D1321298

Encyclopedic Dictionary of Landscape and Urban Planning

Multilingual Reference Book in English, Spanish, French, and German

IFLA

INTERNATIONAL FEDERATION OF LANDSCAPE ARCHITECTS
FEDERATION INTERNATIONALE DES ARCHITECTES PAYSAGISTES

Members/miembros/membres/Mitglieder:

Chairman and Editor-in-Chief/Coordinador y editor/Direction générale du projet et éditeur/Gesamtleitung und Herausgeber:

Dipl.-Ing. Klaus-Jürgen Evert BDLA	Stuttgart, Germany	Deutsch/German

European working group and co-editors/comité europeo y coeditores/groupe de travail européen et coéditeurs/Arbeitsgruppe Europa und Mitherausgeber:

David J. Elsworth MA, Dip LD	Stuttgart, Germany	UK-English
Dipl.-Ing. Icíar Oquiñena	Berlin, Germany	Español/Spanish
Dipl.-Ing. Jean-Marie Schmerber, BDLA	Wolfenbüttel, Germany	Français/French

In collaboration with/con la colaboración de/avec la collaboration de/unter Mitarbeit von:

Lic. Jur. Marga Mielgo	Madrid, Spain	Español/Spanish
Lic. Biol. Begoña Oquiñena	Menorca, Spain	Español/Spanish
Prof. Richard Stiles MA, Dip LD	Wien, Vienna, Austria	UK-English (1978-1990)

US working group/comité americano/groupe de travail américain/Arbeitsgruppe USA:

Team leaders: Edward B. Ballard FASLA,	MLA (1972-2000) (deceased)	US-English
Prof. Nicholas T. Dines FASLA, MLA	Williamsburg, MA, USA	US-English (since 2004)
Beatriz de W. Coffin FASLA MLA	Washington, D.C., USA	Español/Spanish
Dipl.-Ing. Wolfgang Oehme FASLA	Baltimore, MD, USA	US-English
Prof. Robert E. Stipe JD, MRP	(1975-2007) (deceased)	US-English

Klaus-Jürgen Evert (Editor-in-Chief)

Encyclopedic Dictionary of Landscape and Urban Planning
Multilingual Reference Book in English, Spanish, French, and German

Diccionario enciclopédico — Paisaje y urbanismo
Diccionario multilingüe en inglés, español, francés y alemán

Dictionnaire encyclopédique du paysage et de l'urbanisme
Dictionnaire multilingue en anglais, espagnol, français et allemand

Enzyklopädisches Lexikon — Landschafts- und Stadtplanung
Mehrsprachiges Lexikon in Englisch, Spanisch, Französisch und Deutsch

Compiled by the IFLA Committee "Translation of Technical Terms"
Elaborado por el comité de IFLA «Traducción de términos técnicos».
Réalisé par le groupe de travail de l'IFLA « Traduction des termes techniques »
Bearbeitet von der IFLA-Arbeitsgruppe „Übersetzung technischer Begriffe"

Volume 1
A-Z

Editor-in-Chief
Klaus-Jürgen Evert
Rolandstr. 9
70469 Stuttgart
Germany

Editors
Edward B. Ballard (deceased), Alexandria Va., USA
David J. Elsworth, Florian-Geyer-Str. 35, 70499 Stuttgart, Germany
Icíar Oquiñena, Bänschstr. 41, 10247 Berlin, Germany
Jean-Marie Schmerber, Dietrich-Bonhoeffer-Str. 1A, 38300 Wolfenbüttel, Germany
Prof. Robert E. Stipe (deceased), Chapel Hill, N.C., USA

Library of Congress Control Number: 2010924735

ISBN: 978-3-540-76455-7
This publication is available also as:
Electronic publication under ISBN 978-3-540-76435-9 and
Print and electronic bundle under ISBN 978-3-540-76436-6

Springer is part of Springer Science+Business Media

springer.com

Editor: Christian Witschel, Heidelberg, Germany
Development Editor: Sylvia Blago, Heidelberg, Germany
Production: SPI-Publishing, Pondicherry, India
Cover Design: Frido Steinen-Broo, Girona, Spain

Printed on acid-free paper

Foreword

After 23 years of intensive work by the IFLA Working Group in the translation of technical terminology and with the publication in 2001 of the first edition of this comprehensive encyclopedia with German as its 'lead language', it seemed that a large gap in our profession had been closed. The second, corrected reprint was published as early as 2004 and by 2008 almost 1000 copies had been sold worldwide. With continued interest and current strong demand the need arose for the publication of a new, extended edition of the dictionary. And when Springer Publishing House announced its classification as a major reference work, the entire dictionary had to be rewritten so that English would become the 'lead language', enabling an even wider distribution.

In 1978 the chairman of the working group, Klaus-Jürgen Evert formed the IFLA Committee on the Translation of Technical Terms together with Richard Stiles, Zdenek Zvolský, Icíar Oquiñena and Jean-Marie Schmerber. Over the years new colleagues have joined the group: Edward B. Ballard, Bob E. Stipe, Wolfgang Oehme, David Elsworth, Beatriz de W. Coffin as well as Marga Mielgo and Begoña Oquiñena and, following the death of Edward B. Ballard and Bob E. Stipe, Nick Dines.

Now after seven years of intensive work, this improved and updated glossary, supplemented by some 175 additional terms, can be presented, to coincide with the thirtieth anniversary of the establishment of the working group.

A very special circumstance was always of great benefit to the team. It consists of the fact that it brought colleagues together, who shared the same professional idealism. They still share this idealism even after such a long period of time and it has allowed them to continue working together across the continents on such an extensive dictionary.

The glossary takes into account the most important languages of the western world and is a fundamental tool in the hands of planning professions concerned with the environment, urban and landscape planning, science and the entire profession of translators and interpreters. In addition, all of those employed in the sectors of garden and landscape construction and the construction industry itself will also benefit greatly from this work.

Users of the dictionary are welcome to make any suggestions for its improvement, so that with future editions of the work, all of those, who use it, will profit from their comments.

At the International Federation of Landscape Architects World Council meeting in June 2008, Klaus-Jürgen Evert outlined the extraordinary achievement of what might be described as a 'beautiful obsession' on the part of its authors.

The work of the committee received a standing ovation from the World Council delegates. I, too, applaud the publication of this major reference work and now look forward to a good start to this new edition and a great success, worldwide.

Dr. Diane Menzies
President of IFLA
December 2009

Preface

All members of the team are experienced colleagues in the fields of landscape architecture or environmental planning, and especially important is the fact that they are native speakers of the languages into which they translate.

A total of approximately 10,000 terms (including synonyms) have been compiled and translated into five languages (British and American English, Spanish, French and German) as an aid in translation and communication in the very wide, professional fields of landscape architecture, urban planning, nature conservation and environmental protection, as well as for use in private offices, public authorities, research and educational facilities.

One of the dictionary's main aims is to explain the equivalent terms as clearly and comprehensively as possible in each language for the above mentioned professional fields. Collective terms were therefore often selected, under which cross references marked with blue arrows (▶) are leading to a family of similar terms.

The lead language of the dictionary published in 2001 was German (a "brick language" which tends to invent a specific word with its own precise definition for each concept) and intensive research was carried out in German specialist periodicals, textbooks, technical manuals, laws and regulations for the current use of the terminology in order to form a thesaurus as the basis for the dictionary. Each member of the group was not just concerned with providing the simple translation of a term for use in his or her respective language. As often as possible the working group also endeavoured to give a varied picture in the definitions and provided explanations of the terms at the different levels at which they are used in practice. Various connotations of the word, together with an explanation of the different technical and legal backgrounds, not only help to clarify the meaning, but also enable a precise translation of a term.

Over the years, the technical and legislative development of a term was often monitored and updated, as far as was possible, in each of the different language areas. Old termini were retained as synonyms to assist in the evaluation of older literature and legislation.

It was not intended originally to define every term. During the work, however, it became obvious how important short explanations or definitions are, if one is to think in the right direction when searching for the most precise translation. Where this was not possible, the nearest correlation between the term and its concept was devised. With the often very different standards in technology, planning procedures, legislation and environmental consciousness in the various countries in which the languages are spoken, this sometimes seemed to be an impossible task. Seemingly endless, bilateral discussions within the committee about the exact content of most lemmas had therefore to take place. (For example more than 16 journeys to America and many trips to England, France and Spain were also undertaken).

Dialogues with specialists and colleagues or friends of members of the group in both Europe and the U.S.A were held to ascertain how the terms could be best translated and defined. We are most grateful to everybody who helped in this way, because without them our all-encompassing work and the various linguistic possibilities of how to express a term, would not have been achieved.

An attempt was made to formulate the definitions in each language so that a non-specialist would understand the meaning of a term despite the use of technical expressions. In the fields of planning and ecology many terms do not always have the same connotation, because they are derived from various schools of thought. The committee did not venture onto this wide and to a certain extent scientifically insecure terrain. We ask for forgiveness in this respect.

Besides the many planning and construction disciplines, important terms related to fields such as the conservation of ancient monuments, agriculture, forestry, botany, zoology, geography, waste management, building contracts and fee payments as well as surveying and many others were also covered, in order to elucidate the interdisciplinary context of planning terms and their formulation.

For translations we usually turned to specialist literature, technical standards, as well as legislation in each of the languages. In recent years the Internet has also become an invaluable source of information. Multilingual dictionaries were only used in a few cases. Readers, who wish to gain more information, can refer in many instances to the cited sources, often with page numbers.

Many examples of terms are given in context to demonstrate their use and how they are usually expressed, especially when a term is used as a noun, but as a verb in another language. Non-existent or approximate equivalents in one or another language are suggested, and these are indicated by the symbols (≠) and (±).

Variations in the meaning of the same word are, according to their extent, either numbered in rank as lemmas or illustrated in the definition. Words, which originate in Spanish, French, or German, and for which there is no common equivalent in the English language, are underlined in the respective languages.

The basic translation element in the dictionary is the English key word in bold type and these are listed in alphabetical order and numerical sequence, with a symbol for the subject category shown in italics under each term. The term is followed by its definition with applicable references, synonyms and antonyms, and then the equivalent terms and definitions or explanations in the other languages (*s*—Spanish, *f*—French and *g*—German). This unit is called a numbered 'block':

3090 landscape aesthetics [npl *but usually singular*] *land'man. landsc. recr.* (Basic principle of landscape architecture dealing with the allocation of artistic values to the visual appearance of a landscape with regard to composition and visual character; ▶perceived environmental quality, ▶visual quality); *syn. o.v.* landscape esthetics [npl *but usually singular*]; *s* **estética [f] del paisaje** (Ámbito de la planificación del paisaje y de la conservación de la naturaleza que se ocupa de definir y preservar la ▶calidad visual del paisaje; ▶calidad vivencial del ambiente); *f* **esthétique [f] du paysage** (Valeur conceptuelle concernant l'architecture du paysage reposant sur l'étude de la sensibilité visuelle du paysage ; ►qualité perceptuelle, ►valeur esthétique) ; *syn.* esthétique [f] paysagère ; *g* **Landschaftsästhetik [f, o. Pl.]** (Wesentliches Element der Landschaftsarchitektur, das sich mit dem visuellen Erlebniswert der Landschaft befasst; ►Erlebnisqualität, ►Gestaltqualität).

Key words in bold type without numbers, and with ► after them, are references to defined synonyms, e.g.:

lowest tenderer [n] [UK] *contr.* ►lowest bidder [US].

Those terms with asterisks and arrows (∗►) are not synonyms, but indicate definitions and explanations in which the term occurs, e.g.:

administrative personnel [n] [US] *adm. contr.* ∗►office worker.

Blue reference arrows (►) in the blocks are references to terms, which have been defined or to antonyms, which are indicated with *opp.* Compound key words, i.e. those that are composed of several independent words, are usually listed in their normal word order—as the term or idiom [loc] itself—at the beginning of a block and then alphabetically under the other parts of the word, separated from the rest of the word by a comma.

To distinguish between the current usage of terms in the United Kingdom and the United States of America, as well as Canada in certain cases, the terms are designated by [UK], [US], and [CDN], where they differ. The same applies to French for Belgium [B], for Canada [CDN], for Switzerland [CH], and for Luxembourg [L]. Spanish terms used predominantly in Spain are identified by [Es], Latin American variations are also indicated by the name of the country or by [AL]. In German, too, distinction was usually made between Austria [A] and Switzerland [CH].

All abbreviations and symbols are listed in a separate index. The dictionary is organised alphabetically. Compound terms are listed strictly according to the order of the alphabet. Those with a specific difference to the general term follow it, separated by a comma. Compound terms are listed after these, e.g.:

garden [n] *arch. constr. gard.*
garden [n] [UK]**, back** *gard.* ►backyard garden [US].
garden [n]**, baroque** *gard'hist.* ►French classic garden.
garden archaeological excavation [n] *conserv' hist.*
garden art [n] *gard. hist.* ►fine garden design.

British and American ways of spelling, e.g. *colour* and *color* are differentiated, but, for simplicity, the letter 'z' (the American way) has been used instead of 's', in words such as *organize.*

No work is perfect and so it is with this dictionary, although the greatest of care was taken to provide the best translations. The authors know only too well that there are still many, many terms waiting to be dealt with. It is to be noted that a number of terms have yet to be covered in their entirety. In the coming years it is intended to develop the dictionary further by updating and extending its contents. Nevertheless, we have decided to present our work to the public in its current form, with English as the lead language, so that it can already begin to serve as a valuable aid in communication, as a source of alternative ways of formulating sentences and as a basic compendium for day-to-day translations

and study. It should be noted that selected ►terms have been highlighted with a blue color to indicate clickable links in a future electronic document. Users are urged to contribute their suggestions for improvement and additions by email to klaus.j.evert@t-online.de or by post to Klaus-Jürgen Evert, Rolandstrasse 9, 70469 Stuttgart, Germany.

Members of the IFLA Committee "Translation of Technical Terms".

Acknowledgments

Many diligent minds and helping hands have strived together, all on a voluntary basis and over a long period of time, to compile the dictionary. Sincere thanks go here to all of them and also to those colleagues of various disciplines in many different countries, who helped us with great understanding and patience to find the appropriate term in often long discussions.

We give special thanks for the U.S. English in the present version of the dictionary to our co-authors, Edward B. Ballard (1906-2000), leader of the American working group and Prof. Robert E. Stipe (1929-2007), who unfortunately passed away before the publication of the dictionary. As specialists in the fields of design, administration, planning law and conservation, they both worked tirelessly on the definitions and terms which evolved. In 2004 the group was fortunate enough to gain the support of Prof. Nick Dines of Massachusetts, who has continued their intensive work on the dictionary.

Many thanks also go to Wolfgang Oehme, who as a German-American, made a considerable contribution to the first edition of the dictionary with German as the lead language and to Beatriz de Winthuysen Coffin, who as a native Spanish speaker, also greatly contributed to the Spanish part.

My special gratitude goes to the long-serving members of the European working group, who over the years have devoted a great deal of their time to the dictionary: David Elsworth, Stuttgart, responsible for British English, Icíar Oquiñena, Berlin, for Spanish and Jean-Marie Schmerber, Wolfenbuettel for French. Prof. Richard Stiles, Wien, worked upon the UK English part from 1978-1990.

Sincere thanks are extended to my Stuttgart colleagues, who revised critically the German terminology in the following fields: Renate Kübler (nature conservation and impact mitigation regulation), Uwe-Karsten Bruhn (building regulations), Werner Flad (waste management and pollution control), Prof. Dr. Arnulf von Heyl (legal terms on administration, planning regulations and other legal subjects), and Klaus Reissner (surveying and land registering).

Finally, very special thanks go to all our wives and partners for their understanding of how we have spent our free time for the benefit of our profession. An infinite number of leisure hours, weekends and holidays were sacrificed in aid of our dictionary.

Without the financial support of my brother Hans-Friedrich Evert, Krefeld, and Wolfgang Oehme, Baltimore, as well as the financial contribution made by Springer Publishers to this edition, it would certainly have taken many more years to complete the work.

Having said all of this, I am now delighted to be able to hand over the results of our joint efforts to our readers.

Klaus-Jürgen Evert, Chairman of the IFLA Committee on Translation of Technical Terms.

Prólogo

Desde que el Comité Internacional de la Federación Internacional de Arquitectos Paisajistas (IFLA) «Traducción de Términos Técnicos» publicó la primera edición trás veintitrés años de intenso trabajo, se pudo constatar que este amplio diccionario con alemán como «lenguaje guía» había llenado un gran hueco. En 2004 salió una nueva edición corregida. Poco tiempo después se hizo palpable la necesidad de realizar una edición ampliada. La Editorial Springer declaró el diccionario como *major reference work* (obra principal de referencia) y decidió publicarlo con inglés como «lenguaje guía» para posibilitar una mayor divulgación en todo el mundo.

En 1978 Klaus-Jürgen Evert creó el Comité de IFLA con Richard Stiles, Zdenek Zvolský, Icíar Oquiñena y Jean-Marie Schmerber. A los pocos años se adhieron las y los siguientes colegas Edward B. Ballard, Bob E. Stipe, Wolfgang Oehme, David Elsworth, Beatriz de W. Coffin así como Marga Mielgo y Begoña Oquiñena; después del fallecimiento de Edward B. Ballard y Bob E. Stipe, Nick Dines se sumó al grupo como responsable de inglés norteamericano.

Después de siete años de arduo trabajo y coincidiendo con el 32.° aniversario de la existencia del grupo de trabajo, finalmente se puede presentar el diccionario mejorado, actualizado y ampliado en 175 términos.

Esto solo ha sido posible, gracias a que personas con el mismo idealismo profesional se encontraron y trabajaron continuadamente a través de los continentes.

El diccionario incluye los idiomas más importantes del mundo occidental y es un instrumento fundamental en la mano de las profesiones de planificación del medio ambiente, del urbanismo y la ordenación territorial, de la ciencia y de todo el gremio profesional de la traducción escrita y oral. Pero también todos los ramos de la construcción paisajística y de la construcción sacarán ventaja de esta obra.

Sería deseable que los y las usuarias de este diccionario apoyen el desarrollo ulterior de esta obra con correciones y proposiciones.

En la reunión del Consejo Mundial de IFLA en Apeldoom en junio de 2008, el presidente del comité presentó el sobresaliente trabajo que se podría describir como «impresionante pasión» de todas y todos sus autores.

El resultado fue reconocido con una ovación en pie de todas y todos los delegados del Consejo Mundial. También yo quiere dar un gran aplauso a esta obra principal de referencia.

Le deseo un buen inicio y un gran éxito a esta edición a nivel mundial.

Dr. Diane Menzies
Presidenta de IFLA
Diciembre 2009

Introducción

Todas y todos los miembros son colegas con experiencia en el campo de la arquitectura paisajista y la planificación ambiental y —lo que es especialmente importante— traducen a su lengua materna.

Se compilaron y tradujeron aprox. unos 10.000 términos (incluyendo sinónimos) a cinco idiomas (inglés norteamericano – inglés británico – castellano – francés – alemán). El glosario pretende facilitar la comunicación en una amplia gama de campos profesionales como la arquitectura paisajista, el urbanismo, la preservación de la naturaleza y la protección ambiental, así como servir de instrumento para traductores/as e intérpretes independientes, para instituciones públicas, para la investigación y la educación.

Uno de los principales objetivos del glosario es explicar en cada uno de los idiomas, lo más clara y exhaustivamente posible, los términos equivalentes de los campos profesionales nombrados arriba. Por ello, muy a menudo se definen términos genéricos, bajo los cuales se ordenan términos específicos a los cuales se hace referencia a través de una flecha (▶) y marcándolos en azul.

Ya que el idioma guía del diccionario publicado en 2001 es el alemán (una lengua que tiende a formar palabras largas —generalmente sustantivos— con una definición bien precisa), hubo que estudiar detenidamente revistas profesionales, publicaciones científicas, manuales técnicos, leyes y regulaciones alemanas para seleccionar los términos utilizados en su momento y elaborar así un tesauro como base del glosario. Los y las colaboradoras del equipo no sólo se preocuparon de traducir el término a su respectiva lengua, si existe en ella. Más allá de la simple búsqueda del correspondiente término, un objetivo del comité fue siempre presentar por medio de definiciones y explicaciones una visión de los diferentes tipos de práctica de planificación así como de los conceptos legales y técnicos de los correspondientes países, también con la intención de poner en claro las diferencias y los déficits y de esta manera contribuir a aclarar y precisar conceptos. A lo largo de los años, se trató de acompañar lo mejor posible el desarrollo profesional y legal en cada región donde se utiliza cada idioma e incorporar los cambios para así poner al día el diccionario. Los términos antiguos se mantuvieron como sinónimos para posibilitar el entendimiento de textos antiguos.

Originalmente no estaba previsto definir todos los términos. Sin embargo, en el proceso de elaboración se evidenció la importancia que tienen breves explicaciones o definiciones para saber si uno estaba pensando en la dirección correcta a la hora de buscar la traducción más precisa o, si no era posible, por lo menos la correlación más cercana entre el término y el concepto. Debido a los diferentes estándares técnicos, de procedimientos de planificación, legislación y conciencia ambiental en los diversos países donde se hablan los idiomas trabajados, eso a veces nos pareció algo imposible. Debates bilaterales casi interminables sobre el contenido exacto de muchos lemmas fueron ineludibles, tanto dentro del equipo como con múltiples expertos y expertas en cada uno de los países nativos, e implicó más de 16 viajes a los Estados Unidos así como muchos a Inglaterra, Francia y España. Se consultaron a muchos y muchas colegas y amistades de los miembros del equipo para buscar la mejor traducción y definición de los términos. Desde este lugar queremos agradecerles a todos y a todas ellas su colaboración, ya que sin ella nuestro diccionario no habría alcanzado el grado de calidad que creemos que posee.

Se puso esmero en formular las definiciones en un lenguaje entendible por cualquiera, sin prescindir, sin embargo, del uso de los términos técnicos específicos. En los campos de planificación y ecología existen diversos términos que no siempre tienen la misma connotación, dependiendo de la escuela científica a la que pertenezca quien lo utiliza. Esperamos su comprensión por el hecho de que el equipo no se aventuró por ese terreno vasto y a veces inseguro.

Además de los términos de las disciplinas de planificación y construcción, también se incorporaron términos de campos relacionados como la preservación del patrimonio histórico, agricultura, silvicultura, botánica, zoología, geografía, gestión de residuos, contratos de obras, regulación de honorarios o geodesia y muchos otros, con el fin de poner en claro el contexto multidisciplinario de los términos y de las tareas de la planificación.

Para traducirlos se emplearon mayormente publicaciones especializadas, normas técnicas y la legislación en cada uno de los países, en los últimos años crecientemente búsquedas por internet. Solo en casos excepcionales se recurrió a diccionarios multilingües. Quienes deseen más información sobre los respectivos contextos pueden obtenerla en la bibliografía indicada, en muchos casos incluso con el número de página.

Muchos ejemplos de los términos en su contexto demuestran el uso de los mismos y la forma gramatical de expresarlos, especialmente cuando se trata de uno que en alemán se emplea en sustantivo y en los otros idiomas como verbo. Cuando un término no existe o solo tiene equivalente aproximado, se indica con los símbolos (≠) y (±) colocados detrás del mismo.

Variaciones del significado de una misma palabra se definen, dependiendo de la extensión, separadamente como lemmas numerados o dentro de la definición misma. Palabras originarias del castellano, francés o alemán, y para las cuales no existe un equivalente en inglés, están subrayadas en el correspondiente idioma.

Las unidades básicas de traducción en el diccionario son las palabras claves (lemma) en inglés en negrilla numeradas y alistadas en orden alfabético a las que les sigue la indicación en *itálicas* del campo al cual pertenecen. Después del término inglés se encuentra su definición con sus respectivas referen cias, también a términos genéricos o específicos si los hubiere, sinónimos y antónimos. A continuación se encuentran los términos equivalentes y las definiciones o explicaciones en los otros idiomas (*s*—castellano, *f*—francés y *g*—alemán). A esta unidad la denominamos «bloque»:

3090 landscape aesthetics [npl *but usually singular*] *land'man. landsc. recr.* (Basic principle of landscape architecture dealing with the allocation of artistic values to the visual appearance of a landscape with regard to composition and visual character; ►perceived environmental quality, ►visual quality); *syn. o.v.* landscape esthetics [npl *but usually singular*]; *s* **estética [f] del paisaje** (Ámbito de la planificación del paisaje y de la conservación de la naturaleza que se ocupa de definir y preservar la ►calidad visual del paisaje; ►calidad vivencial del ambiente); *f* **esthétique [f] du paysage** (Valeur conceptuelle concernant l'architecture du paysage reposant sur l'étude de la sensibilité visuelle du paysage ; ►qualité perceptuelle, ►valeur esthétique) ; *syn.* esthétique [f] paysagère ; *g* **Landschaftsästhetik [f, o. Pl.]** (Wesentliches Element der Landschaftsarchitektur, das sich mit dem visuellen Erlebenswert der Landschaft befasst; ►Erlebnisqualität, ►Gestaltqualität).

Palabras clave en negrilla sin número y con ► a continuación son referencias a sinónimos definidos en el correspondiente lugar:

lowest tenderer [n] [UK] *contr.* ►lowest bidder [US].

Las que van seguidas de un asterisco y flecha *► no refieren a sinónimos, sino a definiciones o explicaciones en las que se presenta esta palabra como p.ej.:

administrative personnel [n] *adm. contr.* *►office worker.

Flechas azules (►) dentro de los «bloques» son referencias a términos definidos o a antónimos, caracterizados con *opp.*

Voces clave compuestas de varias palabras se encuentran en su orden usual como término o locución [loc] al principio del bloque y también en orden alfabético bajo las demás palabras integrantes, separadas por una coma del resto de la voz.

Para distinguir entre el uso de los términos en los Estados Unidos de América y en el Reino Unido, así como en algunos casos en Canadá, se marcan los términos con [US], [UK] y [CDN], cuando difieren entre sí. Lo mismo ocurre con el francés de Bélgica [B], Canadá [CDN] y Suiza [CH]. Los términos en castellano utilizados predominantemente en España están identificados con [Es], las variantes latinoamericanas con las abreviaturas de los respectivos países o subregiones o con [AL]. En alemán también se diferencian a veces usos típicos de Austria [A] y Suiza [CH].
Todas las abreviaciones y símbolos se encuentran en una lista aparte.

El orden de este diccionario es alfabético. Los términos compuestos están ordenados rígidamente según las letras del alfabeto. Cuando los términos están cualificados por otra palabra detrás de una coma, se ordenan inmediatamente después del término principal y a continuación los términos compuestos sin coma, como p.ej.:

garden [n] *arch. constr. gard.*
garden [n] [UK]**, back** *gard.* ►backyard garden [US].
garden [n]**, baroque** *gard'hist.* ►French classic garden.
garden archaeological excavation [n] *conserv' hist.*
garden art [n] *gard. hist.* ►fine garden design.

Se diferencia la ortografía norteamericana y británica, p.ej. *color* y *colour*. Solamente en el caso de diferencias entre «s» y «z», como p. ej. en *organize*, se utiliza por razones prácticas siempre la «z».

Ninguna obra es perfecta y lo mismo ocurre con este diccionario, aunque pusimos todo el esmero para entregar las mejores traducciones. Sin embargo, los y las autoras saben muy bien que todavía quedan muchos términos por recopilar y elaborar. Se puede constatar que algunos términos aún no han sido definidos suficientemente. A lo largo de los próximos años está previsto continuar actualizando y ampliando el diccionario hasta convertirlo realmente en una obra principal de referencia. A pesar de ello decidimos presentar al público nuestro trabajo en este estado con inglés como lenguaje guía, para que sirva ya de instrumento valioso en la comunicación, de fuente para encontrar alternativas de formulación y como un compendio básico para las traducciones del día a día y para el estudio. Urgimos a los usuarios y usuarias que contribuyan con sus sugerencias a mejorar y ampliar la siguiente edición, y nos las envíen al correo electrónico **klaus.j.evert@t-online.de** o al correo postal Rolandstraße 9, D-70469 Stuttgart.

Los autores y autoras del Comité Internacional de IFLA "Traducción de Términos Técnicos".

Agradecimientos

Muchas mentes despiertas y personas dedicadas han colaborado a lo largo de los años, todas voluntariamente. A todas ellas les quiero agradecer de nuevo su cooperación. También a los muchos y muchas colegas de diferentes disciplinas y países que con gran paciencia y comprensión nos ayudaron, a menudo tras largas discusiones, a encontrar los términos adecuados.

Para esta versión con inglés norteamericano como lenguaje guía agradecemos a nuestros colegas Edward B. Ballard (1906-2000), director del grupo de trabajo estadounidense, y al Prof. Robert E. Stipe (1929-2007), experto en diseño, derecho de planificación y preservación de monumentos, que trabajaron sin descanso en las definiciones y en los términos resultantes de ellas. Afortunadamente en 2004 nuestro comité pudo interesar al Prof. Nick Dines de Massachusetts para continuar el intenso trabajo en el diccionario.

Muchas gracias también a Wolfgang Oehme, que con su dominio del alemán sirvió de nexo con la lengua guía del glosario en la primera edición, y a Beatriz de Windhuysen Coffin que con castellano como lengua materna así mismo aportó importantes colaboraciones a la parte española.

Mi agradecimiento especial corresponde al grupo de trabajo europeo que en parte desde hace más de 30 años dedica mucho de su tiempo libre al diccionario. David Elsworth, Stuttgart, responsable del inglés británico, Icíar Oquiñena, Berlin, para el castellano e Jean-Marie Schmerber, Wolfenbüttel, del francés.

Especialmente agradecido me siento hacia los y las colegas de Stuttgart que revisaron críticamente la terminología alemana en los siguientes campos: Renate Kübler (preservación de la naturaleza y regulación de impactos sobre la naturaleza), Uwe-Karsten Bruhn (derecho de construcción), Werner Flad (gestión de residuos y protección contra inmisiones), Gunter Hägele (planificación de zonas verdes y del paisaje), Prof. Dr. Arnulf von Heyl (términos legales de administración, planificación y otras cuestiones legales), Klaus Reissner (geodesia y catastro).

Finalmente quiero darles gracias especiales a nuestras parejas por su comprensión ante la ocupación de nuestro tiempo libre para el bien de nuestra profesión y el entendimiento internacional. Durante décadas tuvieron que prescindir de muchas horas de tiempo libre, fines de semanas y vacaciones en familia a favor del diccionario. Sin el apoyo financiero de mi hermano Hans-Friedrich Evert, Krefeld, y de Wolfgang Oehme, Baltimore, y sin la contribución financiera de la Editorial Springer habríamos necesitado aún más años hasta terminar el trabajo.

Dicho todo esto, me alegro inmensamente poder entregar ahora los resultados de nuestro trabajo conjunto a sus futuros usuarios y usuarias.

Klaus-Jürgen Evert, Coordinador del Comité de IFLA "Traducción de Términos Técnicos"

Avant-propos

Depuis la parution en 2001 de la première édition du dictionnaire et après 23 ans de travail intensif du groupe de travail de l'IFLA « Traduction de la terminologie technique », il s'avère que cet ouvrage d'envergure, avec l'allemand comme langue de référence, comble une lacune importante. En 2004 paraissait la première réimpression corrigée. Peu après il se révéla nécessaire de publier une deuxième édition. Après avoir classé le dictionnaire en ouvrage de référence, la maison d'édition Springer décidait alors, afin d'assurer ainsi une plus grande diffusion mondiale, d'éditer l'ouvrage avec l'anglais comme langue de référence.

C'est en 1978 que fut constitué par Klaus-Jürgen Evert le Groupe de Travail de l'IFLA avec la participation de Richard Stiles, Zdeněk Zvolský, Icíar Oquiñena et Jean-Marie Schmerber. Au cours des années suivantes se sont joint d'autres collègues : Edward B. Ballard, Bob E. Stipe, Wolfgang Oehme, David Elsworth, Beatriz de W. Coffin ainsi que Marga Mielgo et Begoña Oquiñena ; après le décès de Edward B. Ballard, Nick Dines venait rejoindre le groupe.

Après sept années de travail intensif, une version élargie, améliorée et actualisée avec 175 termes nouveaux a pu être présentée à l'occasion du trentième anniversaire de l'existence du groupe de travail.

Ceci n'a été possible qu'avec la rencontre de personnes partageant le même idéalisme professionnel et déterminées à travailler de manière continue au-delà des continents.

Cet ouvrage prend en compte les principales langues du monde occidental et constitue un outil privilégié dans la main des professionnels de l'aménagement en matière d'environnement, d'urbanisme et de territoire ainsi que dans celle des traducteurs et des interprètes. Mais aussi tous les actifs dans le domaine de l'aménagement de jardin et du paysage ainsi que dans celui du bâtiment tireront profit de cet ouvrage.

Il appartient maintenant aux utilisateurs de ce dictionnaire, d'apporter les corrections et les suggestions nécessaires au perfectionnement de cet ouvrage.

Lors du Congrès Mondial de l'IFLA à Apeldoorn en juin 2008 le président du groupe de travail Klaus-Jürgen Evert a présenté cette œuvre éminente, expression en quelque sorte de cette « impressionnante passion » des auteurs.

Les résultats du groupe de travail furent salués par une ovation debout des délégués du Congrès Mondial. Pour ma part j'applaudis aussi cet ouvrage de référence.

J'espère ainsi que cette édition prenne un bon départ et lui souhaite un grand succès au niveau mondial.

Dr Diane Menzies
Présidente de l'IFLA
Décembre 2009

Préface

Tous les membres sont des spécialistes compétents en matière d'architecture des paysages et de planification environnementale et, élément important, le domaine de traduction est celui de leur langue maternelle.

Ce sont quelques 10.000 termes (y compris les synonymes) qui ont été pris en compte dans cinq langues (anglais américain et britannique, espagnol, français et allemand) afin d'apporter à l'ensemble des spécialistes ainsi qu'aux professionnels de la traduction et de l'interprétation dans les cabinets privés, les administrations publiques et les organismes d'enseignement et de recherche une aide à la traduction et la compréhension dans ce vaste secteur d'activités que sont l'architecture des paysages, l'urbanisme, la protection de la nature et de l'environnement.
Le principal objectif de cet ouvrage est de fournir une explication aussi claire et complète que possible des termes techniques utilisés dans les espaces linguistiques respectifs.

L'allemand fut la langue source de l'ouvrage paru en 2001. La spécificité de celle-ci tient dans sa tendance à créer pour chaque concept un terme approprié — la plupart du temps un mot composé — et à en définir très précisément le contenu. C'est pour cette raison qu'a été réalisée, en vue de l'établissement et le développement du thésaurus, une étude intensive de la terminologie courante dans la littérature spécialisée (ouvrages et périodiques), les manuels techniques, les textes et prescriptions législatifs de langue allemande. Les auteurs du dictionnaire ne se sont pas simplement limités, dans la mesure de leur existence, à une simple traduction des termes dans la langue correspondante. Ils se sont appliqués, aussi souvent que possible, à élaborer dans les définitions et les explications données une présentation différenciée des différents niveaux conceptuels de développement de la terminologie dans la hiérarchie des termes techniques, législatifs et d'aménagement pour, en dernière analyse, exposer les différences et les carences et apporter ainsi les suggestions nécessaires à leur clarification et leur précision. Au cours des années les auteurs ont, dans les divers espaces linguistiques, autant que possible suivi avec attention le développement de la terminologie dans les domaines techniques et législatifs et procédé à son actualisation. Les termes anciens ont été néanmoins conservés comme synonymes afin de pouvoir procurer une information lors de l'étude de littérature surannée ou de titres de loi désuets.

Il n'avait pas été prévu à l'origine de définir la totalité des termes. Au cours des travaux, des explications ou des définitions courtes se sont avérées comme étant la forme la plus appropriée pour cibler le contenu et, si possible, établir une corrélation sans ambiguïté ou, lorsque cela n'était pas possible, une relation approximative entre le terme et son concept. Ceci s'avéra parfois être une tâche quasi insurmontable compte tenu des différents niveaux de développement des standards dans les différents espaces linguistiques en matière de technologies, de procédures d'aménagement, de législation et de conscience environnementale. Pour ces motifs s'imposa rapidement, au cours de très nombreuses discussions bilatérales, non seulement entre les membres du groupe de travail mais aussi avec les experts sur place, la nécessité de définir le contenu exact de la plupart des lemmes (plus de 16 voyages aux États-Unis et de nombreux déplacements en Angleterre, en France et en Espagne). De nombreux spécialistes et

amis des membres du groupe de travail furent consultés en Europe comme aux États-Unis afin de trouver les meilleures traductions et définitions possibles. Nous leur en sommes tous très reconnaissant car sans eux, compte tenu de la multiplicité des expressions linguistiques, nous n'aurions pu obtenir un tel résultat.

Nous nous sommes appliqués à rédiger les définitions dans une langue compréhensible par les profanes, tout en ne renonçant pas à l'utilisation des expressions techniques. Dans les domaines de l'aménagement et de l'écologie certains termes attestent de l'empreinte de différentes écoles. Le groupe de travail ne s'est pas aventuré sur ce vaste terrain d'études, scientifiquement incertain. Qu'il ne nous en soit pas tenu rigueur.

Les concepts d'importance des diverses disciplines voisines telles que p. ex. la protection des monuments historiques, l'agriculture et la sylviculture, la botanique, la zoologie, la géographie, la gestion des déchets, les marchés de travaux, les honoraires, la géodésie ainsi que bien d'autres ont été également traités dans le but de montrer clairement les relations interdisciplinaires existantes avec les concepts et les questions d'aménagement.

La littérature spécialisée, les prescriptions techniques, les textes de loi des espaces linguistiques concernés et plus récemment la recherche sur Internet ont été utilisés en priorité lors de la traduction, très exceptionnellement les dictionnaires multilingues. Les lectrices et lecteurs désirant obtenir des informations complémentaires peuvent se reporter aux sources mentionnées, souvent avec indication du numéro de page. Les termes précédés du signe ▶ de couleur bleue sont des renvois à la même famille de concepts ou à un antonyme.

De nombreux exemples pris dans leur contexte représentent la forme courante d'utilisation de certains termes, spécialement lorsque le substantif est utilisé dans une autre langue sous forme verbale. Les termes inexistants dans une langue, ou pour lesquels les auteurs proposent une traduction, sont respectivement caractérisés par les symboles (≠) et (±).

Les variantes dans la traduction d'un terme sont, selon l'importance, soit classées selon un ordre numérique, soit intégrées à la définition. Les concepts d'origines espagnole, française ou allemande et n'ayant pas de terme correspondant en anglais sont soulignés dans la langue respective.

La plus petite unité de traduction du lexique est caractérisée par un terme-clé anglais (Lemme) alphabétisé et numéroté, à caractère gras, suivi de la mention en italique des domaines d'utilisation. Suivent ensuite la définition et les renvois éventuels, antonymes et synonymes, les références aux termes génériques ou spécifiques ainsi que les traductions et définitions secondaires (*s*—espagnol, *f*—français et *g*—allemand). Cette unité est appelée « bloc ».

3090 landscape aesthetics [npl *but usually singular*] *land'man. landsc. recr.* (Basic principle of landscape architecture dealing with the allocation of artistic values to the visual appearance of a landscape with regard to composition and visual character; ▶perceived environmental quality, ▶visual quality); *syn. o.v.* landscape esthetics [npl *but usually singular*]; **s estética [f] del paisaje** (Ámbito de la planificación del paisaje y de la conservación de la naturaleza que se ocupa de definir y preservar la ▶calidad visual del paisaje; ▶calidad vivencial del ambiente); *f* **esthétique [f] du paysage** (Valeur conceptuelle concernant l'architecture du paysage reposant sur l'étude de la sensibilité visuelle du paysage ; ▶qualité perceptuelle, ▶valeur esthétique) ; *syn.* esthétique [f] paysagère ; *g* **Landschaftsästhetik [f, o. Pl.]** (Wesentliches Element der Landschaftsarchitektur, das sich mit dem visuellen Erlebenswert der Landschaft befasst; ▶Erlebnisqualität, ▶Gestaltqualität).

Les termes-clés en caractère gras non numérotés et précédés du signe ▶ sont des renvois à des synonymes définis, p. ex. :

lowest tenderer [n] **[UK]** *contr.* ▶lowest bidder [US].

Les termes précédés d'un astérisque et du pictogramme *▶ ne sont pas des synonymes mais renvoient aux définitions et explications dans lesquelles le terme est mentionné, p. ex. :

administrative personnel [n] *adm. contr.* *▶office worker.

Les pictogrammes ▶ dans les « blocs » renvoient à des termes définis ou à des antonymes précédés par la mention *opp.*
Les termes composés de plusieurs substantifs indépendants (expression) sont en général répertoriés, tout d'abord dans l'ordre courant de l'utilisation des mots, comme lemme ou comme locution [loc] et se retrouvent ensuite cités par ordre alphabétique par un substantif ou, lorsque celui-ci n'existe pas, par le verbe ou l'adjectif suivi d'une virgule et du reste des mots composant l'expression.

Afin de distinguer les termes couramment utilisés aux États-Unis et en Angleterre ainsi que les différences d'usage entre les concepts américains, anglais, et parfois canadiens, les termes sont caractérisés par les mentions [US], [UK] et [CDN]. Il en est de même pour les termes des pays de langue française, la Belgique [B], le Canada [CDN] la Suisse [CH] et le Luxembourg [L]. Les termes espagnols utilisés couramment en Espagne sont caractérisés par la mention [Es], les variantes des pays d'Amérique latine étant suivies par les diminutifs des pays correspondants ou par la mention [AL]. Concernant les pays de langue allemande, les termes utilisés en Autriche [A] et en Suisse [CH] ont été pris en compte autant que faire ce pouvait.

L'ensemble des abréviations et des symboles est répertorié dans un index séparé.

La classification des termes a été effectuée par ordre alphabétique. Les mots composés sont rigoureusement classés selon l'ordre alphabétique. Les mots composés, dont la spécificité suit la virgule placée après l'élément principal, sont classés directement après le terme déterminant, ensuite viennent les autres mots composés, p. ex. :

garden [n] *arch. constr. gard.*
garden [n]**, baroque** *gard'hist.* ▶French classic garden.
garden [n]**, flower** *gard.* ▶ornamental garden.
garden archaeological excavation [n] *conserv'hist.*
garden art [n] *gard. hist.* ▶fine garden design.

Les différences orthographiques dans la terminologie anglaise américaine et britannique, p. ex. *color* et *colour* a été prise en compte, mais, les mots orthographiés avec le « s » ou le « z »

comme p. ex. *organize,* sont par simplification écrit sans exception avec la lettre « z ».

Aucun ouvrage n'est parfait. Ce dictionnaire n'y fait pas exception, ouvrage dans lequel la pertinence de la traduction se doit de côtoyer la plus grande exactitude. Les auteurs sont conscients qu'un grand nombre de termes supplémentaires demanderait à être traduit. On constatera que de nombreux lemmes ne sont pas encore définis de manière exhaustive. Les auteurs envisagent la réalisation d'un ouvrage encyclopédique dans le cadre de l'actualisation et de l'extension du dictionnaire ces prochaines années. Néanmoins nous avons pris la décision de présenter aux professionnels cet ouvrage en l'état, avec l'anglais comme langue source, afin qu'il apporte, dès aujourd'hui, une aide précieuse à la compréhension, tienne lieu de véritable mine de propositions de traduction et continue d'être une source de données dans le travail quotidien. C'est pourquoi nous prions incessamment les utilisateurs d'apporter toutes corrections, améliorations et suggestions nécessaires pour la prochaine édition et de les adresser par courriel à klaus.j.evert@t-online.de ou à son adresse postale Klaus-Jürgen Evert, Rolandstraße 9, D-70469 Stuttgart.

Les auteurs membres du groupe de travail de l'IFLA « Traduction de la terminologie technique ».

Remerciements

De nombreux cerveaux bien faits ainsi que des mains appliquées ont tous bénévolement, pendant cette longue période, participé à l'élaboration du lexique. À tous mes remerciements les plus sincères et spécialement à tous les collègues provenant de disciplines les plus diverses qui, dans de nombreux pays au cours de discussions souvent longues, nous ont apporté aide et compréhension pour déceler les termes appropriés.
Nous désirons ainsi remercier pour cette édition actuelle avec l'anglais américain comme langue de référence notre collègue Edward B. Ballard (1906-2000), dirigeant du groupe de travail américain ainsi que Prof. Robert E. Stipe (1929-2007), spécialiste dans les domaines de la conception, du droit de l'aménagement et de la protection des monuments historiques, qui, inlassablement, participèrent à l'élaboration des définitions et de leur traduction.

Tous mes remerciements à Wolfgang Oehme qui, pour la première édition, comme « allemand-américain », constitua un lien essentiel pour l'allemand comme langue source ainsi qu'à Beatriz de Windhuysen Coffin qui, avec sa langue maternelle l'espagnol, apporta une contribution importante dans l'élaboration des textes espagnols du dictionnaire.

Mes remerciements tous particuliers vont aux membres européens du groupe de travail qui, pour certains depuis plus de trente ans, ont consacré une grande partie de leur temps à l'élaboration du dictionnaire : David Elsworth, Stuttgart, pour la terminologie anglaise britannique, Icíar Oquiñena, Berlin, pour la terminologie espagnole et Jean-Marie Schmerber, Wolfenbüttel, pour la terminologie française. Professeur Richard Stiles, Vienne, participa aux travaux d'anglais britannique entre 1978 et 1990.

Tous mes remerciements cordiaux aux collègues de la ville de Stuttgart qui, dans différents domaines contrôlèrent avec esprit critique la terminologie allemande ; Renate Kübler (protection de la nature et réglementation relative aux atteintes subies par le milieu naturel), Uwe-Karsten Bruhn (droit de la construction), Werner Flad (gestion des déchets et protection contre les émissions), Gunter Hägele (planification urbaine des espaces verts et planification des paysages), Prof. Dr. Arnulf von Heyl (termes juridiques, administratifs et d'aménagement), Klaus Reissner (géodésie et cadastre).

Pour finir, mes remerciements tous particuliers à l'adresse des épouses et compagnes des auteurs de cet ouvrage pour la compréhension qu'elles apportèrent à ce violon d'Ingres dévoué à notre profession et l'entente internationale. Pendant plusieurs décennies, des heures innombrables prises sur le temps libre, les fins de semaine et les vacances ont été sacrifiées pour la réalisation du dictionnaire. Et sans l'aide financière de mon frère Hans-Friedrich Evert – Krefeld –, de Wolfgang Oehme – Baltimore – et de la maison d'édition Springer de nombreuses années auraient été encore nécessaires pour achever ces travaux.

Ainsi je suis heureux de pouvoir mettre à la disposition d'un public spécialisé l'ouvrage présent, fruit de notre travail commun.

Klaus-Jürgen Evert, Directeur du groupe de travail de l'IFLA « Traduction des termes techniques ».

Vorwort

Seit die erste Auflage nach 23-jähriger intensiver Arbeit durch die IFLA-Arbeitsgruppe „Übersetzung technischer Terminologie" 2001 erschien, zeigte sich, dass dieses umfangreiche Lexikon mit Deutsch als Leitsprache eine große Lücke geschlossen hat. 2004 wurde schon der zweite korrigierte Nachdruck herausgegeben. Schon kurz darauf ergab sich die Notwendigkeit einer erweiterten Auflage. Der Springer-Verlag erklärte das Lexikon zum Hauptstandardwerk, und beschloss für eine noch größere weltweite Verbreitung das gesamte Buch mit Englisch als Leitsprache herauszugeben.

Klaus-Jürgen Evert bildete 1978 die IFLA-Arbeitsgruppe mit den Mitgliedern Richard Stiles, Zdenek Zvolský, Icíar Oquiñena and Jean-Marie Schmerber. Nach wenigen Jahren kamen folgende Kollegen hinzu: Edward B. Ballard, Bob E. Stipe, Wolfgang Oehme, David Elsworth, Beatriz de W. Coffin as well as Marga Mielgo and Begoña Oquiñena; nach dem Tode von Edward B. Ballard und Bob E. Stipe kam Nick Dines als Verantwortlicher des amerikanischen Englisch hinzu.

Nach siebenjähriger intensiver Arbeit kann nun ein mit 175 Termini erweitertes, verbessertes und aktualisiertes Lexikon zum 30-jährigen Jubiläum des Bestehens der Arbeitsgruppe vorgelegt werden.

Dies ist nur möglich, da sich Menschen mit gleichem beruflichen Idealismus begegneten und über die Kontinente hinweg kontinuierlich zusammen weiterarbeiten.

Das Lexikon berücksichtigt die wichtigsten Sprachen der westlichen Welt und ist ein fundamentales Werkzeug in der Hand der planenden Berufe für Umwelt, Stadt und Raum, der Wissenschaft und des gesamten Berufsstandes der Übersetzenden und Dolmetschenden. Aber auch alle in diesem Zusammenhang arbeitenden Sparten des Garten- und Landschaftsbaues und der Bauwirtschaft profitieren von diesem Werk.

Es wäre wünschenswert, wenn Benutzerinnen und Benutzer dieses Lexikons, mit Verbesserungen und Anregungen die Weiterentwicklung dieses Werkes unterstützten.

Auf der Tagung des IFLA-Weltrates in Apeldoorn im Juni 2008 stellte der Vorsitzende der Arbeitsgruppe die herausragende Arbeit, die man als eine „eindrucksvolle Leidenschaft" der ganzen Autorenschaft beschreiben könnte, vor.

Das Ergebnis der Arbeitsgruppe wurde mit einer stehenden Ovation der Delegierten des Weltrates gewürdigt. Auch ich spende diesem Hauptstandardwerk großen Beifall.

So hoffe ich auf einen guten Start dieser Auflage und einen großen Erfolg weltweit.

Dr. Diane Menzies
IFLA-Präsidentin
Dezember 2009

Einführung

Alle Mitglieder sind erfahrene Kolleginnen und Kollegen auf dem Gebiet der Landschaftsarchitektur und Umweltplanung, und was besonders wichtig ist, Muttersprachlerinnen und Muttersprachler für ihren Übersetzungsbereich.

Wir erarbeiteten einen Umfang von ca. je 10.000 Begriffen, incl. Synonyma, in fünf Sprachen (amerikanisches und britisches Englisch – Spanisch – Französisch – Deutsch), um der Fachwelt in dem sehr weiten Spektrum der Tätigkeitsfelder der Landschaftsarchitektur, Stadtplanung, des Natur- und Umweltschutzes sowie den Sprachmittlern und Sprachmittlerinnen in freien Büros, in Behörden, Forschung und Lehre Übersetzungs- und Verständnishilfen vorzulegen.
Ziel dieses Buches ist es, vor allem in den jeweiligen Sprachräumen auftretende Fachtermini der o. g. Fachgebiete, wenn erforderlich, in möglichst klarer und umfassender Form zu erklären. Deshalb wurden oft Sammelstichworte gewählt, bei denen in den Erläuterungen mit Pfeilen (►) blau markierte Verweise zur gleichen Begriffsfamilie hinführen.

Da die Leitsprache des 2001 erschienenen Werkes deutsch ist und man im Deutschen dazu neigt, für fast jeden Begriff ein entsprechendes Wort — meist ein Kompositum — zu schöpfen und das dabei Gedachte entsprechend genau zu definieren, wurde als Grundlage des Thesaurus eine intensive Auswertung der deutschen Fachzeitschriften, Fachbücher, technischen Regelwerke, Gesetze und Rechtsverordnungen hinsichtlich der gebräuchlichen Termini vorgenommen. Es werden von den jeweiligen Bearbeiterinnen und Bearbeitern nicht nur die reinen Begriffsübersetzungen für den betreffenden Sprachraum, soweit vorhanden, erarbeitet. Vielmehr ist es das Ansinnen der Arbeitsgruppe, so oft wie möglich eine differenzierte Darstellung des unterschiedlichen Entwicklungsstandes der planerischen, technischen und juristischen Begriffsebenen in den Definitionen und Erläuterungen zu erarbeiten sowie Unterschiede und Defizite deutlich zu machen, um so Anregungen für begriffliche Klärungen und Präzisierungen zu geben. Im Laufe der Jahre wurde die fachliche und gesetzgeberische Entwicklung in den einzelnen Sprachräumen so weit es möglich war, verfolgt und aktualisierte Ergänzungen vorgenommen. Die alten Termini wurden als Synonyma beibehalten, um bei der Auswertung älterer Literatur und Gesetzesbezeichnungen sich informieren zu können.

Ursprünglich war nicht beabsichtigt, fast alle Begriffe zu definieren. Im Laufe der Arbeit stellte sich jedoch heraus, wie wichtig kurze Erläuterungen oder Definitionen sind, um in den zu übersetzenden Sprachen in die richtige Richtung zu denken und möglichst eindeutige oder wenn dies nicht möglich war, annähernde Zusammenhänge zwischen Wort und Begriff herzustellen — bei dem oft recht unterschiedlichen Entwicklungsstand in Technik, Planungsabläufen, Rechtsentwicklung und Umweltbewusstsein in den einzelnen Sprachräumen eine manchmal fast nicht zu lösende Aufgabe. Deshalb war es dringend geboten, unendlich viele bilaterale Diskussionen über die exakten Inhalte der meisten Stichworte (Lemmata) innerhalb der Arbeitsgruppe aber auch mit Experten „vor Ort" zu führen (z. B. bei über 16 Reisen nach Amerika, viele Reisen nach England, Frankreich und Spanien). Viele Fachleute und Freunde der Arbeitsgruppenmitglieder, sowohl in Europa als auch in den USA, wurden angesprochen, um herauszufinden wie die Begriffe am besten übersetzt und definiert werden sollten. Wir sind ihnen allen sehr

dankbar, weil sich ohne sie bei der Vielfalt der sprachlichen Ausdrucksmöglichkeiten unser umfangreiches Wörterbuch nicht zu diesem Ergebnis entwickelt hätte.

Es wurde versucht, die Definitionen jeweils in einer Sprache zu formulieren, die auch dem Laien verständlich ist, ohne auf fachspezifische Ausdrucksweisen zu verzichten. Im planerischen und ökologischen Bereich sind manche Termini durch die verschiedenen Schulen unterschiedlich besetzt. Auf dieses weite und z. T. wissenschaftlich noch nicht abgesicherte Feld hat sich die Arbeitsgruppe nicht begeben — man möge uns dies nachsehen.

Wichtige Begriffe verschiedener Nachbarfächer wie z. B. Denkmalschutz, Land- und Forstwirtschaft, Botanik, Zoologie, Geografie, Abfallwirtschaft, Bauvertrags- und Honorarwesen sowie Vermessungstechnik und viele andere wurden mitbearbeitet, um den interdisziplinären Zusammenhang planerischer Begriffe und Fragestellungen deutlich zu machen.

Bei der Übersetzung bedienten wir uns vorrangig der Fachliteratur, technischer Regelwerke und Gesetze des jeweiligen Sprachraumes, in letzter Zeit auch der Internetrecherche und nur in ganz wenigen Ausnahmefällen mehrsprachiger Wörterbücher. Für Leserinnen und Leser, die sich weiter informieren möchten, sind in vielen Fällen die Fundstellen, oft mit Seitenzahl, angegeben.

Etliche Kontextbeispiele sollen den Gebrauch von manchen Begriffen so darlegen, wie sie i. d. R. formuliert werden, speziell dann, wenn das substantivierte Stichwort in der jeweiligen Sprache i. d. R. verbal gebraucht wird. Für Termini, die in der einen oder anderen Sprache so (noch) nicht gebraucht werden, macht der/die jeweilige Bearbeiter/-in Übersetzungsvorschläge, die durch die nachgeordneten Symbole (≠) und (±) gekennzeichnet sind.

Bedeutungsvarianten desselben Stichwortes werden je nach Umfang entweder in numerischer Reihenfolge als Lemmata oder in den Definitionen dargelegt. Begriffe, die aus dem Spanischen, Französischen oder Deutschen stammen, und für die es im Englischen kein gebräuchliches Äquivalent gibt, sind in den jeweiligen Sprachen unterstrichen.

Die kleinste selbständige Übersetzungseinheit in dem Lexikon ist das fett gedruckte, nummerierte englische Stichwort (Lemma) mit Angabe des kursiv gedruckten Anwendungsfeldes. Ihm folgen die Definition mit allfälligen Verweisen, Synonymen, Antonymen sowie Hinweise auf Ober- und Unterbegriffe und die nachgeordneten Übersetzungen und Definitionen oder Erläuterungen (*s*—Spanisch, *f*—Französisch und *g*—Deutsch). Diese Einheit bezeichnen wir als ‚Block':

3090 landscape aesthetics [npl *but usually singular*] *land' man. landsc. recr.* (Basic principle of landscape architecture dealing with the allocation of artistic values to the visual appearance of a landscape with regard to composition and visual character; ►perceived environmental quality, ►visual quality); *syn. o.v.* landscape esthetics [npl *but usually singular*]; **s estética [f] del paisaje** (Ámbito de la planificación del paisaje y de la conservación de la naturaleza que se ocupa de definir y preservar la ►calidad visual del paisaje; ►calidad vivencial del ambiente);

f esthétique [f] du paysage (Valeur conceptuelle concernant l'architecture du paysage reposant sur l'étude de la sensibilité visuelle du paysage ; ▶qualité perceptuelle, ▶valeur esthétique) ; *syn.* esthétique [f] paysagère ; *g* **Landschaftsästhetik [f, o. Pl.]** (Wesentliches Element der Landschaftsarchitektur, das sich mit dem visuellen Erlebenswert der Landschaft befasst; ▶Erlebnisqualität, ▶Gestaltqualität).

Fett gedruckte, nicht nummerierte Stichworte mit ▶ sind Verweise auf Synonyme, die dort definiert sind, z. B.

lowest tenderer [n] [UK] *contr.* ▶lowest bidder [US].

Solche mit einem Sternchen und Pfeil ∗▶ sind keine Synonyme, sondern Verweise und deuten auf Definitionen und Erläuterungen hin, in denen dieser Begriff vorkommt, z. B.

administrative personnel [n] *adm. contr.* ∗▶office worker.

Blaue Verweispfeile (▶) in den „Blöcken" verweisen auf definierte Begriffe oder auf Antonyme, gekennzeichnet mit *opp.* Zusammengesetzte, d. h. aus mehreren selbstständigen Worten bestehende Stichwörter sind einmal in ihrer gebräuchlichen Wortfolge als Begriff oder als Redewendung [loc] zu Beginn eines Blockes aufgenommen und dann unter den übrigen Wortteilen mit durch Komma nachgeordnetem Rest der Wortfolge alphabetisch aufgeführt.

Um die in den USA und in Großbritannien heute gebräuchlichen Begriffe und die Unterschiede zwischen amerikanischem und britischem sowie auch manchmal kanadischem Sprachgebrauch deutlich zu machen und ein vielseitig nutzbares Sprachwerk vorzulegen, wurden die entsprechenden Kennzeichnungen mit [US], [UK] und [CDN] vorgenommen. Für das Französische gilt Entsprechendes für Belgien [B], Kanada [CDN] und die Schweiz [CH], Luxemburg [L]. Im Spanischen wurden besonders in Spanien gebräuchliche Termini mit [Es], lateinamerikanische Übersetzungsvarianten mit den entsprechenden Länderbezeichnungen resp. mit [AL] verdeutlicht. Auch im Deutschen wurden teilweise, soweit es bisher möglich war, die Sprachgebiete Österreich [A] und Schweiz [CH] miteinbezogen.
Sämtliche Abkürzungen und Symbole sind in einem gesonderten Verzeichnis aufgeführt.

Die Gliederung des Buches ist alphabetisch. Zusammengesetzte Begriffe werden streng in alphabetischer Buchstabenfolge aufgeführt. Zusammengesetzte Begriffe, bei denen die spezifische Differenz durch Kommata nachrangig dem Hauptbegriff folgt, stehen unmittelbar nach dem Hauptbegriff und anschließend die zusammengesetzten Begriffe, z. B.

garden [n] *arch. constr. gard.*
garden [n] [UK], **back** *gard.* ▶backyard garden [US].
garden [n], **baroque** *gard'hist.* ▶French classic garden.
garden archaeological excavation [n] *conserv' hist.*
garden art [n] *gard. hist.* ▶fine garden design.

Amerikanische und britische Schreibweisen, z. B. *color* und *colour* werden unterschieden. Lediglich das „s" resp. „z", z. B. bei *organize*, wird aus Vereinfachungsgründen durchgängig nur mit „z" geschrieben.

Kein Werk ist vollkommen, so auch nicht dieses Lexikon, bei dem mit größter Sorgfalt versucht wurde, die bestmöglichen Übersetzungen zu erarbeiten. Der Arbeitsgruppe ist auch bewusst, dass noch viele weitere Begriffe erfasst und bearbeitet werden müssen. Es ist festzustellen, dass eine Reihe von Begriffen noch nicht umfassend erläutert wurde. Im Laufe der nächsten Jahre ist im Rahmen weiterer Aktualisierung und Erweiterung beabsichtigt, das Lexikon weiterzuentwickeln. Dennoch haben wir uns entschlossen, den derzeitigen Bearbeitungsstand mit Englisch als Leitsprache in der vorliegenden Form der Fachöffentlichkeit vorzustellen, damit er jetzt schon als wertvolle Verständnishilfe und als Fundgrube für Formulierungsalternativen und weiteres Quellenstudium für die tägliche Arbeit genutzt werden kann. Die Benutzenden werden deshalb dringend gebeten, Verbesserungs- und Ergänzungsvorschläge unter E-Mail **klaus.j.evert@t-online.de** zu machen oder an die Postanschrift Rolandstraße 9, D-70469 Stuttgart zu senden.

Die Autoren der IFLA-Arbeitsgruppe „Übersetzung technischer Terminologie".

Danksagung

Bei einer so langen Bearbeitungszeit haben viele kluge Köpfe und fleißige Hände, alle ehrenamtlich, mitgewirkt. Ihnen allen sei an dieser Stelle noch mal ausdrücklich gedankt. Auch den vielen Kollegen und Kolleginnen unterschiedlicher Fachgebiete aus vielen Ländern, die mit großem Verständnis und viel Geduld geholfen haben, in oft langen Diskussionen die passenden Termini zu finden.
Für die vorliegende Version mit US-Englisch als Leitsprache danken wir unseren Kollegen Edward B. Ballard (1906-2000), Leiter der amerikanischen Arbeitsgruppe, und Prof. Robert E. Stipe (1929-2007), Fachkollege für Gestaltung, Planungsrecht und Denkmalschutz, die unermüdlich an den Definitionen und den sich daraus ergebenden Begriffen gearbeitet hatten. Glücklicherweise konnte unsere Arbeitsgruppe 2004 Prof. Nick Dines aus Massachusetts für die intensive Weiterarbeit an dem Lexikon gewinnen.

Vielen Dank auch an Wolfgang Oehme, der bei der ersten Auflage als Deutsch-Amerikaner ein wesentliches Bindeglied zur deutschen Leitsprache bildete, an Beatriz de Windhuysen Coffin, die mit ihrer spanischen Muttersprache ebenfalls sehr wichtige Beiträge für den spanischen Teil lieferte.

Mein ganz spezieller Dank gebührt den Mitgliedern der europäischen Arbeitsgruppe, die z. T. seit über dreizig Jahren einen großen Teil ihrer Zeit dem Lexikon widmeten: David Elsworth, Stuttgart, der für das UK-Englisch, Icíar Oquiñena, Berlin, die für das Spanische und Jean-Marie Schmerber, Wolfenbüttel, der für das Französische verantwortlich ist. Prof. Richard Stiles, Wien, arbeitete für den UK-englischen Teil 1978-1990.

Ganz herzlichen Dank auch an die Stuttgarter Kollegin und Kollegen, die die deutsche Terminologie in folgenden Gebieten kritisch durchschauten: Renate Kübler (Naturschutz und Eingriffsregelung), Uwe-Karsten Bruhn (Baurecht), Werner Flad (Abfallwirtschaft und Immissionsschutz), Gunter Hägele (Grünordnung und Landschaftsplanung), Prof. Dr. Arnulf von Heyl (planungsrechtliche, sonstige juristische und verwaltungstechnische Begriffe), Klaus Reissner (Vermessungs- und Katasterwesen).

Schlussendlich gilt besonderer Dank allen unseren Ehefrauen und Lebenspartnern für ihr Verständnis für unsere Freizeitbeschäftigung zum Wohle des Berufsstandes und der internationalen Verständigung. Über Jahrzehnte musste auf unendlich viele Freizeitstunden, gemeinsame Wochenenden und Ferien zugunsten des Wörterbuches verzichtet werden. Ohne die finanzielle Unterstützung durch meinen Bruder Hans-Friedrich Evert, Krefeld, Herrn Wolfgang Oehme, Baltimore, und den finanziellen Beitrag des Springer-Verlages zu dieser Version wären sicher noch mehrere Jahre bis zur Fertigstellung ins Land gegangen.

So bin ich sehr froh, unsere gemeinsame Arbeit der Fachöffentlichkeit an die Hand zu geben.

Klaus-Jürgen Evert, Leiter der IFLA-Arbeitsgruppe „Übersetzung technischer Terminologie".

Abbreviations
Abreviaciones
Abréviations
Abkürzungen

[A]	Österreich/in Österreich gebräuchlich.
abbr.	abbreviation.
AbfG	►KrW-/AbfG.
Abk.	Abkürzung.
ABl.	Amtsblatt der Europäischen Union.
abr.	abreviación.
abrév.	abréviation.
AbwAG	Abwasserabgabengesetz.
ADEME	Agence de l'environnement et de la maîtrise de l'énergie.
a. d.	auf dem, auf den
adj	adjective; adjetivo; adjectif, locution adjective ; Adjektiv.
adm.	administration; administración; administration ; Verwaltung.
adv	adverb; adverbio; adverbe ; Adverb.
AFNOR	Association française de normalisation.
agr.	agriculture; agricultura; agriculture, génie rural ; Landwirtschaft.
[AL]	América Latina/utilizado en América Latina;
ANSI	American National Standards Institute.
APD	Avant Projet Détaillé
APS	Avant Projet Sommaire
arb.	arboriculture, tree maintenance; arboricultura, mantenimiento de árboles; arboriculture, travaux d'entretien des arbres ; Baumkunde, Baumpflege.
arch.	architecture; arquitectura; architecture ; Architektur, Gestaltung von Bauwerken.
arr.	Arrêté.
Art./art.	article; artículo; article ; Artikel.
ASLA	American Society of Landscape Architects.
AtAV	Verordnung über die Verbringung radioaktiver Abfälle in das oder aus dem Bundesgebiet (Atomrechtliche Abfallverbringungsverordnung v. 27.07.1998 (BGBl. I Nr. 47, S. 1918);
[AUS]	Australia/used in Australia.
[B]	Belgique/utilisé en Belgique.
BArtSchV	Verordnung zum Schutz wild lebender Tier- und Pflanzenarten (Bundesartenschutzverordnung) vom 14.10.1999 [BGBl. I S. 1955]); zuletzt geändert durch die erste Verordnung zur Änderung der BArtSchV vom 21.12.1999.
BauGB	Baugesetzbuch i. d. Fassung der Bekanntmachung v. 08.12.1986 (BGBl. I S. 2253), i. d. Neufassung vom 27.08.1997 (BGBl. I S. 2141).
BauNVO	Baunutzungsverordnung.
BayNatSchG	Bayerisches Naturschutzgesetz.

BB	Braun-Blanquet.
BBauG	Bundesbaugesetz v. 23.06.1960 (BGBl. I S. 341), novelliert 1976 und 1979, dann abgelöst durch das BauGB.
BBergG	Bundesberggesetz.
BBodSchG	Gesetz zum Schutz vor schädlichen Bodenveränderungen und zur Sanierung von Altlasten — Bundes-Bodenschutzgesetz vom 17.03.1998 (BGBl. I Nr. 16, S. 502-510).
BDA	Bund Deutscher Architekten.
Bd., Bde.	Band, Bände.
BDLA	Bund Deutscher Landschaftsarchitekten.
bes.	besonders.
BGB	Bürgerliches Gesetzbuch.
BGBl.	Bundesgesetzblatt.
BGG	Behindertengleichstellungsgesetz vom 27. April 2002 [BGBl. I S. 1467, 1468], zuletzt geändert durch Artikel 12 des Gesetzes vom 19.12.2007 [BGBl. I S. 3024].
BGH	Bundesgerichtshof.
BImSchG	Bundes-Immissionsschutzgesetz vom 15.03. 1974 (BGBl. I S. 721, 1193) in der Fassung der Bekanntmachung vom 26.0.2002 [BGBl. I S. 3830], zuletzt geändert durch Artikel 1 des Gesetzes vom 23.10.2007 [BGBl. I S. 2470]).
BImSchV	Bundes-Immisssionsschutzverordnung zur Durchführung des Bundes-Immissionsschutzgesetzes.
biol.	biology; biología; biologie ; Biologie.
BNatSchG	Bundesnaturschutzgesetz vom 20.12.1976, gültig in der neuen Fassung v. 25.03.2002 (BGBl. I 1193), zuletzt geändert durch Gesetz vom 22.12. 2008 (BGBl. I 2986).
[BOL]	Bolivia.
bot.	botany; botánica; botanique ; Botanik.
[BR]	Brasil.
BS	British Standards.
[BW]	Baden-Württemberg.
[BY]	Bayern.
[C]	Cuba/utilizado en Cuba.
[CA]	Centroamérica.
Ca	Calcium
ca.	circa.
c.-à-d.	c'est-à-dire.
CAN	Communauté d'agglomération nouvelle
[CAR]	Caribe.
CC.AA.	Comunidades Autónomas.
C.C.A.G.	Cahier des clauses administratives générales des marchés publics.
C.C.A.P.	Cahier des clauses administratives particulières.

CC.EE.   Comunidad Europea.
C.C.H.   Code de la construction et de l'habitation.
C. civil   Code civil.
C.C.T.P.   Cahier des Clauses Techniques Générales
C.C.T.P.   Cahier des clauses techniques particulières.
Cd   cadmium; cadmio; cadmium ; Cadmium, Kadmium.
[CDN]   Canada, used in Canada; Canada, utilisé au Canada.
CEE   Comunidad Económica Europea; Communauté Économique Européenne.
cf.   refer to; véase; conférez, comparez ; [confer] vergleiche.
C. envir.   Code de l'environnement (France).
[CH]   Switzerland, used in Switzerland; Suiza, utilizado en Suiza; [Confédération Helvétique] Suisse, utilisé en Suisse ; [Confoederatio Helveticae] Schweiz, in der Schweiz gebräuchlich.
[CH-GR]   [Confoederatio Helveticae] Schweiz, Graubünden, in Graubünden gebräuchlich.
chem.   chemistry; química; chimie ; Chemie.
circ.   circulaire.
C. marchés publ. Code des marchés publics.
C. min.   Code minier (France).
C.N.F.F.   Comité national pour le fleurissement de la France.
[CO]   Colombia.
conserv.   nature conservation, conservation of landscapes; preservación de la naturaleza; protection du patrimoine naturel ; Natur- und Landschaftsschutz.
conserv'hist.   conservation of historic monuments [UK]/preservation of historic landmarks [US]; protección del patrimonio histórico; protection du patrimoine historique ; Denkmalschutz.
constr.   building material and construction, bioengineering, landscape practice; materiales de construcción, construcción, ingeniería biotécnica, construcción paisajística; matériaux de construction, bâtiment, génie civil, génie biologique, travaux de jardins et paysagers ; Baustoffe und Baukonstruktionen, Ingenieurbiologie, Landschaftsbau.
contr.   contracting; contratos; contrats de marchés publics et de travaux privés ; Vertragswesen.
COS   coefficient d'occupation des sols
CPCU   Code permanent construction et urbanisme.
CPEN   Code permanent environnement et nuisances.
Cr   chromium; cromo; chrome ; Chrom.
[CR]   Costa Rica.
C. rural   Code rural.
[CS]   Cono Sur (Argentina, Chile, Uruguay, Paraguay).
CSPS   Coordinateur de la sécurité et de la protection de la santé
Cu   copper, cuprum; cobre; cuivre ; Kupfer.
cu km   cubic kilometer.
C. urb.   Code de l'urbanisme.
cvs.   cultivars, varieties; cultivares.
D.   Germany; Alemania; Allemagne ; Deutschland.
dB(A)   decibels A-scale; decibelios A; décibel pondéré A ; Dezibel mit Angabe der Bewertungskurve A.
d.b.h.   diameter breast height
décr. n°   décret n°.

D.F.   Distrito Federal (México D.F.).
dir. n°   directive n°.
DIN   Deutsches Institut für Normung; German Institute for Standardization; Institut Allemand de Normalisation ; Instituto Alemán de Normalización.
DIN 18 915   Vegetationstechnik im Landschaftsbau: Bodenarbeiten (Sept. 1990).
DIN 18 916   Vegetationstechnik im Landschaftsbau: Pflanzen und Pflanzarbeiten (Sept. 1990).
Dir. 2006   Directiva 2006/118/CE del Parlamento Europeo y del Consejo de 12 de diciembre de 2006 relativa a la protección de las aguas subterráneas contra la contaminación y el deterioro.
DOG   Document d'orientations générales.
DOM-TOM   Département d'outre-mer — Territoire d'outre-mer : l'ensemble des terres sous souveraineté française situées hors de la métropole.
D.P.L.G.   diplômé par le gouvernement.
DSchG   Landesdenkmalschutzgesetz.
dt./Dt.   deutsch/das Deutsche.
D.T.U.   Documents techniques unifiés.
[EC]   Ecuador.
ECS   eau chaude sanitaire.
ecol.   ecology; ecología; écologie ; Ökologie.
econ.   economics; economía; économie ; Wirtschaft.
EDF   Électricité de France.
EEC   European Economic Community.
EE.UU.   Estados Unidos de América.
EG   Europäische Gemeinschaft (ab 1994).
e.g.   (exempli gratia) for instance.
[EIRE]   Ireland.
eng.   civil engineering; ingeniería civil; génie civil ; Bauingenieurwesen.
envir.   technical protection of the environment; protección técnica del medio ambiente; génie environnemental ; technischer Umweltschutz.
EPCI   Établissement public de coopération intercommunale
erw.   erweitert.
[Es], Es.   España (castellano)/utilizado en España.
[ES]   El Salvador.
etc.   and so forth; et cetera; et cetera ou et cætera ; (et cetera) und so weiter.
et s.   et suivantes.
[EUS]   Euskadi/ País Vasco; l'Euskadi.
EW   Einwohner.
EWG   Europäische Wirtschaftsgemeinschaft (bis 1993).
f   French; francés; français; Französisch.
F.   France; Francia; France ; Frankreich.
[f]   female; feminino; féminin ; feminin.
FAO   Food and Agriculture Organisation of the United Nations; Organisazión de las Naciones Unidas para la Agricultura y la Alimentación; Organisation des Nations Unies pour l'Alimentation et l'Agriculture ; Organisation der Vereinten Nationen für Landwirtschaft, Fischerei und Forsten.
ff   und folgende.
FFH   Fauna-Flora-Habitat gemäß Richtlinie 92/43/EWG des Rates vom 21.05.1992 zur Erhaltung der natürlichen Lebensräume sowie der wild lebenden Tiere und Pflanzen — Fauna-Flora-

	Habitat-Richtlinie (Abl. L 206 vom 22.07.1992, p. 7 und L 363 vom 20.12.2006).
FH	Fachhochschule.
FLL	Forschungsgesellschaft Landschaftsentwicklung Landschaftsbau e. V. mit Sitz in D-53115 Bonn.
FlurbG	Flurbereinigungsgesetz.
for.	forestry; silvicultura; sylviculture ; Forstwesen.
ForstG	(österreichisches) Forstgesetz v. 03.07.1975, BGBl. 1975/440, i. d. F. BGBl. 1977/231, 1978/142, 1987/576 (Forstgesetznovelle 1987), 1993/257 u. 1993/970.
[fpl]	female plural noun; sustantivo femenino en plural; substantif au féminin pluriel ; feminines Substantiv im Plural.
franz.	französisch.
f. S.	frühere Schreibung.
FStrG	Bundesfernstraßengesetz.
g	German; alemán; allemand ; Deutsch.
[GA]	Galicia (Comunidad Autónoma en el NO del Estado Español).
GaLaBau	Garten- und Landschaftsbau.
game'man.	game management; gestión de la caza; biologie de la faune sauvage ; Wildbiologie.
gard.	garden design; paisajismo, diseño de jardines; architecture des jardins ; Gartenarchitektur.
gard'hist.	history of fine garden design; historia del arte de jardinería, patrimonio de jardinería; histoire de l'art des jardins ; Geschichte der Gartenkunst.
GB	Gran Bretaña.
g.b.h.	girth breast height.
GBl. BW	Gesetzblatt für Baden-Württemberg.
GBO	Grundbuchordnung vom 24.03.1897 [RGBl. I S. 139] i. d. F. der Bekanntmachung v. 05.08. 1935 [RGBl. I S. 1073].
GDF	Gaz de France.
GDR	German Democratic Republic.
GG	Grundgesetz.
ggf.	gegebenenfalls.
Ggs.	Gegensatz.
gem.	gemäß.
GenTG	Gesetz zur Regelung der Gentechnik — Gentechnikgesetz in der Fassung der Bekanntmachung vom 16.12.1993 (BGBl. I S. 2066, zuletzt geändert durch Artikel 1 des Gesetzes vom 01.04.2008 (BGBl. I S. 499).
geo.	geography/geology/geomorphology; ciencias de la tierra (geografía/geología/geomorfología); sciences de la terre (géographie/géologie/géomorphologie) ; Geographie/Geologie/Geomorphologie.
ger./gér.	gerund; gerundio; gérondif ; Gerundium.
GRW	Grundsätze und Richtlinien für Wettbewerbe auf den Gebieten der Raumplanung, des Städtebaues und des Bauwesens vom 09.01.1996 (BAnz. Nr. 64 vom 30.03.1996, p. 3922);
ha	hectare; hectárea; hectare ; Hektar.
[HH]	Freie und Hansestadt Hamburg.
hist.	history; historia; histoire ; Geschichte.
HmbNatSchG	Hamburger Naturschutzgesetz.
HOAI	Verordnung über die Honorare für Architekten- und Ingenieurleistungen (Honorarordnung für Architekten und Ingenieure – HOAI); ursprüng-

	liche Fassung vom 17.09.1976 (BGBl. I S. 2805, 3616); Neubekanntmachung vom 04.03. 1991 (BGBl. I S. 533); letzte Neufassung (sechste Novellierung) vom 11.08.2009 (BGBl. I S. 2732), Inkrafttreten am 18.10.2009.
[HON]	Honduras.
hort.	horticulture; horticultura; horticulture ; Gartenbau.
HRLA	Österreichische Honorarrichtlinien für Landschaftsarchitekten; Ende 2006 wurden die HRLA und die Honorarleitlinie der Bundeskammer der Architekten und Ingenieurkonsulenten (HOA) gem. EU-Richtlinien und novelliertem Kartellgesetz von 2005 wegen Wettbewerbsbeschränkung aufgehoben.
H-S-EinbrG	Gesetz über das Verbot der Einbringung von Abfällen und anderen Stoffen und Gegenständen in die Hohe See vom 25.08.1998 [Hohe-See-Einbringungsgesetz] (BGBl. I Nr. 57, S. 2455).
[HS]	Hessen.
hunt.	hunting; caza; chasse ; Jagdwesen.
hydr.	hydrology; hidrología; hydrologie ; Hydrologie.
i.	in, im.
i. A.	im Allgemeinen.
i. d. F.	in der Fassung.
i. d.	in der/die/das.
i. d. R.	in der Regel.
i.e.	[id est] that is.
incl.	inclusive.
INRA	Institut National de la Recherche Agronomique
INSEE	Institut national de la statistique et des études économiques.
i. S. v.	im Sinne von.
ital.	italienisch.
IUCN	International Union for Conservation of Nature and Natural Resources.
i. V. m.	in Verbindung mit.
Jh./Jhs.	Jahrhundert/Jahrhunderts.
K	Kalium.
KG	Gesetz zum Schutz deutschen Kulturgutes gegen Abwanderung [Kulturgüterschutzgesetz].
km²	square kilometer; kilómetro cuadrado; kilomètre carré ; Quadratkilometer.
KrW-/AbfG	Gesetz zur Förderung der Kreislaufwirtschaft und Sicherung der umweltverträglichen Beseitigung von Abfällen (Kreislaufwirtschafts- und Abfallgesetz — KrW-/AbfG) vom 27.09.1994 (BGBl. I S. 2705), zuletzt geändert durch Artikel 10 des Gesetzes vom 3.05.2000 (BGBl. I S. 632).
[L]	Luxembourg, utilisé au Luxembourg.
LAbfG-BW	Landesabfallgesetz von Baden-Württemberg vom 08.01.1990.
land'man.	landscape management; gestión y protección del paisaje; gestion des milieux naturels ; Landschaftspflege.
landsc.	landscape planning; planificación del paisaje; planification des paysages ; Landschaftsplanung.
LBO	Landesbauordnung.
LBO-BW	Landesbauordnung von Baden-Württemberg vom 08.08.1995, zuletzt geändert am 15.12. 1997.

LCAP	Ley 3/1995, de 18 de mayo, de Contratos de las Administraciones Públicas [Es].
leg.	legislation, legal ordinances; legislación, normas legales; législation, normes et obligations règlementaires; Gesetzgebung, Rechtsvorschriften.
LEPro-NRW	Gesetz zur Landesentwicklung Landesentwicklungsprogramm — LEPro von Nordrhein-Westfalen vom 05.10.1089, zuletzt geändert 19.06. 2007.
Ley 16/1985	Ley 16/1985, de 25 de junio, del Patrimonio Histórico Español [Es].
Ley 29/1985	Ley 29/1985, de 2 de agosto, de Aguas [Es].
Ley 22/1988	Ley 22/1988, de 28 de julio, de Costas [Es].
Ley 25/1988	Ley 25/1988, de 29 de julio, de Carreteras.
Ley 4/1989	Ley 4/1989, de 27 de marzo, de Conservación de los Espacios Naturales y de la Flora y la Fauna Silvestres [Es].
Ley 15/1994	Ley 15/1994, de 3 de junio, por la que se establece el régimen jurídico de la utilización confinada, liberación voluntaria y comercialización de organismos modificados genéticamente, a fin de prevenir los riesgos para la salud humana y para el medio ambiente.
Ley 6/1998	Ley 6/1998, de 13 de abril, sobre Régimen del Suelo y Valoraciones [Es].
Ley 10/1998	Ley 10/1998, de 21 de abril, de Residuos
Ley 37/2003	Ley 37/2003, de 17 de noviembre, del Ruido
Ley 62/2003	Ley 62/2003, de 30 de diciembre, de medidas fiscales, administrativas y del orden social que incluye, en su artículo 29, la Modificación del texto refundido de la Ley de Aguas, aprobado por Real Decreto Legislativo 1/2001, de 20 de julio, por la que se incorpora al derecho español la Directiva 2000/60/CE, por la que se establece un marco comunitario de actuación en el ámbito de la política de aguas.
Ley 1/2005	Ley 1/2005, de 9 de marzo, por la que se regula el régimen del comercio de derechos de emisión de gases de efecto invernadero.
Ley 34/2007	Ley 34/2007, de 15 de noviembre, de calidad del aire y protección de la atmósfera.
Ley 42/2007	Ley 42/2007, de 13 de diciembre, del Patrimonio Natural y de la Biodiversidad.
Ley 20.293	Ley N° 20.293 Protege a los cetáceos e introduce modificaciones a la Ley n° 18.882 General de Pesca y Acuicultura, promulgada 14.10. 2008, en vigor desde 25.10.2008; www.bcn.cl
LG-NW	Gesetz zur Sicherung des Naturhaushalts und zur Entwicklung der Landschaft (Landschaftsgesetz von Nordrhein-Westfalen vom 18.02. 1975).
limn.	limnology; limnología; limnologie ; Limnologie.
lit.	letter; letra; lettre ; Buchstabe.
LJagdG	Landesjagdgesetz.
loc	parlance; locución; locution et expression courante ; Ausdrucksweise, Redewendung.
LS	Ley del Suelo; RD 1346/1076, de 9 de abril, por el que se aprueba el texto refundido de la Ley sobre Régimen del Suelo y Ordenación Urbana [Es].
LSG	Landschaftsschutzgebiet.
LUPPAA	Land Use Policy and Planning Assistance Act of June 7, 1973.

LWaldG-BW	Waldgesetz für Baden-Württemberg (Landeswaldgesetz — LWaldG) vom 31.08.1995; GBl.-BW 1995 Nr. 27; p. 685-710).
[m]	male; masculino; masculin ; maskulin.
m	metre/meter; metro; mètre ; Meter.
m²	square meter; metro cuadrado; mètre carré ; Quadratmeter.
m³	cubic meter; metro cúbico; mètre cube ; Kubikmeter.
m³/s	cubic meter per second; metro cúbico por segundo; mètre cube par seconde; Kubikmeter pro Sekunde.
MAB	Man and Biosphere, UNESCO.
Mg	magnesio; Magnesium; magnésium ; Magnesium.
met.	meteorology; meteorología; météorologie ; Meteorologie.
[MEX]	México.
MV	Mecklenburg-Vorpommern
min.	mineral working; recursos minerales/explotación minera ; exploitation des substances minérales ; Abbau, Bergbau.
modif.	modifiziert.
mm	millimeter; milímetro; millimètre ; Millimeter.
[mpl]	male plural noun; sustantivo masculino en plural; substantif au masculin pluriel ; maskulines Substantiv im Plural.
[n]	*English section* noun; *German section* neuter; Neutrum, sächlich.
Na	Natrium; sodio; sodium ; Natrium.
nat'res.	natural resources management; gestión de los recursos naturales; gestion des ressources naturelles ; Ressourcenplanung.
NCSS	National Cooperative Soil Survey in USA.
ndt.	niederdeutsch.
NGF	cotation NGF (Nivellement Général de la France).
N.F.	Norme française publiée par l'Association française de normalisation (AFNOR).
NHG	Natur- und Heimatschutzgesetz der Schweiz.
[NL]	les Pays-Bas.
[NIC]	Nicaragua.
NN	(Höhe über) Normal Null.
n°	numéro ; número.
nordd.	norddeutsch.
[npl]	*English section* noun plural; *German section* sächliches Substantiv im Plural.
Nr.	Nummer.
[NS]	Niedersachsen.
NSG	Naturschutzgebiet.
num.	number.
[NW]	Nordrhein-Westfalen.
[NZ]	New Zealand; Nueva Zelanda; la Nouvelle-Zélande ; Neuseeland.
o.	oder
OB	Oberbegriff.
obs.	obsolete; obsoleto; obsolète ; obsolet.
ocean.	oceanography; oceanografía; océanographie ; Meereskunde.
ÖGLA	Österreichische Gesellschaft für Landschaftsplanung und Landschaftsarchitektur.
o. J.	ohne Jahr (bei Literaturangaben).
ONF	Office National des Forêts.

OPC	Ordonnancement, Pilotage et Coordination.
o. Pl.	ohne Plural.
opp.	opposite; contrario; terme opposé ; Gegenteil.
o. t. st.	of the stands.
o.v.; O. V./o. V.	orthographic variant; orthografische Variante.
p.	page; página; page ; Seite.
[PA]	Panamá.
PADD	Projet d'aménagement et de développement durable.
Pb	lead; plomo; plomb ; Blei.
PBG	Planungs- und Baugesetz des Kantons Zürich: Gesetz über die Raumplanung und das öffentliche Baurecht vom 07.09.1975 [LS 700.1]); nachgeführt bis 1. April 2002.
PDU	Plan de déplacements urbains.
[PE]	Perú.
pedol.	pedology; edafología; pédologie ; Bodenkunde.
p. ej.	por ejemplo.
p. ex.	par exemple.
PflSchG	Gesetz zum Schutze der Kulturpflanzen (Pflanzenschutzgesetz v. 14.05.1998, BGBl. 1998 I Nr. 28, S. 972, zuletzt geändert am 05.03.2008, BGBl I Nr. 8, S. 284).
phys.	physics; física; physique ; Physik.
phyt.	vegetation ecology; fitosociología; phytosociologie ; Pflanzensoziologie.
phytopath.	phytopathology; fitopatología; phytopathologie; Phytopathologie.
plan.	planning science and activities, regional policy; ciencias y actividades de planificación, ordenación del territorio; science et activités de planification, aménagement du territoire et planification spaciale ; Planungswissenschaft, Planungstätigkeit, Raumordnung.
plant.	planting design; utilización de plantas; utilisation des végétaux ; Pflanzenverwendung.
PlanzV 90	Verordnung über die Ausarbeitung der Bauleitpläne und die Darstellung des Planinhalts (Planzeichenverordnung v. 18.12.1990, BGBl. 1991 I, S. 58).
PLH	Programme local de l'habitat.
PLU	Plan local d'urbanisme.
pol.	politics; política; politique ; Politik.
POS	Plan d'occupation des sols.
[pp]	past participle or adjectively used as a past participle; participio perfecto adjetivado; participe passé, participe passé employé comme adjectif ; Partizip Perfekt oder adjektivisch gebrauchtes Partizip Perfekt.
pp.	pages; Seiten.
[ppr]	adjectively used as a present participle; adjektivisch gebrauchtes Partizip Präsenz.
[ppr/adj]	participe présent, adjectif verbal ; adjektivisch gebrauchtes Partizip Präsenz.
PPTOA	Programme ou projet de travaux, d'ouvrages ou d'aménagements
[prep]	predicado nominativo.
[prép]	préposition, locution prépositive.
prof.	professional body of landscape architects, architects and engineers; cuerpos profesionales de arquitectos (paisajistas) e ingenieros; profession des architectes (paysagistes) et ingénieurs ;

	Berufsstand der (Landschafts)architekten und Ingenieure.
pt.	part.
PTT	Postes, Télégraphes, Téléphones remplacé par P. et T. Postes et Télécommunications.
qqn.	quelqu'un.
[RA]	Argentina.
[RCH]	Chile.
RD	Real Decreto [Es].
RD 2159/1978	RD 2159/1978, de 23 de junio, Reglamento de Planeamiento [Es].
RD 2568/1986	Reglamento de Organización, Funcionamiento y Régimen Jurídico de las Entidades Locales, de 28 de noviembre [Es].
RD 1131/1988	Real Decreto 1131/1988, de 30 de septiembre, por el que se aprueba el Reglamento para la Ejecución del Real Decreto Legislativo de Evaluación de Impacto Ambiental
RD 289/2003	Real Decreto 289/2003, de 7 de marzo, sobre comercialización de los materiales forestales de reproducción [Es].
RD 1866/2004	Real Decreto 1866/2004, de 6 de septiembre, por el que se aprueba el Plan Nacional de Asignación de Derechos de Emisión 2005-2007.
RD 314/2006	Real Decreto 314/2006, de 17 de marzo, por el que se aprueba el Código Técnico de la Edificación
RD 125/2007	Real Decreto 125/2007, de 2 de febrero, por el que se fija el ámbito territorial de las demarcaciones hidrográficas.
RDL 1302/1986	Real Decreto Legislativo 1302/1986, de 28 de junio, de Evaluación de Impacto Ambiental [Es].
RDL 1/2001	Real Decreto Legislativo 1/2001, de 20 de julio, por el que se aprueba el texto refundido de la Ley de Aguas.
recr.	facilities for leisure activities, sports; gestión turística, equipamiento de recreo, deporte; aménagement touristique, équipement d'activités de loisirs et de récréation ; Erholungsplanung, Einrichtungen für Freizeit und Erholung, Sport; recreation planning.
réf.	référence.
refl	verbo reflexivo.
Reg.-Bez.	Regierungsbezirk.
rem'sens.	remote sensing; teledetección; télédétection ; Fernerkundung.
resp.	respectively; respectivamente; respectivement ; respektive.
RFA	République fédérale d'Allemagne.
RL 75/440	Richtlinie 75/440/EWG des Rates v. 16. 06. 1975 über die Qualitätsanforderungen an Oberflächenwasser für die Trinkwassergewinnung in den Mitgliedstaaten (ABl. L 194, 25.07.1975, p. 34); zuletzt geändert am 23.12.1991 (ABl. L 377, 31.12.1991, p. 48).
RL 76/160	Richtlinie 76/160/EWG des Rates v. 08.12. 1975 über die Qualität der Badegewässer.
RL 80/778	Richtlinie 80/778/EWG des Rates v. 15. 07. 1980 über die Qualität von Wasser für den menschlichen Gebrauch (ABl. L 229, 30.08. 1980, p. 11); zuletzt geändert durch RL 91/692/

	EWG v. 23.12.1991 (ABl. L 377, 31.12. 1991, p. 48).
RL 84/360	Richtlinie 84/360/EWG des Rates v. 28.06. 1984 zur Bekämpfung der Luftverunreinigung durch Industrieanlagen.
RL 85/337	Richtlinie des Rates 85/337/EWG vom 27.06. 1985 über die Umweltverträglichkeitsprüfung bei bestimmten öffentlichen und privaten Projekten (ABl. L 175, 05.07.1985, p. 40); geändert durch RL 97/11/EG v. 03.03.1997 (ABl. L 73, 14.03.1997, p. 5).
RL 92/43	Richtlinie 92/43/EWG des Rates v. 21.05.1992 zur Erhaltung der natürlichen Lebensräume sowie der wild lebenden Tiere und Pflanzen (ABl. L 206, 22.07.1992, p. 7).
RL 2000/60	Richtlinie 2000/60/EG des Europäischen Parlamentes und des Rates vom 23.10.2000 zur Schaffung eines Ordnungsrahmens für Maßnahmen der Gemeinsachaft im Bereich der Wasserpolitik — Wasserrahmenrichtlinie (WRRL).
RL 2001/18	Richtlinie 2001/18/EG des Europäischen Parlamentes und des Rates über die absichtliche Freisetzung genetisch veränderter Organismen in die Umwelt und zur Aufhebung der RL 90/220/ EWG des Rates vom 12.03.2001 („Freisetzungsrichtlinie").
[RM]	Madagascar.
ROG	Raumordnungsgesetz v. 08.04.1965 (BGBl. I S. 306), geändert am 18.08.1997 (BGBl. I S. 2081), zuletzt novelliert durch das Gesetz zur Neufassung des Raumordnungsgesetzes und zur Änderung anderer Vorschriften (GeROG) v. 22.12.2008 (BGBl. I Nr. 65 S. 2986).
[ROU]	Uruguay.
[RP]	Rheinland-Pfalz.
RSM	Regelsaatgutmischung
s	Spanish; castellano/español; espagnol ; Spanisch.
S.	Seite.
S.A.G.E.	Schéma d'aménagement et de gestion des eaux.
SAN	Syndicat d'agglomération nouvelle
SCA	Syndicat communautaire d'aménagement
SCI	Site of community importance
SCOT	Schéma de cohérence territoriale
[SCOT]	Scotland.
S.D.A.G.E.	Schéma directeur d'aménagement et de gestion des eaux.
S.D.A.U.	Schéma directeur d'aménagement et d'urbanisme.
[SH]	Schleswig-Holstein.
SI	Statutory Instrument [UK].
[SL]	Saarland.
sm	Seemeile.
sociol.	sociology; sociología; sociologie ; Soziologie.
sog.	so genannt.
spätlat.	spätlateinisch.
spec.	species; especie; espèce ; Spezies, Art.
spp.	subspecies; subespecie; sous-espèce ; Subspezies, Unterart.
SRU	Loi n° 2000-1208 du 13 décembre 2000 relative à la solidarité et au renouvellement urbains.
ss	and following pages; seguentes paginas.
stat.	statics and dynamics; estática; statique ; Statik, Tragwerksplanung.
StBauFG	Städtebauförderungsgesetz.
StGB	Strafgesetzbuch.
StLB	Standardleistungsbuch.
StU	Stammumfang.
südd.	süddeutsch.
surv.	surveying, cartography; geodesía, cartografía; géodésie, cartographie ; Vermessungswesen, Kartografie.
syn.	synonym(s); sinónimo(s); synonyme(s) ; Synonym, Synonyma.
tb	también.
TCPA	Town and Country Planning Act.
TFH	Technische Fachhochschule.
trans.	traffic and transportation; tráfico y transportes; transport ; Verkehrswesen.
TrinkwV	Verordnung über Trinkwasser und über Wasser für Lebensmittelbetriebe (Trinkwasserverordnung — TrinkwV) i. d. F. v. 05.12.1990 (BGBl. I S. 2612).
TU	Technische Universität.
Tx	Tüxen.
u.	und.
ü.	über.
u. a.	unter anderem.
UB, UBe	Unterbegriff, ...e.
überarb.	überarbeitet.
UE	Unión Europea.
ugs.	umgangssprachlich.
UICN	Unión Internacional para la Conservación de la Naturaleza; Union internationale pour la conservation de la nature ; ▶IUCN.
UK/U.K.	United Kingdom; Reino Unido; Royaume-Uni ; Vereinigtes Königreich.
UNESCO	United Nations Educational, Scientific and Cultural Organization; Organización de la Naciones Unidades para la Educación, la Ciencia y la Cultura; Organisation des Nations Unies pour l'éducation, la science et la culture ; Organisation der Vereinten Nationen für Erziehung, Wissenschaft und Kultur.
urb.	urban planning; urbanismo; urbanisme ; Stadtplanung.
US/U.S.	United States of America; Estados Unidos de América; États Unis d'Amérique ; Vereinigte Staaten von Amerika.
USDA	United States Department of Agriculture.
UVPG	Gesetz über die Umweltverträglichkeitsprüfung vom 12.02.1990 (BGBl. I S. 205), zuletzt geändert durch Artikel 7 des Gesetzes zur Änderung des Baugesetzbuches und zur Neuregelung des Rechts der Raumordnung (Bau- und Raumordnungsgesetz 1998 — BauROG) (BGBl. I S. 2111) vom 18.08.1997; Gesetz zur Umsetzung der Richtlinie des Rates vom 27.06.1985 über die Umweltverträglichkeitsprüfung bei bestimmten öffentlichen und privaten Projekten (85/337/EWG) vom 12.02. 1990.
v.	von, vom.
v. a.	vor allem.
[vb]	verb; verbo; verbe ; Verb(um).

[vb/intr]	intransitive verb; verbo intransitivo; verbe intransitif ; intransitives Verb(um).
[vb/refl]	reflexive verb; verbo reflexivo; verbe réfléchi ; reflexives Verb(um).
[vb/tr]	transitive verb; verbo transitivo; verbe transitif ; transitives Verb(um).
veröff.	veröffentlicht.
v.o.	variedad ortográfica; variété orthographique.
VO	Verordnung.
VOB	Vergabe- und Vertragsordnung für Bauleistungen; (obs. Verdingungsordnung für Bauleistungen).
VOF	Verdingungsordnung für freiberufliche Leistungen.
VOL	Verdingungsordnung für Leistungen.
Vt	Vermont (USA).
VwV	Verwaltungsvorschrift.
VwVfG	Verwaltungsverfahrensgesetz.
WaStrG	Bundeswasserstraßengesetz von 02.04.1968 i. d. F. der Bekanntmachung vom 23.05.2007 (BGBl. I 962; 2008 I 1980), zuletzt geändert am 31.07.2009 (BGBl. I 2585).
wat'man.	water management, river engineering measures; gestión de recursos hídricos, ingeniería hidráulica de ríos y arroyos; gestion de la ressource en eau, travaux d'aménagement des cours d'eau, génie hydraulique ; Wasserwirtschaft, Gewässerausbau.
WertV	Verordnung über Grundsätze für die Ermittlung der Verkehrswerte von Grundstücken (Wertermittlungsverordnung) i. d. F. vom 06.12.1988 (BGBl. I S. 2209), zuletzt geändert durch Art. 3 des Gesetzes zur Änderung des BauGB und zur Neuregelung des Rechts der Raumordnung (BauROG) v. 18.08.1997 (BGBl I S. 2110).
WFK	Waldfunktionskartierung.
WG-BW	Wassergesetz Baden-Württemberg.
WHG	Gesetz zur Ordnung des Wasserhaushalts (Wasserhaushaltsgesetz) vom 27.07.1957 (BGBl. I S. 1110, 1386); Neubekanntmachung vom 19.08.2002 (BGBl. I S. 3245), zuletzt geändert am 31.07.2009 (BGBl. I S. 2585).
WTO	World Trade Organization; Organización Mundial del Comercio (OMC); Organisation mondiale du commerce ; Welthandelsorganisation: 1995 gegründete Sonderorganisation der UNO zur Förderung und Überwachung des Welthandels.
yr	year.
[YV]	Venezuela.
[ZA]	South Africa.
z. B.	zum Beispiel.
zool.	zoology; zoología; zoologie ; Zoologie.
Zn	zinc; cinc, zinc; zinc ; Zink.
Z.N.I.E.F.F.	zone naturelle d'intérêt écologique, faunistique et floristique.
Z.P.P.A.U.P.	zone de protection du patrimoine architectural, urbain et paysager.
[ZRE]	Zaïre.
z. T.	zum Teil.
ZTVLa-StB	Zusätzliche Technische Vertragsbedingungen und Richtlinien für Landschaftsbauarbeiten im Straßenbau; 1999.
ZVG	Zentralverband Gartenbau zu Bonn.
zw.	zwischen.
±	approximate translation, not exactly synonymous; partially overlapping concept; traducción aproximada, no sinónimo exacto, término que se superpone en parte; traduction approximative, synonyme partiel ; ungefähre Übersetzung, nicht exakt synonym, teilweise überschneidender Begriff.
≠	non-existent in the language concerned or was not found in literature—proposed translation; no existe en el idioma correspondiente o no se encontró en la bibliografía —propuesta de traducción; n'existe pas dans la langue concernée ou na pas été trouvé dans la littérature, proposition de traduction ; existiert nicht in der betreffenden Sprache oder wurde in der Literatur nicht gefunden — Übersetzungsvorschlag.
►	see, cross-reference; véase, referencia a; renvoi à, référence à ; siehe, Verweis.
*►	see, cross-reference, no synonymous term!, indicates a definition or explanation in which the term occurs; véase, referencia a, ¡no es término sinónimo!; se refiere a una definición o una explicación en la que aparece el término; renvoi à, référence à ; n'est pas synonyme de mais renvoie aux définitions et explications dans lesquelles le terme est mentionné ; siehe, Verweis — kein Synonym!; deutet auf eine Definition oder Erläuterung hin, in der dieser Begriff vorkommt).
>	greater than; mayor de; supérieur à ; größer als.
<	less than; menor de; inférieur à ; kleiner als.
†	died; fallecido; décédé ; gestorben.

A

Aapa mire [n] *geo.* ►string bog.

Aapa moor [n] *geo.* ►string bog.

1 abandoned area [n] *plan. urb.* (Derelict or cleared areas of disused industrial plant, parts of a city now uninhabited, because of population declaine or emigration, as well as areas without use, because of the lack of subsidies. Such areas are characteristic of cities all over the world: though they have the potential for the creation of new spaces for urban renewal or recreation; a completely desolate **a. a.** signifies decline, final surrender of its use and the irrevocable loss of social capital and infrastructure; ►derelict land, ►industrial area, ►abandoned land, ►orphan contaminated site); *s* **terreno** [m] **abandonado** (CCEE 1990; en áreas urbanas, superficies de antiguas fábricas cerradas [►ruina industrial] o de zonas de vivienda desocupadas por la reducción de la población o emigración de la misma, que se encuentran generalmente en zonas industriales en declive, así como en áreas rurales por abandono de explotación agrícola [►tierra abandonada]. Estos **t. a.** son un fenómeno extendido en muchas ciudades de los países industrializados y suponen —por un lado— un potencial para la creación de nuevos espacios para el desarrollo urbano y —por otro— si no se recuperan pueden significar abandono definitivo y con ello pérdida irrevocable de capital social e infraestructura. En Es. un ejemplo emblemático de recuperación de **t. a.** y renovación urbana se ha llevado a cabo a partir de la última década del siglo XX en Bilbao Metropolitano); *f* **friche** [f] **(1)** (Espace et/ou bâtiment abandonné temporairement ou définitivement dans l'attente d'une nouvelle occupation ; il peut s'agir de sites d'une activité industrielle et économique désaffectée [►friche industrielle], portuaire, de défense militaire, de transport, de service, d'immeubles d'habitation abandonnés [friche urbaine] ; F., les plus fortes densités de friches industrielles se trouvent en Lorraine, Nord-Pas-de-Calais et région parisienne ; les friches industrielles à cause de la contamination de l'environnement, réelle ou perçue, rendent une expansion ou une reconversion difficile et restent souvent à l'état de friche ; en France, l'ADEME a pris en charge quelques friches orphelines dont la pollution représentait une menace urgente ; ►site pollué orphelin ; **D.**, le phénomène d'abandon total de villages, d'exploitations au point de retourner à l'état sauvage lors de la période de désertification au XIVème siècle est communément désigné par le nom de Wüstung [désolation] ; ►friche sociale) ; *g* **Brache** [f] **(1)** (Flächen stillgelegter Industriebetriebe oder abgeräumte Industrieanlagen, nicht mehr bewohnte Stadtteile, weil die Bevölkerung abnimmt oder abwandert sowie nutzlos gewordene Landstriche, deren Bewirtschaftung durch Wegfall von Subventionen eingestellt wurde. Diese Flächen, die das Erscheinungsbild vieler Städte, besonders im Osten Deutschlands, immer stärker prägen, rühren vom Paradigmenwechsel im Denken und Planen von Expansion auf Schrumpfung. Mit den **B.n** können neue Räume geschaffen werden. Dabei ist im Deutschen zwischen **Brache** und **Wüstung** zu unterscheiden. Die **B.** vereint in sich den Aspekt der Erholung und des Neuen, während Wüstung das endgültige Aufgeben der Nutzung und die Herabstufung von Nutzungsdichte bedeutet, bei der soziales Kapital wie gemeinsam aufgebrachte Infrastruktur unwiederbringlich verloren gehen; cf. FRE 2003; ►Industriebrache, ►herrenloser Altstandort, ►Sozialbrache).

abandoned farmland [n] *sociol.* ►abandoned land.

2 abandoned field [n] *agr.* (Area no longer used for agriculture; ►abandoned area, ►fallow); *s* **parcela** [f] **abandonada** (Campo agrícola que ya no se cultiva; ►terreno abandonado, ►barbecho); *f* **terre** [f] **abandonnée** (Terre non cultivée liée à la déprise agricole, la friche sociale, la friche de spéculation, etc. ; ►friche 2, ►jachère) ; *syn.* terrain [m] en friche ; *g* **aufgelassene Fläche** [f] (Nicht mehr bewirtschaftete landwirtschaftliche Fläche; ►Brache 1, ►Brache 2).

abandoned grassland [n] *agr.* ►abandoned pasture.

abandoned industrial site [n] *plan. urb.* ►derelict land.

3 abandoned land [n] *sociol.* (**1.** *agr.* Agricultural land which has been allowed to go out of cultivation for social or economic reasons; ►abandonment of farmland, ►fallow. **2.** *urb.* Land in a populated area, which has no apparent function; e.g. rest areas of designated open space, that have no real use); *syn. to 1.* abandoned farmland [n]; *s 1* **tierra** [f] **abandonada** (**1.** *agr.* Tierras agrícolas dejadas sin cultivar por cesación de explotación debido a causas económicas y sociales que llevan al éxodo rural; ►abandono de tierras agrícolas, ►barbecho. **2.** *urb.* En zona urbana, porción de terreno sin aparente función, p. ej. resto de espacio libre que por falta de equipamiento o de diseño adecuado no puede cumplir ninguna función); *s 2* **barbecho** [m] **urbano** (Antiguas tierras agrícolas situadas en la zona rururbana abandonadas por sus propietarios que esperan que el precio se eleve debido al cambio de uso del suelo (industrial o residencial); DGA 1986); *syn.* barbecho [m] social; *f* **friche** [f] **sociale** (**1.** Terrains situés dans les campagnes urbanisées, momentanément non travaillés à la suite d'un héritage, d'un changement de propriétaire, d'un morcellement successoral, de l'exode rural, d'un changement de profession de l'exploitant ; MG 1993, 226 ; ►mise en friche de terres agricoles, ►jachère. **2.** *urb.* Espace délaissé en milieu urbain n'ayant pas de fonction apparente, p. ex. une surface restante d'un espace libre et auquel il est difficile d'attribuer une quelconque utilisation) ; *g* **Sozialbrache** [f] (**1.** Aus gesellschaftlichen und wirtschaftlichen Gründen längerfristig nicht genutzte landwirtschaftliche Fläche; ►Auflassung von landwirtschaftlichen Flächen, ►Brache. **2.** *urb.* Flächen im Siedlungsbereich, die mit keiner erkennbaren Funktion versehen sind, z. B. öffentliche Freiflächen oder halböffentliche Flächen im Wohnumfeld von Großsiedlungen, die auf Grund mangelnder Gestaltung und fehlender Nutzungsangebote für Bewohner nicht zugänglich sind oder von diesen nicht angenommen werden sowie übrig gebliebene Freiräume als Restflächen, die objektiv schwer zu funktionalisieren sind; cf. HERLYN 1992).

abandoned land register [n] [US] *adm. urb.* ►derelict land register.

4 abandoned pasture [n] *agr.* (Pasture that has lain fallow and gradually allowed to return to natural grassland or woodland; ►grassland 1); *syn.* fallow grassland [n], abandoned grassland [n], reverted pasture [n]; *s* **herbazal** [m] (Comunidad herbácea que sigue al abandono agrícola y a la intervención degradadora del hombre; ►prado); *f* **prairie** [f] **abandonnée** (►prairie) ; *syn.* prairie [f] délaissée ; *g* **Grünlandbrache** [f] (brachgefallene/brachliegende Grünlandflächen; ►Grünland).

abandoned property register [n] [US] *adm. urb.* ►derelict land register.

5 abandoned quarry [n] *min. landsc.* (Disused quarry, which is no longer worked for the extraction of stone); *syn.* old quarry [n] [also US] (TGG 1984, 186); *s* **cantera** [f] **abandonada**; *f* **carrière** [f] **désaffectée** *syn.* carrière [f] abandonnée ; *g* **aufgelassener Steinbruch** [m] (Stillgelegter Steinbruch, in dem keine Steine mehr gewonnen werden).

K.-J. Evert (ed.), *Encyclopedic Dictionary of Landscape and Urban Planning*, DOI 10.1007/978-3-540-76435-9,
© Springer-Verlag Berlin Heidelberg 2010

A

abandonment [n] *agr.* ►agricultural land abandonment; *econ. envir.* ►industrial plant abandonment; *adm. leg. urb.* ►street abandonment [US]/extinguishment of rights of way [UK].

6 abandonment [n] **of farmland** *agr.* (Ceasing to cultivate agricultural land for economic reasons and/or disinterest in farming; ►abandoned land); *s* **abandono [m] de tierras agrícolas** (en áreas marginales; ►tierra abandonada); *syn.* cesación [f] de explotaciones agrícolas; *f* **mise [f] en friche de terres agricoles** (Terrain foncier improductif laissé à l'abandon ; ►friche sociale) ; *g* **Auflassung [f] von landwirtschaftlichen Flächen** (Aufgabe einer weiteren Bewirtschaftung von landwirtschaftlichen Flächen wegen Unrentabilität, mangelnden Interesses oder Veränderungen im Sozialgefüge der ländlichen Bevölkerung [Strukturwandel]; ►Sozialbrache); *syn.* Nutzungsaufgabe [f] landwirtschaftlicher Flächen.

abatement [n] *envir.* ►noise abatement.

abatement [n]**, flood** *wat'man.* ►flood control.

abatement zone [n] **[UK], noise** *leg. plan.* ►noise zone [US].

abbey close [n] *gard'hist.* ►cloister garden.

abiotic ecological factors [npl] *ecol.* ►ecological factors.

7 abiotic site factor [n] *ecol.* (Physical parameter relevant in determining a location, e.g. data concerning climate [light, temperature, humidity, wind], hydrological values [►measured water level, ►watertable, etc.] or soil condition [►particle size distribution and chemical composition, pH value, etc.]); *s* **factor [m] mesológico abiótico** (Parámetro físico o químico relevante para una determinada estación, p. ej. datos concernientes al clima [luminosidad, temperatura, humedad, vientos], valores hidrológicos [►nivel del agua 2, ►nivel freático] o condiciones edáficas [composición química del sustrato, pH, ►granulometría, etc.]); *f* **facteur [m] stationnel abiotique** (Paramètres physiques, p. ex. facteurs climatiques [lumière, température, humidité de l'air, vent, etc.], facteurs hydrologiques [►niveau d'eau à l'échelle des eaux, ►niveau de la nappe phréatique, etc.] ou facteurs pédologiques [►spectre granulométrique, composition chimique du substrat, valeur pH, etc.]) ; *g* **abiotischer Standortfaktor [m]** (Parameter für einen unbelebten Standort, z. B. Angaben über Klima [Licht, Temperatur, Luftfeuchte, Wind], hydrologische Werte [►Wasserstandshöhe, ►Grundwasserspiegel etc.] oder Bodenbeschaffenheit [►Kornverteilung und chemische Zusammensetzung des Substrates, pH-Wert etc.]).

above-grade masonry [n] *constr.* ►above-ground masonry.

8 above-ground masonry [n] *constr.* (Masonry construction which is built above the finish site grade); *syn.* above-grade masonry [n]; *s* **muro [m] en elevación** (Obra de mampostería sobre el nivel de la superficie); *f* **maçonnerie [f] en élévation** (Ouvrage de maçonnerie située au-dessus du niveau du sol par opposition au mur de soubassement) ; *g* **aufgehendes Mauerwerk [n]** (Mauer[n] oberhalb der Bodenoberkante/oberhalb des Mauersockels sowie die ein Gebäude aussteifende Innenwände).

absorbing power [n] *bot.* ►root absorbing power.

9 absorbing root [n] *bot.* (Final branching of a non-woody root, short in length and covered with ►root hairs, which absorb nutrients and moisture; final part of a ►fibrous root); *syn.* active root [n], feeder root [n]; *s* **raicilla [f] absorbente** (Raíz de último orden, provista de ►pelos radicales; DB 1985; ►raíz fibrosa); *syn.* radícula [f] absorbente, raicilla [f] chupadora, raicilla [f] chupona; *f* **jeune racine [f]** (Dernière ramification de la racine dont l'extrémité est pourvue d'une coiffe et d'une zone pilifère ; ►poil absorbant, ►radicelle) ; *syn.* racine [f] secondaire ; *g* **Saugwurzel [f] (1)** (Letzte, meist kurze, mit ►Wurzelhaaren besetzte Verzweigung der Bodenwurzel, die die Hauptmenge des

Bodenwassers mit den darin gelösten Nährsalzen aufnimmt; Endteil der ►Faserwurzel).

absorbing zone [n] *bot.* ►absorption zone.

absorption [n] *limn. pedol.* ►nutrient absorption.

10 absorption zone [n] *bot.* (**1.** Proximal, unsuberized portion of a root which has not yet developed xylem elements, and which is characterized by ►root hairs. **2.** Main zone of a root for water and nutrient uptake; cf. RRST 1983; ►absorbing root); *syn.* absorbing zone [n], root hair zone [n]; *s* **zona [f] pilífera** (Trozo de la raíz caracterizado por la presencia de ►pelos radicales que se encuentra a pocos milímetros del ápice de la misma; ►raicilla absorbente); *syn.* región [f] pilífera, zona [f] de absorción; *f* **zone [f] pilifère** (LA 1981, 7 ; zone de l'appareil racinaire garni de très nombreux ►poils absorbants, située à quelques millimètres de l'extrémité de la racine ; ►radicelle) ; *syn.* rhizoderme [m] ; *g* **Wurzelhaarzone [f]** (Durch ►Wurzelhaare gekennzeichneter, nicht kutinisierter Teil wenige Millimeter hinter der Wurzelspitze; ►Saugwurzel 1); *syn.* Absorbtionszone [f] (LB 1978, 316).

abundance [n] *phyt. zool.* ►species abundance.

abundance dynamics [npl] *zool.* ►population fluctuation.

abutter [n] **on a street** [US] *leg. urb.* ►adjoining street resident.

abutting lot [n] [US] *leg. urb.* ►neighboring lot [US]/neighbouring plot [UK].

abutting owner [n] **on a street** [US] *leg. urb.* ►adjoining street resident.

abutting resident [n] **on a street** [US] *leg. urb.* ►adjoining street resident.

11 abyssal zone [n] *ocean.* (Oceanic ►benthic zone between ►bathyal zone and ►hadal at a depth of approximately 2,000 to 7,000 m, which is completely without light); *s* **abisal [m]** (Zonas oceánicas más profundas que incluyen las denominadas fosas. Los seres vivos que las habitan se llaman también abisales; ►zona batial, ►zona hadal, ►zona del bentos); *syn.* zona [f] abisal; *f* **zone [f] abyssopélagique** (Dans la zonation verticale du milieu marin, strate hydrosphérique entre les zones bathypélagiques et hadopélagiques, d'une profondeur comprise entre 2000 et 6000 m ; ►zone bathyale, ►zone benthale, ►zone hadopélagique) ; *g* **Abyssal [n]** (Völlig lichtloser Abschnitt des ►Benthals im Ozean zwischen ►Bathyal und ►Hadal in einer Tiefe von ca. 2 000 bis 7 000 m); *syn.* abyssale Region [f].

12 acceleration lane [n] *trans.* (Highway traffic lane designed to permit safe increase of speed while merging onto a highway traffic lane; *opp.* ►deceleration lane); *syn.* merger lane [n] [also US], merger [n] [also US]; *s* **carril [m] de aceleración** (Carril adicional en zonas de acceso a autopista para permitir la aceleración de los vehículos sin dificultar el tráfico fluyente; *opp.* ►carril de deceleración); *f* **bande [f] d'accélération** (Surface en bordure de la chaussée d'une autoroute réservée aux véhicules accédant à cette voie ; *opp.* ►bande de décélération) ; *syn.* voie [f] d'accélération ; *g* **Beschleunigungsspur [f]** (Eine neben der Fahrbahn liegende zusätzliche Fahrspur, die dazu dient, sich einfädelnden Fahrzeugen die Erhöhung der Geschwindigkeit ohne Behinderung des durchgehenden Verkehrs zu ermöglichen; *opp.* ►Verzögerungsspur).

accent grass [n] *hort. plant.* ►specimen grass.

13 accent perennial [n] *hort. plant.* (►Perennial 2 plant which dominates or clearly stands out amongst many other bedding perennials due to its distinct form, foliage texture, or colo[u]r in contrast to its surroundings, e.g. tall grasses such as Miscanthus or plants with a striking colo[u]r such as Delphiniums; an **a. p.** may also be very visible when planted in repetitive groups or to form a basic planting structure; **a. p.s** accentuate the structural composition of a perennial planting, if planted in the foreground and in great numbers, e.g. Aster dumosus or plants which dominate in the spring; even relatively simple perennials can be classified as **a. p.s** if planted repetitively; a sequence of **a. p.s** such as Hollyhocks, followed by Delphiniums, flowering one after the other, will produce a series of effects at various times throughout a season and ensures structural accent over a long period; dense planting of **a. p.s** leads to ►seasonally changing visual dominance); *syn.* theme perennial [n] (HAS 1993, 39), focal point perennial [n]; *s* **vivaz [f] de acentuación** (►Planta vivaz que predomina o sobresale claramente entre muchas otras con sus flores vistosas, la cantidad de las mismas, su forma o el color de sus hojas, de manera que sirve de marco estructurante en un macizo; ►composición estacional de especies); *syn.* perenne [f] de acentuación; *f* **plante [f] vivace dominante** (►Plante vivace occupant une place maîtresse dans une plate-bande de vivaces en raison de sa forme décorative, sa taille, la couleur intense de ses fleurs, sa floraison abondante et constituant l'élément structurant du massif de vivaces ; ►composition florale saisonnière, ►variation/succession saisonnière de la dominance dans la composition florale) ; *g* **Leitstaude [f]** (►Staude, die in einer Beetstaudenpflanzung durch ihre ausgeprägte Form und Gestalt, z. B. Reiherfedergras *[Stipa barbata]*, Größe und Standfestigkeit, z. B. *Miscanthus x giganteus*, ihre auffallende Blütenfarbe, z. B. Rittersporn *[Delphinium]*, und ihren Blütenreichtum aus der Vielzahl der übrigen Beetstauden deutlich hervortritt, gleichsam durch rhythmische Wiederholung in Gruppen das themengerechte „Gerüst" der Staudenpflanzung bildet; bei niedrigen, nur unverdeckt im Vordergrund verwendbaren Arten, z. B. *Aster dumosus* oder im Frühjahrsaspekt vorlaufenden Arten, z. B. Gemswurz *[Doronicum]*, wird die gewünschte Wirkung durch große Stückzahlen erreicht; über ihre Wiederholung und damit verbundene Rangerhöhung kann auch eine eher schlichte Staude thematisiert und zur **L.** erhoben werden. Die Aufeinanderfolge von Leitstauden mit gestaffelten Wirkungszeiträumen [**Leitstaudenfolge**, z. B. Stockmalve *(Alcea)* nach Rittersporn *(Delphinium)*] sichert Strukturen in Pflanzungen langfristig. Eine flächendeckende Verdichtung der **L.n** führt zur Aspektpflanzung/Aspektbildung; ►saisonal wechselnde visuelle Dominanz).

accent plant [n] *hort. plant.* ►specimen plant.

14 accept a bid [loc] [US] *contr.* (►acceptance of a bid [US] 1/acceptance of a tender [UK] 1, ►have a bid accepted [US]/have a tender accepted [UK]); *syn.* accept a tender [UK] [loc]; *s* **dar el remate [loc]** (►conseguir el remate, ►remate); *f* **attribuer le marché à [loc]** (*Contexte* le marché a été attribué à l'entreprise X) ; *syn.* retenir un candidat [vb] ; *g* **jemandem den Zuschlag erteilen [loc]** (►den Zuschlag bekommen, ►Zuschlag).

15 acceptable level [n] **of air pollution** *envir.* (Quantity of contaminants in the air below the level of interference with human health, safety and comfort expressed in parts per million [p.p.m.], which conform to approved standards; cf. RCG 1982; ►adulteration limit, ►air pollution 1); *s* **nivel [m] aceptable de contaminación atmosférica** (Cantidad tolerable de contaminantes para la salud, la seguridad y el bienestar humanos, que se mide en partes por millón [ppm] y que depende de la capacidad del medio de absorber residuos gaseosos; ►contaminación atmosférica, ►valor límite); *f* **niveau [m] acceptable de la pollution atmosphérique** (Du point de vue de la qualité de l'air taux de concentration de polluants atmosphériques, niveau considéré comme satisfaisant pour la santé et l'environnement; ►pollution atmosphérique, ►valeur limite) ; *syn.* taux [m] acceptable à la pollution atmosphérique ; *g* **zulässige Belastbarkeit [f] der Luft** (Zulässige Menge an Luftschadstoffen in einem Gebiet,

die für die Gesundheit und das Wohlbefinden der Menschen unschädlich ist. Die Konzentrationswerte für ►Luftverunreinigungen werden meist in Mikrogramm je Kubikmeter verunreinigter Luft [µg/m³] je Schadstoff gemessen. Die Festlegung der Konzentrationswerte ist in der 23. VO zur Durchführung des BImSchG vom 16.12.1996 nachzulesen; ►Grenzwert).

acceptance [n] *constr. contr.* ►condition for final acceptance [US].

16 acceptance [n] **of a bid** [US] (1) *constr. contr.* (Formal notification of the ►acceptance of a bid/tender, followed normally by ►awarding of contract [US]/letting of contract [UK] which creates a legal obligation between the parties. An amended contract award may arise when the agency/authority or client alters the bid/tender, causing the bidder/tenderer to agree to an adjusted contract sum; ►period for acceptance of a bid [US]/period for acceptance of a tender [UK], ►written notice of contract award); *syn.* acceptance [n] of a tender [UK] (1); *s* **remate [m]** (En concurso-subasta público o restringido, notificación formal de aceptación de la mejor oferta antes del fin del ►periodo/período de remate. Un remate modificado tiene lugar cuando el comitente cambia algo en la oferta del postor y con ello hace por su parte una oferta al postor, que puede aceptarla o no; ►formalización de contrato por escrito, ►adjudicación de contrato, ►adjudicación del remate); *f* **attribution [f] du marché (sur appel d'offres/sur adjudication)** (Acceptation formelle par le maître d'ouvrage, avant la ►limite de validité des offres, de l'offre paraissant la plus intéressante dans le cadre d'un appel d'offres ouvert ou restreint ; ►délai de validité des offres, ►passation d'un marché, ►notification [de l'attribution] du marché à l'entreprise retenue, ►notification d'un marché) ; *syn.* dévolution [f] du marché ; *g* **Zuschlag [m]** (Bei einer öffentlichen oder beschränkten Ausschreibung die Annahme des Angebotes des Bieters mit dem annehmbarsten Angebot durch den Bauherrn [Auftraggeber] vor Ablauf der ►Zuschlagsfrist; cf. § 28 VOB Teil A; ein ‚modifizierter Zuschlag' liegt vor, wenn der Auftraggeber das Angebot des Bieters ändert und somit ein Angebot gegenüber dem Unternehmer abgibt, das dieser annehmen kann; ►Auftragserteilung an eine Firma, ►schriftliche Auftragserteilung, ►Zuschlagserteilung).

17 acceptance [n] **of a bid** [US] (2) *contr.* (Decision to select a particular bid/tender, after bidding/tendering procedure and to award a contract to the ►bidder offering best value for the price [US]/tenderer offering best value for the money [UK]; ►sole source contract award [US]/freely awarded contract [UK], ►written notice of contract award, ►awarding of contract [US]/letting of contract [UK]); *syn.* acceptance [n] of a tender [UK] (2); *s* **adjudicación [f] del remate** (En concurso-subasta público o restringido, declaración escrita enviada al postor de la mejor oferta antes del fin del periodo/período de remate, de que le ha sido adjudicado el remate; al contrario el término ►adjudicación de contrato se utiliza para cualquier tipo de encargo, o sea también para la ►adjudicación negociada; ►formalización de contrato por escrito, ►licitante ganador); *f* **notification [f] (de l'attribution) du marché à l'entreprise retenue** (Dans un marché sur appel d'offres ouvert ou restreint la personne responsable du marché avise par écrit l'entreprise retenue de l'attribution du marché ; ►passation d'un marché, ►attribution d'un marché de travaux, ►notification d'un marché, ►passation des marchés sous forme de marché négocié, ►candidat retenu) ; *syn.* avis [m] d'attribution ; *g* **Zuschlagserteilung [f]** (Bei einer öffentlichen oder beschränkten Ausschreibung an den ►Bieter des annehmbarsten Angebotes ergangene schriftliche Erklärung vor Ablauf der Zuschlagsfrist, dass er den Auftrag erhält; cf. § 28 VOB Teil A; *Kontext* der Zuschlag wird auf das Angebot der Firma X erteilt; das Angebot der Firma X erhält den Zuschlag;

A

Ergebnis erteilter Zuschlag. Das Oberlandesgericht Rostock hat in seiner Entscheidung vom 16.05.2001 [17 W 1/01, 2/01] festgestellt, dass ein Zuschlag dann vorliege, wenn sich der Auftraggeber verbindlich auf einen Vertragspartner festgelegt habe. Der Begriff **Auftragserteilung** gilt unabhängig von Ausschreibungen für jeden Auftrag, d. h. auch für Vergaben ohne förmliches Verfahren [►freihändige Vergabe]; *OB* ►Auftragserteilung an eine Firma. *MERKE*, eine Zuschlagserteilung kann nach deutschem Sprachgebrauch nur nach öffentlichem oder beschränktem Verfahren erfolgen; ►schriftliche Auftragserteilung); *syn.* Erteilung [f] des Zuschlags.

acceptance [n] **of a bid** [US] (3) *contr.* ►period for acceptance of a bid [US].

acceptance [n] **of a tender** [UK] *contr.* ►acceptance of a bid [US] (1).

acceptance [n] **of a tender, period for acceptance** [n] **of a tender** [UK] *contr.* ►period for acceptance of a bid [US].

acceptance period [n] **of a bid** [US] *contr.* ►eriod for acceptance of a bid [US].

accept [vb] **a tender** [UK] *contr.* ►accept a bid [US].

access [n] *plan. trans. urb.* ►delivery access, ►limited vehicular access, ►parking access, ►provision of access for the public; *leg. recr.* ►right of public access.

access basin [n] [US] *constr.* ►cleanout chamber [US].

access driveway [n] **over a pavement** [UK] *urb.* ►access driveway over sidewalk [US].

18 access driveway [n] **over sidewalk** [US] *urb.* (Part of a pavement which is driven over to reach a resident's property by car; ►curb cut [US]/drop kerb [UK]); *syn.* access driveway [n] over a pavement [UK]; *s* **acceso** [m] **de vehículos sobre acera** (Paso a garaje o predio sobre acera; ►bordillo hundido); *f* **bateau** [m] (Partie abaissée d'un trottoir limitée par une bordure basse franchissable par laquelle s'effectue l'accès des véhicules dans une propriété ; ►zone d'abaissement de bordure de trottoir) ; *g* **Überfahrt** [f] (Teilbereich eines [öffentlichen] Gehweges, der zur Überfahrt auf ein Anliegergrundstück dient; ►Bordsteinabsenkung).

19 access eye [n] [US] *constr.* (Pipe fitting with a removable plug which provides access for inspection or cleaning of a pipe run; DAC 1975); *syn.* cleanout [n] [US], cleaning eye [n], rodding eye [n] [UK]; *s* **registro** [m] (Unidad con placa movible o tapón que permite el acceso a la instalación de cañerías u otra instalación de drenaje para efectuar la limpieza; cf. DACO 1988, 345); *f* **regard** [m] **de curage** (Regard placé entre les regards principaux pour permettre le curage hydraulique de la canalisation ou le passage d'une caméra ; VRD 1986, 355) ; *g* **Kontrollschacht** [m] (1) (...schächte [pl]; nicht besteigbare Einrichtung mit niedriger Bauhöhe zur Prüfung, Wartung und zum Durchspülen einer Entwässerungsleitung); *syn.* Reinigungsstutzen [m].

access [n] **for the public** *plan.* ►provision of access for the public.

20 accessibility [n] *plan.* (Ease of approaching or reaching a structure or place; ►barrier-free accessibility); *s* **accesibilidad** [f] (Facilidad para llegar a una instalación, una zona verde o a cualquier otro lugar determinado debido a su conexión a la correspondiente infraestructura viaria); *f* **accessibilité** [f] (Possibilité d'accéder) ; *g* **Erreichbarkeit** [f] (Möglichkeit, dass ein Gebäude, eine Einrichtung oder ein Ort durch Infrastruktureinrichtungen erschlossen und zugänglich ist).

accessible [adj] *plan. sociol.* ►wheel-chair-accessible.

accessible [adj] **for the handicapped/disabled** *leg. plan. sociol.* ►developed for the handicapped/disabled.

access [n] **only for residents** *trans. urb.* ►local street with access only for residents.

21 accessory structure [n] *leg. urb.* (Separate structural unit which, according to its purpose provides support and is subordinate to another structure on the same plot; e.g. pergola, gazebo, swimming pool, trash storage [US]/bin store [UK], telephone box, shelter); *syn.* secondary structure [n], supportive permanent structure [n] [also US], ancillary structure [n] [also UK]; *s* **construcción** [f] **auxiliar** (Unidad estructural separada que —debido a sus fines— está subordinada a otra estructura de la misma parcela, como p. ej. pérgola, piscina, glorieta, cobertizo, etc.); *syn.* construcción [f] complementaria, instalación [f] auxiliar; *f 1* **emplacements** [mpl] **réservés** (F., voies et ouvrages publics, installations d'intérêt général et espaces verts ; surfaces déduites de la surface prise en compte pour le calcul des possibilités de construction sur lequel s'applique le coefficient d'occupation des sols) ; *f 2* **installations** [fpl] **auxiliaires** (D., constructions, installations indépendantes mais d'importance secondaire ou complémentaire (constructions d'approvisionnement en eau, gaz électricité, carport, etc.]) ; *g* **Nebenanlage** [f] (Bautechnisch selbständige, ihrem Zweck und ihrer Nutzung nach aber unselbständige resp. untergeordnete bauliche Anlage; cf. § 14 BauNVO; z. B. Pergola, Gartenlaube, Schwimmbecken, Mülltonnenschrank, Telefonzelle, Schutzdach, Wartehalle; eine Garage ist keine **N.**).

22 access road [n] *plan. trans. urb.* (Road connecting a residential, industrial or recreation area to a larger, public access road or highway network; ►parking access road); *syn.* connecting road [n] [also US]; *s* **carretera** [f] **de acceso** (Carretera que conecta una zona residencial, industrial o de recreo a otra vía pública de mayor importancia o a la red de carreteras; ►calle de acceso a estacionamiento); *syn.* calle [f] de acceso, vía [f] de acceso; *f* **route** [f] **d'accès** (Ouvrage reliant les zones d'habitation, les zones industrielles, les zones de loisirs, etc. au réseau public ; ►voie d'accès à l'aire de stationnement) ; *syn.* voie [f] d'accès, voie [f] de desserte ; *g* **Erschließungsstraße** [f] (Verkehrsweg, der Wohngebiete, Industriebetriebe, Erholungsgebiete etc. an das öffentliche Verkehrsnetz anbindet; ►Parkplatzzufahrtsstraße).

access [n] **to rural land** *leg. recr.* ►public access to rural land.

accident [n] *envir. leg.* ►industrial plant toxic accident, ►oil pollution accident, ►oil tanker accident.

23 accidental [n] *phyt.* (Species that are rare and casual intruders from another plant community or relicts of a preceding community; ►strange species); *s* **accidental** [f] (Especie que se presenta rara o casualmente en una comunidad vegetal a la que no pertenece o que es un relicto de la comunidad precedente; ►especie extraña); *syn.* especie [f] accidental; *f* **espèce** [f] **accidentelle** (Espèce qui se développe habituellement dans d'autres groupements et qui n'apparaît que de façon accidentelle dans l'association décrite ; DEE 1982 ; ►espèce étrangère) ; *syn.* accidentelle [f] ; *g* **Zufällige** [f] (Seltene oder nur ausnahmsweise in eine Gesellschaft eindringende Pflanzenart oder Relikt einer früher dagewesenen Gesellschaft; ►fremde Art); *syn.* zufällige (Pflanzen)art [f].

24 accidental damage [n] **to a tree** *arb.* (Injury to the trunk or branches [►bark wound], as well as the roots of a tree caused for example by impact of a car or by leakage of hazardous substances and contamination of the soil); *s* **daño** [m] **causado a un árbol por accidente** (Herida o daño causado al tronco, las ramas o las raíces por choque de vehículo o por derrame de

sustancias nocivas que contaminan el suelo; ►herida de la corteza); *f* **dégâts [mpl] causés à un arbre (par le choc d'un véhicule accidenté)** (Dommage causé au tronc, au système racinaire ou à la ramure d'un arbre par le choc d'un véhicule ou les substances toxiques s'échappant de celui-ci ; ►blessure de l'écorce) ; *g* **Unfallschaden [m] am Baum** (Autounfallschaden am Stamm und/oder im Astwerk [►Rindenverletzung] sowie durch auslaufende umweltgefährdende Substanzen, die den Boden im Wurzelbereich verseuchen).

acclimation [n] *biol. ecol.* ►acclimatization.

25 acclimatization [n] *biol. ecol.* (**1.** Short term adjustment of organisms to new or changed environmental conditions through phenotypic alteration. **2.** Gradual physiological and behavio[u]ral adaptation or increased tolerance of a species to a changed environment in the course of a few weeks or several generations); *syn.* acclimation [n]; *s* **aclimatación [f]** (Adaptación a corto plazo de los seres vivos a condiciones medioambientales nuevas por medio de cambios fenotípicos); *f* **accoutumance [f]** (des êtres vivants à de nouvelles conditions du milieu au moyen de changements phénotypiques ; le résultat est un accommodat) ; *syn.* accommodation [f] ; *g* **Anpassung [f] (1)** (Gewöhnung von Lebewesen an neue Umweltbedingungen durch z. B. phäno-typische Veränderungen. Im Deutschen wird **Akklimation** im Vergleich zum Englischen nur für die Anpassung von Lebewesen im Labor unter kontrollierten Bedingungen an eine definierte abiotische Umwelt benutzt); *syn.* Akklimatisation [f], phäno-typische Adaptation [f].

accommodation [n] *recr.* ►family accommodation, ►group overnight accommodation, ►hiking accommodation [US], ►individual overnight accommodation, ►rural overnight accom-modation, ►rural vacation accommodation [US], ►vacation accommodation [US] (1), ►vacation accommodation [US] (2).

accommodation [n]**, holiday** [UK] *recr.* ►vacation accommodation [US] (1).

accommodation [n]**, rural holiday** [UK] *recr.* ►rural vacation accommodation [US].

26 accommodation capacity [n] *recr. plan.* (Available number of beds in hotels, guesthouses, etc.; e.g. at a holiday resort); *syn.* holding capacity [n] [also US]; *s* **capacidad [f] hotelera** (Número total de camas en hoteles, pensiones, albergues, etc. p. ej. en un centro turístico); *syn.* capacidad [f] de alojamiento; *f* **capacité [f] d'hébergement** (Quantité disponible des modes d'hébergement, de séjour et d'accueil [hôtels, meu-blés, gîtes, chambres d'hôtes, etc.] dans un lieu de vacances) ; *syn.* capacité [f] d'accueil ; *g* **Übernachtungskapazität [f]** (Menge an vorhandenen Hotels, Pensionen und Gasthäusern, z. B. in einem Ferienort).

accommodation-seeking holiday-makers [npl] [UK] *recr. plan.* ►accommodation-seeking vacationers [US].

accommodation-seeking holiday travellers [npl] [US] *recr. plan.* ►accommodation-seeking vacationers [US].

27 accommodation-seeking vacationers [npl] [US] *recr. plan.* (Persons requiring accommodation in addition to the permanent residential population of a resort community [US]/ holiday resort [UK]); *syn.* accommodation-seeking holiday-makers [npl] [UK], accommodation-seeking vacationists [npl] [US]; accommodation-seeking holiday travellers [npl] [US]; *s* **población [f] turística** (Población que necesita alojamiento temporal en el lugar de estancia); *f* **population [f] touristique** (par opposition à la population permanente) ; *g* **mit Wohnraum zu versorgende Urlaubsreisende [loc]** (Im Gegensatz zur ansäs-sigen Wohnbevölkerung diejenigen Personen, die mit zusätz-lichem Wohnraum versorgt werden müssen).

accommodation-seeking vacationists [npl] [US] *recr. plan.* ►accommodation-seeking vacationers [US].

28 according to certified delivery [loc] [US] *constr. contr.* (*Context* Payment for delivered building material or furniture/equipment on the construction site according to checked delivery schedules [US]/delivery notes [UK]; e.g. bill of lading [US]/delivery note [UK] or weigh bill); *syn.* on proof of verified delivery notes [loc] [UK]; *s* **con justificación [loc]** (*Contexto* liquidación de materiales de construcción suministrados **c. j.** a la obra); *f* **avec justifications à l'appui [loc]** *syn.* justifié par bons d'attachement [loc] ; *g* **auf Nachweis [loc]** (*Kontext* Abrechnung von geliefertem Baustellenmaterial **a. N.**).

account [n]**, advances on** [UK] *constr. contr.* ►interim payment.

29 accounting [n] **of executed work** [US]/**account-ing** [n] **of executed works** [UK] *contr.* (Listing of item costs for work [US]/works [UK] actually executed in accordance with the contract documents; ►bid item [US]/tender item [UK], ►final accounting invoice, ►interim invoice); *s* **liquidación [f] de obras** (Elaboración de la cuenta sobre los servicios realizados según las ►partidas de oferta del contrato; ►liquidación final de obra, ►liquidación intermedia); *f* **règlement [m] des travaux** (Établissement du décompte des travaux à l'entreprise effective-ment exécutés conformément aux articles du détail estimatif ou aux conditions de contrat ; ►article, ►décompte final, ►situation) ; *syn.* règlement [m] des comptes ; *g* **Abrechnung [f] der Bauleistungen** (Prüfbare Aufstellung der Rechnung über tatsächlich ausgeführte Leistungen gemäß ►Positionen der Vertragsunterlagen; ►Schlussabrechnung, ►Zwischenrechnung).

account invoice [n] *contr.* ►final account invoice.

accumulation [n] *biol.* ►bioaccumulation; *met.* ►frost accumulation; *limn. pedol.* ►salt accumulation; *geo. phyt.* ►sand accumulation.

accumulation [n]**, detrital rock** *geo.* ►detrital deposit.

accumulation belt [n] *geo. phyt.* ►sediment accumulation belt.

30 accumulation [n] **of drifted debris** *envir.* (Deposits of plant material or solid waste on gently sloping banks of watercourses or on beaches); *s* **acumulación de maderada** (Concentración de restos de madera o de residuos sólidos al borde de cursos o masas de agua y de playas); *f 1* **embâcles [mpl]** (Accumulation de débris végétaux, de déchets solides, de glaçons, créant un encombrement et faisant obstacle à l'écoule-ment des eaux d'un cours d'eau) ; *f 2* **amas [m] de débris** (Matériaux organiques ou déchets abandonnés par les vagues sur la berge d'un lac) ; *f 3* **laisses [fpl] de mer** (Matériaux d'origine organique ou déchets abandonnés par les vagues soit à la limite supérieure du niveau d'humectation, soit sur l'estran pendant l'étale de marée haute ou marée basse) ; *syn.* dépôts [mpl] organiques des lignes de marées (GEHU 1984, 60) ; *g* **akku-muliertes Treibgut [n, o. Pl.]** (An Rändern von Fließgewässern, Seen oder an Stränden angeschwemmte Ansammlungen von Pflanzenteilen oder festen Abfallstoffen).

accumulation zone [n] *geo. phyt.* ►sediment accumulation zone.

achievement [n] *plan.* ►degree of achievement.

31 acidic grassland [n] *phyt.* (Area with open or closed pioneer plant community, often ephemeral and subject to quickly changing temperature, which is composed chiefly of hemicryp-tophytes and therophytes growing on dry, weakly acid, sandy soils or strongly weathered rocks or substrates of poor to modera-tely good fertility; e.g. Biting Stonecrop pioneer community and siliceous and sand grasslands *[Sedo-Scleranthetea]*; ►nutrient-

A

poor grassland, ►limestone grassland); *s* **pastizal [m] (oligó-trofo) silicícola** (Comunidad vegetal creciendo sobre suelos arenosos pobres en nutrientes y con poca capacidad de retención de agua, por lo que necesitan raíces profundas; p. ej. de la clase de las *Sedo-Scleranthetea*; ►prado oligótrofo, ►pastizal calcícola seco); *syn.* pastizal [m] seco; *f* **pelouse [f] silicicole** (Formation herbeuse naturelle fixée sur un substrat sableux, siliceux, oligotrophe, à végétation ouverte, pauvre en espèces à enracinement profond ; les pelouses alpines silicicoles se répartissent dans le système phytosociologique dans les alliances du *Festucion variae* [pelouse rocheuse silicicole], du *Nardion* et du *Caricion curvulae* ; ►pelouse oligotrophe, ►pelouse calcicole) ; *syn.* pelouse [f] sur sable siliceux, pelouse [f] oligotrophe silicicole ; *g* **Silikatmagerrasen [m]** (Artenarme und tiefwurzelnde, ausgesprochen heliophile Pflanzengesellschaft [Pioniergesellschaft], z. B. Klasse der Mauerpfeffertriften, Sandrasen, Feldgrus- und Felsband-Gesellschaften *[Sedo-Scleranthetea]* auf nährstoffarmen Sandböden mit geringer Wasserhaltekraft; cf. OBER 1978; ►Magerrasen, ►Kalkmagerrasen); *syn.* Sandmagerrasen [m], Sandtrockenrasen [m].

32 acidification [n] *ecol.* (Lowering of the pH value of a soil, body of water, or of precipitation; ►increase in soil acidity, ►acid rain, ►acid soil); *s* **acidificación [f]** (Reducción del pH del suelo, de una masa de agua, de la lluvia, etc.; ►acidificación del suelo, ►lluvia ácida, ►suelo ácido); *f* **processus [m] d'acidification [f]** (Réduction du pH du sol, de la pluie/des pluies, etc. ; ►acidification du sol, ►pluie acide, ►sol acide) ; *g* **Versauerung [f]** (Senkung des pH-Wertes eines Bodens, eines Gewässers oder des Niederschlags; ►Bodenversauerung, ►saurer Boden, ►saurer Regen).

acidity [n] *pedol.* ►increase in soil acidity.

33 acidophilous species [n] *phyt.* (Plant which requires an acid soil [pH < 5] for normal growth. *Generic term* ►indicator plant; *opp.* ►calcareous indicator species); *s* **especie [f] indicadora de acidez** (Planta que necesita un sustrato ácido [pH < 5] para crecer normalmente; *término genérico* ►planta indicadora; *opp.* ►especie indicadora de cal); *syn.* especie [f] acidófila; *f* **espèce [f] indicatrice d'acidité** (Espèce végétale qui requiert pour son développement naturel un substrat acide, pH < 5 ; *terme générique* ►espèce indicatrice ; *opp.* ►espèce indicatrice de calcaire) ; *g* **Säurezeiger [m]** (Pflanzenart, die zu ihrem normalen Gedeihen ein saures Substrat mit einem pH-Wert < 5 benötigt; *OB* ►Zeigerpflanze; *opp.* ►Kalkzeigerpflanze); *syn.* Sauerbodenpflanze [f].

acid precipitation [n] *envir.* ►acid rain.

34 acid rain [n] *envir.* (Rainwater with a concentration of sulphur dioxide and nitrogen oxide, which, when converted to sulphuric and nitric acids, produces a lower pH value than normal rainwater [pH 5.6]; ►particulate deposition); *syn.* acid precipitation [n]; *s* **lluvia [f] ácida** (Precipitación de agua o nieve con contenido de ácido sulfúrico y de óxidos de nitrógeno mayor que por naturaleza. Estos compuestos se depositan en el suelo, en la vegetación o sobre las aguas causando daños ambientales; ►deposición seca); *syn.* precipitación [f] ácida; *f* **pluie [f] acide** (Pollution diffuse transfrontière formée essentiellement de dioxyde de soufre et d'oxyde d'azote subissant pendant leur transfert dans l'atmosphère des oxydations et retombant à terre sous forme d'aérosols de sulfate ou nitrate [dépôts secs] ou d'acides sulfuriques et nitriques [dépôts humides] ; CPEN — Air, Introduction paragraphe 3, 1999. Le terme de **p. a.** été employé pour la 1re fois en 1872 par ROBERT ANGUS SMITH à propos de pluies tombant sur Manchester. En 1961 le Suédois SVANTE ODIN démontra l'importance du phénomène. En 1984, on a reconnu aussi des **p. a.** dans des régions non industrielles

[Afrique et Amérique du Sud] ; QUID 1996, 1472 ; ►déposition atmosphérique sèche) ; *g* **saurer Regen [m,** selten **Pl.]** (Mit Schwefeldioxid und Stickstoffoxid angereicherter Niederschlag, der durch die Umformung in Schwefel- und Salpetersäure einen niedrigeren pH-Wert als normales Regenwasser [pH 5,6] hat. Anfang der 1980er-Jahre lag in Deutschland der durchschnittliche pH-Wert des Regens bei 4,1; cf. AFZ 1982, H. 39; seit Mitte der 1990er-Jahre stieg der pH-Wert im Durchschnitt wieder über 5; der Begriff **s. R.** wurde zuerst 1872 von ROBERT ANGUS SMITH in Manchester benutzt. 1961 wies der Schwede SVANTE ODIN auf die Wichtigkeit dieses Phänomens hin und seit 1984 hat man den **sauren R.** auch in nicht industriealisierten Regionen [Afrika und Südamerika] festgestellt; cf. QUID 1996, 1472; ►trockene Deposition); *syn.* saurer Niederschlag [m], *auf Pflanzen* nasse Deposition [f].

35 acid soil [n] *pedol.* (Soil with pH value less than 7; ►alkaline soil, ►acidification); *s* **suelo [m] ácido** (Suelo en el cual la concentración de iones de hidrógeno es mayor que la de los iones de hidroxilo; MEX 1983; ►suelo alcalino, ►acidificación); *f* **sol [m] acide** (Le pH est inférieur à 7. Cependant, on réserve généralement le terme « sol neutre » au sol dont le pH est compris entre 6,6 et 7,3. Les sols acides ont donc un pH inférieur à 6,6 ; DIS 1986 ; ►sol alcalin, ►processus d'acidification) ; *g* **saurer Boden [m]** (B. mit einem pH-Wert unter 7; ►alkalischer Boden, ►Versauerung).

acoustic screen mound [n] [UK] *constr. envir.* ►noise attenuation mound.

acoustic screen wall [n] [UK] *constr. envir.* ►noise barrier wall [US].

acquisition [n] *adm. plan.* ►land acquisition.

Act [n] *adm. envir. leg.* ►Clean Air Act, ►enabling act [US]; *conserv. leg. nat'res.* ►Endangered Species Act [E.S.A.] [US]; *adm. envir. leg.* ►Federal Clean Air Act [US]; *adm. leg. pol.* ►law, #3; *conserv. leg. nat'res.* ►National Environmental Policy Act [US]/Town and Country Planning [Assessment of Environmental Effects] Regulations 1988 [UK]; *conserv'hist. leg.* ►National Historic Preservation Act 1966 [US]; *agr. for. hort. leg.* ►pest control act.

Act [n]**, Conservation of Wild Creatures and Wild Plants Act 1975** [UK] *conserv. leg.* ►Endangered Species Act [E.S.A.] [US].

Act [n]**, Control of Pollution Act** [UK] *envir. leg.* ►Federal Clean Air Act [US].

Act [n]**, Planning [Listed Buildings and Conservation Areas] Act 1990** [UK] *conserv'hist. leg.* ►National Historic Preservation Act [US].

Act [n]**, Sales of Goods and Services Act** [UK] *constr. contr.* ►General Conditions of the Contract for Furniture, Furnishings and Equipment [US].

Act [n]**, Water Act 1945** [UK] *＊*►water rights.

Act [n]**, Water Resources Act 1963** [UK] *＊*►water rights.

action [n] [US]**, capillary** *pedol.* ►capillary rise.

action group [n] *pol.* ►community action group.

action group [n] [UK]**, anti-motorway** *pol.* ►highway opposition group [US].

action group [n] [UK]**, anti-nuclear** *pol.* ►nuclear power opposition group [US].

36 activated sludge [n] *envir.* *s* **lodo [m] activado** *syn.* fango [m] activado, cieno [m] activado; *f* **boue [f] activée** ; *g* **belebter Schlamm [m]**.

A

37 active bog [n] *geo. phyt.* (Bog in the process of accretion; ►bog growth); *s* **turbera** [f] **viva** (Turbera que se encuentra en proceso de crecimiento; ►crecimiento de turbera); *f* **tourbière** [f] **active** (caractérisée par ►épaississement de la tourbière) ; *syn.* tourbière [f] vivante ; *g* **lebendes Moor** [n] (Ein durch Wachstum gekennzeichnetes Moor; ►Moorwachstum).

38 active play [n] *recr.* (Games played by physical activities); *s* **juegos** [mpl] **activos** *syn.* juegos [mpl] de movimiento; *f* **jeu** [m] **de mouvement** *syn.* activité [f] de mouvement ; *g* **Bewegungsspiel** [n] (Mit körperlichen Bewegungen verbundenes Spiel).

39 active recreation [n] *recr.* (Restorative physical activities, e.g. sports and games, which are necessary for the preservation of human health and joie de vivre; ►local recreation, ►long-stay recreation, ►recreational infrastructure, ►recreational weekend [US]/weekend break [UK], ►short-stay recreation); *s* **recreo** [m] (Recuperación de la fuerza y el espíritu a través del esfuerzo, de la diversión o del juego; DINA 1987; ►descanso de fin de semana, ►equipamiento de recreo, ►recreo local, ►vacaciones cortas, ►vacaciones largas); *syn.* esparcimiento [m]; *f* **loisirs** [mpl] **actifs** (Activités physiques et ludiques permettant la régénération de l'équilibre corporel ; ►équipements collectifs de loisirs, ►loisirs de proximité, ►loisirs de week-end, ►tourisme de passage, ►tourisme de séjour) ; *syn.* récréation [f] active ; *g* **aktive Erholung** [f, o. Pl.] (Zurückgewinnung von Gesundheit und Leistungsfähigkeit durch aktive körperliche oder spielerische Betätigung, die in unregelmäßigen Abständen von den Menschen zur Erhaltung ihres allgemeinen geistigen und körperlichen Wohlbefindens und ihrer Lebensfreude benötigt wird; ►Freizeitinfrastruktur, ►Kurzzeiterholung, ►Langzeiterholung, ►Naherholung, ►Wochenenderholung).

active root [n] *bot.* ►absorbing root.

activity [n] *recr.* ►play activity, ►recreation activity, ►vacation activity [US]/holiday activity [UK].

activity [n] [US]**, free time** *plan. recr.* ►leisure pursuit.

activity [n] [UK]**, holiday** *recr.* ►vacation activity [US].

activity [n] [US]**, leisure time** *plan. recr.* ►leisure pursuit.

activity pattern [n] *plan. recr.* ►leisure activity pattern.

Act [n] **of God** [US] *leg.* ►force majeure.

Act [n] **on Conservation of the Natural Environment** *conserv. leg.* ►Federal Act on Conservation of the Natural Environment [CH].

Act [n] **on Nature Conservation and Landscape Management** *conserv. leg.* ►Federal Act on Nature Conservation and Landscape Management [D].

actual vegetation [n] *phyt.* ►existing vegetation.

40 adaptability [n] *biol. ecol.* (Relative ability of a species to adjust to new environmental conditions; ►ecological tolerance); *s* **adaptabilidad** [f] (Capacidad de las especies de adaptarse a nuevas condiciones ambientales; ►tolerancia ecológica); *syn.* flexibilidad [f]; *f* **capacité** [f] **d'adaptation** (Comportement des espèces végétales et animales basé sur leurs dispositions naturelles et acquises, caractérisé par une adaptation relative aux transformations ou fluctuations des conditions du milieu environnant, les limites d'adaptation variant selon les espèces et leur amplitude écologique ; ►plasticité écologique) ; *syn.* capacité [f] d'accommodation, adaptabilité [f] ; *g* **Anpassungsfähigkeit** [f] (Die relative genetisch festgelegte oder erworbene Fähigkeit von Lebewesen, sich an neue Umweltbedingungen anzupassen; ►ökologische Potenz); *syn.* Adaptabilität [f].

41 adaptation [n] *biol.* (Long term evolutionary adjustment of species to changes of environmental conditions by genetic alteration); *s* **adaptación** [f] **(evolutiva)** (Proceso y resultado de cambios genéticos y sus correspondientes alteraciones morfológicas, fisiológicas y [p]sicológicas de las especies en el marco de las transformaciones del medio ambiente a largo plazo); *syn.* adecuación [f] (genética); *f* **adaptation** [f] **héréditaire** (Processus et résultat de changements génétiques — morphologiques, physiologiques et psychologiques — chez les espèces comme moyen de survie pendant les périodes de transformation des conditions du milieu ; on distingue les adaptations morphologiques, anatomiques et physiologiques) ; *g* **Anpassung** [f] **(2)** (Vorgang und Ergebnis von genetischen — morphologischen, physiologischen, psychologischen — Veränderungen an Arten oder Individuen als Überlebensstrategie); *syn.* Adaptation [f].

42 adaptive use [n] [US] *plan.* (New usage of a plot of land or a building after the present use has been discontinued or had to be abandoned); *syn.* subsequent use [n], after-use [n] [UK]; *s* **uso** [m] **subsiguiente** (Aprovechamiento de un terreno o un edificio para otros fines a los perseguidos hasta el momento); *syn.* uso [m] posterior; *f* **utilisation** [f] **subséquente** (Utilisation d'un fonds, d'un bâtiment qui succède au précédent lorsque celui-ci vient à être abandonné) ; *syn.* usage [m] subséquent ; *g* **Folgenutzung** [f] (Verwendung eines Grundstückes oder eines Gebäudes, wenn die derzeitige Nutzung aufgegeben wird oder aufgegeben werden muss); *syn.* Nachnutzung [f].

43 addendum [n] *constr. contr.* (Modifications to original bills of quantities, often made to reduce the lowest tender figure, if it has exceeded the budget. Variations as well as additional building work should be clearly marked in the bills and drawings and are best negotiated with the contractor before construction commences; ►addendum proposal); *s* **proyecto** [m] **complementario** (Modificaciones de las mediciones y presupuesto originales, presentadas cuando se ha excedido el presupuesto de un artículo en un porcentaje dado. Para este el contratista debe presentar una ►oferta complementaria); *f 1* **avenant** [m] **(au marché)** (Convention écrite conclue entre l'entrepreneur et le client permettant de prendre en compte les modifications nécessaires d'un marché en cours d'exécution [délai, conditions de financement, etc.] ; tout avenant dont le montant dépasse 5 % du montant du marché doit être soumis à l'avis de la commission d'appel d'offres ; l'avenant fait partie intégrante du marché et suit la même procédure [signature, contrôle, etc.] que le marché initial ; un avenant ne doit pas bouleverser l'économie du marché, ni en changer l'objet ; *contexte* **passer un avenant** dans un délai déterminé) ; *f 2* **acte** [m] **spécial** (Acte requérant l'accord des deux parties et revêtant la même forme que l'avenant mais dispensé de transmission a priori à la commission spécialisée des marchés et soumis en revanche, au visa à priori du contrôle financier ; son utilisation est limitée à deux cas précis **1.** l'acceptation d'un sous-traitant à paiement direct, **2.** la fixation du prix définitif résultant d'une modification technique demandée au titre de l'article 19 du CCAG/MI si le montant de la modification ne dépasse pas 10 % du montant initial du marché ou 2 700 000 € HT ; ►proposition d'avenant [au marché]) ; *g* **Nachtrag** [m] (Im Sinne von § 2 VOB Teil B eine [Bau]leistung für eine nicht im Leistungsverzeichnis geforderte Leistung; für diesen Sachverhalt muss der Auftragnehmer in einem ►Nachtragsangebot einen neuen Preis auf Grundlage der Kostenkalkulation des Vertragspreises anbieten; die monetären und terminlichen Auswirkungen sind mitzuvereinbaren); *syn.* Nachtragsleistung [f].

44 addendum [n] **proposal** *constr. contr.* (Price offered for negotiations for changes to the terms and conditions of the signed contract, such as additional building work or additional supply of materials, which has to be procured from the contractor; addenda are issued to correct significant errors, omissions, or conflicts in tender documents/bid documents; ►addendum, ►change order);

A

s **oferta [f] complementaria** (Oferta por escrito firmada por el ►comitente y el ►contratista en la que se acuerdan modificaciones y ampliaciones de la obra después de haberse firmado el encargo inicial; los proyectos complementarios se elaboran para corregir errores significativos, omisiones o conflictos incluidos en los documentos de contrato; ►contrato de modificación, ►proyecto complementario); *f* **proposition [f] d'avenant (au marché)** (Prestation supplémentaire proposée au client par l'entrepreneur et établie sur la base des prix du marché de base et ne constituant pas de modification de l'objet du marché, de bouleversement économique sauf sujétion technique imprévue, le seuil de bouleversement se situant entre 15 et 20 % du montant d'un marché de fournitures, de travaux ou de services ; ►avenant [au marché], ►marché complémentaire) ; *g* **Nachtragsangebot [n]** (Angebot über Änderungen und Ergänzungen von [Bau]leistungen auf Grundlage der Kostenkalkulation des bereits vereinbarten Vertragspreises, das nach Abschluss des Vertrages zwischen Auftraggeber und Auftragnehmer — möglichst vor Ausführung — verhandelt wird; ►Änderungsvertrag, ►Nachtrag).

45 addition [n] [US] *leg. urb.* (**1.** Structure added to the original one some time after initial completion. **2.** Extension or increase in floor area; ►addition of stories, ►zero lot line building [US]; *verb* to build on, to extend); *syn.* extension [n] [UK] (1); *s* **extensión [f] de un edificio** (Superficie añadida a un edificio existente, construida posteriormente a la terminación inicial; *término específico* ala de un edificio; ►sobreedificación, ►construcción sin retranqueo lateral); *f* **extension [f] (1)** (Bâtiment ou local construit ultérieurement et mitoyen à l'ouvrage principal ; ►exhaussement, ►bâtiment mitoyen) ; *g* **Anbau [m] (1)** (**1.** Anbauten [pl]; Zusatzgebäude, das als Ergänzung eines vorhandenen Gebäudes angebaut wurde; ►Aufstockung von Gebäuden, ►Grenzanbau; **2. anbauen [vb]:** eine weitere bauliche Anlage an ein Gebäude seitlich anfügen); *syn. zu 1.* Erweiterungsbau [m].

additional bid documents [npl] [US] *constr. contr.* ►inspection of additional bid documents.

46 additional contractual conditions [npl] *constr. contr.* (Terms which govern, e.g. subcontracts, final acceptance, etc., in addition to the ►General Conditions of Contract for Construction [US]/General Conditions of Government Contracts for Building and Civil Engineering Works [UK]; ►special conditions of [a] contract); *syn.* additional contractual terms [npl]; *s* **pliego [m] de prescripciones administrativas particulares** (Condiciones fijadas en un contrato de obra que reflejan exigencias administrativas específicas del proyecto más allá de las contenidas en el ►pliego de cláusulas administrativas generales para la contratación de obras; ►pliego de prescripciones técnicas particulares); *f* **clauses [fpl] administratives complémentaires (≠)** (D., clauses administratives précisant les stipulations du ►cahier des clauses administratives générales des marchés publics et du ►cahier des clauses administratives particulières, p. ex. les conditions d'exercice de la sous-traitance, délais d'exécution, réception des travaux, garanties, etc. ; en France ces clauses sont inscrites dans le Cahier des clauses techniques particulières CCAP) ; *g* **zusätzliche Vertragsbedingungen [fpl]** (Vertragsbedingungen, die im Vergleich zu den ►allgemeinen Vertragsbedingungen für die Ausführung von Bauleistungen ergänzende Bedingungen wie z. B. Weitervergabe an Subunternehmer, Abnahme etc. regeln; cf. § 10 VOB Teil A; ►besondere Vertragsbedingungen).

additional contractual terms [npl] *constr. contr.* ►additional contractual conditions.

47 additional costs [npl] *contr.* (Supplementary expenditure of money due to the overstepping of a cost estimate or a negotiated contract sum, or an expanded scope of work request; ►cost over-run); *s* **costos [mpl] adicionales** (Fondos adicionales necesitados para un proyecto debido a que los costos totales superan el presupuesto o la suma acordada para el mismo; ►superación del presupuesto); *syn.* costes [mpl] adicionales [Es]; *f* **frais [mpl] supplémentaires** (Moyens financiers nécessaires à la réalisation des travaux après achèvement des ouvrages dépassant les coûts prévisionnels des travaux ; ►dépassement des dépenses de travaux, ►sous-estimation du coût prévisionnel) ; *g* **Mehrkosten [pl]** (Finanzmittel, die zusätzlich zum Kostenanschlag oder zur vereinbarten Vertragssumme benötigt werden; ►Kostenüberschreitung).

additional planning services [npl] [UK] *contr. prof.* ►special professional services.

additional professional services [npl] *contr. prof.* ►supplemental professional services.

48 additional technical contract conditions [npl] **for tree work** *arb. constr. contr.* (In U.S., there are guidelines for tree work, provided by arboricultural organizations; in D. and U.K., supplemental standards for arboricultural practice in carrying out ►tree maintenance work and ►tree treatment work); *s* **condiciones [fpl] técnicas adicionales de contrato para trabajos de mantenimiento de leñosas** (En EE.UU. existen guías para estos trabajos definidas por las organizaciones arboriculturales, en GB y D. se trata de estándares adicionales de arboricultura, específicamente para el ►mantenimiento de árboles y ►saneamiento de árboles); *f* **cahier [m] des clauses techniques générales — entretien des arbres** (D. et U.K., C.C.T.G. s'appliquant aux marchés de ►travaux d'entretien des arbres et de transplantation des arbres ; F., la nature de ces travaux se retrouve dans le fascicule C.C.T.G. n° 35 — Travaux d'espaces verts, d'aires de sports et de loisirs, travaux d'entretien ; ►travaux de restauration arboricole) ; *g* **Zusätzliche Technische Vertragsbedingungen [fpl] — Baumpflege** (*Abk.* ZTV — Baumpflege; **D. u. U.K.**, zusätzliche Vertragsbedingungen für die Ausführung von Baumpflegearbeiten [►Baumpflege] und ►Baumsanierung; **in D.** in der Fassung von 2001 und weiter überarbeitet 2006; diese ZTV ist noch nicht als anerkannte Regel der Technik etabliert, da Erprobung und Bewährung in der Praxis noch ausstehen. Dennoch hat sie zunächst die Vermutung der Richtigkeit für sich, weshalb derjenige, der von den Vorgaben abweicht, dies eingehend und substanziiert begründen muss).

additional tender documents [npl] [UK]**, inspection of** *constr. contr.* ►inspection of additional bid documents [US].

49 additional work [n] **and services** [npl] *constr. contr.* (Services which are an integral part of the contract even though they are not explicitly mentioned); *s* **servicios [mpl] adicionales** (Servicios que son parte integrante del contrato aunque no estén incluidos explícitamente en las ►especificaciones técnicas); *f* **sujétions [fpl] d'exécution** (Nature de prestations non mentionnées explicitement dans la ►description des ouvrages mais faisant partie des travaux à réaliser ; les **s. d'e.** dont les prix ne tiennent pas compte et néanmoins mentionnées dans le cadre du bordereau des prix unitaires sont dénommées sujétions spéciales ; ►spécifications techniques détaillées) ; *g* **Nebenleistungen [fpl]** (Leistungen, die auch ohne Erwähnung in der ►Leistungsbeschreibung 1 zur vertraglichen Leistung gehören; cf. § 2 [1] VOB Teil B in Verbindung mit den ATV der VOB Teil C); *syn.* Nebenarbeiten [fpl].

additional work and services included [loc] *constr. contr.* ►all additional work and services included.

addition [n] **of storeys** [UK] *arch.* ►addition of stories.

50 addition [n] **of stories** [US] *arch.* (Increase of floors on a building); *syn.* addition [n] of storeys [UK]; *s* **sobreedificación**

[f] (Adición de más pisos a un edificio existente); *f* **exhaussement [m]** (de bâtiments d'un ou plusieurs niveaux) ; *syn.* surélévation [f] ; *g* **Aufstockung [f] von Gebäuden** (Erhöhung eines Gebäudes um ein odere mehrere Stockwerke).

additive [n]**, soil** *agr. constr. hort.* ►soil amendment [US].

additives [npl] *constr.* ►gravel additives.

51 add-on item [n] *constr. contr.* (Additional or conditional item in ►list of bid items and quantities [US]/schedule of tender items [UK]; ►alternate specification item, ►basic specification item, ►optional specification item); *s* **partida [f] (de obra) adicional** (En el ►resumen de prestaciones, partida de obra que puede ser necesaria en determinadas condiciones y que tiene como resultado el aumento del precio de la partida básica a la que pertenece que puede ser la ►partida básica, la ►partida discrecional o la ►partida alzada); *syn.* partida [f] (de obra) suplementaria; *f* **numéro [m] de prix pour plus-value** (Numéro d'ordre complémentaire dans un ►descriptif-quantitatif prévoyant une plus-value ou une majoration sur le ►numéro de prix d'une prestation de base ou sur le ►numéro de prix pour une variante ; ►numéro de prix provisoire) ; *syn.* numéro [m] de prix pour majoration ; *g* **Zulageposition [f]** (Im ►Leistungsverzeichnis gekennzeichnete, nicht eigenständige Position, die unter bestimmten, in der Position zu nennenden Voraussetzungen einen Zuschlag zur Vergütung einer Grundposition vorsieht. Denkbar sind **Z.en** zu ►Bedarfspositionen, ►Grundpositionen oder ►Alternativpositionen); *syn.* Zuschlagsposition [f].

52 adhesive [n] *constr.* (Generic term covering any agent that binds two surfaces together; e.g. epoxy, bitumen, glue, mastic, ►rapid-hardening cement); *s* **adhesivo [m]** (Término genérico para denominar cualquier producto que pega dos superficies, p. ej. resina epoxídica, betún, cola, pegamento de poliuretano; ►cemento rápido); *syn.* aglutinante [m], pegamento [m], pegante [m]; *f* **adhésif [m]** (Terme générique pour un matériau permettant de faire adhérer entre elles deux surfaces. Selon la nature des objets à assembler, on utilise différents types d'adhésifs dont la structure chimique va ou non réagir avec la surface des objets assemblés, p. ex. les colles de synthèse telles que les époxydes, les colles vinyliques, acryliques, [famille de résine à deux composants], le bitume, la glue, le mastic, le polyuréthane, ►accélérateur de prise, ►ciment prompt) ; *syn.* colle [f] adhésive ; *g* **Adhäsivkleber [m]** (Oberbegriff für einen Stoff zum Verkleben zweier Oberflächen, z. B. Epoxidharz [Zweikomponentenkleber], Bitumen, Klebstoff, Mastix, Polyurethan-Kleber [PU-Kleber], ►Schnellbinder); *syn.* Kleber [m].

53 adhesive disc/disk [n] *bot.* (Round climbing organ on clinging stems, which attaches itself to vertical supports; e.g. Virginia creeper *[Parthenocissus tricuspidata]*; ►adventitious climbing root); *syn.* adhesive pad [n]; *s* **ventosa [f] adhesiva** (Órgano de fijación de planta trepadora ; ►raíz adventicia); *f* **pelote [f] adhésive** (Organe de fixation de certaines plantes grimpantes sur un support plat telle que la vigne vierge *[Parthenocissus triscupidata]* ; ►crampon) ; *syn.* pelote [f] ventouse, ventouse [f], ventouse [f] adhésive ; *g* **Haftscheibe [f]** (Rundes Kletterorgan an Rankenzweigen bei Kletterpflanzen, das sich an der Oberfläche von Rankhilfen befestigt, z. B. beim Falschen Wein *[Parthenocissus tricuspidata]*; ►Haftwurzel).

adhesive disks [npl] *bot.* ►grasping branch with adhesive disks.

adhesive disks [npl]**, grasping organ with** *bot.* ►grasping branch with adhesive disks.

adhesive disks [npl]**, self-clinging vine with adhesive disks** [UK] *bot.* ►self-clinging vine with adhesive pads [US].

adhesive pad [n] *bot.* ►adhesive disc/disk.

adhesive pads [npl] *bot.* ►self-clinging vine with adhesive pads [US].

adjacent property [n] *leg. urb.* ►neighboring lot [US]/neighbouring plot [UK].

adjoining lot [n] [US] *leg. urb.* ►neighboring lot [US]/neighbouring plot [UK].

adjoining owner [n] **on a street** *leg. urb.* ►adjoining street resident.

54 adjoining street resident [n] *leg. urb.* (One of several adjacent house occupants living on a street); *syn.* abutter [n] on a street [also US], abutting owner [n] on a street [also US], abutting resident [n] on a street [US], adjoining owner [n] on a street; *s* **aledaño/a [m/f]** (de una calle); *syn.* vecino/a [m/f] (de una calle); *f* **riverain [m]** (Personne habitant le long d'une rue) ; *g* **Anlieger [m] (1)** (Anwohner einer Straße); *syn.* Anrainer [m], Anstößer [m] [CH].

adjusting ring [n]**, manhole** *constr.* ►manhole adjustment ring.

55 adjustment [n] *biol.* (Behavio[u]ral response of an organism to different situations or to internal changes to the social organization resulting from experience); *s* **ajuste [m]** (Cambios en el comportamiento de los organismos basados en la experiencia como respuesta a situaciones diferentes o a cambios internos en la organización de la comunidad); *syn.* adaptación [f]; *f* **adaptation [f] éthologique** (Capacité d'une espèce à modifier son comportement ou sa structure dans son milieu naturel en fonction des différentes situations rencontrées ou des changements internes dans l'organisation sociale du groupe) ; *g* **Anpassung [f] (3)** (Durch Erfahrungen bestimmte Verhaltensreaktionen eines Organismus gegenüber unterschiedlichen Situationen oder gegenüber inneren Veränderungen seines Gesellschaftsverbandes).

adjustment course [n] [UK]**, manhole** *constr.* ►manhole adjustment ring.

adjustment ring [n] *constr.* ►manhole adjustment ring.

56 administration [n] **and maintenance** [n] **of public green spaces** *adm. hort. landsc. urb.* (Routine day-to-day or week-to-week operations involved in the management, care and upkeep of public parks, gardens, street planting and other areas of public planting for control and enhancement of plant growth, weeding and cleaning of vegetated areas and maintenance of structures, equipment and amenities; ►establishment maintenance, ►maintenance management, ►maintenance work, ►urban forestry); *syn.* urban landscape management [n] [also US], maintenance regime [n] of public green spaces (GP 2003, Dec., p. 13); *s* **gestión [f] paisajística de espacios verdes** (Todas las medidas necesarias para promover el crecimiento de la vegetación y su estado adecuado, para limpiar las áreas verdes y las instalaciones técnicas y constructivas situadas en espacios libres, así como aquéllas necesarias para desarrollar y mantener en funcionamiento los espacios libres y sus instalaciones o para garantizar la seguridad. Incluyen medidas de rehabilitación, sustitución o replantación resp. resiembra; ►gestión paisajística de espacios verdes, ►mantenimiento de zonas verdes, ►mantenimiento y rehabilitación de zonas verdes, ►silvicultura urbana, ►trabajos de mantenimiento); *f* **gestion [f] des espaces verts publics** (Dans le cadre d'une optimisation des ressources humaines et matérielles tous travaux et méthodes nécessaires à la conservation et au développement normal des plantations [entretien du sol, tuteurage et haubanage, désherbage, tailles et élagages, arrosage, fertilisation, protection phytosanitaire, etc.] ou des semis [tonte, découpage des bordures, aération des pelouses,

A

roulage, etc.] au maintien en bon état de propreté esthétique des espaces verts publics en général, des voiries diverses, des sols sportifs de plein air, des ouvrages, des réseaux et équipements publics divers ; le but de ces opérations est la recherche d'une élévation du standard de qualité des espaces verts de proximité et d'une revalorisation des espaces publics ; ►entretien courant, ►entretien des espaces verts, ►gestion des espaces verts publics, ►maintenance 1, ►travaux d'entretien pendant l'année de garantie) ; *g* **Pflege [f, o. Pl.] und Unterhaltung von öffentlichen Grünflächen** (Unter Optimierung des Ressourceneinsatzes die Anwendung aller Methoden und Maßnahmen zur Lenkung und Förderung des Pflanzenwachstums und zur Säuberung von Vegetationsbeständen, technischen und baulichen Einrichtungen sowie Maßnahmen, die der Entwicklung und Erhaltung der Funktionsfähigkeit der öffentlichen Freianlagen und deren Einrichtungen sowie der Verkehrssicherheit dienen, incl. aller Ersatzbeschaffungen, Ergänzungen oder Nachsaaten und Nachpflanzungen. Ziel ist es, einen möglichst hohen Qualitätsstandard nutzbarer wohnungsnaher Grünflächen vorzuhalten; ►Management öffentlicher Grünflächen, ►Pflege, ►Pflege und Unterhaltung von Grünflächen, ►Unterhaltung).

administrative level [n] of planning *leg. plan.* ►higher administrative level of planning.

administrative personnel [n] [US] *adm. contr.* *►office worker.

57 administrative procedure [n] *adm. leg.* (Management process or action carried out by a public agency/authority as authorized by higher authority, such as a legislative body; e.g. a ►design and location approval process [US]/determination process of a plan [UK], public hearing [US]/public enquiry [UK], public hearing, environmental review, commenting opportunity, a ►preparation of a community development plan [US]/urban development plan [UK]); *s* **procedimiento [m] administrativo** (Procedimiento a seguir por una administración pública para ejecutar una tarea, como p. ej. elaboración de planes parciales o realización de proyectos con ayuda del ►procedimiento de aprobación de proyectos públicos; ►plan urbanístico); *f* **procédure [f] administrative** (Règles et formalités à observer lorsqu'une administration engage une action en vue de réaliser un projet, p. ex. une ►procédure d'instruction de grands projets publics, une procédure d'élaboration d'un PLU/POS ; ►établissement des documents d'urbanisme) ; *g* **Verwaltungsverfahren [n]** (Von einer öffentlichen Verwaltung durchgeführtes Verfahren zur Realisierung einer Aufgabe, z. B. ►Planfeststellungsverfahren, Verfahren zur Aufstellung eines Bebauungsplans, wasserrechtliches Verfahren. **V.** werden nach dem Verwaltungsverfahrensgesetz [VwVfG] des Bundes geregelt; ►Aufstellung eines Bauleitplanes).

administrative staff [n] [US] *adm. contr.* *►office worker.

adoption [n] [UK], tree *conserv.* ►stewardship of trees [US].

adoption [n] of a Proposals Map [UK] *leg. urb.* ►adoption of a zoning map [US].

58 adoption [n] of a zoning map [US] *leg. urb.* (Approval of a zoning map by the local government [US]/local authority [UK], resolution by a district planning authority to carry out a plan [UK]); *syn.* adoption [n] of a Proposals Map [± UK]; *s* **aprobación [f] de un plan parcial** (Decisión del órgano competente de aprobar definitivamente un plan parcial [de ordenación]; **en Es.** los Ayuntamientos de capitales de provincia y ciudades de más de 50 000 habitantes son competentes para la aprobación definitiva de planes parciales y planes especiales que desarrollen y se ajusten a las determinaciones del Plan general; art. 5.1 RD 16/1981; para los municipios de menos de 50 000 habitantes la competencia recae en la Comisión Provincial de

Urbanismo o en el correspondiente órgano de la Diputación Provincial); *f* **approbation [f] du PLU** (Acte approuvant un **p**lan **l**ocal d'**u**rbanisme [PLU] lors de la délibération du conseil municipal ou des organes délibérants des structures intercommunales) ; *g* **Beschluss [m] über den Bebauungsplan (±)** (Satzungsbeschluss der Gemeinde gemäß § 10 BauGB).

59 adret [n] *geo.* (Hillside [US]/hillslope [UK] which faces S. or S.W. in the northern hemisphere, and so receives the maximum available amount of sunshine and warmth; *opp.* ►ubac); *syn.* sunny slope [n]; *s* **ladera [f] solana** (Vertiente de un monte orientada al sur en el hemisferio norte y al norte en el hemisferio sur; *opp.* ►ladera umbría); *f* **adret [m]** (Versant face au sud [ou face au nord dans l'hémisphère austral] par opposition à l'►ubac) ; *g* **Sonnenseite [f]** (Besonnte Hangfläche/Talseite; auf der N-Halbkugel S- und SW-Lagen; *opp.* ►Schattenseite).

60 adulteration limit [n] *adm. envir. leg.* (Maximum measurable level, which is legally allowed for the pollution of the environment, food, etc.; *specific term for drinking water* maximum allowable concentration [= MAC value]; cf. LD 1994 [3], 42; ►emission standard, ►threshold level of air pollution); *syn.* adulteration threshold [n]; *s* **valor [m] límite** (Nivel máximo admisible de emisión: cantidad máxima de un contaminante del aire que la ley permite emitir en la atmósfera exterior; ►valor límite de emisión, ►nivel de referencia de calidad del aire); *f* **valeur [f] limite** (Niveau maximal de concentration de substances polluantes dans l'atmosphère, fixé sur la base des connaissances scientifiques dans le but d'éviter, de prévenir ou de réduire les effets nocifs de ces substances pour la santé humaine ou pour l'environnement ; art. 3 de la loi du 30 décembre 1996 ; ►valeur limite d'émission, ►valeur limite de la qualité de l'air) ; *g* **Grenzwert [m]** (Oberste Grenze, festgelegt in messbaren Werten, bis zu der Belastungen der Umwelt gesetzlich zugelassen sind; ►Emmissionsgrenzwert, ►Immissionsgrenzwert).

adulteration threshold [n] *adm. envir. leg.* ►adulteration limit.

advanced nursery-grown stock [n] [UK] *hort. plant.* ►specimen tree/shrub.

advances on account [n] [UK] *constr. contr.* ►interim payment.

advantage [n] *plan.* ►site advantage.

advective frost [n] *met.* *►radiation frost.

61 adventitious climbing root [n] *bot.* (BOT 1990, 104; small aerial root on stems of climbers, e.g. *Hedera, Hydrangea petiolaris,* and many species of Arum family *[Araceae]*, which attach themselves to solid supports without penetrating); *s* **raíz [f] adventicia** (Raíz aérea de planta trepadora, como *Hedera* y muchas *Araceaes* que se sujeta a otras plantas o a paredes y muros sin penetrar en ellos; cf. DB 1985); *f* **crampon [m]** (Courte racine aérienne, adventive, servant à la fixation superficielle de la plante contre les pierres ou l'écorce d'un arbre, p. ex. le lierre *[Hedera]*, l'*Hydrangea petiolaris* ou en grand nombre des *Araceae*) ; *g* **Haftwurzel [f]** (Kurze Luftwurzel bei Kletterpflanzen, die sich an feste Gegenstände anklammert, ohne in die Stütze einzudringen, z. B. bei Efeu *[Hedera],* Kletterhortensie *[Hydrangea petiolaris]* und vielen Aronstabgewächsen *[Araceae]*).

62 adventitious plant [n] *phyt.* (Plant species which is not indigenous in a certain area, but has recently been introduced by man; *specific terms* ►cultivated plant 1, ►neophyte); *syn.* recently introduced plant [n] (±); *s* **planta [f] adventicia** (Término aplicado a plantas advenedizas que no son propias de la localidad considerada, sino que han sido traídas accidentalmente por el hombre, o por cualquier circunstancia fortuita. Si éstas se aclimatan y resisten la competencia de las demás, se dice que se han naturalizado; ►planta de cultivo, ►neófito); *f* espèce [f] adventice (Espèce introduite accidentellement et capable de se maintenir à l'état sauvage pendant plusieurs années ; *termes spécifiques* ►plante de culture, ►plante néophyte) ; *syn.* espèce [f] subspontanée ; *g* Adventivpflanze [f] (Anthropochore Art, die in einem Gebiet nicht einheimisch ist, sondern erst in jüngster Zeit in dieses Gebiet unter Mitwirkung des Menschen gelangte; cf. BOR 1958; *UB* ►Kulturpflanze, ►Neophyt).

63 adventitious shoot [n] *bot.* (Growth of a dormant bud growing from a cleared trunk or between branches, e.g. a ►coppice shoot); *s* brote [m] adventicio (Retoño que crece desde un lugar no usual, p. ej. ►brote de cepa); *f* pousse [f] adventive (Organe végétal qui apparaît dans une position bien particulière, non classique, en dehors des processus normaux de ramification ; cf. DIB 1988 ; ►rejet de taillis) ; *g* Adventivspross [m] (An nicht gewohnten Stellen einer Pflanze entstehender Zusatzspross aus wieder teilungsfähigem Dauergewebe, z. B. ►Stockaustrieb).

64 adventure playground [n] *recr.* (Non-conventional play area, often under adult supervision, where children may construct huts and other structures with the material provided according to their own imagination); *s* parque [m] infantil de juegos de aventura (Terreno de juegos en zona urbana diseñado y equipado para crear un ambiente de aventura y así permitir a los niños y las niñas jugar activamente y realizar actividades manuales. Generalmente tienen una supervisión pedagógica); *syn.* terreno [m] de juegos de aventura; *f* terrain [m] d'aventure (Espace expérimental en milieu urbain, géré par une collectivité locale, animé par un ou plusieurs permanents et aménagé pour les jeunes enfants de façon à leurs permettre de participer à des activités manuelles et à des jeux éducatifs de nature) ; *syn.* terrain [m] de jeu pour l'aventure ; *g* Abenteuerspielplatz [m] (Unkonventioneller Spielplatz, meist unter Aufsicht, bei dem Kinder mit zur Verfügung gestellten Materialien selbst bauen und hantieren können); *syn.* Aktivspielplatz [m], Robinsonspielplatz [m].

adverse impact [n] *ecol. envir. recr.* ►disturbance, #2; ►environmental stress.

adverse impact [n]**, very** *ecol. envir.* ►severe disturbance.

adverse impact [n] **on the landscape** *conserv. leg. plan.* ►intrusion upon the natural environment.

advertisement board [n] [UK] *urb.* ►commercial sign [US].

65 advertisement structure [n] *landsc. urb.* (Device or fixture in advertizing, which is furnished for electronic means of publicity or for the attachment of notices, placards, posters, or bills; ►commercial sign [US]/advertisement board [UK]); *s* estructura [f] publicitaria (Término genérico para cualquier instalación fija para presentar publicidad electrónica o para fijar anuncios, pósters [Es]/afiches [AL] o carteles, ►cartel publicitario); *f* dispositif [m] publicitaire (Tout moyen matériel destiné à vanter auprès du public par une inscription, forme ou image les mérites d'un produit, d'un bien, d'un service, d'une entreprise ou d'un événement et appliqué sur un support statique, mécanique, lumineux ou électronique, fixe ou mobile, mural, scellé au sol, sur toiture ou terrasse; *termes spécifiques pour un dispositf apposé sur un immeuble* ►enseigne publicitaire ; ►panneau publicitaire) ; *g* Werbeanlage [f] (Vorrichtung, die für das Anbringen von Anschlägen, Transparenten, Folien oder für elektronische Werbemittel eingerichtet ist; *UB* ►Werbetafel).

adviser [n] [UK]**, expert** *prof.* ►expert witness.

advisory services [npl] [UK]**, post-completion** *prof.* ►concluding project review.

aeolian energy [n] *envir.* ►wind energy.

aeolian sand [n] *pedol.* ►wind-blown sand.

66 aeration [n] (1) *agr. limn. pedol.* (Process of introducing air into the ground or into a waterbody by natural or artificial means to increase oxygen availablility); *s* aireación [f] (Proceso de introducir aire en el subsuelo o en un cuerpo de agua por medio natural o artificial para aumentar la disponibilidad de oxígeno); *f* aération [f] (Alimentation en oxygène des cours d'eau, du sol) ; *g* Belüftung [f] (1) (Versorgung eines Gewässers oder Bodens mit Luft, um den Sauerstoffgehalt für die Lebewesen zu erhöhen).

aeration [n] (2) *constr.* ►lawn aeration.

67 aerial photogrammetry [n] *rem'sens. surv.* (Depiction of landscape through systematically reliable measurements of an area by the use of aerial photographs in surveying and mapping; ►topographic mapping); *s* fotogrametría [f] aérea (Método para investigar, describir el paisaje y elaborar mapas por medio de tomas sistemáticas con cámaras fotográficas de medición montadas en aviones; ►cartografía topográfica); *f* photogrammétrie [f] aérienne (Méthode de recherche et de description des paysages consistant dans la couverture photographique aérienne d'une aire prospectée par un quadrillage systématique et dont les opérations comportent l'exécution de photographies d'axe vertical formant des couples stereoscopiques, la restitution graphique ou numérique et la cartographie ; ►phototopographie) ; *syn.* photogrammétrie [f] terrestre ; *g* Luftbildmessung [f] (Methode zur Erforschung und Beschreibung der Landschaft durch systematische Aufnahme eines Gebietes mit einer im Flugzeug eingebauten automatischen Messbildkamera und der Darstellung der Auswertung in ein Kartenwerk; ►Erdbildmessung); *syn.* Aero-Fotogrammetrie [f], *o. V.* Aero-Photogrammetrie [f], Luftfotogrammetrie [f], *o. V.* Luftphotogrammetrie [f].

68 aerial photograph [n] *envir. plan. rem'sens. surv.* (Picture of part of the Earth's surface taken by a camera, usually from an aero-plane; ►infrared color aerial photograph, ►interpretation of aerial photographs, ►oblique aerial photograph, ►satellite photograph, ►vertical aerial photograph); *s* foto(grafía) [f] aérea (►fotografía aérea infrarrojo color, ►fotografía aérea oblicua, ►fotografía aérea vertical, ►imagen de satellite, ►interpretación de foto[grafía]s aéreas); *f* photographie [f] aérienne (Cliché d'une partie de la surface de la terre, pris en général par avion ; ►interpretation des photographies aériennes, ►photographie aérienne infrarouge couleur, ►photographie satellite, ►photographie verticale, ►prise de vue aérienne oblique) ; *g* Luftbild [n] (Fotografische Aufnahme eines Teils der Erdoberfläche, meist von einem Flugzeug aus; ►Infrarotluftbild, ►Luftbildauswertung, ►Luftschrägbild, ►Satellitenaufnahme, ►Senkrechtluftbild); *syn.* Luftbildaufnahme [f].

aerial photograph [n]**, false colo(u)r** *envir. plan. rem' sens. surv.* ►infrared color aerial photograph [US].

aerial photographic interpretation [n] *envir. plan. rem'sens. surv.* ►interpretation of aerial photographs.

69 aerial plant part [n] *bot.* *s* parte [f] aérea de una planta (DA 1975); *f* organe [m] aérien (DAV 1984, 4) ; *syn.* partie [f] aérienne (DA 1975) ; *g* oberirdischer Pflanzenteil [m] *syn.* oberirdischer Teil [m] einer Pflanze.

70 aerial root [n] *bot.* (Above-ground root that for some portion of its length is surrounded by air and, when in contact with soil, it develops a normal root system; not necessarily a breathing root; RRST 1983); *s* raíz [f] aérea (Raíz que no se introduce en el suelo ni en el agua, sino que permanece en el aire, como ocurre con muchas orquídeas epífitas; cf. DB 1985); *syn.* raíz [f] epígea; *f* racine [f] aérienne (Racine adventive qui pend le long d'une tige, pouvant atteindre le sol et se comporter comme une racine ordinaire, p. ex. les racines-piliers, les racines-lianes, etc.) ; *g* Luftwurzel [f] (Oberirdisch entstehende Adven-

A

tivwurzel, die in der Luft meist unverzweigt bleibt, am Boden angekommen, sich jedoch verzweigt und sich wie eine gewöhnliche Wurzel verhält).

aerie [n] *zool. game'man.* ▶eyrie.

aerie observation [n] *conserv. game'man.* ▶protective eyrie observation.

71 aerosol [n] *envir. met.* (Mixture of extremely fine particles either solid (a smoke) or liquid (an aerosol) suspended in a gas. They range in size from less than 10 nanometres to more than 100 micrometres in diameter. Some aerosols occur naturally, in fog, clouds, ash from volcanoes, dust storms, and smoke, especially from burning fossil fuels, forest and grassland fires, living vegetation, and sea spray. On a global average, anthropogenic aerosols arising from industry, traffic and domestic heating currently account for about 10 percent of the total amount of aerosols in the atmosphere); *s 1* **aerosol** [m] (Conjunto de partículas sólidas o líquidas, de tamaño coloidal, suspendidas en un gas, normalmente el aire. Algunos **a.** son p. ej. la niebla, las nubes, cenizas de erupciones volcánicas, tormentas de polvo o el humo, procedente de la combustión de combustibles fósiles, incendios de bosques y savanas y de salpicadura de sal. A nivel global, aerosoles antropógenos procedentes de la industria, el tráfico y la calefacción doméstica son responsables de aprox. 10% del total de los **a.** en la atmósfera); *s 2* **aerosol** [adj] (Término utilizado para designar un gran número de productos envasados en recipientes a presión, que se expelen en forma pulverizada como líquido, como polvo o como espuma); *f* **aérosol** [m] (Ensemble de particules ultramicroscopiques [moins d'un micron], solides [fumées] ou liquides [brouillard] en suspension dans un gaz provenant soit de la suie des feux et des poussières des incendies de forêts, soit des éruptions volcaniques, soit des rejets industriels de poussières [centrales électriques, avions, transports, etc.] ; dans le contexte de la pollution de l'air, un aérosol désigne la matière particulaire fine, plus grosse qu'une molécule, mais assez petite pour rester en suspension dans l'atmosphère pendant au moins plusieurs heures ; le terme « aérosol » est aussi communément utilisé pour désigner un contenant sous pression [pulvérisateur], lequel est créé pour diffuser un léger jet d'une substance telle que de la peinture, insecticides, déodorant, etc. ; les sources naturelles d'aérosols comprennent notamment les particules de sel provenant des embruns marins ainsi que les particules de poussière et d'argile provenant de l'érosion des roches ; les aérosols peuvent également être générés par certaines activités humaines et sont souvent considérés comme des polluants [composés organiques volatils — COV], combinés avec des oxydes d'azote, certains COV peuvent former de l'ozone, qui est un polluant à basse atmosphère ; les aérosols jouent un rôle important dans l'atmosphère, plus précisément dans la condensation des gouttes d'eau et les cristaux de glace, dans différents cycles chimiques, ainsi que dans l'absorption du rayonnement solaire) ; *g* **Aerosol** [n] (...sole [pl.]; aus dem Griech. *a□r* >Luft< und Lat. *solutus* >aufgelöst<; Gemisch von feinst verteilten flüssigen oder festen Stoffen mit Luft. Die Größe der Schwebstoffteilchen liegt zwischen 0,001 und 100 µm. **A.e** sind z. B. Nebel, Wolken, Staub- und Aschewolken sowie Rauch. Die z. T. elektrisch geladenen Teilchen der in der Luft verteilten **A.e** spielen als Kondensationskerne eine wichtge Rolle im Wettergeschehen. Sie werden besonders durch Verbrennungsvorgänge in Industrie und Technik, in Haushalten, durch Verkehrsmittel u. a. künstlich vermehrt).

aerosol protection forest [n] *for. landsc.* ▶dust and aerosol protection forest.

aesthetic landscape assessment [n] *plan. landsc.* ▶visual landscape assessment.

72 aesthetic ordinance [n] [US] *leg. urb.* (Legally-binding local enactment for the preservation of visual amenities, implemented by guidelines and regulations, in order to retain or improve existing buildings and their setting; cf. Civic Amenities Act, 1967 of the U.K.; ▶design ordinance [US], ▶design recommendations); *syn.* building development bye-laws [npl] [UK]; *s* **ordenanza** [f] **municipal de protección de la calidad visual urbana** (En GB y D. norma municipal legalmente vinculante para preservar y mejorar la calidad visual de la construcción y su entorno; ▶reglamento de diseño; ▶guía de diseño); *syn.* ordenanza [f] municipal de protección del aspecto escénico urbano; *f 1* **règle** [f] **(d'urbanisme) concernant l'aspect extérieur des constructions** (≠) (Servitude d'urbanisme spécifique, faisant partie des règles générale d'urbanisme définies dans un document d'urbanisme (PLU/POS), en vue d'améliorer l'esthétique urbain) ; D., prescriptions architecturales communales favorisant la personnalité, l'identité architecturale, paysagère, patrimoniale d'un lieu ainsi que la recherche d'une meilleure qualité architecturale et urbaine en imposant des normes relatives à l'aspect des constructions et de leurs abords; *syn.* prescriptions [fpl] de préservation du paysage urbain, prescriptions [fpl] urbanistiques [B] ; *f 2* **prescriptions** [fpl] **architecturales** (Dispositions architecturales de détail en vue de préserver et mettre en valeur l'environnement architectural dans les secteurs à plan de masse) ; *f 3* **prescriptions** [fpl] **de conservation du patrimoine architectural constitutif d'un ensemble urbain** (Prescriptions et recommandations de conservation des constructions qui, par leurs volumes et leur aspect architectural typique participent, à la qualité architecturale d'une agglomération et s'appliquant à l'intérieur du périmètre d'une zone de protection du patrimoine architectural, urbain et paysager [Z.P.P.A.U.P.] ; *g* **Ortsbildsatzung** [f] (Von einer Gemeinde oder einem Landkreis durch Satzungsbeschluss herbeigeführtes rechtliches Instrumentarium [Rechtsnorm und Gesetz im materiellen Sinne] zur Erhaltung eines bestimmten Ortsbildes durch Festlegung von Gestaltungsdetails an Gebäuden und deren Umgebung; ▶Gestaltungssatzung, ▶Gestaltungsempfehlung).

aesthetic quality [n] *recr. plan.* ▶visual quality.

aesthetics [n] *land'man. landsc. recr.* ▶landscape aesthetics.

aesthetics management [n]**, landscape** *landsc. plan.* ▶visual resource management.

aesthetic value [n] *recr. plan.* ▶visual quality.

aesthetic value [n] **of a landscape** *landsc. plan. recr.* ▶visual quality of landscape.

affidavit [n] [US]**, contractor's** *contr.* ▶bidder's affidavit [US].

73 afforestation [n] *constr. for. land'man.* (Process of planting a new forest in an area, which has not been forested in recent times; ▶high-altitude afforestation, ▶reafforestation); *s* **forestación** [f] (Establecimiento de bosques, mediante siembra o plantación, en terrenos que anteriormente eran agrícolas o estaban dedicados a otros usos no forestales; **en Es.** la planificación y gestión forestal es competencia de las CC.AA.; a la Administración General del Estado le corresponde —entre otras funciones coordinadoras— en colaboración con las CC.AA., la definición de los objetivos generales de la política forestal española, en particular por medio de la aprobación de documentos base como a] la Estrategia forestal española, b] el Plan forestal español, c] el Programa de acción nacional contra la desertificación y d] el Plan nacional de actuaciones prioritarias de restauración hidrológico-forestal. La planificación forestal a nivel de CC.AA. se realiza por medio de los Planes de ordenación de recursos forestales [PORF] que incluyen —si diera el caso— programas de forestación; ▶reforestación); *syn. parcial* repoblación [f] forestal; *f 1* **reboisement** [m] (Constitution artificielle d'un peuplement forestier sur une zone restée longtemps non boisée ; extension d'un peuplement forestier déjà existant en vue de son exploitation ; ▶reboisement d'altitude, ▶reforestation) ; *syn.* afforestation [f] [aussi CDN] ; *f 2* **boisement** [m] (Création d'un jeune peuplement forestier sur un terrain non antérieurement boisé de mémoire d'homme. Au CDN, le terme « boisement » s'oppose à reboisement [d'extension] ; DFM 1975) ; *g* **Aufforstung** [f] (Gründung eines Waldbestandes durch Saat oder Pflanzung auf einer Fläche, die lange nicht bewaldet war; der UB **Ersatzaufforstung** gilt für den Ausgleich des Verlustes an Waldfläche an anderer Stelle; cf. § 18 (1) lit. c ForstG; ▶Hochlagenaufforstung, ▶Wiederaufforstung); *syn.* Bestandesgründung [f], Neuaufforstung [f], Neubewaldung [f] ([A] § 4 ForstG).

afforestation [n] [UK]**, upland** *for. land'man.* ▶high-altitude afforestation.

aficionado [n] [US]**, garden** *gard.* ▶garden enthusiast.

African-Eurasian Waterbirds [npl] *conserv. ecol. pol.* ▶Convention on the Conservation of African-Eurasian Migratory Waterbirds [AEWA].

after care [n] [UK]**, programme of** *constr. hort.* ▶plant maintenance program [US].

after-use [n] [UK] *plan.* ▶adaptive use [US].

74 age class composition [n] *plan. sociol.* (Proportion of population in different age groups; an age group is called *cohort*; ▶demographic pyramid); *syn.* age classification [n], age states [npl] (TEE 1980, 62); *s* **estructura** [f] **demográfica** (Porcentaje de la población de un país en los diferentes grupos de edades, representada gráficamente en la llamada ▶pirámide demográfica); *syn.* distribución [f] demográfica, distribución [f] de población en grupos de edades, estructura [f] de población en grupos de edades; *f* **structure** [f] **par âges de la population** (Représentation graphique de la répartition d'une population par âges ou groupes d'âges ; ▶pyramide des âges ; *g* **Altersstruktur** [f] (Altersmäßige Zusammensetzung einer Bevölkerung resp. einer Population; wird grafisch oft in Form einer ▶Alterspyramide dargestellt. Die **A.** änderte sich im 20. Jh. an der Spitze der Bevölkerungspyramide in zwei Etappen: 1. beruhte zunächst in der ersten Hälfte des 20. Jhs der hohe Zugewinn an Lebenserwartung primär auf dem Rückgang der Säuglings- und Kindersterblichkeit und auf der Abnahme der Sterblichkeit im jüngeren und mittleren Alter als Folge medizinischer Fortschritte bei der Bekämpfung der Infektionskrankheiten. 2. Im Unterschied dazu ist der Zuwachs an Lebenserwartung in den letzten Jahrzehnten des 20. Jhs. primär auf die Abnahme der Sterbewahrscheinlichkeit im höheren Alter durch Fortschritte der Medizin bei der Bekämpfung der Herz- und Kreislauferkrankungen zurückzuführen); *syn.* Altersaufbau [m], Altersgliederung [f], Altersklassenzusammensetzung [f].

age classification [n] *plan. sociol.* ▶age class composition.

agency [n] [US] *adm. contr. leg. pol.* ▶contracting public agency [US], ▶federal government agency [US], ▶land-holding agency [US], ▶law enforcement agency [US], ▶local/regional administrative authority/agency, ▶state government agency [US], ▶water users agency [US].

agency [n] [US]**, approving** *adm. leg.* ▶approving authority.

agency [n] [US]**, operating** *adm. leg.* ▶operating authority.

agency [n] [US]**, public** *adm.* ▶public authority.

75 agency [n] **requesting proposals** *adm. contr.* (Authority requesting submittal of proposals for competitive bidding [US]/competitive tendering [UK] documents; ▶bidding [US]/

tendering [UK], ▶request for qualification [US], ▶request for proposals [US]); *s* **promotor** [m] **de concurso-subasta de obras** (Institución que saca una obra a ▶concurso-subasta de obras, ▶demanda de proposiciones); *f* **service** [m] **responsable de la procédure d'appel d'offres** (Administration qui lance un ▶appel d'offres ; ▶adjudication, ▶concours restreint) ; *g* **ausschreibende Stelle** [f] (Behörde, die eine ▶Ausschreibung organisiert, ▶öffentlicher Teilnahmewettbewerb für Planungsbüros).

agency review [n] [US]**, coordinated public** *adm. leg.* ▶statutory consultation with public agencies [US].

agency [n] **with planning powers** [US]**, public authority or** *adm. plan.* ▶public planning body [US].

agent [n] *agr. chem. envir. for. hort.* ▶defoliating agent, ▶plant protection agent.

agent [n]**, toxic** *envir.* ▶noxious substance.

agent orange [n] *chem.* ▶defoliating agent.

age states [n] *plan. sociol.* ▶age class composition.

76 aggradation [n] *geo.* (Building up of land surface by the importation and accumulation of solid material derived from denudation by a river or the development of a marine beach; DNE 1978; ▶alluvial gravel, ▶land reclamation 1, ▶sediment 1, ▶sedimentation, ▶siltation, ▶silting up 2); *s 1* **acumulación** [f] (Conjunto de materiales sedimentados tras haber sido sometidos a un transporte más o menos largo, a cargo de algunos de los agentes [curso de agua, glaciar, olas, viento]; DGA 1986; ▶acumulación de pedregal, ▶entarquinamiento 1, ▶entarquinamiento 2, ▶ganancia de tierras 1, ▶sedimentación, ▶sedimento); *syn.* aluvionamiento [m], aluvionación [f]; *s 2* **agradación** [f] (Dícese de la superficie generada por acumulación de materiales de origen continental [fluviales, glaciares, etc.]; DGA 1986; ▶acumulación de pedregal, ▶entarquinamiento, ▶ganancia de tierras 1, ▶sedimentación, ▶sedimento); *syn.* sedimento [m] de limo; *f* **alluvionnement** [m] (*Terme générique* apport et amoncellement de sédiments grossiers d'origine marine ou fluviatile ; ▶atterrissage, ▶envasement, ▶épandage alluvial grossier, ▶poldérisation, ▶sédiment, ▶sédimentation; *termes spécifiques* alluvionnement graveleux, alluvionnement limoneux [*syn.* limonage], alluvionnement sableux) ; *syn.* placage [m] alluvionnaire, colmatage [m] alluvionnaire ; *g* **Auflandung** [f] (1) (Anschwemmung und Ablagerung von Lockermaterial im Küsten- oder Fließgewässerbereich; ▶Ablagerung 1, ▶Ablagerung 2, ▶Anlandung 2, ▶Aufschotterung, ▶Landgewinnung, ▶Verlandung).

77 aggregate [n] *constr.* (Granular material, i.e. sand, gravel, crushed stone, slag or cinders, suitable for use in various construction operations and as an additive in the making of mortar, concrete or asphaltic concrete. Aggregate comprises fractions which are roughly the same or of various sizes with a dense structure; ▶exposed aggregate paving block, ▶fine aggregate [US]/stone chippings [UK], ▶mineral aggregate); *s* **árido** [m] (Material mineral o artificial granulado que se utiliza en diferentes operaciones de construcción y de aditivo en la fabricación de mortero, hormigón o asfalto; ▶gravilla, ▶material pétreo, ▶piedra de hormigón con árido visto); *syn.* agregado [m] (pétreo); *f* **granulat** [m] (Matériau grenu concassé ou non, destiné à être aggloméré à un liant, entrant dans la composition du mortier, du béton, de l'asphalte, de masse volumique, porosité et de caractéristiques mécaniques variables ; les **g.** peuvent avoir une granulométrie serrée ou ouverte ; ▶grave, ▶gravillons concassés) ; *syn.* agrégat [m] ; *g* **Zuschlagstoff** [m] (Für die Mörtel-, Beton- und Asphaltbelagsherstellung geeignete ungebrochene und/oder gebrochene Körner aus natürlichen und/oder künstlichen, dichten oder porigen mineralischen Stoffen; der **Z.** besteht aus etwa gleich oder verschieden großen Fraktionen mit dichtem Gefüge; cf. DIN 1045, Pkt. 2.1.3.2 und DIN EN 12 620; ▶Splitt); *syn.* Zuschläge [mpl].

aggregate [n] [US]**, base of crusher-run** *constr.* ▶crushed aggregate subbase.

aggregate [n] [US]**, crusher-run** *constr.* ▶crushed rock.

aggregate [n] [US]**, layer of crusher-run** *constr.* ▶gravel subbase [US]/gravel sub-base [UK].

aggregate stability [n] **of soil** *pedol.* ▶structural stability of soil.

aggregate structure [n] *pedol.* ▶crumb structure.

aggregation [n] *zool.* ▶survival group.

78 agrarian cultural landscape [n] *agr.* (Rural land on which historic traces of cultivation are still visible); *s* **paisaje** [m] **rural (1)** (Paisaje cultural en el que se pueden reconocer las estructuras de producción agrosilvícola del pasado); *f* **paysage** [m] **rural traditionnel** (Paysage rural parcellisé, dont la structure témoignant des méthodes artisanales de l'agriculture traditionnelle) ; *g* **bäuerliche Kulturlandschaft** [f] (Agrarlandschaft, die noch alte Strukturen der Bewirtschaftung erkennen lässt).

agrarian landscape [n] *agr. plan.* ▶agricultural landscape, ▶extensively cleared land (for cultivation).

agrarian landscape [n]**, totally cleared** *land'man.* ▶extensively cleared land (for cultivation).

agrarian structure [n] *agr. plan.* ▶improvement of agrarian structure.

agrarian structure planning [n] *agr. plan.* ▶preliminary agrarian structure planning.

79 agreed measurement [n] **of completed work** [US] *constr. contr.* (Ascertainment of contractually-executed work [US]/works [UK] during final inspection by representatives of the client and the contractor as the basis for issuing a certificate of payment); *syn.* agreed measurement [n] of completed project, agreed measurement [n] of completed works [UK]; *s* **medición** [f] **acordada de los trabajos terminados** (Control de las obras terminadas en la inspección final por parte de representante del cliente como base para la liquidación de cuentas final); *syn.* medida [f] acordada de los trabajos terminados; *f* **métré** [m] **général et définitif** (Situation récapitulative détaillée et complète de tous les ouvrages exécutés et réceptionnés en vue de l'établissement d'un décompte final) ; *g* **örtliches Aufmaß** [n] (Feststellung der vertraglich vereinbarten Leistungen nach Fertigstellung auf der Baustelle als Grundlage für die Abrechnung; wird häufig durch die **Abrechnung nach Ausführungszeichnungen** ersetzt).

agreement [n] *contr. leg. pol. prof.* ▶by mutual agreement, ▶convention 1, ▶fee agreement, ▶oral agreement.

agreement [n]**, conclusion of a binding** *contr. prof.* ▶conclusion of a contract.

agreement [n] [UK]**, form of** *contr. prof.* ▶standard contract form [US].

agreement [n]**, in ~ with** *leg.* ▶by mutual agreement.

agreement [n]**, termination of an** *contr. prof.* ▶termination of a contract [US].

80 agreement [n] **on contract period** *contr.* (Mutual formal or informal accord on the length of execution period between client/owner and contractor; ▶contract time); *s* **fijación** [f] **del plazo de ejecución** (Acuerdo mutuo entre comitente y licitante sobre el periodo/período en el que se ha de realizar una obra; ▶plazo de duración de contrato); *f* **fixation** [f] **des délais d'exécution** (Détermination entre le maître d'ouvrage et l'entrepreneur de la période d'exécution de l'ensemble des

travaux ; ▶délai d'exécution des travaux) ; *syn.* fixation [f] de la date limite d'achèvement des travaux ; *g* **Vereinbarung** [f] **der Ausführungsfrist** (Gemeinsames Festlegen des Ausführungszeitraumes zwischen Auftraggeber und Auftragnehmer; cf. § 11 VOB Teil A und § 5 VOB Teil B; ▶Vertragsfrist).

agreement [n] **on planning proposals** *plan.* ▶obtaining a mutual agreement on planning proposals.

Agreement [n] **on the Conservation of African-Eurasian Migratory Waterbirds [AEWA]** *conserv. ecol. pol.* ▶Convention on the Conservation of African-Eurasian Migratory Waterbirds [AEWA].

agressive plant [n] *hort. phyt.* ▶invasive species.

agricultural and forest land use [n] *agr. for. hort.* ▶intensity of agricultural and forest land use.

81 agricultural area [n] *agr.* (Piece of land devoted to agriculture or concerned with agricultural production and its facilities; ▶agricultural district [US] 1/agricultural land [UK], ▶agricultural production land, ▶priority agricultural area); *syn.* agricultural land [n], farmland [n]; *s* **área** [f] **agrícola** (Zona con infraestructura para el uso agrícola; ▶área priorizada para uso agrícola, ▶suelo agrícola, ▶superficie agrícola útil); *f* **superficie** [f] **agricole** (▶superficie agricole utilisée, ▶zone agricole 1, ▶zone à vocation rurale) ; *syn.* terres [fpl] agricoles ; *g* **landwirtschaftliche Fläche** [f] **(1)** (▶landwirtschaftliche Fläche 2, ▶landwirtschaftliche Nutzfläche, ▶landwirtschaftliche Vorrangfläche); *syn.* Agrarfläche [f], Agrarland [n], landwirtschaftlich genutzte Fläche [f].

agricultural business [n] *agr.* ▶part-time agricultural business.

82 agricultural district [n] [US] **(1)** *agr. leg. plan.* (Zoning category restricted to agricultural use as shown on zoning maps; **in U.S.**, the term "agricultural land" may be also used on a comprehensive plan; in U.S., no zoning term; ▶agricultural area, ▶rural district [US]/rural settlement area [UK]); *syn.* agricultural land [n] [UK]; *s* **suelo** [m] **agrícola** (Categoría de uso del suelo en planes urbanísticos; ▶área agrícola, ▶zona rural urbanizable); *f* **zone** [f] **agricole (1)** (Catégorie d'occupation du sol dans les documents d'urbanisme ; ▶superficie agricole, ▶zone agricole 2) ; *g* **landwirtschaftliche Fläche** [f] **(2)** (In Bauleitplänen dargestellte Nutzungskategorie; ▶Dorfgebiet, ▶landwirtschaftliche Fläche 1); *syn.* Fläche [f] für die Landwirtschaft (§§ 5 [2] 9 u. 9 [1] 18 BauGB), Fläche [f] der Landwirtschaft (§ 4 [1] WertV).

agricultural district [n] [US] **(2)** *leg. urb.* ▶rural district [US]/rural settlement area [UK].

83 agricultural drainage [n] *agr.* (Removal of surface water and soil excess water through the installation of ditches and drainage channels as well as through pipe drains on arable or pasture land; ▶soil drainage 1, ▶field drainage, ▶drainage, ▶land drainage); *s* **drenaje** [m] **agrícola** (Medidas tomadas para reducir el nivel freático en campos de cultivo o prados instalando canales de desagüe o drenes; ▶drenaje, ▶drenaje a gran escala, ▶drenaje agrícola subterráneo, ▶drenaje del suelo); *f 1* **drainage** [m] **agricole** (Ensemble des opérations nécessaires à la suppression des excès d'eau par modelé superficiel, par fossés ouverts ou par canalisations enterrées à l'échelle de la parcelle agricole LA 1981, 424 ; en F. le terme « drainage » désigne l'élimination naturelle et artificielle des eaux excédentaires sur et dans les sols. Le terme allemand « ▶Dränung » désigne seulement l'élimination naturelle et artificielle des excès d'eau dans le sol ; ▶drainage [agricole] à ciel ouvert, ▶drainage agricole souterrain, ▶drainage de sol, ▶évacuation des eaux) ; *syn.* assainissement [m] agricole ; *f 2* **drainage** [m] **superficiel** (Mesures d'élimination de l'engorgement en eau du sol effectué par des labours spéciaux [billons] ou par des petites tranchées, ►fossés de drainage et canaux) ; *syn.* drainage [m] par fossé ; *g* **Entwässerung [f] landwirtschaftlicher Flächen** (Wasserableitung in der bäuerlichen Kulturlandschaft durch offene Gräben oder Dränleitungen; ►Bodenentwässerung, ►Dränung landwirtschaftlicher Flächen, ►Entwässerung, ►Wasserbaumaßnahmen zur Entwässerung landwirtschaftlicher Flächen).

agricultural holdings [npl] *agr. for.* ►scattered agricultural holdings.

agricultural land [n] *agr.* ►agricultural area; *agr. leg. plan.* ►agricultural district [US] (1)/agricultural land [UK].

agricultural land [n]**, consolidation of** *adm. agr. leg.* ►farmland consolidation [US].

agricultural land [n]**, tenant** *agr.* ►agricultural leasehold (property).

84 agricultural land abandonment [n] *agr.* (Leaving formerly cultivated land unused for a long period; ►fallow land, ►leave uncultivated, ►uncultivated farmland); *s* **abandono [m] de tierras de cultivo** (Dejar sin cultivar tierras durante muchos años; ►dejar tierras en barbecho, ►parcela de cultivo abandonada, ►tierra en barbecho); *f* **abandon [m] des cultures** (Mise à l'abandon pendant plusieurs années de sols cultivés avec développement d'une végétation herbacée dense, de hauteur irrégulière, qui, avec un abandon total, évolue progressivement vers un stade buissonnant puis forestier ; ►laisser en friche, ►terre en friche, ►terre en jachère) ; *syn.* abandon [m] de terres agricoles ; *g* **Verbrachung [f]** (Jahrelanges Liegenlassen von vormals kultivierten landwirtschaftlichen Flächen; ►brach fallen, ►Brachfläche, ►Brachland).

agricultural land classification [n] [UK] *agr.* ►agricultural land grade [UK].

85 agricultural land grade [n] [UK] *agr.* (**1**. Land class broadly defining the inherent value for crop production; **In U.K.**, the **Agricultural Land Classification [ALC] System** is a land grading framework first developed in 1966 by the Ministry of Agriculture, aiming to give strategic guidance on land quality for planning purposes. Since then the classification system has been amended and revised many times The maintenance and development of the current ALC system is the responsibility of DEFRA [The Department for Environmental Food and Rural Affairs]. It allows agricultural land to be graded from best [grade 1] to worst [grade 5] and provides a consistent, country-wide system of land classification. **2. In U.S.**, the U.S. Soil Classification Service evaluates land and soils in detail in accordance with their productive capacity for normal crops and grassland, taking into account the long-term risks of soil erosion, the use restrictions for agriculture or other uses. The soil suitability zoning is divided into eight main categories [►land capability classes (US)] and four sub-classes [land capability subclasses]. The main classes indicate the levels of increasing use restrictions and the decline agricultural yield, taking into account the water absorption capacity for irrigation. The sub-classes are equipped with four letters: **E** for the degree of erosion hazard, **W** for the degree of soil moisture, **S** for the limits of root penetration and **C** for climatic limits; ►arable land rank [US], ►land capability class [US], ►land capability classification [US], ►land use capability classification [UK]); *s* **índice [m] de fertilidad de suelos** (**1**. En EE.UU., en el sistema del *U.S. Soil Classification Service* [Servicio de Clasificación de Suelos de los EE.UU.] se valoran los suelos detalladamente según su productividad para el cultivo de frutos del campo y de hierbas de pradera, teniendo en cuenta los peligros de erosión y otras restricciones del uso agrícola. La clasificación se subdivide en 8 clases principales [►clase de aptitud del suelo] y 4 subclases [subclase de aptitud del suelo]. Las clases principales diferencian los niveles crecientes de restricción del uso y la reducción de la productividad agrícola, también considerando la capacidad de absorción del agua de irrigación; las subclases se distinguen con 4 letras: **E** grado de peligro de erosión, **W** grado de humedad del suelo, **S** limitaciones para el enraizamiento y **C** limitaciones climáticas. **2.** En GB, el sistema de **Clasificación de Tierras Agrícolas** [Agricultural Land Classification—ALC] es un marco de graduación de suelos desarrollado por primera vez en 1966 por el Ministerio de Agricultura, con el fin de dar orientación estratégica sobre calidad de los suelos a la planificación. Desde entonces el sistema de clasificación ha sido modificado y revisado muchas veces. El mantenimiento y desarrollo del sistema ALC actual es responsabilidad del Departamento de Asuntos Alimentarios y Rurales Ambientales [Department for Environmental Food and Rural Affairs— DEFRA]. Este permite clasificar las tierras agrícolas del nivel más alto [grado 1] al peor [grado 5] y prevé un sistema consistente de clasificación de suelos para todo el país. **3.** En D. número entre 1 y 100 que indica el valor creciente de los suelos agrícolas; ►clasificación de aptitud de suelos, ►índice de fertilidad de tierras de cultivo); *f 1* **indice [m] de fertilité physique (IFP)** (**1. F.**, nombre variant de 1/20 à 19/20 et qui permet de qualifier la fertilité du sol sur le plan physique, en liaison avec sa profondeur, sa pierrosité, son état calcique et organique. L'IFP correspond au potentiel de captage des réserves nutritives par les racines ; ►classification de l'aptitude culturale des sols. **2. U.S.**, ►classe d'aptitude d'un sol) ; *f 2* **coefficient [m] de fertilité d'un sol (±)** (**D.**, nombre variant de 1 à 100 et définissant la fertilité à attribuer à un sol agricole d'après ses caractères génétiques et fonctionnels ; PED 1984, 206 ; ►classification de l'aptitude des terres [FAO], ►classification des sols et des paysages agricoles de premier choix et marginaux [CDN], ►indice d'estimation des sols) ; *syn.* coefficient [m] de valeur agricole des terres ; *g* **Bodenzahl [f] (±)** (**1**. Ungefähres Maß [relative Wertzahl] für die Ertragsfähigkeit [erzielbarer Reinertrag] der landwirtschaftlich oder gärtnerisch genutzten Böden; in D. wird als fruchtbarster Boden der Schwarzerdeboden der Magdeburger Börde mit der Wertzahl 100 zu Grunde gelegt; ►Ackerzahl, ►Bodenbewertung 2, ►Bodeneignungsbewertung, ►Bodeneignungsklasse, ►Bodenschätzung. **2.** In den USA werden nach dem System des *U.S. Soil Classification Service* Land und Böden detailliert gemäß ihrer Produktionskraft für normale Feldfrüchte und Grünlandgräser beurteilt — bei gleichzeitiger Berücksichtigung der langfristigen Gefahren der Bodenerosion bis zur ►Bodenverheerung, der Nutzungseinschränkung für die Landwirtschaft oder anderer Nutzungen. Die Bodeneignungseinteilung gliedert sich in acht Hauptklassen [►land capability classes (US)] und vier Unterklassen [land capability subclasses]. Die Hauptklassen bezeichnen in Stufen die Zunahme der Nutzungsbeschränkung und die Abnahme landwirtschaftlicher Ertragsfähigkeit, auch unter Berücksichtigung der Wasseraufnahmekapazität bei Beregnung; die Unterklassen sind mit vier Buchstaben gekennzeichnet: **E** Grad der Erosionsgefährdung, **W** Grad der Bodenfeuchtigkeit, **S** Begrenzungen der Durchwurzelbarkeit und **C** für klimatische Einschränkungen).

86 agricultural land improvement [n] *agr. pedol.* (**1**. Technical methods to improve agricultural land by drainage for aeration, or making water available in cultivated soils for better plant growth. **2.** In organic farming, improvement includes also the introduction of beneficial micro-organisms and composted organic amendments as is required to convert to Organic Farming Standards; ►bog cultivation, ►deep plowing [US]/deep ploughing [UK], ►land management measures, ►loosening [of soil], ►soil improvement, ►technical agricultural methods);

A

s **mejoramiento [m] de suelos** (**1.** Conjunto de ▶medidas agro-técnicas para mejorar el suministro de agua y aire en los suelos cultivados con el fin de promocionar el crecimiento de las plantas. **2.** En el ▶cultivo ecológico, las medidas incluyen la introducción de microorganismos beneficiosos y enmiendas de compost orgánico para cumplir con los estándares de cultivo orgánico; ▶arado en profundo, ▶cultivo de turbera, ▶enmienda del suelo, ▶esponjamiento del suelo, ▶medidas de gestión ecológica de zonas agrosilvícolas); *syn.* medidas [fpl] de mejora de suelos; *f 1* **amélioration [f] foncière** (Mesures ou ouvrages tels que la correction de cours d'eau, l'assainissement, l'assèchement, l'irrigation et le drainage, le déboisement ou le reboisement, l'arrachage ou la plantation de haies, la constitution de chemins, le comblement de fossés ou la création de mares, le remaniement parcellaire qui ont pour but de maintenir un d'accroître le rendement des terres agricoles, de faciliter leur exploitation et de les protéger contre les dévastations ou destructions causées par des phénomènes naturels ; cf. LA 1981, 511) ; *f 2* **amélioration [f] du sol (1)** (Ensemble des mesures techniques pour améliorer les propriétés physico-chimiques du sol ; ▶ameublissement, ▶correction de la terre, ▶labour profond, ▶mesures de gestion des territoires agricoles et forestiers, ▶mise en culture de tourbières, ▶techniques culturales) ; *g* **Melioration [f]** (Verbesserung der Luft- und Wasserversorgung in Kulturböden zur nachhaltigen Förderung des Pflanzenwachstums durch ▶kulturtechnische Maßnahmen; ▶Bodenverbesserung 2, ▶landeskulturelle Maßnahmen, ▶Lockerung, ▶Moorkultur, ▶Tiefpflügen).

agricultural land lie fallow [loc]**, let** *agr.* ▶leave uncultivated.

87 agricultural land parcels [npl] *agr. geo.* (Lands of a farming community parceled into a network of fields which may be classified according to their various shapes. In German, the following plot patterns are distinguished, e.g. **1.** *'Blockflur'* [plot pattern of large, contiguous land parcels]; **2.** *'Gewannflur'* [plot pattern of units of parallel parcels scattered in a rural community]; **3.** *'Streifenflur'* [plot pattern of short strip fields with varying ratios of length to width]; **4.** *'Langstreifenflur'* [plot pattern of groups of very long strip fields, 1-10 km long]; *s 1* **esquema [m] de parcelación de tierras agrícolas** (Forma de las parcelas de una comunidad agrícola específica; en alemán se diferencian diferentes modelos de división de las parcelas agrícolas según las tradiciones y condiciones naturales reinantes en los pueblos o comarcas que son **1.** «*Blockflur*» [parcelas grandes], **2.** «Gewannflur» [parcelas dispersas por el territorio comarcal], **3.** «Streifenflur» [parcelas rectangulares en grupos de diferentes tamaños] y **4.** «*Langstreifenflur*» [parcelas muy alargadas de 1-10 km de largo]); *s 2* **finage [m] (1)** (Territorio sobre el que una comunidad de campesinos se instala para roturarlo y cultivarlo y sobre el que ejerce unos derechos agrarios. El concepto es válido para ciertas regiones africanas, en pueblos de nueva planta en España [Plan Badajoz] o en el caso de Israel; cf. DGA 1986); *f 1* **parcellaire [m]** (Ensemble de l'organisation des parcelles cultivées. En D. on distingue différents parcellaires : **1.** « *Blockflur* » : parcellaire à champs massifs, **2.** « *Gewannflur* » : parcellaire divisé en quartiers, **3.** « *Streifenflur* » : parcellaire à longues parcelles, **4.** « *Langstreifenflur* » : parcellaire à très longues lanières, etc. ; DG 1984, 195) ; *f 2* **terroir [m]** (Territoire agricole cultivé n'ayant aucune unité administrative et caractéristique d'une aptitude à la production de certaines denrées alimentaires [terroir viticole, céréalier, de prairies] ; ce terme est utilisé par les géographes français, qui hésitent à élargir le sens strict de « finage » — territoire juridique, administratif, sur lequel la communauté rurale exerce ses droits jusqu'aux limites [fines] de la communauté voisine ; DG 1984, 446) ; *syn.* finage [m] cultural ; *g* **Flur**

[f] (1) (Parzellierte landwirtschaftliche Nutzfläche eines Siedlungs- und Wirtschaftsverbandes des in Kultur genommenen Landes und des im Parzellenverband der Felder eingeschlossenen Waldes; Flurtypisierung erfolgt nach der Grundrissgestaltung der Flurparzellen, z. B. Blockflur, Gewannflur, Streifenflur, Langstreifenflur).

88 agricultural landscape [n] *agr. plan.* (Landscape characterized by agricultural activities; ▶rural land 1); *syn.* agrarian landscape [n]; *s* **paisaje [m] rural (2)** (Paisaje configurado por la actividad agrícola; ▶zona rural); *syn.* paisaje [m] agrícola; *f* **paysage [m] rural** (Lié à l'exploitation agricole ; ▶espace rural) ; *g* **Agrarlandschaft [f]** (Durch landwirtschaftliche Bodennutzung geprägte Landschaft; ▶ländlicher Raum); *syn.* Agrargebiet [n].

89 agricultural land use [n] *agr.* (Form of land usage; e.g. crop farming, pasture, etc.); *s* **tipo [m] de explotación del suelo** (Forma de uso del suelo, p. ej. cultivo o prado); *syn.* tipo [m] de explotación de la tierra; *f* **culture [f] du sol** (Forme d'utilisation du sol à des fins agricoles et art d'exploiter les végétaux cultivés, p. ex. la culture de plein champ, la culture d'herbe, la culture biologique, la culture légumière de pleine terre, etc.) ; *g* **Bodenbewirtschaftung [f]** (Form der landwirtschaftlichen Nutzung, z. B. Ackerbau oder Grünland, Getreide oder Hackfrüchte, Obst-, Wein- und Gemüsebau, Dreifelderwirtschaft).

90 agricultural leasehold (property) [n] *agr.* (Plot of land leased for agricultural use); *syn.* tenant agricultural land [n], tenant farm [n] [also US]; *s* **tierras [fp] (de cultivo) arrendadas** (Terrenos en arriendo para usos agrícolas); *f* **terre [f] louée** (Terrains agricoles soumis à un bail rural à ferme ou à métayage, le **fermage** étant la cession d'usage d'une terre à un fermier moyennant une redevance annuelle indépendante des résultats obtenus et le **métayage** étant la cession de l'usage d'une terre à un métayer moyennant une rétribution représentant une part des produits de l'exploitation ; LA 1981) ; *g* **Pachtland [n]** (Auf bestimmte Zeit gegen Entgelt abgegebene landwirtschaftliche Grundstücke); *syn.* Pachtgrundstück [n] [A nur so].

91 agricultural management [n] *agr.* (Economic use of soil for production of food crops on arable land and ▶grassland farming. Leaving land fallow and policies regulating lands adjacent to agricultural lands for the purpose of watershed management and landuse buffer zones, as well as restrictions on agricultural processes with regard to odors, chemicals, or buildings/processing operations are also part of **a. m. In Europe**, much agricultural land considered to be unsuitable for crop cultivation is directly subsidized in order to protect its visual or aesthetic quality through program[me]s of landscape management. Compensation for such programs, called *nature conservation on a contract basis* [▶contract-based nature conservation] is paid by the state); *s* **gestión [f] agropecuaria** (Aprovechamiento económico del suelo para la producción de alimentos por medio de la agricultura y ganadería. El dejar tierras en barbecho también es una forma de **g. a.** En la UE muchas tierras consideradas inadecuadas para el cultivo se subvencionan directamente con el fin de preservar la calidad ecológica y visual del paisaje a través de programas de gestión del paisaje); *f* **gestion [f] du patrimoine agricole** (Utilisation des sols en vue de la fabrication de produits alimentaires par la culture des végétaux et l'élevage du bétail. La jachère est aussi une forme de **g. du p. a.** ; ▶gestion des cycles naturels. Dans le cas de l'abandon de cultures, l'agriculteur peut conclure un contrat de protection dont l'objet est la protection des espaces naturels [préservation de la qualité des sites et des paysages ainsi que de l'amélioration de la biodiversité animale et végétale] par l'obtention de la maîtrise d'usage de terres agricoles moyennant une subvention de l'État ; on parle ainsi de la ▶protection de la nature par voie contractuelle ;

▶gestion contractuelle de la nature) ; *syn.* gestion [f] agricole ; *g* **landwirtschaftliche Bewirtschaftung [f]** (Wirtschaftliche Nutzung von Böden zur Erzeugung von Lebensmitteln durch Ackerbau und Grünlandnutzung [Viehwirtschaft]. Die Brache ist auch eine Form der **l.n B.** Wird die landwirtschaftliche Nutzung aufgegeben und das Landschaftsbild aus natur- und landschaftspflegerischen Gründen durch den Landwirt mit staatlichen Zuschüssen erhalten, so spricht man von ▶Vertragsnaturschutz).

agricultural methods [npl] *agr.* ▶technical agricultural methods.

92 agricultural planning [n] *agr. plan.* (Planning for the ▶improvement of agrarian structure of an area in connection with the goals of regional land use policy; ▶preliminary agrarian structure planning); *s* **planificación [f] agraria** (Planificación que tiene como meta mejorar la estructura agraria de una comarca, considerando los objetivos de la ordenación del territorio; ▶mejora de la estructura agraria, ▶planificación estructural agraria); *f* **planification [f] rurale** (Ensemble des mesures ayant pour objet l'amélioration des conditions et du cadre de vie des populations rurales par le développement des activités socio-économiques et des services et des équipements, la protection de l'espace naturel ; celles-ci tiennent compte des directives d'aménagement du territoire et des orientations de la planification régionale ; ▶amélioration des structures rurales, ▶étude préopérationnelle d'aménagement foncier et agricole, ▶plan d'aménagement rural) ; *syn.* aménagement [m] rural ; *g* **Agrarplanung [f]** (Planung, die durch Fachpläne der Landwirtschaft der Verbesserung der Agrarstruktur eines Raumes unter Einbeziehung der Ziele der Raumordnung und Landesplanung dient; in einigen Bundesländern besteht die **A.** in Form des *integrierten ländlichen Entwicklungskonzeptes [ILEK]*; ▶agrarstrukturelle Vorplanung, ▶Agrarstrukturverbesserung).

93 agricultural pollution [n] *agr. envir.* (Harmful environmental nuisance affecting soil, air, water bodies, and groundwater caused by liquid and solid wastes from all types of farming, including runoff from pesticides, fertilizers and feedlots, erosion and dust from plowing [US]/ploughing [UK], animal manure and carcasses; RCG 1982; ▶land clearance); *s* **impacto [m] ambiental de la agricultura** (Conjunto de efectos negativos de la producción agropecuaria para el medio, en forma de contaminación de las aguas continentales y subterráneas, el suelo y el aire, fomento de la erosión, etc. que se presentan sobre todo en el cultivo tradicional intensivo con su sobreexplotación del suelo y su tendencia a la transformación del paisaje [▶desmonte total del paisaje] para adecuarlo a sus fines); *f* **pollution [f] agricole** (Atteintes portées par une forme d'agriculture intensive non respectueuse de l'environnement, p. ex. lors de l'application de produits antiparasitaires, par la pollution organique des eaux superficielles et souterraines provoquée ou induite par les engrais, par le déversement ou l'épandage de fumier ou de lisier en provenance de l'élevage industriel, par l'érosion, la formation de nuage de poussières sur les grands espaces sans végétation ; ▶dégagement du paysage) ; *syn.* pollution [f] d'origine agricole ; *g* **Umweltbelastung [f] durch die Landwirtschaft** (Schädliche Einwirkungen auf Boden, Luft, Gewässer und Grundwasser durch eine unsachgemäße, intensive landwirtschaftliche Bewirtschaftungsform, z. B. Abfluss und Versickerung von Pestiziden und übermäßigen Düngergaben, unsachgemäßes Lagern und Ausbringen von Mist und Jauche von Massentierhaltungen sowie Erosion und Staubentwicklung durch ▶Ausräumung der Landschaft).

agricultural production [n] [US]**, curtailment of** *agr. pol.* ▶agricultural reduction program [US]/extensification of agricultural production [UK].

agricultural production [n]**, de-intensification of** *agr. pol.* ▶agricultural reduction program [US]/extensification of agricultural production [UK].

agricultural production [n]**, extensification of** *agr. pol.* ▶agricultural reduction program [US].

94 agricultural production land [n] *agr. plan.* (Farmland used for agricultural production, including both cropland and pasture land); *s* **superficie [f] agrícola útil** (La **s. a. u.** incluye todos los terrenos dedicados a la producción agropecuaria, exceptuando bosques y las zonas no agrícolas [urbanas]); *syn.* superficie [f] agrícola real; *f 1* **superficie [f] agricole utilisée (S.A.U.)** (Notion normalisée dans la statistique agricole européenne ; elle comprend les terres arables [y compris pâturages temporaires, jachères, cultures sous verre, jardins familiaux, etc.], les surfaces toujours en herbe et les cultures permanentes [vignes, vergers, etc.] ; cf. www.insee.fr. Cette expression désigne la dimension d'une exploitation et correspond à la ▶superficie agricole utile moins la superficie des landes improductives et celle des cours et des bâtiments d'exploitation ; LA 1981) ; *syn.* surface [f] agricole ; *f 2* **superficie [f] agricole utile (S.A.U.)** (Cette terme obsolète désigne la superficie agricole totale moins les bois et le territoire non agricole, c.-à-d. les terres labourables, la surface toujours en herbe [S.T.H.] et les cultures permanentes et spéciales, la superficie des bâtiments d'exploitation ainsi que les landes non productives ; LA 1981) ; *g 1* **landwirtschaftliche Nutzfläche [f]** (*Abk.* LN; **1.** der für die pflanzliche Erzeugung genutzte Teil einer Betriebsfläche wie Ackerflächen, Dauergrünland, Brachflächen, Dauerkulturen, z. B. Obst- und Weinbauflächen und Flächen unter Glas); *g 2* **Landwirtschaftsfläche [f]** (*Begriff der Landesverwaltungsämter* Flächen, die dem Ackerbau, der Wiesen- und Weidewirtschaft, dem Gartenbau, incl. Obstbaumanlagen, Hopfenpflanzungen und Baumschulen, und dem Weinbau dienen; dazu zählen auch Moor- und Heideflächen, Brachland sowie unbebaute landwirtschaftliche Betriebsflächen).

95 agricultural reduction program [n] [US] *agr. pol.* (**In U.S.**, a program to limit overproduction in order to protect the income and economic status of farmers; **in Europe**, measures intended to curtail over-production and in consequence to avoid soil and groundwater pollution or threatening animal and plant species due to intensive methods of cultivation; this environment-friendly program[me], which is subsidized by the European Union since 1992, is serving nature conservation goals when on the remaining agricultural land the crop production will be also reduced; ▶setting-aside of arable land [UK] include program[me]s to leave agricultural land uncultivated; ▶organic farming); *syn.* curtailment [n] of agricultural production [US], agricultural yield reduction [n], de-intensification of agricultural production, extensification [n] of agricultural production [UK]; *s* **extensificación [f] de la agricultura** (Sistema de producción agropecuaria aplicado con el fin de reducir la contaminación de suelos y aguas subterráneas y evitar la desaparición de especies de flora y fauna causada por la agricultura industrial, reduciendo la aplicación de insumos artificiales y la densidad de cabezas de ganado por superficie; la política comunitaria existente desde 1992 de fomento del ▶abandono de tierras no contribuye a la protección de la naturaleza si no va acompañada de medidas de extensificación en las superficies que se mantienen en explotación; ▶cultivo alternativo); *f* **extensivité [f] des activités agricoles** (Forme d'agriculture subventionnée par l'union européenne [politique agricole commune — P.A.C.] dans le but de réduire les excédents agricoles, d'empêcher ou de fortement réduire les nuisances causées sur le sol, les eaux superficielles et la nappe phréatique, d'éviter les atteintes portées à la faune et la flore par une agriculture intensive, d'éviter ou de fortement limiter les élevages industriels ainsi que l'utilisation des méthodes de culture

A

industrielle caractérisées par l'emploi intensif d'engrais et de pesticides ; la C.E.E. incite les agriculteurs à introduire des pratiques agricoles compatibles avec les exigences de la protection de l'environnement ainsi que l'entretien de l'espace naturel et du paysage [mesures agri-environnementales, p. ex. ►culture biologique], la mise en jachère [►enfrichement et ►gel environnemental] ou l'abandon de la culture sur certaines parcelles) ; *syn.* extensification [f] ; *g* **Extensivierung [f] der Landwirtschaft** (Landwirtschaftliche Produktionsweise mit dem Ziel, die Belastung des Bodens und des Grundwassers zu verhindern oder stark einzuschränken und die Bedrohung der Tier- und Pflanzenarten durch eine intensive Landwirtschaft zu vermeiden, indem industrielle Kulturmethoden mit künstlicher Düngung und ständigem Pestizideinsatz sowie Massentierhaltungen auf zu kleinen Pro-Kopf-Flächen vermieden oder zumindest stark reduziert werden; die seit 1992 bestehende Förderpolitik der Europäischen Gemeinschaft, Flächen brach liegen zu lassen oder sie aus der Produktion herauszunehmen [►Flächenstilllegung], dient den Zielen des Naturschutzes und der Landschaftspflege nicht, wenn nicht die Nutzung der verbleibenden landwirtschaftlichen Flächen extensiviert wird; ►alternativer Landbau).

agricultural side-line [n] [UK] *agr.* ►part-time agricultural business.

agricultural site [n]**, marginal** *agr. (for.)* ►marginal soil.

agricultural structure [n]**, improvement of** *agr. plan.* ►improvement of agrarian structure.

agricultural tourism [n] [US] *recr.* ►agritourism [US]/agro-tourism [UK].

96 agricultural use [n] *agr. plan. s* **uso** [m] **agrícola del suelo;** *f* **utilisation** [f] **des terres agricoles** *syn.* utilisation [f] des superficies agricoles ; *g* **landwirtschaftliche Bodennutzung** [f] *syn.* landwirtschaftliche Landnutzung [f], Agrarnutzung [f].

97 agricultural wastewater [n] *agr. envir.* (Impure drainage water or precipitation run-off which has been polluted by the leakage of ►liquid manure or silos as well as the concentrated spraying of animal excrement on fields, or high applications of ►chemical fertilizers or by the intensive use of ►plant protection agents for pest and disease control and ►herbicides; ►receiving stream); *syn.* farm runoff [US]; *s* **aguas** [fpl] **residuales agropecuarias** (Agua contaminada, procedente de los usos agrícolas como vertido de ►purín, y cuya alteración se debe a la presencia de ►plaguicidas, ►herbicidas, nitratos, sales y de forna acusada restos de ►fertilizantes [abonos químicos] y una fuerte carga de sólidos que se vierten en el ►curso receptor); *syn.* efluente [m] de explotaciones ganaderas; *f* **eaux** [fpl] **usées agricoles** (Eaux pluviales ou eaux de drainage atteignant un ►émissaire ou ►milieu récepteur préalablement polluées par l'écoulement de ►purin ou de liquide d'égouttage de silos, par des apports excessifs de fumure et d'►engrais artificiel ainsi que par l'utilisation exagérée de ►produits antiparasitaires et d'►herbicides) ; *g* **landwirtschaftliches Abwasser** [n] (...wässer [pl]; durch Abgänge von ►Jauche oder Silosaft belastetes Abwasser sowie durch konzentriertes Ausbringen von tierischen Abfallstoffen, hohen ►Kunstdüngergaben oder durch erhöhten Einsatz von chemischen ►Pflanzenschutzmitteln und ►Herbiziden verunreinigtes, in die ►Vorfluter gelangendes Drän- oder Niederschlagswasser; cf. § 2 AbwAG).

agricultural yield reduction [n] *agr. pol.* ►agricultural reduction program [US]/extensification of agricultural production [UK].

98 agriculture [n] *agr.* (Sector of the economy based upon cultivation of land and characterized by crop husbandry, ►grassland farming, and animal production; ►livestock industry, ►organic farming, ►pasture farming, ►semiproductive cultiva-

tion, ►shifting cultivation, ►traditional farming); *s* **agricultura** [f] (Conjunto de operaciones cuyo objetivo es la producción de alimentos vegetales y animales que supone una alteración casi siempre simplificadora del ecosistema natural. Se pueden distinguir diferentes tipos de **a.**: **a.** de subsistencia, **a.** de mercado, **a.** especulativa; DGA 1986; ►agricultura itinerante, ►cultivo alternativo, ►cultivo extensivo, ►cultivo tradicional, ►ganadería industrial, ►pasticultura); *f* **agriculture** [f] (Culture du sol et, par extension, ensemble des travaux visant à utiliser et à transformer le milieu naturel pour la production de végétaux et d'animaux utiles à l'homme ; LA 1981 ; ►agriculture itinérante [sur brûlis], ►agriculture traditionelle, ►culture biologique, ►culture extensive, ►élevage pastoral, ►industrie liée à l'élevage) ; *g* **Landwirtschaft** [f, o. Pl.] **(1.** Auf Landbewirtschaftung basierender Sektor der Volkswirtschaft, gekennzeichnet durch Pflanzenbau incl. Sonderkulturen wie Garten-, Wein- und Hopfenbau, Grünlandwirtschaft und Tierproduktion; ANL 1984; ►alternativer Landbau, ►extensive Bewirtschaftung, ►konventioneller Landbau, ►Viehwirtschaft, ►Wanderfeldbau, ►Weidewirtschaft. **2.** *leg. gem.* § 201 *BauGB* ... insbesondere der Ackerbau, die Wiesen- und Weidewirtschaft einschließlich Tierhaltung, soweit das Futter überwiegend auf dem zum landwirtschaftlichen Betrieb gehörenden, landwirtschaftlich genutzten Flächen erzeugt werden kann, die gartenbauliche Erzeugung, der Erwerbsgartenbau, der Weinbau, die berufsmäßige Imkerei und die berufsmäßige Fischerei).

agriculture [n]**, grassland** *agr.* ►grassland farming.

agriculture [n]**, traditional** *agr.* ►traditional farming.

agri-environmental contracting [n] *agr. conserv. pol.* ►contract-based nature conservation.

99 agritourism [n] [US]/**agro-tourism** [n] [UK] *recr.* (**In Europe**, popular form of ►countryside recreation, whereby vacationers/holidaymakers become acquainted with the various cultural landscapes, characterized by traditional and regionally different, arable farming systems, such as crops, fruit trees or vines and livestock, as well as enjoying locally produced products; publicity for **a.** includes the marketing of hotels and guest houses in rural areas and vacations/holidays spent on farms; ►farmstay holidays; **in U.S.**, **a.** refers to the act of visiting a working farm or any agricultural, horticultural or agribusiness operation for the purpose of enjoyment, education, or active involvement in the activities of the farm or rural operation; ►ecotourism); *syn.* agricultural tourism [n] [also US]; *s* **agroturismo** [m] (Turismo realizado en zonas rurales. El agroturismo consiste en la prestación de servicios turísticos de alojamiento y restauración por parte de personas agricultoras y ganaderas en sus caseríos e integrados en una explotación agraria. La capacidad del caserío es limitada, en su mayoría 12 personas como máximo; ►turismo ecológico, ►vacación en granja); *f* **agrotourisme** [m] (Activités touristiques proposées par les agriculteurs sur leur exploitation qui permettent aux touristes ou excursionnistes de découvrir les paysages et le milieu agricoles, les différentes formes d'agriculture, proposent une restauration mettant en valeur les produits de la ferme, un hébergement soit en chambre d'hôtes ou en gîte ou cherchent une ressource principale ou d'appoint dans une ferme équestre, dans la création d'un parcours de pêche, d'un parc animalier local ; ►écotourisme, ►vacances à la ferme) ; *g* **Agrotourismus** [m] (Das Reisen resp. der Reiseverkehr zum Kennenlernen bäuerlicher Kulturlandschaften in vielen Teilen Europas, die durch traditionsreiche, regional unterschiedliche landwirtschaftliche Bewirtschaftungsformen wie Acker-, Obst- oder Weinbau und Tierhaltung sowie vor Ort produzierte Produkte gekennzeichnet sind; zum **A.** gehören neben der Vermarktung von Hotels und Pensionen im ländlichen Raum auch die der Angebote über ►Ferien auf dem Bauernhof; ►Ökotourismus).

agroforestry [n]**, silvoarable** *agr. conserv.* ►traditional orchard [UK], #2.

agroforestry system [n] *agr. conserv.* ►traditional orchard [UK], #2.

agropastoral system [n] *agr. conserv.* ►traditional orchard [UK], #2.

agrosilvopastoral system [n] *agr. conserv.* ►traditional orchard [UK], #2.

agro-tourism [n] [UK] *recr.* ►agritourism [US].

A horizon [n] *pedol.* ►topsoil (1).

aims [npl] *plan.* ►provision of data and aims.

100 air bath [n] *recr.* (Hygienic exposure of a body to outdoor air for health reasons); *s* **baño** [m] **de aire** (Exposición del cuerpo al aire como medida de cura); *f* **bain** [m] **d'air** (Exposition du corps aux effets bienfaisants du climat local) ; *g* **Luftbad** [n] (Einwirkenlassen örtlich-klimatischer Einflüsse auf den nackten Körper).

air contamination [n] *envir.* ►air pollution (1).

air corridor [n] *landsc. met.* ►fresh air corridor.

101 air exchange [n] *met. urb.* (Process of substituting existing air masses with fresh air; e.g. for the improvement of urban microclimate by creation of ►fresh air corridors; ►mixing of air masses, ►ventilation); *s* **ventilación** [f] **(1)** (Proceso de sustitución de masas de aire existente por aire fresco, p. ej. para mejorar el microclima urbano creando ►corredores de aire fresco; ►mezcla de masas de aire, ►ventilación 2); *f* **brassage** [m] **de l'air** (Amélioration du climat urbain par brassage de l'atmosphère dans les zones menacées par la pollution de l'air grâce aux couloirs de ventilation ; ►couloir d'air frais, ►échange atmosphérique, ►ventilation) ; *g* **Luftaustausch** [m] (Austausch von Luftmassen, z. B. zur Verbesserung des Stadtklimas durch ►Frischluftschneisen; ►Durchlüftung, ►Durchmischung der Luft).

air flow [n] *geo. met.* ►upslope air flow.

air flow [n]**, barrier effect on frozen** *met.* ►frost accumulation.

102 air layer [n] *met.* (Horizontal layer of the atmosphere; e.g. defined by the height above ground surface, gradient of temperature, gaseous composition; ►ground-level air layer); *s* **capa** [f] **de aire** (Capa horizontal de la atmósfera, definida por la altura sobre el nivel del suelo, el gradiente de temperatura, la composición de gases; ►capa de aire al ras del suelo); *f* **couche** [f] **d'air** (Section horizontale de la couche de l'atmosphère caractérisée par son altitude, sa température, sa composition gazeuse, etc. ; ►couche d'air à proximité du sol) ; *g* **Luftschicht** [f] (Horizontaler Abschnitt der Erdatmosphäre, definiert z. B. durch Höhe über Geländeoberfläche, durch Temperaturverteilung, Gaszusammensetzung etc.; ►bodennahe Luftschicht).

air masses [npl] *met.* ►mixing of air masses.

103 air pollutant [n] *envir.* (Artificial airborne matter—in addition to pollen, fog, and dust of natural origin—in the form of solid particles, liquid droplets, gases, or in combination thereof, and in high enough concentration to harm man, animals, vegetation, or useful material; e.g. particulates, sulphur compounds, volatile organic chemicals, nitrogen compounds, oxygen compounds, halogen compounds, radioactive compounds, and odo[u]rs; *specific term* particulate air pollutant; TGG 1984, 238; ►air pollution 1); *s* **contaminante** [m] **de la atmósfera** (Materia estraña al aire que tiene efectos nocivos sobre los seres vivos; ►contaminación atmosférica); *syn.* contaminante [m] del aire;

f **polluant** [m] **atmosphérique** (Substances physiques ou chimiques émises dans l'atmosphère, dont les principales proviennent des processus industriels [raffinage du pétrole, sidérurgie, métallurgie, industrie des matières plastiques, peintures, laques, solvants, produits de réfrigération ou propulseurs, etc.], de la combustion des combustibles fossiles solides, liquides ou gazeux, des moteurs des véhicules, de l'utilisation des substances radioactives et responsables de la ►pollution atmosphérique) ; *g* **Luftschadstoff** [m] (Jeder Stoff, der direkt oder indirekt vom Menschen in die Luft emittiert wird und schädliche Auswirkungen auf die menschliche Gesundheit und/oder die Umwelt insgesamt haben kann; cf. Art. 2 Richtlinie 96/62/EG und Art. 2 RL 2008/50/EG vom 21.05.2008; ►Luftverunreinigung).

104 air pollution [n] **(1)** *envir.* (Changes in the natural composition and purity of the air, especially due to smoke, soot, dust, gases, aerosol sprays, vapo[u]rs [including water steam], and odorous substances, which affect human health, damage biotic resources and ecosystems, destroy material values and affect environmental amenities and legal uses of the environment. Pollution also includes chemical or physical changes to the air in the atmosphere; e.g. photochemical smog; *definition in the Convention on long-range transboundary air pollution* "introduction by man, directly or indirectly, of substances or energy into the air resulting in deleterious effects of such a nature as to endanger human health, harm living resources and ecosystems and material property and impair or interfere with amenities and other legitimate uses of the environment"; art. 1, Council decision 81/462/EEC; ►air quality monitoring network/system, ►contaminated airspace, ►emission, ►monitoring, ►nonpoint source pollution, ►pollution impact, ►smog, ►transmission); *syn.* air contamination [n], atmospheric pollution [n]; *s* **contaminación** [f] **atmosférica** (Presencia en la atmósfera de materias, sustancias o formas de energía que impliquen molestia grave, riesgo o daño para la seguridad o la salud de las personas, el medio ambiente y demás bienes de cualquier naturaleza; art. 3 b] Ley 34/2007, de calidad del aire y protección de la atmósfera, ►contaminación ubicua, ►emisión, ►inmisión, ►monitoreo ambiental, ►nivel de emisión, ►red de evaluación de la calidad del aire, ►smog, ►transmisión, ►zona de atmósfera contaminada); *syn.* polución [f] atmosférica; *f* **pollution** [f] **atmosphérique** (Introduction par l'homme, directement ou indirectement, dans l'atmosphère et les espaces clos de substances ayant des conséquences préjudiciables de nature à mettre en danger la santé humaine, à nuire aux ressources biologiques et les écosystèmes, à influer sur les changements climatiques, à détériorer les biens, à provoquer des nuisances olfactives excessives ; loi n° 96-1236 du 30 décembre 1996 ; **P. a.** de nature à endommager les ressources biologiques et les écosystèmes, à détériorer les biens matériels et porter atteinte ou nuire aux valeurs d'agrément et aux autres utilisations légitimes de l'environnement ; Dir. n° 84/360/CEE du 28 juin 1984 ; ►émission, ►monitorage, ►pollution superficielle, ►réseau de mesure des pollutions atmosphériques, ►secteur soumis à la pollution de l'air atmosphérique, ►smog, ►transfert) ; *syn.* pollution [f] de l'air, contamination [f] de l'atmosphère ; *g* **Luftverunreinigung** [f] (Veränderung der natürlichen Zusammensetzung der Luft, insbesondere durch Rauch, Ruß, Staub, Gase, Aerosole, Dämpfe [auch Wasserdämpfe] und Geruchsstoffe, die die menschliche Gesundheit gefährden, biotische Ressourcen und Ökosysteme schädigen sowie Sachwerte zerstören und Annehmlichkeiten der Umwelt oder sonstige rechtmäßige Nutzungen in ihr beeinträchtigen; cf. § 3 [4] BImSchG und EG-Richtlinie 89/369, Art. 1; zur Luftverunreinigung gehören auch chemische oder physikalische Veränderungen von Luftverunreinigungen in der Atmosphäre, z. B. fotochemischer Smog; ►Emission, ►flächenhafte Verschmutzung, ►Immission,

A

►lufthygienisches Belastungsgebiet, ►Luftmessnetz, ►Monitoring, ►Smog, ►Transmission); *syn.* Luftverschmutzung [f], Luftbelastung [f], Verschmutzung [f] der Luft.

air pollution [n] (2) *envir.* ►acceptable level of air pollution, ►national or state surveillance system, ►particulate air pollution; *for. landsc.* ►protective forest against air pollution; *hort.* ►resistance to air pollution; *envir. leg.* ►threshold level of air pollution, ►transboundary air pollution.

air pollution [n], **transfrontier air pollution** *envir. leg.* ►transboundary air pollution.

105 air pollution control [n] *envir. leg. nat'res.* (Means of environmental protection, enforced by laws and mechanical equipment, to reduce emissions of noxious substances into the air ; ►air pollution 1, ►clean air plan, ►contaminated airspace, ►pollution control 1); *syn.* air quality management [n] (TGG 1984, 226); *s* **protección [f] del ambiente atmosférico** (Ámbito de la protección ambiental que se dedica a tomar medidas para reducir y evitar la ►contaminación atmosférica, aplicando las leyes y los reglamentos dictados con ese fin; cf. Ley 34/2007, de calidad del aire y protección de la atmósfera, RD 1613/1985, RD 717/1987, RD 1073/2002, RD 653/2003, RD 1796/2003, RD 430/2004; ►plan de protección del ambiente atmosférico, ►protección contra inmisiones, ►zona de atmósfera contaminada); *syn.* control [m] de la polución atmosférica, control [m] de la contaminación atmosférica; *f* **protection [f], surveillance [f] et contrôle [m] de la qualité de l'air** (Activités visant à susciter, animer, coordonner, faciliter, les actions ayant pour objet la prévention, la lutte contre la pollution de l'air, la coordination technique de la qualité de l'air et d'assurer la réalisation d'installations techniques dans le cadre de la réglementation ; cette mission est dévolue en F. à l'Agence de l'environnement et de la maîtrise de l'énergie [A.D.E.M.E. — anciennement Agence pour la qualité de l'air] ; ►lutte contre les émissions, ►plan de protection de l'air, ►pollution atmosphérique, ►secteur soumis à la pollution atmosphérique) ; *syn.* lutte [f] contre la pollution atmosphérique ; *g* **Luftreinhaltung [f, o. Pl.]** (Teilbereich des Umweltschutzes, der sich mit der Verringerung der Schadstoffimmissionen in der Luft befasst und diese mit gesetzlichen Maßnahmen und technischen Einrichtungen gem. Bundes-Immissionsschutzgesetz [BImSchG] und der Verwaltungsvorschrift TA-Luft [Technische Anleitung zur Reinhaltung der Luft] durchsetzt. Nach dem BImSchG ist ein ►Luftreinhalteplan vorgeschrieben, wenn in einem bestimmten Gebiet Luftbelastungen auftreten oder zu erwarten sind; ►Immissionsschutz, ►lufthygienisches Belastungsgebiet, ►Luftverunreinigung).

air pollution control district [n] *envir. leg.* ►air quality control region.

106 air pollution dome [n] *envir. met.* (Concentrated haze canopy covering an admixture of different kinds of air pollutants over an urban agglomeration; ►inversion weather, ►smog); *s* **capa [f] flotante de contaminantes** (DINA 1987; por efecto de tapadera de la ►inversión térmica; ►smog); *syn.* calima [f], calina [f]; *f* **dôme [m] de pollution** (Brume d'inversion constituée de poussières industrielles se formant à basse altitude au-dessus d'une grande agglomération ; ►conditions d'inversion thermique, ►smog) ; *g* **Dunstglocke [f]** (Dunstschicht über Ballungsgebieten, die durch Beimengungen unterschiedlichster Stoffe, die in der Luft schweben, entsteht und nach oben hin scharf abgegrenzt ist; ►Inversionswetterlage, ►Smog).

107 air pollution fee [n] (≠) *adm. envir. leg.* (Charge levied upon industrial firms to encourage them to control air pollution in their production process; ►environmental tax); *syn.* emissions tax [n], emissions fee [n]; *s* **canon [m] de contaminación atmosférica** (Tasa a pagar por los emisores de contaminantes de la atmósfera; ►canon ambiental, ►canon de vertido); *f 1* **taxe [f] sur la pollution atmosphérique** (Taxe sur les émissions de polluants dans l'atmosphère affectée au financement de la lutte contre la pollution de l'air ; F., cette taxe a été remplacée par la ►taxe générale sur les activités polluantes) ; *f 2* **taxe [f] générale sur les activités polluantes (TGAP)** (Taxe créée par la loi de finances pour 1999 et qui se substitue à cinq anciennes taxes fiscales et parafiscales affectées à l'Agence de l'environnement et la maîtrise de l'énergie [ADEME], la **taxe parafiscale sur la pollution atmosphérique** créée en 1985, la **taxe sur le traitement et le stockage de déchets industriels spéciaux** créée par la loi du 2 février 1995, la **taxe parafiscale sur les huiles de base** créée en 1986, la **taxe d'atténuation des nuisances sonores aéroportuaires** créée par la loi du 31 décembre 1992 et la **taxe sur les déchets ménagers et assimilés** créée par la loi du 13 juillet 1992. La création de la TGAP avait pour objectif d'améliorer l'incitation à la protection de l'environnement, en application du principe « pollueur-payeur ». En 2000 le législateur décide d'étendre le champ d'application de la TGAP aux lessives et produits adoucissants ou assouplissants pour le linge, aux grains minéraux naturels, aux produits antiparasitaires à usage agricole et aux installations classées. Soucieux de promouvoir l'utilisation des biocarburants dans le secteur des transports, le Parlement a institué par la loi de finances pour 2005 un prélèvement supplémentaire de TGAP applicable au gazole et supercarburant sans plomb ; ►taxe environnementale) ; *g* **Luftverschmutzungsabgabe [f]** (Aus umweltpolitischen Gründen zur Förderung umweltgerechter Produktionsmethoden orientierte Zahlung, um Verursacher von umwelterheblichen Emissionen so zu lenken, dass Energieverbrauch und Emissionen durch Änderung der Produktionsverfahren reduziert werden. Die Gewährung solcher „Verschmutzungslizenzen" ist sicherlich höchst problematisch, doch ohne sie würde für den Umweltschutz überhaupt nichts oder zu wenig geschehen; ►Umweltabgabe).

108 air pollution monitoring [n] *adm. envir. leg.* (Legally-prescribed repetitive and continued observations, measurements, and evaluation of health and environmental or technical data of air quality to keep track of changes over a period of time; cf. DES 1991, 222; ►monitoring); *s* **evaluación [m] de la calidad del aire** (Monitoreo de la calidad del aire que debe ser realizado por las administraciones competentes en base a la legislación; en Es. es obligación de las CC.AA. y, en su caso, de las entidades locales realizar las mediciones necesarias; art. 28 Ley 34/2007; ►monitoreo ambiental); *f* **surveillance [f] de la qualité de l'air** (Surveillance effectuée par les stations de mesures du réseau de surveillance [organisme agréé] ; décret n° 74-415 du 13 mai 1974 ; ►monitorage) ; *syn.* surveillance [f] de la pollution atmosphérique ; *g* **Überwachung [f] der Luftverunreinigung** (Nach §§ 44-47 a BImSchG vorgeschriebene Überwachung der Luftqualität. Landesregierungen sind durch § 44 BImSchG ermächtigt durch Rechtsverordnung Untersuchungsgebiete festzusetzen. Der Bundesminister für Umwelt, Naturschutz und Reaktorsicherheit erlässt gem. § 45 BImSchG allgemeine Verwaltungsvorschriften über Messobjekte, Messverfahren und Messgeräte, für die Bestimmung der Zahl und der Lage der Messstellen zu beachtenden Grundsätze, die Auswertung der Messergebnisse und die Unterrichtung der Bevölkerung; ►Monitoring); *syn.* Luftüberwachung [f].

109 air pollution-resistant [adj] *hort. plant.* (Characteristic of plants to resist air pollution without noticeable disturbance/harm); *s* **resistente a la contaminación atmosférica [loc]** (Referente a las plantas a las que la contaminación atmosférica no produce daños reconocibles); *f* **résistant, ante à la pollution industrielle [loc]** (Propriété de certains végétaux à réagir à des pollutions atmosphériques, à s'en protéger ou à s'en défendre et ne pas présenter de lésions apparentes) ; *syn.* résistant, ante aux emanations [loc] ; *g* **industriehart [adj] (1)** (Pflanzen betreffend, denen Luftimmissionen keine erkennbaren Schädigungen zufügen); *syn.* immissionsresistent [adj], industriefest [adj], *obs.* rauchhart [adj], *obs.* rauchresistent [adj].

110 air pollution-tolerant [adj] *hort. plant.* (Characteristic of plants to endure air pollution and remain relatively unharmed); *s* **tolerante de la contaminación atmosférica [loc]** (Referente a las plantas a las que la contaminación atmosférica no afecta visiblemente); *f* **tolérant, ante à la pollution atmosphérique [loc]** (Propriété de certains végétaux à ne pas ou peu réagir à la pollution atmosphérique) ; *syn.* tolérant, ante aux emanations [loc] ; *g* **industriehart [adj] (2)** (Pflanzen betreffend, die durch Luftverschmutzungen kaum sichtbar geschädigt werden); *syn.* immissionsverträglich [adj], *obs.* rauchverträglich [adj].

air pollution zone [n] [US] *envir.* ▶contaminated airspace.

air porosity [n] *pedol.* *▶air space ratio, *▶pore volume, *▶porosity.

111 air quality [n] *envir. leg.* (Condition of air as defined by its concentration of admixtures/air pollutants. A. q. is legally defined by limit and assessment values for each noxious substance—i.e. air pollution standards; **in U.S.**, federal and most state regulations prescribe the level of pollutants that may not be exceeded during a given time in a defined area. In 1967 the Air Quality Act required the states to establish regional air quality standards and to control emissions; this was amended by the Clean Air Act of 1970, which empowered the Federal Government to set national air pollution standards; **in U.K.**, the Control of Pollution Act 1974 [COPA], Part IV—atmospheric pollution, prescribes the clean air standards; ▶contaminated airspace); *s* **calidad [f] del aire** (Características físicas del ambiente en cuanto a la ▶contaminación atmosférica; **en Es.** la c. del a. se rige por la ley 34/2007, de calidad del aire y protección de la atmósfera, varias leyes y decretos estatales específicos asi como las normas a nivel de CC.AA., que determinan los objetivos de calidad del aire, los planes y programas para alcanzarlos, así como una serie de medidas de prevención, control de las actividades potencialmente contaminadoras, de ▶evaluación de la calidad del aire y de información pública); *f* **qualité [f] de l'air** (Niveau de concentration d'un nombre déterminé de polluants atmosphériques évalué par des mesures représentatives selon des critères et techniques déterminées [mesures de référence et d'échantillonnage, modélisation] et prenant en compte des valeurs limites, des objectifs de qualité, des seuils de précaution [protection de la santé, protection de la végétation] et d'alerte ; Directive n° 96/62/CE du 27 septembre 1996, loi n° 96-1236 du 30 décembre 1996 ; ▶pollution atmosphérique) ; *g* **Luftqualität [f]** (Beschaffenheit der Luft hinsichtlich ihrer Konzentration von Luftbeimengungen/▶Luftverunreinigung. Die **L.** wird in D. gemäß TA-Luft durch bestimmte Grenz- und Beurteilungswerte je Schadstoff definiert); *syn.* Luftgüte [f] [A].

112 air quality control region [n] *envir. leg.* (DES 1991; area for investigation and control of the air pollution load; **in U.S.**, such federally designated areas are required to meet and maintain federal ambient air quality standards; cf. EPA 1994; such regions are typically large urban conurbations or metropolitan areas which have heavy reliance on motor vehicles and/or smoke stacks [US]/chimney stacks [UK]; ▶air quality control station, ▶air quality monitoring network/system); *syn.* air pollution control district [n]; *s* **zona [f] de evaluación de la calidad del aire** (Parte del territorio delimitada por la Administración competente para la evaluación y gestión de la calidad del aire, seleccionada por sus características de inmisión de contaminación atmosférica para formar parte de la ▶red de evaluación de

la calidad del aire; ▶estación de medición de la calidad del aire aire; cf. art. 3 u] Ley 34/2007); *f* **zone [f] de mesures** (Aire des agglomérations à l'intérieur desquelles sont établis les points d'échantillonnage, emplacement des matériels utilisés pour les mesures de la qualité de l'air ; ▶réseau de mesure des pollutions atmosphériques, ▶station de mesure de la pollution atmosphérique) ; *syn.* zone [f] de surveillance ; *g* **Gebiet [n] zur Luftqualitätsüberwachung** (Gebiet, das mit einem Messnetz zur Untersuchung und Prüfung von Überschreitungen von Immissionsgrenzwerten ausgestattet ist, um die Schadstoffbelastung in der Luft zu dokumentieren; bei Überschreitungen sind Maßnahmenpläne, bei Überschreitungen von Alarmschwellen Aktionspläne aufzustellen und die Bevölkerung so aktuell wie möglich zu informieren; cf. 22. BImSchV; ▶Luftmessnetz, ▶Luftmessstation); *syn.* Luftüberwachungsgebiet [n], lufthygienisches Messgebiet [n], Gebiet [n] für lufthygienische Messungen, Messgebiet [n] für lufthygienische Untersuchungen, Messgebiet [n] für Staub- und Luftmessungen, Messgebiet [n] zur Luftgüteüberwachung.

113 air quality control station [n] *envir.* (Facility for measurement of air pollution; ▶air quality network/system); *s* **estación [f] de medición de la calidad del aire** (▶red de evaluación de la calidad del aire); *f* **station [f] de mesure de la pollution atmosphérique** (Installation de mesure de la qualité de l'air et de mesure des rejets de substances dans l'atmosphère ; ▶réseau de mesure des pollutions atmosphériques) ; *g* **Luftmessstation [f]** (Einrichtung zur Messung der Luftbelastung durch Erfassung luftfremder Stoffe. Je nach Standort der Messeinrichtung werden Stoffe am Ort des Entstehens [Emission] oder am Ort ihrer Wirkung [Immission] erfasst. Gebiets- resp. ortsbezogene Messstationen dienen der Erfassung und Bestimmung der Belastung von Gebieten und damit der dort lebenden Bevölkerung, der Flora und Fauna und von Sachgütern durch luftfremde Stoffe und Schadstoffdepositionen. Die Auswertung ergibt die Beurteilung der Luftqualität; ▶Luftmessnetz); *syn.* Messstation [f] zur Luftqualitätsüberwachung.

air quality management [n] *envir. leg. nat'res.* ▶air pollution control.

114 air quality monitoring network/system [n] *envir.* (Network of automatic measuring facilities for the monitoring of ambient air quality, which every minute continuously transfer data on the concentration of sulfur [US]/sulphur [UK] dioxide, nitrogen oxide, carbon monoxide, ozone and suspended particulate matter to a central computer, where half-hour and daily mean values are calculated; **in U.S.**, federally-designated 'air quality control regions' are required to meet and maintain federal ambient air quality standards; some interstate agreements or compacts exist; **in D.** exist a network of some 250 automatically measuring facilities for the monitoring of air quality. In addition there are measuring points for the monitoring of transboundary air pollution which were agreed upon in international protocol of Helsinki, Montreal and Sofia meetings in order to reduce national annual emissions or their transfrontier transport of sulfur [US]/sulphur [UK] and nitrogen as well as the consumption of chlorofluorocarbons [CFC] and other chemicals which may destroy the ozone layer; ▶air pollution monitoring, ▶air quality control station, ▶monitoring); *s* **red [f] de evaluación de la calidad del aire** (Conjunto de estaciones de medición situadas en lugares estratégicos para el ▶control del ambiente atmosférico en una región. En general se miden los parámetros SO_2, NO_X, CO, ozono y partículas sólidas; ▶estación de medición de la calidad del aire, ▶monitoring, ▶sistema español de información, vigilancia y prevención de la contaminación atmosférica); *syn.* Red [f] Nacional de Vigilancia y Previsión de la Contaminación Atmosférica [Es]; *f* **réseau [m] de mesure des pollutions**

A

atmosphériques (**F.**, le réseau national de surveillance de la pollution de l'air est constitué par une trentaine d'associations agrées ayant pour mission la mesure permanente [mesures centralisées sur ordinateur permettant l'obtention de valeurs moyennes toutes les heures] des concentrations dans l'air de l'anhydride sulfureux, du dioxyde d'azote, des particules fines et des particules en suspension, du plomb, de l'ozone ; ce dispositif de surveillance est mis en place depuis le 1er janv. 1997 dans les agglomérations de plus de 250 000 habitants, depuis le 1er janvier 1998 pour les agglomérations de plus de 100 000 habitants et prévu pour le 1. janvier 2000 pour l'ensemble du territoire national ; conformément à la Convention de Vienne et aux Protocoles d'Helsinki, de Sofia et de Genève sont mis en place des stations de mesure de la pollution transfrontière afin de réduire les émissions d'anhydride sulfureux et de dioxyde d'azote ainsi que la consommation des HCFC et autres substances qui appauvrissent la couche d'ozone ; ▶monitorage, ▶station de mesure, ▶surveillance de la qualité de l'air) ; *syn.* réseau [m] d'alerte, réseau [m] de surveillance ; *g* **Luftmessnetz [n]** (In D. der Luftüberwachung dienendes Netz mit ca. 250 automatisch messenden Stationen, die fortlaufende Messwerte über Schwefeldioxid, Stickstoffoxide, Kohlenmonoxid, Ozon und Schwebstoffe minütlich an zentrale Rechner übermitteln, um dort Halbstunden- oder Tagesmittelwerte zu errechnen. Ferner gibt es Messstationen zur Überwachung grenzüberschreitender Luftverschmutzungen, die in internationalen Protokollen von Helsinki, Montreal und Sofia beschlossen wurden, um die nationalen jährlichen Schwefel- und Stickstoffemissionen sowie den Verbrauch von FCKW und anderer die Ozonschicht zerstörenden Stoffe oder deren grenzüberschreitenden Transport zu reduzieren; ▶Luftmessstation, ▶Monitoring, ▶Überwachung der Luftverunreinigung); *syn.* Immissionsmessungsnetz [n], lufthygienisches Überwachungssystem [n], System [n] zur Überwachung der Luftqualität.

115 air quality standard [n] *adm. envir. pol.* (Prescribed threshold values and indicator values of air pollutants that cannot legally be exceeded during a given time in a specified location. In Europe, in various directives of the European Council are prescribed indicator values for sulphur/sulfur dioxide and suspended material, lead content and nitrogen dioxide in the air); *syn.* air pollution standard [n] (DES 1991, 65); *s* **estándar [m] de calidad del aire** (Valores límite de inmisión y valores indicadores de sustancias contaminantes que no pueden ser excedidos en un periodo de tiempo definido y en un lugar específico. En la UE existen varias directivas que prescriben límites para contaminantes atmosféricos como la Directiva 96/62/CE del Consejo sobre evaluación y gestión de la calidad del aire; la Directiva 1999/30/CE del Consejo que regula los valores límite de SO_2, NO_X, partículas y plomo en el aire ambiente, modificada por decisión de la Comisión 2001/744/CE, la Directiva 2000/69/CE del Parlamento Europeo y del Consejo sobre los valores límite para el benceno y el monóxido de carbono y la Directiva 2002/3/CE del Parlamento Europeo y del Consejo relativa al ozono); *syn.* nivel [m] máximo de inmisión; *f* **valeur [f] limite de qualité de l'air** (Niveau maximal de concentration de substances polluantes dans l'atmosphère, fixé sur la base des connaissances scientifiques, dans le but d'éviter, de prévenir ou de réduire les effets nocifs de ces substances pour la santé humaine ou pour l'environnement [Art. 3 de la loi no 96-1236 du 30 décembre 1996 sur l'air et l'utilisation rationnelle de l'énergie] ; les directives européennes en vigueur, fixent des valeurs limites en matière de protection de la santé humaine, de protection de la végétation et de protection des écosystèmes : dans l'air extérieur ambiant, sont réglementées actuellement, les polluants suivants : SO_2, PS 10, NO_2, O_3, CO, benzène, certains métaux lourds et

HAP ; en F. des arrêtés préfectoraux fixent deux niveaux de pollution : le seuil de recommandation et d'information ainsi que le seuil d'alerte pour le NO_2, le SO_2 et O_3 ; *terme spécifique* valeur [f] limite d'exposition (VLE) : notion utilisée en hygiène du travail qui représente la concentration maximale admissible, pour une substance donnée, dans l'air du lieu de travail, à laquelle le travailleur peut être exposé pour une courte durée [c.-à-d. inférieure ou égale à 15 minutes]) ; *g* **Immissionswert [m]** (In der 22. Verordnung zur Durchführung des Bundes-Immissionsschutzgesetzes [22. BImSchV] vom 26.10.1993 verordneter Grenzwert und Leitwert der Luftqualität, der während einer bestimmten Zeit in einem definierten Gebiet nicht überschritten werden darf. In diversen EG-Richtlinien werden Leitwerte für Schwefeldioxid und Schwebestaub, Bleigehalt, Stickstoffdioxid in der Luft festgesetzt [cf. z. B. Anhang IV, Tabelle A der Richtlinie 80/779/EWG und 89/427/EWG, Tabelle B der Richtlinie 80/779/EWG, 82/884/EWG, 85/203/EWG]. Die **I.e** dürfen zum Schutz vor schädlichen Umwelteinwirkungen nicht überschritten werden. Das Ausmaß der Luftverschmutzung durch Ozon wird durch Schwellenwerte für die Ozonkonzentration festgesetzt; cf. § 1 a 22. BImSchV und Anhang I der Richtlinie 92/72/EWG).

air shaft [n] [US], basement *arch.* ▶basement light well.

air space [n] [US], polluted *envir.* ▶contaminated airspace.

116 air space ratio [n] *pedol.* (Proportion of ▶pore volume of soil filled with air); *syn.* porosity [n]; *s* **contenido [m] de aire** (Porcentaje de aire en los poros del suelo; ▶porosidad total); *f* **capacité [f] en air** (Volume des vides du sol [l'espace poral] occupé par l'air, exprimé en % du volume total et caractéristique de l'état d'aération du sol ; *terme spécifique* **teneur [f] en gaz du sol** : atmosphère contenue dans les pores du sol à l'état gazeux ; PED 1984, 56 et s. ; ▶porosité totale) ; *syn.* porosité [f] non capillaire ; *g* **Luftgehalt [m]** (Anteil des mit Luft gefüllten Porenraumes eines Bodens; ▶Porenvolumen).

air stream [n] [UK] *plant.* ▶traffic-caused wind [US].

air trap [n] *constr.* ▶water seal, #2.

117 aisle [n] *trans.* (Circulation area between parking bays in a parking area; ▶parking maneuver space [US]/parking manoeuvre space [UK]); *s* **pasillo [m] de circulación** (en un área de estacionamiento; ▶espacio de maniobra) ; *syn.* carril [m] de circulación; *f* **voie [f] de circulation** (Allée de circulation des véhicules le long d'une ou entre deux alignements de stationnement dans un parc de stationnement ; ▶bande de manœuvre) ; *syn.* allée [f] de circulation d'un parc de stationnement, desserte [f] d'un parc de stationnement (VRD 1986, 175) ; *g* **Fahrgasse [f]** (Fahrstreifen neben einer oder zwischen zwei Parkbuchten auf einem Parkplatz; ▶Manövrierstreifen).

alder carr [n] [UK] *phyt.* ▶riparian alder stand [US].

alder fen [n] [UK] *phyt.* ▶riparian alder stand [US].

alder stand [n] *phyt.* ▶riparian alder stand [US].

alder swamp woodland [n] [UK] *phyt.* ▶riparian alder stand [US].

algae [npl], subaqueous carpet of chandelier *phyt.* ▶chara vegetation.

alien species [n] *phyt. zool.* ▶introduced species.

alignment [n] *plan. trans.* ▶horizontal alignment, ▶horizontal and vertical alignment, ▶paving alignment, ▶vertical alignment.

alignment [n], interrupted *urb.* ▶interrupted appearance.

118 alignment [n] of a pathway *plan. trans.* (Vertical and horizontal layout of a walkway); *syn.* alignment [n] of a trail [also US]; *s* **recorrido [m] de un camino** (Alineación vertical y horizontal de una vía); *f* **orientation [f] de la voie** (Ligne que suit un chemin sur les plans vertical et horizontal) ; *syn.* tracé [m] de route, direction [f] d'un chemin ; *g* **Wegeführung [f]** (Art und Weise wie ein Weg im Gelände verläuft).

alignment of a trail [n] [US] *plan. trans.* ►alignment of a pathway.

119 alkaline soil [n] *pedol.* (Soil with a pH value greater than 7; *opp.* ►acid soil); *syn.* alkali soil [n]; *s* **suelo [m] alcalino** (Suelo con un pH superior a 7; ►suelo ácido); *syn.* suelo [m] básico; *f* **sol [m] alcalin** (Sol dont le pH est supérieur à 7. Cependant on réserve en général aux « sols neutres » la gamme de pH compris entre 6,6 et 7,3, de sorte qu'on considère comme alcalin un sol dont le pH est supérieur à 7,3 ; DIS 1986 ; *opp.* ►sol acide) ; *syn.* sol [m] basique ; *g* **alkalischer Boden [m]** (B. mit einem pH-Wert über 7; *opp.* ►saurer Boden); *syn.* basischer Boden [m].

alkali soil [n] *pedol.* ►alkaline soil.

120 all additional work and services included [loc] *constr. contr.* *s* **incluídos todos los servicios adicionales [loc]**; *f* **(y) compris toutes sujétions d'exécution [loc]** ; *g* **einschließlich aller Nebenleistungen [loc]**.

121 "All-America City" [n] [US] *hort. plan. urb.* (**In U.S.**, "All-America City" awards are given annually by the National Municipal League for citizen action, effective organization and community improvement; a similar appreciation but restricted to houses and neighborhoods is the Sears-Roebuck "Home and Neighborhood Improvement Award"; and local awards are given by village preservation and improvement societies throughout the country; **in U.K.**, a **"best kept village competition"** in which awards are given for well-maintained villages [or towns] involving the participation of local residents; ►historic preservation measures, ►landscape management; *syn.* best kept village competition [n] [UK]; *s* **concurso [m] de belleza de pueblos** (Competición que galardona a los pueblos más bellos y mejor cuidados, incluyendo la participación de la población en su mantenimiento; ►gestión del Patrimonio Histórico Español, ►gestión y protección del paisaje); *f* **concours [m] national des villes et villages fleuris** (Concours organisé annuellement entre le 15 juin et 15 septembre par le comité national pour le fleurissement de la France [C.N.F.F.] et ouvert à toute les communes. Son but est de développer le fleurissement des espaces urbains et encourager les citadins et les responsables locaux à embellir leur cité ; LEU 1987, 187 ; ►gestion des milieux naturels, ►gestion du patrimoine monumental) ; *g* **Dorfwettbewerb [m]** (D., Anregung der Bewohner von ländlichen Gemeinden oder Ortsteilen durch Wettbewerbe vorhandene Missstände zu beseitigen u. den Wohnwert unter Einbeziehung der bürgerschaftlichen Mitarbeit und Initiative zu verbessern. Seit 1979 wurde der Wettbewerb um die Gesichtspunkte ►Denkmalpflege und ►Landschaftspflege erweitert. In D. gibt es den Wettbewerb „**Unser Dorf soll schöner werden"** für Gemeinden bis 3000 EW, der alle zwei Jahre und den Wettbewerb „**Bürger, es geht um deine Gemeinde"** für Gemeinden mit über 3000 EW, der ca. alle 4 Jahre stattfindet. 1994 nimmt D. erstmals an der „**Entente florale"** als Dorf- und Stadtverschönerungswettbewerb teil. Die Deutsche Gartenbaugesellschaft 1822 e. V. organisiert die Vertretung Deutschlands. Ende der 1990er-Jahre wurde auf Initiative des Zentralverbandes Gartenbau [ZVG], Deutschen Städtetages, Deutsche Städte und Gemeindebundes und des Deutschen Tourismusverbandes der Bundeswettbewerb „**Unsere Stadt blüht auf"** ins Leben gerufen, um die positiven Effekte einer nachhaltigen Stadtbegrünung für Gemeinden mit mehr als 3000 Einwohnern zu fördern, quasi als Ergänzung zum vor genannten Wettbewerb „Unser Dorf soll schöner werden". Der Wettbewerb „**Unsere Stadt blüht auf"** fand erstmals 2001 statt. Ziele dieses

Bundeswettbewerbes sind die Gestaltung und Erhaltung l[i]ebenswerter Städte und Gemeinden, die Förderung einer nachhaltigen Grün- und Freiraumentwicklung im Sinne der Lokalen Agenda 21, die Förderung des Natur- und Umweltschutzes im besiedelten Raum, die Förderung von Handel und Gewerbe durch Schaffung von vielfältigen, lebendigen Innenstädten, die Steigerung der Attraktivität für Touristen und die Förderung des bürgerschaftlichen Engagements. Dieser jährliche Wettbewerb wird seit 2001 auch als Vorentscheid für die europäische **Entente florale** ausgelobt; teilnehmen können auch Städte mit Stadtteilen, die mehr als 15 000 Einwohner und eigene Verwaltungskörperschaften haben).

all-around lawn [n] [UK] *constr.* ►play lawn [US] (1)/all-around lawn [UK].

allee lane [n] [US] *gard. landsc. urb.* ►tree-lined avenue [US].

122 alleviation measure [n] *envir. leg. plan.* (Action to be taken within the scope of an ►environmental impact statement [EIS], aimed at reducing the extent of a project's intrusion upon the environment; **in U.S.**, such a distiction is not made: the generic term is ►mitigation measure; **in D.**, alleviation measures as well as mitigation activities or program[me]s undertaken pursuant to federal or state law; ►prevention measure); *s* **medida [f] mitigante** (En el marco de ►evaluación del impacto ambiental [EIA] medida prevista para reducir los efectos negativos del proyecto sobre el medio; ►medida compensatoria, ►medida correctora); *syn.* medida [f] reductora; *f* **mesure [f] de réduction** (Dans le cadre d'une ►étude d'impact, mesure visant à limiter l'impact [effets directs et indirects temporaires et permanents] des ouvrages sur l'environnement ; ►déclaration environnementale, ►mesure compensatoire, ►mesure de suppression) ; *syn.* mesure [f] de limitation ; *g* **Verminderungsmaßnahme [f] (eines Eingriffs)** (Maßnahme, die der Einschränkung von bedeutenden nachteiligen Auswirkungen eines Projektes auf die Umwelt dient. Während durch ►Vermeidungsmaßnahmen einzelne Beeinträchtigungen von Natur und Landschaft gänzlich unterbleiben, wird durch Maßnahmen zur Minderung nur das Ausmaß einer einzelnen Beeinträchtigung reduziert; ►Ausgleichsmaßnahme [des Naturschutzes und der Landschaftspflege], ►Umwelterklärung); *syn.* Maßnahme [f] zur Minderung (eines Eingriffs).

123 alliance [n] *phyt.* (The most closely related ►associations of plants are united in an alliance. The floristic relationship is shown especially by the presence of a rather large number of species characteristic of the alliance. This floristic relationship is designated by attaching the suffix *-ion*. Where necessary, a species name in the genitive case may be added to the genus name, e.g. *Caricion curvulae*; ►character species of an alliance, ►phytosociological classification); *s* **alianza [f]** (En la sistemática de las asociaciones de BRAUN-BLANQUET, la unidad inmediatamente superior a ►asociación. Se define como «un grupo de asociaciones relacionadas entre sí lo más estrechamente», y en la que «la relación florística se muestra especialmente por la presencia común de un número bastante grande de especies, que caracterizan a la alianza». Se designa por la terminación *-ion* aplicada al nombre latino de las especies más características; cf. DB 1985; ►clasificación fitosociológica, ►especie característica de una alianza); *f* **alliance [f]** (Unité phytosociologique qui regroupe des ►associations ayant des affinités floristiques et sociologiques ; chaque alliance est constituée par des espèces caractéristiques propres se rencontrant dans les diverses associations la constituant, une de ces espèces permettant de dénommer l'alliance [terminaison *-ion*], p. ex. le *Fagion silvaticae,* alliance regroupant les diverses associations où le hêtre est présent ; ►espèce caractéristique d'alliance, ►systématique phytosociologique) ; *g* **Verband [m] (1)** (Die nächste, der ►Assoziation übergeordnete Vegetationseinheit. Sie fasst die

A

floristisch ähnlichen, durch Verbandskennarten charakterisierten Assoziationen zusammen. Verbände werden durch die Endung -*ion*, dem Stamm eines Gattungsnamens angefügt, unterschieden; gegebenenfalls ist dem Gattungsnamen noch ein Artname im Genitiv beizufügen, z. B. *Caricion curvulae*; ►Klassifizierung der Vegetationseinheiten, ►Verbandscharakterart).

124 allocation [n] **of commercial facilities and light industries** *plan. urb.* (Process of developing areas for commercial and industrial use); *s* **instalación [f] de servicios e industrias ligeras** *syn.* implantación [f] de servicios e industrias ligeras; *f* **implantation [f] d'activités commerciales** (Processus par lequel les activités choisissent un lieu d'implantation) ; *syn.* localisation [f] des activités ; *g* **Gewerbeansiedlung [f]** (*Vorgang* Errichten von Gewerbebetrieben in einem dafür vorgesehenen Gebiet).

125 allocation [n] **of disposal cost to waste producer** *envir. pol.* (Legislated inducement for the reduction of waste); *s* **reparto [m] de costes de eliminación entre los generadores de residuos** (Obligación de las empresas e instituciones de pagar los costos para eliminar la basura; uno de los fines de esta medida es promocionar la reducción de residuos); *f* **répercussion [f] des coûts de traitement aux producteurs de déchets** (1. F., montant de la taxe sur le traitement et le stockage des déchets répercuté dans les contrats conclus par les exploitants d'une installation de déchets ménagers et assimilés ou d'une installation d'élimination de déchets industriels spéciaux avec les personnes physiques ou morales dont il réceptionne les déchets. 2. F., taxe reversée à l'Agence de l'environnement et de la maîtrise de l'énergie en vue de la réhabilitation des sites industriels pollués orphelins. 3. D., mesure prévue pour inciter les producteurs à prévenir ou réduire la production de déchets) ; *g* **Kostenumlage [f] auf die Abfallproduzenten** (Gesetzlich geregelte Kostenabwälzung an die Abfallproduzenten als Anreiz für Abfall vermeidendes oder verminderndes Produktionsverhalten).

126 allochthonous [adj] *biol. geo. pedol.* (Descriptive term for rocks or other material formed elsewhere than in the present location; *opp.* ►autochthonous; *noun* allochthon [n]); *syn. o. v.* allochtonous [adj], of foreign origin [n]; *s* **alóctono/a [adj]** (Término aplicado a materiales que se han originado en un lugar distinto del que se encuentran en un momento dado, como los acumulados en el fondo de un valle, procedentes de arrastres en las laderas, o a las rocas constituidas por elementos que no se han originado *in situ*; *opp.* ►autóctono; cf. DINA 1987); *f* **allochtone [adj]** (Qui ne s'est pas formé *in situ* et est constitué d'éléments rapportés ; *opp.* ►autochtone) ; *g* **allochthon [adj]** (Nicht am Fundort entstanden; **a.** sind auch ortsfremde Gesteinskomplexe im Deckengebirge; *opp.* ►autochthon); *syn.* ortsfremd [adj]).

allotment garden (plot) [n] [UK] *urb. leg.* ►community garden [US].

127 allotment garden club [n] [± UK] *sociol.* (Organized society of all members of permanent allotment gardens [UK]; **in U.S.** community garden plot holders are organized in garden clubs; ►permanent community garden area [US]/permanent allotment site [UK]); *syn.* community garden club [n] [US], allotment grower association [n] [± UK], allotment society [n] [± UK]; *s* **asociación [f] de zona de huertos recreativos urbanos** (Organización de los arrendatarios de una ►zona permanente de huertos recreativos urbanos); *f* **association [f] de jardins ouvriers** (Organisme regroupant les adhérents d'un ►lotissement de jardins ouvriers) ; *syn.* organisme [m] de jardins familiaux, société [f] de jardins ouvriers ; *g* **Kleingartenverein [m]** (Organisatorischer Zusammenschluss aller Mitglieder einer ►Dauerkleingartenanlage).

allotment garden development plan [n] [UK] *leg. urb.* ►community garden development plan [US].

allotment garden development planning [n] [UK] *recr. urb.* ►community garden development planning [US].

allotment gardener [n] [UK] *urb. sociol.* ►community gardener [US].

allotment garden hut [n] [UK] *urb.* ►community garden shelter [US].

allotment gardening [n] [UK] *sociol. urb.* ►community gardening [US], ►hobby and allotment gardening [UK].

allotment garden plot [n] [UK] *urb. leg.* ►community garden [US].

allotment garden subject plan [n] [UK] *leg. urb.* ►community garden development plan [US].

allotment grower association [n] [UK] *sociol.* ►allotment garden club [UK].

allotment holder [n] [UK] *urb. sociol.* ►community gardener [US].

allotment land [n] [UK]**, temporary** *urb.* ►temporary community garden area [US].

allotment plot [n] [UK]**, temporary** *leg. urb.* ►temporary community garden plot [US].

allotment site [n] [UK]**, permanent** *leg. urb.* ►permanent community garden area [US].

allotment society [n] [UK] *sociol.* ►allotment garden club [UK].

allowable costs [npl] *contr. prof.* ►overhead expenses.

allowable floor area [n] [US] *leg. urb.* ►floor area ratio (FAR) [US]/floor space index [UK].

all-round fertilizer [n] [UK] *agr. constr. for. hort.* ►complete fertilizer.

128 alluvial cone [n] *geo.* (►Alluvial fan with a steeper angle of slope due to the coarseness of the material: occurs mostly in semiarid regions [US]/semi-arid regions [UK] where usually dry watercourses periodically transport stony material after heavy rainfall); *syn. also* dejection cone [n]; *s* **cono de deyección** (Forma de acumulación de planta cónica, situada en el tramo final de un torrente, cuando éste alcanza el valle principal; DGA 1986; ►abanico aluvial); *f* **cône [m] alluvial** (Relief alluvial qui, en comparaison avec le ►glacis d'épandage, a la forme d'une section de cône comprise entre deux génératrices ; lorsque le **c. a.** construit par un torrent a une forte pente on parle de cône de déjection torrentiels) ; *syn.* cône [m] d'épandage, cône [m] de déjection ; *g* **Schwemmkegel [m]** (Geschiebeakkumulation eines Fließgewässers, die im Vergleich zum ►Schwemmfächer einen steileren Böschungswinkel hat, da das Material gröber ist; entsteht meist in semiariden Gebieten durch den Gesteinstransport zeitweilig fließender Flüsse an Stellen, an denen das Gefälle plötzlich abnimmt).

alluvial deposit [n] *geo.* ►sediment (1).

129 alluvial fan [n] *geo.* (Fan-shaped mass of sediment deposited by a stream where it emerges from an upland valley into a plain; ►alluvial cone); *s* **abanico [m] aluvial** (Acumulación de material detrítico que previamente ha sido transportado por un torrente o curso de agua. Se trata de una formación aluvial de piedemonte, que se localiza en la zona de contacto entre la montaña y la llanura o valle principal; ►cono de deyección); *f* **glacis [m] d'épandage** (Étalement d'alluvions masquant la roche en place, formé par un cours d'eau à forte charge, les alluvions étant déposées sur une étendue plane, unie ou de faible pente au confluent d'une vallée étroite avec une vallée principale ; le terme général de glacis alluvial englobe toutes les formes de recouvrement de la roche en place ; cf. DG 1984 ; ►cône alluvial) ; *g* **Schwemmfächer [m]** (Flache, fächerförmige Aufschüttung von Sand- und Geröllmassen durch ein Fließgewässer, das plötzlich aus einem engen Tal in ein breites Tal oder in eine flache Ebene gelangt; ►Schwemmkegel).

130 alluvial gravel [n] *geo.* (Accumulation of coarse, deposited, waterborne, small stone material; ►gravel river terrace); *s* **acumulación [f] de pedregal** (Deposición de piedras > 2 mm en curso de agua; ►terraza [fluvial] de aluviones); *f* **épandage [m] alluvial grossier** (Dépôt d'alluvions dont la granulation est supérieure à 2 mm ; ►terrasse alluviale) ; *syn.* placage [m] alluvial grossier ; *g* **Aufschotterung [f]** (Ablagerung von Gesteinskörnern > 2 mm durch Fließgewässer; ►Schotterterrasse).

131 alluvial plain [n] *geo.* (Near-level area adjacent to a river or stream, which has been formed by the deposition of material left behind after flooding; ►floodplain [US]/flood-plain [UK], ►regularly flooded alluvial plain, ►upper riparian alluvial plain); *syn.* bottomland [n] [also US] (WET 1993, 13, 32), valley floodplain [n]; *s 1* **vega [f] (1)** (Tierra baja, llana y fértil, generalmente ocupando los fondos de los valles amplios en los cursos medios y bajos de los ríos, donde se depositan los elementos más finos procedentes de la meteorización, descomposición y arrastre de materiales de las zonas más altas de la cuenca del río; DINA 1987; *términos específicos* ►vega de arroyo, ►llanura de inundación, ►vega aluvial ocasionalmente inundada, ►vega aluvial regularmente inundada); *syn.* zona [f] húmeda de aluvión, llanura [f] aluvial; *s 2* **huerta [f] (1)** (En algunas regiones de Es. y de América Latina se aplica a todas las áreas de regadío, aunque no formen valle); *syn.* vega [f] (2); *f* **plaine [f] alluviale** (Zone aplanie par comblement sédimentaire en bordure d'un cours d'eau, caractérisée par des inondations occasionnelles ; *termes spécifiques* plaine alluviale d'un fleuve, plaine alluviale d'une rivière ; ►champ d'inondation, ►plaine alluviale de forêt à bois durs, ►plaine alluviale à bois tendres) ; *syn.* terrains [mpl] alluviaux, vallée [f] alluviale inondable, vallée [f] humide ; *g* **Aue [f]** (Teil des Talbodens, der von Überschwemmungen erreicht werden kann. *UBe* Bachaue, Flussaue, Flussniederung; ►Hartholzaue, ►Überschwemmungsgebiet, ►Weichholzaue); *syn.* Talaue [f], Au [f] [A, CH].

alluvial plain [n] [US]**, lower riparian** *phyt.* ►regularly flooded alluvial plain.

alluvial plain [n] [US]**, lower riverine** *phyt.* ►regularly flooded alluvial plain.

132 alluvial silt [n] *pedol.* (In central Europe, mostly a calcareous wetland soil, similar to loess with an average depth of between 1.5 and 2.5 meters [US]/metres [UK]. The name is not consistent with the definition for ►loam, because alluvial silt partially consists of clay or silt sediments; not used in current U.S. system of soil taxonomy; ►alluvial soil); *s* **limo [m] aluvial** (En Europa Central tipo de ►suelo aluvial generalmente calcáreo, parecido al loess, con profundidad entre 1,5 y 2,5 m. El fango aluvial está constituido por sedimentos arcillosos y limosos; ►limo 2); *f* **limon [m] alluvial** (En Europe centrale il s'agit d'un ►sol alluvial 1 constitué d'alluvions fines argilo-limoneuses de décantation, semblables au lœss et généralement calcaire ; les dépôts ont une épaisseur moyenne de 1,5 à 2,5 m. Cette désignation ne correspond pas toujours à la définition de la fraction granulométrique du ►limon 2 car la texture d'une partie de ces alluvions a plus d'analogie avec l'argile ou le sable) ; *syn.* limon [m] de débordement ; *g* **Auenlehm [m]** (In Mitteleuropa meist kalkhaltiger, lössähnlicher ►Auenboden mit einer durchschnittlichen Mächtigkeit von 1,5-2,5 m. Der Name **A.** stimmt nicht immer mit der Definition der Bodenart ►Lehm überein, da ein Teil der Flussablagerungen mehr einem Ton- oder Schluffsediment entspricht); *syn.* Hochflutlehm [m].

133 alluvial soil [n] *pedol.* (Usually fine-grained soils deposited in floodplains of rivers or streams as a result of the settling out of suspended material on flooded land; ►alluvial plain); *s* **suelo [m] aluvial** (Suelos existentes en los valles de los ríos, formados a partir de sus sedimentos, que no muestran en su perfil acción alguna del agua estancada, pero se encuentran generalmente humedecidos e influenciados fuertemente en su régimen hídrico, su vegetación y biología por la proximidad y el nivel del agua del río; cf. KUB 1953; *término específico* vega [parda y roja]; ►huerta, ►vega 1); *syn.* suelo [m] de aluvión; *f* **sol [m] alluvial (1)** (Sol formé dans une ►plaine alluviale et résultant du dépôt de sédiments transportés par les eaux courantes) ; *syn.* sol [m] alluvionnaire ; *g* **Auenboden [m]** (Dem Fließgewässer zugeführter, erodierter Oberboden, der sich bei Überschwemmungen in der ►Aue abgelagert hat); *syn.* alluvialer Boden [m], *o. V.* Auboden [m], Schwemmlandboden [m].

134 all-weather court [n] [US] *constr. recr.* (Asphaltic concrete or synthetic playing surface which can be used under all weather conditions; ►hard court [US]/hard pitch [UK]); *syn.* all-weather pitch [n] [UK]; *s* **área [f] impermeabilizada para juegos** (Cancha de deportes de bitumen o de material sintético que también se puede usar cuando hace mal tiempo; ►cancha de pavimento duro); *f* **aire [f] tous temps** (Aire de sport équipée d'un revêtement bitumineux ou synthétique permettant son utilisation par mauvais temps ; ►aire à revêtement dur) ; *g* **Allwetterplatz [m]** (Sportfläche aus bitumen- oder kunststoffgebundenem Material hergestellt und deshalb auch bei schlechtem Wetter nutzbar; ►Hartplatz).

all-weather pitch [n] [UK] *constr. recr.* ►all-weather court [US].

135 all-weather surface [n] *constr.* (Covering of sports areas for use under all weather conditions); *s* **suelo [m] (todotiempo) impermeabilizado**; *f* **revêtement [m] tous temps** *syn.* sol [m] tous temps ; *g* **Allwetterdecke [f]** (Belag eines Platzes, der bei jeder Witterung bespielbar ist, z. B. eine Kunststoffdecke, ein Asphaltbelag).

alpine belt [n] *geo. phyt.* ►alpine zone.

136 alpine foreland [n] *geo.* (Highland adjoining the alpine massif; ►foothills); *s* **estribaciones [fpl] de los Alpes** (Zona montañosa situada ante los Alpes; ►estribación); *f* **préalpes [fpl]** (Région de montagnes et de collines de transition entre les massifs alpins et les plaines qui les entourent ; ►contrefort montagneux) ; *g* **Alpenvorland [n]** (Den Alpen vorgelagertes Berg- und Hügelland; ►Vorgebirge).

137 alpine garden [n] (1) *hort.* (In Europe, rock garden in an alpine or in another cold climate area, where alpine plants are cultivated for scientific study and public enjoyment; **in U.S., a. g.** are accessible in national parks and mountainous state parks for scientific study and public enjoyment of native flora; ►alpine garden 2, ►alpine plant); *s* **alpinum [m]** (Jardín botánico especializado en el cultivo, cuidado y estudio de las plantas de alta montaña y del ártico, por lo que en su diseño se utilizan piedras y lechos de grava para simular las condiciones en la que se encuentran las plantas en sus lugares de origen. Puede estar situado en regiones de alta montaña o en otras de clima frío. En países —como GB e Irlanda— donde no se dan las condiciones idóneas para estos jardines, se construyen en las llamadas «casas alpinas» que reproducen lo más fielmente posible sus condiciones ideales de desarrollo; cf. http://es.wikipedia.org/wiki/Alpinum 12.8.08; ►jardín alpino, ►planta alpina); *syn.* alpinarum [m]; *f* **alpinum [m]** (Jardin botanique spécialisé dont l'objectif est la collection, la culture, la climatisation et la mise en valeur des espèces végétales poussant naturellement à de hautes altitudes ; ▶jardin naturel d'altitude, ▶plante alpine, ▶station biologique de la flore de montagne) ; *syn.* jardin [m] botanique alpin, jardin [m] de plantes alpines, jardin [m] alpin) ; *g* **Alpinum [n]** (Alpina [pl], Alpinen [pl]); **1.** Lebensbereich konkurrenzschwacher alpiner und arktischer Pflanzen, meist Stress-Strategen mit speziellen Standortansprüchen an Substrat, Luft- und Bodenfeuchte. **2.** Steingarten, der der Hochgebirgssituation so weit wie möglich angepasst ist, und zur Pflege und zum wissenschaftlichen Studium von Pflanzen der Alpen sowie Pflanzen anderer Hochgebirgsregionen und arktischen Pflanzen dient; ▶Alpengarten, ▶alpine Pflanze).

138 alpine garden [n] (2) *hort.* (**1.** Garden in the Alps, where indigenous alpine plants are cultivated, often for the scientific study of plants in a high alpine situation. **2.** Garden containing plants of the Alps; ▶alpine garden 1; *s* **jardín [m] alpino** (**1.** Jardín situado en los Alpes en el que se cultivan plantas indígenas, a menudo con el fin de estudiar su desarrollo en la ubicación natural de las mismas. **2.** Jardín en el que se cultivan plantas de los Alpes; ▶alpinum); *f 1* **jardin [m] (naturel) d'altitude** (Site naturel de grande envergure, présentant les espèces alpines observées dans leur milieu naturel lors de promenades et de visites guidées, ouvert au public pendant une courte période de l'année, la saison de fermeture allant avec l'enneigement ; ▶alpinum) ; *f 2* **station [f] biologique de la flore de montagne** (Station scientifique d'observations et d'études des plantes alpines dans un grand espace naturel, ouverte aux chercheurs et enseignants-chercheurs) ; *g* **Alpengarten [m]** (**1.** Garten, der in den Alpen angelegt ist, oft zum Zwecke wissenschaftlicher Forschung am natürlichen Hochgebirgsstandort. **2.** Garten, der Pflanzen aus den Alpen enthält; ▶Alpinum).

139 alpine grassland [n] *phyt.* (Grassland vegetation type occurring above tree line, occasionally with dwarf shrubs or gnarled trees, covered with snow for more than six months of the year; ▶nutrient-poor grassland, ▶dry meadow); *syn.* alpine sward [n], alpine meadow [n] [also US]; *s* **pasto [m] alpino** (Vegetación herbácea que crece por encima del límite del bosque, cubierta de nieve más de 6 meses al año; ▶pastizal seco, ▶prado oligótrofo); *f* **pelouse [f] alpine** (Peuplement herbacé occupant la zone sus-jacente à la limite de végétation arborescente, parfois accompagné d'arbrisseaux et de « végétation krummholz » ; la durée du manteau nival est de plus de six mois par an ; le terme spécifique allemand « Urwiese » désigne un peuplement herbacé de l'étage alpin, dense ou même fermé, homogène, reposant sur des sols bien caractérisés, né d'une évolution lente et longue, stable et considéré comme climacique ; VCA 1985, 246 ; ▶pelouse aride, ▶pelouse oligotrophe) ; *g* **alpiner Rasen [m]** (Vorwiegend aus Gräsern bestehende, krautige Pflanzendecke oberhalb der Baumgrenze, nur ausnahmsweise von Sträuchern oder Krummholzkiefern bewachsen, mehr als sechs Monate im Jahr mit Schnee bedeckt; der UB **Urwiese** bezeichnet eine krautige, sehr oft dicht geschlossene, auf gut entwickelten Böden wachsende Pflanzendecke, die sich über eine sehr lange Zeit entwickelt hat und als Klimaxstadium betrachtet wird; ▶Magerrasen, ▶Trockenrasen); *syn.* Alpenmatte [f], alpine Matte [f].

140 alpine hut [n] *recr.* (Small lodging facility of simple construction or simple backcountry shelter in the mountains; smaller than a ▶mountain lodge; ▶mountain cabin); *s* **refugio [m] de alta montaña** (Pequeña construcción de protección o albergue, más pequeño que un ▶refugio de montaña; ▶albergue, ▶cabaña de montaña); *f* **refuge [m] de haute montagne** (Abris et ▶gîte de haute montagne ; ▶gîte de France, ▶refuge de montagne) ; *syn.* gîte [m] de montagne, gîte [m] montagnard, gîte [m] alpin ; *g* **Alpenhütte [f]** (▶Schutzhütte und ▶Beherber-

gungsstätte in den Alpen/im [Hoch]gebirge); *syn.* Hochgebirgshütte [f], Hütte [f] im (Hoch)gebirge.

alpine meadow [n] [US] *phyt.* ▶alpine grassland.

141 alpine pasture [n] *agr.* *syn.* high mountain pasture [n] [also US]; *s* **pastizal [m] de alta montaña** (Pradera que se utiliza en verano para el pastoreo); *syn.* pasto [m] de alta montaña; *f* **alpage [m]** (Pâturage d'altitude sur les pelouses alpines utilisé pour l'estivage ou la transhumance) ; *syn.* haut pâturage [m] ; *g* **Alm [f]** (Der sommerlichen Weidenutzung dienende Wiese im Gebirge); *syn.* Hochweide [f].

142 alpine plant [n] *bot. phyt.* (Plant growing on high mountains above timberline; *s* **planta [f] alpina** (Planta que crece en las altas montañas más allá del límite del bosque); *f 1* **plante [f] alpine** (**1.** Plante poussant naturellement dans les régions montagneuses au-dessus de la limite supérieure des forêts dans l'étage de végétation compris entre 2400 et 3000 m environ. **2.** Plante utilisée en horticulture pour le jardin de rocaille) ; *syn.* plante [f] supraforestière ; *f 2* **plante [f] alpienne** (Terme qualifiant toute plante strictement originaire des Alpes d'Europe Australe alors que le mot alpin désigne toute plante vivant dans une montagne du même type que les Alpes) ; *syn.* plante [f] alpique ; *g* **alpine Pflanze [f]** (Im Hochgebirge oberhalb der Baumgrenze wachsende Pflanze, die spezielle Standortansprüche an Substrat, Luft- und/oder Bodenfeuchte hat; die meisten echten Alpenpflanzen blühen wegen der kurzen Vegetationsperiode im Frühling und Frühsommer. Bei der Pflanzenverwendung für Alpina werden zusätzlich Steingartenpflanzen, die nicht unbedingt aus Gebirgslagen stammen müssen, zur Verlängerung der Blühsaison verwendet); *syn.* Hochgebirgspflanze [f]; *in den Alpen* Alpenpflanze [f].

alpine region [n] *geo.* ▶high mountain region.

alpine sward [n] *phyt.* ▶alpine grassland.

143 alpine zone [n] *geo. phyt.* (Mountainous area below the ▶nival zone down to the tree line; ▶altitudinal belts, ▶páramo); *syn.* alpine belt [n]; *s* **área [f] cacuminal** (Bioma que comprende los territorios situados altitudinalmente por encima de la línea de bosque natural, cualquiera que sea la latitud. En ellos se desarrolla una vegetación del tipo de ▶tundra ártica, hasta llegar a la zona de los suelos desnudos y nieves perpetuas; DINA 1987; ▶páramo, ▶piso nival, ▶zonación altitudinal); *syn.* piso [m] alpino, área [f] alpina, zona [f] oreal; *f 1* **étage [m] alpin** (Étage de végétation compris entre 2400 et 3000 m environ, en général constitué de pelouses rases situées entre la limite supérieure des arbres et l'▶étage nival ; ▶étagement de la végétation 1, ▶páramo) ; *syn.* étage [m] supraforestier ; *f 2* **étage [m] alpien** (Terme qualifiant l'étage de végétation compris entre 2400 et 3000 m environ, caractérisant les Alpes d'Europe Australe alors que le mot « **alpin** » désigne l'étage de végétation d'une montagne du même type que les Alpes ; le terme alpien est aussi utilisé pour désigner les plaines et basses montagnes périalpines) ; *syn.* étage [m] alpique ; *g* **alpine Stufe [f]** (**1.** dem Hochgebirge zugehörig. **2.** Bereich unterhalb der ▶nivalen Stufe bis zur Waldgrenze; die Höhenangaben variieren, je nach Quelle und in Abhängigkeit von der Höhe über NN und geografischer Lage, zwischen 2000-3000 m; ▶Höhenstufung, ▶Páramo); *syn.* Hochgebirgsstufe [f].

alteration [n] *phyt. zool.* ▶anthropogenic alteration; *phyt.* ▶anthropogenic alteration of flora composition; *leg. urb.* ▶building alteration; *leg. plan.* ▶preliminary injunction on alteration.

alteration [n], **plan** *plan.* ▶plan revision.

144 altered construction segment [n] [US] *constr. contr.* (Specified construction work which has to be changed according to an alteration of the design, or caused by order of the

A

client/owner); *syn.* variation [n] [UK]; *s* **trabajos [mpl] modificados** (Cambios en los servicios previstos en un contrato de obra exigidos por el comitente); *f* **travaux [mpl] modificatifs** *syn.* ouvrage [m] modificatif, prestation [f] modificative ; *g* **geänderte Leistung [f]** (Durch Änderung des Entwurfs oder durch andere Anordnungen des Auftraggebers bedingte Änderung für eine im Vertrag vorgesehene Bauleistung; cf. § 2 [5] VOB Teil B).

altered soil [n] [US] *agr. constr. hort. pedol.* ►man-made soil.

145 alternately-jointed [adj] *arch. constr.* (Pattern bond without stack bond joints); *s* **sin juntas cruzadas [loc]** (Aparejo de piedras o ladrillos con juntas no apiladas); *f* **à joints décalés [loc]** (Caractérise l'appareillage d'une maçonnerie, d'un dallage sans joints croisés) ; *syn.* à joints alternés [loc] ; *g* **kreuzfugenfrei [adj]** (So beschaffen, dass im Mauerverband oder Verlegemuster keine Kreuzfugen vorkommen).

146 alternate specification item [n] *contr.* (In competitive contract bidding procedures [US]/tendering procedures [UK], an item which may be totally or partially selected for execution instead of the ►basic specification item. Most **a. s. i.s** specify different materials or methods of execution. Unit prices of **a. s. i.s** are not included in the total figures of a bid [US]/tender [UK]; ►list of bid items and quantities [US]/schedule of tender items [UK], ►optional specification item); *s* **partida [f] alzada** (En el ►resumen de prestaciones partida independiente que puede sustituir total o parcialmente una ►partida básica. Generalmente implica el uso de otros materiales o modos de realización; ►partida discrecional); *syn.* artículo [m] de partidas alzadas; *f* **numéro [m] de prix pour une variante** (Prestation spécifiée dans un ►descriptif-quantitatif pouvant être retenue à la place d'une prestation de base caractérisant en général des produits et une mise en œuvre différents. Pour la remise d'une offre avec variante n'est mentionné que le prix unitaire celui-ci n'influençant pas le prix définitif de l'offre ; ►numéro de prix d'une prestation de base, ►numéro de prix provisoire) ; *g* **Alternativposition [f]** (Im ►Leistungsverzeichnis gekennzeichnete, eigenständige Position, die ganz oder teilweise an Stelle einer ►Grundposition ausgeführt werden kann. In ihr werden meist andere Materialien oder Ausführungsarten beschrieben. Für **A.en** werden im Angebot nur Einheitspreise ausgewiesen, ohne dass diese in die Gesamtwertung der Angebotssumme eingehen; ►Bedarfsposition); *syn.* Wahlposition [f].

147 alternate tree removal [n] *constr. hort.* (Digging out every other tree in an avenue or in a nursery); *s* **aclareo [m] esquemático** (Extracción de árboles de una hilera o de un bosquete o plantación); *syn.* aclareo [m] mecánico; *f* **démariage [m]** (Enlèvement d'un arbre sur deux dans un alignement) ; *g* **Herausnahme [f] eines jeden zweiten Baumes** (Ausgraben [oder Fällen] eines jeden zweiten Baumes in einer Allee oder in einem Baumschulquartier, um den verbleibenden Bäumen bessere Wachstumsbedingungen zu gewähren).

alternative [n] *plan. trans. wat'man.* ►route alternative; *constr. contr.* ►unit price for an alternative.

alternative corridors [npl] *plan. trans.* ►comparison of alternative routes.

alternative power source [n] *envir.* ►renewable power source.

148 altitude [n] *geo.* (►Height above mean sea level, ►difference in altitude, ►high altitude); *s* **altitud [f]** (►altura sobre el nivel del mar, ►diferencia de altitud, ►gran altitud); *f* **altitude [f] absolue** (Hauteur d'un point au-dessus du niveau moyen des mers ; ►altitude absolue) ; *g* **Geländehöhe [f] über NN** (NN: *Abk. für* „Normal Null"; Höhe eines Messpunktes oder einer Fläche über dem Meeresspiegel; ►Höhe über Normalnull).

149 altitudinal belts [npl] *geo. phyt.* (Vertical classification of vegetation zones according to temperature and precipitation on mountain slopes. In northern latitudes the following belts are distinguished: ►colline zone, ►montane zone, ►alpine zone and ►nival zone; in tropical regions of the Andes: *tierra caliente, tierra templada, tierra fria,* and *tierra helada*; ►zonation); *s* **zonación [f] altitudinal** (Distribución de la vegetación en pisos o cinturas en función de la temperatura cambiante con la altitud. En las regiones atlántico-boreales en ►piso basal, ►piso colino, ►piso montano, ►piso alpino y ►piso nival. En las regiones mediterráneas: piso infra-, termo-, meso-, supra-, oro- y crioromediterráneo. *En la región tropical* tierra caliente, tierra templada, tierra fría y tierra helada; cf. DINA 1987); *syn.* zonación [f] de los pisos altitudinales, catenas [fpl] altitudinales, cliseries [fpl] altitudinales; *f* **étagement [m] de la végétation** (Distribution de la végétation, conséquence biogéographique du gradient climatique altitudinal [étages climatiques] ceinturant les montagnes ; dans les zones tempérées, on distingue du bas vers le haut : l'►étage collinéen, l'►étage montagnard, l'►étage alpin et l'►étage nival ; l'étagement de la montagne intertropicale de la cordillère des Andes varie selon la position en latitude, l'exposition aux vents, les courants marins qui longe le littoral pacifique et la configuration du relief ; la nomenclature proposée par Alexander von Humboldt retient a] au littoral à 100 m d'altitude, les terres chaudes, *Tierras calientes*, b] de 1000 à 2000 m, les terres tempérées ou *Tierras templadas*, c] de 2000 m à 3500 m, les terres froides ou *Tierras frias*, d] au-dessus de 3500 m d'altitude, les terres recouvertes de neiges ou *Tierras heladas*, e] l'étage glaciaire ou *Tierras nevadas*, peu développé dans cette région ; en région méditerranéenne on distingue plusieurs types d'étages bioclimatiques : thermo-, méso-, supra-, oro- et cryo-oroméditerranéen [variante méditerranéenne ; selon l'école d'Ozenda], subalpin [variante eurosibérienne] et alpin ; ►zonation de la végétation ; *syn.* zonation [f] altitudinale ; *g* **Höhenstufung [f]** (Durch Temperatur, Niederschlag und Relief bedingte vertikale Vegetationszonierung an einem Gebirgshang. In den gemäßigten Breiten werden von unten nach oben folgende Zonen unterschieden: ►kolline, ►montane, ►alpine und ►nivale Stufe; in den tropischen Anden: *tierra caliente, tierra templada, tierra fria* und *tierra helada*; MEL 1974; *in mediterranen Regonen* infra-, thermo-, meso-, supra- und crioromediterrane Stufen; cf. DINA 1987; ►Vegetationszonierung); *syn.* Höhenstufenfolge [f].

150 altitudinal zone [n] *geo. phyt.* (Elevational zone with distinctive vegetation on a mountain slope; ►altitudinal belt, ►cloud level zone, *geo. phyt.* ►vegetation altitudinal zone); *syn.* elevational zone [n] (TEE 1980, 495); *s* **piso [m] altitudinal** (Piso de vegetación en una ►zonación altitudinal, ►piso de niebla, ►piso de vegetación); *f* **tranche [f] d'altitude** (Zone dans laquelle un ensemble végétal structuré est réuni par une même affinité écologique ; l'►étagement de la végétation qui en résulte est lié au fait que la température décroît régulièrement lorsqu'on s'élève en altitude et dépend de nombreux facteurs écologiques : la pente, l'exposition, la nature du substrat, etc. ; ►étage des brouillards, ►étage de végétation) ; *syn.* tranche [f] altitudinale ; *g* **Höhenstufe [f]** (Vegetationszone an einem Gebirgshang mit einer bestimmten Artenzusammensetzung, die sich auf Grund der ökologischen Gegebenheiten wie Hangneigung, Exposition, Boden, durchschnittliche Jahrestemperatur etc. einstellt; ►Höhenstufung, ►Vegetationsstufe, ►Wolkennebelstufe).

151 altricial animal [n] *zool.* (Nestling that is born naked, blind, and helpless; their eyes and ears are sealed, and they cannot walk, maintain their body temperature, or excrete without assistance; e.g. singing birds, dogs, cats, rabbits, rats; *opp.* ►precocial animal); *s* **animal [m] nidícolo** (Al contrario que el

A

▶animal nidífugo, aquél que nace en un estadio posembrional no completo de manera que necesita del cuidado de sus padres durante algún tiempo, como los pájaros cantores, perros, gatos, conejos, ratas, etc.); *f* **animal [m] nidicole** (Par opposition aux animaux nidifuges, animaux caractérisés par un stade de développement postembryonnaire incomplet, les jeunes se développant en bénéficiant de l'aide de leurs parents [nourrissage et surveillance], p. ex. les oiseaux chanteurs, chiens, chats, lapins, rats ; LAF 1990, 828 ; ▶animal nidifuge) ; *g* **Nesthocker [m]** (Im Gegensatz zum ▶Nestflüchter in einem noch unvollkommenen postembryonalen Entwicklungsstadium geborenes Tier, das noch besonderer Pflege der Eltern bedarf; z. B. Singvögel, Hunde, Katzen, Kaninchen, Ratten).

amateur gardener [n] *gard.* ▶garden enthusiast.

152 ambient noise level [n] *envir.* (▶Noise decibel level, covering the existing [and background] degree of noise intensity; TAN 1975, 89); *syn.* background noise [n] [also US]; *s* **nivel [m] de ruidos ambientales** (Término específico de ▶nivel de ruidos referente a la intensidad de la ▶contaminación acústica medida fuera de los edificios); *f* **niveau [m] sonore ambiant** (L'intensité exprimée en décibels [DB] des émissions sonores naturellement présentes à l'extérieur d'un bâtiment ou dans un milieu naturel ; *terme générique* ▶niveau sonore) ; *syn.* niveau [m] de bruit ambiant ; *g* **Umgebungslärmpegel [m]** (Logarithmisch aufgebaute dB-Skala zur Messung des Schalldruckpegels außerhalb von Gebäuden; ▶Lärmpegel); *syn.* Lärmpegel [m] außerhalb von Gebäuden).

ameliorant [n] [UK]**, soil** *agr. constr. hort. leg.* ▶soil amendment [US].

amended version [n] *plan.* ▶revised version.

amendment [n] *plan.* ▶change in planning, #2, ▶revision; *agr. constr. hort. leg.* ▶soil amendment [US].

amendment [n] **of a plan** *leg. plan.* ▶plan revision.

amenities [npl] *urb.* ▶preservation of local visual amenities.

amenities [npl]**, placement of street** *constr.* ▶installation of street furniture.

amenity [n] [US]**, street** *arch. urb.* ▶piece of street furniture.

153 amenity benefits [npl] **of a forest** *recr.* (Influence of a forest upon the positive physical and psychological well-being of humans); *syn.* amenity value [n] of a forest; *s* **efectos [mpl] psicológicos del bosque** (Efectos del bosque y de las zonas verdes a un estado físico y [p]síquico positivo de la población); *syn.* efectos [mpl] sicológicos del bosque; *f* **rôle [m] de la forêt sur la santé corporelle et l'équilibre psychique** (Influence positive de la forêt sur le bien-être corporel et psychique de l'homme) ; *g* **umweltpsychologische Wohlfahrtswirkungen [fpl] des Waldes** (Einfluss des Waldes auf das positive körperlich-seelische Befinden des Menschen).

amenity facility [n] *arch. constr. urb.* ▶outdoor furniture.

154 amenity grassland [n] [UK] *land'man. recr.* (Generic term covering intensively managed turf, such as golf courses, or grasslands with mainly visual value, as part of a landscape, and only subject to slight management intervention. Grasslands used primarily for grazing or hay production are excluded; **in U.S.** no comparable concept); *s* **pradería [m] de uso recreativo (≠)** (En GB término genérico para zonas verdes bien sometidas a cuidados intensivos como campos de golf, o bien con valor premordialmente escénico, sujetas sólo a cuidados extensivos. No se incluyen superficies dedicadas al pastoreo o a la producción de heno); *f* **prairie [f] d'agrément** (Surface enherbée en vue de son utilisation pour les activités de détente et de loisirs en plein air) ; *g* **Grünlandflächen [fpl] für Freizeit und Erholung (≠)** (U.K.,

nicht wirtschaftlich genutztes Grünland, das entsprechend der Nutzung für Freizeit und Erholung gepflegt wird).

amenity green area [n] *landsc. urb.* ▶green space.

amenity lawn [n] [UK] *constr.* ▶play lawn [US] (1)/all-around lawn [UK].

amenity [n] **of the landscape** [US] *conserv. landsc.* ▶beauty and amenity of the landscape [US]/beauty and amenity of the countryside [UK].

amenity value [npl] **of a forest** *recr.* ▶amenity benefits of a forest.

amenity woodland [n] [UK] *for. plan. recr.* ▶recreation forest.

155 American Landscape Contractors Association [n] **(A.L.C.A.)** [US] *constr. prof.* (National organization of professional builder or installer of soft landscapes, including the construction of pathways, roads, outdoor structures, retaining walls, fences, etc.); *syn. for U.K.* British Association [n] of Landscape Industries (B.A.L.I.); *s* **Sociedad [f] Española de Horticultura (S.E.H.)** (Asociación de empresas y personas dedicadas a la construcción y cuidado de jardines y al cultivo de plantas ornamentales en viveros); *f 1* **Union [f] Nationale des Entrepreneurs du Paysage (U.N.E.P.)** (Organisation professionnelle regroupant à titre volontaire les entrepreneurs de jardins ; *anciennement* Union Nationale des Syndicats d'Entrepreneurs Paysagistes et de Reboiseurs de France — UNSEPRF) ; *f 2* **Comité [m] National Interprofessionnel de l'Horticulture florale, ornementale et des pépinières (C.N.I.H.)** (Organisme interprofessionnel parapublic géré par les professionnels nommés par le Ministre de l'Agriculture et auquel adhèrent obligatoirement les entrepreneurs de jardins) ; *g* **Fachverband [m] des Garten- und Landschaftsbau(e)s (D.,** seit 1967 Verbandsbezeichnung des Berufsstandes der Landschaftsgärtner; *vormals* Fachverband Deutscher Landschaftsgärtner).

156 American Standards [npl] [US] *adm. constr.* (**1.** published by the American National Standards Institute [ANSI]—for all types of construction; Standards are rules of the art, but not laws; they represent the current state-of-the-art at the time of publication, are permanently revised and updated and come into force by publication of the new edition. **2. British Standards** [npl] [UK] [*Abbr.* BS]; published by the British Standards Institute); *s* **normas [fpl] técnicas españolas (UNE) [Es]** (La Asociación Española de Normalización y Certificación [AENOR] es el organismo responsable para elaborar las **n. t. e.** para los diferentes ámbitos de la tecnología, la construcción y también otras áreas comerciales, como el turismo; AENOR es a su vez miembro del Comité Europeo de Normalización [CEN] y del organismo internacional de normalización ISO); *f* **normes [fpl] françaises (N.F.)** (La marque NF est une marque collective de certification confiée à l'association française de normalisation [AFNOR] qui est l'organisme officiel français de normalisation. Ce document de référence apporte la preuve qu'un produit ou service est conforme à des caractéristiques techniques ou commerciales de sécurité et/ou de qualité définies dans le référentiel de certification correspondant au moment de sa publication. Il n'est pas obligatoire de suivre une norme mais elle peut être imposée par un donneur d'ordre lors de l'attribution d'un contrat ; DTB 1985) ; *g* **DIN-Norm [f]** (*Ursprünglich Abk. für* **D**eutsche **I**ndustrie-**N**ormen; heute für **D**eutsches **I**nstitut für **N**ormung e. V.; **DIN-N.en** sind Regeln der Technik, aber keine Gesetze; sie geben den allgemeinen Stand der Technik zum Zeitpunkt ihrer Veröffentlichung wieder, werden daher regelmäßig überarbeitet und als Neufassung verbindlich; das DIN ist gemäß einem Vertrag mit dem Bund die zuständige Normungsorganisation und vertritt die deutschen Interessen in den internationalen

Normengremien ISO [International Organization for Standardization] und CEN [Comité Européen de Normalisation] sowie bei den elektrotechnischen Organisationen IEC [International Electrotechnical Commission] und CENELEC [Comité Européen de Normalisation Electrotechnique]; die Normen kommen Herstellern, der Industrie, dem Handel, der Wissenschaft, den Verbrauchern und der öffentlichen Verwaltung zugute).

amount [n] *constr. contr.* ▶quantity (of an item).

157 amount [n] **of precipitation** *met.* (Quantity of precipitation measured according to the depth in inches/millimeters [US]/millimetres [UK] over a period of time); *s* **pluviosidad [f]** (Cantidad de precipitaciones en un área dada que se mide con ayuda de un pluviómetro, generalmente en milímetros); *syn.* precipitación [f] (anual, mensual, diaria); *f 1* **quantité [f] de précipitations** (CILF 1978, 158 ; mesurée en mm par unité de temps, p. ex. à l'aide d'un pluviomètre) ; *syn.* volume [m] total de l'averse, hauteur [f] de précipitations ; *f 2* **pluviosité [f]** (Rapport de la hauteur des précipitations de l'année considérée à la hauteur de précipitations annuelles moyennes ; ne pas confondre avec pluviométrie [mesure de la pluviosité] ; CILF 1978) ; *g* **Niederschlagsmenge [f]** (Niederschlagsanfall pro Zeiteinheit in mm gemessen).

amphibians [npl] *constr. ecol. zool.* ▶tunnel for amphibians.

amusement [n] *recr.* ▶physical amusement.

158 amusement park [n] *recr.* (Outdoor entertainment complex, e.g, equipped with a roller coaster [US]/switchback railway [UK], big wheel and other amusements such as in Disneyland; ▶leisure park); *syn.* pleasure park [n], fun fair [n] [also UK]; *s* **parque [m] de atracciones** (Centro de recreo con instalaciones de diversión como montaña rusa, noria, tiovivos, etc. o especializado tipo Disneyland; ▶parque recreativo regional); *f* **parc [m] d'attraction** (Parc spécialisé récréatif, thématique et culturelle en milieu périurbain [parc Eurodisneyland] ; ▶parc de loisirs) ; *g* **Vergnügungspark [m]** (Mit Schaustellerbuden und Vergnügungseinrichtungen wie z. B. Achterbahn und Riesenrad oder Disneyland ausgestatteter Park; ▶Freizeitpark).

anabatic wind [n] *geo. met.* ▶upslope air flow.

analysis [n] *plan.* ▶benefit-value analysis; *constr. contr.* ▶bid analysis [US], ▶comparative analysis of bid items [US]; *plan.* ▶comprehensive site survey and analysis, ▶cost-benefit analysis, ▶environmental risk analysis, ▶impact analysis, ▶initial site analysis, ▶landscape analysis; *plan. rem'sens. surv.* ▶satellite data analysis; *constr.* ▶sedimentation analysis; *plan.* ▶site analysis; *pedol.* ▶soil analysis; *plan.* ▶soil quality analysis; *plan. rem'sens. surv.* ▶terrain analysis; *arch. urb.* ▶townscape analysis.

analysis [n]**, community appearance** [US] *arch. urb.* ▶townscape analysis.

analysis [n]**, cost-effectiveness** *plan.* ▶cost-benefit analysis.

analysis [n]**, critical path** *constr. plan.* ▶critical path method.

analysis [n]**, data** *plan.* ▶data evaluation.

analysis [n]**, network** *constr. plan.* ▶critical path method.

analysis [n]**, tender** [UK] *constr. contr.* ▶bid analysis [US].

159 analysis [n] **of environmental impact** *conserv. leg. nat'res. plan.* (Initial phase of an ▶environmental impact assessment [EIA] [US]/environmental assessment [UK] with analysis and investigation to determine according to specified parameters, whether a development project or other landscape intervention will significantly affect the quality of the environment); *s* **estimación [f] de la incidencia significativa sobre el medio** (Primera fase de la ▶evaluación del impacto ambiental en la que se evalúa si un proyecto tendrá efectos notables sobre el medio ambiente de manera que es necesario realizar un estudio de impacto ambiental); *f* **évaluation [f] des incidences significatives sur l'environnement** (Première étape de la ▶procédure d'étude d'impact dans laquelle est effectuée l'analyse de l'état initial du site et de son environnement ainsi que de l'éventualité d'effets dommageables d'un projet sur les milieux naturels et humains) ; *g* **Umwelterheblichkeitsprüfung [f]** (Erste Stufe zur ▶Umweltverträglichkeitsprüfung, bei der zunächst untersucht wird, ob und inwieweit ein geplantes Vorhaben nach Lage, Funktion und den von ihm ausgehenden Auswirkungen auf Gesellschaft, Natur und Landschaft negativ einwirkt. Im Rahmen der Bauleitplanung müssen z. B. alle relevanten Umweltbelange/Schutzgüter [z. B. Boden, Grundwasser, Oberflächengewässer, Luft, Klima, Flora, Fauna, Biotope, Orts- und Landschaftsbild, Belange des technischen Umweltschutzes wie Lärm, Lufthygiene, Erschütterungen, Altlasten, Kampfmittelbeseitigung und des Denkmalschutzes] abgearbeitet und in einem Umweltbericht beschrieben und bewertet werden).

160 analysis [n] **of planning data** *plan.* (Evaluation of basic planning information to determine its completeness, actuality, meaningfulness and usability; ▶evaluation procedure); *syn.* evaluation [n] of planning data; *s* **análisis [m] de datos de planificación** (en cuanto a su calidad para caracterizar la situación de la zona a planificar; ▶método de evaluación); *f* **analyse [f] des données de planification** (Contrôle des documents relatifs à l'état des lieux ou des facteurs caractérisant un aménagement ; ▶procédé d'évaluation, ▶méthode de cotation) ; *g* **Auswertung [f] von Planungsdaten** (Prüfung von Planungsunterlagen/-grundlagen hinsichtlich der Vollständigkeit, Aktualität, Aussagekraft und Verwendbarkeit etc.; Planungsdaten auswerten [vb]; ▶Bewertungsverfahren).

161 analysis [n] **of requirements** *plan.* (Detailed survey of needs and required uses related to a specific planning project); *s* **análisis [m] de la demanda** (Relevamiento exacto de las necesidades y usos a cubrir en el contexto de un proyecto específico); *syn.* análisis [m] de las necesidades; *f* **analyse [f] de la demande** (Techniques d'évaluation, de prévision et de simulation de la demande pour un projet d'aménagement) ; *syn.* analyse [f] des besoins ; *g* **Bedarfsanalyse [f]** (Genaue Erhebung der Bedürfnisse oder Nutzungserfordernisse für ein bestimmtes Planungsvorhaben).

analysis [n] **of tender items** [UK] *constr. contr.* ▶comparative analysis of bid items [US].

162 analysis [n] **of the visual appearance of a landscape** *plan. landsc.* (Analysis of a site based upon the structural perception of the landscape; the study takes into account all physical and natural components, axes and their convergence and equilibrium together with remarks concerning heritage, symbolic values, aesthetics and natural features of the landscape; an analysis employs graphic techniques of expression and is often used when selecting alternative concepts for a particular site); *s* **análisis [m] del aspecto escénico del paisaje** (Estudio de la percepción de un paisaje según sus características morfológicas y escénicas. Este instrumento de planificación se utiliza para elegir posibles alternativas de uso en un emplazamiento específico y consiste en la representación gráfica y descriptiva de la percepción y las interrelaciones visuales de aspectos como la topografía del terreno, la textura de las superficies, los corredores visuales y la presencia y ordenación de elementos singulares en la sección del paisaje estudiada); *f* **étude [f] sitologique** (Approche paysagère d'analyse des sites basée sur la perception structurelle d'un paysage ; l'étude privilégie la prise en compte des points d'appel, des lignes de force, des axes, de leur convergence et équilibre assortie de remarques concernant la valeur patrimoniale, symbolique, esthétique ou scientifique de l'unité paysagère ; cette pratique utilise les techniques d'expression graphiques ; elle est souvent mise en œuvre dans le cadre de l'intégration au site de réalisations prévues par un choix d'aménagement ; GEP 1991, 203-205) ; *g* **Landschaftsbildanalyse [f]** (Untersuchung, wie die Landschaft nach ihren Gestalt- und Strukturmerkmalen durch den Betrachtenden wahrgenommen wird; dieses planerische Instrumentarium stellt grafisch und beschreibend u. a. das Zusammenwirken und visuelle Erleben von Geländekanten, visuellen Bezugspunkten, Geländeformen und deren Oberflächentexturen, Sichtachsen, die Verteilung und den Verbund sowie das Zusammenwirken von Einzelelementen im zu betrachtenden Landschaftsraum dar; die **L.** wird meist angewandt, wenn zwischen unterschiedlichen Gestaltungs- und Nutzungskonzepten für einen bestimmten Landschaftsraum entschieden werden soll).

anchorage net [n] [US]**, nylon** *constr.* ▶root anchoring fabric [US].

anchorage point [n]**, security** *constr. leg.* ▶protection from falling.

anchoring fabric [n] [US]**, root** *constr.* ▶root anchoring fabric.

anchoring membrane [n] [US]**, root** *constr.* ▶root anchoring fabric [US].

163 anchor root [n] *bot.* (Structural root growing obliquely, or horizontally in the case of shallow-rooted plants, which stabilizes a tree in the soil or in the ground; ▶tap root); *s* **raíz [f] de anclaje** (Raíz que crece en diagonal y en las plantas de radicación plana en horizontal, que sirve para fijar el árbol en el suelo; ▶raíz axonomorfa); *f* **racine [f] d'ancrage** (Racine se développant dans une direction oblique dans le sol ou forte racine se développant parallèlement à la surface du sol chez les végétaux à racines traçantes et jouant un rôle de fixation ; ▶racine pivotante) ; *g* **Ankerwurzel [f]** (Schräg in die Tiefe wachsende oder bei Flachwurzlern horizontale Starkwurzel, die für die Befestigung des Baumes im Boden dient [Standsicherheit]; ▶Pfahlwurzel).

anchor support [n] *constr.* ▶metal anchor support.

ancient lake raised bog [n] *geo. phyt.* ▶raised bog on a silted-up lake.

ancillary structure [n] [UK] *leg. urb.* ▶accessory structure.

anemochory [n] *phyt.* ▶wind dispersal.

164 angled stake [n] *arb. constr.* (Tree support to prevent excessive movement of newly planted trees in areas subject to heavy winds; ▶stake 1); *syn.* slanting stake [n] (BS 4428: 1969, 43); *s* **estaca [f] oblicua** (Pilote plantado oblicuamente para fijar árboles; ▶tutor); *f* **tuteur [m] oblique** (Pieu en bois en enfoncé dans le sol en position oblique comme ▶tuteur des arbres nouvellement plantés) ; *syn.* piquet [m] planté en biais ; *g* **Schrägpfahl [m]** (…pfähle [pl]; schräg in die Erde geschlagener Holzpfahl zum Befestigen von gepflanzten Gehölzen, meist so gesetzt, dass die Pfahlköpfe in die Hauptwindrichtung zeigen, in Fahrbahnnähe jedoch stets in Fahrtrichtung; cf. Ziff. 4.7 DIN 18 916; in den *Zusätzlichen Technischen Vertragsbedingungen und Richtlinien für Landschaftsbauarbeiten im Straßenbau [ZTVLa-StB 99]* ist außerdem festgelegt, dass **S.pfähle** an Böschungen oberhalb der Pflanze einzuschlagen sind; ▶Pfahl).

angle [n] **of repose** *constr.* ▶natural angle of repose.

angle [n] **of slope** *constr.* ▶slope angle.

165 angle-parking layout [n] *plan. trans.* (Parking at 30°, 40, 60° angles to traffic flow; ▶in-line parking layout [US]/ parallel parking layout [UK], ▶ninety-degree parking layout); *syn.* echelon parking layout [n] [also US]; *s* **disposición [f] (de aparcamiento) en diagonal** (Orden de colocación de los vehículos en ángulo de 45° a 60° respecto a la dirección del tráfico; ▶disposición [de aparcamiento] en hilera, ▶disposición [de aparcamiento] en perpendicular); *syn.* estacionamiento [m] en diagonal [AL], aparcamiento [m] en posición oblicua; *f* **stationnement [m] en épi** (Disposition des places de stationnement suivant un angle de 45° ou 60° dans le sens de circulation ; ▶stationnement longitudinal, ▶stationnement perpendiculaire) ; *syn.* disposition [f] en épi, rangement [m] en épi, parking [m] en arêtes de poissons, rangement [m] oblique (VRD 1986, 176) ; *g* **Schrägaufstellung [f]** (Aufstellungsanordnung parkender Fahrzeuge im Winkel zwischen 45° und 60° zur Fahrtrichtung; ▶Längsaufstellung, ▶Senkrechtaufstellung).

166 angle-parking row [n] *trans.* (Parking strip of stalls to accommodate parked vehicles in an ▶angle-parking layout); *s* **fila [f] de aparcamiento en diagonal** (▶disposición [de aparcamiento] en diagonal); *syn.* fila [f] de estacionamiento en diagonal, aparcamiento [m] en posición oblicua; *f* **aire [f] de stationnement pour stationnement en épi** (▶stationnement en épi) ; *g* **Schrägparkstreifen [m]** (Parkstreifen für ▶Schrägaufstellung).

angle planting [n] *constr. for.* ▶T-notching.

167 angle step [n] *constr.* (L-shaped precast concrete element used in the construction of a flight of steps or as a step for sitting upon); *s* **peldaño [m] angular** (Pieza prefabricada de hormigón con forma de L para construir escaleras); *syn.* peldaño [m] en L; *f* **marche [f] d'angle** (Élément préfabriqué en béton en forme de L utilisé dans la construction d'un escalier ou d'un banc) ; *syn.* marche [f] (préfabriquée) en L ; *g* **Winkelstufe [f]** (L-förmiges Fertigbetonteil zum Bau einer Treppe oder Sitzstufe); *syn.* L-Stufe [f].

angling [n] *recr.* ▶sport fishing.

animal [n] *zool.* ▶altricial animal; *hunt.* ▶game animal; *zool.* ▶precocial animal; *conserv. leg.* ▶wild animal.

168 animal density [n] *zool.* (**1.** Term for the total number of individuals of all species per unit area—population density of all species in a ▶range 1. **2.** Synonymous with ▶population density 3 of a single species); *s* **densidad [f] de población total** (**1.** Totalidad de los individuos de todas las especies existentes en una unidad espacial; densidad de población de todas las especies en un ▶área de distribution. **2.** Sinónimo de ▶densidad de población 3 de una especie); *f* **densité [f] totale de la population animale** (Nombre total des individus des espèces présentes par unité de surface ; ▶aire de répartition, ▶densité des individus) ; *g* **Wohndichte [f] (1)** (**1.** Bezeichnung für die Gesamtheit der Individuen aller Arten in einer Flächen- oder Raumeinheit — Populationsdichte sämtlicher Arten im ▶Areal. **2.** Synonym zur ▶Individuendichte einer einzigen Art; ÖKO 1983).

169 animal dispersal [n] **of seeds** *phyt. zool.* (Spreading of seeds by animals); *syn.* dissemination [n] by animals, zoochory [n]; *s* **zoocoria [f]** (Dispersión de diásporas por animales. Se diferencia entre la **endozoocoria** en la que las diásporas son transportadas dentro del tubo digestivo del animal y liberadas con los excrementos y la **epizoocoria** en la que las diásporas son diseminadas pegadas con fango a las patas de las aves o adheridas al pelo, pezuñas o plumas de los animales); *syn.* dispersión [f] por los animales, diseminación [f] por los animales; *f* **dissémination [f] par les animaux** (Dispersion des graines assurée par les animaux ; elle concourt autant à la colonisation de sites nouveaux qu'au maintien des espèces végétales dans un environnement donné. On distingue l'**endozoochorie** lorsque les diaspores sont ingérées par l'animal et libérées dans les matières fécales et l'**épizoochorie** lorsque la diaspore dispose de dispositifs d'accrochage (crochets, substance visqueuse) ; *syn.* zoochorie [f] ; *g* **Verschleppung [f] durch Tiere** (Verbreitung von Diasporen durch Tiere. Man unterscheidet zwischen **Endozoochorie** [Ausbreitung der Diasporen durch Kot nach Darmpassage, z. B. Vogelmiere *[Stellaria media]*, Hirtentäschelkraut *[Capsella bursa-pastoris]*, *Chenopodium*-Arten] und **Epizoochorie** [Ausbreitung durch Anheftung an der Oberfläche der Tiere, z. B. bei Pflanzenarten, die mit Klettfrüchten ausgestattet sind, z. B. Hexenkraut *[Circaea lutetiana]* oder Arten mit begrannten oder behaarten Diasporen, z. B. Gräser *[Deschampsia- und Poa-*Arten]); *syn.* Tierverbreitung [f], Zoochorie [f].

170 animal kingdom [n] *zool.* (Whole world of animals as well as term for a large geographical region of animals; ▶wildlife); *s* **reino [m] animal** (Conjunto de todos los animales. También denominación de grandes regiones geográficas animales: holártico, paleotropical, australiano, neotropical, arquinótico; DINA 1987 ; ▶vertebrados endémicos); *f 1* **règne [m] animal** (Ensemble des animaux non domestiques en milieu terrestre et marin, et, plus spécialement, des mammifères, oiseaux, poissons, batraciens, reptiles et de quelques invertébrés supérieurs ; ▶animaux vertébrés indigènes) ; *f 2* **empire [m] faunistique** (Terme utilisé en géographie botanique pour désigner l'unité de premier ordre dans la classification des unités biogéographiques du règne animal en milieu terrestre et marin ; les unités d'ordre suivantes se nomment : région et province ; EBI 1971, 123 et s.) ; *g* **Tierreich [n]** (Gesamtheit aller Tiere. Auch Bezeichnung für tiergeografische Großregionen; ▶heimische Wirbeltiere); *syn.* Tierwelt [f], Faunenreich [n].

171 animal population [n] *zool.* (Term for a group of individuals of various animal species or one species in a defined study area; ▶game population, ▶re-establishment of an animal population); *s* **población [f] animal** (Cantidad de individuos de una o varias especies animales en una zona específica; ▶población cinegética, ▶restauración de poblaciones animales); *f* **peuplement [m] animal** (Ensemble des organismes animaux vivant dans un même milieu ; ▶population cynégétique, ▶restauration des populations) ; *g* **Tierbesatz [m]** (Bezeichnung für die anthropogen beeinflusste Menge von Individuen verschiedener Tierarten oder einer Art in einem bestimmten Bereich; im Deutschen wird unter Besatz immer der vom Menschen direkt beeinflusste Zustand mitgedacht; sonst spricht man von Tierpopulation; ▶Wiederherstellung von Tierpopulationen, ▶Wildbesatz); *syn. z. T.* Tierbestand [m].

animal protection [n] *leg.* ▶domestic animal protection.

animal rest(ing) area [n] *phyt.* ▶vegetation of animal rest area.

172 animal species [n] *zool.* (*Generic term* ▶species 1); *s* **especie [f] animal** (*Término genérico* ▶especie); *syn.* especie [f] de fauna; *f* **espèce [f] animale** (*Terme générique* ▶espèce) ; *g* **Tierart [f]** (*OB* ▶Art).

animal species [npl]**, beneficial** *agr. for. hort.* ▶beneficial species, ▶use of beneficial animal species.

animal species composition [n]**, anthropogenic shift in** *zool.* ▶anthropogenic alteration of the genetic fauna pool.

173 animal species conservation [n] *conserv. leg. zool.* (Specific term of ▶species conservation; ▶nature conservation, ▶domestic animal protection, ▶wildlife conservation; ▶range 1); *s* **protección [f] de especies de fauna en peligro** (Término específico de ▶protección de especies [de flora y fauna]; ▶conservación de la naturaleza, ▶protección de animales, ▶protección de la fauna salvaje, ▶area de distribución); *syn.* protección [f] de especies animales en peligro; *f* **protection [f] des espèces animales** (Terme spécifique pour la ▶protection des espèces [animales et végétales] dans le cadre de la législation de la ▶protection de la nature [10 juillet 1976] ; ▶aire de répartition, ▶protection de la faune sauvage, ▶protection des animaux domestiques et d'expérience) ; *g* **Tierartenschutz [m]** (Aufgabenbereich des Naturschutzes mit dem Ziel, wild lebende Tierarten innerhalb ihres natürlichen Verbreitungsgebietes [▶Areal] so zu erhalten und zu fördern, dass die Evolution der betreffenden Arten gesichert bleibt; nicht zu verwechseln mit ▶Tierschutz; *OB* ▶Artenschutz; ▶Faunenschutz, ▶Naturschutz).

174 animal-tight fence [n] [US] *conserv. for. hunt. trans.* (cf. LAJ 1985, 109; protective structure along busy roads to prevent wild animals [even-hoofed game] from reaching the travelled way [US]/carriage-way [UK]); *syn.* deer-stop fence [n] [UK] (SPON 1986, 274); *s* **cercado [m] de protección de la caza (≠)** (Medida de protección de los animales salvajes —p. ej. a lo largo de carreteras— para evitar accidentes por paso de caza o alrededor de áreas de renuevo de brinzales/latizales para protegerlos de los animales); *f* **clôture [f] gibier** (Installation de protection contre les espèces de gibier le long d'ouvrages routiers ou autour d'un bois en défens pour éviter l'écorçage) ; *g* **Wildschutzzaun [m]** (Schutzeinrichtung gegen wild lebende Tiere, z. B. entlang von Fernstraßen, um besondere Gefährdungen und Unfallhäufungen durch Wildwechsel zu vermeiden oder Einzäunungen um Schonungen, damit vornehmlich Schalenwild keine Verbissschäden verursacht); *syn.* Schutzzaun [m] gegen wild lebende Tiere, Wildsperrzaun [m], Wildgatterzaun [m], Wildzaun [m].

animate material [n] [US] *constr.* ▶living plant material.

175 anmoor [n] *pedol.* (Hydromorphic humic soil, 20-40cm deep, occurring under anaerobic conditions, often with very little decomposed organic material; ▶gleyed anmoor); *syn.* muck [n], saprist [n] [also US]; *s* **anmoor [m]** (Suelo hidromorfo humoso de 20-40 cm de profundidad generado bajo condiciones anaerobias, a menudo con mucha sustancia orgánica sin descomponer; ▶gley turboso); *f* **anmoor [m]** (Horizon organique de surface [humus semi-terrestre], à évolution modifiée par l'hydromorphie temporaire — nappe fluctuante — et ayant une épaisseur de moins de 40 cm et une teneur en matière organique comprise entre 12,5 et 25 % ; cf. DIS 1986, ▶gley tourbeux) ; *g* **Anmoor [n]** (Hydromorpher, unter anaeroben Bedingungen entstandener, 20-40 cm mächtiger Boden, oft mit nur wenig zersetzter organischer Substanz; ▶Anmoorgley).

176 anmoor soil [n] *pedol.* (Hydromorphic humic soil with a humus content of 15-30%, the humus of which is composed primarily of dark-colo[u]red humin compounds—humic acid; ▶anmoor); *s* **suelo [m] paraturboso** (Suelo modificado por hidromorfia temporal que tiene un contenido de humus del 15-30% del cual la mayor parte son compuestos húmicos; ▶anmoor, ▶suelo turboso); *f* **sol [m] paratourbeux** (Sol modifié par une hydromorphie temporaire et ayant une teneur en matière organique comprise entre 15 et 30 % ; cf. DIS 1986, 151 ; ▶anmoor) ; *g* **anmooriger Boden [m]** (Hydromorpher B. mit einem Humusgehalt von 15-30 %, wobei der Humus überwiegend aus Huminstoffen besteht; ▶Anmoor).

177 annex [n] [US] *arch.* (Secondary building, which may be connected to the main building or separate from it); *s* **edificio [m] anexo** (Edificio independiente que cumple las mismas funciones que el edificio principal); *syn.* anexo [m]; *f* **annexe [f]** (Bâtiment indépendant remplissant de multiples fonctions, édifié contre le bâtiment principal ; *syn.* dépendance [f]) ; *g* **Anbau [m] (2)**

A

(Anbauten [pl]; in sich funktionsfähiges Gebäude, das an ein Hauptgebäude angebaut wurde); *syn.* Erweiterungsbau [m] (2).

annexe [n] [UK] *arch.* ▶annex [US].

announce [vb] *leg. plan.* ▶give legal notice.

announcement [n] *adm. prof. leg.* ▶design competition announcement [US]; *adm. leg. plan.* ▶public announcement.

announcement [n] [US], **legal** *adm. leg. plan.* ▶public announcement.

178 annual [n] *bot. hort. phyt. plant.* (**1.** Herbaceous plant that completes its life cycle from germination to flowering and seed production within a single growing season; Its seed survives in the soil—also in areas hostile to vegetation—during the less favo[u]rable growing season. There are two types of annuals: ▶summer annuals [green in summer], which germinate in spring and die back in fall/autumn, and **winter annuals** [green in winter], which germinate in fall/autumn, survive the winter and flower in spring. In addition there are **rainy season annuals** [pluviotherophytes], which germinate at the start of periods of rain. **2.** According to CHRISTEN CHRISTIANSEN RAUNKIAER'S ▶life form categories, annuals are called *therophytes*; ▶biennial plant, ▶casual species, ▶hemicryptophyte, ▶perennial 1); *syn.* annual plant [n], therophyte [n]; *s* **planta** [f] **anual** (**1.** Planta herbácea que completa su ciclo vital en un periodo/período de crecimiento y en las estaciones adversas —también en superficies sin vegetación— sobrevive en el suelo en forma de semilla. Existen ▶plantas anuales estivales [plantas de verde vernal] que germinan en primavera y perduran hasta el otoño y **anuales invernales** [planta de verde invernal] que germinan en otoño, perduran en invierno y florecen en primavera. Finalmente las **plantas pluviifolias** [pluvioterófitos] que germinan al comienzo de la estación de lluvias. **2.** Según el sistema de RAUNKIAER de las ▶formas biológicas, las **p. a.** son denominadas *terófitos*; ▶bienal, ▶efemerófito, ▶hemicriptófito, ▶planta vivaz); *syn.* terófito [m], anual [f]; *f* **plante** [f] **annuelle** (**1.** Plante herbacée ayant un cycle de reproduction très court [quelques mois voire quelques semaines] et un développement rapide passant la mauvaise saison sous forme de graines ; on distingue l'▶annuelle estivale qui germe au printemps et vit jusqu'en automne, l'**annuelle hivernale** [annuelle d'hiver] qui germe à l'automne, lève et se développe avant le début de l'hiver, reste à l'état dormant pendant la saison froide et fleurissent au printemps ainsi que les **thérophytes de saison des pluies**, annuelles qui germent au début de la saison des pluies, fleurissent et grainent en hâte. **2.** Selon la systématique des ▶types biologiques de CHRISTEN CHRISTIANSEN RAUNKIAER les plantes annuelles sont dénommées thérophytes ; ▶espèce fugace, ▶hémicryptophyte, ▶plante bisannuelle, ▶espèce éphémérophyte, ▶plante vivace) ; *syn.* annuelle [f], thérophyte [f] ; *g* **einjährige Pflanze** [f] (**1.** Krautige Pflanze, die ihren Lebenszyklus von der Keimung, zum Blühen bis zur Samenreife in einer Vegetationsperiode abschließt und die ungünstige Jahreszeit — auch in vegetationsfeindlichen Flächen — als Samen im Boden überdauert. E. P.n. sind an nährstoffreiche Standorte gebunden. Es gibt ▶**Sommerannuelle** [sommergrüne Pflanzen], die im Frühjahr keimen und bis zum Herbst einziehen und **winterannuelle** [wintergrüne] **Pflanzen**, die im Herbst keimen, den Winter überdauern und im Frühjahr blühen und fruchten. Ferner gibt es **regengrüne Pflanzen** [Pluviotherophyten], die zu Beginn einer Regenperiode keimen. **2.** *bot.* Nach dem CHRISTEN C. RAUNKIAERschen System der ▶Lebensformen werden **e. P.n Therophyten** genannt — wobei nur sommerannuelle Pflanzen echte Therophyten sind, winterannuelle Pflanzen zählen zu den ▶Hemikryptophyten; ▶Staude, ▶unbeständige Art, ▶Bienne); *syn.* Annuelle [f], annuelle Pflanze [f], Einjährige [f], Therophyt [m].

annual cut [n] **(of a meadow)** [UK] *agr. constr. hort.* ▶single mowing [US].

179 annual periodicity [n] *biol.* (Regularly occurring changes in nature according to the seasons of the year; e.g. bird migration, breeding, moulting, budding, blossom time; ▶circadian rhythm); *syn.* circannual periodicity [n], circannual rhythm [n]; *s* **ritmo** [m] **anual** (Cambios regulares que ocurren en la naturaleza a lo largo del año; ▶ritmo nictemeral); *syn.* periodicidad [f] anual; *f* **périodicité** [f] **annuelle** (Changements de comportement ou répétitivité de phénomènes dans le règne animal et végétal [rythmes biologiques] intervenant à intervalles réguliers au cours de l'année tels que la migration, la nidification, la mue, la pousse des feuilles, la floraison, etc. ; ▶rythme quotidien) ; *g* **Jahresperiodik** [f] (Regelmäßiger Wechsel jahreszeitlicher Lebensvorgänge, z. B. Vogelzug, Brutzeit, Mauser oder Blattaustrieb, Blütezeit etc.; ▶Tagesgang); *syn.* Jahresgang [m], Jahresrhythmik [f], zirkannuale Rhythmik [f].

annual plant [n] *bot. hort. phyt. plant.* ▶annual.

180 annual ring [n] *arb.* (Circle of ▶phloem of woody plants, a growth layer of one year as seen in cross section; ▶cambium, ▶early wood, ▶late wood, ▶growth layer); *s* **anillo** [m] **anual** (▶cámbium, ▶capa de crecimiento, ▶leño tardío, ▶leño temprano); *f* **cerne** [m] **annuel** (En section transversale, ▶couche d'accroissement d'une année dans le bois et l'écorce ; MGW 1964 ; ▶bois initial, ▶bois final, ▶cambium) ; *syn.* couche [f] annuelle ; *g* **Jahresring** [m] (Durch periodische Wachstumstätigkeit des ▶Kambiums entstandene Grenzlinie zwischen dem lockeren, porösen ▶Frühholz und dem vorjährigen dichten ▶Spätholz; ▶Zuwachszone); *syn.* Jahrring [m] [CH].

annual shoot [n] *bot. hort.* ▶one year's growth.

antenna [n], **mobile phone** *urb.* ▶mobile phone antenna.

anthropic soil [n] *agr. constr. hort. pedol.* ▶man-made soil.

anthropochory [n] *phyt. zool.* ▶anthropogenic dispersal.

181 anthropogenic [adj] (1) *ecol. land'man.* (Descriptive term applied to human impact on nature: induced or altered by human presence and activities; ▶man-developed); *syn.* man-made [pp]); *s* **antrópico/a** [adj] (Referente a procesos, acciones, materiales y formas resultantes de la actividad humana; término opuesto a natural o silvestre; cf. DINA 1987; ▶antropógeno/a 2); *syn.* antropógeno/a [adj] (1), antropogénico/a [adj]; *f* **anthropogène** [adj] (1) (Causé par l'homme ; ▶fortement artificialisé) ; *syn. phyt.* anthropique [adj] ; *g* **anthropogen** [adj] (durch den Menschen bedingt/entstanden; ▶naturfern).

anthropogenic [adj] (2) *ecol.* ▶man-developed.

182 anthropogenic alteration [n] *phyt. zool.* (▶Naturalization of animal or plant species in a naturally developed biocoenosis, as a result of which the original characteristics of the ▶biocoenosis are increasingly lost. The term **a. a.** also includes ▶reintroduction of individuals of certain species when they are relocated from geographically distant—taxonomic and genetically different—populations, or when they are introduced into unsuitable areas. Natural immigration of species into an area, on the other hand, does not cause an **a. a.** of fauna or flora. Some critics of animal colonization distinguish between **a. a.** and ▶genetic contamination; ▶anthropogenic alteration of flora composition, ▶anthropogenic alteration of the genetic fauna pool); *s* **contaminación** [f] **biológica** (Introducción de animales o plantas en hábitats donde no viven normalmente. También se entiende por **c. b.** la ▶renaturalización de determinadas especies, cuando se trata de individuos pertenecientes a poblaciones distantes geográficamente [diferentes desde el punto de vista taxonómico y genético] o si han sido introducidas en lugares donde no viven normalmente. En cambio la migración natural de especies no es **c. b.** Algunos autores diferencian entre la **c. b.** [renaturalización de individuos de poblaciones lejanas geográficamente] y la ▶contaminación genética; *términos específicos* **c. b.** de la flora [▶adulteración de la flora endémica], **c. b.** de la fauna [▶adulteración de la fauna endémica]; cf. DINA 1987; ▶biocenosis, ▶naturalización, ▶reintroducción de especies extintas); *syn.* adulteración [f], introducción [f] de especies exóticas; *f* **altération [f] par (l')introduction d'espèces** (Les espèces non indigènes ou envahissantes, introduites intentionnellement ou accidentellement dans une région située en dehors de leur aire de répartition naturelle sont responsables de l'extinction d'autres espèces, de dommages à des populations d'organismes sauvages et domestiqués et de modifications notables dans des ▶biocénoses et écosystèmes ; elles provoquent l'altération du patrimoine biologique ; ▶anthropisation de la composition de la faune sauvage, ▶anthropisation de la composition de la flore, ▶contamination génétique, ▶naturalisation, ▶réintroduction 2 ; *syn.* anthropisation [f] (par influence récente) ; *g* **Verfälschung [f]** (▶Einbürgerung von Tier- oder Pflanzenarten in eine natürlich gewachsene ▶Biozönose mit der Folge, dass diese ihren primären Charakter zunehmend verliert. Außerdem ist auch die ▶Wiederansiedelung bestimmter Arten gemeint, insbesondere wenn die ausgesetzten Individuen aus geografisch fernen [taxonomisch und genetisch anderen] Populationen stammen oder wenn sie an falschen Standorten angesiedelt wurden. Dagegen stellt die natürliche Einwanderung von Arten in ein Gebiet keine **V.** der Fauna oder Flora dar. Einige Kritiker der Tieransiedlungen unterscheiden zwischen **V.** [Wiederansiedelung von Individuen aus geographisch weit entfernten Teilpopulationen] und **Überfremdung;** *UBe* ▶Faunenverfälschung, ▶Florenverfälschung, ▶Überfremdung).

183 anthropogenic alteration [n] **of flora composition** *phyt.* (Major modification of species composition by human introduction of non-native species; ▶ruderalization); *syn.* anthropogenic shift [n] in floristic species composition (TGG 1984, 210), anthropogenic change of flora composition; *s* **adulteración [f] de la flora endémica** (Modificación antropógena de la composición de especies en una zona por introducción de especies no autóctonas; ▶ruderalización); *syn.* contaminación [f] biológica de la flora; *f* **anthropisation [f] de la composition de la flore** (Modification importante de la constitution naturelle d'un peuplement floristique dû à la présence d'espèces végétales introduites ; ▶rudéralisation) ; *g* **Florenverfälschung [f]** (Durch Ausbringen von nicht einheimischen Pflanzenarten [von genetischen Varianten] bedingte wesentliche Veränderung der natürlichen Artenzusammensetzung in einem Gebiet; um zu verhindern, dass nicht heimische Pflanzen den Lebensraum heimischer Pflanzen beschränken, ist eine naturschutzfachliche Begleitung erforderlich; ▶Ruderalisierung); *syn.* Verfremdung [f] der heimischen Flora).

184 anthropogenic alteration [n] **of the genetic fauna pool** *zool.* (Harmful introduction by man of exotic or non-native species which changes the genetic pool or natural population balance in an area); *syn.* anthropogenic shift [n] in animal species composition (TGG 1984, 210); *s* **adulteración [f] de la fauna endémica** (Modificación antropógena de la composición de la fauna de una región); *syn.* alteración [f] de la fauna endémica, contaminación [f] biológica de la fauna; *f* **anthropisation [f] de la composition de la faune sauvage** (Modification dans la constitution de la population animale provoquée par la naturalisation d'animaux non domestiques apparavant non indigènes dans une aire donnée et pouvant provoquer des déséquilibres biologiques) ; *g* **Faunenverfälschung [f]** (Durch Einbürgerung von nicht einheimischen Tierarten bedingte Veränderung der Artenzusammensetzung in einem Gebiet).

anthropogenic change [n] **of flora composition** *phyt.* ▶anthropogenic alteration of flora composition.

185 anthropogenic dispersal [n] *phyt. zool.* (Unintentional spreading of plants and animals caused by human activities, resulting in a change in the original composition of flora and fauna in a defined area. Active **d.** is called ▶introduction of plants or seeds or ▶naturalization); *syn.* anthropochory [n], dispersal [n] by human activity, dissemination [n] by man; *s* **antropocoria [f]** (Dispersión pasiva de animales y plantas por los seres humanos con la consecuencia de que se modifica la flora resp. fauna de la correspondiente zona. La diseminación activa se llama ▶introducción de plantas o semillas resp. ▶naturalización); *syn.* hemerocoria [f], diseminación [f] antropógena, dispersión [f] antropógena; *f* **dissémination [f] anthropogène** (Dispersion passive d'espèces végétales et animales provoquée par l'action humaine entraînant une variation de la répartition des espèces de faune et de flore dans la zone d'origine ; ▶introduction d'espèces végétales, ▶naturalisation) ; *syn.* anthropochorie [f] ; *g* **kulturbedingte Verschleppung [f]** (Durch die bewusste oder unbewusste Tätigkeit des Menschen sowie durch seine Haustiere und Transportmittel bedingte passive Verbreitung von Pflanzen und Tieren, wodurch der ursprüngliche Ausgangszustand von Flora und Fauna in einem bestimmten Gebiet verändert wurde oder wird. Die aktive Verbreitung wird ▶Einbringung von Pflanzen resp. ▶Einbürgerung genannt); *syn.* Anthropochorie [f, o. Pl.], Hemerochorie [f, o. Pl.], kulturbedingte Verbreitung [f], kulturbedingte Ausbreitung [f], Verschleppung [f] durch Menschen.

anthropogenic shift [n] **in animal species composition** *zool.* ▶anthropogenic alteration of the genetic fauna pool).

anthropogenic shift [n] **in floristic species composition** *phyt.* ▶anthropogenic alteration of flora composition.

186 anthropophilous species [n] *biol.* (Plant or animal species which prefers a cultivated habitat due to favo[u]rable growth conditions therein; *opp.* ▶hemerophobic species); *syn.* synanthropic species [n], hemerophilous species [n]; *s* **especie [f] antropófila** (Planta introducida por el hombre, así como la que gracias a él ocupa una estación que no es la suya propia; *opp.* ▶especie hemerófoba); *syn. biol.* especie [f] sinántropa; *syn. phyt.* especie [f] hemerófita, antropófito [m], hemerófito [m]; *f* **espèce [f] anthropophile** (Espèce végétale ou animale vivant au contact de l'homme ou sur les lieux qu'il fréquente et qui bénéficie pour sa dispersion du concours de l'homme, soit que ce dernier le véhicule, soit qu'il crée des conditions propices à son développement ; DIB 1988, 37 ; *opp.* ▶espèce anthropophobe ; *syn.* espèce [f] synanthrope ; *g* **Kulturfolger [m]** (Pflanzen- oder Tierart, die auf Grund der günstigen Lebensbedingungen den Kulturbereich des Menschen als Lebensraum bevorzugt; *opp.* ▶Kulturflüchter); *syn.* Kulturbegleiter [m], hemerophile Art [f], synanthrope Art [f].

187 anti-dazzle planting [n] *landsc. trans.* (Installation of a linear planting of trees and shrubs in medians [US]/motorway central reservations [UK] of major highways [US]/expressways [UK] to reduce reflected daytime glare of cars and buildings and to alleviate headlight glare at night); *syn.* antiglare planting [n] [also US]; *s* **plantación [f] antideslumbramiento** (Instalación de una plantación lineal de leñosas en la banda central de autopistas o autovías para reducir el efecto deslumbrante de automóviles y edificios durante el día y de los faros durante la noche); *syn.* plantación [f] contra el deslumbramiento; *f* **plantation [f] de protection contre l'éblouissement** (Dispositif anti-éblouissement obtenu par la plantation d'écrans végétaux sur les terre-pleins centraux des autoroutes et voies express destiné à éviter ou

A

à atténuer l'effet d'éblouissement provoqué par les feux des véhicules circulant en sens inverse lorsque la largeur du terre-plein central le permet et que les conditions de trafic sont compatibles avec les contraintes d'entretien) ; *syn.* plantation [f] anti-éblouissement ; *g* **Blendschutzpflanzung [f]** (Gehölzpflanzung auf Mittelstreifen von Schnellstraßen zur Verhinderung oder Reduzierung der Blendwirkung entgegenkommender Fahrzeuge bei Tage und zur Verhinderung, dass Scheinwerferlicht entgegenkommender Fahrzeuge blendet).

anti-dazzle screen [n] [UK] *trans.* ▶antiglare screen [US].

antiglare planting [n] [US] *trans.* ▶anti-dazzle planting.

188 antiglare protection [n] *trans.* (Measures or facilities to prevent bewildering brightness; ▶visual screening); *s* **medidas [fpl] (de protección) antideslumbramiento** (▶protección visual); *syn.* protección [f] antideslumbramiento; *f* **protection [f] contre l'éblouissement** (Mesures ou objets de défense contre l'éblouissement causé par les phares de véhicules roulant dans le sens opposé ; ▶protection contre les vues) ; *syn.* protection [f] anti-éblouissement, dispositif [m] anti-éblouissement ; *g* **Blendschutz [m]** (Maßnahmen oder Einrichtungen zur Abwehr der Blendung durch entgegenkommende Fahrzeuge; ▶Sichtschutz); *syn.* Blendabwehr [f].

189 antiglare screen [n] [US] *trans.* (Protection barrier to prevent bewildering brightness; ▶visual screening); *syn.* anti-dazzle screen [n] [UK]; *s* **valla [f] antideslumbramiento** (▶protección visual); *syn.* cortina [f] antideslumbramiento; *f* **écran [m] protecteur de lumière** (Élément de voirie destiné à réduire l'éblouissement causé par les phares des véhicules ; ▶protection contre les vues) ; *syn.* écran [m] anti-éblouissement ; *g* **Blendschutzwand [f]** (Straßenbauliche Einrichtung zur Verringerung der Blendung durch Scheinwerferlicht; ▶Sichtschutz).

anti-motorway action group [n] [UK] *pol.* ▶highway opposition group [US].

anti-nuclear action group [n] [UK] *pol.* ▶nuclear power opposition group [US].

190 anti-transpirant [n] *arb. constr.* (ARB 1983, 233, 352s); (Usually a chemical, foliar spray capable of reducing ▶transpiration); *s* **antitranspirante [m]** (Producto químico que se aplica a las hojas para reducir su ▶transpiración); *f* **anti-transpirant [m]** (Produit chimique réduisant la ▶transpiration des plantes ; *syn.* produit [m] antitranspirant) ; *g* **Verdunstungsschutzmittel [n]** (Chemisches Mittel zur Reduzierung der ▶Transpiration von Pflanzen).

apartment building vacancy (rate) [n] *sociol. urb.* *▶vacancy.

apartment block [n] [UK] *arch. urb.* ▶multistory building for housing [US].

apartment house [n] [US] *arch. urb.* ▶multistory building for housing [US].

apartment tower [n] [US] *arch. urb.* ▶high-rise residential building.

apartment vacancy [n] *sociol. urb.* *▶vacancy.

apex [n] **of a stem** *arb.* ▶stem apex.

191 appeal [n] **of a landscape** *recr.* (Capacity of a landscape to attract a person's interest); *syn.* landscape appeal [n]; *s* **atracciones [fpl] de un paisaje** (Características de un paisaje de atraer el interés de una persona); *f* **attrait [m] d'un paysage** (Caractéristique d'un paysage capable de capter l'attention d'une personne) ; *g* **Attraktivität [f] einer Landschaft** (Eigenschaft einer Landschaft, jemanden in seinen Bann zu ziehen).

appearance [n] *urb.* ▶community appearance [US], ▶interrupted appearance.

appearance [n] **, gap-tooth** *urb.* ▶interrupted appearance.

appearance [n] [US] **, street** *arch. urb.* ▶streetscape.

appearance [n] **of a landscape** *plan. landsc.* ▶analysis of the visual appearance of a landscape.

192 applicant [n] **for bidding documents** [US] *contr.* (Prospective participant in a bid [US]/tender [UK] who has not yet submitted his offer. With the submission of his bid/tender the applicant is known as a ▶bidder [US] 1/tenderer [UK]; ▶bidding [US]/tendering [UK]); *syn.* applicant [n] for tendering documents [UK]; *s* **candidato/a [m/f]** (Persona o empresa que quiere participar en concurso o subasta; al entregar una oferta se convierte en ▶licitante; ▶concurso-subasta de obras); *syn.* empresa [f] candidata; *f* **candidat [m] au marché** (Entreprise ayant exprimé le désir de participer à une consultation [▶adjudication, ▶appel d'offres] ; poser [vb] sa candidature ; ▶concurrent, ▶soumissionnaire) ; *syn.* entreprise [f] candidate, entrepreneur [m] candidat ; *g* **Bewerber/-in [m/f]** (Jemand, der an einer ▶Ausschreibung teilnehmen möchte und noch kein Angebot abgegeben hat. Bei Angebotsabgabe ist dieser Bewerber/diese Bewerberin ein[e] ▶Bieter/-in).

applicant [n] **for tendering documents** [UK] *contr.* ▶applicant for bidding documents [US].

application [n] *adm. leg.* ▶building and site plan permit application [US]/detailed planning application [UK]; *agr. constr. for. hort.* ▶fertilizer application, ▶pesticide application, ▶post-emergent herbicide application, ▶pre-emergent herbicide application.

application [n] [UK] **, detailed planning** *adm. leg.* ▶building and site plan permit application [US].

application [n] [UK] **, outline planning** *adm. plan.* ▶application for preliminary building permit [US].

application documents [npl] *adm. leg. urb.* ▶permit application documents [US].

193 application [n] **for preliminary building permit** [US] *adm. plan.* (Request for building permit [US]/building permission [UK] in outline form); *syn.* outline planning application [n] [UK]; *s* **licencia [f] previa** *syn.* autorización [f] previa, permiso [m] previo; *f* **demande [f] de certificat d'urbanisme** (Procédure permettant de connaître les dispositions d'urbanisme, les limitations administratives au droit de propriété et les équipements publics existants ou prévus sur le terrain) ; *g* **Bauvoranfrage [f]** (Schriftlicher Antrag des Bauherrn an die Baurechtsbehörde zu einzelnen Fragen der Genehmigungsfähigkeit eines Bauvorhabens. Der Bauwillige erhält daraufhin einen bindenden schriftlichen Bescheid [**Bauvorbescheid**], der drei Jahre gültig ist; cf. § 57 LBO-BW).

194 application [n] **of compost** *hort.* (Spreading of a defined quantity of compost to a designated area; ▶fertilizer application, ▶dressing with compost); *s* **aporte [m] de compost** (Aplicación de una cantidad definida de compost en un área determinada; ▶aporte de fertilizante, ▶fertilización con compost); *f* **apport [m] de compost** (Épandage d'une quantité définie de compost sur une unité de surface ; ▶apport d'engrais, ▶fertilisation avec apport de compost) ; *g* **Kompostgabe [f]** (Ausbringen einer definierten Kompostmenge auf eine bestimmte Fläche; ▶Düngergabe, ▶Düngung mit Kompost).

appointment [n] [UK] **, termination of** *contr. prof.* ▶termination of a contract/agreement [US].

appointment [n] **of consultant/designer/planner** *contr. prof.* ▶commissioning a consultant/designer/planner.

approach [n] *trans.* ▶planning approach.

approach [n] **, car park** *trans.* ▶parking access road.

appropriate [adj] **to the site** *landsc. plant.* ►plantation appropriate to the site.

approval [n] *adm. leg.* ►design and location approval [US] (1)/ determination of a plan [UK], ►design and location approval [US] (2)/Secretary of State Approval (of a plan) [UK], ►final approval of a community development plan; *leg. urb. trans.* ►final plan for approval; *constr. contr.* ►inspection and approval of plants.

approval [n] [UK]**, excavation** *adm. conserv'hist. leg.* ►excavation permit [US].

approval [n] **of an urban development plan** *adm. leg.* ►final approval of a community development plan.

approval [n] **of a plan** [UK]**, Secretary of State** *adm. leg.* ►design and location approval [US] (2).

approval [n] **of plants** *constr. contr.* ►inspection and approval of plants.

approval procedure [n] **for route selection** [UK] *plan. trans.* ►corridor approval procedure [US].

195 approve [vb] *adm. leg.* (To formally express agreement with an application, plan or project—e.g. a ►building and site plan permit application [US]/detailed planning application [UK]—by the appropriate certifying authority/agency or permit office that the documents comply with existing regulations, ordinances and standards; ►building permit [US]/building permission [UK]; ►permit; *syn.* give [vb] approval; *s* **autorizar** [vb] (Expresión formal de acuerdo con un plan/una construcción por parte de la autoridad competente; ►licencia de construcción, ►permiso definitivo); *syn.* aprobar [vb] un plan/un permiso de construcción; *syn. parcial* dar [vb] un permiso (de construcción); *f 1* **approuver** [vb] **un document de planification urbaine** *leg. urb.* ; *f 2* **délivrer** [vb] **une autorisation** (*Contexte, p. ex.* délivrer un permis de construire ou un permis de chasse, une autorisation de coupe et d'abattage d'arbres ; ►autorisation, ►permis de construire) ; *syn.* accorder [vb] un permis, octroyer [vb] une autorisation ; *g* **Genehmigung erteilen** [vb] (Die Ausführung eines Antrages oder Gesuches [z. B. eines Baugesuches] amtlich gestatten; ►Baugenehmigung, ►Genehmigung); *syn.* genehmigen [vb].

196 approved development site [n] *leg. urb.* (Land, provided with all services, which has been designated by the authorities for a specific development and is ready for construction according to building regulations; ►land zoned for development, ►undeveloped zoned land); *s* **suelo** [m] **urbanizado (1)** (Suelo supeditado a la normativa de un ►plan parcial [de ordenación] y que ya ha sido equipado con infraestructura; ►suelo edificable, ►suelo urbanizable); *f* **terrain** [m] **viabilisé** (Terrain aménagé en vue d'une construction et desservi par la voirie et déjà pré-équipé en divers réseaux : assainissement, eau potable, électricité, gaz, téléphone ; l'ensemble des travaux de voirie et d'installation des réseaux divers (eau, gaz, électricité, téléphone, égouts) constituent la viabilité ; ►terrain non viabilisé, ►terrain viabilisé, ►zone constructible) ; *syn.* parcelle [f] (constructible) équipée, terrain [m] (constructible) équipé ; *g* **fertiges Bauland** [n, o. Pl.] (►Bauland, das in der Rechtswirksamkeit des Bebauungsplanes [Vollzug] liegt, d. h. nach öffentlich-rechtlichen Vorschriften baulich nutzbar ist, bei dem die Erschließung durch Herstellung von öffentlichen Straßen, Wegen, Plätzen, Parkflächen und Grünanlagen entsprechend den Erfordernissen der Bebauung und des Verkehrs abgeschlossen ist und die Erhebung des Erschließungsbeitrages feststeht; ►Rohbauland); *syn.* baureifes Land [n].

197 approved plan [n] [US] *leg. urb.* (Plan which has been adopted by a public agency/authority); *syn.* approved scheme [n] [UK]; *s* **plan** [m] **aprobado definitivamente** (Plan vigente gracias a la aprobación de la autoridad competente); *f* **plan** [m] **approuvé par arrêté préfectoral** (Document de planification urbaine ayant été soumis au contrôle de la légalité par le préfet, approbation par arrêté du préfet suivie de la publication au Journal Officiel et au recueil des actes administratifs du département) ; *g* **genehmigte Planfassung** [f] (Durch eine Verwaltungsbehörde genehmigter Plan; für einen Bauleitplan cf. §§ 6 u. 11 BauGB).

approved plan [n] **by Secretary of State** [UK] *adm. leg.* ►approved plan for major projects [US].

198 approved plan [n] **for major projects** [US] *adm. leg.* (**In U.S.**, plan approved by a local authority, sometimes by federal agencies—involving federal funds—for major highways, hazardous waste sites, oil pipelines, nuclear power plants, etc.; **in U.K.**, plan of a significant project requiring design and location approval by Secretary of State; ►design and location approval [US] 2/Secretary of State Approval [of a plan] [UK]); *syn.* plan [n] determined by Secretary of State [UK], approved plan [n] by Secretary of State [UK]; *s* **plan** [m] **de proyecto público aprobado** (Plan legalmente vinculante gracias a un ►acuerdo de aprobación definitiva de proyecto público); *f* **plan** [m] **(de grands projets publics) approuvé** (Plan ayant fait l'objet d'une procédure d'instruction de grands projets publics et devenant opposable au tiers par notification d'autorisation ou par la publication de l'arrêté ; ►arrêté d'approbation de grands projets publics) ; *g* **festgestellter Plan** [m] (Plan, der durch einen ►Planfeststellungsbeschluss rechtsverbindlich wird).

approving agency [n] [US] *adm. leg.* ►approving authority.

199 approving authority [n] *adm. leg.* (Agency, board, group or other legally designated individual or authority which has been charged with review and approval of plans, projects, or applications); *syn.* approving agency [n] [also US]; *s* **autoridad** [f] **concesionaria** (Organismo administrativo responsable de otorgar licencias de construcción, de funcionamiento de fábricas, etc.); *f* **autorité** [f] **compétente pour délivrer des autorisations** ; *g* **genehmigende Behörde** [f] (Verwaltungsbehörde, die für genehmigungsbedürftige Planungen, Vorhaben, Maßnahmen und Entscheidungen zuständig ist und entsprechende Urkunden ausstellt); *syn.* Genehmigungsbehörde [f].

200 aqualfs [npl] [US] *pedol.* (**1.** *U.S. Comprehensive Soil Classification System* **a.** is a suborder to 'alfisols', and is saturated with water for periods long enough to limit their use for most crops other than pasture or woodland unless they are artificially drained. **A.** have mottles, low chromas, and Fe-Mn concretions 2mm or more in diameter or gray colors immediately below the A1 or Ap horizons and gray colors in the argillic horizon; cf. SGC 1989, 314 and SST 1997. **2. In Europe**, the nearly correspondent term is **pseudogley**, a hydromorphic soil at a distance from groundwater in which a fluctuation between water saturation and desiccation occurs; **p.** is characterised by a horizon sequence of A_h–B_g–C_{gv}. [**h** for humus, **g** for gleying, **v** for weathered or soft bedrock]; ►aquepts); *s 1* **aqualf** [m] [US] (En el Sistema Integrado de Clasificación de Suelos de EE.UU. **a.** es un suborden de los 'alfisoles'. Está saturado de agua durante periodos/períodos de tiempo suficientemente largos como para limitar el uso agrícola, exceptuando pastos o bosques, a no ser que sea drenado artificialmente. **A.** tiene manchas de colores de baja intensidad, concreciones de Fe/Mn de > 2 mm y manchas grises inmediatamente debajo del horizonte A1 o Ap); *s 2* **pseudogley** [m] (**En Europa**, suelo hidromorfo con capa freática profunda en el que el fuerte cambio entre humedecimiento y sequía superficial lleva a la formación de concreciones y manchas de oxidación sobre todo en el interior de los agregados. El suelo

A

p. tiene la secuencia de horizontes poco diferenciada A_h–B_g–C_{gv}. [h de humus, **g** de tipo gley, v de meteorizado]. En el sistema estadounidense puede ordenarse el **p.** en el suborden *aqualfs* o en el suborden *inceptisols* como ▶*aquepts*; ▶suelo gley); *f* **sol [m] à pseudogley** (Un sol hydromorphe, dont le profil est soumis à la présence d'eau saturant la totalité des pores pendant une période de l'année ; ▶sol à gley) ; *g 1* **Aqualf [m] [US]** (Im US-Compre-hensive-Soil-Classification-System gehört die Unterordnung **A.** zur Ordnung der *Alfisols*. Es ist ein wassergesättigter Boden, der, wenn er nicht dräniert worden ist, auf Grund lang anhaltender Vernässung für die landwirtschaftliche Bodennutzung, außer als Weide oder Wald, nur eingeschränkt nutzbar ist; **A.s** haben eingesprenkelte Farbflecken von schwacher Farbintensität, 2 mm große oder größere Fe/Mn-Konkretionen und graue Farbflecken unmittelbar unter dem A1- oder Ap-Horizont [p für Bodenbear-beitung oder andere Störungen]; *g 2* **Pseudogley [m]** (**Europa**, hydromorpher Boden in Grundwasserferne, in dem ein Wechsel von Stauwasser und Austrocknung Konkretionen und Rostflecken vornehmlich im Aggregatinneren verursacht. **P.e** haben in der am wenigsten differenzierten Ausbildungsweise die Horizontfolge A_h–B_g–C_{gv}. [**h** von Humus, **g** von gleyartig, **v** von verwittert]. Im *US-Comprehensive-Soil-Classification-System* kann der **P.** der Unterordnung *Aqualfs* oder als ▶*Aquept* [▶Gley] den *Inceptisols* zugeordnet werden).

aquatic [n] [US] *bot. phyt.* ▶hydrophyte.

201 aquatic flora [n] *phyt.* (Plants which habitually live in water or inundated areas, whether seawater, brackish or sweet water); *syn.* aquatics [npl] [also US] (TEE 1980, 214 s), aquatic plants [npl]; *s* **flora [f] acuática** (Plantas que crecen en medios acuosos, sean salados, salobres o dulces); *f* **flore [f] aquatique** (Les plantes aquatiques vivant en permanence dans l'eau ou dans les milieux inondés, qu'ils soient marins, saumâtres ou dulça-quicoles [eau douce] ; DEE 1982) ; *g* **Wasserflora [f]** (Pflanzen, die ständig im Wasser oder in überschwemmten Gebieten, sei es im Meer-, Brack- oder Süßwasser, leben); *syn.* Gewässerflora [f].

aquatic plant [n] *bot. phyt.* ▶hydrophyte, ▶emergent aquatic plant (1), ▶emergent aquatic plant (1), ▶submerged aquatic plant.

202 aquatic plant community [n] *bot. phyt.* (Generic term covering all communities living in standing water or water-courses; e.g. ▶chara vegetation, ▶free-floating freshwater com-munity, ▶rooted floating-leaf community); *s* **comunidad [f] acuática** (Término genérico para todo tipo de comunidad de las aguas estancadas y corrientes, p. ej. ▶comunidad de hidrófitos radicantes, ▶comunidad de hidrófitos libremente flotantes; ▶pra-dera acuática sumergida); *f* **groupement [m] d'hydrophytes** (Terme générique englobant tous les herbiers immergés des eaux calmes et courantes, libres en permanence ; ▶groupement d'hydrophytes libres et flottants [des eaux calmes], ▶groupement flottant, fixé, ▶groupement flottant libre, ▶prairie aquatique submergée) ; *g* **Wasserpflanzengesellschaft [f]** (OB zu Gesell-schaften der Still- und Fließgewässer, z. B. ▶wurzelnde Schwimmblattgesellschaft, ▶Schwimmpflanzendecke, ▶Unter-wasserrasen).

aquatic plants [npl] *phyt.* ▶aquatic flora.

aquatics [npl] [US] *phyt.* ▶aquatic flora.

aquatic sports [npl] *recr.* ▶water sports.

203 aquepts [npl] [US] *pedol.* (**1.** *U.S. Comprehensive Soil Classification System* **a.** is a suborder to 'inceptisols', and is saturated with groundwater within 100cm of the mineral soil surface for some time during the year, long enough to limit their use for most crops other than pasture or woodland unless they are artificially drained. **A.** have either a *histic epipedon* [thin organic, water-saturated soil horizon. The thickness of the organic soil

horizon depends on the kind of materials in the horizon and the content of organic carbon]or *umbric epipedon* [surface horizon of mineral soil that is dark colored and relatively thick, contains at least 5.8g/kg organic carbon, is not massive and hard or very hard when dry, has a base water saturation of <50% when measured at pH 7] and gray colors within 50cm of the mineral soil surface, or an *ochric epipedon* [surface horizon of mineral soil that is too light in color, too high in chroma, too low in organic carbon, or too thin to be a *plaggen, mollic, umbric or histic epipedon*, or that is both hard and massive when dry] underlain by a *cambic horizon* [mineral soil horizon that has a texture of loamy very fine sand or finer, contains some weatherable minerals, and is characterized by the alteration or removal of mineral material as indicated by mottling or gray colors, stronger chromas or redder hues than in underlying horizons] with gray colors, or have sodium saturation of 15% or more; cf. SST 1997. **2. In Europe**, the nearly correspondent term for **a.** is gley (soil), a ▶hydro-morphic soil created by oxygen-deficient groundwater at an average depth of 40-80cm; ▶gleyed anmoor; *s 1* **aquept [m] [US]** (En el Sistema Integrado de Clasificación de Suelos de EE.UU., el suborden **a.** pertenece al orden de los *inceptisoles*. Es un suelo saturado de agua subterránea hasta 100 cm de profundidad de la superficie de la capa mineral y que —si no ha sido drenado— no es apto para usos agrícolas, exceptuando pastos y bosques. Los **a.** tienen bien una capa superior orgánica *[histic epipedon]*, cuyo espesor depende del tipo de componentes existentes y del contenido de carbono o una capa superior mineral *[umbric epipedon]* relativamente gruesa y oscura, fuertemente coloreada de gris en los 50 cm superiores y que contiene un mínimo de 5,8 g/kg de carbono orgánico; en caso de sequía el suelo no es ni muy duro ni extremadamente duro y con un pH de 7 está saturado por lo menos en un 50%] o una capa superior mineral de color ocre *[ochric epipedon]* de un color demasiado claro e intenso, con demasiado poco carbono orgánico o dema-siado delgado como para ser un epipedon tipo plaggen, mollic, umbric o histic, o que en el caso de sequía es entre duro y extremadamente duro. Bajo el horizonte A se encuentra uno grisáceo, el horizonte cámbico, con una textura muy limosa de granulación fina, entremezclado con partículas de piedra meteo-rizadas y un contenido de sodio > 15%. Este horizonte está caracterizado por alteraciones o lixiviación de sustancias mine-rales y por manchas de colores; en comparación con el horizonte inmediatamente inferior por tonos de intenso color gris o de rojo oxidado); *s 2* **suelo [m] gley (±)** (Suelo formado debido al anega-miento o la impregnación con agua freática a poca profundidad, que es objeto de procesos químicos especiales, como la reduc-ción. Ésta origina componentes más o menos característicos, como el hierro oxidulado con coloraciones negruzcas y grises, con matices verdosos y azulados, el manganeso acumulado en manchas de color negro, la vivianita azul, etc. que al reoxidarse cambian de color de nuevo. Al horizonte donde ocurren estos procesos se le llama «gley», que es un término de origen ruso; cf. DB 1985; ▶anmoor, ▶gley turboso, ▶suelo hidromorfo regular-mente inundado, ▶suelo semiterrestre); *f* **sol [m] à gley (±)** (Sol qui présente une teneur importante en argiles, essentiellement de la kaolinite et des illites, qui lui donnent une texture plastique. Ce sol formé au niveau des fluctuations de la nappe aquifère se caractérise par une présence constante d'eau en profondeur. Ces fluctuations créent dans le sol des conditions alternativement aérobies et anaérobies. Dans la partie du profil continuellement sous eau l'activité réductrice de l'eau entraîne la différenciation d'un horizon profond dans lequel le fer reste à l'état ferreux, donc réduit et qui se manifeste par une couleur vert bleu [G_r] dans la zone de fluctuation intense où une réoxydation partielle inter-vient, la fluctuation de la nappe entraîne la différenciation d'un horizon marron clair (fer oxydé), le fer se présente alors sous

forme de taches de rouille [G$_o$] ; DIS 1986 ; ►anmoor, ►gley tourbeux, ►sols hydromorphes) ; *g 1* **Aquept [m] [US]** (Im *US-Comprehensive-Soil-Classification-System* gehört die Unterordnung **A.** zur Ordnung der *Inceptisols.* Es ist ein bis 100 cm tief grundwasserwassergesättigter Boden, der, wenn er nicht dräniert worden ist, auf Grund lang anhaltender Vernässung für die landwirtschaftliche Bodennutzung, außer als Weide oder Wald, nur eingeschränkt nutzbar ist; **A.s** haben entweder einen organischen Oberboden [*histic epipedon,* dessen Mächtigkeit von der Art der vorhandenen Bestandteile und vom organischen Kohlenstoffgehalt abhängt] oder einen relativ mächtigen und in den oberen 50 cm stark grau gefärbten mineralischen Oberboden [*umbric epipedon* mit mindestens 5,8 g/kg organischem Kohlenstoff; der Boden ist bei Trockenheit weder steinhart noch sehr hart und hat bei einem pH-Wert von 7 eine Mindestwassersättigung von 50 %] oder einen ockerfarbenen mineralischen Oberboden, den *ochric epipedon* mit sehr heller Farbe, sehr farbintensiv, mit sehr wenig organischem Kohlenstoff oder ein Boden, der bei Trockenheit hart bis steinhart ist. Unter dem A-Horizont folgt ein graugefärbter *cambic horizon* mit einer sehr feinen lehmigen Sandtextur, durchsetzt mit einigen verwitternden Gesteinspartikeln und einem Natriumgehalt von über 15 %. Dieser *cambic horizon* ist durch Veränderung oder Verlagerung mineralischer Stoffe und durch eingesprenkelte Farbflecken, im Vergleich zum darunterliegenden Horizont, durch farbintensive Grau- oder rostrote Farbtöne gekennzeichnet; *g 2* **Gley [m]** (±) (Hydromorpher Boden durch sauerstoffarmes Grundwasser entstanden, mit einem mittleren Grundwasserhorizont in ca. 40-80 cm Tiefe; typische Horizontfolge A$_h$-G$_o$-G$_r$ [h von Humus, o von Oxidation, r von Reduktion]; Auf dem vom Grundwasser unbeeinflussten A$_h$-Horizont folgt der rostartige G$_o$-Horizont [Oxidationshorizont] und darunter der stets nasse, fahlgraue bis graugrüne oder auch blauschwarze G$_r$-Horizont [Reduktionshorizont]. Der mittlere Grundwasserspiegel liegt höher als 80-100 cm, der Kapillarsaum selten höher als 40 cm. Der Name **G.** kommt aus dem Russischen und bedeutet schlammige Bodenmasse. Ältere Synonyma sind u. a. *Bruchboden, Wiesenboden* und *mineralischer Nassboden*; ►Grundwasser- und Stauwasserböden, ►Anmoor); *syn.* Gleiboden [m], Gleyboden [m].

204 aquiclude [n] *hydr. geo.* (Formation which, although porous and capable of absorbing water, does not transmit it at rates sufficient to furnish an appreciable supply for a well or spring; WMO 1974; ►aquitard); *syn.* impermeable stratum [n] (DNE 1978, 16); *s* **acuicludo [m]** (Formación que, aun siendo porosa y capaz de absorber agua, no la transmite en cantidad suficiente para proporcionar una alimentación apreciable a un pozo o manantial; WMO 1974; ►capa de confinación); *f* **aquiclude [m]** (Corps, couche ou massif de roche saturée à porosité si fine que l'eau y est retenue par la tension hygroscopique et qu'aucun flux de drainance ne peut passer [imperméabilité]. Un **a.** est donc improductif en eau et s'oppose à un aquifère et à une couche semi-perméable ; DG 1984) ; ►couche encaissante ; *syn.* formation [f] à microporosité (WMO 1974) ; *g* **Grundwasserstauer [m]** (Untere undurchlässige Schicht eines Grundwasserleiters; ►Grundwassersohlschicht).

205 aquifer [n] *hydr.* (Porous water-bearing formation bed or stratum of permeable rock, sand or gravel capable of yielding significant quantities of water; WMO 1974; ►confined aquifer, ►groundwater reservoir); *syn.* water-bearing stratum [n]; *s* **acuífero [m]** (Formación porosa —capa o estrato— de roca permeable, arena o gravilla, capaz de almacenar y transmitir cantidades apreciables de agua; WMO 1974; ►acuífero confinado, ►embalse de agua subterránea); *f* **formation [f] aquifère** (Formation géologique, sédimentaire pour l'essentiel, à perméabilité d'interstices [sables, graviers] du fond de vallées, de grands

bassins sédimentaires ou de fissures [p. ex. formation karstique], formant un réservoir d'eau souterraine dans lequel la masse d'eau mouvante se déplace par gravité ; ►formation aquifère captive, ►réservoir d'eau souterraine) ; *syn.* aquifère [m], couche [f] conductrice d'eau ; *g* **Grundwasserleiter [m]** (Durchlässige geologische Schicht der Erdrinde aus klüftigem Gestein, Sand oder Kies, in der sich Wasser ansammeln kann und entsprechend der Gefälleneigung fließt; ►Grundwasserstockwerk, ►Grundwasserspeicher); *syn.* Aquifer [m], Wasser führende Schicht [f].

206 aquifer recharge area [n] *hydr.* (TGG 1984, 191; area which contributes water to an aquifer, ►groundwater recharge); *syn.* intake area [n]; *s* **área [f] de alimentación de un acuífero** (►recarga de acuíferos); *syn.* área [f] de toma de un acuífero; *f* **bassin [m] d'alimentation d'un aquifère** (Aire d'apport d'eau externe, de toute origine alimentant une nappe phréatique ; ►alimentation d'une nappe phréatique) ; *syn.* zone [f] d'alimentation d'un aquifère (PPC 1989, 100) ; *g* **Einzugsgebiet [n] eines Grundwasserleiters** (Gebiet, von dem aus einem Grundwasserleiter Wasser zugeführt wird; ►Grundwasseranreicherung); *syn.* Grundwassereinzugsgebiet [n].

aquifer recharge forest [n] *for. wat'man.* ►protected aquifer recharge forest.

207 aquitard [n] *hydr.* (Overlying or underlying [clayey] layer of a nearly impervious and semi-confining nature which permits percolation at a very slow rate, compared to an aquifer; ►pan, ►aquiclude); *syn.* confining bed [n], confining layer [n], confining stratum [n]; *s* **capa [f] de confinación** (Formación que se extiende por encima o por debajo de un acuífero mucho más permeable; WMO 1974, 163; ►capa de cementación, ►acuicludo); *syn.* aquitard [m] (WMO 1974); *f* **couche [f] encaissante** (Formation qui s'étend au-dessus ou au-dessous d'une nappe aquifère beaucoup plus perméable ; WMO 1974 ; ►substratum de la nappe perchée ; ►aquiclude) ; *syn.* aquitard [m] ; *g* **Grundwassersohlschicht [f]** (1. Undurchlässige Schicht, tiefer als 1,3 m unter Flur; 2. liegt diese Schicht oberhalb 1,3 m unter Flur, so wird sie aus bodenkundlicher Sicht **Staunässesohle [f]** oder ►Stauwassersohle genannt; cf. TWW 1982, 133; ►Grundwasserstauer).

208 arable [adj] *agr.* (Capable of being plowed [US]/ploughed [UK] or cultivated); *s* **cultivable [adj]** *syn.* arable [adj], laborable [adj]; *f* **arable [adj]** (Terre pouvant être travaillée et cultivée) ; *g* **ackerbaulich [adj]** (So beschaffen, dass ein landwirtschaftlich nutzbarer Boden gepflügt und bestellt werden kann).

209 arable land [n] *agr.* (Plowable [US]/ploughable [UK] agricultural land, usually under a system of crop rotation; ►arable soil); *s* **tierra [f] de cultivo** (►suelo de cultivo); *syn.* tierra [f] de labranza, tierra [f] laborable, ager [m]; *f* **terre [f] de culture** (Terre cultivée par des outils aratoires produisant en général une récolte annuelle ; ►terre arable) ; *g* **Acker [m] (1)** (Äcker [pl]; mit dem Pflug zu bearbeitende landwirtschaftliche Bodenfläche, die i. d. R. einer Fruchtfolge unterliegt; ►Ackerboden); *syn.* Ackerland [n].

arable land grade [n] [UK] *agr.* ►arable land rank [US].

210 arable land rank [n] [US] *agr.* (Indication of agricultural land quality used in determining a piece of land's suitability for arable use based upon climate and relief factors; ►agricultural land grade [UK], ►Land Capability Classification [US], ►land capability class [US], ►grassland yield index); *syn.* arable land grade [n] [UK]; *s* **índice [m] de fertilidad de tierras de cultivo** (En D., especificación del ►índice de fertilidad de suelos considerando la fertilidad per se y el clima y la topografía; ►clasificación de aptitud de suelos, ►índice de fertilidad de suelos [de prados]); *f* **indice [m] d'estimation des sols** (D., les sols agricoles sont classés en 7 classes de développement prenant en compte les caractères génétiques et fonctionnels des sols ; PED 1984, 206 ; ▸classe d'aptitude d'un sol, ▸classification des aptitudes culturales et pastorales des sols, ▸classification des sols et des paysages agricoles de premier choix et marginaux [CDN], ▸classification de l'aptitude des terres, ▸coefficient de fertilité d'un sol, ▸coefficient de valeur agricole des prairies, ▸indice de fertilité physique) ; *g* **Ackerzahl [f]** (Durch Berücksichtigung von Klima und Relief für die Ackernutzung präzisierte relative Bodenbewertungszahl [7-100], die die Qualität einer Ackerfläche hinsichtlich der Ertragsfähigkeit misst. Zu Grunde gelegt werden optimale Böden, die sog. „Bördeböden" der Magdeburger, Hildesheimer und Soester Börde; Flächen mit einer **A.** unter 20 sind landwirtschaftlich kaum noch nutzbar; ▸Bodeneignungsklasse, ▸Bodenschätzung, ▸Bodenzahl, ▸Grünlandzahl); *syn.* Ackerwertzahl [f].

211 arable parcel [n] *agr.* (Plowable [US]/ploughable [UK] field for growing crops); *s* **campo [m] de cultivo** *syn.* parcela [f] de cultivo; *f* **parcelle [f] de culture** ; *g* **Ackerparzelle [f]** (Stück landwirtschaftliche Fläche, die mit einer Frucht bestellt ist oder wird); *syn.* Ackerschlag [m].

212 arable soil [n] *agr.* (Agriculturally-used and plowed [US]/ploughed [UK] soil suitable for crop production); *s* **suelo [m] de cultivo** (Capa de tierra vegetal arada utilizada para la producción de plantas de cultivo); *f* **terre [f] arable** (Partie superficielle du sol qui reçoit les amendements et les engrais et qui est travaillée par l'agriculteur) ; *syn.* sol [m] arable ; *g* **Ackerboden [m]** (Für die Anzucht von landwirtschaftlichen Kulturpflanzen nutzbarer oder genutzter, umgepflügter Boden).

213 arable use [n] *agr. plan.* (Agricultural use category for soil cultivation); *s* **uso [m] agrícola** (Categoría de uso del suelo para cultivos); *f* **utilisation [f] agricole des terres** (Catégorie d'usage de l'espace à des fins agricoles) ; *syn.* usage [m] agricole des terres ; *g* **Ackernutzung [f]** (Landwirtschaftliche Kultivierung eines Bodens).

214 arable weed community [n] *phyt.* (Plant association of native species on cultivable land belonging to the Secalinetea and Chenopodietea classes; ▸root crop weed community, ▸segetal community, ▸weed community); *s* **comunidad [f] arvense de cultivos** (Término genérico para ▸comunidad arvense de cultivos de cereales, ▸comunidad arvense de los huertos y cultivos de verano; ▸vegetación de malas hierbas; ▸vegetación segetal, ▸comunidad arvense de los huertos y cultivos de verano, ▸comunidad de malas hierbas); *syn.* comunidad [f] de malas hierbas de cultivos; *f* **groupement [m] de plantes adventices des cultures** (Terme générique pour la ▸végétation ségétale, la ▸flore des cultures sarclées, le ▸groupement de cultures sarclées appartenant à la classe des *Secalinetea* et *Chenopodietea* ; ▸groupement anthropique, ▸flore adventice des cultures et des prairies) ; *syn.* groupement [m] de plantes commensales ; *g* **Ackerunkrautgesellschaft [f]** (Klasse der *Secalinetea* [▸Segetalflur] und *Chenopodietea* [▸Hackunkrautflur]; Pflanzengesellschaft, die als Begleiter von landwirtschaftlichen Hauptkulturen auftritt; ihr Vorkommen und die Vielfalt der Artenzusammensetzung sind abhängig von Art und Intensität der Bewirtschaftung; mit zunehmender Bewirtschaftungsintensität gehen etliche Arten in ihrem Bestand zurück, weshalb diese in den ▸Roten Listen enthalten sind; durch Anlage von Ackerrandstreifen, die durch spezielle Ackerrandstreifenprogramme der Länder und Regionen finanziell gefördert werden, kann der Bestand dieser gefährdeten Arten geschützt und aufrechterhalten werden; ▸Unkrautflur); *syn.* Ackerwildkrautflur [f], Ackerwildkrautgesellschaft [f].

215 arbitration proceedings [npl] *constr.* (A hearing granted to disputants to determine the fairness or the accuracy of

conflicting views between client and contractor with regard to quality and quantity; ▸quality control); *s* **procedimiento [m] de arbitraje (≠)** (Repetición de un ▸control de calidad sobre la realización correcta de una obra, en el caso de que el comitente o el contratista hayan cuestionado su calidad); *f* **procédure [f] d'arbitrage** (Renouvellement d'une mesure de ▸vérification qualitative des matériaux et produits effectuée suite à des doutes fondés quant à la conformité, émis par le client ou le contracteur) ; *g* **Schiedsuntersuchung [f]** (Wiederholung einer ▸Kontrollprüfung, an deren sachgerechter Durchführung begründete Zweifel des Auftraggebers oder des Auftragnehmers bestehen); *syn.* Schiedsverfahren [n].

216 arbor [n] [US] *arch. gard.* (Bower formed of vines [US]/climbers [UK] or branches or of latticework covered with climbing plants used as a sitting place in a garden; ▸porch, ▸grapevine arbor [US]); *syn.* arbour [n] [UK], bower [n] [UK]; *s* **cenador [m] (1)** (Pequeño pabellón construido con hierro o cañas bravas, comúnmente redondo, que suele haber en los jardines, cubierto de plantas trepadoras, parras o árboles, y sirve como un lugar de descanso; ▸emparrado, ▸porche); *f 1* **abri [m] de jardin** (Terme générique pour les constructions couvertes ornementales ou d'agrément dans un jardin souvent recouvertes de plantes grimpantes ; *termes spécifiques* pavillon de verdure, ▸pavillon [de jardin], ▸gloriette, ▸kiosque [de jardin], chaumière, etc. ; ▸porche, ▸treille) ; *syn.* abri [m] familial, *en relation avec des plantes grimpantes* pièce [f] de verdure ; *f 2* **fabrique [f]** (Terme générique pour les constructions ornementales dans les jardins au XXIIIème siècle ; terme aujourd'hui encore utilisé pour désigner les abris, pergolas, treillages) ; *g* **Gartenlaube [f]** (Ein Stützgerüst aus Holzlatten oder Metallstäben, das mit Gehölzen oder Kletterpflanzen bewachsen ist und als Sitzplatz dient; ▸Laube, ▸Weinlaube).

217 arborescent [adj] *arb. phyt.* (Descriptive term applied to any plant like a tree in size or form); *syn.* tree-like [adj], dentritic [adj] *geo.*; *s 1* **arbóreo/a [adj]** (De condición parecida a la del árbol, por su desarrollo y sus dimensiones: *planta arbórea*; DB 1985); *s 2* **arborescente [adj]** (Que se hace árbol, que arborece; o también que ha alcanzado el aspecto o la altura de un árbol: helecho arborescente; DB 1985); *f* **arborescent, ente [adj]** (Semblable à un arbre par sa forme ou par sa taille) ; *g* **baumartig [adj]** (So aussehend wie ein Baum; einem Baum ähnlich).

218 arboretum [n] *hort.* (Botanical garden containing a collection of tree species, arranged according to either taxonomic or ecological principles for demonstration and educational purposes); *s* **arboreto [m]** (Parque con colección de leñosas exóticas ordenadas según principios taxonómicos o ecológicos creado con objetivos educativos o de demostración; *término específico para a. especializado en coníferas* pinetum); *syn.* arboretum [m]; *f* **arboretum [m]** (Parc rassemblant une collection variée de végétaux ligneux en général groupés par familles et par genres et comportant souvent des essences étrangères. La croissance naturelle des végétaux fait qu'il est souvent fréquenté par les scientifiques ou les spécialistes) ; *g* **Arboretum [n]** (...ta [pl], ...ten [pl]; Garten mit einer nach taxonomischen oder Standortsansprüchen geordneten Sammlung heimischer und meist auch ausländischer Gehölze zu Ausstellungs- und Studienzwecken); *syn.* Gehölzgarten [m].

219 arboriculture [n] *arb. hort.* (Art and science of growing and care of trees for urban tree preservation and aesthetic purposes, such as specimen trees, street trees, shade trees, town forest, etc.; ▸cultivation of woody plants, ▸tree nursery); *s* **arboricultura [f]** (Conjunto de técnicas y métodos de cultivo que se aplican a los árboles, arbustos y matas, consideradas individualmente, en pequeños grupos o en huertos, en su aprovechamiento agrario; DINA 1987; ▸cultivo de leñosas, ▸vivero

A

[de árboles]); *f* **arboriculture** [f] (Ensemble des techniques de production et d'entretien des végétaux ligneux ; on distingue l'arboriculture fruitière, forestière et ornementale ; ▶élevage des végétaux ligneux, ▶pépinière) ; *g* **Anzucht [f, o. Pl.] und Pflege von Freilandgehölzen** (Gesamtheit der Kulturtechniken, -metho-den und Pflegearbeiten, die bei der Kultivierung und Pflege von Gehölzen für ästhetische, gewerbliche oder wissenschaftliche Zwecke [bis zur verkaufsfertigen Pflanze] angewandt werden — für Solitär- und Straßenbäume, Obstbäume, kleine Baum- oder Zierstrauchgruppen; der englische und französische Begriff *arbo-riculture* beinhaltet sowohl die ▶Gehölzanzucht in der ▶Baum-schule als auch die fachgerechte Pflege und Unterhaltung am Verwendungsort).

220 arboriculturist [n] *arb. hort.* (One who studies the scientific cultivation of trees and shrubs, for economic or ornamental use); *s* **arboricultor, -a [m/f]** (Especialista científico/a en el cultivo de leñosas productivas u ornamentales); *f* **arboriculteur, trice [m/f]** (Spécialiste qui s'occupe de la culture des espèces ligneuses fruitières et ornementales) ; *g* **Gehölzpflanzenzüchter/ -in [m/f]** (Jemand, der sich wissenschaftlich mit der Anzucht und des Anbaues von Nutz- oder Ziergehölzen beschäftigt).

221 arborist [n] *arb. hort.* (Specialist in professional practice of tree maintenance, trained in Europe in the first instance as a tree worker [European Tree Worker—ETW] for 3 months and with further training [5 months] to become a tree technician [European Tree Technician—ETT], recognized by The European Arboriculture Council [EAC]); *s* **arborista [m/f]** (Jardinero especializado en arboricultura); *f* **élagueur [m]** (Personnel qualifié spécialiste de l'entretien des arbres en milieu urbain ; ses activités principales consistent dans la taille et l'élagage des arbres, l'abat-tage des arbres, l'organisation du chantier, la réalisation de traite-ments phytosanitaires et chirurgie arboricole sur le patrimoine arboré, l'entretien courant du matériel) ; *syn.* bûcheron-élagueur [m], arboriste [m], grimpeur-élagueur [m] ; spécialiste [m] des arbres et des arbustes, technicien [m]/technicienne [f] en service d'entretien d'arbres [CDN] ; spécialiste [m] de l'entretien des arbres [CH] ; *g* **Baumpfleger/-in [m/f]** (Jemand, der speziell für die Baumpflege ausgebildet ist — nach der derzeitigen Fachaus-bildungssituation i. d. R. ein Gärtner/eine Gärtnerin mit einer zusätzlichen Ausbildung zum **European Tree Worker [ETW]**; die nächste Fortbildungsstufe mit weiteren Aufbaumodulen ist der **Fachagrarwirt Baumpflege und Baumsanierung [FAW]**; die Qualifikation des FAW ist nahezu identisch mit dem seit 2005 eingeführten und europaweit anerkannten **European Tree Tech-nicien [ETT]**; die Dachorganisation ist der *European Arbori-cultural Council [EAC]*; *syn.* Baumfachmann [m], Baumfachfrau [f].

arbour [n] [UK] *arch. gard.* ▶arbor [US].

archaeological area [n] *conserv'hist.* ▶archaeological excavation area.

archaeological dig [n] [US] *conserv'hist.* ▶archaeological excavation.

archaeological dig area [n] [US] *conserv'hist.* ▶archaeo-logical excavation area.

222 archaeological excavation [n] *conserv'hist.* (Digging up possible existing archaeological or palaeontological findings for study of history and prehistory; ▶garden archaeo-logical excavation; ▶exploratory trenching); *syn.* archaeological dig [n] [also US]; *s* **prospección [f] arqueológica** (art. 42 Ley 16/1985; ▶excavación arqueológica; ▶prospección arqueoló-gica-paisajística); *f 1* **prospection [f] archéologique** (Discipline fondée sur l'analyse de données d'observation ou de mesures effectuées à la surface du sol ayant pour but d'identifier, de décrire et d'interpréter les traces et les vestiges abandonnés à toutes époques par l'homme dans l'espace qui nous environne. Complémentaire de la fouille pour la reconstitution historique du passé, elle s'appuie sur un certain nombre de méthodes telles que l'échantillonnage de vestiges superficiels, la géophysique, etc., spécifiques de chaque cible et adaptées à chaque échelle d'explo-ration ; cf. histoiremesure.rvues.org/index1015.html) ; *f 2* **fouilles [fpl] archéologiques** (Exploration systématique du sol superficiel entreprise pour mettre à jour des objets archéologiques ou paléon-tologiques ; ▶fouilles archéologiques dans un jardin historique, ▶sondage archéologique) ; *syn.* prospection [f] archéologique ; *g* **Grabung [f]** (Freilegung existierender oder vermuteter archäo-logischer oder paläontologischer Funde für die historische und prähistorische Forschung; *UBe* ▶Suchgrabung, ▶gartenarchäo-logische Grabung; *syn.* Ausgrabung [f], Grabungsarbeiten [fpl], *z. T.* archäologische Untersuchung [f].

223 archaeological excavation area [n] *conserv'hist.* (Area which is important for discovery of both prehistoric remains and historic civilizations; ▶exploratory trench); *syn.* archeological dig area [n] [also US]; *s* **zona [f] de prospección arqueológica** (Área arqueológicamente importante; ▶zanja de prospección arqueológica); *f* **site [m] de fouilles archéologiques** (Zone de fouilles d'une grande richesse archéologique et patri-moniale ; ▶tranchée de sondage archéologique) ; *g* **Grabungs-gebiet [n]** (Kulturhistorisch, archäologisch wertvolles Gebiet für Ausgrabungsarbeiten oder in dem Grabungen stattfinden; ▶Such-graben); *syn.* Grabungsort [m].

224 archaeological probability area [n] [US] *conserv' hist. leg.* (Delineated area of archeological interest having a high probability of containing artifacts of historical or prehistoric interest marked on archaeological probability maps); *syn.* protec-ted archaeological area [n] for future digging [UK]; *s* **zona [f] arqueológica** (Zona arqueológica es el lugar o paraje natural donde existen bienes muebles o inmuebles susceptibles de ser estudiados con metodología arqueológica, hayan sido o no extraídos y tanto si se encuentran en la superficie, en el subsuelo o bajo las aguas territoriales españolas; Art. 15 [5], Ley 16/1985); *f* **site [m] archéologique classé** (Zone protégée classée monu-ment historique renfermant ou susceptible de renfermer des monuments, des ruines, substructions, mosaïques, éléments de canalisation antique, vestiges d'habitation ou de sépultures anciennes, des inscriptions ou généralement des objets pouvant intéresser la préhistoire, l'histoire, l'art, l'archéologie ou la numismatique ; cf. loi du 27 septembre 1941) ; *g* **Grabungs-schutzgebiet [n]** (In D. per Rechts-VO zeitlich befristet oder unbefristet festgesetztes Gebiet, das begründete Vermutung nach Kulturdenkmalen von besonderer Bedeutung birgt, weshalb Arbeiten [Änderung der bisherigen Bodengestalt durch Abgra-bung, Auffüllung oder Aufschüttung], die diese Kulturgüter gefährden können, einer behördlichen Genehmigung, meist des Landesdenkmalamtes als der zuständigen Fachbehörde, bedürfen; Gebiete werden zu **G.n** durch untere Denkmalschutzbehörden per Rechts-VO erklärt; cf. z. B. § 22 DSchG-BW).

archaeotope [n] *conserv'hist. land'man.* ▶buried cultural monument.

225 arch culvert [n] *constr. trans.* (Drain conduit with a flat bottom to carry a stream beneath a path or road embankment; ▶culvert); *s* **alcantarilla [f] abovedada** (Apertura perpendicular bajo caminos o carreteras para dejar paso a pequeños cursos de agua; ▶alcantarilla); *syn.* atarjea [f] abovedada [Es], puentecillo [m] abovedado, puente [m] de alcantarilla abovedado, pontón [m] abovedado [Es, YV], tajea [f] abovedada [C], conducto [m] pluvial abovedado [RA]; *f* **passage [m] vouté** (Ouvrage inférieur perpendiculaire à la chaussée, en forme de demi-cercle permet-tant l'écoulement de petits cours d'eau ou le passage de la faune ; ▶passage busé 1) ; *g* **Gewölbedurchlass [m]** (Queröffnung bei

A

Verkehrstrassen für den Durchfluss von kleinen Wasserläufen; ►Durchlass).

archeological area [n] *conserv'hist.* ►archaeological excavation area.

archeological dig [n] [US] *conserv'hist.* ►archaeological excavation.

archeological dig area [n] [US] *conserv'hist.* ►archaeological excavation area.

archeological excavation area [n] *conserv'hist.* ►archaeological excavation area.

archeological probability area [n] [US] *conserv' hist. leg.* ►archaeological probability area [US]/protected archaeological area for future digging [UK].

226 arching [adj] *hort. plant.* (Descriptive term for overhanging of foliage, branches, flower stalks, etc.; ►overhanging growth); *syn.* cascading [adj], overhanging [adj]; *s* **arqueado/a [adj]** (Término descriptivo para ramas, inflorescencias, tallos de gramíneas que crecen formando curvas elegantes; aplicable también a formas obligadas de crecimiento; ►ramaje sobresaliente); *syn.* colgante [adj], llorón, -a [adj]; *f* **courbé, ée [pp/adj]** (p. ex. les longs rameaux souples gracieusement courbés ; ►empiétement de branches) ; *g* **bogig überhängend [ppr/adj]** (1. So beschaffen, dass z. B. Blütenstände, Grashalme über ihre eigene Grundfläche in Form eines Bogens hinausragen. 2. So beschaffen, dass z. B. Zweige, Ranken, Blätter vieler Gräser bogenförmig wachsen; ►Überwuchs).

227 arching ornamental grass [n] *hort. plant.* (Medium to high or extra tall grass); *s* **hierba [f] ornamental llorona** (Gramínea de porte medio a alto con hojas arqueadas o infructescencias colgantes); *f* **graminée [f] ornementale à longues feuilles et hampes retombantes** (Graminée de moyenne et grande taille) ; *g* **Ziergras [n] mit überhängenden Blättern und Blütenständen** (Mittelhohes bis hohes oder Riesengras mit bogig überhängenden Blättern und Blütenständen); *syn.* Schmuckgras [m] mit überhängenden Blättern und Blütenständen.

archipelago [n], **heat** *met. urb.* * ►heat island.

228 architect [n] *prof.* (1. Person skilled and experienced in the art of building and the coordination and supervision of construction sites. 2. Legally-based professional designation for a person who is qualified and licensed to perform architectural services; ►landscape architect, ►Council of Architectural Registration Boards [CARB] [US]/Architects Registration Council [UK] and ►Council of Landscape Architectural Registration Boards [US]); *s* **arquitecto/a [m/f]** (1. Persona formada en el arte de la construcción y de la coordinación y supervisión de obras; ►arquitecto/a paisajista. 2. Término legal para denominar a profesional calificado/a y con licencia para ofrecer servicios de arquitectura; ►Colegio Oficial de Arquitectos); *f* **architecte [m]** (1. Dénomination professionnelle et juridique désignant une personne qualifiée dans l'art de bâtir, qui dessine les plans des bâtiments, dresse les devis et supervise la construction. 2. F., seules les personnes physiques inscrites à un tableau régional d'architectes peuvent porter le titre d'architecte et ont le droit d'exercer sur l'ensemble du territoire national ; [loi n° 77-2 du 3 janvier 1977] ; mission et devoirs de l'architecte cf. décret n° 80-217 du 20 mars 1980 ; ►architecte paysagiste, ►ordre des architectes ; *g* **Architekt/-in [m/f]** (1. OB zu Hochbauarchitekt/-in, ►Landschaftsarchitekt/-in und Innenarchitekt/-in. 2. Jemand, der in dem Fachgebiet der Baukunst ausgebildet ist, Bauentwürfe erstellt und deren Ausführung überwacht. 3. **D.,** gesetzlich geschützter Titel, den nur diejenigen führen dürfen, die in der ►Architektenkammer eingetragen sind); *syn. obs. zu 2.* Baumeister [m].

architect [n] **in a public authority** [UK], **senior landscape** *prof.* ►chief landscape architect in a public authority/agency [US].

architect [n] **in charge of a branch office** [US] *prof.* ►landscape architect in charge of a branch office [US].

architect [n] **in private practice** *prof.* ►landscape architect in private practice, ►senior landscape architect in private practice.

architect principal [n] *prof.* ►landscape architect principal.

229 architect's fee [n] *prof.* (►professional fee 2); *syn.* professional fee [n] (1); *s* **honorarios [mpl] de arquitecto/a** (►honorarios profesionales); *f* **honoraire [m] d'architecte** (►honoraire) ; *syn.* rémunération [f] ; *g* **Architektenhonorar [n]** (►Honorar).

Architects Registration Council [n] [UK] *prof.* ►Council of Architectural Registration Boards (CARB) [US].

230 architectural register [n] *prof.* (In U.S., separate lists of registered architects, landscape architects, engineers, land surveyors and sometimes interior designers in each State; **in U.K.**, the a. r. does not include landscape architects or interior designers; ►Architects Registration Council; **in D.**, list of registered architects, landscape architects, and interior designers in each "Bundesland"); *syn.* register [n] of architects; *s* **registro [m] de arquitectos** (en el ►Colegio Oficial de Arquitectos); *f* **registre [m] des architectes** (F., cette liste ne concerne que les architectes en bâtiment ; **D.,** cette liste inclue les architectes paysagistes et les architectes d'intérieur ; ►ordre des architectes) ; *g* **Architektenliste [f]** (Mitgliederverzeichnis der Hochbau-, Landschafts- und Innenarchitekten bei einer ►Architektenkammer).

231 architecture [n] *arch.* (1. Art and science of building. 2. Sum of architectural structures characterizing an epoch, style, or cultural sphere); *s* **arquitectura [f]** (1. Arte de la construcción como disciplina científica. 2. Conjunto de realizaciones constructivas artísticas de una época, de un estilo, de una cultura; *syn.* estilo [m] arquitectónico. 3. Forma y carácter de un edificio); *f* **architecture [f]** (1. Art de construire. 2. Ensemble des réalisations artistiques d'une époque, d'un style. 3. Forme et caractère d'un édifice) ; *g* **Architektur [f, o. Pl.]** (1. Baukunst als wissenschaftliche Disziplin. 2. Gesamtheit der Erzeugnisse der Baukunst einer Zeitepoche, eines Stils, eines Kulturbereiches; *syn.* Baustil [m]).

arctic prairie [n] *geo. phyt.* ►tundra.

area [n] *recr.* ►agricultural area; *hydr.* ►aquifer recharge area; *conserv'hist.* ►archaeological excavation area, ►archaeological probability area [US]/protected archaeological area for future digging [UK]; *geo. plan.* ►avalanche source area; *recr.* ►bathing area; *urb.* ►blighted area; *phyt.* ►brushy area [US]; *plan. urb.* ►buildable area, ►building area, ►built-up area (1), ►built-up area (2); *adm.* ►burial area; *wat'man.* ►capture area; *rem'sens. surv.* ►check area; *plan.* ►clean air area; *for.* ►clear cutting area [US]; *phyt.* ►climax area; *geo. recr.* ►coastal area; *met.* ►cold air source area; *leg. urb.* ►commercial and light industry area; *conserv. leg.* ►compensation area (1); *prof.* ►competition area; *leg.* ►convention area; *conserv. leg. urb.* ►core area (1), ►core area (2); *plan. recr.* ►countryside recreation area; *wat'man.* ►drinking water catchment area; *conserv. nat'res.* ►Important Birds Area; *adm. pol. urb.* ►local community area; *plan.* ►open space compensation area; *leg. urb.* ►permanent community garden area [US]; *plan. urb.* ►populated area; *agr. plan.* ►priority agricultural area; *plan. recr.* ►rest area; *conserv. ecol.* ►special area of conservation (SAC).

area [n] [US], **allowable floor** *leg. urb.* ▶floor area ratio (FAR) [US]/floor space index [UK].

area [n], **archaeological dig** *conserv'hist.* ▶archaeological excavation area.

area [n] [US], **blowdown** *arb.* ▶windfall area.

area [n] [US], **buffer** *conserv. landsc. urb.* ▶buffer zone, ▶fringe area.

area [n], **burned forest** *for.* ▶forest burn.

area [n], **catchment** *plan.* ▶suburban area; *hydr. nat'res.* ▶drainage basin.

area [n], **catchment** [UK] *recr. urb.* ▶origination area [US].

area [n], **conservation** *arch. conserv' hist. leg.* ▶historic district [US].

area [n], **control** *rem'sens. surv.* ▶check area.

area [n], **clearance** *for.* ▶clear cutting area [US].

area [n], **clear felling** *for.* ▶clear cutting area [US].

area [n] [UK], **development** *leg. urb.* ▶development district [US].

area [n], **exploitable** *agr. for. hort.* ▶area of yield.

area [n], **intake** *hydr.* ▶aquifer recharge area.

area [n], **intensive recreation** *plan. recr.* ▶concentration of recreation facilities.

area [n] [UK], **motorway service** *trans.* ▶service area with all facilities.

area [n], **parking** [US] *trans. urb.* ▶car park (1).

area [n], **peripheral built** *urb* ▶peripheral building development.

area [n], **planting** *constr. landsc.* ▶area of planting.

area [n], **rotted** *arb.* ▶area of rot.

area [n] [UK], **rural settlement** *leg. urb.* ▶rural district [US].

area [n] [UK], **settlement** *plan. urb.* ▶developed area [US].

area [n], **special protection** *conserv. ecol.* ▶special area of conservation (SAC).

area [n], **suburban** *plan.* ▶commuter belt.

area [n] [US], **swimming** *recr.* ▶bathing area.

area [n] [US], **transitional** *conserv. landsc. urb.* ▶buffer zone.

area [n], **vegetated** *constr.* ▶area of vegetation.

area action plan [n] [UK] *leg. urb.* ▶zoning map [US].

area curve [n] *phyt.* ▶species area curve.

232 area [n] **for concentrated leisure activities** *recr.* (Area providing opportunities for concentrated leisure-time or recreational activities); *s* **zona [f] de recreo intensivo** (Áreas a las afueras de las aglomeraciones urbanas equipadas para actividades de recreo intensivas); *f* **espace [m] de loisirs dense** (Espace ludique éloigné des zones de forte concentration urbaine) ; *syn.* zone [f] de loisirs dense ; *g* **Erholungsgebiet [n] für intensive Erholung** (Gebiet außerhalb städtischer Ballungsräume für intensive Erholungsnutzung).

area [n] **for industrial use** *leg. urb.* ▶industrial use zone.

233 area [n] **for throwing the hammer** *recr.* (Sports ground where a metal ball of about 7 kg attached to a wire is thrown in an athletic contest); *syn.* facilities [npl] for throwing the hammer; *s* **área [f] de lanzamiento de martillo**; *f* **installation [f] pour le lancer du marteau** (Installation sportive utilisée pendant une manifestation d'athlétisme pour la discipline du lancer du marteau) ; *g* **Hammerwurfanlage [f]** (Einrichtung auf Sportplätzen zur Durchführung der leichtathletischen Disziplin Hammerwerfen).

234 area [n] **intended for general recreational use** (±) *recr.* (Area with scattered settlement providing opportunities for dispersed leisure-time or recreational activities; ▶recreation in the naturel environment); *s* **zona [f] de recreo extensivo** (Área de poca densidad de poblamiento adecuada para usos recreativos dispersos; ▶recreo extensivo); *f* **espace [m] de récréation diffuse** (Espace ludique éloigné des zones de forte concentration urbaine ; ▶loisirs en milieu naturel) ; *syn.* zone [f] de récréation diffuse ; *g* **Gebiet [n] zur extensiven Erholung** (Gebiet außerhalb städtischer Ballungsräume für extensive Erholungsnutzung; ▶naturnahe Erholung); *syn.* Erholungsgebiet [n] für extensive Freizeitnutzung.

area [n] **of a stream/brook** *geo.* ▶flood area of a stream/brook.

area [n] **of conservation** *conserv. ecol.* ▶special area of conservation (SAC).

area [n] **of decay** *arb.* ▶decayed area.

235 Area [n] **of Outstanding Natural Beauty** [UK] *conserv. leg.* (**1.** *Abbr.* AONB; one of two types of special landscape protection areas created by the National Parks and Access to the Countryside Act 1949; ▶Sites of Special Scientific Interest; an **AONB** is a particularly beautiful landscape, whose distinctive character and natural beauty are so special, that it is in the national interest to protect them. They are designated because of their high qualities of, for example, flora, fauna, historical and cultural associations as well as scenic views and include coastlines, meadows, downland and upland moors.There are 40 **AONBs** in England and Wales: 35 in England, 4 in Wales and one which reaches across the border; together they represent 18% of the finest countryside in England and Wales. There are 9 AONBs in Northern Ireland with a further two proposed. The largest AONB is the Cotswolds with 2038km²; the smallest is the Isles of Scilly, 16 km². Their care is entrusted with local authorities, organisations, community groups and individuals, who live and work within them. The Countryside and Rights of Way Act, 2000—the □CRoW" Act]—added further regulation and protection, ensuring the future of AONBs as important national resources; **in U.S.**, no precise legal equivalent, although somewhat the same result is often achieved through related state or local legislative efforts; ▶natural area preserve, ▶wilderness area); *syn.* Resource Conservation District [n] (RCD) [UK]. **2. In D.**, the legal generic term *Landschaftsschutzgebiet* [landscape protection area] is used for an area of land, in which special protection of natural resources is necessary for the preservation or restoration of the ▶productivity 1 and efficiency of ▶natural systems or the potential use of natural resources, due to the diversity, unique character or beauty of the landscape, or because of its special significance for recreation); *syn.* site [n] of special scientific interest [UK]; *s 1* **Área [m] de Gran Belleza Natural** (En GB, según la Ley de Parques Nacionales y Acceso al Campo de 1949, término legal de una de las dos categorías de protección del paisaje debido a su gran belleza natural, a su carácter o a su significado ecológico, científico, histórico o nacional, siendo la otra ▶parque nacional; otras áreas protegidas son «parajes de especial interés científico»; en EE.UU. no existe un equivalente legal, aunque existen regulaciones a nivel de estados o local que persiguen los mismos fines; ▶régimen ecologico de la naturaleza); *s 2* **paisaje [m] protegido** (En Es. término legal para parte del territorio que las Administraciones competentes, por sus valores naturales, estéticos y culturales, y de acuerdo con el Convenio del paisaje del Consejo de Europa, consideren merecedor de una protección especial; art. 34 Ley 16/2007; ▶espacio natural

A

protegido 1, ►espacio natural protegido 2); *f 1* **site [m] naturel à caractère pittoresque exceptionnel (≠)** (**U.K.**, une des deux catégories de préservation des milieux, des espèces, des paysages ou du patrimoine créée dans le cadre de la législation du National Parks and Access to the Countryside Act de 1949 et caractérisant les sites à sauvegarder pour la valeur de leurs paysages remarquables et originels, pour leur faune et leur flore mais également pour leur patrimoine historique et culturel [site fort de caractère historique et culturel] ; ►monument naturel, ►réserve naturelle de flore et faune sauvage, ►site classé) ; *syn.* ± *littéral* zone [f] d'une beauté naturelle exceptionnelle (≠), site [m] naturel exceptionnel (≠) ; *f 2* **site [m] inscrit** (**F.**, site ou monument naturel protégé au titre de l'art. 4 de la loi du 2 mai 1930 [relève depuis le 20 février 2004 du code du patrimoine] dont la conservation ou la préservation présente du point de vue artistique, historique, scientifique, légendaire ou pittoresque un intérêt général ; l'inscription à l'inventaire des sites a pour effet l'obligation de ne pas procéder à des travaux autres que ceux d'exploitation courante pour les fonds ruraux ou d'aviser l'administration pour les constructions ; ►système naturel, ►site classé, ►zone protégée) ; *g* **Landschaftsschutzgebiet [n] (±)** (*Abk.* LSG; **D.**, rechtsverbindlich festgesetztes Gebiet, in dem ein besonderer Schutz von Natur und Landschaft zur Erhaltung oder Wiederherstellung der Leistungsfähigkeit des ►Naturhaushalts oder der Nutzungsfähigkeit der Naturgüter, wegen der Vielfalt, Eigenart oder Schönheit des Landschaftsbildes, wegen seiner besonderen Bedeutung für die Erholung oder wegen seiner besonderen kulturhistorischen Bedeutung erforderlich ist. Diese Schutzgebietskategorie hat im Vergleich zu anderen Schutzgebieten einen verhältnismäßig schwachen Schutzstatus. Die Regierungspräsidien/Bezirksregierungen als obere Naturschutzbehörden oder andere zuständige Stellen der einzelnen Bundesländer, auch untere Naturschutzbehörden, erlassen Verordnungen, die u. a. die Erklärung zum Schutzgebiet, den Schutzgegenstand und Schutzzweck, Verbote und Erlaubnisvorbehalte resp. zulässige Handlungen beinhalten [Schutzgebietsverordnung]. **2.** Die britische Schutzkategorie *Area of Outstanding Natural Beauty* gibt es in D. nicht; sie kann in etwa mit dem dt. **L.** verglichen werden; ►Naturschutzgebiet 2, ►Schutzgebiet).

236 area [n] **of outstanding scenic beauty** *landsc. recr.* (Area significant for its recognized natural beauty); *s* **paraje [m] natural de gran belleza escénica** (Área destacada por su belleza natural reconocida); *f* **espaces [mpl] les plus remarquables du patrimoine naturel et culturel** (Notamment les gorges, grottes, glaciers, lacs, tourbières, marais, lieux de pratique de l'alpinisme, de l'escalade et du canoë-kayak, etc. ; cf. C. urb. art. L. 145-7) ; *syn.* paysage [m] le plus remarquable du patrimoine naturel et culturel, paysage [m] présentant un intérêt esthétique particulier, paysage [m] à caractère pittoresque, paysage [m] présentant un intérêt visuel particulier ; *g* **landschaftlich besonders schönes Gebiet [n]** (Gebiet, das durch seine Vielfalt, Eigenart und Schönheit von Natur und Landschaft geprägt ist).

237 area [n] **of planting** *constr. landsc.* (**1.** Area designated to receive plants on a planting plan. **2.** *Result of a planting process* planted area; ►plantation 1); *syn.* planting area [n]; *s* **plantío [m]** (**1.** En un ►plan de plantación, área designada para ser plantada. **2.** Resultado de un proceso de plantación; ►plantación 1); *syn.* área [f] de plantación, zona [f] de plantación; *f 1* **aire [f] de plantation** (Espace prévu pour une plantation) ; *syn.* aire [f] à planter, aire [f] des plantations, surface [f] à planter ; *f 2* **aire [f] plantée** (Résultat d'une ►plantation 1) ; *syn.* surface [f] plantée, plantations [fpl] ; *g* **Bepflanzungsfläche [f]** (**1.** Fläche, die bepflanzt werden soll. **2.** *Ergebnis* ►Pflanzung 1; *syn. zu 2.* bepflanzte Fläche [f].

238 area [n] **of regeneration** *conserv. land'man.* (Area set aside for the restoration of a disturbed or destroyed landscape element [e.g. forest or raised bog] or for an impaired population of animals and plants; e.g. by ►reintroduction, limitation or prohibition of certain land uses); *s* **área [f] de regeneración** (Superficie para restaurar elementos del paisaje—como bosques, turberas— destruidos o degenerados o para restablecer poblaciones de especies de fauna por medio de sueltas masivas, prohibición de usos, etc.; ►reintroducción de especies extintas); *f* **site [m] de régénération** (Aire prévue pour le rétablissement d'éléments de paysages gravement menacés ou détruits, p. ex. espaces forestiers, tourbières hautes ou de populations animales et de peuplements végétaux, p. ex. la ►réintroduction 2 d'espèces, restriction ou interdiction de certaines activités) ; *syn.* aire [f] de régénération ; *g* **Regenerationsfläche [f]** (Fläche zur Wiederherstellung von gestörten oder zerstörten Landschaftselementen [z. B. Wald, Hochmoor] oder von gestörten Tier- und Pflanzenbeständen, z. B. durch ►Wiederansiedelung, Nutzungsbeschränkung oder -verbot); *syn.* Regenerationsraum [m].

239 area [n] **of rot** *arb.* (Extensive, well-advanced patch of decay on trunks and main branches of trees; ►decayed area); *syn.* rotted area [n]; *s* **caries [f] húmeda** (Podredumbre avanzada en el tronco o en rama principal; ►zona podrida); *f 1* **pourrissement [m]** (*Processus* décomposition progressive des tissus du bois ; *terme spécifique* **pourrissement interne** : c'est le dépérissement important de l'aubier et du bois parfait pouvant s'étendre à toute la charpentière ou à tout le tronc avec possibilité de formation d'une cavité ; ►point de pourriture [sur le tronc]) ; *f 2* **pourriture [f]** (Résultat du pourrissement des tissus du bois : bois pourri) ; *g* **Morschung [f]** (Umfangreiche, weit fortgeschrittene ►Faulstelle im Stamm- und Starkastbereich. Wenn die **M.** von einer verletzten Rinde ausgeht, spricht man von einer **offenen Fäule**; je nach Art der Holzzersetzung unterscheidet man **Weiß-, Braun- und Moderfäule**; *syn.* Fäule [f], Holzfäule [f].

240 area [n] **of vegetation** *constr.* (Area covered by existing vegetation or proposed to be planted, as opposed to hard-surfaced area; ►sofscape); *syn.* vegetated area [n]; *s* **área [f] de vegetación** (Superficie prevista de vegetación o reservadas para vegetación; ►superficie blanda); *syn.* superficie [f] de vegetación; *f 1* **surface [f] végétale** (Espace planté ou prévu pour une plantation future par opposition aux surfaces minérales ou aux surfaces stabilisées ; ►éléments vivants d'un espace libre) ; *syn.* espace [m] de végétation, espace [m] végétal ; *f 2* **surface [f] végétalisée** (Aire recouverte par une végétation mise en place artificiellement en général résultat des ►techniques de végétalisation, p. ex. pour un talus, une toiture-terrasse) ; *g* **Vegetationsfläche [f]** (Flächen mit vorhandener Vegetation oder für Vegetation vorgesehen; im Gegensatz zu befestigten Flächen; ►vegetationsbestimmte Freianlage).

241 area [n] **of water** *landsc. plan.* (Portion of a land area covered by water bodies); *s* **superficie [f] acuática** (Parte de una zona que está ocupada por cursos o cuerpos de agua); *syn.* superficie [f] de agua; *f* **surface [f] aquatique** (Étendue d'eau recouvrant un territoire) ; *g* **Wasserfläche [f] (1)** (Gewässerflächenanteil in einem Gebiet).

242 area [n] **of yield** *agr. for. hort.* (Area serving for primary production, i.e. agriculture, horticulture or forestry; ►agricultural production land; ►designated land-use area); *syn.* exploitable area [n]; *s* **superficie [f] útil** (Área utilizada para la producción primaria, es decir, agricultura, ganadería y silvicultura; ►superficie agrícola útil; ►area destinada a un uso); *f* **superficie [f] exploitable** (Surface exploitée pour la production primaire ; ►superficie agricole utilisée, ►zone affectée [à un usage]) ; *g* **landwirtschaftliche Produktionsfläche [f]** (Fläche, die der

Urproduktion dient; ►landwirtschaftliche Nutzfläche, ►Nutzungsfläche).

243 area ownership map [n] [US] *adm. surv.* (Cadastral map showing the distribution of parcels of land on a large scale 1:500-1:5,000; ►metes and bounds map [US]/land registration map [UK]); *syn.* land registry map [n] [UK]; *s* **plano [m] catastral** (Representación planimétrica en la que aparecen los límites de las parcelas con su identificación numérica, así como los de las subparcelas y polígonos con indicación de sus aprovechamientos agrarios. Los polígonos son agrupaciones de parcelas cuyos límites siguen elementos permanentes del paisaje rural. En Es., la superficie catastral de un municipio se divide en varios polígonos, oscilando su escala cartográfica entre 1:500-1:5000, según el tamaño y el número de las parcelas donde se representan los límites y los propietarios de los terrenos; DGA 1986; ►mapa catastral); *syn.* plano [m] parcelario; *f* **parcellaire [m] rural** (F., plan à grande échelle [M 1:10 000] sur lequel sont assemblées toutes les parcelles de propriétés du territoire communal ainsi que leur numérotation ; l'informatisation des fichiers du cadastre et la constitution d'une banque de données devrait permettre d'en élargir considérablement son utilisation ; **plan terrier :** plan des propriétés seigneuriales ayant précédé le plan cadastral dans certaines régions ; **D.,** ensemble de la division du sol et représentation cartographique des biens-fonds sous forme d'un plan à grande échelle [de 1:500 à 1:5000] ; ►plan cadastral) ; *syn.* cadastre [m] parcellaire, plan [m] parcellaire ; *g* **Flurkarte [f]** (Kartenmäßige Darstellung der Liegenschaften mit Besitzgrenzen und Eigentumsverhältnissen als Grundrissbild der Flurstücke im Maßstab 1:500 bis 1:5000. Die Gesamtheit der **F.n** eines Gebietes wird **Flurkartenwerk** genannt; ►Katasterkarte).

area planning [n] *plan.* ►metropolitan area planning, ►overall area planning; *landsc. plan. pol. recr.* ►recreation area planning, *plan.* ►rural area planning [US]/countryside planning [UK].

244 area study survey [n] *constr.* (In U.S., investigation and written report of project conditions; in U.K., assessment and recording of conditions within a study area; *s* **investigación [f] sobre terreno** (Relevamiento de un área pequeña p. ej. para preparar un proyecto de construcción); *f* **relevé [m]** (Lever de l'état des lieux de l'état actuel d'un terrain, des éléments au sol rencontrés, du relief constituant le document de base préparatoire aux études d'un projet de travaux) ; *g* **Aufnahme [f] (1)** (Bestandsaufnahme im kleinräumigen Bereich zur Planungsvorbereitung eines Bauprojektes oder eine Bestandserhebung im Rahmen der ►Grundlagenermittlung).

245 area [n] **subject to mining subsidence** *min.* (Caved-in land surface resulting from deep mining; ►ground settlement); *s* **zona [f] de subsidencia [Es, CO]** (Zona que está bajo la influencia de trabajos de minería subterránea; ►asentamiento, ►subsidencia); *syn.* zona [f] de hundimiento [CO, EC, MEX, RA, YV], zona [f] de descenso [RCH]; *f 1* **zone [f] de subsidence minière** (Affaissement progressif du niveau du terrain provoquée par l'activité minière souterraine ; ►subsidence, ►tassement) ; *f 2* **zone [f] d'effondrement minier** (Écroulement brutal du terrain) ; *syn.* zone [f] d'affaissement minier ; *g* **Bergsenkungsgebiet [n]** (Durch Bruchbergbau verursachtes Absinken der Geländeoberfläche; ►Senkung).

area vegetation [n] *phyt.* ►existing area vegetation.

arenicolous [adj] *phyt. zool.* ►psammophilous.

argillaceous [adj] *pedol* ►clayey.

argillic [adj] *pedol* ►clayey.

argillic horizon [n] [US] *pedol.* ►claypan.

argillous [adj] *pedol* ►clayey.

246 Århus convention [n] *adm. envir. leg. plan. pol.* (The Economic Commission for Europe convention, signed by 39 states on 25.06.1998 in Denmark, formulates the requirement of authorities to allow citizens access to environmental information and lays down rules for public participation in individual projects, plans, programs and political guidelines as well as objection procedures and rights of appeal in environmental affairs); *e* **Convenio [m] de Aarhus** (El convenio sobre el acceso a la información, la participación del público en la toma de decisiones y el acceso a la justicia en materia de medio ambiente, firmado el 25 de junio de 1998 en Dinamarca por 39 Estados y la UE, fue ratificado por España en diciembre de 2004 y entró en vigor el 31 de marzo de 2005. A través de la Ley 27/2006 se incorporan las Directivas 2003/4/CE y 2003/35/CE y se establece el deber general de promover la participación real y efectiva del público, sin regular procedimiento alguno, ya que se trata de un ámbito de competencia compartida con las CC.AA. En la ley se definen los derechos de acceso a la información ambiental, de participación pública en asuntos de carácter ambiental y el acceso a la justicia y a la tutela administrativa en asuntos medioambientales); *f* **convention [f] d'Århus** (La **Convention sur l'accès à l'information, la participation du public concerné au processus** décisionnel [plans, programmes et politiques relatifs à l'environnement] et l'accès à la justice en matière **d'environnement** [droit d'engager des procédures administratives ou judiciaires et possibilité dans les procédures de former un recours suffisant et effectif devant une instance judiciaire], signée le 25 juin 1998 au Danemark par 39 États et approuvée par la loi n°2002-285 du 28 février 2002 puis annexée au décret de publication du 12 septembre 2002, est entrée en vigueur le 6 octobre 2002) ; *g* **Århus-Konvention [f]** (Die ECE-Konvention *[Economic Commission for Europe]*, von 39 Staaten am 25.06.1998 in Dänemark unterzeichnet, formuliert den Anspruch auf Zugang der Bürger zu Umweltinformationen gegenüber den Behörden und regelt die Öffentlichkeitsbeteiligung bei einzelnen Vorhaben, Plänen, Programmen und politischen Vorgaben sowie die Widerspruchsverfahren und Klagerechte in Umweltangelegenheiten; BUN 2007).

arid grassland [n] *phyt.* ►dry meadow.

247 aridity indicator plant [n] *phyt.* (Plant which enjoys optimum ecological conditions on dry soils; ►indicator plant *opp.* ►wetness indicator plant); *s* **planta [f] indicadora de sequía** (Especie vegetal cuyas condiciones ecológicas óptimas se encuentran en suelos secos, por lo que se utiliza de ►planta indicadora; *opp.* ►planta indicadora de encharcamiento); *f* **plante [f] indicatrice de sécheresse** (Espèce végétale atteignant son meilleur rendement sur des sols secs ; ►plante indicatrice ; *opp.* ►plante indicatrice de forte humidité) ; *syn.* plante [f] indicatrice de sol sec ; *g* **Trockenheitszeiger [m]** (Bodenpflanze, die auf trockenen Böden ihr ökologisches Optimum hat; ►Zeigerpflanze; *opp.* ►Nässezeiger).

248 arid region [n] *geo.* (Area of the world with an extremely dry climate where average annual total evaporation is greater than the average annual amount of precipitation. **A. r.s.** comprise almost a third of the world's land surface area; ►savanna[h], ►steppe; ►semidesert [US]/semi-desert [UK]); *s* **zona [f] árida** (Parte de la tierra que sufre períodos de sequía, es decir en la que la evaporación anual media excede a la precipitación anual media. Las **z. á.** ocupan un tercio de la superficie de la tierra. *Se dividen en* **z. hiperárida** [desierto], **z. árida** [►semidesierto, subdesierto], **z. semiárida** [►estepa, ►pradera, ►sabana, vegetación mediterránea] y **z. subhúmeda** [sabana tropical, maquis, chaparral, estepa de chernozem]; DINA 1987; ►semidesierto); *syn.* región [f] árida); *f* **région [f] aride** (Région à climat aride, à pluies occasionnelles mais survenant chaque année, sur laquelle

A

l'évaporation totale moyenne annuelle est supérieure aux précipitations moyennes annuelles, comme p. ex. dans les régions de ►savane, de ►steppe ou dans les déserts ; les régions arides représentent 1/3 de la surface des terres du globe ; ►région semi-aride) ; *syn.* zone [f] aride ; *g* **Trockengebiet [n]** (Erdraum mit aridem Klima, d. h. ein Gebiet, bei dem die mittlere jährliche Gesamtverdunstung größer als die mittlere jährliche Niederschlagsmenge ist, wie z. B. in der ►Savanne, ►Steppe oder Wüste. In **T.en** verdunsten Wasserläufe von der Quelle bis zur Mündung und enden oft in abflusslosen Seen oder Salzpfannen; **T.e** nehmen nahezu ein Drittel der Landfläche der Erde ein; ►Halbwüste); *syn.* arides Gebiet [n].

arid sward [n] *phyt.* ►dry meadow.

249 aromatic garden [n] *gard. plant.* (Garden predominantly planted with sweet-smelling herbs and shrubs; ►spices garden); *s* **jardín [m] de plantas aromáticas** (Jardín ornamental sobre todo con hierbas y arbustos aromáticos; ►jardín de hierbas condimenticias); *f* **jardin [m] de senteurs** (Jardin d'agrément privilégiant l'utilisation d'espèces odoriférantes ; ►jardin d'herbes) ; *syn.* jardin [m] d'herbes aromatiques (±) ; *g* **Duftgarten [m]** (Ein vorwiegend mit duftenden Kräutern und Sträuchern gestalteter Ziergarten; ►Gewürzgarten); *syn.* Duftpflanzengarten [m].

aromatic plant [n] *gard. plant* ►fragrant plant.

250 arris rail [n] *constr.* (Horizontal piece of wood upon which the vertical boards are fastened in constructing a ►wooden fence); *s* **traviesa [f]** (Listón longitudinal para sujetar las tablas verticales de una ►valla de madera); *f* **barreau [m]** (Barre de bois horizontale servant de fixation des lattes d'une ►clôture en bois) ; *g* **Riegel [m]** (Querholz bei ►Holzzäunen zur Befestigung der Senkrechtlattung); *syn.* Querriegel [m]).

arterial highway [n] [US] *trans.* ►major highway [US] (1)/major road [UK], ►arterial road.

251 arterial road [n] *trans.* (Main road linking densely-populated cities and towns); *syn.* trunk road [n] (1), arterial highway [n] [also US], major road [n], traffic artery [n] [also UK]; *s* **carretera [f] principal** (Conexión para la circulación de vehículos que enlaza ciudades importantes entre sí); *f* **route [f] à grande circulation** (Voirie à fort trafic en rase campagne) ; *g* **Hauptverkehrsstraße [f] (1)** (Für den Verkehr wichtige Straße außerhalb von Ortschaften).

artificial fertilizer [n] [UK] *agr. constr. for. hort.* ►chemical fertilizer.

artificialization [n]**, level of** *ecol.* ►degree of landscape modification.

252 artificial landform [n] *constr. eng.* (Volume of dumped [US]/tipped [UK] excavated material, building rubble or demolition material sloped and shaped for various visual and physical purposes; ►large spoil area [US]/large tip [UK], ►dumpsite [US]/tipping site [UK]); *s* **montículo [m] artificial** (Forma artificial del terreno resultante de depositar material de desmonte o de escombros de construcción; ►depósito de residuos sólidos, ►escombrera); *f* **dépôt [m] de déchets inertes** (►Amoncellement de terre, de matériaux provenant de chantiers sans mélange avec d'autres déchets ; ►décharge publique) ; *syn.* dépôt [m] de déblai-gravats ; *g* **Auffüllberg [m]** (Künstliche Vollform in der Landschaft als Ergebnis einer Aufschüttung von Aushubmaterial, Bauschutt etc.; ►Deponie, ►Halde 1).

artificial pond [n] *constr.* ►man-made pond.

253 artificial raising [n] **of groundwater table** *hydr.* (Use of influent drainage strategies to increase the presence of groundwater in a designated area: a groundwater recharge strategy); *s* **elevación [f] artificial del nivel freático** (Aplicación de medidas para promover la infiltración de agua en el suelo y de tal manera recargar el correspondiente acuífero); *f* **rehaussement [m] de la nappe phréatique** (Élévation artificielle du niveau de la nappe) ; *g* **Hebung [f] des Grundwasserspiegels** (Künstliche Anhebung des Grundwasserspiegels); *syn.* Grundwasserspiegelhebung [f].

254 artificial streamlet [n] *constr. gard.* (Small man-made stream, which fills a pool in a public open space or private garden); *s* **canalillo [m] de alimentación** (Canal pequeño para alimentar un estanque en parque o jardín); *f* **ruisseau [m] artificiel** (Rigole aménagée dans un espace vert ou un jardin en vue de l'alimentation d'un bassin) ; *g* **künstlich angelegter Wasserlauf [m]** (Künstliches Rinnsal in einer Grünanlage oder im Hausgarten zur Speisung eines Wasserbeckens).

artistic quality [n] *constr. prof.* ►observation of artistic quality.

art [n] **of gardens** *gard. hist.* ►fine garden design.

255 as-built plan [n] *contr. plan.* (Final plan of a project reflecting all changes made during development; ►delivered plan [US]/handover plan [UK]); *s* **plano [m] de realización** (Plano final de un proyecto que representa exactamente todos los cambios realizados durante la obra; ►plan de entrega de obra); *f 1* **plan [m] de récolement (1)** (Document graphique nécessaire à la réception des travaux, fourni par les entreprises et concernant la conformité des travaux réalisés sur lequel sont représentées l'implantation, la destination, la nature, les dimensions des ouvrages exécutés ; ►plan de recolement 2) ; *syn.* plan [m] des ouvrages exécutés ; *f 2* **dessins [mpl] de récolement** (Ensemble des dessins établis dans le cadre des opérations de réception, faisant état des caractéristiques de l'ouvrage réalisé, lorsque celles-ci se trouvent différentes de celles figurant sur les dessins d'exécution) ; *g* **Bestandsplan [m] (1)** (Plan, der nach der Fertigstellung eines Bauvorhabens den gesamten Bestand lagegenau darstellt; ►Übergabeplan).

ascertainment [n] **of the final sum** [UK] *constr.* ►determination of final contract amount [US].

ashlar [n] [US] *arch. constr.* ►cut stone.

ashlaring [n] [US] *arch. constr.* ►ashlar masonry.

256 ashlar masonry [n] *arch. constr.* (Masonry wall or veneer composed of rectangular units of properly-bonded building stones, which have been sawn or dressed and laid in mortar; ►coursed ashlar masonry, ►coursed quarry-faced ashlar masonry, ►irregular coursed ashlar masonry, ►random rough-tooled ashlar masonry, ►rusticated masonry); *syn.* opus [n] quadratum, ashlar walling [n] [also UK], ashlaring [n] [also US]; *s* **sillería [f] de fábrica** (Obra de fábrica formada por sillares labrados, de cara vista, en hiladas con juntas finas. Si los sillares están repujados se llama ►sillería almohadillada; ►fábrica de sillarejo en hiladas, ►mampostería aparejada en hiladas irregulares, ►mampostería aparejada en hiladas regulares, ►sillería ordinaria con mampuestos de labra tosca); *syn.* mampostería [f] de sillares; *f* **maçonnerie [f] en moellons à opus quadratum** (Maçonnerie de moellons de taille à appareillage en opus quadratum et à joints de mortier, pierres de grandes dimensions assemblées en appareil régulier ; lorsque la face vue présente un aspect bosselé, il s'agit d'une ►maçonnerie en pierres bossagées ; ►maçonnerie à appareillage à l'anglaise en pierres bossagées, ►maçonnerie de moellons assisés à parement bossagé, ►maçonnerie de moellons assisés d'égales hauteurs) ; *g* **Quadermauerwerk [n]** (Mauerwerk aus Hau- oder Werksteinen [Quadersteine] in Mörtel mit meist glatten, parallelen Flächen; Lager- und Stoßfugen sind in ihrer gesamten Tiefe bearbeitet. Das **Q.** ähnelt in seinem Aussehen sehr dem ►regelmäßigen Schichtenmauerwerk. Haben die Ansichtsflächen Bossen, so handelt es sich um ein **Bossen-**

quadermauerwerk; ▶Bossenwerk, ▶Schichtenmauerwerk mit bossierten Steinen, ▶unregelmäßiges Schichtenmauerwerk, ▶Wechselmauerwerk mit bossierten Steinen); *syn.* Opus quadratum [n].

ashlar masonry [n]**, coursed dressed** *arch. constr.* ▶coursed dressed ashlar masonry.

ashlar walling [n] *arch. constr.* ▶ashlar masonry.

aspect [n] *met.* ▶orientation; *phyt. plant.* ▶seasonally changing visual dominance.

257 asphalt joint [n] *constr.* (Asphalt-filled space between adjacent pavement units; *generic term* ▶joint 2); **s junta [f] de bitumen** (▶junta); **f joint [m] au bitume** (*Terme générique* ▶joint) ; *syn.* joint [m] de bitume ; **g Bitumenfuge [f]** (Mit flüssigem Bitumen ausgefüllte ▶Fuge zwischen Pflastersteinen oder Belagsdeckenelementen).

assessment [n] *plan. landsc.* ▶visual landscape assessment.

assessment [n]**, aesthetic landscape** *plan. landsc.* ▶visual landscape assessment.

assessment [n]**, site** *plan.* ▶site evaluation.

assessment [n]**, visual tree** *adm. arb. leg.* ▶tree inspection.

assessment criterion [n] *plan.* ▶evaluation criterion.

assessment [n] **of a landscape, visual (impact)** *plan. landsc.* ▶visual landscape assessment.

assessment [n] **of executed works** [UK] *contr.* ▶determination of executed work [US].

assessment [n] **of programme and project impacts in terms of Natura 2000** *conserv. leg. plan.* ▶preliminary assessment of programme and project impacts in terms of Natura 2000.

assessment [n] **of scenic value** *plan. landsc.* ▶visual landscape assessment.

assessment panel [n] [UK] *prof.* ▶competition jury.

assessment protocol [n] *plan.* ▶evaluation method [US].

assessor [n] **of a competition jury** [UK] *prof.* ▶professional member of a competition jury [US].

258 asset [n] *conserv. leg.* (1. Generic term for a resource protected by statutory legislation, legal regulations, administrative decree or contractual agreement. Assets include the health of mankind, whether protected or in need of protection, and such ▶natural resources 1 as water, soils, air, flora and fauna and their habitats, as well as culturally important objects [▶cultural resource (US)/cultural asset (UK)] or mineral resources. The regulations or covenants prescribe measures necessary for the safeguarding, restoration of favo(u)rable conditions or for the sustained existence of an *a*. *NOTE* semantically there are differences between the German term *Schutzgut* and the English *asset*, whereby in the German word the necessity for protection of a resource is emphasized, whilst in English the word simply implies that the resource has a particular quality, which is potentially at risk; *syn.* conservation asset [n], asset [n] to be protected. 2. **Conservation asset base [n] [UK]:** several assets (factors) of a defined built and natural environment valued both for themselves and as a vital part of the quality of life, sense of place and distinctiveness); **s bien [m] a proteger** (Término genérico para un recurso protegido por legislación, decreto administrativo o acuerdo contractual. Los **b. a p.** incluyen la salud humana, protegida o necesitada de protección, y todos aquellos ▶recursos naturales como las aguas, el suelo, el aire, la flora y la fauna y sus hábitats, así como los objetos de valor cultural [bienes culturales] o los recursos minerales. Las regulaciones y convenios prescriben medidas necesarias para preservar,

restaurar las condiciones favorables o para permitir la existencia sostenible de un **b. a p.**); **f composante [f] environnementale** (Terme générique caractérisant les éléments physiques, biologiques et humains fondamentaux susceptibles [▶ressources fondamentales naturelles] d'être touchés par la réalisation d'un projet et qui sont à prendre en compte dans **1.** l'analyse de l'état initial du site et de son environnement et **2.** l'analyse et l'appréciation des effets d'un projet sur l'environnement dans les études environnementales ; les **c. e.** sont précisées dans le deuxième alinéa de l'article 2 du décret du 12 octobre 1977 qui vise nommément la faune et la flore, les sites et les paysages, le sol, l'eau, l'air, le climat, les milieux naturels et les équilibres biologiques, la protection des biens et du patrimoine culturel, la commodité de voisinage, l'hygiène, la santé, la sécurité et la salubrité publique ; cette liste n'est pas exhaustive, elle peut être plus importante mais surtout plus sélective en fonction d'une part de la nature des aménagements et ouvrages projetés et d'autre part des milieux et espaces concernés ; **g Schutzgut [n]** (…güter [pl]; OB für durch Gesetz, Rechts- oder Verwaltungsvorschriften oder vertragliche Vereinbarung geschützte oder zu schützende Gesundheit des Menschen, ▶natürliche Grundgüter wie Wasser, Boden, Luft, Pflanzen und Tiere mit ihren Lebensräumen, kulturell und historisch bedeutende Objekte [Kultur- und Sachgut] oder geschützter Bodenschatz. In den Vorschriften oder Vereinbarungen sind die Maßnahmen, die zur Wahrung, Wiederherstellung eines günstigen Erhaltungszustandes oder zum nachhaltigen Fortbestand dieses Gutes erforderlich sind, dargelegt. *ANMERKUNG* bei einer semantischen Betrachtung des deutschen Terminus *Schutzgut* und des englischen *asset* lassen sich Abweichungen dahingehend feststellen, dass bei **Sch.** der appellative Aspekt des Schützens betont wird, wohingegen *asset* etwas ist, das Qualität hat und gewissen Schadstoffen oder sonstigen Beeinträchtigungen ausgesetzt ist; *UBe* Naturschutzobjekt, Umweltschutzgut); *syn. z. T.* Schutzgegenstand [m], Schutzobjekt [n]).

asset [n] [UK]**, cultural** *conserv'hist.* ▶cultural resource [US].

asset base [n] [UK]**, conservation** *conserv. leg.* ▶asset, #2.

asset [n] **to be protected** *conserv. leg.* ▶asset.

assisted areas [npl] [UK] *plan. leg.* ▶development district [US].

association [n] *adm. plan.* ▶co-operative housing association; *phyt.* ▶ecotonal association; *adm. urb.* ▶nonprofit housing association [US]/non-profit housing association [UK]; *phyt.* ▶plant association.

association [n]**, rock-cleft** *phyt.* ▶rock crevice plant community.

association [n]**, snowpocket** *phyt.* ▶snowbed community.

association [n]**, substitute** *phyt.* ▶substitute community.

association [n] [US]**, water users** *nat'res. wat'man.* ▶water users agency [n] [US].

association [n] **of alpine belt** *phyt.* ▶initial pioneers association of alpine belt.

259 association [n] **of different communities** *phyt.* (Living together of a particular species or group of species in specific plant communities; **s asociación [f] en comunidad vegetal** (Habitación común de una especie o grupo de especies en una determinada comunidad vegetal); **f association [f] de communautés végétales** (Développement harmonieux d'une espèce ou d'un groupe d'espèces au sein d'une association) ; **g Vergesellschaftung [f]** (Zusammenleben einer betreffenden Art oder Gruppe von Arten in bestimmten Pflanzengesellschaften).

association [n] **of pavement joints** *phyt.* ►plant association of pavement joints.

association [n] **of scattered pioneer species** *phyt.* ►initial pioneers association of alpine belt.

at-grade intersection [n] [US] *trans.* ►at-grade junction.

260 at-grade junction [n] *trans.* (►Traffic intersection at the same ground level or elevation; *opp.* ►grade-separated junction); *syn.* at-grade intersection [n] [also US]; *s* **cruce** [m] **a nivel** (►Intersección de tráfico [Es]/intersección de tránsito [AL] a un nivel, de manera que éstas se cruzan directamente; *opp.* ►nudo sin cruces a nivel); *f* **intersection** [f] **routière à un niveau** (►Ouvrage d'intersection routière de différents voiries sur un niveau ; *opp.* ►intersection routière à plusieurs niveaux) ; *syn.* carrefour [m] ; *g* **plangleicher Knotenpunkt** [m] (►Kreuzungsbauwerk, bei dem sich der Verkehr nur in einer Ebene abwickelt; *opp.* ►planfreier Knotenpunkt); *syn.* höhengleicher Knotenpunkt [m], Knotenpunkt [m] in einer Ebene, plangleiches Straßenbauwerk [n].

at ground level [n] *met.* ►ground-level...

261 athletic field [n] *recr.* (Track and field sports area); *s* **pista** [f] **de deportes y atletismo** *syn.* cancha [f] de deportes y atletismo, arena [f] de deportes y atletismo; *f* **terrain** [m] **d'athlétisme** (Équipement sportif utilisé pour les compétitions d'athlétisme etles jeux sur gazon) ; *syn.* stade [m] d'athlétisme ; *g* **Kampfbahn** [f] (Sportliche Wettkampfanlage für Leichtathletik und Rasenspiele. In D. gibt es je nach Ausstattung die drei Typen A, B und C; cf. DIN 18 035, Teil 1).

atlas [n] *plan.* ►regional map atlas [US], ►regional planning atlas.

atlas [n]**, regional planning** *plan.* ►regional map atlas [US].

262 atlas [n] **of aerial maps** *rem'sens. plan.* (Bound compilation of aerial photographs of various parts of the Earth's surface); *s* **atlas** [m] **de foto(grafía)s aéreas** (Compilación encuadernada de fotos aéreas de diversas partes del mundo); *f* **atlas** [m] **de photographies aériennes** (Recueil de photographies aériennes représentant une partie de la surface terrestre établi par l'Institut géographique national [IGN]) ; *g* **Luftbildatlas** [m] (Zusammenstellung von Luftbildern über Teile der Erdoberfläche).

263 atlas [n] **of the environment** *plan.* (Compilation and graphic presentation of basic ecological data as a volume of maps covering existing, natural resources aimed at facilitating planning decisions according to the actual environmental situation of a particular area; ►atmospheric emission inventory, ►lake and river water quality map); *s* **atlas** [m] **ambiental** (Conjunto de datos ecológicos básicos sobre la situación ambiental de un territorio, representado en forma cartográfica, que incluyen los recursos naturales y su grado de contaminación; ►inventario español de emisiones [Es], ►mapa de calidad de las aguas continentales); *f* **atlas** [m] **environnemental** (±) (Jeu de cartes thématiques représentant les données écologiques de base utiles à la prise de décision en matière de planification portant sur la situation réelle de l'environnement [état des ressources naturelles et pollutions] pour un espace défini ; ►carte départementale d'objectifs de qualité, ►carte de pollution des cours d'eau, ►inventaire des émissions des substances polluantes) ; *g* **Umweltatlas** [m] (Atlasförmige, anwendungsorientierte Aufbereitung und Darstellung ökologischer Grunddaten für Planungsentscheidungen über die tatsächliche Umweltsituation für einen bestimmten Raum hinsichtlich der vorhandenen natürlichen Grundgüter und deren Belastung; ►Emissionskataster, ►Gewässergütekarte).

264 atmosphere [n] *met.* (Envelope of gases surrounding the Earth; ►geosphere); *s* **atmósfera** [f] (Capa de aire de unos 100 km de espesor que rodea la superficie de la tierra; ►geosfera); *f* **atmosphère** [f] (Enveloppe gazeuse de la terre ; ►géosphère) ; *g* **Atmosphäre** [f, o. Pl.] (Lufthülle der Erde; ►Geosphäre).

265 atmospheric emission inventory [n] *adm. envir. leg.* (A record set up for the surveillance of emission levels, which contains maps, and continually processes data received from a network of measuring instruments; ►air quality control station, ►air quality monitoring network/system, ►national or state surveillance system for air pollution); *s* **inventario** [m] **español de emisiones** [Es] (►red de evaluación de la calidad del aire, ►estación de medición de la calidad del aire, ►sistema español de información, vigilancia y prevención de la contaminación atmosférica [Es]; art. 5 Ley 34/2007); *f* **inventaire** [m] **des émissions des substances polluantes** (Analyse de la nature, l'origine et l'évolution des rejets des polluants atmosphériques et évaluation de la qualité de l'air ambiant dans les sites à forte pollution pour lesquels sont fixés des objectifs de qualité de l'air, des seuils de précaution, des seuils d'alerte et des valeurs limites ; l'**i. d. e.** est un des éléments de référence des plans régionaux pour la qualité de l'air ou des plans de protection de l'atmosphère, il est publié par l'État chaque année ; loi n° 96-1236 du 30 décembre 1996 ; ►dispositif de surveillance, ►réseau de mesure des pollutions atmosphériques, ►réseau de surveillance, ►station de mesure de la pollution atmosphérique) ; *g* **Emissionskataster** [m, *in A nur so* — oder n] (Auf Grund Landesrecht von zuständigen Behörden erhobene und in einem Register dokumentierte Bestandsaufnahme zur Überwachung der Immissionsentwicklung, das die Immissionsmessdaten [Art, Menge, räumliche und zeitliche Verteilung sowie die Austrittsbedingungen von Luftverunreinigungen bestimmter Anlagen und Fahrzeuge] eines Messstellennetzes enthält, grafisch aufbereitet und fortschreibt; cf. § 46 BImSchG; ►Luftmessnetz, ►Luftmessstation, ►Überwachungsnetz).

266 atmospheric humidity [n] *met. phys.* (Content of vapo[u]r in the air); *s* **humedad** [f] **atmosférica** *syn.* vapor [m] de agua atmosférico; *f* **humidité** [f] **(absolue) de l'air (atmosphérique)** (Masse de vapeur d'eau contenue dans un volume déterminé d'air atmosphérique, non saturé ; CILF 1978, 100) ; *syn.* humidité [f] atmosphérique ; *g* **Luftfeuchtigkeit** [f] (Gehalt von Wasserdampf in der Luft); *syn.* Luftfeuchte [f].

267 atmospheric inversion [n] *met.* (Air layers within which air temperature increases rather than decreases in height—the opposite of the normal meteorological gradient); *s* **inversión** [f] **térmica** (Al contrario que la situación normal en la distribución vertical de la temperatura en la que ésta disminuye con la altura, se habla de **i. t.** cuando aumenta dentro de una cierta altura; su espesor puede ser muy variable; cf. DM 1986); *syn.* inversión [f] de temperatura; *f* **inversion** [f] **thermique** (Anomalie du gradient vertical des températures où l'on observe un accroissement de la température avec l'altitude ; CILF 1978 ; p. ex. l'accumulation par convection de masses d'air froid dans le fond de vallées ou de dépressions [doline] ; en général la température diminue avec l'altitude) ; *syn.* inversion [f] de température ; *g* **Inversion** [f] (Luftschichtung in der Atmosphäre, innerhalb derer die Lufttemperatur mit der Höhe zunimmt. Im Normalfall nimmt die Temperatur mit der Höhe ab); *syn.* Temperaturinversion [f], Temperaturumkehr [f].

atmospheric pollution [n] *envir.* ►air pollution (1).

atrium house [n] *arch.* ►courtyard house.

268 attached building development [n] *leg. urb.* (Uninterrupted building rows; e.g. townhouse/townhome development [US]/row house/terrace development [UK] or row housing [US], ►perimeter block development; *opp.* ►detached building development); *s* **edificación [f] continua** (Edificación orientada a la línea de la calle, en la que los predios están totalmente edificados en el frente de la calle; ►edificación en bloque cerrado; *opp.* ►edificación discontinua); *syn.* edificación [f] alineada (a la calle); *f* **implantation [f] des constructions en ordre continu** (Disposition des bâtiments sur toute la largeur de chaque parcelle, les constructions jouxtant les limites parcellaires ; ►construction en îlot ; *opp.* ►implantation des bâtiments en ordre discontinu, ►implantation des constructions en ordre semi-continu) ; *g* **geschlossene Bauweise [f]** (Bauweise, bei der die Grundstücke zwischen den seitlichen Grenzen in ganzer Breite bebaut werden; ►Blockbebauung; *opp.* ►offene Bauweise 2).

269 attached dwelling [n] **in a cluster** [US] *arch. urb.* (One of a group of two or more individual dwellings connected by common walls; ►row house); *syn.* attached single-family house [n] in a cluster [UK]; *s* **casa [f] unifamiliar agrupada** (≠) (Unidad de grupo de casas individuales construidas en grupo de manera que comparten muros; ►casa adosada); *f* **maison [f] individuelle regroupée** (Construction ayant des murs communs avec les maisons individuelles voisines ; on parle alors de l'habitat groupé (maisons mitoyennes, habitat en bande, petits collectifs, etc.) moins consommateur d'espace et facteur de lien social ; ►maison individuelle construite en ligne) ; *syn.* maison [f] individuelle groupée, logement [m] individuel groupé ; *g* **Einfamilienhaus [n] einer Gebäudegruppe** (Haus in einem Cluster, das mit mehreren Einfamilienhäusern durch gemeinsame Hauswände verbunden ist; ►Reihenhaus); *syn.* Einfamilienhaus [n] einer Hausgruppe.

attached single-family house [n] **in a cluster** [UK] *arch. urb.* ►attached dwelling in a cluster [US].

attack [n] *agr. for. hort.* ►infestation.

attenuation mound [n] *constr. envir.* ►noise attenuation mound.

270 attic [n] *arch.* (Upper part of a house directly under the roof); *s* **ático [m]** (Espacio existente entre el tejado y el techo del piso más alto en los edificios modernos; DACO 1988); *f* **attique [m]** (Étage de moindre importance situé au sommet d'une construction) ; *g* **Attika [f] (1)** (Niedriger Aufbau über dem Hauptgesims eines Bauwerks).

271 attraction point [n] *plan. recr.* (Center [US]/centre [UK] of attraction to deflect visitor pressure away from a sensitive area; ►center of attraction, ►focal point); *syn.* "honey pot" [n]; *s* **centro [m] de gravedad (1)** (Punto de atracción creado en zona sensible para atraer a los visitantes y así evitar daños en ésta; ►centro de atracción, ►centro de gravedad 2); *syn.* aliciente [m]; *f* **pôle [m] d'attraction et d'animation** (Aménagement de centres d'intérêts dans les zones sensibles qui exigent une canalisation de la fréquentation ; ►point d'attrait, ►pôle d'attraction) ; *syn.* point [m] fort d'attraction, pôle [m] de fréquentation ; *g* **Anziehungspunkt [m] (1)** (Planerisches Mittel zur Vermeidung einer flächenhaften Zerstörung von schutzwürdigen Gebieten durch Lenkung der Besucherströme auf einen bestimmten Ort oder einzelne wenige Beobachtungs- oder Aufenthaltspunkte; ►Anziehungspunkt 2, ►Anziehungspunkt 3).

272 attractiveness [n] *recr.* (Possessing pleasant qualities that appeal to the senses, inducing one to approach); *s* **grado [m] de atracción** (Capacidad de atraer el interés de una persona); *f* **attraction [f]** (Capacité d'attirer l'intérêt d'une personne) ; *g* **Attraktivität [f] (1.** Vermögen, jemanden in seinen Bann zu ziehen. **2.** Eigenschaft, anziehend zu sein); *syn.* Anziehungskraft [f].

attractive planting composition [n]**, seasonally** *phyt. plant.* ►seasonally changing visual dominance.

273 autecology [n] *ecol.* (Study of a particular species in relation to its environment and to other species; ►ecology, ►habitat 1, ►population ecology, ►synecology); *s* **autoecología [f]** (Rama de la ►ecología que estudia las especies individuales de flora y fauna en relación con su medio [►hábitat, ►habitación] y en la interacción con otras especies; ►ecología de poblaciones, ►sinecología); *f* **autécologie [f]** (Branche de l'►écologie étudiant les relations existant entre une espèce animale ou végétale et son environnement [►habitat 1, ►habitat naturel] et les autres espèces ; ►démécologie, ►synécologie) ; *g* **Autökologie [f]** (Teilgebiet der ►Ökologie, das sich mit den Beziehungen einer einzelnen Tier- oder Pflanzenart zu ihrer Umwelt [►Habitat] und zu anderen Arten befasst; ►Demökologie, ►Synökologie).

authority [n] *adm. leg.* ►approving authority; *adm. agr. leg.* ►land consolidation authority; *adm.* ►local authority; *adm. leg. pol.* ►local/regional administrative authority/agency; *adm.* ►operating authority, ►public authority (1), ►public authority (2); *adm. plan.* ►public or semipublic authority [US]/public or semi-public authority [UK]; *adm.* ►supervisory authority; *adm. hydr. wat'man.* ►water authority.

authority [n] [US]**, branch** *adm.* ►operating authority.

authority [n] [UK]**, competent** *adm.* ►operating authority.

Authority [n] [England&Wales]**, Central** *adm. leg.* ►law enforcement agency [n] [US].

authority [n] **having jurisdiction** *adm. leg.* ►law enforcement agency [n] [US].

authority [n] **or agency** [n] **with planning powers** [US]**, public** *adm. plan.* ►public planning body [US].

274 author [n] **of a plan** *plan. prof.* (Creator of a plan); *s* **redactor, -a [m/f] de un plan** (Persona que tiene la propiedad artística e intelectual de un plan); *f* **auteur [m] d'un plan** (Celui qui en a la propriété artistique et intellectuelle) ; *g* **Planverfasser/-in [m]** (Urheber/-in eines Planes); *syn.* Plansteller/-in [m/f].

autobahn [n] *trans.* ►freeway [US]/motorway [UK].

275 autochthonous [adj] *geo. pedol.* (Rock or soil that has been formed *in situ*; *opp.* ►allochthonous; *noun* autochthon [n]); *s* **autóctono/a [adj]** (Roca o suelo que se ha formado *in situ*; *sustantivo* autoctonía [f]; *opp.* ►alóctono/a); *f* **autochtone [adj]** (Roche sédimentaire sol non formé sur place ; *substantif* autochtonie [f] ; *opp.* ►allochtone) ; *g* **autochthon [adj]** (Gestein, das oder Boden, der sich am Ort seiner Entstehung befindet; *opp.* ►allochthon; *Substantiv* Autochthonie [f]).

autochthonous plant species [n] *phyt.* ►indigenous plant species.

autochthonous species [n] *phyt. zool.* ►indigenous species.

auto grave yard [n] [US] *envir. urb.* ►car salvage yard [US].

276 automatic irrigation controller [n] *constr.* (Programmed equipment for operation of irrigation system; ►irrigation equipment manhole, ►sprinkler, ►pop-up sprinkler); *s* **dispositivo [m] automático de riego** (Dispositivo montado generalmente en un ►pozo de riego [subterráneo] para la irrigación automática de zonas verdes viarias o azoteas ajardinadas, ►aspersor, ►aspersor escamoteable); *f* **dispositif [m] d'arrosage automatique** (Appareil souvent encastré dans un ►regard d'arrosage et utilisé pour l'arrosage semi-automatique ou automatique des espaces verts d'accompagnement ou des jardins sur dalles ; ►arroseur, ►arroseur escamotable) ; *g* **Bewässerungs-**

A

automat [m] (Meist in einem ►Bewässerungsschacht eingebaute Apparatur zur halb- oder vollautomatischen Bewässerung, z. B. von Straßengrün, intensiv genutzen Rasenanlagen oder Dachgärten; ►Regner, ►Versenkregner).

autoregulation [n] *ecol.* ►self-regulation (1), ►self-regulation (2).

277 auto tourist [n] [US] *recr.* (Driver making a trip with his car in the countryside for recreation activity; ►pleasure driving); *syn.* leisure motorist [n] [UK], pleasure motorist [n] [UK]; *s* **excursionista [m/f]** (Persona que hace una ►excursión en automóvil por una zona atractiva paisajísticamente); *f* **excursionniste [m] motorisé** (Personne utilisant un véhicule pour se promener dans la nature ; *activité* ►randonnée motorisée) ; *syn.* promeneur [m] motorisé ; *g* **Autowanderer [m]/Autowanderin [f]** (Jemand, der mit dem Auto einen Ausflug durch die Landschaft macht; ►Autowandern); *syn.* Autotourist/-in [m/f].

autumnal colo(u)rs/hues [npl] (of leaves) [US] *bot.* ►autumn colors/hues (of leaves) [US]/autumn colours/hues (of leaves) [UK].

autumnal flowering season [n] [UK] *hort. plant.* ►fall-flowering period [US].

278 autumn-blooming plant [n] *hort. plant.* (*Horticultural term* herbaceous or woody plant flowering in autumn; ►late-blooming plant); *syn.* fall-blooming plant [n] [also US], fall-blooming species [n] [also US]; *s* **planta [f] de floración otoñal** (Denominación de horticultura para planta herbácea o leñosa que florece en otoño; ►planta de floración tardía); *f* **plante [f] à floraison automnale** (Terme horticole caractérisant les plantes vivaces et les végétaux ligneux qui fleurissent en automne ; ►plante à floraison tardive) ; *g* **Herbstblüher [m]** (*Gärtnerische Bezeichnung* krautige oder Gehölzpflanze, die im Herbst blüht; ►Spätblüher).

279 autumn circulation period [n] *limn.* (Period of mixing of stratified water masses in a body of water during early autumn: **1.** cooling of surface waters, **2.** density change in surface waters producing convection currents from top to bottom, **3.** circulation of the total water volume by wind action, and **4.** vertical temperature equality; RCG 1982; ►winter stagnation phase, ►thermal stratification); *syn.* fall overturn [n] [also US]; *s* **circulación [f] vertical autumnal/otoñal** (Proceso de mezcla de las capas de agua de un lago templado dimíctico, causado por el descenso de la temperatura de la capa superior y facilitado por las tormentas otoñales, con el resultado de la igualación de densidad y temperatura de <4°C; cf. MARG 1977, 195; ►estratificación invernal, ►estratificación térmica de las masas de agua dulce); *f* **brassage [m] automnal** (Refroidissement et enfoncement des eaux de l'épilimnion dans un lac à mesure qu'elles se rapprochent de 4 °C ; il s'ensuit un mélange des eaux et une égalisation des températures ; après la **b. a.** suit la ►stagnation hivernale ; ►stratification thermique) ; *g* **Herbstzirkulation [f]** (Vorgang und Zustand der Durchmischung der gesamten Wassermasse eines Sees [Meeres] im Herbst, ausgelöst durch die fortschreitende Abkühlung der Oberflächenschicht und durch den Wind [Herbststürme] als Antriebsenergie, so dass der ganze Wasserkörper 4 °C warm ist. Der **H.** folgt die ►Winterstagnation; ►Schichtung 1).

280 autumn colors/hues [npl] (of leaves) [US]/autumn colours/hues [npl] (of leaves) [UK] *bot.* (Changes in foliage colo[u]r of plants from the decomposition of chloroplasts and chlorophyll, caused by the onset of cold weather in temperate climates, with conversion from sugar to starch for storage in shoots and roots; ►discoloration of foliage); *syn.* fall colors [n] [US]; *s* **colores [mpl] otoñales** (Cambios en el color del follaje caduco en otoño debido a la fermentación de los nutrientes almacenados en las hojas que en forma soluble son transportados a las yemas, raíces y otros órganos de almacenamiento. El color amarillento es causado por la esterificación de las carotinas y las xantofilas, el rojizo por la acumulación de antociano en las vacuolas de los tejidos celulares; ►descoloración del follaje); *syn.* colores [mpl] de otoño, tonalidad [f] otoñal; *f* **coloration [f] automnale** (Transformation de la couleur des feuilles en automne provoquée par la migration vers la tige et les racines de protides, d'auxines et des produits de décomposition de la chlorophylle, les feuilles jaunissant ou rougissant par accumulation d'anthocyane dans leurs vacuoles ; la **c. a. jaune** est provoquée par la transformation [estérification] des pigments caroténoides [carotènes et xanthophylles] et la **c. a. rouge** par accumulation d'anthocyane dans les vacuoles des tissus cellulaires ; ►décoloration des feuilles ; cf. MVV 1977, 296) ; *syn.* couleurs [fpl] d'automne ; *g* **Herbstfärbung [f]** (Verfärbung der Laubblätter im Spätsommer oder Herbst durch Fermentierung der in den Blättern angesammelten Nährstoffe, die in gelöster Form zu Knospen, Trieben, Wurzeln und anderen Speicherorganen transportiert werden. Die herbstliche Gelbfärbung der Blätter entsteht zunächst durch den Abbau der grünen Chlorophylle, deren Spaltprodukte durch die Leitungsbahnen abgeleitet werden, so dass allein die Carotinoide zurückbleiben; die herbstliche Rotfärbung wird durch Anthozyane, die den Zellsaft rot färben, hervorgerufen; die spätere Bräunung absterbender Laubblätter beruht auf dem postmortalen [dem Zelltod folgenden] Auftreten wasserlöslicher brauner Farbstoffe; LB 1978, 65; ►Laubverfärbung); *syn.* herbstliche Laubfärbung [f].

281 autumn foliage [n] *bot. hort.* (Term for the leaves of trees, shrubs and perennials which have changed their colo[u]r during the third season of the year; ►autumn colors/hues [of leaves]); *syn.* fall foliage [n] [also US]; *s* **follaje [m] otoñal** (Denominación horticultural para las hojas de árboles, arbustos y vivaces que cambian su color en la tercera estación del año; ►colores otoñales); *f* **frondaison [f] automnale** (Terme horticole caractérisant la couleur du feuillage des végétaux ligneux et plantes vivaces en automne ; ►coloration automnale) ; *g* **Herbstbelaubung [f]** (Gärtnerische Bezeichnung für das durch ►Herbstfärbung charakterisierte Laub von Gehölzen und Stauden).

282 availability [n] (1) *chem. nat'res. pedol.* (Usable existence of nutrients, water, etc.; ►availability of nutrients, ►available soil water); *s* **disponibilidad [f]** (Existencia de agua o nutrientes que puede ser aprovechada por las plantas, ►disponibilidad de nutrientes, ►crescardía); *f* **disponibilité [f]** (p. ex. ressources naturelles dont on a l'usage, qui sont à la disposition ; ►disponibilité en substances nutritives, ►eau utile) ; *g* **Verfügbarkeit [f]** (Nutzbares Vorhandensein von Nährstoffen, Wasser etc.; ►Nährstoffverfügbarkeit, ►pflanzenverfügbares Wasser).

availability [n] (2) *phyt.* ►light availability; *for. phyt.* ►relative light availability.

283 availability [n] of nutrients *pedol.* (Existence of nutrients for plant growth); *s* **disponibilidad [f] de nutrientes** *syn.* disponibilidad [f] de nutrimento; *f* **disponibilité [f] en substances nutritives** (En éléments assimilables conduisant à la nutrition minérale des plantes et satisfaisant leurs besoins immédiats) ; *syn.* possibilité [f] nutritive ; *g* **Nährstoffverfügbarkeit [f]** (Für Pflanzen nutzbares Vorhandensein von Nährstoffen).

available [adj] for development *urb.* ►making available for development.

284 available groundwater resources [npl] *hydr.* (Available quantity of groundwater in saturated soils or aquifers;

►groundwater resources); **s recursos [mpl] acuíferos sustentablemente aprovechables (≠)** (Cantidad máxima de agua que se puede extraer artificialmente de un acuífero, sin hacer que descienda continuamente el nivel de la misma, ni agotar las reservas o alterar la calidad química del agua, es decir que su explotación sea sostenible; ►recursos hídricos subterráneos); *syn.* caudal [m] aprovechable de un acuífero; *f* **ressource [f] en eau souterraine** (Niveau productif d'un aquifère ; ►gîte aquifère) ; *g* **Grundwasserdargebot [n]** (Die für eine nachhaltige Nutzung zur Verfügung stehende Grundwassermenge — i. S. der Ergiebigkeit; der Begriff ►Grundwasservorkommen wird im Vergleich zum Begriff G. nutzungsneutral verstanden).

285 available soil water [n] *pedol.* (Soil water which can be absorbed by plant roots; ►available to plants); **s cresardía [f]** (Término creado por CLEMENTIS para expresar el agua que el suelo puede ceder a la vegetación, que la absorbe para su metabolismo vital; DB 1985; ►disponible para la planta); *syn.* agua [f] útil del suelo (MARG 1980); *f* **eau [f] utile** (Eau contenue dans le sol disponible pour les végétaux ; ►disponible pour la plante) ; *g* **pflanzenverfügbares Wasser [n]** (Wasser im Boden, das durch die Pflanze aufgenommen werden kann; ►pflanzenverfügbar).

286 available [adj] to plants *pedol.* (Descriptive term for water and nutrients which are easily absorbed by plants from the soil; ►mobilization, ►available soil water); **s disponible para las plantas [loc]** (Término descriptivo para agua o nutrientes que son fácilmente absorbidos por las plantas en el suelo; ►mobilización de nutrientes ►cresardía); *f* **disponible pour la plante [loc]** (p. ex. eau utile, eau capillaire absorbable par les racines, mis à la disposition des plantes au cours d'une saison ; cf. PED 1979, 391 ; *contexte* la réserve utile en eau est facilement accessible/utilisable à/par la plante ; DIS 1986, 186 ; ►désagrégation ; ►eau utile) ; *syn. pour des sels nutritifs* assimilable [adj] ; *g* **pflanzenverfügbar [adj]** (Wasser oder Nährstoffe betreffend, die im Boden durch die Wurzelspitzen aufgenommen werden können; ►Aufschließen, ►pflanzenverfügbares Wasser).

avalanche [n] *geo.* ►debris avalanche.

avalanche [n] [UK], course of an *geo.* ►avalanche track [US].

287 avalanche control technique [n] *constr. land'man.* (Method of protecting against avalanches and potential sources of avalanches by engineering works in the form of retaining walls and other retarding structures, as well as afforestation at high altitudes ; ►high-altitude afforestation); **s barrera [f] contra aludes** (Estructuras para controlar aludes, en forma de muros de contención, estructuras de interrupción y forestación a gran altitud; ►forestación de las grandes altitudes); *syn.* protección [f] contra avalanchas; *f* **techniques [mpl] de défense contre les avalanches** (Installations et ouvrages divers de protection contre les avalanches prévus pour réduire les risques dans les secteurs soumis aux avalanches tels que : barrages de pente, barrières de défense, filets paravalanches, ►reboisement d'altitude, plantations en zone de haute montagne) ; *syn.* génie [m] paravalanche, (circ. 88-67 du 20 juin 1988) ; *g* **Lawinenverbauung [f]** (Teilbereich des Landschaftsbaues, Lawinenzüge und potenzielle Lawinenabbruchgebiete meist mit Ingenieurbauten [Hartbauweisen] durch Objektschutzbauten, Ablenkverbau, Stützverbau in der Anrisszone und Bremsbauwerken [Bremshöcker, Fangdämme], aber auch durch Hochlagenaufforstung zu sichern; *UBe* Objektschutz gegen Lawinen, Ablenkverbauung [Galerien, Lenkmauern], Stützverbauung, Bremsverbauung, z. T. ►Hochlagenaufforstung).

avalanche course [n] [UK] *geo.* ►avalanche track [US].

avalanche courses [npl] *plan.* ►map of avalanches courses.

288 avalanche grassland [n] *phyt.* (Non-anthropogenic meadow, occurring as narrow strips of grassland between forests on steep slopes of high mountains, where avalanches descend annually in spring; ELL 1967; ►avalanche track); **s pradera [f] de alud** (Prado no influenciado por el hombre en forma de bandas estrechas de herbáceas entre los rodales en las laderas escarpadas de altas montañas, por las cuales descienden los aludes en primavera; ►curso de avalancha); *syn.* pradera [f] de avalancha; *f* **pelouse [f] de couloir d'avalanches** (Pelouse naturelle développée en bandes étroites sur les ►couloirs d'avalanches entre les espaces de forêts) ; *g* **Lawinenrasen [m]** (Vom Menschen unbeeinflusster Rasen in Form von schmalen Streifen zwischen Waldbeständen an Steilhängen im Hochgebirge, die jährlich im Frühjahr von Lawinen überrollt werden; ►Lawinenbahn); *syn.* Lavinarrasen [m], Lavinarwiese [f].

avalanche hazard map [n] *plan. recr.* ►avalanche location map.

289 avalanche location map [n] *plan. recr.* (**In F.**, a map with a scale of 1:20,000 showing source areas where avalanches are likely to originate. Such a map is elaborated by interpretation of aereal photographs. The Swiss have the longest record of state sponsored avalanche mapping services. **In D.**, instead of such maps there is an "**avalanche warning service**", organized by the German Alpine Association, which may be consulted by telephone; **in U.S.**, Government agencies [local and federal], use on-site avalanche studies for mitigation and protection of infrastructure. Winter resorts typically employ on-site inspectors to patrol and to locate hazard zones, which are frequently reported via internet sites. These studies rely heavily upon empirical observation; ►map of avalanches courses); *syn.* avalanche hazard map [n]; **s mapa [m] de inventario de aludes** (Mapa topográfico a escala 1:20 000 realizado sobre la base de fotos aéreas por el Instituto Geográfico Nacional de Francia en el que se representan las ►zonas de origen de aludes. Suiza es el país con servicio estatal más antiguo de mapeo de avalanchas. En D. no existe un tal mapa, pero sí un servicio de previsión de aludes, organizado por la Asociación Alemana de los Alpes, al que se puede consultar telefónicamente. En EE.UU. agencias gubernamentales (locales y federal), utilizan estudios de aludes in situ para mitigar y proteger la infraestructura; las estaciones de deportes alpinos emplean normalmente inspectores para vigilar y localizar zonas peligrosas, sobre las que normalmente se informa vía internet; ►plan de zonas expuestas a aludes); *f* **carte [f] d'inventaire d'avalanches** (Carte à l'échelle du 1 : 20 000, dessinée à partir de photographies aériennes par l'institut géographique national ; cf. circ. n° 74-201 du 05 décembre 1974 ; ►plan de zones exposées aux avalanches) ; *syn.* carte [f] de localisation probable des avalanches ; *g* **Lawinengefahrenkarte [f]** (**1. F.**, gebräuchliche topografische Karte im M 1:20 000, auf der an Hand von ausgewerteten Luftbildern potenzielle Lawinenentstehungsgebiete und Lawinenbahnen dargestellt werden. **2.** In den Alpenländern dienen diese Karten [M 1:1000 bis 1:10 000] u. a. als lawinentechnische Grundlage für die Erarbeitung von Bauleitplänen [**CH.**, kantonale Richtpläne], für Verkehrs- und Infrastrukturplanungen und zum Schutz des winterlichen Fremdenverkehrs und touristischer Projekte. **3. L.n** werden auch zum aktuellen Lawinenlagebericht für den Wintersport gefertigt; **in D.** gibt es keine offiziellen Lawinenkarten, statt dessen einen **Lawinenwarndienst**, der vom Deutschen Alpenverein [DAV] organisiert wird und telefonisch jederzeit abgerufen werden kann und vom Bayerischen Staatsministerium des Innern in Zusammenarbeit mit dem Bayer. Staatsministerium für Umwelt, Gesundheit und Verbraucherschutz im Internet die Darstellung der aktuellen Lawinenlagen; in CH kann täglich das **Nationale Lawinenbulletin** für alle Regionen, in A. der **Lawinenlagebericht** abgefragt werden;

die Lawinengefahr wird nach der 5-stufigen Europäischen Lawinengefahrenskala beurteilt; ►Lawinenrisikokarte); *syn.* Lawinenkarte [f].

290 avalanche protection forest [n] *for. land'man.* (Forest intended to prevent the formation of avalanches and snowslides and, where possible, to guide, brake, and bring to a halt an avalanche which has been set in motion; ►protective forest for soil conservation purposes); **s bosque [m] protector contra aludes** (Bosque cuya función es prevenir la formación de aludes y deslizamientos de nieve y, dentro de lo posible, desviar, interrumpir y parar aludes ya puestos en movimiento; ►bosque protector del suelo); *syn.* bosque [m] protector contra avalanchas; **f forêt [f] de défense contre les avalanches** (Forêt conçue pour prévenir la formation d'avalanches ou dans le but de détourner, freiner ou arrêter les avalanches ; ►forêt de protection contre l'érosion) ; *syn.* forêt [f] de protection contre les avalanches ; **g Lawinenschutzwald [m]** (Wald, der die Entstehung von Lawinen und Schneerutsch[ung]en verhindern und abgehende Lawinen und Schneerutsche nach Möglichkeit lenken, bremsen und zum Stillstand bringen soll; ►Bodenschutzwald).

291 avalanche source area [n] *geo. plan.* (Area in which a mass of snow begins to descend); **s zona [f] de origen de aludes/avalanchas; f zone [f] d'origine d'avalanche** (À partir de laquelle prend naissance une avalanche) ; *syn.* source [f] d'avalanches (circ. n° 88-67 du 20 juin 1988) ; **g Lawinenentstehungsgebiet [n]** (Fläche, in der eine Lawine abreißt/abbricht); *syn.* Lawinenabbruchgebiet [n], Lawinenanrissgebiet [n].

292 avalanche track [n] [US] *geo.* (LE 1986, 193; course on a slope over which avalanches descend every year to the valley; location of ►avalanche grassland); *syn.* avalanche course [n] [UK], course [n] of an avalanche [UK]; **s curso [m] de avalancha** (Surcos de laderas por los cuales descienden los aludes hasta los valles; ►pradera de alud); *syn.* curso [m] de alud; **f couloir [m] d'avalanches** (Surface d'un versant sur laquelle glissent régulièrement les avalanches ; lieu de développement des ►pelouses de couloir d'avalanches ; *terme spécifique* cône [m] d'avalanche) ; *syn.* zone [f] avalancheuse ; **g Lawinenbahn [f]** (Hangfläche, in der alljährlich Lawinen zu Tal fahren; Standort des ►Lawinenrasens); *syn.* Lahngang [m], Lawinengasse [f], Lawinenrinne [f] (±), Lawinenstrich [m], Lawinenzug [m], Lawinentobel [m].

avenue [n] [UK] *gard. landsc. urb.* ►tree-lined avenue [US].

avenue tree [n] [UK] *hort.* ►standard street tree [US].

average flood level/mark [n] *wat'man.* ►average high-water level/mark.

293 average high-water level/mark [n] *wat'man.* *syn.* average flood level/mark [n]; **s nivel [m] de crecida media** (Cota alcanzada por las aguas de crecida en condiciones normales); **f niveau [m] du débit moyen caractéristique** (Débit d'un cours d'eau égalé ou dépassé pendant 6 mois de l'année) ; *syn.* niveau [m] du débit médian, niveau [m] du débit semi-permanent ; **g mittleres Hochwasser [n]** (...wasser [pl]; *Abk.* mHW); *syn.* mittlerer Hochwasserstand [m].

avifauna [n] *zool.* ►birdlife, ►nidificating avifauna.

avocation [n] [US], **leisure** *plan. recr.* ►leisure pursuit.

award [n] *hort.* ►grand award; *prof.* ►merit award; *contr.* ►recommendation for contract award.

award [n], **contract** *constr. contr.* ►awarding of contract [US].

Award [n], **Home and Neighborhood Improvement** *hort. plan. urb.* *►"All-America City" [US].

awarding authority [n] *adm. prof.* ►initiating authority.

294 awarding [n] **of contract** [US] *constr. contr.* (Act of granting a written or oral binding agreement to award a contract to a construction or maintenance firm, with or without a formal bidding/tendering procedure; let [vb] a contract; ►sole source contract award [US]/freely awarded contract [UK]; ►acceptance of a bid [US] 1/acceptance of tender [UK] 1, ►acceptance of a bid [US] 2/acceptance of a tender [UK] 2, ►contract 1, ►contract awarding procedure); *syn.* contract award [n], letting [n] of contract [UK]; **s adjudicación [f] de contrato** (Encargo a una empresa de forma oral o escrita para realizar trabajos de construcción o mantenimiento; ►adjudicación del remate, ►contrato de obra, ►adjudicación negociada, ►procedimiento de adjudicación, ►remate); *syn. en caso de concurso público* concesión [f] de contrata; **f 1 passation [f] d'un marché** (*Terme générique* ►attribution d'un marché de travaux au titulaire du marché/à l'adjudicataire ; ►contrat, ►marché, ►passation du/des marché[s] sous forme de marché négocié, ►procédure de passation des marchés) ; **f 2 attribution [f] d'un marché de travaux** (à un entrepreneur ; ►attribution du marché [sur appel d'offres/sur adjudication], ►notification [de l'attribution] du marché à l'entreprise retenue) ; *syn.* attribution [f] d'un contrat ; **g Auftragserteilung [f] an eine Firma** (Schriftliche oder mündliche Erklärung über die Übertragung von Bau- oder Pflegeleistungen an eine Firma oder den Bieter mit dem annehmbarsten Angebot nach einem förmlichen Ausschreibungsverfahren; ►freihändige Vergabe, ►Vergabewesen, ►Vertrag, ►Zuschlag, ►Zuschlagserteilung); *syn.* Erteilung [f] des Auftrages an eine Firma, Beauftragung [f] einer Firma, Auftragsvergabe [f] an eine Firma, Vergabe [f] eines Auftrages an eine Firma.

295 awarding [n] **of prizes** *prof.* (Ceremonial handing over or awarding of prizes which were determined by a jury); *syn.* presentation [n] of awards; **s entrega [f] de premios** (Ceremonia de entrega de premios designados por un jurado); **f remise [f] des prix** (L'attribution solennelle des prix aux lauréats d'un concours après avis du jury est prononcée par la personne responsable du marché) ; *syn.* attribution [f] des prix ; **g Preisverleihung [f]** (Feierliche Übergabe oder Zusprechung von Preisen, die ein Preisgericht zuerkannt hat); *syn.* Preisvergabe [f], Überreichung [f] von Auszeichnungen.

awarding period [n], **contract** *contr.* ►period for acceptance of a bid [US].

awarding procedure [n] *constr. contr. prof.* ►contract awarding procedure.

award [n] **of merit** [US] *prof.* ►honorable mention [US]/honourable mention [UK].

award statement [n] [US], **written** *contr.* ►written notice of contract award.

awareness [n] *conserv. sociol.* ►environmental awareness.

296 axillary bud [n] *bot.* (Bud in the angular space between a plant stem and the base of a leaf); **s yema [f] lateral** (Yema normal, que nace en la axila de una hoja; DB 1985); *syn.* yema [f] axilar; **f bourgeon [m] latéral** (Bourgeon situé à l'aisselle de chaque feuille) ; *syn.* bourgeon [m] axillaire ; **g Seitenknospe [f]** (Knospe, die in der Achsel eines Blattes sitzt); *syn.* Achselknospe [f].

ayrie [n] *zool. game'man.* ►eyrie.

B

297 Bachelor [n] of Architecture (B.ARCH) [US] *prof.*
(**In U.S.**, an undergraduate degree given at colleges and universities with programs accredited by the U.S. Department of Education and by the National Architectural Accrediting Board, representing the American Institute of Architects [AIA], Association of Collegiate Schools of Architecture [ACSA], National Council of Architectural Registration Boards [NCARB]. A Bachelor of Science in Architecture [B.S.ARCH] is a non-professional degree; ▶Master of Architecture [M.ARCH] [US]; *syn.* Bachelor [n] of Philosophy in Design (B.Phil.D.) [UK]; *s* **aparejador, -a [m/f]** (En Es. título de técnico superior para personas que se han graduado en arquitectura en una Escuela Técnica Superior después de realizar generalmente cuatro años de estudios); *syn.* licenciado/a [m/f] en arquitectura; *f* **diplôme [m] d'études en architecture (1. F.,** Titre académique, les études d'architecture s'inscrivent dans le schéma européen d'harmonisation des cursus d'enseignement supérieur et sont structurées en trois cycles sur la base du L M D [licence, master, doctorat] ; ces études sont actuellement organisées dans 20 écoles nationales supérieures d'architecture [ENSA — 6 en Île-de-France et 14 en région], placées sous la tutelle du Ministère de la Culture et de la Communication ; un premier cycle d'études d'une durée de 3 ans permettant d'acquérir les bases de la théorie et de la pratique architecturale. Elles mènent au diplôme d'études en architecture conférant le **grade de licence** ou **bachelor** ; un deuxième cycle d'études d'une durée de 2 ans mène au diplôme d'État d'architecte conférant le grade de ▶**Master d'Architecture. 2. Bachelor [m] d'Architecture [U.S.,** grade académique délivré par les collèges et universités sur la base de programmes d'études agrées par le ministère de l'Éducation et la commission d'admission de l'ordre national des architectes]. **3. D.,** grade académique inférieur délivré à la fin des études universitaires d'architecture [formation en 8 semestres minimum, stage pratique de 6 mois] réalisées dans les « Fachhochschulen » et « Gesamthochschulen » [universités de technologie]) ; *g* **Diplomingenieur [m] für Architektur (FH) (±) (1.** In D. seit 1975 an Fachhochschulen und Gesamthochschulen verliehener unterster akademischer Grad nach einem achtsemestrigen praxisorientierten Architekturstudium. Ein sechsmonatiges Praktikum ist in den acht Semestern inbegriffen. **2.** 1999 hatten mit der „Bologna-Erklärung" die Europäischen Bildungsminister eine einheitliche europaweite Umstellung der Studienabschlüsse auf Bachelor und Master bis 2010 festgeschrieben. In Übereinstimmung mit dieser Bologna-Erklärung werden in D. seit 2005 modularisierte, zweistufige Studiengänge eingeführt, die mit einem dreijährigen Bachelorstudiengang und dem Abschluss **Bachelor of Science [B. Sc.] in Architektur** und einen zweijährigen Masterstudiengang mit dem Abschluss **Master of Science in Architektur** [▶Diplomingenieur für Architektur/Master of Science in Architektur] enden; je nach Hochschule werden unterschiedliche Abschlussbezeichnungen verliehen: **B. Sc. in Stadt- und Regionalplanung, B. Sc. in Architektur/Bauingenieurwesen, B. Sc. in Architektur und Städtebau** etc.); *syn. obs.* graduierter Ingenieur [m] für Architektur (bis 1975 verliehener Grad).

298 Bachelor [n] of Landscape Architecture (BLA) [US] *prof.* (**In U.S.**, an undergraduate degree given at colleges and universities accredited by the U.S. Department of Education and by the Council of Landscape Architectural Registration Boards [CLARB] of the American Society of Landscape Architects [ASLA]; ▶Master of Landscape Architecture [MLA]); *syn.* Bachelor [n] of Science in Landscape Architecture (BSLA) [US], Bachelor [n] of Philosophy in Landscape Design (B.Phil.L.D.) [UK]; *s* **ingeniero/a [m/f] técnico/a de planificación del paisaje y de espacios libres** (En A., D., título académico medio concedido después de estudios en Escuela Técnica Superior de cuatro años y seis meses de prácticas; ▶paisajista diplomado/a); *f 1* **ingénieur [m] paysagiste (1. F.,** les **i. p.** sortent des diverses écoles supérieures spécialisées dans l'art et les techniques des jardins et du paysage telles que l'École Nationale d'Ingénieurs de l'Horticulture et du Paysage [E.N.I.H.P.] avec spécialité paysage, l'École Nationale Supérieure de l'Horticulture et d'Aménagement du Paysage [E.N.S.H.A.P] à Angers avec spécialité paysage et aménagement ou d'autres dans lesquelles sont délivrés les grades académiques correspondants, p. ex. **Ingénieur des techniques horticoles et paysagères** [École nationale supérieure des techniques horticoles et paysagères], **paysagiste diplômé en architecture des jardins** [École supérieure d'architecture des jardins] ; durée des études 3 à 4 ans ; **paysagiste ingénieur concepteur** [École nationale supérieure de la nature et du paysage — E.N.S.N.P. — à Angers et Blois] durée des études 5 ans ; **Master Européen de Concepteur en paysage** [École supérieure d'architecture des jardins et des paysages à Paris (E.S.A.J.) — école privée de statut association loi 1901, reconnue par la Fédération Française du Paysage (FFP) et la Fédération Européenne du Paysage (EFLA)] ; en France la profession de paysagiste n'est pas reconnue juridiquement, le titre de paysagiste n'est pas protégé et on trouve parmi les paysagistes des professionnels aux compétences les plus diverses, de l'ingénieur agronome à l'horticulteur-pépiniériste, du spécialiste de bureau d'études d'aménagement à l'installateur de terrains de sport et d'équipements de jeux ; ▶Master en architecture des paysages, ▶Paysagiste DPLG. **2. U.S.,** grade académique délivré par les collèges et universités sur la base de programmes d'études agrées par le ministère de l'Éducation et la commission d'admission de l'ordre national des architectes paysagistes) ; *syn.* bachelier [m] architecte paysagiste [B] ; *f 2* **ingénieur [m] en aménagement des paysages diplômé des universités de technologie (±) (D.,** grade académique inférieur qualifiant depuis 1975 les études suivies dans les « Fachhochschulen » et « Gesamthochschulen » [universités de technologie]) ; études à orientation technique de 4 ans, y compris stage préprofessionnel) ; *g 1* **Bachelor [m] of Engineering (Landschaftsarchitektur)/Bachelor of Science (Landschaftsarchitektur) (1. D.,** erster berufsqualifizierender Abschluss und akademischer Grad, der an Fachhochschulen und Universitäten nach einem — je nach Bundesland — 6-8-semestrigen erfolgreichen Studium verliehen wird [in Bayern 8 Semester]. An den 10 deutschen Fachhochschulen für Landschaftsarchitektur gibt es je nach Studienschwerpunkt unterschiedliche Bachelor-Bezeichnungen: Technische Fachhochschule Berlin *Bachelor of Science Landschaftsarchitektur*, Hochschule Anhalt [FH] in Bernburg *Bachelor of Engineering Landschaftsarchitektur und Umweltplanung*, Hochschule für Technik und Wirtschaft in Dresden *Bachelor of Science Landschafts- und Freiraumentwicklung*, FH Erfurt *Bachelor of Engineering Landschaftsarchitektur*, FH Wiesbaden in Geisenheim *Bachelor of Engineering Landschaftsarchitektur*, Hochschule Ostwestfalen-Lippe [FH] in Höxter *Bachelor of Science Landschaftsarchitektur*, Hochschule Neubrandenburg [FH] *Bachelor of Science Landschaftsarchitektur und Umweltplanung*, Hochschule für Wirtschaft und Umwelt in Nürtingen-Geislingen *Bachelor of Engineering Landschaftsarchitektur*, Fachhochschule Osnabrück *Bachelor of Engineering Landschaftsarchitektur* und Fachhochschule Weihenstephan *Bachelor of Engineering Landschaftsarchitektur*; ▶Master of Science Landscape Architecture. **2.** 1999 hatten mit der „Bologna-Erklärung" die Europäischen

B

Bildungsminister eine einheitliche europaweite Umstellung der Studienabschlüsse auf **Bachelor** und **Master** bis 2010 festgeschrieben. Deshalb wird der **Dipl.-Ing. (FH)** mehrheitlich durch **Bachelor of Engineering**, in einigen Fachhochschulen durch **Bachelor of Science** ersetzt; je nach Fach- und Vertiefungsrichtung können z. B. an der FH Osnabrück ab 2006 folgende akademische Grade verliehen werden: **B. Eng. Landschaftsentwicklung**, **B. Eng. Freiraumplanung** und **B. Eng. Ingenieurwesen im Landschaftsbau**. An der FH Weihenstephan besteht außerdem die Möglichkeit, in einem dualen Studiengang eine abgeschlossene Lehre im Garten- und Landschaftsbau in den Bachelorstudiengang zu integrieren; ▶Master of Science Landscape Architecture); *g 2* **Diplomingenieur [m] für Landschaftsarchitektur (FH)** (±) (*Abk. in D.* Dipl.-Ing. [FH]; seit 1975 an Fachhochschulen [FH] und Gesamthochschulen verliehener akademischer Grad nach einem achtsemestrigen praxisorientierten Studium. Ein sechsmonatiges Praktikum ist in den acht Semestern inbegriffen. In D. wurden an den zehn Fachhochschulen verschiedene Bezeichnungen geführt, z. B. an der TFH Berlin *Dipl.-Ing. [FH] für Landespflege,* an der FH Nürtingen, *Dipl.-Ing. [FH] für Landespflege,* ab 2000 *Dipl.-Ing. [FH] für Landschaftsarchitektur und Landschaftsplanung,* an der FH Erfurt, FH Osnabrück, Gesamthochschule Paderborn [Höxter] und FH Weihenstephan [Freising] *Dipl.-Ing. für Landschaftsarchitektur,* an der FH Wiesbaden-[Geisenheimer Fachbereich] *Dipl.-Ing. [FH] für Landschaftsarchitektur* [bis 2005; ab Herbst 2005 Bachelorausbildung]); *syn. obs.* graduierter Ingenieur [m] für Landespflege (bis 1975 verliehener Grad).

Bachelor [n] of Philosophy in Design (B.Phil.D.) [UK] *prof.* ▶Bachelor of Architecture (B.ARCH) [US].

Bachelor [n] of Philosophy in Landscape Design (B.Phil.L.D.) [UK] *prof.* ▶Bachelor of Landscape Architecture (BLA) [US].

Bachelor [n] of Science in Architecture (B.S.ARCH) [US] *prof.* ▶Bachelor of Architecture (B.ARCH) [US].

Bachelor [n] of Science in Landscape Architecture (BSLA) [US] *prof.* ▶Bachelor of Landscape Architecture (BLA) [US].

299 back-country skiing [n] *recr.* (Form of skiing away from prepared ski runs, on flat or hilly terrain; ▶cross country skiing); *s* **esquí [m] de travesía** (Paseo en esquís por terrenos afuera de las pistas; ▶esqui de fondo); *f 1* **ski [m] de randonnée** (Activité sportive ; ▶ski de fond) ; *f 2* **randonnée [f] à ski** (Effectuer une randonnée avec des skis en dehors des pistes, sur un terrain plat ou montagneux) ; *g* **Skiwandern [n, o. Pl.]** (Wandern auf Skiern abseits der Pisten im flachen oder bergigen Gelände; Ski wandern [vb]; ▶Skilanglauf, Ziff. 3); *syn. o.V.* Schiwandern [n, o.] Pl.].

back court planting scheme [n] [UK] *constr. urb.* ▶courtyard landscaping.

300 backdune [n] [US] *geo. phyt.* (DEN 1971, 13; dune which supports woody vegetation, older than ▶gray dune; ▶dune heath, ▶dune scrub); *syn.* brown dune [n] [UK] (ELL 1988, 369); *s* **duna [f] parda** (Duna muerta más vieja que la ▶duna gris, cubierta de leñosas de la familia de las ericáceas; ▶matorral dunar, ▶landa de dunas); *syn.* duna [f] de brezal; *f* **dune [f] brune** (Ancienne ▶dune grise recouverte de bruyères ; ▶fourré dunaire, ▶lande dunaire) ; dune [f] à bruyères, dune [f] boisée, dune [f] ancienne ; *g* **Heidedüne [f]** (Ältere verheidete ▶Graudüne, deren Sand sich durch Eisenumwandlung braun verfärbt; der Boden entwickelt sich meist zum ▶Podsol; ▶Dünengebüsch, ▶Dünenheide); *syn.* Braundüne [f].

backfill [n] *constr.* ▶fill material.

301 backfilling [n] *constr.* (Refilling of trenches, building excavations, planting pits, etc. in controlled manner by compacting in layers to prevent differential settlement; ▶backfilling behind structures, ▶earth filling 1, ▶earth layer); *s* **relleno [m] de zanjas y hoyos** (Cierre con tierra de zanjas, hoyos de plantación, etc.; *términos específicos* relleno de zanjas, ▶terraplenado; ▶capa de relleno, ▶relleno); *f* **remplissage [m] [d'une tranchée/d'un trou]** (Remblayage en terre propre d'une tranchée après enrobage en sable d'une canalisation, d'une fosse de plantation ; ▶couche de remblai, ▶exécution d'un remblai, ▶remblai) ; *syn.* remblaiement [m] d'une tranchée ; *g* **Verfüllen [n, o. Pl.]** (Zuschütten von Gräben, Baugruben, Pflanzlöchern etc.; *UBe* Grabenverfüllung, ▶Hinterfüllen, Baugrubenverfüllung; ▶Schüttung 1, ▶Schüttung 2).

302 backfilling [n] **behind structures** *constr.* (Refilling of excavated ground behind walls and other structures in controlled manner by compacting in layers to prevent differential settlement; ▶backfilling of on-site excavated material, ▶fill material; *generic term* ▶backfilling); *s* **terraplenado [m]** (Relleno del hueco de cimentación para alcanzar el nivel del suelo; cf. DACO 1988; ▶tierra de relleno, ▶reutilización de material de excavación, ▶relleno de zanjas y hoyos); *f* **remblaiement [m]** (Comblement d'ouvrages avec des ▶matériaux tout-venant, des ▶terres d'emprunt ; ▶matériaux de remblaiement, ▶réemploi de matériaux extraits ; *terme générique* ▶comblement [d'une tranchée/d'un trou]) ; *g* **Hinterfüllen [n, o. Pl.]** (Auffüllen von Arbeitsräumen mit ▶Füllboden; ▶Wiedereinbau von Aushub; *OB* ▶Verfüllen); *syn. z. T.* Wiederverfüllen [n, o. Pl.].

303 backfilling [n] **of on-site excavated material** *constr.* (Reuse of stockpiled excavation material; ▶filling and compacting in layers); *s* **reutilización [f] de material de excavación** (Uso de material sobrante almacenado provisionalmente; ▶relleno y compactación por tongadas); *f* **réemploi [m] de matériaux extraits** (Mise en œuvre de déblais de terrassement stockés temporairement ; ▶terrasser et compacter en couches) ; *g* **Wiedereinbau [m, o. Pl.] von Aushub** (Einbau von zwischengelagertem Aushubmaterial; ▶lagenweiser Einbau); *syn.* Wiederverfüllen [n, o. Pl.].

back garden [n] [UK] *gard.* ▶backyard garden [US].

background level [n] **of environmental pollution** *envir. plan.* ▶existing environmental pollution load.

304 background level [n] **of pollution** [US] *envir.* (Amount and extent of pollution already in an area [soil, water, air] prior to emission of additional pollutants into the corresponding medium or prior to watershed development); *syn.* background load [n] (CRU 1987); *s* **contaminación [f] de fondo** (Nivel de inmisión existente en un área definida, antes de instalar un nuevo foco de contaminación; Orden de 18 Octubre 1976, Anexo 1); *syn.* nivel [m] de contaminación de fondo; *f* **charge [f] polluante existante** (Exprime la quantité et l'étendue de la pollution préalable d'une ressource naturelle de l'environnement [sol, eau, air] avant le rejet de substances nocives dans celui-ci) ; *g* **Grundbelastung [f]** (Menge und Umfang der [ubiquitären] Belastung eines Umweltbereiches [Boden, Wasser oder Luft] bevor zusätzliche Schadstoffe in das entsprechende Medium immittiert werden); *syn.* Vorbelastung [f].

background load [n] *envir.* ▶background level of pollution [US].

background noise [n] [US] *envir.* ▶ambient noise level.

background radiation [n] *envir.* ▶natural background radiation.

background research [n] [UK] *contr.* ▶background work [US].

305 background work [n] [US] *contr.* (Investigations which a bidder [US]/tenderer [UK] has to undertake before he can prepare a competent bid [US]/tender [UK]); *syn.* prior investigation [n], background research [n] [UK]; *s* **trabajo [m] de investigación previa** (Estudios necesarios para que un licitante pueda calcular el precio de su oferta para un concurso-subasta); *syn.* estudios [mpl] preparatorios; *f* **études [fpl] préparatoires** (Études nécessaires à l'établissement d'une offre lors de la consultation des entreprises ; en vue d'obtenir une saine mise en compétition et lorsqu'il est demandé aux entreprises consultées de procéder à des *é. p.*, le nombre de ces entreprises doit être d'autant plus limité que les dépenses d'études sont plus importantes ; CCM 1984, 39) ; *g* **Vorarbeiten [fpl] (1)** (Diejenigen Arbeiten, die für einen Bewerber notwendig sind, um seine Preise für ein Angebotsverfahren sicher zu ermitteln; cf. §§ 8 [2] und 9 [1] VOB Teil A).

306 backless bench [n] *gard. landsc.* (▶Bench 1 without back-rest); *s* **banco [m] sin respaldo** (▶banco sin respaldo ni brazos); *f* **banc [m] sans dossier** (▶Banc sans dossier et accoudoirs) ; *syn.* banc [m] de type banquette (VRD 1994, partie 4, chap. 6.4.3.1, 10) ; *g* **Hockerbank [f]** (▶Sitzbank ohne Rücken- und ohne Armlehne).

backpacking vacation [n] [US] *recr.* ▶hiking holiday.

back slope [n] [US] *constr.* ▶reverse slope [US]/reverse falls [UK].

307 backwater [n] *geo.* (Former arm of a river, which usually remains attached to the main channel only at the downstream end); *syn.* bayou [n] [also US]; *s* **meandro [m] abandonado** (Antiguo brazo de un río que ha sido separado de éste por estrangulamiento por el extremo río arriba); *f* **méandre [m] abandonné** (Provoqué par le recoupement de méandres entraînés dans leur agrandissement et leur migration vers l'aval ; le *m.* est encore en relation, en général en aval, avec le lit mineur) ; *g* **Altwasser [n] (2)** (Abgetrennte Flussschleife, die noch Verbindung zum Fluss hat, meist durch den flussabwärts liegenden Teil der Flussschleife).

308 backyard garden [n] [US] *gard.* (Garden in the rear of a dwelling; *opp.* ▶front yard [US]/front garden [UK]; ▶private garden); *syn.* back garden [n] [UK]; *s* **jardín [m] trasero** (Jardín detrás de la casa; *opp.* ▶jardín delantero, ▶jardín privado); *f* **jardin [m] arrière** (Jardin situé à l'arrière de la maison d'habitation ; il constitue le jardin d'agrément ; ▶jardin particulier, ▶jardin privatif, ▶jardin privé ; *opp.* ▶jardin de façade) ; *g* **Hintergarten [m]** (Garten hinter einem Wohnhaus; ▶Hausgarten, *opp.* ▶Vorgarten).

bacteria [npl] *bot. pedol.* ▶nodule bacteria.

bad drainage indicator plant [n] *phyt.* ▶impeded water indicator plant.

baffle [n] [UK]**, noise** *constr. envir.* ▶noise barrier wall [US].

baked clay paving [n] [UK] *constr.* ▶brick paving.

balanced fertilizer [n] *agr. constr. for. hort.* ▶complete fertilizer.

309 balanced game management [n] *conserv. game' man.* (Maintenance of healthy wild animal populations by reduction of competitive organisms or groups of organisms, including the ▶reintroduction of animals and plants, to preserve genetic and ecological diversity; ▶game thinning); *s* **medidas [fpl] de regulación de poblaciones de fauna** (Incidencia directa en poblaciones de animales para mantener o desarrollar algunas especies o reducir otras con el fin de garantizar la diversidad genética y ecológica. Entre ellas se encuentra también la ▶reintroducción de especies extintas; ▶caza selectiva); *f* **gestion [f] sélective de peuplements d'animaux** (Opérations visant au maintien ou à la reconstitution de certaines espèces ou populations animales y compris la réduction de la population d'espèces concurrentes dans le but de préserver la biodiversité ; ▶réintroduction 2, ▶chasse régulatrice) ; *g* **Bestandslenkung [f]** (Direkte Eingriffe in Bestände von Tieren zur Erhaltung oder Entwicklung bestimmter Organismenarten oder Organismenkollektive einschließlich der Verminderung konkurrierender Organismen mit dem Ziel, eine genetische und ökologische Vielfalt zu sichern. Zur **B.** gehört somit auch die ▶Wiederansiedelung von Tieren oder Pflanzen; ▶Hegeabschuss).

310 balance [n] **of cut and fill** *constr.* (Planned adjustment of earthwork quantities within a construction project, to equalize the amount of excavation and filling operations, so that there is no deficit or surplus of material; ▶volume of cut and fill); *syn.* volume balance [n] of cut and fill; *s* **compensación [f] de desmontes y terraplenes** (Ajuste planificado de las cantidades de tierra en el marco de un proyecto de manera que se consiga un resultado equilibrado entre desmontes y rellenos, y ni sobre ni falte material; ▶volumen de tierras a mover); *syn.* compensación [f] del movimiento de tierras; *f 1* **équilibrage [m] déblai et remblai** (Prise en compte du terrain et de son relief naturel lors des études et de l'établissement du plan de terrassements afin de réduire ou limiter les apports ou les évacuations de terres ; lors de l'exécution des travaux on parle plutôt de *l'ajustage des volumes de déblais et de remblais* ; VRD 1986, 46 ; ▶déblai de masse) ; *f 2* **équilibre [m] des mouvements de terre** (Résultat) ; *g* **Massenausgleich [m]** (Planerische Lenkung von Erdmassenbewegungen innerhalb eines Bauprojektes, um Abtrag und Auftrag auszugleichen, damit kein Fehl oder Überschuss an Erdmassen entsteht; ▶Bodenmasse); *syn.* Erdmassenausgleich [m].

311 balk [n] *constr.* (Any squared timber or log with a cross section of 8-18 cm, finished for use in construction; a squared timber with a larger cross section is called a ▶beam); *s* **madera [f] escuadrada** (Rollizo de madera aserrado de corte transversal cuadrado o rectangular de 6-18 cm de lado que se usa en la construcción. Si el corte transversal es mayor de 20 cm se llama ▶viga de madera); *f* **bois [m] d'équarrissage** (Bois dont les quatre faces d'équerre et planes ont, en principe 15 à 16 cm de largeur minimum ; DTB 1985 ; ▶poutre en bois) ; *syn.* bois [m] équarri (DTB 1985, 36) ; *g* **Kantholz [n]** (...hölzer [pl]) Schnittholz von quadratischem oder rechteckigem Querschnitt mit Querschnittseiten von 6-18 cm. Hölzer mit Querschnittseiten > 20 cm heißen ▶Balken).

ball [n] *constr. hort.* ▶peat ball.

312 balled and burlapped [loc] *hort.* (Descriptive term for woody plants prepared for shipment and wrapped in burlap; *abbr.* B&B; ▶balled and platformed [US]); *s* **envuelto en saco [loc]** (Protección del cepellón de planta para prepararla para el transporte; ▶embalado en malla metálica y enpaletado [del cepellón]); *f* **présenté, ée en motte avec filet [loc]** (Plante livrée avec sa motte enserrée dans un filet à fine ou large maille en plastique ou en jute ; ▶à motte grillagée livrée sur palette) ; *syn.* habillé, ée par un filet [loc], livré, ée avec tontine de coton [loc] ; *g* **mit Ballentuch balliert [loc]** (Gemäß Gütebestimmungen für Gehölze muss das Ballentuch nach spätestens 1,5 Jahren überwiegend verrottet sein; cf. LEHR 1997, 285; ▶balliert und palettiert).

313 balled and platformed [loc] [US] *hort.* (**In U.S.**, descriptive term for a plant in such large size that it must be contained with wire mesh and supported on a wooden platform or base for handling; **in Europe**, this technique is not used because using tree moving machines; ▶ball wiring [of a tree] is sufficient and more economic; ▶balled and burlapped); *s* **embalado en**

B

malla metálica y enpaletado (del cepellón) [loc] (En EE.UU., método de protección del cepellón de leñosas grandes para el transplante según el cual se cubre el cepellón con malla metálica que se soporta en paleta de madera como base para su manejo; en los viveros en Europa este método no se utiliza, sino que, como se usan máquinas de corte del cepellón, es suficiente y además más económico el ►embalaje en malla metálica [del cepellón; ►envuelto en saco); *f* **à motte grillagée livrée sur palette [loc]** (U.S., technique de transport de gros végétaux dont la motte est entourée d'un panier métallique et fixée sur une palette. Cette méthode n'est pas utilisée dans les pépinières européennes car le conditionnement de la motte avec un treillis très résistant lors de la transplantation effectuée à la machine est entièrement satisfaisant et plus économique ; ►emballage en panier métallique, ►présenté, ée en motte avec filet) ; *g* **balliert und palettiert [loc]** (U.S., Beschreibung eines für den Transport vorbereiteten Großgehölzes, dessen Wurzelballen mit einer Drahtballierung versehen und auf einer Holzpalette verpackt ist. Diese Methode wird in europäischen Baumschulen nicht angewendet, da mit den großen Ballenstechmaschinen eine Ballierung mit starken Drahtnetzen ausreichend und wirtschaftlicher ist; ►Drahtballierung, ►mit Ballentuch balliert).

314 balled and potted [loc] *hort.* (Descriptive term for field-grown plants dug with a ball of earth and placed in a container, instead of being wrapped in burlap; *in comparison to* ►bare-rooted plant, ►balled and burlapped); *s* **embalado/a y plantado/a en maceta [loc]** (Término descriptivo para leñosas jóvenes que han crecido al aire libre y son extraídas de la tierra con cepellón y colocadas en contenedores, al contrario que las ►plantas de raíz desnuda; ►envuelto/a en saco); *f* **en motte et présentée en conteneur [loc]** (Mode de présentation de jeunes plants déterrés en pépinière avec une motte de terre et ensuite conditionné pour la vente en conteneur ; par comparaison avec ►plante à racines nues, ►présenté, ée en motte avec filet ; *contexte* plante livrée en conteneur) ; *g* **balliert und in Container gesetzt [loc]** (Qualitätsbeschreibung von Junggehölzen, die im Freiland mit Ballen gestochen werden und anschließend für den Verkauf in einen Container gesetzt werden; *im Vergleich zu* ►Pflanze ohne Ballen, ►mit Ballentuch balliert).

315 balled and potted plant [loc] [US] *hort* (Field-grown plant dug with a ball of earth and placed in a container instead of ►container-grown plant, ►plant boxing); *syn.* balled container plant [n] [UK]; *s* **planta con cepellón embalada y plantada en maceta [loc]** (Leñosa joven que ha crecido al aire libre y se transplanta a maceta; ►planta en maceta, ►encajonamiento del cepellón); *f* **plante en motte et présentée en conteneur [loc]** (►mise en bac, ►plante cultivée en conteneur) ; *g* **ballierte und in Container gesetzte Pflanze [loc]** (►Containerpflanze, ►Einkübeln).

balled container plant [n] [UK] *hort.* ►balled and potted plant [US].

ball stop fencing [n] [UK] *constr. recr.* ►perimeter fencing.

316 ball wiring [n] **(of a tree)** *hort.* (Containment of a dug root ball with wire mesh for shipment; put [vb/pp] in a wire basket [US], secured [pp] with a wire basket; ►wired root ball); *s 1* **embalaje [m] en malla metálica (del cepellón)** (RA 1970; Proceso de sujetar el cepellón con malla para el transporte); *s 2* **embalaje [m] de malla metálica (del cepellón)** (Resultado de 1; ►cepellón con malla metálica); *f* **emballage [m] en panier métallique** (Renforcement de la motte des végétaux ligneux au moyen d'un treillis ou grillage métallique en vue de leur transport ; ►motte avec panier métallique) ; *syn.* exécution [f] d'une motte grillagée, panier [m] grillagé ; *g* **Drahtballierung [f]** (Befestigung der Wurzelballen von Gehölzen für den Transport mit einem unverzinkten, möglichst geglühten Drahtgeflecht [Maschendrahtsicherung oder Drahtkorb]; ►Drahtballen).

317 balustrade planter [n] *gard.* (Raised plant container in place of a railing or ornamental parapet); *syn.* balustrade plant trough [n]; *s* **jardinera [f] para pretil** *syn.* jardinera [f] para parapeto; *f* **jardinière [f] pour balustrade** *syn.* muret-jardinière [m] ; *g* **Pflanzentrog [m] als Mauerbrüstung**.

balustrade plant trough [n] *gard.* ►balustrade planter.

318 banded bird [n] *zool.* (Bird with a band of metal or plastic on its leg, attached by a behavio[u]ral scientist for identification purposes); *syn.* ringed bird [n] [also UK]; *s* **ave [f] anillada** (Ave marcada con anillo de metal o plástico en una pata con el fin de ser identificada para estudio científico de comportamiento o migración); *f* **oiseau [m] bagué** (Bague fixée par les scientifiques à la patte d'un oiseau en vue de son identification lors de l'étude des migrations) ; *g* **beringter Vogel [m]** (Ein von Verhaltensforschern für wissenschaftliche Untersuchen mit einem Ring versehener Vogel.

319 band [n] **of trees and shrubs** *land'man.* (Generic term for border of trees and shrubs along streams, rivers, field rows or field tracks; ►forest edge, ►forest mantle, ►herbaceous edge [US]/herbaceous seam [UK]); *s* **franja [f] de leñosas** (Banda de árboles y/o arbustos a lo largo de cursos de agua, caminos rurales, etc.; ►abrigo del bosque, ►lindero del bosque, ►orla herbácea); *syn.* banda [f] de leñosas, borde [m] arbóreo; *f* **lisière [f] arborescente** (Peuplement ligneux linéaire le long de cours d'eau, de chemins ruraux ; ►lisière forestière, ►manteau [forestier], ►ourlet herbacé) ; *g* **Gehölzsaum [m]** (Gehölzstreifen entlang von Fließgewässern, Ackerschlägen, Feldwegen etc.; ►Krautsaum, ►Waldmantel, ►Waldrand); *syn.* Gehölzstreifen [m].

bank [n] *agr. geo.* ►field bank; *geo.* ►flat bank, ►undercut bank.

bank [n]**, eroding** *geo.* ►undercut bank.

320 bank-caving [n] *geo.* (Lateral undercutting and undermining of riverbanks, especially at the outside of bends; ►bluff 1, ►riverbank collapse [US]/river-bank collapse [UK], ►streambank erosion, ►undercut bank); *syn.* riverbank cave-in [n] [US]/river-bank cave-in [n] [UK]; *s* **socavación [f] de márgenes** (Término específico de ►erosión de márgenes de cursos de agua: socavación lateral de la orilla sobre todo al chocar la corriente contra la ►orilla cóncava [de un rio]; ►arco erosivo de meandro, ►desplome de orillas); *f 1* **affouillement [m] de la berge** (Terme spécifique pour l'►érosion des berges des cours d'eau qui consiste au creusement latéral par les eaux dans les ►berges concaves meubles d'une rivière suivi d'éboulement, d'écroulement du surplomb ou éboulement ; ►berge d'effondrement, ►effondrement d'une berge) ; *syn.* cavitation [f], sapement [m] de la berge, érosion [f] aréolaire, creusement [m] de la berge ; *f 2* **anse [f] d'érosion** (Résultat du processus d'affouillement de la berge d'un cours d'eau) ; *g* **Auskolkung [f] an/von Flussufern** (*UB zu* ►*Erosion an Ufern von Fließgewässern* seitliche Auskolkung und Unterhöhlung des Flussufers, besonders beim Aufprall der Strömung auf einen ►Prallhang; auskolken [vb]; ►Abbruchufer, ►Uferabbruch); *syn.* Unterschneidung [f] von Flussufern.

bank erosion [n] *geo.* ►streambank erosion.

321 bank erosion [n] **by surface runoff** *geo.* (Washing away of surface soils due to excessive storm water, sheet or channel flow velocities; *s* **erosión [f] de orillas por escorrentía (superficial)**; *f* **érosion [f] de berge par ruissellement** (Ruissellement des eaux superficielles provoquant un ravinement du talus de berge à forte pente) ; *g* **Ufererosion [f] durch Oberflächenabfluss** (Zerstörung eines Ufers, das weder befestigt noch

B

mit Vegetation bewachsen ist, durch Abspülung und Erosions-rillen).

322 bank erosion control [n] *constr. landsc.* (Measures to prevent bank erosion of watercourses, or their destruction by other means; ►river bank stabilization); *syn.* bank protection [n]; *s* **protección** [f] **de márgenes** (Medidas para evitar la erosión de orillas de ríos; ►fijación de orillas); *f* **protection** [f] **des berges (contre l'érosion)** (Mesures de prévention contre l'érosion et la destruction des berges des cours d'eau ; cf. circ. n° 88-67 du 20 juin 1988 ; ►confortement des berges) ; *g* **Uferschutz [m, o. Pl.]** (Maßnahmen zur Verhinderung der Erosion oder sonstiger Zer-störungen von Gewässerufern; ►Uferbefestigung).

323 bank guarantee [n] *contr.* (Pledge to pay; *generic term* ►bid bond [US]/tender bond [UK]); *s* **fianza** [f] **bancaria** (*Término genérico* ►aval); *f* **caution** [f] **bancaire** (*Terme générique* ►caution) ; *g* **Bankbürgschaft** [f] (*OB* ►Bürgschaft).

bank protection [n] *constr. landsc.* ►bank erosion control.

324 bank revetment [n] *eng.* (Stabilization of river banks against lateral erosion with, e.g. ►riprap, concrete or rock pavement, ►paving 1; ►streambank erosion; *specific term* rock revetment); *syn.* paved embankment [n]; *s* **revestimiento [m] de orillas** (Término genérico para cualquier tipo de protección contra la ►erosión de márgenes de cursos de agua, como ►escollera, ►adoquinado, asfaltado); *f* **ouvrage [m] de revête-ment de berge** (WIB 1996 ; revêtement de protection des berges contre l'érosion latérale ou l'action des vagues, p. ex. au moyen d'enrochements, du ►cailloutage des berges, ►pavage 2, dallage ; ►érosion des berges des cours d'eau) ; *syn.* revêtement [m] de berge ; *g* **Uferdeckwerk [n]** (Ein uferdeckender Belag zum Schutz gegen ►Erosion an Ufern von Fließgewässern oder Brandungstätigkeit, z. B. durch ►Steinschüttung auf Ufer-böschungen, ►Pflasterung 2, Asphaltbelag); *syn.* Abdeckung [f] von Uferböschungen.

325 bank seepage groundwater [n] *hydr. wat'man.* (Groundwater which is pumped from bank percolation of a river or lake after lateral infiltration); *s* **agua [f] de infiltración de un curso de agua** (Agua subterránea captada de los aluviones de un río o lago); *f* **eau [f] d'infiltration d'un cours d'eau** (Eau captée par un puits alimenté par l'eau en provenance d'un fleuve) ; *g* **uferfiltriertes Grundwasser [n]** (In der Nähe eines Gewässers mit angelegten Brunnen durch seitliche Infiltration gewonnenes Grundwasser); *syn.* Uferfiltrat [n].

bar [n] *geo.* ►barrier-beach.

326 barbed wire fence [n] *constr.* (Fence installed to deter intruders with horizontal wires bearing sharp pointed spikes at close intervals); *s* **valla [f] de alambre de púas** *syn.* valla [f] de alambre espinoso [Es], valla [f] de alambre espigado [CA], valla [f] de alambre de pinchos; *f* **clôture [f] à fils de fer barbelés** ; *g* **Stacheldrahtzaun [m]** (Zaun, der durch miteinander verfloch-tene Drähte, in die in regelmäßigen Abständen Drahtspitzen oder Metallhaken eingearbeitet sind, Tiere und Menschen am Betreten oder Verlassen eines Geländes hindert).

327 barbel zone [n] **[UK]** *limn.* (►River zone of cyprinoid fishes between ►grayling zone [UK] and ►bream zone [UK]. In the relatively fast-moving and oxygen-rich water with changing temperature during the year [in summer-time sometimes above 20 °C] the typical fish species is the barbel *[Barbus barbus]*; ►epipotamon, ►hyporhithron, ►metapotamon, ►potamal); *syn.* chub zone [n], epipotamal [n] (cf. ILL 1961); *s* **zona [f] de barbo** (Zona superior de la ►región ciprinícola con el barbo como pez característico entre la ►zona de umbra y la ►zona de madrilla. Su biocenosis se denomina ►epipótamon; ►hipóriton, ►meta-pótamon, ►zona biológica de un río); *f* **zone [f] à barbeau** (Tronçon d'un cours d'eau entre la ►zone à ombre et la ►zone à

brème caractérisé par un courant relativement fort et des eaux assez bien oxygénées pour une forte variation annuelle de température — en été plus de 20 °C. À côté du barbeau *[Barbus barbus]* on note la présence d'autres cyprins et voraces d'accom-pagnement — gardon et perche. La **z. à b.** et la zone à brème forment la région cyprinicole ou potamale ; ►écologie des eaux courantes, ►épipotamon, ►hyporhithron, ►métapotamon, ►potamal, ►zone écologique des cours d'eau) ; *syn.* épipotamal [m] (cf. ILL 1961) ; *g* **Barbenregion [f]** (Abschnitt von Fließ-gewässern zwischen ►Äschenregion und ►Brachsenregion. In dem noch relativ schnell fließenden und meist sauerstoffreichen Wasser mit im Verlauf des Jahres stark schwankenden Tempe-raturen [im Sommer stellenweise über 20 °C] finden sich neben der Barbe *[Barbus barbus]* als Leitfisch, Flussbarsch und diverse Karpfenfische. In anderen Teilen Europas treten weitere Arten der Gattung *Barbus* als Leitfische hervor; ►Epipotamon, ►Flussregion, ►Hyporhithron, ►Metapotamon, ►Potamal); *syn.* Epipotamal [n] (cf. ILL 1961).

328 Barcelona Convention [n] *conserv. land'man. leg.* (In 1975 under the aegis of the United Nations Environmental Program [UNEP] the Mediterranean neighboring states and the EU member states of that time met to prepare the so-called „Mediterranean Action plan "[MAP] aimed at stopping the progressive destruction of the Mediterranean; "The Convention for the protection of the Mediterranean Sea against pollution" was adopted in 1976 and came into force in1978. In 1995 certain articles were altered, the convention was revised and renamed as the **"The Convention for the Protection of the Marine Environment and the Coastal Regions of the Mediterranean"**; member states are Albania, Algeria, Bosnia-Herzegovina, Croatia, Cyprus, Egypt, the European union, France, Greece, Israel, Italy, Lebanon, Libya, Malta, Monaco, Morocco, Serbia and Montenegro, Slovenia, Spain, Syria, Tunesia, Turkey; the convention consists today of 6 protocols: 1. Land-Based Sources Protocol, 2. Dumping-Protocol, 3. Hazardous Waste Protocol, 4. Offshore Protocol, 5. Emergency Protocol, 6. Special Protection Areas and Biodiversity Protocol, in which detailed measures are set out for the protection of the Mediterranean. At the beginning the emphasis lay particularly on the avoidance of pollutants entering into the Mediterranean; in 1982 further resolutions were made, which define special protection areas of the Mediterranean. The Convention has been revised in its wording several times since 1975 and accepted by the member states, without, however, coming into force. Only two protocols [5 and 6] were ratified so far by the number of member states necessary to ratify the resolution); *s* **Convenio [m] de Barcelona** (En 1975 bajo el auspicio del Programa de las Naciones Unidas para el Medio Ambiente [PNUMA] se reunieron 16 Estados ribereños del Mediterráneo y la Comisión Europea para preparar y adoptar el llamado «Plan de Acción del Mediterráneo» [PAM] con el fin de frenar la destrucción progresiva del Mediterráneo. El «Convenio para la protección del mar Mediterráneo contra la contaminación» fue aprobado junto con dos protocolos [1 y 2] en febrero de 1976 en Barcelona y entró en vigor el 12 de febrero de 1978. En 1995 se modificaron algunos artículos y la convención für revisada y renombrada en **«Convenio para la protección del medio ambiente marino y las regiones costeras del Medi-terráneo»** y sus modificaciones entraron en vigor el 9 de julio de 2004. Estados miembros son: Albania, Algeria, Bosnia-Herze-govina, Croacia, Chipre, Egipto, Eslovenia, España, Francia, Grecia, Israel, Italia, el Líbano, Libia, Malta, Mónaco, Marrue-cos, Serbia y Montenegro, Siria, Túnez, Turquía y la Unión Europea. La convención tiene hoy siete protocolos, en los cuales se definen medidas detalladas para proteger el Mediterráneo: 1. Protocolo sobre la contaminación causada por vertidos desde

B

buques y aeronaves, 2. Protocolo sobre prevención y situaciones de emergencia, 3. Protocolo sobre la contaminación de origen terrestre, 4. Protocolo sobre las zonas especialmente protegidas y la diversidad biológica, 5. Protocolo sobre la contaminación de la exploración y explotación de la plataforma continental, del fondo del mar y de su subsuelo, 6. Protocolo sobre residuos peligrosos, 7. Protocolo sobre gestión integrada de zonas costeras. Al comienzo se puso énfasis en evitar la contaminación del Mediterráneo, en 1982 se acordó sobre el protocolo 4° sobre zonas protegidas, ampliado en 1995. En 1994 fue firmado el Protocolo 5° (llamado "offshore"), en 1996 el 6° sobre residuos peligrosos y a principios de 2008 el 7°. Hasta la fecha [1/2009] han sido ratificados 5 protocolos, de los cuales el 1° fue modificado en 1995, pero aún no está ratificado. La ratificación más reciente fue el 11 de mayo de 2008, por la que entró en vigor la modificación del Protocolo 3° sobre contaminación de origen terrestre. El Protocolo 7° está aún por ratificar. El Estado depositario del Convenio y los Protocolos es España y la sede se encuentra en Atenas, donde el PNUMA ejerce la secretaría para la aplicación del convenio. Éste y sus protocolos junto con el PAM forman parte del Programa de Mares Regionales del PNUMA; cf. www.unepmap.org); *f* convention [f] de Barcelone (Convention pour la protection de la mer Méditerranée contre la pollution [BarCon] à l'initative du programme des Nations unies pour l'environnement visant à protéger et améliorer le milieu marin dans la zone de la mer Méditerranée ; cet accord-cadre adopté à Barcelone le 16 février 1976 et signé à l'heure actuelle par vingt Etats du pourtour méditerranéen [Albanie, Algérie, Bosnie & Herzégovine, Croatie, Chypre, Egypte, France, Grèce, Israël, Italie, Liban, Libye, Malte, Monaco, Maroc, Slovénie, Espagne, Syrie, Tunisie et Turquie] ainsi que la Communauté européenne ; a été ratifié par la France le 11 mars 1978 ; la Convention a été renforcée et amendée en 1995 ; un nouveau Plan d'action pour la protection du milieu marin et le développement durable des zones côtières de la Méditerranée [PAM Phase II] est alors adopté et elle devient la Convention pour la protection de l'environnement marin et des régions côtières de la Méditerranée ; elle a été ratifiée par la France le 16 avril 2001 et est entrée en vigueur le 9 juillet 2004 ; la Convention est constituée de 6 protocoles **1.** protocole relatif à la protection de la mer Méditerranée contre la pollution d'origine tellurique [Protocole LBS], **2.** protocole relatif à la prévention et l'élimination de la pollution de la mer Méditerranée par les opérations d'immersion effectuées par les navires et aéronefs [Protocole sur l'immersion], **3.** protocole relatif à la prévention de la pollution de la mer Méditerranée par les mouvements transfrontaliers de déchets dangereux et leur élimination [Protocole sur les déchets dangereux], **4.** protocole relatif à la protection de la mer Méditerranée contre la pollution résultant de l'exploration et de l'exploitation de la croûte continentale et du fond marin et de ses substrats [Protocole offshore], **5.** protocole relatif à la coopération dans la lutte contre la pollution de la mer Méditerranée par le pétrole et autres substances nocives dans les situations d'urgence [Protocole d'urgence], **6.** protocole relatif aux Aires Spécialement Protégées et à la diversité biologique en Méditerranée [Protocole sur les ASP et la biodiversité] ; bien que les gouvernements aient acceptés sur le papier de protéger la région méditerranéenne, de nombreux gouvernements refusent encore de ratifier tout ou parties de la Convention, d'autres ne l'ont pas encore transposée en lois nationales ; le PAM, s'il était appliqué, permettrait d'inverser radicalement la tendance actuelle de dégradation de l'environnement et d'obtenir une protection efficace de la mer Méditerranée ; la Convention de Barcelone ne deviendra juridiquement contraignante à l'échelle internationale qu'après sa ratification par un quota défini de parties contractantes ; cependant, seuls deux de ces instruments sont entrés en vigueur

[protocoles 5 et 6]) ; *g* Konvention [f] von Barcelona (Auf Initiative des *United Nations Environmental Program [UNEP]* trafen sich 1975 die Mittelmeeranrainerstaaten und die damaligen EU-Staaten, um den sogenannten *Mediterranean Action Plan [MAP]* auszuarbeiten und die fortschreitende Zerstörung des Mittelmeers aufzuhalten. 1976 nahmen die Mitgliedstaaten die „Konvention zum Schutz des Mittelmeers vor Verschmutzung" mit zwei Protokollen [1 & 2] an; sie trat 1978 in Kraft. 1995 wurden einzelne Protokolle geändert, die Konvention wurde ergänzt und schließlich in „**Konvention zum Schutz der marinen Umwelt und der Küstenregionen des Mittelmeers**" umbenannt; Mitgliedsstaaten sind Albanien, Ägypten, Algerien, Bosnien-Herzegowina, Kroatien, Europäische Union, Frankreich, Griechenland, Israel, Italien, Kroatien, Libanon, Libyen, Malta, Monaco, Marokko, Serbien und Montenegro, Slowenien, Spanien, Syrien, Tunesien, Türkei, Zypern. Die Konvention besteht heute aus 7 Protokollen: 1. „Verklappungs-Protokoll", 2. „Prävention und Notfall-Protokoll", 3. „Land-Based Sources"-Protokoll, 4. Protokoll über „Speziell geschützte Gebiete", 5. „Hochsee-Protokoll", 6. „Gefahrgut-Protokoll", 7. „Küstenmanagement"-Protokoll, in denen detaillierte Maßnahmen zum Schutz des Mittelmeers festgehalten sind. Zu Beginn lag der Schwerpunkt vor allem auf der Vermeidung von Schadstoffeinträgen in das Mittelmeer, 1982 wurde ein weiteres Protokoll aufgenommen, das speziell geschützte Gebiete im Mittelmeer behandelt und 1995 modifiziert wurde; 1994 wurden das sog. Offshore-Protokoll, 1996 das Gefahrgut-Protokoll und Anfang 2008 das Protokoll über Integriertes Küstenmanagement unterzeichnet. Bis zum Januar 2009 wurden 5 Protokolle ratifiziert, von denen eines 1995 modifiziert, jedoch bislang nicht ratifiziert wurde. Die jüngste Ratifizierung war im Mai 2008, wodurch die Veränderung des Protokolls über Verschmutzung auf dem Land in Kraft trat. Die Ratifizierung des 7. Protokolls steht noch aus. Die **K. von B.** mit ihren Protokollen sowie dem MAP sind Bestandteil des UNEP-Programms für Regionale Meere; cf. www.unepmap.org); *syn.* Barcelona-Konvention [f].

barchan(e) [n] *geo.* ▶barkhan.

bare [adj] *bot.* ▶leafless.

bare [adj] **in winter** *bot. hort.* ▶winter bare.

329 bare patch [n] *constr. hort.* (Spot of lawn or planting, which has failed to grow after sowing or planting; ▶bare patch of lawn, ▶loss, ▶over-seeding [US]/over-sowing [UK]); *s* **calva** [f] (Parte de césped o plantación que no ha crecido después de siembra o plantación; ▶calvero de césped, ▶marra, ▶resiembra de césped 2); *f1* **pelade** [f] (Partie dénudée d'une surface engazonnée après ensemencement ; ▶pelade de gazon, ▶perte, ▶réensemencement, ▶semis de regarnissage) ; *syn.* zone [f] mal levée ; *f2* **trou** [m] (Zone dépourvue de végétaux après plantation ; *contexte* trou dans la plantation) ; *syn.* manque [m] ; *g* **Fehlstelle** [f] (Bereich im Rasen oder in einer Pflanzung, der trotz Aussaat oder Bepflanzung vegetationslos geblieben ist; ▶Ausfall, ▶Nachsaat, ▶Rasenfehlstelle); *syn.* Bewuchslücke [f], Kahlstelle [f].

330 bare patch [n] **of lawn** *constr. hort.* (▶Bare patch upon which grass will not or will hardly grow); *s* **calvero** [m] **de césped** (Zona del césped en la que prácticamente no crece ninguna hierba; ▶calva); *f* **pelade** [f] **de gazon** (Partie dégarnie ou clairsemée des gazons ; zone dégradée [p. ex. par le piétinement] dans une pelouse peu fournie ou dépourvue de pousses et nécessitant un regarnissage ; ▶pelade, ▶trou) ; *g* **Rasenfehlstelle** [f] (Bereich im Rasen, in dem keine oder fast keine Gräser wachsen; *OB* ▶Fehlstelle).

bare rock [n] *constr.* ▶establishment of vegetation on bare rock.

331 bare-rooted plant [n] *for. hort.* (Plant shipped without a growing medium surrounding the roots, usually applied to young fibrous rooted shrubs and trees; bare-root planting stock [= BR plant]; ▶root-balled plant); *syn.* bare-rooted stock [n] [also US]; **s planta** [f] **de raíz desnuda** (Planta sin cepellón por haber perdido el sustrato de cultura al ser extraida para transplante; *opp.* ▶planta de cepellón); *f* **plante** [f] **à racines nues** (Plante de pépinière dont la terre ou le substrat de culture n'entoure pas les racines ; *opp.* ▶plante en motte) ; *g* **Pflanze** [f] **ohne Ballen** (Pflanze, bei der die Erde oder das Pflanzsubstrat, in dem die Pflanze kultiviert wurde, an den Wurzeln nicht haften blieb oder für den Transport entfernt wurde [= Pflanze mit nackten Wurzeln]; in D. sollen Gehölze, die ohne Ballen verpflanzt werden, ein Wurzelwerk haben, das artspezifisch und bodenbedingt den 10-15-fachen Durchmesser des Stammes haben; *opp.* ▶Ballenpflanze); *syn. for.* wurzelnackte Pflanze [f].

bare-rooted stock [n] [US] *for. hort.* ▶bare-rooted plant.

332 bark [n] *arb.* (Non-technical term used to cover all the tissues outside the xylem cylinder. In older trees usually divisible into inner bark [living], ▶phloem, and ▶outer bark [dead]; MGW 1964); **s corteza** [f] (Término común para denominar el tejido del tronco por fuera del cambium. En los árboles maduros divisible generalmente en corteza secundaria [viva] o ▶floema y corteza externa [muerta] o ▶ritidoma); *f* **écorce** [f] (Terme employé vulgairement pour appeler tous les tissus extérieurs au bois ; dans les arbres âgés, on peut généralement distinguer : l'écorce interne [vivante] et l'écorce externe [morte] ; ▶phloème, ▶écorce ; MGW 1964) ; *g* **Rinde** [f] (Allgemeiner Begriff für alle Gewebe außerhalb des Kambiums: in älteren Bäumen meist unterteilbar in Innenrinde [lebend], leitenden ▶Bastteil und Außenrinde [abgestorben] und ▶Borke; nach MGW 1964).

333 bark chips [npl] *constr. hort.* (Shredded, but not fermented bark, without any additives which is primarily used as mulch. The chips also contain growth-retarding agents such as resin, tannin and especially phenol, which prevent growth of weeds; ▶bark mulch, ▶shredded bark humus); **s astillas** [fpl] **de corteza** (Corteza de árbol triturada sin fermentar y sin aditivos que se utiliza para cubrir el suelo. Las **a. de c.** contienen sustancias que reprimen el crecimiento [resina, taninos y en especial fenoles] y así evitan el crecimiento de hierbas; ▶humus de corteza, ▶mulch de corteza triturada); *f* **copeaux** [mpl] **d'écorce** (Écorce broyée et tamisée utilisée sous forme de grands copeaux [40-60 mm], de moyens copeaux [25-40 mm] ou de petits copeaux [15-25 mm] comme mulch sur les parterres de fleurs, autour des arbres, sur les allées et les aires de jeux, empêchant la pousse des mauvaises herbes par l'action conjuguée de son pH, des substances inhibitrices [résines, tanins, phénols] contenues dans l'écorce et de son opacité ; ▶humus d'écorce, ▶mulch d'écorde fragmentée) ; *syn.* particules [fpl] d'écorce ; *g* **Rindenschrot** [m *oder* n] (Zerkleinerte, nicht fermentierte Rinde ohne Zusätze, die vorwiegend zur Bodenabdeckung verwendet wird. Rindenschrot enthält noch die wachstumshemmenden Wirkstoffe der Rinde [Harze, Tannine und besonders Phenole], die verhindern, dass Wildkräuter aufkommen; ▶Rindenhumus, ▶Rindenmulch).

334 bark habitat [n] *zool.* (Living space for insects in burrows beneath bark or phloem); **s hábitat** [m] **cortícola** (Espacio vital para insectos en conductos de la corteza o del floema); *f* **habitat** [m] **corticole** (Domaine vital des espèces vivant dans des conduits creusés sous l'écorce ou dans le bois des végétaux ligneux) ; *syn.* milieu [m] corticole ; *g* **Rindenhabitat** [m] (Lebensraum für Insekten in Bohrgängen unter der Rinde oder im Bast).

335 barkhan [n] *geo.* (Crescent-shaped sand dune with convex windward side and steeper concave leeside; ▶parabolic dune); *syn.* barchan(e) [n]; **s barkhana** [f] (Duna con planta en forma de media luna, con las puntas orientadas en el sentido del viento, y cuya altura puede alcanzar unos 10 m. Supone la existencia de un viento unidireccional; DGA 1986; ▶duna parabólica); *syn.* barjana [f], duna [f] barjan; *f* **barkhane** [f] (Expression d'origine turque caractérisant une dune libre et mobile dans les zones à déficit d'alimentation sableuse et où le vent souffle modérément dans une direction unique, en forme de croissant, à convexité tournée face au vent ; ▶dune parabolique) ; dune [f] en croissant ; *g* **Sicheldüne** [f] (Dünenform mit konvexer Luv- und konkaver Leeseite; ▶Parabeldüne); *syn.* Barchan [m].

bark humus [n] *constr. hort.* ▶shredded bark humus.

336 bark mulch [n] *constr. hort.* (MET 1985, 10; bark, typically shredded, used to cover soil; ▶bark chips, ▶mulch); **s mulch** [m] **de corteza triturada** (Cobertura del suelo de ▶astillas de corteza; ▶mulch); *f* **mulch** [m] **d'écorce fragmentée** (Revêtement du sol au moyen de ▶copeaux d'écorce ; ▶mulch); *g* **Rindenmulch** [m] (...e [pl]; Bodenabdeckung aus ▶Rindenschrot; ▶Mulch).

337 bark scorch [n] *arb. constr. for.* (**1.** Localized injury to bark and cambium, caused, e.g. by a sudden increase in exposure of a stem or branch to high temperatures from intense sunlight or by fire; cf. SAF 1983. **2.** Damage to a stem or branch caused by infection, e.g. by fungi, bacterias, generating growth anomalies which may cause dieback of bark and cambium; ▶scorch, ▶sunscorch); **s quemadura** [f] **de la corteza** (**1.** Daños locales en la corteza y el cambium causados por el fuego o por radiación solar excesiva en los troncos de especies leñosas con corteza delgada como hayas y tilos; ▶quemadura [del tronco], ▶socarrado. **2.** Daños en la corteza causados por infección, p. ej. de hongos o bacterias, que conllevan anomalías en el crecimiento y pueden conducir a la muerte de la corteza y el cambium); *f 1* **insolation** [f] **d'écorce** (Ensemble des désordres ainsi que des blessures localisées de l'écorce et du cambium causés p. ex. par une exposition excessive au soleil ; cf. DFM 1975 ; ▶brûlure, ▶brûlure solaire) ; *f 2* **lésion** [f] **corticale** (Maladie infectieuse au niveau des tissus corticaux des arbres provoquée par des agents pathogènes [insectes, bactéries et champignons lignivores] entraînant l'apparition de nécroses et une altération importante de l'écorce) ; *g* **Rindenbrand** [m, o. Pl.] (**1.** Thermischer Rindenschaden durch stellenweises Absterben und Ablösen der Rinde und des Kambiums durch übermäßige Sonneneinwirkung bei plötzlich freigestellten glatt- und dünnrindigen Baumarten, z. B. bei Ahorn, Buche, Linde, Vogelbeere, oder durch offenes Feuer; ▶Sonnenbrand, ▶Verbrennungsschaden. **2.** Infektiös verursachte Schädigung im Rindenbereich, z. B. durch Pilze, Bakterien, die zu Wuchsanomalien und zum Absterben von Rinde und Kambium führen kann); *syn. zu 1.* thermischer Rindenschaden [m]; *syn. zu 2.* Rindenkrebs [m].

338 bark wound [n] *arb. hort.* (Injury to the inner bark tissue caused by cutting, hitting or other impact; ▶stem injury); **s herida** [f] **de la corteza** (▶lesion del tronco); *f* **blessure** [f] **de l'écorce** (Dommage causé à l'écorce par une entaille, un coup ; ▶blessure du tronc) ; *g* **Rindenverletzung** [f] (Durch Schnitt, Schlag oder Stoß verursachte Beeinträchtigung oder Schädigung der Rinde, bei der das Kambium freigelegt oder zerstört wird; gelieferte Baumschulware mit **R.en** ist nicht abnahmefähig; ▶Stammverletzung); *syn.* Rindenschaden [m].

barn manure [n] [US] *agr. hort.* ▶straw dung.

339 barochory [n] *phyt. for.* (Natural regeneration of tree and shrub species from heavy wingless seed, dispersed by gravity; ▶seed rain, ▶natural colonization by seed rain); **s 1 barocoria**

B

[f] (Proceso de diseminación que conduce únicamente a la migración de la planta a pequeñas distancias, variables en función de la inclinación del substrato; cf. DB 1985; ▸nuevo vuelo); *s 2* **brinzal [m] natural (de barocoras)** (Resultado de la barocoria; ▸brinzal, ▸colonización natural de «espacio vacio» por nuevo vuelo); *f 1* **barochorie [f]** (Processus de colonisation spontanée né de l'ensemencement naturel de diaspores lourdes ; ▸flux de semences) ; *syn.* apport [m] barochore ; *f 2* **recolonisation [f] barochore** (Résultat de la barochorie ; ▸semis naturel par flux de semences) ; *g* **Aufschlag [m, o. Pl.]** (Aus schweren, nicht flugfähigen Samen durch natürlichen Eintrag entstandener Aufwuchs; ▸Anflug 1, ▸Anflug 2); *syn.* Barochorie [f].

baroque garden [n] *gard'hist.* ▸French classic garden.

barrage [n] *constr. eng. wat'man.* ▸dam, #1.

340 barren land [n] *agr. for.* (Infertile area unsuitable for both agriculture and forestry ; ▸waste land); *s* **terreno [m] improductivo** (Según la FAO, es aquel que aún encontrándose dentro de las superficies agrícolas no es susceptible de ningún aprovechamiento, ni siquiera para pasto, tales como desiertos, pedregales, torrenteras, cumbres nevadas, etc.; DGA 1986; ▸tierra yerma); *syn.* tierra [f] inutilizable (MEX 1983); *f* **terre [f] inculte** (Territoire impropre aux pratiques culturales agricoles ou forestières ; ▸terre improductive) ; *syn.* terrain [m] stérile ; *g* **Unland [n]** (Land- und forstwirtschaftlich nicht kultivierbares und völlig unfruchtbares Land; ▸Ödland).

barrier [n]**, safety** *urb. trans.* ▸pedestrian guardrail [US].

barrier [n]**, sound** *envir.* ▸noise screening facility.

341 barrier-beach [n] *geo.* (Sandy bar above high tide, parallel to the coastline and separated from it by a ▸lagoon; DNE 1978); *syn.* sand spit [n], bar [n]; *s* **cordón [m] litoral** (Acumulación de arena, limo o cantos rodados dispuesta paralelamente a la costa; DINA 1987; ▸albufera, ▸lagoon)); *syn.* lengua [f] de tierra; *f* **flèche [f] littorale** (Accumulation littorale en cordon libre, de forme allongée, disposée plus ou moins parallèlement à la ligne générale du rivage ; DG 1984 ; ▸lagune, ▸lagon) ; *g* **Nehrung [f]** (Dünenrücken oder lang gestreckte Sandinseln als Trennung zwischen Meer und Haff. Eine Verbindung zwischen Meer und Haff/▸Lagune nennt man ‚Tief'); *syn. auch* Lido [m].

342 barrier chain [n] [US] *urb.* (Chain between posts barring entry to a certain area); *syn.* cordon [n] [UK]; *s* **cadena [f] de protección** (Cadena sujeta a postes destinada a impedir el acceso a una zona específica); *f* **chaîne [f] de protection** (Chaîne destinée à interdire l'accès d'un secteur) ; *g* **Absperrkette [f]** (Bewegliche Metallvorrichtung aus ineinandergefügten Gliedern, die an Pfosten oder Pollern befestigt ist und den direkten Zutritt zu einem bestimmten Bereich verhindert).

343 barrier curb [n] [US] *constr.* (Raised rim of concrete or natural stone as edging to vehicular surface, sidewalk [US]/pavement [UK], or planted area; ▸curbstone, ▸drop curb [US]/flush kerb [UK], ▸mountable curb [US]/splayed kerb [UK]); *syn.* curb [n] [US], edge kerb [n] [UK]; *s* **bordillo [m] de acera** (Piedra de separación de la acera y la calzada; ▸bordillo accesible, ▸piedra de bordillo, ▸piedra de bordillo enterrado); *syn.* encintado [m], contra-fuerte [f] de la acera, (piedra [f] de) cordón [m], guardacanto [m], (piedra [f] de) guarnición [f] [MEX], cuneta [CA]; *f* **bordure [f] de trottoir** (Bordure de type T, cf. Norme N.F. P 98 302 ; élément destiné à délimiter la chaussée des autres fonctions et matériaux de la voirie et de récolter les eaux de ruissellement ; ▸bordure arasée d'épaulement, ▸bordure basse franchissable, ▸élément de bordure) ; *g* **Hochbord [m]** (Hohe Steinkante als Gesamtheit aller (Hoch)bordsteine aus Beton oder Naturstein zur Abgrenzung einer Fahrbahnfläche, eines Bürgersteiges oder einer Pflanzfläche; der **H.** wird aus einzelnen Steinen gebaut, wenn er nicht

— wie meist in den USA — aus Ortbeton hergestellt wird; ▸Bordstein, ▸Flachbord, ▸Tiefbordstein); *syn.* Hochbordstein [m].

barrier effect on frozen air flow [n] *met.* ▸frost accumulation.

344 barrier-free [adj] *leg. plan. sociol.* (Designed and built such that buildings, open spaces, methods of transport, items of technical equipment, acoustic or visual information and communication systems, and other facilities are easily accessible for young children, physically disadvantaged or old people and can be used by them without any particular support from others; *specific term* ▸developed for the handicapped/disabled; *noun* ▸barrier-free accessibility); *s* **accesible para personas con movilidad reducida [loc]** (Diseñados y construidos de tal manera que los edificios, espacios libres, medios de transporte, aparatos técnicos, etc. sean fácilmente accesibles para niños pequeños, personas con discapacidad física o personas mayores y puedan ser utilizados por ellas sin ayuda de otras personas; ▸accesibilidad para personas con movilidad reducida, ▸accesible para discapacitados/as); *syn.* accesible para limitados fisico-motores [C] [loc], adaptado para personas con movilidad reducida [loc]; *f* **accessible aux personnes handicapées [loc]** (Conçu et réalisé de telle sorte que l'aménagement favorise l'accessibilité aux personnes handicapées des locaux d'habitation, des lieux de travail et des installations recevant du public ; ▸accessibilité aux personnes handicapées ; *terme spécifique* ▸adapté,.ée aux personnes handicapées à mobilité réduite) ; *syn.* accessible aux personnes à mobilité réduite [loc] ; *g* **barrierefrei [adj]** (**1.** So gebaut und gestaltet, dass eine bauliche Anlage oder Freifläche von kleinen Kindern, behinderten oder alten Menschen ohne fremde Hilfe genutzt werden kann. **2.** So gebaut und gestaltet, dass „bauliche und sonstige Anlagen, Verkehrsmittel, technische Gebrauchsgegenstände, Systeme oder Informationsverarbeitung, akustische und visuelle Informationsquellen und Kommunikationseinrichtungen sowie andere gestaltete Lebensbereiche, wenn sie für behinderte Menschen in der allgemein üblichen Weise, ohne besondere Erschwernis und grundsätzlich ohne fremde Hilfe zugänglich und nutzbar sind"; cf. § 4 BGG; *UB* ▸behindertengerecht; *Substantiv* ▸Barrierefreiheit).

345 barrier-free accessibility [n] *leg. plan. sociol.* (Condition of buildings and other infrastructural facilities, which enables users, especially physically disadvantaged or old people, to move comfortably, ▸barrier-free); *s* **accesibilidad [f] para personas con movilidad reducida** (Condición de edificios y otros equipamientos que permite su uso cómodo sobre todo a personas con discapacidad física o a personas mayores. En Es. según la Ley de Ordenación de la Edificación, uno de los requisitos básicos de la edificación relativos a la funcionalidad es la accesibilidad, de tal forma que se permita a las personas con movilidad y comunicación reducidas el acceso y la circulación por el edificio; art. 3.1.a2] Ley 38/1999; ▸accesible para personas con movilidad reducida); *syn.* accesibilidad [f] para discapacitados/as, accesibilidad [f] para limitados fisico-motores [C]; *f* **accessibilité [f] aux personnes handicapées (à mobilité réduite)** (L'accessibilité au cadre bâti, à l'environnement, à la voirie et aux transports publics ou privés permet leur usage sans dépendance par toute personne qui, à un moment ou à un autre éprouve une gêne du fait d'une incapacité permanente [handicap sensoriel, moteur ou cognitif, vieillissement, etc.] ou temporaire [grossesse, accident, etc.] ou bien encore de circonstances extérieures [accompagnement d'enfants en bas âge, poussettes, etc.] ; cf. décret du 17 mai 2006 mettant en œuvre le principe d'accessibilité généralisée, posé par la loi du 11 février 2005, qui doit permettre à toutes les personnes, quel que soit leur handicap [physique, sensoriel, mental, psychique et cognitif] d'exercer les actes de la vie quotidienne et de participer à la vie sociale ; à compter du 1er juillet 2007, les travaux réalisés sur l'ensemble de la voirie ouverte à la circulation publique et dans les espaces publics en agglomération doivent prendre en compte la nécessité d'assurer progressivement l'accessibilité de la voirie et des espaces publics aux personnes handicapées ; sont également concernés hors agglomération [les zones de stationnement, les emplacements d'arrêt des véhicules de transports en commun, les postes d'appel d'urgence] ; ces prescriptions s'appliquent lors de la réalisation de voies nouvelles, de travaux ayant pour effet de modifier la structure de la voie ou d'en changer l'assiette, de travaux de réaménagement, de réhabilitation ou de réfection des voies, des cheminements existants ou des espaces publics ; de plus un plan de mise en accessibilité de la voirie et des aménagements des espaces publics, prévu par la loi du 11 février 2005, doit être mis en place par les communes ou les établissements publics de coopération intercommunale dans les trois ans suivant la publication de ce décret, en tenant compte d'éventuels plan de déplacement urbain et plan local de déplacements ; ►accessible aux personnes handicapées) ; *g* **Barrierefreiheit [f]** (1. Beschaffenheit von baulichen und sonstigen Infrastruktureinrichtungen, die Nutzern und besonders Behinderten und alten Menschen ein möglichst beschwerdefreies Bewegen ermöglicht. 2. Beschaffenheit von baulichen und sonstigen Anlagen, Verkehrsmitteln, technischen Gebrauchsgegenständen, Systemen oder Informationsverarbeitung, akustischen und visuellen Informationsquellen und Kommunikationseinrichtungen sowie anderen gestalteten Lebensbereichen, die behinderten Menschen in der allgemein üblichen Weise, ohne besondere Erschwernis und grundsätzlich ohne fremde Hilfe zugänglich und nutzbar sind; cf. § 4 BGG; ►barrierefrei).

346 barrier post [n] *urb. trans.* (Barrier to prevent unauthorized access; ►bollard); *s* **poste [m] de cierre de paso** (Término genérico para cualquier tipo de poste para impedir el acceso a un lugar; ►mojón); *f* **poteau [m] d'interdiction de passage** (interdisant l'accès ou le stationnement ; ►borne) ; *syn.* potelet [m] de dissuasion ; *g* **Absperrpfosten [m] (1)** (OB zu jeder Art von Pfosten, der den Zugang unterbindet; ►Poller); *syn.* Sperrpfosten [m].

barrier wall [n] *constr. envir.* ►noise barrier wall.

basal bowing [n] *arb.* ►basal sweep.

347 basal sweep [n] *arb.* (Curvature of the basal part of a tree stem [hence "sabre butt"], generally induced by ►soil creep 2 or snow pressure on slopes; ►contorted growth); *syn.* basal bowing [n], saber butt [n]; *s* **curvatura [f] basal del tronco** (Árbol con el tronco decumbente en la base; ►reptación hídrica del suelo, ►crecimiento achaparrado); *f* **tronc [m] à crosse** (Forme particulière [morphose] d'un tronc d'un arbre causée par l'action de ►reptation d'un versant ou du poids de la neige, la base du tronc étant arqué vers la vallée ; ►croissance tortueuse) ; *g* **Säbelwuchs [m]** (...wüchse [pl]; durch ►Hangkriechen oder Schneedruck am Hang bedingte Baumstammform, die talseitig gebogen ist; ►Krummwuchs).

base [n] *constr.* ►base course (1), ►footing, ►manhole base, ►subbase [US]/sub-base [UK].

base [n] [US]**, crusher-run** *constr.* ►crushed aggregate subbase [US]/crushed aggregate sub-base [UK].

base [n] [UK]**, granular road** *constr.* ►subbase [US]/sub-base [UK].

base [n] [US]**, rock** *constr.* ►crushed aggregate subbase [US]/ crushed aggregate sub-base [UK].

348 baseball backstop [n] [US] (SPON 1986, 278) *constr. recr.* (Enclosure behind the home plate of a baseball field; ►perimeter fencing); *s* **baranda [f] trasera (de campo de béisbol)** (Valla protectora detrás del plato [home plate] en un campo de béisbol; ►enrejado [de pista de juegos de pelota]); *f* **filet [m] d'arrêt de base-ball** (Clôture grillagée entourant un terrain de base-ball ; ►grillage-pareballon) ; *syn.* écran [m] de protection de base-ball ; filet [m] arrière de base-ball ; *g* **Baseball-Fangzaun [m]** (Zaunanlage hinter dem Schlagmal; ►Ballfanggitter); *syn.* Baseball-Fanggitter [n].

349 base course [n] (1) *constr.* (Pavement layer to level or even out irregularities and fill depressions in a rough surface of the ►subbase [US]/sub-base [UK] and ►water-bound surface or wearing course of a flexible surface or an intermediate layer between subbase/sub-base and water-bound surface/wearing course of a flexible surface [UK]; ►blinding, ►pavement structure; in flexible paving construction there are usually three or four layers depending on the bearing load: subbase, base, [base course or levelling layer] and ►surface layer); *syn.* base [n], leveling layer [n] [US]/levelling layer [n] [UK]; *s* **capa [f] de enrase** (En la construcción de caminos y carreteras, capa que sirve para igualar irregularidades. Se encuentra entre la ►capa portante y el ►firme; ►capa de limpieza, ►estructura del cuerpo de carreteras y caminos, ►pavimento 1, ►revestimiento compactado); *syn.* capa [f] de regularización; *f* **couche [f] de base** (Couche mise en place pour corriger les irrégularités lors de la réalisation de travaux de voirie dont la structure est constituée de grave ou comme couche intermédiaire entre la ►couche de fondation et la ►couche de surface en particulier pour les ►sols stabilisés aux liants hydrauliques ; ►couche anticontaminante, ►structure de la voirie) ; *syn.* couche [f] de dressement ; *g* **Ausgleichsschicht [f]** (Schicht zum Glätten von Unebenheiten und Füllen von Zwischenräumen beim Wegeaufbau mit Schottergerüst oder als Zwischenschicht von ►Tragschicht und ►Decke beim Wegeaufbau ►wassergebundener Beläge; ►Sauberkeitsschicht, ►Wegeaufbau).

base course [n] (2) *constr.* ►subbase [US]/sub-base [UK].

base data [npl] *plan.* ►planning data.

350 Basel Convention [n] *envir. leg.* (The Basle Convention established legally-defined, worldwide control of transboundary movements of hazardous wastes and their disposal; cf. decision 93/98/EC of February 01, 1993); *s* **Convención [f] de Basilea** (Acuerdo internacional que regula el control del transporte y la eliminación de residuos tóxicos y peligrosos; cf. Decisión del Consejo n° 93/98/CCEE); *f* **Convention [f] de Bâle** (cf. décision du Conseil n° 93/98/CEE du 1er février 1993 relative à la conclusion, au nom de la Communauté, de la Convention sur le contrôle des mouvements transfrontières de déchets dangereux et de leur élimination) ; *g* **Basler Konvention [f]** (Im Basler Übereinkommen vom 22.03.1989 weltweit geregelte Kontrolle der grenzüberschreitenden Verbringung von gefährlichen Abfällen und ihre Entsorgung; cf. Beschluss 93/98/EWG des Rates v. 01.02.1993 und in D. Abfallverbringungsgesetz vom 30.09.1994, BGBl. I 2771); *syn.* Baseler Übereinkommen [n].

351 base map [n] (1) *plan.* (Graphic representation showing existing physical and cultural data, such as topographical data, utilities, boundary lines, etc. used as a basis for further planning); *s* **mapa [m] básico** (Representación gráfica de los datos topográficos, las fronteras, las instalaciones existentes, etc. que se utiliza como base para la planificación del desarrollo); *syn.* plan [m] básico (1); *f* **plan [m] de l'état initial** (Plan représentant les donnés topographiques et cadastrales, les ouvrages existants, les réseaux divers, etc. et constituant le plan de base sur lequel sont reportées les diverses données nécessaires à la réalisation du projet d'aménagement) ; *g* **Grundlagenkarte [f]** (Karte, die die vorhandenen topografischen Daten, Leitungen, Grenzen etc.

B

darstellt, auf der weitere für die Planungsaufgabe wichtige Daten eingetragen werden können).

352 base map [n] (2) *plan.* (Graphic representation containing information, mostly in the form of survey data, which is used to compile a site plan and form the basis for preliminary planning; ▶planning data); *s* **plan** [m] **básico** (≠) (2) (Informaciones básicas, generalmente contenidas en planos de inventario, que se utilizan de base para elaborar un plano de situación como condición previa para iniciar el proceso de planificación; ▶bases técnicas de la planificación); *f* **fond** [m] **de plan** (Représentation graphique comportant les informations utilisées pour l'élaboration d'un plan de situation et la conception des premières idées d'aménagement ; ▶données d'une opération [d'aménagement]) ; *g* **Planunterlage** [f] (Daten, meist in Form von Bestandsplänen, zur Erstellung eines Lageplanes als Voraussetzung für erste Planungsüberlegungen; ▶Planungsgrundlage).

basement air shaft [n] [US] *arch.* ▶basement light well.

353 basement light well [n] [US] *arch.* (Shaft in front of a basement window with an open top, often covered by a grate, admitting air or light to basement stor[e]y below grade); *syn.* window well [n] [US], basement air shaft [n] [US], light well [n] [UK]; *s* **pozo** [m] **de luz** (Zanja abierta ante las ventanas de sótanos, generalmente cubierta con emparrillado, para permitir que entren luz y aire); *syn.* patio [m] de luz; *f* **cour** [f] **anglaise** (Courette en contrebas du sol environnant sur laquelle débouchent les fenêtres du sous-sol ; DTB 1985) ; *g* **Lichtschacht** [m] (Gemauerte oder betonierte Vertiefung vor Kellerfenstern unterhalb der Geländeoberfläche, oft mit einem lichtdurchlässigen Rost abgedeckt); *syn.* Kellerlichtschacht [m].

base [n] **of crusher-run aggregate** [US] *constr.* ▶crushed aggregate subbase [US]/crushed aggregate sub-base [UK].

base planting [n] [US] *constr.* ▶foundation planting.

base stem [n] *hort.* ▶cane.

basic level [n] **of pollution** [UK] *envir.* ▶background level of pollution [US].

354 basic professional services [npl] *contr. prof.* (Fundamental work generally necessary for the fulfillment of a planning contract. Depending on the field of work, the services are grouped together per individual ▶work phase [US]/work stage [UK]; ▶special planning services); *s* **servicios** [mpl] **profesionales básicos** (Trabajos realizados para la ejecución correcta de un contrato profesional; ▶fase de trabajo, ▶servicios complementarios de planificación); *f* **prestations** [fpl] **élémentaires** (Prestations nécessaires à la réalisation du contenu des divers éléments d'une ▶mission normalisée de maîtrise d'œuvre. Le regroupement d'un certain nombre de prestations élémentaires en une entité logique constitue les ▶étapes de la mission de maîtrise d'œuvre ; ▶prestation complémentaire) ; *g* **Grundleistungen** [fpl] (Diejenigen Leistungen, die zur ordnungsgemäßen Erfüllung eines Planungsauftrages im Allgemeinen erforderlich sind. Sachlich zusammengehörige **G.en** sind zu jeweils in sich abgeschlossenen ▶Leistungsphasen zusammengefasst; § 2 [2] HOAI 2002 u. § 1 [2] HRLA; in der neuen HOAI 2009 wird nicht mehr von **G.**, sondern nur noch von **Leistungen** gesprochen; ▶besondere Leistung 2).

basic specification clause [n] [UK] *contr.* ▶basic specification item.

355 basic specification item [n] *contr.* (Item which may be modified by an ▶add-on item or ▶alternate specification item; ▶separate specification item); *syn.* basic specification clause [n] [also UK]; *s* **partida** [f] **básica** (Descripción independiente de un servicio relacionada con ▶partidas alzadas o

▶partidas [de obra] adicionales; ▶partida normal); *f* **numéro** [m] **de prix d'une prestation de base** (Description d'une spécification technique détaillée référencée par un numéro d'ordre et complétée par une variante [▶numéro de prix pour une variante] ou une plus-value [▶numéro de prix pour plus-value] dans un descriptif de travaux ; ▶numéro de prix de prestation courante) ; *g* **Grundposition** [f] (Eigenständige Leistungsbeschreibung, die jedoch mit ▶Alternativposition oder ▶Zulagepositionen im Zusammenhang steht; ▶Normalposition); *syn.* Hauptposition [f].

356 basin [n] (1) *geo.* (Shallow structural downfold in the Earth's crust; e.g. Paris basin; DNE 1978); *s* **cuenca** [f] **de sedimentación** (Zona de acumulación de grandes cantidades de sedimentos procedentes de áreas circundantes más elevadas, que puede estar enclavada en depresión o cubeta tectónica y —según su localización— puede ser marina, intermontana, pluvial, etc.; cf. DINA 1987, 275); *syn.* cuenca [f] sedimentaria; *f 1* **bassin** [m] **d'effondrement** (Vaste région déprimée, souvent de plusieurs kilomètres de large, délimitée par des failles et correspondant à une aire d'affaissement) ; *syn.* zone [f] dépressionnaire ; *f 2* **bassin** [m] **sédimentaire** (Nom d'ensemble géologique désignant un creux topographique, région déprimée d'une aire continentale sur laquelle se sont entassés des sédiments marins ou continentaux comme p. ex. le Bassin Parisien, le bassin Viennois, le bassin de Mainz) ; *g* **Senkungsbecken** [n] (Im Durchmesser oft viele Kilometer breites Senkungsfeld, z. B. das Pariser Becken, Wiener Becken, Mainzer Becken); *syn.* Sedimentationsbecken [n].

basin [n] (2) *constr.* ▶pool (1).

basin [n] [US], **access** *constr.* ▶cleanout chamber [US].

basin [n], **kettle** *geo.* ▶kettle.

357 basin check method [n] **of irrigation** *agr.* (Agricultural method by which water percolates slowly over mostly-sloping grassland. The water is taken from streams and regulated by small dams); *syn.* controlled flood irrigation [n]; *s* **método** [m] **de riego en tablares** (Técnica de irrigación agrícola en la que se hace fluir el agua lentamente sobre terreno con pendiente suave); *f* **irrigation** [f] **par submersion contrôlée** (Technique d'arrosage agricole par laquelle l'eau circule lentement à l'air libre et s'infiltre progressivement par ruissellement sur une surface en pente ; les eaux en provenance de ruisseaux sont éventuellement stockées au moyen de retenues assurant ainsi sa régulation) ; *g* **Rieselbewässerung** [f] (Landwirtschaftliche Bewässerungsform, bei der das Wasser mit langsamer Geschwindigkeit vor allem über hängiges Grünland läuft. Die Wassermenge wird Bächen entnommen und durch Stauwerke reguliert).

basis [n] **for professional fees** *contr. prof.* ▶construction costs as a basis for professional fees.

358 basket-weave pattern [n] *constr.* (Used in brick or concrete block paving construction; ▶paving pattern); *s* **aparejo** [m] **trenzado a cesta** (Tipo de dibujo en pavimentación de muros, patios o caminos; *término específico* aparejo al tresbolillo/ medio trenzado a cesta; ▶patrón de aparejo); *syn.* aparejo [m] parquet; *f* **appareillage** [m] **à la Bichonière** (▶Calepinage particulier utilisé pour les briques et les pavés longs) ; *syn.* appareillage [m] en opus reticulatum, appareil [m] réticulé ; *g* **Korbflechtverband** [m] (▶Verlegemuster von Klinkern oder länglichen Betonpflastersteinen).

basophilous species [n] *phyt.* ▶calcicolous species.

bastard trenching [n] [UK] *hort.* ▶double digging.

bath [n] *recr.* ▶air bath.

359 bathing area [n] *recr.* (Site or stretch of water for swimming or other recreation activities; e.g. on a beach of a river or lake, on the sea coast; *specific terms* ►bathing beach, ►river bathing beach, ►sunbathing area); *syn.* swimming area [n] [US]; **s zona [f] de baños** (en playa de río o mar; ►playa [de baños], ►playa de río); *f* **lieu [m] de baignade** (p. ex. le long d'un fleuve, d'une rivière, d'un lac ; ►zone de baignade, ►zone de baignade d'une rivière) ; *syn.* aire [f] de baignade ; *g* **Badeplatz [m]** (z. B. am Fluss, am See, an der Küste; *UBe* ►Badestrand, ►Flussbadestrand).

360 bathing beach [n] *recr.* (Pebbly or sandy shore between high and low-water mark of an ocean, gulf, sea, lake, or river for swimming or other recreation activities; ►river bathing beach); *s* **playa [f] (de baños)** (Playa de mar o de río en la que la calidad del agua se controla regularmente y que generalmente es utilizada por muchas personas; ►playa de río); *f* **zone [f] de baignade** (Bord de mer ou des eaux douces, courantes ou stagnantes sur lequel la baignade est expressément autorisée ou n'est pas interdite, habituellement pratiquée par un nombre important de baigneurs ; Directive du Conseil n° 76/160 CEE du 8 décembre 1975 concernant la qualité des eaux de baignade ; ►zone de baignade d'une rivière) ; *g* **Badestrand [m]** (Teil eines Meeresstrandes oder der Rand eines Binnengewässers, an dem das Baden von den zuständigen Behörden [jeweilige Gemeinde im Einvernehmen mit der unteren Gesundheitsbehörde und unteren Wasserbehörde] wegen der hygienischen Gesamtsituation ausdrücklich gestattet oder nicht untersagt ist und an dem üblicherweise eine große Anzahl von Personen badet; cf. RL 76/160/EWG des Rates vom 08.12.1975; gem. Badegewässerverordnung von BW sind ‚Badestellen' Badegewässer oder Teile davon sowie die angrenzenden Landflächen mit den dazugehörigen Einrichtungen, die von Badenden genutzt werden; cf. GBl. BW Nr. 16 v. 15.10.1999, p. 389 ff; ►Flussbadestrand).

bathing complex [n] [US] *recr.* ►indoor bathing complex, ►swimming center [US].

361 bathing water [n] *recr. leg.* (Water of a river, lake or the sea used by a large number of people for swimming. The adjacent land area usually has appropriate facilities. Permission for bathing is granted by a local authority, responsible for controlling the hygenic quality of the water. Bathing in biologically self-cleaning bodies of water is seen as an economic and ecological atlternative to the use of conventional swimming pools. A European Directive on the management of bathing water quality (2006/7/CE) was issued on February 15, 2006; ►bathing beach); *s* **aguas [fpl] de baño** (Aguas de ríos, lagos o del mar utilizadas por número importante de personas para bañarse resp. nadar. Las zonas adyacentes están normalmente equipadas apropiadamente. Las autoridades locales son responsables de dar el permiso para utilizarlas como **a. de b.** y también de controlar la calidad higiénica de las aguas. El 15 de febrero de 2006 se aprobó una «Directiva Europea relativa a la gestión de la calidad de las aguas de baño» [2006/7/CE del Parlamento Europeo y del Consejo] a través de la cual se regulan el control y la clasificación de la calidad de las mismas y la información pública al respecto, con el fin de conservar, proteger y mejorar la calidad del medio ambiente, y de proteger la salud humana. La Directiva debe ser aplicada a más tardar desde el 24 de marzo de 2008; ►playa de baños); *f* **eaux [fpl] de baignade** (*Zone de baignade aménagée ou habituellement fréquentée par le public même sans aménagement particulier* eaux ou parties de celles-ci [zone d'un plan d'eau], douces, courantes ou stagnantes, ainsi que l'eau de mer, dans lesquelles est **1.** soit expressément autorisée par les autorités compétentes dans la mesure où elles satisfont à des normes européennes, **2.** soit n'est pas interdite et habituellement pratiquée par un nombre important de baigneurs ; les eaux de baignade douces et marines constituent un milieu privilégié pour les activités de loisirs pratiquées par un grand

nombre de vacanciers ; leur qualité est étroitement liée au contexte environnemental avoisinant ; des événements, comme les marées noires et des surcharges des réseaux d'assainissement dans les zones touristiques, ont des conséquences importantes sur la qualité des eaux et sur les lieux de baignade ; la qualité des eaux de baignade revêt un double enjeu : préserver la santé des baigneurs et conforter un site touristique et économique ; conformément à la Directive 2006/7/CE du Parlement européen et du Conseil, du 15 février 2006, concernant la gestion de la qualité des eaux de baignade et abrogeant la directive 76/160/CEE et à la réglementation française prise pour son application [Décret n° 81-324 du 7 avril 1981 modifié], les eaux de baignade font l'objet de contrôles sanitaires réalisés pour le Ministère de la santé, de la famille et des personnes handicapées par les DDASS [Direction départementale de l'action sanitaire et sociale] chaque année pendant la saison balnéaire ; ainsi, tous les ans au début de la saison balnéaire, elles publient simultanément des synthèses départementales et nationales analysant les résultats de la saison précédente ; depuis 2002, le site Internet du Ministère chargé de la santé permet d'avoir accès aux cotes de classification bactériologique, au bilan des années antérieures et aux résultats des analyses de la saison en cours ; des actions, comme le pavillon bleu attribué par la FEEE [Fédération européenne d'éducation à l'environnement], reposent sur des démarches volontaires des communes et celles où l'activité touristique se déploie leur accordent une grande importance ; ►zone de baignade ; *termes spécifiques* bassin de baignade naturel, jardin aquatique) ; *g* **Badegewässer [n]** (Gewässer mit angrenzenden Landflächen und dazugehörigen Einrichtungen, das üblicherweise von einer großen Anzahl von Personen zum Baden genutzt wird; das Baden ist von den zuständigen Behörden [jeweilige Gemeinde im Einvernehmen mit der unteren Gesundheitsbehörde und unteren Wasserbehörde] wegen der hygienischen Gesamtsituation ausdrücklich gestattet oder nicht untersagt; cf. GBl. BW Nr. 16 v. 15.10.1999, p. 389 ff; ►Badestrand. Durch umfangreiche Untersuchungen wird das Baden im vollbiologisch gereinigten Schwimmteich als ökonomische und ökologische Alternative zum konventionellen Schwimmbad bestätigt); *syn.* Badeteich [m], Schwimmteich [m], Schwimm- und Badeteich [m]).

362 bathtub effect [n] [US] *constr. arb.* (Waterlogged condition causing death of newly planted trees in compacted soil. In wet seasons, accumulated water sits in tree pits, unable to drain, and tree roots rot; in drought periods, roots cannot penetrate compacted subsoil to reach groundwater; TGG 1984, 192); *syn.* teacup syndrome [n] [US], pot-binding effect [n] [UK]; *s* **efecto [m] de encharcado** (Causa usual de muerte de árboles recién plantados en suelo compacto. En épocas húmedas, el agua acumulada no puede drenar y causa la podredumbre de las raíces; en épocas secas, las raíces no pueden atravesar el subsuelo compacto para llegar al acuífero); *f* **effet [m] fosse de plantation** (Effet négatif d'une fosse de plantation dans des sols fortement compactés, l'enracinement ne s'effectuant que dans le substrat de plantation. La zone de contact des parois de la fosse empêche l'enracinement dans le sol en place par manque d'air ; l'importante variation des forces capillaires entre le sol en place et le substrat à l'intérieur de la fosse entraîne un dessèchement de la motte ; en période de pluie l'engorgement de la fosse peut provoquer le pourrissement des racines et la mort de la plante) ; *syn.* effet [m] lissage des parois ; *g* **Blumentopfeffekt [m]** (1. Negative Wirkung eines Pflanzloches auf frisch gepflanzte Gehölze im stark verdichteten Boden, so dass nur das „lockere" Erdsubstrat innerhalb der Pflanzgrube durchwurzelt wird. Die Pflanzgrubenwand zum anstehendem, festen Boden [Kontaktflächenzone] wird von den Wurzeln wegen des unzureichenden Bodenlufthaushaltes nicht durchdrungen. Ferner wird wegen der

B

B

erhöhten Saugspannung [des höheren Wasserpotenzials] des festen Bodens dem lockeren Pflanzsubstrat Wasser mit der Folge entzogen, dass im ungünstigen Falle der Wurzelballen austrocknet, d. h. in Trockenperioden können die Wurzeln nicht in den Untergrund Richtung Grundwasser wachsen. Bei Regenperioden bildet sich Stauwasser, das Wurzelfäule und den vorzeitigen Tod bewirken kann. **2.** Derselbe Effekt tritt auch bei Oberbodenandeckungen auf, wenn auf Grund mangelnder Verzahnung mit dem Untergrund nur ein flachgründiges Wurzelwachstum möglich ist); *syn.* Blumentopfwirkung [f] (MEY 1982, 269).

363 bathyal zone [n] *ocean.* (Ocean environment and organisms between the ▶littoral and ▶abyssal zones, i.e. between 200 and 4,000m; ▶benthic zone); **s zona** [f] **batial** (Estrato marino entre 200 y 2000 m de profundidad, de oscuridad total y limitado por la temperatura de 4°C, hábitat de los organismos del ▶bentos entre la zona del ▶litoral y del ▶abisal; ▶zona del bentos); **f zone** [f] **bathyale** (Strate hydrosphérique située entre 200 et 2000 m, de nuit totale et limitée par la température de 4 °C ; ▶zone benthique, la ▶zone benthique littorale, la ▶zone abyssopélagique ; cf. DG 1984) ; *syn.* zone [f] bathypélagique ; **g Bathyal** [n] (Lichtloser Lebensraum mariner Organismen des ▶Benthals zwischen ▶Litoral und ▶Abyssal in 200-3000 m Tiefe).

364 batter [n] *arch. constr.* (Backward-leaning slope from the base upwards—as of a wall, typically set by a specified horizontal to vertical unit ratio; ▶battered wall); **s desplome** [m] (Muro o paramento cuyo ángulo exterior con el plano horizontal es mayor de 90°; ▶muro con desplome); **f fruit** [m] (Inclinaison d'un mur ou muret face à la terre retenue et qui lui permet de mieux résister à sa poussée par son propre poids à la poussée de la terre ; ▶mur [à parement] incliné) ; **g Anlauf** [m] **(1)** (...läufe [pl]; Neigung der Ansichtsfläche eines Bauteils in Prozent von unten nach oben; *UB* Maueranlauf; ▶dossierte Mauer); *syn.* Dossierung [f].

365 battered curb(stone) [n] [US] *constr.* (▶batter); *syn.* batter-faced curb [n] [US], battered kerb [n] [UK]; **s piedra** [f] **de bordillo rebajada** (▶desplome); **f bordure** [f] **de trottoir, type T** (▶fruit ; cf. norme N.F. P 98-302) ; **g Bordstein** [m] **mit Anlauf** (▶Anlauf 1).

battered kerb [n] [UK] *constr.* ▶battered curb(stone) [US].

366 battered wall [n] *arch. constr.* (Retaining wall inclined backward from the vertical; ▶batter); **s muro** [m] **con desplome** (Muro con una de sus caras ataludadas; ▶desplome); *syn.* muro [m] ataludado; **f mur** [m] **(à parement) incliné** (Mur auquel on a donné un ▶fruit à un ou aux deux parements ; *contexte* parement extérieur vertical ; cf. MAÇ 1981, 300) ; **g dossierte Mauer** [f] (Schräg von unten nach oben geneigte Mauer); *syn.* Mauer [f] mit ▶Anlauf.

batter-faced curb [n] [US] *constr.* ▶battered curb(stone) [US].

367 battery [n] **of cages** *agr.* (Series of wire enclosures in an intensive and controversially-discussed method of confining up to 10,000 laying hens. During the total duration of their use for laying, the hens do not leave their cramped cages; ▶factory farming); *syn.* multiple henhouse [n] [also US]; **s batería** [f] **de jaulas de puesta** (Instalación industrial con capacidad de hasta 10 000 gallinas ponedoras consistente en sistema de jaulas muy pequeñas [5 dm², 40 cm de altura] montadas en tres o cuatro pisos y en las cuales las gallinas permanecen toda su vida. Este sistema de cría es muy criticado porque no tiene en cuenta las necesidades básicas de estos animales; ▶ganadería intensiva); *syn.* batería [f] de gallinas ponedoras; **f batterie** [f] **de ponte** (Forme la plus intensive et la plus critique [« aviculture rationnelle »] de l'élevage au sol ; les poules pondeuses sont

élevées dans des poulaillers industriels de ponte constitués de batteries comprenant des cages grillagées [5 dm², 40 cm de hauteur] juxtaposées sur trois ou quatre étages qu'elles ne quittent pas pendant toute la durée de l'élevage ; LA 1981, 897 ; ▶élevage intensif) ; *syn.* élevage [m] intensif de la volaille (DUV 1984, 324) ; **g Legebatterie** [f] (Anlage für eine intensivste, sehr fragwürdige Form der Legehennenhaltung in Stallungen mit bis zu 10 000 Tieren, bei der die Hennen die sehr engen Käfige [450 cm² je Tier] während der Gesamtdauer der Legenutzung nicht verlassen. Die artgemäßen Grundbedürfnisse wie Scharren, Picken, geschützte Eiablage, Sandbaden und erhöhtes Sitzen auf Stangen werden nicht berücksichtigt; *opp.* Bodenhaltung, Freilandhühnerhaltung, Freilandhaltung für Legehennen; ▶Massentierhaltung); *syn.* Legehennenbatterie [f].

bayou [n] [US] *geo.* ▶backwater.

368 beach [n] **(1)** *geo.* (**1.** Gently sloping strip of land with accumulation of unconsolidated water-borne material [usually well-sorted sand and pebbles, accompanied by mud, cobbles, boulders, and smoothed rock and shell fragments] that is subject to wave action [swash and backwash] at the shore between the limits of low water and high water; cf. GFG 1997; the limit of a **b.** begins with a characteristic change in material or physiographic form, such as the ▶white dune or cliff; ▶shoreline erosion [US]/shore line erosion [UK]. **2.** Sandy strip above normal water level of a lake or river, and lacking a bare rocky surface. **3.** *Common usage* sandy shore of a sea, lake or river used for recreation; ▶bathing beach, ▶river bathing beach, ▶lake beach, ▶storm beach [US]/storm-beach [UK], ▶barrier beach, ▶shore); *syn.* strand [n] (GFG 1997); **s playa** [f] (Término genérico para ▶playa [de baños], ▶playa de río, ▶playa costera, ▶playa de lago, ▶cordón playero, ▶erosión marina, ▶duna blanca); **f plage** [f] (Zone plate et découverte en bordure d'un plan d'eau marin, fluvial ou lacustre constituée de matériaux de sables ou de galets ; les galets prédominent dans les plages des côtes crayeuses, schisteuses, gréseuses ; sur les littoraux granitiques, les plages sont plutôt sableuses ; terme générique désignant la ▶plage côtière, la berge d'un fleuve, ▶plage d'un lac, ▶zone de baignade, ▶zone de baignade d'une rivière ; ▶érosion côtière, ▶ablation marine, ▶crête de plage, ▶dune blanche) ; *syn.* grève [f] ; **g Strand** [m] (Meist aus Sand aufgebauter flacher Uferbereich eines Meeres, Sees oder Flusses; der **S.**, der am Meer auch aus gröberen Gesteinsfraktionen bestehen kann, wird von der Brandung, dem ständigen Wechsel von Schwall und Sog, zwischen Niedrigwasser und Hochwasser geprägt und auch verändert [▶Küstenerosion]; der **S.** wird an der Küste durch eine ▶Weißdüne oder Steilküste begrenzt; OB zu ▶Badestrand, ▶Flussbadestrand, Flussstrand, ▶Meeresstrand, ▶Seeuferstrand; ▶Strandwall).

beach [n] **(2)** *recr.* ▶developed beach.

369 beach boardwalk [n] *recr.* (Slightly-raised wooden promenade along the shoreline; ▶boardwalk, ▶footbridge); **s pasarela** [f] **a lo largo de la playa** (Camino de madera ligeramente elevado del nivel de la arena que recorre la costa al borde de la playa; ▶pasarela, ▶pasarela de madera); **f passerelle** [f] **de bord de plage** (Pont en bois étroit légèrement surélevé le long d'une plage ; ▶ponceau, ▶passerelle) ; **g Strandsteg** [m] (Leicht erhöhter Holzweg auf dem Sand oder oberhalb des Sandes am Rande eines Strandes; ▶Holzsteg, ▶Steg).

370 beach erosion [n] **(1)** *geo.* (Removal of loose sand or pebbles adjacent to the sea due to wave action; ▶longshore drift); **s erosión** [f] **de playas costeras** (Desplazamiento de sedimentos finos y gruesos de las playas costeras por efecto del oleaje; ▶deriva litoral); **f démaigrissement** [m] **d'une plage** (Enlèvement d'une quantité notable de matériaux superficiels d'une

B

plage, sous l'effet des tempêtes soufflant du large et des courants d'arrachement ; DG 1984 ; ►dérive littorale) ; *g* **Erosion [f] an Strandküsten** (Abtragung der Fein- und Grobsedimente eines Strandes durch Sog und Schwall; ►Küstenversetzung); *syn.* Stranderosion [f].

371 beach erosion [n] (2) *geo.* (Gradual removal of sand from flat banks of rivers or from beaches of calm waterbodies); *s* **erosión [m] de playas de ríos/lagos** (Desplazamiento de la arena en orillas planas de ríos o lagos con formación de playas; *resultado* deriva de playas); *f* **érosion [f] de la berge** (Enlèvement de sable sur une berge des cours d'eau, des lacs et étangs de faible pente avec formation d'une plage de sable) ; *g* **Ufererosion [f]** (Sandabtrag an flachen Ufern von Flüssen oder Stillgewässern mit Strandausbildung).

372 beach lagoon [n] *geo.* (Section of water separated from, but connected with the sea by ►longshore drift on a strip of sand; ►lagoon); *s* **laguna [f] litoral** (Cuerpo de agua de mar formado por ►deriva litoral que, en comparación con la ►albufera se colmata o desaparece de nuevo después de poco tiempo; ►lagoon); *f* **étang [m] littoral** (Étendue d'eau séparée de la mer par un cordon dunaire ou un par cordon littoral suite à une ►dérive littorale, souvent sous forme de dépression marécageuse ; ►lagune, ►lagon) ; *syn.* étang [m] côtier ; *g* **Strandsee [m]** (Vom Meer durch ►Küstenversetzung abgeschnittene Wasserfläche, die im Vergleich zur ►Lagune meist verlandet).

373 beach promenade [n] *recr.* (Wide and attractively laid out pathway running parallel to a beach); *syn.* coastal promenade; *s* **paseo [m] marítimo** (Camino ancho y bien pavimentado que transcurre paralelo a una playa o a la orilla del mar); *syn.* malecón [CAR]; *f* **promenade [f] de bord de mer** (Promenade piétonnière aménagée en bord de mer) ; *g* **Strandpromenade [f]** (Am Strand entlang führender, breit und repräsentativ angelegter Spazierweg).

374 beam [n] *constr.* (Any squared timber, greater than 18cm [UK], which is finished for use as a horizontal bearing or supporting member in construction; ►balk); *syn.* girder [n]; *s* **viga [f] de madera** (Pieza utilizada para soportar una carga sobre un vano, o de un pilar a otro; DINA 1987; ►madera escuadrada); *f* **poutre [f] en bois** (Grosse pièce de bois de section carrée ou rectangulaire supérieure à 20 cm ; ►bois d'équarrissage) ; *g* **Balken [m]** (Schnittholz von quadratischem oder rechteckigem Querschnitt und Seitenlängen > 20 cm; ►Kantholz).

375 beam [n] **of a pergola** *constr.* (Horizontal, round wood beam, balk [squared timber], or steel girder, which connects the posts of a ►pergola; ►hanging beam); *s* **viga [f] horizontal** (Travesaño redondo, cuadrado o de acero que une los postes de una ►pérgola; ►vigueta); *f* **sablière [f]** (Poutre horizontale — rondin, chevron ou profil métallique — réunissant les piliers d'une ►pergola ; ►solive) ; *g* **Pfette [f]** (Horizontaler Balken — Rundholz, Kantholz oder Stahlprofilträger —, der die Pfosten einer ►Pergola miteinander verbindet; ►Auflageholz).

376 bearing capacity failure [n] *constr. eng.* (**1.** Lateral squeezing or shearing of mostly absorptive and otherwise non-loaded soils surrounding a structure, caused by pressure exerted by that structure and resulting in a lateral rotation of the structure's base, usually tending to be below the foot of a slope. Failure is triggered off, when the shear strength of the soil is exceeded; in buildings, **r. s.s** usually occur only where foundations are shallow; ►shear failure of embankment. **2.** In the case of the bearing capacity of a dike, the clay sealing layer behind the dike may be fractured at high tide by the excessive pressure of the water; the strongly flowing water increase **b. c. f.** and causes a strong shifting of the sand and gravel layer below the layer of clay resulting in cavities, which ultimately lead to the collapse of

the dike); *syn. for 1.* rotational slip (failure) [n]; *s* **deslizamiento [m] rotacional [Es, RA, COL, EC, MEX, YV]** (MESU 1977; desplazamiento de una vertiente o parte de ella a través de un plano arqueado, de manera que la masa de tierra se desliza según un eje de círculo imaginario; cf. DGA 1986; ►deslizamiento planar, ►movimientos gravitacionales); *syn.* corrimiento [m] cilíndrico [Es, RA], deslizamiento [m] cilíndrico [RCH] (MESU 1977); *f* **glissement [m] rotationnel** (Basculement lent et continu d'une importante masse de matériaux meubles suivant une surface de cisaillement plus ou moins circulaire, produit généralement dans les sols hétérogènes, la base du cercle de rotation correspondant à une couche plus résistante : **1.** les **cercles de pied** sont les plus courant. **2.** Les **cercles profonds** ne se produisent que dans le cas où le sol situé sous le niveau du pied du talus est de mauvaise qualité ; ►rupture de terrain, ►mouvement de masse) ; *syn.* glissement [m] circulaire ; *g* **Grundbruch [m]** (**1.** Seitliches Ausquetschen meist bindiger Böden des unbelasteten Geländes in der Umgebung von [Erd]bauwerken durch den Druck des Bauwerks, der ein seitliches Wegdrehen des Bauwerksgrundes entlang einer Gleitfläche verursacht, ausgelöst durch die Überschreitung der Scherfestigkeit des Bodens; **G.brüche** ereignen sich bei baulichen Anlagen i. d. R. nur bei Flachgründungen. **2.** *Bei Hochwasser und Deichbau* Aufbrechen der dichtenden Lehmschicht hinter dem Deich durch zu starken Druck des Druckwassers; das stark strömende Wasser weitet den **G.** auf und verursacht eine starke Materialverschiebung in der Sand und Kiesschicht unterhalb der Lehmdecke mit der Folge einer Hohlraumbildung bis zum Absacken des Deiches; ►Geländebruch, ►Massenversatz); *syn. zu 1.* Rotationsrutschung [f].

bearing capacity [n] **of soil** *constr. eng.* ►soil-bearing capacity.

bearing course [n] [UK] *constr.* ►subbase [US]/sub-base [UK].

377 bearing wall [n] *arch.* (Wall supporting a vertical load as well as its own weight); *s* **muro [m] de contención** (En mampostería, cualquier muro que debe soportar otras cargas además de su propio peso); *syn.* muro [m] de carga; *f* **mur [m] portant** (Mur supportant le poids de la charge verticale des différents niveaux d'un ouvrage et dont l'épaisseur augmente d'étage en étage, des combles jusqu'au sol ; MAÇ 1981, 327) ; *g* **Tragmauer [f]** (Mauer, die lotrecht wirkende Lasten aufnehmen muss).

beating [n] [UK] *constr.* ►replacement planting.

beating up [n] [UK] *constr.* ►replacement planting.

beauty and amenity of the countryside [loc] [UK] *conserv. landsc.* ►beauty and amenity of the landscape [US].

378 beauty and amenity of the landscape [loc] [US] *conserv. landsc.* (Aesthetic presence of a natural or cultural environment; the aesthetic perception within a particular culture; ►landscape character); *syn.* beauty and amenity of the countryside [loc] [UK], excellence [n] of landscape, character and beauty of landscape [loc] (UNESCO 1962); *s* **particularidad [f] y belleza [f] del paisaje** (►particularidad del paisaje); *f* **caractère [m] particulier et pittoresque des paysages et des sites** (Appréciation esthétique portée par une personne sur les sites et paysages naturels uniques et emblématiques pour leur caractère visuel, esthétique, environnemental, patrimonial, ludique, légendaire, etc. dans un contexte culturel donné) ; ►caractère particulier des sites et des paysages) ; *g* **Eigenart und Schönheit der Landschaft [loc]** (Besonderheit und Empfinden der Ästhetik einer naturnahen oder Kulturlandschaft durch einen Betrachter eines bestimmten Kulturkreises; der kulturelle Wert von Eigenart

B

und Schönheit ist dem zeitlichen Wandel unterworfen und jeweils vom kulturellen Umfeld abhängig; ►landschaftliche Eigenart).

beauty [n] **of landscape** [US]**, character and** *conserv. landsc.* ►beauty and amenity of the landscape [US]/beauty and amenity of the countryside [UK].

be [vb] **builder for** *constr. contr.* ►be contractor for.

379 become [vb] **established** *constr. hort. for.* (Taking root of a plant, seed, cutting or raft); *syn.* take [vb], take [vb] root [also US]; *s* **arraigar** [vb] (Proceso de inserción de las raíces de una planta en el suelo o en el medio de crecimiento); *syn.* enraizar [vb]; *f* **s'enraciner** [vb] *syn.* prendre [vb] pied, prendre [vb] racine ; *g* **anwachsen** [vb] (Wurzeln schlagen von gepflanzten Pflanzen oder deren Teile).

380 become [vb] **stunted** *hort.* (To cease growing or grow very slowly; *context* the growth of a transplanted mature tree or large shrub is stunted or it has stunted growth/retarded growth); *s* **atrofiarse** [vb] (Proceso de estancamiento del crecimiento o crecimiento muy reducido de árboles grandes transplantados; atrofia [f] [de árboles]); *syn.* crecer [vb] achaparrado (cf. HDS 1987); *f* **végéter** [vb] (S'emploie pour décrire p. ex. le ralentissement ou l'interruption de la croissance chez les sujets transplantés) ; *g* **verhocken** [vb] (Phänomen bei verpflanzten [Groß]gehölzen, die mehrere Jahre keine Triebe bilden oder kümmerlich weiterwachsen).

381 become [vb] **woody** *bot. hort.* (Incorporation of lignine in the cells of young herbaceous stems; lignification [n]; *context* lignified stem); *syn.* lignify [vb]; *s* **lignificarse** [vb] (Almacenamiento de lignina en las paredes celulares de los tallos y las raíces de las plantas leñosas; lignificación [f]); *f* **aoûter** [vb] (Lignification des jeunes pousses des végétaux ligneux vers la fin de l'été ; *substantifs* aoûtement [m], pousse [f] aoûtée, lignification [f] (LA 1981, 86 et PR 1987) ; *syn.* lignifier [vb] ; *g* **verholzen** [vb] (Einlagern von Lignin in Zellwänden krautiger Sprossachsen und in Wurzeln; Verholzung [f], verholzter Trieb); *syn.* holzig werden [vb], lignifizieren [vb].

382 becoming feral [loc] *zool.* (**1.** Introduction of certain domesticated animals into the wild, e.g. house cats, feral pigs, dingos in Australia, mustangs in North America; ►feral species. **2.** *phyt.* Dispersal of cultivated plants in the wild; ►naturalized species, ►overgrowing); *s* **asilvestramiento** [m] (**1.** Proceso por el cual, una especie de animal doméstico, como p. ej. gatos domésticos, cerdos asilvestrados, dingos en Australia o mustangs en Norteamérica, se vuelve silvestre al reproducirse y distribuirse libremente en la naturaleza. **2.** *phyt.* Propagación de plantas cultivadas en la naturaleza; ►especie asilvestrada, ►especie naturalizada, ►invasión de advenedizas, ►invasión de leñosas); *f* **retour** [m] **à l'état sauvage** (**1.** Animaux domestiques vivant en liberté dans la nature, p. ex. les chats, les cochons, les dingos en Australie, les mustangs en Amérique du Nord. **2.** *phyt.* Dissémination dans la nature des espèces de cultures ; ►espèce naturalisée, ►espèce subspontanée, ►espèce retournée à l'état sauvage, ►réapparition des espèces sauvages) ; *g* **Verwilderung** [f] **(1)** (**1.** Leben von bestimmten Haustieren als Wildtier in der freien Natur, z. B. Hauskatzen, verwilderte Hausschweine, Dingos in Australien, Mustangs in Nordamerika. **2.** *phyt.* Verbreitung von Kulturpflanzen in der freien Natur; ►neuheimische Art, ►Verwilderung 2, ►verwilderte Art).

383 becoming interlocked [ppr] *conserv. ecol.* (cf. TGG 1984, 272; process of connecting, e.g. habitats in a ►habitat network, ►hedgerow landscape, ►network 1); *syn.* linkage [n]; *s* **articular** [vb] (Proceso de creación de red p. ej. de biótopos para permitir el intercambio incluido el genético y así asegurar la preservación de las especies a largo plazo; ►paisaje de bocage, ►red de biótopos, ►reticulación, ►reticulación de biótopos);

f **constituer** [vb] **un réseau** (Action de relier entre eux des biotopes identiques ou similaires pour former des refuges et des relais, véritables ponts écologiques ou enchaînement de corridors écologiques dans un espace rural ou urbain, le long des fleuves, etc. ; ►bocage, ►constitution d'un réseau d'habitats naturels, ►réseau de biotopes ; ►formation d'un réseau, ►réseau) ; *syn.* créer [vb] un réseau, former [vb] un réseau ; *g* **vernetzen** [vb] (Verknüpfen von z. B. gleichartigen oder ähnlichen Biotopen zu einem Biotopverbund, Hecken zu einem Heckenverbund [►Heckenlandschaft]; ►Biotopvernetzung, ►Vernetzung).

384 becoming steppe-like [loc] *agr. land'man.* (Transformation to an unforested plain by dehydration of the soil accompanied by alteration in the vegetation; this is caused by climatic change or massive human intervention in the natural environment; e.g. deforestation, overly intensive use of the soil, lowering of groundwater table; ►desertification, ►extensively cleared land [for cultivation]); *s* **estepización** [f] (Sequía del suelo causada por la sobreexplotación de recursos hídricos o por cambios climáticos que conlleva la transformación de la vegetación hacia una que resiste la falta de agua; ►desertificación, ►paisaje desnudo); *syn.* transformación [f] en estepa; *f* **transformation** [f] **en steppe** (Processus d'assèchement du sol et, par corrélation, de transformation de la végétation causé par un changement général des conditions climatiques ou par des atteintes importantes dans un écosystème [dues p. ex. à la déforestation, à une surexploitation agricole] ; *terme spécifique dans les régions tropicales* **savanisation** [f] ; ►désertification, ►paysage ouvert) ; *g* **Versteppung** [f] (Durch Klimaveränderung oder durch massive Eingriffe in Natur und Landschaft — z. B. durch Entwaldung, zu intensive Bodennutzung, Grundwasserabsenkung — verursachte langsame Austrocknung des Bodens und damit verbundene Veränderung der Vegetation; ►Desertifikation, ►ausgeräumte Landschaft); *syn.* Versteppen [n, o. Pl.], zur Steppe werden [vb].

385 be [vb] **contractor for** *constr. contr. syn.* be [vb] builder for [also US]; *s* **ser** [vb] **contratista de**; *f* **exécuter** [vb] **des travaux** ; *g* **ausführen** [vb] *syn.* Ausführung übernehmen [vb].

be [vb] **cut once a year** [UK] *agr. constr. landsc.* to ►be mown once a year [US].

bedding [n] *hort.* ►seasonal bedding.

bedding display [n]**, seasonal** *hort.* ►seasonal bedding.

bedding layer [n] [UK] *constr.* ►laying course for a paved stone surface.

386 bedding material [n] *agr.* (Dry leaves or straw which is used for resting of livestock; *specific terms* bedding straw, ►leaf litter); *s* **camada** [f] (Capa de ►hojarasca o paja para cubrir el suelo en los establos de ganado); *syn.* cama [f], yazija [f]; *f* **litière** [f] (En agriculture, lit de paille ou de feuilles sèches étendu sous les animaux dans les étables ; ►litière feuillue) ; *g* **Einstreu** [f] (In der Landwirtschaft verwendetes trockenes Laub oder Stroh für das Lager von Stallvieh; *UB* ►Laubstreu).

387 bedding plant [n] *hort.* (Plant which is suitable for masses of seasonal rotation planting; *cultivated in nurseries* pot bedding plant; ►seasonal bedding); *s* **planta** [f] **de macizo** (Planta apropiada para plantar en macizo o para ►plantación estacional); *f* **plante** [f] **à massifs** (Plante appropriée aux plantations de bordure ou aux ►plantations saisonnières; *terme spécifique* plante vivace de plate-bande) ; *syn.* plante [f] molle (LA 1981, 653), plante [f] de plate-bande ; *g* **Beetpflanze** [f] (Gärtnerisch kultivierte und gezüchtete Pflanze, die wegen ihrer Blüten- oder Blattwirkung für Rabattenbepflanzungen oder ►Wechselbepflanzung geeignet ist; *UB* Beetstaude).

388 bedding rose [n] *hort. plant.* (Growth form of rose which varies according to the height of the cultivar, 40-100 cm, and according to the panicle or umbel-like inflorescence; e.g. ▶polyantha rose, ▶floribunda rose, cluster rose; ▶hybrid tea); *s* **rosa [f] de macizo** (Tipo de rosa que —según la variedad— alcanza una altura de 40-100 cm y es rica en inflorescencias de forma umbelífera o paniculada; también se le conoce como ▶rosa floribunda o ▶rosa polyantha híbrido; ▶rosa híbrido de té); *f* **rosier [m] buisson à fleurs groupées** (Variétés de rosiers dont la hauteur varie de 40 à 100 cm et caractérisées par des fleurs groupées en bouquets ; elles sont aussi connues sous le nom de ▶rosier floribunda et ▶rosier polyantha ; ▶rosier hybride de thé) ; *syn.* rosier [m] à massifs, rosier [m] polyantha pour massifs et bordures ; *g* **Beetrose [f]** (Rosenwuchsform, die je nach Sorte 40-100 cm hoch wird und zahlreiche Blüten in dolden- oder rispenähnlichen Blütenständen hat; auch als ▶Floribundarose oder ▶Polyantharose bekannt; ▶Edelrose).

389 bedding sand [n] *constr.* (Sand of a ▶laying course for a paved stone surface; ▶sand bed, ▶coarse bedding sand); *s* **arena [f] para adoquinados** (▶Lecho de arena, ▶lecho de asiento [de adoquinado], ▶arena de asiento gruesa); *syn.* arena [f] para empedrados; *f* **sable [m] de pavage** (Sable utilisé pour la forme de pose de pavés sur un ▶lit de sable ; ▶lit de pose du pavage, ▶sable grossier de pose) ; *g* **Pflastersand [m]** (Sand, der als ▶Sandbett für das Setzen von Pflastersteinen dient; ▶Pflasterbett, ▶Pflastergrand); *syn.* Bettungssand [m].

390 bed load [n] *geo. hydr.* (Total of the solid matter which is transported by a watercourse due to ▶tractive force); *syn.* debris load [n]; *s* **carga [f] de fondo** (Cantos y arenas transportadas por un curso de agua sobre el fondo del lecho por rodamiento y saltación; DGA 1986; *término genérico* carga de sedimento; ▶fuerza de tracción); *syn.* arrastre [m] de fondo, acarreo [m] de fondo; *f* **charge [f] de fond** (Galet transporté sur le fond du lit d'un cours d'eau par la ▶force tractive de l'eau) ; *g* **Geschiebe [n] (1)** (Geröll, das in einem Fließgewässer durch die ▶Schleppkraft des Wassers transportiert wird; *UBe* Flussgeschiebe, Bachgeschiebe).

391 bed-load discharge [n] *hydr.* (Amount of solid matter transported per unit of time in watercourses); *s* **descarga [f] de fondo** (Cantidad en peso, masa o volumen de arrastre de fondo transportado, a través de una sección transversal de un curso de agua, por unidad de tiempo; WMO 1974); *f* **charriage [m] de fond** (VOG 1979 ; transport de matériaux sur le fond d'un lit par roulage ou saltation) ; *syn.* transport [m] de la charge (DG 1984, 456) ; *g* **Geschiebeführung [f]** (Geschiebemenge, die ein Fließgewässer je Zeiteinheit transportiert).

392 bed [n] **of a continually-flowing stream** *geo. hydr.* (▶stream bed); *s* **lecho [m] de arroyo de caudal permanente** (Lecho por el cual el agua corre permanentemente; ▶lecho de arroyo); *f 1* **lit [m] d'un ruisseau à écoulement pérenne** (Lit dans lequel l'écoulement des eaux ne cesse jamais) ; *syn.* lit [m] d'une rivière à écoulement pérenne ; *f 2* **lit [m] mouillé** (Partie du ▶lit d'une rivière constamment en eau) ; *g* **immer Wasser führendes Bachbett [n]** (▶Bachbett); *syn.* ständig Wasser führender Bach [m].

bedrock [n] *constr.* ▶massive bedrock.

393 bedrock [n] **[US]/bed-rock** [n] **[UK]** *constr. pedol.* (Solid unweathered rock underlying the superficial layer of topsoil, subsoil, loose sediments and other unconsolidated material; cf. DNE 1978; ▶parent material); *s* **roca [f] in situ** (Roca no alterada que se encuentra bajo el substrato transformado por procesos de edafogénesis; ▶roca madre); *f* **roche [f] en place** (Roche saine située sous la tranche d'altération superficielle ; ▶roche-mère) ; *syn.* roche [f] in situ ; *g* **anstehendes Gestein [n]** (Unter verwitterter Bodenschicht liegendes festes Gestein; ▶Ausgangsgestein).

394 bedroom community [n] [US] *urb.* (Built-up area on the periphery of a large city used mainly for residential purposes; ▶satellite city, ▶housing subdivision [US]/housing estate [UK]); *syn.* bedroom suburb [n] [US], dormitory suburb [n] [UK]; *s* **ciudad-dormitorio [f]** (Aglomeración urbana que tiene una función esencialmente residencial, carece de comercio y servicios y no dispone de puestos de trabajo en número proporcional a la de su población activa. La población activa sólo ocupa estas ciudades después de la jornada laboral; DGA 1986; ▶ciudad-satélite, ▶polígono residencial); *f* **cité-dortoir [f]** (Lieu d'habitation dont la fonction principale est de loger des personnes dont le lieu de travail est dans un centre voisin ; PR 1987 ; ▶ville nouvelle, ▶ensemble résidentiel) ; *syn.* ville-dortoir [f], banlieue-dortoir [f], commune-dortoir [f] ; *g* **Schlafstadt [f]** (Eine auf die Funktion des Wohnens reduzierte Siedlung am Rande einer Großstadt; ▶Trabantenstadt, ▶Wohnsiedlung).

bedroom suburb [n] [US] *urb.* ▶bedroom community [US].

bee forage plant [n]**, woody** *hort. landsc.* ▶woody host plant for bees.

bee plant [n] [US] *hort. landsc.* ▶host plant for bees.

bee plant [n]**, woody** *hort. landsc.* ▶woody host plant for bees.

bees forage plant [n] *hort. landsc.* ▶host plant for bees.

Belgian block pavers [npl] [US] *constr.* ∗▶large-sized paving stone [US]/large-sized granite sett [UK].

395 bell pipe [n] [US] *constr.* (Irrigation, drainage or sewage pipe element which has a widened neck enabling the insertion of the spigot end of a connecting pipe. The joint is sealed by a caulking compound or with a compressible ring); *syn.* socket pipe [n] [UK]; *s* **tubo [m] de enchufe** (Pequeño trozo de tubo que se utiliza para unir dos tramos de tubería; DACO 1988); *syn.* manguito [m]; *f* **tuyau [m] à collet** (Tuyau d'irrigation, d'arrosage ou d'assainissement dont l'extrémité de raccordement est pourvue d'un renforcement tronconique) ; *g* **Muffenrohr [n]** (Bewässerungs- oder Entwässerungsrohr mit einem Rohrverbindungselement in Form eines an einem Ende aufgeweiteten Hohlzylinders).

396 below frostline [loc] *constr. eng.* (UBC 1979, 470; depth below the soil surface that local frost will not penetrate; **in U.S.**, the base of construction footings are typically placed 200-300mm below this line; varies by code; ▶frost heave); *s* **inferior al nivel de helada [loc]** (Referente a la construcción de los cimientos de un edificio que de esa manera no se ven afectados por ▶levantamiento por congelación); *f* **hors gel [loc]** (Ne pouvant être déformé ou dégradé sous l'action du gel, p. ex. une fondation, une chaussée hors gel ; ▶foisonnement par le gel) ; *g* **frostfrei [adj]** (Ein Fundament betreffend, das nicht durch Frosthub verändert werden kann; *Kontext* frostfreie Gründung; eine Mauer frostfrei gründen [vb]; ▶Frosthebung).

below grade [n] [US] *constr.* ▶below ground surface.

397 below-grade construction [n] *urb.* (PSP 1987, 191; installation of an underground development, e.g. underground parking garage or underground train system in order to reduce stationary traffic in the streets, to create underground malls/basement-level shopping concourses [CDN] which are free of traffic noise and fumes); *syn.* subterranean construction [n], underground construction [n]; *s* **construcción [f] subterránea** (de garajes, etc. para reducir la superficie utilizada por vehículos); *f 1* **construction [f] souterraine** (Occupation de l'espace souterrain par différents usages tels que les parcs de stationnement, les aires de stockage des produits énergétiques, les tunnels pour les transports collectifs ou routiers, les centres commerciaux aménagés de zones piétonnes) ; *syn.* construction [f] en souter-

B

rain ; *f 2* **urbanisme [m] souterrain** (Politique d'occupation du sous-sol pour décongestionner les centres urbains, aménager des zones commerciales à l'abri des intempéries et libérées des nuisances de bruit et d'odeurs) ; *g* **Unterbauung [f]** (**1.** *von Gebäuden oder Freianlagen* Anlage von Tiefgaragen oder U-Bahnen in Untergeschossen oder im Untergrund unter öffentlichen Flächen. **2.** Der franz. Terminus *urbanisme souterrain*, der im Dt. mit **städtebaulicher Nutzbarmachung von Untergeschossen und Unterbauungen öffentlicher Flächen** übersetzt wird, ist ein planungspolitisches Konzept zur Verdrängung des ruhenden Verkehrs, zur Schaffung von Verkehrslärm und – schmutz unbelasteten Fußgänger- und Einkaufsbereichen und dient der Entflechtung des Individual- und öffentlichen Personennahverkehrs).

398 below ground surface [n] *constr.* (Term used to measure the depth below ground, e.g. groundwater table, pipeline level; ►depth of groundwater table); *syn.* below grade [n] [also US]; *s* **bajo el nivel [m] del suelo** (►profundidad del nivel freático); *f 1* **sous la surface [f] du sol** (Expression utilisée pour définir l'épaisseur à partir du niveau d'un **terrain**, p. ex. la profondeur d'un aquifère ; ►profondeur de la nappe phréatique) ; *f 2* **sous-sol [m] fini futur** (Définition de la hauteur du fond d'une fouille, d'une tranchée par rapport au niveau du sol fixée par les plans) ; *g* **unter Flur [f]** (*i. S. v. Geländehöhe*, z. B. der Grundwasserstand liegt 2 m unter Flur; ►Grundwasserflurabstand); *syn.* unter Bodenoberfläche [f], unter Erdoberkante [f].

belt [n] *landsc. urb.* ►greenbelt [US]/green belt [UK]; *phyt.* ►riverine reed belt; *geo. phyt.* ►sediment accumulation belt; *landsc.* ►shelterbelt (1); *met.* ►thermal belt.

belt [n]**, alpine** *geo. phyt.* ►alpine zone.

belt [n]**, colline** *geo. phyt.* ►colline zone.

belt [n]**, commuter** *plan.* ►suburban area.

belt [n]**, montane** *geo. phyt.* ►montane zone.

belt [n]**, nival** *geo. phyt.* ►nival zone.

belt [n]**, planar** *geo. phyt.* ►planar zone.

belt [n]**, protective** *landsc.* ►shelterbelt (1).

belt [n]**, screen planting** *gard. landsc.* ►screen planting (2).

belt [n] **[US], stockbroker** *recr. urb.* ►origination area [US].

belt [n]**, subalpine** *geo. phyt.* ►subalpine zone.

belt [n]**, vegetation** *geo. phyt.* ►vegetation altitudinal zone.

belt highway [n] **[US]** *trans.* ►beltway.

belts [npl] *geo. phyt.* ►altitudinal belts.

399 beltway [n] **[US]** *trans.* (Highway encompassing a large city, which can be entered at various points; ►ring road, ►bypass [US]/by-pass [UK]); *syn.* belt highway [n] [US]; *s* **autopista [f] periférica** (Autopista que bordea parcial o totalmente una ciudad; ►anillo periférico, ►carretera de circunvalación); *syn.* variante [f] [Es]; *f* **rocade [f] autoroutière** (Voie de communication autoroutière qui contourne en totalité ou en partie une agglomération urbaine ; ►rocade, ►périphérique, ►route de contournement) ; *syn.* périphérique [m] urbain ; *g* **Autobahnring [m]** (Geschlossene oder nicht geschlossene Umfahrung einer Großstadt; ►Ringstraße, ►Umgehungsstraße).

400 be [vb] **mown once a year** **[US]** *agr. constr. landsc.* (Once-a-year meadow mowing [US]; ►cut a few times per year, ►mowing of meadows [US]/cutting of a meadow [UK], ►once-a-year mown meadow [US]/once-a-year cut meadow [UK], ►single mowing); *syn.* be [vb] cut once a year [UK]; *s* **segado/a [pp] anualmente** (Prado de siega que se corta una vez al año; ►pradera de siega anual, ►segado/a varias veces al año, ►siega, ►siega anual); *f* **fauché,ée une fois par an [loc]** (►Fauche exécutée une fois par an, ►fauche, ►prairie fauchée une fois par

an, ►fauché plusieurs fois par an) ; *g* **einschürig [adj]** (Einmal je Vegetationsperiode gemäht: jährlich einmal zu mähen[de Wiese]; ►einmalige Mahd, ►einschürige Wiese, ►Mahd, ►mehrschürig); *syn.* einmähdig [adj], einmal gemäht [pp], einmal geschnitten [pp].

401 bench [n] (1) *gard. landsc.* (Generic term for outdoor seat [US]/outdoor seating [UK]; ►backless bench, ►circular bench, ►double-back bench, ►free-standing bench/seat, ►park bench, ►permanently installed bench, ►tree bench); *syn.* seat [n]; *s* **banco [m]** (Término genérico para ►bancos adosados, ►banco alrededor de un árbol, ►banco circular, ►banco de jardín, ►banco de parque, ►banco fijo, ►banco móvil, ►banco sin respaldo); *f* **banc [m]** (Terme générique pour ►banc autour d'un arbre, ►banc circulaire, ►banc de jardin, ►banc de parc, ►banc double, ►banc posé, ►banc sans dossier, ►banc scellé) ; *g* **Sitzbank [f]** (OB zu ►Baumbank, ►Doppelbank, ►fest verankerte Bank, ►Gartenbank, ►Hockerbank, ►lose aufgestellte Sitzbank, ►Parkbank, ►Rundbank); *syn.* Bank [f].

bench [n] (2) *min.* ►working bench.

402 benchmark [n] *surv.* (A surveyor's mark on a permanent feature that has a known position and is used as a reference point to determine the relative elevations of a given site; this may be called a 'control point' for smaller surveys; ►datum level, ►datum point); *s* **marca [f] geodésica** (Punto fijo cuya posición se conoce exactamente y que se utiliza como punto de referencia para determinar la posición de otros; ►cota de referencia, ►punto de agrimensura); *f* **repère [m] de niveau (cotation NGF)** (Cote de niveau en général scellée dans une maçonnerie et ramenée au Nivellement Général de la France ; ►cote d'origine, ►point coté) ; *g* **Höhenfestpunkt [m]** (Fest eingebauter Vermessungspunkt; ►Bezugshöhe, ►Vermessungspunkt).

403 benchmark protection [n] *constr.* (Prevention of any changes to datum levels established for a construction site); *s* **protección [f] de cota de referencia** (Prevención de cambios en el nivel de la cota de referencia establecida en el terreno de una obra); *syn.* protección [f] de marca de altura; *f* **conservation [f] des repères de niveau** (Protection des repères de niveau ou des piquets déterminant l'implantation des ouvrages sur un chantier avec l'obligation de les rétablir ou de les remplacer en cas de besoin) ; *g* **Sicherung [f] von Höhenfestpunkten** (Schützen von Höhenfestpunkten auf der Baustelle vor Veränderung).

404 bench [n] **with armrests** *gard. landsc.* (Bench with raised arm supports at each end or at intervals in a continuous strip design); *s* **banco [m] con brazos**; *f* **banc [m] à accoudoirs** ; *g* **Sitzbank [f] mit Armlehne**.

405 bench [n] **with backrest** *gard. landsc.* (*opp.* ►backless bench); *s* **banco [m] con respaldo** (*opp.* ►banco sin respaldo); *f* **banc [m] à dossier** (*opp.* ►banc sans dossier) ; *g* **Sitzbank [f] mit Rückenlehne** (*opp.* ►Hockerbank).

beneficial animal species [npl] *agr. for. hort.* ►beneficial species, ►use of beneficial animal species.

beneficial effects [npl] **of the forest** *land'man.* ►social benefits of the forest.

beneficial influences [npl] **of the forest** *land'man.* ►social benefits of the forest.

406 beneficial species [n] *agr. for. hort.* (Animal species, particularly an insect, which aids cultivation of the land by feeding upon certain injurious species of plants and animals, thus limiting their multiplication; ►biological pest control, ►pest and weed control, ►use of beneficial animal species; *opp.* ►pest); *syn.* pest-eating animal species [n]; *s* **animal [m] beneficioso** (Especie animal aprovechada en la agricultura para combatir especies nocivas para las plantas de cultivo; ►aprovechamiento de animales beneficiosos, ►control biológico, ►control anti-plaguicida; *opp.* ►depredador); *syn.* especie [f] animal benefi-ciosa; *f* **auxiliaire [m]** (Être vivant prédateur ou parasite des espèces animales, néfaste aux cultures ; les auxiliaires sont utilisés en ►lutte biologique pour détruire les ennemis des plantes cultivées ; LA 1981 ; ►protection des végétaux, ►utilisation d'auxiliaires ; *opp.* ►déprédateur) ; *syn.* prédateur [m] pour la lutte biologique, faune [f] utile (LA 1981, 927) ; *g* **Nützling [m]** (Tierart, die durch ihre Ernährungsweise zugunsten wirtschaftlicher Bestrebungen des Menschen zur Eingrenzung bestimmter schädlicher Pflanzen- und Tierarten beiträgt; ►biologischer Pflanzenschutz, ►Nützlingseinsatz ►Pflanzenschutz; *opp.* ►Schädling); *syn.* Nutzorganismus [m].

407 beneficial use [n] **of the environment** *conserv. nat'res.* (Use of the environment or some part of it for the benefit of living organisms, including pollution control and conservation of natural resources; ►natural resources management); *s* **uso [m] usufructuario del medio** (Uso de la naturaleza o de partes de ella para el beneficio de todos los organismos, considerando la protección, conservación y el desarrollo de la capacidad funcional de la misma, ►gestión de los recursos naturales); *f* **gestion [f] économe des espaces, ressources et milieux naturels** (Pratiques d'utilisation des ressources naturelles et des paysages visant à satisfaire les besoins humains et compatibles avec les exigences de protection, de mise en valeur et de développement de l'environnement ; ►gestion des ressources naturelles) ; *syn.* gestion [f] économe du système naturel, gestion [f] respectueuse de l'environnement ; *g* **pflegliche Nutzung [f] der Umwelt** (Nutzung von Natur und Landschaft zum Wohle aller Lebewesen unter Berücksichtigung des Schutzes, der Pflege und Entwicklung der Leistungsfähigkeit des Naturhaushaltes; ►Bewirtschaftung der Naturgüter); *syn.* pflegliche Nutzung [f] des Naturhaushaltes.

benefits [npl] *recr.* ►having recreation benefits; *landsc. plan. urb.* ►long-term benefits.

benefits [npl] **from nature benefits** *ecol.* ►contributions from nature.

benefits [npl] **of a forest** *recr.* ►amenity benefits of a forest.

benefits [npl] **of building materials** *arch.* ►physical benefits of building materials.

benefits [npl] **of open areas** *recr.* ►social benefits of open areas.

benefits [npl] **of the forest** *land'man.* ►social benefits of the forest.

408 benefit-value analysis [n] *plan.* (►Evaluation procedure by which the choice between alternatives is made with reference to several evaluation criteria or principles; ►cost-benefit analysis); *s* **análisis [m] del valor de uso** (►Método de evaluación en el que la elección entre varias alternativas se hace basándose en diferentes criterios o principios de evaluación; ►análisis costo-beneficio); *syn.* análisis [m] costo-eficacia; *f* **analyse [f] coûts-efficacité** (►Procédé d'évaluation pour lequel le choix entre les solutions alternatives s'effectue sur la base d'éléments monétaires et non monétaires qualitatifs tels que l'évaluation de la valeur récréative des forêts, des espaces verts et des paysages naturels, etc. ; ►analyse coûts-avantages) ; *g* **Nutzwertanalyse [f]** (►Bewertungsverfahren, bei dem die zur Wahl stehenden Möglichkeiten bezüglich mehrerer Bewertungsmaßstäbe — oder zumindest bezüglich mehrerer Bewertungsvorschriften — zu bewerten sind; es wird nach Vorteilen klassifiziert und mit Punkten bewertet [Punktierung]; die erreichte Punktzahl gibt Auskunft über den Gesamtnutzwert der zu bewertenden Variante oder des zu bewertenden Gutes; ►Kosten-Nutzen-Analyse).

benthic organisms [npl] *limn. ocean.* ►benthos.

benthic region [n] *limn. ocean.* ►benthic zone.

409 benthic zone [n] *limn. ocean.* (**1.** Bottom of a lake; ►littoral zone [US]/littoral [UK], ►profundal zone. **2.** Bottom of an ocean; ►littoral zone [US]/littoral [UK], ►bathyal zone, ►abyssal zone, ►hadal; ►pelagic zone); *syn. for* 2. benthic region [n]; *s* **zona [f] del bentos** (**1.** Zona del fondo de las aguas continentales que se divide en ►zona litoral y ►zona profunda de un lago. **2.** Zona del fondo de las aguas marinas que se divide en ►litoral, ►zona batial, ►abisal y ►zona hadal; ►zona pelágica); *f* **zone [f] benthique** (Zone correspondant au fond — sol — d'une étendue d'eau. Pour les eaux douces on distingue la ►zone benthique littorale et la ►zone benthique profonde ; en milieu marin on distingue la ►zone marine littorale, la ►zone bathyale, la ►zone abyssopélagique et la ►zone hadopélagique ; ►zone pélagique) ; *g* **Benthal [n]** (Gesamte Bodenregion eines Gewässers. Bei Süßgewässern gliedert man in ►Litoral und ►Profundal; im Meer werden ►Litoral, ►Bathyal, ►Abyssal und ►Hadal unterschieden; ►Pelagial); *syn.* benthischer Bereich [m].

410 benthos [n] *limn. ocean.* (Complete range of forms of plant and animal life living at the bottom of the sea, a lake or river; ►benthic zone); *syn.* benthic organisms [npl]; *s* **bentos [m]** (Organismos ligados a la proximidad del sustrato sólido en los fondos marinos y lacustres; MARG 1981, ►zona del bentos); *syn.* comunidad [f] del bentos; *f* **benthos [m]** (Ensemble des organismes végétaux ou animaux aquatiques — espèces benthiques — vivant sur le substrat solide des fonds marins et lacustres et ainsi que des cours d'eau ; ►zone benthique) ; *g* **Benthon [n]** (Gesamtheit der im ►Benthal lebenden Organismen in Seen, Fließgewässern oder im Meer; benthisch lebend [ppr/adj]); *syn.* Benthos [n].

berm [n] *constr.* ►bund (3); ►earth mound; *trans.* ►road verge.

berm [n] [US] *constr. eng.* ►ledge (1).

berm [n], **noise attenuating** *constr. envir.* ►noise attenuation mound.

411 berm ditch [n] *agr. constr.* (Open trench for conveying water off slopes which are prone to erosion and slippage hazard); *syn.* check swale [n]; *s* **surco [m] en curva de nivel** (Surcos poco profundos que se hacen exactamente a lo largo de las curvas de nivel para recoger las aguas, de modo que se filtren en los terrenos en vez de formar corrientes y provocar la erosión; cf. MEX 1983); *f* **fossé [m] gradin** (Fossé ouvert recoupant très légèrement les courbes de niveau de terrains agricoles en pente comme protection contre le ruissellement et l'érosion) ; *g* **Entwässerungsgraben [m] in einer Böschung** (Offener, hangparalleler Graben an rutsch- oder erosionsgefährdeten Böschungen); *syn.* Rigole [f].

berme [n] *trans.* ►road verge.

berme [n] [US] *constr. eng.* ►ledge (1).

412 Bern Convention [n] *conserv. leg. pol.* ("Convention on the Conservation of European Wildlife and Natural Habitats" adopted at Bern in 1979; wild flora and fauna constitute a natural heritage of great value that needs to be preserved and handed on to future generations. In addition to national protection programmes, the parties to the Convention consider that cooperation should be established at an European level; ►Bonn Convention); *s* **Convención [f] de Berna** («Convención relativo a la conservación de la vida silvestre y del medio natural en Europa» acordada en Berna en 1979; ►Convención de Bonn); *f* **Convention [f] de Berne** (« Convention relative à la conservation de la vie sauvage et du milieu naturel en Europe » ; convention du 19 septembre 1979 fixant la liste des espèces et leur niveau de

B

protection ainsi que la liste des moyens et des méthodes de chasse et autres formes d'exploitation interdites ; ►Convention de Bonn) ; *g* **Berner Konvention [f]** (1979 in Bern verabschiedetes internationales „Übereinkommen zur Erhaltung der wild lebenden Tier- und Pflanzenarten mit ihren Lebensräumen in Europa"; ►Bonner Konvention).

413 berth [n] *recr.* (Place of a ship at a wharf or at anchor in habo[u]r); *s 1* **atracadero [m]** (Lugar en el muelle de un puerto donde se sujetan los botes); *s 2* **fondeadero [m]** (Espacio en puerto o bahía donde los botes pueden echar el ancla); *f* **zone [f] de mouillage** (Emplacement sur le domaine public maritime dans un port, une rade aménagée d'équipements légers ou favorables pour jeter l'ancre assujetti à une autorisation d'occupation temporaire) ; *g* **Liegeplatz [m] (1)** (Ankerplatz im Hafen oder an einer Bootsanlegestelle).

best kept village competition [n] [UK] *hort. plan. urb.* ►"All-America City" [US].

414 best management practices [npl] (BMP) [US] *adm. leg. wat'man.* (In U.S., practices determined by a planning, environmental, or other federal, state, or local government agency to be the most effective and practicable means—after technological, economic and institutional considerations—of preventing ►point source pollution and ►nonpoint source pollution [US]/ non-point source pollution of rivers and streams; cf. RCG 1982. In U.K., the Water Resources Act 1991 governs the control of the pollution of water resources; ►water pollution); *syn.* water quality control [n] [UK]; *s* **protección [f] de la calidad de las aguas continentales** (Todas las medidas técnicas, económicas y administrativas que sirven para evitar la ►contaminación focal y la ►contaminación ubicua de las aguas continentales; ►contaminación de las aguas continentales); *f* **préservation [f] de la qualité des eaux (continentales)** (Toutes mesures techniques, économiques et administratives mises en œuvre par l'agence de bassin en vue de la prévention des risques de pollutions accidentelles ponctuelles et superficielles. Ces mesures trouvent leur application dans les ►schémas directeurs d'aménagement et de gestion des eaux [S.D.A.G.E.] et les ►schémas d'aménagement et de gestion des eaux [S.A.G.E.], les contrats de rivière [instruments contractuels permettant la réalisation des objectifs fixés dans les ►cartes départementales d'objectifs de qualité, le plan décennal de restauration et d'entretien des rivières ainsi que dans le dispositif du S.A.G.E.] ; cf. CPEN ; ►pollution ponctuelle, ►pollution des eaux, ►pollution superficielle) ; *g* **Gewässerreinhaltung [f]** (Alle technischen, ökonomisch und verwaltungsmäßig koordinierten Maßnahmen, die dazu dienen, ►flächenhafte Verschmutzungen und ►punktuelle Verschmutzungen in [Fließ]gewässern zu vermeiden; ►Gewässerbelastung, ►Gewässergütekarte).

415 bestowed [pp] protection status *adm. conserv. conserv'hist. leg.* (Legal condition granted to, e.g. a ►nature reserve, ►national park, ►landscape feature, ►monument, ►special area of conservation [SAC], protected resource, etc. *Context* protective status has been bestowed by law on this monument; ►conservation criterion); *s* **procedimiento [m] de declaración** (de ►espacio natural protegido, de ►bien de interés cultural o de ►zona especial de conservación; ►declarar protegido, ►criterio de protección); *f 1* **procédure [f] de mise sous protection** (*Terme générique,* traduction littérale de l'allemand, terme utilisé en Suisse dans le cadre de la loi sur la protection de la nature et du paysage mais peu usité en France ; on aura tendance à utilisé les l'expressions *f 2* et *f 3* caractérisant les procédures spécifiques mises en œuvre pour la protection des sites) ; *syn.* mise [f] sous protection [B, CH] ; *f 2* **classement [m]** (pour un ►parc national, ►site classé, ►réserve naturelle, ►monument, etc. ; ►classer) ; *syn.* procédure [f] de classement

[m], acte [m] de classement (PPH 1991, 275) ; *f 3* **inscription [f]** (Procédure de mise sous protection de sites et espaces naturels en ►site inscrit ; ►inscrire, ►critère de protection) ; *syn.* procédure [f] d'inscription ; *g* **Unterschutzstellung [f]** (**1.** Verfahren, durch das Flächen oder Einzelobjekte unter Schutz gestellt werden, z. B. ►Naturschutzgebiet 2, ►Naturreservat, ►Nationalpark, ►besonderes Schutzgebiet, ►Landschaftsteil, ►Denkmal etc. **2.** Verfahren, durch das Tiere und Pflanzen z. B. im Washingtoner Artenschutzabkommen, in der EU-Vogelschutzrichtlinie, in der Bundesartenschutzverordnung unter Schutz gestellt werden; ►unter Schutz stellen; ►Schutzkriterium).

best practicable environmental option [n] (BPEO) *envir.* ►state-of-the-art (2).

best practicable means [npl] (BPM) *envir.* ►state-of-the-art (2).

416 best practice [n] *constr.* (Way in which measures are carried out based on state-of-the-art technology, scientific principles as well as practical experience and official recommendations) ; *s* **mejor práctica [f] posible** (Modo de realizar medidas basándose en la ►mejor tecnología asequible, principios científicos así como en la experiencia profesional y en las recomendaciones oficiales al respecto); *f* **bonne pratique [f]** (Mesures exemplaires qui par expérience [résultats et processus suivis] exercent une influence positive sur des systèmes et des pratiques ; ces pratiques méritent d'être exploitées par de nouveaux utilisateurs dans des contextes différents en vue p. ex. d'obtenir l'assurance de bonne qualité, réaliser une bonne fabrication ou de respecter les exigences d'un développement durable ; cette démarche est utilisée dans de très nombreux secteurs économiques, preuve en est la profusion de guides, codes et contrats de bonnes pratiques, p. ex. guide de bonne pratique des déchets de chantiers, guide de bonnes pratiques pour la gestion des véhicules hors d'usage, guide de bonnes pratiques pour les voies vertes en Europe, guides de bonnes pratiques d'hygiène, guide de bonnes pratiques pour le respect de l'environnement, fiche des bonnes pratiques écologiques, etc. ; *contexte* bonnes pratiques de protection de l'environnement ou bonnes pratiques environnementales, bonnes pratiques industrielles, agricoles, etc.) ; *g* **gute fachliche Praxis [f] (GfP)** (Durchführung von Maßnahmen, die in der Wissenschaft als gesichert gelten, die auf Grund praktischer Erfahrungen als geeignet, angemessen und notwendig anerkannt sind und die von amtlichen Beratungen empfohlen werden. Die **g. f. P.** ist im deutschen Recht festgeschrieben und stellt ökologische und naturschutzfachliche Mindestanforderungen an die Bewirtschaftung, z. B. § 5 [4] BNatSchG und § 17 BBodSchG für die landwirtschaftliche Bewirtschaftung, § 2a [1] PflSchG für den Umgang mit Pflanzenschutzmitteln, Verordnungen der Länder über die **g. f. P.** in der Fischerei); *syn.* fachgerechter Umgang [m].

best value for the price bidder [loc] [US] *constr.* ►bidder offering best value for the price [US]/tenderer offering the best value for the money [UK].

beveled [adj] *constr.* ►chamfered.

bevelled [adj] *constr.* ►chamfered.

bg-slab [n] [UK] *constr.* ►grass paver.

B horizon [n] *constr. pedol.* ►subsoil.

bicycle lane [n] [UK] *trans.* ►bike lane [US].

bicycle path [n] [UK] *trans.* ►bikeway [US].

bicycle track [n] [US] *trans.* ►bikeway [US].

bicycle trail system [n] [US] *trans. urb.* ►bikeway network.

417 bid [n] [US] (1) *constr. contr.* (Offer to perform the work described to construct a project for a specified sum; ►bidding procedure); *syn.* tender [n] [UK]; *s* **oferta [f]** (Declaración de una empresa o persona candidata de su intención de realizar un proyecto en el marco del ►procedimiento de convocación de concursos-subasta y de contratación de obras); *syn.* proposición [f]; *f 1* **offre** [f] **(de prix)** (Déclaration d'engagement d'un candidat pour la passation d'un marché sur appel d'offres ; ►organisation des marchés publics) ; *f 2* **soumission** [f] (Déclaration d'engagement d'un candidat pour la passation d'un marché sur adjudication) ; *g* **Angebot** [n] (Willenserklärung einer Partei, die zur Schließung eines Vertrages abgegeben wird. Für die Baubranche wird im Rahmen des VOB-Vergabeverfahrens der Inhalt des **A.s** in § 21 VOB Teil A vorgeschrieben; ►Ausschreibungswesen); *syn.* Offerte [f] [auch CH].

bid [n] [US] **(2)** *constr. contr.* ►accept a bid [US], ►acceptance of a bid [US] (1), ►acceptance of a bid [US] (2), ►have a bid accepted [US], ►invitation for a bid [US], ►subsidiary bid [US]/ subsidiary tender [UK].

bid [n]**, acceptance period of a bid** [US] *constr. contr.* ►period for acceptance of a bid [US].

bid [n] [US]**, supplementary** *contr.* ►subsidiary bid [US]/ subsidiary tender [UK].

418 bid advertisement [n] [US] *adm. constr. contr.* (Public announcement of competitive bidding [US]/competitive tendering [UK] for construction work or delivery of materials—published as an official notice in local media with wide public distribution [US]/published in official gazettes [UK] or in periodicals—with all relevant descriptions about the public developer/authority, subject of bidding/tendering, place of construction, delivery, execution of work[s], or execution of professional services; selective bidding/selective tendering after a ►public pre-qualification bidding notice [US]/public prequalification tendering notice [UK] is also officially published); *syn.* tender notice [n] [UK]; *s* **anuncio** [m] **público de concurso-subasta** (Anuncio publicado en la prensa solicitando ofertas para realizar proyecto de construcción. Se utiliza generalmente para cumplir los requisitos legales de ejecución de obras públicas; ►aviso de candidaturas a concurso público); *f* **avis** [m] **d'appel d'offres** (Publication du lancement de la consultation des entreprises afin que celles-ci puissent participer à l'appel d'offres de marchés de travaux dans un journal habilité à recevoir les annonces légales ou Bulletin officiel des Annonces des Marchés publics ; ►appel public de candidatures pour un appel d'offres) ; *syn.* avis [m] d'appel d'adjudication ; *g* **Ausschreibungsankündigung** [f] (Öffentliche Bekanntmachung von öffentlichen Ausschreibungen über Bauleistungen oder Lieferungen in amtlichen Veröffentlichungsblättern, Tageszeitungen oder Fachzeitschriften mit allen Angaben zum Auftraggeber, Gegenstand der Ausschreibung und Ort der Ausführung, der Lieferung oder der Leistungserbringung; beschränkte Ausschreibungen nach ►öffentlichem Teilnahmewettbewerb werden auch öffentlich bekannt gemacht); *syn.* Bekanntmachung [f] von Ausschreibungen.

419 bid analysis [n] [US] *constr. contr.* (Review and assessment of bids [US]/tenders [UK] to ascertain whether unit prices are reasonable); *syn.* tender analysis [n] [UK]; *s* **control** [m] **de precios de las ofertas** (Verificación de que los precios unitarios de las ofertas son razonables); *f* **analyse** [f] **des prix présentés** (CCM 1984, 37 ; contrôle du sérieux des prix unitaires d'une offre ; les prix doivent paraître compatibles avec une exécution correcte) ; *g* **Preisprüfung** [f] (Prüfung der Angemessenheit der Einheitspreise eines Angebotes oder der Gesamtangebotssumme).

420 bid bond [n] [US] *contr.* (Generic term for surety by a domestic bank, or by government for foreign work; *specific term* ►bank guarantee; ►performance bond, ►surety bond [US]/bond [UK]; *syn.* guarantee bond [n] [UK], tender bond [n] [UK]; *s* **aval**

[m] (Documento a presentar por la empresa licitante que asegure la prestación de fianza por parte de un banco, caja de ahorro u otra institución crediticia autorizada para cubrir la garantía provisional exigida para las empresas que acuden a procedimientos abiertos o restringidos de contratación de obras del Estado; cf. art. 36 [1b] Ley 13/1995 de 18 de mayo, de Contratos de las Administraciones Públicas; ►fianza bancaria, ►garantía de cumplimentio de contrato, ►deposito de fianza); *syn.* fianza [f]; *f* **caution** [f] (Document fournit par l'entrepreneur dans lequel une banque ou tout autre organisme financier garantit envers d'éventuels créanciers l'exécution d'obligations futures en cas d'insolvabilité ; ►caution bancaire, ►garantie de bonne fin des travaux, ►sûretés) ; *g* **Bürgschaft** [f] (In der Baubranche vom Auftragnehmer beizubringende ►Sicherheitsleistung in Form eines Dokumentes, in dem seitens eines Geldinstitutes mit der Höhe der Bürgschaft zeitlich befristet oder unbefristet versichert wird, für die Erfüllung von künftigen Verbindlichkeiten einzustehen, wenn der Auftragnehmer zahlungsunfähig wird; bei Bürgschaften für Bau- und Lieferungsleistungen wird zwischen folgenden **B.en** unterschieden: ►**Vertragserfüllungsbürgschaft**, Vorauszahlungsbürgschaft [muss der Auftragnehmer beibringen, wenn er im Voraus Geld für noch nicht erbrachte Bauleistungen oder noch nicht gelieferte Waren erhalten möchte] und **B. für Mängelansprüche** [tritt nach Abnahme der ausgeschriebenen Leistungen in Kraft, um die Erfüllung der Mängelansprüche incl. Schadensersatz und die Erstattung von Überzahlungen incl. Zinsen sicherzustellen]; ►Bankbürgschaft).

421 bidder [n] [US] **(1)** *contr.* (Firm which has submitted a competitive offer to perform a prime contract, or has responded to a request for a price proposal; ►applicant for bidding documents [US]/applicant for tendering documents [UK], ►bid [US] 1/tender [UK], ►highest bidder [US]/highest tenderer [UK], ►lowest bidder [US]/lowest tenderer [UK], ►selected bidder [US]/selected tenderer [UK]); *syn.* tenderer [n] [UK], submittee [n] of a bid [also US], submitter [n] of a bid; *s* **licitante** [m/f] (Empresa o persona que ha entregado una oferta en concurso o subasta; ►oferta, ►candidato/a); *syn.* concursante [m/f], postor, -a [m/f], licitador, -a [m/f]; *f 1* **concurrent** [m] (Candidat ou entreprise candidate consultée ayant remis une ►offre dans une procédure d'appel d'offres ; ►candidat au marché) ; *syn.* entreprise [f] concurrente ; *f 2* **soumissionnaire** [m] (Candidat ou entreprise candidate consultée ayant remis une ►soumission dans une procédure d'adjudication) ; *g* **Bieter/-in** [m/f] (Firma, die bei einer Ausschreibung oder Preisanfrage ein ►Angebot abgegeben hat; ►Bewerber); *syn.* Erklärende/-r [f/m].

422 bidder [n] [US] **(2)** *contr.* (Responder in competitive bidding [US]/competitive tendering [UK] on contracts for construction projects or supply of goods); *syn.* tenderer [n] [UK]; *s* **participante** [m/f] **en subasta de obras o servicios** (Licitante en una subasta de obras o servicios); *f* **participant** [m] (Concurrent, soumissionnaire à un appel d'offres ou un appel de candidature pour un marché de travaux ou de fournitures) ; *syn.* candidat [m] ; *g* **Teilnehmer/-in** [m/f] (Bieter/-in bei einer Ausschreibung für Bauleistungen oder Lieferungen).

bidder [n] [US]**, best value for the price** *contr.* ►bidder offering best value for the price [US]/tenderer offering the best value for the money [UK].

bidder [n] [US]**, successful** *contr.* ►bidder offering best value for the price [US]/tenderer offering the best value for the money [UK].

bidder [n] [US]**, winning** *contr.* ►selected bidder [US]/ selected tenderer [UK].

B

423 bidder offering best value for the price [n] [US] *contr.* (Successful bidder/tenderer providing the best technical, economic, and design solution for award of the contract; ►acceptance of a bid [US] 1/acceptance of tender [UK]; ►construction contractor); *syn.* successful bidder [n] [US], successful tenderer [n] [UK], best value for the price bidder [n] [US], tenderer [n] offering the best value for the money [UK]; *s* **licitante [m/f] ganador, -a** (Licitante cuya oferta consigue el ►remate por ofrecer las mejores soluciones técnicas, económicas y de diseño; el precio más barato no es decisorio; ►contratista); *f* **candidat [m] retenu** (Après jugement sur la convenance des prix proposés, concurrent ayant remis l'offre jugée la plus intéressante compte tenu du prix des prestations, de leur coût d'utilisation, de leur valeur technique, des garanties professionnelles et financières, du délai d'exécution, le prix le plus bas n'étant pas décisif à lui seul ; le **c. r.** devient attributaire du marché lorsque la personne responsable du marché prend la décision de conclure le marché ; CCM 1984, 36 ; ►entreprise titulaire du marché ; ►attribution du marché [sur appel d'offres/sur adjudication]) ; *syn.* attributaire [m], entreprise [f] retenue, candidat [m] ayant remis l'offre jugée la plus intéressante ; *g* **Bieter/-in [m/f] des annehmbarsten Angebotes** (Person/Firma, die nach Wertung der Angebote hinsichtlich aller technischen und wirtschaftlichen, gegebenenfalls auch gestalterischen und funktionsbedingten Gesichtspunkte den ►Zuschlag erhält; der niedrigste Angebotspreis allein ist nicht entscheidend, d. h. dass Zweitbietenden oder nachrangigen Bietern der Zuschlag erteilt werden kann, wenn die entsprechenden Voraussetzungen dafür gegeben sind; cf. § 25 [2] VOB Teil A; ►Auftragnehmer); *syn.* Bieter/-in [m/f] des wirtschaftlichsten Angebotes, günstigster Bieter [m]/günstigste Bieterin [f].

424 bidder's affidavit [n] [US] *contr.* (Certified statement of the contractor, properly notarized, relating to payment of debts and claims, release of liens, or similar matters requiring specific evidence for the protection of the owner; DAC 1975; ►period for acceptance of a bid [US]/period for acceptance of a tender [UK]); *syn.* contractors affidavit [n] [US], tenderer's confirmation [n] [UK]; *s* **declaración [f] del licitante** (Declaración escrita y legalizada de un licitante de que su oferta es válida hasta la adjudicación del remate, de que conoce las condiciones del lugar de la obra, de que tiene solvencia técnica y económica para ejecutarla y de que puede cumplir las demás condiciones que garantizan el interés del propietario; ►periodo/período de remate); *f* **acte [m] d'engagement** (Document dans lequel le candidat au marché précise les données administratives et financières de son offre, notamment le montant du marché, les prestations confiées aux sous-traitants, le délai d'exécution, déclare être en règle avec ses obligations relatives au paiement des cotisations sociales, de ses impôts et taxes, et fait connaître le délai de validité de son offre ; ►délai de validité des offres) ; *g* **Bietererklärung [f]** (Schriftliche Erklärung des Bieters, dass er bis zum Ablauf der ►Zuschlagsfrist an sein Angebot gebunden ist, dass er sich über die örtlichen Verhältnisse der Baustelle informiert hat, dass er eine Unbedenklichkeitsbescheinigung des Finanzamtes, des Steueramtes sowie des Sozialversicherungsträgers vorlegen wird etc.).

425 bidding [n] [US] *constr. contr.* (Procedure for offering proposals on construction work [US]/works [UK], or the procedure for buying and delivering of, e.g. plants, materials or furnishings, from competitive firms; two main procedures are differentiated: ►selective bidding [US]/selective tendering [UK] and ►open bidding [US]/open tendering [UK]; ►acceptance of the bid 1, ►priced bidding documents [US]/priced tendering documents [UK], ►design competition announcement [US]/ design competition invitation [UK], ►public prequalification bidding notice [US]/public prequalification tendering notice [UK], ►awarding of contract); *syn.* tendering [n] [UK]; *s* **concurso-subasta [m] de obras** (Procedimiento de adjudicación de obras de construcción. Se diferencian ►concurso-subasta restringido y ►concurso-subasta público; ►expediente de concurso-subasta, ►anuncio de un concurso, ►aviso de candidaturas a concurso público; ►remate); *syn.* contratación [f] de obras por concurso-subasta; *f 1* **adjudication [f]** (Procédure de passation des marchés publics pour laquelle la sélection entre les soumissionnaires s'effectue d'après le seul critère du coût global) ; *f 2* **appel [m] d'offres** (Procédure de passation des marchés publics dans laquelle le prix n'est pas l'unique critère d'attribution du marché. Ces critères supplémentaires devront être mentionnés dans le règlement particulier de l'►appel d'offres restreint ou de l'►appel d'offres ouvert. *Termes spécifiques* appel d'offres sans variante, appel d'offres avec variante ; ►adjudication ouverte, ►adjudication restreinte, ►appel public à la concurrence [pour un concours], ►appel public de candidatures pour un appel d'offres, ►attribution du marché [sur appel d'offres/sur adjudication], ►dossier des pièces constitutives du marché) ; *syn.* lancer [vb] un appel d'offres, organiser [vb] un appel d'offres, consulter [vb] les entreprises par appel d'offres ; *g* **Ausschreibung [f]** (**1.** Verfahren, bei dem Unternehmer im Wettbewerb Angebote für die Vergabe von Bauleistungen oder Lieferungen abgeben. Es werden ►beschränkte Ausschreibung 2 und ►öffentliche Ausschreibung unterschieden. **2. In F.** gibt es traditionsgemäß zwei Verfahren: bei der *adjudication* bekommt immer der billigste Bieter den ►Zuschlag. Der *appel d'offres* gibt die Möglichkeit auch nach anderen Kriterien als den Preis, den Zuschlag zu erteilen; jedoch müssen diese Kriterien in der Ausschreibung extra aufgeführt werden; ausschreiben [vb]; ►Angebotsunterlagen, ►Auslobung, ►öffentlicher Teilnahmewettbewerb).

426 bidding and negotiation phase [n] [US] *contr. prof.* (DAC 1975 et HALAC 1985; **in U.S.**, fourth phase of planner's basic services: compilation and distribution of contract documents for invitation and evaluation of bids [US]/tenders [UK], negotiation with bidders [US]/tenderers [UK], and cooperation with owner/client in awarding a contract; **in U.K.**, work stage HJ; ►construction document phase [US]/detailed design phase [UK]); *syn.* tender action [n] and contract preparation [n] [UK] (LI 1996); *s* **preparación [f] y colaboración [f] en la fase de adjudicación** (Fase de planificación o de trabajo después de la ►fase de planificación en detalle en la que se calculan las terracerías para las especificaciones técnicas y se preparan los documentos de contrato para solicitar ofertas, revisar las ofertas, negociar con los licitantes y colaboración en la adjudicación del remate); *f* **dossier [m] de consultation des entreprises (DCE) et assistance aux marchés de travaux (AMT)** (F., constitue un des deux éléments normalisés de la troisième étape de la mission normale de maîtrise d'œuvre d'ingénierie et d'architecture ; HAC 1989, 30 ; l'assistance apportée au maître de l'ouvrage pour la passation du contrat de travaux a pour objet de préparer la sélection des candidats et d'examiner les candidatures obtenues, la consultation des entreprises sur la base d'un dossier permettant à celles-ci de présenter leurs offres en toute connaissance de cause, d'analyser les offres des entreprises, de préparer les mises au point permettant la passation du contrat de travaux par le maître de l'ouvrage ; cf. art. 6 du décret n° 93-1268 du 29 novembre 1993 ; ►projet) ; *g* **Vorbereitung [f] der und Mitwirkung [f] bei der Vergabe** (Planungsabschnitt/Leistungsphase nach der ►Ausführungsplanung, die u. a. die Ermittlung und Zusammenstellung der Massen für die Leistungsbeschreibungen und die Zusammenstellung der Verdingungsunterlagen für das Einholen von Angeboten, das Prüfen und Werten der Angebote, Verhandlungen mit Bietern und Mitwirkung bei der Auftragserteilung beinhaltet, cf. § 15 [1] 6 u. 7 HOAI 2002 resp. §§ 33, 38, 42 und 46 HOAI 2009; in A. werden die Leistungen für diese Phase mit den beiden Teilleistungen *Kostenberechnungsgrundlage* und *Technische und geschäftliche Oberleitung über die Ausführung* erbracht; § 22 lit. d et f HRLA).

bidding documents [npl] [US] *constr. contr.* ▶bid documents [US], ▶priced bidding documents [US], ▶return of bidding documents.

427 bidding period [n] [US] *constr. contr.* (Time during which bids [US]/tenders [UK] have to be submitted; ▶date for submission of bids [US]/date for submission of tenders [UK], ▶bid opening date [US]/submission date [UK]); *syn.* tender period [n] [UK]; *s* **plazo** [m] **de sumisión** (Periodo/período desde el anuncio de concurso/subasta hasta la ▶fecha tope de entrega de ofertas o de trabajos de concurso; ▶fecha de apertura de pliegos); *f* **délai** [m] **de réception des offres/soumissions** (Espace de temps jusqu'à la ▶date limite de réception des offres; ▶date d'ouverture des plis) ; *syn.* délai [m] de remise des offres/soumissions ; *g* **Angebotsfrist** [f] (Zeitraum bis zum ▶Abgabetermin der Angebote; ▶Eröffnungstermin).

428 bidding procedure [n] [US] *adm. contr.* (All activities and measures for preparation of ▶bid documents [US]/tender documents [UK] for projects, the public or selective announcement with invitation to bid [US]/to tender [UK], ▶opening of bids [US]/opening of tenders [UK] and evaluation of ▶bids [US] l/tenders [UK]; ▶project requirements for bidding [US], ▶awarding of contract); *syn.* tendering procedure [n] [UK]; *s* **procedimiento** [m] **de convocación de concursos-subasta y de contratación de obras** (Todos los pasos del proceso de ▶adjudicación de contrato; ▶pliego de condiciones [generales] de concurso, ▶apertura de ofertas, ▶oferta); *f* **organisation** [f] **des marchés publics** (Réglementation régissant les méthodes de passation de commandes publiques d'ouvrages ; ▶adjudication, ▶appel d'offres ; ▶dossier de consultation des entreprises ; ▶dossier d'appel d'offres, ▶ouverture des plis, offre, ▶attribution d'un marché de travaux) ; *g* **Ausschreibungswesen** [n, o. Pl.] (Sämtliche Handlungen und Maßnahmen zur Erarbeitung der ▶Ausschreibungsunterlagen für Bauvorhaben, die öffentliche oder individuelle Bekanntmachung mit Aufforderung zur Bewerbung sowie Einholung, ▶Angebotseröffnung und Auswertung der ▶Angebote; ▶Auftragserteilung an eine Firma).

429 bidding requirements [npl] [US] *contr.* (Prerequisites and participation conditions for competitive bidding [US]/competitive tendering [UK] of contracts); *syn.* requirements [npl] for bidders [US]/requirements [npl] for tenderers [UK], preconditions [npl] for tenderers [UK]; *s* **condiciones** [fpl] **de participación en concurso-subasta** (Prerrequisitos que se exigen y deben cumplirse para participar en un concurso-subasta de obras o servicios); *f* **conditions** [fpl] **de participation (1)** (à un appel d'offres ou un appel de candidature pour un marché de travaux ou de fournitures) ; *g* **Teilnahmebedingungen** [fpl] **(1)** (Voraussetzungen, die für die Teilnahme an Ausschreibungen von Bauleistungen und Lieferungen gefordert werden oder einzuhalten sind).

430 bid documents [npl] [US] *contr. leg.* (All ▶project requirements for bidding [US]/project requirements for tendering [UK], which govern the letting of contracts according to the ▶General Conditions of Contract for Furniture, Furnishings and Equipment [US] and ▶General Conditions of Contract for Construction [US]/General Conditions of Contract for Building and Civil Engineering Works [UK], and which are evaluated after ▶submission of bids [US]/submission of tenders [UK] for award-

ing a contract; ▶inspection of additional bid documents [US]/ inspection of additional tender documents [UK], ▶inspection of bid documents [US]/inspection of tender documents [UK], ▶Standard Form of Building Contract [UK], ▶priced bidding documents [US]/priced tendering documents [UK], ▶return of bidding documents [US]/return of tendering documents [UK]); *syn.* bidding documents [npl] [US], tender(ing) documents [npl] [UK]; *s* **documentos** [mpl] **de contratos de las administraciones públicas** (Regulaciones básicas de la contratación de obras y servicios como los ▶pliegos de cláusulas administrativas generales para la contratación de obras, ▶pliegos de prescripciones técnicas particulares, etc.; cf. Libro I, Título III LCAP; ▶condiciones generales de contratos de obras de las administraciones públicas [Es], ▶condiciones generales de contratos de servicios de las administraciones públicas, ▶pliego de condiciones [generales] de concurso); *f* **pièces** [fpl] **contractuelles constitutives du marché** (Documents constituant le dossier de consultation des entreprises en vue de la conclusion d'un marché, conformément aux ▶clauses générales de passation des marchés publics de travaux et aux ▶clauses générales de passation des marchés publics de fournitures ; ▶cahier des clauses administratives generales [des marchés publics] [C.C.A.G.], ▶dossier de consultation, ▶dossier d'appel d'offres) ; *g* **Verdingungsunterlagen** [fpl] (Alle ▶Ausschreibungsunterlagen, die zum Angebot eines Bieters als Vertragsgrundlage gemäß der ▶Vergabe- und Vertragsordnung für Bauleistungen [VOB], der ▶Verdingungsordnung für Leistungen [VOL] oder der Verdingungsordnung für freiberufliche Leistungen [VOF] gehören und nach ▶Angebotsabgabe für die Vergabe [▶Auftrtagserteilung an eine Firma, ▶Auftragserteilung an ein Planungsbüro] ausgewertet werden; ▶allgemeine Vertragsbedingungen für die Ausführung von Bauleistungen, ▶Ausschreibungsunterlagen).

bid form [n] [US] *constr. contr.* ▶form of bid [US].

431 bid item [n] [US] *contr.* (Single specified clause in a contractor's offer, or a clause in a detailed schedule of items, as part of a numbered sequence in a contract offer; ▶list of bid items and quantities [US]/schedule of tender items [UK], ▶basic specification item, ▶project requirements for bidding, ▶specification item number); *syn.* tender item [n] [UK]; *s* **partida** [f] **de oferta** (Artículo individual en una oferta o en un ▶resumen de prestaciones; ▶número de órdenes, ▶partida básica, ▶pliego de condiciones generales de concurso); *f* **article** [m] (Désigne la description d'un ouvrage ou d'une prestation présentée dans un ordre chronologique sous le chapitre d'un lot de travaux dans un ▶descriptif quantitatif, dans un bordereau des prix ; ▶dossier d'appel d'offres, ▶dossier de consultation, ▶numéro de prix d'une prestation de base, ▶numéro d'ordre) ; *syn.* poste [m] des prix ; *g* **Position** [f] (Einzelposten im Angebot oder in einem ▶Leistungsverzeichnis als Teil einer numerischen Abfolge; ▶Ausschreibungsunterlagen, ▶Grundposition, ▶Ordnungszahl).

bid items [npl] [US] *constr. contr.* ▶comparative analysis of bid items.

432 bid negotiation [n] [US] *contr.* (**1.** Opening and announcement of submitted bids [US]/submitted tenders [UK] on the ▶bid opening date [US]/submission date [UK] at a meeting in which bidders [US]/tenderers [UK] or their authorized representatives are allowed to be present, and the documentation of results, including whether and from whom ▶subsidiary bids [US]/subsidiary tenders [UK] have been submitted, or whether objections were made, in minutes of bid opening [US]/minutes of submission date [UK]. The minutes are signed by the negotiator and the keeper of the minutes, bidders or their representatives present are also entitled to sign the minutes. **2.** Negotiation during the subsequent ▶period for acceptance of a bid [US]/period for

B

acceptance of a tender [UK]; ►contract award negotiation); *syn.* tender negotiation [n] [UK]; *s* **negociación [f] de condiciones de adjudicación** (**1.** Durante la ►fecha de apertura de pliegos o en negociaciones con licitantes. El acta es firmada por el presidente y el secretario de la sesión. **2.** Negociación con licitantes en el ►periodo/período de remate; ►negociación de adjudicación); *f* **discussion [f] des offres** (**1. F.**, les responsables des entreprises concurrentes ne sont pas autorisés à participer à l'ouverture des plis ; *par contre* **D.**, ceux-ci peuvent participer à l'ouverture des plis. **2.** Possibilité qu'a la personne responsable du marché lors d'un appel d'offres ouvert, jusqu'à la limite du ►délai de validité des offres, de discuter avec les candidats pour leur faire préciser ou compléter la teneur de leurs offres [art. 97 ter du Code des marchés publics] ou de procéder en accord avec l'entreprise retenue à une mise au point du marché sans que les modifications entraînées puissent remettre en cause les conditions de l'appel à la concurrence [art. 97 quater du Code des marchés publics] ; la commission dresse un procès-verbal des opérations d'ouverture des plis, qui n'est pas rendu public ; dans le cadre d'un appel d'offres sur performances, les concurrents peuvent, après audition par la commission d'appel d'offres, préciser, compléter ou modifier leur offre ; ►date d'ouverture des plis, ►négociations avec le candidat retenu) ; *g* **Verdingungsverhandlung [f]** (**1.** Öffnung und Verlesung der Angebote für Bauleistungen während des ►Eröffnungstermins, bei dem Bieter oder deren Bevollmächtigte anwesend sein dürfen [§ 22 VOB Teil A]; Fertigung einer Niederschrift über die Ergebnisse, ob und von wem ►Nebenangebote abgegeben wurden, und über eventuelle Einwände; die Niederschrift wird vom Verhandlungsleiter und dem Schriftführer unterschrieben und verlesen; anwesende Bieter oder deren Bevollmächtigte sind berechtigt, mit zu unterzeichnen; in F. findet der Eröffnungstermin stets ohne Anwesenheit der Bieter statt; in D. sind Bieter nur bei der Öffnung der Angebote, die der Verdingungsordnung für Leistungen unterliegen [VOL-Verfahren], nicht zugelassen. **2.** Verhandlungen mit Bietern während der ►Zuschlagsfrist, allerdings nur, um sich z. B. über seine technische und wirtschaftliche Leistungsfähigkeit, über etwaige Nebenangebote, die geplante Art der Durchführung, etwaige Ursprungsorte der Bezugsquellen von Stoffen oder Bauteilen und um sich über die Angemessenheit der Preise Klarheit zu verschaffen; cf. § 24 VOB Teil A; ►Vergabeverhandlung); *syn.* Submissionsverhandlung [f].

433 bid negotiations [npl] [US] *constr. contr.* (Discussions and consultations between employer and contract holder/construction contractor on single prices or on the lump sum in order to reach a mutual basis for awarding a contract; ►sole source contract award [US]/freely awarded contract [UK]); *syn.* tender negotiation [n] [UK]; *s* **negociación [f] sobre precios** (Consulta entre el comitente y el adjudicatario de una obra sobre precios unitarios individuales o sobre el precio total para llegar a un acuerdo aceptable para ambas partes; en Es. los precios fijados al firmar el contrato deben ser ciertos y ser adecuados al mercado. Según Art. 104 de la Ley 13/1995 éstos pueden ser revisados una vez realizado como mínimo el 20% de la obra y transcurridos seis meses desde su adjudicación; en D. en concursos-subastas públicas prohibida según VOB Parte A; ►adjudicación negociada); *f* **négociation [f] des prix** (Consultation menée entre le donneur d'ordre et l'entreprise contractante concernant les prix unitaires ou le montant global d'un marché en vue de trouver un accord acceptable pour les deux parties ; ►passation du/des marché[s] sous forme de marché négocié) ; *g* **Preisverhandlung [f]** (Eingehende Erörterung und Beratung zwischen Auftraggeber und Auftragnehmer über einzelne Einheitspreise oder den Gesamtpreis, um zu einem für beide Seiten annehmbaren Ergebnis zu kommen; in D. bei öffentlichen und beschränkten Ausschrei-

bungen nach VOB Teil A nicht zulässig, es sei denn, die Ausschreibung wird aufgehoben; *∗* ►Verhandlungsverfahren).

bid opening [n] [US] *adm. constr. contr.* ►minutes of bid opening [US], ►opening of bids [US].

434 bid opening date [n] [US] *contr.* (Point in time at which submitted bids [US]/tenders [UK] are opened and announced to any bidders [US]/tenderers [UK] present. This date marks the end of the ►bidding period [US]/tendering period [UK] and the start of the ►period for acceptance of a bid [US]/period for acceptance of a tender [UK]; ►General Conditions of Contract for Construction [US], ►General Conditions of Contract for Furniture, Furnishings and Equipment [US], ►opening of bids [US]/opening of tenders [UK]); *syn.* submission date [n] [UK]; *s* **fecha [f] de apertura de pliegos** (Día en el que se abren todas las ofertas presentadas. Con la **f. de a. de p.** acaba el ►plazo de sumisión de ofertas y empieza el ►plazo de adjudicación; ►apertura de ofertas; ►condiciones generales de contratos de obras de las administraciones públicas, ►condiciones generales de contratos de servicios de las administraciones públicas, ►periodo/período de remate); *syn.* fecha [f] de apertura de proposiciones (económicas), fecha [f] de apertura de plicas; *f* **date [f] d'ouverture des plis** (Date à laquelle a lieu l'ouverture de tous les plis contenant les offres envoyées par les concurrents [conformément aux ►clauses générales de passation des marchés] ; elle correspond à la date limite d'envoi des offres ; ►clauses générales de passation de marchés publics de fourniture, ►délai de soumission, ►délai de réception des offres/soumissions, ►délai de validité des offres, ►ouverture des plis) ; *g* **Eröffnungstermin [m]** (**1.** Zeitpunkt, an dem alle eingegangenen Angebotsunterlagen geöffnet und [nach der ►Vergabe- und Vertragsordnung für Bauleistungen (VOB)] den anwesenden Bietern vorgelesen werden; bei Abwesenheit werden die Ergebnisse auf Anforderung mitgeteilt; der **E.** ist das Ende der ►Angebotsfrist und der Beginn der ►Zuschlagsfrist; ►Angebotseröffnung. **2.** Nach der ►Verdingungsordnung für Leistungen [VOL] werden die Angebote zum **Zeitpunkt der Öffnung** protokolliert, bei der keine Bieter anwesend sein dürfen; das Öffnungsergebnis darf weder den Bietern noch der Öffentlichkeit zugänglich gemacht werden [§ 22 VOL]; Angebote, die der VOL unterliegen, dürfen nicht verhandelt werden, ausgenommen sind Nebenangebote, die dann gewisse Kriterien erfüllen müssen; cf. § 24 VOL); *syn. zu 1.* Submissionstermin [m].

435 bid price [n] [US] *constr. contr.* (Sum that the contractor requires for the execution of the specified work at the ►bid opening date [US]/submission date [UK]); *syn.* tender sum [n] [UK]; *s* **precio [m] de licitación** *syn.* precio [m] ofertado; *f* **montant [m] global de l'offre/la soumission** (►ouverture des plis) ; *syn.* prix [m] global du devis ; *g* **Angebotssumme [f]** (Summe, die ein Bieter zum ►Eröffnungstermin für die ausgeschriebene Leistung verlangt).

bid security [n] [US] *contr.* ►surety bond [US].

bid submission [n] [US] *constr. contr.* ►submission of bids [US].

436 bid summary [n] [US] *contr.* (Compendium of bids sections [US]/tender sections [UK]); *syn.* tender summary [n] [UK]; *s* **lista [f] de costos** (Visión general de los costos/costes al final de una oferta) *syn.* lista [f] de costes [Es]; *f* **récapitulatif [m] du montant des ouvrages** (État présenté en dernière page d'une offre de prix) ; *syn.* récapitulatif [m] des chapitres d'ouvrages ; *g* **Kostenzusammenstellung [f] eines Angebotes** (Auflistung der Kosten am Ende eines Angebotes); *syn.* Kostenübersicht [f] eines Angebotes.

biennial [n] *bot.* ►biennial plant.

437 biennial plant [n] *bot.* (►*Life form* plant that completes its life cycle in two years: in the first year the vegetative part is developed, and in the second year the generative phase ends with the production of seed; ►hemicryptophyte, ►annual, ►perennial 1); *syn.* biennial [n]; *s* **bienal** [f] (►*Forma biológica* planta con ciclo vital de dos años, en el primero desarrolla sus partes vegetativas, en el segundo trascurre la fase generativa [floración y fructificación]; ►hemicriptófito, ►planta anual, ►planta vivaz); *syn.* planta [f] bienal, planta [f] bisanual; *f* **plante** [f] **bisannuelle** (►*Type biologique selon RAUNKIAER* dont le cycle de vie dure deux ans ; pendant la première année se développe l'appareil végétatif [période végétative] et pendant la seconde les appareils floral et de fructification ; ►hémicryptophyte, ►thérophyte, ►plante annuelle, ►plante vivace) ; *g* **Bienne** [f] (►*Lebensform nach RAUNKIAER* Pflanze mit zweijährigem Lebenszyklus: im ersten Jahr wird der vegetative Teil entwickelt, im zweiten die generative Phase [Blühen und Fruchten] abgeschlossen; die Pflanze stirbt nach der Fruchtreife ab; standortbedingt oft Übergänge zu ►Hemikryptophyten; zieht sich die vegetative Phase über mehrere Jahre hin, spricht man von *hapaxanthen Hemikryptophyten* [kurzlebige, nur einmal fruchtende Stauden], z. B. der in den Pyrenäen heimische Steinbrech *Saxifraga longifolia*; ►einjährige Pflanze, ►Staude); *syn.* [f], bienne Pflanze [f], Zweijährige [f], zweijährige Pflanze [f].

438 bike crossing [n] [US] *trans.* (Place where bikers [US]/cyclists [UK] may cross a road); *syn.* cyclist's crossing [n], pedaler's crossing [n] [US], cycle crossing [n] [UK]; *s* **cruce** [m] **para bici(cleta)s** (Banda de cruce de una calzada reservada para ciclistas); *f* **passage** [m] **clouté pour cyclistes** (Équipement de traversée d'une rue, d'une voie de circulation réservée aux cyclistes) ; *g* **Radfahrüberweg** [m] (Übergangsstelle für Radfahrer an einer Fahrbahn/Straße); *syn.* Radfahrerfurt [f].

439 bike lane [n] [US] *trans.* (On inner-city roadways, a specially demarcated lane for bikers [US]/cyclists [UK], in comparison to ►bikeway [US]/cycleway [UK], which is separated from other traffic by a curb [US]/kerbstone [UK] or planted median); *syn.* bicycle lane [n] [UK]; *s* **carril** [m] **de bici(cleta)s** (Vía marcada en la calzada para uso exclusivo de ciclistas sin motor o pequeñas motos, al contrario que el ►camino para bici[cleta]s que está separado del tráfico restante por una banda verde o transcurre al borde de la acera o por parques y jardines); *syn.* carril-bici [m], banda [f] para ciclistas, bicicletero [m]; *f* **bande** [f] **cyclable** (Voie banalisée pour la circulation des deux-roues non motorisés et matérialisée au sol par un dispositif peint sur la chaussée, par opposition à la ►piste cyclable qui est isolée des autres formes de circulation par un trottoir, une bordure en béton ou une bordure végétale ; on peut faire la distinction entre la **b. c.** matérialisée par un dispositif peint continu réservée exclusivement à la circulation des deux-roues et les « couloirs cyclables » matérialisés par une bande peinte discontinue non prioritaire pour les cyclistes ; ►itinéraire cyclable) ; *syn.* couloir [m] cyclable, voie [f] cyclable ; *g* **Radfahrspur** [f] (Innerstädtische, extra abmarkierte Verkehrsspur für Radfahrer auf der Fahrbahn; im Gegensatz zum ►Fahrradweg, der unabhängig von den übrigen Verkehrsarten durch Bordstein oder Trenngrün geführt wird. Es kann auch zwischen **R.** und **Radfahrstreifen** unterschieden werden: Die **R.** ist dann mit einer unterbrochenen Linie abgeteilt und nicht ausschließlich dem Radfahrverkehr vorbehalten. Der Radfahrstreifen ist bei dieser Unterscheidung mit einer durchgezogenen Linie abgetrennt und dient ausschließlich den Radfahrern).

440 bike rack [n] [US] *trans. urb.* (Piece of equipment used for parking bicycles; ►lockable bike rack [US]/lockable cycle rack [UK]); *syn.* bike stand [n] [also UK], cycle rack [n] [UK]; *s* **apoyo** [m] **para bicicleta** (Instalación para estacionar bicicletas; ►apoyo para bicicleta con seguro); *f* **support** [m] **de bicyclettes** (Installation de rangement des cycles ; *termes spécifiques* borne à vélos, étrier d'appui, râtelier à bicyclettes, pince sur potelet, pince murale [terminologie souvent issue de la traduction littérale de produits allemands dans certains catalogues français] ; ►support de bicyclette avec étrier antivol) ; *syn.* porte-vélo [m], range-vélo [m] ; *g* **Fahrradständer** [m] (Vorrichtung zum Abstellen eines Fahrrades; *UBe* Anlehnbügel oder Fahrradgeländer, Einsteigbügel, Fahrradbügelständer, ►Fahrradständer mit Verschlussbügel, Fahrradstützbügel).

441 bike racks [npl] [US] *trans. urb. syn.* cycle racks [npl] [UK]; *s* **aparcamiento** [m] **de bici(cleta)s** (con instalación para amarrarlas); *syn.* parking [m] de bici(cleta)s, parqueo [m] de bici(cleta)s [C, CA], estacionamiento [m] para bici(cleta)s [CS, Es también]; *f* **aire** [f] **de stationnement de bicyclettes** (Aire spécialisée, prévue pour le stationnement de plusieurs bicyclettes, p. ex. dans les cours d'usines, devant les gares, les supermarchés, etc.) ; *g* **Fahrradstandanlage** [f] (Einrichtung zum Abstellen mehrerer Fahrräder); *syn.* Fahrradabstellanlage [f].

bike stand [n] [UK] *trans. urb.* ►lockable bike rack [US].

bike tourism [n] *recr.* ►cycling tourism.

bike trail [n] [US] *trans.* ►bikeway [US]/bicycle path [UK].

442 bikeway [n] [US] *trans.* (Separate trail for use of bicycles. In U.S., bikeway and bike trail are often used synonymously; ►bike lane); *syn.* bicycle path [n] [UK], bicycle track [n] [US], bike trail [n] [US], cycle path [n] [UK], cycleway [n] [UK]; *s* **camino** [m] **para bici(cleta)s** (Vía separada para tránsito de bicicletas; ►carril de bici[cleta]s, ►itinerario de bici[cleta]s); *syn.* pista [f] para bici(cleta)s; *f 1* **itinéraire** [m] **cyclable** (*Terme générique* ouvrage aménagé pour les cyclistes sous forme de piste de marquage au sol ou de couloir indépendant) ; *f 2* **piste** [f] **cyclable** (Élément de voirie en zone urbaine spécialement aménagé pour la circulation des deux-roues soit comme réseau prioritaire comme voie banalisée ; ►bande cyclable) ; *g* **Fahrradweg** [m] (Für Radfahrer gesondert ausgewiesener Fahrweg; ►Radfahrspur); *syn.* Radweg [m], Veloroute [f] [CH].

443 bikeway network [n] *trans. urb.* (System of routes especially designated for the use of bicycles); *syn.* bicycle trail system [n] [also US]; *s* **red** [f] **de pistas para bici(cleta)s** (En zona de recreo o rural, sistema de caminos diseñados y señalizados para uso de ciclistas); *syn.* red [f] de caminos para bici(cleta)s; *f 1* **réseau** [m] **de pistes cyclables** (En milieu urbain système prioritaire continu de couloirs réservés aux deux-roues) ; *syn.* réseau [m] cycliste, réseau [m] cyclable ; *f 2* **réseau** [m] **de randonnée cyclotouristique** (En zone naturelle système prioritaire continu de voies réservées aux deux-roues) ; *g* **Radwegenetz** [n] (Für Radfahrer gesondert gekennzeichnetes Wegesystem); *syn.* Fahrradwegenetz [n], Veloroutennetz [n] [CH].

444 bikeway plan [n] [US] *plan. trans.* (Plan devised for a particular area, showing routes for the use of bicyclists [US]/cyclists [UK]); *syn.* bikeway scheme [n] [UK]; *s* **plan** [m] **de caminos de bici(cleta)s** (Plan elaborado para una ciudad o un área determinada en el que se fijan las rutas prioritarias para bicicletas y se clasifican según el tipo y el estado de los caminos para bicletas. En algunos casos el plan puede incluir medidas previstas para mejorar la infraestructura viaria para bicicletas); *f* **schéma** [m] **directeur de pistes cyclables** (Plan définissant les itinéraires de déplacement à bicyclette dans un secteur précis) ; *g* **Radwegeplan** [m] (Für ein bestimmtes Gebiet erarbeitete Straßen- und Wegekarte für Radfahrer); *syn.* Veloplan [m] [CH].

bikeway scheme [n] [UK] *plan. trans.* ►bikeway plan [US].

445 bi-level grave [n] [US] *adm.* (**In Europe**, extra deep grave for the burial of two coffins or two urns, one on top of the other; **in U.S.**, laws prevent this practice; multiple graves are

B

typically restricted to mausoleum structures; ►burial site); *syn.* two-storey grave [n] [UK]; *s* **tumba [f] de dos niveles** (Tumba de gran profundidad que permite enterrar dos ataudes o urnas; ►tumba); *syn.* tumba [f] de dos pisos; *f* **tombe [f] familiale** (Tombe dont la profondeur de la fosse permet l'inhumation de plusieurs corps ; ►tombe) ; *syn.* tombeau [m] ; *g* **Stockwerks-grab [n]** (Doppeltiefes Grab für die Bestattung von sowohl Särgen als auch Urnen; ►Grabstätte); *syn.* Tiefgrab [n].

446 bill [n] **(1)** *adm. leg.* (Proposed law introduced within a legislative body but not yet adopted); *s* **proyecto [m] de ley** (Propuesta de ley presentada al correspondiente cuerpo legislativo, pero aún no aprobada); *syn.* borrador [m] (de ley), anteproyecto [m] (de ley); *f* **projet [m] de loi** (présenté devant le parlement en vue de son vote) ; *syn.* proposition [f] de loi ; *g* **Gesetzentwurf [m]** (Zur Abstimmung im Parlament erarbeitete Vorlage eines Gesetzes).

bill [n] **(2)** *contr.* ►weigh bill.

bill [n], **interim** *contr.* ►interim invoice.

billboard [n] [US] *urb.* *►commercial sign [US]/advertisement board [UK].

bill [n] **of delivery** [US] *constr. contr.* ►bill of lading [US].

447 bill [n] **of lading** [US] *constr. contr.* (Written and signed schedule specifying kinds and quantity of delivered materials to the construction site; ►weigh bill); *syn.* bill [n] of delivery [US], delivery note [n] [UK]; *s* **recibo [m] de entrega** (Documento sobre el tipo y la cantidad de materiales suministrados a una obra; ►certificado de peso); *f* **bordereau [m] de livraison** (Document certifiant la qualité et la quantité des matériaux livrés ; ►bon de pesage) ; *g* **Lieferschein [m]** (Dokument über Art und Menge von gelieferten Waren/Materialien zur Baustelle; ►Wiegekarte).

448 bill [n] **of quantities** *contr.* (Detailed list with quantities of all items of materials and equipment necessary to construct a project, calculated by a cost estimator [US]/quantity surveyor [UK] according to take-offs from drawings and descriptions of work made by the planner; ►calculation of quantities); *s* **listado [m] de materiales** (Lista detallada de todos los materiales y equipos necesarios para realizar un proyecto; ►cálculo de cantidades); *f* **établissement [m] d'un métré** (Calcul des quantités de travaux ou fournitures effectué par un « métreur » d'après plans et descriptif pour le compte d'un concepteur ; ►avant-métré) ; *g* **Massenaufstellung [f] (1)** (U.K. und F., Auflistung der durch den „Quantity Surveyor"/„métreur" berechneten Bauleistungsmengen nach den Plänen und Leistungsbeschreibungen des Planers; ►Massenberechnung); *syn.* Massenverzeichnis [n], Mengenverzeichnis [n].

449 binding [n] **(1)** *envir.* (Fixation of airborne particulates by plants); *s* **fijación [f]** (de partículas de polvo o de nutrientes por las plantas); *f* **fixation [f]** (de la poussière, des substances nutritives, etc.) ; *g* **Bindung [f] (1)** (1. Ablagerung und Zusammenhalt von Staubpartikeln an der Vegetationsoberfläche; 2. Aufnahme und Festlegung von Nährstoffionen durch Pflanzen oder in Tonmineralen).

450 binding [n] **(2)** *constr.* (Holding loose materials together as in binding agent, such as cement, hydraulic hardening lime, bitumen, polymer plastic, synthetic resin binder); *s* **fraguado [m]** (Endurecimiento de material suelto con un agente ligante como p. ej. cemento, cal hidráulica, bitumen, plásticos polímeros, resina sintética); *f* **prise [f]** (Durcissement de granulats [sable, gravillons] agglomérés par un liant soit hydraulique [ciment] faisant prise par hydratation, un liant hydrocarboné [bitume], un liant synthétique [résines] ; *termes spécifiques* temps de prise, vitesse de prise, accélérateur/retardateur de prise) ; *g* **Abbinden [n, o. Pl.]** (Das Verfestigen von losem Material oder Zuschlag-

stoffen mit Bindemitteln wie z. B. Zement, hydraulisch erhärtende Kalke, Bitumen, polymere Kunststoffe, Kunstharz).

binding force [n] *leg. plan.* ►legal obligation.

binding in law [loc] *adm. leg.* ►legally-binding.

451 binding obligation [n] **to a bid** [US] *contr.* (Agreement/guarantee of a submittee [contractor] to perform the work as outlined in the submitted bid proposal till the end of ►period for acceptance of the bid); *syn.* binding obligation [n] to a tender [UK]; *s* **compromiso [m] a una oferta** (Obligación de un licitante a mantener una oferta hasta la adjudicación del remate; comprometerse [vb] a una oferta, ►periodo/período de remate); *f* **assujettissement [m] à une offre** (Obligation pour un candidat/soumissionnaire qui reste engagé par son offre/sa soumission dans le cadre du ►délai de validité des offres) ; *g* **Bindung [f] (2)** (Verpflichtung des Bieters, zu seinem Angebot bis zum Ablauf der ►Zuschlagsfrist zu stehen; cf. § 19 [3] VOB Teil A).

binding obligation [n] **to a tender** [UK] *contr.* ►binding obligation to a bid [US].

452 binding period [n] *contr.* (Time between ►bid opening date [US]/submission date [UK] and ►awarding of contract [US]/letting of contract [UK] in which a bidder [US]/tenderer [UK] is committed to his offered prices; ►period for acceptance of a bid [US]/period for acceptance of a tender [UK]); *s* **límite [m] de validez de ofertas** (Periodo/período de tiempo entre la apertura de las ofertas y la ►adjudicación en el cual el/la licitante debe mantener los precios de su oferta; ►periodo/período de remate, ►adjudicación de contrato); *syn.* periodo/período [m] de validez de ofertas; *f* **période [f] de validité des offres** (Période entre la remise des offres et l'►attribution d'un marché de travaux pour laquelle le candidat/soumissionnaire est lié à son offre/sa soumission depuis la ►date d'ouverture des plis, jusqu'à la ►passation des marchés ; ►délai de validité des offres) ; *g* **Bindefrist [f]** (Zeitraum vom ►Eröffnungstermin bis zur ►Auftragserteilung an eine Firma, in dem der Bieter an sein Angebot gebunden ist; cf. § 19 VOB Teil A; ►Zuschlagsfrist); *syn.* Angebotsbindefrist [f].

bin store [n] [UK] *urb.* ►trash storage [US].

453 bioaccumulation [n] *biol.* (Term for build-up of organic or inorganic compounds, usually applied to heavy metals, pesticides or metabolites, within the tissues of living organisms; ►food chain); *s* **bioacumulación [f]** (Almacenamiento de sustancias químicas ajenas en un organismo. El grado de **b.** depende fundamentalmente de la persistencia y de la liposolubilidad de las sustancias y así mismo del medio donde se encuentre el organismo; cf. DINA 1987; ►cadena trófica); *f* **bioaccumulation [f]** (Mise en réserve de substances organiques et inorganiques dans les tissus de végétaux ou d'animaux ; ►chaîne trophique) ; *g* **Bioakkumulation [f]** (Anreicherung von organischen und anorganischen Substanzen in Geweben von Pflanzen und Tieren; ►Nahrungskette).

biocenose [n], **forest** *phyt. ecol.* ►forest biocoenosis.

biocenose [n], **river** *limn.* ►river biocoenosis.

454 biocenosis [n] *biol. ecol.* (Biocenoses [pl] or biocoenoses [pl]; distinctive group of animal and plant species living under the same general environmental conditions and having their habitats in the same space or biotope in a relatively stable relationship and evolution; GE 1977; ►habitat 1, ►habitat 2, ►phytocoenosis); *syn. o. v.* biocoenosis [n], *o. v.* biocoenose [n], *o. v.* biocoenose [n], biotic community [n], ecological community [n], life community [n]; *s* **biocenosis [f]** (Comunidad biótica formada por animales y plantas que se condicionan mutuamente y que se mantienen a través del tiempo en posesión de un territorio definido [►biótopo] y en un estado de equilibrio dinámico gracias a la reproducción de los propios organismos que la integran, dependiendo solamente del ambiente exterior inanimado pero no dependiendo, o sólo de manera no esencial, de organismos exteriores a la **b.** El término fue aplicado por primera vez por MÖBIUS en 1877; cf. MARG 1977, DB 1985; ▶hábitat, ▶habitación, ▶fitocenosis); *syn.* comunidad [f] biótica; *f* **biocénose [f]** (Communauté de vie animale et végétale [complexe d'espèces] au sein de laquelle dominent les phénomènes d'interdépendance réciproque et règne un équilibre dynamique. La **b.** représente la composante organique d'un écosystème, l'▶habitat 1 la composante inorganique ; ▶biotope, ▶phytocénose) ; *syn.* communauté [f] biologique, *termes employés dans les pays anglo-saxons* communauté [f] de vie naturelle, communauté [f] biotique ; *g* **Biozönose [f]** (Lebensgemeinschaft als Vergesellschaftung von Pflanzen und Tieren, die durch gegenseitige Beeinflussung und Abhängigkeit in Wechselbeziehungen stehen [dynamisches Gleichgewicht]. Die **B.** ist der organische Anteil eines Ökosystems, während der ▶Habitat [faunistischer oder floristischer „Wohnraum"] dessen anorganische Komponente darstellt; ▶Biotop, ▶Phytozönose); *syn.* Lebensgemeinschaft [f].

455 biochemical oxygen demand [n] *limn.* (*Abbr.* BOD; measure of the oxygen used in meeting the metabolic needs of aerobic microorganisms in water rich in organic matter; RCG 1982; the amount of oxygen used for biochemical oxidation by a unit volume of water at a given temperature and for a given time; the BOD is an index of the degree of organic pollution in water); *s* **demanda [f] bioquímica de oxígeno** (*Abr.* DBO; medida de la capacidad contaminante del agua residual, que indica la cantidad de oxígeno consumido por los microorganismos en la degradación aerobia de la materia orgánica que contiene, en condiciones controladas de tiempo y temperatura; DINA 1987; *f* **demande [f] biochimique en oxygène** (*Abrév.* DBO ; la consommation d'oxygène — exprimée en milligrammes par litre — nécessaire pour que les micro-organismes présents dans les eaux puissent stabiliser les substances organiques existantes dans ce milieu ; LA 1981, 474) ; *g* **biochemischer Sauerstoffbedarf [m]** (*Abk.* BSB; die Messung des BSB; Menge an Sauerstoff, die von Mikroorganismen benötigt wird, um organische Substanz eines definierten Wasserkörpers mit bestimmter Temperatur [20 °C] im Verlaufe eines definierten Zeitraumes [meist fünf Tage] aerob abzubauen; cf. ÖKO 1983; die Messung des BSB gilt als Standardmethode zur Prüfung der Gewässerverschmutzung).

456 biochemical sewage treatment basin [n] *envir.* (Basin for cleansing of domestic and industrial wastewater by a natural biochemical self-purification process; ▶sewage treatment basin); *s* **tanque [m] de tratamiento secundario** (En ▶planta de depuración de aguas residuales gran depósito para el tratamiento biológico aerobio de autodepuración de las aguas. Existen dos procesos: el de lecho bacteriano o filtro percolador y el de fangos activados; en ambos casos se elimina con ayuda de microorganismos (bacterias, protozoos, etc.) gran parte de la carga orgánica que pasa a sedimentarse en los decantadores secundarios en forma de lodo/fango, cf. DINA 1987, 54; ▶decantador); *f* **bassin [m] de lagunage (1)** (Bassin ou étang utilisé pour traiter les eaux usées domestiques ou résiduaires industrielles dans lesquels les effluents sont soumis aux processus biochimiques naturels de l'auto-épuration ; on distingue les bassins de lagunage profond fonctionnant en anaérobiose et les **bassins/étangs de stabilisation** de faible profondeur fonctionnant en aérobiose ; cf. ASS 1987, 40 ; ▶bassin d'épuration) ; *g* **Nachklärbecken [n]** (Becken einer Kläranlage zur Behandlung häuslicher und industrieller Abwässer, in dem sie einem natürlichen biochemischen Prozess der Selbstreinigung ausgesetzt sind und Belebtschlamm abscheiden; ▶Klärbecken).

457 biocide [n] *chem. envir.* (Chemical agent used to kill specific living organisms or groups of them in order to artificially control the population level; ▶herbicide, ▶insecticide, ▶pesticide, ▶plant protection agent); *s* **biocida [m]** (Producto químico aplicado para combatir organismos específicos o grupos de ellos; en *castellano* se utiliza frecuentemente como sinónimo de ▶pesticida, ▶herbicida e ▶insecticida; ▶plaguicida); *f* **biocide [m]** (Produit détruisant les êtres vivants et utilisé en général contre les microorganismes ; ▶herbicide, ▶insecticide, ▶pesticide, ▶produit antiparasitaire [à usage agricole]) ; *g* **Biozid [n]** (*Terminus ursprünglich aus dem Angelsächsischen Schrifttum stammend* Chemikalie, die einzelne Lebewesen oder ganze Populationen, sowohl tierischer wie pflanzlicher Art, abtötet; ▶Herbizid, ▶Insektizid, ▶Pestizid, ▶Pflanzenschutzmittel).

biocoenose [n], forest *phyt. ecol.* ▶forest biocoenosis.

biocoenose [n], river *limn.* ▶river biocoenosis.

biocoenosis [n] *biol. ecol.* ▶biocenosis.

458 biodegradable [adj] *biol. chem.* (Decomposable by bacteria or other living organisms; ▶decomposition 1; *opp.* ▶nonrotting); *s* **biodegradable [adj]** (▶descomposición; *opp.* ▶no degradable); *f* **biodégradable [adj]** (pouvant être décomposé par des êtres vivants ou par les facteurs naturels ; ▶décomposition ; *opp.* ▶imputrescible) ; *g* **abbaubar [adj]** (So beschaffen, dass eine Substanz biologisch abgebaut werden kann; ▶Abbau 1; *opp.* ▶verrottungsfest).

459 biodegradable waste [n] *envir. leg.* (**1.** Recyclable refuse material of animal or plant origin, which can be broken down by micro-organisms, soil-borne organisms or enzymes; waste that cannot be broken down by other living organisms is known as non-biodegradable waste. Forms of **b. w.** include ▶green waste, food waste, paper waste and biodegradable plastics as well as human waste, manure, and sewage. **B. w.** may be converted for re-use by composting, or such processes as anaerobic digestion [e.g. for biogas] and incineration [to generate electricity], but must also be managed, when disposed of in landfills, to reduce the emission of landfill gases [methane] into the atmosphere, which contribute to global warming. In order to limit or stop global warming, the European Union Landfill Directive requires member states to regulate the disposal of biodegradable waste. **2. B. w.** is not to be confused with **biological waste**: any waste, living or deceased, that was used in research, including waste medical equipment and materials such as pipettes, needles, and glassware used in biological research); *s* **residuos [mpl] biodegradables** (Restos reciclables de origen animal o vegetal que pueden ser descompuestos por microorganismos, organismos edáficos o enzimas; residuos que no pueden ser descompuestos por otros organismos se denominan no-biodegradables. Los **r. b.** incluyen ▶residuos verdes, restos de alimentos, papel y plásticos biodegradables, así como excrementos humanos, estiércol y aguas residuales. Los **r. b.** pueden ser transformados para su reutilización por compostaje o por procesos como la digestión anaeróbica [para producir biogás] y la incineración [para generar electricidad], pero también deben ser bien gestionados en caso de deposición en rellenos sanitarios para reducir las emisiones de gases [metano] a la atmósfera, que contribuyen al calentamiento global. Para limitar este efecto, la Directiva de la UE sobre rellenos sanitarios obliga a los Estados miembros a regular la deposición de **r. b.**); *f* **biodéchets [mpl]** (Déchets composés exclusivement de matières organiques biodégradables ; ils regroupent essentiellement **1.** les déchets putrescibles des ménages [fraction fermentescible des ordures ménagères susceptible de se dégrader spontanément dès leur production, déchets verts, papiers-cartons], **2.** les déchets des producteurs non ménagers tels les professionnels [biodéchets des restaurateurs, déchets fermentescibles des commerces alimentaires et supermarchés], les établissements collectifs [déchets alimentaires de cantine, de restaurant, papiers-cartons souillés d'écoles, gymnase, hôpitaux, établissements socioculturels, militaires, pénitentiaires, administrations, etc.], les services communaux [déchets liés aux activités des services espaces verts et nettoiement] ; ils sont susceptibles d'être traités par compostage ou méthanisation ; *terme spécifique* ▶déchets verts) ; *syn.* déchets [mpl] biologiques, déchets [mpl] fermentescibles ; *g* **Bioabfall [m]** (Verwertbarer Abfall tierischer oder pflanzlicher Herkunft, der durch Mikroorganismen, bodenbürtige Lebewesen oder Enzyme abgebaut werden kann; diese Stoffe sind in Anhang 1 Nr. 1 der Bioabfallverordnung vom 21.09.1998 aufgeführt; Bodenmaterial ohne wesentliche Anteile an Bioabfällen gehört nicht hierzu; Pflanzenreste, die auf forst- oder landwirtschaftlich genutzten Flächen anfallen, sind legaldefinitorisch auch keine Bioabfälle; cf. BioAbfV 1998; *UB* ▶Grünabfall).

460 biodiversity [n] *ecol.* (Multiplicity of forms of life and their characteristics and relationships to one another in the ▶biosphere; the full range of variations and variability between systems and organisms can be distinguished on three levels: **1. ecological biodiversity** in the form of multipliple biomes, landscapes and ecosystems including ecological niches; **2. diversity amongst organisms**, i.e. heterogeneous taxonomic groups [e.g. families, genera] including ▶species richness; **3. genetic diversity**, i.e. multiplicity of populations, from individuals to genes. **B.** is distributed very unevenly throughout the world because half of the species occur in the so-called 'hot spots' of the tropics. With deforestation, species deportation and other anthropogenic changes in the environment the loss of **b.** is a global problem. High **b.** requires the long-term stability of ecosystems. ▶Nature conservation and sustainable use of **b.** are the recognized goals of ▶sustainable development. The 1992 Rio de Janeiro U.N. ▶Convention on Biological Diversity represents the most important international instrument in the pursuit of the aims of nature conservation); *syn.* biological diversity [n]; *s* **biodiversidad [f]** (El término «biodiversidad» es una contracción inglesa de la expresión «diversidad biológica», utilizado por primera vez por el biólogo WALTER G. ROSEN y popularizado por EDWARD O. WILSON en el Foro sobre la Diversidad Biológica del Consejo Nacional de Investigación de la Academia de Ciencas de los EE.UU. en 1986. La **b.** es la multiplicidad de formas de vida y sus características e interrelaciones en la ▶biosfera. El rango completo de variaciones y variabilidad entre los sistemas y organismos se puede diferenciar en tres niveles: **1. b. ecológica** en forma de diversos biomas, ecosistemas y paisajes, incluyendo nichos ecológicos; **2. diversidad entre organismos**, es decir la diversidad entre grupos taxonómicos [familias, géneros] y de especies [▶riqueza en especies]; **3. diversidad genética**, es decir la variedad de poblaciones, desde los individuos hasta los genes. La **b.** está muy irregularmente distribuída en la tierra, ya que la mitad de las especies se encuentran en los llamados «hot spots» [puntos calientes] en los trópicos, siendo la región de América Latina y el Caribe con poco más de 2000 mill. de ha. [aprox. 15% de la superficie terrestre], la región con mayor diversidad de especies y de ecorregiones del mundo. Seis países de la región son considerados como megadiversos en cuanto a la diversidad de especies y de endemismos: Brasil, Colombia, Ecuador, México, Perú y Venezuela. Debido a la de[s]forestación, la destrucción de hábitats, la intensificación de la agricultura y otros cambios antropógenos de la naturaleza, la pérdida de **b.** se ha convertido en un problema global. Para promover la preservación de la **b.** se firmó en 1992 en Río de Janeiro el ▶Convenio sobre la Diversidad Biológica [CDB] que es hasta hoy, junto con el Protocolo de Cartagena sobre la seguridad de la biotecnología, el instrumento más importante del derecho internacional para la ▶protección de la naturaleza. A nivel de la UE, la ▶Directiva de Hábitats y la ▶Directiva de las Aves son instrumentos prácticos de implementación de la convención mundial); *syn.* diversidad [f] biológica; *f* **biodiversité [f]** (Le terme « biodiversité » vient de la contraction de l'expression anglaise « biological diversity », soit « diversité biologique » ; il a été introduit pour la première fois par le biologiste WALTER G. ROSEN et popularisé par la suite par EDWARD O. WILSON lors du forum sur la diversité biologique de la National Research Council de l'Académie des Sciences américaine en 1986 ; EDWARD O. WILSON, en donne la définition suivante : « *la totalité de toutes les variations de tout le vivant* » ; selon les scientifiques, la biodiversité est la dynamique des interactions dans des milieux en changement ; dans l'art. 2 de la ▶Convention sur la diversité biologique [Convention de Janeiro], la biodiversité est « la variabilité des organismes vivants de toute origine y compris, entre autres, les écosystèmes terrestres, marins et autres écosystèmes aquatiques et les complexes écologiques dont ils font partie ; cela comprend la diversité au sein des espèces et entre espèces ainsi que celle des écosystèmes » ; en relation avec le ▶réseau Natura 2000 la biodiversité est présentée comme l'expression de la variété de la vie sur la planète à tous ses niveaux d'organisation ; elle comprend notamment les microorganismes, les espèces sauvages végétales et animales ; ce sont aussi des milieux comme les eaux douces, les eaux marines, les forêts, les tourbières, les prairies, les marais, les dunes, etc., c'est encore la diversité des ressources génétiques, des espèces et des écosystèmes ; elle rend compte de la diversité biologique d'un espace donné en fonction notamment de l'importance numérique des espèces animales ou végétales présentes sur cet espace [▶richesse en espèces], de leur originalité ou spécificité, et du nombre d'individus qui représentent chacune de ces espèces ; il est difficile de donner une définition unique et générale de la biodiversité car elle dépend de l'échelle à laquelle on se place [écosystèmes, individus/espèces, gènes] et les disciplines qui l'étudient [écologie, génétique] ; la biodiversité peut être néanmoins définie à trois niveaux différents et complémentaires : **1.** la **diversité écologique [ou diversité écosystémique]**, correspond à la diversité d'écosystèmes évoluant dans le temps, **2.** la **diversité spécifique**, caractérisée par la variations au sein d'une espèces ou entre espèces vivant sur un territoire de dimensions modestes, **3.** la **diversité génétique**, correspondant à la diversité du patrimoine génétique au sein d'une espèce donnée ; ▶biosphère ; la répartition de la biodiversité à la surface du globe est inégale ; tous les milieux sont génétiquement diversifiés, même les milieux extrêmes comme les déserts, les zones de haute altitude ou de grande latitude, mais certains écosystèmes sont plus riches en biodiversité que d'autres, d'autres sont très menacés [récif corallien, mangrove, étang, île, zone humide, tourbière, etc.] ; d'une manière globale, les pays de l'hémisphère sud sont plus riches en biodiversité que ceux du nord ; pour combattre l'appauvrissement considérable de la diversité biologique au niveau global et en Europe par suite des activités humaines a été signée par la Communauté et tous les États membres au cours de la conférence des Nations unies sur l'environnement et le développement, tenue à Rio de Janeiro du 3 au 14 juin 1992 la Convention sur la diversité biologique [CDB] ; en 2002, au sommet mondial sur le développement durable de Johannesburg, les chefs d'États du monde entier se sont mis d'accord sur la nécessité de réduire le taux de perte de diversité biologique de façon significative à l'an 2010 ; la ▶protection de la nature et le ▶développement durable sont les éléments clés de la préservation de la biodiversité) ; *syn.* diversité [f] biologique ; *g* **Biodiversität [f]** (Vielfalt der Lebensformen der ▶Biosphäre in allen ihren Ausprägungen und Beziehungen untereinander; die Bandbreite an

Variation und Variabilität zwischen Systemen und Organismen können auf drei Ebenen festgestellt werden: **1. Ökologische Diversität** in Form der Vielfalt von Biomen, Landschaften und Ökosystemen bis hin zu ökologischen Nischen. **2. Diversität zwischen Organismen**, d. h. Vielfalt zwischen taxonomischen Gruppen [z. B. Familien, Gattungen] bis hin zur ►Artenvielfalt. **3. Genetische Diversität**, d. h. Vielfalt von Populationen über Individuen bis hin zu den Genen. Die **B.** ist weltweit sehr ungleich verteilt, da die Hälfte der Arten auf so genannten „Hot spots" in den Tropen vorkommt. Durch Entwaldung, Artenverschleppung und andere anthropogene Umweltveränderungen ist der Verlust der **B.** ein globales Umweltproblem. Eine hohe **B.** fördert die langfristige Stabilität von Ökosystemen. ►Naturschutz und nachhaltige Nutzung der **B.** sind anerkannte Ziele ►nachhaltiger Entwicklung. Das 1992 in Rio de Janeiro von den Vereinten Nationen verabschiedete ►Übereinkommen über die biologische Vielfalt [Convention on Biological Diversity] stellt das wichtigste völkerrechtliche Instrumentarium für die Verfolgung der naturschutzfachlichen Ziele dar; cf. ÖKO 2003; ►Natura-2000-Schutzgebietssystem); *syn.* biologische Vielfalt [f], biologische Diversität [f].

461 bioengineering [n] *constr.* (Science of engineering and landscape architectural techniques using natural materials for soil reinforcement and stabilization; ►biotechnical construction techniques); *syn.* biotechnical stabilization [n] [also US]; *s* **bioingeniería** [f] (Técnica de estabilización del suelo en zonas propensas a la erosión utilizando material vivo como plantas, ramas o plantación de cesped; ►construcción biotécnica); *f* **génie [m] biologique** (Art de concevoir, de réaliser et de gérer les techniques d'utilisation des végétaux [techniques végétales] dans les ouvrages paysagers limitant ou freinant les processus d'érosion ; ►technique de génie biologique) ; *g* **Ingenieurbiologie [f]** (Lehre und Wissenschaft von der vorwiegenden Verwendung lebenden und toten Pflanzenmaterials zur Sicherung von Landschaftsbauwerken, um durch die Verhinderung oder Bremsung von Abtragungsvorgängen landschaftsbezogene Nutzungsansprüche zu sichern und zu fördern; ►ingenieurbiologische Bauweise); *syn.* Lebendbau [m], Grünverbau(ung) [m/(f)], Lebendverbauung [f].

bioengineering [n] **of rivers and streams** [US] *conserv. landsc. wat'man.* ►natural river engineering measures.

462 biogenetic reserve [n] *conserv.* (Part of a network of nature reserves coordinated since 1976 by the Council of Europe, in which communities of plant and animal species particularly characteristic of European landscapes are conserved, many of which may be rare or under threat; ►protected area); *s* **reserva [f] natural biogenética** (Red de ►espacios naturales protegidos coordinada desde 1976 por el Consejo de Europa como ampliación de la Convención Europea de Protección de la Naturaleza, en la que se encuentran sobre todo aquellos hábitats representativos donde habitan las especies de flora y fauna más típicas, particulares, raras o en peligro de extinción); *f* **réserve [f] biogénétique** (Engagé par le Conseil de l'Europe en 1976, programme de conservation des exemples représentatifs de la flore, de la faune et des zones naturelles d'Europe, garantissant le potentiel et la diversité génétique des biomes européens ; chaque État membre est invité à inventorier les différents types d'habitats, de biocénoses et d'écosystèmes de son pays, afin d'identifier les plus rares et les plus menacés et de proposer en conséquence des sites à inclure dans un réseau de réserves biogénétiques caractérisées par des habitats, des biocénoses et des écosystèmes typiques ou uniques ; ►zone protégée) ; *g* **biogenetisches Reservat [n] (1)** (Als Ergänzung zur Europäischen Naturschutzkonvention ein vom Europarat seit 1976 koordiniertes Netz von ►Schutzgebieten, in denen besonders solche für Natur und

Landschaft in Europa repräsentative Tier- und Pflanzenarten und deren Zönosen erhalten werden, die typisch, einzigartig, selten oder gefährdet sind).

463 biogeochemical cycle [n] *ecol.* (Continual exchange of matter between living and inert components of the ecosystem or its parts; especially important cycles for living organisms are those of oxygen, carbon, mineral [nitrogen and phosphorous]; ►energy flow, ►ecological cycle); *syn.* element cycle [n], material cycle [n]; *s* **ciclo [m] biogeoquímico** (Circulación de los elementos biogenésicos o bioelementos entre el medio y los organismos y viceversa. Son de dos tipos: gaseoso [ciclo del carbono, ciclo del nitrógeno] de rápida circulación y sedimentario [ciclo del fósforo, ciclo del azufre] de circulación lenta; DINA 1987; ►flujo de energia, ►ciclo ecológico [de materia y energía]); *syn.* ciclo [m] de materia; *f* **cycle [m] biogéochimique** (Succession régulière de phénomènes biologiques et physicochimiques au sein de la biosphère entre les composantes biotiques et abiotiques d'un écosystème s'achevant par un retour à la situation initiale ; les principaux **c. b.** sont ceux de l'oxygène, du carbone, des éléments minéraux [azote et phosphore] ; ►flux d'énergie, ►cycle naturel) ; *syn.* cycle [m] de la matière ; *g* **Stoffkreislauf [m]** (Ständiger Austausch von Stoffen zwischen lebenden und unbelebten Komponenten im Ökosystem oder Teilen davon; für Lebewesen besonders wichtige Kreisläufe sind Sauerstoff**k.**, Kohlenstoff**k.**, Mineralstoff**k.** [Stickstoff**k.** und Phosphor**k.**]; ►Energiefluss, ►ökologischer Kreislauf); *syn.* biogeochemischer Kreislauf [m], biogeochemischer Zyklus [m].

464 biogeographical region [n] *geo. phyt. zool.* (Extensive area of land, which exhibits characteristic geogenous spatial structures and associated plant and animal communities. Because of the geographical extent of the EU from Scandinavia to the Mediterranean, nature and landscape and their various cultural differences are extremely diverse. In the European protection area network of NATURA 2000 [►Natura 2000 Protection Area network] there are, therefore, seven **b. r.**: atlantic, continental, alpine, mediterranean, boreal, macaronesian [collective term for the mid-Atlantic region including the Cape Verde islands, the Canary Islands, the Madeira group, the Azores und parts of SW Europe and the Pannonian Region]. Each of these regions has its own peculiar characteristics and typical habitats, as well as animal and plant species; cf. NATURA 2000 region); *s* **región [f] biogeográfica** (Fracción extensa de la tierra que se caracteriza por estructuras geógenas específicas y las comunidades de flora y fauna asociadas a las mismas. Debido a la extensión geográfica de la UE desde Escandinavia al Mediterráneo, su naturaleza y sus paisajes son extremadamente diversos. Por ello, en el sistema europeo de áreas protegidas «Natura 2000» [►Red Ecológica Europea Natura 2000] se diferencian 7 **rr.bb.**: alpina, atlántica, boreal, continental, macronesia [término colectivo para la región centro-atlántica como las Islas Canarias, Madeira, las Azores y partes del SO de Europa], mediterránea y panónica. Cada una de estas regiones tiene sus características específicas, sus particularidades así como sus hábitats y sus especies de flora y fauna características; cf. Natura 2000); *f 1* **région [f] biogéographique** (Dans le cadre du réseau écologique de zones spéciales protégées « Natura 2000 » [►réseau Natura 2000], région géographique et climatique qui peut s'étendre sur le territoire de plusieurs États membres et qui présente une faune, une flore et un milieu biologique conditionnés par des facteurs écologiques homogènes tels que le climat [précipitations, température, etc.] et la géomorphologie [géologie, relief, altitude, etc.] ; l'Union européenne à 27 membres et compte 7 régions biogéographiques : Alpine, Atlantique, Boréale, Continentale, Macaronésienne [Madère, Açores et Canaries], Méditerranéenne et Pannonienne ; l'intégration future de la Rou-

B

manie et de la Bulgarie à l'Union Européenne rajoutera deux nouvelles régions : Steppique et Littoraux de la Mer Noire ; la France est concernée par 4 de ces régions : alpine, atlantique, continentale et méditerranéenne) ; *f 2* **écozone [f]** (Partie de la surface terrestre représentative d'une unité écologique homogène à grande échelle, caractérisée par des facteurs climatologiques, biologiques et physiques particuliers ; la surface terrestre est divisée en 8 écozones [afrotropical, antarctique, australasien, indomalais, paléarctique, néarctique, néotropique et océanien] ; le système des écozones fut proposé en 1975 par MIKLOS UDVARD dans le cadre du programme Homme et Biosphère et est maintenant utilisé internationalement comme système unifié à des fins d'identification biogéographique et de conservation) ; *f 3* **écorégion [f]** (Les analyses de biodiversité régionale réalisées par le Fonds Mondial pour la Nature [WWF] ont abouti à la définition du système mondial des écorégions, fondé sur une association de domaines biogéographiques et de provinces floristiques/zoogéographiques ; il s'agit d'une entité géographique présentant une homogénéité des caractéristiques géologiques, climatiques et topographiques, et par conséquent une homogénéité supposée du fonctionnement écologique ; classiquement utilisée pour les écosystèmes terrestres et la compréhension des associations de végétation, les écorégions peuvent être appliquées aux écosystèmes aquatiques, on parle alors d'hydroécorégions. Le WWF définit 867 écorégions sur la planète ainsi que 238 régions dites « Global 200 » qui sont considérées comme exceptionnelles au niveau biologique et prioritaires en matière de conservation [142 sont des régions terrestres, 53 des régions d'eau douce et 43 des régions marines] ; la classification en écozone est également utilisée dans le cadre des World Heritage Sites ; **CDN**, unité de classification écologique du territoire du Canada, la plus vaste pouvant être subdivisée en unités de plus en plus petites selon les similitudes ou différences des caractéristiques écologiques [climat, sol, eau et espèces sauvages, etc.] ; ce sont par ordre de superficie décroissante l'écoprovince, l'écorégion, l'écodistrict, l'écosection, l'écosite, l'écotope et l'écoélément) ; *g* **biogeografische Region [f]** (Ausgedehntes Gebiet, das in seinem räumlichen Erscheinungsbild charakteristische geogene Strukturen und daran angepasste Pflanzen- und Tiergemeinschaften aufweist. In der geografischen Ausdehnung der EU von Skandinavien bis zum Mittelmeer sind Natur und Landschaft auf Grund der naturräumlichen und landeskulturellen Unterschiede sehr vielfältig. Deshalb werden in Europa im ►Natura-2000-Schutzgebietssystem sieben **b. R.en** unterschieden: die atlantische, kontinentale, alpine, mediterrane, boreale, die makaronesische Region [Sammelbegriff für Gebiete im Mittelatlantik wie Kapverdische und Kanarische Inseln, Madeiragruppe, Azoren und Teile SW-Europas] und die pannonische Region [Ungarische Tiefebene zwischen Balaton und Karpaten]. Jede dieser Regionen hat ihre Besonderheiten und ihre Einzigartigkeit und zeichnet sich durch typische Lebensräume sowie besondere Tier- und Pflanzenarten aus; cf. NATURA-2000-Gebiet).

bio-intensive gardening [n] [US] *agr.* ►organic farming.

biological diversity [n] *ecol.* ►biodiversity, ►Convention on Biological Diversity.

465 biological dying [n] **of a waterbody** *limn. ocean.* (Destructive process of the ecological system of a water body by extreme ►eutrophication, contamination by poisonous matter, or ►acid rain to cause such an imbalance that the system losses its ability to regenerate); *s* **muerte [f] progresiva de un cuerpo de agua** (Proceso destructivo del equilibrio de un ecosistema lótico o léntico por exceso de nutrientes [►eutrofización], de contaminación tóxica o de ►lluvia ácida, de manera que pierde su capacidad de regeneración. Este proceso puede conllevar la ►mortandad acusada de peces por envenenamiento y la reducción del contenido de oxígeno hasta niveles que imposibilitan la vida de la fauna piscícola); *f* **dystrophisation [f]** (Processus de rupture d'équilibre d'un écosystème lotique ou lentique, provoqué par l'►eutrophisation extrême, les ►pluies acides ou la pollution par des substances toxiques, entraînant la mort des organismes végétaux et animaux supérieurs et une altération durable et parfois irréversible — zone morte) ; *syn.* mort [f] progressive d'un écosystème aquatique ; *g* **Umkippen [n, o. Pl.]** (Vorgang der Zerstörung eines Gewässerökosystems durch zu starke ►Eutrophierung, ►sauren Regen und Verschmutzung mit Giftstoffen, so dass es seine Regenerationsfähigkeit verliert).

466 biological indicator [n] *ecol. envir.* (Living organism especially useful in monitoring the level of pollution by virtue of its sensitivity to alterations in environmental conditions, the consequences of which are easily observable as physiological or external changes); *s* **bioindicador [m]** (Organismo vivo utilizado para el control de calidad del aire por sus cambios de aspecto y de composición química dependiendo del grado de contaminación, p. ej. líquenes); *syn.* indicador [m] biógeno; *f* **indicateur [m] biologique** (Organisme vivant [végétal ou animal] ou communauté de vie qui réagit assez spécifiquement à la présence d'une substance toxique ou à une modification du milieu [mortalité, raréfaction, pullulation, etc.] par sa présence ou son absence, par les modifications de son aspect extérieur ainsi que/ou de sa constitution chimique, par son comportement et sert d'indicateur dans le cadre de la surveillance de la pollution de l'environnement, p. ex. les lichens, les abeilles pour la pollution de l'air, les insectes aquatiques, les crustacés, les mollusques pour la pollution de l'eau ; *termes spécifiques* bio-indicateur de l'état des cours d'eau, bio-indicateur marin ; *syn.* bio-indicateur [m] ; *g* **Bioindikator [m]** (Lebewesen, das bei der Überwachung der Umweltverschmutzung auf Grund seiner veränderten äußeren Gestalt oder chemischen Zusammensetzung als ►Zeigerart gilt).

467 biological pest control [n] *agr. for. hort.* (Method of pest control by artificial introduction of the pest's natural enemies instead of applying chemical pesticides, as well as the application of beneficial insects, plants and viruses to supplement a plant's built-in defense system); *s* **control [m] biológico** (Método de lucha antiplaguicida que utiliza los enemigos naturales para reducir las poblaciones de una plaga en vez de aplicar sustancias químicas); *syn.* lucha [f] biológica; *f* **lutte [f] biologique** (Ensemble des méthodes visant à détruire les animaux et les végétaux nuisibles aux plantes cultivées en utilisant judicieusement leurs ennemis naturels [micro-organismes antagonistes, parasites, prédateurs, etc.] ; LAP 1979, 749) ; *g* **biologischer Pflanzenschutz [m, o. Pl.]** (Alle natürlichen und im weiteren Sinne auf biochemischer Basis beruhende Methoden der Schädlingsbekämpfung, die versuchen, die vorhandene natürliche Regulation der Schädlinge zu erhalten und gezielt zu fördern sowie durch Einsatz von Nutzorganismen die fehlenden oder unzureichend vorhandenen Abwehrkräfte aus der Tier- und Pflanzenwelt zu ergänzen; hierzu gehören auch schädlingsabtötende Viren und Pilze. Zur Vermeidung und Substitution von Pflanzenschutzmitteln wurden in D. seit Anfang der 1950er-Jahre zahlreiche Verfahren zum integrierten und **b. P.** erarbeitet, weshalb bis Anfang 2007 vor allem im Gartenbau mehr als 40 Nutzorganismen als Ersatz für Pflanzenschutzmittel in der Praxis zur Verfügung stehen); *syn.* biologische Schädlingsbekämpfung [f].

biological self-cleansing [n] *limn.* ►biological self-purification.

468 biological self-purification [n] *limn.* (Biological and chemical processes which enable bodies of water to break down impurities; aerobic and anaerobic bacteria help in the decomposition of organic pollutants); *syn.* biological self-cleansing [n]; *s* **autodepuración [f]** (Capacidad de cursos y masas de agua de descomponer sustancias orgánicas contaminantes por medio de procesos químicos y biológicos con ayuda de bacterias aerobias y anaerobias); *f* **auto-épuration [f]** (Ensemble des processus physiques, chimiques et biologiques permettant à un milieu naturel pollué de retrouver son état de pureté originelle sans intervention extérieure ; ASS 1987, 255) ; *g* **biologische Selbstreinigung [f]** (Fähigkeit von Gewässern, durch biologische und chemische Vorgänge Verunreinigungen abzubauen; die organischen Schmutzstoffe werden durch aerobe oder anaerobe Bakterien abgebaut); *syn.* natürliche Selbstreinigung [f].

469 biomass [n] *phyt. zool.* (Amount of living material of one or more species, animal or vegetal, in a specified unit; *specific terms* ▶phytomass, ▶zoomass); *syn.* standing biomass [n] (TEE 1980, 143); *s* **biomasa [f]** (Cantidad de materia viva de una o varias especies, animal o vegetal, en una unidad dada del espacio; ▶fitomasa, ▶zoomasa); *f* **biomasse [f]** (Poids de matière vivante, animale ou végétale, par unité de surface ou de volume pour un milieu naturel donné à un moment donné ; *termes spécifiques* **b.** microbienne, **b.** aérienne, **b.** aquatique, **b.** souterraine, ▶phytomasse, ▶zoomasse) ; *g* **Biomasse [f]** (Das Gewicht der zu einem bestimmten Zeitpunkt vorhandenen Lebewesen je Flächen- oder Volumeneinheit einer Lebensstätte; TIS 1975; *UBe* ▶Phytomasse, ▶Zoomasse).

biomass [n]**, standing crop** *bot. ecol.* ▶phytomass.

470 biome [n] *ecol.* (Complex of natural communities or a major biotic unit, adapted to the conditions prevailing within an extensive geographic region, e.g. tundra, tropical rainforest, savanna[h], etc. The vegetative part of a biome is called 'phytome', the animal populations 'zoome'); *s* **bioma [f]** (Comunidad biológica regional importante de plantas y animales que se extiende por grandes regiones naturales, p. ej. tundra, selva tropical, sabana. Se caracteriza por la uniformidad fisionórmica de la climax vegetal y por los animales influyentes y posee constitución biótica característica; DB 1985); *f* **biome [m]** (Unité supérieure d'habitat en écologie correspondant à une zone climatique et un espace vital géographique important sur lequel règne un équilibre biologique. Les biomes sont dénommés d'après la végétation dominante — p. ex. toundra, savane, étage des conifères, etc. La composante végétale d'un biome s'appelle le « **phytom** », la composante animale le « **zoom** ») ; *g* **Biom [n]** (Bioma [pl]; Organismengemeinschaft eines größeren, einer bestimmten Klimazone entsprechenden, geografischen Lebensraumes, in dem sich ein einigermaßen ausgewogenes biologisches Gleichgewicht eingestellt hat. **B.e** werden nach der vorherrschenden Vegetation benannt, z. B. Tundra, tropischer Regenwald, Savanne etc. Der pflanzliche Bestand eines **B.s** heißt **Phytom**, der tierische **Zoom**).

471 biomonitoring [n] *ecol. envir.* (Regular monitoring, evaluation and assessment of the vitality of indicator plants [e.g. trees or lichens or indicator organisms such as fish] to determine environmental quality, including the level of air pollution and changes caused by human activity. On the basis of modifications to indicator species, conclusions can be drawn about the quality and quantity of physical and chemical changes in environmental parameters, and also possible effects of climate change); *syn.* pollution control registering [n]; *s* **bioensayo [m]** (Inspección de plantas y evaluación de los daños, para valorar las condiciones ambientales, la calidad del aire, las posibles pérdidas de cosechas, las fuentes de fitotóxicos, etc.); *f* **surveillance [f] biologique** (Évaluation de la vitalité de certaines espèces végétales indicatrices [p. ex. arbres, lichens] ou d'espèces animales [p. ex. moules, poissons] en vue de la détermination de la qualité de l'environnement ; cette technique de bioindicateurs repose sur l'hypothèse que le contenu en contaminant de la plante ou l'animal reflète la concentration en contaminants biodisponibles

dans l'air, dans l'eau ou le sol sous formes particulaire et/ou dissoute, selon un processus de bioaccumulation ; *terme spécifique* biosurveillance végétale et animale) ; *syn.* biosurveillance [f], biomonitorage [m] ; *g* **Biomonitoring [n]** (Regelmäßige Beobachtung, Erfassung und Bewertung der Vitalität von Indikatorpflanzen [z. B. Bäume oder Flechten] oder Indikatortieren [z. B. Fische in Gewässern] zur Bestimmung der Umweltqualität resp. deren anthropogen bedingten Veränderung. An Hand von Modifikationen an Indikatorarten können Rückschlüsse auf Qualität und Quantität von chemisch-physikalischen Veränderungen der Umweltparameter gezogen und auch mögliche Auswirkungen des Klimawandels beurteilt werden); *syn.* Bewertung [f] der Umweltqualität durch Schadenserfassung bei Pflanzen und Tieren.

472 bioreserve [n] **[US]** *conserv. ecol. leg.* (This term applies to large-scale landscapes that contain outstanding examples of ecosystems, natural communities, and species deserving of protection. They are ecologically and geographically definable entities: a mountain range, a chain of barrier islands, or an entire watershed; **b.s** seek to protect entire ecosystems, rather than single sites for endangered species; ▶biogenetic reserve); *s* **reserva [f] biológica** (≠) (En EE.UU., categoría de protección de espacios naturales de grandes dimensiones que constituyen ejemplos sobresalientes de ecosistemas, biocenosis naturales o de hábitats de especies a proteger. Se trata de entidades territoriales claramente diferenciables ecológica y geográficamente como cordilleras, cadenas de islotes o cuencas hidrográficas, más que hábitats individuales de especies en peligro; ▶reserva natural biogenética); *f* **réserve [f] biologique** (≠) (U.S., *catégorie de zone naturelle protégée* espace naturel de grande étendue présentant des qualités remarquables réunissant des espèces et habitats d'espèces qui méritent d'être conservé ; entité écologique et géographique clairement définie telle qu'une chaîne de montagnes, un archipel ou un bassin hydrographique. La protection a pour objet l'écosystème dans son ensemble et non pas des habitats isolés) ; *g* **biogenetisches Reservat [n] (2)** (≠) (*Gebietsschutzkategorie in den USA* großflächiger Auschnitt einer Natur- und Kulturlandschaft von hervorragender Bedeutung als Ökosystem, natürliche Lebensgemeinschaft und Lebensräume schutzwürdiger Arten. Es ist eine ökologisch und geografisch definierbare Gebietseinheit, z. B. ein Gebirgszug, eine Inselkette oder ein gesamtes Abflussgebiet. Ein solches Gebiet dient dem Schutze eines ganzes Ökosystems und nicht einzelner Habitate).

473 biosphere [n] *ecol.* (Part of the Earth's crust, waters, and atmosphere where living organisms can subsist; RCG 1982; ▶geosphere); *syn.* ecosphere [n]; *s* **biosfera [f]** (Espacio del planeta ocupado por la vida orgánica, en el interior de la corteza terrestre alcanza pocos metros de profundidad; ▶geosfera); *f* **biosphère [f]** (Ensemble des parties de l'atmosphère, hydrosphère et lithosphère occupées par les êtres vivants ; ▶géosphère) ; *syn.* écosphère [f] ; *g* **Biosphäre [f]** (Gesamtheit des von Lebewesen besiedelten Teils der Erde und Erdhülle; ▶Geosphäre); *syn.* Ökosphäre [f].

474 biosphere reserve [n] *conserv. ecol. leg.* (Protected area set up in selected parts of the world under the MAB Program[me] of UNESCO. The multi-purpose of a **b. r.** is to foster sustainable economies of local communities within its borders in harmony with protection of ecosystems and endangered species. **B. r.s** are selected because **[a]** they are a representative of the ▶biome, **[b]** they have outstanding natural characteristics, or **[c]** in the case of an altered or degraded natural system they are suitable for the integration of human activities without interferences of the ecosystem. Long-term research is being carried out with a view to better human understanding of natural systems in order to improve conservation of natural

B

resources. By January 2009, 545 **b. r.s** had been established in 120 countries, 47 of them in the U.S., 13 in U.K., and 15 in Canada; in U.S., some **b. r.s** called 'Great Places' are preserved by *The Nature Conservancy*, an international non-profit organization for similar purposes and many hundreds of state, regional, and local, private, tax exempt non-profit organizations; cf. UNESCO/MAB Website, List of Biosphere Reserves 2009); *s* **reserva [f] de la biosfera** (Programa de acción dentro del programa MAB de la UNESCO iniciado a partir de 1974 y en el que se está creando una red de **r. de la b.** a nivel mundial cuyos objetivos primordiales son: a) la conservación de las áreas naturales y del material genético que contienen [▶bioma], b) la investigación ambiental de las reservas y su entorno y c) la integración de la actividad humana y del desarrollo local en la protección de las mismas. Actualmente existen 545 reservas distribuídas por 120 países del mundo, de las cuales 38 encuentran en Es., en los países latinoamericanos hay un total de 102, de las cuales el mayor número se encuentra en México [11], Argentina y Chile [resp. 7], Cuba [6]; cf. DINA 1987 y sitio web de la UNESCO/MAB, List of Biosphere Reserves, estado de enero 2009); *f* **réserve [f] de la biosphère** (Programme MAB lancé par l'UNESCO en 1970 dont les principaux objectifs sont l'utilisation rationnelle et la conservation des ressources de la biosphère ainsi que la conservation des zones naturelles et du matériel génétique qu'elles contiennent. Les zones constituant le réseau mondial des **r.s de la b.** concernant : [a] des exemples représentatifs de grands ▶biomes naturels, [b] des territoires avec des caractéristiques naturelles inhabituelles d'intérêt exceptionnel et [c] des exemples d'écosystèmes modifiés ou dégradés susceptibles d'être restaurés. La France possède actuellement 10 Réserves de la Biosphère ; cf. CPN 1994 et UNESCO/MAB Website, List of Biosphere Reserves, 2009) ; *g* **Biosphären-reservat [n]** (Ein durch das MAB-Programm der UNESCO unter Schutz gestelltes Gebiet, das [a] für das jeweilige ▶Biom repräsentativ, [b] wegen seiner natürlichen Eigenschaft von herausragendem Interesse ist oder [c] als verändertes oder degradiertes Ökosystem geeignet ist, menschliche Nutzungsansprüche zu integrieren, ohne es zu zerstören. Im Januar 2009 waren weltweit in 120 Ländern bereits 545 **B.e** ausgewiesen. Die Nominierung und rechtliche Festsetzung eines **B.es**, das den Kriterien der UNESCO genügen, aber nicht zwingend von ihr anerkannt werden muss, obliegt der Zuständigkeit des jeweiligen Landes; die Schutzkategorie ist im Bundesnaturschutzgesetz [BNatSchG] verankert; in D. gibt es 13 UNESCO-**B.e**, z. B. das **B.** Berchtesgaden [467 km²], **B.** Spreewald [475 km²], **B.** Flusslandschaft Elbe [3428 km²], **B.** HamburgischesWattenmeer [117 km²], **B.** Niedersächsisches Wattenmeer [2400 km²], **B.** Oberlausitzer Heide- und Teichlandschaft, **B.** Rhön [1850 km²], **B.** Schaalsee [309 km², zwischen Hamburg, Lübeck und Schwerin] und das grenzüberschreitende **B.** Pfälzerwald [3018 km²] nach Frankreich. In der Schweiz gibt es zwei **B.e**: den Schweizer Nationalpark in Graubünden [1724 ha] und das Entlebuch in der Innerschweiz zwischen Bern und Luzern [395 km²]; in **A.** sechs: Gossenköllesee, Tirol [85 ha], Gurgler Kamm in den Ötztaler Alpen, Tirol [1500 ha], Lobau in den Donauauen, Wien [1037 ha], Neusiedler See, Burgenland [250 km²], Großes Walsertal, Vorarlberg [192 km²] und Wienerwald [1050 km²]. Schutzzweck ist neben der Erhaltung repräsentativer Ökosysteme des entsprechenden Naturraumes vor allem die Erhaltung, Entwicklung und Wiederherstellung durch traditionelle Nutzung geprägter Kulturlandschaften sowie die Entwicklung und Erprobung von Wirtschaftsweisen, welche die Naturgüter in besonderer Weise schonen. **B.e** sollen im wesentlichen Teil [in der Kernzone] die Voraussetzungen für ein Naturschutzgebiet mitbringen und im übrigen Teil [Pflegezone] überwiegend den Anforderungen eines Landschaftsschutzgebietes entsprechen. **B.e** bilden weltweit ein Netz von Naturschutzgebieten, in denen zum besseren Verständnis von Ökosystemen Langzeituntersuchungen gemacht werden, um die Systeme und die natürlichen Lebensgrundlagen dann nachhaltig schützen zu können. In D. rechtsverbindlich festgelegte **B.e** sind durch die Bundesländer mit den Schutzkategorien Naturschutzgebiet, Nationalpark oder Landschaftsschutzgebiet zu schützen; cf. § 14a [2] BNatSchG u. UNESCO/MAB Website, List of Biosphere Reserves, Stand Januar 2009; BFN 2006).

475 biotechnical construction technique [n] *constr.* (Method of landscape construction primarily using live or dead plant material to stabilize slopes prone to erosion, to prevent avalanches and to safeguard against the erosion of banks and beds of watercourses; ▶bioengineering, ▶conventional engineering); *syn.* biotechnical treatment [n] [also US]; *s* **construcción [f] biotécnica** (Método de construcción paisajístico, que utiliza sobre todo material vegetal vivo o muerto para fijar el suelo o asegurar zonas propensas a la erosión o a los taludes, orillas y cauces de ríos y arroyos; ▶bioingeniería, ▶construcción convencional); *f* **technique [f] de génie biologique** (Technique paysagère de stabilisation des pentes, de prévention contre les avalanches, de lutte contre l'érosion des berges ou du lit des cours d'eau en utilisant les matériaux végétaux vivant ou mort par opposition au génie conventionnel utilisant des ▶techniques d'ouvrages en dur ; ▶génie biologique) ; *syn.* technique [f] paysagère de stabilisation des sols, technique [f] de génie écologique ; *g* **ingenieurbiologische Bauweise [f]** (▶*Ingenieurbiologie* landschaftsbauliche Konstruktionsweise, bei der durch vorwiegende Verwendung lebenden oder toten Pflanzenmaterials Rutsch- und Erosionshänge, Lawinenabbruchgebiete, Ufer und Sohlen von Fließgewässern etc. gesichert werden; cf. SCHIECHTL 1980; *opp.* ▶Hartbauweise); *syn.* Sicherungsbauweise [f] (DIN 18 918).

476 biotechnical pest control [n] *agr. for. hort. phytopath.* (Measures which limit the multiplication of pests, by using the pests' natural reaction to specific physical [e.g. optical] or chemical stimuli, by e.g. pheromones and thereby keeping the damage at an economically supportable level); *s* **control [m] biotécnico de plagas** (Medidas antiplagas que aprovechan ciertas reacciones naturales a estímulos físicos [p. ej. ópticos] o químicos [p. ej. feromonas] de los depredadores para limitar su reproducción); *f* **lutte [f] raisonnée** (Mesures de lutte antiparasitaire réduisant la multiplication des parasites en utilisant la capacité qu'ont ceux-ci à réagir naturellement à certains stimulants physiques [p. ex. optique] ou chimiques, si bien que les dégâts causés restent à un niveau économiquement supportable) ; *g* **biotechnischer Pflanzenschutz [m, o. Pl.]** (Maßnahmen der Schädlingsbekämpfung, die die Vermehrung der Schädlinge durch Ausnutzung ihrer natürlichen Reaktion auf bestimmte physikalische [z. B. optische] oder chemische Reize [z. B. Lockstoffe] so begrenzt, dass die Schadorganismen unter der wirtschaftlichen Schadensschwelle gehalten werden können).

biotechnical stabilization [n] *constr.* ▶bioengineering.

biotechnical treatment [n] [US] *constr.* ▶biotechnical construction technique.

biotic balance [n] *ecol.* ▶biotic equilibrium.

biotic community [n] *biol. ecol.* ▶biocenosis.

biotic ecological factors [npl] *ecol.* ▶biotic site factor, ▶ecological factors.

477 biotic equilibrium [n] *ecol.* (Dynamic biological balance in a given period of time between different life processes and their interrelationships, and between components of the system, e.g. energy, materials and information flows); *syn.* biotic balance [n]; *s* **equilibrio [m] biótico** (Dentro de un periodo/período determinado de tiempo estado metaestable de los

B

diferentes procesos vitales en un sistema biológico y sus interrelaciones y entre los diferentes componentes del sistema: flujos de energía, materia e información); *f* **équilibre [m] biologique** (État d'équilibre permanent pendant une période donnée entre deux processus vitaux dans un système biologique — interactions entre les composantes du système ; flux d'énergie, de matière et d'information) ; *g* **biologisches Gleichgewicht [n]** (Innerhalb einer bestimmten Zeitspanne konstanter Zustand des Ausgleichs zwischen den verschiedenen Lebensvorgängen in einem biologischen System — Wechselbeziehungen zwischen Systembestandteilen als dynamischer Prozess, z. B. Energie-, Stoff- und Informationsflüsse).

478 biotic site factor [n] *ecol.* (Parameter, which influences the population of a particular ▶location 1 and the interdependencies created there, e.g. population density, competition, shading, mutual influence and interrelationships); *s* **factor [m] mesológico biótico** (Parámetro que influye en una población de una ▶estación particular y en las interdependencias existentes allí, como densidad de población, competencia, grado de sombra, interrelaciones e influencias mutuas de los organismos); *f* **facteur [m] stationnel biotique** (Composantes créés et maintenues par les organismes présents sur la ▶station ainsi que leurs interrelations [p. ex. densité d'une population, compétition, stimulation réciproque]) ; *g* **biotischer Standortfaktor [m]** (Parameter für einen belebten ▶Standort 1 und die dadurch entstehenden Wechselbeziehungen, z. B. Bestandsdichte, Konkurrenz, Belichtung/Beschattung, gegenseitige Beeinflussung und Wechselwirkung).

biotope [n] *phyt. zool.* ▶habitat (2).

biotope mapping [n] *conserv. phyt. zool.* ▶habitat mapping.

bird [n] *zool.* ▶breeding bird, ▶overwintering (migratory) bird, ▶stopover bird, ▶visitor bird.

bird [n]**, summer visitor** *zool.* ▶summer visitor.

bird [n]**, winter visitor** *zool.* ▶winter visitor.

479 bird bath [n] *gard.* (Small, flat bowl in a private garden for birds to drink or bathe); *s* **baño [m] para pájaros** *syn.* estanque [m] para pájaros; *f* **bain [m] d'oiseau** (Élément décoratif du jardin [récipient de faible profondeur et aux bords de faible pente] permettant aux oiseaux de se baigner) ; *g* **Vogeltränke [f]** (Kleines, flaches Becken im Hausgarten, das Vögeln zum Trinken und Baden dient).

bird box [n] [US] *zool.* ▶nesting box.

bird census [n] *conserv. zool.* ▶bird counting.

480 bird counting [n] *conserv. zool.* (Tallying numbers of birds); *syn.* bird census [n]; *s* **conteo [m] de aves** (DINA 1987, 611); *f* **comptage [m] des oiseaux** (p. ex. en vue de l'estimation de la densité de population des espèces d'oiseaux) ; *g* **Vogelzählung [f]** (Erfassung von beobachteten Vogelarten und deren Menge in einem bestimmten Raum in einer definierten Zeitspanne zur wissenschaftlichen Auswertung durch Ornithologen).

bird crash [n] *conserv.* ▶bird strike.

birder [n] [US] *zool.* ▶ornithologist.

bird fauna [n] *zool.* ▶birdlife.

bird forage plant [n] *landsc. plant.* ▶woody host plant for birds.

birdhouse [n] [US] *zool.* ▶nesting box.

481 birdlife [n] *zool.* (**1.** All of the bird species living in a defined area at a specified date or period of time. **2.** Classification of species of birds occurring in a defined area—for species identification); *syn.* avifauna [n], bird fauna [n]; *s* **avifauna [f]** (Fauna de aves circunscrita a un territorio o a un periodo/período de tiempo; DINA 1987); *syn.* fauna [f] de aves; *f 1* **avifaune [f]**

(Ensemble des oiseaux d'une région donnée) ; *syn.* faune [f] ailée, faune [f] des oiseaux, faune [f] avienne [aussi CDN]; *f 2* **monde [m] des oiseaux** (Classification scientifique des diverses espèces d'oiseaux observé sur un territoire déterminé, un continent) ; *g* **Vogelwelt [f]** (**1.** Gesamtheit der zu einem bestimmten Zeitpunkt in einem bestimmten Gebiet lebenden Vögel. **2.** Systematische Zusammenstellung der in einem bestimmten Gebiet vorkommenden Vögel — zur Artbestimmung); *syn.* Avifauna [f].

482 bird migration [n] *zool.* (Departure and return of ▶migratory birds according to the time of year; ▶straggler); *s* **migración [f] de aves** (Marcha y vuelta de las ▶aves migradoras debido a los cambios de estaciones; ▶ave migratoria tardía); *f* **migration [f] des oiseaux** (Mouvement saisonnier de certaines espèces d'oiseaux entre les régions de reproduction au printemps et les zones de repos en hiver ; ▶oiseau migrateur, ▶migrateur tardif) ; *g* **Vogelzug [m]** (Jahreszeitlich bedingtes Fortziehen und Zurückkehren von ▶Zugvögeln; der **V.** in die Winterquartiere wird „Herbstzug" oder „Wegzug" genannt, der Rückzug in die Sommerquartiere/Brutgebiete heißt „Frühjahrszug" oder „Heimzug"; ▶Nachzügler).

bird-nesting box [n] *zool.* ▶nesting box.

483 bird [n] **of passage** *zool.* (NAB 1992, 194; bird species which, seen from the standpoint of a particular area, flies over that area during the course of a longer ▶migration 2 to its ▶breeding range or wintering site [▶hibernation area/region]; ▶partial migrant; *generic term* ▶migratory species; *specific terms* spring migrant [US], fall migrant [US]); *syn.* passage migrant [n] [also UK], passage visitor [n], transient [n]; *s* **ave [f] de paso** (Especie de avifauna que transita por un país o una región al dirigirse a su ▶área de cría o a su ▶área de invernada; *término genérico* ▶especie migratoria; ▶migración 2, ▶ave migradora parcial); *syn.* ave [f] transeunte, ave [f] de pasa y contrapasa; *f* **oiseau [m] de passage** (Du point de vue d'un pays ou d'un secteur déterminé, oiseau survolant ce pays ou ce secteur lors de son vol vers l'▶aire de nidification ou l'▶aire d'hivernage, effectuant une halte sur une aire d'étape et qui n'est visible qu'au printemps et/ou à l'automne ; ▶migration 2, ▶oiseau migrateur partiel, ▶visiteur d'été ; *terme générique* ▶espèce migratrice) ; *g* **Durchzügler [m]** (*Avifauna* vom Standpunkt eines bestimmten Landes oder Gebietes aus gesehen der Vogel, der auf seinem Zug zum ▶Brutareal oder ▶Überwinterungsgebiet dieses Land oder Gebiet überfliegt und während des Zuges Rast macht und somit meistens im frühen Frühjahr und/oder im Herbst dort anzutreffen ist; ▶Migration 2, ▶Teilzieher; *OB* ▶wandernde [wild lebende] Tierart).

bird [n] **of prey** *zool.* ▶predatory bird.

484 bird protection [n] *conserv. leg.* (Species conservation of birds with such measures as habitat conservation and habitat management given priority internationally by ▶conventions on nature conservation; ▶International Council for Bird Protection); *s* **protección [f] de aves** (Preservación de especies de aves por medio de medidas tales como la conservación y el manejo de hábitats que se priorizan internacionalmente en ▶convenciones para la conservación de la naturaleza; ▶Consejo Internacional de Protección de Aves, ▶protección de especies de fauna); *f* **conservation [f] des oiseaux** (Protection des espèces d'oiseaux grâce en particulier aux différentes mesures nationales de protection et de gestion des habitats naturels ainsi qu'internationales [▶convention (internationale) pour la protection de la nature] dont les objectifs sont **1.** l'harmonisation des mesures de protection des espèces et des habitats. **2.** La création de réseaux cohérents d'espaces protégés ; ▶Conseil International pour la Préservation des Oiseaux) ; *g* **Vogelschutz [m, o. Pl.]** (Artenschutz für Vögel durch Maßnahmen des Biotopschutzes und der

Biotopgestaltung — durch ▶Naturschutzkonventionen internatio-
nal aufgewertet. Drei Jahrzehnte vor der Institutionalisierung des
Naturschutzes 1904 gründete 1875 der Geraer Mathematiklehrer
KARL THEODOR LIEBE [1800-1894] den „Deutschen Verein zum
Schutze der Vogelwelt"; cf. N+L 2002 [7], 321; ▶Internationaler
Rat für Vogelschutz).

485 bird protection research center [n] [US] *conserv.*
zool. (Institution concerned with the protection of birds on a
regional or national scale as distinguished from an ▶orntho-
logical station); *syn.* bird protection research centre [n] [UK];
s instituto [m] de ornitología aplicada (Institución pública
existente en D. a nivel de estado federado, que es responsable de
estudiar la situación de las aves en la región y de promocionar los
conocimientos ornitológicos y la protección de aves entre la
población en general; ▶estación ornitológica); *f* **institut [m]
d'ornithologie appliquée (±)** (D., institution publique scientifique des Länder qui, à la différence des ▶stations ornithologiques, se consacre à l'étude de projets de conservation des
oiseaux ainsi qu'à l'ornithologie appliquée) ; *g* **Vogelschutz-
warte [f]** (Staatliches Institut, z. B. in Frankfurt/M, Hamburg,
Hannover, Kiel und Karlsruhe, das sich im Unterschied zur
▶Vogelwarte auf Landesebene dem Vogelschutz und der
angewandten Vogelkunde widmet).

bird protection research centre [n] [UK] *conserv. zool.*
▶bird protection research center [US].

486 bird refuge plant [n] *landsc. plant.* (Tree or shrub
providing birds with important shelter, food, and nesting habitat;
▶woody host plant for birds); *s* **leñosa [f] escondite para aves**
(Árbol o arbusto que sirve a las aves como lugar de protección,
alimento y nidificación; ▶leñosa útil para aves); *f* **plante [f]
ligneuse refuge pour les oiseaux (±)** (Végétaux ligneux procurant aux peuplements d'oiseaux la possibilité de se nourrir, de
se protéger, de nicher ; ▶ligneux nourricier pour les oiseaux) ;
g **Vogelschutzgehölz [n]** (Gehölz, das wild lebenden Vögeln,
Nahrung, Schutz und Nistgelegenheit bietet; ▶Vogelnährgehölz).

487 bird sanctuary [n] *conserv.* (Land set aside primarily as
a refuge for the protection of bird life; ▶natural area preserve);
s **santuario [m] de aves** (Término anglosajón, *sanctuary*, que
designa a una reserva permanente para la protección de ciertas
especies o individuos concretos durante todo o parte de su ciclo
vital; DINA 1987; ▶espacio natural protegido, ▶reserva
natural); *f 1* **réserve [f] ornithologique** (▶Réserve naturelle dont
le but exclusif est de protéger les espèces d'oiseaux ; *terme
spécifique* parc ornithologique ; le terme allemand ne correspon-
dait pas jusqu'à 1979 à une catégorie de protection réglementaire
comme p. ex. les ▶sites classés) ; *f 2* **zone [f] de protection
spéciale [Z.P.S.]** (Territoire classé, institué par la Directive
européenne n° 79/409 du 6 avril 1979 concernant la conservation
des oiseaux et complétée par la directive n° 92/43 du 21 mai 1992
concernant la conservation des habitats naturels ainsi que de la
faune et de la flore sauvages, avec les objectifs **1.** la protection
d'habitats permettant d'assurer la survie et la reproduction des
oiseaux sauvages rares ou menacés. **2.** La protection des aires de
reproduction, de mue, d'hivernage et des zones de relais de
migration pour l'ensemble des espèces migratrices ; ces zones
sont désignées à partir de l'inventaire des **zones importantes
pour la conservation des oiseaux [Z.I.C.O.]**) ; cf. GPE 1998) ;
g **Vogelschutzgebiet [n]** (Ein dem Vogelschutz dienendes Naturreservat; früher kein rechtlicher Begriff wie ▶Naturschutz-
gebiet 2, sondern ein Funktionsbegriff; heute durch EU-Vogelschutzrichtlinie eine Schutzkategorie).

birds area [n] *conserv. nat'res.* ▶Important Birds Area.

488 Birds Directive [n] *conserv. leg.* (European Council
Directive 79/409/EEC on the conservation of wild birds, adopted

in 1979, in response to the 1979 Bern Convention on the
conservation of European Habitats and species. The **B. D.**
provides a framework for the conservation and management of,
and human interactions with, wild birds in Europe. It sets broad
objectives for a wide range of activities, although the precise
legal mechanisms for their achievement are at the discretion of
each Member State. The main provisions of the Directive include:
1. The maintenance of the favourable ▶conservation status 2 of
all wild bird species across their distributional range [Article 2]
with the encouragement of various activities to that end [Article
3]. **2.** The identification and classification of Special Protection
Areas for rare or vulnerable species listed in Annex I of the
Directive, as well as for all regularly occurring migratory species,
paying particular attention to the protection of wetlands of
international importance [Article 4—together with ▶Special
Areas of Conservation [SACs] designated under the Habitats
Directive, SPAs form a network of pan-European protected areas
known as Natura 2000]. **3.** The establishment of a general scheme
of protection for all wild birds (Article 5). **4.** Restrictions on the
sale and keeping of wild birds (Article 6). **5.** Specification of the
conditions under which hunting and falconry can be undertaken
[Article 7]. [Huntable species are listed on Annex II 1 and Annex
II 2 of the Directive]. **6.** Prohibition of large-scale non-selective
means of bird killing [Article 8]. **7.** Procedures under which
Member States may derogate from the provisions of Articles 5-8
[Article 9] — that is, the conditions under which permission may
be given for otherwise prohibited activities. **8.** Encouragement of
certain forms of relevant research (Article 10). **9.** Requirements
to ensure that introduction of non-native birds do not threatened
other biodiversity [Article 11]. **In UK**, a very wide range of other
statutory and non-statutory activities also support the Bird
Directive's implementation. This includes national bird monitoring schemes, bird conservation research, and the UK Biodiversity
Action Plan, which involves action for a number of bird species
and habitats which support them; ▶Habitats Directive);
s **Directiva [f] de las Aves** (Directiva 79/400/CEE del Consejo,
de 2 de abril de 1979, relativa a la conservación de las aves
silvestres, fue aprobada en 1979 como respuesta a la ▶Conven-
ción de Berna, de 1979, sobre la conservación de la vida salvaje y
del medio natural en Europa. La **d. de las a.** proporciona el marco
para la preservación y gestión de las aves silvestres en Europa, así
como de las interacciones humanas con las mismas. Determina
objetivos amplios para una gran gama de actividades, aunque los
mecanismos legales precisos para conseguirlos dependen de la
discreción de cada uno de los Estados miembros. Las principales
previsiones de la directiva incluyen: **1.** El mantenimiento del
estado favorable para todas las especies silvestres en todas sus
áreas de distribución [art. 2]. **2.** La determinación y clasificación
de zonas de protección especial [ZEPA] para las aprox. 200
especies y subespecies amenazadas, clasificadas como tales en el
Anexo I de la Directiva, así como para las especies migratorias,
dando una atención especial a la protección de humedales de
importancia internacional catalogados según la ▶Convención de
Ramsar [art. 4; junto con las Zonas Especiales de Conservación
definidas por la ▶Directiva de Hábitats, las ZEPA forman una
red de áreas protegidas paneuropeas llamada Natura 2000]. **3.** El
establecimiento de un sistema general de protección de todas las
aves silvestres [art. 5]. **4.** Restricciones para la comercialización y
el mantenimiento en cautiverio [art. 6]. **5.** Especificaciones de los
métodos permitidos de caza y falconeo [art. 7], mientras que las
especies que pueden ser objeto de caza figuran en la lista del
Anexo II. **6.** Prohibición de métodos no selectivos de captura y de
sacrificio [art. 8 y Anexo IV]. **7.** Procedimiento bajo el cual los
Estados miembros pueden derogar las previsiones de los arts. 5-8
[art. 9], es decir, las condiciones bajo las cuales se pueden dar
permisos para actividades generalmente prohibidas. **8.** Fomentar

la investigación nombrando temas a los que se debería dar especial atención [art. 10 y Anexo V]. **9.** Requirimientos para asegurar que la introducción de especies no nativas de aves no amenacen la biodiversidad de las otras [art. 11]); *f* **Directive [f] Oiseaux** (La Directive 79/409/CEE — appelée plus généralement Directive Oiseaux — du 2 avril 1979 est une mesure prise par l'Union européenne engageant chaque État à promouvoir et assurer la protection et la gestion de toutes les espèces aviennes sauvages de son territoire ; cette protection s'applique aussi bien aux oiseaux eux-mêmes qu'à leurs nids, leurs œufs et leurs habitats ; les États membres doivent prendre toutes les mesures nécessaires pour préserver, maintenir et rétablir les biotopes et les habitas des oiseaux en créant des zones de protection, entretenant et aménageant les habitats, rétablissant les biotopes détruits, créant des biotopes ; les espèces d'oiseaux mentionnées à l'annexe I de la Directive Oiseaux [qui énumère les espèces les plus menacées de la Communauté, telles que les espèces menacées de disparition, les espèces vulnérables à certaines modifications de leurs habitats, ou encore les espèces considérées comme rares parce que leurs populations sont faibles] doivent faire l'objet de mesures de conservation spéciale concernant leur habitat, afin d'assurer leur survie et leur reproduction ; à cet effet est réalisé un inventaire des ▶zones importantes pour la conservation des oiseaux [ZICO] incluant la protection des zones de reproduction, d'hivernage et de migration ; chaque État classe les territoires les plus appropriés en nombre et en superficie à la conservation des espèces en ▶Zones de Protection Spéciales [ZPS] afin que puissent y être mise en œuvre des mesures de protection et/ou de restauration ; en raison de l'importance des zones humides pour l'avifaune, les zones humides d'importance internationale selon les critères de la ▶Convention de Ramsar sont à prendre en compte dans la constitution des ZPS ; avec l'adoption de la Directive Habitats, les États membres doivent constituer un réseau cohérent de ▶Zones Spéciales de Conservation [ZSC] dénommé ▶réseau Natura 2000 ; ce réseau intègre également les ZPS désignées au titre de Directive Oiseaux, d'où l'appellation commune « Site Natura 2000 » qui est donnée en France aux ZSC comme aux ZPS ; de manière générale, les États membres doivent prendre les mesures nécessaires pour instaurer un régime général de protection de toutes les espèces d'oiseaux en interdisant, quelle que soit la méthode employée, de tuer ou de capturer intentionnellement les espèces d'oiseaux couverts par les directives ; néanmoins la chasse de certaines espèces est autorisée à condition que les méthodes de chasse utilisées respectent certains principes [utilisation raisonnée et équilibrée, chasse en dehors de la migration ou de la reproduction, interdiction de méthodes de mise à mort ou de capture massive ou non sélective], de détruire, d'endommager et de ramasser leurs nids et leurs œufs, de les perturber intentionnellement, de les détenir ; ▶Directive Habitat Faune Flore) ; *g* **Vogelschutzrichtlinie [f]** (Für jedes Mitgliedsland der Europäischen Union [EU] verbindlich umzusetzendes Recht [cf. Richtlinie 79/409/EWG des Rates vom 02.04.1979, geändert durch Richtlinie 97/49/EG des Rates vom 29.07.1997] mit dem vorrangigen Ziel des langfristigen Schutzes und der Erhaltung aller in Europa wild lebenden Vögel [▶Erhaltungszustand] und ihrer natürlichen Lebensräume. Dazu wird ein europaweit vernetztes Schutzgebietssystem ▶Natura-2000-Schutzgebietssystem aufgebaut, da durch den Schutz einzelner, isolierter Gebiete die biologische Vielfalt nicht dauerhaft erhalten werden kann. Die Richtlinie beinhaltet neben dieser Zielsetzung auch naturschutzfachliche Grundlagen und Verfahrensvorgaben zur Errichtung des NATURA-2000-Schutzgebietssystems. Die Mitgliedstaaten müssen für bestimmte, in Anhang I der Richtlinie aufgeführte Vogelarten geeignete Gebiete erhalten und entwickeln. Entsprechendes gilt für alle Zugvogelarten, die nicht in Anhang I aufgeführt sind. Anhang I der **V.** führt rund 180 Vogelarten auf. Auch für die in großer Zahl rastenden Wasser-, Wat- und Greifvögel müssen Schutzgebiete benannt werden. Bei deren Benennung ist gemäß der Richtlinie außerdem der Schutz von Feuchtgebieten und insbesondere von international bedeutsamen nach der ▶Ramsar-Konvention zu berücksichtigen. Da in D. die Schutzvorschriften der ▶Fauna-Flora-Habitat-Richtlinie [FFH-RL] in die §§ 19a bis 19f des Bundesnaturschutzgesetzes übernommen wurden, gelten auch für die EU-Vogelschutzgebiete die §§ 19a bis 19f BNatSchG. Die Vogelschutzgebiete unterliegen auch den Schutzbestimmungen der FFH-RL, da sie nach deren Artikel 7 zu Bestandteilen von NATURA 2000 erklärt worden sind); *syn.* Richtlinie [f] über die Erhaltung der wild lebenden Vogelarten.

489 bird strike [n] *conserv.* (Bird collision with aircraft or severe impact of birds upon glass surfaces, as well as on overhead transmission lines; *specific term* bird-window collision); *syn.* bird crash [n]; *s* **choque [m] de aves contra objetos** (Colisión fuerte de aves contra aviones, ventanales de vidrio o líneas de alta tensión); *f* **heurt [m] avec des oiseaux** (Collision entre des oiseaux et un avion, choc sévère d'oiseaux contre une surface vitrée ou sur une ligne aérienne); *syn.* collision [f] avec des oiseaux (QUID 1996) ; *g* **Vogelschlag [m]** (Zusammenstoß von Luftfahrzeugen mit Vögeln oder heftiger Aufschlag von Vögeln auf Glasflächen sowie Aufschlag auf Überlandleitungen. Nach den Richtlinien zur Verhütung von Vogelschlägen im Luftverkehr des Bundesministers für Verkehr vom 13.02.1974 müssen eigens dafür bestellte **V.beauftragte** Maßnahmen zur V.verhütung organisieren und koordinieren); *syn.* Vogelkollision [f] (N+L 2000 [11], 426).

bird watcher [n] *zool.* ▶ornithologist.

490 black earth [n] *pedol.* (Fertile steppe or prairie soil with A-C profile having an A horizon > 50cm. The U.S. Comprehensive Soil Classification System differentiates between 'udoll' which is characterized by > 8 °C average annual temperature and 'boroll' by < 8 °C average annual temperature); *syn.* chernozem [n]; *s* **chernosem [m]** (Suelo perteneciente a la clase de los suelos de estepa con perfil A-C y un horizonte A de margas de > 50 cm de espesor. Este suelo se da sobre todo en Ucrania, Manitoba y Dakota); *syn.* tierra [f] negra de estepa, chernoziom [m]; *f* **chernozem [m]** (Sol isohumique riche en matière organique profondément incorporée, à structure grumeleuse, avec une couleur noire ou brune très foncée. Ce sol de steppe caractérise un climat continental faiblement aride à pluviosité de 400 à 600 mm ; on le trouve p. ex. en Ukraine, au Manitoba et au Dakota) ; *g* **Schwarzerde [f]** (Zur Klasse der Steppenböden gehörender Boden mit A-C-Profil und > 50 cm mächtigem A-Horizont aus Mergelgestein, vorwiegend anzutreffen in der Ukraine, in Manitoba und in Dakota; cf. SS 1979); *syn.* Tschernosem [m].

491 black spot disease [n] *bot. envir. phytopath.* (Plant disease occurring on roses caused by the parasitic fungi, Massonina rosae [asexual stage] and most commonly, Diplocarpon rosae [sexual stage]. The conidia of the fungi penetrate the cuticle of rose leaves and their haustoria then enter into the epidermis. Infection causes irregular, purple-brown to black spots, up to a diameter of 14mm, on the surfaces of leaves. Spots are circular with a perforated, star-shaped edge and lead to gradual defoliation. Rose stems may also be affected and the whole plant is progressively weakened. Overwintering in leaves as wind-borne mycelia, ascopores and conidia, the disease is spread in the spring during rainy conditions, especially on closely spaced plants and can be brought under control through the careful selection of resistant rose species and varieties as well as good ventilation between the rose stocks. It is also commonly

B

treated by spraying fungicides upon new leaf emergence or first appearance of the disease and by removing the affected leaves); *s* **fumagina [f]** (Costra o polvo negruzco que recubre ramas y hojas, formado por el micelio de diversos hongos, que rara vez se presentan en forma perfecta, por lo que su identificación es muchas veces difícil o imposible. Generalmente su origen es saprofítico, a expensas del melazo exudado o excretado por pulgones y cochinillas, o de origen fisiológico. Se presenta entre otras especies en sauces y chopos, en la vid y en el olivo, así como en las rosas; cf. DB 1985); *syn. popular* negrilla [f], tizne [m]; *f 1* **maladie [f] des tâches noires du rosier** (Maladie cryptogamique causée par des champignons, *Marssonina rosae* et *Diplocarpon rosae*, dont le développement est favorisé par des étés chauds et humides, qui provoquent l'apparition de tâches noires généralement circulaires, le jaunissement des feuilles et leur chute prématurée) ; *f 2* **fumagine [f]** (Maladie crypto-gamique qui se développe sur des dépôts sucrés excrétés à la surface de végétaux par des pucerons. Le mycélium et les fructifications de divers champignons du genre *Apiosporium [Capnodium]* forment un enduit noir sur les parties aériennes des plantes ; cf. LA 1981) ; *g* **Sternrußtau [m, o. Pl.]** (Parasitärer Pilz [asexuelle Form *Massonina rosae*, sexuelle Form *Diplocarpon rosae*], dessen Konidien mit Keimschläuchen die Kutikula der Rosenblätter durchdringen und anschließend Haustorien in die Epidermis senken, so dass oberflächlich unregelmäßige, violettbraune bis schwarze Flecken, die am Rand sternförmig auslaufen, entstehen. Die Krankheit führt zur Schwächung der Pflanzen und zu vorzeitigem Blattabfall; begünstigt wird der Befall durch Feuchtigkeit und engen Pflanzenabstand; durch eine gezielte Auswahl von resistenten Rosenarten und -sorten und durch eine gute Durchlüftung im Bestand sowie durch recht-zeitige Fungizidspritzung kann die Krankheit eingedämmt werden); *syn.* Schwarzfleckenkrankheit [f].

black tide [n] [US] *envir.* ►shore oil pollution.

492 black-top [n] [US&CDN] (1) *constr.* (Bituminous surface material, usually asphalt, used in paving of vehicular, pedestrian, play areas, and airport runways; ►macadam); *syn.* bituminous pavement [n]; *s* **revestimiento [m] bituminoso** (Material bitumi-noso, generalmente asfalto, utilizado para pavimentar superficies; *término específico* revestimiento asfáltico; ►pavimento de macadam); *syn.* pavimento [m] bituminoso; *f* **revêtement [m] bitumineux** (Enduit superficiel à base de bitume, de granulat ou de sable sous forme d'asphalte utilisé pour les voies piétonnes, les aires de jeux, les sols sportifs, ou en émulsion pour la protection hydrique des ouvrages enterrés et l'étanchéité des toiture-terrasses, l'imperméabilisation et le scellement des fissures, la restauration des caractéristiques de surface [rugosité, planéité, aspect visuel, etc.] des couches d'usure légèrement dégradées [chaussées, trottoirs, pistes cyclables, etc.] ; ►revête-ment en macadam) ; *syn.* revêtement [m] noir, enrobé [m] noir ; *g* **Schwarzdecke [f]** (Bitumengebundene, auf eine Tragschicht aufgebrachte schwarze Oberflächenschutzschicht für Verkehrs- und Spielflächen; es gibt je nach Einbauart [heiß, warm oder kalt], Mischgut und Bindemittel unterschiedliche Deckschichten, z. B. **Decke aus Asphalt- und Teerasphaltbeton** [im Heiß- und Warmeinbau], **Gussasphaltdecke**, ►**Makadamdecke, Splitt-mastixasphaltbelag, bituminöse Tragdeckschicht, Edelsplitt-abstreuung auf heißer Bitumenemulsion**); *syn.* bituminöse Decke [f], bituminös gebundene Decke/Deckschicht, bituminöser Belag [m], Schwarzbelag [m].

493 black-top [n] [US&CDN] (2) *arb. for.* (Final phase of ►top-kill [US]/top drying [UK], when all needles of a coniferous tree have fallen; the initial phase is termed ►sorrel-top [US&CDN]; ►dead crown, ►forest decline, ►initial phase of top-kill [US]/initial phase of top drying [UK]); *s* **puntiseco [m]**

en conífera (Fase final de ►puntiseco, cuando todas las agujas de una conifera se han caido; la fase inicial se llama ►puntiseco incipiente; ►copa muerta, ►muerte de bosques); *f* **défoliation [f] de la cime des conifères** (Dégâts occasionnés par des accidents climatiques, des maladies fongiques, des attaques de déprédateurs ou par la pollution et entraînant la perte totale et parfois rapide des aiguilles dans les peuplements de résineux ; une défoliation grave pendant plusieurs années consécutives peut entraîner une diminution de l'accroissement, une descente de cime, des diffor-mités du fût et, dans certains cas, la mort de l'arbre; ►houppe morte, ►mort des forêts, ►phase initiale du dépérissement termi-nal, ►sec, sèche en cime) ; *syn.* dépérissement [m] terminal des conifères, descente [f] de cime des conifères ; *g* **Zopftrockenheit [f] bei Koniferen** (Auf Grund von Schädlingen, Pilzkrankheiten oder Umweltstressfaktoren verursachte völlige Verlichtung/Ent-nadelung eines Nadelbaumes; ►abgestorbene Baumkrone, ►be-ginnende Wipfeldürre, ►Waldsterben, ►wipfeldürr); *syn.* Wip-feldürre [f] bei Koniferen.

blaes [n] [SCOT] *min.* ►red shale.

494 blanket bog [n] *geo. phyt.* (Non-raised oligotrophic bog which is not, or not very markedly, raised above the mineral water table of the surrounding landscape; therefore it is generally wetter and not as oligotrophic as ►raised bogs; cf. ELL 1967); *s* **turbera [f] de cobertura** (Turbera oligótrofa no o muy poco elevada sobre el nivel de la capa freática mineral del paisaje alrededor, por eso, en general, más húmeda y menos oligótrofa que la ►turbera alta); *f* **tourbière [f] de couverture** (Tourbière oligotrophe non bombée ne dépassant pas ou pratiquement pas le niveau de la nappe phréatique des terrains limitrophes. En général plus humide et moins oligotrophe que les ►tourbières hautes à Sphaignes) ; *g* **Deckenmoor [n]** (Nicht gewölbtes oligotrophes Moor, das nicht oder unwesentlich über dem mineralischen Grundwasserstand der umgebenden Landschaft hinausragt; im Allgemeinen nasser als und nicht so oligotroph wie ein ►Hoch-moor; ELL 1967); *syn.* terrainbedeckendes Moosmoor [n].

blast damage [n] *agr. for. hort.* ►storm damage.

blasted tree [n] *agr. for. hort.* ►storm damage.

495 blighted area [n] *urb.* (Urban land with deteriorated or abandoned structures, inadequate or missing community services, vacant land with debris and litter, and impacted by noise, heavy traffic, and odo[u]rs; cf. IBDD 1981; ►run-down housing fit for rehabilitation, ►hazardous old dumpsite); *s* **barrio [f] insalubre (±)** (Zona urbana caracterizada por edificios deteriorados, aban-donados o en ruinas, infraestructura de abastecimiento insufi-ciente, con solares no edificados mal cuidados, a veces afectada por ruido de tráfico pesado o actividades insalubres de origen industrial; ►edificación antigua en mal estado; ►emplazamiento contaminado); *f* **îlot [m] insalubre** (Zone dans laquelle plusieurs immeubles bâtis ou non, attenants ou non à la voie publique et qui constituent soit par eux-mêmes, soit par les conditions dans lesquels ils sont occupés un danger pour la santé des occupants ou des voisins [immeubles abandonnés ou ne correspondant plus aux normes d'hygiène et de sécurité] ; art. 26 et 36 Code de la santé publique ; ►habitat vétuste, ►site pollué ancien) ; *syn.* zone [f] d'habitat insalubre ; *g* **Baugebiet [n] mit städtebau-lichen Missständen** (Baugebiet, das durch Verfall oder verlas-sene Häuser, unzureichende oder fehlende Ver- und Entsorgungs-einrichtungen, durch mit Unrat versehene unbebaute Flächen oder durch Lärm, Schwerverkehr und Geruchsbelästigungen gekenn-zeichnet ist; ►Altbebauung 2, ►Altstandort).

496 blinding [n] *constr.* (Permeable, lower paving course, usually of a granular type material designed to provide a working platform on which subbase construction can proceed with mini-mum interruption from wet weather, also installed to minimize the effect of a weak subgrade [US]/sub-grade [UK] on pavement strength and to prevent clay and silt particles from reaching the ▶subbase [US]/sub-base [UK]; often used as a synonym for ▶frost-resistant subbase [US]/frost-resistant sub-base [UK]); *syn.* capping layer [n]; *s* **capa [f] de limpieza** (En construcción viaria, capa filtrante directamente superior al subsuelo que tiene como fin evitar que partículas de arcilla o limo pasen a la ▶capa portante; el término se utiliza también como sinónimo de ▶capa de protección contra heladas); *f* **couche [f] anticontaminante** (Sous-couche de chaussée, directement placée sur le ▶fond de forme, constituée de matériaux suffisamment fins pour empêcher les remontées d'argile dans le corps de chaussée ; DIR 1977 ; ▶couche antigel, ▶couche de fondation) ; *syn.* couche [f] de propreté (cf. MAÇ 1981, 229) ; *g* **Sauberkeitsschicht [f]** (Dränfähige Schicht im Wegebau direkt über dem Untergrund [▶Erdplanum], die verhindern soll, dass Ton- oder Schluffteile in die ▶Tragschicht gelangen — wird oft syn. für ▶Frostschutzschicht verwendet); *syn.* verbesserter Untergrund [m] (1).

497 blinding layer [n] *constr.* (In sportsground construction, layer between the ▶root-zone layer and the ▶drainage layer of sports fields; ▶filter mat); *s* **capa [f] filtrante (1)** (En la construcción de campos de deporte, capa entre el ▶sustrato de plantación y la ▶capa de drenaje; ▶estera filtrante 1); *f* **couche [f] filtrante** (Dans les travaux d'aires de sports, couche située entre la ▶couche drainante et le ▶support de culture [support végétal d'un gazon, mélange terreux] ; ▶feutre filtrant [anticontaminant]) ; *g* **Filterschicht [f] (1)** (Beim Sportplatzbau oder bei Dachbegrünungen vorgesehene Schicht zwischen ▶Vegetationstragschicht und ▶Dränschicht; ▶Filtermatte).

blinding layer [n] **of sand** *constr.* ▶lay with a (5cm) blinding layer of sand, ▶sand bed.

bloated clay [n] [UK] *constr.* ▶expanded clay.

bloated slate [n] [UK] *constr.* ▶expanded slate [US].

block [n] *constr.* ▶precast paving block [US]/precast concrete paving slab [UK].

block [n] [UK], **apartment** *arch. urb.* ▶multistory building for housing [US].

block [n], **concrete stone** *constr.* ▶cut stone, #3.

block [n] [UK], **tower** *arch. urb.* ▶high-rise residential building.

498 block clearance [n] *urb.* (Removal of buildings or structures within an urban block; ▶perimeter block development, ▶comprehensive redevelopment, ▶center city redevelopment [US]/urban regeneration [UK]); *s* **despeje [m] de patios interiores** (Demolición de edificios en el interior de las manzanas para mejorar las condiciones higiénicas de las viviendas [iluminación, densidad]; ▶edificación en bloque cerrado, ▶reestructuración urbana, ▶rehabilitación urbana); *f* **restructuration [f] d'îlot** (Opération sur une ▶construction en îlot dans les quartiers anciens impliquant la démolition de certains immeubles existants, de bâtiments annexes dans les cours intérieures dans le but de réduire la densité de l'habitat, d'améliorer l'éclairement des logements et d'aménager des espaces pour le jeu ou la détente ; ▶rénovation urbaine, ▶restauration urbaine, ▶restructuration urbaine) ; *g* **Blockentkernung [f]** (Art der städtebaulichen Sanierung durch Beseitigung zu großer Baumassendichte und schlechter Wohnungsbelichtung durch Abbruch von Nebengebäuden und Hofüberbauungen in einer ▶Blockbebauung; ▶Flächensanierung, ▶Stadtsanierung); *syn.* Blockauskernung [f], Entkernung [f].

499 block development [n] *urb.* (**1.** Housing on a block-by-block basis where buildings generally abut one another, enclosing a central open space [courtyard, light well, service area or landscaped area], which is out of view from the public realm; ▶building block. **2.** Buildings which form the edge of a block are referred to as a **perimeter block development**; buildings which surround a semi-public, central open space with greenery, play and sitting areas are referred to as a **courtyard building development**); *syn.* enclosed block development [n], courtyard block development [n]; *s* **edificación [f] en bloque cerrado** (Tipo de construcción en el que los edificios están construidos a lo largo de la línea municipal de fachada y encierran totalmente uno o varios patios en el interior; ▶manzana); *f* **construction [f] en îlot** (Espace urbain clos entièrement entouré par la voirie dans lequel les corps de bâtiment enserrent une cour centrale ou des espaces libres importants [cœur d'îlot] ; l'îlot est lui-même divisé en parcelles, de taille variable mais de forme le plus souvent rectangulaires, et dont les limites sont souvent perpendiculaires aux bâtiments ; la construction à l'intérieur de l'îlot d'immeubles de 2ème, voire de 3ème rang a grandement contribué à l'insalubrité dans les quartiers anciens de cette forme urbaine qui depuis les années 1970 est sujette à des politiques successives de rénovation urbaine [restructuration d'îlot], de réhabilitation du bâti en front de rue et des logements sur cour ainsi que de réorganisation du cœur d'îlot ; ▶îlot) ; *g* **Blockbebauung [f]** (**1.** Geschlossene Bauform mit Gebäuden ohne seitlichen Grenzabstand, die Höfe oder größere Freiflächen allseitig umschließt und den Einblick in die Höfe oder Freianlagen von der Straße verhindert. **2.** Bei der Betrachtung der Baukörper, die einen Block umschließen spricht man von **Blockrandbebauung**; bei der Betrachtung des halböffentlichen und meist begrünten und mit Kinderspielgeräten und Sitzmöglichkeiten ausgestatteten Hofes spricht man von **Hofrandbebauung**; ▶Wohnblock).

block development [n], **courtyard** *urb.* ▶block development.

block development [n], **enclosed** *urb.* ▶block development.

block development [n], **perimeter** *urb.* ▶block development.

block [n] **of flats** [UK] *urb.* ▶multistory building for housing [US].

block [n] **of trees** [UK] *agr. land'man.* ▶coppice [US] (1).

500 blockout [n] *arch.* (Void produced by material attached to formwork within a concrete structure under construction, a space where concrete is not to be placed; used to create a channel or encasement for a fixture; ▶roof blockout); *s* **roza [f]** (Corte hecho en un muro, en forma de canal, para permitir el paso de tuberías y conducciones; DACO 1988; en estructura de hormigón, espacio reservado para otra función como para el ▶paso de tuberías y conducciones; ▶manguito a través de un tejado); *syn.* cavidad [f], escotadura [f], escote [m]; *f* **réservation [f]** (Orifice prévu dans une dalle ou un mur pour faciliter le ▶passage de conduites ; ▶passage en traversée de dalle) ; *g* **Aussparung [f]** (Geplantes Loch oder vorgesehene Vertiefung in einer Wand oder Deckenkonstruktion für die ▶Durchführung von Rohrleitungen oder Kabeln; aussparen [vb]; ▶Dachdurchführung).

block pavement [n] *constr.* ▶interlocking block pavement.

501 block paving [n] [US] *constr. eng.* (▶large-sized paving stone [US]/large-sized paving sett [UK], ▶mosaic block paving [US]/mosaic sett paving [UK]); *syn.* sett paving [n] [UK]; *s* **pavimento [m] de adoquines grandes** (Revestimiento de ▶adoquines grandes con aparejo de juntas a tope); *f* **pavage [m] en gros pavés** (Revêtement en ▶gros pavés posés à joints rompus et calés au marteau) ; *g* **Großsteinpflaster [n]** (Belag aus ▶Großpflastersteinen hammerfest im Verband verlegt); *syn.* Kopfsteinpflaster [n].

B

502 block step [n] *constr.* (Solid tread of whole concrete or stone; ▶roughly-hewn block step); *s* **escalón** [m] **prefabricado** (de una pieza de hormigón o de piedra; ▶escalón de piedras de labra tosca); *f* **bloc marche** [m] (Marche massive en pierre de taille, préfabriquée en béton en bloc, en pierre reconstituée ; VRD 1994, partie 4, chap. 6.4.2.2.1, p. 11 ; ▶dalle-bloc brute ; *terme spécifique* marche [f] en pierre massive) ; *syn.* marche [f] massive ; *g* **Blockstufe** [f] (Aus einem Stück gefertigte Beton- oder Natursteinstufe; ▶grob behauene Natursteinblockstufe).

503 block step [n] **with nosing** *constr.* (Block step with rounded or chamfered tread extension beyond the riser plane; ▶nosing); *s* **escalón** [m] **prefabricado con mampirlán** (▶mampirlán); *f* **bloc marche** [m] **à contremarche biseautée** (▶débordement) ; *g* **Blockstufe** [f] **mit Unterschneidung** (Blockstufe, bei der die vorne leicht gerundete oder gefaste Trittfläche etwas übersteht; ▶Unterschneidung).

blooming period [n] *hort. plant.* ▶flowering period.

blowdown area [n] [US] *landsc.* ▶windfall area.

504 blow-out [n] *geo. landsc.* (Mussel-shaped damage on landforms, e.g. holes on dune crests, which have been torn out by wind on exposed slopes); *s* **excavación** [f] **eólica** (Huecos generalmente con forma de moluscos, causados por el viento en la superficie de formaciones, como p. ej. dunas); *syn.* arranque [m] de pedazos del suelo (por acción del viento; BB 1979, 290); *f* **creux** [m] **de déflation** (Dépression creusée dans le sol par le vent, p. ex. dans les sables dunaires ; DFM 1975) ; *syn.* couloir [m] de souffle-vent ; *g* **Windanriss** [m] (Meist muschelförmige Beschädigung von Oberflächenformen durch Wind, z. B. bei Dünen und windexponierten Hangflächen); *syn.* Ausblasung [f], Windkolk [m].

505 Blue List [n] [US] *conserv. zool.* (**In U.S.**, the National Audubon Society's annual compilation of bird species in decline; ▶Endangered Species List, ▶Red List); *s* **Lista** [f] **Azul** [US] (En EE.UU., la National Audubon Society registra anualmente las especies de aves que han sufrido reducción; ▶Catálogo Español de Especies Amenazadas, ▶lista roja); *f* **Liste** [f] **bleue** [US] (U.S., liste des oiseaux protégés publiée tous les ans par la *National Audubon Society* ; ▶Livre rouge [des espèces menacées], ▶liste des espèces menacées) ; *g* **Blaue Liste** [f] [US] (Zusammenstellung der bedrohten Vögel in den USA, die jährlich von der *National Audubon Society* herausgegeben wird; ▶Rote Liste).

506 bluff [n] (**1**) *geo.* (Steep bank cut by lateral erosion of a watercourse on the outside of a meander; *opp.* ▶slip bank [US]/ slip-off slope [UK]; *specific term* rivercliff [US]/river-cliff [UK]); *s* **orilla** [f] **cóncava** (Orilla abrupta de un meandro que en cursos fluviales naturales frecuentemente está socavada; *opp.* ▶orilla convexa; *término específico* ▶orilla cóncava de un río); *syn.* orilla [f] exterior; *f* **berge** [f] **concave** (Talus abrupt d'un cours d'eau et en constant recul, sapé par le courant formant des affouillements ; *opp.* ▶berge convexe) ; *syn.* rive [f] concave ; *g* **Prallhang** [m] (Außenseite einer Fließgewässerkrümmung, auf die die Strömung auftrifft. Bei natürlichen Wasserläufen wird diese Uferpartie [Unterschneidungsufer] durch Auskolkung unterhöhlt und sukzessive abgetragen; *opp.* ▶Gleithang); *syn.* Prallufer [n], Rivabord [m] [auch CH-GR].

507 bluff [n] (**2**) *geo.* (Steep prominent slope cut by erosion above a sea or lake beach or into the bank of a watercourse on the outside of a meander; ▶bluff 1; *opp.* ▶flat bank); *s* **orilla** [f] **escarpada** (Terraza natural con pediente creciente situada encima del nivel de playa en lago o río o de la línea de la orilla y que se ve afectada por la acción mecánica del agua; ▶orilla cóncava [de un rio]; *opp.* ▶orilla llana); *f* **berge** [f] **escarpée** (Talus vif, généralement abrupt le long des cours d'eau [▶berge concave] ou

en bordure de certains lacs soumis au sapement du courant ou à l'action des vagues ; *opp.* ▶berge à pente douce) ; *syn.* rive [f] abrupte, rive [f] escarpée, rive [f] raide ; *g* **Steilufer** [n] (Oberhalb des See- oder Flussstrandes oder oberhalb der Uferlinie dem Wellenschlag resp. Wasserangriff ausgesetzte, steil ansteigende natürliche Geländestufe; ▶Prallhang; *opp.* ▶Flachufer 1).

board fence [n] *constr.* ▶decorative board fence.

508 boarding [n] **up of tree trunks** *constr.* (Protection of trees on construction sites to prevent cambial tissue damage; ▶protective fencing around trees); *s* **enrejado** [m] **de tablones** (para proteger el tronco durante obras de construcción; ▶valla protectora de árboles); *f* **gaine** [f] **en planches** (Préservation — avec ou sans paillasson protecteur — des troncs d'arbres sur un chantier de construction ; ▶barrière de protection d'arbres) ; *syn.* corset [m] en bois, ceinture [f] de planches ; *g* **Bohlenummantelung** [f] (Senkrechte Bretterverschalung von Baumstämmen zum Schutz der Rinde während den Baubetriebs; ▶Baumschutzzaun).

509 boardwalk [n] *recr.* (**1.** Any pathway made of boards or planks; e.g. over sand, wetlands, or a worn or slippery area. **2.** Raised foot-path construction covered with wood planking, usually a promenade along a beach or shore; generic term covering raised walkways: ▶beach boardwalk, ▶wetland boardwalk [US]/observation boardwalk [UK]; ▶fishing pier, ▶footbridge, ▶landing pier, ▶pier 1); *syn. to 2.* raised trail [n], raised walk [n]; *s* **pasarela** [f] **de madera** (Camino de tablas o tablones elevado unos decímetros sobre el suelo, construido p. ej. sobre arena, humedal o pantano; ▶embarcadero, ▶embarcadero de botes, ▶pasarela a lo largo de la playa, ▶pasarela de observación, ▶pontón de pesca, ▶puente peatonal; *término genérico* ▶pasarela); *syn.* pasadera [f] de madera; *f* **passerelle** [f] **en/de bois** (Étroit sentier en bois fixé dans le sol avec des poteaux courts réservé aux promeneurs utilisé p. ex. le long des bords de plage [▶passerelle de bord de plage], dans les zones protégées [▶passerelle pour la découverte de la nature], ou fixé latéralement sur une falaise au-dessus d'un cours d'eaux ou en bord de mer ; ▶appontement, ▶embarcadère, ▶estacade, ▶ponton de pêche ; *terme générique* ▶passerelle, ▶pont-piéton) ; *syn.* ponton-passerelle [m] ; *g* **Holzsteg** [m] (Schmaler, niedrig aufgeständerter Pfad oder Weg am Strand über Sand [▶Strandsteg], im Feuchtgebiet über sumpfiges Gelände, bei der zumindest die Oberkonstruktion aus Holz besteht; ▶Angelsteg, ▶Beobachtungssteg, ▶Bootssteg, ▶Ufersteg; *OB* ▶Steg).

boardwalk [n] [UK]**, observation** *recr. zool.* ▶wetland boardwalk [US].

boat harbor [n] [US] *recr.* ▶sailboat harbour.

boating [n] [US] *recr.* ▶white-water boating.

boating pier [n] *recr.* ▶landing pier [US].

510 Bodden [n] *geo.* (Irregularly shaped inlet along the South Baltic coast, produced during a previous glacial period by a rise of sea level over a former uneven lowland surface; cf. DNE 1978; ▶lagoon); *s* **bodden** [m] (Bahía abierta y poco profunda localizada en antiguos lóbulos frontales del glaciar escandinavo; se encuentran en el noreste de Alemania en Mecklemburgo-Prepomerania; ▶albufera, ▶lagoon); *f* **bodden** [m] (Rentrant d'un tracé littoral de forme irrégulière fermé par un goulet, constitué lors de l'élévation du niveau de la mer le long des côtes de la Mecklembourg-Poméranie occidentale ; ▶lagune) ; *g* **Bodden** [m] (Während der Eiszeit geformte, ausgedehnte Meeresbucht an der Flachküste Mecklenburg-Vorpommerns mit zahlreichen Untiefen und einem unregelmäßigen Umriss; ▶Lagune).

body [n] **of water** *geo. limn.* ▶waterbody (1).

bog [n] *geo. phyt.* ▶active bog, ▶blanket bog; *land'man.* ▶cutover bog; *geo. phyt.* ▶degraded bog, ▶lens-shaped raised bog;

pedol. phyt. ▶low bog; *geo. phyt.* ▶raised bog; *conserv.* ▶rewatering of a bog; *geo. min.* ▶shallow bog; *geo. phyt.* ▶spring water bog [US]/spring-water bog [UK]; *pedol.* ▶transition bog; *conserv. land'man.* ▶worked bog.

bog [n], **concentric domed** *geo.* ▶lens-shaped raised bog.

bog [n] [UK], **cutaway** *conserv. land'man.* ▶worked bog.

bog [n], **ombrophilous** *geo. phyt.* ▶raised bog.

bog [n], **ombrotrophic** *geo. phyt.* ▶raised bog.

bog [n], **peat** *geo.* ▶peat mass.

bog [n], **quaking** *phyt.* ▶floating sphagnum mat.

511 bog cultivation [n] *agr. land'man.* (Tilling a bog for crop production; in D., there are the following distinct forms: Fehnkultur—fen cultivation; Hochmoorkultur—▶raised-bog cultivation; Sandmischkultur—sand-mix cultivation; Schwarzkultur—▶low-bog cultivation; Sanddeckkultur—sand-cover cultivation; ▶agricultural land improvement, ▶top bog layer); *s* **cultivo** [m] **de turbera** (Dependiendo del tipo de turbera [▶turbera alta, ▶turbera baja], en D. se diferencian diferentes sistemas de cultivo; ▶mejoramiento de suelos, ▶turba bruta); *f* **mise** [f] **en culture de tourbières** (D., on distingue différentes formes de défrichement des tourbières, 1. pour les ▶tourbières hautes, mélange de la ▶tourbe brute avec le substrat sableux sous-jacent après exploitation totale de la tourbière, 2. mise en culture d'une tourbière par drainage suivi de chaulage et d'amendement, 3. mise en culture de tourbières de faible épaisseur [50 cm à 1 m] par labour profond et mélange du substrat sableux avec la tourbe en place, 4. pour les ▶tourbières basses, pâturage intensif après addition d'engrais phosphatés et potassiques sans apport de sable ou recouvrement de la tourbière par une couche de sable ; ▶amélioration du sol) ; *syn.* mise [f] en valeur des tourbières ; *g* **Moorkultur** [f] (Urbarmachung von Mooren. In D. unterscheidet man bei ▶Hochmooren **Fehnkultur** [Vermischung der ▶Bunkerde nach Abtorfung der darunter liegenden Torfschichten mit dem sandigen Untergrund], **Deutsche Hochmoorkultur** [Urbarmachung von nicht abgetorftem Moor durch Entwässerung, Kalkung und Düngung] und **Sandmischkultur** [Urbarmachung von gering mächtigen Mooren (50-150 m) durch Tiefpflügen und Vermischung des Sandes mit dem anstehenden Torf], bei ▶Niedermooren **[Niedermoor]schwarzkultur** [intensive Grünlandnutzung mit Kaliphosphatdüngung ohne Sandauftrag] und **Sanddeckkultur** [Auftrag einer Sandschicht auf die Mooroberfläche]; ▶Melioration); *syn.* Moorkultivierung [f].

512 bog edge [n] *phyt.* (Marginal slope of a ▶raised bog; VIR 1982, 346; ▶lagg); *s* **círculo** [m] **periférico de turbera alta** (Borde de ▶tubera alta, algo más seco que ésta, que cae hacia el ▶lagg periférico); *syn.* anillo [m] periférico de turbera alta, borde [m] periférico de turbera alta; *f* **marge** [f] **sèche de la tourbière ombrogène** (Bord légèrement sec de la ▶dépression périphérique d'une tourbière ombrogène ; ▶tourbière haute) ; *syn.* périphérie [f] de la tourbière ombrogène ; *g* **Randgehänge** [n] (Der etwas trockene, zum ▶Lagg abfallende Rand eines ▶Hochmoores).

513 bog growth [n] *phyt.* (Growth which occurs on ▶raised bogs whose increase in organic material is greater than the loss caused by decomposition, due to permanent wetness and a very low pH; ▶active bog); *s* **crecimiento** [m] **de turbera** (Actividad de ▶turbera alta caracterizada por la acumulación de materia orgánica poco descompuesta debido a la presencia del encharcamiento y un pH muy bajo; ▶turbera viva); *f* **épaississement** [m] **de la tourbière** (Activité de la ▶tourbière haute caractérisée par une accumulation de matière organique — constituée principalement de Sphaignes, de Carex et de Linaigrettes — dont la production est supérieure à la décomposition à cause de l'engorgement permanent et de la très forte acidité du milieu empêchant le

processus de nitrification ; ▶tourbière active) ; *g* **Moorwachstum** [n] (Erscheinung beim ▶Hochmoor, bei dem durch permanente Nässe und durch sehr niedrigen pH-Wert die Anreicherung organischer Substanz größer ist als deren Abbau; ▶lebendes Moor).

514 bog hollow [n] *phyt.* (Inundated or extremly wet depression surrounded by ▶hummocks in a ▶raised bog; ▶bog lake); *s* **pocina** [f] **de turbera** (Depresión inundada o muy húmeda en ▶turbera alta rodeada de ▶mamelones; ▶pocina profunda de turbera); *f* **gouille** [f] **(d'une tourbière)** (▶*Tourbière haute* légère dépression remplie d'eau entourée par des ▶buttes ; ▶œil de la tourbière) ; *g* **Schlenke** [f] (Wassergefüllte oder stark wasserdurchtränkte, seichte Delle im ▶Hochmoor, von höheren ▶Bulten umgeben; ▶Hochmoorkolk).

515 bog lake [n] *geo. phyt.* (Unusually large and deep pool or small pond [▶bog hollow] extending deep down into the peat bog and filled with brown-colo[u]red water); *syn.* soak [n] [EIRE]; *s* **pocina** [f] **profunda de turbera** (▶Pocina de turbera extremadamente profunda o pequeño estanque en el cuerpo de la turbera con agua de color marrón); *syn.* lago [m] de turbera; *f* **œil** [m] **de la tourbière** (▶Gouille [d'une tourbière] particulièrement importante et profonde ou petit étang d'eau brunâtre dans le corps de la tourbière) ; *syn.* mare [f] de la tourbière ; *g* **Hochmoorkolk** [m] (Ungewöhnlich große und tiefe ▶Schlenke oder kleiner Teich im Moorkörper, gefüllt mit bräunlichem Wasser); *syn.* Hochmoorblänke [f], Moorauge [n], Moorsee [m].

516 bog regeneration [n] *conserv.* (Reconstitution of a former raised bog with natural, peat-forming vegetation by rewetting, the removal of the nutrient-rich surface layer and the reintroduction of peat mosses; ▶rewetting of a bog); *syn.* bog restoration [n]; *s* **regeneración** [f] **de turberas** (Rehabilitación de las funciones de una antigua ▶turbera alta por medio de reencharcamiento, eliminación de capas del suelo con alto contenido en nutrientes, reintroducción de especies turfícolas; ▶reencharcamiento de turbera); *f* **régénération** [f] **de la tourbière** (À la suite de l'exploitation industrielle ou traditionnelle, réformation de l'ancienne ▶tourbière haute par réengorgement artificiel [colmatage des fossés], enlèvement de la végétation rudérale, réintroduction des espèces turficoles, provoquant la lente recolonisation totale par les groupements initiaux des stations dégradées ; ▶réengorgement d'une tourbière) ; *syn.* reformation [f] des tourbières, reconstitution [f] de la tourbière ; *g* **Moorregeneration** [f] (Wiederherstellung eines ehemaligen ▶Hochmoores mit natürlicher, Torf bildender Hochmoorvegetation durch Wiedervernässung, Abtransport nährstoffreichen Oberflächenmaterials und Ansiedlung von Torfmoosen; ▶Wiedervernässung eines Moores); *syn.* Moorrenaturierung [f] (±).

bog restoration [n] *conserv.* ▶bog regeneration.

517 bog soil [n] *constr. pedol.* (Peaty soil with surface peat deeper than about 16cm formed under poorly drained conditions. **In U.S.**, 'bog soil' is an obsolete generic term including muck and peat, developed under swamp or marsh types of vegetation; cf. RCG 1982; ▶top bog layer, ▶peat bed 1); *syn.* histosol [n] [US] (±); *s* **suelo** [m] **de turba** (Término no muy bien definido. En el norte de Alemania se denominan así suelos de menos de 30 cm de turba o que contienen < 30% de materia orgánica. En Suiza se emplea el término en construcción paisajística para la ▶turba bruta que se utiliza en ▶macizos de tierra turbosa); *f 1* **sol** [m] **hydromorphe organique (1)** *pedol.* (Sous-classe des sols hydromorphes caractérisée par des sols avec une matière organique de type tourbe et une hydromorphie totale et permanente ; DIS 1986, 111) ; *syn.* sol [m] à tourbe ; *f 2* **terre** [f] **de bruyère** *hort.* (Terre horticole désignant un mélange terreux fortement acide composé de ▶tourbe brute, de sable et de résidus

B

végétaux de la lande tourbeuse non décomposée. Mélange utilisé pour l'aménagement de ►plates-bandes à plantes de terre de bruyère) ; *g* **Moorerde [f]** (**1.** *Kein fest definierter Begriff* in Norddeutschland organische Böden mit < 30 cm mächtigen Torflagen oder Humusgehalten < 30 %); *syn.* Moorgley [m], *auch* Anmoorgley [m] (cf. SS 1979, 351). **2.** In der Schweiz wird im GaLaBau ►Bunkerde, die für ►Moorbeete verwendet wird, als **M.** bezeichnet); *syn.* Moorboden [m].

518 bog subsidence [n] *land'man.* (Diminution of original depth of peat layer caused by drainage, compaction of the surface, and the mineralization of organic material); *s* **hundimiento [m] de turbera** (Dismunición del grosor de una turbera debido al drenaje, la compactación de la superficie y la mineralización de la sustancia orgánica); *f 1* **tassement [m] de la tourbière** (Diminution de l'épaisseur initiale de la tourbe occasionnée par un assèchement de la tourbière entraînant la minéralisation de la matière organique) ; *syn.* affaissement [m] de la tourbière ; *f 2* **tourbière [f] effondrée** (Résultat du tassement de la tourbière : L'arrêt de l'épaississement de la tourbière entraîne sur les tourbières hautes une évolution vers une ►lande humide à *Erica tetralix*) ; *g* **Moorsackung [f]** (Verringerung der ursprünglichen Torfmächtigkeit durch Wasserentzug, statische Belastung und Mineralisation der organischen Substanz. Bei Hochmooren entwickelt sich die Vegetation allmählich zu einer ►Moorheide mit starkem Glockenheidebestand *[Erica tetralix]*).

519 bog vegetation [n] *phyt.* (Total plant cover of a bog); *s* **vegetación [f] turfícola** *syn.* vegetación [f] turfófila; *f* **végétation [f] turficole** (Plantes caractéristiques des tourbières) ; *g* **Moorvegetation [f]** (Gesamtheit der Pflanzen, die für das Moor typisch sind).

520 bollard [n] *trans. urb.* (Wooden, iron, concrete or natural stone post embedded in pavement serving to separate a pedestrian area from vehicular intrusion; *specific term* ►seat bollard, ►traffic bollard); *s* **mojón [m]** (*Mobiliario urbano* pilar bajo y ancho de piedra, hormigón u otro material, que se coloca en aceras, plazas, etc. para cerrar una zona al tráfico o evitar que aparquen vehículos; ►mojón obstáculo, ►mojón de asiento); *syn.* mojón [m] obstáculo, mojón [m] para evitar aparcamiento; *f* **borne [f]** (*Mobilier urbain* bloc de bois, de métal, de béton ou de pierre naturelle planté à l'entrée d'une zone piétonnière, d'une rue à circulation restreinte ou sur une place publique pour empêcher l'accès) ; *termes spécifiques* borne à vélos, borne d'alimentation en eau, ►borne de sécurité, borne de sentier, borne d'interdiction de passage, borne obstacle, ►borne siège, borne thématique) ; *g* **Poller [m]** (Straßen-, Platz- oder Parkmöblierungselement in Form einer niedrigen Säule oder eines dicken Pfostens aus Holz, Metall, Beton oder Naturstein; *UBe* ►Absperrpoller, ►Sitzpoller).

bolting [n] [US] *arb. constr.* ►rod bracing of V crotches.

521 bond [n] *arch. constr.* (Generic term for the pattern created by the juxtaposition of bricks in masonry, as well as the pattern produced by pavers, slabs or tiles in the paving of pathways or squares; ►lay with staggered joints, ►paving pattern; *specific terms* ►basket-weave pattern, ►brick bond, ►cross bond, ►herringbone pattern, ►irregular bond, ►masonry bond, ►ornamental pattern, ►random irregular bond, ►random rectangular bond, ►slab paving pattern, ►stack bond, ►stretcher bond pattern); *syn.* bonding [n]; *s* **aparejo [m]** (Diferentes posiciones en las que se asientan las piedras o ladrillos en una fábrica o pavimento, formando hiladas yuxtapuestas, en prevención de que las llagas o juntas verticales no caigan una sobre otra; cf. DACO 1988; ►colocar con juntas alternadas, ►patrón de aparejo; *términos específicos* ►aparejo a soga, ►aparejo de cruz inglesa, ►aparejo de juntas cruzadas, ►aparejo

de labrillos, ►aparejo de losas, ►aparejo de mampuestos, ►aparejo en espina, ►aparejo ornamental, ►aparejo poligonal, ►aparejo romano, ►aparejo trenzado a cesta, ►opus incertum); *syn.* trabazón [m]; *f* **appareillage [m]** (Terme générique pour les techniques de mise en place des éléments de maçonnerie, dallage, pavage de telle sorte que l'ouvrage soit stable, bien liaisonné et jointoyé, avec des découpes régulières ; ►calepinage, ►poser à joints décalés ; *termes spécifiques* ►appareillage à joints croisés, ►appareillage alterné en croix, ►appareillage à la Bichonière, ►appareillage à panneresses, ►appareillage de dalles, ►appareillage d'une maçonnerie, ►appareillage en chevron, ►appareillage en opus incertum, ►appareillage en opus quadratum, ►appareillage en opus romain, ►appareillage ornemental, ►appareillage polygonal, ►disposition de briques) ; *g* **Verband [m]** (**2**) (Anordnung von Ziegeln im Mauerwerk sowie von Klinkern oder Platten auf Wege- oder Platzflächen; ►im Verband verlegen; ►Verlegemuster; *UBe* ►Fischgrätenmuster, ►Korbflechtverband, ►Kreuzfugenverband, ►Kreuzverband, ►Läuferverband, ►Mauersteinverband, ►Plattenverband, ►polygonaler Verband, ►römischer Verband, ►unregelmäßiger Verband, ►Ziegelverband, ►Zierverband).

bond [n] [UK] *contr.* ►surety bond [US].

bonding [n] *arch. constr.* ►bond.

522 bonding depth [n] *arch. constr.* (Depth of bonding brick or stone in wall construction; ►header); *s* **profundidad [f] de empotramiento** (Profundidad mínima de penetración de un ladrillo o sillar en un muro; ►tizón); *f* **longueur [f] de queue** (Longueur de pénétration d'un moellon dans l'épaisseur d'un mur ; ►boutisse) ; *g* **Einbindetiefe [f]** (**1**) (Abstand, den ein Mauerstein in das Mauerwerk eingreift; ►Binder); *syn.* Einbandtiefe [f], Einbautiefe [f].

bond length [n] *arch. constr.* ►grip length.

bone yard [n] *hist.* ►church yard.

523 Bonn Convention [n] *conserv. leg. pol.* (International convention signed in Bonn 1979, setting goals for the conservation of migratory species of wild animals); *syn.* Convention [n] on the Conservation of Migratory Species of Wild Animals (CMS); *s* **Convención [f] de Bonn** (Convención internacional de 23 de junio de 1979 sobre la conservación de las especies migratorias de la fauna silvestre; ratificada en Es. por Instrumento de 22 de enero de 1985); *f* **Convention [f] de Bonn** (Convention du 23 juin 1979 sur la conservation des espèces migratrices menacées appartenant à la faune sauvage pour lesquelles des mesures de protection devront être prises ou des accords internationaux devront être signés en vue d'assurer le maintien ou le rétablissement des espèces concernées) ; *syn.* convention [f] sur les espèces migratrices ; *g* **Bonner Konvention [f]** (1979 in Bonn verabschiedetes internationales „Übereinkommen zur Erhaltung der wandernden wild lebenden Tierarten").

bonus [n] **for early completion** *contr.* ►bonus payment.

524 bonus payment [n] *contr.* (Provision in a contract of an additional payment to a contractor for completing the work before a stipulated date; this payment may be specified in a **bonus clause** to be included in a contract; *opp.* ►liquidated damages); *syn.* bonus [n] for early completion; *s* **gratificación [f] por entrega adelantada** (Cantidad sobre el presupuesto que se puede acordar como recompensa por contrato si el contratista termina la obra antes de la fecha estipulada; suele hacerse si supone ventaja el adelanto de la entrega de la obra; *opp.* ►multa por retraso); *f* **prime [f] d'avance** (Prime accordée à l'entrepreneur lorsque l'achèvement des travaux a lieu avant le délai d'exécution d'un ouvrage et est source d'économies importantes dans le coût de l'ouvrage ; *opp.* ►pénalité pour retard) ; *syn.* prime [f] pour avance ; *g* **Beschleunigungsvergütung [f]** (Vertraglich vereinbarte Prämie bei Fertigstellung der Bauarbeiten vor Ablauf der Vertragsfrist, wenn dadurch erhebliche Vorteile entstehen; cf. § 12 VOB Teil A; *opp.* ►Vertragsstrafe).

boondocks recreation [n] [US] *recr.* ►recreation in the natural environment.

525 border [n] (1) *constr. gard. hort.* (Planting strip in private garden or public green area with annuals, perennials or shrubs; ►border with shrubs and trees, ►mixed border, ►perennial border, ►shrub border); *s* **arriate/a [m/f]** (Macizo en jardín privado o parque público con flores estivales, perennes y/o leñosas; *términos específicos* ►arriate de flores, ►macizo de leñosas, ►macizo de vivaces); *f* **bordure [f] végétale** (Élément de décoration dans un jardin privatif ou un espace vert public servant à délimiter une allée, une plate-bande, une pelouse, etc., pouvant être temporaire ou fixe et constituée de fleurs à floraison estivale, de plantes vivaces ou d'espèces arbustives ; *termes spécifiques* ►massif d'arbres et d'arbustes [d'ornement], ►massif de plantes vivaces, ►mixed-border) ; *g* **Rabatte [f]** (Niederl./[franz.] *rabat* >Aufschlag an der Kleidung, Kragen<, zu *rabattre* >umschlagen<. **1.** Ursprünglich ein oft vor Gehölzen angelegtes, von Rasen eingefasstes Blumenbeet. **2.** *Heute* langes, schmales Beet entlang von Wegen oder Mauern [Randbeet] in privater oder öffentlicher Grünanlage, mit Sommerblumen, Stauden und/oder Gehölzen bepflanzt; *UBe* ►Blumenrabatte, ►Gehölzrabatte, ►Staudenrabatte).

526 border [n] (2) *constr. gard. hort.* (Linear planting bed; ►perennial border, ►shrub border); *s* **platabanda [f]** (Banda de flores paralela a un camino o casa; ►macizo de vivaces, ►macizo de arbustos); *f* **plate-bande [f]** (Espace de terre cultivé plus long que large, dans un jardin ; PR 1987 ; ►massif de plantes vivaces, ►massif d'arbustes) ; *g* **längliches Pflanzenbeet [n]** (Eine bepflanzte oder zur Bepflanzung vorbereitete längliche Fläche in einem Garten, in einer Parkanlage oder im Straßenraum; ►Staudenrabatte, ►Strauchrabatte); *syn.* längliches Beet [n].

527 border [n] **with shrubs and trees** [n] *constr.* (Planting area of shrubs and specimen trees in public or private green area, often with an underplanting of perennials; ►shrub border); *s* **macizo [m] de leñosas** (Arriate en parque público o jardín privado plantado con especies arbóreas y arbustivas ornamentales, a veces acompañado de vivaces tapizantes; ►macizo de arbustos); *f* **massif [m] d'arbres et d'arbustes (d'ornement)** (Groupement d'espèces arbustives ornementales souvent accompagné de plantes vivaces tapissantes utilisé en bordure de pelouses, comme arrière-plan de décorations florales, comme masque, etc. et constituant des éléments décoratifs importants dans les espaces verts publics ou les jardins privatifs ; ►massif d'arbustes) ; *syn.* massif [m] de ligneux, bordure [f] arbustive ; *g* **Gehölzrabatte [f]** (Beetfläche in öffentlicher oder privater Grünanlage, die mit Sträuchern, einzelnen Bäumen, ggf. mit Stauden unterpflanzt ist; ►Strauchrabatte).

528 bored tunnel [n] *eng. trans.* (Traffic route created by boring through solid rock or sediment layers; ►tunnel); *s* **túnel [m] perforado** (Ruta de tráfico construida perforando por el método del broquel un paso a través de la roca o del correspondiente material geológico; ►túnel); *f* **tunnel [m] de montagne** (Galerie souterraine de communication construite dans le rocher ou sous des couches sédimentaires ; ►tunnel) ; *g* **Bergtunnel [m]** (Im festen Fels oder in Sedimentschichten durch bergmännischen Vortrieb hergestelltes Verkehrsbauwerk; ►Tunnel); *syn.* Gebirgstunnel [m].

529 borrow material [n] *constr.* (Soil excavated from another site; ►backfilling behind structures, ►earthworks, ►fill material, ►hauling to the site [US]/carting away [UK]); *syn.* fill dirt [n] [also US]; *s* **tierra [f] de préstamo** (Tierra excavada de emplazamiento cercano a la obra para rellenar zanjas de conducción, para ►terraplenado o para modelar el terreno; ►terracerías, ►tierra de relleno); *f* **terres [mpl] d'emprunt** (Matériau d'apport de provenance extérieure au chantier mis en ►remblai permettant la mise à niveau d'un terrain dans le cadre des ►terrassements ; ►matériaux de remblaiement, ►matériaux tout-venant, ►terrassements généraux) ; *g* **Füllboden [m] (aus) einer Seitenentnahme** (Mineralisches Material, das aus einer nahe gelegenen Entnahmestelle und für den ►Erdbau und Wegebau verwendet wird; ►Füllboden, *►Schüttboden); *syn.* Auffüllboden [m] (aus) einer Seitenentnahme, Auffüllmaterial [n] (aus) einer Seitenentnahme, Füllmaterial [n] (aus) einer Seitenentnahme, Schüttboden [m] (aus) einer Seitenentnahme, Schüttmaterial [n] (aus) einer Seitenentnahme.

530 borrow pit [n] *min.* (Excavation hole from which soil material has been removed for use in nearby construction projects [US]/civil engineering works [UK]; ►gravel extraction site); *s* **zona [f] de préstamo** (Lugar cercano a una gran obra de construcción del cual se extrae grava o arena para utilizar en aquélla; ►gravera); *syn.* punto [m] de toma; *f* **lieu [m] d'emprunt** (Aire d'extraction de sols, pierres concassées, gravier, sable, etc. dans le cas de gros chantier ; ►site d'extraction de gravier) ; *syn.* aire [f] d'emprunt, banc [m] d'emprunt [CDN], ballastière [f] (DTB 1985) ; *g* **Seitenentnahme [f]** (Ort zur Gewinnung von Steinen und Erden für eine nahe gelegene Großbaustelle; ►Kiesentnahmestelle); *syn.* Entnahmestelle [f] (1).

borrow pit [n]**, flooded** *landsc. recr.* ►flooded gravel pit.

borrow pit [n]**, sand** *min.* ►sand extraction site.

bosk [n] *agr. landsc.* ►grove.

531 boss [n] *arch. constr.* (Roughly shaped stone set to protrude for carving in place); *s* **sillar [m] de labra tosca** (Piedra cuadrada, tal como llega de la cantera, con cara tosca y solamente labrada en las juntas; DACO 1988); *syn.* mampuesto [m] de labra tosca, sillar [m] almohadillado; *f* **pierre [f] bossagée** (Face vue des pierres traitée en taille bossagée : gros éclats formant une saillie bombée ; MAÇ 1981, 181) ; *syn.* moellon [m] de bossage ; *g* **Bossenstein [m]** (An seiner Vorderseite nur roh bearbeiteter Werkstein mit einem oder mehreren Buckeln); *syn.* Bossenquader [m], Buckelquader [m], bossierter Stein [m].

532 bossage [n] [US] *arch. constr.* (Rough tooling the face of natural stones); *syn.* rustication [n] [UK]; *s* **labra [f] tosca** (Trabajo que se da a un sillar con una herramienta de boca dentada; DACO 1988); *f* **bossage [m]** (Action de frapper les pierres naturelles le long des arêtes de la face pour provoquer de gros éclats et donner à la face apparente une structure grossière) ; *g* **Bossierung [f]** (Durch grobes Zuschlagen/Behauen von [Natur]steinen entstandene buckelige Grobstruktur der Ansichtsflächen).

bossage masonry [n] *arch. constr.* ►rusticated masonry.

botanical garden [n] [US] *bot. hort.* ►botanic garden.

533 botanic garden [n] *bot. hort.* (Specialized public or private garden where indigenous and exotic plants are grown and arranged according to taxonomic and systematic criteria for educational and scientific purposes. The oldest European **b. g.s** were developed in Pisa 1543 and Padua in 1545. In the Americas, the **b. g.s** of Huastepec, built by the Aztecs near Tenochititlan between 1440 and 1460, rival the embryonic European gardens in age, complexity, and sophistication; cf. DIL 1987; ►arboretum, ►medicinal herb garden); *syn.* botanical garden [n] [also US]; *s* **jardín [m] botánico** (Parque o jardín privado o público destinado al cultivo de especies vegetales autóctonas o exóticas con un fin científico utilizando criterios sistemáticos taxonómicos o de otro tipo; ►arboreto, ►jardín medicinal); *f* **jardin [m] botanique**

(Parc ou jardin de gestion publique ou privée dans lequel sont cultivées, sur la base de critères taxonomiques ou d'autres critères systématiques, des espèces végétales indigènes et exotiques à des fins scientifiques et éducatives d'initiation à un large public ; ►arboretum, ►jardin médicinal) ; *syn.* jardin [m] des plantes ; *g* **botanischer Garten [m]** (Öffentlicher oder privater, wissenschaftlich geleiteter Garten oder Park, in dem meist nach taxonomischen und anderen systematischen Kriterien heimische und meist auch exotische Pflanzen zu wissenschaftlichen und Schauzwecken kultiviert werden; die beiden ältesten europäischen **b. Gärten** bestehen in Pisa und Padua seit 1543 und 1545; 1542 wurde in Leipzig die erste botanische Sammlung als kleiner *Hortus medicus* [►Heilpflanzengarten] gegründet, der 1577 zum ersten **b. G.** Deutschlands erweitert wurde; ►Arboretum).

bottomland [n] [US] *geo.* ►alluvial plain.

bottomland hardwood forest [n] [US] *phyt.* ►riparian woodland.

534 bottom level [n] *constr.* (Lowest part of a structure, e.g. bottom of wall, bottom of stair; *opp.* ►top level); *s* **borde [m] inferior** (p. ej. de muro, cimiento; *opp.* ►borde superior); *f* **bord [m] inférieur** (Partie inférieure d'une structure ; *contexte* niveau bord inférieur ; *opp.* ►arase) ; *g* **Unterkante [f]** (*Abk.* UK; unterster Teil eines Bauwerks, z. B. UK Fundament; *opp.* ►Oberkante).

535 bottom level [n] **of a wall** *arch. constr.* (**1.** That portion of the wall that rests upon the footing. **2.** The base of the stem segment of retaining wall; ►bottom of wall base, ►wall footing); *s* **borde [m] inferior de muro** (Límite plano entre un muro y su asiento; ►cimentación de muro, ►pie de muro); *f* **fond [m] de forme d'un mur** (Surface du sol d'assise sur laquelle repose la ►fondation d'un mur ; ►pied de mur) ; *syn.* assise [f] de fondation d'un mur, semelle [f] d'un mur ; *g* **Mauersohle [f]** (Untere flächige Abgrenzung einer Mauer zum Fundament; ►Mauerfundament, ►Mauerfuß).

bottom [n] **of slope** *geo.* ►foot of slope.

bottom [n] **of the valley** *geo.* ►valley floor.

536 bottom [n] **of wall base** *arch. constr.* (Wall elevation at ground level or finish grade. This point is marked "b.w." on a drawing elevation or section; ►bottom level of a wall, ►wall base course); *syn.* foot [n] of a wall, masonry wall base [n]; *s* **pie [m] de muro** (Parte inferior de un muro, generalmente sobre asiento, que puede estar conformada como ►zapata de muro; ►borde inferior de muro); *f* **pied [m] de mur** (Partie inférieure reposant sur la fondation et pouvant être réalisée avec un soubassement lorsqu'il s'agit d'un mur portant ; ►fond de forme d'un mur, ►soubassement d'un mur) ; *g* **Mauerfuß [m]** (Unterster, meist auf einem Fundament aufsitzender Teil einer Mauer, der als ►Mauersockel ausgeführt sein kann; ►Mauersohle).

bottom step [n] *arch. constr.* ►step in a flight of steps.

537 boulder [n] *constr. geo.* (**1.** Water-worn or glaciated rounded block of stone of exotic origin, having a diameter greater than 250mm, sometimes used in the design of outdoor spaces. **2.** General term for any rock that is too heavy to be lifted readily by hand; ►perched boulder); *syn. o.v.* bowlder [n] (GFG 1997); *s* **pedrejón [m]** (Trozo de roca o piedra grande colocado en jardines con fines decorativos; ►pedrejón errático); *syn.* pedregón [m] [COL, RCH]; *f* **bloc [m] (de rocher) de forme arrondie** (Rocher de grande taille utilisé dans la composition d'un jardin, p. ex. des « boules de granite » [bloc rocheux volumineux constitué de granite et roches granitoïdes aux formes arrondies] des « meulières » [roche siliceuse irrégulière de structure caverneuse]) ; *g* **Findling [m] (1)** (Ortsfremder, aus der Eiszeit stammender, rundlicher Felsblock oder größerer Stein zur Gestaltung von Außenanlagen; ► Wanderblock).

boulder-clay [n] [UK] *geo. pedol.* ►till [US].

538 boulingrin [n] *gard'hist.* (Sunken ornamental lawn surrounded by sloping banks in French classic gardens. The word is a corruption of 'bowling green'; JEL 1986; ►parterre); *s* **bolingrín [m]** (Del inglés «bowling green». Superficie de césped para juegos de bolas. En jardines barrocos, generalmente una superficie de césped hundida en 50-65 cm, encintada por taludes de césped y con pequeñas decoraciones; ►parterre); *syn.* bulingrín [m]; *f* **boulingrin [m]** (Transposition de l'anglais « bowling green » caractérisant un parterre de gazon légèrement en cuvette encadré par un talus engazonné dans les jardins classiques français ; ►parterre) ; *syn.* parterre [m] de gazon ; *g* **Boulingrin [m]** (Aus dem Englischen *bowling green* >Rasenplatz für Kugelspiele<. In barocken Parkanlagen eine meist 50-65 cm abgesenkte Rasenfläche, die durch Rasenböschungen eingefasst ist und in der Parterrezone oder in Boskets angelegt ist; man unterscheidet einfache **B.s,** die aus reinem Rasen bestehen und mit keinem Zierrat versehen sind, während bei reicher ausgestatteten **B.s** einzelne Rasenstücke mit zusätzlichen Broderien, Skulpturen, Rabatten oder auch Wasseranlagen möglich sind; cf. HEN 1985; ►Parterre); *syn.* abgesenktes Parterre [n] mit Rasenflächen.

boundary line building [n] [UK] *urb.* ►zero lot line building [US].

539 boundary planting [n] *constr. hort.* (Planting around the perimeter of a plot of land, i.e. with a hedge or strip of trees and shrubs); *syn.* peripheral planting [n]; *s* **plantación [f] alrededor de edificio o solar** (Plantación al borde de una propiedad); *f* **plantation [f] d'une clôture végétale** (Entourer le périmètre d'un terrain, d'une parcelle, d'un lotissement par une rangée d'arbres, une haie, une bande boisée) ; *syn.* constitution [f] d'une clôture végétale ; *g* **Umpflanzung [f] (1)** (Anpflanzung, z. B. eines Gehölztreifens um ein Grundstück herum); *syn.* Umpflanzen [n, o. Pl.].

540 boundary survey [n] *surv.* (Plotting of the limits of a property, structures and other items found on it from one or more points of reference, by recording of distances and compass bearings/directions of each enclosing side); *syn.* metes and bounds survey [n] [also US], plot/lot survey [n], property survey [n]; *s* **relevamiento [m] de un solar** (Medición de los límites de un terreno, de los de posibles edificios o construcciones en relación con puntos de referencia); *syn.* relevamiento [m] de lote de terreno [AL]; *f* **relevé [m] planimétrique d'un terrain** (situant les limites parcellaires, l'emplacement des bâtiments et des ouvrages divers, etc. à partir d'un point de référence) ; *g* **Grundstücksvermessung [f]** (Aufmaß der Grundstücksgrenzen, Gebäude und baulichen Anlagen, das auf vermessungstechnische Bezugspunkte bezogen ist); *syn.* Vermessung [f] eines Grundstückes.

bower [n] [UK] *arch. gard.* ►arbor [US]/arbour [UK].

bower [n]**, wine** *gard.* ►grapevine arbor [US]/grapevine arbour [UK].

bowery [n]**, wine** *gard.* ►grapevine arbor [US]/grapevine arbour [UK].

bowlder [n] *constr. geo.* ►boulder.

bowl-shaped basin [n] *geo.* ►location in a basin.

bowl-shaped situation [n] *geo.* ►location in a basin.

541 box culvert [n] *constr. trans.* (►Culvert with a rectangular cross-section); *s* **alcantarilla [f] de cajón** (►alcantarilla); *syn.* tajea [f] de losa, alcantarilla [f] de platabanda [YV]; *f* **passage [m] couvert rectangulaire** (Petit canal [►passage busé] de section rectangulaire ou carrée situé p. ex. sous une route pour permettre l'écoulement des eaux) ; *syn.* dalot [m]

B

rectangulaire ; *g* **rechteckiger Durchlass [m]** (Queröffnung bei Verkehrsstrassen für den Durchfluss von kleinen Wasserläufen; ►Durchlass); *syn.* Plattendurchlass [m].

brace root [n] *arb. bot.* ►stilt root.

bracing [n] [UK] *arb. constr.* ►cabling [US].

542 brackish water area/region [n] *phyt.* (Lowest river section effected by tides, with regionally-typical plant communities; ►glasswort mudflat); *s* **zona [f] de agua salobre** (Tramo inferior de río afectada por las mareas cuyas aguas son mezcla del agua dulce con la marina y que se caracteriza por comunidades vegetales adaptadas a la mayor concentración de sales; ►marisma de salicor); *syn.* medio [m] salobre; *f 1* **milieu [m] saumâtre** (*Terme générique* ; ►marais à salicornes) ; *f 2* **zone [f] de la sansouire** (*Terme spécifique* ensemble des formations végétales ligneuses basses, particulièrement bien adaptées aux milieux salés et humides dans la delta du Rhône) ; *syn.* engane [f] ; *g* **Brackwasserbereich [m]** (Unterster Fließgewässerabschnitt im Einflussbereich der Gezeiten mit regional typischen Pflanzengesellschaften; ►Quellerwatt).

543 braided river [n] *geo.* (River consisting of a tangled network of inter-connected, diverging and converging shallow channels, with banks of alluvial material and shingle between; DNE 1978; ►gravel bank, ►sandbank); *syn.* braided stream [n]; *s* **curso [m] (fluvial) de brazos anastomosados** (Río cuyo lecho está subdividido en varios brazos poco profundos interconectados con bancos de material aluvial; ►banco de arena, ►banco de grava); *f* **cours [m] à méandres anastomosés** (Portion du lit majeur d'un cours d'eau dans lequel par suite de l'accumulation de sédiments après une période de crue se déploient de multiples bras mobiles séparés par des ►bancs de gravier ou des ►bancs de sable) ; *g* **verwilderter Flusslauf [m]** (Teil eines Flussabschnittes, dessen Bett sich infolge ändernder Lockersedimentakkumulationen nach Hochwasser immer wieder in mehrere, miteinander verbundene und durch ►Kiesbänke oder ►Sandbänke getrennte Teilläufe aufspaltet); *syn.* Wildfluss [m].

braided stream [n] *geo.* ►braided river.

544 braiding [n] **of river courses** *geo.* (GGT 1979, 79; middle reaches of a river separated into inter-connected diverging and converging shallow channels, which continually change their course between banks of alluvial material and shingle caused by strong flow velocity and a high rate of sedimentation); *s* **formación [f] de brazos anastomosados** (En los tramos medios de cursos de agua, división de éstos en varios brazos que cambian sus rutas debido a la gran cantidad de carga sedimentaria móvil y a la alta velocidad de flujo del agua); *f* **formation [f] de chenaux anastomosés** (Dans le cours moyen d'un cours d'eau, bras qui divaguent sous l'effet de la sédimentation dans une vallée alluviale fréquemment inondée, se réunissent et se séparent continuellement entre les bancs de galets, graviers ou sable) ; *g* **Verwilderung [f] eines Flusslaufes** (Durch starke Fließgeschwindigkeit und hohe Sedimentzufuhr verursachte Zerteilung von Flussmittelläufen, deren Teilarme sich bei Hochwasserereignissen durch Schotter- und Kiesbänke ständig verlagern).

branch [n] *arb.* ►strong upright branch, ►sturdy branch, ►upright-growing main branch.

branch [n]**, upright scaffold** *arb.* ►upright-growing main branch.

branch attachment [n] [US] *arb.* ►crotch [US], ►cut at the branch attachment.

branch authority [n] [US] *adm.* ►operating authority.

branch collar cut [n] *arb. hort.* ►cut at the branch attachment.

branch cutting [n] *arb.* ►branch pruning.

545 branching [n] **(1)** *arb. hort.* (Overall effect of all the branches/boughs of a tree; the lower branches are usually slanted groundward; cf. GUP 1981, 145; ►branching 2); *s* **ramas [fpl] colgantes** (Conjunto de ramas inferiores y medias de la copa de las leñosas que están inclinadas hacia el suelo, como p. ej. en el caso del tilo; ►ramaje); *f* **branchage [m]** (Ensemble des branches et rameaux d'un arbre ; ►ramure) ; *syn.* lacis [m] de branches (LA 1981, 653) ; *g* **Astbehang [m]** (Im mittleren und unteren Kronenbereich eines Baumes die Gesamtheit der Äste, die abwärts geneigt sind, z. B. bei der Linde; der untere Teil des **A.es**, der bis zum Boden reicht, wird **Astschleppe** genannt; ►Astwerk).

546 branching [n] **(2)** *arb.* (Overall effect of all the branches/boughs of a tree or shrub); *s* **ramaje [m]** (Conjunto de ramas de una leñosa); *f* **ramure [f]** (Ensemble des ramifications principales et secondaires d'un végétal ligneux ; la disposition de celles-ci en constitue la silhouette caractéristique ; ►habitus) ; *g* **Astwerk [n, o. Pl.]** (Gesamtheit aller Äste eines Gehölzes); *syn.* Geäst [n].

branching habit [n] *arb. hort.* ►branching pattern.

branching height [n] *arb. hort.* ►crown base.

547 branching pattern [n] *arb. hort.* (Species-specific form of branching characterizing the ►habit of a tree or shrub); *syn.* branching habit [n]; *s* **ramificación [f] típica** (Forma de distribución de las ramas característica para cada especie de leñosas; ►hábito de crecimiento); *syn.* distribución [f] típica del ramaje; *f 1* **ramification [f]** (Mode de division des branches chez les végétaux ligneux ; ►habitus) ; *f 2* **disposition [f] naturelle des branches** (La répartition des branches sur un tronc est une des caractéristiques morphophysiologiques responsable de l'aspect ou de la silhouette des végétaux ligneux ; ►habitus) ; *syn.* forme [f] caractéristique de la ramure ; *g* **arttypische Astverzweigung [f]** (Artspezifische Form der Astverzweigung, die den ►Habitus eines Gehölzes bestimmt); *syn.* charakteristische Astverzweigung [f].

branch layering [n] *constr.* ►brush layering.

branchlet [n] *arb.* ►small branch, ►twig.

548 branch pruning [n] *arb. hort.* (Cutting off a branch; ►cut at the branch attachment); *syn.* branch cutting [n]; *s* **poda [f] de una rama** (►poda de una rama por la base) *syn.* corte [m] de una rama; *f* **coupe [f] d'une branche** (►coupe à l'empattement) ; *g* **Astabnahme [f]** (**1.** Abschneiden eines Astes am Stamm oder an einem Ast höherer Ordnung. [Zur Schnittführung siehe stammparalleler Schnitt und Astringschnitt bei ►Schnitt an der Astgabel, Ziff. 2]. **2.** *for.* Im forstlichen Sinne bedeutet **Astung** oder **Ästung** das Abschlagen der Äste am liegenden Baum nach der Fällung); *syn.* Astschnitt [m], *for.* Astung [f].

549 branch rail line [n] [US] *plan.* (Line connecting an industrial or other area with the main railway network; ►railroad siding [US]/works siding [UK]); *syn.* branch railway line [n] [UK]; *s* **vía [f] de empalme** (Conexión de ferrocarril a una zona industrial o un puerto; ►vía muerta); *f* **raccordement [m] au réseau ferroviaire** (Raccordement d'équipements portuaires, d'établissements industriels et commerciaux au réseau ferroviaire ; ►voie de garage) ; *g* **Gleisanschluss [m]** (Anbindung von Häfen, Gewerbe- und Industrieeinrichtungen an das Eisenbahnnetz; ►Abstellgleis).

branch railway line [n] [UK] *plan.* ►branch rail line [US].

550 branch stub [n] [US] *arb. for. hort.* (Short portion of a branch left on the tree after breakage or faulty pruning; ►lopping); *syn.* branch stump [n] [UK]; *s* **tocón [m] de rama** (Resto de una rama muerta, quebrada o no podada correctamente; ►trasmocho); *f* **chicot [m] (1)** (Partie d'une branche **[a]** coupée lors d'une taille effectuée trop loin du bourrelet cicatriciel ou du collet ou **[b]** cassée, restant attachée au tronc ou à la charpentière ; ►rapprochement, ►ravalement) ; *g* **Aststummel [m]** (Rest eines abgestorbenen, abgebrochenen oder nicht fachgerecht abgeschnittenen Astes; ►Stummelschnitt).

branch stump [n] **[UK]** *arb. for. hort.* ►branch stub [US].

branch [n] **with adhesive disks** *bot.* ►grasping branch [n] with adhesive disks.

brash [n] **[UK]** *for.* ►slash.

breakdown [n] *biol. chem.* ►decomposition (1).

breakdown [n] **of pollutants** *envir.* ►degradation of pollutants.

breaking [n] **ground [US]** *agr.* ►reversion.

551 breaking [n] **ground and reseeding** [n] **[US]/ breaking ground** [n] **and re-seeding** [n] **[UK]** *agr. constr. landsc.* (of lawns or meadows); *s* **laboreo [m] y resiembra [f] de un césped o una pradera**; *f* **rajeunissement [m] d'une prairie** (Labour et réensemencement d'une prairie ou d'un gazon) ; *g* **Umbruch [m, o.Pl.] und Neueinsaat [f]** (Umpflügen von Rasen- oder Wiesenflächen und Wiederansäen der Flächen).

552 break [n] **of slope** *geo.* (Sudden drop in ground level; ►cuesta, ►edge of a landslip, ►escarpment, ►field dike [US]/ field dyke [UK]); *s* **pequeño declive [m]** (Pequeña y brusca caída en un terreno; ►borde de falla, ►escarpe de falla, ►montadero); *syn.* escalón [m] en el terreno; *f* **rupture [f] de pente** (Petit écart soudain entre deux terrains/niveaux ; ►cuesta, ►escarpement, ►lèvre de faille, ►lèvre de glissement, ►talus embroussaillé) ; *syn.* décrochement [m] de niveau ; *g* **Geländekante [f]** (Plötzlicher, kleiner Abfall eines Geländes; ►Bruchkante, ►Geländeabbruch, ►Hochrain, ►Schichtstufe); *syn.* Geländeknick [m], Geländesprung [m], Geländestufe [f].

553 bream zone [n] **[UK]** *limn.* (Lower section of stream courses that connect downstream to ►barbel zone [UK]. The normally slow-moving water is usually muddy, usually poor in oxygen in deeper water, and in summer very warm [20 °C]. This zone has the richest variety of fish species of all flowing water regions and is seriously threatened by modern wastewater contamination; typical [European] fish species: bream *[Abramis brama]*; ►metapotamon, ►river zone); *s* **zona [f] de madrilla** (Zona media de la ►región ciprinícola con la madrilla como pez característico; su biocenosis se denomina ►metapotamon; ►zona biológica de un río, ►zona de barbo); *f* **zone [f] à brème** (Tronçon inférieur de cours d'eau faisant suite à la ►zone à barbeau et caractérisé par un courant faible et des eaux à faible teneur en oxygène. Cette région cyprinicole a le plus fort peuplement ichtyologique de toutes les zones des cours d'eau. À côté de la brème *[Abramis brama]* on note les cyprins et les voraces d'accompagnement ainsi que les cyprins d'eaux calmes ; cf. MPE 1984 ; ►métapotamon, ►zone écologique des cours d'eau) ; *g* **Brachsenregion [f]** (Unterer Abschnitt von Fließgewässern, der sich stromabwärts an die ►Barbenregion anschließt. Das meist langsam fließende Wasser ist meist trüb, in der Tiefe häufig sehr sauerstoffarm und im Sommer sehr warm [um 20 °C]. Diese Zone hat den artenreichsten Fischbestand aller Fließwasserregionen, der allerdings je nach der Höhe der Abwasserbelastung stark zurückgeht; kennzeichnende Fischart ist die Brasse *[Abramis brama]*; ►Flussregion, ►Metapotamon); *syn.* Brassenregion [f], Bleiregion [f].

breast wall [n] *constr.* ►face wall.

breeding [n] *hort.* ►perennial plant breeding.

554 breeding bird [n] *zool.* (Species of sedentary bird, which breeds and reproduces in a reference area in contrast to ►winter visitor and ►bird of passage. A territory is usually established, where courtship takes place with singing and a nest is built in which young birds are reared; *specific terms* breeding sedentary bird, breeding summer bird; ►migratory bird); *s* **ave [f] nidificante** (Especie de ave que anida y se reproduce siempre en un área de referencia, en contraste con el ►visitante invernal y el ►ave de paso. Generalmente establece un territorio donde tiene lugar el cortejo, construye su nido y cría a sus polluelos; *términos específicos* ave sedentaria nidificante, ►visitante estival nidificante; ►ave migradora); *f* **oiseau [m] nicheur** (Espèce d'oiseau qui niche et se reproduit dans une région considérée — par opposition au ►visiteur d'hiver et ►oiseau de passage ; à cet effet un territoire sera choisi sur lequel a lieu les parades et sera construit un nid pour recevoir les œufs et élever les jeunes oiseaux ; ►oiseau migrateur) ; *syn.* oiseau [m] nidifiant ; *g 1* **Brutvogel [m]** (In einem Bezugsraum brütende und sich fortpflanzende Vogelart — im Gegensatz zum ►Wintergast und ►Durchzügler; es wird i. d. R. ein Revier gegründet, in dem die Balz mit Gesang stattfindet und ein Nest gebaut wird, in dem Jungvögel großgezogen werden); *g 2* **Sommervogel [m]** (UB zu Brutvogel; Vogel, der sein Brutgebiet im Herbst verlässt und in wärmere Gebiete zieht; ►Sommergast, ►Zugvogel).

breeding ground(s) [n(pl)] *zool.* ►breeding site.

555 breeding pair [n] *zool.* (Parent birds which breed the eggs during the incubation period and subsequently raise the offspring); *s* **pareja [f] nidificante** (Pareja de aves que cría los huevos en el periodo/período de incubación y a continuación alimenta a sus polluelos); *f* **couple [m] nidificateur** ; *syn.* couple [m] nicheur, couple [m] nidifiant ; *g* **Brutpaar [n]** (Elterntiere, die während der Brutperiode die Eier ausbrüten und anschließend die Jungvögel aufziehen).

556 breeding range [n] *zool.* (Area where a bird species spends the period during which eggs are laid and the young are reared; for ►migratory birds, breeding range and winter range [►hibernation area] are differentiated; ►breeding territory, ►range 1); *s* **área [f] de cría** (Zona en la que una especie de ave permanece durante el periodo/período de puesta y de cría de los pollitos; en las ►aves sedentarias el **á. de c.** es idéntico al ►área de distribución, mientras que en las ►aves migradoras se diferencia entre el **á. de c.** y el ►área de invernación; ►territorio de cría); *syn.* área [f] de nidada, área [f] de reproducción (de una especie de aves); *f* **aire [f] de nidification** (Aire sur laquelle une espèce d'oiseau passe la période de nidification. Pour les ►oiseaux sédentaires, elle est identique à l'aire de distribution [►aire de répartition] ; pour les oiseaux migrateurs, on distingue une aire de nidification et une ►aire d'hivernage ; ►territoire de nidification) ; *syn.* site [m] de nidification ; *g* **Brutareal [n]** (Gesamtverbreitungsgebiet, in dem eine Vogelart ihre Brutzeit verbringt. Bei ►Standvögeln identisch mit dem Verbreitungsgebiet [►Areal]; bei ►Zugvögeln unterscheidet man Brutareal [Vermehrungsgebiet] und ►Überwinterungsgebiet; ►Brutrevier); *syn.* Brutgebiet [n].

557 breeding season [n] *zool.* (Time of year during which a species breeds); *s* **periodo/período [m] de cría** *syn.* periodo/ período [m] de reproducción; *f* **période [f] de nidification** (Période de l'année durant laquelle une espèce d'oiseau) ; *syn.* période [f] nidicole, temps [m] des nids ; *g* **Brutzeit [f] (1)** (Jahreszeitlicher Abschnitt, in der eine Vogelart brütet).

558 breeding site [n] *zool.* (Specific location where a bird brood and hatch its eggs and rears its young; ►rookery); *syn.* nesting site [n], breeding ground(s) [n(pl)], nesting habitat [n]; *s* **lugar [m] de cría** (Sitio específico elegido por una pareja de aves para criar a sus polluelos; ►colonia de cría); *syn.* lugar [m] de nidada; *f* **emplacement [m] de nidification** (Lieu choisi par les oiseaux pour la ponte des œufs et le développement de leurs petits ; ►colonie nicheuse) ; *syn.* lieu [m] de nidification, aire [f]

de nidification ; *g* **Brutplatz** [m] (...plätze [pl]; Ort, der von Vögeln für die Eiablage und zur Jungenaufzucht gewählt wird; ►Brutgesellschaft); *syn.* Nistplatz [m].

559 breeding territory [n] *zool.* (Area of land or water occupied by a pair of breeding birds; ►breeding range, ►mating territory, ►territory); *s* **territorio** [m] **de cría** (Radio de acción de una pareja de aves en fase de reproducción; ►área de cría, ►territorio, ►territorio de reproducción); *f* **territoire** [m] **de nidification** (Zone recouvrant le rayon d'action d'un couple d'oiseaux nidifiant ; ►aire de nidification, ►territoire, ►territoire de reproduction) ; *syn.* site [m] de nidification, territoire [m] de ponte ; *g* **Brutrevier** [n] (Gebiet, das den Aktionsradius eines brütenden Vogelpaares umfasst; ►Brutareal, ►Fortpflanzungsrevier, ►Territorium).

560 brick bond [n] *arch. constr.* (Pattern of bricks in a wall; *generic term* ►masonry bond); *s* **aparejo** [m] **de ladrillos** (Patrón de ordenación de los ladrillos en obra de fábrica; *término genérico* ►aparejo de mampuestos); *f* **disposition** [f] **des briques** (MAÇ 1981, 191 ; *terme générique* ►appareillage d'une maçonnerie) ; *g* **Ziegelverband** [m] (Anordnung von Ziegeln im Mauerwerk; *OB* ►Mauersteinverband); *syn.* Mauersteinverband [m] aus Ziegeln, Ziegelsteinverband [m].

561 brick paving [n] *constr.* (Surface of dense fired clay unit pavers; ►pavement pattern, ►paver brick [US]/clinkerbrick for paving [UK], ►paving 1); *syn.* baked clay paving [n] [also UK], clinker brick paving [n] [also US]; *s* **pavimento** [m] **de ladrillo clinker** (►adoquinado, ►aparejo de adoquinado, ►ladrillo para pavimento); *f* **pavage** [m] **en terre cuite** (Revêtement de sols ; *termes spécifiques* pavage en brique pleine, pavage en briques, pavage en clinker ; ►brique de pavage, ►calepinage du pavage, ►pavage 2) ; *g* **Klinkerpflaster** [n] (Straßen-, Wege- oder Platzbelag aus Klinkersteinen; ►Pflasterung 1, ►Pflasterung 2, ►Straßenbauklinker); *syn.* Klinkerbelag [m].

brick paving unit [n] [US] *constr.* ►paver brick [US]/clinkerbrick for paving [UK].

brick unit paver [n] [US] *constr.* ►paver brick [US]/clinkerbrick for paving [UK].

562 brick veneer [n] *arch. constr* (Facing of a wall with bricks; ►veneer masonry); *s* **mampostería** [f] **con paramento de ladrillo** (Muro revestido de ladrillos; ►mampostería enchapada de ladrillos); *f* **maçonnerie** [f] **à parement en briques** (►maçonnerie à parement) ; *g* **Ziegelverblendmauerwerk** [n] (►Verblendmauerwerk aus Ziegelsteinen).

563 bridle path [n] [US] *recr. syn.* bridle way [n] [UK], equestrian trail [n] (TGG 1984, 165), riding trail [n]; *s* **sendero** [m] **ecuestre** *syn.* camino [m] de herradura, pista [f] ecuestre, pista [f] para caballos, sendero [m] hípico; *f* **sentier** [m] **équestre** *syn.* allée [f] cavalière, itinéraire [m] équestre, piste [f] cavalière, piste [f] équestre ; *g* **Reitweg** [m] (Zum Reiten angelegter Weg).

564 bridle path network [n] [US] *plan. recr.* (Specially-marked trail system for horseback riding [US]/horse riding [UK]); *syn.* horse riding trail network [n] [UK]; *s* **red** [f] **de senderos ecuestres** *syn.* red [f] de caminos de herradura, red [f] de pistas para caballos; *f* **réseau** [m] **(de randonnée) équestre** (Système de sentiers équestres signalisés) ; *syn.* parcours [m] équestre, randonnée [f] équestre, circuit [m] équestre, réseau [m] de pistes équestres, réseau [m] de chemins équestres, réseau [m] d'itinéraires équestres ; *g* **Reitwegenetz** [n] (Für Reiter gesondert gekennzeichnetes Wegesystem).

bridle way [n] [UK] *recr.* ►bridle path [US].

British Association [n] **of Landscape Industries (B.A.L.I.)** [UK] *constr. prof.* ►American Landscape Contractors Association (A.L.C.A.) [US].

British Standard Codes [npl] **of Practice (C.P.)** [UK] *constr. contr.* ►standards for construction [US].

British Standards [npl] *adm. constr.* *►American Standards.

broached [pp/adj] *constr.* *►tooling of stone [US]/dressing of stone [UK], #8.

B-road [n] [UK] *adm. leg. trans.* ►county road [US].

broadcasting [n] *constr. hort.* ►sowing (2).

565 broadcasting [n] **of seed** *agr. constr. s* **siembra** [f] **al vuelo** *syn.* sembradura [f] al vuelo; *f* **semis** [m] **à la volée** (LA 1981, 1030) ; *g* **Breitwurfsaat** [f] (Das von Hand breifwürfige Ansäen); *syn.* Breitsaat [f].

broadcasting [n] **with a seeder** [US] *agr. constr. hort.* ►mechanical sowing.

566 broad flight [n] **of steps** *arch. constr.* (Grand staircase typically associated with public or formal civic scale design; stairway broadened for greater capacity, visual effects, or both); *s* **escalinata** [f]; *f* **perron** [m] (Escalier d'un petit nombre de marches placé devant l'entrée d'un bâtiment) ; *g* **Freitreppe** [f] (Nicht überdachte, an der Außenseite eines Gebäudes [repräsentativ] vorgelagerte oder sonstige größere, zwei Terrassen/Ebenen verbindende Treppenanlage in Freianlagen, an Flussufern etc. — z. B. die Spanische Treppe in Rom, F. in Versailles „Cent marches", F. am Erfurter Dom; cf. BAR 1968).

broadleaf forest [n] [US] *for. phyt.* ►deciduous forest.

broadleaf tree [n] *bot.* ►deciduous tree.

broadleaf woody plant [n] [US] *bot.* ►broad-leaved woody species.

567 broad-leaved evergreen forest [n] *phyt.* (**1.** Forest with permanently green tree canopies, which are dense or open according to season and humidity. **2.** Generic term covering different types of **b.-l. e. f.**: There are, i.a., 'tropical ombrophilous forests', composed of trees with leaf sizes of 15-40 cm, e.g. in Amazonas, Congo, and on the Sunda Islands; 'temperate evergreen seasonal broad-leaved forests' [with adequate summer rainfall], 'winter-rain evergreen broad-leaved sclerophyllous forests' [often understood as Mediterranean, but present also in south-western Australia, Chile, etc., where there is a pronounced summer drought]; ►broad-leaved, evergreen, sclerophyllous vegetation); *syn.* evergreen broadleaf forest [n]; *s* **bosque** [m] **latifoliado sempervirente** (**1.** Bosque perennifolio en el que la densidad del follaje depende de la estación. **2.** Término genérico para diferentes tipos de **b. l. s.**: Existe el «bosque tropical ombrófilo» [convencionalmente llamado «bosque pluvial tropical»]; «bosque templado latifoliado sempervirente estacional» [con lluvias adecuadas en verano]; «bosque latifoliado sempervirente esclerófilo» con lluvias de invierno [a menudo se toma por mediterráneo, pero se presenta también en el sudoeste de Australia y en Chile, etc., donde el clima tiene una sequía de verano pronunciada]; cf. UNE 1973; ►vegetación esclerófila); *f* **forêt** [f] **sempervirente de feuillus** (**1.** Forêt de feuillus toujours verte, dont la couronne des arbres est persistante et, suivant le régime pluviométrique, épaisse ou éclaircie. **2.** Terme générique caractérisant les formes de **f. s. de f.** sous différentes zones climatiques ; *termes spécifiques* forêt ombrophile tropicale comprenant des arbres aux feuilles de grandeur comprise entre 15 et 40 cm en Amazonie ou dans la cuvette congolaise et sur les îles de la Sonde, les forêts sempervirentes saisonnières tempérées de feuillus [avec pluies d'été suffisantes], les forêts sempervirentes de feuillus sclérophylles [durisilve, forêt laurifoliée] à pluies d'hiver et à été sec prononcé, souvent considérées comme étant méditerranéennes, mais présentes aussi dans le sud-ouest australien, au Chili, etc. ; ►végétation sclérophylle) ; *g* **immergrüner Laubwald** [m] (**1.** Laubwald mit ständig grünen Baum-

kronen, die je nach Jahreszeit und Feuchteregime dicht oder licht sind. **2.** Je nach Klimagebiet unterschiedlich ausgeprägter **i. L.** Es gibt u. a. **immergrüne tropische Regenwälder**, zusammengesetzt aus Baumformen mit 15-40 cm großen Blättern, z. B. im Amazonasgebiet, im Kongobecken und auf den Sunda-Inseln; **immergrüne L.wälder** in winterfeuchten und sommertrockenen Gebieten, die meist im Mittelmeergebiet, in SW-Australien, in Chile etc. vorkommen, werden auch **Hartlaubwälder** genannt; ►Hartlaubvegetation).

568 broad-leaved, evergreen, sclerophyllous vegetation [n] *phyt.* (Plant communities occurring in areas with dry summers and mild winters, in which woody species with leathery, evergreen leaves dominate; ►broad-leaved evergreen forest); *s* **vegetación [f] esclerófila** (Vegetación caracterizada por tener hojas persistentes, pequeñas y duras, capaz de resistir prolongados periodos/períodos de sequía. Plantas de este tipo forman la vegetación característica de la zona mediterránea: encina *[Quercus ilex]*, coscoja *[Q. coccifera]*, alcornoque *[Q. suber]*, acebuche *[Olea oleaster]*, algarrobo *[Ceratonia siliqua]*, etc.; DINA 1987; ►bosque latifoliado sempervirente); *f* **végétation [f] sclérophylle** (Communauté végétale constituée de végétaux ligneux buissonnants pauvres en parenchyme, aux feuilles persistantes petites, dures, souvent épineuses et luisantes ; ►forêt sempervirente de feuillus) ; *g* **Hartlaubvegetation [f]** (In Gebieten mit trockenen Sommern und milden Wintern wachsende Pflanzengesellschaften, in denen Gehölze mit ledrigen immergrünen Blättern dominieren; ►immergrüner Laubwald).

broad-leaved forest [n] *for. phyt.* ►deciduous forest.

broad-leaved tree [n] *bot.* ►deciduous tree.

569 broad-leaved woody species [n] *bot. for.* (Generic term for tree, shrub or ►suffruticose shrub which, in contrast to ►needle-leaved species, bears broad-shaped leaves); *syn.* broadleaf woody plant [n] [also US] (WEB 1993), deciduous woody plant [n]; *s* **frondosa [f]** (Antófito arbóreo dotado de hojas más o menos anchas, por oposición a los de hojas aciculares; DB 1985; ►especie acicufolia, ►sufrútice); *f 1* **feuillu [m]** (**1.** Arbres, arbustes et arbrisseaux portant des feuilles à limbe relativement large le plus souvent caduques ; *opp.* ►espèce à aiguilles. **2. U.S./U.K.**, le terme anglosaxon de sylviculture « hardwood » [bois feuillu] désigne les bois des feuillus car ceux-ci ont habituellement une densité élevée et donc plus dur que les résineux) ; *f 2* **essence [f] (ligneuse) à feuillage caduque** *for.* (Bois provenant d'essences forestières [arbres, arbustes et ►espèces sousfrutescentes], appartenant à la classe des Angiospermes) ; *syn.* essence [f] (ligneuse) à feuilles caduques (PR), bois [m] franc [CDN], bois [m] feuillus (DFM 1975, 128) ; *g* **Laubgehölz [n]** (**1.** OB zu Baum, Strauch und ►Halbstrauch, die im Gegensatz zum ►Nadelgehölz breitflächige Laubblätter ausbilden. **2.** Der forstliche Begriff **Laubholz [n]** bedeutet Holz von Laubbäumen. Im englischen Sprachgebrauch wird **Laubholz** als *hardwood* [Hartholz] bezeichnet, weil das Laubholz ring- oder zerstreutporig ist, d. h., dass es keine Harzkanäle hat und schwerer, fester und härter als das Holz fast aller Nadelbäume ist. *BEACHTE* im Deutschen wird hingegen unter **Hartholz** ein Holz mit einer höheren Rohdichte als 0,55 g/cm³ verstanden. Hierzu gehören neben den harten **Laubhölzern** wie z. B. Eichen-, Eschen-, Robinien-, Bergahorn- und Buchsbaumholz auch das Nadelholz der Eibe [Rohdichte 0,8 à 1,0 g/cm³] sowie Tropenhölzer, z. B. Mahagoni vom Mahagonibaum *[Swieténia mahágoni, S. macrophylla]*, Bangkirai aus Südostasien *[Shorea laevis]*, Bongossi oder Azobé aus dem Senegal bis Kamerun *[Lophira alata]*; cf. WFL 2002, 329).

570 broken range work [n] *arch. constr.* (Squared building stones laid in horizontal courses of different heights, any one course of which may be broken [at intervals] into two courses; cf. DAC 1975; ►cut stone); *syn.* snecked rubble wall [n] [also UK], random work [n] [also US] (DAC 1975); *s* **sillería [f] ordinaria** (Tipo de fábrica en la que los mampuestos no constituyen aparejo alguno. Puede ser de piedra natural o artificial [►sillar]; cf. DACO 1988); *f* **maçonnerie [f] à appareillage à l'anglaise** (Mur à appareillage en assises régulières, en pierre de taille ou en matériaux artificiels assisés, boutisses de rupture et panneresses alternant régulièrement dans chaque assise avec joints creux, provoquant des lits d'inégales hauteurs ; ►matériau de construction minéral) ; *syn.* maçonnerie [f] à appareillage à l'antique ; *g* **Wechselmauerwerk [n]** (Lagerhaftes Mauerwerk aus ►Werksteinen oder Kunststeinen mit rechtwinklig zueinander stehenden, tief bearbeiteten Stoß- und Lagerfugen, bei dem sich die Schichthöhen ständig ändern und lange Horizontalfugen durch größere Steine, sog. *Wechsler* oder *Absetzer*, unterbrochen werden. Gemäß DIN 1053 dürfen Stoßfugen nur über zwei Lagerschichten laufen; weitere handwerkliche Regeln zur Schaffung eines harmonischen Fugenbildes, das auch **schottischer Verband** genannt wird, besagen, dass Folgendes zu vermeiden ist: a] zu kleine Steine, sog. *Streichholzschachteln*, die hin und wieder verwendet werden, um den Wechsler einzubinden; b] sog. *Fallrohre*, d. h. über mehrere Schichten fast senkrechte, meist leicht versetzt verlaufende Stoßfugen; c] sog. *Schubladen*, das sind zwischen zwei großen Steinen einer Schicht eingefügte flache Steine und d] sog. *Treppen*, ein Zickzackfugenbild, das sich ergibt, wenn sich die Steine nicht genügend weit überlappen; cf. EVE 2001).

brooding period [n] *zool.* ►incubation period.

571 brook [n] *geo.* (Small stream, ►torrent); *syn.* creek [n], run [n] (1), kill [n] [north-east US]; *s* **arroyo [m] pequeño** (►aguas vivas, ►torrente); *f* **ruisselet [m]** (Petit ruisseau naturel de largeur moyenne de 0,50 m ; ►torrent, ►gave) ; *g* **Bächlein [n]** (Kleiner natürlicher Bach; ►Wildbach); *syn.* Rinnsal [n].

brooklet [n] *wat'man.* ►streamlet.

brookside flood area [n] [US] *geo.* ►flood area of a stream/brook.

brown dune [n] [UK] *geo. phyt.* ►backdune [US].

572 brown earth [n] *pedol.* (Soil of temperate humid climates with A_h-B_v-C horizons. In the FAO system it is known as 'Cambisols', in U.S. classification system it belongs predominantly to the 'Entisols' or 'Inceptisols'; ►brunification); *syn.* brown forest soil [n]; *s* **tierra [f] parda** (Suelo A [B] C con buen humedecimiento en la región de los bosques de follaje y mixtos húmedos, caracterizada por una desintegración química y formación de arcilla intensas y una cubierta de vegetación exhuberante cerrada; KUB 1953; ►braunificación); *syn.* suelo [m] pardo de bosque; *f* **sol [m] brunifié** (Sol des régions tempérées caractérisé par un profil A [B] C ou A B C. Dans la classification FAO les sols brunifiés sont classés dans l'unité des « cambisols », dans la classification américaine ils appartiennent aux ordres des « entisols » et des « inceptisols » ; ►brunification) ; *syn.* sol [m] brun ; *g* **Braunerde [f]** (Boden des gemäßigt-humiden Klimas mit A_h-B_v-C-Horizont. Im FAO-System werden die Braunerden als ‚Cambisole' bezeichnet, im US-System gehören sie überwiegend zu den ‚Entisolen' oder den ‚Inceptisolen'; ►Verbraunung); *syn.* Brauner Waldboden [m].

573 brownfield [n] *envir. urb.* (Former industrial site that is contaminated or thought to be so, in urban and suburban areas; in U.S., government financial incentives may encourage removal of toxic substances and safe reuse; ►hazardous old dump site [US], ►hazardous waste disposal site, ►orphan contaminated site); *syn.* brownfield site [n]; *s* **finca [f] industrial potencialmente contaminada** (Antigua ubicación industrial cuyo suelo está contaminado o del que se sospecha que lo esté. En Es. la responsabilidad de limpiar y recuperar los suelos contaminados recae sobre los causantes de la contaminación. Sin embargo, las operaciones de limpieza y recuperación pueden realizarse mediante acuerdos voluntarios entre los causantes y la administración pública competente [en general las CCAA]. Estos convenios podrán incorporar también incentivos económicos que puedan servir de ayuda para financiar los costes. En ese caso, las ayudas deben supeditarse al previo compromiso de que las posibles plusvalías que adquieran los suelos revertirán en la cuantía subvencionada en favor de la Administración pública que haya financiado las citadas ayudas; cf. arts. 27.2 y 28 Ley 10/1998; en EE.UU. el gobierno incentiva financieramente la descontaminación de las **f. i. p. c.** y el reciclaje de los mismos para otros usos; ▶depósito de residuos peligrosos, ▶emplazamiento contaminado, ▶emplazamiento contaminado huérfano); **f ancien site [m] industriel potentiellement contaminé par des déchets historiques** (Ancien établissement industriel en zone urbaine ou suburbaine contaminé ou soupçonné de l'être par des déchets historiques ; ▶décharge de déchets industriels [spéciaux], ▶site pollué ancien, ▶site pollué orphelin) ; *syn.* ancien site [m] industriel pollué, site [m] pollué d'un ancien établissement industriel ; **g altlastverdächtige ehemalige Industriefläche [m]** (Frühere Industieanlage im Stadt- oder Vorortbereich, die kontaminiert ist oder bei der Altlasten vermutet werden; ▶altlastverdächtige Fläche, ▶herrenloser Altstandort, ▶Sonderabfalldeponie).

brownfields [n], recycling of *plan. urb.* ▶recycling of derelict sites.

brown forest soil [n] *pedol.* ▶brown earth.

brown-water lake [n] [UK] *limn.* ▶dystrophic lake.

brow [n] of embankment/slope *constr.* ▶crest of embankment/slope.

574 browsing and debarking protection [n] *game' man.* (Measures to prevent damage to trees and shrubs caused by game or grazing livestock; ▶game damage, ▶over-browsing damage); **s protección [f] contra recomido** (Medidas para evitar daños en leñosas causadas por animales salvajes o de ganado; ▶daños causados por la caza, ▶recomido); **f protection [f] contre les morsures** (Protection individuelle du tronc contre les rongeurs et les cervidés ; *terme spécifique* manchon de protection perforé ; ▶dégâts causés par le gibier, ▶abroutissement) ; **g Verbissschutz [m, o. Pl.]** (Vorrichtung zur Verhinderung von Schäden an Gehölzen durch verbeißendes Wild oder verbeißende Weidetiere; ▶Wildschaden, ▶Wildverbiss).

575 bruise [n] *arb. bot.* (Damage to bark tissues and ▶cambium, caused by compression); **s magulladura [f]** (Daños en la corteza y el ▶cambium causados por compresión); *syn.* contusión [f]; **f meurtrissure [f]** (Lésion causée par une pression mécanique exercée sur l'écorce et le ▶cambium) ; **g Quetschung [f]** (Durch mechanischen Druck verursachter Schaden an Borke und ▶Kambium).

576 brunification [n] *pedol.* (Brown-colo[u]ring process in clay soils caused by the release of iron from 2-valent iron silicates or pyroxides, usually after a drop in pH value; ▶brown earth); **s braunificación [f]** (Proceso de formación de suelos en zonas de clima templado-frío y precipitación regular, en el que la hidrólisis de los silicatos no es muy intensa, aunque se llegan a individualizar óxidos de hierro. Las migraciones de arcilla y óxidos de hierro son mínimas e insuficientes para formar un horizonte de acumulación. Los suelos resultantes son los suelos pardos; DINA 1987; ▶tierra parda); **f brunification [f]** (Processus climatique dans les zones à climat tempéré à végétation de forêt feuillue observé sur les matériaux bien drainés, non calcaires avec libération des oxydes de fer libres dans l'horizon B

[avec coloration des argiles] après diminution du pH ; DA 1977 ; ▶sol brunifié) ; *syn.* brunissement [m] ; **g Verbraunung [f]** (Prozess der Braunfärbung von tonhaltigen Böden durch Freisetzung von Eisen aus Fe[II]-haltigen Silikaten oder Pyroxenen, meist nach einem Absinken des pH-Wertes; ▶Braunerde).

brush [n], mat of live hardwood *constr.* ▶brush mat.

577 brush and rock dam [n] *constr. wat'man.* (BIS 1982, 184; *bioengineering* construction method whereby a ▶loose rock dam is reinforced with live branches and twigs which are driven into the subsoil); *syn.* vegetated rock sill [n]; **s solera [f] de piedras y ramas vivas** (Método de construcción biotécnica en el que se insertan ramas vivas entre los huecos de una ▶solera de piedras sueltas); **f seuil [m] en enrochements et branchages (±)** (Ouvrage de génie biologique dans lequel des branches ou des broussailles sont enfoncées dans le sol entre les pierres constituant l'enrochement ; ▶seuil en enrochement) ; *syn.* seuil [m] en enrochement végétalisé ; **g Steinschwelle [f] mit Ast- oder Zweigbesatz** (Ingenieurbiologisches Bauwerk, bei dem in die lose aufgesetzte ▶Steinschwelle lebende Äste oder Zweige bis in den Untergrund gesteckt werden).

brush jetty [n] [US] *constr.* ▶live brushlayer barrier.

578 brush [vb] joints with sand *constr.* (Process); **s rellenar [vb] las juntas con arena** (Proceso); **f garnir [vb] les joints par balayage de sable** (Processus) ; *syn.* figer [vb] les joints au sable ; **g Fugen mit Sand einfegen [vb]** (Vorgang; Fugen eines Pflasters oder Plattenbelages mit Sand ausfüllen).

brushkiller [n] *chem.* ▶defoliating agent.

brush layer [n] *constr.* ▶hedge brush layer.

brush layer barrier [n] *constr.* ▶live brushlayer barrier.

579 brush layering [n] *constr.* (Bioengineering technique to stabilize erosion-prone steep banks by embedding green branches of shrub or tree species, preferably those that will root; e.g. willow; ▶channel deadwood reinforcement); *syn.* branch layering [n], contour brush-layering [n] (BIS 1982, 167s), brush-layer method [n], faggoting [n] [US] (BIS 1982, 168); **s estabilización [f] con ramajes** (Técnica de construcción de bioingeniería para estabilizar taludes o pendientes con ▶peligro de deslizamiento de tierras por medio de ramas verdes de árboles o arbustos, preferentemente de especies que enraizan, como p. ej. el sauce; ▶estabilización de regueros con ramas secas); **f stabilisation [f] de talus par lit de plants et plançons** (*Génie biologique* technique de stabilisation de talus à forte pente ; disposition côte à côte et sur plusieurs paliers parallèles (risbermes superposées) de branches de saules vivantes et/ou de plants à racines nues d'autres espèces végétales indigènes et adaptées à la station ; un géotextile biodégradable à base de fibres de coco tissées peut être intercalé entre les rangées de végétaux ; ▶embroussaillement des ravines par entassement de bois mort) ; *syn.* protection [f] de talus par lit de plançons (WIB 1996) ; **g Buschlagenbau [m]** (Ingenieurbiologische Sicherungsbauweise zur Beruhigung und Festlegung von rutschgefährdeten Steilböschungen durch hangparallelen Einbau von ausschlagfähigem Weidenastwerk und anderem Reisig von mindestens 60 cm Länge, das kreuz und quer bis zu einer Stärke von 15 cm eingelegt, mit dem Aushub der nächsthöheren Terrasse verfüllt und gut verdichtet wird; das herausragende Buschwerk wird auf 10 cm zurückgeschnitten; cf. DIN 18 918; ▶Runsenausgrassung).

brush layer method [n] *constr.* ▶brush layering.

brush laying [n] [UK] *constr.* ▶brush matting with live plant material [US].

580 brush mat [n] *constr.* (Protective ▶biotechnical construction technique using a mat of hardwood brush fastened down with stakes and wire; used to protect streambanks and control

B

erosion; there are two different mats: **1. mat of live hardwood brush**, mainly of willow and alder; *syn.* live brush mattress [n] [also US]; **2. brush mat of deadwood**, hardwood and softwood); *syn.* willow mattress [n] (TAN 1975, WRP 1974); *s* **tapiz [m] de ramas** (Técnica de ►construcción biotécnica para estabilizar y proteger superficies, en especial taludes descubiertos, contra la erosión del suelo, orillas de ríos contra la acción del agua y de las olas, que consiste en extender un tapiz de ramas sujetas con estacas y alambre. Existen dos tipos de tapices: **1. tapiz de ramas vivas**, generalmente sauce y aliso; *syn.* tapiz [m] de ramas de sauce vivo; **2. tapiz de ramas muertas**); *f* **garnissage [m] de boutures** (►*Technique de génie biologique* ouvrage de génie biologique — équipement linéaire et surfacique comprenant des boutures en deux dimensions fixées par des piquets vivants, utilisé pour le piégeage et la rétention des sédiments, p. ex. dans le lit de petites ravines ou dans des aménagements hydrauliques pour consolider les rives des cours d'eau ou les berges des lacs contre les affouillements) ; *syn.* tapis [m] de boutures, tapis [m] de branches à rejets, couche [f] de branches à rejets ; *g* **Spreitlage [f]** (►*Ingenieurbiologische Bauweise* schnell wirksame Flächensicherung standfest geschütteter oder angerissener Böschungen, besonders von Böschungsfüßen, gegen Bodenerosion oder bei Fließgewässern gegen den Angriff fließenden Wassers und Wellenschlag mit **1.** austriebsfähigen, möglichst geraden, wenig verzweigten, jungen Weidenruten mit einer Länge nicht unter 120 cm oder **2.** als **tote Spreitlage** oder **Rauwehr [n]** mit 10 bis 20 cm dicken und dicht gepackten Lagen aus beliebigem Reisig in ganzen Ästen, sowohl von Laub- als auch Nadelhölzern für eine kurzfristige, aber schnell wirksame Bodensicherung. Die Befestigung erfolgt mit Astgabeln oder mit Pfählen und Draht; die untersten Ruten werden in den Boden gedrückt und mit Faschine oder Steinpackung gesichert; cf. LEHR 1997, 350 f); *syn. zu 2.* Raupackung [f], Reisigdeckung [f], Astlage [f].

brush mat revetment [n] [US] *constr.* ►brush matting with live plant material [US]/brush laying [UK].

581 brush matting [n] **with live plant material** [US] *constr.* (►*Biotechnical construction* technique slope protection method to prevent soil erosion on newly constructed slopes or those with slope failure or for the stabilization of watercourse edges, ►brush mat); *syn.* brush mat revetment [n] [US] (WRP 1974, 128), brush laying [n] [UK]; *s* **revestimiento [m] con ramas de sauce vivo** (Técnica de ►construcción biotécnica para frenar la erosión en las márgenes de ríos; ►tapiz de ramas); *f* **stabilisation [f] avec tapis de plançons** (►*Technique de génie biologique* procédé de protection du sol des talus en remblais ou des berges des cours d'eau ; ►garnissage de boutures) ; *g* **Spreitlagenbau [m]** (►*Ingenieurbiologische Bauweise* Flächensicherung lagerhaft und standfest geschütteter oder angerissener Böschungen gegen Bodenerosion oder bei Fließgewässern gegen den Angriff fließenden Wassers und Wellenschlag. Ausschlagfähige, mindestens 150 cm lange Weidenruten werden mit dem dicken Ende Richtung Böschungsfuß dicht nebeneinander gelegt, befestigt und 2-5 cm mit Erde überdeckt; ►Spreitlage).

brush mattress [n] [US]**, live** *constr.* ►brush mat.

brush packing [n] [US] *constr.* ►packed fascine-work.

brush placement [n] **in erosion gullies** *constr.* ►channel brush reinforcement.

brushwood [n] *constr.* ►brushwood thatching.

brushwood bundle [n] [UK] *constr.* ►fascine.

582 brushwood fascine [n] *constr.* (►*Biotechnical construction technique* thin bundle of brushwood 10-15cm thick, laid at an angle of 10°-30° in furrows on slope, to prevent erosion and protect topsoiling); *syn.* fascine pole [n] [also UK]; *s* **fajina [f] de ramillas** (►*Construcción biotécnica* manojos de ramillas de 10-

15 cm de espesor colocados en surcos de taludes con ángulos de 10°-30° para evitar la erosión y fijar la tierra vegetal); *f* **fascine [f] en caniveaux** (►*Technique de génie biologique* ouvrage de stabilisation de talus constitué de fascines disposées en arête de poisson ou en Z dans des rigoles avec un angle de 10° à 30°) ; *g* **Hangfaschine [f]** (►*Ingenieurbiologische Bauweise* diagonal in Rillen, 10°-30° schräg verlegtes, dünnes Reisigbundel von 10-15 cm Dicke, das Erosionen verhindert und den Oberbodenauftrag festlegt); *syn.* Faschinenwurst [f].

583 brushwood thatching [n] *constr.* (►*Biotechnical construction technique* dead material used to cover the soil surface as a rapid but short-term protection against erosion—frequently along riverbanks, also a treatment for stabilizing dunes; ►soil stabilization, ►stabilization of sand dunes); *syn.* thatching [n] with brushwood, soil stabilization [n] with brushwood [also US]; *s 1* **capa [f] protectora de desbrozo y leña menuda** (►*Construcción biotécnica* técnica de bioingeniería usada para proteger a corto plazo márgenes contra la erosión); *s 2* revestimiento [m] (del suelo) con ramas secas (p. ej. en dunas; ►fijación de dunas, ►fijación del suelo); *f 1* **couverture [f] de branchages** (►*Technique de génie biologique* technique paysagère de protection d'un talus ou des berges d'un cours d'eau) ; *f 2* **fixation [f] du sol au moyen de ramilles** (►fixation des dunes, ►fixation du sol) ; *syn.* couverture [f] du sol, recouvrement [m] (des dunes) avec des ramilles ; *g* **Abdeckung [f] mit einem Rauwehr** (►*Ingenieurbiologische Bauweise* Reisigdeckung in Form von Astlagen als vorübergehender, aber schnell wirksamer Schutz der Bodenoberfläche gegen Wasserangriff; cf. DIN 19 657; ►Bodenfestlegung, ►Festlegung von Dünen); *syn.* Deckwerk [n], Raupackung [f], Reisigabdeckung [f], Reisigdeckung [f], Astlagen [fpl], Abdeckung [f] mit Reisig, Bodenabdeckung [f] mit Reisig.

584 brushy area [n] [US] *phyt.* (**1.** Generic term for shrublands and thickets, 50cm to 5m high; **2. scrub [n]:** vegetation consisting chiefly of dwarf and stunted trees and shrubs, often thick and impenetrable, growing in poor soil or in sand, e.g. mallee scrub in south Australia, pine scrub; cf. WEB 1993; also, a tract of country covered with such vegetation, especially a palmetto barren of the southern U.S., or the 'bush' of Australia and South Africa; GGT 1979, 435; ►coniferous evergreen thicket/shrubland, ►dune scrub, ►garide, ►hygrophilic scrub, ►oak scrub [US]/oak copse [UK], ►shrubland. **2.** A special type of **b. a.** is **scrub [n]:** vegetation consisting chiefly of dwarf and stunted trees and shrubs, often thick and impenetrable, growing on poor soil or sand, e.g. **a]** *mallee scrub* in South Australia, comprising mainly of various, low-growing eucalyptus species and surviving upon only 250mm rainfall per annum, **b]** *mulga scrub*, named after the dominating mulga bush *[Acacia aneura]*, which covers approximately one quarter of Australia, **c]** ►**oak scrubs** in Florida composed of different oak species, interspersed with bushes and some pine species or a tract of country covered with such vegetation, especially palmetto barrens of the southern U.S.); *syn.* scrub [n] (BS 3975: part 5), scrubland [n]; *s* **matorral [m] (1)** (Término genérico para formación vegetal constituida por leñosas de pequeña talla, ramificadas desde la base [arbustos y matas]; ►matorral achaparrado de coníferas, ►matorral claro, ►matorral denso, ►matorral de sustitución, ►matorral dunar/de dunas); *syn.* maleza [f] [AL] (DINA 1987); *f 1* **formation [f] buissonnante** (*Terme générique* formation végétale naturelle dense ou claire basse, constituée par une végétation buissonnante et arbustive [►fourré buissonnant, ►fruticée, ►brousse] dont la taille est comprise entre 1 et 2 m, aux branches semi-prostrées très ramifiées, p. ex. la saulaie et rhodoraie-saulaie buissonnante, la saulaie basse à Saule helvétique de l'étage subalpin moyen et surtout supérieur ; ►formation buissonneuse, ►garide, ►taillis hygrophile) ; *f 2* **fourré** [m] (1) (**1.** Terme générique qualifiant le fourré préforestier correspondant à l'évolution dynamique d'une lande vers la forêt ; **2.** terme spécifique désignant une végétation arbustive ou sous arbustive souvent très dense, p. ex. dans le Code Corine des habitats naturels déterminants, fourré d'Aulne vert des Alpes, fourré à buis, etc. ; ►fourré buissonnant, ►fourré dunaire) ; *f 3* **fruticée** [f] (Formation végétale naturelle dense ou claire où prédominent les arbustes, arbrisseaux et sous-arbrisseaux, p. ex. la fruticée des stations rocailleuses à cotoneaster et amélanchier) ; *syn.* végétation [f] frutescente, fourré [m] arbustif ; *f 4* **brousse** [f] *geo. phyt.* (**1.** Terme imprécis désignant une formation arbustive xérophile, très souvent épineuse des régions tropicales. **2.** Terme aussi utilisé en phytosociologie, accompagné d'un complément [brousse à épineux, brousse littorale, ►brousse sempervirente de résineux] et pour désigner une végétation de fourrés buissonnant et arbustifs, p. ex. dans le Code Corine des habitats naturels déterminants, brousse des saules bas des Alpes, etc.) ; *g* **Gebüsche** [npl] (**1.** OB zu offene oder dichte Gebüschformation, die hauptsächlich aus zusammenhängenden, holzigen Blütenpflanzen von 0,5-5,0 m Höhe besteht; ►Dünengebüsch, ►Feuchtgebüsch, ►Garide, ►Gebüsch 2, ►Gebüsch 3, ►immergrünes Nadelgebüsch, ►Kratt. **2.** Eine besondere Form der **G.** ist der **Scrub,** eine Gehölzformation, die hauptsächlich aus niedrigen und gedrungen wachsenden Bäumen und Sträuchern besteht, oft dicht und undurchdringbar, die auf nährstoffarmen Böden oder auf Sand wachsen, z. B. **a]** der *Mallee-Scrub* [Trockenbusch] in Südaustralien, der vornehmlich aus verschiedenen, niedrig wachsenden Eukalyptusarten besteht und mit nur 250 mm Regen im Jahr überlebt; **b]** der *Mulga-Scrub,* nach dem dominierenden Mulga-Busch *[Acacia aneura]* benannt, bedeckt ca. ein Viertel Australiens; **c]** ►**Eichengestrüppe** in Florida aus unterschiedlichen Eichenarten, durchsetzt mit Sträuchern und wenigen Kiefernarten); *syn.* Gebüschformation [f], Strauchformation [f], Busch [m].

bryophyte [n] *bot.* ►shrubland moss (1).

bucking off [n] [US] *constr. for.* ►saw into pieces.

bud break [n] [US]**, spring** *bot.* ►leafing out.

bud union [n] [US] *hort.* ►graft union.

buffer [n]**, regional green** *landsc. urb.* ►regional green corridor.

buffer area [n] *conserv. landsc. urb.* ►buffer zone, ►fringe area.

buffer capacity [n] *pedol.* ►buffering capacity.

buffer capacity [n]**, ecological** *ecol. recr.* ►environmental stress tolerance.

585 buffering [n] *pedol.* (Process which takes place in soils whereby existing nutrients or pH concentration are maintained, despite their diminishment or increase, by presence of buffer compounds; i.e. carbonates, phosphates, oxides, phyllosilicates, and some organic materials); *s* **amortiguación** [f] (Capacidad de los suelos de mantener la concentración constante de nutrientes o de pH a pesar de cambios en el suministro); *f* **neutralisation** [f] (Processus de rétablissement de la valeur du pH d'un sol sous l'influence d'un acide ou d'une base) ; *g* **Pufferung** [f] (Vorgang, bei dem in Böden bestehende Nährstoff- oder pH-Konzentrationen trotz Abschwächung oder Erhöhung aufrechterhalten werden).

586 buffering capacity [n] *pedol.* (Ability of soil or water to resist changes in pH value due to the input or formation of acids or bases in soil or water; RCG 1982; ►buffering); *syn.* buffer capacity [n]; *s* **poder** [m] **tampón** (Capacidad de los suelos o cuerpos de agua de no alterar mucho su pH a pesar de la introducción de sustancias ácidas o básicas en ellos;

►amortiguación); *syn.* poder [m] amortiguador; *f* **pouvoir** [m] **tampon** (Résistance qu'oppose un sol ou de l'eau au changement de pH sous l'influence d'un acide ou d'une base. Un sol est un milieu dit « tamponné » car il s'oppose aux variations brutales du pH ; cf. DIS 1986 ; ►neutralisation) ; *syn.* effet [m] tampon (PED 1979, 355) ; *g* **Pufferkapazität** [f] (Fähigkeit eines Bodens oder Gewässers, bei Zuführung von Säuren oder Basen den pH-Wert nur wenig zu ändern; ►Pufferung); *syn.* Pufferungsvermögen [n].

587 buffering [n] **of nutrients** *pedol.* (TEE 1980, 295; capacity of the soil to maintain the concentration of a nutrient in solution during absorption, e.g. by plants, or supply, e.g. from fertilizers); *s* **capacidad** [f] **tampón de un suelo** (Capacidad de un suelo de mantener la concentración de nutrientes en solución durante la absorción por las plantas o suministro por aporte de fertilizantes); *f* **pouvoir** [m] **tampon d'un sol** (Par sa capacité à absorber des ions H^+ ou OH^-, résistance qu'oppose un sol aux variations brutales de pH sous l'influence d'un acide ou d'une base consécutives, p. ex. à une augmentation saisonnière de la teneur en substances nutritives dans les solutions du sol ou à un apport artificiel d'amendement) ; *syn.* capacité [f] tampon d'un sol ; *g* **Nährstoffpufferung** [f] (Fähigkeit des Bodens, die Konzentration eines Nährstoffs in der Bodenlösung bei dessen Entzug [z. B. durch Pflanzen] oder Zufuhr [z. B. über die Düngung] aufrechtzuerhalten).

588 buffer strip [n] *conserv. landsc. urb.* (Long, narrow separation piece of land, small in area and often only a few meters [US]/metres [UK] wide; ►buffer zone); *s* **banda** [f] **de protección** (Superficie pequeña, generalmente de pocos metros de ancho; ►zona de amortiguación); *f* **bande** [f] **de protection** (Analogue à la ►zone tampon, cependant d'une surface en général plus petite, souvent large de quelques mètres) ; *g* **Abstands- und Schutzzone** [f] (1) (Fläche zur Vermeidung oder Reduzierung von Beeinträchtigungen, die auf Grund der Unverträglichkeit benachbarter Flächennutzungen entstehen. Wie ►Abstands- und Schutzzone 2, jedoch meist kleinere Flächen, oft nur wenige Meter breit).

589 buffer zone [n] *conserv. landsc. urb.* (Area of land providing a separation between two different types of land use in order to reduce problems resulting from their incompatibility; ►greenbelt [US]/green belt [UK], ►green space corridor, ►land reserved for environmental purposes; *syn.* transitional area [n] [also US], buffer area [n]; *s* **zona** [f] **de amortiguación** (Superficies grandes reservadas para evitar perturbaciones entre usos diferentes e incompatibles entre sí; ►cinturón verde, ►corredor verde, ►zona no edificable); *syn.* zona [f] buffer; *f 1* **zone** [f] **tampon** (Zone de protection dont le but est d'éviter les nuisances relatives à l'incompatibilité de l'occupation du sol des zones avoisinantes ; ►ceinture verte, ►coupure verte) ; *f 2* **distance** [f] **d'isolement** (Zone de protection règlementant l'implantation des constructions à usage d'habitation ou autres ouvrages, d'étendue distincte selon la proximité d'un aérodrome, d'un monument historique, d'un site classé, etc., p. ex. 100 m cimetière/agglomération, 100 m bande littorale/construction, etc. ; ►zone *non aedificandi*) ; *g* **Abstands- und Schutzzone** [f] (2) (**1.** Größere Fläche zur Vermeidung oder Reduzierung von Beeinträchtigungen, die auf Grund der Unverträglichkeit benachbarter Flächennutzungen entstehen. **2.** Im engeren Sinne oder kleineren Maßstab wird eine solche **A.- u. S.** auch **Grüntrennzone** genannt; ►freizuhaltende Schutzfläche, ►Grüngürtel, ►Grünzug); *syn.* Pufferzone [f], Freihaltezone [f] [CH].

buildable [adj] *constr.* ►stable.

590 buildable area [n] [US] *plan. urb.* (Area of a building lot remaining after the minimum yard and open space requirements of the zoning ordinance have been met; ►mandatory

B

building line, ▶setback line; *s* **planta [f] edificable** (En ▶plan parcial [de ordenación] parte de los terrenos delimitados por las ▶líneas de retranqueo dentro de las cuales está permitida la edificación; ▶línea de fachada); *f 1* **emprise [f] maximale de construction** (F., surface délimitée par ▶polygone d'implantation ou par des ▶lignes d'implantation à l'intérieur de laquelle l'implantation des constructions est autorisée ; l'emprise au sol des constructions est d'une manière générale déterminée par **1.** les règles d'implantation des constructions par rapport aux voies et emprises publiques, par rapport aux limites parcellaires et par rapport aux constructions voisines, **2.** par un pourcentage maximal de terrain réservé à la construction ou par un pourcentage de surface de terrain devant rester libre de toute construction ; **D.**, dans un « Bebauungsplan » [PLU/POS allemand], espace dont le périmètre est de surcroit délimité par les ▶limites de constructibilité ou éventuellement par des ▶marges de reculement obligatoire et à l'intérieur desquelles est autorisée la construction de bâtiments) ; *f 2* **bande [f] constructible** (*PLU/ POS* surface délimitée sur un document graphique à l'intérieur de laquelle l'implantation des constructions est autorisée ; elle est définie d'un côté par l'alignement sur rue et de l'autre par la ▶limite de constructibilité ; au-delà de la bande constructible seuls les constructions annexes, garages, dépendances et les locaux liés à une activité peuvent être implantés) ; *syn.* bande [f] de constructibilité) ; *g* **Baufenster [n]** (Im Bebauungsplan durch ▶Baugrenzen und evtl. durch ▶Baulinien festgelegter Teil eines Grundstückes, auf dem die Errichtung von Gebäuden zulässig ist; es gibt **Einzelbaufenster**, die für jedes Gebäude auf Grundstücken festgesetzt sind und **Baustreifen**, die als Baufenster über mehrere Gundstücke gehen).

builder [US] *constr. contr.* ▶to be contractor for.

building [n] *arch. urb.* ▶high-rise residential building, ▶residential building.

building [n] [UK]**, multistorey residential** *arch. urb.* ▶multistory building for housing [US].

591 building alteration [n] *leg. urb.* (Modification of an existing building or structure; e.g. windows, doors, bearing walls, façades, annexes, etc., not including repair work); *s* **alteración [f] constructiva** (Cualquier tipo de cambio realizado en partes de una construcción, excepcionando los trabajos de mantenimiento); *syn.* modificación [f] constructiva; *f* **modification [f] d'une construction** (Travaux exécutés sur une construction existante ayant pour effet de modifier l'aspect extérieur, intérieur ou le volume mais ne comprenant pas les travaux d'entretien) ; *g* **bauliche Veränderung [f]** (Veränderung von Teilen eines bestehenden Bauwerkes, z. B. Änderungen an Fenstern, Türen, tragenden Wänden, Giebeln oder Anfügen von Anbauten; ausgenommen sind Instandhaltungsarbeiten).

592 building and site plan permit application [n] [US] *adm. leg.* (Documents of an intended construction project which have to be submitted for approval to the ▶Building Permit Office [US]/Building Surveyors Department [UK]; ▶permit application documents); *syn.* detailed planning application [n] [UK]; *s* **solicitud [f] de permiso de construcción** (▶comisión de urbanismo, ▶documentos de solicitud de licencia de construcción); *syn.* solicitud [f] de licencia de construcción; *f* **demande [f] de permis de construire** (L'exemplaire de la demande et du dossier de permis de construire est adressé ou déposé à la mairie et transmis au service chargé de l'instruction de la demande ; ▶dossier [de demande] de permis de construire, ▶service de l'application du droit des sols) ; *g* **Bauantrag [m]** (Der ▶Baurechtsbehörde zur Genehmigung vorzulegende Unterlagen [Bauvorlagen] für ein beabsichtigtes Bauvorhaben; ▶Bauvorlagen); *syn.* Baugesuch [n].

593 building area [n] *plan. urb.* (Area occupied by one or more buildings); *s* **superficie [f] edificada (1)** (A gran escala, área ocupada por uno o más edificios); *syn.* superficie [f] de planta de edificio(s); *f* **emprise [f] de la construction** (Surface sur laquelle sont implantés un ou plusieurs bâtiments) ; *g* **Gebäudefläche [f]** (*Großmaßstäblich* Fläche, auf der ein oder mehrere Gebäude stehen).

594 building block [n] *urb.* (Rectangular area formed by a group of neighbo[u]ring buildings in a city, enclosed by streets); *s* **manzana [f]** (Dentro de la trama urbana cada uno de los espacios generalmente rectangulares en la que está dividida y dentro de los que se encuentran los edificios; el término «cuadra» —común en AL— no es sinónimo, sino que denomina cada uno de los tramos de calle a lo largo de las **m.**, es decir una **m.** tiene cuatro cuadras); *f* **îlot [m]** (Dans un quartier, ensemble d'immeubles d'habitation, en général rectangulaire, délimité par les rues adjacentes) ; *g* **Wohnblock [m]** (Ein von Straßen umgrenztes Viereck von Wohnhäusern in einem Stadtgebiet); *syn. umgangssprachlich* Block [m].

building code [n] [US] *leg. urb.* ▶uniform building code [US].

595 building code control [n] [US] *adm. leg.* (Observance of published building code for construction by the ▶Building Permit Office [US]/Building Surveyors Department [UK] to ensure that building works comply with building regulations and other relevant associated legislation; **U.S.**, in smaller towns inspection services are typically performed by an appointed and certified *building inspector*); *syn.* Local Authority Building Control [n] [UK], site inspection [n] [UK] (1); *s* **inspección [f] de la construcción** (Control del cumplimiento de las normas de construcción por medio de la ▶autoridad de inspección urbanística; en EE.UU., en las ciudades pequeñas no existe una institución, sino solamente un único puesto de *inspector de construcción*); *syn.* control [m] de obras; *f* **contrôle [m] technique des ouvrages** (Surveillance et contrôle technique sur les constructions immobilières ou de génie civil effectués par l'administration chargée de l'instruction des permis de construire [▶service de l'application du droit des sols] ; **F.**, également par un bureau de contrôle ; cf. loi n° 78-1146 du 7 décembre 1978 et loi n° 78-12 du 4 janvier 1978) ; *syn.* contrôle [m] des constructions ; *g* **Bauaufsicht [f]** (Überwachung der Einhaltung öffentlich-rechtlicher Bauvorschriften und die sich daraus ergebenden Anforderungen seitens der ▶Baurechtsbehörde bei Errichtung, Änderung, Abbruch und Unterhaltung baulicher Anlagen; die für die **B.** zuständigen Behörden werden durch die Baugesetze [▶Bauordnungen] der Bundesländer bestimmt).

596 building construction [n] *arch.* (Section of the building industry which builds above-ground structures; ▶ground civil engineering); *s* **edificación [f]** (Rama de la industria de construcción que se dedica a la construcción de edificios; ▶ingeniería civil); *syn.* construcción [f] arquitectónica; *f* **bâtiment [m]** (Activités liées à l'action de bâtir au-dessus du sol, par comparaison avec les ▶travaux publics ; terme utilisé pour désigner un secteur de l'activité économique de la construction [nomenclature française d'activités]) ; *syn.* (dans le) domaine [m] du bâtiment ; *g* **Hochbau [m, o. Pl.]** (Teilbereich des Bauwesens, das sich im Vergleich zum ▶Tiefbau vornehmlich mit der Errichtung von Bauwerken über dem Erdboden befasst. Eine exakte Trennung zwischen **H.** und Tiefbau ist bei einigen Bauvorhaben kaum möglich).

597 building contract [n] *constr. contr.* (Contract for construction operations according to ▶list of bid items and quantities [US]/schedule of tender items [UK]); *s* **contrato [m] de obra** (Contrato para realizar una obra de acuerdo con el ▶resumen de prestaciones); *f* **marché [m] sur devis** (Marché de travaux passé sur présentation d'un devis qui lie l'entrepreneur à des travaux, des quantités et des prix unitaires fixés ; *terme spécifique* marché sur bordereau de prix qui lie l'entrepreneur à des prix unitaires fixés [*syn.* marché [m] sur prix unitaires] ; ►descriptif quantitatif) ; *g* **Leistungsvertrag [m]** (Vertrag, durch den der Unternehmer zur Herstellung des versprochenen [Bau]werkes auf der Grundlage des ►Leistungsverzeichnisses und sonstiger vereinbarter Vertragsbedingungen und der Besteller [Bauherr] zur Entrichtung der vereinbarten Vergütung verpflichtet werden; cf. § 631 BGB; bei Aufträgen der öffentlichen Hand wird bei der Vergabe [►Auftragserteilung an eine Firma] in einem formalisierten Ausschreibungs- und Vergabeverfahren stets die ►Vergabe- und Vertragsordnung für Bauleistungen zugrunde gelegt); *syn.* Bauvertrag [m].

Building Contract [n] [UK] *constr. contr.* ►Standard Form of Building Contract.

Building Control [n] [UK]**, Local Authority** *adm. leg.* ►building code control [US].

Building Control Department [n] [UK]**, Local Authority** *adm. leg.* ►Building Permit Office [US].

building control officer [UK] [n] *constr. prof.* ►site supervisor representing an authority.

598 building coverage [n] [US] *leg. urb.* (Decimal figure in a ►zoning map [US]/Proposals Map [UK] expressing the ratio between built area and plot area; ►accessory structure, ►zoning ordinance [US]/use class order [UK], Land Use Intensity Index [LUI] [US], ►floor area ratio [FAR] [US]/floor space index [UK]; *syn.* plot ratio [n] [UK], factor [n] for calculation of permissible coverage [ZA]; *s* **coeficiente [m] de ocupación del suelo** (≠) (En D., el **c. de oc. del s.** indica en los ►planes parciales en números decimales el porcentaje máximo del suelo edificable, es decir, la relación entre la superficie edificable y la no edificable, o la superficie descubierta en relación con el área total del terreno, las ►construcciones auxiliares no se tienen en cuenta; **en Es.** no existe tal coeficiente; ►coeficiente de edificación, ►ordenanza de zonificación); *f* **coefficient [m] d'emprise au sol (CES)** (Dans un PLU/POS coefficient exprimant le rapport de la surface du terrain occupée par la base d'un bâtiment [la projection au sol des balcons et terrasses ne reposant pas sur des piliers n'est pas prise en compte] à la surface de la parcelle ou d'un îlot [►emplacements réservés déduits] en vue de définir la surface à attribuer aux espaces verts ; cf. BON 1990, 350 ; ►coefficient d'occupation des sols, ►installations auxiliaires, ►plafond légal de densité, ►règles générales d'utilisation du sol) ; *g* **Grundflächenzahl [f]** (*Abk.* GRZ; in D. gibt die GRZ im Bebauungsplan das höchstens zulässige Maß baulicher Nutzung als Dezimalzahl an, die das Verhältnis zwischen überbauter Fläche und Grundstücksfläche ausdrückt; ►Nebenanlagen werden nicht angerechnet; cf. §§ 17 und 19 BauNVO; ►Baunutzungsverordnung; ►Geschossflächenzahl); *syn.* Überbauungsziffer [f] [CH].

building debris [n] [US] *constr.* ►building rubble.

building density [n] *urb.* ►density of development.

building department [n] [US]**, county** *adm. plan. urb.* ►Building Permit Office [US].

599 building development [n] *leg. urb.* (Generic term covering all types of building construction; e.g. ►attached building development, ►detached and semi-detached housing development, ►detached building development, high-rise building development, ►hillside building development, ►low-density building development, ►mid-rise dwellings development, ►perimeter block development, ►peripheral building development, row house development; ►individual house, ►semidetached

house, ►type of building development); *s* **tipo [m] de edificación** (Término genérico para los diferentes tipos de edificios y formas de ordenación de los mismos desde el punto de vista urbanístico, como ►edificación continua, ►edificación de baja densidad, edificación de torres de pisos, edificación de una ►zona de casas uni- o bifamiliares, ►edificación en bloque cerrado, ►edificación en hilera; ►casa individual, ►casa pareada, ►edificación periférica; *f* **constructions [fpl]** (Terme générique pour le type d'implantation des constructions, p. ex. ►construction en îlot, ►construction en ligne, ►constructions individuelles, ►implantation des constructions en orde discontinu, ►implantation des constructions en odre semi-continu, ►urbanisation aérée; ►constructions périphériques, ►habitat à flanc de colline, ►maison individuelle, ►maisons jumelles) ; *g* **Bebauung [f]** (OB zu Anordnung von Gebäudeeinheiten, z. B. ►Blockbebauung, ►Einzelbebauung, Hochhausbebauung, Komplexbebauung, ►lockere Bebauung, ►offene Bauweise 2, ►Zeilenbebauung; ►Bauweise 2, ►Doppelhaus, ►Einzelhaus, ►Hangbebauung, ►Randbebauung); *syn.* Bebauungsweise [f], Überbauung [f] [CH].

building development bye-laws [npl] *leg. urb.* ►aesthetic ordinance [US].

600 building excavation [n] *constr.* (Open cutting of the ground, which is dug for foundation, basement, underground, car park, etc.); *s* **excavación [f] de cimientos** *syn.* fosa [f] de la obra; *f* **fouille [f] de bâtiment (pour sous-sol)** (Fouille réalisée en pleine masse sur toute l'emprise d'une construction) ; *g* **Baugrube [f]** (Ausschachtung für das Fundament eines Bauwerkes).

601 building frost damage [n] *constr.* (Destruction, shifting or cracking of structures by frost; ►frost heave, ►pavement frost damage); *s* **daños [mpl] causados por heladas** (Destrucción, deterioro o alteración causadas en construcciones por la acción de las heladas; ►dislocamiento causado por helada, ►levantamiento por congelación); *f* **dommages [mpl] causés par le gel** (Destruction, détérioration ou modification provoquées par le gel sur les végétaux ou les constructions ; ►foisonnement par le gel, ►gonflement lors du dégel) ; *syn.* dégâts [mpl] causés par le gel ; *g* **Frostschaden [m] (an Bauwerken)** (Zerstörungen, Beschädigungen oder Veränderungen an Bauwerken als Folge von Frosteinwirkungen; ►Frostaufbruch, ►Frosthebung).

building garden [n] [UK]**, stepped** *arch. constr.* ►terraced townhouse garden [US].

building inspector [n] [US] *adm. leg.* ∗►building code control [US].

building law(s) [n(pl)] *leg. urb.* ►zoning and building regulations.

building line [n] *leg. urb.* ►mandatory building line, ►setback line.

602 building maintenance [n] *constr. urb.* (Measures to keep structures and urban complexes in good condition and appearance); *s* **mantenimiento [m] de edificios** (Todo tipo de medidas necesarias para asegurar el buen estado y aspecto de edificios o complejos urbanos); *syn.* obras [fpl] de mantenimiento; *f* **entretien [m] immobilier** (Mesures de conservation d'un ouvrage, bâtiment ou d'un ensemble immobilier bâti) ; *syn.* maintien [m] en l'état ; *g* **Instandhaltung [f] von Gebäuden** (Alle Maßnahmen, die ein Bauwerk oder ein städtebauliches Ensemble in einem geeigneten [Soll]zustand erhalten); *syn.* Pflege [f, o. Pl.] und Unterhaltung [f, o. Pl.] von Gebäuden.

building material [n] *constr.* ►inert construction material.

Building Occupancy and Construction Authority [n] **(BOCA)** [US] *adm. leg.* ►Building Permit Office [US].

B

603 building [n] **of an open space project** *constr.*
(Process of developing an open area; e.g. a park); *s* **realización
[f] de proyecto de parque o jardín** (Proceso de construcción de
zona verde); *f* **implantation [f] d'un projet d'espaces verts** ;
syn. réalisation [f] d'un projet d'espaces verts, création [f] d'un
projet d'espaces verts ; *g* **Anlage [f] eines Freiraumes** (Erstellung z. B. einer Grünfläche); *syn.* Herstellung [f] eines Freiraumes.

604 Building Official [n] [US] *adm. leg.* (Officer or other
designated person charged with the administration and enforcement of the building code or a duly authorized representative;
UBC 1979; ►Building Permit Office [US]/Building Surveyors
Department [UK]); *s* **inspector, -a [m/f] de construcción** (En
EE.UU., funcionario/a u otra persona designada para la administración y el control del cumplimiento de las normas urbanísticas
y de construcción; ►autoridad de inspección urbanística);
f **contrôleur [m] technique** (Personne agréée pour une duré
maximale de 5 ans par le ministre chargé de la construction après
vérification de ses compétences par une commission d'agrément ;
son rôle, essentiellement préventif, est d'effectuer le contrôle des
travaux [dispositifs techniques du projet, contrôle des réglemen-
tations] et d'évaluer les risques de sinistre ; l'intervention du
contrôleur technique est obligatoire pour les établissements
recevant du public [ERP] et les immeubles de grande hauteur
[IGH] ; le contrôle technique obligatoire porte sur deux missions
de base « la solidité des ouvrages des fondations, de l'ossature,
du clos et du couvert et des éléments d'équipements, qui font
indissociablement corps avec ces ouvrages, ainsi que sur les
conditions de sécurité des personnes dans les constructions » ;
Code de la construction et de l'habitation article R. 111.39 ; il
peut avoir des missions complémentaires telles que l'isolation
thermique et acoustique, l'accessibilité pour les personnes
handicapées, le fonctionnement des installations [cf. norme NF P
03-100]) ; ►bureau de contrôle ; *syn.* technicien [m] du droit des
sols) ; *g* **Baukontrolleur/-in) [m/f]** (Vertreter der ►Baurechts-
behörde, der vor Ort die Einhaltung der baurechtlichen Vorschriften überwacht und überprüft); *syn. obs.* Baupolizist.

building permission [n] [UK] *adm. leg.* ►building permit
[US].

605 building permit [n] [US] *adm. leg.* (**In U.S.**, a written
document issued by building permit office for land development
projects that need official sanction to get a 'construction permit';
in U.K., official sanction issued by to carry out a specified
development granted by statutory planning control authority—
Building Surveyors Department; ►application for preliminary
building permit [US]/outline planning application [UK], ►build-
ing and site plan permit application [US]/detailed planning
application [UK], ►Building Permit Office [US], ►conditional
building permit); *syn.* building permission [n] [UK], construction
permit [n] [US], planning permission [n] [UK]; *s* **licencia [f] de
construcción** (Procedimiento de autorización de obras de
construcción; ►licencia de construcción con restricciones,
►licencia previa); *syn.* permiso [m] de construcción, autorización
[f] de construcción; *f* **permis [m] de construire** (**1.** Procédure
d'autorisation d'urbanisme préalable à l'exécution de travaux
ayant pour objet de contrôler et de s'assurer que les constructions
satisfont aux exigences des règles de l'urbanisme et de la
construction ; ►demande de certificat d'urbanisme, ►demande
de permis de construire, ►permis de construire permissionnel.
2. Document administratif délivré par la mairie et la DDE
[Direction Départementale de l'Equipement] qui autorise la
réalisation de travaux de construction sur un terrain [logement,
bâtiment] ou la modification de bâtiments en vérifiant leur
conformité aux règles d'urbanisme en vigueur) ; *syn.* autorisation
[f] de construire, documents [mpl] constituant le dossier du

permis de construire ; *g* **Baugenehmigung [f]** (Schriftliche
Erklärung einer ►Baurechtsbehörde, dass dem beabsichtigten
Bauvorhaben [Errichtung, Änderung — i. d. R. auch Nutzungsänderung — und Abbruch baulicher Anlagen] nach der Landesbauordnung keine öffentlich-rechtlichen Vorschriften entgegenstehen. Die **B.** wird per ►Bauantrag beantragt. Seit einigen
Jahren sehen manche Bauordnungen Freistellungen und Ausnahmen von der Genehmigungspflicht vor [z. B. Art. 63-66 der
Bayer. BO], z. T. aber Anzeigepflichten. In D. ist neben der **B.**
eine **Baufreigabe** erforderlich. Sie wird erteilt — z. B. in BW
durch den „Roten Punkt" —, wenn die in der **B.** für den Baubeginn enthaltenen Auflagen und Bedingungen erfüllt sind. Ein
Bauvorhaben ohne Genehmigung wird **Schwarzbau** genannt. In
den USA wird für genehmigungspflichtige Vorhaben im Außenbereich der Begriff *construction permit* benutzt; ►Baugeneh-
migung mit Auflagen, ►Bauvoranfrage); *syn.* Baubewilligung [f]
[A u. CH], Bauerlaubnis [f].

606 Building Permit Office [n] [US] *adm. leg.* (Designated
authority charged with the administration and enforcement of the
►uniform building code [US]/building regulations [UK] for
construction; ►building official [US]); *syn.* Building Occupancy
and Construction Authority [n] (BOCA) [US], Building
Surveyors Department [n] [UK], County building department [n]
[US], District Surveyor's Office [n] [US], Local Authority
Building Control Department [n] [UK]; *s 1* **autoridad [f] de
inspección urbanística** (Organismo estatal o municipal responsable de supervisar el cumplimiento de las leyes y regulaciones
de construcción); *syn.* inspección [f] urbanística, policía [f]
urbanística; *s 2* **comisión [f] de urbanismo** (Responsable de
otorgar licencias de construcción; ►reglamento de la construcción); *syn.* administración [f] de urbanismo; *f 1* **service [m] de
l'application du droit des sols** (**1.** *Collectif* administration
communale compétente dans le domaine de l'instruction et de la
délivrance de l'ensemble des autorisations et actes relatifs à
l'occupation ou l'utilisation des sols tels que permis de
construire, déclaration de travaux, autorisation de lotir, certificat
d'urbanisme, permis de démolir, et diverses autorisations [clô-
tures, terrains de camping, stationnement des caravanes, coupes
et abattages d'arbres], lorsqu'un PLU/POS a été approuvé dans la
commune. Dans le cas contraire les décisions dans ce domaine
sont prises au nom de l'État soit par le maire, soit par le
commissaire de la République ; en fonction de la nature du
demandeur, de l'objet de la demande, de la commune d'implantation du projet, le service chargé de l'instruction du dossier et
l'autorité délivrant l'autorisation ainsi que leur dénomination
peuvent varier, p. ex. service urbanisme règlementaire, secteur
application du droit des sols. **2.** Le technicien qui est en charge de
vérifier la conformité de la construction avec les dossiers de
permis de construire lorsque le bâtiment est terminé s'appelle
instructeur) ; *f 2* **bureau [m] de contrôle** (**F.**, organisme privé
agrée par l'État ayant pour mission dans le cadre du ►contrôle
technique de contribuer à la prévention des différents alinéas
techniques susceptibles d'être rencontrés dans la réalisation
d'ouvrages. L'activité du contrôleur technique est incompatible
avec l'exercice de toute activité de conception, d'exécution ou
d'expertise d'un ouvrage ; cf. loi n° 78-1146. **D., U.K., Es.**,
l'activité de contrôle technique est assurée par les administrations
territoriales ; ►Code de la construction et de l'habitation) ;
g **Baurechtsbehörde [f]** (Behörde, die darüber zu wachen hat,
dass bei der Errichtung, baulichen Änderung, Nutzungsänderung,
Instandhaltung oder beim Abbruch baulicher Anlagen auf einem
Grundstück die öffentlich-rechtlichen Vorschriften, z. B. die
►Bauordnung des jeweiligen Bundeslandes und die auf Grund
dieser Vorschriften erlassenen Anordnungen eingehalten werden;
►Baukontrolleur); *syn.* Bauaufsichtsbehörde [f], Baugenehmi-

gungsbehörde [f], Bauordnungsamt [n], Baurechtsamt [n], *obs.* Baupolizeiamt [n].

607 building plot [n] *urb.* (Single parcel of property specified for building development that can be identified and referenced to a recorded plat or ►zoning map [US]/Proposals Map [UK]); *syn.* lot [n] [also US], parcel [n], settlement site [n] [also US]; *s* **solar** **[m]** (Terreno construido o construible según la normativa vigente); *syn.* lote [m] de terreno, parcela [f]; *f* **terrain [m] à** **bâtir** (Un terrain peut être bâti si ses équipements existants ou à créer assurent une desserte convenable lorsque l'autorisation est délivrée ; BON 1990) ; *syn.* terrain [m] constructible ; *g* **Bau-** **grundstück [n]** (Grundstück, das nach den öffentlich-rechtlichen Vorschriften [►Bebauungsplan] mit Gebäuden bebaubar oder bebaut ist); *syn.* Bauparzelle [f], Bauplatz [m].

building process [n] *constr. eng.* ►construction (1).

building project [n] *constr. plan.* ►construction project.

building regulations [npl] [UK] *leg. urb.* ►uniform building code [US], ►zoning and building regulations.

building regulations [npl] [UK]**, relaxation of** *adm. leg.* *urb.* ►variance [US].

building regulations [npl] [ZA]**, standard** *leg. urb.* ►uniform building code [US].

608 building rehabilitation [n] *constr. urb.* (The act or process of bringing a substandard building up to legal building or housing code standards; ►center city redevelopment [US]/urban regeneration [UK], ►modernization 1, ►poor quality housing stock, ►run-down housing fit for rehabilitation); *s* **rehabilitación** **[f] de edificios** (Conjunto de medidas de construcción para mejorar la calidad de los edificios, tanto a nivel estructural como funcional, equipándolos con infraestructuras modernas y adaptándolos a las necesidades actuales de los usuarios; ►edificación antigua con mal trazado, ►edificación antigua en mal estado, ►modernización, ►rehabilitación urbana); *f 1* **restauration [f]** **immobilière** (Ensemble des mesures visant soit à mettre en valeur, remettre en état, à moderniser un immeuble en raison de son caractère historique ou architectural ; ►habitat ancien, ►habitat vétuste, ►modernisation 1, ►rénovation urbaine, ►restauration urbaine) ; *f 2* **réhabilitation [f] immobilière** (Mise aux normes minimales d'habilité de logements dans un immeuble ancien ou récent dégradé, ne comportant que des travaux de réparation, d'assainissement, d'aménagement et d'équipement de confort sans toucher à la structure même du bâtiment ; C. urb., L 313-4, 303-1) ; *g* **Gebäudesanierung [f]** (**1.** Beseitigung mangelhafter sanitärer Einrichtungen, des schlechten, unwirtschaftlichen Grundrisszuschnittes etc. und Anpassung an neue Nutzeransprüche; ►Altbebauung 1, ►Altbebauung 2, ►Modernisierung, ►Stadtsanierung. **2.** Abriss und Wiederaufbau eines oder mehrerer Gebäude); *syn.* Bausanierung [f], Umbau [m] und Verbesserung [f] von Gebäuden.

building restriction line [n] [US] *leg. urb.* ►setback line.

609 building rubble [n] *constr.* (Cast-off masonry, etc.; ►demolition material, ►reuse of building rubble); *syn.* building debris [n] [also US]; *s* **escombros [mpl] (1)** (►material de derribo, ►reutilización de material de derribo); *f* **gravats [mpl]** (Le terme regroupe tous les débris de matières minérales issus des travaux de construction, démolition, rénovation effectués sur un bâtiment ou de terrassement ; on peut distinguer les gravats en mélange et les gravats propres ; les premiers désignent un mélange de gravats avec d'autres déchets non dangereux pouvant être stockés en décharge de classe III et les seconds des déchets inertes non mélangés ou purs pouvant, dans le cadre de la valorisation, être réutilisés après broyage comme grave de recyclage ; ►matériaux de démolition, ►réutilisation de matériaux de démolition) ; *syn.* décombres [fpl] ; *g* **Bauschutt [m]** (Reststoff, der bei Abbruch-, Sanierungs-, Umbau- oder Neubaumaßnahmen anfällt, z. B. Abbruchmaterial von Gebäuden wie Mauersteine, Beton, Fliesen, Mörtelreste, Trümmerreste, Straßenaufbruch und sonstige Baustellenabfälle; durch eine sortenreine Trennung der Stoffe auf der Baustelle und durch Zerkleinerung in Brechanlagen können diese später als Recyclingbeton oder als Schotter und bei Asphaltaufbruch zu Asphaltgranulat im Straßen- und Wegebau wiederverwertet werden; ►Wiedereinbau von Abbruch-/Aufbruchmaterial; *OB* Bauabfall/Baureststoff; ►Abbruchmaterial).

610 building setback regulation [n] *leg. urb.* (Building regulation, which defines minimum distances or ►building setbacks between buildings that have to be kept free of other built objects to ensure healthy, sunny, well-lit and ventilated dwellings and workplaces); *s* **regulación [f] de la distancia legal de** **construcción** (Norma que define el espaciamiento mínimo entre edificios para garantizar viviendas y lugares de trabajo saludables, bien iluminados y aireados; ►distancia legal de construcción); *f* **règles [fpl] d'implantation des constructions les unes** **par rapport aux autres sur une même propriété** (Disposition réglementaire d'occupation et d'utilisation du sol d'une zone définie dans le règlement graphique d'un PLU/POS [article 8] précisant la distance à respecter entre les constructions implantées sur une même unité foncière [Art. R.123-9 du Code de l'Urbanisme] ; ces règles ont pour but de garantir un minimum d'ensoleillement et d'hygiène aux constructions et imposent des normes le plus souvent relatives selon la hauteur et le volume des constructions ; ►espacement minimal entre les constructions, ►espaces des constructions, ►prospect) ; *g* **Abstandsvorschrift** **[f]** (Bauordnungsrechtliche Regelung, die zur Sicherung eines gesunden Wohnens und Arbeitens in gut belichteten, besonnten und belüfteten Gebäuden definierte ►Abstandsflächen vorschreibt, die von oberirdischen Gebäuden und anderen baulichen Anlagen freigehalten werden müssen; cf. Bauordnungen der Bundesländer); *syn.* Abstandsregelung [f].

611 building setbacks [npl] *leg. urb.* (Open space between a lot line and the buildable area within which no structure can be located except as provided in the zoning ordinance; IBDD 1981; ►bulk plane [US], ►minimum building spacing, ►setback line); *syn.* minimum yard [n] [also US], required yard [n] [also US]; *s* **distancia [f] legal de construcción** (Espaciamiento mínimo entre edificios según determinaciones fijadas por reglamentos para la construcción. En D. debe existir un área no construida entre edificios en zonas residenciales cuya longitud es igual a la altura del edificio correspondiente; ►línea de retranqueo); *syn.* retiro [m] [también YZ], separación [f] mínima entre edificios; *f 1* **espacement [m] minimal entre les constructions** (F., une distance minimale réglementaire peut être imposée entre deux bâtiments ou être mesurée horizontalement entre les façades arrières des constructions se faisant vis à vis sur une même propriété ; ►limite de constructibilité, ►polygone d'implantation, ►servitude de vue) ; *f 2* **espace [m] des constructions** (Espace réglementé inconstructible entre les façades de deux bâtiments non contigus, compté par rabattage des façades à l'horizontale) ; *f 3* **prospect [m]** (*Droit de l'urbanisme* distance horizontale autorisée entre un bâtiment et le bâtiment voisin ou la limite de parcelle ou l'alignement opposé d'une voie publique. Les règle de prospect ont pour objet d'assurer un éclairement minimal et de réduire les vues entre voisins ; DUA 1996 ; ►gabarit de prospect) ; *g* **Abstandsfläche [f]** (Fläche, die zwischen gegenüberliegenden Gebäudefronten u. a. zur Sicherung einer ausreichenden Belüftung und Besonnung der Gebäude sowie zur Sicherung nötiger Freiflächen bebauungsfrei bleiben muss und die durch Umklappen der Gebäudefront in die Waagerechte ermittelt wird. In den USA werden Abstandsflächen durch die *bulk plane*, einer imaginären schiefen Ebene mit einem

B

definierten Winkel, und durch Mindestabstände *[minimum building spacing]* festgesetzt; ▶Baugrenze, ▶Mindestabstände von Gebäuden).

612 building site facilities [npl] and equipment [n] *constr.* (Structures and material for site huts, offices, storage compounds, etc. to be installed on a construction site; ▶machinery stock); *syn.* building site installations [n]; *s* **instalaciones [fpl] auxiliares de obra** (Conjunto de instalaciones y aparatos como oficina de obra, zonas de almacenaje, barracones, vallas, máquinas, etc. que se necesitan durante una obra; ▶equipo de máquinas); *f* **installations [fpl] de chantier** (Installations provisoires relatives aux matériels et matériaux nécessaires pendant la durée des travaux ; ▶parc des matériels) ; *g* **Baustellenein-richtung [f] (1)** (Gesamtheit der technischen Ausstattung und fliegenden Bauten, die für den Betrieb einer Baustelle notwendig ist, z. B. Baubüro, Unterkünfte, Baracken, Toiletten, Einzäunung, Maschinen und Geräte; ▶Maschinenpark).

building site installations [npl] *constr.* ▶building site facilities and equipment, ▶dismantlement of building site installations.

building size regulations [npl] [UK] *leg. plan.* ▶bulk regulations [US].

building spacing [n] *leg. urb.* ▶minimum building spacing.

buildings regs sign off [n] [UK] *adm. leg.* ▶occupancy permit [US].

Building Surveyors Department [n] [UK] *adm. leg.* ▶Building Permit Office [US].

613 building terrace [n] *arch.* (Paved, flat area, usually without a roof, immediately adjacent to the ground floor of a building; ▶roof terrace); *s* **terraza [f]** (Plataforma pavimentada adyacente a la planta baja de un edificio, generalmente sin tejado; ▶terraza en azotea); *f* **terrasse [f]** (Surface minérale en général non couverte, située au rez-de-chaussée d'un bâtiment ; ▶toiture-terrasse) ; *g* **Terrasse [f]** (Befestigte Plattform meist ohne Überdachung am Erdgeschoss eines Gebäudes; ▶Dachterrasse).

614 building use category [n] [US] *leg. urb.* (Kind and extent of use on plot[s] as designated in the zoning ordinance; ▶maximum bulk, ▶zoning district category); *syn.* use [n] as defined by Use Class Order [UK]; *s* **uso [m] del suelo** (segun la clasificación de ▶categoría de zonificación; ▶volumen edificable); *f* **utilisation [f] des sols pour la construction** (Nature de l'affectation dominante et dispositions d'occupation des sols fixées par zones dans les documents d'urbanisme ; ▶nature de l'affectation des sols, ▶conditions de l'occupation du sol [et de l'espace]) ; *syn.* affectation [f] des sols, occupation [f] des sols ; *g* **bauliche Nutzung [f]** (Art und Maß der Bebauung von Grundstücken gem. BauNVO; ▶Art der baulichen Nutzung, ▶Maß der baulichen Nutzung).

615 building waste dump [n] *constr. envir.* (Area used on a long-term basis for the orderly disposal of ▶building rubble, construction site waste, soil excavation, road demolition material and other inert waste; ▶rubble disposal site, ▶sanitary landfill [US] 2/landfill site [UK]); *syn.* construction waste tip [n] [UK], construction and demolition waste disposal facility [n]; *s* **vertedero [m] de escombros** (Área destinada a largo plazo para la deposición ordenada de ▶escombros, residuos de emplazamientos de obras, suelos excavados, material de demolición de carreteras y otros residuos inertes; ▶escombrera 2, ▶relleno sanitario); *f* **centre [m] de stockage de déchets inertes** (Installation nouvelle instaurée par la loi du 13 juillet 1992 dite de classe III recevant les gravats et déblais inertes ; ▶décharge contrôlée, ▶dépôt de déchets inertes, ▶gravats) ; *syn.* centre [m] de stockage de classe 3, décharge [f] de gravats ; *g* **Bauschuttdeponie [f]** (Fläche zur zeitlich unbegrenzten geordneten

Ablagerung von ▶Bauschutt, Baustellenabfällen, Bodenaushubmaterial, Straßenaufbruch, Tunnelausbruch und sonstigen Inertabfällen; ▶geordnete Deponie, ▶Schuttablageplatz).

616 building work [n] *constr.* (Tasks involved in building above-ground structures as distinguished from ▶civil engineering works and ▶landscape construction); *s* **obras [fpl] de edificación [f] (±)** (Tareas relacionadas con la construcción de edificios, en contraposición a la de infraestructuras y paisajística. En alemán, el término *«Hochbau»* —entendido como construcción por encima de la tierra— no está claramente delimitado. Se utiliza para las obras arquitectónicas construidas sobre nivel o en altura, mientras que el término *«Tiefbau»* [construcción bajo tierra o a nivel] se utiliza para las ▶trabajos de obras públicas, siendo la delimitación entre ambos prácticamente imposible; ▶construcción paisajística 1); *syn.* construcción [f] arquitectónica; *f* **travaux [mpl] de bâtiment** (Action de construire un bâtiment par comparaison avec les travaux publics [▶travaux de voirie et de réseaux divers] et les travaux d'espaces verts [▶création et entretien des parcs et jardins]) ; *g* **Hochbauarbeiten [fpl]** (Arbeiten zur Herstellung eines Gebäudes — im Vergleich zu ▶Tiefbauarbeiten und Arbeiten des ▶Garten-, Landschafts- und Sportplatzbaues).

building works [npl] [UK], contractor's supervision of the *constr.* ▶contractor's supervision of the project [US].

built area [n], peripheral *urb.* ▶peripheral building development.

617 built portion [n] of a plot/lot *leg. urb.* (Ground surface portion of a plot/lot that is covered by buildings or structures); *syn.* footprint [n] of a building, ground coverage [n], lot coverage [n] [also US]; *s* **superficie [f] construída** (de un solar); *syn.* superficie [f] edificada (2); *f* **surface [f] construite d'un terrain** *syn.* superficie [f] bâtie d'une parcelle, superficie [f] bâtie d'un terrain ; *g* **bebaute Grundstücksfläche [f]** (Der Teil eines Grundstücks, auf dem ein Gebäude oder eine bauliche Anlage steht); *syn.* bebaute Fläche [f] eines Grundstückes, überbaute Grundstücksfläche [f].

618 built-up area [n] (1) *plan. urb.* (An area in which most of the land is developed; ▶building area, ▶populated area); *syn.* developed land [n] [also UK]; *s* **zona [f] edificada (1)** (A pequeña escala, área o zona construida; ▶superficie edificada 1, ▶zona edificada 2); *f* **surface [f] bâtie** (Espace occupé par plusieurs bâtiments ; *terme relatif à un seul bâtiment* ▶emprise de la construction ; ▶zone urbanisée) ; *syn.* surface [f] édifiée (BON 1990, 352), zone [f] bâtie ; *g* **bebaute Fläche [f]** (Fläche oder Gebiet, das mit Gebäuden überstellt ist; ▶besiedelter Bereich; *auf ein Gebäude bezogen* ▶Gebäudefläche); *syn.* überbaute Fläche [f] [CH].

619 built-up area [n] (2) *leg. urb.* (Contiguously developed area as opposed to ▶open land [US]/greenfield site [UK]; ▶undeveloped area, ▶core area 1); *s* **núcleo [f] de población agrupada** (Zona urbanizada dentro del perímetro urbano delimitable funcionalmente. Término utilizado en el contexto de diferenciación entre el ▶perímetro edificado y zonas no edificables [▶suelo no urbanizable y ▶suelo rústico] de una ciudad o municipio; ▶área no edificada); *f* **secteur [m] aggloméré (1. F.,** terme de l'urbanisme définissant à l'intérieur d'une zone urbaine [U] la zone classée dans le cadre du zonage du PLU comme pouvant normalement être bâtie, parce que déjà desservie par des équipements publics et infrastructures urbaines de base; elle est généralement divisée en « secteurs », affectés d'un coefficient d'occupation des sols [COS] et de règles particulières d'implantation, de hauteur, d'aspect extérieurs des bâtiments et d'aménagement de leurs abords ; cf. C. urb., L. 123-1 ; la limite des secteurs agglomérés constitue le périmètre d'agglomération, situé à la limite de la ►zone non urbanisable ; ►zone urbaine 1 ; **2. D.**, zone urbanisée caractérisée par une délimitation fonctionnelle par rapport à l'►zone non urbanisable et qui malgré la présence de dents creuses suscite l'impression du bâti continu et d'un tissu urbain à forte structure organique) ; **g im Zusammenhang bebauter Ortsteil** [m] (**1.** Funktionell abgrenzbarer Siedlungsbereich im Gegensatz zum ►Außenbereich; cf. § 34 BauGB; ►unbebaute Fläche, ►Innenbereich. **2.** In F. ist die Grenze des **i. Z. b.en O.s** der *périmètre d'agglomération*, der die ►*zone non urbanisable* abgrenzt).

620 bulb [n] *bot. hort. plant.* (Underground storage organ with a short, flattened stem and fleshy layers, scales or thickened leaf bases, where nutrients are stored. When seeds ripen, the bulb dies, but it has propagated itself by vegetative reproduction of one or more new bulbs—"baby bulbs"; ►bulbil. Onion, tulip and daffodil bulb have layers; lily bulbs have scales; ►flowering bulb, ►top-size bulb); *s* **bulbo** [m] (Yema subterránea con los catáfilos o las bases foliares convertidos en órganos reservantes, y la porción axial reducida y, generalmente, disciforme, el llamado platillo del bulbo. Existen diferentes formas: **b.** tunicados [cebolla], reticulares, escamosos [azucena]; el **bulbo sólido** o **macizo** [azafrán] no conserva del verdadero **b.** más que su forma redondeada u ovoide, ya que es el tallo o el eje el que actúa como reservante; cf. DB 1985; ►bulbo floreciente); *f* **bulbe** [m] (Organe souterrain d'accumulation de réserves, dont la tige très courte, de forme aplatie supporte un ou plusieurs bourgeons entouré[s] de pièces foliaires [tunique externe protectrice, écailles charnues formant la tunique interne contenant les réserves nutritives] ; avec la maturation des graines le bulbe meurt la plante se renouvelant pendant la période de végétation par multiplication végétative ; ►bulbille. En fonction de la morphologie des feuilles charnues et de l'importance de la tige on distingue les bulbes caulinaires [bulbe solide du Crocus, bulbe rhizomateux du glaïeul], et les bulbes foliaires [bulbe tuniqué de l'oignon, de la tulipe et autres liliacées, bulbe écailleux des Lis et des Jacinthes] ; *dénomination commune* oignon ; ►bulbe à fleur) ; *g* **Zwiebel** [f] (Zu einem Speicher- und Überdauerungsorgan umgewandelter unterirdischer Spross, dessen Sprossachse kegel- oder scheibenförmig als sog. ,Zwiebelkuchen' abgeflacht ist und dessen fleischige Niederblätter [Zwiebelschuppen oder Zwiebelschalen] mit Nährstoffen angereichert sind. Mit der Samenreife stirbt die **Z.** ab, erneuert sich aber zuvor in der Vegetationsperiode durch eine oder mehrere Tochterzwiebeln. *Je nach Einteilungskriterium gibt es folgende UBe* ►Brutzwiebel, Schalenzwiebel [z. B. Tulpe], Schuppenzwiebel [z. B. Madonnenlilie] oder nach dem Verwendungskriterium ►Blumenzwiebel, Küchenzwiebel, Speisezwiebel); *syn.* Bulbus [m].

bulb geophyte [n] *bot. hort.* ►bulbous plant.

621 bulbil [n] *bot.* (Small ►bulb usually formed in a leaf axil that separates from the parent plant and functions in vegetative reproduction; ►flowering bulb); *syn.* bulblet [n]; *s* **bulbillo** [m] (Bulbo de pequeñas dimensiones que nace en la axila de un catáfilo del bulbo padre; DB 1985; ►bulbo, ►bulbo floreciente); *f* **bulbille** [m] (Bulbe axillaire, formé sur le plateau d'oignonets permettant à la plante de se reproduire par multiplication végétative ; ►bulbe, ►bulbe à fleur) ; *syn.* caïeu [m], cayeu [m] ; *g* **Brutzwiebel** [f] (Sich vom Zwiebelkuchen der Mutterpflanze loslösendes, kleines Zwiebelchen, das sich bewurzelt und zum neuen Individuum heranwächst, z. B. Zwiebelchen bei Feuerlilie *[Lilium bulbíferum]*; ►Blumenzwiebel, ►Zwiebel); *syn.* Bulbille [f], Knospenzwiebel [f].

bulblet [n] *bot.* ►bulbil.

622 bulbous plant [n] *bot. hort.* (Perennial plant with underground storage organs for survival during severe climatic

conditions, such as drought or winter cold; ►flowering bulb, ►stem tuber, ►tuberous-rooted plant); *syn.* bulb geophyte [n]; *s* **planta** [f] **de cebolla** (Planta que sobrevive gracias a sus órganos reservantes subterráneos; ►bulbo floreciente, ►planta tuberosa, ►tubérculo caulinar); *syn.* planta [f] cebolluda, planta [f] de bulbo, bulbosa [f]; *f* **plante** [f] **à bulbe** (Plante pluriannuelle qui persiste aux mauvaises conditions climatiques [sécheresse, froid] grâce à un organe souterrain d'accumulation de réserves [bulbe, rhizome ou tubercule] ; ►bulbe à fleur, ►plante à tubercule, ►tubercule caulinaire) ; *syn.* plante [f] à oignon, plante [f] bulbeuse ; *g* **Zwiebelgewächs** [n] (Mehrjährige Pflanze mit unterirdischen knollenförmigen Speicherorganen aus konzentrisch angeordneten, dicken, fleischigen, meist weißen Blättern, mit dünner, trockener, oft schuppiger Schale umgeben, zur Überbrückung von ungünstigen Witterungsbedingungen wie Trockenheit und winterliche Kälte; ►Blumenzwiebel, ►Knollengewächs, ►Sprossknolle); *syn.* Zwiebelgeophyt [m], Zwiebelpflanze [f].

bulk [n] [US] *leg. urb.* ►maximum bulk [US].

623 bulk density [n] *constr.* (Mass of dry soil per unit bulk volume, determined in the laboratory before drying to constant weight at 105 °C, for measurement of soil compaction; ►bulking, ►soil compaction); *s* **compacidad** [f] **(del suelo)** (Medida del grado de compactación de un suelo; se calcula en el laboratorio según DIN 1054; antes del secado hasta un peso constante a 105°C; ►compactación del suelo, ►estado descompactado del suelo); *f* **densité** [f] **en place (d'un sol)** (Masse d'un sol en place par unité de volume ; essais de sol en place en fond de fouille ou à partir de la surface du sol ; ►compactage du sol, ►foisonnement 1, ►tassement du sol) ; *g* **Lagerungsdichte** [f] (Maßstab für den Verdichtungszustand eines Bodens; Berechnungen durch Laboruntersuche nach DIN 1054; ►Bodenauflockerung, ►Bodenverdichtung).

bulk grade [n] [US] *constr.* ►rough grade.

bulk grading [n] [US] *constr.* ►rough grading [US].

624 bulking [n] *constr.* (Increase in volume when rock or undisturbed soil is broken up and excavated; may reach 50%, but averages between 10 and 15%; TANDY 1975); *syn.* swelling [n]; *s* **estado** [m] **descompactado del suelo** (Incremento del volumen de un suelo despues de operación de relleno o acopio); *f 1* **foisonnement** [m] (1) *constr.* (Augmentation de volume de la terre végétale remuée lors de sa reprise et de sa mise en place. Il est estimé selon la nature du sol entre 15 % et 25 %. La terre mise en place sans compactage artificiel se tasse avec le temps sans toutefois retrouver sa densité initiale ; c'est le **foisonnement après tassement** ou foisonnement résiduel) ; *f 2* **foisonnement** [m] (2) *geo.* (Augmentation de volume d'une roche ou d'un sol par hydratation, par absorption d'eau ou par modification de la texture des agrégats et des colloïdes dans un sol ou dans une roche argileuse ; DG 1984) ; *g* **Bodenauflockerung** [f] (1) (Zustand des durch Lösen gelockerten Bodens, der je nach Korngefüge beim Transport und beim Wiedereinbau einen größeren Raum einnimmt [**anfängliche Lockerung**]. Ohne künstliche Verdichtung setzt sich der Boden, ohne den ursprünglichen Lagerungsgrad [Lagerungsdichte] wieder zu erreichen [**bleibende Lockerung**]); *syn.* Bodenlockerung [f].

625 bulk plane [n] [US] *leg. urb.* (An imaginary inclined plane, rising over a lot, drawn at a specified angle from the vertical, the bottom side of which either coincides with the lot line[s] or yard line[s] of the lot, or directly above them. This imaginary plane, together with other bulk regulations and lot size requirements, delineates the maximum bulk or height of a building which may be constructed on the lot. The angle of **b. p.** is established by local regulations; cf. UDH 1986, 24; ►building set-

B

backs, ►cubic content ratio, ►minimum building spacing); *syn.* light plane [n] [US], setback plane [n] [US], sky exposure plane [n] [US]; *s* **ángulo [m] de retranqueo** (≠) (En EE.UU., método utilizado para definir la ►distancia legal de construcción. Un plano inclinado *[bulk plane]* en un ángulo determinado desde la vertical, cuyo extremo se encuentra sobre el límite del solar o el límite de construcción. Este plano imaginario, junto con otras regulaciones del ►volumen edificable y del tamaño de los solares [►retranqueo legal entre edificios], define el volumen o la altura máximas del edificio que se prevé construir en el solar. El **á. de r.** es determinado por reglamentos locales); *f 1* **angle [m] de servitude de vue** (±) (U.S., angle maximum déterminé entre le plan horizontal et la partie supérieure d'un bâtiment fixant l'implantation de bâtiments de propriété privées contiguës ou entre les bâtiments d'un lotissement faisant vis à vis ; *f 2* **servitude [f] de vue** (F., recul que toute vue [vue droite : 1,90 m entre mur et la propriété voisine ; vue oblique : 0,6 m depuis l'angle de la fenêtre ou le rebord du balcon jusqu'au point le plus proche de la limite de la propriété voisine] doit respecter par rapport au terrain du voisin dès lors qu'elle s'exerce a partir d'une fenêtre, d'un balcon ou d'un remblai. **2.** Le règlement du PLU ou du POS peut fixer des règles plus contraignantes : la servitude de vue est l'espace minimal règlementé fixant l'implantation des constructions multiples les unes par rapport aux autres sur un même terrain — art. R. 111-16 — et déduit d'un angle maximum à l'appui des baies éclairant les pièces principales déterminé entre le plan horizontal et la partie d'un immeuble faisant vis à vis ; ►coefficient d'utilisation du sol, ►espacement minimal entre les constructions, ►gabarit de prospect, ►prospect) ; *g* **Winkel [m] zur Berechnung der Abstandsfläche** (≠) (In den USA wird die **A.** u. a. durch eine angenommene schiefe Ebene *[bulk plane]* über einem Baugrundstück, die in einem definierten Winkel von der Senkrechten über der Grundstücksgrenze in das Grundstück hineinreicht, bestimmt. Diese imaginäre schräge Ebene, über die kein Teil eines Gebäude hinausragen darf, bestimmt mit anderen Vorschriften über ►Mindestabstände von Gebäuden *[►minimum building spacing]*, Grundstücksgrößen und das Maß der baulichen Nutzung *[►maximum bulk]* die Ausmaße eines Gebäudes. Die Berechnungsart dieser ►Abstandsfläche 1 wird durch örtliche Vorschriften geregelt; in F. wird der Winkel zur Abstandsflächenberechnung zwischen zwei Gebäuden bis zu den obersten Wohn-/Esszimmerfenstern *[pièces principales]* eines Gebäudes berechnet; ►Baumassenzahl).

626 bulk regulations [npl] [US] *leg. plan.* (Standards controlling density, intensity and location of structures which are expressed by ►floor area ratio [FAR] [US]/floor space index [UK], ►building coverage [US]/plot ratio [UK], number of stories [US]/number of storeys [UK], building height, or other dimensional standards. **In U.S.**, where the Land Use Intensity [LUI] rating approach to zoning is not applicable, **b. r.** prescribe dimensional regulations, such as minimum lot size, height, maximum bulk, front-, rear-, sideyard setbacks, etc. Such regulations are almost universally expressed in zoning regulations, with occasionally some minor overlap with building codes and rarely with subdivision regulation. For all practical purposes, there is no difference between setback, and bulk regulations to be found in the table of district regulations and the limitations expressed in Land Use Intensity [LUI] charts; ►maximum bulk); *syn.* building size regulations [npl] [UK]; *s* **regulación [f] del volumen edificable** (Prescripción urbanística que determina el volumen máximo de edificación a través de diferentes coeficientes como el ►coeficiente de edificación, ►coeficiente de ocupación del suelo, ►número de plantas completas, ►volumen edificable, altura de edificios, etc.); *f* **règles [fpl] relatives à l'implantation et les dimensions des constructions** (**F.**, ces règles sont préci-

sées dans le règlement du PLU ; cf. Règles générales d'urbanisme définissant la localisation et la desserte des constructions [Articles R111-2 à R111-15] l'implantation et le volume des constructions [Articles R111-16 à R111-20] ; l'aspect des constructions [Articles R111-21 à R111-24] ; prescription définissant les règles d'urbanisme ou d'architecture applicables aux constructions comprises dans une zone déterminée d'un telles que le ►coefficient d'emprise au sol, le ►coefficient d'occupation des sols, la hauteur des bâtiments, le ►nombre des niveaux de la construction, le ►plafond légal de densité, etc. — contenu des plans locaux d'urbanisme [Articles R123-1 à R123-14] ; **D.**, prescription définissant les règles d'urbanisme ou d'architecture applicables aux constructions comprises dans une zone déterminée d'un « PLU/POS allemand » telles que le ►coefficient d'occupation des sols, le ►coefficient d'emprise au sol, le ►nombre des niveaux de la construction, la hauteur des bâtiments, etc. ; ►conditions de l'occupation du sol [et de l'espace]) ; *g* **Vorschriften [fpl] zur Bestimmung des Maßes der baulichen Nutzung** (Planungsrechtliche Regelungen, nach denen gemäß §§ 5 [2] 1 und 9 [1] 1 BauGB in Verbindung mit den §§ 16 und 17 BauNVO in Flächennutzungsplänen und Bebauungsplänen der Umfang der baulichen Nutzung [►Maß der baulichen Nutzung] als Verhältnis von überbauter Fläche [GRZ], Geschossfläche [GFZ] oder Baumasse [BMZ] zur Fläche des Baugrundstückes festgelegt werden kann sowie Vorgaben zur Gebäudehöhe gemacht werden können; ►Baumassenzahl, ►Geschossflächenzahl, ►Grundflächenzahl, ►Zahl der Vollgeschosse).

bulk trash [n] [US] *envir.* ►bulky waste.

627 bulky waste [n] *envir.* (Large discarded household items, such as furniture, electric appliances, large car parts); *syn.* bulk trash [n] [US]); *s* **basura [f] voluminosa** (Muebles y enseres domésticos que se quieren tirar, pero que no caben en los contenedores de basura, por lo que en muchas ciudades se recogen en días especiales o según la demanda de los usuarios); *syn.* residuos [mpl] molestos; *f* **encombrants [mpl]** (Déchets volumineux) ; *syn.* déchets [mpl] encombrants ; *g* **Sperrmüll [m, o. Pl.]** (Abfall aus Haushalten, der auf Grund seiner Abmessungen resp. seiner Sperrigkeit als großvolumiger Hausmüll nicht in die Mülleimer passt und in vielen Gemeinden durch Sonderaktionen des Stadtreinigungsamtes entsorgt wird).

628 bulrush swamp [n] [US] *phyt.* (**1.** ►Reed swamp, according to location, predominantly composed of river bulrushes, saltmarsh bulrushes and other rushes [*Scirpus* species]. **2.** Location of reed swamp plant community); *syn.* bulrush wetland [n] [US], club-rush swamp [n] [UK]; *s* **juncar [m] de scirpus** (**1.** Comunidad vegetal de juncos de la familia *Scirpetum.* **2.** Superficie cubierta de scirpus); *f 1* **scirpaie [f]** (Scirpaie à *Scirpus lacustris [Scirpetum lacustris]* ou scirpaie à *Scirpus maritimi [Scirpetum maritimi]* : végétation dominée par espèces de Scirpes [Jonc des tonneliers, Triangle] occupant la partie interne d'une ►roselière immergée en permanence) ; *f 2* **marais [m] à Scirpes** (*Localisation du peuplement*, étendue d'eau stagnante généralement peu profonde et envahie par les Scirpes) ; *syn.* prairie [f] à Scirpes ; *g* **Simsenröhricht [n]** (**1.** Pflanzengesellschaft des Teichbinsenröhrichts *[Scirpetum lacustris]* oder Meerbinsenröhrichts *[Scirpetum maritimi]*, das je nach Standort vorwiegend aus Simsen, Teichsimsen oder Strandsimsen *[Scirpus-* und *Schoenoplectus*-Arten] besteht. **2.** Mit **S.** bestandene Fläche; *OB* ►Röhricht).

bulrush wetland [n] [US] *phyt.* ►bulrush swamp [US].

bunch [n] [US] *phyt. plant. hort.* ►tuft (1).

bunched root system [n] *bot.* ►fascicular root system.

bunch plant [n] [US] *bot. hort.* ►tussock plant.

bund [n] [UK] (1) *constr.* ►earth mound.

B

629 bund [n] (2) *eng. wat'man.* (**1.** *Anglo-Indian term of persian origin* artificial embankment used especially in India to control the flow of water as on a river or irrigated land; WEB 1993. **2.** In Anglo-Chinese ports an embanked quay or thoroughfare along the sea); *s* **dique** [m] (1) (**1.** En inglés término anglo-hindú de origen persa para terraplén artificial utilizado especialmente en la India para controlar el flujo de agua en un río o en campos de irrigación. **2.** En puertos del Lejano Oriente, muelle o vía pública protegida por terraplén); *f* **digue** [f] (**1. En Inde**, endiguement artificiel en vue de la régulation d'un cours d'eau pour l'irrigation des champs. **2. En Extrême-Orient**, talus protégeant un port ou une promenade en bord de mer) ; *g* **Deich** [m] (1) (**1.** In Indien eine künstliche Eindeichung zur Regulierung eines Flusslaufes oder des Wasserflusses auf zu bewässernden Feldern. **2.** Im Fernen Osten ist *bund* eine durch einen Damm geschützte Hafenstraße oder eine Promenade am Meer); *syn.* Damm [m] (1).

630 bund [n] (3) *constr.* (Elongated mound of excavated material temporarily placed adjacent to a trench; *context* trench and bund); *syn.* berm [n]; *s* **montón** [m] **lateral (de material de excavación)** (Montón alargado de tierra situado temporalmente a lo largo de una zanja); *f* **déblai** [m] **de fouille en tranchée, terres jetées sur le côté en cordon** (*Terme spécifique* déblai de fouille en tranchée, terres jetées sur un seul côté en cordon) ; *g* **seitlich gelagerter Grabenaushub** [m].

bund [n]**, noise** [UK] *constr. envir.* ▶noise attenuation mound.

bunding [n] **of a body of water** [UK] *eng. wat'man.* ▶impounding of a watercourse [US].

bundle [n] **of wattles** [US] *constr.* ▶fascine.

bunker [n] [UK] *constr. recr.* ▶sand trap [US].

631 burial area [n] *adm.* (Interment section of a cemetery); *syn.* burial ground [n]; *s* **área** [f] **de tumbas** (Sección de un cementerio con muchas tumbas individuales); *f* **section** [f] **de tombes** (±) (Section d'un cimetière occupée par plusieurs alignements de tombes) ; *g* **Grabfeld** [n] (Fläche mit vielen Einzelgrabstätten).

burial ground [n] *adm.* ▶burial area.

burial marker [n] [US] *conserv'hist. landsc.* ▶burial monument.

632 burial monument [n] *conserv'hist. landsc.* (Large imposing structure on a burial site); *syn.* burial marker [n] [also US]; *s* **monumento** [m] **funerario** (Panteón monumental, monumento o gran lápida conmemorativa dedicada a una persona ilustre); *syn.* tumba [f] conmemorativa; *f* **monument** [m] **funéraire** (*Terme générique* monument destiné à rappeler le souvenir des morts ; on distingue les monuments recouvrant les tombes en pleine terre et ceux recouvrant les tombes avec caveau ; ils peuvent être des monuments simples, de dimension classiques comme les ▶dalles funéraires, les ▶pierres tombales, les ▶plaques tombales, ou être des édicules ayant la taille d'un véritable édifice grandiose et somptueux, construit pour recevoir une ou plusieurs tombes, richement orné de sculptures et quelquefois de mosaïques, de pilastres, comme les chapelles funéraires ou les mausolées ; ▶stèle funéraire) ; *syn.* monument [m] sépulcral ; *g* **Grabdenkmal** [n] (...mäler [pl], in A. nur so; auch ...male [pl]; Bauwerk, Monument oder großer Gedenkstein zu Ehren eines Toten; ▶Grabstele); *syn.* Grabmal [n].

burial place [n] *adm.* ▶burial site.

burial plot [n] *adm.* ▶burial site.

633 burial site [n] *adm.* (Designated plot for an interment in a cemetery; ▶bi-level grave [US]/two storey grave [UK], ▶coffin burial site, ▶family burial site, ▶memorial grave, ▶prominent urn burial site, ▶single row grave); *syn.* burial place [n], burial plot [n], grave site [n]; *s* **tumba** [f] (*Términos específicos* ▶pan-

teón familiar, ▶tumba conmemorativa, ▶tumba commemorativa para urnas, ▶tumba de dos niveles, ▶tumba de tierra, ▶tumba individual en hilera); *syn.* sepultura [f]; *f* **sépulture** [f] (Emplacement funéraire représentant la concession de terrain où est enseveli un défunt et s'il y a lieu, le monument qui y est posé ou la dalle qui le recouvre ; *termes spécifiques* ▶caveau de famille, ▶columbarium, ▶jardin d'urnes, ▶jardin du souvenir, ▶tombe, ▶tombeau commémoratif, ▶tombe à la ligne, ▶tombe familiale) ; *g* **Grabstätte** [f] (Ein für Bestattungen oder Beisetzungen vorgesehener, genau bestimmter Teil eines Friedhofsgrundstückes mit dem darunter liegenden Erdreich; cf. GAE 2000; *UBe* ▶Ehrengrabstätte, ▶Erdgrab, ▶Reihengrab, ▶Sondergrabstätte, ▶Stockwerksgrab, ▶Urnengrabstätte); *syn.* Grab [n], Grabstelle [f].

634 burial stele [n] *adm. conserv'hist.* (Greek, *stēle* >grave pillar<; monolith of modest size, usually with one face only decorated with cut-away carving or low relief sculpture; in antiquity, a free-standing column with an embossed inscription. Marking a tomb, a **b. s.** bears the name of the deceased person. A relief may also depict the family or scenes from the life of the dead; ▶burial monument); *s* **estela** [f] **funeraria** (Originariamente, pequeño monumento, generalmente monolítico, erigido en conmemoración o recuerdo de algo, decorado en general sólo por un lado, sea con bajorrelieve o con inscripciones talladas; ▶monumento funerario); *f* **stèle** [f] **funéraire** (Monument monolithe, en forme de pierre dressée, ayant pour fonction de conserver, par l'image et le texte, le souvenir de personnes ; ▶monument funéraire) ; *g* **Grabstele** [f] (Griech. *stēle* >Grabsäule<; seit der griechischen Antike ein freistehender, mit Relief oder Inschrift versehener Pfeiler; als Grabmal trägt die **G.** den Namen des/der Toten und im Falle von Reliefs können auch die Familie oder Szenen aus dem Leben der Verstorbenen dargestellt sein; ▶Grabdenkmal).

burial vault [n] *adm.* ▶tomb.

635 buried cultural monument [n] *conserv'hist. land'man.* (Archeological buildings and artifacts from historic or prehistoric human activity; ▶cultural monument); *syn.* archaeotope [n], soil-covered cultural monument [n]; *s* **restos** [mpl] **arqueológicos cubiertos** (Elementos geológicos y paleontológicos relacionados con la prehistoria del hombre y sus orígenes que se encuentran en el subsuelo; ▶monumento); *f* **terrain** [m] **qui renferme des stations ou gisements préhistoriques** (Immeuble nu pouvant intéresser la préhistoire, l'histoire, l'art ou l'archéologie, présentant un intérêt public et classé en totalité ou partie comme ▶monument historique ; cf. lois du 31 décembre 1913 et du 25 février 1943) ; *syn.* gisement [m] préhistorique, station [f] préhistorique, vestige [m] archéologique ; *g* **Bodendenkmal** [n] (...mäler [pl], in A nur so, sonst auch ...male [pl]; erdbedecktes ▶Kulturdenkmal aus ur- oder frühgeschichtlicher Zeit, das sichtbare Überreste menschlicher Tätigkeit, keine Kunst- oder Baudenkmäler, dargestellt; Regelungen des Umgangs mit **B.**en enthalten die Denkmalschutzgesetze der Länder und Gemeinden); *syn.* archäologisches Denkmal [n], Archaeotop [m] (N+L 2000, 284).

burlap [n] [US] *constr. arb.* ▶wrapping with burlap.

burlapped [loc] *hort.* ▶balled and burlapped.

636 burlapped root ball [n] *hort.* (Round bundle of soil and undisturbed plant roots bound up in burlap and used in transplanting trees and shrubs; ▶root-balled plant); *syn.* hessianwrapped root ball [n] [also UK]; *s* **cepellón** [m] **sujeto con yute** (▶planta de cepellón); *syn.* cepellón [m] con tela/con saco; *f* **motte** [f] **livrée en tontine de jute** (Motte d'un végétal destiné à la plantation entourée par un filet en jute ; ▶plante en motte) ; *syn.* motte [f] en filet ; *g* **Juteballen** [m] (Ballen eines zu

B

verpflanzenden Gehölzes, der mit Sackleinen umwickelt ist; das Gewebe wird aus der tropischen Rundkapseljute *[Corchorus capsularis]* oder Langkapseljute *[Corchorus olitorius]* hergestellt; ►Ballenpflanze); *syn.* Ballen [m] mit Ballentuch.

637 burlapping [n] *hort.* (Wrapping of dug rootballs in jute mesh/hessian fabric; ►burlapped root ball); *syn.* hessian wrapping [n]; *s* **embalaje [m] de yute** (►cepellón sujeto con yute); *f* **emballage [m] en filet de jute** (En pépinière, pratique courante de traitement des mottes des végétaux sortis de terre en vue de la plantation directe ; ►motte livrée en tontine de jute ; *syn.* emballage [m] en tontine de jute) ; *g* **Juteballierung [f]** (In Baumschulen übliches Einwickeln von ausgegrabenen Erdballen mit Jutegewebe; ►Juteballen).

burn [n]**, leaf** *bot. hort.* ►leaf necrosis.

638 burn clearing [n] *agr.* (Felling and subsequent burning of trees and shrubs; mainly done for ►shifting cultivation and ►land rotation, may be also applied in the context of clear felling to get permanent crop land); *s* **tala [f] a fuego** (Técnica de de[s]forestación que consiste en cortar los árboles y quemarlos a continuación; se utiliza sobre todo como forma de ganancia de tierras para ►agricultura itinerante y ►cultivo en rotación); *syn.* roturación [f] mediante el fuego, roza [f], roza-tumba-quema [f] [MEX], roa [m] [BR]; *f* **brûlis [m] de défrichement** (Technique utilisée dans l'►agriculture itinérante sur brûlis consistant au défrichement et à l'incendie de la végétation suivis d'un travail sommaire du sol. Plusieurs années de récolte alternent avec une période plus ou moins longue de jachère ; ►assolement) ; *syn.* défrichement [m] par écobuage, masole [f] [ZRE], tavy [m] [RM] ; *g* **Brandrodung [f]** (Rodung durch Fällen und anschließendes Abbrennen der Bäume und Sträucher, wobei die Wurzelstöcke vielfach im Boden verbleiben; wird hauptsächlich im Rahmen des ►Wanderfeldbaues und der ►Landwechselwirtschaft angewandt, kann aber auch im Zusammenhang mit einer Neurodung zur Gewinnung von Dauerfeldland betrieben werden; MEL 1972, Bd. 4).

burned forest area [n] *for.* ►forest burn.

bush-hammered [pp/adj] *constr.* *►tooling of stone [US]/dressing of stone [UK], #5.

bush rose [n] [US] *hort. plant.* ►shrub rose.

business [n] *agr.* ►part-time agricultural business.

business [n]**, tourist** *recr. trans.* ►tourism (1).

business district [n] *leg. urb.* ►central business district.

business park [n] [UK] *plan. urb.* ►industrial park [US]/industrial estate [UK].

business space vacancy [n] *sociol. urb.* *►vacancy.

bus lay-by [n] [UK] *trans.* ►bus pullout [US].

bus pull-off strip [n] [US] *trans.* ►bus pullout [US]/bus lay-by [UK].

639 bus pullout [n] [US] *trans.* (Stopping or waiting space for busses adjacent to traffic lanes; ►pullout [US]/lay-by [UK], ►public transit shelter [US]/public transport shelter [UK]); *syn.* bus lay-by [n] [UK], bus pull-off strip [n] [US]; *s* **apartadero [m] de autobús** (Área al borde de la calzada reservada para parada de autobuses o trolebuses; ►apartadero, ►garita de espera, ►refugio); *f* **créneau [m] d'arrêt de bus** (Point d'arrêt de bus en bordure de la chaussée ; ►abri, ►abribus, ►créneau d'arrêt) ; *g* **Bushaltebucht [f]** (Verkehrsfläche neben der Fahrbahn zum Halten von Omnibussen oder Obussen; ►Haltebucht, ►Wartehalle).

butt [n]**, saber** *arb.* ►basal sweep.

640 butt joint [n] *arch. constr.* (Joint between two construction units, where the contact surfaces are tight-jointed; ►butt-jointed,

►coursing joint, ►heading joint); *s* **junta [f] prensada** (En ►mampostería cuando las ►juntas de asiento y las ►juntas a tope se hacen lo más estrechas posible; ►con juntas prensadas); *f* **joint [m] serré** (Intervalle très étroit dans la façon des joints filants et ►joints montants dans une ►maçonnerie ; ►à joints serrés, ►joint délit) ; *g* **Pressfuge [f]** (Enger Zusammenschluss auf der ganzen Länge von ►Lager- und ►Stoßfugen beim ►Mauerwerk; **P.n** sind in Pflastersteinflächen nicht zugelassen; ►fugeneng).

641 butt-jointed [adj/adv] *arch. constr.* (Descriptive term for a close connection of two structural members); *syn.* tight-jointed [adj/adv], tight-butted [adj/adv], close-jointed [adj/adv]; *s* **con juntas prensadas [loc]** (*Contexto* tender [vb] con juntas prensadas); *syn.* con juntas apretadas [loc]; *f* **à joints serrés [loc]** (*Contexte* pose à joints serrés) ; *g* **fugeneng [adv]** (*Kontext* fugeneng/knirsch/press verlegen [vb]); *syn.* knirsch [adv], press [adv].

642 butt [n] **of a tree** *arb.* (Lowest part of a tree trunk where it enters the ground; ►root collar; *opp.* ►stem apex); *s* **pie [m] del tronco** (La base del fustal de un árbol; ►cuello de la raíz; *opp.* ►ápice del tronco); *f* **base [f] du tronc** (La base du fût d'un arbre ; ►collet [racinaire] ; *opp.* ►apex du fût) ; *syn.* collet [m] du tronc, patte [f] d'arbre ; *g* **Stammfuß [m]** (Unterer Teil eines Stammes im Bereich der Wurzelanläufe; ►Wurzelhals; *opp.* ►Stammkopf).

buttress flare [n] *arb. bot.* ►root buttress.

bye-law [n] [UK&CDN]**, city** *adm. leg. urb.* ►city ordinance [US].

bye-laws [npl]**, building development** *leg. urb.* ►aesthetic ordinance [US].

643 by mutual agreement [loc] *adm. leg.* (Reaching reciprocal accord; *context* acting in agreement with; to reach [an] agreement; ►position statement [US]/information notice [UK]); *syn.* in agreement with, in consent with; *s* **por mutuo acuerdo [loc]** (►consultar a [alguien]); *syn.* ponerse de acuerdo con (alguien) [loc]; *f* **en accord avec [loc]** (d'un accord mutuel, unanime ; ►communication pour avis) ; *g* **im Einvernehmen mit [loc]** (*Juristen- und Verwaltungssprache* im gegenseitigen Einverständnis, in Übereinstimmung miteinander; das **E.** ist von einem anderen Gesetzgebungsorgan resp. von einer anderen Behörde [Dienststelle] vor einem Gesetzes- oder Verwaltungsakt durch die federführende Behörde herbeizuführen; unterbleibt die nach einer Rechtsvorschrift erforderliche Mitwirkung einer anderen Behörde, wurde das Einvernehmen also nicht hergestellt, so ist der Verwaltungsakt deswegen nicht nichtig, aber fehlerhaft; cf. § 44 III Nr. 4 VwVerfG; *Kontext* einvernehmlich handeln; ein **E.** erzielen/herstellen; einvernehmlich [adj/adv], einmütig [adj/adv]; nicht zu verwechseln mit ►im Benehmen mit).

644 bypass [n] [US]/**by-pass** [n] [UK] *trans.* (**1.** Road designed to redirect through traffic around the center [US]/centre [UK] of a community. **2.** When referring to a local thoroughfare the **b.** is also called **loop street [US] or loop highway [US]**; ►beltway [US], ►relief road, ►ring road); *s* **carretera [f] de circunvalación** (Carretera diseñada para recoger el tráfico de paso a través de una ciudad y evitar así la congestión dentro de ella; ►anillo periférico, ►autopista periférica, ►calle de desvío, ►carretera); *syn.* circunvalente [f] [C]; *f 1* **route [f] de contournement** (Route à grande circulation déviée en vue du contournement d'une agglomération [contournement routier/autoroutier] ; ►itinéraire de délestage, ►rocade, ►rocade autoroutière, ►route de délestage) ; *syn.* itinéraire [m] de contournement, voie [f] de contournement ; *f 2* **périphérique [m]** (Terme utilisé dans l'aménagement des routes et désignant une route à grande circulation dont le tracé est tangent à la périphérie d'une agglomération ; à

Paris, malgré sa forme de ►rocade autoroutière, on utilise le terme de périphérique ou boulevard périphérique ; *terme spécifique* autoroute périphérique ; ►route de contournement) ; *syn.* boulevard [m] périphérique, route [f] périphérique, ceinture [f] périphérique, voie [f] périphérique, anneau [m] périphérique ; *g* **Umgehungsstraße [f] (1.** Straße, die um einen Ort oder dessen Teil zur Ableitung des Durchgangsverkehrs führt. **2.** Eine Hauptverkehrsstraße, die Städte oder Stadtteile an ihrer Peripherie berührt/„tangiert" wird **Tangente** genannt; ►Autobahnring, ►Entlastungsstraße, ►Ringstraße).

645 bypass channel [n] [US]/by-pass channel [n] [UK] *hydr. wat'man.* (Canal adjacent to a river constructed for diversion and retention of floodwater); *s* **cauce [m] de derivación** (Canal paralelo a un río para captar y desviar las aguas de crecida); *f* **canal [m] latéral** (Canal construit parallèlement à un fleuve à lit encombré afin de retenir les eaux de crues) ; *syn.* canal [m] de délestage ; *g* **Seitengerinne [n]** (Kanal neben einem Fluss zur Aufnahme und Ableitung von Hochwasser); *syn.* Seitenkanal [m], Entlastungsgerinne [n].

646 bypassed meander [n] [US]/by-passed meander [n] [UK] *wat'man.* (Left aside arm of a river; e.g. ►oxbow lake, as a result of river engineering works); *s* **brazo [m] muerto (de origen artificial)** (Resultado de obra de ingeniería; ►brazo muerto); *syn.* caño [m] cegado; *f* **bras [m] mort restructuré** (réhabilité par diverses mesures hydrauliques ; ►bras mort) ; *g* **wiederhergestellter Altwasserarm [m]** (Stillgewässer [ehemaliger verschlammter und versandeter Altarm], das durch Wasserbaumaßnahmen neben einer Flussschleife wieder angelegt wurde; ►Altwasser 1).

C

cable-bracing [n] [UK] *arb. constr.* ►cabling [US].

647 cable television line [n] *envir.* (Communication line either underground or overhead; coaxial or fiber optic cable); *s* **cable [m] de televisión;** *f* **câble [m] de télévision** (Réseau aérien enterré de câbles ou de conducteurs pour la télécommunication) ; *g* **Fernsehkabel [n]** (Ober- oder unterirdisch geführte Leitung zur Übertragung von Fernsehsendungen).

648 cabling [n] [US] *arb. constr.* (Attachment of flexible steel cables to provide mechanical support to the crown of an old or damaged tree; cable-brace [vb]; ►scaffold branch, ►cavity bracing, ►rod bracing); *syn.* bracing [UK], cable-bracing [n] [UK]; *s* **sujeción [f] de las ramas principales (o primarias)** (con cable; ►reforzamiento del tronco con barras metálicas; ►rama primaria) *f* **haubanage [m] de la charpente** (Renforcement de la stabilité des branches ►charpentières par la mise en place de câbles ; ►renforcement métallique des troncs) ; *g* **Kronenverankerung [f] (1.** Veraltete Kronensicherungsmaßnahme, bei der bruchgefährdete ►Starkäste oder Stämmlinge in Bäumen mit Hohlkronen, Zwillingskronen oder breit ausladenden Kronen durchbohrt, mit Hilfe von Stahlgewindestangen, die in den Holzkörper eingebaut und verschraubt wurden, verankert und mit Drahtseilen verstrebt wurden. **2.** Die moderne **Kronensicherung** erfolgt mit verletzungsfreien Kunststoffbändern [Hohltauen] oder Haltegurten; ►Stammverstärkung durch Stabanker); *syn. zu 2.* Seilverankerung [f].

cadastral map [n] [US] *adm. surv* ►metes and bounds map [US]/land registration map [UK].

649 cadastral map [n] of trees *adm. landsc. surv.* (Map at the scale of 1:500-1:1,000, recording the exact location of each tree with species name and label[l]ed tree number for either an entire community or separate jurisdiction therein; e.g. parks department and/or highway department, school department; ►tree data bank, ►metes and bounds map [US]/land registration map [UK]); *s* **catastro [m] de árboles** (En D. registro y plan a escala de 1:500-1:1000 en el que están representados los árboles [especie y n° de registro], clasificados según estén en calles, parques o jardines públicos, o en terrenos administrados por las diferentes instituciones públicas; ►inventario de árboles urbanos, ►mapa catastral); *f* **cadastre [m] des arbres plantés** (Document graphique à l'échelle du 1:500 ou 1:1000 représentant l'inventaire des arbres urbains ; la nature et le nombre des arbres d'alignement sont répertoriés par voie, la fiche d'identification permettant une programmation de l'entretien une véritable gestion du patrimoine arborescent urbain ; *termes spécifiques* répertoire des arbres d'alignement, répertoire des arbres remarquables ; ►fichier informatique de description des arbres, ►plan cadastral) ; *g* **Baumkataster [m,** *in A nur so — oder* **n]** (Im Maßstab 1:500-1:1000 geführtes Karten- oder Planwerk zur lagegenauen Darstellung von Bäumen [Baumart und Baumnummer], getrennt nach Straßenbäumen, Bäumen in öffentlichen Grünanlagen oder Bäumen auf Grundstücken, die in einer Gemarkung oder Teilen davon von einzelnen Ämtern verwaltet werden; ►Baumdatei, ►Katasterkarte).

caducous [adj] *bot.* ►deciduous.

650 calcareous indicator species [n] *phyt.* (Plant species normally characteristic of calcareous [high pH] soil conditions; ►calcicolous species; *opp.* ►acidophilous species); *s* **especie [f] indicadora de cal** (Planta o sinecia que necesita substrato rico en calcio se utiliza por tanto como indicadora de cal; ►especie calcícola; *opp.* ►especie indicadora de acidez); *syn.* planta [f] indicadora de cal; *f* **plante [f] indicatrice de calcaire** (Espèce végétale nécessitant pour sa croissance la présence d'un substrat calcaire ; ►espèce calcicole ; *opp.* ►espèce indicatrice d'acidité) ; *g* **Kalkzeigerpflanze [f]** (Pflanzenart, die zu ihrem normalen Gedeihen ein Ca-reiches Substrat benötigt; ►kalkholde Art; *opp.* ►Säurezeiger); *syn.* Kalkanzeiger [m], Kalk (an)zeigende Art [f], Karbonatzeiger [m], kalkstete Art [f].

calcicole species [n] *phyt.* ►calcicolous species.

651 calcicolous species [n] *phyt.* (Plant growing on limestone or chalky soil; ►calcareous indicator species); *syn.* calcicole species [n], calciphile species [n], lime-loving species [n], basophilous species [n]; *s* **especie [f] calcícola** (Especie o sinecia que necesita substrato rico en calcio, se utilizan como ►especies indicadoras de cal); *syn.* especie [f] calcófila; *f* **espèce [f] calcicole** (Végétal affectionnant les sols calcaires au pH > 7 ; ►plante indicatrice de calcaire) ; *syn.* espèce [f] calciphile ; *g* **kalkholde Art [f]** (Pflanze, deren Verbreitung vorzugsweise auf kalkreichen, warmen und trockenen Böden mit alkalischer Reaktion, pH-Wert > 7, beschränkt ist; ►Kalkzeigerpflanze); *syn.* Kalk liebende Art [f], basiphile Art [f].

calcifuge species [n] *phyt.* ►lime-avoiding species.

calciphile species [n] *phyt.* ►calcicolous species.

calciphobe species [n] *phyt.* ►lime-avoiding species.

652 calculation [n] *contr.* (Advance estimation of prices for a bid [US]/tender [UK] or professional fees, etc.; calculate [vb]); *s* **cálculo [m] (de precios)** (Estimación de precios, honorarios, etc. realizada antes de hacer una oferta); *f* **calcul [m] (des prix)**

C

(En vue de l'établissement d'une offre, d'un devis, d'une proposition d'honoraires, etc.) ; *g* **Kalkulation [f]** (Vorausberechnung von Preisen für ein Angebot, Berechnung eines Honorars etc.; kalkulieren [vb]).

calculation [n] **of contract bid quantities** [US] *constr. plan.* ►calculation of quantities.

653 calculation [n] **of costs** *contr.* (Generic term covering ►total construction cost estimate [US]/final construction cost estimate [UK], ►estimate of probable construction cost, ►establishment of total construction cost, ►preliminary cost estimate); *syn.* costing [n], cost accounting [n]; *s* **cálculo [m] de costos** (Término genérico para ►presupuesto de gastos, ►estimación del costo, ►control de costos, ►estimación aproximada de costos); *syn.* cálculo [m] de costes [Es]; *f* **estimation [f] financière des travaux** (*Termes spécifiques* ►détermination du coût prévisionnel des travaux, ►devis, ►évaluation détaillée du coût des travaux, ►estimation détaillée des dépenses, ►constatation des coûts et ►estimation sommaire des travaux) ; *g* **Kostenermittlung [f]** (OB zu ►Kostenanschlag, ►Kostenberechnung, ►Kostenfeststellung und ►Kostenschätzung).

654 calculation [n] **of quantities** *constr. plan.* (Exact measurement of total quantities required for construction work, according to drawings, as the basis for ►list of bid items and quantities [US]/schedule of tender items [UK] or ►calculation of costs; **in U.S and U.K.**, usually calculated by a cost estimator [US]/quantity surveyor [UK] according to take-offs from drawings and descriptions of work made by the planner; **in D. and F., c. of q.** is normally the responsibility of a planner; ►bill of quantities; ►estimate of quantities); *syn.* calculation [n] of contract bid quantities [also US], volumetrics [npl] [also US], quantity take-off [n]; *s* **cálculo [m] de cantidades** (Calculación exacta de las cantidades totales de materiales y servicios necesitados para realizar un proyecto que se realiza sobre la base de los planes de localización para elaborar el ►resumen de prestaciones o el ►cálculo de costos; ►estimación preliminar de cantidades); *syn.* cómputo [m] de cantidades; *f* **avant-métré [m]** (Calcul par le concepteur de la quantité et de la qualité des travaux ou fournitures demandées par le maître d'ouvrage d'après les plans, énumérant les diverses unités d'œuvre employées dans la construction et indiquant la quantité nécessaire de chacune d'elles, en vue de l'établissement d'un ►descriptif-quantitatif ou d'une ►estimation financière des travaux ; ►avant-métré sommaire) ; *g* **Massenberechnung [f]** (Genaue Ermittlung der gesamten Mengen an Bauleistungen und Lieferungen für ein Bauprojekt anhand von Lageplänen als Grundlage für die Aufstellung eines ►Leistungsverzeichnisses oder zur ►Kostenermittlung; ►Massenüberschlag); *syn.* Massenermittlung [f].

calculations [npl] [US]**, cut/fill quantity** *constr. eng.* ►earthworks calculation.

655 caliper growth [n] *hort. for.* (*also* calliper growth; growth in plant stem thickness; ►trunk diameter); *s* **crecimiento [m] en espesor** (Aumento anual del grosor del tronco o de las ramas; ►diámetro del tronco); *f* **croissance [f] en épaisseur** (IDF 1988, 36 ; augmentation annuelle de la circonférence du tronc ou d'une branche de végétaux ligneux résultant de l'activité des cambiums ; ►diamètre [à hauteur d'homme], ►diamètre de la tige) ; *syn.* croissance [f] en diamètre ; *syn. for.* accroissement [m] (du diamètre) (DFM 1975, 2) ; *g* **Dickenwachstum [n]** (Die jährliche Vergrößerung des Umfanges eines Stammes oder Astes bei Gehölzpflanzen; ►Stammdurchmesser); *syn.* Dickenzuwachs [m].

caliper tree [n] *hort. arb.* ►large-caliper tree/large-calliper tree.

call [n] **for bids** [US] *constr. contr.* ►invitation for a bid [US].

call [n] **for a tender** [UK] *constr. contr.* ►invitation for a bid [US].

call [n] **for entries** [US]**, issuing a** *adm. prof. leg.* ►design competition announcement [US].

calliper growth [n] *hort. for.* ►caliper growth.

calliper tree [n] *hort. arb.* ►large-calliper tree.

656 callus [n] *arb. bot.* (Tissue which is created by the ►cambium on the edges of a wound to form a ►callus cushion; ►callusing); *s* **callo [m] cicatricial** (Tejido formado por el ►cambium alrededor de una herida y que crece formando una ►protuberancia callosa; ►cicatrización); *syn.* tejido [m] cicatrizal, callosa [f] cicatricial, callosidad [f] cicatricial; *f* **cal [m] cicatriciel** (Tissu secondaire de cicatrisation de blessure formé par le ►cambium adjacent recouvrant une plaie chez les végétaux ; ►bourrelet cicatriciel ; ►cicatrisation) ; *syn.* callus [m] ; *g* **Wundkallus [m]** (Gewebe, das am Rande einer Wundfläche durch das angrenzende ►Kambium entsteht und mit einem ►Überwallungswulst die Wunde überwächst; ►Wundheilung); *syn.* Callus [m], Kallus [m], Wundgewebe [n].

657 callus cushion [n] *arb. bot.* (Wound margin/surface covered by ►callus); *syn.* callus margin [n]; *s* **protuberancia [f] callosa** (►Callo cicatricial que se forma paulatinamente sobre una herida del tronco o de las ramas); *syn.* protuberancia [f] cicatricial; *f* **bourrelet [m] cicatriciel** (►Cal cicatriciel circulaire formé sur le pourtour d'une plaie du tronc ou des branches qui pourra être progressivement recouverte) ; *syn.* bourrelet [m] de cicatrisation (annulaire) ; *g* **Überwallungswulst [m]** (Eine Wundfläche schließender ►Wundkallus; LB 1978, 170); *syn.* Überwallungsleiste [f], Callusring [m], Kallusring [m], Wundholz [n].

658 callusing [n] *arb. bot.* (Process of healing over a wound by creating new wood tissue [phellogen] from the cambium on the edge of a wound; ►cambium); *syn.* occlusion [n] [also UK] (ARB 1983, 386, 506; BS 3975: part 5, 53032); *s* **cicatrización [f]** (Formación de nuevo tejido [felógeno] a través del ►cambium vecino a una herida en la corteza o en la madera para cerrarla); *f* **cicatrisation [f]** (Phénomène par lequel une plaie de taille se referme progressivement au rythme des saisons ; il s'élabore des bourrelets cicatriciels de plus en plus rapprochés qui se rejoignent et protègent définitivement les tissus sousjacents ; DIB 1988, 97 ; ►cambium) ; *g* **Wundheilung [f]** (Bildung von neuem Holzgewebe [Phellogen] durch das an den Wundrand angrenzende ►Kambium zur Schließung einer Wundfläche); *syn.* Kallusbildung [f].

callus margin [n] *arb. bot.* ►callus cushion.

camber [n] [UK] *constr.* ►crowning of paved surfaces [US].

cambisol [n] (FAO system) *pedol.* ►gray-brown podzolic soil.

659 cambium [n] *arb.* (**1.** Thin layer of living cells between the bark layer and the wood of a tree. **2.** Actively-dividing layer of cells that lies between, and gives rise to, secondary xylem and ►phloem [vascular cambium]; MGW 1964); *s* **cámbium [m]** (Capa de células en activo proceso de división entre el líber y el leño de las dicotiledóneas, que yace entre el xilema secundario y el ►floema secundario, tejidos a los cuales da origen [cámbium vascular o verdadero]; cf. MGW 1964); *syn.* cambio [m]; *f* **cambium [m]** (Couche cellulaire en voie de division active placée entre le phloème primaire et le xylème primaire, génératrice des tissus secondaires conducteurs, le ►liber [phloème secondaire] et le bois [xylème secondaire]) ; *syn.* assise [f] génératrice (interne) (MGW 1964), zone [f] génératrice ; *g* **Kambium [n]** (Schicht von sich aktiv teilenden Zellen, die zwischen sekundärem Xylem und ►Bastteil liegt und aus der diese beiden Gewebe hervorgehen [vaskulares Kambium]; MGW 1964); *syn. o. V.* Cambium [n].

camp and caravan site [n] [UK] *recr.* ▶camper with trailer site [US].

660 camper van [n] [UK] *recr. syn.* motor caravan [n] [UK]; *s* **camioneta** [f] **de camping**; *f* **camping-car** [m] ; *g* **Campingbus** [m] (Ein zum Freizeitwohnen umgerüsteter Kleinbus).

661 camper with trailer site [n] [US] *recr.* (Generic term for one tent/trailer camping place or one recreation vehicle site); *syn.* camp and caravan site [n] [UK], camping site [n], campsite [n] [US] (1); *s* **parcela** [f] **de camping** (Plaza individual para una tienda de campaña o una ▶autocaravana en un camping); *f* **emplacement** [m] **de camping** (Emplacement individuel pour la tente ou le camping-car) ; *g* **Campingplatzparzelle** [f] (Markierter Einzelstandplatz für eine Campingeinheit).

662 campground [n] [US] *recr.* (Whole area of land designated for short-stay camping by transportable accommodations, such as tents, caravans or trailers, with common facilities; e.g. toilet building, campstore, etc.; ▶camper with trailer site [US]/camp and caravan site [UK], ▶camp out, ▶farm camping, ▶permanent campground [US]/permanent caravan site [UK], ▶short-stay campground [US]/short-stay camping ground [UK], ▶tent camping [US]/camping [UK]); *syn.* campsite [n] [UK] (2); *s* **camping** [m] (1) (Término genérico para lugar de ▶campismo consistente en zona acotada en la que está permitido acampar y que cuenta con diversas instalaciones y servicios; ▶acampar en granja o caserío, ▶acampar por libre, ▶camping de tránsito, ▶camping permanente, ▶hacer camping, ▶parcela de camping); *syn.* campamento [m] turístico [Es], terreno [m] de campismo; *f 1* **terrain** [m] **de camping** (Terme générique pour toute installation réglementée ou terrain aménagé pour la pratique temporaire du camping ; ▶camper en gîte rural, ▶terrain de camping de transit, ▶terrain de camping permanent; **F.**, on distingue selon leur vocation les terrains de camping tourisme, terrains de camping loisirs, terrains de camping aire naturelle [*syn.* camp de tourisme aire naturelle], ▶camping sauvage, ▶emplacement de camping, ▶terrains de camping saisonnier) ; *f 2* **terrain** [m] **de camping-caravaning** (Terme générique pour tout terrain mixte aménagé pour le camping de caractère permanent ou non avec caravane ou camping-car et pour le camping temporaire sous tente) ; *f 3* **aire** [f] **naturelle d'accueil pour le camping-caravaning** (Espaces loués pour une durée limitée, ayant pour but de contrôler la ▶fréquentation touristique sauvage dans les zones fragiles ; ▶camping à la ferme) ; *f 4* **terrain** [m] **de camping libre** (Espace non aménagé sur lequel, aucun arrêté d'interdiction de camping n'a été prononcé par le maire et, avec l'accord du propriétaire, peut être librement pratiqué le camping) ; *g* **Campingplatz** [m] (1. Gelände für einen kurzfristigen ortsgebundenen Aufenthalt im Freien unter Verwendung von transportablen Unterkünften, z. B. Zelte, Wohnwagen oder andere Landfahrzeuge und sonstige Behelfsunterkünfte; in F. werden Flächen ohne Infrastruktureinrichtungen vorgehalten *[terrain de camping libre]*, auf denen frei gezeltet werden kann. **2.** OB zu ▶Dauercampingplatz und ▶Durchgangscampingplatz; cf. ▶Campingplatzparzelle, ▶wildes Lagern und Zelten, ▶Zelten, ▶Zelten auf dem Bauernhof); *syn.* Zelt- und Wohnwagenplatz [m].

camping [n] [UK] *recr.* ▶tent camping [US].

camping [n] [US] *recr.* ▶tent and motor camping.

camping [n] [ZA], **squatter** *urb.* ▶overnight land squatting.

camping [n] **and caravaning** [n] [UK] *recr.* ▶tent and motor camping [US].

camping ground [n] [UK]**, short-stay** *recr.* ▶short-stay campground [US].

camping site [n] *recr.* ▶camper with trailer site [US].

camping trailer [n] [US] *recr.* ▶travel trailer [US].

663 camping vehicle [n] *recr.* (Generic term for a vehicle used for vacation accommodation; ▶travel trailer [US]/caravan [UK], ▶mobile home, ▶recreation vehicle [RV] [US]; *s* **remolque-vivienda** [m] (Vehículo sin motor utilizado como vivienda por gentes de circo o similares, o utilizado durante las vacaciones; ▶remolque de [camping], ▶casa móvil, ▶autocaravana); *syn.* caravana [f], roulotte [f]; *f* **remorque** [f] **de camping** (±) (*Terme générique* véhicule tracté non motorisé, aménagé pour le camping ou roulotte pour le séjour des forains ; ▶caravane, ▶résidence mobile, ▶camping-car) ; *g* **Wohnwagen** [m] (OB zu Anhänger als Camping- [▶Caravan] oder Schaustellerwagen, der für vorübergehendes Wohnen eingerichtet ist; cf. BROCK 1994, Bd. 24; ▶Mobilheim, ▶Wohnmobil); *syn.* Wohnanhänger [m].

664 camp out [vb] *recr.* (Spontaneous, unorganized camping in the countryside with or without shelter); *s* **acampar** [vb] **por libre** (Acampada espontánea, fuera de los límites de un ▶camping 1 oficial); *syn.* acampada [f], camping [m] salvaje; *f* **camping** [m] **sauvage** (Pratique du camping en dehors des terrains aménagés ou sur les zones sur lesquelles cette pratique est interdite par arrêté) ; *syn.* fréquentation [f] touristique sauvage ; *g* **wildes Lagern** [n, o. Pl.] **und Zelten** [n, o. Pl.] (Spontanes, ungeordnetes Zelten in der Landschaft außerhalb von genehmigten Campingplätzen); *syn.* abseits campieren [vb] [A, CH], wildes Campen [n, o. Pl.].

campsite [n] [US] (1) *recr.* ▶camper with trailer site [US].

campsite [n] [UK] (2) *recr.* ▶campground [US].

665 canal [n] (1) *hydr. trans.* (Artificially-created and navigable stretch of flowing or standing water); *s* **canal** [m] (Cauce artificial navegable); *f* **canal** [m] (Cours d'eau aménagé artificiellement pour assurer l'adduction d'eau [canal d'amenée], l'arrosage [canal d'irrigation], pour permettre la navigation fluviale) ; *g* **Kanal** [m] (Künstlich angelegtes, linienförmiges, schiffbares Gewässer größerer Breite, fließend oder stehend).

canal [n] (2) *agr.* ▶irrigation canal.

666 canalization [n] **of a waterbody for navigability** *trans.* (Construction measures carried out to make a stretch of water navigable); *s* **hacer** [vb] **navegable un curso de agua**; *f* **mise** [f] **en navigabilité de cours d'eau** (Aménagement hydraulique permettant la navigation sur un cours d'eau) ; *g* **Schiffbarmachung** [f] **von Gewässern** (Wasserbauliche Maßnahmen an einem Gewässer, um es für Schiffe befahrbar zu machen).

667 canalized stream [n] *wat'man.* (▶river engineering measures); *syn.* channelized stream [n]; *s* **arroyo** [m] **encauzado** (▶ingeniería hidráulica de ríos y arroyos); *syn.* arroyo [m] canalizado; *f* **ruisseau** [m] **canalisé** (▶travaux d'aménagement des cours d'eau) ; *syn.* rivière [f] canalisée, rivière [f] calibrée ; *g* **kanalisierter Bach** [m] (Bach, dessen natürliches Bett im letzten Jahrhundert technisch so ausgebaut wurde, dass der Hochwasserabfluss und die Bodennutzungsverhältnisse auf Kosten der Gewässerökologie verbessert wurden; durch Bachrenaturierungsmaßnahmen werden mit viel Aufwand naturnahe Gewässerläufe wiederhergestellt; ▶Gewässerausbau, ▶Gewässerrenaturierung).

668 cancellation of a construction contract [n] *constr. contr.* (Client's revocation of a contract undertaken with a construction contractor); *s* **resolución** [f] **de contrato (de obra)** (Finalización anticipada de un contrato de obra por parte del comitente; cf. arts. 112-114 LCAP); *syn.* rescisión [f] de contrato (de obra); *f* **résiliation** [f] **du marché de travaux par le maître d'ouvrage** (A l'initiative du maître d'ouvrage il peut être mis fin à l'exécution des travaux avant leur l'achèvement) ; *g* **Entzie-hung [f] des Auftrages (1)** (Wegnehmen eines Auftrages, der einem Unternehmer durch den Auftraggeber erteilt wurde; cf. § 8 VOB Teil B); *syn.* Auftragsentziehung [f] (1), Auftragsentzug [m] (1), Kündigung [f] durch den Auftraggeber.

cancellation |n| **of contract** |US| *contr. prof.* ▶termination of a contract [US].

candelabra espalier [n] [US] *hort.* ▶double-U espalier.

candidate species [n] *conserv. phyt. zool.* ▶near threatened species.

669 cane [n] *hort.* (**1.** Primary stem of a shrub which starts from the ground or close to the ground. Term used in the U.S. in nursery stock specification for quality definition of deciduous shrubs. **2.** Slender, hollow, jointed, usually flexible stem of certain plants—e.g. bamboo—or the stem of a small fruit plant such as blackberry, raspberry, etc.; ▶young shrub transplant, ▶shrub); *syn.* base stem [n], ground shoot [n]; *s* **rama [f] basilar** (Tallo primario de un arbusto que comienza al ras del suelo o cerca del mismo. El término se utiliza para plantas de vivero como criterio de calidad de ▶arbustos [decíduos] y para ▶arbustos jóvenes para transplante; en alemán, en las rosas para los brotes por encima de la soldadura de injerto); *f 1* **tige [f] centrale** (Critère définissant une des caractéristiques végétatives applicable aux jeunes plants et ▶jeunes touffes d'arbustes et d'arbrisseaux ainsi qu'aux ▶arbustes cultivés en pépinières ; ▶arbrisseau) ; *syn.* tige [f] principale ; *f 2* **branche [f]** (Chez les rosiers greffés buissons ou tiges, pousses partant du point de greffe) ; *g* **Grundtrieb [m]** (**D.**, Sortierkriterium zur Gütebestimmung von in Baumschulen gezogenen leichten Sträuchern [▶leichter Strauch] und Sträuchern [▶Strauch 2]; bei Rosen sind es Triebe oberhalb der Veredelung); *syn.* Bodentrieb [m].

670 canopy [n] (**1**) *arch. urb.* (Nearly flat rooflike cover which extends from the wall of a building; e.g. protecting an entrance; ▶porte cochère); *s* **marquesina [f] (1)** (Toldo permanente que se proyecta hacia el exterior de una entrada de edificio o sobre sus ventanas; ▶marquesina de garaje-portal); *f* **auvent [m]** (Aire couverte devant l'entrée d'un bâtiment permettant la descente à pied sec des passagers d'un véhicule ; ▶porche d'entrée de/du/au garage) ; *g* **Schutzdach [n] (1)** (Vordach an Gebäuden über einem Eingang; *syn.* Vordach [n].

canopy [n] (**2**) *phyt.* ▶woodland canopy.

671 canopy closure [n] *for. phyt.* (Progressive reduction of space between crowns as they spread laterally, increasing canopy density; ▶degree of canopy cover); *s* **cierre [m] del dosel (del bosque)** (La reducción progresiva de los espacios libres entre las copas de los árboles de un rodal; ▶expansión horizontal de las copas); *f* **fermeture [f] du couvert** (La réduction progressive des espaces libres entre les cimes d'un peuplement forestier ; ▶indice de cime) ; *g* **Kronenschluss [m] (1)** (...schlüsse [pl]; das allmähliche Zusammenwachsen von Baumkronen in einem Bestand; ▶Deckungsgrad der Baumkronen).

672 canopy cover [n] *for. phyt.* (Completeness of the tree canopy, i.e the ▶canopy closure and therefore an aggregate expression of crown cover. The ▶degree of canopy cover ranges from 'congested' and 'closed', through 'broken' and 'open' to 'gappy'; cf. SAF 1983; ▶crown closure, ▶crown density, ▶tree canopy shading); *syn.* canopy density [n] [also US]; *s* **cobertura [f] del dosel de un rodal** (Cobertura del suelo por la suma de las copas en un rodal/bosque; ▶cierre del dosel [del bosque], ▶densidad de copa, ▶sombreado por árboles); *syn.* vuelo [m] forestal; *f* **couvert [m] d'un peuplement forestier** (Écran formé par l'ensemble plus ou moins continu des branches et du feuillage des arbres d'un peuplement forestier/d'un groupe d'arbre. Quant à sa structure l'▶indice de cime peut être « serré », « fermé/

complet/plein », « clairiéré », « interrompu/discontinu » ou « lacuneux » ; cf. DFM 1975 ; ▶degré d'ouverture, ▶densité du couvert d'une cime, ▶fermeture du couvert) ; *syn.* recouvrement [m] des couronnes, recouvrement [m] de la strate foliaire ; *g* **Kronenschluss [m] (2)** (...schlüsse [pl]; Überschirmung des Bodens durch die Summe der Kronen in einem Waldbestand/Baumbestand; der Kronenschlussgrad [▶Deckungsgrad der Baumkronen] wird unterschieden in 'gedrängt', 'geschlossen', 'locker', 'licht' und 'räumdig'; ▶Beschirmung, ▶Kronenschluss 1, ▶Kronenschluss 3); *syn.* Schlussgrad [m].

canopy density [n] [US] *for. phyt.* ▶canopy cover.

canopy edge [n] **of a tree** *arb. hort.* ▶drip line.

673 cantilever steps [npl] *arch. constr.* (Outdoor stairway with steps inserted in and supported by one end of a wall); *s* **escalera [f] volante** (Escalera en la que los peldaños están sujetados solamente por un lado a la zanca); *f* **escalier [m] suspendu** (Escalier dont les marches sans limon se soutiennent d'elles mêmes en porte-à-faux, grâce à l'encastrement de la queue qui empêche toute rotation ; MAÇ 1981, 497) ; *g* **freitragende Treppe [f]** (Stufen einer Treppe, die freitragend in eine Treppenwange eingespannt sind).

674 cantilever wall [n] *constr.* (DAC 1975; reinforced concrete ▶retaining wall, whereby the weight of the soil rests upon the L-shaped footing behind the face of the wall to prevent it from overturning); *s* **muro [m] de sostenimiento en L** (BU 1959; ▶muro de sostenimiento de hormigón armado en el que el peso de la tierra cae sobre el pie del muro en forma de L por lo que no puede tumbarse); *syn.* muro [m] de apoyo con aleta de retorno (BU 1959); *f* **mur-chaise [m]** (▶Mur de soutènement en béton armé de forme en T renversé ou en L dont le poids de la terre chargeant la semelle arrière assure la stabilité du mur ; VRD, 1986, 64 ; pour de faibles hauteurs le mur de soutènement peut être réalisé avec des éléments préfabriqués) ; *g* **Winkelstützmauer [f]** (▶Stützmauer aus Stahlbeton, bei der das Erdgewicht auf dem L-förmigen Mauerfuß lastet und somit ein Kippen verhindert).

capacity [n] *recr. plan.* ▶accommodation capacity; *constr. eng.* ▶compaction capacity; *pedol.* ▶field capacity; *wat'man.* ▶flood storage capacity; *agr. game'man.* ▶grazing capacity; *constr. pedol.* ▶infiltration capacity; *adm. plan.* ▶interment capacity; *eng.* ▶load-bearing capacity; *pedol.* ▶nutrient storage capacity; *plan. recr.* ▶recreation capacity; *biol. ecol.* ▶regenerative capacity; *limn.* ▶self-cleansing capacity; *pedol.* ▶storage capacity; *pedol.* ▶water-holding capacity.

capacity [n]**, erosion** *geo.* ▶erosion potential.

capacity [n]**, holding capacity** [US] *recr. plan.* ▶accomodation capacity.

capacity [n]**, moisture** *hort.* ▶field capacity.

capacity [n]**, sprouting** *for. hort.* ▶capacity of making new shoots.

capacity [n]**, water storage** *pedol.* ▶water-holding capacity.

capacity [n] **for survival** *biol. ecol.* ▶capacity to survive.

675 capacity [n] **of a contracting firm** *constr.* (Ability of a firm to carry out construction contracts efficiently and on time, based upon the necessary expertise, sufficient specialized personnel and appropriate use of equipment); *s* **capacidad [f] de una empresa** (Requisito mínimo para la contratación de personas naturales o jurídicas por parte de la administración pública; incluye la plena capacidad de obrar y la solvencia económica, financiera y técnica o profesional; cf. arts. 15-19 Ley 13/1995, de Contratos de las Administraciones Públicas); *f* **capacité [f] d'une**

entreprise (Qualité d'une entreprise apportant les garanties professionnelles et financières nécessaires à l'exécution des ouvrages dans le respect des délais et des coûts prévisionnels) ; *g* **Leistungsfähigkeit [f] eines Unternehmens** (Vermögen einer Firma, die zu erbringenden Bauleistungen mit dem erforderlichen Fachwissen, einem ausreichenden Fachpersonal und mit einem der Bauaufgabe gerecht werdenden Maschineneinsatz termingerecht und wirtschaftlich abwickeln zu können); *syn.* Kapazität [f] eines Unternehmens.

capacity [n] **of an area, saturation** *plan. ecol.* ▶carrying capacity of landscape.

capacity [n] **of landscape** *ecol. plan. recr.* ▶carrying capacity of landscape.

676 capacity [n] **of making new shoots** *for. hort.* (Tendency of woody plants to make new shoots when they are severely pruned or ▶coppiced and capacity of ▶dormant cuttings or heel cuttings to generate roots and shoots); *syn.* new growth capability [n] [also US], sprouting capacity [n], sprouting vigo(u)r [n]; *s* **capacidad [f] de rebrote** (Aptitud de leñosas para regenerarse después de un ▶trasmocho o de ser podadas hasta el tocón [▶podar hasta el tocón] y de ▶esquejes leñosos de formar raíces y vástagos); *syn.* capacidad [f] de regeneración; *f* **aptitude [f] de rejeter** (1. Capacité des végétaux ligneux après une taille importante, après recépage, à développer de nouvelles pousses à partir de la tige ou de la souche ; ▶recéper. 2. Capacité des boutures dormantes à développer des racines et des tiges ; ▶bouture de rameau aoûté, ▶bouture de rameau semi-aoûté) ; *syn.* aptitude [f] de repousser, capacité [f] de rejeter ; *g* **Ausschlagfähigkeit [f]** (Fähigkeit von Gehölzen wieder auszutreiben, wenn sie stark zurückgeschnitten oder auf den Stock gesetzt wurden oder das Vermögen von Steckhölzern, Wurzeln und Triebe zu bilden; ▶auf den Stock setzen, ▶Steckholz); *syn.* Ausschlagvermögen [n].

capacity [n] **of natural resources** *conserv. plan.* ▶use capacity of natural resources.

677 capacity [n] **to survive** *biol. ecol.* (Survivability of plants, animals, or ecosystems); *syn.* capacity [n] for survival; *s* **capacidad [f] de supervivencia** (de las plantas, los animales y los ecosistemas); *f* **capacité [f] de survie** (Capacité que possèdent les espèces de faune et de flore, les populations d'espèces ou des écosystèmes tout entier à survivre à une modification importante des facteurs du milieu) ; *g* **Überlebensfähigkeit [f]** (Fähigkeit von Pflanzen, Tieren, Populationen oder ganzen Ökosystemen bei veränderten Umweltbedingungen zu bestehen).

678 capillarity [n] (1. *phys.* Phenomenon which is associated with the surface tension of liquids, particularly in capillary tubes and with porous media where gas, liquid and solid interfaces meet; cf. WMO 1974. 2. *pedol.* Moisture migration from wet soil to dry soil, often against gravity. 3. *constr.* Phenomenon of upward migration of moisture into the subgrade of road beds due to "pumping action" of dynamic loading or traffic induced vibration); *s* **capilaridad [f]** (1. *phys.* Fenómeno de ascensión asociado a la tensión superficial de los líquidos, particularmente en tubos capilares, poros o fisuras. 2. *pedol.* Fenómeno de migración del agua del suelo gracias a la diferencia de tensión de la misma entre las partes secas y húmedas, a menudo en contra de la gravedad); *f* **capillarité [f]** (Phénomène associé à la tension de surface d'un liquide, particulièrement dans les tubes capillaires et les milieux poreux où les interfaces gaz, liquide et solide sont en contact ; WMO 1974) ; *g* **Kapillarität [f, o. Pl.]** (1. *phys.* Phänomen der durch die Oberflächenspannung bestimmte Verhaltensweise von Flüssigkeiten in Kapillaren, engen Spalten oder Poren. 2. *pedol.* Lageveränderung des Bodenwassers vom nassen zum trockenen Bereich auf Grund von Spannungsunterschieden, oft

gegen die Schwerkraft. 3. *constr.* Phänomen der Aufwärtsbewegung des Wassers im Untergrund, verursacht durch die Vibration des Kraftfahrzeugverkehrs).

capillary action [n] [US] *pedol.* ▶capillary rise.

679 capillary break [n] *phys. pedol.* (Interruption of ▶capillary rise); *s* **interrupción [f] de la capilaridad** (Perturbación de la ▶ascensión capilar); *f* **interruption [f] de capillarité** (Perturbation du phénomène de la ▶remontée capillaire) ; *g* **Unterbrechung [f] der Kapillarität** (Störung des ▶kapillaren Aufstieges).

capillary diffusion [n] *pedol.* ▶capillary rise.

680 capillary fringe [n] *pedol.* (Belt of subsurface water, held above the zone of saturation by capillarity; WMO 1974; ▶capillary water); *syn.* capillary zone [n]; *s* **franja [f] capilar** (Cinturón de agua subterránea mantenido, por capilaridad, por encima de la zona de saturación; WMO 1974; ▶agua capilar); *syn.* zona [f] capilar; *f* **frange [f] capillaire** (Zone située immédiatement au-dessus de la surface de la nappe phréatique dans laquelle l'eau est retenue par capillarité au-dessus de cette surface ; WMO 1974 ; ▶eau capillaire) ; *g* **Kapillarsaum [m]** (Zone im Boden, die unmittelbar oberhalb der Grundwasseroberfläche anschließt und durch ▶Kapillarwasser gekennzeichnet ist).

681 capillary interstice [n] *pedol. phys.* (Tube small enough for water to be held in it against gravity above a water table; WMO 1974); *syn.* capillary pore [n], capillary tube [n]; *s* **intersticio [m] capilar** *syn.* poro [m] capilar; *f* **interstice [m] capillaire** (Interstice suffisamment petit pour que l'eau s'y maintienne contre la force de gravité ; WMO 1974) ; *syn.* tube [m] capillaire (PED 1979, 299) ; *g* **Kapillare [f]** (Röhrchen oder langgestreckter Hohlraum mit sehr kleinem Innendurchmesser; MEL 1975, Bd. 13); *syn.* Kapillarröhrchen [n].

capillary migration [n] *pedol.* ▶capillary rise.

capillary pore [n] *pedol. phys.* ▶capillary interstice.

capillary porosity [n] *pedol.* ▶porosity.

682 capillary rise [n] *pedol.* (Upward movement of moisture from ground water through soil or ▶impeded water toward[s] the surface; *opp.* ▶infiltration); *syn.* capillary action [n] [also US], capillary migration [n], capillary diffusion [n]; *s* **ascenso [m] capilar** (Ascensión del agua en el suelo no saturado por encima de la superficie freática, debida a la acción de la capilaridad; cf. WMO 1974; ▶capillarity, ▶agua capilar; *opp.* ▶infiltración); *f* **remontée [f] capillaire** (Mouvement d'eau à travers les interstices d'un milieu non saturé, sous l'effet d'un gradient de potentiel capillaire, ou de tension ; CILF 1978 ; ▶nappe perchée, *opp.* ▶infiltration) ; *syn.* ascension [f] capillaire ; *g* **kapillarer Aufstieg [m]** (Aufwärtsbewegung des Wassers im Boden vom Grund- oder Stauwasser zur Erdoberfläche; ▶Staunässe; *opp.* ▶Versickerung); *syn.* Kapillaranstieg [m], Kapillarbewegung [f].

capillary tube [n] *pedol. phys.* ▶capillary interstice.

683 capillary water [n] *pedol. phys.* (Water which moves upward[s] in the ▶capillary fringe or which is retained against the action of gravity as opposed to downward ▶seepage water); *s* **agua [f] capilar** (Agua retenida por ▶capilaridad por encima del nivel freático en la ▶franja capilar); *f* **eau [f] capillaire** (Eau retenue par capillarité dans le sol au-dessus de la nappe phréatique ; WMO 1974 ; ▶frange capillaire, ▶capillarité) ; *g* **Kapillarwasser [n]** (Wasser, das im Boden im ▶Kapillarsaum durch ▶Kapillarität gehalten wird).

capillary zone [n] *pedol.* ▶capillary fringe.

cap [n] **of a wall** [US] *arch. constr.* ▶crown of a wall.

capping [n] [UK] *envir* ▶sealing.

C

capping layer [n] *constr.* ▶blinding.

684 capping [n] **with soil material** *constr. min.* (LD 1991 [11], 5; placing a less permeable clay soil cover over constructed ▶sanitary landfills [US] 2/landfill sites [UK] to reduce moisture infiltration into refuse cells; ▶dumpsite [US]/tipping site [UK]); *syn.* covering [n] with soil [also US]; *s* **cobertura** [f] **con tierra vegetal** (▶depósito de residuos sólidos); *f* **couverture** [f] **de terre végétale** (Couche de terre végétale faisant partie de la couverture finale mise en place sur les parties comblées en fin d'exploitation lors de la remise en état d'une ▶décharge publique, d'un site de stockage) ; *g* **Über-erdung** [f] (Andecken von Boden auf ▶Deponien, um diese landschaftlich zu gestalten oder beim Auftrag tonigen Bodens, den Wassereintritt in das Abfallgut zu verhindern).

cap stones [n] [US] *arch. constr.* ▶coping.

685 capture area [n] *wat'man.* (Protected acquisiton area around the source of potable water, which may be a watershed [US] around a reservoir; ▶soil and water conservation district [US]/water conservation area [UK]); *s* **zona** [f] **de ganancia** (Área de extracción de aguas subterráneas; ▶perímetro de protección de un acuífero); *f* **zone** [f] **d'appel du captage** (Zone dans laquelle l'ensemble des lignes de courant d'eau se dirigent vers l'ouvrage de prélèvement ; ▶périmètre de protection d'un captage d'eau potable) ; *syn.* terrain [m] de puisage d'eau potable, terrain [m] de captage d'eau potable ; *g* **Entnahmeraum** [m] (Gebiet, das der Trinkwassergewinnung dient; ▶Wasserschutzgebiet).

caracole [n] *arch.* ▶spiral stair(case).

caravan [n] [UK] *recr.* ▶travel trailer [US].

care [n] [US]**, lawn** *constr.* ▶lawn maintenance.

care [n] [US]**, tree** *arb. constr.* ▶tree maintenance.

caretaker [n] **of a tree** [US] *conserv.* ▶voluntary caretaker of a tree.

686 car park [n] (1) *trans. urb.* (Any public or private land designed and used for parking motor vehicles, including parking lots, garages and legally designated areas of public streets; IBBD 1981; ▶multideck car park [US]/multistorey car park [UK], ▶overflow parking area, ▶parking bay, ▶parking garage, ▶parking space, ▶space covered by a parked vehicle); *syn.* parking area [n] (1); *s* **aparcamiento** [m] (1) (Superficie pública o privada reservada para dejar vehículos; ▶aparcamiento 2, ▶aparcamiento de reserva, ▶área de aparcamiento, ▶edificio de aparcamiento 1, ▶edificio de aparcamiento 2, ▶superficie ocupada por un vehículo); *syn.* parking [m], estacionamiento [m] (1) [CS], aparcadero [m] (1) [CA], parqueo [m] (1) [CAR]; *f* **parc** [m] **de stationnement** (Terme générique désignant des emplacements privés ou publics [y compris la voie de desserte] propre à assurer le stationnement de véhicules automobiles ; ▶baie de stationnement, ▶emplacement de stationnement, ▶parking de délestage, ▶parc de stationnement à étages, ▶parc de stationnement couvert, ▶surface occupée par un véhicule) ; *syn.* parking [m], parc [m] à voitures ; *g* **Parkplatz** [m] (1) (Zum Abstellen von mehreren Fahrzeugen vorgesehene private oder öffentliche Fläche, incl. der gegebenenfalls vorhandenen Fahrgassen; ▶Ausweichparkplatz, ▶Parkbucht, ▶Parkhaus 1, ▶Parkhaus 2, ▶Parkplatz 2, ▶Stellfläche); *syn.* Parkfläche [f], Parkierungsfläche [f] [BW].

car park [n] (2) *leg. urb.* ▶communal car park.

car park [n]**, overflow** *plan.* ▶overflow parking area.

car park approach [n] *trans.* ▶parking access road.

687 car park lawn [n] *constr.* (▶Lawn type in landscape construction which is capable of bearing constant, light or periodically heavy traffic in car parks or driveways [US]/house drives [UK]; ▶crushed aggregate lawn, ▶grass-filled modular paving [US]/grass setts paving [UK], ▶reinforced turf); *syn.* parking lot lawn [n] [also US]; *s* **césped** [m] **de aparcamiento** (▶Tipo de césped muy resistente capaz de aguantar tráfico ligero o periódico fuerte; ▶césped armado, ▶césped sobre piedra partida, ▶pavimento de ladrillo de césped); *f* **gazon** [m] **d'aires de stationnement** (▶Type de gazon de composition choisie pour une utilisation sur les aires de stationnement et leur voies d'accès ; **D.**, mélange conforme à la norme DIN 18 917 et souvent utilisé pour l'engazonnement des aires en graves concassées ; ▶gazon armé, ▶gazon sur grave [concassé], ▶pavage engazonné) ; *syn.* gazon [m] pour parking, gazon [m] d'aires stabilisées ; *g* **Parkplatzrasen** [m] (Strapazierfähiger ▶Rasentyp des Landschaftsbaues mit einer besonderen Saatgutmischung für die Verwendungszweck Parkplatz oder Zufahrt, der eine ausreichende Belastbarkeit bei ständig mäßigem oder periodisch starkem Verkehrsaufkommen sicherstellt; cf. DIN 18 917; dieser Rasentyp wird oft für die Anlage von ▶Schotterrasen verwendet; ▶armierter Rasen, ▶Rasenpflaster).

688 car park post [n] *urb. trans.* (Barrier on private parking space to prevent unauthorized use; ▶parking space); *s* **poste** [m] **de cierre de paso de aparcamiento privado** (▶aparcamiento 2); *syn.* poste [m] de cierre de paso de estacionamiento privado; *f* **dispositif** [m] **de condamnation de garage individuel** (Arceau ou potelet rabattable pour un ▶emplacement de stationnement privé) ; *syn.* dispositif [m] de condamnation de parking individuel, barrière [f] pour réservation d'emplacement de stationnement individuel ; *g* **Absperrpfosten** [m] (2) (Pfosten, der Parkplätze für bestimmte [private] Nutzer freihält); *syn.* Sperrpfosten [m].

689 carpentry [n] *constr.* (Wood-working art or craft with wooden construction materials); *s* **carpintería** [f] (Arte de construcción que utiliza la madera como materia prima); *f* **construction** [f] **en bois** (Travaux de fabrication utilisant le bois comme matériau de construction dans le bâtiment ; *termes spécifiques* **menuiserie de bois** [ouvrages en bois de second œuvre assurant le clos intérieur et extérieur des bâtiments] ; **charpenterie** [support des matériaux de couverture de la toiture en pente des bâtiments ; *syn.* menuiserie [f] de charpente en bois]) ; *g* **Holzbau** [m, o. Pl.] (1. Konstruktion, die Holz als Baustoff im Bauwesen verwendet, um z. B. ein Bauwerk in Massivbauweise oder Skelettbauweise [Fachwerkhaus], einen Dachstuhl oder eine Pergola zu errichten; man unterscheidet den konventionellen **zimmermannsmäßigen Holzbau** und den **Ingenieurholzbau** [Konstruktionstechnik mit Brettschichtholz zur Schaffung von Leimbindern, um z. B. großdimensionierte Holzkonstruktionen wie Hallen und Brücken zu bauen]. 2. handwerkliche Leistungen, die im **H.** zu erbringen sind, werden **Zimmermannsarbeiten** genannt).

690 carpet-forming ground cover plant [n] *hort. plant.* (Very flat growing plant, which spreads out by means of stolons above or below the ground, to form a dense carpet of foliage; ▶ground cover perennials, e.g. Waldsteinia *[Waldsteinia ternata]*, Sedum species *[Sedum floriferum 'Weihenstephaner Gold', S. spurium, S. sexangulare]* and very flat-growing suffruticose shrubs, e.g. Periwinkle *[Vinca minor]*, and dwarf shrubs such Partridge Berry *[Gaultheria procumbens]*, Ivy *[Hedera helix]*, Pachysandra *[Pachysandra terminalis]*); the term '**carpet plant**' is often used in connection with aquarium plants such as Brazilian microsword *[Liliaeopsis brasiliensis]*, Pygmy chain sword *[Echinodorus tenellus]*, Baby tears *[Micranthemum umbrosum]*, which are planted to 'carpet' the floor of a tank); *s* **tapizante** [f] (Término horticultor para ▶plantas vivaces tapizantes que crecen al ras de la tierra y se multiplican por medio de estolones epígeos o subterráneos, como la waldsteinia *[Waldsteinia ternata]* y las especies de siempreviva *[Sedum floriferum 'Weihenstephaner Gold', S. spurium, S. sexangulare]*, subarbustos como el tomillo *[Thymus praecox]* y plantas reptantes como la hiedra *[Hedera helix]*); *syn.* planta [f] tapizante; *f* **plante [f] tapissante** (Désigne les végétaux caractérisé par un port, un développement, ras, plaqué au sol, en tapis ; le Waldsteinia *[Waldsteinia ternata]*, les espèces d'Orpins *[Sedum floriferum 'Weihenstephaner Gold', S. spurium, S. sexangulare]* et les sous-arbrisseaux à port très étalé ou rampant, p. ex. la petite Pervenche *[Vinca minor]*, et des arbustes nains comme la Gaulthérie couchée *[Gaultheria procumbens]*, le Lierre *[Hedera helix]*, le Pachysandre terminale *[Pachysandra terminalis]* ont un port tapissant ; en français, une plante tapissante n'est pas forcément « couvre-sol » ; ▶plante vivace couvre-sol) ; *syn.* espèce [f] tapissante ; *g* **teppichbildender Bodendecker [m]** (Gärtnerische Bezeichnung für sehr flach wachsende, sich durch ober- oder unterirdische Ausläufer ausbreitende ▶bodendeckende Stauden, z. B. Golderdbeere *[Waldsteinia ternata]*, Sedumarten *[Sedum floriferum 'Weihenstephaner Gold', S. spurium, S. sexangulare]* und sehr flach wachsende Halbsträucher, z. B. Immergrün *[Vinca minor]*, und Zwergsträucher wie z. B. Scheinbeere *[Gaultheria procumbens]*, Efeu *[Hedera helix]*, Ysander *[Pachysandra terminalis]*).

691 carpet-forming perennial [n] *hort. plant.* (Perennial herbaceous plant forming a lawn-like carpet by rhizomes, stolons or horizontal stems; ▶rhizomatous, ▶stoloniferous, ▶stem-spreading; ▶ground cover perennials); *syn.* dense mat-forming perennial [n]; *s* **planta [f] vivaz cespitosa** (Perenne de crecimiento muy bajo, que generalmente es ▶rizomatosa o ▶estolonífero/a [1 y 2] y cubre completamente el suelo como un césped. En castellano, el término «cespitosa» se utiliza tanto para plantas que forman macollas, como para aquéllas que cubren el suelo como un césped; etimológicamente los dos términos «césped» y «cespitoso» tienen el mismo origen en el latin *caespes* resp. *caespitosus*; cf. DB 1985; ▶planta vivaz tapizante); *syn.* vivaz [f] cespitosa, perenne [f] cespitosa; *f* **plante [f] vivace gazonnante** (Espèce vivace très basse ▶rhizomateuse, ▶stolonifère, ▶traçante) ; *g* **rasenbildende Staude [f]** (Sehr niedrige, meist ▶rhizombildende, ▶Stolonen bildende oder ▶Ausläufer treibende Staude; ▶bodendeckende Staude, ▶teppichbildender Bodendecker).

carpet [n] **of chandelier algae, subaqueous** *phyt.* ▶chara vegetation.

692 car pool [n] *trans.* (Group of people who daily drive together to their place of work instead of in separate vehicles; ▶car pooler); *s* **comunidad [f] de conmutadores (≠)** (Grupo de personas que comparte vehículo a diario para ir a trabajar; ▶miembro de comunidad de conmutadores); *syn.* carpool [m]; *f* **covoiturage [m]** (Augmentation de la capacité de transport des véhicules privés par multiplication du nombre de passagers utilisant un seul véhicule pour se rendre sur leur lieu de travail ; ▶utilisateur de covoiturage) ; *syn.* car pool [m] ; *g* **Fahrgemeinschaft [f]** (Zusammenschluss von Berufstätigen, die täglich mit einem, statt jeder mit seinem Auto zur Arbeitsstelle fahren; ▶Mitglied einer Fahrgemeinschaft).

693 car pooler [n] *trans.* (▶commuter); *s* **miembro [m] de comunidad de conmutadores** (▶conmutador, -a); *syn.* participante [m] de viajes compartidos; *f* **utilisateur [m] de covoiturage** (▶migrant alternant) ; *g* **Mitglied einer Fahrgemeinschaft [f]** (▶Pendler).

694 carport [n] [US]/**car port** [n] [UK] *urb.* (One-sided, roofed structure, attached to the side of a house or sometimes a separate structure with covered breezeway to the dwelling, for

parking or storage of motor vehicles); *s* **garaje [m] sin puertas** (Garaje abierto lateralmente construido en el conjunto de una casa, provisto de tejado, pero sin paredes exceptuando las propias de la casa; cf. DACO 1988; *f* **abri [m] de voiture** (Parc à voiture couvert et ouvert sur les côtés, attenant à une construction ou en position isolée à proximité de l'habitation) ; *syn.* garage/abri [m] [bois] pour voiture, carport [m] ; *g* **Carport [m]** (An ein Haus angebauter überdachter Stellplatz, jedoch ohne Seitenwände); *syn.* überdachter Einstellplatz [m].

carr [n] [UK] *phyt.* ▶fen wood.

carriage porch [n] *arch.* ▶porte cochère.

carriage-way [n] [UK] *trans.* ▶travelled way [US].

carriage-way embankment [n] *eng. trans.* ▶transportation embankment.

carrier channel [n] *constr. eng. wat'man.* ▶receiving stream.

carrier ditch [n] *agr. constr.* ▶drainage ditch.

carrier drain [n] *agr. constr.* ▶collector drain.

carrying [n] **away** [UK] *constr.* ▶hauling away [US].

695 carrying capacity [n] **of landscape** *ecol. plan. recr.* (1. Capability of a landscape to accommodate competing uses, seen from economic and environmental points of view; ▶competing land-use pressure, ▶environmental stress tolerance, ▶recreation load, ▶recreational precedence. 2. The limit in number of humans or animal species which can occupy an area without degrading the environment is called **saturation capacity of an area**; ▶peak visitor use, ▶visitor pressure); *s* **capacidad [f] de usos (de un paisaje)** (1. Posibilidad de un paisaje de acomodar diversos usos que compiten entre sí, considerada normalmente desde un punto de vista económico; ▶resiliencia de un ecosistema, ▶vocación recreativa preferente. 2. **Saturación de una zona:** explotación total de las posibilidades de aprovechamiento de una zona paisajística; ▶cantidad máxima de visitantes, ▶presión por competencia entre usos, ▶presión recreional, ▶presión turística); *f 1* **capacité [f] d'accueil des sites** (1. ▶Capacité de charge d'un écosystème face à des usages concurrents de l'espace ; ▶pression exercée par les usages du sol. 2. Potentiel d'un paysage à tolérer différentes occupations du sol) ; *syn.* capacité [f] d'accueil d'un paysage ; *f 2* **capacité [f] de charge de tourisme (CCT)** (Méthodologie expérimentale, fondée sur l'évaluation de la vulnérabilité spécifique d'une zone [p. ex. îles, qui par leur faible niveau de risque se prêtent à des projets de développement touristique] ; cette approche vise à limiter, devant l'essor du tourisme, les dégradations environnementales et de prévenir les conflits d'usage dans le cadre d'un développement durable ; ▶capacité de charge d'un écosystème, ▶pointe de fréquentation [touristique], ▶pression exercée par les usages des sols, ▶pression exercée par les visiteurs, ▶pression touristique, ▶vocation touristique) ; *f 3* **capacité [f] de charge touristique [CH]** (Concept, instrument de gestion visant à calculer l'intensité d'exploitation touristique ou la fréquentation que peut supporter un site rural ou de montagne compte tenu de ses caractéristiques économiques, écologiques et sociales) ; *g* **Aufnahmekapazität [f] der Landschaft** (1. Mögliche ▶Belastbarkeit 1 durch konkurrierende Nutzungsansprüche, meist unter ökonomischem Aspekt. 2. Potenzial einer Landschaft zur Aufnahme von Nutzungen. 3. Vollständige Ausnützung von Nutzungsmöglichkeiten in einem Landschaftsraum, z. B. von Erholungsuchenden und den damit verbundenen Infrastrukturen; ▶Besucherdruck, ▶Erholungsdruck, ▶Erholungseignung, ▶Nutzungsdruck, ▶Spitzenbesucheraufkommen); *syn.* Kapazität [f] einer Landschaft, Tragfähigkeit [f] einer Landschaft hinsichtlich des Nutzungsdruckes.

696 car salvage yard [n] [US] *envir. urb.* (Confined area for deposit of abandoned inoperable cars; ▶scrapyard); *syn.* auto grave yard [n] [US], car scrap-yard [n] [UK]; *s* **cementerio [m] de coches [Es]** (Lugar de almacenaje de automóviles viejos fuera de uso; ►depósito de chatarra); *syn.* cementerio [m] de carros [AL]; *f* **cimetière [m] de voiture** (Lieu de dépôt d'épaves, de carcasses de véhicules hors d'usage ; ►dépôt de ferraille) ; *syn.* cimetière [m] d'automobiles, cimetière [m] d'épaves automobiles ; *g* **Autofriedhof [m]** (Lagerplatz für ausgediente, schrottreife Autos; ►Schrottplatz); *syn.* Autowrackplatz [m].

car scrap yard [n] [UK] *envir. urb.* ►car salvage yard [US].

697 car sharing [n] *sociol. trans. urb.* (Joint use of one or more cars by members of a community [a company, association, or cooperative] with the aim of reducing car traffic and the number of vehicles on the road and thus the burden on the environment, particularly air pollution. Economically, **c. s.** makes sense for motorists, who drive their cars less than 7,000 km per year; cars may also be rented for an hourly fee. The locations for **c. s.** are distributed throughout a city near connections to public transport [train stations, subway stops, central bus stations, etc.] and are therefore easily accessible); *s* **carsharing [m]** (Aprovechamiento común de uno o más automóviles por la membresía de una comunidad [empresa, asociación o cooperativa] con el objetivo de reducir el tráfico individual y el número de vehículos en las carreteras y con ello la carga ambiental, en especial la contaminación atmosférica. Económicamente **c.** tiene sentido para motoristas que recorren con su vehículo menos de 7000 km al año. Los sistemas de **c.** funcionan de la siguiente manera: la correspondiente comunidad adquiere vehículos de diferentes tipos y tamaños [coches pequeños y medianos, camionetas] que cada participante en la comunidad puede utilizar previo aviso, pagando una cuota por hora o día que varía según el tipo de vehículo, además de la cuota de entrada. La organización se encarga de asegurar y reparar los vehículos. Los vehículos se aparcan en plazas reservadas repartidas por la ciudad cerca de puntos con buena comunicación de transporte público o cerca de los lugares de residencia de la membresía. Para acceder a ellos hay instalados buzones en determinados lugares, donde se depositan las llaves y los documentos de los automóviles); *f* **autopartage [m]** (Service de mise à disposition de voitures pour une courte durée [1 heure ou plus]. L'inscription à ce service permet aux abonnés de réserver facilement un véhicule situé sur un parking à proximité de chez eux. La réservation et l'accès aux véhicules sont facilités par les nouvelles technologies. Les voitures sont ainsi disponibles 24 heures sur 24 et 7 jours sur 7. L'autopartage répond à un vrai besoin puisque 70 % des trajets effectués par ce moyen ne pourraient pas être faits autrement qu'en voiture, la desserte en transports publics n'étant pas assez complète, la course en taxi et la location traditionnelle de voiture n'étant pas adaptées pour des trajets de moins d'une demi-journée. La cible de l'autopartage est principalement urbaine, les abonnés devant avoir la possibilité de se déplacer en transports publics ou par des modes doux [marche, vélo] pour accéder facilement aux véhicules. L'intérêt environnemental de l'autopartage est double. Il induit pour ses utilisateurs une diminution progressive du nombre de kilomètres parcourus en voiture et donc réduit la consommation d'énergie et les émissions de polluants. En outre, il permet de libérer de l'espace urbain utilisé auparavant pour le stationnement des véhicules) ; *g* **Carsharing [n]** (Gemeinschaftliche Nutzung eines oder mehrerer Autos durch Mitglieder einer Interessengemeinschaft [GmbH, Verein, Genossenschaft] mit dem Ziel, den Autoverkehr durch Reduzierung des Kfz-Bestandes und damit die Umweltbelastung, insbesondere die Luftverschmutzung, zu verringern. Wirtschaftlich sinnvoll ist das **C.** für Autofahrer, die jährlich weniger als 7000 km fahren; die Autos sind gegen eine Gebühr auch stundenweise mietbar. Die Standorte der Autos sind über eine Stadt in der Nähe von Knotenpunkten des öffentlichen Nahverkehrs [Bahnhöfe, S-Bahn-Haltestellen, zentralen Busbahnhöfe etc.] verteilt und sind deshalb leicht erreichbar); *syn.* genossenschaftliche Autonutzung [f].

carting [n] **away** [UK] *constr.* ►hauling away [US].

carting [n] **away without tipping certificate** [UK] *constr.* ►hauling away without dumping certificate [US].

carting [n] **to the site** [UK] *constr.* ►hauling to the site [US].

cascade burial [n] [US] *adm.* ►coffin burial.

cascading [adj] *hort. plant.* ►arching.

cast-in-place concrete [n] [US] *constr.* ►poured-in-place concrete [US]/in-situ (cast) concrete [UK].

cast iron channel [n] **with grating** [UK] *constr.* ►cast iron drain with grating [US].

698 cast iron drain [n] **with grating** [US] *constr.* (Drainage channel with U-shaped cross section, often covered by a grill[e] or lid; ►trench drain); *syn.* cast iron channel [n] with grating [UK] (SPON 1974, 250, 319), covered trench drain [n] [US], strip drain [n] (with a metal grill[e]) [US] (OSM 1999, 168), trench frame [n] with grated or solid cover [US]; *s* **canalón [m] de desagüe cubierto con rejilla** (►canal de drenaje [superficial] con sección con forma de U, generalmente cubierto con reja o tapa); *f* **caniveau [m] à grille** (►Caniveau en général préfabriqué en forme de U et équipé de différentes couvertures, p. ex. grille plate ou convexe, grille caillebotis ; cf. norme N.F. P 98 302 pour les caniveaux en béton) ; *syn.* caniveau [m] avec caillebotis ; *g* **Kastenrinne [f]** (►Entwässerungsrinne mit U-förmigem Querschnitt, meist mit Steg-, Gitterrosten oder Deckeln abgedeckt; cf. DIN 19 580).

699 casual species [n] *phyt.* (Casuals are those species, according to the ►degree of naturalization which, while growing wild, have no enduring position in the vegetation, because they are not capable of maintaining themselves without the renewed, direct human intervention; e.g. annuals or cultivated plants which are not hardy, occurring wild on waste sites, but not producing seeds, or species introduced by bird food, whose regular occurrence on particular sites is only made possible by continuous ►reintroduction of their mode of propagation); *syn.* ephemeral plant [n], ephemerophyte [n]; *s* **efemerófito [m]** (Aplícase a los epecófitos que aparecen y desaparecen de manera irregular y accidental, sin instalarse de manera persistente en el país; DB 1985; ►grado de naturalización); *syn.* especie [f] fugaz; *f 1* **espèce [f] fugace** (Dans le processus de naturalisation, espèce existant momentanément à l'état sauvage mais incapable de se reproduire sans une aide directe de l'homme, p. ex. les plantes annuelles ou plantes de région à climat doux ou chaux existant sur les décharges mais dont les graines n'arrivent pas à maturation ; ou bien toutes espèces introduites dont la régularité de leur apparition en des lieux particuliers ne tient qu'a leur introduction constamment renouvelée par l'homme ; ►degré de naturalisation ; *f 2* **espèce [f] éphémérophyte** (Espèce végétale annuelle des climats arides accomplissant tout son cycle [floraison et fructification] en quelques jours après la pluie ; *contexte* éphémérophytes du désert) ; *syn.* éphémérophyte [f] (à ne pas confondre avec un « éphémère » qui est un insecte névroptère de la famille des archiptères qui vit aux abords des plans d'eau et ne vit qu'un jour, voire même que quelques heures) ; *g* **unbeständige Art [f]** (Nach dem ►Grad der Einbürgerung diejenige Pflanzenart, die zwar hier und da wild wachsend auftritt, aber keinen festen Platz in der Vegetation hat, da sie nicht fähig ist, sich aus eigener Kraft ohne erneute direkte Hilfe des Menschen auf längere Zeit zu erhalten; z. B. Einjährige oder nicht winterharte Kulturpflanzen, die auf Schuttplätzen verwildert vorkommen, aber nicht zur Samenreife gelangen oder mit Ölsaaten, Vogelfutter etc. eingebrachte Arten, deren regelmäßiges Auftreten an bestimmten Orten nur dadurch ermöglicht wird, dass ihre Verbreitungseinheiten immer wieder von neuem eingebracht werden); *syn.* Ephemerophyt [m].

casualty [n] *envir. leg.* ►maritime casualty.

casualty [n], **oil pollution** *envir.* ►oil pollution accident.

700 catastrophic rockslide [n] **[US]/catastrophic rock-slide** [n] **[UK]** *geo.* (Sudden, devastating sliding or falling of huge rock volumes en masse over a bedding or faultplane; ►mass slippage [US]/mass movement [UK], ►rockslide [US]/rock-slide [UK]); *s* **avalancha [f] de rocas** (Tipo de ►movimiento gravitacional rápido en el cual las masas de roca descienden con gran velocidad pendiente abajo causando los correspondientes daños; si se derrumba un gran trozo de roca se denomina **desplome [m]**; ►deslizamiento de rocas); *syn.* caída [f] masiva de rocas, derrumbamiento [m] [Es], derrumbe [m] de montaña [AL]; *f* **éboulement [m] catastrophique** (DG 1984, 471 ; *processus d'érosion des versants* chute instantanée en grandes masses provoquant un chaos de rochers ; ►glissement de montagne, ►mouvement de masse) ; *g* **Bergsturz [m]** (Im Vergleich zum ►Bergrutsch plötzlicher, katastrophenartiger ►Massenversatz von Locker- und meist Festgesteinen an Steilhängen oder Felswänden).

701 catch basin [n] **[US]** *constr.* (Inlet structure that receives surface water runoff from vehicular/paved surfaces. **In U.S.**, it consists of a concrete base unit or footing, precast rings or masonry block courses [to specified depth], riser unit, tapered ring [typically 1 200mm to 600mm] to receive inlet grate frame, brick levelling course or adjustment ring, ►cast iron drain with grating and ►storm drain grate [US]/gully grating [UK]; it typically accommodates at least one inflow pipe and an outflow pipe); **in U.K. road gully**, usually comprising precast concrete units as follows: base unit, ►manhole adjustement ring, intermediate ring, ►riser unit [US]/chamber section [UK], sleeve unit with ogee joints, shaft, adjustment ring and gully grating [UK]; **in D.**, it consists of the following components [from bottom to top]: base unit, ►manhole adjustement ring, ►riser unit [US]/chamber section [UK], sleeve unit with ogee joints, shaft, adjustment ring and ►storm drain grate [US]/gully grating [UK], ►sediment basin; ►surface inlet [US]/gully [UK], ►yard drain [US]); *syn.* road gully [n] [UK]; *s* **sumidero [m] de calle** (Dispositivo de evacuación de aguas pluviales de las superficies pavimentadas; ►arqueta de recogida con filtro, ►anillo de ajuste de registro, ►rejilla de entrada [de alcantarilla], ►sumidero de patio); *f* **regard [m] à grille** (Dispositif d'évacuation des eaux pluviales sur les surfaces de circulation qui suivant sa fonction peut comprendre les éléments suivants : ►radier de décantation, élements intermédiaires, raccord sur canalisation, ►rehausse de regard, ►grille de regard [d'évacuation], un regard de petit diamètre d'en général 300 mm est dénommé ►siphon de cour) ; *g* **Straßenablauf [m]** (Zur Entwässerung des Oberflächenwassers auf Verkehrsflächen eingebaute Vorrichtung, meist aus Betonfertigteilen, je nach Funktion bestehend aus einem Bodenteil mit Ablaufrohr oder ►Sumpf 1, Zwischenteil, Muffenteil, Schaft, ►Schachtauflagering und ►Einlaufrost; ein kleinerer Ablauf mit einem Innendurchmesser von 300 mm heißt ►Hofablauf); *syn.* Gully [m], Straßeneinlauf [m], Regeneinlauf [m], Trumme [f] (SUT 1993, 162).

702 catch basin inlet [n] **[US]** *constr.* (Generic term covering a surface opening of a catch basin [US]/road gully [UK], typically on the road or in a road gutter; ►curb inlet [US]/kerbinlet [UK], ►grate inlet [US]/gutter inlet [UK]); *syn.* drain inlet [n] [US], gutter inlet [n] [UK]; *s* **boca [f] de alcantarilla** (Aper-

tura de alcantarilla por donde pasan las aguas pluviales de las calles a la alcantarilla; ►boca de alcantarilla al ras del suelo, ►sumidero en bordillo); *f* **bouche [f] d'égout** (Ouvrage assurant la collecte en surface des eaux pluviales et éventuellement des eaux de lavage des chaussées ; DOFB 1979, 8 ; suivant la façon dont les eaux pluviales sont collectées, on distingue les bouches à accès latéral [avaloir] et les bouches à accès par le dessus [►bouche (d'égout) à grille] ; ►bordure avaloir) ; *g* **Straßenablauföffnung [f]** (Einlaufstelle für Oberflächenwasser an Straßenbordrinnen; ►ebenerdige Straßenablauföffnung, ►Straßenablauf im Bordstein).

703 catch basin [n] **with sump** [US] *constr.* (Inlet structure with a sediment basin at the structure's bottom which collects heavy sediment material before it can enter the pipe; TSS 1997, 330); *syn.* road gully [n] with sump [UK]; *s* **sumidero [m] (de calle) con decantador** (Cisterna o depresión en el punto que una bajante descarga, para evitar que cualquier cuerpo extraño pase a la alcantarilla; DACO 1988); *f* **bouche [f] à décantation** (Bouche dont le radier est en contre-pente par rapport à la conduite aval, de façon à favoriser la rétention des boues) ; *syn.* regard [m] à grille avec cuvette de décantation ; *g* **Straßenablauf [m] mit Sumpf** (Entwässerungsvorrichtung an Straßen mit geschlossenem Bodenteil für das Zurückhalten des Nassschlammes).

catchment area [n] *plan.* ►suburban area; *geo. hydr. nat'res.* ►drainage basin; *recr. urb.* ►origination area [US]; *wat'man.* ►drinking water catchment area.

catchment basin [n] *geo. hydr. nat'res.* ►drainage basin.

category [n] *plan. recr.* ►recreation time category.

704 caterpillar removal [n] **by hand** *hort.* (F., manual removal of caterpillars to prevent damage to trees carried out between 1st December to 1st March; **in U.S.**, such removal is done between similar dates); *s* **recogida [f] de orugas** (En Francia norma sobre el desalojo manual de orugas de los árboles entre el 1º de diciembre y el 1º de marzo); *f* **échenillage [m]** (F., fait partie des travaux d'entretien pendant l'année de garantie et consiste au ramassage des chenilles pendant la période du 1er décembre au 1er mars et pendant la pousse des feuilles) ; *g* **Absammeln [n, o. Pl.] von Raupen** (Pflege- und Unterhaltungsarbeiten, die in F. im Rahmen der Gewährleistung in der Zeit vom 1. Dezember bis 1. März durchgeführt werden).

cathedral forest [n] **[US]** *for.* ∗►old-growth forest.

705 cattail swamp [n] **[US]** *phyt.* (Reed formation of predominantly cattail flag [US]/reedmace [UK] species *[Typha angustifolia* or *Typha latifolia]* and belonging to the association of *Typhetum angustifoliae* or *Typhetum latifoliae*; ►reed swamp); *syn.* reedmace swamp [n] [UK]; *s* **espadañal [m]** (Lugar húmedo en que se crían con abundancia las espadañas, también llamadas aneas, eneas o bohordos *[Typha esp.]*, en las orillas de arroyos y lagunas; ►carrizal de fragmitea); *f* **typhaie [f]** (Groupement végétal composé principalement des espèces de Massettes *[Typha]* et formant les associations du *Typhetum angustifoliae* ou du *Typhetum latifoliae* ; ►roselière, ►roselière a roseaux, ►phragmitaie) ; *syn.* roselière [f] à Massettes, roselière [f] à *Typha* ; *g* **Rohrkolbenröhricht [n]** (Röhrichtbestand, der vorwiegend aus *Typha*-Arten besteht und in Europa zur Assoziation *Typhetum angustifoliae* oder *Typhetum latifoliae* gehört; *OB* ►Röhricht).

cattle grazing areas [npl] *phyt.* ►vegetation of cattle grazing areas.

706 cattle trampling [n] *agr. landsc.* (Heavy treading on meadows by cattle hoofs; ►trampling damage, ►tracks of grazing animals); *s* **pisoteo [m] de animales** (►pisoteo de la vegetación, ►camino de ganado); *f* **piétinement [m] de bétail**

C

(Piétinement causé par le bétail en pâture ; ▶piétinement, ▶pieds de vaches) ; *g* **Viehtritt [m]** (▶Trittbelastung durch Weidevieh, ▶Viehgangeln).

707 causal factor [n] *ecol. plan.* (Determining circumstance, which can have one or more impacts; in the context of an environmental assessment under the EU ▶Habitats Directive; a series of circumstances, which cause a specific effect on the environment, are referred to as the **sum of the influences** acting on a natural habitat; cf. Article 1 of the EU Habitats Directive); *s* **factor [m] causante** (Circunstancia determinante que puede tener un o más efectos; en el contexto de la evaluación ambiental bajo la ▶Directiva de Hábitats de la UE, una serie de circunstancias que causan un efecto específico sobre el medio ambiente son denominadas conjunto de las influencias que actúan sobre el hábitat natural ; cf. art. 1 Directiva de Hábitats); *f* **effet [m]** (Décrit une conséquence d'un projet indépendamment du territoire affecté ; dans le contexte de l'évaluation des incidences sur les sites Natura 2000 ce terme désigne les influences spécifiques d'un projet générées sur l'environnement par les travaux, les ouvrages et leur exploitation ; les changements environnementaux sont sans conteste imputables à ces effets, sont induit par certaines caractéristiques du projet ou sont liés à celui-ci ; ▶Directive Habitat Faune Flore) ; *g* **Wirkfaktor [m]** *ecol. plan.* (Ursache, die eine oder mehrere Auswirkungen auslöst; im Kontext der FFH-Prüfung werden als **W.en** bau-, anlage- und betriebsspezifische Einflüsse bezeichnet, die Umweltveränderungen verursachen; die einzelnen Veränderungen sind ursächlich auf diese Faktoren, d. h. bestimmte Projektmerkmale zurückzuführen bzw. hängen mit diesen zusammen; cf. LEIT 2004; ▶Fauna-Flora-Habitat-Richtlinie).

708 causality research [n] *envir.* (Investigation and possible clarification of a complex of factors which demonstrates a cause-effect connection); *s* **investigación [f] de causas** (Análisis de las relaciones causa-efecto de un complejo de factores); *f* **recherche [f] des causes** (Analyse et clarification éventuelle du complexe des facteurs représentatif de la relation de cause à effet) ; *g* **Ursachenforschung [f]** (Untersuchung und eventuelle Klärung des Faktorenkomplexes, der einen Ursache-Wirkung-Zusammenhang darstellt).

causation [n]**, principle of** *envir. pol.* ▶polluter-pays principle.

709 causer liability [n] *envir. leg.* (Legal duty of the initial causer of environmental damage to pay a penalty, implement measures to repair the damage or for example to introduce environmentally safe production processes; ▶causer of an impact/intrusion, ▶polluter pays principle; *specific term* pollution causer liability); *s* **responsabilidad [f] del causante de un daño** (Responsabilidad legal del ▶causante de un daño al medio ambiente de pagar una multa, sanear el ambiente dañado o de cambiar el procedimiento de producción para evitar daños futuros; ▶principio de causalidad); *f* **responsabilité [f] de l'auteur** (Obligation légale qu'a l'▶auteur d'une atteinte à l'environnement de réparer les dommages commis soit par une amende, soit par des travaux de réhabilitation ou encore par la mise en œuvre de nouvelles technologies respectueuses de l'environnement ; ▶principe pollueur-payeur) ; *g* **Verursacherhaftung [f]** *pol.* (In der Rechtsprechung die Verantwortlichkeit des ▶Verursachers von Umweltschädigungen mit der Folge, Abgaben zu zahlen, Sanierungsmaßnahmen durchzuführen oder z. B. neue umweltschonende Produktionsverfahren anzuwenden; ▶Verursacherprinzip).

710 causer [n] **of an impact/intrusion** *envir. pol.* (Person, company, group, or public authority, who or which causes an adverse impact or ▶intrusion upon the natural environment or creates the preconditions for environmental nuisances); *syn.* initiator [n] of an impact/intrusion; *s* **causante [m/f]** (de una ▶intrusión en el entorno natural); *syn.* inductor, -a [m/f]; *f* **auteur [m]** (Personne physique, maître d'ouvrage, société, établissement public portant ▶atteinte au milieu naturel ; *contexte* auteur de l'atteinte/la pollution) ; *g* **Verursacher/-in [m/f]** (Person, Firma, Gruppe oder öffentliche Hand, die z. B. Umweltschäden oder einen ▶Eingriff in Natur und Landschaft bewirkt oder eine Voraussetzung für die Umweltbelastung schafft).

cave-in [n] **[US], riverbank** *geo.* ▶riverbank collapse [US].

711 cavity [n] *arb.* (Hole in a tree caused naturally by the impact of lightning, breakage in a storm or frost damage, or caused by inadequate or incorrect tree maintenance resulting in decay; ▶area of rot, ▶cleaning of a cavity); *s* **cavidad [f]** (Agujero con ▶caries húmeda, generalmente en tronco de árbol de madera blanda, causado por rayo, tormenta o helada o por falta de cuidado que puede llevar al vaciado del tronco; ▶limpieza de la caries húmeda); *f* **cavité [f] (naturelle)** (Formée naturellement [en général chez les végétaux à bois tendre] sous l'impact de l'éclair, bris de vent ou gélivure, ou causée par des opérations de taille mal exécutées occasionnant le ▶pourrissement du tronc ou de charpentières ; ▶pourriture, ▶raclage des tissus pourris) ; *g* **Höhlung [f]** (Natürlich entstandene [meist bei Weichhölzern], durch höhere Gewalt — Blitzschlag, Sturmbruch oder Frostriss — oder durch unterlassene sowie unsachgemäße Pflege verursachte ▶Morschung an Bäumen, die den Stamm oder Starkäste aushöhlen; *UB* Stammausfaulung; ▶Aushauung).

712 cavity bracing [n] *arb. constr.* (Obsolete term for the mechanical strengthening of a cavity opening in a tree stem by ▶threaded rods. A now seldom used technique, which is without structural effect. It may, in fact, be harmful in that it can create additional wounds, in which diseases can become established); *syn.* solid bracing [n] [also UK]; *s* **reforzamiento [m] del tronco con barras metálicas** (Estabilización de troncos de árboles maduros con cavidades grandes por medio de ▶cinchas roscadas; medida muy utilizada a pesar de que no tiene efectos estructurales y además puede causar heridas adicionales por la cuales pueden penetrar parásitos); *f* **renforcement [m] métallique des troncs** (Pour rétablir la cohésion des troncs creux ou des grosses charpentières en plaçant des ▶tiges filetées environs tous les 40 cm ; méthode souvent utilisée, mais sans aucun effet sur la stabilité de l'arbre. C'est une méthode qui reste incertaine quand à la conséquence sanitaire du tronc — introduction probable de pourriture et de parasites suite à une mauvaise cicatrisation) ; *g* **Stammverstärkung [f] durch Stabanker** (*Obsolete Technik* mit ▶Gewindestäben ausgesteifte Aushauungen von Faulstellen in Baumstämmen. Häufig angewendete, aber statisch nutzlose und für die Gesundheit des Stammes fragwürdige Methode, da zusätzliche, potentielle Krankheitsherde geschaffen werden).

cavity-nester species [n] **[US]** *zool.* ▶cavity-nesting bird [US].

713 cavity-nesting bird [n] **[US]** *zool.* (Bird which breeds in a hole or ▶nesting box); *syn.* cavity-nester species [n] [US], hole-nesting bird [n] [UK]; *s* **ave [f] que anida en cavidades** (Ave que anida en nichos o en cavidades de troncos o en ▶nidales); *f* **oiseau [m] nichant dans des cavités** (Oiseau qui niche dans des trous ou des cavités d'arbres, dans des nids creusés ou dans un ▶nichoir ; LAF 1990) ; *g* **Höhlenbrüter [m]** (In Höhlen oder ▶Nistkästen brütender Vogel); *syn.* höhlenbrütender Vogel [m].

cement [n] *constr.* ▶rapid-hardening cement.

cement [n]**, quick-drying cement** *constr.* ▶rapid-hardening cement.

C

cement [n]**, rapid-setting cement** *constr.* ►rapid-harden-ing cement.

714 cemented soil [n] *agr. hort.* (Soil stabilized with cementing material, i.e. soil cement; ►soil cementing, ►stabi-lized subgrade [US]/stabilised sub-grade [UK]); *s* **suelo** [m] **cimentado** (►subsuelo estabilizado, ►estabilización del suelo con mortero); *f* **sol-ciment** [m] (Technique utilisée pour le renforcement des sols d'un site avec un mélange de terre, de ciment-Portland, d'eau puis compacté ; le sol-ciment est employé pour la stabilisation des sols des chantiers, pour les ouvrages de régulation des cours d'eau [protection des berges, de chenaux], en revêtement d'étangs, etc. ; ►fond de forme stabilisé, ►traitement de sol avec un liant) ; *syn.* sol [m] stabilisé au ciment ; *g* **vermörtelter Boden** [m] (►verbesserter Untergrund, ►Vermörtelung des Bodens).

715 cemetery [n] *adm. leg. urb.* (Enclosed piece of land serving at the present or in the past as a burial ground for the remains of deceased human beings, belonging to a particular political or religious community. A **c.** always contains a set number of graves, often of different kinds, independent of whether they are occupied or not. **Cemeteries** are places of remembrance, mourning, contemplation and reflection and fulfil important urban, cultural, social, and ecological functions as well as those of a recreational character. There are **communal ceme-teries**, which are built, maintained and managed by a munici-pality and **denominational cemeteries**, which are usually established by a certain religious community for the exclusive interment of members of the same confession. Denominational cemeteries are differentiated into Christian, Jewish, Moslem cemeteries or 'mixed' cemeteries, which are used for the burial of members of various confessions and non-believers alike; cf. GAE 2000; ►church yard, ►forest cemetery, ►park-like cemetery, ►pet cemetery); *syn.* graveyard [n]; *s* **cementerio** [m] **(1)** (Terreno cercado para enterrar a los muertos, en algunos países europeos y en los EE.UU. están diseñados como parques de estilo inglés al contrario que en los países iberoamericanos, en los cuales predominan los panteones y las estructuras de piedra. Los **c.** pueden ser comunales o religiosos, es decir estar dedicados a enterrar a miembros de una religión como la católica, protestante, judía, musulmana, etc.; ►cementerio 2; ►cementerio boscoso, ►cementerio de animales, ►cementerio paisajístico); *syn.* camposanto [m]; *f* **cimetière** [m] (Parcelle publique et sacrée enceinte sur laquelle, après une cérémonie, on enterre les morts sans distinction de confession dans des tombes individuelles ou lignagières ; les cimetières sont divisés en concessions cadastrées qui peuvent être louées ou vendus à une personne ou une famille ; selon les cultures et les époques, les cimetières et les sépultures sont plus ou moins monumentalisés et sacralisés ; en France les cimetières sont devenus propriétés communales dont la gestion et l'entretien lui incombent ; dans certains pays occidentaux (Allemagne, Italie, etc.) les cimetières peuvent être **confes-sionnels** tels que les **cimetières catholiques, protestants, juifs/ israélites, musulmans** ; ►cimetière ancien, ►cimetière en ambiance forestière, ►cimetière paysager, ►cimetière pour animaux, ►enclos paroissial) ; *g* **Friedhof** [m] (Räumlich abge-grenztes, eingefriedetes Grundstück, das zur Bestattung der irdischen Reste von Menschen einer bestimmten politischen oder religiösen Gemeinschaft dient oder gedient hat. Ein **F.** umfasst immer eine bestimmte Anzahl von Grabstellen, häufig ver-schiedener Art, unabhängig davon, ob sie belegt sind oder nicht. **F.höfe** sind Orte des Gedenkens, der Trauer, Besinnung und Einkehr und erfüllen wichtige städtebauliche, kulturelle, soziale, ökologische und der Erholung dienende Funktionen. Es gibt **kommunale F.höfe**, die von einer politischen Gemeinde ange-legt, unterhalten und bewirtschaftet werden und **konfessionelle**

F.höfe, die i. d. R. von Religionsgemeinschaften für die aus-schließliche Bestattung von Angehörigen ihres Bekenntnisses errichtet und bewirtschaftet werden. In Baden-Württemberg z. B. werden diese **F.höfe** von den Kommunen getragen. Konfes-sionelle **F.höfe** werden in **kirchliche, jüdische, muslimische** und **Simultanfriedhöfe**, die für die Bestattung auch von Angehörigen anderer Bekenntnisse und Bekenntnislosen zur Verfügung stehen, unterschieden;   ►Kirchhof,   ►Parkfriedhof,   ►Tierfriedhof, ►Waldfriedhof).

716 cemetery department [n] *adm. plan. urb.* (Local authority responsible for burial of the dead; **in U.S.,** cemeteries are privately-owned and -managed, except for National Ceme-teries which are federally owned and managed by the Veterans Administration for military personnel and public personages. Federally-owned Arlington Cemetery in Arlington, County Virginia, is reserved for burial of military dead and prominent personages; **in U.K.,** department of a local authority responsible for burial of the dead); *s* **administración** [f] **de cementerios** *syn.* departamento [m] de cementerios; *f* **service** [m] **des pompes funèbres et cimetières** (Administration communale responsable des opérations funéraires) ; *g* **Friedhofsamt** [n] (Behörde einer Gemeinde oder eines Landkreises, die für die hoheitlichen Aufgaben der Bestattung von Toten zuständig ist).

717 cemetery gardener [n] *hort. prof.* (**In D.,** term for a professionally-trained gardener, who is specialized in mainte-nance and planting of grave sites; **in U.K.,** c. g.s do not receive particular training, but may, for example, take the optional unit on cemeteries and graveyards in the Amenity Horticulture [Land-scaping] Courses offered by National Vocational Qualifications [NVQ] or at horticultural training schools. The job includes measuring, marking out and digging graves, keeping them tidy, checking headstones for damage and instability, looking after grass and plants around the cemetery; **in U.S.,** no equivalent exists; ►cemetery gardening); *syn.* cemetery groundsman [n], cemetery worker [n]; *s* **jardinero** [m] **de cementerios** (D., título profesional para jardineros que se han especializado en el diseño y cuidado de tumbas; ►trabajos de jardinería de cementerios); *f* **jardinier** [m] **de cimetière (±)** (D., dénomination pour un jardinier responsable de l'entretien des tombes et des espaces verts dans les cimetières) ; ►travaux d'aménagement paysager de cimetière) ; *g* **Friedhofsgärtner/-in [m/f]** (*Gärtner der Fachrich-tung Friedhofsgärtnerei* in A., CH und D. staatlich anerkannter Ausbildungsberuf und Berufsbezeichnung für gelernte Gärtner/ -innen, die die Pflege von Rabatten, Rasen- und Gehölzflächen auf Friedhöfen und die Anlage, Betreuung und Erneuerung der Grabanlagen sowie Trauerbinderei, die Dekoration von Trauer-hallen und Särgen durchführen; ►friedhofsgärtnerische Leis-tungen).

718 cemetery gardening [n] *constr.* (D., work of professionally-trained ►cemetery gardeners, including display of sample gravesite designs at garden festivals; no U.S. equivalent); *syn.* cemetery landscaping [n] [also US]; *s* **trabajos** [mpl] **de jardinería de cementerios** (Servicios de jardinería prestados por ►jardineros de cementerios en los mismos cementerios o en exhibiciones de horticultura); *f* **travaux** [mpl] **d'aménagement paysager de cimetière** (Travaux d'aménagement et d'entretien des sépultures exécutés par un ►jardinier de cimetière) ; *g* **fried-hofsgärtnerische Leistungen** [fpl] (Gartenbauliche Leistungen der ►Friedhofsgärtner auf Friedhöfen und als Leistungsschau auch auf Gartenschauen); *syn.* friedhofsgärtnerische Arbeiten [fpl].

cemetery groundsman [n] *prof.* ►cemetery gardener.

cemetery landscaping [n] [US] *constr.* ►cemetery gardening.

C

cemetery worker [n] *prof.* ►cemetery gardener.

center [n] [US] *recr.* ►health resort center, ►leisure center, ►nature study center; *plan.* ►population growth center.

center city [n] [US] *leg. urb.* ►core area (1).

719 center city redevelopment [n] [US] *urb.* (Sum of renewal measures for the improvement of living conditions in deteriorated housing districts, initiated and supervised by local or state government; **in U.S.**, state or federally funded programs for this purpose aimed at low income population, primarily housing, may include mixed uses to stimulate economic development; historically known as *urban renewal*; ►neighborhood redevelopment [US]/district redevelopment [UK], ►block clearance, ►village renewal, ►comprehensive redevelopment, ►urbam redevelopment); *syn.* town centre redevelopment [n] [UK], urban renewal [n], urban regeneration [n] [UK] (LD 1988 [9], 58), regeneration [n] of city cores, downtown regeneration [n] [US]; **s rehabilitación** [f] **urbana** (Conjunto de actividades urbanísticas realizadas con el fin de mejorar las condiciones de habitación y vida de la población de los barrios antiguos de la ciudad. Como en muchos casos la rehabilitación se lleva a cabo barrio por barrio, se habla también de ►rehabilitación de barrios; ►despeje de patios interiores, ►reestructuración urbana, ►remodelación urbana, ►renovación rural); *syn.* renovación [f] urbana; **f 1 restauration** [f] **urbaine** (Ensemble des mesures d'urbanisme visant soit à restaurer et à mettre en valeur des îlots opérationnels situés dans des secteurs sauvegardés, soit à remettre en état, à moderniser et, éventuellement à démolir un ensemble d'immeubles à l'intérieur d'un périmètre défini [zone de restauration urbaine] et impliquant la conservation du cadre immobilier en raison de son caractère historique ou architectural ; ►réaménagement des quartiers, ►réhabilitation des quartiers, ►remodelage urbain, ►rénovation de village, ►restructuration d'îlot, ►restructuration urbaine) ; **f 2 rénovation** [f] **urbaine** (Ensemble des dispositions administratives, juridiques et financières prises pour réaliser une opération d'urbanisme de remise à neuf d'un quartier impliquant la destruction complète de l'habitat existant, le remodelage du plan d'ensemble et la reconstruction des bâtiments et d'équipements nouveaux, la réinstallation temporaire et définitive de la population ; opération à l'initiative de la commune et le plus souvent confiée à un organisme aménageur privé ou public pour laquelle est déterminée une **zone de rénovation urbaine [Z.R.U.]**) ; *syn.* renouvellement [m] urbain ; **g Stadtsanierung** [f] (Staatlich initiierter und gesteuerter Prozess, der sich auf die Gesamtheit aller städtebaulichen Maßnahmen, die der nachhaltigen Verbesserung der Wohn- und Lebensbedingungen städtischer Bevölkerungsgruppen in älteren Stadtvierteln dienen, bezieht. Da meist nur Teile der Stadt saniert werden, spricht man auch von ►Stadtteilsanierung; ►Blockentkernung, ►Dorferneuerung, ►Flächensanierung, ►Stadtumbau. Der heute öfter benutzte Begriff der **Stadterneuerung** drückt mehr ein umfassendes Aufgabenverständnis aus, um sich von rein baulichen oder auf Bodenordnung konzentrierten Ansätzen zu unterscheiden); *syn.* Altstadtsanierung [f].

720 center [n] **of attraction** [US]/**centre** [n] **of attraction** [UK] *recr.* (Spatial setting or use area that serves as a primary focus; ►attraction point, ►focal point); **s centro** [m] **de atracción** (Lugar al que se dirigen muchos visitantes por su atractivo; ►centro de gravedad 1, ►centro de gravedad 2); **f pôle** [m] **d'attraction** (Lieu qui polarise les activités, les touristes vers une zone ; ►point d'attrait, ►pôle d'attraction et d'animation) ; *syn.* lieu [m] d'attraction, espace [m] d'attraction ; **g Anziehungspunkt** [m] **(2)** (Ort, zu dem sich viele Menschen hinbegeben; ►Anziehungspunkt 1, ►Anziehungspunkt 3).

Central Authority [n] [England&Wales] *adm. leg.* ►law enforcement agency [US].

central business core [n] *leg. urb.* ►central business district.

721 central business district [n] *leg. urb.* (*Abbr.* CBD; 1. Legal term describing the boundary of the commercial core of a master plan or a zoning district [US]/planning zone [UK]. 2. Generic term describing the commercial core of any urban area: major shopping area within a city usually containing, in addition to retail uses, governmental offices, service outlets, professional, cultural, recreational and entertainment establishments and uses, residences, hotels and motels, appropriate commercial activities, and transportation facilities; cf. IBDD 1981); *syn.* central business core [n] [also US] (GGT 1979, 135); **s zona** [f] **central** (En D. término legal de la Ley Federal de Construcción [art. 9 (1) BauGB] y del Reglamento de Construcción [art. 1 (2 ss) BauNVO] que denomina a las áreas centrales de las ciudades con usos mayoritariamente comerciales, de servicios y de administración, práticamente sin uso residencial); **f zone** [f] **urbaine consacrée aux activités économiques (Zone UE) (±)** (F., zonage du PLU/POS ; zone d'implantation d'activités secondaires dans laquelle sont autorisés les établissements industriels, surfaces commerciales, les bureaux, les bâtiments et installations d'équipements collectifs et pouvant être partiellement mêlée à l'habitat existant ; cette zone est suivant les besoins divisée en plusieurs secteurs, p. ex. secteur UEa pouvant accueillir des établissements classés soumis à déclaration ou à autorisation, secteur UEb correspondant à l'accueil d'activités industrielles, artisanales, et de service, secteur UEc correspondant à l'accueil d'activités industrielles, artisanales et de service, secteur UEd correspondant à l'accueil d'activités artisanales, tertiaires et commerciales) ; **g Kerngebiet** [n] (D., *Abk.* MK; gemäß § 9 [1] BauGB und § 1 [2 ff] BauNVO nach der besonderen Art ihrer baulichen Nutzung im Bebauungsplan darzustellendes Baugebiet, das vorwiegend der Unterbringung von Handelsbetrieben sowie der zentralen Einrichtungen der Wirtschaft, der Verwaltung und der Kultur dient; § 7 BauNVO).

722 central city environs [npl] [US] *urb.* (Built-up area immediately surrounding the core area of an ►inner city; **in U.K.**, a part of the central area of a town or city characterized by its concentration of one particular type of activity; in UK one talks about a shopping precinct, a town hall precinct, a pedestrian precinct, etc.); *syn.* central city surroundings [npl] [US], city centre precincts [npl] [UK], inner suburbs [npl] [UK]; **s zona** [f] **residencial céntrica (≠)** (En las ciudades centroeuropeas en las que el centro de la ciudad apenas cumple funciones residenciales, barrios residenciales, generalmente de construcción antigua, situados formando un cordón alrededor del ►centro urbano); **f quartiers** [mpl] **péricentraux** (Quartiers anciens entourant immédiatement le ►centre-ville) ; *syn.* quartier [m] périphérique ; **g innerstädtisches Verdichtungsgebiet** [n] (Die unmittelbar um den eigentlichen Stadtkern gelegenen [Alt]baugebiete; ►Innenstadt); *syn.* Cityrandgebiet [n].

central city surroundings [npl] [US] *urb.* ►central city environs [US].

central government body [n] [UK] *adm.* ►federal government agency [US], ►state government agency [US].

central growth point [n] *plan. pol.* ►growth point.

723 central heat [n] *urb.* (Hot air from a ►central heating plant [US]/district heating plant [UK]; in Germany it is not hot air, but hot water; **in U.S.**, it may be either hot air or hot water; ►district heating system); **s calor** [m] **a distancia** (►calefacción a distancia, ►central de calefacción a distancia); **f chauffage** [m] **urbain** (Système de chauffage collectif [en général public] qui permet d'alimenter un certain nombre de bâtiments [privés, publics, industriels] d'une ville, d'un quartier ou d'un ensemble avec de la chaleur [eau chaude sanitaire, vapeur basse pression, eau surchauffée] produite par une ►centrale de production de chaleur et transportée et distribuée par un réseau de canalisations enterrées ; ►réseau de chaleur, ►distribution d'énergie par réseau de chaleur) ; *syn.* chaleur [f] à distance [CH]) ; *g* **Fernwärme [f]** (Thermische Energie, die in einem wärmegedämmten, überwiegend erdverlegten Rohrsystem vom ►Fernheizwerk zum Endabnehmer/zur Verbrauchsstelle geführt wird. Durch Förderung der **F.**, besonders in Ballungsgebieten, können viele emissionsreiche Einzelfeuerungsanlagen wirtschaftlich durch ein Fernheizwerk ersetzt werden und somit einen Beitrag zum Klimaschutz leisten; ►Fernheizung).

724 central heating plant [n] [US] *urb.* (Centralized power plant for the production of heating; ►central heat); *syn.* district heating plant [n] [UK]; *s* **central [f] de calefacción a distancia** (Central de energía que se utiliza además para la producción de ►calor a distancia); *f* **centrale [f] de production de chaleur** (Unité de production de chaleur de très grande puissance générée par toute les énergies existantes [renouvelables, de récupération et conventionnelles] permettant d'alimenter le ►chauffage urbain ; *terme spécifique* centrale biomasse) ; *syn.* chaufferie [f] ; *g* **Fernheizwerk [n]** (Zentrale Stelle [Heiz-(kraft)werk] zur Erzeugung von ►Fernwärme für die Warmwasserversorgung und Beheizung von Räumen; das Übertragungsmedium ist Wasser oder Dampf).

central island [n] [UK]**, turnaround with** *trans. urb.* ►loop turnaround.

central place [n] [UK] *plan.* ►population growth center [US].

central reservation [n] [UK] *trans.* ►median strip [US].

central reservation [n] [UK]**, motorway** *trans.* ►freeway median strip [US].

centre [n] [UK] *sociol. urb.* ►day care centre.

centre [n] [UK]**, commercial** *urb.* ►commercial center [US].

centre [n] [UK]**, health resort** *recr.* ►health resort center [US].

centre [n] [UK]**, leisure** *plan. recr.* ►leisure center [US].

centre [n] [UK]**, nature study** *recr.* ►nature study center [US].

centre [n] [UK]**, tourist** *recr.* ►tourist resort [US].

centre [n] **of attraction** [UK] *recr.* ►center of attraction [US].

cereal weed community [n] *phyt.* ►segetal community.

certificate [n] *agr. for. hort. leg.* ►official phytosanitary certificate.

certificate [n] [US]**, dumping** *constr.* ►dumping certificate [US].

certificate [n] [UK]**, tipping** *constr.* ►dumping certificate [US].

certificate [n] **of completion** *contr.* ►provisional certificate of completion.

certificate [n] **of completion with reduction in payment for contract item** *contr.* ►provisional certificate of completion with reduction in payment for contract item.

certificate [n] **of final completion** *constr.* ►issue of a certificate of final completion.

725 certificate [n] **of practical completion** *constr. contr.* (Planner's written recognition and acceptance by the

planner and contractor that work [US]/works [UK] carried out by the contractor has been satisfactorily completed in accordance with the contract); *s* **acta [f] de recepción/aceptación (de obra)** *syn.* certificado [m] de finalización (de obra); *f* **procès-verbal [m] de réception** (Les opérations préalables à la réception font l'objet d'un **p. v. d. r.** signé par le maître d'œuvre et l'entrepreneur en fin de travaux lorsque l'ouvrage est utilisable avant la prise de possession des ouvrages) ; *syn.* certificat [m] de réception ; *g* **Abnahmeniederschrift [f]** (Bei förmlicher Abnahme schriftlich niedergelegter Befund der [Landschafts]bauarbeiten, der vom bevollmächtigten Planer und Unternehmer unterzeichnet wird); *syn.* Abnahmeprotokoll [n].

certified delivery [n] [US] *constr. contr.* ►according to certified delivery [US].

726 certifying [n] **of seeds** *for. hort.* (Verification of the authenticity of seed by official testing agencies according to international standards); *s* **control [m] de mezcla de semillas** (Verificación de la autenticidad de semillas por parte de agencias oficiales de control de acuerdo con los estándares internacionales); *f* **contrôle [m] des mélanges de semences** (Contrôle et certification des semences effectués par le Service officiel de contrôle des semences et des plants [S.O.C.] selon des règles internationales) ; *g* **Saatgutprüfung [f]** (Prüfung und anschließende Zertifizierung von Sämereien nach international einheitlichen Regeln durch amtliche Prüfstellen; Saatgut prüfen [vb]).

727 chain-link fence [n] *constr.* (Woven wire fabric supported by [metal] posts); *s* **valla [f] de malla metálica** *syn.* cercado [m] de malla metálica; *f* **clôture [f] grillagée** (Grillage en rouleau ou en nappe fixé entre des poteaux) ; *syn.* clôture [f] en grillage (GEN 1982-II, 163) ; *g* **Maschendrahtzaun [m]** (An Pfosten befestigter Zaun aus flächigem Drahtgeflecht); *syn.* Maschengeflechtzaun [m].

728 chain link metal mesh [n] *constr.* (Woven wire fabric for fencing; ►galvanized chain link metal mesh, ►plastic-coated chain link metal mesh); *s* **malla [f] metálica** (►malla metálica galvanizada, ►malla metálica recubierta de plastico); *f* **grillage [m]** (*Termes spécifiques* rouleau, nappe de fils assemblés [torsadé, ancré, soudé] formant des mailles régulières ; ►grillage galvanisé, ►grillage plastifié) ; *g* **Maschendraht [m]** (Flächiges Drahtgeflecht für Zäune; ►kunststoffummantelter Maschendraht, ►verzinkter Maschendraht).

729 chain [n] **of dunes** *geo.* (Long, narrow sand dune or series of dunes, generally oriented parallel to the prevailing wind or in a direction resulting from two or more winds blowing at acute angles to each other. The crest consists of a series of peaks and gaps, and wind direction determines the steep, or slip, face position; ►barrier-beach); *s* **cordón [m] dunar** (Cadena de dunas alargadas, generalmente paralelas al viento dominante u orientadas según la dirección resultante de los vientos predominantes; ►cordón litoral); *f* **cordon [m] dunaire** (Ensablement littoral dont les points hauts, toujours émergés, sont occupés par de nombreuses dunes modelées par le vent en général parallèles à sa direction ; *terme spécifique* lido : cordon dunaire qui sépare la lagune de la mer ; ►flèche littorale) ; *g* **Dünenkette [f]** (Aneinanderreihung verschiedener Düneneinzelformen, die in der Regel parallel zur Hauptwindrichtung oder in der Resultierenden von zwei oder mehreren Windrichtungen in einem entsprechenden Winkel ausgerichtet sind; der Dünenkamm besteht aus mehreren Höhen und Senken; die Steilheit und der Rutschbereich im Lee der Dünen hängt von der Windstärke ab; ►Nehrung).

chair lift [n] *recr.* ►ski lift.

chalk grassland [n] [UK] *phyt.* ►limestone grassland.

chamaephyte [n] *bot.* ►surface plant.

C

chamber cover [n] [UK] *constr.* ▶manhole cover.

chamber cover slab [n] [UK] *constr.* ▶cover slab.

chamber section [n] [UK] *constr.* ▶riser unit [US].

730 chamfer [n] *constr.* (Oblique surface produced by beveling an edge); *s* **chaflán** [m] (Superficie oblicua resultante del corte transversal de una esquina); *syn.* bisel [m], arista [f] muerta; *f* **chanfrein** [m] (Biseau formé par abattement d'une arête sur un matériau de construction) ; *g* **Fase** [f] (Abschrägungsfläche an Kanten von Baumaterialien).

731 chamfered [pp/adj] *constr.* (Descriptive term for a sloped edge in carpentry, concrete and stonework); *syn.* bevel(l)ed [adj]; *s* **achaflanado/a** [pp/adj] *syn.* con esquina biselada [loc]; *f* **à arête chanfreinée** [loc] ; *g* **mit Fase** [loc] (So beschaffen, dass Kanten von Bauelementen mit schmalen Abschrägungsflächen versehen sind); *syn.* gefast [pp/adj].

champion tree [n] [US] *conserv. leg.* ▶monarch tree [US].

chandelier algae [npl], **subaqueous carpet of** *phyt.* ▶chara vegetation.

change [n], **climate** *met.* ▶climatic change.

732 change [n] **in cross section** *constr.* (Increase or decrease in the inside diameter of a pipe, either to a larger or smaller cross section); *syn.* change [n] in diameter (of a pipe); *s* **cambio** [m] **de la sección transversal** (Aumento o reducción del diámetro de una tubería; ▶vano); *f* **changement** [m] **de section** (Variation [augmentation ou diminution] du diamètre nominal de tuyaux le long d'un réseau ; ▶largeur dans-œuvre) ; *g* **Querschnittswechsel** [m] (Übergang der ▶lichten Weite in eine größere oder kleinere Abmessung bei Rohrleitungen).

change [n] **in diameter (of a pipe)** *constr.* ▶change in cross section.

change [n] **in grade** *constr.* ▶change in gradient.

733 change [n] **in gradient** *constr.* (Change in the degree of slope of a road, pathway or paved surface to another one, either similar, flatter or steeper); *syn.* change [n] in grade, change [n] in slope; *s* **cambio** [m] **de pendiente** (Transición entre una pendiente fuerte y una suave, p. ej. de la superficie de caminos o plazas; ▶cambio de rasante); *f* **changement** [m] **de pente** (Point de transition dans une inclinaison continue entre une pente plus faible ou plus forte, p. ex. d'une voie routière, d'une place ; ▶inversement de pente) ; *g* **Gefälleänderung** [f] (Wechsel der Oberflächenneigung auf Straßen, Wegen oder Plätzen von einem Gefälle in ein gleichsinniges geringeres oder größeres; ▶Gefällewechsel); *syn.* Neigungsänderung [f].

734 change [n] **in planning** *plan.* (**1.** Modification of a planning project as a result of new planning objectives; ▶plan revision. **2.** *leg. plan.* **Planning amendment:** revision of a bill or ordinance); *syn.* to 1. planning alteration [n]; *s 1* **revisión** [f] **de la planificación** (Proceso de actualización de un planeamiento debido a cambios sustanciales en los criterios respecto de la estructura general y orgánica del territorio o de la clasificación del suelo; art. 154.3 RD 2159/1978; ▶revisión de un plan); *s 2* **modificación** [f] **de la planificación** (Alteración de algunos elementos de una planificación, sin adoptar nuevos criterios en la estructura general y orgánica del territorio, aunque puede llevar consigo cambios aislados en la clasificación o calificación del suelo; cf. art. 154 RD 2159/1978; ▶modificación de un plan); *f* **modification** [f] **du projet d'aménagement** (Changements apportés au contenu ou à la forme du dessin d'un plan, d'un document graphique ou d'un projet d'aménagement ; ▶révision du plan, ▶modification du plan) ; *g* **Planungsänderung** [f] (Überarbeitung eines Planungsergebnisses oder Planungsvorhabens auf Grund neuer Vorgaben; ▶Überarbeitung eines Planes); *syn.* Änderung [f] einer Planung.

change [n] **in slope** *constr.* ▶change in gradient.

change [n] **of flora composition** *phyt.* ▶anthropogenic change of flora composition.

change [n] **of level** *constr.* ▶level difference.

change [n] **of pitch** [UK] *constr.* ▶gradient change.

change [n] **of slope** *constr.* ▶gradient change.

735 change order [n] [US] *constr. contr.* (Written order to the contractor signed by the owner and the architect, issued after the execution of the contract, authorizing a change in the work or an adjustment in the contract sum or the contract time as originally defined by the contract documents; DAC 1975. *CONTEXT* to issue an instruction to the contractor; ▶addendum proposal); *syn.* instruction [n] to the contractor; *s* **contrato** [m] **de modificación** (Revisión de contratos ya firmados con un contratista, en la que el propietario de obra y el arquitecto firman una orden autorizando cambios en la obra o ajustes de la suma del contrato o del plazo de ejecución respecto a los definidos en el contrato original; ▶oferta suplementaria) ; *syn.* orden [f] de modificación; *f* **marché** [m] **complémentaire** (Contrairement à l'avenant, le **m. c.** constitue un marché ayant une existence juridique propre et ne s'incorpore pas au marché initial, même s'il en est, à certains égards, le prolongement ; il doit répondre aux conditions de l'article 35 III 1a ou 1b du code des marchés publics ; en particulier il faut que les prestations soient devenues nécessaires, à la suite d'une circonstance imprévue, à la réalisation de l'ouvrage tel qu'il y est décrit, à condition que l'attribution soit faite à l'entreprise qui réalise l'ouvrage lorsque ces travaux complémentaires ne peuvent être techniquement ou économiquement séparés du marché principal sans inconvénient majeur pour la personne publique ; enfin, le montant cumulé des marchés complémentaires ne doit pas dépasser 33 % du montant du marché principal ; ces marchés peuvent être passés sans publicité préalable et sans mise en concurrence ; Fiche MINEFI Collectivités locales, 24.04.2003 ; ▶proposition d'avenant [au marché]) ; *g* **Änderungsvertrag** [f] (Änderung eines mit einem Unternehmer bereits abgeschlossenen Bauvertrages, die vom Bauherrn und Architekten unterschrieben wird und eine Änderung der Bauleistung, eine Anpassung der Vertragssumme oder einen neuen Fertigstellungstermin festschreibt; ▶Nachtragsangebot).

736 channel [n] **(1)** *constr. geo.* (Generic term for natural and artificial drainage channel; ▶erosion gully, ▶erosion rill, ▶natural drainage way [US]/natural hillside drainage channel [UK]); *s* **canal** [m] **de desagüe** (Término genérico para desagües naturales y artificiales; ▶canal de drenaje natural, ▶cárcava, ▶reguero); *syn.* albañal [m]; *f* **rigole** [f] **d'écoulement** (Terme générique pour le chenal naturel et artificiel d'écoulement ; ▶ravin, ▶ravine, ▶ravineau, ▶rigole) ; *g* **Abflussrinne** [f] (Natürliche, lineare Mulde/Vertiefung oder linienförmige, oberirdische Entwässerungseinrichtung, die der Wasserabführung dient; *UBe* natürliche und künstliche Abflussrinne; ▶Erosionsgraben, ▶Erosionsrille, ▶Siepen).

channel [n] **(2)** *hydr. wat'man.* ▶bypass channel; *eng. hydr.* ▶constructed channel; *geo.* ▶divergent channel [US]; *agr. for.* ▶drainage channel; *eng. wat'man.* ▶flood relief channel.

channel [n], **cold air drainage** *landsc.* ▶cold air drainage corridor.

channel [n], **flood** *wat'man.* ▶floodway.

channel [n], **flood bypass** *eng. wat'man.* ▶flood relief channel.

channel [n] [UK]**, natural hillside drainage** *geo.*
►natural drainage way [US].

channel [n]**, tidal** *geo.* ►tidal creek.

737 channel brush reinforcement [n] *constr.* (*Bioengineering* Installation of live branches in shallow erosion gullies—up to 3m deep and 8m wide—which only drain off water occasionally and are not filled by mudflow deposits more than 50cm thick; ►channel deadwood reinforcement); *syn.* brush placement [n] in erosion gullies (WIB 1996); *s* **estabilización** [f] **de regueros con ramas vivas** (*Construcción biotécnica* método de control de la erosión en regueros poco profundos [de hasta 3 m de profundidad y 8 m de anchura] por los que sólo fluye agua temporalmente y que no contienen depósitos de corrientes terrosas > 50 cm de grosor; ►estabilización de regueros con ramas secas); *syn.* fijación [f] de regueros con ramas vivas; *f* **embroussaillement** [m] **des ravines par branchages à rejets** (*Technique de génie écologique* protection des talus ou des berges des cours d'eau, stabilisation des « ravines » à fond plat [3 m de haut et 8 m de large], à régime intermittent et à dépôts de lave torrentielle > 50 cm, réalisée par la mise en place, le long du thalweg, d'obstacles répétés constitués par des lits de branchages à rejets; ►embroussaillement des ravines par entassement de bois mort ; WIB 1996, n° 170) ; *g* **Runsenausbuschung** [f] (*Ingenieurbiologie* Einbau von lebendem Astwerk in flache Erosionsgräben — bis 3 m tief und 8 m breit —, die nur zeitweise Wasser führen und keine Murschübe mit Ablagerungen > 50 cm aufweisen; ►Runsenausgrassung).

738 channel deadwood reinforcement [n] *constr.* (Bioengineering method for erosion control in gullies with dead branch material. Installation similar to ►channel brush reinforcement, but using dead branch material); *syn.* dead brushwood construction method [n]; *s* **estabilización** [f] **de regueros con ramas secas** (*Construcción biotécnica* método de control de la erosión en regueros que —al contrario que el de ►estabilización de regueros con ramas vivas — utiliza material vegetal muerto); *syn.* fijación [f] de regueros con ramas secas, estabilización [f]/fijación [f] de regueros con material vegetal muerto; *f* **embroussaillement** [m] **des ravines par entassement de bois mort** (*Technique de génie écologique* se différencie de la stabilisation des ravines par embroussaillement de branchages à rejets, par l'utilisation de branchages morts ; ►embroussaillement des ravines par branchages à rejets) ; *syn.* stabilisation [f] des ravines par tapis de branchage) ; *g* **Runsenausgrassung** [f] (Ingenieurbiologiche Bauweise mit nicht ausschlagfähigem Astwerk oder Reisig als Erosionsschutz in ausgespülten Rinnen an Hängen; ►Runsenausbuschung); *syn.* Grassbau [m].

739 channel discharge cross-section [n] *hydr.* (Cross-sectional area of a watercourse measured to determine the maximum discharge capacity); *s* **capacidad** [f] **máxima de desagüe** (Sección transversal del cauce máximo de un río, incluyendo las zonas inundadas temporalmente, que determina la capacidad máxima del mismo); *f* **section** [f] **(transversale totale) d'écoulement** (Profil de la surface d'écoulement d'un cours d'eau avec son lit mineur et majeur) ; *syn.* profil [m] en travers du lit fluvial, profil [m] transversal du lit fluvial ; *g* **Abflussquerschnitt** [m] (1) (Profildurchflussfläche, die ein Fließgewässer im Bett und bei Hochwasser noch zusätzlich im Auenbereich beansprucht); *syn.* Durchflussprofil [n], Durchflussquerschnitt [m].

channel drain [n] [UK] *constr.* ►trench drain [US].

channel erosion [n] *geo.* ►gully erosion; ►scour.

740 channel invert [n] *constr.* (Lower inside surface [= bottom] of an open drain or channel; ►invert elevation); *s* **lecho** [m] **de canal (de desagüe)** (►nivel de fondo); *syn.* fondo

[m] de canal (de desagüe); *f* **fil** [m] **d'eau** (Génératrice inférieure d'un caniveau, d'une canalisation, etc. ; ►cote du fil d'eau d'un caniveau, ►cote de la génératrice inférieure du tuyau, ►cote du fond de tranchée) ; *g* **Rinnensohle** [f] (Tiefste Linie einer Abflussrinne; ►Sohlenhöhe); *syn.* Sohle [f] einer Abflussrinne.

741 channel irrigation [n] *agr. hort.* (►furrow irrigation); *s* **irrigación** [f] **por canales** (►riego por infiltración); *syn.* riego [m] por canales; *f* **irrigation** [f] **par canaux** (►irrigation à la raie) ; *g* **Bewässerung** [f] **mit Kanälen** (►Furchenbewässerung).

channelization [n]**, stream** *wat'man.* ►stream straightening.

channelized stream [n] *wat'man.* ►canalized stream

742 channel lining [n] *wat'man.* (Layer of heavy stones on bottom and sides of a channel to reduce the flow rate and prevent ►watercourse bed erosion; ►riprap); *syn.* dumped stone lining [n] [also US]; *s* **escollera** [f] **de fondo** (Capa pesada de bloques de piedra sobre el lecho de un río para reducir la velocidad del agua y la ►erosión del lecho; ►escollera); *f* **empierrement** [m] **du lit [d'une rivière]** (Réalisation de rampes en enrochement au niveau plancher du lit mineur d'un cours d'eau, de pente douce pour dissiper l'énergie des eaux, stabiliser le profil en long par réduction de la capacité de transport solide ; ►caillourtage des berges, ►érosion du lit d'un cours d'eau) ; *g* **Berollung** [f] (Eingebaute Steinlage aus schweren Blöcken in ein Flussbett zur Verlangsamung der Fließgeschwindigkeit und gleichzeitigen Bremsung der ►Sohlenerosion; ►Steinschüttung an Uferböschungen).

743 channel realignment [n] *landsc.* (Redirecting of the course of a stream or river, e.g. due to a building project; cf. RIW 1984, 142); *s* **desviación** [f] **de un curso de agua** (p. ej. por la construcción de una autopista); *f* **déviation** [f] **du tracé d'un cours d'eau** (Modification du tracé d'une rivière, d'un fleuve causé p. ex. par un projet autoroutier) ; *g* **Verlegung** [f] **eines Fließgewässers** (Umleitung eines Bach- oder Flusslaufes, z. B. wegen eines Bauprojektes).

channel unit [n] *constr.* ►dished channel unit.

chaparral [n] [MEX&SW-USA] *phyt.* ►garide.

character [n] **and beauty** [n] **of landscape** [US] *conserv. landsc.* ►beauty and amenity of the landscape [US]/beauty and amenity of the countryside [UK].

744 characteristic combination [n] **of species** *phyt.* (Typical group of species in a plant community in comparison to all species in such a community; ►character species); *s* **combinación** [f] **característica de especies** (Está formada por las ►especies características junto con las especies constantes, que constituyen el armazón básico de la comunidad; cf. BB 1979); *f* **ensemble** [m] **caractéristique** (Base floristique typique d'une association végétale par opposition à la liste complète des espèces ; on distingue l'ensemble spécifique complet d'une association, soit la totalité des ►espèces caractéristiques et des espèces compagnes ainsi que l'ensemble spécifique normal d'une association, soit le cortège floristique habituel pour l'étude d'une association) ; *syn.* ensemble [m] spécifique, bloc [m] caractéristique ; *g* **charakteristische Artenkombination** [f] (Typische Artengruppe einer Pflanzengesellschaft im Vergleich zur Gesamtartenliste; ►Kennart).

characteristic delineation [n] [US] *plan.* ►characteristic presentation.

745 characteristic of a landscape [loc] *conserv. landsc. plan. recr.* (Descriptive term for a typical landscape feature incorporating cultural and physiographic aspects; ►landscape character); *s* **característico del paisaje** [loc] (►particularidad del paisaje); *syn.* típico del paisaje [loc]; *f* **caractéristique**

C

d'un paysage [loc] (▶Caractère particulier des sites et des paysages) ; *g* **landschaftstypisch [adj]** (So beschaffen, dass es für eine Landschaft charakteristisch ist; ▶landschaftliche Eigenart); *syn.* landschaftscharakteristisch [adj], landschaftsprägend [ppr/adj].

746 characteristic of the townscape [loc] *urb.* (Descriptive term applied to a typical feature or quality of a town or village); *s* **característico/a del aspecto escénico urbano [loc]** *syn.* característico/a de la calidad visual urbana [loc]; *f* **caractéristique de l'aspect de l'agglomération [loc]** (Qualité descriptive reflétant l'ensemble des caractéristiques esthétiques remarquables du paysage architectural d'une agglomération) ; *syn.* typique de l'aspect des lieux [loc] ; *g* **ortsbildbestimmend [ppr/adj]** (So beschaffen, dass es typisch für einen Ort ist resp. den Charakter eines Ortes prägt).

747 characteristic plant [n] *plant.* (▶accent perennial, ▶character species, ▶indicator plant); *syn.* representative plant [n]; *s* **planta [f] característica** (Planta típica de una región elegida para representarla en el marco p. ej. de una exposición de horticultura; ▶vivaz de acentuación; *phyt.* ▶especie característica, ▶planta indicadora); *f* **plante [f] représentative** (▶plante vivace dominante ; *phyt.* ▶espèce caractéristique [exclusive], ▶plante indicatrice) ; *g* **Leitpflanze [f]** (Pflanze, die im Rahmen eines Ausstellungsthemas, z. B. bei Gartenschauen, für eine bestimmte Region charakteristisch ist; ▶Leitstaude; *phyt.* ▶Kennart, ▶Zeigerpflanze).

748 characteristic presentation [n] *plan.* (Mode of graphic expression); *syn.* characteristic delineation [n] [also US]; *s* **presentación [f] gráfica** (Tipo de expresión gráfica); *f* **graphisme [m]** (Caractères particuliers d'un croquis) ; *g* **grafische Darstellung [f] (1)** (Charakteristische Art einer künstlerischen resp. zeichnerischen Gestaltung).

characteristic species [n] *phyt.* ▶character species.

character [n] **of the ground** *geo.* ▶nature of the terrain.

749 character species [n] *phyt. plant.* (Plant species, whose occurrence is almost completely associated with a certain plant community or closely related vegetational unit; ▶exclusive species, ▶degree of fidelity); *syn.* characteristic species [n]; *s* **especie [f] característica** (Especie de flora que se presenta casi exclusivamente en una asociación por lo que sirve de indicadora de ésta; ▶especie exclusiva, ▶grado de fidelidad); *f* **espèce [f] caractéristique (exclusive)** (Espèce strictement localisée à un groupement végétal déterminé ; ▶espèce exclusive, ▶degré de fidélité) ; *g* **Kennart [f] (1.** *phyt.* Pflanzenart, die in ihrem Vorkommen weitgehend an bestimmte Pflanzengesellschaften oder an Gruppen von nahe verwandten Vegetationseinheiten gebunden ist; ▶treue Art, ▶Treuegrad. **2.** *plant. Terminus für die gestalterische Bedeutung einer Bepflanzung hinsichtlich ihres „Themas"* standortprägnante Pflanzenart, die durch ihre Verwendung die Physiognomie des Lebensbereiches verdeutlicht, weil sie weitgehend an bestimmte Pflanzengesellschaften oder an Gruppen von nahe verwandten Vegetationseinheiten erinnert, z. B. Großblattstauden aus Feuchtbereichen, graulaubige oder sukkulente Pflanzen aus trockenen Freiflächen; der Begriff der **K.** ist der Pflanzensoziologie entlehnt); *syn.* Charakterart [f].

750 character species [n] **of a habitat** *phyt. zool.* (Species, whose dominance, is typical for a particular habitat type); *s* **especie [f] característica de un hábitat** (Especie de flora o fauna cuya dominancia es típica para un tipo de hábitat específico); *f* **espèce [f] caractéristique d'un habitat** (Espèce principalement inféodée à un type d'habitat se trouvant dans un état de conservation favorable et correspondant à un faciès représentatif d'un milieu naturel); *g* **charakteristische Art [f] eines Lebensraums** (Pflanzen- oder Tierart, die ihren eindeutigen und

ausgeprägten Vorkommensschwerpunkt in einem definierten Lebensraumtyp hat).

751 character species [n] **of an alliance** *phyt.* (Typical or characteristic plant species identified with a particular ▶alliance); *s* **especie [f] característica de una alianza** (Especie vegetal que está supeditada a una ▶alianza específica); *f* **espèce [f] caractéristique d'alliance** (Espèce permettant de reconnaître et d'identifier une ▶alliance ; elle trouve les conditions optimales de développement dans les conditions écologiques correspondant à cette alliance ; DEE 1982) ; *g* **Verbandscharakterart [f]** (Pflanzenart, die an einen bestimmten ▶Verband 1 gebunden ist); *syn.* Verbandskennart [f].

752 chara vegetation [n] *phyt.* (Plant community living continually under water e.g. chandelier algae in a standing waterbody); *syn.* subaqueous meadow [n] (ELL 1988, 291), subaqueous carpet [n] of chandelier algae (ELL 1988, 291); *s* **pradera [f] subacuática** (Comunidad vegetal que habita bajo el agua, p. ej. en aguas quietas praderas de algas [*Characeae*] o en el mar; *término específico para el Mediterráneo* pradera de posidonia *[Potamogetonáceas]*); *f* **prairie [f] aquatique submergée** (AEP 1976, 64 ; terme générique englobant les associations végétales du fond des plans d'eau vivant en permanence sous l'eau, p. ex. le groupement à Charas/à characées, le groupement à Potamots ou des milieux marins, dans l'étage infralittoral [p. ex. herbier à posidonie] ; *termes spécifiques* potamaie, charaie) ; *syn.* prairie [f] submergée, fixée, herbier [m] immergé ; *g* **Unterwasserrasen [m]** (Ständig unter Wasser lebende Pflanzengesellschaft, z. B. in Stillgewässern Rasen von Armleuchteralgen [*Characeen*-Rasen] und Laichkraut-Unterwasserwiesen *[Potamogetonion]*); *syn.* Unterwasserwiese [f].

charge [n], **hookup charge** [n] [US] *adm. urb.* ▶utility connection charge.

753 chargeable fee [n] *contr. leg. prof.* (**In Europe**, taken from ▶fee chart, based on the value of the building contract and the contractual differential for the type of work; **in U.S.**, professional fees are individually negotiated; ▶complexity rating classification); *s* **tasa [f] de remuneración** (En D., cantidad fijada en la ▶tabla de honorarios que se deduce de los costos/costes [Es] de la obra y la ▶clase de complejidad); *f* **taux [m] élémentaire de rémunération** (**F.**, le tableau des taux de rémunération fixe le taux des honoraires en pourcentage du coût d'objectif suivant les divers contenus de la mission normalisée, la détermination de l'honoraire étant le produit du taux de rémunération par le montant hors taxes des travaux ; **D.**, montant fixé dans le ▶tableau des taux de rémunération [des honoraires] et calculé en prenant compte de la ▶classe de complexité 1 et le coût d'objectif) ; *syn.* taux [m] d'honoraires [B], valeur [f] du taux de rémunération ; *g* **Honorarsatz [m]** (Der ▶Honorartafel in der HOAI entnommener, mengenabhängiger Betrag, der sich nach den anrechenbaren [Bau]kosten und der ▶Honorarzone richtet; im U.K. und in den USA gibt es beispielsweise keine **H.sätze**; dort werden Honorare jeweils einzeln verhandelt); *syn.* Honoraransatz [m].

charges [npl] [US] *adm. envir. leg.* ▶garbage collection charges.

charges [npl], **overtime** *contr.* ▶overtime payment.

charges [npl], **payment of professional fees** *contr. prof.* ▶remuneration of professional fees.

charges [npl] [UK], **waste collection** *adm. envir. leg.* ▶garbage collection charges [US].

chart [n] *plan.* ▶flow chart.

chasmophytic plant [n] *phyt.* ▶rock plant.

chasmophytic vegetation [n] *phyt.* ▶vegetation of rock clefts.

754 check area [n] *rem'sens. surv.* (Terrestrial control area for verification of objects and identification of surface textures on aerial photographs or satellite images); *syn.* control area [n]; *s* **área** [f] **de entrenamiento** (*En teledetección* área terrestre de control de datos ganados por medio de imágenes de satélite); *syn.* área [f] de control; *f* **aire** [f] **de contrôle** (Aire sur le terrain utilisée pour le contrôle de l'interprétation des photographies aériennes et des photos satellites) ; *g* **Testgebiet** [n] (Terrestrische Kontrollfläche mit Vergleichsdaten zur Überprüfung der Luftbild- oder Satellitendatenauswertung); *syn.* Trainingsgebiet [n].

755 checkdam [n] *wat'man.* (Generic term for the control of fast-flowing rivers [▶torrent control] or streams by a cross-section structure, to decrease stream flow velocity, minimize channel scour, check erosion, promote deposition of sediment, and tooccasionally slow down ▶mud flows; e.g. ▶ground sill, ▶drop structure installation, ▶stream ramp); *s* **dique** [m] **transversal** (Término genérico para estructuras de estabilización de cauces de torrentes como ▶solera de fondo, ▶solera, ▶rampa de solera; ▶corrección de torrentes; ▶corriente terrosa); *f* **ouvrage** [m] **de confortement transversal** (Terme générique pour les ouvrages de stabilisation tels que le ▶seuil transversal d'un cours d'eau, ▶ouvrage de chute ou la ▶rampe en enrochement dans la ▶stabilisation des torrents ; ceux-ci s'opposent à l'érosion régressive du lit du torrent, maintiennent le plan d'eau à l'étiage, régulent le débit solide et ralentissent le dépôt de ▶lave torrentielle) ; *g* **Querwerk** [n] (Oberbegriff für Querschnittsbefestigungen wie ▶Sohlenschwelle, ▶Sohlenabsturz und ▶Sohlenrampe bei der ▶Wildbachverbauung, die eine Tiefenerosion des Bachbettes verhindern und evtl. ▶Muren bremsen); *syn.* Querbauwerk [n].

checking [n] *contr. prof.* ▶ready for checking.

756 checking [n] **and evaluation** [n] **of bids** [US] *constr. contr.* (Critical examination of each bid [US]/tender [UK] for arithmetic mistakes, calculation errors and adaequacy of unit prices in order to be able to award a contract to the contractor who has submitted the most acceptable bid); *syn.* checking [n] and evaluation [n] of tenders [UK]; *s* **análisis** [m] **de ofertas** (Estudio crítico de todas las ofertas entregadas en cuanto a errores aritméticos, de cálculo y a la adecuación de los precios unitarios para posibilitar la adjudicación del contrato al licitante que ha presentado la oferta más aceptable); *syn.* evaluación [f] de ofertas; *f* **jugement** [m] **des offres** (Choix de l'entreprise retenue effectué selon plusieurs critères [prix des prestations, coût d'utilisation, valeur technique, garanties professionnelles et financières] sur l'étude des propositions remises avant la date limite fixée ; l'examen des offres comprend en outre la verification de la concordance des prix entre les indications en lettres et en chiffres, la rectification des erreurs de multiplication, d'addition ou de report dans le détail estimatif ; CCM 1989) ; *syn.* dépouillement [m] des offres, jugement [m] de la consultation ; *g* **Prüfung** [f] **und Wertung** [f] **der Angebote** (Kritische Durchsicht aller eingegangenen Angebote nach Rechenfehlern, Kalkulationsirrtümern, Angemessenheit der Preise, Erkennen von taktischen oder Spekulationspreisen etc., um den Zuschlag auf das annehmbarste Angebot erteilen zu können; cf. §§ 23 u. 25 VOB Teil A); *syn.* Angebotsprüfung [f].

checking [n] **and evaluation** [n] **of tenders** [UK] *constr. contr.* ▶checking and evaluation of bids [US].

checking [n] **of bills** [US] *prof.* ▶checking of invoices.

757 checking [n] **of invoices** *prof.* (Professional obligation to verify that the contractor's work has been completely and satisfactorily executed and that the corresponding invoices have been correctly submitted; in U.S. and U.K., planners are not normally involved in checking invoices; this work is done by accountants and auditors); *syn.* checking [n] of bills [also US]; *s* **revisión** [f] **de cuentas** (Obligación del arquitecto hacia el comitente de controlar si los trabajos realizados se ejecutaron según contrato y correctamente y si fueron puestos en cuenta correctamente; en EE.UU. y GB los arquitectos normalmente no se ocupan de revisar las facturas; este trabajo lo realizan contables o auditores); *f* **contrôle** [m] **des situations** (Obligation de l'architecte dans le cadre de sa mission de vérifier les situations et les décomptes établis par les entreprises et d'établir les propositions de paiement correspondantes) ; *g* **Rechnungsprüfung** [f] (Prüfungsverpflichtung des Planers gegenüber dem Bauherren, ob die erbrachten Leistungen vertragsgemäß und fachtechnisch einwandfrei ausgeführt und rechnerisch richtig berechnet wurden. Im U.K. und in den USA werden Rechnungsprüfungen i. d. R. nicht von Planern, sondern von Buchhaltern oder Rechnungsrevisoren vorgenommen).

check swale [n] *agr. constr.* ▶berm ditch.

758 cheek wall [n] *arch. constr.* (HALAC 1988-II; wall built alongside a series of outdoor steps to retain abutting earth); *s* **zanca** [f] (Viga inclinada sobre la que se apoyan las huellas y contrahuellas de una escalera; DACO 1988); *f* **limon** [m] (Muret de forme variable accompagnant latéralement les marches d'un escalier) ; *g* **Treppenwange** [f] (Seitliche Mauer einer Treppe).

759 cheek wall [n] **with built-in planter** *arch. constr.* (Planted construction element built to contain the edge of outdoor steps; typically requires footings below frost depth as per local conditions); *s* **zanca** [f] **con jardinera**; *f* **limon-jardinière** [m] ; *g* **Treppenwange** [f] **mit eingebautem Pflanzentrog**.

760 chemical fertilizer [n] *agr. constr. for. hort.* (Chemically-manufactured ▶mineral fertilizer. This is a term often used in the context of alternatives to ▶traditional farming in order to be able to distinguish chemical fertilizers from fertilizers produced from natural minerals); *syn.* artificial fertiliser [n] [also UK]; *s* **fertilizante** [m] **(mineral) artificial** (▶fertilizante mineral, ▶cultivo tradicional); *syn.* abono [m] mineral; *f* **engrais** [m] **artificiel** (Terme utilisé par les adeptes de la culture biologique pour différencier les ▶engrais minéraux obtenus par transformation industrielle de matériaux naturels tels phosphates naturels, poudre de roche éruptive broyée et les engrais obtenus par synthèse et utilisés dand l'▶agriculture traditionnelle) ; *syn.* engrais [m] chimique (DUV 1984, 165), engrais [m] organique de synthèse ; *g* **Kunstdünger** [m] (**1.** Synthetisch hergestellter mineralischer sowie organischer Dünger. **2.** *Im allgemeinen Sprachgebrauch* chemisch hergestellter ▶Mineraldünger; oft im Rahmen der Alternativen zum ▶konventionellen Landbau benutzter Begriff, um chemisch hergestellte Düngemittel von mineralischem Dünger aus Gesteinsmehlen zu unterscheiden).

761 chemical pest control [n] *agr. for. hort. phytopath.* (Chemical methods for the prevention and control of diseases and pests which affect the development of plants; the German term 'chemischer Pflanzenschutz' also includes weed control; ▶chemical plant disease prevention and pest control, ▶pest and weed control); *syn.* chemical pest management [n]; *s* **lucha** [f] **antiplaguicida química** (Aplicación de productos químicos para combatir plagas y enfermedades en las plantas y la presencia de las llamadas «malas hierbas»; ▶control antiplaguicida, ▶lucha química contra parásitos); *syn.* control [m] químico de plagas; *f* **lutte** [f] **chimique** (Utilisation de substances ou préparations antiparasitaires [pesticides et produits phytosanitaires] destinées à combattre les ennemis des grandes cultures, des cultures légumières et ornementales ; ▶lutte chimique contre les parasites,

►protection des végétaux) ; *g* **chemischer Pflanzenschutz [m, o. Pl.]** (Chemische Maßnahmen zur Verhütung und Bekämpfung von Krankheiten, Schädlings- *und* Unkrautbefall, um die Entwicklung von Pflanzenbeständen zu fördern; ►chemische Schädlingsbekämpfung, ►Pflanzenschutz, ►Schädlingsbekämpfung).

chemical pest management [n] *agr. for. hort. phytopath.* ►chemical pest control, ►chemical plant disease prevention and pest control.

762 chemical plant disease prevention and pest control [n] *agr. for. hort. phytopath.* (Chemical methods for the prevention and control of diseases and pests which affect the development of plants; ►chemical pest control, ►pest and weed control); *syn.* chemical pest management [n]; *s* **lucha [f] química contra parásitos** (Aplicación de productos químicos para combatir plagas y enfermedades en las plantas causadas por depredadores; ►control antiplaguicida, ►lucha antiplaguicida química); *syn.* control [m] químico de organismos nocivos; *f* **lutte [f] chimique contre les parasites** (Utilisation de substances ou préparations antiparasitaires [pesticides et produits phytosanitaires] destinées à combattre les ennemis des grandes cultures, des cultures légumières et ornementales ; ►lutte chimique, ►protection des végétaux) ; *g* **chemische Schädlingsbekämpfung [f]** (Chemische Maßnahmen zur Verhütung und Bekämpfung von Krankheiten und Schädlingen, die die Entwicklung von Pflanzenbeständen stören; ►chemischer Pflanzenschutz ist nicht synonym, da auch noch Herbizideinsätze mitgedacht werden können; ►Pflanzenschutz).

763 chemical properties [npl] **of soil** *pedol. chem.* (►soil characteristics); *s* **características [fpl] químicas del suelo** (►características del suelo); *f* **caractéristiques [fpl] chimiques du sol** (►propriétés du sol, ►propriétés physiques du sol) ; *g* **bodenchemische Eigenschaften [fpl]** (►Bodeneigenschaften).

chernozem [n] *pedol.* ►black earth.

Chief Executive of English Heritage [n] [UK] *conserv' hist.* ►State Historic Preservation Officer [US].

764 chief landscape architect [n] **in a public authority/agency** [US] *prof.* (Section or branch head in a public agency); *syn.* senior landscape architect [n] in a public authority [UK]; *s* **arquitecto/a [m/f] paisajista en posición dirigente** (en la administración); *f* **ingénieur [m/f] en chef** (Architecte paysagiste en chef cadre supérieur dans les services publics ; Ingénieur principal, e ou en chef dirigeant un ou plusieurs divisions) ; *syn. obs.* ingénieur divisionnaire ; *g* **Garten- und Landschaftsarchitekt/-in [m/f] in leitender Stellung (1)** (in einer Behörde).

child care center [n] [US] *sociol. urb.* ►kindergarten.

765 child care centre [n] [UK] *sociol. urb.* (Facility directed by teachers in which children aged 6 to 12 are supervised at times when they are not attending regular school classes. **In U.S.,** such a facility for school-aged children does not normally exist; this type of arrangement is not part of a national policy. It may be found in the US as a local initiative; in U.S., a **c. c. center** is a private establishment enrolling four or more children between 2-5 years of age and where tuition, fees, or other forms of compensation for the care of the children is charged; IBDD 1981, 49. After school supervision for 6-12 years old children is usually offered in connection with specific music, theatre, athletics, or extra academic study programs, all of which require faculty supervision and is usually associated with the particular school, as opposed to dedicated "centre"; ►day care center [US]/day care centre [UK], ►kindergarten); *s* **guardería [f] infantil pedagógica (≠)** (Centro dirigido por pedagogos/as donde niños y niñas en edad escolar permanecen en las horas libres de clase y reciben apoyo para hacer las tareas escolares; este tipo de servicios son comunes en países como D. donde en la mayoría de los colegios públicos la jornada escolar no dura todo el día; ►guardaría infantil, ►jardín de infancia); *f 1* **garderie [f] d'élèves (≠)** (D., établissement pédagogique assurant la garde des jeunes élèves du primaire en dehors des heures d'école ; ►centre de loisirs maternels, ►garderie d'enfants, ►halte garderie, ►jardin d'enfants) ; *f 2* **salle [f] d'études [f]** (Salle d'école dans laquelle les élèves font leurs devoirs après les cours) ; *g* **Kinderhort [m]** (Pädagogisch geleitete ►Kindertagesstätte, in der Kinder von 6-12 Jahren in der schulfreien Zeit betreut werden; ►Kindergarten); *syn.* Hort [m].

766 children's farm [n] *recr.* (Children's recreation area managed for display of livestock and animal husbandry; ►petting zoo); *s* **granja [f] juvenil** (con fines recreativos y educativos; ►zoo[lógico] para niños/as); *f* **ferme [f] pour enfants** (Équipement situé en général en zone périurbaine souvent intégré dans des structures de loisirs plus importantes, utilisé comme complément d'activités scolaires ou d'éveil à la nature permettant aux enfants de développer leur relation avec le monde animal grâce à la prise en charge personnelle de l'alimentation et des soins procurés aux animaux ; ►zoo pour enfants) ; *g* **Jugendfarm [f]** (Pädagogisch betreuter Spielplatz mit dem Ziel, durch ein Spiel- und Lernangebot kontinuierliche Beziehungen von Kindern untereinander und zu den Tieren sich entwickeln zu lassen und Verantwortung für diese Einrichtung und für die Versorgung und Pflege der Tiere zu übernehmen; ►Streichelzoo).

767 children's playground [n] *urb. recr.* (Public or private area designed to accommodate the active and passive play of children and youth consisting of specialized equipment, hard surface areas and gardens usually in public residential and recreation areas and in private residential green areas; children's playgrounds are laid down by development plan, building permit or contract [within private properties] and legally protected; cf. the new European standard, in which safety requirements are regulated, is DIN EN 1176 and 1177; ►supervised children's playground; *generic term* ►playground); *s* **área [f] de juegos infantiles** (Instalación pública o privada especialmente equipada para juegos y actividades de recreo de niños y niñas, ►área de juegos educativos; *término genérico* ►área de juegos y recreo); *syn.* parque [m] infantil; *f* **aire [f] de jeux (pour enfants)** (Tout espace de petite dimension à usage public de proximité aménagé pour le jeux des enfants ou des adolescents ; l'autorité délivrant le permis de construire ou l'autorisation de lotir, peut exiger du constructeur la réalisation d'une aire de jeux et de loisirs pour les enfants ; des obligations d'aires de jeux peuvent être inscrites dans le règlement du PLU/POS ; DUA 1996, 31 ; ►terrain de jeux éducatifs ; *terme générique* ►terrain de jeux ; *leg.* ►aire collective de jeux) ; *syn.* espace [m] de jeux pour enfants ; *g* **Kinderspielplatz [m]** (Öffentliche oder private Anlage für das Spiel der Kinder und Jugendlichen in Wohn- und Erholungsgebieten, meist in öffentlichen Grünflächen und im privaten Wohngrün; Kinderspielplätze sind Flächen, die durch Bebauungsplan, Baugenehmigung oder Vertrag [innerhalb privater Grundstücke] eigens zum Spielen ausgewiesen und rechtlich abgesichert sind; cf. DIN 18 034; neue Europanormen, die sicherheitstechnische Anforderungen regeln sind DIN EN 1176 und 1177; ►betreuter Kinderspielplatz; *OB* ►Spielplatz).

chinook [n] *met.* ►foehn.

chip [vb] [US] *agr. constr. hort.* ►shred.

chipping bark [n] *arb. bot.* ►outer bark.

chippings [npl] [UK]**, stone** *constr.* ►fine aggregate [US].

chopping-out [n] [UK] *constr. for.* ►stand thinning.

C

768 chorology [n] *phyt. zool.* (Geographical research into individual plant or animal ▶ranges 1); *s* **corología [f]** (Rama de la biogeografía que estudia los fenómenos biológicos en su manifestación espacial, siendo el espacio de dimensión relativamente grande; ▶área de distribución); *f* **chorologie [f]** (Branche de la biogéographie étudiant l'aire de répartition et de dispersion géographique des organismes vivants à la surface du globe. On distingue les unités chorologiques suivantes : empire, région, domaine, secteur, sous-secteur, district ; ▶aire de répartition) ; *syn.* géonémie [f] (DEE 1982) ; *g* **Arealkunde [f]** (Geografische Erforschung der einzelnen Pflanzen- oder Tierareale als Zweig der Biogeografie; ▶Areal); *syn.* Chorologie [f].

chub zone [n] *limn.* ▶barbel zone [UK].

769 church [n] [US] *leg. urb.* (Planning and legal term on ▶community development plans [US]/urban development plan [UK] for religious facilities, which by design and construction are primarily intended for the conducting of organized religious services and accessory uses associated therewith; IBDD 1981); *syn.* place [n] of public worship [UK]; *s* **edificios [mpl] e instalaciones [fpl] con destino religioso** (Superficie asignada en plan parcial [de ordenación] para equipamiento colectivo religioso; ▶plan urbanístico); *f* **zone [f] à vocation d'édifices et équipements de culte (±)** (U.S. et D., dans un ▶document de planification urbaine, espace réservé à un établissement ou un édifice de culte ; F., les locaux destinés à des manifestations cultuelles peuvent être réalisés sur une zone d'aménagement concerté [ZAC]) ; *g* **Kirchen [fpl] und kirchlichen Zwecken dienende Gebäude und Einrichtungen** (Im Bebauungsplan auszuweisende Flächen für den Gemeinbedarf; Bezeichnung gem. Anlage zur PflanzV 90, Kap. 4; ▶Bauleitplan).

770 church yard [n] *adm. hist.* (Area of land adjacent to a church used for the burial of the dead, often enclosed by a wall; predecessor of the modern ▶cemetery); *syn.* grave yard [n], *in the vernacular* bone yard [n]; *s* **cementerio [m] (2)** (Área de enterramiento adjunta a iglesias protestantes o anglicanas, no a iglesias católicas; ▶cementerio 1); *f* **cimetière [m] ancien** (Cimetière qui avec le développement du christianisme au début du Moyen-âge, est implanté à proximité immédiate d'un lieu de culte [chapelle, église, temples, etc.], en général entouré d'une enceinte constituée d'un mur de pierre et considéré comme un espace sacré ; ▶cimetière) ; *syn.* cimetière [m] paroissial ; *f 2* **enclos [m] paroissial** (Ensemble architectural religieux clos, qu'on rencontre principalement en Bretagne ; il se compose généralement d'une porte monumentale ou arche triomphale, d'un mur d'enceinte, d'un ossuaire ou d'une chapelle funéraire, d'un calvaire, d'une église et parfois entouré d'un très petit cimetière) ; *g* **Kirchhof [m]** (Meist mit einer Mauer eingefriedeter, christlicher Kultraum zur Bestattung der Toten, direkt einer katholischen Kirche zugeordnet; ein **K.** zeichnet sich im Vergleich zu Stadtteil-, Haupt- oder Waldfriedhöfen durch einen sehr geringen Grünanteil [allenfalls Alleen, Randbepflanzungen und Schmuckpflanzungen] aus; Vorgänger des heutigen ▶Friedhofes); *syn. obs.* Gottesacker [m].

771 cinder [n] *constr. envir. geo. min.* (**1.** *min.* **1.a** Fused residue—waste product—separated from ores in the process of smelting; **1.b** waste material from making pig iron in a blast furnace. **2.** *geo.* Small vesicular fragmentary lava material which is projected from an erupting volcano); *syn.* slag [n]; *geo. also* volcanic scoria [n]; *s 1* **escoria [f] de alto horno** (Material sobrante de la fundición de acero en lingotes; ▶escoria); *syn.* escoria [f] de fundición, carbonilla [f]; *s 2* **escoria [f] volcánica** (Fragmentos de lava porosa de formas irregulares procedentes de las erupciones volcánicas o que se forman en los bordes superiores o inferiores de ríos de lava); *f* **scories [fpl]** (**1.** *min.* Résidus non métalliques incombustibles, obtenus lors de la réduction des minerais lors des opérations d'élaboration métallurgique, de l'affinage des métaux et de la combustion du charbon ; certains d'entre eux sont revalorisés et utilisés dans la construction routière. **2.** *geo.* **scorie volcanique :** fragment de lave bulleuse de faible densité, à surface irrégulièrement poreuse, rude au toucher, apparaissant dans les projections volcaniques [bombe volcanique] ou sur des coulées dont la surface est craquelée) ; *g* **Schlacke [f]** (**1.** *min.* Bei der Gewinnung von Eisen im Hochofen anfallendes Schmelzprodukt [**Hochofenschlacke**], das als Baumaterial verwendet wird. **2.** *geo.* Glasig-porös geformte Lavabrocken unregelmäßiger Gestalt, die sich an der Unter- und Oberseite von Lavaströmen bilden oder Auswürflinge [**Wurfschlacke**] eines Vulkans).

772 circadian rhythm [n] *biol.* (Approximately 24-hourly pattern of various metabolic activities seen in most organisms, including man, a natural rhythm which is thought to be controlled by an endogenous biological clock); *syn.* diurnal periodicity [n]; *s* **ritmo [m] nictemeral** (En la vida de los organismos, tipo de ritmo endógeno y de corto periodo/período causado por la alternancia de día y noche según el cual los animales [sean diurnos o nocturnos] y las plantas [asimilación] pasan de la actividad a la inactividad o viceversa ante la presencia resp. ausencia de la luz y de otros factores ecológicos [humedad, temperatura]; cf. MARG 1977, 709s.); *syn.* ritmo [m] diario, ritmo [m] diurno, ritmo [m] circadiano, ritmo [m] circadiario; *f* **rythme [m] quotidien** (Alternance des périodes d'activité et de repos des organismes avec une périodicité de 24 heures, déterminée par une montre biologique endogène. Cette montre physiologique possède un mécanisme rythmique autonome réglant chez les végétaux, les animaux et l'homme le métabolisme, les périodes de croissance et les comportements) ; *g* **Tagesgang [m]** (Wechsel zwischen Aktivität und Ruhe von Organismen mit einer 24-Stunden-Periodik, bestimmt durch eine endogene biologische Uhr. Diese physiologische Uhr steuert einen rhythmisch ablaufenden, autonomen Mechanismus, der bei allen Pflanzen, Tieren und beim Menschen Stoffwechselprozesse, Wachstumsleistungen und Verhaltensweisen festlegt); *syn.* zirkadiane Rhythmik [f], Tagesperiodik [f], Tagesrhythmik [f].

circannual periodicity [n] *biol.* ▶annual periodicity.

circannual rhythm [n] *biol.* ▶annual periodicity.

circuit [n] *recr.* ▶riding circuit.

773 circular bench [n] *gard. landsc.* (circular seating; e.g. around a tree); *s* **banco [m] circular** (Banco de forma circular total o parcial que se puede colocar p. ej. alrededor de un árbol); *f* **banc [m] circulaire** (Banc ou banquette de forme circulaire complète ou partielle) ; *syn.* banc [m] en arc de cercle ; *g* **Rundbank [f]** (Kreisförmige Sitzbank in Form eines Voll- oder Teilkreises); *syn.* runde Sitzbank [f].

774 circular path [n] [US] *plan. recr.* (Path designed for walking or rambling returning to its point of origin); *syn.* circular trail [n] [US], circular pathway [n] [UK], loop trail [n] [US] (2); *s* **circuito [m] de excursión** (Camino diseñado para senderismo o paseos que conduce —después de su itinerario— al lugar de origen); *syn.* camino [m] circular; *f* **itinéraire [m] de promenade et de randonnée (pédestre)** (Parcours organisé au terme duquel on revient généralement au point de départ) ; *syn.* circuit [m] (de randonnée) pédestre ; *g* **Rundwanderweg [m]** (Zum Wandern oder Spazierengehen angelegter Weg, der wieder an seinen Ausgangspunkt zurückführt); *syn.* Rundweg [m].

circular pathway [n] [UK] *plan. recr.* ▶circular path [US].

circular pathway system [n] [UK] *plan.* ▶circular walk system [US].

circular stair [n] *arch.* ▶spiral stair(case).

circular trail [n] [US] *plan. recr.* ►circular path [US]/circular pathway [UK].

circular trail system [n] [US] *plan.* ►circular walk system [US].

775 circular walk system [n] [US] *plan.* (System of pathways for pedestrians or cyclists in parks, recreation areas, urban heritage trails, etc., whereby one can return to the point of origin; ►circular path [US]/circular pathway [UK]); *syn.* circular trail system [n] [US], circular pathway system [n] [UK]; *s* **sistema** [m] **de circuitos de excursión** (Sistema de caminos para peatones o ciclistas en parques, áreas de recreo, áreas urbanas, etc. que conducen —después de sus itinerarios— al lugar de origen; ►circuito de excursión); *syn.* sistema [m] de caminos circulares; *f* **circuit** [m] **de plein air** (Parcours pédestre ou cycliste dans les parcs, les zones de loisirs et de plein air, les quartiers urbains, parcours équestre en forêt et en rase campagne, etc. sur lesquels les promeneurs, les cyclistes, les cavaliers reviennent à leur point de départ ; *terme spécifique* ►itinéraire de promenade et de randonnée) ; *syn.* parcours [m] de plein air ; *g* **Rundwegesystem** [n] (Erschließungssystem für Fußgänger oder Radfahrer in Parkanlagen, Erholungsgebieten, Stadtgebieten etc. sowie für Reiter in Wäldern und in der Feldflur, das zum Ausgangspunkt zurückführt; ►Rundwanderweg).

circulation [n] *urb. trans.* ►pedestrian circulation.

circulation network [n] *plan. urb.* ►circulation system; ►transportation system.

circulation plan [n] *urb.* ►pedestrian circulation plan.

776 circulation system [n] *plan. urb.* (1. Whole road and path network by which vehicular and pedestrian access is provided within a defined area. 2. General term for a network of various transportation routes throughout an extensive area; ►transportation system); *syn.* circulation network [n]; *s* **red** [m] **de circulación** (Conjunto de vías e infraestructuras que sirven para el tránsito de vehículos en una zona determinada. La **r. de c.** es la base física sobre la cual se desarrolla la ►red de transportes); *syn.* sistema [m] de circulación; *f* **réseau** [m] **de voirie** (1. Ensemble des routes et des chemins sur un territoire donné. 2. Lignes assurant la liaison et la desserte des aires d'emploi, des principales villes et des zones rurales sur un territoire donné ; ►réseau de transport) ; *syn.* voirie [f] ; *g* **Erschließungssystem** [n] (Gesamtheit des Straßen- und Wegenetzes in einem Betrachtungsraum; ►Verkehrssystem).

circumferential road [n] *trans.* ►ring road.

citizen participation [n] [US] *leg. plan. pol.* ►public participation.

citizen pressure group [n] [US] *pol.* ►community action group.

city [n] *hist. urb.* ►garden city; *landsc. urb.* ►greening of the city; *urb.* ►old part of a city; *plan.* ►surrounding region of a city.

city [n]**, fringe of a** *urb.* ►urban fringe.

city [n]**, outskirts of a** *urb.* ►urban fringe.

city beautification [n] [UK] *landsc.* ►Garden City Movement [UK].

City Beautiful Movement [n] [US] *landsc.* ►Garden City Movement [UK].

city bye-law [n] [UK] *adm. leg. urb.* ►city ordinance [US].

city center [n] [US] *urb.* ►inner city.

city centre precincts [npl] [UK] *urb.* ►central city environs [US].

city cores [n]**, regeneration of** *urb.* ►center city redevelopment [US]/urban regeneration [UK].

city edge [n] *urb.* ►city periphery.

city hygiene department [n] *adm. envir.* ►waste management department.

777 city ordinance [n] [US] *adm. leg. urb.* (Enactment by a municipal authority [►law] as part of its prerogative in governing its own affairs; e.g. ►design ordinance [US], ►zoning map [US], ►aesthetic ordinance [US]/building development bye-laws [UK], ►tree preservation ordinance [US]/tree preservation order [UK], city rates, and use regulations for streets and squares [UK]); *syn.* city/town statutes [npl] [UK], town ordinance [n] [US], bylaw [n] [UK&CDN] (PSP 1987, 191), city bye-law [n] [UK]; *s* **ordenanza** [f] **municipal** (Instrumento de regulación con carácter vinculante creado por decisión de la junta municipal [►ley] como ►plan parcial [de ordenación], ►ordenanza de protección de árboles, ►reglamento de diseño, ►ordenanza municipal de protección de la calidad visual urbana, etc.); *f* **arrêté** [m] **municipal** (Prérogative acte/décision écrit de l'autorité administrative communale [►loi] permettant l'imposition des prérogatives communales en matière d'urbanisme [►plan local d'urbanisme (PLU), ►plan d'occupation des sols (POS), ►carte communale], de prescriptions relatives à l'aspect esthétique des constructions [►réglementation concernant la protection des arbres, ►prescriptions communales architecturales et paysagères, ►règles [d'urbanisme] concernant l'aspect extérieur des constructions, ►prescriptions de conservation du patrimoine architectural constitutif d'un ensemble urbain], de taxes diverses [taxe communale sur la publicité, taxe d'enlèvement des ordures ménagères, taxe locale d'équipement, etc.]) ; *g* **Ortssatzung** [f] (Von einer Gemeinde oder einem Landkreis durch Satzungsbeschluss herbeigeführtes rechtliches Instrumentarium [Rechtsnorm und ►Gesetz im materiellen Sinne] zur Durchsetzung gemeindlicher Selbstverwaltungsaufgaben, z. B. ►Bebauungsplan, ►Baumschutzverordnung, ►Gestaltungssatzung, ►Ortsbildsatzung], Friedhofssatzung, Gebührensatzung für Müllabfuhr, Kanalanschluss oder Erschließungsbeiträge, Sondernutzung von Straßen und Plätzen. Die **O.en** gelten für das Gebiet der Gemeinde oder des Landkreises; sie erfassen innerhalb dieses Gebietes nicht nur Gemeinde- oder Kreiseinwohner, sondern jeden, der in diesem Gebiet einen in der **O.** oder Verordnung geregelten Tatbestand erfüllt; die Beschränkung des Geltungsbereiches auf einen Teil des Gebietes oder auf bestimmte Personenkreise ist aus sachgerechten Gründen möglich; cf. RWB 2007); *syn.* Ortsstatut [n].

city park [n] [US] *urb.* ►inner city park.

778 city periphery [n] *urb.* (Area of land on the boundary of a city; the **c. p.** of a built-up area usually applies to the edge of a city in the sense of a perimeter, i.e. it is the area touching the circumference and not the larger region surrounding a city; *not to be confused with* ►edge of a settlement, ►urban fringe); *syn.* city edge [n] *or* village edge [n], peripheral zone [n] of a city; *s* **periferia** [f] **urbana** (Espacio predominantemente rural situado alrededor de una ciudad, pero que supera sus límites y no se debe confundir con el ►borde de la ciudad); *syn. coloquial* afueras [fpl] de la ciudad; *f* **périphérie** [f] **urbaine** (1) (Espace en général à dominance rurale situé au-delà des limites de la ►frange urbaine d'une agglomération ou de deux communes voisines ; ne pas confondre avec l'►espace périurbain ; ►limite de l'agglomération, ►limite du site construit) ; *syn.* périphérie [f] de l'agglomération ; *g* **Stadtrand** [m] (Äußerer Bereich an einer Stadtgrenze oder in deren Nähe, der durch im Zusammenhang bebaute Siedlungen zur freien Landschaft hin abgegrenzt ist;

nicht zu verwechseln mit ▶Siedlungsrand); *syn.* Peripherie [f] einer Stadt.

city plan [n] **for sports fields** [UK] *plan. recr. urb.* ▶overall master plan for sports and physical activities [US]/ strategic plan for sports and physical activities [UK].

city plan [n] **for sports pitches** [UK] *plan. recr. urb.* ▶overall master plan for sports and physical activities [US]/ strategic plan for sports and physical activities [UK].

city planner [n] [US] *prof.* ▶urban planner [US]/town planner [UK].

city planning [n] [US] *urb.* ▶urban planning.

city plotholder [n] [UK] *urb. sociol.* ▶community gardener [US].

779 cityscape [n] *arch. urb.* (DIL 1987; man's perception of a city in its entirety, including its large-scale landscape, spatial definition, urban elements, urban surroundings and the city's outward appearance determined by uses and functions; ▶community appearance [US], ▶streetscape, ▶townscape); *syn.* city scenery [n] (CWF 1996), urban fabric [n] (LD 1988 [9], 59); *s* **fisionomía [f] de la ciudad** (Percepción visual del conjunto de los elementos constitutivos y estructurantes de la ciudad; ▶aspecto escénico urbano, ▶fisionomía del hábitat residencial, ▶imagen de la calle); *f* **morphologie [f] urbaine** (Concept caractérisant la forme physique et spatiale de la ville fondée sur l'analyse de l'évolution du tissu urbain et du rôle de chacune de ses caractéristiques [site, réseau viaire, trame parcellaire, espaces libres et espaces bâtis ; DUA 1996, 507-509 ; ▶physionomie de l'habitat, ▶aspect d'une agglomération, ▶image de la rue) ; *syn.* forme [f] urbaine, physionomie [f] urbaine ; *g* **Stadtgestalt [f]** (Durch den Menschen wahrgenommene Gesamtheit der raumgliedernden und raumbegrenzenden Elemente einer urbanen Umgebung incl. der durch die äußere Erscheinungsform bedingten Nutzungsfunktionen; ▶Gestalt des Wohnumfeldes, ▶Ortsbild, ▶Straßenbild); *syn.* Stadtbild [n].

city scenery [n] *arch. urb.* ▶cityscape.

city square [n] *urb.* ▶urban square.

city statutes [n] [UK] *adm. leg. urb.* ▶city ordinance [US].

780 city tree [n] *landsc. urb.* (Tree for urban planting which is suited to the micro-climate of a city; ▶street tree); *s* **árbol [m] urbano (≠)** (Árbol adecuado para plantar en ciudades ya que resiste mejor las condiciones del medio ambiente urbano; ▶árbol de calle); *syn.* árbol [m] de ciudad; *f* **arbre [m] urbain** (Arbre adapté aux conditions climatiques en milieu urbain en particulier pour les parcs, les rues et les places ; en F. existe un organisme central, *l'Agence de l'arbre et des espaces verts*, responsable de la protection des arbres urbains ; en D. existe un groupe de travail « l'arbre urbain » de la Conférence permanente des directeurs des services des espaces verts auprès du *« Deutscher Städtetag »* [± association des maires d'Allemagne] ; ▶arbre d'alignement) ; *g* **Stadtbaum [m]** (Für die Anpflanzung im städtischen Klimabereich geeignete Baumart, besonders für Parkanlagen, Straßen und überwiegend befestigte Plätze; in D. gibt es einen Arbeitskreis „Stadtbäume" der Ständigen Konferenz der Gartenamtsleiter beim Deutschen Städtetag; in F. gibt es eine entsprechende zentrale Stelle, die *Agence de l'arbre et des espaces verts*; ▶Straßenbaum).

civic site [n] [US]**, public and** *leg. urb.* ▶site for public facilities.

781 civil engineering [n] *eng.* (Branch of professional engineering relating to the design and construction of buildings, highways, bridges, waterways, harbo[u]rs, railroads, canals, artificial water supply, sewage disposal and other kinds of fixed public works as well as farm drainage; cf. WEB 1993; ▶ground

civil engineering, ▶structural engineering 2); *syn.* construction engineering [n]; *s* **ingeniería [f] de caminos, canales y puertos [Es]** (Ramo profesional de ingeniería dedicado al diseño y la construcción de carreteras, canales, puertos, ferrocarriles, conducciones de suministro de agua, de alcantarrillado y de otras infraestructuras públicas, así como de sistema de drenaje agrícola; ▶ingeniería de construcción); *syn.* ingeniería [f] civil [AL]; *f* **génie [m] civil** (Ensemble des techniques traitant des travaux relatifs à la réalisation d'ouvrages de construction non classés en bâtiments et des infrastructures routières, ferroviaires, fluviales, aériennes, maritimes, à la production et le transport d'énergie, à l'établissement des réseaux d'alimentation en eau et d'assainissement ; COB 1984, 9 ; ▶ingénierie du bâtiment) ; *g* **Bauingenieurwesen [n, o. Pl.] (1)** (Arbeitsgebiet der Planung, Gestaltung, Berechnung, Ausführung und Bauunterhaltung im Tiefbau, Grundbau [Gründungen von Bauwerken aller Art], Wasserbau und Ingenieurbau; ▶Bauingenieurwesen 2).

civil engineering department [n] [UK] *adm.* ▶public works department.

782 civil engineering works [npl] *constr. eng.* (Construction such as roads, railroads [US]/railways [UK], earthworks, hydraulic engineering, laying of sewer and other underground utility lines; ▶ground civil engineering); *syn.* highway construction [n] and maintenance [n] [also US]; *s* **trabajos [mpl] de obras públicas** (Actividades constructivas en infraestructuras; ▶ingeniería civil); *f* **travaux [mpl] de voirie et de réseaux divers** (Travaux en général réalisés pour le compte d'une administration territoriale et concernant la voirie et les réseaux divers ; ▶travaux publics) ; *syn.* travaux [mpl] de V.R.D. (voirie et réseaux divers) ; *g* **Tiefbauarbeiten [fpl]** (Arbeiten des Straßen-, Eisenbahn-, Erd- und Grundbaues, des Wasserbaues, des Kanal- und sonstigen unterirdischen Leitungsbaues; ▶Tiefbau).

783 civilization ecology [n] *ecol.* (Branch of ▶ecology, which studies the adaptation process undergone by plants and animals to their living conditions in human civilization; e.g. in cities, industrial areas, etc.; ▶urban ecology); *s* **ecología [f] de la civilización** (Término que denomina la rama de la ▶ecología que se dedica a estudiar la adaptación de las especies animales y vegetales a las condiciones vitales de la civilización humana, p. ej. en las ciudades, asentamientos industriales, etc.; ▶ecología urbana); *f* **écologie [f] des civilisations (±)** (Branche de l'▶écologie qui étudie le processus d'adaptation du monde animal et végétal aux conditions de vie dans la civilisation humaine ; ▶écologie urbaine) ; *g* **Zivilisationsökologie [f]** (Teilbereich der ▶Ökologie, der die Anpassung von Pflanzen und Tieren an die Lebensbedingungen in der menschlichen Zivilisation, z. B. in Städten, Industrieansiedlungen etc., untersucht; ▶Stadtökologie).

cladding [n] **with panels** [UK] *arch. constr.* ▶veneering with panels.

claim [n] [US] *constr. contr.* ▶remedial construction claim.

claim [n] [UK]**, remedial works** *constr. contr.* ▶remedial construction claim [US].

claims [npl] *constr. contr.* ▶further claims.

clarification [n] **of design brief** [UK] *prof. plan.* ▶determination of planning context [US].

clarification [n] **of planning brief** [UK] *prof. plan.* ▶determination of planning context [US].

784 class [n] (1) *phyt.* (Phytosociological unit superior to an ▶order 1, having the widest range of ▶ecological amplitude. **C.es** are differentiated by the ending *-etea*, which is added to the name of the genus; e.g. *Fagetea*; ▶phytosociological classification); *s* **clase [f]** (Unidad fitosociológica de rango superior a la ▶orden y que tiene una mayor ▶amplitud ecológica que éstas.

Se reconocen por su terminación *-etea* que se añade al nombre genérico; ►clasificación fitosociológica, ►valencia ecológica amplia); *f* **classe** [f] (Dans la détermination des groupements végétaux, unité phytosociologique de rang supérieur, possédant une grande ►valence écologique qui regroupe les ►ordres ayant des affinités floristiques et sociologiques ; le niveau hiérarchique de la **c.** dans la systématique est définit par la terminaison *-etea* ; DEE 1982 ; ►systématique phytosociologique) ; *g* **Klasse** [f] (*Phytosoziologie* die nächste der ►Ordnung übergeordnete Vegetationseinheit mit Klassenkennarten, die eine sehr weite ►ökologische Amplitude haben. Klassen werden durch die Endung *-etea* gekennzeichnet, die dem Stamm eines Gattungsnamens angefügt wird, z. B. *Fagetea*; ►Klassifizierung der Vegetationseinheiten).

class [n] (2) *constr. pedol.* ►particle-size class; *constr. eng.* ►quality class [US]; *constr. pedol.* ►soil textural class.

class [n] [UK], **soil** *constr.* ►class of soil materials [US].

classic garden [n] *gard'hist.* ►French classic garden.

classification [n] *phyt.* ►phytosociological classification.

classification [n], **age** *plan. sociol.* ►age class composition.

classification [n] [UK], **agricultural land** *agr.* ►agricultural land grade [UK].

classification [n] **of biotopes** *ecol. plan.* ►hierarchical classification of biotopes.

785 classification [n] **of rivers and streams** *leg. wat'man.* (Division of all watercourses and canals into categories according to governmental responsibility for performing river maintenance and improvement: Category 1 for navigable rivers under federal or state jurisdiction and Category 2 for non-navigable streams, which are under local jurisdiction; ►river maintenance, ►river engineering measures); *s* **clasificación** [f] **de los cursos de agua** (División de las corrientes fluviales en categorías según la responsabilidad administrativa para su mantenimiento y mejora; en D. se clasifican según su navegabilidad en cursos de agua de 1er orden [hidrovías federales y cursos principales de los estados federados] y cursos de agua de 2do orden [todos los demás]. Esta clasificación tiene el fin de definir la autoridad responsable del ►mantenimiento de cursos de agua y de las obras de ►ingeniería hidráulica de ríos y arroyos); *f* **classement** [m] **des eaux douces** (Classification **1.** des eaux courantes suivant leur calibre : ruisselets, ruisseau, rivière, fleuve. **2.** Des eaux douces suivant leur régime juridique [droit de propriété et droit à l'usage] — les eaux superficielles non domaniales [biens sans maître ou biens collectifs] tels que les cours d'eau ou parties de cours d'eau exclus du domaine public, lacs et étangs situés sur un terrain privé, — les eaux superficielles domaniales appartenant à l'État tels que les cours d'eau navigables ou flottables, les lacs navigables ou flottables, les rivières canalisées, canaux de navigation, étangs et réservoirs d'alimentation ; ►entretien des cours d'eau, ►travaux d'aménagement des cours d'eau) ; *g* **Gewässerklassifizierung** [f] (**D.**, gemäß Wassergesetze des Bundes und der Länder Einteilung der Flüsse und Kanäle in Gewässer erster Ordnung [Bundeswasserstraßen und Landesgewässer] und in Gewässer zweiter Ordnung [alle anderen Gewässer] in Bezug auf die gewässerrechtliche Zuständigkeit der ►Gewässerunterhaltung und des ►Gewässerausbaues; cf. WaStrG und Landeswassergesetze); *syn.* Einteilung [f] der Gewässer.

classification [n] **of soil materials** [US], **use** *constr.* ►soil classification (1).

786 classified highway [n] [US] *adm. trans.* (**In U.S.**, generic term covering interstate highway, federal highway, state highway, beltway; classification by federal and state govern-

ments; **in U.K.**, classification by the Department of Transport); *syn.* classified road [n] [UK]; *s* **carretera** [f] **clasificada** (Término genérico para las vías de dominio y uso público proyectadas y construidas fundamentalmente para la circulación de vehículos automóviles. **En Es.** se clasifican por sus características en autopistas, autovías, vías rápidas y carreteras convencionales. Entre ellas son **carreteras estatales** las integradas en un itinerario de interés general o cuya función en el sistema de transporte afecte a más de una CC.AA. y constituyen la **Red de carreteras del Estado.** Las demás carreteras son competencia de las CC.AA.; cf. arts. 2 y 4 Ley 25/1988; **en EE.UU.** son clasificadas por el gobierno federal y los estatales en carreteras interestatales, carreteras federales, carretera estatales y carreteras de circunvalación); *f* **voie** [f] **classée** (Terme générique désignant la voirie du domaine public routier national [autoroutes et routes nationales], la voirie du domaine public routier départemental [routes départementales] et la voirie du domaine public routier communal [voies communales] ; cf. Code de la voirie routière ; **D.**, terme générique englobant les routes fédérales et les routes de Land [en Bavière, route d'État] et les routes de Kreis) ; *g* **klassifizierte Straße** [f] (**D.**, OB zu Bundesautobahn, Bundesstraße, Landesstraße — *in BY und Sachsen* Staatsstraße — und Kreisstraße; **A.**, alle übergeordneten Straßen, außer jenen, die von der *Autobahnen- und Schnellstraßen- Finanzierungs- Aktiengesellschaft [ASFiNAG]* verwaltet werden, sind Landesstraßen, auch die ehemaligen Bundesstraßen, da sie von den Bundesländern unterhalten werden; **CH**, es werden nationale und kantonale Autostraßen unterschieden; **U.S.**, *Highway* ist der generelle Begriff für öffentliche überörtliche Straßen).

classified road [n] [UK] *adm. trans.* ►classified highway [US].

787 class [n] **of soil materials** [US] *constr.* (Division of soils into groups for earthworks on the basis of their workability or load-bearing characteristics according to their granular or colloidal properties; ►construcrion soil classification, ►soil classification 1); *syn.* soil class [n] [UK]; *s 1* **clase** [f] **de suelo aceptable** (≠) (En Es., definición de las características de composición granulométrica de la tierra fina, de la granulometría, del índice de plasticidad y de la composición química de los suelos modificados según la función prevista para plantaciones, para superficies a encespedar o como suelos estabilizados; cf. RA 1970); *s 2* **grupo** [m] **de suelos** (En D., clasificación de suelos en 10 grupos según las características definidas en las Normas DIN 18 915 para fines de plantación y DIN 18 196 para fines constructivos, sobre la base de su manejabilidad práctica y su capacidad portante; ►clasificación granulométrica, ►clasificación de suelos); *f 1* **classe** [f] **de terrains** (►Classification des terrains d'après leur nature pour l'exécution de travaux de terrassement, l'extraction, le chargement, le transport ou la mise en remblai) ; *f 2* **catégorie** [f] **de sols** (**1.** d'après leur portance comme support de chaussée. **2.** D'après leur aptitude aux travaux culturaux ; ►classification granulométrique des sols) ; *syn.* groupe [m] de sols ; *g* **Bodengruppe** [f] (**D.**, Einordnung der Böden [1] für vegetationstechnische Zwecke in 10 Bodengruppen gem. DIN 18 915 und [2] für bautechnische Zwecke nach DIN 18 196; [3] die Einordnung von Boden und Fels nach ihrem Zustand beim Lösen wird gem. DIN 18 300 in sieben Klassen vorgenommen: ►Bodenklassifikation 1, ►Bodenklassifikation 2); *syn. für den dritten Anwendungsbereich* Bodenklasse [f] (1) (nach DIN 18 300).

class one urn grave [n] [UK] *adm. landsc.* ►prominent urn burial site.

clause [n] *constr. contr.* ►escalation clause

clause [n], **price-revision** *constr. contr.* ►escalation clause.

788 clay [n] **(1)** *constr. eng.* (Mineral soil separate with individual particles less than 0.002mm in diameter [UK and US]; requires moisture for structural ionic bonding and optimum shear resistance, but will liquify if overly moist or subject to significant vibration; ▶silt 1); *s* **arcilla [f]** (Fracción de las partículas del suelo cuyo grosor es menor de 0,005 mm de diámetro; DGA 1986; ▶limo 2); *f 1* **argile [f]** (Terme désignant **1.** un minéral [minéral argileux], **2.** une roche composée pour l'essentiel de ces minéraux, **3.** en granulométrie la fraction minérale du matériau constitué de particules dont le diamètre est inférieur à 2 microns ou 0,002 mm ; DIS 1986 ; ▶limon) ; *f 2* **glaise [f]** (Terre argileuse grasse et plastique pouvant être utilisée pour la fabrication de tuiles, de briques et de la poterie ; *contexte* ▶carrière de glaise) ; *syn.* terre glaise ; *g* **Ton [m]** (**D.**, mineralische Bodenart im Feinkornbereich 0,002 mm; *Abk.* T; cf. DIN 4022 und DIN 4023; Feinkornbereich > 0,002 mm; ▶Schluff).

clay [n] **(2)** *pedol.* ▶translocation of clay.

789 clay content [n] *pedol.* (Constituent amount of clay within a soil); *s* **contenido [m] de arcilla** (de un suelo); *f* **teneur [f] en argile** (d'un sol) ; *g* **Tongehalt [m]** (Mengenanteil an Ton eines Bodens).

790 clayey [adj] *pedol* (Consisting of or relating to clay or clay minerals: like clay); *syn.* argillic [adj], argillaceous [adj], argillous [adj]; *s* **arcilloso/a [adj]** (Que contiene arcilla); *f* **argileux, euse [adj]** (p. ex. sol argileux, horizon argilique) ; *syn.* argilique [adj] ; *g* **tonig [adj]**. (Einen Boden betreffend, der Ton enthält); *syn.* tonhaltig [adj].

791 clay-humus complex [n] *pedol.* (Complex of particles of fine clay and humus, which has the property of attracting and holding ions. In a balanced productive soil, the **c.-h. c.** thus holds sufficient cations—positive charges—to provide essential nutrients for plants); *s* **complejo [m] húmico-arcilloso** (Complejo de minerales de arcilla y humus que genera una textura de agregados estable y liga iones en el suelo que sirven de nutrientes a las plantas); *f* **complexe [m] argilo-humique** (Association en grumeaux construits, de structure stable, résultant de liaisons électrostatiques ou non entre la matière minérale [cations-argiles] et les composés humiques végétaux ou de biosynthèse bactérienne du sol ; grâce à son pouvoir absorbant le **c. a.-h.** met en réserve les éléments nutritifs de la solution du sol nécessaires à la vie des racines et des micro-organismes ; cf. PED 1979, 192 et DIS 1986, 53) ; *syn.* complexe [m] organo-minéral (PED 1979, 205), agrégat [m] argilo-humique (cf. PED 1979, 205), complexe [m] absorbant (LA 1981, 339) ; *g* **Ton-Humuskomplex [m]** (Organo-mineralische Verbindung aus Tonmineralen und Humus, die ein stabiles Aggregatgefüge im Boden erzeugt und Ionen bindet. Z. B. werden in einem fruchtbaren Boden mit hoher biologischer Aktivität genügend Kationen an die **T.-H.e** angelagert, die den Pflanzen die nötigen Nährstoffe liefern).

792 clay lining [n] *constr.* (Compacted clay layer for water retention in a man-made pond or canal; ▶clay sealing, ▶clay sealing layer); *syn.* clay packing [n]; *s* **recubrimiento [m] de limo arcilloso** (Capa de arcilla limosa o limo arcilloso colocada para aislar estanques hacia el subsuelo; ▶capa aislante de arcilla, ▶sellado de arcilla); *f* **corroyage [m] d'argile** (Apport d'une couche d'argile d'environ 20 cm d'épaisseur et placage sur la fond d'une pièce d'eau, constituant son étanchéité ; ▶étanchéisation par couche d'argile, ▶couche argileuse d'étanchéisation) ; *syn.* recouvrement [m] de limon [CH] (WIB 1996) ; *g* **Lehmschlag [m]** (Zur Dichtung von Teichen eingebaute Schicht aus lehmigem Ton oder tonigem Lehm mit einem Wasserdurchlasskoeffizienten von < 1 x 10^{-7} m/s; ▶Tondichtung 1, ▶Tondichtung 2); *syn.* Lehmdichtung [f], Tonlage [f].

793 clay mineral [n] *pedol.* (Crystallized silicates in the form of lamina with a diameter of < 0.002mm, and 2-50nm thick. To a certain extent, clay mineral will absorb and expand in water and is also capable of ion exchange); *s* **mineral [m] de la arcilla** (Silicatos cristalizados generalmente en láminas [cuarzo, clorita, micas], de tamaño de grano inferior a 1/256 mm; cf. DINA 1987); *syn.* mineral [m] arcilloso; *f* **minéral [m] argileux** (Matériau cristallisé inorganique, de structure en général en feuillets [silicates d'Al encadré par un groupement (OH) d'un diamètre < 2 µm et d'une épaisseur comprise entre 7 et 20 Å], rencontré à l'état naturel dans les sols et dans les roches ou formé par néosynthèse, certains ayant la propriété d'absorber de l'eau ou de retenir des cations échangeables ; DIS 1986) ; *syn.* argiles [mpl] ; *g* **Tonmineral [n]** (Kristallisiertes, meist plättchenförmig ausgebildetes OH-haltiges [Schicht]silikat mit einem Blättchendurchmesser von < 2 µm, Dicke 2-50 nm. Es ist z. T. in Wasser aufweitbar und hat die Fähigkeit des Ionenaustausches).

clay packing [n] *constr.* ▶clay lining.

794 claypan [n] *pedol.* (Stiff compact subsoil stratum consisting of translocated clay, forming an impermeable layer below the surface of the soil, causing impeded drainage and waterlogging; cf. DNE 1978; ▶hardpan, ▶translocation of clay); *syn.* argillic horizon [n] [also US], illuvial horizon [n] [also US]; *s* **horizonte [m] de iluviación (1)** (Horizonte compacto, poco permeable, rico en arcilla y separado más o menos toscamente de la capa superior del suelo; MEX 1983; ▶eluviación de arcilla, ▶fragipán, ▶horizonte petrificado); *syn.* horizonte [m] de acumulación de arcilla, horizonte [m] arcillo-compactum [MEX]; *f* **horizon [m] d'accumulation d'argile** (Horizon argileux formé par l'entraînement de particules argileuses et le ralentissement des filets liquides au sein des horizons profonds, plus denses, dont la porosité est presque exclusivement capillaire ; PED 1983, 90 ; ▶horizon d'accumulation durci, ▶horizon de fragipan, ▶processus d'accumulation d'argile) ; *g* **Tonband [n]** (Einlagerungsverdichtung durch ▶Tonverlagerung in Unterböden, z. B. bei Braun- und Parabraunerden, wodurch periodisch Wasserstau verursacht wird. Tonbänder in Marschböden werden auch **Knicks** genannt; ▶Verdichtungshorizont).

795 clay pipe [n] *agr. constr.* (Generic term for earthenware pipe, either laid with open joints for underground drainage, or vitrified with sealed joints for sanitary sewers; ▶vitrified-clay pipe [US]/glazed stoneware pipe [UK]); *s* **tubo [m] de arcilla** (Término genérico para conducto de arcilla para arcantarilla o drenaje subterráneos; ▶tubería de drenaje vitrificada); *syn.* tubería [f] de arcilla; *f* **drain [m] en poterie** (Terme générique pour les tuyaux en terre cuite utilisés pour le drainage des sols ; ▶tuyau en grès ; VRD 1986, 113) ; *syn.* drain [m] en terre cuite, drain [m] en argile ; *g* **Tonrohr [n]** (OB zu Entwässerungs- oder Dränrohr aus Ton; ▶Steinzeugrohr).

796 clay pit [n] *min.* (Extraction site for clay); *s* **gredal [m]** (Lugar de extracción de arcilla); *syn.* yacimiento [m] de tierra arcillosa; *f* **carrière [f] de glaise** (Lieu d'extraction de terre argileuse pour la fabrication des tuiles et des poteries) ; *syn.* carrière [f] d'argile ; *g* **Tongrube [f]** (Entnahmestelle für Ton).

797 clay sealing [n] *constr.* (Non-rigid method of lining the sides and floors of pools and ponds using puddled clay; ▶clay lining); *s* **sellado [m] de arcilla** (Método de impermeabilización de instalaciones de aguas estancadas o pequeños cursos de agua como estanques, filtros verdes, tanques de almacenamiento de aguas pluviales y también para depósitos de residuos; ▶recubrimiento de limo arcilloso); *f* **étanchéisation [f] par couche d'argile** (Méthode utilisée pour l'étanchéisation de bassins de rétention ou de pièces d'eau ; ▶corroyage d'argile) ; *g* **Tondichtung [f] (1)** (Unstarre/labile Bauweise zur Dichtung von stehenden Wasseranlagen oder fließenden Kleingewässern wie Wasserbecken, Teichanlagen, künstlich angelegten Feuchtbiotopen, Badegewässern, Pflanzenkläranlagen, Regenwassersammelbecken oder Bachläufen, aber auch zur Abdichtung von Deponien; ▶Lehmschlag); *syn.* Tonabdichtung [f].

798 clay sealing layer [n] *constr.* (Layer of clay installed to seal the floor of an artificial body of water; ▶clay lining); *s* **capa [f] aislante de arcilla** (Capa de arcilla tendida al fondo de estanques de tipo natural, biótopos húmedos o arroyos renaturalizados para mantener el agua dentro de ellos; ▶recubrimiento de limo arcilloso); *f* **couche [f] argileuse d'étanchéisation** (Couche d'argile utilisée comme étanchéité sur le fond d'un bassin de rétention ou d'une d'eau ; ▶corroyage d'argile) ; *g* **Tondichtung [f] (2)** (Eingebaute Tonschicht zur Abdichtung eines künstlichen Gewässers, Feuchtbiotopes oder renaturierten Bachlaufes; ▶Lehmschlag); *syn.* Tonabdichtung [f].

799 clay slurry [n] *hort.* (Liquid soil mixture into which the root systems of bareroot planting stock are dipped; ▶root puddling); *syn.* root dip [n]; *s* **papilla [f] de barro** (Mezcla líquida de tierra en la que se impregnan las raíces desnudas de las plantas antes de plantarlas; ▶pralinaje de raíces); *syn.* salsa [f] bordelesa; *f* **praline [f]** (Mélange de terre argileuse, d'eau, de fumier ou d'hormones de reprise dans lequel on laisse tremper les racines des végétaux avant la plantation ; ▶pralinage des racines) ; *syn.* pralin [m] ; *g* **Lehmschlämme [f]** (Mit Wasser vermischter, dünnbreiiger Lehm; ▶Tauchen der Wurzeln in Lehmschlämme); *syn.* Lehmbrei [m].

800 clay soil [n] *agr. for. hort. pedol.* (Usually a nutrient-rich dense soil with >30% clay content and unfavo[u]rable physical attributes, thus also known as a **heavy soil**; *opp.* ▶sandy soil); *s* **suelo [m] arcilloso** (Suelo rico en nutrientes de minerales arcillosos con contenido de arcilla superior al 30% y con características físicas desfavorables por lo que también se le llama suelo pesado o fuerte; *opp.* ▶suelo arenoso); *syn.* suelo [m] de adobe [MEX]; *f* **sol [m] argileux** (Sol en général riche en substances nutritives, constitué de roches sédimentaires argileuses dont la teneur en argile est > à 30 % et possédant des caractéristiques physiques défavorables et dénommé « sol lourd » ; *opp.* ▶sol sableux) ; *g* **Tonboden [m]** (Meist nährstoffreicher Boden aus Tongesteinen mit > 30 % Tonanteil und ungünstigen physikalischen Eigenschaften, weshalb er auch **schwerer Boden** genannt wird; *opp.* ▶Sandboden).

801 Clean Air Act [n] *envir.* (Law controlling air pollution; ▶Federal Clean Air Act [US]/Control of Pollution Act [UK]; ▶pollution control 1); *s* **ley [m] de calidad del aire y protección de la atmósfera [Es]** (Ley que rige la ▶protección del ambiente atmosférico, en España Ley 34/2007, de 15 de noviembre, junto con diferentes decretos y órdenes; ▶protección contra inmisiones); *f* **loi [f] relative à la lutte contre les pollutions atmosphériques** (F., il existe de nombreux textes législatifs dirigés contre la pollution atmosphérique dont les plus importants sont : la loi n° 61-842 du 2 août 1961 ▶loi relative à la lutte contre les pollutions atmosphériques et les odeurs, la loi n° 76-663 du 19 juillet 1976 sur les installations classées pour la protection de l'environnement, la loi n° 96-1236 du 30 décembre 1996 sur l'air et l'utilisation rationnelle de l'énergie [transposition en droit français de la Directive n° 96/62/CE du 27 septembre 1996] et d'une manière générale tous les décrets portant transposition des lois et directives du Conseil du Parlement Européen) ; *g* **Immissionsschutzgesetz [n] (±)** (Gesetz, das den ▶Immissionsschutz regelt; D., ▶Bundes-Immissionsschutzgesetz [BImSchG] vom 15.03.1974, i. d. Fassung der Bekanntmachung vom 26.0.2002 [BGBl. I 3830], zuletzt geändert durch Artikel 1 des Gesetzes vom 23.10.2007 [BGBl. I 2470]).

802 clean air area [n] *plan.* (District without air pollution, sometimes used to compare with long-term changes of pollution levels in other areas having contaminated air); *s* **área [f] de aire puro** (Zona casi libre de inmisiones que sirve como base comparativa para estudiar los cambios a largo plazo en zonas más contaminadas); *syn.* área [f] de aire limpio; *f* **territoire [m] de bonne qualité de l'air** (Zone totalement ou presque exempte de pollution atmosphérique et permettant, par comparaison, l'évaluation des changements à long terme de l'air ambiant dans les zones soumises à une forte pollution atmosphérique) ; *g* **Reinluftgebiet [n]** (Gebiet ohne oder fast ohne Immissionsbelastung. Raum, dessen Immissionssituation als Vergleichsbasis für die Bewertung langfristiger Veränderungen in Belastungsgebieten dient).

803 clean air plan [n] *envir. leg.* (Pollution control measures proposed by competent authorities/agencies for an ▶contaminated airspace on the basis of an atmospheric emission inventory for minimizing existing or expected air pollution); *s* **plan [m] de protección del ambiente atmosférico** (Plan de acción para reducir la contaminación en ▶zonas de atmósfera contaminada); *f* **plan [m] de protection de l'air** (Établissement de différents plans visant à prévenir ou à limiter la pollution de l'air ; *on distingue les plans suivants* **1. Plan régional pour la qualité de l'air :** il fixe, sur la base d'un inventaire des émissions et une évaluation de la qualité de l'air et de ses effets sur l'environnement et la santé publique, des orientations destinées à prévenir ou réduire la pollution atmosphérique ou d'en atténuer les effets ainsi que les objectifs de qualité de l'air spécifique à certaines zone. **2. Plan de protection de l'atmosphère :** élaboré dans les agglomérations de plus de 250 000 habitants, vise à ramener la concentration à un niveau inférieur aux valeurs limites admises et à définir les modalités de procédure d'alerte, il peut renforcer les objectifs de qualité ainsi que les mesures techniques relatives à la limitation des sources d'émission de substances polluantes nocives pour la santé humaine et l'environnement ; *syn.* plan [m] de protection atmosphérique. **3. Plan de déplacements urbains :** plan définissant les principes d'organisation des transports des personnes et des marchandises de la circulation et du stationnement ; la diminution du trafic automobile et le développement des transports collectifs et des moyens de déplacement économes et les moins polluants comptent parmi ses principales orientations ; loi n° 96-1236 du 30 décembre 1996 ; ▶secteur soumis à la pollution de l'air atmosphérique) ; *g 1* **Luftreinhalteplan [m]** (In ▶lufthygienischen Belastungsgebieten von den dafür zuständigen Behörden bei auftretenden oder zu erwartenden schädlichen Luftverunreinigungen auf der Grundlage eines Emissionskatasters aufzustellende Maßnahmenpläne; diese Pläne beschreiben Art und Umfang der zu erwartenden Luftverunreinigungen und zeigen Vorgehensweisen zu deren Verminderung auf; solch ein L. dient auch der Vorsorge); *g 2* **Luftqualitätsplan [m]** (Luftreinhalteplan gem. Richtlinie 2008/50/EG vom 21.05.2008 in dem Maßnahmen zur Erreichung der Grenzwerte oder Zielwerte festgelegt sind. **L.pläne** enthalten geeignete Maßnahmen, damit der Zeitraum der Nichteinhaltung der in den Anhängen XI und XIV festgesetzten Grenzwerten und im Anhang XII Abschnitt B festgelegten Alarmschwelle für Ozon so kurz wie möglich gehalten werden kann; cf. Art. 23 RL 2008/50/EG).

clean felling [n] [UK] *for.* ▶clear felling.

clean gravel [n] *constr.* ▶washed gravel.

cleaning eye [n] [UK] *constr.* ▶access eye [US].

804 cleaning [n] **of a cavity** *arb. constr.* (ARB 1983; tree surgeon's removal of rotted material back to healthy wood on trees; ▶cavity, ▶decayed area); *s* **limpieza [f] de la caries húmeda** (Vaciado de las ▶zonas podridas de un árbol para su saneamiento por medio de cirugía; ▶cavidad); *f* **raclage [m] des tissus pourris** (IDF 1988, 116 ; élimination des foyers de pourriture qui se sont installés dans les bois d'un tronc ; ▶cavité [naturelle], ▶point de pourriture) ; *syn.* élimination [f] des tissus pourris ; *g* **Aushauung [f]** (Von einem Baumchirurgen bis in das gesunde Holz ausgeschälte ▶Faulstelle bei lebenden Bäumen; aushauen [vb]; ▶Höhlung).

805 cleaning [n] **of sand areas** *constr. recr.* (Periodic cleansing of sand areas by ▶sand replacement); *s* **limpieza [f] de superficies con arena** (Limpieza regular de áreas de juego sobre arena sea por extracción de las basuras, vidrios y otros residuos o por ▶cambio de la arena); *f* **désinfection [f] des aires sablées** (▶remplacement du sable) ; *g* **Reinigen [n, o. Pl] von Sandflächen** (Regelmäßiges Entfernen von Müll, Glasscherben und anderem Unrat auf Sandspielflächen; ▶Sandaustausch); *syn.* Sandflächenreinigung [f].

cleaning out [n] [UK] *arb. constr.* ▶dead wooding [US] (1).

806 cleaning up [n] **planted areas** [US] *constr.* (Removal of stones, perennial weeds and rubbish; ▶ston picking, ▶landscape maintenance); *syn.* picking up [n] litter and weeding [n] on planted areas [UK]; *s* **limpieza [f] de áreas a plantar** (Conjunto de operaciones de ▶despeje de piedras y desperdicios y desbroce de raíces de malas hierbas realizadas antes de plantar; ▶gestión paisajística de espacios verdes); *f* **désherbage [m] et nettoyage [m] des massifs** (Enlèvement des pierres, de détritus et des racines de plantes adventices sur les zones destinées à être engazonnées ou plantées ; ▶épierrage, ▶entretien des espaces verts) ; *g* **Säubern [n, o. Pl.] von Pflanzflächen** (▶Entfernen von Steinen, Unrat und Wurzelunkräutern in Flächen, die für eine Bepflanzung vorgesehen sind; Pflanzflächen säubern [vb]; ▶Pflege und Unterhaltung von Grünflächen); *syn.* Pflanzflächensäuberung [f].

cleanout [n] [US] *constr.* ▶access eye [US].

807 cleanout chamber [n] [US] *constr.* (Shaft which is installed at intervals and at points of junctions of ▶drainage systems 1 to facilitate control and cleaning; typically with a minimum 450mm diameter); *syn.* inspection chamber [n] [UK], access basin [n] [US]; *s* **pozo [m] de registro (1)** (Unidad de la instalación de drenaje para efectuar alguna limpieza; ▶sistema de drenaje); *f* **regard [m] de visite** (Ouvrage en maçonnerie ou en matériaux divers pratiqué dans le sol permettant la visite et éventuellement le nettoyage de la canalisation ; ▶système de drainage) ; *syn.* regard [m] visitable ; *g* **Kontrollschacht [m] (2)** (...schächte [pl]; Schacht an künstlichen ▶Entwässerungssystemen zur Überwachung und Reinigung der Leitungsstränge); *syn.* Prüfschacht [m], Wartungsschacht [m].

cleanout pipe [n] [US] *constr.* ▶drainage cleanout pipe [US]/drain(age) inspection shaft [UK].

808 clean technology [n] *envir.* (Production method for creating products without generating waste or toxicity; ▶low-waste technology); *s* **tecnología [f] limpia** (Proceso productivo en el que se ha reducido considerablemente las cargas ambientales del mismo, sea por reducción de emisión de contaminantes y de residuos, sea por reducción del consumo energético, de materias primas o de agua; ▶proceso productivo pobre en residuos); *syn.* tecnología [f] de producción limpia; *f 1* **technologie [f] propre (1)** (F., procédé innovant permettant le recyclage de l'eau, des polluants dans les industries consommatrices de matières premières, ou techniques n'engendrant peu ou pas de déchets ou permettant un valorisation maximale par réemploi dans l'entreprise ; ▶technologie propre 2 ; lois n° 75-633 du 15 juillet 1975 et n° 76-663 du 19 juillet 1976) ; *f 2* **technologie [f] non génératrice de déchets (±)** (Au sens littéral du mot allemand, procédé de fabrication permettant la réduction à la source

de la quantité et de la toxicité des déchets produits pouvant atteindre un « niveau 0 » ; ▶technologie faiblement génératrice de déchets) ; *syn.* technologie [f] non productrice de déchets ; *g* **abfallfreies Verfahren [n]** (Fertigungsmethode, Produkte so herzustellen, dass keine Abfallstoffe entstehen; im Franz. wird **a. V.** *und* ▶abfallarmes Verfahren mit dem OB *technologie propre* übersetzt).

clean-up [n] [US], **late** *constr.* ▶late cut.

Clean Water Act [n] [US] *envir. leg.* ▶water rights.

clearance [n] *constr. for.* ▶woody plant clearance.

clearance [n] [UK], **town centre** *plan. urb.* ▶comprehensive redevelopment.

clearance area [n] *for.* ▶clear cutting area [US].

809 clearance height [n] *constr.* (Height limit for passage, e.g. through door openings, arches, bridges; ▶clearance space, ▶clearance width); *s* **altura [f] libre** (Espacio vertical del vano de una puerta o de un construcción; DACO 1988; ▶medida interior, ▶vano); *syn.* altura [f] de paso, altura [f] de franqueo, altura [f] de despeje, espacio [m] libre vertical, franqueo [m] superior, franqueo [m] vertical (BU 1959); *f* **hauteur [f] libre** (Hauteur utile ou minimale dans la construction comprise entre le sol fini et un passage cocher, un plafond [hauteur sous plafond], une porte, etc. ; ▶mesure dans-œuvre, ▶largeur dans-œuvre) ; *syn. souvent utilisé pour les ouvrages d'art routiers* tirant [m] d'air, hauteur [f] sous ouvrage ; *g* **lichte Höhe [f]** (Nutzbare Durchgangs-, Durchfahrts- oder Aufenthaltshöhe bei Türen, Torbögen, Brücken, Räumen etc.; ▶lichtes Maß, ▶lichte Weite).

810 clearance herb formation [n] *phyt.* (Formation of short-lived herb vegetation in a forest clearing or ▶clear cutting area [US]/clear felling area [UK]; ▶tree-fall gap community, ▶forb vegetation in woodland clearing); *syn.* clearing forb formation [n], clearance herb vegetation [n], woodland clearing community [n]); *s* **vegetación [f] de bosques talados** (Formación temporal de vegetación herbácea en claros creados por tala a matarrasa; ▶vegetación de claros, ▶formación de perennes de bosques talados); *syn.* formación [f] de bosques talados; *f* **flore [f] des coupes à blanc(-étoc)** (Groupements temporaires s'installant sur une ▶aire de coupe rase, constitués au départ d'espèces annuelles ou bisannuelles remplacées successivement par les espèces herbacées et sous-ligneuses de la forêt et par les jeunes plants de la future strate arborescente ; ▶flore de clairière, ▶flore herbacée géante de coupe) ; *syn.* végétation [f] de coupes à blanc(-étoc) ; *g* **Kahlschlagflur [f]** (Kurzlebige Waldlichtungskrautflur, die sich nach Fällung oder sonstiger Entnahme aller Bäume einer Waldfläche von selbst einstellt; ▶Lichtungsflur, ▶Staudenschlagflur); *syn.* Schlagflora [f], Schlagflur [f].

clearance herb vegetation [n] *phyt.* ▶clearance herb formation.

811 clearance [n] **of unwanted site material** *constr. hort.* (Removal of stones, perennial weeds and rubbish for ▶hauling away [US]/carting away [UK]; ▶stone picking); *s* **despeje [m] de piedras y desperdicios y desbroce [m] de raíces de malas hierbas** (▶despeje de piedras, ▶transporte afuera de obra); *f* **enlèvement [m] des pierres, de détritus et des racines de plantes adventices** (▶épierrage, ▶évacuation) ; *g* **Entfernen [n, o. Pl.] von Steinen, Unrat und Wurzelunkräutern** (▶Abfahren, ▶Entfernen von Steinen).

812 clearance [n] **of unwanted spontaneous woody vegetation** *conserv. land'man.* (Removal of unwanted tree or shrub material by mechanical or chemical means; ▶spontaneous colonization by scrub); *s* **desbroce [m] (1)** (Proceso de eliminación mecánica o química de la vegetación leñosa, p. ej. abedules, en landas y turberas para mantener un tipo de formación vegetal deseada o regenerar turberas; ▶invasión de leñosas); *f* **débroussaillage** [m] (Enlèvement des repousses arbustives par des moyens mécaniques ou parfois chimiques ; ▶embroussaillement) ; *syn.* mesure [f] de débroussaillement ; *g* **Entkusselung** [f] (*Norddeutscher Begriff* Arbeiten, die mit mechanischen oder manchmal auch chemischen Mitteln zum Entfernen von Gehölzaufwuchs, speziell in Heiden und Mooren, durchgeführt werden, z. B. Birkenentkusselung; entkusseln [vb]; ▶Verbuschung); *syn.* Entbuschung [f].

813 clearance space [n] *constr.* (Generic term for the space between peripheral limits, e.g. internal diameter of pipes, door openings, underpasses, etc.; ▶clearance height, ▶clearance space of trees, ▶clearance width, ▶internal diameter); *syn.* clear opening [n]; *s* **medida** [f] **interior** (Término genérico para espacios libres normalizados según el uso en urbanismo, arquitectura, tecnología, industria, como diámetro interior de tubería, *términos específicos* ▶altura libre, ▶vano, ▶anchura de paso, ▶espacio libre para vehículos); *syn.* luz [f] libre; *f* **mesure** [f] **dans-œuvre** (Mesure prise à la face intérieure des murs, d'une fouille ou d'un tube ; opposée à « hors-œuvre ». *Abrév.* D/O ; DTB 1985 ; *terme spécifique* diamètre intérieur d'un tube ; ▶diamètre nominal [d'écoulement], ▶hauteur libre, ▶hauteur sous couronne, ▶largeur dans-œuvre) ; *g* **lichtes Maß** [n] (Kürzeste Entfernung zwischen zwei gegenüberliegenden Begrenzungen, z. B. Innendurchmesser von Rohren, lichte Höhe einer Unterführung [= nutzbare Durchfahrtshöhe]; *UBe* ▶lichte Höhe, ▶lichte Weite, ▶Lichtraumprofil, ▶Nennweite); *syn.* im Lichten [loc].

814 clearance space [n] **of trees** *trans.* (Space to be kept free of branches for the safe movement of vehicles on tree-lined travelled ways [US]/carriage-ways [UK]. **In D.**, trees must have their branches removed up to a height of 4.5m vertically above the edge of a roadway; ▶clearance height, ▶clear opening); *s* **espacio** [m] **libre para vehículos** (Espacio mínimo libre de ramas de árboles a mantener abierto en avenidas para permitir el paso de vehículos grandes; ▶altura libre, ▶medida interior); *f* **hauteur** [f] **sous couronne** (IDF 1988, 131 ; la dimension minimale à respecter pour obtenir une circulation facile est 4,30 m sous passage public ; cf. arrêté du 13 septembre 1966 ; **D.**, l'élévation de la couronne doit être effectuée jusqu'à une hauteur de 4,5 m mesurée au-dessus du bord de la chaussée ; ▶hauteur libre, ▶mesure dans-œuvre) ; *syn.* gabarit [m], tirant [m] d'air (VRD 1986, 185) ; *g* **Lichtraumprofil** [n] (Freizuhaltender Durchfahrtsraum für Verkehrsteilnehmer bei bepflanzten Straßen und sonstigen Verkehrsflächen zur Sicherung einer ungehinderten und gefahrlosen Nutzung. In D. müssen je nach Verkehrsbedeutung Bäume mindestens bis 4,50 m hoch senkrecht über dem Fahrbahnrand aufgeastet werden; ▶lichte Höhe, ▶lichtes Maß).

815 clearance width [n] *constr.* (Horizontal clearance distance, e.g. the inside diameter of a pipe or width of a door opening; ▶clearance height, *generic term* ▶clearance space); *s* **vano** [m] (Espacio entre apoyos de un arco, viga, apoyos de muros, etc.; ▶altura libre; *término genérico* ▶medida interior); *f* **largeur** [f] **dans-œuvre** (Mesure prise à la face intérieure, p. ex. d'une fouille ou d'un tuyau ; *opp.* mesure hors œuvre ; ▶hauteur libre ; *terme générique* ▶mesure dans-œuvre) ; *g* **lichte Weite** [f] (Kürzeste Entfernung zwischen zwei gegenüberliegenden Begrenzungen, z. B. bei Torpfosten, Innendurchmesser von Rohren; *OB* ▶lichte Höhe, lichtes Maß).

clear cutting [n] [US] *for.* ▶clear felling.

816 clear cutting area [n] [US] *for.* (Area on which the entire standing crop is removed); *syn.* clearance area [n], clear felling area [n] [UK]; *s* **claro** [m] **provocado por la tala a**

matarrasa *syn.* calvero [m]; *f* **aire** [f] **de coupe à blanc** (Peuplement forestier dont la totalité des arbres a été coupé) ; *syn.* aire [f] de coupe rase ; *g* **Kahlschlagfläche** [f] (Waldfläche, auf der sämtliche Bäume gefällt wurden); *syn.* Schlagfläche [f].

clear-cutting method [n] [US] *for.* ▶clear-cutting system [US].

817 clear-cutting system [n] [US] *for.* (Silvicultural system in which the old crop is cleared over a considerable area at one time; regeneration is generally artificial, sometimes raised in conjunction with an agricultural crop; cf. SAF 1983; ▶clear felling, ▶forestry 1); *syn.* clear-cutting method [n] [US], clear felling system [n] [UK]; *s* **explotación** [f] **a matarrasa** (Sistema de ▶selvicultura artificial que utiliza la ▶tala a matarrasa y repobla generalmente con sólo una especie las superficies así taladas); *f* **exploitation** [f] **forestière par coupes rases** (Type d'exploitation comportant exclusivement l'exécution de ▶coupes à blanc enlevant en une seule fois la totalité du peuplement sur des surfaces relativement importantes ; la régénération est souvent obtenue artificiellement, mais la régénération naturelle est aussi possible ; cf. DFM 1975 ; ▶ligniculture, ▶sylviculture) ; *syn.* système [m] des coupes à blanc [CDN], exploitation [f] par coupe à blanc-étoc ; *g* **Kahlschlagswirtschaft** [f, o. Pl.] (Forstliche Wirtschaftsweise, die mit dem ▶Kahlschlag arbeitet und die abgetriebene Fläche wieder aufforstet; ▶Forstwirtschaft); *syn.* Kahlschlag(s)betrieb [m], o. V. Kahlschlagwirtschaft [f].

cleared agrarian landscape [n], **totally** *agr. land'man.* ▶extensively cleared land (for cultivation).

818 clear felling [n] *for.* (Strictly, the removal of the entire standing crop. In practice this may refer to exploitation that leaves much unsaleable material standing; SAF 1983; ▶deforestation, ▶tree clearing and stump removal [operation], ▶slash); *syn.* clean felling [n] [also UK], complete cutting [n] [also US], complete felling [n] [also UK], complete exploitation [n], clear cutting [n] [also US]; *s* **tala** [f] **a matarrasa** (Corta por la cual la totalidad de la población arbórea del sector explotado se corta de una vez; ▶de[s]forestación, ▶despeje de árboles y tocones, ▶residuo forestal, ▶roturación); *syn.* tala [f] rasa, corta [f] total, corta [f] a hecho, corta [f] a matarrasa; *f* **coupe** [f] **à blanc** (*Sensu stricto* coupe de la totalité des arbres d'un peuplement. En pratique, peut inclure des opérations qui laissent sur pied la plus grande partie du matériel non commercialisable ; DFM 1975 ; ▶abattage par extraction de souche, ▶déforestation, ▶défrichement, ▶rémanent[s] de coupe) ; *syn.* coupe [f] blanche, coupe [f] rase, coupe [f] à blanc-étoc ; *g* **Kahlschlag** [m] (Fällung aller Bäume einer Waldfläche unter Beseitigung oder Liegenlassen des ▶Schlagabraumes; ▶Entwaldung, ▶Rodung); *syn.* Kahlhieb [m].

clear felling area [n] [UK] *for.* ▶clear cutting area [US].

clear felling system [n] [UK] *for.* ▶clear cutting system [US].

clearing [n] *agr.* ▶burn clearing; *phyt.* ▶forb vegetation in woodland clearing.

819 clearing [n] **and removal** [n] **of tree stumps** *constr. hort.* (Extraction or grinding away of rootstock [US]/rootstock [UK], as well as its disposal. *Specific terms* ▶stump-chipping and ▶stump out); *s* **destoconado** [m] (Arranque de tocones de árboles con raíz desenterrándolos o triturándolos y transporte de los mismos al lugar de depósito o aprovechamiento; *términos específicos* ▶trituración de tocones y ▶extracción de tocones); *f* **élimination** [f] **de la souche** (Terme générique pour toute opération d'enlèvement du sol de souches soit par détourage de la motte soit par fraisage du sol y compris son évacuation ; *termes spécifiques* ▶dessoucher, ▶broyage de la souche) ; *syn.* enlèvement [m] de la souche ; *g* **Stubbenbeseitigung** [f]

(Entfernen von Wurzelstöcken durch Ausgraben oder Fräsen sowie Abfuhr; OB zu ►Stubben fräsen und ►Stubben roden; Stubben beseitigen [vb]; *syn.* Beseitigung [f] von Stubben, Beseitigung [f] von Wurzelstöcken, Entfernen [n, o. Pl.] von Stubben, Entfernung [f] von Wurzelstöcken, Stubbenentfernung [f], Wurzelstockentfernung [f].

clearing community [n] [UK], **forest** *phyt.* ►tree-fall gap community [US].

clearing forb formation [n] *phyt.* ►clearance herb formation.

clearing work [n] [US], **tree** *constr. for.* ►tree clearing and stump removal (operation) [US]/grub felling [UK].

820 clearly-defined [pp] *plan.* (Term used to describe a planning scheme which delineates the exact boundaries of all plots of land involved; e.g. within a housing subdivision); *s* **claramente delimitado** [pp] (Término utilizado para describir los tipos de planes que consideran exactamente los límites de los predios para definir los usos previstos); *f* **défini, ie au niveau de la parcelled** [loc] (±) (Plan d'aménagement dont la précision graphique est définie par les limites de la parcelle) ; *g* **parzellenscharf** [adj] (Eine Planung betreffend, die die genauen Grundstücksgrenzen berücksichtigt).

clear opening [n] *constr.* ►clearance space.

821 clear-water lake [n] *hydr.* (Waterbody absent of, or hardly discolo[u]red by or without humic matter; mostly oligotrophic deep lakes with a low supply of nutrients; ►distrophic lake); *s* **lago** [m] **de aguas claras** (Lago generalmente oligótrofo que contiene pocas sustancias húmicas; ►lago distrófico); *f* **lac** [m] **d'eau claire** (Lac en général oligotrophe, de forte transparence, non troublé ou coloré par les matières humiques ; ►lac distrophique) ; *syn.* lac [m] d'eau oligotrophe/cristalline ; *g* **Klarwassersee** [m] (See, dessen Wasser nicht oder kaum durch Huminstoffe gefärbt ist; meist ein oligotropher See; ►Braunwassersee).

822 clear-water river [n] *hydr.* (River with relatively clear, transparent water, light to olive green in colo[u]r and with little suspended particulate matter); *s* **río** [m] **de aguas claras** (Río con aguas relativamente transparentes de color claro a verde olivo y con poca materia en suspensión); *f* **fleuve** [m] **d'eau claire** (Fleuve dont l'eau est relativement claire et transparente, de couleur vert clair à vert olive et de faible concentration en particules en suspension) ; *syn.* fleuve [m] d'eau oligotrophe ; *g* **Klarwasserfluss** [m] (Fluss mit relativ klarem, hell- bis olivgrünem, durchsichtigem Wasser und geringem Schwebstoffgehalt).

823 cleft rock [n] *geo.* (Rock characterized by various-sized joints, fissures, crevices, and splits; ►joint 1); *s* **roca** [f] **fisurada** (Roca caracterizada por fracturas, grietas, fisuras y juntas de diferentes tamaños; ►diaclasa); *f* **roche** [f] **fissurée** (Roche divisée par un réseau de fentes lié à sa genèse ou provoqué par les efforts tectoniques et la météorisation subis ; ►diaclase) ; *g* **klüftiges Gestein** [n] (Gestein mit vielen unterschiedlich großen Rissen, Fugen und Spalten; ►Kluft); *syn.* Kluftgestein [n].

clerical personnel [n] *adm. contr.* ►office worker.

clerk [n] **of the works** [UK] *constr. prof.* *►client's project representative [US].

824 client [n] *contr.* (Public authority or private client commissioning and entering into a construction contract with a building contractor; ►developer 1, ►developer 3); *s* **propiedad** [f] **de obra** (Autoridad pública o cliente privado que ordena y contrata una obra a un contratista; ►comitente, ►promotor de obra); *f* **maître** [m] **d'ouvrage** (Personne physique ou morale pour le compte de laquelle est exécuté un ouvrage. Celui-ci commande et finance l'ouvrage et se fait souvent assister dans sa réalisation par un ►maître d'œuvre ; ►donneur d'ordre, ►promoteur-constructeur) ; *syn.* client [m], maître [m] de l'ouvrage ; *g* **Bauherr/-in** [m/f] (►Auftraggeber/-in, ►Bauträger/-in; *UBe* öffentlicher **B.**, privater **B.**).

client [n], **public** *adm. contr.* ►contracting public agency [US].

825 client's project representative [n] [US] *constr. prof.* (An inspector employed by the ►client, architect or landscape architect on the works to ensure compliance with the contract provisions with regard to standards of materials and workmanship. The **c. p. r./clerk of the works** [UK] [CoW] keeps records of time and delivered materials, arrival and departure of workers, and reports in U.K. to the client on architectural and landscape projects, **in U.S.** only on architectural projects or large site construction projects); *syn.* clerk [n] of the works [UK], independent site inspector [n] [UK]; *s* **representante** [m/f] **de la propiedad** (En GB, EE.UU. y Es., inspector, -a representante de la ►propiedad de obra y sólo responsable ante ésta cuya función es asegurar el cumplimiento de las previones del contrato en cuanto a estándares de materiales y calidad profesional; puede dar órdenes a la empresa constructora si la propiedad lo desea); *f* **personne** [f] **responsable du marché** (±) (U.K., représentant légal du ►maître d'ouvrage ne donnant aucune instruction à l'entrepreneur et rendant compte de l'avancement des travaux et de la livraison des matériaux) ; *g* **Überwacher/-in** [m/f] **der Bauarbeiten** (≠) (U.K., als Vertreter des ►Bauherrn ist der *clerk of works* nur diesem Rechenschaft schuldig, darf dem Unternehmer keine Anweisungen geben. Im U.K. gibt es den *clerk of works* bei Hochbau- und Landschaftsbauprojekten, in den USA nur bei Hochbauprojekten. Er rapportiert die gearbeiteten Stunden der Arbeiter und das gelieferte Material; in D. gibt es für den *Clerk of the Works* keine Entsprechung).

826 cliff [n] *geo. min.* (Vertical face of earth or more often rock, usually but not always along the sea-coast, formed by wave action. Other meanings are differentiated by a prefixed term, e.g. sea-cliff, lake-cliff, ►rivercliff; ►coastal cliff); *s* **acantilado** [m] (Abrupto rocoso litoral, generado por la actividad erosiva del mar; DGA 1986; ►costa acantilada); *syn.* costal [m] de abrasión; *f* **falaise** [f] (Escarpement situé sur les côtes et qui est dû à l'érosion marine ; PR 1987 ; ►côte à falaise) ; *g* **Kliff** [n] (**1.** *geo.* durch die Tätigkeit der Brandung an Steilküsten erzeugte Steilwand in Locker- oder Festgesteinen; ►Steilküste. **2.** *min.* Sehr schroffer oder senkrechter Abbruch in einem Steinbruch); *syn. zu 1.* Abbruchufer [n] am Meer; *syn. zu 2.* Steilwand [f].

cliff face [n] *geo.* ►rock face (1).

827 cliff quarry [n] *min.* *s* **cantera** [f] **escarpada**; *f* **carrière** [f] **de roches massives à flanc de coteau** ; *g* **Steinbruch** [m] **an einer Steilwand**.

climate [n] *met. urb.* ►urban climate.

climate [n], **topographic** *met.* ►topoclimate.

climate [n] [UK], **town** *met. urb.* ►urban climate.

climate change [n] *met.* ►climatic change.

828 climate [n] **near the ground** *met.* (Meteorological phenomena in the first two meters [US]/metres [UK] above the surface of the ground; ►ground-level air layer, ►microclimate, ►site climate, ►topoclimate); *s* **clima** [m] **de las capas cercanas a la superficie terrestre** (Estado físico de la atmósfera en una pequeña zona muy próxima a la superficie del suelo; ►capa de aire al ras del suelo, ►mesoclima, ►microclima, ►microclima de estación); *f* **climat** [m] **de la couche d'air à proximité du sol** (Ensemble des qualités de la couche atmosphérique située dans les 2 premiers mètres au-dessus du sol ; ►climat local, ►climat stationnel, ►couche d'air à proximité du sol, ►microclimat) ;

g **Klima [n] der bodennahen Luftschicht** (...klimata [pl], ...klimate [pl]; meteorologische Erscheinungen in den ersten zwei Metern über Geländeoberfläche; die Lehre vom Klima in Bodennähe wird als **Standortsklimatologie** oder **Ökoklimatologie** [GEI 1961, 2] bezeichnet; ►bodennahe Luftschicht, ►Geländeklima, ►Mikroklima, ►Standortklima).

829 climate protection [n] *envir. pol.* (World-wide attempts to reduce emissions of carbon dioxide [CO_2] and ►greenhouse gas emissions and thus limit the ►greenhouse effect and ►climatic change. Almost 20% of the world's population in affluent countries is responsible for half of anthropogenic CO_2 emissions. The USA is the largest climate polluter contributing a quarter of all emissions. With 3.1 billion tonnes China is already in second place and will probably be the greatest emitter of CO_2 within a few years. The International Energy Agency [IEA] expects, that in 2020 the developing countries will be responsible for half of CO_2 emissions. The decision at the first Climate Summit of Kyoto [Japan] in December 1997 to establish binding goals for the reduction of emissions is therefore understandable. Their implementation was the main theme of international conferences in 1998: the Environment Ministers of the EU agreed in June 1998 to a joint programme for **C.P.** and a policy to combat climate change according to the Kyoto Protocol. By the year 2012 CO_2 emissions are to fall by a total of 8.1%. The quota for the individual EU States lie between a reduction of 28% for Luxemburg, 21% for Germany und a plus of 27% for Portugal; cf. ALG 2000); *s 1* **lucha [f] contra el cambio climático** (Política internacional con el fin de reducir la emisiones de CO_2 y de otros ►gases de efecto invernadero y así limitar el ►cambio climático. Consciente de la necesidad de reducir las emisiones, la comunidad internacional aprobó en 1992 en Río de Janeiro la «Convención Marco de las Naciones Unidas sobre Cambio Climático» [UNFCCC] que fue ratificada en 1994. El instrumento práctico de la convención, el Protocolo de Kioto acordado en diciembre de 1997 en la ciudad japonesa, según el cual los países industrializados [responsables de más de 50% de las emisiones de gases de efecto invernadero] deberían reducir éstas en 5,2% entre 2008 y 2012, partiendo de los niveles de emisión de 1990, entró en vigor el 16 de febrero de 2005, trás la ratificación de Rusia, al superar en aprox. 6 por ciento, el mínimo de naciones parte, cuyas emisiones alcanzan 55% del total mundial. Ya en 2002 se había alcanzado el mínimo de 55 Estados signatarios. Hasta principios de 2009 y aunque son responsables de aprox. 25% de las emisiones mundiales en cuestión, los EE.UU. son el único país industrializado que no han ratificado el Protocolo y no han tomado medidas a nivel federal para reducir sus emisiones. Los instrumentos principales definidos en el Protocolo son los llamados «mecanismos flexibles»: el comercio de derechos de emisión de gases de ►efecto invernadero [emissions trading], la aplicación conjunta [joint implementation], el mecanismo de desarrollo limpio [clean development mechanism] y el compartimiento de cargas [burden sharing]. Desde la Conferencia de Kioto en 1997 han tenido lugar 14 Conferencias de las Partes [COP] de la UNFCCC para avanzar en los acuerdos y políticas internacioles de **l. contra el c. c.**, la última de las cuales fue en Poznan [Polonia] en diciembre de 2008, en la que no se consiguió avanzar hacia acuerdos sustanciales más allá de los de Kioto. El dilema que se presenta está entre la postura dura de los países industrializados, sobre todo de los EE.UU. [llamados países del Anexo I], de no aplicar medidas fuertes ante sus industrias resp. poblaciones para reducir las emisiones, y los países emergentes [como Brasil, China, la India, etc.] que no quieren comprometerse oficialmente a tomar medidas por su parte, mientras los primeros no sean consecuentes en la **l. contra el c. c.** El tercer grupo de países, los llamados subdesarrollados, son las víctimas de esta situación, ya que son los más afectados por las consecuencias del cambio climático y no tienen recursos para tomar medidas de adaptación. En diciembre 2009 tendrá lugar la COP-15 en Copenague, en la que se deberá decidir sobre un nuevo acuerdo internacional de **l. contra el c. c.** más allá de Kioto. Los temas más conflictivos son: reducción de emisiones también por parte de los países emergentes, incluidas las emisiones causadas por de[s]forestación; adaptación a los efectos del cambio climático; cuestiones financieras; transferencia de tecnología limpia del Norte al Sur global y otros mecanismos de cooperación. La UE acordó en 1998 internamente una reducción de las emisiones de CO_2 hasta 2012 de 8,1% y un sistema de reducción/aumento de las emisiones por países, de acuerdo con las condiciones de desarrollo, que varía entre 28% para Luxemburgo, 21% para Alemania y Dinamarca, y aumento de 27% para Portugal. Para Es. estaba previsto un aumento del 15% que hasta finales de 2006 ha sido muy superior: 35,6%. La actual estrategia de la UE está definida en su Programa Europeo sobre el Cambio Climático y tiene como meta primordial reducir el aumento de la temperatura media a 2°C con respecto a los niveles preindustriales); *s 2* **protección [f] del clima (±)** (Traducción literal incorrecta que se suele utilizar al traducir textos de algunos idiomas extranjeros, en particular del alemán); *f 1* **lutte [f] contre le changement climatique** (Mesures au niveau mondial en vue de réduire les émissions de GES [►gaz à effet de serre] afin de prévenir de l'effet de l'aggravation de l'►effet de serre et du ►changement climatique ; la part humaine des changements climatiques prend sa source dans les émissions de GES émis par les pays riches depuis la révolution industrielle ; les pays riches avec 20 % de la population sont actuellement responsables de plus de 50 % des émissions de GES ; les USA, avec près de 25 %, est le plus grand émetteur/pollueur au monde, évolution croissante ; l'agence internationale pour l'énergie prévoit que d'ici 2020 les pays en développement émettront près de 50 % de la quantité de CO_2 dans l'atmosphère ; néanmoins la responsabilité des pays développés reste historique, car le réchauffement climatique risque de devenir « une injustice climatique » dans les régions et les sociétés dont les moyens d'adaptation et de déplacement sont limités et ou les populations vivent en grande partie de l'agriculture ; le 10 décembre 1997 la conférence de Kyoto [Japon] a aboutit à l'adoption d'un protocole prévoyant dans les pays industrialisés la réduction de 5 % des émissions de GES par rapport au niveau de 1990 ; les États-Unis ont rejeté ce traité en 2001, peu de temps après l'arrivée de GEORGE W. BUSH à la Maison-Blanche et son administration refuse à ce jour d'entamer de véritables négociations multilatérales sur les suites à donner au protocole de Kyoto. Les pays européens lors du conseil des ministres de l'environnement du 17 juin 1998 ont décidé en fonction de leurs niveaux d'émission de 1990, de leur démographie ou de leurs besoin économiques de réduire les émissions de 8 % réparti sur les divers pays, -21 % pour l'Allemagne, -12,5 % pour le Royaume-Uni, 0 % pour la France, +15 % pour l'Espagne, + 25 % pour la Grèce) ; **F.**, le Plan Climat 2004-2012 permet la transposition des directives européennes en matières de lutte contre le changement climatique et définit différents axes pour améliorer la protection de l'environnement) ; *syn.* prévention [f] du changement climatique ; *f 2* **protection [f] climatique (±)** (Traduction littérale incorrecte souvent utilisée dans la traduction de textes de langues étrangères en particulier en langue allemande) ; *g* **Klimaschutz [m]** (Weltweite Bemühungen, den Ausstoß von Kohlendioxid [CO_2] u. a. ►Treibhausgasen zu verringern, um den ►Treibhauseffekt und ►Klimaveränderungen zu begrenzen. Knapp 20 % der Weltbevölkerung in den reichen Ländern sind für gut die Hälfte der anthropogenen Kohlendioxidemissionen verantwortlich. Dabei sind die USA rund mit einem

Viertel der Emissionen der größte Klimaverschmutzer. China liegt mit seinen über 3,1 Mrd. Tonnen bereits auf Platz 2 der Verschmutzerländer und wird aller Voraussicht nach in wenigen Jahren der größte Emittent von CO_2 sein. Die Internationale Energie-Agentur [IEA] erwartet, dass im Jahre 2020 die Entwicklungsländer für die Hälfte des CO_2-Ausstoßes verantwortlich sein werden. Von daher wird nachvollziehbar, dass im Dezember 1997 der Klimagipfel von Kyoto [Japan] erstmals verbindliche Reduktionsziele festgelegt hatte: Die Industriestaaten reduzieren ihren Ausstoß von Treibhausgasen bis 2012 gegenüber 1990 um 5,5 % [gemessen im Durchschnitt der Jahre 2008-2012], und zwar Japan um 6 %, die USA um 7 %, die 15 Staaten der EU um 8 %. Die Umweltminister der EU-Staaten einigten sich im Juni 1998 auf die interne Verteilung der **K.**-Anstrengungen nach dem Kyoto-Protokoll. Am 16. Februar 2005 trat das Kyoto-Protokoll in Kraft und enthält Regelungen, dass bis zum Jahr 2012 der CO_2-Ausstoß der EU insgesamt um 8,1 % sinken soll. Die Quoten für einzelne EU-Staaten liegen zwischen 28 % Reduktion für Luxemburg, 21 % für Deutschland und Dänemark und 27 % Zuwachs für Portugal sowie 15 % Zuwachs für Spanien, der sich dann auf 35,6 % erhöhte. Mit dem von 174 Staaten getragenen Kyoto-Protokoll gibt es eine völkerrechtlich verbindliche Begrenzung des Treibhausgasausstoßes. Im Dezember 2005 bekannte sich die Staatengemeinschaft auf der Vertragsstaatenkonferenz in Montreal, den Weg mit verbindlichen Reduktionszielen und Obergrenzen des Ausstoßes von Treibhausgasen über das Jahr 2012 fortzusetzen. 2007 beschlossen die europäischen Staats- und Regierungschefs in Brüssel, die CO_2-Emissionen bis 2020 um 20 % zu reduzieren und die Energieeffizienz um 20 % zu erhöhen. Beim G-8-Gipfel im Juni 2007 in Heiligendamm wurde Einigkeit erzielt, dass die Erderwärmung nicht mehr als 1,5-2,5 °C betragen darf, weshalb die Treibhausgasemissionen bis 2050 weltweit halbiert werden sollen).

830 climatic amelioration forest [n] *for. landsc.* (Wood or forest planted to protect residential, recreation or agriculturally-used areas, etc. against cold air damage or harmful winds [local protective forest]. The climate of extensive residential areas or open spaces may be improved by air-mass exchange [regional protective forest]; ▶protective planting, ▶windbreak planting); *s* **bosque** [m] **protector de mejora del microclima** (▶cortina rompevientos, ▶plantación protectora); *syn.* monte [m] protector de mejora del microclima; *f* **forêt** [f] **de protection climatique** (Bois et forêt situés à la périphérie des grandes agglomérations, des zones de loisirs, des cultures agricoles ayant une action favorable sur le mésoclimat ou le climat local ; ils ont pour effet de limiter les dommages causés par les gelées, de réduire l'influence des vents [forêt de protection climatique d'influence locale], d'améliorer les conditions climatiques des zones urbaines voisines : augmentation de la pluviosité, diminution des écarts de température et abaissement des moyennes annuelles [forêts de protection climatique d'influence régionale] ; ▶bande boisée brise-vent, ▶plantation brise-vent, ▶plantation de protection) ; *g* **Klimaschutzwald** [m] (Wald, der Wohnstätten, Erholungsanlagen, landwirtschaftliche Nutzflächen oder Sonderkulturen vor Kaltluftschäden oder vor nachteiligen Windeinwirkungen bewahrt [lokaler Klimaschutzwald] ; er verbessert das Klima benachbarter, großer Siedlungsbereiche oder Freiflächen durch Luftaustausch [regionaler Schutzwald] ; ▶Schutzpflanzung, ▶Windschutzpflanzung).

831 climatic change [n] *met.* (Changes in climate over a long period of time, the causes can be natural, e.g. fluctuations in solar activity, a change in the angle of the sun's rays due to a shift in the inclination of the earth's axis: or the onset or end of an Ice Age cycle. Other natural causes are continental drift and volcanic activity. The unusually rapid warming of the Earth's climate since the Industrial Revolution, especially during the last four decades, has been caused by human activities; ▶global warming, ▶greenhouse effect); *syn.* climate change [n]; *s* **cambio** [m] **climático** (Cambio observado en el clima, a una escala global, regional o subregional que tiene lugar en un periodo/ período largo de tiempo y cuyas causas pueden ser naturales o resultantes de la actividad humana. Desde el comienzo de la industrialización hacia 1850 el contenido de CO_2 en la atmósfera ha aumentado de 280 ppm a más de 380 ppm debido a la combustión de combustibles fósiles y en menor medida a la de[s]forestación. El aumento de CO_2 conlleva el aumento de la temperatura de las capas de aire cercanas a la superficie. Desde 1900 la temperatura global ha aumentado en unos 0,8°C, siendo la principal causa de ello la creciente concentración de CO_2 y de otros gases de efecto invernadero. Si no se consigue frenar el aumento de las emisiones de estos gases, se pronostica un aumento de la temperatura en 5°C hasta el 2100. Una de las consecuencias es el incremento del nivel del mar —causado por el aumento del volumen del agua y por derretimiento de las capas de hielo y de glaciares— que en el último siglo subió entre 15-20 cm y puede aumentar en este siglo entre 50 y 150 cm. Otras posibles consecuencias serían la desaparición de ecosistemas y de especies de flora y fauna, por cambios climáticos a gran escala, que ya se están manifestando en el crecimiento de catástrofes naturales como inundaciones, huracanes, fases largas de sequía e incendios forestales, y la reducción de la disponibilidad de alimentos debida a la falta de agua y a daños en la agricultura causados por catástrofes. Para evitar las peores consecuencias del **c. c.** es necesario reducir el aumento de la temperatura global a un máximo de 2°C, para ello se debe mantener a largo plazo una concentración de gases de efecto invernadero inferior al efecto de 450 ppm de CO_2, lo cual supone que las emisiones de gases de efecto invernadero deben reducirse a nivel mundial hasta mediados del siglo XXI a la mitad de las emisiones del año 1990; cf. PIK 2009; ▶calentamiento global, ▶efecto invernadero); *f* **changement** [m] **climatique** (Modification durable du climat de la Terre à une échelle globale, régionale ou subrégionale résultant de processus intrinsèques naturels de la Terre [déplacement des continents vers les pôles et vulcanisme], d'influences extérieures [variations de l'intensité du rayonnement solaire dues aux variations de l'orbite terrestre ou aux variations de l'activité solaire] et/ou dans les dernières décennies de l'intense activité humaine [industrialisation et utilisation massive d'énergies fossiles] et ayant pour effet le réchauffement de la planète ; celui-ci se traduit par **1.** une augmentation des températures de l'air affectant les huit premiers kilomètres de la basse atmosphère, **2.** une diminution de la fréquence des froids extrêmes et une augmentation des vagues de chaleur, **3.** une influence directe de la diminution de la couche d'ozone de la stratosphère, **4.** une amplification du phénomène ▶El Niño, responsable d'un renforcement des pluies et des sécheresses dans diverses régions des tropiques, **5.** une réduction de l'extension de la couverture neigeuse et de la durée de gel des lacs et des rivières, **6.** le recul des glaciers de montagne, **7.** une réduction de l'étendue de la glace de mer [banquise] au printemps et en été dans l'hémisphère Nord, **8.** une élévation du niveau de la mer de 10 à 20 cm au cours du XXème siècle, **9.** une augmentation des précipitations pendant tout le vingtième siècle dans les hautes et moyennes latitudes de l'hémisphère nord, **10.** un accroissement de la proportion et de la fréquence des cyclones tropicaux de niveau 4 et 5 [les plus élevés de l'échelle de Saffir-Simpson] observé dans le golfe du Mexique et aux Etats-Unis ; Conseil économique et social, Académie des sciences, Académie des technologies, Académie des sciences morales et politiques ; cf. Partager la connaissance et ouvrir le dialogue : le changement climatique, Groupe de réflexion Académies et Conseil économique et social,

C

24 janvier 2006 ; ▶effet de serre, ▶réchauffement global) ; *g* **Klimaveränderung [f]** (Veränderung des Klimas über einen langen Zeitraum; die Ursachen können natürlicher Art sein, z. B. durch Schwankungen der Sonnenaktivitäten, Veränderung des Einstrahlwinkels der Sonnenstrahlen durch Veränderung der Erdachsenneigung: Erklärung für den Eiszeitenzyklus. Weitere natürliche Ursachen sind die Kontinentaldrift und der Vulkanismus. Die ungewöhnlich rasche Erwärmung des Erdklimas seit der Industrialisierung, vor allem in den letzten vier Jahrzehnten ist durch den Menschen verursacht; ▶globale Erwärmung, ▶Treibhauseffekt); *syn.* Klimawandel [m].

832 climatic health resort [n] *recr.* (Place characterized by special climatic conditions which offers health facilities to visitors; ▶health resort, ▶spa); *s* **sanatorio [m] climático** (Lugar caracterizado por sus condiciones bioclimáticas especiales que tiene instalaciones para huéspedes de salud; ▶estación balnearia, ▶estación hidrotermal); *f* **station [f] climatique** (Lieu caractérisé par des conditions bioclimatiques particulièrement favorables et possédant les installations d'accueil des curistes nécessaires ; ▶établissement thermal, ▶station thermale) ; *g* **Luftkurort [m]** (Ort, der sich durch besondere bioklimatische Verhältnisse auszeichnet und über entsprechende Einrichtungen für Kurgäste verfügt; ▶Heilbad, ▶Kurort).

833 climatic map [n] *met.* (Thematic, cartographic representation of climatic elements, showing long-term meteorological observations. There are analytic maps, showing temperature, distribution of precipitation, duration of insolation, etc., and synthetical maps, showing types of climate and climatic zones); *s* **mapa [m] climatológico** (Representación gráfica de elementos del clima que muestra los resultados de las observaciones meteorológicas a largo plazo. Existen mapas analíticos [representación de la distribución de la temperatura, las precipitaciones, duración de la insolación, etc.] y mapas sintéticos que muestran diferentes tipos de clima y diferentes zonas climáticas); *f* **carte [f] climatologique** (Représentation cartographique thématique des éléments météorologiques observés sur plusieurs années ; on distingue les cartes analytiques [représentation des températures, de la pluviosité, de l'ensoleillement, etc.] et les cartes synthétiques [types de climat, zones climatiques, etc.]) ; *g* **Klimakarte [f]** (Thematische, kartografische Darstellung von Klimaelementen, die langjährige meteorologische Beobachtungen aufzeigt. Es werden analytische [Darstellung von Temperatur, Niederschlagsverteilung, Sonnenscheindauer etc.] und synthetische **K.en** [Darstellung von z. B. Klimatypen, Klimazonen] unterschieden).

climatic zone [n] *geo. phyt.* ▶vegetation climatic zone.

climatology [n] *met.* ▶urban climatology.

834 climax [n] *phyt.* (Relatively stable final stage in the successive development of vegetation; ▶climax community); *s* **clímax [f]** (Etapa de mayor madurez de la sucesión con considerable diversificación local. Es en realidad un concepto teórico, ya que existe una gran heterogenidad local de la clímax; MARG 1977; ▶asociación climácica); *f* **climax [m]** (La biocénose la plus stable et la plus complexe, la plus « diversifiée » d'une succession dynamique primaire ; DUV 1984 ; l'adjectif dérivé de climax est « climacique » ; ▶peuplement climax) ; *g* **Klimax [f]** (Klimaxe [pl *selten*]; relativ stabiles Endstadium einer Vegetationsentwicklung; ▶Klimaxgesellschaft).

835 climax area [n] *phyt.* (Territory covered by a ▶climax complex. Large climax areas correspond to ▶biomes); *s* **dominio [m] climácico** (Territorio dominado por un ▶complejo climácico. Los **dd. cc.** grandes corresponden a los ▶biomas); *f* **région [f] climacique** (Territoire géographique de grande étendue caractéristique d'un ▶complexe climacique et correspondant à un ▶biome) ; *g* **Klimaxgebiet [n]** (Das von einem ▶Klimax-

komplex umspannte Gebiet. Die großen **K.e** entsprechen den ▶Biomen).

836 climax community [n] *phyt.* (Final stage in the natural succession of a plant community capable of perpetuation under the prevailing climatic and edaphic conditions; cf. RCG 1982; ▶location 1, ▶climax forest); *s* **asociación [f] climácica** (Comunidades en las cuales la tendencia al aumento de madurez conduce a comunidades de organización más compleja que, por lo menos fisionómicamente y, frecuentemente también desde el punto de vista florístico, presentan una considerable uniformidad de características; MARG 1977; ▶estación, ▶bosque climácico); *syn.* comunidad [f] climácica; *f 1* **peuplement [m] climax** (Peuplement caractérisant le stade final stable de l'évolution progressive des groupements végétaux pour une station donnée, atteint sous des conditions naturelles sans action humaine directe ou indirecte ; ▶station, ▶forêt climacique) ; *syn.* groupement [m] climacique final ; *f 2* **peuplement [m] paraclimacique** (Une association végétale — pelouse ou lande — mais non forestière observée dans une région dont le climax normal est la forêt) ; *syn.* groupement [m] paraclimacique ; *g* **Klimaxgesellschaft [f]** (Endstadium der natürlichen Vegetationsentwicklung auf einem ▶Standort 1; ▶Klimaxwald); *syn.* klimatische Schlussgesellschaft [f].

837 climax complex [n] *phyt.* (Culmination of all successional conditions that lead to one definite climatic terminal community; cf. BB 1965; ▶sere); *s* **complejo [m] climácico** (Culminación de todas las condiciones de sucesión que lleva a la comunidad climácica definitiva; ▶serie de sucesión); *f* **complexe [m] climacique** (Ensemble des séries de végétation qui conduisent au même ▶climax ; DEE 1982 ; ▶série de végétation) ; *g* **Klimaxkomplex [m]** (Gesamtheit aller Entwicklungsserien, die einer bestimmten klimatischen Schlussgesellschaft zusteuern; ▶Sukzessionsserie).

838 climax fluctuation [n] *phyt.* (Notable ▶displacements of species [composition or balance] through human or natural agencies, such as wind, snow, flooding or the advance of sand dunes in mature ▶climax communities); *s* **oscilaciones [fpl] de la clímax** (Cambio brusco de la composición de especies en una ▶asociación climácica debido a fenómenos naturales como avalanchas, crecidas, incendios, etc.); *f* **fluctuation [f] du climax** (Importante ▶évolution de la composition en espèces et de l'équilibre relatif et temporaire d'un ▶peuplement climax créé par des perturbations naturelles comme des avalanches, des incendies ou des inondations) ; *g* **Klimaxschwankung [f]** (Wesentliche ▶Artenverschiebungen bei ausgereiften ▶Klimaxgesellschaften durch Naturereignisse, wie z. B. Windwurf, Lawinen, Brand, Überschwemmungen).

839 climax forest [n] *phyt.* (Final stage in the natural succession of a forest community in a defined ▶location 1; ▶climax community); *s* **bosque [m] climácico** (Estado final en la sucesión natural de una comunidad forestal en una ▶estación determinada; ▶asociación climácica); *f* **forêt [f] climacique** (Stade ultime de l'évolution spontanée progressive d'un peuplement forestier se trouvant dans un équilibre temporaire stable avec son milieu sur une ▶station donnée ; ▶peuplement climax) ; *syn.* forêt [f] climax ; *g* **Klimaxwald [m]** (Endstadium der natürlichen Vegetationsentwicklung eines Waldes auf einem bestimmten ▶Standort 1, der sich im Gleichgewichtszustand mit seiner Umwelt befindet; ▶Klimaxgesellschaft).

climber [n] *bot.* ▶climbing plant, ▶root climber, ▶tendril climber, ▶twining climber.

climber [n]**, hooked** *bot.* ▶scandent plant.

climber [n] **[UK], house wall** *hort. plant.* ▶wall-climbing plant [US].

climber-covered wall [n] [UK] *landsc. urb.* ▶vine-covered wall [US].

climber [n] **for façade planting** *hort. plant.* ▶wall-climbing plant [US].

840 climber planting [n] *constr.* (Vegetative cover with climbers or vines); *s* **cobertura [f] de plantas trepadoras**; *f 1* **plantation [f] des plantes grimpantes** (processus) ; *f 2* **verdissement [m] par plantes grimpantes** (résultat) ; *syn.* verdoiement [m] par plantes grimpantes, végétalisation [f] par plantes grimpantes ; *g* **Berankung [f]** (Bedeckung mit Kletterpflanzen).

841 climber support [n] *constr. gard.* (Climber trellis or support wires for ▶scandent plants, ▶twining climbers and ▶tendril climbers; ▶lattice, ▶tying to a climber support); *s* **enrejado [m] de soporte de trepadoras** (Ayuda artificial para trepar para ▶enredaderas, ▶plantas sarmentosas y ▶plantas zarcillosas; ▶enrejado, ▶sujeción a enrejado de soporte de trepadoras); *f* **support [m] de plantes grimpantes** (Support artificiel sur lequel s'accrochent les ▶plantes grimpantes à vrilles, ▶plantes sarmenteuses, ▶plantes volubiles, tel que des fils tendus, un treillage, une pergola, un arceau, une colonne ; ▶palissage, ▶treillis) ; *g* **Rankgerüst [n]** (Künstliche Kletterhilfe für Gerüstkletterpflanzen wie ▶Rankenpflanzen, ▶Spreizklimmer und ▶Windepflanzen; ▶Befestigen an ein Rankgerüst, ▶Gitter); *syn.* Kletterhilfe [f], Klettervorrichtung [f], *gard'hist.* Treillage [f].

842 climbing perennial [n] *bot.* (▶Hemicryptophyte with annual climbing stems and renewal buds at the base; especially common in warmer parts of the temperate zone; ▶scandent plant; *generic term* ▶climbing plant); *s* **hemicriptófito [m] trepador** (▶Hemicriptófito; ▶planta sarmentosa; *término genérico* ▶planta trepadora); *f* **plante [f] grimpante vivace** (▶Hémicryptophyte développant une tige annuelle, périssant en automne et des bourgeons hivernaux au ras du sol ; ▶plante sarmenteuse, *terme générique* ▶plante grimpante) ; *syn.* liane [f] herbacée ; *g* **Klimmstaude [f]** (*Hemikryptophyta scandentia* ▶Hemikryptophyt mit klimmendem, alljährlich absterbendem Stängel und grundständiger Erneuerungsknospe, z. B. Bittersüß *[Solanum dulcamara]*, Zaunwinde *[Calystegia sepium]*, Japanische Kaiserwinde *[Ipomoea imperialis]*; cf. BOR 1958); ▶Spreizklimmer; *OB* ▶Kletterpflanze).

843 climbing plant [n] *bot. hort.* (Generic term for a climber which employs one of five strategies for climbing acquired during the evolutionary process by using other plants for support or by climbing up another plant in order to reach the light; the following types of climbers may be distinguished: ▶root climber, ▶scandent plant, ▶tendril climber, ▶twining climber; of these there are two main groups: ▶self-clinging climber, ▶trellis climber; **c. p.s** can be annuals or perennials; they thrive especially in humid forests, such as tropical rainforests, riparian woodlands; specific terms are ▶climbing perennial, ▶liana, ▶wall-climbing plant); *syn.* climber [n], vine [n] [also US]; *s* **planta [f] trepadora** (*Término genérico* planta que tiene la capacidad de adherirse a otras plantas o trepar gracias a ellas o algún soporte artificial para alcanzar la luz. Se diferencian las siguientes estrategias: ▶planta sarmentosa, ▶planta trepadora con raíces adventicias, ▶planta voluble, ▶planta zarcillosa y ▶planta zarcillosa con ventosas. Éstas se pueden ordenar en dos grupos principales: ▶trepadora autoadherente y ▶trepadora sobre soporte. En relación a la duración de su vida existen trepadoras anuales y perennes. Las **pp. tt.** crecen especialmente en bosques húmedos [bosque ripícola, pluvisilva]. *Otros términos específicos* ▶bejuco, ▶hemicriptófito trepador, ▶trepadora cubremuros); *syn.* trepadora [f], enredadera [f]; *f* **plante [f] grimpante** (Terme générique

désignant une plante qui, au cours de l'évolution, parmi cinq stratégies différentes a réussi à grimper ou s'accrocher à d'autres végétaux afin d'atteindre la lumière ; on distingue ainsi : la ▶plante grimpante à vrilles, la ▶plante sarmenteuse, la ▶plante volubile, la ▶plante grimpante à racines-crampons et la ▶plante grimpante à ventouses ; ces plantes peuvent être rassemblées en deux groupes : la ▶grimpante pour treillage et la ▶plante grimpante avec des organes d'accrochage ; les plantes grimpantes peuvent être annuelles ou pérennes selon leur durée de vie, être ligneuse [▶liane], herbacée ou vivace [▶plante grimpante vivace] et sont particulièrement abondantes dans les forêts humides [forêt alluviale, forêt tropicale humide] ; ▶plante grimpante couvre-mur) ; *g* **Kletterpflanze [f]** (Pflanze, der es im Laufe der Evolution gelungen ist, sich mit einer von fünf Strategien an anderen Pflanzen zu stützen oder an ihnen hoch zu klettern, um zum Licht zu gelangen. Es werden folgende Strategen unterschieden: ▶Rankenpflanze, ▶Spreizklimmer, ▶Windepflanze, ▶Wurzelkletterer und ▶Haftscheibenranker. Diese können zu zwei Hauptgruppen zusammengefasst werden: ▶Gerüstkletterer und ▶Selbstklimmer; hinsichtlich der Lebensdauer gibt es annuelle **K.n** und perennierende **K.n**. Besonders reichlich gedeihen **K.n** in feuchten Wäldern [Auenwald, Regenwald]; *weitere UBe* ▶Kletterpflanze zur Wandbegrünung, ▶Klimmstaude, ▶Liane); *syn.* Klimmer [m].

climbing plant [n]**, tendril-** *bot.* ▶tendril climber.

climbing root [n] *bot.* ▶adventitious climbing root.

844 climbing rose [n] *hort.* (Generic term covering ▶climbing rose, large-flowered climber or creeping rose; ▶scandent plant; vigorous climbing rose from 3-5 (-8)m high with small clustered flowers on up to 5 meter long supple shoots, derived from *Rosa luciae*, grown over pergolas, arbo[u]rs and pillars; at the outset of the 19th century **c. r.s** were only to be found in Europe on forest edges on moist, rich soils very similar to the native species of *Rosa arvensis*, mostly in Central Europe and *R. sempervirens*, in Southern Europe. In the United States, **Noisette Roses** emerged at the beginning of the 19th century. John Champney bred the variety 'Champney's Pink Cluster' from the repeat-flowering rose, *Rosa chinensis* 'Old Blush' and the musk rose, *Rosa moschata*, both less hardy in Central Europe. His neighbour, Philippe Noisette, selected seedlings of these varieties in 1817 to create 'Blush Noisette', which lends its name to the group. According to English classification, **Rambler Roses** are distinguished from *climbing roses*, by having long, soft, thin and often overhanging shoots, which reach to the ground, mostly with small flowers in clusters or panicles, blooming only once, but modern hybrids are sometimes repeat-flowering or even continuously in flower. As a scandent plant they reach up to 10m in height and repeated hybridization has now made them difficult to classify; varieties include 'Albertine' [soft pink], 'American Pillar' [magenta with a white eye] and 'Bobby James' [cream]. Most Rambler Roses usually bloom only once a year with many relatively small flowers in large clusters. Some varieties develop decorative rosehips in autumn. Regular flowering ramblers are, e.g. 'New Dawn' [pale pink], 'Momo' [dark red/scarlet] and 'Super Dorothy' [pink]. *Climbing roses* are typical scandent plants with thick, hard shoots; flowers are usually large and regularly in bloom, from 2-6 metres high they usually need to be tied, the flowers indicate their origin as tea, floribunda or polyantha roses; e.g. 'Coral Dawn' [coral-pink], 'Gold Star' [deep yellow], 'Sympathy' [deep scarlet]. Mutations of ▶bedding roses are known as *'Climbing sports'*, differing only in their climbing habit with very long shoots, for example, 'Climbing Sarabande' [orange-red]); *syn.* rambling rose [n]; *s* **rosal [m] trepador** (▶Planta sarmentosa que crece de 3 a 5 y hasta 8 m de altura. A principios del siglo XIX en Europa sólo existían las especies silvestres de la *Rosa arvensis,* más usual en Europa Central, y de la *R. sempervirens* en Europa Meridional; ▸rosa de macizo); *f* **rosier [m] grimpant** (▸Plante sarmenteuse de 3 à 5 m de haut, développant des longues tiges de 3 à 5 m qui doivent être palissées sur un support ; avant le début de la culture des roses, il n'existait que des Églantines vivant à l'état sauvage en bordure des champs ou dans les bois telles que l'Églantier des champs *[Rosa arvensis],* le Rosier à feuilles de pimprenelle *[Rosa pimpinellifolia]* ou le Rosier toujours vert à feuillage persistant *[Rosa sempervirens]* ; jusqu'au début du 19ème siècle la culture des roses n'évolua guère, les roses appartenant surtout à trois espèces comprenant la Rose de France *[Rosa gallica],* la Rose cent-feuilles *[Rosa centifolia],* la Rose d'Orient *[Rosa damascena]* ; c'est avec l'importation d'Extrême — Orient du Rosier du Bengale *[Rosa semperflorens]* et du Rosier de Chine *[Rosa chinensis],* fleurissant plusieurs fois l'été, que la culture des rosiers connus une évolution rapide, pour arriver aux formes actuelles grâce à l'obtention des hybrides remontants, croisement entre le rosier à odeur de thé, originaire de Chine et d'Inde, et divers types acclimatés ; EFJ 1969, 1768 ; les races de rosiers sarmenteux comprennent des variétés à grandes fleurs provenant d'une mutation apparue sur un rosier buisson [dénomination anglaise de *climbing*] et des variétés vigoureuses à petites fleurs [dénomination anglaise de *rambler*] à très longs rameaux, appelées **rosiers liane** ; ▸rosier buisson à fleurs groupées) ; *syn.* rosier [m] sarmenteux ; *g* **Kletterrose [f]** (Als 3-5 (-8) m hoher ▸Spreizklimmer emporwachsende Rose oder Rose mit bis zu 5 m langen Trieben, die einer Rankhilfe bedürfen. Anfang des 19. Jhs. gab es bei **K.n** nur die in Europa an Waldrändern auf feuchthumosen Böden heimischen und einander recht ähnlichen Wildarten der *Rosa arvensis,* die mehr im mittleren und der *R. sempervirens,* die mehr im südlichen Europa vorkommt. In Amerika entstanden am Anfang des 19. Jhs. die mehrmals blühenden **Noisetterosen** aus *Rosa chinensis* und *R. moschata,* die in Mitteleuropa wenig winterfest sind. Nach der englischen Sortierung werden **Rambler-Rosen** und **Climbing-Rosen** unterschieden. Die meisten **Rambler-Rosen** bilden lange, weiche, dünne und oft stark überhängende Triebe mit meist kleinen Blüten in Büscheln, Trugdolden oder Rispen, fast immer einmal blühend — jedoch durch moderne Hybriden z. T. auch nachblühend oder sogar öfter blühend —, die den Boden bedecken oder als ▸Spreizklimmer bis zu 10 m in die Höhe wachsen; durch die Hybridisierung werden die Grenzen jeder Klassifizierung immer wieder überschritten; Sorten wie 'Albertine' [zartrosa], 'American Pillar' [karminrosa mit weißem Auge] und 'Bobby James' [cremeweiß] zählen dazu. Einzelne Sorten entwickeln bis zum Herbst schmückende Hagebutten. Mehrfach blühende *Rambler* sind z. B. 'New Dawn', 'Momo' [dunkel- bis karminrot] und 'Super Dorothy' [rosa]. **Climbing-Rosen,** typische Spreizklimmer mit dicken, harten Trieben, i. d. R. großblumig und öfter blühend, die 2-6 m hoch werden und in den meisten Fällen angebunden werden müssen, lassen schon durch ihre Blüten die Herkunft aus Tee-, Floribunda- oder Polyantharosen erkennen: z. B. 'Coral Dawn' [korallenrosa], 'Goldstern' [goldgelb], 'Sympathie' [tief scharlachrot]. Mutanten von ▸Beetrosen, sog. *Climbing sports,* unterscheiden sich von den Ursprungssorten nur durch den langtriebigen, kletternden Wuchs, z. B. 'Climbing Sarabande' [rotorange]).

845 clinging dead leaves [npl] *bot.* (Brown leaves which remain on branches during the winter, mostly until spring, e.g. hornbeam *[Carpinus]* and oak *[Quercus]*); *s* **hoja [f] marcescente** (Hojas que se secan en la planta sin desprenderse. Los quejigos, p. ej., *[Quercus lusitanica]* no pierden las hojas hasta muy avanzado el invierno, porque sus hojas son «marcescentes»; DB 1985; *f* **feuilles [fpl] marcescentes** (Feuilles qui se dessèchent normalement à l'automne mais qui persistent sur les

végétaux ligneux en hiver jusqu'à l'éclatement des nouveaux bourgeons, p. ex. Charme *[Carpinus]* et Chêne *[Quercus]* ; on parle de **feuillage marcescent**) ; *g* **abgestorbene haftende Laubblätter [npl] (≠)** (Nach der Herbstfärbung braun und trocken werdende Blätter, die während des Winters zeitweise bis zum Austrieb an den Ästen einiger weniger Baumarten hängen bleiben, z. B. bei Hainbuche *[Carpinus]* und Eiche *[Quercus]*); *syn.* braunes haftendes Laub [n, o. Pl.] (≠), dürre hängende Laubblätter [npl] (≠), trockenes haftendes Laub [n, o. Pl.] (≠), Winterlaub [n] (GAR 2004, H. 6, 23).

clinker brick [n] **for paving** [UK] *constr.* ▸paver brick [US].

clinker brick paving [n] [US] *constr.* ▸brick paving.

clint vegetation [n] [UK] *phyt.* ▸limestone groove vegetation [US].

846 clipped hedge [n] *constr. gard.* (A narrow sheared hedge is called a 'living fence' [US]); *syn.* trimmed hedge [n]; *s* **seto [m] podado** (Arbustos plantados muy juntos en hileras que se podan regularmente, cumplen funciones divisorias o de delimitación); *f* **haie [f] taillée** (Plantation arbustive d'alignement régulièrement taillée, de hauteur régulière, souvent constituée d'une seule espèce et utilisée en clôture) ; *g* **geschnittene Hecke [f]** (In dichter Reihe angepflanzte, bis zum Boden verzweigte Sträucher oder Heister, die als Trenn- oder Abgrenzungselement regelmäßig geschnitten werden); *syn.* Formhecke [f], Kulturhecke [f], Schnitthecke [f].

847 clipped lawn [n] *constr.* (Thick, regularly mown, carpet-like turf, comprising a fine sward of bright green grass, which is cultivated primarily for its impressive and uniform appearance or for a special design aim and is therefore primarily used for enjoyment and not for active recreation; ▸landscape practice, ▸lawn type); *syn.* pleasure lawn [n]; *s* **césped [m] decorativo** (En la clasificación de la ▸construcción paisajística 2, ▸tipo de césped denso, de hierbas de hojas finas y color intenso que se corta regularmente y da a las superficies un aspecto de alfombra. Su función principal es ornamental, no es adecuado para usos recreativos activos); *f* **gazon [m] d'ornement** (▸Type de gazon haut de gamme composé d'espèces sélectionnées pour leur aspect esthétique, leur pousse lente, la finesse et la densité de leur feuillage permettant d'obtenir un tapis végétal très serré, utilisé pour les aménagements de grand standing, la création des greens de golf ; ▸travaux de paysagisme) ; *syn.* gazon [m] ornemental ; *g* **Zierrasen [m]** (Dichter, regelmäßig geschnittener, teppichartiger ▸Rasentyp des ▸Garten- und Landschaftsbaues aus feinblättrigen, farbintensiven Gräsern, der überwiegend der Repräsentation oder einer anderen besonderen Gestaltungsaufgabe, also im Wesentlichen der passiven, nicht der aktiven Nutzung dient; DIN 18 917; **Z.** verlangt eine hohe bis sehr hohe Pflege, lässt sich fast auf jedem Standort und in jedem Klimaraum anlegen und erhalten und verträgt nur eine geringe Trittbelastung).

848 clipping [n] **of hedges** *constr. hort.* (Regular removal, by hand or mechanical means, of unwanted growth to bring hedges [or shrubs] to the desired shape); *syn.* shearing [n] of hedges [also US] (ARB 1983, 387); *s* **poda [f] de setos** (Operación regular de corte o remoción de las ramas para darles la forma deseada a los setos); *syn.* corte [m] de setos; *f* **taille [f] de haies** (Opération régulière de coupe des pousses de l'année sur une haie afin de lui donner la forme désirée) ; *g* **Heckenschnitt [m]** (Regelmäßiges Zurücknehmen des jährlichen Austriebes von Hecken, um sie in einer gewünschten Form zu halten; den sehr starken Rückschnitt bis kurz über der Bodenebene, damit die Hecke wieder neu aufgebaut werden kann, nennt man „eine Hecke auf den Stock setzen"); *syn.* Rückschnitt [m] von Hecken.

849 clippings [npl] *agr. constr. hort.* (Parts of grass leaves removed from turf or meadows by mowing; ▶lawn clippings); *syn.* cuttings [npl]; *s* **hierba** [f] **segada** (Partes de las hojas de hierba extraidas al cortar un césped o una pradera; ▶hierba de césped cortado); *f 1* **produit** [m] **de tonte/de coupe** *hort.* (obtenu par la coupe d'un gazon d'une surface d'agrément ; ▶produit de tonte de gazon) ; *f 2* **produit** [m] **de fauche** *agr.* (obtenu par la fauche de l'herbe d'un pré) ; *g* **Mähgut** [n, o. Pl.] (Material an Blättern und Sprossachsen, das beim Mähen von Rasen- und Wiesenflächen oder Getreidefeldern anfällt; ▶Rasenschnittgut; *OB* ▶Schnittgut).

clippings [npl] [US], **grass** *constr. hort.* ▶lawn clippings.

850 cloister garden [n] *gard'hist.* (Garden inside a monastery courtyard; ▶monastery garden); *syn.* abbey close [n]; *s* **jardín** [m] **de claustro** (Jardín situado en el interior de un claustro de convento o monasterio; ▶jardín monástico); *f 1* **jardin** [m] **de cloître** (Préau carré circonscrit par les bâtiments d'un monastère bordé par une allée de portiques couverts destinée à la méditation ; espace en général planté de gazon et caractérisé par deux allées perpendiculaires à l'intersection desquelles figurait souvent un puits ou une fontaine ; ▶jardin monastique) ; *f 2* **jardin** [m] **de curé** (Terme utilisé aujourd'hui dans la conception des jardins) ; *g* **Kreuzgarten** [m] (Von einem Kreuzgang umgebener Gartenhof in einer Klosteranlage; ▶Klostergarten).

close [n], **abbey** *gard'hist.* ▶cloister garden.

851 close-boarded fence [n] *constr.* (SPON 1986, 285; fencing which is completely fitted with vertical boards, fixed on ▶arris rails, and having no spaces between them; ▶close board panel fence, ▶visual screen with vertical boards, ▶wooden fence); *syn.* close-sheeted fence [n] [also US] (DAC 1975, 109), solid board fence [n] [also US] (TSS 1988, 450-4), tight board fence [n] [also US]; *s* **valla** [f] **de tablones** (Cerca en la que los tablones están colocados juntos sin espaciamiento y sujetados con ▶traviesas; ▶valla de madera, ▶valla de protección visual de tablones verticales); *syn.* cerca [f] de tablas; *f* **clôture** [f] **en planches jointives** (Clôture dont les lames verticales sont fixées côte à côte sur les ▶barreaux, lices ; ▶clôture en bois, ▶mur brise-vue à lattis vertical) ; *syn.* clôture [f] en plats verticaux (HEC 1985, 66) ; *g* **Bretterzaun** [m] (Einfriedung für Blick- und Windschutz, bei der die meist senkrechten Bretter fugeneng an ▶Riegeln befestigt sind; um ein luftigeres und angenehmeres Erscheinungsbild zu gestalten, werden die vertikalen Bretter oft wechselseitig auf die Querriegel aufgenagelt. In analoger Ausführung sind Konstruktionsformen mit horizontal angeordneten Brettern üblich, bei denen die Bretter wechselseitig direkt auf die Pfosten genagelt werden; ▶Holzzaun, ▶Sichtschutzwand mit senkrechter Verbretterung); *syn.* Bretterwand [f].

close board panel fence [n] *constr.* ▶visual screen with vertical boards.

852 closed canopy [n] *phyt. for syn.* continuous canopy [n], complete canopy [n]; *s* **dosel** [m] **cerrado del bosque**; *f* **couvert** [m] **fermé** *syn.* couvert [m] complet ; *g* **geschlossenes Kronendach** [n].

853 closed recycling system [n] *envir.* (Natural or artificial method of reusing materials derived from natural resources, which have occurred as by-products or are recycled from residues and then used for production of the same materials or substances; ▶recycling); *s* **ciclo** [m] **cerrado de materias** (Sistema natural o método antropógeno de reutilizar materiales producidos a partir de recursos naturales y sus derivados y residuos para producir el mismo tipo de material; ▶reciclado); *f* **cycle** [m] **de la matière fermé** (Système naturel ou artificiel de réemploi de produits obtenus à partir des ressources naturelles et dont les produits dérivés ou résiduels sont utilisés dans la production des mêmes substances, p. ex. dans la production du chlorure de vinyle ; ▶recyclage) ; *g* **geschlossener Stoffkreislauf** [m] (Natürliches oder künstliches System der Wiederverwendung von Materialien, die aus natürlichen Ressourcen hergestellt wurden, sowie von entstehenden Nebenprodukten und Rückständen zur Produktion derselben Stoffe, z. B. Herstellung von Vinylchlorid; ▶Recycling).

854 closed season [n] *conserv. game'man. hunt. leg.* (Prescribed period of time during which certain species of game cannot be hunted, or when certain fish species cannot be caught; ▶hunting season); *syn.* nonhunting/nonfishing season [n] [also US]; *s* **veda** [f] (Época del año en la que no está permitida la caza de especies cinegéticas que varía entre unas especies y otras dependiendo del periodo/período de reproducción de cada una de ellas; *opp.* ▶período de caza); *syn.* periodo/período [m] de veda; *f* **période** [f] **de fermeture de la chasse** (Période déterminée et fixée par arrêté du ministre chargé de la chasse, réglementant l'exercice de la chasse pour prévenir la destruction, favoriser le repeuplement ou assurer la survivance d'espèces d'oiseaux ou de gibier menacées ; *opp.* ▶période d'ouverture de la chasse) ; *syn.* temps [m] de fermeture de la chasse, période [f] de clôture de la chasse ; *g* **Schonzeit** [f] (Durch die Jagdzeitenverordnungen festgesetzte Zeit, in der die einzelnen Wildarten nicht gejagt werden dürfen; in der Fischerei die Zeitperiode, in der nach Festsetzung durch das Fischereirecht bestimmte Fischarten nicht gefischt werden dürfen; *opp.* ▶Jagdzeit); *syn.* Jagdschonzeit [f].

close-jointed [adj/adv] *arch. constr.* ▶butt-jointed.

855 closely united layer [n] *phyt.* (Vegetation layer in a plant community which only occurs together with another layer, e.g. the moss and shrub layers in certain types of raised bog; ▶fusion of layers); *s* **estratos** [mpl] **íntimamente relacionados** (En la estructura vertical de una comunidad [forestal], uno de los subtipos de ▶estratos relacionados en el que aparecen sólo juntos entre sí, como es el caso de los ▶estratos herbáceo y muscinal del *Centunculo-Anthoceretum* y algunas comunidades de turbera alta; cf. BB 1979; ▶estratificación de la vegetación); *syn.* estratos [mpl] íntimamente ligados; *f* **strate** [f] **solidaire** (Strate dépendante de la strate voisine et caractérisant la ▶solidarité écologique des strates ou groupements, p. ex. la strate muscicole de certaines tourbières ne peut exister qu'en dépendance étroite avec la strate arbustive à laquelle elle est subordonnée) ; *syn.* strate [f] fusionnée ; *g* **eng verbundene Schicht** [f] (Bei der ▶Schichtenbindung einer Pflanzengesellschaft vorkommende Schicht, bei der zwei Schichten, z. B. die Moos- und Strauchschicht in gewissen Hochmoortypen, stets nur vereint auftreten).

close-sheeted fence [n] [US] *constr.* ▶close-boarded fence.

close [adj] **to the ground** *met.* ▶ground-level ...

closing date [n] **for design submission** *plan. prof.* ▶submission deadline for design competition.

856 cloud forest [n] *for. phyt.* (Altitudinal zone of tropical rainforest and humid savanna[h]; generally in altitudes of [1,000-] 2,000-3,000m, at the edge of the Andes up to 4,000m above sea level, always on slopes of cloud-covered mountains with a high precipitation rate. The c. f. is characterized by a year-round or nearly all year fog, drizzle, and heavy dew precipitation. In the mountains of Canary Islands the c. f.s are called 'Laurisilva' which are characterized by *Lauráceae* as *Laurus canariensis* and *Laurus platyphylla* and myrtles *[Myrtáceae]*); *syn.* tropical ombrophilous cloud forest [n] (UNE 1973); *s 1* **bosque** [m] **nublado tropical** (Bosques tropicales de la zona altitudinal entre [1000-] 2000 y 3000 m, al borde de los Andes hasta 4000 m, que se presentan en las laderas de montañas nubladas y con altas precipitaciones. Están constituidos por dos capas arbóreas de hasta 20 m de altura con gran abundancia de formas epífitas y troncos cubiertos de musgos y helechos; cf. SILV 1979); *syn.* laurisilva [f] de montaña tropical, bosque [m] tropical ombrófilo nublado; *s 2* **laurisilva [f] canaria** (Bosque cerrado, umbroso y termófilo de frondosas perennifolias de talla variable que forman un solo estrato de suelo a copas con hojas lustrosas y coriáceas, de tipo laurel. Se extiende por las laderas norte y NE de las islas Canarias occidentales y de Las Palmas, entre 600 y 1000 m de altura, asociado a la zona de nieblas producida por el ascenso de los vientos alisios húmedos marinos. Las especies más características pertenecen a las familias de las laureáceas: el loro *[Laurus azorica]*, el viñatigo *[Pérsea indica]*, el barbusano *[Apollónias burbujana]* y el til *[Ocótea phoetens]*. Otras especies características son además el acebiño *[Ilex canariensis]*, el naranjero salvaje *[Ilex platyphylla]*, el madroño *[Arbutus canariensis]*, etc.; cf. DINA 1987); *f* **forêt [f] de brouillard** (Formation végétale tropicale dense se développant sur certaines montagnes soumises à une très forte nébulosité, à une altitude entre 2000 et 3000 m, pouvant atteindre 4000 m sur la bordure occidentale des Andes [forêt dénommée « ceja »] ; sur les versants correspondant au maximum de nébulosité la végétation est caractérisée par l'abondance d'épiphytes ; les **f. d. b.** dans les montagnes de l'archipel des Canaries sont caractérisées par la présence d'espèces laurifoliées [**forêt laurifoliée/laurisilve**] comme *Laurus canariensis, Laurus platyphylla* et différentes Myrtes *[Myrtáceae]* ; *syn.* forêt [f] ombrophile tropicale de brouillard (UNE 1973), forêt [f] de brumes, nebelwald [m] ; *g* **Nebelwald [m]** (...wälder [pl]; Höhenstufe des tropischen Regenwaldes und der Feuchtsavanne; im Allgemeinen in Höhen von [1000-] 2000-3000 m, am O-Rand der Anden bis 4000 m ü. NN, stets an Hängen niederschlagsreicher Gebirge im Wolkengürtel. Der **N.** ist ganzjährig oder die überwiegenden Teile des Jahres von ständigem Nebel, Sprühregen und starkem Taufall geprägt; cf. WAG 1984. In den Bergen der Kanarischen Inseln gibt es auch **N.wälder**, die **Laurisilva** heißen und durch Lorbeergewächse *[Lauráceae]* wie *Laurus canariensis* und *Laurus platyphylla* und Myrtengewächse *[Myrtáceae]* gekennzeichnet sind).

857 cloud level zone [n] *geo. phyt.* (ELL 1988, 466; ►altitudinal zone of ►cloud forest); *s* **piso [m] de niebla** (►Piso altitudinal del ►bosque nublado tropical; ►laurisilva canaria); *f* **étage [m] des brouillards** (►Tranche d'altitude de la ►forêt de brouillard) ; *g* **Wolkennebelstufe [f]** (►Höhenstufe des ►Nebelwaldes).

858 cloudy water [n] *limn. syn.* unclear water [n]; *s* **agua [f] turbia**; *f* **eau [f] trouble** ; *g* **trübes Wasser [n]**.

859 cloverleaf interchange [n] [US] *trans.* (Type of intersection of two highways at different levels, typically employing curved access ramps [US]/access slip roads [UK] on all four sides; ►grade-separated junction, ►freeway interchange [US]/motorway intersection [UK], ►interchange); *syn.* cloverleaf junction [n] [UK]; *s* **trébol [m]** (Denominación de ►nudo sin cruces a nivel; ►cruce de autopistas, ►nudo de enlace); *syn.* cruce [m] de autopista trébol; *f* **trèfle [m]** (Terme caractérisant un dispositif particulier de croisement à niveaux séparés de plusieurs voies autoroutières avec raccordements courbes. On distingue les trèfles à deux, trois ou quatre feuilles ; ►échangeur autoroutier, ►diffuseur) ; *syn.* échangeur [m] en trèfle, croisement [m] en trèfle ; *g* **Kleeblatt [n]** (Bezeichnung für einen ►planfreien Knotenpunkt zweier sich in verschiedenen Ebenen kreuzender Autobahnen; es gibt Kleeblätter mit zwei, drei oder vier Schleifen; ►Autobahnkreuz, ►Anschlussstelle).

cloverleaf junction [n] [UK] *trans.* ►cloverleaf interchange [US].

860 cloverleaf ramp [n] [US] *trans.* (Part of a ►cloverleaf interchange [US]/cloverleaf junction [UK], often constructed using a spiral curve to accommodate required speed transitions; ►grade-separated junction, ►highway ramp [US]/slip road [UK], ►interchange); *syn.* cloverleaf slip road [n] [UK]; *s* **curva [f] de trébol** (Rampa dc acccso o salida en un ►trébol ; ►nudo de enlace, ►nudo sin cruces a nivel, ►rampa de acceso); *f* **bretelle [f] d'échangeur en trèfle** (Voie de raccordement de l'échangeur à niveau séparé ; ►bretelle d'accès, ►diffuseur, ►échangeur, ►intersection routière à plusieurs niveaux, ►trèfle) ; *g* **Kleeblattschleife [f]** (Auf- oder Abfahrtsrampe eines ►Kleeblatts; ►Anschlussstelle, ►Auffahrtsrampe, ►planfreier Knotenpunkt); *syn.* Schleife [f], Schleifenrampe [f].

cloverleaf slip road [n] [UK] *trans.* ►cloverleaf ramp [US].

club-rush swamp [n] [UK] *phyt.* ►bulrush swamp [US].

clump [n] *phyt. plant. hort.* ►tuft (1).

clump [n] of trees *landsc.* ►group of trees.

cluster [n] of trees *landsc.* ►group of trees.

cluster rose [n] *hort. plant.* ►polyantha rose.

clutch [n] of eggs *zool.* ►recurrent clutch of eggs.

coal bing [n] [SCOT] *min.* ►spoil bank [US] (2)/spoil heap [UK].

861 coal mine shale [n] [US] *min.* (Mining spoil of a fine-grained detrital sedimentary rock, originally formed by the compaction of clay, silt or mud); *syn.* colliery shale [n] [UK]; *s* **zafras [fpl] de esquisto** (Residuos de minería de arcilla esquistosa); *f* **schistes [mpl] stériles** (Roche restante produit de l'exploitation du schiste) ; *g* **Berge [pl] (1)** (Taubes Gestein, das bei der Gewinnung von Schiefertonen anfällt).

862 coal mine spoil [n] [US] *min.* (Coal mining waste); *syn.* colliery spoil [n] [UK]; *s* **zafras [fpl] de carbón** (Restos de mineral estéril resultantes de la explotación de minas de carbón); *f* **houille [f] stérile** (Roche restante produit de l'exploitation du charbon) ; *g* **Berge [pl] (2)** (Taubes Gestein, das beim Kohleabbau anfällt).

863 coarse bedding sand [n] *constr.* (Granular particles ranging from 0.5mm to 5.0mm within various classification systems; TSS 1996, 810; ►bedding sand); *s* **arena [f] de asiento gruesa** (Tipo de arena gruesa utilizada de asiento para pavimentos o para superficies de ►revestimiento compactado; ►arena para adoquinados); *f* **sable [m] grossier de pose** (Un ►sable de pavage à forte granulométrie) ; *g* **Pflastergrand [m]** (*Niederdeutscher Ausdruck* grober ►Pflastersand zur Bettung).

864 coarse gravel [n] *constr. pedol.* (Stony material between 20 and 64mm in diameter; ►gravel 1, ►soil textural class); *s* **grava [f] gruesa** (Piedras de un diámetro entre > 20,0 y 63,0 mm; ►clase textural, ►grava); *f* **gros gravier [m]** (Matériau pierreux dont la granularité des grains est comprise entre 20 mm et 63 mm ; ►classe de texture des sols, ►gravier) ; *syn.* gravier [m] grossier ; *g* **Grobkies [m]** (D., *Abk.* gG; ►Bodenart mit Korngrößen von > 20,0 bis 63,0 mm Durchmesser gemäß DIN 4022 und 4023; ►Kies).

coarse root [n] *arb.* ►medium root.

865 coarse sand [n] *constr. pedol.* (**1.** *constr.* ►Soil textural class consisting of mineral particles with diameter from 0.2 to 2.0mm [UK], from 0.5 to 1.0mm [US]; RCG 1982, 159. **2.** *pedol.* 25% or more very coarse, coarse, and medium sand and less than 50% at any other one grade of sand; RCG 1982, 161); *s* **arena [f] gruesa** (Partículas minerales de un diámetro entre 0,6 y 2,0 mm; ►clase textural); *syn.* sablón [m] (BU 1959); *f 1* **sable [m] grossier** (F., en terme de granulométrie, matériau meuble formé de quartz [grains de sable] dont les dimensions sont comprises entre 0,2 à 2 mm ; cf. DIS 1986, 192 ; ►classe de texture des sols) ; *f2* **sable [m] très grossier** (**F.**, en terme de granulométrie, matériau meuble formé de quartz (grains de sable) dont les dimensions sont comprises entre 1 à 2 mm ; cf. DIS 1986, 192 ; ►classe de texture des sols) ; *g* **Grobsand [m]** (**D.**, *Abk.* gS; ►Bodenart mit Korngrößen von > 0,6 bis 2,0 mm Durchmesser; cf. DIN 4022 und 4023).

866 coarse silt [n] *constr. pedol.* (According to the American National Cooperative Soil Survey [NCSS] diameter from > 0.02 to 0.05mm; ►fine silt); *s* **limo [m] grueso** (Fracción granulométrica de partículas minerales de 0,01 a 0,06 mm de diámetro; ►limo fino); *syn.* tarquín [m] grueso; *f* **limon [m] grossier** (Fraction granulométrique d'un sol constitué de particules minérales dont les dimensions sont comprises entre 0,02 et 0,05 mm ; TSF 1985 ; ►limon fin) ; *g* **Grobschluff [m]** (...e [pl] u. ...schlüffe [pl]; **D.**, *Abk.* gU; ►Bodenart mit Korngrößen von > 0,02 bis 0,06 mm Durchmesser gemäß DIN 4022 und 4023, ►Feinschluff).

867 coarse spoil [n] *min.* (Mining waste material of large particles; ►mining spoil); *s* **escoria [f] gruesa** (►Zafras de más de 50 mm de espesor); *f* **stériles [mpl] grossiers** (Déblai d'exploitation minière de granulométrie supérieure à 50 mm ; ►stérile) ; *g* **Grobberge [pl]** (Bergematerial mit Fraktionen > 50 mm; ►Berge 3); *syn.* Grubenberge [pl].

868 coast [n] *geo.* (Strip of land of indefinite width—up to many kilometers—that extends from the sea-shore inland to the first major change in terrain features; the most seaward part of the **c.** is called 'shore'; DOG 1984; ►flat coast, ►littoral zone [US]/littoral [UK], ►water's edge); *s* **costa [f]** (Zona estrecha de contacto entre el mar y los continentes, e influida por aquél; en sentido geomorfológico, zona de contacto entre dos medios morfogenéticos diferentes, y cuyo modelado es resultado tanto de la actividad marina como de la de los agentes continentales, así como también de las características estructurales de la márgen del continente; DGA 1986; ►costa llana, ►litoral, ►orilla); *syn.* costa [f] del mar, costa [f] marina; *f1* **côte [f]** (Au sens géomorphologique partie du ►littoral où le continent entre en contact avec la mer et soumise indirectement au actions marines ; DG 1984, 272 ; ►côte basse, ►zone marine littorale ; *comparaison* ►berge) ; *f2* **rivage [m]** (1) (Bande de terre subissant directement les actions marines) ; *g* **Küste [f]** (Unmittelbar ans Meer angrenzender Streifen des Festlandes — bis zu vielen Kilometern breit; ►Flachküste; *zum Vergleich* ►Ufer; ►Litoral); *syn.* Meeresküste [f].

869 coastal area [n] *geo. recr. syn.* coastal region [n], coastland [n]; *s1* **zona [f] costera** (Término genérico para las áreas situadas a lo largo de la costa); *s2* **ribera [f] del mar** (En Es. la r. del m. es parte del dominio público marítimo-terrestre e incluye el espacio comprendido entre la línea de bajamar escorada o máxima viva equinoccial, y el límite hasta donde alcanzan las olas en los mayores temporales conocidos o, cuando lo supere, el de la línea de pleamar máxima viva equinoccial. Se consideran incluidas en esta zona las marismas, albuferas, marjales, esteros y, en general, los terrenos bajos que se inundan como consecuencia del flujo y reflujo de las mareas, de las olas o de la filtración del agua del mar. También incluye las playas o zonas de depósito de materiales sueltos, incluyendo escarpes, bermas y dunas, tengan o no vegetación, formadas por la acción del mar o del viento marino, u otras causas naturales o artificiales; cf. art. 3.1 Ley 22/1988); *f* **zone [f] côtière** *syn.* espace [m] littoral ; *g* **Küstengebiet [n]**.

870 coastal cliff [n] *geo.* (High steep rockface along a coast line; *specific terms* ocean cliff, sea cliff; ►flat coast, ►cliff); *s* **costa [f] acantilada** (Frente escarpado de roca a lo largo de la costa generado por la actividad erosiva del mar; ►acantilado; *opp.* ►costa llana); *syn.* costa [f] escarpada; *f1* **côte [f] escarpée** (Bande littorale souvent rocheuse, à abrupts rapides et à relief plongeant ; *opp.* côte basse) ; *f2* **côte [f] à falaises** (Côte très escarpée soumise à l'érosion marine et à la dénudation, précédée d'une plate-forme d'abrasion ; ►falaise ; *opp.* côte basse) ; *syn.* côte [f] d'abrasion ; *g* **Steilküste [f]** (Steil zum meist tiefen Wasser abfallende Küste; bei Einwirkung von mariner Erosion und Denudation spricht man von einem ►Kliff; *opp.* ►Flachküste).

coastal cliffs *geo.* ►undercutting of coastal cliffs.

871 coastal dike [n] *constr. wat'man.* (Coastal bank or dam to prevent flooding; ►sea defenses [US]/sea defences [UK], ►sea wall 1); *syn.* coastal dyke [n] [also UK]; *s* **dique [m] costero** (Término específico para ►construcción de protección del litoral costero contra posibles inundaciones; ►dique marítimo); *f* **digue [f] côtière** (Ouvrage maritime de protection constitué d'un remblai longitudinal artificiel, le plus souvent composé de terre dont la fonction principale est **1.** soit de contenir les eaux de la mer et d'empêcher de submerger les basses-terres situées le long de ce dernier, **2.** soit de protéger le littoral contre les actions mécaniques de l'eau [brise-lame, jetée] ; ►ouvrage de protection côtière, ►ouvrage de protection des plages) ; *g* **Küstendeich [m]** (Aufgeschüttetes Erdbauwerk am Meer, das das Gelände zum Festland vor Sturmfluten, Tidehochwässer und Überschwemmungen schützen soll. Seit der Industrialisierung gehört der Deichbau in den industrialisierten Ländern zu den Staatsaufgaben, zuvor wurde er von den Bewohnern der Küstengebiete selbst gebaut; ►Küstenschutzbauwerk, ►Strandmauer); *syn.* Meeresdeich [m], Seedeich [m].

coastal drift [n] *geo.* ►longshore drift.

872 coastal dune [n] *geo.* (Hill of sand bordering a seacoast generally built up by the sea and by eolian [wind driven] processes; ►back dune, ►drifting sand dune, ►embryo dune, ►gray dune, ►white dune); *s* **duna [f] litoral** (►duna blanca, ►duna embrionaria, ►duna gris, ►duna parda, ►duna viva; *f* **dune [f] littorale** (Formation sableuse générée par l'accumulation de sables fournis par l'estran ; en fonction de l'exposition aux influences marines et de l'influence du vent, la zonation dunaire des côtes sédimentaires sableuses du littoral permet de distinguer, de la mer vers l'intérieur, la ►dune embryonnaire, la dune blanche, la ►dune grise et la ►dune brune ; par suite de la destruction du couvert végétal la dune fixée peut se transformer en ►dune mouvante) ; *syn.* dune [f] bordière ; *g* **Küstendüne [f]** (Durch äolische Akkumulation aus Meeressand am Strand aufgehäufte Hügel. Die Sukzessionsentwicklung beginnt von der ►Primärdüne, über die ►Weißdüne zur ►Graudüne und schließlich zur ►Heidedüne; bei Zerstörung der Vegetation kann eine ►Wanderdüne entstehen).

coastal dyke [n] [UK] *constr. wat'man.* ►coastal dike.

coastal erosion [n] *geo.* ►shoreline erosion [US]/shore line erosion [UK].

coastal fishery [n] *agr.* ►fishery.

coastal floodmark vegetation [n] *phyt.* ►driftline community.

873 coastal heath [n] *geo.* (Landscape of open, low growing vegetation of heather *[Calluna]*, grasses, herbs and mosses, typically distributed in oceanic regions; ►heath vegetation); *s* **landa [f] costera** (Paisaje abierto con vegetación rastrera de brezo *[Calluna]*, hierbas y musgos, típica de la región oceánica de zonas entre templadas y frías; ►landa); *f* **lande [f] littorale** (Végétation herbacée et arbustive basse des zones côtières de l'océan atlantique, de la Méditerrannée, de la mer du Nord et de la mer Baltique ; ►lande 3) ; *g* **Küstenheide [f]** (Offene Landschaft an den Küsten der Nord- und Ostsee aus relativ vielen niedrig wachsenden Arten mit ozeanischer Verbreitungstendenz bestehend; ▶Heide).

874 coastal heritage [n] *conserv.* (Total extent of remaining unspoiled coastline; ▶coastal reserve [US]/heritage coast [UK]); *s* **patrimonio [m] natural litoral** (Paisaje litoral no alterado; ▶zona de servidumbre de protección del litoral, ▶zona de reserva); *f* **patrimoine [m] naturel littoral** (Ensemble des zones côtières ayant gardé leur caractère naturel originel ; ▶espace et milieu littoral protégé) ; *syn.* patrimoine [m] côtier ; *g* **erhaltene ursprüngliche Küstenlandschaft [f]** (Gesamtheit der noch weitgehend im ursprünglichen Zustand vorhandenen Küstenabschnitte; ▶geschützter Küstenabschnitt).

875 coastal impounding [n] *eng. wat'man.* (Construction of dikes in the foreland for coastal land reclamation; ▶foredike, ▶polder); *s* **construcción [f] de diques costeros** (Técnica utilizada sobre todo en los Países Bajos para ganar terrenos al mar; ▶dique anterior, ▶polder); *f* **endiguement [m] côtier** (Méthode de conquête sur la mer de terrains immergés en zone littorale ; ▶digue avancée, ▶polder) ; *syn.* endigage [m] ; *g* **Eindeichung [f] an Küsten** (Bau von Deichen auf dem Vorland zur Landgewinnung; eindeichen [vb], einpoldern [vb]; ▶Vordeich, ▶Polder); *syn.* Einpolderung [f].

876 coastal landscape [n] *geo.* (Land or region geomorphologically including the shoreland and the adjacent strip of land as well as the hinterland with its living environment and economic sphere and all of its abiotic resources and biota); *s* **paisaje [m] litoral** (Zona de terreno a lo largo de las costas de mares u océanos que incluye geomorfológicamente el litoral y la franja adyacente de tierra); *syn.* paisaje [m] costero; *f* **paysage [m] littoral** *syn.* paysage [m] côtier ; *g* **Küstenlandschaft [f]** (Landschaftsraum, der geomorphologisch den Randbereich des Meeres, den angrenzenden Festlandstreifen und auch das Hinterland mit seinem gesamten Lebensumfeld und Wirtschaftsraum einschließt).

coastal marsh [n] *geo.* ▶coastal marshland.

877 coastal marshland [n] *geo.* (GGT 1979; low-lying shoreland of a tidal mudflat or in tidal areas of estuaries; ▶salt marsh, ▶tidal mudflat); *syn.* coastal marsh [n], tidal marsh [n]; *s* **marisma [f] marítima** (Terreno bajo y pantanoso que se inunda por las aguas del mar durante las mareas altas. A menudo se encuentra en zonas de desembocadura fluvial; ▶llanura de fango, ▶marisma salina); *f* **marais [m] maritime** (Étendue intertidale d'accumulations récentes de sédiments fins vaseux des secteurs abrités des côtes souvent en voie de colmatage surtout là où les marées sont fortes et les profondeurs faibles ; le **m. m.** est constitué **1.** de la ▶slikke étendue littorale basse, souvent inondée et **2.** du ▶schorre, partie supérieure de l'estran au-dessus des hautes mers moyennes ; les parties basses des marais maritimes sont souvent protégées par des digues) ; *syn.* vasière [f] littorale, cordon [m] vaseux littoral, côte [f] à marais ; *g* **Marsch [f]** (In den letzten 7500 Jahren durch Ablagerung von Schlick entstandene, sehr fruchtbare Flachlandschaft in Höhe des Meeresspiegels an einer Wattenküste oder im Tidebereich der Flüsse; tief liegende **M.en** sind durch Deiche gegen Überflutungen geschützt; die Küsten- und Flussmarschen an der deutschen Nordseeküste umfassen ca. 5500 km²; ▶Salzmarsch, ▶Watt); *syn.* Marschland [n], Seemarsch [f].

878 coastal marsh soil [n] *pedol.* (Soil originating from ▶sediments 2 along a low-lying shore or in the sheltered part of an estuary); *s* **suelo [m] de marsch** (Suelo gley más o menos anmooriforme, de las costas planas, formado por terrestrificación natural o artificial del ▶légamo; cf. KUB 1953); *syn.* suelo [m] de marisma; *f* **sol [m] salin à vase marine** (PED 1983, 470, 483 ;

sol salin à sulfato-réduction formé le long des côtes [formant des polders après endiguement], des estuaires ou des les lagunes côtières. Le matériau initial [« schlick » des auteurs allemands] est un mélange de limons argileux et de matière organique provenant des organismes marins, doué de propriétés réductrices très marquées ; PED 1983, 370 ; ▶boue sédimentaire, ▶vase marine) ; *syn.* sol [m] de polder ; *g* **Marschboden [m]** (Aus ▶Schlick des Küstenwatts oder der Flussmündungsbereiche entstandener Boden); *syn.* Kleiboden [m], Koogboden [m], Polderboden [m].

coastal mudflat [n] *geo. ocean* ▶tidal mudflat.

coastal promenade [n] *recr.* ▶beach promenade.

879 coastal protection [n] *conserv. land'man.* (Measures to prevent or restore coast land subject to wind or wave erosion as well as to protect properties and residents residing in coastal areas. Such protection includes also the construction of residences on stilts, replenishing sand and sand dunes washed away by storms, etc.; **in U.S.**, there are many state and local government programs for **c. p.** and necessary measures are taken by the U.S. Army Corps of Engineers. National Seashores have been established by federal legislation for management by the National Park Service; ▶coastal reserve [US]/heritage coast [UK], ▶conservation of coastal landscapes); *syn.* shoreline protection [n]; *s* **protección [f] del litoral** (Conjunto de medidas que sirven para prevenir la erosión marina o restaurar el litoral de los efectos causados por la misma; *término específico* regeneración de playas; ▶preservación del paisaje litoral, ▶zona de servidumbre de protección del litoral, ▶zona de reserva); *syn.* protección [f] de la costa; *f* **protection [f] du littoral** (Ensemble des mesures susceptibles de protéger directement ou indirectement la côte contre les actions de la houle et du vent et d'éviter un recul des terres ; ▶espace et milieu littoral protégé, ▶protection et mise en valeur des milieux naturels du littoral) ; *g* **Küstenschutz [m, o. Pl.]** (Alle Maßnahmen, die direkt oder indirekt die Küste vor Brandungstätigkeit und Windangriff schützt, um Landverluste zu vermeiden; ▶Erhaltung der Küstenlandschaft, ▶geschützter Küstenabschnitt).

coastal protection structure [n] *conserv. eng.* ▶sea defenses [US]/sea defences [UK].

880 coastal range [n] *geo.* (Chain of mountains along the sea coast; ▶coastal area, ▶coastal strip); *s* **cordillera [f] litoral** (Cadena de montañas a lo largo de la costa; ▶franja costera, ▶zona costera); *syn.* cadena [f] litoral); *f* **chaîne [f] côtière** (Unité montagneuse très allongée qui s'étire le long d'une côte ; ▶frange côtière, ▶zone côtière) ; *g* **Küstengebirgskette** (Bergkette entlang einer Küste; ▶Küstengebiet, ▶Küstenstreifen).

coastal region [n] *geo. recr. syn.* ▶coastal area.

Coastal Regions [npl] of the Mediterranean, Convention for the Protection of the Marine Environment and the *conserv. land'man. leg.* ▶Barcelona Convention.

881 coastal reserve [n] [US] *conserv. landsc.* (Protective designation for the conservation and enhancement of specific sections of coastline area having historic significance; e.g. historic island settlement [US]; ▶coastal heritage); *syn.* heritage coast [n] [UK]; *s 1* **reserva [f] natural marítimo-terrestre (≠)** (En EE.UU. categoría de protección de espacios naturales costeros con significado histórico, como p. ej. asentamientos históricos en islas; ▶patrimonio natural litoral); *syn.* zona [f] del litoral protegida; *s 2* **zona [f] de servidumbre de protección del litoral** (Espacio comprendido entre el límite interior de la ▶ribera del mar y 100 m tierra adentro, en el que, en general, están restringidos los usos a aquéllos relacionados con el uso del dominio público marítimo-terrestre, exceptuando los usos agrícolas; arts. 23 y 25 Ley 22/1988 de Costas); *f* **espace [m] et milieu littoral protégé (±)** (Terme désignant les espaces littoraux à préserver sur les communes littorales en application de l'article L. 146-6 du C. urb. ; ►patrimoine naturel littoral) ; *g* **geschützter Küstenabschnitt [m]** (Schutzkategorie für einen Küstenabschnitt von besonderer Bedeutung für Naturschutz und Landschaftspflege; ►erhaltene ursprüngliche Küstenlandschaft).

882 coastal strip [n] *geo.* (Long narrow area of land and sea bordering the shoreline; ►coastal range [US], ►coastal area); *s* **franja [f] costera** (Zona estrecha y alargada de tierra y mar a lo largo del litoral; ►cordillera litoral, ►zona costera); *f* **frange [f] côtière** (►chaîne côtière, ►zone côtière) ; *syn.* linéaire [m] de côte ; *g* **Küstenstreifen [m]** (Langer, schmaler Gebietsstreifen entlang einer Küste; ►Küstengebiet, ►Küstengebirgskette).

coastal waters [npl] *geo. leg* ►territorial waters.

coastal woodland [n] *conserv. land'man.* ►protective coastal woodland.

coastland [n] *geo.* ►coastal area.

883 coastline [n] *geo.* (Boundary between land and water of a sea or ocean; the term 'shoreline' is not synonymous and frequently used in the sense of "high-water shoreline" or the landward limit of the intermittently exposed shore; cf. DOG 1984); *s* **línea [f] de la costa** (Límite entre la tierra y el mar); *syn.* borde [m] costero; *f* **trait [m] de côte** (Limite entre la bande de terre qui confine à la mer et la zone soumise directement à l'action des vagues en fonction des marées ; DG 1984, 453) ; *syn.* ligne [f] de rivage ; *g* **Küstenlinie [f]** (Grenzsaum zwischen Land und Meer).

coated macadam surface [n] [UK] *constr.* ►macadam [US].

coating [n] **of roots** *constr. hort.* ►root puddling.

cobble [n] [US] *constr.* ►cobblestone [US].

884 cobble paving [n] *constr.* (Paved surface of split or unsplit pebbles, 6-25cm in diameter); *syn.* cobblestone paving [n]; *s* **pavimento [m] de guijarros** *syn.* revestimiento [m] de guijarros, pavimento [m] de peladilla; *f* **pavage [m] en galets** (Revêtement du sol effectué par pose de galets bruts ou d'éclats) ; *g* **Kieselsteinpflaster [n]** (Bodenbelag aus ungespaltenen oder gespaltenen Kieselsteinen); *syn.* Flusskieselpflaster [n].

885 cobblestone [n] [US] *constr.* (Generic term for a roughly-cut rectangular or squared natural stone [usually granite] between 50 and 250mm long used for paving; standard sizes are: 100mm cubes, 100 x 125 x 225mm and 100 x 100 x 200mm **standard**, and 100 x 175 x 250mm **jumbo**; *NOTE* Some concrete paving stones are also called "cobblestone"; ►larg-sized paving stone, ►mosaic paver, ►small paving stone, ►paving stone [US]/paving sett [UK], ►random cobblestone paving [US]/random sett paving [UK]); *syn.* cobble [n] [US], natural stone sett [n] [UK], random sett [n] [UK]; *s* **adoquín [m] natural** (Piedra —generalmente de granito— rectangular o cuadrada de acabado basto entre 50 y 250 mm de largo, utilizada para pavimento; *términos específicos* ►adoquín mosaico, ►adoquín pequeño y ►adoquín grande; ►adoquín, ►adoquinado sin orden); *f* **pavé [m] en pierre naturelle** (Terme générique pour ►pavé mosaïque de pierre naturelle, ►petit pavé et ►gros pavé) ; ►pavé, ►pavage désordonné en pierre ; *syn.* pavé [m] en roche naturelle ; *g* **Natursteinpflasterstein [m]** (OB für ►Mosaikpflasterstein [Kantenlänge/Nenndicke bis 60 mm], ►Kleinpflasterstein [Kantenlänge 60-120 mm] und ►Großpflasterstein [Kantenlänge über 120 mm]; cf. DIN EN 1342; ►Pflasterstein, ►Wildsteinpflaster); *syn.* Pflasterstein [m] aus Naturstein.

cobblestone [n] [US]**, large-sized** *constr.* ►large-sized paving stone [US]/large-sized paving sett [UK].

cobblestone paving [n] *constr.* ►cobble paving, ►random cobblestone paving US].

cockle stair [n] *arch.* ►spiral stair(case).

Codes [npl] **of Practice (C.P.)** [UK] *constr. contr.* ►standards for construction [US].

codominant stem [n] *arb.* ►removal of codominant stem.

codominant stems [npl] *arb.* ►forked growth.

coefficient [n] *constr. hydr.* ►runoff coefficient.

coefficient [n]**, shrinkage** *constr.* ►shrinkage crack.

886 coffin burial [n] *adm.* (Act or ceremony of burying a corpse in a box; ►cremation burial, ►coffin burial site); *syn.* cascade burial [n] [also US]; *s* **entierro [m]** (Acto de enterrar a un muerto en un ataud; ►cremación [de los muertos], ►tumba [de tierra]); *syn.* inhumación [f]; *f* **inhumation [f]** (Action de la mise en terre d'un cercueil ; ►tombe, ►incinération funéraire) ; *syn.* ensevelissement [m], enterrement [m] ; *g* **Erdbestattung [f]** (Absenken eines Sarges in ein Grab in Anwesenheit der Angehörigen; i. d. R. transportiert bei der **E.** ein Bestattungsunternehmer den Sarg mit der Leiche zum Friedhof zur Aufbewahrung in einer Kühlzelle, womit die Leiche in die Verwaltungshoheit des Friedhofsträgers [politische Gemeinde] übergeht; ►Erdgrab, ►Feuerbestattung); *syn.* Beerdigung [f], Begräbnis [n].

887 coffin burial site [n] *adm.* (Grave for ►coffin burial; ►urn burial site); *s* **tumba [f] (de tierra)** (Hoyo en la tierra donde se coloca el ataud; ►entierro, ►tumba para urnas); *f* **tombe [f]** (Fosse dans laquelle est déposé le défunt souvent recouverte d'une dalle ; *terme spécifique en archéologie* fosse sépulcrale ; ►inhumation, ►jardin d'urnes) ; *g* **Erdgrab [n]** (Grab für ►Erdbestattung; ►Urnengrabstätte).

888 coherence [n] *conserv. ecol. pol.* (In the context of conservation within the European Union, individual protection areas connected by linear landscape elements and habitat stepping stones/biotopes are of crucial significance for wild animals and plants, due to their important network function for genetic exchange. The Habitats Directive (formally known as Council Directive 92/43/EEC on the conservation of natural habitats and of wild fauna and flora) requires the preservation of a *coherent* network of protected areas under ►Natura 2000. The overriding feature of this network is the preservation of biodiversity; i.e. the specific diversity of habitats. This central feature of the directive has been permanently secured with the introduction of various measures. These include not only the designation of protection areas and the implementation of coordinated management plans for specific habitats and species, but also other, equally important requirements of the Habitats Directive, such as the protection of species listed in Annex IV [under Article 12 and 13] and the preservation, promotion and maintenance of these landscape and network elements [under Article 10]. The **c.** of Natura 2000 is, therefore, not simply determined by a system of interconnected protection areas characterized by landscape elements, but by a systematic network of permanently viable areas); *syn.* ecological coherence [n]; *s* **coherencia [f] ecológica** (Con la Directiva de Hábitats se creó una red ecológica europea coherente de zonas especiales de conservación, denominada ►Natura 2000, formada por hábitats naturales de interés comunitario, hábitats de especies animales y vegetales de interés comunitario cuya conservación hace necesario designar zonas especiales de conservación, así como por las zonas de protección especial de las aves, definidas según la Directiva 79/409/CEE, de Aves. La función primordial de esta red es la preservación de la biodiversidad, en forma de la

C

diversidad específica de hábitats. Para mejorar la **c. e.** de la red Natura 2000, la directiva urge a los Estados miembros a mantener y, en su caso, desarrollar los elementos del paisaje que revistan primordial importancia para la fauna y flora silvestres. Elementos que, por su estructura lineal y continua [como los ríos con sus correspondientes riberas o los sistemas tradicionales de deslinde de los campos], o por su papel de puntos de enlace [como los estanques o los sotos] resultan esenciales para la migración, la distribución geográfica y el intercambio genético de las especies silvestres. La **c. e.** de la red Natura 2000 no consiste, por tanto, solo en un sistema de áreas protegidas interconectadas caracterizadas por elementos específicos del paisaje, sino en un sistema de áreas viables en sí mismas que contribuyen a preservar la biodiversidad a largo plazo. El cumplimiento de todas las obligaciones de protección previstas en la directiva es así la garantía de salvaguarda de la **c. e.** de la red Natura 2000; cf. art. 3, 10, 12 y 13 Directiva de Hábitats 92/43/CEE); *f* **cohérence [f]** (La directive « Habitats » prescrit la sauvegarde de la cohérence du réseau des sites Natura 2000, donc le maintien des interrelations entre ses divers éléments constitutifs [sites protégés, réseau de corridors biologiques, zones de relais ou pôles relais qui revêtent une importance majeure pour la faune et la flore sauvages] avec leurs relations fonctionnelles nécessaires à la migration, à la distribution géographique et à l'échange génétique d'espèces sauvages ; le « réseau » est cohérent, si la conservation des fonctions de chaque site pris isolément est durable ; la fonction principale du réseau ►Natura 2000 est le maintien de la biodiversité, donc de la diversité spécifique des habitats ; cette fonction centrale de la directive doit être garantie durablement par des mesures diverses ; y contribuent non seulement le classement des sites y compris la mise en œuvre des plans de gestion coordonnés pour les habitats et les espèces déterminés, mais encore d'autres obligations non moins importantes de la directive « Habitats » comme la protection des espèces de l'annexe IV [conformément aux articles 12 et 13] ainsi que le développement d'éléments du paysage [conformément à l'article 10] ; sous le terme de cohérence du réseau Natura 2000 il ne faut pas entendre un système concret de sites reliés par des éléments déterminés de paysage, mais un système de sites viables en eux-mêmes contribuant à la conservation durable de la biodiversité ; le respect de toutes les obligations de protection prévues par la directive « Habitats » est ainsi le garant de la sauvegarde de la cohérence du réseau Natura 2000, comme l'un des éléments central de la directive) ; *g* **Kohärenz [f]** (*Naturschutz innerhalb der EU* Zusammenhang einzelner Schutzgebiete durch verbindende lineare Landschaftselemente und Biotoptrittsteine, die wegen ihrer Vernetzungsfunktion für den genetischen Austausch von ausschlaggebender Bedeutung für wild lebende Tiere und Pflanzen sind. Die Fauna-Flora-Habitat-Richtlinie [FFH-RL] schreibt die Wahrung der **K.** des Schutzgebietsnetzes ►Natura 2000 vor. Die übergeordnete Funktion dieses Netzes ist der Erhalt der Biodiversität, also der spezifischen Vielfalt der Lebensräume und Habitate. Diese zentrale Funktion der Richtlinie soll durch verschiedene Maßnahmen dauerhaft gesichert werden. Dazu tragen nicht nur die Schutzgebietsausweisungen einschließlich der Umsetzung abgestimmter Managementpläne für bestimmte Lebensräume und Arten bei, sondern auch andere, nicht minder wichtige Vorgaben der FFH-RL wie der Schutz der Arten des Anhangs IV [gemäß Art. 12 und 13] sowie den Erhalt, die Förderung und Pflege von o. g. Landschafts- und Verbindungselementen [gemäß Art. 10]. Unter der **K.** des Netzes 2000 ist also nicht ein konkretes, durch bestimmte Landschaftselemente verbundenes System aus Schutzgebieten zu verstehen, sondern ein systemares Verbundensein von in sich dauerhaft lebensfähigen Gebieten); *syn.* ökologische Kohärenz [f].

cohesionless soil [n] [US] *pedol.* ►granular soil [US].

cohesive soil [n] [UK] *pedol.* ►colloidal soil.

cold [n] *landsc. phyt.* ►protection from cold.

889 cold air damage [n] *agr. hort.* (Frost damage to plants caused by accumulation of cold air in depressions [frost pockets]); *s* **daños [mpl] causados por corrientes de aire frío** (Congelación de plantas por concentración de aire frío en hondonadas del terreno); *f* **dommages [mpl] causés par un flux d'air froid** (Destruction des tissus végétaux provoquée par le refroidissement rapide des nappes d'air froid accumulées dans les dépressions de terrain [creux de gel]) ; *syn.* dégâts [mpl] provoqués par les flux d'air froid ; *g* **Kaltluftschäden [mpl]** (Erfrierungen an Pflanzen, die durch Ansammlung von Kaltluft in Geländemulden [Frostlöcher] entstehen).

890 cold air drainage [n] *met.* (LE 1986, 328; **1.** accumulation of cold air, mostly from ►cold air source areas, which flows down from upper slopes. **2.** Flow of dense, often moist, stratified chilled air mass that moves by gravity to low points in the terrain usually within defined topographic or vegetated corridors); *s* **corriente [m] de aire frío** (Concentración de aire frío, generalmente procedente de ►zonas de aparición de corrientes de aire frío); *f* **courant [m] d'air froid** (Masses d'air froid, souvent en provenance des zones ►source d'alimentation en air froid se déplaçant le long des pentes ou vers une dépression de terrain) ; *syn.* flux [m] d'air froid ; *g* **Kaltluftfluss [m]** (Ansammlung kalter Luft, meist aus ►Kaltluftentstehungsgebieten, die hangabwärts oder in eine Geländevertiefung fließt); *syn.* Kaltluftabfluss [m], Kaltluftstrom [m].

cold air drainage channel [n] *landsc.* ►cold air drainage corridor.

891 cold air drainage corridor [n] *landsc.* (Longitudinally-shaped area allowing the unobstructed flow of cold air ranging from the ►cold air source area to the ►frost pocket); *syn.* cold air drainage channel [n]; *s* **canal [m] de drenaje del aire frío** (Franja de terreno sin barreras por la cual puede correr el aire frío de la ►zona de aparición de corrientes de aire frío hasta el punto más bajo del terreno; ►hondonada de aire frío); *f* **couloir [m] d'air froid** (Bande allongée le long de laquelle se déplace une masse d'air froid à partir de la zone ou elle se développe jusqu'au point le plus bas ; ►lac d'air froid ; ►source d'alimentation en air froid) ; *g* **Kaltluftschneise [f]** (Hindernisfreier, langgestreckter Abflussbereich für Kaltluft vom ►Kaltluftentstehungsgebiet bis zum tiefsten Punkt; ►Kältesee); *syn.* Kaltluftabfussbahn [f], Kaltluftbahn [f].

892 cold air source area [n] *met.* (Meadows or agricultural fields which become a nocturnal source of ground-cooled cold air, especially during cloudless ►nights of ground radiation); *s* **zona [f] de aparición de corrientes de aire frío** (Prados o campos de cultivo sobre los cuales, especialmente en las ►noches radiativas, se produce aire frío); *f* **source [f] d'alimentation en air froid** (Espaces naturels [prairies, champs, terrains enherbés] qui après le début du refroidissement nocturne produisent des masses d'air froides en particulier lors des ►nuits en situation radiative) ; *g* **Kaltluftentstehungsgebiet [n]** (Wiesen- oder Ackerflächen, die nach Beginn der nächtlichen Abkühlung, besonders in ►Strahlungsnächten bodenabgekühlte Kaltluft entstehen lassen); *syn.* Kaltluftproduktionsgebiet [n], *auch* Frosteinzugsgebiet [n] (GEI 1961).

893 cold hardiness [n] *agr. bot. for. hort.* (BOT 1990, 144; adaptation of plants to freezing occurrence, so that they can withstand the withdrawal of water during ice formation in the cells by the presence of high sugar concentration acting as 'antifreeze' in the protoplasm; ►frost-resistent); *s* **resistencia [f] a las heladas** (Capacidad de algunas plantas de sobrevivir a

heladas. Éstas resisten la falta de agua por medio de una concentración más alta de azúcares en el líquido celular; ►insensible a heladas); *f* **résistance [f] au gel** (Faculté des végétaux pendant les gelées de résister à la formation de cristaux de glace ou à une perte d'eau par une élévation du gradient de concentration du cytoplasme ; ►résistant, ante au gel) ; *g* **Frosthärte [f]** (Widerstandsfähigkeit gegen Frost: Fähigkeit von Pflanzen, die tödliche Eiskristallbildung im Zytoplasma [viskose bis gallertartige Grundmasse einer Zelle] zu verhindern oder dem Entzug des Wassers bei niedrigen Temperaturen durch eine erhöhte Zellsaftkonzentration zu widerstehen; ►frostbeständig); *syn.* Frostresistenz [f].

894 cold roof [n] *arch. constr.* (Normally flat, sometimes pitched roof where the insulation is placed either between the rafters only or at ceiling joist level. A double-ply, cold roof can be non-ventilated, but is usually ventilated in the space between the outer roof construction and the insulated ceiling. As a rule construction sequence from bottom to top is as follows: **1.** load-bearing construction [roof ceiling rafters], **2.** thermal insulation layer, **3.** ventilated roof space, **4.** outer roof construction, **5.** ►roof membrane, **6.** surface protection [e.g. gravel, vegetation layer]; ►insulated roof membrane assembly; *opp.* ►warm roof); *syn.* ventilated roof [n]; *s* **tejado [m] frío** (Normalmente tejado plano, a veces inclinado, en el que el aislamiento se coloca bien solo entre los cabios o bien al nivel de las viguetas del techo. Un **t. f.** de doble capa puede ser no ventilado, pero generalmente lo es por el espacio entre la construcción externa del tejado y el techo aislado. Generalmente la secuencia de capas, de abajo a arriba, es la siguiente: **1.** capa portante [vigas y viguetas del techo]; **2.** capa de aislamiento térmico; **3.** espacio ventilado del tejado; **4.** construcción externa del tejado; **5.** ►membrana impermeabilizante de tejado; **6.** protección de superficie [p. ej. gravilla, capa portante de vegetación]; ►tejado invertido; *opp.* ►tejado caliente); *syn.* tejado [m] ventilado; *f* **toiture [f] froide** (Toit en pente à double ventilation ; la toiture présente une lame d'air ventilée entre la couverture et la sous toiture, ainsi qu'entre la sous toiture et l'isolation thermique par où circule l'air extérieur ; le système de **t. f.** est susceptible d'occasionner d'importants dégâts à la suite de condensations ; la **t. f.** fut beaucoup utilisée autrefois et est aujourd'hui fortement déconseillée, voire interdite car elle peut provoquer des dégâts importants suite aux problèmes de condensation et de mouvements du support qu'elle entraîne ; la **t. f.** comporte sur le plafond **1.** un pare-vapeur **2.** une isolation thermique, **3.** éventuellement un vide d'air [espace de ventilation, lame d'air ventilée] **4.** un plancher support, **5.** le ►revêtement d'étanchéité et **6.** la couche de lestage, le principe de la **t. f.** occasionne une perte d'inertie thermique et l'éventualité d'un risque de fissuration du support [la variation de température au centre du support peut atteindre une différence de 75 C° entre les périodes les plus chaudes et les plus froides] ; la **t. f.** ne sera utilisée que sur des lieux de stockage sans présence d'humidité et correctement ventilés [carports, annexes] ; ►toiture inversée ; ►toiture chaude) ; *syn.* toit [m] froid ; *g* **Kaltdach [n]** (Zweischalig belüftetes Flachdach; die Schichtenfolge besteht i. d. R. von unten nach oben: **1.** Tragschicht [Dachdecke], **2.** Wärmedämmschicht, **3.** durchlüfteter Dachraum, **4.** Dachhautträger, **5.** ►Dachhaut, **6.** Oberflächenschutz [z. B. Bekiesung, Vegetationstragschicht]; ►Umkehrdach; *opp.* ►Warmdach); *syn.* belüftetes Dach [n].

collapse [n] *geo.* ►riverbank collapse [US]/river-bank collapse [UK].

collapsible post [n] *urb. trans.* ►hinged post.

collar [n] *hort.* ►root collar.

collection charges [npl] [US] *adm. envir. leg.* ►garbage collection charges.

collection charges [npl] [UK]**, waste** *adm. envir. leg.* ►garbage collection charges [US].

collection [n] **of survey information** *plan.* ►data collection.

895 collection [n] **of wild plants or animals** *conserv. phyt. zool.* (Collecting of plants or catching of animals in their habitats); *syn.* taking [n] of species (BEX 1991, 11); *s* **captura [f] de (especies) animales o recogida [f] de plantas** *syn. phyt.* recogida [f] de especies vegetales; *f* **enlèvement [m] des espèces** (Cueillette des espèces végétales ou capture des espèces animales rares ou en voie de disparition ; cf. art. 56 de la loi n° 95-101 du 2 février 1995 relative au renforcement de la protection de l'environnement) ; *g* **Entnahme [f] von Arten** (Fangen oder Sammeln von Tieren oder Pflanzen in freier Natur; Arten entnehmen [vb]); *syn.* Entnehmen [n, o. Pl.] von Arten.

896 collective specification item [n] *contr.* (Part of work in specifications containing several different work items, which are grouped together for a lump-sum price; ►specification item number, ►list of bid items and quantities [US]/schedule of tender items [UK]); *syn.* lump-sum item [n] [also US]; *s* **partida [f] global** (En ►resumen de prestaciones, posición que incluye varios trabajos de diferentes tipos; ►número de órdenes); *f* **poste [m] forfaitaire** (±) (Article dans un ►descriptif-quantitatif regroupant différentes prestations ; ►numéro d'ordre) ; *g* **Sammelposition [f]** (Position im ►Leistungsverzeichnis, die mehrere ungleichartige Teilleistungen, die in Einzelpositionen spezifiziert werden sollten, zusammenfasst; ►Ordnungszahl).

897 collector [n] *conserv.* (Person who takes specimens of flora and fauna; *specific terms* capturer of animals, ►mushroom picker); *s* **coleccionista [m/f] (de plantas o animales)** (Persona que se dedica a recoger muestras de plantas o animales con fines científicos o educativos; ►colector, -a de setas); *f 1* **ramasseur [m]** (de plantes ; ►chercheur de champignons) ; *f 2* **collectionneur [m]** (Personne passionnée par la collection de plantes, d'insectes) ; *g* **Sammler/-in [m/f]** (Jemand, der Pflanzen oder Tiere sammelt; *UBe* z. B. Pflanzensammler/-in, ►Pilzsammler/-in, Schmetterlingssammler/-in).

898 collector drain [n] *agr. constr.* (LAD 1986, 146; **1.** drainage pipe into which individual ►lateral drains discharge; ►main collector drain. **2.** Pipe which collects the sewage or surface water of a plot of land); *syn.* carrier drain [n], collector drainage line [n], leader drain [n], main drainage line [n] (LAD 1986, 146); *s* **colector [m]** (Conducto de recogida de aguas residuales o de drenaje en el que desembocan los ►drenes secundarios; ►colector principal); *f* **collecteur [m]** (**1.** Tuyau d'évacuation collectant les eaux usées, les eaux pluviales, les eaux de drainage en provenance des ►drains adducteurs. **2.** Conduite ou fossé destiné à recevoir les écoulements captés par différents types d'ouvrages et à les acheminer jusqu'à l'exutoire ; GHDA 1993) ; ►collecteur principal) ; *g* **Sammler [m]** (**1.** Entwässerungsleitung, in die einzelne ►Sauger einmünden. **2.** Entwässerungsleitung, in die die Abwässer der Grundstücke und/oder das Oberflächenwasser befestigter Flächen eingeleitet werden; ►Hauptsammler).

collector drainage line [n] *agr. constr.* ►collector drain.

899 collector road [n] *trans. urb.* (Urban road which connects minor ►local street with access only vor residents; ►local feeder road [US]); *syn.* collector street [n] [also US]; *s* **calle [f] de distribución** (Calle que enlaza el tráfico entre las ►calles de vecindario y las ►arterias [secundarias] de tráfico y que sirve también para facilitar el acceso a los distintos edificios y viviendas; cf. DGA 1986); *f* **voie [f] de distribution** (Assure la

C

collecte du trafic issu des ►rues riveraines et permettant la circulation à l'intérieur d'un quartier et reliée à la ►voirie secondaire ; BON 1990, 358) ; *g* **Sammelstraße [f]** (Hauptsächlich den Verkehr zwischen ►Anliegerstraßen und ►Verkehrsstraßen vermittelnde Stadtstraße).

900 collector sewer [n] *envir. urb.* (Main sewer which receives sewage and/or storm water flow from many tributaries covering a large area; ►effluent discharge pipe); *syn.* trunk sewer [n] [also US]; *s* **alcantarilla [f] (1)** (Colector subterráneo que conduce las aguas residuales de las casas o los terrenos a la planta de depuración o al lugar de vertido; ►conducto de desagüe [de aguas residuales]); *f* **collecteur [m] d'eaux usées et d'eaux pluviales** (Égout recueillant les eaux ménagères, les eaux vannes et les eaux pluviales pour les évacuer vers la station d'épuration ; ►conduite d'assainissement) ; *g* **Abwasserkanal [m]** (Öffentlicher Sammler, der Haus- und Grundstücksabwässer aufnimmt und einer Kläranlage zuführt; ►Entsorgungsleitung).

collector street [n] [US] *trans. urb.* ►collector road.

colliery shale [n] [UK] *min.* ►coal mine shale [US].

colliery spoil [n] [UK] *min.* ►coal mine spoil [US].

colline belt [n] *geo. phyt.* ►colline zone.

901 colline zone [n] *geo. phyt.* (►Altitudinal belt between lowlands and ►montane zone); *syn.* colline belt [n]; *s* **piso [m] colino** (Piso de vegetación que se desarrolla de los 0 a los 500 m en una ►zonación altitudinal; ►piso montano); *syn.* piso [m] basal; *f* **étage [m] collinéen** (Étage de végétation entre la vallée et l'►étage montagnard ; ►étagement de la végétation) ; *syn.* étage [m] basal ; *g* **kolline Stufe [f]** (Hinsichtlich der ►Höhenstufung der Bereich zwischen Flachland und ►montaner Stufe; die Höhenangaben variieren, je nach Quelle und in Abhängigkeit von der Höhe über NN und geografischer Lage, zwischen 150-300 m in den Mittelgebirgen und zwischen 300-800 m in den Alpen); *syn.* Hügelstufe [f].

902 colloidal soil [n] *constr. pedol.* (**1.** Soil with particles of molecular aggregates, with a diameter of 0.1 to 0.001 μm. **2.** ►Soil textural class capable of being molded or deformed continuously and permanently, by relatively moderate pressure, into various shapes; cf. SST 1997; *opp.* ►granular soil [US]/non-cohesive soil [UK]); *syn.* cohesive soil [n] [UK], plastic soil [n] [US] (RCG 1982); *s* **suelo [m] cohesivo** (►Clase textural de suelo capaz de almacenar agua, de alto contenido en arcilla o limo; *opp.* ►suelo suelto); *syn.* suelo [m] adhesivo, suelo [m] plástico, suelo [m] tenaz; *f* **sol [m] plastique** (Traduit la capacité d'un sol à l'état humide de se déformer sous la pression sans se briser — teneur en argile au moins 25 % ; ►classe de texture de sol ; *opp.* ►sol à structure particulaire) ; *g* **bindiger Boden [m]** (Wasser haltende ►Bodenart, z. B. Ton, Schluff, Lehm; *opp.* ►nicht bindiger Boden).

903 collusion by bidders [n] [US] *contr. leg.* (Secret, illegal agreement amongst bidders [US]/tenderers [UK] to respect certain prices and not to offer lower prices); *syn.* collusion by tenderers [n] [UK]; *s* **colusión [f] entre licitantes** (Acuerdo secreto ilegal entre licitantes sobre precios a ofrecer para no competir a la baja); *syn.* colusión [f] entre concursantes/postores, connivencia [f] entre licitantes/concursantes/postores; *f* **entente [f] entre entreprises consultées** (Accord exprès ou tacite passé entre plusieurs soumissionnaires de respecter certains prix pour ne pas être moins-disant) ; *g* **Preisabsprache [f] unter Bietern** (Wettbewerbsbeschränkende, rechtswidrige Vereinbarung unter mehreren Unternehmen bei Angebotsabgaben bestimmte Preise einzuhalten und nicht zu unterbieten); *syn.* Absprache [f] bei Angebotsabgabe, Absprache [f] von Angebotspreisen, Absprache [f] unter Bietern, Absprechen [n, o. Pl.] von Angebotspreisen, Preisabrede [f] unter Bietern, Abrede [f] von Angebotspreisen unter Bietern.

collusion by tenderers [n] [UK] *contr. leg.* ►collusion by bidders [US].

904 colonization [n] **(1)** *phyt. zool.* (Introduction of plants or animals leading to the creation of an independent population, capable of reproducing themselves; ►introduction 1, ►naturalization, ►reintroduction, ►resettlement 2, ►restocking of population); *syn.* introduction [n]; *s* **introducción [f] de especies (de flora o fauna)** (►Sueltas [masivas] que llevan a la creación de una población con capacidad de regeneración. Término genérico de ►naturalización, ►reintroducción de especies extintas, ►reposición de especies de fauna y flora, ►reubicación de especies); *f* **1 réintroduction [f] (1)** (►Introduction de végétaux ou d'animaux dans une station qui en est dépourvu et provoquant la formation de populations se reproduisant naturellement en vue de la restauration des populations et de recolonisation des habitats ; terme générique pour ►naturalisation, ►réintroduction 2, ►renforcement de la population, ►transplantation 3) ; *f* **2 recolonisation [f] d'un habitat** (Résultat de la ►réintroduction 1) ; *g* **Ansiedelung [f] (1)** (Erfolgreiche ►Aussetzung von Pflanzen oder Tieren, die zur Bildung einer frei lebenden, fortpflanzungsfähigen Population geführt hat. OB zu ►Bestandsstützung, ►Einbürgerung, ►Umsiedlung, ►Wiederansiedelung; ansiedeln [vb]); *syn. o.V.* Ansiedlung [f].

905 colonization [n] **(2)** *phyt. zool.* (Settlement in a particular area by animals or plants; ►natural colonization, ►primary colonization, ►recent colonization); *s* **colonización [f] natural** (Ocupación de un área específica por una especie de flora o fauna; ►colonización natural de «espacio vacío», ►écesis, ►primera colonización); *f* **colonisation [f] naturelle** (Espèces de faune ou de flore ayant trouvé des conditions favorables de développement sur un territoire donné ; ►apparition d'une espèce nouvelle, ►colonisation primaire, ►colonisation spontanée) ; *g* **Besiedelung [f] (1)** (Das Heimischsein von Tieren oder das Wachsen von Pflanzen in einem bestimmten Gebiet; ►Erstbesiedlung, ►Neubesiedlung, ►spontane Besiedlung); *syn. o. V.* Besiedlung [f].

colonization [n]**, spontaneous** *phyt. plant.* ►natural colonization.

colonization [n] **by scrub** *conserv. land'man.* ►spontaneous colonization by scrub.

colonization [n] **by seed rain** *phyt. for.* ►natural colonization by seed rain.

color aerial photograph [n] [US]**, infrared** *envir. plan. rem'sens. surv.* ►infrared color aerial photograph [US].

colo(u)r aerial photograph [n]**, false** *envir. plan. rem' sens. surv.* ►infrared color aerial photograph [US].

coloured paving block [n] [UK] *constr.* ►tinted concrete paver [US].

colo(u)r photograph [n]**, false** *envir. plan. rem'sens. surv.* ►infrared color photograph [US].

906 columbarium [n] *adm.* (from latin >dovecote<; a particular ►urn burial site with a wall structure of doored compartments to store cinerary urns. Originally a common, underground, Roman, burial site with vaults each containing two urns); *s* **columbario [m]** (Tipo particular de ►tumba para urnas que consiste en una estructura de muro con nichos con pequeñas puertas, donde se depositan las urnas cinerarias. *Arqueología* en los antiguos cementerios romanos, serie de nichos para guardar las urnas cinerarias; CAS 1985); *f* **columbarium [m]** (Bâtiment constitué de niches dans lesquelles sont déposées les urnes cinéraires ; ►jardin d'urnes) ; *g* **Kolumbarium [n]** (Lateinisch *columbarium* >Taubenschlag<; Sonderform einer ►Urnengrab-stätte in Form einer Urnenwand mit angeordneten Nischen resp. Gefachen zum Hineinstellen der Aschenurnen. *Ursprünglich* im 1. und 2. Jh. übliche unterirdische römische Grabstätte mit über- und nebeneinander angeordneten Wandnischen zur Aufnahme von je zwei Urnen); *syn.* Urnenwand [f], Urnennischenwand [f].

907 column [n] *constr.* (Vertical, usually cylindric[al] pillar, often tapering slightly, supporting an entablature or arch, or a free-standing, decorative element); *s 1* **pilar** [m] (Elemento vertical de piedra, mármol, ladrillo u otro material, relativamente esbelto en comparación con su altura. Se utiliza principalmente para soportar superestructuras, pero también puede emplearse como monumento; DACO 1988); *s 2* **columna** [f] (Elemento estructural vertical de soporte con sección circular o rectangular que recibe la carga según la dirección de su eje longitudinal. Las partes de una **c.** son: base, fustel y capitel; DACO 1988); *f 1* **pilier** [m] (Support vertical large et en général maçonné d'une construction) ; *f 2* **colonne** [f] (Support vertical cylindrique d'une construction) ; *g* **Säule** [f] (Senkrechte, meist walzenförmige Stütze eines Bauwerks oder ähnlich geartetes freistehendes, dekoratives Element).

908 columnar [adj] *hort. plant.* (Straight habit of growth of plants with nearly parallel upright branches or shoots; ►fastigiate); *s* **columnar** [adj] (Dícese de las plantas, cuyas ramas se aproximan de tal manera al eje que el conjunto remata en punta, como en el ciprés y el chopo lombardo; DB 1985; ►fastigiado); *f* **columnaire** [adj] (Port de végétaux dont la silhouette érigée évoque une colonne ; ►fastigié) ; *syn.* colonnaire [adj] ; *g* **säulenförmig** [adj] (Schmale, aufrechte Wuchsform von Pflanzen; ►schmal-pyramidal).

combination [n] **of species** *phyt.* ►characteristic combination of species.

combination [n] **of various land uses** *plan.* ►layering of different uses.

909 combined curb [n] **and grate** [n] **[US]** *constr.* (TSS 1988, 330-15; drainage facility consisting of a ►storm drain grate [US]/gully grating [UK] and an adjacent ►curb inlet [US]/kerb-inlet [UK] which has the same length as the grating/grate); *syn.* combined inlet [n] [UK] (SPON 1986, 169); *s* **sumidero** [m] **combinado** (Tipo de sumidero utilizado en F., GB y EE.UU. que tiene acceso en la calzada y en el bordillo; ►sumidero en bordillo); *f* **regard** [m] **à grille-avaloir** (Le dispositif d'entrée des eaux est situé dans la bordure de trottoir et la chaussée ; ►bordure avaloir) ; *syn.* regard à avaloir ; *g* **kombinierter Straßen-Bordsteinablauf** [m] (Z. B. in F., U.K. und USA verwendeter Straßenablauf, bei dem die Einlauföffnung sowohl eben auf der Straße als auch in den Bordstein integriert ist; ►Straßenablauf im Bordstein).

combined inlet [n] **[UK]** *constr.* ►combined curb and grate [US].

910 combined open space pattern/system [n] *landsc. urb.* (Network of concentric, radial, linear or interlinking open space system parts); *s* **trama** [f] **verde combinada** (Sistema de espacios libres urbanos que combina partes circulares, radiales, en bandas o en franjas entre sí); *syn.* red [f] combinada de espacios libres, sistema [m] combinado de espacios libres; *f* **trame** [f] **verte composée** (Ensemble d'éléments d'une trame verte de disposition circulaire, radiale, linéaire ou en bandes par rapport au centre urbain) ; *g* **kombiniertes Freiraumsystem** [n] (Räumlicher Verbund von ringförmigen, radialen, band- oder kammförmigen Freiraumsystemteilen).

911 combined sewer [n] *envir. urb.* (Sewer pipe that carries both sanitary sewage and storm water runoff, typically employed to accommodate emergency overflow conditions and is not recommended for general application; ►sewage treatment plant, ►combined sewerage system); *s* **alcantarilla** [f] **unitaria** (Conducción subterránea perteneciente a un ►sistema de alcantarillado unitario que transporta a la ►planta de depuración de aguas residuales tanto las aguas residuales como la escorrentía superficial de precipitaciones); *f* **collecteur** [m] **unitaire** (Canal recueillant à la fois les eaux ménagères, les eaux vannes, les eaux pluviales et éventuellement certains effluents industriels et les dirigent vers la ►station d'épuration des eaux usées ; ►réseau [d'assainissement] unitaire) ; *g* **Mischkanal** [m] (Entwässerungs-einrichtung für die Abführung aller Arten von Abwasser [zur ►Kläranlage]; ►Mischverfahren).

912 combined sewer system [n] *envir.* (Sewerage system that carries both sanitary sewage and storm water runoff. During dry weather, this system carries all wastewater for treatment to the ►sewage treatment plant. During storm events, part of the load may be intercepted to prevent overloading of the processing facility. In this case, the untreated portion is allowed to enter the receiving stream; RCG 1982; ►separate sewerage system); *s* **sistema** [m] **de alcantarillado unitario** (Sistema conjunto para recoger las aguas residuales y pluviales en una ciudad. En épocas de pluviosidad normal o baja, todo el caudal pasa a la ►planta de depuración de aguas residuales; en fases de grandes precipitaciones, la capacidad de depuración generalmente no es suficiente para tratar toda el agua, por lo cual una parte de ella tiene que ser vertida sin tratamiento al cauce receptor. Para evitar esto en algunos lugares se construyen tanques de retardo subterráneos que permiten almacenar el agua excedente hasta que la capacidad de la planta de tratamiento sea suficiente para depurarlas; ►sistema de alcantarillado separativo); *f* **réseau** [m] **(d'assainissement) unitaire** (Un seul système d'assainissement qui assure l'évacuation de l'ensemble des eaux usées et pluviales ; ►station d'épuration des eaux usées, ►réseau d'assainissement séparatif, ►réseau pseudo-séparatif) ; *syn.* système [m] unitaire d'assainissement ; *g* **Mischverfahren** [n] (Abführung aller Arten von Abwasser in einem Leitungssystem zur ►Kläranlage; ►Trennverfahren); *syn.* Mischkanalisation [f].

913 coming [n] **into effect** *leg.* (Passage of a bill, an ordinance, or a statute into law; *context* at the effective date of the law, at the time when the statute comes/came into force); *syn.* coming [n] into force [also UK], enactment [n], entering [n] into force, taking [n] effect [also US]; *s* **entrada** [f] **en vigor** (de leyes, decretos, reglamentos, etc.; entrar [vb] en vigor); *f* **entrée** [f] **en vigueur** (de lois, de décrets, etc.) ; *g* **In-Kraft-Treten** [n, o. Pl.] (Wirksamwerden von Gesetzen, Satzungen, Verordnungen etc.; *Kontext* bei **I.** des Gesetzes; das in Kraft getretene Gesetz; in Kraft treten [vb]).

coming [n] **into force** **[UK]** *leg.* ►coming into effect.

coming [n] **together** *constr.* ►junction (1).

commencement [n] **of works** **[UK]** *constr. contr.* ►start of construction [US].

comments [npl] *urb. plan.* ►submission of planning comments.

914 commercial and light industry area [n] *leg. urb.* (Area defined on a legally-binding land-use plan or a ►zoning map [US]/Proposals Map [UK]; ►industrial land use, ►industrial park, ►industrial use zone); *syn.* light industry zone [n]; *s* **zona** [f] **de actividades comerciales e industriales no contaminantes** (D., categoría de planificación urbana; ►industrial park, ►parque industrial, ►suelo industrial, ►zona de uso industrial); *f 1* **zone** [f] **à vocation économique** (Zone pouvant accueillir diverses activités, bureaux, commerce, artisanat, industries non polluantes ; F., il est possible de procéder à cette affectation au moyen d'une zone d'aménagement concerté [ZAC] procédure d'urbanisme opérationnel utilisée entre autres pour la mise en œuvre de zones d'activités telles que ZAC industrielle, ZAC commerciale ; ►zone à vocation d'activités économiques, ►zone d'activités industrielles, ►zone d'aménagement concerté à usage industriel, ►parcs d'activités ; *f 2* **zone [f] d'entreprises** (Zone d'activités, créée dans un secteur géographique prioritaire, en général en reconversion, ou des avantages, notamment fiscaux, sont accordés aux entreprises ; cette formule existe dans plusieurs pays et en F. trois zones ont été créées à Dunkerque, La Seyne et La Ciotat ; DUA, 858) ; *f 3* **zone [f] d'investissement privilégié (ZIP)** (Zone sur laquelle les entreprises nouvelles, industrielles ou commerciales, créées depuis le 1er octobre 1988 bénéficient d'une exonération totale puis dégressive d'impôt sur le revenu ou d'impôt sur les sociétés) ; *g* **Gewerbegebiet [n] (1)** (*Abk. im Bebauungsplan* GE; gemäß § 9 [1] 1 BauGB und § 1 [2 ff] BauNVO nach der besonderen Art ihrer baulichen Nutzung im ►Bebauungsplan darzustellendes Baugebiet, das vorwiegend der Unterbringung von nicht erheblich belästigenden Gewerbebetrieben dient; § 8 BauNVO; ►gewerbliche Bauflächen, ►Industriegebiet 2, ►Industriepark).

915 commercial center [n] [US]/**commercial centre** [n] [UK] *urb.* (**1.** Generic term for any urban or rural area in which the predominant use of land is business or commercial enterprises. **2.** Specific term for the commercial portion of a large city core; ►central business district); *s* **centro [m] comercial y financiero** (de una ciudad; ►zona central); *f* **centre [m] des affaires** (Quartier urbain dans lequel se concentre la vie commerciale, caractérisé par sa situation centrale, une grande concentration de bâtiments élevés, une forte intensité du trafic, les valeurs élevées du sol et des impôts payés, le mélange de tous les groupes ethniques et de toutes les classes sociales ; DUA 1996 ; ►zone urbaine consacrée aux activités économiques ; *g* **Geschäftszentrum [n]** (Viertel einer Stadt mit einer Massierung von Handelsunternehmen und Einkaufsläden; ►Kerngebiet).

commercial corridor [n] [US] *urb.* ►strip development [US] (2).

commercial facilities [npl] **and light industries** [npl] *plan. urb.* ►allocation of commercial facilities and light industries, ►location of commercial facilities and light industries.

916 commercial fertilizer [n] *agr. constr. for. hort.* (General term for industrially manufactured ►organic fertilizers and ►mineral fertilizers); *s* **fertilizante [m] industrial** (Término utilizado en general para denominar a abonos minerales u orgánicos fabricados industrialmente; ►fertilizante mineral, ►fertilizante orgánico); *f* **engrais [m] du commerce** (Terme commun utilisé pour les engrais minéraux, organiques ou organo-minéraux produits industriellement ; ►engrais minéral, ►engrais organique) ; *g* **Handelsdünger [m]** (Allgemein benutzter Begriff für industriell hergestelltes mineralisches und organisches Düngemittel; cf. Typenliste der Düngemittelverordnung vom 09.07. 1991, BGBl. I 1450, letzte Neufassung v. 16.12.2008, BGBl. I 2524; ►Mineraldünger, ►organischer Dünger).

commercial forestry [n] *for.* ►forestry (1).

917 commercial horticulture [n] *hort.* (Production of flowers, fruit, vegetables, as well as herbs, perennials, trees, and shrubs for sale; ►fruit growing, ►landscape contracting industry, ►market gardening, ►perennial nursery, ►plant nursery, ►tree nursery); *syn.* nursery trade [n] [also US]; *s* **horticultura [f] comercial** (Rama de la horticultura dedicada a la producción de flores, frutos y verduras así como de árboles, arbustos, perennes y hierbas para su venta; ►empresa de horticultura, ►fruticultura, ►industria de construcción paisajística, ►jardinería, ►vivero [de árboles], ►vivero de vivaces); *f* **horticulture [f]** (À la différence de la ►profession des entrepreneurs paysagistes, branche de

l'agriculture regroupant la production et la reproduction des fruits [►culture fruitière], des légumes, des champignons, des plantes médicinales, condimentaires et aromatiques, des fleurs [►pépinières spécialisées en plantes vivaces], des plantes ornementales ; on distingue l'horticulture vivrière et l'horticulture ornementale [►pépinière] ; ►entreprise horticole, ►jardinerie) ; *g* **Erwerbsgartenbau [m, o. Pl.]** (Im Gegensatz zum ►Garten- und Landschaftsbau ein Wirtschaftszweig des Landbaues/der Agrarwirtschaft mit dem Ziel der Produktion und Vermehrung von Obst [►Obstbau], Gemüse, Pilzen, Heil- und Gewürzpflanzen [Gemüsebau] sowie Blumen und Zierpflanzen [Zierpflanzenbau], Stauden [►Staudengärtnerei], Gehölzen [►Baumschule] und die dazugehörenden Dienstleistungen zu erbringen. Zum **E.** gehört auch die Friedhofsgärtnerei. Die Innenraumbegrünung ist teilweise im Produktionsbereich Zierpflanzenbau enthalten; ►Betrieb einer Endverkaufsgärtnerei, ►Gartenbaubetrieb).

918 commercial sign [n] [US] *urb.* (Board or sign which displays advertising posters, and frequently electronic devices; in U.S., signs on thoroughfares are called 'billboards'; ►advertisement structure); *syn.* advertisement board [n] [UK]; *s* **cartel [m] publicitario** (Marco grande para colocar carteles de publicidad o pantalla electrónica utilizada con fines publicitarios en lugares públicos; el término estadounidense *«billboard»* se utiliza para vallas publicitarias a lo largo de carreteras; ►estructura publicitaria); *f 1* **enseigne [f] publicitaire** (Toute inscription, forme ou image apposée sur un immeuble et relative à une activité qui s'y exerce directement ; ►dispositif publicitaire) ; *f 2* **panneau [m] publicitaire** (Support présentant une information à but publicitaire ; les panneaux publicitaires constituent bien souvent une nuisance visuelle pour le paysage) ; *g* **Werbetafel [f]** (►Werbeanlage, die für das Ankleben von Plakaten oder für elektronische Werbemittel vorgesehen ist); *syn.* Werbeschild [n].

commercial strip development [n] [US] *urb.* ►strip development [US] (2).

commission [n] *contr. prof.* ►earn a commission; *adm. conserv'hist.* ►historic district commission [US]/Historic Buildings and Monuments Commission for England [UK].

commission [n], **planning** *contr. prof.* ►planning contract.

commission [n] [UK], **termination of** *contr. prof.* ►termination of a contract [US].

commissioned planning office [n] [UK] *contr. prof.* ►contract holder [US].

919 commissioning [n] **a consultant/designer/planner** *contr. prof.* (Written or verbal intention to commission a design firm [US]/planning office [UK] for a project, which is assumed to be able to give the best performance on the basis of negotiated contractual conditions; ►earn a commission); *syn.* appointment [n] of consultant/designer/planner; *s* **adjudicación [f] (de proyecto)** (Encargo de forma oral o escrita a un gabinete de arquitectura/planificación para realizar un proyecto de planificación; adjudicar [vb] un proyecto, encargar [vb] un proyecto; ►conseguir un encargo); *syn.* encargo [m] de un proyecto; *f* **attribution [f] de la mission (d'études)** (Notification orale ou écrite à un bureau d'études ; ►obtenir un marché d'étude) ; *syn.* missionnement [m] d'un(e) concepteur, trice ; *g* **Auftragserteilung [f] an ein Planungsbüro** (Schriftliche oder mündliche Erklärung über die Übertragung von Planungsleistungen für ein Planungsprojekt an ein Büro, das auf Grund der ausgehandelten Auftragsbedingungen die bestmögliche Leistung erwarten lässt; ►Auftrag erhalten); *syn.* Erteilung [f] eines Planungsauftrages, Beauftragung [f] eines Büros, Vergabe [f] eines Planungsauftrages.

common [n] [UK] *agr. for. (leg. hist.)* ►town commons [US].

920 common boundary wall [n] *constr. leg.* (Wall centered on a property line; ▶property boundary wall); *syn.* "party wall" [n] [also US]; *s* **muro** [m] **medianero** (Muro que separa dos propiedades adyacentes y pertenece a ambas; cf. DACO 1988; ▶muro de deslinde); *f* **mur** [m] **mitoyen** (Mur séparant deux bâtiments, deux propriétés privées qui, à la diférence du ▶mur séparatif de propriété, appartient à deux propriétaires s'il n'y a ni titre, ni marque du contraire ; art. 653 du Code civil ; la « **mitoyenneté** » est un droit de copropriété immobilière [deux personnes] et ne s'applique qu'aux murs ou clôtures [barrières, fossés, haies, palissades] qui constituent la séparation de deux propriétés privées) ; *g* **Grenzmauer** [f] (Mauer auf der Mitte einer Grundstücksgrenze; ▶Grundstücksmauer).

921 common grazing [n] *agr.* (Extensively grazed pasture without boundary fences, usually infertile and irregularly stocked; mostly on steep hills, predominantly public land. The pasture has a very high number of species in comparison with fodder pastures, often with spiny or thorny herbs; ▶pasture of low productivity; *specific term* grazing commons [US]/grazing common [land] [UK]); *syn.* land pasture [n]; *s 1* **pasto** [m] **común** (±) (Pasto aprovechado conjuntamente por los ganaderos de un pueblo, generalmente de mala calidad y situado en zonas altas y escarpadas; ▶pasto pobre); *s 2* **pastizal** [m] **malo** (Caracterizado por la presencia de un 25% de especies aprovechables como forraje del ganado y un 75% de plantas invasoras no deseables, y posee una capacidad mínima de pastoreo equivalente a una cabeza de ganado por hectárea en las regiones tropicales pluviales; MEX 1983); *syn.* pastizal [m] pobre; *s 3* **dula** [f] *obs.* (Cada una de las tierras en que por turno pacen los ganados de los vecinos de un pueblo); *f* **parcours** [m] (Terrains, non clôturés, de très faible productivité, sur sols souvent rocailleux sur pentes fortes et pauvres en matière organique ; terrains en majorité communaux parfois de vaine pâture parcourus par des troupeaux ; peuplement herbacé riche en espèces et floristiquement diversifié en comparaison avec les prairies de fauche, souvent parsemé d'espèces épineuses ; ▶prairie pâturée maigre) ; *syn.* terres [fpl] de parcours, pacage [m] communal, pâturage [m] communal ; *g* **Hutung** [f] (Nicht eingezäuntes, gewöhnlich wenig ertragfähiges und unregelmäßig bestocktes Weideland auf gering wertigen Böden, meist in steilen Gebirgslagen; überwiegend als Gemeinschaftsbesitz genutzt [Allmendweide] oder verpachtet; Weide mit sehr hoher Artenzahl im Vergleich zur Futter- und Streu[e]wiese, oft mit stacheligen oder dornigen Weideunkräutern bestockt; ▶Hutewald, ▶Magerweide); *syn. o. V.* Allmendweide [f], Hutweide [f], Trift [f], Triftweide [f].

922 common grazing forest [n] *agr. phyt.* (Forest or wood formerly designated as ▶common grazing land for cattle: the undergrowth consists of plant communities corresponding to the degrading of the wood, where its ancient stands of solitary trees have assumed today the character of a park-like landscape; ▶town commons [US]/common [land] [UK], ▶grazed woodland); *s 1* **monte** [m] **patrimonial de pastoreo** (≠) (Monte utilizado antiguamente como tierra de pasto común para el ganado bovino. El sotobosque consiste en comunidades vegetales que coresponden a la fase de degradación del bosque. Hoy en día tienen un cierto carácter de parque, por sus grandes árboles solitarios; ▶bosque de pastoreo, ▶pastizal malo, ▶pasto común, ▶tierras comunales); *s 2* **dehesa** [f] (En el área mediterránea, terrenos acotados de bosque aclarado de encinas, rebollos o fresnos en el que los terrenos libres se utilizan de pastos, bien en dominio de algún pueblo o bien particular existiendo diferentes denominaciones según el uso. Actualmente se aprovechan de esta forma grandes extensiones de tierras en Extremadura y Andalucía

que no sirven para cultivar. El término proviene del latín «defensa»; cf. DINA 1987; *f* **forêt** [f] **pâturée communale** (Forêt appartenant à la commune, à faciès de dégradation du tapis végétal très appauvri correspondant à une pratique ancienne du pacage libre et caractérisée aujourd'hui par un peuplement forestier ouvert ancien, à l'aspect d'une futaie jardinée ; ▶communaux, ▶forêt pâturée, ▶parcours) ; *g* **Hutewald** [m] (Früher in Gemeinschaftsbesitz als Viehweide genutzter und meist degradierter Wald, der entsprechende Pflanzengesellschaften als Unterwuchs aufwies und heute durch einen alten Einzelbaumbestand oft einen parkähnlichen Charakter hat; ▶Allmende, ▶Hutung, ▶Weidewald); *syn.* Hudewald [m].

common land [n] [UK] *agr. for. (leg. hist.)* ▶town commons [US].

923 common open space [n] [US] *plan.* (IBDD 1981; open land within or related to a block development, not individually owned or dedicated for public use, which is intended for the common use of residents in the development); *syn.* community open space [n] [UK]; *s* **espacio** [m] **libre colectivo** (Espacio verde dentro de manzana de viviendas o relacionado a un grupo de éstas dedicado al uso común de los residentes en ellas); *f* **espaces** [mpl] **libres collectifs** (Espaces privés placés aux abords immédiats des bâtiments d'habitations, accessibles aux habitants concernés et réservés à un usage de voisinage) ; *syn.* espace [m] semi-public ; *g* **kollektiv nutzbarer Freiraum** [m] (Gemeinschaftsanlage in Form von Freiflächen innerhalb eines Häuserblocks oder um ein Haus, die vor allem von den Bewohnern benutzt werden); *syn.* halb-öffentlich nutzbarer Freiraum [m].

commons [npl] *agr. for. (leg. hist.)* ▶town commons [US].

924 communal car park [n] *leg. urb.* (Joint parking area [US]/car park [UK] within communal open space used by members of a housing development but not by the general public); *s* **aparcamiento** [m] **de comunidad de vecinos** (Superficie dedicada a aparcamiento de automóviles de los miembros de una comunidad y no para el público en general); *syn.* estacionamiento [m] de comunidad de vecinos [CS], parqueo [m] de comunidad de vecinos [CA, C]; *f* **parc** [m] **de stationnement collectif** (Emplacement réservé au stationnement des véhicules appartenant aux usagers de bâtiments d'habitation et d'équipements collectifs) ; *g* **Gemeinschaftsstellplatz** [m] (In einer Gemeinschaftsanlage zusammengefasste Stellplätze für Autos der Mitglieder einer Wohnanlage; nicht für die allgemeine Öffentlichkeit bestimmt; cf. § 9 [1] 22 BauGB und Bauordnungen der Länder).

communal cemetery [n] *adm. leg. urb.* *∗*▶cemetery.

communal facilities [npl] *leg. urb.* ▶social service facilities.

925 communal facility [n] *leg. urb.* (Installation serving several users and not intended for the general public; e.g. ▶communal garage, common play area; ▶public facility 1); *s* **instalación** [f] **colectiva** (Equipamiento construido para una comunidad de vecinos o una zona pequeña, p. ej. zona de juegos infantiles, garaje, etc.; ▶instalación de equipamiento comunitario, ▶garaje de comunidad de vecinos); *f* **installation** [f] **collective** (F., équipements publics réalisés par les constructeurs dans le cadre d'un programme d'aménagement d'ensemble tels que parc publics, équipements de services publics industriels ou commerciaux ; D, documents d'urbanisme, installation privée prévue pour équiper plusieurs parcelles ou logements au lieu d'une installation par parcelle, p. ex. aire de jeux, emplacements de parking, garages ; ▶équipements collectifs, ▶équipement urbain d'intérêt public, ▶garage collectif, ▶installation d'intérêt général; *g* **Gemeinschaftsanlage** [f] **(1)** (Eine Anlage für mehrere Baugrundstücke oder Wohnungen, anstelle einer gleichartigen Anlage für jedes einzelne Grundstück, z. B. Kinderspielplatz, Freizeiteinrichtungen, Stellplätze und Garagen; nicht für die allgemeine Öffentlichkeit bestimmt; cf. § 9 [1] 22 BauGB; ▶Gemeinbedarfseinrichtung, ▶Gemeinschaftsgarage); *syn.* Gemeinschaftseinrichtung [f].

926 communal garage [n] *leg. urb.* (Adjoining garages which are part of communal housing facilities and are not intended for the general public; ▶underground parking garage [US]/underground car park [UK]); *s* **garaje [m] de comunidad de vecinos** (▶aparcamiento de residentes); *syn.* garaje [m] de bloque de pisos, garaje [m] de residentes; *f* **garage [m] collectif** (Parcs de stationnement regroupés réservés aux habitants d'un immeuble d'habitation ; ▶parking souterrain résidentiel) ; *g* **Gemeinschaftsgarage [f]** (In einer Anlage zusammengefasste Garagen der Mitglieder einer Wohnanlage; nicht für die allgemeine Öffentlichkeit bestimmt; ▶Anwohnertiefgarage).

927 communal use [n] *urb.* (Community use of a structure or land for public purpose; e.g. public building, traffic systems, public utilities, social and cultural institutions; *specific term* institutional use of buildings; ▶public facility 1); *s* **equipamiento [m] comunitario** (En Es. la ley incluye en el **e. c.** los espacios libres de dominio y uso público, los servicios de interés público y social [parques deportivos, equipamiento comercial y social], los centros culturales y docentes, los aparcamientos y una red de itinerarios peatonales; Art. 1, Anexo al Reglamento de Planeamiento RD 2159/1978; ▶instalación de equipamiento comunitario); *f* **services [mpl] collectifs** (Ensemble des activités et des équipements réalisés par les collectivités publiques nécessaires à la satisfaction des besoins de la population urbaine dans les domaines sociaux, médicaux, culturels, de transport, de loisirs, etc. ; *aménagement du territoire* les **Schémas de Services Collectifs** ont été institués par la loi d'orientation pour l'aménagement et le développement durable du territoire du 25 juin 1999, dite loi « Voynet ». Ils sont destinés sur une période de 20 ans à planifier l'action de l'état et à orienter celle des collectivités locales. Ces schémas coordonnent les interventions publiques dans neuf domaines déterminants pour l'aménagement et l'organisation du territoire ; enseignement supérieur et recherché, culture, santé, information et communication, transport de marchandises, transport de personnes, énergie, espaces naturels et ruraux, sports ; *terme spécifique* ▶équipement collectif) ; *g* **Gemeinbedarf [m]** (Öffentliche Einrichtungen und Anlagen, die erforderlich sind, um eine ordnungsgemäße und angemessene Versorgung und Befriedigung der der Allgemeinheit dienenden Ansprüche [verkehrstechnische, kulturelle und zivilisatorische] in einem Gebiet zu gewährleisten; ▶Gemeinbedarfseinrichtung).

928 communication corridor [n] *plan.* (Linear consolidation of transport facilities, energy utilities and telecommunications; ▶growth axis); *s* **banda [f] de infraestructura** (Concentración lineal de infraestructura de transportes, suministro de energía, gas, etc. y de conducciones de telecomunicación a lo largo de un eje; ▶eje de desarrollo); *f* **couloir [m] d'infrastructures de communications** (Concentration linéaire de divers équipements d'infrastructures [route, chemin de fer, télécommunication, transport d'énergie, etc.] ; ▶axe de développement) ; *syn.* corridor [m] d'infrastructures de communications ; *g* **Infrastrukturband [n]** (Lineare Bündelung von Anlagen der Verkehrseinrichtungen, Energieversorgung, Telekommunikation etc.; ▶Entwicklungsachse).

communication route [n] *plan. trans* ▶connecting route.

929 community [n] [US] *adm. pol.* (**1.** Generic term for jurisdictional unit which is the area of local self-government ; ▶section of a community area [US]. **2.** A social group of any size whose members reside in a specific locality, share government, and have a common cultural and historical heritage); *syn.* municipality [n] [UK]; *s 1* **municipio [m]** (Término genérico para unidad territorial básica de un Estado; ▶finage 2); *syn.* municipalidad [f] [AL], comuna [f] [RCH]; *s 2* **comunidad [f]** (Un grupo social de número indefinido de miembros que residen en un lugar específico y comparten un patrimonio social, cultural e histórico común); *f* **commune [f]** (Terme générique pour l'unité territoriale politique et administrative locale ; ▶finage) ; *g* **Gemeinde [f]** (OB zu politisch-administrative Gebietseinheit/Gebietskörperschaft, die die Grundeinheit kommunaler Selbstverwaltung darstellt und i. d. R. aus einer ▶Gemarkung 2 besteht, aber auch aus mehreren Gemarkungen bestehen kann. Im letzteren Falle sind die Gemarkungen Stadtteile/Stadtbezirke).

community [n] *phyt.* ▶edge community, ▶fire plant community, ▶fragment community, ▶plant community 2; *recr.* ▶recreation resort community; *leg. urb.* ▶residential community [US]; *phyt.* ▶riverine community.

930 community action group [n] *pol.* (Established or *ad hoc* organization furthering interests of a particular group in favo[u]r or in opposition to plans or policies pursued by local, state or federal government); *syn.* citizen pressure group [n] [also US]; *s 1* **grupo [m] ciudadano** (Grupo de personas independientes de partidos u organismos políticos o económicos que trabajan en común para impedir la realización de proyectos no deseados o para mejorar la situación de los/las ciudadanos/as en cuestiones puntuales); *syn.* iniciativa [f] ciudadana, «Bürgerinitiative» [f]; *s 2* **asociación [f] de vecinos [Es]** (Grupo de ciudadanos surgidos en los años 60 y 70 en las grandes ciudades españolas para conseguir mejoras infraestructurales y sociales en los barrios; en muchos países latinoamericanos existen también organizaciones vecinales con fines similares); *syn.* junta [f] de vecinos [RCH]; *f* **association [f] de défense** (Groupement de personnes, indépendant des partis politiques ou d'organismes publics, s'unissant spontanément en vue de remédier à certains préjudices) ; *g* **Bürgerinitiative [f]** (Zusammenschluss von Personen außerhalb von Parteien und Interessenverbänden, die Missstände beseitigen resp. Verbesserungen erreichen wollen).

931 community appearance [n] [US] *arch. urb.* (**1.** Overall visual impression, which contributes to a community's sense of place or gestalt. **2.** Visual aspect of the various sections of a city, town, or village, which may encompass indigenous areas, revitalization, historic preservation, integration of conservation and development; some cities issue a Community Appearance Manual which includes examples of plan requirements, landscaping recommendations, and illustrations of elements of architectural design. It provides developers, builders, engineers, architects, elected officials and town staff with a clear understanding of what is expected in development; no UK equivalent; ▶cityscape, ▶townscape); *s* **fisonomía [f] del hábitat residencial** (Impresión visual del conjunto de los elementos constitutivos y estructurantes de un barrio o un pueblo que le dan un carácter propio; ▶aspecto escénico urbano, ▶fisonomía de la ciudad); *f* **physionomie [f] de l'habitat** (Perception visuelle des différentes fractions du territoire d'une ville, d'un village et caractérisées par des traits distinctifs façonnés au fil du temps qui sont un jeu entre les maisons et les paysages, les bâtiments et les espaces urbains, les rues et les places, le matériau et la couleur et qui lui confèrent son individualité ; AVSP; ▶aspect d'une agglomération, ▶morphologie urbaine) ; *syn.* forme [f] urbaine, silhouette [f] d'une cité, apparence [f] de l'habitat ; *g* **Gestalt [f] des Wohnumfeldes** (Wahrgenommene Gesamtheit der raumgliedernden und raumbegrenzenden Elemente einer nachbarschaftlichen Umgebung innerhalb eines Stadtgebietes incl. der durch die äußere Erscheinungsform bedingten Nutzungsfunktionen; ▶Ortsbild, ▶Stadtgestalt); *syn.* Gestaltung [f] des Wohngebietes.

community appearance analysis [n] [US] *arch. urb.* ►townscape analysis.

community area [n] *adm. pol. urb.* ►local community area.

community beautification [n] [US] *landsc.* ►Garden City Movement [UK].

community development [n]**, residential** *urb.* ►housing subdivision [US].

932 community development plan [n] [US]/**urban development plan** [n] [UK] *leg. urb.* (Long-range policy plan indicating the proposed development and types of land use for a community; **in U.S.**, generic term for ►comprehensive plan and ►zoning map; **in U.K.**, the corresponding term **urban development plan** is the generic term covering ►Local Development Framework [LDF] and ►Proposals Map; the **u. d. p.** guides and informs within a system of development control, together with public consultation. The **u. d. p.** forms the statutory basis for planning decisions and is not a single document. Its preparation gives the community the opportunity to participate in policies and proposals during the preparation process. Since the Planning and Compulsory Purchase Act 2004, the former **Structure Plan** has been replaced by Regional Spatial Strategies and the Local Plans and Unitary Development Plans have been substituted by the **Local Development Framework** [LDF]—a non-statutory term used to describe a folder of documents, which includes all of a planning authority's local development documents. These collectively deliver the spatial planning strategy for the local planning authority's area. An LDF comprises Development Plan Documents, Supplementary Planning Documents, the Statement of Community Involvement, the Local Development Scheme, the Annual Monitoring Report, and any Local Development Orders or Simplified Planning Zones that may have been added); *s* **plan** [m] **urbanístico** (≠) (Término genérico para los planes de uso del suelo urbano que se definen a nivel municipal y que en Es. son: el ►plan general municipal de ordenación urbana, los ►planes parciales de ordenación y los ►programas de actuación urbanística); *f* **documents** [mpl] **de planification urbaine** (F., terme générique désignant les règles successives de planification urbaine établies depuis la loi du 14 mars 1919 et du 19 juillet 1924 ; l'ensemble des documents graphiques et écrits [►schéma de cohérence territoriale (SCOT), ►schéma directeur d'aménagement et d'urbanisme (SDAU), ►schéma de secteur, ►plan local d'urbanisme (PLU), ►plan d'occupation des sols (POS), ►carte communale] règlent à long terme les orientations et les règles d'affectation et d'occupation des sols sur les différentes parties du territoire dans le but d'assurer aux habitants de la commune les meilleures conditions de vie ainsi que de sauvegarder et développer l'environnement ; **D.**, terme générique pour le « *Flächennutzungsplan* » et le « *Bebauungsplan* ») ; *syn.* documents [mpl] d'urbanisme, *obs.* plan [m] d'urbanisme, *obs.* plan [m] directeur d'urbanisme [PDU] ; *g* **Bauleitplan [m]** (**1.** OB zu ►Flächennutzungsplan und ►Bebauungsplan, die die bauliche und sonstige Nutzung der Grundstücke langfristig vorbereiten und leiten; **B.pläne** sind gem. § 1 [4] BauGB den Zielen der Raumordnung anzupassen; diesem Anpassungsgebot unterliegen sie bei der Aufstellung, Änderung und Ergänzung. **2.** Planungsinstrument der Gemeinden, das nach Maßgabe des BauGB [§§ 1 ff] eine nachhaltige städtebauliche Entwicklung und eine dem Wohl der Allgemeinheit entsprechende sozialgerechte Bodennutzung gewährleistet und dazu beiträgt, eine menschenwürdige Umwelt zu sichern und die natürlichen Lebensgrundlagen zu schützen und zu entwickeln, auch in Verantwortung für den allgemeinen Klimaschutz, sowie die städtebauliche Gestalt und das Orts- und Landschaftsbild baukulturell zu erhalten und zu entwickeln; § 1 [5] Satz 1 BauGB); *syn.* Richtplan [m] [CH].

933 community development planning [n] [US] *adm. leg. urb.* (Complete range of planning measures introduced by a comprehensive plan [US]/Local Development Framework [LDF] [UK] and the resulting ►zoning map [US]/Proposals Map [UK]. This planning is prepared for the control of land use, particularly buildings, within a community and to ensure sustainable urban development; it includes the preparation of a graphic and written statement depicting an orderly pattern of community development based upon an assessment of environmental, cultural, social, and economic processes and alternative outcome projections; **in U.K.**, the Town and Country Planning system provides the main framework of land use and aims to secure the most efficient and effective use of land in the public interest. At the community level, local authorities decide whether to allow proposals to build on land or to allow a change of use and draw up development plans setting out the authority's policies in its area. Since the Planning and Compulsory Purchase Act 2004, the system of spatial planning in England has been completely revised. New regional planning bodies [RPBs], established for each region, will draw up a statutory Regional Spatial Strategy [RSS], for up to 20 years in advance. The RSS identifies the scale and distribution of provision for new housing, priorities for the environment and considers elements like transport infrastructure, economic development, and the provision of open green spaces, waste treatment and disposal. It is also prepared in the context of the new requirements for sustainability appraisals, to meet European Union rules on strategic environmental assessment [SEA]. In local spatial planning, the development planning system has effectively been replaced with a 'folder' approach to policy making. The main folders are those of the ►Local Development Framework [LDF] and the set of Local Development Documents, which are produced by the local planning authority, collectively delivering the spatial planning strategy for its area in general conformity with the RSS. The Core Strategy is the key plan. **In D.**, urban land-use planning is also a two-tier process involving two types of plan: the preparatory land-use plan [Flächennutzungsplan] and the legally-binding land-use plan [Bebauungsplan]); *syn.* urban land-use planning [n] [UK], development control [n] [UK], urban development planning [n], *obs.* town and country planning [n] [UK]; *s* **planeamiento [m] urbanístico** (En Es., acción de la administración pública competente de definir los usos del territorio de uno o varios términos municipales por medio del ►plan de general municipal de ordenación urbana, de los ►planes parciales de ordenación y los ►programas de actuación urbanística; en alemán, término legal que se refiere a la ordenación del desarrollo urbanístico en la ciudad y en el campo); *syn.* planificación [f] urbana; *f* **planification [f] urbaine** (Action conjointe des collectivités publiques et de l'administration permettant par des documents prospectifs de déterminer les orientations fondamentales d'aménagement urbain ainsi que de fixer les règles générales et les servitudes d'utilisation du sol ; elles s'expriment par les documents suivants : ►schéma de cohérence territoriale [SCOT], — anciennement ►schéma directeur d'aménagement et d'urbanisme [SDAU], ►plan local d'urbanisme [PLU] — anciennement ►plan d'occupation des sols [POS]) ; *g* **Bauleitplanung [f]** (Alle Maßnahmen, die der Vorbereitung und Leitung der baulichen und sonstigen Nutzung der Grundstücke einer Gemeinde dienen und mit Hilfe des ►Flächennutzungsplanes und des daraus zu entwickelnden ►Bebauungsplanes eine nachhaltige städtebauliche Entwicklung vorgeben und eine dem Wohl der Allgemeinheit entsprechende sozialgerechte Bodennutzung gewähren und dazu beitragen, eine menschenwürdige Umwelt zu sichern und die natürlichen Lebensgrundlagen zu schützen und zu entwickeln, auch in Verantwortung für den allgemeinen Klimaschutz, sowie die städte-

C

bauliche Gestalt und das Orts- und Landschaftsbild baukulturell zu erhalten und zu entwickeln; cf. § 1 [5] BauGB).

934 community expansion [n] [US] *urb.* (Planned development measures, which lead to an increase in new ►zoning districts [US]/building zones [UK] and conservation districts within established environmental constraints, or where existing urban districts are developed to form new municipal units; ►satellite city, ►urban land use category, ►urban development); *syn.* town expansion [n] [UK]; *s* **expansión** [f] **urbana (1)** (Medidas de desarrollo urbano para ampliar las áreas urbanizables que pueden conllevar la creación de nuevos barrios; ►categoría de suelo urbano, ►ciudad satélite, ►desarrollo urbano, ►sector urbano); *f 1* **expansion** [f] **urbaine (1)** (Terme général décrivant l'avancement de l'urbanisation ver la périphérie) ; *f 2* **extension** [f] **urbaine (1)** (Mesures de développement des espaces à urbaniser par fixation de zones à urbaniser [►zone urbaine 3, ►secteur urbain] en prenant en compte les besoins en matière d'habitat, d'emploi, de services et de transport des populations futures ; ►développement urbain, ►ville nouvelle, ►zone urbaine 2) ; *syn.* extension [f] de l'urbanisation ; *g* **Stadterweiterung** [f] (Entwicklungsmaßnahmen, die dazu dienen, vorhandene Orte um neue ►Bauflächen zu vergrößern oder vorhandene Orte zu neuen Siedlungseinheiten zu entwickeln; ►Stadtentwicklung, ►Trabantenstadt, ►Baugebiet).

community facility [n] [US] *urb.* ►public facility (1).

935 community garden [n] [US] *urb. leg.* (**1.** Garden plot usually owned by a city, university or like non-profit organization and situated in an urban, suburban or rural setting. **c. g.s** provide gardening opportunities for the physical and social benefit of the people and neighborhoods as well as for environmental awareness, social interaction and community education; **c. g.s** are rented for a small fee and can also be a series of plots dedicated to "urban agriculture". **In U.S.,** the beginning of **c. g.s** was in the early 1970s when engaged citizens in New York, acting on their own initiative, have promoted temporarily planted garden patches as community gardens by occupying unused gaps between buildings in the poorest quarters of the city, in order to fulfil an urgent need for green space in residential areas; in 1999 there were over 1,000 large "allotments" or community gardens between 100 and 700m² in size in the USA. **2. In Europe,** plot of garden land situated away from the home and rented or assigned for leisure use in growing flowers, vegetables, and fruit, though not as a means of living; often closely associated with urban development. In view of its appearance, legal status and incorporation in the planning process, the allotment garden may be permanently used as a garden; ►temporary community garden plot [US]/temporary allotment plot [UK], ►community gardening [US]/allotment gardening [UK], ►tenant garden, ►rented garden [UK], ►leasehold garden); *syn.* allotment garden (plot) [n] [UK]; *s 1* **jardín** [m] **comunitario** (En EE.UU., denominación de pequeño jardín de 100-1000 m² situado entre edificios en zonas céntricas de ciudades grandes diseñado, creado y utilizado por el vecindario como jardín de uso común para paliar el déficit de zonas verdes existente. El movimiento de los **j. c.** comenzó en Nueva York en los años 1970 y se extendió por todo el país; ►jardinería comunitaria); *s 2* **huerto** [m] **recreativo urbano** (Pequeño jardín arrendado por el correspondiente municipio para usos recreativos privados, que no está localizado junto a la vivienda del arrendatario, sino en zona especial más o menos cercana a ella. Su promotor fué D. SCHREBER [1808-1861], Leipzig, Alemania, que los impulsó para mejorar las condiciones de vida de los trabajadores de las grandes ciudades alemanas. Hoy en día las zonas de **h. r. u.**; son una categoría de uso en los planes de ordenación urbana; ►gestión de huertos recreativos urbanos, ►jardín alquilado, ►jardín arrendado, ►jardín de inquilinos, ►parcela de huerto urbano temporal); *syn.* jardín [m] para obreros, huerto [m] familiar urbano; *f 1* **jardin** [m] **collectif** (*Terme générique* catégorie de jardin offrant un loisir très demandé : le jardinage qui contribue à la sauvegarde de la biodiversité des plantes cultivées, fruits, légumes, fleurs, en favorisant leur connaissance, leur culture et leur échange non lucratif entre jardiniers, à sensibiliser à l'environnement, à animer la vie collective d'un quartier et renforcer les liens sociaux de proximité ; *terme spécifique* nouvelle catégorie de jardins collectifs, s'inspirant d'une expérience originale menée à New York en vue de l'appropriation de terrains délaissés pour en faire des jardins gérés collectivement, dont l'objectif est moins le jardinage que l'amélioration du cadre de vie, la création d'un lieu d'échange et de rencontre entre habitants du quartier, l'éducation à l'environnement, l'organisation d'activités culturelles ; ce jardin de proximité à l'initiative d'habitants qui désirent se retrouver dans un lieu convivial pour jardiner ; il est planté et entretenu par les riverains regroupés au sein d'une association ; conçu et planté sur des friches abandonnées suite à des destructions d'immeubles insalubres, c'est un lieu ouvert sur le quartier qui favorise les rencontres entre les générations et les cultures ; dans ce jardin, respect de l'environnement et développement de la biodiversité sont de mise ; éventuellement, il permet de tisser des relations entre les différents lieux de vie de l'arrondissement : écoles, maisons de retraite, hôpitaux ; un tel jardin est confié à une association par convention pour une durée limitée (1 an renouvelable jusqu'à 5 ans) ; Mairie de Paris Direction des Parcs, Jardins et Espaces Verts Sous Direction de l'Animation et de l'Éducation à l'Écologie Urbaine Paris-Nature) ; *syn.* jardin [m] communautaire ; *f 2* **jardin** [m] **d'insertion** (*Terme spécifique* forme de jardin collectif reconnu et réglementé par la loi d'orientation du 29 juillet 1998 relative à la lutte contre les exclusions dont l'objectif, davantage que la production maraîchère, est la réinsertion de personnes en difficulté sociale ou professionnelle [chômeurs, RMIstes, handicapés, personnes isolées, jeunes en difficulté scolaire, anciens détenus, etc.], avec le concours d'un animateur bénévole ou salarié, d'une association chargés de leur accompagnement. Les jardins d'insertion « par le social » cherchent avant tout à réintégrer socialement les publics les plus stigmatisés ; l'activité de production y est relativement accessoire en tant que telle, sa fonction thérapeutique étant privilégiée ; les produits récoltés sont soit conservés par les jardiniers, soit destinés à des structures caritatives [Banque alimentaire, Secours populaire, Restaurants du cœur, etc.]. Les jardins d'insertion « par l'économique » font également de l'activité de jardinage un moyen de réinsertion dans la société. Ils sont un préalable à la réintégration du monde économique, l'objectif à terme étant de retrouver un emploi ; l'activité de production y occupe donc une place plus importante ; en consequence, les produits récoltés sont commercialisés et permettent aux jardiniers de se constituer des revenus d'appoint ; Le Sénat : www.senat.fr) ; *f 3* **jardin** [m] **ouvrier** (±) (*Terme spécifique* parcelles groupées sur des zones d'une superficie supérieure à un hectare, situées soit à proximité de zones à forte densité de population soit en périphérie urbaine, sur lesquelles se réalisent les activités de loisirs en plein air et de jardinage ; ►activités concernant les jardins ouvriers, ►jardin affermé, ►jardin locatif, ►jardin loué, ►parcelle affermée temporairement) ; *syn.* jardin [m] familial ; *g* **Kleingarten** [m] (Für die Freizeit gärtnerisch genutzte, jedoch nicht gärtnerisch erwerbsmäßig bewirtschaftete, wohnungsferne Pachtgartenform, die in D. eng mit der städtebaulichen Entwicklung gekoppelt ist. **K.flächen** sind in Bebauungsplänen ausgewiesen und gesichert. Durch äußeres Erscheinungsbild, rechtliche Sicherung und planerische Behandlung ist der **K.** ein Dauergarten; cf. RICH 1981. Kleingartenrechtliche Regelungen sind im Bundeskleingartengesetz [BKleingG vom

C

28.03.1983] dargelegt. Seit Ende der 1990er-Jahre zeichnet sich eine Entwicklung ab, bei der die Bereitschaft der Städte, insbesondere in den ostdeutschen Bundesländern, gesunken ist, **K.flächen** zu Lasten höherwertiger Nutzungen in Bebauungsplänen auszuweisen; ►Grabelandparzelle, ►Kleingartenwesen, ►Mietergarten, ►Mietgarten, ►Pachtgarten); *syn.* Dauerkleingarten [m], Familiengarten [m] [CH], Heimgarten [m], Laubengarten [m], Schrebergarten [m], *obs.* Arbeitergarten [m].

community garden area [n] [US] *(leg.) urb.* ►permanent community garden area [US], ►temporary community garden area [US].

community garden club [n] [US] *sociol.* ►allotment garden club [UK].

936 community garden development plan [n] [± US] *leg. urb.* (Overall plan which analyzes the requirement for community gardens [US]/allotment garden plots [UK] and shows their spatial distribution as a basis for the ►community development planning [US]/urban land-use planning [UK] of a whole community; ►comprehensive plan [US]/Local Development Framework [LDF] [UK], ►community garden development planning [US]/allotment garden development planning [UK]); *syn.* allotment garden subject plan [n] [± UK], allotment garden development plan [n] [± UK]; **s plan [m] de desarrollo de huertos recreativos urbanos** (En el marco del ►planeamiento urbanístico, estudio de demanda y de localización de ►huertos recreativos urbanos a nivel municipal, como base para la realización de ►Plan General Municipal de Ordenación Urbana; ►planificación de huertos recreativos urbanos); *f* **schéma [m] d'aménagement des jardins ouvriers** (±) (►*Planification urbaine en D.* document d'urbanisme présentant les besoins en lotissements de jardins ouvriers, définissant les hypothèses d'aménagement et les propositions de localisation et expression d'une planification sectorielle dans le cadre du « *Flächennutzungsplan* » [►schéma de cohérence territoriale allemand] ; ►planification des jardins ouvriers) ; *g* **Kleingartenentwicklungsplan [m]** (Im Rahmen der ►Bauleitplanung die Erarbeitung eines Planwerkes, das den Bedarf an Kleingartenanlagen und die räumliche Zuordnung als Grundlage für den ►Flächennutzungsplan für eine gesamte Gemeinde darstellt; ►Kleingartenleitplanung).

937 community garden development planning [n] [US] (≠) *recr. urb.* (Process accompanying ►community development planning [US]/urban land-use planning [UK] involving an analysis of the requirement and space allocation of ►community gardens [US]/allotment garden [plots] [UK] within a built-up area; ►community garden development plan [US]/allotment garden development plan [UK]); *syn.* allotment garden development planning [n] [± UK]; **s planificación [f] de huertos recreativos urbanos** (En el marco del ►planeamiento urbanístico, determinación de la demanda y de la distribución espacial de las superficies asignadas para ►huertos recreativos urbanos dentro de la ciudad; ►plan de desarrollo de huertos recreativos urbanos); *f* **planification [f] des jardins ouvriers** (**D.**, instrument de la ►planification urbaine prenant en considération les besoins en jardins ouvriers ainsi que leur répartition et leur localisation dans la composition urbaine ; ►schéma d'aménagement des jardins ouvriers) ; *g* **Kleingartenleitplanung [f]** (Planungsinstrument der Stadtplanung im Rahmen der ►Bauleitplanung den Bedarf und die räumliche Anordnung von Kleingartenanlagen innerhalb des Stadtgebietes zu ermitteln; ►Kleingartenentwicklungsplan).

938 community gardener [n] [US] *urb. sociol.* (Person renting an community garden plot [US]/allotment garden plot [UK]; ►community garden); *syn.* allotment gardener [n] [UK], allotment holder [n] [UK], city plotholder [n] [UK], urban gardener [n] [US] (TGG 1984, 121); **s arrendatario/a [m/f] de huerto recreativo urbano** (Persona que dispone de un ►huerto recreativo urbano); *f* **exploitant [m] d'un jardin ouvrier** *syn.* utilisateur [m] d'un ►jardin ouvrier ; *g* **Kleingärtner/-in [m/f]** (Besitzer eines ►Kleingartens); *syn.* Kleingartenbesitzer/-in [m/f], *obs.* Schrebergärtner/-in [m/f].

939 community gardening [n] [US] *sociol. urb.* (**1.** Term covering all possible legal, planning, and organizational measures associated with the provision and maintenance of community gardens [US]/allotment garden (plots) [UK]. **In U.S.,** the first community gardens were established in the early 1970s by committed citizens who took over abandoned lots in the poorest quarters of New York; by 1999 their number had reached over 1,000. Usually from 100 to 700m² in size they are looked after by 25,000 to 30,000 gardeners and are sanctioned by the authorities as a temporary use. Nowadays there are over 38 cities in the USA with community gardens. The community gardeners have their own periodical "American Community Gardening Association". In comparison to Germany, less than 5% of all community gardens are statutorily designated. **2. In Europe,** term covering all possible legal, planning, and organizational measures associated with the provision and maintenance of community gardens [US]/allotment garden [plots] [UK]; originally in the middle of the 19th century allotment gardens were derived from the "garden for the poor" and "labo[u]r gardens" after the ideas of the German doctor Dr. SCHREBER [1808-1861], Leipzig, and the Red Cross with the aim of providing garden produce to poor people. In addition the gardens were intended to improve the physical fitness of the children and to make them familiar with nature as part of public health policy; ►guerilla gardening); *syn.* allotment gardening [n] [UK]; **s 1 jardinería [f] comunitaria** (Movimiento ciudadano surgido en los años 1970 en Nueva York de ocupar terrenos baldíos existentes entre los edificios y crear pequeños jardines de uso común del vecindario; a partir de la gran metrópolis se expandió a otras ciudades de los EE.UU.); **s 2 gestión [f] de huertos recreativos urbanos** (Conjunto de medidas legales, de planificación e implementación y de actividades relacionadas con el uso de huertos recreativos urbanos, sean públicas o colectivas); *f* **activités [fpl] concernant les jardins ouvriers** (Terme désignant l'ensemble des mesures juridiques, réglementaires, opératoires et de gestion en faveur de la préservation, de la restauration et du développement des lotissements de jardins ouvriers. Le développement des jardins ouvriers a correspondu à l'avènement de la sociéte industrielle et d'une certaine conception paternaliste de LE PLAY et des Hygiénistes, d'abord en Angleterre [les « champs des pauvres » institués en 1819 destinés aux indigents] puis en Allemagne [à l'initiative du médecin SCHREBER — d'où le terme les « jardins de Schreber » comme syn. de « ►*Kleingarten* »] ensuite en France ou le patronat minier et sidérurgique imposa un urbanisme horizontal en associant logement et jardin [coron, cité ouvrière]. Les lotissements-jardins se développèrent un peu plus tard à la fin du XIXème et au début du XXème siècle, époque où sont promulguées les premières lois sociales en matière de logement ; vers 1890, la Ligue du coin de terre et du foyer, animée par l'abbé LEMIRE, développera les jardins désormais appelés jardins ouvriers par le Docteur LANCRY qui lance le mouvement du « **terrianisme** » ; DUA 1996, 433 ; la gestion des jardins ouvriers est réalisée par des associations privées déclarées et reconnues d'utilité publique ; l'usage commercial du terrain est proscrit par la loi en vigueur depuis 1901 ; *syn.* gestion [f] des jardins familiaux) ; *g* **Kleingartenwesen [n, o. Pl.]** (Sämtliche juristischen, planerischen und organisatorischen Betriebsmaßnahmen, die die Ausstattung und Haltung von Kleingartenanlagen ermöglichen. Der Ursprung des **K.s** geht zurück auf die „Armengärten" und „Arbeitergärten" aus der Mitte des 19. Jhs., die nach den Ideen des Leipziger Arztes Dr. SCHREBER [1808-1861] und des Roten Kreuzes eingerichtet wurden. Das damalige Ziel war eine bessere Versorgung armer Bevölkerungsschichten mit Gartenfrüchten, andererseits sollte die körperliche Ertüchtigung und die Heranziehung der Kinder an die Natur gefördert sowie die allgemeine Gesundheitspolitik unterstützt werden. In den USA gibt es seit Anfang der 1970er-Jahre durch die Besetzung von innerstädtisch brach gefallenen Flächen oder Trümmergrundstücken durch engagierte Bürger in den ärmsten Stadtteilen New Yorks Kleingärten, sog. *community gardens*, die bis 1999 auf über 1000 angestiegen sind. Diese Form der i. d. R. 100-700 m² großen „Kleingärten" [Nachbarschaftsgärten] werden durch ca. 25 000-30 000 *gardeners* betreut und werden offiziell nur für eine temporäre Nutzung geduldet; cf. G+L 1999, 36 ff; mittlerweile gibt es in über 38 amerikanischen Städten *community gardens* und eine Verbandszeitschrift *American Community Gardening Association*. Im Vergleich zu D. sind in den Vereinigten Staaten nur verschwindend wenige *community gardens* [< 5 %] in der Bauleitplanung dauerhaft gesichert).

community garden plot [n] [US] *leg. urb.* ►temporary community garden plot.

940 community garden shelter [n] [US] *urb.* (Small, mostly wooden utility shed within a ►community garden [US]/allotment garden [plot] [UK] which usually serves as a summer house and for the storage of tools. A small structure used only for storage of tools is a 'garden shed'); *syn.* allotment garden hut [n] [UK]; *s* **casita [f] de huerto recreativo urbano** (►huerto recreativo urbano); *f* **abri [m] familial** (Petite structure souvent en bois dans un lotissement de ►jardins ouvriers constituée en général d'une pièce de repos et d'un espace de rangement des outils de jardinage) ; *syn.* cabane [f] de jardin, cabanon [m] ; *g* **Kleingartenlaube [f]** (Kleines Häuschen in einem ►Kleingarten, meist aus Holz, das in der Regel aus einem Aufenthaltsraum und einem Geräteraum besteht).

community importance [n] *conserv. ecol.* ►site of community importance.

community interest [n] *conserv. ecol.* ►natural habitat of community interest.

community involvement [n] *leg. plan. pol.* ►public participation.

941 community land [n] [US] *plan.* (Public property belonging to local authority, community, charitable or nonprofit organization; **in U.S.**, to **c. l.** belongs also land of charitable or nonprofit organizations); *syn.* local authority land [n] [UK] (±); *s* **patrimonio [m] municipal de suelo y edificios** (*Términos específicos* ►reserva de suelo, patrimonio inmobiliario municipal); *syn.* propiedad [f] municipal de suelo y edificios; *f* **propriété [f] de la commune** *termes spécifiques* patrimoine foncier communal, patrimoine immobilier communal ; *syn.* propriété [f] communale, patrimoine [m] communal ; *g* **Gemeindebesitz [m]** (Liegenschaften und Gebäude, die einer Gemeinde gehören); *syn.* Gemeindeeigentum [n].

community open space [n] [UK] *plan.* ►common open space [US].

community planner [n] [US] *prof.* ►urban planner [US]/town planner [UK].

community planning [n] [US] *urb.* ►urban planner [US]/town planner [UK].

942 commuter [n] *sociol. trans. urb.* (Person living in one area and travelling to another area to work or for other purposes [►commuting]; generic term for ►outgoing commuter and

►incoming commuter); *s* **conmutador, -a [m/f]** (Persona que se traslada diariamente a otro lugar para trabajar o estudiar [►migración pendular]; término genérico para ►conmutador, -a saliente y ►conmutador, -a entrante); *f* **migrant [m] alternant** (Personne effectuant des déplacements périodiques [►migration alternante] entre son lieu de domicile et son lieu de travail, de formation ou d'études ; ►migrant alternant considéré du point du vue de la collectivité émettrice, ►migrant alternant considéré du point du vue de la collectivité réceptrice ; *syn.* pendulaire [m] [CH] ; *g* **Pendler, -in [m/f]** (Person, die sich täglich zwischen zwei Orten hin- und herbewegt [►Pendeln], besonders zwischen dem Wohnort und dem Arbeits-, Schul- oder Studienort; OB zu ►Auspendler und ►Einpendler).

commuter belt [n] *plan.* ►origination area [US].

943 commuter flow [n] *sociol. trans. urb.* (Large number of ►commuters moving in one direction, either in public transportation vehicles or in private cars); *syn.* commuter movement [n]; *s* **corriente [f] de migración pendular** (Gran cantidad de personas que se trasladan en transporte público o en automóviles particulares a diario de un lugar a otro; ►conmutador, -a); *syn.* corriente [f] de migración habitual; *f* **flux [m] migratoire journalier** (Population importante se déplaçant d'un point à un autre au moyen de transports individuels ou collectifs ; ►migrant alternant) ; *g* **Pendlerstrom [m]** (Eine sich in eine Richtung bewegende Menschenmenge, sei es mit öffentlichem Nahverkehrsmittel oder Auto; ►Pendler).

commuter lane [n] *trans.* ►offside lane, #2.

commuter movement [n] *sociol. trans. urb.* ►commuter flow.

944 commuter pattern [n] [US] *sociol. trans. urb.* (Daily movements of working people, students or pupils from their home to their working places, universities or schools; ►commuter flow); *syn.* journey-to-work pattern [n] [UK]; *s* **movimientos [mpl] migracionales pendulares** (Estructura de migración pendular de la población en una zona a sus lugares de trabajo o estudio; ►corriente de migración pendular); *syn.* movimientos [mpl] migracionales habituales; *f* **déplacements [mpl] domicile-travail** (Déplacements quotidiens des salariés, étudiants et élèves vers les lieux de travail, les établissements universitaires et scolaires ; ►flux migratoire journalier) ; *syn.* migrations [fpl] quotidiennes, migrations [fpl] alternantes journalières (PR) ; *g* **Pendlerbewegungen [fpl]** (Tägliche Bewegungen der Berufstätigen [Berufspendler], Studenten und Schüler vom Wohnort zur Arbeitsstelle/Hochschule/Schule; ►Pendlerstrom).

945 commuting [n] *sociol. trans. urb.* (Movement of people going to work or to study from their place of residence to another community; ►commuter); *s* **migración [f] pendular** (Desplazamiento de una persona entre su lugar de residencia y su lugar de trabajo; DGA 1986; ►conmutador, -a); *syn.* desplazamiento [m] pendular; *f* **migration [f] alternante** (Ensemble des déplacements journaliers entre le domicile et le lieu de travail ou de formation d'un territoire à un autre ; ►migrant alternant) ; *syn.* déplacement [m] pendulaire, navette [f] domicile-travail, migration [f] régulière du travail, migration [f] quotidienne ; *g* **Pendeln [n, o. Pl.]** (Tägliches Hin- und Herfahren der arbeitenden oder zur Ausbildung gehenden Menschen von ihrem Wohnort in eine andere Gemeinde; ►Pendler).

compacted granular surface [n] [US] *constr.* ►waterbound surface.

compacting [n] **in layers** *constr.* ►filling and compacting in layers.

946 compact [vb] **in layers** *constr.* (Process of consolidating successive soil layers or lifts (150-300mm) to achieve a specified density for a desired bearing or permeability rating);

s **compactar [vb] por tongadas** (Proceso de consolidar sucesivamente el suelo por capas para conseguir una densidad específica para alcanzar la capacidad de carga o la permeabilidad deseadas); *f* **compacter [vb] en couches** ; *g* **lagenweise verdichten [vb]** (Erdbauwerke in Schichten bis zu 30 cm Dicke in einen Zustand größerer Dichte bringen, um die ausgeschriebene Tragfähigkeit zu erreichen).

compaction [n] *constr.* ▶ground compaction, ▶proctor compaction, ▶soil compaction, ▶subsoil compaction.

947 compaction capacity [n] *constr. eng.* (The degree to which a soil may be compressed using mechanical means to achieve a specified density, bearing strength or stability factor—dependent upon ▶particle-size distribution and water content; ▶granular soils require vibration and weight, while colloidal soils [clays] require non-vibrating pronged [sheep's foot] rollers to achieve a specified uniform density; ▶soil structure, particle-size distribution and moisture content determine both the capacity for compaction as well as the most suitable compaction method); *s* **compactabilidad [f] (de suelos)** (Grado máximo de consolidación de suelos que depende de la ▶granulometría y el contenido de agua); *f* **compacité [f] maximale** (Compactage maximal d'un sol support de chaussée ou d'un remblai compte tenu de sa teneur en eau et de son ▶spectre granulométrique ; permet de déterminer la portance d'un sol, sa capacité à supporter les charges qui lui sont appliquées) ; *g* **Verdichtbarkeit [f]** (Von ▶Kornverteilung und vom Wassergehalt abhängige höchst mögliche Verdichtung von Bodenarten).

948 companion plant [n] (1) *phyt.* (▶Indifferent species without pronounced affinities for any plant community; e.g. an ubiquist, which flourishes in quantity and is able to compete in very different plant communities); *s* **especie [f] acompañante (1)** (▶Especie indiferente sin especial afinidad con ninguna comunidad vegetal, como las ubicuistas, que se presenta en grandes cantidades y es capaz de competir en diferentes comunidades vegetales); *syn.* acompañante [f]; *f* **espèce [f] compagne** (Espèce croissant plus ou moins abondamment dans plusieurs groupements végétaux ; espèce qui n'est caractéristique d'aucun groupement végétal ; BB 1928 ; ▶espèce indifférente) ; *syn.* espèce [f] adventice, compagne [f] ; *g* **Begleiter [m] (1)** (Pflanzenart ohne ausgesprochenen Gesellschaftsanschluss, die aber öfter oder in Menge in unterschiedlichsten Pflanzengesellschaften vorkommt; ▶vage Art).

949 companion plant [n] (2) *agr. hort.* (Mutually beneficial plant in vegetable gardens, allowing maximum light, moisture and soil, and sometimes with insect-attracting or repellent properties); *s* **planta [f] acompañante (2)** (Planta utilizada en cultivo hortícola que favorece el crecimiento de otra por sus exigencias edáficas o porque repele a insectos nocivos); *f* **plante [f] associée** (Plante cultivée en association avec d'autres plantes de culture dans un jardin potager ou dans les cultures maraîchères et attirant les insectes utiles ou repoussant les déprédateurs) ; *g 1* **Begleiter [m] (2)** (Pflanze, die zusammen mit anderen Kulturpflanzen in einem Gemüsegarten wächst und nützliche Insekten anzieht oder Schädlinge abweist); *g 2* **Begleitstaude [f]** *hort. plant.* (In einer Rabatte der ▶Leitstaude oder der ranghöchsten Dominante zugeordnete Pflanze, die für den Wechsel des Erscheinungsbildes sorgt, um somit die Aufmerksamkeit des Betrachters wach zu halten).

950 comparative analysis [n] **of bid items** [US] *constr. contr.* (Tabular comparison of bid items [US]/tender items [UK] upon receipt of the contractors bids [US]/tenders [UK] in order to be able to make a comprehensive analysis of unit prices; ▶price list); *syn.* comparative analysis [n] of tender items [UK]; *s* **estudio [m] comparativo de precios** (Listado de los precios

unitarios de diferentes ofertas con ayuda de una tabla para facilitar la evaluación comparativa de las mismas; ▶lista de precios); *f* **étude [f] comparative des prix** (Présentation sous forme de tableau des prix des différents lots d'ouvrages [par corps d'état] remis par les entreprises concurrentes et classement des offres afin de proposer les offres susceptibles d'être retenues et de dégager l'offre la plus intéressante ; ▶série des prix de l'administration publique, ▶série des prix) ; *g* **Preisspiegel [m]** (Tabellarische Gegenüberstellung der Preise für Teilleistungen [einzelner Gewerke] nach Eingang der Unternehmerangebote, um einen umfassenden Kostenvergleich zu ermöglichen; ▶Preisverzeichnis).

comparative analysis [n] **of tender items** [UK] *constr. contr.* ▶comparative analysis of bid items [US].

951 comparison [n] **of alternative routes** *plan. trans.* (Planning investigation of various transportation routes with the aim of selecting the best option within a specified assessment framework); *syn.* comparison [n] of alternative corridors [also UK]; *s* **comparación [f] de trazados** (Comparación de alternativas de rutas de carretera o autopista); *syn.* comparación [f] de alineaciones; *f* **étude [f] de variantes** (Analyse globale des variantes d'un projet routier au niveau de l'environnement afin d'expliciter et de justifier le choix de la solution retenue) ; *g* **Variantenvergleich [m]** (Planerische Untersuchung von verschiedenen Verkehrstrassen oder einer Trassenalternative).

952 compartmentalization [n] **of a landscape** *agr.* (Structuring or enclosure of fields by, e.g. ▶elevated hedgerows, ▶ditch corridor tree planting, wood patches, etc.); *syn.* patterning [n] of a landscape; *s* **compartimentación [f] del paisaje (≠)** (Estructuración tridimensional del paisaje agrario por medio de ▶setos vivos elevados y plantaciones a lo largo de las riberas de cursos de agua [▶plantación-orla de canal]); *f* **parcellisation [f] des paysages** (Fragmentation des grands espaces agricoles par des ▶haies sur talus, des ▶plantations de lisière de fossés) ; *syn.* compartimentation [f] des paysages, compartimentage [m] des paysages ; *g* **Landschaftskammerung [f]** (Durch ▶Knicks und ▶Grabensaumpflanzungen verursachte dreidimensionale Gliederung der bäuerlichen Kulturlandschaft).

compass brick [n] [US] *constr.* ▶concentric manhole brick.

953 compensating function [n] *ecol.* (Role played by ecosystem components or subsystems in reestablishing the ecological balance within a larger ecosystem which has been damaged by local alterations); *s* **función [f] compensatoria** (Función a jugar por los componentes de un ecosistema para reestablecer el balance ecológico dentro de un ecosistema mayor que ha sido dañado por alteraciones locales); *syn.* función [f] reequilibrante; *f* **fonction [f] rééquilibrante** (Rôle joué par certains facteurs d'un écosystème afin de rétablir l'équilibre écologique perturbé par des atteintes locales) ; *g* **Ausgleichsfunktion [f]** (Aufgabe/Rolle von Systemteilen oder Subsystemen, um einen Zustand, der innerhalb eines Gesamt[öko]systems durch beeinträchtigte Bereiche gestört wurde, [wieder]herzustellen).

954 compensation [n] *leg.* (Payment made to owner for property subject to condemnation [US]/compulsory purchase [UK] or to make up for reduction in land value resulting from planning decisions; **in U.S.**, "compensation for planning blight" is referred to as "inverse condemnation compensation," and the value is determined by judicial proceedings; ▶property devaluation caused by planning, ▶eminent domain [power of condemnation]); *s* **indemnización [f]** (Pago en efectivo de un justiprecio, o permuta de parcelas, a propietarios particulares sujetos a ▶expropiación forzosa de terrenos para la realización de un plan; ▶daños y prejuicios de planificación); *f* **indemnité [f]** (Dédom

C

magement financier, transaction de biens de même valeur ou obtention de droits équivalents pour réparation du préjudice direct, matériel et certain causé par une ►expropriation, une réglementation d'une activité ou une diminution de valeur d'un bien ou d'un fond provoquées par des aménagements publics ; ►préjudice causé par des documents d'urbanisme) ; *g* **Entschädigung** [f] (Gesetzlich geregelte Zahlung oder Gewährung von wertgleichen Rechten bei ►Enteignung oder [Nutzungs]beeinträchtigungen durch öffentliche Planungen; ►Planungsschaden).

955 compensation area [n] (1) *conserv. leg.* (Area established to remedy unavoidable residual adverse impacts achieved by replacement of the same resource values at another location: compensation can also be achieved by rehabilitation of the affected site or environment or restoration of the affected site or environment to its previous state or better. **In U.K.**, the Town and Country Planning [Environmental Impact Assessment (EIA)] Regulations, including the provision for **c. a.**, were introduced in 1999, to govern environmental considerations in land use planning and development consent procedures; ►intrusion upon the natural environment, ►impact mitigation, ►alleviation measure, ►mitigation measure); *s* **área** [m] **de compensación** (En el contexto de la ►evaluación de impacto ambiental, superficie localizada cerca del emplazamiento del proyecto estudiado, definida como ►medida compensatoria de los impactos negativos del mismo y en la que se pueden realizar acciones de restauración, o de la misma naturaleza y efecto contrario a los del proyecto planeado ; cf. art. 11 RD 1131/1988; ►compensación [de un impacto], ►intrusión en el entorno natural, ►medida correctora, ►medida mitigante, ►medida sustitutiva, ►regulación de impactos sobre la naturaleza); *f 1* **surface** [f] **compensatoire** (1. Aire, site d'un projet sur lequel des travaux, des pratiques de gestion [conservatoire et/ou restauratoire] visant à offrir une contrepartie, une contrebalance aux atteintes et effets négatifs non réductibles pour l'environnement ou créateurs de nuisances pour l'homme provoqués par un projet ; cette disposition s'applique lorsqu'on a échoué à supprimer ou atténuer les impacts négatifs directs ou indirects d'un projet [►atteinte au milieu naturel] ; sur cette aire doit donc théoriquement être rétabli une situation d'une qualité globale proche de la situation antérieure ou un état de l'environnement jugé fonctionnellement *normal* ou *idéal* [La directive cadre sur l'Eau évoque par exemple le « bon état écologique » des milieux aquatiques] ; *terme spécifiques* boisement compensatoire, ►mesure de réduction, ►mesure de suppression ; *termes spécifiques* site compensatoire, boisement compensatoire [CDN] ; ►mesure de réduction, ►compensation des conséquences dommageables sur l'environnement, ►réglementation relative aux atteintes subies par le milieu naturel. **2.** À ne pas confondre avec le terme suivant ►surface de compensation écologique [CH]) ; *f 2* **surface** [f] **de compensation écologique** [CH] (*Abrév.* SCE ; afin de compenser la perte d'espaces naturels et de stopper la disparition des espèces lors des 50 dernières années, la Suisse a choisi depuis 1993 d'introduire la notion de compensation écologique dans le cadre d'une nouvelle politique agricole ; les exploitants agricoles sont tenu de mettre en place sur leurs terres des SCE telles que prairies extensives, pâturages extensifs, pâturages boisés, prairies peu intensives, ►praires à litières, bandes culturales extensives, ►jachères fleuries, ►jachères tournantes, fossés humides, mares, étangs, arbres fruitiers haute tige, ►praires complantées d'arbres fruitiers, arbres isolés indigènes adaptés au site et allées d'arbres, haies, ►rideaux, ►talus embroussaillés, bosquets champêtres et berges boisées, surfaces rudérales, tas d'épierrage et affleurements rocheux, murs de pierres sèches, chemins naturels non stabilisés, surfaces viticoles présentant une biodiversité naturelle ;

depuis 1999 les agriculteurs consacrant au moins 7 % de leur exploitation à la compensation écologique pour bénéficier de paiements directs ; avec l'ordonnance sur la qualité écologique de 2001, la mise en réseau des SCE peut donner droit à des contributions ; cf. DZV 1998 et ÖQV 2001) ; ►mesure de substitution ; *g 1* **Ausgleichsfläche** [f] (**D.**, Fläche, auf der gem. § 8a BNatSchG i. V. m. §§ 5 [2] 10 und 9 [1] 20 BauGB ein ►Eingriff in Natur und Landschaft, der negative Wirkungen auf die Umwelt verursacht [z. B. Versiegelung, Biotopverlust, nachhaltige Veränderung des Landschaftsbildes] durch Maßnahmen des Naturschutzes und der Landschaftspflege ausgeglichen werden muss; ►Ausgleich [eines Eingriffes], ►Eingriffsregelung, ►Ersatzmaßnahme [des Naturschutzes und der Landschaftspflege], ►Vermeidungsmaßnahme, ►Verminderungsmaßnahme) ; *g 2* **ökologische Ausgleichsfläche** [f] [CH] (*Abk.* ÖAF; in der Schweiz gibt es in Ergänzung zu ausgewiesenen Naturschutzflächen ÖAFs, die Tieren und Pflanzen in landwirtschaftlich intensiv genutzten Gebieten zusätzliche Lebensräume zur Förderung der Artenvielfalt, Vermeidung weiterer Artenverluste und Wiederausbreitung bedrohter Arten schaffen; seit 1993 wurde im Rahmen einer reformierten Agrarpolitik der Begriff ÖAF eingeführt; seitdem sind Landwirte gehalten, extensiv genutzte Wiesen, extensiv genutzte Weiden, Waldweiden, wenig intensiv genutzte Wiesen, ►Streuwiesen/Streuflächen [CH], Ackerkrautstreifen, Feldränder, ►Buntbrachen, ►Wanderbrachen, Gräben, Tümpel, Teiche, ►Streuobstwiesen, standortgerechte Feld- und Ufergehölze, Baumalleen, Hecken, ►Hochraine, ►Stufenraine, Ruderalflächen, Lesesteinhaufen, Trockenmauern, natürliche wassergebundene Wege, Rebflächen mit hoher Artenvielfalt etc. auf ihren Flächen als ÖAF zu pflegen; seit 1999 werden Landwirte, die mindestens 3,5 % der mit Spezialkulturen belegten landwirtschaftlichen Nutzflächen und 7 % der übrigen landwirtschaftlichen Nutzflächen des Betriebes als ökologische Ausgleichsflächen bewirtschaften, vom Gesetzgeber mit Direktzahlungen entgolten; cf. Art. 7 DZV 1998. Seit dem Inkrafttreten der Öko-Qualitätsverordnung 2001 sollen die ökologischen Ausgleichsflächen durch Vernetzung aufgewertet werden und können unter bestimmten Bedingungen gefördert werden; cf. ÖQV 2001); *syn. zu 1.* Kompensationsfläche [f].

compensation area [n] (2) *plan.* ►open space compensation area.

compensation measure [n] *conserv. leg.* ►environmental compensation measure.

956 compensation payment [n] [UK] *conserv. leg.* (Fixed sum of money to be paid to a legally defined public authority, if permission has been granted for an ►intrusion upon the natural environment of a construction project with greater priority, and if an ►impact mitigation or an ►environmental compensation measure of equal value is not possible; **in U.S.**, a developer is required to pay for mitigation on a case by case basis; ►serious intrusion); *s* **indemnización** [f] **compensatoria** (Cantidad de dinero a pagar en el caso de que al realizar una obra, considerada impactante al medio no sean posibles ►medidas correctoras o ►medidas sustitutivas; cf. Ley de Protección de la Naturaleza de Baden-Württemberg § 11 [5] y de Hamburgo § 9 [6]; ►compensación [de un impacto], ►intrusión en el entorno natural, ►intrusión significativa); *syn.* tasa [f] de compensación); *f* **redevance** [f] **compensatoire** (≠) (**F.**, inexistant. **D.**, une somme fixée par l'administration sur toute opération constituant une ►atteinte notable à l'environnement correspondant aux dispositions de protection de l'environnement ou aux ►mesures de substitution ne pouvant être réalisées dans les limites fixées dans le cadre de l'aménagement ou du plan ; cf. loi de protection de la nature des Länder de Baden-Württemberg et de Hambourg ; ►compensation des conséquences dommageables sur l'environ-

nement) ; *syn.* taxe [f] compensatoire (≠) ; *g* **Ausgleichsabgabe [f]** (Ein mit der Gestattung eines ►erheblichen Eingriffs festgesetzter Geldbetrag, der an eine im Gesetz näher bestimmte Behörde zu entrichten ist, wenn der ►Ausgleich [eines Eingriffes] oder ►Ersatzmaßnahmen nicht möglich sind; cf. § 11 [5] NatSchG-BW, § 9 [6] NatSchG-HH; die Gelder sind zweckgebunden für Maßnahmen des Naturschutzes und der Landschaftspflege zu verwenden; ►Eingriff in Natur und Landschaft); *syn.* Ersatzzahlung [f].

competent authority [n] [UK] *adm.* ►operating authority.

957 competing land-use pressure [n] *plan.* (LE 1986, 303; existing or presumed ►land-use requirements and opposing land use demands on the land; *specific terms* ►recreation load, ►development pressure, ►visitor pressure); *syn.* competitive land-use pressure [n], pressure [n] of competitive land-use; *s* **presión [f] por competencia entre usos** (Alta ►demanda de suelo [para un uso] o diferentes usos en una misma área; *términos específicos* ►presión recreacional, ►presión turistica, ►presión urbanizadora); *f* **pression [f] exercée par les usages du sol** (►Exigence d'utilisation du sol existante ou prévisible sur tout ou parties de paysages ; *termes spécifiques* ►pression de l'urbanisation, ►pression exercée par les visiteurs, ►pression récréative, ►pression touristique) ; *g* **Nutzungsdruck [m, o. Pl.]** (Vorhandene oder voraussichtliche ►Nutzungsansprüche an die Landschaft oder deren Teile; *UBe* ►Besucherdruck, ►Erholungsdruck, ►Siedlungsdruck).

958 competition [n] *ecol.* (Rivalry for food, space or other ecological requirements between organisms, whose demands correspond with one another without being sufficiently *met*. Their ability to reproduce or to survive is not impaired by mutually disruptive intrusions. C. can be ►interspecific or ►intraspecific; ►life form); *s* **competencia [f]** (Tipo de interacción negativa entre varias especies o poblaciones que compiten en la utilización de los mismos recursos [alimentarios o espaciales], que, o bien son escasos, o bien los organismos compiten en su búsqueda. La **c.** puede manifestarse de dos formas: por interferencia directa entre las poblaciones concurrentes, en las que una especie impide a la otra el acceso al alimento, o por explotación, en la que una especie, al explotar un determinado recurso, disminuye la disponibilidad de éste para la otra; cf. DINA 1987; ►forma biológica ►interespecífico/a, ►intraespecífico/a); *f* **compétition [f]** (Concurrence qui s'établit pour une même source d'énergie et de matière soit entre individus d'une même espèce [compétition ►intraspécifique], soit entre individus de plusieurs espèces [compétition ►interspécifique] ; DEE 1982 ; la nature du ►type biologique [caractéristiques de développement qui conduisent à la forme naturelle d'une espèce] peut être décisif dans la compétition des espèces) ; *g* **Konkurrenz [f]** (**1.** Wettbewerb um Nahrung, Licht, Raum oder andere ökologische Erfordernisse zwischen zwei Organismen, die in ihren Lebensansprüchen übereinstimmen, ohne dass diese sich in ausreichendem Maße erfüllen lassen oder dass diese durch gegenseitige Störeffekte Fortpflanzung und Überleben beeinträchtigten. **2.** *bot. phyt.* Bei Pflanzen kann sich im natürlichen **K.kampf** in einem Bestand langfristig nur diejenige durchsetzen [durchsetzungsstarke Art], die auf einem Standort rasch eine große Wuchshöhe und Breite gewinnt und somit dauerhaft besteht; die unterlegene Art wird verdrängt. Für den Kampf ums Dasein kann die ►Lebensform entscheidend sein, die sich durch Besonderheiten des Wachstums und der Wuchsform äußert. Die **K.** kann ►interspezifisch oder ►intraspezifisch sein; cf. ÖKO 1983).

competition [n] [US] *prof.* ►final design competition; *ecol.* ►pressure of competition; *plan. prof.* ►submission deadline for design competition.

competition [n] [UK]**, realization** *prof.* ►final design competition [US].

competition [n]**, root** *phyt.* ►rooting stress.

959 competition area [n] *prof.* (Project area for an ►ideas competition or ►final design competition [US]/realization competition [UK]); *syn.* competition lands [npl] [also US]; *s* **área [f] de concurso** (Zona definida sobre la cual se lleva a cabo un ►concurso de ideas o ►concurso de realización); *syn.* recinto [m] de concurso; *f* **zone [f] de réalisation du concours** (±) (Aire définie faisant l'objet d'un ►concours d'idées ou de ►concours de conception-réalisation) ; *g* **Wettbewerbsgebiet [n]** (Definiertes Gebiet für eine gestellte Planungs- oder Bauaufgabe im Rahmen eines ►Ideenwettbewerbes oder ►Realisierungswettbewerbes).

960 competition brief [n] *prof.* (Descriptive project summary of an ►initiating authority or ►sponsor of a competition who wishes to obtain design ideas in the fields of physical planning, urban planning, architecture or landscape design, or solutions which can be implemented); *syn.* competition task [n]; *s* **programa [m] de concurso** (Descripción de proyecto hecha por un ►promotor, -a de concurso 1 sobre cuya base desea obtener proposiciones de solución. Los concursos se aplican en los campos de ordenación del territorio, urbanismo, construcción, planificación del paisaje y paisajismo; ►promotor, -a de concurso 2); *f* **programme [m] du concours** (Description du projet de concours par le ►pouvoir adjudicateur d'un concours dans les domaines de l'aménagement du territoire, l'urbanisme, la construction, la planification des paysages, qui définit les données recueillies, les besoins à satisfaire, les contraintes et les exigences à respecter ; CCM 1984 ; ►promoteur du concours) ; *g* **Wettbewerbsaufgabe [f]** (Definierte Absicht eines Auslobers [►auslobende Behörde/Organisation oder ►privater Auslober] auf den Gebieten der Raumplanung, des Städtebaus, des Bauwesens oder der Landschaftsarchitektur/Landschaftsplanung Gestaltungsideen oder baureife Lösungen zu erhalten).

competition documents [npl] [UK] *prof.* ►competition program package [US].

961 competition entrant [n] *prof.* (Design firm which takes part in a design competition); *syn.* competitor [n], competition participant [n]; *s* **concursante [m/f]** *syn.* participante [m/f] en concurso; *f* **concurrent [m] (d'un concours)** (Personne participant à un concours d'idée ou à un concours de conception-construction) ; *syn.* candidat [m] (sélectionné) au concours ; *g* **Wettbewerbsteilnehmer/-in [m/f]** (Jemand, der am Ideen- oder Realisierungswettbewerb teilnimmt).

962 competition entry [n] *prof.* (Submitted work of a participant in a design competition or ►final design competition [US]/design competition for implementation [UK]. The act of delivering a **c. e.** is known as a 'submission'; ►submittal requirements); *syn.* submission material [n] [also UK]; *s* **trabajo [m] de concurso** (Proyecto entregado para ►concurso de ideas o ►concurso de realización); *f* **prestation [f] de concours** (Document conformes aux exigences du règlement concours remis par les participants à un ►concours d'idées ou un ►concours de conception-construction ; ►rendus de concours) ; *syn.* projet [m] de concours ; *g* **Wettbewerbsarbeit [f]** (Zu einem ►Ideenwettbewerb oder ►Realisierungswettbewerb abgegebene Arbeit; ►geforderter Wettbewerbsleistungen).

963 competition jury [n] *prof.* (Body of usually 7-11 persons, appointed by the ►initiating authority or the ►sponsor of a competition with the job of judging the competition entries. The group is composed of assessors competent in a relevant professional field and ►officials of a competition jury—may include relevant citizen representation; ►competition jury,

►final selection); *syn.* assessment panel [n] [also UK]; *s* **jurado [m]** (Gremio elegido por ►promotor, -a de concurso que tiene la función de evaluar las propuestas entregadas. Está constituido generalmente por ►miembros profesionales de jurado y ►miembros de jurado de concurso); ►selección final); *f* **jury [m]** (Commission chargée par le ►pouvoir adjudicateur d'un concours [composée en règle générale de 7 à 10 membres, les ►personnalités qualifiées membre du jury et les ►membres du jury] de juger les travaux des participants ; ►sélection définitive) ; *g* **Preisgericht [n]** (Vom Auslober bestelltes Gremium, i. d. R. 7-11 Personen, das die Aufgabe hat, Wettbewerbsarbeiten zu beurteilen. Es besteht aus ►Fachpreisrichtern und ►Sachpreisrichtern; ►auslobende Behörde/Organisation, ►engere Wahl 1, ►privater Auslober); *syn.* Preisrichtergremium [n].

competition lands [npl] [US] *prof.* ►competition area.

competition participant [n] *prof.* ►competition entrant.

competition pressure [n] *ecol.* ►pressure of competition.

964 competition program package [n] [US] *prof.* (Plans, texts, photographs, videos, or other representations and simulations necessary for the solution of a competition task, which are provided to design competition participants); *syn.* competition documents [npl] [UK]; *s* **documentación [f] básica de concurso** (Conjunto de textos, planos, material audiovisual, etc. necesario para realizar un trabajo de concurso y que son puestos a disposición de los concursantes); *f* **dossier [m] du concours** (Plans, textes, photos et vidéos mis à la disposition des participants à un concours) ; *g* **Wettbewerbsunterlagen [fpl]** (Texte, Pläne und ggf. Fotos und Videos oder andere Darstellungen und Simulationen, die für die Lösung der Wettbewerbsaufgabe notwendig sind und den Wettbewerbsteilnehmern zur Verfügung gestellt werden).

965 competition rules [npl] [US] *prof.* (Regulations which govern the holding of a design competition, the composition of the jury and the procedures for judging the competition entries); *syn.* competition standing orders [npl] [UK]; *s* **reglamento [m] de concurso** (Conjunto de reglas que rigen la realización de concursos que incluyen la composición del jurado y el proceso de evaluación); *f* **règlement [m] du concours** (Document fixant la description du projet, le dossier de candidature, la composition de jury, les critères de jugement des prestations, les paiements, etc.) ; *g* **Wettbewerbsordnung [f]** (Regularien, die die Durchführung eines Wettbewerbes auf den Gebieten der Raumplanung, des Städtebaues und des Bauwesens, die Zusammensetzung des Preisgerichtes und den Vorgang der Bewertung der abgegebenen Arbeiten bestimmen. In D. gibt es dafür die „Grundsätze und Richtlinien für Wettbewerbe" [GRW 1995, novellierte Fassung vom 22.12.2003, gültig seit 30.01.2004]).

competition standing orders [npl] [UK] *prof.* ►competition rules [US].

competition stress [n] *ecol.* ►pressure of competition.

966 competition submittal [n] *prof.* (Delivery of a ►competition entry [►submittal requirements] on a specific date); *s* **sumisión [f] de trabajo de concurso** (Entrega de ►trabajo de concurso [►requisitos de concurso] al finalizar el ►plazo de sumisión); *syn.* entrega [f] de trabajo de concurso; *f* **remise [f] du projet du concours** (Dépôt des ►prestations de concours [►rendus de concours] à une date prévue) ; *syn.* dépôt [m] du projet de concours ; *g* **Abgabe [f] der Wettbewerbsarbeit** (Einreichen einer ►Wettbewerbsarbeit [►geforderte Wettbewerbsleistungen] zu einem festgelegten Termin); *syn.* Ablieferung [f] des Wettbewerbbeitrages.

967 competition swimming facilities [npl] *recr.* (Swimming venue designed to meet official Olympic or designated competition dimensional standards, public access, and administrative requirements); *syn.* swimming meet installation [n] [also US]; *s* **instalación [f] de deportes de natación** (Centro para actividades de natación equipado con piscina olímpica y todas las demás instalaciones necesarias para cumplir lo requisitos de competiciones irnternacionales); *f* **centre [m] de natation** (Centre sportif accessible au public comprenant différents bassins olympiques homologués ou conçus pour l'organisation de compétitions de natation [bassin de compétition, bassin d'entraînement, bassin de nage synchronisée et de water-polo, bassin de plongeon] ; cet équipement est souvent accompagné d'aires de conditionnement physique ou d'une salle omnisport) ; pour les compétions internationales la piscine de compétition doit avoir de 8 à 10 couloirs et faire de 21 à 25 m x 50 m pour une profondeur minimale de 1,80 m) ; *g* **Schwimmsportanlage [f]** (Schwimmbad, das für Wettkämpfe geeignet ist. Für internationale Wettkämpfe sind Schwimmerbecken mit 8-10 Bahnen nötig, die Abmessungen von 21 bis 25 x 50 m und eine Mindesttiefe von 1,80 m haben).

competition task [n] *prof.* ►competition brief.

competitive land-use [n]**, pressure of** *plan.* ►competing land-use pressure.

competitive land-use pressure [n] *plan.* ►competing land-use pressure.

968 competitiveness [n] *ecol.* (Capacity of a species to compete for survival successfully with other species in an area); *s* **competitividad [f]** (Capacidad de una especie de imponerse ante otras en una estación); *f* **compétitivité [f]** (Potentialité de compétition d'une espèce, déterminant sa capacité d'exploiter les ressources du milieu ; *g* **Konkurrenzkraft [f]** (Vermögen einer Art, sich an einem Standort anderen Arten gegenüber durchzusetzen).

competitive tendering [n] [UK]**, compulsory** *adm. constr. contr.* ►open bidding [US].

competitor [n] *prof.* ►competition entrant.

complete canopy [n] *phyt.* ►closed canopy.

complete cutting [n] [US] *for.* ►clear felling.

969 completed work [n] [US]/**completed works** [npl] [UK] *constr. contr.* (Construction work [US]/works [UK] executed in accordance with the contract documents/►project requirements for bidding; ►issue of a certificate of compliance [US]/issue of a certificate of practical completion [UK], ►measurement of completed work [US]/measurement of completed works [UK]); *s* **trabajo [m] finalizado (en condiciones de aceptación)** (Ejecución de trabajos de construcción [paisajística] que correspondden al documento de contrato hasta la ►aceptación de la obra; ►medición de los trabajos terminados); *f* **ouvrage [m] apte à la réception** (Travaux d'espaces verts achevés conformément aux termes du ►dossier de consultation/►dossier d'appel d'offres et pour lesquels il peut être établi un certificat de ►réception ; ►attachement, ►métré définitif) ; *g* **abnahmefähige Arbeiten [fpl]** ([Landschafts]bauarbeiten, die entsprechend den ►Ausschreibungsunterlagen zur ►Abnahme fertig gestellt sind; ►Aufmaß 1); *syn.* abnahmefähige Bauleistungen [fpl], fertig gestellte Bauleistungen [fpl].

complete exploitation [n] *for.* ►clear felling.

complete felling [n] [UK] *for.* ►clear felling.

970 complete fertilizer [n] *agr. constr. for. hort.* (DIL 1987, 68; inorganic fertilizer containing the four most important macro-nutrients N, P, K, Ca as well as ►trace elements); *syn.* all-round fertilizer [n] [also UK], balanced fertilizer [n]; *s* **abono [m] complejo** (Fertilizante inorgánico mezcla de los nutrientes principales N, P, K y de ►oligoelemento; DINA 1987, 436); *syn.* abono [m] NPK; *f* **engrais [m] complet** (Par opposition aux engrais simples et binaires, engrais composé associant quatre composants [N, P, K, Ca avec magnésie et ►oligo-éléments], en général à longue durée d'action) ; *syn.* engrais [m] complexe ; *g* **Volldünger [m]** (Anorganisches Mehrnährstoffdüngemittel, das die vier wichtigsten Makronährstoffe N, P, K und Ca sowie ►Spurenelemente enthält; *opp.* Einzelnährstoffdünger); *syn.* Mehrnährstoffdünger [m] (cf. VwV DüngeVO vom 16.12.1996-BW).

completion [n] *constr. contr.* ►certificate of practical completion; *constr.* ►issue of a certificate of final completion, ►project completion [US]/**completion of works [US]**, ►site clearance after project completion; *contr.* ►time of completion.

completion [n], **bonus for early** *contr.* ►bonus payment.

completion [n] [UK], **practical** *constr.* ►project completion [US].

completion [n] [UK], **situation of practical** *constr. contr.* ►condition for final acceptance [US].

completion certificate [n] *contr.* ►issue of a sectional completion certificate.

completion [n] **of a contract** [UK] *contr. prof.* ►conclusion of a contract.

completion [n] **of work** [US]/**works** [UK] *constr.* ►project completion [US].

971 complex environmental protection [n] *conserv. pol.* (Whole range of measures and policies for nature conservation and environmental management as well as provisions for recreation, which are established to protect, maintain, and develop natural resources and compatible life conditions for plants, animals, and human beings by preventing disturbances to the ecological equilibrium on a long-term basis); *syn.* integrated environmental protection [n]; *s* **protección [f] integrada del medio ambiente** (Sistema complejo de medidas tendente a obtener el empleo racional, el mantenimiento y la restauración de los recursos naturales y la protección del medio ambiente natural contra la contaminación y otros deteriorios; DINA 1987, 233); *f* **protection [f] intégrée de l'environnement** (±) (Ensemble des opérations et efforts réalisés en faveur de la protection de la nature et des paysages ainsi que du tourisme en espace rural ayant pour objectif la préservation, la gestion et le développement durables de l'environnement naturel de la flore, la faune et de l'homme par des mesures de prévention et de lutte contre les déséquilibres écologiques) ; *g* **komplexer Umweltschutz [m, o. Pl.]** (Gesamtheit der Maßnahmen und Bestrebungen des Naturschutzes und der Landschaftspflege sowie der landschaftsbezogenen Erholungsvorsorge, um die natürlichen Lebensgrundlagen von Pflanze, Tier und Mensch durch die Verhinderung von Störungen des ökologischen Gleichgewichtes nachhaltig zu sichern, zu pflegen und zu entwickeln).

972 complexity rating classification [n] [UK] *contr. leg. prof.* (For buildings and site works, the fee to be charged is first categorized according to the degree of complexity or scope of work; in U.S. no equivalent; ►basic professional services, ►fee chart); *s* **clase [f] de complejidad** (En D. los proyectos de construcción arquitectónica o paisajística se ►tasas de remuneración para el cálculo de honorarios para los ►servicios profesionales básicos están clasificadas en cinco categorías según el grado de dificultad y la envergadura de los trabajos a realizar; ►tabla de honorarios); *f* **classe [f] de complexité (1)** (F., le montant de la rémunération de la mission de maîtrise d'œuvre tient compte de l'étendue de la mission, du coût prévisible des travaux et du degré de complexité de la réalisation des ouvrages ; les ►prestations élémentaires des missions d'études dans le domaine de l'urbanisme et du paysage sont réparties dans les ►tableaux des taux de rémunération [des honoraires] en trois **c.**

de c. — complexité de réalisation faible [1ᵉʳᵉ classe], moyenne [2ᵉᵐᵉ classe] et forte [3ᵉᵐᵉ classe], la première classe comprenant la note de complexité de 0 à 3, la deuxième les notes de 4 à 6 et la troisième 7 à 10 ; HAC 1989, 14, 42) . **D.**, les tableaux des taux de rémunération comprennent cinq **c. de c.** ; ►taux élémentaire de rémunération) ; *syn.* catégorie [f] d'honoraire [B] ; *g* **Honorarzone [f]** (Schwierigkeitsgrad und Komplexität eines Objektes oder einer Flächenplanung [städtebauliche oder landschaftsplanerische Leistung]; **D.**, für ►Grundleistungen bei Gebäuden und Freianlagen sind die ►Honorarsätze entsprechend der Schwierigkeit resp. dem Umfang der zu erbringenden Leistungen in fünf Zonen gegliedert; § 11 ff HOAI 2002 resp. §§ 34, 39, 43, 47 u. 50 HOAI 2009. **A.**, die Honorarermittlung für Leistungen der Landschaftsarchitekten erfolgt über zwei „Gestaltungsklassen"; cf. §§ 17-19 HRLA ; ►Honorartafel); *syn.* Gestaltungsklasse [f] [A].

973 compliance [n] **with plans and specifications** *constr. contr.* (Effective execution of specified construction work by a contractor); *s* **cumplimiento [m] de los trabajos** (Realización de las obras por parte del contratista); *f* **exécution [f] des prestations** (Réalisation des prestations de travaux par un entrepreneur) ; *g* **Erfüllung [f] der Bauleistungen** (Vollständige Ausführung der vertraglich festgelegten Bauarbeiten durch einen Unternehmer).

composite wall [n] *constr.* ►two-tier masonry [US].

974 composition [n] **(1)** *arch. plan.* (Arrangement of a group of elements according to artistic principles); *s* **composición [f] (1)** (Ordenación de los elementos de diseño según reglas artísticas); *f* **composition [f]** (Recherche et arrangement des éléments constitutifs selon certaines lois ; les lois générales de la composition en espaces verts sont entre autres l'échelle, l'unité et la dominante, la hiérarchisation des centres d'intérêt, les proportions, les contrastes et les harmonies, les rythmes) ; *g* **Gestaltung [f] (1)** (Verteilung von Gestaltungselementen nach formal-ästhetischen Gesichtspunkten).

composition [n] **(2)** *plan. sociol.* ►age class composition; *phyt.* ►anthropogenic change/alteration of flora composition; *constr.* ►layer composition; *phyt. zool.* ►species composition.

composition [n], **anthropogenic shift in animal species** *zool.* ►anthropogenic alteration of the genetic fauna pool.

composition [n], **seasonally attractive planting** *phyt. plant.* ►seasonally changing visual dominance.

compositional shift [n] **of faunal or floral communities** *phyt. zool.* ►displacement of species (composition or balance).

975 composition [n] **of a plant community** *phyt.* (Component species and their distribution within a particular ecological setting or plant association); *s* **composición [f] de una comunidad vegetal** (Conjunto de especies de plantas y su distribución en una determinada estación); *f* **organisation [f] d'un groupement végétal** (BB 1928); *syn.* structure [f] d'un groupement végétal ; *g* **Bestandsaufbau [m] (2)** (Zusammensetzung und Verteilung der Arten an einem bestimmten Standort oder in einer bestimmten Pflanzengesellschaft); *syn.* Bestandsstruktur [f].

976 composition [n] **of water features** *arch. gard. gard'hist.* (Designed arrangement of water jets, fountains in differing heights, pools of different shapes, sometimes with cascades; ►water features, ►water theater [US]/water theatre [UK]); *s* **juego [m] de aguas** (Composición de figuras por medio de agua de fuentes o surtidores; ►hidroarte, ►teatro de aguas); *f* **jeu [m] d'eau** (Effets d'animation par l'eau réalisés dans un jardin, sur une place par un ensemble de jets d'eau, fontaines,

C

statues, cascades ; ►art des jeux d'eau, ►théâtre d'eau) ; *g* **Wasserspiel [n]** (Komposition von Springbrunnen und Wasserstrahlfiguren [Fontänen] in unterschiedlichen Dimensionen oder differierenden Höhen, ggf. mit Kaskadenbildern; ►Wasserkunst, ►Wassertheater); *syn.* Wasseranlage [f].

977 compost [n] *hort.* (Product of controlled decomposition of vegetable waste frequently mixed with soil and mineral fertilizers; ►dressing with compost, ►refuse compost, ►sewage sludge); *s* **compost [m]** (Mezcla de materia orgánica descompuesta utilizada para fertilizar y acondicionar suelos. Proviene normalmente de los desechos, basuras, residuos orgánicos, excrementos de animales y lodos de desagüe urbanos; DINA 1987; ►compost de residuos, ►fertilización con compost, ►lodo de depuración); *f* **compost [m]** (Mélange de résidus divers d'origine végétale ou animale, mis en fermentation lente afin d'assurer la décomposition des matières organiques, et utilisé comme engrais et comme amendement, souvent avec incorporation de terre végétale et d'engrais minéraux ; DAV 1984 ; à ne pas confondre avec les ►boues d'épuration qui contiennent des contaminants chimiques et biologiques et dont l'épandage doit faire l'objet d'une vigilance particulière ; ►compost de déchets ménagers, ►fertilisation avec apport de compost) ; *g* **Kompost [m]** (Stabiles und humusartiges, durch den Prozess der aeroben Behandlung entstandenes und den hygienischen Anforderungen entsprechendes Verrottungsprodukt aus pflanzlichen Abfällen [Bioabfällen], das reich an organischer Masse ist, häufig unter Beimischung von Erde und Mineraldüngern. Es gibt in der Vermarktung die Tendenz, den Eindruck zu erwecken, dass K. und ►Klärschlamm das Gleiche sei. K. ist im Vergleich zum Klärschlamm nach Aufarbeitung ein wertvolles biologisches Material, Klärschlamm hingegen eine extrem schadstoffhaltige Mixtur; ►Düngung mit Kompost, ►Müllkompost); *syn.* Komposterde [f].

compost [n]**, mulching with** *constr. hort.* ► dressing with compost.

compost [n]**, waste** *envir. pedol.* ►refuse compost.

compost heap [n] [UK] *hort.* ►compost pile [US].

978 composting [n] *hort.* (**1.** Process of converting waste organic matter into humus; ►refuse composting. **2. Backyard composting:** diversion of organic food waste and yard trimmings from the municipal waste stream by composting them in one's yard through controlled decomposition of organic matter by bacteria and fungi into a humus-like product. It is considered a source reduction, not recycling, because the composted materials never enter the municipal waste stream); *s 1* **compostaje [m]** (Proceso de descomposición aerobia en caliente de los componentes orgánicos de los residuos, hasta obtener un producto sólido, relativamente estable, similar al humus, que se conoce como ►compost; DINA 1987; ►compostaje de residuos orgánicos domésticos); *s 2* **compostaje [m] in situ** (Separación de los residuos orgánicos de alimentos y de plantas en el hogar en el que se producen para producir compost en el jardín propio); *f 1* **compostage [m]** (Procédé d'élimination des déchets : broyage, fermentation et transformation des déchets organiques et minéraux domestiques ou des résidus d'entretien, des produits de tonte et de taille en un produit utilisable comme amendement organique en culture ; *terme spécifique* lombricompostage) ; *f 2* **compostage [m] au jardin** (Transformation des déchets organiques domestiques afin d'améliorer la fertilité du jardin, réduire le volume d'ordures ménagères à traiter par la collectivité et participer à la protection de l'environnement ; ►compostage de déchets ménagers ; *syn.* compostage [m] domestique, compostage [m] individuel, compostage [m] autonome ; ►compostage de/en surface) ; *g 1* **Kompostierung [f]** (Verfahren, aus

organischen und mineralischen Abfallstoffen [Pflanzenreste, Klärschlamm etc.] durch kontrollierte aerobe Rottevorgänge [mit Sauerstoffzufuhr] Kompost entstehen zu lassen); *g 2* **Gartenkompostierung [f]** (Kompostierung von organischen Abfällen aus Küche und Garten im eigenen Garten zur Eigenverwertung. Dadurch wird die durch die Stadtverwaltung zu entsorgende Müllmenge erheblich reduziert; *syn.* Heimkompostierung [f]; *g 3* **eingehauste/eingekapselte Kompostierung [f] (K.** von biologisch abbaubaren Stoffen in einem geschlossenen Reaktor, wobei der Wärmeaustausch mit der Atmosphäre auf ein Mindestmaß reduziert und der Prozess der **K.** durch die Optimierung des Luftaustausches, des Wassergehaltes und der Temperatur beschleunigt wird); *g 4* **Vor-Ort-Kompostierung [f] (K.** von biologisch abbaubaren Stoffen sowie die Verwendung des fertigen Kompostes an dem Standort der Entstehung der Abfälle zur Eigenverwertung; kompostieren [vb]; ►Flächenkompostierung, ►Müllkompostierung).

composting [n]**, surface** *agr. constr. hort.* * ►mulching.

compost management [n] *envir.* ►compost recycling.

979 compost pile [n] [US] *hort. syn.* compost heap [n] [UK]; *s* **montón [m] de compost**; *f* **tas [m] de compost** ; *g* **Komposthaufen [m]** (...haufen [pl]); *syn.* Kompostmiete [f].

compost production plant [n] [US]**, waste** *envir.* ►refuse compost production plant.

980 compost recycling [n] *envir.* (All planning and technical measures under both public and private auspices, for the reutilisation of organic waste matter [vegetable waste, sewage sludge] as a means of ►soil amendment); *syn.* compost management [n]; *s* **gestión [f] de residuos aptos para compostaje** (Todas las medidas de planeamiento y técnicas realizadas por la administración pública o por empresas privadas para reutilizar la materia orgánica de los residuos [residuos vegetales, lodo residual] como ►material de enmienda de suelos); *f* **industrie [f] du compost** (Ensemble des mesures prises par les collectivités locales ou les exploitant d'unités de compostage en vue de la production de compost utilisé comme ►amendement organique des cultures) ; *g* **Kompostwirtschaft [f, o. Pl.]** (Summe aller Maßnahmen öffentlicher oder privater Betreiber, vorwiegend organische Abfallstoffe [Pflanzenreste, Klärschlamm etc.] einer Wiederverwertung als ►Bodenverbesserungsstoff 1 zuzuführen).

981 compound curve [n] [US] *trans.* (Curve described by the front overhang of a vehicle, and the track of the inner rear wheels in motion. The area covered by these curves is called 'swept area'; ►vehicle turning circle); *syn.* sweep circle [n] [UK] (TAN 1978, 179), turning clearance circle [n] [UK] (SPON 1986, 300), vehicle turning radii [npl] [US]; *s* **curva [f] de arrastre** (Curva descrita por el remolque de un camión; ►radio de giro); *f* **épure [f] de giration** (Courbe dessinée en plan, dépendante de la largeur et longueur hors tout d'un véhicule, dépassant le rayon de braquage, permettant de vérifier l'évaluation des manœuvres et l'emprise de balayage de la carrosserie et utilisée pour déterminer le rayon de courbure d'un rond-point, d'une aire de retournement ; cf. BON 1990, 358 ; ►rayon [extérieur] de braquage) ; *g* **Schleppkurve [f]** (Kurve, die bei Bogenfahrten eines Fahrzeuges in der Kurveninnenseite durch die nachfahrenden Hinterräder entsteht und nicht der Spur der Vorderräder entspricht, weshalb eine entsprechende Fahrflächenverbreiterung vorgesehen werden muss; ►Wendekreis).

982 comprehensive [adj] *plan.* (Covering the whole area or subject of study); *s* **exhaustivo [adj]** (Que cubre o afecta a toda un área o un tema de estudio); *f* **couvrant un territoire [loc]** ; *g* **flächendeckend [ppr/adj] (1)** (Ein bestimmtes Gebiet vollständig erfassend).

983 comprehensive plan [n] [US] *leg. urb.* (Long-range preparatory land use plan to guide the policies, growth and development of a community, which includes analysis, recommendations and proposals for the community's population, economy, housing, transportation, utilities, public facilities, and land use, implemented by zoning; cf. IBDD 1981; **in U.S.**, the **comprehensive plan** includes a land use plan [and the policies stated therein would correspond with the more generalized land use projections or intentions shown on the regional plan] and other specialized plan documents, dealing with streets and highways, public works, recreation and open spaces, housing, urban renewal, urban design plans [specifying the major architectural buildings], and historic landmark preservation at a local level; **in U.K.**, the approximate equivalent is the **Local Development Framework [LDF]** and the most important instrument in land use planning from which all future development control stems. In preparing land use plans, planning authorities produce the following documents: **1. Local Development Orders** which extend permitted rights for certain forms of development, with regard to a relevant local development document. **2.** A **Simplified Planning Zone**, an area in which a local planning authority wishes to stimulate development and encourage investment. It operates by granting a specified planning permission in the zone without the need for a formal application or the payment of planning fees. **3. Supplementary Planning Documents** [SPDs] to expand or add details to policies laid out in development plan documents or a saved policy in an existing development plan. These may take the form of design guides or codes, area or site development briefs, a master plan or issue-based documents. The documents use illustrations, text and practical examples to expand on how the authority's policies can be taken forward. **4.** A **Local Development Scheme** or public 'project plan' to identify which local development documents will be produced, in what order and when, according to a timetable for the production of all documents that make up the Local Development Framework over a three-year period; ▶zoning map [US]/Proposals Map [UK], ▶community development plan [US]/urban development plan [UK]; **in D.**, the approximate equivalent is the **preparatory land-use plan** which shows the basic types of land uses envisaged for the entire municipal territory in accordance with anticipated needs for the next ten years. It establishes preliminary development use of plots and identifies general and specific land-use areas including land for public amenities, green spaces, agricultural and woodland areas. The plan obliges the municipality to implement it as adopted, and is put into effect by **legally-binding land-use plans** [Bebauungspläne]); *syn.* preparatory land-use plan [n] [D] (FBC 1993), *obs.* unitary development plan [n] [UK]; *s* **Plan** [m] **General Municipal de Ordenación Urbana** (En Es., instrumento de ordenación integral del territorio que puede abarcar uno o varios términos municipales completos. El **P. G. M. de O.** clasifica el suelo para la aplicación del régimen jurídico correspondiente, define los elementos fundamentales de la estructura general adoptada para la ordenación urbanística del territorio, establece el programa para su desarrollo y ejecución, así como el plazo mínimo de su vigencia; cf. art. 10 Ley del Suelo, RD 1345/1976; ▶plan urbanístico, ▶plan parcial [de ordenación]); *syn.* Plan [m] General de Ordenación Urbana, Plan Director [C], Plan [m] Regulador [RCH]; *f1* **schéma** [m] **de cohérence territoriale (SCOT)** (F., Depuis le 1er avril 2001, le SCOT remplace le ▶schéma directeur d'aménagement et d'urbanisme selon la loi 2000-1208 du 13 décembre 2000 relative à la solidarité et au renouvellement urbains, [loi SRU] ; le SCOT est un document d'urbanisme qui fixe, à l'échelle de plusieurs communes ou groupement de communes, les organisations fondamentales de l'organisation du territoire et de l'évolution des zones urbaines, afin de préserver un équilibre entre zones urbaines, industrielles, touristiques, agricoles et naturelles ; il fixe les objectifs des diverses politiques publiques en matière d'habitat, de développement économique, de déplacements ; le SCOT est élaboré par un ou plusieurs EPCI [Établissement **P**ublic de **C**oopération **I**ntercommunale] et doit couvrir un territoire continu et sans enclaves ; il comprend **a]** un rapport de présentation qui contient un diagnostic du territoire et un état initial de l'environnement et qui explique les choix retenus pour établir le PADD et le DOG ; **b]** le projet d'aménagement et de développement durable [PADD], document obligatoire dans lequel l'EPCI exprime de quelle manière il souhaite voir évoluer son territoire dans le respect des principes de développement durable ; **c]** un document d'orientations générales [DOG] qui est la mise en œuvre du PADD ; **d]** des documents graphiques ; **e]** des dispositions facultatives relatives au transport tels que l'ouverture de nouveaux secteurs à l'urbanisation si création de dessertes en transports collectifs ou la définition de grands projets d'équipement et de service ; le SCOT est opposable au ▶plan local d'urbanisme [PLU] et à la ▶carte communale, aux programmes locaux de l'habitat [PLH], aux plans de déplacements urbains [PDU], aux opérations foncières et d'aménagement, aux schémas de développement commercial et aux autorisations d'urbanisme commercial) ; *f2* **schéma** [m] **directeur d'aménagement et d'urbanisme (SDAU)** (▶Document de planification urbaine antécédant au SCOT et fixant les orientations de l'aménagement de l'espace, et en particulier les choix d'usage du sol, la nature et le tracé des grands équipements d'infrastructure ; DUA 1996 ; il est établi pour des communes, parties ou ensembles de communes comprises dans des agglomérations où des ensembles géographiques présentant une communauté d'intérêts économiques et sociaux et dont les perspectives d'évolution, de mise en valeur et de protection requièrent la définition d'orientations fondamentales d'aménagement, compte tenu de l'équilibre qu'il convient de préserver entre l'extension urbaine, l'exercice des activités agricoles, des autres activités économiques et la préservation des sites et paysages naturels ou urbains ; loi d'orientation pour la ville, n° 91-662 du 13 juillet 1991 ; ▶plan d'occupation des sols) ; *f3* **schéma** [m] **de secteur** (Plan pouvant être établi en complément ou pour la révision d'un SCOT/schéma directeur et concernant une partie du territoire couvert par celui-ci) ; *g* **Flächennutzungsplan** [m] (**1. D.**, ein behördenverbindliches Planwerk, das die beabsichtigte städtebauliche Entwicklung und die Art der Bodennutzung für eine ganze Gemeinde nach den vorausseh-baren Bedürfnissen in den Grundzügen, i. d. R. für die nächsten 10 Jahre, darstellt; gesetzliche Grundlage ist das Baugesetzbuch [BauGB]; ▶Bauleitplan, ▶Bebauungsplan. **2. CH**, das schweizerische Äquivalent ist der **Zonenplan**; gesetzliche Grundlagen sind die Bau- und Zonenordnungen der Städte auf Grundlage der jeweiligen kantonalen Planungs- und Baugesetze [PBG]; **3. F.**, der Geltungsbereich eines **F.es** erstreckt sich über mehrere Kommunen. Deshalb gibt es die Möglichkeit, mit dem *schéma de secteur* Ansprüche von einzelnen Kommunen gesondert zu berücksichtigen resp. Änderungen des SCOT [▶schéma de cohérence territoriale (SCOT)] durchzuführen); *syn.* Flächenwidmungsplan [m] [A], Nutzungszonenplan [CH, Bern], vorbereitender Bauleitplan [m], Zonenplan [m] [CH, Zürich]).

984 comprehensive redevelopment [n] *plan. urb.* (Demolition of whole building complexes, blocks or run-down housing areas, and subsequent reconstruction instead of the preservation and modernization of existing buildings; also the removal of primitive structures which have outlived their usefulness, such as 'pre-fabs'; ▶center city redevelopment [US]/urban regeneration [UK]); *syn.* total redevelopment [n] [US], town centre clearance [n] [UK], urban clearance [n] and redevelopment [n] [US]; *s* **reestructuración [f] urbana** (Método ya poco usual de ▶rehabilitación urbana por el que se demolían manzanas/bloques enteros de casas para dar espacio a construcciones nuevas, en vez de rehabilitar los edificios existentes); *f* **restructuration [f] urbaine** (Opération de transformation immobilière dans les quartiers anciens impliquant la démolition des immeubles existants pour construire à la place de nouveaux logements habités par d'autres occupants ou des locaux affectés à d'autres usages ; ▶restauration urbaine, ▶rénovation urbaine) ; *g* **Flächensanierung [f]** (Abreißen ganzer Baublöcke und Neuaufbau statt Erhaltung und Modernisierung vorhandener Bausubstanz, oft im Zusammenhang mit notwendiger Änderung der Verkehrsflächen und der Notwendigkeit, Gemeinbedarfseinrichtungen zu bauen oder zu kleinteiligen Parzellierungen in neue Grundstückszuschnitte umzuwandeln; gelegentlich auch zur Beseitigung einer nicht mehr vertretbaren Primitivbebauung, z. B. in Laubenkolonien, Barackenvierteln; ▶Stadtsanierung); *syn.* Planerjargon Kahlschlagsanierung [f].

985 comprehensive site survey [n] and analysis [n] *landsc. plan.* (All inclusive survey, review and analysis of basic planning and site data; ▶initial site analysis, ▶survey phase); *syn. also* site investigations [npl], survey [n] of existing site conditions; *s* **elaboración [f] de inventario ambiental** (Investigación integrada de las condiciones naturales y de las bases legales y programáticas de una zona a planificar; ▶análisis del estado inicial, ▶capítulo de análisis de un plan); *f* **étude [f] de reconnaissance** (Prise en compte des caractéristiques du milieu naturel [sol, eau, air, climat, faune, flore] recueil des données socio-économiques, etc. nécessaires à l'élaboration d'un programme, la recherche du parti d'aménagement ou des actions prévues ; ▶analyse de l'état initial, ▶étude préliminaire d'un plan de paysage) ; *g* **Grundlagenuntersuchung [f]** (Untersuchung von Boden, Wasser, Klima, Luft, Vegetations- oder Faunenbeständen, Zusammenstellung der planerischen Vorgaben etc. als Grundlage zur Erarbeitung von Entwicklungszielen oder konkreten Maßnahmen; ▶Bestandsanalyse, ▶Grundlagenteil).

986 compressive strength [n] *constr. stat.* (**1.** Essential structural property of building material to withstand maximum required pressure. **2.** Capacity to resist crushing forces); *s* **resistencia [f] a la compresión** (Máximo grado de compresión que puede resistir un material sin romperse; cf. DACO 1988); *f* **résistance [f] à la compression** (Caractéristique technique des matériaux de construction) ; *g* **Druckfestigkeit [f]** (Eine wesentliche bautechnische Eigenschaft von Werkstoffen, einem geforderten maximalen Druck zu widerstehen).

compulsory competitive tendering [n] [UK] *adm. constr. contr.* ▶open bidding [US].

compulsory purchase [n] [UK] *adm. leg.* ▶eminent domain (power of condemnation) [US].

compulsory purchase order [n] (CPO) [UK] *adm. leg.* ▶condemnation order [US].

compulsory purchase proceedings [npl] [UK] *adm. leg.* ▶condemnation proceedings [US].

computation [n] *contr. prof.* ▶unit of computation.

computation [n] of full contract value [UK] *constr.* ▶determination of final contract amount [US].

987 concave slope [n] *constr.* (Rounded foot or toe of an embankment/slope employed to create a smooth transition to adjacent grades; ▶rounding the top of an embankment/slope); *s* **pie [m] de talud cóncavo** (Resultado de ▶redondeado de talud); *f 1* **arrondi [m] de pied de talus** (Raccordement progressif de pente réalisé entre la partie inférieure d'un talus en

remblai et la surface plane adjacente; *terme spécifique d'aménagement routier* **fossé [m] de pied :** fossé établi au pied de tout talus de remblai ; ▶adoucissement de talus) ; *f 2* **doucine [f] [B]** (Fossé qui recueille les eaux de ruissellement tombant sur ce talus ou provenant de la plate-forme ou de la berme de pied pour les diriger vers les exutoires) ; *g* **ausgerundeter Böschungsfuß [m]** (Ergebnis der Ausrundung des unteren Teiles einer Böschung, die i. d. R. als Entwässerungs- oder Versickerungsmulde dient; ▶Böschungsausrundung).

concealment planting [n] *gard. landsc.* ▶screen planting (2).

988 concentration [n] *chem.* (Amount of a substance per unit volume); *s* **concentración [f]** (Cantidad de una sustancia por unidad de volumen); *f* **concentration [f]** (d'une substance par unité de volume) ; *g* **Anreicherung [f] (1)** (Gespeicherte/angesammelte Menge einer Substanz je Raumeinheit); *syn.* Konzentration [f].

989 concentration [n] of noxious substances *envir.* (Substance per unit volume measurement to determine levels of toxicity); *s* **concentración [f] de sustancias contaminantes** *syn.* concentración [f] de sustancias nocivas/tóxicas; *f* **concentration [f] des substances nuisibles** *syn.* concentration [f] des substances nocives ; *g* **Schadstoffkonzentration [f]** (Gehalt an Schadstoffen in einem Medium je Messeinheit).

990 concentration [n] of recreation facilities *plan. recr.* (Place of aggregated recreation development in a community, on an urban fringe or in a rural area; ▶recreation area 1, ▶leisure park, ▶regional park [US]/country park [UK] (2), ▶leisure center [US]/leisure centre [UK]); *syn.* intensive recreation area [n], recreation center [n] [US]/recreation centre [n] [UK]; *s* **nucleo [m] recreativo** (Lugar en el que se concentran una serie de facilidades de recreo, deportes o culturales al aire libre. En F. estos nucleos existen a nivel urbano, periurbano, rural y natural; ▶zona de uso priorizado para el recreo, ▶parque recreativo regional, ▶centro de recreo); *f* **base [f] de plein air et de loisirs** (Espace libre, animé ouvert à l'ensemble de la population dont les équipements offrent à ses usagers les activités les plus variées permettant la détente et la pratique d'activités sportives, culturelles de plein air et de loisirs dans un cadre naturel préservé du bruit ; *on distingue en F.* les bases urbaines, les bases périurbaines, les bases rurales, les bases de nature ; cf. circ. du 24 mars 1975 ; ▶espace touristique, ▶parc de loisirs, ▶centre ludique, ▶centre de sports et de loisirs) ; *syn.* site [m] récréatif ; *g* **Erholungsschwerpunkt [m]** (Konzentration von Erholungs- und Freizeiangeboten im inneren Stadtgebiet, Stadtrandgebiet oder im ländlichen Raum/in der freien Landschaft; ▶Erholungsgebiet, ▶Freizeitpark, ▶Freizeitzentrum); *syn.* Erholungsstätte [f], Freizeitschwerpunkt [m], Freizeitbereich [m], Freizeitstätte [f].

991 concentration [n] of urban recreation facilities *plan. recr.* (Place of concentrated leisure infrastructure in an urban area); *s* **nucleo [m] recreativo en zona urbana** (Lugar de concentración de equipamiento recreativo en una ciudad); *f* **pôle [m] de loisirs urbain** (Concentration locale d'équipements de loisirs en zone urbaine) ; *syn.* centre [m] de loisirs ; *g* **städtischer Erholungsschwerpunkt [m]** (Ort der Konzentration freizeitrelevanter Infrastruktur in einem Stadtgebiet); *syn.* städtische Freizeitstätte [f].

concentric domed bog [n] *geo.* ▶lens-shaped raised bog.

992 concentric manhole brick [n] *constr.* (Frost- and acid-resistant wedge-shaped brick for the construction of manholes or for brickpaving of concentric patterns); *syn.* radial header [n] [also UK], compass brick [n] [also US, Vt]; *s* **clinker [m] adovelado** (Ladrillo cuneiforme, en D. resistente a heladas y a la acidez, que se utiliza para construir arcos o para pavimentar en formas circulares); *syn.* ladrillo [m] en cuña; *f* **pavé [m] tronqué** (Pavé non gélif et résistant aux acides, de forme conique, utilisé pour la construction de voûtain dans les canaux d'écoulement des eaux pluviales [EP] ou pour la réalisation d'arc de cercle dans les pavages) ; *g* **Kanalschachtklinker [m]** (Frost- und säurebeständiger Klinker mit konischem Format zur Herstellung von Bögen in gemauerten Kanalschächten und Belägen).

993 concentric open space pattern/system [n] *landsc. urb.* (Urban open space system forming a circular pattern around a common center [US]/common centre [UK], that is also supplemented in modern times by an outer ▶greenbelt; ▶earthen wall fortification [US]/town fortification [UK]); *s* **trama [f] verde concéntrica** (Sistema de espacios libres urbanos formados en muchos casos por zonas verdes históricas [▶fortificación de tierra] y hoy en día ampliado por un ▶cinturón verde exterior); *syn.* red [f] concéntrica de espacios libres, sistema [m] concéntrico de espacios libres; *f* **trame [f] concentrique d'espaces libres** (Système de disposition des espaces libres urbains en forme de couronne, souvent expression de la présence d'espaces historiques [▶fortifications de terre] et complété par une ▶ceinture verte périphérique) ; *g* **ringförmiges Freiraumsystem [n]** (Konzentrisches Freiraumsystem im Siedlungsgebiet, das vielfach von historisch vorgegebenen Freiräumen [▶Wallanlagen] geprägt und heute in vielen Städten durch einen äußeren ▶Grüngürtel ergänzt ist).

concept design [n] *plan. prof.* ▶sketch design.

concept site design [n] *plan. prof.* ▶sketch design.

conceptual design phase [n] [US] *contr. plan.* ▶schematic design phase [US].

994 concluding project review [n] [US] *prof.* (Final construction review after completion of the work[s], and after the ▶issue of a certificate of compliance [US]/issue of a certificate of practical completion [UK], to determine any defects and to monitor their correction, as well as to provide further consultant services for a suitable period of time before ▶delivry to the client; ▶construction phase—administration of the construction contract [US]/site management services [UK]); *syn.* post-completion advisory services [npl] [UK], work stage [n] "L" [UK]; *s* **servicios [mpl] en plazo de garantía** (Obligación del contratista de supervisar la eliminación de defectos o reposición de marras después de la ▶aceptación de obra y ▶entrega de obra y de asesorar al cliente en una medida razonable; ▶dirección facultativa de las obras); *f* **mise [f] en service des ouvrages** (Étape de la maîtrise d'œuvre d'une mission normalisée, suivant la ▶réception des travaux et la ▶prise de possession des ouvrages, contenant la constitution et la remise au maître d'ouvrage, en vue de l'exploitation des ouvrages, des notices d'utilisation et des dessins, conformes à l'exécution de chaque ouvrage [dossier des ouvrages exécutés] ; ▶mise à disposition de certains ouvrages, ▶contrôle des travaux, ▶contrôle des entrepreneurs, ▶contrôle général des travaux, ▶réception et décompte des travaux) ; *g* **Objektbetreuung [f] und Dokumentation** (Leistungspflicht des Planers gegenüber dem Bauherrn nach ▶Abnahme und ▶Übergabe des Objektes Mängel vor Ablauf der Verjährungsfristen und Gewährleistungsansprüche festzustellen und diese bei den bauausführenden Unternehmen geltend zu machen, die Überwachung der Mängelbeseitigung sowie eine angemessene weitere Beratung, Mitwirkung bei der Freigabe von Sicherheitsleistungen sowie die systematische Zusammenstellung der zeichnerischen Darstellungen und rechnerischen Ergebnisse des Objektes durchzuführen; cf. § 15 [2] 9 HOAI 2002 resp. §§ 33, 38, 42 und 46 HOAI 2009; ▶Objektüberwachung); *syn.*

Dokumentation [f] über Bauwerk und Pflege sowie Leitung der Garantiearbeiten [CH].

conclusion [n] **of a binding agreement** *contr. prof.* ▶conclusion of a contract.

995 conclusion [n] **of a contract** *contr. prof.* (**1.** Finalization of contract negotiations, validated with the signing of the contract documents; ▶fee agreement. **2. Completion [n] of a contract** [UK]: term often used in property transactions); *syn.* conclusion [n] of a binding agreement; *s* **conclusión [f] de contrato** (Fin de las negociaciones de contrato a las que se les da validez con la firma del mismo; ▶acuerdo sobre honorarios); *syn.* cierre [m] de contrato; *f* **conclusion [f] du marché** (Arrangement final entre le maître d'œuvre et l'architecte aboutissant à la notification du marché ; ▶accord sur les honoraires) ; *g* **Vertragsabschluss [m]** (**1.** Ende einer Vertragsverhandlung, bei der sich die Willenserklärungen beider Seiten über die zu erbringenden Leistungen inhaltlich vollständig decken; durch die beiderseitige Unterzeichnung des Vertrages wird dieser rechtsgültig. Mündliche Abreden [Verbalkontrakt] sind grundsätzlich aureichend, wenn nicht die Schriftform gesetzlich vorgeschrieben ist, sind jedoch bei allen Leistungen für Bauvorhaben und Pflegeleistungen wegen möglicher späterer Streitigkeiten abzulehnen. **2.** Erfüllung eines Vertrages durch Erbringung sämtlicher vereinbarter Leistungen; ▶Honorarvereinbarung).

996 concrete [n] *constr.* (Composite material with Portland cement, ▶aggregates and water; ▶concrete slurry, ▶dry-mix concrete, ▶exposed aggregate concrete, ▶finished concrete, ▶poured-in-place concrete [US]/in-situ [cast] concrete [UK], ▶ready-mix concrete, ▶porous concrete); *s* **hormigón [m]** (Mezcla de material pétreo de diferente granulación [▶árido] con cemento Portland y agua; ▶hormigón amasado durante el transporte, ▶hormigón con árido visto, ▶hormigón de consistencia de tierra húmeda, ▶hormigón fluido, ▶hormigón poroso, ▶hormigón vertido «in-situ», ▶hormigón visto); *syn.* concreto [m] [AL]; *f* **béton [m]** (Matériau de construction constituant un mélange homogène, fabriqué à partir de ▶granulats [sable, gavier, etc.] agglomérés par un liant [p. ex. ciment], l'eau et les adjuvants ; ▶béton coulé sur place, ▶béton de gravillons lavés, ▶béton de parement, ▶béton fluide, ▶béton gâché consistance terre humide, ▶béton poreux, ▶béton prêt à l'emploi) ; *g* **Beton [m]** (**1.** Künstliches Materialgemisch aus grober und feiner Gesteinskörnung [▶Zuschlagstoff], das durch Zement und Wasser zu einem Konglomerat zusammengekittet wird; ▶erdfeuchter Beton, ▶Flüssigmörtel, ▶Ortbeton, ▶Sichtbeton, ▶Transportbeton, ▶Waschbeton, ▶wasserdurchlässiger Beton. **2. Selbstverdichtender Beton [SVB]:** neuartige Betonmischung, dem für erweiterte Gestaltungsmöglichkeiten mit ungewohnten geometrischen Formen besondere Fließmittel zugesetzt werden. Um z. B. dünnere Platten und geschwungene Pfeiler gießen zu können, werden kettenförmige Moleküle, die sich an die Zementpartikel heften und diese auf Abstand zueinander halten, zugesetzt, wodurch die Teilchen nicht verklumpen und die Reibung zwischen ihnen gering ist; somit fließt der Beton wie Honig und muss nicht mit Rüttelgeräten verdichtet werden, weil er von selbst in alle Ecken der Gussform gelangt und kaum Luftblasen einschließt. **3. Ultrafester Faserbeton:** neuartige Betonmischung, bei der zahllose 1-2 cm lange Stahlfasern direkt mit dem Zement in die Mischtrommel gegeben werden und verhindern, dass der Beton, der durch einen Zuschlag aus Kieselpulver sehr fest wird, zerbröseln kann. **4. Textilbeton: a.** An den TU Aachen und Dresden werden Betonformen mit grobmaschigem Bewehrungsgewebe, das von einer Betonmischung umschlossen wird und zu ein oder zwei Zentimeter dicken Platten aushärten kann, entwickelt; Anwendungsbereich für dünnwandige Gerüste für Betonrohre und leichte T-Träger für z. B.

die Fußgängerbrücke über die Döllnitz, ein Nebenfluss der Elbe. **b.** An der Universität Stuttgart wurde die Technik zur Herstellung **vorgespannter textilbewehrter Betonplatten** entwickelt; der Vorteil von textilbewehrten Betonbauteilen besteht darin, dass es keinen Korrosionsangriff im carbonatisierten oder im chlorid-angereicherten Beton gibt und dass Bauteile sehr dünn gefertigt werden können. **5.** *Historie* **B.** ist keine Erfindung der Neuzeit. Die Phönizier stellten schon vor 3000 Jahren ein Gemisch aus Mörtel und vulkanischem Gestein her, das unter Wasser aus-härtete. Über die Griechen gelangte die Kenntnis der Betonher-stellung im 3. Jh. v. Chr. in das damalige Römische Reich und wurde *Opus Caementitium* genannt. Der berühmte Architekt MARCUS VITRUVIUS POLLIO und Ingenieure des antiken Roms priesen die Vorzüge und Vielseitigkeit des **B.s**, weil man damals wusste, dass **B.** unter Wasser abbindet, und verwendeten ihn beim Bau der Cloaca Maxima, Roms historischer Abwasserkanal. Während des Mittelalters ist die alte Technik in Vergessenheit geraten, bis 1804 in Frankreich erste Betonfertigteile wieder hergestellt wurden).

concrete [n] *constr.* ▶dry-mix concrete, ▶exposed aggregate concrete, ▶facing concrete, ▶finished concrete, ▶haunched concrete, ▶lean-mixed concrete, ▶no-fines concrete, ▶porous concrete, ▶prestressed concrete, ▶ready-mixed concrete, ▶sandblasted concrete, ▶smooth-faced concrete, ▶spun concrete, ▶tooled concrete.

concrete [n], **lean** *constr.* ▶lean-mixed concrete.

concrete [n] [US], **premixed** *constr.* ▶ready-mixed con-crete.

concrete [n], **pretensioned** [n] *constr.* ▶prestressed con-crete.

concrete [n] [US], **transit-mixed** *constr.* ▶ready-mixed concrete.

concrete [n], **veneer** *constr.* ▶facing concrete.

concrete block pavement [n] [UK] *constr.* ▶pavement of concrete pavers [US].

concrete compound unit [n] [UK], **prefab(ricated)** *constr.* ▶precast concrete unit.

997 concrete cover [n] *constr.* (Minimum thickness of concrete between the surface of reinforcing rods or mats and the outer surface of the concrete for protection against corrosion and differential expansion and contraction of steel and concrete; cold regions may require 50mm); ***s* cobertura [f] mínima de hormigón** (En hormigón armado, el espesor mínimo de la capa que cubre a la armadura para evitar la oxidación y diferente expansión y contracción entre el acero y el hormigón; en zonas frías puede requirir hasta 50 mm); ***ƒ* enrobage [m] (de béton)** (En parement extérieur, recouvrement du ferraillage d'un minimum de 3,5 cm afin de préserver les aciers de la corrosion ; cf. COB 1984, 149) ; *syn.* recouvrement [m] de béton ; ***g* Betondeckung [f]** (Beim Stahlbeton das Mindestmaß [von 3,5 cm] der Überdeckung von Bewehrungsstählen zum Schutz gegen Korrosion); *syn.* Betonüberdeckung [f], Deckung [f].

998 concrete crib wall [n] *constr.* (Interlocking system of precast concrete elements which create a spatially stable grid structure. When filled with compacted earth, comprising the greater part of the total structure, its exposed side can be planted to maximize soil stability; ▶timber crib wall, ▶free-standing concrete crib wall); ***s* muro [m] de entramado de piezas de hormigón (≠)** (Sistema de entrelazado de elementos de hormigón prefabricados que constituyen una estructura entramada. Se pueden rellenar con tierra y plantar en su lado expuesto; *términos específicos* ▶muro de entramado de rollizos, ▶muro libre de entramado de piezas de hormigón); ***ƒ* système [m] de construc-**

tion en caissons (±) (Empilement d'éléments préfabriqués en béton, formant un maillage rigide, constituant sur la plus grande partie de son profil un important volume de terre compactée et dont la face externe peut être végétalisée ; construction utilisée comme retenue de terrain en remblai, protection de talus, mur fleuri, plantation en sol rocheux ; ▶mur de caissons en béton, ▶mur de/en caissons en bois) ; ***g* Raumgitterkonstruktion [f]** (Verbundsystem aus aufeinander gelagerten Betonfertigteilen, die mit einem verdichteten Erdfüllkörper, der den größten Teil des Gesamtquerschnitts ausmacht, ein räumlich geschlossenes Gitter bilden; die Luftseite kann gärtnerisch begrünt werden; *UBe* ▶Raumgitterwall, Raumgitterwand; ▶Krainerwand).

999 concrete deck [n] *constr.* (Broad reinforced piece of concrete covering a garage or building stor[e]y; e.g. for a flat roof; ▶concrete slab 1, ▶roof garden); *syn.* concrete roof slab [n]; ***s* cubierta [f] de hormigón** (p. ej. de aparcamiento subterrá-neo, de un garaje; ▶azotea ajardinada, ▶losa de cimentación [de hormigón armado]); *syn.* piso [m] de concreto [AL]; ***ƒ* dalle [f] support** (▶jardin sur dalle, ▶radier) ; ***g* Betondecke [f] (1)** (Aus Beton hergestellter oberer Abschluss einer Garage oder eines Stockwerkes, z. B. für ein Flachdach oder einen ▶Dachgarten; ▶Betonplatte).

concrete edge bedding [n] [UK] *constr.* ▶concrete edge footing [US].

1000 concrete edge footing [n] [US] *constr.* (Foundation and backing of in-situ concrete on which an edging unit is laid; ▶haunched concrete, ▶step strip foundation, ▶strip foundation); *syn.* concrete trough [n] [also US] (OEH 1990, 240), concrete edge bedding [n] [UK]; ***s* solera [m] de hormigón** (Cimentación y ▶refuerzo de hormigón para piedras de bordillo; ▶cimiento de escalera, ▶zapata corrida); *syn.* capa [f] de asiento de hormigón; ***ƒ* assise [f] en béton** (Fondation en béton et ▶butée arrière en béton pour la pose de bordures ; ▶fondation d'ancrage, ▶longrine en béton, ▶semelle filante) ; *syn.* blocage [m] béton (VRD 1986, 196) ; ***g* Betonstuhl [m]** (Betonfundament und ▶Betonrückenstütze für Randsteine; ▶Streifenfundament 1, ▶Streifenfundament 2).

concrete element [n], **precast** *constr.* ▶precast concrete unit.

concrete panel construction [n] *urb.* ▶prefabricated concrete panel construction.

1001 concrete pavement [n] *constr.* (Rigid road surfacing which is made up of Portland cement concrete); *syn.* rigid pavement [n]; ***s* pavimento [m] de hormigón** *syn.* pavimento [m] de concreto [AL]; ***ƒ* chaussée [f] en béton** (Surface de la route servant de circulation aux véhicules, réalisée en béton et constituant une chaussée rigide ne pouvant supporter de défor-mations importantes) ; ***g* Betondecke [f] (2)** (Aus Beton herge-stellte Fahrbahndecke).

1002 concrete paver [n] [US] *constr.* (Generic term for any prefabricated portland cement unit paving element; *specific terms* large-sized concrete paver [US], ▶precast paving block [US]/ precast concrete paving slab [UK], ▶standard non-interlocking concrete paver [US]/standard concrete paving block [UK], ▶tinted concrete paver [US]/coloured paving block [UK]); *syn.* concrete unit paver [n] [US], concrete paving stone [n] [UK], concrete paving unit [n] [UK], concrete sett/pavio(u)r [n] [UK]; ***s* piedra [f] de hormigón** (Término genérico; *términos especí-ficos* ▶losa de hormigón, ▶piedra de hormigón coloreada, ▶piedra estándar de hormigón); ***ƒ* pavé [m] en béton** (Terme générique ; *termes spécifiques* ▶dalle préfabriquée en béton pour voirie de surface, ▶pavé en béton coloré dans la masse, ▶pavé en béton ordinaire) ; ***g* Betonpflasterstein [m] (1)** (OB zu ▶Betonwerkstein; Pflasterstein aus Beton zum Befestigen von

Wege-, Platz- und Straßenflächen; *UBe* ►Betonpflasterstein 2, ►eingefärbter Betonpflasterstein, ►Gehwegplatte aus Beton).

concrete pavers [npl] [US] *constr.* ►pavement of concrete pavers.

1003 concrete paver [n] **with exposed crushed basalt** [US] *constr. syn.* exposed basalt aggregate paving block [n] [UK]; *s* **piedra [f] de hormigón con revestimiento de basalto**; *f* **pavé [m] en béton avec parement de basalte** ; *g* **Betonpflasterstein [m] mit Basaltsplittvorsatz.**

1004 concrete paver [n] **with protective coating** [US] *constr.* (Paving block/paver with an additional wearing course for heavy traffic); *syn.* concrete paving block [n] with hard-wearing surface layer [UK]; *s* **piedra [f] de hormigón de superficie de alta resistencia** *syn.* pieza [f] de pavimentación de hormigón de superficie de alta resistencia; *f* **pavé [m] en béton avec couche anti-usure** ; *g* **Betonpflasterstein [m] mit Verschleißvorsatz** (Normalstein mit einer zusätzlichen Verschleißschicht gegen verstärkten Abrieb durch Schwerlastverkehr); *syn.* Betonschwerlastpflasterstein [m].

concrete pavestone [n] [US]**, large-sized** *constr.* ►large-sized paving stone [US]/large-sized paving sett [UK].

concrete paving block [n] **with hard-wearing surface layer** [UK] *constr.* ►concrete paver with protective coating [US].

concrete paving sett [n] [UK]**, large-sized** *constr.* ►large-sized paving stone [US].

concrete paving slab [n] *constr.* ►patterned concrete paving slab.

concrete paving slab [n] [UK]**, precast** *constr.* ►precast paving block [US].

concrete paving stone [n] [UK] *constr.* ►concrete paver [US].

concrete paving unit [n] [UK] *constr.* ►concrete paver [US].

concrete paving unit [n] [AUS]**, interlocking** *constr.* ►interlocking paver [US].

concrete pavior [n] [UK] *constr.* ►concrete paver [US].

concrete paviour [n] [UK] *constr.* ►concrete paver [US].

concrete protection layer [n] *constr.* ►protective screed.

concrete protective slab [n] [US] *constr.* ►protective screed.

1005 concrete reinforcement [n] *constr.* (Strengthening of a concrete structure with cold bendable steel mesh or rods; the product is reinforced concrete, in some countries ferroconcrete); *s* **armadura [f]** (Conjunto de varrillas de acero ultilzadas para dar más estabilidad al hormigón); *syn.* refuerzo [m] (metálico); *f* **ferraillage [m]** (Ensemble des armatures en fer dans le béton pour augmenter la résistance d'un ouvrage) ; *syn.* armature [f] ; *g* **Betonbewehrung [f]** (In Beton verlegte, kalt verformbare Stahleinlagen in Form von Betonstahlmatten oder Stabstählen; das Ergebnis ist Stahlbeton); *syn.* Armierung [f], Bewehrung [f], Stahleinlagen [fpl].

concrete roof slab [n] *constr.* ►concrete deck.

concrete sett [n] [UK] *constr.* ►concrete paver [US].

1006 concrete slab [n] (1) *constr.* (Broad, reinforced piece of concrete; ►poured-in-place concrete slab [US]/in-situ concrete slab [UK]); *syn.* structural slab [n] (OSM 1999); *s* **losa [f] de cimentación (de hormigón armado)** (►losa [de hormigón/cemento] hecha a pie de obra); *syn.* placa [f] de cimentación (de hormigón armado), zapata [f] de cimentación; *f* **radier [m]** (Sol épais et résistant en béton armé, constituant fondation et base

sous un bâtiment, etc. ; ►dalle en ciment coulee sur place) ; *g* **Betonplatte [f]** (Armierte Fundamentplatte, z. B. für Gartenterrassen; ►Betonplatte in Ortbeton).

1007 concrete slab [n] (2) *constr.* (Flat layer of reinforced concrete constructed as the base of, e.g. a building stor[e]y, underground parking garage [US]/underground car park [UK]); *syn.* roof deck [n] [also US]; *s* **placa [f] de hormigón/concreto** (Capa de hormigón armado, p. ej. entrepiso o cubierta de un garaje subterráneo); *syn.* losa [f] de hormigón/concreto; *f* **dalle [f] (en béton)** *syn.* dalle [f] support, plancher [m] dalle ; *g* **Betondecke [f] (3)** (Armierte Betondecke, z. B. als Abschluss eines Stockwerkes, einer Tiefgarage; ►Betondecke 1).

concrete slab [n] [UK]**, in-situ** *constr.* ►poured-in-place concrete slab [US]/in-situ concrete slab [UK].

concrete slab [n] **with imitation stone** *constr.* *►patterned concrete paving slab.

1008 concrete slurry [n] *constr.* (Low viscosity concrete mixture suitable for pouring and pumping; ►shotcrete [US]); *s* **hormigón [m] fluido** (BU 1959; hormigón muy líquido que puede ser bombeado por tuberías; ►gunita); *syn.* hormigón [m] colado, concreto [m] fluido [AL], concreto [m] colado [AL]; *f* **béton [m] fluide** (Béton transporté par pompage et, pour cela, rendu plus fluide; ►béton projeté) ; *syn.* béton [m] coulé, béton [m] pompé ; *g* **Flüssigbeton [m]** (Mit viel Wasser angemachter Beton, um ihn mit Pumpen durch Rohrleitungen zur Einbaustelle zu fördern; ►Spritzbeton).

concrete stone block [n] *constr.* ►cut stone, #3.

1009 concrete tree grate [n] [US] *constr.* (Specific term for ►tree pit cover; ►tree grate [US]/tree grid [UK]; ►tree pit surface); *syn.* concrete tree grid [n] [UK], concrete tree grille [n] [UK]; *s* **cobertura [f] de alcorque (prefabricada) de hormigón** (para proteger el ►alcorque; ►cobertura de alcorque, ►rejilla [metálica] para alcorque); *syn.* pieza [f] (prefabricada de hormigón) para cobertura de alcorque; *f* **panneau [m] de béton perforé pour la protection de pied d'arbre** (*Terme spécifique pour* ►revêtement de protection de pied d'arbre ; ►grille [de protection] de pied d'arbre, ►tour d'arbre) ; *g* **Betonbaumscheibe [f]** (*OB* ►Baumscheibenabdeckung; ►Baumscheibenrost).

concrete tree grid [n] [UK] *constr.* ►concrete tree grate [US].

concrete tree grille [n] [UK] *constr.* ►concrete tree grate [US].

concrete tree vault [n] [US] *constr.* ►precast concrete tree vault.

concrete tree well [n] [UK]**, precast** *constr.* ►precast concrete tree vault [US].

concrete trough [n] [US] *constr.* ►concrete edge footing [US].

concrete unit [n] *constr.* ►precast concrete unit.

concrete unit paver [n] [US] *constr.* ►concrete paver [US]/concrete paving unit [UK].

1010 concrete [n] **with wood board finish** *constr.* (showing rough lumber formwork); *s* **hormigón [m] bruto de encofrado** (Hormigón con marcas del encofrado); *f* **béton [m] brut de décoffrage** (Béton sans traitement de surface laissant apparaître la structure des planches de bois de coffrage) ; *g* **schalungsrauer Sichtbeton [m]** (Beton, dessen sichtbare Oberfläche nach dem Ausschalen die Maserung der Holzbretter ziert); *syn.* schalungsrauer Beton [m].

C

condemnation [n] [US] *adm. leg.* ▶eminent domain (power of condemnation) [US]; ▶land assembly policy [US]/land acquisition policy [UK].

1011 condemnation order [n] [US] *adm. leg.* (In U.S., court order for compulsory taking of private property for a public use, following failure of purchase negotiation and/or independent appraisal to determine amount of just compensation; in U.K., first stage in a compulsory purchase procedure, whereby the appropriated property rights for a certain project have been approved by the authorities in charge. Compulsory purchase powers can also be awarded according to law; ▶condemnation proceedings US]/compulsory purchase proceedings [UK]); *syn.* compulsory purchase order [n] (CPO) [UK]; *s* **orden [f] judicial de expropiación** (según la cual se inicia el procedimiento de expropiación; ▶procedimiento administrativo de expropiación); *f* **ordonnance [f] d'expropriation** (Acte administratif opérant le transfert de propriété ou de droits réels immobiliers lors d'une expropriation lorsque celui-ci n'a pu être effectué par un accord à l'amiable ; ▶procédure d'expropriation) ; *syn.* arrêté [m] d'expropriation ; *g* **Enteignungsanordnung [f]** (Verwaltungsakt als erste Stufe im Enteignungsverfahren, bei der das Eigentumsrecht zu Gunsten eines bestimmten Vorhabens von der zuständigen Verwaltungsbehörde für zulässig erklärt wird. Die Enteignungsermächtigung oder -voraussetzung kann aber auch schon von Gesetzes wegen verliehen sein, z. B. durch einen Planfeststellungsbeschluss; ▶Enteignungsverfahren).

1012 condemnation proceedings [npl] [US] *adm. leg.* (Legal procedure executed by authorities by which private property is expropriated for public purposes; ▶condemnation order [US]/compulsory purchase order [COP] [UK]); *syn.* compulsory purchase proceedings [npl] [UK]; *s* **procedimiento [m] administrativo de expropiación** (▶orden judicial de expropiación); *f* **procédure [f] d'expropriation** (Procédure administrative en vue de déposséder une personne d'un bien immobilier. L'expropriation ne peut être prononcée que si elle a été précédée d'une déclaration d'utilité publique intervenue à la suite d'une enquête préalable de droit commun ; ▶ordonnance d'expropriation) ; *g* **Enteignungsverfahren [n]** (Verfahren, das durch legalen behördlichen/staatlichen Eingriff privates Eigentum für öffentliche, dem Allgemeinwohl dienende Zwecke entzieht; nach Art. 14 III GG ist eine **E.** nur zum Wohle der Allgemeinheit zulässig. Sie darf nur durch Gesetz oder auf Grund eines Gesetzes erfolgen, das Art und Ausmaß der Entschädigung regelt; die Entschädigung muss unter gerechter Abwägung der Interessen der Allgemeinheit und der Beteiligten erfolgen. Das Enteignungsrecht ist weitgehend Landesrecht. Dort sind Enteignungszweck, Gegenstand der Enteignung, Zulässigkeit, Umfang, Art und Maß der Entschädigung sowie das **E.** eingehend geregelt; Bundesgesetze, z. B. §§ 85 ff BauGB für baurechtliche Grundstücksenteignungen, §§ 19 ff FStrG für den Bau von Bundesfernstraßen oder § 11 des Energiewirtschaftsgesetzes für Freileitungen sehen auch Enteignungstatbestände vor; ▶Enteignungsanordnung).

condiment garden [n] *gard.* ▶spices garden.

condition [n] *constr.* ▶existing condition before planning starts, ▶maintenance condition.

condition [n]**, friable soil** *pedol.* ▶granular soil structure.

condition [n]**, oligotrophic** *limn. pedol.* ▶oligotrophy.

1013 conditional building permit [n] [US] *adm.* (Permit to work under special restrictions or requirements; ▶building permit, ▶zoning and building regulations); *syn.* conditional planning permission [n] [UK]; *s* **licencia [f] de construcción con restricciones** (▶licencia de construcción, ▶legislación urbanística); *syn.* permiso [m] de construcción con restricciones, autorización [f] de construcción con restricciones; *f* **permis [m] de**

construire permissionnel (Permis accordé sous réserve de l'observation de prescriptions spéciales ; ▶permis de construire, ▶droit de construire) ; *g* **Baugenehmigung [f] mit Auflagen** (▶Baugenehmigung mit Nebenbestimmungen, die ein Dulden, Unterlassen oder ein bestimmtes Tun, z. B. die extensive Begrünung eines Daches, vorschreiben; ▶Baurecht).

1014 conditional development permit [n] [US] *adm. leg.* (Action by public authority/agency to authorize planning for a specified project development with stipulated requirements); *syn.* conditional planning permission [n] [UK]; *s* **permiso [m] con condiciones** *syn.* licencia [f] con condiciones, licencia [f] con medidas correctoras o de atenuación; *f* **autorisation [f] accordée sous réserve de l'observation de prescriptions spéciales** (Prescriptions spéciales accompagnant l'octroi d'une autorisation, du permis de construire, etc. ; cf. C. urb., art. R. 111-14-1, 111-15, 111-21) ; *syn.* permis [m] accordé sous réserve de l'observation de prescriptions spéciales ; *g* **Genehmigung [f] mit Auflagen** (*Verwaltungsakt* im öffentlichen Recht eine amtliche Gestattung eines Antrages mit auferlegten Bestimmungen, die z. B. eine Befristung, einen Vorbehalt, Bedingungen oder ein Tun, Dulden oder Unterlassen beinhalten; cf. § 36 VwVfG).

conditional planning permission [n] [UK] *adm.* ▶conditional building permit [US]; ▶conditional development permit [US].

condition [n] **before planning starts** [US] *ecol. plan.* ▶existing condition before planning starts.

1015 condition [n] **for final acceptance** [US] *constr. contr.* (Condition of a construction project in accordance with the contract specifications before the issue of a ▶certificate of practical completion); *syn.* situation [n] of practical completion [UK]; *s* **estado [m] de obra en condiciones de recepción** (Estado de obra [paisajista] que corresponde a lo estipulado en el contrato y que se debe mantener por medio de los ▶trabajos de mantenimiento pre-entrega hasta la aceptación de la obra; ▶acta de recepción/aceptación [de obra]); *syn.* estado [m] de obra en condiciones de aceptación; *f* **ouvrage [m] conforme au marché** (État des travaux d'espaces verts avant la réception pour lequel des travaux d'entretien appropriés en garantissent le développement ; ▶entretien [d'espaces verts] jusqu'à la réception des travaux, ▶procès-verbal de réception) ; *g* **abnahmefähiger Zustand [m]** (Zustand von [Landschafts]bauarbeiten, der den vertraglich zugesicherten Eigenschaften der Bauleistungen bis zur Abnahme durch die ▶Fertigstellungspflege entspricht und eine sichere Weiterentwicklung bei sachgerechter Pflege garantiert; cf. DIN 18 917; ▶Abnahmeniederschrift).

conditions [npl] **of engagement and scale of professional charges** [UK] (obs. since 1986) *contr. leg. prof.* ▶Guidance for Clients on Fees [UK].

1016 conditions [npl] **of entry** *prof.* (Requirements for participation in an ▶design competition 1 or ▶final design competition [US]/realization competition [UK]); *s* **bases [fpl] de concurso y pliego [m] de condiciones** (Condiciones de participación en un ▶concurso de ideas o en ▶concurso de realización); *f* **dossier [m] de candidature d'un concours** (Document faisant partie du règlement de la consultation pour un ▶concours d'idées ou un ▶concours de conception-construction) ; *g* **Wettbewerbsbedingungen [fpl]** (Teilnahmebedingungen für einen ▶Ideenwettbewerb oder ▶Realisierungswettbewerb).

1017 condominium [n] *urb.* (Multi-unit building containing individual apartment ownerships and partial ownerships of the property and communal facilities of the building); *s* **condominio [m] [AL]** (Edificio multifamiliar en el que las viviendas son de propiedad individual); *syn.* casa [f] de pisos de propiedad horizontal [Es], propiedad [f] horizontal; *f* **immeuble [m] de copro-**

priété (Immeuble dont la propriété est répartie entre plusieurs personnes, par lots comprenant chacun une partie privative et une quote-part de parties communes) ; *g* **Wohnblock [m] mit Eigentumswohnungen** (Mehrgeschossiges Gebäude, bei dem jeder Wohnungseigentümer einen Miteigentumsanteil am Grundstück und an den gemeinsam genutzten Teilen des Gebäudes hat).

conduction [n] **of surface water/storm water** *constr. urb. wat' man.* ▶surface water drainage.

conductivity [n] *pedol.* ▶hydraulic conductivity.

1018 cone [n] (1) *constr.* (SPON 1986, 171; conical, excentric ▶riser unit [US]/chamber section [UK] used to reduce the diameter of a drain basin to receive an inlet grate at pavement surface—inclined on one side to allow entry for inspection; ▶manhole); *syn.* taper unit [n]; *s* **cono [m] asimétrico de pozo de registro** (▶Anillo de pozo de registro que va disminuyendo paulatinamente adquiriendo forma cónica; ▶pozo de registro 2); *syn.* ahusamiento [m] de pozo de registro; *f* **hotte [f] de regard** (▶Virole conique ou pyramidale avec couronnement incorporé ou non pour recevoir cadre et tampon d'obturation, les ▶regards visitables ayant une hotte conique sur un côté) ; *syn.* cône [m] d'entrée (de regard) ; *g* **Schachthals [m]** (Konisch zulaufender ▶Schachtring. Bei einem besteigbaren Schacht [Einsteigschacht] ist der **S.** einseitig konisch zulaufend; ▶Einstiegsschacht); *syn.* Schachtkonus [m].

cone [n] (2) *geo.* ▶alluvial cone.

cone [n]**, dejection** *geo.* ▶alluvial cone.

1019 configuration [n] **of an area** *arch. constr.* (Topographic design of an area/terrain; ▶ground modeling [US]/ground modelling [UK], ▶landscape design); *syn.* topographic layout [n]; *s* **diseño [m] topográfico (de un terreno)** (Topografía prevista para un área de un proyecto de construcción paisajista; ▶diseño paisajístico, ▶modelado del terreno); *syn.* configuración [f] topográfica; *f 1* **modelage [m] (topographique) du terrain** (*Travaux de terrassement* élaboration du relief sur lequel s'inscrivent les formes et s'édifient les volumes d'un terrain ; ▶création de paysages, ▶mise en forme des terres, ▶modelé de sols) ; *f 2* **configuration [f] du terrain** (Caractéristiques topographiques d'un terrain existant) ; *g* **höhenmäßige Gestaltung [f] eines Geländes** (Topografische Formgebung einer Fläche; ▶Bodenmodellierung, ▶Landschaftsgestaltung); *syn.* Ausformung [f] eines Geländes, Modellierung [f] eines Geländes.

configuration [n] **of a plot** *surv.* ▶plot shape.

1020 confined aquifer [n] *hydr.* (Aquifer overlain and underlain by impervious or almost impervious formations; WMO 1974); *s* **acuífero [m] confinado** (Acuífero encerrado por formaciones impermeables o casi impermeables; WMO 1974); *syn.* acuífero [m] artesiano (±); *f* **formation [f] aquifère captive** (Formation aquifère limitée en dessus et en dessous par une couche imperméable ; WMO 1974) ; *syn.* aquifère [m] à nappe captive ; *g* **Grundwasserstockwerk [n]** (Ein von oben und unten durch eine undurchlässige Schicht abgegrenzter Grundwasserleiter. **G.e** werden von oben nach unten gezählt).

confined planter [n] [US] *constr.* ▶planter with a raised edge.

confining bed [n] *hydr.* ▶aquitard.

confining layer [n] *hydr.* ▶aquitard.

confining stratum [n] *hydr.* ▶aquitard.

confirmation [n] [UK]**, tenderer's** *contr.* ▶bidder's affidavit [US].

conflict [n] *plan.* ▶land use conflict.

1021 congelifraction [n] *constr. geo.* (Disintegration of pavers, rocks, walls, or fragmentation by frost action of moisture contained in splits, cracks, or pores); *syn.* frost splitting [n]; *s* **gelifracción [f]** (Proceso de fragmentación de roca ligado a las alternancias de hielo-deshielo a que es sometida el agua que colmata los poros y fisuras de aquélla; DGA 1986, 83); *syn.* crioclastia [f], gelivación [f]; *f* **gélifraction [f]** (Fragmentation d'une pierre, d'une roche, d'un mur sous l'effet de l'alternance du gel et du dégel de l'eau contenue dans ses fissures ou ses pores. Les fragments détachés sont les « gélifracts ») ; *syn.* cryoclastie [f], gélivation [f] ; *g* **Frostsprengung [f]** (Sprengung von Steinen, Platten, Felsen sowie Mauern oder Abplatzen von Teilen davon durch in Spalten oder Rissen gefrorenes Wasser); *syn.* Frostspaltung [f], Frostverwitterung [f].

1022 conifer [n] *bot.* (Lat. conus >cone<; ferre >bearing<; needle-leaved plants occurring world-wide, classified under *Coniferopsida* [▶needle-leaved species], as gymnosperms with unisexual flowers and cones of various forms, they belong to 7 families: **1. Pinaceae**, e.g. fir *[Abies]*, spruce *[Picea]*, pine *[Pinus]*, larch *[Larix]*, cedar *[Cedrus]*, Douglas fir *[Pseudotsuga]*; **2. Araucariaceae**, e.g. monkey-puzzle tree *[Araucaria]*; **3. Podocarpaceae**, e.g. podocarpus *[Podocarpus]*, celery pine *[Phyllocladus]*, Maniu *[Dacrydium]*; **4. Cupressaceae** incl. Taxodiaceae, e.g. cypress *[Cupressus]*, false cypress *[Chamaecyparis]*, bald cypress *[Taxodium]*, thuja/arbor-vitae *[Thuja]*, juniper *[Juniperus]*, mammoth tree *[Sequoiadendron]*, metasequoia *[Metasequoia]*, Japanese cedar *[Cryptomeria]*; **5. Sciadopityaceae**, with one species, umbrella pine *[Sciadopitys verticillata]*; **6. Cephalotaxaceae** with only one genus, plum yew *[Cephalotaxus]* and **7. Taxaceae** with one genus, yew *[Taxus]*: in Europe Taxus baccata, in North America Taxus canadensis; cf. LB 2008 et GAY 1993); *syn.* coniferous species [n]; *s* **especie [f] conífera** (Término genérico para árboles y arbustos gimnospermos de floración unisexual, con hojas aciculares y piñas de diferentes formas, que pertenecen a la clase de las *Coniferópsida* [▶especie acicufolia] formada por 7 familias: **1. Pinaceae**, p. ej. el abeto *[Abies]*, la picea *[Picea]*, el pino *[Pinus]*, el alerce *[Larix]*, el cedro *[Cedrus]*, el abeto Douglas *[Pseudotsuga]*; **2. Araucariaceae**, p. ej. la araucaria *[Araucaria]*; **3. Podocarpaceae**, p. ej. *[Podocarpus]*, *[Phyllocladus]*, *[Dacrydium]*; **4. Cupressaceae** incl. Taxodiaceae, p. ej. el ciprés *[Cupressus]*, el falso ciprés *[Chamaecyparis]*, la tuya *[Thuja]*, el enebro *[Juniperus]*, la secuoya gigante *[Sequoiadendron]*, la metasequoya *[Metasequoia]*, el cedro japonés *[Cryptomeria]*; **5. Sciadopityaceae**, con una especie *[Sciadopitys verticillata]*; **6. Cephalotaxaceae** con un solo género *[Cephalotaxus]* y **7. Taxaceae** también con un solo género, el tejo *[Taxus]*: en Europa Taxus baccata, en Norteamérica Taxus canadensis; ▶especie acicufolia); *syn.* conífera [f]; *f* **conifère [m]** (lat. conus « cône » ; ferre « porter » ; classe d'espèces ligneuses appartenant à l'orde des Coniférales, embranchement des pinophytes, sous-embranchement des Gymnospermes, que caractérisent un feuillage en aiguilles ou en écaille, des inflorescences femelles en cônes à ovules, des graines nues. L'anglais « softwood » désigne aussi les essences conifèrales elles mêmes ; cf. DFM 1975 ; dans la classification phylogénétique [NCBI] on distingue 7 familles : **1.** la famille des Pinacées [**Pinaceae**] ou abiétacées dans laquelle on rencontre le sapin *[Abies]*, l'épicea *[Picea]*, le pin *[Pinus]*, le mélèze *[Larix]*, le cèdre *[Cedrus]*, le douglas *[Pseudotsuga]* ; **2.** la famille de Araucariacées [**Araucariaceae**] avec l'araucaria *[Araucaria]* ; **3.** la famille des Podocarpacées [**Podocarpaceae**] qui compte entre autre le podocarpus *[Podocarpus]*, le phyllocladus *[Phyllocladus]*, le rimu *[Dacrydium]*; **4.** la famille des Cupressacées ou Cupressinées [**Cupressaceae**] y compris la famille des Taxodiacées qui est devenue la sous-famille des Taxodiacées comprenant entre autres, le cyprès *[Cupressus]*, le faux-cyprès *[Chamaecyparis]*, le cyprès-chauve *[Taxodium]*, le thuya *[Thuja]*, le genévrier *[Juniperus]*, le séquoia géant *[Sequoiadendron]*, séquoïa *[Metasequoia]*, le cèdre du Japon *[Cryptomeria]*; **5.** la famille des Sciadopityacées [**Sciadopityaceae**] comprenant un seul genre et une seule espèce : le pin parasol du Japon *[Sciatopitys verticillata]*; **6.** la famille des Céphalotaxacées [**Cephalotaxaceae**] comprenant un seul genre le céphalotaxus *[Cephalotaxus]* et **7.** la famille des Taxacées [**Taxaceae**] comprenant un seul genre l'if *[Taxus]* : en Europe l'if commun *[Taxus baccata]*, en Amérique du Nord l'if du Canada *[Taxus canadensis]* ; ►espèce à aiguilles) ; *g* **Konifere** [f] (Lat. conus >Zapfen<; ferre >tragen<; Nadelgehölz, der weltweit verbreiteten Klasse der ***Coniferópsida*** [Nadelbäume, ►Nadelbaum], mit getrenntgeschlechtlichen Blüten und unterschiedlich geformten Zapfen, zu der sieben Familien gehören: **1. Pinaceae**, u. a. mit Tanne *[Abies]*, Fichte *[Picea]*, Kiefer *[Pinus]*, Lärche *[Larix]*, Zeder *[Cedrus]*, Douglasie *[Pseudotsuga]*; **2. Araucariaceae**, z. B. mit der Araukarie *[Araucaria]*; **3. Podocarpaceae**, u. a. mit Steineibe *[Podocarpus]*, Blatteibe *[Phyllocladus]*, Maniu *[Dacrydium]*; **4. Cupressaceae** incl. Taxodiaceae, u. a. mit Zypresse *[Cupressus]*, Scheinzypresse *[Chamaecyparis]*, Sumpfzypresse *[Taxodium]*, Thuja *[Thuja]*, Wacholder *[Juniperus]*, Mammutbaum *[Sequoiadendron]*, Urweltmammutbaum *[Metasequoia]*, Sicheltanne *[Cryptomeria]*; **5. Sciadopityaceae** mit der einzigen Art Schirmtanne *[Sciadopitys verticillata]*; **6. Cephalotaxaceae** mit nur einer Gattung der Kopfeibe *[Cephalotaxus]* und **7. Taxaceae** mit der einzigen Gattung Eibe *[Taxus]*: z. B. in Europa *Taxus baccata*, in Nordamerika *Taxus canadensis*; cf. LB 2008 et GAY 1993); *syn.* ►Nadelgehölz [n].

conifer litter [n] *pedol.* ►neddle straw [US]/needle litter [US].

1023 coniferous evergreen thicket/shrubland [n] *phyt.* (Community of plants primarily composed of creeping or dwarf needle-leaved shrubs, growing on the periphery of the ►tree line or ►timberline; ►krummholz community); *s* **matorral** [m] **achaparrado de coníferas** (Comunidad de coníferas achaparradas y de arbustos enanos que crecen al borde del ►límite del bosque, ►límite de los árboles; ►matorral de altura); *syn.* matorral [m] denso sempervirente aciculifoliado; *f* **brousse** [f] **sempervirente de résineux** (Formation arborée peuplée d'espèces de résineux prostrés ou rabougris localisée à la limite naturelle de la forêt/zone de combat ; ►formation de Krummholz, ►limite des forêts, ►limite [supérieur] des arbres et des arbustes) ; *syn.* brousse [f] subalpine, fourré [m] de résineux ; *g* **immergrünes Nadelgebüsch [n]** (Vorwiegend aus kriechenden oder durch Schneemassen niedergedrückten Nadelhölzern bestehende Pflanzenformation am Rande der ►Waldgrenze/►Baumgrenze; ►Krummholzgesellschaft).

1024 coniferous forest [n] *for. phyt.* (Forest consisting entirely or mainly of softwood trees, usually needled evergreens, ranging from temperate to subtropical latitudes; ►conifer, ►needle-leaved species); *s* **bosque** [m] **de coníferas** (Población vegetal constituida mayormente por ►especies coníferas que se presenta de las latitudes templadas a las subtrópicas; ►especie acicufolia); *syn.* bosque [m] acicufolio, bosque [m] acicular; *f* **forêt** [f] **de conifères** (Peuplement végétal constitué principalement de ►conifères dans les régions de latitude tempérée à subtropicale ; ►espèce à aiguilles) ; *syn.* forêt [f] de résineux, forêt [f] coniférienne, forêt [f] résineuse ; *g* **Nadelwald [m]** (Vorwiegend aus ►Koniferen zusammengesetzter Vegetationsbestand in den gemäßigten bis subtropischen Breiten; ►Nadelgehölz); *syn.* Koniferenwald [m].

coniferous species [n] *bot. for.* ►conifer.

1025 coniferous tree [n] *bot. for.* (Tree bearing seeds in a cone, mainly needled evergreens, except Larch *[Larix]*, Bald

Cyprus *[Taxoddium distichum]*, Metasequoia *[Metasequoia glyptostroboïdes]* and Golden Larch *[Pseudolarix amabilis]*; ►conifer, ►needle-leaved species); *syn.* neadle-leaved tree [n], *common usage* conifer [n]; *s* **árbol** [m] **acicufolio** (Término genérico para especie arbórea con hojas aciculares y sempervirente, con excepción del alerce *[Larix]*, el falso alerce *[Pseudolarix]*, el glyptostrobus *[Glyptostrobus]*, la metasecuoya *[Metasequoia]* y el ciprés calvo *[Taxodium]* que son de aguja caediza; ►especie conífera, ►especie acicufolia); *f* **arbre** [m] **conifère** (Terme désignant un arbre riche en résine, à feuillage persistant à l'exception du Mélèze *[Larix]*, du faux Mélèze *[Pseudolarix]*, du Cyprès chauve *[Taxodium distichum]*, du Glyptostrobus *[Glyptostrobus]*, et du Métaséquoia *[Metasequoia glyptostroboïdes]* et dont les organes reproducteurs sont le plus souvent des chatons et des cônes [les réputées pommes de pin ou pignes] ; la plupart des ►conifères possèdent des cellules sécrétrices de résine dans leurs écorces, leurs feuilles ou leur bois, d'où l'appellation courante de **résineux** ou **arbre résineux** ; le terme de résineux, très souvent utilisé dans le commerce [arbre forestier], désigne aussi le bois de ces arbres ; ►espèce à aiguilles) ; *syn.* espèce [f] de conifères, bois [m] résineux, bois [m] de conifères, bois [m] mou [CDN], *dans le langage courant* conifère [m] ; *g* **Nadelbaum [m]** (Meist immergrüner, nadelblättriger Baum — mit Ausnahme von Lärche *[Larix]*, Sumpfzypresse *[Taxodium distichum]*, Urweltbaum *[Metasequoia glyptostroboïdes]* und Goldlärche *[Pseudolarix amabilis]*, die Laub abwerfend sind; ►Nadelgehölz, ►Konifere).

connecting green finger [n] *landsc. urb.* ►green finger connection.

connecting road [n] [US] *plan. trans.* ►access road.

1026 connecting route [n] *plan. trans.* (Vehicular or pedestrian way connecting two places or communities); *syn.* communication route [n]; *s* **vía** [f] **de conexión** (Calle o camino para vehículos o peatones que conecta dos lugares entre sí); *f* **voie** [f] **de liaison** (Chemin/route reliant deux quartiers, deux agglomérations) ; *syn.* voie [f] de raccordement ; *g* **Verbindungsweg [m]** (Fuß- oder Fahrweg, der zwei Orte miteinander verbindet).

1027 connecting section [n] *trans.* (Roadway segment serving as a link to an existing road; ►interchange); *s* **tramo [m] de enlace** (Trozo de carretera que empalma con una vía de transporte existente; ►nudo de enlace); *f* **section** [f] **de raccordement** (sur une voie existante ; ►diffuseur) ; *g* **Anschlussstück [n]** (Verbindungsstück zu einer bestehenden Straße; ►Anschlussstelle).

1028 connection [n] (1) *trans. urb.* (Linkage of an area to a transportation system); *syn.* linkage [n]; *s* **conexión** [f] **(1)** (Unión de una zona al sistema de transporte público); *syn.* empalme [m]; *f* **desserte** [f] (d'une ville, d'une région par les transports en commun ; *termes spécifiques* desserte routière, desserte ferroviaire) ; *g* **Anbindung** [f] **(1)** (Anschluss eines Gebietes an Verkehrseinrichtungen); *syn.* Anschluss [m].

1029 connection [n] (2) *constr. urb.* (Linkage of an area to the public utility system); *s* **conexión** [f] **(2)** (Empalme de una zona a las redes de abastecimiento y de saneamiento); *f* **1 raccordement** [m] (1) *urb.* (Ensemble des mesures de raccordement d'une installation nouvelle ou d'une construction aux divers réseaux publics : voirie réseau d'assainissement ou de distribution EDF, GDF, PTT, eau de ville — fait partie du règlement du PLU/POS) ; *f* **2 branchement** [m] *constr.* (des réseaux d'alimentation en fluides divers et des évacuations des eaux usées sur les réseaux publics correspondants) ; *syn.* jonction [f] ; *g* **Anbindung** [f] **(2)** (Anschluss eines Gebietes an das Versorgungs- und Entsorgungsnetz); *syn.* Anschließen [n, o. Pl.].

connection [n] **to** *constr.* ►junction (1).

connection [n] **to a transportation system** *trans. urb. plan.* ▶traffic linkage.

consent with [loc]**, in** *leg.* ▶by mutual agreement.

conservation [n] *conserv. leg.* ▶habitat conservation; *conserv. nat'res.* ▶heathland conservation; *agr. conserv. nat'res.* ▶hedgerow conservation; *conserv. ecol.* ▶special area of conservation (SAC); *hydr. nat'res.* ▶water resources conservation; *conserv. zool.* ▶wildlife conservation.

conservation agency [n] [US] *adm. conserv.* ▶conservation/recreation agency [US].

conservation area [n] [UK] *arch. conserv' hist. leg.* ▶historic district [US].

conservation asset [n] *conserv. leg.* ▶asset.

conservation asset base [n] [UK] *conserv. leg.* ▶asset, #2.

conservation body [n] [UK]**, nature** *adm. conserv.* ▶conservation/recreation agency [US].

conservation contracts [n] *agr. conserv. pol.* ▶contract-based nature conservation.

1030 conservation criterion [n] *conserv'hist. leg.* (Legal or technical standard used as a basis of comparison in evaluating the conservation value of individual buildings, structures, features or whole areas, or in judging the nature conservation value of natural elements in relation to their degree of naturalness, endangerment, diversity, rarity, representation, ecological stability, use for research, type and extent of human influences); *syn.* conservation standard [n]; *s* **criterio** [m] **de protección** (Estándar legal o científico utilizado para evaluar si un espacio natural o monumento histórico cumple las características para ser declarado como protegido); *f* **critère** [m] **de protection** (Critère légal ou scientifique appréciant la valeur d'un immeuble au titre des monuments historiques ou celle d'un site, paysage ou milieu naturel au titre des zones protégées, comme p. ex. le degré d'artificialisation, la menace, la diversité, la rareté, la stabilité écologique, l'intérêt pour la recherche, le caractère et l'importance de l'influence humaine) ; *g* **Schutzkriterium** [n] (...kriteria [pl], ...kriterien [pl]; rechtliches [Legalkriterium] oder fachspezifisches Kriterium [Fachkriterium] für die Beurteilung der Denkmalschutzwürdigkeit von Einzel- oder Gesamtanlagen oder für die Beurteilung der Naturschutzwürdigkeit von Erscheinungsformen oder Komplexen wie z. B. hinsichtlich der Natürlichkeit, Gefährdung, Vielfalt, Seltenheit, Repräsentanz, ökologischen Stabilität, Forschungsnutzung, der Art und des Ausmaßes menschlichen Einflusses).

conservation crop [n] *conserv. land'man.* ▶soil conservation crop.

conservation grass mix [n] [US] *constr.* ▶low-maintenance grass type.

1031 conservationist [n] *conserv.* (1. Person who supports environment-minded thinking. 2. Person who is actively engaged in nature conservation; ▶ecologist, ▶environmentalist); *s* **conservacionista** [m/f] (Persona que apoya la conservación de la naturaleza activa o pasivamente; ▶ambientalista, ▶ecólogo/a, ▶ecologista); *f* **écologiste** [m] (1) (1. Personne adhérent à l'idée de protection de la nature et de l'environnement. 2. Personne engagée activement dans la lutte pour la protection de la nature et de l'environnement, militant de l'écologisme ; le terme allemand « *Naturschützer* » est syn. de « *Umweltschützer* » ; ▶écologue, ▶écologiste 2, ▶environnementaliste) ; *syn.* défenseur [m] de la nature ; *terme plutôt utilisé dans les publications internationales* conservationiste [m] ; *peu usité* protecteur [m] de la nature ; ▶écologue, ▶écologiste) ; *g* **Naturschützer/-in** [m/f] (1. Anhän-

ger des Naturschutzgedankens. **2.** Jemand, der sich für den Naturschutz einsetzt; ▶Ökologe, ▶Umweltschützer/-in).

1032 conservation measure [n] *conserv. urb.* (Measure to conserve townscape, countryside, historical buildings, parks, etc.); *s* **medidas** [fpl] **de mantenimiento** (de edificios, de jardines históricos, etc.); *f* **mesure** [f] **de sauvegarde** *syn.* mesure [f] conservatoire ; *g* **Erhaltungsmaßnahme** [f] (Tätigkeiten, die der Erhaltung von Bauwerken, des Stadtbildes, von historischen Freiraumanlagen etc. dienen).

Conservation [n] **of African-Eurasian Waterbirds** *conserv. ecol. pol.* ▶Convention on the Conservation of African-Eurasian Migratory Waterbirds [AEWA].

1033 conservation [n] **of coastal landscapes** *conserv. landsc.* (All administrative, planning and technical measures for the protection of coastal areas in order to preserve the cultural heritage); *s* **preservación** [f] **del paisaje litoral** (Conjunto de medidas administrativas, de planificación y técnicas empleadas para proteger las áreas costeras y de esa manera preservar el patrimonio natural y cultural); *f* **protection** [f] **et mise en valeur des milieux naturels du littoral** (Mesures diverses de préservation des paysages littoraux [dunes et landes, plages, zones humides, etc.], p. ex. application des réglementations aux espaces de haute qualité, classement en réserves naturelles ou en forêt de protection, extension des espaces naturels sensibles, élaboration de schémas de mise en valeur de la mer) ; *syn.* protection [f] de l'espace littoral, sauvegarde [f] des paysages côtiers, protection [f] des zones côtières ; *g* **Erhaltung** [f] **der Küstenlandschaft** (Maßnahmen mit dem Ziel, erhaltenswerte Abschnitte einer Küste meist unter Landschafts- oder Naturschutz zu stellen).

Conservation [n] **of European Wildlife and Natural Habitats** *conserv. leg. pol* ▶Convention on the Conservation of European Wildlife and Natural Habitats.

1034 conservation [n] **of flora and fauna** *conserv. phyt. zool.* (All measures which are necessary to preserve the natural habitats of populations of wild animal and plant species for long-term population dynamics, so that these populations may remain livable elements of their natural environment; ▶species conservation, ▶wildlife conservation); *s* **preservación** [f] **de la flora y fauna silvestres** (Todas las medidas necesarias para preservar a largo plazo los hábitats naturales de las poblaciones de animales y plantas silvestres, de manera que éstas puedan sobrevivir como elementos vivos de sus hábitats naturales; ▶protección de especies [de flora y fauna], ▶protección de la fauna salvaje); *f* **conservation** [f] **de la faune et de la flore sauvages** (Ensemble de mesures requises pour maintenir ou rétablir les habitats naturels et les populations d'espèces de faune et de flore sauvages dans un état favorable afin que, dans la dynamique de la population des espèces, celles-ci soient susceptibles de continuer à long terme à constituer un élément viable des habitats naturels auxquels elles appartiennent ; cf. Art. 1 Directive du Conseil n° 92/43/CEE du 21.05.1992 ; ▶protection des espèces [animales et végétales], ▶protection de la faune sauvage) ; *syn.* préservation [f] du patrimoine biologique, conservation [f] des espèces animales et végétales, préservation [f] de la faune et de la flore (sauvages) ; *g* **Erhaltung** [f, o. Pl.] **der Tier- und Pflanzenwelt** (Alle Maßnahmen, die erforderlich sind, um die natürlichen Lebensräume und die Populationen wild lebender Tier- und Pflanzenarten in einem günstigen Erhaltungszustand so zu erhalten, damit sie sich in ihrer Populationsdynamik langfristig so entwickeln können, dass sie lebensfähige Elemente ihrer Lebensräume bleiben können; cf. Art. 1 RL 92/43/EWG des Rates v. 21.05.1992; ▶Artenschutz, ▶Faunenschutz).

C

1035 conservation [n] **of forests** *landsc. conserv.* (All measures for sustainable protection and management of forests in accordance with the principles that assure retention of its capacity for regeneration, its optimum economic use and social enjoyment; cf. SAF 1983); *syn.* forest protection [n]; *s* **gestión [f] de recursos forestales** (Conjunto de medidas destinadas a manejar la conservación y el aprovechamiento integrados y sistémicos de los bosques que considera la totalidad de los posibles usos del mismo, desde la madera al paisaje; cf. DINA 1987); *f 1* **conservation [f] des forêts** (L'aménagement et la gestion des forêts conformément aux principes qui garantissent qu'elles procureront indéfiniment un optimum de bien-être économique et social ; DFM 1975) ; *syn.* sauvegarde [f] des forêts ; *f 2* **protection [f] forestière** (L'ensemble des activités entreprises en vue d'éviter, de limiter ou de supprimer les dommages causés par l'homme, par les organismes vivants nuisibles ou par les troubles atmosphériques divers ; cf. DFM 1975) ; *g* **Walderhaltung [f, o. Pl.]** (Gesamtheit der Maßnahmen, die einen nachhaltigen Schutz und eine nachhaltige Nutzung für die ökonomischen und sozialen Belange sicherstellen); *syn.* Schutz [m, o. Pl.] des Waldes.

1036 conservation [n] **of historic gardens** *conserv' hist. gard'hist.* (Administrative, planning and technical measures for the protection or restoration of municipal or private gardens in order to preserve the cultural heritage of ►fine garden design); *s* **conservación [f] del patrimonio de jardines históricos** (Conjunto de medidas administrativas, técnicas y de planificación necesarias para preservar y restaurar los jardines históricos; ►arte de jardinería) ; *f* **gestion [f] et mise [f] en valeur des jardins historiques** (Mesures administratives d'urbanisme et techniques prises au titre des monuments historiques en vue de la protection et de la mise en valeur de jardins publics ou privés, expression du patrimoine culturel de l'►art de jardins) ; *g* **Gartendenkmalpflege [f, o. Pl.]** (Administrative, planende und technische Maßnahmen zur Erhaltung oder Wiederherstellung von erhaltenswürdigen staatlichen, gemeindlichen oder privaten Gartenanlagen zur Bewahrung des kulturellen Erbes der ►Gartenkunst).

conservation [n] **of historic monuments** [UK] *adm. conserv'hist. leg.* ►preservation of historic landmarks [US].

Conservation [n] **of Migratory Species of Wild Animals (CMS), Convention on the** *conserv. leg. pol.* ►Bonn Convention.

1037 conservation [n] **of natural habitats and of wild fauna and flora** *conserv. leg.* (The main aim of this EC Habitats Directive is to promote the maintenance of biodiversity in the European territory of the Member States to take measures to maintain or restore natural habitats and wild species at a favo(u)rable ►conservation status 2, introducing robust protection for those habitats and species of European importance, taking account of economic, social and cultural requirements and regional and local characteristics; cf. Article 2 Council Directive 92/43/EEC of 21 May 1992; amended by 97/62 EEC of 27.10. 1997; ►conservation of flora and fauna, ►Habitats Directive); *s* **conservación [f] de los hábitats naturales y de la fauna y flora silvestres** (Objetivo de la directiva europea de contribuir a conservar la biodiversidad europea, mediante el establecimiento de una red ecológica y un régimen jurídico de protección de las especies silvestres. Ésta identifica alrededor de 200 tipos de hábitats, unas 300 especies animales y casi 600 especies vegetales como de interés comunitario, y establece la necesidad de protegerlos, para lo cual obliga a que se adopten medidas para mantenerlos o restaurarlos en un estado favorable de conservación [►estado de conservación de una especie]. Corresponde a los Estados miembros de la UE determinar sus zonas especiales

de conservación y establecer, en su caso, planes de gestión que combinen su conservación a largo plazo con las actividades económicas y sociales; cf. Directiva 92/43/CEE del Consejo, de 21 de mayo de 1992, relativa a la Conservación de los Hábitats Naturales y de la Fauna y Flora Silvestres; ►Directiva de Hábitats, ►preservación de la flora y fauna silvestres); *f* **conservation [f] des habitats naturels ainsi que de la faune et de la flore sauvages** (Ensemble de mesures requises pour maintenir ou rétablir les habitats naturels et les populations d'espèces de faune et de flore sauvages dans un état favorable afin que dans la dynamique de la population des espèces celles-ci soient susceptibles de continuer à long terme de constituer un élément viable des habitats naturels auxquels elles appartiennent ; cf. Art. 1 Directive du Conseil n° 92/43/CEE du 21.05.1992 ; ►conservation de la faune et de la flore sauvages, ►Directive Habitat Faune Flore, ►état de conservation) ; *syn.* préservation [f] du patrimoine biologique, conservation [f] des espèces animales et végétales, préservation [f] de la faune et de la flore (sauvages) ; *g* **Erhaltung [f, o. Pl.] der natürlichen Lebensräume sowie der wild lebenden Tiere und Pflanzen** (Sicherung der Artenvielfalt durch die Erhaltung der natürlichen Lebensräume sowie der wild lebenden Tiere und Pflanzen im europäischen Gebiet der Mitgliedstaaten, für die der Vertrag gilt; die Maßnahmen sollen darauf abzielen, einen günstigen ►Erhaltungszustand der natürlichen Lebensräume und wild lebenden Arten von gemeinschaftlichem Interesse zu bewahren und wiederherzustellen, wobei die Anforderungen von Wirtschaft, Gesellschaft und Kultur berücksichtigt sowie auf regionale und örtliche Besonderheiten Rücksicht genommen werden soll; cf. Art. 2 der Richtlinie 92/43 EWG des Rates vom 21.05.1992, geändert durch RL 97/62/EG des Rates vom 27.10.1997; ►Erhaltung der Tier- und Pflanzenwelt, ►Fauna-Flora-Habitat-Richtlinie).

1038 conservation [n] **of nature and natural resources** *conserv. land'man. landsc.* (Administrative, planning and technical measures for the protection or restoration of land areas containing significant physiographic and ecological features that require special protection and management to serve the public health, safety, and welfare and to preserve species or unique biolgical habitats); *s* **preservación [f] del patrimonio natural** (Conjunto de medidas administrativas, técnicas y de planificación para proteger y restaurar áreas naturales con características fisiográficas y ecológicas significativas que necesitan protección y gestión especial con el fin de preservar los ecosistemas, paisajes y las especies de flora y fauna. En España existe el instrumento del Plan Estratégico Estatal del Patrimonio Natural y de la Biodiversidad; arts. 12 y 13 Ley 42/2007); *f* **sauvegarde [f] des espaces, ressources et milieux naturels** (Les espaces, ressources et milieux naturels, les sites et paysages, la qualité de l'air, les espèces animales et végétales, la diversité et les équilibres biologiques auxquels ils participent font partie du patrimoine commun de la nation ; leur protection, leur mise en valeur, leur restauration, leur remise en état et leur gestion sont d'intérêt général et concourent à l'objectif de développement durable qui vise à satisfaire les besoins de développement et la santé des générations présentes sans compromettre la capacité des générations futures à répondre aux leurs ; cf. Code envir. art. 1 L110-1) ; *g* **Erhaltung [f, o. Pl.] der Natur und der natürlichen Hilfsquellen** (Gesamtheit der den Zielen von Naturschutz und Landschaftspflege entsprechenden Maßnahmen, die dazu dienen, vorhandene, natürliche und anthropogene Landschaften, Flora und Fauna mit ihren Lebensräumen sowie ökologische Prozesse zum Wohle der Allgemeinheit zu bewahren und aufrechtzuerhalten).

C

1039 conservation [n] **of plant and animal communities and habitats** *conserv. land'man.* (Administrative, planning and technical measures for the protection or restoration of specified flora and fauna or unique biological habitats; ▶habitat conservation); *s* **preservación [f] del hábitat de especies silvestres de flora y fauna** (Conjunto de medidas administrativas, técnicas y de planificación para proteger las especies de flora y fauna y sus hábitats; cf. Título II Ley 42/2007; ▶protección de biótopos); *f* **préservation [f] des espèces animales ou végétales et d'habitats** (Ensemble des dispositions juridiques et mesures de gestion du patrimoine naturel assurant la protection des espèces animales et végétales contre toutes les causes de dégradation qui les menacent ainsi que le maintien des équilibres biologiques des habitats naturels auxquels ils participent ; cette protection s'inscrit depuis les années 60 dans le classement de grands sites naturels [parcs nationaux, réserves naturelles, zones humides, réserves biologiques, réserves nationales de chasse et de faune sauvage, zones naturelles d'intérêt écologique, faunistique et floristique] ; art. 16 [2] loi n° 76-629 relative à la protection de la nature, article L.411-1 du Code de l'Environnement ; ▶préservation des biotopes) ; *g* **Erhaltung [f, o. Pl.] der Lebensgemeinschaften oder Lebensstätten** (Gesamtheit der den Zielen von Naturschutz und Landschaftspflege entsprechenden Gesetze, Rechtsvorschriften und Maßnahmen, die dazu dienen, Lebensgemeinschaften und Biotope zu schützen. Seit den 1970er-Jahren werden nicht nur kleine und isolierte Biotope geschützt, sondern durch großräumigen Flächenschutz [Naturschutzgebiete, Nationalparke, Biosphärenreservate, Landschaftsschutzgebiete und Naturparke] sowie durch Prozessschutz versucht, den gesamten Naturhaushalt in den Schutzgebieten zu erhalten; cf. §§ 13-18 BNatSchG; ▶Biotopschutz; *syn.* Erhaltung [f, o. Pl.] von Lebensgemeinschaften oder Lebensstätten).

1040 conservation [n] **of the natural and cultural heritage** [CH] *conserv. hist. land'man.* (In Switzerland, measures to protect the visual quality of the countryside, the historic sites and the natural and cultural monuments from damage and destruction; ▶conservation regulations of natural habitats etc. 1994 [UK]); *s* **conservación [f] del patrimonio natural y cultural [CH]** (En Suiza, medidas para proteger la calidad visual del entorno rural, los sitios históricos y los monumentos naturales y culturales; ▶Ley del Patrimonio Natural y de la Biodiversidad [Es]); *f* **protection [f] du patrimoine naturel [CH]** (En Suisse, mesures de mise en valeur, de restauration et de protection contre la destruction de milieux naturels et du patrimoine culturel ; ▶loi relative à la protection de la nature) ; *g* **Natur- und Heimatschutz [m, o. Pl.] [CH]** (Maßnahmen und Handlungen, die die ohne Einfluss des Menschen entstandene Natur und die immateriellen Schöpfungen der Heimat vor Beeinträchtigung und Zerstörung bewahren und gegebenenfalls wiederherstellen; *gesetzliche Grundlage* Bundesgesetz über Natur- und Heimatschutz [NHG] von 1966; ▶Bundesnaturschutzgesetz [D]).

1041 conservation [n] **of the natural environment** *conserv. land'man. landsc.* (**1.** Administrative, planning and technical measures for the protection or restoration of life-sustaining physical and biological systems essential to human life. **2.** Measures to enhance the quality of local or regional environments); *s* **preservación [f] del medio natural** (**1.** Conjunto de medidas administrativas, de planificación y técnicas necesarias para proteger o restaurar la capacidad funcional de los ecosistemas y para posibilitar un aprovechamiento sostenible de los recursos naturales. **2.** Medidas tomadas para mejorar la calidad del medio ambiente a nivel local o regional); *f* **préservation [f] des milieux naturels** (Outils d'aménagement, de développement et de protection, de nature réglementaire et contractuelle [acqui-

sition, aménagement, entretien], issus des échelons européens, nationaux, locaux [DOCUP, CPER, Natura 2000, OLAE/CTE/CAD, réserves, arrêtés de protection de biotope, SDAGE, etc.] contribuant à la préservation et la réhabilitation des espaces et ressources naturelles) ; *syn.* préservation [f] du milieu naturel ; *g* **Erhaltung [f, o. Pl.] der natürlichen Umwelt** (**1.** Alle Planungen und Maßnahmen, die die Erhaltung oder die Wiederherstellung der Verfügbarkeit der natürlichen Ressourcen für eine nachhaltige Nutzung ermöglichen und die Tragfähigkeit der Ökosysteme fördern. **2.** Planungen und Maßnahmen, die die materielle und immaterielle Leistungsfähigkeit der örtlichen und regionalen Umwelt fördern und erhalten).

Conservation [n] **of Wild Creatures and Wild Plants Act 1975** [UK] *conserv. leg.* ▶Endangered Species Act [E.S.A.] [US].

conservation rank [n] *conserv. leg. pol.* ▶conservation status (1).

1042 conservation/recreation agency [n] [US] *adm. conserv.* (Agency/authority established to implement nature conservation laws at local, county and state levels; **in U.S.,** there are several agencies at local, state and federal levels which are legally-established to administer conservation and recreation areas; e.g. local, county or regional authority/board/commission, state park, forest or conservation/recreation agency and federal park, forest or fish and wildlife service; **in U.K.,** the generic term **nature conservation agency** covers the Nature Conservancy Council for England, the Countryside Council for Wales and the Scottish Natural Heritage); *syn.* nature conservation body [n] [UK]; *s* **administración [f] de conservación de la naturaleza** (Término genérico para las autoridades competentes en la protección de la naturaleza a cada nivel de jurisdicción: **1.** Agencia [f] Europea del Medio Ambiente, a nivel de la Europa Comunitaria. **2.** En Es. debido a la competencia de las CC.AA. en cuestiones de protección de la naturaleza, a nivel estatal el Ministerio de Medio Ambiente solo tiene funciones coordinadoras y de diseño general de las políticas, pero no administrativas. **3.** Comisión [f] Estatal para el Patrimonio Natural y la Biodiversidad, en Es. a nivel estatal órgano consultivo y de cooperación entre el Estado y las Comunidades Autónomas. **4.** [Vice-]Consejería [f] de Protección de la Naturaleza, a nivel de Comunidades Autónomas [Es]); *f* **administration [f] pour la protection de la nature** (*Terme générique* **1.** agence [f] européenne de l'environnement, au niveau communautaire. **2.** F., a] direction [f] de la nature et des paysages : administration centrale du ministère de l'environnement. b] Direction [f] régionale de l'environnement (DIREN) : Administration régionale. c] Direction [f] départementale de l'équipement : administration départementale chargée de l'environnement) ; *syn. obs.* direction [f] régionale à l'architecture et à l'environnement (DRAE) ; *g* **Naturschutzbehörde [f]** (**D.,** zum Vollzug des Naturschutzrechtes der Länder eingerichtete Behörde auf Kreisverwaltungsebene [= **untere N.**], Bezirksregierungs- resp. Regierungspräsidentenebene [= **höhere N.**] und Landesministeriums- resp. Senatsebene [= **oberste N.**], nicht auf Bundesebene, weil Naturschutz gemäß Verfassung Zuständigkeit der Länder ist).

1043 conservation regulations [npl] **of natural habitats etc. 1994** [UK] *conserv. leg.* (In U.K., there are several acts and regulations which together provide the basis for nature conservation and comply with the Council Directive 92/43/EEC, such as National Parks and Access to the Countryside Act, Countryside Act 1968 [Areas of Special Scientific Interest], ▶Conservation of Wild Creatures and Wild Plants Act 1975, Wild ▶Birds Directive 79/409/EEC, Wildlife and Countryside Act 1981, Environmental Protection Act 1990;

C

in U.S., there is not one single federal law but a large number of federal and state nature conservation enactments such as the National Park Service Act [1916], Migratory Bird Conservation Act [1929], Fish and Wildlife Act [1956], Fish and Wildlife Conservation Act [1980], Wilderness Act [1964], Wild and Scenic Rivers Act [1968], ►Endangered Species Act [E.S.A.] [US] [1973], National Forest Management Act [1976], ►National Environmental Policy Act [US] as amended [NEPA], Federal Water Pollution Control Act [Clean Water Act, 1972], Clean Air Act [1955], Estuaries and Clean Waters Act [2000], and similar state laws which govern nature conservation; most states, as sovereign units of government, have significant legislation either implementing or supplementing federal laws; **In D.**, ►Federal Act on Nature Conservation and Landscape Management [D]; ►nature conservation enactment); *s 1* **legislación [f] de protección de la naturaleza** (Regulación estatal en forma de leyes y ordenanzas decretadas para alcanzar los objetivos de conservación de la naturaleza y de salvaguarda del paisaje. En GB, al contrario que en Es y D., no hay una jerarquía de leyes, sino varias leyes y ordenanzas sectoriales individuales que son la base para proteger la naturaleza y el paisaje. En EE.UU. tampoco hay una ley federal marco de protección de la naturaleza, sino un sinnúmero de leyes federales y de los estados que, basándose en sus competencias legislativas, amplían la legislación federal según sus necesidades regionales; ►ley de protección de la naturaleza); *s 2* **Ley [f] del Patrimonio Natural y de la Biodiversidad [Es]** (La ley 42/2007, de 13 de diciembre, establece el régimen jurídico básico de la conservación, uso sostenible, mejora y restauración del patrimonio natural y la biodiversidad española, como parte del deber de conservar y del objetivo de garantizar los derechos de las personas a un medio ambiente adecuado para su bienestar, salud y desarrollo. Igualmente recoge normas y recomendaciones internacionales, del Consejo de Europa y de las Naciones Unidas, así como aquéllas dictadas por la Comisión de las Comunidades Europeas y dispone la creación de mecanismos de coordinación y cooperación entre la Administración General del Estado y las Comunidades autónomas como la Comisión Estatal para el Patrimonio Natural y la Biodiversidad. La ley contiene siete títulos que articulan los objetivos e instrumentos previstos: Título preliminar contiene los objetivos, principios básicos, definiciones y los instrumentos básicos; Título I Instrumentos para el conocimiento y la planificación del patrimonio natural y de la biodiversidad; Título II Catalogación, conservación y restauración de hábitats y espacios del patrimonio natural; Título III Conservación de la biodiversidad; Título IV Uso sostenible del patrimonio natural y de la biodiversidad; Título V Fomento del conocimiento, la conservación y restauración del patrimonio natural y la biodiversidad; Título VI De las infracciones y sanciones; cf. Ley 42/2007; ►ley de protección de especies, ►Ley Federal de Conservación de la Naturaleza y de Gestión del Paisaje [D]); *f 1* **loi [f] relative à la protection de la nature** (**F.**, la loi sur la protection de la nature du 10 juillet 1976 porte sur les études d'impact, la protection de la faune et de la flore, la protection de l'animal et les réserves naturelles ; les dispositions diverses définissent en outre les droits reconnus aux associations de protection de la nature et de l'environnement ; ►arrêté réglementant la protection de la faune et de la flore sauvages) ; *f 2* **loi [f] relative au renforcement de la protection de l'environnement** (La loi n° 95-101 du 02.02.1995) dite « loi Barnier » insère dans le code rural les principes généraux issus de la déclaration de Rio de 1992 et précise que les espaces, ressources et milieux naturels, les sites et paysages, les espèces animales et végétales, la diversité et les équilibres biologiques auxquels ils participent font partie du patrimoine commun de la nation. Elle réaffirme d'importantes dispositions du droit de l'environnement, reconnaît

d'intérêt général la protection de l'environnement, défini l'objectif de développement durable, intègre en droit français les principes de précaution, de prévention, de réduction à la source, de responsabilité, de pollueur-payeur et de participation ; la loi renforce la protection de l'environnement dans les domaines de **1.** *la participation du public et des associations* [commission nationale du débat public, enquête publique, procédure d'agrément pour les associations, nouvelles commissions concernant l'environnement], **2.** *la prévention des risques naturels* [procédure d'expropriation, plan de prévention des risques naturels prévisibles (PPR), gestion des cours d'eau non domaniaux], **3.** *la connaissance, protection et gestion des espaces naturels* [inventaire départemental du patrimoine naturel, limitation des abattages d'arbres et de haies, l'extension de la politique des espaces naturels sensibles, la réglementation des boisements, la gestion des parcs naturels régionaux, des disposition concernant les grandes voies de circulation routière, des disposition concernant l'affichage publicitaire, la protection des espèces animales et végétales], **4.** *la gestion des déchets et prévention des pollutions* [transparence des actions publiques dans les dom aines de l'eau et des déchets et délégation de services publics, taxes sur les déchets, Plans d'élimination des déchets ménagers et industriels, la prévention des pollutions, les installations classées] et **5.** *diverses dispositions* [transport de gibier, la protection des paysages, friches]) ; *f 3* **code [m] de l'environnement** (Avec la création du code de l'environnement par l'ordonnance n° 2000-914 du 18 septembre 2000 est codifié à droit constant l'ensemble des textes législatifs du droit de l'environnement et a pour conséquence juridique l'abrogation de l'ensemble des lois antérieures au 21 septembre 2000 pour les organiser en sept livres dans un seul Code) ; *f 4* ►**loi fédérale pour la protection de la nature** [**D**]) ; *g 1* **Naturschutzgesetzgebung [f]** (Staatliche Rechtsetzung in Form erlassener Gesetze und Verordnungen, die den Zielen des Naturschutzes und der Landschaftspflege dienen. Im UK gibt es im Vergleich zu Deutschland keine Gesetzeshierarchie, sondern mehrere Einzelgesetze und Verordnungen, die die Grundlage für den Schutz von Natur und Landschaft sind; auch in den USA gibt es kein übergeordnetes Bundesnaturschutzgesetz, sondern eine Anzahl von Bundesgesetzen und Gesetzen der einzelnen Staaten, die in eigener Zuständigkeit und Gesetzgebungskompetenz die Bundesgesetze entsprechend der regionalen Notwendigkeiten ergänzen); *g 2* **Bundesnaturschutzgesetz [n] [D]** (*Abk.* BNatSchG; Gesetz über Naturschutz und Landschaftspflege vom 20.12.1976; Rahmengesetz des Bundes mit in weiten Teilen sehr detaillierten Vorgaben i. d. neuen Fassung vom 25.03.2002, zuletzt geändert am 22.12.2008 mit dem Oberziel, Natur und Landschaft auf Grund ihres eigenen Wertes und als Lebensgrundlage für die Gesellschaft im besiedelten und unbesiedelten Bereich so zu schützen, dass 1. die Leistungs- und Funktionsfähigkeit des Naturhaushaltes, 2. die Regenerationsfähigkeit und nachhaltige Nutzungsfähigkeit der Naturgüter, 3. die Tier- und Pflanzenwelt einschließlich ihrer Lebensstätten und Lebensräume sowie 4. die Vielfalt, Eigenart und Schönheit sowie der Erholungswert von Natur und Landschaft gesichert sind. Zur Entwicklung optimaler und nachhaltiger, materieller wie immaterieller Leistungen der Naturausstattung von Landschaftsräumen für die Gesellschaft sind u. a. vorgeschrieben resp. geregelt: die Umweltbeobachtung und Landschaftsplanung, der Umgang mit und das Verfahren bei Eingriffen in Natur und Landschaft [►Eingriffsregelung], die Ausweisung von Schutzgebieten, deren Pflege und Entwicklung, die Umsetzung der EU-Richtlinie 92/43 EWG zum Aufbau eines kohärenten Europäischen Netzes besonderer Schutzgebiete „Natura 2000", die Überprüfung der Verträglichkeit oder Unzulässigkeit von Projekten in Natur und Landschaft mit den Erhaltungszielen des Eingriffsgebietes von gemeinschaftlicher Bedeutung, den Schutz und die Pflege wild lebender Tier- und Pflanzenarten und der besonders geschützten Arten, z. B. auch durch den Erlass einer ▶Artenschutzverordnung, das Betreten der Flur zum Zwecke der Erholung, die Mitwirkung von Naturschutzverbänden; das BNatSchG wird durch Ländernaturschutzgesetze weiter konkretisiert. Durch die Einführung der Paragrafen zur Landschaftsplanung wird ein *aktiver* Naturschutz möglich. Der Vorgänger war das **Reichsnaturschutzgesetz** vom 26.06.1935 [RNG], das die gesetzliche Basis für einen *rein konservierenden* Schutz von Natur und Landschaft schuf, ohne jedoch großflächig ein ausreichendes Gegengewicht zu den Belastungen der Landschaft durch Industrie und Landwirtschaft sowie durch großräumige wesentliche Eingriffe in Struktur und Naturhaushalt der bäuerlichen Kulturlandschaft in Form von Meliorationsmaßnahmen [Arbeitsbeschaffung und Arbeitsdienst] zu bilden; *Entsprechung in der Schweiz* **Bundesgesetz über den Natur- und Heimatschutz [NHG]**; in Österreich gibt es kein Bundesgesetz, sondern nur Naturschutzgesetze der Bundesländer).

conservation standard [n] *conserv'hist. leg.* ▶conservation criterion.

1044 conservation status [n] (1) *conserv. leg. pol.* (Legal status of a particular species, site or ▶cultural monument; e.g. based upon a law or convention; ▶threatened species); *syn.* conservation rank [n]; *s* **estatus** [m] **de protección** (Situación legal de una especie de flora o de fauna, de un espacio o un conjunto; ▶especie amenazada, ▶monumento); *f* **statut** [m] **de conservation** (Statut juridique 1. d'une espèce de faune ou de flore afin de les maintenir ou rétablir dans un état favorable, 2. d'une zone, d'un site ou d'un ensemble définis par une loi, une convention ; ▶espèce dont la population n'a pas sensiblement diminué, mais dont les effectifs sont faibles, donc en danger latent ; ▶monument historique) ; *g* **Schutzstatus** [m] (Rechtsstellung einer Tier- oder Pflanzenart, eines Gebietes oder einer Gesamtanlage, z. B. auf Grund eines Gesetzes oder Übereinkommens; ▶bedrohte Art, ▶Kulturdenkmal).

1045 conservation status (2) *conserv. ecol.* (**1.** Condition of a protected species or habitat in a **Natura-2000-Region**, compared with standards laid down for nature conservation and species protection and the criteria for distribution, structure and function as well as the survival of characteristic species. **2. Conservation status of a species:** according to ▶Habitats directive 92/43/EEC of the European Council of 21 May 1992, the sum of the influences acting on the species concerned that may affect the long-term distribution and abundance of its populations within the territory referred to in Article 2; the **c. s.** will be taken as "favourable" when a] population dynamics data on the species concerned indicate that it is maintaining itself on a long-term basis as a viable component of its natural habitats, and b] the natural range of the species is neither being reduced nor is likely to be reduced for the foreseeable future, and c] there is, and will probably continue to be, a sufficiently large habitat to maintain its populations on a long-term basis. **3. Conservation status of a natural habitat:** the sum of the influences acting on a natural habitat and its typical species that may affect its long-term natural distribution, structure and functions as well as the long-term survival of its typical species; the conservative status of a natural habitat will be taken as "favourable" when a] its natural range and areas it covers within that range are stable or increasing, and b] the specific structure and functions which are necessary for its long-term maintenance exist and are likely to continue to exist for the foreseeable future, and c] the conservation status of its typical species is favourable); *s 1* **estado** [m] **de conservación** (Condiciones vitales de las especies y los hábitats protegidos en un área perteneciente a la red **Natura 2000**, en relación a los estándares fijados para la protección de la naturaleza y de las especies y a los criterios de distribución, estructura y función así como de supervivencia de las especies características); *s 2* **estado** [m] **de conservación de una especie** (Conjunto de influencias que actúen sobre la especie y puedan afectar a largo plazo a la distribución e importancia de sus poblaciones en el territorio a que se refiere el artículo; el estado de conservación se considerará "favorable" cuando **1.** los datos sobre la dinámica de las poblaciones de la especie en cuestión indiquen que la misma sigue y puede seguir constituyendo a largo plazo un elemento vital de los hábitats naturales a los que pertenezca, y **2.** el área de distribución natural de la especie no se esté reduciendo ni amenace con reducirse en un futuro previsible, y **3.** exista y probablemente siga existiendo un hábitat de extensión suficiente para mantener sus poblaciones a largo plazo; cf. art. 1 ▶Convención de Bonn); *s 3* **estado** [m] **de conservación de un hábitat** (Conjunto de las influencias que actúan sobre el hábitat natural de que se trate y sobre las especies típicas asentadas en el mismo y que pueden afectar a largo plazo a su distribución natural, su estructura y funciones, así como a la supervivencia de sus especies típicas en el territorio; el **e. de c. de un h.** natural se considerará "favorable" cuando **1.** su área de distribución natural y las superficies comprendidas dentro de dicha área sean estables o se amplíen, **2.** la estructura y las funciones específicas necesarias para su mantenimiento a largo plazo existan y puedan seguir existiendo en un futuro previsible, y **3.** el estado de conservación de sus especies típicas sea favorable); *f 1* **état** [m] **de conservation** (Notion qui rend compte de « l'état de santé » des espèces et des habitats ; l'état de conservation est déterminé à partir de critères d'appréciation en fonction de l'aire de répartition, de la surface occupée, des effectifs des espèces et du bon fonctionnement des habitats ; cette évaluation sert d'une part à définir des objectifs et des mesures de gestion dans le cadre du ▶document d'objectifs [DOCOB], qui vise au maintien ou au rétablissement d'un état de conservation équivalent ou meilleur, et d'autre part, de suivre l'évolution des espèces et des habitats à long terme ; l'état de conservation peut être favorable, pauvre ou mauvais ; dans la pratique, le bon état de conservation vise un fonctionnement équilibré des milieux par rapport à leurs caractéristiques naturelles et avec un impact modéré des activités humaines) ; *f 2* **état** [m] **de conservation d'une espèce** (*Terme spécifique* effet de l'ensemble des influences qui, agissant sur l'espèce, peuvent affecter à long terme la répartition et l'importance de ses populations sur le territoire européen des Etats membres ; l'état de conservation d'une espèce sera considéré comme « **favorable** » [ou un bon état de conservation] dans une zone donnée [administrative : département, Région, Etat, ou biogéographique] lorsque **1.** les données relatives lorsque la répartition et les effectifs de cette espèce ou de cette population [reproduction et hivernage] sont conformes aux caractéristiques biologiques de l'espèce ainsi qu'aux potentialités d'accueil des milieux présents dans cette zone à la dynamique de la population de l'espèce en question indiquent que cette espèce continue et est susceptible de continuer à long terme à constituer un élément viable des habitats naturels auxquels elle appartient et **2.** l'aire de répartition naturelle de l'espèce ne diminue ni ne risque de diminuer dans un avenir prévisible et **3.** il existe et il continuera probablement d'exister un habitat suffisamment étendu pour que ses populations se maintiennent à long terme ; cf. ▶Directive Habitat Faune Flore [DHFF] Art. 1, ▶Convention de Bonn Art. 1) ; *f 3* **état** [m] **de conservation d'un habitat naturel** (*Terme spécifique* effet de l'ensemble des influences agissant sur un habitat naturel ainsi que sur les espèces typiques qu'il abrite, qui peuvent affecter à long terme sa répartition naturelle, sa structure et ses fonctions ainsi que la survie à long terme de ses espèces typiques sur le territoire européen des États membres ; l'état de conservation d'un habitat naturel sera considéré comme « **favorable** » lorsque **1.** son aire

C

de répartition naturelle ainsi que les superficies qu'il couvre au sein de cette aire sont stables ou en extension, **2.** la structure et les fonctions spécifiques nécessaires à son maintien à long terme existent et sont susceptibles de perdurer dans un avenir prévisible, **3.** et l'état de conservation des espèces qui lui sont typiques est favorable ; cf. Directive Habitat Faune Flore [DHFF]) ; *g 1* **Erhaltungszustand [m]** (Zustand schutzzweckrelevanter Arten und Lebensraumtypen in einem **Natura-2000-Gebiet**, gemessen an einem natur- resp. artenschutzfachlichen Leitbild, das sich auf Verbreitung, Struktur und Funktion sowie das Überleben der charakteristischen Arten als Kriterien bezieht. *g 2* **Erhaltungszustand [m] einer Art** (▶Fauna-Flora-Habitat-Richtlinie 92/43/EWG des Rates vom 21. Mai 1992 zur Erhaltung der natürlichen Lebensräume sowie der wild lebenden Tiere und Pflanzen, Art. 1; die Gesamtheit der Einflüsse, die sich langfristig auf die Verbreitung und die Größe der Populationen der betreffenden Arten in dem in Artikel 2 bezeichneten Gebiet auswirken können; der Erhaltungszustand wird als „günstig" betrachtet, wenn **a.** auf Grund der Daten über die Populationsdynamik einer Art anzunehmen ist, dass diese ein lebensfähiges Element des natürlichen Lebensraumes, dem sie angehört, bildet und langfristig weiterhin bilden wird, und **b.** das natürliche Verbreitungsgebiet dieser Art weder abnimmt noch in absehbarer Zeit vermutlich abnehmen wird und **c.** ein genügend großer Lebensraum vorhanden ist und wahrscheinlich weiterhin vorhanden sein wird, um langfristig ein Überleben der Populationen dieser Art zu sichern); *g 3* **Erhaltungszustand [m] eines natürlichen Lebensraumes** (Gesamtheit der Einwirkungen, die den betreffenden Lebensraum und die darin vorkommenden charakteristischen Arten beeinflussen und die sich langfristig auf seine natürliche Verbreitung, seine Struktur und seine Funktion sowie das Überleben seiner charakteristischen Arten auswirken können. Der Erhaltungszustand eines natürlichen Lebensraums wird als „günstig" erachtet, wenn **a]** sein natürliches Verbreitungsgebiet sowie die Flächen, die er in diesem Gebiet einnimmt, beständig sind oder sich ausdehnen und **b]** die für seinen langfristigen Fortbestand notwendige Struktur und spezifischen Funktionen bestehen und in absehbarer Zukunft wahrscheinlich weiterbestehen werden und **c]** der Erhaltungszustand der für ihn charakteristischen Arten günstig ist; Art. 1 Buchstabe e der Richtlinie 92/43/EWG des Rates vom 21.05.1992); *syn.* Erhaltungssituation [f].

1046 conse**rvatory** [n] (1) *hort.* (Large glasshouse for the exhibition of mostly exotic plants [sometimes animals], often erected in botanical or zoological gardens or garden shows); *s* **invernadero [m] (±)** (Invernadero grande para exhibir plantas generalmente exóticas, a menudo en jardín botánico o en exposición de horticultura); *f* **serre [f] d'acclimatation** (Grande serre abritant des plantes ou animaux en général exotiques ou tropicales, très souvent implantée dans les jardins botaniques, zoologiques ou dans les floralies) ; *g* **Schauhaus [m]** (Großes Gewächs- oder Tierhaus zur Schaustellung meist exotischer Pflanzen oder Tiere, oft in botanischen und zoologischen Gärten oder in Gartenschauen anzutreffen).

1047 conse**rvatory** [n] (2) *arch. gard.* (Sunny room or part of a room which can be heated and has large windows or walls of glass, or an extension to a house with large expanses of glass for the overwintering of plants sensitive to winter cold); *s* **jardín [m] de invierno** (Habitación grande con calefacción y grandes ventanales o paredes de vidrio que se utiliza para guardar las plantas sensibles a heladas en invierno); *f* **jardin [m] d'hiver** (Installation en serre adossée à une maison très souvent utilisée comme jardin privé, pouvant être chauffée et abritant pendant la saison froide des plantes rares ou exotiques [gélives]) ; *syn.* véranda [f] ; *g* **Wintergarten [m]** (Heller, heizbarer Raum mit großen

Fenstern oder Glaswänden oder ein Glasanbau für die Haltung von nicht winterharten Pflanzen).

1048 considerations [npl] **of nature conservation and landscape management** *conserv. nat'res.* (Requirements or necessities which should be taken into account when altering a landscape or parts thereof in order to achieve and implement the goals of nature protection and landscape management); *syn.* prerequisites [npl] of nature conservation and landscape management; *s* **exigencias [fpl] de protección de la naturaleza y del paisaje** (Requisitos que hay que cumplir o necesidades a tener en cuenta para no causar daños a la naturaleza o a un paisaje); *f* **préoccupations [fpl] de protection de la nature et des paysages** (Exigence de prise en compte de la nature et des paysages et d'élaboration de méthodes qui permettent de réduire les impacts sur l'environnement) ; *g* **Belange [mpl] des Naturschutzes und der Landschaftspflege** (Erfordernisse, die zur Erreichung der Ziele von Naturschutz und Landsschaftspflege zu berücksichtigen und durchzusetzen sind); *syn.* Belange [mpl] von Naturschutz und Landschaftspflege.

1049 consistency [n] *pedol. phys.* (Soil characteristic with regard to its plasticity, which indicates the ▶workability of a soil at various soil moisture contents and degrees of cementation; ▶plastic limit [of soil], ▶liquid limit of soil); *s* **consistencia [f] del suelo** (Resistencia del suelo a su compactación o rotura. El término *c. del s.* se refiere a los atributos del material del suelo que se expresan por el grado y la clase de su cohesión y adhesión [plasticidad], o por su resistencia a la deformación o rotura; DINA 1987; ▶friabilidad del suelo, ▶límite de plasticidad [del suelo], ▶límite de liquidez [del suelo]); *f* **consistance [f] d'un sol** (Désignation de la structure d'un sol traduisant la cohésion et la résistance à la pression des unités structurales. On emploie les désignations telles que : constance meuble, plastique, friable, cohérente, compacte, durcie ; ▶limite de plasticité, ▶limite de liquidité, ▶possibilités de travail du sol) ; *g* **Konsistenz [f]** (Beschaffenheit von Boden hinsichtlich des Zusammenhaltes resp. der Verformbarkeit [Plastizität], die Aufschluss über die ▶Bearbeitbarkeit des Bodens gibt; ▶Ausrollgrenze, ▶Fließgrenze); *syn.* Zustandsform [f].

consolidated soil [n] *constr.* ▶stabilized soil.

1050 consolidated subgrade [n] *constr. eng.* (Stabilized ground surface at the base of an excavated site, which remains in place and is stable for construction measures; ▶subbase grade [US]/formation level [UK], ▶undisturbed subgrade [US]/undisturbed subsoil [UK]); *syn.* stable soil level [n] [also US]; *s* **subsuelo [m] estable** (Superficie estabilizada del suelo al fondo de una zanja de fundación; ▶subrasante, ▶subsuelo inalterado); *syn.* terreno [m] de cimentación estable; *f* **sol [m] stable** (Sol de fondation — naturel ou rapporté — de bonne homogénéité ayant une capacité portante suffisante pour des fondations normales ; ▶fond de forme, ▶terrain naturel, ▶bon sol) ; *syn.* sol [m] stabilisé ; *g* **fester Baugrund [m]** (Unbearbeiteter oder standfest verdichteter, anstehender Boden am Grunde einer Baugrube; ▶gewachsener Boden, ▶Erdplanum); *syn.* standsicherer Baugrund [m].

1051 consolidate [vb] **scattered land holdings** *adm. agr. leg.* (To reorganize dispersed land holdings of farmers to create more compact lots/plots for obtaining a greater productive efficiency. **In US parlance**, the general sense of this phrase also refers to the reorganisation of dispersed farm lands for the purposes of creating, e.g. a state park, an airport, or any other public or private development project); *s* **llevar [vb] a cabo la concentración parcelaria** (Reorganización del reparto de parcelas de agricultores para crear lotes mayores y así mejorar la productividad; en EE.UU. el término se utiliza también para otros

procesos como p. ej. crear parques nacionales, grandes infraestructuras como aeropuertos u otros grandes proyectos de
desarrollo públicos o privados); *syn.* proceder [vb] a la concentración parcelaria; *f* **remembrer [vb] les terres** *syn.* remembrer
[vb] les parcelles ; *g* **Flurbereinigung durchführen [vb]** *syn.*
flurbereinigen [vb].

consolidation [n] *adm. agr. leg.* ►farmland consolidation
[US].

consolidation [n] **[UK], land** *adm. agr. leg.* ►farmland
consolidation [US].

consolidation of agricultural land [n] *adm. agr. leg.*
►farmland consolidation [US].

1052 consortium [n] (1) *contr.* (Partnership between contractors usually for the purpose of undertaking a large construction
project); *s* **consorcio [m] de contratistas** (Sociedad comanditaria
provisional creada por un grupo de empresas que realizan una
obra conjuntamente); *f 1* **groupement [m] d'entreprises solidaires** (Chaque entreprise est engagée pour la totalité du marché
et doit pallier une éventuelle défaillance de ses partenaires ; l'une
d'entre elles, désignée comme mandataire, représente pour l'exécution du marché tous les entrepreneurs du groupement ; CCM
1984, 46) ; *syn.* entrepreneurs [mpl] groupés solidaires (CCM
1984, 11), groupement [m] solidaire ; *f 2* **groupement [m]
d'entreprises conjointes** (Chacune n'est engagée que pour le ou
les lots qui lui sont assignés, l'une d'entre elles désignée comme
mandataire, étant solidaire de chacune des autres et les représente
pour l'exécution du marché, en ayant en outre la charge et la
responsabilité des tâches de coordination ; CCM 1984, 46) ; *syn.*
entrepreneurs [mpl] groupés conjoints, groupement [m] conjoint ;
g **Arbeitsgemeinschaft [f] (1)** (*Abk.* ARGE; Zusammenschluss
von zwei oder mehreren Baufirmen zur gemeinschaftlichen
Ausführung von Bauleistungen, meist bei Großprojekten); *syn.*
Unternehmergemeinschaft [f].

1053 consortium [n] (2) *prof.* (Partnership between consultants usually for the purpose of undertaking a large planning or
design project); *s* **consorcio [m] de planificación** (Sociedad
comanditaria provisional formada por arquitectos y planifica-
dores, generalmente para realizar conjuntamente un gran
proyecto de planificación); *f* **équipe [f]** (Association de plusieurs
bureaux d'études ou de spécialistes afin de réaliser un projet
important) ; *syn.* consortium [m] ; *g* **Arbeitsgemeinschaft [f] (2)**
(Zusammenschluss von zwei oder mehreren Planungsbüros zur
gemeinschaftlichen Projektierung eines großen Bauvorhabens);
syn. Planungsgemeinschaft [f].

constance [n] *phyt.* ►constancy.

1054 constancy [n] *phyt.* (Synecological term which signifies
in how many separate populations of the same plant community a
certain species occurs within a larger area, compared with the
unit area. There is no difference in principle between ►presence
and **c.** Practically, however, it is advisable to designate as studies
of **c.** those investigations of presence which are made with plots
of sharply limited area; ►frequency, ►exclusive species); *syn.*
constance [n] (TEE 1980, 161); *s* **constancia [f]** (Entre las diferentes acepciones empleadas, la más general es la utilizada en el
método de BROCKMANN-JEROSCH: la recurrencia de las especies
en inventarios de diferentes individuos sineciales. Según este
método son constantes las especies que se encuentran en la mitad
o más de los inventarios de una asociación; DB 1985. Según
BRAUN-BLANQUET [1979] en principio no existe ninguna diferencia entre **c.**, ►frecuencia y ►presencia. Las investigaciones de
presencia referidas a superficies del mismo tamaño reciben el
nombre de determinaciones de **c.** La diferencia entre **c.** y frecuencia reside en que la segunda se refiere a una sola población,
mientras que la determinación de **c.** es válida en la comparación

de poblaciones distintas; BB 1979, 74 s; ►especie exclusiva);
f **constance [f]** (Terme synécologique caractérisant la présence
d'une espèce végétale dans différents relevés d'une même association végétale pour une superficie donnée à l'intérieur d'un
grand territoire. En principe il n'existe pas de différences entre
les termes ►présence, ►fréquence et **c.** ; ►espèce exclusive) ;
g **Konstanz [f]** (Synökologischer Begriff, der besagt, an wieviel
getrennten Beständen der gleichen Pflanzengesellschaft, bezogen
auf eine Flächeneinheit, eine bestimmte Art innerhalb eines größeren Gebietes vorkommt. Prinzipiell besteht zwischen ►Gesell-
schaftsstetigkeit, ►Frequenz und **K.** kein Unterschied. Stetig-
keitsuntersuchungen, die sich auf Probeflächen von ganz be-
stimmtem Umfang beziehen, können als Konstanzbestimmungen
bezeichnet werden; ►treue Art).

1055 constant on-site supervision [n] *constr.* *s* **inspección [f] permanente de obra**; *f* **surveillance [f] permanente de
chantier** ; *g* **ständige Bauleitung [f]**.

constant species [n] *phyt.* ►exclusive species.

1056 construct [vb] **an embankment/slope** *constr.*
(►steepen an existing slope, ►create a slope); *s* **taluzar [vb]**
(►igualar terreno con talud, ►ataluzar); *f* **dresser [vb] un talus**
(Confection et réglage d'un talus ; *termes spécifiques* dresser en
déblai, dresser en remblai ; ►taluter en déblai, ►taluter en remblai) ; *g* **Böschung herstellen [vb]** (*UBe* Abtragsböschung herstellen, Auftragsböschung herstellen; ►abböschen, ►anböschen).

1057 constructed channel [n] *eng. hydr.* (Artificially-
created and confined, linear stretch of water; e.g. ►bypass
channel [US]/by-pass channel [UK], ►canal 1, ►irrigation
canal); *syn.* flume [n]; *s* **reguero [m] (1)** (Cauce excavado por el
hombre para la distribución de agua en el campo con poca o
ninguna fábrica de revestimiento; DGA 1986; ►canal, ►canal de
riego, ►cauce de derivación); *f* **canal [m] artificiel d'écoulement** (Conduit linéaire permettant le passage des eaux, p. ex. des
canaux latéraux pour l'évacuation des hautes eaux ou ►canal
d'irrigation dans les zones arides ; ►canal, ►canal latéral) ; *syn.*
chenal [m] d'écoulement ; *g* **Gerinne [n]** (Künstliches Bauwerk
zur Wasserableitung, z. B. von Hochwassermengen in seitlichen
Kanälen, zur Wasserzuführung für Bewässerungen in ariden
Gebieten; ►Bewässerungskanal, ►Kanal, ►Seitengerinne).

1058 construction [n] (1) *constr. eng.* (Process of building,
e.g. superstructures, infrastructure and landscape work[s]); *syn.*
building process [n]; *s* **construcción [f]** (Proceso de construir
edificios o infraestructuras); *f* **construction [f]** (Processus) ;
g **Bau [m] (1)** (Vorgang des Bauens); *syn.* Erstellung [f] von
Bauwerken.

construction [n] (2) *constr.* ►load-bearing construction,
►masonry construction, ►multicourse construction, ►palisade
construction; *urb.* ►prefabricated concrete panel construction;
constr. eng. ►road construction (1); *eng. trans.* ►road construction (2); ►road-widening construction [US]/road-widening works
[UK]; *constr. contr.* ►start of construction; *constr.* ►stone
construction.

construction [n]**, inert** *constr. eng.* ►conventional engineering.

construction [n]**, subterranean** *urb.* ►below-grade construction.

construction [n]**, underground** *urb.* ►below-grade construction.

1059 construction contract [n] *constr. contr.* (Agreement or covenant with a construction firm to execute a project
usually within a legal framework governing terms, scope and
procedural protocols); *s* **contratación [f] de obras o suministro**
(Acuerdo con empresa para realizar una obra o proyecto, o para

C

suministrar materiales); *f* **contrat [m] de mission de travaux** (Contrat oral ou écrit passé avec l'entrepreneur portant sur des prestations de travaux ou sur une commande) ; *g* **Auftrag [m] an den Unternehmer** (Mündliche oder schriftliche Übertragung einer Aufgabe [Herstellung eines im Angebot versprochenen Werkes in Form einer Bauleistung oder Lieferung von Materialien] seitens des Bestellers [Bauherrn] an einen Unternehmer nach Auswertung des Angebotes resp. des Submissionsergebnisses einer Ausschreibung mit der Verpflichtung, nach erfolgreicher Fertigstellung die vereinbarte Vergütung zu entrichten).

1060 construction contractor [n] *constr. contr.* (Firm or person who guarantees to supply certain materials or do certain work for a stipulated sum or fee; ▶bidder offering best value for the price bidder [US]/tenderer offering the best value for the money [UK]); *s* **contratista [m]** (Empresa o persona que ha recibido el ▶remate; ▶licitante ganador); *syn.* adjudicatario [m]; *f* **entreprise [f] titulaire du marché** (Entreprise attributaire lorsqu'elle reçoit la notification du marché ; ▶attribution du marché [sur appel d'offres/sur adjudication], ▶candidat retenu) ; *syn.* titulaire [m] du marché (CCM 1984, 36), adjudicataire [m] ; *g* **Auftragnehmer [m] (1)** (Ausführender Betrieb/ausführende Firma, der/die den ▶Zuschlag bekommen hat; ▶Bieter/-in des annehmbarsten Angebotes).

construction cost [n] *contr.* ▶establishment of total construction cost, ▶estimate of probable construction cost.

construction cost estimate [n] [UK]**, final** *contr.* ▶total construction cost estimate [US].

1061 construction costs [npl] *constr. contr.* (Monetary costs of a construction project expressed in lump sum, composite square area, or unit costs; ▶estimated construction costs, ▶calculation of costs, ▶overall costs); *s* **costo [m] de construcción/obra** (▶costo estimado de construcción/obra, ▶costo global, ▶cálculo de costos/costes [Es]); *syn.* coste [m] de construcción/obra [Es]; *f* **coût [m] de travaux [F]** (▶coût estimatif des travaux, ▶coût global, ▶estimation financière des travaux) ; *syn.* frais [m] des travaux [B], frais [m] d'exécution [B], montant [m] du coût de travaux [F] ; *g* **Baukosten [pl]** (Kosten eines Bauwerkes; cf. DIN 276; ▶Gesamtkosten, ▶geschätzte Baukosten, ▶Kostenermittlung); *syn.* Kosten [pl] des Bauwerk(e)s, Kostenaufwand [m] eines Bauwerk(e)s.

1062 construction costs [npl] **as a basis for professional fees** *contr. prof.* (A method for calculating a design fee based upon a specified percentage of the total estimated project construction cost including materials, labor, profit and overhead); *s* **costos [mpl] imputables** (Método para el cálculo de honorarios basado en un porcentaje específico del costo total del proyecto incluyendo materiales, mano de obra, gastos generales y ganancia); *syn.* costes [mpl] imputables [Es]; *f* **montant [m] du coût d'objectif [F]** (Montant prévisionnel des travaux de construction défini dans les marchés d'ingénierie incluant la rémunération de l'architecte, généralement assorti d'un taux de tolérance de l'ordre de 10 %. L'architecte s'engage à respecter le coût des travaux à l'intérieur des limites hautes et basses de tolérance définies au contrat sous peine de l'application de réfactions sur sa rémunération) ; *syn.* frais [mpl] totaux d'aménagements [B] ; *g* **anrechenbare Kosten [pl]** (Kosten, die zur Herstellung, zum Umbau, zur Modernisierung, Instandhaltung oder Instandsetzung von Objekten sowie den damit zusammenhängenden Aufwendungen unter Zugrundelegung der Kostenermittlungsarten nach DIN 276-1: 2008-12 zu ermitteln und Grundlage für die Honorarberechnung sind; die Umsatzsteuer ist nicht Bestandteil der **a.n K.**).

construction courses [npl] *constr. eng.* ▶pavement construction courses.

1063 construction document phase [n] [US] *contr. prof.* (In U.S., ▶work phase [US]/work stage [UK] of professional [landscape] architecture services comprising the preparation of drawings and documents necessary to implement a project; in U.K., according to the Landscape Institute Guidance for Clients on Fees; this phase is known as work stage FG [production information]; ▶final design stage, ▶detailed design); *syn.* detailed design phase [n] [UK]; *s* **fase [f] de planificación en detalle** (▶Fase de trabajo de un proyecto después del ▶anteproyecto detallado; ▶planificación detallada); *f* **projet [m]** (▶Étape de la mission de maîtrise d'œuvre comprenant les ▶spécifica-tions techniques détaillées [S.T.D.] et les Plans d'Exécution des Ouvrages [P.E.O.], permettant de lancer la consultation des entreprises ; ▶études de détail) ; *syn.* étude [f] de projet ; *g* **Ausführungsplanung [f]** (Vom Planer zu erbringender Planungsabschnitt [▶Leistungsphase 5], der die Durcharbeitung der Ergebnisse der ▶Entwurfsplanung und Genehmigungsplanung bis zur ausführungsreifen Lösung, d. h. die zeichnerische Darstellung eines Objektes mit allen für die Ausführung notwendigen Einzelangaben auf Ausführungs-, Detail- und Konstruktionszeichnungen, Bepflanzungsplänen, ▶Leistungsbeschreibung etc., beinhaltet; ▶Detailplanung [D]; cf. § 15 [1] 5 HOAI 2002 resp. §§ 33, 38, 42 und 46 HOAI 2009); *syn.* Detailplanung [f] [A] (§ 22 HRLA).

1064 construction drawing [n] *constr.* (Graphic documents showing details of a design project with dimensions, elevations, cross sections for the location on the site or for off-site fabrication; ▶detail drawing, ▶detailed design); *syn.* working drawing [n], detailed design sheet [n]; *s* **plano [m] de ejecución** (Dibujo según el cual se pueden realizar los trabajos en la obra; ▶dibujo en detalle, ▶planificación detallada); *syn.* dibujo [m] de serie o detalle; *f* **plan [m] d'exécution (des ouvrages)** (Document technique destiné à la réalisation du jardin ou d'un aménagement. En général on distingue une série de plans d'exécution des ouvrages [P.E.O.] : plan de piquetage, de nivellement, des circulations, des maçonneries paysagères, de plantations, des réseaux divers, d'équipement ; ▶études de détail, ▶plan de détail) ; *g* **Ausführungszeichnung [f]** (Zeichnung, nach der auf der Baustelle oder in Fertigungsstätten gearbeitet werden soll; ▶Detailplan, ▶Detailplanung); *syn.* Ausführungsplan [m], Bauplan [m], Bauzeichnung [f], Werkplan [m].

construction engineering [n] *eng.* ▶civil engineering.

1065 construction equipment [n] *constr.* (**1.** pecific apparatus designed to carry out a single construction operation. **2.** All the apparatus or device designed to carry out a construction project); *s* **maquinaria [f] de construcción** (**1.** Máquina específica diseñada para realizar una operación de construcción determinada. **2.** Equipamiento total de máquinas necesitado para realizar una obra); *f* **matériel [m] (de chantier)** (**1.** Engin de chantier ; *contexte* nécessitant l'emploi d'un matériel lourd. **2.** Ensemble des outils, instruments, machines utilisés sur un chantier ; terme souvent utilisé au pluriel ; *g* **Gerät [n] (1)** (**1.** Baumaschine; *Kontext* mit schwerem G. arbeiten. **2.** [o. Pl.] Gesamtheit der technischen Ausrüstung, die zur Durchführung [landschafts]gärtnerischer Arbeit benötigt wird); *syn. zu 2.* Geräte [npl] und Maschinen [fpl]).

construction foreman [n] *constr.* ▶contractor's project representative.

1066 construction [n] **for handicapped/disabled persons** *leg. plan. sociol.* (Design and realization of buildings and structures as well as open spaces which can accommodate handicapped people without help of other persons. It redefines design standards and eliminates gratuitous stairways and grade changes for purely visual reasons, by considering the needs of all

users); ▶barrier-free, ▶developed for the handicapped/disabled, ▶handicapped person, ▶universal design); **s construcción [f] adecuada para discapacitados/as** (Diseño y realización de edificios, estructuras y espacios libres de manera que puedan ser utilizados por personas con limitaciones físico-motoras sin ayuda de otras personas; ▶accesible para discapacitados/as, ▶accesible para personas con movilidad reducida, ▶diseño universal, ▶persona con discapacidad [física]); *syn.* construcción [f] adecuada para limitados físicos-motores [C], *obs.* construcción [f] adecuada para minusválidos/disminuidos físicos; **f construction [f] adaptée aux personnes à mobilité réduite** (Bâtiments publics ou privés, espaces de vie et environnement mis en accessibilité et dont les caractéristiques techniques permettent de répondre à la diversité des besoins des usagers, ayant subi une incapacité temporaire ou un handicap permanent ; ▶personne à mobilité réduite, ▶accessible aux personnes handicapées, ▶adapté, ée aux personnes handicapées, ▶conception universelle ; *syn. B, CDN, CH* construction [f] adaptée aux handicapés, construction/construire [f/vb] sans barrieres, construction/construire [f/vb] sans obstacle ; **g behindertengerechtes Bauen [n, o. Pl.]** (Herstellen von baulichen Anlagen oder Freiflächen, die zweckentsprechend von Behinderten ohne fremde Hilfe genutzt werden können; ▶barrierefrei, ▶Behinderte/-r, ▶behindertengerecht, ▶behindertengerechte Planung; *OB* barrierefreies Bauen).

construction items [npl] *constr. contr.* ▶execution of construction items.

construction management services [npl] [UK] *prof.* ▶construction phase—administration of the construction contract [US].

construction manager [n] [UK]**, contractor's** *constr. prof.* ▶contractor's agent.

construction material [n] *constr.* ▶inert construction material, ▶workability of construction material.

1067 construction [n] **of a project** *contr. constr. eng.* (Execution of a landscape or building project); **s ejecución [f] (de un proyecto)** (de edificación o construcción paisajista); **f réalisation [f] d'un projet** ; **g Ausführung [f] eines Projektes** (Realisierung eines Bauprojektes; ausführen [vb]); *syn.* Durchführung eines Projektes [f], Projektdurchführung [f], Projektausführung [f].

1068 construction [n] **of flower beds** *constr. hort.* (Preparation of planting areas to receive annuals and herbaceous perennials with regard to subgrade drainage, required soil or planting medium composition and depth; ▶seasonal flowering plants); *syn.* preparation [n] of flower beds; **s construcción [f] de macizos de flores** (Preparación de áreas de plantación para ▶flores de estación o herbáceas perennes en cuanto al drenaje del subsuelo, la composición y la profundidad de la tierra necesitadas); **f exécution [f] de massif floral** (Mise en œuvre d'un motif ornemental de ▶plantes saisonnières avec une bonne terre drainante bien préparée ; ▶plate-bande) ; **g Herstellen [n, o. Pl.] von Blumenbeeten** (Gärtnerische Anlage von Flächenn mit einem gut vorbereiteten, wasserdurchlässigen Boden[gemisch], die mit ▶Wechselflor bepflanzt werden).

1069 construction [n] **of hard surfaces** *constr.* (Generic term covering construction of roads, pathways, squares, and plazas, including such work associated with pavement grading and drainage; ▶road construction 1); **s trabajos [mpl] de construcción viaria** (Todas las medidas necesarias para construir carreteras, calles, caminos y plazas incluidas las instalaciones auxiliares como taludes, canales de desagüe, etc.; ▶construcción de carreteras); **f travaux [mpl] de voirie** (Terme générique recouvrant tous les travaux afférents à la construction des routes, chemins, places, y compris les travaux annexes tels que les talus,

les fossés de drainage, etc. ; ▶construction routière) ; **g Wegebau [m]** (**1.** [o. Pl.] Teilbereich des Tiefbau[e]s. **2.** ...bauten [mpl]; alle Maßnahmen zur Fertigung von Straßen, Wegen, Plätzen einschließlich der Nebenanlagen wie Böschungen, Entwässerungsgräben etc.; ▶Straßenbau).

1070 construction [n] **of steps** *constr.* **s construcción [f] de escaleras;** **f construction [f] d'escalier** ; **g Treppenbau [m].**

1071 construction period [n] *constr.* (Time span from the beginning of construction to the ▶issue of a certificate of final completion; ▶contract period); **s periodo/período [m] de construcción** (Tiempo de duración de una obra hasta la aceptación final de la misma; ▶aceptación final de obra, ▶plazo de ejecución); **f durée [f] du chantier** (Période comprise entre le début du chantier et la réception de l'ensemble de l'ouvrage dans le cadre de la ▶réception définitive des travaux ; ▶délai d'exécution) ; **g Bauzeit [f]** (Zeitspanne von Baubeginn bis zur ▶Schlussabnahme; ▶Ausführungsfrist).

construction permit [n] [US] *adm. leg.* ▶building permit [US].

1072 construction phase [n] *constr.* (Distinct period of time in the construction process that marks the actual execution of the proposed design from site preparation to final inspection; ▶construction project section, ▶specialist trade); *syn.* construction stage [n]; **s etapa [f] de construcción** (Fase de una obra que se realiza o se debe realizar en un periodo/período de tiempo específico; ▶tramo de obra, ▶lote de oficio); *syn.* fase [f] de construcción; **f phase [f] de travaux (d'un chantier)** (Partie d'un ouvrage à réaliser dans un temps fixé préalablement ; ▶lot technique, ▶tranche de travaux) ; *syn.* phase [f] d'une opération, phase [f] de l'intervention ; **g Bauabschnitt [m]** (Teil eines Bauvorhabens, der in einer bestimmten Zeitspanne durchgeführt wird oder werden muss; ▶Baulos, ▶Fachlos); *syn.* Baustufe [f], Ausbaustufe [f], Bauphase [f].

1073 construction phase—administration [n] **of the construction contract** [US] *prof.* (HALAC 1985, 276; consultant's professional services which ensure the smooth execution of a contract according to the contract documents from the commencement of the work [US]/works [UK] until ▶delivery [US] 1/handover [UK] to the client, as well as the remedying of defects, as established by ▶issue of a certificate of compliance [US]/issue of certificate of practical completion [UK]; tasks include a] preparation and coordination of the time schedule; b] records in the site log; c] acceptance of work and deliveries and the issue of certificates of practical completion; d] measurement of the completed works with the contractor; e] issue of notice of default to the contractor; f] acceptance of the construction services; g] determination of defects and supervision of their removal; h] checking of all invoices for the release of payments by the owner/client; i] monitoring of budget expenditures; j] cost control and evaluation of contractor services in comparison with the contract prices and the ▶total construction cost estimate [US]/final construction cost estimate [UK]; k] schedule of guarantee periods; ▶site supervision, ▶concluding project review [US]/post-completion advisory services [UK]); *syn.* site management services [npl] [UK], construction management services [npl] [UK]; **s dirección [f] facultativa de las obras** (Conjunto de servicios de planificación necesarios para realizar con éxito todas las fases de un proyecto de construcción, teniendo en cuenta el permiso de construcción o la aprobación del comitente, los planes de realización, el resumen de prestaciones así como las normas vigentes y las técnicas reconocidas y aceptadas, desde el comienzo hasta la ▶entrega de obra y su posterior control de la corrección de defectos constatados en la ▶aceptación de obra; ▶inspección de la construcción,

C

►dirección en pie de obra, ►servicios en plazo de garantía); *f 1* **contrôle [m] des travaux** (Contrôle de l'exécution et de l'achèvement des travaux de construction ou d'aménagement ; terme générique pour *f 2, f 3* et *f 4*) ; *f 2* **contrôle [m] des entrepreneurs** (Étape de la mission normalisée de maîtrise d'œuvre comprenant la surveillance de l'exécution des travaux, du début de ceux-ci jusqu'à la ►prise de possession des ouvrages ; **F.,** celle-ci constitue l'étape 4 de la maîtrise d'œuvre et possède deux éléments de mission normalisés comprenant le ►contrôle général des travaux [G.G.T.] et la ►réception et décompte des travaux [R.D.T.] ; ►surveillance permanente de chantier, ►mise à disposition de certains ouvrages) ; *f 3* **contrôle [m] général des travaux (G.G.T.)** (Élément de mission normalisée de maîtrise d'œuvre comprenant l'organisation et la direction des réunions de chantier, le contrôle de la conformité de l'exécution des travaux) ; *f 4* **réception [f] et décompte [m] des travaux (R.D.T.)** (Élément de mission normalisée de maîtrise d'œuvre comprenant la vérification des états quantitatifs, l'établissement des propositions de paiement, la ►réception des travaux et le contrôle de la réfection des ouvrages défectueux) ; *g* **Objektüberwachung [f]** (1. *Nach § 15 [2] Leistungsphase 8 HOAI 2002 resp. §§ 33, 38, 42 u. 46 HOAI 2009* Gesamtheit der Planerleistungen, die einen reibungslosen Ablauf der Ausführung eines Bauvorhabens nach den Vertragsunterlagen [Übereinstimmung mit der Baugenehmigung oder Zustimmung, den Ausführungsplänen und den Leistungsbeschreibungen sowie mit den allgemein anerkannten Regeln der Technik und den einschlägigen Vorschriften] von Baubeginn bis zur ►Übergabe des Objektes sicherstellen; ferner das a] Aufstellen und Überwachen eines Zeitplanes; b] Führen eines Bautagebuches; c] Abnahme von Leistungen und Lieferungen unter Fertigung einer Abnahmeniederschrift; d] gemeinsames Aufmaß mit den bauausführenden Unternehmen; e] Inverzugsetzen der ausführenden Firmen; f] ►Abnahme der Bauleistungen; g] Feststellen von Mängeln und Überwachen sowie Sicherstellung der Beseitigung der festgestellten Mängel; h] Prüfung aller Rechnungen für die Freigabe der Zahlungen durch den Bauherrn; i] Überwachung des Abflusses der zur Verfügung stehenden Finanzmittel; j] Kostenkontrolle durch Überprüfung der Leistungsabrechnung der bauausführenden Firmen im Vergleich zu den Vertragspreisen und zum ►Kostenanschlag; k] Auflistung der ►Gewährleistungsfristen; nach der **O.** folgt als nächste Leistungsphase die ►Objektbetreuung und Dokumentation; 2. *für Ingenieurbauwerke und Verkehrsanlagen gemäß § 55 [2] Leistungsphase 8 HOAI 2002 resp. §§ 42 und 46 LPh 8 HOAI 2009* ►Bauoberleitung [f]).

1074 construction phasing [n] *constr.* (Time schedule for a sequence of construction operations each of which typically prepares for the next operation or completes a particular design element; ►division of a project into sections); *s* **división [f] en fases de construcción** (División temporal de una obra de construcción; ►división de un proyecto en lotes); *syn.* división [f] en etapas de construcción; *f* **phasage [m]** (Division temporelle d'une opération ; ►décomposition en tranches de travaux) ; *g* **Einteilung [f] in Bauabschnitte** (Vorgabe, in welchen Zeiträumen Teile eines Bauvorhabens durchgeführt werden müssen; ►Teilung in Lose).

1075 construction program [n] [US]/**construction programme** [n] [UK] *constr. plan.* (Series of progressive steps for construction work [US]/works [UK] typically associated with a PERT Chart or a Time/Task Bar Chart [*NOTE* PERT: Program Evaluation Review Technique, which is a probability model that estimates and calculates the Critical Path, or the series of interconnected tasks that provide the least slippage in time of execution]; ►progress chart); *syn.* construction progress schedule [n], construction schedule [n], time chart [n]; *s* **calendario [m] de**

ejecución de obras** (►diagrama de etapas de construcción); *syn.* plan [m] de trabajo, plan [m] de ejecución de las obras, gráfico [m] de trabajos (BU 1959); *f* **programme [m] d'une opération** (Établissement d'un programme définissant les différentes phases d'une opération de travaux ; ►planning d'avancement des travaux, ►planning d'exécution des travaux) ; *g* **Bauablaufplan [m]** (Plan, der als generelle Linie die organisatorische und zeitliche Abfolge der einzelnen Projektabschnitte und den Einsatz der sowie die Abhängigkeit zwischen den einzelnen Gewerke aufzeigt; daraus wird dann ein ►Bauzeitenplan entwickelt).

1076 construction progress [n] *constr. contr.* (Incremental advancement of a construction program as measured by agreed bench marks such as percentage of completion or execution of specified tasks); *s* **avance [m] de las obras** *syn.* progreso [m] de las obras, avance [m] de la construcción; *f* **avancement [m] des travaux** (Réalisation de; *g* **Baufortschritt [m]** (Erreichung von vorgegebenen Fertigstellungsterminen einzelner Bauabschnitte oder Gewerke während der Abwicklung eines Bauvorhabens).

construction progress schedule [n] *constr. plan.* ►construction program [US].

1077 construction project [n] *constr. plan.* (Proposal for the implementation of a specific plan or design to build, restore, or redevelop a particular structure or piece of land ; *specific term* ►public construction project); *syn.* building project [n]; *s* **proyecto [m] de construcción** (►proyecto de obras públicas); *syn.* proyecto [m] de edificación; *f 1* **opération [f] immobilière** (pour le génie civil; *termes spécifiques* opération de restauration immobilière, ►opération immobilière publique) ; *syn.* projet [m] de construction ; *f 2* **projet [m] d'aménagement de construction** (pour des espaces libres) ; *g* **Bauvorhaben [n]** (1. OB zu Vorhaben, um eine oder mehrere bauliche Anlagen zu errichten, zu ändern oder abzubrechen; *UB* Sanierungsvorhaben. 2. Landschaftsgärtnerisches Projekt, eine neue Freianlage zu erstellen, eine vorhandene zu ändern oder zu erweitern; *UB* ►öffentliches Bauvorhaben).

1078 construction project section [n] *constr.* (Portion of a large major construction project; ►division of a project into sections, ►specialist trade); *s* **tramo [m] de obra** (Sección de una obra de grandes dimensiones; ►lote de oficio, ►división de un proyecto en lotes); *syn.* sección [f] en construcción; *f* **tranche [f] de travaux** (Fractionnement des ouvrages dans une opération faisant l'objet de marchés successifs ; ►décomposition en tranches de travaux, ►lot technique) ; *g* **Baulos [n]** (Teilabschnitt eines großen Bauvorhabens; ►Fachlos, ►Teilung in Lose); *syn.* Los [n].

construction requirements [npl] *contr. constr.* ►special construction requirements.

construction schedule [n] *constr. plan.* ►construction program [US].

1079 construction segment [n] [US] *constr.* (Individual portion of work in a construction project; ►construction work [US]/construction works [UK], ►altered construction segment [US]/variation [UK]); *syn.* itemised work [n] [UK]; *s* **servicios [mpl] de construcción** (Obra que incluye tanto los ►trabajos de construcción mismos como el suministro de materiales de construcción; ►trabajos modificados); *f* **travaux [mpl]** (Prestations nécessaires à la construction, réparation et entretien de bâtiments ou nécessaires à l'aménagement, la réparation et l'entretien d'espaces verts, y compris toutes fournitures nécessaires à la bonne exécution des travaux ; ►travaux modificatifs) ; *syn.* prestation [f] fournie par l'entreprise ; *g* **Bauleistung [f]** (In einem Leistungsverzeichnis näher beschriebene bauhandwerkliche Maßnahmen mit oder ohne Lieferung von Baustoffen und Bauteilen, die zur Herstellung, Instandsetzung oder -haltung,

Änderung oder Beseitigung eines Bauwerkes notwendig sind oder Arbeiten zur Durchführung landschaftsbaulicher Maßnahmen. Reine Wartungsarbeiten an Bauwerken oder Teilen davon stellen keine **B.** dar, solange nicht Teile verändert oder ausgetauscht werden; Lieferung und Montage maschineller Einrichtungen sind keine **B.en** [cf. § 1 VOB Teil A]. Der aus dem Vertragsrechtlichen stammende Begriff **B.** beinhaltet im Vergleich zu den physischen ▶Bauarbeiten zuzüglich noch die planerischen, organisatorischen und dispositiven Leistungen, die zur Abwicklung von Bauarbeiten notwendig sind; ▶geänderte Leistung).

construction services [npl] **by client/owner** [US] *constr. contr.* ▶services and materials by client/owner.

construction services [npl] **or supplies** [npl] *constr. contr.* ▶supplemental construction services or supplies.

1080 construction site [n] *constr.* (Demarcated area where construction is taking place; ▶transport to the construction site); **s terreno** [m] **de obra** (▶transporte a obra); *syn.* lugar [m] de intervención, lugar [m] de la obra; **f chantier** [m] (▶acheminement sur le chantier) ; *syn.* zone [f] d'intervention ; **g Baustelle** [f] (Fläche, auf der ein Bauvorhaben ausgeführt wird; ▶Transport zur Baustelle).

1081 construction site inspection [n] *constr.* (Examination of a construction site; ▶final site inspection); **s inspección** [f] **de obra** *in situ* (Visita de control de una obra; ▶inspección final de obra); **f reconnaissance** [f] **du chantier** (▶inspection du site) ; *syn.* reconnaissance [f] du site ; **g Baustellenbegehung** [f] (Intensive Besichtigung einer Baustelle, um den Baufortschritt, die Einhaltung der einschlägigen Sicherheitsvorkehrungen und die Qualität der ausgeführten Leistungen zu überprüfen; ▶Objektbegehung; *syn.* Baustellenbesichtigung [f].

1082 construction soil classification [n] *constr.* (In U.S., there are five ▶classes of soil materials [US]/soil class [UK]: 1. massive crystalline bedrock, 2. sedimentary and foliated rock, 3. sandy gravel and/or gravel [GW and GP], 4. sand, silty sand, clayed sand, silty gravel and clayed gravel [SW, SP, SM, SC, GM and GC], 5. clay, sandy clay, silty clay and clayed silt [CL, ML, MH and CH]; in U.K., division of soils into classes according to its characteristics for earth works [cf. BSCP 2003] or for use in [soft] landscape work; cf. BS 3382); **s clasificación** [f] **de suelos (1)** (D., división de suelos en siete clases de suelos según la friabilidad y subdivisión en diez grupos de suelos según la aptitud para plantaciones; ▶grupo de suelos); **f 1 classification** [f] **des terrains** (Classification d'après la difficulté d'exécution des terrassements : [a] terre végétale, sables meubles, [b] terre argileuse, pierreuse ou caillouteuse, tuf, marnes fragmentées, sables agglomérés par un liant argileux, remblais de gravois, [c] argile plastique, glaise franche, marne compacte, sables fortement agglomérés, [d] roche moyennement dure, naissance de masse non compacte, poudingues agglomérés exploitables à la pioche, [e] roche dure exploitable au coin, à la pointerolle ou au marteau-piqueur, [f] roche dure nécessitant l'emploi de la mine ; SDP 1998 ; **D.,** d'après la DIN 18 915 on distingue 10 groupes de sols) ; **f 2 classification** [f] **des sols (1)** (1. D'après la portance pour l'exécution de la voirie on distingue six types de sols : [a] argiles fines saturées, [b] limons plastiques, [c] sables argileux, [d] sables alluvionnaires avec moins de 5 % de fines, [e] matériau insensible à l'eau, grave propre, ancienne chaussée, [f] grave propre et compacte, roche, ancienne chaussée ; VRD 1994, partie 4, chap. 5.2.2. p. 3. **2.** D'après l'aptitude aux travaux culturaux ; ▶catégorie de sols, ▶classe de terrains) ; **g Bodenklassifikation** [f] **(1)** (Untergliederung der Böden **1.** nach ihrem Zustand beim Lösen in **sieben Bodenklassen** gem. DIN 18 300: [1] Oberboden, [2] fließende Bodenarten, [3] leicht lösbare Bodenarten, [4]

mittelschwer lösbare Bodenarten, [5] schwer lösbare Bodenarten, [6] leicht lösbarer Fels und vergleichbare Bodenarten, [7] schwer lösbarer Fels oder **2.** Untergliederung in **zehn Bodengruppen** nach Eignung für vegetationstechnische Zwecke gem. DIN 18 915: [1] organischer Boden; [2] nicht bindiger Boden; [3] nicht bindiger, steiniger Boden; [4] schwach bindiger Boden; [5] schwach bindiger, steiniger Boden; [6] bindiger Boden; [7] bindiger, steiniger Boden; [8] stark bindiger Boden; [9] stark bindiger, steiniger Boden; [10] stark steiniger Boden; ▶Bodengruppe).

construction stage [n] *constr.* ▶construction phase.

construction supervisor [n] [UK]**, contractor's** *constr. prof.* ▶contractor's agent.

construction technique [n] *constr.* ▶biotechnical construction technique.

1083 construction type [n] *constr.* (Generic term for building construction using predominantly one material, e.g. wood, stone, brick, or one method, e.g. prefabrication or *in situ*); **s tipo** [m] **de construcción** (Término genérico para denominar la construcción que utiliza un tipo de material predominante, p. ej. madera, piedra, ladrillo); **f type** [m] **de construction** (Terme générique désignant le caractère principal d'une construction, p. ex. en bois, en pierre, en lamellé-collé) ; **g Bauweise** [f] **(1)** (Art der Konstruktion, z. B. vorwiegend in Holz, Stein, mit Leimbindern, oder ob mit Fertigteilen oder vor Ort gefertigt wird).

construction unit [n] *constr.* ▶living construction unit.

1084 construction [n] **without mortar/cement** *constr.* (Dry laid stone construction that relies on the weight and friction between elements and surfaces for stability); **s construcción** [f] **en seco** (En mampostería, técnica de construcción que no utiliza cemento o mortero); **f construction** [f] **en pierres sèches** (En maçonnerie technique de construction d'ouvrages en pierres naturelles assemblées sans le concours de liants [mortier]) ; **g Trockenbauweise** [f] (Technik, ohne Mörtel Steinbauwerke zu erstellen, die auf Grund ihres Gewichtes und der Reibung zwischen den Bauelementen statisch stabil sind).

1085 construction work [n] [US]**/construction works** [npl] [UK] *constr.* (1. General term related to the application of labor and machines to the building process. 2. Work associated with building activities required to implement a design; ▶scope of construction work, ▶public works); *syn.* construction project [n] [US]; **s trabajos** [mpl] **de construcción** (Todos los trabajos manuales o mecanizados realizados en una obra, no incluye los de planificación o diseño; ▶ámbito de trabajo, ▶obras públicas); **f travaux** [mpl] **de chantier** (▶consistance et étendue des travaux, ▶travaux des chantiers publics) ; *syn.* opérations [fpl] de chantier ; **g Bauarbeiten** [fpl] (Arbeiten auf einer Baustelle; der Begriff **B.** beinhaltet im Vergleich zum Begriff ▶Bauleistung nur die manuellen und maschinellen Tätigkeiten; ▶Art und Umfang der Leistung, ▶öffentliche Bauarbeiten).

1086 construction worker [n] *constr. contr.* (1. Company employee working on a construction site. 2. Employee component of a bid item [US]/tender item UK]); *syn.* laborer [n] [also US], workpeople [n] [also UK]; **s mano** [f] **de obra** (1. Empleado de una empresa que está trabajando en una obra. 2. Partida de cálculo en ofertas); *syn. de 1.* obrero [m] de construcción, trabajador [m] de construcción; **f personnel** [m] **ouvrier** (Main d'œuvre d'entreprise à laquelle est confiée divers travaux manuels pouvant être élémentaires, spécialisés ou nécessitant des connaissances professionnelles sur un chantier ; terme utilisé comme base de calcul d'une offre) ; **g Arbeitskraft** [f] **(1)** (*Abk.* AK; **1.** beschäftigte Person einer Firma auf der Baustelle. **2.** AK als Kalkulationsgröße bei Angeboten; *UB* Bauarbeiter/-in).

C

construction works [npl] [UK] *constr.* ►construction work [US].

construction works [npl] [UK]**, suspension of** *constr. contr.* ►suspension of project [US].

1087 consultant's report [n] **on landscape planning** *landsc.* (Presentation of ecological and visual impacts of, e.g. industrial land uses, infrastructures or urban development, etc. upon natural systems; ►environmental impact assessment [EIA] [US]/environmental assessment [EA] [UK], ►environmental impact statement [EIS], ►environmental impact study [EIS] [US], ►environmental risk analysis); *s* **estudio [m] de los efectos ecológicos y paisajísticos** (Presentación de los impactos ecológicos y visuales de un proyecto; los estudios de impacto ambiental son un tipo de ellos; ►análisis del riesgo ecológico, ►declaración de impacto, ►estudio de impacto ambiental, ►evaluación del impacto ambiental [EIA]); *f* **expertise [f] paysagère** (Description et évaluation des impacts écologiques d'aménagements urbains ou d'infrastructures sur le milieu naturel ou les paysages ; ►analyse des risques écologiques, ►déclaration environnementale, ►étude d'impact, ►notice d'impact, ►procédure d'étude d'impact) ; *g* **landschaftsplanerisches Gutachten [n]** (Darstellung der ökologischen Auswirkungen bestehender oder geplanter wirtschaftlicher Nutzungen, Infrastrukturen oder Siedlungsentwicklungen auf Naturhaushalt und Landschaftsbild resp. auf andere Nutzungen; ►Risikoanalyse, ►Umwelterklärung, ►Umweltverträglichkeitsprüfung, ►Umweltverträglichkeitsstudie).

1088 consultant's report [n] **on noise** *plan.* (**1.** Descriptive analysis of the noise situation to be expected after new construction or substantial changes to a facility, or an investigation of noise levels caused by an existing facility, typically expressed in decibel units. **2.** The description of the noise situation, expected after the establishment of a new or substantially modified facility, is called a **noise forecast**; ►noise pollution); *s* **peritaje [m] de la contaminación acústica** (**1.** Análisis de la ►contaminación acústica de una zona causada por una instalación determinada. **2.** Descripción del grado de contaminación acústica previsible en caso de instalación de una fuente de ruido considerable, que se podría denominar **pronóstico de ruido**); *f 1* **expertise [f] acoustique** (Mesure de niveaux d'émission sonore d'une installation classée) ; *f 2* **contrôle [m] des niveaux de bruit** (Méthode de contrôle des ►nuisances sonores moins exigeante que la précédente, quant aux moyens à mettre en œuvre et à l'appareillage de mesure à utiliser ; arrêté du 23 janvier 1997) ; *f 3* **étude [f] prévisionnelle acoustique** (Évaluation et prévision des niveaux sonores lors de la construction de bâtiments ou d'infrastructures terrestres nouvelles ainsi que de la transformation d'infrastructures existantes ; arrêté du 23 janvier 1997) ; *g* **Lärmgutachten [n]** (**1.** Darstellung der Lärmimmissionen einer Anlage auf ein Beurteilungsgebiet. **2.** Die Beschreibung der Lärmsituation, die bei einer neu zu errichtenden oder wesentlich zu ändernden Anlage zu erwarten sein wird resp. die durch eine Lärm verursachende Anlage gegeben ist, wird **Lärmprognose** genannt; ►Lärmbelastung).

consultation [n] **with public agencies** [US] *plan.* ►statutory consultation with public agencies [US].

1089 consulting contract [n] *contr. prof. s* **contrato [m] de asesoría** *syn.* contrato [m] de consultoría; *f* **contrat [m] pour une mission de conseil** ; *g* **Beratervertrag [m]**.

1090 consulting engineer [n] *prof.* (Professional engineer in private practice who is called for consultation on projects requiring particular engineering expertise); *s* **ingeniero/a [m/f] asesor, -a** (Profesional de ingeniería independiente que es contratado/a para asesorar en proyectos que exigen conocimientos especiales de ingeniería); *f* **ingénieur [m] conseil** (Ingénieur libéral assistant tout client pour l'élaboration, le planning, l'expertise de projets techniques) ; *g* **beratender Ingenieur [m]** (Freischaffender Ingenieur für die Planung, Entwicklung und sachverständige Begutachtung technischer Projekte).

1091 consumer [n] *ecol.* (Heterotrophic life form, which subsists upon other forms of life or parts of them. **Primary consumers** feed upon plants or parts of plants [herbivores, phytophagous species]. **Secondary or tertiary consumers** live upon animals or animal tissue [carnivores, zoophagous species]. **C.s** which obtain nourishment from both plant and animal food [omnivores] cannot be included under these categories; ►food chain, ►food web, ►producer); *s* **consumidor [m]** (En un ecosistema, el conjunto de organismos que consumen la energía de los productores, directamente o mediante intermediarios más próximos a los productores. En la cadena de herbívoros se pueden diferenciar varios eslabones, siendo el primero de ellos el de los *productores* capaces de fijar la energía solar mediante la fotosíntesis y de sintetizar materia orgánica a partir de compuestos inorgánicos. Esta materia es fuente de alimento para los restantes organismos: los **consumidores**. Dentro de ellos cabe diferenciar: **consumidores primarios** —segundo eslabón de la cadena—, que incluye a los animales herbívoros y a los parásitos de las plantas verdes. El tercer eslabón está formado por los **consumidores secundarios** o carnívoros en primer grado, predadores y parásitos que utilizan la materia acumulada por los hervíboros y sirven a su vez de fuente de alimento a los **consumidores terciarios** o carnívoros de segundo grado, que ocupan el cuarto eslabón de la cadena y son alimento, a su vez, de eventuales **consumidores cuaternarios**; cf. DINA 1987; ►cadena trófica, ►red trófica, ►productor); *syn.* heterótrofo [m]; *f* **consommateur [m]** (Organisme hétérotrophe qui se nourrit soit de végétaux soit d'animaux. On distingue **1.** les **consommateurs primaires** ou **c.** de premier ordre qui sont soit des phytophages/herbivores soit des parasites de plantes, **2.** les **consommateurs secondaires**, parmi lesquels les **c.** de second ordre subsistent aux dépens des herbivores alors que les **c. de troisième ordre** se nourrissent de carnivores/zoophages. Les **c.** secondaires peuvent être soit des prédateurs qui se nourrissent de proies vivantes soit des nécrophages qui se nourrissent de cadavres ; DEE 1982 ; ►chaîne trophique, ►producteur, ►réseau trophique) ; *g* **Konsument [m]** (Heterotrophes Lebewesen, das sich von anderen Lebewesen oder deren Teile ernährt. **Primärkonsumenten** ernähren sich von Pflanzen- oder Pflanzenteilen [Herbivore, Phytophage]. **Sekundär-, Tertiär-** usw. **-konsumenten** ernähren sich von Tieren oder tierischem Gewebe [Carnivore, Zoophage]. Konsumenten, die sich sowohl von pflanzlicher als auch tierischer Nahrung ernähren [Omnivore], können diesen Kategorien nicht zugeordnet werden; ANL 1984; ►Nahrungskette, ►Nahrungsnetz, ►Produzent).

1092 consumer behavior [n] [US]**/consumer behaviour** [n] [UK] *sociol.* (Human attitudes, when buying goods and services. The **c. b.** is determined by several influences, such as advertizing, cultural and social origin, orientation to particular social groups, as well as by psychological, demographic and economic conditions); *syn. o.v.* consumer behaviour [n] [UK]; *s* **comportamiento [m] consumista** (Actitud de las personas a la hora de comprar bienes y servicios que depende de diferentes influencias, como la publicidad, el origen cultural y social, orientación a un grupo social asi como de las condiciones sicológicas, demográficas y socioeconómicas); *f* **pratique [f] de consommation** (Comportement des personnes lors de l'achat de biens et de prestations de services [consommation de masse] largement déterminé par l'influence des acteurs économiques [publicité, mercatique], l'origine sociale et culturelle, la référence

à des groupes socioculturels, les donnés démographiques et économiques) ; *g* **Konsumverhalten** [n] (Verhalten der Menschen beim Kauf von Waren und Dienstleistungen. Das **K.** wird u. a. durch mehrere Einflüsse wie Werbung, kulturelle und soziale Herkunft, Orientierung an soziale Bezugsgruppen sowie durch psychologische, demografische und ökonomische Gegebenheiten bestimmt).

consumer behaviour [n] [UK] *sociol.* ►consumer behavior [US].

1093 consumer society [n] *sociol.* (Society whose aim it is to promote consumption, and is characterized by prosperity in most sections of the population. Such a society ignores the need to preserve and reuse finite natural resources in a sustainable economy; ►throw-away-society); *s* **sociedad** [f] **de consumo** (Sociedad que tiene como una de sus metas promover el consumo. La **s. de c.** ignora la necesidad de preservar y reutilizar los recursos naturales en una economía sostenible. En los países industrializados pueden participar de ella una gran mayoría de la población. Sin embargo, en los países en vías de desarrollo o subdesarrollados de economía de mercado, en los que generalmente también se propaga el consumismo, la mayoría no tiene acceso a los bienes de consumo por falta de ingresos; ►sociedad del despilfarro); *f* **société** [f] **de consommation** (Société dont les objectifs primordiaux sont de rechercher, s'assurer et accroître la consommation et caractérisée par l'accès au bien-être de larges couches de la population ; une telle société ignore ou méconnaît la nécessité de préserver les ressources naturelles et de s'engager dans la voie d'un développement économique soutenable ; ►société de gaspillage) ; *g* **Konsumgesellschaft** [f] (Gesellschaft, die die Sicherung und Steigerung des Konsums verfolgt und durch hohen Wohlstand breiter Bevölkerungsschichten gekennzeichnet ist; eine solche Gesellschaft erkennt nicht oder nur unzureichend die Notwendigkeit, die endlichen natürlichen Ressourcen zu schonen und im Sinne einer nachhaltigen Wirtschaftsweise wieder- oder weiterzuverwenden; ►Wegwerfgesellschaft).

1094 consumption [n] **of natural resources** *ecol. nat'res. sociol.* (Human use of air, water, mineral and biotic resources; ►wasteful exploitation, ►sustainable management); *s* **consumo** [m] **de los recursos naturales** (►expoliación de los recursos naturales, ►gestión económica sostenible); *f* **consommation** [f] **des ressources naturelles** (►surexploitation, ►gestion durable) ; *g* **Ressourcenverbrauch** [m] (*Fachsprachlich* ...verbräuche [pl], *sonst o. Pl.*; ►Raubbau, ►nachhaltiges Wirtschaften).

1095 contact herbicide [n] *agr. envir. for. hort.* (Weed killer designed to kill on contact by destroying foliage; *generic term* ►herbicide); *s* **herbicida** [m] **de contacto** (Biocida que se aplica a las partes aéreas de las plantas y que provoca la muerte rápida de las mismas; ►herbicida); *f* **herbicide** [m] **de contact** (Substance appliquée sur la partie aérienne des plantes et provoquant la mort rapide des végétaux ; *terme générique* ►herbicide) ; *g* **Kontaktherbizid** [n] (Unkrautbekämpfungsmittel, das krautige Teile bei Berührung absterben lässt; *OB* ►Herbizid).

1096 container ball [n] *hort.* (Root system of a plant grown in a container filled with growing medium; ►container-grown plant); *s* **cepellón** [m] **en tiesto** (Sistema radical de leñosa precultivada en contenedor relleno de sustrato de cultivo; ►planta en maceta); *f 1* **motte** [f] **d'une plante cultivée en pot** (Système racinaire d'une plante préparée pour la plantation dans un contenant rempli d'un substrat de culture approprié ; terme à ne pas utiliser pour les plantes mises en pot peu de temps avant d'être commercialisées ; ►plante cultivée en conteneur) ; *f 2* **motte** [f] **d'une plante cultivée en godet** (Caractérise le

chevelu des plantes cultivées dans un récipient de forme ronde ou carrée d'un diamètre ou d'un côté compris entre 5 et 13 cm, dont le volume est inférieur à 2 litres) ; *g* **Topfballen** [m] (Wurzelgeflecht einer Pflanze, die in einem mit Kultursubstrat gefüllten Gefäß [vor]kultiviert wurde; ►Containerpflanze).

1097 container garden [n] [US] *gard.* (►Miniature garden in a ►planter, basin or similar container; ►flower planter, ►window box); *syn.* trough garden [n] [UK]; *s* **jardín** [m] **miniatura en jardinera** (Jardín en pequeña escala en un recipiente tipo ►jardinera, ►jardinera para balcón, artesa, ►tiesto de flores o similar; ►jardín miniatura); *f* **jardin** [m] **en forme de jauge** (►Jardin miniature sous forme d'un ►bac de plantation, d'un ►bac à fleurs utilisant des plantes vivaces ou ligneuses à développement réduit ou le cas échéant des plantes nanifiées ; ►jardinière pour balcon) ; *g* **Troggarten** [m] (►Miniaturgarten in einem ►Pflanzenkübel, ►Blumenkübel etc. mit geeigneten, klein bleibenden Stauden und Gehölzen, ggf. mit Bonsaipflanzen; ►Balkonkasten); *syn.* Miniatursteingarten [m].

1098 container-grown [pp] *hort.* (Descriptive term for a plant dug and then grown in a semi-rigid receptacle until established with new fibrous roots to retain shape of root ball when shipping and transplanting; in comparison to ►bare-rooted plant); *s* **cultivada en maceta** [loc] (Término descriptivo de catálogo de plantas de vivero para planta cultivada en contenedor en comparación con la ►planta de raíz desnuda o la planta ►envuelta en saco); *syn.* cultivada en contenedor [loc]; *f 1* **motte** [f] **(livrée) en conteneur** (Désignation sur catalogue de la méthode de culture indiquant que la plante a été cultivée dans un conteneur ; *contexte* plante livrée en conteneur ; *syn.* plante [f] en conteneur ; *f 2* **motte** [f] **(livrée) en pot** (Désignation sur catalogue de la méthode de culture indiquant que la plante a été cultivée dans un pot ; *contexte* plante livrée en pot ; *par comparaison avec les* ►plantes à racines nues et les plantes ►présentées en motte avec filet ; *syn.* plante [f] en pot ; *f 3* **motte** [f] **(livrée) en godet** (Désignation sur catalogue de la méthode de culture indiquant que la plante a été cultivée ou conditionnée dans un godet [petit pot en terre, en fibre de tourbe ou en matière synthétique utilisé pour la culture temporaire des jeunes plants]) ; *syn.* plante [f] en godet ; *g* **mit Topfballen** [loc] (Qualitätsbeschreibung in Pflanzenkatalogen, die besagt, dass die kultivierte Pflanze in einem Topf gezogen wurde; *im Vergleich zu* ►Pflanze ohne Ballen, ►mit Ballentuch balliert); *syn.* mit Containerballen [loc], im Topf kultiviert [loc], mit [ein]getopftem Ballen [loc].

1099 container-grown plant [n] *hort* (Plant dug and then grown in a semi-rigid container until established with new fibrous roots to retain shape of rootball when shipping and transplanting; ►plant boxing, ►burlapping, ►balled and potted plant [US]/balled container plant [UK]); *s* **planta** [f] **en maceta** (Planta cultivada en contenedor semirrígido que sirve para proteger el cepellón a la hora del transplante; ►encajonamiento del cepellón, ►embalaje de yute, ►planta con cepellón embalado y plantada en maceta); *f* **plante** [f] **cultivée en conteneur** [loc] (Plante cultivée dans un milieu de culture ; ►emballage en filet de jute, ►mise en bac, ►plante en motte et présentée en conteneur) ; *syn.* plante [f] en conteneur, végétal [m] en conteneur ; *g* **Containerpflanze** [f] (In D. muss gem. ►Gütebestimmungen für Baumschulpflanzen die Größe des Behälters mindestens 2 Liter betragen, sonst spricht man von *Pflanzen mit Topfballen*; die Größe des Behälters muss in einem angemessenen Verhältnis zur Pflanzengröße stehen, der Behälterinhalt gut durchwurzelt sein; *cf.* auch LEHR 1997, p. 285; ►Einkübeln, ►Juteballierung, ►ballierte und in Container gesetzte Pflanze).

1100 container-grown stock [n] *hort.* (Plants cultivated in containers, for ease of transport and year-round planting, until they reach shipping size. In D. the term is used for containers of

C

2 litres and more. Smaller receptacles are referred to as pots); **s material [m] en maceta** (Planta cultivada en contenedor para facilitar el transporte y la plantación durante todo el año hasta que alcanza el tamaño de plantación definitiva. En D. el término es utilizado para macetas de 2 litros o más de volumen); *f* **fourniture [f] en conteneur** *syn.* plants [mpl] en conteneur ; *g* **Containerware [f]** (Zum Verkauf in mindestens 2 Liter großen Gefäßen kultivierte Pflanzen).

1101 container plant [n] *gard. hort.* (Plant suitable for growing in a pot, tub or planter); **s planta [f] para jardinera** (Planta que se puede plantar en tiesto o jardinera); *f* **plante [f] pour bac** (Végétal particulièrement adapté à la croissance en bac, en vasque ou autres conteneurs) ; *g* **Kübelpflanze [f]** (Pflanze, die sich für die Bepflanzung von Kübeln und Trögen eignet).

contaminant [n] *envir.* ▶noxious substance; *chem. envir.* ▶toxic substance.

1102 contaminated airspace [n] *envir. leg.* (*also* contaminated air space [n]; area with oversight of existing or suspected air pollution. Each state or metropolitan area government may continually record and analyze air contamination in an ▶air quality control region as required by law); *syn.* polluted airspace [n] [US]/polluted air space [n] [US], air pollution zone [n] [US]; **s zona [f] de atmósfera contaminada** (En Es. la ley de calidad del aire y protección de la atmósfera prevé la zonificación del territorio por parte de las CC.AA. según los niveles de contaminantes para los que se hayan establecido objetivos de calidad del aire. Sin embargo, no tipifica las **z. de a. c.** como tales, debido a que la competencia legisladora en el ámbito ambiental, más allá de la legislación básica, recae en las CC.AA.; cf. art. 11 Ley 34/2007; ▶zona de evaluación de la calidad del aire); *f* **secteur [m] soumis à la pollution atmosphérique** (▶zone de mesures) ; *syn.* zone [f] soumise à la pollution de l'air atmosphérique ; *g* **lufthygienisches Belastungsgebiet [n]** (Gebiet, in dem Luftverunreinigungen auftreten oder zu erwarten sind; § 44 BImSchG. Jede Landesregierung kann durch Rechtsverordnung **B.e** festlegen, in denen Luftverunreinigungen fortlaufend festzustellen und zu untersuchen sind; ▶Gebiet zur Luftqualitätsüberwachung); *syn.* Luftbelastungsgebiet [n]).

contaminated land [n] *envir. urb.* ▶hazardous old dumpsite [US].

contaminated land registry [n] [UK] *adm. envir.* ▶toxic site inventory [US].

contaminated site [n] *envir.* ▶orphan contaminated site.

contaminated waste site [n] *envir. urb.* ▶hazardous old dumpsite [US].

1103 contamination [n] (1) *chem. envir. phys.* (Pollution or impairment of the environment [air, water, soil, and living beings] by emission such as dust, waste gas or radioactive radiation, unmanaged refuse dumps [US]/unmanaged tipping sites [UK] as well as micro-organisms; ▶environmental pollution, ▶hazardous waste material); **s contaminación [f]** (1. Liberación artificial de emisiones gaseosas, polvo, sustancias tóxicas, energía o radioactividad en la atmósfera, vertido de residuos en masas y cursos de agua, deposición inadecuada de residuos [▶suelo contaminado], etc. que causan daños al medio ambiente, a los ecosistemas y a los seres vivos que habitan en ellos ▶contaminación ambiental. 2. Contaminación radiactiva. 3. Contaminación lumínica: resplandor luminoso nocturno o brillo producido por la difusión y reflexión de la luz en los gases, aerosoles y partículas en suspensión en la atmósfera, que altera las condiciones naturales de las horas nocturnas, en menoscabo de la fauna, la flora y los ecosistemas en general, y dificulta las observaciones astronómicas de los objetos celestes; cf. art. 3 f y disposición adicional cuarta Ley 34/2007); *syn. de 1* polución [f],

envenenamiento [m]; *f* **contamination [f]** (Introduction dans le milieu naturel [air, eau, sol, organismes vivants] de matières usées, résidus fermentescibles d'origine végétale ou animale, substances solides liquides ou gazeuses, toxiques, dangereuses, transmission de microorganismes dans les aliments ou chez les êtres vivants susceptibles de nuire à la salubrité publique ; ▶déchets historiques, ▶pollution de l'environnement) ; *g* **Kontamination [f]** (1. *envir.* Verschmutzung resp. Vergiftung der Umwelt [Luft, Wasser Boden und Lebewesen] durch schadstoffhaltige Stäube, Abgase oder radioaktive Strahlung, unsachgemäße Abfalllagerung [▶Altlast], Einleitungen in Gewässer sowie durch apathogene oder pathogene Mikroorganismen und Viren mit nachteiligen Auswirkungen für die betroffenen Ökosysteme oder deren Teile; zur **K.** gehört auch die Veränderung von Lebens- und Futtermitteln der Biobauern durch unbeabsichtigte Einkreuzung während der Vegetationszeit oder Vermischung während der Ernte, des Transportes oder im Handel mit gentechnisch veränderten Organismen [GVO]; ▶Umweltverschmutzung; *syn.* Kontaminierung [f], Verschmutzung [f] mit umweltgefährdenden Stoffen, *bei hohem Grad der K.* Verseuchung [f]. 2. *phys.* Verunreinigung durch Kernenergienutzung von z. B. Boden, Luft, Wasser, Menschen, Tieren, Pflanzen und Lebensmitteln); *syn.* radioaktive Kontaminierung [f], radioaktive Verunreinigung [f], radioaktive Verseuchung [f].

contamination [n] (2) *ecol.* ▶danger of contamination; *phyt. zool.* ▶genetic contamination; *envir.* ▶heavy metal contamination.

contamination [n]**, air** *envir.* ▶air pollution (1).

contamination [n]**, groundwater** *envir. wat'man.* ▶groundwater pollution.

contamination [n]**, subsurface** [US] *envir. urb.* ▶hazardous old dumpsite [US].

1104 content [n] **of the contract** *contr.* (Requirements and conditions contained in the contract; ▶contractual terms); **s contenido [m] de contrato** (▶condiciones de contrato); *f* **termes [mpl] du contrat** (▶clauses du contrat) ; *g* **Vertragsinhalt [m]** (▶Vertragsbedingungen).

contingency [n] [UK] *contr.* ▶contingency fund [US].

1105 contingency fund [n] [US] *contr.* (Item included in a cost estimate for construction, services or supplies which cannot be exactly ascertained at the time of the estimate, and which usually serves as a financial reserve); *syn.* contingency [n] [UK]; **s partida [f] de imprevistos** (Posición en plan de costes [Es]/ costos [AL] de cualquier tipo en la que se incluyen los gastos no previsibles al hacer el presupuesto); *syn.* imprevistos [mpl]; *f* **somme [f] provisionnelle** (Prix inclus dans une estimation pour des prestations ou des fournitures qui ne peuvent pas encore être prises en compte à l'époque de l'établissement des calculs et qui constitue en général une réserve financière en cas d'imprévus) ; *syn.* montant [m] provisionnel ; *g* **Unvorhergesehenes [n, o. Pl.]** (Position in Kostenermittlungen jedweder Art für Leistungen oder Lieferungen, die zur Zeit der Berechnung noch nicht festgestellt werden können und in der Regel als Finanzierungspuffer dienen).

1106 continuing education [n] [US] *prof.* (1. Education for career advancement and training in a career or profession; **in U.S.**, required for maintenance of professional license in most professions; **in U.K.**, better known as **Continuing Professional Development [CPD]**); *syn.* education [n] by mid-career course, further education [n] [UK]. **2. on-the-job training** [n] [US]: Guidance in office and field tasks for professional improvement; ▶practical training); **s perfeccionamiento [m] profesional** (**1.** Medidas de capacitación para mejorar y actualizar los conocimientos profesionales; **2.** Formación directamente en el trabajo;

►prácticas profesionales); *syn. de 1.* formación [f] continuada, ampliación [f] de estudios; *syn. coloquial* reciclaje [m] profesional [Es]; *f* **formation [f] (professionnelle) continue** (**1.** Elle permet le maintien dans un emploi et favorise le développement des compétences et l'accès dans l'entreprise aux différents niveaux de qualification professionnelle ; **2.** favorise l'insertion ou la réinsertion des demandeurs d'emploi ; ►stage [de formation]) ; *syn.* formation [f] permanente ; *g* **berufliche Weiterbildung [f]** (**1.** Erweiterung der fachlichen Ausbildung der im Beruf Stehenden. **2.** Berufsbegleitende Fortbildung, z. B. am Arbeitsplatz resp. während der täglichen Arbeit; ►Praktikum); *syn.* berufliche Fortbildung [f], berufsbegleitende Fortbildung [f].

continuing professional development (CPD) [n] [UK] *prof.* ►continuing education [US].

continuous canopy [n] *phyt.* ►closed canopy.

1107 contorted growth [n] *arb. for. hort.* (►basal sweep); *syn.* twisted growth [n]; *s* **crecimiento [m] achaparrado** (►curvatura basal del tronco); *f* **croissance [f] tortueuse** (Caractérise la croissance d'un arbre dont le tronc présente une rectitude irrégulière ; ►tronc à crosse) ; *g* **Krummwuchs [m]** (...wüchse [pl]) Wuchseigenschaft eines Baumes, der keinen geraden Stamm hat; ►Säbelwuchs).

1108 contour [n] *geo. surv.* (Line connecting points of equal elevations on a map; many cartographers differentiate between true contours and ► form lines, the latter being merely intended to convey a diagram[m]atic impression of the topographical situation. The term *contourline* is a tautology; ►hachures); *s* **curva [f] de nivel** (Isolínea utilizada en la representación del relieve en los mapas que une puntos situados a la misma altitud; DGA 1986; ►curva de configuración, ►hachuras); *syn.* curva [f] hipsométrica [RA] (BU 1959); *f* **courbe [f] de niveau** (Ligne reliant tous les points de même altitude ; plus les courbes de niveau sont rapprochées, plus la pente est forte ; ►hachures, ►ligne de forme) ; *syn.* ligne [f] de niveau, isohypse [m], ligne [f] isohypse ; *g* **Höhenlinie [f]** (Gedachte Linie, die Punkte gleicher Höhe miteinander verbindet und in Karten oder Plänen dargestellt wird; ►Höhenstrukturlinie, ►Schraffen); *syn.* Höhenschichtlinie [f].

1109 contour alignment [n] *plan. surv.* (Particular pattern of contours on a survey map indicating the form or geomorphology of a surveyed land area); *s* **curso [m] de las curvas de nivel**; *f* **contour [m] des courbes de niveau** ; *g* **Höhenlinienverlauf [m]**; *syn.* Höhenschichtlinienverlauf [m].

contour brush layering [n] *constr.* ►brush layering.

1110 contour farming [n] *agr. land'man.* (Field operations, such as plowing [US]/ploughing [UK], planting, cultivating and harvesting on the contour to guard against erosion of top soil; cf. RCG 1982; ►contour plowing [US]/contour ploughing [UK], ►contour strip cropping); *s* **cultivo [m] en contorno** (Aquellos cultivos que se hacen arando siguiendo la configuración del terreno, es decir, las curvas de nivel, para evitar la erosión; ►arado en contorno, ►cultivo en fajas); *syn.* cultivo [m] horizontal; *f* **agriculture [f] en lignes de niveaux** (Technique culturale de réduction de l'érosion de sols arables en pente pour laquelle le labour est effectué sans discontinuité en suivant les lignes de niveau ; ►culture en bandes, ►labour suivant les lignes de niveau) ; *syn.* agriculture [f] en bandes [aussi CDN] ; *g* **hangparalleler Anbau [m, o. Pl.]** (Landwirtschaftliche Kultivierungsmethode entlang der Höhenschichtlinien zur Minderung von Bodenabträgen an Hanglagen; ►Konturpflügen, ►Streifenanbau).

contouring [n] **and terrain modelling** [n] [US] *constr.* ►ground modeling [US]/ground modelling [UK].

1111 contour map [n] (1) *geo.* (IBDD 1981; map depicting land elevations in graphic form; ►spot elevation); *syn.* topomap [n] [also US]; *s* **mapa [m] topográfico** (►cota); *syn.* mapa [m] con curvas de nivel; *f* **carte [f] topographique** (Carte représentant tous les éléments visibles d'un terrain ou d'une région avec leur position altitudinale ; ►cote en altimétrie) ; *g* **Höhenkarte [f]** (Darstellung eines Geländes oder eines Gebietes mit Höhenangaben und Höhenlinien; ►Kote); *syn.* topografische Karte [f], *o. V.* topographische Karte [f].

contour map [n] (2) *plan.* ►noise contour map.

1112 contour plowing [n] [US]/**contour ploughing** [n] [UK] *agr. land'man.* (Cultivating crops approximately along the contours of land to control erosion; ►contour farming, ►contour strip cropping); *s* **arado [m] en contorno** (Cultivo siguiendo aproximadamente las curvas de nivel; ►cultivo en contorno, ►cultivo en fajas); *syn.* labranza [f] en contorno, arado [m] en curvas de nivel, labranza [f] en curvas de nivel; *f* **labour [m] suivant les lignes de niveau** (LA 1981, 477 ; méthode culturale des sols à forte pente par laquelle le labour est effectué en suivant les lignes de niveaux afin de limiter l'érosion des sols ; ►agriculture en lignes de niveau, ►culture en bandes) ; *syn.* labour [m] en bandes de niveau, labour [m] en contour [CDN] ; *g* **Konturpflügen [n, o. Pl.]** (In den USA entwickelte landwirtschaftliche Kultivierungsmethode als Erosionsschutzmaßnahme, bei der die Ackerflächen entlang den Höhenlinien gepflügt und bestellt werden; ►hangparalleler Anbau, ►Streifenanbau); *syn.* hangparalleles Pflügen [n, o. Pl.].

contour smoothing [n] [US] *constr.* ►ground modeling [US]/ground modelling [UK].

1113 contour strip cropping [n] *agr. land'man.* (Agricultural method of cultivating crops in a systematic arrangement of strips or bands. Such strips are approximately on the contours or at right angles to the prevailing wind for soil protection reasons and are mainly used in tropical and subtropical regions. Usually strips of grass or close-growing crops are alternated with less close-growing crops or fallow ground; ►contour farming, ►contour plowing [US]/contour ploughing [UK]); *syn.* strip farming [n]; *s* **cultivo [m] en fajas** (Práctica de cultivo arreglado en fajas o bandas en el que se alternan los cultivos para proteger el suelo y la vegetación contra avenidas o vientos. Aquéllas se disponen en contorno en suelos erosionados o en ángulos rectos más o menos a la dirección predominante del viento, cuando existe peligro de erosión eólica; cf. MEX 1983; ►arado en contorno, ►cultivo en contorno); *syn.* cultivo [m] en bandas; *f* **culture [f] en bandes** (Méthode culturale pratiquée dans les régions tropicales et subtropicales afin de protéger les sols contre l'érosion et alternant les cultures protectrices [prairie, friche] avec les cultures non protectrices [champs] disposées en forme de bande le long des courbes de niveaux ou perpendiculairement à la direction principale du vent ; ►agriculture en lignes de niveau. ►labour suivant les courbes de niveau) ; *g* **Streifenanbau [m, o. Pl.]** (Vor allem in tropischen und subtropischen Gebieten praktizierte landwirtschaftliche Anbaumethode zum Erosionsschutz, die abwechselnd gut und schlecht bodenschützende Kulturen [Grünland und Ackerland] oder bestelltes und brachliegendes Land in Streifen entlang den Höhenlinien oder quer zur Hauptwindrichtung anlegt; ►hangparalleler Anbau, ►Konturpflügen); *syn.* Streifenkultur [f].

1114 contract [n] (1) *contr. prof.* (Written agreement between two or more parties; *specific terms* ►building contract, ►construction contract, ►consulting contract, ►daywork contract, ►design contract [US]/letter of appointment [UK], ►lump-sum contract, ►planning contract, ►subcontract); *s* **contrato [m]** (Acuerdo suscrito entre dos o más partes; *términos específicos*

C

►contrato de asesoría, ►contrato de obra, ►contrato de subcontrata, ►contrato global, ►contrato por horas, ►contrato profesional; ►contratatación de obras o suministro, ►contratación de un/a planificador, -a); *f 1* **contrat** [m] (Convention passée entre deux ou plusieurs personnes ; ►contrat d'architecte, ►contrat de mission de travaux, ►contrat de sous-traitance, ►contrat pour une mission de conseil) ; *f 2* **marché** [m] (1) (Convention d'achat ou de vente passée en vue de la réalisation de travaux, fournitures et services on distingue le marchés de travaux, les marchés d'études ; ►marché au forfait, ►marché au temps passé, ►marché sur devis, ►mission d'études, ►mission de maîtrise d'œuvre) ; *g* **Vertrag** [m] (Rechtsgültige schriftliche Abmachung zwischen zwei oder mehreren Partnern; *UBe* ►Architektenvertrag, ►Beratervertrag, ►Leistungsvertrag (Bauvertrag), ►Pauschalvertrag, Planungsvertrag, ►Nachunternehmervertrag, ►Stundenlohnvertrag ►Auftrag an den Unternehmer, ►Planungsauftrag).

contract [n] (2) *constr. contr. prof.* ►awarding of contract [US], ►conclusion of a contract, ►content of the contract, ►fulfillment of the contract [US]/fulfilment of the contract [UK]; *contr. leg.* ►general conditions of contract; *contr.* ►party to a contract; *constr. contr.* ►special technical requirements of contract; *contr. prof.* ►termination of a contract.

contract [n] [US]**, cancellation of** *contr. prof.* ►termination of a contract [US].

contract [n] [UK]**, completion of a** *contr. prof.* ►conclusion of a contract.

contract [n] [UK]**, letting of** *constr. contr.* ►awarding of contract [US].

contract award [n] *contr.* ►prepare a recommendation for contract award, ►recommendation for contract award, ►sole source contract award [US]/freely awarded contract [UK], ►written notice of contract award.

1115 contract award documents [npl] *adm. constr. contr.* (All ►bid documents [US]/tender documents [UK] with prices bidded/tendered for work[s], deliveries, and services, which become contractually binding, if not previously modified within the specified ►period for acceptance of a bid [US]); *s* **documentos** [mpl] **de adjudicación de contrato** (Todos los ►documentos de contrato de las administraciones públicas con indicación de los precios para las obras, sumistros o servicios previstos que son vinculantes si no se modificaron las condiciones en el ►periodo/período de remate); *f* **pièces** [fpl] **du projet de marché** (Toutes les ►pièces contractuelles constitutives du marché détaillées dans le règlement de consultation et remises par le candidat comprenant, pour tous les marchés, l'acte d'engagement et en fonction du marché le mémoire technique, la décomposition du prix global et forfaitaire, le détail estimatif, le bordereau de prix unitaires, et autres pièces qui deviendront les pièces du marché si aucune modification n'intervient pendant la période du ►délai de validité des offres) ; *g* **Auftragsunterlagen** [fpl] (Alle ►Verdingungsunterlagen mit Angabe der Vergütung für die zu erbringende Leistung, die, wenn innerhalb der ►Zuschlagsfrist keine Änderungen vorgenommen werden, zu Vertragsunterlagen werden).

contract awarding period [n] *contr.* ►period for acceptance of a bid [US].

1116 contract awarding procedure [n] *constr. contr. prof.* (Comprehensive regulations which govern the process of awarding of contracts, e.g. as in the ►General Conditions of the Contract for Construction [AIA Document A201] [US]/General Conditions of Contract for Building and Civil Engineering Works [UK] and for construction and furniture contracts between employers/owners and contractors including arbitration rules);

s **procedimiento** [m] **de adjudicación** (Totalidad del proceso de adjudicación y realización de contratos de las administraciones públicas según la Ley de Contratos de las Administraciones Públicas [LCAP]; ►condiciones generales de contratos de obras de las administraciones públicas, ►condiciones generales de contratos de servicios de las administraciones públicas); *syn.* formas [fpl] de adjudicación (BU 1959); *f* **procédures** [fpl] **de passation** [f] **des marchés** (Modalités de dévolution des marchés de travaux, de fournitures ou de services depuis le lancement de la consultation jusqu'au moment de la conclusion de marché fixées dans les ►clauses générales de passation des marchés publics de travaux) ; *syn.* procédure [f] de dévolution des marchés ; *g* **Vergabewesen** [n, o. Pl.] (Das ganze auf Grund der ►Vergabe- und Vertragsordnung für Bauleistungen [VOB] und andere Leistungen [VOL] sowie auf Grund der Verdingungsordnung für freiberufliche Leistungen [VOF] vertraglich geregelte Verfahren, das den Einkauf von Bauleistungen und Lieferungen sowie die Vergabe von Architekten- und Ingenieurleistungen bis zu deren Abrechnung regelt).

1117 contract award negotiation [n] *contr.* (Negotiation with the ►bidder offering best value for the price [US]/tenderer offering the best value for the money [UK], who is expected to be awarded the contract; ►acceptance of a bid 1); *s* **negociaciones** [fpl] **de adjudicación** (Negociación con el ►licitante ganador con la mejor oferta a quien le corresponde conseguir el ►remate); *f* **négociations** [fpl] **avec le candidat retenu** (Mise au point de l'offre retenue dans le but d'améliorer la valeur technique et de réduire les prix de l'offre, le maître d'ouvrage négocie avec le candidat ayant remis l'offre jugée la plus intéressante et auquel doit être attribué le marché ; ►candidat retenu, ►attribution du marché [sur appel d'offres/sur adjudication]) ; *syn.* mise [f] au point de l'offre ; *g* **Vergabeverhandlung** [f] (Verhandlung mit dem/der ►Bieter/-in des annehmbarsten Angebotes, der/die den ►Zuschlag bekommen soll. In D. sind bei öffentlichen Auftraggebern gemäß VOB keine Preisverhandlungen bei öffentlichen Verfahren statthaft).

1118 contract-based nature conservation [n] [UK] *agr. conserv. pol.* (Such contracts form a part of national nature conservation strategies. They are usually agreed upon between nature conservation authorities and land owners and farmers, who for a corresponding amount, voluntarily allow that a certain part of their property [a field, meadow, river bank] is protected and managed according to the goals of nature conservation. The value of the contract is based upon the scope of the work required in maintaining the area (e.g. preservation of a habitat or part of a cultural landscape. Since the Council Regulation (EEC) No 797/85, 12 March 1985 on Improving the Efficiency of Agricultural Structures, so-called agri-environment measures have become the principal instrument for achieving environmental objectives. About one fifth of the EU's agricultural land is covered by agri-environment contracts. Central to the new approach are the concepts of **cross compliance**, **direct income support**, **good farming practice** and **modulation**. Agenda 2000 maintained the nature of the agri-environment schemes as being obligatory for Member States, whereas they are optional for farmers and implemented by contract. Beneficiaries of direct payments will also be obliged to maintain all agricultural land in good agricultural and environmental condition. The EU applies agri-environmental measures which support specifically designed farming practices that help to protect the environment and maintain the countryside. Farmers commit themselves, for a five-year minimum period, to adopt environmentally-friendly farming techniques that go beyond usual good agricultural practice. In return they receive payments that compensate for additional costs and loss of income that arise as a result of altered farming practices.

Examples of commitments covered by national/regional agri-environmental schemes are: **1.** environmentally favourable extensification of farming. **2.** Management of low-intensity pasture systems. **3.** Integrated farm management and organic agriculture. **4.** Preservation of landscape and historical features such as hedgerows, ditches and woods. **5.** Conservation of high-value habitats and their associated biodiversity. In a conservation auction, farmers are asked to bid competitively for a limited number of conservation contracts. There are various types of auction used to allocate conservation contracts of agri-environmental contracts in general; *syn.* conservation contracts, agri-environmental contracting); *s* **custodia [f] del territorio** (Conjunto de estrategias e instrumentos que pretenden implicar a los propietarios y usuarios del territorio en la conservación y el buen uso de los valores y los recursos naturales, culturales y paisajísticos. Para conseguirlo, promueve acuerdos y mecanismos de colaboración continua entre propietarios, entidades de custodia y otros agentes públicos y privados. Para ello se firman contratos individuales entre propietarios de tierras o agricultores y entidades públicas o privadas de conservación de la naturaleza mediante los cuales aquéllos aceptan voluntariamente por una cantidad de dinero determinada que una parte de su propiedad [una pradera, la ribera de un río, un campo de cultivo] sea protegida y gestionada de acuerdo a los objetivos de conservación. El valor del acuerdo de **c. del t.** se basa en el trabajo necesitado para mantener el área natural en cuestión. A partir de la Regulación del Consejo EEC 797/1985, de 12 de marzo, sobre la Mejora de la Eficiencia de las Estructuras Agrarias, se introdujeron las llamadas **medidas agroambientales** que se convirtieron en el principal instrumento para alcanzar los objetivos de protección ambiental. Alrededor de la quinta parte de las tierras agrícolas de la UE están sometidas actualmente a contratos agroambientales. La agenda 2000 prevé la obligatoriedad de aplicación de este instrumento por parte de los Estados miembros, mientras que son opcionales para los agricultores. Todos los beneficiarios de pagos directos están obligados a mantener todas las tierras agrícolas en buenas condiciones agrícolas y ambientales. Más allá de eso, a través de los acuerdos agroambientales, los agricultores se comprometen para un mínimo de cinco años a adoptar prácticas agrarias sostenibles que van más allá de la buena práctica agrícola usual. Ejemplos de contendidos de los acuerdos son: **1.** extensificación de la agricultura; **2.** gestión de sistemas de pastos de baja intensidad; **3.** agricultura orgánica y gestión integrada de granjas; **4.** preservación del paisaje y de componentes históricos del mismo, como setos interparcelarios, acequias, manchas de bosque; **5.** conservación de hábitats de alto valor y de su biodiversidad. **En Es.** las entidades de **c. del t.** pueden ser instituciones públicas u organizaciones privadas sin ánimo de lucro, como fundaciones, organizaciones conservacionistas, ayuntamientos, consorcios u otros tipos de entes públicos. Una experiencia especialmente innovadora e interesante es la de Menorca, isla que en su totalidad está declarada como Reserva de la Biosfera [programa MAB de la UNESCO] y en la que la convergencia entre la administración [Consell Insular], el sector agrario representado por la Unió de Pagesos y los ambientalistas del Grupo Ornitológico Balear [GOB] ha permitido llevar a cabo el Contracte Agrari de la Reserva de la Biosfera [CARB] que considera el mantenimiento del paisaje —principal atractivo turístico de la isla— como servicio o externalidad positiva del modelo agrario. En el CARB el agricultor asume responsabilidades y compromisos obligatorios como la no utilización de transgénicos, un código de buenas prácticas, cumplir un plan de gestión de abonos y residuos de la ganadería, conseguir una calidad y una diversificación del producto y trabajar en proyectos de custodia del territorio. A cambio, el Consell Insular da ayudas hasta un máximo de 6000 euros por finca y ofrece un servicio de asesoramiento ambiental. Además en estos convenios participan también empresas hoteleras; cf. GOB 2009); *syn.* acuerdo [m]/contrato [m] de custodia del territorio, acuerdo[m]/contrato [m] de conservación de un espacio natural; *f 1* **gestion [f] contractuelle de la nature** *agr. pol.* (**1.** Mesures agro-environnementales mise en œuvre dans le cadre de la politique agricole commune [PAC] avec l'article 19 du règlement CEE n° 797-85 du 12 mars 1985 et modifié par les articles 21 à 24 du règlement 2328/91 ainsi que les règlement 2078/92 et 1765/92 du 30 juin 1992, qui tente de concilier l'agriculture tente avec les exigences de protection de l'environnement ; ces règlements peuvent se traduire concrètement par : **a.** des opérations locales agro-environnementales sur des biotopes rares ou des zones de déprises agricole pour lesquelles des contrats de 5 ans sont passés avec les agriculteurs volontaires qui s'engagent à respecter un cahier des charges de pratiques culturales et qui perçoivent une prime à l'hectare, **b.** des opérations de protection et de gestion de la faune et de la flore pour maintenir et reconstituer des biotopes favorables à la conservation de la diversité biologique ; des contrats sont passés avec les agriculteurs volontaires pour l'arrêt de l'activité agricole productive, le retrait à long terme des terres et qui perçoivent une prime à l'hectare, ces mesures agro-environnementales mises en œuvre en France sont regroupées dans un catalogue national. **2.** Démarche d'association des collectivités, propriétaires et usagers à la délimitation de zones de protection, aux règles qui y seront applicables voire à leur gestion ultérieure ; démarche utilisée en particulier lors de la mise en place des zones de protection Natura 2000 et trouvant son expression dans le « contrat Natura 2000 » doit permettre, dans le respect du cahier des charges inclut au DOCOB, la participation des acteurs du monde rural à la gestion durable des habitats naturels et des espèces ayant justifié la désignation des sites figurant dans le réseau Natura 2000 ; il peut prendre la forme **a.** d'un contrat agro-environnemental, faisant l'objet d'un financement du ministère de l'agriculture et de la pêche [MAP] comme contrat territorial d'exploitation [CTE] principalement ou mesures agro-environnementales hors CTE ; **b.** d'un contrat spécifique destiné aux gestionnaires de milieux non agricoles, faisant l'objet d'un financement du Ministère de l'aménagement du territoire et de l'environnement [MATE], mobilisé sur le Fonds de gestion des milieux naturels [FGMN] ; cf. décret n° 2006-922 du 26 juillet 2006 relatif à la gestion des sites Natura 2000) ; *f 2* **protection [f] de la nature par voie contractuelle** *conserv.* (Démarche volontaire d'une personne publique ou privée, physique ou morale possédant des droits sur un terrain méritant d'être préservés au regard de l'intérêt que présentent les espèces faunistiques ou floristiques qu'il abrite, en considération de ses qualités paysagères, etc., de confier la gestion et la préservation de la faune et de la flore s'y trouvant à une autre personne ; cette démarche aboutit à un contrat de protection ; cette convention de maîtrise d'usage prise par voie contractuelle est généralement entreprise par des conservatoires régionaux d'espaces naturels et des associations de protection de la nature ; la plupart des contrats sont conclus pour une durée déterminée, qui peut cependant être particulièrement longue [concession immobilière, bail emphytéotique] ; *syn. traduction littérale incorrecte rencontrée sur le portail du Land de Bavière* protection [f] de la nature contractuelle) ; *g* **Vertragsnaturschutz [m, o. Pl.]** (Strategie und Instrument des Naturschutzes und der Landschaftspflege, das die Umsetzung durchzuführender Maßnahmen, i. d. R. auf landwirtschaftlichen Flächen, ermöglicht, um Kulturlandschaften, naturnahe Landschaften oder deren Teile zu erhalten und weiterzuentwickeln und Erfordernisse des Naturschutzes umzusetzen. Zwischen Naturschutzbehörden und i. d. R. Landwirten werden die naturschutzfachlichen und landschaftspflegerischen Bewirtschaftungsweisen oder die Pflege von Biotopen auf ihren Flächen vertraglich geregelt und gegen entsprechendes Entgelt vergütet. Auf diese Weise können in Schutzgebietsverordnungen formulierte Schutzziele verwirklicht, Pflege- und Entwicklungspläne umgesetzt, spezielle Artenschutzmaßnahmen realisiert und ein Biotopverbundsystem aufgebaut werden; Rechtsgrundlage des **V.es** bilden das Bundesnaturschutzgesetz und Vorschriften der für Naturschutz und Landschaftspflege zuständigen Landesbehörden).

contract basis [n]**, nature conservation on a** *agr. conserv. pol.* ▶contract-based nature conservation.

contract completion [n] *contr.* ▶extension of time for contract completion.

contract conditions [npl] **for tree work** *arb. constr. contr.* ▶additional technical contract conditions for tree work.

1119 contract documents [npl] *constr. contr.* (Complete set of documents which include the contractor's bid [US]/tender [UK] for work[s] to be performed or materials to be delivered, and are necessary for the accomplishment of a contract. The documents comprise ▶project requirements for bidding [US]/project requirements for tendering [UK] with specifications, contractual plans, ▶bidder's affidavit [US], ▶special conditions of [a] contract, project descriptions, ▶special technical requirements of contract for project execution, if applicable, a lump-sum bid [US]/lump sum tender [UK], and finally a ▶written award statement); *s* **documentos [mpl] de contrato** (Conjunto de documentos necesarios para formalizar un contrato incluyendo los precios como ▶pliego de condiciones [generales] de concurso con resumen de prestaciones y los correspondientes planes, normas, informes explicativos, la ▶declaración del licitante, los ▶pliegos de prescripciones técnicas particulares, los ▶pliegos de prescripciones administrativas particulares y la ▶formalización de contrato por escrito); *f* **pièces [fpl] constitutives du marché** (Elles comprennent : l'▶acte d'engagement, le ▶cahier des Clauses Administratives Particulières [C.C.A.P.], le ▶cahier des clauses techniques particulières [C.C.T.P.], l'état des prix forfaitaires, le bordereau des prix unitaires, le ▶cahier des clauses administratives générales [C.C.A.G.], le ▶cahier des clauses techniques générales [C.C.T.G.] et le cas échéant la ▶notification du marché ; ▶dossier d'appel d'offres, ▶dossier de consultation) ; *g* **Vertragsunterlagen [fpl]** (Sämtliche Dokumente mit Angabe der Vergütung für die zu erbringenden Leistungen und Lieferungen, die zur Erfüllung eines Auftrages notwendig sind. Dazu gehören z. B. die ▶Ausschreibungsunterlagen mit Leistungsverzeichnis und den dazugehörenden Plänen, ▶Bietererklärung, ▶Besonderen Vertragsbedingungen, Baubeschreibung, ▶Besonderen Technischen Vertragsbedingungen, ggf. ein Pauschalpreisangebot und die ▶schriftliche Auftragserteilung).

Contract [n] **for Construction (AIA Document A201)** [US]**,** *contr. leg.* ▶General Conditions of the Contract for Construction [AIA Document A201] [US].

Contract [n] **for Construction (AIA Document A201)** [US]**, General and Federal Supplementary Conditions of the** *contr. leg.* ▶General Conditions of the Contract for Construction [AIA Document A201] [US].

Contract [n] **for Furniture, Furnishings and Equipment** [US] *constr. contr.* ▶General Conditions of the Contract for Furniture, Furnishings and Equipment.

contract form [n] [US] *contr. prof.* ▶standard contract form.

contract [n] **for project execution** [UK]**, general conditions of** *constr. contr.* ▶Standard Form of Building Contract [UK].

1120 contract holder [n] [US] *contr. prof.* (Planning/design/engineering firm); *syn.* commissioned planning office [n] [UK]; *s* **gabinete [m] adjudicatario** (Empresa de arquitectura o ingeniería encargada de realizar un proyecto); *syn.* gabinete [m] de arquitectura encargado de proyecto; *f* **bureau [m] d'études contractant** *syn.* contractant [m] ; *g* **Auftragnehmer [m] (2)** (ausführendes Planungs- oder Ingenieurbüro).

contracting member state [n] *contr. pol.* ▶party to a contract.

contracting party [n] *contr. pol.* ▶party to a contract.

1121 contracting public agency [n] [US] *adm. contr.* (Any local, state or federal department/agency which organizes, manages and expends public funds for a project); *syn.* public client [n]; *s* **administración [f] comitente** (Cualquier departamento o agencia local, regional o estatal que tiene a disposición fondos públicos para realizar proyectos); *syn.* autoridad [f] comitente, comitente [m] público; *f* **donneur [m] d'ordre public** *syn.* autorité [f] publique contractante ; *g* **öffentlicher Auftraggeber [m]/öffentliche Auftraggeberin [f]** (Kommunale, Landes- oder Bundesbehörde, die im Rahmen vorgegebener Entscheidungen für die Daseinsfürsorge der Bürger im räumlichen, sozialen und kulturellen Bereich öffentliche Gelder bereitstellt und bewirtschaftet); *syn.* auftraggebende Verwaltung [f].

contraction [n] *constr.* ▶shrinkage.

contraction [n]**, degree of** *constr.* ▶shrinkage crack.

contraction joint [n] *constr.* ▶shrinkage crack.

contract maintenance [n] *constr.* ▶separate contract maintenance.

contractor *constr. contr.* ▶to be contractor for, ▶construction contractor, ▶general contractor, ▶speciality contractor.

contractor [n]**, instruction to the** *constr. contr.* ▶change order [US].

contractor's affidavit [n] [US] *contr.* ▶bidder's affidavit [US].

1122 contractor's agent [n] *constr. prof.* (Contractor's representative responsible on site for the ▶supervision of works. Tasks include overseeing of the works in accordance with the contract documents, in economic terms and according to the time schedule and state-of-the-art technology. The **c. a.** has the duty to ensure that the necessary health and safety regulations are complied with and the measures to protect the environment at the site are implemented. The **c. a.** is responsible for the fulfillment/fulfilment of the legal, local authority and professional association requirements; a **c. a.** is usually a civil engineer, horticulturist or a landscape engineer depending on the project; ▶contractor's project representative); *syn.* contractor's project leader [n]; contractor's construction supervisor [n] [UK], contractor's construction manager [n] [UK]; *s* **jefe/a [m/f] de obra** (Arquitecto/a o aparejador, -a representante de la empresa constructora a pie de obra, que es responsable de la supervisión de un proyecto. Entre las tareas se encuentran la realización de los trabajos oportuna, económica y de acuerdo a la mejor tecnología de construcción posible. Es responsable de que se cumplan las prescripciones de seguridad y protección de la salud y de que se pongan en práctica medidas de protección ambiental en la obra. También es responsable de que se cumplan las normas legales, administrativas y de la mutua de accidentes de trabajo; ▶capataz, ▶dirección de obra, ▶supervisión de obra); *f* **conducteur [m] de travaux** (Homme de terrain représentant le chef d'entreprise ou le ▶directeur des travaux qui, dans le bâtiment et les travaux publics, est responsable de la ▶direction de l'exécution du contrat de travaux ; il planifie, organise et contrôle les travaux de construction, d'aménagement ou d'équipement des chantiers. Il encadre par l'intermédiaire des ►chefs de chantier une ou plusieurs équipes de techniciens, d'ouvriers ou de compagnons. Il est responsable de la qualité des travaux effectués, de la tenue des délais et du respect du budget. Il contrôle toutes les étapes du chantier, depuis l'étude du dossier jusqu'à la réception des travaux ; il fait le lien entre tous les intervenants, décideurs ou exécutants. Il étudie le dossier concernant le projet de construction, les plans d'architecte, les différents cahiers des charges, les devis. Il effectue les démarches administratives et techniques d'ouverture du chantier. Il assure la gestion financière, organise les approvisionnements. Il intervient dans le choix des matériaux et des équipements utilisés ainsi que dans les négociations avec les sous-traitants. Il surveille l'avancement des travaux, étudie avec les ingénieurs les problèmes rencontrés et veille aussi au respect des dispositifs de sécurité. Il gère les plannings de travail. Il peut participer au recrutement des personnels. Le conducteur de travaux rédige les rapports de chantier et rencontre les clients. Il n'existe pas de formation initiale préparant au métier de **c. de t.** Néanmoins, un diplôme de technicien supérieur [BTS bâtiment ou travaux publics, DUT génie civil, travaux publics et aménagement] constitue le niveau minimum requis. L'acquisition d'un diplôme d'ingénieur du bâtiment et des travaux publics, suivie d'une première expérience en tant qu'ingénieur des travaux s'avère être une bonne préparation à la fonction. Les chefs de chantier peuvent accéder à ce poste après quelques années d'expérience professionnelle ; www.cidj.com ; *syn.* conducteur [m] de chantier [B], maître [m] de chantier ; *g* **Bauleiter/-in [m/f] (1)** (Verantwortlicher Vertreter des Unternehmers auf der Baustelle, der für die ►Bauleitung 1 eines Projektes verantwortlich ist. Zu den Aufgaben gehört die termingerechte, wirtschaftliche und nach den allgemein anerkannten Regeln der Baukunst durchzuführende Abwicklung der Arbeiten; er hat dafür zu sorgen, dass die nötigen Vorschriften für die Sicherheit und den Gesundheitsschutz und Maßnahmen zum Umweltschutz auf der Baustelle umgesetzt werden. Der **B.** ist für die Erfüllung der gesetzlichen, behördlichen und berufsgenossenschaftlichen Auflagen verantwortlich; der **B.** ist i. d. R. je nach Projekt ein Bauingenieur oder Gartenbauingenieur; ►Bauleitung 3, ►Polier); *syn.* Bauführer [m].

contractor's construction manager [n] [UK] *constr. prof.* ►contractor's agent.

contractor's construction supervisor [n] [UK] *constr. prof.* ►contractor's agent.

contractor's project leader [n] *constr. prof.* ►contractor's agent.

1123 contractor's project representative [n] *constr.* (Term used in the construction industry for the contractor's on-site representative, who is responsible for the technical completion of the works on time and the coordination between the various trades and contractor management; his superior is the ►contractor's agent of the construction company. Each individual trade has its own ►foreman. **In D.**, the *Polier* is examined by the Chamber of Commerce and Industry before he qualifies, for example, as a master builder; the word originates from *Parlier*, an outdated title for supervisor or spokesperson [French: *parler* >to talk<]; *syn.* construction foreman [n]; *s* **capataz [m]** (Encargado del contratista como responsable de que se lleve a cabo la obra correctamente desde el punto de visto técnico y dentro del plazo previsto y se encarga de coordinar entre los diferentes oficios y la gestión del proyecto por parte del contratista. Su jefe/a es el/la ►jefe/a de obra; ►encargado de obra); *f* **chef [m] de chantier** (GEN 1982-I, 40 ; personne nommée par l'entrepreneur [moyenne et grande entreprise] qui, selon la taille du chantier, dirige en partie ou en totalité les travaux ; il supervise l'installation du chantier, la livraison et la réception des engins et des matériaux ; il seconde le ►conducteur de travaux dans le contrôle des approvisionnements et la gestion du personnel ; présent en permanence sur le chantier, il organise le travail à partir des plans qui lui ont été confiés et coordonne l'action des différents corps de métiers présents simultanément ou successivement sur le chantier. Il est responsable des délais d'exécution et définit les volumes d'heures et de main d'œuvre nécessaires ; il établit les comptes rendus et préparations de réunions de chantier ; il veille également à l'hygiène et au respect des règles de sécurité du travail. Il n'existe pas de formation spécifique initiale préparant au métier de chef de chantier. Néanmoins, certains diplômes du bâtiment et des travaux publics [brevets professionnels (BP), les brevets de techniciens (BT), les bacs professionnels, les bacs technologiques, les BTS ou les DUT du génie civil ou du bâtiment] sont bien adaptés à l'exercice de cette responsabilité, accessible seulement après expérience professionnelle ; www.cidj.com) ; ►chef d'équipe) ; *g* **Polier [m]** (*Berufsbezeichnung im Bauwesen* vom Bauunternehmer mit der fach- und termingerechten Baudurchführung beauftragter Verantwortlicher auf der Baustelle; er ist der Koordinator zwischen den einzelnen Gewerken und der Leitung des Unternehmens; sein Vorgesetzter ist der Bauleiter der Baufirma. Ihm in der Hierarchie nachgeordnet sind ►Vorarbeiter der einzelnen Gewerke. Der deutsche *Geprüfte Polier*, z. B. Fachmeister Hochbau, wird von der Industrie- und Handelskammer [IHK] geprüft; ethymologisch stammt **P.** von *Parlier*, eine veraltete Berufsbezeichnung, für ►Bauleiter/-in 1 oder Sprecher [franz. *parler* >sprechen<] der am Bau beteiligten Maurer, Zimmerleute, Steinbildhauer etc. Der **P.** im Tief- und Straßenbau wird **Schachtmeister** genannt).

contractor's site management [n] *contr. constr.* ►supervision by the person-in-charge.

contractor's supervision [n] **of (building) works** [UK] *constr.* ►contractor's supervision of the project [US].

1124 contractor's supervision [n] **of the project** [US] *constr.* (Project management by the contractor; ►supervision by the person-in-charge, ►supervision of works); *syn.* contractor's supervision [n] of (building) works [UK]; *s* **supervisión [f] de obra** (Control y coordinación de los trabajos a través del **delegado de obra** encargado por el contratista de la obra; ►dirección de obra, ►jefatura de obra); *f* **direction [f] du chantier** (Surveillance, contrôle et coordination du chantier par le responsable de chantier ; ►direction de l'exécution du contrat de travaux, ►maîtrise de chantier) ; *g* **Bauleitung [f] (1)** (Überwachung der Bauarbeiten seitens des auftragnehmenden Unternehmens, das für eine termingerechte, wirtschaftliche und nach den allgemein anerkannten Regeln der Baukunst durchzuführende Abwicklung der Arbeiten verantwortlich ist und die nötigen Vorschriften für die Sicherheit und den Gesundheitsschutz und Maßnahmen zum Umweltschutz auf der Baustelle umsetzt. Die **B.** hat der Erfüllung der gesetzlichen, behördlichen und berufsgenossenschaftlichen Auflagen sicherzustellen; ►Bauleitung 2, ►Bauleitung 3); *syn.* Bauführung [f] (1).

1125 contract period [n] *constr. contr.* (Time span during which a project is implemented; ►agreement on contract period, ►extension of time for contract completion, ►contract time); *s* **plazo [m] de ejecución** (Periodo/período fijado por contrato dentro del cual han de terminarse las obras; ►fijación del plazo de ejecución, ►prolongación del plazo de ejecución, ►plazo de duración de contrato); *syn.* plazo [m] de realización; *f* **délai [m] d'exécution** (Planning détaillé d'exécution établi par l'entrepreneur pouvant devenir contractuel et servir de base pour la coordination des travaux ; ►fixation des délais d'exécution, ►prolongation du délai d'exécution, ►délai d'exécution des travaux) ; *syn.* délai [m] de réalisation ; *g* **Ausführungsfrist [f]** (Vertraglich festgelegter Zeitraum, in dem Bauarbeiten fertig gestellt sein müssen; der Zeitpunkt [Datum] bis zu dem ein Auftrag fertiggestellt werden muss, heißt **Ausführungstermin**; ▶Vereinbarung der Ausführungsfrist, ▶Verlängerung der Ausführungsfrist, ▶Vertragsfrist).

contract preparation [n] [UK] *contr. prof.* ▶negotiation phase [US].

1126 contract price [n] *constr. contr.* (Sum quoted at bid [US]/tender [UK] which is contractually agreed upon, after taking into account surcharges or discounts); *s* **precio [m] de contrato** (Cantidad de dinero estipulada en oferta sobre la que se firma el contrato, después de tener en cuenta los recargos o descuentos); *f* **montant [m] du marché** (Montant remis par l'entrepreneur dans son offre et comprenant toute les dépenses directement ou indirectement nécessaires à l'exécution de tous les ouvrages, soumis aux réajustements et révisions prévus au marché) ; *g* **Vertragspreis [m]** (Die im Angebot angeführte, den vertraglich vorgesehenen Zuschlägen oder Abzügen unterworfene Summe).

1127 contract subletting [n] *contr.* (▶Awarding of contracts [US] by the main contractor to sub-contractors); *syn.* letting [n] to sub-contractors [also UK]; *s* **concesión [f] de subcontrata** (Contratación de una empresa subordinada por parte del contratista principal; ▶adjudicación de contrato); *f* **sous-traitance [f]** (Exécution, par une entreprise [sous-traitant], d'une partie de la production de biens ou de services pour le compte d'une entreprise titulaire du marché [entreprise générale, groupement d'entrepreneurs conjoints ou solidaires] ; DUA 1996, 750 ; ▶passation d'un marché) ; *g* **Weitervergabe [f] an Nachunternehmer** (Die Beauftragung eines Subunternehmers durch den Hauptauftragnehmer ; ▶Auftragserteilung an eine Firma); *syn.* Weitervergabe [f] an Subunternehmer.

1128 contract time [n] *constr. contr.* (Contractually binding dates set for the progress of work [US]/works [UK], which can also apply to a single deadline; ▶contract period); *s* **plazo [m] de duración de contrato** (▶plazo de ejecución); *f* **délai [m] d'exécution des travaux** (Délai maximum [global] des travaux fixé dans le cadre du marché à partir de l'ordre de service de commencement des travaux pouvant mentionner des délais partiels pour l'exécution de certaines tranches de travaux ou de certains ouvrages, ou un planning enveloppe ; cf. C.C.A.G. art. 19 ; ▶délai d'exécution) ; *syn.* délai [m] contractuel ; *g* **Vertragsfrist [f]** (Vertraglich verbindlich festgelegte Frist für den Fortgang der Gesamtarbeit, die auch aus Einzelfristen bestehen kann; cf. § 11 VOB Teil A; ▶Ausführungsfrist).

contractual conditions [npl] *contr. prof.* ▶contractual terms, ▶additional contractual conditions.

1129 contractual terms [npl] *contr. prof. syn.* contractual conditions [npl]; *s* **condiciones [fpl] de contrato** *syn.* cláusulas [fpl] de contrato; *f* **clauses [fpl] du contrat** (Conditions fixées dans un contrat) ; *syn.* conditions [fpl] du contrat ; *g* **Vertragsbedingungen [fpl]** (Konditionen, die in einem Vertrag festgelegt sind).

contractual terms [npl]**, additional** *constr. contr.* ▶additional contractual conditions.

1130 contributions [npl] **from nature** *ecol.* (Attributes of nature which offer human benefit; ▶natural potential, ▶productivity 1 and efficiency of a natural system); *syn.* benefits [npl] from nature; *s* **beneficios [mpl] naturales** (Atributos de la naturaleza que son beneficiosos para la humanidad; ▶potencial natural); *f* **apports [mpl] des éléments constitutifs du milieu naturel** (Contribution apportée par les éléments biotiques et abiotiques du milieu naturel au bénéfice des usages humains ; ▶potentialité du milieu naturel) ; *g* **Leistungen [fpl] der Naturausstattung** (Leistungen der biotischen und abiotischen Natur

gegenüber menschlichen Nutzungsansprüchen; ▶Naturpotenzial).

control [n] *envir. leg. nat'res.* ▶air pollution control; *constr. landsc.* ▶bank erosion control; *agr. for. hort. phytopath.* ▶biological pest control, ▶biotechnical pest control; *adm. leg.* ▶building code control [US]; *agr. for. hort. phytopath.* ▶chemical pest control, ▶ecological pest control; *constr. land'man.* ▶erosion control; *wat'man.* ▶flood control, ▶high-water control; *agr. for. hort. phytopath.* ▶integrated pest control, ▶mechanical pest control; *envir. leg.* ▶noise control; *agr. for. hort. phytopath.* ▶pest control, ▶pest and weed control, ▶physical pest control; *adm.* ▶planning development control; *envir. leg.* ▶pollution control (1); *plan.* ▶project control; *constr.* ▶quality control; *leg.* ▶right-of-passage for fire control; *constr.* ▶torrent control; *plan. trans.* ▶traffic control [US]; *agr. for. hort.* ▶weed control.

Control [n] [UK]**, Local Authority Building** *adm. leg.* ▶building code control [US].

control [n] [US]**, sound** *envir.* ▶noise control.

control [n]**, water quality** [US] *adm. leg. wat'man.* ▶best management practices (BMP) [US].

controlled flood irrigation [n] *agr.* ▶basin check method of irrigation.

controlled landfill site [n] [US] *envir.* ▶sanitary landfill [US] (2)/landfill site [UK].

controlled-release fertilizer [n] [US] *agr. constr. for. hort.* ▶slow-release fertilizer.

controlling [n] *adm. urb.* ▶urban controlling.

control officer [n] [UK]**, building** *constr. prof.* ▶site supervisor representing an authority.

Control of Pollution Act [n] [UK] *envir. leg.* ▶Federal Clean Air Act [US], ▶waste disposal ordinance [US].

control region [n] *envir.* ▶air quality control region.

control station [n] *envir.* ▶air quality control station.

1131 control structure [n] *constr.* (Generic term for ▶avalanche control structure, ▶riverbank stabilization, ▶torrent control; ▶biotechnical construction technique); *s* **estructura [f] estabilizante** (Término genérico de ▶barrera contra aludes, ▶construcción biotécnica, ▶corrección de torrentes, ▶fijación de orillas); *f* **travaux [mpl] de protection (contre les risques naturels)** (Ouvrages de prévention et de protection contre les glissements de terrain, les éboulements et les crues torrentielles, les inondations et les avalanches susceptibles de mettre en péril les constructions et leurs occupants ; ▶confortement des berges, ▶stabilisation d'un torrent, ▶techniques de défense contre les avalanches, ▶technique de génie biologique) ; *g* **Verbauung [f] (1)** (Teilbereich des Landschaftsbaus, Tiefbaus und Wasserbaus, mit Ingenieurbauten [Hartbauweisen] und ▶ingenieurbiologischen Bauweisen Baugruben und -gräben, Hanganschnitte, Rutsch- und Erosionshänge, Lawinenabbruchgebiete, Ufer und Sohlen von Fließgewässern etc. zu sichern; OB zu Baugrubenverbau, ▶Lawinenverbauung, ▶Uferbefestigung, ▶Wildbachverbauung; verbauen [vb]); *syn.* Verbau [m] (1).

1132 conurbation [n] *plan. urb.* (Spatial concentration of population and production centers; generally cities with more than 500,000 inhabitants or several communities and towns that have grown together, and are characterized by a population density of over 1,000 inhabitants/km²; an agglomeration of large cities is a ▶metropolitan region. Whereby a **megalopolis** is a great city or its way of life; an urban complex consisting of a city and its environs [cf. Oxford Dictionary] and is more a 'journalistic' than a specialist term; ▶population growth center);

C

s **aglomeración [f] urbana** (Conjunto urbanizado formado por una ciudad principal a la que se le van uniendo otros núcleos que jurídica, pero no funcionalmente, constituyen municipios autónomos; si se trata de dos o más ciudades que crecen paralelamente hasta fundirse se denomina **conurbación** [f]; a las conurbaciones muy grandes se les denomina **megalópolis** [f]. Según P. HALL (1973) se consideran como tales conurbaciones con un mínimo de 20 millones de habitantes, como Boswash [de Boston a Washington], Chipitts [de Chicago a Pittsburg], megalópolis de Tokio, megalópolis renana [eje del Rín desde Amsterdam hasta Stuttgart] y megalópolis londinense; cf. DGA 1986; ▶área metropolitana, ▶luger central); *f 1* **zone [f] de concentration urbaine** (Grosse agglomération constituée à la suite de l'adjonction de plusieurs villes avoisinantes ; ▶aire métropolitaine, ▶agglomération, ▶région urbaine) ; *syn.* agglomération [f] urbaine, conurbation [f] ; *f 2* **aire [f] urbaine** (Ensemble de communes d'un seul tenant et sans enclave, constitué par un pôle urbain, et par une couronne périurbaine formée de communes rurales ou d'unités urbaines dont au moins 40 % de la population résidente ayant un emploi travaille dans le pôle ou dans des communes attirées par celui-ci ; ▶bassin de vie, ▶lieu central, ▶métropole d'équilibre, ▶pôle urbain) ; *g* **Verdichtungsraum [m]** (In der Regionalplanung und in der Raumordnungsgesetzgebung eine räumliche Konzentration von mehreren Gemeinden oder Städten, die durch eine Mindesteinwohnerzahl von 150 000 und einer Bevölkerungsdichte von über 1000 EW/km² gekennzeichnet ist; cf. BROCK 1986, Bd. 2, 524. Vor der politischen Wende in Europa im Jahre 1989 und der deutschen Wiedervereinigung 1990 war die „Stadtlandschaft" Deutschlands in Politik, Planung und Begrifflichkeit traditionell von ▶zentralen Orten, **V.räumen** und Ballungszentren geprägt. Bei einer stark verdichteten Großstadtregion oder der Konzentration mehrerer Großstädte, z. B. das Ruhrgebiet, die Region um Frankfurt am Main, spricht man vom ▶**Ballungsraum** oder wenn dieser von hoher nationaler und internationaler Bedeutung ist, von einer ∗▶**Metropolregion**/[CH auch Metropolitanregion]); *syn.* Agglomerationsraum [m], Verdichtungsbereich [m], Verdichtungsgebiet [n].

1133 convention [n] (1) *leg. pol.* (Formal agreement between two or more nations); *syn.* agreement [n]; *s* **convenio [m]** (Acuerdo formal entre dos o más países); *syn.* acuerdo [m], convención [f]; *f* **convention [f]** (Accord officiel passé entre des États) ; *syn.* pacte [m], entente [f], accord [m] ; *g* **Abkommen [n]** (Vertragliche Übereinkunft zwischen Staaten); *syn.* Konvention [f], Übereinkommen [n].

Convention [n] (2) *adm. conserv. envir. leg. plan. pol.* ▶Århus Convention, ▶Barcelona Convention, ▶Basel Convention, ▶Bern Convention, ▶Bonn Convention, ▶MARPOL Convention, ▶Paris Convention, ▶Ramsar Convention, ▶Washington Convention.

Convention [n], Desertification *conserv. leg. pol.* ▶Convention to combat desertification.

conventional construction methods [npl] *constr. eng.* ▶conventional engineering.

1134 conventional engineering [n] *constr. eng.* (Application of engineering construction methods, e.g. building of walls, dams, ▶bank revetments instead of ▶biotechnical construction techniques); *syn.* conventional construction methods [npl], hard system [n] [also UK], inert construction [n] (LD 1991 [11]); *s* **construcción [f] convencional** (Uso de métodos de construcción tradicional como muros, taludes, ▶revestimiento de orillas en vez de opp. ▶construcción biotécnica); *syn.* construcción con materiales inertes; *f* **technique [f] d'ouvrages en dur** (Utilisation d'ouvrages de génie civil [endiguement, enrochement, ▶ouvrage de revêtement de berge] au lieu de

▶techniques de génie biologique) ; *g* **Hartbauweise [f]** (Verwendung von Ingenieurbauwerken, z. B. Mauerbauten, Dammschüttungen, ▶Uferdeckwerken, statt ▶ingenieurbiologischer Bauweisen).

1135 conventional river engineering measures [npl] *wat'man.* (Customary methods of river control in disregard of natural conditions of rivers or streams; ▶culverting, ▶natural river engineering measures); *s* **canalización [f] de cursos de agua** (Medidas técnicas de construcción hidráulica en cursos de agua que —al contrario que en las de ▶renaturalización de cursos de agua— no se tienen en cuenta las condiciones naturales del entorno ni se aplican métodos de construcción biotécnica; ▶entubado de un arroyo); *syn.* ingeniería [f] hidráulica convencional de ríos y arroyos; *f* **aménagement [m] des cours d'eau** (Utilisation des techniques de génie civil qui à l'encontre de la restauration en l'état naturel [▶réhabilitation paysagère des cours d'eau] contribue à l'artificialisation d'un cours d'eau ; ▶busage) ; *syn.* aménagement [m] des cours d'eau de type urbain, aménagement [m] dur ; *g* **technischer Gewässerausbau [m, o. Pl.]** (Umgestaltung von Fließgewässern mit dem Ziel der Förderung der Landwirtschaft, der Sicherstellung der Schiffbarkeit der Flüsse, der Besiedelung der Auenbereiche, der Anlage von [landschaftszerstörenden] Verkehrswegen, der Trinkwasserversorgung, der Energiegewinnung und dem Hochwasserschutz. Die dazu nötigen Maßnahmen waren je nach Situation und Nutzungsanspruch die Gewässerverlaufsbegradigung, oft durch Mäanderdurchstich, der Bau von trapezförmigen Regelprofilen mit der Folge der Einschränkung des Gewässerbettes auf minimale Fläche, Bau von Deichen, Rückhaltebecken, Poldern etc., Querbauwerken mit Abstürzen zur Gefälleminimierung oder Schiffsschleusen, meist ohne Aufstiegshilfen für Tiere [Fischtreppen], Eintiefung der Gewässersohle oder Entfernung von Untiefen zum Ausbau der Fahrrinne, Absenkung des Grundwassers etc.; im Vergleich zum naturnahen Gewässerausbau wurden natürliche Gegebenheiten der Ufer und landschaftseingliedernde ingenieurbiologische Verfahren in der ersten Hälfte des letzten Jhs. kaum berücksichtigt; in zunehmendem Maße wurden seit den 1970er-Jahren durch ▶Gewässerrenaturierungen die technischen Einbauten und Befestigungen — wo möglich — unter gewässer- und landschaftsökologischen Gesichtspunkten wieder zurückgebaut resp. verändert. Der **G.** ist ein Eingriff in Natur und Landschaft und es müssen entsprechend der ▶Eingriffsregelung in einem Genehmigungsverfahren in einem Fachplan entsprechende Ausgleichs- und Ersatzmaßnahmen des Naturschutzes und der Landschaftspflege dargelegt werden; ▶Verdolung); *syn.* technischer Ausbau [m, o. Pl.] von Fließgewässern.

1136 convention area [n] *leg.* (Applicable geographical area covered by an international agreement or treaty); *s* **área [f] de vigencia de un convenio internacional;** *f* **zone [f] d'application d'une convention** *syn.* champ [m] d'application territoriale, zone [f] de la convention ; *g* **Geltungsbereich [m] eines Übereinkommens** (Gebiet, für das der Inhalt eines Übereinkommens gilt); *syn.* Übereinkommensbereich [m].

Convention [n] for the Protection of the Marine Environment and the Coastal Regions of the Mediterranean *conserv. land'man. leg.* ▶Barcelona Convention.

1137 Convention [n] on Biological Diversity *conserv. leg.* (Convention of 29 December 1993, providing for the protection of biological diversity in natural habitats, where feasible, and for its sustainability by carefully-limited consumption of natural resources; ▶sustainable development; ▶principle of sustained yield); *s* **Convención [f] sobre la Diversidad Biológica** (Con-

C

vención del 29.12.1993 que prevé la protección mundial de la diversidad biológica, a ser posible en sus hábitats naturales y la ▶sostenibilidad en el uso de especies y ecosistemas; ▶desarrollo sostenible); *f* **convention** [f] **sur la diversité biologique** (Convention Rio de Janeiro entrée en viguer le 29.12.1993, ayant pour but **1.** la protection de la diversité et la richesse des écosystèmes, **2.** le ▶développement durable des espèces et des écosystèmes ; ▶principe d'exploitation durable) ; *g* **Übereinkommen** [n] **über die biologische Vielfalt** (Konvention von Rio de Janeiro vom 29.12.93, die weltweit **1.** den Erhalt der biologischen Vielfalt möglichst in ihren natürlichen Lebensräumen und **2.** die Nachhaltigkeit bei pfleglicher Nutzung von Bestandteilen und Ökosystemen sowie **3.** eine ausgewogene und gerechte Aufteilung der sich aus der Nutzung der genetischen Ressourcen ergebenden Vorteile vorsieht; Deutschland ist seit 1994 Vertragspartei. Der Europäische Rat hat im Jahre 2001 in Göteborg das Ziel vorgegeben, den Verlusten an biologischer Vielfalt in der Union bis zum Jahr 2010 Einhalt zu gebieten; 2002 haben sich die Staats- und Regierungschefs der ganzen Welt auf dem Weltgipfel für nachhaltige Entwicklung in Johannesburg dazu bekannt, dass bis 2010 der Verlust an biologischer Vielfalt entscheidend verringert werden muss [2010-Biodiversitätsziel]. ▶nachhaltige Entwicklung, ▶Nachhaltigkeit der Nutzung); *syn.* Biodiversitätskonvention [f].

1138 convention [n] **on nature conservation** *conserv. leg. pol.* (Generic term for an international agreement on nature conservation; ▶Bonn Convention, ▶Convention on the Conservation of European Wildlife and Natural Habitats, ▶Ramsar Convention, ▶Washington Convention); *s* **convención** [f] **para la conservación de la naturaleza** (Término genérico para acuerdos internacionales de conservación de la naturaleza; ▶Convención de Bonn, ▶Convenio relativo a la conservación de la vida silvestre y del medio natural en Europa, ▶Convención de Ramsar, ▶Convención de Washington); *f* **convention** [f] **(internationale) pour la protection de la nature** (Terme générique pour les accords internationaux relatifs à la protection des espèces ; ▶Convention de Bonn, ▶Convention relative à la conservation de la vie sauvage et du milieu naturel de l'Europe, ▶Convention de Ramsar, ▶Convention de Washington) ; *g* **Naturschutzkonvention** [f] (OB für internationales Übereinkommen für den Naturschutz; ▶Bonner Konvention, ▶Europäische Naturschutzkonvention, ▶Ramsar-Konvention, ▶Washingtoner Artenschutzabkommen).

1139 Convention [n] **on Nature Protection and Wildlife Preservation in the Western Hemisphere** *conserv. pol.* (With this convention, agreed to in 1940 and operational in the U.S. since 28.04.1941, the governments of the American Republics have determined to protect and preserve in their natural habitat representatives of all species and genera of native flora and fauna, including migratory birds, in sufficient numbers and over areas extensive enough to assure them from becoming extinct through any agency within human control; and to protect and preserve scenery of extraordinary beauty, unusual and striking geologic formations, regions and natural objects of aesthetic, historic or scientific value); *s* **Convención** [f] **para la protección de la flora, de la fauna y de las bellezas escénicas naturales de los países de América** (Convención firmada por Argentina, Ecuador, El Salvador, Haití, México, Nicaragua, Perú, República Dominicana, EE.UU. y Venezuela en 1940 y que entró en vigor el 28 de abril de 1941 con el fin de **1.** proteger y conservar en su medio ambiente natural, ejemplares de todas las especies y géneros de su flora y su fauna indígenas, incluyendo las aves migratorias, en un número suficiente y en regiones lo bastante vastas [reservas nacionales] para evitar su extinción por cualquier medio al alcance del hombre, **2.** proteger y conservar

los paisajes de incomparable belleza, las formaciones geológicas extraordinarias, las regiones y los objetos naturales de interés estético o valor histórico o científico [monumentos naturales], **3.** proteger los lugares donde existen condiciones primitivas [reservas de regiones vírgenes]); *f* **Convention** [f] **pour la protection de la flore, de la faune et des beautés panoramiques naturelles des pays de l'Amérique [Convention de Washington du 12.10.1940]** (Convention entrée en vigueur le 28.04.1941, signée par les États-Unis d'Amérique, le Guatemala, le Venezuela, le Salvador, Haïti, la République Dominicaine, le Mexique, l'Équateur, le Nicaragua, l'Argentine, le Pérou, maintenant dépassée et largement ignorée portant sur la protection et conservation dans leur ambiance naturelle **1.** des spécimens de tous les espèces et genres de la flore et de la faune indigènes, y compris les oiseaux migrateurs, en nombre suffisant et dans des régions assez étendues [réserves nationales], pour prévenir leur extinction par quelque moyen que ce soit ou par la main de l'homme [liste très courte, jamais mise à jour], **2.** des paysages d'une beauté rare, des formations géologiques frappantes, des régions et les objets naturels ayant une valeur esthétique, historique ou scientifique [monuments naturels], et des endroits où se rencontrent des conditions primitives [réserves de régions vierges], **3.** de la flore, de la faune et des beautés panoramiques naturelle); *g* **Übereinkommen** [n] **zur Erhaltung der wild lebenden Tier- und Pflanzenarten sowie der Eigenart und Schönheit von Natur und Landschaft auf dem amerikanischen Kontinent** (Mit dieser Konvention, die am 12.10.1940 in Washington, D.C. beschlossen wurde und am 28.04.1941 in Kraft trat, verpflichten sich die amerikanischen Staaten Argentinien, Dominikanische Republik, El Salvador, Ekuador, Guatemala, Haïti, Mexiko, Nicaragua, Peru, USA, Venezuela Vertreter aller Arten und Gattungen der heimischen Flora und Fauna incl. Zugvögel in ausreichender Anzahl und für deren Fortbestand in ausreichend großen Gebieten sowie Landschaften in ihrer außergewöhnlichen Schönheit, besondere und herausragende geologische Formationen, Regionen und Einzelschöpfungen der Natur wegen ihres ästhetischen, historischen oder wissenschaftlichen Wertes und Gebiete in ihrem vom Menschen unbeeinflussten Zustand, so sie von dieser Konvention erfasst werden, zu schützen und zu erhalten).

1140 Convention [n] **on the Conservation of African-Eurasian Migratory Waterbirds [AEWA]** *conserv. ecol. pol.* (The agreement was concluded on 16 June 1995 in the Hague and entered into force on 1 November 1999. The contracting State shall take coordinated measures to maintain migratory waterbird species in a favourable conservation status or to restore them to such a status. The convention has the following purposes **1.** to accord a strict protection for endangered migratory waterbird species, **2.** to identify sites and habitats for migratory waterbirds occurring within their territory and encourage the protection, management, rehabilitation and restoration of these sites, **3.** to coordinate their efforts to ensure that a network of suitable habitats is maintained or, where appropriate, re-established throughout the entire range of each migratory waterbird species concerned, in particular where wetlands extend over the area of any State, **4.** to investigate problems that are posed or are likely to be posed by human activities and endeavour to implement remedial measures, including habitat rehabilitation and restoration, and compensatory measures for loss of habitat, **5.** to cooperate in emergency situations requiring international concerted action and in identifying the species of migratory waterbirds which are the most vulnerable to these situations as well as cooperate in developing appropriate emergency procedures to provide increased protection to these species in such situations and in the preparation of guidelines to assist individual parties in

C

tackling these situations, **6.** to prohibit the deliberate introduction of non-native waterbird species into the environment and take all appropriate measures to prevent the unintentional release of such species if this introduction or release would prejudice the conservation status of wild flora and fauna, **7.** to initiate or support research into the biology and ecology of migratory waterbirds including the harmonization of research and monitoring methods and, where appropriate, the establishment of joint or cooperative research and monitoring programmes, **8.** to develop and maintain programmes to raise awareness and understanding of migratory waterbird conservation; Parties to the Agreement are called upon to engage in a wide range of conservation actions, which are to be drawn up in a detailed **Action Plan** [species and habitat conservation, management of human activities, research and monitoring, education and information, and implementation]. **Conservation Guidelines** are to be developed in relation to priority species and issues; amendments to the Action Plan and to the Conservation Guidelines shall be regularly reviewed); *syn.* Agreement [n] on the Conservation of African-Eurasian Migratory Waterbirds [AEWA]; *s* **Acuerdo [m] para la Conservación de las Aves Acuáticas Migratorias de África y Eurasia (AEWA)** (Convenio acordado el 16 de junio de 1995 en La Haya con el fin de tomar medidas coordinadas para mantener o restablecer un estado de conservación favorable para las especies de aves acuáticas migratorias. El acuerdo tiene los siguientes objetivos: **1.** conceder protección estricta a las especies de aves acuáticas migratorias en peligro; **2.** identificar lugares y hábitats para aves acuáticas migratorias dentro de su territorio y promover la protección, ordenación, rehabilitación y restauración de estos lugares; **3.** coordinar sus esfuerzos para garantizar el mantenimiento o, en su caso, el restablecimiento de una red de hábitats apropiados en toda el área de distribución de cada especie de aves acuáticas migratorias, especialmente cuando los humedales se extiendan en un área que comprenda más de una Parte; **4.** investigar los problemas que plantean o pueden plantear las actividades humanas y esforzarse en aplicar medidas correctoras, incluidas la rehabilitación y la restauración de hábitats, así como medidas compensatorias de la pérdida de hábitats; **5.** cooperar en situaciones de emergencia que requieran acción internacional concertada y en la identificación de especies de aves migratorias acuáticas que son las más vulnerables en esas situaciones, así como en el desarrollo de procedicimientos de emergencia apropiados para proveer creciente protección a esas especies en tales situaciones y preparar directrices para asistir a Partes individuales para manejar esas situaciones; **6.** prohibir la introducción deliberada en el entorno de especies de aves acuáticas no nativas, y tomar las medidas apropiadas para evitar la puesta en libertad no intencionada de estas especies si dicha introducción o puesta en libertad fuera perjudicial para el estado de conservación de la fauna y la flora silvestres; **7.** emprender o apoyar investigaciones sobre la biología y la ecología de las aves acuáticas migratorias, incluida la armonización de los métodos de investigación y seguimiento y, cuando proceda, el establecimiento de programas conjuntos o cooperativos de investigación y seguimiento; **8.** desarrollar y mantener programas para crear concienciación y entendimiento sobre la conservación de las aves acuáticas migratorias. El Acuerdo prevé desarrollar un **Plan de Acción** [conservación de especies, conservación de hábitats, ordenación de las actividades humanas, investigación y seguimiento, educación e información y aplicación] y **Directrices de Conservación** en relación a especies y temas prioritarios. El Plan y las Directrices deben ser revisadas regularmente); *f* **Accord [m] sur la Conservation des Oiseaux d'Eau Migrateurs d'Afrique-Eurasie (AEWA)** (Accord signé le 16 juin 1995 à Den Haag visant la mise en place de mesures coordonnées pour maintenir ou rétablir les espèces d'oiseaux d'eau migrateurs dans un état de conserva-

tion favorable ; l'accord prévoit entre autre **1.** d'accorder une protection stricte aux oiseaux d'eau migrateurs en danger, **2.** d'identifier les sites et les habitats des oiseaux d'eau migrateurs situés sur leur territoire et favoriser la protection, la gestion, la réhabilitation et la restauration de ces sites, **3.** de maintenir ou, lorsque approprié, de rétablir un réseau d'habitats adéquats sur l'ensemble de l'aire de répartition de chaque espèce d'oiseaux d'eau migrateurs concernée, en particulier dans le cas où des zones humides s'étendent sur le territoire de plusieurs États, **4.** mettre en œuvre des mesures correctrices, y compris des mesures de restauration et de réhabilitation d'habitats, et des mesures compensatoires pour la perte d'habitats provoquées par les activités humaines, **5.** de coopérer dans les situations d'urgence qui nécessitent une action internationale concertée et pour identifier les espèces d'oiseaux d'eau migrateurs qui sont les plus vulnérables dans ces situations ainsi que de coopérer également à l'élaboration de procédures d'urgence appropriées permettant d'accorder une protection accrue à ces espèces dans ces situations ainsi qu'à la préparation de lignes directrices ayant pour objet d'aider chacune des Parties concernées à faire face à ces situations, **6.** d'interdire l'introduction intentionnelle dans l'environnement d'espèces non indigènes d'oiseaux d'eau, et de prendre toutes les mesures appropriées pour prévenir la libération accidentelle de telles espèces si cette introduction ou libération nuit au statut de conservation de la flore et de la faune sauvages, **7.** de lancer ou d'appuyer des recherches sur la biologie et l'écologie des oiseaux d'eau, y compris l'harmonisation de la recherche et des méthodes de surveillance continue et, le cas échéant, l'établissement de programmes communs ou de programmes de coopération portant sur la recherche et la surveillance continue, **8.** d'analyser les besoins en formations notamment en ce qui concerne les enquêtes, la surveillance continue et le baguage des oiseaux migrateurs ainsi que la gestion des zones humides. L'accord prévoit l'élaboration d'un **Plan d'action** [concernant la conservation des espèces et des habitats, la gestion des activités humaines, la recherche et la surveillance continue, l'éducation et l'information ainsi que la mise en œuvre] et de **Lignes directrices de conservation**) ; *g* **Abkommen [n] zur Erhaltung der afrikanisch-eurasischen wandernden Wasservögel (AEWA)** (Abkommen, geschlossen am 16. Juni 1995 in Den Haag, zur Ergreifung von koordinierten Maßnahmen, um wandernde Wasservogelarten in einem ausreichenden Bestand zu erhalten oder wieder in einen solchen zu bringen. Das Abkommen sieht unter anderem vor, **1.** gefährdete wandernde Wasservogelarten unter strengen Schutz zu stellen, **2.** Lebensstätten und Habitate für vorkommende wandernde Wasservögel zu bestimmen und den Schutz, das Management, die Sanierung und die Wiederherstellung dieser Stätten zu fördern, **3.** innerhalb des gesamten Verbreitungsgebietes der jeweiligen wandernden Wasservogelart ein Netz geeigneter Habitate zu erhalten oder gegebenenfalls wiedereinzurichten, insbesondere dort, wo Feuchtgebiete sich über das Territorium eines jeden betreffenden Staates erstrecken, **4.** bei menschlichen Eingriffen Abhilfemaßnahmen einschließlich der Sanierung und Wiederherstellung von Habitaten und Ausgleichsmaßnahmen für Habitatverluste zu ergreifen, **5.** in Notsituationen, die international konzertierte Aktionen erfordern, zusammenzuarbeiten und Arten wandernder Wasservögel, die in diesen Situationen am gefährdetsten sind, zu bestimmen sowie in der Entwicklung von Notfallverfahren zusammenzuarbeiten, um einen erhöhten Schutz für diese Arten in solchen Situationen zu gewähren und Richtlinien zur Meisterung dieser Situationen für einzelne Parteien vorzubereiten, **6.** das absichtliche Aussetzen nicht heimischer Wasservogelarten in die Umwelt zu verbieten und alle geeigneten Maßnahmen vorzunehmen, um eine unbeabsichtigte Auswilderung solcher Arten zu verhindern, falls das Aussetzen oder die Ausbringung den Schutzstatus der Wildflora und Fauna beeinträchtigen würde, **7.** die Biologie und Ökologie wandernder Wasservögel zu erforschen einschließlich der Harmonisierung der Forschungs- und Monitoringmethoden und gegebenenfalls der Einrichtung gemeinsamer oder kooperativer Forschungs- und Monitoringprogramme zu unterstützen, **8.** Programme aufzustellen und zu unterhalten, um ein Bewusstsein und Verständnis für wandernde Wasservogelarten zu entwickeln. Das Abkommen sieht vor, einen **Aktionsplan** zum Artenschutz, Habitatschutz, zur Steuerung menschlicher Tätigkeiten, Forschung und Monitoring, Bildung und Information und zur Umsetzung aufzustellen sowie **Leitlinien für Erhaltungsmaßnahmen** zu entwickeln; Änderungen des Aktionsplans und der Leitlinien werden in regelmäßigen Abständen überprüft.

1141 Convention [n] **on the Conservation of European Wildlife and Natural Habitats** *conserv. leg. pol.* (Program[me] for the conservation of endangered plant and animal species, as well as natural habitats in the form of ▶biogenetic reserve, established by the Council of Europe in 1979); *s* **Convenio [m] relativo a la conservación de la vida silvestre y del medio natural de Europa** (Acuerdo europeo de 19 de septiembre de 1979 para desarrollar programas de protección de especies de flora y fauna en peligro y de sus hábitats naturales, creando ▶reservas naturales biogenéticas; en España ratificado por Instrumento de 13 de mayo de 1986); *f* **Convention [f] relative à la conservation de la vie sauvage et du milieu naturel de l'Europe** (Programme de préservation des espèces animales et végétales et d'habitats en voie de disparition réalisé par l'institution de ▶réserves biogénétiques, établi par le Conseil de l'Europe en 1979) ; *g* **Europäische Naturschutzkonvention [f]** (Vom Europarat 1979 aufgestelltes Programm zur Erhaltung gefährdeter Tier- und Pflanzenarten mit ihren natürlichen Lebensräume [Biotopschutz] sowie eine Zusammenarbeit der Europäischen Staaten im Naturschutz, u. a. durch die Ausweisung ▶biogenetischer Reservate und den Aufbau eines Natura-2000-Netzes).

Convention [n] **on the Conservation of Migratory Species of Wild Animals (CMS)** *conserv. leg. pol.* ▶Bonn Convention.

1142 Convention [n] **on the Protection of the Alps** *conserv. land'man. leg. pol.* (Framework agreement for the protection and sustainable development of the Alpine region, signed on 07.11.1991 in Salzburg [Austria] by Austria, France, Germany, Italy, Liechtenstein, Monaco, Switzerland and the EU. Slovenia signed the convention on 29.03.1993; the Convention entered into force on 06.03.1995. In order to achieve the objectives, the Contracting Parties shall take appropriate measures in particular in the following areas: 1] regional planning, 2] soil conservation, 3] conservation of nature and the countryside, 4] mountain farming, 5] mountain forests, 6] tourism and recreation, 7] transport, 8] energy, 9] population and culture, 10] prevention of air pollution, 11] water management, 12] waste management. Until 2008 eight protocols have been agreed upon); *syn.* Alpine Convention [n]; *s* **Convenio [m] para la Protección de los Alpes** (Acuerdo marco para la protección y el desarrollo sostenible de la región alpina, firmado el 7 de noviembre de 1991 en Salzburgo [Austria] por Alemania, Austria, Francia, Italia, Liechtenstein, Mónaco, Suiza y la UE; Eslovenia se adhirió el 29 de marzo de 1993; el Convenio entró en vigor el 6 de marzo de 1995. Para alcanzar los objetivos, las Partes en el Convenio deben adoptar medidas en los siguientes ámbitos: 1] ordenación territorial, 2] protección del suelo, 3] protección de la naturaleza y cuidado del paisaje, 4] agricultura de montaña, 5] bosques de montaña, 6] turismo y actividades recreativas, 7] transportes,

8] energía, 9] población y cultura, 10] calidad del aire, 11] régimen hídrico y 12] residuos. Hasta 2008 se ratificaron 8 protocolos); *syn.* Convenio [m] de los Alpes; *f* **Convention [f] alpine** (Convention cadre signée le 7 novembre 1991 par les pays de l'arc alpin [Allemagne, France, Italie, Principauté du Liechtenstein, Principauté de Monaco, Autriche, Suisse, Slovénie] visant la sauvegarde de l'écosystème naturel ainsi que la promotion du développement durable des Alpes, en protégeant les intérêts économiques et culturels des populations qui y habitent. Pour atteindre ces objectifs, les parties contractantes devront adopter des mesures appropriées [protocoles] dans les douze domaines : 1] aménagement du territoire et développement durable territorial, 2] protection des sols, 3] protection de la nature et entretien des paysages, 4] agriculture de montagne, 5] forêts de montagne, 6] tourisme et loisirs, 7] transports, 8] énergie, 9] population et culture, 10] qualité de l'air, 11] régime des eaux, 12] déchets ; jusqu'à présent les huit premiers protocoles ont été adoptés) ; *g* **Übereinkommen [n] zum Schutz der Alpen** (Ein am 7. November 1991 von den Alpenstaaten [Deutschland, Frankreich, Italien, Fürstentum Liechtenstein, Fürstentum Monaco, Österreich, Schweiz] und der EU in Salzburg, und 1993 von Slowenien, unterzeichnetes internationales Übereinkommen zum Schutze des Naturraums und zur Förderung der nachhaltigen Entwicklung in den Alpen sowie die Sicherung der wirtschaftlichen und kulturellen Interessen der einheimischen Bevölkerung. Um diese Ziele zu erreichen, sind die Vertragsparteien aufgefordert, geeignete Maßnahmen [Protokolle] in zwölf Sachbereichen zu ergreifen: 1] Raumplanung, 2] Bodenschutz, 3] Naturschutz und Landschaftspflege, 4] Berglandwirtschaft, 5] Bergwald, 6] Tourismus und Freizeit, 7] Verkehr, 8] Energie, 9] Bevölkerung und Kultur, 10] Luftreinhaltung, 11] Wasserhaushalt, 12] Abfallwirtschaft; bis 2008 wurden die ersten acht Protokolle beschlossen; das Übereinkommen trat im März 1995 in Kraft); *syn.* Alpenkonvention [f].

1143 Convention [n] **to combat desertification** *conserv. leg. pol.* (*Abbrev.* UNCCD; objective of the convention is to combat desertification and to reduce the effects of drought, particularly in Africa. It promotes efficient sustainable activities at all levels, through international cooperation and partnership agreements between developed countries and developing countries. The Desertification Convention is one of the results of the UN Conference on Environment and Development [UNCED] in 1992 in Rio de Janeiro. It laid the ground for an internationally binding framework for cooperation between the states affected by desertification—especially in Africa—and the industrialized countries. The UNCED requires the affected developing countries to give priority to desertification control in the context of their strategies for sustainable development. Industrialized countries, for their part, are committed to support these efforts through substantial financial aid within the framework of existing bilateral and multilateral development cooperation); *syn.* Desertification Convention [f]; *s* **Convención [f] de las Naciones Unidas de lucha contra la desertificación** (*Abr.* UNCDD; Convenio internacional, elaborado tras la Cumbre de la Tierra 1992 en Río de Janeiro, que fue acordado en París el 17 de junio de 1994 y entró en vigor el 26 de diciembre de 1996. Hasta ahora ha sido ratificado por más de 170 países y es un instrumento jurídicamente vinculante. Tiene como objetivo luchar contra la desertificación y reducir los efectos de la sequía en los países gravemente afectados por sequía o desertificación, en particular en África, mediante medidas eficaces a todos los niveles. La UNCDD obliga a los países en desarrollo afectados, a dar especial atención a la lucha contra la desertificación en el marco de sus estrategias de desarrollo sostenible y concebir los programas de acción nacionales como parte integrante de éllas.

Los países industrializados, por su parte, se comprometen a apoyar estos esfuerzos con fondos sustanciales en el marco de las relaciones de cooperación bi y multilaterales existentes, sobre todo por medio de programas de acción nacionales, subregionales y regionales. La Convención consta de 40 artículos y de 5 anexos con disposiciones de aplicación para África, Asia, América Latina y el Caribe, el Mediterráneo septentrional y Europa Central y Oriental. La sede de su Secretaría reside en Bonn [Alemania]; *NOTA* en publicaciones y documentos españoles y europeos se utiliza del término «desertización» en vez de «desertificación», de uso internacional, que es un anglicismo); *syn.* Convención [f] de las Naciones Unidas de lucha contra la desertización; *f* **Convention [f] des Nations Unies sur la lutte contre la désertification** (Convention conclue à Paris le 17 juin 1994 [faisant suite à la Conférence des Nations Unies sur l'environnement et le développement [CNUED] tenue en 1992 à Rio de Janeiro] ayant pour objectif de lutter contre la désertification et d'atténuer les effets de la sécheresse dans les pays gravement touchés par la sécheresse et/ou la désertification, en particulier en Afrique, grâce à des mesures efficaces à tous les niveaux, appuyées par des arrangements internationaux de coopération et de partenariat en vue de la conception et de l'exécution de programmes de lutte contre la désertification et/ou d'atténuation des effets de la sécheresse pris avec la participation des populations et des collectivités locales, en appliquant des stratégies intégrées à long terme axées simultanément, dans les zones touchées, sur l'amélioration de la productivité des terres ainsi que sur la remise en état, la conservation et une gestion durable des ressources en terres et en eau, et aboutissant à l'amélioration des conditions de vie) ; *g* **Konvention [f] zur Bekämpfung der Wüstenbildung** (Ziel der 1994 in Paris verabschiedeten Konvention ist es, die Wüstenbildung zu bekämpfen und die Auswirkungen von Dürrekatastrophen in Trockengebieten der Welt zu reduzieren. Dazu fördert die Konvention effiziente nachhaltige Aktivitäten auf allen Ebenen, die durch internationale Zusammenarbeit und Partnerschaftsabkommen zwischen den Industriestaaten und den Entwicklungsländern unterstützt werden. Die Desertifikationskonvention ist eines der Ergebnisse der UN-Konferenz für Umwelt und Entwicklung [UNCED] 1992 in Rio de Janeiro. Mit ihr wurde ein international verbindlicher Handlungsrahmen für die Zusammenarbeit der von Desertifikation betroffenen Staaten — insbesondere in Afrika — und den Industrieländern geschaffen. Die UNCED verpflichtet die betroffenen Entwicklungsländer, der Desertifikationsbekämpfung im Rahmen ihrer Strategien zur nachhaltigen Entwicklung einen besonderen Stellenwert einzuräumen und sie damit für alle Unterzeichnerstaaten ein völkerrechtlich verbindliches Abkommen. Die Industrieländer verpflichten sich ihrerseits, diese Bemühungen durch substanzielle finanzielle Leistungen im Rahmen der bestehenden bi- und multilateralen Entwicklungszusammenarbeit zu unterstützen und das Abkommen vor allem mit Hilfe nationaler und regionaler Aktionsprogramme umzusetzen. Am 26.12.1996 tritt die Konvention in 50 Staaten in Kraft; im Dezember 1998 ist sie von 144, 2001 von 170 Staaten ratifiziert; von 1997 bis 2009 fanden weltweit neun Vertragskonferenzen statt); *syn.* Desertifikationskonvention [f], UN-Wüstenkonvention [f].

1144 conversion [n] (1) *leg. plan.* (Change in zoning district category [US]/use class [UK] of land or an existing building; e.g. **c.** of highway into footpath or bridlepath [US]/bridleway [UK]; cf. TCPA 1971, 212; ▶conversion 2, ▶rezoning [US]/rescheduling [UK]); *s* **cambio [m] de uso(s)** (Modificación de la función de áreas o de edificios existentes; ▶conversión, ▶redefinición de uso[s]); *f* **modification [f] de l'usage** (Changement dans l'affectation des sols touchant l'usage principal ou la nature des activités dominantes et les règles relatives à l'implantation et la

dimensions des constructions ; ▶conversion, ▶déclassement) ; *syn.* modification [f] de l'utilisation (des lieux) ; PPH 1991, 66 ; *g* **Nutzungsänderung [f]** (Änderung von Art und Maß der baulichen oder sonstiger Nutzung von Grundstücken oder bestehenden Bauwerken; ▶Umnutzung, ▶Umwidmung).

1145 conversion [n] (2) *urb.* (**1.** Urban planning concept, worked out by local authorities whereby abandoned areas or buildings are earmarked for an adaptive use [US]/subsequent use [UK]; e.g. the transformation and rehabilitation of derelict sites and railroad areas [US]/railway areas [UK], or formerly used military bases into residential and commercial areas, or green open spaces; ▶recycling of derelict sites, ▶urban redevelopment. **2.** *agr. for. plan.* Change of land use in open land [US]/greenfield site [UK]; e.g. from agricultural to forest use); *s* **conversión [f]** (Concepto de planificación urbana de utilizar superficies o edificios abandonados para otros usos después de haber sido saneados y rehabilitados, p. ej. de superficies de uso militar a uso civil, de suelos industriales a residenciales o comerciales, de antiguas fábricas a centros culturales; ▶recuperación de ruinas industriales, ▶remodelación urbana); *f* **conversion [f]** (Politique d'aménagement du territoire et de l'urbanisme de restructuration [changement d'usage] des friches urbaines, industrielles, militaires ; **F.**, les programmes de redéveloppement ont amené à la création de pôles de conversion ; ▶réhabilitation des friches industrielles, ▶remodelage urbain) ; *syn.* reconversion [f] ; *g* **Umnutzung [f]** (**1.** Städtebauliches Konzept, das Kommunen im Rahmen ihrer kommunalen Planungshoheit ausarbeiten, bei dem nicht mehr genutzte Flächen und Gebäude für eine andere Folgenutzung vorgesehen werden, z. B. die Umwandlung und Rekultivierung alter Industrie- und Eisenbahnflächen/-brachen sowie ehemals militärisch genutzter Liegenschaften in Wohn-, Gewerbeflächen oder öffentliche Freiräume; ▶Flächenrecycling, ▶Stadtumbau. **2.** *agr. for. plan.* Änderung der Bodennutzung im Außenbereich, z. B. landwirtschaftliche Fläche in Forstnutzung); *syn. zu 1* Konversion [f].

1146 conveyance [n] **(of land)** *leg.* (Formal transfer of ownership of a plot of land, maintained in a ▶land register); *s* **cesión [f]** (Transmisión formal de propiedad de un terreno y registro en el ▶catastro); *syn.* transmisión [f] de derechos de propiedad; *f* **cession [f]** (Transmission ou transfert de la propriété foncière et enregistrement de l'acquéreur sur le livre foncier ; ▶cadastre) ; *g* **Auflassung [f]** (Formgebundene Einigung, die zusammen mit der Eintragung ins ▶Grundbuch den Eigentumsübergang eines Grundstückes bewirkt; die Eintragung ist nicht Teil der Auflassung, sondern tritt hinzu. Zu unterscheiden von diesem sachenrechtlichen Vorgang ist wiederum das zugrunde liegende schuldrechtliche Rechtsgeschäft [Kaufvertrag]).

1147 co-operative housing association [n] *adm. plan.* (Private housing organization which builds and rents or sells dwellings to members of an association in the form of a stock company, limited stock partnership, or limited company; established as a voluntary and independent organisation to provide good quality affordable homes for rent by local people, the tenants own and manage their housing collectively and re-invest profits in improving the property. There are roughly 600 housing cooperatives in England *syn.* housing co-op [n]; *s* **cooperativa [f] de viviendas** (Empresa sin fines de lucro que construye viviendas para alquilar o vender a sus miembros cooperativistas; Uruguay es el país latinoamericano con una larga tradición en cooperativismo y el mayor número de ellas); *f* **société [f] coopérative de construction** (Société à capital et personnel variable ayant pour objet la construction d'immeubles à vocation d'habitation ou à vocation commerciale destinés à être attribués ou vendus aux associés ; lois du 24 juillet 1867 et du 10 septembre 1947) ; *g* **Wohnungsbaugenossenschaft [f]** (Gemeinnütziges Woh-

C

nungsunternehmen, das Wohnungen zwecks Vermietung oder Verkauf an Mitglieder der Genossenschaft herstellt und in der Rechtsform einer Kapitalgesellschaft — AG, Kommanditgesellschaft auf Aktien oder GmbH — geführt wird).

coordinated public agency review [n] [US] *adm. leg.* ▶statutory consultation with public agencies [US].

coordination [n] [US]**, planning** *plan.* ▶obtaining a mutual agreement on planning proposals.

coordination [n] **of a planning project** [US]/**co-ordination** [n] **of a planning project** [UK] *plan.* ▶obtaining a mutual agreement on planning proposals.

1148 coping [n] *arch. constr.* (Finishing course of a masonry wall or parapet, e.g. with coping stones, or bricks on edge; ▶crown of a wall); *syn.* cap stones [npl] [also US]; *s* **piedra [f] de cubierta** (Cierre de la ▶coronación de un muro o de los bordes de un estanque); *syn.* albardilla [f] de un muro; *f 1* **chaperon [m]** (Élément de ▶couronnement de mur ayant pour but de le protéger contre les effets de la pluie ou des infiltrations ; MAÇ 1981, 317 ; il est en général réalisé un ou deux égouts pour l'écoulement des eaux) ; *syn.* couvertine [f] ; *f 2* **margelle [f]** (Élément de couronnement en pierre d'un puits d'un bassin, d'une piscine, d'une fontaine) ; *g* **Mauerabdeckung [f]** (Abschluss einer ▶Mauerkrone oder von gebauten Teich-/Beckenrändern [Teich-/Beckenrandabdeckung/-randausbildung], z. B. mit Abdeckplatten, Formsteinen, Rollschicht, Überlaufrinnensteinen bei Beckenrändern; *UB* Teichmauerabdeckung).

1149 coppice [n] [US] **(1)** *agr. land'man.* (**1.** Small wood in the rural landscape regularly cut back to encourage regrowth. **2.** Area of undergrowth and small trees grown for periodic cutting to stimulate growth of shoots; ▶residual woodland); *syn.* block [n] of trees [UK], coppice [n] [UK], woodlot [n] [US] (1); *s* **bosquecillo [m] (1)** (Bosque pequeño en paisaje campestre; ▶bosquete residual); *f* **petit bois [m]** (Petite forêt dans un paysage rural ; ▶bosquet restant) ; *g* **Gehölz [n] (1)** ([Kleines] Wäldchen in der bäuerlichen Kulturlandschaft; ▶Restgehölz).

1150 coppice [vb] **(2)** *constr. for. hort.* (Cutting trees close to ground level in order to produce ▶coppice shoots; ▶cutting back of trees and shrubs, ▶pollarding); *syn.* cut [vb] flush with the ground; *s* **podar [vb] hasta el tocón** (Cortar leñosas casi hasta el nivel del suelo para que den ▶brotes de cepa; ▶desmoche); *syn.* truncar [vb], cortar [vb] al nivel del suelo; *f* **recéper [vb]** (Réduction de la longueur des plants d'essences feuillues ou de brins d'un taillis au-dessus du collet pour favoriser le developpement de nouvelles pousses ; *activité* **recépage** ; ▶élagage, ▶étêtage, ▶rejet de taillis) ; *g* **auf den Stock setzen [vb]** (Ausschlagfähige Gehölze oberhalb des Stammfußes abschneiden, damit sie wieder neu durchtreiben; das Austriebsverhalten von Laubbäumen und Sträuchern nach einem Stockhieb ist art- und altersabhängig sehr unterschiedlich. Der Wiederaustrieb ist an eine ausreichende Lichtzufuhr gebunden; *Substantive* Auf-den-Stock-Setzen [n, o. Pl.], Stockhieb [m]; ▶auf den Kopf setzen, ▶Rückschnitt von Gehölzen, ▶Stockaustrieb).

coppice [n] **(3)** *for.* ▶coppice forest.

coppice [n] **(4)** *phyt. for.* ▶thicket.

coppice [n] [UK] **(5)** *landsc.* ▶woodlot [US&CDN] (2).

coppice [n] [UK]**, stored** *for.* ▶coppice with standards.

1151 coppice forest [n] *for.* (Forest originating from ▶coppice shoots, ▶root suckers or both. In comparison with ▶high forest and ▶coppice with standards; a **c. f.** uses a forest management system where continuously rejuvenating shoots are harvested within a short period of ▶rotation 1 for fire wood, tanbark, fascine material or peeled bark); *syn.* sprout forest [n] [also US]; *for a small area* cops [n] [UK], coppice [n] (3);

s 1 **monte [m] bajo** (Masa forestal compuesta por árboles o arbustos originados y perpetuados por ▶brotes de cepa o ▶brotes de raíz, provocados por un tratamiento forestal con cortas periódicas de «rejuvenecimiento» [fases cortas de ▶rotación]. La producción de maderas de pequeñas dimensiones, leñas y cortezas es el aprovechamiento típico de estos montes; DINA 1987; ▶monte alto, ▶monte medio); *syn.* tallar [m]; *s 2* **bosque [m] frutescente** (Formación de talla arbustiva compuesta por especies potencialmente arbóreas, que sobre todo debido a la tala para la obtención de leñas y carbones se mantiene como forma de ▶monte bajo. Están muy extendidas en España. Los más frecuentes son los de encina, mataparda, quejigo y roble [*Quercus pubescens*]; DINA 1987, 590); *f* **taillis [m]** (Par comparaison à la ▶futaie et au ▶taillis-sous-futaie, traitement forestier consistant dans des coupes à blanc répétées [recépage] de courte ▶révolution, donnant naissance à des rejets [▶rejet de taillis et ▶drageons] et une strate arborescente formée de ▶cépées ; traitement appliqué pour obtenir du bois de chauffage, le bois pour la confection des fascines, le tanin et le tan ; *terme spécifique* **taillis fureté** : pratique ancienne du traitement irrégulier, dit de furetage, par lequel ne sont prélevés que les brins d'une dimension préfixée ; ▶durée de renouvellement, ▶révolution) ; *g* **Niederwald [m]** (Im Vergleich zum ▶Hochwald oder ▶Mittelwald eine forstwirtschaftliche Betriebsform, bei der das stets wieder neu austreibende Holz [▶Stockausschlag und ▶Wurzelschösslinge] mit kurzer Umtriebszeit zur Gewinnung von Brennholz, Gerberlohe, Faschinenreisig oder Schälrinde geschlagen wird; ▶Umtrieb 1).

1152 coppice shoot [n] *arb.* (Any shoot arising from an adventitious or dormant bud near the base of a woody plant that has been cut back; SAF 1983); *syn.* sap shoot [n], stool shoot [n]; *s* **brote [m] de cepa** (Rama que crece a partir de yema adventicia en la zona inferior del tronco o del tocón); *syn.* brote [m] adventicio; *f 1* **rejet [m] de taillis** (Tige provenant d'un bourgeon adventice ou dormant à la base d'une plante ligneuse qui a été recépée ; DFM 1975) ; *f 2* **recrû [m]** (Ensemble des repousses issues de semis, rejets de souche, drageons, qui apparaissent après une coupe dans un taillis) ; *g 1* **Stockaustrieb [m] (1)** (1 Ast, der sich aus Adventivknospen im Bereich eines Stammfußes/Baumstumpfes bildet; *Kontext* ein Stock schlägt/treibt wieder aus); *syn.* Stockausschlag [m]; *g 2* **Wiederaustrieb [m]** (**1.** Gesamtheit der neuen Schösslinge, die nach dem Auf-den-Stock-Setzen entstehen. **2.** Vorgang des Wiederaustreibens); *syn.* Stockaustrieb [m] (2).

1153 coppice [n] **with standards** *for.* (Forest in which selected stems are retained, as standards, at each felling to form an uneven-aged overstor[e]y which is removed selectively on a rotation constituting some multiple of the coppice rotation; SAF 1983; sprout forest [US]/▶coppice forest [UK], ▶high forest, ▶hold-over); *syn.* stored coppice [n] [also UK]; *s* **monte [m] medio** (Masa arbórea que procede tanto de chirpiales como de brinzales. Este tipo de monte participa de las características del ▶monte alto y del ▶monte bajo y generalmente procede de la transformación de este último mediante resalveo; DINA 1987; ▶árbol padre); *syn.* tallar [m] bajo; *f* **taillis-sous-futaie [m]** (Par comparaison à la ▶futaie et au ▶taillis, traitement forestier consistant dans des coupes répétées d'une révolution comprise entre 10 et 40 ans, pendant que certains arbres, ▶survivants sont conservés jusqu'à leur exploitation) ; *g* **Mittelwald [m]** (Im Vergleich zum ▶Hochwald und ▶Niederwald eine forstwirtschaftliche Betriebsform, bei der das stets wieder neu austreibende Unterholz [zur Brennholzgewinnung] alle 10-40 Jahre geschlagen wird, während gut gewachsene Stämme [Oberholz zur Wertholzerzeugung] als ▶Überhälter 1 bis zur Hiebsreife stehen bleiben).

coppicing [n] *constr. for. hort.* ►shrub clearing.

cops [n] [UK] *for.* ►coppice forest.

copse [n] [UK] *agr. land'man.* ►coppice [US] (1).

1154 cordgrass marsh [n] *phyt.* (Coastal mudflat covered primarily by cordgrass *[Spartina townsendii]*); *s* **marisma [f] de espartina** (Llanura de fango costera caracterizada por *Spartina townsendii*); *f* **slikke [m] à Spartine** (Espace du marais maritime recouvert par des espèces halophiles pionnières, en particulier la Spartine *[Spartina townsendii]* ou la Salicorne *[Salicornia]* — [slikke à Salicorne] ; DEE 1982, 81) ; *syn.* estran [m] vaseux à Spartine ; *g* **Schlickgraswatt [n]** (Vorwiegend mit Schlickgras *[Spartina townsendii]* bewachsene Wattfläche).

1155 cordon [n] (1) *hort.* (►Espaliered fruit tree which is trained to form one straight trunk without lateral branches, or to two main branches trained in opposite horizontal directions); *s* **cordón [m]** (En fruticultura, ►frutal en espaldera cultivado en forma aplanada; el esqueleto de la planta está constituído por un eje central en el que se insertan las ramitas fructíferas; según la inclinación de éste los **c.** pueden ser horizontales [simples o dobles], verticales y oblicuos); *f* **cordon [m]** (►Arbre fruitier taillé en espalier et palissé avec une seule branche de charpente rectiligne [= **c.** vertical], une seule branche de charpente inclinée d'environ 35° par rapport à la verticale [= **c.** oblique] ou avec une ou deux branches de charpente horizontales [= **c.** horizontal simple et **c.** horizontal à deux bras]) ; *g* **Schnurbaum [m]** (*Obstbau* ►Formobstbaum mit einer Stammhöhe von 30-40 cm und einem einzigen Leittrieb ohne seitliche Tragäste. Durch gezielten Schnitt wird nur seitliches Fruchtholz erzogen; man unterscheidet waagerechte und senkrechte **S.bäume**, bei den waagerechten einarmige und zweiarmige Formen); *syn.* Cordon [m].

cordon [n] (2) [UK] *urb.* ►barrier chain [US].

core [n] *urb.* ►historic core.

1156 core area [n] (1) *leg. urb.* (Planning term for a ►built-up area 2 containing commercial, residential and industrial use zones near the center [US]/centre [UK] of a city or community; ►inner city, ►open land); *syn.* center city [n] [also US]; *s* **perímetro [m] edificado** (En D. término de planificación para denominar los ►núcleos de población agrupada dentro del perímetro urbano y las áreas no edificadas del mismo que se encuentran entre los núcleos. Se utiliza para diferenciar de las zonas no edificables [►suelo no urbanizable y ►suelo rústico]; ►centro urbano); *f* **zone [f] urbaine (1)** (F., *PLU/POS* zone dans laquelle les capacité des équipements publics permettent immédiatement d'admettre des constructions (C. urb. Article R.123-5) ; la typologie des diverses zones urbaines n'est pas réglementée si bien que leur appellation va de UA à UZ, leur contenu variant d'une commune à l'autre ; si néanmoins les zones UA à caractère central d'habitat à majorité ancien, de service et d'activités artisanales et parfois commerciales construit en ordre continu, UB à caractère d'habitat dense [secteurs d'extension du centre-ville], UC extension urbaine à caractère d'habitat de densité moyenne, UD à caractère d'habitat de densité faible en zone périphérique, UE à caractère d'activités semblent garder un contenu comparables dans la majorité des PLU, par contre, suite à une différenciation nouvelle des occupations des sols, certaines dénominations des zones peuvent avoir des contenus très différents, p. ex. UH ensemble urbain, architectural et paysagé à forte valeur historique/zone urbaine réservée aux activités hippiques/zone urbaine du pôle de santé de l'hôpital ; **D.**, terme utilisé en urbanisme pour désigner dans une commune les ►secteurs agglomérés et la zone comprise dans le périmètre d'agglomération ; ►centre-ville ; *opp.* ►zone non urbanisable) ; *g* **Innenbereich [m]** (Planungsbegriff, der die ►im Zusammenhang be-

bauten Ortsteile und gegebenenfalls die nicht überbauten Bereiche einer Gemeinde, die innerhalb im Zusammenhang bebauter Ortsteile liegen, umfassen; ►Innenstadt; *opp.* ►Außenbereich).

1157 core area [n] (2) *conserv.* (Natural Areas Journal, 11 [1] 7; sensitive area of a national park, nature reserve, etc.; ►fringe area, ►marginal protection area); *syn.* core reserve [n]; *s* **zona [f] central** (En parque nacional, espacio protegido, etc., núcleo en donde se encuentran los hábitats por los cuales ha conseguido el estatus de protección; ►zona tampón, ►zona periférica, ►zona periférica de protección); *f* **zone [f] centrale** (Secteur protégé au sens strict dans un parc national soumis à une règlementation particulière de différentes activités telles que la chasse, la pêche, les activités industrielles et commerciales, l'extraction des matériaux, l'utilisation des eaux, la circulation du public et plus généralement toute action susceptible de nuire au développement naturel et d'altérer le caractère du parc ; ►périmètre de protection, ►réserve naturelle nationale « intégrale », ►zone de protection, ►zone périphérique) ; *g* **Kernzone [f]** (Innerer, dem eigentlichen Schutzzweck dienender Teil eines Nationalparks, Naturparks, Naturschutzgebietes etc.; ►geschützte Randzone, ►Randzone).

core reserve [n] *conserv.* ►core area (2).

corkscrew stair [n] *arch.* ►spiral stair(case).

corm [n] *bot.* ►bulbous plant, ►cryptophyte, ►geophyte, ►stem tuber.

corrasion [n] *geo.* ►natural erosion, ►wind erosion.

corridor [n] *landsc.* ►cold air drainage corridor; *plan.* ►communication corridor; *for.* ►forest corridor; *plan. trans.* ►freeway corridor [US]; *landsc. met.* ►fresh air corridor; *landsc. urb.* ►green space corridor; *ecol.* ►native plants and wildlife corridor; *envir. plan.* ►noise corridor; *landsc. urb.* ►regional green corridor; *plan. trans.* ►road corridor, ►transportation corridor; *landsc.* ►wind corridor.

corridor [n] [UK], **motorway** *plan. trans.* ►freeway corridor [US].

corridor [n] [US], **open space** *landsc. urb.* ►linear open space pattern/system.

corridor [n], **planting** *constr. hort.* ►planting trench.

corridor [n] [US], **stream valley** *landsc. urb.* *►linear open space pattern/system.

corridor [n], **ventilation** *landsc. met.* ►fresh air corridor.

corridor [n], **view** *hist. landsc. gard.* ►vista.

corridor [n], **visual** *hist. landsc. gard.* ►vista.

1158 corridor approval procedure [n] [US] *plan. trans.* (Statutory procedure for determining the route of highways, railroads [US]/railways [UK], canals, power transmission lines, sewer system lines and associated linear infrastructures; ►design and location approval process [US]/determination process of a plan [UK]); *syn.* route approval procedure [n] [US], approval procedure [n] for route selection [UK] (TAN 1975, 136); *s* **procedimiento [m] de fijación de trazado** (Procedimiento obligatorio por ley necesario a la hora de planificar carreteras, líneas de ferrocarril o canales, en el que se fija a grosso modo el trazado con la participación de las autoridades e instituciones de carácter público afectadas; ►procedimiento de aprobación de proyectos públicos); *f* **procédure [f] de définition de tracé (≠)** (**D.**, procédure réglementaire par laquelle tout projet de tracé d'équipements d'infrastructure routière, ferroviaire et navigable doit faire l'objet d'une consultation des services de l'État et des personnes publiques associées préalable à la ►procédure d'instruction de grands projets publics) ; *g* **Linienbestimmungsverfahren [n]** (Gesetzlich vorgeschriebenes Verwaltungsver-

C

fahren, bei dem die Planung der Linienführung neu zu bauender überörtlicher Straßen, Eisenbahn-, Kanal- und Hochspannungsleitungstrassen mit den Trägern öffentlicher Belange abgestimmt wird; das **L.** ist kein Genehmigungsverfahren. In einem dann folgenden ►Planfeststellungsverfahren wird mit dem Planfeststellungsbeschluss die Zulassung des Vorhabens beschlossen).

corridors [npl] *plan. trans.* ►comparison of alternative corridors.

corrugation method [n] *agr. for.* ►furrow irrigation.

1159 cosmetic green [n] *plan.* (In planning jargon, a derogatory term expressing **1.** minimal planting design, e.g. in foundation planting, and **2.** minimal allocation of small green open spaces in a development area; ►landscaping of built-up areas, ►provision of green spaces); *s* **decoración [f] verde** (En terminología de planificación, palabra despectiva que denomina la falta de verde en y la poca utilidad de zonas verdes; ►dotación de espacios verdes, ►tratamiento paisajístico 1); *f* **décoration [f] de verdure (±)** (Terme à connotation négative exprimant dans le jargon des aménageurs un déficit dans l'équipement en espaces verts de certaines zones urbaines ou une insuffisance dans l'utilisation de ces mêmes espaces ; ►équipement en espaces verts, ►traitement paysager) ; *g* **Grüngarnierung [f]** (Im Planerjargon negativ besetzter Begriff für eine unzureichende Ausstattung mit Vegetation und für unzureichende Nutzungsmöglichkeiten von zu kleinen Grünflächen; ►Durchgrünung, ►Grünausstattung).

1160 cosmopolitan species [n] *bot. zool.* (A species of world-wide distribution where suitable habitats occur; *opp.* ►endemic species); *s* **especie [f] cosmopolita** (Es aquella cuya área de distribución abarca toda la superficie terrestre; *opp.* ►especie endémica); *syn.* cosmopolita [f]; *f* **espèce [f] cosmopolite** (Espèce répartie dans la majeure partie de la surface du monde ; *opp.* ►espèce endémique) ; *g* **Kosmopolit [m]** (In geeigneten Lebensstätten weltweit verbreitete Art; *opp.* ►endemische Art).

cost accounting [n] *contr.* ►calculation of costs.

1161 cost-benefit analysis [n] *plan.* (Analytical method of assessing and evaluating the costs and benefits of a project; used in calculating net value and when considering the pros and cons of a project. This method has not yet included consumption costs of non-renewable natural resources or incalculable monetary costs of environmental degradation); *syn.* cost-effectiveness analysis [n] (CEA); *s* **análisis [m] de coste-beneficio** (Método analítico utilizado para evaluar costos y beneficios de un proyecto, programa o acción que se basa en técnicas y principios especializados que proceden de la teoría económica y tienen como resultado una valoración monetaria, lo cual limita su aplicabilidad a proyectos en los que los costos y beneficios son fácilmente reconocibles y mensurables; cf. DINA 1988, 977); *f* **analyse [f] coûts-avantages** (Procédure par laquelle l'ensemble des coûts, présents et futurs, d'un ou de plusieurs projets est confronté avec l'ensemble des avantages qui en découlent pendant toute la durée de vie des projets ; on a reproché à cette analyse de ne considérer que les coûts et les avantages évaluables en monnaie et de négliger les éléments qualitatifs en particulier dans l'évaluation des nuisances et atteintes à l'environnement de certains projets. Néanmoins certaines méthodes d'analyses multicritères permettent de prendre en compte les éléments monétaires et non monétaires [analyse coûts-efficacité] ; DUA 1996, 2209) ; *g* **Kosten-Nutzen-Analyse [f]** (Analytisches Verfahren, in dem versucht wird, alle Kosten und Nutzen von Investitionsprojekten zu erfassen und gegenüberzustellen; dient für Rentabilitätsrechnungen und als Entscheidungsgrundlage für das Für und Wider eines Vorhabens, auch hinsichtlich des Nutzens für die Wohlfahrt der Allgemeinheit. Indem der unmittelbare Nutzen

monetarisiert wird, orientiert man sich an früheren Marktpreisen und leitet daraus die zukünftige Marktentwicklung ab).

cost-effective grassing [n] *constr.* ►low-cost grassing.

cost-effectiveness analysis [n] **(CEA)** *plan.* ►cost-benefit analysis.

cost estimate [n] *contr.* ►preliminary cost estimate.

cost estimate [n]**, rough** *contr.* ►preliminary cost estimate.

cost excess [n] *contr.* ►cost over-run.

1162 cost increase [n] *contr. syn.* cost increment [n]; *s* **incremento [m] de costos** *syn.* incremento [m] de costes [Es]; *f* **augmentation [f] des dépenses** (des travaux) ; *g* **Kostenerhöhung [f]**.

cost increment [n] *contr.* ►cost increase.

costing [n] *contr.* ►calculation of costs.

1163 cost [n] **of materials** *constr. contr.* (Commodity value of construction related materials specified for a project); *s* **costo [m] de materiales** *syn.* coste [m] de materiales [Es]; *f* **dépenses [fpl] de matières premières et consommables** ; *g* **Materialkosten [pl]** *syn.* Stoffkosten [pl]).

1164 cost [n] **of materials and equipment** *constr. contr.* (Monetary value of construction materials and equipment or machinery required for installation according to project specifications); *s* **costos [mpl] de equipamiento y maquinaria** *syn.* costes [mpl] de equipamiento y maquinaria [Es]; *f* **dépenses [fpl] de matériels** ; *g* **Kosten [pl] für Geräte und Maschinen**.

1165 cost over-run [n] *contr.* (Cost exceeding the contract amount; *opp.* ►cost saving); *syn.* cost excess [n]; *s* **superación [f] del presupuesto** *opp.* ►ahorro de gastos; *f 1* **sous-estimation [f] du coût prévisionnel** (Dépassement de l'estimation du coût des travaux [avec dépassement du seuil de tolérance], suite à une insuffisance de quantités de nature d'ouvrage prévues ; *opp.* ►surestimation du coût prévisionnel, ►diminution des dépenses de travaux) ; *f 2* **dépassement [m] des dépenses des travaux** (Dépassement du montant du marché, suite à une augmentation de la masse des travaux, à une modification de programme) ; *g* **Kostenüberschreitung [f]** (Mehrausgaben, die über die veranschlagten oder im Vertrag vereinbarten Kosten hinausgehen; *opp.* ►Kostenunterschreitung); *syn.* Kostenüberziehung [f], Überschreitung [f] der Kosten, Überziehung [f] der Kosten.

1166 cost reduction [n] *contr.* (Decrease in construction costs due to lower material or implementation costs, a change in project scope, or a number of other variables; ►cost saving, ►price reduction); *syn.* decrease [n] in costs [also UK]; *s* **reducción [f] de costos** (►ahorro de gastos, ►reducción de precio); *syn.* reducción [f] de costes [Es], disminución [f] de costos/costes [Es]; *f* **diminution [f] des dépenses** (►réfaction sur les prix, ►surestimation du coût prévisionnel, ►diminution des dépenses de travaux) ; *g* **Kostenminderung [f]** (Verringerung der Baukosten auf Grund von günstigeren Angebotspreisen als veranschlagt, baulicher kostensparender Änderungen, durch Standardreduzierung in der Bauausführung oder anderer Parameter; ►Kostenunterschreitung, ►Preisminderung); *syn.* Kostenreduktion [f].

costs [npl] *contr.* ►additional costs.

1167 cost saving [n] *contr.* (Cost decrease in the contract amount; *opp.* ►cost over-run); *s* **ahorro [m] de gastos** *opp.* ►superación del presupuesto; *f 1* **surestimation [f] du coût prévisionnel** (Surévaluation de l'estimation du coût des travaux, suite à un excès de quantités de nature d'ouvrage prévues) ; *f 2* **diminution [f] des dépenses de travaux** (Diminution du montant du marché, suite à une diminution de la masse des travaux, à une modification de programme ; *opp.* ►dépassement

des dépenses de travaux, ►sous-estimation du coût prévisionnel) ; *g* **Kostenunterschreitung [f]** (Minderausgaben, so dass die
volle Höhe der veranschlagten oder im Vertrag vereinbarten
Kosten nicht benötigt wird; *opp.* ►Kostenüberschreitung); *syn.*
Kosteneinsparung [f].

cottage [n] [UK]**, weekend** *recr. urb.* ►weekend house
[US].

cottage garden [n] [UK] *gard. hort.* ►country garden [US].

council housing [n] [UK] *sociol. plan.* ►low-income housing.

1168 Council [n] **of Architectural Registration**
Boards (CARB) [US] **and Council** [n] **of Landscape**
Architectural Registration Boards (CLARB) [US]
prof. (**In U.S.**, most architects belong to the American Institute of
Architects and use the abbreviation A.I.A. or FAIA [Fellow of
A.I.A] after their names; many landscape architects belong to the
American Society of Landscape Architects, and use the abbre-
viation ASLA or FASLA after their names. In order to use pro-
fessional titles, members of each planning and design profession
must pass a uniform official examination to become either 'certi-
fied' [title law] or 'registered' [practice law] professional. In
North America the registration committees of most states of the
USA, together with Puerto Rico and the Canadian provinces of
British Columbia and Ontario have joined together to form the
supreme Council of Landscape Architectural Registration Boards
[CLARB]; cf. www.clarb.org. **In U.K.**, nearest equivalent is
Architects Registration Council [ARCUK], which however does
not cover the professions of landscape architecture and interior
design; **in D.**, organization responsible for the registration of and
representing the professional interests of architects, landscape
architects, and interior designers, without membership of which
the above legally protected professional titles may not be used;
►professional organization); *syn.* Architects Registration Council
[n] [UK]; *s* **Colegio** [m] **Oficial de Arquitectos (En Es.**
corporación de derecho público creada por RD en 1929, como
heredera de la Sociedad Central de Arquitectos, fundada en 1849.
El marco legislativo actual lo constituyen los Estatutos Generales
de los **CC. OO. de AA.** y su Consejo Superior, aprobados por
RD 327/2002 de 5 de abril. Están organizados generalmente por
CC.AA. o por provincias, en el caso de las CC.AA. de Navarra y
el País Vasco existe solo uno, el **C. O. de A.** Vasco-Navarro. El
objetivo primordial de los **CC. OO. de AA.** es servir al interés
general de la sociedad promoviendo la mejor realización de las
funciones propias de los/as arquitectos/as, ordenando y vigilando
el correcto ejercicio profesional de sus colegiados/as, lo cual
sirve a la ciudadanía como garantía de la adecuación del trabajo a
las normativas vigentes. Además ofrece a los/las arquitectos/as
toda una gama de servicios de asesoría técnica, publicaciones,
formación continuada y especialización profesional. En Es. en los
CC. OO. de AA. están integrados tantos profesionales de arquitectura como de arquitectura paisajista; cf. www.coam.org y
www.coavn.org. **En D.**, organización de registro y representación
de los intereses profesionales de arquitectos, arquitectos paisajistas y arquitectos decoradores, sin cuya membresía no está
permitido utilizar los títulos profesionales protegidos legalmente,
aunque no es obligatorio ser miembro para ejercer; ►organiza-
ción profesional); *syn.* Colegio [m] de Arquitectos, Colegio [m]
Territorial de Arquitectos; *f 1* **ordre** [m] **des architectes**
(Organisme des architectes inscrits placé sous la tutelle du
ministre chargé de la culture ; le conseil national et le conseil
régional de l'ordre des architectes concourent à la représentation
de la profession auprès des pouvoirs publics, à la formation
permanente et à la promotion sociale, agissent en faveur de la
protection du titre, etc. ; cf. loi n° 77-2 du 3 janvier 1977 ;
f 2 **Union [f] nationale des syndicats français d'architectes (F.,**

►Organisme professionnel regroupant à titre volontaire les
architectes et chargé de les défendre et de les représenter) ;
g **Architektenkammer [f]** (...n [pl]; **D.**, Organisation der Hoch-
bau-, Landschafts- und Innenarchitekten, die für ihre eingeschrie-
benen Mitglieder vornehmlich berufsständische, aber neben den
Architektenvereinigungen [BDA, BDLA] auch berufsinhaltliche
Interessen wahrnimmt; es besteht keine Pflichtmitgliedschaft; in
D. entstanden seit 1947 **A.n** in den Bundesländern; 1969 wurde
die Bundesarchitektenkammer gegründet; in **Nordamerika**
haben sich die Eintragungsausschüsse der meisten Staaten der
USA, die kanadischen Provinzen British Columbia und Ontario
sowie Puerto Rico unter einem übergeordneten *Council of Landscape Architectural Registration Boards [CLARB]* zusammen-
geschlossen; cf. www.clarb.org; ►Berufsorganisation).

Council [n] **of Landscape Architectural Registra-**
tion Boards (CLARB) [US] *prof.* ►Council of Architectural Registration Boards (CARB) [US].

country [n] *landsc.* ►open country.

1169 country garden [n] [US] *gard. hort.* (Compact garden
close to a rural dwelling; ►monastery garden); *syn.* cottage
garden [n] [UK], rural garden [n]; *s* **jardín** [m] **campesino**
(Jardín rural con combinación tradicional de plantas decorativas,
medicinales, condimenticias y alimenticias que en sus caracte-
ríticas funcionales y la combinación de especies se puede considerar heredero de los antiguos ►jardines monásticos); *syn.* jardín
[m] de granja o caserío; *f* **jardin** [m] **paysan** (Jardin du paysage
rural traditionnel qui pendant des centenaires a procuré une
autosubsistance partielle et parfois totale aux populations rurales.
Ce type de jardin qui fait maintenant partie du patrimoine culturel
et naturel est caractérisé par une clôture faite de palissades, un
compartimentage régulier des parterres avec ses bordures de buis,
le croisement des chemins et la diversité des cultures composées
de fleurs annuelles et vivaces, de roses plantées de façon traditionnelle qui agrémentent les diverses plantes potagères telles que
légumes, herbes, plantes condimentaires et buissons à baies ;
►jardin monastique) ; *syn.* jardin [m] campagnard ; *g* **Bauern-**
garten [m] (Garten der bäuerlichen Kulturlandschaft, der auf
Grund der z. T. jahrhundertealten traditionellen Nutz- und Zierpflanzen ein Teil des kulturellen und natürlichen Erbes bildet.
B.gärten lassen sich in fast allen ihren Funktionsmerkmalen und
Arten-/Sortenbeständen auf alte klösterliche Gärten zurück-
führen; cf. N+L 2002 [7], 289 ; ►Klostergarten).

1170 country park [n] **(1)** *recr.* (Rural recreational facility
within easy reach of a conurbation; ►leisure park); *s* **parque** [m]
rural (Centro de recreo situado en las cercanías de una gran ciudad con instalaciones al aire libre; ►parque recreativo regional);
f **équipement** [m] **récréatif en milieu rural** (Équipement de
récréation et d'hébergement en milieu rural) ; *g* **Erholungsein-**
richtung [f] im ländlichen Raum (Einrichtung meist in der
Nähe von Ballungsräumen, vorwiegend mit landschaftsbezogener
Erholungsmöglichkeit; ►Freizeitpark).

country park [n] [UK] **(2)** *conserv. land'man. recr.* ►regional
park [US].

1171 countryside [n] *agr. landsc.* (Rural area, away from the
influence of large cities and outside their development boundaries, often covered by protective legislation and characterized by
small-scale agriculture, farmland and old landscape elements
such as woodland, hedgerows, or ancient trees in fields; *opp.*
►extensively cleared land; ►hedgerow landscape); *s* **paisaje** [m]
rural (3) (Paisaje estructurado con bosquetes, setos y ejemplares
de árboles y con campos de cultivo y prados relativamente
pequeños, resultado de la explotación por parte de agricultores
pequeños y medianos; *opp.* ►paisaje desnudo; ►paisaje de bocage); *f* **paysage** [m] **rural patrimonial** (Campagne caractérisée

C

par la diversité de ses éléments naturels, bois, haies et groupes d'arbres, témoignage de la civilisation rurale des générations antérieures ; ▶bocage) ; **g kleinbäuerliche Kulturlandschaft [f]** (Kleinteilige Agrarlandschaft abseits des unmittelbaren Einflussbereichs der Großstädte — mit alten Strukturen der Bewirtschaftung —, die stark mit Wäldern, Hecken oder Einzelbäumen gegliedert ist; *opp.* ▶ausgeräumte Landschaft; ▶Heckenlandschaft).

countryside [n] [UK]**, beauty and amenity of the** *conserv. landsc.* ▶beauty and amenity of the landscape [US].

countryside [n]**, nature** [n] **and** *conserv. landsc.* ▶natural landscape.

countryside [n]**, open** *landsc.* ▶open country.

countryside [n] **and nature** *conserv. landsc.* ▶natural landscape.

countryside character [n] [UK] *conserv. landsc. plan. recr.* ▶landscape character.

countryside conservation [n] [US] *conserv. land'man.* ▶rural conservation [US].

1172 countryside management [n] [UK] *plan. recr.* (Planning measures aimed at the conservation and development of areas of countryside, which are subject to heavy recreation pressure with the objective of resolving small-scale conflicts between competing needs for land, and improving the visual appearance such that recreation use will be sustainable in the long term; ▶recreational area management); **s gestión [f] de zonas rurales con fines recreativos** (Medidas de planificación, preservación y desarrollo de zonas rurales que son objeto de gran presión recreacional, con el fin de solucionar conflictos de usos y de mejorar el aspecto escénico del paisaje y así hacer posible el uso recreativo a largo plazo; ▶gestión recreacional); **f gestion [f] des espaces récréatifs** (1. Création, développement et entretien des espaces touristiques, des bases de plein air et de loisirs et des parcs naturels par des acteurs tels que l'État, les collectivités territoriales et locales, les associations ou les opérateurs privés. **2.** Mise en œuvre par les collectivités locales d'actions de réorganisation de certains espaces touristiques ayant pour objectif la maîtrise du développement touristique dans le respect des identités culturelles et des milieux naturels fortement sollicités par les visiteurs afin d'assurer à long terme le fonctionnement des sites et la pratique touristique, notamment par la gestion des flux, le modification du zonage des espaces en fonction de leur sensibilité, l'acquisition de réserves immobilières et foncières, en favorisant l'aménagement d'hébergements banalisés, etc. ; cf. TP 1997, 161 et s. ; ▶gestion de équipements de loisirs, ▶gestion des sites touristiques) ; *syn.* gestion [f] des espaces naturels de loisirs ; **g Pflege [f, o. Pl.] und Erhaltung [f, o. Pl.] der freien Landschaft zum Zwecke der Erholung** (Zielorientierte planerische Maßnahmen zur Sicherung, Pflege und Entwicklung von außerstädtischen Bereichen, in denen der Erholung eine besondere Bedeutung zukommt, um Konflikte durch konkurrierende Nutzungsansprüche so abzubauen, dass eine nachhaltige Erholungsnutzung durch Sicherung oder Verbesserung des Landschaftsbildes gegeben ist; ▶Pflege und Unterhaltung von Erholungsgebieten).

countryside planning [n] [UK] *plan.* ▶rural area planning [US].

1173 countryside recreation [n] *recr.* (Recreation and enjoyment outside of urban settlements on open land, which may comprise the following activities: visiting historic buildings, museums or gardens, parks, zoos, safari parks or bird sanctuaries; going on long walks, hikes or rambles of at least two miles; going on drives, outings, picnics, etc. [including visits to villages]; going fishing; horseback riding [US]/horse riding [UK] or pony

trekking; taking part in any other sport or watching any sport; ▶agritourism [US]/agro-tourism [UK], ▶outdoor recreation); *syn.* leisure [n] in the countryside [also UK]; **s 1 recreo [m] en la naturaleza** (Actividades de esparcimiento realizadas en espacios naturales acondicionados para ese fin con instalaciones de diferentes tipos como aparcamientos, caminos señalizados, itinerarios de deportes o barbacoas, fuentes, mesas para picnics, etc. La concentración de los excursionistas en ciertas zonas es favorable para el desarrollo de los trabajos normales de explotación y para la salvaguarda de las zonas más frágiles; ▶recreo al aire libre); *syn.* esparcimiento [m] en la naturaleza; **s 2 turismo [m] rural** (Término no claramente definido que se utiliza para todas las formas de turismo que se realizan en el medio rural y en la naturaleza y que —según la perspectiva— incluye muchas modalidades como el turismo alternativo y blando, el turismo verde, el ▶agroturismo, el turismo de deporte, el turismo cinegético, el turismo de aventura, el turismo cultural, etc.; cf. CAL 2005; ▶turismo ecológico); **f 1 loisirs [mpl] (de) nature** (Ensemble des activités de détente et de loisirs exercées sur les zones naturelles éloignés des villes telles que visites de parcs et jardins publics, d'espaces verts urbains et périurbains, parcs animaliers, activités nautiques, balades et randonnées, cyclotourisme ; ▶loisirs de plein air) ; **f 2 tourisme [m] vert** (Activités liées à la fréquentation des espaces naturels et ruraux, se référant aux valeurs de nature et de paysage et organisées autour de la découverte, l'observation et l'appréciation de la nature, p. ex. les randonnées pédestres ou équestres, l'hébergement en milieu rural, les activités d'éveil à la nature [découverte de grands sites naturels], le tourisme ornithologique, le tourisme fluvial, le tourisme cynégétique et piscicole, etc. ; ▶agrotourisme, ▶écotourisme) ; *syn.* tourisme [m] de nature, tourisme [m] rural); **g 1 Erholung [f, o. Pl.] in der freien Landschaft** (1. Erholung in der freien Natur wird bei den Menschen immer beliebter und erstreckt sich z. B. auf folgende Tätigkeiten: sportliche Aktivitäten wie Laufen, Reiten, Angeln, Wanderungen, Ausflüge mit Picknick, Ponyfahrten, auch Besichtigungen von Dörfern, historischen Gebäuden, Museen, Parks und Gärten, Tiergehegen, Safariparks; besonders stark werden die Wälder besucht. In D. besteht außer für Bayern [in Art. 141 verfassungsrechtlich garantiert] kein allgemeines Grundrecht auf Erholung; durch die in den einzelnen Landesgesetzen [z. B. Wald- und Forstgesetzen, Jagd- und Naturschutzgesetzen sowie im Wasser-, Straßen- und Verkehrsrecht] festgelegten Regelungen kann die Erholung in verschiedenen Gebieten eingeschränkt oder mit Auflagen verbunden werden, v. a. dann, wenn die Interessengegensätze zwischen Grundstückseigentümern, Jägern, Fischern Reitern, Sportlern, Forstleuten u. a. sowie die Belange von Naturschutz und Landschaftspflege beeinträchtigt sind; ▶Agrotourismus, ▶Erholung im Freien, ▶Ökotourismus); *syn.* Erholung [f] in Natur und Landschaft (§§ 27 und 28 BNatSchG), landschaftsbezogene Erholung [f, o. Pl.], landschaftsgebundene Erholung [f, o. Pl.], Erholung [f, o. Pl.] in freier Natur.

1174 countryside recreation area [n] *plan. recr.* (Landscape of open land suitable or used for recreation due to its variety, unique character or outstanding beauty; ▶agritourism [US]/agro-tourism [UK], ▶countryside recreation); **s paisaje [m] apto para usos recreativos** (Paisaje que debido a su diversidad, su carácter y su belleza es apto para usos recreativos, ▶agroturismo, ▶recreo en la naturaleza); *syn.* paisaje [m] con vocación recreativa; **f paysage [m] récréatif** (Paysage pittoresque à vocation récréative ou touristique ; *terme spécifique* site récréatif ; ▶loisirs de nature, ▶tourisme vert, ▶agrotourisme) ; *syn.* paysage [m] touristique ; **g Erholungslandschaft [f]** (Landschaft, die aufgrund ihrer Vielfalt, Eigenart und Schönheit für die Erholung geeignet ist oder genutzt wird; ▶Agrotourismus, ▶Erholung in der freien Landschaft).

countryside recreation planning [n] [UK] *landsc. plan. pol. recr.* ▶rural recreation planning [US].

county building department [n] [US] *adm. leg.* ▶Building Permit Office [US].

county recorders office [n] [US] *adm.* *▶land-holding agency [US].

1175 county road [n] [US] *adm. leg. trans.* (Borough road which is county-funded and usually in rural areas; ▶classified highway [US]/classified road [UK]; *syn.* B-road [n] [UK]; **s carretera [f] provincial [Es]** (▶carretera clasificada); **f route [f] départementale** (F., terme usuel ; ▶voie classée et devant répondre à des caractéristiques techniques précises faisant partie du domaine public départemental) ; *syn. terme juridique* chemin [m] départemental ; **g Kreisstraße [f]** (D., eine ▶klassifizierte Straße, bei der die Unterhaltungslast vom Landkreis getragen wird).

course [n] [UK]**, bearing** *constr.* ▶subbase [US]/sub-base [UK].

coursed ashlar [n] [US] *arch. constr.* ▶regular coursed masonry of natural stones.

1176 coursed ashlar masonry [n] *arch. constr.* (Masonry composed of rectangular units of properly bonded building stones having sawn, dressed or squared beds and joints laid in courses of various heights, whereby the heights of stones within a course are the same; ▶broken rangework); *syn.* range masonry [n], rangework [n]; **s mampostería [f] aparejada en hiladas regulares** (Tipo de mampostería en la que los mampuestos están unidos con mortero, las caras visibles ordenadas en ángulos rectos con juntas de asiento y a tope de unos 15 cm de profundidad y en la que los mampuestos de una hilada tienen la misma altura; ▶sillería ordinaria); *syn.* sillería [f] aparejada en hiladas regulares; **f maçonnerie [f] de moellons assisés d'égales hauteurs** (Maçonnerie en pierres naturelles à moellons assisés sur mortier, assemblés de telle manière que les joints montants et filants de la face de parement possèdent sur les assises d'épaisseurs diverses une profondeur minimale de 15 cm, les lits de moellons assisés étant d'égales hauteurs ; ▶maçonnerie à appareillage à l'anglaise) ; *syn.* maçonnerie [f] à assises réglées, maçonnerie [f] de moellons assisés (MAÇ 1981, 175) ; **g regelmäßiges Schichtenmauerwerk [n]** (Natursteinmauerwerk aus Ansichtsflächensteinen mit rechtwinklig zueinander stehenden > 15 cm tief bearbeiteten Stoß- und Lagerfugenflächen mit unterschiedlichen Schichthöhen, wobei die Höhe der Steine innerhalb einer Schicht nicht wechseln darf; ▶Wechselmauerwerk).

1177 coursed dressed ashlar masonry [n] *arch. constr.* (Wall made of natural stones, the faces of which are cut to give the wall a smooth surface); **s mampostería [f] en hiladas** (Tipo de mampostería de piedras de labra fina colocadas en hiladas de diferentes alturas); **f maçonnerie [f] en moellons assisés** (Mur permettant d'utiliser des matériaux de pierre très variés assemblés sur une même ligne [rang d'assise] ; *termes spécifiques* maçonnerie de moellons assisés à parement scié, maçonnerie de moellons assisés à parement lité ou éclaté) ; **g Schichtenmauerwerk [n]** (Natursteinmauerwerk aus lagerhaften Steinen mit ebenen Ansichtsflächen).

1178 coursed quarry-faced ashlar masonry [n] *arch. constr.* (Square-cut stones which are dressed to give a rustic or embossed surface); **s fábrica [f] de sillarejo en hiladas** (Tipo de mampostería con ▶sillares de labra tosca colocados en hiladas); **f maçonnerie [f] de moellons assisés à parement bossagé** *syn.* maçonnerie [f] de moellons à parement bossagé

(MAÇ 1981, 181) ; **g Schichtenmauerwerk [n] mit bossierten Steinen**.

coursed rubble [n] *constr.* ▶squared rubble masonry.

1179 course height [n] *constr.* (Height of a row of headers or stretchers; ▶thickness of a layer); **s altura [f] de hilada** (Altura de una capa de sogas o tizones; ▶profundidad de capa); **f hauteur [f] d'assise** (d'un lit dans les maçonneries de moellons assisés ; ▶épaisseur de couche) ; **g Schichtstärke [f] (1)** (Höhe einer Binder- oder Läuferreihe; ▶Schichtstärke 2).

course [n] **of an avalanche** [UK] *geo.* ▶avalanche track [US].

1180 coursing joint [n] *arch. constr.* (Horizontal or arched [mortar] joint between two courses of masonry in a wall or arch; DAC 1975; ▶heading joint); **s junta [f] de asiento** (Junta horizontal entre hiladas; ▶junta a tope); *syn.* junta [f] horizontal; **f joint [m] délit** (DTB 1985, 105 ; joint horizontal sur lequel repose une pierre d'un mur ; ▶joint montant) ; *syn.* joint [m] d'assise ; **g Lagerfuge [f]** (Horizontale oder die Längsrichtung betonende Fuge in einem Mauerwerk an der Ober- und Unterseite der Mauersteine; ▶Stoßfuge).

court [n] *constr. recr.* ▶all-weather court, ▶granular playing court [US]/hoggin playing surface [UK], ▶hard court [US]/hard pitch [UK], ▶tournament tennis court.

court [n] [UK]**, garden** *arch. landsc.* ▶garden courtyard.

court [n] [UK]**, hoggin playing** *constr.* ▶granular playing court [US].

court [n]**, water-bound porous** *arch. landsc.* ▶granular playing court [US]/hoggin playing court [UK].

1181 courtship [n] *zool.* (Mating behavio[u]r during copulation period of birds); *syn.* mating ritual [n]; **s cortejo [m]** (Conjunto de comportamientos típicos de los animales resp. las aves durante su periodo/período de emparejamiento); *syn.* parada [f] nupcial; **f rituel [m] nuptial** (Totalité des comportements chez les oiseaux pendant la période d'accouplement [pariade]) ; **g Balz [f]** (Gesamtheit der Verhaltensweisen während der Paarungszeit bei Vögeln).

1182 courtship territory [n] *zool.* (Area, which many male birds defend against their rivals during the mating period); **s área [f] de cortejo** (Zona defendida por algunas aves macho ante rivales de la misma especie durante la fase de emparejamiento); **f aire [f] de pariade** (Zone défendue par certains oiseaux mâles contre les rivaux de la même espèce pendant la période d'accouplement) ; **g Balzrevier [n]** (Gebiet, das Männchen mancher Vögel gegenüber artgleichen Rivalen während der Paarungszeit verteidigen).

1183 courtyard [n] *arch.* (Enclosed space partly or wholly surrounded by buildings to which it gives access; ▶garden courtyard); *syn.* patio [n]; **s patio [m]** (Espacio libre cerrado total o parcialmente por edificios, ▶patio-jardín); **f 1 cour [f] intérieure** (Espace compris entre les ailes d'un bâtiment ou situé entre plusieurs bâtiments ; ▶cour-jardin) ; **f 2 patio [m]** (▶Cour jardin centrale, fermée, à ciel ouvert, souvent accompagnée de pièces ou de jeux d'eau, traditionnellement bordée d'arcades, d'une maison espagnole ou de style espagnol et dont l'origine remonte à l'atrium des villas de la Rome antique) ; **g Innenhof [m]** (Von einem Gebäude oder von einem Gebäudekomplex umschlossener Außenraum; ▶Gartenhof 2).

courtyard block development [n] *urb.* ▶block development.

1184 courtyard house [n] *arch.* (Usually a one-stor[e]y, angle-shaped house used as a single-family dwelling, where the court is contained within boundary walls, or by adjoining buildings; ▶patio house); *syn.* atrium house [n]; **s casa [f] de patio (1)**

C

(Casa unifamiliar generalmente de una planta construida en forma de L de manera que forma dos límites del patio-jardín mientras que los otros dos están limitados por muros o por las paredes de las casas vecinas; ►casa de patio 2); *f* **maison [f] atrium** (Habitation individuelle d'architecture caractéristique, en général d'un seul niveau, la cour intérieure étant formée par la façade de l'habitation et trois murs limitrophes ou les murs borgnes des habitations voisines ; ►maison patio) ; *g* **Gartenhofhaus [n]** (In der Regel eingeschossiger Winkelbau als Einfamilienhaus, bei dem zwei Wände des Gartenhofes vom Haus, die zwei anderen von Grenzmauern oder von fensterlosen Wänden der Nachbargebäude gebildet werden; ►Patiohaus); *syn.* Atriumhaus [n], Gartenhofwohneinheit [f].

1185 courtyard landscaping [n] *constr. urb.* (Design of court situations within blocks of buildings by introduction of planting instead of hard-surfacing; ►courtyard rehabilitation scheme); *syn.* back court planting scheme [n] [also UK]; *s* **tratamiento [m] paisajístico de patios centrales** (Reestructuración de usos de patios centrales y creación de zonas verdes colectivas para el uso de los residentes de los bloques conlindantes en el marco de la rehabilitación urbana; ►saneamiento de patios centrales); *f* **aménagement [m] végétal d'une cour intérieure** (Conception nouvelle ou réfection de la cour d'un immeuble d'une construction en îlot, remplacement du revêtement imperméable, réorganisation des espaces libres et des espaces plantés ; ►rénovation d'une cour intérieure) ; *syn.* végétalisation [f] d'une cour intérieure, aménagement [m] paysager d'une cour intérieure, traitement [m] paysager d'une cour intérieure ; *g* **Hinterhofbegrünung [f]** (Vegetationsbestimmte Neugestaltung oder vegetationsbetonter Umbau eines versiegelten Hinterhofes einer Blockbebauung; ►Hinterhofsanierung).

1186 courtyard living area [n] *arch.* (Inner area of a ►block development or courtyard which is designed as an outdoor living space; ►outdoor living space [US]/garden living space [UK]); *s* **patio [m] habitación** (Espacio entre bloques o patio diseñado con el fin de ser utilizado como parte de la vivienda; ►jardín habitación; ►edificación en bloque cerrado); *f* **cour [f] d'habitation intérieure** (Espace intérieur d'une ►construction en îlot, cour intérieure ou patio qui en raison de leurs qualités agrémentent l'habitation ; ►jardin d'habitation) ; *g* **Wohnhof [m]** (...höfe [pl]; Innenbereich einer ►Blockbebauung oder Innenhof in einer Teppichhausbebauung, der auf Grund seiner Gestaltung dem Wohnen dient; ►Wohngarten).

1187 courtyard planting [n] *gard. plant.* (Establishment of vegetation within or amongst paved surfaces surrounded by buildings; ►courtyard); *s* **plantación [f] de patios** (►patio); *syn.* ajardinamiento [m] de patios, tratamiento [m] paisajístico de patios; *f* **aménagement [m] végétal d'une cour** (►cour intérieure, ►patio) ; *syn.* plantation [f] d'une cour, garniture [f] végétale d'une cour ; *g* **Hofbegrünung [f]** (Ausstattung mit Pflanzen eines befestigten Platzes zwischen Häusern; ►Innenhofes).

1188 courtyard rehabilitation scheme [n] *constr. urb.* (Renovation of predominantly hard-surfaced or built-up inner space within building blocks; ►courtyard landscaping); *syn.* redevelopment [n] of courtyards; *s* **saneamiento [m] de patios centrales** (Reestructuración de usos en los grandes patios existentes dentro de las manzanas en zonas de edificación antigua de finales del siglo XIX hasta los treinta en ciudades de Europa Central, sobre todo con el fin de eliminar usos y estructuras molestas o antiestéticas y de mejorar la calidad del entorno; ►tratamiento paisajístico de patios centrales); *syn.* clareo [m] de patios centrales; *f* **rénovation [f] de cour intérieure** (Réaménagement de la cour intérieure d'une construction en îlot à revêtement imperméable ou occupée par des bâtiments annexes et

autres appentis ou remises ; ►aménagement végétal d'une cour intérieure) ; *syn.* réfection [f] de cour d'immeuble, réhabilitation [f] de cour d'immeuble, rénovation [f] de cour d'immeuble ; *g* **Hinterhofsanierung [f]** (Umgestaltung eines versiegelten oder mit Nebengebäuden resp. Schuppen zugebauten Hinterhofes einer Blockbebauung; ►Hinterhofbegrünung).

cover [n] *for. phyt.* ►canopy cover; *constr.* ►concrete cover; *arb. phyt.* ►crown cover; *phyt.* ►degradation of vegetative cover, ►degree of species cover, ►degree of total vegetative cover, ►density of vegetative cover, ►duckweed cover; *for. geo. phyt.* ►forest cover; *phyt.* ►forest floor cover, ►grass cover; *constr.* ►grill cover; *phyt.* ►ground cover; *constr.* ►manhole cover; *phyt.* ►plant cover; *for. landsc.* ►tree cover; *constr.* ►tree pit cover; *landsc. urb.* ►vegetation cover.

cover [n] [UK], chamber *constr.* ►manhole cover.

cover [n], forest soil *phyt.* ►forest floor cover.

cover [n], grate *constr.* ►grille cover.

cover [n], grid *constr.* ►cover slab.

cover [n], moss ground *phyt.* ►moss floor.

cover [n], percentage of vegetative *phyt. plant.* ►degree of total vegetative cover.

cover [n] [US], vegetal *phyt.* ►plant cover.

cover [n], vegetative *phyt.* ►plant cover.

cover [n], woodland *for. geo. phyt.* ►forest cover.

coverage [n] *phyt.* ►degree of species cover.

1189 covered parking deck [n] *arch. urb.* (Car park which is not underneath a building but sheltered by a roof and may be planted as a green roof or otherwise used); *s* **aparcamiento [m] enterrado** (Estacionamiento subterráneo que no está situado bajo un edificio, sino bajo una calle o plaza y cuya cubierta puede estar vegetalizada); *syn.* parking [m] enterrado; *f* **parking [m] sous terrasse** (Parking couvert non situé sous un bâtiment mais muni d'une toiture pouvant être végétalisée) ; *syn.* dalles-parking [fpl] ; *g* **überdachte Tiefgarage [f]** (Tiefgarage, die nicht unter einem Gebäude liegt, sondern mit einem begrünten oder anders nutzbaren Dach ausgestattet ist).

1190 covered parking space [n] *urb.* (Generic term for all roofed-over ►parking spaces; ►carport [US]/car port [UK]); *s* **plaza [f] de aparcamiento cubierto** (Término genérico para cualquier tipo de ►aparcamiento 2 cubierto; ►garaje sin puertas); *syn.* plaza [f] de estacionamiento cubierto [CS], aparcamiento [m] cubierto, estacionamiento [m] cubierto, plaza [f] de parqueo cubierto [C, CA]; *f* **abri [m] de voiture fermé** (►Emplacement de stationnement fermé ou couvert, ►abri de voiture) ; *syn.* parking [m] couvert ; *g* **Einstellplatz [m]** (Geschlossener oder überdachter öffentlicher oder privater ►Parkplatz 3; ►Carport).

covered trench drain [n] [US] *constr.* ►cast iron drain with grating [US].

covering [n] *hort.* ►protective covering.

covering [ppr] *phyt.* ►totally covering.

1191 covering [n] of a manhole *constr.* (Closing of a manhole opening to a drain or well with a lid or ►cover slab; ►manhole cover); *s* **cierre [m] de pozo de registro** (Acción de cerrar una apertura a un dren o pozo con una ►placa de cobertura de pozo de registro; ►tapa de pozo de registro); *f* **fermeture [f] d'un regard** (Fermeture de l'entrée d'un regard ou d'une fosse réalisée par un tampon ou une plaque de recouvrement ; ►rehausse sous cadre, ►tampon) ; *syn.* obturation [f] d'un regard ; *g* **Abdecken [n, o. Pl.] eines Schachtes** (Schließen der Schachtöffnung einer Entwässerungs- oder Brunnenanlage mit einem Deckel oder einer Abdeckplatte; ►Schachtabdeckplatte

mit Einstiegsöffnung, ►Schachtdeckel); *syn.* Abdeckung [f] eines Schachtes.

covering [n] **with soil** [US] *constr. min.* ►capping with soil material.

covering [n] **with topsoil** *constr.* ►topsoil spreading.

covering [n] **with vines** [US]**, wall** *gard. urb.* ►façade planting.

1192 cover slab [n] *constr.* (The top of a usually large manhole, which provides the base for the lid of the chamber and carries the ►manhole cover. Typically manufactured in high-strength, precast reinforced concrete, the minimum opening size of 600x600mm for the manhole is created within the cover slab during the casting process. It is also sometimes referred to as a **reducing slab**, because it reduces the apparent opening size or the chamber dimensions, e.g. a circular 1800mm diameter chamber or a rectangular 1200x750mm chamber can be fitted with a standard 600x600mm cover at the surface. **C. s.s** are normally mortar-bedded onto the top of the chamber with the surround concrete brought up to the top of the cover slab. Regulating brickwork is then built on top of the **c. s.** and the frame and manhole cover are installed to finished level); *syn.* shaft cover slab [n] (SPON 1986, 171), chamber cover slab [n] [also UK] (SPON 1986, 171); *s* **placa [f] de cobertura de pozo de registro (≠)** (Pieza prefabricada de hormigón reforzado de gran resistencia y 120-200 mm de grosor utilizada para cubrir la entrada de un ►pozo de registro 2 y que es la base sobre la cual se asienta la ►tapa de pozo de registro. La apertura mínima es de 600 mm); *f* **rehausse [f] sous cadre** (Élément du dispositif de fermeture d'un regard permettant l'ajustement de la hauteur de la cheminée du regard de visite avec le niveau de la chaussée ; élément circulaire de hauteur comprise entre 100 et 250 mm, positionné sur une dalle ou tête réductrice du ►regard visitable sur lequel repose le cadre et son ►tampon ; l'ouverture excentrée est assurée par une tête réductrice de 500 à 700 mm de hauteur en remplacement d'une hotte ou d'une virolle) ; *syn.* dalle [f] de fermeture borgne, dispositif [m] de fermeture d'un regard, tampon [m] de fermeture de regard ; *g* **Schachtabdeckplatte [f] mit Einstiegsöffnung** (120-200 mm dickes Fertigteil aus Stahlbeton- oder Stahlfaserbeton für die oberste Abdeckung eines ►Einstiegsschachtes, auf der der Schachtdeckelrahmen mit Deckel eingebaut wird. Die Schlupfweite der Einstiegsöffnung beträgt mindestens 625 mm; je nach Bauart und Dimensionierung des Gesamtschachtes wird durch eine exzentrische Lage der Öffnung eine solche Platte als Ersatz für einen Schachtkonus eingebaut; ►Schachtdeckel); *syn.* Abdeckplatte [f], Betonauflagerahmen [m] eines Schachtes, *im Brunnenbau* Brunnenabdeckplatte.

crack [n] *pedol.* ►desiccation crack.

crash barrier [n] [UK] *trans.* ►guardrail [US] (1), ►guardrail [US] (2).

1193 create [vb] **a slope** *constr.* (Make a gradient by filling against a structure or road; ►construct an embankment/slope, ►flatten a slope, ►steepen an existing slope); *s* **ataluzar [vb]** (Relleno alrededor de estructuras o en taludes existentes para reducir la pendiente; ►reducir la pendiente de un talud, ►taluzar, ►igualar terreno con talud); *f* **taluter [vb] en remblai** (Confection d'un remblai en pente lors du raccordement d'un bâtiment au niveau du fond de forme ; *action de taluter* **talutage [m] en remblai** ; ►dresser un talus, ►adoucir un talus, ►taluter en déblai) ; *g* **anböschen [vb]** (Ein Bauwerk oder Teile davon schräg mit Boden anfüllen oder eine vorhandene Böschung an Planungshöhen durch Bodenauftrag anpassen; ►Böschung herstellen, ►Böschung abflachen, ►abböschen).

creation [n] **of defined spaces** *gard. landsc. urb.* ►division of spaces.

1194 creation [n] **of rolling hills** *constr. syn.* creation [n] of smoothly-connected landforms; *s* **ondulación [f] del terreno** (Modelación de terreno para dar la impresión de movimiento); *f* **modelage [m] du terrain avec formation de buttes** (Création de mouvements de terrain); *g* **Geländemodellierung [f] mit Hügeln**.

creation [n] **of smoothly-connected landforms** *constr.* ►creation of rolling hills.

crèche [n] [UK] *sociol. urb.* ►day nursery.

creek [n] *geo.* ►brook.

creek [n]**, tidal** *geo.* ►tidal creek.

creep [n] *geo.* ►rock creep; *geo. pedol.* ►soil creep (1), ►soil creep (2), ►talus creep.

1195 creeping dwarf shrub carpet [n] [US] *phyt.* (Spreading dwarf shrub cover, several centimeters high, with intertwined branches prostrate on the ground with species, such as Mountain Avens *[Dryas octopetala]*, Netted Willow *[Salix reticulata]*, Prostrate Willow *[Salix serpyllifolia]*; ►prostrate shrub); *syn.* matted dwarf shrub carpet [n] [US]/matted dwarf-shrub thicket [n] [UK], creeping dwarf-shrub thicket [n] [UK]; *s* **formación [f] de caméfitos reptantes** (En los pisos altitudinales alpino y subalpino inferiores, cobertura de matas rastreras con ramas entrecruzadas de pocos centímetros de altura como formaciones de *Dryas octopetala* o de sauces *[Salix reticulata et S. serpyllifolia]*; ►arbusto en espaldera); *f* **tapis [m] d'arbrisseaux nains (±)** (Groupements d'arbustes nains de l'étage alpin inférieur et de l'étage subalpin supérieur [continuum] constitués p. ex. de la ryade *[Dryas]*, du Saule arbusculeux, du Saule réticulé *[Salix reticulata]*, du Saule prostré *[Salix serpyllifolia]* ; ►arbuste espalier) ; *g* **Zwergstrauchteppich [m]** (Ein Bestand aus Zwergsträuchern mit meist am Boden kriechenden, wenigen Zentimeter hohen Zweigen in der unteren alpinen und subalpinen Stufe, z. B. Bestände aus Silberwurz *[Dryas]*, Netzweide *[Salix reticulata]*, Quendelblättrige Kriechweide *[Salix serpyllifolia]*; ►Spalierstrauch).

creeping dwarf-shrub thicket [n] [UK] *phyt.* ►creeping dwarf shrub carpet [US].

1196 creeping ground cover plant [n] *hort. plant.* (Low prostrate shrub that grows horizontally or a herbaceous perennial that spreads by roots, rhizomes, or entwining stems, typically 50-150mm tall); *s* **rastrera [f]** (Planta vivaz, pequeña leñosa o rosa rastrera que crece horizontalmente por medio de rizomas o estolones y que —según la densidad de plantación— llega a cubrir el suelo en un plazo de 1-3 años); *syn.* planta [f] rastrera; *f* **plante [f] rampante** (Désigne les végétaux dont les tiges s'étalent entièrement sur le sol et pouvant développer naturellement des racines à son contact [stolons]) ; *syn.* plante [f] à port rampant ; *g* **kriechender Bodendecker [m]** (Niedrige, sich dicht an der Erde durch Rhizome oder Ausläufer ausbreitende Staude, Kleingehölz oder Bodendecker-Rose, die den Boden je nach Pflanzdichte in 1-3 Vegetationsperioden schließt und das Unkraut unterdrückt).

1197 creeping soil [n] *geo. pedol.* (Slow mass movement of soil and soil material down relatively steep slopes, primarily under the influence of gravity but facilitated by saturation with water and by alternate freezing and thawing; RCG 1982; ►creeping soil on frozen ground, ►landslide, ►soil creep 2, ►solifluction); *s* **plancha [f] de solifluxión** (Tipo de ►solifluxión que consiste en el deslizamiento de una masa fangosa según un plano inclinado paralelo a la vertiente, determinado bien por la existencia de una formación impermeable, bien por la presencia de láminas de hielo; DGA 1986; ►corrimiento de tierras, ►reptación hídrica del suelo, ►selifluxión, ►tierra fangosa reptante);

C

f **boue [f] reptante** (Glissement lent sur le support compact d'un versant de masses de terre saturées en eau ; ▶gélifluxion, ▶glissement de terrain, ▶reptation d'un versant, ▶terre fluente) ; *g* **Gekriech [n]** (Langsam abgleitende, wassergesättigte Bodenmasse auf einer festen Hangunterlage; im Vergleich zur ▶Fließerde ohne gefrorenen Untergrund; ▶Hangkriechen, ▶Hangrutschung, ▶Solifluktion).

1198 creeping soil [n] on frozen ground *geo. pedol.* (Slow mass downhill movement of soil facilitated by saturation with water or alternate freezing and thawing; ▶creeping soil, ▶soil creep 1, ▶solifluction); *s* **tierra [f] fangosa reptante** (Tipo de movimiento gravitacional lento de sedimentos no consolidados que en declives [> 2°] descienden lentamente sobre el subsuelo helado. Este movimiento se ve facilitado por la saturación de agua de la **t. f. r.** o por procesos de helada y deshielo alternantes; ▶plancha de solifluxión, ▶selifluxión, ▶solifluxión); *syn.* suelo [m] deslizante; *f* **terre [f] fluente** (Sédiment meuble engorgé d'eau se déplaçant à l'état pâteux souvent sur un sous-sol gelé des versants ; ▶boue reptante, ▶gélifluxion, ▶solifluxion) ; *syn.* couche [f] solifluée, coulée [f] boueuse ; *g* **Fließerde [f]** (Lockersediment, das sich in Hanglagen [> 2°] auf gefrorenem Untergrund als wassergesättigter Brei bewegt; *cf. hierzu* ▶Gekriech; ▶Bodenfließen, ▶Solifluktion).

1199 cremation burial [n] *adm.* (Disposal of the ash remains [or ashes] of a cremated human corpse by burial in the ground or scattering in the air from an aeroplane or by burial at sea; ▶coffin burial, ▶urn burial); *s* **cremación [f] (de los muertos)** (Incineración de los cuerpos en crematorio; generalmente a continuación se depositan las cenizas en una urna que se entierra; ▶entierro, ▶entierro en urna); *syn.* incineración [f] (de los muertos); *f* **incinération [f] funéraire** (Action de brûler le corps des morts dans un crématorium et de placer les cendres dans une urne cinéraire ; ▶inhumation, ▶inhumation cinéraire) ; *syn.* crémation [f] des corps ; *g* **Feuerbestattung [f]** (Form der Totenbestattung, bei der nach der Trauerfeier und anschließenden Einäscherung der Leiche im Krematorium meistens einige Tage später die sterblichen Überreste in einer Urne beigesetzt werden; Neben der ▶Urnenbeisetzung gibt es auch anonyme Bestattungen wie das Verstreuen der Asche auf einem Aschestreufeld, das Versteuen der Totenasche auf See [Seebestattung] und das Verstreuen aus einem Flugzeug; ▶Erdbestattung); *syn.* Aschenbeisetzung [f].

1200 crest [n] *geo.* (Upper edge of a slope; ▶cuesta, ▶escarpment 2); *s* **cresta [f]** (Borde superior de una ladera; ▶cuesta, ▶escarpe de falla); *f* **sommet [m] d'un versant** (DG 1984, 341 ; ligne culminante d'un versant de colline, de montagne ; ▶cuesta, ▶escarpement) ; *g* **Hangkante [f]** (Oberes deutliches Ende eines Hanges; ▶Geländeabbruch, ▶Schichtstufe); *syn.* Hangschulter [f].

1201 crest [n] of embankment/slope *constr.* (Transition from sloping face to embankment crown); *syn.* brow [n] of embankment/slope; *s* **frente [m] de talud** (Zona de transición entre el borde y la cima de un talud); *f* **épaulement [m] (de talus)** (Zone de transition entre la partie supérieure du talus et la tête de talus) ; *g* **Böschungsschulter [f]** (Übergang von der oberen Böschungsfläche zur Böschungskrone).

1202 crest vertical curve [n] [US] *trans.* (Formation of a parabolic curve [▶road crest] at a roadway or path [high point] to create a smooth transition from one longitudinal slope to another; *opp.* ▶sag curve); *syn.* vertical curve [n] of a road crest [UK]; *s* **curva [f] vertical de culminación** (En carretera, curva convexa de un gradiente vertical [▶cambio de rasante] que une dos orientaciones diferentes; *opp.* ▶redondeo de depresión); *syn.* redondeo [m] de culminación; *f* **rayon [m] de raccordement convexe**

(Constitution d'une courbe parabolique qui, à partir du point le plus haut d'un gradient d'altitude, forme une transition progressive entre une rampe et une pente sur le profil en long d'une route) ; *opp.* ▶rayon de raccordement concave) ; *g* **Kuppenausrundung [f]** (Bei der Planung einer Wege- oder Straßentrasse die Ausbildung einer vertikalen Kurve [▶Kuppe 2], die am höchsten Punkt eines Gradientenabschnittes einen allmählichen Übergang von der steigenden auf die fallende Längsneigung bildet; *opp.* ▶Wannenausrundung).

criminality [n] *envir. leg.* ▶environmental criminality.

critical area [n] [US] *ecol. plan.* ▶sensitive area.

1203 critically endangered species [n] *conserv. phyt. zool.* (*Faunistic and floristic* ▶*status of endangerment* species, which is threatened with extermination or extinction and is likely to become extinct within the foreseeable future throughout all, or a significant portion of its range. Survival is impossible, when factors causing the danger of extinction are not eliminated by human activities through the introduction of adequate protection measures, which will sustain the population; *according to IUCN categories of threat* CR [Critically Endangered]; ▶endangered species, ▶Red List); *syn. obs.* critical species [n]; *s 1* **especie [f] en peligro crítico (CR)** (Según las categorías de la ▶lista roja de la UICN, un taxón está en peligro crítico [Critically Endangered] cuando la mejor evidencia disponible indica que cumple cualquiera de los cinco criterios definidos para esta categoría y, por consiguiente, se considera que está enfrentando un riesgo extremadamente alto de extinción en estado silvestre; cf. UICN 2001); *s 2* **especie [f] en peligro de extinción** (*En Es. y D., una de las* ▶*categorías de amenaza de extinción de especies* taxones o poblaciones cuya supervivencia es poco probable si los factores causantes de su actual situación siguen actuando; cf. art. 55 Ley 42/2007; ▶Catálogo Español de Especies Amenazadas, ▶lista roja); *f 1* **espèce [f] amenée par sa régression à un niveau critique des effectifs** (▶*Catégorie de régression des populations et de distribution des espèces de faune et de flore* espèce irrémédiablement en voie d'extinction, si, par des mesures de conservation et de préservation, il n'était pas mis un terme à la persistance des influences responsables de son état ; ▶espèce affectée d'une régression forte et continue et qui a déjà disparue de nombreuses région, ▶inventaire local/régional du patrimoine faunistique et floristique, ▶liste des espèces menacées, ▶Livre rouge [des espèces menacées]) ; *f 2* **espèce [f] en danger critique d'extinction (CR)** (Une des neufs catégories du système de la Liste Rouge de l'UICN : espèce remplissant un des critères suivants [texte non exhaustif] : **a.** réduction des effectifs constatée, estimée, déduite ou supposée depuis 10 ans ou 3 générations égale ou supérieure à 80 % …, **b.** espèce dont la répartition géographique dans la zone d'occurrence [estimée inférieure à 100 km²] et la zone d'occupation [estimée à moins de 10 km²] indique un déclin continu, constaté, déduit ou prévu ainsi qu'une fluctuation extrême de la zone d'occurrence, de la zone d'occupation, de l'étendue et/ou de la qualité de l'habitat, du nombre de localités ou de sous-populations, du nombre d'individus matures, **c.** une population estimée à moins de 250 individus matures dont le déclin est estimé à 25 % au moins en trois ans ou une génération, **d.** une population estimée à moins de 50 individus matures et **e.** taxon confronté à une probabilité d'extinction à l'état sauvage de 50 % en l'espace de 10 ans ou de trois générations ; cf. Catégories et Critères de l'UICN pour la Liste Rouge, Version 3.1, UICN 2001 ; ▶espèce menacée) ; *syn.* espèce [f] en voie d'extinction ; *g* **vom Aussterben bedrohte Art [f]** (*Höchste faunistische und floristische* ▶*Gefährdungskategorie* Art, deren Überleben unwahrscheinlich ist, wenn das die das Aussterben verursachenden Faktoren weiterhin einwirken oder bestandserhaltende Schutz- und Hilfsmaßnahmen des Menschen unterbleiben; in D.

zur Gefährdungskategorie 1, gemäß IUCN zur Kategorie CR [Critically Endangered] gehörig: diejenige Population, deren beobachtete, geschätzte, hergeleitete oder vermutete Größe sich [lt. einem von fünf Kriterien] in den letzten 10 Jahren oder in drei Generationen über 90 % verringert hat oder sich in den nächsten 10 Jahren über 80 % verringern wird; ►stark gefährdete Art, ►Rote Liste).

critical path analysis [n] *constr. plan.* ►critical path method.

1204 critical path method [n] *constr. plan.* (*Abbr.* C.P.M.; method of analysis and efficient scheduling of time-saving operations for a complicated project with the aim of completing all specified tasks before a projected due date); *syn.* critical path analysis [n], network analysis [n], network planning technique; *s* **método [m] del «camino crítico» (≠)** (Método de planificación temporal de procesos laborales complejos en el marco de un proyecto); *f* **analyse [f] de chemin critique** (Procédé d'établissement et d'analyse du planning de réalisation d'un projet) ; *syn.* analyse [f] de réseau ; *g* **Netzplantechnik [f]** (Verfahrenstechnik zur Analyse und zeitlichen Planung von komplexen Arbeitsabläufen innerhalb eines Projektes).

1205 critical population size [n] *phyt. zool.* (Minimum number of individuals necessary for the survival of a population); *s* **población [f] mínima** (Cantidad mínima de individuos necesaria para la supervivencia [Es]/sobrevivencia [AL] de una población animal o vegetal); *syn.* población [f] crítica; *f* **nombre [m] minimum d'individus d'une population** (Nombre minimum nécessaire à la survie de la population) ; *g* **kritische Populationsgröße [f]** (Mindestindividuenzahl, die nötig ist, damit eine Population sich erhalten kann).

critical species [n] *conserv. phyt. zool. obs.* ►critically endangered species.

crop [n] *agr. constr. hort.* ►green manure catch crop, ►green manure crop; *agr. for. land'man.* ►nurse crop; *conserv. land' man.* ►soil conservation crop; *for.* ►standing forest crop.

crop [n]**, short term** *agr. hort.* ►green manure catch crop.

crop [n]**, timber** *for.* ►stand of timber.

crop [n]**, tree** *for.* ►stand of timber.

1206 crop area [n] *agr. for. hort.* (Cultivated land suitable for crop production; ►crop growing region/area, ►perennial nursery, ►sod nursery [US]/turf nursery [UK], ►tree nursery); *syn.* cropland [n]; *s* **superficie [f] de cultivo** (Terreno de producción agrícola, silvícola u hortícola; ►plantel de césped, ►región de cultivo, ►vivero [de árboles], ►vivero de vivaces); *f* **surface [f] culturale** (Surface pour la production agricole, sylvicole et maraîchère ; ►gazonnière, ►pépinière, ►pépinière spécialisée en plantes vivaces, ►région culturale) ; *g* **Anbaufläche [f]** (Land- und forstwirtschaftliche sowie gartenbauliche Produktionsfläche; ►Anbaugebiet, ►Anzuchtfläche für Fertigrasen, ►Baumschule, ►Staudengärtnerei); *syn.* Anzuchtfläche [f], Kulturfläche [f].

crop culture [n] *agr. hort.* ►crop growing.

crop farming [n] [US] *agr.* ►cultivation of arable land.

1207 crop growing [n] *agr. hort.* (Planting and harvesting of cultivated plants; ►crop husbandry); *syn.* crop culture [n]; *s* **cultivo [m]** (Plantación y cosecha de plantas de consumo humano o animal; ►producción agrícola); *f* **culture [f]** (de plantes culturales; ►production végétale) ; *g* **Anbau [m, o. Pl.] (1)** (Anpflanzen von Kulturpflanzen, bei denen im Vergleich zur Anzucht nur die Früchte oder andere Pflanzenteile verkauft werden, z. B. Getreide, Kartoffeln, Obst, Blüten — bei der **Anzucht** hingegen werden Pflanzen bis zur verkaufsfertigen Ware kultiviert und als Ganzes verkauft; ►Pflanzenbau).

1208 crop-growing region [n] *agr. for. hort.* (►crop area); *s* **región [f] de cultivo** (Área caracterizada por un cultivo específico, p. ej. de cereales, caña de azucar, etc. ►superficie de cultivo); *syn.* área [f] de cultivo; *f* **région [f] culturale** (►surface culturale) ; *syn.* zone [f] culturale ; *g* **Anbaugebiet [n]** (Größere Flächenverbände als bei einer ►Anbaufläche).

1209 crop husbandry [n] *agr.* (Farm management to create favo[u]rable conditions for the cultivation of plants and the production of crops; ►crop growing, ►cultivation of arable land); *syn.* crop production [n]; *s* **producción [f] agrícola** (Medidas para crear condiciones favorables para el cultivo de plantas; ►cultivo, ►cultivo de tierras); *f* **production [f] végétale** (Mesures prises en vue de créer les conditions favorables à une bonne récolte de produits d'une culture ; ►culture, ►culture des champs) ; *g* **Pflanzenbau [m, o. Pl.]** (Maßnahmen zur Herstellung günstiger Bedingungen für die Ertragsbildung von Kulturpflanzen; ►Anbau 3, ►Ackerbau).

crop [n] **of woody plants** *for. landsc. phyt.* ►nurse crop of woody plants; ►pioneer crop of woody plants.

crop production [n] *agr.* ►crop husbandry; *agr. for.* ►mixed crop production.

crop production [n]**, intensity of** *agr. for. hort.* ►intensity of agricultural and forest land use.

crop production, [n] **intensive** *agr. hort.* ►intensive farming.

crop raising [n] *agr.* ►cultivation of arable land.

1210 crop residues [npl] *agr. for. hort.* (Remains of plants above and below ground after harvesting; ►slash); *s* **residuos [mpl] de cosecha** (Restos de las plantas sobre y bajo la tierra después de la cosecha; ►residuo forestal); *f* **déchets [mpl] de récolte** (Ensemble des parties aériennes et souterraines des plantes après la récolte ; ►rémanent[s] de coupe) ; *g* **Ernterückstand [m]** (Summe der bei der Ernte auf dem Schlag verbleibenden ober- und unterirdischen Pflanzenteile; ►Schlagabraum).

1211 crop rotation [n] *agr.* (Succession of different crops on a piece of ground, so that one type of crop does not occupy the same ground continually and deplete soil fertility; cf. BS 3975: part 5; ►land rotation, ►shifting cultivation); *s* **rotación [f] de cultivos** (Sucesión de diferentes cultivos en un campo agrícola; ►agricultura itinerante, ►cultivo en rotación); *syn.* rotación [f] de frutos; *f* **rotation [f] des cultures** (DG 1984 ; succession des cultures dans le temps sur un même sol pour en conserver la fertilité ; ►agriculture itinérante, ►assolement) ; *syn.* rotation [f] culturale (PR 1987) ; *g* **Fruchtfolge [f]** (Zeitliche Aufeinanderfolge von Kulturpflanzen auf einer landwirtschaftlichen Nutzfläche; ►Landwechselwirtschaft, ►Wanderfeldbau), *syn.* Fruchtfolgesystem [n], Rotation [f].

crop system [n] *agr.* ►single crop system.

1212 crop tree [n] *constr. for.* (Tree within an afforested area or spontaneous stand selected for its special or unique qualities and left standing during cropping or thinning of the forest, until it reaches maturity; ►improvement cut, ►stand thinning); *s* **árbol [m] de élite** (Árbol situado en una población natural o un rodal reproducido artificialmente, elegido por sus características cualitativas para ser preservado a la hora de realizar operaciones de ►limpia, ►raleo o ►cortes de mejora); *f1* **arbre [m] d'élite** (Arbre présentant, pour des caractères sylvicoles importants, une supériorité génétique transmissible prouvée par voie expérimentale ; TSF 1985) ; *f2* **arbre [m] plus** (Arbre présentant en forêt, pour des caractères sylvicoles importants, une supériorité phénotypique nette par rapport à ses voisins de même âge, mais dont la supériorité génétique n'est que présumée et pas toujours réelle et laissé sur pied lors d'une coupe d'éclaircie ; TSF 1985 ;

▶éclaircissage 1, ▶opération d'amélioration) ; *syn.* arbre [m] d'avenir ; *g* **Zukunftsbaum** [m] (Ausgewählter Baum in einem aufgeforsteten oder spontan entstandenen Bestand, der auf Grund seines Wuchses, seiner Qualität oder wegen seiner Einmaligkeit zur Elite des Bestandes gehört und bei Auslichtungshieben oder Durchforstungen [▶Pflegehieb] stehen bleibt; ▶Auslichten im Bestand).

1213 cross bond [n] *arch. constr.* (DAC 1975; bond used in bricklaying with alternating courses of stretchers and headers, whereby headers are placed symmetrically on the joint between the stretchers on the course beneath); *s* **aparejo** [m] **de cruz inglesa** (En mampostería, colocación de los ladrillos alternando hiladas de sogas y tizones y en la que los tizones se colocan simétricamente sobre la junta de las sogas de la capa inferior. Se utiliza cuando se requiere resistencia; cf. DACO 1988; *f* **appareillage** [m] **alterné en croix** (DTB 1985, 13 ; appareillage à la française, disposition des briques dans une maçonnerie avec alternance entre lit de boutisses et lit de panneresses, dans laquelle la boutisse est disposée de part et d'autre de l'axe médian des panneresses des lits inférieurs et supérieurs) ; *g* **Kreuzverband** [m] (Klinkermauersteinverband im aufgehenden 1-Stein-Mauerwerk mit abwechselnder Läufer- und Binderschicht, bei dem zu beiden Seiten der Mittelachse eines Läufers ein Binder angeordnet ist).

1214 cross-country skier [n] *recr.* (Person who pursues ▶cross-country skiing either as a competitive sport or for recreation); *syn.* langlaufer [n] [also US]; *s* **esquiador, -a** [m/f] **de fondo** (Persona que practica profesionalmente o en su tiempo libre ▶esquí de fondo); *f* **fondeur, euse** [m/f] (Personne qui pratique le ▶ski de fond ; PR 1987) ; *g* **Skilangläufer/-in** [m/f] (Sportler oder Erholungsuchender, der ▶Skilanglauf betreibt); *syn.* Langläufer/-in [m/f], *o.V.* Schilangläufer/-in [m/f].

1215 cross-country skiing [n] *recr.* (Form of skiing on prepared ▶loop trails 1; ▶back-country skiing, ▶cross-country skier, ▶down mountain skiing, ▶ski touring); *syn.* langlauf [n] [also US]; *s* **esquí** [m] **de fondo** (Actividad de esquiar realizada en ▶pistas de esquí de fondo preparadas; ▶esqui alpino, ▶esquiador, -a de fondo, ▶esquí de travesía, ▶esquí de travesía en alta montaña); *f* **ski** [m] **de fond** (Sport d'hiver pratiqué sur des parcours préparés de faible dénivellation ; ▶fondeur, euse, ▶piste de ski de fond, ▶randonnée à ski, ▶randonnée à ski en altitude, ▶ski de piste, ▶ski de randonnée) ; *syn.* ski [m] nordique ; *g* **Skilanglauf** [m, o. Pl.] (**1.** Skisport auf präparierten ▶Loipen. **2. S.** wird neuerdings mit einer neuen Generation von Skibrettern durchgeführt, den sog. Nordic-Trend-Skiern, auch **Nordic Cruiser** genannt, bei denen dank eines Mikroschuppenbelages das Behandeln mit Steigwachs entfällt. **3. S.** mit den neuen Trendbrettern, das **Nordic Cruising**, ist die Wintervariante zum Nordic Walking, die sowohl auf Loipen als auch abseits präparierter Loipen betrieben werden kann und bei Letzterem eine Form des ▶Skiwanderns ist; ▶Abfahrtslauf, ▶Skilangläufer/-in, ▶Tourenskilauf) *syn. o. V.* Schilanglauf [m], Skilanglaufen [n, o. Pl.], Nordic Cruising [n].

cross cutting [n] *constr. for.* ▶saw into pieces.

cross fall [n] [UK] *constr. trans.* ▶cross slope [US].

cross joint [n] [US] *arch. constr.* ▶heading joint.

1216 cross section [n] (1) *constr. trans.* (Graphically-shown cut on a drawing at right angles to the longitudinal axis; ▶longitudinal section); *syn.* transverse section [n]; *s* **sección** [f] **transversal** (Corte en ángulo recto al eje longitudinal; ▶sección longitudinal); *syn.* corte [m] transversal, perfil [m] transversal; *f 1* **coupe** [f] **en travers** (En génie routier on utilise le terme de **profil en travers** ; ▶coupe longitudinale) ; *syn.* profil [m] transversal (d'un lit mineur), coupe [f] transversale, section [f] en

travers ; *f 2* **profil** [m] **en dévers** (Caractérisant la déclivité du bord extérieur de la courbe d'une chaussée) ; *g* **Querschnitt** [m] (Im rechten Winkel zur Längsachse dargestellter Schnitt; im Straßenbau wird meistens von einem *Querprofil* gesprochen; ▶Längsschnitt); *syn.* Querprofil [n].

cross section [n] (2) *constr.* ▶pipe discharge cross-section; *eng. wat'man.* ▶river-flow cross section; *constr. eng.* ▶typical cross section.

1217 cross slope [n] [US] *constr. trans.* (Transverse slope of a pathway or road surface; ▶crowning of paved surfaces [US]/camber [UK], ▶reverse slope [US]/reverse falls [UK]); *syn.* cross fall [n] [UK]; *s* **pendiente** [f] **transversal** (Inclinación de superficies de carreteras o caminos en perpendicular al eje longitudinal; ▶bombeo de superficies, ▶contrapendiente); *f 1* **pente** [f] **transversale** (Inclinaison transversale de la voie perpendiculairement à l'axe de celle-ci ; ▶contre-pente, ▶profil en forme de toit) ; *f 2* **dévers** [m] (La pente transversale donnée au profil en travers d'une chaussée dans les courbes ; DIR 1977 ; *g* **Quergefälle** [n] (Neigung rechtwinklig zur Längsachse von Wege- oder Straßenoberflächen; ▶Dachgefälle, ▶Gegengefälle); *syn.* Querneigung [f].

1218 cross-slope furrow irrigation [n] *agr. for.* (Irrigation method in which water flows in parallel channels, usually 75-500m long, with a very small gradient, laid out in a transverse direction to that of the slope; ▶furrow irrigation); *s* **método** [m] **de irrigación del surco oblicuo a la pendiente** (El agua es aplicada por surcos derechos adyacentes de 75-500 m de longitud, situados oblícuamente a la pendiente predominante a fin de reducir el gradiente del surco; FAO 1960; ▶riego por infiltración); *f* **méthode** [f] **d'irrigation par ruissellement avec épis** (L'eau épandue par des rigoles droites de 75 à 500 m de longueur dont le tracé est oblique par rapport au versant et coulant sur une faible pente ; cf. FAO 1960 ; ▶irrigation à la raie) ; *g* **Querfurchenbewässerung** [f] (Bewässerungsmethode, bei der das Wasser von aneinander angrenzenden Furchen aus, die 75-500 m lang und quer zur Hangneigung angelegt sind, mit sehr geringem Grabengefälle fließt; cf. FAO 1960; ▶Furchenbewässerung).

cross-town artery [n] [US]**, major** *trans.* ▶major cross-town artery.

cross-town street [n] [US] *trans.* ▶major cross-town artery.

1219 crotch [n] [US] *arb.* (Point at which a trunk, stem or branch divides); *syn.* branch attachment [n] [US], fork [n] [UK]; *s* **cruz** [f] (Punto de ramificación del tronco); *syn.* axila [f], horcadura [f]; *f* **enfourchement** [m] (Point de division d'une branche) ; *syn.* fourche [f] d'une branche ; *g* **Astverzweigung** [f] (Stelle, an der ein Ast abzweigt); *syn.* Astgabel [f].

crotch [n]**, tight** *arb.* ▶forked growth.

1220 crowding effect [n] *zool.* (ÖKO 1983, 137; selfregulation of animal population by decreased fertility, cannibalism of the young, or stress among mammals); *syn.* mass effect [n]; *s* **efecto** [m] **de masificación de una especie faunística** (Autoregulación de población animal en el caso de sobrepoblación de su área de habitación, por medio de la reducción de la fertilidad, por canibalismo hacia los animales jóvenes o por estrés en las especies de mamíferos); *f* **effet** [m] **de concentration massive d'une espèce animale** (≠) (Régulation de la population par des comportements d'agression, de cannibalisme, de stress provoquant des activités de substitution [diminution de la fertilité], comme moyen de diminuer la forte densité de population ou les natalités importantes [surpopulation], et éviter la surexploitation des ressources alimentaires) ; *g* **Massensiedlungseffekt** [m] (Populationsregulation bei beginnender oder sich abzeichnender Überbevölkerung durch eine herabgesetzte Fruchtbarkeit, durch das Auffressen der eigenen Brut, durch Erhöhung der Mortalität oder Krankheitsanfälligkeit, Auslösung von Wanderungsbewegungen oder durch Stresssituationen [Drängefaktor] bei Säugetieren); *syn.* Kollisionseffekt [m], Masseneffekt [m], Überbevölkerungseffekt [m].

1221 crown base [n] *arb. hort.* (Connecting point between clear stem and a tree crown); *syn.* branching height [n]; *s* **base [f] de la copa** (Zona de transición entre el tronco y la copa); *f* **base [f] de la cime** (Niveau du tronc où commence le houppier et s'arrêtent les branches les plus basses suite à l'élagage) ; *g* **Kronenansatz [m]** (Bereich, in dem der [aufgeastete] Stamm eines Baumes endet und die Krone mit den untersten Astansätzen anfängt; der unterste Bereich der Ast- und Zweigspitzen wird **Kronenbasis** genannt).

crown canopy [n] *for. phyt.* ▶tree canopy.

1222 crown closure [n] *for. phyt.* (**1.** Crown "thickening" in terms of both depth and the closer growth of branches and foliage, which together determine the overall density of an individual tree; SAF 1983. **2. Crown density** [n]: thickness, both spatially [i.e. depth] and in closeness of growth [i.e. compactness] of an individual crown [i.e. its opacity], as measured, e.g. by its shade density; SAF 1983); *s* **densidad [f] de copa** (Grado de espesamiento de la copa tanto en su crecimiento espacial como en la compacidad de los elementos que la componen); *f* **densité [f] du couvert d'une cime** (Consistance d'une cime individuelle, compte tenu à la fois de son développement spatial et de la compacité des éléments qui la composent, c.-à-d. son opacité ; DFM 1975) ; *syn.* couvert [m] de la cime d'un arbre, densité [f] du couvert d'une couronne ; *g* **Kronenschluss [m] (3)** (...schlüsse [pl]; Grad der Dichte des Laub- und Astwerkes einer einzelnen Baumkrone).

1223 crown cover [n] *arb. phyt.* (Ground area covered by a ▶tree crown, as delimited by the vertical projection of its outermost perimeter; SAF 1983); *syn.* crown projection area [n], crown spread [n], crown plan [n]; *s* **cubierta [f] de copa** (Superficie del suelo incluida en la proyección horizontal del perímetro externo de la ▶copa de un árbol; ▶expansión horizontal de las copas, ▶dosel del bosque); *syn.* espesura vertical de copa; *f* **couvert [m] vertical au sol** (Surface de sol comprise à l'intérieur de la projection horizontale, c.-à-d. sur un plan horizontal, de la ▶couronne d'arbre ; DFM 1975 ; ▶indice de cime, ▶couvert continu des cimes) ; *syn.* abri [m] vertical au sol (±) ; *g* **Kronenschirmfläche [f]** (Bodenoberfläche, die durch die senkrechte Projektion der ▶Baumkrone abgedeckt wird; ▶Deckungsgrad der Baumkronen, ▶Kronendach des Waldes).

crown density [n] *for. phyt.* ▶crown closure.

1224 crown form [n] *arb. hort.* (Shape of a tree crown; e.g. protruding, bizarre, conical, round, fastigiate, columnar, umbrella-formed; ▶crown renewal, ▶flat cown [US]/shallow crown [UK], ▶open crown); *syn.* crown shape [n]; *s* **forma [f] de la copa** (▶copa aparasolada, ▶copa hueca, ▶restauración de la copa); *f* **forme [f] de la cime** (Résultat de la taille de formation en vue de l'obtention d'une charpente solide et correspondant à la forme désirée ; ▶houppe au port aéré, ▶houppe au port étalé, ▶restauration de la cime) ; *syn.* forme [f] du houppier, forme [f] de la couronne ; *g* **Kronenform [f]** (Art und Weise wie eine Krone ausgebildet ist, z. B. ausladend, bizarr, flach, konisch, rund, säulenförmig, schirmförmig; ▶flache Baumkrone, ▶lichte Baumkrone, ▶Wiederherstellung der Baumkrone).

crowning off [n] [AUS] *arb.* ▶topping (3).

1225 crowning [n] **of paved surfaces** [US] *constr.* (Slight convexity upon paved areas in cross section; e.g. crowned ¼"-l/2"/ft [2-3%] [US]; ▶cross slope [US]/cross falls [UK]); *syn.*

camber [n] [UK]; *s* **bombeo [m] de superficies** (Forma convexa de superficies revestidas, como plazas y caminos, tejados planos, etc.; ▶pendiente transversal); *syn.* combadura [f] de superficies; *f* **profil [m] en forme de toit** (Convexité d'une chaussée ; DTB 1985, 37 ; ▶dévers, ▶pente transversal) ; *syn.* bombement [m], formation [f] bombée (d'une route), pente [f] en toiture (VRD 1994, 4/6.4.2.2.2) ; *g* **Dachgefälle [n]** (Im Querschnitt konvex geformte Ausbildung von befestigten Platz- oder Wegeflächen; ▶Quergefälle).

1226 crown lifting [n] *constr. for. hort.* (Pruning of lower branches to increase clearance under a tree; BS 3975; Part 5); *syn.* lift-pruning [n] [also US], drop pruning [n] [also US] (DIL 1987), raising [n] the crown/canopy/head [also US]; *s* **limpia [f] del tronco** (Tipo de ▶poda de formación en la que se cortan las ramas inferiores del tronco para aumentar la altura de la copa); *syn.* poda [f] alta; *f* **élagage [m] (1)** (Opération culturale horticole et forestière consistant à couper certains rameaux ou branches d'un arbre afin d'élever la couronne et de former un fût ; cf. IDF 1988, 19 et s.) ; *g* **Aufasten [n, o. Pl.]** (Wegschneiden der unteren Stammäste, damit ein Hochstamm oder Halbstamm entsteht); *syn.* Aufastungsschnitt [m]; *for.* Astreinigung [f], Ästung [f], Aufästung [f].

1227 crown [n] **of a wall** *arch. constr.* (▶coping); *syn.* cap [n] of a wall [also US]; *s* **coronación [f] de un muro** (▶piedra de cubierta); *f* **couronnement [m] de mur** (L'arase ou ce qui termine ou coiffe le sommet d'un mur ; ▶chaperon, ▶margelle) ; *syn.* faîtage [m] d'un mur ; *g* **Mauerkrone [f]** (obere Mauerabgrenzung; ▶Mauerabdeckung).

crown plan [n] *arb. phyt.* ▶crown cover.

crown projection area [n] *arb. phyt.* ▶crown cover.

1228 crown pruning [n] *arb. constr.* (Removing superfluous live growth in a tree crown so as to admit light, reduce weight, and lessen wind resistance; SAF 1983; ▶crown pruning for public safety, ▶crown reduction, ▶crown thinning, ▶drop crotching, ▶routine crown pruning, ▶stand thinning, ▶thinning, ▶topping 3, ▶trainaing of young trees); *s* **poda [f] «en copa»** (Corte de las ramas superfluas en la copa para mejorar las condiciones de crecimiento. Durante el desarrollo del árbol se puede diferenciar entre la ▶poda de formación, la poda de fructificación y la de rejuvenecimiento; ▶aclareo de la copa, ▶ahuecado de la copa, ▶corta de mejora, ▶descopado, ▶limpia, ▶poda de corrección de la copa, ▶poda de mantenimiento de la copa, ▶poda de reducción de la copa, ▶raleo); *f* **taille [f] en cime** (Action d'éliminer une partie des rameaux et branches de la cime d'un arbre, de façon à y faire mieux pénétrer la lumière, ou à en diminuer le poids ou la prise au vent ; DFM 1975 ; ▶éclaircissage 1, ▶éclaircissage 2, ▶éclaircissage de la couronne, ▶éhoupage, ▶opérations d'entretien courant de la cime, ▶réduction de couronne, ▶taille de formation de la charpente, ▶taille d'allégement, ▶taille de correction) ; *g* **Kronenschnitt [m]** (Schnittmaßnahme in der Krone eines Baumes. Während der Entwicklung des Baumes sind folgende Schnittmaßnahmen zu unterscheiden: ▶Erziehungsschnitt in der Baumschule, *Kronenaufbauschnitt* bis zum endgültigen Erreichen der gewünschten Form und der *Kronenpflegeschnitt*, bei dem z. B. unerwünschte Entwicklungen verhindert, Totholz und reibende Äste entfernt werden sowie das Lichtraumprofil hergestellt wird; ▶Auslichten, ▶Auslichten im Bestand, ▶Kronenentlastungsschnitt, ▶Kronenkappung, ▶Kronenlichtungsschnitt, ▶Kronenpflege, ▶Kronenrückschnitt, ▶Kronensicherungsschnitt).

1229 crown pruning [n] **for public safety** *arb. constr.* (Pruning of trees with limited life expectancy and branches prone to breakage, often a temporary precautionary measure); *s* **poda [f] de corrección de la copa** (Poda de árboles de espectativa de vida reducida y con ramas que tienden a quebrarse para evitar daños a terceros. A menudo se realiza sin respetar el porte típico de la especie); *f* **taille [f] de correction** (Opérations de coupe effectuées sur des arbres en mauvaise santé, inadaptés à de nouvelles conditions du milieu, ou présentant des risques pour la sécurité souvent sans respecter le port naturel de l'espèce) ; *g* **Kronensicherungsschnitt** [m] (Einkürzung im Grob- und Starkastbereich an Bäumen mit begrenzter Lebenserwartung und bruchgefährdeten Kronen, meist zur befristeten Herstellung der Verkehrssicherheit, oft ohne Berücksichtigung eines artgerechten Kronenaufbaues).

1230 crown reduction [n] *arb. constr.* (Cutting back the crown of a tree for safety and other reasons, while retaining its characteristic shape; ▶drop-crotching, ▶severe pruning, ▶pollarding); *s* **ahuecado [m] de la copa** (Recorte de la copa para mejorar la estabilidad, suprimiendo ramas muertas o en mal estado, pero manteniendo el porte típico de la especie; ▶desmoche, ▶poda de reducción de la copa, ▶trasmocho); *f* **taille [f] d'allégement** (▶Rapprochement de la couronne d'un arbre dont les branches sont menacées de se rompre tout en gardant son port naturel ; ▶étêtage, ▶rabattage d'un végétal ligneux, ▶réduction de couronne) ; *g* **Kronenentlastungsschnitt [m]** (Einkürzen wesentlicher Teile in bruchgefährdeten, übergewichtigen oder durch Standortmangel geschädigten Kronen unter Beibehaltung des artspezifischen Habitus; ▶auf den Kopf setzen, ▶Kronenrückschnitt, ▶starker Rückschnitt); *syn.* Kronenentlastung [f], Kronengewichtsentlastungsschnitt [m].

1231 crown renewal [n] *arb. hort.* (Pruning of the crown of a poorly growing or damaged tree to create a more natural habit, to improve its health and to give greater structural strength); *syn.* crown restoration [n]; *s* **restauración [f] de la copa** (Poda para dar la forma más natural a la copa, mejorar su salud y darle más fuerza estructural al árbol); *f* **restauration [f] de la cime** (Taille en cime pour obtenir une silhouette caractéristique d'arbre) ; *g* **Wiederherstellung [f] der Baumkrone** (Gezielte Schnittmaßnahme bei einem geschädigten oder deformierten Baum, um eine artgerechte Krone aufzubauen).

crown restoration [n] *arb. hort.* ▶crown renewal.

crown shape [n] *arb. hort.* ▶crown form.

crown spread [n] *arb. phyt.* ▶crown cover.

1232 crown thinning [n] *arb. hort.* (Opening up the top of a tree for deeper light penetration which benefits inner leaves and branches; ▶routine crown pruning); *s* **aclareo [m] de la copa (1)** (Más allá de la ▶poda de mantenimiento, intervención de poda seca consistente en eliminar ramas desde la base para aligerar la copa, pero manteniendo el porte típico de la especie; cf. PODA 1994); *f* **éclaircissage [m] de la couronne** (Une taille à alléger la charpente d'une partie de ses ramifications faisant partie des ▶opérations d'entretien courant de la cime ; le volume de l'arbre n'est généralement pas modifié, mais plutôt sa transparence) ; *g* **Kronenlichtungsschnitt [m]** (Zusätzlich zur ▶Kronenpflege erforderliches Ausdünnen an zu dichten, aber nicht bruchgefährdeten Kronen im Schwachastanteil zur Erhaltung stabiler habitusgerechter Kronen [Verringerung der Windlast] und zur Beseitigung einfacher Fehlentwicklungen, z. B. Zwieselbildung und Ausdünnung überzähliger Wasserreiser; *syn.* Kronenauslichtung [f], Kronenauslichtungsschnitt [m], Auslichten [n, o. Pl.] der Krone.

1233 crumb [n] *pedol.* (Rounded, porous and soft aggregate up to 10mm in diameter, being formed under conditions of frequent wetting and drying in the presence of high organic matter content and organisms activity; cf. ULD 1992, 19; ▶crumb structure, ▶soil mellowness); *syn.* granular ped [n] [also US] (ULD 1992, 19), soil granule [n] (RCG 1982); *s* **grumo [m]** (Unión entre los

agregados elementales del suelo en forma de granos más o menos redondos y porosos de hasta 10 mm de diámetro; cf. DINA 1987; ▶estructura grumosa, ▶sazón); *syn.* agregado [m] compacto; *f* **agrégat [m] (terreux) élémentaire** (Unité structurale du sol ; grain de petite taille [jusqu'à 10 mm de diamètre], plus ou moins arrondi et poreux résultant de l'assemblage de particules de sol ; LA 1981, 339, DIS 1986, 99, 212, DFM 1975 ; ▶bon état de structure grumeleuse, ▶structure fragmentaire) ; *syn.* agrégat [m] du sol (PED 1979, 188), grumeau [m] (PED 1983, 29) ; *g* **Krümel [m]** (Sehr poröses, rundliches Bodenaggregat mit einem Durchmesser von 1-10 mm, das unter dem Einfluss hoher biologischer Aktivität und intensiver Durchwurzelung entsteht; cf. SS 1979; ▶Bodengare, ▶Krümelstruktur 1).

crumbing [n]**, soil** *pedol.* ▶crumb structure.

crumbly soil [n] [US] *agr. for. hort.* ▶friable soil.

1234 crumb structure [n] *pedol.* (Particular state of aggregation of soil particles in which the primary particles are united to form compound particles or ▶crumbs; BS 3975: Part 5; ▶soil mellowness, ▶tilth); *syn.* aggregate structure [n], soil crumbing [n]; *s* **estructura [f] grumosa** (Estado de agregación de las partículas del suelo en el cual éstas están unidas formando ▶grumos; ▶sazón); *syn.* contextura [f] grumosa; *f* **structure [f] fragmentaire** (Mode d'assemblage spatial et d'organisation physique des différentes particules du sol, résultant de processus naturels variés, biologiques, chimiques et mécaniques, dépendant de l'activité biologique de la classe de texture de sol et du travail du sol ; ▶bon état de structure grumeleuse, ▶agrégat [terreux] élémentaire ; cf. DIS 1986); *syn.* structure [f] des agrégats, structure [f] en grumeaux, structure [f] grumeleuse (PED 1979, 236, 241), structure [f] grenue (DIS 1986, 98) ; *g* **Krümelstruktur [f] (1)** (Form der unterschiedlichen Aggregatzusammensetzung von Krümeln entsprechend der biologischen Aktivität im Boden, der Bodenart und der Bodenbearbeitung; ▶Bodengare, ▶Krümel); *syn.* Aggregatgefüge [n].

1235 crushed aggregate lawn [n] *constr.* (Grassed area used occasionally as a parking lot, with a 15-20cm deep layer of crushed rock mixed with suitable soil and so compacted that the crushed material can withstand light or periodically heavy traffic load in car parks or driveways [US]/house drives [UK]; ▶car park lawn, ▶lawn type); *syn.* grass [n] on crushed aggregate; *s* **césped [m] sobre piedra partida** (▶Tipo de césped de mezcla de semillas de ▶césped de aparcamiento usado en superficies utilizadas ocasionalmente como aparcamiento y sembrado sobre capa de piedra partida, capaz de soportar directamente la carga, rellenada con tierra); *f* **gazon [m] sur grave (concassée) (±)** (Aire engazonnée constituée par une couche de grave mélangée à de la terre végétale en général de 15 à 20 cm, utilisée pour une circulation occasionnelle telle que certaines aires de stationnement, les voies pompiers ; ▶gazon d'aires de stationnement, ▶type de gazon) ; *syn.* gazon-gravier [m] [B, CH], gazon [m] empierré [CH] ; *g* **Schotterrasen [m]** (Einfachbauweise des Wegebaus für gelegentlich überfahrene Rasenfläche auf einem 15-20 cm dicken hohlraumarmen Schotterkörper, in den geeignetes Bodensubstrat so eingearbeitet wird, dass der Schotter die Verteilung der Verkehrslast bei ständig mäßigem oder zeitlich begrenztem starkem Verkehrsaufkommen unmittelbar aufnehmen kann, z. B. bei gelegentlich genutzten Parkplätzen, Zufahrten; als strapazierfähiger ▶Rasentyp wird meist die Saatgutmischung für ▶Parkplatzrasen verwendet; cf. FLL 2008).

1236 crushed aggregate subbase [n] [US]/**crushed aggregate sub-base** [n] [UK] *constr.* (Unbound ▶subbase [US]/sub-base [UK] composed of well-graded ▶crushed rock mixture; ▶gravel subbase [US]/gravel sub-base [UK]); *syn.* base [n] of crusher-run aggregate [also US], crusher-run base [n] [also

US], rock base [n] [also US], granular subbase [US]/granular sub-base [UK], layer [n] of graded crushed aggregate; *s* **capa [f] de asiento de piedra partida** (►Capa portante de mezcla de ►piedras partidas de diferentes tamaños sin ligante; ►base de grava); *f* **couche [f] de base en grave** (►Couche de fondation sans liant constituée d'une ►grave de granulométrie serrée ; ►couche de fondation en grave non traité) ; *g* **Schottertragschicht [f]** (Ungebundene ►Tragschicht aus einem korngestuften Schottergemisch im Korngrößenbereich 0/32, 0/45, 0/56, 0/63 oder 0/80; ►Kiestragschicht, ►Schotter); *syn.* Mineraltragschicht [f], Schotterkörper [m] (2).

1237 crushed rock [n] *constr.* (Natural or mechanically-broken stone from a crusher usually graded into sizes varying from 0-65mm in diameter, and commonly used as a surfacing or a ►subbase [US]/sub-base [UK] for hard surfaces; ►fine aggregate [US]/stone chippings [UK]); *syn.* crushed stone [n], crusher-run aggregate [n] [also US]; *s* **piedra [f] partida** (Piedra natural o roca partida en tamaños entre 0-65 mm de diámetro y usada para la construcción viaria; ►capa de asiento de piedra partida, ►gravilla); *syn.* lastre [m], pedregullo [m], cascajo [m] [CO], chancado [m] [RCH], lajilla [f] [RCH], piedra [f] machacada [Es], piedra [f] picada [YZ], piedra [f] triturada [CO, EC, MEX], piedrín [m] [CA], roca [f] partida [RCH]; *f* **grave [f]** (Mélange de matériaux naturels ou concassés de granularité 0-65 mm [cailloux, gravier, sable], utilisé dans la construction des chaussées, p. ex. dans certaines couches de fondation comme la ►couche de base en grave ; ►gravillons concassés) ; *syn.* déchets [mpl] de carrière (ARI 1986, 43) ; *g* **Schotter [m]** (Grobes, ein- oder mehrfach gebrochenes natürliches Locker- oder Festgestein mit den üblichen Lieferkörnungen von 32/45, 45/56, 56/80 mm. Beim Straßen-, Platz- und Wegebau wird **S.** für die Frostschutz- oder Tragschicht [►Schottertragschicht] als Stützkorn und ein Splitt-Sandgemisch als Füllmaterial verwendet, so dass je nach Belastung die Fraktionen 0/32, 0/45, 0/56 oder 0/63 eingebaut werden. Für ungünstige Untergrundverhältnisse im Straßen-, Wege- und Gleisanlagenbau wird als unterste Schicht noch gröberer **S.**, so genannte **Schroppen**, in den Gesteinskörnungen 32/200 und 65/x mm eingebaut. **S.** wird auch als Zuschlagstoff bei der Betonherstellung, als Zuschlagstoff in bituminösen Deck- und Tragschichten, als Bettungsstoff im Eisenbahngleisbau [Schotterbett], als Dränmaterial und als ►Mineralbeton in den Körnungen 0/16, 0/32, 0/45, 0/56 mm verwendet; ►Splitt).

1238 crushed rock top course [n] *constr.* (Layer of crushed stone upon coarse gravel or ►crushed aggregate sub-base); *s* **capa [f] superior de grava** (Revestimiento de piedra partida colocado sobre una capa de grava o ►capa de asiento de piedra partida); *f* **revêtement [m] en grave** (Couche de grave mise en place sur une ►couche de base en grave ou de gravier) ; *g* **Schotterdecke [f]** (1) (Auf eine Kies- oder Schottertragschicht aufgebrachte, oberste Schotterschicht).

crushed stone [n] *constr.* ►crushed rock.

crusher-run aggregate [n] [US] *constr.* ►crushed rock.

crusher-run aggregate [n] [US]**, base of** *constr.* ►crushed aggregate subbase [US]/crushed aggregate sub-base [UK].

crusher-run aggregate [n] [US]**, layer of** *constr.* ►gravel subbase [US]/gravel sub-base [UK].

crusher-run base [n] [US] *constr.* ►crushed aggregate subbase [US]/crushed aggregate sub-base [UK].

1239 cryptophyte [n] *bot.* (Greek *kryptos* >hidden< and *phyton* >plant<: **1.** In RAUNKIAER'S nomenclative classification this is a term for an herbaceous plant [►perennial 1] whose buds, coming from geophytes [tuber, corm or bulb] below the ground,

or in the case of helophytes and hydrophytes, from underwater buds, survive during unfavo[u]rable seasons. **2.** Generic term for ►geophyte, ►helophyte, ►hydrophyte); *s* **criptófito [m]** (División de primer orden en la clasificación biotípica y simorfial de RAUNKIAER: conjunto de formas en que la parte persistente del organismos [►planta vivaz] puede quedar completamente protegida bajo el nivel del suelo o bajo el agua. Comprende tres grupos: ►geófitos o plantas terrestres, ►helófitos o plantas anfibias e ►hidrófitos o plantas acuáticas propiamente dichas; DB 1985); *f* **cryptophyte [m]** (**1.** Classification des végétaux vasculaires d'après RAUNKIAER : végétal particulièrement adapté aux rigueurs de la mauvaise saison puisqu'il ne conserve que ses parties souterraines [bulbe, rhizome ou tubercule] pourvue de bourgeons ; DIB 1988, 115 ; ►plante vivace ; **2.** *termes spécifiques* ►géophyte, ►hélophyte, ►hydrophyte) ; *g* **Kryptophyt [m]** (Griech. *kryptos* >verborgen< and *phyton* >Pflanze<: **1.** *Lebensform nach RAUNKIAER* krautige Pflanze [►Staude], bei der die Erneuerungs-/Überdauerungsknospen der sich periodisch erneuernden/absterbenden, meist nur kurzlebigen oberirdischen Organe während der ungünstigen Jahreszeit bei ►Geophyten in Form von Erdsprossen [Rhizomen, Erdknollen, Sprossknollen, Wurzelknollen und Wurzelknospen] oder Zwiebeln unter der Erdoberfläche und bei Helophyten [►Sumpfpflanze] zeitweilig und bei ►Hydrophyten [Wasserpflanzen] ständig unter der Wasseroberfläche überdauern. **2.** *UBe* Geophyt, Helophyt, Hydrophyt).

1240 cubic content ratio [n] **(of a building)** [US] *leg. urb.* (**In U.S.**, ratio indicating the permissible or existing bulk of a building per m^2 plot area; **in U.K.**, no such equivalent, but there is the term 'cumulative total cubic content' of a building, which is a measure of the total volume of a building); *syn.* cubic index [n] (of a building) [≠ UK] (FBC 1993, 189); *s* **edificabilidad [f]** (Relación que indica cuántos m^3 de espacio se pueden construir por m^2 de superficie del predio y que se fija en el ►plan parcial [de ordenación]); *f* **coefficient [m] d'utilisation du sol** (Le rapport entre le volume du bâtiment à la surface de la parcelle ; BON 1990, 350. F., règle d'urbanisme n'étant plus en vigueur ; décret n° 62-460 du 15 avril 1962) ; *g* **Baumassenzahl [f]** (*Abk.* BMZ; die BMZ gibt an, wieviel m^3 umbauter Raum je m^2 Grundstücksfläche maximal zulässig oder vorhanden sind; cf. § 21 BauNVO. Die **Baumasse** ist das Volumen, das sich aus den bauaufsichtlichen Grundstücksauflagen [GRZ, GFZ, BMZ] ergibt und wird angegeben, wenn die Geschosshöhe nicht vorhersehbar ist, die Erscheinungsform jedoch definiert werden soll, z. B. bei Kirchen, Hallenbauten, gewerblichen Bauten; statt der Verhältniszahl BMZ kann auch die Baumasse [BM] in Kubikmetern als höchstzulässiges Maß der Nutzung angegeben werden; KOR 2005); *syn.* Baumassenziffer [f] [CH].

cubic index [n] **(of a building)** [UK] *leg. urb.* ►cubic content ratio [of a building] [US].

1241 cuesta [n] *geo.* (*Name of Spanish origin* asymmetrical ridge, with gentle slope on one side conforming with the dip of the underlying strata, and a steep or cliff-like face on the other side formed by the outcrop of resistant beds; originally the term applied to a steep slope or scarp that terminates a gently sloping plain at its upper end. Cuestas with dip slopes of 40°-45° or higher are usually called **hogback ridges**. **In US**, there are several examples of **c.** in western New York, the Onondaga escarpment, in S. Ontario, the Niagara escarpment; **in UK,** the Jurassic chalk White Horse Hills and the Cotswolds Edge; **in F.,** Côte d'Or and Côtes du Rhône, the Paris basin; **in D.,** the Swabian Alb; ►break of slope, ►scarp slope); *syn.* scarped ridge [n]; *s* **cuesta [f]** (Relieve estructural cuya génesis requiere, por una parte, la existencia de unas series litológicas con alternancia de niveles resistentes e incoherentes y, por otra, el que dichas

C

series presenten un buzamiento monoclinal. Topográficamente la **c.** forma un relieve disimétrico; la vertiente más suave coincide con un nivel resistente, y se denomina **reverso de cuesta**; la vertiente abrupta, o ►**frente de cuesta**, consta de un talud en la base, en el que afloran materiales incoherentes, y una **cornisa** de roca resistente en la cumbre; DGA 1986; ►pequeño declive); *f* **cuesta** [f] (Rebord de plateau en structure sédimentaire monoclinale comportant la superposition de couches résistantes à des couches tendres ; DG 1984 ; ►rupture de pente, ►versant anaclinal) ; *syn.* côte [f] ; *g* **Schichtstufe** [f] (Unter, geomorphogenetischen, geologischen und tektonischen Voraussetzungen entstandene landschaftliche Geländestufengroßform im Bereich flach lagernder, wenig geneigter Gesteinsschichten entstanden; die eigentliche Stufe ist an widerständige [harte] hangende Gesteine gebunden und der Stufenunterhang befindet sich im Bereich der weniger widerständigen [weicheren] Gesteine. Der flache obere Teil der **S.** ist die **Landterrasse** [Dachfläche] der Stufenfläche, der steile Abhang/die steile Stufenwand die **Stufenstirn** und die obere Hangkante der **Trauf. In Europa** gibt es mehrere Landschaften mit einer Abfolge von **S.n**, so genannte **Schichtstufenlandschaften**, z. B. das Schwäbisch-Fränkische Stufenland, das Thüringer Becken, Pariser Becken, die Südenglische Stufenlandschaft; cf. WAG 2001; ►Geländekante, ►Stufenhang).

1242 cul-de-sac [n] *trans. urb.* (Road which is only open at one end, used primarily by abutting owners); *syn.* dead end [n] [also UK], dead-end street [n] [also US]; *s* **fondo** [m] **de saco** *syn.* callejón [m] sin salida, cul-de-sac [m]; *f* **voie** [f] **sans issue** (Voie accessible par une seule extrémité, n'ayant qu'une issue) ; *syn.* impasse [f], voie [f] en impasse, *langage familier* cul [m] de sac) ; *g* **Sackgasse** [f] (Nur von einem Ende her zugänglicher Verkehrsweg, der für Anlieger bestimmt ist); *syn.* Sackstraße [f], Stichstraße [f].

1243 culinary herb [n] *hort. plant.* (Plant from which specific parts, such as buds [Clove—*Syzygium aromaticum*], bark [Cinnamon—*Cinnamomum zeylanicum*, White Sassafras—*Sassafras albidum*], stems [Rosemary—*Rosmarinum officinalis*, Summer Savory—*Satureja hortensis*], flowers [Indian Cress—*Tropaeolum majus*, Lavender—*Lavandula angustifolia*], seeds [Anise—*Pimpinella anisum*, Pepper—*Piper nigrum*], berries [Common Juniper—*Juniperus communis*, Pimento tree—*Pimenta dioíca*], bulbs [Onion—*Allium ascalonicum*, Garlic—*Allium sativum*], or roots [Horseradish—*Armoracia rusticana*, Curcuma—*Cúrcuma longa*] are taken for use in cooking); *s* **planta** [f] **condimenticia** (Planta de la cual se aprovechan determinadas partes aromáticas como especias en la cocina, p. ej. yemas [clavo—*Syzygium aromaticum*], cortezas [canela—*Cinnamomum zeylanicum*, sasafrás—*Sassafras albidum*], tallos [romero—*Rosmarinum officinalis*, ajedrea—*Satureja hortensis*], flores [capuchina—*Tropaeolum majus*, espliego—*Lavandula angustifolia*], semillas [anís—*Pimpinella anisum*, pimienta—*Piper nigrum*], bayas [enebro—*Juniperus communis*, pimienta de Jamaica—*Pimenta dioíca*], bulbos [cebolla—*Allium ascalonicum*, ajo—*Allium sativum*] o raíces [rábano—*Armoracia rusticana*, cúrcuma—*Cúrcuma longa*]); *f* **plante** [f] **condimentaire** (Plante ayant la particularité d'améliorer la saveur des aliments auxquels elle est incorporée ; les parties aromatiques peuvent être un bourgeon [le Giroflier (clou de girofle) — *Caryophyllus aromaticus* L., *syn. Syzygium aromaticum* Merr. et L. M. Perry, *Eugénia aromática* Baill. non Berg], le bulbe [l'Ail — *Allium sativum*, l'Oignon], l'écorce [le Cinnamone — *Cinnamomum zeylanicum*, le Sassafras — *Sassafras albidum*], la tige [l'Angélique — *Angelica archangelica*, le Romarin — *Rosmarinum officinalis*, la Sariette commune — *Satureja hortensis*], la fleur [la Camomille — *Matricaria chamomilla*, la Grande capucine —

Tropaeolum majus], la feuille [la Sauge officinale — *Salvia officinalis*, la Menthe poivrée — *Mentha piperita*, le laurier sauce — *Laurus nobilis*], la graine [l'Anis — *Pimpinella anisum*, le Poivre — *Piper nigrum*], le fruit [le Genévrier (baies de genièvre) — *Juniperus communis*, le Piment — *Pimenta dioíca*] ou la racine [le Raifort — *Armoracia rusticana*, Curcuma (Safran des Indes) — *Cúrcuma longa*] ; dans le langage courant on utilise souvent le terme de ►**plantes aromatiques** alors que de nombreuses plantes exhalant une odeur agréable n'ont pas de valeur culinaire) ; *g* **Gewürzpflanze** [f] (Pflanze, von der bestimmte aromatische Teile wie Blütenknospe [Gewürznelkenbaum — *Syzygium aromaticum* (verwendet als getrocknete Knospen, Pulver oder Öl)], **Rinde** [Ceylonzimtbaum — *Cinnamomum zeylanicum*, Fenchelholzbaum — *Sassafras albidum*], Stängel [Rosmarin — *Rosmarinum officinalis*, Bohnenkraut — *Satureja hortensis*], **Blüten** [Kapuzinerkresse — *Tropaeolum majus*, Lavendel — *Lavandula angustifolia*], **Samen** [Anis — *Pimpinella anisum* (verwendet als pulverisierte Samenkörner oder Öl), Pfeffer — *Piper nigrum*], Beeren [Wacholder — *Juniperus communis*, Pimentbaum (Nelkenpfefferbaum) — *Pimenta dioíca*], **Zwiebeln** [Schalotte — *Allium ascalonicum*, Koblauch — *Allium sativum*] oder **Wurzeln** [Meerrettich — *Armoracia rusticana*, Kurkuma (Gelbwurzel) — *Cúrcuma longa*] zum Würzen menschlicher Speisen dienen. **Gewürze** werden immer wieder mit **Kräutern** gleichgestellt, obwohl Kräuter zum Würzen streng genommen nur die Blätter von frischen oder getrockneten Pflanzen sind).

1244 cultivar [n] *hort.* (*Abbr.* cv.; technical term derived from the words *cultivated* and *variety*; strain of plant resulting from ►plant breeding, which exhibits new morphologic, physiological, cytological, chemical or other characteristics. The cultivated form retains its typical characteristics after sexual or asexual reproduction. According to the order of rank in botanical nomenclature, a distinction is made between 'varietas' and 'variety' [the cultural form]; **c.** and horticultural variety are used interchangeable terms; cf. Article 10, International Code of Nomenclature for Cultivated Plants [ICNCP 2004]); *syn.* variety [n]; *s* **cultivar** [m] (*Abr.* cv; término técnico derivado de *cultivated variety*; linaje de planta agrícola u hortícola resultante de selección [►fitogenética], que muestra nuevas características morfológicas, fisiológicas, citológicas, químicas o de otro tipo, y que mantiene sus características típicas después de la reproducción sexual o asexual); *syn.* variedad [f] agrícola/hortícola; *f* **cultivar** [m] (Terme international provenant de la contraction du terme anglais *cultivated variety* et désignant une variété d'origine horticole provenant d'une ►sélection végétale, d'une hybridation, du croisement ou de mutation naturelle ou provoquée ; la différence entre un cultivar et une **variété** réside dans le fait que les caractéristiques uniques d'un cultivar ne sont généralement pas transmises d'une génération à l'autre par la semence ; par conséquent, ces plantes doivent être reproduites végétativement) ; *g* **Kultivar** [n] (*Wissenschaftlich* Cultivar; ein von *cultivated variety* abgeleiteter Fachausdruck; *Abk.* cv.; eine sich durch morphologische, physiologische, cytologische, chemische oder andere Merkmale deutlich auszeichnende Varietät einer Pflanze als Ergebnis gärtnerischer ►Pflanzenzüchtung. Diese Kulturform behält seine sortentypischen Merkmale bei geschlechtlicher oder ungeschlechtlicher Fortpflanzung bei. Man sollte zwischen „varietas" als botanische nomenklatorische Rangfolge und der „Varietät" i. S. v. Kulturform/Sorte unterscheiden; cf. Art. 10 Internationaler Code der Nomenklatur der Kulturpflanzen [ICNCP]); *syn.* Cultivar [n], Kulturvarietät [f], Sorte [f], Kulturform [f].

cultivated forest [n] *for.* ►production forest.

1245 cultivated land [n] *agr. for. hort.* (Area used for growing crops, including arable land, grassland, gardens, vineyards, areas where specialized cropping takes place and planted forest); *s* **tierra [f] de cultivo** (Incluye ager, silva y saltus, toda zona de utilización humana); *syn.* tierras [fpl] cultivadas; *f* **terres [fpl] cultivées** (Sol exploité en vue de la production de plantes tels que les champs, prairies, jardins, vignes, vergers, cultures spécialisées et plantations forestières) ; *syn.* terres [fpl] exploitées ; *g* **Kulturland [n, o. Pl.]** (Zum Anbau von Kulturpflanzen [Kulturen] kultivierter Boden. Zum **K.** zählen Ackerland, Grünland, Gärten, Weinberge, Standorte für Sonderkulturen und Forste).

1246 cultivated plant [n] (1) *agr. for. hort.* (Plant species which has been grown and cultivated systematically by man); *s* **planta [f] de cultivo** (Especie vegetal que ha sido cultivada sistemáticamente por la humanidad); *syn.* planta [f] cultivada (1); *f* **plante [f] de culture** (Produit génétique artificiel sélectionné pour sa production et dont l'intérêt agronomique, sylvicole ou horticole est primordial pour l'homme) ; *syn.* plante [f] cultivée (1) ; *g* **Kulturpflanze [f] (1)** (Vom Menschen planmäßig angebaute und der Züchtung unterworfene Pflanzenart; ANL 1984); *syn.* kultivierte Art [f] (1).

1247 cultivated plant [n] (2) *hort. phyt.* (That species, according to the ▶degree of naturalization, with the least naturalness which is grown in ▶plant nurseries for sale, and in parks, gardens and indoors for ornamental use); *syn.* ergasiophyte [n]; *s* **ergasiófito [m]** (Planta exótica con bajo ▶grado de naturalización que se cultiva en ▶empresa de horticultura para su venta, o en parques, jardines o interiores como planta ornamental; cf. DB 1985); *syn.* planta [f] cultivada (2); *f 1* **plante [f] cultivée (2)** (Espèce qui, par son ▶degré de naturalisation, a le taux d'artificialisation le plus élevé et ne se rencontre que sous forme cultivée, se rencontre comme plante ornementale dans les parcs et les jardins ou pour sa production dans une ▶entreprise horticole ; ▶cultivar) ; *syn.* espèce [f] cultivée ; *f 2* **plant [m] cultivé** *hort.* (Par opposition au plant prélevé, plante élevée en pépinière) ; *g* **kultivierte Art [f] (2)** (Nach dem ▶Grad der Einbürgerung die Pflanzenart mit der geringsten Natürlichkeit des Vorkommens; eine Art, die nur in angebautem Zustand vorkommt, meist in ▶Gartenbaubetrieben herangezogen); *syn.* Ergasiophyt [m], kultivierte Pflanzenart [f], Kulturpflanze [f] (2).

1248 cultivated shrub rose [n] *hort. plant.* (Rose, which is cultivated due to its ornamental value as a shrub and not just for its flowers, as in the case of Hybrid Tea roses, which are usually cut down to two or three buds each year; ▶wild rose); *syn.* ornamental rose bush [n] [also UK]; *s* **rosal [m] arbustivo cultivado** (Tipo de rosal que se cultiva por su aspecto decorativo como arbusto y no sólo por sus flores; ▶rosa silvestre); *f* **rosier [m] arbuste d'ornement** (Rosier, qui à cause de son caractère ornemental est entretenu comme un arbuste se développant librement, par comparaison avec les rosiers buissons à fleurs groupées et les rosiers hybrides de thé à taille traditionnelle courte à trois yeux ; ▶rosier sauvage) ; *syn.* rosier [m] buisson d'ornement ; *g* **Zierstrauchrose [f]** (Rose, die wegen ihres Zierwertes als ganzer Strauch habitusgerecht gepflegt wird — anders als Beet- und Teehybriden, die i. d. R. jedes Jahr auf 2-3 Augen zurückgeschnitten werden; ▶Wildrose).

1249 cultivated soil [n] *agr. hort.* (Worked ▶top soil, by digging or tilling, used in crop and plant nursery production); *s* **suelo [m] de cultivo** (Capa de ▶tierra vegetal arada utilizada para la producción de plantas de cultivo); *f* **sol [m] cultivé** (Couche de terre travaillée par les instruments aratoires ; ▶terre végétale) ; *syn.* sol [m] agricole, terrain [m] cultivé ; *g* **Kulturboden [m]** (Landwirtschaftlich oder gärtnerisch bearbeiteter ▶Oberboden).

1250 cultivated terrace [n] *agr.* (Constructed, gently sloping or horizontal agricultural fields on steep slopes of hills and mountains which are stabilized by retaining walls or steep narrow dirt banks; *specific terms* field terrace, terraced paddy rice field, vineyard terrace; ▶terrace cultivation, ▶terracing); *s* **campo [m] en bancales** (Cultivos en bancales; ▶abancalado, ▶cultivo en terrazas); *f* **terrasse [f] de culture** (Large plateforme agricole plane ou légèrement en pente taillée dans un versant et soutenue par une maçonnerie ou délimitée par un talus ; la terrasse peut être parfaitement horizontale surtout si elle est irriguée ; DG 1984, 446 ; ▶aménagement en terrasses, ▶culture en terrasses) ; *g* **Terrassenfeld [n]** (Vom Menschen geschaffenes, flach geneigtes oder waagerechtes, landwirtschaftlich genutztes Feld am steilen Hang, durch Stützmauer oder Steilböschungsabschnitt gegliedert; *UBe* Ackerterrasse, Reisterrasse, Weinbergterrasse; ▶Hangterrassierung, ▶Terrassenanbau).

1251 cultivation [n] *agr. for.* (Management of land by planting and harvesting of cultivated plants; ▶bog cultivation, ▶economic land management, ▶organic farming, ▶permanent cultivation, ▶semiproductive cultivation [US]/semi-productive cultivation [UK], ▶shifting cultivation, ▶soil cultivation, ▶surface cultivation, ▶terrace cultivation, ▶topsoiling and cultivation, ▶traditional farming); *s* **cultivo [m] de tierras** (Parte de la agricultura que se dedica a la explotación de la tierra para producir y cosechar vegetales; ▶cultivo alternativo, ▶cultivo tradicional, ▶gestión agrícola); *f* **cultures [fpl végétales** (Branche de l'agriculture utilisant le travail de la terre pour les productions végétales [céréales, plantes sarclées, oléagineux, légumes, fruits et raisins] ; ▶agriculture traditionnelle, ▶culture biologique, ▶gestion agricole) ; *g* **Landbau [m, o. Pl.]** (Teilbereich der Landwirtschaft, die die regelmäßige Nutzung des Bodens durch Anbau und Ernte von Nutzpflanzen des Acker-, Obst- und Weinbaus betreibt; ▶alternativer Landbau, ▶konventioneller Landbau, ▶Landbewirtschaftung).

1252 cultivation depth [n] *agr. hort. constr.* (Depth to which earth is loosend, either for cultivation of topsoil or for ▶subsoiling); *syn.* loosening depth [n]; *s* **profundidad [f] de desfonde** (Profundidad hasta la cual se descompacta resp. se prevé descompactar el suelo; ▶desfonde); *f* **profondeur [f] de travail du sol** (Travail du sol, profondeur de ▶sous-solage ou de ▶scarifiage ; *termes spécifiques* profondeur de décompactage, profondeur de scarifiage, profondeur de sous-solage) ; *g* **Bearbeitungstiefe [f]** (Tiefe, bis zu der ein Boden gelockert wird resp. gelockert werden soll; ▶Tiefenlockerung); *syn.* Lockerungstiefe [f].

1253 cultivation [n] **of arable land** *agr.* (Tilling to produce favo[u]rable soil conditions for the growth of crop plants); *syn.* crop farming [n] [also US], crop raising [n]; *s* **cultivo [m] de tierras** (Todas las medidas agrícolas para conseguir condiciones de crecimiento favorables para las plantas); *f* **culture [f] des champs** (Techniques culturales en vue de l'établissement de conditions favorables au développement des plantes cultivées) ; *g* **Ackerbau [m, o. Pl.]** (Maßnahmen des Pflanzenbaus auf dem Acker zur Herstellung günstiger Wachstumsbedingungen für Kulturpflanzen); *syn.* Agrikultur [f], Feldbau [m].

1254 cultivation [n] **of woody plants** *arb. hort.* (Commercial or scientific cultivation of trees and shrubs in nurseries by seeds, vegetative propagation, and selection; ▶arboriculture); *s* **cultivo [m] de leñosas** (Cultivo de plantas leñosas para fines comerciales o científicos en viveros por medio de siembra, multiplicación vegetativa o selección; ▶arboricultura); *f* **élevage [m]**

des végétaux ligneux (Ensemble des techniques de production [semis, multiplication végétative et sélection] des végétaux ligneux en pépinière à des fins commerciales ou scientifiques ; ►arboriculture) ; *syn.* culture [f] des végétaux ligneux ; *g* **Gehölzanzucht [f, o. Pl.]** (Das Heranziehen von Bäumen und Sträuchern für gewerbliche oder wissenschaftliche Zwecke in Baumschulen bis zur verkaufsfertigen Ware durch Aussaat, vegetative Vermehrung und Auslese; ►Anzucht und Pflege von Freilandgehölzen); *syn.* Anzucht [f, o. Pl.] von Baumschulware, Anzucht [f, o. Pl.] von Gehölzen.

cultural asset [n] [UK] *conserv'hist.* ►cultural resource [US].

1255 cultural facilities [npl] *leg. urb.* (Such public establishments as museums, libraries, art galleries, botanical and zoological gardens of historic, educational, or artistic interest); *s* **instalaciones [fpl] culturales** (Equipamiento público para actividades culturales, como museos, salas de exposiciones, bibliotecas, cines, teatros, auditorios, centros culturales, centros de jóvenes, jardines botánicos o zoológicos. En Es. los templos de culto no forman parte de las **i. c.**, sino del equipamiento religioso); *f* **zone [f] réservée aux équipements culturels** (Espace réservé à des établissements, équipements ou édifices culturel tels que théâtres, musées, bibliothèques, galeries d'art, jardins botaniques et zoologiques, édifices de cultes; **F.**, dans le plan local d'urbanisme [PLU]/plan d'occupation des sols [POS] ces espaces sont en général prévus pour les équipements sociaux, scolaires, sportifs, de loisirs ; leur dénomination varie d'une commune à l'autre ; *terme spécifique* **zone réservée aux édifices de culte [B]**) ; *syn.* zone [f] à vocation d'équipements sociaux-culturels, zone [f] réservée aux équipements culturels et sportifs ; *g* **kulturellen Zwecken dienende Gebäude [npl] und Einrichtungen [fpl]** (Im Bebauungsplan auszuweisende Flächen für den Gemeinbedarf; cf. Anlage zur PlanzV 90, Kap. 4).

cultural features [npl] *land'man. landsc.* ►small cultural features.

1256 cultural heritage [n] *conserv'hist.* (►Monument or historic area of outstanding value from the point of view of history, art or science, or ►cultural site of outstanding value from historical, aesthetic, ethnological, or anthropological points of view; cf. Paris Convention of 1972; ►conservation of the natural and cultural heritage [CH], ►cultural resource [US]/cultural asset [UK], ►historic district [US]/conservation area [UK], ►preservation of cultural resources); *syn.* cultural patrimony [n]; *s 1* **patrimonio [m] cultural** (Monumento o conjunto histórico de gran valor histórico, artístico o científico, o lugar de interés cultural de gran valor histórico, estético, etnológico o antropológico; cf. Convención de París de 1972); *s 2* **Patrimonio [m] Histórico Español** (Todos los bienes de valor histórico, artístico, cultural o técnico que conforman la aportación de España a la cultura universal. El **PHE** está constituido por los inmuebles y objetos de interés artístico, histórico, paleontológico, arqueológico, etnográfico, científico o técnico, así como por el patrimonio documental y bibliográfico, los yacimientos y zonas arqueológicas y los sitios naturales, jardines y parques, que tengan valor artístico, histórico o antropológico. Ley 6/1985, preámbulo y art. 1 [2]; ►bién de interés cultural, ►Conjunto Historico [Es], ►lugar de interés cultural, ►protección del patrimonio histórico [español], ►Registro General de Bienes de Interés Cultural; ►Convención de París); *f* **patrimoine [m] culturel** (Un ►monument ou un ensemble présentant une valeur exceptionnelle du point de vue de l'histoire, de l'art ou de la science, ou ►site culturel présentant une valeur exceptionnelle du point de vue historique, esthétique, ethnologique ou anthropologique ; cf. Convention de Paris de 1972 ; ►biens culturels, ►protection du patrimoine culturel national, ►secteur sauvegardé, ►zone de protection du patrimoine architectural, urbain et paysager

[Z.P.P.A.U.P.]) ; *syn.* patrimoine [m] historique et esthétique ; *g* **Kulturerbe [n, o. Pl.]** (Geschichtlich, künstlerisch oder wissenschaftlich bedeutsames ►Denkmal oder Ensemble [►flächenhaftes Kulturdenkmal] sowie geschichtlich, ästhetisch, ethnologisch oder anthropologisch bedeutsame ►Kulturstätte; cf. Pariser Übereinkommen von 1972; Kulturdenkmale, die auf der Liste der UN-Kulturorganisation in Paris stehen, gehören zum Welterbe — 2005 ca. 790 Denkmale in 134 Ländern; ►Kulturgut, ►Kulturgüterschutz); *syn.* kulturelles Erbe [n].

cultural heritage landscape [n] *conserv.* ►protected cultural heritage landscape.

1257 cultural heritage [n] **of historic gardens** *conserv'hist. gard'hist.* (Significant historical, artistic or scientific value of ►historic gardens); *s* **patrimonio [m] de jardines históricos** (Conjunto de jardines de importancia histórica, cultural o científica; cf. Ley 16/1985, ►jardín histórico); *f* **patrimoine [m] culturel des parcs et jardins** (Ensemble des ►jardins historiques d'ordre historique, esthétique, scientifique) ; *g* **gartenkulturelles Erbe [n, o. Pl.]** (Geschichtlich, künstlerisch oder wissenschaftlich bedeutsames ►Gartendenkmal oder die Summe solcher Denkmale); *syn.* kulturelles Erbe [n] der Gartenkunst.

1258 cultural landscape [n] *geo.* (**1.** Landscape area which has been largely or completely transformed from its natural state by human intervention. Generic term covering ►agrarian cultural landscape, ►agricultural landscape, ►historic landscape, ►industrial landscape, ►near-natural landscape, ►urban landscape. **2.** Geographic area [including both cultural and natural resources and the wildlife or domestic animals therein] associated with a historic event, activity, or person or exhibiting other cultural or aesthetic values. There are four general types of **c. l.s**, not mutually exclusive: historic site, historic designed landscape, historic vernacular landscape, and ethnographic landscape; LA 1997 [5], 81); *syn.* man-made landscape [n]; *s* **paisaje [m] humanizado** (Paisaje cuyo aspecto se debe a transformaciones causadas por el hombre a lo largo del tiempo. Término genérico para ►paisaje humanizado histórico, ►paisaje industrial, ►paisaje rural 1, ►paisaje rural 2, ►paisaje urbano); *syn.* paisaje [m] cultural, paisaje [m] antropógeno; *f* **paysage [m] culturel** (**1.** Terme utilisé par les géographes allemands et anglo-saxons pour décrire un paysage dont les composantes naturelles ont été modifiées par l'action modélatrice de l'homme, considéré dans sa totalité et intégrant la perception des éléments sensibles de l'environnement, rendant compte des incidences qu'ont pu avoir sur les espaces naturels les diverses pratiques de l'homme [agriculture, urbanisme, etc.]. **2.** Le **p. c.** désigne un paysage qui, à un moment donné de son histoire, a servi de modèle ou de source d'inspiration à des artistes ; TP 1997, 52 ; terme générique englobant le ►paysage culturel faiblement artificialisé, le ►paysage culturel historique, le ►paysage historique, le ►paysage industriel, le ►paysage patrimoine, le ►paysage rural, le ►paysage rural traditionnel, et le ►paysage urbain ; *opp.* paysage naturel) ; *syn.* paysage [m] créé, paysage [m] de création humaine, paysage [m] humanisé ; *g* **Kulturlandschaft [f]** (Landschaft, deren natürlicher Zustand, hervorgeangen aus Geologie, Topographie, Klima und Boden, weit gehend oder völlig durch den Menschen in Jahrhunderten geprägt wurde. Verändert wurden die Landschaften vor allem durch die Landwirtschaft, die für immer mehr Menschen Nahrung bereitstellen musste. Trotz aller Veränderungen wurde Mitteleuropa zu einer der ökologisch stabilsten Regionen der Welt. Fast nirgendwo sonst wird schon seit 7000 Jahren ohne größere Krise Ackerbau betrieben; in D. wird schon seit langem überlegt, durch finanzielle Förderung einer extensiven, nachhaltigen landwirtschaftlichen Nutzung, den Schutz, die Pflege und

Entwicklung der **K.** zu fördern, statt die den naturschutzfachlichen Zielen zuwiderlaufenden Produktionsweisen mit Agrarüberschüssen zu finanzieren. *UBe* ▶Agrarlandschaft, ▶Industrielandschaft und ▶Stadtlandschaft sowie ▶bäuerliche Kulturlandschaft, ▶historische Kulturlandschaft 1, ▶historische Kulturlandschaft 2 und ▶naturnahe Kulturlandschaft).

1259 cultural monument [n] *conserv. hist.* (Generic term for a built object of historical, cultural, architectural or aesthetic interest, typically a building or structure; **in U.S.**, objects of sufficient public interest to be worthy of grant or other form of public protection would normally be listed in a local landmark ordinance, a state registry or the National Register of Historic Places—perhaps all three; ▶historic structure, ▶buried cultural monument, ▶listed landmark building [US]/listed building [UK], ▶historic district [US]/conservation area [UK], ▶Registry of Historic Landmarks [US]); *syn.* historic landmark [n] [also US]; *s* **monumento** [m] (Bien inmueble que constituye una realización arquitectónica o de ingeniería, u obra de escultura colosal siempre que tenga interés histórico, artístico, científico o social; art 15 [1] Ley 16/1985. La declaración como **m.** implica su declaración como ▶bien de interés cultural; art 14 [2] Ley 16/1985; ▶Registro General de Bienes de Interés Cultural [Es]; *términos específicos* ▶Conjunto Histórico, ▶edificio declarado monumento, ▶edificio histórico-artístico; ▶Inventario General de Bienes Muebles, ▶restos arqueológicos cubiertos); *f* **monument** [m] **historique (1)** (Immeuble dont la conservation présente du point de vue de l'histoire ou de l'art un intérêt public et placé sous le régime de protection instauré par la loi du 31 décembre 1913 sur les monuments historiques, modifiée par les lois du 25 février 1943 et du 30 décembre 1966 ; les immeubles sont soit inscrits sur la ▶liste générale des monuments classés [monument classé] soit inscrit à l'▶inventaire supplémentaire des monuments historiques [monument inscrit] ou à l'▶inventaire général des monuments et richesses artistiques de la France ; le classement et l'inscription sont désormais régis par le titre II du Code du patrimoine, qui remplace, la loi du 25 février 1943 ; ▶bâtiment classé ou inscrit au titre des monuments historiques, ▶bâtiment désigné, ▶bâtiment protégé, ▶secteur sauvegardé, ▶terrain qui renferme des stations ou des gisements préhistoriques, ▶zone de protection du patrimoine architectural, urbain et paysager) ; *syn.* édifice [m] classé ou inscrit au titre des monuments historiques ; *g* **Kulturdenkmal** [n] (...mäler [pl], in A nur so, sonst auch ...male [pl]; Objekt öffentlichen Interesses der Architektur [einschließlich ganzer Ensembles], Geschichte, Kunst, Technik, Vorgeschichte etc.; meist durch ▶Denkmalbuch inventarisiert und durch Denkmalschutzgesetze der Bundesländer geschützt; *UBe* ▶Baudenkmal, ▶Bodendenkmal, ▶denkmalgeschütztes Gebäude, ▶flächenhaftes Kulturdenkmal).

cultural monument [n]**, soil-covered** *conserv'hist. land' man.* ▶buried cultural monument.

cultural patrimony [n] *conserv'hist.* ▶cultural heritage.

1260 cultural resource [n] [US] *conserv'hist.* (Valuable item of historic significance belonging to ▶cultural heritage, such as places, buildings, sites, districts and objects of architectural, historical, landscape, ethnic, or archaeological interest; increasingly considered to include intangible as well as tangible resources); *syn.* cultural asset [n] [UK]; *s* **bien** [m] **de interés cultural** [Es] (Bien integrante del ▶Patrimonio Histórico Español declarado como tal por ministerio de la Ley 16/1985 de 25 de Junio, del Patrimonio Histórico Español o mediante Real Decreto de forma individualizada; cf. art. 9 [1]); *f* **biens** [mpl] **culturels** (Biens présentant un intérêt artistique, préhistorique, archéologique ou historique, scientifique, ethnologique, légendaire ou pittoresque et appartenant au ▶patrimoine culturel ; les **b. c.** sont enregistrés sur l'inventaire des monuments et richesses

artistiques de la France) ; *g* **Kulturgut** [n] (...güter [pl]; bewegliches oder unbewegliches Objekt, das für das kulturelle Erbe [▶Kulturerbe] aller Völker von großer Bedeutung ist [z. B. Bauund Kunstdenkmäler, archäologische Stätten], Baulichkeiten, die der Erhaltung oder Ausstellung beweglichen Kulturgutes dienen [z. B. Museen, Bibliotheken, Archive, archivalische Sammlungen, Nachlässe und Briefsammlungen mit wesentlicher Bedeutung für die jeweilige nationale politische, Kultur- und Wirtschaftsgeschichte], Bergungsorte, in denen im Falle bewaffneter Konflikte das bewegliche **K.** in Sicherheit gebracht werden soll sowie Denkmalorte, die in beträchtlichem Umfang **K.**güter aufweisen; cf. Art. 1 Haager Konvention [HK]; **K.** kann mit dem Kulturgutschutz-Kennzeichen versehen werden; Art. 6, 16 f HK; in D. wird hochrangiges **K.** von den Ländern in ein „Verzeichnis national wertvollen Kulturgutes" eingetragen; § 1 (1) KG. Der Bundesminister des Innern führt ein aus den Verzeichnissen der einzelnen Ländern gebildetes „Gesamtverzeichnis national wertvollen Kulturgutes"; § 6 [2] Gesetz zum Schutz deutschen Kulturgutes gegen Abwanderung vom 08.07.1999; BGBl. I S. 1754; zuletzt geändert durch Artikel 2 des Gesetzes vom 18.05.2007; BGBl. I S. 757).

cultural resources [npl] *conserv' hist.* ▶preservation of cultural resources.

1261 cultural site [n] *conserv'hist.* (Object or area of public interest such as a ▶cultural monument in harmony with its manmade and natural environment; ▶cultural heritage, ▶historic site); *s* **lugar** [m] **de interés cultural** (▶monumento, ▶patrimonio cultural, ▶Patrimonio Histórico Español, ▶sitio histórico); *f* **site** [m] **culturel** (Immeuble, ensemble et espaces présentant un intérêt général tels que ▶monuments historiques et leurs abords naturels ou bâtis, sites archéologiques ; ▶patrimoine culturel, ▶site historique culturel) ; *syn.* lieu [m] culturel ; *g* **Kulturstätte** [f] (Objekt oder Fläche öffentlichen Interesses wie ▶Kulturdenkmal oder gemeinsames Werk von Mensch und Natur und die dazugehörende Fläche sowie jede archäologische Stätte; ▶Kulturerbe, ▶kulturhistorische Stätte).

1262 culvert [n] *constr. trans.* (Generic term for any drain, which carries small streams or ditch water beneath a path or road embankment. A ▶pipe culvert has a segmental, full circular, or elliptical arch cross-section, while a ▶box culvert has a rectangular cross-section, and an ▶arch culvert has a flat bottom); *s* **alcantarilla** [f] (Término genérico para la parte de arroyo entubado o de dren que transcurre debajo de un camino, carretera o dique. Dependiendo del tipo de construcción se diferencian en ▶alcantarilla abovedada, ▶alcantarilla tubular y **a.** cubierta con placas [▶alcantarilla de cajón]; cf. BU 1959); *syn.* puentecillo [m], puente [m] de alcantarilla, atarjea [f] [Es], pontón [m] [Es, YV], tajea [f] [C], conducto [m] pluvial [RA]; *f* **passage** [m] **busé** (Partie enterrée d'un ruisseau, d'un canal d'assainissement, d'une voie sous un chemin, une route, une digue constitué d'un dalot rectangulaire ou d'une buse circulaire, etc. ; suivant la nature de l'ouvrage on distingue les termes de ▶passage voûté, ▶passage busé circulaire, ▶passage couvert rectangulaire ; un tunnel de circulation piétonne sous une voie de communication est dénommé « **passage souterrain** ») ; *syn.* dalot [m] ; *g* **Durchlass** [m] (Verdolter Teil eines offenen Baches oder Entwässerungsgrabens unter einem Weg, einer Straße oder einem Dammbauwerk. Je nach Bauart werden ▶**Gewölbedurchlass**, **Rohrdurchlass** [▶runder Durchlass] oder **Plattendurchlass** [▶rechteckiger Durchlass] unterschieden. Ein **D.** für Fußgängerund Radfahrverkehr wird i. d. R. **Unterführung** genannt).

culverted stream [n] [UK] *wat'man.* ▶piped stream [US].

1263 culverting [n] *wat'man.* (Replacement of an open watercourse by canalization in a closed, underground pipe; ▶piped stream; *opp.* uncovering of a culverted stream); *syn.* pipe [vb] ditches or watercourses, piping [n] of ditches or watercourses; *s* **entubado [m] de un arroyo** (Medida de contrucción hidráulica de canalizar un curso de agua pequeño haciéndole cursar por una tubería; ►arroyo entubado); *f* **busage [m]** (Déplacement d'un cours d'eau ouvert et passage des eaux dans une canalisation enterrée de gros diamètre ; ►rivière busée ; *opp.* remise à jour d'un cours d'eau) ; *g* **Verdolung [f]** (Gewässerbaumaßnahme, bei der ein offenes Bachgerinne in eine geschlossene, unterirdische Leitung verlegt wird; ►verdolter Bach; *opp.* Offenlegung eines Baches); *syn.* Bachverrohrung [f], Verrohrung [f] eines Baches.

curb [n] [US] *constr. eng.* ►barrier curb [US], ►battered curb(stone) [US], ►drop curb [US], ►mountable curb.

curb [n] [US]**, batter-faced** *constr.* ►battered curb(stone) [US].

curb [n] [US]**, depressed** *constr.* ►drop curb [US].

curb [n] [US]**, flared** *constr. eng.* ►mountable curb [US]/ splayed kerb [UK].

curb [n] [US]**, flush** *constr.* ►drop curb [US].

curb [n] **and grate** [n] *constr.* ►combined curb and grate [US].

1264 curb [n] **and gutter** [n] [US] *constr.* (Integral component of a monolithic system for curb [US]/kerb [UK] and ►gutter installation; ►dished channel unit); *syn.* kerbstone [n] with gutter [UK]; *s* **pieza [f] de cuneta angular** (Pieza prefabricada en forma de ángulo para la construcción de ►cunetas; ►pieza de cuneta); *f* **bordure [f] caniveau d'épaulement** (Élément de maçonnerie préfabriqué en forme de L destiné à conduire les eaux de ruissellement du ►caniveau vers des exutoires ; selon les normes NF, ces bordures sont désignées par l'appellation AC1 et AC2 ; ►bordure caniveau, ►bordure fil d'eau) ; *syn.* caniveaucoudé [m] ; *g* **Winkelrinnenstein [m]** (Winkelförmiges, Industriell gefertigtes Bordsteinelement; ►Rinnenstein, ►Rinnstein).

1265 curb cut [n] [US] *constr.* (Section of depressed curbstones; e.g. across an entrance drive or at road intersections; ►mountable curb [US]/splayed curb [UK], ►driveway cut [US]/ sunken driveway [UK]); *syn.* drop kerb [n] [UK]; *s* **bordillo [m] hundido** (Tramo de bordillo rebajado en acceso a garaje o a un terreno; ►bordillo accesible, ►acera rebajada); *syn.* zona [f] de bordillo [m] enterrado, zona [f] de cordón enterrado [CS]; *f* **zone [f] d'abaissement de bordure de trottoir** (Abaissement des trottoirs au niveau zéro [raccordement zéro] pour les rendre franchissable pour les cyclistes ou les personnes à mobilité réduite [PMR] au moyen de ►bordures basses franchissables ; ►bateau [d'entrée]) ; *g* **Bordsteinabsenkung [f]** (Bereich der abgesenkten Bordsteine an einer Grundstückseinfahrt oder an Straßenkreuzungen für Rollstühle und Radfahrer; ►Flachbord, ►Gehwegabsenkung); *syn.* Absenkung [f] von Bordsteinen, Wiege [f] [BW].

1266 curb height [n] [US] *constr. syn.* gutter depth [n] [also US]; *syn.* kerb height [n] [UK]; *s* **altura [f] del bordillo**; *f* **hauteur [f] de la bordure** (Différence de hauteur entre la chaussée et le trottoir) ; *g* **Bordsteinhöhe [f] (1)** (Höhendifferenz zwischen Fahrbahn und Gehweg).

1267 curbing [n] [US] *constr.* (Kerbs/curbs used as enclosure for planting beds or ►tree pit edgings [US]/tree circle edging [UK]); *syn.* kerb edging [n] [UK]; *s* **encintado [m] con piedra de bordillo** (Delimitación de alcorques, macizos de flores, etc. con piedras de bordillo; ►recintados de alcorque); *f* **rebord [m] d'entourage** (Délimitation de plate-bande, ►entourage d'arbre au moyen de bordures de voirie) ; *syn.* bordure [f] d'entourage ;

g **Bordsteineinfassung [f]** (Abgrenzung von Beeten, Baumscheiben etc. mit Bordsteinen; ►Baumscheibeneinfassung).

1268 curb inlet [n] [US] *constr.* (TSS 1988, 330-15; opening of a drainage inlet structure integrated with curb [US]/kerbstone [UK]; ►combined curb and grate [US]/combined inlet [UK]); *syn.* kerb-inlet [n] [UK] (SPON 1986, 169), storm water curb opening [n] [US]; *s* **sumidero [m] en bordillo** (Entrada de alcantarilla incrustada en el bordillo de la acera; ►sumidero combinado); *f* **bordure [f] avaloir** (Bouche d'évacuation des eaux pluviales, absorption par ouverture verticale dans la bordure trottoir ; ►regard à grille-avaloir) ; *g* **Straßenablauf [m] im Bordstein** (In den Bordstein eingearbeitete Einlauföffnung; ►kombinierter Straßen-Bordsteinablauf); *syn.* Bordsteinablauf [m].

1269 curb level [n] [US] *constr. syn.* kerb level [n] [UK], top [n] of the curb [US]; *s* **nivel [m] del bordillo**; *f* **niveau [m] fini à arase bordure** ; *g* **Bordsteinhöhe [f] (2)** (Obere Kante eines Bordsteines als Anschlussniveau); *syn.* Oberkante [f] Bordstein, Bordsteinoberkante [f].

curb opening [n] [US]**, storm water** *constr.* ►curb inlet [US]/kerb-inlet [UK].

1270 curbstone [n] [US] *constr.* (Generic term covering ►mountable curb [US]/splayed kerb [UK], ►barrier curb [US]/ edge kerb [UK], ►drop curb [US]/flush kerb [UK]; ►edging stone); *syn.* kerbstone [n] [UK]; *s* **piedra [f] de bordillo** (Término genérico para elemento individual de piedra natural u hormigón para separar el pavimento de la calzada del de la acera o del carril de bicis más elevado; *términos específicos* ►bordillo accesible, ►bordillo de acera, ►piedra de bordillo enterrado; ►piedra de encintado); *syn.* piedra [f] de encintado (de acera), piedra [f] de cordón [RA]; *f* **élément [m] de bordure** (Terme générique caractérisant un élément en pierre naturelle ou en béton destiné à séparer la chaussée des voies spécialisées, îlots, terrepleins, trottoirs, pistes, allées ou à contenir [dispositif de retenue] les matériaux qu'il borde ; CCTG, fascicule 31 ; *termes spécifiques* ►bordure arasée d'épaulement, ►bordure basse franchissable, ►bordure de jardin [type P], ►bordure de trottoir) ; *syn.* bordure [f] de voirie ; *g* **Bordstein [m]** (Element aus Naturstein oder Beton zur Abgrenzung von Straßenbelag und Bürgersteig oder zum höher gelegenen Radweg; Die Bordhöhe beträgt je nach Funktion 5-20 cm; ►Flachbord; *UBe* ►Hochbord, ►Kantenstein 2, ►Tiefbordstein); *syn. ndt.* Kantstein [m], *syn. süddt. auch* Randstein [m], *ostdt. auch* Kantenstein [m] (1).

current [n] *envir. ecol.* ►solar-generated current.

current season's shoot [n] *hort.* ►flowers on current season's shoot.

1271 current state [n] **of a landscape** *landsc.* (G+L 1988 [8]; existing conditions and structure of a landscape, i.e. at the time when a ►landscape plan is being prepared); *syn.* existing conditions [npl] of landscape; *s* **estado [m] actual de un paisaje** (Condiciones existentes y estructura de un paisaje p. ej. cuando se prepara un ►plan de gestión del paisaje); *f* **état [m] d'un paysage** (Ensemble de caractéristiques naturelles, culturelles et symboliques s'appliquant ou appliquées à un paysage déterminé constituant la base d'une évaluation du paysage dans le cadre d'une ►étude paysagère ; cf. observatoire du paysage, www.catpaisatge.net ; *g* **Landschaftszustand [m]** (Beschaffenheit und Struktur einer Landschaft, z. B. bei Aufstellung eines ►Landschaftsplanes; cf. § 11 LG-NW).

curtailment [n] **of agricultural production** [US] *agr. pol.* ►agricultural reduction program [US]/extensification of agricultural production [UK].

curtilage [n] *plan. urb.* ►unbuilt yard.

curve [n] *constr. pedol.* ►particle-size distribution curve.

curvilinear [adj] *gard. plan.* ►designed in an informal manner.

1272 cushion plant [n] *bot. hort. plant.* (Plant with negatively geotropic shoots which form a compact dome-shaped shield, growing close to the ground; ►surface plant); *s* **planta** [f] **pulviniforme** (►Caméfito *[Chamaephyta pulvinata]* con brotes de geotropismo negativo que crecen muy cerca del suelo formando cojinetes compactos); *syn.* planta [f] en almohadillada; *f* **plante** [f] **(vivace) en coussin(et)** (►Chaméphyte dont les bourgeons ont un géotropisme négatif et forment un épais bouclier au centre de la plante) ; *syn.* plante [f] pulviniforme, plante [f] coussinante ; *g* **Polsterpflanze** [f] (*Chamaephyta pulvinata* ►Oberflächenpflanze mit negativ geotropischen und dicht zusammenschließenden Trieben, von niedrigem, kugeligem oder halbkugeligem Wuchs, meist ohne Ausläufer und oft immergrün. Bevorzugte Verwendung in der Bepflanzungsplanung für Steinanlagen und Stein-/Mauerfugen [oft konkurrenzschwache Stress-Strategen], wie z. B. Blaukissen *[Aubrietia]*, Steinbrech *[Saxifraga ferdinandi-coburgii]*); *syn.* kissenbildender Bodendecker [m], *o. V.* Kissen bildender Bodendecker [m], Polsterstaude [f].

1273 cushion shrubland [n] *phyt.* (More or less isolated clumps of dwarf shrubs forming dense cushions, often equipped with thorns; e.g. Astragalus- and Acantholimon 'porcupine' heath in the East-Mediterranean mountains (ELL 1967), in subalpine areas of Sinai and in Jordan the class of *Astragaletea mediterranea* with *Astragalus bethlehemiticus, A. deinacanthus, A. hermoneus, Noaea mucronata* and *Ononis antiquorum* as well as the spiny Tragacanthic dwarf-shrub vegetation in the Middle East dominated by *Acantholimon libanoticum* and *Onobrychis cornuta*; ZOH 1973); *s* **formación** [f] **pulvinular** (Grupos de arbustos enanos, a menudo con espinas, que forman almohadas densas, irregularmente repartidas; tipo de vegetación de las montañas de la región oriental del Mediterráneo; *término específico para formaciones de Erinacetalia* formación pulvinular de erizos); *f* **formation** [f] **d'arbustes nains en coussinets** (±) (Groupes de petits arbustes, souvent épineux, à répartition irrégulière formant, dans les régions désertiques ou de steppes, des coussinets serrés, p. ex. dans l'Est de la Méditerranée l'association de l'Erinacea [ordre des *Erinacetalia*], la classe de l'*Astragalia mediterranea* dans l'étage subalpin dans le Sinai et en Jordanie avec *Astragalus bethlehemicus, A. deinacanthus, A. hermoneus, Noaea mucronata* et *Ononis antiquorum* ainsi qu'au Moyen Orient la formation d'arbustes du *Tragacantha* avec *Acantholimon libanoticum* et *Onobrychis cornuta*) ; *g* **Zwergstrauchpolster-Formation** [f] (Unregelmäßig verteilte, oft dornige Zwergstrauchgruppen, die in Halbwüsten und Steppen dichte Polster bilden, z. B. im östlichen Mittelmeergebiet die Igelginstergesellschaft der Ordnung *Erinacetalia*, die Klasse der *Astragaletea mediterranea* in subalpinen Gebieten im Sinai und in Jordanien mit *Astragalus bethlehemiticus, A. deinacanthus, A. hermoneus, Noaea mucronata* und *Ononis antiquorum* sowie im Mittleren Osten die dornige *Tragacantha*-Zwergstrauchgesellschaft mit z. B. *Acantholimon libanoticum* und *Onobrychis cornuta*; ZOH 1973).

1274 cut [n] (1) *constr. eng.* (Process of soil or rock removal; ►cut material, ►cut shelf, ►earth-moving, ►excavation 1, ►linear excavation [US]/linear cutting [UK], ►overburden 2); *s* **desmonte** [m] (1) (Proceso de excavar para una obra de construcción; ►cascote, ►corte en el terreno, ►excavación de tierras o rocas, ►material de desmonte, ►movimiento de tierras, ►perfil de desmonte [en una pendiente]); *syn.* excavación [f]; *f* **déblaiement** [m] (Action de mouvement des matériaux ; ►affouillement, ►déblai, ►excavation linéaire, ►fouille,

►matériaux de recouvrement, ►mouvement de terre, ►profil de déblai) ; *syn.* décapage [m], dégagement [m] ; *g* **Abtrag** [m] (1) (Vorgang des Abtragens; ►Abgrabung, ►Abraum, ►Abtrag 2, ►Anschnitt, ►Bodenbewegung, ►Einschnitt).

cut [n] (2) *constr.* ►curb cut [US]/drop kerb [UK]; *constr. urb.* ►driveway cut [US]/sunken driveway [UK]; *constr. for.* ►improvement cut; *constr.* ►late cut.

cut [n] [UK], **annual** *agr. constr. hort* ►single mowing [US].

cut [n], **branch collar** *arb. hort.* ►cut at the branch attachment.

cut [n] [US], **exposed** *constr.* ►incised slope.

cut [n], **flush** *arb. hort.* ►cut at the branch attachment.

cut [n], **rejuvenation** *arb. constr. hort.* ►rejuvenation pruning.

1275 cut a few times per year [loc] *agr. constr.* (Descriptive term applied to meadows, which are mown on two or three occasions annually; ►be mown once a year, ►meadow mown several times per year, ►more frequent mowing); *s* **segado/a varias veces al año** [loc] (*Contexto* ►pradera de siega múltiple; ►segado/a anualmente, ►siega múltiple); *f* **fauché, ée plusieurs fois par an** [loc] (Pré, prairie dont le régime de fauche prévoit plusieurs coupes sur la même surface pendant la période de végétation ; ►fauche exécutée plusieurs fois par an, ►fauché, ée une fois par an, ►prairie fauchée plusieurs fois par an) ; *g* **mehrschürig** [adj] (Einen Rasen oder eine Wiese betreffend, die mehrmals pro Vegetationsperiode gemäht wird: mehrmals zu mähender Rasen/zu mähende Wiese; ►einschürig, ►mehrmalige Mahd, ►mehrschürige Wiese).

1276 cut and cover excavation [n] [US] *eng.* (Construction method for installation of a tunnel by open cut rather than by underground boring; ►cut and cover tunnel); *syn.* cut and cover tunnelling [n] [UK], open cut technique [n]; *s* **construcción** [f] **de túneles a zanja abierta** (Método de construcción de túneles en el cual se construye a cielo abierto y se cubre posteriormente después de haber sido terminado con un techo de hormigón; ►túnel construido a zanja abierta); *syn.* técnica [f] (de construcción de túneles) de abrir y tapar; *f* **méthode** [f] **de construction de tunnel en tranchée ouverte** (±) (Technique de construction de tunnels ; ouvrage réalisé en tranchée ouverte et recouverte par une dalle en béton ; ►tunnel en tranchée ouverte) ; *g* **Deckelbauweise** [f] (Verfahren beim Tunnelbau, bei dem statt mit bergmännischem Vortrieb die Baugrube nach Fertigstellung mit einer Betondecke geschlossen wird; ►Unterpflastertunnel); *syn.* offene Bauweise [f] (1).

1277 cut and cover tunnel [n] *trans.* (Tunnel which is built in the open and covered after completion); *s* **túnel** [m] **construido a zanja abierta** *syn.* túnel [m] de abrir y tapar (BU 1959); *f* **tunnel** [m] **en tranchée ouverte** (Tunnel construit en excavation ou fouille linéaire recouverte après exécution) ; *g* **Unterpflastertunnel** [m] (Im Einschnitt oder in offener Baugrube gebauter Tunnel, der nach Fertigstellung überdeckt wird).

cut and cover tunnelling [n] [UK] *eng.* ►cut and cover excavation [US].

1278 cut and fill [n] *constr.* (Grading to change the existing ground surface through excavation and placement of soil to achieve new elevations; ►balance of cut and fill; ►cut material, ►earth fill, ►volume of cut and fill); *s* **desmonte** [m] **y terraplén** [m] (Modificación del nivel original del terreno por ►desmonte o ►relleno de tierra; ►compensación de desmontes y terraplenes, ►material de desmonte, ►volumen de tierras [a mover]); *syn.* corte [m] y terraplén [m], excavación [m] y terraplén [m]; *f* **déblai** [m] **et remblai** [m] (Modification du profil d'un terrain par enlèvement de terres lors de la réalisation de

C

fouilles, par apport de terres en comblement de fouilles ; ►déblai, ►déblai de masse, ►équilibrage déblai et remblai, ►équilibre des mouvements de terre, ►remblai de terres, ►terrassement en remblai) ; *g* **Auftrag** [m] **und Abtrag** [m] (*Abk.* Auf- und Abtrag; Veränderung des Oberflächenprofils durch Abschieben und Aushub von Boden sowie durch ►Bodenauftrag 1 an anderer Stelle; ►Abtrag 2, ►Bodenmasse, ►Massenausgleich).

cut area [n] [US] *constr.* ►linear excavation [US].

1279 cut [n] **at the branch attachment** *arb. hort.* (Two cuts are distinguished: **1. flush cut:** Cut directly at the trunk without leaving a branch collar; ARB 1983, 399. [*NOTE* it is more commonly recommended to avoid flush cuts and to use cuts at the branch collar for ease of healing], and **2. cut at the branch collar:** Cut just beyond outer portion of a branch collar or branch bark ridges; ARB 1983, 398); *syn.* branch collar cut [n]; *s* **poda** [f] **de una rama por la base** (Se distinguen dos tipos: **1. poda** [f] **al ras del tronco:** corte directamente al nivel del tronco sin dejar espolón. Este tipo de corte es generalmente 2-3 veces mayor en superficie que en la espolonadura, por lo que a menudo no se forma bien el callo cicatricial. Esto puede conducir a que no se cierre bien la herida y la planta sufra la invasión de hongos **2. espolonadura** [f]**:** corte de una rama casi por la base, dejando un espolón de dos o tres yemas); *f* **coupe** [f] **exécutée à l'empattement** (Suppression d'un rameau ou d'une branche ; *suivant l'emplacement de la coupe on distingue* **1. la coupe à ras de tronc** : Suppression d'un rameau ou d'une branche sans laisser de chicot ou de ride de l'écorce ; cette coupe est en générale 2 à 3 fois plus importante que la coupe exécutée à l'aisselle, le cal se forme de façon incomplète occasionnant en règle général des plaies sur les côtés de la blessure ; *syn.* coupe [f] au ras du tronc). **2. La coupe exécutée à l'aisselle :** Suppression d'une branche au niveau d'un empattement et réalisée sur un plan allant du bord externe de la ride de l'écorce au sommet du col de la branche et permettant, en comparaison avec la coupe à ras de tronc, la formation plus rapide du bourrelet cicatriciel ; IDF 1988) ; *g* **Schnitt** [m] **an der Astgabel** (Es werden zwei Schnitte unterschieden: **1. stammparalleler Schnitt:** Abschneiden eines Astes mit glattem Schnitt parallel zum Stamm, ohne einen Zapfen oder Astring stehen zu lassen. Ein **s. S.** ist i. d. R. 2-3-mal so groß wie ein Astringschnitt und es erfolgt meistens eine unvollständige Kallusbildung mit ausgeprägten Wundholzleisten an den Seiten der Verletzung; die Folge eines **s. S.es** ist eine schlechtere Abschottung gegen eindringende Pilze in das Stammholz und somit eine Förderung des Fäulnisprozesses. **2. Astringschnitt:** Schnittführung beim Entfernen eines Astes, die am Stamm oder Ast einen Astring stehen lässt, ohne diesen zu verletzen. Methode, die im Vergleich zum stammparallelen Schnitt die Schnittstelle optimal abschotten lässt und eine schnellere Wundüberwallung [Kallusbildung] ermöglicht); *syn.* Schnitt [m] am Astansatz, Schnitt [m] auf Astkragen, Schnitt [m] auf Astring).

cut [n] **at the branch collar** *arb. hort.* ►cut at the branch attachment, #2.

cutaway bog [n] [UK] *conserv. land'man* ►worked bog.

cut face [n] [US] *constr.* ►incised slope.

cut/fill quantity calculations [npl] [US] *constr. eng.* ►earthworks calculation.

1280 cut flower [n] *hort.* (**1.** Stem with flower suitable for a vase. **2.** Flower cut for floral arrangements); *s 1* **flor** [f] **para cortar**; *s 2* **flor** [f] **cortada**; *f 1* **fleur** [f] **à couper** (Plante dont les tiges florales ou le feuillage sont utilisés pour la décoration) ; *f 2* **fleur** [f] **coupée** (Partie de plante portant fleur ou feuillage préalablement coupée en vue de la décoration intérieure) ; *g* **Schnittblume** [f] (**1.** Krautige Pflanze, die geeignete Blüten-

stängel für Vase oder Dekorationszwecke liefert. **2.** Eine bereits für Dekorationszwecke geschnittene Blume).

1281 cut flower garden [n] *gard.* (A garden for harvesting flowers used to create floral arrangements in vases and other containers); *s* **jardín** [m] **de flores para cortar**; *f* **jardin** [m] **bouquetier** (Jardin ou partie de jardin utilisé pour la culture des fleurs à couper) ; *g* **Schnittblumengarten** [m] (Garten oder ein Teil davon, der vorwiegend zum Blumenschneiden dient).

cut flowers [npl] *hort.* ►usefulness for cut flowers.

cut [vb] **flush with the ground** *constr. for. hort.* ►coppice (2).

cut [n] **in fees/charges** *contr. prof.* ►reduction in fees/charges.

1282 cut material [n] *constr.* (Soil or rock material removed in the course of construction work; ►cut 1, ►cut shelf, ►earth moving, ►excavation 1, ►linear excavation [US]/linear cutting [UK], ►overburden 2); *s* **material** [m] **de desmonte** (Suelo o roca que se retira en obra de construcción; ►cascote, ►corte en el terreno, ►excavación de tierras o rocas, ►desmonte, ►movimiento de tierras, ►perfil de desmonte [en una pendiente]); *f* **déblai** [m] (Sols ou roches enlevés pendant des travaux de terrassement ; ►déblaiement, ►excavation linéaire, ►fouille, ►matériaux de recouvrement, ►mouvement de terre, ►profil de déblai) ; *g* **Abtrag** [m] **(2)** (Boden oder Fels, der bei Bauarbeiten abgetragen wird; ►Abgrabung, ►Abraum, ►Abtrag 1, ►Anschnitt, ►Bodenbewegung, ►Einschnitt).

cut [n] **of a meadow** [UK]**, annual** *agr. constr. hort.* ►single mowing [US].

cut-off [n] *geo.* ►oxbow lake.

cutoff spur [n] *geo.* ►meander lobe.

cut [pp] **once a year** [UK]**, to be** *agr. constr. hort.* ►be mown once a year [US].

1283 cut-over bog [n] *land'man.* (Exploited raised bog; ►worked bog); *s* **turbera** [f] **alta explotada** (►turbera explotada); *f* **tourbière** [f] **partiellement exploitée** (Tourbière haute anciennement exploitée, dégradée par l'exploitation de la tourbe ; ►tourbière entièrement exploitée) ; *g* **Leegmoor** [n] ([Teil]abgetorftes Hochmoor; ►abgetorftes Moor).

1284 cut shelf [n] *eng.* (Creation of a berm or short terrace on a slope; ►cut material, ►linear excavation [US]/linear cutting [UK]); *s* **perfil** [m] **de desmonte (en una pendiente)** (Ataluzar un desmonte en bancales; ►corte de terreno, ►material de desmonte); *f* **profil** [m] **de déblai** (Profil d'un terrassement en déblai formant talus ; taluter [vb] ; ►déblaiement, ►excavation linéaire) ; *g* **Anschnitt** [m] (Meist dreieckförmiges Bodenabtragsprofil im Hang; anschneiden [vb]; ►Abtrag 2, ►Einschnitt).

1285 cut site [n] *constr.* (Place where soil or rock material has been removed; ►excavation site); *s* **área** [f] **de desmonte** (Area de terreno donde la tierra o roca ha sido removida; ►superficie de excavación); *syn.* área [f] de excavación; *f* **aire** [f] **de déblaiement** (Surface sur laquelle sont effectués des travaux de terrassement par décapage superficiel du sol ou de matériaux divers ; ►emprise de fouille) ; *g* **Abtragsfläche** [f] (Fläche, auf der Boden abgeschoben oder abgebaggert wird; ►Aufgrabungsfläche); *syn.* Abtragsbereich [m].

1286 cut slope [n] *constr.* (Artificially created incline in a slope by means of excavation; *specific term* rock cut section of a road; ►incised slope; *opp.* ►fill slope); *s* **talud** [m] **de corte** (Talud artificial en una pendiente o cuerpo de tierra; ►cara de corte; *opp.* ►talud de terraplén); *f* **talus** [m] **en déblai** (Versant artificiel raide et dressé selon une pente qui évite tout éboulement ; ►talus de fouille ; *opp.* ►talus en remblai) ; *g* **Anschnittsböschung** [f] (Künstlich angelegte Böschung in einen Hang oder

Erdkörper; ►Einschnittböschung; *opp.* ►Schüttböschung); *syn.* Abtragsböschung [f], *o. V.* Anschnittböschung [f].

1287 cut stone [n] *arch. constr.* (Generic term for any tooled concrete block or natural stone used in construction); **2. tooled natural stone** [n] (►Natural stone trimmed with stone tools to specified dimensions in comparison with ►rubble); *syn.* ashlar [n] [also US]; **3. concrete stone block** [n] (Pre-fabricated reinforced or non-reinforced concrete product of the construction industry with the face of the stone designed for the purpose intended); quarry stone [n]; *s 1* **bloque** [m] (Término genérico para cualquier pieza de construcción de muros de forma paralelepípeda y de mayores dimensiones que un ladrillo, que puede ser de piedra natural, cerámica, hormigón o de otros materiales); *s 2* **sillar** [m] (►Piedra natural labrada utilizada en cimentaciones y para cara vista de ciertos tipos de muro en mampostería; DACO 1988; ►piedra sin labrar); *syn.* piedra [f] de sillería, piedra [f] tallada, piedra [f] labrada, sillar [m] labrado, piedra [f] de talla [ROU] (todos BU 1959); *s 3* **bloque** [m] **de hormigón** (Pieza de construcción fabricada industrialmente de hormigón armado o sin armadura, con cara vista de diferentes texturas y colores según el uso previsto); *f 1* **matériau** [m] **de construction minéral** (Terme générique désignant les matériaux minéraux naturels ou transformés de construction en ►pierre naturelle et en pierres reconstituées ; *syn.* matériau [m] de maçonnerie minéral) ; *f 2* **pierre** [f] **de taille** (Pierre naturelle dure et solide extraite du sol, façonnée au moyen d'un outil de taille en carrière ou à pied d'œuvre suivant les formes et les dimensions du calepin d'appareil ; MAÇ 1981, 179 ; *pour comparaison* ►pierre brute [d'extraction]) ; *syn.* bloc [m] de pierre appareillé (MAÇ 1981, 177), moellon [m] ; pierre [f] d'œuvre ; *f 3* **pierre** [f] **reconstituée** (Éléments naturels broyés et recomposés par moulage avec des liens hydrauliques qui en font un produit proche du béton dans sa structure et offrant une multiplicité de modèles, de finitions et de teintes qui peuvent être utilisés aussi bien en terrasse, près de la piscine, en murets et piliers ou encore en bordure de massifs) ; *syn.* pierre [f] artificielle ; *g 1* **Werkstein** [m] (OB für Naturwerkstein und Betonwerkstein); *g 2* **Naturwerkstein** [m] (UB für einen ►Naturstein, der mit Steinwerkzeugen nach vorgegebenen Abmessungen zugerichtet ist — im Vergleich zum ►Bruchstein); *syn.* Haustein [m]; *g 3* **Betonwerkstein** [m] (Industriell vorgefertigtes Produkt aus bewehrtem oder unbewehrtem Beton mit dem Verwendungszweck entsprechend gestalteten Ansichtsflächen).

cut-through traffic [n] [US] *trans.* ►shortcut traffic [US]/short-cut traffic [UK].

1288 cut timber [n] *constr.* (Wood cut for building purposes in a sawmill); *syn.* sawn timber [n], rough lumber [n]; *s* **madera** [f] **rústica** (Madera que no ha sido cepillada en su superficie, pero que ha sido aserrada, canteada y recortada, hasta por lo menos suprimir las marcas de la sierra en todas sus superficies; DACO 1988); *f* **bois** [m] **de sciage** (Bois découpé dans une scierie) ; *g* **Schnittholz** [n] (Im Sägewerk für Bauzwecke zugeschnittenes Holz).

1289 cutting [n] *hort.* (Generic term for a part of a plant detached for the purpose of vegetative propagation; ►dormant cutting, ►grass cutting, ►hay cutting, ►peat cutting [US]/peat-cutting [UK], ►reed cutting, ►regeneration cutting, ►root cutting); *s* **esqueje** [m] (Fragmento de pequeño tamaño de cualquier parte de un vegetal que se planta para que emita raíces y se desarrolle; RA 1970; ►esqueje leñoso); *f 1* **bouture** [f] (Fragment prélevé sur une plante — rameau, racine, feuille ou écailleux — utilisé à des fins de reproduction végétative ; *terme spécifique* plançon — bouture de tige de saule ; ►bouture de rameau aoûté, ►bouture de rameau semi-aoûté) ; *f 2* **bouture** [f] **de rameau herbacée** (Bouture de jeune tiges prélevée au début

de la période croissance [fin mai] lorsque les tiges sont souples et encore en sève) ; *syn.* bouture [f] feuillée, bouture [f] herbacée ; *g* **Steckling** [m] (Abgetrennter oberirdischer, i. d. R. krautiger Pflanzenteil zur vegetativen Vermehrung; *UBe nach dem Reife- und Verholzungsgrad* krautiger S., nicht verholzter S. [eines Gehölzes] oder halbharter S. [eines Gehölzes], harter S. [eines Gehölzes = ►Steckholz]; *UBe nach dem Pflanzenteil, von dem er abgeschnitten wurde* Augensteckling [z. B. Weinrebe], Blatt-[augen]steckling [z. B. bei Begonien], Kopfsteckling, Triebsteckling, Wurzelsteckling [Wurzelschnittling]).

cutting [n] [CDN] *arb. constr. for.* ►felling (1).

cutting [n], **branch** *arb.* ►branch pruning.

cutting [n], **clear** *for.* ►clear felling.

cutting [n] [US], **complete** *for.* ►clear felling.

cutting [n], **cross** *constr. for.* ►saw into pieces.

cutting [n], **edge** *constr. hort.* ►lawn edge clipping.

cutting [n], **improvement** [US] *constr. for.* ►improvement cut [US].

cutting [n], **lawn edge** *constr. hort.* ►lawn edge clipping.

cutting [n], **lawn edge spade** *constr. hort.* ►lawn edge clipping.

cutting [n] [UK], **linear** *constr.* ►linear excavation [US].

cutting [n], **long grass** *agr. constr. hort.* ►hay cutting.

cutting [n], **reproduction** *for.* ►regeneration cutting [US].

cutting [n], **rough** *agr. constr. hort.* ►scythe.

1290 cutting [n] **a meander** *wat'man.* (Process used to straighten a stream); *s* **separación** [f] **de un meandro** (Proceso de eliminación de sinuosidades de ríos al crear canales entre meandros, dejando a éstos aislados total o parcialmente del curso de agua); *f* **scindement** [m] **d'un méandre** (Elimination d'une sinuosité lors du recalibrage d'un cours d'eau) ; *syn.* rescindement [m] d'un méandre ; *g* **Kappen** [n, o. Pl.] **eines Mäanders** (Abtrennen einer Fließgewässerschleife zur Begradigung eines Wasserlaufes); *syn.* Abhängen [n, o. Pl.] eines Mäanders, Abtrennen [n, o. Pl.] eines Mäanders.

1291 cutting [n] **back of branches** *arb. hort.* (►cutting back of trees and shrubs); *s* **recorte** [m] **de ramas** (►rebaje); *f* **raccourcissement** [m] **des branches** (►élagage 2) ; *g* **Einkürzen** [n, o. Pl.] **der Äste** (►Rückschnitt von Gehölzen); *syn.* Asteinkürzung [f].

1292 cutting back [n] **of perennials** *hort.* (1. Severe cutting back of the vegetative part of perennials to encourage repeat blooming in late summer, removal of faded flower heads and lignified stalks in late autumn or early spring. 2. Cutting back of spent flowers is called ►deadheading of perennials); *s* **rebaje** [m] **de perennes** (1) (Fuerte corte de las partes aéreas de plantas vivaces para promover una nueva floración al final del verano o para eliminar las flores secas y los tallos lignificados en otoño o al principio de la primavera; ►rebaje de perennes 2); *syn.* rebaje [m] de vivaces, poda [f] de perennes/vivaces; *f* **rabattage** [m] **des vivaces** (1) (Les vivaces capables de remonter sont rabattues ou nettoyées après leur floraison principale, afin de stimuler l'émission de nouvelles pousses et d'une floraison automnale ; ►rabattage des vivaces 2) ; *g* **Staudenrückschnitt** [m] **(1)** (1. Starker Rückschnitt der oberirdischen Teile von Stauden, damit sie im Spätsommer noch mal blühen oder die Entfernung der abgeblühten, verholzten oberirdischen Teile im Spätherbst oder vor Frühlingsbeginn. 2. Wegschneiden der verwelkten Blüten, bei manchen Stauden, z. B. bei der Schafgarbe *[Achillea]*, des ganzen Schaftes; ►Staudenrückschnitt 2); *syn.* Rückschnitt [m] von Stauden.

C

1293 cutting back [n] **of trees and shrubs** *constr. hort.* (Removal of unwanted growth to reduce the length of stem, branch or twig, or to stimulate new growth; BS 3975: part 5; ►tree and shrub pruning); *syn.* shortening [n] of trees and shrubs, heading [n] [also US], heading-back [n] [also US] (ARB 1983, 387); *s* **rebaje** [m] (Corte de ramillas, ramas o del tronco para inducir nuevo crecimiento; HDS 1987; rebajar [vb], entresacar [vb]; ►poda de leñosas); *f* **élagage** [m] **(2)** (Opération consistant à éliminer certaines parties de branches d'un arbre afin de maintenir une charpente équilibrée, de lui imposer une forme et de contrôler les projections des branches ; cf. IDF 1988 ; ►taille des végétaux ligneux) ; *g* **Rückschnitt** [m] **von Gehölzen** (Einkürzen oder Wegschneiden von Zweigen, Ästen oder Stämmen, um neues Wachstum anzuregen; ►Gehölzschnitt); *syn.* Gehölzrückschnitt [m], Zurückschneiden [n, o. Pl.] von Gehölzen.

1294 cutting down [n] **a hill** *constr. eng.* (Removal of part or all of a hill in connection with a construction project); *s* **desmonte** [m] **de una colina** (o parte de ella para una obra de construcción); *syn.* corte [m] de una colina; *f* **fouille** [f] **en déblai sur une montagne** (Affouillement réalisé sur une montagne ou partie de montagne en vue de l'implantation d'ouvrages) ; *syn.* abattage [m] d'un pan de montagne/colline, enlèvement [m] d'un pan de montagne/colline ; *g* **Abtrag** [m] **eines Berges** (Abgrabung eines Berges oder Teile davon für Bauvorhaben).

cutting [n] **from inside to outside** *ecol. land'man.* ►meadow cutting from inside to outside.

1295 cutting height [n] *constr.* (Height to which a grass sward is to be cut, as prescribed in a list of bid items and quantities [US]/schedule of tender items [UK]); *s* **altura** [f] **de corte** (p. ej. de césped); *f* **hauteur** [f] **de tonte** (Hauteur des herbes déterminée par le descriptif à obtenir après une tonte) ; *g* **Schnitthöhe** [f] (Im Leistungsverzeichnis vorgegebene Halmlänge, die nach der Mahd erreicht werden muss).

1296 cutting [n] **material to fit** *constr.* *s* **cortar** [vb] **a medida;** *f* **ajustage** [m] (de la pièce à adapter) ; *g* **Einschneiden** [n, o. Pl.] (Maßgenaues Zurichten, z. B. von Passstücken); *syn.* passgenaues Zurichten [n, o. Pl.].

1297 cutting [n] **of heath sods** *agr.* (Removal/digging of heath sods used for animal bedding; ►heath sod); *s* **corte** [m] **de tepes de turba** (Levantamiento de ►tepes de turba para utilizarlos como camada de ganado); *f* **étrépage** [m] **(2)** (Enlèvement manuel des horizons organiques du ►podzol ou de la couche superficielle des landes tourbeuses servant en guise de litière pour le bétail ; GGV 1979, 317 ; ►plaque tourbeuse) ; *g* **Plaggenhieb** [m] (Stechen von rechteckigen, durchwurzelten Soden mit Gras- oder Heidekrautbewuchs zur Einstreu; ►Plagge); *syn.* Plaggenstechen [n, o. Pl.].

cutting [n] **of meadows** [UK] *agr. constr. hort.* ►mowing of meadows [US].

1298 cutting [n] **of sods** [US] *constr. hort.* (Vertical cutting or slitting to a pre-set depth of sods [US]/turves [UK] prior to lifting; cf. BS 3975: part 5; ►sod removal site [US]/turf removal site [UK]); *syn.* scoring [n] of turves [UK]; *s* **levantamiento** [m] **de tepes** (Corte de tepes precultivados para colocarlos en otro sitio; ►zona de préstamo para tepes); *f* **déplacage** [m] **d'un gazon** (Enlèvement des plaques de gazon précultivé pour mise en place à un autre endroit ; ►lieu de prélèvement de gazon) ; *g* **Stechen** [n, o. Pl.] **von Rasensoden** (Entnehmen von vorkultivierten Rasensoden für den Einbau an anderer Stelle; Rasensoden abschälen [vb], Rasensoden stechen [vb]; ►Entnahmestelle für Fertigrasen).

cuttings [npl] *agr. constr. hort.* ►clippings.

1299 cybernetic cycle [n] *ecol.* (*Term adopted from cybernetics* a closed circle of information. The cycle is self-regulating by way of ►feedback; ►self-regulation 2); *s* **ciclo** [m] **cibernético** (Término de la cibernética que se aplica a los ciclos cerrados de información, sistemas parciales que se autoregulan por ►retroalimentación; ►autoregulación 2); *f* **système** [m] **finalisé** (*Terme en provenance de la cybernétique* système évoluant vers un nouvel état antérieurement défini et maintenant sa structure par un mécanisme de ►rétroaction ; ►autorégulation 2) ; *syn.* système [m] régulé ; *g* **Regelkreis** [m] (*Begriff aus der Kybernetik* **1.** ein in sich geschlossener Kreislauf von Informationen. **2.** Ein Teilsystem, das sich durch ►Rückkoppelung selbst regelt; ►Regelung).

cycle crossing [n] [UK] *trans.* ►bike crossing [US].

cycle path [n] [UK] *trans.* ►bikeway [US].

cycle rack [n] [UK] *trans. urb.* ►lockable bike rack [US].

cycle rack [n] [UK]**, lockable** *trans. urb.* ►lockable bike rack [US].

cycle racks [npl] [UK] *trans. urb.* ►bike racks [US].

cycleway [n] [UK] *trans.* ►bikeway [US].

cycling pool [n] *ecol.* ►ecological cycle.

1300 cycling tourism [n] *recr.* (Touring on bicycles while on vacation [US]/holiday [UK]; ►national scenic trail [US]/long distance footpath [UK], ►tourism 1); *syn.* bike tourism [n]; *s* **cicloturismo** [m] (Viaje turístico realizado en bicicleta; ►pasillo verde de largo recorrido, ►turismo); *f* **cyclotourisme** [m] (Forme de ►tourisme utilisant le vélo comme moyen de locomotion ; ►sentier de grande randonnée) ; *syn.* tourisme [m] à bicyclette ; *g* **Radtourismus** [m, o. Pl.] (Das Reisen mit dem Fahrrad, um fremde Gegenden und Länder kennen zu lernen und zur Erholung; ►Fernwanderweg, ►Fremdenverkehr).

cyclists' crossing [n] [UK] *trans.* ►bike crossing [US].

cyclone fence [n] [US trademark] *constr.* ►metal fence.

1301 cyclopean wall [n] *arch.* (**1.** Prehistoric masonry wall made of huge stone blocks laid without mortar. **2.** ►Two-tier masonry [US]/two-leaf masonry [UK] made of big, irregular, stone blocks, which are dressed polygonally, come mostly from a quarry, and are not laid in courses; ►polygonal masonry); *s* **mampostería** [f] **ciclópea** (**1.** Mampostería prehistórica de mampuestos gigantes, sin mortero y generalmente con juntas a tope. **2.** ►Mampostería de dos capas de mampuestos gigantes irregulares de superficie poligonal, colocados sin formar hiladas y sin mortero; ►mampostería poligonal); *f* **maçonnerie** [f] **à appareillage cyclopéen** (**1.** Énormes blocs de pierre de taille unis à joint vif caractérisant les édifices préhistoriques ; **2.** ►maçonnerie à deux parements constituée d'énormes pierres brutes de carrière ou taillées en très gros blocs irréguliers [parfois plus de deux mètres de long] ; ►maçonnerie à appareillage polygonal) ; *g* **Zyklopenmauerwerk** [n] (**1.** Prähistorisches Mauerwerk aus riesigen Steinblöcken, ohne Mörtel, meist fugeneng verlegt. **2.** ►Zweischaliges Mauerwerk aus großen, unregelmäßigen, polygonal behauenen, nicht lagerhaften, meist gesprengten Bruchsteinen, im Verband aufgesetzt. Abweichend von der DIN 1053 können auch Findlinge vermauert werden; ►Polygonalmauerwerk).

1302 cyclopean wall [n] **with hammer-dressed joints** *arch.* (Wall made of big, polygonal stone blocks, which are dressed by a hammer for butt-jointed construction); *s* **mampostería** [f] **ciclópea con juntas a tope** (Mampostería de piedras gigantes dispuestas con juntas a tope); *f* **maçonnerie** [f] **à appareillage cyclopéen à joints étroits** (Pierres cyclopéennes taillées les unes en fonction des autres afin de réaliser des joints d'une épaisseur de 2 mm maximum) ; **g fugenenges Zyklopen-mauerwerk [n]** (Mauerwerk aus großen, polygonalen, hammer-rechten Steinen, die mit Pressfuge verlegt sind).

D

1303 daily billing rate [n] [US] *contr. prof.* (Remuneration for a day's work; ►hourly billing rate [US]); *syn.* daily fee [n], daily rate [n] [UK]; **s tasa [f] por día** (Honorario por día de trabajo; ►tasa de honorarios por hora); **f montant [m] jour-nalier** (Honoraire journalier forfaitaire ; ►prix horaire, ►prix horaire de vacation) ; **g Tagessatz [m]** (Zeithonorar pro Tag; ►Stundensatz).

daily fee [n] *prof.* ►daily billing rate [US].

daily rate [n] [UK] *prof.* ►daily billing rate [US].

1304 daily recreation [n] *plan. recr.* (Short period of leisure activity or relaxation usually at or close to home, mostly within an urban area); **s recreación [f] diaria** (Forma de recreo/descanso cotidiano que se realiza en general en la vivienda o en sus alrededores); **f détente [f] et récréation quotidienne** (Forme de détente courte ayant principalement lieu à proximité de l'habitation en zone urbaine) ; **g Tageserholung [f, o. Pl.]** (Kurzzeiterholungsform, die innerhalb der Siedlungsbereiche vorwiegend in der Wohnung oder in Wohnungsnähe erfolgt und der Entmüdung dient).

1305 dam [n] *constr. eng. wat'man.* (**1.** Barrier constructed across a valley watercourse for impounding water and creating a reservoir; with a series of gates and other control mechanisms the upstream water-level can be controlled and the flow of water supply into a canal can be regulated or diverted; ►loose rock dam; *syn.* barrage [n]. **2.** Body of water held back by such a dam; ►check dam, ►damming 2, ►impoundment, ►spillway); **s presa [f]** (Obra hidráulica destinada a retener el agua de una corriente fluvial para crear un ►embalse; ►aliviadero, ►cons-trucción de presas, ►dique transversal); *syn.* dique [m] (2), represa [f] [AL]; **f barrage [m] de vallée** (Ouvrage hydraulique qui a pour objet de relever le niveau d'un plan d'eau naturel ou d'accumuler l'eau d'une rivière par l'obstruction d'une vallée sur toute sa largeur ; ►lac de retenue, ►construction d'ouvrages de rétention, ►déversoir de crue, ►ouvrage de confortement transversal) ; *syn.* barrage-réservoir [m] ; **g Talsperre [f]** (**1.** Stauanlage, die zur Wasserspeicherung ein Tal in seiner ganzen Breite abschließt und überfluten lässt. **2. T.** wird auch synonym für ►Stausee gebraucht; ►Hochwasserentlastungs-anlage, ►Querwerk, ►Stauanlagenbau); *syn.* Staudamm [m]).

damage [n] *constr.* ►building frost damage; *agr. hort.* ►cold air damage; *agr. constr. hort. for.* ►drought damage; *envir. leg.* ►environmental damage; *geo. pedol.* ►erosion damage; *for.* ►felling and logging damage; *wat'man. land'man.* ►flood damage; *agr. for. hunt.* ►game damage, *landsc.* ►landscape damage; ►over-browsing damage; *constr. eng.* ►pavement frost damage; *arb. hort.* ►root damage; *envir.* ►salt damage; *game' man.* ►scaling damage; *constr. geo.* ►slope damage; *agr. constr. phyt.* ►trampling damage; *landsc. met.* ►weather damage; *agr. for.* ►wind damage.

damage [n]**, blast damage** *agr. for. hort.* ►storm damage.

damage [n]**, dessication** *agr. constr. hort. for.* ►drought damage.

damage [n] [US]**, hauling** *for.* ►felling and logging damage.

damage [n]**, scarred bark** *game'man.* ►scaling damage.

damage [n] [US&CDN]**, skidding** *for.* ►felling and logging damage.

damage [n] [UK]**, stem** *arb. hort.* ►stem injury.

1306 damage [n] **caused by environmental pollu-tion** *envir.* (Harm or injury from air pollution, noise, radiation, etc. that lessens value or usefulness); **s daños [mpl] por conta-minación** (Perjuicios producidos a la salud humana o al medio ambiente por polución atmosférica, acústica, radi[o]activa, etc.); **f dommage [m] causé par les polluants atmosphériques** (Conséquence préjudiciable de l'action de la pollution atmo-sphérique sur les biens matériels, la santé de l'homme et l'envi-ronnement) ; **g Immissionsschaden [m]** (Durch Einwirkung von Luftverunreinigungen, Lärm, Strahlen etc. auf Menschen, Tiere und Pflanzen verursachte starke körperliche/gesundheitliche Beeinträchtigung oder an nicht belebten Sachgütern bedingte Beschädigung oder Zerstörung); *syn.* Immissionsfolgeschaden [m], *obs. UB* Rauchschaden [m].

damaged area [n] *constr. landsc.* ►damaged site.

damaged forest [n] *envir.* ►emission-damaged forest.

1307 damaged site [n] *constr. landsc.* (Destruction caused by erosion to slopes, riverbanks [riverbank collapse (US)/river-bank collapse (UK)] or damage to tree trunks [►trunk wound (US)/stem wound (UK)] caused by mechanical influences, etc.; ►collapse); *syn.* damaged area [n]; **s punto [m] dañado** (Destrucción causada por la erosión o por otra causa mecánica, p. ej. en taludes, orillas [►desplome de orillas], o en troncos de árboles [►herida del tronco], etc.); **f 1 partie [f] endommagée** (Destruction ou dommage provoqué par l'érosion ou une action mécanique, p. ex. sur un talus, les berges, etc. ; ►effondrement d'une berge) ; **f 2 plaie [f]** (Dommage causé sur le tronc d'un arbre [plaie du tronc] ; on utilisera de préférence le mot plaie à la traduction littérale « partie blessée » ; ►plaie du tronc) ; *syn.* partie [f] blessée ; **g Schadstelle [f]** (**1.** Durch Erosion oder mechanische Einwirkung verursachte Zerstörung wie z. B. an Böschungen, Ufern [►Uferabbruch]. **2.** Beschädigung am Baum-stamm [►Stammwunde]).

1308 damage [n] **due to mining subsidence** *min.* (Harm to buildings and land caused by deep mining; ►ground settlement); **s daño [m] minero** (Desperfectos en terrenos y edificios causados por ►subsidencia debida a explotación minera); **f dommage [m] minier** (Dégâts causés par l'exploi-tation minière aux constructions et aux terrains; ►subsidence) ; **g Bergschaden [m]** (Durch ►Senkung, die durch den Bruch-bergbau bedingt sind, entstandene Schäden an Gebäuden und Grundstücken).

1309 damage inventory [n] *envir.* (Compilation of data, photographic records, mapping, etc. related to damage caused to trees, water bodies, or buildings by pollution or other environ-mental influences as well as natural catastrophes; ►forest decline); **s inventario [m] de daños** (Compilación de datos, fotografías, mapas, etc. relacionada con daños causados por la contaminación o por catástrofes naturales p. ej. en árboles, ríos y lagos o en edificios; ►muerte de bosques); **f constat [m] du dommage** (Dans le cadre de l'établissement de la preuve du dommage : constat, analyses prélèvements et le cas échéant inventaire cartographique de l'état sanitaire des arbres, de la qualité de l'eau des cours d'eau où de l'état de bâtiments touchés par des nuisances [pollution de l'air, abattement de nappe, etc.] ou à des catastrophes naturelles ; ►mort des forêts) ; **g Scha-**

D

denserfassung [f] (*Im Rahmen der Beweissicherung* **1.** Aufnahme und ggf. Kartierung von z. B. geschädigten Baumbeständen oder anderen Vegetationseinheiten, Gewässern oder Bauwerken, verursacht durch Belastungen der Umwelt [Luftverunreinigung, Grundwasserabsenkung etc.] oder Naturkatastrophen; **2.** Aufnahme von Schäden, die durch Setzungen beim Tunnelbau unter besiedeltem Gebiet entstanden sind; ▶ Waldsterben).

damages [npl] *constr. contr.* ▶ liquidated damages.

damage [n] to a body of water *envir. leg.* ▶ environmental damage.

damage [n] to a tree *arb.* ▶ accidental damage to a tree.

damage [n] to crops *agr. envir.* ▶ damage to land.

1310 damage [n] to existing vegetation *contr. constr.* (Impairment of plants or parts thereof caused by mechanical action, which may cause deterioration in their vitality); *s* **deterioro [m] de la vegetación** (Destrucción de plantas o partes de ellas con la consecuente reducción de su vitalidad); *f* **détérioration [f] des végétaux** (Destruction de tout ou parties de plantes pouvant être occasionnée par une réduction de la vitalité) ; *g* **Beschädigung [f] von Pflanzen(beständen)** (Mechanische Schäden an Pflanzen oder deren Teile, die vitalitätsmindernd wirken können; der Begriff 'Schädigung' bedeutet in diesem Zusammenhang Verminderung der Vitalität).

1311 damage [n] to land *agr. envir.* (Large-scale destruction of farmland, e.g. by game, or military manoeuvres; ▶ game damage); *syn.* damage [n] to crops; *s* **daños [mpl] agrosilvícolas** (Destrucción a gran escala de superficies agrícolas o silvícolas debido p. ej. a animales de caza o a maniobras militares; ▶ daños causados por la caza); *f* **dommage [m] causé à une surface agricole/forestière** (Destructions à grande échelle de surfaces agricoles ou forestières provoquées p. ex. par des intempéries, le gibier ou lors de manœuvres militaires ; ▶ dégâts causés par le gibier) ; *g* **Flurschaden [m]** (Großflächige Zerstörung von land- und forstwirtschaftlichen Flächen, z. B. durch militärische Übungen oder durch Wild; ▶ Wildschaden).

damage [n] to species and natural habitats *envir. leg.* ▶ environmental damage.

1312 damming [n] (1) *eng. hydr.* (Obstructing or restraining a watercourse by means of a ▶ dam; ▶ impoundment); *s* **embalsado [m]** (Obstrucción de un curso de agua por medio de una ▶ presa; ▶ embalse; embalsar [vb], estancar [vb], rebalsar [vb], represar [vb]); *f* **retenue [f]** (d'un cours d'eau ; ▶ barrage de vallée, ▶ lac de retenue) ; *g* **Aufstauung [f]** (Sammeln von zu Tal fließendem Wasser eines Fließgewässers; aufstauen [vb], anstauen [vb]; ▶ Stausee, ▶ Talsperre); *syn.* Anstauen [n, o. Pl.].

1313 damming [n] (2) *constr. eng.* (Construction of a barrier in a watercourse to form a lake or ▶ impoundment; ▶ dam); *s* **construcción [f] de presas** (Construcción de barreras en cursos de agua [▶ presas] para almacenar agua en ▶ embalses); *f* **construction [f] d'ouvrages de rétention** (Construction de ▶ barrage de vallée, de bief destinés à retenir et orienter les eaux à des fins agricoles, industrielles ou touristiques en augmentant le profil transversal des cours d'eau et de la vallée alluviale ; ▶ lac de retenue) ; *g* **Stauanlagenbau [m]** (Bau von ▶ Talsperren und Wehren zur Speicherung von Wasser, indem neben dem Flussquerschnitt auch Teile des Talquerschnitts abgesperrt werden; ▶ Stausee); *syn.* Staubauwerksbau [m].

1314 damp meadow [n] *phyt.* (Primarily moist to relatively wet grasslands including rushes and tall forb vegetation; ▶ wet meadow); *syn.* moist soil meadow [n] [UK] (ELL 1988, 572); *s* **prado [m] húmedo** (Pradera natural de húmedad moderada que incluye juncares y formaciones megafórbicas; ▶ prado higrófilo);

f **prairie [f] mésophile** (Prairie naturelle à engorgement modéré appartenant aux alliances de l'*Arrhenatherion* pour la prairie de fauche et du *Cynosurion* pour les prairies de pâture ; ▶ prairie hygrophile) ; *syn.* prairie [f] humide, prairie [f] fraîche ; *g* **Feuchtwiese [f]** (Überwiegend feuchtes bis mäßig nasses Grünland einschließlich binsenartiger Bestände und Hochstaudenfluren; ▶ Nasswiese); *syn.* Feuchtgrünland [n].

1315 damp mortar [n] *constr.* (Moist mortar with fine aggregate for laying of pavers, setts, slabs, etc.; *related terms* dry-packed mortar [US] and semi-dry mortar [UK]); *s* **mortero [m] húmedo** (Mortero con consistencia de tierra húmeda para colocar adoquines o losas); *f* **mortier [m] gâché à consistance de la terre humide** (Produit de mise en œuvre de matériaux utilisé comme lit de pose de pavés, dalles, klinker, etc.) ; *g* **erdfeuchter Mörtel [m]** (Mit wenig Wasser angemachter Mörtel für ein Pflasterbett zum Verlegen von Klinkern, Pflastersteinen, Platten etc.).

1316 dampproofing [n] [US]/damp-proofing [n] [UK] *constr.* (Protection against moisture using impervious materials; e.g. additives in dampproof concrete [US]/damp-proof concrete [UK], coating for surface applications); *s* **aislamiento [m] antihumedad** (Tratamiento de superficies contra la humedad, p. ej. aplicando pintura aislante); *f* **isolation [f] contre l'humidité** (Protection contre l'humidité [infiltrations et remontée d'eau] par l'emploi de substances ou matériaux protecteurs [isolant hydrophile], p. ex. enduits hydrauliques d'imperméabilisation, enduits de parement plastiques ou incorporation à la fabrication d'adjuvant tels que les rétenteurs d'eau, les entraîneurs d'air les hydrofuges) ; *g* **Isolierung [f] gegen Feuchtigkeit** (Schutz gegen Feuchtigkeit durch Dichtungsstoffe, z. B. durch Betonzuschlagstoffe bei Sperrbeton, Isolieranstrich bei Oberflächenbehandlungen).

1317 dampproof mortar [n] [US]/damp-proof mortar [n] [UK] *constr.* (Mortar treated by use of a suitable admixture to make the passage or absorption of water impossible; ▶ waterproofing); *s* **mortero [m] impermeabilizante** (Mortero que, gracias a su contenido de aditivos químicos, no deja pasar la humedad; ▶ sellado); *f* **mortier [m] hydrofuge** (Mortier rendu imperméable par l'apport d'un adjuvant imperméabilisant par la formation d'un film imperméable continu à la surface du mortier ; ▶ étanchéisation) ; *syn.* mortier [m] étanche, mortier [m] hydrophobe ; *g* **Sperrmörtel [m]** (Mörtel, der durch besondere chemische Zusätze keine Feuchtigkeit durchlässt; ▶ Dichtung).

danger [n] *landsc. met.* ▶ frost danger.

danger [n], flood *hydr. wat'man.* ▶ flood hazard.

1318 danger [n] of contamination *ecol.* (Risk of an element of the environment becoming polluted; e.g. air, groundwater, soil; ▶ risk of groundwater pollution); *s* **riesgo [m] de contaminación** (▶ peligro de contaminación de acuífero); *syn.* peligro [m] de contaminación; *f* **vulnérabilité [f] à la pollution** (Risque de pollution des éléments de l'environnement [sol, nappe, air, etc.] par des substances indésirables ; ▶ vulnérabilité de la nappe aux pollutions) ; *g* **Verschmutzungsgefährdung [f]** (Die Möglichkeit, dass ein Umweltmedium mit unerwünschten Stoffen verschmutzt wird, z. B. Boden, Grundwasser, Luft; ▶ Grundwassergefährdung).

1319 danger [n] of high water *wat'man.* (Threat that a high-water level will occur; ▶ flood hazard); *s* **peligro [m] de crecida** (▶ peligro de inundación); *f* **risque [m] de hautes eaux** (Eventualité de la survenance d'une crue ; ▶ risque d'inondation) ; *g* **Hochwassergefahr [f]** (Möglichkeit, dass ein Hochwasserereignis eintritt; ▶ Überschwemmungsgefahr).

1320 danger [n] of trees falling down *arb. leg. syn.* threat [n] of trees toppling over; *s* **peligro [m] de caída de**

árboles; *f* danger [m] de chute d'arbres ; *g* Umsturzgefahr [f] von Bäumen (Möglichkeit, dass Bäume wegen mangelnder Standfestigkeit, z. B. wenn Ankerwurzeln durch Bauarbeiten im Wurzelschonbereich gekappt wurden, umfallen).

dangerous waste [n] *envir. leg.* ▶hazardous waste.

data [npl] **and aims** *plan.* ▶provision of data and aims.

data bank [n] *adm. landsc.* ▶green space data bank; ▶landscape data bank.

data bank [n]**, tree** *adm. arb.* ▶tree data bank.

1321 data base [n] *adm. plan.* (Generic term for collected information in written or computerized form; ▶geographic information system); *s* base [f] de datos (1. Término genérico para el conjunto de información recopilada, sea en forma escrita o computarizada, en cuyo caso puede ser interrelacionada; ▶sistema de información geográfica. 2. Informaciones recopiladas en terreno o de documentos existentes utilizadas como base para el análisis del estado de un lugar específico); *syn.* de 1. banco [m] de datos; *f 1* base [f] de données (Informatique ensemble de données alphanumériques stockées sur un support identique et reliées entre elles au sein d'une structure cohérente, traitables par ordinateur permettant d'organiser et d'archiver des données dans le but d'être facilement retrouvées et accessibles ; ▶système d'information géographique) ; *f 2* données [fpl] (Informations collectées sur le terrain ou à partir de documents utilisées pour l'analyse) ; *g* Datengrundlage [f] (1. Gesammelte Informationen in geschriebener Form oder elektronisch gespeichert und verwaltet, die miteinander verknüpft werden können und auf die man mit entsprechenden Suchprogrammen zugreifen kann; ▶Geografisches Informationssystem. 2. *plan. ecol.* Die für eine Analyse gebrauchten Informationen, die aus Geländeerhebungen oder vorhandenen Dokumenten stammen); *syn. zu 1.* Datenbank [f].

data capture [n] *plan.* ▶data collection.

1322 data collection [n] *plan.* (Gathering and compilation of basic information to provide the contextual background for planning or scientific studies; ▶data collection for planning purposes); *syn.* data capture [n], collection [n] of survey information; *s* recogida [f] de datos (Compilación de datos básicos para desarrollar el planeamiento; ▶colección de información básica); *syn.* recopilación [f] de datos, colecta [f] de datos, toma [f] de datos; *f* relevé [m] des données (▶Inventaire des documents existants relatif à la mission) ; *syn.* collecte [f] des données, recueil [m] des données, recensement [m] des données ; *g* Datenerhebung [f] (Gewinnung von Daten durch Untersuchungen als Grundlage für den Planungsprozess; ▶Ermittlung der Planungsgrundlagen); *syn.* Datengewinnung [f], Datenermittlung [f], Datenerfassung [f].

1323 data collection [n] **for planning purposes** *plan.* (Gathering of detailed information on which to base planning solutions; ▶data collection); *s* colección [f] de información básica (Recopilación de conjunto de datos y documentos necesarios para desarrollar un trabajo de planificación; ▶recogida de datos); *f* inventaire [m] des documents existants relatif à la mission (▶relevé des données) ; *syn.* collecte [f] de l'ensemble des données disponibles, collecte [f] de l'ensemble des informations disponibles ; *g* Ermittlung [f] der Planungsgrundlagen (Bestandsaufnahme, ▶Datenerhebung, deren Darstellung und Bewertung als Grundleistungen für landschaftsplanerische Leistungen; cf. §§ 45 a [1] und 46 [1] HOAI 2002, §§ 23-24 und 26-27 HOAI 2009).

1324 data evaluation [n] *plan.* (Complex process of assigning values to fundamental data and assessing relationships and impacts as well as implications of various patterns or outcomes as they apply to a planning project); *syn.* data analysis

[n]; *s* evaluación [f] de datos (Proceso complejo de valoración de las informaciones básicas, sus interrelaciones y consecuencias para una planificación específica); *syn.* ponderación [f] de datos; *f* évaluation [f] des données ; *g* Datenauswertung [f] (Komplexer Prozess der Bewertung von notwendigen Grundlagen, deren Zusammenhänge und Auswirkungen für eine bestimmte Planung); *syn.* Datenanalyse [f].

data form [n] *adm. conserv. ecol.* ▶standard data form.

1325 data processing [n] *plan.* (Series of operations on data, especially by a computer, to retrieve or classify information); *s* procesamiento [m] de datos *syn.* elaboración [f] de datos; *f* traitement [m] des données ; *g* Datenverarbeitung [f].

date [n] **for design submission, closing** *plan. prof.* ▶submission deadline for design competition.

1326 date [n] **for submission of bids** [US] *constr. contr.* (Time limit for submitting bids [US]/tenders [UK] to the authority/agency sending requests for proposals [R.F.P.]; the submission of tenders/bids is followed by ▶checking and evaluation of bids [US]/tenders [UK], and by the ▶awarding period); *syn.* date [n] for submission of tenders [UK], deadline [n] for submission of bids [US]/deadline [n] for submission of tenders [UK]), submission deadline [n]; *s* fecha [f] tope de entrega de ofertas (Límite temporal para entregar las ofertas a la autoridad/agencia que las ha solicitado. A la entrega le siguen el ▶análisis de ofertas y el ▶periodo/período de remate); *f* date [f] limite de réception des offres (Date limite à laquelle doivent être remises les offres/soumissions ; après cette date suit le ▶jugement des offres et le ▶délai de validité des offres) ; *syn.* date [f] limite de remise des offres/soumissions, date [f] limite d'envoi des offres/soumissions, date [f] d'expiration du délai de réception des offres/soumissions, date [f] limite de dépôt, date [f] limite de réception des soumissions [CDN] ; *g* Abgabetermin [m] (Termin, bis zu dem Angebote bei der ausschreibenden Stelle eingehen müssen. Nach dem **A.**/Eröffnungstermin folgt die ▶Prüfung und Wertung der Angebote und die ▶Zuschlagsfrist); *syn.* Angebotstermin [m].

date [n] **for submission of tenders** [UK] *constr. contr.* ▶date for submission of bids [US].

1327 datum [n] *constr. geo. surv.* (**1.** Fixed or assumed point, line or surface in relation to which others are determined. **2.** A level surface, such as fixed ground level or mean sea level, used as a reference from which elevations are computed; ▶benchmark, ▶datum point. **3.** The **d.** of a given site is a *site datum*); *syn.* datum level [n], datum plane [n]; *s* cota [f] de referencia (Nivel de altura fijado en obra al cual se refieren todas las demás alturas; ▶marca geodésica, ▶punto de agrimensura); *syn.* altura [f] de referencia; *f* cote [f] d'origine (Point fixé servant de référence à tous les travaux de nivellement sur le chantier ; ▶point coté ; ▶repère de niveau [cotation NGF]) ; *syn.* cote [f] de référence, point [m] d'origine, point [m] de niveau de base, repère [m] ; *g* Bezugshöhe [f] (Auf der Baustelle festgesetzte Höhe [Höhenbezugspunkt], auf die andere Höhen bezogen werden; ▶Höhenfestpunkt, ▶Vermessungspunkt); *syn.* Ausgangshöhe [f].

datum level [n] *constr. geo. surv.* ▶datum.

datum plane [n] *constr. geo. surv.* ▶datum.

1328 datum point [n] *surv.* (Ground survey point established as a reference for further survey work); *s* punto [m] de agrimensura (Punto de la tierra medido en longitud y latitud que sirve de referencia para mediciones en terreno); *syn.* banco [m] de nivel; *f* point [m] coté (Point matérialisé sur le terrain et coté par un géomètre ; on distingue le point planimétrique [élément de base de la projection plane horizontale de la topographie] et le point altimétrique [expression numérique de l'altitude d'un point préalablement défini en tant que point planimétrique] ; cf. HAC 1989, 527) ; *syn.* repère [m] coté ; *g* **Vermessungspunkt** [m] (In Höhe und Lage eingemessener Punkt im Gelände).

day care center [n] [US] *sociol.* ►day care centre [UK].

1329 day care centre [n] [UK] *sociol. urb.* (Private or public facility for supervised care of children in different ages during the day by a person other than the child's parents. Once a child reaches the age of twelve, he or she is no longer covered by day care legislation; **In UK** a wide range of childcare is offered, including childminders, day nurseries, and playgroups as well as pre-school education in schools. It is regulated by OFSTED, [Office for Standards in Education, Children's Services and Skills] which came into being on 1 April 2007. It brings together the wide experience of four formerly separate inspectorates and regulates the care of children and young people, as well as inspecting education and training in the sector; ►child care center [US]/child care centre [UK], ►day nursery, ►kindergarten); *syn. o.v.* day care center [n] [± US]; *s* **guardería** [f] **infantil (1)** (Centro de atención de niños y niñas de diferentes edades generalmente durante todo el día; término genérico para ►jardín de infancia y ►guardería infantil pedagógica; ►casa-cuna); *f* **garderie** [f] **d'enfants** (≠) (**D.**, établissement socio(-)pédagogique assurant la garde régulière des enfants pendant la journée ; terme générique englobant le ►jardin d'enfants, la ►garderie d'élèves et la ►crèche ; ►centre de loisirs maternels, ►école maternelle, ►halte garderie) ; *g* **Kindertagesstätte** [f] (*Abk.* Kita; soziale Einrichtung, in der Kinder verschiedener Altersstufen regelmäßig, z. T. ganztägig betreut werden; *UBe* ►Kindergarten, ►Kinderhort, ►Kinderkrippe und Krabbelstuben [für 1-2-jährige Kleinkinder]).

1330 day nursery [n] *sociol. urb.* (Daytime child care facility for children from two months to two years); *syn.* crèche [n] [also UK]; *s* **casa-cuna** [f] (Establecimiento de cuidado de niños y niñas de menos de 3 años de edad en días laborables); *syn.* hogar [m] infantil [AL]; *f* **crèche** [f] (Établissement assurant la garde régulière pendant la journée des enfants de moins de trois ans ; cf. PR 1987) ; *g* **Kinderkrippe** [f] (Pädagogisch geleitete Tagesstätte für Kinder/Säuglinge, die älter als zwei Monate sind).

1331 day trip [n] [US]/**day-trip** [n] [UK] *recr.* (Excursion such as rambling or an outing in the car lasting one day); *s* **excursión** [f] **de un día**; *f* **excursion** [f] **journalière** (Promenade pédestre ou motorisée d'une durée d'un jour) ; *syn.* excursion [f] d'un jour, sortie [f] journalière ; *g* **Tagesausflug** [m] (Sich über einen Tag erstreckende Wanderung oder Spazierfahrt).

1332 day tripper [n] [US]/**day-tripper** [n] [UK] *recr.* (Person making a day excursion by foot or motor vehicle; ►day trip [US]/day-trip [UK]); *s 1* **excursionista** [m/f] (*Término genérico* ►excursión de un día); *s 2* **dominguero** [m] [Es] (Término peyorativo utilizado para persona que desarrolla actividades de tiempo libre mediante desplazamientos de corta duración y predominantemente en domingo); *f* **excursionniste** [m] (Personne qui fait une promenade seul ou en groupe, en voiture ou à pied ; ►excursion journalière) ; *g* **Ausflügler** [m] (Jemand, der eine Wanderung oder eine Spazierfahrt macht; ►Tagesausflug).

day-trippers [npl]**, recreation area for** [UK] *recr.* ►day-use recreation area [US] (1).

day-trip recreation area [n] [UK] *plan. recr.* ►day-use recreation area [US] (2).

1333 day-use recreation area [n] [US] **(1)** *recr.* (Area which is frequented by ►day trippers [US]/day-trippers [UK]; ►hiking area [US]/rambling area [UK]); *syn.* excursion area [n] [UK], recreation area for day-trippers [UK]; *s* **zona** [f] **de excursiones** (Zona atractiva para el esparcimiento en la natura-

leza frecuentada por ►excursionistas; ►zona de montañismo/senderismo); *syn.* zona [f] de esparcimiento; *f* **lieu** [m] **d'excursion** (►Zone de promenade et de randonnée qui accueille les ►excursionnistes) ; *syn.* zone [f] d'excursion ; *g* **Ausflugsgebiet** [n] (Gebiet, das von Erholungsuchenden aufgesucht wird; ►Ausflüglern ►Wandergebiet).

1334 day-use recreation area [n] [US] **(2)** *plan. recr.* (Area, which is developed for recreation activities of one day or less, where overnight accommodation is usually not provided; ►local recreation area, ►short-stay recreation); *syn.* day-trip recreation area [n] [UK], day-use recreation zone [n] [US]; *s* **área** [f] **de recreación diaria** (Zona cercana a la ciudad que sirve para actividades de recreo sin ofrecer generalmente posibilidades de alojamiento; ►puente, ►vacaciones cortas, ►zona de recreo local); *f* **base** [f] **urbaine de plein air et de loisirs** (Aménagement de loisirs et de détente très proche des centres villes pouvant satisfaire les besoins quotidiens de détente de plein air ; ►base périurbaine de plein air et de loisirs, ►tourisme de passage) ; *g* **Tageserholungsgebiet** [n] (Gebiet in der Nähe städtischer Siedlungsgebiete, das der Tageserholung dient und i. d. R. keine Übernachtungsmöglichkeiten vorhält; ►Kurzzeiterholung, ►Naherholungsgebiet).

day-use recreation zone [n] [US] *plan. recr.* ►day-use recreation area [US] (2)/day-trip recreation area [UK].

1335 daywork [n] *constr. contr.* (Minor construction operations which are primarily paid according to negotiated ►hourly billing rates [daywork pay] and recorded on ►daywork sheets to verify the amount of time taken); *s* **trabajos** [mpl] **por salario horario** (Trabajos secundarios en obra que generan mayormente costos de salario y que se liquidan según tasa de salario horario acordada y que deben ser registrados en ►lista de trabajo por horas; ►tasa de honorarios por hora); *syn.* trabajos [mpl] por horas; *f* **travaux** [mpl] **exécutés en régie** (Travaux limités effectués au temps passé, réglés sur la base d'un prix horaire fixé à l'avance et sur présentation de factures justifiées par les bons d'attachement signés par le maître d'ouvrage ; ►prix horaire, ►prix horaire de vacation, ►rapport de chantier journalier) ; *g* **Stundenlohnarbeiten** [fpl] (Überwiegend Lohnkosten verursachende Bauleistungen geringeren Umfanges, die nach vereinbarten ►Stundensätzen und auf Nachweis durch ►Stundenlohnzettel abgerechnet werden); *syn.* Regiearbeiten [fpl].

1336 daywork contract [n] *constr. contr.* (Written agreement for the payment of wages for construction projects [US]/building works [UK] according to accepted hourly rates; ►remuneration on a time basis); *s* **contrato** [m] **por horas** (Acuerdo según el cual la mano de obra necesaria para realizar los trabajos de construcción se paga a una tasa por hora prefijada; ►honorarios por tiempo empleado); *f* **marché** [m] **au temps passé** (Marché pour lequel les travaux réalisés correspondent dans leur majorité à des prestations de main d'œuvre et sont réglées sur la base de tarifs horaires contractuels; ►honoraire au temps consacré) ; *g* **Stundenlohnvertrag** [m] (Vertrag, bei dem die erbrachten Bauleistungen, die meist überwiegend Lohnkosten und [fast] keine Materialkosten verursachen, nach vereinbarten Stundensätzen vergütet werden; ►Honorar nach Zeitaufwand).

1337 daywork sheet [n] *constr. contr.* (Record of ►daywork of labo[u]r and on machines according to an agreement which must be approved daily or weekly by a project supervisor; ►according to certified delivery [US]/on proof of verified delivery notes [UK], ►proof of executed work); *syn.* time sheet [n]; *s* **lista** [f] **de trabajo por horas** (Registro de los ►trabajos por salario horario o de las horas de empleo de maquinaria para presentar al jefe de obra y para liquidar las cuentas; ►con justificación, ►verificación de trabajos cumplidos); *f* **rapport**

[m] **de chantier journalier** (Pièce sur laquelle figure l'effectif du personnel, le relevé des heures effectuées, la description des travaux réalisés, les ▶travaux exécutés en régie ; celle-ci est contrôlée journalièrement ou hebdomadairement par le maître d'œuvre ; ▶avec justifications à l'appui, ▶justification des prestations réalisées) ; *g* **Stundenlohnzettel [m]** (Liste, auf der ▶Stundenlohnarbeiten und Maschinenstunden aufgeführt sind und je nach Vereinbarung werktäglich oder wöchentlich vom Bauleiter anerkannt werden müssen; ▶auf Nachweis, ▶Nachweis der erbrachten Leistungen); *syn.* Stundenrapportzettel [m].

d. b. h. *hort.* ▶girth; ▶trunk diameter.

dead brushwood construction method [n] *constr.* ▶channel deadwood reinforcement.

1338 dead crown [n] *arb. for.* (**1.** Generic term for the dead upper part of deciduous and evergreen trees; **2. in U.S.**, the dead tip of conifers is called **spike-top** [n]; the advanced colo[u]r phase of the foliage of a coniferous tree killed by pests, e.g. bark beetles, or pathogens is termed **red-top** [n] [US&CDN]; *NOTE* the initial phase is termed **sorrel-top**; the final phase, when all the needles have fallen, ▶**black-top** [n] [US&CDN] 2; cf. SAF 1983; ▶dead crown, ▶dieback, ▶top-kill [US]/top drying [UK]); *syn. for conifers* spike-top [n] [US], red-top [n] [US&CDN]; *s* **copa** [f] **muerta** (▶muerte regresiva de plantas, ▶puntiseco, ▶puntiseco en conífera); *f* **houppe** [f] **morte** (▶asphyxie, ▶défoliation de la cime des confères, ▶descente de cime) ; *g* **abgestorbene Baumkrone [f]** (Toter oberirdischer Teil eines Baumes; ▶Absterben, ▶Wipfeldürre, ▶Zopftrockenheit bei Koniferen).

dead end [n] [UK] *trans. urb.* ▶cul-de-sac.

dead-end street [n] [US] *trans. urb.* ▶cul-de-sac.

1339 deadheading [n] **of perennials** *hort.* (Cutting back of spent flowers; ▶cutting back of perennials); *s* **rebaje [m] de perennes** (Fuerte corte de las partes aéreas de plantas vivaces para eliminar las flores secas y los tallos lignificados en otoño o al principio de la primavera; ▶rebaje de perennes 1); *f* **rabattage [m] des vivaces (2)** (Les vivaces capables de remonter sont nettoyées à la fin de l'automne ou au début du printemps ; ▶rabattage des vivaces 1) ; *g* **Staudenrückschnitt [m] (2)** (Komplettes Entfernen der abgeblühten und verholzten oberirdischen Staudenteile im Spätherbst oder vor Frühlingsbeginn; ▶Staudenrückschnitt 1); *syn.* (spät)winterlicher Rückschnitt [m] der Stauden.

1340 dead leaves [npl] *arb. bot.* (of deciduous woody species; ▶clinging dead leaves, ▶foliage); *s* **hojarasca** [f] **(1)** (Conjunto de las hojas caídas de los árboles; ▶follaje, ▶hoja marcescente); *f* **feuilles** [fpl] **mortes** (▶feuillage, ▶feuilles marcescentes, ▶frondaison) ; *g* **abgestorbenes Laub [n, o. Pl.]** (**1.** Von selbst von Gehölzen fallende und auf der Erde liegende, abgestorbene Blätter. **2.** Dürre, braune Laubblätter, die an Bäumen während des Winters hängen; ▶abgestorbene haftende Laubblätter, ▶Belaubung, ▶Laub); *syn.* abgestorbene Blätter [npl], tote Blätter [npl], totes Laub [n, o. Pl.]; *syn. zu 1.* Falllaub [n].

dead-level roof [n] *arch. constr.* ▶flat roof.

deadline [n] [US] *contr. plan.* ▶expiration of a deadline.

deadline [n] [UK]**, expiry of a** *contr. plan.* ▶expiration of a deadline [US].

deadline [n] **for design competition** *plan. prof.* ▶submission deadline for design competition.

deadline [n] **for submission of bids** [US] *constr. contr.* ▶date for submission of bids [US]/date for submission of tenders [UK].

1341 deadman [n] *constr.* (Wooden block embedded horizontally in the ground to support guy wires anchoring transplanted trees; ▶guying, ▶wire guys); *syn.* ground anchor [n]; *s* **rollizo** [m] **de anclaje** (Trozo de madera que se entierra horizontalmente para fijar los ▶vientos de anclaje de árboles transplantados; ▶fijación con vientos [metálicos]); *f* **rondin** [m] **d'ancrage** (Pieu enfoui à l'horizontale dans le sol et utilisé comme fixation du haubanage sur un arbre transplanté ; ▶câble de haubanage, ▶haubanage) ; *g* **Rundholzanker [m]** (Im Boden horizontal eingegrabenes Rundholz zur Befestigung der ▶Drahtseilverankerung von verpflanzten Bäumen; ▶Drahtanker).

1342 dead plant material [n] *constr.* (Non-living vegetative material used in ▶biotechnical construction techniques; e.g. brushwood, ▶fascines, wickerwork, as well as construction timber, etc.; ▶inert construction material); *s* **material** [m] **vegetal muerto** (Partes secas de plantas que se utilizan en la ▶construcción biotécnica, como ramas secas, ▶fajinas secas, madera de construcción, etc.; ▶material de construcción inerte); *f* **matériau** [m] **organique mort** (Végétaux morts utilisés dans les ▶techniques de génie biologique, tels que branches mortes, ▶fascines mortes, rondins, perches, pieux en bois mort ; ▶matériau inerte) ; *g* **nicht lebender Baustoff [m]** (Trockenes pflanzliches Baumaterial für ▶ingenieurbiologische Bauweisen, wie z. B. tote Äste [Grass], tote ▶Faschinen, nicht ausschlagfähiges Reisig, Bauholz; ▶toter Baustoff); *syn.* nicht lebender Bauteil [m] (DIN 18 320).

dead (standing) tree [n] *ecol.* ▶snag [US&CDN] (3).

dead top [n] [AUS] *arb. for.* ▶top-kill [US].

dead valley [n] *geo.* ▶dry valley.

deadwood [n]**, brush mat of** *constr.* ▶brush mat.

1343 dead wooding [n] [US] **(1)** *arb. constr.* (ARB 1983; removing of broken, diseased, dying, and dead limbs, those that cross, are weakly attached, or are of low vigo[u]r; ▶tree and shrub pruning, ▶tree maintenance); *syn.* cleaning out [n] [UK]; *s* **ramoneo** [m] (Eliminación de ramas débiles, secas o dañadas irremediablemente por enfermedades; PODA 1994; ▶poda de leñosas, ▶mantenimiento de árboles); *syn.* monda [f]; *f* **taille** [f] **d'entretien courant** (Une ▶taille des végétaux ligneux de faible envergure et essentiellement préventive comportant **1.** limination des branches mortes ou cassées, des drageons ou des gourmands, la suppression des chicots et **2.** ablation des charpentières mal disposées ; cf. IDF 1988, 45 ; ▶travaux d'entretien des arbres) ; *g* **Erhaltungsschnitt [m]** (▶Gehölzschnitt im Rahmen der regelmäßigen Pflegearbeiten, bei dem trockene, abgebrochene, beschädigte, scheuernde oder nach innen wachsende Triebe herausgenommen werden; ▶Baumpflege).

1344 dead wooding [n] [US] **(2)** *arb. hort.* (ARB 1983, 420; removal of broken, diseased, dying and dead wood from a tree; ▶removal of woody debris); *syn.* cutting out [n] [UK] (BS 3975: Part 5), cleaning out [n] [UK] (BS 3975: Part 5), dry pruning [n] [UK]; *s* **escamonda** [f] (Supresión de ramas muertas en copas de árboles también para garantizar la seguridad pública; ▶eliminación de restos de leñosas); *f* **élagage** [m] **en sec** (Action de couper les branches mortes ; DFM 1975, 107 ; ▶enlèvement du bois mort) ; *syn.* taille [f] de toilette [aussi B] ; *g* **Totholzentfernung [f] (1)** (**1.** Herausschneiden der toten, absterbenden oder gebrochenen Äste im Rahmen der Kronenpflege und zur Erfüllung der Verkehrssicherungspflicht. **2.** Wegnahme von toten Stämmen, Ästen [Trockenäste/Dürräste] und Zweigen, die auf dem Waldboden liegen; ▶Totholzentfernung 2); *syn.* Totholzbeseitigung [f].

deadwood reinforcement [n] *constr.* ▶channel deadwood reinforcement.

debris [n] *envir.* ▶accumulation of drifted debris; *bot.* ▶detritus (1); *geo.* ▶glacial debris, ▶rock debris.

debris [n] [US], **building** *constr.* ▶building rubble.

debris [n], **drifted** *envir.* ▶accumulation of drifted debris.

1345 debris avalanche [n] *geo.* (Rapid flowage type of ▶mass slippage [US]/mass movement [UK]; flowing slide of a big quantity of rock debris in narrow tracks down steep slopes; GGT 1979; ▶rockfall); *s* **avalancha** [f] **de rocas** (Tipo de ▶movimiento gravitacional rápido en el que —al contrario que en el caso de la ▶caída de rocas— gran cantidad de éstas se desplazan a mucha velocidad valle abajo); *syn.* alud [m] de rocas; *f* **avalanche** [f] **de pierres** (Forme du ▶mouvement de masses par lequel pierres et blocs de pierres, par comparaison avec les ▶chutes de pierres, dévalent en grand nombre subitement et au même moment une pente) ; *g* **Steinlawine** [f] (Form des ▶Massenversatzes, bei dem sich Steine und Blöcke im Vergleich zum ▶Steinschlag in großer Zahl gleichzeitig und plötzlich talwärts bewegen; cf. WAG 1984).

debris flow [n] *geo.* ▶mud flow, #2.

debris load [n] *geo. hydr.* ▶bed load.

1346 debris slide [n] *geo.* (*Form of a mass movement* rapid rolling or sliding of unconsolidated earth debris without backward rotation of the mass; GGT 1979; ▶mass slippage [US]/mass movement [UK]); *s* **deslizamiento** [m] **de detritos** (Tipo de movimiento gravitacional rápido por el que los detritos descienden por una pendiente de terreno, bajo la influencia de la gravedad y accionados por flujos de agua superficial o subterránea; DINA 1987; ▶movimientos gravitacionales); *f* **glissement** [m] **d'éboulis** (▶Mouvement de masse rapide de débris meubles en direction d'une vallée sans formation d'un plan de glissement incurvé concave) ; *g* **Schuttrutschung** [f] (Form des ▶Massenversatzes, bei dem lockerer Schutt schnell talwärts gleitet, ohne eine wannenförmige Gleitfläche zu bilden).

decantation test [n] [UK] *constr.* ▶sedimentation analysis.

decay [n] *biol.* ▶root decay; *envir.* ▶stone decay.

decay [n], **leaf litter** *pedol.* ▶leaf litter decomposition.

1347 decayed area [n] *arb.* (Woody part of tree or shrub which has been attacked/destroyed by bacterial, viral or fungal agents; ▶area of rot, ▶wound area); *syn.* area [n] of decay; *s* **zona** [f] **podrida** (Tejido lignificado de árbol o arbusto que ha sido atacado por bacterias, virus u hongos; ▶caries húmeda, ▶herida); *syn.* zona [f] de putrefacción (en la madera); *f* **point** [m] **de pourriture (sur le tronc)** (Tissu de l'aubier ou du bois parfait atteint et détruit par les micro-organismes ou les champignons ; ▶plaie, ▶pourrissement) ; *syn.* cavité [f] de pourriture, trou [m] de pourriture, bois [m] cari (DFM 1975, 43), foyer [m] de pourriture (IDF 1988, 116) ; *g* **Faulstelle** [f] (Durch Krankheitserreger befallene und zerstörte Gewebeteile der Splintholzschicht oder der oberen Kernholzschicht bis ca. ¼ des Ast- oder Stammdurchmessers; ▶Morschung, ▶Wundstelle); *syn.* Faulherd [m].

1348 deceleration lane [n] *trans.* (Additional lane adjacent to a traffic lane, which allows vehicles to leave the road at a reduced speed without hindering the rest of the through traffic; *opp.* ▶acceleration lane); *s* **carril** [m] **de desaceleración** (Carril adicional adyacente a la calzada de autopista que permite a los vehículos salir de ésta sin impedir al resto del tráfico; *opp.* ▶carril de aceleración); *f* **bande** [f] **de décélération** (Surface en bordure de chaussée d'une autoroute réservée aux véhicules quittant cette voie ; *opp.* ▶bande d'accélération) ; *syn.* voie [f] de décélération ; *g* **Verzögerungsspur** [f] (Eine neben der Fahrbahn liegende zusätzliche Fahrspur, die dazu dient, sich ausfädelnden Fahrzeugen die Herabsetzung der Geschwindigkeit ohne Behin-

derung des durchgehenden Verkehrs zu ermöglichen; *opp.* ▶Beschleunigungsspur); *syn.* Ausfädelungsspur [f].

decibel level [n] [UK] *constr. envir. leg.* ▶sound pressure level.

1349 deciduous [adj] *bot.* (Characteristic of ▶broad-leaved woody species to shed their leaves during a climatically unfavo[u]rable season; ▶leafless, ▶summer green, ▶winter bare, ▶winter green; *opp.* ▶evergreen); *syn.* caducous [adj]; *s* **caducifolio/a** [adj] (Así se llaman los árboles y arbustos que no se conservan verdes todo el año, porque se les cae la hoja al empezar la estación desfavorable; DB 1985; *syn. para latitudes templadas* ▶desnudo en invierno, ▶sin hojas, ▶verde vernal; *opp.* ▶perennifolio); *syn.* deciduo/a [adj], de hoja caduca [loc], de hojas caedizas [loc]; *f 1* **à feuilles caduques** [loc] (Caractère des végétaux ligneux qui perdent toutes leurs feuilles chaque année pendant la saison froide ou sèche ; *terme spécifique pour les arbres* **caducifolié** ; ▶à feuillage vert pendant l'été, ▶dénudé, ▶dépourvu de feuillage pendant l'hiver, ▶semipersistant, ante ; *opp.* ▶à feuilles persistantes) ; *syn.* à feuillage caduc [loc], à feuillage décidu [loc] ; *f 2* **tropophile** [adj] (Végétation de type forêts sèches caducifoliées des zones tropicale et subtropicales qui perdent leurs feuilles pendant la saison sèche) ; *g* **Laub abwerfend** [ppr/adj] (1. Eigenschaft von Laubgehölzen, in klimatisch ungünstigen Jahreszeiten das Laub abzuwerfen; *opp.* ▶immergrün; *syn. für gemäßigte Breiten* ▶winterkahl, ▶sommergrün; **2.** Eine Zwischenform von **L. a.** und immergrün ist ▶wintergrün, da nach dem Winter das Laub vollständig erneuert wird; ▶unbelaubt).

1350 deciduous forest [n] *for. phyt.* (Forest consisting entirely or mainly of broad-leaved trees which, with a few exceptions, lose all their leaves seasonally; BS 3975, Part 5; ▶broad-leaved woody species, ▶coniferous forest, ▶mixed forest, ▶needle-leaved deciduous forest); *syn.* broad-leaved forest [n], broad-leaved woodland [n], broadleaf forest [n] [also US] (TEE 1980, 249); *s* **bosque** [m] **de frondosas** (Formación vegetal en la que predominan árboles caducifolios al contrario que en el ▶bosque mixto y el ▶bosque de coníferas; ▶bosque aciculifoliado deciduo, ▶frondosa); *syn.* bosque [m] de hojas latifoliadas, bosque [m] planocaducifolio, bosque [m] planifolio, bosque [m] de hoja caduca, bosque [m] de hoja caediza; *f* **forêt** [f] **(à essences) feuillue(s)** (Par opposition à la ▶forêt mélangée, la ▶forêt mixte et la ▶forêt de conifères, groupement végétal dans lequel dominent les ▶essences [ligneuses] à feuillage caduque ; ▶feuillu, ▶forêt de résineux caducifoliée) ; *syn.* forêt [f] feuillue (caducifoliée), forêt [f] (feuillue) décidue [aussi CDN] ; *g* **Laubwald** [m] (Pflanzenformation, in der ▶Laubgehölze im Vergleich zum ▶Mischwald und ▶Nadelwald vorherrschen; ▶sommergrüner Nadelwald).

1351 deciduous tree [n] *bot.* (Angiospermous tree which bears broad-shaped leaves as opposed to those of ▶coniferous trees; ▶macrophanerophyte); *syn.* broad-leaved tree [n], broadleaf tree [n]; *s* **árbol** [m] **frondoso** (Árbol o arbusto que no se conserva verde todo el año, porque se le caen las hojas al empezar la estación desfavorable [estación fría o seca]; DB 1985; *opp.* ▶árbol acicufolio; ▶macrofanerófito); *syn.* árbol [m] caducifolio; *f 1* **arbre** [m] **feuillu** (Qualifie un arbre ligneux qui par opposition aux ▶arbres conifères portant des feuilles à large limbe ; ▶macrophanérophyte) ; *f 2* **bois** [m] **de feuillus** (Bois provenant d'essences forestières, appartenant à la classe des Angiospermes) ; *syn.* bois [m] franc [CDN], bois [m] feuillus (DFM 1975, 128) ; *g* **Laubbaum** [m] (Bedecktsamiger Baum, der im Gegensatz zum ▶Nadelbaum breitflächige Blätter ausbildet; ▶Makrophanerophyt).

D

deciduous woody plant [n] *bot.* ►broad-leaved woody species.

decked car park [n] [US] *trans. urb.* ►multideck car park [US]/multi-storey car park [UK].

decline [n]**, population** *sociol.* ►depopulation.

1352 decline [n] **in number of species** *phyt. zool. syn.* decline [n] in species diversity, reduction [n] in species diversity; *s* **merma [f] de especies** (Reducción de la diversidad de especies en un área); *syn.* regresión [f] de especies; *f* **dégénérescence [f] des espèces** *syn.* régression [f] des espèces ; *g* **Artenrückgang [m]** (Verringerung der Artenvielfalt); *syn.* Artenverarmung [f], Artenschwund [m, o. Pl.].

decline [n] **in species diversity** *phyt. zool.* ►decline in number of species.

1353 decomposer [n] (1) *biol. ecol.* (Organism at the base of a food chain which breaks down organic matter into simpler inorganic substances); *syn.* reducer [n] (1); *s* **descomponedor [m]** (Organismo heterótrofo, normalmente un hongo o una bacteria, que descompone, para alimentarse, los cuerpos de las plantas o animales muertos. Estos organismos llevan a cabo la mineralización de la materia orgánica; DINA 1987); *syn.* reductor [m], desintegrador [m]; *f* **décomposeur [m]** (Microorganismes fongiques et bactériens assurant la minéralisation des substances organiques) ; *syn.* réducteur [m], bioréducteur [m] ; *g* **Destruent [m]** (...en [pl]; organische Substanz abbauende und zu anorganischen Stoffen umwandelnder Organismus [Bakterium, Pilz]); *syn.* Reduzent [m].

decomposer [n] (2) *pedol. phyt.* ►raw humus decomposer.

1354 decomposition [n] (1) *biol. chem. pedol.* (Molecular breakdown of organic matter resulting in physical/chemical separation into its elements or simpler constituents; ►leaf litter decomposition, ►litter decomposition); *syn.* breakdown [n], degradation [n]; *s* **descomposición [f]** (Disgregación molecular de la materia orgánica; ►descomposición de la hojarasca, ►descomposición de litter); *syn.* degradación [f], alteración [f] [YZ también]; *f* **décomposition [f]** (Dislocation moléculaire et décomposition chimique et physique des matières organiques ; ►décomposition de la litière des feuilles, ►décomposition de la litière) ; *syn.* altération [f] (1), dégradation [f] ; *g* **Abbau [m, o. Pl.] (1)** (Molekulare Zersetzung und physikalisch-chemische Veränderung [Vereinfachung] der organischen Substanz; ►Laubstreuzersetzung, ►Streuzersetzung); *syn.* Zersetzung [f].

decomposition [n] (2) *envir.* ►waste decomposition.

1355 decomposition dump [n] [US] *envir.* (Loosely dumped [US]/tipped [UK] organic material in a heap up to 2m high, which decomposes aerobically, whereby the detritus or rotting matter is reduced to roughly 50% of its original volume); *syn.* decomposition tip [n] [UK]; *s* **depósito [m] de putrefacción** (Almacenamiento de residuos orgánicos hasta un máximo de 2 m que se putrifican en proceso aeróbico, de manera que se reduce el volumen en aprox. un 50%); *f* **dépôt [m] de décomposition** (±) (Matières ou résidus organiques déposés en tas de hauteur maximale de 2 m, décomposés par un processus aérobe à la fin duquel les matières décomposées auront été réduites à 50 % de leur volume d'origine) ; *g* **Rottedeponie [f]** (Bis maximal 2 m hoch, locker geschüttete Deponiestoffe, die durch aerobe Zersetzungsprozesse abgebaut werden, wodurch sich das Schüttgut bis auf ca. 50 % seines ursprünglichen Volumens reduziert).

decomposition tip [n] [UK] *envir.* ►decomposition dump [US].

1356 decontamination [n] **of soil** *envir.* (Purification of toxic soil; ►hazardous waste material, ►recycling of derelict sites); *syn.* remediation [n] of contaminated soil; *s* **saneamiento**

[m] de suelos (Conjunto de medidas para descontaminar un ►suelo contaminado con el resultado de la recuperación del mismo para un nuevo uso; ►recuperación de ruinas industriales); *f* **décontamination [f] d'un sol pollué** (►déchets historiques, ►réhabilitation des friches industrielles, ►réparation de dommages causés à l'environnement) ; *syn.* dépollution [f] du sol, réhabilitation [f] des sols pollués ; *g* **Bodensanierung [f]** (Entgiftung eines schadstoffbelasteten Bodens; ►Altlast, ►Flächenrecycling, ►Sanierung); *syn.* Bodendekontaminierung [f], Bodenreinigung [f], Dekontamination [f] des Bodens.

decontamination [n] **of toxic waste sites** *envir. plan.* ►toxic site reclamation.

1357 decontamination plant [n] *envir.* (Plant which causes an accelerated breakdown of organic substances in combination with micro-organisms in contaminated soils or in wastewater; *specific term* aquatic plant for wastewater treatment (WWT 1989, 506); ►soil decontamination with plants, ►wastewater treatment wetland, ►wetland wastewater treatment); *s* **planta [f] descontaminante** (±) (Especie vegetal que —con ayuda de microorganismos— puede descomponer sustancias orgánicas contaminantes en suelos, masas de agua o ►filtros verdes más rápidamente de lo normal; ►fitosaneamiento de suelos, ►tratamiento de aguas residuales por filtros verdes); *f* **plante [f] décontaminante** (Plante responsable de la décomposition des substances organiques toxiques dans les sols contaminés ou les eaux usées par l'intermédiaire de micro-organismes ; l'utilisation de plantes pour dépolluer les milieux souillés, p. ex. par des métaux lourds [zinc, mercure, plomb, arsenic, sélénium, etc.] s'appelle la **phytorémédiation** et les travaux correspondants la **phytorestauration** ; ►décontamination du sol par les plantes, ►lagunage à macrophytes, ►lagunage par roselière) ; *syn.* plante [f] phytorémédiatrice ; *g* **Repositionspflanze [f]** (Bodenfilterbildende Pflanze, die einen beschleunigten Abbau von organischen Schadstoffen in kontaminierten Böden, im Abwasser, in Pflanzenkläranlagen oder in Badeteichen zusammen mit Mikroorganismen bewirkt. Effektive **R.n** sind z. B. Süßgräser *[Phragmites australis, Phalaris arundinacea]*, Sauergräser *[Carex acutiformis, C. gracilis, C. riparia]*, Rohrkolben *[Thypha angustifolia, T. minima]* und Sumpfschwertlilie *[Iris pseudacorus]*; **R.n** werden wegen ihrer Reinigungsfähigkeit auch zur Rekultivierung von Uferrändern und Feuchtgebieten gepflanzt; ►Pflanzkläranlage, ►Phytosanierung, ►Wurzelraumentsorgung).

decontamination [n] **with plants** *envir.* ►soil decontamination with plants.

1358 decorative board fence [n] *constr.* (**1.** Fence built of vertical wood members fixed at close regular intervals to ►arris rails, secured between posts. The pickets may be sawn or planed from hardwood or softwood; of rectangular, triangular or half-rounded section. **2.** Wooden fence with pointed vertical pales; ►wooden fence; *generic term* ►fence); *s* **valla [f] de tablas perfiladas** (Cerca construida con listones o tablas perfiladas colocadas a cierta distancia entre sí y fijadas por un lado a las ►traviesas horizontales; *término genérico* ►valla; ►valla de madera); *f* **clôture [f] de planches ajourée** (Planches ou lattes verticales fixées avec espacement sur un côté des lices ; ►clôture 2, ►clôture en bois) ; *syn.* clôture [f] de planches à claire-voie ; *g* **Profilbretterzaun [m]** (Einfriedigung mit Profilbrettern oder -latten, die senkrecht und mit Abstand an den [waagerechten] ►Riegeln einer Seite befestigt sind; ►Holzzaun; *OB* ►Zaun).

1359 decorative foliage plant [n] *hort. plant.* (Decorative perennial, e.g. with a coarse-textured foliage pattern, such as Elephant's Ears [US]/Bergenia [UK] *[Bergenia cordifolia]*,

Plantain-Lily *[Hosta spp. and cvs.]*, Rodgersia *[Rodgersia spp., Astilboides tabularis]*, Giant Reed *[Arundo donax]*); ▶ornamental perennial); *s* **vivaz [f] de hojas decorativas** (Perenne que se cultiva por su follaje colorista, más que por sus flores, como p. ej. la bergenia *[Bergenia cordifolia]*, el llantén *[Hosta spp. y cvs.]*, la rodgersia *[Rodgersia spp., Astilboides tabularis]*; ▶vivaz ornamental); *syn.* perenne [f] de hojas decorativas, planta [f] vivaz de hojas decorativas; *f* **plante [f] vivace à feuilles ornementales** (Plante vivace dont l'utilisation culturale est due principalement à la valeur décorative de son feuillage ; ▶plante vivace d'ornement) ; *syn.* plante [f] vivace décorative par son feuillage ; *g* **Blattschmuckstaude [f]** (*Terminus für das Wuchsverhalten einer Staude hinsichtlich ihrer ästhetischen Wirkung und gärtnerischen Verwendung*: Staude, die primär durch ihr Laub ziert; es wirkt meist länger als die Blüten, die manchmal unauffällig sind, und ist in seiner auffälligen und attraktiven Blattgestalt präsent: [großflächig, linear, farbig, Textur etc. wie z. B. Bergenie *(Bergenia cordifolia)*, Funkie *(Hosta in Arten und Sorten)*, Schaublatt *(Rodgersia spec., Astilboides tabularis)*, Pfahlrohr *(Arundo donax)*]; ▶Schmuckstaude); *syn.* Blattstaude [f].

decrease [n] in costs [UK] *contr.* ▶cost reduction.

1360 decrease [n] in population *biol.* (Reduction in the number of individuals of a species); *s* **disminución [f] de una población** (Reducción del número de individuos de una especie); *f* **régression [f] d'un peuplement** (du nombre d'individus d'une population) ; *g* **Abnahme [f] eines Bestandes** (Verminderung der Individuen einer Population).

deed register [n] [US] *adm. surv.* ▶real property identification map [US].

1361 deep [adj] *pedol.* (Descriptive term referring to a thick upper soil profile; *opp.* ▶shallow 2); *s* **profundo/a [adj]** (Término descriptivo referente al grosor de la capa vegetal [horizonte A] de un suelo; *opp.* ▶poco profundo/a 2); *f* **profond, onde [adj]** (Relatif à l'épaisseur de la couche de terre végétale ; *opp.* ▶superficiel, elle) ; *syn.* épais, aisse [adj] (LA 1981, 1048) ; *g* **tiefgründig [adj]** (Einen Oberboden mit starker Mächtigkeit betreffend; *opp.* ▶flachgründig 2).

1362 deep mining [n] *min.* (Underground exploitation of mineral resources; ▶mining; *opp.* ▶open-pit mining [US]/opencast mining [UK]); *syn.* underground mining [n] [also US]; *s* **minería [f] en galería** (Explotación subterránea de recursos minerales; ▶minería; *opp.* ▶minería a cielo abierto); *syn.* minería [f] subterránea; *f* **exploitation [f] minière souterraine** (Extraction de minerais métallurgiques, de minéraux industriels et de métaux natifs et dont l'exploitation a lieu dans le sous-sol ; ▶exploitation minière ; *opp.* ▶exploitation [minière/de carrière] à ciel ouvert) ; *g* **Untertagebau [m, o. Pl.]** (Unterirdischer Abbau von Steinen, Erden und anderen Bodenschätzen; ▶Bergbau; *opp.* ▶Tagebau); *syn.* unterirdischer Abbau [m, o. Pl.], untertägiger Abbau [m, o. Pl.].

1363 deep pit mining [n] *min.* (Form of ▶mining by which the ▶deposits are extracted from several hundred meters [US]/metres [UK] below the earth's surface, e.g. brown coal lignite operations in the German Rhineland and the exploration of copper deposits in Bingham Canyon, Utah, USA); *s* **explotación [f] a cielo abierto profunda** (Forma de ▶minería superficial en la que se excavan los ▶yacimientos hasta varios kilómetros cuadrados de extensión y más de cien metros de profundidad, como p. ej. la minería de lignitos en Renania o de cobre en Utah [USA]); *f* **exploitation [f] minière à ciel ouvert** (Forme d'▶exploitation minière à ciel ouvert, pour laquelle les ▶gisements de minéraux et de fossiles peuvent être exploités à plusieurs centaines de mètres de profondeur, p. ex. le bassin

houiller de Rhénanie) ; *g* **Tieftagebau [m]** (Form des ▶Bergbaus, bei dem die ▶Lagerstätten von der Erdoberfläche aus mehrere hundert Meter tief abgebaut werden können, z. B. Rheinisches Braunkohlenrevier oder der Kupferabbau im Bingham Canyon in Utah, USA).

deep ploughed soil [n] [UK] *pedol.* ▶deep plowed soil [US].

deep ploughing [n] [UK] *agr. for. hort.* ▶deep plowing [US]/deep ploughing [UK].

1364 deep plowed soil [n] [US] *pedol.* (▶Great soil group [US]/soil group [UK] of ▶man-made soil, created by a ▶deep plowing [US]/deep ploughing [UK]. The inverted horizon is > 40cm deep; ▶trenching); *syn.* deep ploughed soil [n] [UK]; *s* **suelo [m] subsolado en profundo** (▶Tipo de suelo de los ▶cultosoles originado por operaciones de ▶arado en profundo. El horizonte de subsolado es > 40 cm; ▶subsolar en profundo); *syn.* suelo [m] de arado profundo; *f* **sol [m] de culture défoncé** (▶Groupe de sol appartenant aux ▶sols de culture caractérisé par le retournement en profondeur [▶défonçage] des couches ; la profondeur de l'horizon de défonçage est > à 40 cm ; ▶labour profond) ; *g* **Rigosol [m]** (▶Bodentyp 1 der Kultosole, der durch tiefgründige Bodenumschichtung [▶Rigolen] entstanden ist. Der Rigolhorizont ist > 40 cm tief; ▶anthropogener Boden, ▶Tiefpflügen); *syn.* rigolter Boden [m].

1365 deep plowing [n] [US]/**deep ploughing [n]** [UK] *agr. for. hort.* (Mechanical working of the soil below the usual level of a normal plow [US]/plough [UK] to break up compacted soil horizons; ▶subsoiling); *s* **arado [m] en profundo** (Consiste en dar a la tierra una labor más profunda de lo normal para romper la compacidad del suelo; ▶desfonde); *f* **labour [m] profond** (Opération de travail profond du sol ayant pour objet de provoquer la fracturation d'un sol compacté à une profondeur supérieure à celle des horizons habituellement labourés ; DAV 1984 ; ▶décompactage) ; *syn.* défoncement [m] profond (LA 1981, 396) ; *g* **Tiefpflügen [n, o. Pl.]** (Unter die übliche Pflugtiefe hinabreichende mechanische Bodenbearbeitung mit dem Ziel, verdichtete Bodenhorizonte aufzureißen; ▶Tiefenlockerung); *syn.* Tiefenmelioration [f], Tiefenbruch [m].

deep ripping [n] [UK] *constr.* ▶subsoiling.

deep-rooted plant [n] [US] *bot. hort.* ▶deep-rooting plant.

1366 deep-rooting plant [n] *bot. hort.* (Plant whose root system extends below the upper soil layers; ▶tap-rooted plant, *opp.* ▶shallow-rooting plant); *syn.* deep-rooted plant [n] [also US]; *s* **planta [f] de radicación profunda** (▶Planta con raíz axonomorfa; *opp.* ▶planta de radicación profunda); *f* **plante [f] à racines profondes** (Se dit des plantes dont l'ensemble des racines s'enfonce profondément dans la sol ; ▶plante à racine pivotante ; *opp.* ▶plante à racines traçantes) ; *g* **Tiefwurzler [m]** (Pflanze mit tief in den Boden eindringender Haupt- oder Pfahlwurzel; ▶Pfahlwurzler; *opp.* ▶Flachwurzler); *syn.* tief wurzelnde Art [f], tief wurzelnde Pflanze [f].

deep sea fishing [n] *agr.* ▶fishery.

deep soil roof planting [n] [US] *constr.* ▶intensive roof planting.

deer-stop fence [n] [UK] *conserv. for. hunt. trans.* ▶animaltight fence [US].

1367 defect [n] *contr. leg.* (Deficiency in a piece of executed work, which does not correspond to accepted standards due to the non-compliance with specifications as laid down by a contract; also any condition or characteristic which detracts from the appearance, strength or durability of an object); *syn.* defective work [n]; *s* **defecto [m] (de obra)** (Deficiencia en un trabajo ejecutado por no corresponder a los estándares o a las especificaciones fijadas en el contrato); *syn.* fallo [m] (de construcción), trabajo

[m] defectuoso; *f 1* **imperfection** **[f]** (*Dans le cadre de la réception des travaux* caractéristique d'un ouvrage n'étant pas entièrement conforme aux spécifications du marché, aux prescriptions, normes et documents techniques généraux et dont les éléments défectueux ne sont pas de nature à porter atteinte à la sécurité, au comportement ou à l'utilisation de l'ouvrage) ; *syn.* déficience [f] ; *f 2* **malfaçon** **[f]** (*Dans le cadre de la réception des travaux* caractéristique d'un ouvrage ne respectant pas la qualité exigible ou non conforme aux spécifications du marché, aux prescriptions, normes et documents techniques généraux) ; *f 3* **vice** **[m]** **de construction** (Malfaçon importante pour laquelle le décèlement du vice, le rétablissement de l'intégrité de l'ouvrage, ou sa mise en conformité avec les règles de l'art et les stipulations du marché peut comprendre la démolition partielle) ; *g* **Mangel** **[m]** (Mängel [pl]; Nichtentsprechung von vertraglich zugesicherten Eigenschaften [geschuldeter Sollzustand] der ausgeführten Bauleistungen, da diese nicht den anerkannten Regeln der Technik entsprechen. Kein **M.** ist die Falschleistung, bei der eine ganz andere als die vereinbarte Leistung erbracht wurde; hier hat der Vertragspartner nicht schlecht, sondern im Rechtssinne überhaupt keine Leistung erbracht); *syn.* Sachmangel [m].

defective work [n] *contr. leg.* ▶defect.

1368 defects liability period [n] *contr. leg.* (Period of time after ▶issue of a certificate of compliance [US]/issue of a certificate of practical completion [UK], specified in contract, usually one year, during which the contractor must replace dead or diseased plants and repair defective building or landscape construction; ▶defect); *syn.* guarantee period [n] [also US]; *s* **plazo** **[m]** **de garantía** (Periodo/período de dos o más años que comienza después de la ▶aceptación de obra y en el cual se han de corregir los posibles ▶defectos que surjan o reponer las posibles marras); *syn.* periodo [m] de garantía; *f* **délai** **[m]** **de garantie** (Période de 2, 5 ou 10 ans après ▶réception des travaux pendant laquelle le titulaire d'un marché doit la réfection des ▶malfaçons constatées sur les ouvrages défectueux réalisés par ses soins ; ▶imperfection, ▶vice de construction) ; *g* **Gewährleistungsfrist** **[f]** (Zeitraum von fünf oder zwei Jahren nach ▶Abnahme, den ein Auftragnehmer gem. BGB oder VOB zur Behebung von ▶Mängeln an Bauleistungen einräumen muss); *syn.* Rügefrist [f] [CH].

1369 deficiency [n] **in implementation** *envir. pol.* (Insufficient implementation of environmental goals into action, due to difficulties and obstacles in carrying out the law or legally binding plans); *s* **déficit** **[m]** **de implementación** (Cumplimiento insuficiente de una ley o un programa ambiental debido a problemas y obstáculos para poner en práctica sus objetivos); *syn.* deficiencia [f] de implementación; *f* **insuffisances** **[fpl]** **dans l'application** (Faiblesses dans la mise en œuvre d'objectifs en matière de politique environnementale causées par des difficultés et des obstacles dans l'application des lois) ; *g* **Vollzugsdefizit** **[n]** (Zustand unzureichender Verwirklichung umweltpolitischer und anderer, insbesondere gesetzlicher Zielsetzungen durch Schwierigkeiten und Hindernisse im Gesetzesvollzug).

deficit [n] *plan.* ▶shortfall.

1370 defined elevation [n] *constr.* (Specified level on the construction site, to which the ground has to be graded; ▶meeting finished grade [US]/marrying with existing levels [UK]; ▶make a finished grade flush with); *s* **cota** **[f]** **de nivel** (En obra, nivel del suelo proyectado hasta el cual hay que rellenar o cortar el terreno; ▶enrase, ▶equiparar alturas); *f* **cote** **[f]** **de raccordement** (Cote projetée ou cote existante sur un chantier à laquelle doivent se raccorder les bâtiments ou les matériaux mis en place ; ▶raccordement au niveau existant, ▶se raccorder au

niveau fini) ; *g* **Anschlusshöhe** **[f]** (Höhe auf der Baustelle, an die mit einzubauenden Baustoffen oder Konstruktionen angeschlossen wird; ▶Höhenanschluss, ▶höhengerecht anschließen).

1371 definition [n] **of spatial units** *plan.* (Outlining of areas when establishing ▶spatial distribution 2); *s* **delimitación** **[f]** **de unidades de estructura ecológica del territorio** (Definición de unidades espaciales en el marco de la ▶estructuración ecológica del paisaje); *f* **délimitation** **[f]** **spatiale** (Définition des unités spatiales lors de la division en types de territoires naturels ; ▶division spatiale) ; *g* **Raumabgrenzung** **[f]** (Definierung von Raumeinheiten bei der Durchführung von ökologischen ▶Raumgliederungen).

1372 deflation [n] *geo.* (Removal of dry unconsolidated particles, especially dust and sand, from the land surface by the wind, mostly in dry climates with poor vegetative cover; cf. DNE 1978; ▶extreme soil erosion, ▶truncated soil profile, ▶wind erosion); *s* **deflación** **[f]** (▶Erosión eólica producida al ponerse en movimiento partículas sueltas; DINA 1987; ▶destrucción del suelo. ▶suelo decapitado); *f* **déflation** **[f]** **éolienne** (Enlèvement des sols affleurants par l'action du vent ; ▶dévastation des sols, ▶érosion éolienne 2, ▶sol décapé) ; *syn.* érosion [f] éolienne (1) ; *g* **Deflation** **[f]** (Flächige Abtragung von Boden durch ausblasende Tätigkeit des Windes, meist in Klimazonen mit geringer Vegetationsbedeckung; ▶Bodenverheerung, ▶geköpftes Bodenprofil, ▶Windabtragung); *syn.* äolische Abtragung [f], Auswehung [f].

defoliant [n] *chem.* ▶defoliating agent.

1373 defoliate [vb/intr] (1) *hort.* (To remove leaves from a growing plant partially or completely as a cultural practice; ▶defoliating agent); *s* **defoliar** **[vb]** (1) (1. Remoción manual de las hojas de árboles y arbustos como medida horticultural. 2. Destrucción de la vegetación de grandes superficies por medio de agentes desfoliantes, tal como la practicó el ejército norteamericano durante la guerra de Vietnam utilizando el llamado *agent orange* que contenía dioxinas; ▶defoliador) ; *syn. de 1.* deshojar [vb], desfoliar [vb], desfollonar [vb] (DB 1985); *f* **défolier** **[vb]** (1. Opération manuelle d'arboriculture consistant à supprimer les jeunes feuilles entourant le bourgeon terminal en activité afin de régulariser la croissance des arbres et des arbustes ; l'action de défolier ou le résultat est la « défoliation » ; ▶phytocide. 2. Destruction de forêts entières par l'usage de ▶défoliants [agent orange] pendant la guerre du Viêt-Nam) ; *g* **entlauben** **[vb]** (1. Anuelles Entfernen von Blättern bei Sträuchern und Bäumen als gärtnerische Kulturmaßnahme. 2. Zerstörung ganzer Wälder durch den Einsatz von ▶Entlaubungsmitteln [Agent orange] während des Vietnamkrieges); *syn. zu 1.* entblättern [vb].

1374 defoliate [vb/tr] (2) *bot.* (To shed leaves usually due to extremes in climatic or other environmental conditions; ▶defoliation); *s* **defoliar** **[vb/tr]** (2) (Desprendimiento natural de las hojas, principalmente de los árboles y arbustos, cuyo resultado es la ▶defoliación; cf. DB 1985); *f* **perdre** **[vb]** **les feuilles** (À la fin de la période de végétation les végétaux à feuilles caduques des zones climatiques froide et tempérée perdent leurs feuilles ; *contexte* en automne commence la chute des feuilles ; ▶défeuillaison) ; *g* **Laub abwerfen** **[vb]** (Abstoßen der Blätter bei sommergrünen Gehölzen am Ende der Vegetationsperiode in kalten und gemäßigten Klimazonen; *Kontext* die Gehölze werfen das Laub ab; im Herbst fällt das Laub [von den Bäumen]; ▶Laubfall).

1375 defoliating agent [n] *chem.* (Chemical substance for removing woody plant leaves; arboricides are called **brushkillers** in U.S.; cf. SAF 1983, 4350; ▶herbicide); *syn.* defoliant [n], 'agent [n] orange'; *s* **defoliador** **[m]** (Sustancia química que sirve para destruir las hojas de leñosas; ▶herbicida); *syn.* agente [m]

D

defoliante; *f 1* **phytocide** [m] (Produit chimique utilisé pour détruire les végétaux ligneux) ; *f 2* **défoliant** [m] (Terme désignant les produits chimiques [phytocides] utilisés en traitement aérien par les américains au Viêt-Nam ; terme impropre parfois utilisé pour désigner les phytocides ; ►herbicide) ; *syn.* produit [m] défoliant ; *g* **Entlaubungsmittel** [n] (Chemisches Mittel zur Entfernung von Blättern an Gehölzen; ►Herbizid); *syn.* Entblätterungsmittel [n].

1376 defoliation [n] *bot.* (**1.** Natural process, whereby trees and shrubs loose their leaves in autumn, triggered by various plant hormones. The leaves are shed from plants, which are green during the summer, at the end of the growing season or as a cleaning process during the growing season by congestion of the cells in the abscission tissue at the end of the petiole. Before **d.** takes place nutrients are transported from the leaves into the buds, stems, roots and other storage organs, as reserve substances; ►senescence 1 of the foliage of ►hemicryptophytes and ►cryptophytes; **d.** of ►summer green plants is concentrated for a short period before winter [**autumn, annual d.**], though summer drought can also lead to a premature shedding of leaves [**summer d.**], for example in the small-leaved lime tree *[Tilia cordata]*. In ►evergreen plants **d.** is more or less [often rhythmically] distributed throughout the year; between the two are those trees and shrubs, which are ►winter green whereby leaves remain for a maximum of 12 months before completely defoliating in or after the winter; ►deciduous; **2. to shed leaves** [vb]: *context* in the autumn the leaves are shed from the trees, the shrubs shed their leaves; ►defoliate 2, ►leaf litter; *syn.* shedding [n] of leaves, leaf fall [n] [also US]; *s* **defoliación** [f] (Desprendimiento natural de las hojas en otoño; ►caducifolio/a, ►caída de las hojas, ►criptófito, ►defoliar, ►hemicriptófito, ►hojarasca, ►perennifolio/a, ►verde invernal, ►verde vernal); *f* **défeuillaison** [f] (Chute naturelle des feuilles des arbres et arbustes en automne ; ►disparition des feuilles chez les hémicryptophytes et les kryptophytes ; ►à feuilles caduques, ►à feuilles persistantes, ►à feuillage vert pendant l'été, ►litière feuillue, ►perdre les feuilles, ►semi-persistant) ; *syn.* chute [f] des feuilles, défoliation [f], effeuillement [m], effeuillaison [f] ; *g* **Laubfall [m, o. Pl.]** (Natürlicher Vorgang, ausgelöst durch verschiedene Pflanzenhormone, bei dem die Blätter der sommergrünen Holzgewächse am Ende einer Vegetationsperiode oder als Reinigungsprozess während der Vegetationsperiode durch Verschleimung oder Abrundung der Zellen des parenchymatischen Trennungsgewebes am Grunde des Blattstieles von den Bäumen und Sträuchern abfallen. Vor dem **L.** werden noch Nährstoffe aus den Blättern in die Knospen, Triebe, den Stamm und in Wurzeln und sonstige Speicherorgane transportiert und als Reservestoffe eingelagert; ►Einziehen des Laubes bei ►Hemikryptophyten und ►Kryptophyten; der **L.** konzentriert sich bei ►sommergrünen Gehölzen auf eine kurze Periode *vor* dem Winter [**Herbstl., alljährlicher L.**], wobei besonders bei Trockenheit schon im Sommer ein Teil der Blätter abgeworfen werden kann [**Sommerlaubfall**], z. B. bei Winterlinde *[Tilia cordata]*. Bei ►immergrünen Gehölzen ist der **L.** mehr oder weniger stark [meist rhythmisch] über das Jahr verteilt [**Treiblaubfall**]; dazwischen stehen ►wintergrüne Gehölze, deren Blätter maximal 12 Monate leben und die einen vollständigen **L.** *im* oder *nach* dem Winter vollziehen; ►Laub abwerfend, ►Laub abwerfen, ►Laubstreu); *syn.* Blattfall [m, o. Pl.]).

1377 deforestation [n] *for. land'man.* (**1.** Removal of forest cover without subsequent replanting; ►burn clearing, ►clear felling, ►reforestation, ►shifting cultivation, ►tree clearing and stump removal [operation]; *opp.* ►afforestation. **2** Processes of tree felling in the Mediterranean which encouraged soil erosion and laid bare calcareous or other rock surfaces is also called

denudation); *s* **de(s)forestación** [f] (Eliminación de las masas forestales sin restituirlas, antiguamente como resultado de la ►agricultura itinerante, en la actualidad en América Latina más como resultado de la explotación del suelo para el cultivo de soja como planta forrajera y de soja, caña de azúcar o palmas para producir agrocombustibles o, como pasto para el ganado vacuno. La ganancia de maderas preciosas es un negocio adicional en este contexto; ►despeje de árboles y tocones, ►reforestación, ►roturación, ►tala a fuego, ►tala a matarrasa); *syn. vb.* desboscar [vb] (DB 1985, 173); *f 1* **défrichement** [m] (Opérations ayant pour effet de détruire l'état boisé d'un terrain par abattage des arbres et arrachage des souches et de mettre fin à sa destination forestière ; cf. DFM 1975 ; défricher [vb], déboiser [vb] ; ►abattage par extraction de souche, ►agriculture itinérante [sur brûlis], ►brûlis de défrichement, ►coupe à blanc) ; *syn.* défrichage [m], essartage [m], déboisement [m], travaux [mpl] de défrichement ; *f 2* **déforestation** [f] (Résultat de la suppression [destruction durable de la destination forestière sans reboisement ultérieur] sur une certaine surface de terrain de la forêt qu'elle portait ; cf. DFM 1975 ; *opp.* ►reforestation) ; *syn.* déboisement [m], déforestage [m]) ; *g* **Entwaldung** [f] (**1.** Verminderung des Waldbestandes als Ergebnis von ►Rodung oder Abbrennen ohne Wiederaufforstung; ►Brandrodung, ►Kahlschlag, ►Wanderfeldbau, ►Wiederaufforstung); **2.** *leg. for.* [A] Verwenden von Waldboden [Waldflächen] zu anderen Zwecken als solchen der Waldkultur; § 17 [1] ForstG; bloßes Unbrauchbarmachen von Waldboden für Zwecke der Waldkultur wird in Österreich **Waldverwüstung** genannt; cf. § 16 [2] ForstG); *syn. zu 1.* Rodung [f] ohne Wiederaufforstung, Ausreutung [f] ohne Neuaufforstung [CH].

degradation [n] (1) *biol. chem.* ►decomposition (1).

degradation [n] (2) *envir.* ►deterioration (1).

degradation [n] (3) *geo.* ►denudation.

degradation [n] [US] *geo.* ►riverbank degradation.

degradation [n], **environmental** *envir.* ►environmental deterioration.

1378 degradation [n] **of pollutants** *envir.* (Chemical reaction in soil, air or water, where complex contaminant molecules are broken down into simpler components; ►decomposition 1); *syn.* breakdown [n] of pollutants; *s* **degradación** [f] **biológica** (Reacción química en la tierra, el aire o el agua gracias a la cual se descomponen las moléculas complejas de sustancias contaminantes; ►descomposición); *syn.* biodegradación [f]; *f* **biodégradation** [f] (de matières polluantes dans les milieux pédologiques, aériens et aquatiques ; ►décomposition) ; *g* **Abbau [m, o. Pl.] von Schadstoffen** (Chemische Reaktion im Boden, in der Luft oder im Wasser, bei der komplizierte Moleküle nun Schmutzstoffen in einfacher Strukturen zerlegt werden; ►Abbau 2); *syn.* Schadstoffabbau [m].

1379 degradation [n] **of soil** *pedol.* (Morphological alteration of topsoil [A horizon] into simpler compounds, mostly by human influences; e.g. reduction of ►structural stability of soil, fading of upper A_h horizon, decrease of lime, washing out of fine materials, etc.; ►truncated soil profile); *s* **degradación** [f] **del suelo (vegetal)** (Alteraciones de la capa vegetal del suelo [horizonte A] en general por influencias antropógenas p. ej. de[s]forestación con resultado de pérdida de la fertilidad del mismo; ►estabilidad de la grumosidad, ►suelo decapitado); *f* **dégradation** [f] **du sol** (Modification de l'évolution naturelle des sols par lessivage ou altération souvent provoquée par l'action de l'homme généralement par une intensification culturale du sol et ayant pour conséquence une diminution de la ►stabilité structurale du sol. Elle s'exerce soit directement — défrichement — et brutalement, soit indirectement — coupes, pâturages — par l'intermédiaire de la végétation ; ►sol décapé) ;

g **Degradation [f] des Bodens** (Morphologische Veränderung des Oberbodens [A-Horizontes], meist durch anthropogene Einflüsse, z. B. Verringerung der ►Gefügestabilität, wodurch die Verschlämmungs- und Erosionsneigung erhöht wird, Aufhellung des oberen A$_h$-Horizontes, Entkalkung, Auswaschung von Teilen der Verwitterungsprodukte; ►geköpftes Bodenprofil); *syn.* Degradierung [f] des Bodens.

1380 degradation [n] **of vegetative cover** *phyt.* (Wearing out of vegetative cover; e.g. by trampling); *s* **degradación [f] de la cubierta vegetal** (Destrucción de la vegetación p. ej. por pisoteo); *syn.* deterioro [m] de la cubierta vegetal; *f* **dégradation [f] du couvert végétal** (p. ex. causée par le piétinement) ; *g* **Beeinträchtigung [f] der Vegetation** (Zerstörerische Veränderung der Vegetation[sdecke], z. B. durch Trittbelastung); *syn.* Degradation [f] der Vegetation, Degradation [f] des Bewuchses.

1381 degraded bog [n] *geo. phyt.* (Bog downgraded by human activities); *s* **turbera [f] degradada** (Turbera que ha sufrido deterioro por la actividad del hombre; *f* **tourbière [f] anthropisée** (Tourbière fortement dégradée par l'activité de l'homme) ; *g* **degradiertes Moor [n]** (Durch Nutzungsansprüche des Menschen stark beeinträchtigtes Moor).

degree [n] *ecol. envir* ►environmental stress degree.

1382 degree of achievement [n] *plan.* (Measure by which a goal has been reached in the eyes of the evaluator; often expressed on a scale of 1 to 10); *s* **grado [m] de realización** (Desde la perspectiva del evaluador, medida cuantitativa del alcance de una meta, a menudo expresada en una escala de 1 a 10); *syn.* grado [m] de logro; *f* **degré [m] de réalisation de l'objectif** (Du point de vue de la personne effectuant l'évaluation, valeur donnant le résultat d'un objectif atteint, souvent présenté sur une échelle de 1 à 10) ; *g* **Zielerreichungsgrad [m]** (Maß für die Erreichung eines Zieles aus der Sicht des Bewerters; die Bewertungsdimension wird häufig in einer 10-Punkteskala ausgedrückt); *syn.* Zielerfüllungsgrad [m].

1383 degree [n] **of canopy cover** *phyt.* (Ratio of area covered by all tree crowns; ►canopy closure); *syn.* degree [n] of crown cover; *s* **expansión [f] horizontal de las copas** (Porcentaje del suelo cubierto por todas las copas; ►cierre del dosel [del bosque]); *syn.* índice [m] de proyección de las copas, índice [m] del área de copa, grado [m] de cobertura del dosel; *f* **indice [m] de cime** (Valeur indiquant la continuité du couvert végétal, le degré de plénitude de couvert d'un peuplement forestier ; ►fermeture du couvert) ; *syn.* indice [m] couronne ; *g* **Deckungsgrad [m] der Baumkronen** (Anteil der Fläche, die durch die Baumkronen im Grundriss beschirmt ist; ►Kronenschluss 1); *syn.* Beschirmungsgrad [m].

degree [n] **of contraction** *constr.* ►shrinkage crack.

degree [n] **of crown cover** *phyt.* ►degree of canopy cover.

1384 degree [n] **of fidelity** *phyt.* (Extent to which a species is affiliated with a particular plant community: five grades may be distinguished: exclusive, selective, preferential, indifferent and strange; ►accidental, ►exclusive species, ►preferential species, ►indifferent species, ►selective species and ►strange species; ►site fidelity); *s* **grado [m] de fidelidad** (Mayor o menor limitación de las especies a determinadas comunidades vegetales. Se diferencian cinco grados: exclusivo, electivo, preferente, indiferente y extraño; ►especie electiva, ►especie exclusiva, ►especie extraña, ►especie indiferente y ►especie preferente; BB 1979; ►accidental, ►fidelidad); *f* **degré [m] de fidélité** (Rend compte de la manière dont une espèce est localisée exclusivement ou préférentiellement dans les diverses communautés biologiques ; on distingue suivant un **d. de f.** décroissant les ►espèces exclusives, les ►espèces électives, les ►espèces

préférantes [préférentielles], les ►espèces indifférentes et les ►espèces étrangères ; ►espèce accidentelle, ►fidélité [au domaine/lieu]) ; *g* **Treuegrad [m]** (Grad des engen Gebundenseins von Arten an bestimmte Pflanzengesellschaften; es werden fünf Abstufungen unterschieden: treu, fest, hold, vag[e] und fremd; ►feste Art, ►fremde Art, ►holde Art, ►treue Art, ►vage Art, ►Zufällige; ►Standorttreue).

1385 degree [n] **of landscape modification** *ecol.* (Level of combined human influence on ecosystems or landscapes; may be graded according to FORMAN and GODRON: ►virgin landscape, ►natural landscape, ►semi-natural, ►man-developed landscape, managed landscape, cultivated landscape, suburban landscape, urban landscape; LE 1986, 286 ; ►near natural landscape); *syn.* level [n] of human-caused landscape modification (LE 1986, 286), level [n] of artificialization (LE 1986, 286), gradient [n] of landscape modification; *s* **grado [m] de modificación (de ecosistemas y paisajes)** (Nivel de transformación antropogénica sufrido por los paisajes naturales en relación a sus estados originales. Se puede diferenciar en ►virgen, ►poco alterado, ►seminatural, ►antropógeno); *syn.* grado [m] de artificialización (de ecosistemas y paisajes); *f* **degré [m] d'artificialisation** (La modification du milieu, conséquence de la pression humaine [présente et passée] sur un écosystème ou différents espaces de paysages, est caractérisée par un gradient du plus faible au plus fort qui caractérise les milieux/systèmes où l'influence de l'homme est pratiquement nulle à ceux où l'homme a presque totalement modifié le milieu naturel ; on distingue les degrés : ►d'artificialisation nulle, ►faiblement artificialisé, ée, ►d'artificialisation moyenne, ►fortement artificialisé ; cf. VRD 1994, tome I, 2/1.4.2 page 41) ; *g 1* **Natürlichkeitsgrad [m]** (Grad unterschiedlicher menschlicher Beeinflussung von Ökosystemen oder Landschaftsräumen. Es kann unterschieden werden in ►natürlich 2, ►naturnah, ►halbnatürlich und ►naturfern); *g 2* **Hemerobie [f]** (Griech. *hémeros* >gezähmt< und *bíos* >Leben<; Grad der menschlichen Beeinflussung von Landschaften, Biotoptypen, Biozönosen oder ganzen Ökosystemen mit folgenden Hemerobiestufen: **ahemerob** [unbeeinflusst], **mesohemerob** [gering beeinflusst, z. B. sehr gering besiedelte Gebiete der Hochgebirge, Wüsten und Tundra], **oligohemerob** [stark/mittel beeinflusst, z. B. dünn besiedelte Kulturlandschaften] und **euhemerob** [stark beeinflusst wie Agrarlandschaften und Siedlungsgebiete], **polyhemerob** [sehr stark beeinflusst, z. B. versiegelte Flächen, Deponien] und **metahemerob** [naturfremd, z. B. Kerngebiete der Innenstädte, Industrieflächen]; cf. FREY 2004; in der Praxis werden die Termini *Natürlichkeitsgrad* und *Hemerobie* uneinheitlich und z. T. syn. verwendet).

1386 degree [n] **of naturalization** *phyt. zool.* (Capability of plants or animals to survive in areas in which they did not live previously and to adapt to abiotic and biotic site conditions by genetic modifications occurring over long periods of time, or by behavio[u]ral modifications in response to changing conditions. Based on varying degrees of naturalization, five groups are recognized according to SCHROEDER 1969: 1. ►indigenous species, 2. ►naturalized species, 3. ►epecophytes, 4. ►casual species, 5. ►cultivated species); *s* **grado [m] de naturalización** (Capacidad de las plantas y los animales de adaptarse a un lugar desconocido anteriormente. Se divide en cinco escalas según la naturalidad decreciente: 1. ►especie indígena, 2. ►especie naturalizada, 3. ►epecófito, 4. ►efemerófito y 5. ►ergasiófito); *syn.* grado [m] de adaptación; *f* **degré [m] de naturalisation** (Capacité des espèces animales et végétales à s'adapter à un milieu dans lequel elles n'existaient pas auparavant et à s'y maintenir pour se comporter comme une espèce indigène ; on distingue 5 catégories de naturalisation : 1. ►espèce indigène, 2. ►espèce naturalisée, 3. ►espèce acclimatée, 4. ►espèce

D

D

fugace, 5. ►espèce cultivée) ; **g Grad [m] der Einbürgerung** (Fähigkeit von Pflanzen oder Tieren, sich den am früher nicht bewohnten Ort herrschenden abiotischen und biotischen Standortfaktoren gegenüber zu behaupten; Einteilung nach abnehmender Natürlichkeit des Vorkommens in fünf Gruppen [nach SCHROEDER 1969]: 1. [eigentlich] ►einheimische Art, 2. ►neuheimische Art, 3. ►kulturabhängige Art, 4. ►unbeständige Art, 5. ►kultivierte Art 2).

1387 degree [n] **of presence** *phyt.* (A six-degree scale upon which the ►determination of presence is expressed: **5** or **VI** = constantly present [in 80 to 100% of the stands]; **4** or **V** = mostly present [in 60 to 80% o. t. st.]; **3** or **IV** = often present [in 40 to 60% o. t. st.]; **2** or **III** = not often present [in 20 to 40% o. t. st.]; **1** or **II** = seldom present [in 2 to 20% o. t. st.]; **ss** or **I** = rare [in less than 2% of the stands]); *s* **grado [m] de presencia** (Escala de valores mediante la cual se realiza la ►determinación de la presencia: **5** ó **VI** = casi siempre presente [entre el 80 y 100% de los representantes de asociaciones (individuos de asociaciones)]; **4** ó **V** = presente en la mayoría de las veces [entre el 60 y 80% de los representantes]; **3** ó **IV** = presente frecuentemente [entre el 40 y 60%]; **2** ó **III** = presente con poca presencia [entre el 20 y 40%]; **1** ó **II** = rara [entre el 2 y 20% de los representantes]; **ss** ó **I** = presente de forma muy esporádica [en menos del 2% de los representantes]; BB 1979); *f* **coefficient [m] de présence** (La présence de chaque espèce, lors de la ►détermination de la présence d'une espèce, est exprimée par 5 classes : **V** = espèces présentes dans 80-100 % des relevés ; **IV** = espèces présentes dans 60-80 % des relevés ; **III** = espèces présentes dans 40-60 % des relevés ; **II** = espèces présentes dans 20-40 % des relevés ; **I** = espèces présentes dans 1-20 % des relevés ; *syn.* classe [f] de fréquence, catégorie [f] de fréquence, classe [f] de constance ; *g* **Stetigkeitsgrad [m]** (Sechsteilige Skala nach der die ►Stetigkeitsbestimmung vorgenommen wird: **5** oder **VI** = stets vorhanden [in 80-100 % der Einzelbestände (Assoziationsindividuen)]; **4** oder **V** = meist vorhanden [in 60-80 % der Einzelbestände]; **3** oder **IV** = öfter vorhanden [in 40-60 % der Einzelbestände]; **2** oder **III** = nicht oft vorhanden [in 20-40 % der Einzelbestände]; **1** oder **II** = selten vorhanden [in 2-20 % der Einzelbestände]; **ss** oder **I** = ganz vereinzelt [in < 2 % der Einzelbestände]).

1388 degree [n] **of species cover** *phyt.* (Ratio of area covered by all individuals of one species in a single layer of a plant community, projected vertically to the ground; e.g. ►tree layer, ►shrub layer, ►herb layer, ►moss layer, and lichen layer; ►dominant species); *syn.* dominance [n], range [n] of species cover (TEE 1980, 161), coverage [n] (TEE 1980, 165); *s* **dominancia [f]** (Hegemonía y máximo biológico representados por una o más unidades sistemáticas en una asociación. Por el concepto de hegemonía, las especies dominantes pueden definirse como aquellas a cuyas necesidades está supeditada la ecesis [= proceso de germinación, crecimiento y reproducción de la planta] de las demás. Como máximo biológico, las dominantes forman la masa vegetal de la colectividad; DB 1985; ►estrato arbóreo, ►estrato arbustivo, ►estrato herbáceo, ►estrato muscinal; ►especie dominante); *f* **dominance [f]** (Tranche de valeurs — de 1 à 5 — exprimant en pourcentage la surface [projetée dans le plan horizontal] de recouvrement des individus de chaque espèce d'un groupement végétal précis ; pour les groupements pluristrates, la **d.** est évaluée pour chaque strate [►strate arborescente, ►strate arbustive, ►strate herbacée, ►strate muscinale] ; ►espèce dominante) ; *g* **Deckungsgrad [m] einer Art** (Mengenanteil der Individuen einer Art in einer bestimmten Pflanzengesellschaft als senkrechte Projektion des gesamten Spross- und Blattwerkes einer betreffenden Schicht [►Baumschicht, ►Strauchschicht, ►Krautschicht, ►Moosschicht und

evtl. Flechtenschicht] auf dem Boden; ►dominante Art); *syn.* Dominanz [f].

1389 degree [n] **of total vegetative cover** *phyt. plant.* (Ratio of plant cover to bare ground; ►plant sociability, ►total estimate); *syn.* percentage of vegetative cover [n] (HASL 1981); *s* **expansión [f] horizontal** (Sección determinada en la superficie del suelo por el haz de proyección horizontal del cuerpo de la planta. Aplicando este concepto a las formaciones, éstas se dividen en formación cerrada, formación semicerrada, formación abierta, formación subdesértica, formación desértica; cf. DB 1985; ►estima [global] del grado de cobertura, ►sociabilidad); *f* **degré [m] de recouvrement** (1. *phyt. plant.* Rapport entre la surface du sol recouverte par la végétation et l'étendue restée nue) ; *syn.* surface [f] occupée (DB 1985, 450 ; 2. *phyt.* ►estimation globale, ►sociabilité) ; *g* **Deckungsgrad [m]** (1. In der Vegetationskunde das Verhältnis der den Boden bedeckenden Pflanzen zu den unbewachsenen bzw. von anderen Arten bedeckten Flächen, ausgedrückt in Prozenten oder in Ziffernstufen 1-5 [►Gesamtschätzung, ►Häufungsweise]. 2. In der Bepflanzungsplanung die Beschreibung der Flächenanteile von Arten [Artmächtigkeit] in einer definierten Pflanzung als Ist- oder Sollzustand. Stärker als von den gepflanzten Stückzahlen wird der **D.** von der Wuchsform [niederliegende mit höherem **D.** als aufragende] und Ausbreitungsstrategie [Strategietypen] der eingebrachten Arten bestimmt); *syn.* Artmächtigkeit [f], Bedeckungsgrad [m], Bodenbedeckung [f], Bodendeckung [f], Flächendeckung [f].

degree [n] **of use** *recr.* ►frequency of use.

1390 degree [n] **of vitality** *phyt.* (Level of plant species survival in a community; the following four grades are used: **a.** well-developed, regularly completing the life circle; **b.** strong and increasing but usually not completing the life cycle [many mosses]; **c.** feeble but spreading, never completing the life cycle; **d.** occasionally germinating but not increasing; many ephemeral adventive plants; ►vitality); *s* **grado [m] de vitalidad** (Escala de clasificación de la vitalidad de especies florísticas: **a.** plantas bien desarrolladas, que cumplen regularmente su ciclo vital completo; **b.** plantas que se desarrollan débilmente, pero se multiplican, o con buen desarrollo, pero sin cumplir regularmente con su ciclo vital; **c.** plantas que vegetan miserablemente y se multiplican, pero no presentan su ciclo vital completo y **d.** plantas que germinan esporádicamente, pero que no se multiplican; BB 1979, 60s; ►vitalidad); *f* **degré [m] de vitalité** (Degré de vigueur et de prospérité atteint par les différentes espèces ; on distingue les **d. de v.** suivants : **a.** plantes bien développées, accomplissant régulièrement leur cycle évolutif complet ; **b.** plantes à cycle évolutif généralement incomplet, développement végétatif vigoureux ; **c.** plantes à cycle évolutif incomplet, développement végétatif restreint ; **d.** plantes germant accidentellement, ne se multipliant pas [beaucoup d'espèces adventices] ; ►vitalité) ; *g* **Vitalitätsgrad [m]** (Ausmaß des Vorkommens einer Art in einem Bestand/einer Pflanzengesellschaft auf Grund ihrer Lebenskraft und Durchsetzungsfähigkeit/Konkurrenzstärke. Der **V.** einer Art kann durch folgende Einteilung vorgenommen werden: **a.** gut entwickelte, regelmäßig ihren vollständigen Lebenskreislauf abwickelnde Pflanzen; **b.** schwächer entwickelte, sich vermehrende Pflanzen [verminderte Soziabilität, spärliches Auftreten, Nanismus etc.] oder üppig entwickelte, die aber ihren Lebenskreislauf nicht regelmäßig abwickeln; **c.** kümmerlich vegetierende, sich vermehrende Pflanzen, die ihren Lebenskreislauf nicht vollständig abwickeln; **d.** zufällig gekeimte, sich nicht vermehrende Pflanzen; ►Vitalität).

dehorning [n] [US]/**de-horning** [n] [UK] *arb.* ►topping (3).

1391 dehydration [n] *biol.* (Loss of the total water content of a living organism; ▶frost-desiccation); *syn.* desiccation [n]; *s* **deshidratación** [f] (Pérdida total del contenido de agua en organismos vivos; ▶desecación invernal); *f 1* **déshydratation** [f] (Perte d'eau contenue dans les tissus des organismes vivants) ; *f 2* **dessèchement** [m] (Perte totale de l'eau contenue dans les tissus des végétaux ; ▶dessèchement par le gel) ; *g* **Vertrocknen** [n, o. Pl.] (Verlust des gesamten Wassergehaltes eines lebenden Organismus; ▶Frosttrocknis).

1392 dehydration [n] **of sulphur emissions** *envir.* (Process which eliminates hydrogen sulphide [H₂S] from manufactured gases, either by chemical fixing to iron oxide to form iron sulphide or by absorption of activated charcoal; ▶humid desulphurization); *s* **desulfuración** [f] **en seco** (Procedimiento de eliminación de anhídrido sulfúrico [SH₂] de gases de emisión, bien por fijación a óxido de hierro o bien por adsorción con carbono activado; ▶desulfuración húmeda); *f* **désulfuration** [f] **par voie sèche** (Procédé de récupération du soufre par traitement des effluents gazeux en vue de l'extraction de l'hydrogène sulfuré. Les processus réactionnels de récupération s'opèrent au moyens de différents matériels tels que des décanteurs, des régénérateurs, des absorbeurs, etc. ; ▶hydro-désulfuration) ; *g* **Trockenentschwefelung** [f] (Verfahren, durch das Schwefelwasserstoff [H₂] aus technischen Gasen meist durch chemische Bindung an Eisenoxid unter Bildung von Eisensulfid oder auch durch Adsorption an Aktivkohle entfernt wird; ▶Nassentschwefelung).

1393 de-icing salt [n] *envir.* (Salt that is spread for thawing of ice and snow on traffic areas, by lowering the freezing point of water, principally sodium chloride [NaCl], also sodium acetate/formate, calcium magnesium acetate, and 'Ice Ban' [US]); *syn.* road salt [n] [also US] (HOU 1984, 115), rock salt [n] [also US]; *s* **sal** [f] **anticongelante** (Sal que se echa en carreteras para evitar heladas debido a la reducción de la termperatura de congelación. Hasta hace unos años se utilizaba mayormente cloruro sódico [ClNa], pero debido a sus efectos nocivos sobre la vegetación, últimamente se usan bien una mezcla de sales de amonio, cloruro cálcico [CaCl₂] y cloruro magnésico [MgCl₂] bien arena, gravilla o granulados de zafras); *syn.* sal [f] para deshelar carreteras; *f* **sel** [m] **de déneigement** (Produit chimique utilisé pour abaisser le point de transformation de l'eau en glace sur les trottoirs et la voirie. Reconnu pour sa toxicité sur les végétaux et son action corrosive sur les matériaux, le NaCl est peu utilisé et remplacé par des mélanges de sels tels que les sels d'ammonium, le chlorure de calcium [CaCl₂], le chlorure de magnésium [MgCl₂] et des produits de substitution comme le sable, le gravier, les granulats de scories, etc.) ; *syn.* sel [m] de déverglaçage, fondants [mpl] de déneigement ; *g* **Streusalz** [n] (Auftaumittel aus gemahlenem Steinsalz, hauptsächlich Natriumchlorid [NaCl], zur Beseitigung von Eis und Schnee auf Verkehrsflächen durch Herabsetzen des Gefrierpunktes von Wasser. Wegen der für Pflanzen sehr schädlichen und auf Materialien korrodierenden Wirkung wird NaCl nur noch sehr beschränkt eingesetzt und durch Salzgemische wie z. B. Ammoniumsalze, Calciumchlorid [CaCl₂], das stark hygroskopische Magnesiumchlorid [MgCl₂ — Feuchtsalz] und durch Sand, Splitt, Schlackengranulate u. Ä. ersetzt. 1982 hat Berlin als erste deutsche Stadt die Verwendung von NaCl auf allen Straßen mit wenigen festgelegten Ausnahmen verboten; der Begriff **Streumittel** ist der OB zu Streusalz und mineralischem Streugut wie Sand, Splitt etc. Während NaCl sich für Temperaturen bis –8 °C eignet, sind CaCl₂ oder MgCl₂ auch noch bei tieferen Temperaturen wirksam; cf. MKT 2002, 51. In D. haben sich beim Winterstreudienst zwei Varianten der Feuchtsalzstreuung herausgebildet: FS 5 und FS 30. Bei den beiden Verfahren enthalten die ausgebrachten Salzmengen fünf resp. 30 Gewichtsprozent Feuchtigkeit. Die benetzende Flüssigkeit [meist eine 20-25-prozentige Calciumchloridlösung] wird kurz vor dem Ausbringen mit einer speziellen Geräteausstattung auf den Streuteller gebracht. Feuchtsalze haften besser auf der Straße und werden dadurch weniger stark verweht. Während auf deutschen Bundesautobahnen, wo von jeher verkehrsbedingt die größten Mengen ausgebracht wurden, in den 1960er- und 70er-Jahren in Unkenntnis der phytotoxischen Wirkung der Auftaumittel durchschnittlich 15-50 t Salz/km gestreut wurden (ca. 700-2 500 g/m²), verminderten sich die Salzmengen bereits in den 1980er-Jahren deutlich auf unter 10 t Salz/km². Heute hat sich vielerorts durch die Einführung eines Winterdienstes in Einsatzstufen der Verbrauch noch weiter reduziert und liegt in Abhängigkeit von den jährlichen Erfordernissen z. B. in Berlin zwischen 0,05-0,35 kg/m²; cf. BAL 2004; *syn.* Auftausalz [n], Auftaumittel [n], Tausalz [n].

de-intensification [n] **of agricultural production** *agr. pol.* ▶agricultural reduction program [US]/extensification of agricultural production [UK].

dejection cone [n] *geo.* ▶alluvial cone.

delaying [n] **storm runoff** [US] *envir. urb. wat'man.* ▶retention of rainwater.

delineation [n] [US]**, characteristic** *plan.* ▶characteristic presentation.

1394 delivered plan [n] [US] *constr. contr.* (Plan showing the status of actually executed work, drawn up for the ▶delivery [US] 1/handover [UK]; ▶as-built plan, ▶project documentation); *syn.* handover plan [n] [UK]; *s* **plan** [m] **de entrega de obra** (Plan que contiene todos los trabajos ejecutados en la obra y se prepara para la ▶entrega de obra; ▶documentaciónde un proyecto, ▶plano de realización); *syn.* plan [m] de entrega de proyecto; *f* **plan** [m] **de récolement (2)** (Plan des ouvrages conforme à l'exécution et remis au maître d'œuvre y compris notices de fonctionnement et entretien des ouvrages ; ▶dessins de recolement, ▶dossier des ouvrages exécutés, ▶mise à disposition de certains ouvrages, ▶plan de recolement 1, ▶prise de possession des ouvrages) ; *g* **Übergabeplan** [m] (Plan, der den Stand der tatsächlich ausgeführten Bauarbeiten darstellt und für die ▶Übergabe an den Bauherrn angefertigt wird; ▶Bestandsplan 1, ▶Dokumentation).

1395 delivery [n] [US] **(1)** *contr. prof.* (Transfer of a completed project to a client); *syn.* handover [n] [UK]; *s* **entrega** [f] **de obra** *syn.* entrega [f] de proyecto; *f 1* **prise** [f] **de possession des ouvrages** (par le maître de l'ouvrage après les opérations de réception, lorsque l'ouvrage est utilisable) ; *f 2* **mise** [f] **à disposition de certains ouvrages** (Concerne certains ouvrages ou parties d'ouvrages non achevés lorsque le maître de l'ouvrage veut faire exécuter des travaux autres que ceux qui font l'objet du marché) ; *g* **Übergabe** [f] (Freigeben eines fertiggestellten Bauobjektes an den Bauherren zu dessen Nutzung).

delivery [n] **(2)** *constr. contr.* ▶according to certified delivery [US].

delivery [n]**, on proof of verified delivery notes** [UK] *constr. contr.* ▶according to certified delivery [US].

1396 delivery access [n] *trans. urb.* (Property driveway for conveying materials and packages to the consignee; ▶service road); *s* **acceso** [m] **de suministro** (Vía asfaltada dentro de un predio que sirve para el suministro de bienes; ▶vía de servicio); *f* **voie** [f] **de desserte** (*Voirie tertiaire* voie réservée au trafic ayant une de ses extrémités le long de la voie [parking et immeubles] ; ▶contre-allée de service) ; *syn.* voie [f] de service ; *g* **Anlieferungsweg** [m] (Fahrweg auf einem Grundstück, auf dem Güter und Sendungen dem Empfänger zugestellt werden; ▶Nebenfahrbahn).

D

delivery and haulage traffic [n] *plan. trans.* ▶stationary traffic.

1397 delivery [n] **at the construction site free of charge** *constr. contr.* (▶transport to the construction site); *s* **suministro** [m] **a la obra sin recargo** (▶transporte a obra); *f* **fourniture** [f] **à pied d'œuvre** (Acheminement des matériaux sur le chantier sans frais de livraison ; ▶acheminement sur le chantier) ; *g* **Lieferung** [f] **frei Baustelle** (▶Transport zur Baustelle ohne besondere Berechnung der Frachtkosten).

delivery note [n] [UK] *constr. contr.* ▶bill of lading [US].

delivery notes [npl] [UK]**, on proof of verified** *constr. contr.* ▶according to certified delivery [US].

1398 delivery [n] **of materials** *constr.* (Transport to and unloading of materials at a construction site; ▶supply of materials); *s* **entrega** [f] **de materiales** (▶suministro de materiales); *f* **fourniture** [f] **de/des matériaux** (▶approvisionnement de matériaux) ; *g* **Materialanlieferung** [f] (Der direkte Transport zur und Abladen von Baustoffen auf der Baustelle; ▶Baustofflieferung); *syn.* Baustoffanlieferung [f].

1399 delivery [n] **of plants** *constr.* (Transport of an ordered supply of plants to a client or construction site); *s* **suministro** [m] **de plantas** (Transporte de plantas a una obra o a un cliente); *f 1* **fourniture** [f] **de végétaux** (Quantité des végétaux livrés ou à livrer) ; *syn.* approvisionnement [m] en végétaux ; *f 2* **fourniture** [f] **à pied d'œuvre des végétaux** (Végétaux à livrer sur le chantier) ; *g* **Pflanzenlieferung** [f] (**1.** Bestimmte Menge der gelieferten oder zu liefernden Pflanzen. **2.** Belieferung einer Baustelle oder eines Kunden mit bestellten oder gekauften Pflanzen).

1400 delta [n] *geo.* (Fan-shaped alluvial deposit of sediment at the mouth of a river); *s* **delta** [m] (Forma de acumulación aluvial localizada en la desembocadura de un río, y cuyo origen está en la pérdida de potencia de éste como consecuencia de una disminución de su pendiente y velocidad en el tramo inferior del recorrido; DGA 1986); *f* **delta** [m] (Accumulation littorale de sédiments créée par un cours d'eau à son embouchure dans la mer ou dans un lac, souvent anastomosée, lorsque sa capacité d'alluvionnement est supérieure à la capacité d'érosion du milieu dans lequel il se jette) ; *g* **Delta** [n] (Fächerförmige Aufschüttung fluvialer Lockersedimente vor einer Flussmündung in ein Meer oder in einen See).

demarcated forest [n] [ZA] *for. leg.* ▶reserved forest.

demecology [n] *ecol.* ▶population ecology.

1401 demographic pyramid [n] *plan. sociol.* (Diagram[m]atic representation of the age class distribution of a population, usually also illustrating the sex ratio); *syn.* population pyramid [n] [also US]; *s* **pirámide** [f] **demográfica** (Representación gráfica en forma de pirámide escalonada, en la que se anchura de cada bloque expresa el porcentaje de individuos [diferenciando sexos] existente para cada clase de edad; DINA 1987, 407); *syn.* pirámide [f] de edad; *f* **pyramide** [f] **des âges** (Représentation graphique de la distribution des classes d'âges d'une population) ; *g* **Alterspyramide** [f] (Grafische Darstellung der Zusammensetzung einer Bevölkerung nach Alter und meist auch nach Geschlecht).

1402 demolition material [n] *constr.* (Dismantlement substances originating from demolishment of a structure; ▶demolition work [US]/demolition works [UK], ▶building rubble); *s* **material** [m] **de derribo** (▶escombros, ▶trabajos de derribo); *syn.* material [m] de demolición; *f* **matériaux** [mpl] **de demolition** (Débris provenant de la démolition d'un bâtiment ; ▶démolition de constructions existantes, ▶gravats, ▶travaux de démolition) ; *g* **Abbruchmaterial** [n] (Beim Abreißen eines

Bauwerkes anfallender Bauabfall, wie z. B. Straßenaufbruch, ▶Bauschutt von Gebäuden und sonstige Baustellenabfälle sowie Bodenaushub. Diese Stoffe müssen gem. Ziff. 5.2.6 TA Siedlungsabfall v. 14.05.1993 an der Abfallstelle getrennt erfasst und einer Verwertung zugeführt werden; ▶Abbrucharbeiten).

demolition permission [n] [UK] *leg.* ▶demolition permit.

1403 demolition permit [n] [US] *leg.* (Authorization for the partial or complete, demolishment of a building or structure); *syn.* demolition permission [n] [UK]; *s* **permiso** [m] **de derribo** *syn.* permiso [m] de demolición; *f* **permis** [m] **de démolir** (Obligation imposée à toute personne privée, collectivité ou établissement public désirant démolir tout ou partie d'un bâtiment) ; *g* **Abbruchgenehmigung** [f] (Von einer Behörde erteilte Erlaubnis, ein Bauwerk abreißen zu dürfen).

1404 demolition work [n] [US]/**demolition works** [npl] [UK] *constr.* (Dismantlement of structures, and related operations); *s* **trabajos** [mpl] **de derribo** *syn.* trabajos [mpl] de demolición; *f 1* **travaux** [mpl] **de démolition** ; *f 2* **démolition** [f] **de constructions existantes** *syn.* démolition [f] d'ouvrages existants ; *g* **Abbrucharbeiten** [fpl] (Beim Abbrechen oder Abtragen von Bauwerken anfallende Arbeiten).

1405 demonstration farm [n] *agr. sociol.* (Agricultural facility intended as a place of learning for school classes, kindergartens and other children's organizations. Stocked with a wide range of animals and various cultivated plants; the farm's purpose is to re-establish the link with nature and natural processes for interested children living in towns and cities; ▶children's farm); *s* **granja** [f] **pedagógica** (Granja con cría de diferentes especies animales y cultivo de diferentes plantas que tiene como fin principal servir fines educativos para niños/as de ciudades; ▶granja juvenil); *f* **ferme** [f] **pédagogique** (Exploitation agricole spécialisée dans l'élevage d'un large éventail d'animaux de ferme ainsi que dans la culture de nombreuses plantes et proposant un lieu d'apprentissage ou de connaissances aux classes scolaires, aux jardins d'enfants, aux institutions travaillant dans le domaine de l'enfance afin de procurer aux enfants des villes un éveil à la nature et ses processus naturels ; ▶ferme pour enfants) ; *syn.* ferme [f] modèle ; *g* **Schaubauernhof** [m] (Landwirtschaftlich geführter Betrieb, der als anschaulicher Lernort mit einem möglichst vielseitigen Tierbestand und dem Anbau von unterschiedlichen Kulturpflanzen für Schulklassen, Kindergärten und anderen mit Kindern arbeitenden sozialen Institutionen eingerichtet ist, um interessierten Großstadtkindern die Verbindung zur Natur und zu natürlichen Vorgängen wiederherzustellen; ▶Jugendfarm).

1406 demonstration garden [n] (1) *gard. hort. plant.* (Garden planted with certain plants or according to a particular theme for show purposes); *s* **jardín** [m] **tipo** (Jardín que sirve para presentar determinadas plantas o un tema de diseño determinado); *f* **jardin** [m] **de démonstration** (Jardin aménagé pour la présentation de certaines espèces végétales ou de compositions particulières) ; *g* **Schaugarten** [m] (Garten, der der Demonstration eines bestimmten Pflanzen- oder Gestaltungsthemas dient); *syn.* Schauanlage [f].

1407 demonstration garden [n] (2) *gard.* (Garden designed for educating the public in botany, gardening, and applied ecology); *s* **jardín** [m] **pedagógico** (Jardín diseñado para enseñar botánica, horticultura y ecología aplicada al público en general); *syn.* jardín [m] educativo; *f* **jardin** [m] **pédagogique** (Jardin conçu pour l'enseignement au grand public de la science botanique, des pratiques horticoles et paysagères et de l'écologie appliquée) ; *syn.* jardin [m] de démonstration ; *g* **Lehrgarten** [m] (Gärtnerische Anlage, die einer breiten Öffentlichkeit für den

Unterricht in Botanik, Gartenbau- und Gartenpflegearbeiten sowie angewandter Ökologie dient).

demonstration study [n] *ecol. landsc. plan.* ▶pilot study.

1408 dendrologist [n] *arb.* (Person who specializes primarily in the scientific study of trees; ▶dendrology, ▶plantsman); *s* **dendrólogo/a [m/f]** (Persona especializada en el estudio de las leñosas; ▶dendrología, ▶experto/a en plantas); *f* **dendrologue [m]** (Personne spécialisée dans l'étude des arbres et des plantes lignifiées ; ▶dendrologie, ▶spécialiste des plantes) ; *g* **Dendrologe [m]/Dendrologin [f]** (Experte, der sich vornehmlich mit dem Studium der Bäume befasst; ▶Gehölzkunde, ▶Pflanzenfachmann); *syn.* Gehölzkundler/-in [m/f].

1409 dendrology [n] *arb.* (Branch of science dealing with identification and systematic classification of trees and other woody species; ▶dendrologist); *s* **dendrología [f]** (Rama de la botánica que estudia las plantas leñosas; *término específico* dendrocronología: Estudia los anillos de crecimiento de los árboles y a través de ellos las características de los procesos que han vivido, como sequías, infecciones de parásitos, etc., así como las variaciones climáticas del pasado; ▶dendrólogo/a); *f* **dendrologie [f]** (Branche de la botanique appliquée traitant de l'identification et de la classification systématique des espèces arborescentes ; ▶dendrologue) ; *g* **Gehölzkunde [f]** (Wissenschaftszweig der angewandten Botanik, der sich mit dem Studium der Bäume und anderen Holzgewächsen beschäftigt; ▶Dendrologe); *syn.* Dendrologie [f].

1410 denitrification [n] *chem. pedol.* (Biochemical reduction of nitrate ions [NO_3^-] by ▶microorganisms into nitrogen dioxide [NO_2], nitrous oxide [N_2O], elementary nitrogen [N_2] and only a little ammonia [NH_3]; *opp.* ▶nitrification); *s* **desnitrificación [f]** (Reducción de nitratos a nitritos, a óxido nitroso o a nitrógeno elemental; *opp.* ▶nitrificación); *f* **dénitrification [f]** (Réduction de nitrates [ions NO_3^-] en NO_2 puis en N moléculaire ; par contre réduction en faible quantité en gaz NH_3 ; SS 1979 ; *opp.* ▶nitrification) ; *g* **Denitrifizierung [f]** (Reduktion von Nitrationen [NO_3^-] zu Stickstoffdioxid [NO_2], Distickstoffoxid [N_2O] und elementarem Stickstoff [N_2], dagegen nur in geringer Menge zu Ammoniak [NH_3]; SS 1979; *opp.* ▶Nitrifizierung); *syn.* Denitrifikation [f].

denominational cemetery [n] *adm. leg. urb.* ∗▶cemetery.

1411 dense foliage [n] **of woody plants** *bot. hort. syn.* heavy foliage [n] of woody plants; *s* **follaje [m] frondoso** *syn.* frondosidad [f]; *f* **frondaison [f] dense** ; *g* **dichte Belaubung [f, o. Pl.]** *syn.* dichtes Laubwerk [n, selten Pl.].

dense mat-forming perennial [n] *hort. plant.* ▶carpet-forming perennial.

density [n] *zool.* ▶animal density; *constr.* ▶bulk density; *for.* ▶forest stand density; *game'man.* ▶game density; *agr.* ▶increase in stocking density; *plan. sociol. urb.* ▶population density (1); *biol.* ▶population density (2); *phyt. zool.* ▶population density (3); *sociol. urb.* ▶residential density; *ecol.* ▶species density; *trans.* ▶traffic density;

density [n], **building** *urb.* ▶density of development.

density [n] [US], **canopy** *for. phyt.* ▶canopy cover.

density [n], **crown** *for. phyt.* ▶crown closure.

density [n] [US], **land use** *urb.* ▶density of development.

density [n], **reduction in housing** *urb.* ▶wide spacing in a housing development.

density [n], **settlement** *plan. sociol. urb.* ▶population density (1).

density [n], **turf** *constr. hort.* ▶density of sward (1).

1412 density [n] **of development** *urb.* (General term covering the number and all types of buildings in relation to unit of land area per hectare or acre on which they are situated and depending on a particular land use type [residential, office, etc.]. **In U.S., d. o. d.** is indicated on zoning maps by average lot size in each zoning district, and specified in ▶zoning ordinances as to ▶cubic content ratio [of a building] [US], ▶floor area ratio [FAR] [US] and ▶building coverage [US]; **in Europe, d. of d.** is indicated on a ▶Proposals Map [UK] or ▶Site Allocations Development Plan Document [UK] by ▶floor space index [UK] and ▶plot ratio [UK]); *syn.* building density [n], land use density [n] [US]; *s* **densidad [f] de edificación** (Término genérico que denomina la relación entre la superficie de planta de los edificios y la del área sobre la cual están construidos; **en D.** se determina en los ▶planes parciales [de ordenación] por medio del ▶volumen edificable, el ▶coeficiente de edificación, el ▶coeficiente de ocupación del suelo y el número de pisos [Es]/plantas [AL]; ▶ordenanza de zonificación); *f* **densité [f] de construction** (Le rapport entre la surface de plancher hors œuvre nette de cette construction et la surface du terrain sur laquelle elle est ou doit être implantée ; BON 1990, 350 ; pour limiter la densité de construction les règlements de ▶plan local d'urbanisme [PLU] ou de plan d'occupation des sols [POS] utilisent diverses valeurs telles que le ▶coefficient d'emprise au sol, ▶coefficient d'occupation des sols [COS], ▶coefficient d'utilisation du sol, ▶plafond légal de densité [PLD] ; ▶règles générales d'utilisation du sol) ; *g* **Bebauungsdichte [f]** (Im Wesentlichen durch ▶Baumassenzahl, ▶Geschossflächenzahl, ▶Grundflächenzahl und/oder Zahl der Vollgeschosse in ▶Bebauungsplänen vorgegebene Dichte der Bauentwicklung auf einer definierten Fläche; ▶Baunutzungsverordnung).

1413 density [n] **of existing vegetation** *for. phyt.* (Thickness of a forest stand or plant community expressed in the number of plants per unit area); *s* **densidad [f] de población vegetal** (Cantidad de plantas en un rodal o una comunidad vegetal); *f* **densité [f] d'un peuplement végétal** ; *g* **Bestandsdichte [f]** (Anzahl von Pflanzen je Flächeneinheit eines Waldbestandes oder einer Pflanzengesellschaft); *syn.* Dichte [f] eines Pflanzenbestandes, *o. V.* Bestandesdichte [f].

1414 density [n] **(of species)** *phyt.* (Number of individuals of one species per unit area within a plant community; ▶species abundance); *s* **densidad [f] (de especies)** (Número de individuos de una misma especie vegetal en un territorio dado; ▶abundancia [de una especie]); *f* **densité [f] (d'une espèce)** (Rapport exact entre le nombre des individus d'une même espèce végétale observés sur un territoire donné et l'étendue de ce territoire ; ▶abondance) ; *g* **Dichtigkeit [f] (einer Art)** (Auf eine Pflanzenart bezogenes quantitatives Merkmal über den mittleren Abstand ihrer Individuen in einer Pflanzengesellschaft; ▶Abundanz).

1415 density [n] **of sward (1)** *constr. hort.* (Number of blades of grass of a lawn per unit area); *syn.* turf density [n]; *s* **densidad [f] de tapiz de césped** (Cantidad de tallos de hierba por unidad de superficie); *f* **densité [f] du feuillage (d'un gazon)** (Nombre des brins des graminées pour un gazon par unité de surface) ; *g* **Narbendichte [f]** (Anzahl der Grashalme eines Rasens je Flächeneinheit).

1416 density [n] **of sward (2)** *constr. phyt.* (Closeness of turf as a ratio of bare to dense ▶turf sward; ▶density of vegetative cover); *s* **densidad [f] de tápiz herbáceo** (Despúes de los ▶trabajos de mantenimiento pre-entrega, relación entre las superficies cerradas y abiertas de césped o hierba; ▶densidad de la cobertura vegetal, ▶tapiz de césped); *f* **taux [m] de recouvrement du couvert herbacé** (Valeur caractérisant le rapport

D

d'occupation spatiale entre la végétation herbacée et le sol nu ; ▶densité du couvert végétal, ▶tapis de graminées) ; *g* **Besto-ckungsdichte [f] einer Grasnarbe** (**1.** *constr.* Bei der Abnahme von Ansaaten nach der ▶Fertigstellungspflege: Verhältnis von offener zu geschlossener Rasen- oder Wiesenfläche; ▶Besto-ckungsdichte, ▶Rasennarbe. **2.** *constr. phyt.* Verhältnis der Gesamtdeckung zu einzelnen Artengruppen: z. B. Gräser, Legu-minosen, andere Kräuter, Anflug aus der Umgebung); *syn.* Besatzdichte [f] von Kräuter- und Grasarten, Bestockungsgrad [m] von Kräuter- und Grasarten, Bestockungsgrad [m] einer Grasnarbe.

1417 density [n] **of use** (1) *plan.* (Prescribed ratio of built-up area in relation to the lot size in a settlement area, defined by ▶floor area ratio [FAR] [US]/floor space index [UK] and ▶building coverage [US]/plot ratio [UK]); *s* **densidad [f] de usos** (Relación fijada por medio del ▶coeficiente de edificación y ▶coeficiente de ocupación del suelo entre la superficie edificada y la total de un solar); *f* **densité [f] d'occupation du sol** (Définition des droits ou possibilités de construction associés à la disponibilité d'une certaine surface de sol en zone urbaine exprimés par le ▶coefficient d'occupation des sols [COS], le ▶plafond légal de densité [PLD], le ▶coefficient d'emprise au sol [CES]) ; *g* **Nutzungsdichte [f]** (Durch ▶Geschossflächenzahl [GFZ] und ▶Grundflächenzahl [GRZ] vorgegebenes Verhältnis von bebauter Fläche zur Grundstücksfläche in einem Siedlungs-gebiet).

density [n] **of use** (2) *recr.* ▶frequency of use.

1418 density [n] **of vegetative cover** *for. phyt.* (Calcu-lated or estimated ratio of bare to dense; there are three distinct forms: 1. bare [adj], 2. sparsely vegetated [pp/adj], 3. densely vegetated [pp/adj]); *s* **densidad [f] de la cobertura vegetal** (Relación estimada o calculada de la cubierta vegetal entre desnuda y cerrada; se diferencian tres grados de **d.**: 1. desnuda [pp/adj], 2. rala [adj], 3. densa [adj]); *f* **densité [f] du couvert végétal** (Valeur évaluée ou calculée caractérisant le rapport d'occupation spatiale entre le couvert et le sol ; 1. dénudée [pp/adj], 2. clairsemée [pp/adj], discontinue [pp/adj], 3. dense [adj], fermée [pp/adj], continue [adj]) ; *syn.* densité [f] de peuplement végétal ; *g* **Bestockungsdichte [f]** (**1.** Geschätztes oder errech-netes Verhältnis von offener zu geschlossener Vegetationsdecke; es werden drei Dichtegrade unterschieden: 1. nicht bestockt [pp/adj], 2. dünn bestockt [pp/adj], offen [adj], 3. dicht bestockt [pp/adj], geschlossen [adj]. **2.** Mengenbezeichnung für das Auftreten einer Art in einem bestimmten Vegetationsbestand, z. B. Rasen, Wald, Wiese); *syn.* Bestockungsgrad [m], *phyt.* Besatzdichte [f] von Pflanzenarten.

1419 density [n] **of wood** *for. phys.* (Weight per unit volume, generally expressed in g/cm³ or lb/ft³, dependent upon the content of humidity; in forestry there are three different densities: **1a. green density** based on the weight and volume of wood when green [standing or freshly felled]; **1b. nominal density:** density based on oven-dry weight and the volume when tested-commonly when the wood is green or at 50 or 12% moisture content; **2. bulk density:** kg per unit of bulk volume; **3. oven-dry density:** measured in g/cm³; density based on the weight and volume of wood when dried to constant weight in a ventilated oven at a temperature above the boiling point of water-generally 103 ± 2 °C; SAF 1983; ▶hardwood, ▶softwood); *s* **densidad [f] de la madera** (Peso por unidad de volumen, generalmente expresada en g/cm³, que depende del contenido de humedad; en silvicultura se diferencian tres densidades: **1. densidad aparente** [medida g/cm³]; densidad en estado original o recién talada; **2. densidad en volumen** [medida kg/m³ de madera sólida]; **3. compactibilidad en estado desecado** [medida g/cm³]; ▶madera dura, ▶madera blanda); *f* **densité [f]**

du bois (En F. les bois sont classés suivant leur densité moyenne, la densité pouvant être variable suivant les conditions de climat, de sol, de luminosité, donc d'exposition ; dans les diverses parties d'un même bois : le cœur est plus dense que l'aubier ; le bois de la base de l'arbre est plus dense que celui du sommet. Mais surtout pour un même bois la densité varie beaucoup en fonction de la teneur en eau. On ne peut donc faire de comparaisons utiles que si l'on convient d'un pourcentage commun à toutes les éprouvettes. Le taux de 15 % a été retenu ; **a)** parmi les feuillus : les **bois très légers** [densité 0,4 à 0,5] : le peuplier *[Populus]*, le saule *[Salix]*, le tilleul *[Tilia]*, le tremble *[Populus tremula]* ; les **bois légers** [densité 0,5 à 0,65] : l'aulne *[Alnus]*, le bouleau *[Betula]* ; les **bois demi-lourds** [densité 0,65 à 0,8] : le charme, le châtaignier, le chêne, l'érable, le frêne, le hêtre, le merisier, le noyer, l'orme, le platane, le poirier ; **les bois lourds** [densité 0,8 à 0,95] : le buis, le chêne vert, le cornouiller, le sorbier, l'olivier, le houx ; **b)** pour les bois résineux : les **bois très légers** [densité moins de 0,4] Pin Weymouth ; les **bois légers** [densité de 0,4 à 0,5] : le sapin, l'épicéa ; les **bois mi-lourds** [densité de 0,5 à 0,6] le pin sylvestre, le pin maritime, le pin du Nord ; les **bois lourds** [densité de 0,6 à 0,8] : le pin Laricio ; les **bois très lourds** [densité de 0,8 à 1] : l'if ; ▶bois dur, ▶bois tendre ; *termes spécifiques* densité à l'état vert ; densité anhydre [*syn.* densité à 0 %]) ; *g* **Holzdichte [f]** (Verhältnis der Holzmasse zum Holzvolumen in Abhängigkeit des Feuchtigkeitsgehaltes, da feuchtes Holz schwerer als trockenes Holz ist und deshalb eine höhere Dichte, d. h. ein höheres Gewicht je Volumeneinheit, hat. In der Forstwirtschaft sind folgende Messgrößen üblich: **1. Roh-dichte:** [Maßeinheit g/cm³]; Dichte bei beliebigem Feuchtig-keitsgehalt; bei nassem Holz höher als bei trockenem und somit bei derselben Holzart stark schwankend; man unterscheidet R. im Frischezustand und R. im Darrzustand. **2. Raumdichte:** Maß-einheit kg/Festmeter; sie bezeichnet das Verhältnis von Darrge-wicht zu Frischvolumen und gibt an, wie viel Kilogramm Trockensubstanz in einem Frischfestmeter einer Holzart enthalten sind. **3. Darrdichte:** Maßeinheit g/cm³; sie ist die Dichte des völlig trockenen Holzes und gibt das Verhältnis von Darrgewicht zu Darrvolumen an, d. h. wie viel Gramm wasserfreier Holzstoff in einem Kubikzentimeter resp. kg/m³ trockenem Holz enthalten ist; WFL 2002; ▶Laubholz, ▶Weichholz).

dentritic [adj] *geo.* ▶arborescent.

1420 denudation [n] *geo.* (Generic term for geomorphologic processes of progressive lowering of the Earth's surface by weathering with resultant leveling [US]/levelling [UK] of all land; originally the 'laying bare' of rocks by removal of material covering them; some authors regard **d.** as the actual process, and 'degradation' as the collective result on the landscape; cf. DNE 1978; ▶extreme soil erosion, ▶karstification, ▶natural erosion, ▶sheet erosion, ▶shoreline erosion [US]/shore line erosion [UK]); *syn.* degradation [n]; *s* **denudación [f]** (*Cambio morfo-lógico por erosión* ▶erosión ejercida por los diversos agentes externos fundamentalmente sobre áreas interfluviales, provo-cando un desgaste superficial; DGA 1986; ▶destrucción del suelo, ▶erosión laminar, ▶erosión marina, ▶karstificación); *f* **érosion [f] continentale** (CILF 1978 ; terme générique pour tout processus géomorphologique qui occasionne un aplanis-sement des reliefs continentaux ; ▶abrasion marine, ▶ablation marine, ▶dévastation des sols, ▶érosion, ▶érosion côtière, ▶érosion pelliculaire, ▶karstification) ; *syn.* équiplanation [f], dénudation [f] ; *g* **Abtragung [f]** (OB zu geomorphologischem Vorgang, der die Einebnung der Oberflächenformen des Fest-landes bewirkt; der Begriff 'Denudation' beinhaltet im Deutschen nicht die ▶Küstenerosion; ▶Bodenverheerung, ▶Erosion, ▶Flä-chenerosion, ▶Verkarstung); *syn.* Denudation [f], Destruktion [f].

deoxidization [n] *chem. limn.* ▶deoxidation.

deoxygenation [n] *chem. limn.* ▶deoxidation.

1421 deoxidation [n] *chem. limn.* (Decrease in oxygen of a body of water caused by ▶eutrophication); *syn.* deoxidization [n], deoxygenation [n]; *s* **desoxigenación [f]** (Disminución del contenido de oxígeno disuelto en el agua por efecto de la ▶eutrofización; DINA 1987); *syn.* desoxidación [f]; *f* **désoxygénation [f]** (Diminution de la teneur en oxygène dans les eaux douces provoquée par ▶eutrophisation) ; *g* **Sauerstoffentzug [m]** (Verminderung des Sauerstoffgehaltes in einem Gewässer durch ▶Eutrophierung).

department [n] *adm.* ▶cemetery department, ▶park department [US]/parks department [UK], ▶public works department, ▶waste management department.

department [n] [UK]**, building surveyors** *adm. plan. urb.* ▶Building Permit Office [US].

department [n]**, city hygiene** *adm. envir.* ▶waste management department.

department [n] [UK]**, civil engineering** *adm.* ▶public works department.

department [n] [US]**, County building** *adm. plan. urb.* ▶Building Permit Office [US].

Department [n] [UK]**, Local Authority Building Control** *adm. leg.* ▶Building Permit Office [US].

department [n]**, local government** *adm.* ▶local authority.

department [n]**, municipal hygiene** *adm. envir.* ▶waste management department.

department [n] [UK]**, parks and cemetery** *adm. urb.* ▶park department [US]/parks department [UK].

department [n] [US]**, parks and recreation** *adm. urb.* ▶park department [US]/parks department [UK].

department [n]**, public hygiene** *adm. envir.* ▶waste management department.

department [n] [US]**, sanitation** *adm. envir.* ▶waste management department.

department [n] **of planning and development** [CDN] *adm. urb.* ▶local planning agency/office [US].

department [n] **of public works** [CDN] *adm.* ▶public works department.

depauperation [n] **of fauna/flora** [US] *phyt. zool.* ▶depletion of fauna/flora.

1422 dependent layer [n] *phyt.* (Plant community connected with a certain vegetative layer; e.g. the highly developed broad-leaved sclerophyll shrub layer in the Evergreen Oak/Holm Oak *[Quercus ilex]* virgin forest of the Atlas mountains; ▶fusion of layers, ▶stratification of plant communities); *s* **estrato [m] dependiente** (En la estructura vertical de una comunidad [forestal], uno de los subtipos de ▶estratos relacionados que depende y está condicionado por otro determinado estrato, como es el caso del estrato arbustivo de laurifolios, muy desarrollado en el bosque virgen de *Quercus ilex* en el Atlas, o el de helechos en los bosques de *Fraxinus* y *Corylus* de Irlanda Occidental; cf. BB 1979; ▶estratificación de la vegetación); *f* **strate [f] dépendante** (Dans la structure verticale d'une association végétale ; strate liée à l'existence d'une autre strate, p. ex. la strate arbustive du Laurier et des végétaux sclérophiles dans la forêt vierge du Chêne vert *[Quercus ilex]* dans l'Atlas ; ▶solidarité écologique des strates, ▶stratification de la végétation) ; *g* **abhängige Schicht [f]** (Hinsichtlich der ▶Schichtenbindung einer Pflanzengesellschaft vorkommende Schicht, die von einer bestimmten Vegetationsschicht abhängig ist, z. B. die loorbeerblättrige Strauch-

schicht des *Quercus-ilex*-Urwaldes im Atlas; ▶Vegetationsschichtung).

1423 depletion [n] **of fauna/flora** *phyt. zool.* (Reduction in species diversity in the plant and animal kingdoms); *syn.* depauperation [n] of fauna/flora [also US] (TEE 1980, 209); *s* **empobrecimiento [m] de la fauna/flora** (Reducción de la diversidad de especies o de poblaciones); *f* **appauvrissement [m] faunistique/floristique** (Réduction de la diversité des espèces ou des peuplements) ; *g* **Verarmung [f] der Fauna/Flora** (Verringerung der Artenvielfalt).

1424 depletion [n] **of natural resources** *ecol. envir.* (Decrease in natural resources through consumption or ▶exhaustion; ▶wasteful exploitation); *s* **merma [f] de los recursos naturales** (Reducción de los recursos naturales por consumo humano o ▶agotamiento; ▶expoliación de los recursos naturales); *f* **régression [f] des ressources naturelles** (Diminution des réserves en ressources naturelles, ▶épuisement des ressources naturelles, ▶pillage des ressources naturelles) ; *syn.* raréfaction [f] des ressources naturelles, appauvrissement [m] en ressources naturelles ; *g* **Abnahme [f] natürlicher Grundgüter** (Verminderung natürlicher Grundgüter durch Verbrauch oder ▶Abbau 2; ▶Raubbau).

1425 depopulation [n] *sociol.* (Decrease in population number in a defined area; ▶migration 1 from an area, ▶rural exodus); *syn.* population decline [n]; *s* **despoblamiento [m]** (Reducción de la población humana en un área definida; ▶éxodo rural, ▶migración 1); *f* **dépeuplement [m]** (Diminution du nombre d'habitants [declin démographique] dans une région donnée ; ▶exode rural, ▶migration 1) ; *syn.* désertification [f] humaine, dépopulation [f]; *g 1* **Entvölkerung [f]** (Ständige Abnahme der Bevölkerungszahl in einem bestimmten Gebiet mit der Folge, dass dieses in letzter Konsequenz menschenleer wird. Früher fand die Abwanderung wegen Epidemien, Hungersnöte oder Kriege statt [z. B. Dreißigjähriger Krieg]; heute sind meist hohe Arbeitslosigkeit in strukturschwachen Gebieten und ein starker Geburtenrückgang die Ursachen; ▶Landflucht, ▶Migration 1); *g 2* **Entsiedelung [f]** (Entvölkerung eines bestimmten strukturschwachen Gebietes und die Aufgabe der Wohnhäuser und Betriebsgebäude sowie der Landwirtschaft; Häuser und Infrastrukturen verfallen oder werden abgerissen und die Natur überwuchert alles Anthropogene [Verbrachung der Landschaft mit ortstypischer ▶Sukzession bei ungelenkter Landschaftsdynamik]; *opp.* Siedlungsverdichtung).

1426 deposit [n] *geo. min.* (Natural accumulations of organic or inorganic materials in solid, liquid or gaseous form, which make up parts of the earth's crust and can be commercially exploited; ▶detrital deposit, ▶loose sedimentary deposit, ▶natural stone deposit); *s* **yacimiento [m]** (Acumulación natural de materias primas orgánicas o anorgánicas en forma sólida, líquida o gaseosa que forma parte de la corteza terrestre y puede ser aprovechada comercialmente; ▶yacimiento de recursos petrográficos); *f* **gisement [m] de minéraux et de fossiles** (Portion de l'écorce terrestre [interne ou de surface] où se trouvent accumulées des substances minérales ou organiques sous forme solide, liquide ou gazeuse et économiquement utiles ; DG 1984 ; ▶gisement de pierres naturelles) ; *syn.* gîte [m] de substances minérales ou fossiles ; *g* **Lagerstätte [f]** (Natürliche Ablagerungen oder Ansammlungen anorganischer oder organischer Rohstoffe in fester, flüssiger oder gasförmiger Form, die Bestandteile der Erdkruste bilden und für Wirtschaftszwecke verwertet werden können; ▶Natursteinvorkommen); *syn.* Rohstoffvorkommen [n].

deposit [n]**, alluvial** *geo.* ▶sediment (1).

deposit [n]**, surface** *geo.* ▶sediment (1).

D

D

deposit [n]**, unconsolidated sedimentary** *geo.* ▶loose sedimentary deposit.

deposit draft [n] [UK] *leg. urb. trans.* ▶preliminary planning proposal.

1427 deposition [n] (1) *biol.* (Accumulation of organic or inorganic matter in plant cells, standing water or on vegetation, or as leaf litter on the forest floor; ▶bioaccumulation); *s* **almacenamiento** [m] **de materia** (Acumulación de materia orgánica o anorgánica en células, en las aguas estancadas o sobre las plantas o el suelo del bosque; ▶bioacumulación); *syn.* depósito [m] de materia, acumulación [f] de materia; *f* **dépôt** [m] (Action et résultat de l'accumulation d'éléments organiques ou inorganiques dans les cellules vivantes, les eaux calmes ou sur les plantes, les sols forestiers etc. ; ▶bioaccumulation) ; *g* **Ablagerung [f] (1)** (**1.** *Vorgang* das Anhäufen von organischen oder anorganischen Stoffen in Pflanzenzellen, Stillgewässern oder auf Pflanzen, Waldböden etc. **2.** *Ergebnis* das Abgelagerte in Zellen, Stillgewässern oder auf Pflanzen, Waldböden etc.; ▶Bioakkumulation).

deposition [n] (2) *envir.* ▶particulate deposition.

deposition [n]**, final** *envir.* ▶final disposal.

deposition [n]**, sediment** *geo.* ▶sedimentation.

1428 deposition site [n] *geo.* (Location of surface deposit of solid particles by action of water, ice or wind); *s* **zona [f] de aluvión** (Lugar de sedimentación de material edáfico erosionado o de productos de la meteorización); *syn.* zona [f] de sedimentación; *f* **zone [f] sédimentaire** (Aire sur laquelle s'entassent les dépôts produits des activités érosives ; *terme spécifique* plaine d'épandage : zone sur laquelle se déposent les alluvions fluviatiles ou fluvio-glaciaires) ; *syn.* aire [f] de sédimentation ; *g* **Ablagerungsfläche [f]** (Ort der Akkumulation/Anhäufung von erodiertem Bodenmaterial/von Verwitterungsprodukten).

deposits [npl] *geo.* ▶detrital deposits; *min.* ▶extraction of deposits; *geo.* ▶loose sedimentary deposits.

deposits [npl]**, superficial** *geo.* ▶sediment (1).

depressed curb [n] [US] *constr.* ▶drop curb [US].

1429 depression [n] *geo.* (Saucer-like concavity, oblong or round in shape on the surface of the ground with slightly sloping sides; ▶location in a basin); *s* **depresión [f] (1)** (Cualquier parte del relieve de la superficie terrestre hundida respecto a su entorno, especialmente cuando no posee salidas para el drenaje superficial; DINA 1987; ▶localización en valle cerrado); *syn.* hondonada [f]; *f* **dépression [f]** (Forme de terrain en creux dont la taille se mesure en mètres [une petite doline], en hectomètres [dépression monoclinale en structure plissée], en kilomètres [dépression périphérique du massif ancien], et qui ne peut être assimilée une vallée ; DG 1984 ; ▶situation en cuvette) ; *g* **Geländemulde [f]** (Längliche oder rundliche, ringsum von leicht ansteigenden Hängen begrenzte flache Hohlform der Erdoberfläche; ▶Kessellage); *syn.* Mulde [f] (1).

depth [n] *arch. constr.* ▶bonding depth; *agr. hort. constr.* ▶cultivation depth; *constr.* ▶drainage depth; *wat'man* ▶drawdown area depth; *constr.* ▶fill depth, ▶frost-penetration depth (in soil); *agr. hort. constr.* ▶installation depth; *surv.* ▶lot depth; *constr. hort.* ▶planting depth.

depth [n]**, gutter** *constr.* ▶curb height.

depth [n]**, loosening** *agr. hort. constr.* ▶cultivation depth.

depth [n] [UK]**, plot** *surv.* ▶lot depth [US].

1430 depth [n] **of groundwater table** *hydr.* (Distance between groundwater level and ground surface); *s* **profundidad [f] del nivel freático** (Distancia entre la superficie de la capa freática y la del suelo); *f* **profondeur [f] de la nappe phréatique** (Distance comprise entre la surface de la nappe et la surface au sol) ; *g* **Grundwasserflurabstand** [m] (Abstand zwischen Grundwasserspiegel und Geländeoberkante).

dereliction [n] **of land** *plan. urb.* *►* ▶derelict land.

1431 derelict land [n] *plan. urb.* (**1.** Area or tract of land occupied by a disused industrial plant or which has been cleared of its former industrial uses, an industrial blighted area, or even a disused spoil pile [US]/spoil heap [UK] in a mining region; ▶hazardous old dumpsite; **2.** *process* dereliction of land; ▶brownfield, ▶reclamation of derelict land, ▶recycling of derelict sites, ▶waste land, ▶barren land); *syn.* abandoned industrial site [n], blighted industrial land [n] [US], brownfield [n] (±), disused land [n] [also US], industrial waste land [n], disturbed (industrial) land [n]; *s* **ruina [f] industrial** (Áreas caracterizadas por instalaciones industriales que ya no están en el proceso de producción o antiguas zonas de depósitos de zafras mineras; ▶emplazamiento contaminado, ▶finca industrial potencialmente contaminada, ▶recuperación de ruinas industriales, ▶tierra yerma, ▶terreno improductivo); *syn.* zona [f] industrial abandonada, suelo [m] industrial abandonado; *f* **friche [f] industrielle** (**1.** Espaces abandonnés par suite de la cessation d'activités industrielles, du démantèlement d'installations industrielles, d'une relocalisation des activités ; ▶site pollué ancien ; **2.** dépôts de stériles non recultivés dans le cadre des exploitations minières, nécessitant en général une réhabilitation du site ; ▶ancien site industriel potentiellement contaminé par des déchets historiques, ▶réhabilitation des friches industrielles, ▶terre improductive, ▶terre inculte) ; *g* **Industriebrache [f]** (Flächen, die durch stillgelegte Industriebetriebe oder durch abgeräumte Industrieanlagen gekennzeichnet sind sowie durch den Bergbau verursachte, nicht genutzte Bergedeponieflächen; ▶altlastverdächtige ehemalige Industriefläche, ▶Altstandort, ▶Flächenrecycling, ▶Ödland, ▶Unland).

1432 derelict land register [n] *adm. urb.* (Computerized data bank with detailed information on ruins, industrial wasteland and derelict properties, some of which may be contaminated; **in U.S.**, all derelict land files are kept at the local level within cities and states. The Federal government keeps track of "Super Fund" sites designated to receive federal funds to remove toxic materials, but a national computer data bank containing all "derelict" land parcels is not maintained at the Federal level. **In UK & Ireland**, the official Land Registry comprises a computerised data bank with detailed nation-wide information on ruins, industrial wasteland and derelict properties, some of which may be contaminated. **In D.**, the register is applied in the planning of new uses for brownfield sites with the aim of limiting new development of greenfield sites within the scope of an overall sustainable land management policy; ▶recycling of derelict sites); *syn.* abandoned land register [n] [also US], abandoned property register [n] [also US]; *s* **inventario [m] de suelos industriales abandonados** (Banco de datos computarizado con información detallada de las ruinas Industriales y suelos abandonados, que pueden estar contaminados. En **Es.** no existe tal inventario, solamente se registran a nivel de CC.AA. los emplazamientos de ▶suelos contaminados. **En EE.UU.** se registran en las ciudades y/o en los estados. A nivel federal solo están registrados los emplazamientos contaminados del «Super Fund» designados para recibir fondos federales para su saneamiento. **En GB e Irlanda** el registro de la propiedad de la tierra es un banco de datos de cobertura nacional con información detallada sobre ruinas industriales y suelos abandonados. **En F.** no existe un inventario estatal, sino que solamente se registran los emplazamientos abandonados en las grandes ciudades, aglomeraciones urbanas y en algunos departamentos, generalmente sobre la base del inventario histórico de antiguos emplazamientos industriales y de actividedes de servicios [BASIAS].

En D. el registro se realiza a nivel municipal con el fin de planificar la reutilización de los suelos y así evitar el consumo de suelo; ►recuperación de ruinas industriales); *f* **inventaire [m] des friches industrielles (F.**, recensement des sites industriels et d'activités de services désaffectés ou sous-utilisés, caractérisés par une contamination potentielle, l'obsolescence des bâtiments et/ou une infrastructure inadéquate ; il est très souvent établi au cas par cas suivant les nécessités, par les grandes villes, agglomérations ou un département en vue d'une planification urbanistique économique et environnementale ; il est en général réalisé sur la base de l'inventaire historique d'anciens sites industriels et activités de service dont les données sont regroupées dans la base de données d'anciens sites industriels et activités de service [BASIAS] ; ►réhabilitation des friches industrielles) ; *g* **Brachflächenkataster [m,** *in A nur so* — oder **n]** (Rechnergestützte, flächendeckende Erhebung und lagegenaue Darstellung aller erfassten brachliegenden Industrie- und Gewerbeflächen, oft belastet, zwecks Wiederverwendung für städtebauliche Nutzungen [Folgenutzung von Brachen] mit dem städtebaulichen Ziel, den Landschaftsverbrauch in einem Gemeindegebiet im Rahmen eines übergeordneten Flächenmanagements nachhaltig einzuschränken; ►Flächenrecycling).

1433 derelict land [n] **resulting from mining operation** *min.* (**In U.S.,** abandoned surface-mined land that was inadequately reclaimed by the operator and for which the operator no longer has any fixed responsibility; **d. l.** usually refers to lands mined before the passage of comprehensive reclamation laws; cf. RCG 1982); *syn.* orphan land [n] [also US]; *s* **paisaje [m] minero abandonado** (Mina a cielo abierto que ha sido abandonada sin recultivar); *f* **friche [f] minière (U.S.,** paysage industriel engendré avant l'application de la réglementation instituant l'obligation générale pour l'exploitant de remise en état des sites industriels pollués) ; *g* **nicht rekultivierte Bergbaufolgelandschaft [f] (U.S.,** Industrielandschaft, die durch den Bergbau vor In-Kraft-Treten der Rekultivierungsbestimmungen entstand und für die der Betreiber hinsichtlich der Wiederherstellung eines geordneten Landschaftsgefüges keine Auflagen erhielt).

1434 description [n] **of professional services** *contr. prof.* (Work performed by architects, engineers, landscape architects, or other specialists who are involved in planning a project); *s* **descripción [f] de servicios profesionales** (Definición de los trabajos realizados por arquitectos/as, consultores/as, ingenieros/as y planificacadores/as); *f 1* **fixation [f] de l'étendue de la mission** (La mission du prestataire est soit de conception soit de maîtrise d'œuvre) ; *f 2* **définition [f] de la forme de la mission** (Mission complète avec respect du coût d'objectif ou mission partielle sur la base d'une estimation prévisionnelle des travaux) ; *f 3* **définition [f] du contenu de la mission** (Fixation de la catégorie et du contenu de la mission normalisée confiée aux architectes, ingéniers, paysagistes etc. membres d'une équipe pluridisciplinaire de conception ou de maîtrise d'œuvre participant à une opération) ; *g* **Leistungsbeschreibung [f] (1)** (Beschreibung der Architekten-, Berater-, Ingenieur- oder Leistungen anderer Projektbeteiligter).

1435 de-sealing [n] *constr. envir.* (Removal of impermeable surfaces and application of permeable surfacing or vegetation; *opp.* ►sealing of soil surface); *s* **desimpermeabilización [f] del suelo** (Remoción de pavimentos impermeables y sustitución por pavimentos permeables o por tierra vegetal; *opp.* ►sellado del suelo); *syn.* sustitución [f] de pavimentación impermeable por permeable; *f* **reperméabilisation [f] des sols** (Opération visant à retrouver la perméabilité des supports originels par enlèvement du revêtement existant et son remplacement par un revêtement perméable ou par la reconstitution d'un couvert végétal ; perméa-

biliser [vb] ; *opp.* ►imperméabilisation du sol) ; *g* **Entsiegelung [f] des Bodens** (Entfernung von undurchlässigen Bodenbelägen sowie der Abriss baulicher Anlagen und Ersatz durch wasser- und luftdurchlässige Beläge oder Wiederherstellung von Vegetationsflächen; *opp.* ►Versiegelung des Bodens); *syn.* Bodenentsiegelung [f], Entsiegelung [f] befestigter Flächen.

1436 desertification [n] *geo.* (Transformation process of formerly fertile land into desert; ►Convention to combat desertification); *s* **desertificación [f]** (Proceso causado por cambios bioclimáticos y/o la acción antropógena de pérdida de fertilidad de áreas y paulatina formación de desierto. El progreso de la desertificación va acompañado, en general, de la degradación física de los ecosistemas del desierto, que puede tomar la forma de movimientos de arenas ó invasión de dunas, de crecidas repentinas, de entarquinamiento o formación de cárcavas en las cuencas degradadas, todo lo cual ocasiona una perdida de tierras productivas; DINA 1987; ►Convención de las Naciones Unidas de lucha contra la desertificación); *syn.* desertización [f]; *f* **désertification [f]** (Évolution bioclimatique [facteurs physiques et biologiques] et action anthropique [facteurs politiques, sociaux, culturels et économiques] entraînant la progression des déserts au dépend des terres cultivées suite à la dégradation irréversible des sols ; cf. décision du Conseil n° 98/216/CE relative à la conclusion, au nom de la Communauté européenne, de la convention des Nations unies sur la lutte contre la désertification dans les pays gravement touchés par la sécheresse et/ou la désertification, en particulier en Afrique ; ►Convention des Nations Unies sur la lutte contra la désertification) ; *g* **Desertifikation [f]** (Aus dem Lat. *deserta, -orum [n]* >Einöden<, >Wüsten<; Verwüstung und Verödung von Landstrichen und Verschlechterung deren Böden durch Vormarsch vorhandener Wüsten. Die Wüstenbildung ist ein schwerwiegendes Umweltproblem, das durch vielschichtige Wechselwirkungen zwischen physikalischen, biologischen, politischen, sozialen, kulturellen und wirtschaftlichen Faktoren verursacht wird; cf. Beschluss 98/216/EG des Rates v. 09.03.1998 über den Abschluss — im Namen der Europäischen Gemeinschaft — des Übereinkommens der Vereinigten Nationen zur Bekämpfung der Wüstenbildung in den von Dürre und/oder Wüstenbildung schwer betroffenen Ländern, insbesondere in Afrika; ►Konvention zur Bekämpfung der Wüstenbildung); *syn.* Ausbreitung [f] der Wüsten, Wüstenausbreitung [f], Wüstenbildung [f].

Desertification Convention [n] *conserv. leg.* ►Convention to combat desertification.

1437 desert reclamation [n] *agr.* (Reclaiming of arid areas by irrigation); *s* **ganancia [f] de tierras desérticas** (Posibilidad de aprovechamiento de tierras en zonas áridas gracias a la irrigación); *f* **colonisation [f] agricole en zone désertique** (Conquête de terres agricoles par irrigation des régions sèches ou des zones désertiques) ; *syn.* conquête [f] de terres vierges, gain [m] de terres par irrigation ; *g* **Landgewinnung [f] in Wüsten** (Gewinnung von landwirtschaftlichen Bodenflächen durch Bewässerung in ariden Gebieten).

1438 desiccation crack [n] *pedol.* (DNE 1978; open fissure, often in a polygonal pattern, caused by drying out of wet clay soils; ►seasoning check, ►shrinkage crack); *s* **fisura [f] de contracción** (Grieta en suelo arcilloso, a menudo de forma poligonal, causada al secarse éste; ►fenda de merma, ►grieta de contracción); *syn.* fisura [f] de desecación, grieta [f] de contracción, grieta [f] de desecación; *f* **fente [f] de retrait (1)** (Fissure ouverte dessinant souvent un réseau polygonal dans des matériaux argileux par suite de la dessiccation du sol ; ►fente de retrait 1, ►fissure de retrait [de dessiccation] dans le béton) ; *syn.* fente [f] de dessiccation, fissure [f] de retrait ; *g* **Trockenriss [m]** (Offener, meist durch Austrocknung entstandener, polygonal ange-

D

ordneter Bodenspalt in wasserhaltigen Lehm-, Ton- oder Mergelböden; ►Schwindriss in Beton, ►Schwindriss in Holz); *syn.* Schrumpfungsriss [m] (1).

1439 design [n] (1) *arch.* (**1.** Result or process of an intentional organization or composition of structural elements of a work of art or object. **2.** Shape, delineation, pattern or motif of an artistic work or object); *s 1* **forma [f]** (Resultado); *s 2* **diseñar [vb]** (Proceso de dar forma a un objeto); *f 1* **forme [f]** (Résultat) ; *f 2* **mise [f] en forme** (Processus) ; *g* **gestalterische Ausbildung [f]** (Formgebung; ausbilden [vb]); *syn.* Ausformung [f], Ausgestaltung [f], Design [n], Gestaltung [f] (2).

1440 design [n] (2) *arch. plan.* (Schematic plan aimed at organizing a physical, functional and aesthetic arrangement of a variety of structural elements in a project to achieve desired social, cultural and ecological outcomes: the concept is often illustrated in two or three-dimensional forms; ►layout 1, ►planning 1, ►planning 2); *s* **diseño [m] (de un proyecto)** (Arreglo funcional y estético de estructuras y elementos paisajísticos en un proyecto; ►plan de diseño, ►planificación, ►proyectar); *syn.* concepción [f] (de un proyecto); *f* **conception [f]** (1) (Mise en forme d'un projet selon des critères fonctionnels, techniques, esthétiques et écologiques ; ►études, ►mise en page, ►planification) ; *g* **Gestaltung [f]** (3) (**1.** Erarbeitung eines Entwurfs nach funktionalen, technischen, formal-ästhetischen, gesellschaftlichen und ökologischen Gesichtspunkten. **2.** Zwei- oder dreidimensionales Entwurfskonzept einer planerischen Idee. **3.** Anordnung von Text und Bildelementen sowie weiterer Darstellungsformen auf einem Plan, Poster, Plakat etc.; ►Gestaltung 3, ►Planung 1, ►Planung 2); *syn. zu 2.* Design [n]; *syn. zu 3.* Layout [n] (1).

design [n] (3) *plan.* ►detailed design; *arch. constr.* ►edge design; *plan.* ►final project design; *gard. hist.* ►fine garden design; *gard. plan.* ►garden design; *constr. plan.* ►grading design; *adm. landsc.* ►grave design; *adm.* ►gravestone design; *land'man. landsc.* ►landscape design; *gard. landsc.* ►planting design; *plan. prof.* ►preliminary design, ►sketch design; *leg. plan. pol. sociol.* ►universal design; *arch. urb.* ►urban design.

design [n]**, concept** *plan. prof.* ►sketch design.

design [n]**, concept site** *plan. prof.* ►sketch design.

design [n] [US]**, grading** *constr. plan.* ►ground modeling [US]/ground modelling [UK].

design [n]**, guarding** *constr. leg.* *►protection from falling.

design [n]**, open space** *landsc. recr. urb.* ►open space planning.

design [n]**, schematic** *plan. prof.* ►sketch design.

design [n]**, site** *plan. prof.* ►site planning.

design and build competition [n] [UK] *constr. contr.* ►design/build competition [US].

1441 design and build firm [n] *constr. prof.* (Professional firm doing both design and construction of [landscape] projects); *s* **gabinete [m] de diseño y construcción de jardines** (En EE.UU. una empresa que se dedica tanto a planificar y diseñar parques y jardines como a la construcción paisajística); *f* **entreprise [f] paysagiste de conception et de réalisation de jardins** (En général moyenne et grande entreprise disposant d'un bureau d'études et d'une direction des travaux) ; *g* **Betrieb [m] für Gartenplanung und Ausführung** (Zusammenschluss [sog. ARGE, meist in der Form einer GmbH] eines ausführenden Garten- und Landschaftsbaubetriebes mit Landschaftsarchitekten, um sich direkt für mögliche Aufträge zu bewerben mit der Folge, dass es keine funktionale Trennung mehr von Landschaftsarchitekten und Garten- und Landschaftsbaubetrieben gibt. Nach deutscher Berufsauffassung ist einem planenden Architekten die gewerbliche Tätigkeit im Baugewerbe untersagt, da dies nicht

dem Standesrecht entspricht. In den USA üblich als Firma, die ein Planungsbüro und einen Ausführungsbetrieb unterhält); *syn.* Arbeitsgemeinschaft [f] für Planung und Ausführung.

1442 design and build program [n] [US]/**design and build programme** [n] [UK] *constr.* (Description of project and services to be provided in planning and execution of major work [US]/works [UK]. Selection of a construction contractor for optimum execution of a project from technical, economic, and design viewpoints is then determined by competitive bidding [US]/competitive tendering [UK]; ►design/build competition [US], ►project description); *s* **memoria [f] descriptiva de obra con programa de trabajo** (En el caso de obras de gran envergadura y complejidad y a diferencia de la ►memoria descriptiva normal, representación de la obra prevista en forma de programa de desarrollo de los trabajos, de manera que para expertos terceros sea posible reconocer las diferentes fases y las condiciones importantes de la obra. Esto sirve además para posibilitar el desarrollo y la elección de las mejores soluciones técnicas, económicas, funcionales y estéticas; ►concurso con programa establecido); *f* **dossier [m] de consultation avec programme** (Un dossier du concours qui comporte, en annexe au règlement particulier de l'appel d'offres, le « programme » qui définit les données recueillies, les besoins à satisfaire, les contraintes et les exigences à respecter ; CCM 1984, 78 ; ►appel d'offres avec concours. À la différence des ►spécifications techniques détaillées [S.T.D.] le programme définit les données recueillies, les besoins à satisfaire, les contraintes et les exigences à respecter permettant aux concurrents de présenter un projet et une offre pour lesquels les propositions techniques, financières, architecturales adoptées constituent la meilleure solution pour la réalisation des ouvrages ; cf. CCM 1984, 78 ; ►description des ouvrages) ; *g* **Leistungsbeschreibung [f] mit Leistungsprogramm** (Leistungsbeschreibung bei großen und technisch schwierigen Vorhaben, bei der abweichend zur ►Baubeschreibung die Leistungen programmatisch dargestellt werden, aus der die Bewerber alle für die Entwurfsbearbeitung und für ihr Angebot maßgebenden Bedingungen und Umstände erkennen können, um die technisch, wirtschaftlich und gestalterisch beste sowie eine funktionsgerechte Lösung der Bauaufgabe zu ermitteln und vorzuschlagen; cf. § 9 Nr. 10 u. 11 VOB Teil A; ►Ausschreibung mit Leistungsprogramm).

design and build tendering [n] [UK] *constr. contr.* ►design/build competition [US].

1443 design and location approval [n] [US] (1) *adm. leg.* (Authorization of a project, which gives planning consent, including necessary subsequent mitigation measures to accommodate affected public interests. The permissibility of a project is determined/ascertained in the ►design and location approval process [US]/determination process of a plan [UK]. All public, legal/judicial relationships between the submitting authority and those affected by the planning proposals are legally defined by the **d.**; ►approved plan; ►approved plan for major projects [US]); *syn.* determination [n] of a plan [UK]; *s* **aprobación [f] de proyectos públicos** (Comprobación de la licitud de un proyecto público, incluyendo las medidas necesarias en otras instalaciones, en cuanto a todos los intereses públicos. A través de la **a.** se regulan todas las relaciones de derecho público entre el responsable del proyecto y los afectados por el mismo; ►plan de proyecto público aprobado, ►procedimiento de aprobación de proyectos públicos); *f* **approbation [f] de grands projets publics (≠)** (**D.**, contrôle de la recevabilité d'une opération publique y compris celles des mesures induites pouvant toucher des intérêts publics ou privés dans le cadre d'une ►procédure d'instruction de grands projets publics ; la procédure prévoit la consultation des personnes concernées et le déroulement d'une ►enquête publique ;

▶plan [de grands projets publics] approuvé) ; **g Planfeststellung [f]** (Festlegung der Zulässigkeit eines öffentlichen Vorhabens einschließlich der notwendigen Folgemaßnahmen an anderen Anlagen im Hinblick auf alle von dem Vorhaben berührten öffentlichen Belange im Rahmen eines ▶Planfeststellungsverfahrens. Durch die **P.** werden alle öffentlich-rechtlichen Beziehungen zwischen dem Träger des Vorhabens und den durch den Plan Betroffenen rechtsgestaltend geregelt; cf. § 75 VwVfG; ▶festgestellter Plan).

1444 design and location approval [n] [US] (2) *adm. leg.* (Final decision which gives planning consent to an important project after a ▶design and location approval process [US]/ determination process of a plan [UK] according to the requirement of statutory consultation with public authorities/agencies and ▶public hearing [US]/public inquiry [UK]); *syn.* Secretary of State Approval [n] (of a plan) [UK]; *s* **acuerdo [m] de aprobación definitiva de proyecto público** (Decisión definitiva sobre un proyecto en el marco del ▶procedimiento de aprobación de proyectos públicos después del proceso de ▶audiencia pública); *f* **arrêté [m] d'approbation de grands projets publics** (Décision définitive consacrant l'autorisation d'un projet après ▶enquête publique dans le cadre d'une ▶procédure d'instruction de grands projets publics et le rendant imposable aux tiers) ; *syn.* arrêté [m] portant approbation ; **g Planfeststellungsbeschluss [m]** (Endgültige Entscheidung über ein Vorhaben im Rahmen des ▶Planfeststellungsverfahrens nach dem Anhörverfahren; ▶öffentliche Anhörung).

1445 design and location approval process [n] [US] *adm. leg.* (Administrative procedure prescribed by law for the development of such momentous public projects as major highways, pipelines, hazardous waste disposal sites, nuclear power plants, etc. The permissibility of such projects and necessary subsequent measures for mitigation etc., are determined/ ascertained in this procedure. All public, legal/judicial relationships between the submitting authority, department and citizen representatives affected by the planning proposals are settled after a ▶public hearing [US]/public inquiry [UK]); *syn.* determination process [n] of a plan [UK]; *s* **procedimiento [m] de aprobación de proyectos públicos** (Procedimiento administrativo para llevar a cabo la planificación de proyectos públicos, como p. ej. autopistas o carreteras, aeropuertos, oleoductos o gaseoductos, depósitos de residuos, centrales nucleares, etc. En él se comprueba la legalidad del proyecto y se fijan todo tipo de medidas necesarias para evitar daños al bien público. Todas las relaciones legales entre la autoridad comitente y las personas e instituciones afectadas por el proyecto se regulan en él con ayuda de la audiencia de autoridades e instituciones públicas y semipúblicas; ▶audencia pública. En D es un procedimiento claramente regulado y obligatorio por ley); *f* **procédure [f] d'instruction de grands projets publics** (Procédure administrative prescrite pour la réalisation d'aménagements ou d'ouvrages entrepris par les collectivités publiques telles que les grandes infrastructures [routes et autoroutes, lignes haute tension, etc.], installations classées [décharges, centrales nucléaires, etc.] ; par cette procédure est engagé le contrôle de la recevabilité d'une opération publique y compris celles des mesures induites pouvant toucher des intérêts publics ou privés ; la procédure prévoit la consultation des personnes concernées et le déroulement d'une ▶enquête publique) ; **g Planfeststellungsverfahren [n]** (**D.**, Ein durch Rechtsvorschrift angeordnetes Verwaltungsverfahren zur Verwirklichung von staatlichen Planungen wie z. B. Fernstraßen, Fernleitungen, Deponien, Kernkraftwerke etc. Dabei wird die Zulässigkeit eines staatlichen Vorhabens einschließlich der notwendigen Folgemaßnahmen an anderen Anlagen im Hinblick auf alle von dem Vorhaben berührten öffentlichen Belange festge-

legt. Durch das **P.** werden alle öffentlich-rechtlichen Beziehungen zwischen dem Träger des Vorhabens und den durch die Planung Betroffenen unter Einbeziehung eines Anhörungsverfahrens rechtsgestaltend geregelt; cf. § 75 VwVfG; ▶öffentliche Anhörung).

1446 designated land-use area [n] *plan.* (Area identified for a certain use by a ▶community development plan, ▶landscape plan, etc.); *s* **área [f] destinada a un uso** (Superficie identificada en un ▶plan urbanístico, ▶plan parcial, ▶plan de gestión del paisaje, etc. para un determinado uso); *f* **zone [f] affectée** (Espaces, zones dont l'affectation est fixée dans les documents d'urbanisme, un ▶plan de paysage, etc. ; ▶documents de planification urbaine) ; **g Nutzungsfläche [f]** (Fläche, die in ▶Bauleitplänen, ▶Landschaftsplänen etc. ausgewiesen ist).

designated services [npl] [US] *contr. prof.* ▶special professional services.

designation [n] *leg. plan.* ▶statutory designation.

designation [n]**, legal** *leg. plan.* ▶statutory designation.

designation [n] **by legally binding land-use plan** *leg. plan.* ▶designation by zoning map [US].

1447 designation [n] **by zoning map** [US] *leg. plan.* (Legally-binding delineation of areas and objects worthy of retention and preservation, etc.; **in U.K.**, designations are shown on the **Proposals Map,** which shows the location of proposed schemes in all current Development Plan Documents and the **Site Allocations Development Plan Document,** which allocates land for new development, identifies the characteristics and requirements of such development and defines the development boundaries for further growth; ▶zoning map [US]/Proposals Map/Site Allocations Development Plan Document [UK]); *syn. in Europe* designation [n] by a legally binding land-use plan; *s* **prescripción [f] en plan parcial** (Determinación legal vinculante de usos del suelo, declaraciones de edificios protegidos, etc. por medio del ▶plan parcial [de ordenación]); *f* **prescription [f] édictée par un plan local d'urbanisme/plan d'occupation des sols** (Dispositions architecturales ou d'urbanisme spécifiques opposables au tiers et destinées à maintenir ou à créer une forme urbaine caractéristique d'une construction ou d'un ensemble de constructions ; cf. C. urb., art. R. 123-21) ; **g Festsetzung [f] durch den Bebauungsplan** (Rechtsverbindliche Ausweisung gegen jedermann von Flächen, schutzwürdigen Objekten etc. im ▶Bebauungsplan); *syn.* Festsetzen [n, o. Pl.] durch den Bebauungsplan.

designation [n] **of landscape planning requirements** *leg. landsc.* ▶statutory designation of landscape planning requirements.

designation [n] **of land uses** *adm. leg. plan.* ▶statutory land use specification.

1448 design/build competition [n] [US] *constr. contr.* (▶design and build program [US]/design and build programme [n] [UK]); *syn.* design and build competition [n] [UK], design and build tendering [n] [also UK]; *s* **concurso [m] con programa establecido** (▶memoria descriptiva de obra con programa de trabajo); *f* **appel [m] d'offres avec concours** (Cette variante d'appel d'offres est recommandée dans tous les cas où il est manifeste que le maître de l'ouvrage n'a pas intérêt à définir, ou n'est pas en mesure de définir sans une grande part d'arbitraire, les grandes lignes de la conception technique des ouvrages ; CCM 1984, 79 ; ▶dossier de consultation avec programme) ; **g Ausschreibung [f] mit Leistungsprogramm** (Ausschreibungsvariante mit einer Programmbeschreibung, die i. d. R. zu einem

D

Pauschalangebot führt; ▶Leistungsbeschreibung mit Leistungs-programm).

1449 design competition [n] **(1)** *prof.* (Competitive process between planning and design firms, requested by a public agency to submit written and graphic proposals for a development project, often used merely to generate ideas or concepts. Often such proposals are then evaluated by a professional jury which makes recommendations to the agency regarding the award of a contract. The agency may also select a few firms to prepare refined and detailed plans, based on their concepts, or to participate in a ▶final design competition [US]/realization competition [UK] for selection of a firm for contract award; ▶public prequalification bidding notice, ▶request for proposals [US]); *s* **concurso [m] de ideas** (Procedimiento de competencia entre empresas de planificación y diseño que son llamadas por una administración pública a presentar proposiciones para proyectos específicos. Las proposiciones son evaluadas por un jurado profesional que hace las recomendaciones sobre los premios; ▶aviso de candidaturas a concurso público; ▶concurso de realización); *f* **concours [m] d'idées** (Mise en compétition par avis d'appel public de candidatures de bureau d'études, d'aménageurs, d'architectes, etc. dans l'intention de définir les grandes lignes de conception d'ouvrages, d'aménagements [prestations intellectuelles] dans le domaine de l'aménagement du territoire, de l'urbanisme ou de la construction, sans promesse de réalisation ; les candidats ayant remis un projet sont jugés et primés par un jury qui motive son choix permettant à l'organisateur du concours d'allouer des primes, récompenses ou avantages aux auteurs des prestations les mieux classées ; ▶appel public de candidatures pour un appel d'offres, ▶concours de conception-construction) ; *g* **Ideenwettbewerb [m]** (Auslobungsverfahren, mit dem ein Auftraggeber eine Vielfalt von Ideen für die Lösung einer Aufgabe auf dem Gebiet der Raumplanung, des Städtebaues, der Freiraumplanung resp. Landschaftsarchitektur oder des Bauwesens anstrebt, ohne dass eine Absicht zur Realisierung der Aufgabe besteht und das von einem Preisgericht auf Grund vergleichender Beurteilungen bewertet und mit oder ohne Verteilung von Preisen entschieden wird; cf. Teil 1 Nr. 2.1.1 GRW 1995 u. § 20 [1] VOF; es gibt den offenen anonymen, den begrenzt offenen [beschränkten] und den Einladungswettbewerb; ▶öffentlicher Teilnahmewettbewerb für Planungsbüros, ▶Realisierungswettbewerb. **In USA** ist das öffentliche Wettbewerbswesen im Vergleich zu Europa unterrepräsentiert, da vorwiegend betriebswirtschaftliche Gesichtspunkte es einschränken. **I.e** mit experimentellen Ansätzen, die eine kreative Plattform zur Meinungs- und Gestaltungsfindung bieten, werden von etablierten Planungsbüros kaum wahrgenommen. **I.e** werden national ausgeschrieben und vor allem Lehrende, Studierende und junge Büros beteiligen sich an ihnen); *syn.* Gestaltungswettbewerb [m], Planungswettbewerb [m] (§ 25 VOF).

design competition [n] **(2)** *adm. plan. prof.* ▶limited design competition, ▶open design competition, ▶submission deadline for design competition.

1450 design competition announcement [n] [US] *adm. prof. leg.* (Public announcement of ▶design competition 1 or ▶final design competition [US]/realization competition [UK] and amount of cash prizes; ▶merit award, ▶prize, ▶honorable mention [US]/honourable mention [UK]); *syn.* design competetion invitation [n] [UK], issuing [n] a call for entries; *s* **anuncio [m] de un concurso (de ideas)** (Anuncio público del tema, las condiciones y los premios previstos de un ▶concurso de ideas o de un ▶concurso de realización; ▶mención, ▶mención honorífica, ▶premio); *f* **appel [m] public à la concurrence (pour un concours)** (Mise en compétition de maîtres d'œuvre, présentation publique du dossier de consultation, du mode de désignation des

lauréats et de l'allocation des ▶prix ou récompenses attribuées dans le cadre d'un ▶concours d'idée, d'un ▶concours de conception-construction, d'un concours d'architecture et d'ingénierie, de l'attribution d'un marché par appel d'offres avec concours d'idées ; ▶projet mentionné, ▶travaux primés) ; *g* **Auslobung [f]** (Öffentlich bekannt gemachtes, einseitiges Versprechen einer Belohnung für die Vornahme, z. B. eines ▶Ideenwettbewerbes oder ▶Realisierungswettbewerbes, insbesondere für die Herbeiführung eines Erfolges; cf. §§ 657 ff BGB u. Teil 1, Nr. 1.3 GRW 1995; ausloben [vb]; ▶Ankauf, ▶engere Wahl, ▶Preis); *syn.* Ausschreibung [f] eines (Ideen)wettbewerbes.

1451 design competition documents [npl] *plan. prof.* (Description of design requirements provided by requesting authority/client to prospective offerers of professional services, as a basis for submission of design solutions; ▶design competition invitation, ▶competition rules [US]/competition standing orders [UK]); *s* **dossier [m] de bases del concurso de ideas** (Documentación de los requerimientos de un concurso de ideas puesta a disposición por la administración que llama a concurso y que sirve de base para elaborar las propuestas de diseño; ▶anuncio de un concurso [de ideas]; ▶reglamento de concurso); *f* **dossier [m] de concours d'idées** (Documents constituant le dossier permettant la consultation des candidats et composé d'une présentation exhaustive du concours et de son règlement, de l'état des lieux, du cahier des charges, du dossier cartographique et photographique, du fond de plan de travail et des gabarits de rendu y compris tous documents statistiques ou pré-études techniques ; ▶appel public à la concurrence [pour un concours], ▶règlement du concours) ; *g* **Ausschreibungsunterlagen [fpl] für einen Ideenwettbewerb** (Auslobungstext, der die Wettbewerbsaufgabe umfassend und eindeutig beschreibt und alle Anforderungen, die von den Wettbewerbern erfüllt werden müssen, klar definiert incl. aller erforderlichen Karten, Luftbilder, Grundlagenpläne und sonstigen Unterlagen sowie Ergebnisse etwaiger Voruntersuchungen, Statistiken und Prognosen; ▶Auslobung, ▶Wettbewerbsordnung).

design competition invitation [n] [UK] *adm. prof. leg.* ▶design competition announcement [n] [US].

1452 design contract [n] [US] *prof.* (Contractual agreement between the planner and his client; ▶consulting contract, ▶Guidance for Clients on Fees [UK], ▶planning contract); *syn.* letter [n] of appointment [UK], professional contract [n]; *s* **contrato [m] profesional** (Acuerdo contractual del arquitecto [paisajista] con el cliente; ▶contratación de un/a planificador, -a, ▶contrato de asesoría, ▶reglamento de honorarios); *f* **contrat [m] d'architecte** (Contrat passé entre l'architecte et le maître d'ouvrage ; ▶contrat pour une mission de conseil, ▶mission d'études, ▶mission de maîtrise d'œuvre, ▶rémunération des prestations de maîtrise d'œuvre) ; *syn.* marché [m] d'études, contrat [m] d'études ; *g* **Architektenvertrag [m]** (Vertrag eines [Landschafts]architekten mit einem Bauherrn; ▶Planungsauftrag, ▶Beratervertrag, ▶Honorarordnung für Architeketen und Ingenieure).

design development phase [n] [US] *contr. prof.* ▶final design stage.

design drawing [n] *constr. plan.* ▶detailed design drawing.

1453 designed in an informal manner [loc] *gard. plan.* (Informal as opposed to a ▶formal garden/landscape design. *NOTE* also may include asymmetrical or non-axial compositions; rural or rustic nature); *syn.* curvilinear (±) [adj]; *s* **de diseño paisajístico [loc]** (Diseñado siguiendo las pautas de un [determinado] paisaje o de una idea de paisaje, al contrario que el diseño geométrico de parques y jardines; ▶jardín geométrico);

syn. de diseño irregular [loc]; *f* **à la conception paysagère [loc]** (Conception motivée par le souci de respecter le relief naturel du terrain, la croissance libre des végétaux et de représenter un paysage ou une scène naturelle, par opposition à une conception régulière des jardins classiques ; ►jardin régulier) ; *syn.* de conception paysagère [loc], de dessin irrégulier [loc] ; *g* **landschaftlich gestaltet [loc]** (So gestaltet, dass es den Eigenschaften und Gegebenheiten einer [bestimmten] Landschaft oder Landschaftsvorstellung entspricht, im Gegensatz zum ►formalen Garten und zur formalen Parkgestaltung).

design element [n] *arch.* ►design feature.

1454 design elements [npl] [US] *constr.* (Physical features incorporated in the environmental design of gardens, parks, and urban open spaces, including furnishings, play equipment, and associated accommodations); *syn.* design equipment [n] [UK]; *s* **equipamiento [m] de espacios libres** (Instalación de elementos de mobiliario urbano y construcciones en espacios libres; equipar [vb] espacios libres); *f 1* **équipement [m] de plein air** (Terme générique pour un espace libre avec ses objets ou les dispositifs qui le composent) ; *f 2* **accessoires [mpl] de plein air** (*Terme spécifique pour* agrès, bancs, éclairage, etc.) ; *syn.* mobilier [m], meubles [mpl] ; *g* **Ausstattung [f] in Freiräumen** (Einrichtung von Möblierungselementen und baulichen Anlagen in Freianlagen; ausstatten [vb]).

design equipment [n] [UK] *constr.* ►design elements [US].

1455 designer [n] *plan. prof.* (Person who creates plans, drawings, or models according to aesthetic or functional principles aimed at achieving a desired outcome; e.g. a sustainable setting suitable for human habitation ; ►planner, ►site designer); *s* **diseñador, -a [m/f]** (Persona que planifica, dibuja o modela creativamente según conceptos estético-formales y funcionales; ►planificador/a); *f* **concepteur [m], conceptrice [f]** (Personne à laquelle est confiée une mission d'ingénierie et d'architecture et assurant la conception et le contrôle d'exécution d'un ouvrage ; ►aménageur, ►concepteur, trice [paysagiste] d'opération, ►projeteur) ; *g* **Gestalter [m]** (Jemand, der kreativ etwas z. B. nach formal-ästhetischen und funktionalen Vorstellungen plant oder formt und seine Denkprozesse und Forschungsergebnisse, die teils deduktiv, teils intuitiv gewonnen wurden, sichtbar und kommunizierbar macht; ►Objektplaner/-in, ►Planer/-in); *syn.* Entwerfer/-in [m/f].

designer [n]**, project** *prof.* ►site designer.

1456 design feature [n] *arch.* (Outstanding element or artistic component of a design); *syn.* design element [n]; *s* **elemento [m] de diseño** (Componente artístico o llamativo de un diseño); *syn.* elemento [m] de configuración; *f* **élément [m] de la composition** (Composante formelle ou esthétique caractérisant la conception d'une construction, d'un aménagement) ; *g* **Gestaltungselement [n]** (Formaler Bestandteil oder künstlerische Komponente einer Planung).

1457 design firm's representative [n] [US] *constr. prof.* (Representative of a planning or engineering office responsible for ►supervision of works of a project. His tasks include supervision of execution of the work according to the time schedule and generally accepted state-of-the-art technology; the **d. f. r.** has to ensure that the ►contractor's agent or ►foreman provides necessary, on-site safety and health protection as well as measures to protect the environment. The **d. f. r.** is also responsible for the fulfillment/fulfilment of the legal, local authority and professional association requirements. He is usually a civil engineer, architect or landscape architect); *syn.* inspector [n] [UK], resident engineer [n], site engineer [n] [UK], supervisor [n] representing a design practise [UK], supervising officer [n] [US]; *s* **arquitecto/a [m/f] director de las obras** (Arquitecto/a o

ingeniero/a representante del gabinete de arquitectura que está continuamente a pie de obra y es responsable de la ►dirección de obra. Entre sus tareas se encuentra la vigilancia de la realización de los trabajos puntual, económica y con la mejor tecnología de contrucción asequible. Es responsable de que el jefe de obra o el capataz de la empresa contratista ponga en práctica las normas de seguridad y de protección de la salud y las medidas de protección ambiental. El **a. d. de las o.** también es responsable —como representante del gabinete de arquitectura— del cumplimiento de las prescripciones legales, administrativas y de la mutua de accidentes de trabajo); *syn.* director, -a [m/f] de las obras; *f* **représentant [m] du maître d'œuvre** (Ingénieur présent sur le chantier et responsable pour un bureau d'études de la ►direction de l'exécution du contrat de travaux) ; chantier ; il planifie, organise et contrôle les travaux de construction ; il s'assure par l'intermédiaire des ►chefs de chantier des entreprises du respect des normes de qualité, d'hygiène et de sécurité ; il est responsable de la tenue des délais et du respect du budget, assure la gestion financière, organise les approvisionnements, surveille l'avancement des travaux et gère les plannings de travail ; le représentant du maître d'œuvre est en général un ingénieur, un architecte [paysagiste] ou un ingénieur du bâtiment ; *g* **Bauleiter/-in [m/f] (2)** (Vertreter eines Planungsbüros/beratenden Ingenieurs, der für die ►Bauleitung 3 eines Projektes verantwortlich ist. Zu seinen Aufgaben gehört die termingerechte, wirtschaftliche und nach den allgemein anerkannten Regeln der Baukunst durchzuführende Abwicklung der Arbeiten; er hat dafür zu sorgen, dass der Bauleiter/Polier der bauausführenden Firma die nötigen Sicherheitsvorkehrungen trifft und Maßnahmen zum Gesundheitsschutz und Umweltschutz auf der Baustelle vorsieht. Der **B.** ist als Vertreter des Büros für die Erfüllung der gesetzlichen, behördlichen und berufsgenossenschaftlichen Auflagen verantwortlich; der **B.** ist i. d. R. ein Bauingenieur, Hochbauarchitekt, Baumeister oder Landschaftsarchitekt); *syn.* örtlicher Bauleiter [m]/örtliche Bauleiterin [f], bauüberwachende(r) (Landschafts)architekt/-in [m/f]/Ingenieur/-in [m/f], bauleitende(r) (Landschafts)architekt/-in [m/f]/Ingenieur/-in [m/f].

1458 design guideline [n] *plan.* (Planning criteria for guiding and directing the design of roads, play areas, conservation and historic areas, etc.); *s* **directriz [f] de diseño** (Instrucciones para la planificación p. ej. de carreteras, zonas de juegos de niños, etc.); *syn.* normas [fpl] técnicas de diseño; *f* **directive [f] de présentation des dessins** (Recommandations ou prescriptions définissant la terminologie et la représentation graphique d'ouvrages dans le bâtiment, le génie civil, etc. ; **F.**, celles-ci sont définies dans le cadre de la norme NF) ; *g* **Entwurfsrichtlinie [f]** (Planungsanleitung resp. -vorgaben, z. B. für die Anlage von Straßen, Spielplatzanlagen etc.).

design guidelines [npl] *leg. urb.* ►design regulations.

1459 design load [n] *eng.* (In structural engineering the calculated load-bearing capacity for the dimensioning and design of building elements; distinct forms are: permanent load [dead load], live load, wind load, snow load and catastrophic load—e.g. earthquake); *s* **hipótesis [f] de carga** (En planeamiento de estructuras portantes, cantidad y dirección calculadas de las posibles cargas, p. ej. sobre un edificio, como base para el dimensionamiento de la construcción y la selección de los materiales. Se diferencia entre carga muerta, carga móvil, carga debida al viento y a la nieve y carga catastrófica); *f* **surcharge [f] présumée** (*Calcul des structures* valeur des charges, des forces et des coefficients utiles à la définition des états limites dans la conception des bâtiments ; on distingue les charges permanentes et les charges d'exploitation, les surcharges climatiques [neige, vent] et les déformations dues aux mouvements du sol ; cf. BON 1990, 156) ; *g* **Lastannahme [f]** (*Tragwerksplanung* rechnerisch ange-

D

nommene Größe und Richtung möglicher Kräfte, die z. B. auf ein Bauwerk als Grundlage für Dimensionierung und Ausgestaltung von Bauteilen wirken. Es werden u. a. ständige Last [Eigenlast], Verkehrslast, Wind- und Schneelast sowie Katastrophenlast [z. B. Erdbeben] unterschieden).

design loading [n] *eng.* ▶ load-bearing capacity.

design [n] **of landscape around buildings** [UK] *plan. prof.* ▶ site planning.

1460 design [n] **of material surfaces** *arch.* (**1.** Specification of texture, color, form, size, and structural capacity of exterior construction units. **2.** Shaping and exterior surfacing of construction units); *s* **diseño** [m] **de superficie de materiales** (Resultado del trabajado de mampuestos o piezas de construcción); *f* **traitement** [m] **des matériaux** (Idées conceptionnelles dans l'utilisation et la mise en forme des matériaux divers ainsi que résultats du travail des matériaux industriels ou de parties d'ouvrages ; *terme spécifique* traitement des sols) ; *g* **Materialgestaltung** [f] (Formgebung und Ergebnis der Oberflächenbearbeitung von Werksteinen oder Bauteilen).

design [n] **of the environment** *conserv. nat'res. plan.* ▶ planning and design of the environment.

1461 design ordinance [n] [US] *leg. urb.* (Local legislative enactment [for cities and counties] for compliance with design stipulations within the scope of ▶ community development planning, cemetery planning, etc.; ▶ design recommendations, ▶ aesthetic ordinance [US]/building development bye-laws [UK]); *syn.* design requirements [npl] [UK], design regulations [npl] [UK]; *s* **reglamento** [m] **de diseño** (Normas municipales que fijan el aspecto exterior de los edificios y de los espacios libres correspondientes; ▶ guía de diseño, ▶ ordenanza municipal de protección de la calidad visual urbana); *f* **prescriptions** [fpl] **communales architecturales et paysagères** (±) (Dispositions réglementaires régissant l'aspect extérieur des constructions et la réalisation des espaces verts dans le cadre d'aménagements [cahier des prescriptions architecturale et paysagère] ou des plans d'urbanisme ou p. ex. dans certains articles du règlement du PLU/POS ou dans un « cahier des prescriptions architecturales » comportant l'implantation de la construction, le recul par rapport aux voies publiques et privées, par rapport aux limites de propriété, l'emprise au sol, le coefficient d'occupation des sols, l'aspect extérieur concernant les matériaux autorisés, afin de préserver les particularités du paysage bâti et de garder une harmonie entre les bâtiments à construire ou à aménager et l'environnement bâti existant ; ▶ prescriptions de conservation du patrimoine architectural constitutif d'un ensemble urbain, ▶ recommandations architecturales et paysagères, ▶ règle [d'urbanisme] concernant l'aspect extérieur des constructions) ; *syn.* cahier [m] des prescriptions techniques et environnementales ; *g* **Gestaltungssatzung** [f] (Juristisch festgelegte Vorschriften als kommunales 'Gesetz' zur Durchsetzung von Gestaltungsvorschlägen gegen jedermann im Rahmen der Stadtplanung, der Friedhofsgestaltung etc.; ▶ Gestaltungsempfehlung, ▶ Ortsbildsatzung); *syn.* Gestaltungsverordnung [f], Gestaltungsvorschriften [fpl]).

design process [n] *plan.* ▶ planning process.

1462 design recommendations [npl] *urb.* (Design suggestions of a local authority for urban building projects and sites without regulatory or statutory power; ▶ design ordinance); *s* **guía** [f] **de diseño** (Propuestas de diseño sin obligatoriedad hechas por la administración competente para proyectos de construcción y paisajismo; ▶ reglamento de diseño); *f* **recommandations** [fpl] **architecturales et paysagères** (±) (Dispositions techniques ou esthétiques satisfaisant aux exigences réglementaires en matière d'aménagements urbains, souvent précisées

dans un cahier de recommandations ; ▶ prescriptions communales architecturales et paysagères) ; *g* **Gestaltungsempfehlung** [f] (Gestaltungsvorschläge für städtebauliche Objekte und Anlagen ohne rechtliche Durchsetzungsmöglichkeit; ▶ Gestaltungssatzung).

design regulations [npl] [UK] *leg. urb.* ▶ design ordinance [US].

design requirements [npl] [UK] *leg. urb.* ▶ design ordinance [US].

design sheet [n]**, detailed** *constr.* ▶ construction drawing.

1463 design speed [n] *eng. trans.* (Value expressed in km/hr to determine road design parameters for vertical curves, radii, superelevations and road surfacing, according to set traffic speeds); *s* **velocidad** [f] **directriz** (Velocidad guía empleada para el cálculo de los elementos principales de un proyecto de carretera como radios, peraltes, superficies, etc.); *syn.* velocidad [f] de régimen, velocidad [f] básica, velocidad [f] tomada por base del proyecto; *f* **vitesse** [f] **de référence** (Vitesse sur la base de laquelle sont définis le tracé et les caractéristiques géométriques limites [valeur des rayons, des dévers, de la déclivité, etc.] d'une voie routière permettant d'offrir de bonnes conditions de sécurité sur des sections homogènes ; prescriptions ICTAAL) ; *g* **Entwurfsgeschwindigkeit** [f] (Richtwert bei der straßenbaulichen Detailplanung hinsichtlich der Verkehrsgeschwindigkeit zur Ermittlung von Kurvenradien, Überhöhungen, Ausbaubreiten etc.).

design submission [n]**, closing date for** *plan. prof.* ▶ submission deadline for design competition.

1464 desilting [n] *constr. land'man.* (Removal of fine sediments from lakes or ponds; ▶ dredging 2); *s* **dragado** [m] **de lagos colmatados** (Extracción de sedimentos finos de lagos o estanques; ▶ dragado de lodo/fango); *f 1* **curage** [m] **d'un lac** (Terme générique pour les travaux d'entretien curatif d'extraction mécanique d'une partie des sédiments [▶ désenvasement, désensablement enlèvement de la végétation] d'un plan d'eau en voie de comblement et donc de vieillissement) ; *f 2* **dragage** [m] **d'un lac** (Enlèvement mécanique de sable de gravier ou de vase au fond d'un lac au moyen d'une drague qui opère soit sur l'eau [flotteur en forme de péniche pour la drague à cuiller, à grappin ou élévateur], soit à partir de la berge pour la drague hydraulique qui aspire les déblais dans un tuyau) ; *g* **Ausbaggerung** [f] **verlandeter Seen** (Mechanische Entfernung von Schlamm/Sedimenten aus einem Stillgewässer mit Sandbagger, der entweder vom Ufer oder von einer Schwimmplattform aus operiert und mit Hilfe von Strahlpumpen und mechanischen Fräsköpfen ein Wasser-Sediment-Gemisch im Verhältnis 1:3 bis 1:25 erzeugt — je nach Transportstrecke; ▶ Schlammräumung); *syn.* Entschlammung [f] verlandeter Seen.

dessication [n] *biol.* ▶ dehydration; *pedol.* ▶ drying out (1).

dessication damage [n] *agr. constr. hort. for.* ▶ drought damage.

1465 destination traffic [n] *trans.* (Proportion of traffic which moves to another particular area; ▶ recreation destination area, ▶ origin traffic); *s* **tráfico** [m] **de destino** (Tránsito que se dirige a una zona o a un lugar específico; ▶ zona de destino de recreacionistas, ▶ tráfico de origen); *syn.* tránsito [m] de destino; *f* **circulation** [f] **de destination** (Trafic qui se dirige vers une zone ou un lieu précis ; ▶ bassin touristique, ▶ destination touristique, ▶ circulation de provenance) ; *syn.* trafic [m] de destination ; *g* **Zielverkehr** [m] (Verkehr, der in ein bestimmtes Gebiet fließt oder einen bestimmten Ort ansteuert; ▶ Erholungszielgebiet 1, ▶ Quellverkehr).

1466 destruction [n] **of vegetation** *conserv. land'man*; *s* destrucción [f] de la vegetación; *f* destruction [f] de la végétation ; *g* Vegetationszerstörung [f].

1467 desulphurization [n] *envir.* (Generic term for ►humid desulphurization, ►flue gas desulphurization and ►dehydration of sulphur emissions); *s* desulfuración [f] (Término genérico para ►desulfuración en seco, ►desulfuración de gas de combustion y ►desulfuración húmeda); *f* désulfuration [f] (Terme générique pour l'►hydro-désulfuration, le ►désulfuration des gaz combustibles et le ►procédé de récupération du souffre) ; *g* Entschwefelung [f] (OB zu ►Nassentschwefelung, ►Rauchgasentschwefelung und ►Trockenentschwefelung).

1468 detached [pp/adj] *urb.* (e.g. a single free standing house); *syn.* isolated [pp/adj] (1), separate [adj]; *s* separado/a [pp/adj] (1) (En inglés término utilizado para casas o edificios individuales en un predio); *syn.* independiente [adj/pp]; *f 1* isolé, ée [pp/adj] (Caractérise une habitation, une résidence éloignée d'une zone urbanisée); *f 2* détaché, ée [pp/adj] (Désigne une construction située à l'écard du bâtiment principal) ; *syn.* secondaire [adj], accessoire [adj] ; *g* frei stehend (1) [ppr/adj] (Bezeichnung für ein Gebäude, das auf einer Fläche einzeln steht).

1469 detached and semidetached housing [n] [US]/ **detached and semi-detached housing** [n] [UK] *urb.* (Residential development of ►individual houses, ►single-family detached houses and ►semidetached houses; ►detached building development, ►dispersed building development); *s* zona [f] de casas uni- o bifamiliares (►casa individual, ►casa pareada, ►casa unifamiliar, ►casa bifamiliar, ►edificación discontinua, ►edificación dispersa); *syn.* área [f] de casas uni- o bifamiliares; *f* constructions [fpl] individuelles (Zone d'habitations constituée de ►maisons individuelles ou ►maisons jumelles ; ►habitat dispersé, ►implantation des constructions en ordre discontinu, ►implantation des constructions en ordre semi-continu) ; *g* Einzelbebauung [f] (Aus ►Einzelhäusern oder ►Doppelhäusern bestehendes Siedlungsgebiet; ►offene Bauweise, ►Streubebauung).

1470 detached building development [n] *leg. urb.* (Interrupted building arrangement; e.g. detached or semidetached housing subdivision [US]/detached or semidetached house development [UK]; ►low-density building development, ►zero lot line development [US]); *s* edificación [f] discontinua (≠) (Tipo de ordenación de edificios en un área de casas individuales, mellizas o grupos de casas con una longitud máxima de 50 m y con distanciamiento mínimo entre los límites de los predios; ►androna, ►edificación de baja densidad, ►edificación discontinua sin retranqueo); *f* implantation [f] des constructions en ordre discontinu (Disposition caractérisant des maisons individuelles, maisons jumelles ou groupes d'habitations non contiguës dans un lotissement, une distance minimale pouvant être imposée entre la construction et la limite parcellaire [►recul de constructibilité par rapport aux limites séparatives] ; ►urbanisation aérée) ; *f 2* implantation [f] des constructions en odre semi-continu (Disposition caractérisant des maisons individuelles, maisons jumelles ou groupes d'habitations non contiguës dont l'implantation du bâti se situe sur une des deux limites séparatives ; ►exception aux règles d'implantation des constructions [par rapport aux limites séparatives] avec édification en limite latérale) ; *g* offene Bauweise [f] (2) (Gebäudestellung von Einzelhäusern, Doppelhäusern oder Hausgruppen mit einer Länge von max. 50 m in einem Baugebiet mit jeweiligem seitlichem Mindestabstand [►Bauwich] zur Nachbargrenze; im Bebauungsplan können Flächen festgesetzt werden, auf denen nur Einzelhäuser, nur Doppelhäuser, nur Hausgruppen oder zwei dieser Hausformen zulässig sind; cf. § 22 BauNVO; ►abweichende Bauweise mit einseitiger Grenzbebauung, ►lockere Bebauung).

detached dwelling [n] [US] *arch. urb.* ►single-family house.

detached dwelling [n] [US], **single-family** *arch. urb.* ►single-family detached house.

detached house [n] *arch. urb.* ►single-family detached house.

1471 detail drawing [n] *plan.* (Detailed graphic presentation of a specific portion of a project which is prepared by the designer, typically at an enlarged scale; ►construction drawing, ►shop drawing); *s* dibujo [m] en detalle (Representación gráfica elaborada por el/la diseñador, -a de una parte específica de un proyecto; ►dibujo de construcción, ►plano de ejecución); *syn.* proyecto [m] en detalle; *f* plan [m] de détail (L'ensemble des plans de détail souvent annexés au descriptif-quantitatif est désigné par un « carnet de détails » ; ►fiche de détail d'un produit, ►plan d'exécution [des ouvrages] [P.E.O.]) ; *g* Detailplan [m] (Zeichnerische Darstellung eines Objektes oder Teile davon in einem größeren Maßstab bis 1:1 mit allen notwendigen Einzelangaben für die Konstruktions- und Bauausführung; ►Ausführungszeichnung, ►Produktzeichnung); *syn.* Detailzeichnung [f], Einzelzeichnung [f].

1472 detailed design [n] *plan.* (General term covering working drawings, cross sections, planting details, etc.; ►construction document phase [US]/detailed design phase UK]); *syn.* detail planning [n]; *s* planificación [f] detallada (Término genérico para la elaboración de los planos de trabajos específicos como de instalaciones técnicas, caminos, plantación en el marco del diseño de un proyecto; ►fase de planificación en detalle); *f* études [fpl] de détail (*Terme général* exécution des plans relatifs aux divers lots techniques ou travaux des diverses spécialités dans le cadre de l'élément normalisé des plans d'exécution des ouvrages [P.E.O.] ; ►projet) ; *g* Detailplanung [f] (1. Allgemeiner Begriff für das Fertigen von Arbeits-, Werks- und Einzelzeichnungen sowie von Bepflanzungsplänen resp. Bepflanzungsangaben. 2. In A. Bezeichnung der Leistungsphase ►Ausführungsplanung gem. § 22 lit. c HRLA).

1473 detailed design drawing [n] *constr. plan.* (Graphic presentation of a ►final project design often referred to as a design development drawing); *s 1* diseñar [vb] un proyecto (Proceso de elaborar un ►anteproyecto detallado); *syn.* dibujar [vb] los planes de un proyecto; *s 2* plano [m] (detallado) de proyecto (Resultado de 1); *f* plan [m] d'avant-projet détaillé (d'A.P.D.) (Documents graphiques nécessaires à la bonne compréhension du projet ; ►avant-projet détaillé) ; *g* Entwurfszeichnung [f] (*Objektplanung* zeichnerische Ausarbeitung eines ►Entwurfs).

detailed design phase [n] [UK] *contr. prof.* ►construction document phase [US].

detailed design sheet [n] *constr.* ►construction drawing.

detailed planning application [n] [UK] *adm. leg.* ►building and site plan permit application [US].

detail planning [n] *plan.* ►detailed design.

detention basin [n] *hydr. wat'man.* ►retarding basin, ►storm water detention basin.

detention reservoir [n] *wat'man.* ►flood control reservoir.

detention structure [n] *hydr. wat'man.* ►storm water detention basin.

1474 deterioration [n] (1) *envir.* (Process of worsening or degradation, e.g. of an environment, water quality, soil structure; ►environmental deterioration, ►forest deterioration); *syn.* degra-

D

dation [n]; **s deterioro [m]** (Empeoramiento de las condiciones p. ej. ambientales, de calidad del agua, de la estructura del suelo; ▶deteriorio forestal, ▶degradación del medio ambiente); **f 1 détérioration [f]** (Modification déséquilibrante des caractéristiques de l'environnement, des conditions de vie, du temps) ; **f 2 dégradation [f] (2)** (Terme plus souvent utilisé pour qualifier les modifications de la qualité des eaux, de la stabilité structurale d'un sol ; ▶destruction de l'environnement, ▶maladie des forêts) ; **g Verschlechterung [f]** (Verminderung der Qualität einer Situation/eines Zustandes oder eines Umweltmediums, z. B. der Umweltbedingungen, der Wasserqualität, des Bodengefüges; ▶Umweltzerstörung, ▶Walderkrankung).

deterioration [n] (2) *conserv. ecol.* ▶impoverishment.

1475 deterioration [n] **of plant growth** *bot. ecol.* (Reduction in vitality of plant growth; e.g. by air pollution, lowering of groundwater table, drought); **s perturbación [f] del crecimiento vegetal** (Reducción de la vitalidad del crecimiento de las plantas causada p. ej. por contaminación, reducción del nivel freático, sequía); **f perturbation [f] de la croissance des végétaux** (Phénomène qui s'oppose au développement de la végétation, p. ex. la pollution, un rabattement de la nappe phréatique, la sècheresse) ; **g Beeinträchtigung [f] des Pflanzenwachstums** (Reduzierung des Wachstums, z. B. durch Luftverschmutzung, Grundwasserabsenkung, Trockenheit).

deterioration resistant [adj] *biol.* ▶nonrotting [US]/nonrotting [UK].

determination [n] **of a plan** [UK] *adm. leg.* ▶design and location approval [US] (1).

1476 determination [n] **of executed work** [US] *contr.* (Acceptance of scope of work[s] performed, which is often necessary for the final accounting according to the specifications, especially for underground facilities. This can be done, e.g. by an ▶interim agreed measurement of complete work [US]/works [UK]); *syn.* assessment [n] of executed works [UK]; **s revisión [f] intermedia de las obras realizadas** (Para poder confirmar la calidad de realización de trabajos que son dificiles de controlar una vez terminada la obra; ▶medición intermedia de los trabajos realizados); **f 1 détermination [f] quantitative des travaux exécutés** (Terme générique définissant toutes les activités nécessaires au paiement d'acomptes en cours de travaux et l'établissement d'un décompte général en fin de travaux) ; **f 2 prise [f] d'attachements** (Reconnaissance commune des ouvrages effectuée par l'architecte et l'entreprise afin de constater contradictoirement sur le chantier les travaux effectués en vue de l'établissement par l'entrepreneur du projet de décompte mensuel ; ▶métré provisoire ; ATJ 1973, Tome II, 193) ; **g Feststellung [f] von ausgeführten Leistungen** (Für die Abrechnung meist notwendige gemeinsame Baustellenbegehung, um erbrachte Leistungen, die bei Weiterführung der Arbeiten nur schwer feststellbar sind, z. B. Verlegung von Leitungen, durch ein ▶Zwischenaufmaß zu bestätigen; cf. § 14 [2] VOB Teil B).

1477 determination [n] **of final contract amount** [US] *constr.* (Final calculation of all costs incurred in order to ascertain prices and provide cost control); *syn.* ascertainment [n] of the final sum [UK], computation [n] of full contract value [UK]; **s cálculo [m] definitivo** (Determinación final de la suma de contrato sobre la base de los costes [Es]/costos [AL] reales); **f décompte [m] général et définitif** (Détermination du coût constaté de tous les ouvrages exécutés et réceptionnés dans le cadre du contrôle de coût) ; **g Nachkalkulation [f]** (Ermittlung der bei der Bauausführung effektiv angefallenen Kosten zur Preisermittlung [für zukünftige Berechnungen] und Kostenkontrolle des laufenden Projektes).

1478 determination [n] **of planning context** [US] *prof. plan.* (Basic investigations and establishment of the preconditions for the solution of a planning task as the first phase [▶work phase] of the whole ▶scope of professional services); *syn.* fact-finding data collection [n], clarification [n] of design/planning brief [UK]; **s 1 inventario [m] preliminar de un proyecto** (Investigaciones básicas para preparar las propuestas de planificación que son parte de la primera ▶fase de trabajo; ▶descripción de servicios de planificación); **s 2 inventario [m] ambiental** (Término utilizado en el caso de un ▶estudio de impacto ambiental); **f 1 recueil [m] des données** *constr.* (Première partie de l'Avant-projet sommaire [A.P.S.] ayant pour but de rassembler tous les documents nécessaires à la définition d'un programme d'aménagement ; **en D.**, cette prestation constitue le premier élément normalisé d'une mission normalisée [▶étape de la mission de maîtrise d'œuvre] ; ▶forme et étendue de la mission) ; *syn.* recherches [fpl] et études, études [fpl] préliminaires [B] ; **f 2 inventaire [m] des données** *plan. envir. ecol.* (Relevé des données quantitatives et qualitatives de base dans le cadre d'études de paysages ou d'environnement) ; **g Grundlagenermittlung [f]** (**D.**, Erste ▶Leistungsphase des ▶Leistungsbildes Objektplanung für Gebäude, Freianlagen und raumbildende Ausbauten, bei der die Voraussetzungen zur Lösung der Bauaufgabe durch die Planung ermittelt werden; cf. § 15 [1] HOAI 2002 resp. §§ 33, 38, 42 u. 46 HOAI 2009; **CH.**, in der schweizerischen Ordnung für Leistungen und Honorare der Landschaftsarchitekten entspricht die **G.** der „Problemanalyse" und dem „Studium von Lösungsmöglichkeiten" als 1. und 2. Leistungsabschnitt der „Vorprojektphase").

1479 determination of presence [n] *phyt.* (An indication of the number of stands under investigation in which a particular species occurs as a competitor; ▶degrees of presence, ▶constancy); **s determinación [f] de la presencia** (Determinación de en cuántas de las diferentes poblaciones investigadas aparece una especie; ▶grado de presencia, ▶constancia); **f détermination [f] de la présence (d'une espèce)** (Rapport entre le nombre des aires-échantillons contenant une espèce et le nombre total des aires analysées dans le même individu d'association ; BB 1928 ; ▶coefficient de présence, ▶constance) ; *syn.* détermination [f] de la fréquence, détermination [f] de la constance ; **g Stetigkeitsbestimmung [f]** (Angabe, in wie vielen der untersuchten Einzelbestände einer Pflanzengesellschaft eine bestimmte Art als Mitbewerber auftritt; ▶Stetigkeitsgrad, ▶Konstanz).

1480 determination [n] **of requirements** *plan.* (Analysis of a site or design program[me] to determine the scope and complexity of a proposed project); **s determinación [f] de la demanda** *syn.* determinación [f] de las necesidades; **f détermination [f] de la demande** *syn.* détermination [f] des besoins ; **g Bedarfsermittlung [f]** (Untersuchung einer Fläche oder Analyse einer Planungsaufgabe, um den Umfang und die Anforderungen eines Projektes zu ermitteln); *syn.* Bedarfsfeststellung [f].

determination process [n] **of a plan** [UK] *adm. leg.* ▶design and location approval process [US].

1481 dethatching [n] *constr.* (Vertical cutting out of superfluous, brown-colo[u]red, dead grass leaves, including the surficial roots, between the soil surface and grass leaves of a lawn, in order to ventilate the sward; ▶lawn aeration, ▶separate contract maintenance, ▶thatch, ▶thatching); **s escarificación [f]** (Labrado del terreno que corta verticalmente la tierra y las raíces y sirve para eliminar los residuos de plantas muertas en el suelo para airearlo; ▶aireación del césped, ▶fieltro, ▶formación de fieltro, ▶trabajos regulares de mantenimiento; escarificar [vb]); **f 1 scarification [f]** (Opération de cisaillage par des couteaux sur une profondeur de 10 cm du feutre superficiel des pelouses [▶feutre de gazon] situé entre la surface du sol et la base d'un gazon au moyen d'un scarificateur dans le cadre de ▶travaux d'entretien courant ; ▶aération de pelouse, ▶feutrage [d'un gazon]) ; *syn.* scarification [f] au verticut, défeutrage [m] ; *f 2* **verticutage [m]** (Opération de régénération d'un gazon au moyen d'une machine [verticuteur] dont les lames rotatives griffent le sol sur environ 1 à 2 mm de profondeur éliminant la mousse, les mauvaises herbes et le feutre du gazon ; le sol est ainsi rendu perméable à l'air, à l'eau et aux nutriments) ; *g* **Vertikutieren [n, o. Pl.]** (Spätlat. *verticalis* >senkrecht< und franz. *couteau* >Messer<; senkrechtes, wurzeltiefes Herausschneiden des im Überfluss gebildeten, braun gefärbten, abgestorbenen Pflanzenmaterials [▶Rasenfilz] zwischen Bodenoberfläche und Blattzone eines Rasens mit einem Vertikutiergerät, um die Grasnarbe im Rahmen der ▶Erhaltungspflege zu belüften; der Rasen wird auf 2 cm zurückgeschnitten; optimaler Zeitpunkt für das **V.** ist der Zeitpunkt vor den Wachstumsschüben — je nach Jahr und Region — im April/Mai oder Anfang August; im zeitigen Frühjahr lässt sich Moos gut entfernen; ▶Belüftung des Rasens, ▶Verfilzen).

1482 detraction [n] **from visual quality** *landsc.* (Reduction in ▶visual quality; ▶visual disturbance, ▶impairment of a landscape); *syn.* visual detriment [n] (MET 1985, 4), reduction [n] in visual quality, visual defacement [n] [also US]; *s* **desfiguración [f] visual** (Cambio negativo de los recursos visuales causado por un proyecto o acción de la sociedad; ▶calidad visual, ▶desfiguración del paisaje [rural o urbano], ▶deterioro del paisaje); *syn.* desvalorización [f] visual; *f* **nuisance [f] visuelle** (Tout facteur visuel à caractère permanent, continu ou discontinu, qui constitue une gêne, une entrave, un préjudice immédiat ou différé pour la santé, l'environnement et qui rend la vie malsaine ou pénible ; cf. sites-pollues. ecologie.gou.fr ; ▶altération visuelle du caractère [des sites et des paysages], ▶défiguration des sites et des paysages, ▶valeur esthétique) ; *g* **visuelle Beeinträchtigung [f] (1)** (Negative Wirkung auf den Betrachter beim Anblick von störend empfundenen „Bildern"; ▶Gestaltqualität, ▶Verunstaltung der Landschaft, ▶visuelle Beeinträchtigung 2).

detrimental [adj] **to health** *envir.* ▶injurious to health.

1483 detrital deposit [n] *geo.* (Coarse rock debris produced by weathering and erosion processes which has been transported from its original site; ▶rock debris); *syn.* detrital rock accumulation [n]; *s* **depósito [m] detrítico** (Acumulación de ▶detrito procedente de procesos de erosión); *syn.* derrubios [mpl]; *f* **accumulation [f] détritique** (Masse importante de débris accumulés lors de processus érosifs ; (DG 1984, 135) ; ▶éboulis) ; *g* **Schotterkörper [m] (1)** (Durch Erosionsvorgänge angelagerte, große Schottermassen; ▶Gesteinsschutt).

detrital rock accumulation [n] *geo.* ▶detrital deposit.

detrital sediment [n] *geo.* ▶loose sedimentary deposit.

1484 detritus [n] **(1)** *bot.* (Organic remains resulting from the decomposition of vegetable matter on land or in water; ▶premature shedding of leaves); *syn.* debris [n] [also US]; *s* **detrito [m]** (Partículas de materia orgánica procedentes de la descomposición de organismos muertos; ▶defoliación prematura); *syn.* detritus [m] (1); *f* **détritus [m] (1)** (Substances organiques en voie de décomposition en milieu terrestre ou aquatique ; ▶chute prématurée des feuilles) ; *g* **Detritus [m]** (In Zersetzung befindliche organische Substanz auf dem Boden [z. B. unter älterem Falllaub] oder in Gewässern; ▶frühzeitiger Laubfall).

detritus [n] **(2)** *geo.* ▶rock debris.

detritus [n]**, organic** *pedol. phyt.* ▶litter (1).

detritus tank [n] **(for sewage treatment plants)** (DET 1976, 128) *constr. hydr.* ▶sand arresting trap [US].

1485 developed area [n] [US] *plan. urb.* (Land made suitable for human habitation and provided with various types of buildings, facilities together with the appropriate infrastructure such as roads and utilities; ▶populated area, ▶urban land use category, ▶vehicular and pedestrian infrastructure); *syn.* settlement area [n] [UK]; *s* **área [f] de asentamiento** (Término genérico para superficies apropiadas para el asentamiento y en las que se encuentran todo tipo de construcciones y las correspondientes infraestructuras como las ▶superficies viarias, instalaciones de suministro de electricidad, etc.; ▶categoría de suelo urbano, ▶zona edificada 2); *f* **espace [m] urbanisé** (Ensemble de constructions accompagnées des bâtiments de service et des infrastructures correspondantes appartenant par nature à une agglomération ou à un village existant, ▶circulations, ▶voirie, etc. ; ▶zone urbaine 2, ▶zone urbanisée) ; *syn.* site [m] urbanisé, aire [f] urbanisée ; *g* **Siedlungsfläche [f]** (Anthropogen überformtes, für die Besiedlung geeignetes Land mit baulichen Anlagen aller Art und den dazugehörenden Infrastruktureinrichtungen wie ▶Verkehrsflächen, Anlagen der Energieversorgung etc.; ▶Baufläche, ▶besiedelter Bereich 2); *syn.* Siedlungsgebiet [n], Siedlungsland [n], Siedlungsraum [m].

1486 developed beach [n] *recr.* (Stretch of land along a sea coast or an inland lake used for recreation; ▶bathing beach); *s 1* **línea [f] de playa** (Zona aprovechable para el tiempo libre y el recreo; ▶playa [de baños]); *syn.* costa [f], litoral [m]; *s 2* **playa [f] equipada** (Playa costera en la que existen servicios de temporada para uso de la población); *f 1* **linéaire [m] de plage** (Espace géographique d'accueil très recherché pour le tourisme en bordure de mer, des étangs ou des lacs intérieurs considéré comme paramètre représentatif de l'attractivité du littoral ; *terme spécifique en façade maritime* linéaire [m] côtier naturel, ▶zone de baignade) ; *syn.* plage [f] touristique, ligne [f] de plage ; *f 2* **plage aménagée** (Espace littoral public pour lequel l'aménagement, l'exploitation et l'entretien font l'objet d'une concession accordée en vertu du décret du 26 mai 2006, la durée de la concession ne pouvant excéder douze ans ; le pourcentage de superficie et de linéaire de plage, qui devront rester libres en permanence de tout équipement et de toute installation, est au minimum de 80 % pour les plages naturelles et de 50 % pour les plages artificielles ; les équipements et installations de plage autorisés devront être démontables et effectivement démontés durant la période hivernale, sauf exceptions justifiées par la fréquentation hivernale de certaines plages et l'accueil d'activités permanentes sur celles-ci) ; *g* **Freizeitstrand [m]** (Für Freizeit und Erholung nutzbarer Bereich am Meer oder am Ufer eines Inlandgewässers; ▶Badestrand).

developed [pp] **by man**-*ecol.* ▶man-developed.

1487 developed [pp] **for the handicapped/disabled** *leg. plan. sociol.* (Designed for the needs of ▶handicapped persons; **in U.S.**, according to the requirements of the Americans with Disabilities Act of 1990, and may include aspects of ▶Universal Design; **in U.K.**, provisions for the benefit of the disabled are contained in the Chronically Sick and Disabled Persons Act 1970 and prescribed in the Code of Practice for Access of the Disabled to Buildings [British Standards Institution code of practice BS 5810: 1979]; ▶wheel-chair-accessible); *syn.* accessible [adj] for the handicapped/disabled; *s* **accesible [adj] para discapacitados/as** (Diseñado sin barreras para permitir el acceso a ▶personas con discapacidad [física] o con movilidad reducida; ▶accesible con silla de ruedas, ▶accesible para personas con movilidad reducida); *syn.* accesible [adj] para limitados físico-motores [C]; *f* **adapté, ée aux personnes handi-**

D

capées **[loc]** (Qualité d'un espace, d'un bâtiment conçu pour les ►personnes à mobilité réduite [PMR] ; ►accessible aux fauteuils roulants ; *terme générique* ►accessible aux personnes handicappées) ; *syn.* conçu, ue en fonction des handicaps [loc], aménagé, ée pour personnes handicapées [loc], adapté, ée aux handicaps [loc], adapté, ée aux personnes à mobilité réduite [loc] ; *g* **behindertengerecht [adj]** (Den Bedürfnissen von ►Behinderten entsprechend barrierefrei gebaut und gestaltet. In den Landesbauordnungen [z. B. § 39 LBO-BW] sind die baulichen Anlagen aufgeführt, die barrierefrei herzustellen sind; cf. DIN 18 024: „Barrierefreies Bauen", DIN 18 025: „Barrierefreie Wohnungen", Ergänzung zur DIN 18 034: „Barrierefreie Spielplätze und Freiflächen zum Spielen"; ►rollstuhlgerecht; *OB* ►barrierefrei).

developed land [n] [UK] *plan. urb.* ►built-up area (1).

developed with roads [loc] *urb.* ►land developed with roads.

1488 developer [n] **(1)** *adm. urb.* (Legal or beneficial owner[s] of property included in a proposed development of 'ready-built' dwellings for purchase and occupancy; ►client, ►developer 2, ►development corporation; *s* **promotor [m] de obra** (En general, empresa que construye viviendas para venderlas con llave en mano; ►compañía promotora, ►sociedad pública de promoción urbanística, ►propiedad de obra); *syn.* constructor [m], empresa [f] promotora de obra; *f 1* **promoteur-constructeur [m]** (Dans la plupart des cas, société construisant pour le compte de tiers et commercialisant clés en mains ; ►maître d'ouvrage, ►établissement public d'aménagement, ►société d'économie mixte, ►société de développement) ; *f 2* **développeur [m] territorial** (Personne qui maîtrise les techniques de management de projet et dont les activités principales consistent dans la conduite de diagnostics territoriaux ou thématiques ; la contribution à la construction d'une stratégie de développement territorial, conception, formalisation, conduite des programmes, contrats, procédures ou des projets opérationnels ; mobilisation des acteurs locaux, animation des réseaux, conseil et/ou réalisation des prestations de service et d'accompagnement des porteurs de projet, construction et mise en œuvre une stratégie de communication interne et externe, contribution à l'évaluation des politiques et les actions de mise en œuvre, réalisation d'une ville sur la réglementation, les outils et méthodes du développement territorial, l'environnement économique, social, culturel, politique) ; *g* **Bauträger/-in [m/f]** (Einzelperson, Konsortium oder Gesellschaft, die Wohngebäude plant, erstellt und dann schlüsselfertig verkauft oder vermietet und verwaltet; manche **B.** sind kommunale Wohnungsunternehmen; ►Bauherr/-in, ►Entwicklungsgesellschaft 1, ►Entwicklungsgesellschaft 2); *syn.* Projektentwickler/-in [m/f], Projektträger/-in [m/f], Vorhabensträger/-in [m/f].

1489 developer [n] **(2)** *adm. urb.* (Private company involved in the construction, development and sale of residential, commercial and industrial property, usually in a managerial capacity); *syn.* development company [n] [also UK], property developer [n] [also UK]; *s* **compañía [f] promotora** (Empresa privada que coordina y financia obras urbanísticas [en Es programas de actuación urbanística], incluyendo la construcción de edificios e infraestructuras); *f* **société [f] de développement** (Société de droit privée ayant pour objet le financement et la coordination d'opérations de construction, d'équipement, de viabilisation et de vente de bâtiments à vocation d'habitation ou à usage commercial) ; *g* **Entwicklungsgesellschaft [f] (1)** (Privatwirtschaftliches Unternehmen [Kommandit- oder Aktiengesellschaft], das die Finanzierung und Koordination für die Erschließung, den Bau und den Verkauf von Wohn- und Geschäftsgebäuden oder Industrieanlagen betreibt); *syn.* Entwickler/-in [m/f], Investor/-in [m/f], Projektentwickler/-in [m/f], Projektträgergesellschaft [f].

1490 developer [n] [US] **(3)** *contr.* (►Contracting public agency [US], ►developer 1 or private client commissioning and entering into a construction contract with a building contractor; ►client, ►contracting public agency [US]); *syn.* owner [n] [UK]; *s* **comitente [m]** (Persona física o jurídica que encarga una obra o un proyecto; ►administración comitente, ►promotor de obra, ►propiedad de obra); *f* **donneur [m] d'ordre** (Personne morale ou physique qui se propose de conclure un contrat ; ►donneur d'ordre public, ►maître d'ouvrage, ►promoteur-constructeur) ; *syn.* client [m], contracteur [m], personne [f] contractante ; *g* **Auftraggeber/-in [m/f]** (Juristische oder natürliche Person, die einen Auftrag vergibt; der Begriff **A.** wird meist in Ausschreibungen und Verträgen im Begriffspaar „Auftraggeber/Auftragnehmer" verwandt; ►Bauherr/-in, ►Bauträger/-in, ►öffentlicher Auftraggeber/öffentliche Auftraggeberin).

1491 developing [n] **of plant parts** *bot.* (Emergence of new plant growth in the form of shoots or buds); *s* **desarrollo [m] de partes de plantas** (Nuevo crecimiento de las plantas en forma de yemas, ramas, órganos de reproducción y raíces); *f* **naissance [f] de parties de végétaux** (Donner naissance à, émettre des plantes nouvelles ou des parties de végétaux) ; *syn.* émission [f] de parties de végétaux, développement [m] de parties de végétaux ; *g* **Bildung [f] von Pflanzenteilen** (Entstehen von Knospen, Trieben, Fortpflanzungsorganen und Wurzeln).

development [n] *leg. urb.* ►attached building development, *plan. urb.* ►block development; *leg. urb.* ►building development, ►detached building development; *urb.* ►dispersed building development, ►established industrial development; *for. land'man.* ►forest development; *plan. urb.* ►haphazard development, ►hillside building development, ►infill development; *plan. trans. urb.* ►infrastructure development; *plan.* ►landscape development; *urb.* ►land zoned for development, ►low-density building development, ►mid-rise dwellings development, ►mixed use development, ►new housing development, ►odd-lot development [US]; *plan. urb.* ►open land needed for development; *recr.* ►park-like vacation development; *plan. urb.* ►perimeter block development, ►peripheral building development, ►planned industrial development, ►ready for development; *recr. urb.* ►residential leisure development; *eng. wat'man.* ►river development; *bot. hort.* ►root development; *constr. eng.* ►shear development; *plan.* ►state development; *urb.* ►strip development (1), ►strip development [US] (2); *envir. nat'res. sociol.* ►sustainable development, ►unzoned land ripe for development [n] [US]/white land [UK]; *urb.* ►urban development; *recr. urb.* ►weekend house development; *leg. urb.* ►zero lot line development [US].

development [n] [US]**, commercial strip** *urb.* ►strip development [US] (2).

development [n] [UK]**, continuing professional** *prof.* ►continuing education [US].

development [n]**, courtyard block** *urb.* ►block development.

development [n]**, enclosed block** *urb.* ►block development.

development [n]**, humus** *pedol.* ►humification.

development [n]**, perimeter block** *urb.* ►block development.

development [n]**, real estate** *urb.* ►subdivision [US].

development [n]**, residential community** *urb.* ►housing subdivision [US].

development [n] [UK]**, ribbon** *urb.* ►strip development [US] (1), ►strip development [US] (2).

development [n]**, root system** *hort.* ►root penetration (1).

development [n]**, slip-plane** *hort.* ▶shear development.

development [n]**, soil** *pedol.* ▶soil formation.

development area [n] [UK] *plan. leg.* ▶development district [US].

development company [n] [UK] *plan.* ▶developer (2).

development control [n] *adm.* ▶planning development control.

1492 development corporation [n] *adm. plan.* (**In U.S.**, quasi-public or private non-profit organization, with some government support, giving loan and equity funding for economic development; **in U.K.**, governmental agency set up to create new towns or revitalize parts of existing cities, being empowered amongst other things to hold land, carry out building and provide necessary services; ▶co-operative housing association, ▶low-income housing); *s* **sociedad** [f] **pública de promoción urbanística** (En el Reino Unido, agencia gubernamental que tiene como fin crear nuevas ciudades o revitalizar barrios de ciudades existentes, con derecho de poseer suelo, construir y proveer los servicios necesarios; ▶construcción de viviendas de protección oficial, ▶cooperativa de viviendas); *f 1* **établissement** [m] **public d'aménagement (E.P.A.)** (Établissement créé pour réaliser pour son compte ou pour le compte de l'État ou d'une collectivité locale toutes interventions foncières ou opérations d'aménagement ; ▶habitat social, ▶société coopérative de construction) ; *f 2* **société** [f] **d'économie mixte (S.E.M.)** (Société pouvant réaliser des interventions foncières et exerçant une activité de construction ou de gestion de logements sociaux et détenue à plus de 50 % sans pouvoir excéder 65 % par des collectivités territoriales ou des groupements de ces collectivités) ; *g* **Entwicklungsgesellschaft** [f] **(2)** (Behördenähnliche, dem Gemeinwohl verpflichtete Planungs- und Verwaltungsgruppe, die neue Städte oder Stadtteile entstehen lässt; ▶sozialer Wohnungsbau, ▶Wohnungsbaugenossenschaft).

1493 development district [n] [US] *plan. leg.* (Term used to describe a region, area within a state or locality [US], or part of a country or county [UK], in which the level of economic development is considerably lower than that of comparable areas. Such areas are nationally recognized as in need of assistance and are supported by public funds, in order to promote their economic growth; **in U.S.**, there are federal, state, and local economic development districts; **in U.K.**, term used in sense of regeneration of an urban area; since 1972 three classes of assisted areas are recognized: in order of increasing aid they are designated as 'intermediate areas', 'development areas', and 'special development areas'; ▶area action plan, ▶old industrial region, ▶regional funds of the European Community); *syn.* assisted areas [npl] [UK], development area [n] [UK]; *s* **región** [f] **con debilidades estructurales** (Según la Carta Europea de Ordenación del Territorio de 1983, región con ingresos inferiores a la media del estado correspondiente, cuyas condiciones de vida y de trabajo han progresado poco o que tienen riesgo de quedar atrasadas como consecuencia de cambios en su base económica y necesitan una ayuda particular; cf. DINA 1987, 676; en D. zonas poco desarrolladas que reciben ayuda estatal para promover la creación de puestos de trabajo y elevar el PSB per cápita; ▶zona industrial en declive, ▶fondo regional de las CC.EE. [FEDER]); *f* **zone** [f] **d'investissement prioritaire** (*Aménagement du territoire* terme désignant les régions, parties de territoire en déclin ou menacées de dépérissement dans lesquelles le revenu moyen par habitant est inférieur à celui du reste du territoire national ; afin de corriger ce déséquilibre, celles-ci bénéficient prioritairement d'aides diverses de l'État [aides aux investissements industriels, création de zones d'activités, amélioration des infrastructures de communication, etc.] ; ▶fond européen de développement régional (F.E.D.E.R.), ▶région industrielle en déclin) ; *syn.* zone [f] d'aides, zone [f] de développement ; *g* **Entwicklungsgebiet** [n] (**1.** *plan. Raumordnung* Bezeichnung für eine Region innerhalb eines Landes oder Landesteiles, deren ökonomischer Entwicklungsstand gemessen am Pro-Kopf-Einkommen hinter dem des übrigen Landes wesentlich zurückgeblieben ist. Deshalb entstanden staatlich anerkannte und mit öffentlichen Mitteln geförderte Gebiete, um deren wirtschaftliches Wachstum zu fördern. **2.** *leg.* gemäß § 2 [2] 7 ROG von 1997 und § 2 [2] 4 ROG von 2008 ein Raum, in dem die Lebensbedingungen in ihrer Gesamtheit im Verhältnis zum Bundesdurchschnitt wesentlich zurückgeblieben sind oder ein solches Zurückbleiben zu befürchten ist; ▶altindustrialisierte Region, ▶Regionalfond der EG); *syn.* Förderungsgebiet [n], Ausbaugebiet [n], strukturschwaches Gebiet [n], strukturschwacher Raum [m] (§ 2 [2] 7 ROG von 1997, § 2 [2] 4 ROG von 2008), strukturschwache Region [f], zurückgebliebenes Gebiet [n]).

Development Document [n] [UK]**, Local** *leg. urb.* ▶community development plan.

development maintenance [n] [US] *contr.* ▶project development maintenance.

1494 development mitigation plan [n] [US] *leg. landsc. plan.* (Plan which must be prepared or approved by a government department, whenever development proposals for, e.g. a highway project, mineral extraction, industrial development, etc. will have an impact on natural systems, and which lays down ▶mitigation measures necessary with regard to nature conservation. **In U.S.**, a corrective plan with detailed mitigation measures, which is prepared by the department responsible for planning the development proposals; ▶impact mitigation regulation); *syn.* landscape envelope plan [n] [UK] (LD 1990 [1], 59); *s* **plan** [m] **paisajístico complementario (≠)** (Contribución de especialistas del paisaje a la planificación de otras disciplinas que tienen efectos sobre la naturaleza, como la planificación viaria, planificación hidrológica, etc.; ▶medida compensatoria, ▶regulación de impactos sobre la naturaleza); *syn.* contribución [f] paisajística a la planificación sectorial; *f 1* **schéma** [m] **directeur paysager** (F., études et plans prenant en compte les enjeux du paysage réalisées dans le cadre de l'avant-projet sommaire [APS] aboutissant à l'établissement de recommandations et de lignes d'action de protection des paysages ainsi que la formulation, à l'issue de l'enquête publique, d'aménagements concrets sur les sites concernés en vue d'une meilleure insertion dans le paysage des infrastructures ferroviaires, des aménagements routiers, des carrières, des champs d'éolienne, etc.) ; *f 2* **plan** [m] **de protection et de mise en valeur des paysages (±)** (D., plan établi dans la cadre du dossier technique d'une opération d'aménagement d'infrastructure au stade de l'avant projet détaillé ; cette étude d'environnement normalisée s'inspire des principes et des orientations de la loi relative à la protection de la nature, prend en compte les résultats de l'étude d'impact, elle fixe la nature et l'ampleur des mesures susceptibles de limiter ou compenser les incidences de l'aménagement et de celles tendant à compenser les impacts non réductibles et dommageables du projet sur l'environnement ; ▶mesure compensatoire, ▶réglementation relative aux atteintes subies par le milieu naturel) ; *g* **Landschaftspflegerischer Begleitplan** [m] (*Abk.* LBP; D., Planwerk, das eine Fachbehörde zur Umsetzung der ▶Eingriffsregelung bei einem Eingriff in Natur und Landschaft, der auf Grund eines nach öffentlichem Recht vorgesehenen Fachplanes vorgenommen werden soll, erstellen muss, um nach Bestandsaufnahme und Bewertung von Natur und Landschaft einmal die zum Ausgleich dieses Eingriffes erforderlichen Maßnahmen oder Ersatzmaßnahmen aus der Sicht von Naturschutz

D

und Landschaftspflege in Text und Karte darzustellen und zum anderen Gestaltungsmaßnahmen festzulegen; cf. § 8 [4] BNatSchG. I. d. R. beauftragt der Vorhabensträger ein Planungsbüro mit der Erarbeitung eines LPBs; die Kosten dafür sowie die Kosten für die Umsetzung der Maßnahmen hat der Vorhabenträger zu tragen. Der LBP ist Bestandteil des Fachplans und erlangt mit diesem allgemeine Rechtsverbindlichkeit; cf. BFN 2006; ▶Ausgleichsmaßnahme); *syn.* landespflegerischer Fachbeitrag [m].

1495 development mitigation planning [n] [≠ US] *leg. landsc. plan.* (Process of preparing supplemental/auxiliary plans which must be developed and approved by government departments to accompany specific planning projects, such as land consolidation, traffic planning, mineral extraction, industrial development, etc. with a view to mitigating any negative effects upon the environment, which may arise from these projects; ▶development mitigation plan [US]/landscape envelope plan [UK]); *syn.* landscape envelope planning [n] [UK]; *s* **planificación [f] paisajística complementaria** (≠) (**D.**, la planificación sectorial, p. ej. concentración parcelaria, planificación viaria, etc., es complementada por esta planificación paisajista ya que los causantes de interferencias en la naturaleza están obligados por ley a minimizar y compensar y, a ser posible, evitar los impactos negativos; ▶plan paisajístico complementario); *f* **planification [f] de protection et de mise en valeur paysagère** (≠) (**D.**, planification sectorielle soumise à l'approbation de l'administration de la protection de la nature [§ 8 (4) de la loi fédérale sur la protection de la nature], fournie par un maître d'ouvrage concernant tout aménagement, ouvrage et travaux dans les domaines d'activités tels que les infrastructures de transport, le remembrement, les carrières, les cours d'eau etc. et présentant les mesures envisagées pour éviter, réduire ou remédier aux effets négatifs du projet sur l'environnement ; en F. pourrait être assimilé au terme **volet paysager**, instauré par la loi 93-24 du 8 janvier 1993 instituant la prise en compte du paysage dans la procédure du permis de construire [document graphique précisant l'insertion d'un projet dans l'environnement avec ses accès et ses abords, son impact visuel] ; ▶plan de protection et de mise en valeur des paysages ; ▶schéma directeur paysager) ; *g* **landschaftspflegerische Begleitplanung [f]** (**D.**, Fachplanung des Naturschutzes und der Landschaftspflege, z. B. zur Flurbereinigungsplanung, Straßenplanung etc., als ergänzender landschaftspflegerischer Planungsprozess, zu dem ein Planungsträger [Fachbehörde] zwecks Ausgleichs eines Eingriffes in Natur und Landschaft gemäß § 8 [4] BNatSchG verpflichtet ist; ▶Landschaftspflegerischer Begleitplan).

1496 development [n] **of a meadow** *phyt.* (Growth of a seeded or existing meadow under actual or future site conditions); *s* **desarrollo [m] de pradera** (Desarrollo de una pradera sembrada o de una existente bajo determinadas condiciones mesológicas); *f* **développement [m] prairial** (Développement d'une prairie ensemencée ou existante sous les conditions actuelles ou futures ; dans l'évaluation de la qualité des fourrages on parle du **développement phénologique d'une prairie**) ; *g* **Wiesenentwicklung [f]** (Entwicklung einer angesäten oder vorhandenen Wiese unter bestehenden oder zukünftigen Standortbedingungen).

1497 development [n] **of green spaces** *constr.* (**1.** Planning and design of green spaces for ecological restoration or human uses. **2.** Planning for or construction of facilities within existing ▶green spaces; ▶landscape practice); *s* **ejecución [f] de proyectos de parques y jardines** (Realización de planes de diseño paisajístico incluyendo todas las medidas constructivas o la remodelación de ▶zonas verdes o elementos de diseño y el correspondiente equipamiento; ▶construcción paisajística 2);

f **travaux [mpl] d'espaces verts** (Études et aménagements de création ou de remodelage d'▶espaces verts, de parcs et jardins ; ▶travaux de paysagisme) ; *g* **Grünflächenbau [m]** (Planungen und Baumaßnahmen zur Herstellung oder Umgestaltung von ▶Grünflächen und Grünelementen incl. der dazugehörigen Einrichtungen; ▶Landschaftsbau).

1498 development [n] **of mineral resources** *min.* (Planning and technical measures which are necessary for mineral working); *s* **ganancia [f] de recursos geológicos** (Conjunto de medidas de planificación y técnicas necesarias para la explotación de recursos geológicos); *f* **recherche [f] et exploittation [f] des ressources minérales** (Ensemble des mesures de planification et techniques réglementant la recherche et l'exploitation des gîtes de substances minérales ou fossiles) ; *g* **Erschließung [f] von Bodenschätzen** (Gesamtheit der planerischen und technischen Maßnahmen, die die Voraussetzung für Gewinnung von Bodenschätzen sind).

1499 development [n] **of prefabricated panel buildings** *urb.* (**1.** Subdivision area which is predominantly to be found in eastern parts of Germany and former Communist countries in eastern Europe, the buildings of which are made of ▶prefabricated concrete panel construction. **2. In U.S.**, residential or commercial subdivision employing prefabricated structural components and/or whole structures; may also include onsite tilt slab or unitized construction methods); *syn.* precast concrete panel apartment buildings, development of industrialized apartment blocks; *s* **polígono [m] residencial de edificios prefabricados con grandes placas** (Tipo de asentamiento residencial de bloques de viviendas prefabricadas existente en los antiguos países socialistas y en Cuba, construidas utilizando grandes placas de hormigón; ▶construcción prefabricada de placas de hormigón/concreto); *f* **grand ensemble [m] construit en plaques de béton préfabriqué** (Quartier composé d'immeubles érigés sur le principe d'une ▶construction en plaques de béton préfabriqué ; terme utilisé pour décrire les immeubles standartisés en béton préfabriqué dans le cadre de la politique d'industrialisation du bâtiment en RDA et les anciens pays socialistes) ; *g* **Plattenbausiedlung [f]** (Vorwiegend in ostdeutschen Gemeinden und in Osteuropa anzutreffender Wohnbaugebietstyp, dessen Gebäude vorwiegend aus Fertigbetonteilplatten hergestellt wurden; ▶Plattenbauweise); *syn.* Plattenbauquartier [n].

development permit [n] [US] *adm. leg.* ▶conditional development permit.

development plan [n] *leg. urb.* ▶community development plan, ▶community garden development plan [US]; *landsc. leg.* ▶mandatory landscape development plan; *plan.* ▶master plan (2); *leg. plan.* ▶state development plan [US].

development plan [n] [UK]**, allotment garden** *leg. urb.* ▶community garden development plan [US].

development plan [n] [UK]**, unitary** *obs. leg. urb.* ▶comprehensive plan [US]/Local Development Framework (LDF) [UK].

development plan [n]**, urban** *leg. urb.* ▶community development plan.

1500 development plan [n] **for a rural area** (≠) *leg. plan.* (**1. In France**, a legally binding plan for the preservation of ▶rural land 1, agricultural activities, and the countryside, especially for villages which have no comprehensive plan. **2. In U.S.**, such a plan may involve planning to protect rural character and agricultural productivity by delimiting areas for non-agricultural uses, as in a land use plan. It may also refer to infrastructure plan to bring utilities to an underdeveloped or impoverished area); *s* **plan [m] especial** (±) (En Es., plan para la ordenación de recintos y conjuntos artísticos, protección del paisaje y de las vías de comunicación, conservación del medio rural, reforma interior, saneamiento de poblaciones; cf. art. 17 [1] LS; ▶zona rural. En Francia, plan de protección del espacio natural rural); *f 1* **plan [m] d'aménagement rural (P.A.R.)** (Règles d'urbanisme et les servitudes d'utilisation des sols établies sur les territoires à vocation agricole en vue d'assurer un développement harmonieux d'unités géographiques et économiques [communes rurales dont la population totale est inférieure à 10 000 habitants]) ; *syn. obs.* plan [m] de zone d'environnement protégé (Z.E.P.) ; *f 2* **charte [f] intercommunale de développement et d'aménagement** (Les **c. i. de d. et d'a.** définissent les perspectives à moyen terme du développement économique, social et culturel des communes concernées, déterminent les programmes d'action correspondants, précisent les conditions d'organisation et de fonctionnement des équipements et services publics ; en zone rurale la **c. i. de d. e. d'a.** se substitue au plan d'aménagement rural ; ▶espace rural) ; *g* **Bauleitplan [m] für den ländlichen Raum (≠)** (**In F.** ein gegen jedermann verbindlicher Plan für Gemarkungen im ▶ländlichen Raum mit dem Ziel des Schutzes der Agrarlandschaft und von Natur und Landschaft, speziell für solche Gemeinden, die keinen Bauleitplan haben).

development planning [n] [UK]**, urban** *adm. leg. urb.* ▶community development planning.

1501 development pressure [n] [US] *urb.* (Population increase and social pressures which lead to the expansion of built-up areas into open or ▶open land [US]/greenfield sites [UK]; ▶competing land use pressure); *syn.* settlement pressure [n] [UK]; *s* **presión [f] urbanizadora** (Crecimiento de población y/o tendencia a la disgregación social a nivel residencial que conllevan la urbanización de zonas rurales en la periferia de las urbes con el resultado de la pérdida creciente de áreas naturales; ▶suelo no urbanizable, ▶suelo rústico, ▶presión por competencia entre usos); *f* **pression [f] de l'urbanisation** (Pression exercée par différents acteurs économiques et politiques entraînant le grignotage des espaces ruraux en marge des périmètres construits à la périphérie des agglomérations, des ▶zones non urbanisables dans certaines régions littorales favorisées par l'attrait de leurs paysages ; ▶pression exercée par les usages des sols) ; *g* **Siedlungsdruck [m, o. Pl.]** (Gesellschaftliche Kräfte, die das Ausufern von Ortslagen in das weitere Umland der Städte [in den ▶Außenbereich] oder Neugründungen von Siedlungsgebieten „auf der grünen Wiese" mit der Folge des weiteren Rückgangs naturnaher Flächen bewirken; ▶Nutzungsdruck).

development site [n] *leg. urb.* ▶approved development site.

development zone [n] [UK] *leg. urb.* ▶zoning district [US].

development zone [n] **with maximum building height** [UK] *leg. urb.* ▶zone with an easement limiting the height of a building.

1502 dewatering [n] *constr.* (Removal of groundwater from a construction site by sump pumps, well points, or drainage systems); *s* **desagüe [m] de fundación** (BU 1959; instalaciones y medidas para evacuar el agua subterránea que aparece al excavar un terreno para construir los cimientos de un edificio); *syn.* agotamiento [m] [ES, MEX], desagote [m] [RA], achique [m] [RA, EC, YV], desecación [f] [CO] (1), abatimiento [m] [EC] (MESU 1977); *f* **épuisement [m]** (Travaux et techniques permettant l'excavation, le traçage et l'exécution des fondations d'un ouvrage sur un fond de fouille sec lors de l'afflux d'eau souterraine ou lorsque celui-ci se trouve au-dessous du niveau de la nappe ; on procède soit par enceinte étanche, soit à un rabattement de nappe par pompage) ; *g* **Wasserhaltung [f]** (Einrichtungen und Maßnahmen, die das beim Baugrubenaushub zu Tage tretende Grundwasser abführen. Man unterscheidet 'offene Wasserhaltung' und 'Wasserhaltung durch Grundwasserabsenkung').

dew formation [n] *met.* ▶dew production.

1503 dew production [n] *met.* (Process by which very small drops of water are formed on surfaces by the condensation of water vapo[u]r); *syn.* dew formation [n]; *s* **formación [f] de rocío** (Proceso de condensación del vapor de agua de la atmósfera y su deposición sobre las plantas en forma de gotas de agua muy pequeñas); *f* **formation [f] de rosée** (Processus de condensation de la vapeur d'eau et de son dépôt sous forme de fines gouttelettes d'eau) ; *g* **Taubildung [f]** (1. Vorgang, bei dem durch Kondensation des Wasserdampfes sich auf bodennahen Oberflächen, die meist durch Strahlungskälte unter den Taupunkt abgekühlt sind, feintropfiger Niederschlag absetzt. Der Taupunkt ist erreicht, wenn die relative Luftfeuchtigkeit von 100 % erreicht ist. **2.** Bei nachträglichem Gefrieren der Tautropfen entsteht **weißer Tau**; eine **T.** unter dem Gefrierpunkt ergibt **Reif**).

diaclase [n] *geo.* ▶joint (1).

diaclasis [n] *geo.* ▶joint (2).

1504 diagonal wattlework [n] [US]/**diagonal wattlework** [n] [UK] *constr.* (Vertically-installed fence of live willows, alders or the like, which is diagonally embedded on steep slopes, as a bioengineering technique; ▶wattlework); *syn.* diagonal wickerwork [n] [US]/diagonal wicker-work [n] [UK] (with willow branches); diamond-shaped wattle-work [n] [UK], diamond-shaped wattlework [n] [US]; *s* **malla [f] reticular (de ramas)** (*Construcción biotécnica* ▶varaseto de ramas que se coloca en cruz en taludes de 30°- 45° de pendiente); *f* **treillis [m] à maillage croisé** (*Génie biologique* assemblage de branchages — à rejets — dont l'entrecroisement est effectué en diagonale, angle de 30° à 45° et mis en œuvre sur les terrains en pente ; ▶treillis de branches [de saules] à rejets, ▶claie de branches à rejets ; *technique* clayonnage en losange) ; *syn.* claie [f] à mailles croisées, treillage [m] à maillage croisé, treillis [m] à maillage en losange, claie [f] à mailles en losange, clayon [m] à mailles en losange ; *g* **Diagonalgeflecht [n]** (*Ingenieurbiologie* ▶Flechtwerk, das an Böschungen kreuzweise mit 30°-45° Neigung eingebaut wird); *syn.* Rautenflechtwerk [n], Rautengeflecht [n].

diagonal wickerwork [n] [US]/**diagonal wicker-work** [n] [UK] *constr.* ▶diagonal wattlework [US]/diagonal wattlework [UK].

1505 diagram [n] (1) *constr. plan.* (Simplified two-dimensional schematic representation of relationships intended to explain the functioning of a system of construction; *specific term* diagrammatic cross-section); *s* **esquema [m] representativo** (Representación gráfica bidimensional); *syn.* diagrama [m] representativo; *f* **schéma [m] du procédé** (Figure donnant une représentation simplifiée et fonctionnelle d'un processus ou d'un objet ; cf. PR 1987) ; *syn.* schéma [m] de fonctionnement d'un système ; *g* **Systemskizze [f]** (Vereinfachte zeichnerische Darstellung von systemaren Zusammenhängen oder eines Konstruktionsaufbaues).

diagram [n] (2) *pedol.* ▶soil texture diagram.

diagram [n]**, flow** *plan.* ▶flow chart.

diameter breast height [n] *hort.* *▶girth, ▶trunk diameter.

diamond lane [n] *trans.* ▶offside lane, #2.

1506 diamond-shaped interchange [n] *trans.* (Rhombus type of traffic junction at two different levels with 4 parallel ramps; ▶interchange); *s* **nudo [m] de enlace con forma de rombo** (Cruce de tráfico a dos niveles con cuatro rampas paralelas; ▶nudo de enlace); *f* **raccordement [m] de type losange** (Intersection routière à deux niveaux et échangeur à quatre branches parallèles ; ▶diffuseur) ; *syn.* raccordement [m]

D

en losange ; *g* **Rautenanschluss [m]** (Verkehrsknotenpunkt in zwei Ebenen an Autobahnen oder sonstigen Schnellstraßen mit vier Parallelrampen; ▶Anschlussstelle).

diamond-shaped wattlework [n] [US]/**diamond-shaped wattle-work** [n] [UK] *constr.* ▶diagonal wattlework [US]/diagonal wattle-work [UK].

1507 dieback [n] *envir. hort.* (Progressive dying from the extremity of any part of a plant; ▶dieback of trees [US]/die-back of trees [UK], ▶top-kill [US]/top drying [UK]); *s* **muerte [f] regresiva de plantas** (Senescencia prematura de las plantas por exceso o falta de humedad o por otras causas ambientales; ▶muerte de árboles, ▶puntiseco incipiente); *f* **asphyxie [f]** (par excès d'humidité ou sécheresse ; ▶dépérissement des arbres, ▶phase initiale du dépérissement terminal) ; *syn.* mort [f] de végétaux ; *g* **Absterben [n, o. Pl.]** (Langsames Sterben von Pflanzen, das meist an den äußeren Teilen beginnt; ▶Baumsterben, ▶beginnende Wipfeldürre).

1508 dieback [n] **of firs** [US]/**die-back** [n] **of firs** [UK] *for. envir.* (Premature senescence, lack of growth [▶stork's nest] or dieback in crowns of fir trees caused by persistently harmful growing conditions due to decades of air pollution and its synergetic consequences for forest ecosystems; ▶forest decline); *s* **muerte [m] de abetos** (Senescencia prematura, recesión del crecimiento de la punta [▶nido de cigüeña] y ahuecamiento de la copa causados por el efecto persistente de la contaminación atmosférica con todas sus consecuencias sinergéticas para los ecosistemas forestales; ▶muerte de bosques, ▶nid de cigogne); *f* **dépérissement [m] des Sapins** (Vieillissement prématuré, arrêt subit de la croissance de la cime et éclaircissement de la ramure provoqués par une altération des conditions de croissance due à la persistance pendant plusieurs décennies des nuisances et de leur effets synergétiques sur l'écosystème de la forêt ; ▶mort des forêts, ▶nid de cigogne) ; *g* **Tannensterben [n]** (Frühzeitige Vergreisung, Wachstumsstau im Wipfel [▶Storchennest] und Kronenverlichtung auf Grund nachhaltiger Beeinträchtigungen der Wachstumsbedingungen durch jahrzehntelange Luftverschmutzungen mit allen ihren synergetischen Folgen auf das Waldökosystem; ▶Waldsterben).

1509 dieback [n] **of trees** [US]/**die-back** [n] **of trees** [UK] *for. envir.* (Premature aging and death of trees as a result of negative environmental influences; ▶dead crown, ▶black-top [US&CDN] 2, ▶dead crown, ▶Dutch elm disease, ▶forest decline, ▶top-kill [US]/top drying [UK]); *s* **muerte [f] de árboles** (Muerte prematura de los árboles debida a la contaminación del medio; ▶copa muerta, ▶grafiosis del olmo, ▶muerte de bosques, ▶puntiseco, ▶puntiseco en conífera); *f* **dépérissement [m] des arbres** (Vieillissement précoce et mort prématurée des arbres en milieu rural et urbain causés par les influences négatives sur l'environnement ; ▶défoliation de la cime des conifères, ▶descente de cime, ▶graphiose de l'orme, ▶houppe morte, ▶mort des forêts) ; *g* **Baumsterben [n, o. Pl.]** (Frühzeitige Vergreisung und vorzeitiges Absterben von Bäumen in Landschaft und Stadt durch negative Umwelteinflüsse mit all ihren Folgewirkungen; ▶abgestorbene Baumkrone, ▶Ulmensterben, ▶Waldsterben, ▶Wipfeldürre, ▶Zopftrockenheit bei Koniferen).

1510 difference [n] **in altitude** *geo.* (Difference in elevation between two geographically different points); *s* **diferencia [f] de altitud** (Diferencia de elevación entre dos puntos diferentes del terreno); *f* **dénivellation [f]** (Différence de niveau, d'altitude entre deux points) ; *syn.* dénivelé(e) [m/f] ; *g* **Höhenunterschied [m] (1)** (Differenz zwischen zwei geografisch verschiedenen Höhenpunkten).

1511 differential species [n] *phyt. zool.* (Plant or animal species which, while occurring in different life communities, is confined to certain subunits [US]/sub-units [UK], and may thus be used in their identification); *s* **especie [f] diferencial** (Son taxones que, sin poseer una limitación sociológica estrecha, se presentan sólo en una de dos o más comunidades afines, poniendo así de manifiesto determinadas diferencias bióticas, edáficas, microclimáticas, corológicas o genéticas; BB 1979); *f* **espèce [f] différentielle** (Espèce animale ou végétale, plus ou moins exclusivement cantonnée dans l'une de deux ou plusieurs associations ou sous-associations affines. Elle aide surtout à caractériser surtout les unités inférieures) ; *g* **Trennart [f]** (Pflanzen- oder Tierart, die in verschiedenen Lebensgemeinschaften auftritt, dort aber jeweils auf bestimmte Untereinheiten beschränkt ist und zu deren Unterscheidung dient); *syn.* Differentialart [f], Differenzialart [f].

different uses [n] *plan.* ▶layering of different uses.

1512 diffused particulate matter [n] *envir.* (Finely-divided solid or liquid particles in the air or in an emission; RCG 1982, 119; ▶suspended particulate matter); *syn.* diffused particulates [npl]; *s* **partículas [fpl] en suspensión en el aire** (Contaminantes atmosféricos sólidos de tamaño entre 1 y 100 micras que sedimentan lentamente bajo la acción de la gravedad; ▶materias en suspensión); *f* **particules [fpl] en suspension** (Particules liquides ou solides très fines en suspension dans l'air ; ▶matières en suspension) ; *g* **Schwebstaub [m]** (...stäube [pl]; in der Luft in feinster Verteilung vorkommende kleine, feste oder flüssige Partikelchen, die auf Grund ihrer geringen Größe und ihres geringen Gewichtes fast homogen verteilt sind und somit für einige Zeit in der Schwebe bleiben; gesundheitliche Schädigungen entstehen, wenn feine Partikel mit einem Durchmesser von < 10 µm in die Lunge eindringen und dort durch Reizungen Entzündungen hervorrufen, die zu Bronchitis oder Lungenkrebs führen können. Noch kleinere Staubteilchen, < 2,5 µm, können aus der Lunge in den Blutkreislauf gelangen und Veränderungen im Blutbild hervorrufen; ▶Schwebstoffe); *syn. o. V.* Schwebestaub [m], Feinstaub [m].

diffused particulates [n] *envir.* ▶diffused particulate matter.

diffuse root system [n] *arb. hort.* ▶fibrous root system.

diffusion [n]**, capillary** *pedol.* ▶capillary rise.

1513 dig [vb] *hort.* (Turning the soil with a spade or fork; ▶trenching, ▶double digging); *s* **cavar [vb]** (Remover la tierra con laya/pala u horca; ▶subsolar en profundo, ▶desfonde a dos palas); *f* **bêcher [vb]** (Retourner la terre avec une bêche ; ▶défonçage, ▶bêcher à deux fers) ; *g* **umgraben [vb]** (Boden mit einem Spaten oder einer [Grabe]gabel wenden; ▶Rigolen, ▶holländern).

dig [vb] **a planting hole** [US] *constr. hort.* ▶excavate a planting hole.

dig area [n] [US]**, archaeological** *conserv'hist.* ▶archaeological excavation area.

1514 digested sludge [n] *envir.* (Solid product of sewage treatment resulting from the bacterial breakdown of the organic components; ▶activated sludge, ▶sewage sludge, ▶sewage sludge digestion); *s* **lodo [m] digerido** (Lodo secundario resultante de la depuración de aguas residuales que se produce por ▶digestión anaerobia de lodo de depuración en cámaras de digestión produciéndose metano y CO_2; ▶lodo activado, ▶lodo de depuración); *syn.* fango [m] digerido, cieno [m] digerido; *f* **boues [fpl] digérées** (Matière épaisse [boue fraîche] issue du traitement des eaux usées traitées par ▶digestion anaérobe des boues [flore bactérienne] dans une cuve de fermentation fermée [digesteur] et produisant du méthane et du gaz carbonique ; ▶boue activée, ▶boues d'épuration) ; *g* **Faulschlamm [m] (1)** (Bei der Abwasserreinigung anfallender Frischschlamm, der durch anaerobe Bedingungen in Faulkammern bei Abgabe von

Methan und CO_2 verfault; ►belebter Schlamm, ►Klärschlamm, ►Schlammfaulung).

digging [n] **with an excavator** *constr. eng. min.* ►excavation (1).

dike [n] *constr. wat'man.* ►coastal dike, *agr. constr.* ►drainage ditch, *conserv. land'man.* ►field dike; *eng. wat'man.* ►foredike, *wat'man.* ►main dike [US]/main dyke [UK], ►summer dike, ►winter dike.

dike [n] [UK] *agr.* ►dirt bank [US].

1515 dimension [n] *constr.* (Spatial measure of a single line, such as length, breadth, height, thickness, caliper, and circumference); *s* **dimensión** [f] (Extensión de un objeto en una dirección determinada como largo, ancho, altura, espesor, perímetro); *syn.* medida [f]; *f* **dimension** [f] (Grandeur d'un objet [longueur, largeur, hauteur] mesuré) ; *syn.* mesure [f] ; *g* **Abmessung** [f] (Maß eines Gegenstandes: das mit einem Längenmaß Gemessene wie Länge, Breite, Höhe, Dicke/Stärke und Umfang).

1516 dimensioning [n] *constr. plan.* (**1.** Process of noting dimensions such as lengths or diameters on a working drawing. **2.** Result of noting dimensions onto a plan); *s 1* **medición** [f] (Proceso de tomar medidas de un lugar); *s 2* **medida** [f] (Resultado de 1); *syn.* tamaño [m], dimensión [f]; *f* **dimensionnement** [m] (**1.** Processus de définition des longueurs, diamètres etc. sur des plans d'études. **2.** Résultat des mesures transcrites sur un plan) ; *syn.* donnée [f] de mesures ; *g* **Vermaßung** [f] (**1.** Vorgang der Bestimmung der Längenabmessungen, Durchmesser etc. auf Werkplänen. **2.** Ergebnis der eingetragenen Abmessungen auf Werkplänen); *syn. zu 2.* Maßangaben [fpl].

1517 diminishing step [n] *arch. constr.* (Step on sloping ground with decreasing riser height); *s* **escalón** [m] **menguante (±)** (Peldaño de escalera en terreno en pendiente cuyas contrahuellas se van reduciendo con la altura); *f* **marche** [f] **en sifflet** (Escalier extérieur dont la hauteur des contremarches diminue avec la pente) ; *g* **Schleppstufe** [f] (Stufe in hängigem Gelände mit einer ständig geringer werdenden Auftrittshöhe); *syn.* einschleifende Stufe [f].

1518 diminution [n] **of water resources** *wat'man.* (Lessening in the amount of available water resources); *syn.* lowering [n] (of water resources), lessening [n] (of water resources) [also US]; *s* **degradación** [f] **de los recursos hídricos** (Efectos negativos sobre la cantidad de agua disponible para los usos humanos); *syn.* reducción [f] de los recursos hídricos; *f* **altération** [f] **de la qualité des ressources en eau** (Impact négatif sur les ressources en eau disponibles) ; *g* **Beeinträchtigung** [f] **der Wasserressourcen** (Negative Auswirkung auf die verfügbare Menge der Wasserressourcen).

1519 dimple spring [n] *hydr.* (Seepage spring from point source); *s* **manantial** [m] **puntual** (Surgencia de aguas de un acuífero en un solo punto); *f* **source** [f] **ponctuelle** (Émergence isolée des eaux de la nappe phréatique) ; *syn.* source [f] isolée ; *g* **Punktquelle** [f] (Punktueller Grundwasseraustritt).

1520 dip [n] *geo.* (Angle of inclination of a geological stratum, measured in degrees from the horizontal); *s* **inclinación** [f] **(1)** (Ángulo de un estrato geológico, medido en grados a partir de la horizontal); *f* **pendage** [m] (d'une couche géologique, d'une strate, d'un plan de faille) ; *g* **Neigung** [f] **(1)** (Winkel einer geologischen Schicht von der Horizontalen gemessen).

dip curve [n] [UK] *trans.* ►sag curve [US].

Diploma [n] **in Landscape Architecture (DipLA)** [UK] *prof.* ►Master of Landscape Architecture (MLA).

Diploma [n] **in Landscape Design (DipLD)** [UK] *obs. prof.* ►Master of Landscape Architecture (MLA).

1521 dip slope [n] *geo.* (Sloping ground surface running approximately parallel to dip in the underlying strata, usually in connection with a corresponding ►scarp slope); *s* **reverso** [m] **de cuesta** (Vertiente más suave de una cuesta que coincide topográficamente con un nivel resistente; cf. DGA 1986; ►frente de cuesta); *f* **versant** [m] **cataclinal** (Versant doux, dont la pente suit à peu près parallèlement la stratification sous-jacente et en général en relation avec un versant escarpé ; DG 1984, 118 ; ►versant anaclinal) ; *syn.* revers [m] ; *g* **Stufenfläche** [f] (Flach abgedachte Hochfläche einer Schichtstufe, deren Gefälle ungefähr parallel zu den darunterliegenden Schichten verläuft; i. d. R. im Zusammenhang mit einem ►Stufenhang, einem steilen Abhang auf der anderen Seite des Berges); *syn.* Stufenlehne [f].

1522 direct discharge [n] *envir. leg.* (Ejection of untreated effluent into the sea or a ►receiving stream); *s* **vertido** [m] **directo** (Descarga directa de efluentes en el ►curso receptor); *f* **rejet** [m] **direct** (Apport d'effluents non épurés dans un milieu récepteur ; LA 1981, 473 ; ►émissaire, ►milieu récepteur) ; *g* **Direkteinleitung** [f] (Entsorgung von ungeklärten Abwässern ins Meer oder in einen ►Vorfluter; ►Direkteinleiter).

1523 directional lane [n] *trans.* (Lane in which traffic may only flow in one direction; three types of lanes are distinguished: right turn lane, straight lane, and left turn lane); *s* **carril** [m] **unidireccional** (Vía de carretera o autopista que sólo se puede utilizar para una dirección. Se diferencia entre carril de giro a la derecha, carril de giro a la izquierda o carril recto); *syn.* calzada [f] de una dirección; *f* **chaussée** [f] **unidirectionnelle** (Chaussée ne comportant qu'une direction de circulation) ; *g* **Richtungsfahrbahn** [f] (Verkehrsraum mit einem oder mehreren Fahrstreifen, auf der der Verkehr nur in einer Richtung fließen darf; es werden Rechtsabbiegespur, Geradeausfahrstreifen und Linksabbiegespur unterschieden; *fachsprachlich veraltet* Fahrspur).

Directive [n] *conserv. leg.* ►Birds Directive; *adm. leg.* ►European Union Directive; *wat'man. pol.* ►EU Water Framework Directive; *conserv. leg.* ►Habitats Directive; *constr. contr.* ►rectification directive [US].

Directive [n] **for the conservation of natural habitats, wild animals and plants** *conserv. leg.* ►Habitats Directive.

1524 directive [n] **on planting requirements** [US] *leg. urb.* (Formal legally-binding instruction issued by an administrative authority/agency requiring planting of prescribed vegetation; **in U.S.**, planting requirements are usually determined during **site plan review** of existing conditions and proposed development, prior to review by the planning commission and approval by the administrative authority; **in D.**, specified e.g. in legally-binding land-use plans [►Bebauungsplan]; ►plant preservation requirement [US]/obligation to preserve existing plants [UK]); *syn.* enforcement notice [n] on planting requirements [UK]; *s* **prescripción** [f] **de plantar** (Orden administrativa [p. ej. contenida en ►plan parcial [de ordenación] dirigida a los propietarios de los correspondientes predios de llevar a cabo medidas de plantación en un plazo determinado o de cumplir con la ►prescripción de preservar la vegetación); *f* **obligation** [f] **de réaliser des plantations** (Décision administrative communale obligeant le propriétaire d'un terrain à bâtir à effectuer des plantations [arbre de haute tige, haie d'arbustes etc.] dans le cadre d'un ►plan local d'urbanisme [PLU]/plan d'occupation des sols [POS]; **D.**, l'ampleur de l'obligation de plantation, déterminée sur la base d'un procédé d'évaluation spécifique, constitue une des mesures compensation des conséquences dommageables [imperméabilisation du sol, suppression de la végétation etc.] de la construction sur l'environnement [sol, eau, climat, flore, faune] ; ►obligation de préservation de la végétation existante) ;

D

g Pflanzgebot [n] (Eine durch behördlichen Bescheid ergangene Anordnung einer Gemeinde, die den Eigentümer eines Grundstücks innerhalb einer bestimmten Frist verpflichtet, das Vollzugsdefizit einer im ▶Bebauungsplan festgesetzten Pflanzverpflichtung [Pflanzzwang] oder ▶Pflanzbindung nach § 9 [1] 25 BauGB zu beseitigen; cf. § 178 BauGB; in den USA gibt es die Form des **P.es** aus bauleitplanerischer Sicht nicht, jedoch in anderer Form: Baugesellschaften müssen i. d. R. im Rahmen des Genehmigungsverfahrens einen Freiflächengestaltungsplan mit der zu schützenden und zu pflanzenden Vegetation vorlegen. Dieser Planungsschritt wird *site plan review* genannt).

directive [n] to rectify [US] *constr. contr.* ▶rectification directive [US].

1525 direct runoff [n] *constr. geo.* (That part of precipitation reaching channels or other ▶receiving streams immediately after rainfall or snow melting); *syn.* direct surface runoff [n], immediate runoff [n], storm flow [n], storm runoff [n], storm water runoff [n]; **s escorrentía [f] superficial (1)** (Parte de la precipitación que se desplaza en la superficie del terreno); *syn.* escurrimiento [m] superficial; *f* **ruissellement [m]** (Circulation initiale des eaux de pluie suivant un tracé diffus sur un domaine restreint, alimentant dans les talwegs l'écoulement des eaux de surface ; ▶milieu récepteur) ; **g Oberflächenwasserabfluss [m] (1)** (Regenwasser, das kleinräumig oberirdisch unmittelbar nach dem Regenereignis oder der Schneeschmelze zur ▶Vorflut 1 wird und in natürliche oder künstliche Wasserläufe [▶Vorfluter] abfließt); *syn.* Abfluss [m] des Oberflächenwassers, Regenwasserabfluss [m].

direct selection [n] and negotiation [n] [US] *contr.* ▶sole source contract award [US]/freely awarded contract [UK].

direct surface runoff [n] *constr. geo.* ▶direct runoff.

1526 dirt bank [n] [US] *agr.* (Narrow dirt bank of a rice field); *syn.* dike [n] [UK], dyke [n] [UK], levee [n]; **s dique [m] (3)** (Talud estrecho para estancar el agua en una arrocera); *f* **levée [f] (1)** (Bourrelet de terre encadrant des rizières) ; *syn.* diguette [f] (DG 1984, 398) ; **g Damm [m] (2)** (Aufschüttung zur Eindeichung eines Reisfeldes).

disabled [pp] *leg. plan. sociol.* ▶developed for the handicapped/disabled.

disabled person [n] *sociol.* ▶handicapped person.

disabled persons [npl] *leg. plan. sociol.* ▶construction for disabled persons.

disabled persons [npl], garden for *gard.* ▶garden for the handicapped.

disappeared species [n] *conserv. phyt. zool.* ▶extinct species.

disc [n] *bot.* ▶adhesive disc/disk.

discharge [n] *envir. leg.* ▶direct discharge; *wat'man.* ▶flood discharge; *constr.* ▶outflow; *hydr. wat'man.* ▶peak discharge.

discharge [n], maximum *wat'man.* ▶flood discharge; ▶peak discharge.

discharge cross-section [n] *hydr.* ▶channel discharge cross-section.

discharge fee [n] *adm. envir. leg.* ▶effluent discharge fee.

1527 discharge [n] of effluents *envir.* (1. Release of treated and untreated liquid waste into the sewerage system, a watercourse or the sea. 2. Release of hazardous liquid waste and ▶toxic substances either into environmentally-safe separation facilities, or illegally into watercourses or the sea; ▶disposal by dilution, ▶dumping at sea, ▶effluent disposal, ▶underwater effluent discharge into the sea, ▶water polluting firm); **s vertido [m]** (1. Eliminación de aguas o productos residuales depurados o sin depurar en un cauce de agua o en el mar; cf. art. 92 Ley de Aguas. 2. Eliminación de ▶sustancias tóxicas líquidas en plantas de tratamiento o ilegalmente en el sistema de desagüe público, en un curso receptor o en el mar; ▶flujo de efluente [sin depurar], ▶descarga de residuos en alta mar, ▶vertido directo, ▶vertido por dilución, ▶emisario submarino); *syn.* descarga [f] de efluentes; *f* **rejet [m] des effluents** (Déversement ou écoulement de substances quelconques liquides [en général de ▶substances toxiques] dans les eaux superficielles ou souterraines ; ▶auteur d'un déversement direct [des effluents], ▶clapage, ▶élimination des eaux usées, ▶émissaire en mer, ▶immersion de déchets en mer, ▶milieu récepteur) ; *syn.* déversement [m] des eaux usées (LA 1981, 473), rejet [m] des eaux résiduaires ; **g Einleitung [f] (1.** Entsorgung von geklärten oder ungeklärten Abwässern in die Kanalisation, in einen ▶Vorfluter oder ins Meer. **2.** Entsorgung von flüssigen ▶Giftstoffen entweder in eigens dafür hergerichteten Aufbereitungsanlagen oder illegalerweise in die Kanalisation, in Vorfluter [Oberflächengewässer], unterirdische Gewässer oder ins Meer; ▶Abwasserbeseitigung, ▶Abwasserleitung ins Meer, ▶Direkteinleiter, ▶Verklappen von flüssigen Abfallstoffen, ▶Verklappung).

discharge pipe [n] into the sea *envir.* ▶underwater effluent discharge pipe into the sea.

1528 discoloration [n] of foliage [US]/**discolouration [n] of foliage** [UK] *bot.* (**1.** Colo[u]r alteration which is caused by disease or change in a plant's growing conditions. **2.** ▶autumn colors/hues [of leaves] [US]/autumn colours/hues [of leaves] [UK]); **s descoloración [f] del follaje** (**1.** Alteración del color de las hojas por enfermedad o cambio de condiciones de crecimiento. **2.** ▶colores otoñales); *f* **décoloration [f] des feuilles** (**1.** Changement de la coloration naturelle d'une feuille provoqué par une maladie ou des changements stationnels. **2.** ▶coloration automnale) ; **g Laubverfärbung [f] (1.** Veränderung der natürlichen Blattfarbe eines Laubgehölzes durch Krankheit oder Standortveränderung/Veränderung der Wachstumsbedingungen. **2.** ▶Herbstfärbung).

discontinuity layer [n] *limn. ocean.* ▶thermocline.

disease [n] *bot. envir. phytopath.* ▶black spot disease; *arb. phytopath.* ▶Dutch elm disease; *agr. for. hort.* ▶fungus disease; *constr. hort.* ▶plant disease; *hort.* ▶viral disease.

disfigurement [n] of landscape *conserv. land'man.* ▶impairment of landscape.

1529 dished channel unit [n] *constr.* (Shaped stone or prefabricated concrete unit used in road construction to form a ▶gutter; ▶channel 1, ▶curb and gutter [US]/kerbstone with gutter [UK]); *syn.* spoon drain [AUS]; **s pieza [f] de cuneta** (Elemento prefabricado de piedra u hormigón utilizado para construir ▶cunetas o ▶canales de desagüe en pavimentos de caminos; ▶pieza de cuneta angular); *f* **1 bordure [f] fil d'eau** (Élément de maçonnerie préfabriqué ou en pierre naturelle à double dévers — bordure CC ; cf. Norme N.F. P 98 302 — destiné à conduire les eaux de ruissellement vers des exutoires ; ▶bordure caniveau d'épaulement, ▶caniveau, ▶rigole d'écoulement) ; *f* **2 bordure [f] caniveau** (Élément de maçonnerie préfabriqué ou en pierre naturelle à simple dévers — bordure C ; cf. Norme N.F. P 98 302 — destiné à conduire les eaux de ruissellement vers des exutoires ; ▶caniveau, ▶bordure caniveau d'épaulement) ; **g Rinnenstein [m]** (Vorgefertigtes Steinelement für den Straßenbau, das zum Bau eines ▶Rinnsteins oder einer ▶Abflussrinne im Wege-/Platzbelag dient; *nicht zu verwechseln mit* Rinnstein; ▶Winkelrinnenstein).

1530 disjunctive range [n] *phyt.* (Widely separated islands of a former continuous range, in which plant distribution may not occur under recent conditions because seeds and spores cannot

spread over long distances; ►relict area); *s* **área [f] disyunta** (En geobotánica, área fragmentada, es decir, constituida por dos o más áreas secundarias muy separadas unas de otras; DB 1985; ►área relíctica); *syn.* área [f] discontinua; *f* **aire [f] disjointe** (Une aire de répartition géographique est disjointe lorsque les zones de répartition de l'espèce sont séparées par des distances supérieures à celles que leurs diaspores peuvent franchir ; DEE 1982 ; ►aire relictuelle) ; *syn.* aire [f] discontinue ; *g* **disjunktes Areal [n]** (Areal, das aus Teilen besteht, zwischen denen eine Verbreitung von Pflanzen unter den heutigen Bedingungen unmöglich erscheint; BOR 1958; ►Reliktareal).

disk [n] *bot.* ►adhesive disk.

disks [npl] **[UK], self-clinging vine with adhesive** *bot.* ►self-clinging vine with adhesive pads [US].

1531 dislodgement [n] *zool.* (Forcing of animals to leave their habitats, because of human activities); *s* **expulsión [f] de animales** (Hacer huir a animales de sus hábitats por perturbaciones causadas por actividades humanas); *syn.* ahuyentar [vb] animales; *f* **exode [m]** (Abandon de l'aire de répartition par les animaux soumis à des perturbations intentionnelles ou involontaires causées par les activités humaines) ; *g* **Vertreibung [f]** (Durch menschliche Aktivitäten verursachter Zwang auf Tiere, ihren Lebensraum zu verlassen).

1532 dismantlement [n] **of building site installations** *constr.* (Removal of ►building site facilities and equipment; *opp.* ►site facilities installation); *s* **desmontaje [m] de las instalaciones auxiliares de obra** (después de terminada ésta; *opp.* ►montaje de las instalaciones auxiliares de obra); *f* **repli [m] d'installations de chantier** (à la fin des travaux ; *opp.* ►mise en place d'installations de chantier) ; *g* **Abbau [m, o. Pl.] von Baustelleneinrichtungen** (Entfernen von aufgebauten ►Baustelleneinrichtungen 1; *opp.* ►Aufbau von Baustelleneinrichtungen).

1533 dispersal [n] *phyt. zool.* (**1.** Active or passive movement of organisms from one location to another where they may produce offspring; e.g. plant seeds, fungal spores by wind; insect adults and birds by flight; other animals by walking or swimming; cf. NRM 1996; *specific terms* ►animal dispersal of seeds, ►anthropogenic dispersal, ►seed dispersal, ►water dispersal, ►wind dispersal. **2.** ►spread 1; ►distribution); *s* **dispersión [f] (1)** (Desplazamiento activo o pasivo de organismos a otras áreas en las que pueden desarrollar su ciclo vital completo y dejar descendencia. En los vegetales, la **d.** tiene lugar por medio de las diásporas. El hecho del desplazamiento de las diásporas, cuando éstas son frutos o semillas, se llama ►diseminación, mientras que el término **d.** queda para el efecto que origina. El resultado de la **d.** es la ►expansión de la correspondiente especie; cf. DINA 1987; términos específicos para la ►diseminación de semillas o frutos y de organismos por diferentes medios [viento, agua, animales y seres humanos]: ►anemocoria, ►antropocoria, ►hidrocoria, ►zoocoria); *f* **dissémination [f]** (Terme générique pour le dispersion naturelle des graines par le vent ou l'eau ; ►anémochorie, ►anthropochorie, ►dispersion, ►dissémination des semences, ►dissémination par des animaux, ►hydrochorie) ; *g* **Verbreitung [f] (1)** (Vorgang des Ausbreitens eines Organismus von einem Ort zu einem anderen Gebiet, wo er sich vermehrt. Die passive **V.** von Diasporen durch Wasser oder Wind und z. B. die von Insekten durch starken Wind wird auch **Verdriftung**, die **V.** durch Menschen oder Tiere **Verschleppung** genannt; die aktive **V.** der Vögel und Insekten erfolgt i. d. R. durch Flug, die anderer Tiere durch Wanderung und Schwimmen; ►Ausbreitung, ►kulturbedingte Verschleppung, ►Samenverbreitung, ►Verbreitung 2, ►Verdriftung durch Wasser, ►Verdriftung durch Wind, ►Verschleppung durch Tiere).

dispersal [n] **[UK], pollution** *envir.* ►pollution dispersion [US].

dispersal [n] **by human activity** *phyt. zool.* ►anthropogenic dispersal.

dispersal model [n]**, emission** *envir.* ►emission distribution model.

dispersal [n] **of seeds** *phyt. zool.* ►animal dispersal of seeds.

1534 dispersed building development [n] **[US]** *urb.* (Pattern of low density urban growth with scattered groups of housing and service facilities in ►open country; ►uncontrolled proliferation of settlements); *syn.* dispersed settlement [n] [UK]; *s* **edificación [f] dispersa** (Estructura de agrupación de edificios en la que éstos no tienen muros comunes, sino que se encuentran repartidos individualmente o en pequeños grupos en el campo; ►campiña, ►desfiguración del paisaje; *syn.* construcción [f] dispersa, urbanización [f] dispersa; *f* **habitat [m] dispersé** (Structure d'implantation caractérisant l'isolement de l'habitat en zone rurale ; *contexte* proscription de l'éparpillement des constructions ; ►mitage du paysage) ; *syn.* urbanisation [f] dispersée (R. 111-14-1) ; *g* **Streubebauung [f]** (Siedlungsgefüge in Form von nicht aneinandergrenzenden, bebauten Grundstücken oder lockeren Gruppensiedlungen in der ►freien Landschaft; *syn. mit negativem Sinn* Splittersiedlung [f] [§ 35 (3) BauGB]; ►Zersiedelung [der Landschaft]); *syn.* Streusiedlung [f].

dispersed fruit tree planting [n] **[US]** *agr. conserv.* ►traditional orchard [UK].

dispersed settlement [n] **[UK]** *urb.* ►dispersed building development [US].

1535 dispersion [n] (1) *phyt.* (Spreading of individuals of a species within a ►plant association/population. There are three patterns of spatial distribution: 1. random pattern, 2. clumped/aggregated pattern, and 3. regular/overdispersed pattern; TEE 1980, 54s; ►dispersal, ►distribution); *syn.* spatial distribution [n] (1), dissemination pattern [n] (TEE 1980, 54) ; *s* **dispersión [f] (2)** (Distribución espacial de los individuos de una especie en una ►asociación. Se distinguen tres tipos: 1. Subdispersión [unidades ordenadas más regularmente], 2. normal y 3. supradispersión [plantas agrupadas]. Esta última es la normal en las comunidades vegetales naturales; cf. BB 1979; ►sociabilidad, ►dispersión 1); *f* **dispersion [f]** (**1.** Distribution spatiale des individus d'une espèce à l'intérieur d'une ►association [végétale] en relation avec la ►sociabilité. On distingue la dispersion de hasard, la dispersion en groupe et la dispersion régulière ; ►répartition [des espèces]. **2.** *zool.* Répartition dans l'espace des individus d'une population ainsi que de leurs mouvements à l'intérieur de leur habitat naturel, déterminée par les facteurs topologiques et le comportement des individus entre eux ; cf. LAF 1990, 293 ; ►dissémination) ; *g* **Verteilung [f]** (Räumliches Vorhandensein von Individuen einer Art in einer Pflanzengesellschaft [►Assoziation]. **V.** steht mit der ►Häufigsweise im Zusammenhang. Es werden drei unterschiedliche Verteilungsmuster unterschieden: 1. zufällige **V.**, 2. gehäufte/aggregative **V.** und 3. regelmäßige **V.**; ►Verbreitung 1, ►Verbreitung 2); *syn.* Dispersion [f], Streuung [f].

dispersion [n] **[US] (2)** *envir.* ►pollution dispersion, ►traffic noise dispersion.

1536 displacement [n] **of species (composition or balance)** *phyt. zool.* (Alteration of the species balance within a plant or animal community as a result of changes in environmental conditions); *syn.* compositional shift [n] of faunal or floral communities; *s* **modificación [f] de la composición de especies** (Alteración del balance de especies constituyentes de una comu-

D

nidad como resultado de cambios de las condiciones ambientales); *syn.* desplazamiento [m] de especies, evolución [f] de la composición de especies (correspondiente a la sucesión); *f* **évolution [f] de la composition en espèces** (Changement naturel de la constitution d'une population animale ou végétale au cours du cycle évolutif ou modification de la constitution d'un peuplement végétal provoqué par l'action humaine [p. ex. pâturage intensif dans les forêtes avec en conséquence l'apparition d'un puplement secondaire) ; *syn.* modification [f] de la composition des espèces ; *g* **Artenverschiebung [f]** (Veränderung der Artenzusammensetzung einer Tier- oder Pflanzengesellschaft auf Grund der Sukzessionsentwicklung); *syn.* Verschiebung [f] im Artengefüge, Verschiebung [f] des Artenspektrums.

display [n] *hort.* ▶floral display, ▶summer bloom display, ▶summer plant display.

display [n]**, seasonal bedding** *hort.* ▶seasonal bedding.

disposal [n] *envir.* ▶effluent disposal, ▶waste treatment and disposal [US]/public cleansing and waste disposal [UK].

1537 disposal [n] **by dilution** *envir.* (Disposal of liquid waste or excavation material in an inland body of water or on the high seas; ▶discharge of effluents, ▶dumping at sea, ▶marine pollution); *s* **vertido [m] por dilución** (Disposición de residuos líquidos en alta mar, como p. ej. ácidos diluidos de la producción de dióxido de titano, lodos de depuración o residuos resultantes de la limpieza de los tanques de barcos o petroleros. Tras la ratificación de los Convenios Internacionales de **Oslo** para el ámbito del Mar del Norte y el Atlántico [Convenio de 15 de febrero de 1972 para la prevención de la ▶contaminación marina provocada por vertidos desde buques y aeronaves, ratificado en Es. por Instrumento de 19 de febrero de 1973] y de **Londres** para todos los mares del mundo [Convenio de 29 de diciembre de 1972 sobre prevención de la contaminación del mar por vertimiento de desechos y otras materias, ratificado en Es. por Instrumento de 13 de julio de 1974], se dictaron la Orden de 26 de mayo de 1976 y la Ley 21/1977, derogada en lo que se refiere a la aplicación de sanciones en los casos de contaminación marina provocada por vertidos desde buques por medio de la Ley 27/1992 de Puertos del Estado y de la Marina Mercante y el RD 258/1989, según el cual se fijan las normas de vertido en las aguas interiores y en el mar territorial de sustancias peligrosas, clasificadas en dos categorías según el grado toxicidad, persistencia y bioacumulación, se establece el procedimiento de control y se regula el establecimiento de programas para evitar y reducir la contaminación por sustancias peligrosas; cf. Art. 1.3 RD 258/1989; por medio de la Ley 16/2002, de prevención y control integrados de la contaminación fue derogado el Art. 4 del RD 258/1989 que en determinadas circunstancias permitía sobrepasar los límites de emisión; ▶descarga de residuos en alta mar, ▶vertido); *f* **clapage [m]** (Vidange en mer de substances souvent non polluées [généralement, déchets ou produits de dragage prélevés dans les ports, leurs chenaux d'accès ou darses — produit de type vase] en un lieu réservé à cet effet [zone de clapage en mer], en principe à l'aide d'un navire dont la cale peut s'ouvrir par le fond ; par extension, ce terme désigne toute opération de rejet en mer de boues ou de solides [p. ex. par refoulement à l'aide de pompes] ; les rejets en mer par clapage des produits de dragage des ports sur des sites autorisés relèvent de la **Convention d'Oslo** du 15.02.1972 pour le domaine maritime de la mer du Nord et de l'Atlantique et de la **Convention de Londres** du 29.12.1972 pour toute les mers du globe et plus précisément des lignes directrices de cette convention adoptées à Berlin les 14 et 19 juin 1993 ; la Convention pour la protection du milieu marin de l'Atlantique du Nord-Est, dite **Convention OSPAR**, a été signée à Paris le 22 septembre 1992 ; elle est née de la fusion de la Convention d'Oslo [1972] traitant de la prévention de la pollu-

tion marine par les opérations d'immersion, et de la Convention de Paris [1974], traitant des rejets d'origine tellurique ; ▶pollution marine, ▶immersion des déchets en mer) ; *g* **Verklappen [n, o. Pl.] von flüssigen Abfallstoffen** (Verbringung von flüssigen Abfallstoffen auf See, z. B. Dünnsäure aus der Titandioxidproduktion, Klärschlämme oder Rückstände, die bei Schiffs- oder Tankreinigungen anfallen. Nach der Ratifizierung der internationalen **Abkommen von Oslo** [15.02.1972 für den Geltungsbereich Nordsee und Atlantik] und **London** [29.12.1972 für alle Meere der Welt] und nach In-Kraft-Treten des „Hohe-See-Einbringungsgesetzes" von 1977 ist die ▶Verklappung in Deutschland erlaubnispflichtig, und zwar durch das Deutsche Hydrographische Institut; 1988 einigten sich die EG-Umweltminister auf die Richtlinie zur Beendigung der ▶Meeresverschmutzung durch Abfälle aus der Titandioxidproduktion, nach der die Dünnsäureverklappung bis zum 31.12.1989 eingestellt werden sollte. Für die Mitgliedstaaten Großbritannien, Frankreich und Spanien wurde unter bestimmten Voraussetzungen eine Übergangsfrist vorgesehen. Am 22.09.1992 wurde in Paris das „Übereinkommen zum Schutze der Meeresumwelt und des Nordatlantiks", die **OSPAR-Konvention**, benannt nach den beiden Vorläufern Oslo-Konvention und Pariser Konvention, mit dem Ziel unterzeichnet, Verschmutzungen zu verhindern und Verschmutzungen zu beseitigen sowie den Geltungsbereich des Übereinkommens vor schädlichen Auswirkungen menschlicher Tätigkeiten zu schützen; ▶Einleitung; *OB* Abfallbeseitigung auf See); *syn.* Verklappung [f] flüssiger Abfallstoffe.

disposal cost [n] **to waste producer** *envir. pol.* ▶allocation of disposal cost to waste producer.

disposal industry [n] *envir.* ▶waste treatment and disposal industry.

disposal [n] **of mining gob** [US] *min.* ▶disposal of mining spoil.

disposal [n] **of mining gob** [US]**, underground** *min.* ▶underground disposal of mining spoil.

1538 disposal [n] **of mining spoil** *min.* (Disposition of mining waste by spreading over and shaping the ground surface, or returning to the worked galleries or adits; ▶mining spoil, ▶underground disposal of mining spoil); *syn.* disposal [n] of mining gob [also US]; *s* **depositar [vb] zafras** (Transporte y almacenamiento definitivo de ▶zafras a cielo abierto o en mina abandonada [▶relleno de minas (subterráneas) con zafras]); *syn.* depositar [vb] estériles; *f* **stockage [m] de stériles** (Transport et mise en place [stockage] des masses constituant les haldes et les terrils ainsi que le ▶stockage souterrain de stériles dans les galeries désaffectées ; ▶stériles) ; *g* **Bergeverbringung [f]** (Transport und profilgerechter Einbau [Endlagerung] von ▶Berge 3 in die Landschaft oder Rückführung der Berge in die Stollen — ▶Bergeversatz).

1539 disposal [n] **of pollutants in water bodies** *envir.* (▶Disposal by dilution or discharge of ▶noxious substances in rivers or the sea; ▶direct discharge, ▶dumping at sea); *s* **vertido [m] de sustancias contaminantes** (Descarga de ▶sustancias nocivas en las masas de agua; ▶vertido directo, ▶descarga de residuos en alta mar, ▶vertido por dilución); *f 1* **rejet [m] de polluants** (▶Immersion de déchets en mer ou déversement de substances polluantes dans les eaux intérieures ; ▶clapage, ▶rejet direct, ▶substance génératrice de nuisances) ; *f 2* **déversement [m] de rejets dilués** (Rejet continu dans les eaux douces superficielles continentales ou en haute mer des eaux de refroidissement des installations nucléaires, des eaux épurées par les stations d'épuration, d'effluents domestiques dilués) ; *g* **Schadstoffeinleitung [f]** (Verklappen oder Aus-

D

bringing von ▶Schadstoffen in Gewässer; ▶Direkteinleitung, ▶Verklappen von flüssigen Abfallstoffen, ▶Verklappung).

1540 disposal [n] **of radioactive waste** *envir. leg.* (Legally-authorized **d. of r. w.**; ▶final disposal site for radioactive waste, ▶final disposal, ▶nuclear fuel reprocessing plant); *s* **disposición** [f] **de residuos radi(o)activos/nucleares** (▶cementerio nuclear, ▶almacenamiento definitivo de residuos, ▶planta de reciclaje de residuos radiactivos); *f* **élimination [f] des déchets radioactifs** (La gestion des déchets radioactifs comprend le traitement, l'entreposage et le stockage des substances radioactives dangereuses ainsi que la décontamination, la valorisation et le conditionnement des déchets faiblement radioactifs ; ▶centre de stockage de déchets radioactifs, ▶stockage définitif, ▶station de traitement des déchets radioactifs) ; *g* **Entsorgung [f] des Atommülls** (Geordnete Beseitigung des Atommülls; ▶Endlager für radioaktive Stoffe, ▶Endlagerung, ▶Wiederaufbereitungsanlage für Kernbrennstoffe); *syn.* Atommüllentsorgung [f]).

disposal [n] **of surplus material** *constr.* ▶final disposal of surplus material, ▶off-site disposal of surplus material.

disruption [n]**, environmental** *ecol. envir. recr.* ▶disturbance.

1541 disruptive conditions [npl] **for fauna** *conserv. nat'res.* (Changes in the environment adversely affecting fauna); *s* **perturbación [f] de la fauna** (Factores que ponen en peligro a la fauna o causan molestias que pueden conducir a agravar su estado de conservación); *f* **perturbation [f] affectant la faune** (Facteur mettant des espèces animales en danger ou susceptible d'en aggraver le statut) ; *syn.* effets [mpl] adverses affectant les espèces animales ; *g* **Beeinträchtigung [f] der Tierwelt** (Negative Auswirkungen auf wild lebende Tiere durch Eingriffe in ihre Habitate, durch Umweltverschmutzung oder z. B. indem ihnen nachgestellt wird, sie mutwillig beunruhigt oder wenn deren Nist-, Brut-, Wohn- oder Zufluchtsstätten durch Aufsuchen, Fotografieren, Filmen oder ähnliche Handlungen gestört werden).

dissemination [n] **by animals** *phyt. zool.* ▶animal dispersal of seeds.

dissemination [n] **by man** *phyt. zool.* ▶anthropogenic dispersal.

dissemination pattern [n] *phyt.* ▶dispersion (1).

distance [n] *zool.* ▶individual distance; *constr. hort.* ▶planting distance; *plan. recr.* ▶walking distance.

distributary [n] [UK] *geo.* ▶divergent channel [US].

1542 distribution [n] *phyt. zool.* (**1.** Geographic spread of plant or animal species. **2.** The manner, pattern, or relative frequency with which organisms are dispersed among a population or over a specified area/region [pattern of spatial arrangement of individuals]; different **d.** patterns, such as regular/uniform, random, or aggregated/clustered are distinguished; NRM 1996, LE 1986, 50. **3.** The process of spreading of an organism from one place to another is called ▶**dispersal**; ▶anthropogenic dispersal, ▶animal dispersal of seeds, ▶range 1, ▶water dispersal, ▶wind dispersal); *syn.* dispersion [n] (2); *s* **distribución [f] biogeográfica** (Distribución de las especies de flora y fauna en el espacio y en el tiempo, determinada por variaciones geográficas, climáticas y ecológicas debidas a diversos procesos desde la deriva continental y tectónica, las glaciaciones, la evolución, las condiciones climáticas y orográficas causantes de barreras o la existencia de puentes intercontinentales, por nombrar los más importantes. En su conjunto la **d. b.** permanece inexplicada, aunque existen diversas hipótesis al respecto; cf. MARG 1977, DINA 1987; ▶anemocoria, ▶antropocoria, ▶área de distribución, ▶diseminación de semillas, ▶dispersión 1,

▶hidrocoria, ▶zoocoria); *f* **répartition [f] (des espèces)** (Disposition dans l'espace des organismes vivants ; la **r.** peut être géographique ou écologique [stationnelle] ; ▶aire de répartition, ▶anémochorie, ▶dissémination, ▶dissémination anthropogène, ▶dissémination des semences, ▶dissémination par les animaux, ▶hydrochorie) ; *syn.* dispersion [f] ; *g* **Verbreitung [f] (2)** (Geografisch gekennzeichnetes Vorkommen von Pflanzen- oder Tierarten als Folge und Ergebnis geologischer, orografischer, klimatischer und sonstiger ökologischer Bedingungen sowie des Zeitfaktors; die Art und Weise eines Verbreitungsmusters resp. die relative Häufigkeit des Vorkommens von Arten kann regelmäßig, vereinzelt/zufällig oder aggregativ/gehäuft sein; ▶Areal, ▶kulturbedingte Verschleppung, ▶Verbreitung 1, ▶Samenverbreitung, ▶Verdriftung durch Wasser, ▶Verdriftung durch Wind, ▶Verschleppung durch Tiere).

distribution curve [n] *constr. pedol.* ▶particle-size distribution curve.

1543 distribution limit [n] *phyt. zool.* (Maximum bounds of a species or community); *syn.* range limit [n] (TEE 1980, 110); *s* **barrera [f]** (Límite de distribución de una especie de flora o fauna); *f* **limite [f] de l'aire de répartition (d'une espèce)** (Limite de l'aire naturelle d'extension d'une espèce, d'une association à l'intérieure de laquelle celles-ci sont présentes) ; *g* **Verbreitungsgrenze [f]** (Ende eines Gebietes, bis zu dem eine Art oder Gesellschaft vorkommt).

distribution model [n] *envir.* ▶emission distribution model.

1544 distribution [n] **of plants** *geo. phyt.* (Spread of plant species in a large area; ▶range 1, ▶dispersal, ▶plant geography, ▶synchorology); *s* **distribución [f] de especies vegetales** (Presencia de una especie en una región grande; ▶área de distribución, ▶dispersión 1, ▶fitogeografía, ▶sincorología); *f* **répartition [f] des espèces** (Disposition dans l'espace des organismes vivants. Cette distribution peut dépendre de leur ▶aire de répartition [= répartition géographique] ou de leurs caractéristiques écologiques qui les inféodent plus ou moins à un type de milieu donné [= répartition stationnelle] ; ▶dissémination, ▶phytogéographie, ▶synchorologie) ; *g* **Pflanzenverbreitung [f]** (Vorkommen einer Pflanzenart in einem definierten Gebiet; ▶Areal, ▶Ausbreitung, ▶Pflanzengeografie, ▶Synchorologie).

distribution [n] **of spatial patterns** *ecol.* ▶ecological distribution of spatial patterns.

distributor road [n] [UK]**, local** *plan. trans. urb.* ▶local feeder road [US].

distributor road [n] [UK]**, peripheral** *plan. trans. urb.* ▶peripheral highway [US].

district [n] *leg. urb.* ▶agricultural district [US] (1)/agricultural land [UK], ▶central business district [US], ▶development district [US], ▶historic district [US], ▶residential zoning district [US]/residential zone [UK]; *hydr. leg. wat'man.* ▶river basin district; *leg. urb.* ▶rural district [US]/rural settlement area [UK]; *hydr. leg. wat'man.* ▶soil and water conservation district [US]/water conservation area [UK]; *leg. urb.* ▶zoning district [US].

district [n] [US]**, air pollution control** *envir. leg.* ▶air quality control region.

District [n]**, Lake** *geo. hydr. landsc.* ▶lake landscape.

district [n] [UK]**, local** *leg. urb.* ▶section of a community area [US].

District [n] [UK]**, Resource Conservation** *conserv. leg.* ▶Area of Outstanding Natural Beauty [UK].

district commission [n] *adm. conserv'hist.* ▶historic district commission [US]/Historic Buildings and Monuments Commission for England [UK].

district heating plant [n] [UK] *urb.* ▶central heating plant [US].

1545 district heating system [n] *urb.* (TGG 1984, 251; centralized provision of heating from a heating plant for residential or industrial areas; ▶central heating plant); *s* **calefacción** [f] **a distancia** (Suministro de calor y agua caliente a zonas residenciales e industriales aprovechando el calor residual de centrales eléctricas tradicionales. También existen «centrales electrotérmicas de bloque» que suministran una zona simultáneamente con energía eléctrica, calor y agua caliente; ▶central de calefacción a distancia); *f 1* **réseau** [m] **de chaleur** (Installation de production et de distribution d'énergie thermique [souvent en provenance de l'incinération des ordures ménagères ou des biogaz issus de la fermentation organique] produite par une ▶centrale de production de chaleur parfois à cogénération [production mixte de chaleur et d'électricité] pour alimenter des zones d'habitation, zones d'activités et zones industrielles) ; *syn.* réseau [m] de chauffage urbain ; *f 2* **distribution** [f] **d'énergie par réseau de chaleur** (*Contexte* approvisionnement en énergie par réseau de chaleur) ; *g* **Fernheizung** [f] (Einrichtung, die über Rohrleitungen mittels Wasser oder Dampf von einem ▶Fernheizwerk Wohn-, Gewerbe- oder Industriegebiete oder von einem Blockheizkraftwerk eine Wohnsiedlung mit Wärme beliefert); *syn.* Fernwärmeversorgung [f], Versorgung [f] mit Fernwärme.

district park [n] [UK] *recr. urb.* ▶neighborhood park [US].

district redevelopment [n] [UK] *urb.* ▶neighborhood redevelopment [US].

District Surveyor's Office [n] [US] *adm. leg.* ▶Building Permit Office [US].

1546 disturbance [n] *ecol. envir. recr.* (**1.** Negative effects on the functioning of natural systems; e.g. *specific terms* ▶preventable disturbance, ▶nonpreventable disturbance [US]/nonpreventable disturbance [UK], ▶detraction from visual quality, ▶persistent environmental disturbance, ▶severe disturbance; ▶intrusion upon the natural environment, ▶environmental stress); *syn.* disruption [n], adverse impact [n]. **2.** Human pressure that causes stressfull effects on biotic, climatic, edaphic, and topographic conditions, and reduces the sustainable functioning of an area or ecosystem is called **environmental disturbance**); *syn.* environmental disruption [n]; *s* **influencia** [f] **perturbadora** (Efectos negativos sobre el ▶régimen ecológico de la naturaleza; *términos específicos* ▶desfiguración [visual], ▶perturbación significativa, ▶influencia perturbadora persistente, ▶perturbación inevitable, ▶perturbación evitable, ▶perturbación persistente; ▶impacto ambiental, ▶intrusión en el entorno natural); *syn.* distorsiones [fpl], impacto [m] negativo, perjuicio [m] causado al régimen ecológico; *f* **nuisance** [f] (Facteur comportant un risque notable pour la santé et le bien-être de l'homme, ou qui peut atteindre indirectement celui-ci, par des répercussions sur son patrimoine naturel, culturel et économique ; *termes spécifiques* ▶nuisance considérable, ▶nuisance persistante, ▶nuisance inévitable, ▶nuisance évitable, ▶nuisance visuelle, ▶nuisance considérable ; ▶atteinte au milieu naturel) ; *syn.* perturbation [f] ; *g* **Beeinträchtigung** [f] (Negative Auswirkungen auf die Funktionsfähigkeit des Naturhaushaltes oder dessen Teile sowie auf das Wohlbefinden des Menschen; **B.** wird im Vergleich zu ▶Belastung im Allgemeinen als Störung oder graduelle Verschlechterung einer vorhandenen Qualität, Funktion oder eines Erscheinungsbildes verstanden; *Kontext* man spricht z. B. von der **B.** der Gewässergüte, aber von der Belastung eines Gewässers; die **B.** der Erholungsfunktion, aber die Belastung eines Erholungsgebietes; *UBe* ▶erhebliche Beeinträchtigung, ▶nachhaltige Beeinträchtigung, ▶unvermeidbare Beeinträchtigung, ▶vermeidbare Beeinträchtigung, ▶visuelle Beeinträchtigung 1; ▶Eingriff in Natur und Landschaft).

disturbance [n]**, long-lasting** *ecol. envir.* ▶persistent environmental disturbance.

disused land [n] [US] *plan. urb.* ▶derelict land.

ditch [n] *agr. constr.* ▶berm ditch, ▶intercepting ditch.

ditch [n]**, diversion** *constr.* ▶intercepting ditch.

ditch cleaning [n] *wat'man.* ∗▶ditch clearing.

ditch cleanup [n] [US] *wat'man.* ∗▶ditch clearing.

1547 ditch clearing [n] *wat'man.* (Maintenance of ditches, including cleaning out, deepening and widening; *specific terms* ditch cleaning, ditch cleanup [also US]; ▶sediment removal); *syn.* ditching [n] [also UK]; *s* **limpieza** [f] **de zanjas de avenamiento** (Trabajo de mantenimiento de canales de drenaje que incluyen la remoción de fango, basuras y vegetación; ▶drenaje de cursos de agua); *f* **curage** [m] **des fossés** (Travaux d'entretien des fossés de drainage ou d'assèchement consistant dans l'enlèvement des dépôts de vases, des déchets et de la végétation qui pourraient entraver le bon écoulement des eaux ; ▶curage des cours d'eaux) ; *g* **Grabenräumung** [f] (Im Rahmen der Instandhaltung von Entwässerungsgräben das Entfernen von Schlammablagerungen, Unrat und Bewuchs, die den Abflussquerschnitt vermindern; ▶Gewässerräumung).

1548 ditch corridor tree planting [n] [US] *agr. conserv.* (Linear tree planting along ditches, usually of fruit trees on the edges of improved agricultural land, which were planted to compensate for the usable land loss caused by excavation of drainage ditches; *s* **plantación-orla** [f] **de canal** (±) (Plantación de frutales a lo largo de canales de desagüe de praderas desecadas, originalmente para contrarrestar la pérdida de tierra); *f* **plantation** [f] **de lisière de fossés** (Formation ligneuse linéaire plantée en particulier en bordure des fossés de drainage afin de compenser les surfaces agricoles perdues lors des travaux d'amélioration foncière) ; *syn.* plantation [f] de bord de fossés ; *g* **Grabensaumpflanzung** [f] (Linearer [Obst]baumbestand, vor allem am Rande meliorierter Grünlandflächen; ursprünglich meist angelegt, um den Landverlust, der durch Entwässerungsgräben verursacht wurde, auszugleichen).

ditching [n] [UK] *wat'man.* ▶ditch clearing.

diurnal periodicity [n] *biol.* ▶circadian rhythm.

1549 divergent channel [n] [US] *geo.* (River branch which reaches the sea independently, as in a delta); *syn.* distributary [n] [UK]; *s* **brazo** [m] **de desembocadura** (Brazo de un río que llega al mar o a un lago independientemente del resto); *f 1* **bras** [m] **d'embouchure** (Partie d'un fleuve qui se jette dans la mer ou dans un lac) ; *f 2* **grau** [m] (*Terme spécifique dans le Midi de la France* chenal par lequel un cours d'eau, un étang débouche dans la mer ; petit lac saumâtre) ; *g* **Mündungsarm** [m] (Teil eines aufgefächerten Flusses an der Küste oder bei der Mündung in einen See, z. B. in einem Delta).

diversion ditch [n] *constr.* ▶intercepting ditch.

1550 diversity [n] *ecol.* (Generic term for the number of different species present and their relative abundance in an ecosystem or parts thereof, as well as distribution of individuals among the species; there are five principal components of diversity: [1] ▶genetic diversity [inherited varability of all living organisms], [2] ▶taxonomic diversity [variety of organisms = species diversity], [3] ▶structural diversity [heterogeneity of three-dimensional spatial elements], [4] functional **d.** [ecological services that organisms and ecosystems do for each other], [5] abiotic matrix **d.** [unity of soil, water, and air, within which organisms coexist]; ▶Convention on Biological Diversity, ▶landscape diversity, ▶richness, ▶species richness); *s* **diversidad** [f]

ecológica (Número de especies o de estructuras de un ecosistema o parte de él; se distingue entre la diversidad de especies y la ▶diversidad estructural; ▶biodiversidad, ▶Convención sobre la Diversidad Biológica, ▶diversidad genética, ▶espectro de diversidad, ▶heterogeneidad del paisaje, ▶riqueza en especies, ▶variedad); *syn.* diversidad [f] biológica; *f* **diversité [f]** (Importance numérique des espèces [diversité spécifique] et structures d'un écosystème ou parties d'écosystèmes [diversité écologique] rencontrées dans une unité structurale paysagère ; on distingue la ▶diversité des espèces et la ▶diversité structurale ; ▶convention sur la diversité biologique, ▶diversité des paysages, ▶diversité génétique, ▶variété, ▶richesse en espèces) ; *g* **Diversität [f]** (Auf Raumeinheiten bezogenes Maß für die Menge an Erscheinungsformen [Arten und Strukturen] in einem Ökosystem oder Teilen davon; cf. ANL 1991; es wird zwischen ▶Artendiversität und ▶Strukturdiversität unterschieden; ▶Artenvielfalt, ▶genetische Vielfalt, ▶landschaftliche Vielfalt, ▶Übereinkommen über die biologische Vielfalt, ▶Vielfältigkeit); *syn.* Mannigfaltigkeit [f].

diversity [n], spatial *plan.* ▶spatial heterogeneity.

diversity [n], species *ecol.* ▶taxonomic diversity.

diversity [n] amongst organisms *ecol.* ▶biodiversity.

divide [n] [US] *geo. hydr. wat'man.* ▶watershed.

diving [n] *recr.* ▶scuba diving, ▶skin diving.

1551 diving pool [n] *recr.* (Facility mostly used for high-diving competitions); *s* **piscina [f] de salto** *syn.* piscina [f] con trampolín; *f* **fosse [f] à plongeon** (Bassin profond dans une piscine réservé au saut à partir d'un plongeoir) ; *syn.* bassin [m] à plongeon ; *g* **Sprungbecken [n]** (Wasserbecken im Sportbad für Turmspringen).

division [n] *geo.* ▶physiographic division.

divisional planting [n] [US] *gard. landsc.* ▶structure planting.

division [n] of a project into partial tenders [UK] *constr. contr.* ▶division of a project into sections.

1552 division [n] of a project into sections *constr. contr.* (Wherever possible, complex construction projects are properly divided into lots and awarded separately as subcontracts. Contracts for work [US]/works [UK] involving various trades are usually awarded for each separate ▶trade; ▶construction project section, ▶construction phasing, ▶specialist trade); *syn.* division [n] of a project into partial tenders [UK]; *s* **división [f] de un proyecto en lotes** (Los proyectos de construcción complejos normalmente se subdividen en lotes y se adjudican a empresas diferentes. Cuando se trata de obras que implican a diferentes ▶oficios, éstas se adjudican individualmente a cada uno de ellos; ▶lote de oficio, ▶división en fases de construcción, ▶tramo de obra); *f* **décomposition [f] en tranches de travaux** (Fractionnement des travaux en tranches opérationnelles ; les ouvrages sont en général découpés en ▶lots techniques et confiés à differentes entreprises ; ▶tranche de travaux, ▶corps d'état, ▶phasage) ; *syn.* partition [f] des travaux en tranches (CCM 1984, 45), fractionnement [m] d'une opération en plusieurs tranches (CCM 1984, 60) ; *g* **Teilung [f] in Lose** (Untergliederung umfangreicher Bauleistungen in Bauabschnitte und deren Vergabe nach Losen [Teillose]; cf. § 4 VOB Teil A; Bauleistungen verschiedener Handwerks- oder Gewerbezweige sind in der Regel nach Fachgebieten oder Gewerbezweigen getrennt zu vergeben [▶Fachlose]; § 4 [3] VOB Teil A; ▶Baulos, ▶Gewerk, ▶Einteilung in Bauabschnitte).

division [n] of solid waste [US], collecting and recycling *adm. envir.* ▶waste management department.

1553 division [n] of spaces *gard. landsc. urb.* (Intentionally planned spatial structuring by employing plants, buildings or other structures; ▶structure planting); *syn.* creation [n] of defined spaces; *s* **creación [f] de espacios** (Por medio de plantas, edificaciones u otros elementos constructivos; ▶plantación estructurante); *f* **structuration [f] spatiale** (Délimitation ou définition d'un espace par le jeu de plantations, de bâtiments ou de la structure du bâtis ; ▶plantation structurante) ; *syn.* structuration [f] de l'espace ; *g* **Raumbildung [f]** (Abgrenzen oder Definieren eines Raumes durch Pflanzen, Gebäude oder Bauteile; ▶Rahmenpflanzung).

Document [n] [UK], Local Development *leg. urb.* ▶community development plan.

documentation [n] *prof.* ▶project documentation.

dog park [n] [US] *landsc. urb.* ▶pet exercise area [US]/dogs' loo [n] [UK].

dogs' loo [n] [UK] *landsc. urb.* ▶pet exercise area [US].

dome grate [n] [US] *constr.* ▶strainer.

1554 domestic animal protection [n] *leg.* (Governmental measures and/or program[me]s for the protection of the lives and health of animals, especially with respect to laboratory experiments and the keeping of house pets; **in UK**, under the auspices of the Royal Society for the Prevention of Cruelty to Animals (RSPCA); ▶animal species conservation); *s* **protección [f] de animales** (A diferencia de la ▶protección de especies de fauna en peligro, la **p. de a.** se dedica a proteger las condiciones de vida de los animales de laboratorio, de los de ganadería y de los domésticos); *f* **protection [f] des animaux domestiques et d'expérience** (Mesures réglementant la protection des animaux domestiques et les expériences ou recherches pratiquées sur les animaux vivants ; cf. décret n° 87-848 du 19 octobre 1987 relatif aux expériences pratiqués sur les animaux par opposition à la ▶protection des espèces animales) ; *g* **Tierschutz [m]** (Bestrebungen zum Schutze von Leben und Wohlbefinden des Tieres aus der Verantwortung des Menschen für das Tier als Mitgeschöpf. Der **T.** ist in D. durch das Tierschutzgesetz vom 25.05. 1998 geregelt, insbesondere die Tierhaltung, das Töten von Tieren, Eingriffe an Tieren, Tierversuche sowie Zucht, Halten von Tieren und Handel mit Tieren. Mit der Änderung des Artikels 20a des Grundgesetzes vom 27.07.2002 wurde der **T.** zum Staatsziel erhoben; cf. BGBl. I Nr. 53, p. 2862 v. 31.07. 2002; *zum Unterschied* ▶Tierartenschutz).

domestic refuse [n] *envir.* ▶household waste.

1555 domestic sewage [n] *envir.* (Sewage containing human excrement and household wastes, which originates from sanitary conveniences of dwelling, office building, or institution; ▶toilet wastewater [US]/foul sewage [UK]); *syn.* sanitary sewage [n]; *s* **aguas [fpl] residuales domésticas** (Conjunto de residuos líquidos precedentes de domicilios particulares, oficinas y actividades comerciales; ▶aguas negras); *f* **eaux [fpl] ménagères et eaux-vannes** (Eaux usées en provenance de la cuisine, de la salles d'eau et cabinets d'aisance [eaux ménagères ou eaux grises] et des W.-C. [▶eaux vannes ou eaux noires]) ; *g* **Haushaltsabwasser [n]** (...wässer [pl]; Abwasser aus Küche, Bad und Toilette; ▶Toilettenabwässer).

domestic waste [n] *envir.* ▶household waste.

domestic waste collection [n] [UK] *envir.* ▶household waste collection [US].

dominance [n] *phyt.* ▶degree of species cover, ▶seasonally changing visual dominance.

1556 dominant species [n] *phyt. zool.* (Species which, by their activity, behavio[u]r, or number, have considerable influence or control upon the conditions of existence of

D

associated species; a species which "controls" its habitat and ►food web; RCG 1982; ►degree of total vegetative cover); *s* **especie [f] dominante** (Especie que ejerce mayor influencia en la caracterización de una comunidad; ►expansión horizontal); *f* **espèce [f] dominante** (Espèce qui a la densité la plus élevée [règne animal] ou le taux de recouvrement le plus élevé [règne végétal]. Les espèces végétales dominantes déterminent la physionomie du groupement ; DEE 1982 ; ►degré de recouvrement) ; *g* **dominante Art [f]** (Art, die durch ihre relativ hohe Anzahl [hoher ►Deckungsgrad], ihr Verhalten, ihre Dynamik bedeutenden Einfluss auf oder Kontrolle über das Vorkommen benachbarter Arten hat. Dominante Pflanzen bestimmen z. B. das Erscheinungsbild einer ganzen Pflanzengesellschaft); *syn.* vorherrschende Art [f].

door grate [n] [US] *constr.* ►shoe-scraper.

1557 dormancy [n] *bot. hort.* (Condition of a plant or seed in which life-functions are virtually at a standstill. This is found in non-germinating seeds and non-growing buds; ►growing season, ►senescence 1); *s* **período [m] de reposo** (En los tropófitos, dícese de la época en que el vegetal pasa al estado de vida latente, durante el cual sus funciones vitales se reducen al mínimo; DB 1985; *opp.* ►periodo/período de crecimiento; ►caída de las hojas; *syn.* dormancia [f]; *f* **dormance [f]** (1. Période de vie ralentie d'un végétal qui a momentanément suspendu son développement sans perdre ses facultés végétatives ; la levée de dormance se fera grâce à un choc climatique — froid ou sécheresse ; DIC 1993 ; ►disparition des feuilles ; *opp.* ►période de végétation. 2. Passer l'hiver pour les organes des végétaux) ; *syn.* période [f] de repos végétatif ; *g* **Vegetationsruhe [f]** (Teil des Jahres, in dem Wachstumsvorgänge bei Pflanzen oder deren Diasporen stillstehen oder kaum merklich vorangehen; ►Einziehen des Laubes; *opp.* ►Vegetationsperiode); *syn.* Wachstumsruhe [f], Ruheperiode [f], Ruhezeit [f], Ruhestadium [f].

1558 dormant cutting [n] *constr. hort.* (**1.** Woody shoot severed from a parent plant during autumn, winter or early spring for vegetative propagation purposes; ►cutting, ►slip 1. **2.** A cutting including a small amount of the older wood at the base—usually of a two-years-old shoot—is called **heel cutting**); *s* **esqueje [m] leñoso** (Brote lignificado de una leñosa que se utiliza para la reproducción vegetativa; ►esqueje arrancado, ►esqueje); *f 1* **bouture [f] de rameau aoûté** (►Bouture de tige de bois mur prise sur du bois de l'année après la saison de croissance, au début de l'automne ou pendant la période de repos végétatif ; *terme spécifique* bouture d'œil en sec ; DFM 1975 ; ►bouture arrachée) ; *syn.* bouture [f] de rameau ligneux, bouture [f] de tige ligneuse (LA 1981, 191), bouture [f] ligneuse ; *f 2* **bouture [f] de rameau semi-aoûté** (Bouture de tiges au début de leur lignification, encore pourvues de feuilles prélevée en fin d'été ; *terme spécifique* bouture [f] d'œil en vert) ; *syn.* bouture [f] semi-ligneuse, bouture [f] demi-ligneuse, bouture [f] demi-aoûtée ; *g* **Steckholz [n]** (1. Ausgereifter, d. h. verholzter, meist einjähriger Trieb eines Gehölzes in unbelaubtem Zustand zur vegetativen Vermehrung. 2. Der Trieb eines Gehölzes [z. B. Lavendel, *Potentilla*], der am Ende einer Vegetationsperiode als **Kopfsteckling** zur Vermehrung geschnitten wird, heißt **nicht ganz ausgereiftes Steckholz** oder **halbhartes S.** und wird deshalb auch ►Steckling genannt; Letzteres entspricht dem Französischen *f 2*; ►Rissling); *syn.* harter Steckling [m].

1559 dormant state [n] *bot. zool.* (TEE 1980, 69; condition of organisms during ►dormancy or ►hibernation [►winter sleep]; genetically determined pause in the development of plants and animals to ensure survival during adverse environmental conditions; ►overwintering 1, ►overwintering 2); *s* **estado [m] de dormancia** (Condición de los organismos durante el ►periodo/período de reposo [flora] o la ►invernación 1 [fauna]);

f **état [m] de repos végétatif** (État caractérisant les organismes pendant la ►dormance ; ►hibernation) ; *g* **Ruhezustand [m]** (Genetisch fixierte Entwicklungsruhe bei Pflanzen und Tieren zum Überdauern ungünstiger Umweltbedingungen, z. B. als ►Vegetationsruhe [durch Keimhemmung] oder ►Überwinterung 1 [►Winterschlaf]; *UB bot.* Samenruhe); *syn.* Dormanz [f], Latenzzustand [m].

dormitory suburb [n] [UK] *urb.* ►bedroom community [US].

dotard [n] [UK] *ecol.* ►snag [US&CDN] (3).

1560 double-back bench [n] *gard. landsc.* (Bench with common back and seats facing in opposite directions); *s* **bancos [mpl] adosados** *syn.* banco [m] doble; *f* **banc [m] double** *syn.* banc [m] adossé ; *g* **Doppelbank [f]**.

1561 double digging [n] *hort.* (Method of manually cultivating soil to a depth of spades [US]/two spits [UK] without reversing the order of topsoil and subsoil; BS 3957: part 5; ►trenching, ►dig); *syn.* bastard trenching [n] [also UK]); *s* **desfonde [m] a dos palas** (Método de arado manual hasta una profundidad de dos palas sin voltear el suelo; ►subsolar en profundo, ►cavar); *f* **bêcher [vb] à deux fers** (MBG 1972 ; ►défonçage, ►bêcher) ; *syn.* défoncer [vb] à deux fers ; *g* **holländern [vb]** (zwei Spaten tief umgraben; ►Rigolen, ►umgraben); *syn.* zweischichtig umgraben [vb].

double house [n] [US] *arch. leg. urb.* ►semidetached house.

1562 double-sided masonry wall [n] *constr.* (Wall with a fair face on both sides; ►wall face); *s* **mampostería [f] con paramento de cara vista bilateral** (Muro con dos lados visibles; ►cara vista de un muro); *syn.* mampostería [f] de cara vista bilateral, mampostería [f] de dos caras; *f* **maçonnerie [f] à deux parements** (Maçonnerie de pierres dont le remplissage entre les deux faces vues appareillées est réalisé le plus souvent avec un béton maigre ou des éclats de pierres dans un mortier maigre ; VRD 1994, tome 2, partie 4, chap. 6.4.2.3, p. 36 ; il s'agit souvent de murs en béton banché de pierres ; ►parement) ; *g* **zweihäuptiges Mauerwerk [n]** (Mauer mit zwei Ansichtsflächen; ►Mauerhaupt); *syn.* doppelhäuptiges Mauerwerk [n].

1563 double-U espalier [n] *hort.* (►Espaliered fruit tree with a trunk height of 40 cm and candelabra-like, U-shaped branches, 40 cm apart, whereby the central leader has been removed; ►palmette espalier); *syn.* candelabra espalier [n] [also US]; *s* **palmeta [f] en U doble** (*En fruticultura* ►frutal en espaldera con horquilla a unos 40 cm del suelo, despunte de la flecha y ramificación en U en la que las ramas crecen a una distancia de 40 cm entre sí; ►palmeta regular [de ramas horizontales/oblicuas]); *syn.* épsilon [m]; *f 1* **palmette [f] en U double** (Palmette pour laquelle le tronc se sépare en deux puis chacune des deux branches forme un U simple ; *f 2* **palmette [f] Verrier** (Forme dans laquelle les deux U simples sont imbriqués l'un dans l'autre ; LA 1981, 480 ; ADT 1988, 49 ; ►arbre fruitier taillé en espalier, ►palmette à branches horizontales/obliques); *g 1* **Doppel-U-Spalier [n]** (*Obstbau* ►Formobstbaum mit einer Stammhöhe von 40 cm und einer U-förmigen Astverzweigung, an deren beiden Enden wiederum eine U-förmige Astverzweigung gezogen wurde; die beiden stammverlängernden Mitteltriebe wurden entfernt; alle vier Senkrechttriebe haben einen Abstand von ca. 40 cm; *g 2* **Verrierpalmette [f]** (*Obstbau* ►Formobstbaum mit einer Stammhöhe von 40 cm und einer U-förmigen, jeweils 40 cm auseinander liegenden Astverzweigung, bei dem der stammverlängernde Mitteltrieb entfernt wurde; ►Spalier mit waagerechten/schrägen Ästen).

downgrading [n] [UK] *leg. urb.* ►downzoning [US]/reduction of permissible building volume [UK].

1564 downhill ski run [n] *recr.* (Descent on skis down a ►ski run of compacted snow ['piste'], or in deep snow);

s **descenso [m] en esquís** (Bajada con esquís por una ▶pista de esquí preparada o sobre la nieve profunda); *f* **descente [f] à ski** (Descendre une pente avec des skis que ce soit sur une piste ou en neige profonde ; ▶piste de ski alpin) ; *g* **Skiabfahrt [f]** (Das Abfahren mit Skiern, sei es auf einer Piste oder im Tiefschnee; ▶Skipiste); *syn.* Hangabfahrt [f], *o. V.* Schiabfahrt [f].

1565 down mountain skiing [n] *recr.* (**1.** Alpine ski sport on compacted snow for ski runs in mountain areas. **2.** Ski sport on short ski runs down a prepared snow-covered hill is called **downhill skiing**; ▶cross-country skiing, ▶ski touring); *s* **esquí [m] alpino** (▶esquí de fondo, ▶esquí de travesía en alta montaña); *f* **ski [m] de piste** (Sport de ski alpin sur des pistes préparées ; ▶ski de fond, ▶randonnée à ski en altitude) ; *syn.* ski [m] alpin ; *g* **Abfahrtslauf [m, o. Pl.]** (Disziplin des alpinen Skisports auf präparierten Pisten in Gefällstrecken; ▶Skilanglauf, ▶Tourenskilauf).

downpour [n] [US] *met.* ▶heavy rainfall.

1566 downslope wind [n] *geo. met.* (▶mountain wind, ▶slope wind); *syn.* katabatic wind [n]; *s* **viento [m] catabático** (Viento descendiente a lo largo de las laderas, causado por el enfriamiento del suelo; ▶brisa de montaña, ▶viento de ladera); *f* **vent [m] catabatique** (Vent descendant le long d'une pente. Ce vent est associé à un refroidissement de la surface de la pente ; ▶brise de montagne, ▶vent de versant) ; *g* **Hangabwind [m]** (Den Hang herunter wehender Wind; ▶Bergwind, ▶Hangwind); *syn.* katabatischer Wind [m].

downtown [n] [US] *urb.* ▶inner city.

downtown park [n] [US] *urb.* ▶inner city park [US]/public garden [UK].

downtown regeneration [n] [US] *urb.* ▶center city redevelopment [US]/urban regeneration [UK].

1567 downzoning [n] [US] *leg. urb.* (Legal act of rezoning a tract of land for a less intensive use than the existing or permitted use; e.g, reduction in permitted number of buildings in an area, or otherwise reducing of bulk regulations, or the change in the zoning classification of higher use to one that is lower, e.g. from commercial to residential; *opp.* upzoning); *syn.* downgrading [n] [UK], reduction [n] of permissible building volume [UK]; *s* **reducción [f] del volumen edificable permitido** (Reducción legal del volumen edificable de suelo urbanizado o urbanizable); *f* **réduction [f] de la densité de construction** (Réduction des valeurs de coefficient des règles d'urbanisme définissant la densité de construction à l'intérieur d'une zone ou secteur d'un PLU/POS ; *opp.* augmentation/réévaluation de la densité de construction [F., par transfert ou report de coefficient d'occupation des sols]) ; *g* **Verringerung [f] des Maßes der zulässigen baulichen Nutzung** (Reduzierung der nach bisherigen gesetzlichen Bestimmungen gültigen Nutzungsmöglichkeiten von Bauflächen/Baugebieten; *Planerjargon* Herabstufung [f]; *OB* Änderung der zulässigen Nutzung; cf. § 42 BauGB; *opp.* Erhöhung des Maßes der baulichen Nutzung); *syn.* Abzonung [f] [CH] (RBU 1999, 131).

draft planning proposal [n] *leg. urb. trans.* ▶preliminary planning proposal.

drain [n] *agr. constr.* ▶collector drain; *constr.* ▶drop drain [US]/tumbling bay [UK], ▶fascine drain, ▶French drain, ▶lateral drain, ▶main collector drain, ▶ring drain, ▶slit drain, ▶slot trench drain, ▶trench drain [US]/channel drain [UK], ▶yard drain.

drain [n]**, carrier** *agr. constr.* ▶collector drain.

drain [n] [UK]**, channel** *constr.* ▶trench drain [US].

drain [n] [US]**, covered trench** *constr.* ▶cast iron drain with grating [US].

drain [n]**, leader** *agr. constr.* ▶collector drain.

drain [n]**, open** *agr. for.* ▶drainage channel.

drain [n] [UK]**, slit** *constr.* ▶slot trench drain.

drain [n] [AUS]**, spoon** *constr.* ▶dished channel unit.

drain [n] [UK]**, storm** *urb. wat.'man.* ▶storm sewer [US].

drain [n] [US]**, strip** *constr.* ▶cast iron drain with grating [US].

1568 drain [vb] **a fish pond** *agr.* (To remove the water from a fish pond); *s* **vaciar [vb] un estanque (de vivero)**; *f* **vider [vb] un étang** (Ensemble des opérations ayant pour objet la mise à sec d'un plan d'eau. Le résultat est un « étang en assec » ; cf. AEP 1976) ; *syn.* vidange [m] ; *g* **Fischteich ablassen [vb]** (Das Wasser aus einem Fischteich herauslaufen lassen).

1569 drainage [n] (1) *constr.* (**1.** Generic term for the removal of excess water from the ground surface or subsurface. **2.** The English term 'drainage' is different from the German term 'Drainage/Dränung', because the former is the generic term for collection and removal of both excess surface and subsurface water from land areas, whereas the German term 'Drainage/Dränung' covers only ▶subsurface drainage; ▶drainage pattern, ▶sewerage, ▶surface water drainage); *s* **drenaje [m]** (Término genérico para todas las medidas constructivas tomadas para reducir la cantidad de agua en la superficie o en el perfil del suelo; el término «drenaje» tiene significado diferente del término alemán «Dränung» ya que éste sólo se refiere al ▶drenaje subterráneo; drenar [vb]; ▶canalización, ▶drenaje superficial, ▶red de drenaje); *syn.* avenamiento [m]; avenar [vb] [Es también] [RA], desagüe [m]; *f* **évacuation [f] des eaux** (▶Évacuation des eaux superficielles. Le terme anglais « drainage » englobe l'élimination naturelle et artificielle de tous les excès d'eau tant en milieu rural qu'en milieu urbain. F., le terme « drainage » désigne l'élimination naturelle et artificielle des eaux excédentaires sur et dans les sols ; l'évacuation des eaux superficielles dans les agglomérations est définie par le terme « assainissement ». D., le terme allemand « Dränung » désigne seulement l'élimination naturelle et artificielle des excès d'eau dans le sol ; ▶drainage souterrain, ▶réseau [public] d'assainissement, ▶réseau hydrographique) ; *g* **Entwässerung [f]** (**1. D.**, *OB zu* Rohrleitungsbau, ▶Oberflächenentwässerung und ▶Dränung zur Abführung von Regen- oder Bodenwasser; ▶Entwässerungssystem; ▶Kanalisation. **2. U.K.**, im Englischen wird im Vergleich zum Deutschen unter *drainage* sowohl 'Dränung' als auch die oberirdische Ableitung des Regenwassers und die natürliche Entwässerung der Bodenschichten verstanden; **F.**, im Französischen wird unter *drainage* die ober- und unterirdische, natürliche und künstliche Entwässerung offener Böden verstanden. Die Entwässerung versiegelter Flächen in Siedlungsgebieten [Regen- und Schmutzwasser] wird durch den Begriff ▶*assainissement* abgedeckt).

drainage [n] (2) *agr.* ▶drying out (2).

drainage [n] (3) *agr.* ▶agricultural drainage; *met.* ▶cold air drainage; *agr.* ▶field drainage; *constr.* ▶grid pattern drainage, ▶herringbone pattern drainage; *agr. eng. wat'man.* ▶land drainage; *agr. constr.* ▶mole drainage, ▶subsurface drainage; *constr. urb. wat.' man.* ▶surface water drainage.

drainage [n]**, impeded** *pedol.* ▶impeded water.

drainage [n]**, soil** *agr. constr.* ▶subsurface drainage.

drainage [n] [US]**, storm water** *constr. urb. wat.' man.* ▶surface water drainage.

drainage area [n] *geo. hydr. nat'res.* ▶drainage basin.

1570 drainage basin [n] *geo. hydr. nat'res.* (Area of land which drains water with organic matter, dissolved nutrients and sediments into a collecting watercourse or waterbody; ▶water-

D

shed); *syn.* drainage area [n], catchment basin [n], catchment area [n], river basin [n] [US]/river-basin [n] [UK], watershed [n] [also US]; *s* **cuenca [f] hidrográfica** (Territorio en el cual las aguas fluyen al mar a través de una red de cauces secundarios que convergen en un cauce principal único; ►divisoria de aguas); *syn.* cuenca [f] vertiente, cuenca [f] de drenaje, cuenca [f] de desagüe, cuenca [f] receptora, cuenca [f] fluvial; cuenca [f] de captación [RA, CO, MEX], cuenca [f] imbrífera [RA], área [f] de drenaje [EC, PE] (MESU 1977); *f* **bassin [m] versant** (Aire d'alimentation à l'intérieur de laquelle les précipitations se dirigent vers un cours d'eau et ayant pour limite la ligne de crête ou ►ligne de partage des eaux le séparant du **b. v.** voisin) ; *syn.* bassin [m] hydrographique, bassin [m] d'alimentation, bassin [m] de drainage, impluvium [m] ; *g* **Abflussgebiet [n]** (Einzugsgebiet eines Fließgewässers mit seinem gesamten Oberflächenabfluss und all seinen organichen Substanzen, gelösten Nährstoffen und Sedimenten, die an einer einzigen Flussmündung, einem Ästuar oder Delta ins Meer gelangen; ►Wasserscheide); *syn.* Entwässerungsgebiet [n], Flussgebiet [n], Stromgebiet [n], Wassereinzugsgebiet [n].

1571 drainage channel [n] *agr. for.* (Open drain, with mostly flowing water; ►drainage ditch, ►receiving stream); *syn.* open drain [n]; *s* **canal [m] de avenamiento** (Canal artificial a cielo abierto que generalmente contiene agua; ►curso receptor, ►zanja de avenamiento); *syn.* canal [m] de desagüe; *f* **canal [m] de drainage** (fossé ouvert, à plafond, en général à écoulement pérenne ; ►émissaire, ►milieu récepteur) ; *g* **Entwässerungsgraben [m] (1)** (Offener Graben, meist mit fließendem Wasser; ►Entwässerungsgraben 2; ►Vorfluter).

drainage channel [n]**, cold air** *landsc.* ►cold air drainage corridor.

drainage channel [n] [UK]**, natural hillside** *geo.* ►natural drainage way [US].

1572 drainage cleanout pipe [n] [US] *constr.* (Shaft for inspection of subsurface drain[age] lines); *syn.* drain cleanout pipe [n] [US], drain(age) inspection shaft [n] [UK]; *s* **registro [m] (de drenaje)** (Pozo para inspeccionar las condiciones subterráneas de un drenaje); *f* **regard [m] de drainage** ; *g* **Dränschacht [m]** (*o. V.* Drainschacht).

drainage corridor [n] *landsc.* ►cold air drainage corridor.

drainage course [n] [UK] *constr.* ►drainage layer.

1573 drainage depth [n] *constr.* (Distance from invert level to ground surface); *s* **profundidad [f] de drenaje** (Distancia entre la solera del tubo de drenaje y la superficie del suelo); *f* **profondeur [f] de drainage** (Mesure prise entre le niveau inférieur du drain fond de tranchée — et le niveau supérieur du terrain) ; *syn.* profondeur [f] de drain ; *g* **Dräntiefe [f]** (Das Maß von der Rohrunterkante [Grabensohle] bis zur Erdoberfläche); *syn. o. V.* Draintiefe.

1574 drainage ditch [n] *agr. constr.* (**1.** Trench for draining the soil; ►French drain, ►berm ditch); *syn.* drainage trench [n]; **2. dyke [n] [UK]** (Artificial watercourse for draining marshy land and moving surface water. Usually 2-4m wide; found beside roads and between fields; draining into larger drains, sewers, rivers, or the sea. Named originally in East Anglia [Norfolk and Suffolk] and Fenland; HASL 1981, 146); *syn.* carrier ditch [n], dike [n] [US]; *s 1* **zanja [f] de avenamiento** (Canal artificial a cielo abierto de pequeñas dimensiones, construido en tierra o roca, para desagüe del suelo; ►dren francés, ►surco en curva de nivel); *s 2* **canal [m] de desagüe de zona pantanosa** (Un curso de agua artificial que sirve para drenar zonas pantanosas y aguas superficiales. Normalmente de 2-4 m de ancho, se encuentra a menudo al lado de carreteras y entre los campos de cultivo; descarga en drenes grandes, en alcantarillas, ríos o en el mar. El

término inglés *«dyke»* se aplica a los **c. de d.** existentes en zonas pantanosas de East Anglia y Fenland); *f* **fossé [m] de drainage** (Tranchée ouverte peu profonde sans plafond ou à plafond réduit, servant à l'assainissement des sols trop humides ; ►tranchée drainante, ►fossé gradin) ; *g* **Entwässerungsgraben [m] (2)** (Offener, meist kein Wasser führender, flacher Graben zur Entwässerung des Bodens; ►Entwässerungsgraben 3, ►Entwässerungsgraben in einer Böschung).

drainage grate [n] [US] *constr.* ►storm drain grate [US].

drainage inlet structure [n] [US] *constr. eng.* ►surface inlet [US].

drainage inspection shaft [n] [UK] *constr.* ►drainage cleanout pipe [US].

drainage inspection shaft [n] [UK]**, earth-covered** *constr.* ►earth-covered drainage cleanout pipe [US].

1575 drainage layer [n] *constr.* (Subsurface layer of permeable material to collect and lead away surface water percolating from above, through a pipe system into a receiving stream, where the ground beneath is impermeable, or a water receiving layer on rooftop gardens; ►filter layer); *syn.* drainage course [n] [also UK]; *s* **capa [f] de drenaje** (Capa subterránea de material permeable para recoger y evacuar las aguas de percolación en el caso de que el subsuelo o el tejado portante sean impermeables; ►capa filtrante 2); *syn.* manto [m] de drenaje, capa [f] permeable, capa [f] de desagüe; *f* **couche [f] drainante** (Terme générique pour toute couche utilisée pour le captage des eaux d'infiltration et leur évacuation vers un collecteur, mise en œuvre sur sous-sol imperméable ; ►couche filtrante, ►milieu récepteur) ; *g* **Dränschicht [f]** (**1.** Schicht über einem nicht ausreichend durchlässigem Bau- oder Untergrund zur Aufnahme von Sickerwasser und zur Weiterleitung des Überschusswassers durch Dränleitungen in einen ►Vorfluter. **2.** Wasser aufnehmende Schicht auf Dachgärten; ►Filterschicht 2; *o. V.* Drainschicht).

drainage line [n] *constr.* ►subsurface drainage line.

drainage line [n]**, collector** *agr. constr.* ►collector drain.

drainage line [n]**, main** *agr. constr.* ►collector drain.

drainage lines [n]**, spacing of** *constr.* ►spacing of lateral drains.

drainage net [n] [UK] *geo.* ►drainage pattern.

drainage [n] **of soil** *pedol.* ►water permeability.

1576 drainage pattern [n] *geo.* (Arrangement of rivers and their tributaries within a ►drainage basin; *specific terms* river system, stream system); *syn.* drainage system [n] (2) [also UK], drainage net [n] [also UK]; *s* **red [f] de drenaje** (Disposición de los cursos fluviales y sus afluentes en una ►cuenca hidrográfica; *término específico* red fluvial); *f* **réseau [m] hydrographique** (Ensemble de canaux de drainage naturels, permanents ou temporaires, par où s'écoulent les eaux provenant du ruissellement soit par restitution continue le long du lit d'un cours d'eau soit par restitution par les nappes souterraines sous forme de sources ; CILF 1978 ; ►bassin versant) ; *g* **Entwässerungssystem [n] (1)** (Form der Fließgewässeranordnung zueinander in einem ►Abflussgebiet; *UB* Flusssystem).

drainage permeability [n] **of soil** *pedol.* ►water permeability.

drainage piping pattern [n] [US] *constr.* ►subsurface drainage system.

drainage-poor soil [n] *agr. hort.* ►poorly drained soil.

1577 drainage swale [n] *constr. eng.* (Drainageway with a smooth, shallow, and relatively wide, often parabolic cross-section, either paved or usually grass-covered which allows surface water run-off to percolate into the soil. A grass **d. s.** often has a gravel-filled trench underneath; ►infiltration swale, ►ponding area, ►receiving stream); *s* **canal [m] de drenaje (subterráneo)** (Conducto de drenaje de poca profundidad y frecuentemente de sección parabólica que puede estar relleno de grava o pavimentado; cf. DIS 1993, 90; ►acequia de infiltración, ►curso receptor); *f 1* **fossé [m] de pied de talus** (Fossé large et peu profond en pied de talus de faible pente à fond pavé ou à couvert enherbé sur sous-couche de gravier pour l'évacuation des eaux de ruissellement ; ►noue d'infiltration, ►cuvette d'infiltration, ►milieu récepteur) ; *f 2* **noue [f] de drainage** (Large fossé à faible pente, élément constitutif de tout un réseau souvent relié à un ou plusieurs petits bassins paysagers d'infiltration contribuant à soulager le réseau d'assainissement et à dépolluer naturellement les eaux d'écoulement de la voirie dans les lotissements ; fait souvent partie des prescriptions paysagères du *« Bebauungsplan »* [PLU/POS allemand]) ; *g* **Entwässerungsmulde [f]** (Ausgepflasterte oder mit einer Kiesschüttung unter der Grasnarbe versehene muldenartige Rinne entlang einer Böschung zur Abführung von Oberflächenwasser in einen ►Vorfluter; ►Versickerungsmulde).

1578 drainage system [n] (1) *constr.* (Designed network of pipes, channels or ditches for the removal of excess surface and/or soil water from a specific area; ►subsurface drainage system); *s* **sistema [m] de drenaje** (Red de tubos o canales construidos para evacuar las aguas superficiales o edáficas de un área específica; ►sistema de drenaje subterráneo); *syn.* sistema [m] de avenamiento, sistema [m] de desagüe; *f* **système [m] de drainage** (Terme générique pour l'ensemble des dispositifs de drainage superficiels ou enterrés d'une parcelle ou d'un ensemble de parcelles ; ►système de drainage souterrain) ; *syn.* réseau [m] de drainage ; *g* **Entwässerungssystem [n]** (2) (Gesamtheit der zur Oberflächen- oder Bodenentwässerung eines Gebietes oder einer Fläche angelegten, miteinander verbundenen Leitungen oder Gräben; ►Dränsystem).

drainage system [n] (2) [UK] *geo.* ►drainage pattern.

drainage system [n] (3) *envir. urb.* ►storm water drainage system.

drainage system [n] [US]**, underground storm** *envir. urb.* ►storm water drainage system.

drainage trench [n] *agr. constr.* ►drainage ditch.

drainage way [n] *geo.* ►natural drainage way [US].

drain cleanout pipe [n] [US] *constr.* ►drainage cleanout pipe.

drained granular material [n] [US] *constr. pedol.* ►well-drained soil.

drain fill [n] *constr.* ►granular drain fill.

drain grate [n] [US] *constr.* ►storm drain grate [US].

drain hole [n] *constr.* ►weep hole.

drain infiltration [n] *constr.* ►trench drain infiltration.

drain inlet [n] [US] *constr.* ►catch basin inlet [US].

drain inspection shaft [n] [UK] *constr.* ►drainage cleanout pipe [US]/drain(age) inspection shaft [UK].

drain inspection shaft [n] [UK]**, earth-covered** *constr.* ►earth-covered drainage cleanout pipe [US].

1579 drain invert level [n] *constr.* (Elevation of the bottom of a drain pipe or a drain ditch/swale; ►invert elevation); *s* **invert [m]** (Parte más baja de una tubería; ►nivel de fondo); *syn.* solera [f] [EC], cubeta [f] [MEX] (MESU 1977); *f* **génératrice [f] inférieure extérieure du tuyau** (*Contexte* niveau génératrice inférieure extérieure) ; *g* **Rohrsohle [f]** (Unterster Bereich eines äußeren Rohrquerschnittes; ►Sohlenhöhe).

drain line [n]**, subsurface** *constr.* ►subsurface drainage line.

1580 drain outfall [n] *constr.* (Final outlet for storm water into a ►receiving stream); *syn.* drain outlet pipe [n]; *s* **emisario [m] (de drenaje)** (Salida final de un dren al ►curso receptor); *f* **bouche [f] du collecteur** (dans l'►émissaire ; ►milieu récepteur) ; *g* **Dränausmündung [f]** (Öffnung eines Dränstranges in den ►Vorfluter; *o. V.* Drainausmündung); *syn.* Dränabflussstelle [f].

drain outlet pipe [n] *constr.* ►drain outfall.

1581 drain pipe [n] *constr.* (Generic term for ►clay pipe, perforated plastic pipe or perforated concrete pipe for the removal of subsurface water; ►subsurface drainage line); *syn.* interceptor pipe [n] [also US]; *s* **tubo [m] de drenaje** (Término genérico para ►tubo de arcilla, tubo de hormigón poroso y tubo de plástico perforado o ranurado; ►dren); *syn.* tubo [m] de avenamiento, caño [m] de avenamiento; *f* **tuyau [m] de drainage** (►Drain en poterie, tuyau perforé en béton ou en P.V.C. utilisé pour la réalisation d'un ►drain) ; *syn.* tuyau [m] drainant ; *g* **Dränrohr [n]** (►Tonrohr, perforiertes Beton- oder Kunststoffrohr zum Verlegen einer ►Dränleitung; *o. V.* Drainrohr).

drain pipe with woven fabric soil separator sleeve [n] *constr.* ►perforated drain pipe with woven fabric soil separator sleeve.

drain [n] **with a metal grill(e)** [US]**, strip** *constr.* ►cast iron drain with grating [US].

1582 drawdown area depth [n] *wat'man.* (Zone on banks of watercourses or lake shores which is characterized by ►fluctuating water levels); *s* **amplitud [f] máxima de la zona de nivel variable de agua** (Altura de las partes de las márgenes de ríos, arroyos y lagos afectada por ►nivel variable del agua); *f* **amplitude [f] de la zone des variations du niveau des eaux** (Zone de la berge d'un plan d'eau ou d'un cours d'eau caractérisée par des ►plans variables du niveau d'eau [entre les hautes et les basses eaux]) ; *syn.* amplitude [f] de la zone de balancements des eaux ; *g* **Höhe [f] der Wasserwechselzone** (Uferbereich an Seen und Flüssen, der durch einen ►wechselnden Wasserspiegel gekennzeichnet ist); *syn.* Schwankungshöhe [f] der Wasserwechselzone.

drawdown [n] **of groundwater table** [US] *constr. wat'man.* ►lowering of groundwater table.

1583 drawdown strip [n] *limn.* (Portion of a stream bank or lake shore which occasionally dries up in summer and has different water levels according to the time of year); *s* **zona [f] de nivel variable de agua** (Parte de las márgenes de ríos, arroyos o lagos inundada durante gran parte del año que en verano se deseca por reducción del caudal afluente); *f* **zone [f] à plan d'eau variable** (Zone des bordures des étangs et des cours d'eau inondées ou engorgées durant une partie importante de l'année avec des sols plus ou moins asséchés en été) ; *syn.* zone [f] de balancement des eaux ; *g* **Wasserwechselzone [f]** (Im Sommer zeitweilig austrocknender Uferbereich von Still- oder Fließgewässern mit im Jahresverlauf unterschiedlichen Wasserhöhen).

1584 drawing [n] *plan.* (Graphic presentation; e.g. picture, sketch, delineation, or design; ►construction drawing, ►detail drawing, ►detailed design drawing, ►isometric drawing, ►perspective drawing); *s* **dibujo [m]** (►dibujo en detalle, ►dibujo en perspectiva, ►plano de ejecución, ►plano [detallado] de proyecto, ►proyección isométrica); *syn.* croquis [m]; *f* **dessin [m]** (►isométrie, ►perspective, ►plan d'avant-projet détaillé [d'A.P.D.], ►plan de détail, ►plan d'exécution [des ouvrages]) ; *syn.* croquis [m] ; *g* **zeichnerische Darstellung [f]** (i. S. v.

Zeichnung; ▶Ausführungszeichnung, ▶Detailplan, ▶Entwurfs-zeichnung, ▶Isometrie, ▶Perspektive).

drawing [n]**, working** *constr.* ▶construction drawing.

1585 dredge and fill permit [n] [US] *adm. leg. wat'man.* (Issue of an order giving a right, according to water regulations, to the use of water, alterations to a body of water, its channel, banks, or abutting land as well as to interference within the natural water cycle. **In U.S.,** a **d.&f. p.** is issued jointly by the E.P.A.—or their State designee—and the Army Corps of Engineers); *s 1* **permiso** [m] **de construcción hidráulica o de explotación de las aguas continentales** (Aprobación de una medida constructiva en un curso de agua o de un uso específico según la ▶legislación de las aguas); *s 2* **concesión** [f] **de uso privativo de aguas** [Es] (Art. 72 Ley de Aguas); *s 3* **concesión** [f] **de agua para riego** [Es] (Art. 60 Ley de Aguas); *f* **autorisation** [f] **au titre de la police des eaux** (Autorisation administra-tive relative aux installations, ouvrages, travaux ou activités soumis à la police des eaux et à autorisation ou déclaration en vertu de la nomenclature Eau ; décret n° 93-743 du 29 mars 1993 qui fixe les seuils d'autorisation et de déclaration suivant les dangers qu'ils représentent et la gravité de leurs effets sur la ressource en eau et les systèmes aquatiques ; ▶droit de l'eau) ; *g* **wasserrechtliche Genehmigung** [f] (Nach dem ▶Wasserrecht geregelte behördliche Genehmigung für die Nutzung oder Ver-änderung eines Gewässers oder einen Eingriff in den Wasser-kreislauf); *syn.* wasserrechtliche Bewilligung [f] [A].

1586 dredging [n] (1) *min.* (**1.** Extraction of gravel or sand from a pit below ground-water level. **2.** Extraction of gravel or sand from a pit by pipes, whereby the water also acts as a means of transport, is called **hydraulic excavation**; *specific term* ▶gravel dredging; *opp.* ▶dry excavation); *s* **dragado** [m] **sub-acuático** (Extracción de arena y grava por dragado bajo la capa freática; *término específico* ▶extracción de grava por dragado; *opp.* ▶extracción en seco [de rocas y arenas]); *f* **extraction** [f] **de matériaux alluvionnaires en nappe** (Extraction de gravier ou de sable en nappe alluviale ou en dessous du niveau de la nappe phréatique ; *termes spécifiques* extraction de matériaux alluvion-naires dans le lit mineur d'un cours d'eau/dans un plan d'eau, ▶dragage de gravier, extraction de sable/gravier en mer ; ▶extraction de matériaux alluvionnaires hors nappe) ; *syn.* exploitation [f] de roches alluvionnaires en eau, extraction [f] en site immergé, extraction [f] en eau ; *g* **Nassbaggerung** [f] (Ge-winnung von Kies oder Sand in einer Entnahmestelle unterhalb des Grundwasserspiegels; *UB* ▶Nassauskiesung; *opp.* ▶Tro-ckenabbau); *syn.* Nassabbau [m].

1587 dredging [n] (2) *wat'man.* (Removal of mud or sedi-ments from settling basins, ponds, lakes, river channels, canals, or harbo[u]rs; ▶desilting); *s* **dragado** [m] **de lodo/fango** (Remoción del lodo del fondo de ríos, lagos, canales, puertos o tanques de sedimentación; ▶dragado de lagos colmatados); *f* **désenvasement** [m] (Enlèvement des dépôts sédimentaires d'un étang, bassin de décantation, d'un lac comblé par des atterris-sements ; *terme spécifique* dragage des atterrissements ; ▶curage d'un lac, ▶dragage d'un lac) ; *g* **Schlammräumung** [f] (Entfer-nung von Schlamm/Sedimenten aus Absetzbecken, Teichen, verlandeten Seen oder zur Erhaltung der Solltiefe für die Schiff-fahrt auf Flüssen oder in Hafenbecken; ▶Ausbaggerung verlandeter Seen); *syn.* Entschlammung [f], Entlandung [f].

1588 dredging [n] **of navigable channels** *wat'man.* (Removal of mud and debris to create or maintain deeper channels in waterways or harbo[u]rs; *s* **dragado** [m] **en profundo** (Medida técnica de remoción de los sedimentos del fondo de cauces para mejorar su navigabilidad); *f* **approfondis-sement** [m] **du chenal** (Action de surcreusement du lit des

fleuves, de nettoyage des fonds des chenaux fluviaux pour la mise à gabarit afin d'assurer la navigation fluviale ; *termes spécifiques* dégravoiement, dessablement) ; *g* **Eintiefung** [f] **von Gewässerfahrrinnen** (Technische Maßnahmen, die der Anpas-sung von tiefer gelegten Fahrrinnen an die Schifffahrt dienen); *syn.* Ausbaggerung [f] von Gewässerfahrrinnen, Fahrrinnen-eintiefung [f], Gewässerfahrrinnenausbaggerung [f], Gewässer-fahrrinneneintiefung [f], Vertiefung [f] von Fahrrinnen.

dressed and matched boards [npl] *constr.* ▶tongue and groove.

dressed flagstone [n] *constr.* ▶flag cut to shape.

dressing [n] **for turfing** *constr.* ▶topdressing for turfing [US]/top-dressing for turfing [UK].

dressing [n] **of stone** [UK] *constr.* ▶tooling of stone [US].

1589 dressing [n] **with compost** *agr. constr. hort. syn.* mulching [n] with compost; *s* **fertilización** [f] **con compost**; *f* **fertilisation** [f] **avec apport de compost** ; *g* **Düngung** [f] **mit Kompost** (mit Kompost düngen [vb]); *syn.* Abdecken [n, o. Pl.] mit Kompost, Düngen [n, o. Pl.] mit Kompost.

drift [n] *geo.* ▶sediment (1).

drifted debris [n] *envir.* ▶accumulation of drifted debris.

1590 drifting sand dune [n] *geo.* (Dune which moves in the direction of the wind by the shifting of sand from the windward side of the dune to the leeward side due to the absence of vegetation cover; ▶stabilization of sand dunes, ▶parabolic dune, ▶barkhan); *syn.* shifting dune [n]; *s* **duna** [f] **viva** (Duna que avanza en el espacio bajo la acción del viento por falta de cubierta vegetal que la estabilice; cf. DINA 1987; *opp.* duna muerta; ▶fijación de dunas, ▶duna parabólica, ▶barkhana); *syn.* duna [f] funcional; *f* **dune** [f] **mouvante** (Dune en mouvement sous l'action du vent par suite de l'absence d'un couvert végétal ; ▶fixation des dunes, ▶dune parabolique, ▶barkhane) ; *syn.* dune [f] mobile, dune [f] vive ; *g* **Wanderdüne** [f] (Düne, die sich wegen fehlender Vegetationsbedeckung in Windrichtung bei starkem Sandtransport über die Luv- zur Leeseite bewegt; *opp.* ortsfeste Düne; ▶Festlegung von Dünen, ▶Parabeldüne, ▶Sicheldüne).

1591 driftline [n] *phyt.* (Organic flood line deposition [▶drift-line litter] on sea beaches or on flat banks of a river, forming an irregular, low mound); *s* **orla** [f] **de cúmulos de residuos orgánicos** (Franja de materiales orgánicos depositados en playas o en orillas de ríos; ▶residuos de deriva de la orla de cúmulos); *f 1* **ourlet** [m] **halonitrophile** (Amas de matériaux organiques déposés par les marées sur le haut de plage et formant un alignement de tas) ; *f 2* **haut** [m] **de plage** (Partie de plage qui n'est recouverte par la mer qu'aux grandes marées et sur laquelle se déposent des amas de matériaux organiques [▶laisses de mer] ; la ligne jalonnée par les débris abandonnés par la mer sur le haut de plage est souvent appelée laisse de mer ; VOG 1979, 114) ; *syn.* zone [f] de dépôts organiques des lignes de marée (GEHU 1984, 60) ; *g* **Spülsaum** [m] (Unregelmäßiger, niedriger Materialwall aus angeschwemmten [▶Spülsandgetreibsel], meist organischen Stoffen, die vom Meer an den Strand oder vom Fluss an flache Ufer gespült wurden); *syn.* Flutmarke [f].

driftline colonizer [n] *phyt.* ▶driftline plant.

1592 driftline community [n] *phyt.* (Lush-growing annual forb formation on organic high tide deposits rich in proteins at the flood line; e.g. sea-rocket community *[Cakiletalia maritimae]* on beaches and beggar-ticks community *[Bidentetalia]* on river-banks*)*; a **strandline community** [n] is a forb formation on a strand floodmark); *syn.* coastal floodmark vegetation [n]; *s 1* **vegetación** [f] **halófila de los cúmulos de residuos orgá-nicos** (Comunidad vegetal generalmente de plantas anuales que se presenta sobre acumulaciones orgánicas ricas en proteínas y bien húmedas en las orlas de las playas y que pertenece a la clase *Cakiletalia maritimae*; cf. BB 1979); *s 2* **vegetación [f] de las acumulaciones de restos orgánicos** (Comunidad vegetal, rica en especies anuales, que se presenta sobre la orla de acumulaciones orgánicas en las orillas de ríos y que pertenece a la clase *Bidentetalia*); *f 1* **flore [f] d'ourlet halonitrophile** (Végétation à forte proportion d'annuelles se développant sur le sable enrichi en nitrates provenant de la décomposition des algues et des débris rejetés par la mer, périodiquement inondé en bordure des grèves littorales maritimes, p. ex. le Cakile maritime *[Cakiletalia maritimae]*) ; *syn.* flore [f] du haut de plage, groupement [m] d'ourlet halonitrophile ; *f 2* **flore [f] des grèves (alluviales)** (Végétation des vases argilo-calcaires des berges de grandes rivières et des grèves d'étangs eutrophes à niveau variable, riche en espèces annuelles comme le Chanvre d'eau *[Bidens tripartita]* appartenant à l'ordre du *Bidentetalia tripartitae*) ; *syn.* végétation [f] des berges, groupement [m] des grèves alluviales à Bidens, groupement [m] à Bidens ; *g* **Spülsaumflur [f]** (Oft üppig wachsende, annuelle Pflanzengesellschaft auf meist stark eiweißhaltigen und gut durchfeuchteten Hochwasserablagerungen von Pflanzen- und Tierresten an den Flutmarken der Strände, z. B. Ordnung der Meersenfgesellschaften *[Cakiletalia maritimae]* oder an Ufern von Fließgewässern [z. B. die halbruderale Zweizahnflur — *Bidentetalia])*; *syn.* Spülsaumgesellschaft [f].

1593 driftline litter [n] *phyt.* (Organic deposition on sea beaches or on flat river banks forming a floodmark. The material, rich in proteins and moisture, originates from plants and animals and is a nutrient-rich growing medium for the germination of annuals and a habitat for specialized small fauna); *syn.* floodmark litter [n] (VIR 1982, 310), driftline material [n]; *s* **residuos [mpl] de deriva de la orla de cúmulos** (Deposición orgánica en playas marinas u orillas de ríos formando una línea de marea. El material de origen animal o vegetal, rico en proteínas y humedad, es un medio rico en nutrientes para la germinación de plantas anuales y como hábitat para muchas especies de fauna menor [microorganismos y crustáceos] especializadas en los medios arenosos); *f* **laisses [fpl]** (Débris organiques et d'origine anthropique laissés le long des plages ou des berges à pente faible de lacs et de cours d'eau ; ceux-ci lorsqu'ils ne sont pas polluées, constituent un medium abritant et nourrissant de nombreuses espèces vivant dans le sable [micro-organismes et crustacés] ; une fois dégradé ils favorisent la germination de plantes annuelles [Natura 2000 type d'habitat naturel 1210 — végétation annuelle des laisses de mer] et contribuent à la fixation des plages ; essentiellement ; *termes spécifiques* laisses [fpl] de mer, laisses [fpl] des hautes eaux, laisses [fpl] côtières, laisses [fpl] de cours d'eau) ; *g* **Spülsaumgetreibsel [n]** (An den Flutmarken flacher Küsten und Ufer der Seen und Flüsse angeschwemmte, meist stark eiweißhaltige und gut durchfeuchtete Hochwasserablagerungen von Pflanzen und Tieren, die nach der Zersetzung ein nährstoffreiches Medium für die Keimung einjähriger Pflanzenarten und Lebensstätten für spezielle Kleintiere sind; an Meeresstränden bilden sich z. B. Meersenf-Gesellschaften *[Cakiletea maritimae]* und Bestände verschiedener Meldenarten *[Atriplex]* — Natura-2000-Code 1210).

1594 driftline plant [n] *phyt.* (Usually an annual plant growing on the ►driftline); *syn.* driftline colonizer [n]; *s* **planta [f] de línea de playa** (Planta generalmente anual que crece en la ►orla de cúmulos de residuos orgánicos); *f 1* **plante [f] d'ourlet halonitrophile** (Plante basse, annuelle, halophile, souvent nitrophile se maintenant sur le ►ourlet halonitrophile) ; *syn.* plante [f] de haut de plage ; *f 2* **plante [f] des grèves alluviales** (Plante basse en général annuelle croissant sur le ►haut de plage) ;

g **Spülsaumpflanze [f]** (Meist annuelle Pflanze des ►Spülsaums); *syn.* Spülsaumbesiedler [m].

1595 driftwood [n] *envir.* (Floating wood on a watercourse, or wood debris washed ashore on a beach or riverbank; ►driftline, ►snag 4); *s* **madera [f] flotante** (Restos de árboles o trozos de madera que flotan en los ríos o en el mar y pueden ser depositados en las orillas por efecto de las olas; ►orla de cúmulos de residuos orgánicos); *syn.* maderada [f]; *f* **bois [m] flottant** (Bois dérivant sur l'eau ; ►amas de bois, ►embâcle de bois) ; *g* **Treibholz [n, o. Pl.]** (Auf dem Wasser schwimmendes oder an Rändern von Fließgewässern, Seen oder an Stränden angeschwemmtes Holz; ►Spülsaum, ►Holzhindernis); *syn.* Schwemmholz [n] [CH].

drilling [n] *min.* ►off-shore drilling.

drilling [n]**, prospective** *geo. min.* ►test boring.

1596 drinking water [n] (1) *wat'man.* (Water which is suitable for drinking. **In U.S.**, the federal Drinking Water Act of 1973, as amended, and the Clean Water Act of 1976 govern all potable water supplies; **in Europe**, the potable water supplied to the public is controlled by the E.E.C. Directive on The Quality of Water Intended for Human Consumption 80/778/EEC. OJ No L 229. Water quality standards to be met relate to: 1. organoleptic factors, e.g. colo[u]r, turbidity, taste, odo[u]r, pH, conductivity, etc. 2. Physico-chemical factors, e.g. chloride, sulphate, calcium, magnesium, dissolved oxygen, etc. 3. Substances undesirable in excessive amounts, e.g. nitrate, ammonia, fluoride, iron, manganese, etc. 4. Toxic substances, e.g. arsenic, mercury, lead, cadmium, etc. 5. Microbiological parameters. Till 1990, the EEC had 13 Directives on water quality); *syn.* potable water [n] [also US]; *s* **agua [f] potable** (Agua adecuada para el consumo humano procedente de acuíferos, de ríos, embalses o de aguas salobres o saladas desalinizadas y mejoradas que debe reunir condiciones mínimas de calidad e higiene; en la Europa Comunitaria éstas están reguladas por la Directiva 80/778/EEC; en España por la Reglamentación técnico-sanitaria sobre la calidad de las aguas potables para consumo público; cf. RD 1423/1982); *syn.* agua [f] bebible [RCH]; *f* **eau [f] potable** (Eau destinée à la consommation humaine prélevée dans le milieu naturel à une profondeur suffisante et à partir de couches de bonne capacité filtrante, de parfaite qualité et à l'abri de toute contamination [eaux souterraines ou eaux de source]. Les prélèvements dans les eaux superficielles doivent satisfaire après traitement à des exigences de qualité précises [décret du 3 janvier 1989] en ce qui concerne les paramètres organoleptiques, physico-chimiques [magnésium, sodium, chlorures, etc.], microbiologiques, les paramètres concernant les substances indésirables [nitrates, nitrites, etc.], les substances toxiques [arsenic, cadmium, cyanure, chrome, etc.], les pesticides et produits apparentés ; la gestion de la distribution d'eau potable est assurée en général soit par la collectivité, soit par un organisme privé lié à la commune par un contrat de concession ou d'affermage) ; cf. dir. du Conseil n° 80/778/CEE relative à la qualité des eaux destinées à la consommation humaine ; cette directive sera abrogée le 26 décembre 2003 (cf. dir. n° 98/83 du 3 novembre 1998) ; *g* **Trinkwasser [n]** (Für den menschlichen Gebrauch geeignetes Wasser aus genügender Tiefe und ausreichend filtrierenden Schichten von einwandfreier Beschaffenheit, das dem natürlichen Wasserkreislauf entnommen [Grund- oder Quellwasser] und in keiner Weise beeinträchtigt wurde. Anderes Wasser [Oberflächenwasser von Seen und Flüssen, Uferfiltrat, entsalztes Meerwasser] wird zu Trinkwasser, wenn durch Aufbereitungsmaßnahmen, von Ausnahmen abgesehen, die entsprechende Zusammensetzung erreicht wird [bestimmte Mindestkonzentration an Calcium-, Magnesium- und Hydrogencarbonat-Ionen, einen geringen Gehalt an organischen Verbindungen und mit Sauerstoff gesättigt]; **in D.** sind die

D

Anforderungen an Beschaffenheit und Aufbereitung des Trinkwassers sowie die Pflichten des Unternehmers oder eines sonstigen Inhabers einer Wasserversorgungsanlage und die Überwachung durch das Gesundheitsamt in hygienischer Hinsicht in der Trinkwasserverordnung [TrinkwV] in der seit 16.02.2001 dem Bundesrat vorgelegten und seit dem 01.01.2003 in Kraft getretenen Fassung geregelt; durch die RL 80/778/EWG des Rates v. 15.07.1980, geändert durch die RL 98/83/EG des Rates vom 03.11.1998, über die Qualität von Wasser für den menschlichen Gebrauch müssen die Rechtsvorschriften der einzelnen Länder angeglichen werden).

drinking water [n] (2) *chem. envir.* ▶fluoridation of drinking water.

1597 drinking water catchment area [n] *wat'man.* (Bottom runoff area draining into an aquifer accessible for pumping potable water; ▶soil and water conservation district [US]/water conservation area [UK]); *s* **zona [f] de captación de agua potable** (Área de alumbramiento de aguas subterráneas o de ganancia de aguas superficiales en la que existen restricciones de uso con el fin de proteger los recursos acuíferos; ▶perímetro de protección de acuífero); *f* **champ [m] captant** (Zone d'alimentation en eau potable sur laquelle sont situés les captages d'eau potable ; les **c. c.** sont limités en vue de la prévention de la pollution des eaux souterraines ou des eaux de sources destinées à l'alimentation par des zones de protection de captage d'eau destinée à la consommation humaine [périmètres de protection] ; ▶périmètre de protection d'un captage d'eau potable) ; *syn.* zone [f] de captage d'eau potable ; *g* **Trinkwasserentnahmegebiet [n]** (Auf dem Verordnungswege ausgewiesenes Trinkwasserschutzgebiet, das entsprechend den jeweiligen Landeswassergesetzen in mehrere Schutzzonen gegliedert ist, in denen Auflistungen mit gestaffelten Vorschriften für verbotene und beschränkt zulässige Handlungen und Einrichtungen gelten; ▶Wasserschutzgebiet).

drinking water distribution [n] *plan. urb.* ▶public water supply.

1598 drinking water extraction and treatment plant [n] *wat'man.* (Facility for the processing of water after removal from ground or surface water, before it is treated chemically or physically and provided to the consumer as drinking water; standards within the European Community for water removal have been laid down by Directive 75/440/EEC; ▶water treatment plant); *s* **planta [m] de extracción y potabilización de agua** (Instalación técnica para la ganancia y el tratamiento de agua superficial o subterránea para adecuarla al consumo humano por medio de procesos físicos o químicos; ▶planta de tratamiento de aguas); *f* **station [f] de pompage et de traitement d'eau potable** (Installation de prélèvement d'eau brute dans les aquifères et les eaux superficielles et de traitement pour potabilisation au moyen de procédés physico-chimiques et physiques ; ▶usine de traitement des eaux) ; *g* **Trinkwassergewinnungs- und -aufbereitungsanlage [f]** (Einrichtung zur Förderung von Rohwasser aus Grund- und Oberflächenwasser, um es mit chemisch-physikalischen und physikalischen Aufbereitungsverfahren [Grob- und Feinfilter] für die Trinkwasserversorgung aufzubereiten; der Rat der Europäischen Gemeinschaft hatte mit der RL 75/440/EWG Normen für die Trinkwassergewinnung festgelegt; diese Richtlinie wurde durch die RL 2000/60/EG des Rates vom 23.10.2000 ab 22.12.2007 ersetzt; ▶Wasseraufbereitungsanlage).

1599 drinking water treatment [n] *wat'man.* (TGG 1984, 234; processing of drinking water from a stored supply of natural water in a ▶water treatment plant; ▶drinking water extraction and treatment plant); *s* **potabilización [f] del agua** (Procesamiento del agua para adecuarla al consumo humano por

medio de ▶planta de extracción y potabilización de agua o en el caso de las aguas salobres o saladas en plantas de desalinización que utilizan diferentes métodos físico-químicos como la destilación térmica, la ósmosis inversa y la electrodiálisis; cf. DINA 1987; ▶planta de tratamiento de aguas); *f* **traitement [m] d'eau potable** (Prélèvement d'eau brute dans les aquifères sédimentaires, les eaux superficielles, les aquifères alluviaux ou dans la mer et traitement par des procédés physico-chimiques et physiques dans une installation de traitement de potabilisation ; ▶station de pompage et de traitement d'eau potable, ▶usine de traitement des eaux) ; *g* **Trinkwasseraufbereitung [f]** (Gewinnung von Trinkwasser aus gefasstem Rohwasser aus Grund- oder Oberflächenwasser, Uferfiltrat oder Meerwasser durch chemisch-physikalische und physikalische Behandlungsverfahren mittels einer ▶Wasseraufbereitungsanlage; ▶Trinkwassergewinnungs- und -aufbereitungsanlage).

1600 drip irrigation [n] *agr. hort.* (Water-saving method of ▶irrigation especially devised for farms, gardens, greenhouses and in semiarid [US]/semi-arid [UK], arid and sub-tropical countries, whereby water is emitted in drip form from low-pressure pipe lines at a constant and specified rate; ▶sprinkler irrigation); *syn.* trickle irrigation [n] [also UK]; *s* **irrigación [f] por goteo** (Sistema de riego de gran eficacia aplicado en la producción de frutas y hortalizas y de plantas ornamentales; DINA 1987; ▶irrigación por aspersión, ▶irrigación); *f* **arrosage [m] au goutte à goutte** (Procédé d'▶arrosage caractérisé par une grande économie d'eau utilisé sous les serres de production, en pépinières ou dans les régions arides et subtropiques par lequel l'eau s'écoule goutte à goutte par des goutteurs placés à intervalles réguliers le long d'une canalisation souple ou par un tuyau poreux fonctionnant sous pression ; ▶arrosage par aspersion, ▶irrigation) ; *syn.* goutte à goutte [m] ; *g* **Tröpfchenbewässerung [f]** (Wasser sparendes Bewässerungsverfahren, besonders für Gewächshäuser sowie in semiariden, ariden und subtropischen Gebieten, bei dem das Wasser ober- oder unterirdisch tröpfchenweise aus Rohr- oder Schlauchleitungen heraustritt; *OB* ▶Bewässerung; ▶Beregnung); *syn.* Tropfbewässerung [f].

1601 drip line [n] *arb. hort.* (Outer limit of a tree canopy); *syn.* canopy edge [n] of a tree (TEE 1980, 166); *s* **borde [m] del dosel** (Perímetro externo de la copa de un árbol); *f* **égouttement [m]** (Périphérie d'une couronne d'arbre) ; *g* **Kronentraufe [f]** (Äußere, seitliche Begrenzung einer Baumkrone); *syn.* Kronentrauf [m].

1602 drip line area [n] *bot. constr.* (CPF 1986-III, 126; rooted area defined by canopy coverage projected vertically onto the ground. In arid and semiarid regions [US]/semi-arid regions [UK], the **d. a.** extends well beyond the tree canopy edge; ▶rhizosphere, ▶rooting zone); *syn.* root spread [n] (1), root system spread [n]; *s* **zona [f] de desarrollo de la raíz** (Área ocupada por las raíces, pensada como proyección vertical de la parte aérea de la planta. En zonas áridas o semiáridas la **z. de d. de la r.** supera por mucho el borde del dosel del árbol; ▶zona de penetración [de la raíz], ▶rizosfera); *f* **zone [f] de développement des racines** (Zone d'emprise en projection horizontale du système racinaire d'un arbre, correspondant en général à la surface de la couronne ; dans les climats arides et semi-arides cette zone peut être nettement supérieure et atteindre quatre fois le diamètre de la couronne pour les arbres à forme pyramidale ; ▶horizon racinaire, ▶rhizosphère) ; *syn.* aire [f] d'étalement des racines ; *g 1* **Wurzelbereich [m]** (Vom Wurzelsystem durchdrungene Fläche, die als senkrechte Projektion des oberirdischen Pflanzenteils gedacht ist und eine art- und standortbedingte unterschiedlich große Ausdehnung hat. In ariden und semiariden Gebieten geht der **W.** von Gehölzen weit über die Traufkante

D

hinaus; bei säulenförmigen Bäumen kann dieser mehr als den vierfachen Kronendurchmesser betragen); **g 2 Kronentraufbereich [m]** (Wurzelbereich, der innerhalb der senkrechten Projektion der Baumkrone liegt; ▶Rhizosphäre, ▶Wurzelraum).

drip line protection [n] [US] *arb.* ▶root zone protection.

1603 driveway [n] *urb.* (Vehicular access to a property; ▶access driveway over sidewalk [US]/access driveway over a pavement [UK]); *syn.* driveway entrance [n] [also US], vehicle crossing [n] [also UK], entryway [n] [also US], entrance drive [n] [also US], entry drive [n] [also US], vehicle entrance [n] [also UK]; *s* **vía [f] de acceso (de vehículos)** (Entrada de vehículos a una parcela de terreno; ▶acceso de vehículos sobre acera); *f* **voie [f] d'accès (pour véhicules)** (Entrée d'une parcelle de terrain accessible pour les véhicules ; ▶bateau) ; *g* **Einfahrt [f]** (Zugang für Kraftfahrzeuge auf ein Grundstück; ▶Überfahrt); *syn.* Zufahrt [f].

driveway [n] [UK], **sunken driveway** *constr. urb.* ▶driveway cut [US].

1604 driveway cut [n] [US] *constr. urb.* (Sunken sidewalk [US]/sunken pedestrian pavement [UK] across a driveway; ▶curb cut [US]/drop kerb [UK]); *syn.* sunken driveway [n] [UK]; *s* **acera [f] rebajada** (Zona rebajada de la acera y el bordillo en la zona de acceso a un predio; ▶bordillo hundido); *syn.* zona [f] de acera rebajada; *f* **bateau [m] [d'entrée]** (Zone d'abaissement du trottoir au droit de l'entrée d'une voie permettant l'accès des véhicules ; ▶zone d'abaissement de bordure de trottoir) ; *syn.* abaissé [m] de trottoir ; *g* **Gehwegabsenkung [f]** (Tieferlegen eines Bürgersteiges und der Bordsteine im Bereich einer Grundstückseinfahrt; ▶Bordsteinabsenkung).

driveway entrance [n] [US] *urb.* ▶driveway.

drop chute [n] *constr.* ▶paved rubble drop chute [US]/rough bed channel.

1605 drop-crotching [n] *arb. constr.* (ARB 1983, 390; heavy reduction in size of a crown by thinning the terminal or major branches to a large lateral while retaining the characteristic shape of a tree; ▶topping 3, ▶crown reduction); *syn.* reducing [n] of a crown [UK], drop crotch pruning [n] [also US], severe crown pruning [n] [also US]; *s* **poda [f] de reducción de la copa** (Fuerte poda de la copa de un árbol para reducir su tamaño manteniendo el porte típico de la especie. Se suele realizar cuando los alrededores inmediatos de la ubicación han sufrido algún cambio —como reducción del nivel freático— que perjudica la estabilidad o el crecimiento del árbol, o cuando éste tiene poca estabilidad estática o sufre mucha presión del viento; ▶ahuecado de la copa, ▶descopado); *f* **réduction [f] de couronne** (*Méthode anglaise* les charpentières sont rabattues à l'aisselle d'une de leurs ramifications jouant le rôle de tire-sève située dans la direction souhaitée ; la silhouette obtenue reste proche de la silhouette naturelle, la charpente n'étant pas affectée par la taille ; IDF 1988, 172-178 ; ▶taille d'allégement, ▶éhoupage) ; *syn.* raccourcissement [m] de la cime ; *g* **Kronenrückschnitt [m]** (Baumschnitt, bei dem die gesamte Krone in ihrer Höhe und ihrer seitlichen Ausdehnung entsprechend den Erfordernissen des Baumumfeldes oder der Verkehrssicherheit [z. B. mangelnde Standsicherheit bei zu großer Windlast] eingekürzt wird; meist nach schweren Eingriffen in die unmittelbare Umgebung eines Baumes wie z. B. Grundwasserabsenkung, Bodenverdichtung, Versiegelung des Baumstandortes oder wegen Bruchgefahr. Die Grob- und Starkäste werden auf Saft führendes Holz unter Berücksichtigung eines arttypischen Kronenaufbaues zurückgenommen; bei Rückschnitt von Teilbereichen der Krone spricht man von **Einkürzung von Kronenteilen**; ▶Kronenentlastungsschnitt, ▶Kronenkappung); *syn.* Einkürzen [n, o. Pl.] von

Baumkronen, Kroneneinkürzung [f], Kroneneinkürzungsschnitt [m].

drop crotch pruning [n] [US] *arb. constr.* ▶drop-crotching.

1606 drop curb [n] [US] *constr.* (Curbstone [US]/kerbstone [UK] with rectangular cross-section, the front edge of which is usually chamfered and laid flush with the adjacent area of paving as opposed to an ▶barrier curb [US]/edge kerb [UK]; ▶mountable curb [US]/splayed kerb [UK]); *syn.* depressed curb [n] [US], drop-kerb [n] [UK], flush curb [n] [US], flush kerb [n] [UK]; *s* **piedra [f] de bordillo [m] enterrado** (Bordillo con perfil rectangular, generalmente biselado, cuyo borde superior —al contrario que el ▶bordillo de acera común— se encuentra al mismo nivel que la calzada adyacente; ▶bordillo accesible); *syn.* bordillo [m] enterrado, cordón [m] enterrado [CS]; *f* **bordure [f] arasée d'épaulement** (Type CR ; cf. norme N.F. P 98 302 ; élément de profil rectangulaire, en général chanfreiné, dont le bord supérieur, à la différence des ▶bordures de trottoir, est au même niveau que le revêtement adjacent ; cf. VRD 1994, Tome 2: 5.4, p. 4 ; ▶bordure basse franchissable) ; *g* **Tiefbordstein [m]** (1. Bordstein aus Beton oder Naturstein mit rechteckigem Profil, meist gefast, der im Gegensatz zum ▶Hochbord höhengleich mit der Belagsfläche abschließt. 2. Die Gesamtheit aller **T.e** bildet den **Tiefbord**; ▶Flachbord).

1607 drop drain [n] [US] *constr.* (Vertical drop in a ▶collector drain, with ▶cleanout chamber [US]/inspection chamber [UK], to offset an abrupt change in level and avoid an excessive slope [US]/fall [UK]; ▶access eye [US]/rodding eye [UK]); *syn.* tumbling bay [n] [UK]; *s* **dren [m] vertical** (Parte vertical de un ▶colector para superar diferencias de pendiente, generalmente provisto de una tapa de ▶registro); *f* **chute [f] de drainage** (Élément vertical d'un ▶collecteur [▶regard de visite] mis en œuvre afin de surmonter les différences de niveaux importantes et muni en général d'un tampon) ; *g* **Dränabsturz [m]** (Senkrecht stehendes Steinzeug- oder Betonrohr in einem ▶Sammler, in dem der Drän zur Überwindung großer Gefälleunterschiede durchgeleitet wird, meist mit einem Schachtdeckel versehen; *o. V.* Drainabsturz).

drop inlet [n] [US] *constr.* ▶yard drain [US].

drop-kerb [n] [UK] *constr.* ▶curb cut [US]; ▶drop curb [US].

drop [n] **of a slope** [US] *constr. geo.* ▶slope gradient.

dropper [n] [UK] *bot. hort.* ▶sinker root.

drop pruning [n] [US] *constr. for. hort.* ▶crown lifting.

1608 drop structure [n] *constr. envir.* (Protective construction around an ▶outfall pipe into a receiving stream; ▶outfall [of sewage]); *syn.* outfall structure [n] [also US]; *s* **acometida [f] a colector principal** (Construcción protectora del lugar de entrada de una tubería de descarga en el curso receptor; ▶tubería de salida, ▶punto de vertido); *f* **ouvrage [m] de rejet** (Ouvrage hydraulique de déversement d'effluents dans un émissaire ; ▶point de rejet [des effluents], ▶tuyau d'arrivée) ; *syn.* déversoir [m] ; *g* **Einlaufbauwerk [n] (1)** (Baulich gefasste ▶Einleitungsstelle an einem Vorfluter; ▶Einlaufrohr).

1609 drop structure installation [n] *wat'man.* (WRP 1974, 116; artificially-created step in the bed of a watercourse to reduce its longitudinal gradient; *generic term* ▶checkdam); *s* **solera [f]** (Grada artificial construida en cursos de agua para reducir la pendiente y así disminuir la velocidad de la corriente; *término genérico* ▶dique transversal); *f* **ouvrage [m] de chute** (Ouvrage de stabilisation du profil en long d'un cours d'eau par diminution de la pente longitudinale ; *terme générique* ▶ouvrage de confortement transversal) ; *g* **Sohlenabsturz [m]** (Sohlenstufe in einem Fließgewässer zur Verringerung des Längsgefälles; *OB* ▶Querwerk).

D

1610 drought damage [n] *agr. constr. hort. for.* (**1.** Damage to plants in the form of die-back or wilting caused by lack of water). **2.** Consequence of a drought period, wind action, etc.; ►frost-desiccation); *syn.* desiccation damage [n]; *s* **daño** [m] **por sequía** (Pérdida producida por la sequía; ►desecación invernal); *f* **dommages** [mpl] **causés par la sécheresse** (Phénomènes de fanaison [flétrissement provoqué par déshydratation] suite au dessèchement du sol ou au ►dessèchement par le gel) ; *g* **Trockenschaden** [m] (Dürre- oder Welkeerscheinung an Pflanzen durch Austrocknung des Bodens oder ►Frosttrocknis).

1611 drought-resistant [adj] *phyt.* (Characteristic of a plant that adapts to periodic low moisture conditions); *s* **resistente a la sequía** [loc] (Característica de plantas que se pueden adaptar a largos periodos/períodos de sequía); *f* **résistant, ante à la sécheresse** [loc] (Qualifie les végétaux susceptibles de supporter une longue période de sécheresse) ; *g* **trockenresistent** [adj] (Pflanzen betreffend, die eine längere Trockenheit ertragen).

1612 dry excavation [n] *min.* (Extraction of stone and soil material above ground water level; *opp.* ►dredging 1); *s* **extracción** [f] **en seco (de rocas y arenas)** (Extracción de arena y grava por encima del nivel freático; *opp.* ►dragado subacuático); *f* **extraction** [f] **de matériaux alluvionnaires hors nappe** (Exploitation à ciel ouvert de substances minérales ou fossiles au-dessus du niveau de la nappe phréatique ; l'extraction peut avoir lieu par le haut du gisement [exploitation en fouille] ou par le bas [exploitation en butte] ; *opp.* ►extraction de matériaux alluvionnaires en nappe) ; *syn.* extraction [f] en terrain sec, extraction [f] à sec, exploitation [f] de roches alluvionnaires hors nappe ; *g* **Trockenabbau** [m] (Gewinnung von Steinen und Erden oberhalb des Grundwasserspiegels; *opp.* ►Nassbaggerung); *syn.* Trockenbaggerung [f].

1613 dry gravel excavation [n] *min.* (*opp.* ►gravel dredging, ►dry excavation); *s* **extracción** [f] **de grava en seco** (Término específico de ►extracción en seco [de rocas y arenas]; *opp.* ►extracción de grava por dragado); *syn.* extracción [f] seca de ripio [RA], extracción [f] seca de granza [CO], extracción [f] seca de granzón [YV]; *f* **extraction** [f] **de gravier hors nappe** (Extraction à ciel ouvert de gravier d'origine fluvio-glaciaire au-dessus de la nappe phréatique ; *terme générique* ►extraction de matériaux alluvionnaires hors nappe ; *opp.* ►dragage de gravier) ; *g* **Trockenauskiesung** [f] (Gewinnung von Kies oberhalb des Grundwasserspiegels; *OB* ►Trockenabbau; *opp.* ►Nassauskiesung); *syn.* Trockenkiesabbau [m].

1614 dry habitat [n] *phyt. zool.* (►Habitat associated with a location's dry conditions; ►xeric site); *syn.* xeric habitat [n]; *s* **biótopo** [m] **seco** (►Biótopo que crece bajo condiciones mesológicas de sequía; ►estación xerofítica); *syn.* biótopo [m] xerofítico; *f* **biotope** [m] **sec** (►Biotope caractérisé par des conditions stationnelles sèches ; ►station sèche) ; *syn.* biotope [m] xérique ; *g* **Trockenbiotop** [m, *auch* n] (►Biotop, der an trockene Standortbedingungen gebunden ist; ►Trockenstandort).

dry habitat roof planting [n] [US] *constr.* ►extensive roof planting.

drying [n] [UK]**, top** *arb. for.* ►top-kill [US].

1615 drying out [n] (1) *geo.* (Desiccation of waters, humid meadows, ponds, etc.; ►drying up, ►parching); *s* **desecación** [f] (2) (Proceso natural por el cual arroyos, humedales, etc. van perdiendo agua hasta secarse completamente y desaparecer; ►aridecimiento, ►desecación 3; *término específico* desecación estival); *f* **dessèchement** [m] (Processus naturel pendant lequel un sol se dessèche, p. ex. dessèchement de prairies humides, de marais, de lacs ; *terme spécifique* dessèchement estival ; ►dessiccation, ►tarissement) ; *g* **Trockenfallen** [n, o. Pl.] (►Austrock-

nung 2 von Fließgewässern, Feuchtwiesen, Teichen etc.; ►Austrocknung 3).

1616 drying out [n] (2) *pedol.* (Draining of soils, damp meadows, bogs, or swamps to make them usable for human activities; ►parching); *syn.* drainage [n] (2); *s* **desaguado** [m] (Acción de sacar activamente el agua de prados húmedos, turberas, pantanos, etc.; desecar [vb]; ►aridecimiento); *syn.* desecación [f] (activa), desecamiento [m] activo, drenaje [m] (de prados húmedos, turberas, pantanos); *f* **assèchement** [m] (Action anthropique, drainage en vue de la « mise en valeur » de prairies humides, de tourbières, de marais, d'un lac, etc., assainissement de terrains marécageux ; ►dessiccation) ; *g* **Trockenlegung** [f] (Dränung von Böden, Feuchtwiesen, Mooren, Sümpfen etc.; ►Austrocknung 3); *syn.* Entwässerung [f] (von Mooren/Sümpfen), Austrocknung [f] (1), Trockenlegen [n, o. Pl.].

1617 drying shrinkage [n] **of concrete** *constr.* (Reduction in volume due to setting and curing dependent upon the humidity of the surrounding air, the water and cement content of the concrete and the outer dimensions of construction; deformation of concrete caused by hydration of cement generating heat during curing/setting is known as **thermal shrinkage** or **thermal cracking**; ►shrinkage, ►shrinkage crack); *s* **merma** [f] **del hormigón** (Reducción del volumen del hormigón al fraguarse que puede producir fisuras; ►grieta de contracción, ►merma); *syn.* merma [f] del concreto [AL], contracción [f] del concreto [AL], contracción [f] del hormigón; *f* **retrait** [m] **du béton** (Le retrait correspond à des variations dimensionnelles mettant en jeu des phénomènes physiques avant, pendant ou après la prise des bétons ; on distingue **1.** le retrait plastique [ou avant prise] résulte de l'évaporation de l'eau utile à la prise et génère une faible contraction du matériau et est une cause fréquente de fissurations précoces ; **2.** le retrait de dessication ou hydraulique, qui résulte de l'évaporation de l'eau excédentaire, se manifeste en décroissant sur plusieurs années, et provoque une contraction importante ; **3.** le retrait thermique lié au retour à température ambiante [refroidissement] des pièces en béton ayant au préalable subi une élévation de température due aux réactions exothermiques d'hydratation du ciment ; **4.** le retrait d'auto-dessication lié à la contraction du béton en cours d'hydratation et protégé de tout échange d'eau avec le milieu environnant ; la conséquence essentielle du retrait est l'apparition de phénomènes de fissuration pouvant diminuer la durabilité des structures en béton armé ou précontraint, et/ou limiter leur capacité portante ; cette fissuration peut conduire à limiter l'adhérence entre un matériau rapporté en surface et le support en béton ; ►retrait, ►fissure de retrait [de dessication] dans le béton) ; *syn.* contraction [f] du béton ; *g* **Schwinden** [n, o. Pl.] **von Beton** (Verkleinerung des Rauminhalts/der Längenveränderungen, die beim Beton als Folge des Abbindens und Erhärtens [Austrocknens] entsteht, abhängig von der Feuchtigkeit der umgebenden Luft, dem Wasser- und Zementgehalt des Betons und den äußeren Abmessungen des Bauteils; Verformungen des Betons unter dem Einfluss der Abbindwärme fallen nicht unter den Begriff des **S.s** und werden als **Schwinden durch Hydratation/Hydration** bezeichnet; schwinden [vb]; ►Schwinden, ►Schwindriss in Beton).

1618 drying up [n] *hydr.* (of a river, spring or well); *s* **desecación** [f] (3) (de río, arroyo, fuente, etc.); *f* **tarissement** [m] (Cessation de l'écoulement d'un cours d'eau, d'une source, d'un puits) ; *g* **Austrocknung** [f] (2) (Das Trockenwerden von Flussbetten, Quellen, Brunnen; austrocknen [vb]).

1619 dry masonry [n] *arch. constr.* (Stone wall laid without mortar; ►gravity retaining wall, ►masonry construction); *syn.* dry (stone) wall [n] (TSS 1988-410-8), dry stane dyke [n] [SCOT], dry rubble construction [n] (MEA 1985); *s* **mampostería** [f] **en seco** (Cualquier tipo de ►mampostería en la que no se ha utilizado mortero; DACO 1988; ▶muro de gravedad); *f* **maçonnerie [f] à sec** (Construction décorative montée avec des pierres naturelles, sans liant de mortier ; ▶maçonnerie, ▶mur poids) ; *syn.* maçonnerie [f] de pierres sèches (MAÇ 1981, 6), mur [m] à joints vifs ; *g* **Trockenmauerwerk [n]** (Ein ohne Mörtel errichtetes ▶Mauerwerk; Trockenmauern werden als ▶Schwergewichtsmauern — Stützmauern oder als niedrige, frei stehende Mauern gebaut; Fundamentierung auf gewachsenem Boden 40 cm unter Geländeoberkante mit Schotter oder Kiestragschicht, Fundamentdicke am Mauerfuß ca. ein Drittel der Mauerhöhe); *syn.* Trockenmauer [f].

1620 dry meadow [n] *phyt.* (TPE 1987, 493; VIR, 1982, 336s; nutrient-poor grassland community growing on humus-deficient, shallow or deep sandy soils, which requires relatively warm conditions and tolerates greater aridity than other grassland associations; ▶acidic grassland, ▶limestone grassland, ▶nutrient-poor grassland, ▶semidry grassland [US]/semi-dry grassland [UK], ▶xerothermous meadow; *opp.* ▶wet meadow); *syn.* dry sward [n], arid sward [n] (ELL 1988, 476), arid grassland [n] (ELL 1988, 467ss); *s* **pastizal [m] seco** (Grupo de comunidades de vegetación que crece en suelos pobres, poco profundos, necesita bastante calor y puede soportar la sequía; BB 1979; ▶pastizal calcícola seco, ▶pastizal [oligótrofo] silicícola; ▶pastizal semiseco, ▶pastizal xerotérmico, ▶prado oligótrofo; *opp.* ▶prado higrófilo); *f* **pelouse [f] aride** (Formation herbacée naturelle rase, floristiquement très diversifiée et très riche en espèces herbacées et graminées thermophiles et xérophiles appartenant à l'alliance du *Xerobromion* ; localisée sur pentes rocailleuses, versants abrupts ou affleurements rocheux calcaires caractérisés par un sol généralement squelettique, pauvre en éléments nutritifs et une mauvaise alimentation en eau ; ▶pelouse calcicole maigre, ▶pelouse héliophile dense, ▶pelouse oligotrophe, ▶pelouse silicicole, ▶pelouse xérophile et thermophile ; GGV 1979 ; *opp.* ▶prairie hygrophile) ; *syn.* pelouse [f] sèche ; *g* **Trockenrasen [m]** (Magerrasenpflanzengesellschaft auf humusarmen, flachgründigen oder tiefgründigen, sandigen Böden wachsend, die ein relativ großes Wärmebedürfnis hat und zugleich mehr Trockenheit erträgt als die übrigen; ▶Halbtrockenrasen, ▶Kalkmagerrasen, ▶Magerrasen, ▶Silikatmagerrasen, ▶Xerothermrasen; *opp.* ▶Nasswiese); *syn.* Volltrockenrasen [m].

1621 dry-mix concrete [n] *constr.* (Composite material with Portland cement, aggregate and little water in relation to its other components; *related terms* 'dry-packed concrete' [also US], 'dry-tamp concrete' [also US] and 'dry-shake concrete' [also US]); *s* **hormigón [m] de consistencia de tierra húmeda** *syn.* concreto [m] de consistencia de tierra húmeda [AL]; *f* **béton [m] gâché consistance terre humide** ; *g* **erdfeuchter Beton [m]** (Mit wenig Wasser angemachter Beton).

dry-packed concrete [n] [US] *constr.* ▶dry-mix concrete.

dry pruning [n] [UK] *arb. hort.* ▶dead wooding [US] (2).

dry rubble construction [n] *arch. constr.* ▶dry masonry.

1622 dry seeding [n] *constr.* (▶Bioengineering technique of sowing seeds in a dry state—if necessary with addition of fertilizer and ▶soil amendments—on the area to be protected. The surfaces are usually covered with a mulch layer of straw or similar material either before or after seeding and then fixed with a plastic emulsion; *opp.* ▶hydroseeding); *s* **siembra [f] en seco** (Procedimiento de **s.** en la ▶bioingeniería en el que las semillas, a veces junto con abono y enmiendas, son esparcidas en seco sobre la superficie a recubrir. Las superficies son cubiertas antes o después de la operación con una capa de mulch y a continuación fijadas con pegamento; *opp.* ▶riego de semillas por emulsión; ▶material de enmienda de suelos); *f* **ensemencement [m]**

à sec (▶génie biologique consistant dans un apport à sec de semences, parfois avec incorporation d'engrais et d'▶amendements ; les surfaces ensemencées sont recouvertes préalablement ou à posteriori par de la paille ou des produits similaires et sont ensuite traitées par projection d'un fixateur ; *opp.* ▶procédé d'enherbement par projection hydraulique) ; *g* **Trockensaat [f]** (▶*Ingenieurbiologie* Saatverfahren, bei dem das Saatgut, gegebenenfalls unter Beigabe von Dünger und ▶Bodenverbesserungsstoffen 1, trocken auf die zu sichernden Flächen aufgebracht wird. Die Flächen werden meist entweder zuvor oder danach mit einer Mulchschicht aus Stroh oder ähnlichen Stoffen abgedeckt und anschließend mit Klebern festgelegt; *opp.* ▶Anspritzverfahren).

dry stane dyke [n] [SCOT] *arch. constr.* ▶dry masonry.

dry stone wall [n] *arch. constr.* ▶dry masonry.

dry sward [n] *phyt.* ▶dry meadow.

1623 dry valley [n] *geo.* (Valley which has no or almost no permanent stream. A **d. v.** can occur, e.g, 1. on porous rocks [karst landscapes] where the groundwater level has dropped so deep that surface flow is no longer possible. 2. As a result of climate change and 3. due to drainage, deforestation or industrial water extraction and thus changes in the ecosystem such that the river runs dry; a **d. v.** in desert or semi-arid areas of North Africa or the Middle East is called a ▶wadi, in South Africa, a **rivier**); *syn.* dead valley [n]; *s* **valle [m] seco** (Valle sin curso de agua permanente, típico de regiones kársticas, en el que las aguas se han infiltrado para alimentar cursos subterráneos. Los **v. s.** en zonas desérticas o semiáridas del norte de África o del Medio Oriente se llaman ▶wadis; cf. DGA 1986); *f* **vallée [f] sèche** (Vallée parcourue par un cours d'eau intermittant ou n'évacuant plus d'eau, p. ex. dans les vallées des régions karstiques ou désertiques ; ▶oued) ; *g* **Trockental [n]** (Tal, das zeitweise oder dauernd kein Wasser führt. Ein **T.** kann z. B. entstehen, wenn 1. in durchlässigen Gesteinen [Karstlandschaften: Karsttal] der Grundwasserspiegel soweit abgesunken ist, dass kein oberirdisches Fließen mehr möglich ist, 2. infolge einer Klimaveränderung die Wasserführung verändert oder eingestellt ist und 3. bei Trockenlegung, Waldrodung oder industrieller Wasserentnahme der Wasserhaushalt der Landschaft so verändert wird, dass das Fließgewässer trockenfällt; **T.täler** in Wüsten oder semiariden Gebieten Nordafrikas oder des Vorderen Orients heißen ▶**Wadis**, die in Südafrika **Riviere**); *syn. für den Großraum Südafrika* Rivier [n] [ZA].

dry wall [n] *arch. constr.* ▶dry masonry.

1624 dry wall plant [n] *gard. hort. plant.* (Suitable plant for a dry habitat; e.g. dwarf woody shrubs and perennials for a ▶rock garden); *s* **planta [f] mural** (Planta que crece en lugar seco, p. ej. planta de ▶rocalla) ; *f* **plante [f] des murs secs** (Végétal adapté ou approprié aux milieux secs, p. ex. divers végétaux ligneux nains ou les plantes vivaces de rocailles ; ▶jardin de rocaille) ; *g* **Trockenmauerpflanze [f]** (Für trockene Standorte geeignete Pflanze, z. B. diverse Zwerggehölze, Steingartenstauden aus dem Lebensbereich Steinanlage; ▶Steingarten).

1625 drywell sump [n] [US] *constr.* (Pit or catch basin [US] filled with gravel or pebbles for the collection and percolation of surface runoff in poorly draining soils); *syn.* seepage pit [n] (DET 1976, 171), soakaway [n] [UK]; *s* **sumidero [m]** (Registro relleno de piedras o grava cuya solera y cuyas paredes inferiores son permeables para permitir la infiltración del agua en el suelo); *syn.* pozo [m] filtrante; *f* **puisard [m]** (Puits creusé dans les sols imperméables pour recevoir les eaux superficielles et rempli de gravier ou de galet) ; *syn.* puits [m] perdu, trou [m] perdu ; *g* **Sickerschacht [m]** (Mit Kies oder Geröll gefüllter Schacht in

D

schwer wasserdurchlässigen Schichten zum Auffangen und Versickern von Oberflächenwasser); *syn.* Sickergrube [f].

1626 duckweed cover [n] *phyt.* (Free-floating *Lemna* community; *specific term* ▶free-floating freshwater community); *s* **lemnal** (Término específico de ▶comunidad de hidrófitos libremente flotante); *f* **lemnaie [f]** (Association d'hydrophytes libres et flottants des eaux dormantes de la classe des *Lemnetea* ; *terme générique* ▶groupement d'hydrophytes libres et flottants [des eaux calmes]) ; *syn.* peuplement [m] à *Lemna* ; *g* **Wasserlinsendecke [f]** (Freischwimmende Wasserpflanzengesellschaft; *OB* ▶Schwimmpflanzendecke).

duff mull [n] [US&CDN] *pedol.* ▶moder.

dummy-jointed [pp/adj] *constr.* *▶tooling of stone [US]/dressing of stone [UK], #10.

dump [n] [US] *envir.* ▶dumpsite [US]/tipping site [UK].

dump [n] [US]**, overburden** *min.* ▶overburden pile.

dumped fill [n] *constr.* ▶fill material.

dumped stone lining [n] [US] *wat'man.* ▶channel lining.

dumping [n] *constr.* ▶earth filling (1), ▶front end dumping, ▶side dumping.

dumping [n] [US]**, elongated pile** *constr.* ▶front end dumping.

1627 dumping [n] **at sea** *envir.* (Disposal of waste or other substances, from ships, aircrafts, platforms or other facilities built in the sea or any other disposal, mostly toxic, on the bottom or under the bed of a high sea; ▶discharge of effluents, ▶disposal by dilution, ▶off-site disposal of surplus material); *s* **descarga [f] de residuos en alta mar** (Eliminación de residuos, generalmente tóxicos, en alta mar; ▶disposición de excedentes de terracerías; ▶vertido, ▶vertido por dilución); *syn.* vertido [m] de residuos en alta mar; *f* **immersion [f] des déchets en mer** (Rejet et déversement délibéré dans la mer de déchets d'origine industrielle ou domestique [substances liquides ou solides nocives] et autres matières à partir de navires, aéronefs, engins ou plates-formes, et ne résultant pas de la marche normale de ceux-ci ; *termes spécifiques* dégazage des hydrocarbures, déballastage ; ▶clapage, ▶enlèvement des excédents, ▶rejet des effluents) ; *g* **Verklappung [f]** (Beseitigung von Abfällen oder sonstigen, meist giftigen Stoffen von Schiffen, Luftfahrzeugen, Plattformen oder sonstigen auf See errichteten Anlagen oder jede Lagerung von Abfällen oder sonstigen Stoffen auf dem Meeresboden und im Meeresuntergrund; cf. H-S-EinbrG; ▶Einleitung, ▶Massenverbringung, ▶Verklappen von flüssigen Abfallstoffen); *syn. leg.* Einbringung [f] von Abfällen, Versenken [n, o. Pl.] im Meer (Art. 1 [2] lit. c RL 78/176/EWG des Rates v. 20.02.1978), Tiefversenkung [f] (Art. 1 RL 82/883/EWG des Rates v. 03.12.1982).

1628 dumping certificate [n] [US] *onstr.* (Record of trips from the construction site to a dump [US]/tip [UK] as verification of completed work for payment; ▶dumping record [US]/tipping record [UK]); *syn.* tipping certificate [n] [UK]; *s* **justificación [f] de descarga** (Documentación de los viajes de transporte de la obra a vertedero para la liquidación de servicios de construcción; ▶registro de descarga en vertedero); *f* **justificatif [m] de décharge** (Document justifiant p. ex. l'enlèvement des excédents [terres ne pouvant être utilisées sur place, gravats, etc.] aux décharges et légitimant les propositions de paiement de prestations de travaux ; ▶établissement de la liste des justificatifs de décharge) ; *g* **Kippennachweis [m]** (Bescheinigung über Transportfahrten von der Baustelle zur Deponie resp. Ablagerungsstelle für die Abrechnung von Bauleistungen bezüglich der Entsorgung von Aushubmaterial, Bauschutt und sonstigen Bau- und Grüngutabfällen; ▶Auflistung der Kippennachweise).

1629 dumping layer [n] [US] *in.* (Levelled course of a ▶large spoil area [US]/large tip [UK]); *syn.* tipping layer [n] [UK]; *s* **piso [m] de escombrera** (Capa nivelada de una ▶escombrera grande); *f* **tage [m] d'un terril** (▶amoncellement) ; *g* **Schüttscheibe [f]** (Etage einer ▶Halde 1).

1630 dumping [n] **of earth** [US] *constr. eng.* (Process of depositing excavated earth, rubble material, etc.; e.g. to create street or railroad embankments [US]/railway embankments [UK], large spoil areas [US]/large tips [UK] or to fill depressions; ▶dumping of refuse [US]/tipping of refuse [UK], ▶artificial landform, ▶infilling); *syn.* tipping [n] of earth [UK]; *s* **terraplenado [m]** (Proceso de depositar áridos, escombros, tierra, etc. p. ej. para construir un terraplén de carretera; ▶descarga de residuos, ▶montículo artificial, ▶relleno); *syn.* construcción [f] de un terraplén; *f 1* **mise [f] en remblai** (Action d'exhausser le niveau d'un terrain, d'une chaussée, d'une voie de chemin de fer ; ▶dépôt de déchets, ▶dépôt de déchets inertes, ▶comblement) ; *f 2* **exhaussement [m]** (Résultat de l'action d'exhausser le niveau d'un terrain) ; *syn.* remblai [m] ; *f 3* **terre-plein [m]** (Surface horizontale construite par remblaiement ; cette levée de terre est souvent soutenue par une maçonnerie ; par extension, toute surface plane en terre) ; *g* **Aufschüttung [f]** (1. Vorgang des Aufschüttens von Aushubmaterial, Bauschutt etc., z. B. bei [Straßen-, Bahn]dämmen, Halden oder Geländemulden; ▶Ablagerung von Abfällen, ▶Auffüllen. 2. Ergebnis der Aufschüttung; ▶Auffüllberg); *syn. zu 1.* Aufschütten [n, o. Pl.].

1631 dumping [n] **of refuse** [US] *envir. min.* (Unloading of waste material; ▶artificial landform, ▶dumping of earth [US]/tipping of earth [UK], ▶dumpsite, ▶large spoil area, ▶refuse dump [US]/refuse tip [UK], ▶waste disposal); *syn.* tipping [n] of refuse [UK]; *s* **descarga [f] de residuos** (Entrega de desechos en un ▶depósito de residuos sólidos, ▶depósito de residuos sólidos urbanos, ▶eliminación de residuos sólidos, ▶escombrera, ▶montículo artificial, ▶terraplenado); *syn.* vertido [m] de residuos; *f* **dépôt [m] de déchets** (de déchets ménagers, industriels ; ▶amoncellement, ▶décharge de déchets menagers, ▶décharge publique, ▶dépôt de déchets inertes, ▶élimination des déchets, ▶exhaussement, ▶mise en remblai) ; *g* **Ablagerung [f] von Abfällen** (Abladen von Abfällen, die nicht vermeidbar und nicht rezyklierbar sind, auf behördlich bewilligten Deponien; gesetzliche Grundlage ist das Kreislaufwirtschafts- und Abfallgesetz; ▶Abfallbeseitigung, ▶Abfalldeponie, ▶Auffüllberg, ▶Aufschüttung, ▶Deponie, ▶Halde 1); *syn.* Deponieren [n, o. Pl.] von Abfällen.

dumping price [n] [UK] *constr. contr.* ▶predatory price [US].

1632 dumping record [n] [US] *constr.* (List with attached ▶dumping certificates [US]/tipping certificates [UK]); *syn.* tipping record [n] [UK]; *s* **registro [m] de descarga en vertedero** (Lista de desechos a eliminar con los correspondientes certificados; ▶justificación de descarga); *f* **établissement [m] de la liste des justificatifs de décharge** (▶justificatif de décharge) ; *g* **Auflistung [f] der Kippennachweise** (▶Kippennachweis).

dumping site [n] *min.* ▶overburden dumping site.

1633 dumpsite [n] [US] *envir.* (Area for the deposit of waste materials; *specific terms* ▶landfill to original contour, ▶large spoil area [US]/large tip [UK], ▶hazardous waste disposal site, ▶overburden dumping site, ▶refuse dump [US]/refuse tip [UK], ▶sanitary landfill [US] 2/landfill site [UK], ▶unmanaged dumpsite, ▶unauthorized dumpsite [US]/fly tipping site [UK]); *syn.* dump [n] [US], land disposal site [n] [US] (TGG 1984, 233), spoil area [n] [US], spoil bank [n] [US] (1), tip [n] [UK], tipping site [n] [UK]; *s* **depósito [m] de residuos sólidos** (Área de deposición de basuras y material sobrante; ▶depósito de cascote, ▶depósito de residuos peligrosos, ▶depósito de residuos sólidos

D

urbanos, ▶escombrera, ▶vertedero clandestino, ▶vertedero controlado, ▶vertedero en hondonada, ▶vertedero uncontrolado); *syn.* vertedero [m]; *f* **décharge [f] publique** (Lieu public ou sont déversés les ordures, remblais, et déchets divers ; la réglementation sur la récupération des déchets voudrait que seul les déchets ultimes puissent être mis en ▶décharges contrôlées ; *termes spécifiques* ▶amoncellement, ▶centre de décharge des déchets ultimes, ▶centre de stockage de déchets spéciaux ultimes et stabilisés, ▶décharge [des déchets] sauvage, ▶décharge de déchets industriels [spéciaux], ▶décharge de déchets ménagers, ▶décharge de matériaux de recouvrement, ▶décharge non contrôlée, ▶décharge par enfouissement, ▶installation de stockage des déchets ménagers et assimilés) ; *syn.* installation [f] de stockage ; *g* **Deponie [f]** (Fläche zur zeitlich unbegrenzten oberirdischen Ablagerung von Abfällen oder Bodenmaterial; deponieren [vb]; *OB* Abfallbeseitigungsanlage; *UBe* ▶Abfalldeponie, ▶Abraumdeponie, ▶Flurkippe, ▶geordnete Deponie, ▶Halde 1, ▶Sonderabfalldeponie, ▶ungeordnete Deponie, ▶wilde Deponie); *syn.* Ablagerungsfläche [f] für Abfälle, Kippe [f], Materialablagerung [f] [CH]).

dumpsite [n], **rubble** *envir.* ▶rubble disposal site.

1634 dumpster [n] [US] *constr.* (Movable waste container which can be lifted hydraulically into a dump truck); *syn.* skip [n] [UK]; *s* **contenedor [m]** (Caja móvil de camión para transporte de basuras, escombros, etc.); *syn.* container [m]; *f* **bac [m]** (Conteneur utilisé pour l'évacuation des déchets ou des matériaux de démolition au moyen d'un camion multi-bennes) ; *syn.* conteneur [m], benne [f] ; *g* **Mulde [f] (2)** (Container für den Abtransport von Abfällen, Schutt- und Bodenmassen mit einem Schwenkarm-Muldenkipper).

1635 dune [n] *geo.* (Sand hill formed by wind action; ▶backdune [US]/brown dune [UK], ▶barkhan, ▶coastal dune, ▶drifting sand dune, ▶embryo dune, ▶gray dune [US]/grey dune [UK], ▶inland dune, ▶parabolic dune, ▶shrub dune, ▶white dune); *s* **duna [f]** (Acumulación individual de arena propia de zonas litorales y desérticas. Se origina a partir de las llanuras arenosas, sin vegetación, remodeladas por acción eólica. Las d. que poseen una forma mejor definida son las ligadas a vientos constantes unidireccionales, como es el caso de la ▶barkhana, la ▶duna parabólica, el rebdou y la nebka. Las dunas tienden a acumularse en campos dunares [ergs] en los que predomina un determinado tipo de acumulación. Existen ▶dunas vivas o funcionales y dunas muertas que están estabilizadas; ▶duna arbustiva, ▶duna blanca, ▶duna embrionaria, ▶duna gris, ▶duna interior, ▶duna litoral, ▶duna parda); *f* **dune [f]** (Accumulation de sable d'importance variable, formée par le vent ; ▶dune arbustive, ▶barkhane, ▶dune blanche, ▶dune brune, ▶dune continentale, ▶dune embryonnaire, ▶dune grise, ▶dune littorale, ▶dune mouvante, ▶dune parabolique) ; *g* **Düne [f]** (Durch Wind aufgehäufte[r] Sandhügel; Feinsedimentablagerung in unterschiedlichsten Formen und Entwicklungsstadien wie ▶Binnendüne, ▶Graudüne, ▶Heckendüne, ▶Heidedüne, ▶Küstendüne, ▶Parabeldüne, ▶Primärdüne, ▶Sicheldüne, ▶Wanderdüne, ▶Weißdüne).

dune [n] **fixed** *geo. phyt.* ▶gray dune [US]/grey dune [UK].

dune [n] [US], **gray** *geo. phyt.* ▶gray dune [US]/grey dune [UK].

dune [n] [UK], **grey** *geo. phyt.* ▶gray dune [US].

dune [n], **primary** *geo. phyt.* ▶embryo dune.

dune [n], **upsiloidal** *geo.* ▶parabolic dune.

1636 dune destruction [n] *agr. land'man.* (Elimination of plant cover on dunes causing wind ▶blowouts, which may result in ▶drifting sand dunes); *s* **destrucción [f] de dunas** (Deterioro y destrucción de la vegetación dunaria que posibilita la ▶excava-

ción eólica que puede llevar hasta la creación de ▶dunas vivas); *syn.* deterioro [m] de dunas; *f* **destruction [f] des dunes** (Dégradation du couvert végétal de la dune entraînant la formation de ▶creux de déflation pouvant aller jusqu'à la constitution de ▶dunes mouvantes) ; *syn.* détérioration [f] des dunes ; *g* **Dünenzerstörung [f]** (Zerstörung der Pflanzendecke auf Dünen, wodurch ▶Windanrisse und dadurch ausgelöst im ungünstigsten Falle ▶Wanderdünen entstehen können).

1637 dune face [n] *geo.* (Steep front of a dune eroded by wave and wind action); *s* **frente [m] de duna erosionada** (Cara escarpada de duna costera causada por la acción de las olas); *f* **front [m] de dunes érodées** (Escarpement provoqué sur une dune littorale par l'activité érosive de déferlement de la houle) ; *g* **Dünenkliff [n]** (Durch Brandungstätigkeit verursachter steiler Abfall einer Küstendüne).

dune forest [n] *phyt.* ▶dune woodland community.

1638 dune grassland [n] *phyt.* (Xeromorphic grassland community characterized by lack of nitrogen, salt spray, wind stress and full sun exposure; ▶extreme habitat); *s* **herbazal [m] dunar/de dunas** (Comunidad herbácea xerofítica adaptada a la estación extrema caracterizada por pobreza en materia orgánica y nitratos y por la acción del viento y la fuerte insolación; ▶biótopo extremo); *f* **pelouse [f] dunaire** (Association végétale herbacée xérophyte adaptée à un milieu caractérisé par des facteurs très sélectifs comme la pénurie en matières organiques, l'action du vent et des embruns salés, la déshydratation due à une forte insolation ; ▶biotope aux conditions stationnelles extrêmes) ; *g* **Dünenrasen [m]** (Durch Stickstoffmangel, mechanische Windbeanspruchung, salzhaltige Gischt und starke Sonneneinstrahlung gekennzeichnete, meist xeromorphe Rasengesellschaft; ▶Extrembiotop); *syn.* Dünengrasflur [f].

1639 dune heath [n] *phyt.* (Dwarf shrub community on a fixed sand dune such as ▶back dune and ▶gray dune); *s* **landa [f] de dunas** (Formación vegetal arbustiva característica de las ▶dunas grises y ▶dunas pardas); *syn.* landa [f] dunar; *f* **lande [f] dunaire** (Formation végétale d'arbustes nains caractérisant les ▶dunes grises et les ▶dunes brunes) ; *g* **Dünenheide [f]** (Zwergstrauchformation der ▶Graudüne und ▶Heidedüne).

1640 dune planting [n] *constr. landsc.* (Installation of dune vegetation in order to hold sand dunes in place); *s* **plantación [f] de dunas** (Fijación de la superficie de las dunas plantando especies adaptadas a las condiciones mesológicas dominantes); *syn.* plantío [m] de dunas; *f* **plantation [f] dunaire** (Mesure de fixation de dunes et de lutte contre l'ensablement par la reconstitution de la végétation locale [mise en défens] suivie de l'installation d'une couverture végétale [reboisement]) ; *g* **Dünenbepflanzung [f]** (Befestigung der Dünenoberfläche mit standortgerechter Vegetation, die dem Entwicklungsstand der Düne [Weißdüne, Graudüne, Heidedüne] entspricht, um den Sand festzulegen.

1641 dune protection [n] *constr. landsc.* (Measures to prevent or repair dune erosion, such as ▶blow-out or undercutting of dune faces, as well as measures for stabilization of ▶drifting sand dunes, typically by redirecting pathways, planting, or installing wind fence barriers); *s* **protección [f] de dunas** (Todo tipo de medidas dedicadas a frenar o parar la destrucción de dunas, p. ej. estabilización de ▶dunas vivas; ▶excavación eólica); *f* **protection [f] des dunes** (Mesures pour empêcher la dégradation des dunes [comme p. ex. le déchaussement d'un front dunaire] par l'interdiction de coupe des plantes aréneuses conservatrices de dunes ou la remise en état des dunes détériorées, p. ex. fixation de ▶dunes mouvantes ; ▶creux de déflation) ; *g* **Dünenschutz [m, o. Pl.]** (Alle Maßnahmen zur Beseitigung und Verhinderung von Dünenschäden, wie z. B. ▶Windanriss oder Unterspülung eines Dünenkliffs sowie Maßnahmen zur Festlegung von ▶Wanderdünen).

1642 dune scrub [n] *phyt.* (Shrub community on a ▶back dune forming a transition zone to a ▶dune woodland community); *s* **matorral** [m] **dunar/de dunas** (Comunidad de matorral en la ▶duna parda como transición al ▶bosque dunar); *f* **fourré** [m] **dunaire** (Groupements arbustifs de la dune à bruyère caractérisant le stade préforestier de la colonisation des dunes littorales ; *terme spécifique* désigne un groupement arbustif méditerranéen tel que la brousse à olivier et à lentisque ; ▶dune brune, ▶forêt dunaire) ; *syn.* brousse [f] dunaire ; *g* **Dünengebüsch** [n] (Buschgesellschaft in der ▶Heidedüne als Übergang zum ▶Dünenwald); *syn.* Dünengestrüpp [n].

1643 dune slack [n] *phyt.* (Damp or wet hollow in ▶back dunes or ▶gray dunes, which is underlain by impervious deposits within a large dune system); *syn.* dune trough [n] [also US] (DEN 1971, 13); *s* **valle** [m] **interdunar** (Hondonada húmeda entre ▶dunas grises o ▶dunas pardas sobre lentejas de agua freática dulce; ubicación potencial de ▶bosque dunar); *syn.* corral [m] (FGB 1981); *f* **dépression** [f] **dunaire** (Dépression humide, marécageuse, pouvant contenir de l'eau douce ou saumâtre [▶mardelle], située en pied de la retombée de la dune vive ou à l'intérieur d'un ensemble de ▶dunes grises ou ▶dunes brunes ; station potentielle de la forêt dunaire ; *terme spécifique* **souillère** [f] : dépression localisée entre les dunes littorales) ; *syn.* mare [f] dunaire, panne [f] (dunaire) ; *g* **Dünental** [n] (Feuchte Vertiefung in ▶Graudünen oder ▶Heidedünen, die auf süßen Grundwasserkissen liegen; potenzieller Dünenwaldstandort; bildet sich ein stehendes Gewässer aus, so spricht man von einem ▶Heidetümpel).

dune stabilization [n] *constr.* ▶stabilization of sand dunes.

1644 dune system [n] *geo.* (An aggregate of active or old dunes in a given area); *s* **sistema** [m] **dunar** (Conjunto de dunas vivas o muertas en un área dada que —debido a su origen eólico común— conforman un paisaje diferenciable del de los alrededores); *f* **système** [m] **dunaire** (actif ou ancien) ; *syn.* appareil [m] dunaire ; *g* **Dünensystem** [n] (Gesamtheit von z. T. unterschiedlichen Dünenformen, die auf Grund ihrer äolischen Entstehung einen von der Umgebung abgrenzbaren, ganzheitlichen landschaftlichen Zusammenhang bilden).

dune trough [n] [US] *phyt.* ▶dune slack.

1645 dune woodland community [n] *phyt.* (Final stage in the development of dune vegetation of ▶gray dunes, and especially of ▶backdunes [US]/brown dunes [UK], in a humid location preceded by ▶dune scrub; ▶dune slack); *syn.* dune forest [n] (TEE 1980, 212); *s* **bosque** [m] **dunar** (Estado final del desarrollo de la vegetación dunar, sobre todo en lugares húmedos [▶valle interdunar] de ▶dunas grises y sobre todo de ▶dunas pardas; ▶matorral dunar/de dunas); *syn.* monte [m] dunar; *f* **forêt** [f] **dunaire** (Dernier stade de l'évolution de la végétation sur les sables des ▶dunes grises et des ▶dunes brunes, fortement exposée au vent et poussant en bordure des dépressions humides ; dans les endroits les plus abrités du vent pousse une végétation arborescente avec diverses espèces de pins ; *terme spécifique* bosquet dunaire ; ▶dépression dunaire, ▶fourré dunaire) ; *syn.* forêt [f] littorale ; *g* **Dünenwald** [m] (Meist windgeschorener Waldbestand vor allem in feuchten Lagen [▶Dünental] im Grau- und besonders im Heidedünenbereich; ▶Dünengebüsch, ▶Graudüne, ▶Heidedüne).

duplex [n] [US] *arch. leg. urb.* ▶semidetached house.

1646 durability [n] *ecol. econ. envir.* (**1.** Long-lasting quality of manufactured or processed products to obtain a favorable lifecycle cost ratio aimed at decreasing the consumption of natural resources and maintenance over time. **2.** Concept of prolonging the viability of materials and energy in economic processes); *s* **durabilidad** [f] (**1.** Larga permanencia de los productos en el ciclo económico para ahorrar recursos naturales. **2.** Planteamiento con el fin de prolongar el periodo/período de aprovechamiento de materia y energía en la economía); *f* **long cycle** [m] **de vie** (Caractéristique des produits à longue durée de vie occasionnant une réduction de la consommation des ressources naturelles et ayant une incidence moindre sur l'environnement) ; *syn.* durabilité [f] ; *g* **Langlebigkeit** [f] (1) (**1.** lange Verweildauer von produzierten Gütern im Wirtschaftskreislauf, um den Verbrauch der Naturgüter nachhaltig zu reduzieren. **2.** Denkansatz mit dem Ziel, die Verweildauer von Materie und Energie im Wirtschaftssystem zu verlängern).

1647 duration [n] **of insolation** *met.* (Period of time, i.e. number of hours of sunshine, in which ▶insolation occurs); *s* **periodo/período** [m] **de exposición solar** (Número de horas de sol; ▶insolación); *f* **insolation** [f] (Quantité de rayonnement solaire directe ; ▶ensoleillement) ; *syn.* durée [f] d'ensoleillement ; *g* **Sonnenscheindauer** [f] (Zeitspanne der ▶Besonnung).

1648 dust and aerosol protection forest [n] *for. landsc.* (Type of ▶protective forest absorbing air pollution; the species of trees planted are chosen according to their tolerance either of non-phytotoxic or strongly-harmful emissions); *s* **bosque** [m] **protector contra la inmisión de polvo y aerosoles** (Tipo de ▶bosque protector contra inmisiones cuya composición de especies leñosas se elige teniendo en cuenta su resistencia ante inmisiones no fototóxicas ni fuertemente dañinas para la vegetación); *f* **forêt** [f] **de protection contre les poussières** (Forme de la ▶forêt de protection contre la pollution atmosphérique ; le choix des espèces ligneuses la constituant dépend exclusivement de leur tolérance à la nature et la concentration des particules en suspension dans l'atmosphère) ; *g* **Staubschutzwald** [m] **und Aerosolschutzwald** (Form eines ▶Immissionsschutzwaldes; die Auswahl der anzupflanzenden Gehölzarten richtet sich nach deren Verträglichkeit gegenüber nicht phytotoxischen oder stark pflanzenfeindlichen Immissionen).

dust bin [n] [UK] *envir.* ▶trash can [US] (1).

1649 dust filtering [n] *envir.* (Removal of dust from the air by screening through foliage; ▶dust settlement); *syn.* filtration [n] of dust from the air, filtering [n] of dust; *s* **filtración** [f] **del polvo** (Remoción del polvo del aire por medio de cortinas de vegetación que lo hacen sedimentar; ▶fijación de polvo); *syn.* filtración [f] de las partículas sólidas; *f* **dépoussiérage** [m] (Filtration des fines particules solides retenues par les tissus des végétaux lors du passage d'air chargé de poussières à travers un couvert végétal ; ▶fixation des poussières) ; *g* **Staubfilterung** [f] (Zurückhalten von Staubpartikeln in Vegetationsflächen, durch die die staubbelastete Luft streicht; ▶Staubbindung).

dust particles [npl] *envir.* ▶particulate matter (1).

1650 dust precipitation [n] *envir. limn. pedol.* (Transport and deposition of dust particles from the air onto a certain area; *generic term* ▶matter import); *s* **deposición** [f] **de polvo** (Transporte atmosférico y sedimentación del polvo en un lugar determinado; *término genérico* ▶aporte de sustancias); *syn.* deposición [f] de partículas sólidas; *f* **émission** [f] **particulaire** (Transport et dépôt de particules apportées par le vent dans une région définie ; *terme spécifique* ▶apport d'éléments minéraux) ; *syn.* émanation [f] de poussières ; *g* **Staubeintrag** [m] (Transport von Staub durch die Luft in ein bestimmtes Gebiet; *OB* ▶Stoffeintrag); *syn.* Staubniederschlag [m], Staubdeposition [f].

1651 dust screening [n] *landsc.* (Protection from blowing dust; e.g. by planting of a shelter belt or ▶protective forest absorbing air pollution; ▶dust and aerosol protection forest); *s* **protección** [f] **contra el polvo** (Conjunto de medidas e instalaciones, como plantaciones protectoras y ▶bosques protectores contra inmisiones, que sirven para proteger ante las inmisiones de polvo y aerosoles; ▶bosque protector contra inmisión de polvo y aerosoles); *f* **protection [f] contre les poussières** (Ensemble des mesures et ouvrages de protection [plantation de protection, ▶forêt de protection contre la pollution atmosphérique, zone de protection etc.] réduisant la concentration ou l'exposition aux particules en suspension dans l'air ; ▶forêt de protection contre les poussières) ; *g* **Staubabschirmung [f]** (Summe der Maßnahmen und Einrichtungen, z. B. Schutzpflanzungen, ▶Immissionsschutzwälder, die Schutz vor Staubbelastung gewähren; ▶Staubschutzwald und Aerosolschutzwald).

1652 dust settlement [n] *envir.* (Deposition of ▶particulate matter 1 from the air, after its passing through a belt of vegetation; ▶dust filtering, ▶particulate deposition); *s* **fijación [f] de polvo** (Deposición de ▶partículas sólidas y fijación de las mismas p. ej. gracias a la capacidad de la vegetación de filtrar el aire; ▶filtración del polvo; ▶deposición seca); *f* **fixation [f] des poussières** (Dépôt et rétention des ▶poussières sur une surface, p. ex. ▶dépoussiérage sur les surfaces végétales ou par ▶déposition atmosphérique sèche) ; *g* **Staubbindung [f]** (Ablagerung und Festhalten der ▶Stäube auf Oberflächen, sei es durch ▶Staubfilterung in Vegetationsflächen oder ▶trockene Deposition).

1653 Dutch elm disease [n] *arb. phytopath.* (Infectious agent, *Ceratocystis ulmi*, causing a pathological obstruction of the vascular system in elm trees, and leading to their death; this agent is usually accompanied by a simultaneous infestation of elm bark beetles—*Scolytus scolytus*; ▶dieback of trees [US]/dieback of trees [UK]); *s* **grafiosis [f] del olmo** (Enfermedad que sufre esta especie y que es causada por un hongo *[Ceratocystis ulmi]* que produce la obstrucción patológica del sistema vascular, conlleva su muerte y está poniendo en peligro su supervivencia. Esta enfermedad va frecuentemente acompañada de la infección con escarabajos de la corteza del olmo *[Scolytus scolytus]*; ▶muerte de árboles); *f* **graphiose [f] de l'orme** (Maladie due à un champignon *[Ceratocystis ulmi]*, catastrophique pour cette espèce ; les spores sont disséminées par le vent, la pluie, le contact du grand Scolyte *[Scolytus scolytus]* et véhiculées par la sève, germent et donnent un mycélium qui se développe dans les vaisseaux de l'aubier ; le champignon secrète des toxines et forme des bouchons gommeux [thyles] qui empêchent la circulation de la sève ; les arbres malades débourrent tardivement et incomplètement, les feuilles jaunissent en mai-juin, meurent et tombent, l'écorce du tronc se décolle et l'arbre meurt ; ▶dépérissement des arbres) ; *syn.* dépérissement [m] de l'orme ; *g* **Ulmensterben [n, o. Pl.]** (Absterben der Ulmen durch pathologische Verstopfung der Wasserleitbahnen durch den Pilz *Ceratocystis ulmi* mit meist gleichzeitigem Befall durch den Ulmensplintkäfer *[Scolytus scolytus]*. In den 1930er-Jahren fiel z. B. die Erstpflanzung des Berliner Boulevards, „Kurfürstendamm", dieser Krankheit vollständig zum Opfer; ▶Baumsterben).

duty [n] of occupier to make land or premises safe for persons or vehicles *leg.* ▶public responsibility.

duty to consent [loc], owner's *leg.* ▶private owner's obligation.

1654 dwarf fruit tree [n] [US] *hort.* (**In U.S.**, a fruit tree which is measured from 240-300cm [8-10 feet] of full height; a **miniature fruit tree [n]** is measured from 180-240cm [6-8 feet] of full height. **In U.K.**, a **short standard** is a fruit tree with a stem height of 100-120cm according to the British Standards for Nursery Stock: BS 3936: part 1. The equivalent for small stems is 'short stem' [40-60cm] and for rose trees 'short standard' with a stem height of 30-40cm; in ▶semi-dwarf fruit tree [US]/half

standard [UK]; ▶standard rose); *syn.* short standard [n] [UK]; *s* **árbol [m] de pie bajo** (Altura del tronco de 40-60 cm en frutales, de unos 40 cm en ▶rosal de pie alto; ▶árbol de pie medio); *syn.* pie [m] bajo; *f 1* **arbre [m] mini-tige** (Suivant la Norme N.F. V 12 052 terme qui désigne les arbres fruitiers dont la hauteur de la tige du collet jusqu'à la branche la plus basse doit être égale à 50 cm avec un écart de formation de ± 10 cm sur la hauteur de la tige ; ▶arbre demi-tige) ; *syn.* mini-tige [m] ; *f 2* **arbre [m] courte-tige** (Suivant la Norme N.F. V 12 055 terme qui désigne les arbres dont la hauteur du tronc mesurée du collet jusqu'à la première branche est égale à 0,80 m, l'écart de formation sur la hauteur de la tige est de ± 5 cm ; ▶rosier-tige) ; *syn.* courte-tige [m] ; *g* **Niederstamm [m]** (Gemäß Gütebestimmungen für Baumschulpflanzen ein Obstgehölz mit einer Stammhöhe von 40-60 cm [Kern- und Steinobst] und 40-50 cm bei Beerenobst. Bei Stammrosen muss die Stammhöhe ca. 40 cm betragen [Niederstammrose]; ▶Halbstamm, ▶Rosenhochstamm); *syn.* Fußstamm [m], *obs.* Viertelstamm [m].

1655 dwarfish [adj] *hort. plant.* (Term applied to understock with low vigo[u]r on which cultivars are grafted; ▶vigorous growing; *opp.* ▶vigorous); *s* **de crecimiento débil [adj]** (Término horticultor aplicado p. ej. a patrón sobre el que ha sido injertada una planta; *opp.* ▶de crecimiento vigoroso; ▶vigoroso/a); *f* **à croissance lente [loc]** (Par exemple les porte-greffes à croissance lente qui sont déterminant pour le développement végétatif de l'arbre ainsi que la mise à fruits ; ▶vigoureux ; *opp.* ▶à croissance rapide) ; *g* **schwach wüchsig [adj]** (Eine Pflanze betreffend, die in ihrem Wachstum im Vergleich zu anderen [der selben Art oder Gattung] mit geringerer Vitalität ausgestattet ist und deshalb auch in ihrer Größe kleiner bleibt, z. B. eine **s. w.e** Veredlungsunterlage, **s. w.e** Ziergehölze; ▶wüchsig; *opp.* ▶stark wüchsig); *syn.* schwach wachsend [ppr/adj].

1656 dwarfism [n] *bot. zool.* (**1.** Dwarf or stunted growth of plants caused by dry sites, lack of nutrients or trace elements in the soil. Dwarf forms of ▶fruit trees are originated by grafting on ▶dwarfish ▶understock. **2.** Genetically dwarf growth of animals or a consequence of endocrinic or other metabolic disturbances, sometimes also caused by lack of food); *s* **enanismo [m]** (**1.** En plantas, crecimiento reducido o malformado causado por sequía, falta de nutrientes u oligoelementos. Formas enanas en ▶árboles frutales se crían por injertos sobre ▶patrones que crecen poco [▶crecimiento débil]. **2.** En animales, fenómeno de origen genético o como consecuencia de alteraciones endócrinas u otras enfermedades del metabolismo, algunas veces por mala nutrición); *f* **nanisme [m]** (Ensemble des caractères présentés par les organismes vivants dont la taille est inférieure à la normale ; l'obtention de ces caractères peut être due à la sélection chez les espèces animales et végétales ou bien **1.** pour les végétaux à des conditions stationnelles ou climatiques défavorables [sécheresse, manque en éléments nutritifs ou en éléments trace — formes naines ou rabougries], pour les ▶arbres fruitiers à la qualité du ▶porte-greffe ▶à croissance lente [porte-greffe nanifiant] ou encore **2.** pour les animaux à des troubles physiologiques) ; *g* **Zwergwüchsigkeit [f]** (**1.** Bei Pflanzen durch trockenen Standort, Mangel an Nährstoffen oder Spurenelementen verursachte Zwerg- oder Kümmerformen. Zwergformen bei ▶Obstbäumen entstehen durch Veredlung auf ▶schwach wüchsigen ▶Unterlagen. **2.** Bei Tieren genetisch bedingt oder die Folge von endokrinen oder anderen Stoffwechselstörungen, manchmal auch durch Nahrungsmangel verursacht; *syn.* Nanismus [m], Zwergwuchs [m].

1657 dwarf shrub [n] *bot.* (Small woody plant, up to 50cm in height, frequently growing prostrate on the ground, which lives predominantly in the oceanic, arctic, subarctic, or alpine regions. Dwarf shrubs and ▶suffruticose shrubs belong to the lifeform

D

class *Chamaephytae*; ▶small shrub, ▶surface plant); *s* **mata** [f] (Arbusto de poca altura, a lo sumo de 50 a 60 cm, que crece sobre todo en regiones oceánicas, árticas o subárticas y en las altas montañas; pertenecen junto a las ▶sufrútices a los ▶caméfitos; ▶pequeño arbusto); *syn.* matilla [f]; *f* **arbuste** [m] **nain** (Plante ligncusc bassc pouvant attcindrc 50 cm dc hautcur, de forme prostrée, couchée sur le sol, répandue dans les régions océaniques, arctiques ou subarctiques ainsi qu'en haute montagne ; les **a. n.** appartiennent avec les ▶espèces sous-frutescentes à la forme biologique des ▶chaméphytes ; ▶petit arbuste ; *syn.* buisson [m] nain ; *g* **Zwergstrauch** [m] (Bis 50 cm hohes Gehölz, oft spalierförmig am Boden ausgebreitet, das vornehmlich in ozeanischen, arktischen oder subarktischen Gebieten und im Hochgebirge vorkommt; Zwergsträucher zählen mit den ▶Halbsträuchern zur Lebensform der ▶Oberflächenpflanzen *[Chamaephytae]*; ▶Kleinstrauch).

dwarf-shrubland [n] *phyt.* ▶dwarf-shrub plant community.

1658 dwarf-shrub plant community [n] *phyt.* (Dense or loose dwarf-shrub formation, rarely exceeding 50cm in height, sometimes called heaths or heathlike formations, living in the lower alpine and subalpine zones, e.g. Alpenrose heaths *[Rhododendro-Vaccinietum]*, Heather heaths on acid soils, e.g. sandy heaths *[Genisto-Callunetum typicum]*, Crowberry vegetation *[Empetrion boreale]*, wet-heath vegetation [e.g. *Oxycocco-Sphagnetea]*, etc. According to the varying densities of dwarf-shrub cover, there are 'dwarf-shrub formations' or 'cryptogamic formations' with patches of lichens, mosses and scattered dwarf shrubs; ▶wet-heath vegetation); *syn.* dwarf-shrubland [n]; *s* **matorral** [m] **bajo** (Conjunto de formaciones vegetales de leñosas de pequeño tamaño [< 50 cm] que habitan la zona alpina y subalpina como p. ej. las landas de rosa de los Alpes *[Rhododendro-Vaccinietum et Vaccinio-Pinetum rhododendretosum hirsuti]*, landas de enebro enano *[comunidades de Juniperus alpina]*, landas arenosas *[Calluno-Genistetum typicum]* y en la zona mediterránea: *términos específicos* **romeral** [m] [sobre suelo calcáreo], **brezal** [m] [de *Calluna* y *Erica* sobre suelos ácidos], **jaral** [m] [de especies de *Cistus* sobre suelo de caliza o silíceo], **cantuesal** [m] [sobre suelo ácido cuando domina *Lavandula*], **jaguarzal** [m] [sobre suelo ácido cuando domina *Halimium*], **tomillar** [m] [formación de *Thymus vulgaris*], entre otros muchos; ▶landa pantanosa); *f* **groupement** [m] **d'arbrisseaux nains** (A. Fruticées basses des landes subalpines, 1. rhodoraies et vacciniaies sous forêt claire *[Rhododendro-Vaccinietum et Vaccinio-Pinetum rhododendretosum hirsuti]*, sur terrain *calcaire* le groupement extra-sylvatique à éricacées *[Rhodoreto-Vaccinietum extrasilvaticum]* de la zone de combat ; *syn.* rhodoraie [f] extra-sylvatique ; *sur terrain siliceux* groupement à Genévrier nain ; *syn.* lande [f] extra-sylvatique. **2.** Les juniperaies, groupements à Genévrier nain *[Juniperus alpina]*. **3.** Les formations frutescentes rases, à arbustes prostrés, des stations exposées au vent *[Empetro Vaccinietum]* ; VCA 1985, 211-212. **4.** Fruticées basses des landes atlantiques, les groupements à Éricacées sur sables siliceux telles que les landes sèches à Callune *[Genisto-Callunetum typicum]*, les landes mésophiles et les landes humides *[Ericion tetralicis]*, les groupements à Camarine *[Empetrion boreale]*, les ▶landes marécageuses sur sol tourbeux *[Oxycocco-Sphagnetea]*) ; *syn.* lande [f] d'arbrisseaux nains, lande [f] à fruticées basses ; *g* **Zwergstrauchgesellschaft** [f] (Dichte oder offene, selten über 50 cm hohe Formationen aus niedrig wachsenden Gehölzen der unteren alpinen und subalpinen Stufen, z. B. Alpenrosenheiden *[Rhododendro-Vaccinietum et Vaccinio-Pinetum rhododendretosum hirsuti]*, Zwergwacholderheiden *[Juniperus alpina*-Gesellschaften], *Calluna*-Heiden auf sauren Böden des Tief- und Berglandes, z. B. Sandheiden *[Genisto-Callunetum typicum]*, Krähenbeerenheiden *[Empetrion

boreale]*, ▶Sumpfheiden [z. B. *Oxycocco-Sphagnetea*] etc. Entsprechend der Dichte der Zwergstrauchdecke wird zwischen dichten und offenen Zwergstrauchformationen oder Kryptogamen-Formationen mit Zwergsträuchern unterschieden; cf. ELL 1967); *syn.* zwergstrauchreiche Formation [f], *z. T.* Zwergstrauchheide [f].

1659 dwarf standard rose [n] *hort.* (Low bengalese or polyantha rose which is grafted onto a 40cm ▶standard rose); *s* **rosal** [m] **enano de pie alto** (▶Rosal de pie alto de 40 cm de altura injertado con rosa bengala o rosa polyantha); *f* **rosier** [m] **mini-tige miniature** (Rosier du Bengale ou polyantha greffé sur un ▶rosier tige de 40 cm de hauteur) ; *g* **Zwergstammrose** [f] (Zwergbengalrose oder Polyantharose, die auf einen 40 cm hohen ▶Rosenhochstamm veredelt wurde).

1660 dwarf woody species [n] *arb. hort.* (Tree or shrub which remains naturally small or stunted in growth in contrast to other stronger growing species of its kind); *s* **leñosa** [f] **enana** (Árbol o arbusto que no crece hasta la misma altura que sus congéneres); *f* **espèce** [f] **ligneuse naine** (Plante ligneuse basse qui biologiquement reste basse ou espèce dont la faible croissance est due à l'action des facteurs du milieu) ; *syn. hort.* plante [f] ligneuse naine, *phyt.* arbrisseau [m] nain ; *g* **Zwerggehölz** [n] (Eine von Natur aus klein bleibende oder eine wuchsschwache Abweichung einer stark wüchsigen Gehölzart).

dwelling [n]**, multifamily dwelling (unit)** *arch. urb.* ▶multifamily housing.

dwelling [n]**, terraced** *arch.* ▶stepped house.

dwelling building [n] *arch. urb.* ▶residential building.

1661 dwelling environs [n] *sociol. urb.* (Surroundings of a residential building or residential area which have a direct influence upon the quality of human life; factors include the size and design of open spaces, the convenient location of communal facilities, children's playgrounds, as well as facilities for noise screening and reduction in exhaust gas emissions, if these are necessary; ▶dwelling quality); *syn.* residential environment [n], dwelling surroundings [npl]; *s* **hábitat** [m] **(residencial)** (Totalidad de condiciones físicas que influyen sobre una vivienda o un vecindario, como el tamaño y calidad de las zonas verdes, la ubicación de servicios, de zonas de juegos infantiles, la presencia de medidas de protección contra ruidos o contaminación atmosférica, etc.; ▶calidad residencial); *syn.* ámbito [m] residencial; *f* **1 environnement** [m] **de l'habitat** (Espace extérieur urbain proche du logement accessible aux habitants sous forme d'équipements commerciaux et sociaux, d'emplois, de services ou d'espaces récréatifs ou sportifs pour les enfants, de terrains de jeux ; ▶qualité de l'habitat) ; *syn.* environnement [m] du logement, cadre [m] de vie ; *f* **2 bassin** [m] **de vie** *plan.* (Terminologie de l'aménagement du territoire ; espace de proximité au sein duquel s'organise le lien social ; le regroupement de plusieurs **b.s de v.** permet de former un pays pouvant conduire à un projet de développement économique et d'organisation des services ; circ. du 21 avril 1995) ; *g* **Wohnumfeld** [n] (Nahbereich als Gesamtheit aller äußeren Umgebungsfaktoren einer Wohnung oder eines Wohngebietes mit den auf das Wohnen unmittelbar einwirkenden Lebensbedingungen, wie z. B. Größe und Ausstattung der Freiflächen, Lage der Gemeinbedarfseinrichtungen, der Kinderspielplätze sowie, wenn nötig, Anlagen gegen Lärm und Abgasbelastungen. Der Begriff **Lebensraum**, der syn. verwendet werden könnte, ist im Deutschen durch den politischen Sprachgebrauch seit den 1870er-Jahren und besonders während der Nazizeit [1933-1945] im Sinne von Gebietsforderungen und territorialer Expansion immer noch negativ besetzt; ▶Wohnqualität); *syn.* Nahumwelt [f], Wohnumwelt [f], gebaute Umwelt [f] (±).

1662 dwelling quality [n] *sociol.* (Attribute of a residential area characterized by its attractive location, and other accommodating amentities such as, convenient shopping facilities, proximity to local public transportation system [US]/local transport system [UK]); *s* **calidad** [f] **residencial** (Valoración cualitativa de las características de barrios residenciales como su ubicación, la oferta de servicios, la conexión al transporte público, la calidad de zonas verdes, etc.); *syn.* calidad [f] del hábitat; *f* **qualité** [f] **de l'habitat** (Valeur de l'habitat caractérisée par une situation paysagère privilégiée, la proximité des équipements, services publics et des transports en commun, par l'existence d'une entité communautaire, etc.) ; *g* **Wohnqualität** [f] (Durch bestimmte Merkmale wie z. B. Lage, Einkaufsmöglichkeiten, Anbindung an das öffentliche Personennahverkehrsnetz, Versorgung mit öffentlichen Grünflächen gekennzeichnete Wohnbedingungen).

dwelling surroundings [npl] *sociol. urb.* ▶dwelling environs.

dwelling unit [n]**, multifamily** *arch. urb* ▶multi-family housing.

1663 dy [n] *pedol.* (Decomposed or little decomposed, organic acid deposits in oxygen-poor and usually nutrient-poor water bodies; ▶gyttja, ▶sapropel, *generic term* ▶subhydric soil); *s* **dy** [m] (Sedimento muy húmico, turboso propio de las aguas ácidas; MARG 1977; ▶gyttja, ▶sapropel, *término genérico* ▶suelo subacuático); *f* **dy** [m] (Couche organique meuble d'aspect vaseux dans les eaux brunes, constituée presque exclusivement de flocules humiques amorphes ; les eaux sont pauvres en oxygène et la plupart du temps également pauvres en matières nutritives ; DIS 1986 ; ▶gyttja, ▶sapropèle ; *terme générique* ▶sol subaquatique) ; *g* **Dy** [m] (In sauerstoff- und meist auch nährstoffarmen Gewässern häufig saure, wenig zersetzte organische Ablagerungen; ▶Gyttja, ▶Sapropel; *OB.* ▶Unterwasserboden); *syn.* Braunschlammboden [m].

dying [n]**, top** *arb. for.* ▶top-kill [US].

dying [n] **of a waterbody** *limn. ocean.* ▶biological dying of a waterbody.

dyke [n] [UK] *constr. wat'man.* ▶coastal dike; *agr. constr.* ▶drainage ditch; *agr.* ▶dirt bank [US].

dyke [n] [UK]**, field** *conserv. land'man.* ▶field dike [US].

dyke [n] [UK]**, main** *wat'man.* ▶main dike [US].

dynamic equilibrium [n] *ecol.* ▶ecological equilibrium.

dynamics [npl] *ecol.* ▶population dynamics.

dynamics [npl] *constr. eng. phys.* ▶statics [npl] and dynamics.

dynamics [npl] **of relationships** *ecol.* ▶interdependency within an ecosystem.

1664 dynamic species equilibrium [n] *ecol.* (Species balance maintained with gradual alteration of species composition within an area resulting from influx of ▶migratory species, ▶extinction, or departing species, but being controlled overall by the number of [available] ecological niches); *s* **equilibrio** [m] **dinámico de especies** (Alteración gradual de la composición de especies de un área como resultado del flujo de ▶especies migratorias, de la ▶extinción o emigración de especies. Ambos procesos se ven influenciados por el número de especies o de nichos ecológicos a disposición); *f* **équilibre** [m] **dynamique des espèces** (Variation permanente de la constitution de la population provoquée par l'afflux d'espèces immigrantes et la perte d'espèces en voie de disparition ou émigrantes, maintenue en équilibre par le nombre d'espèces intégrées aux niches écologiques ; ▶espèce migratrice, ▶extinction) ; *g* **dynamisches Artengleichgewicht** [n] (Kontinuierlicher Wandel der Artenzusammensetzung durch Zustrom einwandernder Arten und Verlust ausster-

bender oder abwandernder Arten, wobei beide Variablen in gegenläufiger Weise durch die Anzahl vorhandener Arten resp. besetzter ökologischer Nischen beeinflusst werden; ▶wandernde [wild lebende] Tierart, ▶Aussterben).

1665 dystrophic [adj] *limn.* (Characteristic of rivers or ponds with brown acidic water, low nutrient content and low plant production); *s* **dístrofo** [adj] (Aplícase a los medios acuáticos muy pobres en sustancias disueltas aprovechables para la nutrición de las algas, pero ricos en ácidos húmicos, que dan al agua una acidez más o menos considerable; DB 1985); *syn.* distrófico [adj]; *f* **dystrophe** [adj] (Caractéristique des eaux douces brunes, riches en acides humiques, pauvres en calcaire et à faible production végétale) ; *g* **dystroph** [adj] (Braune, huminsäurereiche Gewässer betreffend, die einen sehr geringen Kalkgehalt haben und eine sehr geringe Pflanzenproduktion aufweisen).

1666 dystrophic lake [n] *limn.* (Lime- and nutrient-poor lake, the water of which is colo[u]red yellow to brown by organic matter); *syn.* brown-water lake [n] [also UK]; *s* **lago** [m] **dístrófico** (Lago pobre en nutrientes, pero rico en humus y teñido de colores amarillo o pardo); *f* **lac** [m] **dystrophique** (Lac pauvre en calcaire et en éléments nutritifs, dont l'eau est de teinte jaune à brune par la présence des humus) ; *g* **Braunwassersee** [m] (Kalk- und sehr nährstoffarmer [dystropher] See, dessen Wasser durch Humusstoffe gelb bis braun gefärbt ist); *syn.* dystropher See [m].

early completion [n]**, bonus for** *contr.* ▶bonus payment.

early foliage [n] [US] *bot. hort.* ▶spring foliage.

1667 early frost [n] *met.* (Premature occurrence of frost in autumn; ▶radiation frost; *opp.* ▶late frost); *s* **helada** [f] **blanca de otoño** (Helada prematura que se presenta en otoño; ▶helada de irradiación; *opp.* ▶helada blanca de primavera); *f* **gelée** [f] **hâtive** (Gelée apparaissant très tôt en automne ; ▶gelée de rayonnement ; *opp.* ▶gelée tardive) ; *syn.* gelée [f] d'automne ; *g* **Frühfrost** [m] (Vorzeitig im Herbst eintretender Frost; ▶Strahlungsfrost; *opp.* ▶Spätfrost).

1668 early wood [n] *arb. bot.* (Less dense, large-celled/open-textured, first-formed part of an ▶annual ring; cf. MGW 1964; ▶late wood); *syn.* spring wood [n]; *s* **leño** [m] **temprano** (Madera nueva del ▶anillo anual actual; ▶leño tardío); *syn.* leño [m] de primavera; *f* **bois** [m] **initial** (Partie d'un ▶cerne annuel dont le bois est le moins dense, à cellules plus grosses et le premier formé ; MGW 1964 ; ▶bois final) ; *syn.* bois [m] de printemps ; *g* **Frühholz** [n, o. Pl.] (Das weniger dichte, erstgebildete Holz eines ▶Jahresringes mit großlumigen Zellen; MGW 1964; ▶Spätholz).

1669 earn [vb] **a commission** *contr. prof.* (To gain or acquire a professional design or planning contract based upon merit, a successful technical and financial proposal or a competition entry. *CONTEXT* to be awarded with a contract); *s* **conseguir** [vb] **un encargo** (Obtener un contrato profesional de diseño o planificación sobre la base de méritos, de una oferta técnica y financiera, por medio de adjudicación negociada o por haber ganado un concurso); *f* **obtenir** [vb] **un marché d'étude** (*Contexte* être missionné[e] pour une étude après avoir remis une offre, avoir gagner un concours ou été retenu lors de la passation d'un marché sous forme de marché négocié) ; *syn.* contracter [vb]

E

un marché d'étude avec qqn. ; *g* **Auftrag erhalten [vb]** (Mit der Ausführung einer Planung auf Grund eines annehmbaren technischen und finanziellen Angebotes, eines gewonnenen Wettbewerbes oder durch eine freihändige Vergabe betraut werden).

earth [n] *constr. eng.* ►dumping of earth; *agr. constr. hort. pedol.* ►soil (1).

earth [n] [UK]**, tipping of** *constr. eng.* ►dumping of earth [US].

1670 earth-covered drainage cleanout pipe [n] [US] *constr. syn.* earth-covered drain cleanout pipe [n], earth-covered drain(age) inspection shaft [n] [UK]; *s* **registro [m] de drenaje cubierto de tierra**; *f* **regard [m] borgne de drainage** ; *g* **erdüberdeckter Dränschacht [m]**.

earth-covered drain(age) inspection shaft [n] [UK] *constr.* ►earth-covered drainage cleanout pipe [US].

1671 earthen wall fortification [n] [US] *urb.* (**In Europe**, defensive system of ramparts and outworks with wide, ramped, earthen walls that surrounded medi[a]eval cities. Most of these walls were destroyed after World War I, and they were replaced by public open green spaces or by broad tree-lined streets [US]/avenues [UK]; **in U.S.**, earthen entrenchments were built during wars in earlier times; e.g. in a ring of forts around Washington, D.C., during the 'Civil War' or 'War Between the States', sometimes preserved in historic battlefield parks; ►removal of fortfications); *syn.* town fortification [n] [UK] (JEL 1986, 457); *s* **fortificación [f] de tierra** (±) (Antiguas fortificaciones de ciudades medievales hechas de terraplenes de tierra, que en Alemania fueron desmontadas después de la 1ª Guerra Mundial y convertidas en parques o amplias avenidas de ronda arboladas; ►desmantelamiento de fortificaciones); *f* **fortification [f] de terre** (Ouvrage militaire défensif, enceinte entourant un grand nombre de villes européennes constitués de levées de terre et d'un glacis ; la plupart des **f.s** furent détruites après la première guerre mondiale et transformées en allées ou espaces verts ; ►arasement de[s] fortifications) ; *g* **Wallanlage [f]** (Historische Befestigungsanlage europäischer Städte mit geböschten Wällen, die im Vergleich zu Befestigungsmauern nicht mehr „eingeworfen" werden konnten. Die meisten **W.n** wurden nach dem Ersten Weltkrieg geschleift und als Grünanlagen oder zu Baumalleen umgebaut; ►Schleifen von Wallanlagen).

1672 earth excavation [n] *constr.* (Removal of earth from the ground to the subbase grade [US]/formation level [UK] and the resulting depression; ►excavated material, ►topsoil stripping, ►volume of stripped topsoil); *s* **excavación [f] de tierra** (Remoción de tierra para llegar al nivel de subrasante; ►material de desmonte, ►remoción de la capa de tierra vegetal, ►volumen de acopio de tierra vegetal); *f 1* **terrassement [m] en pleine masse** (Afin de dégager des ouvrages enfouis ou d'exécuter en déblai une plate-forme pour la voirie ou la fondation d'ouvrages ; ►matériau extrait, ►retroussement de terre végétale, ►retroussis de terre végétale) ; *f 2* **terrassement [m] en déblai** (Action de terrasser) ; *syn.* mouvement [m] de terre en déblai ; *g* **Bodenabtrag [m]** (zwecks Freilegung tiefer gelegener Objekte oder Erstellung eines tiefer liegenden Planums; ►Aushub, ►Oberbodenabtrag 1, ►Oberbodenabtrag 2).

1673 earth fill [n] *constr.* (**1.** Piece of land raised artificially to a required level. **2.** Extent, condition and result of ►earth filling 1); *s* **relleno [m] de tierra (1)** (Resultado de ►relleno de tierra 1); *f* **remblai [m] de terres** (Résultat d'un ►remblaiement) ; *g* **Bodenauftrag [m] (1)** (Umfang, Zustand, Ergebnis von ►Bodenauftrag 2).

earth fill [n] **around a tree** *arb. constr.* ►groundfill around a tree.

1674 earth filling [n] (1) *constr.* (*Generic term for three methods* ►filling and compacting in layers, ►front end dumping, ►side dumping; ►dumpsite [US]/tipping site [UK], ►backfilling); *syn.* dumping [n]; *s* **relleno [m] (1)** (*Término genérico para tres tipos de rellenos* ►relleno y compactación por tongadas, ►terraplenado frontal, ►descargamiento lateral; ►depósito de residuos sólidos, ►relleno de zanjas y hoyos); *syn.* terracerías [fpl] [MEX]; *f* **exécution [f] d'un remblai** (*Terme générique désignant deux formes du remblaiement* ►terrasser et compacter en couche ainsi que le ►remblai par avancement ou ►déversement latéral ; ►décharge publique, ►remplissage [d'une tranchée/d'un trou]) ; *g* **Schüttung [f] (1)** (*OB zu drei Bodeneinbauverfahren* ►lagenweiser Einbau und ►Kopfschüttung sowie ►Seitenschüttung; ►Deponie, ►Verfüllen).

1675 earth filling [n] (2) *constr.* (Process of adding soil on a site; ►topsoil spreading, ►replacement of topsoil); *s* **relleno [m] de tierra (2)** (Proceso; ►extender de tierra vegetal, ►reextensión de tierra vegetal); *f* **terrassement [m] en remblai** (Processus de mise en place de terres ; ►mise en place de la terre végétale, ►remise en place de la terre végétale) ; *syn.* mouvement [m] de terre en remblai ; *g* **Bodenauftrag [m] (2)** (Vorgang des Einbaus von Bodenmaterial; ►Andecken von Oberboden, ►Wiederandecken von Oberboden); *syn.* Boden(auf)schüttung [f], Erdaufschüttung [f].

1676 earthflow [n] *geo.* (Relatively quick flowage movement of water-saturated clayey or silty earth material down low-angle terraces or hillsides, forming a circular cut, or shear plane and a bulging accumulation mound; cf. GGT 1979; however, dry flows of granular material are also possible; ►mass slippage [US]/mass movement [UK]); *s* **alud [m] de tierras [Es, YV]** (Tipo de ►movimiento gravitacional rápido causado por la saturación del suelo arcilloso o limoso de terrazas ligeramente inclinadas o en pendientes, caracterizado por un plano de corte circular sobre el que se produce y un montículo de acumulación); *syn.* flujo [m] de tierras [CO, EC], flujo [m] de barro [RCH] (MESU 1977); *f* **coulée [f] de terre [f]** (Mouvement rapide de descente sur un versant de matériaux fortement imbibés d'eau en masse et mettant en place des accumulations bosselées ou en forme de langue ; à l'état de liquidité, on parle de « coulée boueuse » ; cf. DG 1984, 419 ; ►mouvement de masse) ; *syn.* écoulement [m] de sol ; *g* **Erdschlipf [m]** (Schnell fließende Massenbewegung in wasserdurchtränkten Lockersediment- und Verwitterungsdecken an Hängen, bei der eine halbkreisförmige Abrissnische und muldenförmige Gleitbahn sowie eine wulst- oder zungenartige Akkumulationsform entsteht. Dieser gravitative ►Massenversatz entsteht auch an Schichten geringer Mächtigkeit und wird in der Fachsprache auch **Frana** [ital.] genannt); *syn.* Bergschlipf [m], Frana [f].

earth forces [n] [US] *constr. eng.* ►earth pressure.

1677 earth formwork [n] *constr.* (Sides of a ►foundation trench consisting of stable soil used to contain foundation material; ►formwork); *s* **encofrado [m] de tierra** (Bordes de ►zanja de cimentación de tierra compacta que se utilizan en vez del ►encofrado de madera); *f* **coffrage [m] en terre** (►Tranchée de fondation dont les parois en terre sont assez stables pour servir de ►coffrage pour la mise œuvre des matériaux de fondation ; (MAÇ 1981, 149) ; *syn.* coffrage [m] dans le sol ; *g* **Erdschalung [f]** (Standfeste Bodenwände eines ►Fundamentgrabens zur seitlichen Begrenzung des einzubauenden Fundamentschüttgutes statt einer Holzschalung; diese wird nur im oberen Teil nötig, um einen einwandfreien Übergang vom Fundament zum Mauerfuß herzustellen; ►Schalung).

1678 earth layer [n] *constr.* (Layer created by filling to a certain thickness; ►fill material); *s* **capa [f] de relleno** (Resultado; ►tierra de relleno); *f* couche [f] de remblai (Résultat des travaux de mise en remblai sur une certaine hauteur des ►matériaux de remblaiement) ; *g* Schüttung [f] (2) (*Ergebnis* Schicht eines in bestimmter Höhe eingebauten ►Füllbodens); *syn.* Schicht [f].

1679 earth mound [n] *constr.* (Dumped [US]/tipped [UK], longitudinal pile of soil [US]/heap of soil [UK], or earthen sculptural form); *syn.* bund [n] [also UK], berm [n]; *s* terraplén [m] (Montón alargado de tierra de 3-4 m de altura que puede ser objeto de tratamiento paisajístico o tener funciones de protección contra caída de rocas, contra el ruido, etc.); *syn.* berma [f], albarrada [f]; *f* merlon [m] (*En général* ouvrage de terre en remblai longitudinal, de 3 à 4 m de hauteur, constitué de matériaux récupérés lors de déblais, recouverts de terre, faisant l'objet d'un traitement paysager et utilisé pour la protection contre les chutes de blocs rocheux, contre le bruit, etc., implanté en lisière d'agglomération ou le long d'infrastructures de transport) ; *g* Erdwall [m] (Geschütteter, länglicher Erdkörper).

1680 earth-moving [n] *constr.* (Movement of large volumes of [usually] subsoil in connection with ground model[l]ing, including excavation, transport and intermediate storage; ►earthworks, ►excavation 1, haulage, loading and transport of *in situ* material); *s* movimiento [m] de tierras (Movimiento de grandes volúmenes de subsuelo para conseguir un nivel de subrasante o para modelar el terreno; ►excavación, movimiento, carga y transporte de tierra, ►terracerías); *syn.* movimiento [m] de suelos; *f* mouvement [m] de terre (Terme général désignant le décapage, l'extraction ou la reprise sur stock, le chargement et le transport au lieu d'emploi, de dépôt ou de décharge de déblais de toute nature ; ►extraction, déplacement, chargement et transport des terres, ►terrassements) ; *syn.* déblais [mpl] généraux, mouvement [m] des terres ; *g* Bodenbewegung [f] (►Lösen, Fördern, Laden und Transportieren von Bodenmassen; ►Erdbau); *syn.* Bodenförderung [f], Erdbewegung [f], Erdmassenbewegung [f], Massenbewegung [f].

earth opening [n] *constr.* ►earth void.

1681 earth pressure [n] *constr. eng.* (Force of a body of earth, with certain soil characteristics such as soil structure, density, and cohesion or the angle of internal friction upon a structure; e.g. behind a ►retaining wall; there are two types: the **active e. p.**, which occurs when the structure can give way to a certain extent — e.g. a few centimetres; if the load is higher, there is an increased **e. p.**; **passive e. p.** is the greatest pressure which a body of earth can exert upon a structure without the structure being moved; *syn.* soil pressure [n], earth forces [npl] [also US]; *s* empuje [m] de tierra (Esfuerzo oblicuo debido a las tierras en un ►muro de sostenimiento. Hay dos tipos de **e. de t.**: **empuje activo** se presenta cuando la construcción puede ceder p. ej. unos pocos centímetros; cuando la carga es superior existe un **e. de t. activo incrementado.** El **e. de t. pasivo** es la presión máxima con la que un cuerpo de tierra puede apoyar una construcción sin que ésta se mueva; cf. DACO 1988); *f* poussée [f] des terres (Poussée exercée par le poids propre d'une butée de terre, p. ex. contre un ►mur de soutènement, dépendant des caractéristiques mécaniques [nature du sol, cohésion, etc.] et géométriques du sol [pente, angle de frottement] ; parmi les forces qui déterminent la cinématique d'un ouvrage de soutènement [rigide ou souple] on distingue **1.** la **poussée** ou **pression active**, c.-à-d. la force agissante qui tente de pousser ou de faire pivoter un mur et qu'il va falloir contrer, **2.** la **butée** ou **pression passive** des terres constituant l'appui opposé par le sol à la partie en fiche de l'ouvrage) ; *syn.* poussée [f] du sol ; *g* Erddruck [m] (Druck eines Erdkörpers mit definierten Eigenschaften wie Bodenart, Dichte, Konsistenz resp. innerer Reibungswinkel auf ein Bauwerk, z. B. auf eine ►Stützmauer; es gibt zwei Arten: der aktive Erddruck tritt ein, wenn das Bauwerk in gewissen Grenzen nachgeben kann — in wenigen Zentimetern; wenn die Belastungen höher sind, liegt ein **erhöhter aktiver Erddruck** vor. Der **passive Erddruck** ist der größte Druck, mit dem ein Erdkörper ein Bauwerk stützen kann, ohne es zu verschieben).

earth root-ball [n] [UK] *hort.* ►soil root ball.

earth shaping works [n] [UK] *constr. plan.* ►ground modeling [US]/ground modelling [UK].

1682 earth void [n] *constr.* (SPON 1986, 373ss; interstitial space in a concrete ►grass paver to be filled with growing medium); *syn.* earth opening [n], hole [n] [UK], interstice [n] (TGG 1984, 150); *s* alveolo [m] (Agujeros en ►adoquín pavicésped que se rellenan de tierra como sustrato vegetal); *f* alvéole [f] (Cavité d'une ►dalle gazon remplie de terre) ; *g* Erdkammer [f] (Hohlraum in ►Rasengittersteinen oder Lochziegeln zur Aufnahme von Erdsubstrat. Wenn diese **E.n** mit Rasen bestockt sind, werden sie auch **Rasenkammern** genannt); *syn.* Vegetationskammer [f].

1683 earthworks [npl] *constr. eng.* (Generic term for excavation and earth grading involved in earthwork construction for paths, roads, embankments, building foundations, retaining walls, etc., and in spreading of topsoil for planting; ►cut and fill, ►earth moving); *syn.* groundworks [npl] [also UK]; *s* terracerías [fpl] (Término genérico para todo tipo de trabajo de excavación y aplanado de tierras en el contexto de la construcción de caminos, carreteras, cimientos de edificios, muros de contención, taludes, etc.; ►desmonte y terraplén, ►movimiento de tierras); *f 1* terrassements [mpl] (Terme générique pour tous travaux de modelage du sol ou du terrain existant pour lui faire prendre le profil projeté ; on distingue parfois les terrassements préliminaires — [retroussements, déblais, remblais, fouilles], les terrassements secondaires [encaissement, nivellement superficiel et le terrassements particuliers [déblais pour réalisation de fosses, mise à gabarit de fosses, fouilles en tranchées pour conduites et réseaux divers, décapage de drains, remise sous profil d'accotements, déblais pour purges, terrassements pour ouvrages d'art, etc] ; ►déblai et remblai, ►mouvement de terre) ; *syn.* travaux [mpl] de terrassement ; *f 2* terrassements [mpl] généraux (Mouvements de terre dans les travaux d'espaces verts [enlèvement, chargement, transport, évacuation, stockage] comprenant le décapage du terrain ou des emprises de terrassement [déblais de terre de retroussement] avec mise en dépôt de la terre végétale, les remblais généraux avec essais à la plaque et la construction de la couche de forme pour la réalisation des cheminements ou de voies de circulation, les déblais généraux, la réalisation de fouilles, de tranchées ou de fosses pour tout ouvrage d'art, murs de soutènement, etc., ainsi que les remblais de terre végétale pour gazonnements et plantations ; cf. C.C.T.C. fascicule n° 3 ; ►travaux préliminaires, ►terrassements particuliers) ; *g* Erdbau [m, o. Pl.] (Gesamtheit der Baumaßnahmen, die Boden jeglicher Art in seiner Lage und Lagerungsdichte durch Lösen, Laden, Fördern, Einbauen und Verdichten verändert, z. B. beim Bau von Erdkörpern für Wege, Straßen, Plätze, Dämme und Erdmodellierungen, bei Bodenverfüllungen, bei der Herstellung von Gründungen und Baugruben von Bauwerken, Stützmauern etc. sowie beim Einbau von Oberboden für vegetationstechnische Zwecke; ►Auftrag und Abtrag, ►Bodenbewegung); *syn.* Erdarbeiten [fpl], Erdbaumaßnahmen [fpl].

1684 earthworks calculation [n] *constr. eng.* (Estimation of the volume of earth-moving at the final design stage, which is usually broken down into amounts of ►cut and fill; ►no cut-no fill line, ►real earthworks measurement, ►rough estimate of earthworks quantities); *syn.* cut/fill quantity calculations [npl] [also US]; *s* cálculo [m] del movimiento de tierras (Estimación

del movimiento de tierras hecha [▶desmonte y terraplén] para determinar precios para una oferta; ▶cómputo final del movimiento de tierras, ▶estimación de volúmenes de movimiento de tierras, ▶línea de excavación); *syn.* cálculo [m] de terracerías; *f* **évaluation [f] des cubatures** (Calcul du mouvement des terres à remuer aussi bien en déblai qu'en remblai exécuté en vue de l'établissement d'un descriptif et présenté au métré linéaire, au m³ pour une épaisseur > 25 cm ou au m² pour une épaisseur < 25 cm ; ▶calcul des cubatures [en déblai et/ou en remblai], ▶estimation sommaire des cubatures, ▶tracé du projet) ; *g* **Erdmassenberechnung [f]** (Berechnung der zu bewegenden Erdmassen [▶Auftrag und Abtrag] im Rahmen der Massenermittlung für Ausschreibungen; ▶Erdmassenabrechnung, ▶Erdmassenüberschlag, ▶Nulllinie); *syn.* Erdmassenermittlung [f], Erdmengenberechnung [f], Erdmengenermittlung [f].

earthworks measurement [n] *constr. contr.* ▶real earthworks measurement.

earthworks quantities [npl] *constr. eng.* ▶rough estimate of earthworks quantities.

1685 easement [n] *leg.* (**1.** Grant of permitted or prohibited use and access over private property to a neighbo[ur]ing piece of land; ▶private owner's obligation. **2.** Grant for specified access and use over private property for such public purposes as, e.g. ▶utility easement, ▶vehicular easement, ▶vehicular and pedestrian easement: Easements may be both affirmative and negative. An affirmative **e.** entitles the holder to do affirmative acts on the land of another: a right of passage, ditch, drain or other servitude. Negative **e.** prohibits or restricts a landowner from doing something he would have a right to do on his own land; e.g. put up a building of a certain design, etc. They may be created by express grants, reservation, implication, estopel, dedication, prescription, etc.; ▶private owner's obligation, ▶right-of-way); *s* **servidumbre [f] inmobiliaria** (En general, privilegio intrínseco adquirido que una persona puede tener sobre el terreno de otro; ▶obligación de tolerar servidumbres, ▶servidumbre de acceso, ▶servidumbre de acceso para vehículos, ▶servidumbre de paso de conducciones, ▶servidumbre de servicios públicos); *f* **servitude [f]** (Charge ou restriction imposée à l'usage et au droit de la propriété immobilière et foncière. La servitude peut imposer **1.** des interdictions au propriétaire du fonds servant [p. ex. servitude « non altius tollendi » — de bâtir au-dessus d'une hauteur déterminée] ou bien de tolérer certains droits au bénéfice du propriétaire du fond dominant [p. ex. droit de passage], **2.** des restrictions au droit de propriété immobilière pour raison d'intérêt général [servitude de mitoyenneté, de vue, etc.] ou d'utilité publique [servitude de passage, de voire, d'urbanisme, etc.] ; ▶obligation de tolérer les servitudes d'utilité publique, ▶servitude de passage, ▶servitude de passage de véhicules, ▶servitudes d'utilité publique, ▶servitudes relatives à l'utilisation des réseaux d'alimentation et de canalisation) ; *g* **Grunddienstbarkeit [f]** (Das dem jeweiligen Eigentümer eines Grundstückes oder der Allgemeinheit zustehende Recht zur begrenzten Nutzung eines anderen Grundstücks, z. B. ▶Fahrrecht, ▶Geh- und Fahrrecht, ▶Leitungsrecht. Inhalt kann auch sein, dass der Eigentümer des dienenden Grundstücks bestimmte, ihm sonst mögliche Handlungen unterlassen [z. B. Bebauungsbeschränkungen] oder vom herrschenden Grundstück ausgehende Beeinträchtigungen [z. B. übermäßige Immissionen], die er sonst verbieten könnte, dulden muss; cf. MEL, Bd. 11, 93; ▶Duldungspflicht, ▶Geh-, Fahr- und Leitungsrecht).

easement [n] **limiting the height of a building** *leg. urb.* ▶zone with an easement limiting the height of a building.

1686 easily soluble [adj] *chem. pedol.* (Descriptive term applied to the easy dissolvability of soil nutrients in water; *opp.*

▶poorly soluble); *s* **de gran solubilidad [loc]** (Término descriptivo aplicado a la facilidad de los nutrientes del suelo de disolverse en el agua; *opp.* ▶poco soluble); *syn.* fácilmente soluble [adj]; *f* **facilement soluble [loc]** (PED 1979, 376, 388 ; dont les caractéristiques permettent une dissolution rapide, p. ex. les substances nutritives dans le sol ; *opp.* ▶difficilement soluble) ; *syn.* de bonne solubilité [loc], de solubilité facile [loc] (PED 1979, 180) ; *g* **leicht löslich [adj]** (So beschaffen, dass es sich in Flüssigkeit leicht auflöst, z. B. Nährstoffe im Boden; *opp.* ▶schwer löslich).

1687 easily workable soil [n] *agr. hort.* (Coarse-textured soil with a low drawbar pull, and hence easy to cultivate; cf. RCG 1982; ▶crumb structure, ▶fine-textured soil, ▶firm soil, ▶granular soil structure, ▶sandy soil); *syn. obs. in scientific use* light soil [n] [also US]; *s* **suelo [m] ligero** (Término antiguo que se usaba para ▶suelos arenosos o de textura gruesa. Se aplica a suelos de textura entre gruesa y media con un contenido bajo de limo y arcilla o de estructura de un solo grano de consistencia suelta. Es un suelo fácil de cultivar; MEX 1983; ▶estado grumoso, ▶suelo consistente, ▶suelo pesado); *f* **terre [f] douce** (Terre caractérisée par des propriétés physiques [▶état structural grumeleux] assurant une bonne réserve en eau et un travail facile ; LA 1981, 1093 ; ▶sol ferme, ▶sol sableux, ▶terre forte) ; *syn.* terre [f] franche, terre [f] légère, sol [m] meuble ; *g* **leicht bearbeitbarer Boden [m]** (Boden, der auf Grund der gut entwickelten und stabilen ▶Krümelstruktur 2 oder auf Grund des hohen Sandanteils, z. B. anlehmiger Sand [Sl], lehmiger Sand [lS], stark sandiger Lehm [SL], mit Maschinen und Geräten ohne zusätzlichen Aufwand bearbeitet werden kann; im praktischen Landbau gibt es die nicht exakt definierten Begriffe wie ,leichter Boden', ▶mittelschwerer Boden und ▶schwerer Boden; ▶Krümelstruktur 1, ▶Sandboden); *syn. agr. hort.* leichter Boden [m].

1688 east-facing slope [n] *geo. met.* (Side of a hill facing east); *s* **pendiente [f] este** (Ladera de montaña orientada hacia el este); *f* **versant [m] est** (Versant d'une montagne orienté vers l'est) ; *g* **Osthang [m]** (Ein nach Osten exponierte Seite eines Berges).

echelon parking layout [n] *plan. trans.* ▶angle-parking layout.

eco-audit [n] *envir. leg.* ▶environmental audit.

ecoclimate [n] *for. met. phyt.* ▶site climate.

ecofactors [npl] *ecol.* ▶ecological factors.

1689 ecological [adj] *ecol. envir.* (**1.** Relating to the science of ▶ecology. **2.** Descriptive term applied to the environment of living organisms or with their pattern of interactions and relationships. **3.** *NOTE* often used incorrectly as an attribute for processes or objects, the existence of which have little or no impact on natural resources or systems; in this sense the terms environmentally safe, environment-friendly or environmentally sound or appropriate would be better); *s* **ecológico/a [adj]** (**1.** Relacionado con la ciencia de la ▶ecología. **2.** Término descriptivo aplicado al medio ambiente de los organismos vivos o a sus interacciones e interrelaciones. **3.** *NOTA* frecuentemente el término es usado incorrectamente como atributo de procesos u objetos cuya existencia no tiene ningún o muy poco impacto en los recursos naturales o en los ecosistemas; en este sentido sería mejor utilizar apropiado al, no dañino al, respetuoso con o compatible con el medio ambiente); *f* **écologique [adj]** (**1.** Qui a trait à la science de l'▶écologie. **2.** Terme utilisé à mauvais escient désignant les processus ou objets dont l'existence ont une influence négative très limitée sur les écosystèmes naturels et dont les cycles et les flux énergétiques se rapprochent des écosystèmes non dégradés ; dans cette acception il est préférable d'utiliser p. ex. les termes de respectueux de la nature/de l'environ-

nement, compatible avec l'environnement, critères environnementaux) ; **g ökologisch [adj]** (**1.** Die ►Ökologie betreffend. **2.** Beschreibender Begriff für die Wechselwirkung und das Beziehungsgeflecht der Organismen mit ihrer Umwelt. **3.** Als Attribut begrifflich häufig falsch verwendet für Vorgänge oder Objekte, deren Existenz die natürlichen Lebensgrundlagen möglichst wenig belasten und sich den Funktionsweisen, den Stoffkreisläufen und Energieflüssen unbeeinflusster Ökosysteme annähern. In diesem Sinne sollten besser Termini wie naturverträglich, umweltverträglich, umweltschonend oder „nach umweltrelevanten Kriterien" benutzt werden).

1690 ecological amplitude [n] *ecol.* (Maximum extent of genetically established tolerance or limits within which a species reacts to changes in one or more environmental factors by adapting, dying [plants], adapting or migrating [animals]; ►ecological factors, ►ecological tolerance); *syn.* ecological range [n] [also US] (TEE 1980, 30), ecological valence [n]; **s amplitud [f] ecológica** (Expresión de la tolerancia ambiental de una especie, normalmente medida a través de la diferencia entre valores máximos y mínimos de uno o varios parámetros ecológicos que la especie es capaz de soportar sin menoscabo de su desarrollo; DINA 1987; ►factores ecológicos, ►tolerancia ecológica); *syn.* valencia [f] ecológica (MARG 1977); **f valence [f] écologique** (Tolérance d'un organisme envers l'action d'un ou de plusieurs facteurs du milieu ; un individu possédant une grande v. é., est qualifié d'euryèce, un individu de faible v. é. est qualifié de sténoèce ; ►facteurs écologiques, ►plasticité écologique ; DEE 1982, 91) ; *syn.* amplitude [f] écologique, domaine [m] écologique ; **g ökologische Amplitude [f]** (Genetisch festgelegte maximale Reaktionsbreite einer Art, innerhalb derer Veränderungen eines oder mehrerer Umweltfaktoren [►Ökofaktoren] ertragen werden; ►ökologische Potenz); *syn.* ökologische Valenz [f].

ecological balance [n] *ecol.* ►ecological equilibrium.

ecological biodiversity [n] *ecol.* ►biodiversity.

ecological buffer capacity [n] *ecol. recr.* ►environmental stress tolerance.

ecological capacity [n] *ecol. recr.* ►environmental stress tolerance.

ecological coherence [n] *conserv. ecol. pol.* ►coherence.

ecological community [n] *biol. ecol.* ►biocenosis.

1691 ecological cycle [n] *ecol.* (Circulation of matter and energy in natural systems transmitted by animals and plants; ►biogeochemical cycle, ►energy flow, ►hydrologic cycle); *syn.* cycling pool [n] (TEE 1980, 248); **s ciclo [m] ecológico (de materia y energía)** (Circulación de materia y energía en sistemas naturales realizada por los organismos vivos; ►ciclo biogeoquímico, ►ciclo hidrológico, ►flujo de energía); **f cycle [m] naturel** (Cycles biogéochimique et d'énergie dans un système naturel dans lequel la distribution des bioéléments est assurée par la faune et la flore ; ►cycle biogéochimique, ►cycle de l'eau, ►flux d'énergie) ; **g ökologischer Kreislauf [m]** (...läufe [pl]) Stoff- und Energiekreislauf im Naturhaushalt, dessen Träger und Übermittler Tiere und Pflanzen sind; ►Energiefluss, ►Stoffkreislauf, ►Wasserkreislauf).

1692 ecological distribution [n] **of spatial patterns** *ecol. plan.* (LE 1986, 204s; spatial organization of one or more ►landscape factors, as a means of evaluating individual spatial units with a view to their function as habitats or use by society; ►physiographic division, ►spatial distribution 2); *syn.* ecological landscape pattern [n]; **s estructuración [f] ecológica del territorio** (Término genérico para todas las estructuraciones del paisaje o de uno o varios ►factores del paisaje, que se

realizan con el fin de valorar las diferentes unidades espaciales para las necesidades de uso de la sociedad; ►tipología del paisaje, ►estructuración del territorio); **f division [f] de l'espace en territoires naturels** (Division spatiale des paysages établie sur la base d'un ou plusieurs ►facteurs paysagers dans le but d'une évaluation des unités spatiales pour les besoins de la société ; ►division spatiale, ►typologie physiographique des paysages) ; *syn.* division [f] en espaces naturels ; **g ökologische Raumgliederung [f]** (Räumliche Gliederung einer Landschaft oder einzelner resp. mehrerer ►Landschaftsfaktoren, die mit dem Ziel einer Wertung der einzelnen Raumeinheiten für Nutzungsansprüche der Gesellschaft und hinsichtlich der Funktion als Biotop durchgeführt wird; ►naturräumliche Gliederung, ►Raumgliederung).

1693 ecological equilibrium [n] *ecol.* (Functional state of a system or parts thereof which retains its identity despite a continual flow of material and energy. Many regulation mechanisms allow the system to oscillate in a steady state); *syn.* dynamic equilibrium [n], flow equilibrium [n], ecological balance [n], steady state [n]; **s equilibrio [m] dinámico** (Estado funcional de un sistema caracterizado por un flujo continuo de materia y energía, sin que se altere la identidad del mismo. Muchos mecanismos de regulación permiten al sistema variar alrededor de un estado de equilibrio *[steady state]*); *syn.* equilibrio [m] ecológico; **f équilibre [m] écologique** (État fonctionnel d'un système caractérisé par une circulation permanente d'énergie et de bioéléments sans que l'identité du système soit remise en question ; la nature de l'é. é. est telle que celui-ci est constamment modifié par un grand nombre de mécanismes de régulation et oscille autour d'un état moyen d'équilibre) ; *syn.* équilibre [m] dynamique ; **g ökologisches Gleichgewicht [n]** (Funktionszustand eines Systems oder Teilen davon, der sich durch einen ständigen Material- und Energiefluss auszeichnet, ohne dass die Identität des Systems verlorengeht. Zur Natur des ö. G.es gehört es, dass es nie ganz eingehalten werden kann, sondern durch eine Vielzahl von Regelmechanismen um einen mittleren Gleichgewichtszustand *[steady state]* pendelt); *syn.* Fließgleichgewicht [n], dynamisches Gleichgewicht [n].

1694 ecological factors [npl] *ecol.* (Conditions of life upon which all living organisms are dependent [US]/dependant [UK], which are differentiated as follows: **abiotic** [e.g. soil, water, topography, climate] and **biotic e. f.s** [plants, animals, humans]. Due to the importance of food for organisms, **trophic factors** form a third group of e. f.s; ►site factors); *syn.* ecofactors [npl], environmental factors [npl]; **s factores [mpl] ecológicos** (Todas las características del medio susceptibles de actuar directamente sobre los seres vivos. Se distingue entre factores bióticos y abióticos [climáticos, edáficos, químicos]; PARRA 1984; ►factores mesológicos); *syn.* factores [mpl] ambientales; **f facteurs [mpl] écologiques** (Tout caractère du milieu physique ou biotique susceptible d'agir sur la répartition géographique et stationnelle et sur la vie des organismes vivants dans la nature. Les facteurs écologiques sont nombreux et peuvent être classés selon **1.** les milieux où ils s'exercent [facteurs édaphiques, topographiques, biotiques], **2.** leurs modalités d'action sur le milieu [facteurs énergétiques, chimiques, hydriques, abiotiques], **3.** l'importance pour l'alimentation des organismes [facteurs trophiques] ; cf. DEE 1982 ; ►facteurs de la station) ; *syn.* facteurs [mpl] du milieu, facteurs [mpl] environnementaux ; **g Ökofaktoren [mpl]** (Wirkgrößen, von denen die Lebensmöglichkeiten aller Lebewesen abhängen; man unterscheidet **abiotische** [z. B. Boden, Wasser Relief, Klima] und **biotische** Ökofaktoren [Pflanzen, Tiere, Menschen]. Wegen der Bedeutung der Nahrung für Organismen unterscheidet man als dritte Gruppe

die **trophischen Faktoren**; ▶Standortfaktoren); *syn.* ökologische Faktoren [mpl], Umweltfaktoren [mpl].

1695 ecological interactions [npl] **and interrelationships** [npl] *ecol.* (Reciprocal actions and interconnections with interdependency and spheres of influence among living organisms within a community, as well as their relationship with the abiotic environment); *s* **sistema [m] de interrelaciones e interacciones ecológicas** (Efectos, dependencias e interrelaciones entre los organismos de una biocenosis entre sí, y entre éstos y el medio abiótico); *f* **tissus [mpl] d'interactions et de rapports interspécifiques écologiques** (Système d'actions réciproques et de dépendances des espèces d'une communauté entre elles ainsi qu'avec les facteurs abiotiques du milieu de vie); *g* **ökologisches Gefüge [n]** (Wirk-, Abhängigkeits- und Verflechtungsbeziehungen sowohl zwischen den Organismen einer Lebensgemeinschaft als auch zwischen diesen und ihrer abiotischen Umwelt).

ecological intercompatibility [n] *landsc. plan.* ▶matrix on ecological intercompatibility.

ecological landscape pattern [n] *ecol.* ▶ecological distribution of spatial patterns.

1696 ecologically-based land management [n] *agr. for. nat.res. obs.* (Preservation and improvement of conditions in a landscape ecosystem, which ensures the sustained productivity and efficiency of ▶agriculture and ▶forestry 1); *s* **gestión [f] ecológica de zonas agrosilvícolas** (±) (D. por el término «Landeskultur» casi obsoleto se entienden todo tipo de medidas para promover el buen funcionamiento ecológico y productivo de las áreas rurales con usos agropecuarios [▶agricultura] y silvícolas [▶silvicultura artificial]; a veces se utiliza como sinónimo de «Landespflege»); *f* **gestion [f] du terroir** (±) (D., Terme quasi-obsolète dont l'usage était courant au 19ème et au début du 20ème siècle désignant la conservation et la valorisation durable des systèmes écologiques assurant un développement des potentialités du milieu naturel en faveur des activités de l'▶agriculture et de la ▶sylviculture ; cette notion a pour synonyme le terme « Landespflege » [cf. ▶architecture des paysages] ; F., approche géographique et économique du développement patrimonial durable dans les zones rurales ainsi que dans les zones périurbaines adjacentes, associée à des démarches de qualité par la mise en place de projets de développement respectueux des cadres de vie et des cultures locales proposant des solutions économiques viables ; *syn.* gestion [f] des terroirs) ; *g* **Landeskultur [f, o. Pl.]** (**1.** Erhaltung und Förderung eines Zustandes des Natur- oder Landschaftshaushaltes, durch den die nachhaltige land- und forstwirtschaftliche Leistungsfähigkeit der Kulturlandschaft gesichert wird. **2.** Der Begriff **L.** trat Ende des 18. Jhs. auf und umfasst nach einer etwas anderen Definition des landwirtschaftlichen Arbeitskreises kulturtechnische Forschung „alle Maßnahmen der grundlegenden und nachhaltigen Verbesserung des land- und forstwirtschaftlich genutzten Standortes als Voraussetzung für eine gesunde Ordnung im ländlichen Raum". In der Gesetzgebung wurde der Begriff **L.** zum ersten Mal in dem preußischen Edikt zur Beförderung der Land-Cultur vom 14.09. 1811 erwähnt. Der Begriff **Landespflege** [siehe ▶Landschaftsarchitektur] ist mit **L.** gleichsinnig oder sinnverwandt; ▶Forstwirtschaft, ▶Landwirtschaft).

1697 ecologically-sound [adj] *ecol.* (Descriptive term applied to human activities which do not adversely affect nature or parts thereof and are in accordance with ecological principles; ▶environmentally safe); *s* **respetuoso/a con la naturaleza [loc]** (Término descriptivo para actividades humanas que no tienen efectos adversos para la naturaleza; ▶respetuoso/a con el medio ambiente); *f* **respectueux, euse de la nature [loc]** (Terme quali-

fiant toute activité humaine impliquant une prévention des nuisances et une gestion rationnelle des ressources naturelles ; ▶respectueux, euse de l'environnement) ; *g* **naturschonend [ppr/adj]** (So beschaffen, dass Tätigkeiten des Menschen die Natur oder deren Teile nicht nachhaltig belasten; ▶umweltfreundlich).

1698 ecological niche [n] *ecol.* (Role of a species within an ecosystem); *s* **nicho [m] ecológico** (Término muy controvertido. Desde el punto de vista ambiental se considera como «nicho» el papel funcional de un organismo en su comunidad o ecosistema; DINA 1987); *f* **niche [f] écologique** (Ensemble de caractères écologiques [spatiaux et biologiques] permettant à une espèce donnée de s'intégrer à une biocénose ; théoriquement il n'existe qu'une espèce par **n. é.**, deux espèces ayant les mêmes besoins ne pouvant cohabiter, une espèce viendra à être éliminer dans le cadre de la sélection naturelle ; DEE, 1982) ; *g* **ökologische Nische [f]** (Rolle oder Funktion einer Art in einem Ökosystem; *MERKE*, **ö. N.** bezeichnet keinen Raum, in dem eine Art lebt, sondern das Beziehungsgefüge, das sich im Verlaufe der Evolution einer Art entwickelt hat und das ihr die Möglichkeit gibt, in Wechselbeziehung zu den biotischen und abiotischen Gegebenheiten in ihrer Umwelt zu überleben); *syn.* ökologische Funktion [f] einer Art, ökologische Rolle [f] einer Art.

ecological patch type [n] *ecol. land'man. plan.* ▶ecological spatial unit.

1699 ecological pest control [n] *agr. for. hort. phytopath.* (Further development of ▶biological pest control by re-establishment of a species-rich ▶biocenosis, including the restriction of monocultures, by the protection of indigenous ▶beneficial species and improvement of their living conditions, and by improving areas where ▶pests have found shelter, as well as introduction of beneficial arthropods; e.g. ladybugs [US]/ ladybirds [UK]); *s* **control [m] ecológico antiplagas** (Desarrollo ulterior del ▶control biológico por medio de la restauración de las ▶biocenosis ricas en especies —incluyendo la restricción de monocultivos—, la protección de especies ▶animales beneficiosas endémicas y la mejora de sus condiciones de vida, el saneamiento de zonas que dan cobijo a los ▶depredadores y la introducción de antrópodos beneficiosos, p. ej. mariquitas); *f* **lutte [f] écologique** (±) (Pratique récente développée à partir de la ▶lutte biologique consistant dans le rétablissement d'une ▶biocénose riche en espèces [limitation de la monoculture], la protection des ▶auxiliaires indigènes et amélioration de leurs conditions de vie, assainissement des lieux favorables à la propagation des ▶déprédateurs ou des parasites, introduction d'insectes utiles [entomophages] comme la coccinelle) ; *g* **ökologischer Pflanzenschutz [m, o. Pl.]** (*Weiterentwicklung des ▶biologischen Pflanzenschutzes* Wiederherstellung einer artenreichen ▶Biozönose [auch Einschränkung von Monokulturen], Schutz einheimischer ▶Nützlinge und Verbesserung ihrer Lebensbedingungen, Sanierung der Areale, die ▶Schädlingen Unterschlupf gewährten sowie der Import von Nutzgliederfüßlern, z. B. Marienkäfer); *syn.* ökologische Schädlingsbekämpfung [f]).

1700 ecological planning [n] *conserv. nat.res. plan.* (Planning which gives priority to the protection and development of the ▶natural potential, and at the same time takes into consideration the compatibility of competing interests/land uses; ▶landscape planning); *s* **planificación [f] ecológica** (Tiene el fin de informar al gestor de cuáles son, desde el punto de vista de los condicionantes ambientales del territorio [▶potenciales naturales], las mejores opciones para localizar las actuaciones humanas; cf. DINA 1987; ▶planificación del paisaje); *f* **planification [f] écologique** (Approche permettant d'évaluer dans la planification spatiale les possibilités d'aménagement et leurs consé-

quences sur l'environnement grâce à la prise en compte des ▶potentialités du milieu naturel, c.-à-d. les aptitudes et contraintes relatives à divers usages ainsi que les incompatibilités entre eux ; ▶planification des paysages) ; *g* **ökologische Planung [f]** (Vorbereitung von Maßnahmen für räumliche Planungen mit dem vorrangigen Ziel der Sicherung und Entwicklung des ▶Naturpotenzials unter Berücksichtigung der Verträglichkeit konkurrierender Nutzungsansprüche; ▶Landschaftsplanung); *syn.* naturverträgliche Planung [f], Umweltplanung [f], umweltverträgliche Planung [f].

ecological race [n] *ecol.* ▶ecotype.

ecological range [n] *ecol.* ▶ecological amplitude, ▶narrow ecological range, ▶wide ecological range.

ecological resistance [n] *ecol. recr.* ▶environmental stress tolerance.

1701 ecological risk [n] *ecol. plan.* (Evaluated threat of intense reduction in productivity and efficiency of natural systems or parts thereof, because of modified or harmful environmental conditions; *s* **riesgo [m] ecológico** (Evaluación cualitativa de la probabilidad de que un ecosistema o partes de él pierdan y en qué intensidad lo hagan su capacidad de funcionamiento debido a los efectos ecológicos significativos y previsibles de determinado uso previsto, a los parámetros de impactos espaciales y a la estructura de uso existente en la zona afectada); *f* **risque [m] écologique** (Jugement de valeur par lequel il est possible de présumer la probabilité et l'intensité d'une diminution de la productivité et de l'efficience de tout ou parties d'un écosystème au vu des effets supposés, significatifs du point de vue écologique, des paramètres spatiaux et de la structure de l'usage des sols considéré) ; *g* **ökologisches Risiko [n]** (Qualitatives Werturteil darüber, mit welcher Wahrscheinlichkeit und Intensität auf Grund der angenommenen ökologisch bedeutsamen Effekte, der raumwirksamen Parameter und der Struktur der betroffenen Flächennutzung mit einer Minderung der Leistungs- und Funktionsfähigkeit eines betrachteten Ökosystems oder seiner Teile gerechnet werden muss).

1702 ecological significance [n] *ecol.* (Relevance to ecological conditions); *s* **significado [m] ecológico** *syn.* importancia [f] ecológica; *f* **importance [f] écologique** ; *g* **ökologische Bedeutung [f]**.

1703 ecological spatial unit [n] *ecol. land'man. plan.* (Part of the ▶ecological distribution of spatial patterns); *syn.* ecological patch type [n] (LE 1986, 207); *s* **unidad [f] ambiental** (Porción del territorio definido en el marco de la ▶estructuración ecológica del territorio); *f* **écotope [m] territorial naturel** (Unité de paysage définie dans le cadre de la ▶division de l'espace en territoires naturels ; GEP 1991, 29) ; *g* **ökologische Raumeinheit [f]** (Im Rahmen der ▶ökologischen Raumgliederung ausgeschiedene Landschaftsbereiche).

1704 ecological stability [n] *ecol.* (Characteristic condition of natural ecosystems not only in species and habitat-rich systems but also in those poor in species, e.g. woodrush-beech woodland *[Luzulo-Fagetum]*, raised bogs, etc. where trophic conditions are constant; ▶elasticity); *syn.* persistence [n] (LE 1986, 433); *s* **estabilidad [f] ecológica** (Condición característica de ecosistemas naturales que se da no sólo en aquéllos ricos en especies y en hábitats, sino también en los pobres en diversidad, como las turberas, siempre y cuando se mantengan las condiciones tróficas relativamente constantes; ▶elasticidad de un ecosistema); *f* **stabilité [f] écologique** (Faculté des écosystèmes naturels, des communautés biologiques à demeurer proche d'un équilibre ou à retrouver celui-ci après des perturbations ; MPE 1984, 17 ; ▶résilience d'un écosystème) ; *g* **Stabilität [f]** (Dauerhaftigkeit für natürliche Ökosystembereiche. S. ist nicht nur in

arten- und lebensformreichen, sondern auch in artenärmeren Systemen, z. B. im Hainsimsen-Buchenwald *[Luzulo-Fagetum]*, Hochmoor etc., bei entsprechender Trophiekonstanz gegeben; ▶Funktionsfähigkeit eines Ökosystems).

ecological system [n] *ecol.* ▶ecosystem.

1705 ecological tolerance [n] *ecol.* (Capacity of an organism to tolerate certain conditions or changes to its environment, due to physiological and morphological properties. A **large e. t.** signifies a wide range of adaptability; a **low e. t.** signifies a narrow range; ▶adaptability, ▶ecological amplitude); *syn.* environmental tolerance [n] [also US]; *s* **tolerancia [f] ecológica** (Capacidad de un organismo de tolerar cambios o condiciones adversas en su medio gracias a sus características morfológicas y fisiológicas; ▶adaptabilidad, ▶amplitud ecológica); *f* **plasticité [f] écologique** (Aptitude d'une espèce à tolérer des variations importantes des conditions de milieu. Les espèces dites plastiques ont de grandes facultés d'adaptation ; DEE 1982, 51 ; ▶capacité d'adaptation, ▶valence écologique) ; *syn.* tolérance [f] écologique ; *g* **ökologische Potenz [f]** (Die Fähigkeit eines Organismus auf Grund physiologischer und morphologischer Ausstattung bestimmte Verhältnisse oder Veränderungen seiner Umwelt zu ertragen. **Große ö. P.** bedeutet weite, **geringe ö. P.** enge Anpassungsgrenzen. In der angelsächsischen Literatur wird für ö. P. häufig der Begriff *tolerance* verwendet; ▶Anpassungsfähigkeit, ▶ökologische Amplitude); *syn.* ökologische Plastizität [f], ökologische Reaktionsbreite [f], ökologische Toleranzbreite [f].

ecological valence [n] *ecol.* ▶ecological amplitude.

1706 ecologist [n] *ecol. envir. sociol.* (Scientist who studies and practices ▶ecology; ▶conservationist, ▶landscape ecologist, ▶plant ecologist); *s* **ecólogo/a [m/f]** (Científico/a que estudia la ▶ecología; ▶conservacionista, ▶ecólogo/a del paisaje, ▶ecólogo/a vegetal); *f* **écologue [m/f]** (Spécialiste de l'▶écologie ; *terme spécifique* ▶ingénieur-écologue en paysages ; ▶écologiste 1, ▶phyto-écologue) ; *g* **Ökologe [m]/Ökologin [f]** (Jemand der sich wissenschaftlich mit der ▶Ökologie beschäftigt; ▶Landschaftsökologe, ▶Naturschützer/-in).

1707 ecology [n] *ecol.* (Science dealing with the relationships and interactions between living organisms and their environment; ▶autecology, ▶civilization ecology, ▶ecosystem, ▶human ecology, ▶landscape ecology, ▶plant ecology, ▶population ecology, ▶synecology, ▶urban ecology); *s* **ecología [f]** (Ciencia que estudia las relaciones de los seres vivos entre sí y con el medio; *términos específicos* ▶autoecología, ▶ecología de la civilización, ▶ecología de paisajes, ▶ecología de poblaciones, ▶ecología humana, ▶ecología urbana, ▶ecología vegetal, ▶ecosistema, ▶sinecología); *f* **écologie [f]** (Étude des relations réciproques entre les organismes et le milieu dans lequel ils vivent et se reproduisent ; ▶autécologie, ▶démécologie, ▶écologie des civilisations, ▶écologie du paysage, ▶écologie humaine, ▶écologie urbaine, ▶écologie végétale, ▶écosystème, ▶synécologie) ; *g* **Ökologie [f]** (Griech. *oikos* >Haus<, *logos* >Lehre<, d. h. Lehre vom Haushalt; der 1866 von dem Biologen ERNST HAECKEL geprägte Begriff beinhaltet die Wissenschaft von den Wechselwirkungen der Lebewesen untereinander und mit ihrer Umwelt; mit dem wachsenden Umweltbewusstsein in der zweiten Hälfte des 20. Jhs. entwickelte sich der Begriff weit über den engen naturwissenschaftlichen Rahmen der Biologie hinaus; ▶Autökologie, ▶Demökologie, ▶Humanökologie, ▶Landschaftsökologie, ▶Ökosystem, ▶Pflanzenökologie, ▶Stadtökologie, ▶Synökologie, ▶Zivilisationsökologie).

ecology [n]**, vegetation** *phyt.* ▶plant ecology; ▶phytosociology.

1708 ecology [n] **of habitat islands** *ecol.* (Branch of ecosystem research, which investigates the existence and

structure of natural [true islands] and man-made isolated habitats, as for example ►remnant habitat, ►habitat island within the cultural landscape, ►environmental resource patch, ►process of habitat fragmentation); *s* **ecología [f] de biótopos aislados** (Rama de la investigación de ecosistemas que se ocupa de las relaciones de interdependencia de espacios vitales aislados [«islas»], sean naturales o artificiales, como ►biótopos relícticos y ►hábitats residuales; ►islote de vegetación natural, ►aislamiento de biótopos); *f* **écologie [f] des biocénoses relictuelles** (Branche de la recherche en écologie étudiant les phénomènes d'isolement [insularisation] de nombreux habitats naturels [isolement/morcellement/disjonction spatial naturel ou anthropogène des aires de nombreuses espèces] comme p. ex. les ►biotopes relictuels, les ►habitats relictuels, [territoires refuges], ►habitat relictuel, espèces à aires disjointes ; ►îlot de ressources naturelles, ►isolement des biotopes) ; *g* **Inselökologie [f, o. Pl.]** (Forschungsrichtung der Ökosystemforschung, die den Bestand und das Wirkungsgefüge von natürlichen [echte Inseln] und anthropogen isolierten Lebensräumen wie z. B. ►Biotopreste und ►Habitatinseln [ökologische Inseln] inmitten der Kultur- resp. Zivilisationslandschaft untersucht; ►Ökozelle, ►Verinselung von Biotopen).

1709 economic forest management system [n] *for.* (Efficient operation of forests for production of timber and other forest products; ►clear cutting system [US]/clear felling system [UK], ►selection system); *syn.* silvicultural management system [n]; *s* **explotación [f] forestal** (Bosque explotado con fines económicos; ►entresaca, ►explotación a matarrasa); *f* **exploitation [f] sylvicole de production** (Sylviculture orientée vers des objectifs économiques ; ►exploitation forestière par coupes rases, ►jardinage) ; *syn.* gestion [f] de la forêt ; *g* **forstwirtschaftlicher Betrieb [m]** (Ein auf wirtschaftliche Ziele ausgerichtetes Forstbetriebssystem; ►Kahlschlagswirtschaft, ►Plenterwaldbetrieb); *syn.* waldbauliche Bewirtschaftung [f].

economic forestry [n] *for.* ►forestry (1).

1710 economic land management [n] *agr. for.* (Measures taken for the production of regular crop yields according to economic principles; ►agricultural land use); *s* **gestión [f] agrícola (±)** (Conjunto de medidas en el uso agrosilvícola de la tierra para conseguir buenos rendimientos económicos; ►tipo de explotación del suelo); *f* **gestion [f] agricole** (LA 1981, 575 ; forme et méthodes d'exploitation des sols en vue d'obtenir des rendements réguliers et un haut niveau de productivité par l'intégration des progrès techniques sur la base de critères économiques ; ►culture du sol) ; *g* **Landbewirtschaftung [f]** (Maßnahmen der Bodennutzung zur Erzielung regelmäßiger Erträge nach ökonomischen Grundsätzen; ANL 1984; ►Bodenbewirtschaftung).

economic plant species [n] *agr. for. hort.* ►useful plant species.

economy [n] *agr. envir. for. hort.* ►recycling economy.

eco-roof system [n] [US] *constr.* ►roof garden system.

ecosphere [n] *ecol.* ►biosphere.

1711 ecosystem [n] *ecol.* (Functional unit of the biosphere with complex interactions and interrelationships between living organisms and natural or man-made abiotic elements and structures involving the flow of energy, material and information; ►biotope, ►biocoenosis, ►ecological interactions and interrelationships, ►forest ecosystem, ►interdependency within an ecosystem, ►landscape ecosystem, ►natural systems); *syn.* ecological system [n]; *s* **ecosistema [m]** (Término creado por TANSLEY en 1936 que describe la unión funcional entre los componentes biológicos —organismos, materia orgánica— e inertes —energía, materia inorgánica— que coexisten en un lugar

manteniendo unas relaciones recíprocas y en las que se pueden cuantificar «entradas» y «salidas» de materia y energía. Los sistemas de este tipo constituyen, desde el punto de vista del ecólogo, las unidades básicas de la naturaleza; DINA 1987; ►biótopo, ►biocenosis, ►ecosistema forestal, ►régimen ecológico de la naturaleza, ►sistema de interrelaciones e interacciones ecológicas); *f* **écosystème [m]** (Champ d'action formé par le milieu vivant, les éléments naturels abiotiques avec leurs transformations anthropogènes, leurs structures techniques y compris leurs relations matérielles, énergétiques et d'information ; ►biotope, ►biocénose, ►écosystème forestier, ►système naturel, ►tissus d'interactions et de rapports interspécifiques écologiques) ; *g* **Ökosystem [n]** (Funktionelle Einheit der Biosphäre als Wirkungsgefüge aus Lebewesen, unbelebten natürlichen und vom Menschen geschaffenen Bestandteilen, die untereinander und mit ihrer Umwelt in energetischen, stofflichen und informatorischen Wechselwirkungen stehen; ANL 1984; ►Biotop, ►Biozönose, ►Naturhaushalt, ►ökologisches Gefüge, ►Waldökosystem).

ecosystems [npl] *ecol.* ►interactions of ecosystems.

ecosystems [npl]**, effective functioning of natural** *ecol.* ►natural potential.

ecosystems [npl]**, feasibility of recreating** *conserv.* ►recoverability.

1712 ecotechnology [n] *envir.* (Construction methods and other processes which have been technologically adapted according to ecological criteria aimed at eliminating toxic materials and enhancing conditions that support life; e.g. a sewage treatment plant operating on the basis of biological self-purification, the ►recycling of raw materials according to ecological processes; ►low-waste technology); *s* **ecotécnica [f]** (Tecnología que tiene en cuenta los principios ecológicos, como procesos naturales de depuración para plantas de tratamiento de aguas residuales o para el ►reciclado; ►proceso productivo pobre en residuos); *syn.* tecnología [f] apropiada, tecnología [f] blanda, tecnología [f] suave; *f* **génie [m] écologique** (Techniques de construction et procédés de production permettant le ►recyclage de l'eau [procédés sur la base de l'auto-épuration], ►récupération des matières premières [prise en compte des processus écologiques ; ►technologie faiblement génératrice de déchets, ►technologie propre 2) ; *syn.* technologie [f] écologique, technologie [f] douce ; *g* **Ökotechnik [f]** (Ökologischen Strukturen und Entwicklungen angepasste technische Bauweisen und Verfahren mit dem Ziel, giftige Stoffe zu entfernen und die Lebensbedingungen zu verbessern, z. B. Kläranlagen auf der Grundlage der biologischen Selbstreinigung; auch Wiederverwertungsverfahren [►Recycling] von Rohstoffen entsprechend den ökologischen Kreisläufen; ►abfallarmes Verfahren); *syn.* angepasste Technologie [f], sanfte Technologie [f].

1713 ecotonal association [n] *phyt.* (TEE 1980, 123; transitional strip of vegetation between two plant communities, having characteristics of both; ►ecotone 1, ►ecological factors); *syn.* ecotonal plant community [n]; *s* **comunidad [f] de ecotono** (Comunidad de transición entre dos comunidades bióticas o tipos de vegetación. Es una zona de unión o cinturón de transición que normalmente es más estrecha que las áreas de las comunidades inmediatas; DINA 1987, 352; ►ecotono, ►factores ecológicos); *syn.* comunidad [f] ecotónica; *f* **groupement [m] de transition (1)** (Groupement végétal caractéristique d'une modification de végétation [mélange de végétaux d'écologie opposée] liée au changement d'un ou plusieurs ►facteurs écologiques biotiques ou abiotiques [gradient hydrique, variations chimiques du sol, etc.] entre deux groupements contigus ; ►écotone) ; *syn.* association [f] de transition, groupement [m] transitoire, groupement

[m] intermédiaire ; *g* **Übergangsgesellschaft [f] (1)** (Pflanzengesellschaft, die auf Grund von einem oder mehreren sich ändernden ►Ökofaktoren den räumlichen Wechsel von einer Pflanzengesellschaft zur anderen anzeigt; ►Ökoton).

ecotonal plant community [n] *phyt.* ►ecotonal association.

1714 ecotone [n] (1) *ecol.* (Natural transition of an ecosystem between various landscapes or their components or between two ecosystems, in which the supply of life's necessities [food, shelter, heterogeneity of microclimatic conditions] is often greater than in both of the adjacent uniform landscape units; ►transitional biotope, ►seam biotope); *s* **ecotono [m]** (Una transición entre dos o más comunidades bióticas o tipos de vegetación. En general, las comunidades ecotónicas son más ricas florística y faunísticamente que las comunidades adyacentes; cf. DINA 1987; ►biótopo de la orla herbácea, ►cinturón de transición); *f* **écotone [m]** (Zone de transition entre deux communautés biologiques et donc entre deux milieux. D'une façon générale, les écotones sont toujours plus riches floristiquement et faunistiquement que les biocénoses en contact. Ce sont aussi des zones plus hétérogènes dans lesquelles on retrouve des espèces liées à chacune des communautés ; DEE 1982 ; ►biotope d'ourlet herbacé, ►ligne de démarcation) ; *g* **Ökoton [n]** (Natürlicher Ökosystemübergang, Grenzbereich zwischen verschiedenen Landschaften resp. Landschaftsteilen oder Landschaftsbestandteilen, in denen oft das Angebot an Lebenserfordernissen [Nahrung, Deckung, Mannigfaltigkeit kleinklimatischer Bedingungen] größer ist als in den beiden sich anschließenden einförmigen Landschaftsräumen; cf. ÖKO 1983; ►Grenzlinienbereich, ►Saumbiotop).

ecotone [n] (2) *phyt.* ►timberline ecotone.

1715 ecotope [n] *geo.* (In ►landscape ecology the smallest possible land unit—that is still a holistic [environmentally complete] unit; ►habitat 2); *syn.* tessera [n] (LE 1986, 11); *s* **ecotopo [m]** (Hábitat diferenciado dentro de otro mayor; DINA 1987; ►biótopo, ►ecología del paisaje); *f* **écotope [m]** (Unité cartographique en ►écologie du paysage définie à la fois par des caractères de station et de localité ; aire occupée par une biocénose caractérisée par l'ensemble des facteurs climatiques [climatope] et édaphiques [édaphotope] ; ces facteurs sont dits stationnels ; VOC 1979 ; ►biotope) ; *g* **Ökotop [m]** (...tope [pl]; in der ►Landschaftsökologie die kleinste Einheit einer Landschaft in ihrer natürlichen räumlichen Ausprägung, bestimmt durch klimatische und bodenkundliche Gegebenheiten, die stets eine bestimmte Pflanzengesellschaft bedingen; *syn. in der allgemeinen Bioökologie* ►Biotop); *syn.* Landschaftszelle [f], Physiotop [m].

ecotopes [npl] *ecol.* ►pattern of ecotopes.

1716 eco-tourism [n] *recr.* (Environmentally sound tourism specially focused on limiting the consumption of natural resources, as well as reducing the production of waste. E. also recognizes the human dignity and cultural characteristics of indigenous people in the countries visited); *syn.* sustainable tourism [n]; *s* **turismo [m] ecológico** (Alternativa al turismo de masas con sus graves impactos ambientales. El concepto de **t. e.** tiene como *leitmotiv* el respeto a la naturaleza, el reducir al mínimo el consumo de recursos naturales y la producción de residuos así como el respeto de la dignidad humana y de las culturas de los pueblos indígenas); *syn.* ecoturismo [m], turismo [m] alternativo; *f* **écotourisme [m]** (Terme générique qui rassemble toutes les formes récentes du voyage organisé ou non de petits groupes de personnes dont les activités de détente favorisent le patrimoine culturel, ne recourant pas aux hébergements banalisés et aux moyens classiques de transport touristiques mais faisant large place au tourisme axé sur la nature, la randonnée,

l'hébergement mobile [circuits aventures], l'observation de zones naturelles ; cette forme du tourisme durable respecte une certaine éthique dont les principales composantes sont une contribution à la préservation de l'environnement, la protection de la nature et le soutien à l'économie locale dont les retombées directes contribuent au bien-être des populations locales ; de nombreux labels existent qui permettent de choisir sa destination et son hébergement prenant en compte les critères de développement durable ; *termes spécifiques* **tourisme solidaire et responsable :** forme de tourisme qui met au centre du voyage l'homme, la rencontre et l'implication des populations locales dans les différentes phases d'un projet touristique, l'implication du voyageur dans la nécessité sociale du lieu de sa destination, le respect de la personne, des cultures et de leur environnement ; **tourisme équitable :** forme de tourisme favorisant la consommation de produits de commerce équitable et les échanges culturels et sociaux avec leurs producteurs ; **tourisme durable :** forme de tourisme qui s'implique dans une démarche globale de développement durable qui exploite au maximum les ressources mettant en valeur un territoire touristique tout en préservant les processus écologiques essentiels et la biodiversité qui génèrent ce même tourisme) ; *syn.* tourisme [m] écologique ; *g* **Ökotourismus [m, o. Pl.]** (Tourismus als Alternative zum umweltzerstörenden Massentourismus, der v. a. die Schonung der Umwelt, besonders den sehr sparsamen Verbrauch natürlicher Ressourcen, eine starke Reduzierung von Abfällen und sonstiger Hinterlassenschaften sowie die Achtung der Menschenwürde und kulturellen Eigenständigkeit sowie der Besonderheiten der jeweiligen einheimischen Bevölkerung zum Leitbild hat. Das Konzept des **Ö.** geht davon aus, dass das Reisen zu den Grundlagen und attraktiven Seiten des Lebens in modernen Industriegesellschaften gehört); *syn.* Alternativtourismus [m], sanfter Tourismus [m], nachhaltiger Tourismus [m].

1717 ecotoxicology [n] *ecol. envir.* (Science of the spread and influence of harmful materials upon ►ecosystems or their components; ►bioindicator, ►noxious substance); *s* **ecotoxicología [f]** (Ciencia que estudia la distribución y los efectos de sustancias tóxicas en el medio ambiente; ►bioindicador, ►ecosystema, ►sustancia nociva); *f* **écotoxicologie [f]** (Science étudiant la propagation et l'impact de ►substances génératrices de nuisances, de substances toxiques sur les populations d'organismes vivants et les ►écosystèmes ou leurs composantes ; ►indicateur biologique) ; *g* **Ökotoxikologie [f]** (Wissenschaft von der Verbreitung und den Wirkungen schädlicher Stoffe auf ►Ökosysteme oder deren Teile; ANL 1984; ►Bioindikator, ►Schadstoff).

1718 ecotype [n] *ecol.* (Genetic variation of a species adapted to a particular environment and characterized by a recognizably different morphology or physiology. Sometimes used; *syn.* race [n], genecotype [n], ecological race [n] (TEE 1980); *s* **ecotipo [m]** (Estirpe o subdivisión de una especie que está genéticamente adaptada al hábitat y clima local. Estos grupos genéticos son más simples que un biotipo y menos amplios que una especie; DINA 1987); *f* **écotype [m]** (Variation génétique au sein d'une espèce adaptée à un milieu particulier et caractérisée par des différences morphologiques et physiologiques nettes) ; *g* **Ökotyp [m]** (Durch natürliche Selektion entstandene Teilpopulation einer Tier- oder Pflanzenart mit erblich bedingter Anpassung an die Standortbedingungen in ihrem Verbreitungsgebiet; ANL 1984).

ectotrophic mycorrhiza [n] *bot. pedol.* ►mycorrhiza.

1719 edaphic [adj] *agr. biol. constr. hort. pedol.* (**1.** Term applied to chemical, physical or biological conditions of a soil or substrate that affects the growth of plants or other organisms: acidity, alcalinity, water content, texture. **2.** Term related to soil science. **3.** Descriptive term meaning 'of the soil'); *s* **edáfico/a**

[adj] (**1.** Término aplicado a las características biológicas, químicas y físicas del suelo o de un substrato que afectan el crecimiento de las plantas y de otros organismos: acidez, alcalinidad, contenido de agua, textura, etc. **2.** Término relacionado con el estudio del suelo. **3.** Término descriptivo que significa «del suelo»); *f* **édaphique** [adj] (Relatif au sol) ; *g* **bodenkundlich** [adj] (Den Boden oder ein Bodensubstrat hinsichtlich biologischer, chemischer und physikalischer Eigenschaften betreffend); *syn.* Boden... [als Präfix], edaphisch [adj].

edaphic properties [npl] *pedol.* ▶soil characteristics.

edge [n] *phyt.* ▶bog edge; *ecol.* ▶field woodland edge [US]; *for.* ▶forest edge; *phyt.* ▶herbaceous edge; *constr.* ▶laid on edge, ▶lawn edge; *geo.* ▶water's edge.

edge [n], **city** *urb.* ▶city periphery.

edge [n], **lake** *geo. landsc.* ▶lakeshore.

edge [n], **mowing** *constr.* ▶mowing strip.

edge [n], **upper** *arch. constr.* ▶top level.

edge [n], **village** *urb.* ▶city periphery.

1720 edge community [n] *phyt.* (Vegetation unit which regularly occurs in the close proximity of woodland); *s* **comunidad** [f] **vecinante** (En zonación local de comunidades, aquéllas que se presentan siempre en contacto entre sí, como *Scirpo-Phragmitetum* y *Magnocaricion*; cf. BB 1979); *f* **groupement** [m] **de contact** (Association végétale de transition rencontrée régulièrement entre un peuplement forestier et les autres formations végétales voisines) ; *syn.* peuplement [m] de contact ; *g* **Kontaktgesellschaft** [f] (Vegetationseinheit, die mit dem Walde regelmäßig eng benachbart vorkommt).

1721 edge design [n] *arch. constr.* (**1.** Form of separation between two adjacent surfaces; e.g. between different planting areas, the edges of paths or ponds; common edging materials are brick, stone, concrete or wood; ▶roof parapet). **2.** Configuration of open space boundary to accommodate various plant, and animal species or human activities in the design of parks and arboreta; *s* **delimitación** [m] (**1.** Diseño del borde de un camino o arriate a la superficie colindante para lo que generalmente se utilizan ladrillos, piedras, hormigón o madera; *en tejados planos* ▶parapeto de tejado plano. **2.** Diseño de los bordes de parques o bosques para posibilitar usos vicinales o para crear condiciones adecuadas para la flora y la fauna); *syn.* de l. borde [m]; *f 1* **délimitation** [f] (Bordure végétale ou non, mise en place pour marquer la limite d'un chemin, d'un massif, d'une pelouse d'un étang, etc. ; pour les toitures terrasses se référer à ▶acrotère ; *contexte* l'allée est délimitée par une rangée de pavés ; *terme spécifique* création [f] de lisière fleurie) ; *syn.* constitution [f] d'une bordure, bande [f] de rive (VRD 6/21, 42) ; *f 2* **aménagement** [m] **des lisières** (Travaux de structuration paysagère en milieu urbain ou en zone de cultures agricoles intensives par la réalisation de cordons boisés le long des cours d'eau, de haies et de lisières forestières assurant le maillage de différentes zones et le refuge pour les espèces animales et végétales) ; *g* **Randausbildung** [f] (**1.** Gestaltung der Abgrenzung zu benachbarten Flächen, z. B. bei Wegen, Rabatten, Teichen; *auf Flachdächern* ▶Attika 2. **2.** Gestaltung von Park- oder Waldrändern, um angrenzende Nutzungen zu ermöglichen oder für Flora und Fauna entsprechende Lebensbedingungen zu schaffen); *syn. zu 1.* Kante [f], Randeinfassung [f].

1722 edge effect [n] *ecol. landsc.* (Favo[u]rable effect upon increased diversity of biocoenosises in border areas between various types of landscape patches, biotopes, etc., in comparison with the homogenous vegetation of the neighbo[u]ring biotopes; ▶ecotone 1); *s* **efecto** [m] **de borde** (Tendencia de los organismos a aparecer en diferentes variedades y densidades en las comunidades de enlace de ▶ecotonos); *f 1* **effet** [m] **de lisière** (**1.** en écologie, situation favorisant la diversité des biocénoses dans la zone de transition de différentes formations végétales, de divers biotopes ou de formes distinctes de développement, etc. par comparaison avec les peuplements végétaux homogènes des biotopes voisins. **2.** En sylviculture l'**e. de l.** se produit lorsque les arbres après déboisement se retrouvent brusquement en lisière et peuvent être déracinés par le vent, ce qui peut provoquer un lent recul de la lisière de la forêt ; LA 1981, 695 ; ▶écotone) ; *f 2* **effet** [m] **de bordure** (En agriculture, influence du milieu environnant sur les plantes situées aux bords des parcelles ; LA 1981, 185) ; *g* **Randeffekt** [m] (Günstige Wirkung auf die erhöhte Diversität von Biozönosen im Grenzbereich zwischen verschiedenen Landschaftsteilen, Bewuchsformen, Biotopen etc. im Vergleich zu den homogenen Vegetationsbeständen der benachbarten Biotope; ▶Ökoton); *syn.* Grenzlinienwirkung [f], Randzoneneffekt [m].

edge kerb [n] [UK] *constr.* ▶barrier curb [US].

1723 edge [n] **of a landslip** *geo.* (Upper edge of a slope failure caused by a ▶shear failure of embankment; ▶slip plane); *s* **borde** [m] **de falla** (Borde superior de una pendiente provocado por un ▶deslizamiento planar; ▶plano de deslizamiento); *syn.* borde [m] de dislocación; *f 1* **lèvre** [f] **de glissement** (Bords des terrains formé par un ▶rupture de terrain ; ▶plan de glissement) ; *f 2* **lèvre** [f] **de faille** (Bords des terrains formés par une cassure ; ▶plan de faille) ; *g* **Bruchkante** [f] (Unterbrechung des kontinuierlichen Verlaufes eines Hanges, z. B. ausgelöst durch einen ▶Geländebruch; ▶Gleitfläche); *syn.* Abbruchkante [f].

1724 edge [n] **of a settlement** *urb.* (General term for the outer edge of a village, town or city demarcated by the buildings of the urbanised area; ▶city periphery, ▶peripheral building development, ▶urban fringe); *s* **borde** [m] **de la ciudad** (Límite exterior de un asentamiento/una ciudad hacia el campo abierto; ▶edificación periférica, ▶franja rururbana, ▶periferia urbana); *syn.* borde [m] urbano; *f 1* **limite** [f] **de l'agglomération** (Terme général désignant la limite des zones d'urbanisation d'une ville, d'une agglomération ; ▶franje urbaine) ; *f 2* **limite** [f] **du site construit** (Limite déterminée de l'espace bâti, du périmètre d'agglomération par rapport à l'espace agricole ou naturel ; ▶constructions périphériques, ▶périphérie urbaine l) ; *syn.* limite [f] du (domaine) bâti, limite [f] du périmètre bâti/construit ; *g* **Siedlungsrand** [m] (**1.** Abschluss einer Siedlungsfläche zur freien Landschaft. **2.** Äußere Begrenzung eines Ortes/einer Stadt durch Siedlungsbauten); ▶Randbebauung, ▶Stadtrand); *syn.* Ortsrand [m].

edge piece [n] *constr.* ▶edge restraint.

1725 edge restraint [n] *constr.* (SPON 1986, 371; **1.** specially-formed ▶interlocking paver [US]/interlocking paving block [UK] integrated within the bond of the paving, which acts a] as a **half-size paver** along the side, or b] as a **starting** or **finishing paver** when paving an area. Pavers for normal use are known as **standard pavers**; ▶edge unit. **2.** A linear metal or plastic device placed at the edge of unit pavers and fastened to the granular base by means of metal spikes to resist lateral pressure exerted under normal use); *syn. for 1.* edge piece [n]; *s* **semiadoquín** [m] **de borde** (▶Adoquín de ensamblaje de forma especial que se utiliza para los bordes laterales [**semiadoquín de borde lateral**] o frontales [**semiadoquín de final de hilada**] de un pavimento; ▶ladrillo de arranque); *f* **pavé** [m] **d'extrémité** (Type de pavé utilisé pour les finitions de pavage en ▶pavés autobloquants ; suivant la marque ou la forme du pavé on distingue le **demi-pavé** ou le **pavé de rive** [parfois petit pavé de rive] pour les finitions longitudinales et le ▶**pavé de départ**, les autres pavés ayant la dénomination de

pavés standards) ; *g* **Randstein [m]** (Besonders geformter, in den Verbund eingreifender ▶ Verbundpflasterstein. Je nach Lage, Ausformung oder Fabrikat unterscheidet man den **seitlichen Randstein** oder **Halbstein** für den seitlichen Abschluss und den **Anfang-** oder **Schlussstein** für den Verlegeanfang oder für das Verlegeende. Die übrigen Steine werden **Normalsteine** genannt; ▶ Anfängerstein).

1726 edge trees [npl] *for.* (SAF 1983; trees on the edge of a crop or stand growing under conditions of light and exposure; ▶ forest mantle); *syn.* forest border trees [npl], marginal trees [npl] of a forest; *s* **árboles [mpl] marginales** (DFM 1975; conjunto de árboles de ramaje bajo que crecen al borde de un bosque por lo que las condiciones de iluminación y exposición al sol difieren de las de los árboles del interior del mismo; ▶ abrigo del bosque); *f* **arbres [mpl] de lisière** (Arbres situés à la lisière d'un peuplement forestier et croissant dans les conditions de lumière et d'exposition différentes de celles qui régissent l'intérieur du peuplement ; DFM 1975 ; ▶ manteau forestier) ; *g* **Trauf [m] eines Waldes** (Tief beasteter, standfester Baumbestand, der den Rand eines Waldes bildet; ▶ Waldmantel 1); *syn.* Traufrand [m], Waldtrauf [m] (WT 1980, 191).

1727 edging [n] *constr.* (Border of a bed, planting area, path, etc.; ▶ curbing [US]/kerb edging [UK], ▶ edge design, ▶ flush edging, ▶ wooden edging); *s* **encintado [m]** (▶ Delimitación de un macizo o arriate de flores; ▶ acabado plano de los bordes, ▶ encintado con piedra de bordillo, ▶ encintado de madera); *syn.* recintamiento [m]; *f* **bordure [f]** (Alignement végétal ou minéral destiné à délimiter un massif, une allée, une plantation, etc. [▶ délimitation] ; ▶ bordurette de voliges, ▶ butée en rive, ▶ rebord d'entourage) ; *g* **Einfassung [f]** (▶ Randausbildung eines Beetes, einer Pflanzfläche, eines Weges etc.; ▶ Bordsteineinfassung, ▶ höhengleiche Einfassung, ▶ Holzeinfassung).

edging [n], **lawn** *constr. hort.* ▶ lawn edge clipping; ▶ lawn edge spade cutting.

edging [n] [UK], **tree circle** *constr.* ▶ tree pit edging [US].

edging [n] **of a flight of steps** *arch. constr.* ▶ side edging of a flight of steps.

1728 edging stone [n] *constr.* (Tooled natural stone or prefabricated concrete stone used for borders of planting beds, paths, etc.; ▶ curbstone, ▶ edge restraint); *s* **piedra [f] de encintado** (Piedra natural o prefabricada para bordear macizos, caminos, etc.; ▶ piedra de bordillo, ▶ semiadoquín de final de hilada); *f* **bordure [f] de jardin (type P)** (pour la délimitation de plates-bandes, chemins, de pistes, etc. ; cf. N.F. P 98 302 ; ▶ élément de bordure, ▶ pavé d'extrémité) ; *syn.* bordurette [f] ; *g* **Kantenstein [m] (2)** (Vorgefertigter Natur- oder Betonstein zur Einfassung von Beeten, Wegen etc.; ▶ Bordstein, ▶ Randstein).

1729 edging unit [n] *constr.* (▶ Interlocking paver [US]/ interlocking paving block [UK] in the first row, i.e. along the outer edges of a paved area; *specific terms* half edging stone, full edging stone); *s* **ladrillo [m] de arranque** (▶ Adoquín de ensamblaje con el que se comienza a tender un pavimento); *syn.* arranque [m]; *f* **pavé [m] de départ** (KRO 1989 ; ▶ pavé autobloquant avec lequel débute la rangée) ; *g* **Anfängerstein [m]** (▶ Verbundpflasterstein, mit dem die erste Reihe begonnen wird; **A.e** sind Ergänzungssteine zu den Hauptsteinen/Normalsteinen); *syn.* Anfänger [m], Anfangstein [m].

edging up of a lawn [n] *constr. hort.* ▶ lawn edge clipping.

edging with a spade [n], **lawn** *constr. hort.* ▶ lawn edge spade cutting.

education [n] *prof.* ▶ professional education.

education [n] **by mid-career course** *prof.* ▶ continuing education [US].

effect [n] *constr. arb.* ▶ bathtub effect [US]; *zool.* ▶ crowding effect; *ecol. landsc.* ▶ edge effect; *met.* ▶ El Niño effect; *ecol.* ▶ filtering effect of vegetation; *ecol. envir.* ▶ greenhouse effect; *biol.* ▶ isolation effect; *landsc.* ▶ venturi-effect.

effect [n], **forest** *for. nat'res. recr.* ▶ forest function.

effect [n], **mass** *zool.* ▶ crowding effect.

effect [n], **negative** *ecol.* ▶ negative impact.

effect [n] [UK], **pot-binding** *constr. arb.* ▶ bathtub effect [US].

effective functioning [n] **of natural (eco)systems** *ecol.* ▶ natural potential.

effective length [n] *constr.* ▶ laying length.

1730 effective management [n] **of migratory species of wild animals** *conserv.* (*Context* conservation and **e. m. of m. s. of w. a.** require the concerted action of all States within the national jurisdictional boundaries of which such species spend any part of their life cycle; cf. Bonn Convention); *s* **gestión [f] eficaz de especies migratorias** (*Contexto* la preservación de las especies migratorias precisa de acción concertada de todos los Estados en los cuales las especies pasan parte de su ciclo vital; cf. Convención de Bonn); *f* **gestion [f] efficace des espèces migratrices** (*Contexte* une conservation et une **g. e. des e. m.** appartenant à la faune sauvage requièrent une action concertée de tous les États à l'intérieur des limites de juridiction nationale desquelles ces espèces séjournent à un moment quelconque de leur cycle biologique ; cf. Convention de Bonn) ; *g* **wirksame Hege [f] und Nutzung [f] wandernder Tierarten** (*Kontext* Erhaltung sowie **w. H. und N. w. T.** erfordern gemeinsame Maßnahmen aller Staaten, in deren nationalen Zuständigkeitsbereich diese Arten einen Teil ihres Lebenszyklus verbringen; cf. Präambel 6. Abs. der Bonner Konvention).

1731 effect [n] **of environmental pollution** *envir.* (Tangible or intangble influence of air pollution, noise, noxious substances, radiation, etc. on living beings or inert material); *s* **efectos [mpl] de la contaminación** (Influencias tangibles o intangibles de la polución atmosférica, del ruido, sustancias tóxicas, radiación, etc. sobre organismos vivos o materia inerte); *f* **action [f] (nocive) de la pollution atmosphérique** (Ayant des conséquences préjudiciables, visibles ou pernicieuses sur les êtres vivants et la matière inerte) ; *syn.* influence [f] (dommageable) de la pollution atmosphérique ; *g* **Immissionseinwirkung [f]** (Sichtbare oder unsichtbare Beeinflussung durch Umwelteinwirkungen wie Luftverunreinigung, Geräusche, Strahlen etc. auf Lebewesen oder tote Materie).

effect [n] **on frozen air flow, barrier** *for. nat'res. recr.* ▶ frost accumulation.

1732 effects [npl] *ecol.* (Intended or not intended results: good, bad or indifferent); *syn.* impact [n]; *s* **efecto [m]** (En sentido neutral de la palabra: consecuencia deseada o no deseada de algo); *syn.* incidencia [f], impacto [m]; *f* **effet [m]** (au sens neutre ; p. ex. incidences sur l'environnement) ; *syn.* impact [m], incidence [f], conséquence [f] ; *g* **Auswirkung [f]** (1. Bezweckte oder nicht bezweckte [wertneutrale] Wirkung im Sinne einer Beeinflussung oder Veränderung durch ein Ereignis oder Vorgang; auswirken [vb], wirken [vb]. 2. Ergenis von 1.); *syn.* Effekt [m], Wirkung [f].

efficiency [n] *ecol. plan.* ▶ functional efficiency.

1733 efflorescence [n] *constr. pedol.* (1. Deposition of salts on masonry. 2. Deposition of crystalized salts on the surface of the soil through the evaporation of capillary water); *s* **eflorescencia [f] (1)** (1. Deposición de sal sobre muros. 2. Deposición de sal en la superficie del suelo en zonas áridas y semiáridas causada por la evaporación del agua capilar); *f* **efflorescence [f]** (**1.** Dépôt de sel sur les murs. **2.** Formation de sels pulvérulents en particulier sur les sols des régions arides et semi-arides par suite de l'évaporation de l'eau de cristallisation) ; *g* **Ausblühung [f]** (**1.** Salzkruste an der Oberfläche einer Mauer. **2.** Salzbildung durch Verdunstung der Bodenfeuchtigkeit vor allem auf Böden arider und semiarider Gebiete; ausblühen [vb]); *syn.* Auswitterung [f].

1734 effluent [n] *envir.* (Generic term pertaining to any outflowing, polluted water or wastewater discharge, sometimes referred to also as wastewater effluent, covering outflow from a sewage treatment facility, or industrial facility, ▶domestic sewage, ▶industrial sewage and/or storm water. *NOTE*, in English, effluent merely means, "outflowing," and does not necessarily refer to the nature of the ▶outflow; effluent in sanitary engineering often means sanitary wastewater or "sewage effluent"; **sewer effluent** is merely outflowing sewage; outflow from a wastewater treatment plant is referred to as **secondary effluent** or **treated effluent**; **in U.S.**, storm water is normally collected in separate drainage system for discharge into watercourses; ▶flow of liquid waste, ▶mixed effluent, ▶reuse of effluent, ▶sewage 1, ▶wastewater); *s* **efluente [m]** (Término genérico para la descarga de agua contaminada o aguas residuales a un curso receptor, sea directamente desde la canalización o desde una ▶planta de depuración de aguas residuales; *términos específicos* ▶aguas residuales 1, ▶aguas residuales 2, ▶aguas residuales domésticas, ▶aguas residuales industriales, ▶aguas residuales mixtas, ▶caudal efluente [de tubería]; ▶flujo de efluente); *f* **effluent [m]** (Terme générique désignant une eau résiduaire urbaine, [▶effluent urbain] ou industrielle [▶effluent industriel], et plus généralement tout rejet liquide susceptible de contenir une certaine charge polluante [dissoute, colloïdale ou particulaire] en suspension [matières organiques oxydables, des nitrates ou des sulfates, etc.] nuisible à l'environnement et rejeté par le réseau d'assainissement dans l'environnement ; *termes spécifiques* ▶eaux ménagères et eaux vannes, ▶eaux usées, ▶eaux usées et eaux pluviales ; ▶débit sortant, ▶eaux usées, ▶écoulement des effluents) ; *syn.* eaux [fpl] usées, eaux [fpl] résiduaires, eaux [fpl] d'égout ; *g* **abfließendes Abwasser [n]** (Das durch häuslichen, gewerblichen und industriellen Gebrauch verunreinigte ▶Schmutzwasser, das in die Kanalisation gelangt oder gereinigt aus Abwasserkläranlagen in die Vorflut fließt. Der englische Begriff *effluent* bezeichnet nur das aus einem Leitungsrohr [in den Vorfluter] auslaufende Abwasser und sagt zunächst nichts über den Verschmutzungsgrad aus; *UB* ▶Abfluss 1, ▶Abfluss flüssiger Emissionen, ▶Haushaltsabwasser, ▶Industrieabwasser, ▶Mischwasser).

1735 effluent discharge fee [n] *adm. envir. leg.* (Legally-fixed fee for the discharge of untreated effluent into a ▶receiving stream; the amount of fee depends on the degree of harmful pollution of the effluent); *s* **canon [m] de vertido** (Exacción destinada a la protección y mejora del medio receptor de cada cuenca hidrográfica, cuya cantidad depende de la carga contaminante valorada según unidades de contaminación; arts. 93 y 105, Ley de Aguas 29/1986); *f 1* **redevance [f] relative à la détérioration de la qualité de l'eau** (Taxe perçue par les agences financières de bassin pour les pollutions dues aux usages domestiques et industriels produites par les personnes publiques ou privées ; taxe perçue sur la base de la consommation en eau potable du réseau public ou des installations de captage ; art. 14-1 de la loi n° 64-1245 du 16 décembre 1964 relative au régime et à la répartition des eaux et à la lutte contre leur pollution) ; *f 2* **redevance [f] d'assainissement** (Taxe prélevée par les services publics d'assainissement communaux ou intercommunaux pour tout rejet dans un réseau d'assainissement) ; *g* **Abwasser-**

abgabe [f] (**D.**, nach dem Abwasserabgabengesetz [AbwAG] von 1976, i. d. F. der Bekanntmachung vom 18.01.2005, festgelegter Beitrag für die Einleitung ungeklärter Abwässer von Kommunen, Industrie und Gewerbe in die ▶Vorfluter. Die **A.** richtet sich nach der Schädlichkeit des Abwassers, die unter Zugrundelegung der oxidierbaren Stoffe, des Phosphors, des Stickstoffs, der organischen Halogenverbindungen, der Metalle Quecksilber, Cadmium, Chrom, Nickel, Blei, Kupfer und ihrer Verbindungen sowie der Giftigkeit des Abwassers gegenüber Fischeiern nach der Anlage zu diesem Gesetz in Schadeinheiten bestimmt wird; cf. § 3 AbwAG. Ziel ist die Vermeidung von Abwasser, die Reduzierung der Verschmutzung und die Abwasserbehandlung vor Einleitung. Bei Einhaltung strenger Mindestanforderungen wird die Abgabe pro Schadstoffeinheit um 75 % günstiger. Damit besteht ein zusätzlicher finanzieller Anreiz, schonendere Verfahren zu entwickeln; cf. ÖKO 2003).

1736 effluent discharge pipe [n] *envir. urb.* (Drainage line for the disposal of sullage, sewage, and precipitation water; ▶collector sewer); *syn.* sewage discharge pipe [n], sewage disposal pipe [n] [also UK]; *s* **conducto [m] de desagüe (de aguas residuales)** (Tubería de descarga de aguas residuales o de lluvia; ▶alcantarilla); *f* **conduite [f] d'assainissement** (Conduite utilisée pour l'évacuation des eaux usées ; ▶collecteur d'eaux usées et d'eaux pluviales) ; *g* **Entsorgungsleitung [f]** (Rohrleitung, die der Entfernung von Abwässern dient; ▶Abwasserkanal); *syn.* Abwasserleitung [f], Entwässerungsleitung [f].

effluent discharge pipe [n] **into the sea** *envir.* ▶underwater effluent discharge pipe into the sea.

1737 effluent disposal [n] *envir.* (Disposing of purified or unpurified sewage; ▶discharge of effluents); *s* **eliminación [f] de aguas residuales** (Vertido de aguas residuales depuradas o sin depurar; ▶vertido); *f* **élimination [f] des eaux usées** (Opération d'évacuation des eaux usées traitées ou non et de rejet dans un exutoire naturel ; ▶rejets d'effluents) ; *syn.* élimination [f] des eaux résiduaires, *par extension* assainissement (2) ; *g* **Abwasserbeseitigung [f]** (Abführung von geklärtem oder ungeklärtem Abwasser; die **A.** obliegt den Gemeinden. Sie haben das Abwasser insbesondere zu sammeln, den Abwasserbehandlungsanlagen zuzuleiten, zu reinigen und die hierfür erforderlichen Kanäle, Rückhaltebecken, Pumpwerke, Regenwasser- und Abwasserbehandlungsanlagen herzustellen, zu unterhalten und zu betreiben; ▶Einleitung).

1738 effluent pollution [n] *envir.* (Pollution of a watercourse or water body by the discharge of sewage; ▶effluent pollution load, ▶discharge of effluents); *s* **contaminación [f] del agua** (Polución de un curso de agua causada por ▶flujo de efluente; ▶carga de aguas residuales); *f* **pollution [f] due aux effluents** (Pollution des fleuves et des eaux de ruissellement produites par les ▶rejets d'effluents ; ▶charge de pollution 1) ; *g* **Abwasserbelastung [f]** (Grad der Verschmutzung eines Gewässers durch Abwasser; die Bemessung der Belastung durch definierte Substanzen erfolgt durch Angaben der auftretenden Konzentrationen, jedoch wird sie auch pauschal in Form von Saprobiestufen definiert; neuere Bestimmungsverfahren benutzen auch limnologische und autökologische Kriterien und stellen die Größe einzelner ökologischer Wirkungsfaktoren fest; ▶Abwasserlast, ▶Einleitung, ▶Saprobie).

1739 effluent pollution load [n] *envir.* (Quantity of effluent pollution in a watercourse or waterbody); *s* **carga [f] de aguas residuales** (Cantidad de contaminantes vertidos en un curso de agua); *syn.* carga [f] de contaminación, carga [f] de polución; *f* **charge [f] de pollution (1)** (Taux de pollution d'un cours d'eau engendré par un effluent pendant un temps défini) ; *syn.* charge [f] polluante (ASS 1987, 256) ; *g* **Abwasserlast [f]**

(Umfang der Verunreinigung eines [fließenden] Gewässers mit Abwasser).

effluent seepage [n] *geo. hydr.* ▶groundwater seepage.

effluent treatment [n] *envir.* ▶industrial effluent treatment.

1740 effluent volume [n] *envir. syn.* sewage volume [n]; *s* **caudal** [m] **de aguas residuales** (Volumen de aguas residuales que fluyen a la red del alcantarillado); *f* **débit** [m] **d'eaux usées** (Volume des effluents collectés dans la canalisation) ; *g* **Abwasseranfall** [m] (Das in die Kanalisation gelangende Wasser).

1741 elasticity [n] *ecol.* (Capability of an ecosystem to create a stable cyclical system by linkage and feedback of individual characteristics; ▶ecological stability, ▶environmental stress tolerance, ▶natural potential); *syn.* resilience [n]; *s* **elasticidad** [f] **de un ecosistema** (Velocidad a la que un sistema retornaría al punto estable tras cesar el efecto de una perturbación; DINA 1987; ▶estabilidad ecológica, ▶potencial natural, ▶resiliencia de un ecosistema); *f* **résilience** [f] **d'un écosystème** (Aptitude d'un écosystème par interactions entre divers facteurs à générer des cycles stables c.-à-d. en s'autorégulant ; la résilience d'un écosystème forestier est p. ex. sa capacité à se reconstituer après sa destruction par un incendie et à revenir à son état d'équilibre ; ▶capacité de charge d'un écosystème, ▶potentialité du milieu naturel, ▶stabilité écologique) ; *syn.* élasticité [f] d'un écosystème, pouvoir [m] de résistance d'un écosystème ; *g* **Funktionsfähigkeit** [f, o. Pl.] **eines Ökosystems** (Fähigkeit eines Ökosystems durch Kopplung und Rückkopplung von Einzelfaktoren in sich stabile, d. h. sich selbst regulierende Regelkreise zu bilden; ▶Belastbarkeit 1, ▶Naturpotenzial, ▶Stabilität).

elastic layer [n] *constr.* ▶resiliant layer.

elbow [n] *constr.* ▶pipe elbow, ▶reducing elbow.

electricity [n] *envir. ecol.* ▶solar-generated current.

1742 electric power line [n] *envir.* (Generic term for ▶overhead power line and ▶underground power line); *s* **conducción** [f] **eléctrica** (*Términos específicos* ▶línea aérea, ▶cable eléctrico subterráneo); *f* **ligne** [f] **électrique** (*Termes spécifiques* ▶ligne aérienne de distribution d'électricité, ▶ligne électrique souterraine) ; *syn.* conduite [f] électrique ; *g* **Stromleitung** [f] (*UBe* ▶Freileitung und ▶unterirdisches Stromkabel).

electric power plant [n] *envir.* ▶power plant.

1743 electric power supply [n] *plan.* (Provision of electricity in a particular area); *s* **suministro** [m] **de electricidad** *syn.* suministro [m] de energía eléctrica, abastecimiento [m] de electricidad; *f* **alimentation** [f] **en électricité** (Approvisionnement d'une zone déterminée en courant électrique) ; *g* **Stromversorgung** [f] (Bereitstellung von elektrischem Strom in einem definierten Gebiet).

electric power transmission line [n] [US] *constr. envir.* ▶high tension power line.

1744 electromagnetic radiation [n] *envir.* (Electromagnetic phenomenon emanating from a high tension power line, television, radar and microwaves, as well as household electrical equipment, which may cause "electrostress" to people); *s* **contaminación** [f] **electromagnética** (Radiación electromagnética de las líneas de alta tensión, de aparatos de televisión, radares, antenas de telefonía móvil, microondas y otros aparatos domésticos que puede tener efectos nocivos para la salud de la población); *f* **ondes** [fpl] **électromagnétiques** (Terme commun pour désigner le rayonnement électromagnétique émit par les lignes à haute tension, les radars et les appareils électriques ménagers pouvant avoir d'effets négatifs sur la personne humaine par « l'électrostress » qu'il produit) ; *g* **Elektrosmog** [m] (Umgangssprachliche Bezeichnung für elektromagnetische Strahlung, die von künstlichen elektrischen Quellen, z. B. Hochspannungs-

leitungen, Fernseh-, Radar- und Mikrowellen sowie elektrischen Haushaltsgeräten, aber auch von Senderantennen [Mobilfunksender] im hochfrequenten Bereich, erzeugt wird. Diese Strahlungen übertreffen um mehrere Größenordnungen die Stärke des natürlichen Magnetfeldes und des so genannten Schönwetterfelds der Erde, sehr schwache Gleichfelder ohne Richtungswechsel, an die sich Mensch und Natur seit Jahrtausenden angepasst haben. Beim Menschen beeinflusst der **E.** seine Befindlichkeit und kann sich zum sog. Elektrostress schädlich auf die Gesundheit auswirken. Die Beeinträchtigungen reichen von einer Beeinflussung der Gehirnströme über Schlafstörungen [gebremste Melatonin-Ausschüttung] und lokale Überwärmung bis hin zum vermehrten Auftreten von Leukämie bei Kindern, die unter Hochspannungsleitungen leben; die beobachteten Wirkungen werden von der Industrie und einigen Wissenschaftlern bezweifelt; cf. ÖKO 2003; speziell die Unbedenklichkeit der Mobilfunkstrahlen der Antennenmasten ist zurzeit schwerlich zu beweisen, da der Zeitraum der Nutzung noch viel zu kurz ist, um Langzeitauswirkungen festzustellen).

electrostress [n] *envir.* * ▶electromagnetic radiation.

element [n] *phyt.* ▶floristic element; *chem.* ▶trace element.

element [n], **design** *arch.* ▶design feature.

element [n], **landscape** *conserv. land'man. landsc.* ▶landscape feature.

element [n], **planter** *gard. landsc.* ▶planter unit.

element [n], **precast concrete** *constr.* ▶precast concrete unit.

element [n], **street design** *arch. urb.* ▶piece of street furniture.

element [n] [US], **streetscape** *arch. urb.* ▶piece of street furniture, ▶piece of street furniture element.

element [n], **visual** *conserv. land'man. landsc.* ▶landscape feature.

element cycle [n] *ecol.* ▶biogeochemical cycle.

1745 elevated hedgerow [n] [US] *geo. land'man.* (Raised vegetative buffer, originally planted as a protective barrier around fields, which may be on a 50cm high ridge or berm, constructed of stones or earth; ▶coppice 2, ▶hedgerow); *syn.* quickset hedge [n] [UK]; *s* **seto** [m] **vivo elevado** (▶Seto vivo interparcelario típico del norte de Alemania; está formado por un montículo de tierra y piedras de 0,5 m de altura sobre el cual crece la vegetación arbórea y arbustiva; ▶podar hasta el tocón); *f* **haie** [f] **sur talus** (Haie bocagère plantée sur un talus pouvant atteindre 1,50 m de haut, utilisée autrefois en protection des champs ; une haie sans talus est appelée « haie de plein pied » ou « haie à plat » ; LA 1981, 596 ; ▶haie champêtre, ▶recéper) ; *g* **Knick** [m] (SH, ▶Feldhecke, die ursprünglich als Schutzeinfriedung der Felder auf ca. 0,5 m hohen Erd- oder Steinwällen angelegt wurde und im Rahmen der Feldheckenpflege etwa alle 7-12 Jahre am Boden abgeschnitten wird. Das Wort **K.** kommt von ‚knicken': Ursprünglich wurden in SH junge Gehölze, damit sie dichter werden, angeschnitten und auf den Boden zur Bewurzelung und Neuverzweigung abgesenkt. Diesen uralten Brauch gibt es heute nicht mehr. Stattdessen wird ‚knicken' heute i. S. v. ▶auf den Stock setzen verstanden; cf. WE 1967. Zur Zeit des 30-jährigen Krieges gab es in Mitteleuropa viele auf dem Stock gesetzte, schwer durchdringbare [Hainbuchen]wehrhecken, die man *Knickicht, Wehrholz, Landheeg* oder *Gebück* nannte; cf. LAU 2002); *syn.* Wallhecke [f].

1746 elevated highway [n] [US] *trans.* (Road built overhead to reduce the amount of traffic on downtown streets by creating a second tier; especially for inner-city express highways); *syn.* elevated road [n] [UK]; *s* **calle** [f] **elevada** (Calle construida sobre pilares para evitar cruces y así acelerar el tránsito dentro de la ciudad); *syn. popular* excalectrix [m] [Es]; *f* **voie [f] suspendue** (Route construite sur des piliers en créant un deuxième niveau de circulation [voie rapide] favorisant le décongestionnement des centre-villes) ; *g* **Hochstraße [f]** (Hochgelegte oder aufgeständerte Straße auf Brückentragwerken in Innenstädten oder Ballungsgebieten zur Entlastung des Verkehrs in den Stadtstraßen resp. zur Entflechtung von Verkehrsknoten durch Schaffung einer zweiten Ebene, besonders für Stadtschnellstraßen).

elevated road [n] [UK] *trans.* ▶elevated highway [US].

1747 elevation [n] *plan.* (Generic term covering all vertical views of a structure; *specific terms* ▶defined elevation, ▶finished elevation, ▶front elevation, ▶invert elevation, ▶side elevation, and ▶rear elevation, ▶spot elevation, ▶TW-elevation [US]; ▶top view); *s* **alzado [m]** (Vista de frente; ▶vista de planta, ▶alzado posterior, ▶alzado lateral, ▶alzado frontal); *syn.* elevación [f], plano [m] de elevación; *f* **élévation [f]** (Représentation graphique d'un objet en projection sur un plan vertical ; ▶vue en plan, ▶élévation vue arrière, ▶vue latérale, ▶élévation avant) ; *g* **Ansicht [f]** ([Zeichnerische] Darstellung eines Objektes in einer Projektionsebene; ▶Draufsicht, ▶Rückansicht, ▶Seitenansicht, ▶Vorderansicht).

elevation [n] [UK], **establishment of final** *constr.* ▶establishment of finished elevation.

elevation [n], **existing** *constr. eng. plan.* ▶existing ground level.

elevation [n] [US], **finished floor (FFE)** *constr.* ▶finished floor level.

elevation [n] [US], **TW-** *arch. constr.* ▶top level of wall.

elevation [n] **above mean sea level** *surv.* ▶height above mean sea level.

elevational zone [n] *geo. phyt.* ▶altitudinal zone.

elfin forest [n] *phyt.* ▶krummholz community.

elfinwood [n] *phyt.* ▶krummholz community.

elfinwood [n] [US], **pinus-dominated** *phyt.* ▶pinus-dominated krummholz formation.

1748 elimination [n] **of disturbing impacts** *ecol. land' man.* (Mitigation of physical, visual, or ecological hazards through various means); *s* **eliminación [f] de incidencias perturbadoras**; *f* **élimination [f] des nuisances** ; *g* **Beseitigung [f] störender Einflüsse.**

ell [n] [US] *constr.* ▶pipe elbow.

elm disease [n] *arb.* ▶Dutch elm disease.

1749 El Niño effect [n] *met.* (El Niño is an abnormal warming of waters in the eastern Pacific which climatologists link to worldwide weather extremes, from flood to drought. **In U.S.**, it is likely to create higher-than-normal storm activity, particularly on the west coast in January, February and March. The current El Niño is being compared to a similar pattern in 1982-83 which caused $2 billion damage nationwide, $100 million of that along the coast of California. An anomalous warming of ocean water resulting from the oscillation of a current in the South Pacific, usually accompanied by heavy rainfall in the coastal region of Peru and Chile, and reduction of rainfall in equatorial Africa and Australia); *s* **efecto [m] de El Niño** (Calentamiento anómalo del agua del océano a gran distancia de las costas de América del Sur debido a la oscilación de una corriente del Pacífico del Sur, usualmente acompañado por fuertes lluvias en la región costera de Perú y Chile, y la reducción de lluvia en África ecuatorial y Australia); *f* **effet [m] d'El Niño** (Perturbation océanique et atmosphérique provoquée par le réchauffement

irrégulier et anormal de l'eau océanique dans la zone tropicale de l'océan Pacifique au large des côtes d'Amérique du Sud ; en temps normal, les alizés forts et prédominants soufflent vers l'ouest et poussent les eaux chaudes de l'océan vers l'Australie et l'Indonésie, faisant remonter en surface, le long des côtes de l'Amérique du Sud, des eaux plus froides, riches en éléments nutritifs et donc très poissonneuses ; tous les trois à sept ans, les alizés faiblissent et changent de direction ; les eaux chaudes qui sont normalement poussées vers l'ouest reviennent vers l'Amérique du Sud ; cette énorme masse d'eau chaude réchauffe [jusqu'à 5 °C] une partie des eaux de l'océan Pacifique et modifie notamment la pression atmosphérique et le régime des pluies tropicales en provoquant des phénomènes extrêmes de sècheresse sur l'Afrique équatoriale et l'Australie ainsi que des fortes précipitations et inondations dans les régions côtières du Pérou et du Chili ; le nom El Niño a été donné à ce phénomène par les Péruviens, car il se manifeste souvent autour de Noël. **La Niña** est le phénomène inverse d'El Niño, au cours duquel on assiste à un réchauffement anormal des eaux de l'ouest du Pacifique et à des alizés plus forts ; La Niña apporte des hivers plus froids dans l'ouest du Canada et dans l'Alaska, ainsi qu'un temps plus sec et plus chaud dans le sud-est de l'Amérique) ; *g* **El-Niño-Effekt [m]** (Span. *niño* >der Junge<, >das Christkind<; ein etwa alle drei bis sieben Jahre im Pazifik in der Weihnachtszeit auftretendes Klimaphänomen, das Witterungsextreme mit verheerenden Folgen wie sintflutartige Starkregenfälle, Stürme, Überschwemmungen in den Küstenregionen von Peru und Chile, Dürre und Waldbrände in Äquatorialafrika und Australien hervorruft. In normalen, nicht von El Niño beeinflussten Jahren bläst der Südostpassat, welcher von den subtropischen Hochdruckgürteln zur äquatorialen Tiefdruckrinne weht und durch die Erdrotation abgelenkt wird, im Bereich des Äquators von Osten nach Westen. Er treibt somit kühles Oberflächenwasser von der südamerikanischen Küste nach Westen und setzt durch die Verschiebung der Wassermassen einen Kreislauf in Gang. Dem in Südostasien ankommenden, inzwischen erwärmten Oberflächenwasser weicht kaltes Wasser in genau umgekehrter Richtung aus. So bewegt sich kaltes, nährstoffreiches und dadurch auch fischreiches Wasser, welches sich wegen seiner größeren Dichte in tieferen Regionen des Pazifiks befindet, von Westen nach Osten. Vor Südamerikas Westküste gelangt dieses Wasser in den Auftriebsgebieten an die Oberfläche. Deshalb befindet sich dort der kalte und nährstoffreiche Humboldtstrom. In einem El-Niño-Jahr, in dem sich die Passatwinde abschwächen, erwärmt sich in Äquatornähe der östliche Pazifik um ca. 5-8 °C [vor den Westküsten Amerikas] und schwächt damit erheblich die Wirkung des kalten Humboldtstromes ab. Der westliche Pazifik vor Südostasien und Australien kühlt sich ab und ändert entsprechend die Meeresströmung im Pazifik [Zurückfließen der Warmwassermassen von Indonesien in Richtung Südamerika] und den atmosphärischen Luftdruck. Die so entstehenden Zirkulationsanomalien lösen die oben beschriebenen folgenschweren Naturkatastrophen aus; cf. www.elnino.info).

1750 elongated garden [n] *gard. plan.* (Long and narrow-shaped garden of a town house [US]/row house; ▶row house garden); *syn.* town house garden [n] [also US]; *s* **jardín [m] de bolsillo** (≠) (Jardín pequeño de forma rectangular situado generalmente detrás de casa familiar; ▶jardín de casa adosada); *f* **jardin [m] en mouchoir de poche** (Jardin étroit et long, en général situé en arrière d'une habitation en ligne ; ▶jardin d'une construction en ligne) ; *g* **Handtuchgarten [m]** (Schmales, langes Gartengrundstück, meist auf der Rückseite von Einfamilienreihenhäusern angelegt; ▶Reihenhausgarten).

elongated pile dumping [n] [US] *constr.* ▶front end dumping.

1751 embankment [n] *constr.* (**1.** Slope of a filled, linear mound of compacted soil or other material. **2.** Artificial structure of earth, gravel, crushed aggregate or rock [e.g. from tunnel excavation], or long artificial mound of stone with steep slopes, usually of uniform gradient constructed primarily to retain water, or to carry a roadway or railroad, as well as for noise and sight protection. **3.** Term is sometimes loosely applied to the steep, artificial side of a river, e.g. Thames embankment; ►levee [US] 2/flood bank [UK], ►natural slope 1, ►river embankment, ►road embankment, ►standard slope, ►transportation embankment, ►subgrade [US] 1/sub-grade [UK]); *s* **talud** [m] (**1.** Ladera de un terraplén de tierra u otros materiales. **2.** Estructura artificial de tierra, grava, agregado triturado o roca [p. ej. de excavación de túnel] o montículo de piedras alargado con laderas muy pendientes, generalmente de la misma inclinación, construido primordialmente para soportar una vía de transporte [carreter, ferrocarril] o como ►dique de contención para retener agua en un estanque artificial; ►dique fluvial, ►subbase, ►talud estándar, ►talud natural, ►terraplén para carretera. **3.** En inglés se utiliza también para denominar orillas muy pedientes estabilizadas artificialmente); *f 1* **talus** [m] (Pente naturelle ou artificielle prononcée et par extension relief présentant un plan allongé ou deux versants raides opposés par leur sommet ou leur terre-plein commun ; *f 2* **ouvrage** [m] **linéaire en remblai** (**1.** Talus d'un ouvrage linéaire exécuté par remblaiement ; **2.** remblai de route, de voie ferroviaire constitué de terre, de graves, de rochers pouvant constituer la ►couche de forme d'infrastructures de transport, une ►digue de protection contre les crues, une retenue d'eau ou un barrage ; **3.** en anglais *embankment* signifie aussi le talus stabilisé artificiellement des berges d'un cours d'eau ; ►dique fluviale, ►profil type de talus, ►remblai d'infrastructure de transport, ►remblai routier, ►talus naturel) ; *syn.* remblai [m] linéaire ; *g 1* **geschüttete Böschung** [f] (Durch Aufschüttung entstandene Böschung eines linearen Erdkörpers); *g 2* **Damm** [m] (**3**) (Durch Baumaßnahmen entstandener länglicher Erdkörper aus Boden, Kies, Schotter oder Fels [z. B. Tunnelausbruch]. Ein Damm kann der ►Unterbau 1 von Verkehrswegen [Straßendamm, Bahndamm] incl. Kanälen sein, als ►Deich 2 vor Hochwasser schützen, als Umrandung von künstlichen Speicherbecken oder als Staudamm dienen. **3.** Im Englischen bedeutet *embankment* auch die anthropogen veränderte/stabilisierte Uferböschung eines Fließgewässers; ►Damm für Verkehrsbauten, ►Flussdeich, ►natürliche Böschung, ►Regelböschung).

embankment [n]**, brow of** *constr.* ►crest of embankment/slope.

embankment [n]**, carriage-way** *eng. trans.* ►transportation embankment.

embankment [n]**: failure** *constr.* ►shear failure of embankment.

embankment [n]**, highway** *constr. eng.* ►road embankment.

embankment [n]**, paved** *eng.* ►bank revetment.

embankment [n]**, rounding the crest of** *constr.* ►rounding the top of an embankment/slope.

embankment [n]**, toe of an** *constr.* ►foot of an embankment/slope.

1752 embankment crown [n] *constr.* (Top of an embankment/slope); *s* **cima** [f] **de talud** *syn.* cumbre [m] de talud; *f* **tête** [f] **de talus** *syn.* crête [f] de talus, épaule [f] de talus ; *g* **Böschungskrone** [f].

embankment gradient [n] *constr.* ►slope angle.

embankment slope [n] *constr.* ►unstable embankment slope.

1753 embankment stabilization [n] *constr.* (Measures to minimize the risk of erosion or embankment movement;

►slope stabilization, ►unstable embankment slope); *s* **estabilización** [f] **de talud** (Medidas de protección contra la erosión o el deslizamiento; ►consolidación de taludes, ►talud inestable 1); *f* **stabilisation** [f] **de talus** (Mesure de protection de la surface d'un talus contre l'érosion ; ►stabilisation de versant, ►talus instable) ; *syn.* consolidation [f] de talus ; *g* **Böschungssicherung** [f] (Maßnahme zum Erosionsschutz und zur Rutschsicherung von Böschungen; ►Hangsicherung, ►instabile Böschung).

embossed [pp/adj] *constr.* *►tooling of stone [US]/dressing of stone [UK], #1.

embossed natural stone wall [n] *arch. constr.* ►rough-tooled natural stone wall.

1754 embroidered parterre [n] *gard'hist.* (Level garden of various ornamentally designed flowerbeds, without trees or tall shrubs, usually bordered by clipped boxwood—originating from the French baroque period; ►parterre); *s* **parterre** [m] **bordado** (Tipo de parterre barroco desarrollado en Francia a finales del siglo XVI. formado por una zona plana de jardín sin leñosas con diferentes macizos de flores generalmente «bordados» [en francés *broderie*] por pequeños setos de boj; ►parterre); *syn.* parterre [m] de broderie; *f* **parterre** [m] **de broderie** (Type de ►parterre développé en France à la fin du XVIème siècle par Claude Mollet, formé de rinceaux de fleurs ou de buis taillés se détachant sur fond sablé, leur dessin curviligne s'apparente aux broderies ; DIC 1993, 519) ; *syn.* compartiment [m] de broderie, parterre [m] en broderie ; *g* **Broderieparterre** [n] (Franz. *broderie* >Stickerei<; ebene, repräsentativ-zierende Gartenanlage ohne hohe Gehölze, mit ornamental gestalteten Beeten, die entweder mit unterschiedlich farbigen Gesteinsmaterialien resp. gemahlenem terracottafarbigem Ziegelsplitt ausgelegt oder mit Sommerblumen bepflanzt und vorwiegend mit geschnittenem *Buxus* [Buchsbaum] eingefasst sind — entstanden zur Zeit des französischen Barocks, z. B. Parkanlagen des Schlosses zu Versailles, Augustusburg zu Brühl bei Köln; ►Parterre 2); *syn.* Parterre [n] mit starker Ornamentierung.

1755 embryo dune [n] *geo. phyt.* (Lowest beach dune between ►driftline and ►white dune. In Central and Eastern Europe its formation is fostered by beach grass *[Agropyron junceum]* where sand is in sufficient supply); *syn.* foredune [n], primary dune [n]; *s* **duna** [f] **embrionaria** (Primera fila de dunas de poca altura entre la ►orla de cúmulos de residuos orgánicos y la de ►dunas blancas); *syn.* embrión [m] de duna; *f* **dune** [f] **embryonnaire** (Dune sur laquelle s'observent les premières accumulations subhorizontales de sable, située entre le ►haut de plage et la ►dune blanche dont la fixation s'effectue avec le Chiendent jonciforme *[Agropyron junceum]* et s'élèvent par captage éolien ; stade initial de formation de la dune vive ; ►ourlet halonitrophile) ; *syn.* avant-dune [f] (DG 1984, 145), banquette [f], dune [f] jeune ; *g* **Primärdüne** [f] (Niedrigste Düne am Strand zwischen ►Spülsaum und ►Weißdüne. Die Dünenbildung wird in Mittel- und Osteuropa bei ausreichender Sandzufuhr durch die Strandquecke *[Agropyron junceum]* eingeleitet); *syn.* Embryonaldüne [f], Vordüne [f].

emerged peat moss hollow community [n] *phyt.* ►floating sphagnum mat.

1756 emergency lane [n] *trans.* (Lane between travelled way [US]/carriage-way [UK] and ►road verge reserved for emergencies); *syn.* stopping lane [n]; *s* **carril** [m] **de emergencia** (En autopista, carril adicional entre la calzada y la ►berma 2 cuya función es permitir la parada de vehículos en caso de emergencia); *f* **bande** [f] **d'arrêt d'urgence** (Élément de chaussée en bordure de l'►accotement non stabilisé constitué par une surlargeur portant la bande de guidage ainsi qu'une partie stabilisée et revêtue présentant si possible un aspect différent de

E

celui de la chaussée, permettant l'arrêt d'urgence d'un véhicule en panne, l'intervention des véhicules de secours et d'entretien) ; *g* **Standstreifen [m]** (An hochbelasteten Außerortsstraßen [Autobahnen und autobahnähnlichen Bundesstraßen und Schnellstraßen mit Richtungsfahrbahnen] zwischen Fahrbahnen und ▶Bankett liegende Spur zum Anhalten eines Kraftfahrzeuges bei Pannen, Not- oder Unfällen); *syn.* Seitenstreifen [m], Pannenstreifen [m] [A, CH], *umgangssprachlich* Standspur [f] *und* Notspur [f].

emergency preservation notice [n] [UK] *conserv. leg. plan.* ▶temporary restraining order [US].

emergent aquatic [n] [US] *phyt.* ▶emergent aquatic plant (1), ▶emergent aquatic plant (2).

1757 emergent aquatic plant [n] (1) *phyt.* (Helophyte of reed and sedge swamp; ▶emergent aquatic plant 2); *syn.* reed plant [n] [also UK], emergent aquatic [n] [also US], emergent macrophyte [n] [also US]; *s* **planta [f] semiacuática** (Gramínea que crece en las orillas de masas de agua con niveles de agua cambiantes, entre las cuales se presentan a menudo las fragmiteas y la hierba cinta; ▶planta acuática emergente); *f* **plante [f] des roselières** (Plante semi-aquatique à puissants rhizomes croissant dans la vase, dont l'appareil végétatif est aérien et émettant chaque année des pousses verticales qui meurent au premier hiver et subsistent sèches jusqu'au printemps, localisée en marge des cours d'eau ou dans l'eau dans les zones à niveaux variables ; le Roseau commun et le Roseau à balai *[Phragmites communis* et *P. australis]* ou l'Alpiste *[Phalaris arundinacea]* constituent les plantes principales des roselières ; les plantes des roselières sont des ▶hélophytes ; ▶amphiphyte) ; *syn.* plante [f] semi-aquatique ; *g* **Röhrichtpflanze [f]** (An Wasserrändern oder im Wasser mit schwankendem Wasserstand wachsendes Rohr, oft als Schilfrohr *[Phragmites communis* und *P. australis]* oder Rohrglanzgras *[Phalaris arundinacea]* anzutreffen; ▶Überwasserpflanze).

1758 emergent aquatic plant [n] (2) *bot.* (Tall, rooted, herbaceous angiosperm in ▶reed swamps, stagnant or slightly running water, that may be temporarily or permanently flooded at the base but does not tolerate prolonged inundation of the entire plant; ▶helophyte); *syn.* tall helophyte [n] (VIR 1982), emergent hydrophyte [n] (CWD 1979), emergent aquatic [n] [also US]; *s* **planta [f] acuática emergente** (Planta herbácea que crece en carrizales, cuyas raíces están permanentemente sumergidas en el agua, y que tolera la inundación incluso permanente de la parte inferior del tallo, pero no de todas las partes aéreas; ▶carrizal de fragmitea, ▶helófito); *f* **amphiphyte [m]** (Plante herbacée *[Angiospermae]* des milieux à plan d'eau variable des bords de mares [zone de balancement du niveau d'eau] ou sur sols inondés ou engorgés durant une partie importante de l'année, enracinée sur les sédiments, susceptibles de s'installer et de se développer au-dessous et au-dessus de la surface de l'eau [certaines parties — feuilles et fleurs — poussent à l'extérieur de l'eau], croissant en majorité dans les différentes ▶roselières ; les principales espèces en France sont les jussies et le myriophylle du Brésil ; ▶hélophyte) ; *syn.* plante [f] aquatique émergente, plante [f] amphiphyte, plante [f] amphibie ; *g* **Überwasserpflanze [f]** (Meist im ▶Röhricht lebende, krautige höhere Pflanze *[Angiospermae]*, die mit Wurzeln und den untersten Sprossteilen im Wasser steht und eine zeitlich begrenzte bis permanente Überflutung nur der unteren Pflanzenteile verträgt; ▶Sumpfpflanze).

emergent hydrophyte [n] *bot.* ▶emergent aquatic plant (2).

emergent macrophyte [n] [US] *phyt.* ▶emergent aquatic plant (1), ▶emergent aquatic plant (2).

1759 emerging leaves [npl] *bot. hort.* (Leaves opening on woody plants and perennials); *syn.* new leaves [npl]; *s* **foliación**

[f] (Desarrollo de las yemas folíferas en los árboles y arbustos caducifolios; cf. DB 1985); *syn.* brotadura [f] de las hojas; *f* **feuillaison [f]** (Période pendant laquelle commencent à se développer les bourgeons et à apparaître les nouvelles feuilles des végétaux ligneux et des plantes vivaces) ; *syn.* foliation [f], frondaison [f] ; *g* **Laubaustrieb [m]** (Sprießen der Blätter bei Gehölzen und Stauden im Frühling resp. nach einer Periode der Vegetationsruhe; *allgemeinsprachliches Syn. für Bäume und Sträucher* Grünwerden [n, o. Pl.] der Gehölze); *syn.* Blattaustrieb [m].

1760 emigration [n] *sociol. zool.* (Outward ▶migration 2 of individual animals within a population from their home range never to return, usually due to shortage of food, as well as due to economic hardship, political, or religious reasons of people; ▶migration 1; *opp.* ▶immigration); *s* **emigración [f]** (Migración de individuos de fauna al exterior de su biótopo por falta de alimentos o de seres humanos a otras regiones o países por falta de posibilidades de vida en su región/país original; ▶migración 1, ▶migración 2; *opp.* ▶inmigración); *f* **émigration [f]** (Mouvement d'individus ou de populations quittant leur lieu d'origine pour aller s'établir dans un autre milieu de vie ; pour la population humaine l'émigration est en général un phénomène économique et politique ; ▶migration 1, ▶migration 2 ; *opp.* ▶immigration) ; *g* **Emigration [f]** (Dauerhaftes Sichentfernen einzelner bis vieler Individuen aus ihrer Population und ihrem angestammten Lebensraum, meist wegen Nahrungsmangel oder beim Menschen i. d. R. wegen wirtschaftlicher Not, politischer oder religiöser Gründe; ▶Migration 1, ▶Migration 2; *opp.* ▶Immigration); *syn.* Auswanderung [f].

1761 eminent domain [n] **(power of condemnation)** [US] *adm. leg.* (In U.S., legal authority of a government to take or to authorize the taking, of private property for public use, subject to strict rules of procedure and payment of fair compensation to the owner according to the Vth Amendment of the Federal Constitution and equivalent provisions in state constitutions; **in U.K.**, legal procedure whereby property is acquired by public authorities for the realization of planning projects, when the owner is unwilling to sell of his own accord; compulsory purchase and compensation is governed by the *Planning and Compensation Act, 1991;* ▶administrative procedure); *syn.* condemnation [n] [US], compulsory purchase [n] [UK]; *s* **expropiación [f] forzosa** (▶Procedimiento administrativo para la enajenación de terrenos privados y la indemnización de sus propietarios para realizar obras de utilidad pública; cf. Ley de Expropiación Forzosa, de 16 de diciembre de 1954 y Ley sobre Régimen del Suelo y Valoraciones, RD 6/1998, arts. 33-40); *syn.* enajenación [f] forzosa; *f* **expropriation [f]** (Pour cause d'utilité publique, abandon par un propriétaire de la propriété d'un bien immobilier en faveur de l'administration moyennant indemnité dans le cadre d'une ▶procédure administrative en vue de la réalisation d'une opération d'aménagement ou d'urbanisme) ; *g* **Enteignung [f]** (Mit Hilfe eines förmlichen ▶Verwaltungsverfahrens bewirkte zwangsweise Überführung von Privateigentum in das Eigentum eines Gemeinwesens zur Realisierung von Planungsvorhaben; die **E.** ist in zahlreichen bundes- und landesrechtlichen Gesetzen, z. B. §§ 85 ff BauGB, geregelt; **im UK** werden Belange der Enteignung und Entschädigung im *Planning and Compensation Act von 1991* geregelt); *syn.* förmliche Enteignung [f].

1762 emission [n] *envir.* (Release of fumes, noise, vibrations, light, heat, radiation, etc. from a certain source that causes pollution; ▶emission standard, ▶excess gaseous emission, ▶pollution impact, ▶transmission); *s 1* **emisión [f]** (*Sensu generale* expulsión a la atmósfera, al agua o al suelo de sustancias, vibraciones, calor o ruido procedentes de forma directa o indirecta de fuentes

fuentes puntuales o difusas de una instalación; cf. Art. 3 k Ley 16/2002, de Prevención y Control Integrados de la Contaminación; ▶emisión gaseosa, ▶inmisión, ▶transmisión, ▶valor límite de emisión); *s 2* **nivel [m] de emisión [Es]** (*Sensu juris* cantidad de un contaminante emitido a la atmósfera por un foco fijo o móvil, medido en una unidad de tiempo; cf. Prevención y Corrección de la Contaminación Atmosférica de Origen Industrial, Orden de 18 de Octubre de 1976, Anexo 1); *f* **émission [f]** (Rejet [processus et son résultat] dans l'atmosphère de gaz, de particules solides ou liquides, corrosifs, toxiques ou odorants, bruit, vibration et trépidation, lumière, chaleur et rayonnement de nature à compromettre la santé publique ou la qualité de l'environnement, à nuire au patrimoine agricole, forestier ou bâti ; décret n° 74-415 du 13 mai 1974 ; **D.**, terme définissant les émissions considérées du point de vue de la source émettrice ; ▶immission, ▶transfert, ▶rejet gazeux, ▶valeur limite d'émission) ; *g* **Emission [f]** (Von einer festen oder beweglichen Anlage ausgehende Luftverunreinigungen, Geräusche, Gerüche, Erschütterungen, Wärme, [Licht]strahlen und ähnliche Erscheinungen oder die Abgabe solcher Beeinträchtigungen an die Umwelt; auch von Produkten abgegebene Schadstoffe zählen dazu; cf. § 3 [3] BImSchG. Die maximal zulässige Abgabe von **E.en** wird durch Gesetze, Verordnungen oder sonstige Regelwerke [Technische Anleitungen] mit Richtwerten und/oder Grenzwerten festgelegt; im englischen und französischen Sprachraum deckt der Begriff **E.** den Ausstoß und die ▶Immission ab; ▶Abgas 1, ▶Emissionsgrenzwert, ▶Transmission).

emission-avoiding [ppr] *envir. pol.* ▶environmentally safe.

1763 emission certificate [n] *envir. pol.* (Environmental policy instrument in the form of an authorized permit fee, which allows the holder to emit environmentally harmful gases, especially CO_2, into the atmosphere. The aim of issuing **e. c.s** is to ensure that industry no longer discharges billions of tons of CO_2 into the atmosphere, without paying for their disposal. In this way, the emission of pollutants can be reduced and brought under control with minimal cost to the economy; in the initial pilot phase, 2005 to 2007, CO_2 licenses were issued, for example, to electricity companies free of charge); *s* **derecho [m] de emisión** (Instrumento de política ambiental acordado en el Protocolo de Kioto como uno de los llamados mecanismos de flexibilidad, junto a los basados en proyectos de inversión en tecnología limpia en países terceros [mecanismo de desarrollo limpio y aplicación conjunta] para favorecer la reducción de emisiones de efecto invernadero. Consiste en el derecho subjetivo a emitir, desde una instalación incluida en el ámbito de aplicación de la correspondiente ley, una tonelada equivalente de dióxido de carbono, durante un periodo/período determinado. La Directiva 2003/87/CE establece un régimen para el comercio de derechos de emisión de gases de efecto invernadero en la Comunidad Europea. **En Es.** esta directiva fue traspuesta por medio de la Ley 1/2005, que regula los mecanismos de aplicación, las autorizaciones de emisión de gases de efecto invernadero y define la naturaleza y el contenido del Plan Nacional de asignación de **d. de e.** así como su procedimiento de aprobación. En la primera fase de aplicación de cinco años a partir del 1 de enero de 2008, la asignación de derechos es gratuita, salvo para los nuevos entrantes. En el ▶régimen de comercio de derechos de emisión de gases de efecto invernadero están incluidas hasta el momento solamente las instalaciones industriales de generación de energía térmica y de combustión de más de 20 MW de potencia, todas las refinerías de hidrocarburos, todas las coquerías así como las instalaciones de producción y transformación de metales férreos, las fábricas de cemento sin pulverizar, de cal en hornos rotatorios, de vidrio y fibra de vidrio, de productos cerámicos mediante horneado, de pasta de papel y de papel y cartón a partir de una

capacidad de producción definida en la ley; cf. art. 2 y anexo I Ley 1/2005); *f* **certificat [m] d'émission (de gaz à effet de serre)** (Instrument de politique environnementale instauré dans le cadre du protocole de Kyoto instituant un système d'échange de quotas d'émission de gaz à effet de serre [certificat — unité de quantité en tonnes de CO_2] afin de réduire les émissions de gaz à effet de serre ; les entreprises se voient attribuer des autorisations d'émission et sont tenues d'avoir un nombre équivalent de certificats [quotas] représentant le droit d'émettre une quantité donnée de gaz ; les entreprises dépassant leurs plafonds d'émissions de gaz à effet de serre sont autorisées à acheter des certificats [à un prix fixé à tout moment par l'offre et la demande] auprès d'entreprises plus performantes sur le plan environnemental qui polluent moins ; l'acquéreur paie pour polluer plus alors que le vendeur est récompensé parce qu'il lui reste des certificats ; plusieurs bourses d'échange ont vu le jour en Europe, afin de permettre la vente et l'achat des certificats d'émission en interne aux pays industrialisés, et entre pays industrialisés et pays émergents ; *syn.* certificat [m] de réduction d'émission ; *g* **Emissionszertifikat [n]** (Umweltpolitisches Instrument in Form einer kostenpflichtigen Bewilligungsbescheinigung, mit der der Inhaber das Recht erwirbt, umweltschädliche Gase, besonders CO_2, in die Atmosphäre ausstoßen zu dürfen. Mit den **E.en** soll erreicht werden, dass die Industrie nicht mehr Milliarden Tonnen des Klimagiftes CO_2 kostenlos in die Atmosphäre endlagern kann, und dass Schadstoffemissionen mit minimalen volkswirtschaftlichen Kosten verringert werden; in der ersten Pilotphase von 2005-2007 wurden CO_2-Lizenzen, z. B. an Stromkonzerne noch kostenlos abgegeben).

1764 emission-damaged forest [n] *envir.* (Forest adversely affected by air pollutants, e.g. by ▶acid rain, ▶forest decline); *s* **bosque [m] dañado por la contaminación** (▶lluvia ácida, ▶muerte de bosques); *f* **forêt [f] contaminée par la pollution atmosphérique** (Forêt exposée à une forte pollution de l'air et particulièrement altérée, p. ex. par les ▶pluies acides ; ▶mort des forêts) ; *syn.* forêt [f] affectée par la pollution atmosphérique ; *g* **immissionsgeschädigter Wald [m]** (Wald, der durch starke Luftverschmutzung, z. B. durch ▶sauren Regen, erheblich beeinträchtigt ist; ▶Waldsterben).

emission dispersal model [n] *envir.* ▶emission distribution model.

1765 emission distribution model [n] *envir.* (Model for the calculation of areas, affected by pollution emissions); *syn.* emission dispersal model [n]; *s* **modelo [m] de difusión (de emisiones)** (Esquema para calcular las superficies posiblemente afectadas por emisiones); *syn.* modelo [m] de propagación (de emisiones); *f* **modèle [m] de diffusion** (Modèle permettant le calcul des surfaces pouvant être atteint par les émissions) ; *syn.* modèle [m] de propagation ; *g* **Ausbreitungsmodell [n]** (Modell zur Berechnung von Flächen, die von Emissionen beeinträchtigt werden können).

emission inventory [n] *adm. envir. leg.* ▶atmospheric emission inventory.

1766 emission [n] **of noxious substances** *envir.* (Release of harmful matter from factories, machinery, and other equipment, as well as from sites with stored toxic materials which pollute air, soil, waterbody or groundwater; ▶emission, ▶noxious substance); *s* **emisión [f] de (sustancias) contaminantes** (Salida de sustancias nocivas de fábricas, máquinas u otros aparatos técnicos así como de lugares con materiales tóxicos almacenados que contaminan las aguas, el aire o el suelo; ▶emisión, ▶sustancia nociva); *syn.* emisión [f] de sustancias nocivas/tóxicas; *f* **émission [f] de substances nuisibles** (Rejet, introduction de substances polluantes pour l'air, l'eau et le sol, en

E

E

provenance de machines, installations, équipements ainsi que de sites pollués par des dépôts contenant des substances nocives ; ►émission, ►substance génératrice de nuisances) ; *syn.* émission [f] de produits nocifs ; *g* **Schadstoffemission [f]** (Von Betriebsstätten, Maschinen oder anderen technischen Geräten sowie von Grundstücken mit schadstoffhaltigem Lagermaterial ausgehende Luft-, Boden- oder Gewässer-/Grundwasserverunreinigung; ►Emission, ►Schadstoff).

1767 emission [n] of odo(u)r *envir.* (Discharge of stench or offensively smelling gaseous emissions); *s* **inmisión [f] de malos olores**; *f* **émission [f] de substances malodorantes** ; *g* **Geruchsimmission [f]** (Auf Menschen einwirkende übelriechende Gasemission).

1768 emission point [n] *envir.* (Source of emission which contaminates the environment with undesirable matter and can be exactly located, e.g. chimney, sewage pipe; ►point source pollution; *opp.* ►nonpoint source pollution [US]); *s* **foco [m] de contaminación** (Fuente de contaminación puntual; ►contaminación focal, *opp.* ►contaminación ubicua); *syn.* foco [m] de polución; *f* **source [f] de pollution ponctuelle** (Installations exactement localisées qui rejettent en permanence ou accidentellement des substances polluantes indésirables telles que les cheminées des grandes installations de combustion, points de rejet d'effluents industriels ; ►pollution ponctuelle, *opp.* ►pollution superficielle) ; *g* **punktuelle Verschmutzungsquelle [f]** (Genau lokalisierbare Emissionsstelle, wie z. B. Schornstein, Abwassereinleitungsrohr, die die Umwelt mit unerwünschten Stoffen verunreinigt; ►punktuelle Verschmutzung, *opp.* ►flächenhafte Verschmutzung).

emissions [npl] *envir.* ►industrial gaseous emissions.

emissions fee [n] *adm. envir. leg.* ►air pollution fee.

emission source [n] *envir.* ►pollution source.

1769 emission standard [n] *envir. leg.* (Legally defined maximum level of specific pollutants that may be emitted from a certain source in a specified period of time; ►adulteration limit, ►threshold level of air pollution); *s* **valor [m] límite de emisión** (Cantidad máxima de un contaminante del aire o del agua que la ley permite emitir al medio exterior. Se establece un límite para la emisión instantánea y otros para los valores medios en diferentes intervalos de tiempo; cf. Prevención y Corrección de la Contaminación Atmosférica de Origen Industrial, Orden de 18.10.1976, Anexo 1 y Titulo II Ley 16/2002; ►valor límite, ►nivel de referencia de calidad del aire); *syn.* nivel [m] máximo admisible de emisión, límite [m] de emisión; *f 1* **valeur [f] limite d'émission** (Concentration et/ou masse de substances polluantes dans les émissions en provenance d'installations pendant une période déterminée, à ne pas dépasser ; dir. n° 84/360/CEE du 28 juin 1984, arrêté du 1er mars 1993 ; ►valeur limite, ►valeur limite de la qualité de l'air) ; *syn.* valeur [f] limite de rejet d'effluents gazeux, taux [m] de produits polluants à ne pas dépasser ; *f 2* **valeur [f] guide de qualité atmosphérique** (Niveau de concentration dans l'atmosphère vers lequel l'anhydride sulfureux, les particules en suspension, l'ozone, le plomb et le dioxyde d'azote doivent tendre dans la mesure du possible sur une période donnée et en dessous duquel ils ne devrait avoir aucun effet préjudiciable sur la santé ; décr. n° 74-415 du 13 mai 1974) ; *syn.* valeur [f] guide de la qualité de l'air ; *f 3* **objectif [m] de qualité** (Niveau de concentration de substances polluantes dans l'atmosphère, fixé sur la base des connaissances scientifiques dans le but d'éviter, de prévenir ou de réduire les effets nocifs de ces substances pour la santé humaine ou pour l'environnement à atteindre dans une période donnée) ; *syn.* valeur [f] cible (dir. n° 96/62/CE du 27 septembre 1996), norme [f] de qualité de l'air ; *g* **Emissionsgrenzwert [m]** (Gesetzlich festgelegter

Höchstwert für näher definierte Emissionen, die eine Emissionsquelle in einem festgelegten Zeitraum abgeben darf; ►Grenzwert, ►Immissionsgrenzwert).

emissions tax [n] *adm. envir. leg.* ►air pollution fee.

1770 emissions trading system [n] *envir. pol.* (As permitted by the Kyoto Protocol of 1997, the sale of unused quotas of carbon dioxide [CO_2] emissions to other countries, which can then use them as credits for their own accounts. The protocol stipulates that by 2008 these emission rights will be traded worldwide for carbon dioxide emissions. The EU introduced the multinational **European Union Emission Trading System** (EU ETS) as early as 01.01.2005, but initially only until 2007 [Phase I]. It is the world's largest multinational, emissions trading scheme and is the foundation of EU climate policy covering more than 10,000 installations in the energy and industrial sectors. The European **legal basis** for emissions trading is the **Emissions Trading Directive** 2003/87/EG of the European Parliament and the Council of 13.10.2003. The EU scheme allows a regulated operator to use **carbon credits** in the form of **Emission Reduction Units** [ERU] to comply with its obligations. Issuance of certificates and monitoring of emissions is the responsibility of each of the countries; for each phase, the total quantity to be allocated by each Member State is defined in the Member State National Allocation Plan [NAP — equivalent to its UNFCCC-defined carbon account]. The European Commission oversees the NAP process and decides if the NAP fulfills the 12 criteria set out in the Annex III of the Emission Trading Directive [EU Directive 2003/87/EC]. Currently, the EU does not allow CO_2 credits under ETS to be obtained from **sinks**; e.g. reducing CO_2 by planting trees); *syn.* Emissions trading [m]; *s* **régimen [m] de comercio de derechos de emisión de gases de efecto invernadero** (Partiendo del Protocolo de Kioto, según el cual se permite el comercio internacional de cuotas de emisión de dióxido de carbono [CO_2] con el fin de reducir emisiones con el costo menor posible, en la UE se estableció por medio de la Directiva 2003/87/CE el **r. de c. de d. de e. de g. de e. i.** para fomentar reducciones de las emisiones de estos gases de una forma eficaz y de manera económicamente eficiente. La directiva prevé que las instalaciones sometidas a su ámbito de aplicación deben tener autorización de emisión a partir del 1 de enero de 2005. Para esto los Estados miembros debían transponer la Directiva en leyes nacionales y elaborar los llamados Planes Nacionales de asignación [PNA]. **En Es.** esta directiva fue traspuesta por medio de la Ley 1/2005, que regula los mecanismos de aplicación, las autorizaciones de emisión de gases de efecto invernadero y define la naturaleza y el contenido del Plan Nacional de asignación de derechos de emisión así como su procedimiento de aprobación. Una primera versión del Plan Nacional de asignación 2005-2007 fue aprobada anteriormente por medio del RD 1866/2004. En 2008 estaban incluidas más de 1000 instalaciones de generación de energía, de combustión o industriales con una asignación promedio anual de 175,8 millones de ►derechos de emisión. En la totalidad de la UE más de 10 000 instalaciones están incluidas en el sistema de comercio de derechos. La Comisión Europea vigila el proceso de los PNA y decide si éstos cumplen los 12 criterios previstos en el anexo III de la Directiva. Actualmente ésta permite el comercio de derechos dentro de cada país miembro o de la UE, con terceros países con compromiso de reducción o limitación de emisiones que sea parte del Protocolo de Kioto, siempre que exista previo reconocimiento en un instrumento internacional, así como con reducciones certificadas de emisión procedentes de los mecanismos de desarrollo limpio o de aplicación conjunta, que excluye instalaciones nucleares, actividades de uso de tierra, cambio de uso de tierra o selvicultura y proyectos de hidroeléctricas de capacidad superior a 20 MW que no cumplan con los criterios y directrices aprobadas por la Comisión Mundial de Presas; cf. Ley 1/2005 y MMA 2008); *syn.* comercio [m] de derechos de emisión; *f* **système [m] d'échange de quotas d'émissions de gaz** (Mécanismes de « projets » pour réduire le coût de la réduction des émissions prévus dans le protocole de Kyoto autorisant la vente de quotas excédentaires de gaz à effet de serre [GES] à d'autres pays ; par la directive 2003/87/CE du 13 octobre 2003 la communauté européenne pour satisfaire aux obligations du protocole de Kyoto a mis en place depuis le 1er janvier 2005 des quotas d'émissions de dioxyde de carbone [CO₂] pour un certain nombre d'entreprises du secteur de la production d'énergie, de l'industrie manufacturière et des services ; ces quotas sont échangeables et négociables ; une quantité initiale de quotas est allouée à ces entreprises au titre de leurs installations concernées ; le plan national d'allocation de quotas élaboré par chaque État doit préciser pour la période définie **1. le montant total d'émissions** sur lequel porte le marché d'échange de quotas d'émissions [en tonnes de dioxyde de carbone], **2. la répartition de ce total** entre les différents secteurs d'activité couverts par le marché, puis à l'intérieur de ces secteurs entre les installations couvertes ; le 30 avril de chaque année au plus tard, les exploitants des installations restituent un nombre de quotas correspondant au total de leurs émissions au cours de l'année précédente ; ces quotas restitués sont ensuite annulés) ; *g* **Emissionshandel [m]** (Politikinstrument der Europäischen Umweltpolitik auf marktwirtschaftlicher Basis, mit dem das im Kyoto-Protokoll von 1997 festgelegte Klimaschutzziel der Reduktion von Treibhausgasemissionen durch den Verkauf von nicht ausgeschöpften Kontingenten an Kohlendioxid[CO₂]-Ausstoß an andere Länder, die diese gekauften Differenzmengen ihrem eigenen Konto gutschreiben können, erlaubt ist. Das Protokoll sieht vor, dass ab 2008 weltweit mit diesen Emissionsrechten für den Kohlendioxidausstoß gehandelt werden soll. Die EU führte das multinationale **E.rechtehandelssystem** bereits zum 01.01.2005, zunächst nur für 2005-07 [Phase I], ein. In D. legt ein „Nationaler Allokationsplan" [NAP] in der Phase II für 2008-2012 fest, wieviele Kohlendioxidzertifikate die Branchen der Wirtschaft vom Staat erhalten; die Regeln für die Phase III, die ab 2013 beginnt, werden die gesetzlichen Grundlagen erarbeitet und verhandelt, die im Jahre 2009 im Gesetzgebungsverfahren verabschiedet werden sollen. Europäische **Rechtsgrundlage** des Emissionshandels bildet die **Emissionshandelsrichtlinie** 2003/87/EG des Europäischen Parlaments und des Rates vom 13.10.2003, geändert durch RL 2004/101/EG am 27.10.2004. Ausgabe von Zertifikaten und Überwachung der Emissionen übernimmt in Deutschland die Deutsche Emissionshandelsstelle des Umweltbundesamtes auf Grundlage des Treibhausgas-Emissionshandelsgesetzes [Gesetz zur Umsetzung der Richtlinie 2003/87/EG]. In Österreich regelt die Umsetzung des Emissionsrechtehandels das Emissionszertifikategesetz [EZG] vom 30.04.2004); *syn.* Emissionsrechtehandel [m].

employee [n] [US] *adm. constr. contr.* ▶staff person; *adm. contr.* ▶technical employee.

empty plot [n] [UK] *urb.* ▶vacant lot.

1771 enabling act [n] [US] *adm. leg.* (Legal order necessary for the implementation of specific legislation; **in U.S.**, this is usually state legislation, which authorizes or permits local governments to undertake various regulatory, spending and other programs and activities. A few so-called 'home rule' states, under their state constitutions, do not need such legislation; the term is sometimes incorrectly used to refer to acts of Congress; ▶order [US] 2); *syn.* enabling statute [n] [UK]; *s* **decreto [m]** (Conjunto de reglas legisladas para la aplicación de una ley; ▶reglamento jurídico); *syn.* real decreto [m] [Es] (*Abr.* RD); *f* **décret [m]**

d'application (Règle juridique déterminant les modalités d'application des textes législatifs ; ▶ordonnance légale) ; *syn.* ordonnance [f] d'exécution [CH] ; *g* **Durchführungsverordnung [f]** (*Abk.* DVO; ▶Rechtsverordnung, die der Durchführung einer Gesetzesbestimmung dient); *syn.* Vollziehungsverordnung [f] [CH].

enabling statute [n] [UK] *adm. leg.* ▶enabling act [US].

enactment [n] *leg.* ▶coming into effect.

enclosed block development [n] *urb.* ▶block development.

1772 enclosed medieval garden [n] *arch. hist. gard'hist.* (In Europe, an utilitarian garden of the Middle Ages in a walled courtyard; ▶cloister garden); *s* **jardín [m] medieval** (Jardín amurallado de origen medieval, en muchos casos se trata de jardines en conventos o monasterios; ▶jardín de claustro); *f* **jardin [m] clos** (Jardin entouré de murs, propre à l'époque de moyen-âge ; cf. DIC 1993 ; ▶jardin de cloître, ▶jardin de curé) ; *syn.* hortus [m] conclusus ; *g* **Gartenhof [m] (1)** (Von Mauern umgebener mittelalterlicher Garten; ▶Kreuzgarten); *syn.* Hortus [m] conclusus.

enclosing wall [n] *arch.* ▶perimeter wall.

1773 enclosure [n] *constr. landsc.* (Generic term covering fence, property boundary wall, hedge or other surrounding barrier; ▶fencing, ▶wild animal enclosure); *s* **cercado [m]** (Término genérico para todo tipo de construcción de deslinde; cercar [vb]; ▶envallado); *syn.* cerca [f]; *f 1* **clôture [f] (1)** (Terme générique pour toute enceinte délimitant un espace, p. ex. ▶clôture 2 métallique, clôture en maçonnerie, clôture de haies vives ; ▶enclore) ; *f 2* **enclosure [f]** *hist.* (Terme utilisé par les propriétaires anglais du XVIeme au XVIIIeme siècle qualifiant la technique de clôture de terrains acquis par eux lors du partage des communaux) ; *g* **Einfriedung [f]** (*UBe* Zaun, Grundstücksmauer, Hecke, ▶Umzäunung; einfrieden [vb], einfriedigen [vb]); *syn.* Einfriedigung [f].

enclosure [n], **game** *conserv. hunt.* ▶shooting preserve [US]/shooting reserve [UK].

1774 endangered species [n] *conserv. phyt. zool.* (*Faunistic and floristic* ▶*status of endangerment* species the survival of which may be uncertain when factors causing the danger of extinction are not eliminated by man through protection measures sustaining the population; in U.S. law, any species which is in danger of extinction throughout all or a significant portion of its range, other than a species of the Class Insectar determined by the U.S. Department of Environmental Protection [EPA] to constitute a pest; *according to IUCN categories of threat* EN [Endangered]; ▶Red List, ▶threatened species); *s* **especie [f] en peligro (EN)** (Según las categorías de la ▶lista roja de la UICN, un taxón está en peligro [Endangered] cuando la mejor evidencia disponible indica que cumple cualquiera de los cinco criterios para esta categoría y, por consiguiente, se considera que está enfrentando un riesgo muy alto de extinción en estado silvestre; cf. UICN 2001. **En EE.UU.**, en esta categoría se incluye cualquier especie en peligro de extinción en su totalidad o en parte significativa de sus hábitats, con excepción de las especies de la clase de los insectos que han sido declaradas por la Agencia de Protección Ambiental [EPA] como insecto nocivo. **En Es.** no existe tal categoría a nivel estatal, sino que corresponde a las CC.AA. establecer categorías específicas que se consideren necesarias; cf. art. 55 Ley 42/2007; ▶Catálogo Español de Especies Amenazadas. **En D.** corresponde a la categoría 2; ▶categoría de amenaza de extinción de especies, ▶especie amenazada); *f 1* **espèce [f] affectée d'une régression forte et continue et qui a déjà disparue de nombreuses régions** (▶*Catégorie de régression des populations et de distribution des*

E

espèces toute espèce animale ou végétale dont le peuplement ou l'habitat naturel est directement ou indirectement menacé par l'homme et par suite de son importante régression est en danger de disparition dans son aire de répartition naturelle ; ▶espèce dont la population n'a pas sensiblement diminuée, mais dont les effectifs sont faibles, donc en danger latent, ▶liste des espèces menacées, ▶Livre rouge [des espèces menacées]) ; *syn.* espèce [f] fortement menacée, espèce [f] en danger de disparition ; *f 2* **espèce [f] en danger (EN)** (Une des neufs catégories du système de la Liste Rouge de l'UICN : espèce remplissant un des critères suivants [texte non exhaustif] : **a.** réduction des effectifs constatée, estimée, déduite ou supposée depuis 10 ans ou 3 générations égale ou supérieure à 70 % lorsque les causes de la réduction sont clairement réversibles et ont cessé ou de 50 % lorsque les causes de la réduction n'ont pas cessé ou ne sont peut-être pas réversibles…, **b.** espèce dont la répartition géographique dans la zone d'occurrence [estimée inférieure à 5000 km²] et la zone d'occupation [estimée à moins de 500 km²] indique un déclin continu, constaté, déduit ou prévu ainsi qu'une fluctuation extrême de la zone d'occurrence, de la zone d'occupation, de l'étendue et/ou de la qualité de l'habitat, du nombre de localités ou de sous-populations, du nombre d'individus matures, **c.** une population estimée à moins de 2500 individus matures dont le déclin continu est estimé à 20 % au moins en cinq ans ou deux génération, **d.** une population estimée à moins de 250 individus matures et **e.** taxon confronté à une probabilité d'extinction à l'état sauvage d'au moins 20 % en l'espace de 20 ans ou de cinq générations ; *cf.* Catégories et Critères de l'UICN pour la Liste Rouge, Version 3.1, UICN 2001 ; ▶espèce menacée) ; *g* **stark gefährdete Art [f]** (*Faunistische und floristische* ▶ *Gefährdungskategorie* jede durch direkte oder indirekte menschliche Einwirkungen nahezu in ihrem gesamten einheimischen Verbreitungsgebiet signifikant verringerte oder regional verschwundene Tier- und Pflanzenart; **in D.** zur Gefährdungskategorie 2, gemäß IUCN zur Kategorie EN *[Endangered]* gehörig: diejenige Population, deren beobachtete, geschätzte, hergeleitete oder vermutete Größe sich [it. einem von fünf Kriterien] in den letzten 10 Jahren oder in drei Generationen über 70 % verringert hat oder verringern wird; in der U.S.-Gesetzgebung zählen hierzu alle Arten, die in ihrer Gesamtheit oder zu einem bemerkenswerten Teil vom Aussterben bedroht sind mit Ausnahme der Arten der Klasse der Insekten, die von der U.S.-Naturschutzbehörde als Schädlinge deklariert werden; ▶bedrohte Art, ▶Rote Liste).

1775 Endangered Species Act [n] (E.S.A.) [US] *conserv. leg.* (**In U.S.**, 1973 legislation, amended 1978 and 1982, for the protection of endangered and threatened animal and plant species, including lists of those affected. It also provides a flexible framework for resolving conflicts between the interests of industry and survival of species. **In the U.K.** exists also the Wild Life and Countryside Act 1981; ▶critically endangered species, ▶Red List, ▶threatened species, ▶Washington Convention); *syn.* Conservation [n] of Wild Creatures and Wild Plants Act 1975 [UK]; *s* **ley [f] de protección de especies (En US**, ley de 1973, modificada en 1978 y 1982, para proteger ▶especies en peligro de extinción o ▶especies amenazadas de flora y fauna, que incluye una lista de las especies afectadas. También prevé un marco para la resolución de conflictos entre los intereses de la industria y los de la supervivencia de las especies. **En Es.** no existe tal ley, sino que la protección de especies está regulada en los arts. 52 a 58 de la Ley 42/2007, del Patrimonio Natural y de la Biodiversidad. **En D.** el correspondiente decreto nombra las especies amenazadas de flora y fauna y regula su protección; ▶Catálogo Español de Especies Amenazadas, ▶Convención de Washington, ▶lista roja); *f* **arrêté [m]**

réglementant la protection de la faune et de la flore sauvages (*Préservation du patrimoine biologique* les ▶listes des espèces menacées végétales et animales protégées sur l'ensemble ou une partie du territoire national ont été pour la première fois publiée par arrêté du ministre de l'Environnement du 20 janvier 1982 et sont régulièrement complétées et modifiées. Différents arrêtés ministériels ou préfectoraux fixent, en général par classe, la liste des espèces animales et les territoires, notamment maritime, sur lesquels les interdictions s'appliquent. [mammifères terrestres et marins, mollusques, insectes, poissons, grenouilles et crustacés, amphibiens et reptiles, espèces végétales, etc.]. Le C. rural, art. R. 211 et s. réglemente la protection **1.** intégrale des spécimens sauvages des espèces les plus menacées, **2.** artielle des espèces rares mais faisant l'objet d'une certaine utilisation ou **3.** temporaire des espèces pas nécessairement rares, mais dont l'exploitation peut devenir préoccupante dans certains départements et en certaines périodes de l'année. Il fixe les conditions dans lesquelles sont fixées les listes limitatives des espèces protégées, la durée des interdictions, et l'étendue des territoires sur lesquels elles s'appliquent ainsi que les différentes réglementations qui touchent à la coupe, la mutilation l'arrachage, la cueillette ou l'enlèvement de végétaux de ces espèces ainsi que la destruction ou l'enlèvement des œufs ou des nids, la mutilation, la destruction ou la capture, l'enlèvement, la naturalisation d'animaux de ces espèces ; **D.**, réglementation prise par décret fédéral et décrets des Länder. **CH.**, la protection des espèces est réglementée par les articles 23 à 27 de l'ordonnance d'exécution de la loi fédérale sur la protection de la nature et du paysage ; ▶convention de Washington, ▶espèce amenée par sa régression à un niveau critique des effectifs, ▶espèce dont la population n'a pas sensiblement diminué, mais dont les effectifs sont faibles, donc en danger latent, ▶inventaire local/régional du patrimoine faunistique et floristique, ▶Livre rouge [des espèces menacées]) ; *syn.* arrêté [m] relatif à la protection des espèces menacées, arrêté [m] réglementant la protection des espèces animales et végétales ; *g* **Artenschutzverordnung [f]** (Rechtsverordnung zum Schutze wild lebender Tier- und wachsender Pflanzenarten; z. B. in D. die **Bundesartenschutzverordnung** [BArtSchV], erstmals 1986 erlassen und seitdem mehrfach fortgeschrieben, die auf der Basis der §§ 39-55 des Bundesnaturschutzgesetzes den Artenschutz der EG-Artenschutzverordnung von 1984, durch die in allen EG-Mitgliedstaaten das ▶Washingtoner Artenschutzabkommen in Kraft gesetzt wurde, umsetzt und den Schutz wild lebender Tier- und Pflanzenarten durch Verbote und Beschränkungen für Entnahmen aus der freien Natur für Handel, Haltung etc. regelt; die BArtSchV enthält gem. § 1 in der Anlage 1 Spalte 2 und 3 ein Verzeichnis ▶besonders geschützter Arten und streng geschützter Arten. **CH.**, in der Schweiz ist der Artenschutz durch Art. 18-23 NatHSchG sowie durch die Art. 23-27 der Vollziehungsverordnung zum NatHSchG geregelt; ▶bedrohte Art, ▶Rote Liste, ▶vom Aussterben bedrohte Art); *syn.* Artenschutzgesetz [f] [A].

Endangered Species List [n] [US] *conserv. phyt. zool.* ▶Red List, #2.

endangerment [n] *conserv. phyt. zool.* ▶status of endangerment.

1776 endangerment [n] of species *conserv. phyt. zool.* (Conditions or acts that threaten existing species with extinction); ▶critically endangered species); *syn.* threat [n] of extinction; *s* **amenaza [f] de extinción** (Criterio de protección de especies de flora y fauna, ▶especie en peligro crítico, ▶especie en peligro de extinción); *f* **menace [f]** (Critère utilisé pour l'appréciation de la diminution du nombre des espèces d'un peuplement ou de la réduction de l'étendue du territoire occupé par ce peuplement ; ▶espèce amenée par sa régression à un niveau critique des effectifs) ; *syn.* risque [m] menaçant ; *g* **Gefährdung [f] der Arten** (Kriterium des Verbreitungs- und Bestandsrückganges von Tier- und Pflanzenarten; ►vom Aussterben bedrohte Art).

1777 endemic species [n] *phyt. zool.* (Organism living only in a limited natural area since the last glacial epoch, but which, in comparison to an ►indigenous species, may not originate in this range); *s* **especie [f] endémica** (Especie propia exclusivamente de un determinado país, de una cordillera, de una isla, etc. El término se utiliza también como sinónimo de ►especie indígena, aunque sería conveniente limitarlo a la primera definición; cf. DB 1985); *f* **espèce [f] endémique** (Caractérise une espèce ou autres unités de classification du règne animal ou végétal dont l'aire de répartition géographique est liée à une localité ou une région bien définie qui, en comparaison avec les unités taxonomiques autochtones, n'en sont pas obligatoirement originaire ; ►espèce indigène) ; *g* **endemische Art [f]** (Art, Gattung oder andere systematische Kategorie, die nur in einem natürlich begrenzten Areal vorkommt, jedoch im Vergleich zur autochthonen taxonomischen Einheit nicht unbedingt dort entstanden ist; ►einheimische Art); *syn.* endemischer Organismus [m], Endemit [m].

end face [n] **of a wall** [US] *arch. constr.* ►wall head.

end moraine [n] *geo.* ►terminal moraine.

energy [n], **aeolian** *envir.* ►wind energy.

energy [n] [US], **flux of** *ecol.* ►energy flow.

energy [n], **wave-wash** *envir.* ►wave action.

1778 energy conservation program [n] [US]/**energy conservation programme** [n] [UK] *envir. leg. pol.* (Public/governmental and private measures to reduce energy consumption through legal enactments, technical devices or construction methods, as well as measures intended to influence or change consumer behaviour); *s* **programa [m] de ahorro energético** (Paquete de medidas legislativas, técnicas y constructivas orientadas a la reducción del consumo energético y/o a influenciar el comportamiento de la población); *f* **programme [m] d'économie d'énergie** (Mesure politique et dispositions légales en faveur de la limitation de la consommation d'énergie, de l'utilisation rationnelle de l'énergie et de l'augmentation de l'efficacité énergétique au travers d'une réglementation thermique des bâtiments, l'utilisation d'énergies renouvelables, la promotion d'installations ou de matériels destinés à économiser l'énergie et des actions pour influencer le comportement des consommateurs) ; *syn.* programme [m] de limitation de la consommation d'énergie, programme [m] de réduction de la consommation d'énergie ; *g* **Energiesparprogramm [n]** (Politische Maßnahmen zur Senkung des Energieverbrauches durch gesetzliche Vorschriften, technische und bauliche Vorrichtungen sowie dadurch verursachte oder beeinflusste Verhaltensänderungen der Nutzer).

1779 energy flow [n] *ecol.* (Transfer of energy within an ecosystem); *syn.* flux [n] of energy [also US]; *s* **flujo [m] de energía** (En un ecosistema transferencia de energía [absorción y restitución] por medio del ciclo biogeoquímico); *syn.* flujo [m] energético; *f* **flux [m] d'énergie** (Dans un écosystème, transfert d'énergie [absorption et restitution] par le jeu du cycle biogéochimique) ; *g* **Energiefluss [m]** (Die mit dem Stoffkreislauf verbundene Energieaufnahme und -weitergabe in einem Ökosystem); *syn.* Energieumsatz [m], Energietransfer [m].

energy source [n] *envir.* ►power source.

1780 energy supply [n] *plan.* (Provision of energy from power sources in connected networks for conveyance and distribution of electricity, fuel and heating mediums); *s* **suministro [m] de energía** (Equipamiento y procesos para proveer de energía eléctrica, gas y calor a nivel estatal o regional); *syn.* abastecimiento [m] de energía; *f* **alimentation [f] en énergie** (Installations et processus de production et de distribution d'électricité et de chaleur) ; *syn.* approvisionnement [m] en énergie ; *g* **Energieversorgung [f]** (Vorhalten von Einrichtungen und Vorgänge als staatliche Daseinsvorsorge, die der Erzeugung und Verteilung von Elektrizität, Wärme und Kraftstoff dienen).

enforcement agency [n] *adm. leg.* ►law enforcement agency.

enforcement notice [n] **on planting requirements** [UK] *leg. urb.* ►directive on planting requirements [US].

engagement [n] [UK]**, termination of** *contr. prof.* ►termination of a contract [US].

1781 engineer [n] *prof.* (Generic term for a member of the engineering profession, covering persons in, e.g. civil engineering, structural engineering, mechanical engineering, sanitary engineering, etc. with a professional academic degree. Members of the other planning and design vocations [architecture, landscape architecture, urban planning, etc.] have separate academic training with overlapping competence and cooperating relationships with engineering and other professional expertise. In U.S., most members of the engineering profession belong to the American Society of Professional Engineers [NSPE] and use the abbreviation P.E. after their names; in U.K., engineers in professional practice belong to the Institute of Civil Engineers; in D. general term for a member of the engineering vocation who has the professional academic degree ►Diplomingenieur. 'Diploma engineer' is a title which has been used in D. by all members of the allied professions of engineering such as architecture and landscape architecture; since 1999, however, in accordance with the **Bologna accords,** the intention has been to make academic degree standards and quality assurance in European higher education more comparable throughout Europe, and since 2005 modular, two-tier study courses have been introduced in D., which lead, for example in architecture, to a Bachelor's degree after three years of study, e.g. **Bachelor of Science [B. Sc.]** and a Master's degree after two years, e.g. **Master of Science [M. Sc.].** As such the German term 'engineer' for an architect is likely to become obsolete; ►Bachelor of Architecture [B.ARCH] [US], ►consulting engineer, ►structural engineer); *s* **ingeniero/a [m]** (Término genérico para personas que pertenecen al cuerpo profesional de la ingeniería. En Es. el título de ingeniero [para el nivel universitario] y de ingeniero técnico [para el nivel de Escuela Superior Técnica] se utiliza para las profesiones técnicas [ingeniería civil, electrónica, mecánica, etc.] y para las de ciencias prácticas [montes, agronomía, etc.]; en el contexto del proceso de equiparación de títulos [Proceso de Bologna] de la UE se introdujeron los títulos de Bachelor y Master de Arquitectura. En D. denominación utilizada tanto para profesiones técnicas [ingeniería civil, aeronáutica, electrónica, mecánica, etc.] como para los estudios de arquitectura y arquitectura paisajista; ►ingeniero/a asesor, -a, ►ingeniero/a de estructuras); *f* **ingénieur [m]** (Titre délivré par un établissement à toute personne disposant d'une solide culture scientifique et d'un ensemble de connaissances et savoir-faire techniques, économiques, sociales et humaines, le rendant apte à résoudre des problèmes de nature technologique, concrets et souvent complexes, liés à la conception, à la réalisation et à la mise en œuvre de produits, de systèmes ou de services ; F., le titre d'**ingénieur diplômé** sanctionne en France une formation [longue en principe de 5 ans] dispensée par un établissement d'enseignement supérieur habilité par une instance placée auprès ministère de l'Enseignement supérieur, la Commission des Titres d'Ingénieur, créée par la loi du 10 juillet 1934 ; D., terme générique utilisé pour dénommer les ingénieurs diplômés d'État ; dénomination professionnelle protégée, attribuée aux étudiants sortant des universités techniques et des facultés des sciences ainsi que des universités de technologie [Fachhochschulen] ainsi qu'aux professionnels habilités par la loi à porter ce titre académique; ▶ingénieur conseil, ▶ingénieur [en] structure[s]) ; *g* **Ingenieur [m]** (*OB zu* ▶*Diplomingenieur,* ▶*Diplomingenieur [FH] und graduierter Ingenieur (obs)* **1.** geschützte Berufsbezeichnung für Absolventen einer technisch-naturwissenschaftlichen Hochschule oder Fachhochschule sowie für Fachleute, die kraft Ingenieurgesetzes zur Führung des Ingenieurtitels berechtigt sind. **2.** Berufspraktiker, der technische Probleme auf Grund naturwissenschaftlicher Grundlagen, des aktuellen technischen Wissens und wirtschaftlicher Vorgaben empirisch-analytisch für einen bestimmten Ort lösen muss; ▶beratender Ingenieur, ▶Statiker).

engineer [n]**, resident** *constr.* ▶design firms' representative [US].

engineer [n] [UK]**, site** *constr.* ▶design firms' representative [US].

engineering [n] *constr.* ▶bioengineering; *constr. eng.* ▶civil engineering, ▶ground civil engineering, ▶hydraulic engineering, ▶structural engineering (2).

engineering [n]**, construction** *constr. eng.* ▶civil engineering.

engineering [n]**, structural** *constr. eng. phys.* ▶statics and dynamics.

engineering [n] **for agriculture** *agr.* ▶hydraulic engineering for agriculture.

English Heritage [n] [UK]**, Chief Executive of** *conserv'hist.* ▶State Historic Preservation Officer [US].

1782 enhancement [n] **of soil fertility** *agr. hort.* (Cultivation measures to maintain and improve soil quality for plant growth; ▶soil fertility, ▶soil mellowness, ▶tilth); *s* **mejora [f] de la fertilidad del suelo** (Medidas para mantener o mejorar la ▶fertilidad del suelo; ▶sazón); *f* **mise [f] en valeur du sol par la culture** (Création et maintien de conditions favorables du sol au développement des racines ; ▶fertilité du sol, ▶bon état de structure grumeleuse ; PED 1984, 208) ; *g* **Bodenpflege [f, o. Pl.]** (Kulturmaßnahmen, die die ▶Bodenfruchtbarkeit erhalten oder verbessern; ▶Bodengare).

1783 enjoyment [n] **of nature** *recr.* (Physical and psychological pleasure derived from the immediate experience of nature or landscape); *s* **disfrute [m] de la naturaleza** (Placer físico y psíquico derivado del goce y la percepción directa de la naturaleza); *f* **jouissance [f] de la nature** (Émotion physique et psychique que l'on retire de la perception immédiate de la nature ou d'un paysage) ; *syn.* délice(s) [m(pl)] de la nature, plaisir(s) [m(pl)] de la nature ; *g* **Naturgenuss [m]** (Aus dem unmittelbaren Erlebnis der Natur oder naturbetonten Landschaft herrührendes physisches und psychisches Wohlbefinden).

1784 enlargement [n] *plan. urb.* (▶Expansion of existing facilities); *syn.* extension [n] (2); *s* **extensión [f]** (Expansión de una instalación existente; ▶ampliación); *f* **extension [f] (2)**; *g* **Ausbau [m, o. Pl.]** (i. S. v. Vergrößerung; ▶Erweiterung).

1785 enlargement [n] **of a plan** *plan.* (Modification of a plan; e.g. by increasing the scale; *opp.* ▶reduction of a plan); *s* **ampliación [f] de un plano** (Modificación de un plano aumentando la escala; *opp.* ▶reducción de un plano); *f* **agrandissement [m] de plan** (Modification de l'échelle d'un plan à une échelle plus grande ; *opp.* ▶réduction de plan) ; *g* **Planvergrößerung [f]** (Änderung eines Planes, z. B. durch Vergrößern des Maßstabes; *opp.* ▶Planverkleinerung).

1786 enlargement [n] **of land holdings** *agr. for. urb.* (Augmentation of land ownership); *s* **aumento [m] del tamaño**

de las parcelas agrícolas; *f* **agrandissement [m]** (Augmentation de la superficie des parcelles, p. ex. d'une exploitation agricole) ; *syn.* accroissement [m] ; *g* **Aufstockung [f] der Grundstücksflächen** (Vergrößerung/Erweiterung der Grundstücksflächen); *syn.* Grundstücksvergrößerung [f].

enrichment [n] *chem. biol.* ▶increase; *limn. pedol.* ▶nutrient enrichment.

1787 enrichment planting [n] *for.* (**1.** Method of afforestation in tropical rain forests, whereby cultivated seedlings are planted in forests, which have been cut down or on derelict and non-forested land to accelerate natural regeneration. There are two methods: **forest corridor planting** whereby seedlings are planted in strips with the necessary distance to existing tree canopies, and **cluster planting** in natural or artificially created clearings. This planting method was developed by foresters in Southeast Asia, with the intention of increasing future logging in production forests. **2.** Method of restoring species-rich forests by planting additional deciduous tree species or the introduction of additional species in succession areas to enhance species diversity); *s* **plantación [f] de enriquecimiento** (**1.** Método de reforestación en selvas tropicales que consiste en plantar plántulas de semilla en zonas taladas o yermas para acelerar la regeneración natural. Existen dos métodos: la **plantación en corredores forestales**, por el cual se plantan las plátulas en franjas con la distancia necesaria al bosque existente, y la **plantación en grupos**, en claros naturales o artificiales. Este método de plantación fue desarrollado por silvicultores en el Sureste de Asia, con el fin de incrementar la cosecha de futuras talas en bosques de producción. **2.** Método para restaurar bosques mixtos ricos en especies plantando especies de árboles caducos o introduciendo especies adicionales en áreas de sucesión para aumentar la diversidad de especies); *f* **plantation [f] d'enrichissement** (Technique sylvicole d'exploitation des forêts tropicales dans lesquelles les plants et les gaulis d'espèces à conserver sont insuffisants pour la régénération naturelle, soit en raison du taux de survie intrinsèquement bas des plants soit comme conséquence d'opérations forestières destructrices ; elle consiste dans la plantation de plants « éduqués » produits par semis en pépinière traditionnelle de semences récoltées au pied des arbres semenciers sélectionnés ou la réalisation de semis direct en plantation afin d'accélérer la régénération naturelle ; il existe deux méthodes de plantation d'enrichissement : **1.** la **plantation en layons**, dans laquelle des plants sont placés dans des corridors, généralement débarrassés dans une bonne mesure de leur partie supérieure, et **2.** la **plantation groupée** [plantation sur trouée], dans laquelle des groupes de plants sont plantés dans des ouvertures créées naturellement ou artificiellement dans le couvert forestier ; ces méthodes ont été développées par les forestiers en Asie du sud-est désireux d'enrichir leurs forêt de production en vue d'une coupe plus bois future ; cf. Les unités pilotes d'aménagement, de reboisement et d'agroforesterie ; nouvelle approche de gestion durable des forêts congolaises, Bois et forêts des tropiques, 2005, N° 285 ; *syn.* reboisement [m] d'enrichissement ; *g* **Anreicherungspflanzung [f]** (**1.** Bei Aufforstungsprojekten in tropischen Regenwäldern das Auspflanzen von vorkultivierten Sämlingen auf abgeholzte Flächen, um die natürliche Regeneration zu beschleunigen. Es gibt zwei Methoden: die **Schneisenpflanzung**, bei der die Pflanzung bandartig mit entsprechendem Abstand vom bestehenden Kronendach eingebracht wird, und die **Gruppenpflanzung**, die in natürlich oder künstlich entstandenen Lichtungen erfolgt. Diese Pflanzmethode wurde von Förstern in Südostasien entwickelt, um mit ihr geeignete, standortgerechte Waldbaukonzepte zu entwickeln und ihre Produktionswälder für einen späteren Holzeinschlag aufzustocken; cf. FAS 2008. **2.** Ein Lenkungsinstrument zur

Wiederherstellung artenreicher Mischwälder durch das Pflanzen von zusätzlichen Laubbaumarten oder das Einbringen zusätzlicher Arten in Sukzessionsflächen, um die Artenvielfalt zu fördern).

entering [n] **into effect** *leg.* ▶coming into effect.

entering [n] **into force** *leg.* ▶coming into effect.

enthusiast [n] *gard.* ▶garden enthusiast.

1788 entisol [n] **[US]** **(1)** *pedol.* (Soil that has no diagnostic pedogenetic horizons, e.g. on very recent geomorphic surfaces, either on steep slopes that are undergoing active erosion, or on fans and floodplains where the recently eroded materials are deposited. They may also be on parent materials which are resistant to alteration; cf. RCG 1982; ▶soil suborder); *syn.* immature soil [n] [UK]; *s* **suelo** [m] **mineral bruto** (▶Clase de suelos terrestres poco evolucionados caracterizados por la falta de evolución de la materia mineral, de manera que tienen un horizonte A muy poco desarrollado directamente encima de la roca madre; *términos específicos* regosol [m] [**s. m. b.** blando] y litosol [m] [**s. m. b.** macizo y compacto]; EDAFO 1982); *f* **sol** [m] **peu évolué** (Sol à profil AC caractérisé essentiellement par la faible altération du milieu minéral et, dans la majorité des cas, la faible teneur en matière organique du profil : la classification distingue trois sous-classes : **1.** sols peu évolués climatiques, **2.** sols peu évolués d'érosion et **3.** sols peu évolués d'apport ; cf. PED 1983, 201 ; ▶sous-classe pédologique) ; *syn.* sol [m] minéral brut (PED 1983, 201) ; *g* **Rohboden** [m] (▶*Bodenklasse der Landböden* Initialstadium der Bodenbildung, mit einem nahezu humusfreien, wenig belebten, gering mächtigen [A-]Horizont, der unmittelbar dem unverwitterten Gestein aufliegt; SS 1979).

entisol [n] **[US]** **(2)** *land'man.* ▶vegetation establishment on entisol.

1789 entrance area [n] *arch.* (Exterior space in front of a building, property gate, etc.); *s* **zona** [f] **de entrada** (Zona exterior de la portada de un edificio, de una puerta de acceso, etc.); *f* **zone** [f] **d'entrée** (devant un bâtiment, un portail) ; *syn.* aire [f] d'entrée ; *g* **Eingangsbereich** [m] (Außenbereich vor einem Gebäude, vor einem Grundstückstor etc.).

entrance drive [n] **[US]** *urb.* ▶driveway.

entrant [n]**, competition** *prof.* ▶competition entrant.

entrydrive [n] **[US]** *urb.* ▶driveway.

entryway [n] **[US]** *urb.* ▶driveway.

enumeration survey [n] **[UK]** *for.* ▶forest stand inventory [US].

1790 environment [n] **(1)** *conserv. ecol. nat'res. urb.* (All conditions, circumstances, and influences surrounding and affecting the development of an organism or a group of organisms; ▶cultural landscape, ▶dwelling environs, ▶natural environment, ▶urban environment); *s* **medio ambiente** [m] (Conjunto en un momento dado, de los agentes físicos, químicos, biológicos o de los factores sociales susceptibles de causar un efecto directo o indirecto, inmediato o a plazo, sobre los seres vivientes y las actividades humanas; cf. DINA 1987; ▶hábitat [residencial], ▶medio ambiente urbano, ▶medio natural, ▶paisaje humanizado); *syn.* medio [m]; *f* **environnement** [m] (Ensemble des éléments physiques, chimiques, biologiques et sociaux et culturels qui caractérisent un espace et influent le développement d'un organisme ou un groupe d'organisme ; le terme **e.** est souvent utilisé dans un sens restreint pour ne désigner que les agents physiques ayant une influence sur l'homme ; VOC 1979 ; ▶environnement de l'habitat, ▶environnement urbain, ▶milieu naturel, ▶paysage culturel) ; *g* **Umwelt** [f, selten Pl.] (**1.** Ursprünglich von JAKOB J. BARON VON UEXKÜLL 1921 geprägter Begriff für die spezifisch lebenswichtige Umgebung

einer Tierart. **2.** Gesamtheit der äußeren Lebensbedingungen [ökologische Faktoren], die auf eine bestimmte Lebenseinheit, auf ein Individuum oder auf eine Lebensgemeinschaft an dessen Lebensstätte einwirken; zu diesen Bedingungen gehören abiotische Faktoren wie Klima [Temperatur, Luftfeuchte, atmosphärisches CO_2], Emissionen [Lärm, Schadstoffe], Boden, Gewässer und biotische Faktoren [Pflanzen, Tiere, Menschen]. **3.** auf die menschliche Gesellschaft bezogen, ist **U.** ein sehr komplexes, vielschichtiges Wirkungs- und Beziehungsgefüge, das die Menschen umgibt, also alle natürlichen, gebauten und sozialen Umweltfaktoren, die die Menschen beeinflussen und auf die sie Einfluss nehmen können; diese sehr komplexe **U.** wird als ▶natürliche Umwelt, gebaute Umwelt und soziale Umwelt wahrgenommen. In der dt. Gesetzgebung herrscht der restriktive Umweltbegriff, der die ▶natürlichen Grundgüter und die Kulturlandschaften sowie deren Teile dem Naturschutz und der Landschaftspflege unterstellt, vor; erweitert wird dieser Begriff durch § 1 Bundeswaldgesetz, in dem zudem die Leistungsfähigkeit des Naturhaushaltes und die Agrar- und Infrastruktur genannt werden; die Gleichsetzung von **U.** mit natürlicher Umwelt im Sinne des restriktiven Umweltbegriffs ist in einer modernen Gesellschaft unangemessen, weil es dort nur noch selten um den Schutz natürlicher Ursprünglichkeit, sondern i. d. R. um vom Menschen gestaltete Räume, sog. ▶Kulturlandschaften, geht; cf. ÖKO 2003; ▶städtische Umwelt, ▶Wohnumfeld).

environment [n] **(2)** *plan.* ▶atlas of the environment.

environment [n]**, protection of the** *sociol. urb.* ▶technical protection of the environment.

environment [n]**, residential** *sociol. urb.* ▶dwelling environs.

environment [n]**, root** *arb. hort.* ▶rooting zone.

environmental assessment [n] **(EA)** **[UK]** *conserv. leg. nat'res. plan.* ▶environmental impact assessment [EIA] [US].

1791 environmental audit [n] *envir. leg.* (Investigation and assessment as to whether the purchase, production and management of goods and services have caused any environmental impacts, and subsequently an investigation to determine whether the examined firm complies with the ISO certificate "environmentally-controlled firm"); *syn.* eco-audit [n], environmental review [n]; *s* **ecoaudit** [m] (En la UE procedimiento voluntario de analizar las empresas del sector industrial en cuanto a sus impactos ambientales y en caso de obtener resultados positivos de otorgarles un «certificado de empresa ecoauditada». Estos análisis sólo los pueden realizar consultoras reconocidas oficialmente); *syn.* ecoauditoría [f], auditoría [m] ambiental; *f* **éco-audit** [m] (Procédure volontaire des entreprises du secteur industriel de s'engager dans un politique de management environnemental ; la déclaration environnementale et le programme d'action sont validés par un vérificateur environnemental indépendant selon des normes reconnues [ISO] ; les entreprises enregistrées peuvent se prévaloir d'une déclaration de participation ; cf. CPEN, feuillets 101, p. 3980 B) ; *syn.* audit [m] environnemental ; *g* **Öko-Audit** [n] (Verfahren zur Überprüfung des Umweltverhaltens eines Betriebes beim Einkauf, bei der Produktion und Verwaltung mit anschließender Untersuchung, ob der nach ISO-Normen [Europäische Norm EN ISO 14001:2004] validierte Unternehmensstandort die Bedingungen für das Zertifikat „umweltgeprüfter Betrieb" erfüllt; mit diesem Verfahren soll eine kontinuierliche Verbesserung des betrieblichen Umweltschutzes erreicht werden. **In D.** wurden erstmals durch das Bayerische Staatsministerium 1996 und 1997 Schritte eines EG-Ö.-A.s nach der EG-Verordnung Nr. 1836/93 vom 29.06.1993,

E

mittlerweile ersetzt durch EG-Verordnung Nr. 196/2006 der Kommission vom 03.02.2006, durchgeführt).

1792 environmental awareness [n] *conserv. sociol.* (Informed comprehension and pronounced responsibility for environmental problems on the part of citizens and those responsible in administration and industry. It is based upon an understanding of the threats imposed upon fundamental natural resources and the willingness to implement remedial measures); *syn.* environmental consciousness [n]; *s* **conciencia** [f] **ecológica** (Actitud de parte de la población, de responsables de instituciones públicas y privadas de conocimiento y responsabilidad ante los problemas ambientales y de la naturaleza con la consecuente disponibilidad a contribuir a solucionarlos); *syn.* conciencia [f] ambiental; *f* **conscience** [f] **écologique** (Attitude de compréhension éclairée et de responsabilisation en face aux atteintes à l'environnement se traduisant pour un certain nombre de citoyens et de responsables politiques, administratifs et économiques par la conviction de la nécessité d'entreprendre des actions, d'imposer ou de prendre des mesures contre la dégradation des équilibres biologiques et ses causes) ; *g* **Umweltbewusstsein** [n] (Das aufgeklärte Verständnis und ausgeprägte Verantwortungsbewusstsein für die Umweltproblematik bei Bürgern und Verantwortlichen in Verwaltung und Wirtschaft, bestehend aus der Einsicht in die Gefährdung der natürlichen Lebensgrundlagen und deren Ursachen und der Bereitschaft zur Abhilfe).

1793 environmental biology [n] *ecol.* (Branch of biology and ►environmental research concerned with reactions of organisms to disturbances in their environment as well as biological problems of ►environmental protection and ►environmental hygiene); *s* **biología** [f] **ambiental** (Parte de la biología y de la ►investigación ambiental que estudia las reacciones de los organismos ante distorsiones de su medio natural y los problemas biológicos de la ►protección del medio ambiente y de ►higiene ambiental); *f* **biologie** [f] **environnementale** (Branche de la biologie et de la ►recherche sur l'environnement qui traite des réactions des organismes face aux désordres subis par leur environnement naturel ainsi que des problèmes de biologie dans le domaine de la ►protection de l'environnement et de l'►hygiène environnementale) ; *g* **Umweltbiologie** [f] (Teilgebiet der Biologie und der ►Umweltforschung, das die Reaktion der Organismen auf Störungen ihrer natürlichen Umwelt sowie die biologischen Probleme des ►Umweltschutzes und der ►Umwelthygiene umfasst; MEL 1979, Bd. 24).

1794 environmental compatibility [n] *conserv. nat'res. plan.* (Harmonious existence promoted by implementation of congruous policies, action program[me]s, building projects, or the manufacture of a product in such a way that any detrimental effects upon society and the natural environment are ruled out or reduced to a minimum); *s* **compatibilidad** [f] **con el medio ambiente** (El hecho de que un proyecto previsto no tenga consecuencias negativas graves para el medio, de manera que su realización suponga una ganancia para la sociedad. Se comprueba a través de la evaluación del impacto ambiental); *f* **compatibilité** [f] **avec l'environnement** (Concordance entre les concepts politiques, les programmes d'action, les projets de construction, la production des biens et produits de consommation avec l'objectif de limiter au strict minimum les atteintes portées aux éléments du système naturel et les effets néfastes sur la santé humaine, de satisfaire aux préoccupations d'environnement) ; *g* **Umweltverträglichkeit** [f] (Die Übereinstimmung einer bestimmten Politik, eines bestimmten Aktionsprogrammes, Bauvorhabens oder die Herstellung bzw. Beschaffenheit eines Produktes mit dem Erfordernis, dass sämtliche Beeinträchtigungen auf

Gesellschaft, Natur und Landschaft ausgeschlossen oder auf ein Minimum reduziert sind).

environmental compatibility matrix [n] *landsc. plan.* ►matrix on ecological intercompatibility.

1795 environmental compensation measure [n] *conserv. leg.* (Replacement of disturbed habitat; e.g. wetland, by a mitigation project for landscape management and nature conservation, where a development area cannot be restored to its natural condition. Such a measure is then to be undertaken as an improvement of ecological conditions on another site, i.e. there is usually no functional connection between the place of interference or ►intrusion upon the natural environment and the location of a compensation measure. In U.S., this measure results in 'no net loss'; ►impact mitigation, ►impact mitigation regulation, ►mitigation measure, ►compensation payment [UK]); *s* **medida** [f] **sustitutiva** (≠) (**En D.**, en el marco de la ►regulación de impactos sobre la naturaleza, medidas a tomar en el caso de que no sean posibles ►medidas compensatorias en el contexto de una ►intrusión en el entorno natural. Éstas se llevan a cabo en otro lugar y tienen como fin contribuir a mejorar la situación ecológica general. **En Es.** legalmente no existe este instrumento de compensación; ►compensación [de un impacto], ►indemnización compensatoria); *f* **mesure** [f] **de substitution** (≠) (Terme utilisé en D. dans le cadre de la protection de la nature dont le contenu juridique n'est pas exactement défini ; des **m. d. s.** doivent être envisagées pour tous travaux ou ouvrages qui, par leurs effets, portent ►atteinte au milieu naturel et dont les conséquences dommageables ne peuvent être compensées ; les **m. d. s.** sont considérées comme une amélioration des éléments du milieu naturel pouvant remplir d'autres fonctions ou objectifs de protection et de sauvegarde de l'environnement ; la réalisation des **m. d. s.,** par opposition aux ►mesures compensatoire, n'est donc pas directement liée au site du projet ; par contre les aires de compensation doivent toujours être des espaces dégradés et aptes à une réhabilitation, leur état final par rapport à l'état initial devant constituer une réelle augmentation de la valeur écologique ou une amélioration notable des fonctions du milieu ainsi qu'être similaire au milieu naturel ou aux paysages auxquels le projet porte une atteinte irréversible ; ►compensation [des conesquences dommageables], ►redevance compensatoire, ►réglementation relative aux atteintes subies par le milieu naturel) ; *g* **Ersatzmaßnahme** [f] **(des Naturschutzes und der Landschaftspflege)** (Maßnahme des Naturschutzes und der Landschaftspflege, die bei Eingriffen in Natur und Landschaft durchgeführt werden müssen, wenn ►Ausgleichsmaßnahmen im Bereich eines Eingriffs nicht möglich sind. **E.** ist ein in den Naturschutzgesetzen nicht, aber in Verwaltungsrichtlinien näher definierter Begriff. Bundesrechtlich ist **E.** nur in der Weise bestimmt, dass sie nach § 8 [9] BNatSchG sich auf nicht ausgleichbare, aber vorrangige Eingriffe beziehen. Da **E.n** die Nichtausgleichbarkeit eines ►Eingriffs in Natur und Landschaft voraussetzen, muss Ersatz etwas qualitativ anderes sein als ►Ausgleich. **E.n** werden dann als Verbesserung anderer Leistungen des Naturhaushaltes für andere Ziele und Objekte von Naturschutz und Landschaftspflege verstanden, d. h. in der Regel besteht kein örtlicher/räumlicher funktionaler Wirkungszusammenhang mit dem Eingriffsgeschehen. Die Naturschutzgesetze der Bundesländer verlangen auch, dass **E.n** auf den Eingriffsort zurückwirken müssen. Es kommen nur solche Flächen in Betracht, die aufwertungsbedürftig und -fähig sind. Diese Voraussetzung erfüllen sie, wenn sie in einen Zustand versetzt werden können, der sich im Vergleich mit dem früheren als ökologisch höherwertig einstufen lässt. Dazu muss zudem ein Zustand geschaffen werden, der den durch das geplante Vorhaben beeinträchtigten Funktionen des Naturhaushaltes oder des Landschaftsbildes gleichwertig oder zumindest ähnlich ist; cf. BAUR 1999, 484 f; ►Ausgleichsabgabe, ►Eingriffsregelung).

1796 environmental conditions [npl] *landsc. land'man.* (State of a landscape based on its climate, vegetation, fauna as well as geomorphological elements, human settlements, etc. which affects the quality and stability of natural systems); *s* **características [fpl] de un paisaje** (Estado de un paisaje determinado por los elementos geomorfológicos, el clima, el suelo, el agua, la flora y la fauna, así como por el tipo de asentamiento humano); *f* **composantes [fpl] d'un paysage** (État d'un paysage relatif au climat, faune, flore ainsi que les éléments géomorphologiques, la structure du bâti, etc.) ; *syn.* caractéristiques [fpl] d'un paysage ; *g* **Gegebenheiten [fpl] einer Landschaft** (Zustand in einer Landschaft hinsichtlich des Bodens, Wassers, Klimas, der Vegetation, Fauna sowie der geomorphologischen Elemente, der Siedlungs- und Infrastruktur etc.).

environmental consciousness [n] *conserv. sociol.* ►environmental awareness.

1797 environmental conservation [n] *conserv. envir.* (Conservation of the natural or human cultural environments; ►environmental protection); *s* **preservación [f] del medio ambiente** (►protección ambiental); *f* **sauvegarde [f] de l'environnement** (►protection de l'environnement) ; *syn.* préservation [f] du milieu naturel ; *g* **Erhaltung [f, o. Pl.] der Umwelt** (►Umweltschutz).

1798 environmental considerations [npl] *conserv. nat'res.* (►considerations of nature conservation and landscape management, ►environmental relevance); *s* **exigencias [fpl] del medio ambiente** (►exigencias ambientales, ►exigencias de protección de la naturaleza y del paisaje); *syn.* factor [m] medio ambiente; *f* **préoccupations [fpl] d'environnement** (►préoccupations de protection de la nature et des paysages, ►exigence environnementale) ; *syn.* préoccupations [fpl] environnementales ; *g* **Belange [mpl] der Umwelt** (►Belange des Naturschutzes und der Landschaftspflege, ►Umweltbelang).

1799 environmental criminality [n] *envir. leg.* (Violations of laws concerning environmental protection, e.g. environmentally harmful waste disposal, illegal shipping of waste products, pollution of air and water, contamination of soil); *s* **criminalidad [f] ambiental** (Infracción contra las leyes de protección del medio ambiente, como p. ej. eliminación clandestina de residuos tóxicos, exportación ilegal de residuos, vertido ilegal de contaminantes en cursos de agua, etc.); *f* **criminalité [f] dans le domaine de l'environnement** (±) (Violations des dispositions de la réglementation, infractions à la réglementation sur la protection de l'environnement portant atteinte à l'équilibre du milieu naturel et soumises à sanctions pénales, p. ex. dépôt et abandon d'ordures, rejet illicite dans les eaux, non-respect des prescriptions imposées sur les niveaux d'émission dans l'atmosphère) ; *g* **Umweltkriminalität [f]** (Verstöße gegen strafbewehrte Rechtsvorschriften zum Schutze der Umwelt, z. B. umweltschädliche Abfallbeseitigung, illegaler Mülltourismus, Luft- und Gewässerverunreinigung, Bodenverschmutzung).

1800 environmental damage [n] *envir. leg.* (**1.** Often nationwide and with global impact, an observable adverse change that occurs directly or indirectly, causing damage, pollution or severe impairment of objects, natural habitats or natural resources [species, habitats, water and soil]. **E. d.** includes air pollution, greenhouse effect, oil slick on the seas, rapid soil erosion, ozone depletion, deforestation ►forest decline; due to the synergism effect, the causes of modern **e. d.s** are no longer individually identifiable. **2.** Legal definition **in D., a]** damage to species and natural habitats as classified under Section 21a BNatSchG is any damage that has a significant adverse effect on attaining or maintaining the favorable conservation status of habitats or species; **b] damage to a body of water** as per Section 2a of the Water Resources Act is any damage that significantly adversely affects the ecological or chemical state of surface waters or coastal waters, and the chemical or quantitative status of groundwater, **c] degradation of soil** leading to a deterioration of soil function within the meaning of Section 2, Federal Soil Conservation law, caused by direct or indirect transfer of substances, organisms or micro-organisms on, in or under the soil leading to risks to human health; cf. Act of 10.05.2007 implementing EU Directive 2004/35/EG of the European Parliament and the Council of 21.04.2004 regarding environmental liability and the prevention and remediation of environmental damage; ►EU Water Framework Directive); *s* **daño [m] ambiental** (Degradación grave directa o indirecta, frecuentemente de grandes dimensiones territoriales o incluso globales, de espacios o recursos naturales [especies, ecosistemas, aguas y suelos] o perturbación grave de la función del medio natural por accidente o por acumulación, como contaminación atmosférica, contaminación radi[o]activa, efecto invernadero, manchas de petróleo en los océanos, erosión del suelo acelerada, destrucción de la capa de ozono o ►muerte de bosques; ►Directiva Marco del Agua); *syn.* daño [m] al medio ambiente; *f* **dommage [m] à l'environnement** (Toute action ayant une incidence négative [dégradation, destruction, modification ou perte] résultant d'une atteinte aux ressources naturelles et incluant les atteintes matérielles et immatérielles. **1.** *Article L 161-1 du Code de l'environnement* les **dommages causés à l'environnement** [terminologie légale française] sont constitués par les détériorations directes et indirectes mesurables de l'environnement qui : **a.** créent un risque d'atteinte grave à la santé humaine du fait de la contamination des sols, **b.** affectent gravement l'état et le potentiel écologique des eaux, **c.** affectent gravement le maintien ou le rétablissement dans un état de conservation favorable les espèces et habitats d'espèces visés au 2 de l'article 4, à l'annexe I de la directive 79/409/CEE du Conseil du 2 avril 1979 et à l'annexe I et II de la directive 92/43/CEE du Conseil du 21 mai 1992, les sites de reproduction et aires de repos des espèces énumérées à l'annexe IV de la directive 92/43/CEE du Conseil du 21 mai 1992, **d.** affectent les fonctions [services écologiques] assurées par les sols, les eaux et les espèces et habitats au bénéfice d'une de ces ressources naturelles ou au bénéfice du public ; les dommages causés à l'environnement comprennent également ceux générés par les activités professionnelles. **2.** *En droit communautaire dans le cadre de la responsabilité environnementale pour la réparation d'un dommage selon le principe du pollueur-payeur* : par dommage environnemental on entend **a.** les dommages causés aux espèces et habitats naturels protégés, à savoir tout dommage qui affecte gravement la constitution ou le maintien d'un état de conservation favorable de tels habitats ou espèces, **b.** les dommages affectant les eaux, à savoir tout dommage qui affecte de manière grave et négative l'état écologique, chimique ou quantitatif ou le potentiel écologique des eaux concernées, **c.** les dommages affectant les sols, à savoir toute contamination des sols qui engendre un risque d'incidence négative grave sur la santé humaine du fait de l'introduction directe ou indirecte en surface ou dans le sol de substances, préparations, organismes ou micro-organismes ; cf. Directive 2004/35/CE du Parlement Européen et du Conseil du 21 avril 2004 ; on distingue trois catégories de dommages ; **a.** dommages à la biodiversité, **b.** dommages aux sites contaminés, **c.** dommages aux ressources environnementales à travers leur application économique ; ►mort des forêts, ►Directive pour la protection des eaux intérieures de surface, des eaux de transition, des eaux côtières et des eaux souterraines) ; *syn.* dommage [m] causé à l'environnement, dommage [m] environnemental, dommage [m] écologique ; *g* **Umweltschaden**

E

E

[m] (**1.** Oft flächendeckende und global wirkende, direkt oder indirekt eintretende, feststellbare nachteilige Veränderungen, Beschädigungen oder starke Beeinträchtigungen von Objekten, Lebensräumen oder natürlichen Grundgütern [Arten, Biotope, Gewässer und Boden] oder Beeinträchtigung der Funktion einer natürlichen Ressource durch große Belastungen und Verschmutzungen, z. B. Luftverschmutzung, Treibhauseffekt, Ölteppich auf Meeren, beschleunigte Bodenerosion, Abbau der Ozonschicht, ▶Waldsterben; durch Wirkungssynergismen sind die Ursachen moderner **U.schäden** nicht mehr eindeutig identifizierbar. **2.** *Legaldefinition* nach der Umwelthaftungsrichtlinie [RL 2004/35/EG] und den relevanten Bundesgesetzen gibt es drei Typen von **U.schäden: 1. Schädigung geschützter Arten und natürlicher Lebensräume:** jeder Schaden, der erhebliche nachteilige Auswirkungen auf die Erreichung oder Beibehaltung des günstigen Erhaltungszustandes der Lebensräume oder Arten hat; cf. auch § 21a BNatSchG; **2. Schädigung der Gewässer:** jeder Schaden, der erhebliche nachteilige Auswirkungen auf den ökologischen oder chemischen Zustand eines oberirdischen Gewässers oder Küstengewässers, das ökologische Potential oder den chemischen Zustand eines künstlichen oder erheblich veränderten oberirdischen Gewässers oder Küstengewässers oder den chemischen oder mengenmäßigen Zustand des Grundwassers hat; cf. auch § 22a Wasserhaushaltsgesetz und Definition in der ▶Wasserrahmenrichtlinie; **3. Schädigung des Bodens:** Beeinträchtigung der Bodenfunktion im Sinne des § 2 [2] Bundes-Bodenschutzgesetz, die durch eine direkte oder indirekte Einbringung von Stoffen, Zubereitungen, Organismen oder Mikroorganismen auf, in oder unter den Boden hervorgerufen wird und Gefahren für die menschliche Gesundheit verursacht; cf. Gesetz v. 10.05.2007 zur Umsetzung der Richtlinie 2004/35/EG des Europäischen Parlaments und des Rates vom 21.04.2004 über Umwelthaftung zur Vermeidung und Sanierung von Umweltschäden).

environmental degradation [n] *envir.* ▶environmental deterioration.

environmental design [n] *conserv. nat'res. plan.* ▶planning and design of the environment.

1801 environmental deterioration [n] *envir.* (Worsening of the environment by massive ▶environmental nuisances, so that sustained productivity and efficiency of natural systems for human land-use requirements as well as life conditions of fauna and flora are no longer possible or only possible to a limited extent); *syn.* environmental degradation [n]; *s* **degradación [f] del medio ambiente** (Daños causados en el ambiente por intrusiones negativas que reducen o empeoran la capacidad funcional y las condiciones vitales de flora y fauna; ▶impacto ambiental); *syn.* deterioro [m] del medio ambiente, destrucción [f] del medio ambiente; *f* **destruction [f] de l'environnement** (Dégradation intensive provoquée par des nuisances massives ayant pour conséquence de réduire totalement ou presque la production en ressources biologiques des systèmes naturels pouvant satisfaire les exigences des usages humains ainsi que de limiter durablement les conditions de vie de la flore et de la faune ; ▶nuisance d'environnement) ; *syn.* détérioration [f] de l'environnement ; *g* **Umweltzerstörung [f]** (Schädigung der Umwelt durch massive ▶Umweltbelastungen, so dass die nachhaltige Leistungsfähigkeit für menschliche Nutzungsansprüche sowie die Lebensbedingungen für Fauna und Flora nicht mehr oder nur sehr eingeschränkt gegeben ist).

environmental disruption [n] *ecol. envir. recr.* ▶disturbance.

environmental disturbance [n] *ecol. envir. recr.* ▶disturbance.

environmental factors [npl] *ecol.* ▶ecological factors.

1802 environmental hygiene [n] *conserv. envir.* (Branch of environmental medicine concerned with measures to prevent pollution of air, water, soil, plants, and food via noxious materials such as chemicals, micro-organisms, industrial emissions, and radioactive materials); *s* **higiene [f] ambiental** (Ámbito de la medicina ambiental que se ocupa de las medidas preventivas de la contaminación atmosférica, del agua, suelo, flora y fauna y alimentos por medio de contaminantes ambientales: productos químicos, microorganismos, polvo, emisiones industriales y urbanas y sustancias radiactivas); *f* **hygiène [f] environnementale** (Branche de la médecine environnementale qui traite des mesures de prévention de la pollution de l'air, l'eau, le sol, les végétaux et les aliments par des substances nocives telles que les produits chimiques, les micro-organismes, les poussières industrielles, les gaz d'échappement et les substances radioactives) ; *g* **Umwelthygiene [f]** (Teilgebiet der Umweltmedizin, das sich mit vorsorgenden Maßnahmen zur Vermeidung der Verunreinigung von Luft, Wasser, Boden, Pflanzen und Lebensmitteln durch Umweltnoxen — Chemikalien, Mikroorganismen, Industriestaub, Abgase und radioaktive Stoffe — befasst; MEL 1979, Bd. 24).

1803 environmental illness [n] *envir.* (Occupational or other sicknesses caused directly or indirectly—e.g. via infected/contaminated animals or plants—by substances in the environment which are noxious to human organisms); *s* **enfermedad [f] ambiental** (Dolencia causada por contaminación ambiental o por exposición a productos nocivos durante el trabajo); *f* **maladies [fpl] environnementales** (Maladies générales ou professionnelles contractées directement ou indirectement par infection ou contamination et provoquées par des substances nocives telles que les produits chimiques, les micro-organismes, les poussières industrielles, les gaz d'échappement et les substances radioactives) ; *g* **Umweltkrankheit [f]** (Durch unmittelbare oder mittelbare [z. B. über kontaminierte Tiere oder Pflanzen] Einwirkung von Umweltnoxen [Chemikalien, Mikroorganismen, Industriestaub, Abgase und radioaktive Stoffe] auf den menschlichen Organismus verursachte allgemeine oder Berufskrankheit; MEL 1979, Bd. 24).

1804 environmental impact [n] *conserv. nat'res. plan.* (Existence of negative influences of a planned project upon the environment which necessitates an ▶environmental impact assessment [EIA] according to legal regulations; ▶analysis of environmental impact); *s* **incidencia [f] significativa sobre el medio ambiente** (El análisis de la importancia de los efectos sociales y ambientales de un proyecto. Es el primer paso al realizar una ▶evaluación de impacto ambiental; ▶estimación de la incidencia significativa sobre el medio); *syn.* repercusión [f] significativa sobre el medio ambiente; *f* **incidence [f] notable sur l'environnement** (Existence d'effets négatifs d'un projet sur les milieux naturels et humains qui nécessite que le projet soit soumis à une ▶procédure d'étude d'impact ; ▶évaluation des incidences significatives sur l'environnement) ; *g* **Umwelterheblichkeit [f]** (Vorhandensein von negativen Auswirkungen eines geplanten Projektes auf die biotische und abiotische Umwelt, so dass die Durchführung einer ▶Umweltverträglichkeitsprüfung nach dem ▶Umweltverträglichkeitsprüfungs-Gesetz zwingend erforderlich wird; ▶Umwelterheblichkeitsprüfung).

1805 environmental impact assessment [n] **(EIA)** [US] *conserv. leg. nat'res. plan.* (Legally prescribed comprehensive procedure of analysis and evaluation for public and private projects before they are given approval, if it is expected that they will have highly significant effects upon the environment. Technical aspects required by law are usually investigated in an

►environmental impact study [EIS] [US]. **In the U.S.**, the **EIA** is usually required under many state environmental impact laws for large-scale Federal and similar projects significantly affecting the environment, according to the National Environment Policy Act [NEPA] of 1970; **in Europe**, the **EA** [environmental assessment] is governed by the Directive of the European Council 85/337/EU of 27.06.1985 [Council Directive on the assessment of the effects of certain public and private projects on the environment] and by national legislation; ►environmental impact statement [EIS]; **in U.K.**, the **environmental assessment [EA]** is a procedure that must be followed for certain types of development before they are granted development consent. It is required by the European Directive [85/33/EEC as amended by 97/11/EC] in which it is known as the **E.I.A.** An **E.A.** is the whole process required to reach a decision on whether or not to allow a project to proceed. It involves the presentation, collection and assessment of information on the environmental effects of a project and also the final judgement upon it. An important part of that process is the submission of an **Environmental Statement [ES]** which provides information put forward by the applicant in his application for project planning permission. An **E.S.** outlines the likely effects of the development on the environment and proposes mitigation measures. It is circulated to statutory bodies and made available to the public for comments, which must be taken into account by the competent authority [e.g. local planning authority] before granting consent. The *Town and Country Planning [Environmental Impact Assessment—England] Regulations 2008* and the Explanatory Memorandum contain an informal consolidation of the **EIA** Regulations as amended); *syn.* environmental assessment [n] (EA) [also UK]; *s* **evaluación [f] del impacto ambiental (EIA)** (Procedimiento administrativo de toma de decisión sobre proyectos, planes y programas con incidencia importante en el medio ambiente utilizando un conjunto de estudios y sistemas técnicos que permiten estimar los efectos que la ejecución de un determinado programa, plan, proyecto, obra o actividad causa sobre el medio ambiente y, por medio de la ►declaración de impacto, fijar las condiciones en las que se deben realizar aquéllos así como las prescripciones pertinentes para la correspondiente vigilancia ambiental; cf. RD Legislativo 1302/1986, de evaluación de impacto ambiental, arts. 5 y 18 RD 1131/1988 y Ley 6/2006, sobre evaluación de los efectos de determinados planes y programas en el medio ambiente; en la Unión Europea regulada por la directiva 85/337/CEE del 27 de junio de 1985 [revisada por directiva del 3 de marzo de 1997] y las leyes nacionales de aplicación; en los EE.UU. es obligatoria para todos los proyectos federales de envergadura según el National Environment Policy Act 1970; ►estudio de impacto ambiental); *f* **procédure [f] d'étude d'impact** (Procédure préalable à la réalisation d'aménagements ou d'ouvrages publics ou privés qui, par leur nature, leur localisation, leurs dimensions ou leurs effets, peuvent porter atteinte au milieu naturel ; l'évaluation des incidences sur l'environnement est en général réalisée dans l'►étude d'impact ; la **p. d'é. d'i.** est réglementée par la directive du Conseil n° 85/337/CEE du 27 juin 1985 [révisée par la directive du Conseil du 3 mars 1997] et traduit en droit interne par diverses réglementations ; ►déclaration environnementale, ►notice d'impact) ; *g* **Umweltverträglichkeitsprüfung [f]** (*Abk.* UVP; gemäß Gesetz über die Umweltverträglichkeit [UVPG] vorgeschriebener unselbständiger Teil verwaltungsbehördlicher Verfahren nach einem definierten Ablaufschema, das bei der Prüfung über die Zulässigkeit von öffentlichen oder privaten Projekten angewendet wird, wenn diese erhebliche Auswirkungen auf die Umwelt haben. Die UVP umfasst die Ermittlung, Beschreibung und Bewertung der Auswirkungen eines Vorhabens auf **1.** Menschen, Fauna, Flora, Boden, Wasser, Luft, Klima und Landschaft mit ihren Wechselwirkungen und

2. Kultur- und sonstige Sachgüter. Die UVP wird unter Einbeziehung der Öffentlichkeit und unter Beteiligung der Träger öffentlicher Belange durchgeführt. Die nach dem Gesetz erforderlichen fachlichen Grundlagen und Einschätzungen werden i. d. R. in der ►Umweltverträglichkeitsstudie abgehandelt. Die UVP wird in der Europäischen Gemeinschaft durch die Richtlinie des Europarates 85/337/EWG vom 27.06.1985 [geändert durch Richtlinie 97/11/EG vom 03.03.1997] und durch nationale Gesetze geregelt; in den USA wird sie für alle großen öffentlichen Maßnahmen des Bundes gemäß National Environment Policy Act 1970 verlangt; ►Umwelterklärung); *syn.* Prüfung [f] der Umweltverträglichkeit.

1806 environmental impact statement [n] (EIS) *conserv. leg. nat'res. plan.* (Report containing the results of a comprehensive analysis of the environmental consequences of policies, program[me]s or development projects, with special regard to their ecological effects. Required in the U.S. for all major Federal actions significantly affecting the quality of the human environment. *Statutory basis in the U.S.* National Environment Policy Act 1970); *s* **declaración [f] de impacto** (Formulación del órgano ambiental en relación a la autorización de una obra, instalación o actividad en la que —sobre la base de los resultados del estudio de impacto ambiental— se determinan **1.** la conveniencia o no de realizar el proyecto, y en caso afirmativo, **2.** fijará las condiciones que deben establecerse en orden a la adecuada protección del medio ambiente y los recursos naturales y **3.** incluirá las prescripciones pertinentes sobre la forma de realizar el seguimiento de las actuaciones, de conformidad con el programa de vigilancia ambiental. Las **d. de i.** se harán públicas en todo caso; cf. Art. 4 (1) RD 1302/1986, de 28 de junio, de Evaluación de Impacto Ambiental y Reglamento RD 1131/1988); *f* **déclaration [f] environnementale** (Déclaration annuelle établie volontairement par les entreprises participant au système communautaire de management environnemental et d'audit, ayant pour but à partir de la description des activités de l'entreprise et de l'évaluation des problèmes environnementaux [production de déchets, consommation d'énergie et d'eau, le bruit, etc.] de présenter la politique, le programme et le système de management que l'entreprise a mise en œuvre sur le site considéré ; cf. règlement du Conseil n° 1836/93/CEE du 29 juin 1993) ; *g* **Umwelterklärung [f]** (Bericht über die Ergebnisse einer umfassenden Untersuchung über Folgen der Politik, von Programmen oder Projekten unter besonderer Berücksichtigung der ökologischen Auswirkungen. In Europa ist die gesetzliche Grundlage die Verordnung [EWG] Nr. 1836/93 des Rates vom 29.06.1993 über die freiwillige Beteiligung gewerblicher Unternehmen an einem Gemeinschaftssystem für das Umweltmanagement und die Umweltbetriebsprüfung).

1807 environmental impact study [n] (EIS) [US] *conserv. leg. nat'res. plan.* (Comprehensive analysis and evaluation of the environmental consequences of policies, program[me]s or development projects, with special regard to their ecological effects, with the results incorporated in a study report. The objective of an **e. i. s.** is to ensure consideration of all key issues as early as possible. ►Scoping is used to determine the scope of the study. The **e. i. s.** is also intended to provide an effective medium by which the project design can be amended at an early stage to take account of environmental effects and regulatory requirements; LD 1994 [6], p. 16. **U.S.**, for all major federal actions, an initial ►environmental impact assessment [EIA] [US]/environmental assessment [UK] is made to determine if the proposal is a major action significantly affecting the quality of the human environment requiring an ►environmental impact statement [EIS]. Statutory basis is the National Environment Policy Act [NEPA] 1970; **U.K.**, statutory basis is the EC

E

Directive 85/337; ►environmental risk analysis); **s estudio [m] de impacto ambiental** (Estudio realizado para identificar, predecir e interpretar así como prevenir las consecuencias o efectos ambientales que determinadas acciones, programas o proyectos pueden causar a la salud y al bienestar humanos y al entorno. El contenido está regulado en España por Reglamento en el RD 1131/1988; ►análisis del riesgo ecológico, ►consulta sobre los contenidos de un estudio de impacto ambiental, ►evaluación del impacto ambiental); ***f 1* étude [f] d'impact** (F., études préalables à la réalisation d'aménagements effectuées dans le cadre de la ►procédure d'étude d'impact permettant d'apprécier les conséquences d'ouvrages publics ou privés qui, par leur importance, leurs dimensions ou leurs incidences, peuvent porter atteinte au milieu naturel et de proposer les mesures prises pour supprimer, limiter ou compenser ces incidences ; loi relative à la protection de la nature du 10 juillet 1976, modifiée par la loi n° 96-1236 du 30 décembre 1996 ; ►analyse des risques écologiques, ►cadrage préalable) ; ***f 2* notice [f] d'impact** (Étude préalable à la réalisation d'aménagements permettant d'apprécier les conséquences d'ouvrages publics ou privés dont l'incidence sur l'environnement a été jugée moindre et devant indiquer les conditions dans lesquelles l'opération projetée satisfait aux préoccupations d'environnement ; cf. décr. n° 77-1141 du 12 octobre 1977) ; ***g* Umweltverträglichkeitsstudie [f]** (*Abk.* UVS; umfassendes ökologisches Gutachten, das den Umfang der Auswirkungen eines Großprojektes auf Mensch, Natur und Landschaft sowie auf Sachgüter und das kulturelle Erbe identifiziert, beschreibt und bewertet, die Wechselwirkungen zwischen den genannten Faktoren aufzeigt und Abhilfen oder Alternativen bei schädlichen Auswirkungen darlegt. Die **UVS** bildet die fachliche Grundlage für die ►Umweltverträglichkeitsprüfung; ►Risikoanalyse, ►Scoping); *syn.* Umweltverträglichkeitsuntersuchung [f], gesamtökologisches Gutachten [n] (hinsichtlich der Nutzungsverträglichkeit).

1808 environmentalist [n] *conserv.* (Proponent of environmental protection; ►conservationist); **s 1 ambientalista [m/f]** (Persona que se ocupa activamente del medio ambiente. Es versión del inglés «*environmentalist*»; DINA 1987); **s 2 ecologista [m/f]** (Término que designa a las personas y entidades que se preocupan activamente de defender la naturaleza. Es versión del inglés «ecologist», propiamenete ecólogo; DINA 1987; ►conservacionista); ***f 1* écologiste [m] (2)** (1. Personne adhérent à l'idée de protection de la nature et de l'environnement. 2. Personne engagée activement dans la lutte pour la protection de la nature et de l'environnement, militant de l'écologisme ; le terme allemand « Naturschützer » est synonyme de « *Umweltschützer* » ; ►écologiste 1) ; *syn.* défenseur [m] de l'environnement ; ***f 2* environnementaliste [m/f]** (Spécialiste de l'environnement. Terme rarement utilisé au profit du mot écologiste) ; ***g* Umweltschützer/-in [m/f]** (Anhänger des Umweltschutzgedankens; ►Naturschützer/-in).

1809 environmentally safe [adj] *envir. pol.* (Descriptive term for an ecologically-desirable effect with regard to well-being of the environment; ►ecologically-sound); *syn.* environment-friendly [adj], environmentally sound [adj], emission-avoiding [ppr] (±) ; **s respetuoso/a con el medio ambiente [loc]** (Término descriptivo para actividades humanas que no tienen efectos adversos sobre el medio ambiente; ►respetuoso/a con la naturaleza); *syn.* no dañino/a al medio ambiente [loc], no contaminante [adj] (±); ***f* respectueux, euse de l'environnement [loc]** (Terme décrivant toute activité peu consommatrice d'énergie et de matières premières ou qui engendre peu ou pas de dégradations des facteurs biotiques ou abiotiques de l'environnement ; ►respectueux de la nature) ; *syn.* de manière écologiquement acceptable [loc], respectueux, euse des qualités du milieu [loc] ;

***g* umweltfreundlich [adj]** (So beschaffen, dass Tätigkeiten des Menschen oder Ergebnisse von Handlungen die Umwelt mit allen biotischen und abiotischen Komponenten nicht nachhaltig belasten; ►naturschonend); *syn.* umweltverträglich [adj], umweltschonend [adj].

environmentally sound [adj] *envir. pol.* ►environmentally safe.

1810 environmental medicine [n] *conserv.* (Branch of medicine concerned with the influence and effects of environmental pollution upon human organisms; **s medicina [f] ambiental** (Parte de la medicina que se ocupa de las influencias y los efectos de la contaminación ambiental sobre el organismo humano; *término específico* toxicología ambiental); ***f* médecine [f] environnementale** (Branche de la médecine qui traite des influences et des effets de la pollution environnementale sur l'organisme humain ; *terme spécifique* toxicologie environnementale : étude des contaminants présents dans nos différents environnements) ; ***g* Umweltmedizin [f]** (Teilbereich der Medizin, der sich mit den Einflüssen und Auswirkungen der Umweltverschmutzung auf den menschlichen Organismus befasst; MEL 1979, Bd. 24; *UB* **Umwelttoxikologie [f]:** Wissenschaft, die sich mit den Gesundheitswirkungen von Schadstoffen in der Umwelt befasst; die Aufnahme der Stoffe kann über das Wasser, die Nahrung, durch die Atemluft oder über die Haut erfolgen).

1811 environmental nuisance [n] *envir.* (Conditions adversely affecting upon the environment, caused either directly or indirectly by humans, which result in such changes to the ecosystem that the sustainability of resources for human use—both material and immaterial—is diminished, or that the living conditions for fauna and flora deteriorate; ►environmental deterioration); **s impacto [m] ambiental** (Efectos o consecuencias negativas de la incidencia humana sobre el medio, en especial cuando se producen modificaciones de los sistemas naturales. La trascendencia depende de la vulnerabilidad, *sensu lato*, del territorio; cf. DINA 1987; ►degradación del medio ambiente); *syn.* distorsión [f] ambiental; ***f* nuisance [f] d'environnement** (Tout effet produit directement ou indirectement par les usages humains et qui modifie sensiblement des stations ou des écosystèmes et qui a pour conséquence la diminution de la l'utilisation durable des ressources pour les usages humains [matériels et immatériels] ou la détérioration des conditions de vie de la faune et de la flore ; ►destruction de l'environnement) ; *syn.* nuisance [f] environnementale 1 ; ***g* Umweltbelastung [f]** (Jeder Einfluss, der mittelbar oder unmittelbar von menschlichen Nutzungen ausgeht und Standorte oder Ökosysteme so verändert, dass deren nachhaltige Leistungsfähigkeit für menschliche [materielle und immaterielle] Nutzungen gemindert wird oder die Lebensbedingungen der Fauna und Flora oder deren Teile verschlechtert werden; ►Umweltzerstörung); *syn.* Umweltbeeinträchtigung [f].

environmental option [n] **(BPEO), best practicable** *envir.* ►state-of-the-art (2).

environmental planning [n]**, design** [n] **and conservation** [n] *landsc. recr.* ►landscape architecture.

1812 environmental planning input [n] (±) *landsc. plan.* (Planning statements, dealing with the comprehensive spatial development on a national, state, regional, or local scale, which investigate the compatibility of the various ecological, economic, and social conditions; ►multidisciplinary planning [US]/cross-sectional scope of planning [UK]); **s contribución [f] paisajística a la planificación general** (Dictamen sobre la planificación general a nivel nacional, regional o comunal de las exigencias de uso del espacio de las planificaciones sectoriales, teniendo en cuenta los aspectos ecológicos, económicos y sociales; ►planificación integrada); ***f* volet [m] environnemental**

d'une planification globale (≠) (D., document de planification justifiant la compatibilité des différentes planifications sectorielles au niveau national, d'un Land, d'une région ou de la commune en tenant compte des aspects écologiques, économiques et sociaux ; ►planification intégrée) ; *g* **landespflegerischer Beitrag [m] zur Gesamtplanung** (Landschaftsplanerische Aussagen zur räumlichen Gesamtentwicklung auf Bundes-, Landes-, regionaler oder kommunaler Ebene, die die Verträglichkeit der Nutzungsansprüche sektoraler Fachplanungen unter ökologischen, ökonomischen und sozialen Aspekten überprüfen; ►Querschnittsplanung).

1813 environmental policy [n] *conserv. pol.* (Measures which are necessary at a political level to ensure a healthy environment befitting human beings, to protect natural resources from harmful human activities and to remove the consequences of such activities. These measures are to be taken in keeping with ecological principles and a sustainable economic system); *s* **política [f] de medio ambiente** (Conjunto de medidas legislativas, administrativas, científicas, técnicas y de planificación que se consideran necesarias para preservar el medio ambiente sano o para restaurarlo de manera que sirva de base vital para la sociedad); *syn.* política [f] ambiental; *f* **politique [f] environnementale** (Ensemble des objectifs globaux, programmes, actions et prescriptions réglementaires qui, au niveau politique, sont nécessaires pour garantir un environnement sain et humain, pour sauvegarder et protéger le milieu naturel des effets néfastes des atteintes causées par les activités humaines, compte tenu de l'objectif d'un développement durable à un coût économiquement acceptable des ressources naturelles) ; *g* **Umweltpolitik [f]** (Gesamtheit aller Maßnahmen als konflikthafte Reaktion auf der Ebene politischer Problemlösungen, die notwendig sind, um eine gesunde und menschenwürdige Umwelt nachhaltig zu sichern, um Naturgrundlagen vor nachteiligen Wirkungen menschlicher Eingriffe zu schützen und Folgen schädlicher Eingriffe zu beseitigen — vor dem Hintergrund der Steuerung des ökologisch-ökonomischen Umweltsystems auf der Basis eines nachhaltigen Wirtschaftens).

Environmental Policy Act [n] [US] *conserv. leg. nat'res. plan.* ►National Environmental Policy Act.

1814 environmental pollution [n] *envir. leg.* (Direct or indirect introduction of deleterious substances or emissions into an environmental medium [air, water or soil], by human actions with resulting harmful effects, e.g. human health hazards, deterioration of flora and fauna or ecosystems, damages to objects [structures, monuments, artwork, etc.], reduction in the comfort-level of the environment or its lawful uses; ►contamination 1, ►damage caused by environmental pollution, ►effect of environmental pollution); *s* **contaminación [f] ambiental** (Liberación artificial, en el medio ambiente, de sustancias o energía, que causa efectos adversos sobre el hombre o sobre el medio, directa o indirectamente; DINA 1987; ►contaminación); *syn.* polución [f] ambiental, contaminación [f] del medio ambiente; *f* **pollution [f] de l'environnement** (Introduction par l'homme, directement ou indirectement, de substances ou d'énergie dans le milieu naturel [sol, eau, air] et ayant une action nocive de nature à mettre en danger la santé de l'homme, à endommager la flore, la faune et les écosystèmes, à détériorer les biens matériels [édifices, monuments historiques, objets d'art, etc.] et à porter atteinte ou à nuire aux valeurs d'agrément et aux autres utilisations légitimes de l'environnement tel que le sol, l'eau, l'air, les aliments ; cf. art. 2 directive du Conseil n° 84/360/CEE du 28 juin 1984 ; ►contamination) ; *g* **Umweltverschmutzung [f]** (Unmittelbare oder mittelbare Zuführung von Stoffen oder Energie durch den Menschen in ein Umweltmedium [Boden, Wasser, Luft], aus der sich abträgliche Wirkungen wie eine

Gefährdung der menschlichen Gesundheit, eine Schädigung der Tier- und Pflanzenwelt oder Ökosysteme sowie von Sachwerten [Bauwerke, Denkmale, Kunstobjekte etc.] und eine Beeinträchtigung der Annehmlichkeiten der Umwelt oder sonstiger rechtmäßiger Nutzungen der Umwelt ergeben. Einschlägige Rechtsnormen regeln auf nationaler und europäischer Ebene, welches Verhalten der Gesetzgeber zur Verhinderung oder Eindämmung der **U.** gebietet und welche Sanktionen für verbotswidriges Verhalten drohen; cf. Bundes-Bodenschutzgesetz [BBodSchG], Kreislaufwirtschafts- und Abfallgesetz [KrW-/AbfG], RL 84/360/EWG etc.; ►Kontamination); *syn.* Umweltkontamination [f], Umweltkontaminierung [f], Verschmutzung [f] der Umwelt.

environmental pollution [n]**, background level of** *envir.* ►existing environmental pollution load.

1815 environmental pollution load [n] *envir.* (Level of pollution caused by air pollution, noise, noxious substances, radiation, etc.); *s* **nivel [m] de inmisión** (Cantidad de contaminación sólida, líquida y gaseosa, existente entre 0-2 m de altura sobre el suelo; cf. Orden de 18 de octubre de 1976 —Anexo 1); *f* **étendue [f] de la pollution atmosphérique** (Dimension des dommages engendrés par les rejets de polluants dans l'atmosphère) ; *syn.* importance [f] de la pollution atmosphérique, intensité [f] de la pollution de l'air, niveau [m] de pollution atmosphérique ; *g* **Immissionsbelastung [f]** (Umfang schädlicher Umwelteinwirkungen, die durch Immissionen verursacht sind).

1816 environmental precautions [npl] *conserv. nat'res. plan.* (Anticipatory measures and techniques to ward off possible environmental damage; ►precautionary principle, ►preventive measures); *s* **prevención [f] ambiental** (Medidas y técnicas aplicadas de antemano para evitar daños ambientales, posibles peligros, desventajas graves o influencias perturbadoras considerables; ►prevención, ►principio de prevención [de deterioros ambientales]); *syn.* precaución [f] ambiental; *f* **prévention [f] des atteintes à l'environnement** (Pratiques et techniques environnementales utilisées dans la lutte, par priorité à la source, contre les atteintes à l'environnement ; ►prévention, ►principe d'action préventive et de correction, ►principe de prévention ; *syn.* action [f] de prévention en faveur de l'environnement ; *g* **Umweltvorsorge [f]** (Frühzeitige Maßnahmen und Techniken, die Umweltschäden und vor allem sonstige Gefahren, erhebliche Nachteile und erhebliche Belästigungen gar nicht erst aufkommen lassen; cf. RL 96/61/EG des Rates v. 24.09.1996 über die integrierte Vermeidung und Verminderung der Umweltverschmutzung [IVU-Richtlinie]; ►Vorsorge, ►Vorsorgeprinzip).

1817 environmental program [n] [US]/**environmental programme** [n] [UK] *conserv. pol.* (Procedural plan and priority actions to be taken by a government in accordance with principles and aims of a long-term ►environmental policy); *s* **programa [m] de medio ambiente** (Líneas y objetivos generales de la ►política de medio ambiente a largo plazo y correspondientes programas de acción para implementarla); *syn.* programa [m] ambiental; *f 1* **programme [m] de défense de l'environnement** (Action globale, établie par un gouvernement, de protection du patrimoine naturel définissant les principes et objectifs d'une politique à long terme de préservation de l'environnement ; ►politique environnementale) ; *f 2* **programme [m] environnemental** (Description des objectifs et des activités spécifiques de l'entreprise destinés à assurer une meilleure protection de l'environnement sur un site donné, y compris une description des mesures prises ou envisagée pour atteindre ces objectifs et, le cas échéant, les échéances fixées pour leur mise en œuvre ; art. 2, règlement du Conseil n° 1836/93/CEE du 29 juin 1993) ; *g* **Umweltprogramm [n]** (Plan, der die Grundlinien und Ziele einer langfristig angelegten ►Umweltpolitik einer Regie-

E

rung und Aktionsprogramme vorrangig vorgesehener Maß-nahmen vorsieht).

1818 environmental protection [n] *conserv. pol.* (Generic term now divided into two categories: ▶complex environmental protection and ▶technical protection of the environment; ▶legislation on environmental protection, ▶sustainable development); *s* **protección [f] ambiental** (Conjunto de programas políticos y medidas para la preservación a largo plazo de la naturaleza y de los recursos naturales vitales para la humanidad. El término fue acuñado en los años 1960, cuando se descubrió la **p. a.** como un nuevo campo de acción política. Término genérico para ▶protección integrada del medio ambiente y ▶protección técnica del medio ambiente; ▶desarrollo sostenible); *syn.* protección [f] del medio ambiente; *f* **protection [f] de l'environnement** (Terme générique utilisé pour désigner deux notions de protection des ressources naturelles, la ▶protection intégrée de l'environnement et la ▶génie environnemental) ; ▶développement durable) ; *g* **Umweltschutz [m, o. Pl.]** (Gesamtheit der Maßnahmen und politischen Bestrebungen zur dauerhaften Erhaltung der natürlichen Lebensgrundlagen des Menschen und die der Natur insgesamt, um die Auswirkungen früherer, gegenwärtiger und zukünftiger Tätigkeiten des Menschen so zu gestalten, dass die Qualität der Umwelt ein gesundes Leben in einer sich weiter vielfältig entwickelnden Natur ermöglicht. In den 1960er-Jahren wurde der Begriff geprägt, als die Umweltproblematik als neues Politikfeld entdeckt wurde. Der OB 'Umweltschutz' wird u. a. in zwei Konsequenzen als ▶komplexer Umweltschutz und als ▶technischer Umweltschutz verstanden. Es waren nicht zuletzt die alternativen Forschungsinstitute — auf der Grundlage von Bürgervereinen gegründet —, die einen auf Vorsorge orientierten **U.** einforderten. Das Konzept des präventiven **U.es**, d. h. in Kreislaufprozessen zu denken und bei Eintrag von Energien und Stoffen durch Nutzungsoptimierung die Belastung der Umwelt zu minimieren, entwickelten das Öko-Institut e. V. in Freiburg/Breisgau, Darmstadt und Berlin [seit 1977] und das Wuppertaler Institut für Klima, Umwelt, Energie — seit 1991; in den 1990er-Jahren kam der Begriff der Nachhaltigkeit [▶nachhaltige Entwicklung] resp. die Zukunftsfähigkeit in die **U.**-Diskussion. Viele geltende Verträge und Programme für den **U.** kranken heute daran, dass sie im Zusammenhang einer weltweit einseitig ökonomischen Praxis eingefordert werden, die sich nur nach der Eigendynamik des Marktes und der Finanzströme weltweit richtet und deshalb umweltpolitische Rahmenbedingungen noch weitgehend ignoriert. In D. teilen sich die Verantwortung für den **U.** der Bund, die Länder und die Gemeinden; die Gesetzgebungskompetenz des Bundes ist auf die Vorgabe von Rahmenregelungen beschränkt; der **U.** wird zunehmend durch europäisches Recht und internationale Übereinkommen geprägt); *syn.* Schutz [m, o. Pl.] der Umwelt.

environmental protection [n]**, integrated** *conserv. pol.* ▶complex environmental protection.

1819 environmental protection technology [n] *envir.* (Technical measures for the protection of the environment; ▶technical protection of the environment); *s* **tecnología [f] de protección ambiental** (▶protección técnica del medio ambiente); *syn.* tecnología [f] ambiental; *f 1* **technologie [f] de protection de l'environnement** (Ensemble des techniques utilisées dans le domaine de la protection de l'environnement ; ▶génie environnemental) ; *f 2* **technique [f] de protection de l'environnement** (Procédé utilisé dans le domaine de la protection de l'environnement) ; *g* **Umweltschutztechnik [f]** (Technische Maßnahmen, Geräte und Verfahren, die auf Grund der Analyse und Bewertung von bestehenden Umweltschäden entwickelt wurden, um bereits entstandene Schäden zu sanieren; es gibt **nachgeschaltete Technologien**, die z. B. Abgase und

Abwässer reinigen sowie verseuchte Böden dekontaminieren und **produktionsintegrierte U.en**, die bereits im *statu nascendi* mit Hilfe von *Clean technologies* Produktionsverfahren so steuern, dass [fast] keine Umweltschäden entstehen; ▶technischer Umweltschutz).

environmental purposes [npl] *leg. urb.* ▶land reserved for environmental purposes.

1820 environmental quality [n] (1) *envir. pol.* (Result of successful integration of ecology and design to achieve functional, social, and ▶sustainable development); *s* **calidad [f] ambiental** (Balance y evaluación de la situación del medio ambiente resultante de todos los cambios y las influencias antropógenas. Los informes periódicos de las diferentes agencias del medio ambiente dan una visión general de la **c. a.** de cada uno de los países o de las regiones resp. CC.AA. [en Es.]; ▶desarrollo sostenible); *f* **qualité [f] de l'environnement** (Résultat de l'intégration ou de la symbiose réussie entre les fonctions écologiques et les concepts d'aménagement garantissant un ▶développement durable du milieu naturel et de la vie sociale ; terme à rapprocher de l'expression « cadre de vie » des architectes et des urbanistes) ; *syn.* qualité [f] du milieu, *par extension* qualité [f] du cadre de vie ; *g* **Umweltqualität [f]** (Beschaffenheit des Zustandes der Umwelt [Umweltsituation] als Bilanz und Bewertung aller anthropogenen Veränderungen und Beeinflussungen. Mit Hilfe von **U.szielen** werden Parameter geschaffen, die die Grenzen aufzeigen, innerhalb derer das Wirtschaften mit immer knapper werdenden Ressourcen noch möglich ist resp. möglich sein sollte, um eine ▶nachhaltige Entwicklung von Wirtschaft und Gesellschaft zu ermöglichen. Die vom Umweltbundesamt in Berlin, das seit 1974 besteht, veröffentlichten Jahresberichte „Daten zur Umwelt" liefern einen Gesamtüberblick zur **U.** der Bundesrepublik Deutschland).

environmental quality [n] (2) *plan. recr. sociol.* ▶perceived environmental quality.

1821 environmental quality standards [npl] *envir. pol.* (Relative level of availability of natural resources measured by success in integrating ecology and design, based on ▶environmental awareness, to attain functional, social, and sustainable development, and to lower the strain on ▶ecological factors and natural resources); *s* **estándar [m] de calidad ambiental** (Parámetro cuantificable fijado por medio de ordenanza o reglamento como p. ej. los niveles máximos de emisión e inmisión, las normas DIN, las normas de eliminación de residuos sólidos, etc. Los **e. de c. a.** se fijan para cada uno los diferentes campos de protección ambiental como la protección del ambiente atmosférico, del suelo, de las aguas continentales, la gestión de residuos. En evaluaciones de impacto ambiental se fijan estándares interdisciplinariamente. Cuanto más altos los **e. de c. a.** menor es la contaminación de los ▶factores ecológicos); *syn.* standard [m] de calidad ambiental; *f* **standards [mpl] de qualité de l'environnement** (Niveau relatif de disponibilité des ressources naturelles dépendant du niveau de ▶conscience écologique et de la capacité d'une société à garantir un développement durable du milieu naturel et de la vie sociale ; ce niveau de qualité est défini par des paramètres précisés par des dispositions législatives, normes ou règlements spécifiques relatives aux différents domaines de la protection de l'environnement tels que la qualité de l'air, la protection du sol, la qualité des eaux, les déchets, etc. ; plus le niveau des standards est élevé, moins les dégradations des facteurs environnementaux sont importantes ; ▶facteurs écologiques) ; *g* **Umweltqualitätsstandard [m]** (Operationalisierter resp. quantifizierbarer/quantifizierter Parameter oder Indikator, der nach Durchführungsverordnungen und Verwaltungsvorschriften, z. B. Technische Anleitung Luft, DIN-Normen, VDI-Vorschriften, diverse Verordnungen für die

Abfallbeseitigung, definiert wird. **U.s** werden für die einzelnen Bereiche des Umweltschutzes wie z. B. Luftreinhaltung, Bodenschutz, Gewässerschutz und Kreislaufwirtschaft festgelegt und sind konkrete Bewertungsmaßstäbe, die Umweltqualitätsziele oder unbestimmte Rechtsbegriffe operationalisieren, indem sie für einen bestimmten Parameter oder Indikator die angestrebte Ausprägung, das Messverfahren und die Rahmenbedingungen festlegen. Sie können kardinal [z. B. Grenzwert für SO₂], ordinal [z. B. Gefährdung nach den Roten Listen] oder nominal [z. B. schutzwürdige Biotope nach § 30 BNatSchG und Gebiete, die dem Aufbau und dem Schutz des Europäischen ökologischen Netzes „Natura 2000" dienen] skaliert sein. Umweltverträglichkeitsprüfungen legen **U.s** fachbereichsübergreifend fest. Je höher die Standards, desto unbelasteter sollen die Umweltfaktoren [▶Ökofaktoren] sein); *syn.* Umweltstandard [m].

1822 environmental relevance [n] *envir.* (Important consequences for the environment of a particular activity, which have to be taken into consideration due to their significant size or extent; ▶environmental considerations, ▶environmental impact); *syn.* environmental significance [n]; *s* **exigencias [fpl] ambientales** (Cuestiones relevantes para el medio ambiente, ▶exigencias del medio ambiente y del paisaje, ▶incidencia significativa sobre el medio ambiente); *f* **exigence [f] environnementale** (Constat de l'importance de la prise en compte ou de la signification des activités en faveur de l'environnement ; exigence essentielle dans le domaine de la protection de l'environnement ; ▶incidence notable sur l'environnement, ▶préoccupations d'environnement) ; *g* **Umweltbelang [m]** (Ein für die Umwelt wichtiger Tatbestand oder eine zu berücksichtigende Erheblichkeit; ▶Belange der Umwelt, ▶Umwelterheblichkeit).

1823 environmental relief [n] *envir.* (Result of steps taken to relieve or avoid pressures upon the environment or to introduce ▶mitigation measures; *opp.* ▶environmental nuisance); *s* **minimización [f] de impacto sobre el medio** (Resultado de todas las medidas para evitar, reducir, y si es posible, compensar los efectos negativos importantes de un proyecto sobre el medio ambiente; ▶impacto ambiental, ▶medida compensatoria); *f* **réduction [f] de l'importance des impacts sur l'environnement (±)** (Résultat de toutes les mesures et actions ayant pour but d'éviter de réduire ou de compenser une dégradation de l'environnement ; ▶mesure compensatoire; *opp.* ▶nuisance d'environnement) ; *g* **Umweltentlastung [f]** (Ergebnis aller Maßnahmen und Handlungen, die darauf abzielen, eine weitere Belastung der Umwelt zu vermeiden, zu mindern oder ▶Ausgleichsmaßnahmen [des Naturschutzes und der Landschaftspflege] vorzusehen; *opp.* ▶Umweltbelastung); *syn.* Entlastung [f] der Umwelt.

1824 environmental research [n] *ecol. sociol.* (**1.** In a biological sense as in ▶ecology. **2.** In a sociological sense the study and research of changes in the environment caused by human intervention and the complex relationship between this "artificial" or cultural environment and the natural ecosystem); *s* **investigación [f] ambiental** (En sentido biológico, en principio ▶ecología, en sentido sociológico la investigación de las transformaciones causadas por la humanidad en su medio y las interacciones complejas entre el medio ambiente antropógeno y los ecosistemas naturales); *f* **recherche [f] sur l'environnement** (Au sens biologique du terme correspond à ▶écologie ; dans l'acceptation sociologique, analyse et recherche sur les changements causés par les activités humaines et les interrelations complexes entre ce monde « artificiel » et les écosystèmes naturels) ; *syn.* recherche [f] environnementale ; *g* **Umweltforschung [f]** (**1.** Im biologischen Sinne soviel wie ▶Ökologie. **2.** Im sozialwissenschaftlichen Sinne die Untersuchung und Erforschung der durch die Tätigkeit des Menschen auftretenden Veränderungen seiner Umwelt und der komplexen Wechselwir-

kungen zwischen dem künstlichen Geschaffenen und dem natürlichen Ökosystem; cf. MEL 1979, Bd. 24; zur **U.** gehören Teilgebiete wie Umweltökonomie, Umweltplanung, Umweltpolitologie, Umweltpsychologie, Umweltsoziologie sowie Umweltkommunikation im weitesten Sinne mit Bürgerbeteiligung, Öffentlichkeits- und Informationsarbeit).

1825 environmental resource patch [n] *ecol.* (LE 1986, 93; near-natural area of varying size differing completely from the surrounding matrix, e.g. an intensively used landscape, as part of a ▶habitat network, has the function of supporting the ecological equilibrium of a man-made landscape; ▶ecology of habitat islands, ▶succession area); *s* **islote [m] de vegetación natural** (Área natural poco alterada de tamaño variable que se encuentra entre las parcelas cultivadas en zonas de agricultura intensiva; puede formar parte de una red de biótopos [▶reticulación de biótopos] con la función de contribuir al equilibrio ecológico de los paisajes humanizados; ▶área de sucesión, ▶ecología de biótopos aislados); *f* **îlot [m] de ressources naturelles** (Parcelle de surface variable entre les parcelles cultivées en zones de cultures intensives en vue de la ▶constitution d'un réseau d'habitats naturels et le maintien des équilibres biologiques dans les paysages culturels ; ▶aire de succession, ▶écologie des biocénoses relictuelles, ▶réseau de biotopes) ; *g* **Ökozelle [f]** (Naturnahe Fläche unterschiedlicher Größe innerhalb der intensiv genutzten Kulturlandschaft, die mit anderen Flächen in Beziehung steht [▶Biotopvernetzung] und zum ökologischen [Fließ]gleichgewicht in der Kulturlandschaft beitragen soll; ▶Inselökologie, ▶Sukzessionsfläche); *syn.* ökologische Zelle [f].

environmental review [n] *envir. leg.* ▶environmental audit.

1826 environmental risk analysis [n] *ecol. landsc. plan.* (Evaluation of risk of likely environmental hazard—within an ▶environmental risk assessment—caused by the impact of competitive demands of a particular development. The analysis is made to further determine the likelihood of damage in an ecological sense to the functioning of ▶natural systems, to the use capacity of natural resources and to plants and animals, and thus to protect sustainably the diversity of the environment. When costs are applied to the analysis it is termed risk-benefit analysis; cf. WES 1985; ▶environmental impact statement [EIS], ▶environmental risk assessment); *s* **análisis [m] del riesgo ecológico** (Estimación del riesgo en el marco del ▶procedimiento de análisis del riesgo ecológico, por medio del cual se analiza si los usos del suelo en una zona específica son compatibles entre sí y con el medio y permiten mantener las funciones del ecosistema natural y de su flora y fauna y el aprovechamiento del mismo de forma sostenible; ▶declaración de impacto, ▶evaluación del impacto ambiental, ▶régimen ecológico de la naturaleza); *f* **évaluation [f] des risques environnementaux** (Dans le cadre d'un ▶procédé d'évaluation des risques écologiques, constatation de la compatibilité des demandes spatiales concurrentes à l'intérieur d'un même système d'utilisation des sols afin de sauvegarder durablement, au sens écologique, l'efficience du ▶système naturel, la capacité d'utilisation des ressources fondamentales naturelles, la flore et la faune ainsi que la diversité de l'environnement spatial ; ▶déclaration environnementale, ▶procédure d'étude d'impact) ; *syn.* analyse [f] des risques écologiques, analyse [f] des risques environnementaux ; *g* **Risikoanalyse [f]** (Abschätzung des Risikos im Rahmen des ▶Risikoverfahrens, ob die Verträglichkeit von konkurrierenden Raumansprüchen innerhalb eines Flächennutzungsmusters gegeben ist, um im ökologischen Sinne die Funktionsfähigkeit des ▶Naturhaushaltes die Nutzungsfähigkeit der Naturgüter, die Pflanzen- und Tierwelt und die Vielfalt der räumlichen Umwelt nachhaltig zu sichern; ▶Umwelterklärung, ▶Umweltverträglichkeitsprüfung); *syn.* ökologische Risikoanalyse [f].

1827 environmental risk assessment [n] *ecol. landsc. plan.* (Evaluation technique involving an ▶impact analysis and the ensuing ▶environmental risk analysis during comprehensive landscape planning in order to devise a suitable land use pattern. A matrix of questions is evolved whereby the significant probable effects of a use are measured in terms of impact, benefits and acceptability—risk perception—to the natural environment and human welfare. In this context interrelationships between neighbo[u]ring and superimposing land use requirements are investigated according to a cause-effect relationship); *s* **procedimiento** [m] **de análisis del riesgo ecológico** (Procedimiento de evaluación ecológica que, en el marco de la planificación del paisaje y sirviéndose del ▶análisis de efectos y del ▶análisis del riesgo ecológico, tiene como fin la definición de los usos del suelo. En este procedimiento se estudian las relaciones de causa-efecto existentes entre los diferentes usos); *f* **procédé** [m] **d'évaluation des incidences environnementales** (±) (Procédé d'évaluation écologique établi sur la base de l'▶analyse des impacts et de l'▶évaluation des risques environnementaux et utilisé au cours d'une des phases du processus d'élaboration d'une étude paysagère afin d'établir une matrice d'usages spatiaux traitant des questions posées par les interactions écologiques provoquées par la superposition d'usages ou la proximité d'usage concurrentiels, prenant donc en compte les relations de cause à effet dans l'atteinte au milieu naturel) ; *g* **Risikoverfahren** [n] (Ökologisches Bewertungsverfahren, durch das mit Hilfe der ▶Wirkungsanalyse und der dann in einem nächsten Schritt folgenden ▶Risikoanalyse während eines Planungsvorganges im Rahmen einer landschaftlichen Gesamtplanung ein Flächennutzungsmuster erarbeitet wird, bei dem die wesentliche Fragestellung nach den ökologischen Wechselwirkungen zwischen einander benachbarten oder sich überlagernden Nutzungsansprüchen untersucht, also von einem Verursacher-Folgewirkung-Betroffener-Zusammenhang ausgegangen wird).

1828 environmental sensitivity [n] *envir. land'man.* (Degree to which a particular environment copes with either natural or human intrusion upon the natural environment, as with polluting substances which cause disturbances, deterioration, or defacement); *syn.* environmental susceptibility [n]; *s* **sensibilidad** [f] **ante impactos** (Característica de organismos y del medio ambiente natural de reaccionar negativamente ante injerencias externas negativas o inmisión de contaminantes); *syn.* vulnerabilidad [f] ante impactos, fragilidad [f] ante impactos; *f* **sensibilité** [f] **à la pollution des milieux** (Caractéristique que possèdent certaines ressources naturelles ou humaines à constituer des cibles particulièrement vulnérables aux polluants atmosphériques [population humaine, faune, flore, patrimoine historique]) ; *syn.* sensibilté [f] des milieux à la pollution, fragilité [f] à la pollution, vulnérabilité [f] à la pollution ; *g* **Empfindlichkeit** [f] **gegen Umweltbelastungen** (Eigenschaft, auf bestimmte Eingriffe oder Umweltbelastungen sensibel oder mit Beeinträchtigungen zu reagieren).

1829 environmental sensitivity level [n] *plan. land' man.* (Limit of elasticity or tolerance for disruption or change within the biotic and abiotic environment, as assessed and evaluated by use of given criteria; ▶environmental sensitivity); *s* **grado** [m] **de sensibilidad** (Nivel de impacto de una injerencia humana sobre un parámetro ambiental específico; ▶sensibilidad ante impactos); *syn.* grado [m] de vulnerabilidad, grado [m] de fragilidad; *f* **degré** [m] **de sensibilité** (Résultat d'une évaluation lors de l'analyse des nuisances préjudiciables aux paramètres environnementaux ; ▶sensibilité à la pollution des milieux) ; *syn.* degré [m] de fragilité, degré [m] de vulnérabilité ; *g* **Empfind-**

lichkeitsstufe [f] (Bewertungsergebnis der Untersuchung von Beeinträchtigungen der biotischen und abiotischen Umwelt auf Grund nachvollziehbarer Kriterien, z. B. mit Hilfe eines Bewertungsbaumes; ▶Empfindlichkeit gegen Umweltbelastungen).

environmental significance [n] *envir.* ▶environmental relevance.

1830 environmental stress [n] *ecol. envir. recr.* (Direct or indirect alteration or disruption of environmental media caused by human activities, so that sustainable functioning of an area or ecosystem is diminished; ▶background level of pollution, ▶disturbance, ▶environmental pollution, ▶environmental stress tolerance, ▶intrusion upon the natural environment, ▶traffic load, ▶visitor pressure); *syn.* adverse impact [n]; *s* **estrés** [m] **ambiental** (Efecto nocivo, generalmente resultado directo o indirecto de actividades humanas sobre el medio natural o alguno de sus factores; a diferencia de la ▶perturbación el **i. a.** implica siempre el empeoramiento material de un medio, como el suelo o las aguas por sustancias tóxicas o nutrientes, etc.; ▶carga de tráfico, ▶contaminación de fondo, ▶contaminación ambiental, ▶intrusión en el entorno natural, ▶presión turística, ▶resiliencia de un ecosistema); *syn.* incidencia [f] ambiental, distorsión [f], impacto [m] negativo, perjuicio [m] causado al medio ambiente; *f* **stress** [m] **environnemental** (Effet provoqué directement ou indirectement par une activité humaine transformant les stations ou écosystèmes si profondément, que leur résilience en est altérée pour longtemps ; ▶atteinte au milieu naturel, ▶capacité de charge d'un écosystème, ▶charge polluante existante, ▶intensité du trafic, ▶nuisance, ▶résilience d'un écosystème, ▶pollution de l'environnement, ▶pression exercée par les visiteurs) ; *syn.* nuisance [f] environnementale (2) ; *g* **Belastung [f]** (Jeder Einfluss, der meist direkt oder indirekt von menschlichen Nutzungen ausgeht und Standorte oder Ökosysteme stofflich so verändert, dass die nachhaltige Funktionsfähigkeit gemindert wird; im Vergleich zur ▶Beeinträchtigung bezeichnet **B.** immer die materielle Verschlechterung eines Mediums, z. B. des Bodens, Gewässers durch Gifte, Eutrophierung etc.; beim Menschen drückt **B.** die physische und seelische Beanspruchung aus; ▶Belastbarkeit 1, ▶Besucherdruck, ▶Eingriff in Natur und Landschaft, ▶Grundbelastung, ▶Umweltverschmutzung, ▶Verkehrsbelastung).

1831 environmental stress degree [n] *ecol. envir.* (Intensity of strain on the environment; ▶disturbance); *s* **intensidad [f] de influencia perturbadora** (Grado de impacto negativo desde el punto de vista cualitativo; ▶influencia perturbadora); *syn.* grado [m] de carga; *f* **intensité [f] de pollution** (Intensité d'une ▶nuisance déterminée par la concentration, la teneur d'un polluant) ; *g* **Belastungsgrad** [m] **(1)** (Ausmaß einer ▶Beeinträchtigung, bestimmt durch die Belastungsstärke); *syn.* Belastungsintensität [f].

1832 environmental stress indicator [n] *biol. ecol. envir.* (Symptom of strain on the environment; e.g. ▶indicator plant, deteriorated building material, and discolo[u]red or pitted sculptures; ▶biological indicator); *s* **indicador** [m] **del nivel de contaminación** (Planta utilizada para demostrar el grado de contaminación debido a su sensibilidad a la misma; también puede utilizarse el término en el contexto del estado de conservación de edificios, es decir cuando éstos reflejan deterioro causado por la contaminación; ▶bioindicador, ▶planta indicadora); *f* **indicateur** [m] **(du degré) de pollution** (Facteur chimique [p. ex. les polluants], physique [p. ex. la dégradation des bâtiments], ou biologique [organisme végétal et animal] qui fait l'objet de mesure et rend compte de l'importance de la pollution d'un milieu ; il est un outils d'évaluation de la qualité de l'environnement ; ▶indicateur biologique, ▶plante indicatrice) ; *syn.* indicateur [m] de l'intensité de pollution ; *g* **Belastungs-**

indikator [m] (**1.** Merkmal, das eine Umweltbelastung anzeigt, z. B. Verwitterung von oder Farbveränderungen an Bau- oder Kunstwerken oder deren Teile. **2.** Organismus, der Luftverschmutzung oder organische Gewässerverunreinigung anzeigt; ▶Bioindikator, ▶Zeigerpflanze).

1833 environmental stress tolerance [n] *ecol. recr.* (Endurance under pressure from different levels of human use with required management, which can be sustained in a natural, semi-natural or man-made ecosystem without a consequent deterioration in its structure or function; ▶carrying capacity of landscape, ▶elasticity, ▶environmental stress, ▶grazing capacity, ▶recoverability from trampling); *syn.* ecological (buffer) capacity [n], ecological resistance [n] (LE 1986, 433); *s* **resiliencia [f] de un ecosistema** (Capacidad de absorción de impactos; ▶capacidad de pastoreo, ▶capacidad de usos [de un paisaje], ▶elasticidad de un ecosistema, ▶impacto ambiental, ▶resistencia al pisoteo); *syn.* capacidad [f] ecológica, capacidad [f] puffer; *f 1* **capacité [f] de charge d'un écosystème** *ecol.* (Faculté d'un écosystème naturel ou anthropique à combler le plus vite possible et sans perte totale ou partielle des éléments principaux le caractérisant tout vide provoqué par des agents naturels ou anthropogènes ; ▶capacité au piétinement, ▶capacité biotique d'un pâturage, ▶capacité d'accueil des sites, ▶stress environnemental, ▶résilience d'un écosystème) ; *syn.* élasticité [f] d'un écosystème, pouvoir [m] de résistance d'un écosystème, seuil [m] de tolérance d'un milieu ; *f 2* **capacité [f] de charge d'un site touristique** *recr.* ; *g* **Belastbarkeit [f] (1)** (Fähigkeit eines natürlichen, naturnahen, halbnatürlichen oder anthropogenen Ökosystems, natürliche und durch den Menschen bedingte Einflüsse auszugleichen, ohne Verlust oder dauerhafte Funktionsminderung seiner das System bestimmenden Bestandteile; ▶Aufnahmekapazität der Landschaft, ▶Belastbarkeit durch Tritt, ▶Belastung, ▶Funktionsfähigkeit eines Ökosystems, ▶Weidebelastbarkeit); *syn.* ökologische Pufferkapazität [f], ökologische Tragfähigkeit [f].

1834 environmental survey [n] *ecol. plan.* (Collection of data on climatic, air pollutional, pedologic, phytosociological, or groundwater conditions relating to the existing environment; ▶vegetation impact study); *s* **inventario [m] ambiental** (Descripción del medio ambiente que existe y en qué estado en un área determinada [= estado cero] en la que se piensa localizar una acción o proyecto; ▶procedimiento de comprobación de efectos); *f* **relevé [m] de l'état de l'environnement** (Analyse des éléments naturels, sol, eau, air, climat, faune et flore ; ▶procédure de suivi environnemental par relevé de végétation) ; *syn.* relevé [m] de l'état d'un milieu ; *g* **Bestandsaufnahme [f] des Umweltzustandes (in einem Raum)** (Je nach Fragestellung die Erfassung der aktuellen klimatischen, lufthygienischen, bodenkundlichen, pflanzen- und tierökologischen Gegebenheiten oder der Grundwasserverhältnisse eines Untersuchungsraumes; ▶Beweissicherungsverfahren durch Vegetationskartierung).

environmental susceptibility [n] *envir. land'man.* ▶environmental sensitivity.

1835 environmental tax [n] *adm. envir. leg. pol.* (Payment of a levy for an act affecting the environment, especially the causing of emissions. In Germany, the rate is assessed according to regulations contained in environmental legislation. **In U.S.**, instead of an **e. t.**, local or state governments are reducing other taxes levied on polluters who decrease their pollution levels; ▶air pollution fee); *s* **canon [m] ambiental** (Tasa a pagar obligatoriamente por actividad con impacto ambiental como p. ej. vertido de aguas depuradas en ríos o en el mar, emisiones contaminantes; ▶canon de contaminación atmosférica); *f 1* **taxe [f] environnementale** (Taxe perçue au titre de diverses législations pour des atteintes à l'environnement ; *en F. on distingue* **1.** la

▶taxe générale sur les activités polluantes [TGAP], **2.** la redevance au titre de la détérioration de la qualité de l'eau, **3.** les taxes sur l'énergie [Taxe Intérieure de Consommation sur le Gaz Naturel — TICGN, Taxe Intérieure de consommation sur les Produits Pétroliers, — TIPP] et la taxe sur le CO_2 mise en place en 2006 ; ▶écotaxe ; *f 2* **écotaxe [f]** (Type d'impôt écologique, qui taxe un produit commercialisé provoquant des dommages sur l'environnement, p. ex. l'écotaxe voiture [loi de finances pour 2007] pour les véhicules particuliers réceptionnés CE, sur le taux de CO2 émis dès lors qu'il est supérieur à 160g/km ; ▶taxe sur la pollution atmosphérique) ; *g* **Umweltabgabe [f]** (Aus umweltpolitischen Gründen zur Beeinflussung eines umweltgerechten Verhaltens [Verhaltenslenkung] orientierte Zahlung für umwelterhebliche Handlungen, insbesondere für die Verursachung von Emissionen nach Maßgabe der Bestimmungen einschlägiger Umweltgesetze; eine Abgabe ist als **U.** zu betrachten, wenn sich die als Besteuerungsgrundlage dienenden Eigenschaften eindeutig umweltschädigend auswirken. Eine Abgabe kann jedoch auch als **U.** betrachtet werden, wenn sie eine vielleicht weniger deutliche, doch klar feststellbare positive Umweltauswirkung hat. In der EU obliegt der Nachweis der erwarteten Umweltauswirkung einer Abgabe generell den Mitgliedsländern; cf. Mittlg. der Kommission 2001/C37/03 v. 03.02.2001; **U.n** sind z. B. die Abfallabgabe, naturschutzrechtliche Ausgleichsabgaben, die Ressourcennutzungsgebühr im Rahmen der Wasserentnahmeentgelte, das Zwangspfand auf Einwegverpackungen, Mineral- und Stromsteuer, die ▶Luftverschmutzungsabgabe); *syn.* Umweltlenkungsabgabe [f].

environmental tolerance [n] *ecol.* ▶ecological tolerance.

environment-friendly [adj] *envir. pol.* ▶environmentally safe.

1836 environment-friendly construction practices [npl] *constr.* (Building measures, which take good care of nature and make sustainable use of its resources, e.g. use of renewable energies; exploiting the insulation properties of a structure for the optimal usage of energy [energy-efficient building]; catchment of storm water in a system of infiltration swales and trench drains, thereby minimizing the burden on piped systems, rivers and streams and thus the environment. **In D.**, the term *passive house* is used as a quality standard for houses which are environment-friendly, because they are built with regard to the climate; **in CH.**, they are known as *Minergie (minimum energy)* houses); *s* **construcción [f] respetuosa al medio ambiente** (Obra nueva o de rehabilitación en la que se tienen en cuenta los problemas ambientales por medio de diferentes medidas entre las que se encuentran el uso de materiales renovables, de larga duración y no contaminantes, medidas de ahorro energético y de agua, de protección acústica, de instalación de sistemas de energías renovables [fotovoltaica, solar térmica, etc.], etc., además de reducir la cantidad de residuos de obra producidos); *syn.* construcción [f] ecológica; *f 1* **construction [f] respectueuse de l'environnement** (Ouvrage neuf ou en réhabilitation prenant en compte les préoccupations écologiques à travers différentes mesures qui, au départ se limitaient à la réduction du volume de déchets de chantier mis en décharge, le développement de l'utilisation de matériaux renouvelables ; on parle désormais de démarche Haute Qualité Environnementale [HQE] dans le cadre de la mise en œuvre du Système de Management Environnemental [SME] du maître d'ouvrage selon la série des normes internationales ISO 14 000 qui inclue entre autre l'intégration du bâtiment avec son environnement immédiat, des préoccupations d'efficacité énergétique, de durabilité des matériaux, de gestion de l'eau et des déchets, de confort hygrothermique et acoustique, de qualité sonore ou lumineuse, de qualité sanitaire, etc.) ; *f 2* **construction [f] écologique (et durable)** (Expression, certes à la mode mais peu satisfaisante, désignant un mode de construction qui intervient de manière la plus préventive possible dans le cycle de la nature et pour lequel le bilan énergétique constitue l'objectif principal ; **D.**, les constructions peuvent obtenir le label *Passivhaus* [consommation d'énergie de chauffage + ECS + ventilation inférieure à 30 kWh/m²/an], en **CH.** existe un label *Minergie Habitat Passif* [consommation d'énergie de chauffage + ECS + ventilation inférieure à 42 kWh/m²/an] ; **F.**, les constructions sont régies par la règlementation thermique RT 2000 qui n'impose pas de valeurs de consommation d'énergie finale mais des valeurs maximales de déperdition et il n'existe aucune norme basse énergie ou très basse énergie [habitat passif]) ; *g* **umweltverträgliche Bauweise [f]** (Art des Bauens durch schonendem Umgang mit der Natur und ihren Ressourcen, Einsatz erneuerbarer Energien, Ausnutzung der Speicherwirksamkeit der Baukonstruktion zur optimalen Energienutzung [energiesparendes Bauen], Versickerung des anfallenden Regenwassers in Mulden-/Rigolensysteme, so dass eine möglichst geringe Belastung der Umwelt erfolgt. **D.**, das Qualitätslabel *Passivhaus* zeigt eine umweltverträgliche, klimafreundliche Bauweise an; in **CH.** heißt dieses Gütezeichen *Minergie*. Ein Fehlname wäre *ökologische Bauweise*, da der Wissenschaftsbegriff „ökologisch" nur beschreibt, nicht aber wertend im Sinne von „naturnah" benutzt werden sollte); *syn.* naturverträgliche Bauweise [f], Bauweise [f] nach umweltrelevanten Kriterien.

environs [npl] *urb.* ►central city environs [US].

1837 epecophyte [n] *phyt.* (Species according to the ►degree of naturalization, which has become adapted to varying conditions with an enduring place in the ►existing vegetation, but not in the ►potential natural vegetation, and which would disappear with a cessation of human activity); *s* **epecófito [m]** (Planta naturalizada, que se desarrolla en las tierras de labor, en las proximidades de las habitaciones humanas, en los muros, etc., como dependiente de las actividades humanas; ►grado de naturalización, ►vegetación natural potencial, ►vegetación real); *f* **espèce [f] végétale acclimatée** (Espèce introduite dont l'installation et le maintien nécessitent des soins renouvelés de l'homme ; ►degré de naturalisation, ►végétation potentielle naturelle, ►végétation réelle) ; *g* **kulturabhängige Art [f]** (Nach dem ►Grad der Einbürgerung diejenige Pflanzenart, die einen festen Platz in der heutigen ►realen Vegetation hat, nicht aber in der potentiellen natürlichen; beim Aufhören menschlicher Tätigkeit würde sie verschwinden; ►potentielle natürliche Vegetation); *syn.* Epökophyt [m].

ephemeral plant [n] *phyt.* ►casual species.

ephemerophyte [n] *phyt.* ►casual species.

epicormic [n] [US] *arb. bot. hort.* ►water sprout.

epicormic branch [n] *arb.* ►epicormic shoot of stem.

epicormic growth [n] *arb.* ►epicormic shoot of stem.

epicormic shoot [n] *arb. bot. hort.* ►water sprout.

1838 epicormic shoot [n] **of stem** *arb.* (Branch which grows from a latent bud on a woody stem; cf. ARB 1983, 19; ►adventitious shoot, ►coppice shoot, ►water sprout); *syn.* epicormic branch/growth [n]; *s* **chupón [m] del tronco** (Ramificación vertical que surge en troncos, provista de yemas de leño; cf. PODA 1994; ►brote adventicio, ►brote de cepa, ►hijuelo); *f* **gourmand [m] de tige** (Branche apparaissant tardivement sur des tiges déjà âgées, à partir de bourgeon dormant situé sous l'écorce ; cf. IDF 1987, 164 ; ►gourmand, ►pousse adventive, ►rejet de taillis) ; *g* **Stammaustrieb [m]** (Zweig oder Ast, der sich aus einem schlafenden Auge am Stamm bildet; ►Adventivspross, ►Stockaustrieb 1, ►Wasserreis).

epigeal [adj] *ecol.* ►epigeous.

epigean [adj] *ecol.* ►epigeous.

1839 epigeous [adj] *ecol.* (Descriptive term applied to organisms which live on or close to the ground); *syn.* epigeal [adj], epigean [adj]; *s* **epígeo/a [adj]** (Referente a organismo que vive sobre o cerca del suelo); *syn.* terrícola [adj]; *f* **épigé, ée [adj]** (Vivant à la surface du sol) ; *g* **auf dem Boden lebend [loc]** *syn.* epigäisch [adj].

epilithic [adj] *bot. phyt.* ►rupicolous.

epilithic plant [n] *phyt.* ►rock plant.

epipotamal [n] *limn.* ►barbel zone [UK].

1840 epipotamon [n] *limn.* (Life community of the ►barbel zone [UK]); *s* **epipótamon [m]** (Biocenosis de la ►zona de barbo); *f* **épipotamon [m]** (Communauté animale de la ►zone à barbeau) ; *g* **Epipotamon [n]** (Fließgewässerlebensgemeinschaft der ►Barbenregion).

1841 epirhithron [n] *limn.* (Life community of the ►trout zone; ►potamon, ►rhithron); *s* **epiritrón [m]** (Biocenosis de la zona superior de la ►región salmonícola [►zona de trucha]; ►potamium, ►ritrón); *f* **épirhithron [m]** (Communauté animale de la zone supérieure de la ►région salmonicole [►zone à truite], de la source jusqu'au premier confluent ; ►potamon, ►rhithron) ; *syn.* crénon [m], zone [f] supérieure de la truite ; *g* **Epirhithron [n]** (Fließgewässerlebensgemeinschaft der obersten ►Bergbachregion [►Forellenregion]; ►Potamon, ►Rhithron).

equator-facing slope [n] [US] *geo. met.* ►south-facing slope.

equestrian riding [n] [US] *recr.* ►horseback riding [US].

equestrian sport [n] [UK] *recr.* ►horseback riding sport [US].

equestrian trail [n] *recr.* ►bridle path [US]/bridle way [UK].

equilibrium [n] *ecol.* ►dynamic species equilibrium, ►ecological equilibrium.

equilibrium [n]**, dynamic** *ecol.* ►ecological equilibrium.

equilibrium [n]**, flow** *ecol.* ►ecological equilibrium.

equilibrium [n]**, landscape** *ecol.* ►landscape ecosystem.

equipment [n]**, fall arrest** *constr. leg.* ►protection from falling, #2.

equipment [n]**, yard** *constr.* ►machinery stock.

ergasiophyte [n] *phyt.* ►cultivated plant (2).

1842 ericaceous plant [n] **(1)** *bot.* (belonging to the heath family *Ericaceae*); *s* **ericácea [f]** (Perteneciente a la familia de los brezos: *Ericaceae*); *f* **Éricacée [f]** (Plante appartenant à la famille des *Ericaceae*) ; *g* **Heidekrautgewächs [n]** (Pflanze, die zur Familie der *Ericaceae* gehört); *syn.* Ericacea [f], Ericacee [f].

ericaceous plant [n] **(2)** *hort.* ►peat garden plant [UK].

1843 erodability [n] *agr. constr. geo. pedol.* (Characteristics of soils which are susceptible to erosion, when such stabilizing surface material as vegetation is removed; ►erosion risk); *s* **erosionabilidad [f]** (Facilidad o capacidad con la que un determinado tipo de suelo se erosiona bajo unas condiciones específicas de pendiente, en comparación con otros suelos bajo las mismas condiciones; DINA 1987; ►peligro de erosión); *f* **vulnérabilité [f] à l'érosion** (Caractéristique des sols facilement attaqués par les agents érosifs par manque de couverture végétale ; ►risque d'érosion) ; *g* **Erodierbarkeit [f]** (Eigenschaft von Böden, die leicht abgeschwemmt werden können, wenn keine schützende Vegetationsdecke mehr besteht; ►Erosionsgefahr).

eroding bank [n] *geo.* ►undercut bank.

erosion [n] *geo.* ►beach erosion (1), ►beach erosion (2), ►extensive erosion, ►extreme soil erosion, ►flash erosion, ►gully erosion, ►natural erosion, ►rill erosion, ►sheet erosion, ►shoreline erosion, ►soil erosion, ►splash erosion, ►streambank erosion, ►stream erosion, ►vertical erosion, ►watercourse bed erosion, ►wind erosion.

erosion [n]**, bank** *geo.* ►streambank erosion.

erosion [n]**, channel** *geo.* ►scour.

erosion [n]**, coastal** *geo.* ►shoreline erosion [US]/shore line erosion [UK].

erosion [n] **[US], geologic** *geo.* ►natural erosion.

erosion [n] **[UK], mass** *geo.* ►mass slippage [US].

erosion [n] **[UK], raindrop** *geo. pedol.* ►splash erosion.

erosion [n]**, riverbank** *geo.* ►streambank erosion.

erosion [n]**, streambed** *geo.* ►watercourse bed erosion.

erosion [n]**, wide-scale** *geo.* ►extensive erosion.

erosion capacity [n] *geo.* ►erosion potential.

1844 erosion control [n] *constr. land'man.* (Measures to minimize the effects of ►natural erosion); *syn.* erosion protection [n]; *s* **protección** [f] **contra la erosión** (Medidas para reducir la ►erosión); *f* **lutte** [f] **contre l'érosion** (Mesures contre les ►érosions et l'envahissement des cours d'eau) ; *syn.* défense [f] du sol ; *g* **Erosionsschutz [m, o. Pl.]** (Maßnahmen zur Eindämmung der ►Erosion).

1845 erosion control facility [n] *constr. land'man.* (Temporary installation on construction sites such as ►silt fence, ►row of straw bales or ►sediment retention basin for trapping silt particles to prevent runoff into receiving streams or drains); *s* **instalación** [f] **de protección contra la erosión** (Medida temporal de control de la erosión durante obra de construcción como ►interceptor de fango, ►fila de pacas de paja o ►decantador de fango/lodos para evitar el vertido de material edáfico en la alcantarilla o en un curso receptor); *f* **équipement** [m] **de protection contre l'érosion** (Installations temporaires sur les chantiers, p. ex. ►clôture [de protection] contre les écoulements de boue, une ►rangée de bottes de paille, un ►bassin de décantation [des boues] qui évitent le transport de matériaux fins dans le réseau d'eau pluviale) ; *g* **Erosionsschutzeinrichtung [f]** (Temporäre Anlage auf Baustellen wie ►Schlammfangzaun, ►Strohballenreihe oder ►Schlammabsetzbecken, die das Abschwemmen von Boden in die ►Vorflut 2 verhindern).

1846 erosion damage [n] *geo. pedol.* (Destruction caused by soil erosion); *s* **daños** [mpl] **de erosión**; *f* **dommage** [m] **causé par l'érosion** ; *g* **Erosionsschaden [m]**.

1847 erosion gully [n] *geo. pedol.* (Channel or ravine resulting from erosion caused by rushing precipitation water during and immediately following heavy rains; the distinction between gully and rill is one of depth; ►erosion rill, ►gully erosion, ►natural erosion, ►rill erosion, ►stream erosion); *s* **cárcava [f]** (Barranca que van formando en la tierra las avenidas impetuosas de los torrentes, consiste en una profunda incisión en el terreno originada generalmente cuando existe una gran concentración de escorrentía en alguna zona determinada; cf. DINA 1987; *término específico para formaciones de arenisca o conglomerados sueltos* torre [f] o chimenea [f] de erosión; cf. GEO 1990; ►erosión, ►erosión en regueros, ►erosión en cárcavas, ►reguero); *syn.* carcavón [m], barranco [m] (1), cañada [f] (WMO 1974); *f 1* **ravin [m]** (Entaille étroite et importante sur un profil transversal en V de plusieurs hectomètres de long et jusqu'à 50 m de large, fixée ou non par la végétation ; DG 1984 ; ►érosion, ►ravinement 1, ►ravinement 2, ►rigole, ►ruissellement en filets) ; *f 2* **ravine [f]** (Ravin en voie d'évolution et non stabilisé

sur un versant de matériaux meubles, provoqué par les eaux de ruissellement ; le « ►ravineau » désigne une petite ravine) ; *syn. pour un groupement de ravines en pays méditerranéen* roubine [f] ; *g* **Erosionsgraben [m]** (...gräben [pl]; durch stoßweise linear fließendes Wasser entstandene, über 50 cm bis mehrere Meter tiefe Furche im geneigten Gelände — verursacht durch heftige Niederschläge; im Vergleich zu ►Erosionsrillen können **E.gräben** durch eine Bodenbearbeitung nicht mehr ausgeglichen werden; ►Erosion, ►Grabenerosion 1, ►Grabenerosion 2, Rillenerosion); *syn.* Runse [f].

erosion hazard [n] *geo. pedol. plan.* ►erosion risk.

1848 erosion potential [n] *geo.* (Amount of eroded soil that can be calculated using wind or water, soil-loss equations which measure the volume that may be dislodged and moved from its place of origin, depending on the slope of a terrain and its degree of vegetative cover; cf. RCG 1982); *syn.* erosion capacity [n]; *s* **capacidad** [f] **erosiva** (Cuantificación de la pérdida potencial de suelo sobre la base de fórmula empírica en la que se combina el índice de erosionabilidad pluvial, la erosionabilidad del suelo, la longitud de pendiente y su gradiente y otros factores relacionados con el tipo y la práctica de cultivo; cf. DINA 1987, 380); *syn.* potencia [f] erosiva; *f* **puissance** [f] **d'érosion** (PED 1983, 144 ; pouvoir que possède le vent et l'eau d'entraîner une certaine quantité de sol par unité de temps en fonction de l'inclinaison du terrain et de la nature du couvert végétal) ; *g* **Erosionskraft [f]** (Vermögen des Windes oder des Wassers eine gewisse Menge Boden pro Zeiteinheit abzutragen — u. a. abhängig von der Neigung eines Geländes und vom Grad der Vegetationsbestockung).

erosion protection [n] *constr. land'man.* ►erosion control.

1849 erosion rill [n] *geo. pedol.* (Shallow and narrow channel on a slope caused by flowing precipitation water; usually only several inches/centimeters deep; ►erosion gully, ►natural erosion, ►rill erosion); *s* **reguero** [m] **(2)** (Canal de escorrentía estrecho, bien diferenciado y relativamente grande; DINA 1987; ►cárcava, ►erosión, ►erosión en regueros); *syn.* surco [m] de erosión; *f* **rigole** [f] (Entaille stabilisée importante sur un profil transversal en V de plusieurs hectomètres de long et de profondeur centimétrique à décimétrique ; VOG 1979 ; ►érosion, ►ravin, ►ravine, ►ruissellement en filets) ; *g* **Erosionsrille [f]** (Durch linear fließendes Niederschlagswasser entstandene flache Furche am Hang; ►Erosion, ►Erosionsgraben, ►Rillenerosion); *syn.* Erosionsrinne [f].

1850 erosion risk [n] *geo. pedol. plan.* (Risk of soil being eroded when stabilizing vegetation is non-existent; ►erodability [of soil]); *syn.* erosion hazard [n], erosion susceptibility [n]; *s* **peligro** [m] **de erosión** (Riesgo de erosión del suelo cuando no está o está poco cubierto de vegetación; ►erosionabilidad); *syn.* riesgo [m] de erosión; *f* **risque** [m] **d'érosion** (Risque d'enlèvement de matière provoqué par les agents d'érosion [eau, vent] sur les sols et roches à faible ou dépourvus de couverture végétale ; ►vulnérabilité à l'érosion) ; *syn.* danger [m] d'érosion, vulnérabilité [f] à l'érosion ; *g* **Erosionsgefahr [f]** (Gefahr, dass bei vegetationslosen oder spärlich bestockten Flächen der Oberboden durch Wasser oder Wind verstärkt abgetragen wird; ►Erodierbarkeit); *syn.* Erosionsgefährdung [f], Erosionsanfälligkeit [f], Erosionslabilität [f].

erosion susceptibility [n] *geo. pedol. plan.* ►erosion risk.

1851 erratic block [n] *geo.* (erratics [pl]; rock fragment of any size which is carried by glacial ice and deposited at some distance from the outcrop from which it was derived; ►boulder, ►exotic block, ►perched boulder); *s* **bloque** [f] **errático** (Bloque de gran tamaño que ha sido transportado por un glaciar y posteriormente depositado sobre un sustrato de diferente

E

naturaleza a la de aquél del que el bloque procede; DGA 1986; ►bloque alóctono, ►pedrejón, ►pedrejón errático); *f* **bloc [m] erratique** (Rocher de dimension métrique et parfois décamétrique, d'origine étrangère aux roches sur lesquelles il repose et dans une position telle qu'il n'a pu être apporté que par un courant de glace ; DG 1984 ; ►bloc allochthonc, ►bloc [de rocher] de forme arrondie, ►bloc perché) ; *g* **Findling [m] (2)** (Großer, ortsfremder Felsblock, der meist durch Gletscher oder Inlandeismassen von seinem weit entfernten Ursprungsgebiet — in Mitteleuropa aus Skandinavien — zu seinem heutigen Fundort transportiert worden ist; ►allochthoner Block, ►Findling 2, ►Wanderblock); *syn.* erratischer Block [m] (KFL 1962), Irrblock [m] (WAG 2001, 184).

1852 escalation clause [n] *constr. contr.* (Contractual agreement between parties to a contract not to alter the ratio negotiated between services offered and the bid prices [US]/tendered prices [UK], for a specified period, because of general price changes or inflation rates [cost increases in wages and materials] to the disadvantage of one of the parties); *syn.* price-revision clause [n]; *s* **cláusula [f] de precio fijo** (Acuerdo entre las dos partes de un contrato de no cambiar los precios acordados a pesar de que hayan cambios generales de precios o de la inflación); *f 1* **clause [f] d'actualisation des prix** (Les prix fermes sont actualisés dans les conditions prévues par la réglementation en vigueur) ; *f 2* **clause [f] de révision des prix** (Les prix révisables sont révisés dans les conditions prévues par la réglementation en vigueur. La variation dans les prix concerne l'actualisation et la révision des prix et représente la possibilité pour l'entrepreneur de répercuter sur les prix du marché les variations des éléments constitutifs du coût des travaux ; sauf si le marché les exclut ou s'il ne contient pas les éléments nécessaires, l'actualisation ou la révision des prix se fait sur la base des conditions économiques du mois de référence [mois zéro], en appliquant des coefficients établis à partir de l'index de référence fixés par le marché pour le lot concerné ; la valeur initiale de l'index à prendre en compte est celle du mois d'établissement des prix ; décret n° 76-87 du 21 janvier 1976 — C.C.A.G.) ; *g* **Gleitklausel [f]** (1. OB zu Lohn- und Stoffpreisklausel. 2. Vereinbarung zwischen Vertragsparteien mit dem Ziel, das vereinbarte Verhältnis der [Bau]leistungen zum angebotenen Preis durch allgemeine Preisveränderungen oder Inflationsraten [Kostenerhöhungen bei Löhnen und Baustoffen] nicht zum Nachteil einer Vertragspartei zu verändern; eine derartige Klausel muss ausdrücklich im Bauvertrag festgelegt werden, da es im allgemeinen Werkvertragsrecht und auch in der VOB keine Bestimmung über einen Preisvorbehalt gibt); *syn.* Gleitpreisklausel [f], Preisgleitklausel [f], Wertsicherungsklausel [f].

1853 escarpment [n] *geo.* (Steep face of a ►cuesta; a line of cliffs produced by faulting or erosion is called 'scarp'; ►break of slope, ►scarp slope); *s* **escarpe [f] de falla** (Desnivelación topográfica entre dos bloques fallados; DGA 1986; ►cuesta, ►frente de cuesta, ►pequeño declive); *syn.* escarpadura [f]; *f* **escarpement [m]** (Versant de forte pente [abrupt] souvent formé par une faille [escarpement de faille]) ; ►cuesta, ►rupture de pente, ►versant anaclinal ; *g* **Geländeabbruch [m]** (Meist durch Verwerfung bedingter, plötzlicher großer Abfall eines Geländes; ►Geländekante, ►Stufenhang); *syn.* geomorphologische Großform ► Schichtstufe [f].

1854 espalier [n] *hort.* (Wooden or wire trellis on which trees or shrubs can be trained in a vertical plane, either free-standing or against a wall; ARB 1983; *specific terms* ►double-U espalier, ►fan espalier, ►palmette espalier, ►rose espalier, ►espaliered fruit tree); *s* **espaldera [f]** (Soporte de metal, alambre, plástico o madera, libre o fijado a muros o fachadas para el cultivo de ►frutales en espaldera o que sirve de sujección a plantas

trepadoras; *término específico* ►espaldera de rosas; ►palmeta en abanico, ►palmeta en U doble, ►palmeta regular [de ramas horizontales/oblicuas]); *f* **treillage [m]** (Armature de lattes et de fil de fer fixée le long d'un mur ou tendue verticalement dans le sol sur laquelle sont palissés les ►arbres fruitiers taillés en espalier ; ADT, 1988, 43 ; ►espalier de rosiers, ►palmette à branches horizontales/obliques, ►palmette en éventail, ►palmette Verrier) ; *g* **Spalier [n] (1)** (Freistehendes oder an Mauern befestigtes Gerüst aus Holz-, Plastik- oder Metalllatten oder Pfosten mit Drahtbespannung für ►Formobstbäume, Weinreben oder Ziergehölze; *UBe* ►Fächerspalier, freistehendes Spalier, ►Rosenspalier, ►Spalier mit waagerechten/schrägen Ästen, Wandspalier; ► Verrierpalmette); *syn.* Spaliergerüst [n].

1855 espalier [vb] *hort.* (To train ►espaliered fruit trees or shrubs in a vertical plane on a wire or wooden trellis, either freestanding or against a wall; ►espalier row); *s* **espaldar [vb] árbol frutal** (Podar y sujetar un frutal a su tutor para darle una forma obligada; ►frutal en espaldera, ►hilera de plantas en espaldera); *f 1* **palisser [vb] des arbres fruitiers** (Fixation des branches d'arbres fruitiers à un mur ou à un autre support ; palissage [m] des arbres fruitiers ; ►arbre fruitier taillé en espalier, ►contre-espalier, ►espalier) ; *f 2* **accolage [m]** (Fixation de jeunes pousses, p. ex. d'arbres, de vigne, sur un support ; accoler [vb]) ; *g* **Formobstbaum befestigen [vb] und schneiden [vb]** (Erziehen eines Obstgehölzes zum gewünschten ►Formobstbaum; ►Spalier 2).

espalier. [n] [US]**, candelabra** *hort.* ►double-U espalier.

1856 espaliered fruit tree [n] *hort.* (Fruit tree trained to grow flat against a wall or trellis in a desired shape or form; ►cordon 1, ►double-U espalier, ►fan espalier, ►palmette espalier); *s* **frutal [m] en espaldera** (En fruticultura, árbol frutal con forma de cultivo obligada, sea aplanada o apoyada en muro u otro tipo de sujeción de madera o metal. Puede ser de diferentes formas: ►cordón, épsilon, ►palmeta en abanico, ►palmeta en U doble, ►palmeta regular [de ramas oblicuas/horizontales], piramidal); *f* **arbre [m] fruitier taillé en espalier** (Pratique culturale sur les arbres fruitiers plantés et palissés le long d'un mur, de treillis en bois, de barreaux en métal ou du fil de fer tendu pour obtenir des formes telles que la forme pyramidale [fuseau, quenouille, gobelet], le ►cordon, la ►palmette à branches horizontales/oblique, la ►palmette en éventail, la ►palmette en U double, la palmette en U simple et la ►palmette Verrier; ADT 1988, 42-60) ; *syn.* arbre [m] fruitier palissé (LA 1981, 533) ; *g* **Formobstbaum [m]** (*Obstbau* auf Zwergunterlage veredelte Obstsorte, die an Mauern oder an frei stehenden, senkrechten Gerüstwänden aus Holz, Metallstäben oder Spanndrähten zu folgenden Formen erzogen werden: ►Fächerspalier, Pyramidenform, ►Schnurbaum, ►Spalier mit waagerechten/schrägen Ästen, U-Form, ►Verrierpalmette); *syn.* Spalierobstbaum [m].

1857 espalier row [n] *hort.* (One or more rows of low, fruiting or flowering trees, grape vines, roses, raspberries, etc. trained to grow on an espalier. The fruits are called espalier-trained fruits); *s* **hilera [f] de plantas en espaldera** (Una o más filas de árboles frutales o decorativos, rosales o vides cultivadas en forma obligada utilizando estructura de apoyo; *términos específicos* emparrado [m], espaldera [f] de frutales, espaldera [f] de rosas); *f 1* **plante [f] palissée** (*Terme générique* végétaux dont les branches sont attachées à un treillage) ; *f 2* **espalier [m]** (Arbres fruitiers ou d'agrément, plants de vigne, rosiers, framboisiers palissés le long d'un mur et disposés sur un seul plan) ; *f 3* **contre-espalier [m]** (Une suite d'arbres fruitiers ou d'agrément, de plants de vigne, de rosiers, de framboisiers, etc. plantés en plate-bande et non plus contre un mur. Les arbres conduits sur une armature de lattes et de fil de fer tendue verticalement sont taillés sévèrement de façon à leur donner une forme plate ; ADT 1988, 43) ; *g* **Spalier [n] (2)** (Anlage von an Spaliergerüsten befestigten, niedrigstämmigen Zier- oder Obstbäumen, Weinreben, Rosen, Himbeeren etc.; *UBe* Obstspalier, Rosenspalier, Wandspalier).

1858 espalier shape [n] *hort.* (Narrow vertical form of a trained woody plant on a trellis or fence; e.g. ►cordon 1, ►double-U espalier, ►fan espalier, ►palmette espalier, ►rose espalier, U-shape e.; ►espaliered fruit tree); *s* **forma [f] de espaldera** (Talla vertical y estrecha de leñosa obligada, p. ej. ►cordón, ►espaldera de rosas, ►palmeta en abanico, ►palmeta en U doble, ►palmeta regular [de ramas oblicuas/horizontales]; ►frutal en espaldera); *f* **forme [f] palissée** (Taille spéciale de végétaux ligneux sous les formes suivantes : ►cordon, ►espalier de rosiers, forme pyramidale, ►palmette à branches horizontales/oblique, ►palmette en éventail, ►palmette en U double, palmette en U simple, ►palmette Verrier ; ►arbre fruitier en espalier) ; *g* **Spalierform [f]** (Unterschiedliche, in aufrechter Scheibenform geschnittene Gehölze wie z. B. ►Fächerspalier, ►Rosenspalier, ►Spalier mit waagerechten/schrägen Ästen, ►Schnurbaum, U-Form, ►Verrierpalmette; ►Formobstbaum).

esplanade [n] *urb.* ►waterfront promenade.

1859 established industrial development [n] *urb.* *s* **industria [f] existente** (en una zona); *f* **établissement [m] industriel** (Installation industrielle existante) ; *syn.* équipement [m] industriel ; *g* **vorhandene Industrieansiedlung [f]** (Bestehende Industrieanlage; im Vergleich zur ►geplanten Industrieansiedlung die realisierte **I.**).

establishing [n] **a finished grade** *constr.* ►fine grading.

1860 establishment [n] *plan. urb.* (**1.** The act of bringing into existence; e.g. of industrial firms in an area. **2.** Result of 1.); *s* **establecimiento [m]** (Proceso y resultado de instalación p. ej. de nuevas industrias en una zona); *syn.* instalación [f]; *f* **implantation [f]** (par exemple d'industries nouvelles) ; *g* **Ansiedelung [f] (von Betrieben)** (**1.** Der Vorgang des Ansiedelns von z. B. Industriebetrieben; [sich] ansiedeln [vb], sich niederlassen. **2.** Ergebnis von 1.); *syn. o. V.* Ansiedlung [f] (von Betrieben), *syn. zu* 2. Niederlassung [f].

1861 establishment maintenance [n] *constr.* (Maintenance operations carried out during the contractual maintenance period of usually one year by the landscape contractor after practical completion of the planting work [US]/planting works [UK] until the end of the ►defects liability period; ►issue of a certificate of compliance [US]/issue of a certificate of practical completion [UK], ►separate contract maintenance, ►project development maintenance [US]/pre-handover maintenance [UK]); *s* **mantenimiento [m] inicial (±)** (Cuidados posteriores a la ►aceptación de obra de plantación para asegurar el desarrollo típico de la vegetación hasta el final del ►plazo de garantía; ►trabajos regulares de mantenimiento, ►trabajos de mantenimiento pre-entrega); *syn.* trabajos [mpl] de mantenimiento inicial; *f* **travaux [mpl] d'entretien pendant la période de garantie** (Travaux de confortement effectués après la ►réception des travaux et pendant le ►délai de garantie afin d'assurer la consistance des aménagements paysagers, des aires de sports et de loisirs, le développement conforme aux capacités éco-physiologiques des plantations ; ►travaux d'entretien courant, ►entretien [d'espaces verts] jusqu'à la réception des travaux) ; *g* **Entwicklungspflege [f, o. Pl.]** (**1.** Pflegemaßnahmen nach ►Abnahme der vegetationstechnischen Arbeiten zur Sicherung einer bestandstypischen Weiterentwicklung bis zum Ende der ►Gewährleistungsfrist; die **E.** schließt nach DIN 19 916, DIN 18 917 resp. DIN 18 918 an die ►Fertigstellungspflege an. Die Dauer der **E.** reicht bis zum Flächenschluss des Vegetations-

bestandes. Im UK-Englischen sind *establishment maintenance* und *contract maintenance* sich überschneidende Begriffe, wobei sich *establishment m.* mehr auf den Zweck der Pflege und *contract m.* sich auf die Vertragsbasis bezieht. **2.** *adm. conserv.* Summe aller Maßnahmen zur nachhaltigen Lenkung und Förderung der Funktionsfähigkeit einer Garten- und Parkanlage oder einer Biotopfläche mit ihren angestrebten oder weiterzuentwickelnden bestandstypischen Vegetationsbildern; diese über 1. hinaus reichenden Maßnahmen sind ggf. in einem Pflegeplan darzustellen. Der Zeitraum für die **E.** ergibt sich aus den festgesetzten Funktionszielen, Pflegemaßnahmen und insbesondere aus den Entwicklungszeiten und Pflegeerfordernissen; ►Erhaltungspflege).

establishment [n] **of final elevation** [UK] *constr.* ►establishment of finished elevation.

1862 establishment [n] **of finished elevation** *constr.* (General term used to cover **1.** construction of final grade in earthworks; **2.** construction, e.g. of a wall to a planned finished level, often notated on a drawing as TW [top of wall]; ►subbase grade preparation [US]/grading of formation level [UK]); *syn.* establishment [n] of final elevation [also UK]; *s 1* **relleno [m] a subrasante** (Relleno para establecer el nivel de superficie necesitado; ►excavar una [sub]rasante); *s 2* **ejecución [f] de obra hasta la altura final** (Construcción de un muro o un edificio hasta su altura definitiva); *f 1* **mise [f] à niveau définitif** (Exécution du règlement définitif ; ►dressement du fond de forme) ; *syn.* nivellement [m] définitif ; *f 2* **exécution [f] au niveau fini** *syn.* réalisation [f] (d'un ouvrage) à hauteur finie ; *g 1* **Einbau [m, o. Pl.] auf fertige Höhe** (*Erdbau* Herstellen eines Feinplanums, ►Herstellen des Erdplanums); *g 2* **Errichten [n, o. Pl.] (einer Mauer) auf fertige Höhe** *syn.* Hochziehen [n, o. Pl.] (einer Mauer) auf geplante Höhe.

1863 establishment [n] **of grass/lawn areas** *constr.* (Seeding, sodding, plugging or sprigging to establish a green sward, in arid regions often by ►stolonizing); *s* **establecimiento [m] de superficies de césped** (►establecimiento de césped con estolones); *f* **engazonnement [m] d'une pelouse** (►engazonnement par jet de stolons de graminées) ; *syn.* gazonnage [m] d'une pelouse, établissement [m] d'une pelouse ; *g* **Herstellung [f] von Rasenflächen** (Anlegen einer Pflanzendecke mit Gräsern durch Ansaat, Verlegen von Rollrasen, in ariden und semiariden Gebieten durch Auspflanzen von Grasstolonen etc., so dass eine fest verwurzelte Grasnarbe entsteht; ►Begrünung mit Grasstolonen); *syn.* Rasenflächenherstellung [f].

establishment [n] **of industries, planned** *plan. urb.* ►planned industrial development.

1864 establishment [n] **of total construction cost** *contr.* (Recalculation of the actual total construction cost at the issue of the final certificate of completion: the basis for the final account and project documentation; ►calculation of cost); *s* **control [m] de costos** (Nuevo ►cálculo de costos de construcción reales al elaborar el acta de aceptación de obra como base para la liquidación final y la documentación del proyecto); *syn.* control [m] de costes [Es]; *f* **constatation [f] des coûts** (Établissement du décompte général à la fin des travaux ; *résultat* coûts constatés ; ►estimation financière des travaux) ; *g* **Kostenfeststellung [f]** (Nachweis der tatsächlich entstandenen Kosten; Voraussetzung für Vergleiche und Dokumentation nach DIN 276 in der Fassung 12/2008 oder nach dem wohnungsrechtlichen Berechnungsrecht; *Ergebnis* festgestellte Kosten, Aufstellung der festgestellten Kosten; ►Kostenermittlung); *syn.* Feststellung [f] der Kosten.

1865 establishment [n] **of vegetation** *constr.* (**1.** Planting or seeding on bare ground to control erosion, or for land-

E

scaping of buildings, including their façades and roofs, and of structures. **2.** Planting measures taken to encourage plant cover on recently disturbed soil in either new construction or reclamation circumstances; ►courtyard planting, ►establishment of vegetation on bare rock, ►façade planting, ►hydroseeding, ►intermediate planting, ►landscaping of built-up areas, ►planting of a streambank, ►pre-construction planting, ►roof planting, ►spoil reclamation [US]/tip reclamation [UK], ►vegetation establishment on entisols [US]/vegetation establishment on immature soil [UK]); *s* **establecimiento [m] de vegetación** (Creación de cobertura vegetal en superficies desnudas o en edificios; ►establecimiento de vegetación en suelo mineral bruto, ►plantación anticipada, ►plantación de escombreras, ►plantación de fachadas, ►plantación de patios, ►plantación intermedia, ►revestimiento vegetal de tejados, ►riego de semillas por emulsión, ►tratamiento paisajístico 1, ►vegetalización [artificial] de rocas, ►vegetalización de riberas); *f* **végétalisation [f]** (Création d'un couvert végétal sur une surface nue ; ►aménagement de terrasses-jardins, ►aménagement végétal d'une cour, ►colonisation des rochers par les végétaux, ►ensemencement intermédiaire d'engrais vert, ►préverdissement, ►procédé d'enherbement par projection hydraulique, ►traitement paysager, ►végétalisation des berges, ►végétalisation des façades, ►végétalisation des toitures, ►végétalisation de sol brut, ►végétalisation de terril ; verduriser [vb] (DUV 1984, 311) ; *syn.* verdissement [m] (TJ 1988, 128) ; *g* **Begrünung [f]** (Schaffung einer Pflanzendecke auf vegetationslosen Flächen oder die Begrünung von Bauwerken; ►Anspritzverfahren, ►Begrünung vor Baubeginn, ►Dachbegrünung, ►Durchgrünung von Siedlungsgebieten, ►Fassadenbegrünung, ►Felsbegrünung, ►Haldenbegrünung, ►Hofbegrünung, ►Rohbodenbegrünung, ►Uferbegrünung, ►Zwischenbegrünung).

1866 establishment [n] of vegetation on bare rock *constr.* (Natural colonization, or planting to protect against erosion on rock faces or rocky slopes, often using hydro-seeding methods); *s 1* **colonización [f] de rocas** (Proceso natural); *s 2* **vegetalización [f] (artificial) de roca rasa** *syn.* establecimiento [m] de vegetación en roca rasa; *f 1* **colonisation [f] des rochers par les végétaux** (Processus naturel) ; *f 2* **végétalisation [f] des rochers** (Processus artificiel de protection par la végétation des parois ou des talus rocheux) ; *g* **Felsbegrünung [f]** (Natürliche oder durch Sicherungsmaßnahmen beschleunigte Pflanzenansied[e]lung an Felswänden oder Felshängen).

estate [n] [UK], housing *urb.* ►housing subdivision [US].

estate [n] [UK], industrial *plan. urb.* ►industrial park [US], ►location of commercial facilities and light industries.

1867 estate garden [n] [US] *gard.* (Much larger ►ornamental garden than a ►private garden, often with a park-like character, usually situated on the periphery of a city or in the countryside; *syn.* villa garden [n] [UK]; *s* **jardín [m] de villa** (►Jardín privado mucho más grande que los de las casas unifamiliares generalmente con parte ornamental y con parte tipo jardín inglés, ►jardín ornamental); *f* **jardin [m] de banlieue** (Par comparaison avec le ►jardin particulier, le ►jardin privatif et le ►jardin privé, ►jardin d'agrément à usage privatif de superficie plus importante et souvent à caractère paysager, situé en général en limite d'agglomération urbaine) ; *g* **Villengarten [m]** (Im Vergleich zum ►Hausgarten ein meist wesentlich größerer ►Ziergarten mit oft parkartigem Charakter, meist am Stadtrand liegend).

estate parcel [n] [US], real *surv.* ►real property parcel [US]/plot [UK] (2).

esthetics [n] *land'man. landsc. recr.* ►landscape aesthetics.

1868 estimated construction costs [npl] *constr. contr.* (►Preliminary cost estimates that may include approximate materials, equipment, and labor costs expressed as a lump sum, a unit cost or a square area cost; i.e. cost per square meter); *s* **costo [m] estimado de construcción/obra** (►estimación aproximada de costos); *syn.* coste [m] estimado de construcción/obra [Es]; *f* **coût [m] estimatif des travaux** (►estimation sommaire du coût des travaux) ; *g* **geschätzte Baukosten [pl]** (Überschlägig ermittelte Kosten eines Bauprojektes, meist auf der Grundlage von Erfahrungswerten, z. B. €/m² Freianlage, € je km Verkehrsinfrastruktur; ►Kostenschätzung).

estimated volume [n] *contr.* ►increase in estimated volume, ►reduction in estimated volume.

estimate [n] of earthworks quantities *constr. eng.* ►rough estimate of earthworks quantities.

1869 estimate [n] of probable construction cost *contr.* (Ascertainment of approximate total cost used as a prerequisite in the decision whether a project should be executed as planned and also as a basis for the necessary financing); *s* **estimación [f] del costo** (Cálculo aproximado de los costos totales de un proyecto como prerequisito para la toma de decisión si éste se puede realizar o no); *syn.* estimación [f] del coste [Es]; *f 1* **évaluation [f] détaillée du coût des travaux** (Évaluation des dépenses afférentes à l'exécution des ouvrages fondée sur les avant-métrés, établie pendant la phase d'Avant-Projet Détaillé et tenant comte des particularités des ouvrages ; cette évaluation devra être cohérente avec le coût d'objectif ; HAC 1989, 34) ; *syn.* estimation [f] du coût prévisionnel des travaux, *pour l'architecte* estimation [f] globale du coût des travaux ; *f 2* **estimation [f] détaillée des dépenses** (F., évaluation des dépenses relatives à l'exécution des ouvrages contenue dans les spécifications techniques détaillées [S.T.D.] établie pendant la phase du Projet s'appuyant sur le devis quantitatif et fournie au maître d'ouvrage dans le cadre d'une mission normalisée permettant au maître d'ouvrage d'arrêter le coût prévisionnel de la solution d'ensemble et d'évaluer les coûts d'exploitation et de maintenance ; HAC 1989, 34 ; arrêté du 21 décembre 1993, annexe III ; D., évaluation établie suivant la norme DIN 276) ; *g* **Kostenberechnung [f]** (Ermittlung der angenäherten Gesamtkosten im Rahmen der Entwurfsplanung als Voraussetzung für die Entscheidung, ob ein Bauvorhaben, wie geplant, durchgeführt werden soll und als Grundlage für die erforderliche Finanzierung; die Berechnung erfolgt nach DIN 276 oder nach dem wohnungsrechtlichen Berechnungsrecht).

1870 estimate [n] of quantities *constr. plan.* (Rough ►calculation of quantities for construction used as a basis for a ►preliminary cost estimate); *s* **estimación [f] preliminar de cantidades** (Cálculo a grosso modo de las cantidades necesarias para una obra para la ►estimación aproximada de costos; ►cálculo de cantidades); *syn.* cálculo [m] preliminar de cantidades, cálculo [m] aproximativo; *f* **avant-métré [m] sommaire** (Évaluation sommaire des travaux ou fournitures en vue de l'établissement de l'►estimation sommaire du coût des travaux ; ►avant-métré) ; *g* **Massenüberschlag [m]** (Überschlägige Erfassung der Mengen an Bauleistungen und Lieferungen für eine ►Kostenschätzung; ►Massenberechnung); *syn.* überschlägige Massenermittlung [f].

1871 estuary [n] *geo.* (Inlet or arm of the sea, especially the wide mouth of a river, where the tide meets the current); *s* **estuario [m]** (Brazo de mar o desembocadura fluvial ensanchada donde se produce un contacto entre agua dulce y agua salada y mareas evidentes. En estas condiciones se establece una dinámica entre las dos masas de agua y las mareas dan lugar a condiciones ambientales muy específicas que producen fenó-

E

menos sedimentarios característicos; DINA 1987); *f* **estuaire [m]** (Embouchure d'un fleuve en général en forme d'entonnoir) ; *g* **Ästuar [n]** (Trichterförmige Flussmündung); *syn.* Trichtermündung [f], Mündungstrichter [m].

1872 etiolation [n] *bot. hort.* (Pale-green colo[u]ring, yellowing or whitening, and drawn, rank look of plants or their structurally weak parts due to shortage of natural light); *s* **ahilamiento [m]** (Anomalía del crecimiento en las plantas, que se han desarrollado en la oscuridad, caracterizada, en los tallos, por el alargamiento de los entrenudos, y la decoloración por falta de clorofila. El término «etiolamiento» es un galicismo; cf. DB 1985); *syn.* etiolamiento [m]; *f* **étiolement [m]** (État des plantes, insuffisamment éclairées, et qui de se fait, n'acquièrent ni la couleur, ni la robustesse, ni la brièveté des entre-nœuds, des sujets correctement traités ; DIB 1988) ; *g* **Etiolierung [f]** (Durch Lichtmangel verursachtes schnelles Wachsen von Pflanzen oder deren Teile. Das Ergebnis sind blass-grüne, gelbliche oder weißliche, hochgeschossene, weiche [geile] Triebe und dünnere Blätter. Schon bei kurzzeitiger Belichtung etiolierter Pflanzen setzt die Deetiolierung ein [Photomorphogenese]; cf. LB 2002); *syn.* Etiolement [n] (LB 2002), Geilwuchs [m], Vergeilen [n, o. Pl.].

European Community [n] *adm. plan.* ▶regional funds of the European Community (EC).

1873 European Diploma [n] *pol.* (Award instituted in 1965 by the Council of Europe in recognition of important areas of landscape, usually nature reserves, national parks, etc.; **in U.S.**, the federal government maintains a National Register of Natural Areas, and many states have equivalent registers or program[me]s); *s* **diploma [m] europeo** (Distinción creada por el Consejo de Europa en 1965 para galardonar áreas naturales importantes, que suelen ser espacios naturales protegidos, parques nacionales o parques naturales); *f* **diplôme [m] européen (des espaces protégés)** (Distinction instituée par le Conseil de l'Europe en 1965 et attribuée à tout ou partie de territoires présentant des qualités remarquables, en général à des réserves naturelles, parcs nationaux ou parcs naturels) ; *g* **Europadiplom [n]** (Vom Europarat seit 1965 verliehene Auszeichnung für bedeutende einmalige Landschaften oder Teile davon, meist Naturschutzgebiete, Nationalparke oder Naturparke).

1874 European Diploma Area [n] *conserv. pol.* (Area of landscape which has been awarded the ▶European Diploma); *s* **área [f] de diploma europeo** (Espacio natural protegido de importancia europea dotado de un ▶diploma europeo); *f* **zone [f] d'un diplôme européen d'espaces protégés** (Territoire protégé d'importance européenne qui en raison de qualités remarquables du point de vue scientifique, culturel ou esthétique s'est vu attribué un ▶diplôme européen [des espaces protégés] ; on parle d'une **zone diplômée**) ; *g* **Europadiplomgebiet [n]** (Schutzgebiet von europäischer Bedeutung, das mit einem ▶Europadiplom ausgezeichnet wurde).

1875 European Information Centre [n] for Nature Conservation *conserv. pol.* (Organization set up in 1967 by the Council of Europe to further nature conservation in the member states); *s* **Centro [m] Europeo de Información para la Conservación de la Naturaleza y el Medio Ambiente** (Organismo del Consejo de Europa creado en 1967 con el fin de promover la conciencia sobre los problemas ambientales en los países miembros; CIMA 1978, 875); *f* **Centre [m] européen d'information pour la conservation de la nature** (Organisme du Conseil de l'Europe créé en 1967 pour la promotion de la protection de la nature dans les états membres) ; *g* **Europäische Informationszentrale [f] für Naturschutz** (Seit 1967 beste-

hende Einrichtung des Europarates zur Förderung des Naturschutzes in den Mitglied[s]staaten).

1876 European Landscape Convention [n] *adm. conserv. leg. pol.* (Signed in Florence October 20, 2000; the aims of this Convention are to promote landscape protection, management and planning, and to organize European co-operation on landscape issues; each Party undertakes: **a]** to recognise landscapes in law as an essential component of people's surroundings, an expression of the diversity of their shared cultural and natural heritage, and a foundation of their identity; **b]** to establish and implement landscape policies aimed at landscape protection, management and planning through the adoption of the specific measures; **c]** to establish procedures for the participation of the general public, local and regional authorities, and other parties with an interest in the definition and implementation of the landscape policies mentioned in paragraph b above; **d]** to integrate landscape into its regional and town planning policies and in its cultural, environmental, agricultural, social and economic policies, as well as in any other policies with possible direct or indirect impact on landscape); *s* **Convenio [m] Europeo del Paisaje** (Convención firmada en Florencia el 20.10.2000 que tiene por objetivo promover la protección, gestión y ordenación de los paisajes, así como organizar la cooperación europea en ese campo; cada Parte se compromete a: **a]** reconocer jurídicamente los paisajes como elemento fundamental del entorno humano, expresión de la diversidad de su patrimonio común cultural y natural y como fundamento de su identidad; **b]** definir y aplicar en materia de paisajes políticas destinadas a la protección, gestión y ordenación del paisaje mediante la adopción de las medidas específicas; **c]** establecer procedimientos para la participación del público, las autoridades locales y regionales y otras partes interesadas en la formulación y aplicación de las políticas en materia de paisaje mencionadas en la anterior letra b; **d]** integrar el paisaje en las políticas de ordenación territorial y urbanística y en sus políticas en materia cultural, medioambiental, agrícola, social y económica, así como en cualesquiera otras políticas que puedan tener un impacto directo o indirecto sobre el paisaje); *syn.* Convención [f] Europea del Paisaje; *f* **Convention [f] européenne du paysage** (Convention signée à Florence le 20.10.2000 ayant pour objet de promouvoir la protection, la gestion et l'aménagement des paysages, et d'organiser la coopération européenne dans ce domaine ; chaque Partie s'engage : **a]** à reconnaître juridiquement le paysage en tant que composante essentielle du cadre de vie des populations, expression de la diversité de leur patrimoine commun culturel et naturel, et fondement de leur identité ; **b]** à définir et à mettre en œuvre des politiques du paysage visant la protection, la gestion et l'aménagement des paysages par l'adoption des mesures particulières ; **c]** à mettre en place des procédures de participation du public, des autorités locales et régionales, et des autres acteurs concernés par la conception et la réalisation des politiques du paysage mentionnées à l'alinéa b ci-dessus ; **d]** à intégrer le paysage dans les politiques d'aménagement du territoire, d'urbanisme et dans les politiques culturelle, environnementale, agricole, sociale et économique, ainsi que dans les autres politiques pouvant avoir un effet direct ou indirect sur le paysage) ; *g* **Europäisches Landschaftsübereinkommen [n]** (In Florenz am 20.10.2000 unterzeichnetes Übereinkommen, das das Ziel verfolgt, den Schutz, die Pflege und die Gestaltung der Landschaft zu fördern und die europäische Zusammenarbeit in Landschaftsfragen zu organisieren; jede Vertragspartei verpflichtet sich, **a]** Landschaften als wesentlichen Bestandteil des Lebensraums der Menschen, als Ausdruck der Vielfalt ihres gemeinsamen Kultur- und Naturerbes und als Grundstein ihrer Identität rechtlich anzuerkennen; **b]** durch Ergreifen von spezifischen Maßnahmen eine auf den Schutz, die Pflege und die

Gestaltung der Landschaft ausgerichtete Landschaftspolitik zu erarbeiten und umzusetzen; c] Verfahren für die Beteiligung der Öffentlichkeit, der Kommunal- und Regionalbehörden und anderer Parteien, die ein Interesse an der Festlegung und Umsetzung der unter Buchstabe b genannten Landschaftspolitik haben, einzuführen; d] die Landschaft in ihre Regional- und Stadtplanungspolitik und in ihre Kultur-, Umwelt-, Agrar-, Sozial- und Wirtschaftspolitik sowie in andere, sich möglicherweise unmittelbar oder mittelbar auf die Landschaft auswirkende politische Tätigkeiten und Maßnahmen aufzunehmen).

1877 European Union Directive [n] *adm. leg.* *s* **Directiva** [f] **de las Comunidades Europeas (CC.EE.);** *f* **Directive** [f] **de la Communauté Européenne (C.E.E.)** ; *g* **EG-Richtlinie** [f] *syn.* Richtlinie [f] der Europäischen Gemeinschaft.

1878 European Wilderness Reserve [n] *conserv. pol.* (International title of hono[u]r for an area in Europe awarded by the International Council for Bird Protection, which has advisory status with the Council of Europe. Such areas have a great diversity of migrating birds for periods of time on breeding, feeding, or resting grounds and wintering areas. They are mostly ▶wetlands with special significance for wading and aquatic birds); *s* **reserva** [f] **natural europea** (Título honorífico creado por el Consejo Internacional de Protección de Aves —una organización supraestatal con status de asesor en el Consejo de Europa— a zonas europeas utilizadas, por lo menos durante una época del año, por un porcentaje importante de las especies de aves como territorio de reproducción, de alimentación, de descanso o de invernación; generalmente se trata de ▶zonas húmedas de gran importancia para aves limícolas y acuáticas y áreas de descanso para gansos y grullas); *f* **réserve** [f] **européenne** (Titre international délivré par le Conseil international pour la protection des oiseaux — organisme conseil auprès du Conseil de l'Europe — et attribué aux territoires européens constituant pour de nombreuses espèces d'oiseaux des aires de nidification, de nourriture, d'escale ou d'hivernage ; il s'agit en général de ▶zones humides pour les espèces limicoles et l'avifaune aquatique) ; *g* **Europareservat** [n] (Ein vom Internationalen Rat für Vogelschutz — einer überstaatlichen Organisation mit Beraterstatus beim Europarat — verliehener internationaler Ehrentitel für Gebiete in Europa, die einer Vielzahl von Vögeln mindestens eine Zeitlang als Brut-, Nahrungs-, Rast- oder Überwinterungsräume dienen; meist ▶Feuchtgebiete mit besonderer Bedeutung für Wat- und Wasservögel).

eurybathic [adj] *ecol.* ▶euryecious.

1879 euryecious [adj] *ecol.* (Descriptive term applied to a species which tolerates a wide range of environmental conditions, usually species with a wide geographical distribution; *specific terms* **1. eurybathic:** descriptive term for organisms living in a body of water at different depths, **2. euryhaline:** able to live in waters of a wide range of salinity; e.g. eel *[Anguilla anguilla]*, salmon *[Salmo salar]*, stickleback *[Gasterosteus aculeatus]*, **3. euryionic:** able to live in soil or water with a wide range from acidity to alkalinity, **4. euryphotic:** able to live under conditions with a wide range of light intensity, **5. eurythermal/eurythermic/eurythermous:** able to live in a wide range of temperatures; ▶ubiquitous; *opp.* ▶stenoecious); *syn.* eurytopic [adj]; *s 1* **eurioico** [adj] (Aplícase a la planta que en determinado clima se halla en las más diversas estaciones, es decir que posee una gran amplitud ecológica; *término específico* **altídomo** [adj]: aplícase a la planta que en determinado país vive en localidades extremas separadas por desniveles de 2000 m o más; las plantas altídomas son altitudinalmente eurítopas; *opp.* brevídomo; DB 1985 und DINA 1987); *s 2* **eurítopo** [adj] (Se dice de las plantas o sinecias que no son exclusivas de un área geográfica reducida,

sino que viven en países diferentes. Este concepto tiene toda clase de grados. En los máximos se expresa más correctamente con el término cosmopolita; *términos específicos* **1. euribático:** que se extiende a través de un gran espesor de agua, **2. eurihalino:** que tolera amplias oscilaciones en la concentración de sales haloídeas, **3. euriiónico:** poco sensible a las variaciones del pH del suelo, **4. eurífoto:** que tolera la intensidad de luz variable, **5. euritermo:** poco sensible a las variaciones de temperatura; cf. DB 1985; ▶ubicuo/a; *opp.* ▶estenoico/a); *syn.* euricoro [adj], euriálico [adj]; *f* **euryèce** [adj] (Terme générique caractérisant les organismes capables de supporter une grande amplitude d'action des facteurs écologiques ; *termes spécifiques* **1. eurybathique** [capable de vivre en milieu marin à des profondeurs variables], **2. euryhalin :** capable de supporter des variations importantes variation de salinité, **3. euryionique** [pH compris entre 5 et 8], **4. euryphote :** intensité lumineuse variable, **5. eurytherme :** capable de supporter des variations importantes de température ; ▶ubiquiste ; *opp.* ▶sténoèce) ; *syn.* eurytope [adj]) ; *g* **euryök** [adj] (Tiere oder Pflanzen betreffend, die Schwankungen lebenswichtiger Umweltfaktoren innerhalb weiter Grenzen ertragen; meist ubiquitäre Organismen; *UBe* **1. eurybath:** Organismen betreffend, die im Wasser in unterschiedlichen Tiefenzonen leben können und nicht an eine bestimmte Tiefenzone gebunden sind, **2. euryhalin:** Organismen betreffend, die unterschiedliche Salzgehalte im Wasser ertragen können, z. B. Aal *[Anguilla anguilla]*, Lachs *[Salmo salar]*, Dreistacheliger Stichling *[Gasterosteus aculeatus]*, **3. euryion:** Organismen betreffend, die in Medien [Boden, Wasser] von saurer bis alkalischer Reaktion leben können, **4. euryphot:** Organismen betreffend, die mit einer weiten Lichtamplitude leben können, **5. eurytherm:** Organismen betreffend, die große Temperaturdifferenzen ertragen können; ▶ubiquitär; *opp.* ▶stenök; *syn.* eurytop [adj], euryözisch [adj].

euryhaline [adj] *ecol.* ▶euryecious.

euryionic [adj] *ecol.* ▶euryecious.

euryphotic [adj] *ecol.* ▶euryecious.

eurythermal [adj] *ecol.* ▶euryecious.

eurythermic [adj] *ecol.* ▶euryecious.

eurythermous [adj] *ecol.* ▶euryecious.

eurytopic [adj] *ecol.* ▶euryecious.

1880 eutrophic [adj] *ecol.* (Descriptive term applied to soil and waterbodies containing a high concentration of nitrogen and phosphorous, or having an excessive concentration, mostly of nitrogen [N] and phosphorous [P_2O_5] to create a dead area of soil or dead waterbody; cf. SST 1997; ▶distrophic, ▶oligotrophic; ▶mesotrophic, ▶nutrient-rich, ▶trophic level of a waterbody; *syn.* nutrient-rich [adj]; *s* **eutrófico** [adj] (Término relacionado con medios ▶ricos en nutrientes, que favorecen un gran desarrollo de la vegetación; se aplica también a los organismos que viven en ese tipo de ambientes; *opp.* ▶dístrofo, ▶oligótrofo; ▶mesótrofo, ▶nivel trófico de masa de agua); *syn.* éutrofo [adj]; *f* **eutrophe** [adj] (Qualifie une eau ou un milieu ▶riche en substances nutritives ; *opp.* ▶dystrophe, ▶oligotrophe ; ▶mésotrophe, ▶niveau trophique des eaux douces) ; *g* **eutroph** [adj] (**1.** An nährstoffreiche Böden oder Gewässer gebunden. **2.** Zu viel Nährstoffe, meist N und P_2O_5 enthaltend [in Gewässern]; **3.** *Gewässergüte bei Stillgewässern* so beschaffen, dass eine starke Nährstoffbelastung und hohe Algenproduktion vorliegen, zeitweise Algenblüten möglich sind, eine geringe Sichttiefe und zeitweise ein totaler Sauerstoffschwund im Tiefenwasser vorherrscht; ▶nährstoffreich; ▶mesotroph; *opp.* ▶dystroph, ▶oligotroph; ▶Trophiestufe eines stehenden Gewässers); *syn. limn.* überdüngt [adj].

1881 eutrophication [n] *limn. pedol.* (Natural or artificial increase in the nutrient content of an ecosystem upsetting the ecological balance and leading to a long-term alteration in the ▶characteristic combination of species. In limnic ecosystems plant and animal plankton accumulate, and at great depths oxygen becomes depleted; ▶biochemical oxygen demand, ▶deoxidation, ▶trophic state); *s* **eutrofización [f]** (Enriquecimiento de las aguas con nutrientes a un ritmo tal que no puede ser compensado por su eliminación definitiva por mineralización total, de manera que la descomposición del exceso de materia orgánica producida hace disminuir enormemente la concentración de oxígeno en las aguas profundas y por ende altera a largo plazo la ▶combinación característica de especies; cf. MARG 1977, 759; ▶demanda bioquímica de oxígeno, ▶desoxigenación, ▶estado trófico); *f* **eutrophisation [f]** (Enrichissement naturel ou artificiel des écosystèmes limniques par augmentation de la concentration en substances nutritives provoquant une prolifération du plancton animal et végétal et par la même une ▶désoxygénation des eaux profondes, une altération durable de l'▶ensemble caractéristique ; ▶degré trophique, ▶demande biochimique en oxygène) ; *syn.* eutrophication [f] ; *g* **Eutrophierung [f]** (Natürliche oder künstliche Zunahme des Nährstoffgehaltes [hauptsächlich Stickstoff (N) und Phosphor (P$_2$O$_5$)] im Ökosystem oder Teilen davon, die das vorhandene ökologische Gleichgewicht stört und die ▶charakteristische Artenkombination nachhaltig verändert. In limnischen Ökosystemen tritt z. B. pflanzliches und tierisches Plankton massiert auf und in den tieferen Schichten wird Sauerstoff entzogen; ▶biochemischer Sauerstoffbedarf, ▶Sauerstoffentzug, ▶Trophie); *syn.* Gewässerbelastung [f] durch Nährstoffeintrag, Gewässereutrophierung [f].

1882 eutrophy [n] *limn. pedol.* (Quality or state of a nutrient-rich soil or water body; ▶oligotrophy, ▶trophic state); *s* **eutrofia [f]** (Estado o calidad de un suelo o un cuerpo de agua rico en nutrientes; *opp.* ▶oligotrofia; ▶estado trófico); *syn.* riqueza [f] en nutrientes; *f* **eutrophie [f]** (Enrichissement excessif des sols et des eaux par les matières nutritives ; *opp.* ▶oligotrophie ; ▶degré trophique) ; *syn.* richesse [f] trophique ; *g* **Eutrophie [f]** (**1.** Sehr hoher Versorgungsgrad von Böden oder Gewässern mit Nährstoffen. **2.** *limn. Klassifizierung der Gewässergüte* Gewässer mit einer starken Nährstoffbelastung, hohen Algenproduktion [Algenblüten zeitweise möglich], mit einer geringen Sichttiefe und einem zeitweise totalen Sauerstoffschwund; ▶Trophie; *opp.* ▶Oligotrophie); *syn.* Nährstoffreichtum [m].

1883 EU Water Framework Directive [n] *wat'man. pol.* (*Abbr.* WFD; Directive 2000/60/EC, which came into force on 22nd December 2000 established a new framework for Community action in the field of water policy as part of a substantial restructuring of EU water policy and legislation. The Directive rationalises and updates existing water legislations and provides for water management on the basis of River Basin Districts (RBD's). The main activities for the implementation of the WFD will take place in the context of River Basin Management Projects led in the UK by local authorities. The overall objective of river basin projects is to establish an integrated monitoring and management system for all waters within a RBD, to develop a dynamic programme of management measures and to produce a River Basin Management Plan, which will be continually updated. Central to the Water Framework Directive is a requirement for Member States to encourage the active involvement of all interested parties in its implementation. The WFD sets a framework for comprehensive management of water resources in the European Community, within a common approach and with common objectives, principles and basic measures. It addresses inland surface waters, estuarine and coastal waters and groundwater. The fundamental objective of the Water Framework Directive aims at maintaining "high status" of waters where it exists, preventing any

deterioration in the existing status of waters and achieving at least "good status" in relation to all waters by 2015. Member States shall ensure that a co-ordinated approach is adopted for the achievement of the objectives of the WFD and for the implementation of programmes of measures for this purpose. The objectives of the WFD are: **1.** to protect and enhance the status of aquatic ecosystems [and terrestrial ecosystems and wetlands directly dependent on aquatic ecosystems], **2.** to promote sustainable water use based on long-term protection of available water resources, **3.** to provide for sufficient supply of good quality surface water and groundwater as need for sustainable, balanced and equitable water use, **4.** to provide for enhanced protection and improvement of the aquatic environment by reducing/phasing out of discharges, emissions and losses of priority substances, **5.** to contribute to mitigating the effects of floods and droughts, **6.** to protect territorial and marine waters, **7.** to establish a register of 'protected areas'; e.g. areas designated for protection of habitats or species); *s* **Directiva [f] Marco del Agua (DMA)** (La Directiva 2000/60/CE del Parlamento Europeo y del Consejo, de 23 de octubre de 2000, que establece un marco comunitario de actuación en el ámbito de la política de aguas, entró en vigor el 22 de diciembre del 2000. Con ella se introduce un nuevo marco para la acción comunitaria en el campo de la política del agua como parte de una reestructuración sustancial de la política y legislación del agua. La directiva racionaliza y actualiza la legislación existente y prevé la gestión hidrológica sobre la base de las cuencas hidrográficas [CC.HH.]. Las actividades principales para la implementación de la DMA tendrán lugar en el contexto de proyectos de gestión de CC.HH. El objetivo general de los proyectos de CC.HH. es establecer un sistema integrado de monitoreo y gestión, y crear un Plan de Gestión de Cuencas Hidrográficas que será actualizado regularmente. La DMA se ocupa de las aguas superficiales interiores, estuarios y aguas costeras, así como de las aguas subterráneas. El objetivo fundamental de la DMA aspira mantener un «alto estado» de las aguas donde exista, evitando cualquier deterioro del estado existente para alcanzar por lo menos un «buen estado» de todas las aguas hasta 2015. Los Estados miembros asegurarán que se adopte un enfoque coordinado para alcanzar los objetivos y para implementar los programas de medidas para este fin. Los objetivos concretos de la DMA son: **1.** Proteger y realzar el estado de los ecosistemas acuáticos [y de los ecosistemas terrestres y humedales directamente dependientes de ecosistemas acuáticos]. **2.** Promover el uso sostenible del agua basado en la protección a largo plazo de los recursos de agua disponibles. **3.** Proveer suministro suficiente de agua superficial o subterránea de buena calidad como necesidad para el uso sostenible, equilibrado y egalitario del agua. **4.** Asegurar la protección y mejora del medio ambiente acuático reduciendo y eliminando descargas de emisiones y pérdidas de sustancias prioritarias. **5.** Contribuir a mitigar los efectos de inundaciones y sequías. **6.** Proteger las aguas territoriales y marinas. **7.** Establecer un registro de «áreas protegidas», p. ej. áreas designadas para la protección de hábitats o especies. En Es. la transposición de la Directiva Marco del Agua se realizó mediante la Ley 62/2003, de 30 de diciembre, de medidas fiscales, administrativas y del orden social que incluye, en su artículo 129, la modificación del texto refundido de la Ley de Aguas, aprobado por el RD Legislativo 1/2001 por la que se incorpora al derecho español la Directiva 2000/60/CE, estableciendo un marco comunitario de actuación en el ámbito de la política de aguas; cf. mma.es/secciones/acm); *f* **Directive [f] pour la protection des eaux intérieures de surface, des eaux de transition, des eaux côtières et des eaux souterraines** (Dir. n° 2000/60/CE du 23 octobre 2000 ; directive-cadre établissant un cadre pour une politique communautaire dans le domaine de l'eau ayant pour but **1.** d'identifier les eaux européennes et leurs caractéristiques, recensées par bassin et district hydrographiques [analyse des caractéristiques de chaque district hydrographique, étude de l'incidence de l'activité humaine sur les eaux, analyse économique de l'utilisation de celles-ci et registre des zones qui nécessitent une protection spéciale], **2.** d'organiser la gestion des eaux intérieures de surface, souterraines, de transition et côtières, de prévenir et de réduire leur pollution, de promouvoir leur utilisation durable, de protéger leur environnement, d'améliorer l'état des écosystèmes aquatiques et d'atténuer les effets des inondations et des sécheresses ; le plan de gestion et les programmes de mesures doivent **a.** prévenir la détérioration, améliorer et restaurer l'état des masses d'eau de surface, atteindre un bon état chimique et écologique de celles-ci, ainsi que réduire la pollution due aux rejets et émissions de substances dangereuses ; **b.** protéger, améliorer et restaurer les eaux souterraines, prévenir leur pollution, leur détérioration et assurer un équilibre entre leurs captages et leur renouvellement; **c.** préserver les zones protégées ; ces mesures doivent être accompagnées **1.** d'une politique de tarification incitant les consommateurs à utiliser les ressources de façon efficace et à ce que les différents secteurs économiques contribuent à la récupération des coûts des services liés à l'utilisation de l'eau, y compris les coûts pour l'environnement et les ressources **2.** de régimes assortis de sanctions effectives, proportionnées et dissuasives en cas violations de la directive-cadre ; une liste de substances polluantes prioritaires sélectionnées parmi celles qui constituent un risque important pour ou via le milieu aquatique a été élaborée, via une procédure associant surveillance et modélisation ; par ailleurs des mesures de contrôle relatives à ces substances prioritaires, ainsi que des normes de qualité applicables aux concentrations de celles-ci sont également proposées ; la loi n° 2004-338 du 21 avril 2004 transpose la directive 2000/60/CE du Parlement européen et du Conseil du 23 octobre 2000 ; cette dernière confortant le dispositif français, introduit une obligation de résultat, en fixant un objectif de « bon état des masses d'eau » à l'horizon 2015 ; elle complète les dispositions concernant les documents de planification [schémas directeurs et schémas d'aménagement et de gestion des eaux — SDAGE et SAGE] tant pour leur élaboration, avec des éléments de calendrier et la consultation obligatoire du public, que pour leur mise en œuvre, en organisant la compatibilité des documents d'urbanisme) ; *syn.* directive [f] cadre dans le domaine de l'eau ; *g* **Wasserrahmenrichtlinie [f]** (RL 2000/60/EG v. 23.10.2000; zum 22.12.2000 in Kraft getretene Europäische Richtlinie [WRRL] mit dem Ziel, bis zum Jahre 2015 für alle Gewässertypen und das Grundwasser einen „guten Zustand" zu erreichen — und dies mit einem gesamtheitlich ökologischen Anspruch. Dieser orientiert sich für die Oberflächengewässer maßgeblich an biologischen Qualitätsmerkmalen, die über bestimmte Organismengruppen definiert werden. Als räumliche Bezugseinheiten sind Einzugsgebiete und somit die hydrologischen Bedingungen maßgebend, nicht mehr Verwaltungs- oder Staatsgrenzen. Die Richtlinie bezieht sich nicht nur auf die Beschaffenheit der Gewässer selbst, sondern nennt als Ziel ausdrücklich auch den Schutz und die Zustandsverbesserung der von den aquatischen Ökosystemen abhängigen Landökosystemen und Feuchtgebiete).

evaluation [n] *plan.* ▶data evaluation, *plan. landsc. recr.* ▶landscape evaluation; *plan.* ▶site evaluation, ▶suitability evaluation; *arb. conserv.* ▶tree and shrub evaluation

1884 evaluation criterion [n] *plan.* (Quantitative or qualitative standard of judgement used to assess relative values within an array of physical or cultural factors); *syn.* assessment criterion [n]; *s* **criterio [m] de evaluación** (Característica cualitativa o cuantitativa utilizada en una evaluación) *syn.* indicador [m] de evaluación; *f* **critère [m] d'évaluation** (Notions quantitatives et qualitatives définies dans un système d'évaluation, une échelle d'évaluation/estimation) ; *g* **Bewertungsmaßstab [m]**

(Quantitative oder qualitative Eigenschaft als Leitlinie für zu bewertende Objekte); *syn.* Bewertungskriterium [n]).

1885 evaluation factor [n] *plan.* (Specific quantitative or qualitative numerical value applied in an assessment process; usually expressed as a multiplication factor); *s* **factor [m] de evaluación**; *f* **facteur [m] d'évaluation** (Valeur, coefficient multiplicateur utilisé dans les métodes normées de mise en évidence ou de reconnaissance d'un résultat qui ne peut pas être mesuré directement ; *terme spécifique* facteur de pondération) ; *syn.* facteur [m] d'estimation ; *g* **Bewertungsfaktor [m]** (Die in die Bewertung eingehende Zahl als Multiplikator).

1886 evaluation method [n] [US] *plan.* (Practical program[me] of action to assess the relative values of various characteristics of an area; ▶evaluation procedure); *syn.* assessment protocol [n], evaluation scheme [n] [UK]; *s* **esquema [m] de evaluación** (Regla aplicable en la práctica según la cual se pueden valorar alternativas en un proceso de decisión o evaluación: ▶método de evaluación); *syn.* matriz [f] de evaluación; *f* **schéma [m] d'évaluation** (Utilisation de facteurs permettant dans le cadre d'une évaluation ou d'une prise de décision de comparer entre elles des solutions alternatives pratiques) ; *syn.* matrice [f] d'évaluation ; *g* **Bewertungsvorschrift [f]** (Praktisch anwendbare Regel, nach der die in einer Entscheidungs- oder Bewertungssituation vorgegebenen Alternativen bewertet werden können; ▶Bewertungsverfahren).

evaluation [n] **of bids** *constr. contr.* ▶checking and evaluation of bids.

1887 evaluation [n] **of environmental factors** *plan.* (Assessment or classification of the significance and sensitivity of abiotic [e.g. soils, water, topography, climate] and biotic factors [flora, fauna and human beings] for nature and landscape within a particular area under investigation; the major criteria are e.g. rarity, degree of naturalness, size, degree of eutrophy. Evaluation procedures include ▶cost-benefit analysis, ▶benefit-value analysis, ▶environmental risk assessment); *s* **evaluación [f] de los factores ambientales** (Estimación resp. clasificación de la importancia y la sensibilidad de factores abióticos [p. ej. agua, suelo, relieve, clima] y bióticos [flora, fauna y seres humanos] sobre la naturaleza y el paisaje en un espacio a estudiar; criterios de evaluación son p. ej. rareza, grado de naturalidad, tamaño del área, grado de eutrofización. Como procedimientos de evaluación se utilizan p. ej. el ▶análisis de coste-beneficio, el ▶análisis del valor de uso o el ▶análisis del riesgo ecológico); *f* **évaluation [f] des facteurs environnementaux** (Évaluation des projets et des programmes qui détermine la vulnérabilité sur un site donné des facteurs abiotiques [sol, eau, air et climat] et biotiques [homme, flore, faune] sur la base de critères d'évaluation tels que p. ex. raréfaction, degré d'artificialisation, degré d'eutrophisation, etc. et en ayant recours à des procédés d'évaluation : ▶analyse coûts-avantages, ▶analyse coûts-efficacité, ▶évaluation des risques environnementaux) ; *g* **Bewertung [f] von Umweltfaktoren** (Einschätzung resp. Einstufung der Bedeutung und Empfindlichkeit von abiotischen [z. B. Boden, Wasser Relief, Klima] und biotischen Faktoren [Pflanzen, Tiere und Mensch] auf Natur und Landschaft in einem Untersuchungsraum; Bewertungsgrößen sind z. B. Seltenheit, Natürlichkeitsgrad, Flächengröße, Eutrophierungsgrad. Als Bewertungsverfahren dienen z. B. die ▶Kosten-Nutzen-Analyse, die ▶Nutzwertanalyse, die ökologische ▶Risikoanalyse).

evaluation [n] **of planning data** *plan.* ▶analysis of planning data.

1888 evaluation procedure [n] *plan.* (▶evaluation method [US]); *s* **método [m] de evaluación** (▶esquema de evaluación); *syn.* método [m] de valoración, procedimiento [m] de evaluación, procedimiento [m] de valoración; *f 1* **procédé [m] d'évaluation** (▶schéma d'évaluation) ; *syn.* méthode [f] d'estimation ; *f 2* **méthode [f] de cotation** (Fixation d'une valeur ou dimension du système cardinal) ; *g* **Bewertungsverfahren [n]** (▶Bewertungsvorschrift); *syn.* Bewertungsmethode [f].

evaluation scheme [n] [UK] *plan.* ▶evaluation method [US].

1889 evaporation [n] *ecol. met.* (Vaporization of water from non-vegetated soil and water surfaces); *s* **evaporación [f]** (Agua transferida a la atmósfera a partir de superficies de agua, hielo o nieve o del suelo no cubierto de vegetación); *f* **évaporation [f]** (de surfaces aquatiques et minérales exemptes de végétaux) ; *g* **Evaporation [f]** (Verdunstung der vegetationsfreien Erd- und Wasseroberfläche).

1890 evapotranspiration [n] [US]/**evapo-transpiration** [n] [UK] *ecol. met.* (Combined loss of water by ▶evaporation from the soil surface and ▶transpiration from plants); *s* **evapotranspiración [f]** (Pérdida total de agua, en forma de vapor, de la superficie del suelo y de las aguas [▶evaporación] y de la vegetación [▶transpiración]); *f* **évapotranspiration [f]** (Hauteur totale d'eau rejetée par la surface d'un sol végétalisé et composée de l'▶évaporation du sol et de la ▶transpiration du couvert végétal) ; *g* **Evapotranspiration [f]** (Gesamtverdunstung einer bewachsenen Bodenoberfläche, die sich aus der Verdunstung der Bodenoberfläche [▶Evaporation] und der Wasserabgabe der Pflanzen [▶Transpiration] zusammensetzt).

1891 even-hoofed game [n] *hunt.* (Hunting term for artiodactyls; e.g. roe deer, red deer, chamois, boar; ▶scaling damage); *s* **caza mayor [f] artiodáctila** (Animales con doble uña como el venado, ciervo, rebeco, jabalí; en alemán es un término utilizado por cazadores; ▶daño de descortezamiento); *f* **gibier [m] artiodactylique** (Dans le langage des chasseurs, gibier de la famille des ongulés ayant un nombre pair de doigts à chaque patte comme le chevreuil, le cerf, le chamois, le sanglier ; ▶dégâts d'écorçage) ; *g* **Schalenwild [n]** (Waidmännische Bezeichnung für Paarhufer wie z. B. Reh, Hirsch, Gemse, Wildschwein; ▶Schälschaden).

even spacing [n] *constr.* ▶regular spacing.

1892 evergreen [adj] *bot.* (Descriptive term for vegetation with foliage persisting throughout the year; ▶sommer green; *opp.* ▶deciduous; ▶defoliation; ▶winter green); *s* **perennifolio/a [adj]** (Así se designan los árboles y arbustos verdes todo el año, como las encinas, los laureles, los pinos, las araucarias, etc. En todos los árboles perennifolios las hojas viejas no se caen antes de haberse desarrollado otras nuevas, y aun en muchas especies se conservan las de varias brotaduras [en los abetos y las araucarias se conservan en el árbol diez años o más], DB 1985; ▶verde invernal; *opp.* ▶caducifolio/a; ▶defoliación, ▶verde invernal); *syn.* siempre verde [adj]; *f* **à feuilles persistantes [loc]** (Espèces végétales dont les feuilles ne tombent pas à la fin de la période de végétation et restent toujours vertes pendant plusieurs années ; la ▶défeuillaison est masquée par la présence constante de feuilles vivantes ; le feuillage est persistant bien que les feuilles soient caduques ; *les végétaux ligneux sont aussi appelés* feuillu persistant/sempervirent, conifère persistant/sempervirent ; ▶feuillage vert pendant l'été ; *opp.* ▶à feuilles caduques ; ▶semi-persistant, ante ; *syn. pour un végétal ligneux* sempervirente [adj], à feuillage persistant/sempervirent [loc] ; feuillu persistant [loc], feuillu sempervirent [loc]; *g* **immergrün [adj]** (Eine Pflanze betreffend, die während des ganzen Jahres grüne Blätter hat und deren Chlorophyllapparat [im Ggs. zu ▶wintergrün] nach der ungünstigen Periode wieder reaktiviert wird, z. B. Buchsbaum *[Buxus sempervirens]*; bei **i.en** Gehölzen

ist der ▶Laubfall mehr oder weniger stark [meist rhythmisch] über das Jahr verteilt [WIESNER (1904): Treiblaubfall], das einzelne Blatt lebt länger als zwölf Monate; ▶sommergrün, *opp.* ▶Laub abwerfend); *syn.* eusempervirent [adj].

evergreen broadleaf forest [n] *phyt.* ▶broad-leaved evergreen forest.

1893 examination [n] **of unit prices** *constr. contr.* (Review of unit prices for possible modification when considerable changes in specified items are sought after contract award, which may involve addenda; ▶unit price); *s* **inspección [f] del cálculo de precios de oferta** (Control de los ▶precios unitarios de una oferta en el caso de modificación considerable de ítems específicos después de adjudicado un contrato); *f* **examen [m] du mode d'établissement des prix unitaires** (Contrôle des prix unitaires du bordereau pour des travaux en plus ou en moins dans le cadre de l'établissement d'un avenant au marché ; ▶prix unitaire) ; *g* **Einsicht [f] in die Preiskalkulation** (Überprüfung eines kalkulierten Einheitspreises seitens der ausschreibenden Stelle bei erheblicher Massenmehrung oder Massenminderung im Rahmen von Nachtragsverhandlungen; ▶Einheitspreis).

1894 excavate [vb] **a planting hole** *constr. hort. syn.* excavate [vb] a planting pit, dig [vb] a planting hole [also US]; *s* **apertura [f] de hoyo (de plantación)** (Excavar [vb] un hoyo [de plantación]); *syn.* excavación [f] de hoyo (de plantación); *f* **creuser [vb] le trou de plantation** *syn.* creuser [vb] la fosse de plantation ; *g* **Pflanzloch ausheben [vb]**.

excavate [vb] **a planting pit** *constr. hort.* ▶excavate a planting hole.

1895 excavate [vb] **by hand** *constr.* (▶manual excavation); *s* **excavar [vb] manualmente** (▶excavación manual); *f* **exécuter [vb] une excavation à la main** (▶excavation exécutée à la main) ; *syn.* exécuter [vb] une fouille manuellement ; *g* **in Handschacht [m] ausführen [vb]** (im Sinne von „von Hand aufgraben"; ▶Handschachtung).

1896 excavated material [n] *constr.* (Soil which is removed as a result of earthworks operations; ▶surplus excavated material); *s* **material [m] de desmonte** (Material resultante de movimiento de tierras para construir zanjas, cortes del terreno o fosas de cimentación; ▶material sobrante); *syn.* material [m] excavado, tierra [f] excavada; *f* **matériau [m] extrait** (Volume de terre déplacée lors de la réalisation de fouilles en rigole, en tranchée, en puits ou en pleine masse] ; ▶déblai excédentaire) ; *syn.* matériaux [mpl] déblayés, déblai [m] de fouilles ; *g* **Aushub [m] (1)** (Bodenmasse, die bei Erdarbeiten zur Herstellung von Gräben, Einschnitten oder Baugruben anfällt und bewegt wird; ▶Überschussmasse); *syn.* Aushubmaterial [n], Erdaushub [m], Bodenaushub [m].

1897 excavated muck [n] **and plant material** [n] *constr. land'man. wat'man.* (Material excavated during maintenance of ditches, streams or lakes; ▶ditch clearing); *s* **material [m] de desbroce** (Material resultante de los trabajos de mantenimiento de canales y acequias; ▶limpieza de zanjas de avenamiento); *f* **produit [m] de curage** (Matériaux enlevés du lit d'un cours d'eau ou d'une eau stagnante lors de travaux d'entretien ; ▶curage des fossés) ; *syn.* produit [m] de dégagement ; *g* **Räumgut [n, o. Pl.]** (Material, das während einer Pflegemaßnahme Gräben, Fließ- oder Stillgewässer entnommen wird; ▶Grabenräumung).

1898 excavation [n] **(1)** *constr. eng.* (**1.** Process of cutting into an existing ground surface and removing the rock and soil material. **2.** Digging of trenches, ▶building excavations, etc. with the use of an excavator; ▶excavation process, ▶linear excavation [US]/linear cutting [UK], ▶cut 1, ▶cut material, ▶cut shelf); *syn.* cut [n], digging [n] with an excavator;

s excavación [f] de tierras o rocas (1. ▶Corte en el terreno existente. 2. Proceso de hacer una fosa de cimentación o una zanja con ayuda de una excavadora; excavar [vb] tierras o roca; ▶desmonte, ▶excavación de cimientos, ▶excavación de tierra, ▶material de desmonte, ▶perfil de desmonte); *f* **fouille [f]** (Entaille, ▶excavation linéaire effectuée lors de travaux de terrassements destinés à la pose de canalisations, la réalisation de fondations et du dégagement du volume des sous-sols ; on distingue suivant leur forme les ▶fouilles ponctuelles, linéaires et en excavation ; ▶affouillement 2, ▶déblaiement, ▶profil de déblai) ; *g* **1 Abgrabung [f]** (*Vorgang und Ergebnis* ▶Einschnitt in eine vorhandene Geländeoberfläche; ▶Anschnitt); *g* **2 Ausbaggerung [f]** (1) (Mit Hilfe eines Baggers eine Baugrube oder einen Graben ausheben; ▶Abtrag 1, ▶Aushub 2); *syn. zu 2.* Aushub [m] mit einem Bagger.

excavation [n] (2) *conserv'hist.* ▶archaeological excavation; *constr.* ▶building excavation; *eng.* ▶cut and cover excavation [US]/cut and cover tunnelling [UK]; *min.* ▶dry excavation, ▶dry gravel excavation; *constr.* ▶earth excavation; *conserv'hist.* ▶garden archaeological excavation; *constr.* ▶linear excavation, ▶manual excavation, ▶pathway excavation, ▶soil excavation.

excavation [n], **footpath** *constr.* ▶pathway excavation.

excavation [n], **gravel** *min.* ▶dry gravel excavation.

excavation [n], **hydraulic** *min.* ▶dredging (1).

excavation [n], **mineral** *min.* ▶mineral extraction.

excavation approval [n] [UK] *adm. conserv'hist.* ▶excavation permit [US].

1899 excavation [n] **for exposure** *constr.* (Digging of soil to reveal underground features; ▶manual excavation); **s excavación [f]** (Poner al descubierto conducciones u otros objetos enterrados bajo tierra; ▶excavación manual); *f* **affouillement [m]** (1) (Action de creusement d'une fouille pour mise à jour d'objets enterrés ; ▶excavation exécutée à la main) ; *g* **Aufgrabung [f]** (1. Freilegen von unter der Bodenoberfläche liegenden Objekten. 2. Graben oder Loch als Ergebnis von 1.; ▶Handschachtung).

1900 excavation, haulage, loading and transport of in-situ material [loc] *constr.* (Typical on-site earthwork operations; ▶earth-moving); **s excavación, movimiento, carga y transporte de tierra [loc]** (▶movimiento de tierras; *verb* excavar, mover, cargar y transportar tierra); *f* **extraction, déplacement, chargement et transport des terres [loc]** (▶mouvement de terre) ; *g* **Lösen, Fördern, Laden und Transportieren von Bodenmassen [loc]** (▶Bodenbewegung).

1901 excavation [n] **of services trenches** *constr.* (Digging of linear trenches for utility or drainage lines); **s excavación [f] de zanjas (de conducción)**; *f* **ouverture [f] des tranchées** (ouvrir [vb] des tranchées, creuser [vb] des tranchées) ; *g* **Ausheben [n, o. Pl.] von Leitungsgräben**.

1902 excavation permit [n] [US] *adm. conserv'hist.* (Sanction for an archaeological investigation); *syn.* excavation approval [n] [UK]; **s autorización [f] de excavaciones o prospecciones arqueológicas** *syn.* permiso [m] para excavaciones o prospecciones arqueológicas; *f* **autorisation [f] de fouiller** (Autorisation délivrée par le ministère de la Culture nécessaire à la réalisation de fouilles à l'effet de recherche de monuments ou d'objets pouvant intéresser la préhistoire, l'histoire, l'art ou l'archéologie ; art. 1 de la loi du 27 septembre 1941) ; *syn.* autorisation [f] de fouilles ; *g* **Grabungsgenehmigung [f]** (Genehmigung eines Landesdenkmalamtes, Grabungen in einem Gebiet mit begründeter Vermutung nach Kulturdenkmalen von besonderer Bedeutung durchführen zu dürfen).

1903 excavation process [n] *constr.* (Operation of digging out or removal of earth material; the result is a pit or ▶building excavation; ▶excavation 1, ▶excavation of service trenches, ▶mineral extraction by an excavator); **s excavación [f] de tierra** (Proceso; excavar [vb] tierra; ▶excavación de cimientos, ▶excavación de tierras o rocas, ▶excavación de zanjas [de conducción], ▶extracción mineral con excavadora); *f* **affouillement [m]** (2) (Action d'extraire des matériaux en terre ferme au moyen d'une pelle, d'un bulldozer, etc. ; ▶extraction réalisée avec une pelle, ▶fouille, ▶fouille de bâtiment [pour sous-sol], ▶ouverture des tranchées) ; *syn.* creusement [m], excavation [f] ; *g* **Aushub [m]** (2) (Arbeitsvorgang des Ausschaufelns oder Ausbaggerns von Erde; das Ergebnis ist ein Loch oder eine ▶Baugrube; ausheben [vb]; ▶Aushub 1; *UBe* ▶Ausbaggerung 1, ▶Ausbaggerung 2, ▶Ausheben von Leitungsgräben); *syn.* Aushubarbeiten [fpl], Ausschachtung [f], Ausschachtarbeiten [fpl].

1904 excavation site [n] *constr.* (Location specified for the removal of soil to install utilities [US]/services [UK], construct foundations, etc.; ▶building excavation); **s superficie [f] de excavación** (Localización específica para remover tierra para instalar conducciones o fundar cimientos; ▶excavación de cimientos); *syn.* área [f] de excavación; *f* **emprise [f] de fouille** (Aire d'excavation destinée à la pose de canalisations, la réalisation de fondations des bâtiments, le dégagement du volume des sous-sols, etc. ; ▶fouille de bâtiment [pour sous-sol]) ; *syn.* aire [f] d'excavation ; *g* **Aufgrabungsfläche [f]** (Fläche, die z. B. zur Verlegung von Leitungen, zum Einbau von Bauwerken ausgehoben wird; *UB* ▶Baugrube).

excavation [n] **to formation grade** [UK] *constr.* ▶excavation to subbase grade [US].

excavation [n] **to formation level** [UK] *constr.* ▶subbase grade preparation [US].

1905 excavation [n] **to subbase grade** [US] *constr.* (Removal of soil and rock to subbase level for construction of roads and hard surfaces; ▶pathway excavation, ▶subbase grade [US]/formation level [UK], ▶subgrade [US] 1/sub-grade [UK]); *syn.* excavation [n] to formation grade [UK]; **s excavación [f] (para una carretera)** (Excavar hasta el nivel de ▶subbase; ▶subrasante, ▶vaciado para carretera); *f* **décaissement [m]** (*Travaux de voirie* fouille exécutée pour l'implantation de la ▶couche de forme ; ▶décaissement des circulations, ▶fond de forme) ; *g* **Auskofferung [f]** (*Wegebau* Aushub für den ▶Unterbau 1; auskoffern [vb]; ▶Wegekoffer, ▶Erdplanum).

excavator [n] *min.* ▶mineral extraction by an excavator.

excavator [n], **digging with an** *min.* ▶excavation (1).

excellence [n] **of landscape** *conserv. landsc.* ▶beauty and amenity of the landscape [US]/beauty and amenity of the countryside [UK].

1906 excess gaseous emission [n] *envir.* (Generic term for surplus gas emanating from technical or chemical processes, especially by combustion; ▶air pollution 1); **s emisión [f] gaseosa** (Témino genérico para gases emitidos durante procesos técnicos o químicos, especialmente por combustión; ▶contaminación atmosférica); *f* **rejet [m] gazeux** (Terme générique pour les gaz non réutilisables introduits dans l'atmosphère, résultant des processus techniques ou chimiques en provenance des installations industrielles [en particulier la combustion] ; ▶pollution atmosphérique) ; *syn.* émission [f] gazeuse, effluent [m] gazeux ; *g* **Abgas [n]** (1) (OB zu einem meist nicht mehr nutzbaren Gas, das bei technischen oder chemischen Prozessen [besonders bei Verbrennungsvorgängen] entsteht; ▶Luftverunreinigung).

1907 excessive price [n] *constr. contr.* (Cost figure submitted in a bid [US]/tender [UK], which is calculated too high for a specified material or construction work; ►reasonable price); **s precio [m] excesivo** (En entrega de oferta, precio de ítem específico demasiado alto; ►precio razonable); **ƒ prix [m] excessif** (Valeur démesurée d'une marchandise ou d'une prestation calculée ou fixée dans une offre ; ►prix raisonnable) ; **g überhöhter Preis [m]** (Zu hoch kalkulierter/angesetzter Preis für eine bestimmte Ware oder Leistung bei Angebotsabgaben; ►angemessener Preis).

1908 excess rainwater [n] *constr. wat'man.* (Precipitation which does not completely percolate into the soil); **s aguas [fpl] pluviales excedentes** (Aguas que por la cantidad en la que se presentan no pueden infiltrarse/percolar en el suelo y, por tanto, fluyen superficialmente); **ƒ eaux [fpl] pluviales excédentaires** (Partie des précipitations qui ne peuvent pas être absorbées par sol) ; **g Überschusswasser [n]** (Wasser, das nicht vom Boden aufgenommen werden kann).

exchange [n] *met. urb.* ►air exchange, ►voluntary land exchange.

1909 exclusive species [n] *phyt.* (Species completely or almost completely confined to one community; ►character species, ►constancy, ►degree of fidelity, ►fidelity); *syn.* faithful species [n], constant species [n]; **s especie [f] exclusiva** (Especie limitada exclusivamente o casi exclusivamente a una comunidad determinada; BB 1979; ►constancia, ►especie característica, ►fidelidad, ►grado de fidelidad); *syn.* especie [f] constante, especie [f] fiel; **ƒ espèce [f] exclusive** (►Espèce caractéristique [exclusive], qui du point de vue de son ►degré de fidélité, est liée à peu près strictement à un groupement végétal déterminé ; ►constance, ►fidélité) ; *syn.* espèce [f] constante, espèce [f] fidèle ; **g treue Art [f]** (Art, die hinsichtlich des ►Treuegrades bei Pflanzengesellschaften ausschließlich oder nahezu ausschließlich an eine bestimmte Gesellschaft gebunden ist; ►Gesellschaftstreue, ►Kennart, ►Konstanz); *syn.* absolute Charakterart [f], gesellschaftstreue Art [f].

excursion area [n] [UK] *recr.* ►day-use recreation area [US] (1).

executed work [n] [US] *contr.* ►accounting of executed work [US]/accounting of executed works [UK], ►determination of executed work [US]/assessment of executed works [UK], ►proof of executed work.

executed works [npl] [UK]**, assessment of** *contr.* ►determination of executed work [US].

execution [n] **of a plan** *plan.* ►implementation of a plan.

1910 execution [n] **of building works** *constr.* (Implementation phase of a project on site carried out with construction equipment. The individual steps are laid down in the ►construction programme. E. of b. w. begins with the ►preparation of site facilities and ends with ►issue of a certificate of compliance; in U.S., construction specifications and general conditions documents establish parameters of scope and sequence of construction); **s ejecución [f] de obra** (Fase de realización de un proyecto de construcción. Los pasos individuales se definen en el ►calendario de ejecución de obras. La **e. de o.** comienza con la ►preparación de las instalaciones de obra y termina con la emisión del certificado de ►aceptación de obra); *syn.* realización [f] de obra; **ƒ exécution [f] des travaux** (Phase de réalisation des travaux dont les différentes étapes sont fixées dans le ►programme d'une opération ; la première étape constitue la ►préparation du chantier, la dernière étant la ►réception des travaux) ; *syn.* réalisation [f] des travaux ; **g Bauausführung [f]** (Phase der Realisierung eines Objektes auf der Baustelle mit Hilfe bautechnischer Maßnahmen. Die einzelnen Schritte sind im ►Bauablaufplan fest-

gelegt. Die **B.** beginnt mit der ►Baustelleneinrichtung 2 und endet mit der ►Abnahme); *syn.* Baudurchführung [f].

1911 execution [n] **of construction items** *constr. contr.* (**In U.K.** and **F.**, the construction of a particular item is not explicitly mentioned in the specifications, but is contained implicitly by means of headings under 'scope of work' before the itemized descriptions); *syn.* execution [n] of piece of work; **s ejecución [f] de unidades de obra** (En F. y GB no se nombra directamente en las especificaciones técnicas de obra, sino que se implican al ser mencionadas como títulos dentro de la lista de trabajo); **ƒ exécution [f] de travaux** (D., terminologie utilisée dans les textes de descriptif-quantitatif ; F. et U.K., ce terme n'est pas explicitement mentionné dans les spécifications techniques mais est implicitement contenu dans le titre de description des ouvrages) ; *syn.* confection [f] d'ouvrages ; **g Herstellen [n, o. Pl.] von Bauleistungen** (U.K. und F., in den Leistungsbeschreibungen wird das Herstellen der Leistungen explizit nicht erwähnt, sondern ist implizit durch die Überschriften der Gegenstände der Bauleistungen mit dem nachfolgenden Leistungsbeschrieb enthalten); *syn.* Ausführung [f] von Bauleistungen, Herstellung [f] von Bauleistungen.

execution [n] **of piece of work** *constr. contr.* ►execution of construction items.

1912 execution [n] **of planning services** *plan.* (All of the services provided by a planner: ►determination of planning context [US]/clarification of design/planning brief [UK], ►schematic design phase [US]/outline and sketch scheme proposals [UK], ►final design stage, ►construction document phase [US]/detailed design phase [UK], ►bidding and negotiation phase [US]/tender action and contract preparation [UK], ►construction phase—administration of the construction contract [US]/site management services [UK], ►concluding project review [US]/post-completion advisory services [UK]); **s ejecución [f] de trabajos de planificación** (Conjunto de trabajos realizados por un/a planificador, -a : diseño, planificación, organización, realización y supervisión así como monitoreo de las partes técnicas de un proyecto de construcción; ►inventario preliminar de un proyecto, ►fase de anteproyecto, ►fase de anteproyecto detallado, ►fase de planificación en detalle, ►preparación y colaboración en la fase de adjudicación, ►superintendencia de obras, ►servicios en plazo de garantía); **ƒ maîtrise [f] d'œuvre** (Tout ou parties des éléments de mission de conception et d'assistance confiés par le maître d'ouvrage au ►maître d'œuvre permettant d'apporter une réponse architecturale, technique et économique au programme d'ouvrages qu'il a défini et comprenant le ►recueil des données, les études d'esquisse, les études d'►avant-projet, la ►phase d'avant projet détaillé, les études de ►projet, la phase d'instruction d'autorisations si nécessaire, l'assistance apportée au maître de l'ouvrage pour la passation du contrat de travaux [►dossier de consultation des entreprises (DCE) et assistance aux marchés de travaux (AMT)], les études d'exécution ou l'examen de la conformité au projet et le visa de celles qui ont été faites par l'entrepreneur, la direction de l'exécution du contrat de travaux, l'ordonnancement, le pilotage et la coordination du chantier, l'assistance apportée au maître de l'ouvrage lors des opérations de réception et pendant la période de garantie de parfait achèvement [►mise en service des ouvrages] ; article 7, loi MOP n° 85-704 du 12 juillet 1985 modifiée par l'ordonnance 2004-566 du 17 juin 2004, Décret n° 93-1268 du 29 novembre 1993) ; **g Projektbearbeitung [f]** (Gesamtheit der objektplanerischen Leistungen eines Planers: Planung, organisatorische Durchführung und Überwachung des technischen Teils eines Bauvorhabens mit folgenden Grundleistungen: 1. ►Grundlagenermittlung, 2. ►Vorplanung, 3. ►Entwurfsplanung, 4. Genehmigungsplanung, wenn erforderlich, 5. ►Ausführungsplanung,

6. und 7. ▶Vorbereitung der und Mitwirkung bei der Vergabe, 8. ▶Objektüberwachung, 9. ▶Objektbetreuung und Dokumentation gemäß § 15 [2] 1-9 HOAI 2002 resp. §§ 33, 38, 42 u. 46 HOAI 2009).

exercise trail [n] [US] *recr.* ▶fitness trail.

exhaust fumes [npl] *envir.* ▶exhaust gas.

1913 exhaust gas [n] *envir.* (Gaseous emission from internal combustion engines); *syn.* traffic fumes [npl], exhaust fumes [npl]; *s* **gas** [m] **de escape** (Salida de gas de combustión interna de motores); *f* **gaz** [m] **d'échappement** (Gaz émis par les moteurs à combustion) ; *g* **Abgas** [n] **(2)** (Auspuffgas von Verbrennungsmotoren).

1914 exhaustion [n] *ecol. land'man.* (Total consumption of natural resources; ▶depletion of natural resources, ▶utilization of natural resources, ▶wasteful exploitation); *s* **agotamiento** [m] (Consumo total de un recurso natural; ▶explotación de los recursos naturales, ▶merma de los recursos naturales, ▶utilización de los recursos naturales); *f* **épuisement** [m] **des ressources naturelles** (Consommation intégrale des ressources naturelles ; ▶régression des ressources naturelles, surexploitation, ▶utilisation des ressources naturelles) ; *g* **Abbau** [m, o. Pl.] **(2)** (Vollständiger Verbrauch der natürlichen Grundgüter; ▶Abnahme natürlicher Grundgüter, ▶Raubbau, ▶Nutzung natürlicher Ressourcen).

exhibition [n] **of floristry** *hort.* ▶florist's show.

1915 existing area vegetation [n] *phyt.* (Floristic species inventory of a defined study area; ▶tree cover, ▶standing forest crop); *syn.* plant species stock [n]; *s* **población** [f] **(1)** (Reunión de individuos vegetales cualquiera. El término no prejuzga el valor fitosociológico de dicha reunión. Lo mismo se puede emplear para designar un representante de una asociación bien definida que un herbazal heterogéneo o un grupo de líquenes establecido sobre un tejado; DB 1985; *términos específicos* ▶cobertura arbórea, ▶población forestal); *f* **peuplement** [m] **végétal** (Inventaire floristique des espèces croissant sur une aire déterminée ; *termes spécifiques* ▶peuplement arborescent, ▶peuplement forestier 2) ; *g* **Bestand** [m] (Floristisches Arteninventar einer bestimmten Aufnahmefläche oder eines Betrachtungsgebietes; *UBe* ▶Baumbestand 2, Gehölzbestand, ▶Waldbestand 2); *syn.* floristischer Artenbestand [m], floristischer Bestand [m], Pflanzenbestand [m], Vegetationsbestand [m].

1916 existing condition [n] **before planning starts** [US] *ecol. plan.* (Conditions which exist within an area at a given moment in time with regard to planning parameters, e.g. land use, topography, flora and fauna, soils and climatic conditions); *syn.* existing situation [n] before planning starts [UK]; *s* **estado** [m] **cero** (Situación en un área antes de comenzar algún tipo de planificación); *syn.* situación [f] inicial, estado [m] inicial; *f* **état** [m] **initial du site et de l'environnement** (État existant d'une zone avant aménagement ou planification caractérisé par une série de paramètres tels que l'usage du sol, la topographie, la flore et la faune, le sol, le climat, etc.) ; *syn.* état [m] initial du paysage et de l'environnement ; *g* **Zustand** [m] **vor Planungsbeginn** (Augenblicklich bestehende Gegebenheiten in einem Gebiet hinsichtlich aller planungsrelevanter Parameter wie z. B. Bodennutzung, Topografie, vorhandene Flora und Fauna, Boden- und Klimaverhältnisse); *syn.* Bestand [m] vor Planungsbeginn.

existing elevation [n] *constr. eng. plan.* ▶existing ground level.

1917 existing environmental pollution load [n] *envir. plan.* (Existing extent of pollution within a planning area or part of it, before further increases in pollution are taken into account, which may be caused by regionally important planning projects); *syn.* background level [n] of environmental pollution;

s **nivel** [m] **de contaminación de fondo** (La que existe en un área definida, antes de instalar un nuevo foco de contaminación. Se indica separadamente para cada contaminante; cf. Orden 18/10/76 Anexo 1); *f* **niveau** [m] **de pollution dans l'air ambiant (±)** (Concentration des polluants dans l'air extérieur [à l'exclusion des lieux de travail] dans un territoire déterminé permettant l'évaluation de la qualité de l'air) ; *g* **Immissionsvorlast** [f] (Immissionsgegebenheiten in einem Planungsraum oder Teilen davon vor Berücksichtigung weiterer Immissionszuwächse infolge von raumbedeutsamen Planungsvorhaben); *syn.* Immissionsgrundlast [f].

existing grade [n] [US] *constr. eng. plan.* ▶existing ground level.

existing ground height [n] *constr. eng. plan.* ▶existing ground level.

1918 existing ground level [n] *constr. eng. plan.* (**1.** Height above a given datum level, shown on plans as the level of the present ground surface. **2.** Elevation prior to grading; ▶natural ground surface, ▶top level); *syn.* existing grade [n] [also US], existing elevation [n], existing ground height [n], existing level [n] [UK] (SPON 1986, 156); *s* **nivel** [m] **existente del terreno** (**1.** Indicación en plano de elevación de la altura existente en un terreno. **2.** Nivel del terreno antes de comenzar una obra; ▶borde superior, ▶superficie natural del terreno); *syn.* línea [f] del terreno, nivel [m] del suelo; *f* **niveau** [m] **existant de terrain naturel** (**1.** Donnée altimétrique sur une coupe de profil traduisant le niveau du terrain existant. **2.** Donnée altimétrique du ▶terrain naturel avant le début des travaux ; ▶arase) ; *g* **Geländeoberkante** [f] (**1.** Höhenangabe bei Profilzeichnungen für vorhandenes Niveau eines Geländes. **2.** Höhe des unverritzten Geländes vor Baubeginn; ▶Gelände 1, ▶Oberkante); *syn.* Geländehöhe [f], vorhandene Höhe [f].

existing ground surface [n] *constr. plan.* ▶natural ground surface.

existing land uses [npl]**, mapping of** *plan.* ▶survey of existing uses.

existing level [n] [UK] *constr. plan.* ▶existing ground level.

existing levels [loc] [UK]**, marrying with** *constr.* ▶meeting finished grade [US].

existing situation [n] **before planning starts** [UK] *ecol. plan.* ▶existing condition before planning starts [US].

existing slope [n] *constr.* ▶steepen an existing slope.

existing trees [npl] *arb. constr. landsc. urb.* ▶preservation of existing trees.

existing use plan [n] [US] *plan.* ▶survey of existing uses.

1919 existing vegetation [n] *phyt.* (Vegetation present at the time of vegetation mapping; ▶damage to existing vegetation, ▶density of existing vegetation, ▶preservation of existing vegetation); *syn.* actual vegetation [n] (TEE 1980, 198); *s* **vegetación** [f] **real** (Vegetación existente en el momento del inventario de vegetación); *syn.* vegetación [f] actual, vegetación [f] existente); *f* **végétation** [f] **réelle** (Végétation existant réellement sur une aire déterminée) ; *g* **reale Vegetation** [f] (Vegetation, die tatsächlich in einem Betrachtungsraum resp. auf einem Standort wächst).

exodus [n] *agr. sociol.* ▶rural exodus.

1920 exotic block [n] *geo.* (Rock body that is unrelated to the rocks with which it is associated. Exotic masses of tectonic origin are *allochthonous*; those of glacial origin are erratics; cf. DOG 1984, 174; ▶erratic block); *s* **bloque** [m] **alóctono** (Bloque de roca de gran tamaño procedente de un lugar distinto al emplazamiento en el que encuentra. Los bloques de origen

tectónico son *alóctonos*, los de origen glacial *erráticos*; ▶bloque errático); *f* **bloc [m] allochtone** (**1.** Rocher qui n'appartient pas originellement au lieu ou il se trouve ; les rochers transportés pendant la période glaciaire sont nommés ▶blocs erratiques ; *syn.* bloc [m] exotique, rocher [m] exotique. **2.** Ensemble de couches déposées ailleurs lors d'une poussée latérale et chevauchant une couche formée localement [bloc autochtone]) ; *g* **allochthoner Block [m]** (Ein von einem anderen Ursprungsgebiet stammender Steinblock oder stammendes Felsgestein; in der Glazialzeit transportierte Felsblöcke werden ▶Findlinge 2, **erratische Blöcke** oder **Irrblöcke** genannt).

1921 exotic species [n] *phyt. zool.* (exotics [npl]; nonnative species coming from another geographic region, which were introduced from abroad; ▶strange species; *opp.* ▶indigenous species); *s* **especie [f] exótica** (Especie de flora o fauna de otras regiones geográficas que ha sido introducida en una región en la cual no existía anteriormente; ▶especie extraña; *opp.* ▶especie indígena; *f* **espèce [f] exotique** (Organisme introduit directement ou indirectement par suite des activités humaines et vivant à l'état libre hors de son milieu naturel d'origine ; ▶espèce étrangère ; *opp.* ▶espèce indigène) ; *g* **exotische Art [f]** (Meist aus fernen Ländern stammende Art; ▶fremde Art; *opp.* ▶einheimische Art).

1922 expanded clay [n] *constr.* (Clay fired in rotary kilns at 1200 °C to form round pellets that are porous, inert, pH neutral and have no nutrient value. **E. c.** is used as a lightweight aggregate in concrete or a planting medium, or as a drainage layer; the baked clay granules are called *Haydite* in the US after S. J. HAYDE, who invented the material in 1917 and are also known by the trademarks 'Hydroton' or LECA [light expanded clay aggregate]; **e. c.** is suitable for hydroponic systems in plant cultivation, where they substitute for soil, the nutrients being controlled in water solution. It is considered to be an ecologically sustainable and re-usable growing medium, it can be cleaned and sterilized; *syn.* bloated clay [n] [also UK], haydite [n] [also US]; *s* **arcilla [f] expansiva** (Material ligero y de gran porosidad que se obtiene a partir de arcilla sometida a grandes temperaturas. Se utiliza como material de drenaje en las ▶azoteas ajardinadas, como árido para hormigón ligero o como sustrato para plantas en hidrocultura); *f* **argile [f] expansée** (Matériau léger et de forte porosité constitué à partir d'argile soumise à de hautes températures. Il est utilisé comme matériau de drainage sur les ▶jardins sur toitures-terrasses ou ▶jardins sur dalle, comme élément constitutif du mortier/béton allégé ou comme support dans l'hydroculture) ; *g* **Blähton [m]** (Feucht angemachter, fetter Ton, der bei niedrigen Temperaturen dichtsintert, im Drehrohrofen bei hohen Temperaturen [ca. 1100-1350 °C] gebrannt wird und durch Wasserverdampfung feine Poren bildet, wodurch eine niedrige Rohdichte [450-1300 kg/m³] entsteht; wird z. B. als Zuschlagstoff für Leichtbeton, als Dränschicht bei ▶Dachgärten oder bei Hydrokulturen verwendet. **B.** wurde in den USA 1917 von S. J. HAYDE erfunden, weshalb dieses Material dort auch *Haydite* genannt wird. Das Synonym **Leca** ist die Abkürzung für *lightweight expanded clay aggregates*); *syn.* Leca [n].

1923 expanded slate [n] [US] *constr.* (Slate which has been expanded by heating [1,000 °C] to many times its original volume as a result of exfoliation; suitable as a lightweight aggregate for concrete, or as a drainage layer for ▶roof gardens); *syn.* bloated slate [n] [UK]; *s* **pizarra [f] expansiva** (Granulado artificial de textura porosa que se obtiene al cocer restos de pizarra a 1000°C. Se utiliza como material de drenaje en las ▶azoteas ajardinadas, como árido para hormigón ligero o como sustrato para plantas en hidrocultura); *syn.* pizarra [f] esponjada; *f* **schiste [m] expansé** (Granulat artificiel à texture alvéolaire

obtenu par cuisson — porté à une température d'environ 1000 °C. Est utilisé comme matériau de drainage dans les ▶jardins sur toitures-terrasses ou ▶jardins sur dalle, comme élément constitutif du mortier/béton allégé ou comme support dans l'hydroculture) ; *g* **Blähschiefer [m]** (Bei ca. 1000 °C gebrannte Dachschieferabfälle, die sich unter Bildung von Poren stark aufblähen; wird als Zuschlagstoff für Leichtbeton oder als Dränschicht bei Dachterrassen oder ▶Dachgärten verwendet).

1924 expansion [n] *plan. urb.* (Enlargement of an area, e.g. a subdivision, park or open space; ▶community expansion, ▶urban expansion); *s* **ampliación [f]** (de un parque o un jardín; ▶expansión urbana 1, ▶expansión urbana 2); *f* **extension [f] (3)** (d'un parc, d'un jardin public, d'un espace vert ; ▶expansion urbaine 1, ▶expansion urbaine 2, ▶extension urbaine) ; *g* **Erweiterung [f]** (Vergrößerung eines Baugebietes, Parks, einer Grünanlage etc.; ▶Stadterweiterung, ▶Ausbreitung einer Stadt).

1925 expansion joint [n] *constr.* (DAC 1975; gap between construction units in a structure or concrete work, which permits a relatively slight movement due to temperature changes or other conditions without rupture or damage; cf. DAC 1975; ▶isolation joint; DACO 1988); *s* **junta [f] de dilatación** (Separación entre unidades de construcción en una estructura o trabajo de hormigón que permite pequeños movimientos, de manera que cuando hay cambios de temperatura o de otras condiciones no se produzcan roturas ni daños; ▶junta de separación); *syn.* junta [f] de expansión; *f* **joint [m] de dilatation** (Joint prévu dans un ouvrage afin d'éviter la formation de fissures suite à des variations thermiques ou des mouvements de terrain ; on distingue le joint de retrait et le ▶joint de rupture) ; *g* **Dehnungsfuge [f]** (Fuge zur Aufnahme von möglichen Rissbildungen zwischen zwei Bauteilen, Bauwerken oder Materialien, deren Ausdehnungs- und Schwindverhalten unterschiedlich sind, die bei geringfügigen Ausdehnungen durch Temperatureinflüsse [Erwärmung und Abkühlung] oder Bodensetzungen entstehen können. Man unterscheidet dabei ▶Trennfugen oder Raumfugen und Scheinfugen); *syn.* Dehnfuge [f], Bewegungsfuge [f].

expansive soil [n] *pedol.* ▶swelling soil.

expenses [npl] *contr.* ▶list of expenses, ▶overhead expenses, ▶travel expenses.

expenses [npl] [US]**, reimbursable** *contr. prof.* ▶overhead expenses.

1926 experiencing nature [n] *recr.* (The subjective, communicative, intellectual and emotional impression made by nature upon human beings, and the perception of images of the natural environment); *s* **vivencia [f] de la naturaleza** (Encuentro subjetivo y comunicativo con la naturaleza e impresión causada en el interior de las personas por las imágenes reales de la naturaleza con sus cualidades de producir encanto, gracias al incentivo de autorealización y de autodeterminación); *f* **perception [f] de la nature** (Rencontre subjective, communicative, intellectuelle et émotionnelle avec la nature qui est le fait de l'observation par les sens et de la prise en compte d'images matérielles de l'environnement naturel ainsi que de la perception des qualités de beauté et de charme, reflet des impressions imagées enregistrées, stimulée par l'aspiration à l'épanouissement personnel et la quête de l'autoformation de la personnalité morale); *g* **Naturerlebnis [n]** (Die subjektive, kommunikative, intellektuelle und emotionale Begegnung mit und das Bemerken und Aufnehmen von gegenständlichen Bildern der natürlichen Umwelt sowie das Innewerden von Anmutungsqualitäten als Widerschein der aufgenommenen bildhaften Eindrücke, ausgelöst durch den Antrieb zur Selbstentfaltung und Selbstgestaltung); *syn.* Naturerleben [n, o. Pl.], Erleben [n, o. Pl.] der Natur.

experimental area [n] *agr. for. hydr. phyt.* ▶test area.

E

1927 expert [n] *prof.* (Generic term for a person who has special skill or knowledge [precise expertise] in some particular field. *NOTE* expert and ▶specialist 1 are not synonymous!); *s* **especialista [m/f] (1)** (Persona altamente calificada y con mucha experiencia en un campo profesional; ▶especialista 2); *syn.* perito/a [m/f] (1), experto/a [m/f] (1); *f* **professionnel [m]** (Personne de métier spécialisée dans un secteur précis ; ▶spécialiste) ; *syn.* expert [m] ; *g* **Fachmann [m]** (Fachleute [pl], *selten* ...männer [pl], Fachfrau [f], ...en [pl]; jemand, der auf einem bestimmten Gebiet besonders ausgebildet, spezialisiert, sehr erfahren und entsprechend sachverständig ist; ▶Sonderfachmann); *syn.* Experte [m], Expertin [f], Sachverständige/-r [f/m].

expert [n]**, plant** *bot. hort.* ▶plantsman.

expert adviser [n] [UK] *prof.* ▶expert witness.

expert on historic garden management [n] *conserv' hist.* ▶historic garden expert [US].

1928 expert opinion [n] *prof. plan.* (Legally recognized expertise by a consultant in a particular case contained in a written report); *s* **dictamen [m] (pericial)** (Informe escrito de un/a experto/a sobre una cuestión profesional); *f* **expertise [f]** (Instruction et rapport écrit d'un expert) ; *g* **Gutachten [n]** (Schriftliche Darlegung eines Sachverständigen über einen bestimmten Sachverhalt; der Begriff '**gutachtlich**' bedeutet *in der Form eines Gutachtens* oder *ein Gutachten betreffend*); *syn.* gutachterliche Stellungnahme [f].

1929 expert opinion [n] **on the value of trees and shrubs** *arb. conserv.* (▶tree and shrub evaluation); *s* **peritaje [m] (escrito) del valor de leñosas** (Documento de peritaje elaborado sobre la base del ▶cálculo del valor de leñosas); *f* **expertise [f] en végétaux ligneux** (*Terme spécifique* expertise en arbre ; ▶calcul de la valeur des végétaux ligneux) ; *g* **Gehölzwertgutachten [n]** (Schriftliche Ausführungen eines Sachverständigen mit Hilfe einer ▶Gehölzwertermittlung).

1930 expert witness [n] [US] *prof.* (Specialist in a particular profession or field, certified by a locally responsible Chamber of Industry or Handicrafts [UK], whose services may be enlisted to resolve a conflict between owner and designer or contractor and, if necessary, to prepare ▶expert opinions for use in legal proceedings; not certified in U.S.); *syn.* expert adviser [n] [UK]; *s* **experto/a [m/f] (2)** (D. especialista de una profesión particular, reconocido oficialmente por la Cámara de Comercio e Industria u otra institución similar, cuyos servicios pueden ser solicitados para preparar opiniones expertas en pleitos legales; ▶dictamen [pericial]); *syn.* perito/a [m/f] (2); *f* **expert [m] agréé** (Spécialiste agréé par certaines administrations, les tribunaux, les chambres de commerce et de l'industrie, les chambres syndicales de diverses professions, habilité à effectuer des contrôles techniques, des examens, à établir des ▶expertises pour certaines catégories d'activités ou d'installations) ; *g* **öffentlich bestellte Sachverständige [f]/öffentlich bestellter Sachverständiger [m]** (D., von der zuständigen örtlichen Industrie- und Handelskammer oder Handwerkskammer fachlich beurteilter und vereidigter Fachmann eines Berufsstandes oder eines Fachgebietes, der für das Verfassen von gerichtlichen ▶Gutachten herangezogen wird).

1931 expiration [n] **of a deadline** [US] *contr. plan.* (Termination of a time limit), *syn.* expiry [n] of a deadline [UK]; *s* **expiración [f] de una fecha tope**; *f* **expiration [f] d'un délai** ; *g* **Ablauf [m, o. Pl.] einer Frist** (Erlöschen einer Zeitspanne, die für einen bestimmten Zweck festgelegt wurde); *syn.* Verstreichen [n, o. Pl.] einer Frist.

expiry [n] **of a deadline** [UK] *contr. plan.* ▶expiration of a deadline [US].

1932 explanatory board [n] *recr.* (Any flat piece of material, as wood, stone, metal, or plastic, designed to convey descriptive and educational information); *s* **tablón [m] de información** (en parque natural o zona de recreo al aire libre); *f* **panneau [m] de signalisation** (Élément de signalétique en bois, pierres, métal ou plastique matérialisant des signes, un texte, dessin ou des images) ; *syn.* panneau [m] d'information ; *g* **Hinweistafel [f]** (Holz-, Stein-, Metall- oder Kunststoffplatte mit anzeigendem oder erklärendem Text und Bildern); *syn.* Hinweisschild [n], Informationstafel [f], Bild- oder Schrifttafel [f].

1933 explanatory report [n] *plan.* (Written statement or narrative text, often with graphic material, which describes and substantiates planning proposals; ▶progress report); *syn.* written statement [n]; *s* **informe [f] descriptivo** (Texto explicativo de fundamentación de un plan o un concepto de planificación; ▶avance del estado de los trabajos) *syn.* memoria [f] descriptiva (1); *f* **rapport [m] de présentation** (Texte et représentation graphique présentant les objectifs et mesures d'aménagement exposés dans les documents graphiques ; ▶rapport d'avancement) ; *g* **Erläuterungsbericht [m]** (Schriftliche oder grafische Aussagen zur Begründung einer Planfassung resp. Planungskonzeption; ▶Bericht über den Stand der Arbeiten/Planung).

1934 explanatory statement [n] *plan.* (Written report to accompany, e.g. a legally-binding land-use plan); *syn.* explanatory text [n] [also US]; *s 1* **memoria [f] justificativa de la ordenación** (Texto que debe acompañar a los planes generales municipales de ordenación urbana y a los planes parciales; cf. arts. 37 y 57 RD 2159/1978); *s 2* **memoria [f] descriptiva (2)** (Texto que debe acompañar los proyectos de urbanización en el que se describen las características de las obras a realizar; Art. 69 [1] RD 2159/1978); *f* **rapport [m] justificatif** (p. ex. pour un plan d'urbanisme) ; *g* **Begründung [f]** (Erläuternder Bericht zu einer Planfassung, z. B. zum Bebauungsplan).

explanatory text [n] [US] *plan.* ▶explanatory statement.

exploitable area [n] *agr. for. hort.* ▶area of yield.

1935 exploitation [n] *ecol. min. recr. sociol.* (Use of natural resources or the natural environment for commercial profit; ▶wasteful exploitation); *s* **explotación [f]** (Uso de recursos naturales o de la naturaleza con fines de lucro y sin tener en cuenta su capacidad de regeneración; ▶expoliación de los recursos naturales); *syn.* aprovechamiento [m]; *f* **surexploitation [f]** (Exploitation abusive des ressources naturelles nonobstant la prise en compte de leur renouvellement et ayant des conséquences dommageables sur la biodiversité, les ressources non renouvelables et les équilibres naturels ; ▶pillage des ressources naturelles) ; *g* **Ausbeutung [f, Pl. selten]** (Exzessiver Abbau oder Verbrauch von natürlichen Ressourcen, ohne auf ihre Erneuerbarkeit zu achten; ▶Raubbau).

exploitation [n] **of mineral resources** *min.* ▶mineral working.

exploration capacity [n] *conserv. plan.* ▶use capacity of natural resources.

1936 exploration [n] **for foundation** *constr. eng.* (Geotechnical investigation of the ground by soil borings and laboratory analysis to define the ▶load-bearing capacity, the height of groundwater level, groundwater flow, frost resistance, workability characteristics, and to determine possible soil contamination; ▶soil investigation, ▶soil mechanics); *syn.* geotechnical investigation [n]; *s* **reconocimiento [m] del terreno** (Análisis del subsuelo p. ej. en cuanto a la profundidad de la capa freática, al flujo del agua freática, a su calidad estática, su resistencia a las heladas, friabilidad para construcción paisajística, etc. para determinar su ►capacidad de carga; *término genérico* ►prospección de suelos; ►mecánica de suelos); *syn.* reconocimiento [m] del subsuelo, investigación [f] del terreno; *f 1* **recherche [f] géotechnique** (Étude des sols et des ouvrages en granulat naturel du point de vue de leur résistance mécanique ; DIR 1977, 59) ; *f 2* **reconnaissance [f] du sol** (Essais et mesures effectués dans le cadre de l'élaboration du projet des fondations d'un ouvrage en vue de définir la nature et l'épaisseur des couches, de préciser les caractéristiques de la nappe aquifère, d'étudier les caractéristiques physiques et mécaniques des couches rencontrées et éventuellement de déceler la présence de déchets historiques ; *terme spécifique* sondage des sols de fondation ; cf. COB 1984 ; *terme générique* ►prospection du sol ; ►capacité de charge [d'exploitation], ►mécanique des sols) ; *g* **Baugrunduntersuchung [f]** (Geologische und bodenmechanische Untersuchung des Baugrundes z. B. hinsichtlich der ►statischen Belastbarkeit des Baugrundes, der Grundwasserhöhe, des Grundwasserflusses, der Frostbeständigkeit, Bearbeitbarkeit für Erdbauarbeiten, evtl. Prüfung von Altlasten/Kontaminationen; *OB* ►Bodenuntersuchung; ►Bodenmechanik).

1937 exploratory trench [n] *conserv'hist.* (First narrow excavation in an archaeological investigation area to determine the more important places to dig; ►archaeological investigation area, ►exploratory trenching); *s* **zanja [f] de prospección (arqueológica)** (Pequeña zanja en terreno de prospección arqueológica para determinar por dónde seguir la investigación; ►excavación arqueológica, ►zona de prospección arqueológica); *syn.* zanja [f] de excavación arqueológica; *f* **tranchée [f] de sondage archéologique** (Tranchée étroite sur un ►site de fouilles archéologiques ; ►fouille archéologique préventive, ►sondage archéologique) ; *g* **Suchgraben [m]** (Schmaler Graben in einem archäologischen Untersuchungsgelände, um herauszufinden, wo im Gelände mit den Grabungsarbeiten begonnen werden soll; ►Grabungsgebiet, ►Suchgrabung).

1938 exploratory trenching [n] *conserv'hist.* (Systematic layout of narrow trenches [►exploratory trench] in an archaeological excavation, in order to establish whether the remains of an historic garden or old structure still exist, or whether historic plans were actually carried out; ►garden archaeological excavation); *syn.* trial trenching [n] [also UK] (GP 2003; Dec. p. 34); *s* **excavación [f] arqueológica** (Trabajos de excavación en zona de prospección arqueológica para buscar posibles restos de edificios o jardines históricos; ►zanja de prospección [arqueológica], ►prospección arqueológico-paisajística) ; *f 1* **sondage [m] archéologique** (Exécution de tranchées étroites sur un terrain de fouilles afin de constater si des restes de monuments ou de jardins pouvant intéresser l'archéologie, sont encore présents ou ont été véritablement réalisés d'après les plans historiques existants ; PPH 1991, 223 ; ►tranchée de sondage archéologique, ►fouilles archéologiques dans un jardin historique) ; *f 2* **fouille [f] archéologique préventive** (Fouille déclenchée à l'initiative des archéologues de l'administration compétente à l'occasion de chantiers extérieurs à l'archéologie et qui permet, avant d'entamer des travaux d'aménagement ou d'urbanisation sur un terrain susceptible de receler un site archéologique, d'éviter que le patrimoine ne soit détruit lors de la réalisation de ces travaux) ; *g* **Suchgrabung [f]** (Gezieltes Anlegen von ersten schmalen Gräben [►Suchgraben] in einem geschichtsträchtigen Untersuchungsgelände, um z. B. herauszufinden, ob Reste alter Bauwerke oder Gartenanlagen noch vorhanden sind oder ob historische Pläne tatsächlich ausgeführt wurden; ►gartenarchäologische Grabung).

1939 expose [vb] **a soil profile** *pedol.* (Reveal soil layering characteristics through excavation); *s* **abrir [vb] una calicata**

(Excavar una fosa para exponer el perfil de un suelo); *f* **établir [vb] une fosse pédologique** ; *g* **Bodenprofil freilegen [vb]**.

1940 exposed aggregate concrete [n] *constr.* (Concrete finish achieved by removing the surface layer of cement to reveal specified stone aggregate through washing or brushing during intitial curing stage, or sand blasting after curing; ►finished concrete); *s* **hormigón [m] con árido visto** (Tipo de acabado de hormigón en el que por medio de lavamiento y/o frotación queda visto el árido con objeto de crear un efecto agradable; cf. DACO 1988; ►hormigón visto); *f 1* **béton [m] de gravillons lavé** (Béton de granulométrie soignée à pour lequel le dégagement partiel des grains est obtenu par enlèvement superficiel de la couche de ciment par lavage après le début de prise du mortier de surface sous l'action d'un jet d'eau ; DTB 1985 ; ►béton de parement) ; *f 2* **béton [m] lavé** (Béton à granulats apparents obtenus par enlèvement eau jet d'eau du mortier de surface) ; *g* **Waschbeton [m]** (Betonoberfläche, bei der während des Abbindens durch Abwaschen oder Abbürsten Zuschlagstoffe, z. B. Kieselsteine, Basaltsplitt, in der obersten Schicht freigelegt werden; ►Sichtbeton).

1941 exposed aggregate paving block [n] *constr. syn.* paver [n] with exposed aggregate [also US]; *s* **piedra [f] de hormigón con árido visto**; *f* **pavé [m] avec béton de parement** (Avec parement sur la face vue, p. ex. gravillon rond ou concassé) ; *g* **Betonpflasterstein [m] mit Natursteinvorsatz** (z. B. Splitt, Riesel).

exposed basalt aggregate paving block [n] [UK] *constr.* ►concrete paver with exposed crushed basalt [US].

exposed cut [n] [US] *constr.* ►incised slope.

exposed face [n] [US] *constr.* ►incised slope.

1942 exposed mud [n] *limn. pedol.* (VIR 1982, 310; wet organic soil in a water body temporarily left uncovered); *s* **fango [m] expuesto** (Fango al descubierto por desecación de una laguna o por vaciado de un estanque); *f* **vase [f] exondée** (Boue reposant sur le fond d'un plan d'eau résultant de sa vidange et de son assèchement) ; *g* **offener Schlammboden [m]** (Schlamm, der durch Austrocknen eines Stillgewässers oder durch Ablassen des Wassers aus Teichen offen daliegt); *syn.* trockenfallender Schlamm [m].

1943 exposed mud vegetation [n] *phyt.* (Nitrophilous vegetation mostly of summer annuals, especially on wet ground or in shallow water with a fluctuating watertable: Burmarigold Class: *Bidentetea* ; VIR 1982; ►littoral vegetation, ►riparian vegetation); *s* **vegetación [f] de bordes de estanques** (Vegetación nitrófila perteneciente a la clase *Bidentetea* que crece en bordes de estanques; ►vegetación del litoral y de las riberas de las aguas continentales, ►vegetación ripícola); *f* **groupement [m] des grèves d'étang** (Végétation annuelle estivale nitrophile, soumise à une inondation temporaire de faible amplitude en dehors de la période hivernale, appartenant à la classe des *Bidentetea* ; ►végétation des grèves, ►végétation du littoral et des rives des cours d'eau, ►végétation ripicole) ; *g* **Teichuferflur [f]** (Nitrophile, sommerannuelle Vegetation an Teichrändern, die zeitweise überschwemmt wird und zur Klasse der Zweizahnfluren *[Bidentetea]* gehört; ►Uferbewuchs, ►Vegetation entlang von Küsten und Ufern).

1944 exposed soil horizons [npl] *pedol.* (Exposed section through soil horizons; ►soil profile, ►outcrop); *s* **perfil [m] aflorado** (►Perfil del suelo expuesto a la luz; ►afloramiento natural); *syn.* afloramiento [m] de perfil de suelo; *f* **profil [m] pédologique tranché** (►affleurement, ►profil pédologique) ; *g* **Bodenaufschluss [m]** (Freigelegtes ►Bodenprofil; ►Aufschluss).

E

1945 exposed stone formation [n] *geo. pedol.* (Visible, relatively unweathered bedrock [C horizon] revealed by ▶soil erosion; ▶truncated soil profile); *s* **formación [f] de suelo decapitado** (Proceso de pérdida de suelo hasta dejar al descubierto el horizonte C de roca o material geológico poco meteorizado debido a la ▶erosión del suelo; ▶suelo decapitado); *f* **formation [f] de pavage** (*Formation superficielle* concentration de cailloux grossiers et jointifs de l'horizon C d'un sol ou mise à nu d'un matériau rocheux peu météorisé constituant l'horizon C d'un sol par érosion superficielle [enlèvement des éléments fins par érosion éolienne ou ruissellement diffus] ; dans le cas des roches en plaques on parle de « formation de dallage » ; cf. DIS 1986 ; ▶érosion des sols, ▶sol décapé) ; *g* **Steinpflasterbildung [f]** (Freilegung des wenig verwitterten Gesteinsmaterials eines C-Horizontes durch ▶Bodenerosion; ▶geköpftes Bodenprofil).

exposure [n] *constr.* ▶excavation for exposure; *met.* ▶light exposure; *envir.* ▶noise exposure; *met.* ▶orientation; *bot.* ▶poor light exposure, ▶strong light exposure.

exposure [n]**, low level of light** *bot.* ▶poor light exposure.

exposure [n]**, slope** *geo. met.* ▶slope aspect.

exposure [n]**, solar** *met.* ▶insolation.

exposure [n]**, weak light** *bot.* ▶poor light exposure.

expressway [n] [UK] *trans.* ▶major highway [US] (2).

expressway [n] [UK]**, four-lane undivided** *trans. urb.* ▶four-lane undivided highway [US].

expressway [n]**, inner-city** [UK] *trans.* ▶inner-city freeway [US] (1).

expropriation [n] [US] *adm. leg.* ▶land assembly policy [US]/land acquisition policy [UK].

extensification [n] **of agricultural production** [UK] *agr. pol.* ▶agricultural reduction program [US].

extension [n] [UK] (1) *leg. urb.* ▶addition [US].

extension [n] (2) *plan. urb.* ▶enlargement.

1946 extension [n] **of a wall line** *arch. constr.* (Imaginary extension of a straight line projected from a wall, often accomplished through supporting spatial design elements); *s* **alineación [f] de muro** (Línea recta imaginaria que resulta de la prolongación de la línea sobre la que se asienta un muro); *f* **alignement [m] d'un mur** (Ligne que dessine le corps d'un mur) ; *g* **Mauerflucht [f]** (Richtung, gerade Linie, die ein Mauerkörper anzeigt, auf die oft mit ergänzenden Gestaltungselementen Bezug genommen wird).

1947 extension [n] **of time for contract completion** *contr.* (▶contract period); *s* **prolongación [f] del plazo de ejecución** (▶plazo de ejecución); *f* **prolongation [f] du délai d'exécution** (▶délai d'exécution) ; *g* **Verlängerung [f] der Ausführungsfrist** (▶Ausführungsfrist).

1948 extensive erosion [n] *geo.* (Wearing away of soil over broad areas which occurs mainly in arid and semiarid regions and which is caused by rare and heavy storms; those areas are characterized by lack of vegetation; ▶sheet erosion); *syn.* wide-scale erosion [n]; *s* **erosión [f] aerolar** (Término específico para denominar un proceso erosivo que afecta a grandes superficies; ▶erosión laminar); *f* **érosion [f] aréolaire** (Processus s'exerçant sur de vastes étendues ; ▶érosion pelliculaire) ; *g* **massive, großflächige Erosion [f]** (Massive Bodenabtragungen nach heftigen Niederschlägen in weiten Gebieten arider und semiarider Regionen, die durch Vegetationsmangel geprägt sind; ▶Flächenerosion).

1949 extensively cleared land [n] **(for cultivation)** *land'man.* (Rural land largely cleared of natural vegetation to facilitate the use of agricultural machinery; ▶hedgerow clear-

ance); *syn.* totally cleared agrarian landscape [n]; *s* **paisaje [m] desnudo** (Paisaje agrícola desarbolado; ▶tala de setos interparcelarios/setos vivos elevados); *syn.* ager [m]; *f* **paysage [m] ouvert** (1. Terme désignant un paysage d'openfield étendu ou le regard porte à perte de vue sans rencontrer de contrastes physiques. 2. Paysage agraire caractérisé par l'absence souvent totale d'arbres et d'arbustes suite à l'▶enlèvement des haies, des haies sur talus) ; *syn.* paysage [m] agricole nu, paysage [m] dénudé ; *g* **ausgeräumte Landschaft [f]** (Über weite Flächen durch menschlichen Einfluss von Bäumen und Sträuchern oft völlig freie Landschaft, z. B. durch ▶Schleifen von Feldhecken/ Knicks, Roden von kleinen Wäldchen; in der **a.n L.** werden vorwiegend großflächig); *syn.* ausgeräumte Agrarlandschaft [f], Agrarsteppe [f], Kultursteppe [f].

extensive open land [n] [US] (TGG 1984, 247) *landsc.* ▶open country.

extensive planting [n] [UK] *constr.* ▶low maintenance planting.

extensive recreation [n] *recr.* ▶recreation in the natural environment.

1950 extensive roof planting [n] *constr.* (Establishment of vegetation on relatively dry habitat roofs, within a growing medium [▶root-zone layer] that typically does not exceed 10cm deep or upon ▶substrate mats. The plants selected to withstand such extreme conditions do so despite the lack of available water and nutrients and require hardly any maintenance; ▶intensive roof planting); *syn.* shallow soil roof planting [n] [also US], functional roof planting [n], low maintenance roof planting [n], dry habitat roof planting [n] [also US]; *s* **revestimiento [m] vegetal extensivo de tejados** (Establecimiento de vegetación en tejados en zonas relativamente secas sobre capa de ▶sustrato de plantación o estera de sustrato de menos de 10 cm de espesor. Las plantas utilizadas para una ubicación externa tal pueden sobrevivir con poca agua y pocos nutrientes y prácticamente no necesitan mantenimiento; ▶estera de sustrato sintético, ▶revestimiento vegetal intensivo de tejados); *syn.* plantación [f] extensiva de tejados; *f* **végétalisation [f] extensive des toitures** (Terrasse-jardin en général inaccessible, d'un entretien réduit, constituée d'un ▶support de culture inférieur à 10 cm et n'autorisant que l'implantation d'un couvert végétal xérophile ; *opp.* ▶végétalisation intensive des toitures, ▶natte de substrat synthétique) ; *g* **extensive Dachbegrünung [f]** (Bauwerksbegrünung auf Dächern mit dünnen, meist unter 10 cm mächtigen Substratschichten [▶Vegetationstragschicht] oder ▶Substratmatten. Die für solche Extremstandorte ausgewählten Pflanzen können mit geringsten Wasser- und Nährstoffreserven dauerhaft überleben und benötigen kaum Pflege; *opp.* ▶intensive Dachbegrünung); *syn.* dünnschichtige Dachbegrünung [f] (DGA 9/94, 575), Extensivbegrünung [f] eines Daches.

extent [n] **of planning services** *contr. prof.* ▶scope of professional services.

1951 extermination [n] *conserv.* (Destruction of total population of a plant or animal species by human activities; ▶extinction; **extirpation** is the destruction of a local population of a plant or animal species by human activities; ▶extinction); *s* **exterminio [m]** (Eliminación por la acción humana de una especie de flora o de fauna o de una población de aquéllas en regiones de la tierra; ▶extinción); *syn.* exterminación [f]; *f* **extermination [f]** (Destruction anthropogène d'une espèce végétale ou animale toute entière ou d'un peuplement naturel correspondant ; ▶extinction) ; *g* **Ausrottung [f]** (Völlige Vernichtung einer ganzen Pflanzen- oder Tierart resp. die Vernichtung einzelner Populationen in Gebieten der Erde durch den Menschen; ausrotten [vb]; ▶Aussterben).

1952 external costs [npl] *econ. sociol.* (Expenses generated by negative, external effects, e.g. air and water pollution, which are not calculated by the polluter as part of production costs but are passed on to the public); *s* **costos** [mpl] **ambientales y sociales** (Costos generados por efectos externos negativos, p. ej. de la contaminación del aire o del agua, que no se tienen en cuenta por el contaminante en sus costos de producción, por que se convierten en costos a llevar por la sociedad en general); *syn.* externalidades [fpl], costes [mpl] ambientales y sociales [Es]; *f* **coûts** [mpl] **externes** (Effets négatifs engendrés par le processus de production [p. ex. pollution de l'air ou des eaux] et que le pollueur ne prend pas en compte dans ses coûts de production et qui sont donc supportés par le contribuable) ; *g* **externe Kosten** [pl] (Durch Produktion verursachte negative externe Effekte [z. B. Luft- oder Gewässerverschmutzung], die der Verursacher nicht in seine betriebliche Kostenrechnung aufführt und somit gesamtgesellschaftlich abwälzt).

1953 external space [n] *constr. urb.* (Open space within the site of a specific construction project, usually subject to landscape treatment; ▶landscape planting in built-up areas); *syn.* outdoor area [n]; *s* **zona** [f] **ajardinada** (Espacio libre dentro de los límites de un predio o proyecto, generalmente sujeto a tratamiento paisajístico; ▶zonas verdes en áreas residenciales); *f* **espace** [m] **extérieur** (Espace laissé libre par les constructions et la voirie sur les terrains déjà aménagés ou à aménager en zone urbaine ; ▶espaces verts d'une zone urbanisée) ; *syn.* espace [m] d'accompagnement (d'immeuble/de bâtiments) ; *g* **Außenanlage** [f] (Meist grünplanerisch zu gestaltender oder gestalteter Freiraum innerhalb von Grundstücksgrenzen in bebauten Gebieten; ▶Siedlungsgrün); *syn.* in der HOAI Freianlage [f] (cf. § 13 u. 14 HOAI 2002 resp. § 38 HOAI 2009), Freifläche [f] bei öffentlichen oder privaten Bauwerken (§ 14 Ziff. 3 und 4 HOAI 2002).

1954 extinction [n] *conserv. phyt. zool.* (Complete disappearance of a plant or animal species from the world; ▶extermination, ▶extinct species, ▶Red List, ▶threatened with extinction); *s* **extinción** [f] (Desaparición de una especie de flora o fauna o de una de sus poblaciones en una región de la Tierra; ▶Catálogo Español de Especies Amenazadas, ▶en peligro de extinción, ▶especie extinta, ▶exterminio, ▶lista roja); *f* **extinction** [f] (d'une espèce animale ou végétale tout entière ; ▶espèce en voie d'extinction, ▶espèce disparue ou éteinte, ▶extermination, ▶liste des espèces menacées, ▶Liste rouge, ▶Livre rouge [des espèces menacées]) ; *g* **Aussterben** [n, o. Pl.] (Das Erlöschen einer ganzen Pflanzen- oder Tierart resp. von einzelnen Populationen; aussterben [vb]; ▶ausgestorbene Art, ▶in freier Wildbahn ausgestorben Art, ▶verschollene Art, ▶Ausrottung, ▶Rote Liste, ▶vom Aussterben bedroht).

extinction [n]**, threat of** *conserv. phyt. zool.* ▶endangerment of species.

1955 extinct species [n] *conserv. phyt. zool.* (**1.** Species which lived more than 100 years ago, and has now with certainty or great probability disappeared worldwide. The existence of **e. s.** could not be verified despite investigations; therefore, it is suspected that the population has been annihilated; *according to IUCN categories of threat* EX [extinct]; a taxon is Extinct when there is no reasonable doubt that the last individual has died and when exhaustive surveys in known and/or expected habitat, at appropriate times [diurnal, seasonal, annual], throughout its historic range have failed to record an individual; ▶extinction, ▶Red List); **2. extinct in the wild species** [n] (*According to the IUCN categories of threat* EW [Extinct in the Wild]; a taxon is Extinct in the Wild when it is known only to survive in cultivation, in captivity or as a naturalized population [or populations] well outside the past range. A taxon is presumed

Extinct in the Wild when exhaustive surveys in known and/or expected habitat, at appropriate times [diurnal, seasonal, annual], throughout its historic range have failed to record an individual. The species now only exists in culture, in enclosures or zoos or has survived as a ▶naturalized species in populations outside the original distribution area; *syn.* species extinct in the wild); **3. disappeared species** [n] (Species not found or recorded for at least 10-20 years despite search and is therefore suspected to be annihilated or extinct; extinction cannot be exactly determined, nor whether it is likely to reappear or be rediscovered); *s 1* **especie** [f] **extinta (EX)** (**1.** Un taxón de flora o fauna se considera extinto [Extinct] cuando no queda ninguna duda razonable de que el último individuo existente ha muerto. Esto se presume cuando prospecciones exhaustivas de sus hábitats, conocidos y/o esperados, en los momentos apropiados [diarios, estacionales, anuales], y a lo largo de su área de distribución histórica, no han podido detectar un sólo individuo. Las prospecciones deberán ser realizadas en periodos/períodos de tiempo apropiados al ciclo de vida y formas de vida del taxón; UICN 2001; ▶extinción, ▶lista roja); *s 2* **especie** [f] **extinta en estado silvestre [EW]** (Según las categorías de la ▶lista roja de la UICN, un taxón está extinto en estado silvestre [Extinct in the Wild] cuando sólo sobrevive en cultivo, en cautividad o como población [o poblaciones] naturalizadas completamente fuera de su distribución original. Se presume esto cuando prospecciones exhaustivas de sus hábitats, conocidos y/o esperados, en los momentos apropiados [diarios, estacionales, anuales], y a lo largo de su área de distribución histórica, no han podido detectar un sólo individuo. Las prospecciones deberán ser realizadas en periodos/períodos de tiempo apropiados al ciclo de vida y formas de vida del taxón; UICN 2001, ▶especie naturalizada); *s 3* **especie** [f] **desaparecida** un taxón cuya extinción no puede ser determinada exactamente o que podría reaparecer o ser descubierta de nuevo); *f 1* **espèce** [f] **éteinte** (**1.** ▶*Catégorie de régression des populations et de distribution des espèces* espèce vivant il y a plus d'un siècle et dont l'▶extinction a été depuis constatée avec certitude ou grande vraisemblance. La présence d'▶espèces disparues n'a pu être prouvée malgré de longues recherches, c'est pourquoi il est justifié de supposer que ces populations sont définitivement éteintes ; **2.** selon le système de classement de la ▶Liste rouge de l'UICN correspond à la catégorie EX [espèce éteinte] ; un taxon est *dit* éteint lorsqu'il ne fait aucun doute que le dernier individu est mort ; une espèce est **présumée** *éteinte* lorsque des études exhaustives menées dans ses habitats connus et/ou probables, à des périodes appropriées [rythme diurne, saisonnier, annuel], et dans l'ensemble de son aire de répartition historique n'ont pas permis de noter la présence d'un seul individu. Les études doivent être faites sur une durée adaptée au cycle et aux formes biologiques du taxon ; cf. Catégories et Critères de l'UICN pour la Liste Rouge, Version 3.1, UICN 2001 ; ▶liste des espèces menacées, ▶liste rouge, ▶Livre rouge [des espèces menacées] ; *f 2* **espèce** [f] **éteint à l'état sauvage (EW)** (Une des neufs catégories du système de la Liste Rouge de l'UICN : une espèce est **dite** *éteinte à l'état sauvage* lorsqu'elle ne survit qu'en culture, en captivité ou dans le cadre d'une population naturalisée nettement en dehors de son aire de répartition ; ▶espèce naturalisée ; une espèce est **présumée** *éteinte à l'état sauvage* lorsque des études détaillées menées dans ses habitats connus et/ou probables, à des périodes appropriées [rythme diurne, saisonnier, annuel], et dans l'ensemble de son aire de répartition historique n'ont pas permis de noter la présence d'un seul individu. Les études doivent être faites sur une durée adaptée au cycle et aux formes biologiques du taxon ; cf. Catégories et Critères de l'UICN pour la Liste Rouge, Version 3.1, UICN 2001 ; ▶espèce menacée); *f 3* **espèce**

[f] disparue (Catégorie de régression des populations et de distribution des espèces : espèce pour laquelle aucune observation n'a été faite depuis plusieurs dizaines d'années et réputée pour avoir localement ou régionalement entièrement disparu) ; *g 1* **ausgestorbene Art [f]** (Noch vor ca. 100 Jahren lebende und in der Zwischenzeit — z. T. weltweit — mit Sicherheit oder großer Wahrscheinlichkeit erloschene Art; in D. zur Gefährdungskategorie 0, gemäß IUCN zur Kategorie EX [extinct] oder EW [Extinct in the Wild — in freier Wildbahn ausgestorben] gehörig; *g 2* **in freier Wildbahn ausgestorbene Art [f]** (*Gemäß IUCN zur Gefährdungskategorie EW [Extinct in the Wild — in freier Wildbahn ausgestorben] gehörig* Art, die nur in Kultur, in Gehegen oder Zoos oder als ▶neuheimische Art in Populationen außerhalb ihres ursprünglichen Verbreitungsgebietes überlebt; eine Art wird dieser EW-Kategorie zugeordnet, wenn nach umfangreichen Untersuchungen in bekannten und/oder vermuteten Lebensräumen zu entsprechenden Zeiten [täglich, saisonal, jährlich] kein Individuum in seinem angestammten Areal mehr gefunden wurde; in D. zur Gefährdungskategorie 0 gehörig); *syn.* Art [f], die in freier Wildbahn ausgestorben ist; *g 3* **verschollene Art [f]** (Seit längerer Zeit [mindestens 10-20 Jahre] trotz Suche nicht nachgewiesene Art und es besteht deshalb der Verdacht, dass die Art erloschen/ausgestorben ist; ▶Aussterben, ▶Rote Liste).

extinguishment [n] of rights of way [UK] *adm. leg. urb.* ▶street abandonment [US].

extirpation [n] *conserv.* ▶extermination.

1956 extraction [n] **(1)** *min.* (Generic term covering the excavation of gravel, sand or clay; ▶gravel extraction, ▶mineral extraction, ▶sand extraction); *syn.* sand and gravel working [n]; *s* **préstamo [m]** (Término genérico para la explotación de yacimientos de grava, arena o arcilla; ▶extracción de arena, ▶extracción de grava, ▶extracción de minerales superficiales o rocas industriales); *syn.* extracción [f]; *f* **extraction [f]** (Terme générique pour l'exploitation de gravier, de sable ou d'argile ; ▶extraction de gravier, ▶extraction de sable, ▶extraction des substances minerals) ; *g* **Entnahme [f]** (OB zu Ausbaggerung von Kies, Sand oder Ton; entnehmen [vb]; ▶Kiesabbau, ▶Materialentnahme, ▶Sandabbau).

extraction [n] **(2)** *wat'man.* ▶groundwater extraction [US]/ground water extraction [UK]; *for.* ▶timber extraction; *hydr. nat'res. wat'man.* ▶water extraction.

extraction [n]**, mineral** *min.* ▶mineral working.

extraction area [n] *min. plan.* ▶future extraction area.

1957 extraction [n] **of deposits** *min.* (Mining of usable mineral resources); *s* **explotación [f] de yacimientos mineros**; *f* **exploitation [f] d'un gisement** (Exploitation réalisée sur terre, sous terre ou en milieu aquatique des ressources utiles) ; *syn.* exploitation [f] d'un gîte ; *g* **Lagerstättenabbau [m, o. Pl.]** (Gewinnung nutzbarer Bodenschätze).

1958 extraction [n] **right** *min. leg.* (Lawful claim for mineral extraction; ▶permit for mineral extraction [US]/planning permission for mineral extraction [UK], ▶mineral resources); *s* **derechos [mpl] extractivos** (Derecho de explotación de ▶recursos minerales de dominio público sean sujetos de la legislación de minas o no; ▶permiso de explotación); *f* **droit [m] d'exploitation** (Tout droit du propriétaire du sol ou du concessionnaire à exploiter des gîtes de substances minérales [mines ou carrières — ▶richesses naturelles du sous-sol] sous réserve des dispositions légales ; ▶autorisation d'exploitation, ▶autorisation d'extraction) ; *g* **Gewinnungsberechtigung [f]** (Jedes Recht zur Gewinnung von bergfreien [= dem Bergrecht unterstellten] oder grundeigenen [= im Eigentum des Grundeigentümers stehenden]

▶Bodenschätzen; cf. §§ 3 [2 u. 3] und 4 [6] BBergG; ▶Abbaugenehmigung).

1959 extraction site [n] *min.* (Area from which soil material such as gravel, sand, or clay is extracted; ▶mineral working site; *specific terms* ▶borrow pit, ▶gravel extraction site, ▶sand extraction site; ▶proposed extraction site); *s* **zona [f] de extracción** (Término genérico para las superficies en las que se extraen grava, arena o arcilla; ▶área de explotación minera [mina o cantera]; *términos específicos* ▶zona de préstamo, ▶gravera, ▶lugar de extracción de arenas); *f* **site [f] d'extraction de matériaux** (Lieu d'où on extrait des matériaux rocheux servant à la construction ; ▶emprise d'extraction ; *termes spécifiques* ▶lieu d'emprunt, ▶site d'extraction de gravier, ▶site d'extraction de sable) ; *syn.* carrière [f] d'emprunt, zone [f] d'extraction de matériaux ; *g* **Entnahmestelle [f] (2)** (Fläche zur Gewinnung von Steinen und Erden [speziell Schüttgüter] im Tagebau; ▶Abbaufläche; *UBe* ▶Seitenentnahme, ▶Kiesentnahmestelle 1, ▶Sandentnahmestelle; ▶Abbaugebiet).

extra thick sod [n] *constr. hort.* ▶sod [US] (1), #2.

1960 extra work [n] *constr. contr.* (Additional item of work as an addendum to a main item in a schedule of quantities; ▶add-on item, ▶list of bid items and quantities [US]/schedule of tender items [UK]); *s* **trabajo [m] adicional** (Trabajo de construcción adicional que se describe en el ▶resumen de prestaciones después de la partida principal y que se paga independientemente; ▶partida [de obra] adicional); *f 1* **plus-value [f]** (Travaux supplémentaires et à rémunérer, prévus dans le ▶descriptif-quantitatif ; ▶numéro de prix pour plus-value ; CCM 1984, 189) ; *f 2* **majoration [f]** (Travaux supplémentaires et à rémunérer, prévus dans la description des travaux et les spécifications techniques ; *contexte* majoration sur le prix pour façons complémentaires, façons accessoires ; CCM 1984, 189) ; *g* **Zulage [f]** (Zusätzlich zu erbringende und zu vergütende Bauleistung, die in einer Folgeposition zur Hauptposition im ▶Leistungsverzeichnis beschrieben ist; ▶Zulageposition).

1961 extreme habitat [n] *ecol.* (Very stressful habitat in which only certain natural or culturally altered conditions have considerable influence on the organisms which live there [▶biocenosis]. Natural extreme sites include, e.g. deserts, cliffs, alpine regions, etc.; man-made extreme sites include, e.g. cliffs in quarries, piles [US]/tips [UK] with toxic waste, paving joints, a big city, etc.; ▶isolated patch habitat, ▶location 1); *syn.* hostile habitat [n] (TGG 1984, 175), severe habitat [n] (TEE 1980, 84), extreme site [n] (TEE 1980, 84); *s* **biótopo [m] extremo** (Hábitat en el que una o pocas condiciones ambientales extremas determinan las posibilidades de crecimiento de las ▶biocenosis; pueden ser naturales, como los desiertos cálidos o fríos, o antropógenos como las aglomeraciones urbanas, los vaciaderos de gangas tóxicas o muy ácidas, juntas entre las losas de pavimento, paredes de canteras, etc.; ▶estación, ▶hábitat natural aislado); *syn.* residencia [f] ecológica extrema, estación [f] extrema, hábitat [m] extremo; *f* **biotope [m] aux conditions stationnelles extrêmes** (On distingue les milieux naturels [▶biocénoses] aux conditions extrêmes comme le désert, les falaises, les régions alpines, les vasières, ainsi que les milieux anthropogènes comme la falaise dans une carrière, un mur, les joints dans un pavage, le terril, la ville, etc. dans lesquels seuls quelques facteurs agissent sur les organismes vivants ; ▶station, ▶habitat relictuel isolé) ; *syn.* station [f] aux conditions extrêmes ; *g* **Extrembiotop [m,** *auch* **n]** (Lebensraum, in dem nur sehr wenige Umweltfaktoren, aber dafür um so ausgeprägter auf die dort lebenden Organismen [▶Biozönosen] wirken. Es gibt **natürliche E.e** wie Wüsten, Sanddünen, Felswände, alpine Regionen, Wattenschlick etc. und **anthropogene E.e** wie z. B. Felswände in Steinbrüchen, Halden mit giftigem oder sehr saurem Bergematerial, Fugen im Straßenpflaster, die Wärmeinsel Großstadt etc.; ►Sonderstandort, ►Standort 1); *syn.* Extremstandort [m].

extreme site [n] *ecol.* ►extreme habitat.

1962 extreme soil erosion [n] *conserv. pedol.* (Severe soil washing caused by excessive cultivation; ►denudation, ►truncated soil profile); *syn.* soil obliteration [n] [also US]; *s* **destrucción** [f] **del suelo** (Forma extrema de erosion provocada por la forma de cultivar la tierra; ►denudación, ►suelo decapitado); *f* dévastation [f] des sols (Forme extrême de l'érosion du sol causée par l'excès de travail du sol, le surpâturage, le défrichement, etc. ; ceux-ci provoquent la destruction de la cohésion du sol, la diminution de la perméabilité, etc., autant d'éléments qui accélèrent les phénomènes d'érosion ; ►érosion continentale, ►sol décapé) ; *syn.* ravages [mpl] causés à la terre végétale ; *g* **Bodenverheerung** [f] (Extremes Ausmaß der Bodenerosion durch Maßnahmen der Bodenbewirtschaftung; ►Abtragung, ►geköpftes Bodenprofil); *syn.* Bodenzerstörung [f].

eye [n] [US] *constr.* ►access eye.

eye [n]**, rodding** [UK] *constr.* ►access eye [US].

1963 eyrie [n] *zool. game'man.* (Nest of a bird of prey, e.g. hawk or eagle typically found in tall trees or high rock formations); *syn.* aerie [n], eyry [n], ayrie [n]; *s* **nido** [m] **de ave rapaz** (Nido de rapaz normalmente situado en árboles altos o en formaciones rocosas de gran altura; *término específico* aguilera [f]); *f* aire [f] d'un oiseau de proie (Site de nidification des espèces-proies ; très souvent cavité rupestre ou espace plat sur les corniches rocheuses ou les falaises escarpées, grands arbres sur lequel est installé le grand nid des rapaces) ; *g* **Horst** [m] **(eines Greifvogels)** (Meist auf unzugänglichen Felsen oder in großer Höhe gebautes Nest eines Greifvogels, z. B. Falke oder Adler. Im Deutschen werden Nester von Stelzvögeln [Reiher, Storch] auch **H.** genannt); *syn.* Greifvogelhorst [m].

eyrie observation [n] *conserv. game'man.* ►protective eyrie observation.

eyry [n] *zool. game'man.* ►eyrie.

F

fabric [n] **of the landscape** *landsc.* ►landscape structure.

façade [n] *gard. urb.* ►planted façade.

façade greenery [n] *gard. urb.* ►façade planting.

1964 façade planting [n] *gard. urb.* (*Also* facade planting; **1.** Covering of a building wall with ►climbing plants, window boxes or espaliered woody plants which may be shaped by trimming and fastened to the wall surface; ►foundation planting; *generic term* building green. **2.** *Result* ►planted façade); *syn.* façade greenery [n], vertical garden [n] [also CDN], vertical green space [n], vertical greenery [n], vertical landscaping [n], wall covering [n] with vines [also US]; *s* **plantación** [f] **de fachadas** (**1.** *Proceso* cobertura de paredes de edificios por medio de ►plantas trepadoras, jardineras en ventanas o leñosas en espaldera; ►jardinería lineal bordeando fachadas; *término genérico* vegetalización [f] de edificios. **2.** *Resultado* ►fachada vegetalizada); *f* végétalisation [f] des façades (**1.** Implantation d'une draperie de ►plantes grimpantes ou de plantes palissées ; ►ceinture végétale sur rez-de-chaussée. **2.** *Résultat* ►façade tapissée [de]) ; *syn.* jardinage [m] des surfaces verticales, jardinage [m] vertical, création [f] des jardins verticaux, jardinage [m] des façades ; *g* **Fassadenbegrünung** [f] (**1.** *Vorgang* Gestaltung von Gebäudewänden mit ►Kletterpflanzen, Balkonkästen oder mit Gehölzen, die durch Formschnitt Wände flächig bedecken; in D. wurde in der zweiten Hälfte der 1970er-Jahre die **F.** als Instrument zur Abmilderung innerstädtischer Gründefizite erkannt und in vielen Städten durch Gemeinderatsbeschlüsse finanziell gefördert; ►Pflanzung am Gebäudesockel; *OB* Bauwerksbegrünung. **2.** *Ergebnis* ►begrünte Fassade, vertikales Grün); *syn.* Vertikalbegrünung [f], vertikale Gebäudebegrünung [f] (S+G 1995 [11], 736).

face [n] [US]**, exposed** *constr.* ►incised slope.

face concrete [n] *constr.* ►facing concrete.

faced wall [n] [US] *constr.* ►one-sided masonry.

1965 face wall [n] *constr.* (Wall constructed on excavated ground in front of stable, existing soil or rock which, in contrast with a ►retaining wall, is not subject to excessive earth pressure); *syn.* breast wall [n] (ARB 1983, 174); *s* **muro** [m] **cortina** (Muro exterior de cerramiento, más o menos ligero, que no posee función resistente; DACO 1988; *opp.* ►muro de sostenimiento); *f* mur [m] de revêtement (Mur placé devant un front de déblai ou contre le terrain naturel d'un talus stable et qui à la différence d'un ►mur de soutènement n'a pas à contenir la poussée de terres qui lui sont adossées) ; *syn.* perré [m] ; *g* **Futtermauer** [f] (Mauer im Abtragsplanum vor gewachsenem, standfestem Boden, die im Gegensatz zur ►Stützmauer keinen oder nur unbedeutenden ►Erddruck aufnehmen muss).

1966 facies [n] **(1)** *phyt.* (Physiognomic predominance of a species composition within a ►plant association or ►subassociation); *s* **facies** [f] **(1)** (Voz empleada en geobotánica con diferentes significados. H. de VILLAR propone esta definición: «La variante o el número de variantes cuyos límites sociológicos coinciden con los límites topológicos». Para que haya diferencia de **f.** debe, pues, haber diferencia florística, pero ésta debe ofrecer paralelismo o relación con la del medio. Así, el encinar, en el centro de España, ofrece una **f.** silicícola y una **f.** calcícola, que están acusadas por diferencias florísticas. En la escuela de Zürich-Montpellier, **f.** es la unidad de menor categoría de la sistemática fitosociológica; cf. DB 1985 ; ►asociación, ►subasociación); *f* faciès [m] **(1)** à (*Aussi* facies [m] ; individualisation physionomique d'un étage, d'une ►association [végétale] ou d'une ►sous-association liée à des conditions stationnelles particulières et caractérisée localement par l'abondance ou la dominance d'une espèce) ; *g* **Fazies** [f] **(1)** (Physiognomisch stark hervortretende Artenkombination innerhalb einer ►Assoziation oder ►Subassoziation).

1967 facies [n] **(2)** *geo.* (Appearance, nature or character, applied especially to descriptions of the composite character of sedimentary rock); *s* **facies** [f] **(2)** (Conjunto de caracteres litológicos [litofacies] y paleontológicos [biofacies] que presenta una roca sedimentaria. La **f.** es indicativa del medio, marino o continental, en el que la roca se formó así como de los procesos de erosión y transporte previos a su deposición; DGA 1986; *f* faciès [m] **(2)** (*Aussi* facies [m] ; ensemble des caractères pétrographiques et paléontologiques définissant un dépôt ou une roche ; DG 1984) ; *g* **Fazies** [f] **(2)** (Gesamtheit aller Merkmale eines Sedimentgesteins hinsichtlich seines petrografischen Aufbaues).

facilities [npl] **for throwing the hammer** *recr.* ►area for throwing the hammer.

facility [n] *leg. urb.* ►communal facility; *constr. land'man.* ►erosion control facility; *recr.* ►indoor tennis facility; *envir.* ►noise screening facility; *recr.* ►pole vault facility; *plan.*

F

►power distribution facility; *envir.* ►pretreatment facility; *urb.* ►public facility (1); *plan.* ►public facility (2); *recr.* ►recreation facility, ►simply-provided recreation facility, ►sports facility; *trans.* ►traffic facility, ►traffic guidance facility; *recr.* ►well-provided recreation facility.

facility [n]**, amenity** *arch. constr. urb.* ►outdoor furniture.

facility [n] [US]**, community** *urb.* ►public facility (1).

facility [n]**, construction and demolition waste disposal** *constr. envir.* ►building waste dump.

facility [n]**, poorly-provided recreation** *recr.* ►simply-provided recreation facility.

1968 facing concrete [n] *constr.* (Final surfacing of exposed aggregate applied to a structural concrete wall surface to provide a finished and more attractive appearance); *syn.* face concrete [n], veneer concrete [n]; *s* **hormigón** [m] **de acabado** (Capa de agregado de mejor calidad que se aplica a una superficie de hormigón estructurado); *f* **béton** [m] **à granulats apparents** (Surface d'un béton de parement sur laquelle on fait apparaître les granulats constitués en général de matériaux de qualité supérieure à ceux contenus dans la masse du béton) ; *g* **Vorsatzbeton** [m] (Sichtbetonoberfläche, die im Vergleich zur dahinter liegenden Betonmasse aus höherwertigem Gesteinsmaterial gefertigt ist).

fact-finding data collection [n] *prof. plan.* ►determination of planning context [US].

factor [n] **for calculation of permissible coverage** [ZA] *leg. urb.* ►building coverage [US].

1969 factory-built house [n] [US] *arch.* (**1.** Dwelling unit constructed and assembled completely at a factory, transported to the building site, and placed on a permanent foundation); *syn.* pre-assembled house [n] [UK]; **2. prefabricated house** [n] (Dwelling assembled and erected at the site primarily from standardized manufactured components; DAC 1975); *syn. for 1. and 2.* prefab [n] [also US]; *s* **casa** [f] **prefabricada** (**1.** Unidad de vivienda construida y montada completamente en una fábrica, que es transportada al lugar de emplazamiento y montada sobre cimientos permanentes. **2.** Casa montada en el emplazamiento utilizando mayormente elementos estandarizados prefabricados); *f* **maison** [f] **préfabriquée** (**1.** Maison, logement préassemblés en usine et transportés sur le site. **2.** Plus généralement système constructif utilisant des composants fabriqués en usine et assemblés sur le chantier) ; *g* **Fertighaus** [n] (In einer Fabrik witterungsunabhängig vorgefertigtes [Wohn]haus, das aus meist serienmäßig hergestellten, typisierten, geschosshohen und raumbreiten Bauteilen zusammengesetzt wird. Vorteile des Fertigbaus im Vergleich zu konventionellen Baumethoden sind v. a. kürzere Bauzeiten und niedrigere Kosten).

1970 factory farming [n] *agr.* (Breeding and rearing of animals on a large, intensive, and industrial scale within a confined space); ►battery of cages); *s* **ganadería** [f] **intensiva** (Gran empresa de cría de ganado bajo condiciones cuasiindustriales; *término específico* ►batería de jaulas de puesta); *syn.* ganadería [f] industrial; *f* **élevage** [m] **intensif** (Exploitation visant à la production quasi-industrielle d'animaux domestiques ou utiles ; *terme spécifique* poulailler industriel ; ►batterie de ponte) ; *syn.* élevage [m] industriel, élevage [m] hors sol (LA 1981, 443 et s.) ; *g* **Massentierhaltung** [f] (Großbetrieb zur intensiven, industriemäßigen Produktion von Nutztieren auf engem Raum; *UB* ►Legebatterie).

faggoting [n] [US] *constr.* ►brush layering.

failure [n] *constr.* ►bearing capacity failure, ►shear failure of embankment, ►slope failure.

failure [n]**, riverbank** *geo.* ►riverbank collapse [US]/riverbank collapse [UK].

fair-faced concrete [n] [UK] *constr.* ►smooth-faced concrete.

fair game [n] *leg. hunt.* ►huntable game.

faithful species [n] *phyt.* ►exclusive species.

fall [n] [UK] *constr.* ►slope [US] (2).

fall [n] [UK]**, longitudinal** *constr.* ►longitudinal gradient.

fall arrest equipment [n] *constr. leg.* ►protection from falling, #2.

fall-blooming plant [n] [US] *hort.* ►autumn-blooming plant.

fall-blooming species [n] [US] *hort.* ►autumn-blooming plant.

fall colors [npl] [US] *bot.* ►autumn colors/hues (of leaves) [US].

1971 fall-flowering period [n] [US] *hort. plant.* (**1.** ►Flowering period in September, October and/or November, not to be confused with a **fall/autumn floral display:** horticultural expression for those plants, cultivated in flower beds for fall/autumn flowering, until the spring floral display begins. **2. Fall/autumn flowering** means a noticeable abundance of one or several flowering plant species in the garden or in the countryside in the fall/autumn); *syn.* autumnal flowering season [n] [UK]; *s* **periodo/período** [m] **de floración otoñal/autumnal** (**1.** En el hemisferio norte periodo/período de floración en septiembre, octubre y noviembre; no se debe confundir con el **florecimiento otoñal:** término horticultor para el conjunto de plantas que se cultivan en macizos para la floración otoñal y se cuidan hasta el cambio a la floración prevernal. **2. Floración** [f] **otoñal/autumnal** denomina la abundacia [visual] de flores de una o varias especies en jardines o en el campo durante el otoño); *f* **floraison** [f] **automnale** (**1.** Processus de développement des fleurs pendant la saison de l'automne [►floraison]. **2.** Période d'épanouissement des fleurs située entre septembre et novembre ; *contexte* arbuste, vivace, bulbe à floraison automnale ; à ne pas confondre avec le terme horticole des **fleurs automnales** ou **décoration florale automnale** qui regroupe les plantes cultivées en plate-bande ou en massifs et entretenues jusqu'à leur remplacement par les fleurs printanières) ; *g* **Blütezeit** [f] **im Herbst** (**1.** ►Blütezeit im September, Oktober und/oder November; nicht zu verwechseln mit **Herbstflor:** gärtnerischer Ausdruck für die Gesamtheit der Pflanzen, die für die Herbstblüte in Blumenbeeten kultiviert und bis zum Wechsel zum Frühjahrsflor gepflegt werden. **2. Herbstblüte** bezeichnet die [visuell] wahrnehmbare Fülle einer Pflanzenart oder mehrerer blühender Arten im Garten oder in der Landschaft im September bis November); *syn.* Herbstblüte [f, o. Pl].

fall foliage [n] [US] *bot. hort.* ►autumn foliage.

fall [n] **of a slope** [UK] *constr. geo.* ►slope gradient.

fallout [n] *envir.* ►radioactive fallout.

fall overturn [n] [US] *limn.* ►autumn circulation period.

1972 fallow [n] *agr.* (Condition of land left uncropped for 1 or 2 years as part of an agricultural cropping system; ►abandonment of farmland, ►abandoned land, ►abandoned pasture, ►fallow land, ►green fallow, ►setting-aside of arable land [UK]); *s* **barbecho** [m] (Tierra sin cultivar durante un tiempo como parte del sistema de cultivo; ►abandono de tierras, ►abandono de tierras agrícolas, ►barbecho a corto plazo, ►herbazal, ►tierra abandonada, ►tierra en barbecho); *f 1* **jachère** [f] (Phase improductive temporaire d'une terre labourable dans la technique culturale ; ►enfrichement, ►friche sociale, ►gel environnemental, ►jachère verte, ►prairie abandonnée, ►terre en jachère) ;

f **2 friche [f] (2)** (Terrain laissé momentanément ou durablement à l'abandon, résultat du dépeuplement des territoires agricoles ou du désintéressement des propriétaires dans les campagnes urbanisées ; DG 1984, 203 ; ▶enfrichement, ▶friche sociale, ▶mise en friche de terres agricoles) ; *g* **Brache [f] (2)** (OB zu Grün-, Teil-, Dauer-, Kurzzeit- und Schwarzbrache. Nutzungsfreie Phase eines landwirtschaftlichen Bodennutzungssystems; im Französischen wird mit *jachère* die **Kurzzeitbrache** als Bestandteil der Kulturfolge und mit *friche* eine **Langzeitbrache** als Folge von z. B. ▶Flächenstilllegung, Desinteresse der Landwirte, die lieber in der Stadt arbeiten wollen, unterschieden; ▶Auflassung von landwirtschaftlichen Flächen, ▶Brachland, ▶Grünbrache, ▶Grünlandbrache, ▶Sozialbrache).

fallow [adj], let agricultural land lie *agr.* ▶leave uncultivated.

fallow grassland [n] *agr.* ▶abandoned pasture.

1973 fallow land [n] *agr.* (Unused open arable land as a part of a ▶crop rotation to restore its fertility; ▶fallow); *s* **tierra [f] en barbecho** (Tierra no cultivada temporalmente como parte de la ▶rotación de cultivos en una zona; ▶barbecho); *f* **terre [f] en jachère** (Terre laissée temporairement improductive dans le cadre de la ▶rotation des cultures sur un même terrain ; ▶jachère) ; *g* **Brachland [n]** (Im Rahmen einer ▶Fruchtfolge nicht genutztes, offen gehaltenes Ackerland; ▶Brache 2).

falls [npl] [UK] *constr.* ▶slopes [US].

falls [n, only pl.] [UK], long *constr.* ▶longitudinal gradient.

falls [npl] and gradients [npl], surveyed to proposed *constr.* ▶graded to proper line and levels.

false colo(u)r aerial photograph [n] *envir. plan. rem'sens. surv.* ▶infrared color aerial photograph [US].

false colo(u)r photograph [n] *envir. plan. rem'sens. surv.* ▶infrared color photograph [US].

1974 family accommodation [n] *recr.* (Place of lodging for one or more families; ▶hostelry); *s* **albergue [m] familiar** (Hospedaje de vacaciones para una o más familias; ▶albergue); *f* **gîte [m] familial** (Habitation de vacances meublée, souvent labelisée, prévu pour l'accueil d'une ou plusieurs familles ; ▶gîte, ▶gîte de France) ; *g* **Ferienunterkunft [f] für (eine oder mehrere) Familien (≠)** (▶Beherbergungsstätte).

1975 family burial site [n] *adm. landsc.* (▶Burial site which is distinguished from ▶single row graves by its size and favo[u]rable location and therefore offers the opportunity to erect a larger burial monument); *s* **panteón [f] familiar** (▶Tumba de mayores dimensiones que las ▶tumbas individuales en hilera, de manera que hay espacio para construir un monumento conmemorativo); *f* **caveau [m] de famille** (▶Sépulture qui se distingue d'une ▶tombe à la ligne par sa grandeur et son emplacement privilégié et qui donne la possibilité d'édifier un monument funéraire) ; *syn.* caveau [m] familial ; *g* **Sondergrabstätte [f]** (▶Grabstätte, die sich durch Größe, bevorzugte Lage und durch die Möglichkeit zur Bestattung mehrerer Verstorbener einer Familie oder eines durch Verwandschaft verbundenen Personenkreises vom ▶Reihengrab unterscheidet; ein anderes weitergehendes Nutzungsrecht ist die Errichtung eines größeren Grabdenkmales. Die Nutzungsdauer ist nicht auf die Dauer der Ruhezeit beschränkt, sondern wird wesentlich länger bemessen [40-60 Jahre]; cf. GAE 2000; es gibt ein-, zwei- und dreistellige **S.n**); *syn.* Familiengrab [n], Familiengrabstätte [f], Familiengrabstelle [f], Sondergrab [n], Sondergrabstelle [f], Vorzugsgrab [n], Vorzugsgrabstätte [f], Vorzugsgrabstelle [f], Wahlgrab [n], Wahlgrabstätte [f], Wahlgrabstelle [f].

1976 family or youth camp [n] *recr.* (Residential recreation facility in rural areas or ▶tourist area; *specific terms in U.S.* organization or group camp for low income persons or groups; *also* camps for Boy Scouts, Girl Scouts, Campfire Girls, etc.; ▶resort community [US]/holiday village [UK]); *s* **colonia [f] de vacaciones** (Agrupación de casas sencillas para acoger a jóvenes o a familias en época de veraneo, situado en zona rural o en ▶zona de veraneo; ▶pueblo de veraneo); *f* **colonie [f] de vacances** (Habitation ou centre utilisé par des enfants, des adolescents, des familles en zone rurale ou dans les unités touristiques nouvelles en zone de montagne ; *termes spécifiques* camp d'adolescents, centre aéré [dans le centre ou les abords d'agglomérations urbaines pour les couches sociales défavorisées avec migration journalière entre la maison d'habitation et le centre de vacances] ; art. 10, décret 92-273 du 23 mars 1992 relative aux plans de zones sensibles aux incendies de forêts ; ▶zone de tourisme, ▶village de vacances) ; *g* **Ferienkolonie [f]** (Siedlung im ländlichen Raum, die vorwiegend aus Ferienhäuser für Kinder, Jugendgruppen oder ganze Familien besteht oder ein kleines ▶Ferienerholungsgebiet; ▶Feriendorf).

fan [n] *geo.* ▶alluvial fan, ▶rock fan; *plan.* ▶vista fan.

1977 fan espalier [n] *hort.* (Fruit tree or other woody plant trained to grow flat against a wall or trellis in a fan shape; ▶espaliered fruit tree, ▶palmette espalier); *s* **palmeta [f] en abanico** (En fruticultura, ▶frutal en espaldera obligado a crecer en forma aplanada apoyado contra muro o celosía de madera, con la horquilla a unos 40 cm del suelo y ramas en forma de abanico; ▶palmeta regular [de ramas oblicuas/horizontales]); *syn.* espaldera [f] en abanico); *f* **palmette [f] en éventail** (Arbre fruitier, à forme plate, possédant deux ou plusieurs branches partant d'un tronc commun ; ADT 1988, 48 ; ▶arbre fruitier taillé en espalier, ▶palmette à branches horizontales/obliques) ; *g* **Fächerspalier [n]** (*Obstbau* ▶Formobstbaum mit einer Stammhöhe von 40 cm und einer fächerförmigen Astverzweigung; ▶Spalier mit schrägen/waagerechten Ästen).

fan pattern [n] [UK] *constr.* ▶fish scale paving [US].

fan-trained fruit tree [n] *hort.* ▶palmette espalier.

farm [n] *recr.* ▶children's farm; *agr. sociol.* ▶demonstration farm; *envir.* ▶wind farm.

farm [n], tenant [US] *agr.* ▶agricultural leasehold (property).

1978 farm camping [n] *recr.* (Type of vacation [US]/holiday [UK] with overnight lodging in a tent on farm property; ▶camp out [vb]); *s* **acampado [m] en granja o caserío** (Forma de vacaciones rurales en la que se acampa en el terreno de una granja; acampar [vb] en granja o caserío, hacer [vb] camping en granja o caserío; ▶acampar por libre); *f* **camping [m] à la ferme** (Forme de l'hôtellerie de plein-air caractérisant la pratique du camping sous tente, effectuée sur une exploitation agricole, souvent de capacité limitée pour éviter à l'agriculteur d'être soumis au statut d'entreprise commerciale ; forme de camping parfois dénommée camping diffus ; ▶camping sauvage) ; *syn.* camping [m] fermier, camping [m] en gîte rural ; *g* **Zelten [n, o. Pl.] auf dem Bauernhof** (Urlaubsform, bei der bei einem Bauern gezeltet wird; ▶wildes Lagern und Zelten).

farmer [n] *agr.* ▶organic farmer.

farm holding [n] [UK], resited *agr.* ▶relocated farmstead [US].

farm holdings [npl] [UK], relocation of *agr. urb.* ▶farmstead resettlement [US].

farm holidays [npl] [UK] *recr.* ▶farm vacation [US]/farmstay holidays [UK].

farming [n] *agr. land'man.* ▶contour farming; *agr.* ▶factory farming, ▶grassland farming, ▶intensive farming, ▶organic farming, ▶pasture farming, ▶pond farming, ▶traditional farming.

farming [n], **crop** *agr.* ▶cultivation of arable land.

farming [n], **livestock** *agr.* ▶livestock industry.

farming [n], **mixed** *agr. for.* ▶mixed crop production.

farming [n], **one-crop** *agr.* ▶single crop system.

farming [n], **sheep** *agr* ▶sheep raising.

farming [n], **strip** *agr. land'man.* ▶contour strip cropping.

farmland [n] *agr.* ▶abandonment of farmland, ▶agricultural area; *plan. agr.* ▶uncultivated farmland.

farmland [n], **abandoned** *agr. sociol.* ▶abandoned land.

farmland [n], **re(-)allocation of** *adm. agr. leg.* ▶farmland consolidation [US]/land consolidation [UK].

farmland arrangement [n] *agr. land'man.* ▶farmland structure.

1979 farmland consolidation [n] [US] *adm. agr. leg.* (Grouped ownership of scattered agricultural or forest plots in order to create more compact holdings, thereby promoting a greater productive efficiency; ▶rearrangement of lots [US]/re-organisation of plot boundaries [UK]); *syn.* consolidation [n] of agricultural land, reallocation [n] of farmland [US]/re-allocation [n] of farmland [UK], land consolidation [n] [UK]; *s* **concentración** [f] **parcelaria** [Es] (Política agraria de concentrar las tierras repartidas de diferentes dueños con el fin de mejorar las condiciones de producción agro-silvícola; ▶reordenación de parcelas); *f* **remembrement** [m] **rural** (Réorganisation de la propriété rurale grâce à un regroupement et une nouvelle répartition parcellaire assurant une structure des exploitations agricoles propre à une meilleure utilisation et rentabilité des terres ; ▶remembrement parcellaire) ; *g* **Flurbereinigung** [f] (Zusammenlegung von zersplittertem oder unwirtschaftlich geformtem land- und forstwirtschaftlichem Grundbesitz zur Verbesserung der Arbeits- und Produktionsbedingungen; *gesetzliche Grundlagen* Flurbereinigungsgesetz von 1976, zuletzt geändert 19.12. 2008, und Ausführungsgesetze der Bundesländer [D], Flurverfassungs-Grundsatzgesetz von 1951 [A], zuletzt geändert 2000; der Begriff **Zweckflurbereinigung** bezeichnet die Flurneuordnung für Großbauprojekte von z. B. Straßen, Flughäfen, Messegelände; die **Verkoppelung** war eine einfache Art der **F.** im 18. Jh., bei der die Gemengelage des Grundbesitzes durch eine Neugliederung der Flur aufgehoben wurde, z. B. die Knicklandschaft in Schleswig-Holstein; ▶Umlegung); *syn.* Flurneuordnung [f], Kommassierung [f] [auch A], Güterzusammenlegung [f] [CH], *obs.* Verkoppelung [f].

farmland consolidation act [n] *adm. agr. leg.* ▶land consolidation act [UK].

1980 farmland consolidation area [n] [US] *agr. leg.* *syn.* land consolidation area [n] [UK]; *s* **área** [f] **de concentración parcelaria**; *f* **périmètre** [m] **de remembrement** ; *g* **Flurbereinigungsgebiet** [n] (Gesamtgebiet, das der Flurbereinigung unterliegt).

1981 farmland consolidation procedure [n] [US] *adm. agr. leg.* (System whereby the execution of land consolidation is authorized for the area concerned according to the legislation governing its implementation; **in U.S.**, this is done by private enterprise without governmental involvement, except for agricultural zoning, which only restricts land to agricultural use; ▶farmland consolidation [US]/land consolidation [UK], ▶farmland consolidation area, ▶land consolidation act); *syn.* land consolidation procedure [n] [UK]; *s* **procedimiento** [m] **de concentración parcelaria** (Procedimiento administrativo de planificación y ejecución de la ▶concentración parcelaria llevado a cabo por la correspondiente administración sobre la base de la correspondiente legislación, en D. sobre la base de la ▶ley de

concentración parcelaria; ▶área de concentración parcelaria); *f* **procédure** [f] **de remembrement** (Ensemble des règles d'organisation d'ordre administratif pour lequel chaque phase de la procédure de ▶remembrement rural est portée par voie d'enquête publique à la connaissance des intéressés, qui peuvent présenter leurs réclamations ; ▶périmètre de remembrement, ▶loi relative au remembrement des propriétés rurales) ; *g* **Flurbereinigungsverfahren** [n] (Behördlich geleitetes Verfahren zur Durchführung der ▶Flurbereinigung innerhalb eines ▶Flurbereinigungsgebietes nach dem ▶Flurbereinigungsgesetz [D]/Flurverfassungs-Grundsatzgesetz [A] und dessen Ausführungsgesetzen; cf. Zusammenlegungsverfahren [§ 91 FlurbG]).

1982 farmland structure [n] *agr. land'man.* (Structuring of farmland by hedgerow, woodland, drainage channels and wayside planting, mostly to protect against wind and water erosion); *syn.* farmland arrangement [n]; *s* **estructuración** [f] **de los campos agrosilvícolas** (Creación de caminos, zanjas de avenamiento, aterrazamiento de pendientes, plantaciones pantalla o de bosquecillos en zona agrícola teniendo en cuenta las exigencias de protección del suelo de la erosión); *f* **structuration** [f] **de l'espace rural** (Ouvrages ou plantations à caractère de protection contre l'érosion, implantés le long des chemins, fossés de drainage, terrasses ; *syn.* organisation [f] des espaces ruraux) ; *g* **Flurgliederung** [f] (Meist dem Erosionsschutz dienende zweckmäßige Anlage der Wirtschaftswege, Entwässerungsgräben, Hangterrassierungen, Windschutzpflanzungen und Waldparzellen).

farm pond [n] [UK] *agr.* ▶fish pond.

farm runoff [n] [US] *agr. envir.* ▶agricultural wastewater.

farmstay holidays [npl] [UK] *recr.* ▶farm vacation [US].

1983 farmstead [n] *agr.* (Farmhouse and its adjacent buildings and service areas; a yard connected with farm buildings or enclosed by them is called **farmyard**; ▶relocated farmstead [US]/resited farm holding [UK]); *s 1* **granja** [f] (▶granja relocalizada); *syn.* casa [f] de labranza, finca [f], hacienda [f] [AL], rancho [m] [AL], cortijo [m] [Andalucía], masía [f] [Catalunya], caserío [m] [País Vasco], pazo [m] [Galícia]; *s 2* **estancia** [f] **[RA]** (Gran finca de explotación ganadera en la Argentina); *f* **ferme** [f] (Ensemble des bâtiments d'habitation et d'exploitation agricoles ; LA 1981 ; ▶ferme transférée) ; *syn.* habitat [m] rural ; *g* **Gehöft** [n] (Landwirtschaftliches Anwesen mit den dazugehörenden Wohn- und Wirtschaftsgebäuden, die sich um einen **Wirtschaftshof** gruppieren; ▶Aussiedlerhof); *syn.* Hoflage [f], Hofstelle [f].

1984 farmstead resettlement [n] [US] *agr. urb.* (Relocation of farmsteads [US]/farm holdings [UK] and their reconstruction outside of the confines of existing settlements); *syn.* relocation [n] of farm holdings [UK]; *s* **reasentamiento** [m] **de granjas** (Traslado de una empresa agrícola a las afueras de las zonas urbanizadas); *syn.* relocalización [f] de granjas; *f 1* **transfert** [m] **d'exploitations agricoles** (1. Déplacement d'une exploitation agricole ou d'une habitation rurale, p. ex. lorsqu'elle est traversée par un projet entrainant la destruction des bâtiments ou la coupure des élevages de leur espace naturel, ne permettant pas le maintien sur place de l'ensemble exploité ; cf. www.sarthe.pref.gouv.fr ; **2.** transmission d'une exploitation agricole à un repreneur [relève familiale ou acquéreur étranger — reprise partielle ou totale]) ; *syn.* transfert de ferme [CDN] ; *f 2* **restructuration** [f] **d'exploitations agricoles** (Reconstitution d'un ensemble exploité à partir de l'îlot le mieux adapté) ; *terme spécifique* relocalisation de bâtiments agricoles ; *syn.* déplacement [m] de bâtiments d'exploitation agricole, transplantation [f] (1) ; *g* **Aussiedlung** [f] (Verlegung eines landwirtschaftlichen Haupterwerbsbetriebes aus meist beengter Ortslage, in der sich der Betrieb nicht zukunftsorientiert entwickeln kann, und wegen immissionsschutzrechtlicher Gründe in die Flur; aussiedeln [vb]. Nach dem 2. Weltkrieg begannen die **A.en**, da sich für Landwirte die wirtschaftliche Notwendigkeit ergab, größere Felder als vorher zu bewirtschaften und größere Viehbestände zu halten; **A.en** wurden durch die Raumordnung gefördert, um den Lebensstandard der ländlichen Bevölkerung an den der städtischen anzupassen und um moderne Betriebsanlagen zu bauen und die notwendige Infrastruktur bereitzustellen).

farm track [n] [US] *agr.* ▶field track.

1985 farm vacation [n] [US] *recr.* (*Specific term* ▶farm camping; ▶rural overnight accommodation); *syn.* farmstay holidays [npl] [UK], farm holidays [npl] [UK]; *s* **vacación [f] en granja** (Tipo de turismo basado en el alojamiento en casas rurales y donde el usuario participa en las tareas cotidianas del lugar: agrícolas, ganaderas, artesanales; *término genérico* ▶turismo rural; *término específico* ▶acampar en granja; ▶hospedaje en el campo); *f* **vacances [mpl] à la ferme** (*Terme spécifique* ▶camping à la ferme ; ▶hébergement rural) ; *syn.* vacances [fpl] dans un gîte rural ; *g* **Ferien [pl] auf dem Bauernhof** (*UB* ▶Zelten auf dem Bauernhof; ▶Beherbergung auf dem Lande).

farmyard [n] *agr.* ▶farmstead.

1986 fascicular root system [n] *bot.* (Dense root system, consisting of many thin, profusely branched roots, mostly of c[a]espitose grasses, with no ▶tap root; ▶fibrous root); *syn.* bunched root system [n]; *s* **raíz [f] fasciculada** (Sistema radical en el que las raicillas están agrupadas formando un haz, como p. ej. en el ajo; ▶raíz axonomorfa, ▶raíz fibrosa); *f* **racine [f] fasciculée** (LA 1981, 937 ; système racinaire dense et ténu se développant en forme de faisceaux, sans ▶racine pivotante, p. ex. les racines de certaines graminées ; ▶radicelle) ; *g* **Büschelwurzel [f]** (Dichtes, aus gleichartigen Wurzeln [homorrhizes System] bestehendes Wurzelwerk ohne ▶Pfahlwurzel; ▶Faserwurzel).

1987 fascine [n] *constr.* (Bundle of flexible twigs and branches bound together and used to control erosion in gullies, or to stabilize banks of watercourses or slopes; ▶brushwood fascine); *syn.* fascine pole [n] [also UK], bundle [n] of wattles [also US], brushwood bundle [n] [also UK]; *s* **fajina [f]** (Haces de matas dispuestas en hileras para contener la erosión en taludes o en orillas de ríos; ▶fajina de ramillas); *f* **fascine [f]** (*Technique paysagère de stabilisation* terme générique pour tous fagots de branchage constitués de longues ramilles de bois divers, en général mélangés à des boutures de saules utilisés pour la fixation du sol sur des talus à pente de 45° à 75° ou pour la stabilisation des berges d'un cours d'eau ; ▶fascine en caniveaux) ; *g* **Faschine [f]** (*Sicherungsbauweisen* OB zu walzenförmiges Gebinde aus langem, elastischem Reisig beliebiger Gehölze mit Weiden gemischt, falls Weidenausschlag erwünscht, zur Festlegung des Oberbodenauftrages auf nicht zu steilen Böschungen oder zur Ufersicherung von Fließgewässern; ▶Hangfaschine).

1988 fascine construction method [n] *constr.* (Bound bundles of live branches partly buried in trenches for erosion control; ▶fascine, ▶wicker fence construction); *s* **enfajinado [m]** (Método de construcción para controlar la erosion con haces de matas dispuestas en hileras; ▶construcción de valla trenzada, ▶fajina); *syn.* fajinada [f]; *f* **fascinage [m] (de talus)** (Travaux de consolidation au moyen de ▶fascines de branchages à rejets ; ▶clayonnage) ; *syn.* consolidation [f] en fascines (WIB 1996) ; *g* **Faschinenbau [m]** (Sicherungsbauweise unter Verwendung von ▶Faschinen; ▶Flechtzaunbau).

1989 fascine drain [n] *constr.* (Drain constructed of ▶fascines buried in a slope to conduct away seepage water); *s* **dren [m] de fajinas** (Matas de ▶fajinas dispuestas en hileras para facilitar el drenaje en taludes); *f* **cordon [m] drainant de fascines** (Ouvrage de drainage au moyen de ▶fascines permet-

tant l'évacuation de l'eau par percolation ou le long de la pente d'un talus et permettant sa stabilisation ; les fagots utilisés sont appelés **fascines drainantes**) ; *g* **Faschinendrän [m]** (*Sicherungsbauweisen* aus ▶Faschinen hergestellte Dränleitung zur Hang- und Sickerwasserabführung an Böschungen; die Faschinenbündel werden auch **Dränfaschinen** genannt).

fascine pole [n] [UK] *constr.* ▶fascine; ▶brushwood fascine.

1990 fast-growing [adj] *agr. for. hort.* (Descriptive term for plants which increase rapidly in size; *opp.* ▶slow-growing; ▶vigor [US]/vigour [UK]; *s* **de crecimiento rápido [loc]** (*opp.* ▶de crecimiento lento; ▶vigor); *f* **à croissance rapide [loc]** (Relatif aux végétaux ; *opp.* ▶à croissance lente ; ▶vigueur) ; *g* **schnell wüchsig [adj]** (Pflanzen betreffend, die jährlich lange und kräftige Triebe bilden; *opp.* ▶langsam wüchsig, ▶Wüchsigkeit).

1991 fastigiate [adj] *hort. plant.* (Narrow, erect habitat of growth with close vertical branching; *cf.* RCG 1982; ▶columnar); *s* **fastigiado/a [adj]** (Denominación de la forma de crecimiento de árboles cuyas ramas se aproximan al eje de tal manera que el conjunto acaba en punta, como en el ciprés y el chopo lombardo; *cf.* DB 1985; ▶columnar); *f* **fastigié, ée [adj]** (Caractérisant un arbre qui a un tronc unique, droit, continu et vigoureux avec une forte densité de branches et de ramifications régulièrement disposées autour du tronc, parfois dès le niveau du sol ; ▶columnaire) ; *g* **schmal-pyramidal [adj]** (Wuchsformbezeichnung von Bäumen; ▶säulenförmig).

1992 fast-moving stream [n] *geo. limn. wat.man.* (▶intermittent watercourse, ▶torrent); *syn.* fast water [n] [also US]; *s* **arroyo [m] de flujo rápido** (1. Curso de agua o tramo del mismo con fuerte pendiente, rápidos y flujo de agua alto y rápido, que se presta para la práctica de deportes acuáticos como rafting, piragüismo, etc. **2. rabión [m]** se denomina la corriente impetuosa de un río en los parajes estrechos o de mucho declive; CAS 1985; ▶aguas vivas, ▶curso de agua intermitente, ▶torrente); *f* **eau [f] vive** (Terme générique pour les cours d'eau irréguliers à débit rapide [▶torrent, cascade, gorge, canyon] qui donnent souvent lieu à la pratique des sports d'eaux vives [rafting, canyoning, kayak, hydro speed ou nages en eaux vives] ; ▶cours d'eau intermittent ; ▶gave) ; *syn.* eau [f] sauvage ; *g* **schnell fließendes Gewässer [n]** (Gewässerlauf oder ein Abschnitt davon mit stärkerem Gefälle, Stromschnellen und hoher und schneller Wasserabflussrate, der sich für Wassersportarten wie z. B. Rafting, Kanu- und Kajakfahrten eignet; ▶zeitweilig Wasser führendes Fließgewässer, ▶Wildbach).

fast water [n] [US] *geo. limn. wat.man.* ▶fast-moving stream.

1993 fauna [n] (1) *zool.* (All the animals to be found in a specific area); *s* **fauna [f]** (Conjunto de animales de una región determinada); *f* **faune [f]** (Totalité des espèces animales d'une région ou d'un milieu déterminé) ; *syn.* monde [m] animal ; *g* **Fauna [f]** (Gesamtheit der Tierarten eines bestimmten Gebietes); *syn.* Tierwelt [f].

fauna [n] (2) *phyt. zool.* ▶depletion of fauna/flora; *conserv. nat.res.* ▶disruptive conditions for fauna; *phyt. zool.* ▶inventory of fauna and flora; *zool.* ▶small fauna, ▶species of soil fauna.

fauna [n], **bird** *zool.* ▶birdlife.

fauna [n] [US], **depauperation of** *phyt. zool.* ▶depletion of fauna/flora.

fauna pool [n] *zool.* ▶anthropogenic alteration of the genetic fauna pool.

feasibility [n] **of recreating ecosystems/habitats** *conserv.* ▶recoverability.

1994 feasibility study [n] *plan. trans.* (Evaluation of a major project to determine its practicability and environmental impact; e.g. the study of a road alignment by an ▶environmental impact assessment [EIA] [US]/environmental assessment [UK] or an ▶environmental impact statement [EIS]); **s estudio [m] de factibilidad** (Estudio evaluativo de un proyecto para determinar su oportunidad, p. ej. en el marco de una ▶evaluación de impacto ambiental [EIA]); ***f* étude [f] de faisabilité** (Étude préalable dégageant l'opportunité, la localisation, la qualité et la possibilité d'une opération d'aménagement, p. ex. d'un projet routier ; ▶procédure d'étude d'impact) ; ***g* Eignungsuntersuchung [f]** (Bewertung eines größeren Bauvorhabens, z. B. einer Verkehrstrasse im Rahmen einer ▶Umweltverträglichkeitsprüfung hinsichtlich seiner Durchführbarkeit).

feather (branch) [n] [UK] *arb.* ▶slender branch [US].

feathered tree [n] [UK] *arb. hort.* ▶ground-branching tree [US].

feathered tree [n] [UK]**, one-year-old** *arb. hort.* ▶strongly branched liner [US].

1995 Federal Act [n] **on Conservation of the Natural Environment** [CH] *conserv. leg.* **s Ley [f] Federal de Conservación de la Naturaleza y el Paisaje [CH]**; ***f* loi [f] fédérale sur la protection de la nature et du paysage [CH]** ; ***g* Bundesgesetz [n] über den Natur- und Heimatschutz [CH]**.

1996 Federal Act [n] **on Nature Conservation and Landscape Management** [D] *conserv. leg.* (**In D.**, there are federal legal guidelines with certain directives, for example: ▶Endangered Species Act [E.S.A.] [US]/Conservation of Wild Creatures and Wild Plants Act 1975 [UK], participation of nature conservation groups, regulations in landscape preservation planning, regulations on encroachments into the landscape; reinforced and fulfilled in detail by state nature conservation laws; ▶National Environmental Policy Act [US]/Town and Country Planning [Assessment of Environmental Effects] Regulations 1988 [UK], ▶nature conservation enactment); **s Ley [f] Federal de Conservación de la Naturaleza y de Gestión del Paisaje [D]** (En D. ley marco que contiene algunas directrices de vinculación directa como la promulgación del ▶decreto para la protección de especies, la participación de las organizaciones naturistas en la política de protección, la planificación del paisaje, la evaluación de impacto ambiental; por lo demás son los estados federados [Bundesländer] los responsables de promulgar y aplicar leyes de protección de la naturaleza; ▶ley del patrimonio natural y de la biodiversidad, ▶ley de protección de la naturaleza, ▶real decreto legislativo de evaluación de impacto ambiental); ***f* loi [f] fédérale relative à la protection de la nature [D]** (La loi sur la protection de la nature et la conservation des paysages, loi cadre fédérale du 20 décembre 1976 modifiée par la loi du 25 mars 2002 outre son caractère général [§§ 1-2], comporte des dispositions particulières concernant la constitution d'un réseau d'habitats naturels [§ 3], l'institutionnalisation de la ▶planification des paysages [§§ 12-17] au niveau des différentes collectivités territoriales, les dispositions à prendre face aux atteintes portées sur le milieu naturel [études d'impact, mesures compensatoires, etc. — §§ 18-21], la définition des différentes catégories nationales de zones protégées [§§ 22-30], la protection des espèces et habitats d'espèces dans les zones protégées du Réseau Natura 2000 [transposition en droit allemand des directives « Habitats » et « Oiseaux » — §§ 32-34], la protection des espèces de faune et de flore sauvages ainsi que leurs habitats [§§ 39-43], la participation des associations agréées de protection de la nature à l'action des organismes publics ainsi que l'action en justice des associations [§§ 58-61] ; cette loi est concrétisée par les lois relatives à la protection de la nature des différents Länder. **CH.**, la loi s'intitule **« loi fédérale sur la protection de la nature et du paysage ».** **En A.** il n'existe pas de loi fédérale mais des lois des Länder ;

▶code de l'environnement, ▶loi relative à la protection de la nature, ▶loi relative aux études d'impact) ; ***g* Bundesnaturschutzgesetz [n] [D]** (*Abk.* BNatSchG; Gesetz über Naturschutz und Landschaftspflege vom 20.12.1976; Rahmengesetz des Bundes i. d. F. vom 25.03.2002 mit dem Oberziel, Natur und Landschaft auf Grund ihres eigenen Wertes und als Lebensgrundlage für die Gesellschaft im besiedelten und unbesiedelten Bereich so zu schützen, dass **1.** die Leistungs- und Funktionsfähigkeit des Naturhaushaltes, **2.** die Regenerationsfähigkeit und nachhaltige Nutzungsfähigkeit der Naturgüter, **3.** die Tier- und Pflanzenwelt einschließlich ihrer Lebensstätten und Lebensräume sowie **4.** die Vielfalt, Eigenart und Schönheit sowie der Erholungswert von Natur und Landschaft gesichert sind. Zur Entwicklung optimaler und nachhaltiger, materieller wie immaterieller Leistungen der Naturausstattung von Landschaftsräumen für die Gesellschaft sind u. a. vorgeschrieben resp. geregelt: die Umweltbeobachtung und Landschaftsplanung, der Umgang mit und das Verfahren bei Eingriffen in Natur und Landschaft [▶Eingriffsregelung], die Ausweisung von Schutzgebieten, deren Pflege und Entwicklung, die Umsetzung der EU-Richtlinie 92/43 EWG zum Aufbau eines kohärenten Europäischen Netzes besonderer Schutzgebiete „Natura 2000", die Überprüfung der Verträglichkeit oder Unzulässigkeit von Projekten in Natur und Landschaft mit den Erhaltungszielen des Eingriffsgebietes von gemeinschaftlicher Bedeutung, den Schutz und die Pflege wild lebender Tier- und Pflanzenarten und der besonders geschützten Arten, z. B. auch durch den Erlass einer ▶Artenschutzverordnung, das Betreten der Flur zum Zwecke der Erholung, die Mitwirkung von Naturschutzverbänden; das BNatSchG wird durch Ländernaturschutzgesetze weiter konkretisiert. Durch die Einführung der Paragrafen zur Landschaftsplanung wird ein *aktiver* Naturschutz möglich. Der Vorgänger war das **Reichsnaturschutzgesetz** vom 26.06.1935 [RNG], das die gesetzliche Basis für einen *rein konservierenden* Schutz von Natur und Landschaft schuf, ohne jedoch großflächig ein ausreichendes Gegengewicht zu den Belastungen der Landschaft durch Industrie und Landwirtschaft sowie durch großräumige wesentliche Eingriffe in Struktur und Naturhaushalt der bäuerlichen Kulturlandschaft in Form von Meliorationsmaßnahmen [Arbeitsbeschaffung und Arbeitsdienst] zu bilden; **CH.**, *Entsprechung in der Schweiz* **Bundesgesetz über den Natur- und Heimatschutz [NHG]** vom 01.07.1966, zuletzt geändert 06.10.2006; **A.**, in Österreich gibt es kein Bundesgesetz, sondern nur Naturschutzgesetze der Bundesländer; ▶Naturschutzgesetz, ▶Umweltverträglichkeitsprüfungs-Gesetz).

1997 Federal Clean Air Act [n] [US] *envir. leg.* (**In U.S.**, Federal law controlling air pollution nation-wide and adhered to in state laws; The **Air Pollution Act** of 1963 affirmed the authority of the Federal Government in dealing with interstate pollution situations, although it recognized air pollution to be primarily a state and local problem. The Air Pollution Act of 1970 empowered the Federal Government to set national air pollution standards [in place of state standards] and required the states to meet those standards [but to develop their own ways of doing so]. It also set up ▶air quality control regions immediately; **in U.K.**, Control of Pollution Act —*abbr.* COPA—two **Clean Air Acts** passed by the United Kingdom Government. The 1956 Act dealt with the control of smoke from industrial and domestic sources, and was extended by the Act of 1968, particularly to control gas cleaning and heights of stacks of installations in which fuels are burned and also to deal with smoke from industrial open bonfires. Both Acts are now subsumed and extended under the **COPA 1974**. This Act covers a wide range of legislation on pollution issues, and under Part IV—atmospheric pollution; cf. DES 1991; ▶contaminated airspace, ▶emission,

►pollution control 1); *syn.* Control of Pollution Act [n] [UK]; *s 1* ley [f] de calidad del aire y protección de la atmósfera [Es] (La ley 34/2007 tiene por objeto establecer las bases en materia de prevención, vigilancia y reducción de la contaminación atmosférica con el fin de evitar y cuando esto no sea posible, aminorar los daños que de ésta puedan derivarse para las personas, el medio ambiente y demás bienes de cualquier naturaleza. A las prescripciones de esta ley están sujetas todas las fuentes de los contaminantes ya sean de titularidad pública o privada, exceptuando los ruidos y vibraciones, las radiaciones ionizantes y no ionizantes, y los contaminantes biológicos, que se rigen por su normativa específica. La ley se orienta en los principios de cautela y acción preventiva, de corrección de la contaminación en la fuente misma y de quien contamina paga, y abarca además los diferentes problemas de la contaminación atmosférica tales como la ►contaminación transfronteriza, el agotamiento de la capa de ozono o el ►cambio climático. Consta de 7 capítulos: I. Disposiciones generales; II. Evaluación y gestión de la calidad del aire; III. Prevención y control de las emisiones; IV. Planificación; V. Instrumentos de fomento de protección de la atmósfera; VI. Control, inspección, vigilancia y seguimiento y VII. Régimen sancionador, así como de las disposiciones adicionales, derogatorias y finales, y de 4 anexos; cf. Preámbulo y arts. 1 y 2 Ley 34/2007, de 15 de noviembre; ►emisión, ►protección del ambiente atmosférico); *s 2* ley [f] federal de protección del aire puro (±) (En EE.UU. ley federal de control de la contaminación atmosférica en todo el país y a la cual se adhieren leyes de los estados; la ley sobre polución de 1963 [Air Pollution Act] reafirmaba la competencia del Gobierno Federal en cuestiones de contaminación interestatal, aunque reconocía que la contaminación era primordialmente un problema estatal y local. La ley sobre polución de 1970 dió poder al Gobierno Federal para fijar estándares de emisión [en lugar de los estatales] y obligaba a los estados a cumplirlos, aunque desarrollando sus propios instrumentos para conseguirlo. También determinó inmediatamente ►zonas de evaluación de la calidad del aire. **En GB**, la ley de control de la contaminación [**Control of Pollution Act** —*abr.* COPA] consiste en dos leyes aprobadas por el Gobierno del Reino Unido. La ley de 1956 trataba el humo de fuentes industriales y domésticas, y fue ampliada por la ley de 1968, particularmente para controlar la depuración de los gases, la altura de las chimeneas de instalaciones de combustión y también el humo de hornos industriales. Ambas leyes fueron sustituidas y ampliadas por la **COPA 1974**. Esta ley cubre un amplio ámbito de regulaciones sobre cuestiones de contaminación y en su Capítulo IV, la contaminación atmosférica); *f* **Loi [f] relative à la lutte contre les pollutions atmosphériques et les odeurs** (Loi portant modification de la loi du 19 décembre 1917 ; loi n° 61-842 du août 1961 ; art. 1: Les immeubles, établissements industriels, commerciaux, artisanaux ou agricoles, véhicules ou autres objets mobiliers possédés, exploités ou détenus par toutes personnes physiques ou morales, devront être construits, exploités ou utilisés de manière à satisfaire aux dispositions prises afin d'éviter les pollutions de l'atmosphère et les odeurs qui incommodent la population, compromettant la santé ou la sécurité publique, ou nuisent à la production agricole, à la conservation des constructions et monuments ou au caractère des sites ; ►émission, ►lutte contre les émissions, ►secteur soumis à la pollution atmosphérique, ►zone de mesures) ; *g* **Bundes-Immissionsschutzgesetz [n]** (±) (*Abk.* BImSchG; Gesetz v. 15.03.1974 i. d. Fassung der Änderungsgesetze, zuletzt geändert durch Artikel 1 des Gesetzes vom 23.10.2007 [BGBl. I 2470], zum Schutz vor schädlichen Umwelteinwirkungen durch Luftverunreinigungen, Geräusche, Erschütterungen und ähnliche Vorgänge; es trifft Regelungen zum Schutz von Menschen, Tieren, Pflanzen und anderen Sachen vor schädlichen Umwelt-

einflüssen und beugt solchen Einflüssen vor. Das **B.** gliedert sich in sieben Teile und regelt im Wesentlichen: Errichtung und Betrieb von Anlagen — wobei zwischen genehmigungspflichtigen und nicht genehmigungsbedürftigen Anlagen unterschieden wird —, Beschaffenheit von Anlagen, Stoffen, Erzeugnissen, Brenn- und Treibstoffen, Beschaffenheit und Betrieb von Kraftfahrzeugen, Bau und Änderung von Straßen und Schienenwagen; ►Emission, ►Gebiet zur Luftqualitätsüberwachung, ►Immissionsschutz, ►lufthygienisches Belastungsgebiet).

1998 federal government agency [n] [US] *adm.* (Federal department/division responsible for executing government program[me]s; ►state government agency [US]); *syn.* central government body [n] [UK], federal government authority [n] (±); *s* **autoridad [f] estatal (federal)** (Instituciones de administración central de un Estado [federal]; ►autoridad estadual); *syn.* administración [f] estatal (federal); *f* **autorité [f] centrale [fédérale]** (►autorité régionale) ; *syn.* autorité [f] administrative fédérale ; *g* **Bundesbehörde [f]** (Oberste staatliche Ressortbehörde; ►Landesbehörde).

federal government authority [n] *adm.* ►federal government agency [US].

1999 federal government construction project [n] *plan.* (Project of a federal, state or local authority/agency [with national significance]; ►state government construction project); *s* **proyecto [m] de obras públicas estatal (1)** (►proyecto de obras públicas estatal 2); *f* **projet [m] immobilier de l'État (1)** (Opération de construction d'un ensemble immobilier de logements ou d'équipements publics réalisée et financée par l'État ; ►projet immobilier de l'État 2) ; *g* **staatliches Bauvorhaben [n] (1)** (Bauprojekt, das von einer staatlichen Behörde [Bundesbehörde (Oberfinanzdirektion, Finanzbauamt) oder Landesbehörde (Staats- resp. Landesbauämter)] durchgeführt wird; ►staatliches Bauvorhaben 2); *syn. für die Bundesbauverwaltung* Bauvorhaben [n] des Bundes.

federal highway [n] [US] *adm. leg. trans.* ►interstate highway [US].

Federal Supplementary Conditions [npl] **of Contract for Construction (AIA Document A201)** [US], **General and** *contr. leg.* ►General Conditions of Contract for Construction [US].

fee [n] *adm. envir. leg.* ►air pollution fee; *prof.* ►architect's fee ; *contr. leg. prof.* ►chargeable fee; *adm. envir. leg.* ►effluent discharge fee, ►hazardous waste disposal fee [US]; *contr. leg. prof.* ►lump-sum fee; *contr. prof.* ►lump-sum fee, ►retainer fee.

fee [n], **daily** *prof.* ►daily billing rate [US].

fee [n], **emissions** *adm. envir. leg.* ►air pollution fee.

fee [n] [UK], **hazardous substances** *adm. envir. leg.* ►hazardous waste disposal fee [US].

2000 fee agreement [n] *contr. prof.* (Verbal agreement or written contract governing the ►remuneration of professional fees); *s* **acuerdo [m] sobre honorarios** (Contrato oral o escrito utilizado como base para el ►pago de honorarios); *f* **accord [m] sur les honoraires** (Accord oral ou contrat écrit fixant les conditions de ►rémunération des honoraires d'une mission de maîtrise d'œuvre) ; *syn.* convention [f] sur les honoraires ; *g* **Honorarvereinbarung [f]** (Mündliche Absprache oder schriftlicher Vertrag über die Grundlagen der ►Honorarvergütung); *syn.* Vereinbarung [f] eines Honorars.

fee [n] **based on an hourly rate, professional** *contr. prof.* ►time charges.

2001 fee calculation [n] *contr. prof.* (Determination of design and planning fees based upon lump sum estimate, cost plus profit, or labor times a specified overhead and profit multiplier; ▶Guidance for Clients on Fees [UK]); **s cálculo [m] de honorarios** (Determinación de la suma a pagar por los servicios de los/las arquitectos/as e ingenieros/as según el ▶baremo de honorarios o en forma de honorarios globales o por tiempo empleado); *f* **calcul [m] des honoraires** (▶rémunération des prestations de maîtrise d'œuvre [exercée pour le compte de maîtres d'ouvrage publics]) ; *g* **Honorarberechnung [f]** (Ermittlung des Entgeltes für Leistungen der Architekten und Ingenieure nach der ▶Honorarordnung für Architekten und Ingenieure [HOAI] resp. nach den Honorarrichtlinien ÖGLA in Österreich oder in Form eines Pauschal- oder Zeithonorars; *syn.* Berechnung [f] des Honorars.

2002 fee chart [n] *contr. leg. prof.* (Summary table of fee rates according to the degree of difficulty or complexity of the work [▶complexity rating classification]; **In U.S.**, a fixed fee schedule is considered 'price-fixing' and is unlawful. Instead of a table of fixed fees such as the German ▶Honorarordnung für Architekten und Ingenieure [HOAI], which is mandatory, **the UK** pamphlet 'Guidance for Clients on Fees' shows a percentage fee graph for standard services from which a percentage fee can be read against the contract sum and according to four categories of complexity; ▶construction costs as a basis for professional fees, ▶Guidance for Clients on Fees [UK]); *syn.* fee scale [n] [UK], fee table [n] [also UK], scale [n] of charges [also UK], scale [n] of fees [also UK]; **s tabla [f] de honorarios** (Lista de las tasas de honorarios que corresponden a los diferentes servicios profesionales que se encuentra en el ▶reglamento de honorarios de D., F., GB y otros países de la UE. En los EE.UU. la fijación de ▶baremos de honorarios se considera «fijación de precios» y es ilegal; ▶clase de complejidad, ▶costos imputables); *syn.* escala [f] de honorarios; *f* **tableau [m] des taux de rémunération (des honoraires)** (Recueil de tableaux indiquant les valeurs du taux de rémunération des honoraires en fonction des classes et des notes de complexité et des différents ▶montants du coût d'objectif ; ▶classe de complexité, ▶rémunération des prestations de maîtrise d'œuvre [exercée pour le compte de maîtres d'ouvrage publics]) ; *syn.* barème [m] des honoraires, tarif [m] d'honoraires [aussi B] ; *g* **Honorartafel [f]** (Tabellarische Übersicht über Honorarsätze, abhängig von der Schwierigkeit und Komplexität der Aufgabe [▶Honorarzone] und den ▶anrechenbaren Baukosten, die in der ▶Honorarordnung für Architekten und Ingenieure [HOAI] dargestellt sind); *syn.* Aufstellung [f] von Honorarsätzen, Honorartabelle [f].

2003 feedback [n] *ecol.* (*Term in cybernetics* the correction of processes or behavio[u]r of individuals, who do not conform to the norm in a self-regulating system, whereby the continued existence of the system is guaranteed; a loop in which one component affects a second one that in turn affects the first component; cf. LE 1986; ▶cybernetic cycle, ▶self-regulation 2); *syn.* loop [n]; **s retroalimentación [f]** (Término utilizado en varias disciplinas científico-técnicas, entre ellas la cibernética; mecanismo que actúa en un proceso de modo que las salidas inciden sobre las entradas. Puede ser negativo, con acción reguladora o inhibidora, o positivo, con acción aceleradora; aunque existen voces castellanas el término inglés "feedback" es de uso generalizado en español; cf. DINA 1987, DPD 2005; ▶autoregulación 2, ▶ciclo cibernético); *syn.* feedback [m], realimentación [f], retroacción [f]; *f* **rétroaction [f]** (*Terme utilisé en cybernétique* processus correctifs agissant sur des mécanismes ou des comportements d'individus non conformes à la norme à l'intérieur d'un système autorégulant et garantissant ainsi l'existence du système ; ▶autorégulation 2, ▶système finalisé) ; *syn.* feed-back [m] ; *g* **Rückkoppelung [f]** (*Begriff aus der Kybernetik* Korrektur von Vorgängen oder Verhaltensweisen von Individuen [Regelgrößen], die

nicht der „Norm" entsprechen, in einem sich selbst regelnden System, wodurch der Bestand des Systems garantiert ist; ▶Regelkreis, ▶Regelung).

feeder road [n] [US] *plan. trans. urb.* ▶local feeder road.

feeder root [n] *bot.* ▶absorbing root.

feeding [n] *bot. hort.* ▶plant nutrition; *arb. constr.* ▶tree feeding.

feeding [n], root *arb. constr.* ▶tree feeding.

2004 feeding territory [n] *zool.* (Limited area, part of the home range of a species, which is regularly visited for food uptake; ▶home range 2); **s área [f] trófica** (Zona limitada dentro del ▶área de habitación de una especie animal, que éste frecuenta regularmente para conseguir su alimento; *f* **territoire [m] alimentaire** (Espace limité, secteur de l'▶aire d'habitation d'une espèce occupé régulièrement en vue de l'acquisition de nourriture) ; *syn.* territoire [m] de nutrition ; *g* **Nahrungsrevier [n]** (Begrenztes Gebiet, Teil des Wohngebietes einer Tierart, das von dieser zur Nahrungsaufnahme regelmäßig aufgesucht wird ▶Wohngebiet 1).

2005 feed pipe [n] *constr.* (Pipe which conveys water or gas to the point of consumption); *syn.* supply line [n]; **s tubería [f] de alimentación** (Tubería principal que suministra el líquido, bien directamente al punto de su utilización, bien a una tubería secundaria; DACO 1988; *término específico* tubería [f] de alimentación a pisos); *syn.* alimentador [m]; *f* **conduite [f] d'alimentation (1)** (Assurant l'adduction d'eau ou de gaz jusqu'à son lieu de consommation) ; *g* **Versorgungsleitung [f] (1)** (Leitung, in der Wasser oder Gas bis zur Gebrauchsstelle transportiert wird).

2006 fee grade [n] *contr. leg. prof.* (Class of charges specified according to the degree of project scope and complexity; ▶fee chart); **s nivel [f] de complejidad de proyecto (≠)** (*En el reglamento de honorarios de D* clasificación en dos niveles de los proyectos de planificación del paisaje o de estudios de impacto según su tamaño y su grado de dificultad; ▶tabla de honorarios); *f* **1 unité [f] d'intervention** (F., Mode de rémunération des missions d'urbanisme réglementaire ou des études préalables, préopérationnelles et opérationnelles calculées sur la base d'**u. d'i.** [UI] déterminées par tranches d'importance de population [entre 400 et 1000 habitants, etc.] pour les premières ou par tranches de superficie [de 5 à 10 ha, etc.] pour les secondes) ; *syn.* catégorie [f] du projet [B] ; *f* **2 classe [f] de complexité (2)** (D., les missions d'études d'expertise ou exploratoires dans le domaine de l'urbanisme et du paysage sont réparties dans les ▶tableaux des taux de rémunération [des honoraires] en deux **c. de c.** — normale et complexe) ; *g* **Honorarstufe [f]** (Für Gutachten und Wertermittlungen, für städtebauliche und landschaftsplanerische Leistungen sind der Schwierigkeit resp. dem Umfang der zu erbringenden Leistung entsprechend in ▶Honorartafeln zwei Vergütungsrubriken — Normalstufe und Schwierigkeitsstufe — vorgesehen; cf. Teil V und VI HOAI 2002 resp. Teil 2 HOAI 2009).

2007 fee installment [n] [US]/fee instalment [n] [UK] *contr. prof.* (Portion of remuneration of professional fees or charges, as one of several usually equal payments spread over an agreed period of time); **s pago [m] a plazos (1)** (Pago parcial de los honorarios profesionales); *syn.* pago [m] parcial; *f* **paiement [m] par acompte** (Paiement partiel sur un honoraire) ; *syn.* acompte [m] ; *g* **Abschlagszahlung [f] (1)** (Teil eines zu zahlenden Honorars, abhängig von den bereits erbrachten Leistungen); *syn.* Akontozahlung [f], Teilzahlung [f].

fee invoice [n] *contr. prof.* ▶final fee invoice.

2008 fee negotiation [n] *contr. prof.* (Discussion of remuneration for contract services between client and planner);

s **negociación [f] de honorarios** (Intercambio entre la propiedad y el/la arquitecto/a para fijar los honorarios a pagar resp. cobrar por el servicio a contratar); *f* **négociation [f] des honoraires** (entre le maître d'ouvrage et le concepteur en vue de fixer l'étendue de la mission et sa rémunération) ; *g* **Honorarverhandlung [f]** (Verhandlung zwischen Bauherrn und Planer mit dem Ziel, eine Vereinbarung über zu erbringende Leistungen und die dafür zu zahlende Vergütung abzuschließen).

fees [npl] *contr. leg. prof.* ►Guidance for Clients on Fees [UK]; *contr. prof.* ►increase in fees/charges, ►reduction in fees/charges, ►remuneration of professional fees.

fees [npl], **cut** [n] **in ~/charges** *contr. prof.* ►reduction in fees/charges.

fees [npl], **payment of professional** *contr. prof.* ►remuneration of professional fees.

fee scale [n] **[UK]** (obs. since 1986) *contr. leg. prof.* ►Guidance for Clients on Fees [UK].

fee table [n] **[UK]** *contr. leg. prof.* ►fee chart.

2009 felling [n] **(1)** *arb. constr. for.* (Act of cutting down a single tree or a group of trees; ►tree felling work, ►tree clearing and stump removal [operation] [US]/grub felling [UK]); *syn.* cutting [n] [CDN]; *s* **tala [f] (1)** (Corte de árboles individuales o en grupos; cortar [vb], talar [vb]; ►despeje de árboles y tocones, ►trabajos de tala y despeje de árboles); *syn.* apeo [m], corta [f]; *f* **abattage [m]** (Action de coupe d'un arbre sur pied, ou une groupe d'arbres ; ►travaux d'abattage d'arbres ; ►abattage par extraction de souche) ; *g* **Fällen [n, o. Pl.]** (Umsägen von Bäumen; ►Baumfällarbeiten, ►Rodung); *syn.* Fällarbeiten [fpl], Fällung [f], Schlägern [n, o. Pl.] [A].

2010 felling [n] **(2)** *constr. for.* (**1.** Cutting down trees; ►clear felling, ►deforestation, ►tree clearing and stump removal [operation] [US]. **2.** The process of tree felling in the Mediterranean which encouraged soil erosion and laid bare calcareous or other rock surfaces is called **denudation**; ►karstification); *s* **tala [f] (2)** (Corte de árboles en masa para dar al suelo un destino distinto al de monte arbóreo; ►despeje de árboles y tocones, ►karstificación, ►tala a matarrasa, ►roturación, ►de[s]forestación); *syn.* desmonte [m] (2); *f* **coupe [f]** (En sylviculture **1.** secteur d'une surface forestière exploitée. **2.** L'abattage des arbres dans un peuplement forestier ; on distingue la ►coupe à blanc, les coupes d'abri, claires, sombres, les coupes de régénération et d'ensemencement ; ►abattage par extraction de souche, ►déforestation, ►défrichement, ►karstification) ; *g* **Abholzung [f]** (**1.** Fällen von Bäumen auf einer Fläche. **2.** *for.* Baumfällungen in einem Wald zur Nutzholzgewinnung; die Fläche der **A.** [des Hiebes] wird **Schlag**, die geerntete Holzmenge **[Holz]einschlag, Ernteertrag, Abtriebsmasse** oder **Hiebsanfall** genannt. **3.** Entfernung des Baumbestandes durch ►Kahlschlag in einem Gebiet. **4.** Rodungsprozess im mediterranen Raum, der die Bodenerosion fördert und kahle Kalkfelsen und Felspartien anderer Gesteine freilegt; ►Rodung, ►Entwaldung, ►Verkarstung); *syn. zu 1.* Abholzen [n, o. Pl.]; *syn. zu 2.* Einschlag [m] (1), Hieb [m], Schlägerung [f] [A].

felling [n] **[UK], grub** *constr. for.* ►tree clearing and stump removal (operation) [US].

felling [n] **[UK], improvement** *constr. for.* ►improvement cut [US].

2011 felling and logging damage [n] *for.* (Damage within forests; e.g. to tree trunks and butts and compaction of soil caused by sliding felled timber along the ground); *syn.* skidding damage [n] [also US&CDN], hauling damage [n] [US]; *s* **daño [m] de tala y arrastre** (Detrimento en bosques, p. ej. en troncos o pies de árboles o compactación del suelo causada por el arrastre de troncos talados sobre el suelo); *syn.* daño [m] de explotación forestal; *f* **dégâts [mpl] causés par l'abattage ou le débardage** (Dommages produits sur les arbres lors de l'abattage et du débusquage du bois, p. ex. sur le tronc, la base du tronc, le sol [ornières, compactage du sol]) ; *g* **Rückeschaden [m]** (Schäden im Waldbestand an Stämmen, Stammfüßen und am Boden [Fahrspuren, Verdichtung] durch unsachgemäßes Herausnehmen gefällter Bäume); *syn.* forstlicher Nutzungsschaden [m].

2012 fen [n] **(1)** *geo.* (Low marshy land with shallow water and soil of an inorganic [mineral] basis caused by terrestrialization of waterbodies or by paludification of bad drainage areas; ►low bog; *opp.* ►raised bog); *syn.* minerotrophic peatland [n], soligenous peatland [n] (WET 1993, 374), rheophilous peatland [n] (WET 1993, 374), rheotrophic peatland [n] (WET 1993, 374); *s* **turbera [f] baja (1)** (Denominación de paisaje formado por colmatación de lagos poco profundos o por falta de desagüe de tierras, dando lugar a acumulación de turba sobre suelos minerales. Ésta se compone de restos de vegetales poco descompuestos, que se almacenan en aguas ricas en nutrientes, pero probres en oxígeno; cf. BB 1979, 413; ►turbera baja 2; *opp.* ►turbera alta); *f* **tourbière [f] plate** (Tourbière alimentée par les eaux de la nappe phréatique ou les eaux de ruissellement superficielles, bien oxygénée et à pH élevé ou neutre ; par opposition à la ►tourbière haute, la pluviosité est insuffisante pour assurer un engorgement permanent, la végétation ne s'élève pas au-dessus du niveau de la nappe [absence de Sphaignes] ; *terme spécifique* tourbière topogène alcaline, tourbière à pluviosité déficitaire alimentée par des eaux de ruissellement alcalines; GGV 1979, 240-262 ; *D., paysage marécageux* zone à engorgement permanent et drainage extrêmement faible, colonisée par une végétation turficole ; ►tourbière basse) ; *syn.* tourbière [f] à Pleurocarpes, tourbière [f] alcaline, tourbière [f] mésotrophe, tourbière [f] infra-aquatique, tourbière [f] de vallée alcaline, basmarais [m], marais [m] verts (*car les végétaux sont verts*) ; *g* **Flachmoor [n]** (*Landschaftsbezeichnung* durch Verlandung von Gewässern oder durch Versumpfung von Flächen mit gehemmtem Wasserablauf entstandene vegetationsbedeckte Ablagerung von Torf im Mineralbodenwasserbereich; ►Niedermoor; *opp.* ►Hochmoor); *syn.* Fehn [n], Moos [n] (1), Ried [n].

fen [n] **(2)** *pedol. phyt.* ►low bog.

fen [n] **[UK], large-sedges** *phyt.* ►tall sedge swamp.

fen [n] **[UK], small-sedges** *phyt.* ►low sedge swamp.

2013 fence [n] *constr.* (Enclosure of a plot of land using wire netting, wooden boards or laths, round section timber, metal posts, etc.; *specific terms* ►animal-tight fence [US]/deer-stop fence [UK], ►barbed wire fence, ►chain-link fence, ►close-boarded fence, ►decorative board fence, ►interwoven wood fence, ►metal fence, ►metal grid fence, ►palisade fence, ►picket fence, ►silt fence [US], ►snow fence, ►stockade fence [US], ►trellis fence, ►wicker fence, ►wire fence, ►wooden fence, ►wooden rail fence); *s* **valla [f]** (Término genérico para cercado de un terreno construido de metal, madera [planchas, paneles, listones, estacas], bambú u otros materiales; *términos específicos* ►cerca de parcela de pasto, ►cercado de protección de la caza, ►valla de alambre de púas, ►valla de empalizada, ►valla de enrejado de madera, ►valla de enrejado de metal, ►valla de estacas partidas, ►valla de listones, ►valla de madera, ►valla de malla metálica, ►valla de metal, ►valla metálica, ►valla de tablas perfiladas, ►valla de tablones, ►valla de trenzado, ►valla paranieves, ►valla trenzada; ►interceptor de fango); *f* **clôture [f] (2)** (Terme générique pour tout élément en béton [lissage, dalles pleines, panneaux ajourés, mixtes], métallique [grillage noué ou soudé], en bois [rondins, bois débités, bois usinés], en plastique [lissage ou barreaudage en PVC], en pierre,

F

F

brique et maçonnerie, utilisé pour remplir diverses fonctions telles que délimiter un terrain, une propriété, assurer la sécurité, cacher les vues désagréables ou mettre en valeur un terrain ; *termes spécifiques* ►clôture à fils de fer barbelés, ►clôture à lamelles tressées, ►clôture à lattes, ►clôture à treillis, ►clôture de lices/lisses en bois, ►clôture de planches ajourée, ►clôture [de protection] contre les écoulements de boue, ►clôture en bois, ►clôture en échalas, ►clôture en planches jointives, ►clôture gibier, ►clôture grillagée, ►clôture jointive en rondins, ►clôture métallique, ►clôture métallique à caillebotis, ►clôture pare-neige, ►palissade tressée de branchages à rejets) ; *g* **Zaun [m]** (Einfriedung eines Grundstückes aus Drahtgeflecht, Brettern, Latten, Rundholz- oder Metallstäben; etymologisch ist **Z.** als äußere Begrenzung verwandt mit dem englischen *town* [Stadt] und niederländischen *tuin* [Garten]; *UBe* ►Bretterzaun, ►Draht-zaun, ►Flechtzaun, ►Holzflechtzaun, ►Holzzaun, ►Koppel-zaun, ►Lattenzaun, ►Maschendrahtzaun, ►Metallgitterzaun, ►Palisadenzaun, ►Profilbretterzaun, ►Scherenzaun, ►Schlammfangzaun, ►Schneezaun, ►Sichtschutzholzzaun aus Latten/Staketen, ►Stacheldrahtzaun, ►Stahlzaun, ►Wildschutz-zaun).

fence [n]**, close board panel** *constr.* ►visual screen with vertical boards.

fence [n] **[UK], deer-stop** *conserv. for. hunt. trans.* ►animal-tight fence [US].

fence [n]**, interwoven lattice** *constr.* ►interwoven wood fence.

fence [n] **[UK], lath** *constr.* ►picket fence.

fence [n] **[US], living** *constr. hort.* *►clipped hedge.

fence [n] **[UK], pale** *constr.* ►picket fence.

fence [n] **[US], paling** *constr.* ►picket fence.

fence [n]**, post and rail** *constr. agr.* ►wooden rail fence.

fence [n]**, vertical slatted** *constr.* ►picket fence.

fence [n]**, wooden rail** *constr. agr.* ►wooden rail fence.

fence construction [n] *constr.* ►wicker fence construction.

fence [n] **of lattice work** *constr.* ►interwoven wood fence.

2014 fence post [n] *constr.* (Round or square-shaped, verti-cally-positioned structural member as a part of a fence); *s* **poste [m] de valla** (Pieza estructural en la estabilización de cercas que se introduce verticalmente en el suelo); *syn.* poste [m] de cerca; *f* **poteau [m] de clôture** (Élément vertical stabilisateur de la clôture ; *termes spécifiques* poteau droit, poteau à renvoi, poteau à bavolet) ; *syn.* piquet [m] de clôture ; *g* **Zaunpfahl [m]** (Stabi-lisierendes, vertikales Element eines Zaunes); *syn.* Zaunpfosten [m], *o. V.* Zaunspfahl [m].

fencerow [n] **[US]** *agr. land'man.* ►hedgerow.

2015 fencing [n] *landsc.* (**1.** Act of enclosing a piece of land with a fence. **2.** ►Enclosure by a continuous structure); *s* **en-vallado [m]** (**1.** Acción de cercar una parcela de terreno con valla. **2.** ►Cercado por medio de una estructura continua); *f 1* **enclore [vb]** (Action de délimiter une parcelle par une ►clô-ture 2 ; *syn.* clôturer [vb] ; *f 2* **clôture [f]** (3) (Résultat ; l'espace entouré d'une clôture est dénommé l'**enclos**) ; *g* **Umzäunung [f]** (**1.** Vorgang der ►Einfriedung eines Grundstückes mit einem Zaun. **2.** Zaun als Ergebnis).

fencing [n] **[UK], ball stop** *constr. recr.* ►perimeter fencing.

fencing [n] **around trees** *constr.* ►protective fencing around trees

2016 fen [n] **covered with pine krummholz** *phyt.* (►pine krummholz, ►raised bog, ►transition bog); *s* **paular [m] con pinos mugo** (►Humedal de transición o ►turbera alta cubierta de pinos mugo); *f* **brousse [f] à Pins rampants** (►Tour-

bière haute ou ►tourbière de transition peuplée par le Pin mugo ; ►pin rampant) ; *syn.* tourbière [f] arbustive à Pins rampants, tourbière [f] boisée de Pins rabougris ; *g* **Latschenfilz [m]** (Mit strauchartig gewachsenen Bergkiefern *[Pinus mugo]* [►Latsche] bestocktes ►Übergangsmoor oder ►Hochmoor).

fen peat [n] **[UK]** *constr.* ►sedge peat [US].

2017 fen wood [n] *phyt.* (**1.** Generic term covering a forest on a permanent wet, organic soil. **2. In Europe**, a deciduous wood-land community [alder fen *(Alnetum glutinosae)*] developed on a minimum 10-20cm thick peat layer in which the groundwater stands permanently near the surface. Fluctuation of water table is rarely more than 1 m, usually flooded in early spring. In contrast to ►riparian woodland, not exclusively associated with mineral sediments; *specific term* cedar swamp [US]; ►riparian alder stand [US]/alder carr [UK], ►swamp forest); *syn.* carr [n] [UK], swamp wood [n]; *s* **bosque [m] turbícola** (Comunidad vegetal, p. ej. *Alnetum glutinosae*, que se desarrolla sobre un sustrato turboso de 10-20 cm de espesor como mínimo. El nivel freático se encuentra muy cerca de la superficie y sus fluctuaciones raramente sobrepasan 1 m. Normalmente inundada en primavera. Al contrario que el ►bosque ripícola, no está asociado a sustrato mineral; ►aliseda turbícola/turfófila, ►bosque pantanoso, ►soto [ripícola]); *syn.* bosque [m] sobre suelo orgánico; *f* **forêt [f] tourbeuse** (Association végétale, p. ex. *Alnetum glutinosae*, établie sur un substrat tourbeux d'une épaisseur minimale de 10-20 cm. La nappe phréatique affleure constamment ou temporaire-ment la surface du sol. Une variation de 1 m du niveau de la nappe est exceptionnelle, l'inondation printanière est courante. La circulation de la nappe est faible ou importante. En compa-raison avec la ►forêt alluviale, ce groupement n'est pas exclu-sivement associé à un substrat minéral ; ►aulnaie marécageuse, bétulaie marécageuse, pinède naturelle humide ; ►forêt maré-cageuse) ; *syn.* forêt [f] humide à Sphaignes ; *g* **Bruchwald [m]** (Auf mindestens 10-20 cm mächtiger [Bruchwald]torfschicht stehende Waldgesellschaft [z. B. *Alnetum glutinosae*], in der das Grundwasser dauernd nahe der Oberfläche steht. Schwan-kungen des Wasserspiegels ausnahmsweise mehr als 1 m, Überschwemmungen gewöhnlich im zeitigen Frühjahr; im Vergleich zum ►Auenwald <u>nicht</u> ausschließlich an mineralische Sedimente gebunden; *UBe* Birkenbruchwald, ►Erlenbruchwald, Fichtenbruchwald, Kiefernbruchwald, Schwarzerlenbruchwald; ►Sumpfwald).

fen woodland [n] **[EIRE]** *geo. phyt.* ►swamp forest.

fen woodland [n] **[US]** *phyt.* ►fen wood.

2018 feral species [n] *phyt. zool.* (Domestic, free-ranging/ free-roaming animals or cultivated plants uncontrolled by man which have multiplied in the wild; ►naturalized species, ►over-growing); *s* **especie [f] asilvestrada (1)** (**1.** Se aplica a las espe-cies de fauna doméstica que se reproducen y distribuyen libre-mente en la naturaleza. **2.** Se aplica a especies de plantas silves-tres que proceden de semillas de plantas cultivadas. **3.** En silvicultura, especie que fue introducida de otro país, pero que se reproduce naturalmente, sin cultivo; en esta acepción equivale a ►especie naturalizada, ►invasión de advenedizas); *syn. phyt.* especie [f] cimarrona [AL]); *f* **espèce [f] retournée à l'état sauvage** (Animal domestique vivant à l'état sauvage ou plante de cultures ayant formé des populations se reproduisant naturel-lement ; ►espèce naturalisée, ►espèce subspontanée, ►réappa-rition des espèces sauvages) ; *g* **verwilderte Art [f]** (Frei lebende Haustier- oder Kulturpflanzenart, die der menschlichen Kontrolle nicht mehr unterliegt und eine vermehrungsfähige Population gebildet hat; ►neuheimische Art, ►Verwilderung 2).

fertilization [n] *agr. constr. for. hort.* ►fertilizing, ►mineral fertilization, ►over-fertilization.

fertilized hay field [n] [US] *agr. phyt.* ▶fertilized meadow.

2019 fertilized meadow [n] *agr. phyt.* (Mostly intensively-used agricultural land, mown once or more per year on slightly moist to moist soils; ▶fertilized pasture, ▶hay meadow, ▶pasture); *syn.* fertilized hay field [n] [also US]; *s* **prado** [m] **jugoso** (Pradera abonada explotada intensamente, de una o más siegas al año, generalmente sobre suelos fértiles, de húmedos a semihúmedos; ▶pastizal jugoso, ▶pradera fertilizada, ▶pastizal); *f* **prairie** [f] **(mésophile) de fauche** (Prairie naturelle permanente sur un substrat bien pourvu en eau [parfois très humide] caractéristique de l'alliance de l'*Arrhenatherion elatioris,* souvent fertilisée, fauchée plusieurs fois par an et utilisée pour la production de foin ou de silage ; GGV 1979, 285 ; ▶herbage, ▶pâturage, ▶prairie à foin) ; *syn.* pré [m] (mésophile) de fauche ; *g* **Fettwiese** [f] (Meist intensiv genutztes, gedüngtes ein- oder mehrschüriges Wiesengrünland — in Mittel- und Westeuropa i. d. R. zum Verband der Glatthafer-Talfettwiesen *[Arrhenatherion elatioris BB 1925]* gehörig —, auf meist frischen bis feuchten und fruchtbaren Böden mit einer ganzjährigen guten Wasserversorgung; ▶Fettweide, ▶Futterwiese, ▶Weide); *syn.* Mähwiese [f], Schnittwiese [f].

2020 fertilized pasture [n] *agr.* (Intensively grazed grassland, mostly on variably moist soils; ▶fertilized meadow, ▶hayfield, ▶pasture); *s* **pastizal** [m] **jugoso** (Campo de pastoreo abonado de uso intensivo, generalmente sobre suelos de semihúmedos a húmedos; ▶pastizal, ▶prado de siega, ▶prado jugoso); *syn.* pasto [m] jugoso; *f* **herbage** [m] (Prairie naturelle de qualité supérieure, à pâture et fertilisation intensive ; ▶pâturage tournant, ▶prairie [mésophile] de fauche ; LA 1981, 902) ; *syn.* prairie [f] d'embouche (DA 1977, 60), pré [m] d'embouche (PR 1987), prairie [f] grasse ; *g* **Fettweide** [f] (Intensiv genutztes und gedüngtes Weidegrünland auf meist frischen bis feuchten Böden; ▶Fettwiese, ▶Mähweide, ▶Weide).

2021 fertilizer [n] *agr. constr. for. hort.* (Generic term covering any substance—may be organic: straw manure, ▶liquid manure, ▶semiliquid manure [US]/semi-liquid manure [UK], plant residues, ▶refuse compost, ▶sewage sludge, green manure, bonemeal, blood as well as fishmeal, or may be inorganic: compounds of nitrogen, phosphorous, potassium, magnesium, etc., which can be added to the soil and which provides essential nutrients for plant growth and to maintain or improve soil fertility; ▶chemical fertilizer, ▶commercial fertilizer, ▶complete fertilizer; ▶humus fertilizer, ▶mineral fertilizer, ▶organic fertilizer, ▶slow-release fertilizer; ▶plant nutrient); *s* **fertilizante** [m] (Término genérico para cualquier sustancia orgánica [estiércol, ▶licuame, ▶purín, compost, ▶compost de residuos, ▶lodo de depuración, etc.] o anorgánica [compuestos de nitrógeno, fósforo, potasio, magnesio, etc.] que se aplica al suelo para mejorar el crecimiento de las plantas; ▶abono complejo, ▶abono húmico, ▶fertilizante de acción controlada, ▶fertilizante industrial, ▶fertilizante mineral, ▶fertilizante [mineral] artificial, ▶fertilizante orgánico, ▶nutriente [de plantas]); *syn.* abono [m]; *f* **engrais** [m] (Substance fertilisante organique [▶lisier, ▶purín, compost, ▶compost de déchets ménagers, engrais vert, ▶boues d'épuration], ou minérale [azote, phosphore, potassium, magnésium, etc.] dont la fonction principale est d'apporter les éléments nutritifs aux végétaux ; ▶amendement humifère, ▶engrais à décomposition lente, ▶engrais artificiel, ▶engrais complet, ▶engrais du commerce, ▶engrais minéral ; ▶engrais organique, ▶élément nutritif) ; *syn.* fertilisant [m] ; *g* **Dünger** [m] (OB; organischer [Mist, ▶Gülle, ▶Jauche, Kompost, ▶Müllkompost, Gründünger, ▶Klärschlamm, Knochen-, Blut-, Fischmehl] oder anorganischer Stoff [Stickstoff, Phosphor, Kalium, Magnesium etc.], der dem Boden zur Erhöhung der Erträge von Kulturpflanzen und zur Verhinderung der Nährstoffverarmung des Bodens zugeführt wird; ▶Handelsdünger, ▶Humusdünger, ▶langsam fließender Dünger, ▶Kunstdünger, ▶Mineraldünger, ▶organischer Dünger, ▶Volldünger; ▶Pflanzennährstoff); *syn.* Düngemittel [n].

fertilizer [n] [UK]**, all-round** *agr. constr. for. hort.* ▶complete fertilizer.

fertilizer [n] [UK]**, artificial** *agr. hort.* ▶chemical fertilizer.

fertilizer [n]**, balanced** *agr. constr. for. hort.* ▶complete fertilizer.

fertilizer [n] [US]**, controlled-release** *agr. constr. for. hort.* ▶slow-release fertilizer.

2022 fertilizer application [n] *agr. constr. hort. for.* (Spreading of a specified quantity of fertilizer per plant or unit area; ▶fertilizing, ▶plant nutrition); *s* **aporte** [m] **de fertilizante** (Aplicación de una cantidad específica de abono por unidad de superficie o por planta; ▶fertilización, ▶nutrición de [las] plantas); *syn.* aporte [m] de abono; *f* **apport** [m] **d'engrais** (Épandage d'une quantité précise d'engrais par végétaux de m^2 de végétation ou de culture ; ▶fertilisation, ▶fumure, ▶nutrition des plantes) ; *g* **Düngergabe** [f] (Das Ausbringen einer bestimmten Düngermenge je Pflanze oder Flächeneinheit; ▶Düngung, ▶Pflanzenernährung).

2023 fertilizer program [n] [US]/**fertilizer programme** [n] [UK] *agr. hort.* (Schedule prescribing the kind and quantity of fertilizer to be used during a specific period of time); *s* **plan** [m] **de abonado** (▶plan [m] de fertilización); *f* **plan** [m] **de fumure** *syn.* programme [m] de fertilisation ; *g* **Düngeplan** [m] (Plan, der für eine Pflanzenkultur die Düngeart und -menge während eines bestimmten Zeitraumes vorgibt).

2024 fertilizing [n] *agr. constr. for. hort.* (Enrichment of soil with prepared organic or inorganic ▶fertilizers; BS 3975: part 5; ▶fertilizer application, ▶green manuring, ▶inorganic fertilizing, ▶organic fertilizing, ▶plant nutrition, ▶over-fertilization); *syn.* manuring [n], fertilization [n]; *s* **fertilización** [f] (▶Aporte de fertilizantes al suelo para mejorar sus propiedades físico-químicas; abonar [vb]; ▶abonado excesivo, ▶abonado verde, ▶fertilización mineral, ▶fertilización natural, ▶fertilizante, ▶nutrición de [las] plantas); *syn.* abonado [m]; *f 1* **fertilisation** [f] (Apport de matières fertilisantes [▶amendements, ▶d'engrais, ▶apport d'engrais] qui modifient les propriétés physico-chimiques et biologiques d'un sol ; ▶fertilisation excessive, ▶fertilisation minérale, ▶fertilisation par [apport d']engrais vert, ▶fumure organique, ▶nutrition des plantes) ; *f 2* **fumure** [f] (Étymologiquement, entretien ou amélioration de la fertilité du sol par enfouissement de fumier ; par extension, ensemble des apports de matières fertilisantes, sous formes organique et minérale, ou quantités d'engrais ou d'unités fertilisantes fournies au sol ou aux cultures — c'est dans ce dernier sens que le mot « fumure » est le plus employé ; LA 1981, 554) ; *g* **Düngung** [f] (Das Ausbringen von Düngemitteln [▶Düngergabe] und das Einarbeiten in den Boden, um für das Pflanzenwachstum die Bodenfruchtbarkeit zu erhalten und das Nährstoffangebot zu erhöhen; düngen [vb]; ▶Dünger, ▶Gründüngung, ▶anorganische Düngung, ▶organische Düngung, ▶Pflanzenernährung, ▶Überdüngung); *syn.* Ausbringen [n, o. Pl.] von Dünger, Aufbringen [n, o. Pl.] von Düngemitteln, Düngeraufbringung [f].

fertilizing [n] [US]**, tree** *arb. constr.* ▶tree feeding.

2025 fertilizing hole [n] *arb. constr.* (Specially prepared hole for the fertilizing and watering of an older tree repeatedly around the area of its roots; ▶tree feeding); *s* **hoyo** [m] **de alimentación** (Agujero hecho con máquina especial en la zona radical de árboles maduros para mejorar el suministro de agua y de nutrientes; ▶fertilización de árboles); *syn.* hueco [m] de

F

alimentación; *f* **orifice [m] d'alimentation** (Méthode de ▶nutrition des arbres [alimentation en eau et fumure] consistant à pratiquer des trous d'environ 60 cm de profondeur et espacés de 60 à 80 cm l'un des autres sur une circonférence minimale correspondant au diamètre de la couronne) ; *g* **Fütterungsloch [n]** (...löcher [pl]; im Wurzelbereich eines älteren Baumes mit Spezialgeräten eingebrachtes Loch für die Nährstoff- und Wasserversorgung; ▶Baumfütterung).

2026 fertilizing [n] **with straw** *agr.* (Leaving remaining cereal straw and its under-plowing [US]/under-ploughing [UK] after harvest); *s* **abonado [m] con paja** (Aprovechamiento de los residuos de cosecha de cereales como fertilizantes mezclándolos con el suelo al ararlo); *syn.* fertilización [f] con paja; *f* **fertilisation [f] par enfouissement de la paille** (Utilisation de la paille comme engrais qui est enfouie dans le sol [déchaumage] et restitue à celui-ci des éléments nutritifs et fournit un substrat organique aux micro-organismes) ; *syn.* fertilisation [f] par déchaumage ; *g* **Strohdüngung [f]** (Belassen und Unterpflügen des zerkleinerten und verteilten Getreidestrohs auf Ackerflächen).

2027 fibric peat [n] **(1)** *pedol.* (Weakly-humified bog peat, i.e. more than 66% of its bulk is composed of undecomposed plant fibers [US]/fibres [UK]; ▶humified raised bog peat); *s* **turba [f] blanca** (Turba poco humificada en la que se reconocen muy bien los restos de las plantas; ▶turba negra); *f 1* **tourbe [f] fibreuse** (Tourbe, dont la matière organique est peu décomposée ; la structure des plantes est encore visible ; ▶tourbe noire) ; *syn.* tourbe [f] feuilletée ; *f 2* **tourbe [f] blanche** (Tourbe prélevée sur la couche suppérieure d'une tourbière haute, peu décomposée, constitue un état intermédiare entre la sphaigne morte et la tourbe blonde) ; *g* **Weißtorf [m] (1)** (In kühleren Klimaverhältnissen des Subatlantikums entstandener Hochmoortorf, der schwach bis kaum humifiziert ist, so dass Pflanzenrückstände sehr gut erkennbar sind; ▶Schwarztorf).

2028 fibric peat [n] **(2)** *constr. hort.* (Lightly-decomposed *Sphagnum* peat); *s* **turba [f] clara** (Turba de esfagno poco descompuesta y sin humificar); *syn.* turba [f] rubia; *f* **tourbe [f] blonde** (Tourbe à Sphaigne peu décomposée de la couche inférieure de la tourbière) ; *g* **Weißtorf [m] (2)** (Nach DIN 11 542 ist als Qualitätsbezeichnung statt **W.** nur „wenig zersetzter Hochmoortorf" zu verwenden); *syn.* wenig zersetzter Hochmoortorf [m].

fibrous bark [n] *arb. bot.* ▶outer bark.

2029 fibrous root [n] *arb. hort.* (Thin, profusely branched, thread-like root, occupying a large volume of shallow soil around the plant's base; the root system is called ▶fibrous root system or "diffuse root system" (BOT 1990, 35); ▶fine root, ▶absorbing root); *s* **raíz [f] fibrosa** (Raíz prolongada y de poco grosor, semejante a una fibra; cf. DB 1985; ▶raicilla absorbente, ▶raíz fina, ▶sistema de raíces fibrosas); *f* **radicelle [f]** (Ébauche de la ▶jeune racine provenant de la ramification de la racine principale d'un diamètre < 5 mm ; la radicelle a souvent un aspect fibreux, c'est la raison pour laquelle elle est parfois dénommée **racine fibreuse** ; ▶racine fine, ▶chevelu racinaire) ; *g* **Faserwurzel [f]** (Zur Aufnahme von Wasser und Nährstoffen fein verzweigte, dünne Wurzel mit einem Durchmesser < 1 mm im Oberbodenbereich; ▶Faserwurzelwerk, ▶Feinwurzel, ▶Saugwurzel 1); *syn.* Feinstwurzel [f], Kurzwurzel [f].

2030 fibrous root system [n] *arb. hort.* (▶fibrous root); *syn.* diffuse root system [n]; *s* **sistema [m] de raíces fibrosas** (Conjunto de ▶raíces fibrosas de una planta); *syn.* sistema [m] radical fibroso; *f* **chevelu [m] racinaire** (Ensemble des jeunes racines d'une plante ; ▶radicelle) ; *g* **Faserwurzelwerk [n]** (Sämtliche ▶Faserwurzeln einer Pflanze); *syn.* Feinwurzelwerk [n].

2031 fidelity [n] *phyt.* (Degree of restriction of a species to a plant community; species with a most significant fidelity for the [total] range of an association are called ▶character species; ▶degree of fidelity); *s* **fidelidad [f] (1)** (Constancia de la presencia de una especie en una comunidad vegetal. Las especies de mayor **f.** se denominan ▶especies características; ▶grado de fidelidad); *f* **fidélité [f]** (Caractérise le coefficient de présence d'une espèce dans une association végétale [caractéristiques, compagnes, accidentelles] ; une espèce localisée exclusivement dans une association végétale est une ▶espèce caractéristique [exclusive] ; ▶degré de fidélité) ; *g* **Gesellschaftstreue [f]** (Grad der Bindung einer Art an eine Pflanzengesellschaft; Arten stärkster Treue werden ▶Kennarten genannt; ▶Treuegrad).

field [n] *agr.* ▶abandoned field; *recr.* ▶athletic field; *envir. urb.* ▶brownfield; *constr. recr.* ▶grass sports field; *agr. hort.* ▶nursery field.

field [n], **fertilized hay** *agr. phyt.* ▶fertilized meadow.

field [n], **grass playing** *constr. recr.* ▶grass sports field.

field [n], **open** *landsc.* ▶open country.

2032 field bank [n] *agr. geo.* (Sloping terrace margin created by erosion in agrarian landscapes: the bottom end of a sloping field parallel to the contours and covered by bushy vegetation); *s* **borde [m] de bancal de cultivo** (Pequeña terraza paralela a las curvas de nivel cubierta de matas que sirve de límite entre las parcelas de cultivo); *f* **talus [m] embroussaillé** (Petit talus buissonnant créé par ablation d'un versant et parallèle à la pente séparant deux parcelles cultivées étagées) ; *g* **Stufenrain [m]** (Hangparallel vergraste oder verbuschte Geländestufe in der bäuerlichen Kulturlandschaft, die, ausgelöst durch hangabwärtiges Pflügen, durch Abspülung bis zur Ackerparzellengrenze an geneigten Hängen entstanden ist; je nach Stärke der Hangneigung können mehrere **S.**e die Feldflur gliedern).

2033 field border [n] *agr. ecol.* (Uncultivated strip at the edge of an arable field; ▶field dike [US]/field dyke [UK]); *syn.* headland [n] [also US]; *s* **lindero [m] de campo de cultivo** (Banda no cultivada al borde de un campo de cultivo o entre dos campos; ▶montadero); *syn.* linde [m] de un campo de cultivo; *f* **lisière [f] d'un champ** (Bande de terrain inculte en bordure des champs ; ▶rideau) ; *syn.* marge [f] d'un champ ; *g* **Rain [m] (1)** (Unbewirtschafteter Streifen an der Ackergrenze oder zwischen zwei Äckern; ▶Hochrain); *syn.* Ackerrain [m], Ackerrandstreifen [m], Feldrain [m].

2034 field capacity [n] *pedol.* (Amount of water retained in a soil after it has become saturated and has drained freely; this is usually expressed as a percentage of the oven-dry [105 °C] weight of the soil; ▶water holding capacity); *syn. hort.* moisture capacity [n]; *s* **capacidad [f] de campo** (La cantidad de agua retenida por un suelo que ha perdido su agua gravitacional define la **c. de c.** Ésta marca un límite entre el agua gravitacional y el agua capilar, e indica la máxima cantidad de agua que puede retener el suelo tras dos días de aporte de la misma; cf. DINA 1987; ▶capacidad de retención de agua); *f* **capacité [f] (de rétention) au champ** (État d'équilibre de la teneur maximale en eau d'un sol — capillaire et liée —, exprimé en % du poids ou du volume ; ▶capacité de rétention [maximale] pour l'eau) ; *g* **Feldkapazität [f]** (Gleichgewichtszustand des Bodenwassergehalts, ausgedrückt in Gewichts- oder Volumenprozenten und bezogen auf Boden von 105 °C; ▶Wasseraufnahmefähigkeit); *syn. hort.* Wasserkapazität [f].

2035 field check [n] *rem'sens. surv.* (Ground check for evaluation of aerial photography. Checking reliability of ▶interpretation of aerial photographs is often called 'ground truth';

TEE 1980, 198; ▶check area); *syn.* ground check [n]; *s* **contra-stación [f] del terreno** (Proceso de control *in situ* de los resultados del procesamiento de datos de satélite o de aerofotos; ▶área de entrenamiento, ▶interpretación de foto[grafía]s aéreas); *f* **contrôle [m] des données sur le terrain** (Contrôle sur le site des résultats de l'interprétation des images satellites ; ▶aire de contrôle, ▶interprétation des photographies aériennes) ; *g* **Geländevergleich [m]** (Kontrolle und Ergänzungen von Aus-wertungsergebnissen der Luftbilder oder Satellitendaten im Gelände; ▶Luftbildauswertung, ▶Testgebiet); *syn.* Feldvergleich [m].

field check [n] of soil [US] *constr.* ▶on-site soil investigation.

2036 field dike [n] [US] *conserv. land'man.* (Earth berm along edges of fields and pathways of sloping agricultural areas to prevent erosion; ▶field bank, ▶field border); *syn.* field dyke [n] [UK]; *s* **montadero [m]** (Normalmente pared sin argamasa con ligera inclinación hacia arriba de contención de bancal en ladera aterrazada; cf. DGA 1986; ▶borde de bancal de cultivo, ▶lindero de campo de cultivo); *syn.* dique [m] (2); *f* **rideau [m]** (Ressaut de terrain à l'aval d'un champ : à la limite entre deux parcelles ou sur les bords de chemin ruraux, formation, selon la pente, d'un bourrelet ou d'un talus par accumulation de terre labourée au-dessus et descente de celle-ci au-dessous ; le terme a été employé en Picardie, puis généralisé ; DG 1984 ; ces éléments structurant du paysage rural, plus ou moins embroussaillés, sont des facteurs de limitation de l'érosion et de stabilisation de la terre ; les rideaux et les ▶talus embroussaillés en bordure des chemins de terre anciens, dont certains sont à l'état de chemins creux constituent souvent les derniers espaces naturels et refuges faune flore en zone rurale ; ▶lisière d'un champ ; *g* **Hochrain [m]** (Wallartig erhöhter Feld- oder Wegrand an geneigten landwirtschaftlichen Flächen als Erosionsschutz. **H.e** sind oft aus Lesesteinwällen, die an Feldrändern zusammengetragen wurden, entstanden; ▶Rain 1, ▶Stufenrain).

2037 field drainage [n] *agr.* (Measures taken to reduce the water table of agricultural land, usually through the installation of pipe drains; ▶agricultural drainage, ▶drainage 1, ▶land drainage, ▶soil drainage 1, ▶subsurface drainage); *s* **drenaje [m] agrícola subterráneo** (Medidas tomadas para reducir el nivel freático en campos de cultivo, normalmente instalando drenes; ▶drenaje, ▶drenaje agrícola, ▶drenaje a gran escala, ▶drenaje del suelo, ▶drenaje subterráneo); *f* **drainage [m] agricole souterrain** (Mesures d'assainissement des sols agricoles au moyen de drains pour corriger les mauvaises propriétés physiques, améliorer les conditions de croissance des cultures par du sol : ▶drainage de sol au moyen de tuyaux pour l'▶évacuation des eaux ; ▶drainage agricole, ▶drainage [agricole] à ciel ouvert, ▶drainage souterrain, ▶drainage superficiel) ; *syn.* assainissement [m] des sols (LA 1981, 113), drainage [m] agricole enterré ; *g* **Dränung [f] landwirtschaftlicher Flächen** (▶Entwässerung durch Dränrohre auf landwirtschaftlich genutzten Flächen zur Verbesserung der Wachstumsbedingungen; ▶Bodenentwässerung, ▶Dränung, ▶Entwässerung landwirtschaftlicher Flächen, ▶Wasserbaumaßnahme zur Entwässerung landwirtschaftlicher Flächen).

field dyke [n] [UK] *conserv. land'man.* ▶field dike [US].

field features [npl] [UK] *land'man. landsc.* ▶small landscape features.

2038 field-grown plant [n] *hort.* (As distinguished from a plant grown in a greenhouse; ▶nursery field); *s* **planta [f] cultivada al aire libre** (Al contrario que una planta cultivada en invernadero; ▶campo de cultivo al aire libre); *f* **plante [f] de pleine terre** (Plante cultivée en plein air par distinction avec la

plante de serre ou la plante en pots ; ▶pleine terre) ; *syn.* plante [f] cultivée en plein air, plante [f] de plein champ ; *g* **Freiland-pflanze [f]** (Pflanze, die im Gegensatz zur Gewächshauspflanze im Freien kultiviert wird; ▶Freiland).

field hedge [n] *agr. land'man.* ▶hedgerow.

2039 field investigation [n] *ecol. geo. pedol.* (Scientific study carried out on site; ▶field survey 1, ▶topographic survey 1); *syn.* field research [n]; *s* **investigación [f] de campo** (Estudios científicos realizados sobre terreno; ▶levantamiento de planos, ▶trabajo de campo); *f* **étude [f] sur le terrain** (Travaux scientifiques effectués sur le terrain ; ▶étude préliminaire de terrain, ▶reconnaissance du site) ; *syn.* prospection [f] sur le terrain, étude [f] de terrain ; *g* **Felduntersuchung [f]** (Wissenschaftliche Arbeiten im Gelände; ▶Feldaufnahme, ▶Geländeaufnahme 1).

2040 field nursery [n] *for. hort.* (Loose term for a tree nursery, generally not permanent, established in or near the forest rather than near administrative or executive headquarters; SAF 1983; ▶forest tree nursery); *s* **vivero [m] forestal volante** (Vivero forestal temporal ubicado en lugar cercano a los montes a repoblar; ▶vivero forestal); *f* **pépinière [f] volante** (Terme général désignant une pépinière non permanente, établie p. ex. en forêt ou dans son voisinage plutôt qu'à proximité d'un centre administratif ou de gestion ; DFM 1975 ; ▶pépinière forestière) ; *g* **temporäre Forstbaumschule [f]** (In der Nähe eines Waldes angelegte Baumschule zur Heranzucht von Jungpflanzen; ▶Forstbaumschule).

field research [n] *ecol. geo. pedol.* ▶field investigation.

2041 fields [npl] *agr.* (General term covering agricultural land with few or no trees and shrubs; ▶open country; *in comparison* ▶hedgerow landscape); *s* **campo [m] abierto** (Término común para el paisaje humano sin bosque; ▶bocage, ▶campiña); *f* **openfield [m]** (Paysage agraire caractérisé par des parcelles sans enclos souvent géométriques et en forme de lanières ; *en comparaison* ▶bocage, ▶campagne 2, ▶paysage d'openfield, espace naturel ; LA 1981, 792) ; *syn.* campagne [f] (1), champagne [f], champs [mpl] ouverts ; *g* **Flur [f] (2)** (*Allgemeiner Begriff* offenes, nicht bewaldetes Kulturland; die in der ▶freien Landschaft enthaltenen Gewässer, Wälder und Felsareale werden bei dem Begriff **F.** nicht mitgedacht; *zum Vergleich* ▶Heckenlandschaft); *syn.* Feldflur [f].

field shrub [n] *agr. land'man.* ▶field tree or shrub.

2042 field survey [n] (1) *biol. geo. landsc. phyt. surv.* (General term for a site inventory to collect and delineate information by a specialist; ▶topographic survey 3); *s* **trabajo [m] de campo** (Término común para colecta de datos sobre terreno realizada por especialista; inventariar [vb] sobre terreno; ▶medición topográfica); *f* **étude [f] préliminaire de terrain** (Visite du terrain, examen de l'état général du terrain et recensement de toutes les informations pouvant avoir un effet sur le projet ; ▶levé, ▶arpentage) ; *syn.* inventaire [m] de terrain ; *g* **Feldaufnahme [f]** (Aufzeichnung des Geländeinventars oder Teile davon; ▶Vermessung).

field survey [n] (2) *surv.* ▶topographic survey (3).

2043 field track [n] *agr.* (**1.** Maintained, paved or unpaved road in the rural landscape. **2. In U.S.,** a dirt road made with a grass median and used by farm vehicles with little or no maintenance is called **farm track [US]**; *s 1* **camino [m] rural** (Vía de tránsito en zona agrícola); *s 2* **camino [m] carretero** (Vía de tránsito de vehículos de tracción animal; cf. DGA 1986); *f 1* **chemin [m] rural** (Voie carrossable appartenant à la commune en zone rurale et affectée à l'usage du public) ; *f 2* **chemin d'exploitation** (Chemin appartenant au propriétaire riverain et servant exclusivement à la communication entre les divers fonds ou à leur exploitation) ; *g* **Feldweg [m]** (Befahrbarer Weg in der Agrarlandschaft); *syn.* landwirtschaftlicher Weg [m].

2044 field tree [n] *agr. land'man.* (Single tree left standing in a field or hedgerow; *specific term* hedgerow tree [US]/hedgerow timber [UK]); *s* **solitario [m] campestre** (Árbol solitario en el campo); *f* **arbre [m] champêtre** (en zone rurale agricole) ; *g* **Feldbaum [m]** (Baum in der bäuerlichen Kulturlandschaft).

2045 field tree [n] **or shrub** [n] *agr. land'man.* (Indigenous woody plant in the rural landscape; in pasture land: pasture tree; a group of such trees or shrubs is known as copse; ▶coppice [US] 1/cops [UK], ▶woody clump); *s* **leñosa [f] campestre** (**1.** Árbol o arbusto que se encuentra en el campo. **2.** En alemán, «Feldgehölz» es un término genérico que cubre tanto leñosas solitarias como grupos o hileras de éstas; ▶bosquecillo 1, ▶bosquecillo 3); *f 1* **ligneux [m] champêtre** (Espèces ligneuses indigènes isolées, groupées ou alignées dans les régions agricoles) ; *f 2* **bosquet [m] champêtre** petit massif forestier constituée par l'association d'arbres et d'arbustes locaux en zone rurale ; ▶bosquet 2, ▶bouquet, ▶haie champêtre, ▶petit bois) ; *g* **Feldgehölz [n]** (**1.** Einzelner bodenständiger Baum oder Strauch sowie Gruppe/Reihe von Gehölzen in der bäuerlichen Kulturlandschaft. **2.** Ein F. ist auch eine Gruppe von Bäumen und Sträuchern oder ein kleines Wäldchen von ein bis mehreren Hektar Größe in der bäuerlichen Kulturlandschaft, meist in steilen Hangfluren, feuchten Senken und Hangfußlagen, auf erodierten Kuppen und in frostgefährdeten Talverengungen, oft als Relikt eines ehemaligen Bauernwaldes; kleine Flächen, die mit **F.en** bestockt sind, oder kleine Wäldchen werden auch **Feldholzinseln** genannt; ▶Gehölz 1, ▶Gehölz 2); *syn.* Flurgehölz [n].

2046 field-woodland edge [n] [US] *ecol.* (TGG 1984, 195; transition area of low growth between woods and fields; ▶ecotone 1, ▶edge effect, ▶forest edge); *syn.* prairie-woodland edge [n] [US], forest-meadow boundary [n] [UK]; *s* **límite [m] entre bosque y pradera** (Estación de transición entre bosque y pradera o campo de cultivo que se caracteriza por una gran diversidad de especies de las comunidades conlindantes; ▶ecotono, ▶efecto borde, ▶lindero del bosque); *f* **frange [f] de contact champ-forêt** (Zone de contact entre les groupements forestiers et les groupements voisins de culture ou de prairie et caractérisée par un nombre d'espèces et d'individus important des groupements de lisières ; ▶écotone, ▶effet de bordure, ▶effet de lisière, ▶lisière forestière) ; *syn.* limite [f] de boisement ; *g* **Feld-Wald-Grenze [f]** (Lebensraum von Saumbiozönosen mit großer Arten- und Individuenzahl einer Population; ▶Ökoton, ▶Randeffekt, ▶Waldrand).

fill [n] *constr.* ▶cut and fill, ▶dumped fill, ▶earth fill, ▶fill material, ▶general fill [US], ▶granular drain fill, ▶granular fill, ▶land filling, ▶structural fill [US].

fill [n], **dumped** *constr.* *▶fill material.

fill [n] [US], **general** *constr.* *▶fill material.

fill [n] [US], **structural** *constr.* *▶fill material.

2047 fill depth [n] *constr.* (Magnitude of ▶fill material backfilled to a specific height); *syn.* fill height [n]; *s* **altura [f] de relleno** (Grosor de una capa de ▶tierra de relleno); *f* **hauteur [f] de remblaiement** (Épaisseur des ▶matériaux de remblaiement, ▶matériaux tout-venant ; *terme spécifique* hauteur d'empierrement) ; *syn.* hauteur [f] des matériaux remblayés, hauteur [f] des matériaux en remblai ; *g* **Schütthöhe [f]** (Mächtigkeit eines in bestimmter Höhe eingebauten ▶Schüttbodens).

fill dirt [n] [US] *constr.* ▶fill material.

2048 filler [n] *plant.* (Single tree or shrub to fill a space in screening); *s* **leñosa [f] para protección visual** (Árbol o arbusto

que se utiliza para evitar la visibilidad de un objeto); *f* **végétal [m] ligneux de remplissage** (Arbre ou arbuste jouant le rôle d'écran visuel) ; *syn.* végétal [m] ligneux de masse ; *g* **Deckgehölz [n]** (Baum oder Strauch, der dazu dient, den Blick auf ein Objekt zu verhindern; *UB* Deckstrauch).

fill height [n] *constr.* ▶fill depth.

filling [n] *constr.* ▶backfilling, ▶earth filling (1), ▶earth filling (2), ▶hydraulic filling, ▶infilling, ▶land filling; *min.* ▶refilling.

2049 filling [n] **and compacting** [n] **in layers** *constr.* (of soil material; ▶backfilling of on-site excavated material); *s* **relleno [m] y compactación [f] por tongadas** (Relleno, allanado y compactación de tierra en capas; ▶reutilización de material de excavación); *syn.* terraplenado [m] y compactación [f] por tongadas; *f* **terrasser [vb] et compacter [vb] en couches** (▶réemploi de matériaux extraits) ; *syn.* mise [f] en place et compactage en couches, remblai [m] par couches alternées ; *g* **lagenweiser Einbau [m, o. Pl.]** (Lagenweises Schütten, Planieren und Verdichten von Bodenmaterial; ▶Wiedereinbau von Aushub); *syn.* Lagenbauweise [f], Lagenschüttung [f].

filling [n] **fail places** [US] *for.* ▶replacement planting.

2050 filling [n] **of paver holes** [US] *constr.* (Filling the grid of holes in ▶grass pavers with growing medium); *syn.* soiling [n] of holes, soiling [n] of voids [UK] (cf. SPON 1974, 329); *s* **relleno [m] de huecos de adoquines pavicésped** (Relleno con tierra de los alveolos de los ▶adoquines pavicésped); *f* **remplissage [m] des alvéoles** (Apport de terre végétale dans les vides des ▶dalles gazon) ; *g* **Verfüllen [n, o. Pl.] der Erdkammern** (Füllen der Hohlräume von ▶Rasengittersteinen mit einem Erdsubstrat).

2051 fill material [n] *constr.* (**1.** Suitable material for filling up of trenches, working pits or low-lying areas; ▶backfilling behind structures, ▶borrow material, ▶land filling); **2.** Any type of loose subsoil including sand, gravel or rock fragments, used in ▶earthworks); *syn.* fill [n] [also US], fill dirt [n] [also US]; **3. structural fill** [n] [US] (*Specific term* granular soil that can be prepared to bear weight and to avoid subsidence); *syn.* granular fill [n]; **4. general fill** [US] (*Specific term* non-toxic overburden that can safely be used to bring a non-structural portion of a site up to specified grades for purposes of slope management and re-contouring for drainage and planting, **5. backfill** [n] (Material used for ▶backfilling of utility trenches and around new buildings or over underground construction); **6. dumped fill** [n] (Soil that has been deposited, usually by truck, and which has not been spread or compacted; MEA 1985); *s* **tierra [f] de relleno** (**1.** Tierra u otro material adecuado, incluyendo material de construcción reciclable utilizado para rellenar zanjas de conducciones, bordes de edificios, hondonadas u otras depresiones; ▶relleno de zanjas y hoyos. **2.** Cualquier tipo de subsuelo suelto, incluyendo arena, grava o fragmentos de roca, utilizable para ▶terracerías; ▶terraplenado, ▶tierra de préstamo, ▶relleno del terreno); *syn.* material de relleno; *f 1* **matériaux [mpl] tout-venant** (Terres ou granulats de remplissage de fouilles des réseaux divers, de recouvrement ou de comblement d'ouvrages ; *termes spécifiques* grave [f] tout-venant, sable [m] tout-venant ; *f 2* **matériaux [mpl] de remblaiement** (**1.** Matériau approprié [p. ex. terre de déblai ou terre des fouilles purgée de tout détritus et pierrailles] utilisé pour le remblaiement des tranchées pour canalisation ; ▶terres d'emprunt). **2.** Sols et matériaux rocheux, matériaux de substitution, sous-produits industriels dont le comportement justifie l'utilisation pour la réalisation de remblais [remblai en terre fine pour le ▶remplissage [d'une tranchée/d'un trou], le remplissage de fouilles des réseaux divers, remblai en terre saine pour le recouvrement ou de comblement d'ouvrages] et de couches de forme ; ▶comblement du terrain, ▶terrassement,

▶terrassement généraux, ▶terres d'emprunt) ; *syn.* matériaux [mpl] de construction des remblais, matériaux [mpl] pour remblais ; *g 1* **Füllboden [m]** (Geogenes mineralisches Bodenmaterial, z. B. Unterboden, höherwertiges Schüttgut, z. B. Wandkies, oder rezyklierte Baustoffe zum Verfüllen von Arbeitsräumen, Leitungsgräben, Überschütten oder Hinterfüllen von Bauwerken sowie für den ▶Erdbau zum Auffüllen/Zuschütten tiefer liegender Flächen [z. B. Geländemulden]; ▶Füllboden [aus] einer Seitenentnahme, ▶Hinterfüllen, ▶Geländeauffüllung); *g 2* **Schüttboden [m]** (OB zu Füllboden. *MERKE*, im Dt. kennzeichnet die Vorsilbe ,Schütt-' das ▶Verfüllen/Zuschütten von Hohlformen und das Aufschütten von Hügeln und Wällen; hingegen die Vorsilbe ,Füll-' bedeutet stets nur das Verfüllen/ Schließen/Anfüllen von Hohlräumen oder Auffüllen/Erhöhen niedrig liegender Flächen); *syn. zu 1.* Auffüllboden [m], Auffüllmaterial [n], Füllmaterial [n]; *syn. zu 2.* Schüttbaustoff [m], Schüttgut [n], Schüttmaterial [n], Schüttstoff [m] (S+G 1995 [11], 737), Schüttsubstrat [n].

2052 fill slope [n] *constr.* (Slope created by earthwork operations; ▶road embankment); *s* **talud [m] de terraplén** (Pendiente creada por trabajo de terracerías; ▶terraplén para carretera); *syn.* talud [m] de relleno (BU 1959); *f* **talus [m] en remblai** (Versant artificiel raide créé par terrassement en remblai ; *termes spécifiques* talus de route, talus de voie ferrée ; ▶remblai routier) ; *g* **Schüttböschung [f]** (Böschung eines künstlich geschütteten Erdkörpers; ▶Straßendamm); *syn.* Auftragsböschung [f].

filter blanket [n] [US] *constr.* ▶filter mat.

filter cloth [n] *constr.* ▶filter mat.

filter fabric [n] [US] *constr.* ▶filter mat.

2053 filtering effect [n] **of vegetation** *ecol.* (Absorption of air pollutants by plants); *s* **efecto [m] filtrante de la vegetación** (Capacidad de las plantas de absorber o servir de depósito de sustancias contaminantes de la atmósfera); *f* **effet [m] de filtration de la végétation** (Action purificatrice des végétaux face à la pollution atmosphérique ; sédimentation des poussières consécutive à la réduction de la vitesse du vent, absorption, dépôt, fixation sur le feuillage, précipitation au sol par formation de brouillard condensant) ; *syn.* action [f] filtrante des végétaux ; *g* **Filterwirkung [f] von Pflanzenbeständen** (Absorption oder oberflächliche Deposition der aus der Luft aufgefangenen Stoffe durch Pflanzen).

filtering [n] **of dust** *envir.* ▶dust filtering.

2054 filter layer [n] *constr.* (Layer which prevents the infiltration of washed-out soil particles into the subbase [US]/subbase [UK] after a period of frost or persistent rain; ▶filter mat); *syn.* filter stopping layer [n] [also US]; *s* **capa [f] filtrante (2)** (Capa que evita el deslavado del suelo hacia la capa de drenaje después de heladas o de lluvias fuertes; ▶estera filtrante); *f* **couche [f] anticapillaire** (Sous-couche de chaussée dépourvue d'éléments fins empêchant les remontées capillaires en provenance du fond de forme ; DIR 1977 ; ▶feutre filtrant) ; *g* **Filterschicht [f] (2)** (Schicht, die verhindert, dass nach Frostperioden oder anhaltenden Regenfällen aufgeweichter Boden des Baugrundes in die darüberliegende Tragschicht eindringt; ▶Filtermatte).

2055 filter mat [n] *constr.* (Prefabricated pervious sheet of, e.g. glass fiber [US]/glass fibre [UK], polyester, polypropylene or a geotextile between topsoil and the ▶drainage layer to prevent movement of soil particles due to infiltrating water; ▶wash fine material); *syn.* soil filter [n] [also US], filter blanket [n] (OSM 1999) [also US], filter fabric [n] [also US], soil separator [n], filter cloth [n], filter membrane [n]; *s* **estera [f] filtrante** (Lámina permeable de fibra de vidrio, poliéster u otros materiales que se coloca sobre la ▶capa de drenaje para evitar la ▶colmatación en

revestimientos vegetales de tejados); *f* **feutre [m] filtrant (anti-contaminant)** (Feutre non tissé en fibres polyester imputrescibles disposé sur la ▶couche drainante et relevé le long des parois des remontées afin d'éviter le ▶colmatage de la couche drainante tout en assurant une perméabilité nécessaire à l'évacuation des eaux pluviales excédentaires) ; *syn.* feutre-jardin [m], filtre [m] drainant, écran [m] filtrant, filtre [m] anticontaminant, géotextile [m] anticontaminant, (feutre) bidim [m] ; *g* **Filtermatte [f]** (Aus Glasfaser, Polypropylen, Polyester etc. hergestelltes Vlies, das zwischen Oberboden und ▶Dränschicht zur Verhinderung des ▶Einschlämmens von Feinanteilen, z. B. bei Dachbegrünungen, eingebaut wird); *syn.* Filtervlies [n].

filter membrane [n] *constr.* ▶filter mat.

filter stopping layer [n] [US] *constr.* ▶filter layer.

filtration [n] **of dust from the air** *envir.* ▶dust filtering.

filtration spring [n] *hydr. geo.* ▶seepage spring.

final acceptance [n] [US] *constr. contr.* ▶condition for final acceptance.

2056 final account invoice [n] *contr.* (Complete project bill prepared by the contractor for payment by the client; ▶final fee invoice, ▶interim invoice); *s* **liquidación [f] final de obra** (Listado completo de todos los servicios prestados para realizar una obra para liquidar los gastos no abonados hasta el momento; ▶liquidación final de honorarios, ▶liquidación intermedia); *f* **décompte [m] final** (Établissement par l'entrepreneur du montant total des travaux auquel il peut prétendre après achèvement de ceux-ci ; ▶projet de décompte général, ▶situation) ; *syn.* projet [m] de décompte final, décompte [m] général ; *g* **Schlussabrechnung [f]** (Durch den Bauunternehmer aufgestellte, vollständige Auflistung aller erbrachten Leistungen für ein Bauvorhaben zwecks Begleichung der noch ausstehenden Zahlungen; ▶Honorarschlussrechnung, ▶Zwischenrechnung); *syn.* Schlussrechnung [f].

2057 final approval [n] **of a community development plan** [US] *adm. leg.* (by a governing authority); *syn.* approval [n] of an urban development plan [UK]; *s* **aprobación [f] definitiva** (Decisión administrativa o de un consejo municipal de aprobar un plan de ordenación urbana o un plan parcial); *f* **approbation [f] d'un document d'urbanisme** (C. urb., art. L. 123-5 ; par délibération de l'autorité compétente, p. ex. approbation du plan par le préfet ; cf. art. 129-9) ; *g* **Genehmigung [f] eines Bauleitplanes** (Verwaltungsakt einer übergeordneten Verwaltungsbehörde [Regierungspräsidium/Bezirksregierung], z. B. einen Flächennutzungsplan oder Bebauungsplan zu genehmigen).

final completion [n] *constr.* ▶issue of a certificate of final completion.

final construction cost estimate [n] [UK] *contr.* ▶total construction cost estimate [US].

final contract amount [n] [US] *constr.* ▶determination of final contract amount.

final deposition [n] *envir.* ▶final disposal.

2058 final design competition [n] [US] *prof.* (Often a second phase in a competition between selected firms chosen to prepare detailed plans and sometimes models for a fixed and defined program[me], based on their preliminary concepts. All participants receive fees for their competition entries. The plans are evaluated by the owner/developer or professional jury, and the winner is designated for contract award; ▶design competition 1); *syn.* realization competition [n] [UK]; *s* **concurso [m] de realización** (Concurso anunciado por un promotor de concurso que —sobre la base de un programa concreto y de unas condiciones a cumplir— tiene como fin mostrar las posibilidades y vías de realización de un proyecto; ▶concurso de ideas);

F

f **concours [m] de conception-construction** (Concours pour la passation d'un contrat de conception-réalisation portant à la fois sur l'établissement des études et l'exécution des travaux, le maître d'ouvrage arrêtant un programme et des prestations précises ; le recours au **c. de c.-c.** ne peut avoir lieu que si l'association de l'entrepreneur aux études est nécessaire pour réaliser l'ouvrage, en raison de motifs techniques liés à sa destination ou à sa mise en œuvre technique ; décret n° 93-1270 du 29 novembre 1993 ; ►concours d'idées) ; *g* **Realisierungswettbewerb [m]** (Von einem Auslober ausgeschriebener Wettbewerb, der auf der Grundlage eines fest umrissenen Programmes und bestimmter Leistungsanforderungen die planerischen Möglichkeiten für die Durchführung eines Projektes aufzeigt; cf. Teil 1 Nr. 2.1.2 GRW 1995; bei großen Projekten oft als zweite Phase nach einem ►Ideenwettbewerb, bei dem dann alle Teilnehmer für ihre Entwürfe Honorar erhalten. **In USA** sind eingeladene Wettbewerbe ohne angemessenes Bearbeitungshonorar für die abzuliefernde Arbeit nicht üblich. Aus betriebswirtschaftlicher Sicht gibt es dort nicht die Idealisierung der europäischen Wettbewerbskultur); *syn.* Bauwettbewerb [m].

2059 final design stage [n] *contr. prof.* (Third phase of the design process, as part of a planner's services, which is based on approved ►schematic design phase [US]/outline and sketch scheme proposals [UK] including the preparation of the design development documents [US] consisting of drawings, outline specifications, ►estimate of probable construction cost and other documents to fix and describe the scope of the whole project); *syn.* design development phase [n] [also US]; *s* **fase [f] de anteproyecto detallado** (Fase definitiva de diseño de proyecto sobre la base del ►anteproyecto aprobado en la que se incluye la elaboración de planes, memoria descriptiva, la ►estimación del costo y otros documentos necesarios para describir el proyecto; ►fase de anteproyecto); *f* **phase [f] d'avant-projet [m] détaillé (d'A.P.D.)** (Étape de conception secondaire dans une mission de maîtrise d'œuvre pour des recherches et études relatives aux ouvrages dans le cadre d'une solution d'ensemble retenue sur la base de l'►avant-projet sommaire [A.P.S.] accepté par le maître d'ouvrage. Ces recherches et études ont pour buts essentiels l'approfondissement de la solution d'ensemble au niveau des ouvrages considérés, la présentation des choix architecturaux et techniques comprenant les plans des ouvrages, le descriptif sommaire et les caractéristiques des ouvrages ainsi que l'établissement d'un estimatif sommaire par corps d'état ou par ouvrage d'exécution ; HAC 1989 ; ►avant-projet, ►estimation détaillée des dépenses, ►évaluation détaillée du coût des travaux) ; *g* **Entwurfsplanung [f]** (Durcharbeiten des Planungskonzeptes der ►Vorplanung bis zu einer endgültigen Lösung in Form einer zeichnerischen Darstellung des Gesamtentwurfes, ggf. mit Detailplänen incl. der Objektbeschreibung und einer ►Kostenberechnung nach DIN 276-1: 2008-12 oder nach dem wohnungsrechtlichen Berechnungsrecht; cf. § 15 [1] 3 HOAI 2002 resp. §§ 33, 38, 42 u. 46 HOAI 2009).

2060 final disposal [n] *envir.* (Final deposing of waste material; ►final disposal site for radioactive waste); *syn.* final deposition [n]; *s* **almacenamiento [m] definitivo de residuos** (Disposición final de residuos urbanos o industriales en un vertedero tras su tratamiento para que no causen daños al medio ambiente o de residuos radiactivos en ►cementerio nuclear); *syn.* confinamiento [m] definitivo de residuos; *f* **stockage [m] définitif** (p. ex. de déchets industriels ultimes ; ►centre de stockage de déchets radioactifs) ; *g* **Endlagerung [f]** (Methode der Beseitigung von Abfallstoffen auf Deponien durch geregelte Ablagerung; die endgültige Lagerung wird in der Technischen Anleitung zur Verwertung, Behandlung und sonstigen Entsorgung von Siedlungsabfällen [TA Siedlungsabfall] und der Technischen Anleitung zur Lagerung, chemisch-physikalischen, biologischen Behandlung, Verbrennung und Ablagerung von ►besonders überwachungsbedürftigen Abfällen [TA Abfall] geregelt; ÖKO 2003; die Entsorgung und Endlagerung von Kernbrennstoffen und sonstigem radioaktiven Material regeln die §§ 9a-9g Atomgesetz [AtG] vom 23.12.1959, zuletzt geändert am 17.03.2009; ►Endlager für radioaktive Stoffe).

2061 final disposal [n] **of surplus material** *constr.* (Final placement of ►surplus excavated material which could not be placed *in situ* after excavation or stripping/superficial removal; ►off-site disposal of surplus material); *s* **depósito [m] definitivo de material sobrante** (►disposición de excedentes de terracerías, ►material sobrante); *f* **dépôt [m] définitif de déblais excédentaires** (Stockage définitif de matériaux ne pouvant plus être mis en place sur leur lieu d'enlèvement ; ►déblais excédentaires, ►enlèvement des excédents) ; *g* **Endlagerung [f] von Überschussmassen** (Endgültige Lagerung von Erdmassen, die am Ort ihres Aushubes/Abtrages nicht wieder eingebaut werden konnten; ►Überschussmassen, ►Massenverbringung).

2062 final disposal site [n] **for radioactive waste** *envir.* (Place for final deposition of radioactive waste; ►final disposal); *s* **cementerio [m] nuclear** (Depósito de residuos de características especiales para el ►almacenamiento definitivo de residuos radiactivos); *f* **centre [m] de stockage de déchets radioactifs** (Lieu de dépôt définitif des déchets radioactifs ; ►stockage définitif) ; *g* **Endlager [m] für radioaktive Stoffe** (Endgültige Deponie für radioaktive Stoffe, meist ein Tiefenlager; ►Endlagerung).

final elevation [n] **[UK], establishment of** *constr.* ►establishment of finished elevation.

2063 final fee invoice [n] *contr. prof.* (Final bill of a planner showing the compilation of all relevant charges for contract services after all interim payments have been deducted); *s* **liquidación [f] final de honorarios** (Factura en la que aparecen todos los servicios prestados relevantes para el cálculo de los honorarios definitivos); *f* **projet [m] de décompte général** (Facture établie par le concepteur faisant état du solde des honoraires dus en fonction des phases techniques d'exécution réalisées après prise en compte des acomptes mensuels ; le **p. de d. g.** accepté ou rectifié par le conducteur d'opération devient le décompte général et définitif ; *syn.* décompte [m] définitif des honoraires ; *g* **Honorarschlussrechnung [f]** (Rechnung des Planers, in der abschließend alle vergütungsrelevanten Beträge zusammengestellt werden und nach Abzug der geleisteten Akontozahlungen der restliche Auszahlungsbetrag zur Anforderung ermittelt wird); *syn.* Schlusshonorarnote [f] [auch A].

2064 final grade [n] (1) *constr.* (Result of ►fine grading: the final grade of the site which conforms to the approved plan; ►subbase grade [US]/formation level [UK]); *syn.* finish grade [n] [also US], finished grade [n], finished levels [npl]; *s* **superficie [f] refinada** (Resultado de la ►operación de refino [de superficie]; ►subrasante); *f* **règlement [m] final** (Résultat de la mise en forme définitive au râteau des surfaces à ensemencer ou à planter ; ►exécution du règlement final, ►fond de forme) ; *syn.* réglage [m] final, dressement [m] final ; *g* **Feinplanum [n]** (Für Vegetationsarbeiten hergestellte Ebenflächigkeit der Bodenoberfläche [Vegetationstragschicht]: bei Rasenflächen je nach Ausschreibungstext mit einer Abweichung von der geplanten Höhe auf einer 4-m-Messstrecke von meist ± 1,5 cm, bei Pflanzflächen ± 5 cm. Anschlüsse müssen bündig sein und können nach unten bis 3 cm abweichen; cf. DIN 18 915; ►Erdplanum, ►Herstellen des Feinplanums); *syn.* Feinplanie [f].

2065 final grade [n] (2) *constr.* (Result of modifying the ground surface); *s* **nivel [m] definitivo del terreno** (Resultado de la modelación del terreno para obras de ingeniería civil); *f* **modelé [m] final** (Configuration définitive d'un terrain après travaux) ; *g* **fertige Geländemodellierung [f]** (Ergebnis eines Erdbauwerkes oder z. B. einer in die Landschaft eingepassten Straßentrassierung).

2066 final investigative report [n] *plan.* (Conclusive written or oral statement on a scientific investigation); *syn.* report [n] of results; *s* **informe [m] final de una investigación** (Presentación definitiva de los resultados de una planificación o investigación); *f* **rapport [m] définitif de travaux de recherche** (Présentation des résultats de travaux de recherche) ; *syn.* rapport [m] de synthèse d'une recherche ; *g* **Abschlussbericht [m] einer Untersuchung** (Abschließende Darstellung des Ergebnisses einer Forschungsarbeit oder einer Untersuchung); *syn.* Ergebnisbericht [m] einer Untersuchung, Schlussbericht [m] einer Untersuchung.

2067 final invoice amount [n] *contr.* (Sum of money invoiced by a contractor for services and work performed, which becomes due after the client or his authorized representative has audited and determined the final sum; a final account would normally be paid within two months of its submission); *s* **cuenta [f] final** (Suma a calcular por el contratista o adjudicatario del contrato sobre los servicios prestados que es controlada por el comitente o su encargado; el pago final debe realizarse en el plazo acordado o el regulado por ley); *f 1* **montant [m] du décompte final** (Somme correspondant au décompte établi dans les marchés de travaux par le maître d'œuvre à partir du projet de décompte final éventuellement rectifié présenté par l'entrepreneur ; le décompte final établit le montant total en prix de base hors taxe sur la valeur ajoutée des sommes auxquelles l'entrepreneur peut prétendre du fait des prestations réellement exécutées à l'occasion du marché ; en cas de retard dans la présentation du projet de décompte final, et après mise en demeure restées sans effet, le maître d'œuvre peut le faire établir aux frais de l'entrepreneur ; article 13.33 du C.C.A.G. Travaux ; *f 2* **montant [m] du décompte général** (Somme correspondant au décompte établi dans les marchés de travaux par le maître d'œuvre, signé par la personne publique et notifié à l'entrepreneur par un ordre de service ; le montant du décompte général est égal au montant récapitulatif des acomptes mensuel et du solde ; le décompte général devient le **décompte général définitif** après son acceptation par l'entrepreneur ; article 13.4 du C.C.A.G. Travaux ; *g* **Schlussabrechnungssumme [f]** (Vom Auftragnehmer für seine erbrachten Leistungen zu ermittelnder Betrag, der vom Auftraggeber oder seinem Treuhänder geprüft und festgestellt wird; die Schlusszahlung wird nach § 16 VOB Teil B spätestens innerhalb von zwei Monaten nach Zugang fällig).

2068 final plan [n] *leg. urb.* (Finished project design, which has been developed from a draft version after consideration of arguments and representations of those who are involved in the planning process, and which will be presented as a planning proposal for approval by public authority or agency; ►final plan for approval); *s* **plan [m] definitivo** (Plan desarrollado a partir del ante-proyecto del plan correspondiente después de haber sido consideradas las sugerencias de los participantes en el proceso de planeamiento. Este plan es sometido a información pública; ►plan aprobado provisionalmente); *f* **projet [m] de plan approuvé** (Dans la procédure d'élaboration d'un PLU/SCOT, projet éventuellement modifié suite à l'enquête publique et approuvé par délibération des organes à l'initiative de l'élaboration du document d'urbanisme ; si l'approbation d'un PLU n'intervient pas dans un délai de trois ans il cesse d'être opposable aux tiers ; ►plan approuvé par arrêté préfectoral) ; *g* **endgültige Planfassung [f]** (Planungsergebnis, das aus der vorläufigen Planfassung unter Berücksichtigung der Darlegungen und Erörterungen der am Planungsprozess Beteiligten entwickelt wurde; diese Fassung wird bei Bauleitplänen gem. § 10 [1] BauGB als Satzung beschlossen; ►genehmigungsfähige Planfassung).

2069 final plan [n] **for approval** *leg. urb. trans.* (Final form of a plan [US]/scheme [UK] as part of ►community development planning [US]/urban land-use planning [UK] and in a ►design and location approval process [US]/determination process of a plan [UK] for major projects, modified to take account of ►objections and supporting representations, and submitted by a public agency/authority for approval by a governing body; ►final plan, ►publication of planning proposals); *s* **plan [m] aprobado provisionalmente** (En proceso de ►planeamiento urbanístico o en ►procedimiento de aprobación de proyectos públicos, ►plan definitivo concepcionalmente que —una vez consideradas las ►alegaciones, comentarios y proposiciones en el marco de la ►sumisión a información pública— ha sido aprobado por el municipio correspondiente y que debe ser aprobado definitivamente por la autoridad superior competente); *f* **plan [m] approuvé** (►*Planification urbaine* plan ayant été soumis à la délibération 1. de l'organe délibérant de l'Établissement Public de Coopération Intercommunale [EPCI] chargé de l'élaboration du SCOT après enquête publique, 2. du conseil municipal chargé de l'élaboration du PLU/POS après enquête publique ; ►enquête publique, ►observations, ►mise à la disposition du public, ►procédure d'instruction de grands projets publics, ►projet de plan approuvé) ; *g* **genehmigungsfähige Planfassung [f]** (1. Im Bauleitplanungsprozess die endgültige Fassung eines Planentwurfes, die unter Berücksichtigung der ►Anregungen [und Bedenken], die während der öffentlichen Auslegung eingingen, von der Gemeinde beschlossen wurde und der höheren Verwaltungsbehörde [Regierungspräsidium/Bezirksregierung] zur Genehmigung vorgelegt wird; ►Auslegung eines Planentwurfes, ►Bauleitplanung. 2. Im ►Planfeststellungsverfahren die endgültige Planfassung, die von der zuständigen Verwaltungsbehörde, z. B. Regierungspräsidium/Bezirksregierung, festgestellt wird; ►endgültige Planfassung); *syn.* genehmigungsfähiger Entwurf [m], genehmigungsfähiger Plan [m].

2070 final planning report [n] *plan.* (Conclusive written or oral presentation on the completion of a planning project; *s* **informe [m] final de una planificación** (Presentación definitiva de los resultados de una planificación); *f* **rapport [m] définitif d'une étude** (Présentation des résultats d'une étude) ; *syn.* rapport [m] de synthèse d'une étude ; *g* **Abschlussbericht [m] einer Planung** (Abschließende Darstellung des Ergebnisses einer Planung, einer Forschungsarbeit oder einer Untersuchung); *syn.* Ergebnisbericht [m] einer Planung.

2071 final project design [n] *plan.* (Final development documents for solution of a planning task which has been mutually agreed to by all affected persons and agencies/authorities; ►final design stage, ►preliminary design); *s* **anteproyecto [m] detallado** (Representación gráfica y descripción de la solución propuesta para cumplir una tarea de planificación o realizar un proyecto de construcción paisajística; ►fase de anteproyecto detallado, ►anteproyecto); *f* **avant-projet [m] détaillé (A.P.D.)** (Présentation complète des solutions d'ensemble et des dispositions et caractéristiques techniques, fonctionnelles, dimensionnelles, financières et de gestion des ouvrages ; ►avant-projet sommaire [A.P.S.] ; ►phase d'avant-projet détaillé) ; *syn.* dossier [m] d'avant-projet détaillé (d'A.P.D.) ; *g* **Entwurf [m]** (1. Darstellung eines endgültigen Lösungsvorschlages für eine Planungsaufgabe, der mit allen zu Beteiligenden abgestimmt ist. 2. In A. Bezeichnung der Leistungsphase ►Entwurfsplanung gem. § 22 lit. b HRLA; ►Vorentwurf); *syn.* baureifer Entwurf [m].

2072 final selection [n] [UK] *prof.* (*Process* assessment of competition entries which is made by a jury during an objective evaluation. Qualified entries are considered in the awarding of

F

F

▶prizes or ▶merit awards; ▶honorable mention [US]/honourable mention [UK]); *s* **selección [f] final** (*Proceso* en el marco de un concurso, clasificación por parte del jurado sobre la base de una evaluación objetiva de los trabajos entregados que podrían ser premiados o adquiridos; ▶mención, ▶mención honorífica, ▶premio); *f* **sélection [f] définitive** (Processus de décision par la personne adjudicatrice, sur proposition du jury de concours, des participants sélectionnés susceptibles d'être lauréat ou dont les prestations risquent d'être primées ; ▶prix, ▶projet mentionné, ▶travaux primés) ; *g* **engere Wahl [f, o. Pl.] (1)** (*Vorgang* Festlegung durch ein Preisgericht während einer sachlichen Prüfung derjenigen eingereichten Wettbewerbsarbeiten, die für eine Preisverleihung oder einen ▶Ankauf in Betracht zu ziehen sind; cf. Teil 1, Nr. 5.6.5 GRW 1995, zuletzt novelliert am 22.12.2003; ▶engere Wahl 2, ▶Preis).

2073 final site inspection [n] *prof.* (Concluding site examination to establish satisfactory project completion, prior to the expiration [US]/expiry [UK]of the ▶period of limitation covering the ▶remedial construction claim [US]/remedial works claim [UK]; ▶construction site inspection); *s* **inspección [f] final de obra** (Obligación del jefe de obra de inspeccionar la construcción para constatar que no tenga defectos antes de que termine el ▶plazo de prescripción del ▶derecho a garantía; ▶inspección de obra *in situ*); *f* **inspection [f] du site** (Obligation du maître d'œuvre d'effectuer une visite de chantier avant le ▶délai de prescription de la garantie de parfait achèvement en vue de la constatation éventuelle d'imperfections ou de malfaçons ; ▶droit à l'obligation de garantie, ▶reconnaissance du chantier) ; *g* **Objektbegehung [f]** (Leistungspflicht des Planers vor Ablauf der ▶Verjährungsfristen für ▶Gewährleistungsansprüche an bauausführende Unternehmen das Bauobjekt zur Mängelfeststellung noch einmal zu inspizieren; cf. § 15 [2] 9 HOAI 2002 resp. Leistungsphase 9 in Anlage 11 und 12 HOAI 2009; ▶Baustellenbegehung).

final sum [n] [UK]**, ascertainment of the** *constr.* ▶determination of final contract amount [US].

2074 fine aggregate [n] [US] *constr.* (Fine grade of ▶crushed rock used in pathway and road construction); *syn.* stone chippings [n] [UK]; *s* **gravilla [f]** (En construcción viaria, material pétreo triturado, generalmente de rocas como basalto, granito, caliza, etc., de tamaño menor que la ▶piedra partida); *f* **gravillons [mpl] concassés** (Granulat en destiné à la construction et l'entretien des chaussées, provenance du concassage dans une carrière de roches telles que le basalte, granite, grauwacke, calcaire, porphyre, etc. de granularité 5/11 ou 11/22 mm après concassage primaire, ou de granularité 2/5, 5/8, 8/11, jusqu'à 16/22 mm après concassage successifs ; granularité plus fine que la ▶grave ; cf. DIR 1977, 64) ; *g* **Splitt [m]** (Für den Wegebau verwendetes, einfach gebrochenes Festgestein — meist aus Gesteinsarten wie Basalt, Granit, Grauwacke, Kalkstein, Porphyr, Melaphyr etc. — in den Lieferkörnungen nach einfachem Brechen von 5/11 oder 11/22 und nach mehrfachem Brechen in den Edelsplitt-Fraktionen 2/5, 5/8, 8/11 bis 16/22 mm; kleiner als ▶Schotter).

fine arts [npl] **of garden design** *gard. hist.* ▶fine garden design.

2075 fine garden design [n] *gard. hist.* (Aesthetic composition of open spaces with a comprehensive display of naturally-growing or artistically-trained plants, with pathways, ground modeling, architectural elements, lakes, pools, streams, fountains, as well as sculptural features representing religious or aesthetic themes which reflect the economic and social conditions of the age and region; ▶garden design, ▶garden culture); *syn.* fine arts [npl] of garden design, garden art [n], art [n] of gardens; *s* **arte**

[m/pl. f] de jardinería (Arte de composición estética de espacios libres con plantas de crecimiento natural y plantas con formas obligadas, modelación del terreno, superficies de agua, fuentes, caminos, elementos arquitectónicos, etc. que representa la visión estética o religiosa de la cultura regional y refleja así mismo las condiciones socioeconómicas de una época; ▶diseño de jardines, ▶patrimonio de jardinería y horticultura ornamental); *f* **art [m] des jardins** (Conception de l'organisation paysagère des espaces extérieurs selon un caractère esthétique fonctionnel caractérisée par l'utilisation de végétaux au port naturel taillés artistiquement, de chemins, modelés de sol, éléments de construction architecturés, étangs et jeux d'eau, statues exprimant les conceptions religieuses ou esthétiques caractéristiques des conditions économiques et sociales de l'époque ; ▶architecture des jardins, ▶patrimoine culturel des jardins) ; *g* **Gartenkunst [f]** (1. Gesamtheit der in den einzelnen geschichtlichen Epochen von der Antike bis heute künstlerisch gestalteten und gebauten vegetationsbestimmten Freiräume; 2. Gestalterisches Gesamtkonzept von Freiräumen mit natürlich gewachsenen oder kunstvoll geschnittenen oder besonders kultivierten Pflanzen, mit Wegen, Erdmodellierungen, baulichen Architekturelementen, mit Wasserflächen oder Wasserspielen sowie Plastiken entsprechend den jeweiligen örtlichen, religiösen und ästhetischen Vorstellungen sowie wirtschaftlichen und gesellschaftlichen Bedingungen. Neben dem wirtschaftlichen Garten-Ertragsdenken trat irgendwann der Lustgewinn hinzu, die Freude an den Pflanzen, der Sinn für Formen, Farben, Düfte und das ästhetische Erleben des gestalteten Raumes. Aus dem bescheidenen, aber wohnlichen und ästhetisch ansprechenden *Hortus conclusus* der mittelalterlichen Burgen, entwickelten sich in der Renaissance und im Barock riesige Parkanlagen, deren Gestaltungsprinzip später auch auf die Freiraumstrukturen der Städte übertragen wurde; ▶Gartenarchitektur, ▶Gartenkultur).

2076 fine grading [n] *constr.* (Final stage of ground model[l]ing to achieve an exact ▶final grade 1; ▶subbase grade [US]/formation level [UK], ▶rough grade); *syn.* minor grading [n], regulating [n] (of topsoil), finished grading [n] [also US], establishing [n] a finished grade; *s* **operación [f] de refino (de superficie)** (Aplanado final de la superficie hasta el nivel deseado; en la técnica de vegetalización, antes de la siembra o plantación; en la construcción de carreteras o caminos, antes de la colocación de las capas superiores; ▶nivel de rasante de acabado, ▶subrasante, ▶superficie refinada); *f* **exécution [f] du règlement final** (Étalage et mise en forme [processus] du terrain pour obtenir une surface bien dressée au niveau fini ; ▶fond de forme, ▶nivellement grosso modo, ▶règlement final) ; *syn.* règlement [m] définitif, dressement [m] final, modelage [m] du terrain aux cotes finies (VRD 1986, 326) ; *g* **Herstellen [n, o. Pl.] des Feinplanums** (1. Profilgerechte Einebnung der Bodenoberfläche als Vorgang; in der Vegetationstechnik vor Ansaat und Bepflanzung; 2. *Wegebau* profilgerechte Einebnung vor Einbau der Oberflächenschicht[en]; ▶Erdplanum, ▶Feinplanum, ▶Grobplanum); *syn.* Herstellen [n, o. Pl.] der Feinplanie [f], Herstellung [f] des Feinplanums.

2077 fine gravel [n] *constr.* (Mineral particles with diameter in U.S. from 2.0 to 12.5mm; ▶gravel 1, ▶soil textural class); *s* **gravilla [f]** (Fracción de las partículas del suelo cuyo grosor es de 2 a 6,3 mm de diámetro; ▶clase textural, ▶grava); *syn.* grava [f] menuda; *f* **gravier [m] fin** (▶Classe de texture de sol dont le diamètre des particules est compris entre 2 et 6,3 mm ; ▶gravier) ; *g* **Feinkies [m]** (D., ▶Bodenart mit einem Durchmesser von > 2 bis 6,3 mm; *Abk.* fG; DIN 4022 und DIN 4023; ▶Kies).

2078 fine particles [npl] *constr. pedol.* (Clay and silt fraction of soil; ▶soil structure); *s* **partículas [fpl] finas**

(Fracción de arcilla y limo en un suelo; ►textura del suelo); *f* **particules [fpl] fines** (Particules argileuses et limoneuses ; ►structure du sol) ; *g* **Feinanteile [mpl]** (Meist tonige und schluffige Anteile in einem Boden; ►Bodenstruktur).

2079 fine root [n] *arb. hort.* (cf. RRST 1983, 79; small root with diameter beteen 1 to 5mm; ►fibrous root); *s* **raíz [f] fina** (Pequeña raíz de un diámetro entre 1-5 mm; ►raíz fibrosa); *f* **racine [f] fine** (Caractérise une racine de diamètre compris entre 1 et 5 mm ; ►radicelle) ; *g* **Feinwurzel [f]** (Wurzel mit einem Durchmesser von 1-5 mm; ►Faserwurzel).

2080 fine sand [n] *constr. pedol.* (**1.** *constr.* ►*Soil textural class* mineral particles with diameter from 0.1 to 0.25mm [US]/ 0.02 to 0.2mm [UK]. **2.** *pedol.* 50% or more fine sand, or less than 25% very coarse, more fine sand, or less than 25% very coarse, fine sand; RCG 1982, 161; *s* **arena [f] fina** (Fracción de las partículas del suelo cuyo grosor va de 0,10 a 0,05 mm de diámetro; DGA 1986; ►clase textural); *f 1* **sable [m] fin** (F., en terme de granulométrie, matériau meuble formé de quartz [grains de sable] dont les dimensions sont comprises entre 0,05 à 0,1 mm ; cf. DIS 1986, 192 ; ►classe de texture des sols) ; *f 2* **sable [m] moyennement fin** (F., en terme de granulométrie, matériau meuble formé de quartz [grains de sable] dont les dimensions sont comprises entre 0,1 à 0,2 mm ; cf. DIS 1986, 192 ; ►classe de texture des sols) ; *g* **Feinsand [m]** (D., ►Bodenart mit einem Durchmesser von > 0,06 bis 0,2 mm; *Abk.* fS; DIN 4022 und DIN 4023).

2081 fine silt [n] *constr. pedol.* (*According to BS* mineral particles with diameter from 0.002 to 0.006mm [UK]; according to NCSS: diameter from 0.002 to 0.005mm [US]; ►soil textural class); *s* **limo [m] fino** (Fracción de las partículas del suelo cuyo grosor va de 0,002 de 0,005 mm de diámetro; ►clase textural); *f* **limon [m] fin** (Fraction granulométrique d'un sol dont le diamètre des particules est compris entre 0,002 et 0,005 mm ; ►classe de texture de sol) ; *g* **Feinschluff [m]** (D., ►Bodenart mit einem Durchmesser von > 0,002 bis 0,006 mm; *Abk.* fU; DIN 4022 und DIN 4023).

2082 fine spoil [n] *min.* (Waste material of small particles; ►mining spoil); *s* **escoria [f] fina** (Material sobrante de partículas de 2-8 mm de diámetro; ►zafras); *f* **stériles [mpl] fins** (Matériau produit par l'exploitation minière et inutilisable comme minéral, de granulométrie 2-8 mm; ►stérile) ; *g* **Feinberge [pl]** (Bergematerial in den Korngrößen 2-8 mm; ►Berge 3).

2083 fine-textured soil [n] *agr. hort.* (RCG 1982; soil consisting of or containing large quantities of fine fractions, primarily silt and clay; ►clay soil, ►easily workable soil, ►firm soil); *syn.* very firm soil [n] [also US], *obs. in scientific use* heavy soil [n]; *s* **suelo [m] pesado** (Suelo arcilloso difícil de trabajar; ►suelo arcilloso, ►suelo consistente, ►suelo ligero); *syn.* suelo [m] apelmazado [MEX]; *f* **terre [f] forte** (Sol compact, contenant beaucoup de particules fines, se labourant difficilement ; LA 1981, 1093 ; ►sol argileux, ►sol ferme, ►terre douce) ; *syn.* sol [m] lourd ; *g* **schwerer Boden [m]** (Bezogen auf die Bearbeitbarkeit im praktischen Landbau nicht genau definierter Begriff für ►Tonböden; hierzu gehören z. B. sandiger Lehm [sL], Lehm [L], lehmiger Ton [lT]), des Weiteren anmooriger Boden und Moorboden [Mo]; ►leicht bearbeitbarer Boden, ►mittelschwerer Boden).

finish [n] *constr.* ►surface finish.

2084 finished concrete [n] *constr.* (Concrete surface which will be treated or left untreated after removal of the formwork; e.g. ►concrete with wood board finish, ►exposed aggregate concrete, smooth-faced concrete, ►tooled concrete); *s* **hormigón [m] visto** (Hormigón cuya cara visible no se cubre después de quitar el encofrado, aunque se le puede dar un tratamiento de superficie; se diferencian p. ej. hormigón alisado, ►hormigón bruto de encofrado, ►hormigón con árido visto, ►hormigón trabajado con herramientas); *f* **béton [m] de parement** (Béton qui reste visible après décoffrage, et présente un parement devant répondre à certains critères de qualité et architecturaux ; on distingue p. ex. le béton banché, le ►béton brut de décoffrage, le ►béton de gravillons lavé, ►béton lavé, ►béton à parement éclaté, béton matricé) ; *syn.* béton [m] architectonique ; *g* **Sichtbeton [m]** (Beton, dessen sichtbare Oberfläche nach dem Ausschalen nicht oder nachbehandelt wird; man unterscheidet z. B. glatten, strukturlosen Sichtbeton, ►schalungsrauhen Sichtbeton, ►Sichtbeton mit Schlagbearbeitung, ►Waschbeton).

2085 finished elevation [n] [US] *constr. plan.* (**1.** Height planned for a building or structure or a part of it, in relation to a reference point. **2.** Final elevation of the ground surface according to proposed contours and spot elevations, in relation to a reference point ; ►establishment of finished elevation, ►spot elevation, ►top level); *syn.* finish elevation [n] [US], finished grade [n] (of a land surface) [US], finished level [n] [UK], proposed level [n] [UK]; *s* **altura [f] final** (**1.** Altura final de un edificio o partes del mismo o en exteriores fijada en el proyecto. **2.** Nivel final de la superficie del suelo según plan; ►borde superior, ►relleno a subrasante, ►cota); *syn.* cota [f] final, cota [f] proyectada; *f* **cote [f] finie** (**1.** Valeur attachée à l'altitude d'un point d'un bâtiment, du terrain, définie par un plan. **2.** Hauteur projetée d'ouvrages ou de parties d'ouvrages ; ►arase, ►cote en altimétrie, ►mise à niveau définitif) ; *syn.* niveau [m] final, niveau [m] de sol fini, cote [f] projetée, cote [f] de plan ; *g* **fertige Höhe [f]** (**1.** Geplante Höhe von Bauwerken oder Teilen davon. **2.** Geplante Höhe beim Erdbau und bei der Feinplanie; ►Einbau auf fertige Höhe, ►Kote, ►Oberkante); *syn.* Nennhöhe [f], Planungshöhe [f].

finished floor elevation [n] (FFE) [US] *constr.* ►finished floor level.

2086 finished floor level [n] *constr.* (*Abbr.* FFL; top surface of a floor, used as a reference level); *syn.* finish floor level [n] (FFL) [also US], finish(ed) floor elevation [n] (FFE) [also US]; *s* **nivel [m] final del suelo** (Superficie superior de planta utilizada como nivel de referencia); *f* **niveau [m] plancher fini** (Cote de niveau de référence indiquée sur un plan) ; *g* **Oberkante [f] Fußboden** (*Abk.* OKF; in Plänen angegebene Bezugshöhe [fertige Höhe] des Erdgeschosses); *syn.* Fußbodenoberkante [f] (FOK), Erdgeschoss Fußbodenhöhe [f] (EFH).

finished grade [n] *constr.* ►final grade (1), ►meeting finished grade [US].

finished grade [n], **establishing a** *constr.* ►fine grading.

finished grade flush with [loc] [US] *constr.* ►make a finished grade flush with.

finished grade [n] **(of a land surface)** [US] *constr. plan.* ►finished elevation [US]/finished level [UK].

finished grading [n] [US] *constr.* ►fine grading.

finished level [n] [UK] *constr. plan.* ►finished elevation [US]/finished level [UK].

finished levels [npl] *constr.* ►final grade (1).

finish floor level [n] (FFL) [US] *constr.* ►finished floor level.

finish grade with [loc] [UK], **marry** *constr.* ►make a finished grade flush with [US].

2087 fire blight [n] *agr. hort. for. phytopath.* (Dangerous bacterial disease on woody plants of the Rose family, principally of the *Maloideae* or *Pomoïdeae* with the main hosts such as *Amelanchier*, *Chaenomeles*, *Cotoneaster*, *Crataegus*, *Cydonia*, *Malus*, *Pyracantha*, *Pyrus*, *Sorbus* and *Stranvesia* species, caused

F

by *Erwinia amylovora* whereby leaves and branches wilt and die back. The bacterium spread in the bark and goes over from stems into the leaves. Therefore first of all discolor the stems of leaves and then the ripps of the leaves. In market gardening a precautionary desease treatment is made with streptomycin agents. The affected parts must be removed and burned); *s* **tizón [m] de fuego** (Plaga peligrosa causada por la bacteria *Erwinia amylovora* que se presenta en leñosas de la familia de las rosáceas, sobre todo en las *Maloideae*, debido a la cual las hojas y ramas se secan y mueren. Las partes afectadas de las plantas deben ser podadas y quemadas); *f* **feu [m] bactérien** (Maladie redoutable provoquée par la bactérie *Erwinia amylovora* produisant le dessèchement des feuilles de nombreuses rosacées ; conformément à la législation en vigueur, les plantes attaquées doivent être éliminées et détruites par le feu) ; *g* **Feuerbrand [m, o. Pl.]** (Durch das Bakterium *Erwinia amylovora* verursachte gefährliche Krankheit an Gehölzen der *Rosaceae* — vornehmlich bei Kernobstgewächsen [*Maloïdeae* oder *Pomoïdeae* mit den Hauptwirten *Amelanchier-, Chaenomeles-, Cotoneaster-, Crataegus-, Cydonia-, Malus-, Pyracantha-, Pyrus-, Sorbus-* und *Stranvesia-*Arten] —, bei der Blätter und Zweige welken und verdorren. Das Bakterium breitet sich in der Rinde aus und gelangt über die Stiele in die Blätter. Deshalb verfärben sich zuerst die Blattstiele und die Mittelrippe der Blätter und erst zuletzt Blattränder und Blattspitzen. Die befallenen Pflanzenteile müssen entsprechend den Gesetzesvorschriften unbedingt abgeschnitten und verbrannt werden; zur vorbeugenden Behandlung im Erwerbsgartenbau reicht i. d. R. eine Spritzung während der Blüte mit einem streptomycinhaltigen Präparat. Jedoch in feuchten, starken Befallsjahren müssen bis zu vier Spritzungen erfolgen. Diese Bakteriose wurde erstmals 1780 in den USA im Hudson Tal im Staate New York beobachtet und in den 1950er-Jahren nach Deutschland eingeschleppt).

2088 fire break [n] *for.* (Forest corridor cleared for fire protection); *syn.* fire control line [n] [also US]; *s* **cortafuego(s) [m(pl)]** (Pista sin árboles que sirve para frenar la expansión del fuego en el bosque); *syn.* pasillo [m] cortafuego(s); *f* **pare-feu [m]** (Type d'équipement traditionnel consistant à cloisonner une forêt par des bandes déboisées entretenues [jusqu'à 300 m de largeur] ; pour une protection plus grande contre les grands feux on peut envisager l'aménagement de coupures agricoles ; cf. VRD 1994, partie 4, 9.3.1., 4) ; *syn.* tranchée [f] pare-feu, ligne [f] pare-feu ; *g* **Schneise [f] (1)** (Für den vorbeugenden Feuerschutz angelegter breiter, baum- und strauchfreier Streifen im Wald); *syn.* Brandschneise [f].

fire control [n] *leg.* ▶right-of-passage for fire control.

fire control line [n] [US] *for.* ▶fire break.

2089 fire hydrant [n] *urb. syn.* fire-plug [n] [also UK]; *s* **boca [f] de incendios**; *f* **bouche [f] d'incendie** *syn.* prise [f] d'incendie ; *g* **Hydrant [m] zur Löschwasserentnahme**.

2090 fire lane [n] [US] *constr. urb.* (Paved pathway between buildings, reserved for the use of fire engines in the case of fire); *syn.* fire path [UK]; *s* **vía [f] de bomberos** (Carril asfaltado entre edificios reservado para el acceso de las máquinas de bomberos en caso de incendio); *syn.* entrada [f] de bomberos; *f* **voie [f] pompiers** (Chaussée aménagée exclusivement pour les véhicules de secours des sapeurs pompiers dans un lotissement) ; *syn.* voie [f] de secours incendie, piste [f] de pompier ; *g* **Feuerwehrweg [m]** (Für die Brandbekämpfung in bebauten Gebieten vorgesehener, befestigter Weg); *syn.* Feuergasse [f] [BW].

fire path [n] [UK] *constr. urb.* ▶fire lane [US].

firepath pot [n] [UK] *constr.* ▶grass paver.

2091 fire plant community [n] *phyt.* (Plant association which is morphologically adapted to survive major [wild] fire

events); *s* **comunidad [f] pirófita** (Sabana o matorrales de Australia ricos en protáceas, mirtáceas, etc. que mantienen los frutos cerrados largo tiempo y los abren después del incendio, cuando el suelo está ya frío caen las semillas; las leñosas poseen una corteza muy gruesa que protege el cambium; SILV 1979, 93); *f* **association [f] pyrophyte** (Constituée d'espèces présentant des adaptations morphologiques particulières leurs permettant de résister aux incendies, p. ex. les séries du chêne vert et du chêne liège dans les forêts méditerranéennes) ; *g* **Feuergesellschaft [f]** (Pflanzengesellschaft, die auf Grund ihrer morphologischen Anpassungen Feuerereignisse größeren Ausmaßes überleben kann).

fire-plug [n] [UK] *urb.* ▶fire hydrant.

fire pond [n] [UK] *constr.* ▶fire suppression pond [US].

2092 firescaping [n] *landsc.* (Process of landscape treatment and maintenance composed of fire-resistant plants, structures, and utilities to withstand fire events which threaten buildings in areas prone to natural fires/wildfires. The term 'natural fire' refers to fire that begins via natural ignition [usually lightening]; 'wildfire' means fire that is out of control of human suppression efforts, regardless of ignition source); *syn.* firewise landscaping [n], safescaping [n] (±); *s* **diseño [m] paisajístico anti-incendios (≠)** (En EE.UU., tipo de planificación y diseño de parques y jardines para zonas castigadas frecuentemente por incendios naturales o de origen antropógeno que consiste en utilizar plantas y construir muros y estructuras resistentes al fuego); *f* **étude [f] d'aménagement d'espaces extérieurs dans les zones sensibles aux incendies (≠)** (Plan d'aménagement et d'entretien des espaces verts dans les zones sensibles aux incendies préconisant une organisation de l'espace soucieuse de limiter les dommages éventuels sur les habitations provoqués par le feu grâce à l'utilisation de végétaux et de matériaux résistants au feu ainsi qu'à une implantation adaptée des bâtiments et autres aménagements tels que murs, cheminements et modelé du terrain) ; *g* **Planung und Gestaltung von Feuer abweisenden Außenanlagen [fpl] (≠)** (Planungs- und Pflegekonzept, das in semiariden Gebieten Außenanlagen mit Gebäuden und sonstigen baulichen Anlagen durch Mauern, Wege und Geländemodellierungen sowie mit feuerresistenter Bepflanzung so gestaltet und pflegt, dass sich Feuerereignisse im angrenzenden Umfeld nicht oder nur eingeschränkt zu den Häusern hin ausbreiten können).

2093 fire suppression pond [n] [US] *constr.* (Pool in which water is kept for extinguishing fires); *syn.* fire pond [n] [UK]; *s* **estanque [m] de reserva** (Laguna artificial mantenida para suministrar agua para apagar incendios); *f* **point [m] d'eau de défense contre l'incendie** (Réserve d'eau aménagée et desservie par une piste de défense des forêts contre l'incendie [DFCI] balisée ; VRD 1994, tome 3, 9.3.1., p. 13) ; *syn.* étang [m] de lutte contre l'incendie ; *g* **Feuerlöschteich [m]** (Teich, in dem Wasser zum Löschen eines Brandes vorgehalten wird).

firewise landscaping [n] *landsc.* ▶firescaping.

firm [n]**, specialist** *constr. hort.* ▶speciality contractor.

2094 firm soil [n] *agr. hort.* (Generic term indicating the consistency of a moist silty or clayey soil that offers noticeable resistance to crushing, but can be crushed with moderate pressure between thumb and forefinger; in dry soils the term is **hard soil**; cf. RCG 1982; ▶easily workable soil, ▶fine textured soil); *s* **suelo [m] consistente (≠)** (Término genérico no claramente definido utilizado para suelos limosos relativamente difíciles de cultivar; ▶suelo ligero, ▶suelo pesado); *f* **sol [m] ferme** (Sol résistant à l'émiettement mais pouvant être détruit par une forte pression ; DIS 1986 ; ▶terre douce, ▶terre forte) ; *g* **mittelschwerer Boden [m]** (Bezogen auf die Bearbeitbarkeit im praktischen Landbau nicht genau definierter Begriff für Schluff-

und Lehmböden; ►leicht bearbeitbarer Boden, ►schwerer Boden).

2095 firm [n] **whose bid was accepted** [US] *contr.* *syn.* firm [n] whose tender was accepted [UK]; *s* **empresa** [f] **ganadora del remate;** *f* **entreprise** [f] **retenue** *syn.* entreprise [f] attributaire ; *g* **Firma, die den Zuschlag** [m] **bekommen hat.**

firm [n] **whose tender was accepted** [UK] *contr.* ►firm whose bid was accepted [US].

2096 first floor [n] [US] *arch. constr.* (Floor of a building at ground level, usually slightly above the surrounding terrain; **in U.S.**, the ground floor is referred to as the first floor; **in U.K.** the first floor is the second storey); *syn.* ground floor [n] [UK]; *s* **planta** [f] **baja** (Piso de un edificio al nivel del terreno o ligeramente elevado respecto a él; en algunos países latinomericanos se denomina primera planta); *syn.* piso [m] bajo, bajos [mpl]; *f 1* **rez-de-chaussée** [m] (Partie d'un édifice dont le plancher est sensiblement au niveau de la rue, du sol ; PR 1987) ; *f 2* **rez-de-jardin** [m] (Partie d'un édifice dont le plancher est au niveau d'un jardin) ; *g* **Erdgeschoss** [n] (Das untere Stockwerk eines Gebäudes, das meist etwas höher liegt als die Umgebungsflächen. Im US-Englischen ist dies der *first floor*, wobei im UK-Englischen mit *first floor* das erste Obergeschoss bezeichnet wird); *syn.* Parterre [n] (1).

2097 first mowing [n] [US] *constr.* (First cut on newly sown or newly sodded lawn [US]/turfed lawn [UK]); *syn.* topping [n] [UK] (1); *s* **primer corte** [m] (Primera vez que se corta un césped recién sembrado o uno de tepes); *f* **première tonte** [f] (Effectuée après la germination lorsque les jeunes herbes ont atteint 5 à 8 cm) ; *syn.* première coupe [f] ; *g* **erster Schnitt** [m] (Mähgang nach dem Auflaufen eines Rasens, wenn die Gräser etwa 8-10 cm Höhe erreicht haben); *syn.* Schröpfschnitt [m].

first refusal [n] [US] *adm. leg.* ►right of first refusal.

fish [n] *envir. zool.* ►mass death of fish, ►restocking of fish, ►stocking of fish.

2098 fishery [n] *agr.* (Occupation of catching fish and other aquatic animals in lakes, rivers, or the sea; *specific terms* coastal fishery, deep sea fishing, freshwater fishing; ►sport fishing); *s* **pesca** [f] (*Términos específicos* pesca continental, pesca de altura/alta mar, pesca de bajura, ►pesca recreativa, ►pesca deportiva); *f* **pêche** [f] **(commerciale)** (Activité commerciale de capture des poissons ; *termes spécifiques* pêche en eau douce, pêche en haute mer ; ►pêche [de loisirs]) ; *g* **Fischerei** [f, o. Pl.] (Gewerbsmäßig betriebener Fischfang; *UBe* Süßwasserfischerei, Hochseefischerei, Küstenfischerei; ►Angelsport).

fish habitat area [n] *conserv. leg. zool.* ►protected fish habitat area.

2099 fish hole [n] *conserv. zool.* (Place in a watercourse, lake or pond where fish hide or lay eggs); *s* **refugio** [m] **de freza** (±) (Lugar al borde de un curso fluvial o lago en el que los peces se esconden para el desove); *f 1* **mouille** [f] (Abri naturel ou poche dans le lit d'un cours d'eau constamment en eau et constituant une cache piscicole) ; *f 2* **sous-berge** [f] (Abri artificiel le long de la berge d'un cours d'eau) ; *g* **Fischunterstand** [m] (Bereich am Rande eines Gewässers, das Fischen z. B. Schutz zum Laichen gewährt).

2100 fishing industry [n] *agr.* (Organization, administration and marketing in the business of catching and rearing fish and other aquatic animals for sale); *s* **industria** [f] **pesquera** (Conjunto de actividades industriales y comerciales relacionadas con la pesca industrial); *syn.* industria [f] de la pesca; *f* **industrie** [f] **poissonnière** (Organisation, gestion et promotion de la pêche commerciale) ; *syn.* industrie [f] de la pêche ; *g* **Fischereiwirt-**schaft [f, o. Pl.] (Organisation, Verwaltung und Marketing des gewerbsmäßigen Fischfangs); *syn.* Fischereiwesen [n, o. Pl.].

2101 fishing pier [n] *recr.* *s* **pontón** [m] **de pesca;** *f* **ponton** [m] **de pêche** ; *g* **Angelsteg** [m].

2102 fish ladder [n] *wat'man. zool.* (Installation which allows migrating fish to bypass a dam in a watercourse; ►fish pass); *s* **escala** [f] **de peces** (Construcción que permite a los peces migrantes sobrepasar una presa en un curso de agua; *términos específicos* escala [f] salmonera, paso [m] salmonero; ►paso [natural] de peces); *syn.* paso [m] de peces; *f* **échelle** [f] **à poissons** (Dispositif de franchissement d'obstacles artificiels s'opposant à la migration des poissons ; *termes spécifiques pour ces passes à pente très raide* échelles à bassins successifs, échelles à ralentisseurs ; AEP 1976, 264-269 ; la passe à poissons du pont barrage de Vichy construite en 1996 mesure 120 m de long et est composée de 22 bassins de 8 m de long ; elle équipée d'un système de vidéo-surveillance pour la détermination et le comptage des espèces de poissons. C'est la seule passe à poissons en Europe où le public peut venir voir admirer le passage des poissons migrateurs ; ►itinéraire de migration des poissons) ; *syn.* passe [f] à poissons ; *g* **Fischleiter** [f] (Anlage, die den Wanderfischen künstliche Höhenunterschiede in Fließgewässern überbrücken hilft; ►Fischwechsel); *syn.* Fischaufstiegsanlage [f], Fischpass [m], Fischtreppe [f], Fischweg [m] (1).

fish mortality [n]**, massive** *envir. zool.* ►mass death of fish.

2103 fish pass [n] *zool.* (Natural migration path for fish which can be obstructed by weirs or dams; ►fish ladder); *s* **paso** [m] **(natural) de peces** (Vía natural de migración de peces que puede estar obstaculizada por una obra hidráulica; ►escala de peces); *f* **itinéraire** [m] **de migration des poissons** (Cours d'eau fréquenté par les poissons migrateurs dont le déplacement peut être stoppé par divers ouvrages hydrauliques ; ►échelle à poissons) ; *g* **Fischwechsel** [m] (Natürlicher Fischwanderweg; kann durch Gewässerbauwerke behindert werden; ►Fischleiter); *syn.* Fischweg [m] (2).

2104 fish poaching [n] *leg. hunt.* (Breach of a fishing law by unauthorized fishing, misappropriation, damage or destruction; ►game poaching); *s* **pesca** [f] **furtiva** (Lesión de la ley de pesca por pescar en zona o época del año vedada; ►caza furtiva); *syn.* pesca [f] ilegal; *f* **braconnage** [m] **de poissons** (Infractions au droit de la pêche ; pêcher sans permis ou à une période, en un lieu et avec des engins prohibés, de détruire la faune piscicole ; ►braconnage de gibier) ; *syn.* pêche [f] illicite ; *g* **Fischwilderei** [f] (Verletzung des Fischereirechts durch Fischen oder durch Aneignung, Beschädigung oder Zerstörung einer Sache, die dem Fischereirecht unterliegt; § 293 StGB; ►Jagdwilderei).

2105 fish pond [n] *agr.* (Artificial pond for breeding and production of ►fry and fish for the table; ►drain a fish pond, ►pond farming); *syn.* farm pond [n] [also UK]; *s* **estanque** [m] **de vivero (de peces)** (Estanque artificial utilizado para la cría de ►alevines y de peces de consumo; ►piscicultura de estanques, ►vaciar un estanque); *syn.* estanque [m] de piscifactoría; *f* **étang** [m] **piscicole** (Plan d'eau artificiel destiné à l'élevage et à la production d'►alevins et de poissons adultes ; AEP 1976, 232 ; l'activité économique liée à ces plans d'eau se dénomme la pisciculture d'étang ; ►pisciculture en étangs, ►vider un étang) ; *syn.* étang [m] à gestion piscicole, étang [m] de production piscicole, lac [m] de retenue ; *g* **Fischteich** [m] (Künstlicher Teich zur Zucht und Produktion von Setzlingen und Speisefischen; ►Fischteich ablassen, ►Teichwirtschaft, ►Setzling 1).

2106 fish population [n] *zool.* (Presence of fish in a body of water); ►management of fish population, ►restocking of fish); *syn.* fish stock [n]; *s* **población** [f] **de peces** (Existencia de fauna piscícola en un cuerpo de agua; ►cuidados ictiogénicos, ►suelta

F

de peces, ▶repoblación piscícola); *syn.* población [f] piscícola, población [f] ictícola; *f* **population [f] piscicole** (▶alevinage, ▶aménagement piscicole et halieutique, ▶réempoissonnement, ▶repeuplement de poissons) ; *syn.* population [f] ichtyologique, ressources [fpl] en poissons, peuplement [m] piscicole (AEP l979, 15), stock [m] piscicole, cheptel [m] piscicole ; *g* **Fisch-besatz [m]** (Vorhandensein von Fischen in Gewässern; ▶Fisch-einsatz, ▶Fischhege); *syn.* Fischbestand [m].

2107 fish scale paving [n] [US] *constr.* (TSS 1988, 440-4; pavement pattern in the form of a fan, typical in historic European cities, usually created with cobblestones, granie setts and other natural stone units. The pattern can also be achieved with concrete shaped pavers, or in **stamped concrete**, whereby a special stencil—available on rolls—isoverlaid and pressed into freshly poured concrete; ▶pavement pattern, ▶paving pattern); *syn.* fan pattern paving [n] [UK], European fan paving [n] [UK]; *s* **adoquinado [m] en abanico** (Tipo de ▶aparejo de adoquinado, ▶patrón de aparejo); *f* **appareillage [m] en queue de paon** (*Termes spécifiques* appareillage en vraie/fausse queue de paon ; ▶calepinage, ▶calepinage du pavage) ; *syn.* pose [f] en queue de paon ; *g* **Fächerpflasterung [f]** (Pflasterung mit Klein- oder Mosaiknaturpflastersteinen im schuppenförmigem ▶Verlege-muster, das oft in historischen Städten Europas anzutreffen ist; ▶Pflasterung 1. Das Fächermuster wird, wenn keine Natursteine verwendet werden, mit vorgefertigter Kunststoffschablone [Stem-pel] in einen weichen speziellen Ortbeton nach DIN 1045 [**Prägebeton**] eingedrückt, „eingeprägt" resp. „gestempelt"); *syn.* Pflasterung [f] im Schuppenverband.

2108 fish spawning area [n] *zool.* (Area of a water body in which fish lay their eggs); *s* **área [f] de freza** (Zona de un cuerpo de agua en la que los peces desovan); *syn.* zona [f] de freza, área [f] de desove de peces; *f* **cours d'eau [m] de fraye** (Cours d'eau ou portion de cours d'eau abritant ou susceptible d'abriter des ▶frayères pour les poissons ; AEP 1976, 88-89) ; *syn.* site [m] de fraye, site [m] de frayère, zone [f] de frai(e)/fraye, zone [f] de frayère (ARI 1986, 48) ; *g* **Fischlaichgewässer [n]** (Teil eines Gewässers, in dem Fische laichen).

fish stock [n] *zool.* ▶fish population.

2109 fish winter kill [n] *envir. zool.* (Death of fish caused by lack of oxygen in frozen lake or pond); *s* **mortandad [f] invernal de peces** (Muerte de peces por falta de oxígeno en lago helado); *syn.* asfixia [f] invernal; *f* **asphyxie [f] hivernale** (Mort des poissons par manque d'oxygène dans les lacs gelés) ; *g* **Fischsterben [n]** (1) (Sterben von Fischen durch Sauer-stoffzehrung im zugefrorenen See).

fitness course [n] [US] *recr.* ▶fitness trail.

2110 fitness trail [n] *recr.* (Combining jogging and the per-formance of specific exercises at predetermined locations along an outdoor trail; e.g. 18-station parcourse); *syn.* fitness course [n] [also US], exercise trail [n] [also US], trim trail [n], parcours(e) [n]; *s* **circuito [m] de gimnasia** (En parque o bosque, recorrido con instalaciones de ejercicios físicos, en general en zona muy frecuentada por la población urbana); *f* **parcours [m] d'initia-tion sportive (en forêt)** ; *syn.* parcours [m] d'éducation physique (en forêt) ; *g* **Trimm-dich-Pfad [m]** (Ein mit Turngeräten ausgestatteter, für jedermann frei zugänglicher Weg im Wald für sportliche Aktivitäten der unter Bewegungsmangel leidenden Bevölkerung); *syn.* Waldsportpfad [m].

2111 fitted piece [n] [US] *constr.* (Individually-formed piece used to close a gap after the installation of regularly-dimensioned building elements); *syn.* piece made-to-measure [loc] [UK]; *s* **pieza [f] de empalme** (Pieza de construcción de dimensiones especiales fabricada para conectar elementos de construcción de dimensiones normadas, una vez colocados éstos en obra); *syn.*

pieza [f] de ajuste; *f 1* **découpe [f] d'adaptation** (Élément pré-paré sur mesure afin de refermer un espace libre entre les maté-riaux, p. ex. dans un pavage, un dallage) ; *f 2* **pièce [f] ajustée** (Élément réalisé sur mesure et mis en place en dernier après la pose des pièces standards) ; *syn.* élément [m] sur mesure, pièce [f] sur mesure, pièce [f] d'adaptation, raccord [m] d'adaptation ; *g* **Passstück [n]** (Zur Schließung einer Lücke angefertigtes Bauteil, das nach dem Einbau der regelmäßigen Bauelemente als Reststück eingepasst wird).

2112 fitting [n] *constr.* (Usually a standardized pipe part used for joining two or more sections of pipe together; ▶fitted piece [US]/piece made to measure [UK]; ▶pipe fitting); *s* **accesorio [m]** (Dispositivo manufacturado que se utiliza para unir piezas de construcción como p. ej. tuberías, es decir «eles», «tes», etc.; ▶pieza de empalme); *f* **raccord [m] (de tuyauterie)** (Accessoire préfabriqué standard utilisé pour les tuyaux de canalisation ; ▶pièce ajustée) ; *g* **Formstück [n]** (Vorgefertigtes Bauelement, z. B. für Rohrleitungen; ▶Passstück).

fitting [n]**, junction** *constr.* ▶pipe coupling [US]/pipe junction piece [UK, AUS].

fitting [n]**, pipe joint** *constr.* ▶pipe coupling [US]/pipe junction piece [UK, AUS].

fixation [n] *biol. pedol.* ▶nitrogen fixation; *pedol.* ▶nutrient fixation.

fixed dune [n] *geo. phyt.* ▶gray dune [US]/grey dune [UK].

flag [n] *constr.* ▶natural stone flag.

flag [n]**, raw stone** *constr.* ▶quarry flagstone.

flag [n]**, roughly-hewn** *constr.* ▶quarry flagstone.

2113 flag [n] **cut to shape** *constr.* (Flagstone with finished sides or surfaces); *syn.* dressed flagstone [n]; *s* **losa [f] cortada a medida** (Losa de piedra natural con bordes o cara vista traba-jada); *f* **dalle [f] clivée avec parements travaillés** (Élément de pierre naturelle brute de taille dont les arêtes des faces de pare-ment sont taillées) ; *g* **Natursteinplatte [f] mit bearbeiteten Kanten**.

flagstone [n] *constr.* ▶natural stone flag, ▶quarry flagstone, ▶sawn flagstone.

flagstone [n]**, dressed** *constr.* ▶flag cut to shape.

2114 flagstone step [n] *arch. constr.* (Step slab, usually 5-8cm in thickness, which is laid with its front edge projecting from the face of the riser; ▶block Stepp, ▶nosing); *syn.* slab [n] on riser [also UK], flagstone tread [n] and riser [n] [also US]; *s* **escalón [m] de losa** (Elemento de escalera de unos 5 a 8 cm de espesor que se coloca en horizontal con el borde sobresaliendo con vuelo ligero sobre la contrahuella; ▶escalón prefabricado, ▶mampirlán); *f* **dalle [f] giron** (Élément d'escalier constitué d'une dalle à retour de parement, de 5 à 8 cm d'épaisseur formant giron et posée sur la contre marche avec constitution d'un nez de marche saillant [▶débordement] ; VRD 1994, tome 4, 6.4.2.2.1, p. 15 ; ▶bloc-marche) ; *syn.* dalle [f] de marche, marche [f] plaquée (NEU 1983, 141) ; *g* **Legstufe [f]** (Treppenstufenelement von meist 5-8 cm Dicke, die mit Überstand [▶Unterschneidung] auf eine Unterlage in Form von Unterlegsteinen gelegt wird. Durch den Schattenwurf des Überstandes wirken **L.en** sehr leicht; ▶Blockstufe); *syn.* Plattenstufe [f].

flagstone tread [n] **and riser** [n] [US] *arch. constr.* ▶flagstone step.

flamed [pp/adj] *constr.* *▶tooling of stone [US]/dressing of stone [UK], #11.

flared curb [n] [US] *constr. eng.* ▶mountable curb [US]/splayed kerb [UK].

2115 flash erosion [n] *geo.* (Severe wearing away of soil by sudden rainfall in gullies or ravines, which have been denuded of vegetation by human activities; ►gully erosion); **s erosión [f] repentina brusca (≠)** (En zonas áridas y semiáridas arrastre masivo de material edáfico en cárcavas y surcos profundos desnudos de vegetación por efecto de la interferencia humana; ►erosión en cárcavas); **ƒ érosion [f] brutale** (PED 1984, 118 ; ablation importante de sols en pente défrichés dans les zones arides ou semi-arides provoquée par des pluies violentes ; ►ravinement 1) ; **g massive Bodenabspülung [f] (≠)** (Durch plötzliche, heftige Regenereignisse verursachte starke Erosion in ariden und semiariden Gebieten in vegetationslosen Gräben und tiefen Erosionsschluchten, die durch menschlichen Einfluss vegetationslos wurden; ►Grabenerosion).

2116 flash flood [n] *geo.* (**1.** Sudden but short-lived torrent in a usually dry valley, notably in a[n] [semi]arid area after a rare, brief, but intensive rainstorm; DNE 1978; a **f. f.** has occurred, when an enormous quantity of water has flooded a river bed or wadi within a time period of six hours; **f. f.s** can loosen and move huge boulders, uproot trees, destroy buildings and bridges, create new water channels; the volume of water can cause the water level to rise up to 10m above normal and can cause mud avalanches, when topsoil is washed away. **2.** Release of a great quantity of water after the breach of a dyke, which occurs without any warning and leads to the sudden inundation of a large area; ►heavy rainfall); **s crecida [f] repentina** (**1.** Crecida de corta duración con punta de caudal muy elevada en un valle generalmente seco, muy frecuentemente en áreas [semi]áridas. Se trata de una **c. r.** cuando una gran cantidad de agua inunda un lecho de río o wadi en un plazo de seis horas. Una **c. r.** puede poner en movimiento grandes bloques erráticos, destruir puentes y edificios, desenraizar árboles, crear nuevos canales de agua, y el volumen de agua puede causar el incremento del nivel del agua hasta 10 m sobre el nivel normal y causar avalanchas de barro si se ve afectado el suelo. **2.** Salida de gran cantidad de agua al romperse un dique, con la consecuencia de inundación repentina de un área grande; ►lluvia torrencial); **ƒ crue [f] éclair** (Crue de courte durée et de montée brusque avec un débit de pointe relativement élevé en particulier dans les régions de climat [semi]aride ; ►pluie torrentielle) ; **g plötzliche Überschwemmung [f] (≠)** (**1.** *Kurzzeitiges „Hochwasser" in Trockentälern* plötzliches Hochwasser bei Starkregenfällen in [semi]ariden Gebieten; eine **p. Ü.** liegt vor, wenn innerhalb von sechs Stunden nach einem Starkregenereignis riesige Wassermassen ein Flussbett oder Wadi überfluten; **p. Ü.en** können Felsblöcke bewegen, Bäume entwurzeln, Gebäude und Brückenbauwerke zerstören und neue Wasserrinnen auswaschen; die Wassermassen können bis zu 10 m über dem normalen Wasserstand ansteigen, und durch das Wegspülen von Boden Schlammlawinen erzeugen. **2.** Plötzliches, für die Menschen unvorhersehbares Hochwasser nach einem Deichbruch, wenn riesige Wassermassen über ein Gebiet hereinbrechen; ►Starkregen); *syn.* Sturzflut [f].

2117 flat bank [n] *geo.* (Low slopping land along rivers and streams; ►slip bank [US]/slip-off slope [UK]; *opp.* ►bluff 2); **s orilla [f] llana** (Orilla de poca pendiente en aguas continentales; ►orilla convexa; *opp.* ►orilla escarpada); **ƒ berge [f] à pente douce** (bordant les cours d'eau et les eaux calmes ; ►berge convexe; *opp.* ►berge escarpée) ; *syn.* rive [f] à pente douce ; **g Flachufer [n] (1)** (Flach ansteigendes Gelände am Rande von Fließ- und Stillgewässern; ►Gleithang; *opp.* ►Steilufer).

2118 flat coast [n] *geo.* (Low sloping shore condition associated with shallow tidal or mud flats; ►coastal marshland, ►mangrove stand; *opp.* ►coastal cliff); **s costa [f] llana** (Orilla de poca pendiente en el litoral; ►manglar, ►marisma marítima; *opp.* ►costa acantilada); *syn.* costa [f] baja, litoral [m] llano;

ƒ côte [f] basse (Étendue de largeur variable en bordure de mer constituée de plages de sable souvent bordées d'un cordon dunaires ; parmi les côtes basses, on compte les côtes à étangs, à lagunes, à limans, à ►mangrove, les ►marais maritimes ; DG 1984, 273 ; *opp.* ►côte à falaises, ►côte escarpée) ; **g Flachküste [f]** (Flacher Landstreifen von unterschiedlicher Breite [mehrere Meter bis viele Kilometer], der den allmählicher Übergang des Festlandes ins Meer mit Vorstrand, Strand und Dünen darstellt oder ein durch Brandung planierter Streifen küstennahen Meeresbodens [Abrasionsfläche]; breite **F.n** sind z. B. die durch Gezeiten geprägten Wattenküsten mit ihren ►Marschen und die sandigen Küstenzonen der ►Mangroven; sofern dieser Bereich mit Sand oder Geröll [Materialanschwemmung] bedeckt ist, spricht man auch von ►Strand; meerwärts dehnt sich an vielen Küstenabschnitten die **Schorre** aus; *opp.* ►Steilküste).

2119 flat crown [n] [US] *arb. syn.* shallow crown [n] [UK]; **s copa [f] aparasolada**; **ƒ houppe [f] au port étalé** ; **g flache Baumkrone [f]**.

2120 flat roof [n] *arch. constr.* (Nominally horizontal roof on residential, commercial, or industrial buildings, having zero pitch or slope. A **f. r.** under a roof garden is used to retain water for planting; excess rainwater is removed by means of overflow drains, gutters or external scuppers discharges; most membrane roofs require 1%-2% pitch to drain properly; ►pitched roof, ►planted flat roof, ►roof garden); *syn.* dead-level roof [n]; **s tejado [m] plano** (►azotea ajardinada, ►tejado inclinado, ►tejado plano plantado); *syn.* azotea [f]; **ƒ toiture [f] à pente nulle** (►jardin sur dalle, ►jardin sur toiture-terrasse, ►toiture inclinée, ►toiture-terrasse rampante, ►toiture-terrasse végétalisée/plantée) ; *syn.* toiture [f] plate ; **g Flachdach [n] (1)** (Dach ohne Neigung, meist als begehbares Terrassendach oder als ►Dachgarten ausgestaltet; ►begrüntes Dach, ►geneigtes Dach; *opp.* Steildach); *syn.* Nullgrad-Dach [n], Nullgrad-Gefälle-Dach [n], gefälleloses Dach [n].

2121 flat shore [n] *geo.* (Level land along coasts); **s orilla [f] del mar llana**; **ƒ rivage [m] (2)** (Terrain plat — formé de sables, graviers —, situé au bord de la mer) ; *syn.* grève [f] ; **g Flachufer [n] an Küsten** (Flache, unterschiedlich breite Zone, die sich bis zum obersten Bereich, in dem die Meeresbrandung noch wirkt, erstreckt; sofern dieser Bereich mit Sand oder Geröll [Materialanschwemmung] bedeckt ist, spricht man auch von ►Strand; *opp.* Steilküste; ►Flachküste).

2122 flatten [vb] **a slope** *constr.* (Filling to reduce the gradient of an existing embankment; ►steepen an existing slope); *syn.* lessen [vb] a slope; **s reducir [vb] la pendiente de un talud** (►igualar terreno con talud); **ƒ adoucir [vb] un talus** (Réduction de l'angle d'un talus existant au moyen d'un remblai ; ►taluter en déblai) ; **g Böschung abflachen [vb]** (Die Steilheit einer vorhandenen Böschung reduzieren; ►abböschen); *syn.* Böschung flacher ausformen/gestalten [vb].

2123 flat terrain [n] *landsc.* (Flat piece of land as opposed to a slope or inclined plane); *syn.* level ground; **s terreno [m] plano**; **ƒ terrain [m] plat** ; **g ebene Fläche [f]** (Fläche ohne Gefälle; im Vergleich zum Hang/zu einer hängigen Fläche).

flat undressed stone [n] [UK] *constr.* ►ragstone [US].

flexible surface [n] [UK] *constr.* ►water-bound surface.

flight [n] **from the land** *agr. sociol.* ►rural exodus.

2124 flight [n] **of block steps** *arch. constr.* (Outdoor stairway constructed of precast concrete or stone masonry units); **s escalera [f] de bloques** (Escalera exterior construida de elementos prefabricados de hormigón o de bloques de piedra natural); **ƒ escalier [m] à blocs-marches** ; **g Treppe [f] aus Blockstufen** (Treppe aus vorgefertigten Naturstein- oder Betonblöcken); *syn.* Blockstufentreppe [f].

flight [n] **of stairs** *arch. constr.* ►steps.

flight [n] **of stairs/steps** *arch. constr.* ►stairway.

flight [n] **of steps** *arch. constr.* ►broad flight of steps, ►foundation of a flight of steps, ►slope of a flight of steps, ►side edging of a flight of steps, *►stairway, #1, ►step in a flight of steps, ►steps, ►total rise of a flight of steps, ►width of a flight of steps.

flight of steps [n], length of a *arch. constr.* ►total run of a stairway [US].

flight [n] of steps, steepness of a *arch. constr.* ►slope of a flight of steps.

2125 flight [n] of step slabs on risers *arch. constr.* (►flagstone step); *s* **escalera [f] de losas** (►escalón de losa); *f* **escalier [m] en dalles** (►dalle giron) ; *g* **Treppe [f] aus Legstufen** (►Legstufe); *syn.* Legstufentreppe [f].

2126 flight [n] of steps on a slope *arch. constr.* *s* **escalera [f] en talud**; *f* **escalier [m] en accompagnement de talus** *syn.* escalier [m] en façade de talus ; *g* **Treppe [f] in einer Böschung.**

2127 flight [n] of steps with lateral edging *constr.* (Stairway in which the ground on each side of the steps is contained by an edging stone; ►cheek wall); *s* **escalera [f] incrustada y encintada lateralmente** (Escalera en terreno fijada lateralmente con piedras de encintado; ►zanca); *f* **escalier [m] sur talus rentrant** (Escalier posé sur un talus, dont la ligne de pente passe par les nez de marche, la terre étant maintenue sur les flancs par une bordure ; ►limon) ; *syn.* escalier [m] sur talus incrusté ; *g* **Treppe [f] mit Bekantung** (Treppe, die seitlich zum Erdreich hin mit einem Kantenstein begrenzt ist; ►Treppenwange); *syn.* Treppe [f] mit seitlicher Begrenzung.

flight [n] of steps without cheek walls [UK] *arch. constr.* ►open stairway [US]/flight of steps without cheek walls [UK].

floating aquatics [npl] [US], mat of *phyt.* ►floating meadow.

floating-leaf community [n] *phyt.* ►rooted floating-leaf community.

2128 floating-leaved plant [n] *bot. phyt.* (Rooted, herbaceous hydrophyte with some leaves on the surface of the water, e.g. Fragrant Water lily *[Nymphaea odorata]*; cf. CWD 1979, 42); *syn.* rooted water plant [n], rooted hydrophyte [n]; *s* **hidatófito [m] radicante** (Planta acuática fija, en la que algunas hojas y flores flotan en la superficie del agua, como p. ej. el nenúfar *[Nymphaea odorata]*); *syn.* hidrófito [m] radicante; *f* **plante [f] flottante et fixée** (Plante hydrophile fixée qui étale ses feuilles sur l'eau [à feuilles natantes] ou à inflorescences affleurantes ; *terme spécifique* plante [f] flottante, fixée, à feuilles partiellement submergées) ; *syn.* hydrophyte [m] flottant et fixé, hydrophyte [m] immergé, fixé et affleurant, plante [f] immergée de surface, hydrophyte [m] enraciné et flottant ; *g* **Schwimmpflanze [f] (1)** (Wurzelnder Hydrophyt mit meist auf der Wasseroberfläche liegenden Blättern, z. B. Weiße Seerose *[Nymphaea alba]*); *syn.* Schwimmblattpflanze [f].

2129 floating meadow [n] *phyt.* (Very dense blankets of free-floating and rooted water plants, e.g. in tropical and subtropical bodies of water [herbaceous floating]); *syn.* mat [n] of floating aquatics [also US] (TEE 1980, 214); *s* **pradera [f] flotante** (Comunidad herbácea muy densa de los fondos de aguas marinas en zonas tropicales o subtropicales; *término específico para el Mediterráneo* pradera de posidonia); *f* **herbier [m]** (Végétation phanérogamique, dense, p. ex. fixée sur le fond meuble ou rocheux en milieu marin [l'herbier à posidonie en Méditerranée], ou flottante dans certaines eaux tropicales) ; *g* **Schwingrasen [m] (1)** (Sehr dichte, schwimmende und im Boden wurzelnde Pflanzenmatte aus krautigen Pflanzen, z. B. in tropisch-subtropischen Gewässern); *syn.* schwimmende Wiese [f].

2130 floating sphagnum mat [n] *phyt.* (TEE 1980, 214; *sphagnum* moss layer covering the surface of a ►bog lake, often densely rooted by higher plants with extended rhizomes; ►floating meadow); *syn.* quaking bog [n], schwingmoor [n] [EIRE], emerged peat moss hollow community [n] (ELL 1988, 289); *s* **tremedal [m]** (Zona pantanosa con abundancia de plantas hidrófilas, que llegan a encespedar con una cierta consistencia; DINA 1987; ►pocina profunda de turbera, ►pradera flotante); *syn.* tremadal [m]; *f* **tourbière [f] flottante** (Formation de lentilles superficielles de Sphaignes et de rhizomes flottants qui s'enfoncent en s'épaississant dans une mare de tourbière ; ►herbier, ►œil de la tourbière) ; *syn.* tourbière [f] tremblante, marais [m] tremblant, gazon [m] flottant [aussi CH] ; *g* **Schwingrasen [m] (2)** (Auf der Wasseroberfläche eines Moorsees oder ►Hochmoorkolkes sich bildende Sphagnumdecke, die meist von höheren Pflanzen mit lang ausstreichenden Rhizomen durchwurzelt ist; ►Schwingrasen 1).

2131 flood [n] *geo. wat'man.* (Inundation of any land not usually covered with water, through a temporary rise in the normal level of a river, lake or the sea; cf. DNE 1978; ►flash flood, ►high-water [level] of watercourses, ►hundred-year flood, ►probable maximum flood [PMF]); *syn.* inundation [n], flooding [n]; *s* **inundación [f]** (Anegamiento de tierras que normalmente no están cubiertas de agua por desbordamiento temporal de un curso de agua, lago o del mar; ►avenida, ►crecida repentina, ►crecida máxima del siglo, ►crecida máxima probable); *syn.* anegamiento [m]; *f* **inondation [f]** (DG 1989, 249 ; envahissement limité du lit d'inondation suite au débordement des eaux d'un cours d'eau lors d'une ►crue inondante, de terrains inondables par les eaux d'un lac ou par la mer ; ►crue éclair, ►crue centennale, ►crue maximale probable) ; *syn.* débordement [m], submersion [f] ; *g* **Überschwemmung [f]** (Zeitlich begrenzte Überflutung von Flächen durch Ansteigen des Wasserspiegels eines Fließgewässers, Sees oder des Meeres; ►Hochwasser der Fließgewässer, ►Jahrhunderthochwasser, ►Maximalhochwasser, ►plötzliche Überschwemmung); *syn.* Überflutung [f].

flood [n], sheetflood *geo.* ►sheetwash.

flood abatement [n] *wat'man.* ►flood control.

2132 floodable land [n] *wat'man.* (Additionally provided areas over which flood water may spread more gradually to reduce flow peaks and prevent or limit downstream damage; ►floodplain [US]/flood-plain [UK]); *syn.* flood control pool [n]; *s* **terreno [m] inundable** (Superficie adicional reservada para regular crecidas máximas; ►llanura de inundación); *f* **polder [m] d'inondation** (Technique de ralentissement dynamique des crues par la mise en eau temporaire de zones de surstockage en vue de l'écrêtement des pointes de crue ; ►champ d'inondation ; le terme de *polder* est historiquement associé aux ouvrages similaires du Rhin) ; *syn.* lac [m] réservoir, réservoir [m] de crues, exutoire [m] de crues (ARI 1986, 44), aire [f] de surstockage, unité [f] de surstockage, bassin [m] de rétention, casier [m] de rétention ; *g* **Überschwemmungsbereich [m]** (Zusätzlich vorgehaltene Fläche in den Gewässerniederungen, auf der sich das Hochwasser zur Entschärfung und Regulierung von Hochwasserspitzen ausdehnen kann; ►Überschwemmungsgebiet); *syn.* Hochwasserretentionsraum [m], Hochwasserrückhaltefläche [f], Hochwasserrückhalteraum [m], Retentionsraum [m] für Hochwasser, Überflutungsfläche [f], Überschwemmungspolder [m].

2133 flood area [n] of a stream/brook *geo.* (►alluvial plain); *syn.* streamside flood area [n] [also US], brookside flood area [n] [also US]; *s* **vega [f] de arroyo** (►vega); *f* **plaine [f] alluviale d'une rivière** (►Plaine alluviale d'un cours d'eau de faible importance) ; *g* **Bachaue [f]** (Teil des Talbodens eines kleinen Fließgewässers, das bei Hochwasser überflutet und mit angeschwemmten Sedimenten überlagert wird; ►Aue).

flood bank [n] [UK] *eng. wat'man.* ►levee [US] (2).

flood berm [n] *eng. wat'man.* ►levee [US] 2/flood bank [UK].

flood bypass channel [n] *eng. wat'man.* ►flood relief channel.

flood channel [n] *wat'man.* ►floodway.

2134 flood control [n] *wat'man.* (Measures to reduce or prevent danger of flooding and damages caused by temporary inundations: by ►levees [US] 2/flood banks [UK], by enlarging the streambed, or by providing of additional ►floodable land); *syn.* flood abatement [n], flood prevention [n], flood protection [n]; *s* **protección** [f] **contra crecidas o mareas vivas** (Medidas de construcción orientadas a disminuir los efectos de las crecidas, p. ej. construcción o aumento de la altura de ►diques de contención, designación de ►terrenos inundables, etc.); *syn.* control [m] de crecidas o mareas vivas, protección [f] contra avenidas; *f* **défense** [f] **contre les crues** (Ensemble des dispositions prises pour assurer la protection des terres et des biens contre les inondations, ou pour minimiser les dommages qu'ils peuvent causer grâce à la construction ou l'élévation de ►digues, l'élargissement du lit mineur ainsi que l'affectation de zones fluviale en ►polders d'inondation ; cf. WMO 1974) ; *syn.* protection [f] contre les crues (GCE 1980, 170), défense [f] contre les inondations ; *g* **Hochwasserschutz** [m, o. Pl.] (Maßnahmen zur Minderung oder Verhinderung der Hochwassergefahr, z. B. durch Bau oder Erhöhung von Deichen, durch Speicheranlagen und Verbreiterung von Fließgewässerbetten sowie durch Ausweisung zusätzlicher Mulden und ►Überschwemmungsbereiche; ►Deich 2); *syn.* Überflutungsschutz [m].

flood control pool [n] *wat'man.* ►floodable land.

2135 flood control reservoir [n] *wat'man.* (Natural or artificially-created basin used to retain a certain quantity of flood water, thus helping to reduce the peak flow in a river—usually with uncontrolled outlets; ►rainwater interceptor basin, ►retarding basin, ►storm water detention basin); *syn.* detention reservoir [n], flood detention basin [n]; *s* **embalse** [m] **de retención** (Hondonada artificial o natural que sirve para almacenar las aguas de crecida; ►embalse de regulación de aguas pluviales, ►embalse de retardo, ►embalse de sobrecarga); *syn.* embalse [m] de control de crecidas; *f* **bassin** [m] **de rétention des crues** (Réservoir naturel [ancien champ fluvial] ou artificiel [barrage réservoir] qui écrète les pointes de crue d'un cours d'eau, permettant l'épandage du débit excédentaire pendant un délai limité sans qu'il en résulte d'importants dommages pour les activités agricoles qui pourraient être pratiquées dans ces zones ; ►bassin de retenue des eaux pluviales, ►bassin de rétention, ►déversoir d'orage) ; *syn.* réservoir [m] de contrôle des crues, réservoir [m] d'écrêtement des crues, polder [m] fluvial ; *g* **Hochwasserrückhaltebecken** [n] (Natürliches oder künstlich geschaffenes Becken, das einen Teil des Hochwasserabflusses aufnimmt und dadurch die Spitzenwerte der Hochwasserwellen eines Fließgewässers abmindern hilft; ►Regenrückhaltebecken, ►Regenüberlaufbecken, ►Rückhaltebecken); *syn.* Hochwasserretentionsbecken [n].

flood crest [n] *wat'man.* ►flood peak.

2136 flood damage [n] *wat'man. land'man.* (Destruction caused by flooding); *s* **daños** [mpl] **por inundación**; *f* **dégâts** [mpl] **causés par la crue** ; *g* **Hochwasserschaden** [m] (Durch ein Hochwasserereignis verursachter Schaden).

flood danger [n] *hydr. wat'man.* ►flood hazard.

flood detention basin [n] *wat'man.* ►flood control reservoir.

2137 flood discharge [n] *wat'man.* (Maximum rate of discharge for a given period; ►stream flow, ►peak discharge); *syn.*

maximum discharge [n] (WMO 1974); *s* **desagüe** [m] **máximo** (Valor máximo del caudal de crecida; ►flujo de un cauce, ►caudal de punta); *f* **débit** [m] **caractéristique maximal (de crue)** (*Abrév.* DCM ; plus fort débit observé pendant au moins 10 jours dans l'année ; cf. CILF 1978 ; ►débit de pointe, ►écoulement fluviatile) ; *syn.* débit [m] des crues ; *g* **Hochwasserabfluss** [m] (1. Oberer Grenzwert der Abflüsse bei Hochwasserereignissen; 2. Das bei Hochwasser abfließende Wasser eines Fließgewässers in Abhängigkeit eines hohen Niederschlagaufkommens pro Zeiteinheit, der Schneeschmelze oder eines Rückstaus [z. B. eines Nebenflusses durch den Hauptfluss] und der Größe und Gliederung des Einzugsgebietes; ►Abfluss 2, ►Abflussspitze 1); *syn.* Hochwasserabführung [f].

flooded borrow pit [n] *landsc. recr.* ►flooded gravel pit.

2138 flooded gravel pit [n] *landsc. recr.* (Pond created by excavation; ►borrow pit, ►extraction site, ►wet gravel pit); *syn.* flooded borrow pit [n]; *s* **laguna** [f] **artificial** (Cuerpo de agua artificial creado por extracción de grava o arena, a menudo recultivado para ser utilizado como lugar de recreo; ►gravera subacuática, ►zona de extracción, ►zon de préstamo); *f* **étang** [f] **de fouille** (Eaux closes artificielles résultant de l'extraction de matériaux nécessaires à la construction, telles marnières, sablières, gravières, ballastières remises en état dans le cadre de travaux paysagers ; *terme spécifique en partie synonyme* lac de gravière ; ►gravière en nappe, ►lieu d'emprunt, ►site d'extraction de matériaux) ; *syn.* carrière [f] en eau, carrière [f] dans la nappe ; *g* **Baggersee** [m] (Künstliches Stillgewässer, das durch Sand- oder Kiesgewinnung entstand und oft nach Beendigung der Entnahme durch [Kiesgruben]rekultivierung landschaftspflegerisch gestaltet und als Fisch- oder Badesee genutzt wird; ►Entnahmestelle 2, ►Kiesentnahmestelle 2, ►Seitenentnahme; *UBe z. T. syn.* Kiesgrubengewässer [n], Kiesweiher [m]); *syn.* Abgrabungssee [m].

flooded soil [n] *pedol.* ►seasonally-flooded soil.

2139 flood fringe area [n] [US] *wat'man.* (IBDD 1981; that portion of the flood hazard area outside of the floodway, which is rarely flooded); *syn.* high-water margin [n] [UK]; *s* **zona** [f] **borde de crecida** (≠) (Zona aluvial inundada muy excepcionalmente); *f* **marge** [f] **du champ d'inondation** (Zone de la plaine alluviale exceptionnellement inondée) ; *syn.* bord [m] de la zone affectée par les hautes eaux ; *g* **Hochwasserrandzone** [f] (Selten überfluteter Auenbereich).

2140 flood hazard [n] *hydr. wat'man.* (TGG 1984, 100; possibility that an area or surface may be subject to flooding); *syn.* flood danger [n]; *s* **peligro** [m] **de inundación** *syn.* riesgo [m] de inundación; *f* **risque** [m] **d'inondation** *syn.* danger [m] d'inondation) ; *g* **Überschwemmungsgefahr** [f] (Möglichkeit, dass ein Gebiet oder eine Fläche überschwemmt werden kann).

flood hazard area [n] [US] *hydr. wat'man.* ►floodplain [US]/flood-plain [UK].

flooding [n] *geo. wat'man.* ►flood.

2141 flood irrigation [n] *agr.* (Surface water supply by flooding; e.g. cranberry bog [in U.S.], rice field, etc.); *s* **riego** [m] **por inundación** *syn.* riego [m] a manta; *f* **irrigation** [f] **par submersion** *syn.* irrigation [f] par inondation ; *g* **Bewässerung** [f] **durch oberirdisches Fluten**.

flood level/mark [n], **average** *wat'man.* ►average high-water level/mark.

floodmark litter [n] *phyt.* ►driftline.

floodmark vegetation [n], **coastal** *phyt.* ►driftline community.

flood [n] **of tourists** *recr. plan.* ►tourist flow.

2142 flood peak [n] *wat'man.* (Highest level reached by a flood); *syn.* flood crest [n]; *s* **punta [f] de crecida** (Nivel máximo alcanzado durante una crecida); *syn.* punta [f] de avenida; *f* **pointe [f] de crue** (Niveau du débit maximal de la crue); *g* **Hochwasserspitze [f]** (Höchster Hochwasserstand bei einem Hochwasserereignis).

2143 floodplain [n] [US]/**flood-plain** [n] [UK] *hydr. wat'man.* (Floor of a valley over which a watercourse may spread in time of flood; ▶alluvial plain); *syn.* flood hazard area [n]; *s* **llanura [f] de inundación** (Parte del valle que se inunda cuando hay crecida; ▶vega); *syn.* zona [f] de inundación; *f* **champ [m] d'inondation** (Zone en bordure d'un cours d'eau, de plus ou moins grande étendue, submergée en période de crue inondante ; ▶plaine alluviale) ; *syn.* lit [m] d'inondation (DG 1989, 116, 350), vallée [f] alluviale inondable, zone [f] d'épandage des crues, zone [f] inondable ; *g* **Überschwemmungsgebiet [n]** (Talboden, der von einem Fließgewässer bei Hochwasser überflutet wird; ▶Aue); *syn.* Hochwasserbereich [m], Überflutungsgebiet [n].

floodplain [n], **valley** *geo.* ▶alluvial plain.

floodplain zone [n] *leg. wat'man.* ▶preservation of floodplain zone [US]/preservation of flood-plain zone [UK]; *leg. plan.* ▶statutory floodplain zone [US]/statutory flood-plain zone [UK].

flood prevention [n] *wat'man.* ▶flood control.

2144 flood-prone area [n] *wat'man.* (Area susceptible to flooding; ▶danger of high water); *s* **zona [f] con peligro de inundación** (Terreno que puede ser inundado en caso de crecida; ▶peligro de crecida); *f* **zone [f] exposée aux risques d'inondations** (Zone sujette à des inondations inévitables et imprévisibles ; ▶risque de hautes eaux) ; *syn.* territoire [m] exposé aux risques d'inondations ; *g* **hochwassergefährdetes Gebiet [n]** (Geografisch definierter Bereich, der unabweisbaren und nicht vorhersehbaren Überschwemmungen ausgesetzt ist; ▶Hochwassergefahr).

flood protection [n] *wat'man.* ▶flood control.

2145 flood relief channel [n] *eng. wat'man.* (Artificial watercourse parallel to river or stream to contain peak flows); *syn.* flood bypass channel [n], floodway [n] (WMO 1974); *s* **canal [m] de derivación** (Canal destinado a transportar el exceso de agua procedente de las crecidas); *syn.* canal [m] de desagüe, canal [m] de by-pass (WMO 1974); *f* **canal [m] d'évacuation des crues** (Canal construit pour évacuer l'eau excédentaire des crues ; dépassant le volume d'eau qui peut être charrié sans risque par le cours d'eau ; WMO 1974) ; *syn.* canal [m] de dérivation, canal [m] de décharge ; *g* **Hochwasserseitengerinne [n]** (Seitenkanal eines Fließgewässers zur Abführung von Hochwasserspitzen).

2146 flood savanna(h) [n] *geo. phyt.* (Periodically inundated land in the tropics in various mosaic patterns, with either palms or groups of other trees on uplands; cf. ELL 1967; ▶parkland); *s 1* **savana [f] de inundación** (Tipo de sabana que se inunda en la estación lluviosa y en cuya composición son importantes las ciperáceas. Dentro del tipo general de carácter subxerofítico existen diferentes subtipos; cf. DB 1985; ▶sabana arbolada); *s 2* **llanos [mpl]** (*Término específico* nombre propio y expresión geográfica de las grandes llanuras que se extienden en Venezuela entre la región andina y la orilla izquierda del Orinoco, prolongándose también en Colombia, entre los Andes y los bosques amazónicos. Su vegetación corresponde al grupo de las sabanas: en grandes extensiones carece de árboles, con excepción de las zonas húmedas; en otras partes, la salpican árboles aislados. Extensas zonas se inundan por los desbordes del Orinoco; cf. DB 1985); *f 1* **savane [f] inondable** (▶Savane-parc inondée périodiquement composée par une mosaïque de dépres-sions herbacées et de groupes d'arbres situés sur des promontoires) ; *f 2* **savane [f] galerie** (Savane installée sur les rives caillouteuses ou sableuses des cours d'eau en milieu forestier humide, p. ex. en Amazonie) ; *g* **Überschwemmungssavanne [f]** (Periodisch überflutete Parklandschaft, bestehend aus einem Mosaik von erhöhten Baumgruppen und tiefer liegendem Grasland; ▶Parklandschaft 1; eine Form der Ü. ist die **Termitensavanne**).

2147 flood storage capacity [n] *wat'man.* (TGG 1984, 155; capability of retaining large amounts of water for short periods of time to control highwater levels; e.g. by ▶dams, ▶floodable land, ▶flood control reservoir, ▶storm water detention basin); *s* **capacidad [f] de almacenamiento de aguas de crecidas** (Poder de retención de aguas de avenida que tienen algunas zonas p. ej. ▶terrenos inundables, ▶embalses de regulación de aguas pluviales o ▶embalses de retención, o a través de ▶presas); *f* **capacité [f] de rétention des crues** (Pouvoir que possèdent certaines zones naturelles [▶polders d'inondation] ou ouvrages [barrage-réservoir, ▶barrage de vallée, ▶bassin de retenue d'eaux pluviales, ▶bassin de rétention des crues], à amortir les ondes de crues et à retenir temporairement d'importantes masses d'eau lors des crues) ; *g* **Hochwasserrückhaltevermögen [n]** (Die Fähigkeit, bei Hochwasserereignissen Wassermassen kurzzeitig aufnehmen zu können, z. B. durch ▶Hochwasserrückhaltebecken, ▶Regenrückhaltebecken, ▶Talsperren, ▶Überschwemmungsbereiche).

2148 flood water mark [n] *wat'man.* (Measured level of a flood); *syn.* high-water mark [n]; *s* **nivel [m] de aguas de crecida o marea**; *f* **niveau [m] maximum des crues** ; *g* **Hochwasserstand [m]** (Gemessene Höhe eines Hochwasserereignisses).

2149 floodway [n] *wat'man.* (Channel of natural stream or river and portions of the floodplain adjoining the channel, which are reasonably expected to carry and discharge the floodwater or flood flow; cf. IBDD 1981; ▶flood relief channel (WMO 1974), ▶flood fringe area [US]/high-water margin [UK]); *syn.* flood channel [n]; *s* **lecho [m] mayor** (Espacio por el que circulan las aguas de crecida sin incluir la ▶zona borde de crecida; ▶canal de derivación); *syn.* cauce [m] de desagüe de crecidas, canal [m] de desagüe de crecidas; *f* **lit [m] majeur (1)** (Chenal d'écoulement des eaux envahi par les hautes eaux ; ▶canal d'évacuation des crues, ▶marge du champ d'inondation) ; *g* **Hochwasserbett [n]** (Abflussquerschnitt eines Fließgewässers ohne ▶Hochwasserrandzone; TWW 1982, 468 ; ▶Hochwasserseitengerinne).

floor [n] *arch. constr.* ▶first floor [US]/ground floor [UK]; *for. pedol.* ▶forest floor; *phyt.* ▶moss floor; *leg. urb.* ▶story [US]; *geo.* ▶valley floor.

floor [n] [UK], **ground** *arch. constr.* ▶first floor [US].

floor [n], **manhole** *constr.* ▶manhole base.

2150 floor area [n] *leg. plan. urb.* (Area of each floor in a building; *specific terms* 'gross floor area', which contains the entire area within the perimeter of the outside walls of a floor, and ▶net floor area, which is the actual occupied area of a floor. The sum of all single floor areas is loosely called 'floor area'); *s* **superficie [f] de planta** (Área total de una planta de un edificio; se diferencia entre la **s. de p. bruta** y la **s. de p. neta**; a la suma de todas las **s. de p.** se le llama **s. de p. total**; ▶superficie útil de un edificio); *f 1* **surface [f] de plancher hors œuvre (d'un niveau de construction)** (Dans le cadre de l'instruction d'une demande de permis de construire ou d'une déclaration de travaux on distingue les deux types de surfaces de plancher hors oeuvre ; ▶surface utile) ; *f 2* **surface [f] de plancher hors œuvre brute (SHOB)** (Somme des surfaces de plancher de chaque niveau de construction calculées à partir du nu extérieur des murs de façades et au niveau supérieur du plancher, y compris les combles et sous-sols, aménageables ou non, les balcons, les loggias et toitures-terrasses, non compris les éléments ne constituant pas de surface de plancher, comme les terrasses non couvertes de plain-pied avec le rez-de-chaussée, les saillies à caractère décoratif, les vides [trémies d'ascenseur ou d'escalier, rampes d'accès]) et *f 3* **surface [f] de plancher hors œuvre nette (SHON)** (Celle-ci est égale à la surface hors œuvre brute [SHOB] d'une construction, après déduction 1. des surfaces des combles et des sous-sols non aménageables pour l'habitation ou pour des activités à caractère professionnel, artisanal, industriel ou commercial [notamment hauteur sous plafond ou sous toiture inférieure à 1,80 m], 2. des surfaces des toitures-terrasses, 3. des balcons et des parties non closes situées au rez-de-chaussée, des surfaces des bâtiments ou parties des bâtiments aménagés en vue du stationnement des véhicules [garage], 4. des surfaces des bâtiments affectés au logement des récoltes, des animaux ou du matériel agricole, 5. d'une surface égale à 5 % de la SHON affectée à l'habitation [déduction forfaitaire relative à l'isolation des locaux], d'une déduction spécifique aux opérations de réfection des immeubles d'habitation dans la limite de 5 m^2 par logement pour des travaux tendant à l'amélioration de l'hygiène) ; *g* **Geschossfläche [f]** (Fläche, die durch ein Geschoss eines Gebäudes definiert wird ; es wird zwischen **Bruttogeschossfläche** und **Nettogeschossfläche** unterschieden, wobei als Brutto**g.** die von den Außenflächen eines Gebäudes begrenzten Flächen der Geschosse gelten ; die Summe aller **G.n** bildet die **Gesamtgeschossfläche**; ►Nutzfläche).

floor area [n] [US]**, allowable** *leg. urb.* ►floor area ratio (FAR) [US]/floor space index [UK].

2151 floor area ratio [n] **(FAR)** [US] *leg. urb.* (Ratio of the total floor area of a building to the area of a building plot. FAR provisions may be used in combination with other bulk regulations, such as height limits, open space, and building space requirements to determine the stipulated ►density of development; ►building coverage [US]/plot ratio [UK]; ►Building Permit Office; *syn.* allowable floor area [n] [US] (UBC 1979, 50), floor space index [n] [UK], permissible floor area [n] [ZA]; *s* **coeficiente [m] de edificación (≠)** (El c. de e. fija el máximo de m^2 del total de las plantas de un edificio permitido en relación a la superficie del terreno a construir. En la D. en el cálculo sólo se incluye el solar mismo, en UK se considera también la mitad de la superficie de calle correspondiente. En España el **c. de e.** no existe, sino que la densidad máxima de construcción está regulada por el número de viviendas por hectárea y la altura máxima de edificación; cf. arts. 74, 75 LS Texto refundido y arts. 30 c, 47, 100 RD 2159/1978; ►densidad de edificación, ►coeficiente de ocupación del suelo, ►comisión de urbanismo); *f 1* **coefficient [m] d'occupation des sol (COS)** (Est le rapport exprimant le nombre de m^2 de plancher hors œuvre susceptibles d'être construit par m^2 de sol s'appliquant à la surface d'une parcelle ou d'un îlot [emplacements réservés déduits] et déterminant la densité de construction pour une zone ou partie de zone d'un plan local d'urbanisme [PLU] ou d'un plan d'occupation des sols [POS] ; le dépassement du COS peut être autorisé sous réserve dans la limite de 20 % afin de favoriser la diversité de l'habitat [logements locatifs sociaux] ; dans les autres cas le dépassement du COS est lié au versement d'une participation des constructeurs ; la présentation de la demande est adressée pour autorisation au service chargé de l'instruction des permis de construire ou le service de l'État chargé de l'urbanisme qui fixe les modalités d'établissement et de recouvrement de cette contribution ; C. urb., art. L. 332-1 à 5 ; plafond légal de densité ; ►coefficient d'emprise au sol, ►service de l'application du droit des sols) ; *f 2* **plafond [m] légal de densité (PLD)** (Limite generale de densité de construction définie par le rapport entre la

surface de plancher hors œuvre nette de cette construction et la surface de terrain sur laquelle elle est ou doit être implantée ; au-delà de cette limite, le droit de construire sur un terrain n'appartient plus au propriétaire du sol mais relève de la collectivité ; le PLD est instauré facultativement par les conseils municipaux de communautés urbaines et autres groupements de communes, lequel ne peut légalement être inférieur à 1 et, pour la Ville de Paris, à 1,5. Lorsqu'un dépassement du PLD est nécessaire et autorisé, le constructeur doit opérer le versement à la collectivité d'une somme égale à la valeur de l'acquisition du terrain qui lui manque, la constructibilité du terrain est cependant toujours limitée au ►coefficient d'occupation des sols [COS] ; cf. C. urb, art. R. 112-1, R. 112-2 ; cf. BON 1990, 350 ; ►densité de construction) ; *g* **Geschossflächenzahl [f]** (*Abk.* GFZ; die **G.** gibt an, wieviel Quadratmeter Bruttogeschossfläche der einzelnen Geschosse je Quadratmeter Grundstücksfläche zulässig sind und bestimmt mit die gewünschte Bebauungsdichte. Die Geschossfläche ist nach den Außenmaßen der Gebäude in allen Vollgeschossen zu ermitteln; cf. § 20 BauNVO. **In D.** bezieht sich die GFZ nur auf das zu bebauende Grundstück; **im U.K.** wird noch die angrenzende halbe Straßenfläche mitgerechnet. **In F.** gibt es außerdem gesetzlich generell festgesetzte Mindestgrenzen der Bebauungsdichte, ausgedrückt in PLD *[Plafond légal de densité]*; sie beträgt in Paris 1,5 und außerhalb von Paris 1,0. Eine Überschreitung dieser Höchstgrenze ist auf Antrag des Bauherrn bei der Genehmigung durch die ►Baurechtsbehörde möglich. In diesem Fall muss eine Abgabe an die Gemeinde in Höhe des Betrages abgeführt werden, der nötig wäre, um ein Grundstück zu kaufen, damit die gesetzlich festgelegte PLD nicht überschritten wird; cf. Gesetz v. 31.12.1975, Art. R. 112-1, R. 112-2; ►Bebauungsdichte, ►Grundflächenzahl); *syn.* Ausnützungsziffer [f] [CH].

floor elevation [n] [US]**, finished (FFE)** *constr.* ►finished floor level.

floor level [n] [US]**, finish (FFL)** *constr.* ►finished floor level.

2152 floor plan [n] *arch.* (Plan showing doors, walls, and rooms on each floor of a building, with designated use[s] of each delineated space); ►plan view); *s* **planta [f] de un piso** (Representación gráfica de la superficie de una ►planta de un edificio); *f* **dessin [m] en plan d'un étage/niveau** (►dessin en plan) ; *g* **Geschossgrundriss [m]** (Maßstabsgerechte, horizontale Darstellung eines Gebäudestockwerkes; ►Grundriss).

floor space index [n] [UK] *leg. urb.* ►floor area ratio (FAR) [US].

2153 flora [n] **(1)** *bot.* (1. Systematic classification of all plant species growing in a particular area or region or in a specific environment—as distinguished from the ►vegetation, which includes all of the plant communities of this area. 2. Classified list or book of plants of a certain region, which helps in identification of species); *s* **flora [f]** (1. Conjunto de las especies y variedades de plantas de un territorio dado; DINA 1987; ►vegetación. 2. Manual de determinación de la flora; DINA 1987); *f* **flore [f]** (1. Liste des espèces végétales qui vivent en une région ou en un point donné. 2. Ouvrage donnant le moyen d'identifier des espèces végétales qui vivent dans une région ; ►végétation 3) ; *g* **Flora [f]** (1. Systematisch erfasste Gesamtheit der Pflanzenarten eines bestimmten Gebietes; im Vergleich zur ►Vegetation, die die Gesamtheit seiner Pflanzengesellschaften beinhaltet. 2. Pflanzenbestimmungsbuch für ein bestimmtes Gebiet).

flora [n] **(2)** *phyt. zool.* ►aquatic flora; *phyt. zool.* ►depletion of fauna/flora; *bot.* ►fungal flora; *phyt. zool.* ►inventory of fauna and flora; *bot. pedol.* ►species of soil flora.

F

flora [n] [US]**, depauperation of** *phyt. zool.* ▶depletion of fauna/flora.

flora [n] **and fauna** [n] *conserv. ecol. phyt. zool.* ▶conservation of flora and fauna, ▶inventory of flora and fauna.

flora composition [n] *phyt.* ▶anthropogenic alteration of flora composition.

2154 floral display [n] *hort.* (Entire bloom in a flower bed; ▶flowering period); *s* **florecimiento** [m] (±) (Conjunto de plantas en flor p. ej. en un macizo floral, en un parque, etc.; ▶floración); *f* **fleurissement** [m] (Ensemble des végétaux d'un massif floral ; ▶floraison) ; *g* **Blumenflor** [m] (Gesamtheit der blühenden Pflanzen in einem Beet; ▶Blütezeit).

2155 floribunda rose [n] *hort. plant.* (Garden rose which results from crossing between ▶polyantha rose and ▶hybrid tee rose; ▶bedding rose); *s* **rosa** [f] **floribunda** (Rosal de flores agrupadas en racimos resultante del cruce de la ▶rosa polyantha y la ▶rosa híbrido de té; ▶rosa de macizo); *f* **rosier** [m] **floribunda** (▶Rosier buisson à fleurs groupées en bouquets issu du croisement entre des ▶rosiers polyantha et des ▶rosiers hybrides de thé) ; *g* **Floribundarose** [f] (Vielblütige Gartenrose, die wie die Polyanthahybride aus der Kreuzung zwischen ▶Polyantharose und ▶Edelrosen entstand; ▶Beetrose).

2156 floristic element [n] *phyt.* (Group of species of flora classified according to certain criteria. The following terms are distinguished according to various aspects: 'geographical **f. e.**', 'genetic **f. e.**' [floristic group of species with the same center [US]/centre [UK] of origin and development], 'migratory **f. e.**' [floristic group of species with the same corridor of immigration], 'chronological **f. e.**' [historical floristic element with the same immigration time period]. There are also geographically-termed **f. e.s**, such as 'atlantic **f. e.**', 'central European **f. e.**', 'mediterranean **f. e.**', 'pacific **f. e.**', etc.); *s* **elemento** [m] **fitogeográfico** (Expresión florística de un grupo de plantas caracterizado según determinados criterios: «elemento geográfico» [grupo de plantas de área semejante], «elemento genético» [grupo de plantas originadas en la misma región], «elemento migratorio» [siguiendo el mismo camino] y «elemento histórico» [inmigradas contemporáneamente]. Se puede hablar también de los elementos atlántico, medioeuropeo, mediterráneo, etc.; cf. DB 1985); *syn.* cortejo [m] florístico (DINA 1987, 442); *f* **cortège** [m] **floristique** (Ensemble des espèces végétales d'une flore, qui caractérise certains critères. *On distingue* l'élément géographique l'élément génétique, l'élément migratoire et l'élément historique. On distingue ainsi l'élément atlantique, l'élément médioeuropéen, l'élément méditerranéen, etc.) ; *syn.* éléments [mpl] floristiques ; *g* **Florenelement** [n] (Eine nach bestimmten Kriterien zusammengefasste Artengruppe der Flora. Je nach Einteilungsgesichtspunkten wird unterschieden in ‚geografisches Florenelement', ‚Genoelement' [**F.** mit gleichem Entstehungs- u. Entfaltungszentrum], ‚Migroelement' [**F.** gleicher Einwanderungsrichtung mit gleichem Einwanderungsweg] und ‚Chronoelement' [historisches **F.**, gleiche Einwanderungszeit] oder man unterscheidet die Geoelemente in atlantisches **F.**, mitteleuropäisches **F.**, mediterranes **F.** etc.).

2157 floristic kingdom [n] *phyt.* (Geobotanical term for the largest geographical unit according to the botanical classification and spatial distribution of plants on the earth's surface. There are six units: boreal [holarctic], paleotropic[al], neotropic[al], Australian, Cape [province] and Antarctic region); *s* **reino** [m] **floral** (Jerarquía fitogeográfica superior: Existen siete unidades de **r. f.**: holártico, paleotropical, neotropical, del Cabo, australiano, antártico, oceánico; cf. DINA 1987); *f* **empire** [m] **floristique** (Terme utilisé en géographie botanique pour désigner l'unité de premier ordre dans la classification des unités biogéographiques du règne végétal en milieu terrestre ; les unités d'ordre suivantes se nomment : région, domaine, secteur et district ; d'après LEMÉE on distingue six **e. f.** : « holarctique » ou « boréal », « africano-malgache », « néotropical », « australo-papou », « antarctique », « oriental » ou « indo-malais » ; suivant les écoles scientifiques, il existe différentes classifications, p. ex. le « Capensis » de DIELS est une région de l'empire africano-malgache chez LEMÉE ; EBI 1971, 123-137) ; *g* **Florenreich** [n] (Geobotanischer Begriff für die größte geografische Einheit hinsichtlich einer räumlichen Gliederung der Pflanzendecke der Erdoberfläche auf Grund botanisch-systematischer Einheiten. Es gibt sechs **F.e**: holarktisches [boreal] [Holarktis], paläotropisches [Paläotropis], neotropisches [Neotropis], australisches [Australis], kapländisches [Capensis] und antarktisches **F.** [Antarktis]; cf. MEL, Bd. 9, 1973).

floristic species composition [n]**, anthropogenic shift in** *phyt.* ▶anthropogenic alteration of flora composition.

2158 floristry [n] *hort.* (Florist's art and business); *s* **floristería** [f] (Arte y negocio del/de la florista); *f* **fleuristerie** [f] (**1.** Commerce de fleurs ; **2.** métier de l'horticulture de l'art floral spécialisé dans les techniques de préparation des végétaux, materiaux et accessoires divers en vue de la réalisation de compositions de fleurs) ; *syn. peu usité* fleuristique [f] ; *g* **Floristik** [f, o. Pl.] (**1.** Arbeitsgebiet des Floristen; **F.** ist in D. keine Fachrichtung des Gartenbaus. **2.** *phyt.* Wissenschaft, die sich mit dem Studium der Pflanzenarten und -sippen in natürlichen Verbreitungsgebieten befasst — Zweig der ▶Geobotanik).

2159 floristry work [n] *hort.* (Results of a florist's work); *s* **trabajos** [mpl] **de floristería;** *f* **réalisations** [fpl] **florales** *syn.* réalisations [fpl] des fleuristes ; *g* **floristische Leistungen** [fpl] (Arbeitsergebnisse der Blumenbinder).

2160 florist's show [n] *hort.* (The showing of flower arrangements, floral decorations, etc., primarily using different plant material); *syn.* exhibition [n] of floristry; *s* **exhibición** [f] **de floristería** (Exposición de ramos y arreglos de flores y elementos decorativos hechos mayormente de plantas); *f* **exposition** [f] **florale** (Manifestation périodique destinée à présenter des arrangements, des décorations floraux et utilisant principalement des plantes) ; *g* **Floristenschau** [f] (Ausstellung von Blumengebinden, Gestecken und Dekorationen, vorwiegend aus Pflanzenmaterial).

2161 flounder zone [n] [UK] *limn.* (*Brackish water region* lowest section of a stream course between ▶bream zone [UK] and the sea: characterized by non-uniform currents [tidal and wind action], often variable salt content of the water; recognizable European fish species: flounder *[Platychthis flesus]*, ruffle *[Acerina cernua]*, stickleback *[Gasterosteus aculeatus]*, and migratory fish; ▶river zone, ▶hypopotamon); *syn.* mullet zone [n]; *s* **región** [f] **salobre** (Tramo inferior de los cursos de agua entre la ▶zona de madrilla y el mar: caracterizada por corrientes cambiantes, contenido de sal variable y en la que en Europa predominan la platija *[Platychthis flesus]*, la acerina *[Acerifla cernua]*, el gasteósteo *[Gasterosteus aculeatus]* y diversas especies piscícolas migratorias; ▶zona biológica de un río, ▶hipopótamon); *syn.* tramo [m] salobre; *f* **zone** [f] **à éperlan** (Cours inférieur d'un cours d'eau entre la ▶zone à brème et la mer caractérisée par des courants changeants [milieu intertidal] et de fréquentes variations de salinité des eaux [milieu saumâtre] ; Flet *[Platychthis flesus]*, Grémille *[Acerina cernua]*, Epinoche *[Gasterosteus aculeatus]* et divers poissons migrateurs ; AEP 1976, 13 ; ▶zone écologique des cours d'eau ; ▶hypopotamon) ; *syn.* hypopotamal [m] ; *g* **Brackwasserregion** [f] (Unterster Fließgewässerabschnitt zwischen ▶Brachsenregion und Meer; gekennzeichnet durch uneinheitliche Strömung [Gezeiten und Windeinwirkung], häufig schwankender Salzgehalt des Wassers;

kennzeichnende Fischarten Flunder *[Platychthis flesus]*, Kaulbarsch *[Acerina cernua]*, Stichling *[Gasterosteus aculeatus]* und diverse Wanderfische; ▶Flussregion, ▶Hypopọtamon); *syn.* Kaulbarschregion [f], Kaulbarsch-Flunderregion [f], Hypopotamal [n].

flow [n] *ecol.* ▶energy flow; *hydr.* ▶groundwater flow; *geo.* ▶mud flow; *phyt.* ▶stem flow; *geo. wat'man.* ▶stream flow; *recr. plan.* ▶tourist flow.

2162 flow chart [n] *plan.* (Graphic presentation showing progress of project phases and time periods for a complicated planning procedure); *syn.* flow diagram [n]; *s* **organigrama [m]** (Representación gráfica); *syn.* esquema [m] de organización; *f* **organigramme [m]** (Représentation graphique de la succession des phases de travaux) ; *syn.* schéma [m] du planning, diagramme [m] de succession des opérations ; *g* **Ablaufschema [n]** (Grafische Darstellung über geplante oder vorhandene Organisations-oder Zeitabläufe [Zeitflussplan]); *syn.* Organigramm [n], Organisationsschema [n].

flow diagram [n] *plan.* ▶flow chart.

flow equilibrium [n] *ecol.* ▶ecological equilibrium.

flower [n] *bot.* ▶meadow flower; *gard. hort. plant.* ▶summer flower; *hort.* ▶wild flower.

2163 flower bed [n] *gard.* (Area predominantly planted with ▶annuals; ▶parterre, ▶seasonal bedding, ▶seasonal flower bed); *s* **macizo [m] de flores** (Área plantada mayormente con ▶plantas anuales; ▶macizo de flores estacionales, ▶parterre, ▶plantación estacional); *f 1* **massif [m] floral (d'annuelles)** *syn.* massif [m] fleuri ; *f 2* **plate-bande [f] florale/fleurie** (Surface plane étroite sur laquelle sont plantées des fleurs saisonnières, d'espèces différentes et située en général le long d'un cheminement ou d'un mur ; ▶parterre, ▶plantation saisonnière, ▶plante annuelle, ▶plate-bande [florale] saisonnière) ; *g* **Blumenbeet [n]** (Mit vorwiegend Einjährigen bepflanztes Beet; ▶einjährige Pflanze, ▶Wechselbeet, ▶Wechselbepflanzung, ▶Parterre 2).

flower beds [npl] *constr. hort.* ▶construction of flower beds.

2164 flower bud [n] *bot. s* **capullo [m]** *syn.* yema [f] floral, botón [m] floral; *f* **bouton [m] floral** ; *g* **Blütenknospe [f]**.

flower garden [n] *gard.* ▶ornamental garden.

2165 flowering aspect [n] *phyt.* (Blooming appearance in most plant communities of one or more distinct seasonal waves of flowering; ▶seasonal aspects); *syn.* flowering trait [n]; *s* **aspecto [m] de floración** (En la mayoría de las comunidades vegetales, aspecto cambiante debido a la variación de las plantas en flor; ▶sucesión de aspectos estacionales); *f* **aspect [m] floral saisonnier** (Aspects saisonniers du milieu naturel en particulier d'un peuplement végétal, produit par la variation rythmique du développement des végétaux) ; *syn.* faciès [m] saisonnier ; *g* **Blühaspekt [m]** (Durch die Vegetationsrhythmik bedingte Erscheinung einer oder mehrerer jahreszeitlich abgrenzbarer Blütenwogen einer Pflanzengesellschaft; ▶Aspektfolge).

2166 flowering bulb [n] *bot. hort.* (▶Bulbous plant with spectacular flowers, such as crocus, daffodil, tulip, squill *[Scilla]*; ▶stem tuber, ▶tuberous-rooted plant); *s* **bulbo [m] floreciente** (▶Planta de cebolla con flores llamativas como el crocus, el jacinto, el narciso o el tulipán; ▶planta tuberosa, ▶tubérculo caulinar); *f* **bulbe [f** *et* **m] à fleur** (Terme horticole pour désigner les ▶plantes à bulbe aux fleurs très colorées telles que les narcisses, les tulipes, les scilles. Par contre le perce-neige, le crocus et le colchique ne sont pas des plantes à bulbe au sens botanique mais des plantes à ▶tubercules caulinaires ; LA 1981, 209 ; ▶plante à tubercule) ; *syn.* oignon [m] à fleur ; *g* **Blumenzwiebel [f]** (Gärtnerische Bezeichnung für ein ▶Zwiebelgewächs mit auffallend farbigen Blüten wie z. B. Narzisse, Tulpe, Scilla.

MERKE, Krokus, Herbstzeitlose und Winterling sind im botanischen Sinne keine Blumenzwiebeln, sondern unterirdische ▶Sprossknollen [Erdknollen]; ▶Knollengewächs).

2167 flowering period [n] *hort. plant.* (▶floral display, ▶summer-flowering period); *syn.* blooming period [n]; *s* **floración [f]** (▶florecimiento, ▶floración estival); *syn.* periodo/período [m] de floración; *f* **floraison [f]** (▶fleurissement, ▶floraison estivale) ; *syn.* époque [f] de floraison, période [f] de floraison ; *g* **Blütezeit [f]** (Zeitabschnitt eines Jahres, in dem Pflanzen blühen; nicht zu verwechseln mit ▶Blumenflor, ▶Blütezeit im Sommer); *syn.* Blüte [f], Blühabschnitt [m], Blühdauer [f]; Blühperiode [f], Blühzeit [f].

flowering season [n] [UK]**, autumnal** *hort. plant.* ▶fall-flowering period [US].

flowering shrub [n] *hort. plant.* ▶ornamental shrub.

flowering trait [n] *phyt.* ▶flowering aspect.

flowering tree [n] **or shrub** [n] *hort. plant.* ▶ornamental woody plant.

2168 flowering woody plant [n] *hort. plant. s* **leñosa [f] floreciente** (Árbol o arbusto caracterizado por sus flores llamativas); *f* **plante [f] à fleurs ligneuse** (Arbre, arbuste caractérisés par leurs fleurs remarquables ; *termes spécifiques* arbuste à fleurs, arbre à fleurs) ; *syn.* au pluriel (végétaux) ligneux [mpl] à fleurs ; *g* **Blütengehölz [n]** (Durch auffällige Blüten gekennzeichneter Baum oder Strauch).

2169 flower planter [n] *gard. landsc.* (Container designed to hold flowers in a growing medium, typically large pots or boxe; ▶planter); *s* **tiesto [m] de flores** (Término genérico ▶jardinera); *f* **bac [m] à fleurs** (Terme générique ▶bac de plantation) ; *g* **Blumenkübel [m]** (Großes Gefäß, das vorwiegend der Bepflanzung mit annuellen Pflanzen dient; *OB* ▶Pflanzenkübel).

2170 flower show [n] *hort.* (Usually an indoor ▶horticultural exhibition; ▶indoor [horticultural] exhibition); *s* **exhibición [f] floral** (▶exhibición de horticultura, ▶exhibición [de horticultura] en pabellones); *f* **floralies [fpl]** (Exposition horticole où est présentée une grande variété de fleurs, soit en pots, en bouquets, en compositions ou bien associées à de la céramique, des écorces, des feuillages, etc. ; ▶exposition de jardins, ▶exposition horticole, ▶exposition couverte) ; *syn.* exposition [f] florale ; *g* **Blumenschau [f]** (Ausstellung in einer Halle und/oder im Freien, in der die gärtnerische Leistungsfähigkeit vorwiegend im Produktionsbereich Zierpflanzenbau und Staudenanzucht gezeigt und die Pflanzen zum Verkauf angeboten werden, z. B. im Rahmen einer ▶Gartenbauausstellung oder regionalen Verkaufsschau; ▶gärtnerische Hallenschau).

2171 flowers [npl] **on current season's shoot** *hort.* (Reference to flower buds which form on new growth stems rather than old growth stems); *s* **flor [f] en vástago del año** (Referente a capullos formados en vástagos nuevos); *f* **fleurs [fpl] poussant sur le bois de l'année** ; *syn.* fleurs [fpl] à floraison sur le bois de l'année ; *g* **Blüten [fpl] am treibenden Holz** (Der geschlechtlichen Fortpflanzung dienende Sprosse oder Sprossabschnitte an Ästen, die in der Vegetationsperiode neu gebildet werden).

flowery mead [n] [UK] *gard. landsc. poet.* ▶wildflower meadow.

2172 flowing surface water [n] *hydr.* (Surface water run-off which is or has to be drained into lower-lying watercourses or into storm water pipes; ▶sheet flow); *s* **agua [f] de escorrentía superficial** (≠) (Agua que fluye sobre la superficie hasta el curso de agua más cercano o hasta una alcantarilla; ▶escorrentía laminar); *f* **eau [f] d'écoulement de surface** (Lors des débits de pointe d'un réseau de drainage, eau qui s'écoule rapidement en

F

surface et dans l'horizon perturbé pour être ensuite captée par la tranchée ou le fossé de drainage ; ►ruissellement en nappes) ; *syn.* écoulement [m] superficiel (1) ; *g* **Vorflut [f] (1)** (Abfließßendes Oberflächenwasser [Abflussspende], das von tiefer liegenden natürlichen oder künstlichen Wasserläufen oder in Abwasserleitungen aufgenommen und weitergeleitet wird; ►Flächenabfluss).

2173 flow [n] of liquid waste *envir.* (Discharge of liquid emissions/sewerage; ►effluent, ►wastewater); *s* **flujo [m] de efluente** (Descarga de ►aguas residuales 2; ►efluente); *f* **écoulement [m] des effluents** (Flux de déchets liquides plus ou moins pollués ; ►effluent, ►effluent urbain) ; *g* **Abfluss [m] flüssiger Emissionen** (►abfließendes Abwasser, ►Abwasser).

2174 flow rate [n] *hydr.* (Volume of water flowing per unit of time; ►water resources); *s* **caudal [m]** (Volumen de agua que pasa en la unidad de tiempo a través de una sección transversal de una corriente determinada; ►recursos hídricos); *f* **débit [m]** (Quantité liquide écoulée par unité de temps ; ►ressources hydrauliques) ; *g* **Abflussmenge [f]** (Abgeführtes Wasservolumen pro Zeiteinheit [meist m³/s]; ►Wasserdargebot).

2175 flow slide [n] *constr.* (Type of erosion in ►earthworks whereby part of the topsoil slips off the ►slip plane parallel to the slope in a fluid state of the material; ►bearing capacity failure, ►shear failure of embankment, ►slop failure, ►soil creep 1, ►soil slippage); *s* **corrimiento [m] por flujo [Es]** (Forma de erosión en ►terracerías por la que parte de la tierra vegetal se desliza sobre un ►plano de deslizamiento paralelo al talud debido a su estado fangoso por entrada de agua; MESU 1977; ►caída de un talud, ►deslizamiento del suelo, ►deslizamiento planar, ►deslizamiento rotacional, ►soliflución); *syn.* deslizamiento [m] por flujo [CO, PE] (MESU 1977), corrimiento [m] por liquefacción (BU 1959); *f* **fluage [m]** (Forme de glissement de terrain lent lors de travaux de ►terrassement résultant d'une déformation gravitaire continue d'un remblai ramolli par l'augmentation de sa teneur en eau qui se met en mouvement le long d'un ►plan de glissement [zone de transition avec le massif stable], en général parallèlement au plan du versant ; cf. Mém. Fiche RN4 ; ►glissement de talus, ►glissement rotationnel, ►foirage, ►rupture de terrain, ►solifluxion) ; *g* **Schlammrutschung [f]** (Form der ►Rutschung beim ►Erdbau, bei dem Teile des angedeckten Oberbodens durch Wasserzutritt oder Staunässe verschlammt und auf einer ►Gleitfläche meist parallel zur Böschungsebene abrutschen; ►Bodenfließen, ►Geländebruch, ►Grundbruch, ►Hanggleiten).

2176 flow velocity [n] *hydr.* (Rate and direction of water motion in units of time); *s* **velocidad [f] de flujo** (Camino recorrido por el agua de un río o arroyo por unidad de tiempo); *syn.* velocidad [f] de la corriente; *f* **vitesse [f] du courant** (Vecteur indiquant l'intensité et la direction de la vitesse des particules liquides en mouvement ; WMO 1974) ; *g* **Fließgeschwindigkeit [f]** (In Fließrichtung pro Zeiteinheit zurückgelegter Weg).

2177 fluctuating water level [n] *hydr. syn.* oscillating water level [n]; *s* **nivel [m] variable del agua** (Fluctuación del nivel del agua de un curso de agua, etc.); *f* **plan [m] variable du niveau d'eau** (Fluctuation du niveau de l'eau d'un cours d'eau, d'un plan d'eau ; *contexte* lac de rétention à plan variable) ; *g* **wechselnder Wasserspiegel [m]** (Wechselnde Höhe der Wasseroberfläche); *syn.* wechselnder Wasserstand [m], wechselnder Pegelstand [m].

fluctuation [n] *phyt.* ►climax fluctuation; *zool.* ►population fluctuation.

2178 fluctuation [n] of groundwater level *hydr.* (Periodical variation in level[s] of groundwater); *s* **oscilación [f] del nivel de la capa freática** (Reducción y subida periódicas del

nivel de la capa freática); *syn.* oscilación [f] de la superficie freática; *f* **battement [m] de la nappe phréatique** (Abaissement et remontée périodique de la surface libre d'une nappe phréatique pendant l'année) ; *syn.* oscillation [f] de la nappe phréatique (PED 1983, 390), jeu [m] de la nappe ; *g* **Grundwasserschwankung [f]** (Periodisches Absinken und Ansteigen des Grundwasserspiegels).

2179 fluctuation [n] of species *ecol.* (Continual change —as in size, composition, and geographical distribution of flora and fauna under genetic, environmental, and group influences; may sometimes cause rapid increase of certain organisms; ►population fluctuation, ►population dynamics); *s* **fluctuación [f]** (Alteraciones de la densidad, composición y distribución geográfica de la población de una especie de flora o fauna en una estación específica debido a influencias genéticas, ambientales u otras. Puede conducir a la reproducción masiva de algunos organismos, p. ej. insectos; ►fluctuación de poblaciones, ►dinámica de poblaciones); *f* **fluctuation [f]** (Variation des effectifs d'une population animale ou d'un groupement végétal provoquée par certains facteurs de la biocénose à certaines époques, p. ex. d'une année à l'autre [fluctuation du cycle annuel] ou à chaque génération nouvelle] ou à intervalles réguliers de quelques années [fluctuation cyclique séculaire] ; DUV 1984, 16 ; est parfois à l'origine d'une augmentation considérable du nombre des individus de certains organismes ; ►variation de la densité de la population, ►dynamique des populations) ; *g* **Fluktuation [f]** (Durch Umwelteinflüsse bedingte Schwankungen in der Wohndichte einer Tierpopulation oder in einem Pflanzenbestand in einem definierten Gebiet innerhalb bestimmter Zeiten [z. B. eines Jahres oder von Generation zu Generation]; kann zur Massenvermehrung [Gradation] bestimmter Organismen führen; ►Populationsbewegung, ►Populationsdynamik); *syn.* Massenwechsel [m].

2180 flue gas [n] *envir.* (Waste gas, usually from a combustion process); *s* **gas [m] de combustión** (Emisión gaseosa producida en proceso de combustión); *syn.* gas [m] de humo; *f* **gaz [m] combustible** (Gaz produit lors de la combustion de mélanges air-matière combustible, dans les chambres de combustion des installations d'incinération, de combustion ou de chauffage) ; *syn.* gaz [m] de combustion ; *g* **Rauchgas [n]** (Bei der Verbrennung von Brennstoffluftgemischen in Feuerräumen von Kesselanlagen entstehende Gase).

2181 flue gas desulphurization [n] *envir.* (Purification of ►flue gas to reduce sulphur emissions either by dehydration with activated carbon or by rinsing with an appropriate absorbing fluid [= ►humid desulphurization]; ►dehydration of sulphur emissions); *s* **desulfuración [f] de gas de combustión** (Limpieza del ►gas de combustión para reducir la emisión de dióxido sulfúrico [SO_2] por medio de ►desulfuración en seco con carbono activado o por lavado con líquido absorbente adecuado [►desulfuración húmeda]); *f* **désulfuration [f] des gaz combustibles** (Épuration des ►gaz combustibles en vue de la diminution des rejets sulfureux au moyen d'une installation de désulfuration par séchage [►désulfuration par voie sèche] ; ►désulfuration par voie humide) ; *g* **Rauchgasentschwefelung [f]** (Reinigung der ►Rauchgase zur Verminderung des Schwefelausstoßes durch ►Trockenentschwefelung mit Aktivkohle oder durch Auswaschen mit geeigneten Absorptionsflüssigkeiten [►Nassentschwefelung]).

2182 flue gas purification [n] *envir.* (Process used in thermic waste treatment; ►flue gas desulfurization); *s* **purificación [f] de gas de combustión** (Proceso de reducción por medio de filtros de las sustancias perniciosas para la salud de las personas y para el medio ambiente emitidas en plantas de incineración de residuos y en grandes instalaciones de combustión;

procesos específicos son los de ►desulfuración de gas de combustión y de denitrificación de gas de combustión); *f* **épuration [f] des gaz combustibles** (►désulfuration des gaz combustibles) ; *g* **Rauchgasreinigung [f]** (Maßnahmen bei der thermischen Abfallbehandlung und bei Großfeuerungsanlagen, die die für Umwelt und Mensch schädlichen Bestandteile der Rauchgase durch Filteranlagen verringern; spezielle Verfahren sind die ►Rauchgasentschwefelung und Rauchgasentstickung).

flume [n] *eng. hydr.* ►constructed channel.

2183 fluoridation [n] **of drinking water** *chem. envir.* (Addition of fluorine to drinking water); *s* **fluorización [f] del agua potable** *syn.* enriquecimiento [m] con fluor (del agua potable); *f* **fluoration [f]** (Addition de fluor dans l'eau potable) ; *syn.* fluoruration [f] ; *g* **Anreicherung [f] des Trinkwassers mit Fluor** (Zuführung von Fluor in Trinkwasser); *syn.* Fluoridierung [f] des Trinkwassers, Fluoranreicherung [f].

flush curb [n] [US] *constr.* ►drop curb [US].

flush cut [n] *arb. hort.* ►cut at the branch attachment.

2184 flush edging [n] *constr.* (Flat border at same level, e.g. to a path or paved surface; ►drop curb [US]/flush kerb [UK], ►mowing strip); *syn.* trim [n] [also UK] (1); *s* **acabado [m] plano de los bordes** (Delimitación lisa entre zonas pavimentadas y no pavimentadas; ►borde de corte, ►piedra de bordillo enterrado); *syn.* recintado [m] plano; *f* **butée [f] en rive** (Limite plane, p. ex. rangée de pavés entre une surface stabilisée et non stabilisée ; ►bordure de pelouse, ►bordure arasée d'épaulement) ; *g* **höhengleiche Einfassung [f]** (Ebene Begrenzung, z. B. Pflasterzeile oder Tiefbordband zwischen befestigten und unbefestigten Flächen; ►Mähkante, ►Tiefbordstein; *UB* höhengleiche Wegekante); *syn.* ebenerdige Begrenzung [f], ebenerdige Einfassung [f], höhengleiche Kante [f].

2185 flush-jointed [pp/adj] *constr.* (Descriptive term applied to masonry joints which are pointed up with a mineral-based mortar; ►pointing); *s* **de junta llena [loc]** (Término descriptivo para junta de mortero entre mampuestos, enrasada con el paramento; DACO 1988; ►rejuntado); *f* **de joint plat [loc]** (►jointoiement ; DTB 1985) ; *syn.* de joint plein [loc] (DTB 1985) ; *g* **vollfugig [adj]** (So bearbeitet, dass z. B. Fugen eines Mauerwerkes mit mineralischem Fugenmörtel komplett ausgefüllt sind; ►Verfugen).

flush kerb [n] [UK] *constr.* ►drop curb [US].

flush with the ground, cut [vb] *constr. for. hort.* ►coppice (2).

flux [n] **of energy** [US] *ecol.* ►energy flow.

2186 fly ash [n] *envir.* (Particulate matter entrained in flue gases resulting from combustion of fuel or other material); *s* **cenizas [fpl] volantes** (ERNST 1973; partículas sólidas emitidas en procesos de combustión); *f* **cendres [fpl] volantes** (Produit pulvérulent issu de la combustion du charbon ; DTB 1985) ; *g* **Flugasche [f]** (Aus Verbrennungsvorgängen herrührende und durch Rauchgase mitgeführte, feste, feine Teilchen, die durch Wind verteilt werden).

fly tip [n] [UK] *envir.* ►unauthorized dumpsite [US].

fly tipping site [n] [UK] *envir.* ►unauthorized dumpsite [US].

2187 focal point [n] *plan. recr.* (Unifying feature in a landscape; ►attraction point); *s* **centro [m] de gravedad (2)** (Elemento atractivo de un paisaje o en un parque; ►centro de gravedad 1); *f* **point [m] d'attrait** (Élément de paysages ou de décor d'un jardin qui attire l'attention ; ►pôle d'attraction et d'animation) ; *syn.* point [m] d'intérét, point [m] fort ; *g* **Anziehungspunkt [m] (3)** (Besonders attraktives Element, z. B. in einer Landschaft, in einem Park; ►Anziehungspunkt 1).

focal point perennial [n] *hort. plant.* ►accent perennial.

fodder meadow [n] *agr.* ►hay meadow.

2188 fodder tree [n] *agr.* (Tree from which live branches with leaves are stripped to provide forage or bedding for domestic animals; ►pollarded tree); *syn.* lopped tree [n]; *s* **árbol [m] de forraje** (Árbol cuyas ramas vivas sirven de alimento o de camada para animales; ►árbol desmochado); *f* **arbre [m] d'émonde** (Un arbre émondé sur toute la hauteur de sa tige [suppression des branches latérales du tronc] pour en obtenir des rameaux ou du feuillage ; ►arbre têtard ; DFM 1975, 109) ; *syn.* arbre [m] élagué ; *g* **Schneitelbaum [m]** (Baum, dessen beblätterte Schösslinge zur Futtergewinnung — Laubheugewinnung — für den Winter wiederholt beschnitten werden; ►Kopfbaum); *syn. o. V.* Schneidelbaum [m].

2189 foehn/föhn [n] *met.* (**1.** Local name in Central Europe for a warm and dry descending wind occurring in northern Alpine valleys. **2.** Any descending warm and relatively dry wind on the lee side of a mountain ridge; **in U.S.,** a warm dry foehnlike wind that descends the eastern slopes of the Rocky mountains is called **chinook**; ►mountain wind); *s* **foehn [m]** (Viento catabático, calentado y reseco por efecto de su descenso que, por lo general, se registra en las laderas de montañas. Se produce cuando una corriente de aire húmedo asciende a barlovento de una cadena de montañas y, tras la condensación por enfriamiento adiabático, genera nubes y lluvias de estancamiento y se enfría. Una vez descargado de humedad, el aire desciende a sotavento, calentándose según la adiabática seca. El nombre procede de los Alpes, pero posteriormente se generalizó a cualquier viento del mismo tipo, otros nombres utilizados son **chinook** en las Rocosas, **berg** en el suroeste africano, **bora** en Dalmacia, **sirocco** en Sicilia, **tramontana** en el Languedoc y Rosellón y **zonda** en los Andes argentinos. El **leveche** es un viento cálido y seco de componente sur que sopla sobre la costa sudeste de España por delante de una depresión que avanza; con frecuencia trae gran cantidad de polvo y arena, hasta el punto que su llegada se anuncia por una banda nubosa marrón que se observa hacia el sur del horizonte; cf. DM 1986 y DGA 1986; ►brisa de montaña); *syn.* föhn [m]; *f* **fœhn [m]** (Vent catabatique — descendant sur un versant — chaud et sec avec des caractères thermiques et hygrométriques liés à la subsidence — mouvement descendant de l'air — qui réchauffe l'air et diminue son humidité relative ; sa violence est due à l'effet de canalisation par les vallées ; cf. DB 1984 ; ►brise de montagne) ; *g* **Föhn [m]** (ahd. *phönno*, von lat. *favonius* >lauer Westwind<; **1.** warmer, trockener Fallwind in den Vorländern beidseitig des Alpenkammes. **2.** Allgemeiner Begriff für einen warmen und relativ trockenen Fallwind auf der Leeseite eines Gebirges; F. entsteht, wenn Luft über ein Gebirge strömt und sich dabei beim Aufstieg trocken-adiabatisch bis zum Kondensationsniveau um ca. 1 °C je 100 Höhenmeter und beim weiteren Anstieg feucht-adiabatisch um ca. 0,5-0,7 °C abkühlt. Nach Überqueren des Gebirgskammes und Abregnen erwärmt sich die in die Täler abfallende Luft unter Wolkenauflösung trocken-adiabatisch um ca. 1 °C je 100 m und erreicht deshalb im Lee auf der gleichen Ausgangshöhe wie im Luv eine um einige Grade höhere Tempera-tur; je höher die Gebirge desto ausgeprägter sind die F.winde. Föhnige Winde in abgeschwächter Form treten auch in den Mittelgebirgen auf; cf. MEL 1973; dasselbe Phänomen wird in Argentinien **Zonda**, in Nordamerika auf der Ostseite der Rocky Mountains **Chinook** genannt; der warme bis sehr heiße und staubbeladene Wind aus Nordafrika in Richtung Mittelmeer heißt **Schirokko** [ital. Scirocco], im südlichen Spanien **Solano** oder **Leveche**; in Dalmatien gibt es den trockenen kalten Fallwind **Bora**, in Norditalien die aus den Alpen kommende frischkühle **Tramontana**; ►Bergwind); *syn.* Fallwind [m].

F

2190 fog [n] [UK] (1) *hort.* (CWF 1996; long grass left standing in winter; e.g. Reed Bentgrass *[Calamagrostis x acutiflora]*, Switch Grass *[Panicum virgatum]*; ►perennial with lignified seed stalks); *s* **gramínea** [f] **resistente a heladas** (≠) (Gramínea alta, como p. ej. la calamagrostis *[Calamagrostis x acutiflora]*, que por su aspecto decorativo se mantiene en jardines durante el invierno; ►perenne con infructescencia invernal); *f* **graminée** [f] **vivace résistante aux rigueurs de l'hiver** (Graminée haute recherchée pour son aspect décoratif pendant l'hiver quand elle est sèche, p. ex. le Calamagrostis *[Calamagrostis x acutiflora]*, le panic raide *[Panicum virgatum]*; ►vivace à infrutescence [lignifiée] pérenne) ; *syn.* graminée [f] rustique résistante aux rigueurs hivernales ; *g* **winterständiges Gras** [n] (Hohes Gras, das über den Winter stehen bleibt und seinen Zierwert behält, z. B. Reitgras *[Calamagrostis x acutiflora]*, Rutenhirse *[Panicum virgatum]*; ►Wintersteher).

fog [n] (2) *met.* ►frequency of fog.

2191 foliage [n] *bot. hort.* (Live leaves of monocotyledoneae and eudicotyledoneae; ►clinging dead leaves, ►dead leaves, ►discoloration of foliage, ►foliage of woody plants, ►needles); *syn.* leafage [n]; *s* **follaje** [m] (1) (Conjunto de hojas de los árboles y otras plantas; ►aguja, ►descoloración del follaje, ►follaje 2, ►hoja marcescente, ►hojarasca); *f* **feuillage** [m] (1) (Ensemble des organes foliaires vivants des mono- et dicotylédones ; le feuillage des résineux/conifères est constitué par plusieurs générations d'►aiguilles ; ►discoloration des feuilles, ►feuilles marcescentes, ►frondaison) ; *g* **Laub** [n, o. Pl.] (1. Lebende Blätter aller Bedecktsamer. 2. Auf der Erde liegende, abgestorbene Pflanzenblätter der Bedecktsamer [Einkeimblättler und Zweikeimblättler]: ►abgestorbenes Laub, ►abgestorbene haftende Laubblätter; das **L.** der Koniferen heißt ►**Nadeln** [fpl]; ►Belaubung, ►Laubverfärbung).

2192 foliage color [n] [US]/**foliage colour** [n] [UK] *bot. plant.* (1. Way in which a leaf is colo[u]red: the main colo[u]ring agent of leaves is leaf green *[chlorophyll]*, xanthophyll colo[u]rs the leaves yellow, *carotene* and *anthocyanin* red and yellow. 2. **Variegation** [n] is a special form of **f. c.** caused by colo[u]rless leucoplasts [absence of chlorophyll], which occur in the yellowish parts of leaves of numerous [variegated] varieties; cf. LB 1976, 65; ►autumn colors/hues [of leaves] [US]/autumn colours/hues [of leaves] [UK], ►discoloration of foliage [US]/discolouration of foliage [UK]); *s* **color** [m] **del follaje** (1. Tipo de color de las hojas: el principal agente de color es el verde causado por la clorofila, la xantofila colorea las hojas de amarillo, la carotina y antocianina de rojo y amarillo. 2. La **variegación** [f] es una forma especial del **c. del f.** causada por leucoplastos descoloridos [por ausencia de la clorofila] que tiene lugar en parte de las hojas amarillentas de muchas variedades [variegadas]; ►color otoñal, ►descoloración del follaje); *f* **couleur** [f] **de la feuille** (1. La couleur des feuilles est due à la présence simultanée de pigments colorés dans la texture des cellules végétales qui composent ces feuilles : de couleur verte [chlorophylle], de couleur jaune orangé [carotène et xanthophylle], et de couleur rouge [anthocyane] ; la coloration des feuilles en automne est attribuable à la diminution de la longueur des jours et de l'énergie solaire ; l'obturation des nervures par un bouchon de liège empêche la régénération chlorophyllienne qui est d'abord ralentie puis stoppée ; celle-ci disparaît pour laisser place à des pigments jaunes normalement masqués par le vert ; la variété des teintes automnales est principalement conditionnée par la température, les températures basses favorisant la synthèse des anthocyanes [pigments rouges] qui apparaissent dans la feuille à la fin de l'été chez certains végétaux [érable, chêne, sumac]. 2. Une forme particulière de la couleur d'une feuille est l'**aspect** [m] **panaché des feuilles** de certains végétaux pour

lesquelles la présence de leucoplastes dépourvus de pigments est responsable de l'aspect blanc jaunâtre de certaines parties de la feuille ; ►coloration automnale, ►décoloration des feuilles) ; *syn.* coloration [f] de la feuille ; *g* **Blattfärbung** [f] (1. Art, wie ein Blatt gefärbt ist: der Hauptfarbstoff der Blätter ist das Blattgrün *[Chlorophyll]*, das *Xanthophyll* färbt die Blätter gelb, *Carotin* gelbrot und *Anthocyan* rot. 2. **Panaschierung** [f] als besondere Form der **B.** entsteht durch farblose Leukoplasten [Fehlen von Chlorophyll], die in gelblichweißen Blatteilen zahlreicher [weißbunter] Varietäten vorkommen; cf. LB 1976, 65; ►Herbstfärbung, ►Laubverfärbung).

2193 foliage [n] **of woody plants** *bot. hort.* (All live leaves of deciduous woody plants; ►autumn foliage, ►dense foliage of woody plants, ►foliage, ►new foliage of woody plants, sparse foliage of woody plants, ►spring foliage); *syn.* leafage [n] of woody plants, live deciduous leaves [npl]; *s* **follaje** [m] (2) (Conjunto de hojas vivas de leñosas; ►foliación joven, ►foliación prevernal, ►follaje 1, ►follaje frondoso, ►follaje hueco, ►follaje otoñal); *f* **frondaison** [f] (Le volume constitué par l'ensemble des feuilles d'une plante ; le terme frondaison est aussi utilisé pour désigner la période de l'apparition des feuilles [le temps de la feuillaison] ; ►frondaison automnale, ►frondaison dense, ►frondaison légère, ►frondaison printanière, ►feuillage 1, ►jeune frondaison) ; *syn.* feuillage [m] (2) ; *g* **Belaubung** [f, o. Pl.] (1. Alle lebenden Blätter einer Pflanze als Gesamterscheinungsbild, i. d. R. bei Gehölzen; wenn die **B.** im Herbst zu Boden fällt, wird sie zu ►Laub; im Deutschen wird der Begriff **Laub** [n] für lebende und abgestorbene Blätter verwendet; Blätter der Nadelbäume heißen **Nadeln**; 2. sich belauben [vb]; ►dichte Belaubung, ►frische Belaubung, ►Frühjahrsbelaubung, ►Herbstbelaubung, ►lockere Belaubung); *syn.* Laubwerk [n, selten Pl.], Blattwerk [n, selten Pl.].

follow-up care [n] [US] *constr.* ►separate contract maintenance.

2194 follow-up costs [npl] *adm. plan.* (1. Total costs for the running of a project, comprising those connected with the initial investment, principally interest on loans, and those for its operation, upkeep and administration, which begin after the project's completion. 2. In comparison to **f.-up c.**, so-called **life-cycle costs** comprise all costs incurred throughout the life/duration of a project/product or equipment from creation/production until the date of disposal or complete renovation; the term **l.-c. c.** applies mainly to products or equipment); *s* **costos** [mpl] **ocasionados** (1. Costos posteriores a la terminación de una obra para mantenerla en servicio y que pueden incluir el pago de intereses y del crédito solicitado para realizarla. 2. Los llamados **costes** [Es]/**costos** [AL] **de ciclo de vida** incluyen todos aquéllos causados durante la duración/vida de un proyecto/producto o equipamiento desde su producción hasta la fecha de eliminación o rehabilitación total); *syn.* costes [mpl] ocasionados [Es]; *f* **coût** [m] **d'utilisation et d'entretien** (1. Coûts nécessaires au fonctionnement et à la maintenance des ouvrages achevés ou d'équipements d'infrastructure. 2. Par comparaison les **coûts de cycle de vie** [Life-cycle costs] correspondent aux **coûts** générés sur l'ensemble des étapes de la vie des procédés, services, produits et équipements tels que extraction et traitement des matières premières, fabrication, transport et distribution, utilisation et réemploi, recyclage et gestion des déchets qui ont une incidence sur l'environnement et l'économie) ; *syn.* coût [m] d'exploitation et d'entretien) ; *g* **Folgekosten** [pl] (1. Kosten, die nach Fertigstellung einer Anlage oder Ausstattungseinrichtung zu deren Betrieb, Unterhaltung und Zinszahlungen für aufgenommenes Kapital anfallen. 2. Im Vergleich beinhalten die sogenannten **Life-Cycle-Costs** alle Kosten, die während der gesamten Lebensdauer/Bestandsdauer — von der Erstellung/Produktion bis zum Zeitpunkt der Entsorgung resp. Totalerneuerung — anfallen).

follow-up maintenance [n] [US] *constr.* ▶separate contract maintenance.

food [n] *zool.* ▶supply of food.

2195 food chain [n] *ecol.* (Linear pattern of food intake based upon the connection between plants as ▶producers, animals as ▶consumers, and ▶decomposers 1; a series of organisms in any natural community in which each member feeds on the one before and is in turn eaten by the one after; ▶bioaccumulation, ▶food web, ▶life pyramid, ▶trophic level); *s* **cadena [f] trófica** (Concepto introducido por ELTON [1927] para describir las relaciones tróficas en una sucesión lineal de organismos vivos que se nutren unos de otros en un orden determinado [▶productor, ▶consumidor, ▶descomponedor]. La **c. t.** constituye un caso límite ideal de la ▶red trófica, a la que en realidad no se llega nunca; cf. DINA 1987 y MARG 1977; ▶bioacumulación, ▶nivel trófico, ▶pirámide ecológica); *syn.* cadena [f] alimentaria; *f* **chaîne [f] trophique** (Succession d'organismes vivants dont chacun se nourrit aux dépens du précédent. Les végétaux [▶producteurs primaires] sont mangés par des animaux phytophages [▶consommateurs primaires] qui sont à leur tour mangés par des animaux prédateurs [consommateurs secondaires de 1er ordre, puis de 2e ordre, etc. En réalité, les différents organismes de la **ch. t.** ont entre eux des relations multiples et complexes ; elles sont ramifiées et forment alors un ▶réseau trophique ; DEE 1982 ; ▶bioaccumulation, ▶décomposeur, ▶pyramide écologique, ▶niveau trophique) ; *syn.* chaîne [f] alimentaire ; *g* **Nahrungskette [f]** (Lineares Modell der Verknüpfung von Pflanzen [▶Produzenten], Tieren [▶Konsumenten] und ▶Destruenten durch die Nahrungsaufnahme; ANL 1984; ▶Bioakkumulation, ▶Nahrungsnetz, ▶Nahrungspyramide, ▶Trophiestufe).

2196 food patch [n] *hunt.* (Area sown or planted to improve the supply or provide additional food for small populations of ▶game); *s* **comedero [m] de pastos** (Área sembrada o plantada para proporcionar alimento a la ▶caza 2; una de las posibles medidas en el marco de la ▶gestión de la caza 2); *f* **culture [f] à gibier** (DFM 1975 ; champ semé ou planté pour procurer des aliments au ▶gibier et considéré comme mesure de ▶gestion cynégétique) ; *syn.* prairie [f] de gagnage, pré [m] de gagnage [m], gagnage [m], terrain [m] à viander (DFM 1975, 82) ; *g* **Wildacker [m]** (Mit bestimmten Kulturpflanzen und Kräutern angelegtes Feld zur Nahrungsverbesserung resp. -ergänzung für das ▶Wild als Maßnahme der ▶jagdlichen Hege).

2197 food plant [n] *ecol. landsc. plant.* (Generic term covering ▶host plant for bees, ▶woody host plant, ▶woody host plant for birds); *s* **planta [f] útil para animales salvajes** (Término genérico para ▶planta apícola, ▶leñosa útil para insectos y aves, ▶leñosa útil para las aves); *syn.* planta [f] alimento para animales salvajes; *f* **plante [f] nourricière** (Terme générique pour ▶plante mellifère, ▶ligneux nourricier, ▶ligneux nourricier pour les oiseaux) ; *g* **Nahrungspflanze [f]** (*UBe* ▶Bienennährpflanze, ▶Nährgehölz, ▶Vogelnährgehölz).

food supply [n] *zool.* ▶richness of food supply.

2198 food web [n] *ecol* (Network-like pattern of all interconnected ▶food chains in a natural community; ▶life pyramid); *s* **red [f] trófica** (Conjunto de relaciones tróficas [o alimentarias] existentes entre las especies de una comunidad biológica y que refleja el sentido de flujo de materia y energía que atraviesa el ecosistema. Toda **r. t.** puede considerarse formada por ▶cadenas tróficas; DINA 1987. *Términos específicos* cadena de herbívoros, cadena saprofítica o de detritus; ▶pirámide ecológica); *syn.* red [f] alimentaria; *f* **réseau [m] trophique** (DEE 1982, 77 ; la multi-

plicité de ▶chaînes trophiques dans un écosystème s'anastomosant en un réseau alimentaire ; DUV 1984, 50, 52 ; ▶pyramide écologique) ; *syn.* réseau [m] alimentaire ; *g* **Nahrungsnetz [n]**. (Modell des netzartigen Verbundes von ▶Nahrungsketten; ANL 1984; ▶Nahrungspyramide).

2199 footbridge [n] *constr. trans.* (**1.** Narrow structure, usually wooden, carrying a path across a stream, ravine or ditch; *specific term* wooden footbridge; ▶boardwalk. **2.** Narrow bridge, which allows pedestrians to cross from one side of a road to the other; *syn.* pedestrian bridge; *s 1* **pasarela [f]** (Puente estrecho, generalmente de madera, que atraviesa un río, arroyo o zanja; *término específico* ▶pasarela de madera; si se trata de un puente provisional para atravesar una corriente de agua se denomina **pasadera [f]**; si el puente está formado de maderos o de una sola tabla se denomina **pontón [m]**); *s 2* **puente [m] peatonal** (Puente sobre río o autopista construido para uso exclusivo de peatones y ciclistas); *f 1* **passerelle [f]** (Petit pont rustique en bois, d'une seule arche et souvent vouté, placé pour franchir un fossé ou un ruisseau ; *terme spécifique* ▶passerelle en/de bois ; ▶appontement) ; *syn.* ponceau [m] ; *f 2* **pont [m] piéton** (Passerelle fixe étroite permettant aux piétons le franchissement de voies de circulations) ; *g* **Steg [m]** (**1.** Kleine, schmale Brücke über einen Bach oder Graben; *UB* ▶Holzsteg. **2.** Überführung über Verkehrswege in Form einer kleinen schmalen Brücke. **3.** Kurzform für ▶Bootssteg); *syn. zu 2.* Fußgängerbrücke [f].

2200 foothills [npl] *geo.* (Transitional line of hills, lying between and more or less parallel to a main range of mountains and a plain; DNE 1978; rolling landscape leading from the foot of a mountain to flat land is called 'piedmont'; **in eastern U.S.**, f. generally refers to the marginally hilly country located in the western portion of the 'piedmont' where it borders a mountain region; the entire piedmont generally encompasses all of a region between the mountains and a coastal plain); *s* **estribación [f]** (Ramificación de montañas o colinas que se desprende a uno u otro lado de una cordillera; ▶piedemonte se denomina una región baja situada en contacto con un macizo montañoso; cf. DGA 1986); *syn.* contrafuertes [mpl], cadena [f] secundaria; *f* **contrefort [m] (montagneux)** (Suite d'accidents de relief bas collinaire ou légèrement montagneux de transition s'appuyant sur un massif montagneux) ; *syn.* avant-mont [m], chaîne [f] de montagne secondaire ; *g* **Vorgebirge [n]** (Höhenrücken oder Kette von Hügeln, die dem höheren Gebirge vorgelagert sind).

2201 footing [n] *constr.* (Substructure made of e.g. natural stone, concrete, wood, metal for benches, ▶planters, etc.); *syn.* base [n], substructure [n]; *s* **zapata [f]** (Cimentación o ensanchamiento de la base de un soporte, que tiene como cometido repartir las cargas sobre el terreno; DACO 1988; ▶jardinera); *f* **piédestal [m]** (*Terme ancien* ensemble des pièces support en pierres, béton, bois, métal, constituant le pied ou la base [soubassement] d'un ouvrage, p. ex. d'une colonne, d'une statue, d'un banc, de ▶bacs de plantation, etc.) ; *syn.* base [f], support [m], élément [m] de la base ; *g* **Unterkonstruktion [f]** (Bauteil aus Materialien wie z. B. Naturstein, Beton, Holz, Metall für Bänke, ▶Pflanzkübel, ▶Pflanzenkübel etc.); *syn.* Untergestell [n].

footing [n] [US], **strip** *constr.* ▶strip foundation.

2202 foot [n] **of an embankment/slope** *constr. syn.* toe [n] of an embankment/slope; *s* **pie [m] de talud**; *f* **pied [m] du talus** ; *g* **Böschungsfuß [m]** (Unterster Teil einer Böschung).

foot [n] **of a wall** *arch. constr.* ▶bottom of wall base.

2203 foot [n] **of slope** *geo. syn.* bottom [n] of slope; *s* **pie [m] de ladera**; *f* **pied [m] de versant** ; *g* **Hangfuß [m]** (Unterster Teil eines Hanges).

F

2204 footpath [n] *landsc. urb.* (**1.** Paved or compacted granular track for the use of persons walking in parks and open spaces; **2.** ▶sidewalk [US]/pavement [UK]); *syn.* pathway [n], track [n] [also US], trail [n], walk [n]; *s* camino [m] (**1.** Vía para peatones asfaltada, pavimentada o con revestimiento compactado en parque o paseo, **2.** ▶accra); *syn.* sendero [m]; *f* allée [f] piétonnière (pour espace vert au niveau d'un groupe d'immeubles ou d'espaces publics ; ▶trottoir) ; *syn.* allée [f] piéton, cheminement [m] piétonnier ; *g* Gehweg [m] (**1.** Asphaltierter, gepflasterter oder wassergebundener Fußweg in Grünanlagen und Parks; *im Vergleich zum* Fahrweg; *OB* Weg; **2.** ▶Bürgersteig); *syn.* Fußweg [m].

footpath excavation [n] *constr.* ▶pathway excavation.

footpath network [n] *recr.* ▶pathway network.

2205 footpath paving slab [n] *constr.* (*Generic term* slab for surfacing pedestrian areas made of concrete or natural stone materials; *specific terms* ▶precast paving block [US]/precast concrete paving slab [UK], ▶natural garden stone flag, ▶patterned concrete paving slab); *s* losa [f] de pavimentación (*Término genérico* placa de piedra natural u hormigón utilizada para pavimentar aceras y caminos; *términos específicos* losa de arenisca, losa de granito, ▶losa de hormigón, ▶losa de hormigón con dibujo, ▶losa de piedra natural, losa de pórfiro; ▶losa de jardín [de piedra natural o artificial]); *f* dalle [f] pour circulation piétonne (Terme générique pour les matériaux naturels ou normalisés fabriqués industriellement et utilisés comme revêtement de voirie et de sols extérieurs ; ▶dalle pavée ; *termes spécifiques* ▶dalle préfabriquée en béton pour voirie de surface, dalle gazon en béton pour circulation piétonne, dalle de/en granit, dalle granit, dalle de/en porphyre, dalle porphyre, dalle de/en grès, dalle grès, ▶dalle de jardin) ; *syn.* dalle [f] pour sols extérieurs, dalle [f] pour sentier ; *g* Gehwegplatte [f] (OB für eine im Straßen-, Garten- und Landschaftsbau zur Befestigung von Gehwegen verwendete Beton- oder Natursteinplatte; ▶Gehwegplatte aus Beton; *UBe* Betongehwegplatte, Gehwegplatte aus Beton, Natursteinplatte [z. B. Granitplatte, Porphyrplatte, Sandsteinplatte] für Gehwege und Terrassenbeläge; ▶Gartenplatte).

footprint [n] **of a building** *leg. urb.* ▶built portion of a plot/lot.

forb [n] *phyt.* ▶tall forb.

forb [n]**, perennial** *bot. hort. plant.* ▶perennial (1).

forb community [n] *phyt.* ▶forb vegetation, ▶perennial forb community, ▶tall forb community.

forb reed marsh [n] [US] *phyt.* ▶tall forb reed marsh [US]/tall forb reed swamp [UK].

forb reed swamp [n]**, tall** [UK] *phyt.* ▶tall forb reed marsh [US].

2206 forb-rich [adj] *phyt.* (Descriptive term for an area having mostly broad-leaved herbs with very few grass species); *syn.* rich in forbs [loc]; *s* rico/a [adj] en herbáceas (Término descriptivo para una comunidad vegetal en la que predominan las plantas herbáceas y tiene pocas gramíneas); *f* riche en espèces herbacées [adj] (Terme caractérisant un groupement végétal dans lequel prédominent les plantes herbacées à larges feuilles non graminiformes) ; *g* krautreich [adj] (Mit vorwiegend breitblättrigen, kaum grasartigen Kräutern versehen).

2207 forb vegetation [n] *phyt.* (Perennial, short-lived or ephemeral plant formation composed primarily of non-graminaceous species. Woody plants are rarely present; ▶tall forb community); *syn.* forb community [n], herbaceous vegetation [n]; *s* vegetación [f] herbácea (Comunidad vegetal de herbáceas permanente, de corta vida o efímera, con poca presencia de gramíneas y en la que raramente se presentan leñosas; ▶formación

megafórbica); *f* végétation [f] herbacée (1) (Groupement de végétaux pérennes, à croissance courte ou épisodique en majorité constitué de plantes herbacées non graminiformes dans lequel les végétaux ligneux sont rarement présents ; ▶mégaphorbiaie) ; *syn.* formation [f] herbacée ; *g* Krautflur [f] (Ausdauernde, kurzlebige oder episodisch auftretende Pflanzenformation aus vorwiegend nicht graminoiden Kräutern. Gehölze treten meist nicht auf; ▶Hochstaudenflur).

2208 forb vegetation [n] **in woodland clearing** *phyt.* (▶Clearance herb formation where mainly perennials grow; e.g. a willow-herb community *[Epilobietalia angustifolii]*; ▶tall forb community); *s* formación [f] de perennes de bosques talados (Formación vegetal en la que predominan las plantas herbáceas perennes; en Europa Central pertenece a la clase *Epilobietalia angustifolii*. Término genérico ▶vegetación de bosques talados; ▶formación megafórbica); *f* flore [f] herbacée géante de coupe (*Terme générique* ▶flore de coupes à blanc[-étoc] ; ce groupement temporaire des coupes des forêts mésophiles constitue la classe des *Epilobietalia angustifolii* ; ▶mégaphorbiaie) ; *g* Staudenschlagflur [f] (Vorwiegend aus mehrjährigen krautigen Pflanzen bestehende Vegetationseinheit auf Kahlschlägen; z. B. in mitteleuropäischen Schlag- und Vorwaldgesellschaften die Ordnung der Weidenröschengesellschaft *[Epilobietalia angustifolii]*; *OB* ▶Kahlschlagflur; ▶Hochstaudenflur).

2209 force [n] **majeure** *leg.* (Unforeseeable natural event which cannot be prevented; e.g. earthquake, flood, avalanche, forest fire; ▶major natural event, ▶natural hazard); *syn.* Act [n] of God [also US]; *s* fuerza [f] mayor (Fenómeno natural poco usual e inevitable; ▶catástrofe natural, ▶riesgo natural); *f* force [f] majeure (Évènement naturel, imprévisible, hors du commun, agissant de l'extérieur, inévitable ; ▶catastrophe naturelle, ▶risque naturel, ▶risque naturel majeur) ; *g* höhere Gewalt [f] (Unvorhersehbares, ungewöhnliches, von außen einwirkendes, nicht abwendbares Ereignis, das Leben oder Gesundheit zahlreicher Menschen oder Tiere, die Umwelt, erhebliche Sachwerte oder die lebensnotwendige Versorgung der Bevölkerung in so ungewöhnlichem Maße gefährdet oder schädigt, das auch durch Anwendung äußerster, den Umständen nach möglicher und dem Betreffenden zumutbarer Sorgfalt nicht zu vermeiden war [z. B. schwerer Orkan, Flutkatastrophe]; ▶natürliche Gefahr, ▶schweres Naturereignis).

forecast [n]**, noise** *plan.* ▶consultant's report on noise.

2210 foredike [n] *eng. wat'man.* (Protective embankment designed to reduce the wearing effect of wave action, surf, current or ice drift on the main ▶levee [US] 2/flood bank [UK]); *s* dique [m] anterior (Construcción de contención delante del dique principal que sirve para reducir los ataques de las olas, las corrientes, etc. sobre éste y así protegerlo; ▶dique de contención); *syn.* dique [m] rompeolas; *f* digue [f] avancée (Ouvrage de protection ayant pour fonction de protéger la digue principale contre la houle, les vagues déferlantes, le courant ou la débâcle ; ▶digue) ; *g* Vordeich [m] (Deichschutzwerk, das die Beanspruchung des Hauptdeiches durch Seegang, Brandung, Strömung oder Eisgang wesentlich verringert; ▶Deich 2).

foredune [n] *geo. phyt.* ▶embryo dune.

foreign origin [adj]**, of** *biol. geo. pedol.* ▶allochthonous.

foreign tourism [n]**, promotion of** *econ. plan. recr.* ▶promotion of national tourism, #2.

2211 foreland [n] (1) *wat'man.* (**1.** Low, flat land created by waves and current at the foot of a coastal dike, which is regularly flooded and often used for grazing; sea embankments may be installed for land reclamation); *syn.* foreshore [n]; **2. foreland** [n] **of a river** (Space between the margin of a watercourse [▶shore line] and the embankment, which is often used for grazing and

may cause stream pollution by faecal matter]); **s *1* llanura [f] de marea (±)** (Acumulación de partículas finas transportadas por la pleamar hasta el interior de albuferas o el pie de diques costeros, donde permanecen, ya que la bajamar no tiene suficiente energía para romper la cohesión de aquéllas); **s *2* lecho [m] mayor (de río)** (DGA 1986; terreno adyacente y casi al mismo nivel que el cauce de un curso de agua, que sólo se inunda cuando el caudal excede la capacidad del cauce normal; WMO 1974; ►línea de orilla); *syn.* terreno [m] de fondo, cauce [m] mayor; ***f 1* avant-côte [f]** (Zone littorale régulièrement submergée par les eaux, comprise entre le bas de plage et la digue, correspond en fait à l'estran) ; ***f 2* lit [m] majeur (2)** (Zone non régularisée d'écoulement des eaux d'un fleuve ou d'une rivière à large fond alluvial, ou zone comprise entre la berge [►ligne de rive] et la digue, soumise à l'inondation hivernale, couverte de prairies fauchées ou pâturées ; cf. DG 1984) ; ***g* Vorland [n] (1)** (1. Bereich vor dem Deich einer Küste, der regelmäßig überflutet wird; Lahnungen können zur Landgewinnung angelegt werden. 2. Bewachsenes, der Weidenutzung dienendes Land zwischen ►Uferlinie und Damm/Deich eines Fließgewässers).

2212 foreland [n] (2) *geo.* (Stable line or area marginal to an orogenic belt in front of the line of a mountain chain; e.g. ►alpine foreland, often called Fore-Alp; cf. GGT 1979); *syn.* frontline [n]; **s piedemonte [m]** (Región baja situada en contacto con un macizo montañoso. En medios semiáridos el **p.** se corresponde frecuentemente con una llanura, modelada en glacis o en abanicos aluviales, que en pendiente suave establece la conexión entre la montaña y el fondo del valle; en medios fríos el **p.** puede coincidir con extensas llanuras de acumulación fluvioglaciar; DGA 1986; ►estribaciones de los Alpes); ***f* piémont [m]** (Région basse de plaine ou peu accidentée située au pied d'un relief montagneux, p. ex. les plaines préalpines ; ►préalpes) ; *syn.* avant-pays [m] ; ***g* Vorland [n] (2)** (Gebiet oder Landschaft vor einem Gebirge, z. B. ►Alpenvorland).

foreland [n] **of a river** *wat'man.* ►foreland (1), #2.

2213 foreman [n] *constr.* (Work gang leader of an individual trade on a construction site); *syn.* ganger [n] [also UK]; **s encargado [m] de obra** (Supervisor a cargo de un grupo de trabajadores en una obra); ***f* chef [m] d'équipe** (Personne qualifiée qui dirige un groupe de travailleurs) ; ***g* Vorarbeiter [m]** (Leiter der Arbeitergruppe eines Gewerkes auf einer Baustelle).

foreman [n]**, construction** *constr.* ►contractor's project representative.

foreshore [n] *wat'man.* ►foreland (1).

2214 forest [n] (1) *ecol. for. leg. phyt.* (**1.** *ecol.* Ecosystem characterized by an extensive tree cover. **2.** *ecol.* Plant community, composed principally of trees, shrubs and lianas, and covering a large area; *specific terms* ►natural forest, riparian forest [►riparian woodland], ►second-growth [forest] [US]/secondary forest [UK], ►virgin forest. **3.** *for.* ►production forest; a **wood** has a smaller extend than a forest; ►woods [US]); **s *1* monte [m]** (Extensión de terreno cubierta de formaciones vegetales integradas por asociaciones de plantas leñosas, semileñosas y herbáceas; si se ha originado sin la intervención del hombre, por dispersión natural de la semilla se le llama «bosque natural»; SILV 1979; *términos específicos* ►bosque de explotación forestal, ►bosque de galería, ►bosque natural, bosque pluvial —►pluviisilva—, ►bosque primario, ►bosque protector, ►bosque ripícola, ►bosque secundario); **s *2* bosque [m] (1)** (Denominación utilizada cuando en el monte predominan los árboles; SILV 1979, 133; ►bosque 3, ►monte alto); *syn.* silva [f]; ***f* forêt [f]** (1. Formation végétale ligneuse, de composition variée, occupant une surface importante et de couvert plus ou moins dense ; on distingue la ►forêt primaire, la ►forêt

vierge, la ►forêt naturelle, la ►forêt de substitution, la ►forêt [de production]. **2.** Formation végétale constituée d'arbres plantés ou spontanés, aux cimes jointives ou peu espacées, dominant souvent un sous-bois arbustif ou herbacé. **3.** Espace couvert par ce type de végétation. **4. F.**, espace boisé dont la superficie est d'au moins 4 ha avec une largeur moyenne en cime d'au moins 25 mètres ; des classes de superficie peuvent être distinguées : 4 à 25 ha, 25 à 100 ha, etc. ; définition IFN [Inventaire Forestier National], cf. www.limousin.ecologie.gouv.fr ; **5. CH.**, la définition de la forêt dans l'Inventaire Forestier National suisse repose, à savoir, le degré de recouvrement, la largeur de la forêt et la hauteur dominante des arbres ; une surface boisée est considérée comme forêt à partir du moment où sa largeur est supérieure à 50 mètres et que les couronnes des arbres couvrent plus de 20 % de la surface au sol ; ►bois) ; ***g* Wald [m] (1)** (1. W. bedeutete ursprünglich eine der Bewirtschaftung nicht unterworfene Fläche; in Orts- und Flurnamen haben sich im Laufe der Jahrhunderte für siedlungsferne [Gebirgs]wälder Synonyme wie z. B. Hart, Hard, Haard, Haardt und im Norddeutschen für einen großen Wald das Synonym **Horst** gebildet; ein ►Forst ist ein nach forstwirtschaftlichen Grundsätzen bewirtschafteter Wald. 2. Pflanzengesellschaft, die vorwiegend aus Bäumen besteht und eine größere Fläche bedeckt. Es werden je nach Einteilungskriterium ►Urwald, ►Naturwald, ►Sekundärwald und Wirtschaftswald oder nach den Besitzverhältnissen Staatswald, Körperschaftswald, Kirchenwald und Privatwald unterschieden. 3. *leg.* Jede mit Forstpflanzen [Waldbäume und Waldsträucher] bestockte Fläche, aber auch kahl geschlagene oder verlichtete Grundflächen, Waldwege, Waldblößen und Lichtungen, Waldwiesen, Waldäsungsplätze und Holzlagerplätze; ferner Waldparkplätze, Teiche, Weiher, Gräben und andere Gewässer von untergeordneter Bedeutung sowie Moore, Heiden und Ödflächen, soweit sie zur Sicherung der Funktion des angrenzenden Waldes erforderlich sind; cf. § 2 LWaldG-BW; im Amerikanischen wird ein großflächiger Wald grundsätzlich *forest* [nicht *wood*] genannt; ►Wald 2).

forest [n] (2) *recr.* ►amenity value of a forest, ►amenity benefits of a forest; *for. land'man.* ►avalanche protection forest; *phyt.* ►broad-leaved evergreen forest; *landsc.* ►climatic amelioration forest; *phyt.* ►climax forest; *for. phyt.* ►cloud forest; *phyt.* ►common grazing forest; *for. phyt.* ►coniferous forest; *agr. for.* ►coppice forest; *for. phyt.* ►deciduous forest; *phyt.* ►dune forest, ►dune woodland community; *for. landsc.* dust and aerosol protection forest; *envir.* ►emission-damaged forest; *phyt.* ►gallery forest, *phyt.* ►fringing forest, ►gallery forest; *for.* ►high forest; *for. phyt.* ►lowland forest; *for.* ►mixed forest; *phyt. geo.* ►mountain forest; *for. recr.* ►multiple-use forest; *for. land'man.* ►natural forest; *for. phyt.* ►needle-leaved deciduous forest; *landsc.* ►noise attenuation forest; *for.* ►old-growth forest [US]; *for. recr.* ►open meadow in a forest; *for.* ►pole-stage forest, ►production forest; *for. wat'man.* ►protected aquifer recharge forest; *conserv. leg. nat'res.* ►protected habitat forest; *for. landsc.* ►protective forest (1), ►protective forest (2); *for. phyt.* ►pure stand; *geo. phyt.* ►rain forest [US]/rain-forest [UK]; *phyt.* ►ravine forest; *for.* ►recreation forest, ►recruitment of a forest; *for. leg.* ►reserved forest; *for. hist.* ►royal hunting forest [UK]; *for.* ►sapling-stage forest; *for. landsc.* ►screen forest; *for.* ►selection forest; *adm. for.* ►state forest/national forest [US]/State Forest [UK]; *urb.* ►suburban forest; *geo. phyt.* ►swamp forest; *urb.* ►town forest [UK]; *phyt.* ►virgin forest.

forest [n]**, aerosol protection** *for. landsc.* ►dust and aerosol protection forest.

forest [n]**, aquifer recharge** *for. wat'man.* ►protected aquifer recharge forest.

forest [n]**, bottomland hardwood** *phyt.* ►riparian woodland.

forest [n] [US], **broadleaf** *for. phyt.* ►deciduous forest.

forest [n] [US], **cathedral** *for.* ►old-growth forest.

forest [n], **cultivated** *for.* ►production forest.

forest [n], **damaged** *envir.* ►emission-damaged forest.

forest [n] [ZA], **demarcated** *for. leg.* ►reserved forest.

forest [n], **elfin** *phyt.* ►krummholz community.

forest [n], **evergreen broadleaf** *phyt.* ►broad-leaved evergreen forest.

forest [n], **fringing** *phyt.* ►gallery forest.

forest [n] [US], **hardwood** *phyt.* ►riparian woodland.

forest [n], **hydric** *geo. phyt.* ►swamp forest.

forest [n] [US], **national** *adm. for.* ►state forest/national forest [US]/State Forest [UK].

forest [n] [UK], **ombrophilous** *geo. phyt.* ►rainforest [US]/rain-forest [UK].

forest [n] [NZ], **permanent state** *adm. for.* ►reserved forest.

forest [n], **prim(ev)al** *phyt.* ►virgin forest.

forest [n], **primordial** *phyt.* ►virgin forest.

forest [n], **pristine** *phyt.* ►virgin forest.

forest [n] [US], **protective marine** *conserv. land'man.* ►protective coastal woodland.

forest [n], **pure** *for. phyt.* ►pure stand.

forest [n], **riparian** *phyt.* ►riparian woodland.

forest [n] [UK], **secondary** *phyt. for.* ►second-growth (forest) [US].

forest [n] [US], **sprout** *for.* ►coppice forest.

forest [n], **tropical ombrophilous cloud** *for. phyt.* ►cloud forest.

forest [n], **urban** *urb.* ►town forest [UK].

forest [n] **absorbing air pollution** *for. landsc.* ►protective forest absorbing air pollution.

forest [n] **against air pollution** *for. landsc.* ►protective forest against air pollution.

forest area [n] *for. leg.* ►forest terrain.

forest biocenose [n] *phyt. ecol.* ►forest biocoenosis.

forest biocoenose [n] *phyt. ecol.* ►forest biocoenosis.

2215 forest biocoenosis [n] *phyt. ecol.* (Life community of plants and animals in a forest; ►biocenosis); *syn. o.v.* forest bioc(o)enose [n], forest biocenosis [n]; *s* **biocenosis** [f] **forestal** (Todas aquellas partes del bosque que sean homogéneas sobre un área específica con estructura y composición típicas forman una **b. f.** Ésta viene determinada por la cobertura vegetal del bosque [fitocenosis], la fauna [zoocenosis], los microorganismos [microbiocenosis], la superficie de rocas, las condiciones del suelo [edátopo], las condiciones atmosféricas, microclimáticas e hidrológicas [climátopo], y la acción recíproca entre ellos, el intercambio de sustancias y energía entre sus componentes y otras influencias de la naturaleza; SILV 1979, 113; ►biocenosis); *f* **biocénose** [f] **forestière** (Communauté biotique des êtres vivants végétaux et animaux ainsi que des microorganismes présents dans une forêt ; ►biocénose) ; *syn.* biocénose [f] sylvatique, biocénose [f] forêt ; *g* **Waldbiozönose** [f] (Lebensgemeinschaft von Pflanzen und Tieren in einem Wald; ►Biozönose).

forest biotope [n] *biol. conserv.* ►forest habitat.

forest border trees [npl] *for.* ►edge trees.

2216 forest burn [n] *for.* (Forest area destroyed by fire); *syn.* burned forest area [n]; *s* **quemado** [m] (DFM 1975; superficie de bosque en la cual ha habido un incendio recientemente); *f 1* **forêt** [f] **incendiée** (Espace forestier détruit par le feu) ; *f 2* **brûlis** [m] (Surface parcourue par le feu et non encore régénérée dans le cadre de l'agriculture itinérante ; *terme spécifique* brûlis à blanc [destruction totale du couvert végétal] ; DFM 1975) ; *g* **abgebrannte Waldfläche** [f] *syn.* Waldbrandfläche [f].

forest canopy [n] *for.* ►tree canopy, ►opening up of forest canopy.

2217 forest cemetery [n] *adm. landsc. urb.* (►Park-like cemetery whereby the location of coffin and urn graves within a woodland setting is designed to provide solace and lessen the fear of death); *s* **cementerio** [m] **boscoso** (Tipo de cementerio ideado posteriormente al ►cementerio paisajístico y en el cual se plantan árboles y arbustos imitando a un bosque hueco); *f 1* **cimetière** [m] **en ambiance forestière** (±) (Conception d'aménagement des cimetières du début du siècle dans les pays anglo-saxons consécutive à la période des ►cimetières paysagers consistant dans une composition à dominante paysagère dans laquelle les tombes sont réparties sous une futaie claire ou accompagnées de plantations en forme de bosquets afin de dédramatiser la mort ; CH., cimetière forestier ; cimetière aménagé sur le modèle d'un cimetière-parc et qui rappelle un bois) ; *syn.* cimetière [m] paysager forestier (±) ; *f 2* **cimetière** [m] **vert** (U.S., UK., formule environnementale de l'enterrement, à la mode depuis 1995, ayant lieu dans un cimetière naturel, dans lequel les défunts ne reposent pas sous des pierres tombales, les cercueils étant entièrement biodégradables et peu enfoncés dans la terre (45 à 60 cm) afin de faciliter et d'accélérer le processus de décomposition et ou il est interdit d'enterrer des corps ayant usé de fluides d'embaumement. Ces cimetières sont souvent aménagés en forêt ou en bordure de forêt, l'emplacement de la sépulture étant marqué par un arbre d'une espèce indigène [arbre de mémoire] ; *syn.* cimetière [m] naturel, cimetière [m] écologique, écocimetière [m] ; *g* **Waldfriedhof** [m] (In der Entwicklung der Freiraumplanung dem ►Parkfriedhof folgende Friedhofsvariante, die das Ziel hatte, mittels einer versöhnlicheren Grabfeldgestaltung durch einen lichten Wald und einer hainartigen Bepflanzung dem „Tode seinen Schrecken" zu nehmen. In D. wurde 1905 der erste **W.** in München angelegt; cf. RICH 1981; inspiriert von dem Münchener **W.** wurde 1914 in Stockholm ein internationaler Architektenwettbewerb über 80 ha ausgeschrieben, den die Architekten GUNNAR ASPLUND und SIGURD LEWERENTZ gewannen und zwischen 1920 und 1940 einen in der Architekturgeschichte und Friedhofskultur einmaligen Friedhof, den **Skogskyrkogården** [Waldfriedhof] in Enskede im Süden Stockholms schufen. Er wurde 1994 als Kulturgut auf Grund seines außergewöhnlichen universellen Wertes in die Welterbeliste der UNESCO aufgenommen; cf. ZAD 2002, pp. 53-64).

2218 forest clearing [n] *for.* (Open space of natural or man-made origin in a forest or brushwood stand); *s* **claro** [m] **(del bosque)** (Pequeña área desarbolada dentro de un bosque o matorral. Puede ser de origen natural o artificial); *f 1* **clairière** [f] (Espace de surface naturellement ou artificiellement ouvert dans la forêt ou la brousse ; DFM 1975) ; *f 2* **trouée** [f] **forestière** (Une petite clairière créée temporairement dans un but sylvicole [régénération naturelle et successions végétales], ou provoquée par accident météorologique ; DFM 1975, 58) ; *g* **Waldlichtung** [f] (Auf natürliche oder künstliche Weise entstandene freie Stelle im Wald); *syn.* Lichtung [f].

forest clearing community [n] [UK] *phyt.* ►tree-fall gap community [US].

forest community [n] *phyt.* ►natural forest community.

forest constitution [n] *for.* ►forest structure.

2219 forest corridor [n] *for.* (Stretch of forest clearing used for the removal of logs and to subdivide the forest into units; ►vista, ►fire break, ►fresh air corridor); *s* **pista** [f] **forestal** (En bosque camino utilizado para transportar los troncos de madera talados o para subdividirlo en unidades; ►corredor visual, ►cortafuego[s], ►corredor de aire fresco); *f* **couloir** [m] **forestier** (Terme générique désignant une bande déboisée de terrain entre deux coupes pour le transport des produits forestiers ; *termes spécifiques* layon [passage étroit], laie [large couloir] ; ►couloir de vue, ►cône de visibilité, ►pare-feu, ►couloir d'air frais) ; *g* **Waldschneise** [f] (Zur Forsteinteilung und Holzabfuhr angelegter, waldfreier Streifen; ►Schneise 1, ►Sichtachse, ►Frischluftschneise); *syn.* Schneise [f] (2).

2220 forest cover [n] *for. geo. phyt.* (Whole forest vegetation in a large area predominantly characterized by a dense ►tree cover; *generic term* ►plant cover; ►stand of timber, ►standing forest crop); *syn.* woodland cover [n]; *s* **cobertura** [f] **forestal** (Conjunto de toda la vegetación de un área en la que predominan los árboles; *término genérico* ►cobertura vegetal; ►cobertura arbórea, ►población forestal, ►rodal); *syn.* cobertura [f] vegetal boscosa [SILV 1979, 30], masa [f] forestal, manto [m] forestal; *f* **couverture** [f] **forestière** (►Couverture végétale désignant la surface couverte par la projection verticale des houppiers de l'ensemble des arbres d'un ►peuplement arborescent ou d'une essence donnée ; TSF 1985 ; ►peuplement forestier 1, peuplement forestier 2) ; *syn.* couverture [f] arborescente, couvert [m] arboré, couvert [m] forestier, massif [m] forestier ; *g* **Waldbedeckung** [f] (UB zu ►Bewuchs; *sensu lato* gesamte Vegetation einer großen Fläche, die vorwiegend durch einen dichten Baumbestand gekennzeichnet ist ; ►Baumbestand 1, ►Baumbestand 2, ►Waldbestand 2); *syn.* Bewaldung [f] (1), Waldbestand [m] (1).

forest crop [n], **standing** *for.* ►standing forest crop.

2221 forest decline [n] *envir. for.* (Premature dieback and aging symptoms exhibited in coniferous and broad-leaved woodlands, probably caused by, amongst other things, high levels of atmospheric pollution; ►dieback of trees [US]/die-back of trees [UK], ►Dutch elm disease, ►forest deterioration, ►top-kill [US]/top drying [UK]); *syn.* waldsterben [n] [also UK]; *s* **muerte** [f] **de bosques** (Envejecimiento y muerte prematuras de los bosques de coníferas y frondosas y de ejemplares individuales en parques urbanos que ocurre en especial en el Europa Central, causada por la contaminación atmosférica y por consecuencia del suelo, sobre todo de óxidos sulfúricos y de nitrógeno y ozono que producen, entre otros efectos, la acidificación del suelo; ►muerte de árboles, ►grafiosis del olmo, ►deterioro forestal, ►puntiseco); *syn.* desaparición [f] de bosques; *f* **dépérissement** [m] **des forêts** (Vieillissement précoce et mort prématurée de groupements de résineux ou de feuillus provoqués principalement par la pollution atmosphérique et ses conséquences, ►dépérissement des arbres, ►graphiose de l'orme, ►maladie des forêts, ►descente de cime) ; *syn.* dépérissement [m] forestier, mort [f] des forêts ; *g* **Waldsterben** [n, o. Pl.] (Frühzeitige Vergreisung und vorzeitiges Absterben von großflächigen Nadel- und Laubbaumbeständen, das hauptsächlich durch hohe Luftverschmutzung infolge des Ausstoßes von Schwefeldioxid bei der Verbrennung fossiler Brennstoffe entsteht, aber auch durch in Haushalten und Industrie erzeugte Stickoxidverbindungen, Ammoniak und Schwermetalle mit allen ihren Folgewirkungen. Zur Klärung dieses vielschichtigen Phänomens werden auch nicht immissionsbedingte Schadfaktoren wie Stress durch extreme Witterungs- und Klimaereignisse, mangelnde forstliche Pflege, Schädlings- und Pilzbefall sowie die globale Klimaveränderung herangezogen; cf. ALG 1997; In den 1970er-Jahren bemerkten Umweltschützer eine großflächige Umweltveränderung durch massenhaftes Absterben von Tannen und Fichten, das anfangs als

►Baumsterben und später als **W.** bezeichnet wurde. Mit zunehmender Industrialisierung und wachsendem Autoverkehr breitete sich das **W.** über ganz Europa und Nordamerika aus, begünstigt durch den Bau hoher Schornsteine, so dass die o. g. Schadstoffe weitflächig verbreitet wurden. In D. wurde 1984 das **W.** erstmalig in Form von Schad-/Vitalitätsstufen erfasst und seitdem werden jährlich in den **Waldzustandsberichten** der Bundesländer und im Bericht des Bundeslandwirtschaftsministeriums die Ergebnisse der Auswertungen von Beobachtungsnetzen, mit denen die Landesforstverwaltungen den Zustand der Wälder regelmäßig überwachen, vorgestellt; ►Ulmensterben, ►Walderkrankung, ►Wipfeldürre).

2222 forest deterioration [n] *for. envir.* (Prolonged decline of the normal physiological state and vital functions of a forest caused by pathogenic agents, ►injurious organisms, ►air pollution 1, ►lowering of groundwater table, etc.; ►forest decline); *s* **deterioro** [m] **forestal** (Alteración de los procesos fisiológicos normales y de las funciones vitales de una población forestal o partes de ella debida a ►organismos nocivos, ►contaminación atmosférica, ►reducción del nivel freático, etc.; ►muerte de bosques); *f* **maladie** [f] **des forêts** (Altération du fonctionnement normal des processus physiologiques causée p. ex. par des ►organismes nuisibles, la ►pollution atomsphérique, le ►rabattement de la nappe phréatique, etc. ; ►dépérissement des forêts) ; *g* **Walderkrankung** [f] (Durch ►Schadorganismen, ►Luftverunreinigung, ►Grundwasserabsenkung, Häufigkeit trockener Sommer etc. ausgelöste Veränderung normaler physiologischer Vorgänge im Vegetationsbestand des Waldes; ►Waldsterben).

2223 forest development [n] *for. land'man.* (Process of ►afforestation, ►natural regeneration (1), ►natural regeneration (2), ►reforestation); *s* **desarrollo** [m] **forestal** (*Proceso* ►dasocracia, ►forestación, ►reforestación, ►regeneración natural); *f* **développement** [m] **du couvert forestier** (Processus ; ►reboisement, ►reforestation, ►régénération naturelle) ; *g* **Bewaldung** [f] (2) (*Vorgang* ►Aufforstung, ►Wiederaufforstung; der OB **Neubewaldung** beinhaltet in Österreich Aufforstung, ►Naturverjüngung; cf. § 4 ForstG; sich bewalden [vb]).

2224 forest ecosystem [n] *phyt. ecol.* (►ecosystem, ►forest biocenosis); *s* **ecosistema** [m] **forestal** (►Ecosistema de un bosque compuesto por la ►biocenosis forestal y su medio ambiente abiótico); *f* **écosystème** [m] **forestier** (►Écosystème d'une forêt composé de la communauté biotique et de son environnement abiotique ; ►biocénose forestière) ; *syn.* écosystème [m] forêt ; *g* **Waldökosystem** [n] (►Ökosystem eines Waldes; ►Waldbiozönose).

2225 forest edge [n] *for.* (Boundary between a forest and other landscape elements; e.g. meadow, field, water body or built-up area. The **f. e.** includes the ►forest mantle and ►forest seam; ►ecotone 1, ►edge trees); *syn.* woodland edge [n]; *s* **lindero** [m] **del bosque** (Zona de transición entre el bosque y otros componentes del paisaje como prados, campos de cultivo, superficies de agua o zonas urbanas; el **l. del b.** incluye el ►abrigo del bosque y la ►orla herbácea del bosque; ►árboles marginales, ►ecotono); *syn.* linde [f] del bosque; *f* **lisière** [f] **forestière** (Zone de contact franche ou en progression entre la forêt et les zones non boisées attenantes telles que les champs, un pré, un plan d'eau, une route, etc. ; la dénomination pré-bois est parfois utilisée lorsque l'aspect de la lisière est caractérisé par une futaie dispersée, un taillis discontinu, des buissons bas irréguliers situés au pied des arbres et une strate herbacée dense et haute ; ►arbres de lisière, ►écotone, ►manteau forestier, ►ourlet préforestier herbacé) ; *syn.* bordure [f] de la forêt [CDN] ; *g* **Waldrand** [m] (Grenzbereich zwischen Wald und unbewaldeten Landschaftsteilen wie z. B. Wiese, Acker, Wasser-

fläche oder Baugebiet. Der **W.** schließt den Waldmantel und ►Waldsaum mit ein; ►Ökoton, ►Trauf eines Waldes, ►Waldmantel 2).

forested land [n] *for.* ►forest land.

forest effect [n] *for. nat'res. recr.* ►forest function.

forest enumeration [n] [UK] *for.* ►forest stand inventory [US].

2226 forest floor [n] *for. pedol.* (**1.** Upper soil layer of a forest supporting vegetation. **2.** Layer of non-decomposed and decomposing organic matter lying on top of the mineral forest soil); *syn.* forest soil [n]; *s* **suelo** [m] **forestal** (Capa superior del suelo con capa de litter en una población forestal); *f* **sol** [m] **forestier** (Horizons organiques et minéraux des sols à couvert forestier, p. ex. sol brun forestier) ; *g* **Waldboden** [m] **(1)** (Obere Bodenschicht mit der Streuauflage in einem Waldbestand).

2227 forest floor cover [n] *phyt.* (Herbaceous vegetation and ►litter layer covering the mineral forest soil; ►litter 1); *syn.* forest soil cover [n]; *s* **cubierta** [f] **del suelo del bosque** (Capa de vegetación herbácea y de ►litter que cubre el suelo del bosque; ►capa de litter); *f* **strate** [f] **forestière au sol** (Couverture d'un sol forestier constituée par la strate herbacée et les ►débris végétaux ; ►litière) ; *g* **Waldbodendecke** [f] (Bedeckung des Waldbodens durch Vegetation und ►Streu; ►Streuschicht).

2228 forest floor litter [n] *for. pedol.* (Uppermost layer, the L layer, of organic debris on a forest floor; ►leaf litter, ►litter 1, ►litter layer, ►needle straw [US]/needle litter [US]); *syn.* forest litter [n]; *s* **litter** [f] **del bosque** (Capa superior orgánica del suelo forestal; ►capa de barrujo, ►capa de litter, ►hojarasca, ►litter); *f* **litière** [f] **forestière** (Couche organique de ►débris végétaux de toute nature accumulés sur un sol forestier ; ►couverture morte, ►litière d'aiguilles, ►litière feuillue) ; *syn.* litière [f] de forêt, *par extension* humus [m] forestier; *g* **Waldbodenstreu** [f] (Auf dem Waldboden aufliegender, noch nicht oder kaum zersetzter Bestandesabfall, der die organische Schicht bildet; ►Laubstreu, ►Nadelstreu, ►Streu, ►Streuschicht); *syn.* Streuauflage [f] im Wald.

forest [n] **for soil conservation purposes** *conserv. land'man.* ►protective forest for soil conservation purposes.

forest [n] **for water resources** *nat'res. wat'man.* ►protective forest for water resources.

2229 forest fragmentation [n] *for.* (Process of reducing extensive and continuous forests to a mosaic of smaller patches; cf. NRM 1996); *s* **fragmentación** [f] **espacial del bosque** (Reducción de la extensión de los bosques por tala o cambio de usos hasta convertirse en un mosaico de pequeñas manchas forestales); *f* **morcellement** [m] **de la forêt** (Division de grands massifs forestiers en une mosaïque de petites surfaces boisées suite à des coupes ou brûlis) ; *syn.* fragmentation [f] de la forêt ; *g* **Waldzerstückelung** [f] (Reduzierung ausgedehnter Waldbestände in ein Mosaik von Restwaldflächen durch Rodung und Nutzungsumwandlungen); *syn.* Atomisierung [f] der Waldflächen, *o. V.* Waldzerstücklung [f].

2230 forest function [n] *for. nat'res. recr.* (Role of forests which contributes to the well-being of humans—recreation, climatic protection, groundwater recharge, supply of wood, etc.—as well as the ecological functions resulting from the forest ecosystem and its effect on the surroundings); *syn.* forest effect [n]; *s* **función** [f] **del bosque** (Diferentes destinos del bosque en sentido ecológico y para el bienestar de la sociedad como función recreativa, función protectora del clima, función protectora contra la erosión, función productora de madera, etc.; ►inventario de funciones del bosque); *f* **fonction** [f] **de la forêt** (Faculté de la forêt à satisfaire les aspirations de bien-être de l'homme —

fonction récréative et sportive, fonction climatique, fonction anti-érosive, fonction de production, etc. — ainsi que les actions bio-écologiques propres à l'écosystème forestier sur l'environnement en général ; ►inventaire des fonctions de la forêt) ; *g* **Waldfunktion** [f] (Leistung des Waldes für das Wohlergehen des Menschen — Erholungsfunktion, Klimaschutzfunktion, Erosions-schutzfunktion, Naturschutzfunktion, Holzlieferung etc. — sowie für landschafts- und bioökologische Wirkungen, die sich aus den Vorgängen im Waldökosystem für Standort und Umgebung ergeben; cf. WT 1980; **W.en** können über die Ausweisung von Schutzgebieten geschützt werden; ►Waldfunktionskartierung).

forest functions [npl] *for. nat'res. envir. recr.* ►mapping of forest functions.

forest grazing ground [n] *agr. for.* ►forest pasture.

2231 forest habitat [n] *biol. conserv.* (►habitat 2); *syn.* forest biotope [n]; *s* **biótopo** [m] **forestal** (►biótopo); *syn.* habitat [m] forestal; *f* **biotope** [m] **forestier** (►biotope) ; *syn.* biotope [m] sylvatique ; *g* **Waldbiotop** [m, *auch* n] (►Biotop).

forest habitat [n]**, protected** *conserv. leg. nat'res.* ►protected habitat forest.

forest heritage [n] *conserv. land'man. nat'res.* ►forest patrimony.

2232 forest land [n] *for.* (**1.** Area covered by forest trees or, if totally lacking, non-stocked forest land [US]/unstocked forest land [UK]), bearing evidence of former forest and not now in other use; cf. SAF 1983; ►forest product area, ►shifting cultivation); **2. forested land** [n] (Area afforested by human activity); *s* **área** [f] **forestal** (En las estadísticas de superficies, área cubierta predominantemente por bosque de explotación forestal. Se incluyen también superficies taladas si se vuelven a repoblar o se repueblan por sucesión [p. ej. sucesión secundaria tras ►agricultura itinerante], también sabanas con un grado de cobertura forestal de 0,05; ►suelo forestal); *syn.* área [f] de bosque, área [f] de montes; *f* **terrain** [m] **forestier** (Terrain portant une forêt en croissance ; s'il n'y en existe pas [terrain forestier nu] : terrain dont l'aspect montre à l'évidence qu'il en a porté et n'est pas actuellement mis en valeur d'une autre manière ; DFM 1975 ; ►agriculture itinérante [sur brûlis], ►surface sylvicole) ; *g* **forstliche Landfläche** [f] (In der Flächenstatistik alle Flächen, die vom Baumbewuchs beherrscht werden, unabhängig von der forstwirtschaftlichen Nutzung. Dazu gehören auch Kahlschläge, wenn sie wieder bestockt werden oder sich durch Sukzession wieder bestocken [z. B. sekundäre Sukzession nach ►Wanderfeldbau]; ferner Savannen mit einem Beschirmungsgrad von mindestens 0,05; cf. WT 1980; ►forstwirtschaftliche Nutzfläche).

forest land use *agr. for. hort.* ►intensity of agricultural and forest land use.

forest litter [n] *for. pedol.* ►forest floor litter.

2233 forest management [n] **(1)** *for.* (Planning and practical application of scientific, economic and social principles to the administration and working of a forest area for specified objectives; or, more specifically, that branch of forestry concerned *sensu lato* with the over-all administrative, economic, legal and social aspects, and *sensu stricto* with the essentially scientific and technical aspects, especially ►forestry 1, protection and forest regulation; cf. SAF 1983; ►forest science); *s* **dasocracia** [f] (Arte y ciencia del gobierno de los montes; DB 1985; ►dasonomía, ►silvicultura artificial); *syn.* gestión [f] de montes, gestión [f] forestal; *f* **aménagement** [m] **forestier** (**1.** Branche de la sylviculture qui s'intéresse aux sciences et techniques forestières dans leurs rapports avec les principes d'ordre administratif, législatif, économique et social de la gestion des forêts. **2.** Application pratique des connaissances scientifiques et techniques à la

gestion d'une forêt, à la conduite des exploitations et aux travaux à y exécuter, en vue d'objectifs à attendre ; DFM 1975 ; ▶science forestière, ▶sylviculture) ; *syn.* plan [m] d'aménagement forestier, procès-verbal [m] d'aménagement forestier ; *g* **Forsteinrichtung [f]** (Zweig der ▶Forstwissenschaft und der forstwirtschaftlichen Tätigkeit, der die mittel- und langfristige forstliche Planung und periodische Inventur des Waldbestandes betreibt; ▶Forstwirtschaft).

2234 forest management [n] (2) *for.* (Practical administration and working of a forest area, usually by management units and districts, for economic timber production and ▶provisions for recreation as well as for wildlife preservation; ▶clear-cutting system, ▶economic forest management system, ▶forest management 1, ▶selection system); *syn.* silvicultural management [n]; *s* **gestión [f] de empresa forestal** (Empresa con dos funciones principales: la producción maderera y el uso recreativo del bosque; ▶dasocracia, ▶entresaca, ▶explotación a matarrasa, ▶explotación forestal, ▶previsión de posibilidades de recreo); *f* **gestion [f] forestière** (Mesures d'exploitation forestière pouvant soit favoriser une ▶politique de prévoyance concernant les loisirs et le tourisme, l'équilibre écologique, soit poursuivre principalement des buts économiques ; ▶aménagement forestier, ▶exploitation forestière par coupes rases, ▶exploitation sylvicole de production, ▶jardinage) ; *g* **waldbauliche Bewirtschaftung [f]** (Forstbetrieb, der durch Wirtschaftsziele oder solche der ▶Erholungsvorsorge gekennzeichnet ist; ▶forstwirtschaftlicher Betrieb [= Forstbetrieb], ▶Forsteinrichtung, ▶Kahlschlagswirtschaft, ▶Plenterwaldbetrieb).

forest management system [n] *for.* ▶economic forest management system.

2235 forest mantle [n] *phyt.* (Deciduous shrub growth on the forest edge which together with the ▶forest seam creates a transitional zone between forest and agricultural areas or ▶nutrient-poor grassland; e.g. **in Central Europe**: *Prunetalia spinosae, Salicetum triandro-viminalis*; **in U.S.**, species of the same genera, such as cherry *[Prunus]* and willow *[Salix]*; ▶forest edge, ▶woodland edge scrub community); *s* **abrigo [m] del bosque** (Arbustos de frondosas, en Europa Central p. ej. *Prunetalia spinosae, Salicetum triandro-viminales*, que —junto con la ▶orla herbácea del bosque— forman el ecotono entre el bosque y las áreas colindantes de uso agrícola o de ▶prados oligótrofos; ▶comunidad arbustiva del lindero del bosque, ▶lindero del bosque); *f* **manteau [m] forestier** (Végétation ligneuse linéaire qui avec l'▶ourlet préforestier herbacé forme une zone de transition en ▶lisière forestière située entre la forêt, les champs ou friches, les prairies ou ▶pelouses oligotrophes ; ▶groupement de manteau préforestier buissonnant) ; *g* **Waldmantel [m] (1)** (Schmaler Streifen aus Laubholzsträuchern am ▶Waldrand, z. B. in Mitteleuropa die Schlehengesellschaft [Ordnung der *Prunetalia spinosae*], die mit dem ▶Waldsaum den Übergang vom Waldbestand zu den landwirtschaftlichen Flächen oder ▶Magerrasen bildet; eine weiter **W.**-Gesellschaft ist die Assoziation des Uferweidengebüsches, das *Salicetum triandro-viminalis* in Weichholzauenwäldern; ▶Mantelgesellschaft).

forest-meadow boundary [n] [UK] *ecol.* ▶field-woodland edge [US].

forest nature reserve [n] [UK] *conserv. leg. nat'res.* ▶forest research natural area [US].

2236 forest pasture [n] *agr. for.* (Forest for livestock grazing, as a secondary use, which is detrimental to the development of the forest. Due to compaction of the soil and browsing of young trees, buds and branches; it can lead to the destruction of the forest); *syn.* wood pasture [n], forest grazing ground [n]; *s* **pastizal [m] forestal** (Aprovechamiento adicional del bosque

que es negativo para el desarrollo natural del mismo ya que produce compactación del suelo y pérdida de la capacidad de regeneración natural, porque el ganado muerde las yemas y plántulas jóvenes); *f* **pâturage [m] en forêt** (Pratique ancienne du pacage des animaux dans la forêt encore utilisée en Méditerranée entraînant une dégradation progressive de la végétation arborée des peuplements climaciques sous le couvert de laquelle existe un tapis herbacé productif, celle-ci étant causée par le tassement du sol, le broutement des bourgeons, des jeunes pousses et du feuillage, interdisant toute régénération naturelle et pouvant entraîner à terme la destruction de la forêt ; bois défrichés pour la pratique du pâturage dans lequel il ne reste plus que quelques arbres isolés irrégulièrement disposés ; MPE 1984, 186 ; VCA 1985, 116) ; *syn.* pacage [m] en forêt, pâturage [m] boisé, pâture [f] en forêt ; *g* **Waldweide [f]** (Forstliche Nebennutzung als Viehweide, die für die Entwicklung und den Bestand des Waldes abträglich ist und durch Verdichtung des Bodens, Verbiss des Aufwuchses, der Knospen und Triebe bis zur Zerstörung des Waldes führen kann); *syn.* Wytweide [f] [CH].

forest path [n] [UK] *for.* ▶timber road [US].

2237 forest patrimony [n] *conserv. land'man. nat'res.* (Sum of woods and forests as a natural heritage of a specified area/region); *syn.* forest heritage [n]; *s* **patrimonio [m] forestal** (Conjunto de bosques naturales o cultivados que cumplen funciones ecológicas, climáticas, de protección, económicas, cinegéticas, recreativas o turísticas); *f* **patrimoine [m] forestier** (Ensemble des peuplements forestiers cultivés et naturels participant à diverses fonctions de production, cynégétique, climatique, protectrice, récréative, touristique, etc.) ; *g* **Gesamtheit [f] der Waldressourcen** (als natürliches Erbe).

2238 forest play area [n] *recr.* (Natural play area in a forest); *s* **terreno [m] forestal para juegos infantiles** (Parte acotada de un bosque natural utilizable para juegos relacionados con los árboles); *f* **aire [f] de jeux en forêt** (Equipement récréatif naturel léger en forêt) ; *g* **Waldspielplatz [m]** (Naturspielgelände im Wald).

2239 forest products area [n] [US] *for.* (Area permanently concerned with the production of timber, including the area where production is reduced as a preservation measure or for recreation; ▶forest land); *s* **suelo [m] forestal** (Todas las superficies que sirven para la producción de madera o que cumplen funciones protectoras o de recreación; ▶área forestal); *f* **surface [f] sylvicole** (Toute surface utilisée pour la sylviculture, y compris les forêts de protection et récréatives comprenant l'ensemble des bois [superficie] et le fonds ; ▶terrain forestier) ; *g* **forstwirtschaftliche Nutzfläche [f]** (Sämtliche Flächen, die der dauernden Erzeugung von Holz dienen, incl. der Flächen mit verminderter Produktion [z. B. Schutz- und Erholungswälder]; ▶forstliche Landfläche).

forest protection [n] *landsc. conserv.* ▶conservation of forests.

2240 forest recreation [n] *recr.* *s* **recreo [m] en el bosque** *syn.* recreación [f] en el bosque, esparcimiento [m] en el bosque; *f* **loisirs [mpl] et récréation en forêt** *syn.* activités [fpl] récréatives en forêt ; *g* **Erholung [f, o. Pl.] im Wald.**

2241 forest research natural area [n] [US] *conserv. leg. nat'res.* (Area established, by public or private agency, specifically to preserve a representative sample of an ecological forest community, primarily for scientific or educational purposes. In U.S., commercial exploitation, e.g. of forest or range, is ordinarily not allowed and general public use is discouraged; SAF 1983; ▶protective forest 1, ▶protective forest 2, ▶timber research area, ▶wilderness area); *syn.* natural forest reserve [n] [UK], forest nature reserve [n] [UK] (COU 1978, 197), scientific

purpose area [n] [CDN]; *s* **reserva** [f] **forestal integral** (En EE.UU. área de bosque público o privado protegido en la que generalmente no se permite la explotación comercial y se restringe el uso recreativo. En algunos estados federados de Alemania categoría de protección legal para preservar, desarrollar e investigar las biocenosis forestales en estado poco alterado, por lo que su gestión se rige exclusivamente por criterios ecológicos y no económicos; ▶bosque protector 1, ▶bosque protector 2, ▶reserva forestal experimental, ▶reserva natural 2); *f 1* **réserve** [f] **biologique domaniale** (Milieux forestiers riches, rares ou fragiles faisant partie du domaine forestier de l'État géré par l'Office national des forêts avec comme objectifs **1.** une gestion particulièrement orientée vers la sauvegarde de la faune, de la flore et de toute autre ressource naturelle, **2.** des programmes d'observation scientifiques, **3.** des actions d'éducation du public ; le classement permet la création d'une réserve intégrale [pénétration du public interdite, opérations sylvicoles exclues], une réserve dirigée [ouverture contrôlée du site, intervention sylvicoles limitées dans un but de protection], une zone tampon périphérique ; GPE 1998) ; *f 2* **réserve** [f] **biologique forestière** (Objectifs et effet du classement identiques à la ▶réserve biologique domaniale et s'appliquant aux forêts non domaniales appartenant aux communes, aux départements, aux régions et aux établissements publics ; F., les massifs forestiers, grands ensembles naturels riches et peu modifiés sont très souvent classés en « ▶zone naturelle d'intérêt écologique, faunistique et floristique [Z.N.I.E.F.F.] de type II » ; circ. n° 91-71 du 14 mai 1991 ; ▶espace boisé protégé, ▶forêt de protection, ▶parcelle permanente d'observation, ▶paysages classés à caractère pittoresque) ; *g* **Naturwaldreservat** [n] (In einigen Bundesländern [in BY, HS, MV, NS] rechtlich geschützte Waldfläche, die der Erhaltung, Entwicklung und Erforschung naturnaher Waldökosysteme dient, weshalb wirtschaftsbestimmte Eingriffe ausgeschlossen sind. Es ist einmal eine sich selbst überlassene Waldfläche — ein Urwald von morgen —, zum anderen handelt es sich um einen **Schonwald**, in dem eine bestimmte landschaftstypische Waldgesellschaft resp. ein bestimmter Bestandsaufbau oder der Lebensraum für gefährdete Tier- und Pflanzenarten durch zielgerichtete Pflege zu erhalten oder zu entwickeln ist; **N.e** dienen auch zur Erhaltung und Förderung von Lebensraumtypen und des Vorkommens der Lebensstätten von Arten nach der Fauna-Flora-Habitat-Richtlinie. Ein weiterer Schutzzweck ist in ausgewählten Gebieten der Beitrag zur Erhaltung der biologischen Vielfalt in Europa im Rahmen der Schaffung eines Europäischen Netzes Natura 2000; ▶Bannwald 1, ▶Schutzwald, ▶Wildnis[schutz]gebiet, ▶Weiserfläche); *syn. in BW* Waldschutzgebiet [n], Bannwald [m] (1), Schonwald [m]; *syn. in NW, RP und SL* Naturwaldzelle [f], *syn. z. T. in HS* Naturschutzgebiet [n] (1).

forest reserve [n] *for. leg.* ▶reserved forest.

forest ride [n] [UK] *for.* ▶forest track [US].

2242 forestry [n] (1) *for.* (**1.** Science and professional practice of managing and using for human benefit the natural resources that occur on or in association with ▶forest land; cf. SAF 1983; ▶forest science, ▶forestry techniques, ▶natural silviculture, ▶silviculture 2, ▶tending of a forest land); *syn.* silviculture [n] (1); **2. economic forestry** [n] (Husbandry of tree crops from the producer's point of view or, from the economist's, the profitable exploitation of the resources intrinsic to ▶forest land; SAF 1983); *syn.* commercial forestry [n] [also US]; *s* **silvicultura** [f] **artificial** (Rama de la silvicultura que se dedica al cultivo de bosques mayormente monoespecíficos y de edad uniforme con el único fin de aprovechar su madera u otras partes de los árboles. En realidad se trata en este caso de sistemas agrícolas que imponen limitaciones biocenóticas sobre el medio ambiente que dependen de las especies dominantes; *opp.* ▶silvicultura natural;

DINA 1987; ▶área forestal, ▶dasonomía, ▶dasotomía, ▶silvicultura, ▶tratamiento silvicultural [de mantenimiento, mejora y protección de rodales]); *syn.* selvicultura [f] artificial; *f 1* **sylviculture** [f] **(1)** (Ensemble des sciences, des arts et des activités, qui ont pour objet la conservation, l'entretien, la régénération, le reboisement, l'aménagement et la gestion des forêts et des domaines forestiers ainsi que leur création, en vue de la consommation et du renouvellement de leurs ressources matérielles et immatérielles ; DFM 1975 ; ▶méthodes de sylviculture, ▶science forestière, ▶soins culturaux d'un peuplement forestier, ▶sylviculture 2, ▶sylviculture écologique, ▶terrain forestier) ; *syn.* foresterie [f] [spécifique au CDN] ; *f 2* **ligniculture** [f] (Sylviculture intensive [culture de produits ligneux] appliquée notamment au pin maritime et aux peupliers ; TSF, 1985) ; *syn.* foresterie [f] intensive [spécifique au CDN] ; *g* **Forstwirtschaft** [f] (Im weitesten Sinne Zweig der Landwirtschaft, der die Aufgabe hat, die ▶forstlichen Landflächen der Gesellschaft durch die Bereitstellung materieller [Holz] und immaterieller Leistungen [z. B. Erholungsnutzung, Naturbeobachtung] mit den Tätigkeiten des ▶Waldbaus, der Holzernte und der ▶Waldbestandspflege nutzbar zu machen; ▶Forstwissenschaft, ▶naturnaher Waldbau, ▶Waldbau 1, ▶Waldbau 2).

forestry [n] (2) *adm. for. hort. landsc. urb.* ▶urban forestry.

2243 forestry commission [n] *adm. for.* (Public body of persons formed to guide administration and use of forest land); *s* **administración** [f] **forestal** (Organismo privado o público que se dedica a la administración y el uso de áreas forestales); *syn.* administración [f] de silvicultura; *f* **office** [m] **national des forêts** (*Abrév.* O.N.F. ; établissement public national à caractère industriel et commercial créé en 1964 et chargé de la gestion et de l'équipement des forêts appartenant à l'État ainsi que de la mise en œuvre du régime forestier appartenant à l'État et aux collectivités territoriales) ; *syn.* administration [f] des eaux et forêts ; *g* **Forstverwaltung** [f] (Jede Maßnahme oder Handlung privater oder öffentlicher Körperschaften zur Erhaltung und Pflege des Waldes. In D. und A. wird die **F.** hauptsächlich durch staatliche Forstbehörden ausgeübt; große Privatwaldbesitzflächen haben eigene Forstverwaltungen, kleinere Flächen werden von staatlichen Forstämtern beaufsichtigt).

forestry production land [n] [UK] *for.* ▶forest products area [US].

2244 forestry techniques [npl] *for.* (Techniques used in forest establishment, cultivation, and management; ▶forestry 1); *s* **dasotomía** [f] (Arte y ciencia de dirigir las cortas de los montes para su buen gobierno; DB 1985; ▶silvicultura artificial); *f* **méthodes** [fpl] **de sylviculture** (Interventions de nature à maintenir, mettre en valeur, exploiter, entretenir et améliorer le patrimoine forestier ; ▶sylviculture) ; *syn.* techniques [fpl] de sylviculture, techniques [fpl] forestières ; *g* **Waldbau** [m] **(1)** (Technische Entwicklung von Methoden der Waldbegründung und -pflege; ▶Forstwirtschaft).

2245 forest science [n] *for.* (Science of creating, conserving, and managing the natural resources that occur on and in association with forest land for the continuing use and benefit of society; cf. SAF 1983; ▶forestry 1); *syn.* science [n] of forestry; *s* **dasonomía** [f] (Ciencia de los bosques, que trata principalmente de su conservación y aprovechamiento; DB 1985; ▶silvicultura artificial); *f* **science** [f] **forestière** (Science qui a pour l'objet la création, la conservation, l'aménagement et la gestion des ressources naturelles des forêts et des domaines forestiers en vue d'un rendement soutenu pour le bénéfice de la société ; ▶sylviculture) ; *g* **Forstwissenschaft** [f] (Wissenschaft und Lehre von den Grundlagen, Methoden und Ergebnissen der Forstwirtschaft und von der Abgrenzung und Auslotung aller gesetzlichen Probleme zwischen Mensch und Wald; ▶Forst-wirtschaft).

2246 forest seam [n] *phyt.* (Linear-shaped stand of forb vegetation on the outer edge of a ▶forest mantle; ▶herbaceous edge [US]/herbaceous seam [UK]); *s* **orla [f] herbácea del bosque** (▶Orla herbácea que crece en el borde del bosque junto al ▶abrigo del bosque); *f* **ourlet [m] préforestier herbacé** (Frange linéaire herbacée [▶ourlet herbacé] précédant le manteau forestier en ▶manteau forestier) ; *g* **Waldsaum [m]** (Krautige, hochstaudengeprägte Pflanzengesellschaft [▶Krautsaum], die vor dem ▶Waldmantel wächst und sich deutlich vom niedrigeren Bewuchs der angrenzenden landwirtschaftlichen Flächen und sonstigen Wiesen und Rasen abhebt).

2247 forest seam formation [n] *phyt.* (Herb community on a forest margin; ▶seam community); *s* **comunidad [f] herbácea del lindero del bosque** (Comunidad herbácea en los bordes de los bosques; *término genérico* ▶comunidad de la orla herbácea); *syn.* comunidad [f] de la orla herbácea del bosque; *f* **groupement [m] d'ourlet préforestier à mégaphorbes** (Peuplement de mégaphorbes em lisière de forêt ; *terme générique* ▶peuplement d'ourlet herbacé) ; *g* **Waldsaumflur [f]** (Stauden-flur an Waldrändern; *OB* ▶Saumgesellschaft).

forest soil [n] *for. pedol.* ▶forest floor.

forest soil cover [n] *phyt.* ▶forest floor cover.

2248 forest species [n] *for. phyt.* (Plant species which grows primarily in a forest); *s* **especie [f] forestal** (Especie vegetal que crece mayormente en el bosque); *f* **espèce [f] sylvatique** (Espèce croissant principalement dans un peuplement forestier) ; *syn.* espèce [f] forestière ; *g* **Waldart [f]** (Pflanze, die vor-wiegend im Wald wächst).

forest stand [n] *for.* ▶standing forest crop, ▶tending of a forest stand.

2249 forest stand density [n] *for.* (Closeness of woodland or forest trees); *syn.* stand density [n] of forest/woodland; *s* **densidad [f] de un rodal**; *f* **densité [f] de peuplement forestier** *syn.* matériel [m] relatif, densité [f] relative, recouvrement [m] fores-tier ; *g* **Bestockungsdichte [f] einer Waldfläche** *syn.* Besto-ckungsgrad [m] einer Waldfläche.

2250 forest stand inventory [n] [US] *for.* (Counting of one or more tree species, generally above a specified size limit, and their classification by size, condition, etc. in order to deter-mine the production potential; cf. SAF 1983); *syn.* forest enu-meration [n] [UK], *obs.* enumeration survey [n] [UK]; *s* **inventa-riación [f] de un rodal** (Registro y evaluación de una masa forestal en cuanto a su potencial productivo); *f* **inventaire [m] forestier** (Dénombrement et évaluation du peuplement forestier existant afin d'établir la surface, la répartition et la qualité du bois des essences forestières et leur potentiel de production ; *termes spécifiques* inventaire pied à pied, inventaire statistique ; DAV 1984) ; *g* **forstliche Bestandsaufnahme [f]** (Ermittlung und Bewertung des aktuellen Baumbestandes hinsichtlich seines Produktionspotenzials); *syn.* forstliche Bestandsinventur [f].

2251 forest structure [n] *for.* (Distribution of age or size classes of a ▶standing forest crop, and the ▶stratification of plant communities, degree of canopy closure, ▶ecological stability, etc.); *syn.* structure [n] of a forest stand, forest consti-tution [n]; *s* **estructura [f] de un bosque** (Caracterización de una ▶población forestal según diferentes criterios como **a]** la distri-bución vertical de especies forestales [▶estratificación de la vegetación], **b]** la composición de clases de edades, **c]** la estruc-tura vertical del estrato arbóreo [mono-, bi- o pluriestratificado] o **d]** la distribución de las especies individuales [▶sociabilidad]; ▶estabilidad ecológica); *f* **structure [f] d'une forêt** (Caractéri-sation d'un peuplement selon **a]** la distribution verticale [▶strati-fication de la végétation] et horizontale des espèces forestières, **b]** la composition des classes d'âges, **c]** la structure verticale [p. ex. monostrate, tristrate, pluristrate] ou de la répartition des différentes espèces d'arbres [▶sociabilité, p. ex. isolé, en touffe, en groupe], les classes de dimensions, [diamètres ou circonfé-rences], classes de cimes, de couvert ; les peuplements irréguliers [âges et tailles différents] et parfois mélangés ont un rôle impor-tant dans la protection contre le vent, l'érosion, les avalanches ; ▶peuplement forestier 1, ▶stabilité écologique) ; *syn.* structure [f] d'une station forestière, structure [f] d'un peuplement fores-tier ; *g* **Waldaufbau [m, o. Pl.]** (Strukturformen, Entwicklungs-phasen oder Kennzeichnung eines Bestandes nach **a]** der Art wie Baumarten in horizontaler und vertikaler Verteilung [▶Vegeta-tionsschichtung] angeordnet sind, **b]** der Alterszusammenset-zung/dem Altersgefüge, **c]** dem vertikalen Aufbau [z. B. ein-schichtig, dreischichtig, mehrschichtig] oder nach der Verteilung der einzelnen Baumarten [▶Häufungsweise, z. B. einzeln, horst-weise, truppweise], ▶Stabilität etc.; das unterschiedliche Alter und die dadurch verschieden großen Bäume haben z. B. eine besonders gute aerodynamische Wirkung bei Windschutzpflan-zungen; ▶Waldbestand 2); *syn.* Bestandsaufbau [m], Bestands-struktur [f], Waldstruktur [f].

2252 forest study trail [n] *recr.* (Forest pathway with illustrated texts on signs conveying information about the nature of the forest and its constituents; ▶nature trail [US]/nature study path [UK]); *s* **itinerario [m] forestal** (Camino señalizado por un bosque a lo largo del cual, por medio de tablones con imágenes y textos de explicación y objetos originarios del mismo, se trans-miten conocimientos sobre ese bosque y sus componentes; *tér-mino genérico* ▶itinerario de la naturaleza); *f* **sentier [m] de découverte de la forêt** (Itinéraire de promenade balisé et illustré par des panneaux informatifs ; ▶sentier de découverte) ; *syn.* circuit [m] d'initiation aux particularités forestières (DUV 1984, 288) ; *g* **Waldlehrpfad [m]** (Beschildeter Wanderweg, der an Hand von originalen Objekten und durch Darstellungen in Bild und Schrift Kenntnisse über die Natur des Waldes und seiner Bestandteile vermittelt; ▶Lehrpfad).

2253 forest terrain [n] *for. leg.* (Land covered by an exten-sive stand of trees. There are also non-stocked [US]/unstocked [UK] forest lands; cf. SAF 1983; in comparison to the synony-mous term ▶forest land, **f. t.** refers to topographic features as well); *syn.* forest area [n], forest land [n]; *s* **terreno [m] forestal** (En sentido legal, superficie de uso forestal, que puede estar rodada provisionalmente; ▶área forestal); *syn.* superficie [f] forestal; *f* **surface [f] forestière** (Terrain portant une forêt en croissance ou terrain à vocation forestière après une coupe ; ▶terrain forestier) ; *syn.* étendue [f] boisée, terrain [m] boisé, superficie [f] forestière, superficie [f] boisée, zone [f] forestière ; *g* **Waldfläche [f]** (Durch einen ausgedehnten Baumbestand gekennzeichnete Fläche; es gibt auch nicht bestockte Waldflä-chen, wenn diese als Waldflächen planungsrechtlich ausgewiesen sind; ▶forstliche Landfläche); *syn.* Waldboden [m] (2) [A].

2254 forest track [n] [US] *for.* (Compacted sand or gravel path, track or road through a wood or forest, often of sufficient width to allow vehicles to pass each other; SAF 1983; ▶timber road [US]/forest path [UK]); *syn.* forest ride [n] [UK]; *s* **camino [m] forestal (1)** (Término genérico para cualquier tipo de cami-no, generalmente sin asfaltar, en el bosque; ▶camino forestal 2); *syn.* sendero [m] forestal; *f* **chemin [m] forestier** (Chemin constitué d'un sol ferme bien drainé ; lorsque celui-ci est consti-tué d'une chaussée de matériaux compactés ou à revêtement asphalté on parle de « piste » ou de « route forestière » ; DFM 1975 ; ▶laie, ▶layon) ; *syn.* sentier [m] forestier, sentier [m] sylvestre ; *g* **Waldweg [m]** (**1.** Der nicht dem öffentlichen Ver-kehr gewidmete Weg im Staats-, Körperschafts- und Privatwald.

2. OB zu meist unbefestigter, manchmal auch asphaltierter Weg im Wald; ▶forstwirtschaftlicher Weg).

forest tree liner [n] *for. hort.* ▶forest tree seedling.

2255 forest tree nursery [n] *for. hort.* (**1.** Area set aside in which trees are cultivated for use in forestry. **2.** Commercial activity concerned with the production of seeds for forestry. The source of seed is a specially reserved stand of forest trees, which is documented in an international directory [**seed harvest stock**]. Seed is harvested from genetically high-quality forest plants [**seed producers**]. **In Europe,** four categories of forest seed are commercially recognized: [a] **proven source**—seed, which is only suitable for the cultivation of wild, woody tree species for the open countryside, but not for silviculture; [b] **selected seed**—used for the production of an approved stand of forest, which produces the lowest seed quality; [c] **certified seed**—from an approved stand, which complies with quality standards set for the designation of a forest in which seeds are harvested and the resulting trees are checked with regard to certain criteria; [d] **quality seeds**—seeds of the highest quality for silviculture, whereby the seeds are only harvested from specially selected, above-average phenotypes [slips on understock, several years old] of the same species or from a collection of clones. In order to control their fertilization these trees must be at a distance of more than 400m from the same species. A **Seed harvest stock** for tested and certified forest plant seeds is known as a **seed orchard**; ▶seed stand); *s* **vivero** [m] **forestal** (**1.** Superficie cercada en zona forestal utilizada para cultivar plántulas para regenerar el bosque; *términos específicos* **v. f. permanente/fijo** —destinado a proporcionar plantones para la reproducción de grandes áreas— y **v. f. volante** —ubicado en lugar cercano a los montes a repoblar. Su ventaja es que comparte las características ecológicas de las zonas a repoblar; cf. DINA 1987. **2.** Actividad comercial dedicada a la producción de semillas para la silvicultura. El origen de las semillas son rodales de árboles forestales especialmente acotados [**rodales semilleros**], que se documentan en un Catálogo Nacional de Materiales de Base [CNMB], que a su vez forma parte del Catálogo Común Europeo. En la UE, los **materiales forestales de reproducción** se subdividen en 4 categorías: **1. identificados: mm. ff. de r.** obtenidos de materiales de base que pueden ser bien una fuente semillera, bien un rodal situados dentro de una única región de procedencia y que satisfacen las exigencias establecidas en el anexo II; **2. seleccionados: mm. ff. de r.** obtenidos de materiales de base que se corresponden con un rodal situado dentro de una única región de procedencia, que hayan sido seleccionados fenotípicamente a nivel de población y que satisfacen las exigencias establecidas en el anexo III; **3. cualificados: mm. ff. de r.** obtenidos de materiales de base que se correspondan con huertos semilleros, progenitores de familias, clones o mezclas de clones, cuyos componentes han sido individualmente seleccionados fenotípicamente y satisfacen las exigencias establecidas en el anexo I; **4. controlados: mm. fm. de r.** obtenidos de materiales de base que se corresponden con rodales, huertos semilleros, progenitores de familias, clones o mezclas de clones. La superioridad del material de reproducción debe haber sido demostrada mediante ensayos comparativos o estimada a partir de la evaluación genética de los componentes de los materiales de base. Los materiales de base deberán satisfacer las exigencias establecidas en el anexo V; art. 2, l] y anexos RD 289/2003; cf. Directiva 1999/105/CE del Consejo, de 22 de diciembre; ▶rodal semillero); *syn.* vivero [m] silvicultor; *f 1* **pépinière** [f] **forestière** (Surface consacrée à l'élevage de tout jeunes arbres — matériels de reproduction, c.-à-d. semis, plants, boutures, etc. —, principalement en vue de leur plantation et du reboisement. Une pépinière peut comporter des « planches de semis » ou des « planches de plants » ; une pépinière peut être

permanente ou **temporaire** ; DFM 1975) ; *f 2* **verger** [m] à **graines** (Pépinière de plants forestiers avec un matériel de base admis dans les catégories qualifiées et testées, destinée à la production de semences sélectionnées ; les matériels forestiers de reproduction sont répartis en quatre catégories **a. identifiée** — si la source de graines est située dans une région de provenance de l'essence considérée ; **b. sélectionnée** — lorsque le matériel de base constitue un peuplement qui est situé dans une seule région de provenance et dont la population a fait l'objet d'une sélection phénotypique ; **c. qualifiée** — un matériel de base peut être admis en catégorie « qualifiée », lorsqu'il constitue un verger à graines, des parents de famille, un clone ou un mélange clonal dont les composants ont fait l'objet d'une sélection phénotypique individuelle ; **d. testée** — lorsque le matériel de base constitue un peuplement, un verger à graines, des parents de famille, un clone ou un mélange clonal pour lequel la supériorité des matériels de reproduction par rapport à des matériels témoins doit avoir été démontrée par des tests comparatifs ou par une estimation établie à partir de l'évaluation génétique des composants des matériels de base ; Code forestier Art. R. 552-1-10 ; ▶peuplement porte-graines) ; *g* **Forstbaumschule** [f] (**1.** Baumschule zur Aufzucht von Forstpflanzen. **2.** Forstsaatgutbetrieb mit ausgewiesenen und in einem internationalen Verzeichnis erfassten Waldfläche [▶**Samenerntebestand**], auf der Saatgut zur bedarfsgerechten Erzeugung von genetisch hochwertigen Forstpflanzen [**Saatgutproduzenten**] geerntet werden darf. In D. werden vier Kategorien von forstlichem Saatgut gehandelt: **a. quellengesichert** — Saatgut, das nur für die Anzucht von Landschaftsgehölzen, jedoch nicht für den Waldbau geeignet ist; **b. ausgewählt** — Saatgut für die unterste Qualitätsbezeichnung eines für die waldbauliche Samenernte zugelassenen Bestandes; **c. geprüft** — Saatgut für die Qualitätsbezeichnung eines für die waldbauliche Samenernte zugelassenen Bestandes, bei dem die Nachkommenschaft hinsichtlich bestimmter Kriterien geprüft ist; **d. qualifiziert** — Saatgut für die höchste waldbauliche Samenqualität, bei der die Samen nur von besonders ausgelesenen, überdurchschnittlichen Phänotypen [Pfropfreiser auf mehrjährigen Unterlagen] derselben Art oder von einer Klonsammlung stammen. Diese Bäume müssen wegen der Kontrollierbarkeit der Befruchtung über 400 m von Exemplaren derselben Art entfernt sein. Ein **Samenerntebestand** für geprüftes und qualifiziertes Forstpflanzensaatgut wird **Samenplantage** [engl. *seed orchard*] genannt).

2256 forest tree seedling [n] *for. hort.* (Young plant grown in a ▶forest tree nursery); *syn.* forest tree liner [n]; *s* **postura** [f] (Pie de árbol cultivado en un ▶vivero forestal); *f* **plant** [m] **forestier** (Jeune plante cultivée dans une ▶pépinière forestière, ▶verger à graines) ; *g* **Forstpflanze** [f] (In einer ▶Forstbaumschule kultivierte Jungpflanze).

forest windbreak [n] *landsc.* ▶forest wind shield.

2257 forest wind shield [n] *landsc.* (Deciduous shrub layer which protects the interior of the forest from the wind); *syn.* forest windbreak [n]); *s* **manto** [m] **de amparo** (Rompevientos de un rodal de bosque; SILV 1979, 86); *f* **manteau** [m] **de protection contre le vent** (Bande boisée protégeant l'intérieur de la forêt contre l'action dévastatrice du vent) ; *g* **Waldmantel** [m] **(2)** (Waldmantel, der für das Innere des Waldes den Wind abhält und somit z. B. gegen Windwurf schützt).

fork [n] [UK] *arb.* ▶crotch [US].

2258 forked growth [n] *arb.* (SAF 1983, 153; U- or V-formed branching of a trunk or vertical main branch; ▶crotch); *syn.* tight crotch [n] (ARB 1983, 403, 490), codominant stems [npl] (SHI 1991); twin stems [n] (±); *s* **cima** [f] **en horquilla** (Bifurcación del tronco o de una rama principal vertical en forma de U o V [ahorquillado]; ▶cruz); *f* **fourche** [f] (Branches de force égale issue de la division sous forme de V ou de U d'un tronc ou d'une charpentière verticale ; l'▶enfourchement en V est souvent sujet à un arrachement des tissus internes de la ride de branche de l'écorce [entre-écorce] ou à la forte pression exercée sur le bois à cet endroit) ; *syn.* arbre [m] fourchu [CDN] (±) ; *g* **Zwiesel** [m] (U- oder V-förmige Gabelung eines Stammes oder senkrechten Starkastes in ca. gleich starke Stämmlinge; der V-Zwiesel [Druckzwiesel] ist wegen der einwachsenden Rinde und wegen der Druckholzbildung in diesem Bereich bruchgefährdeter als der U-Zwiesel [Zugzwiesel]; den Längsriss unterhalb eines V-Zwiesels nennt man **Zwieselriss**; je nach Lage der **Z.** spricht man von **Hochzwiesel** oder **Tiefzwiesel**, ▶Astverzweigung; *UBe* U-Zwiesel, V-Zwiesel; cf. QBB 1992); *syn.* U-förmige Stämmlinge [mpl], U-Vergabelung [f], V-förmige Stämmlinge [mpl] (N+L 1988 [6], 382), V-Vergabelung [f], Zwieselwuchs [m].

form [n] [US] *arb. hort.* ▶crown form; *constr. contr.* ▶form of bid [US]; *geo. landsc.* ▶landform; *biol.* ▶life form; *contr. prof.* ▶standard contract form; *adm. conserv. ecol.* ▶standard data form.

2259 formal garden [n] *gard'hist.* (Garden using a symmetrically-balanced layout with one or more axes and a strong geometric form, often with an ornamental floral pattern; ▶landscape garden, ▶pastoral park [UK]; *opp.* ▶informal garden); *s* **jardín** [m] **geométrico** (Jardín diseñado con formas regulares de estructura y de distribución de las plantas y flores. El jardín barroco en un ejemplo de **j. g.**; ▶jardín paisajístico; *opp.* ▶jardín informal); *f* **jardin** [m] **régulier** (Jardin caractérisé par des lignes droites et des formes géométriques existant depuis la plus haute antiquité. Les jardins réguliers à la française, dont Le Nôtre a fourni les plus beaux spécimens, en sont une illustration parfaite ; LA 1981 ; ▶jardin de style libre ; *opp.* ▶jardin paysager) ; *g* **formaler Garten** [m] (*Im Deutschen nicht exakt definierter Begriff* ein meist nach geometrischen oder floral-ornamentalen Mustern intensiv gestalteter Garten oder Park, oft mit Blumenbroderien, geschnittenen Hecken/Boskets oder in Formen geschnittenen Einzelgehölzen [Ars topiaria]; Bauerngärten gehören auch zur Kategorie **f. G.**; unter *architektonischer Garten* wird in der Geschichte der Gartenkunst i. d. R. der französische Barockgarten verstanden; ▶Landschaftspark; *opp.* ▶nicht formaler Garten); *syn.* Regelgarten [m], regelmäßiger Garten [m] (GOT 1926, 450), z. T. architektonischer Garten [m].

formation [n] *geo. pedol.* ▶exposed stone formation; *geo.* ▶peat formation; *phyt.* ▶plant formation, ▶reed swamp, ▶reed swamp formation; *conserv. geo. land'man.* ▶rock formation; *biol.* ▶root formation; *bot. hort.* ▶tussock formation.

formation [n], **tuft** *bot. hort.* ▶tussock formation.

formation grade [n], **excavation to** [UK] *constr.* ▶excavation to subbase grade [US].

formation level [n] [UK] *constr.* ▶subbase grade [US]; *constr. stat.* ▶subsoil level [US].

formation level [n] [UK], **excavation to** *constr.* ▶subbase grade preparation [US].

formation level [n] [UK], **grading of** *constr.* ▶subbase grade preparation [US].

formation level [n] **of a road/path** [UK] *constr.* ▶subbase grade of a road/path [US].

2260 formation [n] **of ice crystals** *constr. pedol.* (Formation of lens-shaped ice crystals in frozen cohesive soils, which grow through being fed by capillary water from the surrounding soil, and result in the deformation of the overlying ground surface; ▶frost heave); *s* **formación** [f] **de lentejas de hielo** (En suelo cohesivo helado creación de cristales de hielo con forma de lenteja que van aumentando de tamaño gracias al agua capilar de

su alrededor y tienen como consecuencia la deformación de la superficie del suelo de encima de ellos; ▶levantamiento por congelación); *f* **formation** [f] **de lentilles de glace** (Action de la glace dans un substrat minéral poreux dans lequel, à la suite d'infiltrations capillaires, a lieu un processus d'individualisation de la glace pour former une **l. d. g.** en soulevant le sol et la végétation environnante. Les collines isolées dont l'intérieur est constitué de glace sont dénommées tertres de toundra, DG 1984, 210 ; ▶foisonnement par le gel) ; *g* **Eislinsenbildung** [f] (Phänomen des Gefrierens von bindigen Böden, in denen sich linsenförmige Eiskristallkörper bilden, die sich durch ständigen kapillaren Wasserzutritt aus den umgebenden Poren vergrößern und den Bodenkörper heben; ▶Frosthebung).

2261 formation [n] **of U or V crotches** *arb.* (1. Growing of a second competing leader or vertical main branch of a tree in the form of an U- or V-shaped bifurcation. 2. Result of 1: ▶forked growth); *s* **formación** [f] **de cima en horquilla** (1. Crecimiento de una segunda rama vertical en competencia con la rama principal de un árbol, dándole forma de horquilla en V o U. 2. Resultado ▶cima en horquilla); *f* **formation** [f] **de fourche** (Division en plusieurs branches d'un tronc d'arbre ; ▶fourche) ; *g* **Zwieselbildung** [f] (1. Entstehung eines zweiten konkurrierenden Leittriebes an einem Baum in Form einer U- oder V-förmigen Verzweigung. 2. Resultat von 1: ▶Zwiesel).

formative pruning [n] *hort.* ▶training of young trees.

2262 form line [n] *plan.* (Line on a map, approximating a ▶contour, but for lack of elevation control, merely sketched in so as to give a general indication of the relief; SAF 1983); *s* **curva** [f] **de configuración** (En la representación del relieve de los mapas, líneas que se utilizan para dar una idea aproximada del mismo, sin precisión métrica alguna. Fue uno de los sistemas cartográficos empleados antes de la introducción de las ▶curvas de nivel; DGA 1986); *f* **ligne** [f] **de forme** (▶Courbe de niveau qui représente le relief général, la configuration d'un terrain ou les lignes de même cote d'un projet sans référence à un point géodésique) ; *g* **Höhenstrukturlinie** [f] (▶Höhenlinie ohne geodätischen Bezug, die lediglich ein generelles Relief oder eine Bodenmodellierung andeutet).

form [n] **of agreement** [UK] *contr. prof.* ▶standard contract form [US].

2263 form [n] **of bid** [US] *constr. contr.* (Outline sheets sent by the authority/agency as a request for proposals [R.F.P.] to document provisions of the ▶bid [US] 1/tender [UK]; ▶priced bidding documents [US]/priced tendering documents [UK]); *syn.* bid form [n] [US], form [n] of tender [UK]; *s* **formulario** [m] **de propuesta** (Formulario elaborado por la autoridad/agencia convocadora de concurso-subasta que debe ser rellenado por los postores; ▶expediente de concurso-subasta, ▶oferta); *syn.* modelo [m] de proposición; *f* **modèle** [m] **de dossier d'appel d'offres/de soumission** (Formulaires délivrés par la personne responsable de la consultation, permettant aux entreprises ou aux concepteurs de remettre une offre ; ▶dossier des pièces constitutives du marché ; ▶offre [de prix], ▶soumission) ; *syn.* modèle [m] de dossier de consultation ; *g* **Angebotsformular** [n] (Von der ausschreibenden Stelle herausgegebener Vordruck zum Ausfüllen der Angebotspreise und sonstiger Vordrucke; ▶Angebot, ▶Angebotsunterlagen).

form [n] **of tender** [UK] *constr. contr.* ▶form of bid [US].

Form Standard Codes [npl] [US] *constr. contr.* ▶standards for construction [US]/Codes of Practice (C.P.) [UK].

2264 formwork [n] *constr.* (Temporary construction to contain wet concrete in the required shape while pouring and setting; ▶earth formwork, ▶permanent formwork); *syn.* shuttering [n]; *s* **encofrado** [m] (Construcción provisional de tablas

F

de madera para retener el hormigón húmedo y darle la forma deseada hasta que se seca; ►encofrado de tierra, ►encofrado perdido); *f* **coffrage [m]** (Dispositif étanche et résistant, en bois ou en métal qui moule et maintient le béton que l'on coule ; PR 1987 ; ►coffrage en terre, ►coffrage perdu) ; *g* **Schalung [f]** (Aus Holz gebildete Hohlform für Betonierarbeiten; ►Erdschalung, ►verlorene Schalung).

fortification [n] [US] *urb.* ►earthen wall fortification.

fortification [n] [UK]**, town** *urb.* ►earthen wall fortification [US].

fortifications [npl] *urb.* ►removal of fortifications.

fossil [n] *biol. geo.* ►indicator fossil.

foul sewage [n] [UK] *envir.* ►toilet wastewater [US].

foundation [n] *constr. eng.* ►exploration for foundation; *constr.* ►post foundation [US], ►stepped foundation, ►step strip foundation, ►strip foundation.

foundation [n] [US]**, pillar** *constr.* ►post foundation [US].

foundation [n] [UK]**, spot** *constr.* ►post foundation [US].

2265 foundation [n] **of a flight of steps** *constr.* (Strip footings for step construction or, in the case of a long flight of steps, an additional reinforced concrete slab; where frost heave is not a problem, a less rigid method of foundation construction for a few steps is also possible); *s* **cimentación [f] de escalera** (Cimento de escalera de zapata corrida o, en el caso de tramos muy largos, reforzado por placa de hormigón armado; en suelos sin peligro de heladas se pueden utilizar fundaciones más ligeras); *f* **fondation [f] d'un escalier** (Ouvrage de fondation constitué d'une semelle filante ou d'une paillasse ; sur un sol très compact et pour un escalier rustique, il est possible de découper les marches dans le sol, le terrain servant de fondation, ou d'employer un type de fondation plus léger) ; *g* **Treppengründung [f]** (Fundamentbauwerk aus Streifenfundamenten oder bei langem Treppenlauf mit zusätzlicher Stahlbetonplatte, die die beiden Streifenfundamente verbindet; bei nicht frostgefährdeten Böden ist für kleine Treppen auch eine labile Bauart möglich); *syn.* Treppenfundament [n].

2266 foundation planting [n] *constr.* (Planting of shrubs, small trees and ground covers—very often neatly clipped—close to a building, sometimes in a continuous strip; ►façade planting); *syn.* base planting [n] [also US] (DIL 1987, 121), mustache planting [n] [also US]; *s* **jardinería [f] lineal bordeando fachadas (≠)** (*Forma de plantación usual en los EE.UU.* Plantación en plantel estrecho de arbustos, tapizantes y árboles pequeños —a menudo podados en topiaria— situado al borde de los edificios; ►plantación de fachadas); *f* **ceinture [f] végétale sur rez-de-chaussée (≠)** (U.S., forme typique de plantation étroite sur le pourtour des façades d'un bâtiment, constituée d'arbustes, de plantes tapissantes et de petits arbres très souvent taillés ; ►végétalisation des façades) ; *g* **Pflanzung [f] am Gebäudesockel (≠)** (*Eine in den USA übliche Bepflanzungsform* schmales Pflanzbeet von Sträuchern, Bodendeckern und kleinen Bäumen — oft in Formschnitt gehalten — am Sockelgeschoss; ►Fassadenbegrünung); *syn.* Sockelgeschosspflanzung [f] (≠), Sockelbepflanzung [f] (≠).

2267 foundation trench [n] *constr.* (Elongated excavation for a supporting part of a wall or structure); *s* **zanja [f] de cimentación** (Excavación alargada para dar cabida al fundamento de un muro o una estructura); *syn.* trinchera [f] de cimentación; *f* **tranchée [f] de fondation** (Fouille plus haute que large destinée à permettre la réalisation des fondations) ; *g* **Fundamentgraben [m]** (Durch Ausschachtarbeiten entstandener Graben für den Bau einer Gründung).

2268 fountain pool [n] *urb. gard.* (Ornamental basin with single or multiple jets of water; ►pool 1, ►water features, ►composition of water features, ►garden pool); *s* **fuente [f]** (►Estanque ornamental con uno o varios surtidores de agua; ►estanque, ►hidroarte, ►juego de aguas); *syn.* surtidor [m]; *f* **fontaine [f] à jet d'eau** (Équipement ornemental dans lequel l'eau jaillit d'un jet pouvant comporter différents ajutages permettant des figures d'eau variées et retombant dans un bassin ; ►art des jeux d'eau, ►bassin ornemental, ►jeu d'eau, ►pièce d'eau) ; *g* **Springbrunnen [m]** (Brunnen, bei dem ein Springstrahl oder mehrere in die Höhe steigen und in ein Becken zurückfallen; ►Wasserbecken, ►Wasserkunst, ►Wasserspiel, ►Zierbecken).

four-lane undivided expressway [n] [UK] *trans. urb.* ►four-lane undivided highway [US].

2269 four-lane undivided highway [n] [US] *trans. urb.* (Road for high-speed traffic, usually without a central median, but with an additional lane on steep inclines for slow-moving, heavy vehicles; ►slow vehicle lane); *syn.* four-lane undivided expressway [n] [UK]; *s* **carretera [f] de cuatro carriles** (Vía de tránsito rápido sin división intermedia que en zonas de pediente tiene un carril adicional para los vehículos lentos; ►carril lento); *f* **voie [f] express à deux fois deux voies** (Route prévue pour une circulation à grande vitesse, souvent dépourvue de terre-plein central, avec une ►voie pour véhicules lents sur les sections en rampe) ; *syn.* voie [f] rapide à 4 voies ; *g* **vierspurige Schnellverkehrsstraße [f]** (Den schnelleren Kraftfahrzeugen dienender Verkehrsweg, meist ohne Mittelstreifen; für den Lastverkehr gibt es an Steigungen oder Gefällstrecken oft eine ►Kriechspur).

fourrés [npl] [UK] *phyt.* ►brushy area [US].

fracture [n] *arb. stat.* ►resistance against fracture.

fragile area [n] [US] *ecol. plan.* ►sensitive area.

fragmentation [n] *for.* ►forest fragmentation; *ecol.* ►process of habitat fragmentation.

2270 fragmentation [n] **of an area** *conserv. plan. urb.* (Separation of a normally contiguous landscape or urban space by linear infrastructure such as a road, railway, etc.; large-scale infrastructure such as roads and railway lines present insurmountable obstacles for terrestial fauna and especially for large mammals in terms of genetic exchange; the situation is often exacerbated by protective fences and only encourages the ►process of habitat fragmentation. Other effects of landscape fragmentation are to be seen in the following six problem areas: **1.** Soil compaction and sealing, **2.** micro-climatic change, **3.** environmental pollution load [noise, fumes, road salt, etc.], **4.** changes in the water resource management, **5.** impairment of the visual quality of the landscape, and **6.** consequences for land use [structural change, namely changes in spatial relationships caused by fragmentation]. The consequences of landscape fragmentation are increasingly serious, because mobility—in the sense of transportation—is an essential tool in society for most activities, and for many people a highly desirable goal. However, to what extent the effects of high mobility are assessed politically and in rational terms remains open; ►undissected area with low traffic intensity); *syn.* splitting-up [n] of an area; *s* **fragmentación [f] de espacios (naturales o urbanos)** (División de espacios naturales o urbanos por infraestructuras de transporte, como autopistas, carreteras o líneas de ferrocarril, que tiene como consecuencia el efecto de corte para la fauna, en especial para la fauna edáfica, los anfibios, los reptiles y los mamíferos; los principales efectos de la **f. de e.** son la separación entre biótopos de diferentes ciclos vitales [anfibios], el aislamiento genético de poblaciones y el ►aislamiento de biótopos. Más allá de los efectos negativos para la fauna, la **f. de e.** conlleva múltiples impactos: **1.** Compactación y sellado del suelo. **2.** Alteración del microclima. **3.** Contaminación atmosférica, acústica y del suelo. **4.** Alteración del régimen hidrológico. **5.** Intrusión visual en el paisaje y **6.** consecuencias para los usos del suelo. Los impactos de la **f. del e.** son cada vez más graves, ya que la movilidad —en el sentido del transporte en el espacio— se considera imprescindible para realizar muchas actividades en la sociedad industrial moderna; ▶área no fragmentada con poca densidad de tráfico); *syn. según el contexto* fragmentación [f] de superficies, fragmentación [f] de paisajes, fragmentación [f] de zonas urbanas; *f 1* **coupure [f] (des espaces)** (Destruction du tissu urbain et des espaces paysagers de leurs fonctions de proximité par les infrastructures linéaires routières et ferroviaires [effet de coupure paysagère et physique] ; les infrastructures routières et en particulier les autoroutes constituent des obstacles pratiquement infranchissables pour les animaux et notamment les grands mammifères sauvages; cette situation est souvent accentuée par les grillages et clôtures de protection des voies ; elles sont des barrières qui empêchent la circulation de divers éléments de l'écosystème : plantes ou animaux, mais aussi éléments non vivants ; les autoroutes induisent un effet de coupure biologique entre les populations animales qu'elles séparent ; elles accentuent les effets provoqués par l'▶isolement des biotopes, et participent à la fragmentation du paysage ; dans la fragmentation des habitats la perte de surface a plus d'impact que la fragmentation en tant que telle ; cela signifie que pour la capacité de survie de populations végétales et animales, la grandeur de la surface de l'habitat au sein d'un paysage est plus importante, du moins à court terme, que la distance spatiale entre les fragments d'habitats ; les effets annexes de la fragmentation des paysages [effet de séparation, effet de barrière] sont : **1.** le compactage et l'imperméabilisation des sols, **2.** le changement du microclimat, **3.** les émissions [bruit, gaz d'échappement, sel de déneigement, etc.], **4.** la dégradation de la qualité des ressources en eau, **5.** la défiguration du paysage, **6.** la dispersion et difficultés d'accessibilité des parcelles ; les conséquences de la fragmentation des paysages sont d'autant plus graves que la mobilité [au sens des transports dans l'espace] est une composante incontournable de l'activité économique actuelle ; la question de l'évaluation rationnelle des impacts paysagers engendrés par le culte de la mobilité dans la société et leur justification par la classe politique reste ouverte ; ▶zone naturelle pauvre en infrastructures) ; *syn. suivant le contexte* fragmentation [f] des espaces, fragmentation [f] des espaces urbains, fragmentation [f] paysagère, fragmentation [f] éco-paysagère, fragmentation [f] du/des paysage(s), fragmentation [f] de l'espace par une infrastructure de transport ; *f 2* **coupure [f] paysagère** (Élément paysager [vallon boisé, haie) qui structure le territoire) ; *g* **Zerschneidung [f] (von Räumen)** (Trennung von Landschafts- und Stadträumen durch linienhafte Verkehrsinfrastrukturen; große Verkehrswege wie überörtliche Straßen und Bahnstrecken stellen für alle am Boden lebenden Tiere und besonders für große Säugetiere ein unüberwindliches Hindernis für den genetischen Austausch dar; dieser Zustand wird oft noch durch Wildzäune verstärkt und fördert die ▶Verinselung von Biotopen. Weitere Auswirkungen der Landschaftszerschneidung sind folgende sechs Problemfelder: **1.** Bodenverdichtung und -versiegelung, **2.** Veränderung des Kleinklimas, **3.** Immissionsbelastung [Lärm, Abgase, Streusalz etc.], **4.** Veränderung des Wasserhaushaltes, **5.** Beeinträchtigung des Landschaftsbildes und **6.** Folgen für die Landnutzung [strukturelle Veränderung, d. h. durch die Fragmentierung verursachte Veränderungen von räumlichen Lagebeziehungen]. Die Folgen der Landschaftszerschneidung werden immer gravierender, da die Mobilität — im Sinne von räumlichen Transporten — ein wesentliches Mittel für viele Aktivitäten in einer Gesellschaft ist und für viele Akteure ein hoch bewertetes Ziel darstellt. Inwieweit aber die

Folgen der hohen Mobilität auch rational bewertet und politisch verantwortet werden, bleibt offen; ▶unzerschnittener verkehrsarmer Raum); *syn. je nach Kontext* Flächenzerschneidung [f], Landschaftszerschneidung [f], Zerschneidung [f] städtischer Räume.

2271 fragment community [n] *phyt.* (Plant association which previously had a diverse ▶species composition and is now reduced to only a few representative species due to change in environmental conditions); *s* **comunidad [f] fragmentaria** (Comunidad vegetal que anteriormente tenía mayor diversidad en su ▶composición de especies pero que la ha perdido debido a influencias ambientales negativas); *f* **groupement [m] fragmentaire** (Association ne possédant pas la totalité de ses constantes, réduite à ses éléments essentiels ; ▶composition des espèces d'une communauté) ; *syn.* association [f] fragmentaire ; *g* **Rumpfgesellschaft [f]** (Eine auf Grund veränderter Umweltbedingungen auf wenige Vertreter reduzierte Pflanzengesellschaft, die ursprünglich eine vielfältige ▶Artenzusammensetzung hatte).

2272 fragrant plant [n] *gard. plant.* (▶culinary herb); *syn.* aromatic plant [n]; *s* **planta [f] aromática** (Especie vegetal cuyas hojas o flores exhalan olores agradables debido a su contenido de aceites aromáticos. Se utilizan como condimento en la cocina y para fines medicinales. Algunas plantas aromáticas son también ▶plantas condimenticias); *f* **plante [f] aromatique** (Plante dont l'arome exhalé ou les huiles produites par les feuilles ou les fleurs leurs permettent d'être utilisées en cuisine et en médecine ; certaines plantes aromatiques sont également des ▶plantes condimentaires ; LA 1981, 103) ; *g* **Duftpflanze [f]** (Pflanze, deren Blüten oder Blätter einen starken aromatischen Duft haben, weswegen sie in der Küche, Medizin, zur Körperpflege und zur Verbesserung des allgemeinen Wohlbefindens verwendet werden; ▶Gewürzpflanze).

Framework Plan [n] *urb.* ▶Urban Design Framework Plan.

Fray damage [n] [UK] *hunt. zool.* ▶rub off the velvet [US].

fray off [vb] **the velvet** [UK] *hunt. zool.* ▶rub off the velvet [US].

2273 free-floating freshwater community [n] *phyt.* (UNE 1973; aquatic plant community on the surface of an oxbow lake, pond, pool or bay in the sea: community cannot exist if the water is turbulent; ▶floating-leaved plant, ▶free-floating waterplant, ▶rooted floating-leaf community; *specific term* ▶Duckweed cover); *syn.* free-floating lacustrine community [n]; *s* **comunidad [f] de hidrófitos libremente flotantes** (Formación vegetal de plantas acuáticas que flotan en la superficie de lagos, estanques o meandros abandonados. Este tipo de comunidad sólo puede crecer en aguas quietas; ▶comunidad de hidrófitos radicantes, ▶hidrófito libremente flotante, ▶hidatófito radicante; *término específico* ▶lemnal); *syn.* comunidad [f] (de agua dulce) que flota libremente; *f* **groupement [m] d'hydrophytes libres et flottants (des eaux calmes)** (Terme générique pour les formations flottantes libres des eaux calmes des latitudes moyennes et hautes [bras morts, parties calmes des rivières, canaux, étangs, mares, larges fossés toujours en eau], disparaissant l'hiver, appartenant à la classe des *Lemnetea* ; *terme spécifique* ▶lemnaie ; ▶plante flottante, fixée, ▶groupement flottant, fixé, ▶plante flottante et libre) ; *syn.* groupement [m] lentique flottant libre, herbier [m] libre et flottant ; *g* **Schwimmpflanzendecke [f]** (Frei schwimmende Wasserpflanzen[gesellschaft] an der Oberfläche von Altwässern, Teichen, Tümpeln oder Seebuchten; auf stärker bewegtem Wasser können sie sich nicht zu Gesellschaften zusammenschließen; ▶wurzelnde Schwimmblattgesellschaft, ▶Schwimmpflanze 1, ▶Schwimmpflanze 2; *UB* ▶Wasserlinsendecke); *syn. zum Teil* Schwimmpflanzengesellschaft [f],

Schwimmdecke [f], frei schwimmende Wasserpflanzenformation [f], frei schwimmende Stillwassergesellschaft [f].

2274 free-floating water plant [n] *bot. phyt.* (CWD 1979, 13; non-anchored plant that floats freely in the water or on the surface; e.g. Water hyacinth *[Eichhornia]*, Duckweed *[Lemna]*; ►free-floating freshwater community, ►hydrophyte); *s* **hidrófito** [m] **libremente flotante** (Planta acuática con los órganos asimiladores flotantes; ►comunidad de hidrófitos libremente flotantes, ►hidrófito); *syn.* hidatófito [m] flotante; *f* **plante** [f] **flottante et libre** (Plante aquatique dont l'appareil foliaire flotte à la surface de l'eau ; ►groupement d'hydrophytes libres et flottants [des eaux calmes], ►hydrophyte) ; *g* **Schwimmpflanze** [f] **(2)** (Freischwimmender ►Hydrophyt, z. B. Wasserhyazinthe *[Eichhornia]*, Wasserlinse *[Lemna]*; ►Schwimmpflanzendecke).

freely awarded contract [n] [UK] *contr.* ►sole source contract award [US].

2275 free-of-charge [n] *trans. pol.* (Use of public transport without payment, as a means of improving the economic situation of poorer sections of the population and reducing the amount of individual traffic within cities); *s* **tarifa** [f] **cero** (Uso de transporte público sin pagar para mejorar la situación de la población de ingresos bajos y reducir el tráfico individual; reivindicación de la izquierda en Alemania desde 1968); *syn.* tarifa [f] gratis; *f* **tarif** [m] **gratuit** (±) (Acheminement gratuit des personnes en zone urbaine par des moyens de transport collectifs publics dans le but d'améliorer la condition des classes sociales défavorisées et de réduire l'importance des moyens de transport individuels) ; *g* **Nulltarif** [m] (Unentgeltliche [oder fast unentgeltliche] Beförderung mit öffentlichen Verkehrsmitteln zwecks wirtschaftlicher Besserstellung ärmerer Bevölkerungsschichten und zur Verdrängung des Individualverkehrs aus den Städten: Forderung der politischen Linken seit ca. 1968).

2276 free play area [n] *recr.* (Multipurpose area for different kinds of informal recreational activities; ►practice game area [US]/kickabout area [UK]); *syn.* unstructured play area [n] [US]; *s* **área** [f] **de juegos libres** (Pequeña superficie libre como campa o superficie de arenilla para juegos informales; ►pista de juegos de pelota); *f* **aire** [f] **de jeux libres** (±) (Équipement de plein air et de loisirs constitué par un espace enherbé ou une aire à sol stabilisé aux liants hydrauliques, utilisé pour différentes formes de jeux de balle, la course, le saut, etc. ; ►espace multisports) ; *g* **Tummelplatz** [m] (Freizeiteinrichtung in Form einer Wiese oder eines wassergebundenen Platzes, vornehmlich für Ballspiele aller Art, Laufen, Springen etc.; ►Bolzplatz).

2277 free-standing [adj] *constr.* (So constructed that a building or part of a building is capable of supporting itself without the additional aid of columns or beams); *s* **autoportante** [adj] (Edificio o parte del mismo construido de tal manera que resulta estable sin soporte adicional); *syn.* autosostenible [adj]; *f* **autoportant, ante** [ppr/adj] (Dont les caractéristiques assurent à un ouvrage une stabilité propre, sans renforcement ou lien à d'autres ouvrages, p. ex. un mur portant libre ; DTB 1985) ; *syn.* autoporteur, euse [adj] (DTB 1985) : *p. ex.* un portail autoporteur, mur autoporteur ; *g* **selbsttragend** [ppr/adj] (So beschaffen, dass ein Bauwerk oder Bauteil ohne zusätzliche Stützen oder Träger statisch stabil bleibt, z. B. eine freistehende Mauer, ein gewölbter Plafond).

2278 free-standing bench/seat [n] *gard. landsc.* (portable or removable); *s* **banco** [m] **móvil**; *f* **banc** [m] **posé** ; *g* **lose aufgestellte Sitzbank** [f].

2279 free-standing concrete crib wall [n] *constr.* (An often symmetrical ►concrete crib wall constructed on grade, with extremely steep-sloping sides—►planted noise attenuation struc-

ture—, which can be planted. Usually used in noise protection schemes); *s* **muro** [m] **libre de entramado de piezas de hormigón** (≠) (Paredón libre, generalmente simétrico que se aplica en terraplenes muy pendientes y que puede plantarse a ambos lados [►muro de jardineras antirruidos]. Se utiliza a menudo como ►pantalla antirruidos; ►muro de entramado de piezas de hormigón); *f* **mur** [m] **de caissons en béton** (Construction libre, en général symétrique, constituée d'un empilement de caissons végétalisables creux en béton avec entretoises s'assemblant par emboîtement, chaque rang étant rempli de terre végétale au montage, de forte inclinaison [►merlon en éléments préfabriqués] pouvant être végétalisée des deux côtés. Le premier rang est posé sur une assise béton, le montage des rangs supérieurs est effectué à sec par double emboîtement des entretoises. Cet équipement est souvent utilisé comme écran antibruit ; les caissons en béton sont aussi utilisables pour la réalisation de berges ou de murs de soutènement [droits, inclinés ou courbes] végétalisés ; *terme générique* ►système de construction en caissons) ; *g* **Raumgitterwall** [m] (Freistehende, meist symmetrische ►Raumgitterkonstruktion aus Beton mit übersteilen Böschungen [►Steilwall], die von beiden Seiten begrünt werden kann. Sie wird für Lärmschutzanlagen eingesetzt; *OB* Raumgitterkonstruktion).

2280 free-standing wall [n] *arch. constr.* (In comparison with ►retaining wall, a wall with a front and a rear elevation, the stability of which being dependent upon its own weight and the relation of its depth to its height. The ►crown of a wall usually has a coping, i. e. a row of stones of the same thickness as those used in the masonry, which projects slightly beyond the face of the wall. Alternatively, the coping may be a rowlock course [US]/layer of headers-on-edge [UK] or a soldier course with a uniformly thick layer of stones; ►rowlock course); *s* **muro** [m] **libre** (En comparación con un ►muro de sostenimiento, muro con caras visibles por ambos lados, cuya estabilidad depende de su propio peso y de la relación entre el grosor y la altura del mismo. La ►coronación de un muro generalmente se realiza con piezas de recubrimiento del mismo material que pueden tener un borde que sobresale, a no ser que se recubra con una ►hilada a sardinel); *f* **mur** [m] **libre** (Par comparaison avec le ►mur de soutènement, désigne un mur constitué par un parement avant et arrière dont la stabilité dépend de son propre poids et de la relation entre son épaisseur et sa hauteur ; le ►couronnement de mur est en général réalisé avec des dalles de recouvrement de même épaisseur [pierres somnitales] avec un débord bilatéral symétrique lorsqu'il n'est pas prévu de lit de matériaux disposés sur chant, de pierres posées sur la tranche ou d'un faîtage particulier) ; ►lit à disposition sur chant ; *syn.* mur [m] isolé ; *g* **frei stehende Mauer** [f] (Im Vergleich zur ►Stützmauer, eine Mauer mit einer Vorder- und Rückansichtsseite, deren Standfestigkeit von ihrem Eigengewicht und der Dicke in Relation zur Höhe abhängig ist. Die ►Mauerkrone schließt i. d. R. mit überstehenden, gleichstarken Abdeckplatten ab, wenn keine ►Rollschicht oder Abdeckschicht mit einheitlicher Schichtstärke vorgesehen ist).

free time [n] *plan. recr.* ►leisure.

free time activity [n] [US] *plan. recr.* ►leisure pursuit.

2281 freeway [n] [US] *trans.* (Grade-separated highway [i.e. without intersections], with at least two lanes in each direction, which is structurally separated by a ►median strip [US]/central reservation [UK], and also has an inside lane for broken-down vehicles or those which need to stop for a short time; ►emergency lane. A minimum speed is prescribed, so that tractors or cyclists are not allowed to use such roads. Until the 1930s the term was not covered by any particular international standard. The first freeway in the world was the Long Island Motor Parkway, 45 miles in length, built from 1908-1911 in the USA by the eccentric billionaire son WILLIAM K. VANDERBILT JR. of New York. **In D.**, the first freeway [AVUS] was built between 1912-1921 as a racing track and road for driving practice in Berlin-Grunewald. **In Italy**, the Autostrada dei Laghi, designed by the engineer PIERO PURICELLI Milan [1883-1951] and inaugurated in 1924, was the first long freeway in Europe, connecting Milan with the Italian lakes [Milano-Varese-Como-Gallarate—known today as the A 8 and part of the A 9]; inspired by the Italians, Germany's first freeway or 'autobahn' as they came to be known, was a 20 kilometres long, four lane freeway constructed between Cologne and Bonn from 1929 to 1932. German 'autobahns' became symbols of the Third Reich during the late Thirties and World War II; **in Great Britain**, motorways were first introduced under the Special Roads Act 1949, with the first motorway, the M6 Preston Bypass, opening in 1958. The first major motorway to open was the M1 between Crick, [Northamptonshire] and Berrygrove [Hertfordshire], opening in 1959. Thereafter, motorways opened on a regular basis well into the 1980s; by 1972 the first 1,000 miles (1,609 km) of motorway had been built. The M25 London Orbital opened in 1986. By 1996 the total length of motorways had reached 2,000 miles (3,219 km); ►interstate highway [US]/motorway and trunk road [UK]; ►major highway [US] 1/major road [UK], ►major highway [US] 2/expressway [n] [UK]); *syn.* superhighway [n] [US], limited-access highway [n] [US], motorway [n] [UK], autobahn [n]; *s* **autopista [f]** (►Vía interurbana sin cruces a nivel y por lo menos dos carriles por dirección separados por una ►franja divisoria central, por la que no pueden circular vehículos lentos. En algunos países sólo se pueden utilizar pagando peaje; ►autovía, ►carretera estatal, ►carretera federal, ►carretera interestatal, ►carril de emergencia); *f* **autoroute [f]** (►Infrastructure routière à grande distance comportant deux chaussées à sens unique, séparées par un ►terre-plein central ou une double ►glissière de sécurité, chaque chaussée comporte sur le côté extérieur une ►bande d'arrêt d'urgence, elle-même en général bordée par une glissière de sécurité ou un terre-plein ; une autoroute ne comporte aucun croisement à niveau ; les croisements entre autoroutes et avec le réseau routier ordinaire se font par des ►échangeurs de liaison à deux chaussées, chacune à un sens de circulation et sans croisement, séparées par un terre-plein central, accessible en des points aménagés à cet effet ; l'accès et la sortie se font par des ►bretelles dont le tracé est tangentiel à celui de la chaussée, appelées « voies d'accélération » ou de « décélération » ; le service aux usagers est assuré par des ►aires de repos et des ►aires de services ; le réseau autoroutier français comporte en janvier 2006 10 843 km [8236 km concédées et 2607 km non concédées] et est en grande partie à péage basé sur un forfait ou sur la distance parcourue ; la vitesse maximum est fixée en rase campagne à 130 km/h [110 km/h par temps de pluie] dans les zones urbaines à 110 km/h ; le concept d'autoroute apparaît en 1909 en Allemagne avec la société AVUS [**A**utomobil-**V**erkehrs-**u**nd **Ü**bungs-**S**traße] ; d'une longueur de 9 km et situé au sud-ouest de Berlin, ce nouveau type de route est mis en service en 1921 ; la première autoroute au monde, voie protégée pour les hommes d'affaires à Long Island et longue de 65 km a été créée en 1914 aux États-Unis par un fils de milliardaire new-yorkais WILLIAM K. VANDERBILT JR ; en Europe « l'autoroute des lacs » [*Autostrada dei Laghi*] inaugurée en 1924 et conçu par l'ingénieur milanais PIERO PURICELLI [1883-1951], relie Milan et Varese [77 km] ; le terme « Autobahn », « autoroute », créé en 1929 pour la première fois par ROBERT OTZEN dans le cadre du « Autobahnprojekt HaFraBa » [Projet d'autoroute Hambourg-Frankfort-Bâle], s'intègre dans le vocabulaire populaire à partir de la création en 1933 du réseau allemand de la « Reichs-autobahn » dans le cadre de la politique agressive et expansion-

niste des nationalsocialistes ; la première liaison autoroutière française a été inaugurée le 9 juin 1946 sur une vingtaine de km entre Saint-Cloud et Orgeval aujourd'hui devenue A13 ; le réseau autoroutier français a été lent à se développer ; la première liaison d'envergure fut celle de l'axe Nord-Sud [Lille-Paris-Lyon-Marseille], inaugurée en octobre 1970 ; ►grand itinéraire, ►voie express) ; *g* **Autobahn [f]** (Kreuzungsfreie Fernstraße mit mindestens zwei Fahrstreifen je Fahrtrichtung, die baulich durch einen ►Mittelstreifen getrennt sind, sowie eines Randstreifens bzw. ►Standstreifens zum Abstellen von Pannenfahrzeugen. Eine Mindestgeschwindigkeit ist vorgeschrieben, so dass Traktoren oder Radfahrer nicht fahren dürfen. Mit dem Begriff **A.** war bis in die 1930er-Jahre kein bestimmter Standard verbunden. Somit ist die erste „Autobahn" der Welt, der zunächst 45 Meilen lange *Long Island Motor Parkway,* der von 1908-1911 in den USA von dem exzentrischen Milliardärssohn WILLIAM K. VANDERBILT JR. aus New York gebaut wurde. **In D.** entstand als erste Autostraße 1912-1921 die Versuchs- und Rennstrecke AVUS [**A**utomobil-**V**erkehrs- und **U**ebungs-**S**traße] in Berlin-Grunewald. In Italien wurde 1924 die erste lange Autobahn Europas eingeweiht, die „Autostrada dei Laghi", konzipiert von dem Mailänder Ingenieur PIERO PURICELLI [1883-1951], von Mailand nach den oberitalienischen Seen [Milano-Varese-Como-Gallarate — heute A 8 und Teil der A 9]. Durch die Italiener inspiriert, wurde von 1929-1932 von Köln nach Bonn eine vierspurige 20 km lange Kraftwagenstraße, damals noch ohne Fahrbahntrennung und als **„Nur-Autostraße"** bezeichnet, heute die 6-streifige A 555, gebaut. Mit dem Beginn des Langzeitprojektes zum Bau der „Reichsautobahnen" im Sommer 1933 wurde der Begriff Autobahn, der 1929 erstmals von ROBERT OTZEN, dem Vorsitzenden des Autobahnprojektes Hamburg-Frankfurt/M.-Basel [HaFraBa] in Analogie zur Eisenbahn eingeführt. Unter der Leitung des Generalinspektors für das deutsche Straßenwesen, Dr. FRITZ TODT, wurde mit dem Landschaftsarchitekten Prof. ALWIN SEIFERT und dem Architekten Prof. PAUL BONATZ die Reichsautobahnen in den 1930er-Jahren auf der Grundlage des Reichsverkehrswegeplanes konzipiert und gebaut; ►Bundesfernstraße, ►Fernstraße, ►Schnellverkehrsstraße).

2282 freeway corridor [n] [US] *plan. trans.* (Strip of land encompassing a highway) *syn.* motorway corridor [n] [UK]; *s* **trazado [m] de autopista**; *f* **tracé [m] d'autoroute**; *g* **Autobahntrasse [f]** (Aus dem Franz. *tracé* >Spur, Umriss<; geplante und später im Gelände ausgesteckte Linienführung einer Autobahn).

2283 freeway interchange [n] [US] *trans.* (Junction of two superhighways [US]/motorways [UK] on different levels; ►interchange, ►grade-separated junction); *syn.* motorway intersection [n] [UK]; *s* **cruce [m] de autopistas** (Empalme entre dos autopistas a diferentes niveles para evitar los cruces a nivel; ►nudo de enlace, ►nudo sin cruces a nivel); *f* **échangeur [m] autoroutier** (Raccordements entre autoroutes effectués sur des niveaux distincts de telle sorte qu'un cisaillement ne se produise sur l'autoroute ; ►diffuseur, ►échangeur) ; *syn.* nœud [m] autoroutier ; *g* **Autobahnkreuz [n]** (Kreuzung zweier Autobahnen auf verschiedenen Ebenen, so dass ein Anschluss in allen Richtungen kreuzungsfrei möglich ist; ►Anschlussstelle, ►planfreier Knotenpunkt).

freeway junction [n] [US] *trans.* ►interchange.

2284 freeway median strip [n] [US] *trans.* (Mostly vegetated ►median strip [US]/central reservation [UK] which separates traffic flowing in opposite directions); *syn.* motorway central reservation [n] [UK]; *s* **franja [f] divisoria central de autopista** (Banda —generalmente verde— de separación entre las dos direcciones de una autopista; ►franja divisoria central); *syn.* banda [f] central de separación de autopista; *f* **terre-plein**

[m] central [T.P.C.] d'une autoroute (Séparation matérielle des deux sens de la circulation d'une autoroute ; ▸terre-plein central) ; g Autobahnmittelstreifen [m] (I. d. R. begrünter ▸Mittelstreifen zwischen den Richtungsfahrbahnen).

2285 freeze [n] in a depression met. (Frost originating in depressions and forming a ▸frost pocket); s helada [f] en hondonada (Helada que se presenta en depresiones del terreno y forma una ▸hondonada de aire frío); f gelée [f] de bas fond (Gelée se produisant dans une dépression et donnant lieu à un ▸lac d'air froid) ; g Muldenfrost [m] (...fröste [pl]; Frost, der in Geländemulden entsteht und einen ▸Kältesee bildet).

2286 French classic garden [n] gard'hist. (Garden characterized by strong axes, divided by *allées* into a regular grid-like plan, using a hierarchy of ornate *parterres de broderie, parterres de gazon* [turf], and *bosquets*. Such baroque gardens were created predominantly in the 17th century; e.g. Vaux-le-Vicomte [1656-1661, André Le Nôtre], Chantilly [1663-1666, A. Le Nôtre], and Versailles [1662-1688, A. Le Nôtre]); *syn.* baroque garden [n]; s jardín [m] francés (Jardín de rígida distribución axial y caracterizado particularmente por el uso de los parterres, siendo típico de la época barroca en la Francia del siglo XVII. El ejemplar más conocido mundialmente es el de Versailles); *syn.* jardín [m] barroco; f jardin [m] à la française (Forme de jardin à architecture géométrique rigoureuse, dont les principaux éléments, allées, pelouses et bosquets, parterres et broderies, fontaines et plans d'eau sont ordonnés dans une symétrie absolue ou relative le long d'un axe central fixé sur le centre de l'édifice ; André Le Nôtre, créateur incontesté du jardin classique, privilégie, à la différence des Italiens, les surfaces planes et les amples niveaux offrant un large panorama ; la perspective à perte de vue, le gigantisme des proportions, la hiérarchie stricte des éléments et la fixation du jardin sur le château symbolisent le rayonnement de la Cour et la gloire du Grand-Siècle ; l'époque des jardins classiques en D. prend naissance à la fin du XVIIᵉᵐᵉ siècle principalement sous l'influence des jardins classiques français tels que Vaux le Vicomte [1656-1661, André Le Nôtre], Chantilly [1663-1666, A. Le Nôtre] et en particulier Versailles [1662-1688, A. Le Nôtre] ; l'utilisation de la dénomination baroque dans la terminologie allemande trouve son origine dans la définition des périodes stylistiques par les historiens d'art allemands à partir de la fin du XIXᵉ siècle ; DUA 1996) ; *syn.* jardin [m] classique français ; g Barockgarten [m] (Aus dem Franz. *baroque* >sonderbar<, >schwülstig<; eigentlich >schief<, >unregelmäßig<; streng axial ausgerichtete architektonische Parkanlage mit Broderien geschmückten Parterres, Bosketts und Wasseranlagen. Die Gestaltung passt sich nicht dem Gelände an, sondern unterwirft es und schließt die Gartenteile mit dem Bauwerk als Ausdruck einer großhöfischen Gesellschaftsordnung zusammen. Der B. symbolisiert eine staatlich legitimierte Herrschaft über Land und Natur. Erste Barockgärten wurden in Italien geschaffen, z. B. Villa Aldobrandini in Frascati [ab 1600], Villa d'Este in Tivoli. In D. begann die barocke Gartenkunst am Ende des 17. Jhs. und wurde wesentlich durch die französischen Gärten in Vaux-le-Vicomte [1656-1661, André Le Nôtre], Chantilly [1663-1666, A. Le Nôtre] und vor allem durch den Schlossgarten von Versailles [1662-1688, A. Le Nôtre] geprägt. Barocke Großanlagen in D. sind z. B. der „Große Garten" von Herrenhausen in Hannover [1665ff], Nymphenburg [1671] und Schleißheim [1683] bei München); *syn.* klassischer französischer Garten [m].

2287 French drain [n] constr. (Gravel-filled ditch, sometimes with porous drainage pipe at the bottom, and which may be covered with a layer of topsoil); s dren [m] francés (Canal pequeño relleno de grava, a veces con tubo de drenaje poroso en el fondo, que puede estar cubierto de tierra); *syn.* dren [m] fil-

trante, dren [m] ciego, zanja [f] de filtrado; f tranchée [f] drainante (Drains posés en fond d'une tranchée remblayée de materiaux filtrants permettant aux eaux recueillies d'être absorbées par le sol ou d'être évacuées) ; *syn.* tranchée [f] absorbante, tranchée [f] de drains ; g Entwässerungsgraben [m] (3) (Kiesgefüllter Graben, manchmal mit Dränrohr auf der Sohle); *syn.* Drängraben [m], Steindränung [f], Sauggraben [m], Sickergraben [m].

2288 frequency [n] phyt. (Syn-ecological term which records how often a certain species occurs in a separate ▸sample plot. The difference between 'frequency' and '▸constancy' is as follows: the former refers to a single population and thus illustrates the population's structure, whereas 'constancy' is determined by comparing various populations; ▸presence); s frecuencia [f] (Término de la sinecología que indica la cantidad de veces que aparece una especie en diferentes ▸superficies de muestreo en una misma población. La diferencia entre f. y ▸constancia es que la primera se refiere a la población individual y contempla así la estructura interna de la misma, mientras que la determinación de la constancia se utiliza para comparar distintas poblaciones entre sí; ▸presencia); f fréquence [f] (Terme synécologique caractérisant le pourcentage des relevés dans un même peuplement contenant une espèce par rapport au nombre total de relevés étudiés. La différence entre la « fréquence » et la « ▸constance » réside dans le fait que la première caractérise un peuplement déterminé par la même structure interne de la population, et que la seconde compare plusieurs peuplements de la même association ; ▸aire d'échantillonnage floristique, ▸présence) ; g Frequenz [f] (Dieser synökologische Begriff gibt Auskunft, an wieviel getrennten Probeflächen desselben Einzelbestandes eine bestimmte Art vorkommt. Der Unterschied zwischen F. und ▸Konstanz besteht darin, dass sich erstere auf den Einzelbestand bezieht und daher die Innenstruktur der Population beleuchtet, während die Konstanzbestimmung dem Vergleich verschiedener Populationen gilt; ▸floristische Probefläche, ▸Gesellschaftsstetigkeit).

2289 frequency [n] of fog met. (Commonness of occurrence of clouds of water droplets); s frecuencia [f] de nieblas *syn.* abundancia [f] de nieblas; f nébulosité [f] (Fraction de ciel couverte à un moment donné par l'ensemble des nuages, brouillards et brumes visibles) ; *syn.* fréquence [f] de brouillard ; g Nebelhäufigkeit [f] (Häufiges Vorkommen von Nebel).

2290 frequency [n] of use recr. (How often recreation facilities are used); *syn.* degree [n] of use, density [n] of use (2); s grado [m] de frecuentación (Frecuencia de uso de una instalación de recreo); f densité [f] de fréquentation (Degré d'utilisation d'un équipement) ; *syn.* taux [m] de fréquentation ; g Nutzungsintensität [f] von Erholungseinrichtungen (Häufigkeit der Nutzung von Erholungseinrichtungen); *syn.* Nutzungshäufigkeit [f].

2291 fresh air corridor [n] landsc. met. (Elongated or wedge-shaped area without buildings and mostly with low vegetation, which allows the passage of fresh air from ▸cold air source areas to climatically-deprived urban areas); *syn.* ventilation corridor [n]; s corredor [m] de aire fresco (Superficie no edificada y con vegetación baja de forma alargada o de cuña que sirve para dejar paso al aire fresco producido en ▸zonas de aparición de corrientes de aire frío y así ventilar zonas urbanas); f couloir [m] d'air frais (Espace non construit, de forme linéaire ou en coin, pourvu d'une couverture végétale le plus souvent basse, favorisant la pénétration de masses d'air frais dans les zones urbaines à partir des ▸sources d'alimentation en air froid) ; *syn.* couloir [m] de ventilation ; g Frischluftschneise [f] (Unbebaute und meist niedrig bewachsene, lineare oder keilförmige Fläche, die Frischluft aus ▸Kaltluftentstehungsgebieten in klimatisch belastete Räume führt).

2292 freshwater [n] *geo. limn.* (Water of a river or lake with less than 0.5% salt content); *s* **agua** [f] **dulce** (Agua que contiene menos de 0,5% de cloruro sódico); *f* **eau** [f] **douce** (Eaux intérieures dont la concentration en sels est < à 0,05 %) ; *g* **Süßwasser** [n] (Wasser der Binnengewässer mit weniger als 0,5 % Salz).

freshwater community [n] *phyt.* ▶free-floating freshwater community.

2293 friable soil [n] *agr. for. hort.* (Easily-crumbled soil with a well-developed and stable ▶crumb structure, which is advantageous for moisture migration, root growth and oxygenation); *syn.* crumbly soil [n] [also US]; *s* **suelo** [m] **friable** (Suelo poroso, granular, blando, fácilmente trabajable sin que se haga compacto; MEX 1983; ▶estructura grumosa); *syn.* suelo [m] grumoso; *f* **sol** [m] **à structure grumeleuse** (Dans la classification génétique des structures de sol, sol à structure construite due à des facteurs biologiques dont les particules sont des agrégats de forme arrondie irrégulière, très poreux de 1 mm à 1 cm et qui possèdent une porosité totale élevée ; PED 1979, 238 ; ▶structure fragmentaire) ; *syn.* sol [m] à structure en grumeaux, sol [m] grumeleux, sol [m] friable ; *g* **krümeliger Boden** [m] (Boden mit einer gut entwickelten und stabilen ▶Krümelstruktur I).

friable soil condition [n] *pedol.* ▶granular soil structure.

friction [n] *constr. stat.* ▶internal friction.

2294 frighten [vb] **away** *conserv. hunt. zool.* (Strike game or birds with sudden fear, so that they move to other perches or grazing spots); *syn.* scare [vb]; *s* **espantar** [vb] (Molestar a animales de caza o aves por lo que cambian su lugar de reposo o de pasto); *f* **effaroucher** [vb] (Déranger fréquemment le gibier ou les oiseaux de telle sorte qu'ils changent de lieu d'escale ou de pâture ; F., utilisation du substantif **dérangement** [m] ; l'espèce abandonne les sites de reproduction, principale cause de la régression ou de la disparition de l'espèce) ; *syn.* déranger [vb] irrémédiablement ; *g* **vergrämen** [vb] (Wild oder Vögel mehrfach stören, so dass sie andere Rast- oder Weideplätze aufsuchen).

fringe [n] *pedol.* ▶capillary fringe; *urb.* ▶urban fringe.

fringe [n], **green space** *landsc. urb.* ▶greenbelt [US]/greenbelt [UK].

fringe [n] **of a city/town** *urb.* ▶urban fringe.

2295 fringe area [n] *conserv. landsc. urb.* (Less-critical habitat surrounding and buffering the core area of, e.g, a nature preserve, but directly affecting its ecological health; ▶core area 2); *syn.* buffer area [n]; *s 1* **zona** [f] **periférica** (Área situada al borde p. ej. de una ciudad); *s 2* **zona** [f] **tampón** (Área adyacente o circundante de una reserva natural integrada por ecosistemas que van desde el natural y seminatural, adyacente a la ▶zona central, hasta el fuertemente modificado; cf. DINA 1987); *f* **zone** [f] **périphérique** (1. *Terme général* espaces situés à la périphérie, p. ex. du centre, de l'agglomération, d'un aménagement. 2. *Protection de la nature en F.* zone autour d'un parc national dans laquelle peuvent être prises toutes mesures permettant un ensemble de réalisations et d'améliorations d'ordre social, économique et culturel tendant à rendre plus efficace la protection de la nature à l'intérieur du parc ; D., zone de protection proprement dite, située autour de la réserve intégrale d'un site classé, d'un parc national, etc. ; ▶zone centrale) ; *syn.* aire [f] périphérique ; *g* **Randzone** [f] (1. Flächen am Rande, z. B. einer Stadt. 2. Bereich um die ▶Kernzone innerhalb eines [Natur]schutzgebietes); *syn.* Randbereich [m].

fringe [n] **of a water body, shallow** *phyt. zool.* ▶shallow water area.

fringing forest [n] *phyt.* ▶gallery forest.

from a leisure perspective [loc] *plan. recr.* ▶leisure-related.

2296 frontage [n] *urb.* (**1.** Front face of a building or a row of buildings. **2.** That side of a lot abutting on a street, [watercourse, or body of water]); *syn.* street frontage [n] (PSP 1987, 194); *s 1* **fachada** [f] **principal** (Cara principal de un edificio o una hilera de edificios); *s 2* **hilera** [f] **de fachadas** (Alineación de las fachadas principales a lo largo de una calle, un río o un lago); *syn.* frente [m] de fachadas; *f 1* **front** [m] **de façade** (Face antérieure d'un bâtiment, exposé à la vue donnant le plus souvent sur la rue ; *f 2* **alignement** [m] **de façades** (Alignement construit le long d'une voie, l'implantation des bâtiments étant en général continu et sans éléments décalés) ; *g* **Gebäudefront** [f] (Vorderansicht eines Gebäudes oder einer Häuserzeile ohne wesentliche Versprünge); *syn.* Vorderhausfassade(n) [f(pl)], Hausfront [f], Häuserfront [f], Straßenseite [f] eines Hauses/von Häusern.

frontage road [n] [US] *trans.* ▶service road.

frontal moraine [n] *geo.* ▶terminal moraine.

2297 front elevation [n] *plan.* (**1.** Vertical view of an object or building from the front. **2.** Front view of an object or building shown in a drawing, typically rendered as a two dimensional plane; *generic term* ▶elevation); *s* **alzado** [m] **frontal** (Vista de frente; *término genérico* ▶alzado); *syn.* elevación [f] frontal, plano [m] de elevación frontal; *f* **élévation** [f] **avant** (Projection de la face avant d'un objet sur un plan vertical parallèle à l'axe de l'objet ; *terme générique* ▶élévation) ; *syn.* vue [f] de face, vue [f] avant, vue [f] frontale ; *g* **Vorderansicht** [f] (Ansicht eines Objektes von vorne; *OB* ▶Ansicht).

2298 front end dumping [n] *constr.* (Construction method in earthworks, whereby the transport vehicle dumps [US]/tips [UK] soil over an exposed face of an oblong-shaped pile [US]/heap [UK] of earth on to the apex slope [the narrow part of the heap]; ▶earth filling 1, ▶side dumping); *syn.* elongated pile dumping [n] [also US]; *s* **terraplenado** [m] **frontal** (Método de construcción en terracerías por el cual el volquete descarga la tierra sobre un borde lateral de un talud alargado; ▶relleno, ▶descargamiento lateral); *f* **remblai** [m] **par avancement** (Méthode d'exécution de remblai par déchargement des matériaux en tête de talus ; les couches successives ainsi rapportées sont en général obliques ; ▶exécution d'un remblai, ▶déversement latéral) ; *g* **Kopfschüttung** [f] (*Erdbau* Einbaumethode, bei der die Transportfahrzeuge den Boden von einem vorhandenen langen Erdkörper auf die Kopfböschung [schmaler Teil eines Erdkörpers] abkippen; ▶Schüttung 1, ▶Seitenschüttung).

front garden [n] [UK] *urb.* ▶front yard [US].

frontline [n] *geo.* ▶foreland (2).

2299 front lot line [n] [US] *surv. syn.* front plot boundary [n] [UK]; *s* **límite** [m] **delantero de solar** *syn.* límite [m] delantero de lote de terreno/predio; *f* **limite** [f] **séparative avec l'espace public** (Limite d'une propriété qui donne sur les voies ou emprises publiques; le règlement d'une zone d'un PLU/POS peut spécifier que l'implantation d'une construction doit être réalisée à l'alignement de l'espace public, c.-à-d. sur la limite parcellaire qui sépare l'espace public de l'espace privé) ; *g* **vordere Grundstücksgrenze** [f] *syn. adm. leg.* vordere Flurstücksgrenze [f].

front plot boundary [n] [UK] *surv.* ▶front lot line [US].

front setback line [n] *leg. urb.* ▶setback line.

2300 front yard [n] [US] *urb.* (TGG 1984, 193; private area of green between street and façade of a house); *syn.* front garden [n] [UK]; *s* **jardín** [m] **delantero** (Jardín delante de la casa); *f* **jardin** [m] **de façade** (Jardin situé entre la voirie et la façade d'une maison d'habitation) ; *syn.* jardin [m] avant, jardin [m] sur rue, jardinet [m] de façade ; *g* **Vorgarten** [m] (Private, meist

F

einsehbare Grünfläche zwischen Straßengrenze und Hausfassade).

frost [n] *met.* ▶early frost, ▶late frost; *constr. hort.* ▶lifting of plants by frost; *met.* ▶radiation frost.

frost [n], **advective** *met.* ▶radiation frost.

frost [n], **radiative** *met.* ▶radiation frost.

2301 frost accumulation [n] *met.* (Nocturnal collection of cold, moisture laden air behind an obstructing barrier that freezes as it condenses on surfaces; *syn.* barrier effect [n] on frozen air flow; *s* **acumulación** [f] **de aire frío** (Concentración nocturna de aire frío detrás de una barrera); *f* **accumulation** [f] **d'air froid** (Masse d'air froid retenue par un obstacle) ; *g* **Kaltluftstau** [m] (Kaltluftansammlung hinter einem Hindernis).

2302 frost crack [n] *arb.* (Longitudinal split—whether internal or external—in the ▶heartwood of a stem or branch of a tree caused by internal stresses set up by freezing); *syn.* frost split [n] [also US]; *s* **grieta** [f] **por congelación** (Fisuras en el tronco o en las ramas que llegan generalmente hasta el ▶duramen, causadas por tensión interna en la madera generada por temperaturas extremadamente frías); *syn.* grieta [f] por helada; *f* **gélivure** [f] (Fente observée sur le tronc ou les branches des arbres, partant de l'écorce et pouvant atteindre le ▶bois parfait, occasionnée par des tensions internes dans le bois sous l'effet du gel) ; *g* **Frostriss** [m] (Riss im Stamm oder an Ästen, der meistens von der Rinde bis in das ▶Kernholz reicht, verursacht durch interne Spannungen im Holz, die bei extremen Kältetemperaturen auftreten).

frost damage [n] *constr.* ▶building frost damage, ▶pavement frost damage.

2303 frost danger [n] *landsc. met.* (Frost hazard typically found in low lying areas subject to nocturnal cold air drainage); *syn.* frost hazard [n]; *s* **peligro** [m] **de heladas** *syn.* riesgo [m] de heladas; *f* **danger** [m] **de gel** *syn.* risque [m] de gel ; *g* **Frostgefahr** [f] (Möglichkeit, dass Frost auftritt, oft in Geländemulden, in denen sich Kaltluft ansammelt); *syn.* Frostgefährdung [f].

2304 frost-desiccation [n] *agr. bot. for. hort.* (Drying out of plants in frozen soil with winterkill, especially of broad-leaved evergreens, caused by transpiration of excessive moisture; ▶drought damage); *syn.* winter desiccation [n], winter drying [n] (SAF 1983), winter burn [n] [also US], parch blight [n] (SAF 1983); *s* **desecación** [f] **invernal** (Desecación de las plantas debida a la falta de agua causada por la congelación del suelo en invierno; ▶daño por sequía; *f* **dessèchement** [m] **par le gel** (Dessèchement du feuillage ou des ramilles des végétaux provoqué par le vent lorsque, dans un sol gelé pendant l'hiver, l'alimentation en eau des parties supérieures n'est plus assurée ; ▶dommages causés par la sécheresse) ; *syn.* dessiccation [f] hivernale (DFM 1975), dessiccation [f] par le gel, dessèchement [m] hivernal ; *g* **Frosttrocknis [f, o. Pl.]** (Dürreerscheinung an Pflanzen, bei der diese oder deren Teile bei Schönwetter-Frostperioden auf Grund intensiver Sonneneinstrahlung verdunsten müssen und vertrocknen oder zurücktrocknen, da durch Bodenfrost oberirdische Teile nicht mehr mit genügend Wasser versorgt werden können; ▶Trockenschaden).

frost hazard [n] *landsc. met.* ▶frost danger.

2305 frost heave [n] *constr. phys.* (Upward movement of soil surface due to the expansion of frozen water stored between particles in the first 50 to 100cm of the soil profile due to saturation or poor or obstructed drainage; ▶frost-susceptible soil, ▶lifting of plants by frost, ▶pavement frost damage); *syn.* frost lifting [n]; *s* **levantamiento** [m] **por congelación** (Alzamiento de la superficie del suelo al formarse cristales, capas, lentejas de hielo en el mismo; ▶dislocamiento causado por helada, ▶levan-

tamiento de plantas por helada, ▶suelo susceptible a la congelación); *f* **foisonnement** [m] **par le gel** (Sous l'effet du gel, mouvement d'un revêtement de sol, d'une fondation provoqué par la formation de cristaux, de couches ou de lentilles de glace dans les couches supérieures du sol ; ▶déchaussement par le gel, ▶gonflement lors du dégel, ▶sol sensible au gel) ; *g* **Frosthebung** [f] (Aufwärtsbewegung von Wegedecken oder Fundamenten durch Bildung von Eiskristallen, Eislagen oder Eislinsen in den oberen Bodenschichten, die meist einen unzureichenden Wasserabzug haben; ▶Frostaufbruch, ▶frostempfindlicher Boden, ▶Hochfrieren von Pflanzen); *syn.* Frosthub [m], Hochfrieren [n, o. Pl.], Auffrieren [n, o. Pl.].

frost-heave [n] **of plants** [US] *constr. hort.* ▶lifting of plants by frost.

frost injury [n] *agr. arb. for. hort.* ▶plant frost injury.

frost lifting [n] *constr. phys.* ▶frost heave.

2306 frost-penetration depth [n] **(in soil)** *constr.* (Vertical distance measured from soil surface to the lowest soil horizon that is affected by freezing for a particular climate zone, which typically determines minimum structural footing depth); *s* **profundidad** [f] **de helada (en un suelo)** *syn.* profundidad [f] de helamiento (del suelo); *f* **profondeur** [f] **atteinte par le gel** *syn.* profondeur [f] du sol gelé ; *g* **Frosttiefe** [f] (Tiefe, bis zu der ein Boden gefroren ist oder auf Grund der Klimazone gefrieren kann).

2307 frost pocket [n] *met.* (Nocturnal collection of cold air in a low-lying area; ▶freeze in a depression); *s* **hondonada** [f] **de aire frío** (Concentración nocturna de aire frío en las zonas más bajas de un terreno; ▶helada en hondonada); *syn.* hondonada [f] helada, cuenca [f] con aire frío estancado, lago [m] de aire frío; *f* **lac** [m] **d'air froid** (*Stratification thermique* écoulement d'air froid et formation d'une nappe dans les parties les plus basses d'un terrain ; ▶gelée de bas fond ; LA 1981, 565 ; *terme spécifique* poche d'air froid) ; *syn.* trou [m] à gelée, trou [m] de gel [aussi CDN], creux [m] de gel ; *g* **Kältesee** [m] (Nächtliche Kaltluftansammlung an der tiefsten Stelle eines Geländes; ▶Muldenfrost); *syn.* Frostloch [n], Kälteinsel [f], Kaltluftsee [m].

2308 frostproof [adj] *constr.* (Characteristic of building materials unsusceptible to freezing; ▶frostproof soil); *s* **resistente a heladas** [loc] **(1)** (Característica de materiales de construcción de no alterarse por helada; ▶suelo no susceptible a la congelación); *f* **non gélif, ive** [loc] **(1)** (Caractéristique de certains matériaux non-sensibles au gel ; ▶sol non gélif) ; *syn.* insensible au gel [loc] ; *g* **frostsicher** [adj] **(1)** (So beschaffen, dass Baumaterialien durch Frost keine Beeinträchtigungen erfahren; ▶frostsicherer Boden).

2309 frostproof soil [n] *constr.* (Non-cohesive soil such as sand or gravel, especially with narrow-graded particle size distribution which is not susceptible to frost heave; *opp.* ▶frost-susceptible soil); *syn.* frost-resistant soil [n]; *s* **suelo** [m] **no susceptible a la congelación** (Suelo no cohesivo, p. ej. de arena o gravilla, que no retiene el agua y por lo tanto no sufre peligro de ▶levantamiento por heladas); *f* **sol** [m] **non gélif** (*opp.* ▶sol sensible au gel) ; *g* **frostsicherer Boden** [m] (Nicht bindiger Boden wie Sand oder Kies, besonders bei enger Stufung; LEHR 1981; *opp.* ▶frostempfindlicher Boden).

2310 frost protection [n] *agr. hort. landsc.* (Shelter from damage by frost); *s* **protección** [f] **contra heladas**; *f* **protection** [f] **contre le gel** ; *g* **Frostschutz** [m, o. Pl.] (Schutz gegen Einwirkungen des Frostes).

2311 frost resistance [n] *constr.* (Capacity of non-cohesive soil or backfill not to form ice crystals and changes in soil structure during frost, despite the existence of water); *s* **resistencia** [f] **a heladas** (Característica de suelo no cohesivo o de material de relleno de no formar lentejas de hielo o transformar su textura al presentarse fases de congelación); *syn.* insensibilidad [f] a heladas, durabilidad [f] al helamiento; *f* **gélivité [f]** (Aptitude des matériaux en présence d'eau à ne pas former de lentilles de glace ou à ne pas modifier leur unité structurale sous l'action du gel) ; *syn.* résistance [f] au gel ; *g* **Frostsicherheit [f]** (Eigenschaft von Schüttmaterial beim Gefrieren trotz vorhandenen Wassers keine Eislinsenbildung und Gefügeveränderungen zu zeigen); *syn.* Frostunveränderlichkeit [f].

2312 frost-resistant [adj] *agr. bot. constr. hort.* (Characteristic of building materials and plants to resist damage by frost; ▶frostproof); *syn. hort.* hardy [adj]; *s* **insensible a heladas [loc]** (Característica de plantas o materiales de construcción de no sufrir daños por heladas; ▶resistente a heladas 1); *syn.* resistente a heladas [loc] (2); *f* **résistant, ante au gel [loc]** (Propriété des matériaux et des végétaux à supporter sans dommages l'action du gel) ; *syn. pour les matériaux* ▶non gélif, ive 1, non sensible au gel [loc], insensible au gel [loc] ; *g* **frostbeständig [adj]** (Eigenschaft von Baustoffen und Pflanzen bei Frosteinwirkungen keine Frostschäden zu zeigen; ▶frostsicher 1); *syn.* frosthart [adj], frostresistent [adj], frostsicher [adj] (2), frostunempfindlich [adj].

frost-resistant soil [n] *constr.* ▶frostproof soil.

2313 frost-resistant subbase [n] [US]/**frost-resistant sub-base** [n] [UK] *constr.* (Drainage layer typically composed of crushed aggregate situated directly under the subbase [US]/sub-base [UK] of roads and pavements to reduce negative effects of moisture in the subgrade by preventing upward migration into the subbase course, e.g. formation of ice crystals, and to increase its bearing capacity; ▶blinding); *s* **capa [f] de protección contra las heladas** (Capa con capacidad de drenaje que se tiende directamente sobre el subsuelo como capa inferior de la estratificación de superficies revestidas para evitar influencias negativas de la humedad y mejorar la capacidad portante; ▶capa de limpieza); *syn.* capa [f] de perfilado, capa [f] de regularización; *f* **couche [f] antigel** (Sous-couche de fondation, non sensible au gel, destinée à isoler thermiquement le sol de fondation afin d'éviter les variations de portance liées à l'alternance gel-dégel ; on parle alors d'une chaussée hors gel ; DIR 1977, 32 ; ▶couche anticontaminante) ; *syn.* couche [f] non-gélive ; *g* **Frostschutzschicht [f]** (*Wegebau* dränfähige Schicht aus frostunempfindlichen Mineralstoffen direkt über dem Unterbau/Untergrund als unterste Schicht des Oberbaus zur Verringerung negativer Feuchtigkeitseinflüsse des Untergrundes auf den Oberbau [z. B. Eislinsenbildung] und zur Erhöhung seiner Tragfähigkeit; verwendet werden Kiese und Kies-Sand-Gemische, Gemische aus Splitt und Brechsand der Lieferkörnungen 0/5 bis 0/32, Gemische aus Schotter, Splitt und Brechsand der Körnungen 0/45 und 0/56 sowie Recycling-Baustoffe); *syn.* ▶Sauberkeitsschicht [f].

frost split [n] *arb.* ▶frost crack.

frost splitting [n] *constr. geo.* ▶congelifraction.

2314 frost-susceptible soil [n] *constr.* (Clay or clay-rich soil and loamy sand that have the capacity to absorb and retain water from below by capillary rise; *opp.* ▶frostproof soil); *s* **suelo [m] susceptible a la congelación** (Suelo arcilloso o arena limosa capaz de absorber y retener agua por acción capilar; *opp.* ▶suelo no susceptible a la congelación); *f* **sol [m] sensible au gel** (Argile ou sol riche en argile, sable argileux ayant la capacité d'absorber de l'eau sous-jacente par action capillaire ; *opp.* ▶sol non gélif) ; *g* **frostempfindlicher Boden [m]** (Ton oder tonreicher Boden, lehmiger Sand [eng gestuft] mit dem Vermögen, von unten Wasser kapillar aufsaugen zu können; *opp.* ▶frostsicherer Boden); *syn.* frostgefährdeter Boden [m].

frozen air flow [n], **barrier effect on** *met.* ▶frost accumulation.

2315 frozen root ball [n] *hort.* (Exposed root ball that is frozen solid in winter, thus facilitating the transplanting of semimature trees); *s* **cepellón [m] congelado** (Método de jardinería que facilita el transplante de leñosas grandes en invierno); *f* **motte [f] gelée** (Motte de gros végétaux mise à nue et exposée au gel ; méthode de phytotechnie ou opération horticole précédant la transplantation) ; *g* **Frostballen [m]** (Zum Durchfrieren frei gelegter Wurzelballen eines Großgehölzes. Gärtnerische Methode, um das Verpflanzen zu vereinfachen).

2316 fruit and vegetable garden [n] *gard.* (Generic term for a garden where produce is not sold commercially. *Specific terms* ▶herb garden, ▶kitchen garden, ▶medicinal herb garden, ▶orchard, ▶spices garden; *opp.* ▶ornemental garden); *s* **huerto [m]** (Término genérico para jardín en el cual —a diferencia del ▶jardín ornamental— se producen plantas aprovechables, pero sin fines comerciales; *términos específicos* huerto de verduras, ▶huerta 2, ▶huerto frutal, ▶jardín de hierbas finas, ▶jardín de plantas condimenticias, ▶jardín medicinal); *f* **jardin [m] de rapport** (Terme générique pour tout jardin de plantes culinaires et d'arbres fruitiers cultivés à des fins non commerciales ; *termes spécifiques* ▶jardin d'herbes, ▶jardin d'herbes condimentaires, ▶jardin médicinal, ▶jardin potager, ▶verger, par opposition au ▶jardin d'agrément) ; *syn.* jardin [m] d'utilité, jardin [m] utilitaire ; *g* **Nutzgarten [m]** (Nicht erwerbswirtschaftlich genutzter Garten; *UBe* Gemüsegarten, ▶Gewürzgarten, ▶Heilpflanzengarten, ▶Küchengarten, ▶Kräutergarten, ▶Obstgarten; *zum Unterschied* ▶Ziergarten).

fruit bush [n] [UK], **pyramidal** *hort.* ▶pyramidal dwarf fruit tree [US].

fruit culture [n] *agr. hort.* ▶fruit growing.

2317 fruit growing [n] *agr. hort.* (Branch of agriculture and horticulture dealing with permanent cultivation of fruit crops); *syn.* fruit culture [n]; *s* **fruticultura [f]** (Rama de la agricultura y de la horticultura que se dedica al cultivo de árboles frutales); *f 1* **culture [f] fruitière** (Terme générique) ; *f 2* **arboriculture [f] fruitière** (Ensemble des techniques appliquées aux arbres et aux arbustes qui produisent des fruits comestibles ; LA 1981) ; *g* **Obstbau [m, o. Pl.]** (Zweig der Landwirtschaft resp. des Gartenbaus, der den Anbau obsttragender Dauerkulturen betreibt); *syn.* Obstanbau [m, o. Pl.].

2318 fruit species [n] *agr. hort.* (Edible, mostly juicy, fleshy cultivated fruits, including the malaceous fruits of apple and pear, such stone fruits as cherry, peach and plum, and such berry fruits as strawberry, redcurrant [US]/red currant [UK], etc.; ▶fruit variety); *s* **especie [f] de fruta** (Fruta comestible, carnosa y generalmente jugosa de variedades cultivadas de frutales como el manzano, el peral, el naranjo, el limón [frutos de pepita], el mango, la cereza, el durazno [AL]/melocotón [Es], la ciruela [frutos de hueso], la grosella, la grosella espinosa [frutos de baya], la fresa/frutilla [CS], etc.; ▶variedad de fruta); *syn.* especie [f] frutal; *f* **espèce [f] fruitière** (LA 1981, 550 ; fruit en général juteux et charnu d'espèces cultivées telles que la pomme, la poire [fruits à pépins], la cerise, la griotte, la pêche, la prune [fruits à noyau], la groseille, la groseille à maquereau [baies et petits fruits], la fraise, etc. ; ▶variété fruitière) ; *g* **Obstart [f]** (Essbare, meist saftreiche, fleischige Frucht von Kultursorten wie z. B. Apfel, Birne [Kernobst], Süßkirsche, Sauerkirsche, Pfirsich, Pflaume [Steinobst], Brombeere und Himbeere [*bot.* Sammelsteinfrüchte], Johannisbeere, Stachelbeere [Beerenobst], Erdbeere [*bot.* Sammelnussfrucht] etc.; ▶Obstsorte).

2319 fruit tree [n] *hort.* (Malaceous or stone fruit trees are measured in Europe by different stem heights: **in U.S.**, fruit trees

F

are measured in mature size at these heights: standard 18-25 ft, semi-dwarf 12-15 ft, dwarf 8-10 ft, miniature 6-8 ft; **D. and U.K.**, tall standard [UK]/►shade tree [US] 2 [1.80-2.00m], half standard [UK]/►semi-dwarf fruit tree [US] [1.20-1.50m], short standard [UK]/►dwarf fruit tree [US] [1.00-1.20m], short stem [0.40-0.60m]; ►espaliered fruit tree, ►miniature fruit tree [US] 2, ►pyramidal dwarf fruit tree [US]/pyramidal fruit bush [UK], ►pyramidal fruit tree); *s* **árbol** [m] **frutal** (Leñosa de diferentes alturas de tronco. En D. se diferencian las siguientes formas: ►árbol de pie alto [1,80-2,00 m], ►árbol de pie medio [1,20-1,50 m], árbol de pie cuarto o de metro [0,80-1,00 m] y ►árbol de pie bajo [0,40-0,60 m]. En EE.UU. no se diferencian por la altura del tronco, sino por la altura final del árbol en edad madura: estándar 18-25 pies, árbol de pie medio 12-15 pies, árbol de pie bajo 8-10 pies, árbol miniatura 6-8 pies; ►arbusto de tipo «Spindel», ►frutal de porte piramidal, ►frutal enano, ►frutal en espaldera); *syn.* frutal [m]; *f* **arbre** [m] **fruitier** (A. f. à pépins ou à noyaux ; on distingue les formes d'**a. f.** suivantes avec les hauteurs de tiges mesurées à partir du collet jusqu'à la branche la plus basse : l'►arbre de haute-tige [180 cm], l'►arbre demi-tige [130 cm], l'►arbre courte-tige [80 cm] l'arbre mini-tige [50 cm] ; cf. norme N.F. V 12 052 ; ►arbre fruitier taillé en espalier, ►buisson [d'arbres fruitiers], ►fuseau [d'arbres fruitiers], ►quenouille [d'arbres fruitiers]) ; *g* **Obstbaum** [m] (Kern- oder Steinobstgehölz mit unterschiedlichen Stammhöhen. Es werden folgende Obstbaumformen unterschieden: ►Hochstamm 1 [1,80-2,00 m], ►Halbstamm [1,20-1,50 m], Viertel- oder Meterstamm [0,80-1,00 m] und ►Niederstamm [0,40-0,60 m]. In den USA wird nicht nach Stammhöhe, sondern nach Höhen unterschieden, die der Baum im Endzustand erreicht; ►Buschbaum, ►Formobstbaum, ►Spindel, ►Spindelbusch).

fruit tree [n]**, fan-trained** *hort.* ►palmette espalier.

fruit tree grove [n] [US] *hort.* ►orchard.

fruit tree planting [n] [US]**, dispersed** *agr. conserv.* ►traditional orchard [UK].

2320 fruit tree pruning [n] *hort.* (Cutting back of undesirable fruit tree shoots and old unproductive limbs. The pruning measures, depending on the age and development condition of a fruit tree are carried out in stages as follows: **1.** Pruning to train the tree generally. **1a.** Pruning during planting [e.g. selection and shaping of the leader, removal of inward-growing competing shoots; **1b.** pruning after the first year [e.g. removal of competitive branches, training of the leader, pruning of the leader to lengthen the trunk]; **1c.** pruning until the first picking of the fruit [e.g. removal of excessive branches, creation of fruiting branches from the leader, pruning of the leader to lengthen the trunk]; **2.** general pruning maintenance [e.g. leader is no longer pruned, thinning of branches and height restriction, lateral branches are subordinated in favour of the leader, thinning out and height restriction]; **3.** fruit branch rejuvenation; **4.** rejuvenation cut [e.g. thinning out, shortening of branches or training of more favorable branch length); *s* **poda** [f] **de (árboles) frutales** (Corte de retoños no deseados o de ramas viejas improductivas, que se realizan dependiendo de la edad y del desarrollo del frutal en las siguientes etapas: **1.** Poda de formación, **2.** poda de mantenimiento, **3.** poda de fructificación y **4.** poda de rejuvenecimiento); *f* **taille** [f] **d'arbres fruitiers** (En arboriculture fruitière la taille a pour fonction d'obtenir une végétation vigoureuse, de favoriser la croissance des boutons à fruits, d'obtenir une fructification en rapport avec les conditions stationnelles ; cf. ADT 1988, 18) ; *g* **Obstbaumschnitt** [m] (Schnittmaßnahmen, die je nach Alter und Entwicklungsstand eines Obstbaumes wie folgt durchgeführt werden: **1.** Erziehungsschnitt: **1a.** Pflanzschnitt [z. B. Leitäste auswählen und formieren, nach innen gerichtete Konkurrenztriebe entfernen; **1b.** Schnitt nach dem ersten Standjahr [z. B.

Konkurrenztriebe entfernen, Leitäste formieren, Rückschnitt der Leitäste und Stammverlängerung]; **1c.** Schnitt bis Ertragsbeginn [z. B. überflüssige Triebe entfernen, Fruchtäste an Leitästen anbauen, Rückschnitt der Leitäste und Stammverlängerung]. **2.** Instandhaltungsschnitt [z. B. Leitäste nicht mehr anschneiden, Auslichten und Höhenbeschränkung, Seitenäste den Leitästen unterordnen, Auslichten und Höhenbeschränkung]. **3.** Fruchtholzverjüngung. **4.** Verjüngungsschnitt [z. B. Auslichten, Einkürzen resp. auf günstige Verlängerungen weiterleiten).

fruit trees [npl] *agr. conserv.* ►plowed land with fruit trees [US]/ploughed land with fruit trees.

2321 fruit variety [n] *agr. hort.* (Fruit cultivar which was developed by cross-breeding, mutation or selection; ►fruit species); *s* **variedad** [f] **de fruta** (►Especie de fruta modificada por cruce, mutación o selección); *f* **variété** [f] **fruitière** (►Espèce fruitière obtenue par croisement, mutation ou sélection) ; *g* **Obstsorte** [f] (Durch Kreuzung, Mutation oder Züchtung veränderte ►Obstart).

2322 fry [n] *agr.* (Young fish raised in ►pond farming; ►restocking of fish); *s* **alevín** [m] (Pequeño pez criado en ►piscicultura de estanques para la repoblación o con fines comerciales; ►suelta de peces); *f* **alevin** [m] (Jeune poisson destiné au repeuplement des eaux ou à l'élevage ; LA 1981 ; ►alevinage, ►pisciculture en étangs, ►réempoissonnement) ; *g* **Setzling** [m] (1) (In der ►Teichwirtschaft herangezogener Jungfisch; ►Fischeinsatz); *syn.* Besatzfisch [m], Setzfisch [m].

fuel reprocessing plant [n] *envir.* ►nuclear fuel reprocessing plant.

fulfillment [n] *prof.* ►planning goal fulfillment.

2323 fulfillment [n] **of planning requirements** [US]/ **fulfilment** [n] **of planning requirements** [UK] *plan.* (Satisfactory achievement of predetermined planning objectives and/or statutory directives); *s* **satisfacción** [f] **de la demanda** *syn.* satisfacción [f] de las necesidades; *f* **couverture** [f] **de la demande** *syn.* satisfaction [f] des besoins ; *g* **Bedarfsdeckung** [f] (Ausreichende Bereitstellung der erforderlichen Notwendigkeiten auf Grund planerischer Ziele oder gesetzlicher Vorgaben).

2324 fulfillment [n] **of professional responsibilities** [US]/**fulfilment** [n] **of professional responsibilities** [UK] *contr. prof.* (Satisfactory completion of professional services as agreed upon in a contract); *s* **cumplimiento** [m] **de prestaciones profesionales;** *f* **accomplissement** [m] **de la mission de maîtrise d'œuvre** (par un concepteur/aménageur) ; *g* **Erfüllung** [f] **der Planungsleistungen** (Vollständige Erbringung/Abwicklung der vertraglich vereinbarten Arbeiten durch einen Planer/ eine Planerin); *syn.* Erfüllung [f] der planerischen Leistungen.

2325 fulfillment [n] **of the contract** [US]/**fulfilment** [n] **of the contract** [UK] *contr.* (Complete execution of all required planning or construction work [US]/construction works [UK] according to a contract); *s* **cumplimiento** [m] **de contrato** (Ejecución completa de los trabajos de construcción y demás condiciones de un contrato); *f* **exécution** [f] **des engagements** (Concerne les engagements contractuels de bonne exécution des travaux pris par l'entrepreneur lors de la signature du marché) ; *g* **Vertragserfüllung** [f] (Vollständige Realisierung der in einem Vertrag dargelegten Leistungen und Bedingungen).

full contract value [n] [UK]**, computation of** *constr.* ►determination of final contract amount [US].

fumes [npl]**, exhaust** *envir.* ►exhaust gas.

fumes [npl]**, traffic** *envir.* ►exhaust gas.

function [n] *for. nat'res. recr.* ►forest function.

2326 functional efficiency [n] *ecol. plan.* (Effective functioning of a land use pattern, of natural systems, etc.; ►elasticity); *s* **capacidad** [f] **funcional** (Funcionamiento efectivo de un mosaico de usos del suelo, del régimen de la naturaleza, etc.; ►elasticidad de un ecosistema); *f* **fonctionnement** [m] (Capacité, p. ex. du zonage dans un document d'urbanisme à permettre un développement harmonieux de l'urbanisation ; ►résilience d'un écosystème) ; *g* **Funktionsfähigkeit** [f, o. Pl.] (Leistung z. B. eines Flächennutzungsmusters hinsichtlich der ausgewiesenen baulichen und sonstigen Nutzungen eine geordnete städtebauliche Entwicklung zu ermöglichen; **F.** des Naturhaushalts: ►Funktionsfähigkeit eines Ökosystems).

2327 functional landscape planning [n] [US] *landsc. leg. plan.* (Independent landscape architetural involvement in planning a development project; e.g. restoration and reuse of a refuse landfill area, or a mineral extraction site; ►development mitigation plan [US]/landscape envelope plan [UK] (LD 1990 [1], 59), ►functional planning [US]/sectoral planning [UK], ►professional contribution); *syn.* sectoral landscape planning [n] [UK]; *s* **planificación** [f] **sectorial del paisaje** (En D., contribución especializada en forma de un plan de gestión del paisaje p. ej. antes de la explotación de yacimientos superficiales; ►contribución profesional, ►planificación sectorial, ►plan paisajístico complementario); *f* **planification** [f] **sectorielle des paysages** (≠) (D., planification spatiale fixant au moyen du ►plan de protection et de mise en valeur des paysages les actions et mesures en faveur de la protection de la nature et des paysages s'exerçant sur l'aire d'un projet d'infrastructure routière, d'une carrière, etc. ; ►contribution spécialisée, ►planification sectorielle, ►schéma directeur paysager) ; *g* **landespflegerische Fachplanung** [f] (Eigenständige Fachplanung in Form eines ►Landschaftspflegerischen Begleitplanes, z. B. beim Bodenabbau gemäß Bodenabbaugesetz oder bei Straßenbauprojekten; ►Fachbeitrag, ►Fachplanung).

2328 functional plan [n] [US] *plan.* (Graphic and written document resulting from the process of ►functional planning [US]/sectoral planning [UK]); *syn.* sectoral plan [n] [UK]; *s* **plan** [m] **sectorial** (Resultado de la ►planificación sectorial); *f* **plan** [m] **sectoriel** (Résultat d'une ►planification sectorielle) ; *g* **Fachplan** [m] (Ergebnis einer ►Fachplanung); *syn.* Sachplan [m] [CH].

2329 functional planning [n] [US] *plan.* (Process whereby planning proposals are developed for a specific sphere of activity, usually governed by a particular authority/department/agency [whether ministerial or federal, state or local]; e.g. transportation, water supply, agriculture. These are then integrated in a comprehensive plan, after which detail plans are prepared at a larger scale for roads, utilities, drainage, etc.; ►multidisciplinary planning, ►overall area planning); *syn.* sectoral planning [n] [UK]; *s* **planificación** [f] **sectorial** (Proceso de planificación de una administración relacionado con un ámbito específico como p. ej. transportes, suministro de aguas o protección de la naturaleza; ►planificación general, ►planificación integrada); *syn.* planeamiento [m] sectorial; *f* **planification** [f] **sectorielle** (Par opposition à la ►planification intégrée, organisation du développement d'un domaine limité d'activités telles que p. ex. les transports, la gestion de l'eau, l'agriculture, la protection de la nature, etc. ; les objectifs de ces diverses **p. s.** sont coordonnés et intégrés par la suite dans plan général ; ►planification globale) ; *g* **Fachplanung** [f] (Im Gegensatz zur querschnittsorientierten Planung die planerische Bearbeitung eines abgegrenzten Sachgebietes, z. B. Verkehr, Wasserwirtschaft, Landwirtschaft, die von Ministerien, Ämtern und sonstigen Verwaltungen von Bund, Ländern und Gemeinden betrieben wird; anschließend werden die Inhalte dieser Pläne in einen Gesamtplan integriert und durch groß-

maßstäbige Pläne weiter detailliert; ►Gesamtplanung, ►Querschnittsplanung).

functional roof planting [n] *constr.* ►extensive roof planting.

functioning [n] **of natural (eco)systems, effective** *ecol.* ►natural potential.

funds [npl] *adm. econ.* ►public funds.

funds [npl] **of the European Community (EC)** *adm. plan.* ►regional funds of the European Community (EC).

fun fair [n] [UK] *recr.* ►amusement park.

2330 fungal flora [n] *bot.* (All fungus species *[Mycophyta]* growing in a particular area or region; *generic term* ►flora 1); *s* **flora** [f] **micológica** (Conjunto de especies de hongos *[Mycophyta]* que crecen en un determinado lugar; *término genérico* ►flora); *f* **flore** [f] **mycologique** (Ensemble des espèces de champignons *[Mycophyta]* en un lieu donné ; *terme générique* ►flore) ; *g* **Pilzflora** [f] (Systematisch erfasste Gesamtheit der Pilzarten *[Mycophyta]* in einem bestimmten Gebiet; *OB* ►Flora; **2.** Pilzbestimmungsbuch für ein bestimmtes Gebiet).

2331 fungicide [n] *chem. hort. phythopath.* (Chemical ►pesticide to control fungal plant infestation; ►biocide, ►pest and weed control, ►plant protection agent); *s* **fungicida** [m] (Producto químico utilizado para eliminar hongos; ►biocida, ►control antiplagicida, ►pesticida); *f* **fongicide** [m] (Substance ou préparation ayant la propriété de tuer les champignons et utilisée pour lutter contre les champignons nuisibles aux cultures et aux produits récoltés. *Remarque* les fongicides et les bactéricides sont souvent groupés sous le nom de « produits anticryptogamiques » ; LAH 1983 ; ►biocide, ►pesticide, ►protection des végétaux) ; *syn.* anticryptogamique [m] ; *g* **Fungizid** [n] (Mittel [des Pflanzenschutzes] zur Bekämpfung von pilzlichen Krankheitserregern; ►Biozid, ►Pestizid, ►Pflanzenschutz).

fungus [n] *agr. for. hort. phytopath.* ►noxious fungus.

2332 fungus disease [n] *agr. for. hort.* (Plant disease caused by specific fungal ►decomposers 1; e.g. rust disease [class of *Urediniomycetes*], blight, powdery mildew *[Erysipha]*, downy mildew *[Peronospora]*, mould [e.g. *Penicillium* species]; ►noxious fungus); *s* **micosis** [f] (Nombre genérico de las infecciones producidas por hongos; *términos específicos* **m.** defoliadora, **m.** endocelular, **m.** lignivora, etc.; ►descomponedor, ►hongo nocivo); *f* **maladie** [f] **cryptogamique** (Affection d'une plante provoquée par un champignon parasite ; LA 1981, 714 ; ►champignon parasite, ►décomposeur) ; *syn.* mycose [f] ; *g* **Pilzkrankheit** [f] (Durch bestimmte ►Destruenten, z. B. Rostpilz, Brandpilz, Echter *[Erysipha]* und Falscher Mehltaupilz *[Peronospora]*, Schimmelpilz [z. B. *Penicillium*-Art], verursachte Pflanzenkrankheit; ►Schadpilz; ARB 1983, 568); *syn.* Pilzbefall [m] (±).

2333 furnish [vb] **and install** [vb] *constr. contr.* (*abbr.* **f & i**, i.e. provide materials and put in place for the execution of the work [US]/works [UK]; ►placement of soil, ►supply of materials); *syn.* provide [vb] and install [vb], supply [vb] and install [vb]; *s* **proveer** [vb] **e instalar** [vb] (Provisión de mano de obra y entrega de materiales para su utilización en una obra; ►relleno de tierra, ►suministro de materiales); *f 1* **fournir** [vb] **et mettre** [vb] **en place** (Traduction littérale. En français on utilisera de préférence l'expression formée à partir de substantifs) ; *f 2* **fourniture** [f] **et mise** [f] **en place** (►approvisionnement de matériaux, ►remblaiement de/en terre) ; *g* **liefern** [vb] **und einbauen** [vb] (►Einbau von Boden, ►Baustofflieferung).

furniture [n] *arch. constr. urb.* ►installation of street furniture, ►movable outdoor furniture, ►outdoor furniture, ►piece of street furniture, ►street furniture.

F

furniture [n]**, site** *arch. constr. urb.* ▶outdoor furniture.

furrow [n] *constr. hort.* ▶planting furrow.

2334 furrow irrigation [n] *agr. for.* (Partial surface flooding method normally used with clean-tilled crops where water is applied in furrows or rows of sufficient capacity to contain the desired flow; cf. RCG 1982; ▶cross-slope furrow irrigation); *syn.* corrugation method [n] (FAO 1960); *s* **riego [m] por infiltración** (Método tradicional de irrigación en el que el agua es conducida a surcos en el terreno, a través de los cuales circula y se infiltra en el suelo; ▶método de irrigación del surco oblicuo a la pendiente); *syn.* irrigación [f] por surcos, irrigación [f] por infiltración, riego [m] por surcos, regado [m] por surcos; *f* **irrigation [f] à la raie** (Infiltration progressive de l'eau d'arrosage par ruissellement le long de rigoles parallèles ; ▶méthode d'irrigation par ruissellement avec épis) ; *g* **Furchenbewässerung [f]** (Verteilen des Wassers auf der zu bewässernden Fläche in parallele Furchen, in denen es versickert; ▶Querfurchenbewässerung).

2335 furrow planting [n] *constr.* (**1.** *Bioengineering* contour planting method on slopes to avoid topsoil slippage by digging shallow trenches [▶planting furrows] in the subsoil stratum, thus interlocking subsoil and topsoil layers. **2.** Setting out plants in a plowed [US]/ploughed [UK] groove); *s* **plantación [f] en surcos** (*Bioingeniería* método de plantación en laderas para evitar el deslizamiento del suelo cavando zanjas pequeñas hasta el subsuelo para mezclar las capas del suelo; ▶surco de plantación); *f* **plantation [f] en sillons** (Technique de génie biologique, méthode de plantation assurant la protection des sols superficiels contre le glissement sur les talus, la couche superficielle étant retenue par le sol sous-jacent grâce à des ▶rigoles de plantations longues, étroites et peu profondes ; DFM 1975) ; *g* **Riefenpflanzung [f]** (*Ingenieurbiologie* Pflanzmethode mit ▶Pflanzriefen auf Normalböschungen, bei der Oberbodenrutschungen durch Verzahnung von Unterboden mit aufgetragenem Oberboden verhindert werden).

2336 further claims [npl] *constr. contr.* (Call for additional payments after the receipt of what was assumed to be the final payment of account); *s* **demanda [f] de nuevos pagos** (Exigir pagos adicionales después de haber recibido lo que se suponía que sería el pago final de la cuenta); *f* **demande [f] d'indemnité** (Réclamation présentée par l'entrepreneur après la signature du décompte général et définitif donnée sans réserve, comportant le versement d'une indemnité ; art. 50 du C.C.A.G.) ; *g* **Nachforderung [f]** (Geltendmachung von zusätzlichen Zahlungen nach vorbehaltloser Annahme der als solche gekennzeichneten Schlusszahlung; cf. § 16 Nr. 3 [2] VOB Teil B).

further education [n] [UK] *prof.* ▶continuing education [US].

2337 fusion [n] **of layers** *phyt.* (Two or more layers of vegetation are often closely united and always appear together in certain communities; there are ▶closely united layers, ▶overlapping layers and ▶dependent layers); *s* **estratos [mpl] relacionados** (Dos o más estratos de la vegetación pueden estar relacionados entre sí; se diferencia entre ▶estratos íntimamente relacionados, ▶estratos transgresivos y ▶estratos dependientes; cf. BB 1979); *syn.* fusión [f] de estratos; *f* **solidarité [f] écologique des strates** (Participation d'une strate aux conditions écologiques de sa voisine ; ▶strate dépendante, ▶strate solidaire, ▶strate transgressive) ; *g* **Schichtenbindung [f]** (Enges Verbundensein von zwei oder mehreren Vegetationsschichten in einer Pflanzengesellschaft; es wird zwischen ▶eng verbundenen Schichten, ▶übergreifenden Schichten und ▶abhängigen Schichten unterschieden).

2338 future extraction area [n] *min. plan.* (Site or area of a possible extraction project); *s* **zona [f] de explotación en** estudio (Área explotable aún sin delimitar); *f* **aire [f] d'exploitation projetée** (Dont les limites d'extraction des substances minérales ou fossiles ne sont pas encore exactement localisées) ; *syn.* aire [f] d'extraction projetée ; *g* **zukünftiges Abbaugebiet [n]** (Noch nicht genau für die Gewinnung von Steinen, Erden oder anderen Bodenschätzen festgelegtes Gebiet).

G

2339 gabion [n] *eng. wat'man.* (From Italian *gabbione [m]*: augmentative of *gabbia [f]* >cage<); wire mesh basket packed with loose rock material, e.g. for retaining or freestanding walls, stabilizing steep slopes or protecting stream banks; ▶reno mattress); *s* **gabión [m]** (Mallas o cestos de alambres llenos de grava que se utilizan para la protección de las orillas contra la erosión lateral; ▶gabión de fondo, ▶gabión de carrizo); *f* **gabion [m]** (Grande corbeille de grillage remplie de pierres pour la protection des rives ou comme éléments de soutènement de murs ; ▶gabion matelas, ▶matelas de gabions végétalisés) ; *g* **Drahtschotterbehälter [m]** (Mit Schotter gefüllter Drahtkorb, z. B. für den Uferschutz und als Konstruktionselement für Stützmauern und freistehende Mauern; ▶Gabionenmatratze, ▶Röhrichtgabione); *syn.* Drahtschotterbox [f], Drahtschotterkörper [m], Drahtskelettkörper [m], Gabione [f] (aus dem Ital. *gabbione [m]*: Vergrößerungsform von *gabbia [f]* >Käfig<), Steinkorb [m].

gabion [n]**, vegetated** *eng. wat'man.* ▶reno mattress.

2340 gallery forest [n] *phyt.* (Trees forming a strip along a watercourse, lake or in ravines within savanna[h] formations; ▶riparian woodland); *syn.* fringing forest [n]; *s* **bosque [m] de galería** (Bosque ribereño que se genera por la humedad edáfica y que contrasta con la masa de vegetación contigua, fisionómicamente diferente, que no goza de dicha humedad. Cuando el río en que se forma es suficientemente estrecho, las copas de ambas orillas se enlazan y de ahí su nombre. El prototipo es el que se desarrolla en los ríos de las sabanas tropicales; DGA 1986; el término se emplea también para denominar el ▶bosque ripícola); *syn.* bosque [m] en galería; *f* **galerie [f] forestière** (Forêts riveraines dans les régions tropicales, constituées par des bandes forestières en lisière des berges des cours d'eau, des lacs intérieurs ou des gorges couvertes par la savane ; ▶forêt alluviale) ; *syn.* forêt [f] galerie, forêt [f] galerie-ripisilve haute ; *g* **Galeriewald [m]** (Waldstreifen in wechselfeuchten Tropen entlang von Fließgewässern, Binnenseen oder Schluchten von Savannenformationen in einer sonst unbewaldeten Auenlandschaft; in ariden Gebieten gibt es entlang von Wadis **G.**wälder, die mit Phreatophyten, d h. Bäumen bestockt sind, die mit ihren Wurzeln das Grundwasser erreichen; ▶Auenwald).

2341 galvanized [pp/adj] *constr.* (Relating to steel, which has been coated with zinc as a protection against rust; zinc-coated); *s* **galvanizado/a [pp/adj]** (Relativo al acero que ha sido protegido contra la corrosión por medio de un baño de zinc); *f* **galvanisé, ée [pp/adj]** (Relatif aux pièces métalliques protégées contre la corrosion par une couche de zinc) ; *g* **feuerverzinkt [pp/adj]** (Stahl betreffend, der durch Eintauchen in ein Zinkbad mit einer Schutzschicht gegen Rost versehen ist).

2342 galvanized chain link metal mesh [n] *constr.* (Zinc-coated woven wire fabric); *s* **malla [f] metálica galvanizada**; *f* **grillage [m] galvanisé** ; *g* **verzinkter Maschendraht [m]**.

2343 game [n] *hunt.* (**1.** Term for ►wild animals covered by hunting laws; ►even-hoofed game, ►game birds, ►huntable game, short-furred game. **2.** Animal classified as game); *syn.* game animal [n], game species [n]; *s 1* **caza [f]** (Denominación común para aquellos ►animales salvajes que son cazados y están contemplados en la legislación de la caza vigente en España; comúnmente se distingue entre las especies de caza mayor, como lobo, ciervo, gamo, jabalí, cabra montés, y las de caza menor, como aves [ánsar, ánades, pato cuchara, perdiz, faisán, codorniz, etc.] y mamíferos pequeños [conejo, liebre, ardilla, comadreja, zorro, marta, etc.]; cf. RD 1095/1989, de 8 de septiembre; ►aves cinegéticas, ►caza mayor artiodáctila, ►especie cinegética); *syn.* fauna [f] cinegética; *s 2* **pieza [f] de caza** (Individuo de especie cinegética que es objeto de caza); *f* **gibier [m]** (**1.** Terme cynégétique qualifiant les espèces sauvages chassables soumises à la réglementation de la chasse ; ►espèce animale non domestique, ►espèce chassable, ►faune sauvage, ►faune sauvagine, ►gibier artiodactylique. **2.** Espèces de gibier) ; *g* **Wild [n]** (**1.** Waidmännische/weidmännische Bezeichnung für alle ►wild lebenden Tiere, die dem Jagdrecht unterliegen; cf. §§ 1 u. 2 BJagdG; ►jagdbares Wild. **2.** Zum Wild gehörendes Tier; das W. wird unterteilt in ►Federwild und Haarwild [Haarnutz- und Haarraubwild]. Alle Paarhufer des Haarwildes werden als ►Schalenwild bezeichnet).

game animal [n] *leg. hunt.* ►game, ►huntable game.

2344 game biologist [n] *game'man. hunt.* (Scientist dealing with gun techniques and ballistics, as well as hunting zoology: i.e. habits and behavio[u]r of ►huntable game, knowledge of footprints, flight silhouettes, identification of animal sounds, and the prevention of game damage; ►wildlife biologist); *s* **biólogo/a [m/f] cinegético/a** (Científico/a que domina tanto las cuestiones técnicas de la caza [armas, balística] como la zoología cinegética: formas de vida y comportamiento de las ►especies cinegéticas, sabe identificarlas por las huellas, las llamadas, etc. y se preocupa por minimizar los daños causados por la caza; ►biólogo/a de fauna salvaje; *syn.* biólogo/a [m/f] de fauna cinegética; *f* **cynégète [m]** (Scientifique spécialisé dans l'étude de l'art de la chasse et en particulier des comportements et formes de vie du gibier dont la chasse est autorisée [traces, chant des animaux] ainsi que de la lutte contre les dégâts causés par la faune sauvage ; ►biologiste spécialiste de la faune sauvage ; ►espèce chassable) ; *g* **Jagdwissenschaftler/-in [m/f]** (Jemand, der sich u. a. mit der Waffentechnik und Ballistik und der Jagdzoologie, d. h. mit den Lebens- und Verhaltensweisen ►jagdbarer Tiere [Fährtenkunde, Flugbilder, ,Ansprechen' von Lautäußerungen der Tiere] und mit der Bekämpfung von Wildschäden beschäftigt; ►Wildbiologe).

2345 game birds [npl] *hunt.* (Comprehensive term covering huntable birds); *s* **aves [fpl] cinegéticas** (Aves declaradas como especies objeto de caza); *f* **faune [f] sauvagine** (*Terme générique* les espèces d'oiseaux chassés ; *terme spécifique* gibier d'eau) ; *syn.* gibier [m] à plumes ; *g* **Federwild [n]** (Sammelbegriff für alle dem Jagdrecht unterliegenden Vogelarten: jagdbare Vögel); *syn.* Flugwild [n], Wildgeflügel [n], *obs.* Geflüg [n].

2346 game crossing [n] *game'man. hunt. trans.* (Passageway of game, especially for movement of game across a travelled way [US]/carriage-way [UK]; ►game trail); *s* **cruce [m] de animales** (Tránsito de animales salvajes o domésticos por una carretera; *término específico* cruce [m] de Ganado; ►paso de caza); *f* **traversée [f] de gibier** (Traversée fréquente d'une route par les animaux sauvages sur une zone de passage habituelle ; les

lieux les plus exposés à la traversée de gibier sont les routes en forêt ou en lisière de bois et les zones agricoles notamment à la tombée de la nuit et au lever du jour ; les périodes de l'année les plus propices aux traversées du gibier se situent en automne lorsque le gibier est à la recherche de nourriture mais est également plus souvent dérangé par la fréquentation de la forêt [champignons, chasse, etc.] et au printemps qui correspond à la période de reproduction et engendre des déplacements importants ; ►piste animale) ; *g* **Wildwechsel [m] (1)** (Das Überwechseln des Wildes, bes. über eine Verkehrsstraße ; ►Wildwechsel 2); *syn.* Wechsel [m].

game culling [n] [US] *hunt.* ►game thinning.

2347 game damage [n] *agr. for. hunt.* (Harm to agricultural crops and forest stands, such as scaling, browsing, selective losses of most preferred plant species, digging up of root crops, rubbing off [US]/fraying off [UK] on young trees, etc. caused by over-population of game; ►game population, ►over-browsing damage, ►scaling damage, ►rub off the velvet [US]/fray off the velvet [UK]); *s* **daños [mpl] causados por la caza** (Detrimento de los cultivos agrosilvícolas por una ►población cinegética demasiado densa, por descortezamiento, mordiscos, pasto selectivo de especies preferidas, etc.; *términos específicos* ►daño de descortezamiento, ►recomido; ►frotar la borra); *syn.* daños [mpl] causados por la fauna cinegética; *f* **dégâts [mpl] causés par le gibier** (Déprédations causées aux cultures agricoles et forestières par une ►population cynégétique trop importante, p. ex. dommages forestiers dus au grand gibier ; ils se manifestent par l'écorçage, l'abroutissement, la disparition sélective d'espèces préférentielles, le défoncement de cultures sarclées, etc. ; *termes spécifiques* dégâts de frotture, ►abroutissement, ►dégâts d'écorçage ; ►frotter le velours) ; *g* **Wildschaden [m]** (Schaden in land- und forstwirtschaftlichen Kulturen und in der Fischereiwirtschaft durch Wild. In der Landwirtschaft entsteht W. hauptsächlich durch Schwarzwild [z. B. Umbruch von Hackfruchtkulturen], gelegentlich auch durch Rot- und Damwild, in Obstgärten und Baumschulen durch Hasen und Kaninchen. In der Forstwirtschaft entstehen W.schäden durch Verbiss in Jungbeständen bis zum Stangenholzalter, Schälen, selektiver Ausfall bevorzugter Arten, Fegen von Jungpflanzen etc., der meist durch zu hohen ►Wildbesatz verursacht wird; in der Fischereiwirtschaft ist der Graureiher und Kormoran Hauptschadensverursacher durch Verzehr und Anpicken der kleinen Fische in Zuchtanlagen; *UBe* Fegeschaden, ►Schälschaden, ►Wildverbiss; ►Bast fegen).

2348 game density [n] *game'man.* (Number of individuals in the summer population of a game species per unit area; in D., the optimum g. d. would be 3-11 roe deer per 100 ha and 1-2 red deer per 100 ha; ►overprotection [US]/over-protection [UK], ►game damage, ►game population); *s* **densidad [f] de población cinegética** (Número de individuos de caza por especie cinegética y unidad de superficie; ►daños causados por la caza, ►población cinegética, ►sobreprotección de la caza); *f* **densité [f] de la population cynégétique** (Nombre d'individus de la population estivale d'une espèce cynégétique par unité de surface ; D., elle représente 3 à 11 chevreuils/100 ha, 1 à 2 cerfs/100 ha ; ►dégâts causés par le gibier, ►population cynégétique, ►réglementation excessive de la chasse) ; *g* **Wilddichte [f]** (Zahl der Individuen des Sommerbestandes einer Wildart pro Flächeneinheit; in D. wäre z. B. das Optimum 3-11 Rehe/100 ha, 1-2 Stück Rotwild/100 ha; ►Überhege, ►Wildschaden, ►Wildbesatz); *syn.* Dichte [f] des Wildbestandes.

game enclosure [n] *conserv. hunt.* ►shooting preserve [US]/ shooting reserve [UK].

2349 game hunting [n] *hunt.* (All activities and measures according to hunting protocol and customs including hunting laws and hunting science); *syn.* venery [n]; *s* **actividades [fpl] cinegéticas** (Conjunto de medidas necesarias para posibilitar la práctica de la caza, incluyendo la correspondiente legislación y la ciencia cinegética); *f* **exercice [m] de la chasse** (Toute mesure et activité conforme aux usages et à la culture des coutumes de la chasse) ; *g* **Jagdwesen [n, o. Pl.]** (Sämtliche Maßnahmen zur Ermöglichung einer waidgerechten/weidgerechten Jagdwirtschaft, die Jagdausübung und alle Aktivitäten zur Pflege des jagdlichen Brauchtums. Zum **J.** gehören ferner das Jagdrecht und die Jagdwissenschaft).

game keeper [n] [UK] *game'man. leg.* ►game warden.

2350 game management [n] *game'man. recr.* (**1.** Measures for the care and regulation of huntable game stock according to a policy of maintaining an animal population compatible with its habitat and to ensure that hunting in the long-term will be possible [e.g. in Africa and India]. In the early 1980s African countries began to change their policies, so that *Game Management* was replaced by ►wildlife management. **2. In D.**, ►hunting management is practised outside nature reserves. **3.** Otherwise, fallow deer are mostly kept as livestock on agricultural land or on farms in enclosures; ►balanced game management); *s* **gestión [f] de la caza (1)** (Conjunto de medidas orientadas al aprovechamiento cinegético de una región. En España la legislación estatal prevé la creación del Inventario Español de Caza y Pesca, con el fin de mantener la información más completa de las poblaciones, capturas y evolución genética de las especies cuya caza o pesca estén autorizadas, con especial atención a las especies migradoras. Más allá de las normas generales la competencia exclusiva de la caza reside en las CC.AA.; cf. art. 64 Ley 42/2007 y leyes orgánicas de Estatutos de Autonomía; ►conservación de la fauna silvestre, ►gestión de la caza 2, ►medidas de regulación de poblaciones de fauna); *f* **gestion [f] de la faune sauvage** (**1.** Gestion et régulation des espèces sauvages dont la chasse sportive est autorisée [p. ex. la chasse au grand gibier en Afrique et en Inde] ; depuis le début des années 80 s'est opéré dans le secteur de la chasse sportive un changement de mentalité avec le développement de principes de gestion durable des ressources naturelles aptes à mieux contribuer à la conservation de la biodiversité ; ►conservation et gestion durable de la faune sauvage. **2. F.,** à l'extérieur des ►réserves de chasse et de faune sauvage la ►gestion cynégétique est mise en œuvre par le Schéma Départemental de Gestion Cynégétique [SDGC] ; ►gestion sélective de peuplements d'animaux) ; *syn.* gestion [f] faunistique ; *g* **Wildbewirtschaftung [f]** (**1.** Pflege und Regulation von jagdbaren Wildbeständen mit einer für den Lebensraum tragbaren Bestandsdichte, so dass eine nachhaltige Jagd [z. B. Großwildjagd in Afrika und Indien] möglich ist; Anfang der 1980er-Jahre fand jedoch in afrikanischen Ländern ein Umdenken statt, so dass das *Game Management* durch das ►Wildlife Management abgelöst wurde; ►Bestandslenkung. **2. In D.** die ►jagdliche Hege außerhalb von Naturschutzgebieten. **3.** Landwirtschaftliche Wildhaltung oder die nutztierartige Aufzucht von vorwiegend Damwild in Gehegen auf Wildfarmen zur Wildfleischproduktion); *syn. für 3.* Wildhaltung [f].

game park [n] (1) *conserv. game'man. hunt. leg. obs.* ►game reserve.

2351 game park [n] (2) *recr.* (Enclosed area of woodland or forest used for recreation where ►even-hoofed game are usually kept; ►zoological garden, ►shooting preserve [US]/shooting reserve [UK]); *s* **parque [m] de animales salvajes** (Área cercada de bosque en la que viven sobre todo animales de caza destinada mayormente a la recreación; ►parque zoológico, ►reserva de

caza); *f* **parc [m] animalier** (Parc de loisirs implanté en milieu périurbain rassemblant en enclos dans un cadre naturel des animaux sauvages, en particulier les cervidés ; ►enclos de réserve de chasse, ►jardin d'acclimatation, ►jardin zoologique) ; *g* **Wildpark [m]** (Der Naherholung dienende eingezäunte Wald- oder parkähnliche Fläche, auf der besonders ►Schalenwild gehalten wird; ►zoologischer Garten, ►Wildgehege).

2352 game pass [n] *constr. ecol. game'man.* (Underpass or overpass of a travelled way [US]/carriage-way [UK] for game animals, in order to mitigate negative effects of a progressive process of habitat fragmentation; ►tunnel for amphibians; *specific term for an overpass* wildlife viaduct); *s* **cruce [m] para animales salvajes** (Pase subterráneo o sobre nivel creado para permitir a los animales cruzar carreteras sin peligro de ser alcanzados por los vehículos que circulan; ►túnel para anfibios); *syn.* paso [m] para animales salvajes; *f* **passage [m] à faune** (Ouvrage de passage supérieur ou inférieur, spécifique ou mixte réalisé afin de faciliter la traversée des routes et autoroutes dans les zones de transit du haut gibier ; ►tunnel pour amphibiens) ; *syn.* passage [m] de gibier, écoduc [m] ; *g* **Wildtierpassage [f]** (**1.** Über- oder Unterführung, die Wildtieren das niveauverschiedene Queren von Fahrbahnen ermöglicht, um negative Auswirkungen der fortschreitenden Verinselung von Lebensstätten in Natur und Landschaft zu mindern; ►Amphibientunnel. **2.** Angelegte Trasse, um Wildtieren die Wanderung in Gebiete zu ermöglichen, in denen sie auf Grund des Klimawandels ausweichen müssen, um bessere Überlebenschancen zu haben); *syn. zu 1.* Grünbrücke [f], *syn. zu 2.* Wildtierkorridor [m].

2353 game poaching [n] *leg. hunt.* (Taking ►game by unlawful means on the property of a public or private owner; ►fish poaching); *s* **caza [f] furtiva** (Infracción del derecho por cazar sin permiso o en zonas vedadas; ►caza, ►pesca furtiva); *f* **braconnage [m] de gibier** (Chasser le ►gibier sans permis à une période ou en un lieu interdit, avec des engins prohibés. En français « braconnage » inclut aussi la pêche sans permis ; ►braconnage de poissons) ; *syn.* chasse [f] illicite ; *g* **Jagdwilderei [f]** (Verletzung fremden Jagdrechtes durch Nachstellen, Fangen oder Aneignung von ►Wild oder durch Aneignung, Beschädigung oder Zerstörung einer Sache, die dem Jagdrecht unterliegt; cf. § 292 StGB; demzufolge ist ohne Erlaubnis des Jagdausübungsberechtigten z. B. das Sammeln von Federwildeiern oder Abwurfstangen **J.**; ►Fischwilderei; in A. StGB §§ 137 ff; in der Schweiz Art. 17 ff Jagdgesetz); *syn.* Jagdvergehen [n], *obs.* Jagdfrevel [m].

2354 game population [n] *game'man.* (Total number of game or individuals of a game species in a defined area; ►game, ►game density, ►restocking of game population); *s* **población [f] cinegética** (Cantidad total de animales de ►caza o de individuos de una especie de caza en una zona determinada; ►densidad de población cinegética, ►repoblación de la fauna cinegética); *f* **population [f] cynégétique** (Totalité du ►gibier ou des individus d'une espèce cynégétique sur un territoire donné ; ►densité de la population cynégétique, ►repeuplement des populations animales sauvages) ; *g* **Wildbesatz [m]** (Anthropogen beeinflusster Gesamtbestand an ►Wild oder an Individuen einer Wildart in einem Gebiet; ►Wilddichte, ►Aufstockung des Wildbestandes); *syn.* Wildbestand [m], Wildstand [m].

2355 game preserve [n] [US] *hunt.* (Tract of land or land and water managed for the production and harvesting of wildlife; RCG 1982, 76; ►hunting district, ►shooting preserve [US]/ shooting reserve [UK]); *syn.* hunting ground [n] [UK]; *s* **territorio [m] de caza** (Término genérico; ►terreno cinegético de aprovechamiento común, ►reserva de caza); *syn.* zona [f] de caza; *f* **territoire [m] de chasse** (**1.** Très grand ►domaine de chasse. **2.** Groupement de petits domaines de chasse constitué par les propriétaires fonciers d'une commune regroupés en une association de chasse communale agrée [A.C.C.A.] ; ▶enclos de réserve de chasse) ; *syn.* zone [f] de chasse ; *g* **Jagdgebiet [n]** (**1.** Ein großes Jagdrevier. **2.** Zusammenfassung mehrerer zusammenliegender kleinerer ▶Jagdbezirke mit land-, forst- oder fischereiwirtschaftlich nutzbaren Flächen, ▶Wildgehege).

game refuge [n] [US] *conserv. game'man. hunt. leg.* ▶game reserve.

2356 game reserve [n] *conserv. game'man. hunt. leg.* (TGG 1984, 157; area designated for the protection of game animals and birds within which hunting and fishing are either prohibited or strictly controlled; RCG 1982, 67; *more generally* wildlife reserve; cf. SAF 1983; ▶game park 2, ▶nature reserve, ▶shooting preserve [US]/shooting reserve [UK]); *syn.* game refuge [n] [also US], game sanctuary [n], *obs.* game park [n]; *s 1* **Refugio [m] de Fauna [Es]** (Zona establecida por decreto cuando por razones biológicas, científicas o educativas sea preciso asegurar la conservación de determinadas especies de la fauna cinegética, en las cuales el ejercicio de la caza está prohibido con carácter permanente, salvo por razones de orden biológico, técnico o científico. La declaración es competencia exclusiva de las CC.AA.; ▶parque de animales salvajes, ▶reserva de caza, ▶reserva natural 1); *s 2* **Reserva [f] Regional de Caza [Es]** (Zona geográficamente delimitada y sujeta a régimen cinegético especial, establecida por la ley con la finalidad de promover, fomentar, conservar y proteger determinadas especies —como cabra montés, rebeco, corzo, oso, urogallo y otras—, subordinando a esta finalidad el posible aprovechamiento de su caza. La declaración es competencia exclusiva de las CC.AA.); *f 1* **réserve [f] de chasse** (*Terme générique* zone dans laquelle est prise toute mesure propre à prévenir les dommages aux activités humaines, à favoriser la protection du gibier et de ses habitats et à maintenir les équilibres écologiques ; loi n° 90-85 du 23 janvier 1990 ; ▶enclos de réserve de chasse, ▶parc animalier, ▶réserve naturelle nationale, ▶réserve naturelle régionale) ; *syn.* réserve [f] cynégétique ; *f 2* **réserve [f] de chasse et de faune sauvage** (Zone dans laquelle est réglementée la protection et la gestion du gibier, la préservation de ses habitats ; le classement a pour effet en particulier l'interdiction de chasse, l'autorisation de captures de gibier et de repeuplement à des fins scientifiques, la préservation des habitats, la réglementation et l'interdiction exceptionnelle de l'accès des véhicules et des piétons) ; *f 3* **réserve [f] nationale de chasse et de faune sauvage** (Territoire concerné par une réserve de chasse et de faune sauvage et présentant un intérêt particulier en raison de son étendue, par la présence d'espèces dont les effectifs sont en diminution sur tout ou partie du territoire national ou d'espèces présentant des qualités remarquables, en fonction des études scientifiques, techniques ou des démonstrations qui y sont poursuivies ; le classement a notamment pour objet, la protection des espèces menacées, le développement du gibier à des fins de repeuplement, les études scientifiques et techniques, la réalisation d'un modèle de gestion du gibier, la formation des personnels spécialisés et l'information du public ; GPE 1998) ; *g* **Wildschutzgebiet [n]** (Durch das Jagdgesetz der Länder bestimmter, abgegrenzter Bereich, in dem während eines festgelegten Zeitraumes eines Jahres aus Gründen der Bestandeserhaltung seltener oder in ihrem Bestand bedrohter Arten nicht oder durch Verordnungen nur eingeschränkt gejagt werden darf; cf. § 20 BJagdG, Art. 21 BayJagdG, § 20 HessJagdG; ▶Naturreservat, ▶Wildgehege, ▶Wildpark); *syn.* Jagdschutzgebiet [n], Jagdschongebiet [n], Wildreservat [n], Wildasyl [n] [auch CH]).

game sanctuary [n] *conserv. game'man. hunt. leg.* ▶game reserve.

game species [n] *leg. hunt.* ▶game, ▶huntable game.

2357 game thinning [n] *hunt.* (Hunting of diseased or enfeebled game and reduction in numbers in order to avoid ▶overprotection [US]/over-protection [UK] and ▶game damage to vegetation; ▶balanced game management, ▶hunting management); *syn.* game culling [n] [also US]; *s* **caza [f] selectiva** (Caza de ejemplares cinegéticos enfermos o débiles y reducción en número para evitar ▶daños causados por la caza; ▶gestión de la caza, ▶medidas de regulación de poblaciones de fauna, ▶sobreprotección de la caza); *f* **chasse [f] régulatrice** (Chasse des animaux sauvages malades ou de faible constitution, régulation de la population sauvage afin de limiter les ▶dégâts causés par le gibier, ▶gestion cynégétique, ▶gestion sélective de peuplements d'animaux, ▶réglementation excessive de la chasse) ; *syn.* chasse [f] sélective ; *g* **Hegeabschuss [m]** (Bejagung von krankem, verletztem, schwachem und überaltem Wild, das sich nicht weiter vermehren soll sowie die Verringerung des Wildbesatzes zur Vermeidung von ▶Wildschäden durch ▶Überhege; ▶Bestandslenkung, ▶jagdliche Hege 2).

2358 game trail [n] *game'man. hunt.* (Regularly-used pathway between the usual living area of game—mostly even-hoofed game—and the area of feeding, drinking, wallowing or salt licking. A path often used by short-furred game is known as a 'pass'; ▶game crossing); *syn.* migratory path/trail [n]; *s* **paso [m] de caza** (Camino que conecta el espacio vital de las especies de caza con otras áreas frecuentadas por ellas que es utilizado regularmente; ▶cruce de animales); *f* **piste [f] animale** (Chemin régulièrement emprunté par le grand gibier entre les remises et les zones de nourriture, abreuvoirs, souille, etc. ; ▶traversée de gibier) ; *syn.* passe [f] (PR 1987), piste [f] de transit ; *g* **Wildwechsel [m] (2)** (Regelmäßig benutzter Pfad zwischen gewöhnlichem Standort des Wildes — meist Schalenwild — und dem Ort der Nahrungsaufnahme, Tränke, Suhle oder Salzlecke, auf dem die Tiere hin und her ziehen. Ein regelmäßig benutzter Pfad des niederen Haarwildes — Dachs, Fuchs, Hase, Marder, außer Rehwild — nennt man **Pass**; ▶Wildwechsel 1).

2359 game warden [n] *game'man. leg.* (**1.** Person whose duty it is to care for game. **2.** Person who enforces game laws); *syn.* game keeper [n] [also UK]; *s* **guardabosques [m] (de caza)** (Persona ocupada del cuidado de la caza y que controla el cumplimiento de la legislación correspondiente); *f* **garde-chasse [m]** (Personne préposée à la garde du gibier et soucieuse de faire respecter la législation sur la chasse) ; *g* **Wildhüter [m]** (Jemand, dem die Hege des Wildes obliegt und der die Einhaltung der Jagdgesetze überwacht).

ganger [n] [UK] *constr.* ▶foreman.

gap community [n] *phyt.* ▶tree-fall gap community [US].

gapping [n] [UK] *constr.* ▶replacement planting.

gap plot [n] [UK] *urb.* ▶vacant lot [US].

gap site [n] [UK] *urb.* ▶vacant lot [US].

gap-stopping [n] [UK] *urb.* ▶odd-lot development [US].

gap-tooth appearance [n] *urb.* ▶interrupted appearance.

garbage [n] [US] *envir.* ▶household waste, ▶kitchen waste.

2360 garbage collection charges [npl] [US] *adm. envir. leg.* (In U.S., local taxes are laid down and levied by municipalities or counties to finance local and regional landfill charges. Collection is both private and public and varies by region. Household waste is typically brought to recycling centers [glass, plastic, paper, electronics, etc.], and waste is transferred to larger regional landfills that are regulated by State and Federal authorities. Income from recycled materials is used to offset waste processing charges at the local level. **In England and Wales** there are 376 Waste Collection Authorities reponsible for collecting waste from nearly 22 million homes. Charges for refuse collection are included in the Council Tax, which is a local tax, set by councils to help pay for local services like policing and refuse collection. It applies to all domestic properties, including houses, bungalows, flats, maisonettes, mobile homes or houseboats, whether owned or rented. When the Council Tax system was introduced, all properties were valued and put into a 'valuation band'. The valuation bands for homes in England are based on their value on 1 April 1991); *syn.* waste collection charges [npl] [UK]; *s* **tasa [f] de recogida de residuos urbanos** (Cantidad a pagar mensual o anualmente por parte de los poseedores de residuos domésticos para cubrir los gastos del servicio municipal de gestión de residuos urbanos. En Es. es competencia de las entidades locales que pueden actuar directamente o delegar sus funciones a empresas públicas o privadas de acuerdo a la legislación sobre régimen local); *syn.* tasa [f] de recogida de basuras domésticas; *f 1* **taxe [f] d'enlèvement des ordures ménagères [TEOM]** (Taxe fiscale additionnelle prélevée sur toute les propriétés assujetties à la taxe foncière sur les propriétés bâties à l'exception des usines et des maisons louées pour un service public relevant des articles 1520 à 1526 du Code général des impôts et perçue avec la taxe foncière par les communes et leurs groupements) ; *f 2* **redevance [f] générale d'enlèvement des ordures ménagères [REOM]** (Redevance relevant des articles L. 2333-76 à L. 2333-80 du code général des collectivités territoriales se substituant à la taxe d'enlèvement des ordures ménagères, à la redevance pour l'enlèvement des déchets des campings et à la redevance spéciale ; celle-ci est calculée en fonction du service rendu pour l'enlèvement de tous les déchets et résidus dont la collecte est assurée ; son montant varie selon le nombre de personnes dans un foyer pour les ménages ou au volume du bac pour les autres usagers. Ce montant est ainsi lié à la quantité moyenne de déchets produits par les différents types d'usagers) ; *f 3* **redevance [f] incitative d'enlèvement des ordures ménagères** (Redevance proposée à l'initiative lancée l'ADEME [Agence de l'Environnement et de la Maîtrise de l'Energie] liée à la quantité de déchets produits ; celle-ci s'inscrit dans les politiques de prévention et de valorisation des déchets en favorisant la réduction de leur production ; elle permet d'optimiser les filières de valorisation, de maîtriser la hausse des coûts du service public déchets et d'en améliorer la transparence ; elle est déterminée par mesure des quantités [poids ou volume] d'ordures ménagères résiduelles présentées à la collecte [bac « gris »] ; plus l'usager produit d'ordures ménagères résiduelles, plus sa redevance est élevée ; les collectivités utilisent trois types de mesure **1.** le comptage du nombre de fois où le bac « gris » [quipé d'une puce électronique] est collecté, **2.** la pesée du bac « gris » [équipé également d'une puce] lors de sa collecte, **3.** le volume du bac « gris », plus l'usager choisit un bac petit, moins il paye, **4.** le sac payant ou la vignette : les ordures ménagères résiduelles ne sont collectées que dans des sacs achetés auprès de la collectivité ou sur lesquels une vignette achetée auprès de la collectivité est apposée ; il est également possible, comme le font certains pays européens, de facturer les différents bacs [ordures ménagères résiduelles, déchets recyclables et déchets fermentescibles] en fonction de tarifs différenciés afin d'inciter l'usager au tri de ses déchets ; les fractions recyclables étant facturées moins chères que la fraction résiduelle) ; *g* **Abfallgebühren [fpl]** (Durch örtliche Satzung der Stadt- und Landkreise festgesetzte Gebühren, die je nach Satzung entweder jeder Grundstückseigentümer für die Entsorgung des Abfalles, der auf seiner Liegenschaft produziert wird, zahlen muss oder die jeder einzelne Haushalt als Gebührenschuldner für die Entsorgung seines Abfalls entrichten muss).

garbage incinerator [n] [US] *envir.* ►waste incineration plant.

2361 garden [n] *arch. constr. gard.* (Area of land planted and cultivated with ornamental plants, fruits, and/or vegetables, usually enclosed by fence, hedge, or other opaque vegetative screen for privacy; generic term for ►alpine garden 1, ►alpine garden 2, ►aromatic garden, ►backyard garden [US]/back garden [UK], ►botanic garden, ►cloister garden, ►community garden [US]/allotment garden (plot) [UK], ►container garden, ►country garden [US]/cottage garden [UK], ►cut flower garden, ►demonstration garden (1), ►demonstration garden (2), ►elongated garden, ►enclosed medieval garden, ►estate garden, ►formal garden, ►French classic garden, ►fruit and vegetable garden, ►historic garden, ►herb garden, ►informal garden, ►kitchen garden, ►landscape garden, ►lapidary garden, ►leasehold garden, ►medicinal herb garden, ►miniature garden, ►model garden, ►monastery garden, ►naturalistic garden, ►old-style garden, ►ornamental garden, ►park-like garden, ►perennials test garden, ►private garden, ►rented garden, ►rock garden, ►roof garden, ►rose garden, ►row house garden, ►school garden, ►special garden, ►spices garden, ►sunken garden, ►tenant garden, ►terraced townhouse garden [US]/stepped building garden [UK], ►test garden, ►theme garden); *s* **jardín [m]** (Terreno cercado y cultivado con plantas ornamentales, frutales y/u hortalizas. El **j.** es una expresión de la cultura, tradición y del estilo de vida de un pueblo y refleja la vida social de cada época y de su correspondiente población. El desarrollo de la ►jardinería estuvo ligado desde sus orígenes a la arquitectura y a la agricultura, y supone la introducción del mundo natural, en especial vegetal, ordenado y gestionado por los seres humanos, en los hábitats urbanos. En España las primeras manifestaciones de jardines surgen en la época romana; los árabes desarrollaron aquí los primeros jardines botánicos de Europa. La mayoría de los jardines destacables son jardines histórico-artísticos, como los de la Alhambra y el Generalife en Granada, los de Aranjuez, la Granja en Segovia, el Retiro en Madrid y los numerosos jardines y huertos de los claustros monacales. Los jardines europeos fueron influenciados por jardines exóticos. Sin el modelo de los jardines árabes, no se habrían desarrollado los magníficos juegos de agua de los jardines del Renacimiento. Sin conocimiento del mundo jardinero chino, no se habría creado el jardín inglés. Y las rocallas japonesas aún siguen influenciando con su genial reduccionismo a lo imprescindible una cultura jardinera de orientación artística; cf. PARRA 1984; término genérico para ►alpinum, ►azotea ajardinada, ►huerta 2, ►huerto, ►huerto recreativo urbano, ►jardín alpino, ►jardín alquilado, ►jardín arrendado, ►jardín botánico, ►jardín campesino, ►jardín clásico, ►jardín comunitario, ►jardín de casa adosada, ►jardín de claustro, ►jardín de flores para cortar, ►jardín de hierbas condimenticias, ►jardín de hierbas finas, ►jardín de inquilinos, ►jardín de plantas aromáticas, ►jardín de villa, ►jardín escolar pedagógico, ►jardín especializado, ►jardín experimental, ►jardín experimental de vivaces/perennes, ►jardín geométrico, ►jardín histórico, ►jardín lapidario, ►jardín hundido, ►jardín informal, ►jardín medicinal, ►jardín medieval, ►jardín miniatura, ►jardín miniatura en jardinera, ►jardín modelo, ►jardín monástico, ►jardín ornamental, ►jardín paisajístico, ►jardín parque, ►jardín pedagógico, ►jardín privado, ►jardín silvestre, ►jardín temático, ►jardín tipo, ►jardín trasero, ►rocalla, ►rosaleda, ►terraza-jardín); *f* **jardin [m]** (Espace, ordinairement clos par une haie, une clôture, un mur sur lequel on cultive des végétaux utiles [légumes, arbres fruitiers] ou d'agrément [fleurs, arbustes ornementaux] ; ►jardin d'agrément. Le jardin est un lieu prédéfini par une culture, il est l'expression de la tradition et de la manière de vivre et reflète la vie sociale de son créateur et de son utilisateur au cours de l'histoire ; terme générique pour ►alpinum, ►jardin affermé, ►jardin à la française, ►jardin arrière,

►jardin botanique, ►jardin bouquetier, ►jardin clos, ►jardin collectif, ►jardin conservatoire, ►jardin creux, ►jardin d'apparat, ►jardin d'essai, ►jardin d'essais comparatifs des plantes vivaces, ►jardin d'herbes, ►jardin d'insertion, ►jardin d'une construction en ligne, ►jardin d'utilité, ►jardin de banlieue, ►jardin de cloître, ►jardin de curé, ►jardin de démonstration, ►jardin de rapport, ►jardin de rocaille, ►jardin de senteurs, ►jardin de style libre, ►jardin en forme de jauge, ►jardin en mouchoir de poche, ►jardin historique, ►jardin lapidaire, ►jardin locatif, ►jardin loué, ►jardin médicinal, ►jardin miniature, ►jardin monastique, ►jardin naturel, ►jardin [naturel] d'altitude, ►jardin ouvrier, ►jardin particulier, ►jardin paysager, ►jardin paysan, ►jardin pédagogique, ►jardin potager, ►jardin privatif, ►jardin régulier, ►jardin scolaire, ►jardin spécialisé, ►jardin spécimen, ►jardin sur dalle, ►jardin sur toiture-terrasse, ►jardin thématique, ►roseraie, ►terrasse-jardin) ; *g* **Garten [m]** (Mit Hecke, Zaun oder Mauer eingefriedetes Stück Land, das zur Anzucht von Nutzpflanzen dient oder als ►Ziergarten genutzt wird. Ein G. ist Ausdruck von Kultur, Tradition und Lebensart und spiegelt das gesellschaftliche und soziale Leben einer jeden Zeit und der Menschen selbst wieder. Europäische Gärten wurden auch von exotischen Gärten beeinflusst. Ohne das Vorbild der maurischen Gärten hätten sich in den Renaissancegärten nicht die prächtigen Wasserspiele entwickelt. Ohne die Kenntnis der chinesischen Gartenwelt wäre nicht der englische Gartenstil entstanden. Und noch immer beeinflussen japanische Steingärten mit ihrer genialen Beschränkung auf das Wesentliche eine künstlerisch orientierte Gartenkultur, die auf ästhetische Reduktion ausgerichtet ist. Der **G.** der Stadt, der öffentliche Park also, ist traditionellerweise der Ort der Begegnung und gelebter Toleranz. Er entstand im 19. Jh. als Versuch, Stadtgesellschaften zu befrieden, d. h. eine Annäherung der Stände zu ermöglichen, woraus sich im Laufe der Zeit ein friedlich freundliches Nebeneinander aller sozialen Gruppen und Kulturen einer Stadtbevölkerung, das auch zu einem Miteinander werden kann, als ungeschriebenes Toleranzgebot über jegliche Parknutzung, entwickelte [►Volkspark]. Gärten sind so unterschiedlich wie die menschlichen Vorstellungen vom Paradies als irdische Andeutung [vom persischen *pairi daeza* zum griechischen *parádeisos* = umzäunter Park] — etymologisch deshalb interessant, da **G.** der eingefriedete Raum [germanische Wurzel *gerd* = gürten] bedeutet [cf. MIL 2002]. Mit dem Hannoveraner EXPO-2000-Slogan „Die Stadt als Garten" wurde der Gartenbegriff wieder in den Mittelpunkt städtebaulichen Denkens gerückt, wie es der preußische Gartenkünstler und Landschaftsarchitekt PETER JOSEF LENNÉ [1789-1866] und der Anhaltiner Fürst LEOPOLD III FRIEDRICH FRANZ [1740-1817] in Wörlitz vorgedacht hatten: der **G.** als Symbol für einen nachhaltigen Städtebau, für eine gesunde Lebensumwelt, für eine nachhaltige Landnutzung. Eine herausragende Aufgabe der Gartenkunst für das 21. Jh. wird sein, den **G.** als Bild für eine Harmonie von Natur und Kultur[en] zu schaffen; *UBe* ►Alpengarten, ►Alpinum, ►Bauerngarten, ►botanischer Garten, ►Duftgarten, ►formaler Garten, ►Gartendenkmal, ►Garten zur Pflege alter Kulturmethoden, ►Handtuchgarten, ►Hausgarten, ►Hintergarten, ►Kleingarten, ►Klostergarten, ►Kräutergarten, ►Kreuzgarten, ►Küchengarten, ►Landschaftspark, ►Lapidarium, ►Lehrgarten, ►Mietergarten, ►Mietgarten, ►Miniaturgarten, ►Mustergarten, ►naturnaher Garten, ►nicht formaler Garten, ►Nutzgarten, ►Pachtgarten, ►parkartiger Garten, ►Reihenhausgarten, ►Schaugarten, ►Schnittblumengarten, ►Schulgarten, ►Senkgarten, ►Sichtungsgarten, ►Sondergarten, ►Staudensichtungsgarten, ►Steingarten, ►Terrassenhausgarten, ►Themengarten, ►Troggarten, ►Villengarten).

garden [n] [UK], **allotment** *urb. leg.* ►community garden [± US].

garden [n] [UK], **back** *gard.* ►backyard garden [US].

garden [n], **baroque** *gard'hist.* ►French classic garden.

garden [n], **condiment** *gard.* ►spices garden.

garden [n] [UK], **cottage** *gard. hort.* ►country garden [US].

garden [n], **courtyard** *arch. landsc.* ►garden courtyard.

garden [n], **flower** *gard.* ►ornamental garden.

garden [n] [UK], **front** *urb.* ►front yard [US].

garden [n], **natural-like** *gard.* ►naturalistic garden.

garden [n] [UK], **perennials trial** *hort.* ►perennials test garden.

garden [n] [UK], **physic** *gard.* ►medicinal herb garden.

garden [n] [UK], **public** *urb.* ►inner city park [US].

garden [n] [UK], **rear** *gard.* ►private garden.

garden [n] [US], **residential** *gard.* ►private garden.

garden [n] [US], **rooftop** *arch. constr.* ►roof garden.

garden [n] [UK], **split-level house** *arch. constr.* ►terraced townhouse garden [US].

garden [n] [UK], **stepped building** *arch. constr.* ►terraced townhouse garden [US].

garden [n] [US], **townhouse** *gard. plan.* ►elongated garden.

garden [n] [UK], **trial** *hort.* ►test garden.

garden [n] [UK], **trough** *gard.* ►container garden [US].

garden [n] [US], **vegetable** *gard.* ►kitchen garden.

garden [n] [CDN], **vertical** *gard. urb.* ►façade planting.

garden [n] [UK], **villa** *gard.* ►estate garden [US].

garden aficionado [n] [US] *gard.* ►garden enthusiast.

garden archaeological dig [n] [US] *conserv'hist.* ►garden archaeological excavation.

2362 garden archaeological excavation [n] *conserv' hist.* (►Exploratory trenching in ►historic gardens, in order to find remnants of the original design or alterations of it; ►archaeological excavation); *syn.* garden archaeological dig [n] [also US]; *s* **prospección [f] arqueológico-paisajística** (Trabajos de excavación en ►jardín histórico para buscar posibles restos del diseño original o cambios del mismo; ►excavación arqueológica, ►prospección arqueológica); *f* **fouilles [fpl] archéologiques dans un jardin historique** (Travaux de prospection archéologique dans un ►jardin historique afin d'en retrouver les restes de la composition originelle ou les transformations réalisées sur le projet initial ; ►fouilles archéologiques, ►prospection archéologique, ►dage archéologique) ; *g* **gartenarchäologische Grabung [f]** (Grabungsarbeiten in einem historischen Garten, um mögliche Reste des Urentwurfs oder Veränderungen des ursprünglich realisierten Entwurfs zu finden; ►Gartendenkmal, ►Grabung, ►Suchgrabung).

garden art [n] *gard. hist.* ►fine garden design.

2363 garden bench [n] *gard. landsc.* (Sturdy seat of wood, stone or concrete, sometimes with a backrest for the use of several persons; ►park bench); *syn.* garden seat [n]; *s* **banco [m] de jardín** (►banco de parque); *f* **banc [m] de jardin** (Meuble de jardin, point de repos naturel et élément esthétique ; ►banc de parc) ; *g* **Gartenbank [f]** (Im Garten aufgestellte Sitzgelegenheit, auf der mehrere Personen Platz haben; ►Parkbank).

2364 garden city [n] *hist. urb.* (Proposal in 1898 by EBENEZER HOWARD, that was published in "Garden Cities for Tomorrow" in 1902. This planning approach aimed to combine the advantage of living in the city with the benefits of rural life.

G

The ground remained community property in order to prevent inappropriate alteration. Letchworth and Welwyn in Britain were the first **g. c.s**); *s* **ciudad [f] jardín** (Tipo de ciudad nueva ideada por EBENEZER HOWARD en 1898, definida por él como «un núcleo planeado para la producción y para la vida saludable, de un tamaño que haga posible la plenitud de la vida social, pero no mayor, rodeada de un cinturón permanente de espacio rural, siendo todos los terrenos de propiedad pública poseídos en comunidad por los ciudadanos»; DGA 1986); *f* **cité-jardin [f]** (Conception d'organisation de l'espace urbain élaborée en 1898 par l'urbaniste anglais EBENEZER HOWARD qui essaye d'unifier les avantages de la vie urbaine et rurale. La **c.-j.** est organisée à partir d'un parc central public autour duquel sont disposées en ceintures les zones d'habitations et industrielles. Chaque habitation possède un jardin d'utilité, le sol restant propriété communale) ; *g* **Gartenstadt [f] (1)** (Von EBENEZER HOWARD 1898 entwickelter, einheitlich geplanter, selbständiger Stadttypus mit Wohn- und Arbeitsstätten, der die Vorzüge des Land- und Stadtlebens miteinander vereinigt. Der Boden blieb im Eigentum der Gemeinde, um eine missbräuchliche Nutzung zu verhindern; zu jedem Familienheim gehörte ein wirtschaftlich nutzbarer Garten; die ersten englischen Gartenstädte waren Letchworth und Welwyn; cf. DIL 1987. In D. entstand 1906 „Hellerau" bei Dresden und 1909 die „Margarethenhöhe" in Essen; cf. RICH 1981).

2365 Garden City Movement [n] [UK] *landsc. obs.* (**In U.K.**, EBENEZER HOWARD founded the **g. c. m.**, which was a forerunner of modern town planning, not only in terms of controlling the layout and design of cities, but also as part of national policy on economic and social planning and the improvement of public health in urban areas; **in U.S.** and D., generic term of the 19th century covering architecture, landscape architecture and civil engineering with the main objective of improving human living conditions by creating a beautiful environment with trees, lawns, gardens, and lakes, and also bridges, houses, etc. The modern term is 'community beautification' [US]/'city/town beautification' [UK]; ▶garden city); *syn.* City beautiful Movement [n] [US]; *s* **movimiento [m] de la ciudad-jardín** (En GB, fundado por EBENEZER HOWARD que fue uno de los precursores del urbanismo moderno, no sólo en cuanto al control del diseño de las ciudades, sino también como parte de una política nacional de planificación económica y social, y para mejorar la salud pública en áreas urbanas. En otros países se utiliza como término genérico del siglo XIX que cubre la arquitectura, el paisajismo y la ingeniería civil con el principal objetivo de crear un medio ambiente bello con árboles, campas de césped, jardines y lagos, y también casas, puentes, etc. El término moderno es el de **embellecimiento [m] de la ciudad**; ▶ciudad jardín); *f* **mouvement [m] de la cité-jardin** (Mouvement de réforme en urbanisme engagé par l'anglais EBENEZER HOWARD qui préconise en 1898 la création de ceintures vertes autour des grandes villes ainsi que la construction à leur périphérie de cités n'excédant pas 30 000 habitants, les soit disant cités-jardins ; conçue pour limiter l'exode rural et éviter la formation de conurbation la ▶cité-jardin donnera une impulsion nouvelle à l'urbanisme et l'aménagement des espaces verts en milieu urbain) ; *g* **Gartenstadtbewegung [f]** (Das Reformbestreben im Städtebau des Engländers EBENEZER HOWARD, der 1898 für Großstädte Grünringe und in ihrem Umkreis die Entstehung von neuen, auf 30 000 Einwohner begrenzte so genannte „Gartenstädte" zur Eindämmung der Landflucht und Begrenzung der räumlich ausufernden Großstädte forderte; die **G.** hat der Stadt- und städtischen Grünplanung in der Folgezeit etliche Impulse gegeben; ▶Gartenstadt 1).

garden club [n] [US], community *sociol.* ▶allotment garden club [UK].

garden conservator [n] [UK] *conserv'hist.* ▶historic garden expert [US].

garden court [n] [UK] *arch. landsc.* ▶garden courtyard.

2366 garden courtyard [n] *arch. landsc.* (Small, open space between buildings, often in shade for much of the day, that is designed with plants; the term *patio* means a small paved courtyard or terrace with or without planting. *NOTE* in US, **yard** means in this context the ground which adjoins or surrounds a house, public building, or structure; ▶courtyard, ▶courtyard planting); *syn.* garden court [n], courtyard garden [n]; (courtyard) patio [n]; *s* **patio-jardín [m]** (Pequeño espacio libre con abundante vegetación en el interior de una casa o de una manzana de casas en zona urbana con gran densidad de construcción; ▶patio, ▶plantación de patios); *syn.* patio [m] ajardinado; *f* **cour-jardin [f]** (Cour intérieure de surface réduite en zone urbaine d'habitat dense, caractérisée par un court ensoleillement et aménagée par un décor végétal et de nombreux éléments minéraux ; ▶aménagement végétal d'une cour, ▶cour intérieure, ▶patio) ; *syn. dans les pays méditerranéens* patio [m] ; *g* **Gartenhof [m] (2)** (Kleiner, von Mauern und Gebäudefassaden begrenzter, begrünter Freiraum im Siedlungsbereich mit hoher Bebauungsdichte, oft nur wenige Stunden am Tag von der Sonne beschienen; ▶Hofbegrünung, ▶Innenhof); *syn.* bepflanzter Innenhof [m]; *syn. in mediterranen Ländern* Patio [m].

2367 garden culture [n] *arch. gard'hist.* (The sum of all aesthetic aspects of gardening including formal design elements, which has developed over a peiod of time within a certain area and as a reaction to the prevailing environment; **g. c.** is often viewed as a part of national or local heritage; contemporary **g. c.** is promoted by city authorities and includes nature conservation and landscape management programmes, often directed at the improvement of urban areas, such as courtyards, rented gardens, residential areas generally, the encouragement of facade greening and roof gardens. Publicity campaigns are carried out to increase public awareness and understanding of nature and **g. c.** Linguistically the usage of the word 'culture' has a positive connotation; holistically speaking, **culture** is the sum of all spiritual, material, intellectual and environmental aspects, which characterize society or a social group. In the context of gardens it is associated with the aesthetic and spiritual experience to be enjoyed there. Basically it is the search for the lost Garden of Paradise or Eden—not a wilderness, but a garden conceived new with every epoch according to the spirit of the times; ▶fine garden design); *s* **patrimonio [f] de jardinería y horticultura ornamental** (Conjunto de trabajos del arte de la jardinería y horticultura de una región o un país; ▶arte de jardinería); *f* **patrimoine [m] culturel des jardins (±)** (Ensemble des travaux artistiques concernant la conception et la réalisation de jardins englobés dans le processus de maîtrise et d'aménagement de l'environnement, héritage d'une époque donnée, ainsi que le processus de recherche et de recréation des formes et différents éléments esthétiques de la composition ; ▶art des jardins) ; *g* **Gartenkultur [f]** (Gesamtheit der zu bestimmten Zeiten in einer abgegrenzten Region auf Grund der Auseinandersetzung mit der Umwelt in ihrer Gestaltung hervorgebrachten gartenkünstlerischen Ergebnisse sowie der Prozess des jeweiligen Hervorbringens und Reproduzierens verschiedener formal-ästhetischer Gestaltungselemente; zur Förderung der **G.** werden heute seitens der Stadtverwaltungen unter Einbeziehung der Ziele von Naturschutz und Landschaftspflege Programme zur Realisierung von Mietergärten, Durchführung von Innenhof- und Wohnumfeld-, Fassaden-, Dachbegrünungen und Vorgartenprogramme erarbeitet und durchgeführt sowie durch Öffentlichkeitsarbeit das Verständnis für die Natur und **G.** gefördert. Im allgemeinen Sprachgebrauch ist der Kulturbegriff im Sinne, dass Kultur Geist und Seele erfreut, positiv besetzt. **Kultur** bedeutet im weiteren

Sinne die Gesamtheit der einzigartigen geistigen, materiellen, intellektuellen und emotionalen Aspekte, die eine Gesellschaft oder eine soziale Gruppe kennzeichnen. Im Gartenkontext kommt es auf den ästhetisch-seelischen Aspekt des Gartenerlebens an. Es ist im Grunde die Sehnsucht nach dem verlorenen Garten Eden, dem Paradiesgarten, der nicht Wildnis, sondern ein nach der Vorstellung einer jeweiligen Epoche gestalteter Gartenraum war; ▶Gartenkunst).

2368 garden design [n] *gard. plan.* (Process of planning for development of enclosed spaces for flowers, fruits or vegetables according to functional or aesthetic principles; ▶fine garden design, ▶landscape architecture); *s* **diseño** [m] **de jardines** (En alemán término actualmente poco usual que implica la planificación y el diseño de espacios libres en zonas urbanas o de parques en haciendas feudales o burguesas según criterios funcionales o estético-formales; ▶arte de jardinería, ▶paisajismo); *f* **architecture** [f] **des jardins** (Terme peu utilisé désignant l'aménagement et la conception des espaces libres urbains, ou des parcs et jardins à différentes périodes de l'histoire d'après des critères fonctionnels ou esthétiques ; ▶architecture des paysages, ▶art des jardins) ; *syn.* conception [f] des jardins ; *g* **Gartenarchitektur** [f] (*Heute wenig verwendeter Begriff* Ergebnis der Planung und Gestaltung von Freiräumen im Siedlungsbereich oder von Parkanlagen im feudalen oder großbürgerlichen Grundbesitz nach funktionalen oder formal-ästhetischen Gesichtspunkten; ▶Gartenkunst, ▶Landschaftsarchitektur); *syn.* Gartengestaltung [f].

2369 garden design theme [n] *hort. gard.* (Main concept which characterizes a garden, expressed with, e.g. perennials, water features, play facilities; ▶theme garden); *syn.* theme [n] of garden design; *s* **tema** [m] **de diseño de jardines** (Concepto central de diseño de un jardín basado, p. ej. en vivaces, elementos acuáticos, instalaciones para juegos, etc.; ▶jardín temático); *f* **thème** [m] **d'un jardin** (autour duquel est organisé un jardin, p. ex. jardin de rocaille, jardin d'herbes, jardin écologiques, etc. ; ▶jardin thématique) ; *g* **Gartenthema** [n] (Thema, das die Gestaltung eines Gartens besonders prägt, z. B. Stauden, Wasserkunst, Spielgeräte; ▶Themengarten).

2370 garden enthusiast [n] *gard.* (Person who loves to garden; ▶garden for connoisseurs); *syn.* garden lover [n], amateur gardener [n]; garden aficionado [n] [also US]; *s* **jardinero/a** [m/f] **aficionado/a** (▶jardín de aficionado); *syn.* horticultor, -a [m/f] aficionado/a; *f* **amateur** [m] **de jardins** (Personne portant un intérêt particulier pour la conception, la plantation et l'entretien de jardins ; ▶jardin d'amateur) ; *g* **Gartenliebhaber/-in** [m/f] (Jemand, der ein besonderes Interesse an der Gestaltung, Pflege und Unterhaltung von Gärten hat; ▶Liebhabergarten); *syn.* Amateurgärtner/-in [m/f], Hobbygärtner/-in [m/f], Freizeitgärtner/-in [m/f].

gardener [n] *hort. prof.* ▶landscape gardener.

gardener [n], **allotment** [UK] *urb. sociol.* ▶community gardener [US].

gardener [n], **amateur** *gard.* ▶garden enthusiast.

gardener [n], **perennial** *hort. prof.* ▶perennial grower, #3.

gardener [n], **urban** *urb. sociol.* ▶community gardener [US].

garden exhibition [n] *hort. urb.* ▶horticultural exhibition.

garden festival [n] [UK] *hort. urb.* ▶horticultural exhibition.

garden festival area [n] *hort. urb.* ▶horticultural exhibition area.

2371 garden [n] **for connoisseurs** *gard.* (Garden characterized by its individual design and especially by its unusual planting; ▶garden enthusiast); *syn.* garden [n] for plant lovers; *s* **jardín** [m] **de aficionado/a** (Jardín caracterizado por su diseño

individual y por su aplicación poco usual de plantas; ▶horticultor, a aficionado/a); *f* **jardin** [m] **d'amateur** (Jardin caractérisé par une conception très individuelle, en particulier par une composition recherchée de végétaux ; ▶amateur de jardins) ; *syn.* connaisseur de jardins ; *g* **Liebhabergarten** [m] (Garten, der durch eine sehr individuelle Gestaltung, insbesondere durch eine ausgefallene Pflanzenverwendung geprägt ist; ▶Gartenliebhaber).

garden [n] **for plant lovers** *gard.* ▶garden for connoisseurs.

2372 garden [n] **for the blind** *arch.* (Garden or part of a park especially equipped with signage in Braille or embossed type and a system of pathway directions for blind people or those with severe visual impairment. Modern planning aims to create public parks and gardens, which cater for the needs of all handicapped people; ▶universal design); *s* **jardín** [m] **para invidentes** (Jardín o parte de uno equipado especialmente con señales en Braille o en letras talladas y un sistema guía de caminos adecuado para personas invidentes o de visión reducida. La planificación moderna de parques y jardines tiene como objetivo crear espacios públicos que sean utilizables para todo tipo de personas con limitaciones físicas; ▶diseño universal); *syn.* jardín [m] para ciegos; *f* **jardin** [m] **pour aveugles** (Jardin ou parc spécialement conçu pour que les non-voyants puissent s'y promener sans aucune aide ; les aménagements se caractérisent par des revêtements de sol distincts pour les différents parcours, plans en relief comme descriptifs tactiles d'un lieu avec plaques descriptives en Braille, plates-bandes rehaussées pour permettre d'appréhender les senteurs, le modelé et la texture des végétaux ; la tendance actuelle dans l'aménagement des espaces verts est de rendre les parcs et jardins accessibles pour les personnes aveugles et malvoyantes ; ▶conception universelle) ; *g* **Blindengarten** [m] (Garten oder Teil eines Parkes, der besonders für blinde und stark sehbehinderte Menschen mit geeigneten Hinweistafeln in Braille- oder reliefartig hervorgehobener Normalschrift und einem blindengerechten Wegeleitsystem ausgestattet ist. Die moderne Projektplanung zielt darauf ab, alle öffentlichen Parks und Gärten in ihrer Gesamtheit so zu gestalten, dass sie für alle Behinderten nutzbar sind; ▶behindertengerechte Planung).

garden [n] **for the disabled** *gard.* ▶garden for the handicapped.

2373 garden [n] **for the handicapped** *gard.* (Garden designed to enable all visitors to fully access the design features through accommodative grading [max. 6%], paving, planting, seating, signage, lighting, and spatial configuration and dimensional standards; ▶universal design); *syn.* garden [n] for the disabled, garden [n] for disabled/handicapped persons; *s* **jardín** [m] **para personas con movilidad reducida** (Jardín diseñado de tal manera que permite el acceso sin impedimentos a todas las personas, por medio de pendientes suaves [max. 6%], pavimentos, plantaciones, asientos, iluminación, configuración espacial y estándares de dimensión adecuados; ▶diseño universal); *syn.* jardín [m] para limitados físico-motores [C]; *f* **jardin** [m] **pour handicapés** (Jardin conçu et aménagé comme un jardin pour tous prévoyant des aménagements simples [allées larges, rampe] l'utilisation diversifiée de matériaux pour les cheminements qui facilitent l'orientation, de repères sensoriels et spatiaux qui par leur texture ou leur odeur sont adaptés aux différentes formes de handicap ; ▶conception universelle) ; *syn.* jardin [m] pour personnes à mobilité réduite ; *g* **Behindertengarten** [m] (Garten, der allen Nutzern einen vollständigen, ungehinderten Zugang zu allen Teilen durch geringe Steigungen in der Wegeführung [max. 6 %], leicht begehbare Oberflächengestaltung, Vorhaltung ausreichender Sitzmöglichkeiten und lesbarer Hinweistafeln sowie durch Beleuchtung ermöglicht; ►behindertengerechte Planung); *syn.* Garten [m] für Behinderte.

gardening [n] *constr.* ►cemetery gardening; *sociol. urb.* ►community gardening; *landsc.* ►guerilla gardening; *hort. recr.* ►hobby and allotment gardening [UK]; *hort.* ►market gardening; *constr. landsc.* ►prairie gardening [US], ►wildflower meadow gardening [US].

gardening [n] [UK], **allotment** *sociol. urb.* ►community gardening [US], ►hobby and allotment gardening [UK].

gardening [n] [US], **bio-intensive** *agr.* ►organic farming.

gardening [n] [US], **market** *hort.* ►commercial horticulture.

gardening [n] [US], **organic** *agr.* ►organic farming.

gardening [n] [US], **truck** *hort.* ►market gardening.

garden living space [n] [UK] *gard.* ►outdoor living space [US].

garden lover [n] *gard.* ►garden enthusiast.

2374 garden maintenance [n] [US] *constr. hort.* (Measures for long-term development and ongoing care to ensure vitality and sustainability of a garden; ►maintenance costs); *syn.* garden upkeep [n] [UK]; *s* **mantenimiento [m] de jardines** (Todos los trabajos e inversiones necesarias [►costo[s] de mantenimiento] para mantener y desarrollar la capacidad funcional de un jardín); *f* **entretien [m] des jardins** (Tous travaux et investissements assurant le développement et à long terme le maintien en bon état d'un jardin ; ►coûts d'entretien) ; *g* **Pflege [f, o. Pl.] und Unterhaltung von Gärten** (Alle Arbeiten und Investitionen, die der Entwicklung und Erhaltung der nachhaltigen Funktionsfähigkeit eines Gartens dienen; ►Pflegekosten).

garden pavilion [n] *gard.* ►gazebo.

garden paving slab [n] [UK] *constr.* ►natural garden stone flag.

garden plant [n] *hort.* ►peat garden plant [UK].

2375 garden plot area [n] *hort.* (Land close to home often rented for production of fruit and vegetables, and occasionally uncultivated; ►community garden [US]/allotment garden [plot] [UK], ►permanent community garden area [US]/permanent allotment site [UK], ►temporary community garden area [US]/temporary allotment land [UK]); *s* **parcela [f] familiar** (Pequeño terreno cultivado por persona o familia para su aprovechamiento privado situada cerca de la vivienda, puede ser un ►área de huertos urbanos temporales o un ►huerto recreativo urbano; ►zona permanente de huertos recreativos urbanos); *f* **terres [fpl] maraîchères** (Parcelle cultivée pour la production personnelle à proximité du lieu d'habitation ou située à la périphérie des zones résidentielles et organisée en ►jardins ouvriers ou en ►lotissement de jardins ouvriers ; ►terres affermées temporairement rairement) ; *g* **Gartenland [n]** (Meist zur Eigenversorgung genutztes, zeitweilig unbestelltes Land in Wohnungsnähe oder auch als ►Grabeland oder ►Kleingarten im näheren Siedlungsbereich gelegen; ►Dauerkleingartenanlage).

2376 garden pool [n] *constr. gard.* (Decorative basin with a very shallow water depth, designed only as an ornament and sometimes equipped with a small fountain; ►composition of water features, ►fountain pool, ►pool 1); *s* **estanque [m] ornamental** (Pequeño estanque de muy poca profundidad diseñado sólo con fines decorativos y a veces equipado con una fuentecilla; ►estanque, ►fuente, ►juego de aguas); *syn.* estanque [m] decorativo; *f* **bassin [m] ornemental** (►Pièce d'eau de faible profondeur, conçue comme élément décoratif [bassin en boulingrin dans le jardin classique] en général équipé ►d'une fontaine à jet d'eau ; ►jeu d'eau) ; *syn.* pièce [f] d'eau décorative, bassin [m] décoratif, bassin [m] d'agrément ; *g* **Zierbecken [n]** (►Wasser-

becken mit sehr geringer Wassertiefe, das lediglich zur Zierde angelegt wurde und meist mit einem kleinen Springstrahl ausgestattet ist; Z. mit Springstrahl werden auch **Zierbrunnen** genannt; ►Springbrunnen, ►Wasserspiel); *syn.* Zierteich [m].

garden preservationist [n]**, professional** *conserv'hist.* ►historic garden expert [US].

garden seat [n] *gard. landsc.* ►garden bench.

garden show [n] *hort. urb.* ►horticultural exhibition.

garden soil [n] [US] *pedol.* ►man-made humic soil.

garden stone flag [n] *constr.* ►natural garden stone flag.

2377 garden suburb [n] *urb.* (Contemporary term for a suburban community outside the city which is rich in green areas and developed for residential use, also called ►bedroom community [US]/dormitory suburb [UK]); *s 1* **suburbio [m] ajardinado** (Tipo de barrio residencial con muchas zonas verdes localizado en las afueras de una ciudad; ►ciudad-dormitorio); *s 2* **urbanización [f] suburbana** (Barrio de chalets con jardín localizados en los alrededores de las grandes urbes); *f* **ville [f] verte (±)** (Terme définissant aujourd'hui les banlieues vertes des grandes agglomérations à usage strictement résidentiel ; ►cité-dortoir) ; *syn.* cité [f] verte (±) ; *g* **Gartenstadt [f] (2)** (Heute Bezeichnung für stark durchgrünte Vorstädte, die meist als reine Wohnbezirke dienen; ►Schlafstadt).

2378 garden tool [n] *hort.* (Implement used in the garden for cultivation and/or cleanup; e.g. spade, hand trowel, hoe, rake, etc.); *s 1* **herramienta [f] de jardinería** (Objeto que sirve para realizar trabajos de jardinería como laya, pala, azada, rastrillo, etc.); *s 2* **equipamiento [m] de jardinería** (Conjunto de herramientas de jardinería); *f 1* **outil [m] de jardinage** (Objet fabriqué à l'usage des travaux de jardinage) ; *f 2* **matériel [m] de jardinage** (Terme englobant l'ensemble des outils de jardinage) ; *g* **Gartengerät [n]** (1. Gegenstand, mit dem gärtnerische Tätigkeiten verrichtet werden können, z. B. Spaten, Pflanzschaufel, Hacke, Harke etc. 2. [o. Pl.] Gesamtheit der Geräte, die zur gärtnerischen Arbeit benötigt werden); *syn.* Gartenwerkzeug [n].

garden upkeep [n] [UK] *constr. hort.* ►garden maintenance [US].

garden waste [n] [UK] *envir.* *►green waste.

garden yard [n] [UK] *arch. landsc.* ►garden courtyard.

2379 garide [n] *phyt.* (Scrub formation in the Mediterranean or similar climatic areas. This type of vegetation is degraded from the original broad-leaved evergreen, sclerophyllous forest with different flora according to geographical location: ►garrigue on calcareous soils [southern France], ►macchia on acid soils [Italy], chaparral in Mexico and SW-USA, phrygana in Greece, trachiotis in Cyprus; *s* **matorral [m] (de sustitución)** (Formación vegetal predominante en la región mediterránea y en regiones del mundo de clima similar en la que se distinguen a «grosso modo» tres grandes tipos en el camino de la regresión: ►maquis, ►garriga y formaciones con predominio de herbáceas; según las regiones donde se presentan tienen diferentes denominaciones como **chaparral** en México y el suroeste de EE.UU.; cf. DINA 1987; ►garida); *f* **garide [f]** (Groupement végétal [fourré] à feuilles caduques typique, formé surtout de plantes supportant la sécheresse — xérophytes —, qui colonise en particulier les dalles calcaires du pied méridional du jura et les étages montagnards et subalpins des régions méditerranéennes ; il s'agit de formations secondaires, de composition floristique différente suivant leur situation géographique : le **tomillar** en Espagne, la ►garrigue sur sol calcaire [sud de la France], le ►maquis sur sol basique dans le sud-est de la France et en Italie, la **phrygana** en Grèce, le **trachiotis** à Chypre, le **chaparral** au Nord-Ouest du Mexique [formation végétale arbustive naturelle] et dans le sud-ouest des États Unis) ; *g* **Garide [f]** (Mediterrane Gebüschformation. Sie ist eine Degradationsstufe des Hartlaubwaldes mit unterschiedlicher floristischer Zusammensetzung je nach geografischer Lage: **Coscojar** in Spanien, ►**Garrigue** auf Kalkböden [Südfrankreich], ►**Macchia** auf sauren Böden [Italien], **Phrygana** auf nährstoffarmen Kalkfelsen und Silikatgestein sowie auf mehrmals abgebrannten Flächen in Griechenland, **Trachiotis** auf Zypern, **Chaparral** [als natürliche, nicht degradierte Pflanzenformation] in Mexiko und SW-USA); *syn.* Felsenheide [f].

2380 garrigue [n] *phyt.* (Low, scattered shrubs on calcareous soils in the Mediterranean region [southern France]; primarily kermes oak *[Quercus coccifera]*, dwarf sclerophyllous shrubs, mastic species, Rock rose *[Cistus]* and aromatic scrub; *generic term* ►garide; ►macchia; *s 1* **garriga [f]** (Formación vegetal de la vegetación mediterránea, normalmente sobre áreas calizas, consistente en un ►matorral [de sustitución] de hoja perenne y coriácea que tiene una disposición espacial abierta y cubierta ocasionalmente en sus claros por gramíneas en macollas; DINA 1987; ►maquí) ; *syn.* coscojar [m] [Es]; *s 2* **garida [f]** Denominación aplicada por R. CHODAT a una vegetación en general fruticosa y frecuentemente abierta, que recuerda la de la garriga mediterránea, pero que crece en clima centroeuropeo de vegetación generalmente mesofítica; DB 1985); *f* **garrigue [f]** (Formations végétales frutescentes ou sous-frutescentes résultant de la dégradation de la chênaie verte sur substrat calcaire ; on distingue la **g.** à chêne kermès *[Quercus coccifera]* sur calcaire dur et la **g.** à romarin *[Rosmarinus officinalis]* sur calcaire tendre ; la dégradation de ces formations donne des pelouses basses à Brachypodes *[erme]* ; *termes spécifiques* matorral, fruticée méditerranéenne ; *terme spécifique à l'extérieur des régions méditerranéennes* garissade ; ►garide, ►maquis) ; *g* **Garrigue [f]** (Niedrige, meist lichte Gebüschformation auf Kalkböden in der Mediterranregion [Südfrankreich], vor allem aus Kermeseiche *[Quercus coccifera]*, Hartlaubzwergsträuchern, Mastixarten [z. B. *Pistacia lentiscus, Thymus mastichina*], Zistrosen *[Cistus* und *Halimium]* und aromatisch duftenden Sträuchern; *OB* ►Garide; ►Macchia); *syn.* Garigue [f].

gas [n] *envir.* ►exhaust gas, ►flue gas; *ecol. envir.* ►greenhouse gas.

gaseous emission [n] *envir.* ►excess gaseous emission.

gaseous emissions [npl] *envir.* ►industrial gaseous emissions.

2381 gaseous pollution impact [n] *envir.* (Input of mostly toxic or unpleasant gases into the environment; ►emission of odo[u]r, ►pollution impact); *s* **inmisión [f] gaseosa** (Sustancia gaseosa, generalmente perjudicial para la salud humana o para la naturaleza, que incide en una determinada zona; ►inmisión, ►inmission de substances malodorantes); *syn.* contaminación [f] gaseosa; *f* rejet [m] à l'atmosphère de polluants gazeux (En général substances gazeuses émises ayant des conséquences préjudiciables de nature à mettre en danger la santé humaine, à nuire aux ressources biologiques et à détériorer les biens matériels ; ►immission, ►émission de substances malodorantes) ; *syn.* rejet [m] de substances gazeuses, rejet [m] d'effluents gazeux ; *g* **Gasimmission [f]** (Meist schädliche Einwirkungen von emittierten Gasen auf Menschen, Tiere, Pflanzen und nicht belebte Sachgüter; ►Immission, ►Geruchsimmission).

2382 gasoline trap [n] [US] *envir.* (Device for separating spilt fuel from surface water before it enters a drainage system); *syn.* petrol trap [n] [UK]; *s* **separador [m] de gasolina** (Dispositivo para separar gasolina y aceites del agua de escorrentía superficial antes de que ésta entre al sistema de alcantarillado; *término genérico* separador de hidrocarburos); *syn.* separador [m]

de nafta [RA]; *f* **séparateur [m] à hydrocarbures** (Appareil placé sur un réseau d'assainissement et destiné à retenir les huiles et essences minérales volatiles contenues dans les eaux résiduaires des garages, parcs de stationnement, installations de lavage, etc.) ; *g* **Benzinabscheider [m]** (Vorrichtung in Bodenabläufen und Kläranlagen zur Trennung von Benzin und Öl, das nicht in die Entwässerungsleitung gelangen soll).

2383 gas propellant [n] *envir.* (Gas contained under pressure; e.g. in spray cans); *s* **agente [m] propulsor de aerosoles** (Gas licuado o comprimido que sirve de agente de arrastre para expeler sustancias de un recipiente a presión en forma pulverizada como líquido, como polvo o como espuma; cf. DINA 1987); *syn.* gas [m] propelente; *f* **gaz [m] propulseur** (Gaz contenu dans les récipients des générateurs d'aérosols, p. ex. dans les atomiseurs ou vaporisateurs) ; *g* **Treibgas [n]** (Verdichtetes oder unter Druck leicht verflüssigbares Gas, das z. B. aus Spraydosen für Kosmetika oder Lacke zerstäubt wird; wegen seiner Ungiftigkeit und Unbrennbarkeit im letzten Jahrhundert häufig verwendetes Treibmittel, das in der Vergangenheit den Ozonabbau in der Stratosphäre beschleunigte und zum Treibhauseffekt in der Atmosphäre beitrug; heute nur noch in Sonderfällen verwendet).

gas purification [n] *envir.* ►flue gas purification.

2384 gate valve [n] *constr.* (Device regulating the flow of water in a channel, sluice, or pipe by means of a flat disc that travels in a slot to open or close the flow orifice); *s* **válvula [f] de cierre** (Dispositivo de cierre de una tubería para regular el flujo de agua); *syn.* válvula [f] de compuerta, válvula [f] de corredera, válvula [f] esclusa [RA], válvula [f] de cortina [CO], compuerta [f] tubular [Es] (BU 1959, DACO 1988); *f 1* **vanne [f] d'arrêt** (Appareil de régulation du débit d'eau dans un tuyau, une canalisation) ; *f 2* **régulateur [m] de débit** (Dispositif intercalé dans une canalisation ou un réseau d'arrosage régulant le passage d'eau) ; *g* **Schieber [m]** (Absperrvorrichtung zur Regulierung des Durchflusses in einer Rohrleitung).

2385 gazebo [n] *gard.* (Small garden house of light construction to provide a cool and shady retreat in summer; ►arbor [US]/arbour [UK], ►community garden shelter [US]/allotment garden hut [UK], ►grapevine arbor [US]); *syn.* garden pavilion [n], summerhouse [n]; *s 1* **glorieta [f]** (Casita de jardín de construcción ligera, generalmente de madera, que sirve como lugar sombreado y fresco para descansar en el verano; ►casita de huerto recreativo urbano, ►cenador, ►emparrado); *syn.* cenador [m] (2), belvedere [m]; *s 2* **quiosco [m]** (Templete o pabellón en parques o jardines, generalmente abierto por todos los lados, que entre otros usos ha servido tradicionalmente para celebrar conciertos populares); *f 1* **cabanon [m] de bois** (Petite maisonnette de construction légère, souvent fixe, en bois, en canne de Provence ou préfabriqué; ►abri familial, ►fabrique, ►treille) ; *f 2* **gloriette [f]** (Petit ►pavillon ornemental non couvert, généralement en fer forgé inspiré des modèles de ferronnerie du XVIème siècle, parfois en treillage en bois, placé en position centrale dans un jardin clos, souvent recouvert de plantes grimpantes ; *terme générique* ►abri de jardin) ; *f 3* **pavillon [m] (de jardin)** (Construction très souvent couverte, légère, de forme ronde ou carrée, édifiée dans un jardin sur un espace ouvert ; *terme générique* ►abri de jardin) ; *syn.* recouvert par de la végétation pavillon [m] de verdure, cabinet [m] de verdure ; *f 4* **kiosque [m] (de jardin)** (Terme provenant du Moyen-Orient ; ►abri de jardin léger, ouvert de tous les côtés, en général de forme octogonale et de décoration orientale) ; *g* **Gartenpavillon [m]** (Kleines, leichtes Gartenhaus; ►Gartenlaube, ►Kleingartenlaube, ►Weinlaube).

g. b. h. *hort.* ►girth; ►trunk diameter.

2386 geest [n] *geo.* (Dry, infertile, glacial, sandy soils in the north-west lowlands of Germany; *opp.* ▶coastal marshland); *s* **geest** [m] (Terreno seco y arenoso de origen glacial en el noroeste de Alemania; *opp.* ▶marisma marítima); *f* **geest** [m] (Paysage du nord-ouest de l'Allemagne correspondant à une partie de la grande plaine nord-européenne fortement modelée par les glaciations quaternaires. Cette plaine intérieure est une zone d'épandage de dépôts proglaciaires principalement constituée de sols sableux secs et peu fertiles ; *opp.* ▶marais maritime) ; *g* **Geest** [f] (Trockene, unfruchtbare glaziale Auf-schüttungslandschaft mit vorwiegend sandigen Böden in der nordwestdeutschen Tiefebene; *opp.* ▶Marsch).

genecotype [n] *ecol.* ▶ecotype.

2387 gene pool [n] *biol.* (Combination of inherited characte-ristics in a species population; ▶inherited genes); *s* **banco** [m] **de genes** (Conjunto de características genéticas de una población dada; ▶patrimonio genético); *f* **pool** [m] **génique populationnel** (Ensemble des caractères génétiques d'une population donnée ; ▶patrimoine génétique) ; *syn.* stock [m] génétique (DUV 1984, 286) ; *g* **Genpool** [m] (Qualitative und quantitative Summierung aller unterschiedlichen Gene, die in einer abgegrenzten Popula-tion vorhanden sind; ▶Erbgut).

General and Federal Supplementary Conditions [npl] **of Contract for Construction (AIA Document A201)** [US] *contr. leg.* ▶General Conditions of Contract for Construction [US].

2388 General Conditions [npl] **of Contract** *contr. leg.* (Generic term for documents governing construction and supply contracts between employers/owners and contractors with legally-binding clauses. **In U.S.**, there are ▶General Conditions of Contract for Furniture, Furnishings and Equipment [AIA Document A271] and ▶General Conditions of Contract for Construction [AIA Document A201]; **in U.K.**, there are regula-tions for the delivery of goods and equipment governed by the *Supply of Goods and Services Act 1982 (amended 2002)*, and for construction *General Conditions of Contract for Building and Civil Engineering Works*; for government contracts GC/Works/1 General Conditions of Contract for Building and Civil Engineer-ing, edition 3, 1997; several organisations issue **g.c. of c**. and local authorities also have their own conditions; ▶bid documents [US]/tender(ing) documents [UK]; ▶general conditions of con-tract for project execution); *s* **condiciones** [fpl] **generales de contratos de las administraciones públicas** (Término genérico para las normas que regulan la ▶adjudicación de contratos entre las administraciones del Estado y las empresas privadas; en Es. la Ley 13/1995, de 8 de mayo, de Contratos de las Administraciones Públicas [LCAP] es la base jurídica para todo tipo de contratos, sean éstos de obras, de gestión de servicios públicos, de sumi-nistro o de consultoría y asistencia [Libro II]. Esta ley incluye así mismo las disposiciones generales [Libro I, Título I], los requisitos para contratar con la Administración [Libro I, Tít. II], las actuaciones relativas a la contratación [Libro I, Tít. III], la revisión de precios en los contratos de la Administración [Libro I, Tít. IV], la extinción de contratos [Libro I, Tít. V], la cesión de contratos y la subcontratación [Libro I, Tít. VI], la contratación en el extranjero [Libro I, Tít. VII], y el Registro Público de Contratos [Libro I, Tít. VIII]; ▶condiciones generales de con-trato de servicios de las administraciones públicas, ▶documentos de contratos de las administraciones públicas, ▶pliego de cláu-sulas administrativas generales para la contratación de obras); *f* **clauses** [fpl] **générales de passation des marchés** (Documents et procédures types normalisés utilisés par les services publics pour la consultation des entreprises et la passation de marchés publics de fourniture ou de commande d'ouvrages ; ▶cahier

des clauses administratives générales [des marchés publics], ▶clauses générales de passation des marchés publics de fourni-tures, ▶pièces contractuelles constitutives du marché) ; *syn.* règlement [m] de passation des marchés, règles [fpl] de passation des marchés ; *g* **Verdingungsordnung** [f] (Regelwerk für forma-lisierte Ausschreibungs- und Vergabeverfahren und Bestimmun-gen für die Ausgestaltung von Bau- und Lieferverträgen zwi-schen Auftraggebern und Auftragnehmern, an das besonders die öffentliche Hand gebunden ist; ▶allgemeine Vertragsbedin-gungen für die Ausführung von Bauleistungen, ▶Verdingungs-unterlagen, ▶Verdingungsordnung für Leistungen, ▶Vergabe-und Vertragsordnung für Bauleistungen).

2389 General Conditions [npl] **of Contract for Con-struction (AIA Document A201)** [US] *contr. leg.* (**In U.S.**, there is an Architect's Handbook of Professional Practice, Volume 2, dealing with owner-contractor agreements issued by The American Institute of Architects [AIA]. The procedure for construction projects is written down in the AIA document A201. For construction work to be done for the federal government there is the **General and Federal Supplementary Conditions of the Contract for Construction** written down in the AIA documents A201 and A201/SC. These comprehensive documents cover all of the requirements applicable to private or public contracts for the most appropriate and economic execution of construction projects; ▶General Conditions of Contract for Furniture, Furnishings and Equipment [US]/Sales of Goods and Services Act [UK]. **In U.K.**, several regulations exist dealing with owner-contractor agreements. The **General Conditions of Government Contracts for Building and Civil Engineering Works** are issued by the Property Services Agency [PSA]; cf. 3rd edition, 1989, quoted in BCD 1990; in U.K., there are other alternative standard contract forms; e.g. those published by the "Joint Contracts Tribunal" [JCT] and those issued by the **National Joint Consultative Council** [n] **(NJCC) for Building Code of Procedure for Single or Two Stage Tendering [UK]**. These are general regulations, which prescribe generally accepted standards for 'open' and 'selective' tendering); *syn.* General Conditions [npl] of Government Contracts for Building and Civil Engineering Works [UK]; *s* **condiciones** [fpl] **generales de con-tratos de obras de las administraciones públicas [Es]** (En Es. la contratación de obras se rige por la Ley de Contratos de las Administraciones Públicas [LCAP]; cf. ▶condiciones generales de contratos de las administraciones públicas); *syn.* reglamento [m] de adjudicación para la ejecución de obras; *f* **clauses** [fpl] **générales de passation des marchés publics de travaux** (Docu-ments et procédures types normalisés utilisés par les services publics pour la consultation des entreprises et la passation de marchés publics de travaux de bâtiment ou de génie civil ; ▶clauses générales de passation des marchés publics de fourni-tures ; **D.**, ces clauses générales constituent la VOB, ouvrage dans lequel figurent selon la terminologie française les cahiers des clauses administratives générales et particulières ainsi que les cahiers des clauses techniques générales) ; *g* **Vergabe- und Ver-tragsordnung** [f] **für Bauleistungen** (*Abk.* VOB; *vor 2002* **Verdingungsordnung für Bauleistungen;** für öffentliche Auf-traggeber ist in D. die VOB seit 1926 ein Regelwerk als Grund-lage für die Ausgestaltung von Bauverträgen, die entsprechend den haushaltsrechtlichen Grundsätzen eine zweckmäßige und wirtschaftliche Deckung des Bedarfs an Bauleistungen durch formalisierte Ausschreibungs- und Vergabeverfahren sicher-stellen soll; die VOB ist gegliedert in die VOB Teil A — Allge-meine Bestimmungen für die Vergabe von Bauleistungen DIN 1960 mit vier Abschnitten, VOB Teil B — Allgemeine Vertrags-bedingungen für die Ausführung von Bauleistungen DIN 1961 und VOB Teil C — Allgemeine Technische Vertragsbedingun-

G

gen für Bauleistungen [ATV] DIN 18 299 bis DIN 18 459. Im Rahmen der EG-weiten Harmonisierung sind in den Teil A diverse Richtlinien des Europäischen Parlamentes und des Rates in die VOB eingearbeitet worden, zuletzt die Richtlinie 2004/18/EG über die Koordinierung der Verfahren zur Vergabe öffentlicher Bauaufträge, Liefer- und Dienstleistungsaufträge und die EU-Richtlinie 2004/17/EG zur Koordinierung der Zuschlagserteilung durch Auftraggeber im Bereich Wasser-, Energie- und Verkehrsversorgung sowie der Postdienste. Der Text der VOB Teil B wurde 2002 intensiv überarbeitet und dabei insbesondere an das Gesetz zur Modernisierung des Schuldrechtes angepasst; in der VOB Teil C wurden insgesamt 40 Allgemeine Technische Vertragsbedingungen [ATV] überarbeitet resp. neu erstellt; ▶Verdingungsordnung für Leistungen); *syn. obs.* Verdingungsordnung [f] für Bauleistungen.

2390 General Conditions [npl] **of Contract for Furniture, Furnishings and Equipment** [US] *constr. contr.* (**In U.S.**, stipulations contained in the Architect's Handbook of Professional Practice, Volume 2, dealing with owner-contractor agreements issued by The American Institute of Architects [AIA]. The procedure for delivery of furniture, furnishings and equipment is set out in the AIA document A271; **in U.K.**, the supply of goods and services including contracts for works and materials is governed by statute; Part I of the 1982 Act deals with the supply of goods and its provisions affect building contracts; Part II deals with the supply of services including professional services; ▶General Conditions of Contract for Construction (AIA Document A201) [US]/General Conditions of Government Contracts for Building and Civil Engineering Works [UK]); *syn.* Sales of Goods and Services Act [n] [UK]; *s* **condiciones** [fpl] **generales de contratos de servicios de las administraciones públicas** [Es] (En Es. la contratación de servicios de todo tipo se rige por la Ley de Contratos de las Administraciones Públicas [LCAP]; ▶condiciones generales de contratos de obras de las administraciones públicas); *syn.* reglamento [m] de adjudicación de servicios; *f* **clauses** [fpl] **générales de passation des marchés publics de fournitures** (**D.**, documents et procédures types normalisés utilisés par les services publics pour la consultation des entreprises et la passation de marchés publics de fournitures ; **F.**, réglementation régie dans le cadre du Code des marchés publics ; la directive du Conseil des communautés européennes n° 88-295 du 22 mars 1988 modifiant la directive n° 77-62 du 21 décembre 1976 porte le montant des fournitures à un seuil de 200.000 Euro ; ▶clauses générales de passation des marchés publics de travaux) ; *g* **Verdingungsordnung [f] für Leistungen** (*Abk.* VOL; für öffentliche Auftraggeber ist in D. die VOL seit 1932 [Erlass des Reichsfinanzministeriums vom 24.09. 1932] ein Regelwerk als Grundlage für Leistungen aller Art aus Kauf-, Werk- und Werklieferungsverträgen, die entsprechend den haushaltsrechtlichen Grundsätzen eine zweckmäßige und wirtschaftliche Deckung des Bedarfs an Lieferungen durch formalisierte Ausschreibungs- und Vergabeverfahren sicherstellen soll; auf Grund der Richtlinie des Rates vom 22.03.1988 [88/295/EWG] zur Änderung der Richtlinie 77/62/EWG ist sie europaweit bei Auftragswerten von mindestens 200.000 Euro anzuwenden; ausgenommen sind Bauleistungen, für die die ▶Vergabe- und Vertragsordnung für Bauleistungen [VOB] und Leistungen freiberuflich Tätiger, für die die VOF gilt. Die VOL ist gegliedert in die VOL Teil A — Allgemeine Bestimmungen für die Vergabe von Leistungen — und VOL Teil B — Allgemeine Bedingungen für die Ausführung von Leistungen).

general conditions [npl] **of contract for project execution** [UK] *constr. contr.* ▶Standard Form of Building Contract [UK].

General Conditions [npl] **of Government Contracts for Building and Civil Engineering Works** [UK] *contr. leg.* ▶General Conditions of Contract for Construction (AIA Document A201) [US].

2391 general contractor [n] *contr.* (Entrepreneur commissioned by a building owner to oversee the execution of the complete building work/works, including subcontractor construction, coordination and supervision; ▶consortium 1); *s* **contratista** [m] **general** (Empresario encargado por el comitente de realizar todos los trabajos correspondientes para construir una obra, que generalmente encarga por su parte a subcontratistas. Tiene por lo tanto la responsabilidad de coordinar y dirigir la obra; ▶consorcio de contratistas); *f* **entreprise** [f] **générale** (Titulaire d'un marché sous-traitant une partie de ses prestations, que celles-ci constituent ou non un lot technique ; ▶groupement d'entreprises solidaires) ; *g* **Generalunternehmer** [m] (Bauausführender Unternehmer, der vom Bauherrn mit sämtlichen zu einem Bauwerk gehörenden Leistungen i. d. R. unter Einschaltung von Subunternehmern, also mit der gesamten Bauausführung, der Koordinierung und Leitung des Bauvorhabens beauftragt wird. Der **Generalübernehmer** hingegen übernimmt wie vor ein gesamtes Projekt als Manager, ohne jedoch selbst Bauleistungen auszuführen; ▶Arbeitsgemeinschaft 1).

general development planning [n] *plan.* ▶overall area planning.

general fill [n] *constr.* ▶fill material.

2392 generalist species [n] *phyt. zool.* (Species with broad food or habitat preferences; ▶indifferent species, ▶ubiquist); *s* **especie** [f] **generalista** (Organismo capaz de vivir bajo condiciones ambientales muy variadas y de aprovechar recursos alimenticios diversos, pero sin utilizarlos al cien por cien. Son los primeros en ocupar los nichos que quedan vacíos; DINA 1987; ▶especie indiferente, ▶especie ubicuista); *f* **espèce** [f] **généraliste** (Espèce de grande amplitude écologique telles que les annuelles et vivaces des lieux perturbés ou pollués [bords de route, terrils, jardins et cultures abandonnés, terres irradiées, etc.] ; ▶espèce indifférente, ▶espèce ubiquiste) ; *syn.* espèce [f] de grande amplitude écologique, généralistes [mpl] ; *g* **Generalist** [m] (Art mit einer großen Amplitude von Nahrungs- oder Biotopansprüchen; ▶Ubiquist, ▶vage Art).

general land-use area [n] [D] *leg. urb.* ▶urban land use category.

2393 general plan [n] *plan.* (Generic term covering the long-range, overall plan to guide the growth and development of an area; ▶community development plan, ▶community garden development plan [US]/allotment garden development plan [UK], ▶overall master plan for play areas [US]/strategic plan for play spaces [UK], ▶overall master plan for sports and physical activities [US]/strategic plan for sports and physical activities [UK], ▶master plan 2); *s* **plan** [m] **director** (Tipo de plan cuya función principal es el desarrollo de líneas generales o directrices de planificación en un campo específico, p. ej. Plan Hidrológico Nacional, ▶plan de ordenación de los recursos naturales o a nivel local plan general municipal de ordenación, ▶plan de desarrollo, ▶plan de desarrollo de áreas de juegos y recreo infantiles, ▶plan de desarrollo de equipamientos deportivos, ▶plan de desarrollo de huertos recreativos urbanos, ▶plan municipal de zonas verdes y espacios libres, ▶plan urbanístico); *f* **schéma** [m] **directeur sectoriel** (±) (Document définissant les grandes orientations, les engagements à long terme en matière d'aménagements sectoriels tels que les déplacements, les sports, les espaces verts, etc. ; *termes spécifiques* schéma directeur de végétalisation des espaces publics urbains, schéma directeur des pistes cyclables, schéma directeur de cheminements piétonniers, schéma directeur départemental des circulations douces, schéma directeur d'aménagement lumière, schéma directeur d'accessibilité de la voie publique aux personnes handicapées, schéma directeur des infrastructures ; ▶documents de planification urbaine, ▶programme d'équipements sportifs, ▶schéma d'aménagement des jardins ouvriers, ▶schéma directeur d'aires de jeux, ▶plan de développement ; *syn.* plan [m] directeur [CDN, CH]) ; *g* **Leitplan [m]** (Plan, der als Orientierungs- und Informationsgrundlage die strategischen Vorstellungen für zukünftige Entwicklungen für ein Fach- oder Tätigkeitsbereich in einem definierten Gebiet darstellt; OB zu z. B. ▶Bauleitplan, ▶Kleingartenentwicklungsplan, ▶Spielflächenleitplan, ▶Sportstättenentwicklungsplan etc.; ▶Entwicklungsplan).

2394 general plan [n] for urban open spaces [US] (±) *landsc. urb.* (Plan and the explanatory report which accompanies the legally-binding land-use plan such as the ▶zoning map [US]/Proposals Map/Site Allocations Development Plan Document [UK] and presents the goals and measures of ▶general urban green space planning; ▶directive on planting requirements [US]/enforcement notice on planting requirements [UK], ▶plant preservation requirements [US]/obligation to preserve existing plants [UK]; ▶comprehensive plan [US]/Local Development Framework [LDF] [UK], ▶landscape plan, ▶plan area of a zoning map, ▶public or semipublic authority [US]/public or semi-public authority [UK]); *syn.* green infrastructure plan [± UK], green open space structure plan [n] [± UK], green space strategy plan [n] [± UK]; *s* **plan [m] municipal de zonas verdes y espacios libres** (Plan con un informe escrito que plasma los fines y las medidas de la ▶planificación de zonas verdes y espacios libres urbanos acompañando a un ▶plan general municipal de ordenación urbana o a un ▶plan parcial [de ordenación]. Puede incluir ▶prescripciones de preservar la vegetación existente o ▶prescripciones de plantar; ▶área de vigencia de un plan parcial, ▶autoridades e instituciones públicas y semipúblicas, ▶plan de gestión del paisaje); *f 1* **volet [m] paysager d'un PLU** (≠) (F., plan établi dans le cadre de l'élaboration du PLU ; il réalise un diagnostic paysager du territoire communal, et propose un projet de paysage avec ses orientations techniques ainsi qu'un programme d'actions traduisant les orientations politiques en matière de paysages ; il a pour but d'intégrer la valorisation et la préservation du patrimoine paysager dans le PLU ; D., expression réglementaire des objectifs et mesures en matière de ▶planification urbaine des espaces verts, ce document d'urbanisme opposable au tiers dans certains Länder constitue l'instrument de protection, de mise en valeur et d'organisation du paysage et des espaces naturels en milieu urbain ; plan à l'échelle du 1:1000 qui « accompagne » les *Bebauungspläne* [PLU/POS allemand] ; les prescriptions proposées en matière de paysages, d'espaces verts et d'écologie urbaine comprennent par principe les facteurs sol [protection et imperméabilisation des sols], eau [limitation de l'imperméabilisation, bassins et aires de rétention/infiltration des eaux pluviales], végétaux [conservation de la végétation existante — ▶obligation de préservation de la végétation existante ou plantations compensatoires — ▶obligation de réaliser des plantations] ainsi que la protection des espèces de faune, de flore et de leurs habitats [interdiction de certaines usages, constitution de biotopes nouveaux et mesures d'entretien et de gestion] ; proposition de traduction : « schéma directeur d'espaces verts » ; ce document constitue en quelque sorte le « volet paysager » du *Bebauungsplan* [PLU/POS allemand] ; ▶périmètre d'un plan local d'urbanisme/plan d'occupation des sols, ▶personnes publiques [associées à l'élaboration du plan], ▶schéma de paysage) ; *f 2* **plan [m] paysage** (±) (F., à l'échelle d'une communauté de communes, d'une grande agglomération ou d'une communauté d'agglomérations il constitue un document-cadre spécifique ayant pour objectif de guider les choix et les orientations d'aménagement en fédérant les approches « paysages » des diverses démarches de planification ; il définit les choix et orientations en matière d'identité paysagère ainsi que les programmes d'actions, en particulier en ce qui concerne la gestion durable du patrimoine vert et des paysages ainsi que la mise en valeur du patrimoine bâti, les maillages verts, la préservation des sites sensibles sur les plans écologique et paysager, la restauration d'une continuité écologique, visuelle et fonctionnelle entre les espaces ; il développe des recommandations pour intégrer au mieux le végétal dans tout projet d'aménagement et assure la mise en cohérence des règles d'urbanisme et des projets d'aménagement ; il s'articule au ▶plan local d'urbanisme [PLU] sur la base de principes reconnus par tous concernant la protection, la réhabilitation et la valorisation des paysages et espaces verts) ; *g* **Grünordnungsplan [m]** (*Abk.* GOP; landschaftspflegerischer Fachbeitrag mit Plan- und Textteil, der selbständig oder begleitend zum ▶Bebauungsplan die Ziele, Erfordernisse und Maßnahmen des Naturschutzes und der Landschaftspflege sowie die Ordnung und Gestaltung der Grünflächen darstellt [▶Grünordnung]. Grünordnungsplanerische Festsetzungsmöglichkeiten umfassen grundsätzlich die Regelungsfelder Boden [z. B. Mutterbodenschutz und Versiegelungsanteil], Wasser [Begrenzung der Versiegelungsflächen, Ausweisung von Niederschlagsversickerungsflächen und Regenrückhalteflächen], Pflanzen [Erhaltung bestehender Vegetation — ▶Pflanzbindung oder Verpflichtung von Neupflanzungen — ▶Pflanzgebot] sowie Arten- und Biotopschutz [Nutzungsausschluss, Erstellung von Biotopflächen und Maßgaben zur Pflege]. Der G. ist nicht in der Bundesgesetzgebung, sondern nur in den meisten Landesnaturschutzgesetzen verankert. Dementsprechend gibt es Unterschiede hinsichtlich der Verpflichtung **G.pläne** zu erarbeiten, der Inhalte und Einarbeitung der Festsetzungen und Darstellungen in die Bebauungspläne; der **G.** wird aus der nächst höheren Planungsebene, d. h. dem ▶Landschaftsplan abgeleitet und vertieft resp. ergänzt dessen Aussagen für den jeweiligen Planungsraum; der Geltungsbereich ist i. d. R. mit dem des Bebauungsplanes identisch. Der **G.** ist eine wichtige Informationsunlage für die Umweltprüfung, die gem. § 2 [4] BauGB zur Aufstellung von Bebauungsplänen durchgeführt werden muss, um die Belange des Umweltschutzes und voraussichtliche erhebliche Umweltauswirkungen zu ermitteln. Während des Planungsprozesses sind die Öffentlichkeit und die ▶Träger öffentlicher Belange zu beteiligen; cf. BFN 2006; ▶Geltungsbereich eines Bebauungsplanes).

2395 general planning [n] [US] *plan.* (Overall planning for ultimate development, which may be executed in stages/phases over a long period of time, and which—with periodic updating— serves as a guide for most of the specialist planners; ▶overall area planning, ▶overall master planning for play areas [US]/play spaces strategic planning [UK], ▶overall master planning for sports and physical activities [US]/strategic planning for sports and physical activities [UK]; ▶scope of professional services); *syn.* strategic planning [n] [UK]; *s* **planificación [f] directriz** (Planteamientos generales de desarrollo a medio y largo plazo en un campo específico; ▶descripción de servicios de planificación, ▶planificación general, ▶planificación integrada de áreas de juegos infantiles, ▶programa de gestión de equipamientos para deportes); *f* **planification [f] référentielle** (±) (Planification représentant pour les spécialistes pendant une certaine période un modèle de référence ; ▶forme et étendue de la mission, ▶planification d'aires de jeux, ▶planification globale, ▶programmation des équipements sportifs) ; *syn.* planification [f] directrice [CH] ; *g* **Leitplanung [f]** (Musterlösung hinsichtlich des ▶Leistungsbildes für und der strategischen Vorgehensweise bei einer Planung, die zumindest für eine gewisse Zeit der Mehrzahl von

Fachleuten als Vorbild dient und periodisch bei Bedarf fortgeschrieben wird; ▶Gesamtplanung, ▶Spielflächenleitplanung, ▶Sportstättenentwicklungsplanung).

general recreational use [n] *recr.* ▶area intended for general recreational use.

general stagnation situation [n] [US] *met.* ▶inversion weather.

2396 general urban green space planning [n] *landsc. urb.* (Goals and measures of open space planning in developed areas to preserve and enhance the spatial and functional relationship of all open spaces and recreational facilities, to each other and to the built-up areas, according to ecological and aesthetic principles for the sustained benefit of the citizens; ▶general plan for urban open spaces [US]/green open space structure plan [UK], ▶impact mitigation regulation, ▶open space planning); *syn.* green space strategy [n] [also UK]; *s* **planificación** [f] **de zonas verdes y espacios libres urbanos** (Planificación del paisaje en zonas urbanas con el fin de asegurar las relaciones espaciales y funcionales de todos los espacios libres con todo su equipamiento de recreo entre sí y con las zonas edificadas, considerando los principios ecológicos y estéticos con el fin de su disfrute por la población. Los resultados se fijan en un plan en el que se muestran los objetivos y las medidas a realizar; ▶planificación de espacios libres, ▶plan municipal de zonas verdes y espacios libres, ▶regulación de impactos sobre la naturaleza); *f* **planification** [f] **urbaine des espaces verts** (F., objectifs et mesures du paysagisme d'aménagement dans le cadre de l'urbanisme qui s'appuie sur plusieurs corps de règles relevant du Code de l'urbanisme, du Code rural, du Code forestier et de la législation sur la protection de la nature. D., mesures réglementaires définissant l'ordonnancement fonctionnel des espaces libres végétalisés ou minéraux par rapport aux constructions dans le développement et la composition urbaine, selon des critères esthétiques, écologiques et de valorisation des activités de loisirs et dans le cadre de la prise en compte de la ▶réglementation relative aux atteintes subies par le milieu naturel [plan et mesures paysagères constituant le volet paysager du « Bebauungsplan » — PLU/POS allemand] ; ces mesures trouvent leur expression dans le ▶volet paysager d'un PLU ; ▶aménagement des espaces libres, ▶plan paysage) ; *g* **Grünordnung** [f] (Ziele und Maßnahmen der Landschaftsplanung im Siedlungsbereich, die der Sicherung und Förderung der räumlichen sowie funktionellen Ordnung aller vegetationsbestimmten und nicht vegetationsbestimmten Freiräume zueinander und zu den baulichen Anlagen im Zusammenhang mit der stadtplanerischen Entwicklung nach ökologischen und gestalterischen Gesichtspunkten und aus Gründen der Erholungsvorsorge nachhaltig dienen [zur Gegensteuerung einer zunehmenden Bebauungsverdichtung], wie es zum geistigen und körperlichen Wohlbefinden des Menschen erforderlich ist. Ziele und Maßnahmen der **G.** werden in einem ▶Grünordnungsplan dargestellt; durch die **G.** fließen die Ziele des Umwelt- und Naturschutzes sowie die naturschutzrechtliche ▶Eingriffsregelung in die Bauleitplanung mit ein; ▶Freiraumplanung); *syn.* Grünordnungsplanung [f], *z. T.* Naturschutz [m, o. Pl.] und Landschaftspflege [f, o. Pl.] im Siedlungsbereich.

genes [npl] *biol.* ▶inherited genes.

2397 genetically modified organism (GMO) [n] *biol. ecol. leg.* (Living creature, with the exception of a human being, in which the genetic material has been altered using molecular biological methods, in a way not naturally possible by reproduction; a genetically modified organism is also an organism created by reproduction processes of genetically modified organisms, or other forms of propagation of a genetically modified organism; cf. Directive 2001/18/EC; genetic engineering uses several methods to introduce foreign DNA into an organism and to create a GMO, also known as a **transgenic organism**, with new properties. The abbreviation DNA stands for the term Desoxyribonucleic Acid); *s* **organismo** [m] **modificado genéticamente (OMG)** (Cualquier organismo, con excepción de los seres humanos, cuyo material genético ha sido modificado de una manera que no se produce de forma natural en el apareamiento o en la recombinación natural, siempre que se utilicen técnicas que reglamentariamente se establezcan; cf. Art. 2 Ley 15/1994. La modificación genética de organismos es objeto de grandes controversias. En el caso de las plantas de cultivo transgénicas, la crítica principal de las organizaciones ecologistas es que estos organismos pueden diseminarse por medio del viento o de animales y contaminar así los cultivos naturales. Así mismo se considera negativo que unas pocas empresas multinacionales puedan dominar el mercado de semillas casi en todo el mundo, dictar los precios e incluso llevar a juicio a agricultores en cuyos campos se encuentren plantas transgénicas, aunque ellos no las hayan plantado. Solo hay 4 cultivos transgénicos a gran escala: soja, algodón y maíz suponen el 95% del total. El 5% restante es colza. **En Es.** el cultivo de maíz transgénico se ha extendido mucho en los últimos años en regiones como Aragón y Catalunya, con la consecuencia de que ya no existe la posibilidad de cultivar maíz ecológico. Ante esta situación una coalición de organizaciones ecologistas [Amigos de la Tierra, Ecologistas en Acción, Plataforma Transgènics Fora, Plataforma Galega Antitransxénicos] y la Coordinadora de Organizaciones de Agricultores y Ganaderos [COAG] están fomentando una campaña de declaración de zonas libres de transgénicos, a la que ya se han sumado varias CC.AA. [Asturias, País Vasco, Baleares, Canarias], la provincia de Málaga y muchos municipios. **En AL**, se ha extendido incluso de forma ilegal el cultivo de soja transgénica sobre todo en la Argentina, Paraguay y el Brasil. Los EE.UU. y Canadá son los otros dos países del mundo en los que se cultivan transgénicos a gran escala. En estos países, la aplicación de herbicidas se ha disparado desde que se introdujeron los transgénicos: En EE.UU. el uso de glifosato se multiplicó por 15, en Brasil aumentó en un 80%. En Europa 7 países han prohibido el cultivo de transgénicos en su territorio en base a pruebas científicas e incertidumbres. La población en Europa rechaza de forma abrumadora los cultivos transgénicos; cf. www.tierra.org consulta 23.8.2009); *syn.* organismo [m] transgénico; *f* **organisme** [m] **génétiquement modifié (OGM)** (Organisme vivant [animal, végétal, bactérien] dont le patrimoine génétique a été modifié par une technique de génie génétique soit pour accentuer certaines de ses caractéristiques ou lui conférer des propriété nouvelles considérées comme désirables, soit au contraire pour atténuer, voire éliminer certaines caractéristiques considérées comme indésirables ; ce processus s'inspire des techniques de sélection ou de mutation, qui existent déjà dans le monde agricole. Le génie génétique permet de transférer des gènes sélectionnés d'un organisme à un autre, y compris entre des espèces différentes. Il offre ainsi potentiellement la possibilité d'introduire dans un organisme un caractère nouveau dès lors que le ou les gène(s) correspondants sont préalablement identifiés ; le génie génétique peut concerner à l'avenir les domaines médicaux, agricoles, de l'alimentation et environnementaux ; les OGM sont controversés car ils peuvent avoir un impact aussi bien positif que négatif sur l'environnement [possibilité et conséquences du transfert du caractère nouveau d'un OGM à une espèce sauvage proche, sur l'impact direct ou indirect sur la faune ou la flore, sur le caractère réversible de modifications éventuelles de la biodiversité, sur les bénéfices attendus pour l'environnement ; site interministériel sur les OGM, www.ogm.gouv.fr) ; *g* **gentechnisch veränderter Organismus** [m] (*Abk.* GVO; Lebewesen, mit Ausnahme des Menschen, bei dem das genetische Material mit Hilfe moleku-

G

G

larbiologischer Methoden in einer Weise verändert worden ist, wie es natürlicherweise durch Kreuzen oder natürliche Rekombination nicht möglich ist; ein genetisch veränderter Organismus ist auch ein Organismus, der durch Kreuzung oder natürliche Rekombination zwischen gentechnisch veränderten Organismen oder durch andere Arten der Vermehrung eines gentechnisch veränderten Organismus entstanden ist; cf. § 3 GenTG, RL 2001/18/EG. Die Gentechnik verfügt über verschiedene Methoden, um fremde DNA in einen Organismus einzuführen und so einen GVO mit neuen Eigenschaften, auch **transgener Organismus** genannt, herzustellen [die Abk. DNA steht für die englische Bezeichnung Desoxyribonucleic Acid]); der Einsatz von GVOs wird sehr kontrovers diskutiert, da die Auswirkungen auf die Umwelt positiv und negativ sein und noch nicht sicher abgeschätzt werden können, z. B. die Möglichkeiten und Konsequenzen der Übertragung neuer Eigenschaften eines GVOs auf eine nahestehende wild lebende Art, die direkten oder indirekten Auswirkungen auf Flora und Fauna, auf mögliche Veränderungen der Biodiversität, auf den Nutzen für die Umwelt, Nahrungskette und Gesundheit. In der EU unterliegt der Vertrieb von GVOs strengen EU-Kennzeichnungsrichtlinien und der Anbau genveränderter Pflanzen ist durch strenge Auflagen sehr stark eingeschränkt. Hingegen nimmt der Anbau von gentechnisch veränderten Pflanzen wie Sojabohne, Mais, Baumwolle, Raps und Reis in den USA und vielen Entwicklungsländern ständig zu; 2005 90 ha, 2006 über 104 Mio. ha Anbaufläche weltweit; cf. ALG 2007, 233 und wikipedia 2009); *syn.* gentechnisch modifizierter Organismus [m].

2398 genetic contamination [n] *phyt. zool.* (Significant adulteration of gene potential of a population caused by interbreeding with individuals of foreign taxonomic and genetic origin; ▶anthropogenic alteration); *s* **contaminación [f] genética** (Alteración significativa del patrimonio genético de una población provocada por la mezcla de individuos de origen taxonómico y genético diferente; ▶contaminación biológica); *f* **contamination [f] génétique (±)** (Altération significative du patrimoine génétique d'une population provoquée par le mélange avec des individus d'origine taxonomique et génétique différente ; ▶altération par [l']introduction d'espèces) ; *g* **Überfremdung [f]** (Beachtliche Veränderung des Genpotenzials einer Population durch Vermischung mit Individuen anderer taxonomischer und genetischer Herkunft; ▶Verfälschung).

2399 genetic diversity [n] *biol. ecol.* (Variety of genetic building blocks among individuals of the same species, providing resilience in the face of environmental stress, and allowing adaptation to changing conditions, which is essential for ultimate survival of a species; ▶biodiversity); *s* **diversidad [f] genética** (Presencia de una gran variedad genética en una población que da a ésta gran resiliencia para resistir ante factores ambientales negativos y le permite adaptarse a las nuevas condiciones, lo cual es esencial para la supervivencia de la especie; ▶biodiversidad); *f* **diversité [f] génétique** (Ensemble de l'information génétique contenue dans tous les êtres vivants et correspondant à la variabilité des gènes et des génotypes entre espèces et au sein de chaque espèce ; LBI 1997 ; le large éventail du matériel génétique des individus d'une même espèce leurs permet de mieux réagir à des stimulations extérieures et leurs procure une meilleure adaptation aux variations des conditions du milieu ; ▶biodiversité) ; *syn.* diversité [f] biochimique ; *g* **genetische Vielfalt [f]** (Großes Erbanlagenspektrum von Individuen derselben Art, das erlaubt, mit Umweltstress besser fertig zu werden und sich an verändernde Lebensbedingungen besser anpassen zu können; ▶Biodiversität); *syn.* genetische Diversität [f].

genetic potential [npl] *biol.* ▶inherited genes.

2400 geobotany [n] *phyt.* (Study of plants and their distribution as indicators of soil composition and depth, rock characteristics, groundwater, and historic conditions; ▶chorology, ▶phytosociology, ▶plant geography); *s* **geobotánica [f]** (La ciencia de la relación entre la vida vegetal y el medio terrestre, es decir que estudia el hábitat de las plantas en la superficie terrestre. El término fue creado por RÜBEL pues no había ninguno cuyo valor etimológico correspondiese a este conjunto de conocimientos: los de geografía botánica y ecología que se usaban, y aún siguen usándose en su lugar, son más adecuados para expresar divisiones del mismo conjunto. Así H. DEL VILLAR divide la **g.** en sinecología, fitoecología y ▶fitogeografía o geografía botánica; cf. DB 1985; ▶corología, ▶fitosociología); *f* **géographie [f] botanique** (Science étroitement liée à la géographie des plantes, étudiant les causes de la répartition géographique des diverses espèces végétales et des différents types de végétation à la surface du globe ainsi que les circonstances qui influent sur le développement des individus et des associations et leurs relations avec ceux des époques géologiques antérieures ; on distingue différents domaines de recherche **1.** la ▶chorologie et la synchorologie, **2.** la géographie botanique historique et génétique, **3.** la synécologie et la ▶phytosociologie ; cf. DEE 1982 et VOC 1979 ; ▶phytogéographie) ; *g* **Geobotanik [f]** (*Teilgebiet der Botanik* Lehre von der Verbreitung von Pflanzen und Pflanzengesellschaften in Abhängigkeit von Umweltfaktoren und historischen Bedingungen; die **G.** ist eng mit der Vegetationsgeografie verbunden und wird mit dieser zur ▶Pflanzengeografie zusammengefasst. Es gibt folgende Forschungsgebiete: **1.** floristische **G.** [▶Arealkunde], **2.** historische und **genetische G.**, **3.** ökologische **G.** und **soziologische G.** als ▶Pflanzensoziologie).

2401 geodesy [n] *geo. surv.* (Science of determining the size and shape of the earth and the precise location of points on its surface by triangulation or ▶topographic survey 3; ▶aerial photogrammetry, ▶topographic mapping); *s* **geodesia [f]** (Ciencia que tiene por objeto la determinación de la forma y dimensiones de la Tierra. La **g.** evalúa las relaciones geométricas existentes entre los diferentes puntos de la superficie terrestre; DGA 1986; ▶cartografía topográfica, ▶fotogrametría aérea, ▶medición topográfica); *f* **géodésie [f]** (Ensemble des recherches et mesures visant à préciser la forme géométrique de la Terre et la position des lieux à la surface [altitude, latitude, longitude] ; DG 1984 ; ▶arpentage, ▶levé, ▶photogrammétrie aérienne, ▶phototopographie) ; *g* **Geodäsie [f]** (Lehre von der Ausmessung und Abbildung der Erdoberfläche; man unterscheidet die ‚niedrige **G.**' im Sinne der ▶Vermessung und die ‚höhere **G.**', die sich mit Mess- und Rechenverfahren zur Bestimmung der Erdfigur sowie mit der Landesvermessung befasst; ▶Erdbildmessung, ▶Luftbildmessung).

geographical research [n] *geo.* ▶landscape geographical research.

2402 geographic information system [n] **(GIS)** *adm. geo. landsc. plan.* (Computer-aided system of information comprising hardware, software, data and organizational methods to collect, administer, process, analyse, model and visualize spatial data; ▶landscape information system); *s* **sistema [m] de información geográfica (SIG)** (Cualquier tipo de sistema de información computarizado que incluye el hardware, software, datos y métodos de organización utilizados para recolectar, administrar, procesar, analizar, modalizar y visualizar datos geográficos; ▶sistema de información paisajística); *f* **système [m] d'information géographique (SIG)** (Système automatisé constitué par le matériel, les logiciels, les données et les savoir-faire liés à l'utilisation de ces derniers et conçu pour enregistrer, organiser, traiter, analyser, modéliser et visualiser des données spatiales géographiquement référencées ; ▶système d'information paysager) ; *g* **Geografisches Informationssystem [n] (GIS)** (Ein aus Hardware, Software, Daten und Organistionsformen bestehendes rechnergestütztes Informationssystem zur Erfassung, Verwaltung, Bearbeitung, Analyse, Modellierung und Visualisierung raumbezogener Daten; ►Landschaftsinformationssystem); *syn.* Geoinformationssystem [n].

2403 geography [n] *geo.* (Science dealing with all aspects of the earth's surface, its natural, social and political divisions, climate, production, population, etc.; ►plant geography); *s* **geografía [f]** (Ciencia que estudia las variaciones de las distribuciones espaciales de los fenómenos de la superficie terrestre [abióticos, bióticos y culturales], así como las relaciones entre el medio natural y el hombre y la individualización y análisis de las regiones en la superficie terrestre; DGA 1986; ►fitogeografía); *f* **géographie [f]** (Science humaine ayant pour objet l'étude **1.** des formes et des dimensions de la terre, **2.** des phénomènes naturels, sociaux et politiques dans des cadres spatiaux de dimensions hiérarchisées, la nature et l'intensité de leurs interactions qui animent la surface de la terre ; cf. DG 1984 ; ►phytogéographie) ; *g* **Geografie [f]** (Wissenschaft, die sich mit der Beschreibung der Erdoberfläche und der Entwicklung der Landschaften bis zum heutigen Zustand, mit den unterschiedlichen politischen und sozialen Gegebenheiten, deren Entwicklung und Wechselwirkungen befasst; ►Pflanzengeografie); *syn. o. V.* Geographie [f].

geohydrology [n] *hydr.* ►hydrogeology.

geologic erosion [n] [US] *geo.* ►natural erosion.

2404 geology [n] *geo.* (Study of materials of which planet earth is made, natural processes acting on these materials, products formed, history of the planet and its life forms since its origin); *s* **geología [f]** (Ciencia que estudia la geosfera, es decir la capa abiótica de la Tierra, su génesis, su constitución, los procesos naturales que la rigen, las transformaciones sufridas por ella a lo largo de la historia del planeta y los recursos minerales existentes en ella); *f* **géologie [f]** (Étude de la composition, de la structure et de l'évolution de la Terre, ainsi que des processus qui s'y déroulent au sein de ses enveloppes gazeuses, liquides et solides ; tandis que la **g.** dynamique étude les phénomènes qui se déroulent dans la lithosphère et sa surface, la **g.** historique retrace l'histoire de l'évolution terrestre et de la vie organique ; DG 1984, 215) ; *g* **Geologie [f]** (Wissenschaft, die durch Untersuchung der durch natürliche oder künstliche Aufschlüsse zugänglichen Teile der Erdkruste mit ihren Gesteinen, deren Lagerungs- und Umwandlungserscheinungen sowie ihrem Fossilinhalt versucht, ein Bild von der Geschichte der Erde und des Lebens zu entwerfen; GEO 1972).

2405 geomorphology [n] *geo.* (Study of the classification, description, nature, origin, and development of land-forms and their relationships to underlying geological structures); *s* **geomorfología [f]** (Ciencia que estudia la génesis y evolución de las formas de relieve terrestres; DGA 1986); *f* **géomorphologie [f]** (Étude scientifique des formes de la surface terrestre et de leur évolution, et en particulier, l'étude du relief et du modelé ainsi que la description des lois régissant la création, l'évolution et la dynamique des formes de la surface terrestre [morphogenèse] ; DG 1984) ; *g* **Geomorphologie [f]** (Lehre von den Oberflächenformen der Erde und von den sie gestaltenden Kräften und Vorgängen. Insbesondere wird die feinere Gestaltung der Erdoberflächenformen hinsichtlich der Grund- oder Leitformen sowie die Gesetzmäßigkeit ihrer Entstehung, Umwandlung und Verbreitung untersucht und beschrieben).

2406 geophyte [n] *bot.* (Greek *gē* >earth< and *phyton* >plant<; nomenclature according to RAUNKIAER'S classification of life forms of an herbaceous plant with perenniating parts which are below ground level, e.g. bulb, corm, tuber, etc. and thus able to survive during unfavo[u]rable seasons [►carpet-forming perennial]. To geophytes belong **bulbs** *[Geophyta bulbosa]*; e.g. *Narcissus* and *Tulipa*; perennial **ground tubers** with annual, underground corms; e.g. autumn crocus *[Colchicum autumnale]*, crocus, winter aconites *[Eranthis]*; **stem tubers** with a strongly thickened hypocotylar such as cyclamen and sugar beet *[Beta vulgaris var. altissima]* and **underground stem tubers** such as the potato *[Solanum tuberosum]*, **root tubers** e.g. foxtail lilies *[Eremurus]*, **rhizome geophytes** *[Geophyta rhizomatosa]* with underground shoots; e.g. Lily of the Valley *[Convallaria]*, Solomon's seal *[Polygonatum]*, Couch Grass *[Agropyron repens]* and **root shoot geophytes** *[Geophyta radicigemma]* on which secondary shoots appear from root shoots; e.g. Lesser Bindweed *[Convolvulus arvensis]*. Geophytes survive in landscapes with long periods of drought with a subsequent period of rain. Leaves of **g.** last from early spring to early summer in Central Europe—*Eranthis*, Lesser Celandine *[Ranunculus ficaria]* or may persist throughout the summer *[Convallaria, Polygonatum]*. Spring **g.** have great importance in garden design due to their early flowering; ►bulbous plant, ►hemicryptophyte, ►sommer green, ►surface plant; *generic term* ►cryptophyte); *s* **geófito [m]** (Son las plantas terrestres según la clasificación biotípica y simorfial de los ►criptófitos de RAUNKIAER, que pierden periódicamente toda su parte aérea y cuyas yemas de sustitución quedan situadas bajo el suelo. Se subdividen según la morfología de su parte perenne en **g. con bulbo** *[Geophyta bulbosa]*, como el narciso y el tulipán; **g. con bulbo subterráneo** anual, como el azafrán, el cólquico *[Colchicum autumnale]* o las merenderas *[Crocus sp.]*; **g. con tubérculo** con porción caulinar engrosada, como el ciclamen *[Cyclamen]* y la remolacha azucarera *[Beta vulgaris var. altissima]*, **g. con tubérculo subterráneo**, como la patata *[Solanum tuberosum]*; **g. con tubérculos radicales**, como el lirio estepario *[Eremurus]*; **g. con rizoma** *[Geophyta rhizomatosa]*, como el lirio de los valles *[Convallaria]* o el helecho real *[Pteridium aquilinum]* y **g. con yemas radicales** *[Geophyta radicigemma]*, cuyas yemas persistentes se localizan en la superficie del suelo, como el acónito *[Aconitum napellus]* y la correhuela *[Convolvulus arvensis]*. Los **g.** sobreviven en estaciones con largos periodos/períodos de sequía y las subsiguientes lluvias. En climas templados, las hojas de los **g.** perduran de principios de primavera a principios de otoño —*[Ranunculus ficaria]*— o durante todo el verano —*[Convallaria, Polygonatum]*. Los **g.** de primavera tienen gran importancia en el diseño de jardines por su floración temprana; cf. DB 1985 y DINA 1987; ►caméfito, ►hemicriptófito, ►planta de cebolla, ►planta vivaz, ►verde vernal); *f* **géophyte [m]** (*Selon la classification des végétaux vasculaires d'après RAUNKIAER* désigne un végétal dont les organes végétatifs visibles face au froid et à l'enneigement sont enfouis durant la mauvaise saison dans le sol non inondé [►plante vivace] ; selon la nature de l'organe de conservation souterrain on distingue : les **géophytes à bulbe** *[Geophyta bulbosa]*, p. ex. le Narcisse *[Narcissus]* et la Tulipe *[Tulipa]*, *syn.* géophyte [m] bulbeux ; les **géophytes à corme** *[Geophyta bulbosa]*, p. ex. le Colchique d'automne *[Colchicum autumnale]*, le Crocus, l'Éranthe d'hiver *[Eranthis]* ; les **géophytes à tubercule** qui possèdent un renflement situé à la base de la tige comme le Cyclamen *[Cyclamen]* et la Betterave à sucre *[Beta vulgaris var. altissima]* ou un organe souterrain comme la pomme de terre *[Solanum tuberosum]* ; les **géophytes à racine tubérisée**, p. ex. le Lis des steppes *[Eremurus]* avec ses racines charnues transformées en organe de réserve ; les **géophytes à rhizome** *[Geophyta rhizomatosa]* avec une tige à développement horizontal, souterraine ou rampante, p. ex. le Muguet *[Convallaria]* et le Sceau de Salomon *[Polygonatum]*, le Chiendent rampant *[Agropyron repens]* ; les

géophytes à bourgeons racinaires *[Geophyta radicigemma]*, comme le Liseron des champs *[Convolvulus arvensis]* chez lequel les racines possèdent des bourgeons internes qui deviennent de nouvelles pousses, de nouvelles racines ou encore restent dormants. Les **g.s** sont capables de survivre dans les régions soumises à de longues périodes sèches et régulières suivies d'une saison de pluie ; les feuilles des **g.s** en Europe méridionale persistent jusqu'au début de l'été comme l'Éranthe d'hiver *[Eranthis]*, la Ficaire fausse-renoncule *[Ranunculus ficaria]* ou pendant l'été comme le Muguet *[Convallaria]*, le Sceau de Salomon *[Polygonatum]* ; les **g.s** à germination printanière prennent une place importante parmi les plantes ornementales de nos jardins ; *terme générique* ▶cryptophyte ; ▶à feuillage vert pendant l'été, ▶chaméphyte, ▶hémicryptophyte, ▶plante à bulbe) ; *g* **Geophyt [m]** (Griech. *gē* >Erde< und *phyton* >Pflanze<; *Lebensform nach RAUNKIAER* ausdauernde, höhere krautige Landpflanze [▶Stauden], bei der die Erneuerungs-/Überdauerungsknospen der sich periodisch erneuernden/absterbenden, meist nur kurzlebigen oberirdischen Organe, unter der Bodenoberfläche als Sprossachsenteile angelegt sind [*OB* ▶Kryptophyt]. Zu den **G.en** gehören **Zwiebel-G.en** *[Geophyta bulbosa]*, z. B. Narzisse *[Narcissus]* und Tulpe *[Tulipa]*, **Erdknollen-G.en** *[Geophyta bulbosa]* als ausdauernde Pflanzen mit vergänglichen, einjährigen, unterirdischen Knollen, z. B. Herbstzeitlose *[Colchicum autumnale]*, Crocus, Winterling *[Eranthis]*, **Sprossknollen-G.en**, die durch starke Verdickung des Hypokotyls entstehen, wie Alpenveilchen *[Cyclamen]* und Zuckerrübe *[Beta vulgaris var. altissima]*, **unterirdische Sprossknollen** wie die Kartoffel *[Solanum tuberosum]*, **Wurzelknollen-G.en**, z. B. Steppenkerze *[Eremurus]*, **Rhizom-G.en** *[Geophyta rhizomatosa]* mit unterirdischen Sprossachsen, z. B. Maiglöckchen *[Convallaria]* und Weißwurz *[Polygonatum]*, Quecke *[Agropyron repens]* und **Wurzelknospen-G.en** *[Geophyta radicigemma]*, bei denen aus Wurzelknospen Tochtersprosse treiben, z. B. bei Ackerwinde *[Convolvulus arvensis]*. **G.en** überdauern auch in Landschaften mit regelmäßigen, langen Trockenzeiten, wenn anschließend eine Regenperiode folgt. Die Blattausdauer von **G.en** in Mitteleuropa ist vorsommergrün — Winterling *[Eranthis]*, Scharbockskraut *[Ranunculus ficaria]* oder ▶sommergrün — Maiglöckchen *[Convallaria]*, Weißwurz *[Polygonatum]*; in der gärtnerischen Verwendung haben die früh im Jahr blühenden Früjahrsgeophyten eine große gestalterische Bedeutung; ▶Hemikryptophyt, ▶Oberflächenpflanze, ▶Zwiebelgewächs); *syn.* Bodenpflanze [f], Erdpflanze [f].

geophyte [n], bulb *bot. hort.* ▶bulbous plant.

2407 geosphere [n] *ecol. geo.* (Solid portion of the earth, the ▶lithosphere, together with the ▶atmosphere and ▶hydrosphere; ▶biosphere); *s* **geosfera [f]** (La parte sólida de la tierra — ▶litosfera— junto con la ▶atmósfera y la ▶hidrosfera; cf. DINA 1987; ▶biosfera); *syn.* geosistema [m]; *f* **géosphère [f]** (Espace disposé autour du globe terrestre, la croûte terrestre [▶lithosphère], les étendues aquatiques [▶hydrosphère], ainsi que la couche d'air [▶atmosphère] dans lequel se développe l'ensemble des organismes vivants ; ▶biosphère) ; *g* **Geosphäre [f]** (Der sich parallel zur Erdoberfläche erstreckende Raum, bestehend aus Gesteinskruste [▶Lithosphäre], Wasser [▶Hydrosphäre] und Lufthülle [▶Atmosphäre], in dem sich organisches Leben als ▶Biosphäre entwickelt).

geotechnical investigation [n] *constr. eng.* ▶exploration for foundation.

germinated plantlet [n] *bot.* ▶germinated young plant.

2408 germinated young plant [n] *bot.* (Plant established by natural seed dispersal; ▶seedling, ▶tree of seedling origin); *syn.* germinated plantlet [n]; *s 1* **plántula [f] germinada** (Planta

que ha crecido de semilla por regeneración natural; ▶plántula de semilla, ▶brinzal 1); *syn.* plantín [m] germinado (HDS 1987); *s 2* **semillón [m]** (En Argentina arbolito nacido espontáneamente, fuera der vivero, por lo general bajo los árboles padres; DB 1985); *f* **plantule [f]** (Jeune plante issue de la germination de la graine et se nourrissant encore aux dépens de celle-ci ; DAV 1984 ; ▶brin de franc-pied, ▶semis 2) ; *g* **Keimling [m]** (**1.** Der in Keimwurzel, Keimblätter und Stengel gegliederte Embryo im Samen. **2.** Durch Aussaat hervorgehende junge Pflanze mit ausgebildeten Keim- und Primärblättern; ▶Sämling, ▶Kernwuchs); *syn.* Keimpflanze [f].

getaway [n] [US] *recr.* ▶vacation accommodation (2).

girder [n] *constr.* ▶beam.

2409 girth [n] *hort.* (Circumference of a tree trunk, usually measured at breast height, 1.30m from the ground, **in D.**, 1m from the ground; *measurement* **g**irth **b**reast **h**eight [g.b.h.] or diameter **b**reast **h**eight: [d.b.h.]; ▶trunk diameter); *s* **circunferencia [f] del tronco** (En D., F. y Es. se mide a un metro del suelo, en GB y EE.UU. a 1,30 m del mismo; ▶diámetro de tronco); *f* **circonférence [f] du tronc** (Caractéristique dimensionnelle des arbres-tiges d'alignement, d'ornement ou fruitiers mesurée à 100 cm du collet ; ▶diamètre [à hauteur d'homme]) ; *syn.* circonférence [f] à hauteur de poitrine ; circonférence [f] de la tige ; *g* **Stammumfang [m]** (Größenmaß eines Stammes, das in D. bei Bäumen 1 m über Grund, in angelsächsischen Ländern in Brusthöhe [**Brusthöhenumfang** bei 1,30 m Stammhöhe] gemessen wird; Stämmlinge mehrstämmiger Bäume werden einzeln gemessen; ▶Stammdurchmesser).

girth breast height [n] (GBH) *hort.* ▶girth.

give [vb] approval *adm. leg.* ▶approve.

2410 give [vb] legal notice *leg. plan.* (**1.** Legally required announcement such as an advertisement in a local newspaper or internet, e.g. for open bidding, adoption of a zoning map, ▶design competition 1, street abandonment, intended to give warning of a pending public meeting or a pending public business, or to notify in advance, the legal rights or implications of a pending matter. **2.** A required **public notice** is a necessary form of **l. n.** as set forth in statutory regulations in local, state and Federal jurisdictions. Public notice is required to obtain legal standing for the process or action advertised; to be in conformance with the law to avoid legal challenges for non-compliance later in the process by disgruntled parties; ▶public announcement); *syn.* announce [vb], give [vb] public notice [US]; give [vb] legal advertisement; *s* **anunciar [vb] públicamente** (Aviso por medio de comunicación de masas de una decisión de una autoridad pública, como p. ej. de la aprobación inicial de un plan parcial [de ordenación], de un ▶concurso de ideas que se somete a información preceptiva ; ▶anuncio preceptivo, ▶publicación de un plan); *syn.* publicar [vb]; *f* **rendre [vb] public** (Phase finale de l'élaboration d'un document d'urbanisme, de la passation d'un marché, de l'organisation d'un ▶concours d'idées ou d'un ▶appel public de candidatures pour un appel d'offres dans les documents sont publiés par affichage, par voie de presse, etc. ; ▶publication ; *g* **öffentlich bekannt machen [vb]** (Ortsüblich in der Presse oder im Internet anzeigen, z. B. öffentliche Ausschreibungen, ▶Ideenwettbewerbe, ▶öffentlichen Teilnahmewettbewerb, Satzungsbeschlüsse von Bebauungsplänen, Einziehungen von Straßen; ▶öffentliche Bekanntmachung); *syn.* öffentlich kundmachen [vb] [A].

2411 glacial debris [n] *geo.* (Rock fragments lying in moraines or at the base or 'talus' of a glacier; the larger stones are known as ▶boulders, ▶erratic block); *syn.* glacial scree [n] [also UK]; *s* **detritus [m] glaciar** (Material rocoso de acarreo por glaciares, cuando se trata de una roca de gran tamaño se denomina ►bloque errático; ►pedrejón); *f* **débris [mpl] glaciaires** (Matériaux rocheux transportés par un glacier ou une calotte glaciaire et dont l'accumulation constitue une moraine ; les rochers de volume important sont appelés ►bloc [de rocher] de forme arrondie, ►blocs erratiques) ; *g* **Geschiebe [n] (2)** (Vom Gletscher oder Inlandeis transportierte und in Moränen abgelagerte Gesteinsbrocken, von denen besonders große als ►Findlinge 2 bezeichnet werden; ►Findling 1).

glacial kettle [n] *geo.* ►kettle.

2412 glacial loam [n] *geo. pedol.* (Loose sediment composed of weathered, decalcified boulder clay; ►weathering); *s* **harina [f] glaciar** (Se dice de las partículas de fracción más fina integrantes de una morrena y que constituyen su matriz; DGA 1986; ►meteorización); *syn.* arcilla [f] glaciar; *f* **limon [m] glaciaire décarbonaté** (Sédiment meuble de la moraine de fond se décalcifiant graduellement au cours de la ►météorisation) ; *syn.* limon [m] glaciaire décalcifié ; *g* **Geschiebelehm [m]** (Lockersediment der Grundmoräne, das durch ►Verwitterung allmählich entkalkt).

2413 glacial relict [n] *phyt. zool.* (Remnant or fragment of flora or fauna that remains from the post-glacial period during which it was more widely distributed; today this species is restricted to some relatively small geographic areas with cold climatic conditions; ►relict species); *s* **reliquia [f] glacial** (Especie de flora o fauna abundante en la época posglacial, que hoy en día está reducida a hábitats relativamente pequeños con climas fríos; ►especie relíctica) ; *f* **relicte [m] glaciaire** (Individu, espèce ou communauté végétale plus largement répandu pendant l'époque postglaciaire, ayant survécu à la disparition ou la modification de la végétation climacique et vivant aujourd'hui grâce à l'existence de conditions stationnelles froides ; ►espèce relictuelle) ; *g* **Glazialrelikt [n]** (In der unmittelbaren Nacheiszeit weitverbreitete Art, die heute auf einige für sie klimatisch günstige [kältere] Standorte beschränkt ist; ►Reliktart); *syn.* Eiszeitrelikt [n].

2414 glacial sand [n] *geo.* (Granular glacial drift deposited directly by ice or transported by meltwater); *s* **arena [f] glaciar** (Sedimentos aluviales de arenas transportadas por el hielo de glaciaciones terrestres, por glaciares o corrientes de agua de deshielo); *f* **sable [m] glaciaire** (Sédiment alluvial meuble transporté lors des glaciations inlandsissiennes, par les glaciers ou les courants d'eau de fonte) ; *g* **Geschiebesand [m]** (Vom Inlandeis, von einem Gletscher oder durch Schmelzwasser transportiertes sandiges Lockersediment); *syn.* Decksand [m], Geschiebedeckensand [m].

glacial scree [n] [UK] *geo.* ►glacial debris.

2415 glasswort mudflat [n] *geo. phyt.* (Coastal mudflats vegetated with glasswort *[Salicornia]*; ►glasswort vegetation); *s* **marisma [f] de salicor** (►Llanura de fango litoral poblada mayormente de salicor; ►vegetación de salicor); *f* **marais [m] à Salicornes** (Étendue de sols limoneux plus ou moins inondés [►slikke] occupée en majorité par les Salicornes *[Salicornia]* ; en Camargue cette zone est dénommée la **sansouire** ; ►peuplement à Salicornes) ; *syn.* slikke [m] à Salicornes, *pour la région du delta du Rhône* sansouire [f] ; *g* **Quellerwatt [n]** (...watten [pl]; Schlickbereich im ►Watt, das vorwiegend mit Queller *[Salicornia]* bewachsen ist; ►Quellerflur).

2416 glasswort vegetation [n] *phyt.* (Pioneer vegetation of ►tidal mudflats, on shores of Red Sea, Arabian Gulf, etc., vegetated mainly with glasswort *[Salicornia]*; this succulent green plant absorbs substantial amounts of atmospheric carbon, helping to counter global warming; ►brackish water area/region, ►glasswort mudflat); *syn.* marine crest vegetation [n], samphire vegetation [n]; *s* **vegetación [f] de salicor** (Vegetación pionera

de las ►llanuras de fango litoral bajo influencia de las mareas, en las orillas del Mar Rojo, en el Golfo de Arabia, etc. formada casi exclusivamente por salicor *[Salicornia]*; esta planta suculenta absorbe grandes cantidades de carbono atmosférico, contribuyendo así a reducir el aumento de la temperatura media a nivel global; ►marisma de salicor, ►zona de agua salobre); *f* **peuplement [m] à Salicornes** (Formation végétale pionnière du ►slikke, de ►milieux saumâtres ou de la ►zone de la sansouire constituée exclusivement de Salicornes *[Salicornia]* ; ►marais à Salicornes) ; *syn. pour la région du delta du Rhône* engane [f] (à Salicornes) ; *g* **Quellerflur [f]** (Pioniervegetation im ►Watt, an den Ufern des Roten Meeres, am Strand des Arabischen Golfes etc., die fast ausschließlich aus Queller *[Salicornia]* besteht; Queller absorbiert größere Mengen an CO_2 der Luft und kann bedingt dazu beitragen, die weltweite Erwärmung zu verlangsamen; ►Brackwasserbereich, ►Quellerwatt).

glazed stoneware pipe [n] [UK] *constr.* ►vitrified clay pipe [US].

gley [n] [UK] *pedol.* ►aquepts [US].

2417 gleyed anmoor [n] *pedol.* (Sticky waterlogged [US]/water-logged [UK] soil with a histic epipedon, developed under conditions of poor drainage in cold climates [tundra and snowbeds of alpine regions], having significant hydromorphic characteristics and a A_h-G_r profile; the shallow, blackish and muddy Ah horizon contains 15-30% of organic matter; in U.S., **g. a.** is not used in the current soil taxonomy system but is similar to ►Aquepts, especially the subgroup *Histic Lithic Cryaquepts*); *s* **gley [m] turboso** (Suelo hidromorfo cuyo horizonte hístico se ha desarrollado bajo condiciones de mal drenaje en climas fríos [tundra y vallecitos nevados de áreas alpinas], tiene un perfil A_h-G_r con un horizonte A_h entre 15 y 30% de sustancia orgánica. En la nomenclatura de suelos de los EE.UU. el **g. t.** no se utiliza, pero es similar al ►aquepts, en especial al subgrupo «cryaquepts líticos hísticos»; ►suelo gley); *f* **gley [m] tourbeux** (Sol à nappe phréatique profonde dans les dépressions ou plaines alluviales, sol de transition caractérisé par un profil A_h-G_r, à nappe phréatique de faible oscillation [moins de 40 cm], l'horizon A_h contenant entre 15 et 30 % de matière organique ; PED 1983, 388-390 ; ►sol à gley) ; *syn.* gley [m] à anmoor ; *g* **Anmoorgley [m]** (Grundwasserbeeinflusster Boden kalter Klimate [in der Tundra und in Schneetälchen alpiner Gebiete] mit deutlichen hydromorphen Merkmalen und A_h-G_r-Profil; der gering mächtige, schwärzliche und meist schlammige Ah-Horizont hat 15-30 % organische Substanz; cf. SS 1979; ►Aquept, ►Gley).

gley soil [n] [UK] *pedol.* ►aquepts [US].

2418 global warming [n] *met.* (Since 1860 and especially since the early 1970s, the progressive increase in the average temperature of the atmosphere close to Earth and of the oceans, as well as the further warming expected in the future; **g. w.** is generally assumed to be caused by an intensification of the ►greenhouse effect; ►climatic change); *s* **calentamiento [m] global** (Aumento progresivo de la temperatura media de la troposfera [capa inferior de la atmósfera hasta una altitud aproximada de 12 km] y de la superficie de los océanos que tiene lugar desde 1860 y en especial desde el comienzo de los años 1970 y que sigue ocurriendo, y cuya causa principal es la intensificación del ►efecto invernadero; ►cambio climático); *f* **réchauffement [m] global** (Augmentation progressive des températures moyennes de l'atmosphère proche de la surface terrestre ainsi que des mers et des océans depuis 1870 et en particulier depuis le début des années 70 provoquée par un ►effet de serre additionnel dû aux activités humaines responsables de l'augmentation de la concentration dans l'atmosphère de plusieurs gaz à effet de serre tels que le dioxyde de carbone [CO_2], le méthane [CH_4] et le

G

protoxyde d'azote [N$_2$O] ; le réchauffement climatique actuel aura des conséquences physiques sur l'environnement [cycle des eaux, océans, nuages et pluies, répartition des espèces végétales et animales], et sur les activités humaines [modes de production et de consommation d'énergie, politique de l'eau, agriculture, élevage et pêche, aménagement du territoire, activités Industrielles, habitat et transports, prévention et couverture des risques, santé humaine] ; ►changement climatique) ; *syn.* réchauffement [m] climatique, réchauffement [m] planétaire ; *g* **globale Erwärmung [f]** (Allmählicher Anstieg der Durchschnittstemperatur der erdnahen Atmosphäre und der Meere seit 1860 und vor allem seit Anfang der 1970er-Jahre sowie die künftig zu erwartende weitere Erwärmung; als Ursache wird die Verstärkung des ►Treibhauseffektes angenommen; cf. ALG 2007; ►Klimaveränderung).

gnarled wood [n] *phyt.* ►krummholz.

goal [n] *plan.* ►planning goal.

goal fulfillment [n] *prof.* ►planning goal fulfillment.

gob [n] [US] *min.* ►mining spoil.

gob [n] [US]**, disposal of mining** *min.* ►disposal of mining spoil.

gob [n] [US]**, underground disposal of mining** *min.* ►underground disposal of mining spoil.

2419 golf course [n] *recr.* (Turf sports facility from 20 to 60 ha [50 to 150 acres] with nine, 18 or up to 36 holes; a course usually includes ponds, trees and shrub planting and sand traps); *s* **campo [m] de golf** (Área de juego de golf de 20-60 ha de extensión, con 9, 18 y hasta 36 hoyos, que incluye generalmente lagos artificiales, plantaciones de árboles y arbustos y obstáculos de arena llamados «bunkers»); *f* **terrain [m] de golf** (Espace de loisirs et sportif à caractère d'espace vert, à la périphérie des zones résidentielles, occupant entre 20 et 60 ha, comprenant 9, 18 ou 36 trous, agrémenté de plantations, d'espaces aquatiques et d'obstacles artificiels [bunker], conçu et exécuté par des spécialistes hautement qualifiés ; *syn.* parcours [m] de golf ; *g* **Golfplatz [m]** (20-60 ha große Rasensportanlage mit neun, 18 oder bis zu 36 Spielbahnen [Löchern], die mit Tümpeln, Gehölzpflanzungen sowie künstlichen Hindernissen [sandgefüllte ,Bunker'] ausgestattet ist und dem Golfspiel dient).

gorge [n] *geo.* ►ravine, #2.

government agency [n] *adm.* ►federal government agency [US], ►state government agency [US].

government authority [n]**, federal** *adm.* ►federal government agency [US].

government authority [n]**, state** *adm.* ►state government agency [US].

government body [n] [UK]**, central** *adm.* ►federal government agency [US].

government construction project [n] *plan.* ►federal government construction project, ►state government construction project.

2420 gradation [n] *zool.* (Stages in the mass reproduction of a species, especially for insects); *s* **metamorfosis [f]** (Conjunto de etapas de reproducción de organismos, en especial de insectos); *f* **processus [m] de reproduction de masse d'une espèce** (Étape complète de la reproduction d'un organisme, en particulier chez les insectes) ; *g* **Gradation [f]** (Gesamtablauf der Massenvermehrung einer Organismenart, besonders bei Insekten).

grade [n] *constr.* ►excavation to subbase grade [US]; *contr. leg. prof.* ►fee grade; *constr.* ►final grade (1), ►final grade (2), subgrade [US] (1)/sub-grade [UK].

grade [n]**, below** [US] *constr.* ►below ground surface.

grade [n]**, change in** *constr.* ►change in gradient.

grade [n]**, establishing a finished** *constr.* ►fine grading.

grade [n]**, excavation to formation** *constr.* ►excavation to subbase grade [US].

grade [n]**, existing** [US] *constr. eng. plan.* ►existing ground level.

grade [n]**, finished** *constr.* ►final grade (1).

grade [n]**, finish** [US] *constr.* ►final grade (1).

grade [n]**, quality** [UK] *constr. eng.* ►quality class [US].

2421 graded sediment group [n] [US] *constr. (pedol.)* (Measured granular or silt-clay material that can be separated by sieve screening or water-grading in a soil classification system; ►particle-size distribution); *syn.* soil separate group [n] [also US], particle-size group [n] [UK]; *s 1* **grupo [m] granulométrico** *pedol.* (En la clasificación de suelos, gradación en milímetros de las partículas minerales del suelo; ►granulometría); *syn.* fracción [f] granulométrica/mineral; *s 2* **tamaño [m] de grano** *constr.* (Dimensión en mm de los materiales inertes utilizados en la construcción); *f* **fraction [f] granulométrique** (Dimension en mm [fixée par convention] des grains de sols minéraux ; on distingue les **f. g.** définies au tamis et les **f. g.** définies par siphonnage ; plusieurs fractions granulométriques déterminent la composition granulométrique d'un sol ou de granulats ; PED 1979, 228 et s. En pédologie les proportions de particules sont classées selon cinq fractions granulométriques ou plus : fraction argile < 2 mm ; fraction limons fins comprise entre 2 et 20 mm ; fraction limons grossiers comprise entre 20 et 50 mm ; ►spectre granulométrique) ; *syn. pedol.* fraction [f] granulaire ; *g* **Korngruppe [f]** (In Millimeter konventionell festgelegte Größen von Mineralkörpern [Gesteinskörnung] der einzelnen Bodenarten innerhalb von zwei gewählten Siebkorn- oder Schlämmkorngrößen [Korngröße, Kornfraktion], z. B. bei Feinsand > 0,06-0,2 mm. Es gibt Siebkorngruppen und Schlämmkorngruppen. Mehrere Korngruppen ergeben die ►Kornverteilung eines Bodens oder Zuschlagstoffes, d. h. das Korngemisch, z. B. 8/16 — Kleinstkorn 8 mm, Größtkorn 16 mm; cf. DIN 4226).

2422 graded to proper line and levels [loc] *constr.* (Descriptive term applied to a terrain which is graded to proposed spot elevations and contours; ►spreading of soil material to true contours); *syn.* surveyed to proposed falls and gradients [loc]; *s* **nivelado hasta rasante de acabado [loc]** (Relativo al terreno que ha sido rellenado hasta los niveles definitivos; ►relleno de tierra hasta nivel de rasante); *f* **conforme/conformément aux plans de profil [loc]** (Profil de terrain exécuté conformément aux pièces contractuelles [plans et détails] ; ►profilage aux pentes indiquées) ; *g* **profilgerecht [adj/adv]** (Ein Gelände betreffend, das nach den geplanten Höhen hergestellt ist; ►profilgerechter Einbau).

grade [n] **of a land surface** [US]**, finished** *constr. plan.* ►finished elevation [US]/finished level [UK].

grade [n] **of a slope** [US] *constr. geo.* ►slope gradient.

grade-separated intersection [n] [US] *trans.* ►grade-separated junction.

2423 grade-separated junction [n] *trans.* (Generic term covering any traffic intersection on different levels with ramps to facilitate directional changes, superimposed on existing land forms; ►freeway interchange, ►cloverleaf interchange, ►diamond-shaped interchange, ►interchange, *opp.* ►at-grade junction); *syn.* grade-separated intersection [n] [also US]; *s* **nudo [m] sin cruces a nivel** (Intersección de carreteras a dos niveles, de manera que éstas no se crucen directamente; ►cruce de autopistas, ►nudo de enlace, ►nudo de enlace con forma de rombo, ►trébol; *opp.* ►cruce a nivel); *f* **échangeur [m]** (Terme géné-

rique carrefour dont les échanges sont séparés les uns des autres et gérés en dehors des axes principaux ; *termes spécifiques* ▶diffuseur, ▶échangeur autoroutier, ▶raccordement de type losange, ▶trèfle ; *opp.* ▶intersection routière à un niveau) ; *syn.* carrefour [m] dénivelé, intersection [f] routière à plusieurs niveaux ; *g* **planfreier Knotenpunkt [m]** (Knotenpunkt, bei dem sich der Verkehr auf verschiedenen Ebenen abwickelt; ▶Anschlussstelle, ▶Autobahnkreuz, ▶Kleeblatt, ▶Rautenanschluss; *opp.* ▶plangleicher Knotenpunkt); *syn.* Knotenpunkt [m] in mehreren Ebenen, kreuzungsfreies Straßenbauwerk [n], planfreies Straßenbauwerk [n], höhenfreier Knotenpunkt [m], höhenfreies Straßenbauwerk [n].

2424 grade stake [n] [US] *constr. surv.* (Peg marking a spot elevation); *syn.* level stake [n] [UK]; *s* **piquete [m] (de nivel)** (Estaca que marca la altura sobre nivel); *f* **piquet [m] de niveau** (Piquet définissant sur le terrain les points de niveau aux cotes du projet) ; *g* **Höhenpflock [m]** (Pflock, der eine eingemessene Höhe im Gelände anzeigt).

2425 gradient [n] (1) *trans.* (▶Slope [US] 2/fall [UK] of a path or road expressed in percentage or degrees; ▶lost gradient); *s* **pendiente [f] en subida** (Porcentaje o relación 1:x de la desviación hacia arriba de una vía de la línea horizontal; ▶gradiente perdido, ▶pendiente); *syn.* rampa [f] [Es, RA, CO]; *f* **déclivité [f] en rampe [f]** (Pente ascendante d'un cheminement, d'une route ; ▶déclivité perdue, ▶pente) ; *syn.* montée [f], déclivité [f] ascendante ; *g* **Steigung [f]** (1) (Das Maß in Prozent oder das Verhältnis 1:x für die Abweichung eines Verkehrsweges von der Horizontalen; im Gleisbau wird die **S.** in Promille angegeben; z. B. das Ansteigen eines Weges, einer Straße, eines Schienenstranges; ▶Gefälle, ▶verlorene Steigung).

gradient [n] (2) *agr. constr. hydr.* ▶hydraulic gradient; *constr.* ▶longitudinal gradient.

2426 gradient change [n] *constr. trans.* (Change in the rate of ascent or descent; ▶change in gradient); *syn.* incline change [n]; *s* **cambio [m] de rasante (1)** (Transición entre una pendiente de subida y una de bajada; ▶cambio de pendiente); *f* **inversement [m] de pente** (Transition dans un profil en long entre une pente et une rampe ; ▶changement de pente) ; *g* **Neigungswechsel [m]** (Übergang von einem Gefälle in eine Steigung oder umgekehrt; ▶Gefälleänderung).

gradient [n] **of a slope** [US] *constr. geo.* ▶slope gradient.

gradient [n] **of landscape modification** *ecol.* ▶degree of landscape modification.

gradient profile [n] [US] *trans.* ▶vertical alignment.

grading [n] *constr.* ▶fine grading, ▶ground modeling [US]/ground modelling [UK], ▶rough grading [US]/major grading [UK].

grading [n] [US], **bulk** *constr.* ▶rough grading [US]/major grading [UK].

grading [n] [UK], **down-** *leg. urb.* ▶downzoning [US].

grading [n] [US], **finished** *constr.* ▶fine grading.

grading [n] [UK], **major** *constr.* ▶rough grading [US].

grading [n], **minor** *constr.* ▶fine grading.

2427 grading [n] **and leveling** [n] [US]/**grading** [n] **and levelling** [n] [UK] *constr.* (SPON 1986, 422; creating a gradient and smoothing of the ground surface; ▶subbase grade [US]/formation level [UK]); *s* **nivelado [m] del terreno** (▶subrasante); *syn.* aplanado [m] del terreno, explanado [m] del terreno, allanado [m] del terreno; *f* **régalage [m]** (Travaux de terrassement consistant à disposer les matériaux mis en tas en un profil en nappe régulière ; ▶fond de forme) ; *syn.* régalement

[m] ; *g* **Planieren [n, o. Pl.]** (Eine Fläche einebnen; einebnen [vb], planieren [vb]; ▶Erdplanum); *syn.* Einebnung [f].

2428 grading design [n] *constr. plan.* (Planning of the future topography of a piece of land by delineation with spot elevations and existing and proposed contours; ▶grading plan, ▶ground modeling [US]/ground modelling [UK]); *s* **estudio [m] de nivelación** (Elaboración del plan para el relieve que se quiere dar a un terreno; ▶modelado del terreno, ▶plano de nivelación); *f* **cotation [f] d'un projet** (Étude déterminant la topographie d'un projet d'aménagement et fixant l'ensemble des cotes en vue de sa réalisation ; ▶mise en forme des terres, ▶plan de nivellement) ; *g* **Planung [f] der Höhenabwicklung** (Planerische Ausarbeitung des auszuführenden Reliefs eines Geländes; ▶Höhenplan für den Erdbau, ▶Bodenmodellierung).

grading [n] **of finished level to all falls and gradients** [UK] *constr.* ▶spreading of soil to true contours.

grading [n] **of finished level to all gradients** [US] *constr.* ▶spreading of soil to true contours.

grading [n] **of formation level** [UK] *constr.* ▶subbase grade preparation [US].

2429 grading plan [n] [US] *constr. plan.* (Plan showing ground modeling [US]/ground modelling [UK] of a land area by means of contours, spot elevations, and slope ratio indications; ▶hachures); *syn.* levelling plan [n] [UK]; *s* **plano [m] de nivelación** (En terracerías, plano que muestra la futura modelación del terreno; ▶hachures); *f* **plan [m] de nivellement** (Plan sur lequel figure l'altitude du terrain existant ou les cotes futures nécessaires à l'exécution des travaux de terrassement ; ▶hachuras) ; *g* **Höhenplan [m] für den Erdbau** (Plan, der [vorhandene und] geplante Höhen für den Erdbau in Form von Höhenpunkten, Höhenlinien oder Steigungsverhältnissen von Böschungen oder Hängen angibt; ▶Schraffen; *UBe* Höhenlinienplan, Schraffenplan).

2430 gradually sloping ground [n] *constr.* (Long gradual slope with significant fall); *s* **terreno [m] en pendiente** (Superficie con una pendiente de más del 15%); *f* **surface [f] pentue** (de plus de 15 %) ; *syn.* terrain [m] en pente (LA 1981, 477) ; *g* **hängiges Gelände [n, o. Pl.]** (Gelände mit deutlichem Gefälle).

2431 grafting [n] *hort.* (Horticultural method of improving the quality of a plant by inserting a scion of the desired variety into a slit of the rooted ▶understock); *s* **injerto [m]** (Método horticultural de mejora de la calidad de ▶patrones insertando una púa o yema de la planta deseada); *f 1* **greffage [m]** (Action de greffer) ; *f 2* **greffe [f]** (Résultat de l'opération horticole de multiplication végétative visant l'obtention d'une meilleure qualité des végétaux bénéficiant des qualités des deux sujets réunis par implantation d'un greffon [bourgeon ou rameau de la variété à reproduire] sur un ▶porte-greffe) ; *g* **Veredelung [f]** (Qualitätssteigerung im Pflanzenbau von geeigneten, weniger edlen ▶Unterlagen durch Transplantieren eines Edelreises/Edelauges einer gewünschten Pflanze resp. Art/Unterart/Sorte); *syn. o. V.* Veredlung [f].

2432 graft union [n] *hort.* (BOT 1990, 91, ARB 1983, 548; **1.** point of ligature at which the scion is inserted into the ▶understock. **2.** Specific term for a swollen or enlarged area where a bud has been grafted to a stock: grafting callus cushion, graft swelling area [cf. ARB 1983, 548], bud head [also US]); *syn.* bud union [n] [also US]; *s* **soldadura [f] (de injerto)** (Punto en el ▶patrón en el que se injerta la púa o yema, que según la compatibilidad puede formar una protuberancia callosa); *f* **point [m] de greffe** (Point d'union du greffon avec le ▶porte-greffe ; *terme spécifique* bourrelet de greffe [légère déformation du tronc d'un arbre au point de greffage dont le fort développement est le signe d'une compatibilité imparfaite entre le porte-greffe et le greffon ; cf. LA 1981) ; *g* **Veredelungsstelle [f]** (Ort der ▶Unterlage, an der das Edelreis/Edelauge eingesetzt wurde; die **V.** ist — je nach Verträglichkeit der Unterlage — durch eine unterschiedlich große Verdickung, den „Veredlungswulst", gekennzeichnet); *syn. o. V.* Veredlungsstelle [f].

grain-size distribution [n] *constr.* ▶particle-size distribution.

graminoid [n] *bot.* ▶grass.

2433 grand award [n] *hort.* (An emblem or medal for outstanding performance during a garden show awarded to participants who have received several gold medals in individual competitions); *syn.* grand award [n] for best overall performance; *s* **gran premio** [m] (Galardón donado en exhibiciones de horticultura a participantes que han recibido medallas de oro en varias competiciones); *f* **grand prix** [m] (Distinction décernée aux participants à une exposition florale ayant obtenu plusieurs médailles d'or lors des concours individuels) ; *g* **Großer Preis** **[m]** (Bei Gartenschauen verliehene Auszeichnung für Teilnehmer, die mehrere Goldmedaillen für Leistungswettbewerbe erhalten haben).

grand award [n] **for best overall performance** *hort.* ▶grand award.

granite sett [n] [UK]**, large-sized** *constr.* ▶large-sized paving stone [US].

2434 granular drain fill [n] *constr.* (Coarse material installed above a drain[age] line); *s* **capa [f] de material filtro** (Capa de grava o áridos que se tiende encima de los drenes); *f* **remblai** [m] **drainant** (Remblai de remplissage de tranchée ou de fouille en galets ou en grave recouvrant un drain) ; *g* **Dränpackung [f]** (Kies- oder Schotterfüllung überhalb einer Dränleitung; *o. V.* Drainpackung).

granular fill [n] *constr.* ▶fill material.

granular ped [n] [US] *pedol.* ▶crumb.

2435 granular playing court [n] [US] *constr.* (Enclosed sports area [US]/sports pitch [UK] surfaced with a self-binding gravel/sand mixture of various grades; ▶hard court [US]/hard pitch [UK], ▶granular playing surface [US]/hoggin playing surface [UK]); *syn.* water-bound porous court [n], hoggin playing court [n] [UK] (cf. SPON 1986, 344); *s* **cancha [f] de arenilla** (Campo de deportes de ▶superficie de arenilla; ▶cancha de pavimento duro); *f* **terrain** [m] **de sport à revêtement perméable** (Aire sportive réalisée avec un revêtement perméable ; ▶aire à revêtement dur, ▶surface de sport à revêtement perméable) ; *g* **Tennenplatz [m]** (Sportplatz als ▶Tennenfläche hergestellt; ▶Hartplatz).

2436 granular playing surface [n] [US] *constr.* (Recreational surface comprising self-binding gravel/sand mixture of various grades. The surface allows water to percolate, can be used in nearly all weathers throughout the year, except during necessary maintenance periods; ▶water-bound surface); *syn.* water-bound playing surface [n] [US], hoggin playing surface [n] [UK]; *s* **superficie [f] de arenilla** (Campo de juegos con revestimiento de mezcla mineral sin ligantes, permeable y utilizable sin limitaciones de las condiciones meteorológicas; ▶revestimiento compactado); *f1* **surface [f] de sport à revêtement** **perméable** (Aire de sport constituée par un revêtement superficiel perméable [sablé ou gravillonné] sur fond de forme minéral, sans liant hydraulique et, à l'exception des périodes d'entretien, pouvant être utilisée par tous les temps ; ▶sol stabilisé aux liants hydrauliques) ; *f2* **surface [f] en terre battue** (**1.** Aire constituée par une terre compactée par pilonnage, par exemple l'aire d'un sol en terre-plein sans revêtement pour un vide sanitaire en sous-sol ; **2.** *courts de tennis* constitué du bas vers le haut d'un empierrage avec réseau de drainage, d'une

couche de calcaire [6 cm] de granulométrie fine [craon], stabilisée au rouleau perméable à l'eau, d'une couche de mâchefer, [3 cm] à granulométrie variable qui stocke l'eau d'arrosage et d'une couche de brique pilée rouge qui sert à glisser et à avoir un bon contraste de couleur avec les balles ; *syn.* surface [f] stabilisée de confort ; *g* **Tennenfläche [f]** (Sportfläche aus mineralischen Korngemengen ohne künstliche Bindemittel. Sie ist wasserdurchlässig, weitgehend wetterunabhängig und mit Ausnahme der notwendigen Pflegezeiten uneingeschränkt nutzbar; LEHR 1981; ▶wassergebundener Belag).

granular road base [n] [UK] *constr.* ▶gravel subbase [US]/gravel sub-base [UK].

2437 granular soil [n] [US] *pedol.* (Soil of sand and gravel incapable of being molded or deformed continuously and permanently by relatively moderate pressure; ▶colloidal soil); *syn.* cohesionless soil [n] [also US], non-cohesive soil [n] [UK], nonplastic soil [n] [US]; *s* **suelo [m] suelto** (Suelo arenoso sin estructura en el cual cada partícula está separada, como en las dunas de arena. Es muy susceptible a los efectos erosivos del viento; cf. MEX 1983; *opp.* ▶suelo cohesivo; *syn.* suelo [m] no cohesivo; *f* **sol [m] à structure particulaire** (Sol dépourvu ou peu abondant en ciments argileux ou limoneux et à texture grossière ; ▶sol plastique) ; *g* **nicht bindiger Boden [m]** (Boden ohne oder mit sehr geringem Ton- und Schluffanteil: Sandboden, Kies; *opp.* ▶bindiger Boden); *syn. o. V.* nichtbindiger Boden [m].

2438 granular soil structure [n] *pedol.* (ULD 1992; physical condition of soil characterized by rounded, porous peds with large macropores between them. It is typically found in surface horizons [A horizons], in most forest soil horizons, and in many well-managed agricultural plow [US]/plough [UK] layers; cf. ULD 1992; ▶crumb, ▶crumb structure, ▶soil mellowness, ▶tilth); *syn.* friable soil condition [n]; *s* **estado [m] grumoso** (Estructura física de un suelo cultivado caracterizada por grumos redondeados porosos con grandes poros entre ellos; ▶estructura grumosa, ▶grumo, ▶sazón); *syn.* estado [m] friable del suelo; *f* **état [m] structural grumeleux** (L'état physique d'un sol caractérisant sa capacité culturale ; ▶agrégat [terreux] élémentaire, ▶bon état de structure grumeleuse, ▶structure fragmentaire) ; *syn.* état [m] structural friable, structure [f] granulaire ; *g* **Krümelstruktur [f] (2)** (Physikalischer Zustand eines kultivierten Bodens, der durch die räumliche Anordnung der Mineralpartikel und organischen Stoffe, die zu Aggregaten verknüpft sind, gekennzeichnet ist und bei geringer biologischer Aktivität nicht stabil sein muss, z. B. bei sandreichen Böden mit vorherrschendem Einzelkorngefüge; ▶Bodengare, ▶Krümel, ▶Krümelstruktur 1).

granular subbase [n] [US]/**granular sub-base** [n] [UK] *constr.* ▶crushed aggregate subbase [US].

2439 granular surface course [n] [US] *constr.* (Top or wearing course composed of a mixture of compacted clay, coarse sand and fine gravel or fine crushed aggregate; ▶granular playing surface [US]/hoggin playing surface [UK]); *syn.* hoggin surface course [n] [UK]; *s* **suelo [m] de arenilla** (Suelo compactado de campo de juegos; ▶superficie de arenilla); *f* **sol [m]** **sportif stabilisé mécaniquement** (Revêtement superficiel/ couche d'usure d'une ▶surface de sport à revêtement hydraulique, ▶surface en terre battue) ; *g* **Tennenbelag [m]** (Wassergebundene Deckschicht/Verschleißschicht einer ▶Tennenfläche); *syn.* Tennendecke [f].

2440 granulometric percentage [n] *constr. pedol.* (Proportional weight in metric measure of a particular sediment grade expressed as a percentage of the total weight of a graded mixture in that measure; ▶graded sediment group [US]/particle size group [UK]); *s* **porcentaje [m] de una fracción granulométrica**

(Peso proporcional de un ▶grupo granulométrico en una mezcla de partículas de diferentes tamaños; ▶grupo granulométrico); *f* **pourcentage [m] de la fraction granulométrique** (Proportion d'une classe granulométrique exprimée en pourcentage du poids total d'un granulat ; ▶fraction granulométrique) ; *g* **Kornanteil [m]** (Prozentualer [Gewichts]anteil einer ▶Korngruppe in einem Körnungsgemisch).

2441 grapevine arbor [n] [US]/**grapevine arbour** [n] [UK] *gard.* (Pergola constructed to support grapevines on a trellis); *syn.* wine bower [n], wine bowery [n]; *s* **emparrado [m]** (Pérgola construida para sostener viñas); *syn.* parral [m], parrón [m] [RCH]; *f* **treille [f]** (Berceau de ceps de vigne soutenus par un treillage ; PR 1987) ; *g* **Weinlaube [f]** (Pergolenartiger Bau als Stützgerüst für Weinranken).

2442 graphic representation [n] *plan.* (Drawing or diagram expressing ideas); *s* **representación [f] gráfica** (Dibujo o diagrama que refleja ideas sobre papel o en pantalla de ordenador); *f* **représentation [f] graphique** (d'une idée par des croquis et diagrammes) ; *g* **grafische Darstellung [f] (2)** (von Gedanken durch Zeichnungen und Diagramme); *syn. o. V.* graphische Darstellung [f].

2443 grasping branch [n] **with adhesive disks** *bot.* (Specialized stem tips that adhere to surfaces to facilitate climbing; e.g. Virginia Creeper—*Parthenocissus veitchii*); *syn.* grasping organ [n] with adhesive disks; *s* **zarcillo [m] caulinar con ventosas** (Órgano trepador p. ej. en la parra virgen *[Parthenocissus esp.]*); *syn.* zarcillo [m] rameal con ventosas; *f* **vrille [f] raméale à disques-ventouses** (Organe aérien de fixation de certaines vignes grimpantes, p. ex. la Vigne vierge *[Parthenocissus triscuspidata 'Veitchii']*) ; *g* **Sprossranke [f] mit Haftscheiben** (Kletterorgan, z. B. beim Wilden Wein *[Parthenocissus veitchii]*).

grasping organ [n] **with adhesive disks** *bot.* ▶grasping branch with adhesive disks.

2444 grass [n] *bot.* (Group of small, green-bladed plants belonging to the family of *gramíneae*, including cereals, reeds, many species of lawn grass, and bamboos; ▶lawn grass, ▶nurse grass, ▶ornamental grasses); *syn.* graminoid [n]; *s* **gramínea [f]** (Planta de la familia de las *gramíneae*, del orden de las glumifloras, de flores, por lo regular, hermafroditas. El fruto es una cariopsis, con abundante tejido nutricio. En general, esta familia comprende plantas herbáceas, con tallos cilíndricos y huecos [cañas], provistos de nudos manifiestos, hojas angostas, largamente envainadoras y con una lígula en el encuentro de la vaina y el limbo; las flores se disponen en espículas, compuestas de varias de ellas, y de dos glumas estériles en la base; las espículas se agrupan a su vez en espigas, racimos o panículas, a veces muy complicadas. En total cuentan con unas 4000 especies, entre ellas los cereales; y toman parte preponderante en la formación de las praderas, estepas y sabanas; DB 1985; ▶gramínea para césped, ▶gramínea nodriza, ▶hierba ornamental); *syn. popular* hierba [f], yerba [f]; *f* **graminée [f]** (Plante vivace de la famille des monocotylédones, à tige cylindrique creuse mais cloisonnée au niveau des nœuds, aux fleurs peu apparentes et groupées en épis, aux fruits farineux réduits à des grains, au port herbacé, et qui comprend les céréales, les herbes des prairies, des steppes et des savanes, les bambous, la canne à sucre ; ▶graminée abri, ▶graminée ornementale, ▶graminée pour gazon) ; *g* **Gras [n]** (Einkeimblättrige, windblütige Pflanze mit Stängeln, die verdickte Knoten [Nodien] und meist hohle Internodien hat und mit schmalen, spitz zulaufenden, parallelnervigen Blättern ausgestattet ist. Sie gehört zu der Familie der Gräser *[Gramíneae]* mit Vertretern wie z. B. Weizen, Roggen, Gerste, Hafer, Schilfrohr, Bambus; ▶Ammengras, ▶Rasengras, ▶Ziergras).

grass [n]**, accent** *hort. plant.* ▶specimen grass.

grass area [n] *constr. recr.* ▶turf area.

grass areas [npl] *constr.* ▶establishment of grass/lawn areas.

grass clippings [npl] [US] *constr. hort.* ▶lawn clippings.

grass concrete slab [n] [UK] *constr.* ▶grass paver.

2445 grass cover [n] *phyt.* (Ground cover with grass; ▶turf sward); *s* **cobertura [f] herbácea** (▶tapiz de césped); *syn.* cobertura [f] de hierba; *f* **couverture [f] herbacée** (▶tapis de graminées) ; *syn.* couvert [m] herbacé, couverture [f] de graminées ; *g* **Grasbewuchs [m, o. Pl.]** (Bedeckung des Bodens mit Gras; ▶Rasennarbe).

grass-crete paver [n] [US] *constr.* ▶grass paver.

2446 grass cutting [n] *constr. hort.* (Mowing of grass to maintain the ▶turf sward at a low height); *syn.* mowing [n] (SPON 1986, 120, 205); *s* **corte [m] del césped** (Corte regular del césped para mantener el ▶tapiz de césped con poca altura); *syn.* siega [f] del césped; *f* **tonte [f] du gazon** (Intervention régulière de coupe pour obtenir un gazon court et un aspect soigné ; ▶tapis de graminées) ; *syn.* coupe [f] de gazon, tonte [f] de la pelouse ; *g* **Rasenschnitt [m]** (Mähgang, der zum Kurzhalten der ▶Rasennarbe durchgeführt wird); *syn.* Rasenmahd [f].

grass cutting [n]**, long** *agr. constr.* ▶hay cutting.

2447 grassed roof [n] *constr. urb.* (Roof covered by turf); *syn.* sod roof [n] [US] (OSM 1999); *s* **tejado [m] de hierba** (Tejado cubierto de hierba); *f* **toit [m] en pré** (TJ 1988, 139) ; *syn.* toit [m] enherbé (TJ 1988, 140) ; *g* **Grasdach [n]** (I. d. R. durch Grasansaat begrüntes Dach).

grassed strip [n] *constr.* ▶grass strip (1).

grass-filled joint [n] *constr.* ▶turf-filled joint.

2448 grass-filled modular paving [n] [US] *constr.* (TSS 1988, 330-7; paved surface for parking lots, fire-engine access, etc., constructed using large-sized paving stones [US]/large-sized paving setts [UK] of concrete or natural stone laid on a ▶subbase [US]/sub-base [UK] with wide joints. These are filled with sandy topsoil and sown with a hard-wearing grass mixture); *syn.* grass setts paving [n] [UK] (LD 1988 [9]); *s* **pavimento [m] de adoquines con césped** (Revestimiento de superficies de aparcamientos, accesos a predios, etc. generalmente de adoquines grandes tendidos sobre la ▶capa portante con juntas anchas que se rellenan de tierra vegetal y arena y se siembran con gramíneas resistentes); *syn.* revestimiento [m] de adoquines con césped; *f* **pavage [m] engazonné** (Revêtement de pavés pour aires de stationnement, voies pompiers, etc. constitué de pavés à joints larges mis en place sur une ▶couche de fondation souvent mélangée à un terreau. Les joints sont remplis d'un mélange terre végétale/sable et le semis est effectué avec un mélange de semences résistantes au piétinement) ; *g* **Rasenpflaster [n]** (Bodenbelag für Parkplätze, Feuerwehrzufahrten etc., meist aus Großpflastersteinen o. Ä. [früher nur mit auf Lücke gesetzten Großpflastersteinen aus Naturstein, seit Anfang der 1990er-Jahre auch aus Betonrasenfugensteinen mit angeformten oder integrierten Abstandhaltern hergestellt], mit den breiten Fugen auf eine ▶Tragschicht verlegt werden. Da die so verlegten Steine nicht in der Lage sind, durch Fahrbetrieb verursachte Scherkräfte aufzufangen, muss die Fugenfüllung mit vergleichsweise grobkörnigen Mineralstoffen, gemischt mit Erdsubstrat, und ausreichendem Porenvolumen die erforderlichen Stabilisierungsaufgaben übernehmen. Das Fugengemisch wird mit strapazierfähigen Gräsern eingesät); *syn.* Rasenfugenpflaster [n].

grass-filled paving block [n] [US] *constr.* ▶grass paver.

grass growth [n]**, uncontrolled** *for. hort.* ▶uncontrolled grass intrusion.

grassing [n] *constr.* ▶lawn seeding, ▶low-cost grassing

grassing [n]**, cost-effective** *constr.* ▶low-cost grassing.

grass intrusion [n] *for. hort.* ▶uncontrolled grass intrusion.

2449 grassland [n] (1) *agr.* (Agrarian land used for hay or fresh fodder, pasturing, field forage growing; *specific terms* ▶natural grassland, ▶permanent grassland, ▶temporary grassland [US]/ley [UK]); *s* **prado** [m] (Formación herbácea producida por el hombre a partir del monte rozado, estercolado, segado y pastado. Es España es costumbre llamar «pradera» al prado sembrado, para distinguirlo así del que salió del monte rozado; DINA 1987; *términos específicos* ▶prado natural, ▶pradera permanente, ▶pradera temporal); *syn.* pradera [f]; *f* **prairie** [f] (Surface enherbée, de composition floristique et de durée d'établissement variable, principalement pâturée, mais dont le fourrage est également récolté mécaniquement pour l'alimentation à l'auge ou pour la conservation du foin — ensilage ; DAV 1984 ; ▶prairie naturelle, ▶prairie permanente, ▶prairie temporaire) ; *g* **Grünland** [n] (Landwirtschaftliche Nutzfläche, die als Mähwiese, Weideland oder Feldfutterfläche genutzt wird; *UBe* ▶natürliches Grünland, ▶Dauergrünland, ▶zeitweiliges Grünland).

grassland [n] (2) *phyt.* ▶acidic grassland, ▶alpine grassland; *land' man. recr.* ▶amenity grassland; *phyt.* ▶avalanche grassland, ▶dune grassland, ▶limestone grassland; *agr.* ▶natural grassland; *phyt.* ▶nutrient-poor grassland; *agr.* ▶permanent grassland, ▶plowing-up of grassland [US]/ploughing-up of grassland [UK]; *phyt.* ▶semidry grassland [US]/semi-dry grassland [n] [UK], ▶steppe-like grassland; *agr.* ▶temporary grassland [US]; *phyt.* ▶tropical grassland.

grassland [n]**, abandoned** *agr.* ▶abandoned pasture.

grassland [n]**, arid** *phyt.* ▶dry meadow.

grassland [n] [UK]**, chalk** *phyt.* ▶limestone grassland.

grassland [n]**, fallow** *agr.* ▶abandoned pasture.

grassland [n]**, old orchard** *agr. conserv.* ▶traditional orchard meadow [UK].

grassland [n] [US]**, open orchard** *agr. conserv.* ▶traditional orchard meadow [UK].

grassland [n] [UK]**, ploughing-up of** *agr.* ▶plowing up of grassland [US]/ploughing-up of grassland [UK].

grassland [n] [US]**, turning-over of** *agr.* ▶plowing up of grassland [US].

grassland agriculture [n] *agr.* ▶grassland farming.

2450 grassland community [n] (1) *phyt.* (Plant association growing on [agricultural] grassland areas); *s* **asociación** [f] **pratícola** (Comunidad herbácea que crece en prados sembrados); *syn.* asociación [f] herbácea; *f* **groupement** [m] **prairial** (Association végétale croissant sur une prairie) ; *g* **Grünlandgesellschaft** [f] (Auf Grünland wachsende Pflanzengesellschaft).

2451 grassland community [n] (2) *phyt.* (Plant community of, e.g. ▶alpine grassland, ▶dry meadow, ▶nutrient-poor grassland, ▶salt meadow); *s* **comunidad** [f] **pratícola** (Comunidad herbácea, p. ej. de ▶pastizal seco, ▶pasto alpino, ▶pasto salado, ▶prado oligótrofo, etc.); *f* **groupement** [m] **de pelouse** (Terme générique relatif aux associations végétales des pelouses, p. ex. ▶pelouse alpine, ▶pelouse aride, ▶pelouse oligotrophe, ▶pré salé) ; *syn.* végétation [f] des pelouses ; *g* **Rasengesellschaft** [f] (Pflanzengesellschaft z. B. des ▶alpinen Rasens, ▶Magerrasens, ▶Trockenrasens, der ▶Salzwiese).

2452 grassland farming [n] *agr.* (Planning, administration, development and cultivation of grassland for fodder. There are various forms of **g. f.** according to the type of farm; meadow farming, pasture farming, ▶pasture farming for hay production

or field forage); *syn.* grassland agriculture [n], grassland management [n]; *s* **praticultura** [f] (Metodología agronómica que se ocupa del cuidado de los prados, los aspectos agronómicos relacionados con su siembra, explotación y mantenimiento; cf. DINA 1987; ▶pasticultura); *f* **exploitation** [f] **des prairies** (Culture fourragère sur prairies permanentes, temporaires ou artificielles ; selon la forme d'affouragement on distingue, le fanage, le pâturage, le pâturage tournant ; ▶élevage pastoral) ; *g* **Grünlandwirtschaft** [f] (Futterbauwirtschaft auf Grünland. Es wird je nach Bewirtschaftungsform zwischen Wiesenwirtschaft, ▶Weidewirtschaft und Mähweidewirtschaft unterschieden); *syn.* Grünlandbewirtschaftung [f], Grünlandnutzung [f].

grassland management [n] *agr.* ▶grassland farming.

2453 grassland species [n] *agr. phyt.* (Herbaceous plant, such as grass, clover and other species of the *Leguminosae* familiy growing on grassland, which is used for fodder and hay production as well as silage purposes); *s* **planta** [f] **pratense** (Planta herbácea, como las gramíneas, el trébol y otras especies de la familia de las leguminosas, que crecen en prados sembrados, y que son utilizadas para la producción de forraje y heno); *syn.* especie [f] pratense; *f* **plante** [f] **prairiale** (Espèces herbacées telles que les graminées, légumineuses et autres papilionacées *[Fabaceae]* qui croissent sur les prairies permanentes ou temporaires et sont utilisées comme fourrage vert ou récoltées pour l'ensilage ou la distribution de foin) ; *syn.* espèce [f] prairiale ; *g* **Grünlandpflanze** [f] (Krautige Pflanze [z. B. Gras, Klee und andere Schmetterlingsblütler *(Fabaceae)*], die auf Dauergrünlandflächen wächst und der Frischfutter-, Heu- oder Silagefuttergewinnung dient); *syn.* Grünlandart [f], Grünlandkraut [n].

grassland vegetation [n] *phyt.* ▶mapping of grassland vegetation.

2454 grassland yield index [n] (≠) *agr. leg.* (**D.**, an index compiled to express numerically the yield capabilities of grasslands according to soil types, hydrological conditions, climate, yield period, relative humidity, topography for land evaluation and tax assessment; ▶agricultural land grade [UK], ▶arable land rank [US]/arable land grade [UK], ▶Land Capability Classification); *s* **índice** [m] **de fertilidad del suelo (de prados)** (≠) (**D.**, valoración cuantitativa del tipo de suelo, el suministro de agua, el clima y otros factores mesológicos; ▶clasificación de aptitud de suelos, ▶índice de fertilidad de suelos, ▶indice de fertilidad de tierras de cultivo); *f* **coefficient** [m] **de valeur agricole des prairies** (▶classification des aptitudes culturales et pastorales des sols, ▶indice d'estimation des sols, ▶indice de fertilité physique) ; *syn.* coefficient [m] de valeur agronomique des prairies ; *g* **Grünlandzahl** [f] (**D.**, an Hand der Reichsbodenschätzung auf der Grundlage des Bodenschätzungsgesetz vom 16.10.1934 die zahlenmäßige Erfassung [relative Wertzahl] der Ertragsfähigkeit von Grünland nach Bodenart [Sand, Schluff, Lehm, Ton, Moorboden/Moorerde] und Zustandsstufe des Bodens, Wasserverhältnissen, Klima sowie Vegetationsdauer, Luftfeuchtigkeit und Geländegestaltung; die Zustandsstufe gibt den Entwicklungsgrad an, den ein Boden bei seiner Entwicklung vom Rohboden über eine Stufe höchster Leistungsfähigkeit bis zur Ausbildung eines Podsols erreicht hat; SS 1979. Die Agrar- und Umweltwissenschaftliche Fakultät der Universität Rostock hat die Methoden der Grünlandbewertung auch aus ökologischer und naturschutzfachlicher Sicht weiter verfeinert; ▶Ackerzahl, ▶Bodenschätzung, ▶Bodenzahl).

grass mix [n] [US]**, conservation** *constr.* ▶low-maintenance grass type.

grass mixture [n] *constr.* ▶sports grass mixture.

grass [n] **on crushed aggregate** constr. ▶crushed aggregate lawn.

2455 grass paver [n] constr. (Hollow concrete blocks which allow vegetative growth to occur in interstitial spaces; ▶grass-filled modular paving [US]/grass setts paving [UK]); syn. turf block [n] [also US] (TGG 1984, 81), turf paver [n] [also US] (HALAC 1992, 133), firepath pot [n] [also UK], lattice (work) concrete block [n] [also US] (TGG 1984, 150), bg-slab [n] [also UK] (SPON 1986, 375), grass concrete slab [n] [also UK] (SPON 1986, 375), grass-filled paving block [n] [also US] (TGG 1984, 81), grass-crete paver [n] [also US], grid block [n] [also US]; *s* **adoquín** [m] **pavicésped** (Ladrillo especial con agujeros que permiten el crecimiento de hierba entre los espacios intersticiales; ▶pavimento de adoquines con césped); syn. ladrillo [m] de hormigón para césped; *f* **dalle** [f] **gazon** (Dalle alvéolée en béton et depuis le début des années 90 en polyéthylène haute densité recyclé [*terme spécifique* dalle nid d'abeilles] utilisée pour les circulations engazonnées, en renforcement de berge, sur couche filtrante dans les bassins de décantation ; ▶pavage gazonné) ; syn. dalle [f] alvéolée pour gazon, dalle [f] BG [béton-gazon], dalle [f] evergreen, dalle [f] perforée ; *g* **Rasengitterstein** [m] (Betonhohlstein oder Rasenziegel aus Ton mit Erdkammern für befahrbare Rasenflächen. Seit Anfang der 1990er-Jahre werden in D. solche Elemente auch aus rezykliertem Kunststoff eingebaut und **Rasenwaben** genannt; *UBe für R. aus Beton* Betongitterstein [m], Beton-Gras-Stein [m], Betonrasenplatte [f]; *aus gebranntem Ton* Rasenlochklinker; ▶Rasenpflaster); syn. Rasenstein [m].

grass playing field [n] constr. recr. ▶grass sports field.

grass playing pitch [n] [UK] constr. recr. ▶grass sports field.

2456 grass plugging [n] constr. (Vegetative establishment of turf grasses by planting small plugs, approximately 5cm wide, containing the top growth, roots, rhizomes or stolons. This planting method is used in semiarid [US]/semi-arid [UK] and arid regions, and for certain grasses in temperate regions; e.g. *Zosia*); *s* **establecimiento** [m] **de césped por tallos rizomatosos** (Método utilizado en zonas áridas y semiáridas para establecer césped, por medio de hierbas precultivadas, generalmente de especies estoloníferas; syn. plantación [f] de césped por motas [C]; *f* **plantation** [f] **d'un tapis de graminées** (±) (Plantation de graminées pré-cultivées pour gazon — en général des espèces à racines traçantes — en vue de la création d'une aire engazonnée. Cette méthode est principalement utilisée dans les régions arides et semi-arides et pour certaines graminées en régions tempérées, p. ex. le *Zoysia*) ; *g* **Pflanzen** [n, o. Pl.] **einer Grasnarbe** (Setzen von vorkultivierten Rasengräsern — meist Ausläufer treibende Arten — für die Anlage einer Rasenfläche. Diese Methode wird vorwiegend in semiariden und ariden Klimagebieten angewendet).

grass prairie [n] [US]**, tall** phyt. ▶tall grass steppe.

grass savanna(h) [n] phyt. ▶tropical grassland.

2457 grass seeding [n] constr. (Process of sowing and rolling of grass seed into the soil; ▶seeding, ▶seeding of grass/lawn areas, ▶seeding of meadows); *s* **siembra** [f] **de césped (1)** (Proceso de extensión y rastrillado de las semillas de hierba en el suelo; ▶siembra, ▶siembra de césped 2, ▶siembra de praderas); *f* **semis** [m] **d'un gazon** (Dans les travaux d'établissement d'un gazon, opération de distribution et d'incorporation des semences ; ▶enherbement, ▶engazonnement d'une pelouse par semis, ▶semis 1) ; *g* **Rasenansaat** [f] (Vorgang des gleichmäßigen Ausbringens und Einarbeitens von Rasensaatgut; ▶Ansaat 1, ▶Herstellung von Rasenflächen durch Ansaat, ▶Wiesenansaat).

2458 grass-seed mixture [n] constr. hort. (▶lawn type, ▶seed mixture, ▶standard grass-seed mixture); syn. turf-seed mixture [n] [also US] (TGG 1984, 238); *s* **mezcla** [f] **de semillas de césped** (▶mezcla estándar de semillas, ▶mezcla de semillas, ▶tipo de césped); *f* **mélange** [m] **de semences pour gazon** (destiné à l'engazonnement ; ▶mélange de graines, ▶mélange standardisé de semences pour gazon, ▶type de gazon) ; *g* **Rasensaatgutmischung** [f] (Zusammenstellung von Gräsern zu einer Grassamenmischung für einen bestimmten Nutzungsanspruch auf einem definierten Standort, der durch seine geografische Lage, Boden und Klima bestimmt ist; für den Garten- und Landschaftsbau sind gem. DIN 18 917 sieben ▶Rasentypen entwickelt worden, die im Handel als ▶Regelsaatgutmischungen erworben werden können; ▶Saatgutmischung); syn. Grassaatmischung [f], Grassamenmischung [f].

grass seeds [n] [UK]**, landscape** constr. ▶low-maintenance grass type.

grass setts paving [n] [UK] constr. ▶grass-filled modular paving [US].

2459 grass sports field [n] constr. recr. syn. grass playing field [n] (SPON 1974, 337), grass playing pitch [n] [also UK], turf playing field [n]; *s* **campo** [m] **de deportes de césped**; *f* **pelouse** [f] **sportive** (Pelouse dans une installation sportive) ; syn. pelouse [f] de jeu ; *g* **Rasenspielfeld** [n] (Rasenfläche in Sportanlagen); syn. Rasenspielplatz [m].

grass steppe [n] phyt. ▶tall grass steppe.

2460 grass strip [n] **(1)** constr. (Narrow vegetated strip predominantly covered with grass: maintained ▶turf strips 1 usually require regular maintenance; in comparison ▶meadow strips receive little maintenance; *generic term* ▶vegetated strip); syn. grassed strip [n]; *s* **franja** [f] **de hierba** (Según el grado de mantenimiento y su estado se diferencian ▶franja de césped y ▶banda de herbáceas; ▶banda verde central, ▶banda de césped); *f* **bande** [f] **engazonnée** (▶Bande verte d'accompagnement ensemencée ; selon l'intensité de l'entretien on distingue l'▶accotement engazonné ou l'▶accotement) ; *g* **Grasstreifen** [m] (Vorwiegend mit Grasarten bestockter ▶Grünstreifen [OB]. Je nach Pflegeaufwand [Schnitthäufigkeit] oder -zustand werden ▶Rasenstreifen 1 oder ▶Wiesenstreifen unterschieden).

grass strip [n] **(2)** constr. ▶turf strip (1).

grass type [n] constr. ▶low-maintenance grass type.

grate [n] constr. ▶steel grate, ▶tree grate, ▶window well grate.

grate [n] [US]**, drainage** constr. ▶storm drain grate [US].

grate [n] [US]**, gutter** constr. ▶storm drain grate [US].

grate [n] [US]**, inlet** constr. ▶storm drain grate [US].

grate cover [n] constr. ▶grille cover.

2461 grate inlet [n] [US] constr. (Surface opening of a ▶catch basin [US]/road gully [UK]); syn. gutter inlet [n] [UK]; *s* **boca** [f] **de alcantarilla al ras del suelo** (▶sumidero de calle); *f* **bouche** [f] **(d'égout) à grille** (Bouche à accès par le dessus ; ▶regard à grille) ; *g* **ebenerdige Straßenablauföffnung** [f] (▶Straßenablauf).

grave [n] adm. landsc. ▶individual urn grave, ▶single row grave, ▶use period of grave [US]/rest period of grave [UK].

grave [n] [UK]**, rest period of** adm. ▶use period of grave [US].

2462 grave cross [n] adm. conserv'hist. (Gravesite marker or emblem of Christianity representing the Cross on which Christ was crucified; *s* **cruz** [f] **funeraria** (Elemento de ornamentación de una tumba en forma de cruz de piedra, madera o metal); *f* **croix** [f] **funéraire** (Élément commémoratif (pierre, bois, métal, etc.) sculptée sous forme de croix ornant une tombe) ;

g **Grabkreuz [n]** (Meist durch einen Bildhauer in Form eines Kreuzes gestaltetes Grabdenkmal).

2463 grave design [n] *adm. landsc.* (Design of each grave-site in a cemetery); *s* **decoración [f] de tumbas** *syn.* ornamentación [f] de tumbas; *f* **ornementation [f] de(s) tombe(s)** (Forme de composition pouvant utiliser des éléments esthétiques naturels [plantations diverses] ou la sculpture de la pierre tombale) ; *syn.* conception [f] de tombe, décoration [f] des sépultures ; *g* **Grabgestaltung [f]** (Ausstattung eines Grabes mit Bepflanzung und Grabmal nach formal-ästhetischen Gesichtspunkten).

2464 grave digger [n] *adm.* (Person who excavates graves in a cemetery); *s* **cavador [m] de tumbas** *syn.* enterrador [m]; *f* **fossoyeur [m]** (Personne qui creuse les fosses dans un cimetière) ; *g* **Totengräber [m]** (Jemand der auf einem Friedhof Gräber aushebt).

grave [n] **in a row** *adm. landsc.* ▶urn grave in a row.

2465 gravel [n] **(1)** *constr. eng.* (Generic term for coarse particles of rock that result from naturally occurring disintegration or that are produced by crushing weakly bound conglomerate; they are normally retained in a No 4 sieve [US]. Some authorities limit gravel to sizes of 2 to 10mm, referring to the larger 10 to 50mm stones as pebbles. One definition in the U.S. uses a diameter of from 4.76mm to 76mm, others only up to 64mm. In fact, there is no real agreement. The Wentworth Scale of particle size in the U.S. excludes the term gravel and sizes a pebble as 4 to 64mm; DNE 1978; ▶coarse gravel, ▶fine gravel, ▶medium gravel); *s* **grava [f]** (Sedimentos no consolidados de piedras y piedrillas de 2 a 63 mm de diámetro, que se utilizan para la construcción; ▶grava gruesa, ▶grava media, ▶gravilla); *syn.* ripio [m] [RA], granza [f] [CO], granzón [m] [YV]; *f* **gravier [m]** (Matériau sédimentaire meuble formé de granulats grossiers dont le diamètre est compris entre 2 et 60 mm ; dans les travaux de jardins on distingue le **g. roulé** du **g. concassé** ; ▶gravier fin, ▶gravier moyen, ▶gros gravier) ; *g* **Kies [m]** (Lockergestein — Sedimente — mit Korndurchmessern von 2 bis 63 mm; DIN 4022 und DIN 4023; im Landschaftsbau wird auch zwischen **Rundkies** und **gebrochenem Kies** [Schotter] unterschieden; ▶Feinkies, ▶Grobkies, ▶Mittelkies).

gravel [n] **(2)** *geo.* ▶alluvial gravel; *constr. pedol.* ▶coarse gravel, ▶fine gravel; *constr. min.* ▶in situ gravel; *constr.* ▶pea gravel, ▶pit gravel, ▶washed gravel.

gravel [n]**, clean** *constr.* ▶washed gravel.

gravel [n]**, quarrying of** *min.* ▶gravel extraction.

2466 gravel additives [npl] *constr.* (Gravel aggregates added to a soil so as to improve drainage, oxygen supply and structural stability); *s* **incorporación [f] de grava** (Añadidura de grava, p. ej. en sustrato de suelo, para aumentar la permeabilidad para el agua y el aire y la estabilidad estructural); *f* **incorporation [f] de gravier** (Addition de gravier, p. ex. dans un sol pour en améliorer la constitution physique en augmentant la perméabilité et la stabilité structurale) ; *g* **Kiesbeimengung [f]** (Zumischung von Kies, z. B. in ein Bodensubstrat zur Erhöhung der Wasser- und Luftdurchlässigkeit und Strukturstabilität).

2467 gravel bank [n] *geo.* (Accumulation of gravel in a river, usually exposed at low water; ▶sandbank); *s* **banco [m] de grava** (Acumulación de grava en un curso de agua, que suele salir a la luz cuando el río tiene nivel de estiaje; ▶banco de arena); *syn.* llanura [f] de grava; *f* **banc [m] de gravier** (Accumulation de gravier formant îles entre des bras de tressage ou d'anastomoses dans la plaine d'inondation d'un cours d'eau ; ▶banc de sable, ▶levée 2) ; *syn.* levée [f] de rive ; *g* **Kiesbank [f]** (Kiesansammlung/Kiesinsel in einem Fließgewässer; ▶Sandbank).

gravel chippings [npl] [UK]**, layer of** *constr.* ▶layer of gravel chips [US].

gravel chips [npl] [US] *constr.* ▶layer of gravel chips [US].

2468 gravel-covered roof [n] *arch. constr.* (▶Flat roof on top of which gravel or stone chips [US]/stone chippings [UK] have been spread; ▶low-pitch roof); *s* **azotea [f] recubierta de grava** (▶Tejado plano sobre el cual se ha extendido una capa de grava o gravilla; ▶tejado plano con pendiente maxima de 5°); *f* **toiture [f] recouverte de gravier** (Dalle comportant un lit de gravillons non liés comme protection lourde de l'étanchéité ; ▶toiture à pente nulle, ▶toiture-terrasse plate) ; *g* **Kiesdach [n]** (Mit einer wenigen Zentimeter hohen Kiesschicht versehenes Flachdach ▶Flachdach 1, Flachdach 2).

2469 gravel dredging [n] *min.* (*Generic term* ▶dredging 1); *syn.* wet gravel workings [npl] [also UK]; *s* **extracción [f] de grava por dragado** (Ganancia de grava en una gravera bajo el nivel del agua freática o en un curso de agua; *término genérico* ▶dragado subacuático); *f* **dragage [m] de gravier** (L'extraction de roches alluvionnaires dans la nappe au moyen d'une drague ; *terme générique* ▶extraction de matériaux alluvionnaires en nappe) ; *syn.* extraction [f] de gravier dans la nappe ; *g* **Nassauskiesung [f]** (Gewinnung von Kies in einer Entnahmestelle [Kiesgrube] unterhalb des Grundwasserspiegels oder in einem Fließgewässer; weil bei einer **N.** in einem Baggersee die das Grundwasser schützende Deck- und Filterschicht abgetragen wird, und es somit auf Dauer zu schädlichen Veränderung der physikalischen oder biologischen Beschaffenheit des Grundwassers kommen kann, bedürfen **N.en** eines wasserrechtlichen Zulassungsverfahrens, das in den meisten Fällen eine Umweltverträglichkeitsprüfung nach dem UVP-Gesetz bedingt; weniger umweltbeeinträchtigend ist die Trockenauskiesung; *OB* ▶Nassbaggerung).

gravel excavation [n] *min.* ▶dry gravel excavation.

2470 gravel extraction [n] *min.* (Generic term for ▶gravel dredging and ▶dry gravel excavation); *syn.* quarrying [n] of gravel (TGG 1984, 101); *s* **extracción [f] de grava** (*Términos específicos* ▶extracción de grava en seco, ▶extracción de grava por dragado); *syn.* explotación [f] de grava; *f* **extraction [f] de gravier** (*Termes spécifiques* ▶dragage de gravier, ▶extraction de gravier hors nappe) ; *g* **Kiesabbau [m, o. Pl.]** (OB zu ▶Nassauskiesung und ▶Trockenauskiesung); *syn.* Abbau [m] von Kies, Auskiesung [f], Kiesbaggerung [f], Kiesentnahme [f], Kiesgewinnung [f].

2471 gravel extraction site [n] *min.* (*Generic term* ▶borrow pit; ▶sand and gravel pit, ▶wet gravel pit); *syn.* gravel pit [n]; *s* **gravera [f]** (Término genérico para ubicación de extracción de grava; ▶gravera subacuática, ▶hoyo de excavación, ▶zona de préstamo); *syn.* cantera [f] de grava; *f* **site [m] d'extraction de gravier** (Lieu d'extraction de graviers ayant une profondeur qui peut atteindre plusieurs mètres et qui est souvent rempli d'eau provenant de la nappe phréatique ; ▶carrière en fosse, ▶gravière en nappe, ▶lieu d'emprunt) ; *syn.* gravière [f] ; *g* **Kiesentnahmestelle [f] (1)** (Abbaufläche zur Kiesgewinnung; ▶Baggergrube, ◀Kiesentnahmestelle 2, ▶Seitenentnahme).

gravel pit [n] *landsc. recr.* ▶flooded gravel pit, *min.* ▶gravel extraction site; *landsc. recr.* ▶wet gravel pit.

2472 gravel pit reclamation [n] *land'man. landsc.* (Landscape planning measures for the creation of a recreation area or a wetland habitat, which may contain rare or endangered species, in and around a disused gravel pit); *s* **recuperación [f] de paisaje de gravera** (Conjunto de medidas paisajísticas para desarrollar un área de recreo o crear un humedal en antigua laguna de extracción de grava); *syn.* recultivo [m] de cantera de grava; *f* **réaménagement [m] de gravière** (Dans le cadre de la remise en état du site d'une gravière en nappe, compte tenu des caractéristiques du milieu environnant, aménagement d'une zone de loisirs ou création d'une zone humide favorisant la vie des espèces rares et menacées) ; *g* **Kiesgrubenrekultivierung [f]** (Landschaftsplanerische Maßnahmen zur Errichtung eines Erholungsgebietes oder zur Schaffung eines Feuchtgebietes für seltene und bedrohte Arten in einer ausgebeuteten Kiesentnahmestelle).

2473 gravel river terrace [n] *geo.* (Flat platform created by scour of a riverbed into previously accumulated ▶detrital deposits; ▶lower river terrace, ▶middle river terrace, ▶upper river terrace); *s* **terraza [f] (fluvial) de aluviones** (Rellano paralelo al cauce de un río formado por la erosión lineal del ▶depósito detrítico; ▶terraza fluvial alta, ▶terraza fluvial baja, ▶terraza fluvial media); *syn.* terraza [f] (fluvial) de cantos rodados; *f* **terrasse [f] alluviale** (Terrasse entaillée dans les alluvions plus anciennes [▶accumulation détritique] résultant de l'enfoncement d'un cours d'eau ; ▶terrasse [fluviatile] inférieure, ▶terrasse [fluviatile] moyenne, ▶terrasse [fluviatile] supérieure) ; *syn.* terrasse [f] emboîtée ; *g* **Schotterterrasse [f]** (Durch das Eintiefen eines Flusses in seinen vorher aufgeschütteten ▶Schotterkörper 1 entstandene Terrasse; ▶Niederterrasse, ▶Mittelterrasse, ▶Hochterrasse); *syn.* Aufschüttungsterrasse [f], Akkumulationsterrasse [f].

2474 gravel subbase [n] [US]/**gravel sub-base** [n] [UK] *constr.* (Densely-packed mixture of granular material, ground in a crusher and not sorted for particle size, which is used as a ▶subbase [US]/sub-base [UK] under roads or pathways; ▶crushed aggregate subbase [US]/crushed aggregate sub-base [UK]); *syn.* granular road base [n] [UK], *regional parlance* layer [n] of crusher-run aggregate [US]; *s* **base [f] de grava** (Mezcla compacta de material granulado de diferentes tamaños usada como ▶capa portante en la construcción viaria; ▶capa de asiento de piedra partida); *f* **couche [f] de fondation en grave non traité** (Mélange de cailloux, de graviers et de sable de granulométrie étalée [calibré et normé], à faible proportion de vides constituant la couche de forme dans les travaux de voirie ; ▶couche de base en grave, ▶couche de fondation) ; *g* **Kiestragschicht [f]** (Hohlraumarmes, frostfreies, korngestuftes Kies-Sandgemisch als Unterbau für den Wegebau; ▶Schottertragschicht, ▶Tragschicht).

gravel workings [npl] [UK]**, wet** *min.* ▶gravel dredging.

2475 grave planting [n] *hort. plant. syn.* gravesite planting [n]; *s* **plantación [f] de tumbas**; *f* **plantation [f] des tombes** ; *syn.* travaux [mpl] de plantations funéraires ; *g* **Grabbepflanzung [f]** (Pflanzenausstattung eines Grabes); *syn.* Grabstättenbepflanzung [f].

grave site [n] *adm.* ▶burial site.

gravesite planting [n] *hort. plant.* ▶grave planting.

2476 grave slab [n] *adm.* (**1.** Generic term for the following kinds of grave slabs made of stone, bronze, etc.; **2. complete grave cover slab [n]** (Wide, flat block covering an entire grave surface area with or without inscription or decoration such as figures); **3. partial grave cover slab [n]** (Slab which covers only part of the grave surface with or without an inscription or decoration; ▶grave tablet); **4. epitaph [n]** (Special form of grave slab covering the whole of the grave to mark the memory of a deceased person, erected on a platform or column, on the inner or outer wall of a church, usually with an inscription describing the life of the deceased); *s* **losa [f] sepulcral** (En cementerios paisajísticos lápida sepulcral colocada en horizontal al nivel del suelo); *f* **dalle [f] tombale** (**1.** Terme générique pour les monuments funéraires horizontaux recouvrant tout ou partie de la surface de la tombe et sur lesquels sont portées des inscriptions et parfois des sculptures méplates ; ▶dalle mortuaire) ; *syn.* dalle [f] sépul-

crale, pierre [f] tombale horizontale); **2.** *Termes spécifiques* **2.1 dalle [f] tombale recouvrant toute la surface de la tombe**; **2.2 plaque [f] funéraire** (Dans un cimetière, plaque de taille restreinte et de formes diverses [livres, métiers, loisirs, etc.] posée sur la tombe ; ▶dalle mortuaire) ; *syn.* plaque [f] sépulcrale, plaque [f] tombale; **2.3 dalle [f] funéraire** (Située à l'intérieur d'un édifice, plaque peu épaisse, en matériaux divers, portant une inscription [**l'épitaphe**], parfois des armoiries et un décor sculpté ou gravé et qui constitue une partie d'un tombeau, souvent encastrée dans un mur près d'une tombe, ou qui peut former à elle seule ce tombeau) ; **2.4 plate-tombe [f]** (Monument funéraire situé à l'intérieur d'un édifice, composé d'une dalle funéraire ou d'une lame funéraire fermant la tombe, encastrée dans le sol ou légèrement surélevée par rapport à ce dernier) ; *g* **Grabplatte [f]** (**1.** OB für folgende Arten von Grabplatten aus Stein oder Bronze; **2.** *UBe* **2.1 Vollgrababdeckung [f]** (Platte, die die gesamte Grabnutzungsfläche abdeckt; mit oder ohne Beschriftung/Inschrift oder figürlichen Schmuck); **2.2 Teilgrababdeckung [f]** (Grabplatte, die nur einen Teil der Grabnutzungsfläche abdeckt; mit oder ohne Beschriftung/Inschrift oder figürlichen Schmuck; ▶Kissenstein); **2.3 Epitaph [n]** (…e [pl] und …ien [pl]; **a.** Grabschrift; **b.** *Bildhauerkunst* eine besondere Form der Grabplatte als Erinnerungsmal für einen Toten in der Größe einer Vollgrababdeckung, an einen Pfeiler, eine Innen- oder Außenwand einer Kirche angebracht, die i. d. R. voll beschriftet ist und das Leben des Verstorbenen beschreibt; im Franz. wird unter *épitaphe* nur der ausführliche Text auf der Grabplatte verstanden); *syn.* Epitaphium [n].

2477 gravestone [n] *adm. conserv'hist.* (Small, upright headstone or flat ▶grave tablet marking location of buried remains of a deceased person; ▶grave slab); *syn.* tombstone [n], headstone [n] [also US]; *s* **lápida [f] sepulcral (1)** (Pequeña losa sepulcral colocada en pie a la cabeza de la tumba o tendida sobre la cabecera de la misma; ▶lápida sepulcral 2); *syn.* piedra [f] sepulcral; *f* **pierre [f] tombale** (▶Monument funéraire vertical, en pierre ou en marbre, sur lequel peut être inscrit le nom, la date de naissance et de mort ainsi qu'une épitaphe ; ▶dalle mortuaire, ▶dalle tombale) ; *syn.* pierre [f] sépulcrale ; *g* **Grabstein [m]** (Gedenkstein auf einem Grab; ▶Grabplatte, ▶Kissenstein).

2478 gravestone design [n] *adm.* (Sculptural design of ▶gravestones, ▶grave crosses or ▶grave slabs; *s* **ornamentación [f] de lápidas sepulcrales** (▶lápida sepulcral, ▶cruz funeraria, ▶losa sepulcral); *syn.* ornamentación [f] de piedras sepulcrales; *f* **ornementation [f] des pierres tombales** (Travaux de sculpture effectués sur les ▶pierres tombales, ▶croix funéraires, ▶dalles tombales ou tombeaux) ; *syn.* conception [f] des pierres tombales ; *g* **Grabmalgestaltung [f]** (Bildhauerische Tätigkeit an ▶Grabsteinen, ▶Grabkreuzen, ▶Grabplatten oder größeren Grabdenkmalen); *syn.* Gestaltung [f] von Grabmalen/Grabsteinen).

2479 grave tablet [n] *adm.* (Small, flat grave marker of metal or stone on a burial site; ▶grave slab); *s* **lápida [f] sepulcral (2)** (Pequeña lápida de metal o piedra que se coloca tendida al ras del suelo sobre la cabecera de la tumba; ▶losa sepulcral); *f* **dalle [f] mortuaire** (Petite plaque, dalle, pierre horizontale gravée en métal ou en matériau de pierre naturelle, placée au ras du sol et ne recouvrant qu'une petite partie de la tombe l'autre partie étant aménagée et entretenue en plantations ; ▶dalle tombale) ; *g* **Kissenstein [m]** (Liegendes Grabmal auf Urnengräber oder am Kopfende eines Erdgrabes mit Inschrift/ Beschriftung, das die Grabstätte nur teilweise bedeckt; auf Rasengrabstätten bodeneben verlegt, um den Stein beim Rasenschnitt zu schützen; ▶Grabplatte); *syn.* Grabliegeplatte [f].

grave yard [n] *adm. urb.* ▶cemetery; *adm. hist.* ▶church yard.

G

gravitational water [n] *pedol.* ▶seepage water.

2480 gravity retaining wall [n] *arch. constr.* (DIL 1987, 139; ▶retaining wall which will withstand earth pressure due to its own weight and proportion); *s* **muro** [m] **de gravedad** (▶Muro de sostenimiento que soporta el empuje lateral gracias a su propio peso); *syn.* muro [m] de sostenimiento por gravedad; *f* **mur** [m] **poids** (▶Mur de soutènement en maçonnerie ou en béton, assurant sa stabilité par lui-même, sans lien à d'autres ouvrages et équilibrant par son propre poids la poussée des terres) ; *syn.* mur [m] autoporteur, mur [m] massif ; *g* **Schwergewichtsmauer** [f] (▶Stützmauer, die auf Grund ihres Eigengewichtes und entsprechend berechneten Querschnittes dem Erddruck standhält).

2481 gray-brown podzolic soil [n] *pedol.* (Soil profile A_h-A_l-B_t-C, up to 60cm deep, the moderately dark A horizon of which has little clay content. In Central Europe, the dark brown B_t horizon, which may be between 40 and 400cm deep, has been enriched with a high percentage of bases and an appreciable quantity of illuviated silicate clay; formed on relatively young land surfaces, mostly glacial deposits, from material relatively rich in calcium, under deciduous forests in humid temperate regions; **in U.S.**, this term is not used in current system of soil taxonomy; in U.S. soil science, a similar type is the *great soil group* of **udalfs**: soils which generally have brownish colors throughout, and are not saturated with water for periods long enough to limit their use for most crops; SST 1997); *syn.* udalf [n] [US], cambisol [n] (FAO system); *s* **tierra** [f] **parda eutrófica** (KUB 1953; suelo con perfil A_h-A_l-B_t-C, caracterizado por un horizonte A de hasta 60 cm, lixiviado de minerales arcillosos y por un horizonte B_t de acumulación de aquéllos, que en los suelos de Europa Central puede tener un espesor entre 40 y 400 cm); *syn. sistema de clasificación de los EE.UU.* udalf [m], *syn. sistema de clasificación de la FAO/UNESCO* cambisol [m]; *f* **sol** [m] **brunifié lessivé** (PED 1983, 300 ; sol caractérisé par un mull avec un horizon [B] ou très pauvre en matière organique et coloré en brun par les oxydes de fer. Le mull ne provoque aucune dégradation des argiles, celles-ci subissent au maximum un simple entraînement mécanique [lessivage] ; DIS 1986) ; *syn.* sol [m] lessivé ; *g* **Parabraunerde** [f] (Terrestrischer Boden mit A_h-A_l-B_t-C-Profil: bis 60 cm mächtiger, an Ton verarmter A-Horizont; er umfasst den krümeligen, humosen, geringmächtigen A_h- und den humusarmen, fahlbraunen, häufig plattigen A_l-Horizont. Im tiefbraunen B_t-Horizont, der in Mitteleuropa zwischen 40 und 400 cm mächtig sein kann, hat eine Tonanreicherung stattgefunden; stark lessivierte **P.n** mit mächtigem, sehr verfahltem A_l-Horizont werden als **Fahlerden** bezeichnet; im FAO-System werden **P.n** als **Cambisols** und die Fahlerden als **Podzoluvisols**, im US-System als **Udalfs** und die Fahlerden als **Boralfs** [insbesondere Glossoboralfs] bezeichnet; cf. SS 1979).

2482 gray dune [n] [US]/**grey dune** [n] [UK] *geo. phyt.* (Dune that has become stabilized by vegetative development, predominantly grasses, mosses, and low herbaceous species and where soil development has begun to take place; ▶backdune); *syn.* fixed dune [n]; *s* **duna** [f] **gris** (Duna muerta de origen organógeno relativamente alejada de la costa, que apenas sufre recubrimiento de arena y que se encuentra poblada de herbáceas, musgos y gramíneas y, en algunas zonas, de monte bajo. Si la **d. g.** está cubierta de brezales, se le llama ▶duna parda o duna de brezal); *f* **dune** [f] **grise** (Zone dunaire côtière éloignée du rivage, pauvre en substances nutritives, rarement ensablée, recouverte d'une végétation constituée de graminées basses, de mousses, de nombreux végétaux herbacés et par endroit par une forêt dunaire ; lorsque des anciennes dunes grises sont envahies par les bruyères, celles-ci sont appelées ▶dunes brunes ou dunes à bruyères) ; *syn.* dune [f] fixée ; *g* **Graudüne** [f] (Strandferner, nährstoffarmer,

selten übersandeter Dünenbereich an der Küste mit einer Pflanzendecke aus niedrigen Gräsern, Moosen und zahlreichen Kräutern, stellenweise von einem Dünenwald bewachsen; wenn ältere Graudünen verheiden, nennt man sie Braundünen oder ▶Heidedünen).

2483 grayling zone [n] [UK] *limn.* (River zone between the upper zone [▶trout zone], and the middle fish zone [▶barbel zone (UK)] where in Europe, the grayling *[Thymallus thymallus]* is the typical fish species. In contrast to the trout zone it is characterized by a moderately fast current and deeper water, which is nevertheless still well oxygenated; ▶hyporhitron, ▶river zone); *s* **zona** [f] **de umbra** (Zona inferior de la region salmonícola con la umbra *[Thymallus thymallus]* como pez característico en los ríos de la Europa templada. Se encuentra entre la ▶zona de trucha y la ▶zona de barbo. La biocenosis que habita en ella se denomina ▶hipóritron; ▶zona biológica de un río); *f* **zone** [f] **à ombre** (Tronçon d'un cours d'eau situé entre la ▶zone à truite et la ▶zone à barbeau — cours inférieur des rivières et fleuves de montagne — portant le nom de l'espèce la plus caractéristique, l'ombre commun *[Thymallus thymallus]*. Par comparaison avec la zone à truite, cette zone est caractérisée par des eaux assez froides, très oxygénées, une vitesse de courant assez rapide et une plus grande profondeur des eaux ; ▶hyporhithron, ▶zone écologique des cours d'eau); *g* **Äschenregion** [f] (Abschnitt von Fließgewässern zwischen ▶Forellenregion und ▶Barbenregion [Unterläufe der Bäche resp. Gebirgsflüsse], in dem die Europäische Äsche *[Thymallus thymallus]* als Charakterfisch lebt. Das im Unterschied zur Forellenregion mäßig schnell fließende Wasser ist durch Sauerstoffreichtum und größere Wassertiefe gekennzeichnet; ▶Flussregion, ▶Hyporhithron); *syn.* Hyporhithral [n].

2484 grazed woodland [n] *agr. for.* (ELL 1988, 14; mostly degraded woodland thinned out by livestock grazing, and leaving isolated irregularly-grouped trees; ▶common grazing forest); *syn.* pasture woodland [n]; *s* **bosque** [m] **de pastoreo** (Generalmente bosque degradado por pastoreo del que sólo quedan algunos árboles aislados agrupados irregularmente; ▶monte patrimonial de pastoreo); *f* **forêt** [f] **pâturée** (En général manteau et tapis végétal forestier dégradé par le pâturage en forêt ; ▶forêt pâturée communale) ; *g* **Weidewald** [m] (Als Viehweide genutzter, meist degradierter Wald; ▶Hutewald).

2485 grazing [n] *agr.* (**1.** Agricultural use of grassland to provide pasture for cattle grazing upon meadows. **2.** In French, the term *pacage* is used for pasturage on nutrient-poor, natural grassland; ▶common grazing, ▶overgrazing); *syn.* pasturing [n]; *s* **pastoreo** [m] (▶pasto común, ▶sobrepastoreo; apacentar [vb]); *syn.* apacentamiento [m]; *f 1* **pâturage** [m] (1) (Action de pâturer ; ▶parcours, ▶surpâturage) ; *syn.* pâture [f] ; *f 2* **pacage** [m] (1) (Pâturage sur prairie naturelle de faible productivité) ; *g* **Beweidung** [f] (**1.** Landwirtschaftliche Nutzung des Grünlandes zur Fütterung des Viehs durch das Abgrasen der Wiesenflächen. **2.** Im Französischen gibt es den UB *pacage* für Beweidung von Magerwiesen; ▶Hutung, ▶Überbeweidung).

grazing [n], **rotational** *agr.* ▶rotation pasture [US].

2486 grazing capacity [n] *agr. game'man.* (Maximum stocking rate possible in a given year without damaging vegetation or related resources; RCG 1982; ▶overgrazing); *s* **capacidad** [f] **de pastoreo** (Cantidad máxima de ganado que puede pastar en una zona sin que ésta se deteriore por ▶sobrepastoreo); *f* **capacité** [f] **biotique d'un pâturage** (LA 1981, 272 ; charge animale, maximale liée aux possibilités de production de matière végétale d'un milieu ; ▶surpâturage) ; *syn.* capacité [f] de charge animale, capacité [f] de contenance biologique ; *g* **Weidebelastbarkeit** [f] (Höchstmöglicher Tierbesatz, der in einem Gebiet

G

leben kann, ohne dass die Vegetation durch ►Überbeweidung nachhaltig beeinträchtigt oder zerstört wird).

grazing common land [n] [UK] *agr. for. (leg. hist.)* ►town commons [US].

grazing commons [n] [US] *agr. for. (leg. hist.)* ►town commons [US].

grazing pressure [n] *agr.* ►grazing stress.

2487 grazing stress [n] *agr.* (Degree of grazing intensity resulting from use of grassland by pasturing livestock; ►grazing capacity); *syn.* grazing pressure [n]; *s* **intensidad [f] de explotación de pastos** (Grado de aprovechamiento de pastizales que puede llevar a superar la ►capacidad de pastoreo); *f* **intensité [f] de l'exploitation pastorale** (Degré d'utilisation des zones de pâture par le bétail ; ►capacité biotique d'un pâturage) ; *g* **Weidedruck [m]** (Grad der Nutzung von Grasfluren durch Weidevieh; ►Weidebelastbarkeit); *syn.* starke Weidebeanspruchung [f].

grazing system [n] *agr.* ►rotational grazing system.

2488 grease separator [n] *constr.* (Device used to remove oily substances from wastewater); *syn.* grease trap [n]; *s* **separador [m] de grasa** (Dispositivo que se utiliza para extraer sustancias aceitosas de las aguas residuales); *f* **bac [m] séparateur à graisses** (Bac destiné à la rétention des graisses difficilement biodégradables ; ASS 1987, 144 ; *syn.* séparateur [m] de graisse (ASS 1987, 150) ; *g* **Fettabscheider [m]** (Vorrichtung zur Trennung von Fettbestandteilen aus Abwässern).

grease trap [n] *constr.* ►grease separator.

great colonies [npl]**, growing in** *phyt.* ►growing in large groups.

great fen-sedge swamp [n] *phyt.* ►saw sedge swamp.

2489 great soil group [n] [US] *pedol.* (Soils with identical developmental origins as defined by a particular horizontal layer combination; the soil type with its distinctive profile is used as the basis of soil mapping); *syn.* soil type [n] [UK]; *s 1* **(gran) grupo [m] de suelos** (En la clasificación de los EE.UU., tercer nivel jerárquico que utiliza como criterios los horizontes y las propiedades de diagnóstico); *s 2* **tipo [m] de suelo** (En D., penúltima subdivisión en la clasificación de suelos que se caracterizan por un desarrollo igual y se define por una combinación específica de horizontes de características similares); *f* **groupe [m] de sols** (F., subdivision des sous-classes dans la nomenclature de classification des sols établie sur la base du degré d'évolution et de différenciation du profil, du facteur temps, de la présence d'horizons spéciaux, des modifications anthropogènes, etc. PED 1983, 195s ; *syn.* classification américaine famille [f] de sols) ; *syn.* type [m] génétique de sols (PED 1984, 206) ; *g* **Bodentyp [m] (1)** (Boden mit gleichem Entwicklungszustand, der durch eine bestimmte Horizontkombination definiert ist).

green [n] *plan.* ►cosmetic green; *recr.* ►putting green.

green [adj] *bot. hort.* ►summer green.

green area [n] *landsc. urb.* ►green space, ►industrial green area.

green area [n]**, amenity** *landsc. urb.* ►green space.

green area [n]**, private** *leg. urb.* ►private green space.

green area [n]**, public** *leg. urb.* ►public green space.

green area [n] [US]**, zoned** *leg. urb.* ►zoned green space.

2490 greenbelt [n] [US]/**green belt** [n] [UK] *landsc. urb.* (Continuous open land, up to several kilometers [US]/kilometres [UK] in width around towns or cities, usually in agricultural use, and often containing parkland, woodland and sportsgrounds, which is maintained free from major developments by means of planning controls, in order to contain urban growth, and to fulfill local recreational requirements and climatic functions; ►buffer

zone, ►greenbelt concept, ►green open space ring, ►green space corridor, ►open country); *syn.* green space fringe [n]; *s* **cinturón [m] verde** (Zona que rodea un lugar densamente construido, en la que hay restricciones a la edificación y predominan los espacios verdes. En Inglaterra suelen tener varios kilómetros de ancho. Se utilizan para la agricultura y para fines recreativos y cumplen funciones climáticas para la correspondiente ciudad y en general tienen como fin evitar el crecimiento indiscriminado de la ciudad y permitir el acceso directo al campo; en D. se crearon cc. vv. más estrechos a partir de las antiguas fortificaciones y ►anillos verdes con funciones ordenadoras dentro de la ciudad; cf. DINA 1987; ►campiña, ►concepto de «cinturón verde», ►corredor verde, ►zona de amortiguación); *f* **ceinture [f] verte** (Zones vertes urbaines ou périurbaines d'un ou plusieurs km de large, en forme d'anneau que les documents d'urbanisme ont laissées libre de construction à des fins sociales et hygiéniques ; *terme spécifique* couronne forestière ; ►boucle verte, ►campagne 2, ►concept de ceinture verte, ►coupure verte, ►zone tampon) ; *g* **Grüngürtel [m]** (In England meist mehrere Kilometer breite Freiraumzone, bestehend aus landwirtschaftlich genutztem Land, Parklandschaften, Wälder und Sportflächen, die im Rahmen einer langfristigen Stadtplanung um Städte herum angelegt wurde und vorgehalten wird, um die städtische Ausdehnung einzudämmen und einen direkten Zugang zur ►freien Landschaft für die Erholung und um stadthygienische Funktionen zu ermöglichen. In D. entstanden wesentlich schmalere **G.** durch Umwandlung von geschleiften Wallanlagen in Grünflächen und als stadtgliedernde Grünringe; ►Green-Belt-Idee, ►Grünring, ►Grünzug, ►Abstands- und Schutzzone 1); *syn.* Grünzone [f].

2491 greenbelt concept [n] [US]/**green belt concept** [n] [UK] *landsc. urb.* (An urban design method of containing the inner city with open space, and of providing recreational facilities; there may be a series of such peripheral green rings with urban development between them; ►greenbelt [US]/green belt [UK]); *s* **concepto [m] de «cinturón verde»** (Concepto urbanístico según el cual las ciudades deberían estar rodeadas por un ►cinturón verde, zonas de recreo y ►anillos verdes estructurantes, como p. ej. las antiguas ►fortificaciones de tierra); *f* **concept [m] de ceinture verte** (Concept de planification urbaine utilisant les ►ceintures vertes, les ►boucles vertes structurantes, ou les vastes couronnes préservées de l'urbanisation, souvent aménagées en zones de loisirs et constituant un élément de structuration d'une agglomération urbaine ; ►fortification de terre) ; *g* **Green-Belt-Idee [f]** (Stadtgestaltungskonzept in Form von stadtbegrenzenden ►Grüngürteln, stadtumgebenden Freizeitzonen und stadtgliedernden ►Grünringen, z. B. ►Wallanlagen).

2492 greenbelt policy [n] *landsc. pol. urb.* (In U.K., governing principle for preservation and use of continuous open land around developed areas; in some parts of U.S., development boundaries are established to maintain mostly agricultural land surrounding urban areas; ►greenbelt, ►regional green corridor); *s* **medidas [fpl] de protección de cinturones verdes** (Política aplicada en Inglaterra de crear grandes ►cinturones verdes alrededor de las grandes ciudades con el fin de evitar que dos ciudades cercanas lleguen a unirse; ►corredor verde regional); *f 1* **politique [f] d'aménagement de ceintures vertes** (Politique d'urbanisme de création et de sauvegarde de larges ►ceintures vertes afin d'éviter le regroupement de grandes agglomérations voisines) ; *f 2* **politique [f] des zones de discontinuité** (F., mesures et actions des collectivités territoriales de protection d'espaces naturels d'intérêt régional soumis aux menaces de la pression urbaine afin de constituer de vastes ►coupures vertes périurbaines entre les axes de développement urbain) ; *g* **Grün-**

G

G

zugpolitik [f] (±) (Auf die Durchsetzung der Schaffung, Erhaltung und Sicherung von breiten ►Grüngürteln und Grünzügen gerichtetes Handeln von z. B. Regionalparlamenten, Nachbarschaftsverbänden, Umweltorganisationen, um große Nachbarstädte nicht flächendeckend mit Baugebieten zusammenwachsen zu lassen; ►regionaler Grünzug).

green buffer [n]**, regional** *landsc. urb.* ►regional green corridor.

green connection [n] [UK]**, roadside** *lands. urb.* ►urban greenway [US].

green corridor [n] *landsc. urb.* ►regional green corridor.

greenery [n]**, vertical** *gard. urb.* ►façade planting.

2493 green fallow [n] *agr.* (**1.** Condition of agricultural land with spontaneous vegetation or seeded with a fodder crop and used as a temporary pasture between harvest and resowing); *syn.* seeded fallow [n]; **2. wildflower fallow** [n] *agr. ecol.* (Specific term for a type of fallow land which has been set aside for several years to spontaneously develop diverse vegetation, or a piece or strip of land, which is seeded with a special mixture of locally cultivated flowering plants, as well as wild and seldom-found, annual and perennial herbs together with host plants, serving as habitats for endangered fauna and flora; type of mitigation measure aimed at creating biodiversity on agricultural land, where intensive use has eradicated wildflowers; technique first introduced in Switzerland. Plants, which are not desired as part of the vegetation, have to be mechanically removed from the vicinity and to prevent colonization by shrubs, the area must be mown in sectional strips from about the third or fourth year and followed by recultivation as an agricultural production area); **3. shifting fallow land** [n] *agr. ecol.* (A special form of ►wildflower fallow practised in a system of shifting cultivation, whereby a part of a field is set aside each year in a strip, while on the other side of the field the same size of land is recultivated for agriculture); **4. bee pasture fallow** [n] *agr. ecol.* (A special form of ►wildflower fallow with many flowering plants especially set-aside for the production of great amounts of high-quality pollen and nectar to support large populations of both pollinators and honey bees); *s* **barbecho** [m] **a corto plazo** (Vegetación espontánea o sembrada de plantas forrajeras en tierra de cultivo no utilizada en ese periodo/período del año que se aprovecha como pasto entre la cosecha y la resiembra); *f 1* **jachère** [f] **verte** (Pratique culturale agricole consistant en la mise en repos d'une surface agricole entre la récolte et la culture suivante et utilisant comme pâturage extensif les plantes fourragères ensemencées ou des plantes améliorantes [jachère cultivée], la végétation spontanée disséminée par le vent ou présentes dans le sol de la jachère [jachère morte]) ; *syn.* jachère [f] partielle, semi-jachère [f], jachère [f] intermittente ; *f 2* **jachère** [f] **fleurie** *agr. ecol.* (*Terme spécifique pour jachère verte,* parcelle à l'abandon couverte d'une végétation spontanée ou ensemencée avec un mélange de plantes culturales, accompagné de plantes annuelles et pérennes à riche floraison dont l'objectif premier est une amélioration de la qualité des sites et des paysages ainsi que la préservation et l'amélioration de la biodiversité animale ; réalisées dans le cadre de la réglementation sur les **Jachères Environnement et Faune Sauvage** [JEFS], les jachères fleuries sont colonisées par de nombreux insectes butineurs et apportent la nourriture à une faune variée tels que papillons, petits mammifères et oiseaux ; les JEFS sont fixes ou rotationnelles [la parcelle non cultivée changeant d'année en année]) ; *syn.* jachère [f] florale ; *f 3* **jachère** [f] **tournante** *agr. ecol.* (Forme particulière de la ►jachère fleurie constituée de bandes installées pour une courte durée (1 à 2 ans) en bordure de surfaces assolées en grandes cultures et plantées de semences d'espèces annuelles et bisan-

nuelles) ; *f 4* **jachère** [f] **apicole** *agr. ecol.* (Forme particulière de la ►jachère fleurie ; il s'agit d'une parcelle semée avec des plantes pollinifères et/ou nectarifères ayant pour objectif de participer au renforcement des populations d'insectes butineurs en leurs assurant des approvisionnements plus réguliers en pollen de bonne qualité et en nectar et ainsi d'améliorer la biodiversité en favorisant les insectes pollinisateurs) ; *syn.* jachère [f] fleurie mellifère, prairie [f] mellifère, jachère [f] à intérêt apicole, jachère [f] pollinique ; *g 1* **Grünbrache** [f] (Phase eines landwirtschaftlichen Bodennutzungssystems mit angesäten Futterpflanzen oder spontaner Vegetation auf Brachland, das zwischen geernteten Kulturpflanzen und einer Neubestellung als extensives Weideland dient); *syn.* Halbbrache [f], Teilbrache [f]; *g 2* **Buntbrache** [f] *agr. ecol.* (*UB zu Grünbrache* eine sich mehrere Jahre selbst überlassene Fläche mit spontaner, vielfältiger Vegetation oder mit einer speziellen Samenmischung aus alten Kulturpflanzen und seltenen ein- und mehrjährigen Wildkräutern sowie Beimengung von entsprechenden Wirtspflanzen eingesäte stillgelegte Fläche, die als Lebensraum für gefährdete Tiere und Pflanzen, die in der intensiv genutzten Agrarlandschaft kaum noch einen Platz finden, dient. Erforderlichenfalls werden Pflanzen, welche in der Umgebung unerwünscht sind, mechanisch entfernt. Um eine Verbuschung zu verhindern, wird die Fläche etwa ab dem dritten oder vierten Jahr abschnittweise gemäht und später wieder als landwirtschaftliche Produktionsfläche bewirtschaftet); *g 3* **Wanderbrache** [f] *agr. ecol.* (Eine besondere Form der ►Buntbrache, bei der auf einem Feld streifenweise jedes Jahr ein Stück brachgelassen wird, während auf der anderen Seite des Feldes die gleichgroße Fläche wieder bewirtschaftet wird); *g 4* **Bienenweide** [f] **auf Stilllegungsflächen** *agr. ecol.* (Eine besondere Form der ►Buntbrache mit vielen Pollen und Nektar produzierenden Pflanzen, die Blütenbestäuber und Honigbienenbestände fördern und für diese eine reichhaltige hochqualitative Pollentracht und genügend Blütennektar vorhalten).

greenfields [npl] [UK] *leg. urb.* ►open land [US].

green-field site [n] [UK] *leg. urb.* ►open land [US].

2494 green finger connection [n] *landsc. urb.* (Generic term for a mostly narrow green area, which may have an avenue or walk, between built-up areas, or from such an area to a special point of interest or to the ►open country; ►green space corridor); *syn.* green wedge [n], connecting green finger [n]; *s* **conexión** [f] **verde (1)** (Zona verde, generalmente estrecha, paseo o avenida con función de comunicación para peatones o ciclistas que une otras zonas verdes o lugares de interés entre sí. Puede encontrarse en la ciudad o en el campo; ►campiña, ►corredor verde); *f* **coulée** [f] **verte** (Espace vert de faible largeur, bande boisée, allée ou espace de circulation pour les deux-roues ou les piétons bordés d'arbres, reliant entre eux des espaces verts, des points de référence urbains avec les zones bâties ou les espaces naturels ; ►campagne 2 ; cf. DIC 1993, 209 ; ►coupure verte) ; *syn.* espace [m] vert de liaison, cordon [m] vert ; *g* **Grünverbindung** [f] (Meist schmale Grünfläche, Allee oder von Bäumen markierter Weg mit vorwiegender Erschließungsfunktion für Fußgänger oder Radfahrer. Sie verbindet Grünflächen und andere städtebauliche Zielpunkte miteinander, mit Baugebieten oder der ►freien Landschaft; ►Grünzug).

2495 greenhouse effect [n] *ecol. envir.* (Increase in temperatures of the troposphere [lowest layer of the terrestrial atmosphere, to approx. 12 km height] and sea surfaces, caused by increasing carbon dioxide content [CO_2], primarily accelerated by the burning of fossil fuels for power production, industry and vehicles. The secondary most important ►greenhouse gas is methane [CH_4], which is released during drilling for oil and natural gas, but also occurs with the cultivation of wet rice. Further greenhouse gases are e.g. laughing gas [N_2O], which is released particularly in agriculture and the chemical industry are set free, as well as ozone [O_3] and chlorofluorocarbons [CFC]. These gases allow unhindered the passage of direct sunlight happen, absorb, however, a large part of the radiant heat reflected by the earth's surface and reflect half of it back again, down to the earth. Without the natural **g. e.** of water vapour and carbon dioxide, it would be more than 30 °C colder on earth. Natural, life-sustaining **g. e.** has gradually increased in strength since the Industrial Revolution due to human activity and comprises 50% CO2, 25% CFC and similar gases, 13% methane, 7% ozone, close to the surface and 5% N_2O; cf. ALG 2000; the term **g. e.** is often equated with the phenomenon of ▶global warming, ▶climatic change); *syn.* glasshouse effect [n]; *s* **efecto [m] invernadero** (Efecto de ciertos gases en la atmósfera que permiten el paso de la radiación solar de onda corta hacia la tierra y detienen, por el contrario, la salida del calor irradiado, en medida proporcional a su concentración. Esto tiene como consecuencia el aumento de la temperatura de la troposfera [capa inferior de la atmósfera hasta una altitud aproximada de 12 km] y de la superficie de los océanos. El principal ▶gas de efecto invernadero es el dióxido de carbono [CO_2], cuyo contenido en el aire ha aumentado desde 1850 de 280 ppm a 380 ppm, debido sobre todo a la combustión de combustibles fósiles para la producción de electricidad, para la industria, la calefacción y los vehículos, y a la de[s]forestación. El segundo grupo de gases de efecto invernadero más importante es el de los gases fluroclorocarbonados [FCC], el tercero es el metano [CH_4] que se emite en la prospección de petróleo y gas natural, así como en el cultivo de arroz y también en la digestión del ganado vacuno. Otros gases de efecto invernadero son el gas hilarante [N_2O], que se emite mayormente en la agricultura y en la industria química, así como el ozono [O_3]. Sin el **e. i.** natural del vapor de agua y del dióxido de carbono, la temperatura media de la tierra sería 30°C más fría. Este **e. i.** natural y favorable a la vida en la tierra ha aumentado desde la industrialización y se debe en un 50% al CO_2, un 25% a los FCC y gases similares, en 13% al metano, en 7% al ozono cercano al suelo y en 5% al gas hilarante. Muy frecuentemente se equipara erróneamente el **e. i.** al ▶calentamiento global; cf. DINA 1987; ▶cambio climático); *f* **effet [m] de serre** (1. Processus naturel par lequel les gaz présents dans l'atmosphère terrestre retiennent la chaleur du soleil, empêchant ainsi la planète de geler. 2. Élévation de la température de la troposphère [couche inférieure de l'atmosphère terrestre jusqu'à 12 km de la surface du globe] et à la surface de la mer provoquée par les activités humaines produisant des ▶gaz à effet de serre comme le dioxyde de carbone [CO_2] provenant principalement de la combustion d'énergies fossiles, le méthane libéré lors du raffinage du pétrole ou par l'élevage du bétail et la culture du riz, les chlorofluorocarbures utilisés dans les systèmes de réfrigération et climatisation et responsables du ▶réchauffement climatique ; les gaz à effet de serre [GES] sont transparents à certaines longueurs d'onde du rayonnement solaire qui pénètrent profondément dans l'atmosphère jusqu'à la surface du globe ; la partie du rayonnement absorbée par la Terre est restituée à son tour en direction de l'atmosphère sous forme de chaleur ; les GES et les nuages empêchent une partie de la chaleur de s'échapper l'emprisonnent ainsi près de la surface du globe, il s'ensuit une augmentation artificielle de l'effet de serre et provoque un réchauffement de l'atmosphère basse ; depuis le début de la révolution industrielle la concentration des GES a considérablement augmenté ; le dioxyde de carbone engendre environ 55 % de l'effet de serre dû à l'homme, le méthane de 15 %, le protoxyde d'azote [N_2O] de 5 %, les halocarbures — CFC 12, HCFC-22 — de 15 % ; si l'augmentation de la température moyenne de notre planète est d'environ 0,5 °C dans la seconde moitié du vingtième siècle, selon le rapport du **G**roupe **i**ntergouvernemental d'**e**xperts sur les

changements climatiques [GIEC] publié en janvier 2001, celui-ci pourrait atteindre 1,4 °C à 5,8 °C au cours du siècle à venir ; ▶changement climatique) ; *g* **Treibhauseffekt [m]** (1. Anstieg der Temperatur in der Troposphäre [unterste Schicht der Erdatmosphäre, bis ca. 12 km Höhe] und der Meeresoberflächen infolge des seit 1850 von 280 ppm auf 380 ppm zunehmenden Kohlendioxid-Gehaltes [CO_2] in der Luft, vornehmlich durch Verbrennung fossiler Brennstoffe für die Energiegewinnung, in der Industrie oder im Verkehr. Die zweitwichtigsten ▶Treibhausgase sind die Fluor-Chlor-Kohlenwasserstoffe [FCKW] gefolgt vom Methan [CH_4], das bei der Förderung von Erdöl und Erdgas freigesetzt wird, aber auch beim Nassreisanbau anfällt und bei der Verdauung in Mägen der Rinder produziert wird. Weitere Treibhausgase sind z. B. das Distickstoffoxid oder Lachgas [N_2O], das vor allem bei der Landwirtschaft und in der chemischen Industrie freigesetzt wird, sowie Ozon [O_3]. Diese Gase lassen das direkte Sonnenlicht ungehindert passieren, absorbieren aber einen großen Teil der von der Erdoberfläche reflektierten Wärmestrahlung und geben die Hälfte davon wieder nach unten ab. Ohne den natürlichen **T.** des Wasserdampfes und des Kohlendioxids wäre es auf der Erde über 30 °C kälter. Der natürliche, lebensbewahrende **T.** wird seit der Industrialisierung zusätzlich vom Menschen verstärkt und beruht zu 50 % auf CO_2, zu 25 % auf FCKW und ähnlichen Gasen, zu 13 % auf Methan, zu 7 % auf bodennahem Ozon und zu 5 % auf Lachgas N_2O; cf. ALG 2000; oft wird der Begriff **T.** verkürzt mit ▶globaler Erwärmung gleichgesetzt. 2. Ursprünglich wurde der Begriff verwendet, um den Effekt zu beschreiben, dass hinter Glasscheiben eines Gewächshauses [Treibhauses] und dadurch im Innenraum die Temperaturen ansteigen solange die Sonne darauf scheint ; ▶Klimaveränderung); *syn.* Glashauseffekt [m].

2496 greenhouse gas [n] *ecol. envir.* (Gas in the troposphere, which allows direct sunlight to pass unhindered, but absorbs a large part of the radiant heat reflected from the earth's surface and returns half of the heat back to the earth, thus contributing to a heating-up of the atmosphere. Anthropogenic **g. g.** which magnify the natural ▶greenhouse effect are primarily carbon dioxide [CO_2], methane [CH_4], laughing gas [N_2O], surface-near ozone [O_3] and chlorofluorocarbons [CFC]; ▶climatic change, ▶climatic protection, ▶global warming); *s* **gas [m] de efecto invernadero** (Gas en la troposfera que permite el paso de la luz solar hacia la tierra, pero absorbe una gran parte de la radiación de calor reflejada de la superficie de la tierra y la devuelve en un 50%, con lo que contribuye al aumento del ▶efecto invernadero natural. Entre los **gg. de e. i.** antropógenos se encuentran el dióxido de carbono [CO_2], el metano [CH_4], el gas hilarante [N_2O], el ozono cercano al suelo [O_3] y los compuestos fluroclorocarbonados [FCC]. La emisión y el empleo de estos gases están regulados en acuerdos internacionales [Protocolo de Montreal para el empleo de los FCC, Protocolo de Kyoto para la emisión de **gg. de e. i.**]; ▶calentamiento global, ▶cambio climático, ▶protección del clima); *f* **gaz [m] à effet de serre [GES]** (Les GES sont des gaz dont les propriétés physiques sont telles que leur présence dans l'atmosphère terrestre contribue à renforcer l'▶effet de serre naturel à la surface de la Terre ; les principaux GES naturels sont la vapeur d'eau [H_2O], le dioxyde de carbone [CO_2], le méthane [CH_4], le protoxyde d'azote [N_2O], l'ozone [O_3] ; les GES industriels sont les chlorofluorocarbonates [CFC] et [HCFC-22, perfluorométhane [CF_4], hexafluorure de soufre [SF_6] ; l'utilisation des GES est réglementée par divers accords internationaux [Protocoles de Montréal, Kyoto, etc.] ; les GES ont une nocivité différente ; c'est pourquoi les émissions de GES sont mesurées avec une unité commune — l'équivalent CO_2 — appelé **potentiel de réchauffement global [PRG]** ; celui-ci permet de déterminer les émissions produites par exemple par une entreprise et de réaliser dans le cycle industriel un bilan global de pollution prenant en compte les émissions directes et indirectes ; ►changement climatique, ►lutte contre la changement climatique, ►réchauffement global) ; *g* **Treibhausgas [n]** (Gas in der Troposphäre, das das direkte Sonnenlicht ungehindert passieren lässt, aber einen großen Teil der von der Erdoberfläche reflektierten Wärmestrahlung absorbiert und die Hälfte davon wieder nach unten abgibt und somit zur Aufheizung der Atmosphäre beiträgt. Zu den anthropogenen **T.en**, die den natürlichen ►Treibhauseffekt zusätzlich verstärken, gehören vornehmlich Kohlendioxid [CO_2], Methan [CH_4], Distickstoffoxid oder Lachgas [N_2O], bodennahes Ozon [O_3] und Fluor-Chlor-Kohlenwasserstoffe [FCKW]; ►globale Erwärmung, ►Klimaschutz, ►Klimaveränderung).

green infrastructure plan [n] [UK] *landsc. urb.* ►general plan for urban open spaces [US]/green open space structure plan [UK].

greening [n] [UK] *landsc.* ►planting treatment.

greening [n] **of built-up areas** *urb. landsc* ►landscaping of built-up areas.

2497 greening [n] **of the city** *landsc. urb.* (Popular slogan used in the promotion of more green in urban areas in order to reduce the urban heat island effect and to create a more habitable and supportive human environment); *s* **más verde [m] en la ciudad** (Slogan popular que reivindica la creación de más zonas verdes y la plantación de árboles, etc. en las aglomeraciones urbanas para mejorar la calidad del ambiente y así la habitabilidad de las mismas para la población); *f* **le végétal [m] dans la ville** ; *g* **Grün [n, o. Pl.] in der Stadt** (Slogan, der dazu aufrufen soll, weitere Grünflächen in Ballungsgebieten anzulegen, um das lokale Kleinklima zu verbessern und das Wohn- und Arbeitsumfeld für die Menschen attraktiver und lebenswerter zu machen).

2498 green island [n] *landsc. urb.* (Isolated green space surrounded by built environment); *s* **islote [m] verde** (Espacio verde en una ciudad que no está integrado en la trama verde de la misma); *f* **îlot [m] vert** (Espace ponctuel en zone urbaine non relié à une trame d'espaces libres) ; *g* **Grüninsel [f]** (Punktuelle Grünfläche in einem Stadtgebiet, die nicht an ein Freiraumsystem angeschlossen ist).

2499 green manure catch crop [n] *agr. hort.* (Interim vegetative method of establishing a quick soil cover between the growing of main crops, mostly with Leguminosae, such as Yellow Lupine *[Lupínus lúteus]*, Alexandrine Clover *[Trifolium alexandrinum]*, Crucíferae, such as White Mustard *[Sinapis alba]* and Radish *[Raphanus sativus var. oleiformis]* or Italian rye grass for stabilization, shading or keeping free of weeds on temporarily open ground as well as for mobilization of nutrients and improvement of soil structure; ►green manuring); *syn.* short term crop [n]; *s* **intercalado [m] de cultivos de abonado verde** (Método de cultivo secundario para establecer rápidamente una cobertura vegetal del suelo entre los cultivos principales, que generalmente se realiza con leguminosas como p. ej. altramuz *[Lupinus albus]*, alfalfa *[Medicago sativa]*, trébol *[Trifolium]*, o con crucíferas como la mostaza blanca *[Brassica hirta]* y el rábano *[Raphanus sativus var. oleiformis]*, y que también sirve para fijar el suelo, dar sombra o evitar el crecimiento de plantas advenedizas, así como para aumentar el contenido de nutrientes y mejorar la estructura del suelo; ►abonado verde, ►plantación intermedia); *f* **culture [f] d'engrais vert en dérobé** (Culture secondaire de plantes de courte durée [fourrages annuels ou engrais verts], intercalée entre deux cultures principales [chou fourrager repiqué entre une céréale d'hiver précoce et une betterave ou un maïs ; colza fourrager d'été ou moutarde semé entre une céréale et une culture fourragère de printemps], ensemencement de légumi-

neuses telles que p. ex. le Lupin jaune *[Lupínus lúteus]*, le Trèfle d'Alexandrie *[Trifolium alexandrinum]* ou de crucifères telles que la moutarde blanche *[Sinapis alba]* et le radis cultivé *[Raphanus sativus var. oleiformis]* afin de protéger la parcelle cultivée contre l'érosion et d'améliorer la fertilité du sol ; cf. LA 1981, 460 ; ►apport d'engrais vert, ►engrais vert, ►engrais vert en interculture) ; *syn.* engrais [m] vert en culture dérobée, engrais [m] vert dérobé, engrais [m] vert en interculture, interculture-CIPAN [f] (= culture intermédiaire piège à nitrates); *g* **Gründüngung [f] als Zwischenkultur** (Begrünung einer Acker- oder gärtnerischen Freilandfläche mit z. B. Leguminosen wie Gelbe Lupine *[Lupínus lúteus]*, Alexandriner Klee *[Trifolium alexandrinum]* oder Kreuzblütlern wie Weißer Senf *[Sinapis alba]* und Ölrettich *[Raphanus sativus var. oleiformis]* zur vorübergehenden Festlegung, Beschattung oder Unkrautfreihaltung von offenen Bodenflächen sowie zur Bodenaufschließung, Nährstoffversorgung und Verbesserung des Bodengefüges; ►Gründüngung, ►Zwischenbegrünung).

2500 green manure crop [n] *agr. constr. hort.* (Growing stand of green manure plants on a cultivated area; ►green manuring); *s* **cultivo [m] para abonado verde** (►abonado verde); *f* **culture [f] d'engrais vert** (Catégorie de végétaux [►engrais vert] cultivés ; ►apport d'engrais vert) ; *g* **Gründüngungspflanzen [fpl]** (Wachsender Bestand von **G.** auf einer Kulturfläche; ►Gründüngung); *syn.* Gründüngerpflanzen [fpl].

2501 green manure plant [n] *agr. constr. hort.* (Plant used for ►green manuring; e.g. luzerne *[Medicago sativa]*, clover *[Trifolium]*, scorpion weed *[Phacelia]*); *s* **planta [f] para abonado verde** (Planta utilizada como abono verde, como p. ej. alfalfa *[Medicago sativa]*, trébol *[Trifolium]*; ►abonado verde); *f* **engrais [m] vert** (1. Plantes fourragères légumineuses [le trèfle violet/incarnat — *Trifolium*, la luzerne — *Medicago sativa*] ou les crucifères [la moutarde, la phacélie — *Phacelia tanacetifolia*, la navette — *Brassica rapa*, le colza, etc.] à croissance rapide semées en arrière saison après une culture d'été, coupées et enfouies avant la culture de printemps pour enrichir le sol et en améliorer sa structure ; 2. le produit d'amendement dénommé **engrais vert** est aussi appelé « **culture intermédiaire piège à nitrates** » [CIPAN] ; ces couverts végétaux permettent d'éviter que les sols restent nus pendant l'hiver et d'empêcher le lessivage des nitrates entre deux cultures classiques ; ►apport d'engrais vert ; les espèces végétales susceptibles d'améliorer la fertilité du sol grâce aux organismes fixateurs d'azote sont souvent dénommées « améliorants » ; DIC 1993) ; *g* **Gründüngungspflanze [f]** (Pflanze, meist Hülsenfrüchtler, z. B. Luzerne *[Medicago sativa]*, Klee *[Trifolium]*, aber auch Büschelschön *[Phacelia]*, die auf zeitweilig unbestellten Flächen die obere Bodenschicht beschattet, den Boden vor Austrocknung und Erosion schützt, verhindert, dass Nährstoffe ausgewaschen werden und nach Einarbeitung in den Boden die Gare verbessert und das Bodenleben aktiviert; ►Gründüngung); *syn.* Gründünger [m], Gründüngerpflanze [f].

2502 green manuring [n] *agr. constr. hort.* (Method of soil improvement by which an herbaceous mass of plants, usually leguminous, is worked into the soil in its green state to improve the soil's structure and to increase nutrients in the upper layers; ►soil mellowness); *s* **abonado [m] verde** (Método de fertilización que consiste en la introducción de la masa herbácea de plantas leguminosas [abono verde o sideral] en el suelo en estado fresco para mejorar la *sazón* del mismo, su estructura y el contenido de nutrientes, y también para proteger el suelo contra la erosión); *f 1* **fertilisation [f] par (apport d')engrais vert** (Forme d'amendement organique dans la gestion de l'interculture utilisant les propriétés de certains végétaux [légumineuses, graminées, crucifères, Phacélie, etc.] afin de couvrir le sol et d'améliorer le ▶bon état de structure grumeleuse d'un sol, de lutter contre les adventices par effet de concurrence au moyen d'une culture [dérobée ou associée à la culture principale] destinée à être enfouie dans le sol après fauchage ou broyage ; *syn.* implantation [f] d'engrais vert, amendement [m] par (apport d')engrais vert ; ▶engrais vert) ; *f 2* **apport [m] d'engrais vert** (Processus) ; *syn.* implantation [f] d'engrais vert, incorporation [f] d'engrais vert ; *g* **Gründüngung [f]** (**1.** Düngungsart und Vorgang, bei der massenwüchsige Krautpflanzen, meist Hülsenfrüchtler *[Leguminosae]*, als Kurzzeitbegrünung eingesät und im grünen Zustand als Ganzes zur Verbesserung der ▶Bodengare und zur Anreicherung der oberen Bodenschichten mit Nährstoffen in den Boden eingearbeitet werden; **G.** aktiviert das Bodenleben, schützt vor Austrocknung, Verschlämmung und unterdrückt das Unkraut, dient oft auch dem Erosionsschutz, speziell im Winter während der Vegetationsruhe. **2.** Die zur Düngung verwendeten Kräuter werden **Gründünger**, ▶Gründüngungspflanzen oder **Gründüngerpflanzen** genannt).

green open space [n] *landsc. urb.* ▶green space.

2503 green open space ring [n] *landsc. urb.* (Narrow ▶greenbelt circling entire or portions of urban areas); *s* **anillo [m] verde** (Elemento de estructuración de la ciudad más estrecho que un ▶cinturón verde); *f* **boucle [f] verte** (Étroite ▶ceinture verte comme élément de structuration en milieu urbain) ; *syn.* anneau [m] de verdure, boucle [f] de verdure ; *g* **Grünring [m]** (Schmale, kreisförmige Freiraumzone als raumwirksames Element zur Stadtgliederung; ▶Grüngürtel); *syn.* Grünflächenring [m].

green open space structure plan [n] [UK] *landsc. urb.* ▶general plan for urban open spaces [US].

2504 green pruning [n] *agr.* (SAF 1983, 207; removal of live branches with leaves from a ▶fodder tree for use as forage or bedding for domestic animals); *syn.* summer pruning [n]; *s* **poda [f] en verde** (Corte de vástagos con hojas para utilizarlos de forraje o camada de animales; ▶árbol de forraje); *f* **élagage [m] en vert (±)** (Action de couper les branches vivantes d'un arbre pour récolter le feuillage utilisé comme litière ou fourrage ; DFM 1975, 107 ; ▶arbre d'émonde) ; *g* **Schneiteln [n, o. Pl.]** (Abschneiden von beblätterten Trieben eines Baumes zur Laubheu- oder Streugewinnung; schneiteln [vb]; ▶Schneitelbaum).

green roof construction [n] [US] *constr.* ▶roof garden system.

green roof installation [n] [US] *arch. constr. urb.* ▶roof planting.

green roof system [n] *constr.* ▶roof planting method.

green roof technology [n] [US] *constr.* ▶roof planting method.

2505 green separation zone [n] *plan. landsc.* (Area of planting which divides different type of land use; ▶buffer zone, ▶screen planting 2, ▶separator); *s* **verde [m] separador** (Banda de vegetación para separar diferentes usos; ▶banda de separación, ▶pantalla vegetal, ▶zona de amortiguación); *syn.* verde [m] amortiguador; *f* **bande [f] végétale de séparation** (Espace planté, de faible largeur, délimitant des utilisations différentes du sol ; ▶bande de séparation, ▶plantation brise-vue, ▶zone tampon) ; *syn.* espace [m] vert linéaire structurant, trame [f] verte structurante ; *g* **Trenngrün [n, o. Pl.]** (Grünstreifen zur Trennung unterschiedlicher Flächennutzungen; ▶Abstands- und Schutzzone 2, ▶Sichtschutzpflanzung; ▶Trennstreifen).

2506 green space [n] *landsc. urb.* (**1.** Generic term used in planning for primarily planted, public and private open spaces in urban areas such as parks, gardens, allotments, roadside planting,

cemeteries, sportsfields, play areas, camping grounds, open-air swimming pools, designated in development plans and green open space structure plans. **2.** Term for a public or private area within a populated area occupied primarily by planted or to be planted vegetation; ▶open space, ▶public green space, ▶residential green space); *syn.* landscaped open space [n] (TGG 1984, 167), green open space [n], green area [n]; **3. amenity green area [n]** (Specific term for the individual, completed facility, e.g. park, cemetery, allotment garden, external space of a public building); *s* **zona [f] verde** (**1.** Término genérico utilizado en planificación para espacios públicos o privados determinados predominantemente por la vegetación, como parques, jardines, cementerios, áreas de juegos y deportes, etc., y que son definidos como tales en los planes de ordenación urbana. **2.** Espacio público o privado que ofrece a los usuarios condiciones óptimas para la práctica del deporte y juegos, para paseos, esparcimiento y reposo, y en el que el elemento fundamental es la vegetación; cf. DINA 1987; ▶espacio libre, ▶parques y jardines públicos, ▶zonas verdes residenciales); *syn.* espacio [m] verde; *f* **espace [m] vert** (**1.** Terme générique qui recouvre les espaces non bâtis urbains, périurbains ou ruraux, privés ou publics, boisés ou non. **2.** Emplacements réservés à vocation de loisirs [parcs et jardins, squares, espaces naturels aménagés et jardins familiaux], plantations d'accompagnement [voiries, bâtiments publics, habitations, industries et commerces, etc.], équipements sportifs, cimetières, camping, sites et paysages naturels, espaces protégés, etc. pouvant apparaître sur les documents graphiques d'un plan local d'urbanisme [PLU]/plan d'occupation des sols [POS] ; ▶espace libre, ▶espace vert public, ▶espaces verts résidentiels) ; *g* **Grünfläche [f]** (**1.** Im planerischen Sprachgebrauch der OB für öffentliche und private vegetationsbestimmte Flächen wie z. B. Parkanlagen, Dauerkleingärten, Straßenbegleitgrün, Friedhöfe, Sport-, Spiel-, Zelt-, Badeplätze, die in Bauleitplänen und Grünordnungsplänen ausgewiesen werden/sind; ▶öffentliche Grünfläche. **2.** Vorwiegend mit Pflanzenbewuchs auszustattende oder ausgestattete öffentliche und private Fläche im Siedlungsbereich; ▶Wohngrün; der UB **Grünanlage** ist die einzelne, fertig gestellte Anlage, z. B. Parkanlage, Friedhof, Kleingartenanlage, Außenanlage eines öffentlichen Gebäudes; ▶Freiraum); *syn. bei Betonung des räumlichen Aspekts* Grünraum [m].

green space [n]**, neighborhood** *urb.* ▶proximity green space.

2507 green space corridor [n] *landsc. urb.* (Usually broad connected, linear green spaces for urban structuring, provision of various recreation facilities, and to improve air quality; ▶green finger connection); *s* **corredor [m] verde** (Zona verde, generalmente con forma de banda, que estructura la ciudad y puede dar espacio a diversos usos recreativos y sirve además para mejorar la calidad del aire; ▶conexión verde 1); *syn.* pasillo [m] verde (1); *f* **coupure [f] verte** (Espaces verts urbains boisés ou non, en général de forme linéaire, dont la fonction est de délimiter l'usage du sol afin d'assurer les différentes fonctions antiseptique et sanitaire, de loisirs, d'équilibre écologique ; ▶coulée verte) ; *syn.* corridor [m] vert ; *g* **Grünzug [m]** (Stadtgliedernde, in der Regel unterschiedlich breite, bandförmige Grünfläche mit diversen Nutzungs- und stadthygienischen Funktionen und unterschiedlichen Erholungsinhalten; ▶Grünverbindung).

2508 green space data bank [n] *adm. landsc.* (Inventory of all public green spaces and trees of a community with a detailed description of facilities by data processing for effective management and ▶open space planning; ▶green space survey, ▶tree data bank); *s* **inventario [m] de zonas verdes y de árboles** (Registro computarizado de todas las zonas verdes y todos los árboles de una ciudad con indicación detallada de su equipa-

G

miento y su estado. Tiene como fin facilitar la organización del trabajo de mantenimiento y sirve además de base para la ►planificación de espacios libres y para la creación de un sistema de reticulación de biótopos; ►inventario de árboles urbanos, ►inventariación de zonas verdes); *f* **fichier [m] informatique de description des espaces verts** (Description de l'ensemble des espaces verts et arbres d'ornement d'une ville comprenant des données détaillées sur les différentes utilisations des espaces et ayant pour objectif grâce à l'informatique d'améliorer l'►aménagement des espaces libres et la gestion des travaux d'entretien par les services techniques ; ►fichier informatique de description des arbres, ►inventaire du patrimoine vert) ; *g* **Grünflächendatei [f]** (Erfassung sämtlicher Grünflächen und Bäume einer Stadt mit detaillierten Angaben über die Ausstattung der Flächen mit Hilfe der Datenverarbeitung zur Steuerung der betriebswirtschaftlichen Organisation des Unterhaltungs- und Pflegebetriebes. Eine Grünflächendatei dient außerdem als Grundlage für die städtische ►Freiraumplanung und zur Erarbeitung eines Biotopverbundsystems; ►Bestandsaufnahme von Grünflächen, ►Baumdatei); *syn.* Grünflächenkataster [**m**, *in A nur so* — oder **n**].

2509 green space feature [n] [US] *landsc. urb.* (Component part of urban open space distinguished by its vegetation); *syn.* green townscape feature [n] [UK]; *s* **elemento [m] natural (de diseño)** (Componente de vegetación de parque o jardín urbano); *f* **composante [f] végétale** (Élément naturel de l'espace urbain) ; *syn.* élément [m] de verdure ; *g* **Grünelement [n]** (Vegetationsbestimmter Einzelbestandteil von städtischen Freiräumen).

green space fringe [n] [US] *landsc. urb.* ►greenbelt [US].

green space planning [n] *landsc. urb.* ►general urban green space planning.

green space policy [n] [UK] *pol. landsc. urb.* ►open space policy [US].

green spaces [npl] *adm. landsc. pol. recr. urb.* ►maintenance of green spaces, ►provision of green spaces, ►requirements for green spaces, ►supply of developed green spaces, ►travelled way green spaces, ►urban green spaces.

green space strategy [n] [UK] *landsc. urb.* ►general urban green space planning.

green space strategy plan [n] [UK] *landsc. urb.* ►general plan for urban open spaces [US]/green open space structure plan [UK].

2510 green space survey [n] *adm. landsc. urb.* (Inventory of public green spaces in a community, which is recorded in the ►green space data bank); *s* **inventariación [f] de zonas verdes** (Registro de las áreas verdes de una ciudad cuyo resultado puede ser un ►inventario de zonas verdes y de árboles); *f* **inventaire [m] du patrimoine vert** (Relevé des différentes catégories d'espaces verts urbains ; ►fichier informatique de description des espaces verts) ; *syn.* inventaire [m] des espaces verts ; *g* **Bestandsaufnahme [f] von Grünflächen** (Erfassung der Grünflächen einer Stadt/Gemeinde. Das Ergebnis ist die ►Grünflächendatei); *syn.* Bestandserfassung [f] der Grünflächen.

green townscape feature [n] [UK] *landsc. urb.* ►green space feature [US].

2511 green waste [n] *envir.* (Recyclable refuse material of vegetable origin occurring in house, allotment and fruit and vegetable gardens, as well as on agriculturally-used land or in the landscape construction business; the material is broken down by micro-organisms, soil-borne organisms or enzymes; *generic term* ►biodegradable waste; *specific term* yard waste [US]/garden waste [UK]); *s* **residuos [mpl] verdes (≠)** (Material residual

reciclable de origen vegetal de jardines privados, huertos o huertos recreativos urbanos, de parques y jardines públicos así como de granjas agrícolas o viveros de plantas, que puede ser descompuesto por microorganismos, organismos edáficos o enzimas; *término genérico* ►residuos biodegradables); *f* **déchets [mpl] verts** (Résidus végétaux issus des activités horticoles, de l'entretien et du renouvellement des espaces verts publics et privés [parcs et jardins, terrains de sport, etc. des collectivités territoriales, des organismes publics et parapublics, des sociétés privées et des particuliers] ; *terme générique* ►biodéchets) ; *g* **Grünabfall [m]** (Verwertbare Reste pflanzlicher Herkunft aus Haus-, Nutz- und Kleingärten, landwirtschaftlichen Betrieben oder Unternehmen des Garten- und Landschaftsbaues, die durch Mikroorganismen, bodenbürtige Lebewesen oder Enzyme während eines Rotteprozesses abgebaut werden; *OB* ►Bioabfall; *UB* Gartenabfall).

greenway [n] [US] *lands. urb.* ►urban greenway.

green wedge [n] *landsc. urb.* ►green finger connection.

gregariousness [n] *phyt.* ►plant sociability.

grey dune [n] [UK] *geo. phyt.* ►gray dune [US]/grey dune [UK].

grid [n] *constr.* ►metal grid, ►planting grid, ►triangular planting grid, ►turf grid, ►wooden grid.

grid [n] [UK]**, concrete tree** *constr.* ►concrete tree grate [US].

grid [n] [UK]**, light well** *constr.* ►window well grate [US].

grid [n]**, quincunx planting** *constr.* ►triangular planting grid.

grid [n]**, square** *constr.* ►square planting grid.

grid [n] [UK]**, tree** *constr.* ►tree grate [US].

grid [n]**, tree planting** *constr. hort.* ►tree planting pattern.

grid [n]**, triangular** *constr.* ►triangular planting grid.

grid block [n] [US] *constr.* ►grass paver.

grid cover [n] *constr.* ►cover slab.

gridiron system [n] [US] *constr.* ►grid pattern drainage.

2512 grid pattern drainage [n] *constr.* (Drainage system, whereby ►lateral drains follow the gradient of a slope and are tied to collectors; ►herringbone pattern drainage, ►subsurface drainage, ►subsurface drainage system); *syn.* gridiron system [n] [also US]; *s* **drenaje [m] longitudinal** (Sistema de ►drenaje subterráneo en el que los ►drenes secundarios siguen la pendiente of la ladera; ►drenaje a espina de pez, ►sistema de drenaje subterráneo); *f* **drainage [m] longitudinal** (►Drainage souterrain dans lequel les ►drains adducteurs sont disposés dans le sens de la plus grande pente ; ►drainage transversal, ►système de drainage souterrain) ; *g* **Längsdränung [f]** (►Dränung, bei der die ►Sauger mit dem Hanggefälle verlaufen; ►Dränsystem, ►Querdränung).

2513 grid pavement [n] [US] *constr.* (TSS 1988, 840-17; surfacing for fire-engine access, garage drives, parking lots. etc. constructed of ►grass-pavers laid on a subbase, voids of which are filled with sandy topsoil and sown with a hard-wearing grass mixture; ►crushed aggregate lawn [US], ►grass-filled modular paving [US]/grass setts paving [UK]); *syn.* surfacing [n] of lattice concrete blocks [also US] (TGG 1984, 150), pavement [n] of grass pavers [UK]; *s* **revestimiento [m] de adoquines pavicésped** (Pavimento de ladrillos con agujeros que se utiliza para revestir accesos de garajes, entradas de bomberos, aparcamientos, etc.; ►césped sobre piedra partida, ►adoquín pavicésped, ►pavimento de adoquines con césped); *f 1* **revêtement [m] de dalles gazon** (Surface stabilisée par ►dalles gazon pour allées résidentielles, parcs de stationnement, accès de garages, voies pompiers, etc.) ; *f 2* **béton [m] gazonné** (Dalle béton munie d'alvéoles permettant de garder un aspect gazonné aux voie d'accès, zone de parking, etc. : ►gazon sur grave concassée, ►pavage engazonné) ; *g* **Rasengittersteinbelag [m]** (Belag aus Betonhohlsteinen oder Tonrasenziegeln mit Erdkammern zur Befestigung von Feuerwehrwegen, Garagenzufahrten, Parkplätzen etc.; in F. gibt es noch den *béton gazonné*: das sind in Ortbeton ausgesparte Erdkammern für die Rasenansaat; ►Rasengitterstein, ►Rasenpflaster, ►Schotterrasen); *syn.* Rasensteinbelag [m].

2514 grid planting [n] **of reed stems** *constr.* (Rows of *Phragmites* reed canes planted in a network for ►stabilization of sand dunes or sandy spoil dumps [US]/spoil tips [UK]); *s* **estabilización [f] con carrizo** (Estabilización de dunas móviles o montes de escoria por medio de hileras cruzadas de *Phragmites* formando redes; ►fijación de dunas); *f* **treillis [m] d'ancrage de Roseaux (±)** (Dispositif de ►fixation des dunes mouvantes ou des surfaces sablonneuses des terrils, réalisé par ancrage sur le sol de rangées de chaumes de Roseaux *[Phragmites]* mis en œuvre pour former un treillis) ; *g* **Reetbesteck [n]** (Netzförmig gesteckte Reetreihen aus Schilfhalmen *[Phragmites]* zur Befestigung von Wanderdünen oder sandigen Bergehalden; ►Festlegung von Dünen).

grill [n] *constr.* ►metal grid.

grill [n], **steel** *constr.* ►steel grate.

grill [n], **tree** *constr.* ►tree grate [US]/tree grid [UK].

grille [n] *constr.* ►metal grid.

grille [n] [UK], **concrete tree** *constr.* ►concrete tree grate [US].

grille [n], **tree** *constr.* ►tree grate [US]/tree grid [UK].

grille [n], **steel** *constr.* ►steel grate.

2515 grille cover [n] *constr.* (Grating of metal bars for shoe-scraper, cast iron drain with grating [US]/cast iron channel with grating [UK]; ►cast iron drain with grating, ►steel grate, ►storm drain grate [US]/gully grating [UK], ►tree grate [US]/tree grid [UK], ►window well grate); *syn.* grate cover [n]; *s* **rejilla [f] de cubierta** (Pequeña reja que sirve para cubrir canalillos de desagüe o de quitabarros; ►canalón de desagüe cubierto con rejilla, ►parrilla de pozo de luz, ►rejilla de acero, ►rejilla de entrada [de alcantarilla], ►rejilla [metálica] para alcorque); *syn.* parrilla [f] de cubierta; *f* **grille [f]** (Élément métallique recouvrant les dispositifs d'assainissement de voirie, les entourages d'arbre, etc. ; ►caniveau à grille, ►grille de cour anglaise, ►grille [de protection de pied] d'arbre, ►grille de regard d'évacuation, ►grille maille métallique) ; *syn.* grille [f] caillebotis ; *g* **Gitterrost [m]** (Abdeckung für Fußabstreifer oder ►Kastenrinnen oder Oberflächenentwässerungseinrichtungen; ►Baumscheibenrost, ►Einlaufrost, ►Lichtschachtrost, ►Stahlgitterrost; *OB* Abdeckrost); *syn.* Gitterabdeckung [f].

grillwork [n] [US] *constr.* ►lattice.

2516 grip length [n] *arch. constr.* (Length of a reinforcing bar or rod to be anchored in concrete); *syn.* bond length [n]; *s* **longitud [f] de anclaje** (Largo de las barras de refuerzo o de las varillas de armadura en el hormigón); *f* **longueur [f] de scellement** (Longueur des ferraillages dans le béton assurant la liaison d'armatures ; cf. COB 1984, 149) ; *g* **Einbindetiefe [f] (2)** (**1.** Maß, das ein Bauelement in einen Baukörper eingebunden ist, z. B. Bewehrungsstäbe in Beton; **2. E.** für Mauersteine ►Einbindetiefe 1); *syn.* Ankerlänge [f], Verankerungslänge [f].

grit chamber [n] *constr. hydr.* ►sand arresting trap [US].

groove vegetation [n] [US] *phyt.* ►limestone groove vegetation.

gross floor area [n] *leg. plan. urb.* ►floor area.

ground [n] *geo.* ►character of the ground; *constr.* ►gradually sloping ground; *recr.* ►playground; *gard'hist.* ►pleasure ground; *agr. constr. hort. pedol.* ►soil 1; *recr.* ►sports ground.

ground [n], **burial** *landsc.* ►burial area.

ground [n], **character of the** *geo.* ►nature of the terrain.

ground [n], **forest grazing** *agr. for.* ►forest pasture.

ground [n] [UK], **hunting** *recr.* ►game preserve [US].

ground [n], **level** *landsc.* ►flat terrain.

ground [n] [UK], **short-stay camping** *recr.* ►short-stay campground [US].

ground [n] [UK], **static caravan** *recr.* ►permanent campground [US].

ground [n], **wintering** *zool.* ►hibernation site.

ground anchor [n] *constr.* ►deadman.

2517 ground-branching tree [n] [US] *arb. hort.* (Young tree with laterals close to the ground. In **U.K.**, such trees [feathered trees] are typically young and container-grown, with an upright central leading shoot and a stem furnished with evenly spread and balanced lateral growths down to near ground level, according to species such as silver birches, poplars and alders. They are supplied in sizes 1.75 to 2 metres and 2 metres to 2.5 metres; it is important that the leader is retained: a forked or double leader would create a point of weakness. A feathered tree will branch naturally and symmetrically on its own. As the tree develops, the lowest laterals can be removed in stages over two or three years, retaining the lower laterals to thicken up the main trunk, finally leaving a short, clear stem; ►strong branched liner [US]/one-year-old feathered tree [UK]); *syn.* feathered tree [n] [UK]; *s* **leñosa [f] ramificada desde la base (≠)** (D. denominación utilizada en horticultura para árbol joven ramificado desde la base y que ha sido transplantado por lo menos dos veces en vivero; ►leñosa joven ramificada desde la base); *f 1* **baliveau [m]** (Plant greffé ou jeune arbre obtenu par semis, bouturage ou marcottage, avec une hauteur supérieure à 150 cm, cultivé à distance et présentant une tige comportant généralement peu ou pas de branches latérales et une flèche verticale. Le végétal ligneux est toujours considéré comme **b.** tant que la circonférence du tronc à 1 m du sol n'atteint pas 6 cm ; au-delà il devient une « tige » ; cf. VRD 1994, 5/2.3, p. 1 ; ►jeune baliveau) ; *f 2* **baliveau [m] ramifié** (Terme correspondant à la catégorie allemande de qualité des végétaux de pépinières ornementales *Stammbusch* ; baliveau présentant une tige centrale bien définie, d'où partent des pousses latérales régulièrement réparties, à partir du sol jusqu'au sommet ; le diamètre de la motte doit représenter au moins trois fois la circonférence du tronc, mesurée au collet ; ENA-F 1996, 5, 35) ; *g 1* **Heister [m]** (D., in der Baumschule mindestens 2 x verpflanzter junger Baum, nach Höhe von 125-150 bis 250-300 cm, Stammumfang in 30 cm Höhe 5-6 cm, ohne Krone und mit Seitenholz vom Boden an garniert; dem natürlichen Wuchs der betreffenden Baumart entsprechend tief bezweigt. Bei Angeboten, Ausschreibungen und Rechnungen gilt seit 09/1995 als Sortierkriterium nur noch die Höhenangabe; Lieferung mit oder ohne Ballen sowie im Container; ►leichter Heister); *g 2* **Stammbusch [m]** (►Heister mit einem Stammumfang [StU] von mindestens 12 cm und einer Höhe von mindestens 250 cm; PGL 2005).

2518 ground breeder [n] *zool.* (Bird nesting on the ground); *syn.* ground-nesting bird [n]; *s* **ave [f] que anida en el suelo** (Ave que construye su nido sobre la tierra); *f* **nicheur [m] au sol** (Oiseau qui construit un nid ouvert sur le sol, cavité creusée dans la terre et tapissée ou non de matériaux) ; *syn.* oiseau [m] qui niche sur le sol ; *g* **Bodenbrüter [m]** (Vogel, der auf dem Boden sein Nest baut).

G

ground check [n] *rem'sens. surv.* ▶field check.

2519 ground civil engineering [n] *constr. eng.* (Section of the building industry predominantly concerned with the construction of roads, paths, sewer lines and other underground utilities, tunnels, railroads [US]/railways [UK], bridges as well as with earthworks and hydraulic engineering; ▶building construction, ▶civil engineering works); *s 1* **ingeniería** [f] **civil** (Parte de la construcción que se dedica en especial a las obras a nivel del suelo o en el subsuelo, es decir construcción de carreteras, canales, puentes, vías de ferrocarril, túneles, conducciones subterráneas, etc.; ▶edificación); *syn.* ingeniería [f] de caminos, canales y puertos; *s 2* **obras** [fpl] **públicas** (Construcciones de infraestructura que generalmente son realizadas por las administraciones públicas; ▶trabajos de obras públicas); *f* **travaux** [mpl] **publics** (Secteur d'activités économiques qui par comparaison au ▶bâtiment est spécialisé dans les ▶travaux de voirie et de réseaux divers) ; *g* **Tiefbau** [m] (**1.** [o. Pl.]; Teilbereich des Bauwesens, das sich im Vergleich zum ▶Hochbau vornehmlich mit Bauarbeiten zu ebener und unter der Erde befasst, z. B. mit der Herstellung von Straßen, Wegen und Plätzen sowie die Verlegung der Kanalisation und unterirdischer Leitungen, den Erd-, Grund-, Tunnel-, Gleis-, Brücken- und Wasserbau plant und durchführt. Eine exakte Trennung zwischen **T.** und Hochbau ist kaum möglich. **2.** ...bauten [pl]; Bauwerk an oder unter der Erdoberfläche; ▶Tiefbauarbeiten. **3.** *min.* Im Bergbau wird der Abbau von Lagerstätten nutzbarer Minerale unter Tage im Gegensatz zum Tagebau auch Tiefbau genannt).

2520 ground compaction [n] *constr.* (Compression of the soil surface, which results in compaction and consolidation of soil particles; doing this with a hand tamper is called 'tamping'; compaction with a plate vibrator or vibrating plate compactor is also carried out mechanically by a machine-operated tamper fitted with a flat base designed for compacting soil, asphalt, and similar materials and typically used when laying foundations for roads, sidewalks, and any other areas that need to be overlaid with material. A plate vibrator is driven by a gas-powered unit that fits on top of the bottom plate); ▶soil compaction); *s* **compactación** [f] **superficial del suelo** (Compresión de los centímetros superiores del suelo para consolidarlo; si se realiza con un pisón a mano se le dice «apisonado», para realizarlo mecánicamente se utiliza un vibrador; ▶compactación del suelo); *syn.* apisonado [m] del suelo; *f* **compactage** [m] **superficiel** (Compression des matériaux de la couche superficielle d'un sol, par pilonnement au moyen d'une pilonneuse, par vibration au moyen d'une plaque vibrante ; ▶compactage du sol, ▶tassement du sol) ; *g* **oberflächliche Verdichtung** [f] (Verdichtung, die nur die oberen Zentimeter des Bodens erfasst; wird diese Tätigkeit mit einem Handstampfer ausgeführt, so wird das 'stampfen', mit einer Rüttelplatte 'abrütteln' genannt; ▶Bodenverdichtung).

2521 ground cover [n] *phyt.* (Low, spreading vegetative cover of the ground; ▶degree of total vegetative cover, ▶vegetation layer); *syn.* surface vegetation [n] (VDE 1980); *s* **cobertura** [f] **vegetal del suelo** (Vegetación rastrera que cubre el suelo; ▶estrato de vegetación, ▶expansión horizontal); *syn.* cubierta [f] vegetal del suelo, tapiz [m] vegetal del suelo; *f* **couvert** [m] **végétal du sol** (Végétation recouvrant le sol ; ▶degré de recouvrement, ▶strate de végétation) ; *syn.* tapis [m] végétal, couverture [f] (végétale) du sol ; *g* **Bodenbedeckung** [f] (Niedriger Bewuchs des Bodens; ▶Deckungsgrad, ▶Vegetationsschicht); *syn.* Bodendecke [f], Pflanzendecke [f].

ground cover [n], **moss** *phyt.* ▶moss floor.

ground coverage [n] *leg. urb.* ▶built portion of a plot/lot.

2522 ground cover perennial [n] *hort. plant.* (Generic term for the prostrate growth habit of herbaceous plants with regard to the required design effect and their usage in planting perennial, which forms a compact vegetation cover and thus hinders the germination of weeds, the intrusion of other unwanted plants and helps to prevent soil erosion. In design terms the impact of large areas of one species or variety is important [colour, texture, etc.]. Although other taller perennials may also cover the soil, **g. c. p.s** are known for their usually low, creeping, carpetforming growth, e.g. New Zealand Burr *[Acaena]* or Waldsteinia *[Waldsteinia ternata]*. Substitutes for **g. c. p.s** are species without stolons or only short stolon growth, when these are planted more densely, e.g. *Waldsteinia geoides*; less suitable are tussocks of grasses, which make the character of the planting less effective, if the grasses are planted densely, e.g. Fountain Grass *[Pennisetum]*, Autumn moor grass *[Sesleria autumnalis]*); *s* **planta** [f] **vivaz tapizante** (Término genérico para plantas perennes que sirven para cubrir el suelo al crear una capa cerrada de vegetación y así impiden la erosión y el crecimiento de plantas advenedizas); *syn.* vivaz [f] tapizante, perenne [f] tapizante; *f* **plante** [f] **vivace couvre-sol** (Plante vivace naine utilisée dans de nombreuses situations dans la composition d'un jardin pour son port étalé et son effet décoratif ; plante à croissance basse qui recouvre le sol de manière dense ; elle a l'avantage de limiter le développement des adventices et de protéger le sol de l'érosion ; *terme spécifique* plante vivace tapissante, [de quelques cm à moins de 20 cm de hauteur], au port tapissant, comme la lampourde de Magellan *[Acaena buchananii]*, la petite pervenche *[Vinca minor]*, la gaulthérie couchée *[Gaulteria procumbens]*, les orpins *[Sedum]*, etc.) ; *syn.* vivace [f] couvre-sol ; *g* **bodendeckende Staude** [f] (*Terminus für das Wuchsverhalten einer Staude hinsichtlich ihrer gestalterischen Wirkung und gärtnerischen Verwendung* Staude, die einen geschlossenen Bewuchs bildet und dadurch keimende Unkräuter, Fremdaufwuchs und die Bodenerosion verhindert. Für die Gestaltung ist der art-/sortenspezifische Flächeneindruck [Färbung, Textur] wichtig. Obwohl auch höhere, nicht aufkahlende Stauden den Boden bedecken, wird eine **b. S.** gewöhnlich als niedrige, durch ober- oder unterirdische Ausläufer teppichbildende Staude verstanden, wie z. B. Stachelnüsschen *[Acaena]*, Golderdbeere *[Waldsteinia ternata]*. Stellvertretend sind nicht oder nur kurze Ausläufer bildende Arten in entsprechend höherer Pflanzdichte geeignet, wie z. B. Golderdbeere *[Waldsteinia geoides]*; weniger geeignet sind Horstgräser, wenn ein wirkungsvoller Formcharakter durch Dichtpflanzung verunklart wird, z. B. Lampenputzergras *[Pennisetum]*, Herbst-Kopfgras *[Sesleria autumnalis]*); *syn.* Flächendeckstaude [f], Staude [f] für Flächendeckung.

2523 ground cover plant [n] *hort. plant.* (Small plant, herbaceous or woody, often prostrate, that covers the ground in place of turf and prevents weed growth and soil erosion; ▶carpetforming ground cover plant, ▶creeping ground cover plant); *syn. abbreviation* ground cover [n]; *s* **coberturas** [fpl] (CMJ 1982; plantas que crecen hasta 50 ó 60 cm y que se usan para cubrir superficies, decorar, bordear canteros y controlar la erosion; ▶rastrera, ▶tapizante); *f* **plante** [f] **couvre-sol** (Désignation des espèces végétales vivaces ou arbustives naines plantées pour former un couvert dense et limiter ainsi le développement des adventices et réduire les riques d'érosion ; suivant leur hauteur on distingue 1. les plantes dites tapissantes ou gazonnantes [moins de 20 cm], 2. les couvre-sols de taille moyenne [20 à 40 cm] et 3. les couvre-sols de grande taille [entre 40 et 60 cm] ; ▶plante rampante, ▶plante tapissante) ; *syn.* couvre-sol [m], végétal [m] couvre-sol ; *g* **Bodendecker** [m] (Gärtnerische Bezeichnung für niedrige flächendeckende Stauden [< 20 cm] und niedrige Gehölze [10-60 cm], die gepflanzt werden, um einen geschlossenen Bewuchs zu bilden und dadurch keimende Unkräuter, Fremdaufwuchs und die Bodenerosion verhindern; ▶kriechender

Bodendecker, ▶teppichbildender Bodendecker); *syn.* bodendeckende Pflanze [f], Flächendecker [m].

2524 ground cover rose [n] *hort. plant.* (Rose which grows compactly and is suitable for large areas of planting, thus preventing weed growth during the main growing season; the word is somewhat misleading, since it implies that the roses are low-growing plants. Depending on their strength they may reach a height of 130cm and they can also be planted individually or in groups. In the 1990s a collection of low bush roses was started in Dresden Pillnitz, which has made a valuable contribution in rose trials); *s* **rosal [m] cubresuelo** (Rosal de crecimiento compacto, adecuado para plantaciones de superficie, ya que cubre todo el suelo y no permite el crecimiento de plantas advenedizas. El adjetivo «cubresuelo» es engañoso, ya que —dependiendo de su vigor— los **rr. cc.** pueden llegar hasta 130 cm de altura); *syn.* rosal [m] rastrero; *f* **rosier [m] couvre-sol** (Dans la classification des rosiers, catégorie de rosier de végétation petite à moyenne en hauteur [0,30 m à 0,80 m], produisant de long rameaux, sans être toutefois sarmenteux ; grâce à leur port tapissant ils recouvrent rapidement le sol et conviennent autant pour l'espace restreint d'une jardinière que pour le recouvrement d'un talus ; ils sont parfois dénommés **rosiers à massifs** quand ils sont utilisés pour des plantations en masse) ; *syn.* rosier [m] rampant ; *g* **Bodendeckerrose [f]** (In der Rosenklassifizierung ein Verwendungstyp kompakt wachsender Rosen, die zur Flächenpflanzung geeignet sind und in der Hauptvegetationszeit kein Unkraut hochkommen lassen sollen; das Wort **B.** ist insofern problematisch, da es den oft zutreffenden Eindruck einer niedrigen oder kriechenden Pflanze erweckt; aber gelegentlich werden auch Pflanzen bis zu 130 cm Wuchshöhe als **B.** angeboten und verwendet. Manche Sorten können auch als Beetrosen, Strauchrosen oder sogar als Kletterrosen verwendet werden. In Dresden Pillnitz gibt es seit den 1990er-Jahren die Kleinstrauchrosenvergleichspflanzung, die einen wichtigen Beitrag zur Qualitätssichtung liefert); *syn.* flächendeckende Rose [f], *Pillnitzer Sprachgebrauch* Kleinstrauchrose [f].

ground cover vegetation [n] *plant.* ▶shrubby ground cover vegetation.

2525 groundfill [n] **around a tree** *arb. constr.* (Raised soil level at the base of an existing tree or a newly-planted one, which for most tree species usually leads to their death; ▶overfilling of root zone); *syn.* earth fill [n] around a tree, raised soil level [n] near a tree trunk (cf. ARB 1983, 172); *s* **relleno [m] con tierra (alrededor del tronco)** (Elevación del nivel del suelo alrededor de un árbol existente que en la mayoría de los casos produce la muerte del mismo; ▶recubrimiento de la zona de raíces); *f* **enfouissement [m] du tronc** (Remblaiement de terre sur plusieurs dm du pied d'un arbre provoquant la mort chez la plupart des espèces ; ▶recouvrement de la zone racinaire) ; *g* **Stammeinschüttung [f]** (Bodenaufschüttung am Stammfuß, die bei vielen Baumarten bei großer Schütthöhe zum Absterben führt; die Rotbuche *[Fagus sylvatica]* verträgt überhaupt keine Anschüttung, auch kein Überfahren des Wurzelbereiches!; ▶Überfüllung des Wurzelbereiches).

ground floor [n] [UK] *arch. constr.* ▶first floor [US].

ground height [n]**, existing** *constr. eng. plan.* ▶existing ground level.

2526 ground-level ... [adj] *met.* (▶ground level air layer); *syn.* close to the ground [loc] [also UK], at ground level [loc]; *s* **al ras del suelo** [loc] (▶capa de aire al ras del suelo); *f* **au ras du sol** [loc] ; (▶couche d'air à proximité du sol) ; *syn.* à proximité du sol [loc] ; *g* **bodennah** [adj] (▶bodennahe Luftschicht).

ground level [n] *constr. eng. plan.* ▶existing ground level.

2527 ground-level air layer [n] *met.* (TGG 1984, 78; air layer up to 2m thick above ground surface. In the dividing line between earth and atmosphere are mostly existing other conditions than above the 2m layer above ground. In this layer the wind velocity is slowed down by friction; the mix of air is reduced, the ground surface absorbing sun rays and radiating warmth is both heat source and cold source for the adjacent air. Furtheron this very restricted layer is characterized by special conditions through dust, water vapor and gases which are escaping the soil; ▶climate near the ground, ▶microclimate); *s* **capa [f] de aire al ras del suelo** (Capa de aire de hasta 2 m sobre la superficie del suelo. En la zona de transición entre la tierra y la atmósfera reinan generalmente condiciones muy diferentes a las del aire más arriba. La fricción del aire con el suelo reduce la velocidad del mismo y por ello la mezcla de aire; la superficie del suelo absorbe la radiación solar y refleja calor y se convierte tanto en fuente de calor como de frío para el aire vecino y además polvo, vapor de agua y gases del suelo crean condiciones especiales en espacios muy reducidos; ▶clima de las capas cercanas a la superficie terrestre, ▶microclima); *f* **couche [f] d'air à proximité du sol** (Couche d'une épaisseur maximale de 2 m située au-dessus du sol ; dans la couche de transition située entre la terre et l'atmosphère règnent en général des conditions différentes que dans les couches basses supérieures ; p. ex. la modification du profil de vitesse du vent ainsi que du mélange vertical de l'air en raison des effets du frottement au sol [sous couche rugueuse], l'absorption ou la perte de chaleur de la surface du sol suite au rayonnement solaire, la dispersion des poussières, la variation de la teneur en vapeur d'eau et les flux de gaz émis par le sol ; ▶climat de la couche d'air à proximité du sol, ▶microclimat) ; *g* **bodennahe Luftschicht [f]** (Luftschicht bis 2 m über der Oberfläche des Erdbodens. In der Grenzschicht von Erde und Atmosphäre herrschen meist wesentlich andere Bedingungen vor, als über 2 m, da z. B. durch die Reibung am Boden die Windgeschwindigkeit und damit auch die Durchmischung der Luft abnimmt, die Sonnenstrahlen absorbierende und Wärme ausstrahlende Bodenoberfläche bald eine Wärmequelle, bald eine Kältequelle für die anliegende Luft ist sowie Staub, Wasserdampf und aus dem Boden austretende Gase die besonderen Verhältnisse auf kleinstem Raum ausmachen; cf. GEI 1961; ▶Klima der bodennahen Luftschicht, ▶Mikroklima).

2528 ground modeling [n] [US]/**ground modelling** [n] [UK] *constr. plan.* (Earth grading to reshape the ground surface to proposed levels and contours; ▶ground restoration); *syn.* contouring [n] and terrain model(l)ing [n] [also US] (LA 10/89, 35), contour smoothing [n] [also US], earth shaping works [n] [also UK] (SPON 1986, 156), grading [n], ground shaping [n] [also US], shaping [n] the ground [also US], grading design [n] [also US], land shaping [n], site contouring [n] [also UK] (SPON 1986, 121); *s* **modelado [m] del terreno** (*Proceso y resultado* transformación del relieve de la superficie de un terreno por obras de construcción; modelar [vb] el terreno; ▶reestablecimiento de la superficie del terreno); *syn.* modelación [f] del terreno; *f 1* **mise [f] en forme des terres** (*Processus* exécution de travaux de terrassement afin d'amener le terrain aux cotes projetées ; ▶rétablissement des lieux) ; *f 2* **modelé [m] de sols** (Résultat du modelage. Le vallonnement se réalise par la mise en forme d'un terrain en relief avec vallons et buttes) ; *g* **Bodenmodellierung [f]** (1. Durchführen von Erdbauarbeiten zur Schaffung/Ausformung einer Geländeoberfläche in geplanten Höhen; ▶Wiederherstellung der Geländeoberfläche. 2. Ergebnis der Erdbauarbeiten zur Geländemodellierung); *syn. zu 1.* Herstellung [f] der Erdmodellierung, Modellieren [n, o. Pl.] eines Geländes; *syn. zu 1. u. 2.* Erdmodellierung [f], Geländeausformung [f], Geländemodellierung [f], Höhenmodellierung [f].

G

2529 ground moraine [n] *geo.* (Debris carried at the base of a glacier and deposited as a horizontal sheet of till and overridden by a glacier; cf. GGT 1979); *s* **morrena [f] de fondo** (Conjunto de materiales que van siendo abandonados por el hielo en su base, formada tapizando toda la base rocosa por descargas continuas y regulares; DINA 1987); *f* **moraine [f] de fond** (Accumulation de débris de roches situées sous la glace au contact du lit du glacier) ; *g* **Grundmoräne [f]** (An der Gletscherbasis bewegter Moränenschutt, der meist als sandig-toniger Mergel mit Geschieben horizontal abgelagert wurde).

ground-nesting bird [n] *zool.* ▶ground breeder.

ground preparation [n] *constr. hort. agr.* ▶soil preparation.

ground radiation [n] *met.* ▶night of ground radiation.

2530 ground relief [n] *geo.* (Actual configuration of the Earth's surface being measured by the difference in elevation between highest and lowest level. It should not be confused with topography, though American usage of 'topographical relief' is permissible; cf. DNE 1978); *s* **relieve [m] (1)** (Este término alude a los contrastes topográficos de la superficie terrestre que tienen carácter destacado, de resalte, frente a aquellas otras zonas de su entorno que se toman como base de referencia. Aquí el relieve es sinónimo de elevación y, por tanto, se enfrenta a términos como llanura, depresión e incluso topografía regular; DINA 1987); *f* **relief [m] (1)** (L'importance d'un relief se mesure à son étendue et à sa hauteur ou commandement, appréciée à la verticale à partir de sa base ; DG 1984) ; *g* **Relief [n] (1)** (Höhenverhältnisse der Erdoberfläche).

ground reshaping [n] *constr.* ▶ground restoration.

2531 ground restoration [n] *constr.* (Earth regrading to restore a ground surface to its original contours; ▶ground modeling [US]/ground modelling [UK]); *syn.* ground reshaping [n]; *s* **reestablecimiento [m] de la superficie del terreno** (Trabajos de terracerías para restaurar la forma original de la superficie de un terreno; ▶modelado del terreno; reestablecer [vb] la superficie del terreno); *f* **rétablissement [m] des lieux** (Travaux de terrassement afin de rétablir un terrain dégradé dans son état antérieur ; C. urb., art. L. 480-4 ; ▶mise en forme des terres) ; *g* **Wiederherstellung [f] der Geländeoberfläche** (Erdarbeiten, die ein zerstörtes Gelände in den alten Zustand versetzen; ▶Bodenmodellierung).

2532 ground settlement [n] *constr. geo. min.* (**1.** Subsidence due to a building's own weight upon the formation level, or due to a sinkage of the ground level in, e.g. an ▶area subject to mining subsidence. **2.** Geomorphological downward displacement of part of the Earth's surface relative to its surroundings. **3.** Sinking of terrain as a result of deep mining; ▶basin 1, ▶karstification); *syn.* subsidence [n] (TGG 1984, 99); *s 1* **asentamiento [m] [AL]** (Hundimiento del terreno debido al peso propio de un edificio o causado por trabajos de minería en una ▶zona de subsidencia); *syn.* asiento [m] [Es, RA]; *s 2* **subsidencia [f] [Es, CO]** (**1.** Fenómeno de hundimiento lento al que son sometidos los geosinclinales y las cuencas de sedimentación como consecuencia de una acumulación progresiva de materiales; DGA 1986. **2.** Hundimiento del terreno causado por trabajos de minería en el subsuelo o por procesos naturales; el término específico «colapso» se aplica cuando el hundimiento es rápido; ▶cuenca de sedimentación, ▶karstificación, ▶zona de subsidencia); *syn.* hundimiento [m] (del terreno) [CO, EC, MEX, RA, YV], descenso [m] [RCH]; *f 1* **subsidence [f]** (**1.** Terme géomorphologique désignant le processus d'affaissement lent d'une partie de la surface terrestre. **2.** Affaissement d'un terrain, provoqué par l'écroulement du plafond d'une cavité souterraine naturelle ou artificielle [subsidence minière] ; ▶bassin d'effondrement, ▶bassin sédimentaire, ▶karstification, ▶zone de subsidence

minière) ; *syn.* affaissement [m], effondrement [m] ; *f 2* **tassement [m]** (Déformation verticale d'un sol par compression sous la charge d'un ouvrage ou d'engins agricoles, par le piétinement) ; *g* **Senkung [f]** (**1.** Durch Eigenlast eines Bauwerkes entstehendes Absinken des Baugrundes oder z. B. das Absinken eines Geländes in einem ▶Bergsenkungsgebiet. **2.** Geomorphografischer Begriff für den Vorgang einer langsamen oder plötzlichen, regionalen oder lokalen Absenkung eines Teils der Erdoberfläche; diese kann durch natürliche geologische/tektonische Vorgänge eine größere oder kleinere Geländevertiefung [Senke] hervorrufen [▶Senkungsbecken], durch chemische Vorgänge im Untergrund mit löslichen oder ausspülfähigem Gestein [▶Verkarstung] oder durch Tauprozesse Hohlformen erzeugen; **3.** Absacken eines Geländes durch Bruchbergbau; sich senken [vb], sich setzen [vb], absacken [vb]; ▶Bergsenkungsgebiet, ▶Senkungsbecken); *syn.* Absenkung [f] (1), Sackung [f], Setzung [f] (1).

ground shaping [n] [US] *constr. plan.* ▶ground modeling [US]/ground modelling [UK].

ground shoot [n] *hort.* ▶cane.

2533 ground sill [n] *wat'man.* (Horizontal structure on the bottom of a watercourse designed to reduce flood velocities by creating stillwater reaches upstream to dissipate energy; ▶checkdam, ▶loose rock dam); *s* **solera [f] de fondo** (Estructura construida al nivel del fondo de un cauce para evitar la socavación; ▶dique transversal, ▶solera de piedras sueltas); *syn.* durmiente [m], rastra [f] [MEX]; *f* **seuil [m] transversal d'un cours d'eau** (Protection d'un cours d'eau par la mise en place d'un exhaussement transversal en bois, en enrochement, en béton sur le fond du lit afin de réguler le débit solide, de recréer une dissipation d'énergie et de protéger le fond contre les affouillements ; ARI 1986 ; ▶ouvrage de confortement transversal, ▶seuil en enrochements) ; *syn.* seuil [m] antiérosion (ARI 1986, 34), gradin [m] d'un cours d'eau, seuil [m] de calage, seuil [m] de fond ; *g* **Sohlenschwelle [f]** (Mit dem Fließgewässerboden abschließendes ▶Querwerk, das ein Eintiefen des Gewässerbettes verhindern soll; ▶Steinschwelle); *syn.* Grundschwelle [f], Sohlschwelle [f].

2534 ground sinkage [n] *constr.* (Subsidence of the surface level due to closer compaction of soil particles or moisture loss; ▶settled soil, ▶soil loosening); *s* **asentamiento [m] del terreno** (Ligero hundimiento de la superficie del suelo debido a la compactación de las partículas del mismo; ▶descompactación del suelo, ▶suelo asentado); *syn.* asiento [m] del terreno; *f* **tassement [m] naturel (du sol)** (Abaissement progressif naturel du niveau des masses de terre meuble rapportées sous leur propre poids ; ▶ameublissement du sol, ▶sol tassé naturellement) ; *g* **Setzung [f] (2)** (Absinken der Einbauhöhe von eingebauten, lockeren Bodenmassen durch natürliche Verdichtung oder durch Verlust der Bodenfeuchte; ▶Bodenauflockerung 2, ▶gesetzter Boden).

ground surface [n] *constr. plan.* ▶natural ground surface.

ground surface [n]**, existing** *constr. plan.* ▶natural ground surface.

2535 groundwater [n] [US]/**ground water** [n] [UK] *hydr. wat'man.* (Subsurface water accumulated up to great depths in pores and holes of loose sedimentary deposits of the earth's upper crust, and forming a natural reservoir of potable water; this water arises from infiltration of precipitation as well as penetration of river water through the channel bed; ▶aquifer, ▶watertable); *syn.* phreatic water [n], subterranean water [n], underground water [n] (GGT 1979, 235); *s* **agua [f] subterránea** (Término genérico; agua almacenada en el subsuelo entre rocas o sedimentos que se diferencia del agua edáfica en que aquélla se encuentra en la zona saturada; ►acuífero, ►nivel freático); *f* **nappe [f] phréatique** (Eau emmagasinée dans des formations géologiques perméables sédimentaires des nappes alluviales [fonds de vallée, grands bassins sédimentaires] ou dans les roches, s'infiltrant dans la zone non saturée du réservoir d'eau souterraine pour atteindre la surface de la nappe phréatique [zone saturée] ; en plaine alluviale une ►formation aquifère peut être aussi alimentée par les eaux fluviales filtrées ; *termes spécifiques* nappe libre, nappe captive ; ►niveau de la nappe phréatique) ; *syn.* nappe [f] souterraine, nappe [f] aquifère ; *g* **Grundwasser [n]** (Das im Untergrund angesammelte, bis in größere Tiefen in Hohlräumen der Lockersedimente und des Gesteins der oberen Erdkruste vorkommende Wasser. Es entsteht durch Versickerung des Niederschlagswassers und durch das Eindringen des Flusswassers aus dem Gerinnebett [Uferfiltrat]; ►Grundwasserleiter, ►Grundwasserspiegel).

groundwater [n] *hydr. wat'man.* ►bank seepage groundwater.

groundwater contamination [n] *envir. wat'man.* ►groundwater pollution [US].

groundwater depletion [n] *wat'man.* ►groundwater exploitation [US], ►groundwater exhaustion [US].

2536 groundwater exhaustion [n] [US]/**ground water exhaustion** [n] [UK] *hydr. nat'res. wat'man.* (Continued withdrawal of water from groundwater at a rate greater than the rate of replenishment; ►groundwater exploitation [US]); *syn.* groundwater depletion [n] [US]/ground water depletion [n] [UK]; *s* **agotamiento [m] del agua subterránea** (Disminución continua del agua de un embalse subterráneo a una velocidad superior a la de alimentación; cf. WMO 1974; ►explotación de las reservas subterráneas); *syn.* extinción [f] del agua subterránea; *f* **tarissement [m] de la nappe** (Exploitation excessive d'eau souterraine dans une formation aquifère sans réalimentation suffisante par infiltration, provoquant une décroissance de débit ; ►surexploitation des réserves en eau souterraine) ; *syn.* épuisement [m] de la nappe phréatique ; *g* **Grundwassererschöpfung [f]** (Exzessives Fördern von Grundwasser ohne Möglichkeit einer natürlichen Regeneration; ►Grundwasserausbeutung); *syn.* Erschöpfung [f] des Grundwassers.

2537 groundwater [n] **exploitation** [US]/**ground water exploitation** [n] [UK] *wat'man.* (Withdrawal from a groundwater reservoir in excess of the average rate of replenishment; WMO 1974; ►lowering of groundwater table, ►groundwater extraction [US]/ground water extraction [UK]); *syn.* groundwater depletion [n] [US]/ground water depletion [n] [UK]; *s* **explotación [f] de las reservas subterráneas** (Extracción de agua de un acuífero más allá de la capacidad de regeneración, de manera que se produce una ►reducción del nivel freático; ►extracción de agua subterránea); *f* **surexploitation [f] des réserves en eau souterraine** (Prélèvement d'eau d'un aquifère en quantité supérieure à sa capacité de réalimentation et provoquant un déséquilibre permanent entre la ressource et les besoins ; ►prélèvement d'eau souterraine, ►rabattement de la nappe phréatique) ; *g* **Grundwasserausbeutung [f]** (Grundwasserentnahme, die das natürliche Regenerationsvermögen in den grundwasserführenden Schichten übersteigt; ►Grundwasserabsenkung, ►Grundwasserförderung); *syn.* Ausbeutung [f] des Grundwassers.

2538 groundwater extraction [n] [US]/**ground water extraction** [n] [UK] *wat'man.* (Pumping of groundwater from saturated soil or aquifer); *syn.* groundwater withdrawal [n] [US]/ground water withdrawal [n] [UK]; *s* **extracción [f] de agua subterránea** *syn.* captación [f] de las reservas subterráneas, alumbramiento [m] de agua subterránea; *f 1* **prélèvement [m] d'eau souterraine** (Terme générique désignant les différents types de prélèvement sur les eaux souterraines tels que l'usage de l'eau des sources [captage de source], les opérations dans les nappes d'eau souterraines [puits, forage] ; *syn.* exploitation [f] (des réserves) d'eaux souterraines ; *f 2* **captage [m] d'eau souterraine** (*Terme spécifique* captage d'une source ou pompage des eaux souterraines destinées à la consommation humaine et soumis à autorisation administrative (prélèvement > 8 m³/h pour les particuliers) ; *g* **Grundwasserförderung [f]** (Hochpumpen von Wasser aus einem Grundwasserleiter); *syn.* Grundwassergewinnung [f], Grundwasserentnahme [f].

2539 groundwater flow [n] *hydr.* (Movement of water in an aquifer); *s* **flujo [m] de aguas subterráneas** (Movimiento del agua en un acuífero); *f 1* **écoulement [m] souterrain** (Mouvement horizontal de l'eau dans une nappe aquifère) ; *f 2* **drainance [f]** (Flux d'eau, à composante essentiellement verticale, passant d'un niveau aquifère à un autre à travers une couche semi-perméable ; DG 1984) ; *g* **Grundwasserfluss [m]** (Horizontale Wasserbewegung in einem Grundwasserleiter); *syn.* Grundwasserstrom [m].

groundwater flow [n]**, velocity of** *hydr.* ►groundwater velocity [US].

groundwater level [n] *hydr.* ►watertable; ►fluctuation of groundwater level, ►natural rise of groundwater level.

2540 groundwater pollution [n] [US]/**ground water pollution** [n] [UK] *envir. wat'man.* (Contamination of groundwater by human activities, mostly occurring in populated areas, due to hazardous waste, leachate from waste dumps and the accumulation of over a century of industrial production as well as agricultural use [pollution caused by nitrates, pesticides and their metabolites — products of metabolism, biodegraded products, etc]. The European Groundwater Directive 2006/118/EC, was introduced on 12.12.2006 for the protection of groundwater against pollution and deterioration; it lays down the minimum requirements for groundwater protection for agriculture, waste management, transport, pharmaceutical industry and plant protection agents with deadlines by which they must be complied with; ►remediation of groundwater pollution, ►risk of groundwater pollution); *syn.* groundwater contamination [n] [US]/ground water contamination [n] [UK]; *s* **contaminación [f] del agua subterránea** (Polución del agua freática causada por actividades humanas, generalmente en relación con la urbanización [suelos industriales contaminados, aguas de percolación de depósitos de residuos y desarrollo industrial de más de 100 años] y por la agricultura [entrada de nitratos, pesticidas y sus sustancias de descomposición]. El 12.12.2006 entró en vigor la Directiva 2006/118/CE relativa a la protección de las aguas subterráneas contra la contaminación y el deterioro, que define los criterios para evaluar el estado químico de las aguas subterráneas, procedimientos de evaluación de las mismas, la determinación de las tendencias significativas y sostenidas al aumento y definición de los puntos de partida de las inversiones de tendencia, así como medidas para prevenir o limitar las entradas de contaminantes en las aguas subterráneas; ►peligro de contaminación de acuífero, ►saneamiento puntual de aguas subterráneas); *syn.* polución [f] del agua subterránea; *f* **contamination [f] de la nappe phréatique** (Présence de dans la nappe phréatique de substances minérales ou organiques nocives ou indésirables comprenant les micro-organismes, les minéraux dissous dont la présence est naturelle, les produits chimiques fabriqués par l'homme et les minéraux radioactifs ; la pollution de la nappe phréatique est principalement liées aux activités humaines et peut être classifiée selon l'origine de la contamination [industrielles, transport, activités agricoles, collectivités] ou leur type : pollutions énergétiques [radioactives], pollutions mécaniques [apports massifs de

G

matières en suspension], les pollutions organiques [eaux usées domestiques, industrielles et épandages agricoles], les pollutions par hydrocarbures, les pollutions chimiques [métaux lourds, phytosanitaires, toxiques divers], les pollutions bactériologiques [contamination lors de crues importantes] ; ▶résorption de la pollution de la nappe, ▶vulnérabilité de la nappe aux pollutions) ; *syn.* pollution [f] de la nappe phréatique ; *g* **Grundwasserserschmutzung [f]** (Verunreinigung des Grundwassers durch menschliche Aktivitäten, meist im Zusammenhang mit der Siedlungsentwicklung [Altlastenthematik, Sickerwässer der Müllkippen und über 100 Jahre Industrieproduktion], und der landwirtschaftlichen Nutzung [Eintrag von Nitraten, Pestiziden und deren Metaboliten — Stoffwechselprodukte, Abbau- und Reaktionsprodukte]. Am 12.12. 2006 wurde zum Schutze des Grundwassers vor Verschmutzung und Verschlechterung die EG-Grundwasserrichtlinie [GWRL] RL 2006/118/EG verabschiedet, die Mindestanforderungen an den GW-Schutz für Landwirtschaft, Abfallwirtschaft, Verkehr, Arznei- und Pflanzenschutzmittelzulassungen beinhaltet resp. zeitliche Vorgaben, bis wann diese erarbeitet sein sollen, vorgibt; ▶Sanierung der Grundwasserverschmutzung); *syn.* Grundwasserkontamination [f].

groundwater pollution [n], remedial measures for *envir. wat'man.* ▶remediation of groundwater pollution.

groundwater recession [n] *hydr.* ▶natural groundwater recession.

2541 groundwater recharge [n] *hydr. wat'man.* (Natural or artificial process, by which water is added from outside to the zone of saturation of an aquifer; cf. WMO 1974; ▶aquifer recharge area); *syn.* groundwater replenishment [n]; *s* **recarga [f] de acuíferos** (Procesos naturales o artificiales por los cuales se aporta agua del exterior a la zona de saturación de un acuífero, bien directamente a la formación o bien indirectamente a través de otra formación; WMO 1974; ▶área de alimentación de un acuífero); *syn.* alimentación [f] de acuíferos, alimentación [f] del agua subterránea; *f* **alimentation [f] d'une nappe phréatique** (Réapprovisionnement en eau de la zone de saturation d'une formation aquifère ou emmagasinement d'eau souterraine par des processus naturels [▶bassin d'alimentation d'un aquifère] ou des méthodes artificielles ; WMO 1974 ; *termes spécifiques* alimentation artificielle d'une nappe, recharge artificielle d'une nappe) ; *syn.* réalimentation [f] d'une nappe phréatique, réapprovisionnement [m] d'une nappe phréatique ; *g* **Grundwasseranreicherung [f]** (Natürliche oder anthropogene Zuführung von Wasser in wasserführende Schichten; ▶Einzugsgebiet eines Grundwasserleiter); *syn.* Grundwasserergänzung [f], Grundwassererneuerung [f], Grundwasserneubildung [f], Grundwasserregeneration [f].

2542 groundwater regime [n] *hydr.* (Natural hydraulic system with variations in alimentation from watercourses, lakes and precipitation); *s* **balance [m] hídrico del agua subterránea** (Cuantificación de los flujos de agua, la capacidad de almacenamiento y las condiciones de tiempo y funcionamiento en relación con el agua subterránea en una zona determinada o en un acuífero; cf. DINA 1987); *syn.* régimen [m] hidrológico del agua subterránea; *f* **régime [m] hydraulique de la nappe phréatique** (Système de variations naturelles entre les débits des apports [infiltration efficace, alimentation par un cours d'eau] et ceux des écoulements dans un système hydrologique délimité [consommation en période de sécheresse] déterminant la quantité d'eau de la nappe disponible) ; *g* **Grundwasserhaushalt [m]** (Natürliches hydraulisches Beziehungsgefüge zwischen der „Rücklage" aus versickerndem Niederschlag und Zuführungsrate aus Uferfiltraten der Oberflächengewässer und dem „Aufbrauch" in Trockenzeiten, das die Menge des vorhandenen Grundwassers in einem Gebiet bestimmt); *syn.* Grundwasserregime [n].

groundwater replenishment [n] *hydr. wat'man.* ▶groundwater recharge.

2543 groundwater reservoir [n] *hydr.* (▶Aquifer where groundwater is stored and may later be extracted and used; cf. WMO 1974); *syn.* groundwater storage [n]; *s* **embalse [m] de agua subterránea** (▶Acuífero en el cual el agua subterránea se almacena en cantidad considerable por lo que puede ser aprovechada posteriormente); *f* **réservoir [m] d'eau souterraine** (▶Formation aquifère dans laquelle est emmagasinée de l'eau souterraine qui pourra être extraite et utilisée ultérieurement ; WMO 1974) ; *syn.* réservoir [m] aquifère ; *g* **Grundwasserspeicher [m] (1)** (▶Grundwasserleiter, der auf Grund seines Hohlraumvolumens Wasser speichert, das später genutzt werden kann).

2544 groundwater resources [npl] *hydr.* (Occurrence of an existing amount of ground water in a defined area; ▶available groundwater resources); *s* **recursos [mpl] hídricos subterráneos** (Total de agua freática disponible en una zona determinada; ▶recursos acuíferos sustentablemente aprovechables); *f* **gîte [m] aquifère** (Quantité d'eau présente dans une couche souterraine dans une zone géologique ou hydrogéologique donnée, susceptible d'être exploitée en vue de l'alimentation en eau de la population ; p. ex. gîte aquifère alluvial ; ▶ressource en eau souterraine) ; *g* **Grundwasservorkommen [n]** (In einem bestimmten Gebiet vorhandene Grundwassermenge; ▶Grundwasserdargebot).

2545 groundwater seepage [n] [US]/ground water seepage [n] [UK] *geo. hydr.* (Diffused, non-point source discharge of ground water to the ground surface); *syn.* effluent seepage [n] (WMO 1974); *s* **filtración [f] efluente** (Caudal de aguas subterráneas que se propaga a la superficie del terreno, WMO 1974); *f* **suintement [m] (1)** (Zone d'écoulement diffus, permanent ou temporaire, des eaux de la nappe sur un versant ou au pied d'un relief ; c'est une zone humide ponctuelle caractérisée par la forte présence de mousses perpétuellement détrempées et de fougères. En milieu calcaire, le dépôt de calcium dissout peut former des concrétions incrustantes appelées « tuf » ou « travertin ») ; *syn.* source [f] de filtration, filtration [f] effluente (WMO 1974, 300), émergence [f] de suintement, sourcin [m] ; *g* **diffuser Grundwasseraustritt [m]** (Heraussickern des Grundwassers an mehreren Stellen, statt eine eindeutige Quelle zu bilden); *syn.* diffuser Austritt [m] des Grundwassers.

2546 groundwater soil [n] [US]/ground water soil [n] [UK] *pedol.* (▶Hydromorphic soil characterized by a high groundwater level or ▶alluvial soil which often is influenced by water pressure from high-water level of a watercourse; ▶water-saturated gley sometimes has its groundwater level up to the ground surface; ▶soil affected by impeded water); *s* **suelo [m] semiterrestre de inundación** (Según la clasificación alemana, suelo caracterizado por el alto nivel de la capa freática o ▶suelo aluvial caracterizado por la presión de agua producida por crecidas frecuentes. En el caso del ▶gley de moder y del gley anmoor el agua freática llega en parte hasta la superficie del suelo; ▶suelo de agua estancada, ▶suelo semiterrestre); *f* **sol [m] à nappe phréatique profonde** (▶Sol hydromorphe caractérisé par un engorgement permanent provoqué par la présence ou la remontée de la nappe phréatique comme pour les ▶sols alluviaux pendant la période des crues. Dans les gley à anmoor, les ▶gley oxydés humifères la nappe peut affleurer par intermittence la surface du sol ; ▶sol à nappe temporaire perchée) ; *syn.* sol [m] à nappe phréatique permanente ; *g* **Grundwasserboden [m]** (Boden, der durch hoch anstehendes Grundwasser oder ▶Auenboden, der oft durch Hochwasser bewirktes Druckwasser geprägt ist. Beim ►Nassgley und Anmoorgley steht das Grundwasser zeitweilig bis zur Bodenoberfläche; ►Grundwasser- und Stauwasserböden, ►Stauwasserboden).

groundwater storage [n] *hydr.* ►groundwater reservoir; ►recharged groundwater storage.

groundwater surface [n] *hydr.* ►watertable.

groundwater table [n] *hydr.* ►depth of groundwater table; *constr. wat'man.* ►lowering of groundwater table.

groundwater table [n] [US]**, drawdown of** *constr. wat' man.* ►lowering of groundwater table.

2547 groundwater velocity [n] [US]/**ground water velocity** [n] [UK] *hydr.* (Rate at which groundwater flows through a particular soil as determined by hydraulic conductivity, effective porosity, and hydraulic gradient calculated according to Darcy's Law); *syn.* velocity [n] of groundwater flow [US]/velocity [n] of ground water flow [UK]; *s* **velocidad [f] de flujo de las aguas subterráneas** (Razón según la cual fluye el **a. s.** a través de un suelo específico y que está determinada por la conductividad hidráulica, la porosidad efectiva y el gradiente hidráulico calculado según la Ley de Darcy); *f* **vitesse [f] d'écoulement de l'eau d'un aquifère** (PCE 1990, 22) ; *syn.* vitesse [f] de déplacement de l'eau d'un aquifère, vitesse [f] de circulation de l'eau d'un aquifère ; *g* **Fließgeschwindigkeit [f] des Grundwassers** (Geschwindigkeit, mit der das Grundwasser durch ein bestimmtes poröses Medium auf einer definierten Fließstrecke auf Grund der hydraulischen Durchlässigkeit, der Porosität und des hydraulischen Gradienten oder Potentialgefälles fließt).

groundwater withdrawal [n] *wat'man.* ►groundwater extraction.

2548 groundwater yield [n] [US]/**ground water yield** [n] [UK] *hydr.* (Groundwater quantity which can be pumped during a given period of time; ►available groundwater resources, ►potential groundwater yield); *s* **rendimiento [m] hídrico** (Cantidad de agua que puede ser extraída de un acuífero en un periodo/período de tiempo definido; ►potencial hídrico de agua subterránea, ►recursos acuíferos sustentablemente aprovechables); *f* **débit [m] d'une nappe** (Volume d'eau traversant par unité de temps, une section définie d'un aquifère, sous l'effet d'un gradient hydraulique donné ; GHDA 1993 ; ►potential hydrique d'un aquifère, ►ressource en eau souterraine ; *syn.* rendement [m] d'une nappe ; *g* **Grundwasserergiebigkeit [f]** (Wassermenge, die in einem bestimmten Zeitraum gefördert werden kann; ►Grundwasserdargebot, ►Grundwasserhöffigkeit).

groundworks [npl] [UK] *constr. eng.* ►earthworks.

2549 group [n] **of trees** *landsc.* (Bunch of trees standing together; ►grove, ►woodlot [US&CDN] 2); *syn.* clump [n] of trees, cluster [n] of trees (TGG 1984, 193); *s 1* **grupo [m] de árboles** (Conjunto de árboles que crecen juntos; ►arboleda herbosa, ►bosquecillo 2); *s 2* **bosquete [m]** (Pequeño bosque artificial en parque o jardín); *f 1* **groupe [m] d'arbres** (Ensemble de plusieurs arbres, p. ex. dans un parc, jardin ou paysage rural ; ►bois, ►boqueteau 1, ►boqueteau 2, ►bosquet, ►parquet) ; *syn.* bouquet [m] d'arbres ; *f 2* **massif [m] d'arbres** (Composition d'un groupe dense d'arbres sur un espace de dimension variable) ; *g* **Baumgruppe [f]** (Einzelne, zusammenstehende Bäume; ►Hain, ►Wäldchen).

2550 group overnight accommodation [n] *recr.* (Lodging for travelling holiday/vacation groups in hotels, inns, etc.; ►hostelry); *s* **hospedaje [m] (colectivo)** (Alojamiento de grupo en cualquier tipo de centro hotelero, ►albergue, hostal de juventud, etc.); *f* **hébergement [m] collectif** (Action d'héberger des groupes, des organismes ou associations de vacanciers

dans des auberges de jeunesse, etc. ; ►gîte) ; *g* **Beherbergung [f] von Gruppen (≠)** (Unterbringung von Reisegruppen, Ferienvereinen; ►Beherbergungsstätte); *syn.* Unterbringung [f] von Gruppen.

2551 grouting [n] *constr.* (Finishing of joints in pavement or tiled surfaces with semi- liquid or a workable mortar; ►pointing); *s* **rejuntado [m] con mortero líquido** (Relleno de juntas de un pavimento de losas o de azulejos; ►rejuntado); *f* **jointoiement [m] avec une barbotine** (Remplissage des joints de dallage avec un matériau de liaison liquide ; ►jointoiement) ; *g* **Verfugen [n, o. Pl.] mit flüssigem Mörtel** (Verfüllen von Fugen eines Plattenbelages oder einer gekachelten Fläche ; ►Verfugen).

2552 grove [n] *agr. landsc.* (Relatively large group of trees within a settled area, without an understor[e]y shrub cover, either planted or natural; a **g.** on agricultural land used to be used for fattening pigs. In the Mediterranean olive trees are cultivated in an 'olive grove'); *syn.* bosk [n]; *s* **arboleda [f] herbosa** (Bosquete sin sotobosque arbustivo situado dentro o cerca de zona poblada, pero también en superficies agrícolas antiguamente utilizadas para la cría de puercos. En el Mediterráneo pertenecen a esta categoría los olivares y los encinares, melojares o las fresnedas utilizadas como ►dehesas; cf. DINA 1987); *f* **boqueteau [m] (1)** (F., au sens de l'►inventaire forestier national, toute surface, d'une largeur d'au moins 25 m, comprise entre 50 ares et 5 ha, où l'état boisé est acquis ; TSF 1985 ; D., espace arboré de faible étendue, dépourvu de strate arbustive, à proximité des zones urbaines ou définissant p. ex. certains espaces boisés agricoles [engraissement des porcs] ou la culture des olives en zone méditerranéenne ; *terme spécifique* oliveraie) ; *syn.* petit bois [m] ; *g* **Hain [m]** (Strauchloser, lichter Baumbestand im Siedlungs- oder siedlungsnahen Bereich, aber auch z. B. eine landwirtschaftlich genutzte Fläche — früher für die Schweinemast — oder für Olivenbäume im Mittelmeergebiet; *UBe* Eichenhain, Obsthain, Olivenhain).

grower [n] *hort.* ►plant nursery.

growing in great colonies [loc] *phyt.* ►growing in large groups.

2553 growing in groups, small patches, or cushions [loc] *phyt.* (According to Braun-Blanquet's classifycation the third category of ►plant sociability pertaining to plants, which grow closely together either as individuals or in groups of shoots on the surface of the ground); *s* **creciendo en grupos mayores** [loc] (En la clasificación de Braun-Blanquet 3ª categoría de ►sociabilidad o modo de apruparse de los individuos de una especie; BB 1979); *syn.* creciendo en pequeños rodales o almohadillas [loc]; *f* **croissant en troupes** [loc] (Unité de l'échelle de ►sociabilité selon Braun-Blanquet concernant la disposition des individus [ou des pousses] d'une même espèce, à l'intérieur d'une population donnée, croissant en troupes/en touffes moyennes, espacées) ; *g* **truppweise wachsend [ppr/adj]** (*Nach Braun-Blanquet Skalierungseinheit der* ►*Häufungsweise* Pflanzen einer Art betreffend, die als Individuen oder in Form der Gruppierung oberirdischer Sprosse kleine Flecken oder Polster bilden).

2554 growing in large groups [loc] *phyt.* (Pure stand of a single species on scale of sociability; ►plant sociability); *syn.* growing in great colonies [loc]; *s* **creciendo en una población continua** [loc] (En la clasificación de Braun-Blanquet 5ª categoría de ►sociabilidad o modo de apruparse de los individuos de una especie; BB 1979); *f* **croissant en peuplements serrés et continus** [loc] (Unité caractérisant l'échelle de ►sociabilité des individus [ou les pousses] d'une même espèce à l'intérieur d'une population donnée) ; *g* **große Herden bildend [loc]** (Skalierungseinheit bei der ►Häufungsweise).

G

2555 growing medium [n] *constr. hort.* (Soil mixture of different soil textural classes and organic matter; e.g. topsoil, peat, sand, ▶leaf mold [US]/leaf-mould [UK], shredded bark humus, forming a root-zone layer; ▶planting soil); *syn.* topsoil mix(ture) [n], planting mix(ture) [n] [also US]; *s* **sustrato** [m] **vegetal** (Mezcla de diferentes materiales, como tierra vegetal, compost, turba, ▶mantillo de hojarasca, arena o lava, que sirve de capa portante a la vegetación; ▶tierra para plantaciones); *f* **mélange** [m] **terreux** (Mélange de divers matériaux, p. ex. terre végétale, humus d'écorce, tourbe, sable, ▶terreau de feuilles, etc. spécifique à une culture particulière ; ▶terre végétale améliorée) ; *syn.* substrat [m] végétal, support [m] de culture ; *g* **Erdgemisch** [n] (Bodenmaterial, das i. d. R. aus mehreren miteinander vermischten Stoffen in Erdenwerken oder auf der Baustelle hergestellt wird, z. B. aus Oberboden, Blähton, Kompost, Lava, ▶Lauberde, Rindenhumus, Sand, Tonziegelsplitt, Torf; je nach Verwendungszweck wird bei einem zweischichtigen Aufbau einer Vegetationstragschicht zwischen **Oberbodensubstrat** [mit organischer Substanz] und **Unterbodensubstrat** [mit sehr geringem Anteil an organischer Substanz] unterschieden; ▶Pflanzerde); *syn.* Bodenmischung [f], Bodensubstrat [n], Erdmischung [f], Erdsubstrat [n], Pflanzsubstrat [n], Substrat [n], Vegetationssubstrat [n], *im Zierpflanzenbau* Kulturerde [f], Kultursubstrat [n].

2556 growing on peat [loc] *phyt.* (Descriptive term for plants which require peat, bog soil or other histic soil for normal growth); *syn.* turficolous [adj]; *s* **turfófilo/a** [adj] (Término descriptivo para plantas que necesitan un sustrato de turba u otros suelos hísticos para crecer normalmente; cf. DB 1985); *f* **turficole** [adj] (qui vit sur une tourbière ou sur sol tourbeux) ; *g* **auf Torf wachsend** [loc] *syn.* auf Torf lebend [loc].

2557 growing season [n] *bot. hort.* (Period of general plant growth in any one year; *opp.* ▶dormancy); *s* **periodo/período** [m] **de crecimiento** (*opp.* ▶período de reposo); *syn.* periodo/período [m] de vegetación; *f* **période** [f] **de végétation** (Période de croissance des végétaux, de la germination jusqu'à la formation des organes reproducteurs [▶floraison], du développement des limbes jusqu'à leur mi-brunissement ; DUV 1984, 92 ; *opp.* ▶dormance) ; *syn.* période [f] végétative (LA 1981) ; *g* **Vegetationsperiode** [f] (Zeitraum des allgemeinen Pflanzenwachstums während eines Jahres; *opp.* ▶Vegetationsruhe); *syn.* Vegetationszeit [f], Wachstumsperiode [f].

2558 growing singly [loc] *phyt.* (First category of ▶plant sociability); *s* **creciendo aislado** [loc] (En la clasificación de BRAUN-BLANQUET 1ª categoría de la ▶sociabilidad de los individuos de una especie ; BB 1979); *f* **croissant isolément** [loc] (Première catégorie de la ▶sociabilité) ; *syn.* dispersé, ée [pp/adj], isolé, ée [pp/adj] ; *g* **einzeln wachsend** [ppr/adj] (Skalierungseinheit der ▶Häufungsweise).

2559 growing stock [n] *for.* (Number of trees within a given area of forest); *syn.* growing timber [n]; *s* **masa** [f] **de un rodal** (Número de árboles en una superficie forestal determinada); *syn.* volumen [m] de un rodal, existencias [fpl]; *f* **matériel** [m] **sur pied existant** (Nombre d'arbres faisant partie d'un peuplement forestier donné ; DFM 1975) ; *g* **Bestockung** [f] **(1)** (Anzahl der Bäume auf einer Bestandsfläche); *syn.* stehender Bestand [m], stehendes Holz [n], Bestandsmasse [f].

growing timber [n] *for.* ▶growing stock.

2560 growth [n] **(1)** *biol.* (Increase in size of an organism or its individual parts until it has reached its genetically programmed final size; ▶vigor [US]/vigour [UK]); *s* **crecimiento** [m] (Aumento de la talla de un organismo hasta alcanzar las dimensiones definitivas determinadas genéticamente; ▶vigor); *f* **croissance** [f] (Augmentation de la taille d'un organisme jusqu'à accéder aux dimensions définitives déterminées génétiquement ; terme à distinguer de la notion de développement ; ▶vigueur) ; *g* **Wachstum** [n] (Größerwerden eines Organismus oder einzelner Teile bis zur genetisch festgelegten Endgröße; ▶Wüchsigkeit).

growth [n] **(2)** *phyt.* ▶bog growth; *arb. for. hort.* ▶caliper growth, ▶contorted growth; *bot. ecol.* ▶deterioration of plant growth; *arb.* ▶forked growth; *biol. hort.* ▶inhibiting (plant growth); *bot. hort.* ▶one year's growth; *constr. hort. leg.* ▶overhanging growth; *bot.* ▶plant growth (1); *plan. sociol.* ▶population growth; *phyt.* ▶primary (plant) growth; *landsc.* ▶regenerated woody growth; *bot.* ▶root growth; *hort.* ▶spontaneous growth; *envir. for.* ▶stork's nest growth; *econ. nat'res. sociol.* ▶sustained growth; *agr. for. hort.* ▶weed growth.

growth [n]**, epicormic** *arb.* ▶epicormic shoot of stem.

growth [n]**, stunting (plant)** *biol. hort.* ▶inhibiting (plant growth).

growth [n]**, twisted** *arb. for. hort.* ▶contorted growth.

growth [n]**, uncontrolled grass** *for. hort.* ▶uncontrolled grass intrusion.

growth [n] [US]**, volunteer** *for. phyt.* ▶natural regeneration (2).

2561 growth axis [n] *plan.* (Planned linear concentration of settlement areas, traffic routes and main services by axial growth; ▶communication corridor, ▶strip development [US] 2/ribbon development [UK]); *s* **eje** [m] **de desarrollo** (Concentración del crecimiento urbano-industrial a lo largo de vías de tránsito superiores, que parten generalmente de las grandes ciudades y unen a éstas entre sí; ▶banda de infraestructura, ▶crecimiento desordenado de la periferia de la ciudad); *syn.* banda [f] de desarrollo; *f* **axe** [m] **de développement** (*Aménagement du territoire* concentration linéaire de zones d'urbanisation, des voies de communication et des réseaux d'approvisionnement ; ▶couloir d'infrastructures de communications, ▶développement urbain périphérique désordonné en bordure des axes d'accès) ; *g* **Entwicklungsachse** [f] (*Raumordnung* bandartige, gegliederte, unterschiedlich dichte Folge von Siedlungsgebieten, Verkehrswegen, Versorgungssträngen und anderen Infrastruktureinrichtungen; ▶Infrastrukturband, ▶Ortsrandauswucherung entlang von Ausfallstraßen); *syn.* Entwicklungsband [n].

growth form [n] *bot. zool.* ▶habit.

growth inhibitor [n] *chem. agr. hort.* ▶growth retardant.

2562 growth layer [n] *arb.* (Layer of wood or bark apparently produced during one growing period, especially in woods of the temperate zones, divisible into ▶early wood and ▶late wood or bark, and seen as growth rings in cross-section; cf. MGW 1964; ▶annual ring, ▶phloem); *syn.* growth ring [n]; *s* **capa** [f] **de crecimiento** (Parte de madera o floema que se ha formado en un periodo/período de crecimiento que —especialmente en las maderas de las zonas de clima templado— se puede diferenciar en ▶leño temprano y ▶leño tardío resp. en floema temprano y tardío; ▶anillo anual, ▶floema); *syn.* incremento [m] de crecimiento; *f* **couche** [f] **d'accroissement** (Couche de bois ou de liber produite apparemment pendant une période de végétation ; pouvant être fréquemment différencié en ▶bois initial ou liber initial et en ▶bois final ou liber final, notamment dans les bois des régions tempérées ; MGW 1964 ; ▶cerne annuel, ▶liber) ; *syn.* cerne [m], anneau [m] ligneux, anneau [m] de croissance ; *g* **Zuwachszone** [f] (Bereich aus Holz oder Bast, der offensichtlich während einer Wachstumsperiode gebildet worden ist; häufig besonders in Hölzern der gemäßigten Klimazonen, einteilbar in ▶Frühholz und ▶Spätholz resp. Früh- und Spätbast; MGW 1964; ▶Bastteil, ▶Jahresring); *syn.* Zuwachsring [m].

2563 growth [n] per year *arb.* (Annual increase of a tree in its girth, height, canopy diameter, etc.; ►annual ring, ►caliper growth, ►growth layer); *s* **crecimiento [m] anual** (Aumento del espesor del tronco, de la altura, del volumen de la copa, etc. de un árbol durante un año; ►anillo anual, ►capa de crecimiento, ►crecimiento en espesor); *f* **accroissement [m] annuel** (du volume du bois — l'accroissement ligneux annuel —, de la circonférence, de la hauteur, etc. ; DFM 1975, 2 ; ►cerne annuel, ►couche d'accroissement, ►croissance en épaisseur) ; *g* **jährlicher Zuwachs [m]** (Größerwerden eines Baumes im Stammumfang, Höhe, Kronendurchmesser etc. innerhalb eines Jahres; ►Dickenwachstum, ►Jahresring, ►Zuwachszone).

2564 growth point [n] *plan. pol.* (Central community with regional supporting functions providing major facilities for smaller outlying settlements; e.g. opera house, central libraries, sport stadiums, entertainment complexes, convention halls, hotel accommodations, etc.; ►population growth center); *syn.* growth pole [n], central growth point [n]; *s* **polo [m] de desarrollo** (Centro que cuenta con un conjunto de actividades integradas y dinámicas que impulsan su crecimiento y pueden llegar a difundirlo en su entorno. Teoría desarrollada por el economista francés F. PERROUX que fue aplicada en los años 1960 en varios países —incluida España— como guía de la planificación regional, en los que el Estado promovió —con más o menos éxito— la creación de diversos polos en zonas atrasadas. Estas actuaciones han sido progresivamente sustituidas por otras de ámbito territorial más amplio; cf. DGA 1986; ►lugar central); *syn.* polo [m] de crecimiento; *f* **pôle [m] de développement** (*Aménagement du territoire,* territoire sur lequel sont mises en œuvre des opérations favorisant la concentration de l'habitat, des activités et des grandes infrastructures de services, p. ex. les métropoles d'équilibre ; ►lieu central, ►pôle urbain) ; *g* **Wachstumspol [m]** (*Raumordnung* Zentrum mit einer Reihe von integrierten und dynamischen Wirtschaftsaktivitäten, die das regionale Wachstum fördern und in die Umgebung ausstrahlen sollen. Diese Theorie von F. PERROUX wurde in den 1960er-Jahren in verschiedenen europäischen Ländern als regionalplanerische Leitlinie angewandt, indem der Staat die Schaffung von Polen in wirtschaftsschwachen Räumen begünstigte; ►zentraler Ort); *syn.* Entwicklungsschwerpunkt [m].

growth pole [n] *plan. pol.* ►growth point.

2565 growth regulator [n] *chem. agr. for. hort.* (Benign substance which controls the growth processes of plants without providing nutrition or damaging them; ►growth retardant); *s* **sustancia [f] reguladora del crecimiento** (Materia destinada a influir en el crecimiento de las plantas, sin ser alimento ni reprimir su crecimiento; ►inhibidor); *f* **régulateur [m] de croissance** (Limitation de la croissance, reduction de la croissance ; ►substance inhibitrice) ; *g* **Wachstumsregler [m]** (Stoff, der dazu bestimmt ist, die Lebensvorgänge von Pflanzen zu beeinflussen, ohne ihrer Ernährung zu dienen und ohne sie zum Absterben zu bringen; ►Hemmstoff).

2566 growth retardant [n] *chem. agr. hort.* (Chemical substance that, when applied to plants in small amounts, retards or inhibits the growth process; ►growth regulator); *syn.* growth inhibitor [n] (TEE 1980, 99); *s* **inhibidor [m]** (Sustancia química que cuando se aplica a las plantas en pequeñas dosis, retrasa o inhibe el proceso de crecimiento; ►sustancia reguladora de crecimiento); *syn.* sustancia [f] inhibidora; *f* **substance [f] inhibitrice** (Substance chimique ralentissant la croissance des végétaux ; ►régulateur de croissance) ; *g* **Hemmstoff [m]** (Chemische Substanz, z. B. in dem Mittel *Cycocel* oder *Alar,* die auf die Bildung der Gibberelinsäure in der Pflanze einwirkt und somit das Streckungswachstum hemmt. In der gärtnerischen Kultur-

technik spricht man vom „Stauchen" der Pflanzen und „Hemmen" des Wachstums; ►Wachstumsregler); *syn.* Retardator [m].

growth ring [n] *arb.* ►growth layer.

2567 growth ring boundary [n] *arb.* (The outer limit of a ►growth layer produced during one growing period; ►annual ring); *s* **límite [m] de anillo de crecimiento** (Borde externo de la ►capa de crecimiento formada en un periodo/período de crecimiento; ►anillo anual); *f* **limite [f] de cerne** (Le bord externe de la ►couche d'accroissement formée pendant une période de végétation ; ►cerne annuel) ; *syn.* limite [f] de l'anneau de croissance ; *g* **Zuwachsringgrenze [f]** (Äußere Grenze einer ►Zuwachszone, die während einer Vegetationsperiode gebildet wird; ►Jahresring).

groyne [n] [UK] *wat'man.* ►jetty [US].

grubbing up [n] *constr. for. hort.* ►stump out.

grub felling [n] [UK] *constr. for.* ►tree clearing and stump removal [operation] [US].

grub out [vb] [US] *constr. for. hort.* ►stump out.

grub up [vb] [UK] *constr. for. hort.* ►stump out.

2568 guarantee [n] *constr. contr.* (Assurance of the contractor that construction work[s] and delivered materials will conform to specification and the current state-of-the-art for a given period of time; ►replacement guarantee); *s* **garantía [f] (1)** (Seguro del contratista de que sus obras de construcción y los materiales utilizados corresponden a lo fijado en el contrato y a las normas de calidad correspondientes al momento; ►garantía de reposición de marras); *f 1* **garanties [fpl]** (Terme générique pour les responsabilités des constructeurs d'ouvrage édictées par les art. 1792 et 2270 du C. civil qu'a l'entrepreneur à exécuter les ouvrages dans les règles de l'art et conformément aux stipulations du C.C.A.P. et le rend responsable de plein droit envers le maître ou l'acquéreur de l'ouvrage, des dommages, même résultant d'un vice du sol, qui compromettent la solidité de l'ouvrage ou qui, l'affectant dans l'un de ses éléments constitutifs ou l'un de ses éléments d'équipement, le rendent impropre à sa destination ; CCH, art. L. 111-13 ; ►garantie de reprise des végétaux, ►garantie de bonne levée) ; *f 2* **garantie [f] de parfait achèvement** (Responsabilité de l'entrepreneur qui est tenu pendant un délai d'un an à compter de la réception d'exécuter les travaux non réalisés lors de la réception des travaux, à réaliser les travaux assortis de réserves, de procéder aux travaux confortatifs ou modificatifs nécessaires suite aux désordres révélés postérieurement à la réception, de fournir au maître d'œuvre les plans des ouvrages exécutés et documents divers [notice de fonctionnement et d'entretien] ; cf. CCH, art. L. 111-19) ; *f 3* **garantie [f] décennale** (Responsabilité décennale des constructeurs d'ouvrages conformément aux exigences de la loi n° 78.12 du 04 janvier 1978, comprenant la garantie dite de bon fonctionnement des éléments d'équipement et celle des dommages immatériels consécutifs) ; *f 4* **garantie [f] de bon fonctionnement** (Responsabilité des constructeurs d'ouvrages d'une durée minimale de deux ans à compter de la date de réception de l'ouvrage pour les éléments d'équipement ne formant pas indissociablement corps avec les ouvrages de viabilité, de fondation, d'ossature, de clos ou de couvert ; cf. CCH, art. L.111-15/16) ; *g* **Gewährleistung [f]** (Vom Auftragnehmer übernommene Gewähr, dass seine Leistung zur Zeit der Abnahme die vertraglich zugesicherten Eigenschaften hat, den anerkannten Regeln der Technik entspricht und nicht mit Fehlern behaftet ist; cf. § 13 VOB Teil B. In der VOB wird seit 29.10.2002 der Begriff **G.** durch **Verjährung der Mängelansprüche** ersetzt. Der **G.** kann sich kein Auftragnehmer im gesetzlichen Rahmen entziehen. Umfang und zeitliche Länge sind abhängig von der zu Grunde gelegten Rechtsgrundlage, z. B. BGB, VOB Teil B oder AGB [Allgemeine Geschäfts-

G

bedingungen]. In D. ist durch das BGB ein grundsätzlicher Ausschluss der **G.** bei auch noch so ausgefeilten Geschäftsbedingungen nicht möglich. Der Begriff **Garantie** ist in D. bau- und vertragsrechtlich ohne Grundlage. Dieser Begriff stellt für den Verwender ein hohes Risiko dar, da er für die Nichteinhaltung in vollem Umfange haftet und zwar neben den sowieso bestehenden gesetzlichen Gewährleistungsansprüchen auch ohne jegliches Verschulden; ▶Anwachsgarantie); *syn.* Verjährung [f] der Mängelansprüche.

guarantee bond [n] [UK] *contr.* ▶bid bond [US].

2569 guarantee fulfillment obligation [n] [US]/**guarantee fulfilment obligation** [n] [UK] *contr.* (The legal duty of a supplier/contractor to deliver a product or a piece of work in a faultless condition; under the warranty the customer/client has the rights of ▶remedying defects, ▶price reduction and payment for damages); *syn.* seller's warranty [n]; *s* **obligación** [f] **(por ley) de cumplir con la garantía** (Compromiso legal de un deudor/contratista de suministrar un producto o una obra sin defectos; los derechos que tiene un acreedor/propietario de obra son ▶corrección de defectos [en obra], reducción de la remuneración [▶reducción de precio] e indemnización); *f* **obligation** [f] **de parfait achèvement** (▶réfaction sur les prix, ▶réfection des imperfections ou malfaçons) ; *syn.* obligation [f] de bon achèvement, *obs.* obligation [f] d'entretien ; *g* **Gewährleistungspflicht** [f] (Die gesetzliche Verpflichtung eines Schuldners/Auftragnehmers, eine Sache oder ein Werk im mangelfreien Zustand abzuliefern; Rechte, die der Gläubiger/Bauherr aus der ▶Gewährleistung [Verjährung der Mängelansprüche] hat, sind ▶Mängelbeseitigung, Minderung der Vergütung [▶Preisminderung] und Schadensersatz).

guarantee period [n] *contr. leg.* ▶defects liability period.

guard [n] *arb. constr.* ▶tree guard.

guarding design [n] *constr. leg.* *∗*▶protection from falling.

2570 guardrail [n] [US] **(1)** *trans.* (Vehicular safety barrier installation adjacent to a travelled way [US]/road carriage-way [UK], which prevents vehicles from leaving the road and aids in the visual direction of traffic); *syn.* crash barrier [n] [UK], safety barrier [n] [UK] (LOV 1973, SPON 1986, 294), vehicle barrier [n] [UK] (LD 1992 [12], 45); *s* **quitamiedo** [m] (Barandilla metálica colocada al borde de las carreteras y autopistas que sirve de guía lateral y para evitar que los vehículos se salgan de las mismas); *syn.* guardacarril [m], valla [f] protectora, baranda [f] de guía; *f* **glissière** [f] **de sécurité** (Dispositif de retenue des véhicules contre les sorties accidentelles de chaussée placé p. ex. **1.** sur le terre plein central d'un autoroute lorsque la largeur de celui-ci est inférieure à 12 m ; **2.** sur l'accotement dans les courbes à l'extérieur de la chaussée de grand rayon lorsque celui-ci est inférieur à 1,5 Rm [rayon minimal], lorsque la hauteur des remblais dépasse 4 m ou 1 m en cas de dénivellation brutale, en présence d'obstacles durs à moins de 10 m du bord de chaussée ; cf. ICTA 1985) ; *g* **Leitplanke** [f] (Seitlich der Fahrbahn eingebaute Sicherheitseinrichtung, die ein Abkommen der Fahrzeuge von der Fahrbahn verhindern soll und der optischen Verkehrsführung dient); *syn.* Leitschiene [f] [A], Schutzplanke [f].

2571 guardrail [n] [US] **(2)**/**guard-rail** [n] [UK] *constr. arch.* (Safety barrier or hand rail installed along ramps, stairways, and at wall edges; ▶pedestrian guardrail [US]/pedestrian guardrail [UK], ▶protection from failling); *s* **barandilla** [f] (Barrera de seguridad o baranda de escalera, rampa, terraza, etc. para evitar caídas; ▶barrera de protección, ▶protección contra el riesgo de caídas); *f 1* **garde-corps** [m] (▶Barrière de protection placée sur les côtés d'un escalier ouvert, le pourtour d'un palier, d'un balcon, d'une mezzanine, d'un toit, d'une toiture, d'une terrasse, d'un pont, etc. afin d'empêcher une chute accidentelle ;

▶protection contre les [risques de] chutes de hauteur) ; *f 2* **rampe** [f] **(1)** (Garde-corps disposé le long d'un escalier) ; *g* **Geländer** [n] (Vorrichtung an Treppen, Balkonen, Terrassen, Brücken etc. zur ▶Absturzsicherung; ▶Abschrankung).

2572 guerrilla gardening [n] *landsc.* (*Also* guerilla gardening [n]; primarily in large cities, the unauthorized appropriation by members of the public of urban open spaces such as road medians, tree pit surfaces, neglected patches, etc., which are then decoratively planted at whim [surprise planting] and tended or the transformation of deserted, waste land in the city into colourful gardens or flowering meadows, which are accessible to everyone. This form of gardening is to be seen as a kind of protest, which began in Great Britain and has now spread in the past few years in particular to the metropolises of the western world—most conspicuously to New York since the beginning of the 1970s; ▶community gardening); *s* **jardinería** [f] **guerrillera** (Movimiento existente en grandes ciudades que consiste en la apropiación no autorizada de espacios libres, como solares descuidados, franjas divisorias de calzadas, alcorques de árboles, etc. y plantación de los mismos por activistas de grupos de la llamada **«guerrilla jardinera»** con el fin de transformar áreas abandonadas en áreas floridas, accesibles a toda la población. Este movimiento, relacionado con la ecología y el cuidado del medio ambiente, comenzó en Gran Bretaña y se extendió a las metrópolis de los países occidentales, en especial a Nueva York desde comienzos de los años 1970; actualmente tiene grupos muy activos en Londres y Berlín; en los últimos años se ha extendido a países como Bélgica, Dinamarca, Brasil, México, Canadá y Sudáfrica; el término **«green guerilla»** fue utilizado por primera vez en Nueva York a mediados de los años 1970, donde se fundó una ONG del mismo nombre; 30 años después británico RICHARD REYNOLDS popularizó el de **«guerrilla gardening»** a través del sitio web www.guerrillagardening.org; cf. JAH 2007; ▶jardinería comunitaria); *f* **guérilla** [f] **jardinière** (Activité militante communautaire pacifique qui consiste dans la réappropriation illicite d'espaces [verts] souvent publics ou de terrains vagues, constitués en priorité dans les grandes agglomération par les terre-pleins centraux de la voirie, les tours d'arbre, les plates-bandes délaissées etc. en vue d'embellir ces espaces au moyen de graines de fleurs ou de légumes, de plantes et de les transformer ainsi en jardins colorés, en potager ou en prés fleuris ; les actions des « guerrilleros jardiniers » ont pour but d'inciter les propriétaires et les municipalités à entretenir leurs espaces verts et de créer ainsi, en réinventant les liens sociaux et la convivialité, un espace de partage et de vie meilleur ; ce mouvement de contestation de groupes de jardiniers activistes menant souvent la nuit « une opération commando » armés de pelles et de râteaux et prenant d'assaut les lieux à l'abandon a pris naissance en Angleterre et s'est étendu depuis le début des années 70 dans les grandes métropoles des États-Unis — avec en tête New York — et du Canada ; ce mouvement semble avoir trouver en France des adeptes depuis le début de 2007 [échanges sur le net] ; ▶activités concernant les jardins ouvriers) ; *syn.* guerilla gardening [m], jardinage [m] illégal, rébellion [f] jardinière ; *g* **Guerilla-Gardening** [n, o. Pl.] (Aus dem Spanischen *guerilla* >kleiner Krieg<, Diminutiv von *guerra*; unautorisierte Inbesitznahme von Teilen öffentlicher [Grün]flächen, vorrangig in Großstädten, bei der Mittelstreifen, Baumscheiben, verwahrloste Beete etc. zur Verschönerung mit Pflanzen eigener Wahl bestückt werden [Überraschungspflanzung] oder triste, innerstädtische Brachflächen in bunte Gärten oder Blumenwiesen, die jedermann zugänglich sind, verwandelt werden. Diese Form der Inbesitznahme ist eine Protestform — sei es aus stadtökologischem Anliegen, mit künstlerischem Anspruch oder aus politischem Protest —, die sich, von Großbritannien ausgehend, seit einigen Jahren insbesondere in den Metropolen der westlichen Welt — allen voran New York — verbreitet. Der Brite RICHARD REYNOLDS hat 30 Jahre später diese Bewegung auf der Webseite www.guerillagardening.org einem breiteren Publikum bekanntgemacht. Die Protestbürger werden **Gartenpiraten** [DIE ZEIT online Nr. 38 v. 18.09.2008], **Garten-Guerilleros** oder **Guerilla-Gärtner** genannt; Näheres über die Besetzung von innerstädtisch brach liegenden Flächen in New York Anfang der 1970er-Jahre ▶Kleingartenwesen); *syn.* Gartenpiraterie [f].

guidance facility [n] *trans.* ▶traffic guidance facility.

2573 Guidance [n] **for Clients on Fees** [UK] *contr. leg. prof.* (In certain European countries, a legally-specified remuneration for planners, designers and structural engineers in planning of facilities and structures, based on project scope and complexity; **in U.S.**, such a legally-specified remuneration is illegal. Instead of this there exists an Architect's Handbook of professional practice, Volume 3, dealing with several standard forms on owner-architect agreements issued by The American Institute of Architects [AIA], such as the **Standard Form of Agreement between Owner and Architect** written down in the AIA document B141/CM, the "Standard Form of Agreement between Owner and Architect for Designated Services"—AIA document B161/CM—, the "Scope of Designated Services"—AIA document B162—, the "Standard Form of Agreement between Owner and Architect for Special Services"—AIA document B727—, etc.; ▶fee chart. Corresponding contract documents for landscape architects were issued in the year 2000 by the American Society of Landscape Architects [ASLA]; **in UK.,** the Landscape Institute withdrew its mandatory **Scale of Professional Charges** in 1986 and in the absence of a formal basis on which to advise clients and landscape consultants on fees, it published a **Guidance for Clients on Fees** in 1996, revised in 2002 with an appendix added in 2003. The purpose of the booklet is to provide information, specifically to clients, to enable a better understanding of how landscape architects' fees are calculated. The fee scale graph in the booklet is indicative and intended as a guide in evaluating offers and negotiating fees. Fees for architectural and engineering services in UK have also been deregulated. There is no fixed fee scale, so an architect charges a fee taking into account the requirements of the project, his or her skills, experience, overheads, the resources needed to undertake the work, as well as profit and competition. The Fees Bureau, recognized by the RIBA, annually provides construction professionals with researched business operating data on the construction sector based on exclusive research by a team of professionals. An architect's fees may thus be charged as a percentage of the total construction cost or at an hourly rate or as a lump sum, depending on the service supplied. Fee rates are analysed by job, building type and contract type. Each practice devises its own method of fee calculation, for example, from inception to planning at a flat rate; tender to mobilisation, at day rate and construction to completion, as percentage of the Contract Sum. The data, published yearly, is shown in a chart, plotting construction cost against percentage fee and showing an average fee line, which is a line of best fit based on an analysis of the raw survey data. For engineers, too, fee data is researched for 20 market sectors to give average 'going rates' currently being charged by consulting civil and structural engineers. Hourly charge-out rates are analysed for different types of staff, by region and by size of firm. The fee decreases with increasing contract value. For both existing and new works, the architect's and the engineer's fee is inversely proportional to the contract sum and follows an exponential curve); *syn. obs. since 1986* fee scale [n] [UK], conditions [npl] of engagement and scale [n] of professional charges [UK]; **s *1* baremo [m] de honorarios** (Guía para fijar honorarios elabo-

rada por los colegios profesionales; ▶tabla de honorarios); *syn.* tarifas [fpl] de honorarios; **s *2* reglamento [m] de honorarios** (En D. existe un reglamento de honorarios para arquitectos, urbanistas, paisajistas e ingenieros de planificación que tiene carácter legal y por el que fijan las tarifas para los diferentes tipos de servicios, los grados de dificultad, etc.); *ƒ* **rémunération [f] des prestations de maîtrise d'œuvre (exercée pour le compte de maîtres d'ouvrage publics)** (Conditions de rémunération des missions d'ingénierie et d'architecture [architectes, urbanistes, paysagistes, architectes d'intérieur, ingénieurs-conseils, bureaux d'études, techniciens économistes de la construction, métreurs, métreurs-vérificateurs, géomètres-experts] remplies pour le compte des collectivités publiques par des prestataires de droit privé principalement régi par le décret du 28 juillet 1973, par la loi du 12 juillet 1985 et la loi du 01 décembre 1988 ; HAC 1989) ; *syn.* réglementation [f] relative à la rémunération des prestations de maîtrise d'œuvre, réglementation [f] relative à la rémunération des prestations d'ingénierie et d'architecture (terme révolu depuis la loi du 12 juillet 1985) ; *g* **Honorarordnung [f] für Architekten und Ingenieure (HOAI) (D.,** seit 01.01.1977 gesetzlich geregelte Vergütung [zwischen Höchst- und Mindestsätzen] der Leistungen für Hochbauarchitekten, Landschaftsarchitekten, Innenarchitekten, Städteplaner und Ingenieure [BGBl. I 2805 vom 17.09.1976], geändert am 10.11.2001 [BGBl. I 2992] und ab 01.01.2002 in Kraft und in zuletzt geänderter Fassung am 18.08.2009 [BGBl. I Nr. 53] in Kraft getreten. Es handelt sich um eine preisrechtliche Vorschrift, welche die Vergütungsansprüche der Architekten und Ingenieure nach oben und unten begrenzt, ohne in die Vertragsfreiheit einzugreifen. Gemäß gängiger Rechtsprechung handelt es sich dabei stets um eine ergebnis- und erfolgsorientierte, nicht aber um eine zeit- oder tätigkeitsbezogene Vergütung. Die neue HOAI 2009 ist eine sog. *Inländer-HOAI,* die nur für Leistungen von Planungsbüros mit Sitz in Deutschland gilt, um Art. 16 der Europäischen Richtlinie über Dienstleistungen am Binnenmarkt vom 12.12.2006 zu entsprechen. *Historie* die HOAI löste am 1. Januar 1977 die GOA 1950 [Gebührenordnung für Architekten aus dem Jahr 1950] und die GOI 1956 [Gebührenordnung der Ingenieure aus dem Jahr 1956] ab; ▶Honorartafel. **A.,** in der Landschaftsarchitektur gibt es in Österreich die **„Honorarrichtlinien für Landschaftsarchitekten",** herausgegeben von der Österreichischen Gesellschaft für Landschaftsplanung und Landschaftsarchitektur [ÖGLA], und in der Schweiz die **„Ordnung für Leistungen und Honorare der Landschaftsarchitektinnen und Landschaftsarchitekten",** herausgegeben vom Bund Schweizer Landschaftsarchitekten [BSLA] und dem Schweizerischen Ingenieur- und Architektenverein [SIA]. Im 19. Jh. entstanden in D. die ersten Honorarregelungen und 1871 die erste **H. f. A.;** cf. DAB 1999 [10], p. XII. **In den USA** ist eine **HOAI** illegal; ob in einem Europa der Deregulierung eine Honorarordnung bestehen bleiben wird, wird die Zukunft zeigen; **im UK** gibt es auch kein Äquivalent zur HOAI. Kalkulationsmethoden zur Ermittlung von Honoraren sind z. B. für Landschaftsarchitekten in dem unverbindlichen Leitfaden *Engaging a Landscape Consultant* vom *Landscape Institute* aufgeführt; Vergütungen werden auf Prozentbasis [▶time charge] oder als Pauschalhonorare [▶lump-sum fee] vereinbart); *syn.* Gebührenordnung [f] für Architekten und Ingenieure, Vergütungsordnung [f] für Architekten und Ingenieure, Verordnung [f] über die Honorare für Leistungen der Architekten und Ingenieure.

2574 guideline [n] **in planning** *plan.* (Criterion for determining a course of action to attain a desired objective); **s directriz [f] de planificación/planeamiento** *syn.* directiva [f] de planificación/planeamiento; *ƒ* **parti [m] d'aménagement** (Choix déterminé par l'architecte ou l'aménageur d'une solution parmi plusieurs) ; *g* **Leitlinie [f] einer Planung** (Bestimmender planerischer Grundgedanke eines Planungsvorhabens).

2575 guideline (value) [n] *plan.* (Number of units aimed for in planning to reach a particular result; ►adulteration limit, ►open space standard); *s* **valor [m] estándar** (Valor utilizado en planificación como orientación para alcanzar un resultado específico; ►estándar de zonas verdes, ►valor límite); *syn.* valor [m] guía, valor [m] de orientación, valor [m] indicativo; *f* **valeur [f] guide** (Valeur utilisée par les aménageurs constituant une orientation ou objectif défini à réaliser ; ►valeur des besoins en espaces libres, ►valeur limite) ; *syn.* valeur [f] indicative ; *g* **Richtwert [m] (für Leitplanungen)** (Zielorientierte, planungsrelevante, überschlägige Bedarfszahl; ►Bedarfszahl für Freiflächen, ►Grenzwert); *syn.* Orientierungswert [m], Richtzahl [f] (für Leitplanungen).

guide post [n] *trans.* ►road edge guide post.

2576 gulch [n] [US] *geo.* (Short, narrow, very steep-sided, nearly vertically rocky walls of a ►ravine or cleft in a hillside, formed and occupied by a torrent; cf. WEB 1993, GFG 1997; term used in the western U.S.; in U.K. rarely used; ►gorge); *s* **barranco [m] (2)** (En alta montaña, ►desfiladero muy angosto, muchas veces de ancho de pocos metros, con paredes muy pendientes, casi verticales, causado por fuerte erosión lineal remontante y formación de hoyos de torbellinos, en cuyo fondo fluye un torrente; en inglés el término se utiliza en el oeste de los EE.UU., raramente en GB); *f* **gorge [f] de raccordement** (Incision fluviatile qui raccorde des vallées suspendues à des vallées glaciaires profondes. Agent d'érosion le plus important dans l'évolution du relief alpin, ces gorges sont encaissées et étroites, aux parois verticales et souvent de quelques mètres de large) ; *g* **Klamm [f]** (Im Hochgebirge eine steile und sehr enge, oft nur wenige Meter breite ►Schlucht mit fast senkrechten, stellenweise überhängenden Felswänden als Folge starker rückschreitender Tiefenerosion und Strudellochbildung, in deren Sohle ein Wildwasserbach fließt).

gully [n] *geo. pedol.* ►gully erosion.

gully [n] [UK] *constr. eng.* ►surface inlet [US], ►yard drain [US].

gully [n] [UK]**, road** *constr.* ►catch basin [US].

2577 gully erosion [n] *geo. land'man.* (Wearing of channels and small ravines by heavy rainfall and concentrated runoff, which, over short periods, removes the soil to considerable depths, typically ranging from 0.5 to as much as 25 to 30m; cf. SST 1997; ►natural erosion, ►rill erosion); *syn.* channel erosion [n]; *s* **erosión [f] en cárcavas (1)** (Consiste en profundas incisiones en el terreno originadas generalmente cuando existe una gran concentración de escorrentía en alguna zona determinada. Normalmente va precedida de fuertes erosiones laminares y en regueros, aunque también existen áreas en las que, sin haber existido con anterioridad fenómenos erosivos, la violencia de las precipitaciones, en conjunción con unas condiciones propicias del suelo, pueden desencadenar la creación de un barranco en el transcurso de un solo aguacero; cf. DINA 1987; ►erosión, ►erosión en regueros); *syn.* erosión [f] en barrancos (1); *f* **ravinement [m] (1)** (Creusement par les eaux de ruissellement de rigoles profondes sur un versant incliné de matériaux meubles ; ►érosion, ►ruissellement en filets) ; *syn.* érosion [f] en rigoles (LA 1981, 477) ; *g* **Grabenerosion [f] (1)** (Erosionsform, bei der durch plötzlichen Regenfall durch konzentrierte Abflusssammlung tiefe Rinnen und Gräben [> 50 cm tief] in die Bodenoberfläche gespült werden. Der Vorgang wird auch **Runsenspülung** genannt; ►Erosion, ►Rillenerosion); *syn.* Furchenerosion [f], Gullyerosion [f], Runsenerosion [f], Runsenspülung [f].

gully grating [n] [UK] *constr.* ►storm drain grate [US].

gun-applied concrete [n] [UK] *constr.* ►shotcrete [US].

guncrete [n] [UK] *constr.* ►shotcrete [US].

2578 gutter [n] *constr.* (Paved shallow channel adjacent to street traffic lane [US]/road carriage-way [UK]; ►channel invert, ►curb and gutter [US]/kerbstone with gutter [UK]); *s* **cuneta [f]** (Canaleta plana de desagüe al borde de calles y carreteras; ►lecho de canal [de desagüe], ►pieza de cuneta angular); *syn.* arroyo [m] de la calle; *f* **caniveau [m]** (Élément de voirie permettant de recueillir et d'évacuer latéralement les eaux pluviales en provenance de la chaussée et du trottoir, de cheminement et d'aires à sol stabilisé ; ►bordure caniveau d'épaulement, ►fil d'eau) ; *syn.* rigole [f] ; *g* **Rinnstein [m]** (Seitliche Abflussrinne entlang von Straßen; ►Rinnensohle, ►Winkelrinnenstein).

gutter [n] [UK]**, kerbstone with** *constr.* ►curb and gutter [US].

gutter depth [n] [US] *constr.* ►curb height [US].

gutter grate [n] [US] *constr.* ►storm drain grate [US].

gutter inlet [n] [UK] *constr.* ►catch basin inlet [US], ►grate inlet [US].

2579 guying [n] *arb. constr. hort.* (Method of securing newly planted, usually semi-mature, trees using ►wire guys or cabling attached to anchoring pegs, stakes, or deadmen; ►cabling [US]/cable bracing [UK], ►root bracing); *s* **fijación [f] con vientos (metálicos)** (Método para asegurar árboles maduros trasplantados; ►fijación del cepellón, ►sujeción de las ramas principales, ►viento de anclaje); *syn.* anclaje [m] con vientos (metálicos), anclaje [m] con cables; *f 1* **haubanage [m]** (Méthode d'immobilisation des gros arbres [Ø du tronc > 18 cm] ou de conifères grâce à un ►hauban pour les protéger contre l'action du vent ; ►haubanage de la charpente, ►haubanage de la motte) ; *f 2* **câble [m] de haubanage** (Câble métallique permettant de tenir un arbre en position verticale après la plantation : ►hauban) ; *g* **Drahtseilverankerung [f]** (1. *Vorgang* Befestigung von meist verpflanzten Großbäumen mit oberirdischen korrosionsbeständigen Drähten oder dünnen Drahtseilen; die Drähte/Seile werden in einem Winkel von maximal 60 Grad an Bodenankern befestigt und mit Spannschlössern gespannt. 2. *Ergebnis* von 1.; eine Befestigung mit dünnen Drahtseilen wird [Draht]seilverankerung genannt. 3. Eine D. kann auch bei älteren Bäumen notwendig werden, wenn Kipp- oder Bruchgefahr besteht und die Lasten im Baum selbst nicht abgetragen werden können; hierzu sind i. d. R. 3-4 Seile aus der Krone im Winkel von 45 ° zum Boden zu führen und dort durch Erdanker oder Fundamente zu sichern; cf. ZTV-Baumpflege; ►Drahtanker, ►Kronenverankerung, ►Wurzelballenverankerung); *syn.* Drahtverspannung [f], Drahtverankerung [f].

2580 gypsum salt swamp [n] *geo.* (Swampy area containing hydrated calcium sulfate [CaSO₄]; large occurrences in Australia); *s* **pantano [m] gipsáceo** (Ecosistema muy frecuente en Australia en cuyo suelo se presenta yeso); *f* **marais [m] gypsifère** (Écosystème de grande étendue en Australie) ; *g* **Gipssumpf [m]** (In Australien in größerer Ausdehnung auf gipshaltigen Böden vorkommendes Feuchtgebiet).

2581 gypsum vegetation [n] *phyt.* (Plant cover growing on gypsum [$CaSO_4 \cdot 2 H_2O$] and anhydrite [$CaSO_4$]); *s* **vegetación [f] gipsícola** (Plantas que crecen en suelos ricos en yeso [$SO_4 Ca + 2 H_2O$] y anhidrita [$SO_4 Ca$]); *f* **végétation [f] de gypsophytes** (Peuplement végétal se développant sur le gypse — $CaSO_4 + 2 H_2O$ — et anhydride — $CaSO_4$) ; *g* **Gipsvegetation [f]** (Pflanzenbestände auf Gips [$CaSO_4 \cdot 2 H_2O$] und Anhydrit [$CaSO_4$]).

2582 gyttja [n] *pedol.* (Lacustrine or palustrine fine humose sludge, consisting mainly of plant and animal residues precipitated from standing water, well-aerated and supplied with nutrients; ▶dy); *s* **gyttja [m]** (Masa, sin estructura determinada, de materia orgánica descompuesta y restos orgánicos en descomposición, que se forma en el fondo de ciertos lagos y otras aguas tranquilas. Es un caso especial de formación de humus; DB 1985; *términos populares* sueco [m], lodo [m] gris; ▶dy); *f* **gyttja [m]** (Dans la classification morpho-chimique des humus, vase noire exondée des eaux calmes bien aérées et riches en matière organique ; ▶dy) ; *g* **Gyttja [f]** (Humoser Unter-wasserboden in gut durchlüfteten, nährstoffreichen Stillge-wässern; ▶Dy); *syn.* Grauschlammboden [m].

H

2583 habit [n] *bot. zool.* (General apearance and characteristics of an individual organism, typical for its species); *syn. bot.* growth form [n]; *s* **hábito [m] de crecimiento** (Aspecto general y características de un individuo, típicas para su especie); *syn. phyt.* porte [m]; *f* **habitus [m]** (1. Totalité des caractères morpho-logiques constituant l'apparence générale d'un végétal permettant de l'identifier. 2. Ensemble des pratiques entrant dans la culture, voire dans la personnalité de l'individu, de la personne ; MG 1993, 250) ; *syn. bot.* silhouette [f] végétale, port [m] ; *g* **Habitus [m]** (Gesamtheit der morphologischen Merkmale, die den Indi-viduen einer Art eigen sind); *syn. bot.* Wuchsform [f], *syn. zool.* Gestalt [f].

2584 habitat [n] (1) *phyt. zool.* (Autecological term for a defined spatial unit providing the living conditions necessary for survival of one organism or group of organisms of the same species; ▶habitat 2); *s 1* **hábitat [m]** (Lugar geográfico en que mora una planta o un animal. El término «hábitat» se emplea en castellano tanto para referirse al lugar en donde crece [▶loca-lización] como a las condiciones mesológicas bajo las cuales se cría una planta [▶estación]; cf. DB 1985; ▶localidad, ▶bió-topo); *s 2* **habitación [f]** (Término usual empleado metafórica-mente en botánica para referirse al lugar geográfico en que mora una planta. En realidad, corresponde al cruce de un meridiano y un paralelo, con especificación de la altitud en que se encuentra sobre el nivel del mar. Las condiciones mesológicas del lugar corresponden a la ▶estación); *syn. parcial* hábitat [m]; *f 1* **habitat [m]** (1) (Ensemble des facteurs écologiques qui caractérisent le lieu où se développe une espèce ou une commu-nauté biologique ; l'habitat peut être constitué de plusieurs ▶biotopes, en particulier pour certains animaux ; cf. DEE 1982) ; *f 2* **habitat [m] naturel** *conserv.* (Zone terrestre ou aquatique délimitée se distinguant par ses caractéristiques géographiques, abiotiques et biotiques, qu'elle soit entièrement naturelle ou semi-naturelle ; ▶localité, ▶station) ; *g* **Habitat [m]** (...e [pl]; autökologischer Begriff für einen Ort, an dem Organismen einer Art regelmäßig anzutreffen sind resp. in dem die Art in einem der Stadien ihres Lebenskreislaufes vorkommt; **H.** wird im Deut-schen im Vergleich zu ▶Biotop immer nur aus der Sicht einer Population einer Art oder ggf. nur eines Individuums definiert; ein Biotop kann mehrere unterschiedliche **H.e** bereitstellen; z. B. hat ein Wald für Insekten Strukturen wie Kronen-, Strauch- und Krautschicht, Astlöcher etc. Anders betrachtet kann ein **H.** auch mehrere Biotope umfassen, sich ergänzende **H.e**, wie bei wan-dernden Tierarten, die z. B. Winter- und Sommerquartiere nur zur Nahrungssuche nutzen).

2585 habitat [n] (2) *phyt. zool.* (Totality of biotic and abiotic living conditions for a population system of organisms of differ-ent animal and plant species [▶biocoenosis], characterizing a defined area; ▶dry habitat, ▶extreme habitat, ▶habitat 1, ▶living space, ▶seam biotope, ▶substitute habitat, ▶wetland habitat; ▶location 1); *syn.* biotope [n]; *s* **biótopo [m]** (Espacio limitado en el que vive una ▶biocenosis. Un biótopo puede ser ecológicamente homogéneo o bien puede constar de una carac-terística agrupación de diferentes residencias ecológicas, como es el caso de un lago; cf. DB 1985; ▶biótopo de la orla herbácea, ▶biótopo extremo, ▶biótopo húmedo, ▶biótopo seco, ▶biótopo sustituyente, ▶dominio vital, ▶hábitat; ▶estación, ▶localiza-ción); *f* **biotope [m]** (*Au sens strict* ensemble [▶domaine d'exis-tence, système] des composantes biotiques et abiotiques caracté-risant une communauté d'organismes végétaux et animaux pour un territoire donné [▶biocénose] ; ▶biotope aux conditions stationnelles extrêmes, biotope de substitution, ▶biotope d'ourlet herbacé, ▶biotope humide, ▶biotope sec, ▶habitat 1 ; ▶localité, ▶station) ; *syn.* écobiocénose [f] ; *g* **Biotop [m,** *auch* n] (1. *Sensu stricto* Gesamtheit [▶Lebensraum, System] der abiotischen und biotischen Lebensbedingungen eines Bevölkerungssystems von Organismen verschiedener Tier- und Pflanzenarten [▶Biozö-nose] in einem bestimmten Gebiet, das in seiner relativ klar abgrenzbaren Beschaffenheit zur Umgebung ablesbar ist; im Vergleich zum ▶Habitat ist der **B.** ein synökologischer Begriff; ▶Ersatzbiotop, ▶Extrembiotop, ▶Feuchtbiotop, ▶Saumbiotop, ▶Trockenbiotop. 2. *Sensu lato* abgrenzbarer Raumabschnitt, der als Wohn- oder Standort die Lebensansprüche eines Organismus oder einer Gruppe von Organismen, z. B. einer Population, er-füllt; dieser weiter gefasste Begriff ist syn. mit ▶Habitat. Der Biotop ist ein wertfreier Begriff und enthält neben natürlich entstandenen Landschaften mit ihren Bestandteilen auch — entgegen dem umgangssprachlichen Gebrauch — vom Menschen geschaffene Landschaftsbestandteile wie „Betonwüsten" in Stadtlandschaften; ▶Standort 1, ▶Fundort).

habitat [n] (3) *zool.* ▶bark habitat; *phyt. zool.* ▶character species of a habitat; *biol. conserv.* ▶forest habitat; *ecol.* ▶isolated patch habitat; *conserv. ecol. leg.* ▶legally protected habitat; *conserv. leg. nat'res.* ▶protected habitat forest; *conserv. ecol. land'man.* ▶remnant habitat; *phyt.* ▶ruderal habitat; *ecol.* ▶transitional habitat.

habitat [n]**, hostile** *ecol.* ▶extreme habitat.

habitat [n]**, isolated patches of** *ecol.* ▶habitat island.

habitat [n]**, mesic** *ecol.* ▶wetland habitat.

habitat [n]**, nesting** *zool.* ▶breeding site.

habitat [n]**, protected forest** *conserv. leg. nat'res.* ▶pro-tected habitat forest.

habitat [n]**, reinstallment of a** *landsc. conserv.* ▶habitat restoration.

habitat [n]**, replacement** *ecol. leg.* ▶substitute habitat.

habitat [n]**, replacement of a** *landsc. conserv.* ▶habitat restoration.

habitat [n]**, severe** *ecol.* ▶extreme habitat.

habitat [n]**, xeric** *phyt. zool.* ▶dry habitat.

2586 habitat conditions [npl] *phyt. zool.* (The total of location prerequisites for plants or animals, such as climate conditions [sun, shadow, temperature, wind], soil [soil textural class, great soil group [US]/soil type [UK], nutrient, oxygen and water supply], spatial conditions [surface and underground spread and living conditions]); *s* **condiciones [fpl] mesológicas** (Condi-ciones que reinan en el lugar de crecimiento de un organismo específico); *f* **conditions [fpl] de la station** (Ensemble des facteurs nécessaires à la vie de la faune ou de la faune sur un site donné) ; *syn.* conditions [fpl] stationnelles, conditions [fpl] écologiques ; *g* **Standortbedingungen [fpl]** (1) (Gesamtheit der

H

an einem Ort vorherrschenden Gegebenheiten, wie Klimaverhält-nisse [Sonne, Schatten, Temperatur, Wind], Boden [Bodenart, Bodentyp, Nährstoff-, Luft- und Wasserversorgung], räumliche Gegebenheiten [ober- und unterirdische Ausbreitungs- und Lebensmöglichkeiten], die das Leben für Pflanzen oder Tiere bestimmen); *syn.* standörtliche Verhältnisse [npl], Standortver-hältnisse [npl].

2587 habitat conservation [n] *conserv. leg.* (Nature conservation policy involving the protection and management of wildlife habitats for stability); *s* **protección [f] de biótopos** (Medidas dedicadas a la preservación y gestión de hábitats naturales para su estabilización. En el contexto de la política europea de protección de hábitats naturales los Lugares de Importancia Comunitaria constituyen la ►Red Ecológica Europea Natura 2000 hasta su transformación en Zonas Especiales de Conservación o Zonas de Especial Protección para las Aves, y son «aquellos espacios del conjunto del territorio nacional o de las aguas marítimas bajo soberanía o jurisdicción nacional...que contribuyen de forma apreciable al mantenimiento o, en su caso, al restablecimiento del estado de conservación favorable de los tipos de hábitats naturales y los hábitats de las especies de interés comunitario...en su área de distribución natural»; arts. 41 y 42 Ley 42/2007; ►Directiva de Hábitats); *syn.* conservación [f] de biótopos; *f* **préservation [f] des biotopes** (Mesures de protection de la nature destinées à la protection, la conservation et la gestion des espaces naturels ; *terme spécifique pour la protection de la faune* préservation des habitats) ; les biotopes peuvent être protégés par arrêté [Arrêté de Biotope] ; ►biotope protégé par arrêté préfectoral ; c'est une mesure de protection établie par le décret du 25 novembre 1977 dans le but de favoriser la conservation des biotopes tels que les mares, les marécages, les marais, les haies, les bosquets, les landes, dunes, pelouses ou toutes autres formes naturelles peu exploitées par l'Homme ; la création est à l'initiative de l'État ; l'arrêté est pris par le préfet qui peut aussi interdire les actions susceptibles de porter atteinte à l'équi-libre biologique des milieux) ; *g* **Biotopschutz [m]** (Maßnahmen von Naturschutz und Landschaftspflege, die der Erhaltung, Sicherung und Pflege von Lebensräumen dienen. In den Landes-naturschutzgesetzen wird außerdem in dem Abschnitt „Allge-meiner Schutz der Pflanzen und Tiere" ein Mindestschutz auch der nicht besonders geschützten Arten gewährleistet, vor allem der Schutz von Lebensstätten, die den Tieren dienen, aber auch der Schutz des Naturhaushaltes und der Naturgüter [Schutz des Bodens und des Kleinklimas]; Naturschutzbehörden können durch Rechtsverordnungen oder Einzelanordnungen für Lebens-stätten bestimmter Arten, insbesondere ihre Standorte, Brut- und Wohnstätten, zeitlich befristet besondere Schutzmaßnahmen, insbesondere Verbote festlegen; cf. § 29 NatSchG-BW; ►gesetz-lich geschützter Biotop).

habitat forest [n] *conserv. leg. nat'res.* ►protected habitat forest.

habitat fragmentation [n] *ecol.* ►process of habitat frag-mentation.

2588 habitat island [n] *ecol.* (Primary habitat of an orga-nism, now isolated or undergoing isolation [= primary state of insularity] due to a natural separation process, e.g. flooding by the sea, topographical movement, climatic fluctuations or to a severing caused by man, e.g. infrastructural measures of a linear nature, the expansion of an intensively used agricultural area. The least favo[u]rable result is that of a biotope remnant [= secondary state of insularity]; ►isolated patch habitat, ►remnant habitat, ►process of habitat fragmentation); *syn.* isolated patches [npl] of habitat; *s* **hábitat [m] residual** (Lugar de crecimiento de una o varias especies que ha sido o está siendo aislado por causas natu-rales [inundación por el mar, movimientos tectónicos, fluctua-

ciones climáticas] o antropógenas [intensificación de la agricul-tura, línea de infraestructura]. El resultado peor es un ►biótopo relíctico; ►habitat natural aislado, ►aislamiento de biótopos); *syn.* enclave [m] residual; *f* **habitat [m] relictuel** (Aire de répar-tition géographique naturelle originaire très restreinte d'une espèce due à des barrières d'isolement naturelles telles que la submersion marine, la surrection d'une chaîne montagneuse, les changements climatiques, etc. [isolement primaire] ou à des effets de coupure anthropogènes telles que les infrastructures linéaires, l'extension de la culture intensive des surfaces agri-coles, etc. [isolement secondaire] ; type d'habitat détruit dont des vestiges occupent des situations favorables à leur maintien (reliefs peu accessibles) ; ►biotope relictuel, ►habitat relictuel isolé, ►isolement des biotopes) ; *syn.* habitat [m] résiduel, habitat [m] isolé, îlot [m] de station disjointe ; *g* **Habitatinsel [f]** (Lebensraum einer Art, der durch natürliche Trennbedingungen, z. B. durch Meeresüberflutung, Hochgebirgszug, biologische Isolierung durch Klimaschwankungen [= primäre oder natürliche Inselstruktur] oder durch anthropogene Zerschneidungseffekte, z. B. durch linienartige Infrastrukturmaßnahmen, Erweiterung intensiver landwirtschaftlicher Nutzflächen, entstanden ist oder entsteht und im ungünstigsten Falle zum ►Biotoprest [= sekun-däre Inselstruktur] wird; ►Sonderstandort, ►Verinselung von Biotopen).

habitat island [n], **remnant** *conserv. ecol. land'man.* ►remnant habitat.

habitat islands [npl] *ecol.* ►ecology of habitat islands.

2589 habitat management [n] *conserv. land'man.* (Mea-sures necessary to sustain a natural ►habitat 2); *s* **gestión [f] de biótopos** (Conjunto de medidas necesarias para preservar o restaurar hábitats naturales y la riqueza estructural del paisaje antropógeno; ►biótopo); *syn.* gestión [f] de hábitats naturales; *f* **gestion [f] des biotopes** (Ensemble de mesures contribuant à préserver, maintenir, restaurer ou mettre en valeur l'intégrité des milieux naturels ; ►biotope) ; *syn.* gestion [f] des habitats naturels ; *g* **Biotoppflege [f, o. Pl.]** (Gesamtheit der Maßnahmen, die dazu dienen, einen ►Biotop in seinem Gesamtgefüge und naturraumtypischen Umfeld zu erhalten und, wenn nötig, zu verbessern oder wiederherzustellen und eine strukturreiche Kul-turlandschaft zu erhalten und zu fördern).

2590 habitat mapping [n] *conserv. phyt. zool.* (Recording, categorizing and plotting the extent and distribution of important wildlife habitats and their species inventory within a specifically defined area; ►habitat 1, ►habitat 2, ►habitat network, ►land-scape feature, ►urban habitat mapping; *genegric term* ►map-ping); *syn.* biotope mapping [n]; *s* **inventario [m] cartográfico de biótopos** (Registro de especies y de ►biótopos con fines de conservación; ►componente del paisaje, ►hábitat, ►inventario de biótopos urbanos, ►reticulación de biótopos; *término gené-rico* ►inventario cartográfico); *f 1* **cartographie [f] des biotopes** (Technique d'élaboration de cartes de ►biotopes, d'►habitats, d'►éléments de paysages ; ►cartographie des biotopes urbains, ►constitution d'un réseau d'habitats naturels ; *terme générique* ►cartographie) ; *syn.* cartographie [f] des habitats naturels ; *f 2* **cartographie [f] écobiocénotique** (Relevé cartographique dynamique des communautés naturelles) ; *g* **Biotopkartierung [f]** (Bestandsaufnahme und grafische Darstellung von ►Bio-topen, ►Habitaten und ►Landschaftsbestandteilen aus der Sicht des Naturschutzes und der Landschaftspflege auf Karten, Plänen oder in einem Atlas; ►Biotopvernetzung, ►Stadtbiotopkar-tierung; *OB* ►Kartierung).

2591 habitat network [n] *conserv. ecol.* (Connected green areas, wetland habitats, hedgerows, etc. for habitat preservation; ►green space corridor, ►process of habitat fragmentation);

s **reticulación [f] de biótopos** (Creación de conexiones entre los diferentes biótopos en una región para garantizar la interconectividad entre ellos y así permitir el intercambio mutuo incluido el genético y asegurar su preservación a largo plazo; ▶aislamiento de biótopos, ▶corredor verde); *syn.* red [f] de biótopos; *f 1* **constitution [f] d'un réseau d'habitats naturels** (Constitution d'un système d'espaces naturels ou de surfaces à pratique agricole extensive reliant entre eux des biotopes isolés sensibles. Objectif de cette mesure : la reconstitution de populations animales et végétales par une augmentation des aires de répartition naturelles des espèces et l'échange génétique des espèces sauvages, la mise en valeur de la qualité des paysages et une augmentation de leur valeur récréative ; ▶coupure verte) ; *syn.* constitution [f] d'un réseau de biotopes, constitution [f] d'un tissu d'habitats naturels, constitution [f] d'un tissu de biotopes, constitution [f] d'un réseau de corridors écologiques, mise [f] en place d'un réseau de biotopes ; mise [f] en réseau des biotopes [B et CH] ; *f 2* **réseau [m] de biotopes** (Système d'espaces vitaux favorisant la conservation et le développement des relations spatiales entre les milieux naturels dans le but d'assurer à long terme la survie des espèces de faune et de flore, des communautés biologiques et de leurs habitats ; résultat du maillage fonctionnel d'éléments de paysage interconnectés participant aux déplacements des espèces, aux échanges génétiques indispensables à la régénération des milieux naturels et diminuant l'▶isolement des biotopes) ; *g* **Biotopvernetzung [f]** (1. Herstellung eines Flächensystems zur Erhaltung und Wiederherstellung der räumlichen Voraussetzungen und funktionalen Beziehungen in Natur und Landschaft mit dem Ziel, Tiere, Pflanzen, ihre Lebensgemeinschaften und Lebensräume langfristig zu sichern. Dabei beziehen sich die räumlichen Voraussetzungen auf die Sicherung und Bereitstellung von Flächen für ein funktional zusammenhängendes „Netz", das landschaftstypische Lebensräume und Lebensraumkomplexe einbindet und das den Auswirkungen räumlicher ▶Verinselung von Biotopen entgegenwirkt; cf. BUR 2004. 2. System naturnaher oder extensiv genutzter Flächen, das verstreut und isoliert liegende, gleichartige oder ähnliche Biotope miteinander verbindet. Ziele von Maßnahmen, die diesen Verbund wiederherstellen, sind einmal die Ermöglichung und Förderung der Ausbreitung von Pflanzen- und Tierarten sowie des Genaustausches zwischen isolierten Populationen und zum anderen sollen von wild lebenden Arten gering besiedelte Landschaftsteile wieder als Lebensraum zur Verfügung stehen; ferner wird dadurch das Landschaftsbild bereichert und meist der Erholungswert der Landschaft erhöht; ▶Grünzug); *syn. zu 2.* Biotopverbund [m], Biotopverbundsystem [n], Netz [n] verbundener Biotope, Biotopnetz [n].

habitat [n] of community interest *conserv. ecol.* ▶natural habitat of community interest.

habitat range [n] [US] *phyt. zool. leg.* ▶living space.

2592 habitat restoration [n] *landsc. conserv.* (Measures to restore a destroyed or degraded habitat to its previous functioning condition); *syn.* reinstallment [n] of a habitat, replacement [n] of a habitat; *s* **restauración [f] de un biótopo** (Medidas que sirven para restaurar el estado original sostenible de un biótopo degradado o destruido; restaurar [vb]/regenerar [vb] un biótopo); *syn.* regeneración [f] de un biótopo; *f* **réhabilitation [f] d'un biotope** (Mesures favorisant le rétablissement des conditions nécessaires à la restauration d'un biotope préalablement détruit ou dégradé ; la réhabilitation doit être en particulier réalisée lorsqu'un habitat naturel, bien qu'ayant disparu dans une grande unité paysagère, était autrefois largement représenté en raison des conditions stationnelles) ; *syn.* restauration [f] d'un biotope ; *g* **Wiederherstellung [f] eines Biotopes** (Maßnahmen, die einen zerstörten oder degradierten Biotop in seinen früheren funktionsfähigen

Zustand zurückversetzen. Die **W.** soll besonders immer dann erfolgen, wenn ein Lebensraumtyp in einer naturräumlichen Haupteinheit nicht mehr vorkommt, aber auf Grund der Standortbedingungen ursprünglich dort verbreitet war).

habitats [npl], feasibility of recreating *conserv.* ▶recoverability.

2593 Habitats Directive [n] *conserv. leg.* (Obligatory for all member countries of the European Union, this is the shortened term for the European Flora, Fauna, Habitats Directive—more formally known as **Council Directive 92/43/EEC on the Conservation of natural habitats and of wild fauna and flora**—dated 21.05.1992, and amended by 97/62 EEC of 27.10.1997 with the primary goal of preserving or re-establishing the biological diversity of Europe. To support this, a European-wide network of conservation areas, ▶Natura 2000, is to be developed, because the protection of individual, isolated areas alone cannot promote biodiversity. Instead individual habitats must be linked with one another via landscape feature ssuch as watercourses, river and road embankments, ribbon-like vegetation structures, etc. The directive also contains basic procedures fornature conservation under the establishment of NATURA 2000. In Appendix I of the FFH directive approximately 250 natural or near-natural habitat types are listed, which are worthy of protection. Appendix II contains approximately 250 animal and 450 plant species in the European Union, which are strongly endangered and in need of protection. Habitats requiring such strict conservation are priority areas and given special attention); *syn.* Directive [f] for the conservation of natural habitats, wild animals and plants; *s* **Directiva [f] de Hábitats** (Directiva 92/43/CEE del Consejo, de 21 de mayo de 1992, relativa a la conservación de los hábitats naturales y de la fauna y flora silvestres, modificada por la Directiva 97/62/CE de 27.10.97 y por Reglamento [CE] n° 1882/2003 del 23.12. 2003, que tiene el objeto de contribuir a garantizar la biodiversidad mediante la conservación de los hábitats naturales y de la fauna y flora silvestres en el territorio europeo de los Estados miembros. Para alcanzar este objetivo crea una red ecológica europea coherente de *zonas especiales de conservación*, denominada ▶Red Ecológica Europea Natura 2000, para garantizar el mantenimiento o el reestablecimiento, en un estado de conservación favorable, de los tipos de hábitats naturales y de los hábitats de las especies de que se trate en su área de distribución natural, que así mismo constituyen ejemplos representativos de características típicas de una o varias de las siete regiones biogeográficas siguientes: alpina, atlántica, boreal, continental, macaronesia, mediterránea y panónica. Esta red incluye también las *zonas de especial protección para las aves* designadas con arreglo a la Directiva de Aves. La **D. de H.** tiene seis anexos. El anexo I [Tipos de hábitats naturales de interés comunitario cuya conservación requiere la designación de zonas de especial conservación] enumera unos 250 tipos de hábitats que merecen protección y determina los llamados «hábitats prioritarios» que son aquéllos amenazados de desaparición. El anexo II [Especies animales y vegetales de interés comunitario para cuya conservación es necesario designar zonas especiales de conservación] contiene unas 250 especies de fauna y unas 450 de flora de la UE que están fuertemente amenazadas y necesitan protección. El anexo III define los criterios de selección de los lugares que pueden clasificarse como lugares de importancia comunitaria y designarse zonas especiales de conservación. El anexo IV nombra las especies animales y vegetales de interés comunitario que requieren una protección estricta. El anexo V define las especies cuya recogida en la naturaleza y cuya explotación pueden ser objeto de medidas de gestión. El anexo VI determina los métodos y medios de captura y sacrificio y modos de transporte prohibidos); *f* **Directive [f] Habitat Faune**

H

Flore [DHFF] (La Directive 92/43/CEE du 21 mai 1992, modifiée par la Directive 97/62/CE du 27 octobre 1997 et le Règlement (CE) n° 1882/2003 du 29 septembre 2003, dénommée « **Directive Habitat** », contribue à assurer la biodiversité par la conservation des habitats naturels ainsi que de la faune et flore sauvage sur le territoire européen ; cette directive complète la directive Oiseaux et en accord avec les exigences de la ►Convention de Berne [1979], dont elle reprend les grandes lignes, les renforce et les amplifie sur le territoire des Etats membres de la Communauté Européenne ; elle donne pour objectif final la constitution et la préservation d'un réseau européen cohérent de zones spéciales de conservation [ZSC], dénommé « Natura 2000 » afin de conserver ou rétablir les habitats et les espèces d'intérêt communautaire dans leur aire de répartition naturelle ; elle prend en compte sept **zones biogéographiques** [alpine, atlantique, boréale, continentale, macaronésienne, méditerranéenne et pannonienne] et est composée de 6 annexes ; les **annexes I** [types d'habitats naturels d'intérêt communautaire] et **II** [espèces animales et végétales d'intérêt communautaire] de la directive fournissent des indications quant aux types d'habitats et d'espèces dont la conservation nécessite la désignation de ZSC [certains d'entre eux sont définis comme des types d'habitats ou des espèces « prioritaires » — en danger de disparition], l'**annexe III** donne les critères de sélection des sites susceptibles d'être identifiés comme d'importance communautaire et désignés comme ZSC, l'**annexe IV** énumère les espèces animales et végétales d'intérêt communautaire qui nécessitent une protection stricte, l'**annexe V** concerne les espèces animales et végétales d'intérêt communautaire dont le prélèvements dans la nature et l'exploitation sont susceptibles de faire l'objet de mesures de gestion et l'**annexe VI** énumère les méthodes et moyens de capture et de mise à mort et modes de transport interdits ; le réseau Natura 2000 est réalisé au travers de la mise en place de mesures de protection ou de gestion des zones concernées, en tenant compte des exigences économiques, sociales, culturelles et des particularités locales, afin de contribuer au développement durable ; la désignation des ZSC se fait en trois étapes [**1.** inventaire des sites et établissement d'une liste nationale de **p**roposition de **s**ite d'**i**ntérêt **c**ommunautaire [PSIC], **2.** la Commission arrête une liste des ►sites d'importance communautaire [SIC] pour chacune des sept régions biogéographiques, **3.** après sélection d'un site comme SIC, l'État membre concerné désigne ce site comme ZSC ; celui-ci est doté d'un ►document d'objectifs [DOCOB], établi en concertation avec les acteurs locaux intéressés et devant fixer les orientations de gestion et les moyens financiers d'accompagnement ; il sera le document de référence au plan régional comme au plan européen pour une gestion équilibrée des territoires mais aussi pour l'obtention des cofinancements nationaux, européens et locaux mis en place pour sa gestion ; la directive prévoie que les états membres prennent les mesures appropriées pour éviter, non seulement à l'intérieur du périmètre des ZSC, la détérioration des habitats naturels et des habitats d'espèces ainsi que les perturbations susceptibles d'avoir un effet significatif sur les espèces pour lesquelles les zones ont été désignées, mais aussi en dehors, quand des activités sont susceptibles d'avoir un effet négatif sur le site ; à cet effet une **étude d'incidence** doit être réaliser pour tous les projets ou plans qui ne sont pas directement liés à la gestion des sites mais susceptible d'affecter ce site de manière significative, individuellement ou en conjugaison avec d'autres plans et projets ; cf. Directive « Habitats » Art. 6) ; *g* **Fauna-Flora-Habitat-Richtlinie [f]** (*Abk.* FFH-RL; für jedes Mitgliedsland der Europäischen Union [EU] verbindlich umzusetzendes Recht [cf. Richtlinie 92/43/EWG des Rates vom 21.05.1992, geändert durch Richtlinie 97/62 EG vom 27.10.1997] mit dem vorrangigen Ziel der Erhaltung der in Europa vorhandenen biologischen Vielfalt resp. deren Wiederherstellung. Dazu wird ein europaweit vernetztes Schutzgebietssystem ►Natura 2000 aufgebaut, da durch den Schutz einzelner, isolierter Gebiete die biologische Vielfalt nicht dauerhaft erhalten werden kann. Vielmehr müssen einzelne Lebensräume miteinander über Landschaftselemente wie z. B. Fließgewässer, Böschungen, bandartige Vegetationsstrukturen etc. verknüpft werden. Die Richtlinie beinhaltet neben dieser Zielsetzung auch naturschutzfachliche Grundlagen und Verfahrensvorgaben zur Errichtung von NATURA 2000. In **Anhang I** der FFH-RL sind rund 250 zu schützende natürliche und naturnahe Lebensraumtypen von gemeinschaftlichem Interesse aufgelistet. Der **Anhang II** beinhaltet rund 250 Tier- und 450 Pflanzenarten von gemeinschaftlichem Interesse, die EU-weit als stark gefährdet gelten und für deren Erhaltung besondere Schutzgebiete ausgewiesen werden müssen; **Anhang III** beinhaltet Kriterien zur Auswahl der Gebiete, die als Gebiete von gemeinschaftlicher Bedeutung bestimmt und als besondere Schutzgebiete ausgewiesen werden könnten; **Anhang IV** beinhaltet eine Liste streng zu schützender Tier- und Pflanzenarten von gemeinschaftlichem Interesse; **Anhang V** beinhaltet Tier- und Pflanzenarten von gemeinschaftlichem Interesse, deren Entnahme aus der Natur und deren Nutzung Gegenstand von Verwaltungsmaßnahmen sein können; **Anhang VI** behandelt verbotene Methoden und Mittel des Fangs, der Tötung und Beförderung. Die Schutzvorschriften der FFH-RL wurden in die §§ 19a-19f des BNatSchG übernommen); *syn.* Richtlinie [f] zur Erhaltung der natürlichen Lebensräume sowie der wild lebenden Tiere und Pflanzen, FFH-Richtlinie [f].

2594 Habitats Directive Management Plan [n] *conserv. ecol.* (Plan for each of the NATURA-2000 special protection areas, which according to Article 6 paragraph 1 of the ►Habitats Directive, require conservation measures to maintain favourable conditions for habitats and species within the context of the European network 'Natura 2000'; similarly management plans for huntable bird species are also being implemented according to the Birds Directive. The standard procedure of a Management Plan contains 6 stages: **1.** Elaboration of the legal, organisational and methodical base data for the Natura-2000 area. **2.** Description of the general characteristics of the area [information on natural site conditions, ownership, past and present usage, conservation status, knowledge of ecology, importance for the European network]. **3.** Conservation measures for the area [aims, habitat type according to Annex I, selection, results, species according to Annex II of the Directive]. **4.** Comprehensive review [discussion of the conservation status of the habitat types according to Annex I and species according to Annex II, complete evaluation, threat analysis, conflicts and synergies, necessary conservation measures, final assessment of function and fulfillment/fulfilment of the function of the area within the natural landscape unit, instruments for implementtation]. **5.** Suggestions for conservation concepts. **6.** Recommendations for monitoring and control of success. The first drafts of management plans for huntable bird species to be considered in unfavourable conservation status were prepared between 1997 and 2000. Subsequently, the plans were updated with new information, expanded to cover EU25 and the layout has been changed to be harmonised with those developed under international agreements. The long-term objectives of the plans are to restore the populations of the species to a favourable conservation status in the EU. Therefore, the plans identify short-term objectives to be achieved during the initial 3-years period. These address the most urgent issues to halt the species declines in the EU. Evaluation and review of the plans after three years includes an assessment of the results achieved. During this process new objectives for the next period are identified to ensure

the recovery of the populations and the achievement of the long-term objectives); *s* **plan [m] de gestión de la directiva de hábitats** (Plan que debe ser elaborado para cada una de las zonas especiales de conservación de la red Natura 2000, que según el art. 6 para. 1 de la ▶Directiva de Hábitats necesitan medidas de conservación para mantener las condiciones favorables de los hábitats y las especies de interés comunitario. El procedimiento estándar de un **p. de g.** incluye seis fases: **1.** Elaboración de la base de datos legales, organizacionales y metodológicos para el área Natura 2000. **2.** Descripción de las características generales del área [condiciones naturales del lugar, régimen de propiedad, usos pasados y presentes, estado de protección, conocimientos de su ecología, importancia para la red europea]. **3.** Medidas de conservación para el área [objetivos, tipo de hábitat de acuerdo al anexo I, selección, resultados, especies de acuerdo al anexo II]. **4.** Análisis de conjunto [valoración del estado de conservación de los tipos de hábitats del anexo I y de las especies del anexo II, evaluación general, análisis de las amenazas, conflictos y sinergias, medidas de conservación necesarias, evaluación final de la funcionalidad y del cumplimiento de la función del área dentro de la unidad de paisaje natural, instrumentos para la implementación]. **5.** Propuestas para planteamientos de conservación. **6.** Recomendaciones para el monitoreo y el control del éxito. Los objetivos a largo plazo son restaurar las poblaciones de las especies hasta un estado de conservación favorable en la UE. Por lo tanto, los planes identifican objetivos a ser cumplidos en el periodo/período inicial de los primeros tres años. Estos apuntan a los problemas primordiales para frenar la desaparición de especies en la UE. La evaluación y actualización de los planes después de tres años incluye la valoración de los resultados alcanzados. Durante este proceso se identifican nuevos objetivos para el siguiente periodo para asegurar la recuperación de la población y el cumplimiento de los objetivos de largo plazo); *f* **document [m] d'objectifs [DOCOB]** (Plan de gestion élaboré pour une durée de 6 ans sur un site Natura 2000 [art. 6 paragraphe 1 ▶Directive Habitat Faune Flore] ; il est établi en concertation avec les acteurs locaux ; il vise à définir les objectifs et les orientations de gestion et préciser les moyens à utiliser pour le maintien ou le rétablissement des habitats et espèces dans un état de conservation favorable ; ces propositions élaborées localement et issues d'une concertation approfondie sont approuvées par l'Etat qui a la responsabilité de l'application des directives ; c'est un document de diagnostic et d'orientation de la gestion des sites Natura 2000 ; il est une aide à la décision pour les acteurs ayant compétence ou droits d'usage sur le site et vise à la mise en cohérence des actions publiques et privées, des activités économiques, sociales et culturelles qui ont une incidence directe ou indirecte sur les habitats ou espèces d'intérêts communautaires ; le DOCOB des zones de protection spéciales [ZPS] est souvent réalisé avec celui des ZSC lorsque les périmètres sont superposés ; chaque site Natura 2000 est géré par un gestionnaire désigné lors de la création du site ; un **comité de pilotage local [COPIL]** élabore le DOCOB et est chargé de la mise en œuvre et du suivi des actions de gestion du site; il est composé de tous les acteurs en présence sur le site [collectivités territoriales, administrations et établissements socioprofessionnels et association de protection de la nature, propriétaires ayant droit et usagers, experts et pouvant être complété par des personnes de droit public ou de droit privé] ; les tâches administratives, techniques et financières afférentes à l'élaboration du DOCOB sont en général assurées en régie par une collectivité territoriale qui peut faire appel à un organisme ou structure tiers que l'on appelle « opérateur », en charge de l'animation du comité de pilotage et de la rédaction du DOCOB ; un « Guide méthodologique » pour la réalisation des documents d'objectifs issu d'un programme européen est utilisé sur 37 sites

pilotes français ; celui-ci est constitué de 6 grandes parties : **1.** un rapport de présentation établissant un diagnostic du territoire centré sur les espèces et habitats d'intérêt communautaire présents [identification, cartographie, évaluation de l'état de conservation], les mesures et actions de protection de toute nature qui s'appliquent au site, les activités humaines présentes sur le site [agriculture, loisirs, sites industriels, infrastructures, etc.] et leurs interactions avec les espèces et habitats d'intérêt leurs effets sur leur état de conservation. **2.** La définition d'objectifs de développement durable du site dont la finalité est de maintenir, voire d'accroître, la quantité et la qualité des habitats et espèces présents en tenant compte des enjeux socio-économiques ou culturels qui s'exercent sur le site. **3.** La définition de mesures permettant d'atteindre ces objectifs avec la priorité dans leur mise en œuvre. **4.** Un ou plusieurs cahiers des charges applicables aux contrats Natura 2000 précisant les bonnes pratiques à respecter et l'indication des dispositifs en particulier financiers destinés à faciliter la réalisation des objectifs. **5.** La liste des engagements faisant l'objet de la charte Natura 2000 du site ; **6.** les modalités de suivi et d'évaluation des mesures proposées et de l'état de conservation des habitats naturels et espèces ; cf. Décret n° 2006-922 du 26 juillet 2006 relatif à la gestion des sites Natura 2000) ; *g* **FFH-Managementplan [m]** (Spezieller Plan für die jeweiligen NATURA-2000-Schutzgebiete, für die nach Artikel 6 Absatz 1 der ▶Fauna-Flora-Habitat-Richtlinie Erhaltungsmaßnahmen notwendig sind, um einen günstigen Erhaltungszustand der Lebensraumtypen und/oder Arten zu gewährleisten, die maßgeblich für die Aufnahme in das Europäische Netz „Natura 2000" waren; analog dazu werden auch für Europäische Vogelschutzgebiete Managementpläne erstellt. In Bayern wird der Managementplan als Leitlinie staatlichen Handelns verstanden und von den Fachbehörden erstellt; er soll Klarheit und Planungssicherheit schaffen, hat jedoch keine rechtliche Bindungswirkung für die bestehende Nutzung durch die Grundeigentümer. Rechtsverbindlich ist nur das gesetzliche Verschlechterungsverbot, das unabhängig vom Managementplan greift; alle Maßnahmen, die zu einer erheblichen Verschlechterung der für das Gebiet maßgeblichen Lebensraumtypen und Arten führen, sind demnach verboten. Der **FFH-M.** soll am Runden Tisch offen und gegenüber den Belangen der Grundeigentümer aufgeschlossen diskutiert werden; die Mustergliederung des Managementplans sieht 6 Kapitel vor: **1.** Darstellung der gesetzlichen, organisatorischen und methodischen Grundlagen für das Natura-2000-Gebiet. **2.** Beschreibung der Allgemeinen Gebietseigenschaften [Allgemeine Gebietsbeschreibung, Besitzverteilung, natürliche Grundlagen, Nutzungsgeschichte und gegenwärtige Nutzung, Schutzsituation, Ökologischer Kenntnisstand, Rolle und Bedeutung des Gebietes im Europäischen Netz Natura 2000]. **3.** Schutzobjekte und Erhaltungsmaßnahmen [Erhaltungsziele, Lebensraumtypen nach Anhang I der FFH-RL, Auswahl, Ergebnisse, Arten nach Anhang II der FFH-RL]. **4.** Zusammenfassende Betrachtung [Darstellung des Erhaltungszustandes der Lebensraumtypen nach Anhang I und der Arten nach Anhang II, Gesamtbewertung, Gefährdungsanalyse, Zielkonflikte und Synergien, Notwendige Erhaltungsmaßnahmen, Gesamtbeurteilung der Funktion und der Funktionserfüllung des Gebietes im Naturraum, Umsetzungsinstrumente]. **5.** Vorschläge für eine Schutzkonzeption. **6.** Empfehlungen für Monitoring und Erfolgskontrolle; cf. LWF Arbeitsanweisung Managementpläne für Waldflächen NATURA 2000, Bayerische Landesanstalt für Wald und Forstwirtschaft, 2004); *syn.* FFH-Gebietsmanagementplan [m].

habitat transition line/zone [n] *ecol.* ▶transitional biotope.

habitat type [n] *conserv. ecol.* ▶natural habitat type, ▶priority natural habitat type.

H

hachure lines [npl] *surv.* ▶hachures.

2595 hachures [npl] *surv.* (Lines on a map depicting steep slopes to give a general indication of the relief; ▶contours, ▶form line); *syn.* hachure lines [npl]; *s* **hachuras** [fpl] (Rayado de líneas paralelas cortas perpendiculares a las cotas de nivel representando topografía; ▶curva de configuración, ▶curva de nivel); *f* **hachures** [fpl] (Lignes disposées suivant les lignes de plus grande pente sur des équidistantes horizontales exprimant suivant l'épaisseur et l'écartement la pente d'un terrain ; ▶courbe de niveau, ▶ligne de forme) ; *syn.* hachure [f] cartographique ; *g* **Schraffen** [fpl] (Senkrecht zu ▶Höhenlinien dargestellte Linien auf Karten, Plänen etc., die Böschungen oder Hänge anzeigen; ▶Höhenstrukturlinie).

2596 hadal [n] *ocean.* (Deepest part of the ocean, beneath the ▶abyssal zone, with ocean trenches over 6,000m in depth); *s* **zona** [f] **hadal** (Espacio submarino situado a más de 6000 m de profundidad más allá de la zona ▶abisal); *f* **zone** [f] **hadopéla-gique** (Dans la zonation verticale du milieu marin, strate hydrosphérique d'une profondeur au-dessous de 6000 m ; ▶zone abyssopélagoqic) ; *syn.* zone [f] hadale ; *g* **Hadal** [n] (Auf das ▶Abyssal folgender Abschnitt der Ozeane unterhalb ca. 6000-7000 m Wassertiefe); *syn.* hadale Region [f], Ultraabyssal [n].

half-shrub [n] *bot.* ▶suffruticose shrub.

half standard [n] [UK] *arb. hort.* ▶semi-dwarf fruit tree [US].

hall [n] *recr.* ▶multipurpose hall.

halophilous [adj] *plant.* ▶salt-tolerant.

2597 halophilous vegetation [n] *phyt.* (**1.** Plant communities on saline soils; e.g. on sea coasts, saline springs. **2.** Vegetation on sodium chloride, sodium sulfate, and sodium carbonate soils; ▶halophyte); *syn.* halophytic vegetation [n]; *s* **vegetación** [f] **halófila** (**1.** Comunidad vegetal con preferencia por los suelos salinos, a los que está especialmente adaptada por medio de ciertas estructuras morfológicas que le permite acumular sales o eliminarlas. **2.** Vegetación de suelos secos con muchas sales; BB 1979, 357; ▶halófito); *syn.* vegetación halófita; *f* **végétation** [f] **halophile** (Groupement végétal d'un terrain salé qui peut supporter un degré élevé de salinité ; ▶halophyte) ; *syn.* végétation [f] salitrale ; *g* **Salzflur** [f] (**1.** Pflanzengesellschaft[en] auf salzreichen Böden, z. B. an Meeresküsten, Brunnensalzstellen, auch im Binnenland, z. B. die Salzgebiete am Neusiedler See oder am Rande trockener Salzpfannen. **2.** Vegetation der Kochsalz-, Sulfat- und Sodaböden; ▶Salzpflanze); *syn.* Halophytenvegetation [f], Salzvegetation [f].

2598 halophyte [n] *bot.* (Greek *halos* >salt<, *phyton* >plant<; plant growing on saline soil—sea coasts, salt pans in arid or semiarid regions [US]/semi-arid regions [UK]—with the ability to store a high salt content in its cells); *s 1* **planta** [f] **halófila** (Planta [o sinecia] que crece solo en medios salinos [tierra o agua]; DB 1985); *s 2* **halófito** [m] (Planta propia de suelo fisiológicamente seco por su dosis excesiva de sales. Comprende las plantas halófilas no acuáticas; DB 1985); *f* **halophyte** [m] (Plante qui vit dans les milieux salés [côte marine, dépressions salées des régions arides ou semi-arides] et dont les tissus peuvent supporter un degré élevé de salinité) ; *g* **Salzpflanze** [f] (An salzreichen Standorten — Meeresküste, Ränder von Salzpfannen in ariden oder semiariden Gebieten — wachsende Pflanze, die durch einen Selektionsvorteil gegenüber anderen Pflanzen Salzwirkungen saliner Standorte ohne Schaden ertragen und einen hohen Salzgehalt in ihren Zellen speichern kann und zum Keimen und optimalen Wachstum mehr als 0,5 % Salzgehalt im Boden benötigt [obligater Halophyt]; cf. HER 2005); *syn.* Halophyt [m].

halophytes [npl] *phyt.* ▶vegetation of perennial halophytes.

halophytic vegetation [n] *phyt.* ▶halophilous vegetation.

hammer [n]**, facilities for throwing the** *recr.* ▶area for throwing the hammer.

hammered [pp/adj] *constr.* *▶tooling of stone [US]/dressing of stone [UK], #6.

2599 hammerhead turnaround [n] *trans. urb.* (Usually a T-shaped widening at the end of a ▶cul-de-sac to enable vehicles to reverse course using a 3-point turn maneuver/manoeuvre; ▶turnaround); *syn.* "T" turn [n] [also US], "Y" turn [n] [also US], T-shaped turnaround [n] [also US]; *s* **martillo** [m] **de (re)vuelta** (≠) (Ampliación de la calzada en forma de T al final de un ▶fondo de saco para que los vehículos puedan dar la vuelta; *término genérico* ▶revuelta); *syn.* martillo [m] de giro (≠); *f* **T** [m] **de retournement** (±) (▶Dispositif de retournement en forme de T à l'extrémité d'une voie en impasse ; ▶voie sans issue) ; *g* **Wendehammer** [m] (Meist T-förmige Fahrbahnausweitung [▶Wendeanlage] am Ende einer ▶Sackgasse für das Wenden von Fahrzeugen).

hammock [n] [US] *phyt.* ▶hummock.

hand [n] *constr.* ▶excavate by hand.

handicapped [pp] *leg. plan. sociol.* ▶developed for the handicapped/disabled.

handicapped [pp]**, accessible** [adj] **for the handicapped/disabled** *leg. plan. sociol.* ▶developed for the handicapped/disabled.

2600 handicapped person [n] *sociol. syn.* disabled person [n]; *s* **persona** [f] **con movilidad reducida** (Cualquier persona con una discapacidad física, sea por enfermedad, edad o accidente, que le limita sus posibilidades de actividad y necesita condiciones de acceso adecuadas a los espacios públicos y privados. En alemán el término «Behinderte/,-r» se aplica también para personas con deficiencias mentales); *syn.* persona [f] con discapacidad (física), limitado/a [m/f] físico-motor/a [C], impedido/a [m/f] físico/a, discapacitado/a [m/f], *obs.* minusválido/a [m/f]; *f* **personne** [f] **à mobilité réduite (PMR)** (Toute personne dont le handicap subi dans son environnement constitue une limitation d'activité ou une restriction de participation à la vie en société en raison d'une altération substantielle, durable ou définitive d'une ou plusieurs fonctions physiques, motrices ou sensorielles, d'une déficience des fonctions mentales, cognitives ou psychiques, d'un poly-handicap ou d'un trouble de santé invalidant, de l'âge et dont la situation requiert une attention particulière et une adaptation spécifique des services proposés) ; *syn.* personne [f] handicapée, handicapé [m], handicapée [f] ; *g* **Behinderte/-r** [f/m] (Mensch, dessen körperliche Funktion, geistige Fähigkeit oder seelische Gesundheit mit hoher Wahrscheinlichkeit länger als sechs Monate von dem für das Alter typischen Zustand abweichen und dessen Teilhabe am Leben in der Gesellschaft daher beeinträchtigt ist; § 2 Sozialgesetzbuch IX).

handicapped persons [npl] *leg. plan. sociol.* ▶construction for handicapped persons.

handicapped persons [npl]**, garden for** *gard.* ▶garden for the handicapped.

2601 hand-laid stone subbase [n] [US] *constr.* (Manual method of installing the ▶subbase [US]/sub-base [UK] of paths using local, crushed natural stone [particle size 50-250mm]; this hand-packed Telford [type] subbase [US] (BU 1973) is uneconomic due to the high proportion of labo[u]r involved; ▶hard base [US]/hardcore [UK]); *syn.* hand-pitched stone sub-base [n] [UK]; *s* **cimiento** [m] **Telford** (Método manual de establecer la ▶capa portante de caminos utilizando piedra gruesa triturada de tamaño entre 50-250 mm de origen local. Por su alto costo de mano de obra en los países industrializados es un método prácticamente obsoleto; ►subbase de piedra compactada); *syn.* cimiento [m] de piedra gruesa, fundación [f] Telford, base [f] Telford (BU 1959); *f* **empierrement [m] manuel** (►Couche de fondation constituée de matériaux concassés extrait d'un gisement local [granularité 5/25 cm] ; dans cette technique ancienne de réalisation d'allées stabilisées, la forme est bloquée à la main, pierre par pierre pour former un hérisson consolidé par du sable tout venant ou du concassé de carrière ; GEN 1982-II, 57-58 ; ►empierrement) ; *syn.* empierrement [m] à l'eau ; *g* **Setzpacklage [f]** (Wegen des hohen Lohnanteils heute unwirtschaftliche, manuell ausgeführte ►Tragschicht im Wegebau aus ortsnahem, gebrochenem Natursteinmaterial in den Körnungen 50-250 mm; ►Schüttpacklage).

handover [n] [UK] *contr. prof.* ►delivery [US] (1).

handover plan [n] [UK] *constr. contr.* ►delivered plan [US].

hand-pitched stone subbase [n] [UK] *constr.* ►hand-laid stone subbase [US].

2602 handrail [n] *arch. constr.* (Security railing held by the hand when using a stairway; ►railing from falling); *s* **pasamanos [m]** (Parte superior de una barandilla a la que van empotrados o apoyadas las cabezas de los barrotes o balaustres de una escalera; DACO 1988; ►protección contra el riesgo de caídas); *f* **maincourante [f]** (►protection contre les [risques de] chutes de hauteur); *g* **Handlauf [m]** (Seitliche Begrenzung einer Treppe in Form eines Geländers aus Stahl, Holz etc., an dem man sich mit der Hand festhalten kann; ►Absturzsicherung).

handrail [n] *gard.* ►planter serving as a handrail.

2603 hand sowing [n] *constr. hort.* (Spreading or broadcasting seeds by hand; ►sowing 2); *s* **siembra [f] manual** (Diseminación manual de semillas; ►siembra 2); *syn.* siembra [f] al vuelo, siembra [f] a voleo; *f* **semis [m] manuel** (Épandage à la main de semences, de fruits ou de jeunes pousses ; *terme générique* ►semis 3) ; *syn.* semis [m] à la main, semis [m] à la volée ; *g* **Handsaat [f]** (Ausstreuen von Samen, Früchten oder Sprossen mit der Hand; *UB zu* ►Aussaat); *syn.* Ansaat [f] von Hand, Handausbringung [f].

hand work [n] [US] *constr.* ►manual labor [US]/manual labour [UK].

2604 hanging beam [n] *constr.* (Wooden cross-piece laid upon ►beams of a pergola); *syn.* overhead timber [n] [also US]; *s* **vigueta [f]** (Listón de madera que se coloca sobre las vigas para conformar una pérgola; ►viga horizontal); *f* **solive [f]** (Élément de couverture d'une pergola posé sur les ►sablières) ; *g* **Auflageholz [n]** (…hölzer [pl]; Holz, das auf den ►Pfetten einer Pergola aufliegt).

2605 haphazard development [n] *urb.* (Disorderly urban growth); *s* **desarrollo [m] urbano anárquico** (Crecimiento urbano sin planificación); *f* **développement [m] anarchique de l'urbanisation** *syn.* développement [m] désordonné de l'urbanisation ; *g* **ungeordnete bauliche Entwicklung [f]**.

2606 harbor mud [n] [US]/**harbour mud** [n] [UK] *envir.* (Mostly contaminated sludge of dredged sediments from ship basins); *s* **légamo [m] portuario** (Sedimentos mostly contaminados dragados de los puertos industriales que necesitan tratamiento especial antes de ser depositados o utilizados; *f* **boue [f] de dragage portuaire** (Vase provenant du dragage d'un bassin portuaire, en général fortement contaminée et considérée comme déchets) ; *g* **Hafenschlick [m]** (Meist hoch kontaminierter Baggerschlamm aus Hafenbecken).

2607 hard base [n] [US] *constr.* (Layer, normally of coarse material, composed of local, broken stone, slag or cinders with particle sizes of 50-250mm used as a subbase in path and road construction; ►subbase [US]/sub-base [UK]); *syn.* hardcore [n]

[UK]; *s* **subbase [f] de piedra compactada** (En construcción viaria, ►capa portante de piedra natural, escorias de altos hornos o industriales de granulometría entre 50-250 mm); *f* **empierrement [m]** (►Couche de fondation ou de base d'une chaussée constituée de matériaux concassés ou de tout-venant extrait d'un gisement local [granularité 5/25 cm] ou de mâchefer industriel apporté en vrac, étalé, fortement cylindré et sablé ; cf. DTB 1985, 35) ; *syn.* empierrement [m] compacté ; *g* **Schüttpacklage [f]** (►Tragschicht im Wegebau aus ortsnahem, gebrochenem Natursteinmaterial, aus Hochofen- oder Industrieschlacken in den Körnungen 50-250 mm).

hardcore [n] [UK] *constr.* ►hard base [US].

2608 hard court [n] [US] *constr. recr.* (Generic term covering ►granular playing court [US]/hoggin playing surface [UK], asphalted concrete, or synthetic playing surfaces; ►all-weather court [US]/all-weather pitch [UK]; ►granular playing surface); *syn.* hard pitch [n] [UK]; *s* **cancha [f] de pavimento duro** (Término genérico para ►superficie de arenilla, superficie agregada con bitumen o superficie de plástico; ►área impermeabilizada para juegos); *f* **aire [f] à revêtement dur** (Terme générique pour les ►surfaces de sport à revêtement perméable, les aires sportives à revêtement bitumineux ou synthétique ; ►aire tous temps, ►surface en terre battue) ; *g* **Hartplatz [m]** (*UBe* ►Tennenfläche, bitumengebundene oder Kunststofffläche für Sportaktivitäten; ►Allwetterplatz).

hardiness [n] *agr. bot. for. hort.* ►cold hardiness.

2609 hard landscaping [n] *constr. plan.* (Planning and execution of non-living components of open spaces, including paved surfaces, steps, walls, overhead structures, and amenities in comparison with ►soft landscaping); *s* **diseño [m] y construcción [f] de componentes duros/inertes en espacios libres (≠)** (Planificación y ejecución de los componentes inertes de espacios libres incluyendo superficies pavimentadas, escaleras, muros, estructuras de cobertura e instalaciones de juego; ►diseño, ejecución y plantación de componentes vivos en espacios libres); *f* **conception [f] des éléments inertes de l'espace** (Élaboration et exécution de dallages, maçonneries paysagères et divers équipements d'un espace libre par comparaison avec la ►conception des éléments vivants de l'espace) ; *g* **technisch-bauliche Gestaltung [f] des Außenraumes (≠)** (Planung und Ausführung von befestigten Flächen, Treppen, Mauern und sonstigen technischen Einrichtungen und Möblierungselementen eines Freiraums; zum Unterschied zur ►vegetationsbestimmten Gestaltung des Außenraumes).

2610 hardpan [n] *pedol.* (**1.** Thin, hard stratum in the lower A horizon or B horizon where soil grains become cemented by such bonding agents as iron oxide and calcium carbonate; when the air spaces in the soil are filled with fin clay particles, the subsoil is called ►claypan; ►iron pan 1, ►orthod, ►ortstein, ►plow pan [US]/plough pan [UK]); *syn.* pan [n]; **2. fragipan** [n] (In American soil taxonomy, an altered subsurface horizon, 15cm or more thick, with very low organic matter, high bulk density, that restricts the entry of water and roots into the soil matrix when dry, but shows a moderate to weak brittleness when moist. Most fragipans have redoximorphic features, have evidence of translocation of clay, and are slowly or very slowly permeable to water; cf. KST 1996); **3. duripan** [n] (Subsurface soil horizon that is cemented by illuvial silica, usually opal or microcrystalline forms of silica, to the degree that less than 50% of the volume of air-dry fragments will slake in water or during prolonged soaking in acid [HCl]. Duripans vary in the degree of cementation by silica and, in addition, they commonly contain accessory cements, chiefly iron oxides and calcium carbonate. They form almost exclusively in arid or Mediterranean climates; soils with a duripan layer are predominantly used for grazing or wildlife habitat, and are seldom cultivated); *s 1* horizonte [m] petrificado (Término genérico para capa endurecida en la parte inferior del horizonte A o en el horizonte B como ▶horizonte de iluviación, ▶orterde, ▶ortstein, ▶piso de arado; ▶capa ferruginosa); *syn.* horizonte [m] endurecido, costra [f] cementada, «hardpan» [m] [Es, EC], tosca [f] [PA], suelo [m] duro [PE] (MESU 1977); *s 2* fragipán [m] (Término específico. En la clasificación de suelos de los EE.UU., horizonte subsuperficial de textura media, densidad aparentemente elevada en relación con los horizontes que lo rodean. Es duro y compacto cuando está seco, pero al humedecerse, las unidades estructurales se rompen bruscamente cuando se las presiona entre los dedos. Es pobre en materia orgánica, la conductividad hidráulica es lenta, el drenaje imperfecto, presenta motas y grietas emblanquecidas, de textura más gruesa y forma poligonal. Su formación parece estar asociada a las glaciaciones cuaternarias, el peso del hielo habría compactado a este horizonte, situado encima de uno permenentemente helado y debajo de otro helado durante el período frío; EDAFO 1994); *s 3* duripán [m] (Término específico. En la clasificación de suelos de los EE.UU., horizonte subsuperficial endurecido por sílice cementada en medio ácido, generalmente ópalo o formas microcristalinas de sílice. Los fragmentos secos no se deshacen en agua, ni en ácido clorhídrico. A veces contiene otros cementos secundarios, como carbonatos y óxidos de hierro. Los **dd.** se presentan principalmente en suelos de áreas con materiales volcánicos recientes, con climas mediterráneos subhúmedos o climas áridos. Durante la estación húmeda puede tener lugar la translocación de la sílice procedente de la meteorización de los vidrios volcánicos. Cuando un **d.** se halla en la superficie se denomina *duricrust*; cf. EDAFO 1994); *f 1* horizon [m] d'accumulation durci (Couche de faible épaisseur et dure formée dans l'horizon B des podzols [DG 1984, 11] ; terme générique pour ▶alios durci, ▶alios faiblement durci, ▶alios ferrugineux, ▶horizon d'accumulation d'argile, ▶semelle de labour) ; *f 2* horizon [m] de fragipan (*Classification américaine* horizon de profondeur fortement compacté, induré ou ayant une très haute teneur en argile peu perméable gorgé d'eau saisonnièrement ou en permanence ; cf. DIS 1986, PED 1983, 312 ; *contexte* sol hydromorphe à fragipan) ; *syn.* fragipan [m] ; *g* Verdichtungshorizont [m] (Dünne, harte Schicht unterhalb der Bodenoberfläche, entstanden durch Bodenentwicklungsprozesse [Vorkonsolidierung aus der Glazialzeit durch hohe Druckeinwirkung oder durch Verlagerung von Tonpartikeln, Eisenoxid oder Kalziumkarbonat in tiefere Bodenschichten] oder durch einseitige Bodenbearbeitung [▶Pflugsohle]; *UBe* ▶Eisenschwarte, ▶Orterde, ▶Ortstein, ▶Tonband; *in der amerikanischen Bodentaxonomie* ▶fragipan und ▶duripan).

hard pitch [n] [UK] *constr. recr.* ▶hard court [US].

hard rock [n] [US] *geo.* ▶solid rock.

2611 hardscape [n] *constr.* (Non-living components of open spaces, including paved surfaces, steps, walls, overhead structures, and amenities in comparison with ▶softscape; ▶hard landscaping, ▶stone construction); *syn.* hard landscape [n]; *s* componentes [mpl] duros de espacios libres (≠) (Conjunto de partes de una zona verde o jardín caracterizadas por sus elementos constructivos de materiales inertes como superficies pavimentadas, escaleras, muros, instalaciones técnicas o mobiliario, al contrario que la ▶superficie blanda caracterizada por la vegetación; ▶diseño y construcción de componentes duros/inertes en espacios libres, ▶obras de fábrica y de revestimiento de suelos); *f* éléments [mpl] inertes d'un espace libre (Circulations et maçonnerie paysagère correspondant aux éléments durs d'un espace tels que les sols durs, dallages, murets, escaliers, rocailles et enrochements, bassins et constructions et accessoires décoratifs divers

par opposition aux ▶éléments vivants de l'espace ; ▶conception des éléments inertes d'un espace, ▶maçonnerie paysagère) ; *g* Technisch-Bauliches [n] einer Freianlage (≠) (Gesamtheit der befestigten Flächen, Treppen, Mauern und sonstigen konstruktiven Einrichtungen und Möblierungselemente einer Außenanlage — zum Unterschied zur ▶vegetationsbestimmten Freianlage; *Kontext* zuerst wird das Technisch-Bauliche realisiert; ▶technisch-bauliche Gestaltung eines Außenraumes, ▶Steinbau im Garten- und Landschaftsbau).

hardscape area [n] [US] *constr. urb.* ▶paved area.

2612 hard shoulder [n] *trans.* (Paved margin adjacent to a travel lane [US]/carriage-way [UK] or parallel to its verge, used by stopping vehicles; in US usage a "pull-off lane" may be paved or unpaved); *syn.* paved pull-off lane [n] [also US], paved shoulder [n] [also US]; ▶road verge; *s* arcén [m] (Borde pavimentado de autopista o carretera donde pueden parar los vehículos en caso de emergencia; ▶berma 2); *f* accotement [m] stabilisé (DIR 1977, 9 ; surface latérale de la plate-forme bordant extérieurement la chaussée susceptible de supporter la charge d'un véhicule victime d'un accident, aménagée entre la chaussée et le fossé ; ▶accotement non stabilisé) ; *g* befestigter Seitenstreifen [m] (Teil der Straßenkrone, der neben der Fahrbahn resp. neben dem äußeren Randstreifen einer Straße liegt und da er kein Teil der Fahrbahn ist, nicht dem fließenden Verkehr dient; er ist im Vergleich zum ▶Bankett in geringem Maße befestigt, jedoch nicht so belastbar wie die Fahrbahn, und dient bei Bedarf der Entmischung des Verkehrs).

hard soil [n] *agr. hort.* ▶firm soil.

hard surface [n] *constr. urb.* ▶paved area.

hard surfaces [npl] *constr.* ▶construction of hard surfaces.

hard systems [npl] [UK] *constr. eng.* ▶conventional engineering.

2613 hard-wearing lawn [n] *constr.* (Grassed area for sports or other heavy use; ▶play lawn [US] 1/all-round lawn [UK]); *syn.* hard-wearing turf [n]; *s* césped [m] resistente (Césped que resiste tanto el pisoteo, como el peso de vehículos; ▶césped normal); *f* gazon [m] rustique (Variété de gazon utilisée pour résister sans protection particulière au piétinement des utilisateurs ou au roulement des véhicules ; *termes spécifiques* ▶gazon d'agrément et de détente, gazon résistant au piétinement, gazon résistant au roulement des véhicules) ; *g* belastbarer Rasen [m] (Häufigen Tritt und Befahren mit Fahrzeugen vertragender Rasen; *UBe* trittbelastbarer **R.**, für Fahrverkehr belastbarer **R.**; ▶Gebrauchsrasen); *syn.* strapazierfähiger Rasen [m].

hard-wearing turf [n] *constr.* ▶hard-wearing lawn.

2614 hardwood [n] *for.* (1. In commerce timber of *Angiospermae* [broad-leaved trees] is conventionally distinguished from that of *Gymnospermae*—needle-leaved trees; ▶density of wood, ▶softwood). 2. In English-speaking countries, timber of broad-leaved ▶trees 1 belonging to the botanical group of *Angiospermae*; *opp.* ▶softwood. 3. In German-speaking countries only very firm, heavy wood, characterized by a high percentage of wood fiber [US]/wood fibre [UK], and small-luminal wood; e.g. boxwood *[Buxus]*, oak *[Quercus]*, mahagony *[Swieténia mahágoni, S. macrophylla]*; ▶density of wood); *s* madera [f] dura (1. En el comercio de la madera, se denomina así la procedente de frondosas [Angiospermas], o sea de árboles de hoja ancha, al contrario que las coníferas o resinosas [gimnospermas] de hoja estrecha [acículas] que —en general— es mucho más blanda; ▶densidad de la madera; *opp.* ▶madera blanda. 2. En países de habla alemana, madera muy densa y pesada caracterizada por el alto contenido de fibra y conductos estrechos como la de boj, roble, caoba; ▶árbol, ▶densidad de la madera; *opp.* ▶madera blanda); *f 1* **bois [m] feuillu** (**U.S./U.K.**, Terme anglosaxon de sylviculture désignant les bois des feuillus car ceux-ci ont habituellement une densité élevée et donc plus dur que les résineux ; ▶arbre) ; *f 2* **bois [m] dur** (**1. F.**, en sylviculture bois caractérisé par une grande résistance, provenant des essences feuillues telles que l'érable, le hêtre, le chêne ou de quelques conifères tel que l'if, etc. ; par opposition au ▶bois tendre provenant en majorité des résineux ; on considère les bois comme durs pour une densité comprise entre 0,75 à 0,90 g/cm³, et comme tendres quand elle est comprise entre 0,55 à 0,65 g/cm³ ; ▶densité du bois. **2.** Bois très dense et lourd, caractérisé par un taux important de fibres ligneuses et d'éléments conducteurs, p. ex. le Buis *[Buxus]*, le Chêne *[Quercus]*, le Mahagoni *[Swieténia mahágoni, S. macrophylla]* ; *opp.* ▶bois tendre) ; *g* **Laubholz [n]** (…hölzer [pl]). **1.** Dieser forstliche Begriff bedeutet Holz von Laubbäumen. Im englischen Sprachgebrauch wird **L.** als *hardwood [Hartholz]* bezeichnet, weil das Laubholz ring- oder zerstreutporig ist, d. h., dass es keine Harzkanäle hat und schwerer, fester und härter als das Holz fast aller Nadelbäume ist. *BEACHTE* im Deutschen wird hingegen unter **Hartholz** ein Holz mit einer höheren Rohdichte als 0,55 g/cm³ verstanden, d. h. es ist ein sehr festes und schweres Holz, das durch hohen Anteil an Holzfasern und enge Gefäße gekennzeichnet ist. Hierzu gehören neben den harten **Laubhölzern** wie z. B. Eichen-, Eschen-, Robinien-, Bergahorn- und Buchsbaumholz auch das Nadelholz der Eibe [Rohdichte 0,8 bis 1,0 g/cm³] sowie Tropenhölzer, z. B. Mahagonibaum *[Swieténia mahágoni, S. macrophylla]*; cf. WFL 2002, 329, 452; ▶Baum, ▶Holzdichte; *opp.* ▶Weichholz).

hardwood brush [n], mat of live *constr.* ▶brush mat.

2615 hardy [adj] *bot. hort.* (Descriptive term applied to plants which stand up to winter cold without protection, and survive outdoors all year round; ▶frost-resistant); *s* **resistente al frío (invernal) [loc]** (Término aplicado a las plantas que aguantan bien el frío, por lo que pueden sobrevivir todo el año a la intemperie; ▶insensible a heladas); *f* **non gélif, ive [loc] (2)** (Caractérise une plante résistante aux températures hivernales ; ▶résistant, ante au gel) ; *g* **winterhart [adj]** (Eine Pflanze betreffend, die die winterliche Witterung gut zu überstehen vermag; ▶frostbeständig).

hardy [adj] [US], root *bot.* ▶perennial (2).

harmful insect [n] *agr. for. hort. phytopath.* ▶injurious insect.

2616 harpooning [n] *game'man. recr.* (Hunting of whale or fish with a barbed missile weapon on a long cord); *s* **pesca [f] con arpón** (Pesca de ballenas o peces con arpones; arponear [vb]); *f* **pêche [f] au harpon** ; *g* **Harpunieren [n, o. Pl.]** (Jagen im Wasser mit Harpune).

haulage [n] *constr.* ▶excavation, haulage, loading and transport of in-situ material.

2617 hauling [n] **away** [US] *constr.* (Removal of [unwanted] material from a site; *opp.* ▶hauling to the site [US]/carting to the site [UK]); *syn.* carrying [n] away [UK], carting [n] away [UK]; *s* **transporte [m] afuera de obra** (Evacuación de material sobrante de una obra que ya no se necesita; *opp.* ▶transporte a obra); *f* **évacuation [f]** (Enlèvement de matériaux et mise en décharge publique ; *opp.* ▶transport sur le chantier) ; *g* **Abfahren [n, o. Pl.]** (Wegfahren von Material, das auf einer Baustelle nicht mehr benötigt wird, oder die Abfuhr von Grüngut; *opp.* ▶Anfahren); *syn.* Abtransport [m], Abfuhr [f].

2618 hauling [n] **away without dumping certificate** [US] *constr. syn.* carting [n] away without tipping certificate [UK]; *s* **transportar [vb] sin justificación**; *f* **transport [m] à la décharge sans justificatif** ; *g* **Abfahren ohne Kippennachweis [m]** (ohne Kippennachweis abfahren [vb]).

hauling damage [n] [US] *for.* ▶felling and logging damage.

2619 hauling [n] **to the site** [US] *constr.* (Transporting construction material to the construction site. If soil material is transported to the site from another place it is called '**borrow material**' from a ▶borrow pit; *opp.* ▶hauling away [US]/carting away [UK]); *syn.* carting [n] to the site [UK], importing [n] [US]; *s* **transporte [m] a obra** (Suministro de una obra con los materiales necesarios; ▶zona de préstamo; *opp.* ▶transporte afuera de obra); *f* **transport [m] sur le chantier** (Transport de matériaux ; ▶lieu d'emprunt ; *opp.* ▶évacuation) ; *g* **Anfahren [n, o. Pl.]** (Transport von Baustoffen auf die Baustelle; ▶Seitenentnahme; *opp.* ▶Abfahren).

2620 haunched concrete [n] *constr.* (Wedge of in-situ concrete providing lateral support for a curb [US]/kerb [UK] or edging unit; ▶concrete edge footing [US]/concrete edge bedding [UK]); *s* **refuerzo [m] de hormigón** (Apoyo lateral de hormigón para bordillo de aceras o encintados; ▶solera de hormigón); *f* **butée [f] arrière en béton** (Renforcement en béton coulé sur place calant la rive arrière des bordures. *Contexte* contribué par un massif en béton ; *syn.* solin [m] de béton ; ▶assise en béton) ; *g* **Betonrückenstütze [f]** (Hinterbeton bei Bordsteinen, Kantensteinen, Randsteinen etc.; ▶Betonstuhl); *syn.* Hinterbetonierung [f].

2621 haustorium [n] *bot.* (Outgrowth from a hypha that penetrates a host cell in order to absorb nutrients); *s* **haustorio [m]** (Término propuesto por DE CANDOLLE, que generalmente equivale a chupador. Se aplica para órganos de morfología muy diversa que utilizan los antófitos parásitos, como el muérdago, para chupar los jugos vitales de la planta hospedante; cf. DB 1985); *f* **racine-suçoir [f]** (Jeune racine d'un végétal parasite puisant sa nourriture dans les tissus d'une plante-hôte) ; *syn.* suçoir [m] ; *g* **Saugwurzel [f] (2)** (Feiner Faden an den Haustorien von Schmarotzern, welche in den Wirt eindringen); *syn.* Haustorie [f].

2622 have [vb] **a bid accepted** [US] *constr.* (*Context* a bid/tender of a firm was accepted; ▶acceptance of a bid 2 [US]/acceptance of a tender 2 [UK]); *syn.* win [vb] the contract [also US], have a tender accepted [UK]; *s* **conseguir [vb] el remate** (Conseguir el contrato por ofrecer la oferta más económica, después de la evaluación de los resultados del concurso-subasta; ▶adjudicación del remate); *f* **être [vb] retenu** (*Contexte* l'offre de l'entreprise X a été retenue, l'entreprise X a été retenue pour le marché de ..., le candidat retenu est l'entreprise X ; ▶notification [de l'attribution] du marché à l'entreprise retenue) ; *syn.* être [vb] attributaire du marché ; *g* **den Zuschlag bekommen [vb]** (Als Bieter des annehmbarsten Angebotes nach Auswertung des Ergebnisses einer öffentlichen oder beschränkten Ausschreibung mit dem Auftrag betraut werden; ▶Zuschlagserteilung); *syn.* den Zuschlag erhalten [vb], den Auftrag bekommen/erhalten [vb].

have [vb] **a tender accepted** [UK] *constr.* ▶have a bid accepted [US].

2623 having recreation benefits [loc] *recr.* (Descriptive term applied to landscapes with favo[u]rable factors for recreation; ▶landscape component for recreation); *s* **con efecto recreativo [loc]** (Término descriptivo aplicado a paisajes con factores favorables para el recreo; ▶componente recreativo del paisaje); *f* **présentant un intérêt récréatif [loc]** (▶Élément du paysage présentant un intérêt récréatif) ; *syn.* à valeur récréative [loc], procurant une valeur récréative [loc] ; *g* **erholungswirksam [adj]** (Eine Landschaft oder deren Teile betreffend, die eine positive Wirkung für die Erholung hat; ▶erholungswirksames Landschaftselement).

H

2624 hay cutting [n] *agr. constr. hort.* (**1.** Mowing of grass for fodder. **2.** Mowing of meadow grass in green open spaces after the peak blooming and natural seeding of wildflowers for ecological and aesthetic reasons); *syn.* long grass cutting [n]; *s* **siega** [f] **del heno** (**1.** Siega de pradera para henificar. **2.** Corte de superficies de hierba en parques después de la floración principal); *syn.* henificación [f]; *f* **coupe** [f] **du fourrage** (**1.** Sectionnement sur pied des tiges des graminées et des plantes herbacées fourragères en vue du fanage. **2.** Sectionnement sur pied des espèces prairiales après la floraison dans les espaces verts urbains pour des raisons écologiques ou esthétiques) ; *syn.* fauche [f] du fourrage, fauchage [m] du fourrage ; *g* **Heuschnitt** [m] (**1.** Schnitt von langem Wiesengras zur Heugewinnung. **2.** Schnitt von langem Gras in Grünanlagen nach der Hauptblüte der Wiesenblumen — aus ökologischen und gestalterischen Gründen sowie zur Förderung der natürlichen Aussaat); *syn.* Heumahd [f], Grasschnitt [m], Langgrasschnitt [m].

haydite [n] [US] *constr.* ►expanded clay.

2625 hayfield [n] [US] *agr. (phyt.)* (►Permanent grassland subdivided into grazing lots, which are alternately grazed to allow for regeneration. Such meadows are periodically cut, in order to remove weeds and lush vegetation patches of animal rest areas [around accumulation of animal droppings], or to obtain winter fodder [hay and silage production]; ►hay meadow, ►rotation pasture, ►intensively grazed pasture, ►fertilized pasture); *syn.* hay meadow [n] [UK] (1); *s* **prado** [m] **de siega** (►Pradera permanente dividida en parcelas que se utilizan alternando como pasto y pradera de siega; ►pastizal en rotación, ►pastizal jugoso, ►pastizal jugoso, ►pradera fertilizada); *syn.* pastizal [m] de siega; *f* **pâturage** [m] **tournant** (**1.** ►Prairie permanente cloisonnée en parcelles ou enclos pâturés successivement en vue de la régénération des herbes broutées et fauchées selon les exigences en vue de la mise en réserve de l'excédent d'herbe pour l'alimentation hivernale du bétail — fourrage et ensilage — ou afin d'éliminer les refus ; LA 1981, 903 ; ►herbage, ►prairie à pâturage rationné, ►prairie à foin, ►prairie à pâturage en paddocks, ►pâturage libre, ►surface toujours en herbe. **2.** Forme d'alimentation des animaux par laquelle le fourrage est récolté sur pied sur des pâturages tournant) ; *syn. de 1.* pâturage [m] en rotation ; *syn. de 2.* système [m] warmbold ; *g* **Mähweide** [f] (In Koppeln aufgeteiltes ►Dauergrünland, das zur mehrwöchigen Erholung des abgeweideten Pflanzgutes abschnittsweise beweidet und zur Gewinnung von Winterfutter — Heu- und Silagefutterherstellung — oder zum Entfernen von Unkrautflächen und Geilstellen hin und wieder gemäht wird; ►Fettweide, ►Futterwiese, ►Standweide, ►Umtriebsweide); *syn.* Mähumtriebsweide [f], Rotationsmähweide [f].

hay field [n] [US]**, fertilized** *agr. phyt.* ►fertilized meadow.

2626 haying [n] *agr.* (Cutting, lifting, drying, and stacking of meadow grasses for use as fodder); *syn.* hay making [n]; *s* **henificación** [f] (Siega y secado de la hierba para su posterior uso de alimento del ganado); *f* **fanage** [m] (Opérations de fauche, de retournement et de ratissage du fourrage pendant la fenaison) ; *g* **Mahd** [f] **zur Heugewinnung** (Schneiden und Zusammenrechen von Gras zwecks Heugewinnung); *syn.* Heuen [n, o. Pl.].

hay making [n] *agr.* ►haying.

2627 hay meadow [n] *agr.* (Forage grasses, fertilized and mown several times per year; ►fertilized meadow); *syn.* fodder meadow [n]; *s* **pradera** [f] **fertilizada** (Prado permanente abonado, de gran calidad, en el que se producen hierbas de forraje; en comparación con el ►prado jugoso, en éste se consideran sobre todo sus condiciones mesológicas de riqueza de nutrientes y humedad); *f* **prairie** [f] **à foin** (Prairie permanente fertilisée, de qualité supérieure, fauchée plusieurs fois par an dont le fourrage

est exploité pour l'alimentation des animaux d'embouche et des fortes laitières ; cf. DAV 1984, 386 , ►prairie [mésophile] de fauche ; *opp.* ►prairie pâturée maigre) ; *syn.* prairie [f] fraîche de fauche, prairie [f] à faner ; *g* **Futterwiese** [f] (Gedüngte, meist mehrschürige Wiese, die der Futtergewinnung dient. Im Vergleich zur ►Fettwiese wird bei letzterer vornehmlich der nährstoffreiche und feuchte Standort angesprochen).

hay meadow [n] [UK] *agr. (phyt.)* ►hayfield [US].

hazard [n] *hydr. wat'man.* ►flood hazard; *land'man.* ►landslide hazard; *leg.* ►natural hazard; *constr.* ►slippage hazard.

hazard [n]**, erosion** *geo. pedol. plan.* ►erosion risk.

hazard [n]**, frost** *landsc. met.* ►frost danger.

hazard area [n] [US]**, flood** *hydr. wat'man.* ►floodplain [US]/flood-plain [UK].

hazard map [n]**, avalanche** *plan. recr.* ►avalanche location map.

2628 hazardous material [n] *envir. leg.* (Substances which, according to their characteristics, may be a danger to people during shipment; such substances may be explosive, combustible, inflammable, very toxic, toxic, corrosive, irritant, carcinogenic, genetically harmful, etc.; ►toxic substance); *syn.* hazmat [n] [also US]; *s* **sustancia** [f] **tóxica y peligrosa** (Sustancias y preparados que debido a sus cualidades conllevan un alto riesgo para la seguridad pública o para la salud; están clasificados en a] explosivos, b] comburentes, c] extremadamente inflamables, d] fácilmente inflamables, e] inflamables, f] muy tóxicos, g] tóxicos, h] nocivos, i] corrosivos, j] irritantes, k] sensibilizantes, l] carcinógenos, m] mutagénicos, n] tóxicos para la reproducción (teratogénicos) y o] peligrosos para el medio ambiente; Cap. I, Art. 2 RD 363/1995, de 10 de Marzo, por el que se aprueba el Reglamento sobre Notificación de Sustancias Nuevas y Clasificación, Envasado y Etiquetado de Sustancias Peligrosas; residuo tóxico y peligroso y **s. t. y p.** no son sinónimos, sino términos que se solapan; ►sustancia tóxica); *f* **substance** [f] **dangereuse** (Substance ou produit qui en raison de ses propriétés chimiques intrinsèques peut constituer un danger pour la population et les travailleurs qui l'emploient lors de sa production, manipulation et transport ; est dangereuse au sens de la directive n° 67/548/CEE du 27 juin 1967 les substances et préparations a] explosives, b] comburantes, c] extrêmement inflammables, d] facilement inflammables, e] inflammables, f] très toxiques, g] toxiques, h] nocives, i] corrosives, j] irritantes, k] sensibilisantes, l] cancérogènes, m] mutagènes, n] toxiques pour la reproduction, o] dangereuses pour l'environnement ; ►substance toxique) ; *syn.* matière [f] dangereuse ; *g 1* **Gefahrstoff** [m] (Stoff und Zubereitung nach § 3a des Chemikaliengesetzes sowie Stoff und Gegenstand, der auf Grund seiner Beschaffenheit bei Transporten auf öffentlichen Verkehrswegen eine Gefahr für die öffentliche Sicherheit darstellen kann; gem. Richtlinie 67/548/EWG des Rates vom 27.06.1967 zählen zu **G.en** Güter und Zubereitungen, die folgende Eigenschaften aufweisen: a] explosionsgefährlich, b] brandfördernd, c] hochentzündlich, d] leicht entzündlich, e] entzündlich, f] sehr giftig, g] giftig, h] mindergiftig, i] ätzend, j] reizend, k] sensibilisierend, [l] krebserzeugend, [m] Erbgut verändernd, [n] fortpflanzungsgefährdend [reproduktionstoxisch] und [o] umweltgefährlich; cf. auch Richtlinie 1999/45/EG des Rates vom 31.05.1999. Gefährlicher Abfall und **G.** sind nicht syn., sondern sich überschneidende Begriffe; ►Giftstoff); *syn.* umweltgefährdender Stoff [m], gefährlicher Stoff [m]; *g 2* **Gefahrgut** [n] (Gefahrstoff, der transportiert wird; Gefahrstoff und **G.** sind nicht syn.; in D. gelten für Gefahrstoffe und **G.güter** unterschiedliche Kennzeichnungen; während bei ersteren die Kennzeichnung Auskunft über den Umgang mit den Stoffen bei der Herstellung, Weiterverarbeitung und (Wieder)verwendung gibt, ist die **G.kennzeichnung** für den Transport vorgesehen, um z. B. Feuerwehr oder Technischem Hilfwerk [THW] die entsprechenden Informationen bei Unfällen zu geben); *syn.* umweltschädliches Gut [n].

2629 hazardous materials [npl] *envir.* (Waste materials dangerous to the environment which belong to legal category of toxic waste; ▶hazardous material, ▶hazardous waste, ▶water-polluting substance); *syn.* hazmat [n] [also US]; *s* **residuos [mpl] tóxicos y peligrosos** (Todos los residuos cuyas características les permiten clasificarlos como tales y que están sometidos a un régimen especial de control en sentido legal como ▶residuos industriales peligrosos; ▶sustancia tóxica para el agua. Las características coinciden mayormente con las aplicadas a las ▶sustancias tóxicas y peligrosas y son las siguientes : H1] explosivo, H2] comburente, H3-A] fácilmente inflamable, H3-B] inflamable, H4] irritante, H5] nocivo, H6] tóxico, H7] carcinógeno, H8] corrosivo, H9] infeccioso, H10] tóxico para la reproducción, H11] mutagénico, H12] sustancias o preparados que emiten gases tóxicos o muy tóxicos al entrar en contacto con el aire, con el agua o con un ácido, H13] sustancias o preparados susceptibles, después de su eliminación, de dar lugar a otra sustancia por un medio cualquiera, por ejemplo un lixiviado, que posea una de las características enumeradas anteriormente y H14] peligrosos para el medio ambiente; Anexo I, Tabla 5 RD 833/1988, de 20 de Julio, por el que se aprueba el Reglamento para la Ejecución de la Ley 20/1986, de 14 de Mayo, Básica de Residuos Tóxicos y Peligrosos); *syn.* desechos [mpl] tóxicos y peligrosos; *f* **déchets [mpl] générateurs de nuisances** (Déchets pouvant, en l'état ou lors de leur élimination, produire des effets nocifs sur le sol, la flore et la faune, dégrader les sites ou les paysages, polluer l'air et les eaux, engendrer des bruits ou des odeurs et porter atteinte à la santé de l'homme et à l'environnement, etc. ; les entreprises qui produisent, importent, exportent, éliminent, transportent ou se livrent à des opérations de courtage ou de négoce de ces déchets sont tenues de fournir à l'administration toutes informations les concernant ; [loi n° 75-633 du 15 juillet 1975] ; *termes spécifiques* ▶déchets dangereux, ▶déchets encombrants et déchets dangereux, déchets industriels spéciaux, ▶substance dangereuse, ▶substance polluante pour les eaux) ; *g* **umweltgefährdender Abfall** [m] (Alle Stoffe, die dem besonders überwachungsbedürftigen Abfall im rechtlichen Sinne zugerechnet werden; *UB* ▶besonders überwachungsbedürftiger Abfall; ▶Gefahrstoff, ▶wassergefährdender Stoff).

2630 hazardous old dumpsite [n] [US] *envir. urb.* (TGG 1984, 103; place with hidden soil contamination in an old industrial area harbo[u]ring toxic chemicals or radioactive materials; ▶hazardous waste disposal site, ▶hazardous waste material); *syn.* contaminated waste site [n], contaminated land [n], subsurface contamination [n] [US], toxic dumpsite [n] [US]; *s* **emplazamiento [m] contaminado** (Superficie, p. ej. de antigua industria o vertedero, peligrosa para la salud y el medio ambiente por un alto grado de contaminación; ▶depósito de residuos peligrosos, ▶suelo contaminado); *f* **site [m] pollué ancien [m]** (Zone contaminée par des ▶déchets historiques, p. ex. friche industrielle, décharge constituant un danger potentiel pour la santé humaine, la faune et la flore et sur laquelle la gravité des risques présentés ne peuvent être évalués ; ▶décharge de déchets industriels [spéciaux]) ; *syn.* site [m] contaminé, dépôt [m] ancien ; *g* **Altstandort [m]** (Von ▶Altlasten geprägte Fläche, z. B. ehemaliger Industriestandort oder abgeschlossene Deponieanlage, auf der mit umweltgefährdenden Stoffen umgegangen wurde und somit für Mensch, Tier und Pflanze Gefahren ausgehen, die zeitlich und in ihrem Umfange nicht genau bestimmbar sind. Nach § 2 [5] Nr. 2 BBodSchG sind hiervon Anlagen

ausgenommen, deren Stilllegung einer Genehmigung nach dem Atomgesetz bedarf. Ein UB ist die **altlastverdächtige Fläche:** Altablagerung oder **A.** [beide Begriffe sind nicht syn.], bei der/dem die Besorgnis besteht, dass das Wohl der Allgemeinheit durch schädliche Bodenveränderungen oder sonstige Gefahren beeinträchtigt ist oder zukünftig beeinträchtigt wird; cf. § 2 [6] BBodSchG; ▶Sonderabfalldeponie).

hazardous slope [n] *geo. pedol.* ▶unstable slope.

hazardous substances fee [n] [UK] *adm. envir. leg.* ▶hazardous waste disposal fee [US].

2631 hazardous waste [n] *envir. leg.* (Refuse matter which cannot be disposed of with ordinary household rubbish due to environmental risks but has to be kept at a ▶hazardous waste disposal site. Such wastes are listed in the appendices of the Council directive 94/31/EEC; additionally the Council decision 94/904/EC on an index of hazardous wastes applies; **h. w.** possesses at least one of four characteristics (ignitability, corrosivity, reactivity, or toxicity); **in U.S.,** such substances or materials appear on special EPA list; cf. EPA 1994; according to the WHO definition **h. w.** has **1.** short-term hazards, such as toxicity by ingestion, inhalation or skin absorption, corrosivity or other skin or eye contact hazards or the risk of fire or explosion; or **2.** long-term environmental hazards including chronic toxicity upon repeated exposure, carcinogenicity—which may in some cases result from acute exposure but with a long latent period—, resistance to detoxification processes such as biodegradation, the potential to pollute underground or surface waters or aesthetically objectionable properties such as offensive smells; cit. in DES 1991; ▶hazardous material); *syn.* toxic waste [n], dangerous waste [n]; *s* **residuos [mpl] industriales peligrosos** (Desechos peligrosos clasificados por Directiva del Consejo 91/689/CEE, modificada por Directiva del Consejo 94/31/CEE, que no pueden ser depositados junto con los residuos domésticos, sino que por su peligrosidad para el medio ambiente deben ser eliminados en ▶depósitos de residuos peligrosos o en plantas de incineración de residuos industriales peligrosos; ▶residuos tóxicos y peligrosos); *syn.* desechos [mpl] industriales peligrosos, residuos [mpl] tóxicos, desechos [mpl] tóxicos, residuos [mpl] especiales (1); *f* **déchets [mpl] dangereux** (Résidus de consommation ou de production, qui par leur nature polluante, ne peuvent être éliminés comme les ordures ménagères ; les circuits d'élimination sont soumis à un contrôle strict, le stockage s'effectue sur des décharges de classe I [▶décharge de déchets industriels (spéciaux)] ; les catégories ou types génériques de **d. d.** sont définis dans la liste établie dans l'annexe I de la directive du Conseil n° 91/689/CEE du 12 déc. 1991, modifiée par la directive 94/31/CE et complétée par la liste des **d. d.** de la directive 94/904/CE ; ▶déchets générateurs de nuisances) ; *syn.* déchets [mpl] toxiques et dangereux, déchets [mpl] spéciaux ; *g* **besonders überwachungsbedürftiger Abfall [m]** (Gefährlicher Abfallstoff, der durch eine Rechtsverordnung nach § 41 [1] oder § 41 [3] Nr. 1 KrW-/AbfG bestimmt ist und der wegen seiner besonderen Umweltgefährdung nicht mit normalem Haus- oder Gewerbemüll beseitigt/entsorgt werden kann, sondern auf ▶Sonderabfalldeponien gelagert werden muss. Diese gefährlichen Abfälle sind in einem auf den Anhängen I und II der RL 91/689/EWG vom 31.12.1991, geändert durch RL 94/31/EWG des Rates, beruhenden Verzeichnis aufgeführt. Ferner gilt die Entscheidung 94/904/EG des Rates vom 22.12.1994 über ein Verzeichnis gefährlicher Abfälle im Sinne von Artikel 1 Abs. 4 der Richtlinie 91/689/EWG über gefährliche Abfälle [ABl. EG Nr. L 356 S. 14]; §§ 43-47 KrW-/AbfG regeln für die Beseitigung und Verwertung von **b. ü. Abfällen** ein obligatorisches Nachweisverfahren; ▶umweltgefährdender Abfall); *syn. ugs. Kurzfassung* Sonderabfall [m], Sondermüll [m].

hazardous waste [n]**, incineration plant for** *envir.*
▶incinerator for toxic/hazardous waste [US].

2632 hazardous waste disposal fee [n] [US] *adm. envir. leg. obs.* (**In U.S.**, permit payment for disposal of dangerous byproducts; **in U.K.**, this fee is governed by the Planning [Hazardous Substances] Act 1990 and Planning [Hazardous Substances] Regulations 1992 [SI 1992 No. 656]; **in D.**, legally-prescribed payment for ▶hazardous waste which has to be controlled by a public agency with the aim of minimizing the amount of ▶hazardous materials or to encourage the producer to reduce or eliminate waste by recycling and limiting the use of natural resources; since 1998 the legal basis was withdrawn by the Federal Constitution Court [Bundesverfassungsgericht]); *syn.* hazardous substances fee [n] [UK]; *s* **canon** [m] **de residuos tóxicos y peligrosos** (**En Es.**, los productores resp. gestores de residuos tóxicos y peligrosos deben correr con todos los gastos de eliminación e inspección del proceso. La administración competente puede imputar a los solicitantes los costos de los servicios de inspección previa a la concesión de autorizaciones para gestión de residuos; art. 30 Ley 10/1998, de Residuos; **en D.**, pago obligatorio a cargo de los productores de ▶residuos industriales peligrosos que se emplea para el control y la vigilancia de la eliminación de los mismos y que tiene además como fin promover la reducción o el reciclaje de los mismos; ▶residuos tóxicos y peligrosos); *f 1* **redevance** [f] **spéciale pour l'enlèvement des déchets assimilés (aux déchets ménagers)** (**F.**, redevance créée dans le cadre de l'enlèvement des emballages industriels et commerciaux afin d'inciter les entreprises commerciales et artisanales à favoriser le recyclage et la valorisation des emballages au détriment de l'abandon en décharge ou de la destruction par incinération simple ; **D.**, taxe relative aux ▶déchets dangereux à et ayant pour but de diminuer la quantité de produits générateurs de déchets ou d'encourager les producteurs à opérer une politique de valorisation des déchets industriels ; ▶déchets encombrants et déchets dangereux, ▶déchets générateurs de nuisances) ; *f 2* **taxe** [f] **sur les déchets industriels spéciaux** (**F.**, taxe versée à l'ADEME et dont le produit est affecté exclusivement au traitement et à la réhabilitation des sites pollués, lorsque cette participation est devenue nécessaire du fait de la disparition ou de l'insolvabilité du producteur ou du détenteur de déchets) ; *g* **Abfallabgabe** [f] (Bei einigen Bundesländern [z. B. BW, Hessen, Sachsen] durch Landesgesetz vorgeschriebene Zahlung für ▶besonders überwachungsbedürftigen Abfall mit dem Ziel, das Aufkommen für überwachungsbedürftigen Abfall zu mindern oder den Erzeuger zur Abfallverwertung sowie zum sparsamen Verbrauch natürlicher Ressourcen zu ermuntern. Mit Urteil des Bundesverfassungsgerichtes vom 07.05.1998 [AZ: 2 BvR 1876/91] ist die Abfallabgabe für rechtswidrig erklärt worden; ▶umweltgefährdender Abfall).

2633 hazardous waste disposal site [n] *envir.* (Area for the deposition of hazardous waste such as clinical waste, industrial waste, nuclear waste, poisonous waste, 'difficult wastes' [in U.K.], which according to their physical properties, composition and amount present handling problems or are known to be harmful to the environment; *s* **depósito** [m] **de residuos peligrosos** (Superficie de almacenamiento de desechos industriales, de talleres, empresas e instituciones que por sus características o su cantidad pueden causar daños a la salud, al medio ambiente, ser explosivos o combustibles, o contener organismos que transmiten enfermedades; ▶residuos industriales peligrosos); *f 1* **décharge** [f] **de déchets industriels (spéciaux)** (Terme désignant une installation de dépôt existante avant la loi du 13 juillet 1992 dite de classe 1 lorsque sont déposés des déchets industriels spéciaux ou des ▶déchets dangereux [explosifs, comburants, facilement inflammables, inflammables, irritants, nocifs, toxiques

infectieux, etc.] et dite de classe 2 lorsque les déchets industriels sont assimilés aux ordures ménagères) ; *f 2* **centre** [m] **de stockage de déchets spéciaux ultimes et stabilisés** (**1.** Installation résultant d'une évolution technique récente, sur laquelle est admis le stockage permanent de déchets industriels spéciaux ultimes et stabilisés et réalisée sans intention de reprise ultérieure des déchets. **2.** Installation nouvelle instaurée par la loi du 13 juillet 1992 dite de classe I recevant des déchets industriels spéciaux, ultimes et stabilisés) ; *syn.* installation [f] de stockage des déchets industriels spéciaux (ultimes et stabilisés), centre [m] de stockage de classe 1 ; *g* **Sonderabfalldeponie** [f] (Fläche zur ordnungsgemäßen Ablagerung von Abfällen aus gewerblichen oder sonstigen wirtschaftlichen Unternehmen oder öffentlichen Einrichtungen, die nach Art, Beschaffenheit oder Menge in besonderem Maße gesundheits-, luft- oder wassergefährdend, explosibel oder brennbar sind oder Erreger übertragbarer Krankheiten enthalten resp. hervorbringen können; cf. § 41 KrW-/AbfG; ▶besonders überwachungsbedürftiger Abfall); *syn. leg.* Abfallbeseitigungsanlage [f] zur Ablagerung von besonders überwachungsbedürftigen Abfällen, *ugs.* Sondermülldeponie [f].

2634 hazardous waste material [n] *envir.* (TGG 1984, 103, 233; identified or unidentified toxic material on former industrial sites, ▶dumpsites [US]/tipping sites [UK], etc., which may continue to contaminate the soil and ground water, poison watercourses or pollute the atmosphere; ▶hazardous old dumpsite, ▶hazardous waste disposal site, ▶toxic site inventory [US]/ contaminated land registry [UK]); *s* **suelo** [m] **contaminado** (Todo aquel cuyas características físicas, químicas o biológicas han sido alteradas negativamente por la presencia de componentes de carácter peligroso de origen humano, en concentración tal que comporte un riesgo para la salud humana o el medio ambiente; Art 3 Ley 10/1998, de Residuos. En general se trata de antiguos ▶depósitos de residuos peligrosos y zonas industriales contaminadas que deben ser saneadas para poder reutilizarse; cf. PDPS 1994, Tomo I; *términos específicos* zona industrial abandonada, zona de depósito de residuos peligrosos; ▶depósito de residuos sólidos, ▶emplazamiento contaminado, ▶inventario de suelos potencialmente contaminados); *syn.* superficie [f] contaminada abandonada, «vieja carga» [f]; *f* **déchets** [mpl] **historiques** (Déchets constitués de substances ayant contaminé le sol, l'atmosphère et les eaux souterraines sur des anciens sites industriels ou des ▶décharges publiques ; ▶centre de stockage de déchets spéciaux ultimes et stabilisés, ▶inventaire historique régional d'anciens sites industriels et activités de service, ▶inventaire [national] des sites et sols pollués, ▶site pollué ancien) ; *syn.* pollution [f] sournoise, pollution [f] inconsciente [CH], pollution [f] à retardement [CH] ; *g* **Altlast** [f] (Altlasten [pl]; OB zu Altablagerung und ▶Altstandort; dem Umfang nach identifizierte, zu identifizierende oder noch unbekannte und meist nicht sichtbare Lagerung von giftigen Stoffen auf [ehemaligen] Industriestandorten, ▶Deponien oder sonstigen Flächen, die vor vielen Jahren oder Jahrzehnten stattfand und geeignet ist, a] die Gesundheit des Menschen zu gefährden und sein Wohlbefinden zu beeinträchtigen, b] Nutz- und wild lebende Tier- und Pflanzenarten zu gefährden, c] Boden, Grundwasser und Gewässerläufe schädlich zu beeinflussen, d] schädliche Umwelteinwirkungen durch Luftverunreinigungen herbeizuführen und e] die Belange von Naturschutz und Landschaftspflege sowie der Stadtplanung nicht zu wahren; cf. § 10 [4] KrW-/AbfG. Bei **Altablagerungen** [fpl] handelt es sich um stillgelegte Abfallbeseitigungsanlagen sowie sonstige Grundstücke, auf denen Abfälle behandelt, gelagert oder abgelagert worden sind; § 2 [5] 1 BBodSchG; In D. begannen viele Städte Anfang der 1980er-Jahre mit der Bearbeitung von Altlasten; ▶Altlastenkataster, ▶Altstandort).

hazardous wastes [npl] *envir. leg.* ►transboundary movements of hazardous wastes and their disposal.

hazmat [n] [US] *envir.* ►hazardous material.

2635 header [n] *arch. constr.* (Masonry unit laid into the thickness of a wall, so as to expose its ends; ►stretcher); *s* **tizón** [m] (Colocación de un sillar o ladrillo con su dimensión mayor perpendicular al paramento; DACO 1988; ►soga); *f* **boutisse** [f] (Pierre dont la plus grande dimension est perpendiculaire à la façade d'un mur ; ►panneresse) ; *g* **Binder** [m] (Der mit seiner Längsseite quer zur Mauerflucht liegende und in die Mauer 'einbindende' Stein; im Vergleich zum ►Läufer); *syn.* Binderstein [m].

2636 header bond pattern [n] *arch. constr.* (►Masonry bond consisting of ►header courses; ►rowlock course [US]/ layer of headers-on-edge [UK], ►stretcher bond pattern); *syn.* heading bond (pattern) [n] [also US]; *s* **aparejo** [m] **a tizón** (Aparejo o disposición de ladrillos o sillares a tizón; DACO 1988; ►aparejo a soga, ►aparejo de mampuestos, ►hilada a sardinel, ►hilada de tizones); *f* **appareillage** [m] **en boutisses** (DTB 1985 ; ►appareillage d'une maçonnerie réalisé à partir de ►lits de boutisses ; ►appareillage à panneresses, ►lit à disposition sur chant) ; *syn. résultat* appareil [m] en boutisses ; *g* **Binderverband** [m] (Aus ►Binderschichten bestehender ►Mauersteinverband; ►Läuferverband, ►Rollschicht).

2637 header course [n] *arch. constr.* (A continuous line of masonry units, laid so that their ends are exposed, as in the face of a wall; ►stretcher course); *syn.* heading course [n] [also US]; *s* **hilada** [f] **de tizones** (Serie horizontal de sillares o ladrillos colocados a tizón; DACO 1988; ►hilada de sogas); *f* **lit** [m] **de boutisses** (Technique de mise en place des lits en maçonnerie pour laquelle une assise n'est constituée que de boutisses c.-à-d. d'éléments dont le côté long est posé perpendiculairement à l'alignement du mur. On parle d'un appareillage à la Française lorsqu'il y a alternance entre un lit de boutisses et un lit de panneresses ; ►assise de panneresses) ; *syn.* assise [f] de boutisses ; *g* **Binderschicht** [f] (Beim Mauersteinverband verwendete Schicht, bei der die Steine mit der Längsseite senkrecht zur Mauerflucht resp. Verlegerichtung gesetzt werden; ►Läuferschicht).

heading [n] *constr. hort.* ►cutting back of trees and shrubs; *hort.* ►deadheading of perennials; *arb.* ►topping (3).

heading back [n] [US] *constr. hort.* ►cutting back of trees and shrubs.

2638 heading back [n] **of tree scaffolds** *arb. hort.* (Severe cutting back of main branches for training characteristic shape of crown; ►topping 3); *syn.* stubbing [n] of main scaffolds; *s* **poda** [f] **de las ramas principales** (Poda fuerte de las ramas primarias para dar forma característica a la copa; ►descopado); *f* **rabattage** [m] **d'une branche forte** (Suppression des branches charpentières dans le but de provoquer la pousse de branches nouvelles ; ►éhoupage) ; *g* **Kappen** [n, o. Pl.] **von Starkästen** (Starkes Zurückschneiden der Starkäste, um einen neuen Kronenaufbau zu erhalten; ►Kronenkappung).

heading bond (pattern) [n] [US] *arch. constr.* ►header bond pattern.

heading course [n] [US] *arch. constr.* ►header course.

2639 heading joint [n] *arch. constr.* (Vertical joint between two masonry units; ►coursing joint, ►butt joint); *syn.* vertical joint [n], cross joint [n] [US] (DAC 1975); *s* **junta** [f] **a tope** (Unión vertical o casi vertical entre las piezas de un muro; ►junta de asiento, ►junta prensada); *syn.* junta [f] vertical, junta [f] recta; *f* **joint** [m] **montant** (Joint vertical dans une maçonnerie ; ►joint délit, ►joint serré ; DTB 1985, 118) ; *g* **Stoßfuge** [f] (In einem Mauerwerk eine senkrechte oder fast

senkrechte Fuge an der Seite eines Mauersteins; ►Lagerfuge, ►Pressfuge).

heading off [n] [NZ] *arb.* ►topping (3).

head joint [n] *arch. constr.* ►heading joint.

headland [n] [US] *agr. ecol.* ►field border.

headstone [n] [US] *adm. conserv'hist.* ►gravestone.

headstream [n] [UK] *geo.* ►headwater stream [US].

2640 headwall [n] **with wing walls** *constr. wat'man.* (Structure which guides a stream into an opening of a culvert); *s* **embocadura** [f] (Estructura que guía a un arroyo hasta la entrada de una tubería de entubado); *syn.* abocinamiento [m]; *f* **bajoyer** [m] (Dans un ouvrage hydraulique, mur limitant latéralement une buse, un pertuis) ; *syn.* mur [m] bajoyer, mur [m] de tête ; *g* **Einlaufbauwerk** [n] **(2)** (Bauwerk, das der Fassung eines Baches in eine Dole dient).

headwater [n] *wat'man.* ►spring water.

headwaters [npl] [US] *geo.* ►upper reaches.

2641 headwater stream [n] [US] *geo.* (TGG 1981, 167; uppermost part of a river or stream near its source; ►lower reaches, ►middle reaches, ►spring water, ►upper reaches); *syn.* headstream [n] [UK]; *s* **cabecera** [f] **(de río)** (Parte superior del curso de un río; ►agua de manantial, ►tramo inferior [de un curso de agua], ►tramo medio [de un curso de agua], ►tramo superior [de un curso de agua]); *f* **cours** [m] **supérieur près de la source** (Partie amont d'un cours d'eau près de sa source dans le profil longitudinal ; ►cours inférieur, ►cours moyen, ►cours supérieur, ►eau de source) ; *g* **Quelllauf** [m] (Oberster Teil eines Fließgewässers; ►Mittellauf, ►Oberlauf, ►Quellwasser, ►Unterlauf; *UBe* Quellbach, Quellfluss).

health [n]**, detrimental to** *envir.* ►injurious to health.

2642 health care facilities area [n] *leg. urb.* (Land designated on a ►zoning map [US]/Proposals Map/Site Allocations Development Plan Document [UK] for installation of a ►communal use designed to meet community health care requirements); *syn.* health facilities area [n] [also US]; *s* **instalaciones** **[fpl] de centros asistenciales y sanitarios [Es]** (Áreas fijadas en ►plan parcial [de ordenación] para ►equipamiento comunitario; cf. art. 13.2.d, LS Texto refundido); *syn.* instalaciones [fpl] de salud pública, equipamientos [mpl] sociales; *f* **secteur** [m] **correspondant à l'implantation d'équipements hospitaliers et sanitaires** (Terme d'urbanisme qualifiant un secteur d'une zone U d'un ►plan local d'urbanisme affecté aux équipements médico-sociaux ; sa dénomination est variable suivant la commune ; ►équipement urbain d'intérêt public, ►services collectifs) ; *g* **gesundheitlichen Zwecken dienende Gebäude** [npl] **und Einrichtungen** [fpl] (Im ►Bebauungsplan auszuweisende Flächen für den ►Gemeinbedarf an Gesundheitseinrichtungen; cf. Anlage zur PlanzV 90, Kap. 4).

health facilities area [n] [US] *leg. urb.* ►health care facilities area.

2643 health resort [n] *recr.* (Place to which people go for treatment of their diseases or for prevention of their occurrence; ►climatic health resort, ►spa); *s* **estación** [f] **balnearia** (Término genérico para lugar de descanso con tratamiento médico; ►estación hidrotermal, ►sanatorio climático); *syn.* balneario [m]; *f* **station** [f] **thermale** (*Terme générique* endroit dans lequel on suit un traitement médical ; *termes spécifiques* station balnéaire [bains de mer], ►établissement thermal, ►station climatique) ; *g* **Kurort** [m] (Ort, in dem Krankheiten behandelt oder nachbehandelt werden oder deren Entstehung vorgebeugt wird. In D. ist die Bezeichnung **K.** gesetzlich geschützt; ►Heilbad, ►Luftkurort).

H

2644 health resort center [n] [US]/**health resort centre** [n] [UK] *recr.* (Medicinal center [US]/medicinal centre [UK] of a health resort equipped with a sanatorium and assembly rooms; ▶spa); *s* **sanatorio** [m] **balneario** (Centro médico de una estación balnearia; ▶estación hidrotermal); *f* **centre** [m] **de cure** (Établissement pour un traitement médical d'une certaine durée ; ▶établissement thermal) ; *g* **Kurzentrum** [n] (...zentren [pl]; medizinisches Zentrum eines Kurortes mit Sanatorien, Kurheimen; ▶Heilbad).

2645 health resort park [n] *recr.* (Well-cared for, enclosed grounds in a health resort or ▶spa intended to benefit the convalescent during his stay and to add to the image of the ▶health resort); *s* **parque** [m] **de balneario** (Parque envallado con funciones decorativas y recreacionales en el recinto de una ▶estación balnearia; ▶estación hidrotermal); *f* **parc** [m] **thermal** (Grands espaces verts d'agrément, de détente et de repos attenant à une ▶station thermale ; ▶établissement thermal) ; *syn.* parc [m] de la station thermale, parc [m] d'une station balnéaire ; *g* **Kurpark** [m] (...parks [pl], *seltener* ...parke [pl], *CH meist* ...pärke [pl]; zur Repräsentation gepflegter und dem Erholungsaufenthalt dienender, eingegrenzter Freiraum in einem ▶Kurort; ▶Heilbad).

2646 health resort recuperation [n] *recr.* (Recovery while receiving medical therapy under the supervision of a doctor); *s* **cura** [f] **de reposo** (Periodo/período de reconvalescencia en un balneario con tratamiento médico); *f* **cure** [f] **de repos** (Repos pris comme méthode thérapeutique sous traitement médical) ; *g* **Kurerholung** [f, o. Pl.] (Erholung mit unter ärztlicher Aufsicht durchgeführten Heilverfahren während eines mehrwöchigen Aufenthaltes in einem Sanatorium oder einer Rehabilitationsklinik).

heap [n] [UK], **compost** *constr.* ▶compost pile [US].

heap [n] [US], **rubble** *envir.* ▶rubble pile [US]/rubble tip [UK].

heap [n] [UK], **silt** *min.* ▶silt pile [US].

heap [n] [UK], **slag** *envir. min.* ▶slag pile [US].

heap [n] [UK], **soil** *constr.* ▶soil pile [US].

heap [n] [UK], **spoil** *min.* ▶spoil bank [US] (2).

heap [n] [UK], **topsoil** *constr.* ▶topsoil pile [US].

2647 heart root [n] *bot. hort.* (Root originating from a lateral root near the rootstock, but descending at an angle rather than growing vertically downward; several lateral roots of about equal size extending obliquely downward from the base of a stem form a **heartroot system**; e.g. European Birch [*Betula pendula*], Beech [*Fagus sylvatica*], Douglas Fir [*Pseudotsuga menziesii*], Black Alder [*Alnus glutinosa*], Hornbeam [*Carpinus betulus*], European Larch [*Larix decidua*], Linden [*Tilia cordata*]; ▶taproot); *s* **raíz** [f] **fasciculada** (La que está constituida por un manojo de raicillas del mismo o parecido grosor como p. ej. en el abedul [*Betula pendula*], la haya [*Fagus sylvatica*], el abeto de Douglas [*Pseudotsuga menziesii*], el aliso [*Alnus glutinosa*], el carpe [*Carpinus betulus*], el alerce [*Larix decidua*], el tilo [*Tilia cordata*]; ▶raíz axonomorfa); *f* **racine** [f] **fasciculée** (Système radiculaire caractérisé par une homogénéité entre la racine principale et les racines secondaires, celui-ci apparaissant comme un faisceau de racines comme p. ex. chez le Bouleau [*Betula pendula*], le Hêtre [*Fagus sylvatica*], le Sapin de Douglas [*Pseudotsuga menziesii*], l'Aulne [*Alnus glutinosa*], le Charme [*Carpinus betulus*], le Mélèze [*Larix decidua*], le Tilleul [*Tilia cordata*]; ▶racine pivotante) ; *g* **Herzwurzel** [f] (Senkrecht und schräg vom Wurzelstock ausgehende Wurzel, die im Verbund mit anderen **H.n** ein kompaktes Wurzelsystem bildet, z. B. bei Birke [*Betula pendula*], Buche [*Fagus sylvatica*], Douglasie [*Pseudotsuga menziesii*], Erle [*Alnus glutinosa*], Hainbuche [*Carpinus betulus*], Lärche [*Larix decidua*], Linde [*Tilia cordata*]; ▶Pfahlwurzel; *OB* Vertikalwurzel).

2648 heartwood [n] *arb.* (Inner layers of wood which, in the growing tree, have ceased to contain living cells and in which the reserve materials [e.g. starch] have been removed or converted into heartwood substances. It is generally darker in colo[u]r than ▶sapwood; MGW 1964); *s* **duramen** [m] (La parte muerta del leño de un árbol, no apta, por lo tanto, para acarrear el agua y las sustancias disueltas en ella absorbidas por las raíces. Ocupa la parte interna del tronco y es más dura y más oscura que las porciones leñosas aún vivas; ▶albura); *syn. término popular* corazón [m]; *f* **bois** [m] **parfait** (Dans l'arbre vivant, couches internes de bois qui ne contiennent plus de cellules vivantes et dans lesquelles les réserves [p. ex. l'amidon] ont disparu ou ont été transformées en substances du bois [gommes, résines, substances tanniques] parfait ; le **b. p.** est généralement plus coloré, plus robuste et plus durable que l'▶aubier ; cf. MGW 1964) ; *syn.* duramen [m], bois [m] de cœur ; *g* **Kernholz** [n] (...hölzer [pl]; die inneren Zonen im Holz, die im stehenden Stamm keine lebenden Zellen mehr enthalten und in denen die Reservestoffe [z. B. Stärke] i. d. R. abgebaut oder in Kernholzsubstanzen umgebaut worden sind; in der Farbe dunkler als das ▶Splintholz; cf. MGW 1964); *syn.* Herzholz [n].

heat [n] *urb.* ▶central heat; *met.* ▶radiant heat, ▶reradiation of heat; *envir.* ▶waste heat;

heat archipelago [n] *met. urb.* ▶heat island.

heath [n] *geo.* ▶coastal heath; *phyt.* ▶dune heath; *geo. phyt.* ▶steppe-heath.

2649 heather [n] *bot.* (Generic term for Scotch Heather [*Calluna vulgaris*] and Heath [*Erica*]); *s 1* **brezo** [m] (Término genérico para especies de la familia de las ericáceas y del género *Erica*); *s 2* **brecina** [f] (Término específico para el brezo de la especie *Calluna vulgaris*); *syn.* biércol [m]; *f* **bruyère** [f] (Appellation courante pour les végétaux regroupés sous les genres *Erica* et *Calluna*); *g* **Heide** [f] (1) (OB zu Besenheide [*Calluna vulgaris*], Glockenheide [*Erica tetralix*] und andere Pflanzen der Gattung *Erica*); *syn.* Heidekraut [n].

2650 heather moor [n] [UK] *phyt.* (ELL 1988, 343; dwarf shrub heath, characterized, e.g. by the order of *Ericetalia tetralicis*, resulting from conditions brought about by the increasingly effective drainage on ▶raised bogs as well as on natural heathland of hydromorphic humic sandy soils, both in regions of Atlantic climate and in central Europe); *s* **landa** [f] **turbosa** (Matorral bajo resultado de la degeneración de ▶turbera alta poco deshidratada y landa natural sobre suelos arenosos anmooriformes que se presentan en los climas templados de Europa Central, caracterizada por el orden *Ericetalia tetralicis/Sphagno-Ericetalia*); *f* **lande** [f] **humide** (Stade de dégénérescence de la ▶tourbière haute partiellement drainée ou se développe une lande naturelle oligotrophe, à *Erica tetralix* et à Sphaignes, sur podzol très humide, à gley superficiel et sable siliceux à faible profondeur : lande : **1.** la **lande tourbeuse** appartenant à l'alliance de l'*Ericion tetralicis*. **2.** La **lande paratourbeuse** appartenant à l'alliance de l'*Ulcinion nani* ; cf. GGV 1979, 316) ; *syn.* lande [f] humide à Ericacées, lande [f] hygrophile ; *g* **Moorheide** [f] (Zwergstrauchheide als Degenerationsstadium wenig entwässerter ▶Hochmoore sowie z. T. natürliche Heiden anmooriger Sandböden, beide in sommerkühlen und wintermilden Klimabereichen Mitteleuropas, die z. B. durch die Ordnung des *Ericetalia tetralicis/Sphagno-Ericetalia* [Glockenheidemoor] geprägt ist); *syn.* Heideanmoor [n], Heidemoor [n].

2651 heath invasion [n] *phyt.* (Gradual incursion of plants, characteristic for raised bogs, predominantly by heather [*Calluna vulgaris*], caused by draining); *s* **invasión** [f] **de brezos** (Incursión gradual resultando en desplazamiento creciente por brezos *[Calluna vulgaris]* de las plantas características de turbera alta en turberas desecadas); *f* **envahissement [m] par la bruyère** (Processus de colonisation progressive par la Callune fausse-bruyère *[Calluna vulgaris]* sur les tourbières asséchées qui repousse les espèces caractéristiques de la végétation des tourbières hautes) ; *g* **Verheidung [f]** (Zunehmende Verdrängung hochmoortypischer Pflanzen bei entwässerten Hochmooren durch die Besenheide *[Calluna vulgaris]*).

2652 heathland [n] *geo. land'man.* (Extensive [pasture] land of uncultivated, podzolic or podzolized soils with a vegetation dominated by dwarf ericaceous shrubs and dwarf oak, juniper, etc.; ▶heath vegetation, ▶steppe-heath, ▶woody heathland); *s* **landa [f]** (Paisaje de pastos acidófilos sobre suelos arenosos en zonas templadas húmedas. En Europa se presenta principalmente en las regiones nórdicas y atlánticas. En la península ibérica en la zona húmeda del norte y noroeste sobre substrato de granito, gneiss y rocas antiguas; cf. DB 1985; ▶brezal, ▶landa esteparia); *f* **lande [f] (1)** (Espaces incultes ou improductifs, à végétation buissonnante, en général stade ultime de la dégradation anthropique des forêt utilisés autrefois comme pâturage extensif, pour la fourniture de litière ou la cueillette tels que p. ex. les forêts claires de pins sur les plaines sablonneuses d'Europe occidentale ou les ▶pelouses calcicoles des régions montagneuses ; ▶fruticée à bruyère, ▶lande 2, ▶landes steppiques) ; *syn.* landes [fpl] ; *g* **Heide [f] (2)** (1. In Nordwestdeutschland unkultivierte, meist nährstoffarme, sandige, vorwiegend mit Zwergsträuchern bewachsene, teilweise extensiv beweidete Fläche/Landschaft; in Süddeutschland dagegen Kalkmagerrasen im Berg- und Hügelland und im östlichen Mitteleuropa auch lichte Kiefernwälder der sandigen Ebenen; ▶Heide 3, ▶Steppenheide); *syn.* Heidefläche [f], Heidelandschaft [f]. 2. In Südschweden auf der Insel Öland eine steppenartige Kalkheide, das **Große Alvar** *[Stora Alvaret]*, eine in der Welt einzigartige steppenähnliche Kulturlandschaft — abhängig von Weidewirtschaft — mit flachem, karstigem Untergrund, teilweise bedeckt mit einer dünnen Schicht Oberboden, die abwechselnd extremer Trockenheit und Überschwemmungen ausgesetzt ist und mit endemischen Pflanzen wie das Öland-Sonnenröschen *[Helianthemum oelandicum var. oelandicum* und *H. oelandicum var. canescens]* und der Alvar-Wermuth *[Artemisia oelandica]* sowie mit solchen Pflanzen, die normalerweise nur in Südeuropa [Überreste einer vorgeschichtlichen Wärmeperiode], im Hochgebirge oder in Sibirien vorkommen, bedeckt ist).

2653 heathland conservation [n] *conserv. nat'res.* (Statutory protection and maintenance of heathland, which warrants conservation of special elements contained in a heath landscape, e.g. Lüneburg Heath in northern Germany and the Great Alvar *[Stora Alvaret]* on the isle of Oeland in southern Sweden. Most large areas of heathland are natural area preserves [US]/nature reserves [UK]); *s* **protección [f] de landas** (Política de declarar como espacios naturales protegidos a parajes de landas, como las Landas de Luneburgo en Alemania, por ser un tipo de paisaje cultural representativo del norte del país y porque desaparecerían sin medidas específicas de conservación —como el pastoreo— que les dió origen y que es necesario para su preservación); *f* **protection [f] de la lande** (Conservation et mise en valeur des landes pour le caractère original des paysages qu'elles constituent ; les landes oligotrophes les plus remarquables demanderaient à être protégées comme l'est la lande de Lüneburg en Allemagne du nord) ; *g* **Heideschutz [m, o. Pl.]** (Gesetzlicher Schutz und Pflege von Heiden als besonders erhaltenswerte Landschaftsbestandteile; größere Heideflächen, z. B. die Lüneburger Heide in Norddeutschland, das Große Alvar *[Stora*

Alvaret] auf Öland in Südschweden, stehen meist unter Naturschutz).

2654 heathland pool [n] *geo. phyt.* (Body of water on boggy heathland; ▶dune slack); *s* **charca [f] de brezal** (Cuerpo de agua en el matorral de brezos; ▶valle interdunar); *f* **mardelle [f]** (Plan d'eau dans les landes paratourbeuses, les dunes grises et brunes ; ▶dépression dunaire) ; *g* **Heidetümpel [m]** (Stehendes Gewässer in vermoorten Heiden, Grau- oder Braundünen; ▶Dünental); *syn.* Heideweiher [m].

2655 heath sod [n] *agr. pedol.* (Stripped-off surface layer of a thoroughly rooted, hum[o]us topsoil with heather and grass vegetation. The sods served as litter in stables and afterwards, when soaked in excrement and urine, they were used to fertilize mostly podzolic soils and brown earth); *syn.* turf sod [n]; *s* **tepe [m] de turba** (Porción de tierra vegetal muy húmica y enraizada de suelos arenosos con vegetación de landa o hierba. Los *tt. de t.* servían de cama para el ganado y a continuación, empapados de excrementos y orina, de abono para los suelos generalmente podsolizados o pardos; *syn.* pan [m] de turba; *f* **plaque [f] tourbeuse** (Couche superficielle provenant de l'étrépage manuel des sables humifères fortement enracinés et à forte teneur en matière organique des sols [podzol] des landes tourbeuses à Éricacées et graminées ; ce matériau était jadis utilisé comme litière à l'étable et ensuite mélangé aux déjections animales solides et liquides pour être en dernier lieu épandu comme engrais sur les sols podzolisés ou sols bruns) ; *g* **Plagge [f]** (Niederdeutscher Ausdruck für Placken >Flecken<; abgeschälte Sode eines stark durchwurzelten und stark humosen Oberbodens von sandigen Böden mit Heide- oder Grasbewuchs. **P.n** dienten früher im Stall zur Einstreu und anschließend mit Kot und Harn durchtränkt als Dünger für meist podsolierte Böden oder Braunerden; *UBe* Grassode, Heidesode; plaggen [vb]).

2656 heath vegetation [n] *phyt.* (Plant formation dominated by a wide range of dwarf shrubs; ▶dwarf-shrub plant community, ▶garrigue, ▶macchia, ▶steppe-heath. In Central European lowlands dwarf shrubs, such as Heather *[Calluna vulgaris]*, Cross-leaved Heather *[Erica tetralix]*, and Black Crowberry *[Empetrum nigrum]* can spread extensively in humid-cool climate on lime-free sand dunes, on hummocks and edges of raised bogs as well as on extreme acid anmoor or peat soils. In the highlands, dwarf-shrub communities, i.e. *Nardus* grassland and *Calluna* heath vegetation are not so extensively distributed as on wide sandy plains; in the south of Germany more Calluna heath associations of the *Vaccinio-Genistetalia* alliance are to be found; cf. ELL 1978 et OBER 1978); *syn.* heathland [n]; *s* **brezal [m]** (Nombre vulgar de las asociaciones vegetales de las ▶landas que se componen principalmente de vacciniáceas *[Vaccinium myrtillus]*, ericáceas *[Erica cinerea, E. ciliaris, E. vagans, Calluna vulgaris]* y ulex *[Ulex europeus, U. minor, U. Galii]*. También se presentan como sotobosque de diferentes tipos forestales p. ej. de *Picea excelsa* y *Pinus silvestris* en el norte de Europa, pasando por *Fagus silvatica* y *Quercus robur* hasta *Pinus pinaster* y *Quercus suber* en el suroeste de Francia y en Portugal. La riqueza en especies leñosas varía geográficamente, encontrándose la mayor variedad en Europa en la Cordillera Cantábrica y en el País Vasco [España] con 13 especies de brezo y 13 de genista y la menor en el noroeste de Alemania y las regiones costeras del oeste del Báltico con sólo dos especies de brezo y tres de genista; cf. DB 1985 y DINA 1987; ▶garriga, ▶landa esteparia, ▶maquis, ▶matorral bajo); *syn.* landa [f] (2); *f 1* **lande [f] (2)** (Formation végétale de grande diversité, où dominent des végétaux ligneux, sous-arbrisseaux et arbrisseaux, à feuilles persistantes et les plantes herbacées, généralement fermée, pouvant comporter des buissons plus élevés ou des arbres isolés, établie sur les sols pauvres et acides ; ▶groupement d'arbrisseaux nains. On distingue suivant leur répartition géographique **1.** les **landes atlantiques** constituées de landes climaciques sur le littoral [landes littorales] et de landes secondaires ou une végétation de substitution a remplacé les peuplements forestiers telles que les landes à éricacées, formation végétale dont la strate la plus importante est formée de végétaux ligneux, sous arbrisseaux et arbrisseaux sempervirents tels que la Callune fausse-bruyère *[Calluna vulgaris]*, la Bruyère tétragone *[Erica tetralix]*, la Camarine noire *[Empetrum nigrum]*, etc., disséminées à partir des landes climaciques suite à un processus de déforestation sur sols acides, dépassant rarement 30 cm de hauteur, appartenant à la classe des *Calluno-Ulicetea* ; on distingue les landes sèches atlantiques *[Ulici-Ericion cinereae]*, les landes sèches médio-européennes *[Calluno-Genistion]*, les landes mésophiles *[Ulicion nani]*, les landes humides *[Ulicion nani et Ericion tetralicis]* ; GGV 1979, 311-315). **2.** Les **landes montagnardes et subalpines** telles que la lande épineuse, la brousse de Pin mugo, la brousse d'Aulne vert, les groupements à saules, les rhodoraies et les vacciniaies, les junipéraies, les landines ; cf. VCA 1985 ; ▶garrigue, ▶lande steppique, ▶maquis) ; *f 2* **fruticée [f] à bruyères** (F., formation ligneuse haute comportant de nombreuses espèces de grandes bruyères : Bruyère arborescente *[Erica arborea]*, Bruyère à balais *[Erica scoparia]*, Bruyère à fleurs terminales *[Erica terminalis]*, etc.) ; *f 3* **brande [f]** (Basse lande où dominent la grande bruyère à balais *[Erica scoparia]* ; la **b.** est aussi recouverte d'autres espèces de bruyères ainsi que d'ajoncs, de fougères et de genêts ; LA 1981, 198) ; *g* **Heide [f] (3)** (Vegetationstyp und Lebensbereich, der ein breites Spektrum von ▶Zwergstrauchgesellschaften umfasst, weitgehend baumlos. Im mitteleuropäischen Flachland können sich Zwergsträucher wie Besenheide *[Calluna vulgaris]*, Glockenheide *[Erica tetralix]* und Krähenbeere *[Empetrum nigrum]* in feucht-kühlem Klima auf entkalkten Flugsanddünen, auf Bulten und Randgehängen der Hochmoore sowie auf extrem sauren Anmoor- oder Torfböden massiv ausbreiten. Im Hügel- und Bergland waren Zwergstrauchheiden, also *Nardus*-Rasen und *Calluna*-Heiden, [Borstgrasheiden der Klasse *Nardo-Callunetea*] nie so stark ausgebreitet wie auf weiten sandigen Ebenen; statt dessen sind im süddeutschen Raum vermehrt Pflanzengesellschaften der Heidekraut-Heiden des Verbandes *Vaccinio-Genistetalia*, früher *Calluno-Ulicetalia*, anzutreffen; cf. ELL 1978 et OBER 1978; Stauden sind in **H.n** eher untergeordnet vertreten, gern auf Offensandflächen [Sandfluren]; ▶Garrigue, ▶Macchia, ▶Steppenheide).

heating plant [n] [US] *urb.* ▶central heating plant.

heating system [n] *urb.* ▶district heating system.

2657 heating up [n] *envir. limn.* (Increase of temperature of a body of water or the air); *syn.* raising [n] the temperature; *s* **calentamiento [m]** (Aumento de la temperatura de un curso de agua o del aire); *f* **échauffement [m] (excessif)** (de la température d'un cours d'eau ou de l'air) ; *g* **Aufheizung [f]** (Erhöhung der Temperatur eines Gewässers oder der Luft); *syn.* Erwärmung [f], Aufheizen [n, o. Pl.].

2658 heat insulation [n] (1) *constr.* (**1.** Protection against cold provided by cold-shielding materials in outer walls of a building to conserve heat and save energy. **2.** In English, the generic term **thermal insulation** is used for both heat and cold resistance; ▶heat protection); *s* **aislamiento [m] térmico (1)** (Medidas constructivas en muros, suelos y techos de los edificios para aislarlos contra el frío reduciendo las pérdidas de calor y así ahorrar energía; ▶protección contra pérdidas de calor); *syn.* aislamiento [m] contra el frío, protección [f] térmica (1), protección [f] contra el frío; *f* **isolation [f] thermique** (Amélioration de l'efficacité énergétique [réalisation d'économie d'énergie] des bâtiments par l'utilisation de solutions techniques ou de matériaux limitant la déperdition thermique des logements, la lame

principale de matériaux isolants pouvant être placée à l'extérieur ou à l'intérieur ; ▶protection thermique) ; *g* **Wärmeisolierung [f] (1)** (Zwecks Einsparung von Wärmeenergie die Abdichtung gegen Kälte durch technische Einrichtungen — wärmedämmende Stoffe — in der äußeren Haut eines Gebäudes mit dem Ziel, den Wärmedurchgang zeitlich zu verzögern, damit die dem Gebäude zugeführte Wärmemenge länger erhalten bleibt; ▶Wärmeschutz); *syn.* Wärmedämmung [f], Wärmeisolation [f].

2659 heat insulation [n] (2) *constr.* (Protection against heat provided by heat-shielding materials in the outer walls of a building to prevent heat build-up in hot regions or in temperate climates during the summer. In temperate latitudes such insulation is a measure of "summer heat protection"); *s* **aislamiento [m] térmico (2)** (Medidas constructivas en muros, suelos y techos con diferentes materiales para prevenir la transmisión de calor hacia el interior de los edificios); *syn.* aislamiento [m] contra el calor, protección [f] térmica (2), protección [f] contra el calor; *f* **protection [f] solaire** (Limitation de la température de l'air et des parois en été par une protection antisolaire des parois vitrées, un revêtement isolant thermique en l'extérieur [isolation par l'extérieur] avec élévation de l'inertie thermique des bâtiments dans les pays chauds) ; *g* **Wärmeisolierung [f] (2)** (Zwecks Einsparung von Energie die Abdichtung durch technische Einrichtungen in der äußeren Haut eines Gebäudes, damit in heißen Regionen oder in gemäßigten Klimaregionen in den Sommermonaten Gebäuden oder Räumen wenig Wärme zufließt. In gemäßigten Breiten ist eine solche Abdichtung eine Maßnahme des „sommerlichen Wärmeschutzes").

2660 heat island [n] *met. urb.* (Accumulation of warm air in the center [US]/centre [UK] of a city due to solar radiation absorption by pavements and building roofs, which in turn warm the night air through re-radiation of stored heat. Heat islands are mitigated by use of canopy tree cover to produce evapotranspiration cooling and use of lighter colored building materials to increase albedo or reflectivity. In large urban agglomerations, such a homogeneous accumulation may not be formed, because large green areas or forests divide the heat island into clusters of separate units, thus forming a so-called **heat archipelago**); *s* **isla [f] de calor** (Acumulación de calor en las ciudades. Si éstas tienen grandes zonas verdes en su interior se puede hablar de «archipiélago de calor»); *syn.* isla [f] térmica; *f* **îlot [m] thermique** (Colonne d'air chaud accumulée pendant la journée et s'élevant au-dessus du centre d'une ville ; dans le cas des grandes agglomérations il est préférable de parler « d'archipel thermique » lorsque des espaces verts ou des zones forestières importantes ne permettent pas la formation d'un îlot thermique homogène) ; *g* **Wärmeinsel [f]** (Ansammlung warmer Luft im Inneren einer Stadt. Bei Großstädten sollte man i. d. R. besser von einem **Wärmearchipel** sprechen, wenn durch großflächige Grünanlagen oder Waldflächen keine homogene Wärmeinsel entstehen kann).

2661 heat load [n] of the air *envir.* (TGG 1984, 247; increase in air temperature caused by reflected radiation of warm air from the surface of pavements and buildings); *s* **carga [f] térmica del aire** (Aumento de la temperatura del aire por radiación de calor de superficies pavimentadas y de edidicios); *f* **nuisance [f] thermique provoquée par le réchauffement de l'air** (Élévation de la température provoquée par le rayonnement solaire sur les surfaces imperméabilisées et les bâtiments) ; *g* **Wärmebelastung [f] der Luft durch Aufheizung** (Erhöhung der Lufttemperatur durch Wärmeabstrahlung von befestigten Flächen und Gebäuden); *syn.* Strahlungshitze [f].

2662 heat protection [n] *constr. leg.* (Measures to prevent or avoid the loss of heat by limiting heat interchange with insulation materials. **H. p.** applies to both protection against heat and

protection against cold; ▶heat insulation 1, ▶heat insulation 2);
s **protección [f] contra pérdidas de calor** (Conjunto de medidas de aislamiento fuerte de edificios para evitar o reducir pérdidas innecesarias de energía. **En Es.** regulada por el art. 15 del Código Técnico de Edificación que determina exigencias básicas para el ahorro de energía en la proyección, construcción, uso y mantenimiento de edificios; cf. RD 314/2006. **En D.** regulada desde 01.02. 2002 por la Ordenanza de Ahorro de Energía [*Antiguamente* Ordenanza de Aislamiento de Edificios contra Pérdidas de Calor]; ▶aislamiento térmico 1, ▶aislamiento térmico 2); *f* **protection [f] thermique** (Mesures contre les dépertes d'énergie [utilisation de matériaux isolants, ventilation ; la **p. t.** est également un élément favorable à l'▶isolation thermique en hiver ; ▶protection solaire) ; *g* **Wärmeschutz [m, o. Pl.]** (Maßnahmen zur Verhinderung resp. zur Vermeidung von Wärmeverlusten durch Einschränkung des Wärmeaustausches durch wärmedämmende Stoffe und Bauweisen; **W.** bewirkt sowohl einen Hitze- als auch einen Kälteschutz; in D. durch die Energieeinsparverordnung seit 01.02. 2002, die die frühere Wärmeschutzverordnung und Heizungsanlagenverordnung ablöst, geregelt; die zweite Fassung erschien 2004 und seit 01.10.2007 gilt eine Neufassung, die die EG-Richtlinie 2002/91/EG umsetzt; ▶Wärmeisolierung 1, ▶Wärmeisolierung 2).

2663 heat stress [n] *biol.* (TEE 1980, 46; impact of intense heat upon a living being, which may lead to death); *s* **insolación [f]** (Conjunto de fenómenos causados a los seres vivos por la exposición prolongada al sol, que puede conducir a la muerte); *syn.* estrés [m] térmico; *f* **insolation [f]** (Ensemble des phénomènes provoqués sur un individu par une exposition prolongée au soleil et pouvant entraîner la mort) ; *g* **Wärmebelastung [f]** (Starke Hitzeeinwirkung auf Lebewesen, die im Einzelfall zum Hitzetod führen kann); *syn.* Hitzestress [m].

heavy foliage [n] **of woody plants** *bot. hort.* ▶dense foliage of woody plants.

heavy maintenance [n] *constr. hort.* ▶high maintenance.

2664 heavy metal contaminated site [n] *phyt.* (Location characterized by the heavy metal toxicity of the soil in which only heavy metal tolerant plants will grow; *specific term* ▶zinc-contaminated site); *s* **ubicación [f] contaminada con metales pesados** (Emplazamiento caracterizado por la toxicidad de metales pesados contenidos en el suelo, por lo que sólo pueden crecer plantas metalofitas en él; *término específico* ▶estación calaminar); *syn.* sitio [m] contaminado con metales pesados; *f* **site [m] métallicole** (*Terme spécifique* ▶station calaminaire) ; *syn.* site [m] contaminé par les métaux lourds, zone [f] contaminée par les métaux lourds ; *g* **Schwermetallstandort [m]** (Standort, der durch Schwermetalltoxizität des Bodens geprägt ist und deshalb nur Schwermetallpflanzen wachsen lässt; *UB* ▶Galmeistandort).

2665 heavy metal contamination [n] *envir. s* **contaminación [f] de metales pesados**; *f* **contamination [f] par les métaux lourds** ; *g* **Schwermetallverseuchung [f]** *syn.* Schwermetallvergiftung [f], Schwermetallkontamination [f].

2666 heavy metal-tolerant [adj] *phyt.* (Ability of certain plants to grow on ▶heavy metal contaminated sites); *s* **resistente a metales pesados [loc]** (Referente a plantas que pueden crecer en ▶ubicación contaminada con metales pesados); *f* **résistant, ante aux métaux lourds [loc]** (Propriété des végétaux se développant sur les ▶sites contaminés par les métaux lourds) ; *g* **schwermetallresistent [adj]** (Eigenschaft von Pflanzen, die auf ▶Schwermetallstandorten wachsen können).

2667 heavy metal-tolerant plant [n] *phyt.* (Plant which grows on soils with toxic concentrations of e.g. Cu, Zn, Pb or Cr; *specific term* ▶zinc-tolerant plant); *syn.* metallophyte [n];

s **planta [f] metalofita** (Planta característica de suelos con metales pesados como Cu, Cr, Pb o Zn; *término específico* ▶planta calaminar); *syn.* metalofita [f]; *f* **espèce [f] métallophyte** (Flore spécifique qui pousse sur des sols riches en métaux lourds tels que Cu, Zn, Pb, Cd ou Cr et donc toxiques ; on distingue les **métallophytes strictes** [ou espèces strictement métallicole] que l'on ne retrouve que sur les terrains contaminés et les **pseudo-métallophytes** qui vivent habituellement sur d'autres sols mais peuvent également s'y adapter et y survivre [pseudométallophytes électifs ou indifférents] ; cf. REM 2006 ; *terme spécifique* ▶plante calaminaire) ; *syn.* métallophyte [f], espèce [f] métallicole, plante [f] (caractéristique) des sols toxiques ; *g* **Schwermetallpflanze [f]** (Pflanze, die auf Böden mit toxischen Konzentrationen von z. B. Cu, Zn, Pb, Cd oder Cr wächst; *UB* ▶Galmeipflanze); *syn.* Metallophyt [m].

2668 heavy metal-tolerant vegetation [n] *phyt.* (Herbaceous vegetation of areas rich in heavy metals—mainly zinc, copper, lead, nickel, cobalt, cadmium, chromium and arsenic—either from undisturbed metal ore near the soil surface or from mining and industrial activities; VIR 1982; *specific terms* ▶zinc-tolerant plant community); *s* **vegetación [f] metalofita** (Vegetación característica de suelos con metales pesados; *términos específicos* vegetación cuprícola, ▶vegetación calaminar); *f* **groupement [m] métallicole** (Association végétale strictement inféodé à un milieu fortement contaminé par les métaux lourds par un apport de cuivre, zinc, plomb, cadmium, etc. ; il s'agit en général de communautés végétales herbacées, [pelouses métallicoles] assez proches de celles des affleurements serpentiniques ; *terme spécifique* ▶groupement calaminaire ; *terme générique* groupement des sols toxiques) ; *syn.* association [f] métallicole, groupement [m] des sols contaminés par les métaux lourds; *g* **Schwermetallflur [f]** (Pflanzengesellschaft auf schwermetalltoxischem Boden, z. B. durch Kupfer-, Zink-, Blei- oder Kadmiumeintrag verursacht; *UBe* ▶Galmeiflur, Schwermetall-Grasnelkenrasen *[Armerietum halleri]* im Harz); *syn.* Schwermetallvegetation [f].

2669 heavy rainfall [n] *met.* (Precipitation rate for storm water of at least 5 l/m^2 in the first five minutes, 10 l/m^2 after 20 minutes and > 17 l/m^2 after one hour); *syn.* heavy storm water [n] [also US], downpour [n] [also US], torrential downpour [n] [also US], torrential pour [n] [also US]; *s* **lluvia [f] torrencial** (Lluvia intensa de más de 4 mm/h; DM 1986); *f* **pluie [f] torrentielle** (Pluie violente, de forte intensité, de longue durée et dont l'intensité minimale est de 5 l/m^2 les cinq premières minutes, 10 l/m^2 après vingt minutes et > 17 l/m^2 après une heure) ; *syn.* pluie [f] orageuse, pluie [f] cyclonique (CILF 1978) ; *g* **Starkregen [m]** (Niederschlag, bei dem innerhalb der ersten fünf Minuten mindestens 5 l/m^2, 10 l/m^2 nach 20 Minuten und > 17 l/m^2 nach 1 Std. fallen müssen).

heavy rainwater run-off [n] [UK] *constr. plan.* ▶heavy storm runoff [US].

2670 heavy recreation use [n] *plan. recr.* (Recreation use characterized by high visitor frequency; e.g. peak weekend recreation use [US]; TGG 1984, 160); *s* **uso [m] recreativo intensivo** (Alto grado de aprovechamiento de una zona, una instalación, etc. para usos recreativos); *syn.* frecuentación [f] turística intensiva; *f* **fréquentation [f] touristique dense** (à fort flux récréatif ou touristique) ; *g* **intensive Erholungsnutzung [f]** (Nutzung, die durch hohes Aufkommen von Erholungsuchenden gekennzeichnet ist).

heavy soil [n] *agr. hort. obs.* ▶fine-textured soil.

2671 heavy storm runoff [n] [US] *constr. plan. syn.* heavy rain-water run-off [n] [UK]; *s* **escorrentía [f] de lluvia**

torrencial; *f* ruissellement [m] torrentiel ; *g* **Starkregenabfluss** [m].

heavy storm water [n] [US] *met.* ►heavy rainfall.

heavy-use lawn [n] *constr.* ►play lawn [US] (1)/all-around lawn [UK].

2672 hedge [n] *gard. landsc.* (Linear planting of closely-growing shrubs or trees, mostly grown for enclosure and privacy screening; ►clipped hedge, ►elevated hedgerow [US]/quickset hedge [UK], ►hedgerow, ►thorny hedge, ►untrimmed hedge [US]/untrained hedge [UK], ►windbreak hedge); *s* **seto** [m] (Plantación lineal de arbustos o de árboles con capacidad de ►brote de cepa. Se diferencian ►setos vivos y ►setos podados; ►seto espinoso, ►seto rompevientos, ►seto vivo elevado, ►seto vivo interparcelario); *f* **haie** [f] (Plantation arbustive linéaire destinée à délimiter ou protéger un espace ; on distingue la ►haie libre, la ►haie taillée et la ►haie vive ; ►haie brise-vent, ►haie champêtre, ►haie défensive, ►haie sur talus) ; *g* **Hecke** [f] (Linear angeordnete Anpflanzung von Sträuchern oder stockausschlagfähigen Bäumen. Es werden ►freiwachsende Hecken und ►geschnittene Hecken unterschieden; ►Dornenhecke, ►Feldhecke, ►Knick, ►Windschutzhecke).

hedge [n], **field** *agr. land'man.* ►hedgerow.

hedge [n] [UK], **quickset** *geo. land'man.* ►elevated hedgerow [US].

hedge [n], **trimmed** *constr. gard.* ►clipped hedge.

hedge [n] [UK], **untrained** *gard. landsc.* ►untrimmed hedge [US].

2673 hedge brush layer [n] *constr.* (►Biotechnical construction technique of inserting dead, coarse branch material, even conifer branches as well as rooted and easy-rooting whips and dormant cuttings into in-situ soil in a stepped formation on unstable slopes; ►brush layering, ►hedge layering); *s* **bancal** [m] **de estabilización con ramajes y esquejes (≠)** (*Construcción biotécnica* pequeños bancales construidos en pendientes muy fuertes, en los que introducen ramas y desbrozos muertos y plántulas y esquejes para fijarlos; ►estabilización con arbustos, ►estabilización con ramajes); *f* **lit** [m] **de plants et plançons** (WIB 1996 ; terrasse exécutée parallèlement à la pente d'un talus instable sur laquelle est mis en place un amas de grosses branches et de ramilles mortes mélangées à des plançons [branche de saule utilisée comme bouture] ; ►procédé de stabilisation de talus par lits de plants, ►stabilisation de talus par lit de plants et plançons) ; *g* **Heckenbuschlage** [f] (In rutschgefährdeten Steilböschungen hangparallel eingegrabene Stufe/Terrasse, in die nicht ausschlagfähiges, grobes Reisig und Astwerk, auch Nadelholzäste sowie Jungpflanzen — Sämlinge und bewurzelte Steckhölzer — eingelegt werden; cf. DIN 18 918; ►Buschlagenbau, ►Heckenlagenbau).

2674 hedge layer [n] *constr.* (►Biotechnical construction technique of inserting live, rooted and easy-rooting whips at least 100cm long, into in-situ soil in a stepped formation on unstable slopes; ►hedge layering); *s* **bancal** [m] **de estabilización con arbustos (≠)** (*Construcción biotécnica* pequeños bancales construidos en pendientes muy fuertes, en los que se plantan arbustos vivos de un mínimo de 100 cm de altura con raíces adventicias y de enraizamiento fácil para fijarlos; ►estabilización con arbustos); *f* **lit** [m] **de plants** (Terrasse exécutée parallèlement à la pente d'un talus instable sur laquelle est mis en place, sur une épaisseur minimale de 100 cm, des jeunes plants capables de former rapidement des racines adventives ; WIB 1996 ; ►procédé de stabilisation de talus par lits de plants) ; *g* **Heckenlage** [f] (In rutschgefährdeten Steilböschungen hangparallel eingegrabene Stufe/Terrasse, in die ausschließlich bewurzelte, leicht Adventiv-

wurzeln bildende Junggehölze von mindestens 100 cm Höhe eingelegt werden; cf. DIN 18 918; ►Heckenlagenbau).

2675 hedge layering [n] *constr.* (►Biotechnical construction technique for establishing a permanent vegetation cover which is suited to site conditions by ►hedge layers; in comparison with ►brush layering, this method can be used only during the planting period); *s* **estabilización** [f] **con arbustos** (Método de ►construcción biotécnica para establecer una cubierta de vegetación permanente, adecuada al emplazamiento, en pendientes fuertes, para lo que se utilizan ►bancales de estabilización con arbustos. Este método sólo se puede aplicar — al contrario que el de ►estabilización con ramajes— en la época de plantación); *f* **procédé** [m] **de stabilisation de talus par lits de plants** (►Technique de génie biologique employée pour l'établissement au moyen de ►lits de plants d'une végétation permanente pour la stabilisation des talus à forte pente menacés de glissement ; technique ne pouvant être mise en place que pendant la période de végétation ; ►stabilisation de talus par lit de plants et plançons) ; *g* **Heckenlagenbau** [m] (►Ingenieurbiologische Bauweise zur Herstellung einer standortgerechten Dauervegetation durch ►Heckenlagen auf rutschgefährdeten Steilböschungen; cf. DIN 18 918. Diese Methode kann im Vergleich zum ►Buschlagenbau nur während der Pflanzzeit angewendet werden).

2676 hedge plant [n] *hort.* (Woody species grown in a nursery, which can be trimmed [US]/trained [UK] to form a thick screen); *syn.* hedging plant [n] [also UK]; *s* **planta** [f] **para setos** (Especie de leñosa cultivada en vivero, bien ramificada en la base y con aptitud para el corte, que sirve en especial para crear setos que se cortan regularmente); *f* **plante** [f] **de haies** (Espèce végétale supportant la taille, cultivée en pépinière et adaptée à la constitution de haies ; *terme spécifique* arbuste pour haies) ; *syn.* espèce [f] pour haies ; *g* **Heckenpflanze** [f] (In Baumschulen kultivierte, von unten an gut verzweigte, schnittverträgliche Gehölzart [Strauch oder Heister], die sich besonders gut für die Anlage einer regelmäßig zu schneidenden Hecke eignet; Sortierung entweder nach Triebzahl 2, 3/4, 5/7, 8/12 oder nach Höhe; cf. LEHR 1997, 289).

2677 hedgerow [n] *agr. land'man.* (Linear planting of usually native shrubs and trees to form a field boundary, often with the purpose of hindering the movement of animals—many **h.s** are also called 'fencerows'; cf. LE 1986, 401; ►elevated hedgerow [US]/quickset hedge [UK], ►hedgerow landscape, ►windbreak planting, ►untrimmed hedge); *syn.* field hedge [n]; *s* **seto** [m] **vivo interparcelario** (Plantación lineal de arbustos que divide los campos de cultivo y que cumple función de pantalla cortavientos; ►cortina rompevientos, ►paisaje de bocage, ►seto vivo, ►seto vivo elevado); *f* **haie** [f] **champêtre** (Haie constituée par l'association d'arbres et d'arbustes locaux en zone rurale agricole ; ►bande boisée brise-vent, ►bocage, ►haie libre, ►haie sur talus, ►haie vive, ►plantation brise-vent) ; *syn.* haie [f] bocagère (dans un paysage bocager) ; *g* **Feldhecke** [f] (Lineare Anordnung von aus standortgerechten, stockausschlagfähigen Sträuchern und Bäumen, oft in Form einer ►Windschutzpflanzung, die in der bäuerlichen Kulturlandschaft Felder und Wiesen umgrenzt; ►freiwachsende Hecke, ►Heckenlandschaft, ►Knick); *syn.* Baum- und Strauchhecke [f], flurschützende Hecke [f].

2678 hedgerow clearance [n] *agr. landsc.* (Grubbing up of hedgerows; ►elevated hedgerow [US]/quickset hedge [UK], ►land clearance [US]/landscape clearance [UK]); *syn.* ripping out [n] of hedgerows [also UK]; *s* **tala** [f] **de setos interparcelarios** (Remoción total de los setos en una zona; ►desmonte total del paisaje, ►seto vivo elevado); *f* **enlèvement** [m] **des haies** (Enlèvement des haies provoquant la destruction partielle ou totale du paysage bocager ; ►haie sur talus; LA 1981, 180 ;

▶dégagement du paysage) ; *syn.* destruction [f] des haies, arasement [m] des haies (LA 1981, 180) ; *g* **Schleifen [n, o. Pl.] von Feldhecken** (Entfernen von Feldhecken, Feldhecken schleifen [vb]; ▶Ausräumung der Landschaft, ▶Knick).

2679 hedgerow conservation [n] *agr. conserv. nat'res.* (Preservation and maintenance of hedges with respect to their important ecological functions in the agriculturally-cultivated countryside); *s* **protección [f] de setos** (Preservación y mantenimiento de setos debido a su importancia ecológica en áreas agrícolas); *f* **protection [f] des haies** (Sauvegarde et entretien des haies réputées pour leurs multiples fonctions écologiques en zone rurale) ; *g* **Heckenschutz [m, o. Pl.]** (Erhaltung und Pflege von Hecken wegen ihrer wichtigen ökologischen Funktionen in der bäuerlichen Kulturlandschaft).

2680 hedgerow landscape [n] *land'man.* (Broad fields divided by linear planting to separate crop areas from pasturing and as a ▶windbreak planting to protect against erosion; ▶hedgerow); *s* **paisaje [m] de bocage** (Paisaje agrario típico de Europa Occidental presente, sobre todo, en la fachada atlántica. Se caracteriza por la presencia de ▶setos vivos interparcelarios o setos vivos elevados en los linderos de los campos; cf. DGA 1986; ▶cortina rompevientos); *syn.* paisaje [m] de campos cercados; *f* **bocage [m]** (Type de paysage caractéristique de l'Ouest de la France, constitué de champs irréguliers limités par des talus et des haies ; DEE 1982 ; ▶haie champêtre, ▶plantation brise-vent) ; *syn.* paysage [m] bocager (LA 1981, 180), paysage [m] de bocage ; *g* **Heckenlandschaft [f]** (Durch ▶Feldhecken geprägte Landschaft; ▶Windschutzpflanzung); *syn. in SH* Knicklandschaft [f].

2681 hedgerow maintenance [n] *agr. land'man. landsc.* *syn.* hedging [n] [also UK]; *s* **cuidado [m] de setos vivos interparcelarios**; *f* **entretien [m] des haies champêtres** *syn.* entretien [m] des haies bocagères ; *g* **Feldheckenpflege [f, o. Pl.]** *syn. in SH* Knickpflege [f, o. Pl.], Wallheckenpflege [f, o. Pl.].

hedgerows [npl] **and woodland patches** [npl] *agr. land'man.* ▶planting of hedgerows and woodland patches.

hedgerow timber [n] [UK] *agr. land'man.* ▶field tree.

hedgerow tree [n] [US] *agr. land'man.* ▶field tree.

hedging [n] [UK] *agr. land'man. landsc.* ▶hedgerow maintenance.

hedging plant [n] *hort.* ▶hedge plant.

heel cutting [n] *constr. hort.* ▶dormant cutting.

2682 heeling-in [n] *constr. hort.* (Temporary digging in of plant material between delivery to site and final planting out; heel in [vb] [US]/lay-in [vb] [UK]; ▶heeling on site, ▶site nursery); *s* **enterrado [m] de plantas** (Cobertura temporal de las raíces de plantas en vivero o en obra antes de su plantación definitiva, para protegerlas contra la sequedad, el calor o las heladas. Los cepellones que se hayan secado durante el transporte deben ser regados resp. mantenidos húmedos hasta su replantación; ▶depósito de material de plantación, ▶enterrado de plantas in situ); *syn.* amurillado [m] de plantas; *f* **mise [f] en jauge** (Mise en terre provisoire sur le chantier ou en pépinière de végétaux prévus pour la plantation ; mettre [vb] en jauge ; ▶jauge, ▶jauge d'attente) ; *g* **Einschlag [m] (2)** (Bedecken der Wurzeln von ausgegrabenen Pflanzen zur Zwischenlagerung auf der Baustelle oder in der Baumschule mit Boden oder geeigneten Materialien, um die Pflanzen gegen Austrocknung, Überhitzung oder Frost zu schützen; beim Transport trocken gewordene Ballen müssen sofort gegossen resp. feucht gehalten werden; Gehölze, die länger als zwei Tage im Einschlag bleiben, sind abseits des Baubetriebes in vorbereitete Gräben [Einschlaggräben] zu stellen, zu befeuchten, an den Wurzeln/Ballen mit lockerem Boden zu umgeben und

gegebenenfalls einzuschlämmen; cf. DIN 18 916; ▶Baustelleneinschlag, ▶Einschlagplatz); *syn.* Einschlagen [n, o. Pl.] von Pflanzen, Pflanzeneinschlag [m].

2683 heeling-in [n] **on site** *constr.* (Method of storing plants temporarily in the ground or other media until they can be planted in their final positions; ▶heeling-in, ▶site nursery); *s* **enterrado [m] de plantas in situ** (Cobertura temporal de las raíces de plantas transportadas a obra, si no pueden ser plantadas en un plazo máximo de 48 horas; ▶depósito de material de plantación, ▶enterrado de plantas); *syn.* amurillado [m] (de plantas) en obra; *f* **jauge [f] d'attente** (Plantation provisoire sur le chantier des végétaux ne pouvant être plantés dans la journée de leur livraison ; ▶jauge, ▶mise en jauge) ; *syn.* mise [f] en jauge provisoire ; *g* **Baustelleneinschlag [m]** (Temporäres Überdecken von Wurzeln einer Pflanzenlieferung, die nicht innerhalb von 48 Stunden nach Eintreffen auf der Baustelle gepflanzt werden kann; cf. DIN 18 916; ▶Einschlag, ▶Einschlagplatz).

2684 height [n] **above mean sea level** *surv.* (Elevation on land with reference to an average level calculated to lie between low and high tides; **in U.S.,** the National Geodetic Vertical Datum [NGVD 88] is referenced to a single datum located at Father Point, Rimouski, Quebec, Canada; **in U.K.,** the Ordnance Datum for mean sea level is at Newlyn in Cornwall; **in D.,** the zero reference level is that of Amsterdam; for Austria, the level of Trieste; for Switzerland, the boulder—Pierre de Niton—on Lake Geneva; for France, the level of Marseille and for Spain, the level of Alicante); *syn.* elevation [n] above mean sea level; *s* **altura [f] sobre el nivel del mar** (En Es. se usa como nivel del mar el nivel medio del Mediterráneo en Alicante, considerándole nivel cero y refiriéndose a ese punto y nivel cualquier altura o profundidad del territorio nacional; DGA 1986); *syn.* altitud [f] absoluta; *f* **altitude [f] absolue** (Élévation d'un point au-dessus de l'ellipsoïde de référence [F., ellipsoïde de Clarke] coïncidant lui-même avec le niveau de la mer défini par un médimarémètre [**F.,** celui de Marseille, **D.,** d'Amsterdam, **A.,** de Trieste, **CH.,** de la Pierre de Niton sur le lac de Genève, **Es.,** d'Alicante] ; DG 1984 ; *g* **Höhe [f] über Normalnull** (*Abk.* Höhe über NN; für D. ist die Bezugshöhe der Nullpunkt des Amsterdamer Pegels, für Österreich eine Höhenmarke am Triester Pegel und für die Schweiz der Felsblock — Pierre de Niton — am Genfer See, für das U.K. der Pegel in Newlyn, Cornwall, in Frankreich der Marseiller Pegel und für Spanien der Pegel von Alicante); *syn.* Höhe [f] über dem Meeresspiegel.

helical stair [n] *arch.* ▶spiral stair(case).

heliophilous species [n] *for. phyt.* ▶light-demanding plant.

heliophilous woody species [n] *for. phyt.* ▶light-demanding woody species.

2685 helophyte [n] *bot.* (Greek *helos* >swamp< and *phyton* >plant<; plant growing in permanently saturated or frequently flooded soils; ▶emergent aquatic plant 2); *syn.* marsh plant [n], swamp plant [n] [US]; *s* **helófito [m]** (Planta acuática que arraiga en el fondo sumergido, atraviesa con su tallo la masa acuática, y desarrolla sus hojas, flores y frutos en el medio aéreo; DB 1985; ▶planta acuática emergente); *syn.* planta [f] anfibia; *f* **hélophyte [m]** (Plante qui croit dans la vase et dont l'appareil végétatif est aérien et dressé ; DEE 1982 ; les principales espèces sont les joncs [*Juncus* spp.], roseaux [*Phragmites* spp.], carex [*Carex* spp.], typhas [*Typha* spp.], iris [*Iris pseudacorus*]; ▶amphiphyte) ; *syn.* plante [f] palustre, plante [f] paludéenne ; *g* **Sumpfpflanze [f]** (Griech. *helos* >Sumpf< und *phyton* >Pflanze<; auf ständig nassen oder jährlich häufig überfluteten Böden wachsende Pflanze; ▶Überwasserpflanze); *syn.* Helophyt [m].

helophyte [n], **tall** *bot.* ▶emergent aquatic plant (2).

hemerophilous species [n] *biol.* ▶anthropophilous species.

2686 hemerophobic species [n] *biol.* (Plant or animal species which cannot live within an area influenced by human occupation; *opp.* ▶anthropophilous species); *s* **especie** [f] **hemerófoba** (Plantas que rehuyen los cultivos, que no medran en los huertos, jardines y campos culturados; DB 1985; *opp.* ▶especie antropófila); *f* **espèce** [f] **anthropophobe** (Espèce de faune ou de flore ne pouvant vivre sous l'influence de l'homme ; *opp.* ▶espèce anthropophile) ; *g* **Kulturflüchter** [m] (Pflanzen- oder Tierart, die nur außerhalb des menschlichen Einflussbereichs leben kann; *opp.* ▶Kulturfolger); *syn.* hemerophobe Art [f].

2687 hemicryptophyte [n] *bot.* (Greek. *hemi* >half<, *kryptos* >hidden< and *phyton* >plant<: Nomenclature according to RAUNKIAER'S classification for herbaceous plant whose buds survive on the soil's surface during the winter. **H.s** are subdivided in ▶**rosette plants** *[Hemikryptophyta rosulata]*; e.g. Dandelion *[Taraxacum officinale]*, Plantain *[Plantago]*, **scape-bearing plants** *[Hemikryptophyta scaposa]* overwintering without rosette, e.g. Wormwood *[Artemisia]*, White Wood Aster *[Aster divaricatus]*, ▶**tussock plants** *[Hemikryptophyta caespitosa]*, e.g. Waldsteinia *[Waldsteinia geoides]* and a lot of tussock grasses; **perennials with runners** *[Hemikryptophyta reptantia]*, e.g. Creeping Buttercup *[Ranunculus repens]*, Woodland Strawberry *[Fragaria vesca]* and ▶**climbing perennials** *[Hemikryptophyta scandentia]*, e.g. Morning-Glory *[Ipomoea imperialis]*; ▶perennial 1); *s* **hemicriptófito** [m] (En la clasificación de RAUNKIAER conjunto de plantas herbáceas en las que muere anualmente la parte aérea, y las yemas de reemplazo quedan próximamente a ras del suelo. Se subdividen en: **1.** ▶**Plantas cespitosas** *[Hemikryptophyta caespitosa]* como los géneros *Festuca, Sesleria, Carex, Poa.* **2.** ▶**Plantas en roseta** o arrosetadas *[Hemikryptophyta rosulata]*, con todas las hojas en roseta basal y tallo sin hojas, a las cuales pertenecen, entre otras, muchas especies de *Anemone, Primula, Gentiana, Plantago, Taraxacum.* **3. Hemicriptófitos escaposos** *[Hemikryptophyta scaposa]* que en su mayoría son hierbas altas, que en las zonas cálidas pueden alcanzar varios metros de altura, y pueden tener roseta basal *[Ranunculus, Geranium, Salvia]* o no *[Aconitum, Epilobium, Gentiana]*. **4.** ▶**Hemicriptófitos trepadores** *[Hemikryptophyta scandentia]*, a los que pertenecen muchas perennes con tallo trepador que muere anualmente y con yemas de renovación basales, como los géneros *Aristolochia, Corydalis, Galium.* **5. Hemicriptófitos decumbentes** *[Hemikryptophyta repentia]*, que permanecen sobre el suelo y no arraigan, como *Mertensia maritima* o especies de *Centaurea* y **6. Hemicriptófitos reptantes** *[Hemikryptophyta reptantia]*, plantas perennes que producen estolones que luego arraigan, como la fresa/frutilla silvestre *[Fragaria vesca]*, especies de *Potentilla*, entre otras; cf. BB 1979; ▶planta vivaz); *f* **hémicryptophyte** [m] (Du grec *hemi* « à demi », *kryptos* « caché » et *phyton* « plante » ; classification des végétaux vasculaires [plantes vivaces et graminées] d'après RAUNKIAER désignant des plantes enracinées dont le bourgeon est situé au ras du sol, et dont la partie aérienne meurt pendant la saison défavorable [froide ou sèche] ; on distingue les **hémicryptophytes à rosette**, *[Hemikryptophyta rosulata]*, à couronne de feuilles au niveau du sol, p. ex. le Pissenlit *[Taraxacum officinale]*, le Plantain *[Plantago]* ; les **hémicryptophytes caulescentes**, à tige développée, généralement feuillue et ramifiée ; les **hémicryptophytes cespiteuses**, *[Hemikryptophyta caespitosa]* formant des touffes, comme beaucoup de graminées, de cypéracées et de joncs, comme le Waldsteinia *[Waldsteinia geoides]* et de nombreuses graminées en touffes ; les **hémicryptophytes grimpantes** *[Hemikryptophyta scandentia]*, comme l'Ipomée impériale *[Ipomoea imperialis]*, à tige volubile **ou décombrantes** *[Hemikryptophyta reptantia]*, comme la Renoncule rampante *[Ranunculus repens]*, la Fraise des bois *[Fragaria vesca]* ;

▶plante à rosette, ▶plante cespiteuse, ▶plante grimpante vivace, ▶plante vivace) ; *g* **Hemikryptophyt** [m] (Griech. *hemi* >halb<, *kryptos* >verborgen< und *phyton* >Pflanze<; nach RAUNKIAER Lebensformklasse der ▶Stauden und Gräser [viele *Poaceae*], bei der die Überwinterungs-/Erneuerungsknospen der sich periodisch erneuernden/absterbenden, meist nur kurzlebigen oberirdischen Organe auf der Höhe der Erdoberfläche — geschützt durch Erde und Laubstreu — liegen; **H.en** werden untergliedert in ▶Rosettenpflanzen [Hemikryptophyta rosulata], z. B. Löwenzahn *[Taraxacum officinale]*, Wegerich *[Plantago]*, ohne Blattrosette überwinternde **Schaftpflanzen** *[Hemikryptophyta scaposa]*, z. B. Beifuß *[Artemisia]*, Waldaster *[Aster divaricatus]*, ▶Horstpflanzen [Hemikryptophyta caespitosa], z. B. Ungarwurz *[Waldsteinia geoides]* und viele Horstgräser; **Stauden mit oberirdischen Ausläufern** [Hemikryptophyta reptantia], z. B. Kriechender Hahnenfuß *[Ranunculus repens]*, Walderdbeere *[Fragaria vesca]* und ▶**Klimmstauden** *[Hemikryptophyta scandentia]*, z. B. Japanische Kaiserwinde *[Ipomoea imperialis]*. Ein Großteil der Staudenarten der gemäßigten Florenzonen gehört zu dieser Lebensform); *syn.* Erdschürfepflanze [f], Erdschürfestaude [f].

herb [n] *hort. plant.* ▶culinary herb; *bot. plant.* ▶medicinal herb.

herb [n], **perennial** *bot. hort. plant.* ▶perennial (1).

herb [n], **wild** *hort.* ▶wild flower.

2688 herbaceous edge [n] [US] *phyt.* (Linear-shaped stand of forb vegetation along the edge of copses, hedges or forests; ▶forest edge); *syn.* perennial herb border [n] [LE 1986, 108], herbaceous seam [n] [UK]; *s* **orla** [f] **herbácea** (Borde de vegetación herbácea a lo largo de sotos, setos y bosques; ▶lindero del bosque); *syn.* franja [f] herbácea; *f* **ourlet** [m] **herbacé** (Frange herbacée linéaire en bordure de taillis, de buissons, de haies vives et de forêts ; ▶lisière forestière) ; *g* **Krautsaum** [m] (...säume [pl]; linearer, krautiger Vegetationsbestand entlang Gebüschen, Hecken und Wäldern; ▶Waldrand); *syn.* Krautrain [m].

herbaceous fringe biotope [n] *phyt. zool.* ▶seam biotope.

herbaceous fringe community [n] *phyt.* ▶seam community.

herbaceous seam [n] [UK] *phyt.* ▶herbaceous edge [US].

herbaceous stratum [n] *phyt.* ▶herb layer.

herbaceous vegetation [n] (1) *phyt.* ▶forb vegetation.

2689 herbaceous vegetation [n] (2) *agr. for. hort.* (**1.** Any form of non-woody plants covering the ground; BS 3975: Part 5. **2.** Leafy growth of herbaceous plants; ▶herb layer); *syn.* herbage [n]; *s* **cobertura** [f] **herbácea** (Vegetación de plantas herbáceas; ▶estrato herbáceo); *syn.* herbaje [m]; *f* **couverture** [f] **herbacée** (Peuplement végétal constitué exclusivement de plantes herbacées ; ▶strate herbacée) ; *syn.* couvert [m] herbacé, tapis [m] herbeux, végétation [f] herbacée (2) ; *g* **Krautbewuchs** [m, o. Pl.] (Vegetationsbestand aus nur krautigen Pflanzen; ▶Krautschicht); *syn.* Krautvegetation [f].

herbage [n] *agr. for. hort.* ▶herbaceous vegetation (2).

herb community [n], **perennial** *phyt.* ▶perennial forb community.

herb community [n], **tall** *phyt.* ▶tall forb community.

2690 herb garden [n] *gard.* (Garden or area thereof in which aromatic, medicinal or kitchen herbs [spices] are cultivated; ▶aromatic garden; ▶culinary herb, ▶kitchen garden, ▶medicinal herb garden, ▶spices garden; *generic term* ▶fruit and vegetable garden); *s* **jardín** [m] **de hierbas finas** (Parte de un jardín en el que se cultivan plantas aromáticas, ▶plantas condimenticias o plantas medicinales; *términos específicos* ▶huerta 2, ▶jardín de hierbas condimenticias, ▶jardín de plantas aromáticas, ▶jardín medicinal; *término genérico* ▶huerto; ▶jardín de plantas aromáticas); *f* **jardin [m] d'herbes** (*Terme générique* jardin utilisé pour la culture des plantes aromatiques, plantes médicinales ou ►plantes condimentaires ; *termes spécifiques* ►jardin de senteurs, ►jardin d'herbes aromatiques, ►jardin d'herbes condimentaires, ►jardin médicinal, ►jardin potager, ; *terme générique* ►jardin de rapport) ; *g* **Kräutergarten [m]** (Teil einer Gartenanlage, in der Duft-, Heil- oder Küchenkräuter [►Gewürzpflanzen] kultiviert werden. ,Kräuter' werden immer wieder mit ,Gewürzen' gleichgestellt, obwohl folgender Unterschied besteht: Kräuter sind die Blätter von frischen oder getrockneten Pflanzen und **Gewürze** deren aromatische Teile wie z. B. Knospen, Samen, Beeren, Wurzeln oder Rinde; ►Duftgarten, ►Gewürzgarten, ►Heilpflanzengarten, ►Küchengarten; *OB* ►Nutzgarten).

2691 herbicide [n] *agr. chem. envir. for. hort.* (Specific term of 'phytocide'; chemical agent used to kill or inhibit the growth of forbs and grasses; ►contact herbicide, ►defoliating agent, ►nonselective herbicide [US]/non-selective herbicide [UK], ►root herbicide, ►selective herbicide); *syn.* weedkiller [n]; *s* **herbicida [f]** (Sustancia química empleada para luchar contra las hierbas indeseadas en los cultivos. Se diferencia entre ►herbicida total y ►herbicida selectivo; ►defoliador, ►herbicida de contacto, ►herbicida de translocación); *f* **herbicide [m]** (Substance chimique de lutte contre les mauvaises herbes ; on fait la distinction entre l'►herbicide sélectif et l'►herbicide total ; ►défoliant, ►défoliant, ►herbicide de contact, ►herbicide racinaire, ►phytocide) ; *syn.* désherbant [m] ; *g* **Herbizid [n]** (Chemisches Mittel zur Bekämpfung von Unkräutern. Es wird zwischen ►Teilherbizid und ►Totalherbizid unterschieden; ►Entlaubungsmittel, ►Kontaktherbizid, ►Wurzelherbizid); *syn.* Unkrautvernichtungsmittel [n], Unkrautbekämpfungsmittel [n], Wildkrautbekämpfungsmittel [n], Unkrautvertilgungsmittel [n].

herbicide application [n] *agr. for. hort. envir.* ►post-emergent herbicide application, ►pre-emergent herbicide application.

2692 herb layer [n] *phyt.* (Herbage and young woody plants, which do not project above the **h. l.** and occur in the stratifycation, e.g. between the ►moss layer and ►shrub layer of a forest. ►Tall forb formations may be distinguished in upper and lower **h. l.s**; ►stratification of plant communities); *syn.* herbaceous stratum [n] (TEE 1980, 279); *s* **estrato [m]. herbáceo** (Capa de vegetación de plantas herbáceas y plántulas arbóreas y arbustivas que se encuentra en el monte alto entre el ►estrato muscinal y el ►estrato arbustivo; ►estratificación de la vegetación, ►formación megafórbica); *f* **strate [f] herbacée** (Groupement dans laquelle prédominent les plantes herbacées, avec présence de petits végétaux ligneux et de jeunes plants d'arbres ou d'arbustes ; dans la structure verticale de la végétation d'un groupement forestier la **s. h.** se situe entre la ►strate muscinale et la ►strate arbustive ; dans les groupements non forestiers dans lesquels se succèdent deux synusies on distingue les **s. h.** inférieure et supérieure ; ►mégaphorbiaie, ►stratification de la végétation) ; *syn.* couche [f] herbacée ; *g* **Krautschicht [f]** (Aus krautigen Pflanzen und Jungpflanzen von Gehölzen bestehender Vegetationsbestand, der nicht über die **K.** hinausragt und in der Schichtenfolge, z. B. in einem Wald, zwischen ►Moosschicht und ►Strauchschicht liegt. Bei ►Hochstaudenfluren kann z. B. zwischen einer oberen und unteren **K.** unterschieden werden; ►Vegetationsschichtung).

herb reed swamp [n] [UK]**, tall** *phyt.* ►tall forb reed marsh [US]/tall forb reed swamp [UK].

heritage [n] *conserv.'hist. land'man. nat'res.* ►conservation of the natural and cultural heritage [CH], ►coastal heritage, ►cultural heritage, ►natural heritage.

Heritage [n] [UK]**, Chief Executive of English** *conserv' hist.* ►State Historic Preservation Officer [US].

heritage [n]**, forest** *conserv. land'man. nat'res.* ►forest patrimony.

heritage [n]**, landscape resource** *conserv. land'man.* ►landscape patrimony.

heritage area [n] [UK] *arch. conserv'hist. leg.* ►historic district [US].

heritage coast [n] [UK] *conserv. landsc.* ►coastal reserve [US].

heritage landscape [n] *conserv.* ►protected cultural heritage landscape.

heritage management [n] [UK] *adm. conserv'hist.* ►historic preservation measures [US]/measures for conservation of historic monuments [and sites] [UK].

heritage [n] **of historic gardens** *conserv'hist. gard'hist.* ►cultural heritage of historic gardens.

heritage site [n] *conserv. leg. pol.* ►world heritage site.

2693 heron rookery [n] [US] *zool.* (Colony where herons breed); *syn.* heronry [n] [UK]; *s* **colonia [f] de garzas** (Colonia de garzas y el lugar donde anida y cría sus polluelos); *f* **héronnière [f]** (Colonie de hérons sur son aire de nidification) ; *g* **Reiherkolonie [f]** (Kolonie mit Reihernistplätzen).

2694 herringbone pattern [n] *constr.* (Arrangement of zigzag lines; ►paving pattern); *syn.* spicatum opus [n] [also UK]; *s* **aparejo [m] en espina** (Patrón de aparejo de pavimentos característico de la época romana cuyos ladrillos clinker o adoquines están ordenados en forma de zig-zag); *syn.* dibujo [m] en espina; *f* **appareillage [m] en chevron** (Mise en œuvre caractéristique de l'époque romane, la pose de briques s'effectuant suivant une ligne en zigzag ; ►calepinage) ; *syn.* appareillage [m] en épi, appareillage [m] en opus spicatum, appareillage [m] en arêtes de poissons (VRD 1994, tome 2, 5.3.3, p. 8) ; *g* **Fischgrätenmuster [n]** (►Verlegemuster von Klinkern oder länglichen Betonpflastersteinen in Zickzacklinien); *syn.* Fischgrätenverband [m].

2695 herringbone pattern drainage [n] *constr.* (LAD 1986, 215; ►subsurface drainage whereby ►lateral drains are laid in a transverse direction to the slope of the ground; i.e. almost parallel or slightly angled/slanted to the ►contours; ►grid pattern drainage); *s* **drenaje [m] a espina de pez** (Sistema de drenaje en el cual los ►drenes secundarios se encuentran en perpendicular a la caída del terreno, es decir más o menos paralelos a las ►curvas de nivel; ►drenaje longitudinal, ►drenaje subterráneo); *syn.* drenaje [m] en forma de espina; *f* **drainage [m] transversal** (►Drainage souterrain dans lequel les drains recoupent légèrement les ►courbes de niveau, le collecteur étant disposé dans le sens de la plus grande pente ; ►drain adducteur, ►drainage longitudinal) ; *g* **Querdränung [f]** (►Dränung, bei der die ►Sauger quer zum Längsgefälle eines Geländes, also annähernd parallel zu den ►Höhenlinien liegen; ►Längsdränung).

Hessian strips [npl] [UK]**, wrapping with** *constr. arb.* ►wrapping with burlap [US].

hessian-wrapped root ball [n] [UK] *hort.* ►burlapped root ball.

hessian wrapping [n] *hort.* ►burlapping.

heterogeneity [n] *plan.* ►spatial heterogeneity.

heterogeneity [n]**, landscape** *land'man. landsc.* ►landscape diversity.

hibernaculum [n] *zool.* ►hibernation site.

H

2696 hibernation [n] *zool.* (Winter dormancy of animals in a state of deep sleep, when metabolic processes are slow and body temperature falls; ▶hibernation area/region, ▶overwintering 1, ▶overwintering 2, ▶winter sleep); *s* **invernación [f] (1)** (1. Dormancia invernal de animales en la que la termperatura corporal desciende y las funciones vitales se reducen al mínimo. **2.** Paso del invierno en ▶área de invernación; ▶hibernación 2, ▶invernación 2, ▶invernación 3); *f* **hibernation [f]** (Passer l'hiver en état d'engourdissement suite à l'abaissement permanent de la température centrale chez certains mammifères et certains oiseaux ; ▶aire d'hibernage, ▶hivernation 1, ▶hivernation 2, ▶sommeil hibernal, ▶territoire d'hibernation) ; *g* **Überwinterung [f] (1)** (1. Verbringung des Winters in einem Zustand des tiefen Schlafes, bei dem die Körpertemperatur abgesenkt wird und alle Lebensfunktionen auf ein Minimum reduziert sind. **2.** Verbringen der kalten Jahreszeit durch jahresperiodische Wanderungen in ein wärmeres Gebiet; Zugvögel im ▶Überwinterungsgebiet werden „Wintergäste" genannt. **3.** Insekten können durch unterschiedliche Entwicklungsstadien, z. B. durch Eier, Larven und Puppen oder durch zusätzliche Bildung von Gefrierschutzproteinen, überwintern; ▶Überwinterung 2, ▶Winterschlaf); *syn.* Hibernation [f].

2697 hibernation area/region [n] *zool.* (Area where animals retreat for the winter, their metabolic rate is reduced to a minimum and the animals enter a deep sleep; ▶hibernation site, ▶overwintering area); *s* **área [f] de invernación** (Zona donde los animales invernantes pasan el invierno, durante el cual su metabolismo se reduce al mínimo y éstos caen en un profundo sueño; ▶área de invernada, ▶hibernáculo 2); *f 1* **territoire [m] d'hibernation** (Aire recherchée par les animaux [émigration] afin de survivre aux menaces que constituent le climat et le manque de nourriture en hiver sur les quartiers d'été ; ▶hibernacle, ▶quartiers d'hiver) ; *syn.* zone [f] d'hivernage, aire [f] d'hiver ; *f 2* **aire [f] d'hivernage** (pour les oiseaux migrateurs) ; *syn.* site [m] d'hivernage ; *g* **Überwinterungsgebiet [n]** (Gebiet, das Tiere aufsuchen, um den Bedrohungen und Widrigkeiten des Winters in ihren Sommergebieten zu entgehen; ▶Winterlager, ▶Winterquartier); *syn.* Überwinterungsareal [n].

2698 hibernation site [n] *zool.* (Place in which an animal hibernates during the winter; ▶hibernation area/region, ▶overwintering area); *syn.* hibernaculum [n], wintering ground [n], winter resting place [n]; *s* **hibernáculo [m]** (Lugar en el que los animales invernantes pasan el invierno; ▶área de invernación, ▶área de invernada); *f* **hibernacle [m]** (Lieu sur lequel un animal hiberne ; ▶aire d'hivernage, ▶quartiers d'hiver, ▶territoire d'hibernation) ; *syn.* gîte [m] d'hibernation, site [m] d'hibernation ; *g* **Winterlager [n]** (1. Platz, an dem ein Tier im Ruhezustand überwintert. **2.** ▶Winterquartier 2; ▶Überwinterungsgebiet); *syn.* Hibernakulum [n], Winterquartier [n] (1).

hiemvirent [adj] *bot. hort.* ∗▶winter green.

2699 hierarchical classification [n] **of biotopes** *ecol. plan.* (Systematic evaluation of biotopes in a given area according to their relative scarcity and state of natural development, etc.; ▶habitat mapping); *s* **clasificación [f] de biótopos** (Evaluación sistemática de los biótopos de una zona determinada según su estado de conservación, su importancia ecológica, etc.; ▶inventario cartográfico de biótopos); *f 1* **hiérarchisation [f] des biotopes** (Évaluation du niveau d'intérêt d'après une notation établie à partir de critères basés sur la description des données biotiques et abiotiques du milieu, de l'intérêt floristique et faunistique des groupements végétaux et animaux ainsi que de leurs habitats selon, leur degré d'évolution et d'altération par l'homme, etc. ; ▶cartographie des biotopes, ▶cartographie écobiocénotique, ▶inventaire des zones naturelles d'intérêt écolo-

gique, faunistique et floristique [ZNIEFF]) ; *f 2* **sélection [f] des sites** (Évaluation de l'importance des sites susceptibles d'être identifiés en tant que d'importance communautaire selon leur valeur relative pour la conservation de différents types d'habitat naturel ou d'espèces) ; *g* **Bewertung [f] von Biotopen** (Einschätzung und Einstufung/Skalierung der biotischen und abiotischen Gegebenheiten eines Untersuchungsbereiches nach der Rangfolge ihrer Wertigkeit, ihres Entwicklungsstadiums etc.; ▶Biotopkartierung); *syn.* Klassifizierung [f] von Biotopen, Bonitierung [f] von Biotopen.

hierarchy [n] **of central places** [UK] *plan.* ▶hierarchy of population growth centers [US].

2700 hierarchy [n] **of population growth centers** [US] *plan.* (Division of a whole country or State into a hierarchy of settlements; **in U.K.**, Regional Centres, Sub-Regional Centres, Key Inland Towns, Sub-Urban Towns, Key Settlements, based on various urban and rural patterns as well as transport networks; ▶population growth center); *syn.* hierarchy [n] of central places [UK]; *s* **jerarquización [f] de lugares centrales** (Estructuración de los asentamientos humanos de todo un país en un sistema jerárquico. **En GB** se utilizan las siguientes categorías: centros regionales, centros subregionales, ciudades claves del interior, ciudades suburbanas y asentamientos claves. **En D.** se subdividen en ▶lugares centrales superiores, medios, inferiores y centros pequeños, a los cuales les corresponden funciones específicas. Dependiendo de la categoría de los centros, éstos están dotados de determinadas infraestructuras y servicos públicos para la población. **En Es.** no existe tal jerarquización); *f* **maillage [m] du territoire selon les lieux centraux** (D., partition d'un territoire fédéral selon les catégories « centre supérieur », « centre intermédiaire », « centre inférieur », « petit centre », établie sur la base d'une différenciation de l'armature urbaine et de la desserte rurale, des centres métropolitains et de leurs zones périphériques ; ▶lieu central, ▶pôle urbain) ; *g* **zentralörtliche Gliederung [f]** (Funktionale Gliederung des gesamten Landesgebietes in einer Stufung von Ober-, Mittel- und Grundzentren entsprechend der unterschiedlichen Siedlungsstruktur und der Verkehrserschließung der ländlichen Bereiche, der Ballungsrandzonen und der Ballungskerne, um eine bestmögliche Versorgung der Bevölkerung in allen Teilen des Landes zu ermöglichen; cf. § 22 Gesetz zur Landesentwicklung Landesentwicklungsprogramm [LEProNRW 1989, zuletzt geändert 2007]; ▶zentraler Ort).

2701 high altitude [n] *geo.* (Upper reaches of highlands; in high mountains, the upper sub-alpine or alpine level between tree line and permanent snow, close to the limit of plant growth caused by cold temperatures; ▶timberline); *s* **gran altitud [f]** (Zona en la parte superior de las [altas] montañas, en los pisos subalpino superior, alpino y subnival, entre el ▶límite del bosque y las nieves perpetuas); *f* **haute altitude [f]** (Domaine d'altitude élevée d'une [moyenne] montagne, en haute montagne étage subalpin, alpin et subnival entre la limite supérieure de la végétation ligneuse [▶limite des forêts] et la limite inférieure des neiges permanentes, proche de la limite thermique de croissance des végétaux) ; *syn.* zone [f] d'altitude, zone [f] de haute altitude ; *g* **Hochlage [f]** (1. Gebiet in den oberen Höhen eines [Mittel]gebirges, im Hochgebirge die obere subalpine, alpine und subnivale Stufe zwischen der ▶Waldgrenze und dem ewigen Schnee — nahe der Kältegrenze des Pflanzenwachstums. **2.** *leg.* Zone innerhalb von 500 Höhenmetern unterhalb der natürlichen Baumgrenze; § 142 (2) ForstG).

2702 high-altitude afforestation [n] *for. land'man.* (Establishment of forests at ▶high altitudes to restore the natural tree cover, to protect against soil erosion, and to prevent avalanches); *syn.* upland afforestation [n] [also UK] (LD 1991 [11], 14); *s* **forestación [f] de las grandes altitudes** (Establecimiento

de montes de nubes o montes frescos [▶gran altitud] para pro-teger contra la erosión, evitar aludes o reestablecer la cobertura arbórea original); *syn.* reforestación [f] de las grandes altitudes; *f* reboisement [m] d'altitude (Repeuplement des zones de ▶haute altitude par des espèces forestières afin de reconstituer les peuplements naturels en vue de la protection des sols contre l'érosion, la protection contre les avalanches, etc.) ; *syn.* reboise-ment [m] en zone de haute montagne ; *g* Hochlagenaufforstung [f] (Gründung von Waldbeständen in ▶Hochlagen zur Wieder-herstellung des natürlichen Waldbestandes, zum Schutze des Oberbodens gegen Erosion, zur Verhinderung von Lawinen etc.).

high branched tree [n] [US] *arb.* ▶high crowned tree.

2703 high crowned tree [n] *arb.* (Standard tree with a long clear stem; ▶standard street tree [US]/avenue tree [UK]); *syn.* high branched tree [n] [also US]; *s* árbol [m] podado en vertical (Árbol estándar con tronco largo y libre de ramificaciones; ▶ár-bol de alineación); *f* houppier [m] (1) (Arbre ébranché auquel on n'a laissé que sa houppe, c.-à-d. la seule extrémité de sa cime ; DFM 1975 ; ▶arbre d'avenue) ; *g* hoch aufgeasteter Baum [m] (Entsprechend der „FLL-Gütebestimmungen für Baumschul-pflanzen" ein Hochstamm, der bei einem Stammumfang von 20/25 cm eine Stammhöhe bis zum Kronenansatz von mindestens 2,20 m aufweist, ab 25 cm Stammumfang mindestens 2,5 m; cf. FLL 2000. Im Straßenraum kann bei älteren Bäumen zur Schaf-fung des nötigen Lichtraumprofils fahrbahnseitig mindestens bis zu 4,50 m aufgeastet werden; ▶Alleebaum).

2704 higher administrative level [n] **of planning** *leg. plan.* (Planning at a legally-prescibed, higher administrative or policy level; e.g. comprehensive community development planning [US]/urban land-use planning [UK] takes precedence over zoning [preparation of zoning maps]); *syn.* planning [n] on a higher statutory level [UK]; *s* planificación [f] superior (≠) (Planeamiento existente a nivel superior, como p. ej. la ordena-ción del territorio está por encima de la planificación urbana); *f* planification [f] à l'échelon (administratif) supérieur (P. ex. le schéma régional d'aménagement et d'urbanisme par rapport au schéma de cohérence territoriale [SCOT] et ce dernier par rapport au plan local d'urbanisme [PLU]/plan d'occupation des sols [POS]) ; *g* übergeordnete Planung [f] (Planung einer höheren Planungsebene: z. B. ist die Flächennutzungsplanung der Bebau-ungsplanung übergeordnet).

2705 highest bidder [n] [US] *contr. syn.* highest tenderer [n] [UK]; *s* licitante [m/f] a la alta; *f* candidat/soumissionnaire [m] le plus-disant ; *g* teuerster Bieter [m]/teuerste Bieterin [f].

highest tenderer [n] [UK] *contr.* ▶highest bidder [US].

2706 high forest [n] *for.* (Forest, generally with ▶trees of seedling origin, that normally develop a high closed canopy; ▶coppice forest, ▶coppice with standards); *s* monte [m] alto (Masa boscosa de ▶árboles generados de semilla que normal-mente desarrolla un dosel cerrado; ▶monte medio, ▶monte bajo); *syn.* bosque [m] (2), fustal [m] (1); *f 1* futaie [f] (Forêt haute formée par ▶brins de franc pied ; ▶taillis) ; *syn.* haute futaie [f] ; *f 2* futaie [f] sur souche (Futaie développée à partir de ▶taillis ou de ▶taillis-sous-futaie) ; *g* Hochwald [m] (Aus ▶Kernwüchsen, Pfropflingen oder Stecklingen sich aufbauender oder hervorgegangener Wald; ▶Mittelwald, ▶Niederwald).

2707 high jump installation [n] *recr.* (Facility for high jump competition on sports grounds); *s* instalación [f] de salto de altura (Equipamiento deportivo para practicar la disciplina de atletismo de salto de altura); *f* installation [f] de saut en hauteur (Équipement sportif permettant la pratique de la discipline d'athlétisme du saut en hauteur) ; *g* Hochsprunganlage [f] (Ein-richtung auf Sportplätzen zur Durchführung der leichtathletischen Disziplin Hochsprung).

2708 high jump landing area [n] *recr.* (Area [3-5m wide and 5-7m long] containing a cushion approx. 1m high, filled with synthetic foam to break the fall at the base of a high jump; in the past, the landing area was a pit filled with sand; ▶high jump installation); *s* foso [m] de salto de altura (Área de 3-5 m de ancho y 5-7 m de largo cubierta por un colchón de 1 m de altura para frenar la caída en los saltos de altura. Antiguamente consis-tía en un foso de arena; ▶instalación de salto de altura); *f* bac [m] de saut en hauteur (Autrefois fosse remplie de sable d'une ▶installation de saut en hauteur ; aujourd'hui aire matelassée [zone de chute] ou s'effectue la réception de l'athlète après le saut et équipée de tapis de saut [éléments creux en mousse synthétique et revêtement en tissu synthétique] de dimensions 600 x 400 x 60 cm ou de 500 x 300 x 60 cm pour le sport scolaire) ; *g* Hochsprunggrube [f] (Früher mit Sand gefüllte Grube einer ▶Hochsprunganlage [3-5 m breit und 5-7 m lang], heute mit einem Sprunghügel aus ca. 1 m hohen Weichboden-Sprungmatten aus Schaumstoff ausgestattet).

2709 highland [n] *geo.* (**1.** Low mountainous or elevated region of any country standing prominently above adjacent low areas; occasionally also in the names of geographical districts; e.g. the Kenya White Highlands, Highlands of Scotland. For small areas 'heights' is sometimes used. **2.** The German term *Mittelgebirge* [Hercynian mountains after the Harz Mountains] is applied to the wooded mountain system of Middle Germany, or to portions of it; ▶high mountain region); *s* montañas [fpl] medias (Un espacio de terreno de montañas de mediana altura, de altitud entre 500-1000 m, contrapuesto al terreno llano o campiña; a nivel de vegetación corresponden con el ▶piso montano; ▶alta montaña); *syn.* serranía [f]; *f* moyenne mon-tagne [f] (Partie saillante ou relief de l'écorce terrestre à la fois élevé de plusieurs centaines de mètres au-dessus de son soubas-sement. La moyenne montagne est entièrement déneigée en été [en dehors des zones polaires] et où le modelé glaciaire hérité reste peu apparent, sans crêtes alpines notamment ; DG 1984 ; ▶haute montagne) ; *g* Mittelgebirge [n] (*Europa* Berg- oder Hügelland mit einem Höhenunterschied von 500-1000 m zwi-schen Gebirgsfuß und gerundeten Gipfeln; in Zentraleuropa erstreckt sich das **M.** vom Rheinischen Schiefergebirge über den Harz, Thüringer Wald bis Böhmer Wald, das Erzgebirge und die Sudeten sowie die Böhmisch-Mährische Höhe in Tschechien als herzynisches [etymolog. von Harz] oder variskisches Falten-gebirge [nach Varisker/Narisker einem alten germanischen Volksstamm benannt]; ▶Hochgebirge).

high level [n] **of light exposure** *bot.* ▶strong light ex-posure.

2710 high maintenance [n] *constr. hort.* (Descriptive term for an area requiring a high degree of maintenance work; e.g. **h. m.** lawn); *syn.* heavy maintenance [n], labo(u)r-intensive main-tenance [n]; *s* de mantenimiento intenso [loc] (Término descrip-tivo referente a una zona [verde] que requiere mucho trabajo de mantenimiento); *f* d'entretien intensif [loc] (qui nécessite un entretien important) ; *g* pflegeintensiv [adj] (Eine Anlage be-treffend, die einen hohen Aufwand an Pflege erfordert); *syn.* mit hohem Pflegeaufwand [loc].

2711 high maintenance planting [n] *constr.* (Creation of dense plant cover or establishment of a ▶clipped lawn, requiring intensive upkeep; ▶intensive roof planting); *syn.* intensive planting [n] [also UK] (2); *s* revestimiento [m] vegetal intensivo (Creación de cubierta vegetal o de césped ornamental costoso y que exige mantenimiento intenso; ▶césped decorativo, ▶revestimiento vegetal intensivo de tejados); *syn.* plantación [f] intensiva; *f* végétalisation [f] intensive (Création d'une planta-tion d'agrément ou d'accompagnement sur une épaisseur de

H

substrat supérieure à 15 cm, d'un ►gazon d'ornement, nécessitant un entretien important et régulier ; ►végétalisation intensive des toitures) ; *g* **intensive Begrünung [f]** (Anlage einer anspruchsvollen und pflegeaufwendigen Bepflanzung oder eines ►Zierrasens; ►intensive Dachbegrünung).

high mountain pasture [n] [US] *agr.* ►alpine pasture.

2712 high mountain plant [n] *bot.* (►montane plant); *s* **orófito [m]** (Planta de montaña; ►planta montana); *syn.* especie [f] de montaña, planta [f] de montaña; *f* **orophyte [m]** (Plante croissant exclusivement en haute montagne et se développe aux étages subalpin et alpin à des altitudes comprises entre 1900 m à 3250 m ; ►plante montagnarde) ; *syn.* plante [f] de haute montagne ; *g* **Gebirgspflanze [f]** (Im Gebirge lebende Pflanze; ►Mittelgebirgspflanze); *syn.* Orophyt [m].

2713 high mountain region [n] *geo.* (Mountains with a difference in height of more than 1,000m between their foot and the summit; their peaks and narrow ridges frequently rise above the snow line; ►mountain range, ►highland); *syn.* alpine region [n]; *s* **alta montaña [f]** (►Montaña en la que la diferencia de altura entre su pie y su cumbre es superior a 1000 m, cuyas cimas y crestas superan frecuentemente el límite de las nieves perpetuas; ►montañas medias); *f* **haute montagne [f]** (Montagne dont la différence d'altitude entre son soubassement et le sommet est supérieure à 1000 m ; très souvent les aiguilles et les arêtes montagneuses dépassent la limite des neiges permanentes ; ►espace montagnard, ►massif montagneux, ►moyenne montagne) ; *g* **Hochgebirge [n]** (Gebirge mit > 1000 m Höhenunterschied zwischen Gebirgsfuß und Gipfel; vielfach ragen die spitzen Gipfel und schmalen Grate über die Schneegrenze hinaus; ►Gebirge, ►Mittelgebirge).

high occupancy vehicle lane [n] *trans.* *►offside lane.

high plains [npl] *geo.* ►plateau (1), #2.

high-rise [n] *arch. urb.* ►multistory building for housing [US].

2714 high-rise building [n] *arch.* (**1.** Especially tall building of more than 23m [in US] or approximately 6 stories [US]/ storeys [UK] typically framed with structural steel and requiring exceptional footing conditions for stability. **2.** With the exception of the high-rise mud buildings of Shiban, Hadhramaut in the Yemen, which are over 300 years old, the multi-story buildings of the Pueblo Indians in the Southwest of the USA and the mediaeval dynasty towers of Tuscany, the development of modern **h.-r. b.s** started in Chicago. From the 1870s onwards, technical progress in construction such as the development of strong, rolled steel, OTIS elevators, which were fitted with safety measures to prevent their falling, and the introduction of central heating enabled the construction of ►office towers of 20 stories and above for the first time. The following decades saw the construction of the first real **skyscrapers,** after additional improvements in the properties of construction materials and methods were made. The term **skyscraper** originated in 1880; the *Empire State Building* in New York reached a height of 381m in 1931; for a quarter of a century the *Sears Tower* [1969-1974] in Chicago with its height of 431m was the highest building in the world; the *Chestnut-DeWitt-Building*, finished in 1962, was the first **h.-r. b.** completely constructed in reinforced concrete; *Burj Dubai*, the world's tallest building in 2009, is in Dubai U.A.E. and has a height of 818 metres; ►high-rise residential building); *s* **edificio [m] alto** (**1.** Edificio de mucha altura; **en Es.** no existe definición legal de la altura a partir de la cual se considera alto un **e.**; **en EE.UU.** a partir de seis plantas; **en D.** a partir de 22 m sobre rasante, altura definida por la accesibilidad de la planta superior con escaleras de bomberos; *términos específicos* ►torre de oficinas, ►torre de pisos. **2.** Aparte de las casa altas de adobe en el sur de Yemen de más de 300 años de antigüedad, las edifi-

caciones de varias plantas de los indígenas Pueblo del SO de los EE.UU. y las torres medievales de la aristocracia urbana, la construcción moderna en altura comenzó en Chicago. La disponibilidad de acero laminado de mayor rigidez, de elevadores seguros desarrollados por OTIS y de sistemas centrales de calefacción posibilitaron a principios dc los años 1870 construir por primera vez torres de oficinas de 20 plantas o más. Las subsiguientes mejoras de la calidad de los materiales y de las técnicas constructivas permitieron construir en las décadas siguientes **rascacielos** —el término inglés *skyscraper* se creó en los EE.UU. hacia 1880— como el *Empire State Building* 1931 en Nueva York de 381 m de altura o la *Sears Tower* 1969-1974 en Chicago de 431 m, que durante un cuarto de siglo fue el edificio más alto del mundo; el *Chestnut-DeWitt-Building* [111 m] terminado en 1962 fue el primer rascacielos construído totalmente de hormigón armado. La torre de Dubai [*en árabe* Burdsch Dubai] en la Unión de Emiratos Árabes es actualmente [2009] con sus 206 pisos y 818 m de altura el edificio más alto del mundo); *syn. de 1* torre [f]; *f* **immeuble [m] de grande hauteur (I.G.H.)** (Tout corps de bâtiment dont le plancher bas du dernier niveau [PBDN] est situé, par rapport au niveau du sol le plus haut utilisable pour les engins des services publics de secours et de lutte contre l'incendie à 50 m pour les immeubles à usage d'habitation et à plus de 28 m pour tous les autres immeubles ; R122-2 et 3, Code de la Construction et de l'Habitation ; plusieurs classes d'I.G.H. sont répertoriées : immeubles à usage d'habitation [G.H.A.], immeubles à usage d'hôtel [G.H.O.], immeubles à usage d'enseignement [G.H.R.], immeubles à usage de dépôts d'archives [G.H.S.], immeubles à usage sanitaire [G.H.U.], immeubles à usage de bureaux 28 m < PBDN ≤ 50 m [G.H.W.1], immeubles à usage de bureaux PBDN > 50 m. [G.H.W.2] ; un **I.G.H.** doit répondre à un ensemble de règles de construction et d'équipement visant à limiter le plus possible la propagation d'un feu d'un étage à l'autre et à permettre l'évacuation des occupants. Avec l'amélioration des caractéristiques des matériaux, des structures ainsi que des méthodes de construction il fut possible d'ériger des grattes-ciel [skyscraper en anglais] dès la fin du XIXème siècle ; l'*Empire State Building* à New York, avec ses 381 m de hauteur en 1931 restera le plus haut bâtiment du monde pendant plus de 40 ans, le *Sears Tower* [1969-1974] à Chicago avec ses 431 m de hauteur dépassera le World Trade Center détruit par les attentats du 11 septembre 2001 et est à ce jour l'immeuble le plus haut des États-Unis ; la *Burj Dubaï* ou *tour de Dubaï* est, depuis le 17 janvier 2009, avec une hauteur de 818 mètres la plus haute structure du monde ; *termes spécifiques* ►tour [à usage] de bureaux, ►tour d'habitations) ; *g* **Hochhaus [n]** (**1.** Entsprechend der deutschen Landesbauordnungen ein Gebäude, bei dem der Fußboden eines Geschosses mindestens mehr als 22 m über der Geländeoberfläche liegt; diese Festsetzung fußt auf der Erreichbarkeit der obersten Geschosse mit Feuerwehrleitern. *UBe* ►Bürohochhaus, ►Wohnhochhaus. **2.** Abgesehen von den über 300 Jahre alten Lehmhochhäusern im südlichen Jemen, den mehrstöckigen Wohnanlagen der Puebloindianer im SW der USA und der Geschlechtertürme des Stadtadels im Mittelalter, z. B. in der Toskana, startete die Entwicklung des modernen **H.baues** in Chicago. Die Entwicklung gewalzter höherfester Stähle, des absturzsicheren Aufzugs durch OTIS und der [Dampf]zentralheizung ermöglichten es zu Beginn der 1870er-Jahre, erstmals 20-geschossige Bürohochhäuser und noch höhere Häuser zu bauen. Nach weiterer Verbesserung der Materialeigenschaften und Konstruktionsmethoden wurden in folgenden Jahrzehnten **Wolkenkratzer** errichtet — das englische Wort *skyscraper* entstand in den USA um 1880; *Empire State Building* in New York mit 381 m Höhe, 1931; für ein Vierteljahrhundert war der 431 m hohe *Sears Tower* [1969-1974] in Chicago das höchste Haus der Welt; das 1962 fertiggestellte *Chestnut-DeWitt-Building*

H

[111 m] in Chicago war das erste ganz aus Stahlbeton gebaute **H.**; cf. SOB 2003 und BROCK 1989, Bd. 10, p. 130; der *Turm von Dubai*, [arab. *Burdsch Dubai*], in den Vereinigten Arabischen Emiraten, ist mit 818 m Höhe und 206 Stockwerken zurzeit [2009] das höchste Gebäude der Welt).

2715 high-rise residential building [n] *arch. urb.* (Multistor[e]y building; **in D.**, the term is applied to a building in which the floor of at least one apartment is situated more than 22m above ground; **in U.S.**, there is no such fixed height; ►multistory building for housing [US]/multistorey building for housing [UK], ►tower building); *syn.* apartment tower [n] [US]/ tower block [n] [UK]; *s* **torre [f] de pisos** (Edificio residencial de gran altura, **en D.** como mínimo con una planta a la altura de 22 m, **en F.** de seis plantas, **en Es. y EE.UU.** no existe altura mínima definida; ►casa de pisos, ►torre de viviendas); *syn.* edificio [f] alto de pisos, torre [f] de viviendas; *f* **tour [f] d'habitations** (Edifice résidentiel de six étages ou plus qui comprend plusieurs logements ; ►habitat collectif, ►tour à noyau central) ; *g 1* **Wohnhochhaus [n]** (Gebäude mit acht und mehr Stockwerken und bei dem der Fußboden mindestens eines Aufenthaltsraumes 22 m über Gelände liegt; cf. IWU 1977, KOR 2005); *g 2* **Wohnturm [m]** (Freistehende, sehr hohe Sonderform des Wohnhochhauses; ►Geschosswohnungsbau; *syn.* ►Punkthaus [n].

2716 high spoil pile [n] [US] *envir. min.* (Mound which rises well above the original ground surface either on an undisturbed site, or where waste material is dumped [US]/tipped [UK] at an extraction site; ►dumpsite, ►landfill to original contour); *syn.* high tip [n] [UK]; *s* **depósito [m] sobre el nivel de superficie** (►Depósito de residuos sólidos en terreno no socavado o en una zona de extracción que se rellena más allá del nivel del terreno formando un montículo; ►vertedero en hondonada); *f* **décharge [f] par dépôt** (Installation d'élimination de déchets sur un terrain naturel, dans une cavité naturelle du sol ou sur un site d'extraction de matériaux pour laquelle les déchets stockés dépassent le niveau du terrain naturel ; ►décharge publique, ►décharge par enfouissement) ; *g* **Hochkippe [f]** (►Deponie auf unverritztem Gelände oder in einer Entnahmestelle, bei der die verkippten Materialien über die ursprüngliche Geländeoberfläche hinaus abgelagert werden; ►Flurkippe); *syn.* Oberflurkippe [f].

2717 high tension power line [n] *envir.* (Above ground heavy cable carrying a high voltage supply of electric power; ►electric power line, ►overhead power line, ►underground power line); *syn.* high voltage transmission line [n] [also US], electric power transmission line [n] [also US]; *s* **línea [f] de alta tensión** (Línea eléctrica superficial que permite el transporte de electricidad a grandes distancias; ►cable eléctrico subterráneo, ►conducción eléctrica, ►línea aérea); *f* **ligne [f] à haute tension** (Ligne électrique permettant à partir de la source de production le transport de l'énergie électrique sur de grandes distances ; ►ligne aérienne, ►ligne électrique, ►ligne électrique souterraine) ; *g* **Hochspannungsleitung [f]** (Elektrische Leitung, die Energie von einer Energiequelle, z. B. einem Elektrizitätswerk, über weite Distanzen transportiert; ►Freileitung, ►unterirdisches Stromkabel, ►Stromleitung).

high tip [n] [UK] *envir. min.* ►high spoil pile [US].

high-veld [n] [ZA] *geo.* ►plateau (1).

high voltage transmission line [n] [US] *constr. envir.* ►high tension power line.

high water [n] *wat'man.* ►danger of high water.

2718 high-water control [n] *wat'man.* (Regulating the discharge of watercourses by building ►by-pass channels [US]/by-pass channel [UK] and by preserving floodable land, ►dams, ►retarding basins, through river engineering; ►damming 2);

s **regulación [f] de aguas de crecidas** (Conjunto de medidas que tienen el objetivo de evitar inundaciones o mitigar los efectos de las mismas por medio de la construcción de ►cauces de derivación, ►embalses de retardo, ►presas, etc. o ampliando el cauce; ►construcción de presas); *f* **régulation [f] des crues** (Toutes les mesures ayant pour objectif la régulation du débit des cours d'eau au moyen de ►barrages de vallée, ►bassins de rétention, ►canaux latéraux et polders d'inondation ou par l'aménagement du cours d'eau [élargissement de la section du lit, endiguement et rectification rectiligne du tracé] ; ►construction d'ouvrages de rétention) ; *syn.* écrêtement [m] des crues, laminage [m] des crues ; *g* **Hochwasserregelung [f]** (Alle Maßnahmen, die dazu dienen, den Abfluss von Wasserläufen durch Anlagen von ►Talsperren, ►Rückhaltebecken, ►Seitengerinnen und Flutmulden/ Poldern oder durch den Ausbau des Gewässers — Vergrößerung des Profils incl. Deich und Bau einer gestreckten Linienführung — zu regeln; ►Stauanlagenbau).

high-water level [n] *wat'man.* ►average high-water level/ mark, ►record high-water level/mark.

high-water level [n]**, peak** *wat'man.* ►record high-water level/mark.

2719 high-water (level) [n] **of watercourses** *wat' man.* (Annual maximum depth of rivers or streams, especially after melting snow and heavy rainfall; ►flood; *opp.* ►low-water level); *s* **avenida [f] (1)** (Nivel alto de aguas en cursos de agua, que se presenta especialmente después de fuertes precipitaciones o tras el deshielo de la nieve en las montañas; ►inundación; *opp.* ►estiaje; *syn.* crecida [f], riada [f]; *f* **crue [f]** (Période ou phénomène de montée du niveau des eaux dans un cours d'eau provoqué par la fonte des neiges [crue nivale] ou des pluies intenses sur un bassin-versant [crue d'averse], la rupture d'un barrage d'objets flottants obstruant le lit d'un cours d'eau [crue d'embâcles] ; ►inondation, *opp.* ►étiage) ; *syn.* hautes eaux [fpl] ; *g* **Hochwasser [n] der Fließgewässer** (...wasser [pl]; Wasserhochstand, besonders nach Schneeschmelze oder nach starken Regenfällen; ►Überschwemmung, *opp.* ►Niedrigwasser[stand]).

high-water margin [n] [UK] *wat'man.* ►flood fringe area [US].

high-water mark [n] *wat'man.* ►average high-water level/ mark, ►flood water mark.

high-water mark [n]**, peak** *wat'man.* ►record high-water level/mark.

high-water mark [n]**, record** *wat'man.* ►record high-water level/mark.

highway [n] [US] *adm. trans.* ►elevated highway, ►classified highway, ►four-lane undivided highway, ►interstate highway, ►major highway (1), ►major highway (2), ►peripheral highway, ►state highway.

highway [n] [US]**, arterial** *trans.* ►arterial road, ►major highway (1).

highway [n] [US]**, belt** *trans.* ►beltway [US].

highway [n] [US]**, federal** *adm. leg. trans.* ►interstate highway [US].

highway [n] [US]**, inner-city express** *trans.* ►inner-city freeway [US] (1)/inner-city expressway [UK].

highway [n] [US]**, limited-access** *adm. leg. trans.* ►freeway [US]/motorway [UK].

highway [n] [US]**, loop** *trans.* ►bypass [US]/by-pass [UK].

highway [n] [US]**, super-** *adm. leg. trans.* ►freeway [US]/ motorway [UK].

highway [n] [US]**, trunk** *trans.* ►major highway [US] (1)/ major road [UK].

highway construction [n] **and maintenance** [n] [US] *constr.* ▶civil engineering works.

highway embankment [n] *constr. eng.* ▶road embankment.

2720 highway opposition group [n] [US] *pol. syn.* anti-motorway action group [n] [UK]; *s* **grupo [m] ciudadano anti-autopista** *syn.* grupo [m] ecologista antiautopista; *f* **association [f] de défense contre une autoroute** ; *g* **Bürgerinitiative [f] gegen den Bau einer Autobahn.**

highway planning [n] *plan. trans.* ▶road planning.

2721 highway ramp [n] [US] *trans.* (Inclined road connecting major highway with other highway at a ▶grade-separated junction; ▶cloverleaf ramp, ▶interchange); *syn.* slip road [n] [UK]; *s* **rampa [f] de acceso** (Tramo en pendiente de conexión entre carretera y autopista ▶nudo sin cruces a nivel; ▶curva de trébol, ▶nudo de enlace); *f* **bretelle [f] d'accès** (Voie de raccordement à niveau séparé d'une autoroute de liaison avec le réseau routier ; ▶bretelle d'échangeur en trèfle (PR 1987), ▶diffuseur, ▶échangeur ; *g* **Auffahrtsrampe [f]** (Verbindungsstück von kreuzungsfrei geführten Schnellstraßen an das übrige Straßennetz; ▶Anschlussstelle, ▶Kleeblattschleife, ▶planfreier Knotenpunkt).

Highways and Transportation Services [npl] [US] *adm.* ▶public works department.

2722 highway tunnel [n] [US] *trans.* (Structure built for an underground road. *Generic term* vehicular traffic tunnel; ▶bored tunnel, ▶cut and cover tunnel); *syn.* road tunnel [n] [UK]; *s* **túnel [m] de carretera** (Construcción para una carretera subterránea; ▶túnel construido a zanja abierta, ▶túnel perforado); *f* **tunnel [m] routier** (Ouvrage souterrain d'infrastructure de transport routier ; ▶tunnel de montagne, ▶tunnel en tranchée ouverte) ; *g* **Straßentunnel [m]** (Verkehrsbauwerk für eine unterirdisch geführte Straße, ▶Bergtunnel, ▶Unterpflastertunnel).

2723 hiking accommodation [n] [US] *recr.* (Lodging facility for short stays by hikers and bikers along ▶national scenic trails [US]/long distance footpaths [UK]; e.g. youth hostel); *syn.* hostel [n] [also US], ramblers' hostel [n] [UK]; *s* **hostal [m] de senderistas** (Lugar de alojamiento a lo largo de un ▶pasillo verde de largo recorrido; *término específico para el Camino de Santiago* hostal de peregrinos); *f* **gîte [m] d'étape** (Lieu d'hébergement pour une étape sur un ▶sentier de grande randonnée) ; *g* **Wanderheim [n]** (Übernachtungshaus auf einem ▶Fernwanderweg).

2724 hiking area [n] *plan. recr.* (**In Europe**, district serving the recreational pursuit of rambling [UK]; **in U.S.**, an extensive area where hiking is enjoyed; ▶day-use recreation area 1 [US]/recreation area for day-trippers [UK]); *syn.* rambling area [n] [also UK]; *s* **zona [f] de montañismo/senderismo** (Área propicia para practicar el montañismo o senderismo; ▶zona de excursiones); *f* **zone [f] de promenade et de randonnée** (Zone favorisant la découverte des sites naturels et des paysages par le développement de la promenade et de la randonnée ; ▶lieu d'excursion) ; *g* **Wandergebiet [n]** (Gebiet, das der Erholungsform 'Wandern' dient; ▶Ausflugsgebiet).

hiking footpath [n] [UK] *recr.* ▶hiking trail [US].

2725 hiking holiday [n] *recr.* (Time spent traveling on foot for recreation; ▶auto tourist); *syn.* backpacking vacation [n] [also US], rambling holiday [n] [also UK], walking holiday [n] [also UK]; *s* **vacaciones [fpl] de senderismo** (Vacaciones dedicadas a realizar recorridos largos a pie o en bicicleta; ▶excursionista motorizado/a); *f* **vacances [fpl] itinérantes** (Activité de détente consacrée principalement à la randonnée à pied ou à vélo ; ▶excursioniste motorisé) ; *g* **Wanderurlaub [m]** (Der Erholung dienende freie Zeit, die primär dem Fuß- oder Radwandern dient; ▶Autowanderer).

2726 hiking network [n] *recr.* (Interconnected hiking trails; TGG 1984, 158; ▶pathway network); *s* **red [f] de pasillos verdes [Es]** (A pesar de la falta de tradición senderista en España, en los últimos años se ha comenzado a crear una **r. de p. v.** partiendo de los trazados abandonados de ferrocarriles, de las vías de servicio de canales y embalses y de partes de la extensísima red de vías pecuarias [cañadas] para convertirlos en **p. v.**, en senderos, en rutas para recorrer a pie, en bicicleta o a caballo. En España existen unos 125 000 [!] km de cañadas); *syn.* red [f] de caminos de senderismo, red [f] de senderos recreativos; *f* **réseau [m] de randonnée pédestre** (Sytème de sentiers pour les randonneurs dans les espaces naturels ; ▶réseau de sentiers de promenade et de randonnée pédestre) ; *syn.* parcours [m] de randonnée pédestre, circuit [m] pédestre ; *g* **Wanderwegenetz [n]** (Ein System von Wegen für Wanderer in der freien Landschaft; ▶Wegenetz).

2727 hiking trail [n] [US] *recr. syn.* hiking footpath [n] [UK], rambling pathway [n] [UK], ramblers' route [n] [UK], rambling footpath [n] [UK], walking path [n] [US], walking trail [n] [US] (TGG 1984, 87; ▶national scenic trail [US]/long distance footpath [UK]); *s* **pasillo [m] verde (2) [Es]** (▶pasillo verde de largo recorrido); *syn.* camino [m] de senderismo, ruta [f] de senderismo; *f 1* **sentier [m] de promenade** (Utilisé pour un déplacement court ; ▶sentier de grande randonnée) ; *syn.* chemin [m] piétonnier, itinéraire [m] de promenade ; *f 2* **sentier [m] de randonnée** (Utilisé pour un déplacement long sur un réseau de randonnée pédestre) ; *syn.* itinéraire [m] de randonnée, parcours [m] pédestre, itinéraire [m] pédestre ; *g* **Wanderweg [m]** (Weg, der durch die Natur führt und dem Wandern dient; ▶Fernwanderweg).

hiking trails [n] *recr. for.* ▶marking of hiking trails.

2728 hill [n] *geo.* (Natural elevation of the land surface rising prominently above the surrounding area, usually considered less than 300m from base to summit; cf. DOG 1984; elevations which are higher than hills are mountains); *s* **colina [f]** (Elevación menor del terreno que destaca aisladamente sobre el territorio que la rodea. Aunque la diferenciación con la montaña es imprecisa, las características más destacadas de un paisaje de colinas son: una altitud de la base entre 150 y 200 m sobre el nivel medio del mar, desniveles relativos entre 300 y 600 m; pendientes pronunciadas; cumbres con tendencia a la horizontalidad y escaso o nulo escalonamiento altitudinal de la vegetación; DGA 1986); *f* **colline [f]** (Relief peu élevé, de faible altitude [jusqu'à 300 m] aux pentes douces et au sommet arrondi, se présentant en famille sinon on parlera d'une butte) ; *g* **natürlicher Hügel [m]** (Sanft ansteigende Erhebung in einer ebenen Landschaft; kleiner Berg).

hillside breeze [n] [US] *met.* ▶slope wind.

2729 hillside building development [n] *urb*; *s* **urbanización [f] en/de ladera** (Construcción de casas en una ladera); *f* **habitat [m] à flanc de colline** (implanté sur un versant) ; *syn.* habitat [m] à flanc de coteau ; *g* **Hangbebauung [f]** (Wohnbebauung in Hanglage).

hillside drainage channel [n] [UK]**, natural** *geo.* ▶natural drainage way [US].

2730 hillside location [n] *agr. urb.* (Place of cultivated land or a built area situated on a slope; ▶slope aspect); *s* **ubicación [f] en pendiente** (Descripción de la situación de tierras de cultivo o de áreas de construcción en una ladera; ▶exposición); *f* **exposition [f] en pente** (Caractérise l'emplacement de bâtiments ou de cultures situées sur un versant ; ▶exposition) ; *g* **Hanglage [f]** (Situationsbezeichnung für Kulturflächen oder Baugebiete am Hang, z. B. untere, mittlere, obere Hanglage; ▶Hangexposition).

2731 hinged post [n] *urb. trans.* (Barrier which can be dropped for vehicular access); *syn.* collapsible post [n]; *s* **poste [m] de cierre de paso unido por bisagra** (Barrera abatible por la que se regula el acceso de vehículos a un predio); *f* **poteau [m] d'interdiction de passage basculable** ; *g* **umlegbarer Absperrpfosten [m].**

histic lithic cryaquepts [n] *pedol.* ▶gleyed anmoor.

historic buildings [npl] *conserv'hist.* ▶preservation of historic buildings.

Historic Buildings and Monuments Commission [n] [UK] *adm. conserv'hist.* ▶State Historic Preservation Office [US].

Historic Buildings and Monuments Commission [n] **for England** [UK] *adm. conserv'hist.* ▶historic district commission [US].

historic community area [n] [UK] *arch. conserv' hist. leg.* ▶historic district [US].

2732 historic core [n] *urb.* (Historic center [US]/historic centre [UK] of a town or city); *syn.* old town [n]; *s* **centro [m] histórico** (Parte antiqua de la ciudad); *syn.* casco [m] antiguo, casco [m] viejo, barrio [m] antiguo; *f* **vielle ville [f]** (Centre historique d'une ville ou d'une agglomération ; *terme spécifique* noyau historique) ; *syn.* centre [m] historique ; *g* **Altstadt [f]** (Historischer Kern einer Stadt; da die **A.** i. d. R. ein Teil einer größeren Gemeinde ist, wird sie auch **Altstadtviertel** genannt); *syn.* Altstadtquartier [n], Altstadtviertel [n].

2733 historic district [n] [US] *arch. conserv' hist. leg.* (**1.** Area of special architectural or historic interest which is designated for protection by local ordinance; a distinction should be drawn between a **local historic district**, which affords a degree of protection, and a **National Register historic district**, which is essentially honorific and ultimately offers no protection; ▶cultural monument, ▶historic district commission [US]/Historic Buildings and Monuments Commission for England [UK], ▶preservation of historic landmarks [US]/conservation of historic monuments [UK]); *syn.* historic zoning district [n] [US]. **2. conservation area** [n] [UK] (Group of buildings, streets, squares or other areas of special architectural or historical interest, the character of which it is desirable to preserve or enhance; Civic Amenity Act 1967 [UK]; ▶cultural monument, ▶préservation of historic landmarks [US]/conservation of historic monuments [UK]); *syn.* heritage area [n], historic community area [n]; *s* **Conjunto [m] Histórico [Es]** (Agrupación de bienes inmuebles que forman una unidad de asentamiento, continua o dispersa, condicionada por una estructura física representativa de la evolución de una comunidad humana por ser testimonio de su cultura o constituir un valor de uso y disfrute para la colectividad. Asímismo es **C. H.** cualquier núcleo individualizado de inmuebles comprendidos en una unidad superior de población que reúna esas mismas características y pueda ser claramente delimitado; cf. art. 15 [3] Ley 6/1985; ▶administración [local] de protección de monumentos, ▶monumento, ▶protección del Patrimonio Histórico Español); *f 1* **secteur [m] sauvegardé** (Zone présentant un caractère historique ou esthétique de nature à justifier la conservation, la restauration et la mise en valeur de tout ou partie d'un ensemble d'immeubles ; document d'urbanisme visant à protéger exclusivement les sites urbains par un plan de sauvegarde et de mise en valeur soumis à enquête publique et se substituant à tout document d'urbanisme existant ; ▶administration [locale] des monuments historiques, ▶protection des monuments historiques) ; *f 2* **zone [f] de protection du patrimoine architectural, urbain et paysager (Z.P.P.A.U.P.)** (Zone instituée par la loi du 8 janvier 1993 autour des monuments historiques et dans les quartiers, sites et espaces à protéger ou à

mettre en valeur pour des motifs d'ordre esthétique, historique ou culturel [p. ex. quartiers anciens, ensembles d'habitat rural, villages, éléments du patrimoine préindustriel ou industriel, architectures du début du siècle, balnéaires ou thermales, espaces paysagers directement liés à un ensemble bâti ou eux-mêmes fortement architecturés, etc.] ; à l'intérieur de la Z.P.P.A.U.P. sont instituées des prescriptions particulières en matière d'architecture et de paysage affectant les travaux de construction, de démolition, de déboisement et de modification de l'aspect des immeubles ; le projet de zone est soumis à enquête publique ; se substitue à la zone de protection du patrimoine architectural, urbain [ZPPAU] instituée par la loi du 7 janvier 1983 ; ▶monument historique) ; *g* **flächenhaftes Kulturdenkmal [n]** (...mäler [pl], in A. nur so; auch ...male [pl]; geschützte Gruppe einzelner oder miteinander verbundener Gebäude, Straßen-, Platz- oder Ortsbilder [Ensemble], die wegen ihrer Architektur, ihrer Geschlossenheit oder ihrer Stellung in der Landschaft aus wissenschaftlichen, künstlerischen oder heimatgeschichtlichen Gründen von besonderem öffentlichem, denkmalschützerischem Interesse sind; ▶Denkmalschutz, ▶Denkmalschutzbehörde, ▶Kulturdenkmal); *syn.* geschützte Gesamtanlage [f], geschützte Sachgesamtheit [f].

2734 historic district commission [n] [US] *adm. conserv'hist.* (Advisory board which recommends to local authorities the designation of historic buildings and districts; **in U.S.,** a local government body created by local ordinance to review application by property owners to change, demolish, or move a locally designated historic landmark or a building within a locally designated historic district. Decisions by a commission are in the form of a "Certificate of Appropriateness", issuance of which is a condition precedent to the granting of other required permits such as a building, zoning, or occupancy permits. Commissions are normally limited to the review of exterior building features; they may occasionally also have survey and decisions of such commissions varies widely according to authorizing legislation in each of the 50 states and the preferences of the local government; **in U.K.,** the conservation of historic monuments and sites is governed by the Planning [Listed Buildings and Conservation Areas] Act 1990; ▶State Historic Preservation Office [US]/ Historic Buildings and Monuments Commission [UK]); *syn.* Historic Buildings and Monuments Commission [n] for England [UK]; *s* **administración [f] (local) de protección de monumentos** (Administración municipal de protección del patrimonio histórico-artístico y cuya denominación varía de un municipio a otro; ▶administración regional de protección de monumentos); *f 1* **administration [f] locale du patrimoine historique** (Administration municipale responsable du recensement, de la protection, la sauvegarde et la mise en valeur du patrimoine historique et culturel de la commune, la mise en œuvre d'actions culturelles d'animation et de promotion du patrimoine historique, d'activités culturelles éducatives ; la dénomination de ces services patrimoniaux en liaison avec les services extérieurs patrimoniaux est variable selon les communes : atelier du patrimoine, service du patrimoine, service patrimoine historique et culturel, etc.) ; *f 2* **service [m] départemental de l'architecture et du patrimoine (SDAP)** (F., service déconcentré du ministère de la Culture et de la Communication à l'échelon départemental, placé sous l'autorité du directeur régional des affaires culturelles, par délégation du préfet de région ; il existe un SDAP dans chaque département. Les SDAP exercent trois grandes missions : **1.** conseil [promotion d'une architecture et d'un urbanisme de qualité], **2.** contrôle [délivrent des avis sur tous les projets qui ont pour effet d'apporter des modifications dans les espaces protégés bâtis et paysagers (sites, abords des monuments historiques, ▶secteurs sauvegardés et ▶zones de protection du patrimoine architectural,

urbain et paysager — ZPPAUP)], et **3.** conservation des monuments [maîtrise d'œuvre des travaux d'entretien des édifices classés au titre des monuments historiques] ; cf. www.culture. gouv.fr ; ▶direction régionale des monuments historiques) ; **g Denkmalschutzbehörde [f]** (Die **untere D.** ist die genehmigende Behörde großer Kreisstädte, kreisfreier Städte, für Gemeinden, Landratsämter und Verwaltungsgemeinschaften. Da in Deutschland der Denkmalschutz der Kulturhoheit der Länder obliegt, ist die Zuständigkeit von Entscheidungen unterschiedlich geregelt. In BW z. B. genehmigt bei kommunalen Objekten und Bauvorhaben das Regierungspräsidium als die nächst **höhere D.**; Entscheidungen in Dissensfällen und bei Landeseigentum trifft das Wirtschaftsministerium als **oberste D.**; das ▶Landesdenkmalamt hat als Fachbehörde keine Genehmigungskompetenz, sondern berät die untere und höhere Denkmalschutzbehörde).

2735 historic garden [n] *conserv'hist. gard'hist.* (Garden with cultural significance displaying a particular aspect of ▶fine garden design); *syn.* heritage garden [n] [also UK]; **s jardín [m] histórico [Es]** (Espacio delimitado, producto de la ordenación humana de elementos naturales, a veces complementado con estructuras de fábrica, y estimado de interés en función de su origen o pasado histórico o de sus valores estéticos, sensoriales o botánicos. Los **jj. hh.** declarados constituyen parte del Patrimonio Histórico Español y son asímismo Bienes de Interés Cultural; cf. Ley 16/1985; ▶arte de jardinería); **f jardin [m] historique** (Espace vert inscrit ou classé parmi les monuments historiques et témoin du patrimoine culturel de l'▶art des jardins) ; **g Gartendenkmal [n]** (Gartenanlage als Zeugnis des kulturellen Erbes der ▶Gartenkunst).

2736 historic garden expert [n] [US] *conserv'hist.* (Professionally-trained person who is in charge of or advises on restoration and future management of historic gardens; ▶conservation of historic gardens); *syn.* garden conservator [n] [UK], expert [n] on historic garden management, professional garden preservationist [n]; **s jardinero/a [m/f] conservador, -a [m/f] del patrimonio de jardines históricos** (Profesional responsable del cuidado, la restauración y/o la gestión y ▶conservación del patrimonio de jardines históricos); *syn.* conservador, -a [m/f] de jardines históricos; **f conservateur, trice [m/f] des jardins historiques** (Personne responsable de la ▶gestion et de la mise en valeur des jardins historiques) ; **g Gartendenkmalpfleger [m]/…pflegerin [f]** (Speziell ausgebildete Person, die sich mit der ▶Gartendenkmalpflege befasst).

historic landmark [n] [US] *conserv. hist.* ▶cultural monument.

2737 historic landscape [n] *conserv'hist. land'man. leg.* (**1.** Landscape associated with a specific date, period or historic person. This might include a battlefield site, a treaty site, or a former president's personal garden, gardens or parks that are associated with important historical figures, or a once-natural landscape transformed by past human interactions into an area bearing witness to a cultural evolution. **2.** Landscape that was consciously designed or laid out by a landscape architect, master gardener, architect, engineer, or horticulturist according to historic design principles, or an amateur gardener working in a recognized style of tradition. The landscape may be associated with a significant person, trend or event in landscape architecture or illustrate an important development in the theory and practice of landscape architecture. Aesthetic values play a significant role in designed landscapes. Examples include parks, campuses, and estates; LA 1997 [5], 81; ▶historic site, ▶historic vernacular landscape); *syn.* heritage landscape [n], historic designed landscape [n] (LA 1997 [5], 81); **s paisaje [m] humanizado histórico (1)** (Lugar o paraje natural vinculado a acontecimientos o recuerdos del pasado, a tradiciones populares, creaciones cultu-

rales o de la naturaleza y a obras del hombre, que posean valor histórico, etnológico, paleontológico o antropológico; art. 15 [4] Ley 6/1985; ▶paisaje humanizado histórico 2, ▶sitio histórico); **f 1 paysage [m] historique** (Tout ou partie de paysage présentant un intérêt particulier en liaison avec un événement ou une personne historique, p. ex. un champs de bataille, la signature d'un traité de paix ou les jardins d'un ancien homme d'état et constituant un objet culturel ; ▶site historique culturel) ; **f 2 paysage [m] patrimoine** (Grands sites remarquables classés au titre de la loi de 1930, expression d'anciennes formes d'exploitation tels que la lande de Lüneburg ; ▶paysage vernaculaire) ; paysage [m] patrimonial ; **f 3 paysage [m] culturel historique** ; **g historische Kulturlandschaft [f] (1)** (Landschaft oder ein Teil davon, die einen engen Bezug zu einem bedeutsamen historischen Ereignis oder zu einer bestimmten Person hat wie z. B. ein Schlachtfeld, ein Ort, an dem ein Friedensvertrag geschlossen wurde oder die [privaten] Gartenanlagen eines früheren Staatsoberhauptes oder heute oft denkmalpflegerischer Obhut unterliegende Gebiete, die durch frühere Bewirtschaftungsweisen entstanden sind, z. B. Reste der Lüneburger Heide — durch extensive Heidschnucken-Weidewirtschaft bis ins 18. Jh. offen gehaltene *Calluna*-Heidegebiete —, die noch weit bis ins 19. Jh. als eine wüstenhafte Landschaft galt, und deren ästhetische Wertschätzung erst entdeckt wurde, als sie bereits im Verschwinden begriffen war; cf. JES 2004. In D. sind seit 1998 die Ziele des Naturschutzes und der Landschaftspflege auch für **h. K.en** gem. § 2 [1] 13 BNatSchG festgeschrieben; ▶historische Kulturlandschaft 2, ▶kulturhistorische Stätte).

Historic Preservation Act [n] **1966** [US] *conserv'hist. leg.* ▶National Historic Preservation Act 1966 [US].

2738 historic preservation measures [npl] [US] *adm. conserv'hist.* (Administrative and practical activities to protect, preserve, restore or reconstruct the fabric of important historic buildings, structures, and the existing form and vegetative cover of a site. It may include initial stabilization work, where necessary, as well as ongoing maintenance of the historic building materials. **In U.S.**, these activities may be supported by a wide variety of regulatory, fiscal, and other incentives; ▶cultural monument, ▶preservation of historic structures, ▶preservation of historic landmarks [US]/conservation of historic monuments [UK], ▶State Historic Preservation Office [US]/historic buildings and monuments commission/council [UK]); *syn.* heritage management [n] [UK], measures [npl] for conservation of historic monuments (and sites) [UK]; **s gestión [f] del Patrimonio Histórico Español [Es]** (Conjunto de medidas destinadas a conservar el Patrimonio Histórico, promover el enriquecimiento del mismo y fomentar y tutelar el acceso de todos los ciudadanos a los bienes comprendidos en él. La **g. del p. h.** concierne a la Administración del Estado y a las Comunidades Autónomas; cf. art. 2.1 Ley 16/1985; ▶administración regional de protección de monumentos, ▶conservación de monumentos histórico-artísticos, ▶monumento, ▶protección del Patrimonio Histórico Español); *syn.* gestión [f] de monumentos histórico-artísticos; **f gestion [f] du patrimoine monumental** (Mesures administratives, opérations et programmes visant à la conservation [entretien, restauration ou réutilisation] des ▶monuments historiques ; ▶direction régionale des monuments historiques, ▶garde et conservation des monuments historiques, ▶protection des monuments historiques) ; **g Denkmalpflege [f, o. Pl.]** (Administrative, planende und technische Maßnahmen zur Sicherung und Pflege oder Wiederherstellung von Kulturdenkmälern als Verpflichtung des jeweiligen Eigentümers. In D. wird die Koordination überwiegend als staatliche Aufgabe verstanden und von den Bundesländern im Rahmen ihrer Kulturhoheit wahrgenommen; ▶Baudenk-

malpflege, ▶Denkmalschutz, ▶Kulturdenkmal, ▶Landesdenkmalamt).

2739 historic range [n] *zool.* (Total geographical limits inhabited in history by an animal species; e.g. range occupied by the elk or wolf. Term mostly used for species about which there is information on the size of their range. Almost all **h. r.s** have developed naturally in contrast to those in modern times which have been established through ▶introduction 1 or reintroduction); *s* **área** [f] **histórica** (Total del área geográfica que una especie de fauna ocupó como máximo en un momento de la historia. Casi todas las **aa. hh.** son de origen natural, al contrario que las actuales creadas o reestablecidas por ▶sueltas [masivas]); *f* **aire** [f] **historique d'une espèce** (Unité biogéographique de premier ordre occupée à une époque donnée par une espèce animale, p. ex. l'aire historique du loup ou de l'élan. Cette appellation est employée en général en relation avec les espèces pour lesquelles une évolution de la répartition de l'espèce a été étudiée. Presque toutes les aires historiques des espèces sont d'origine naturelle par opposition à notre époque récente pendant laquelle d'importantes unités ont été créées ou reconstituées par des ▶introductions) ; *g* **historisches Areal** [n] (Gesamter geografischer Raum [maximale Ausdehnung], welcher in historischer Zeit von einer Tierart bewohnt war, z. B. das historische Areal des Elches, des Wolfes. Zumeist wird dieser Begriff im Zusammenhang mit solchen Arten angewandt, bei denen Angaben über Veränderungen ihrer Arealgröße vorliegen. Fast alle **h. A.e** sind auf natürliche Art und Weise entstanden, im Gegensatz zur Neuzeit, in der große Arealteile durch ▶Aussetzungen begründet resp. wiederhergestellt werden); *syn.* historisches Verbreitungsgebiet [n].

historic rose [n] *hort.* ▶old-fashioned rose.

2740 historic site [n] *conserv'hist.* (Place of historic, architectural, and cultural value or importance; ▶cultural monument; ▶historic landscape); *s* **sitio** [m] **histórico** (Lugar o paraje natural vinculado a acontecimientos o recuerdos del pasado, a tradiciones populares, creaciones culturales o de la naturaleza y a obras del hombre, que posean valor histórico, etnológico, paleontológico o antropológico; art. 15 [4] Ley 16/1985; ▶monumento, ▶paisaje humanizado histórico 1); *f* **site** [m] **historique culturel** (**1.** Lieu reconnu pour ses qualités culturelles en particulier comme témoignage de l'activité humaine au cours de l'histoire [site ethnologique, paléontologique ou anthropologique ou plus récemment site industriel] ; ▶monument historique, ▶paysage historique, ▶paysage patrimoine. **2.** Espace portant l'empreinte d'un personnage illustre, jardin artistique, parc ou champ de bataille) ; *g* **kulturhistorische Stätte** [f] (**1.** Ort kultureller Bedeutung, besonders als Zeugnis menschlichen Schaffens und Wirkens in der Geschichte; ▶Kulturdenkmal. **2.** Durch eine bedeutende Persönlichkeit geprägte Anlage, gestalteter Garten, Park oder durch eine Feldschlacht historisch wichtiger Ort; ▶historische Kulturlandschaft); *syn.* historische Stätte [f] [CH].

2741 historic structure [n] *conserv'hist.* (Cultural artifact, e.g. building, bridge or other edifice, which is verifiably authentic for a specific historical date or period, and for which public interest in its preservation exists; **h. s.s** are covered by ▶historic landmarks [US]/cultural monuments [UK]); *s* **edificio** [m] **histórico-artístico** (Edificación o construcción de origen humano representativo para la arquitectura de una época histórica y cuya preservación es de interés público; ▶monumento); *syn.* edificio [m] histórico; *f* **monument** [m] **historique (2)** (Immeubles, monuments mégalithiques et les terrains qui renferment des stations ou gisements préhistoriques dont la conservation présente du point de vue de l'histoire ou de l'art un intérêt public et sont classés en totalité ou en partie comme ▶monuments historiques ; cf. art. 1 loi sur les monuments historiques du 31

décembre 1913) ; *syn.* monument [m] classé ; *g* **Baudenkmal** [n] (...mäler [pl], in A nur so, sonst auch ...male [pl]; von Menschenhand geschaffenes Objekt, das für einen bestimmten kulturhistorischen Zeitabschnitt im Bereich der Architektur charakteristisch ist und an dessen Erhaltung aus künstlerischer oder geschichtlicher Sicht ein öffentliches Interesse besteht; **B.mäler** gehören zu den ▶Kulturdenkmälern); *syn.* Baukulturdenkmal [n].

2742 historic vernacular landscape [n] *conserv'hist. land'man. leg.* (Landscape that evolved through use by the people whose activities or occupancy shaped it. Through social or cultural attitudes of an individual, a family, or community, the landscape reflects physical, biological, and cultural characters of everyday lives. Function plays a significant role in a **h. v. l.** This can be a farm complex or a district of historic farmsteads along a river valley. Examples include rural historic districts and agricultural landscapes; LA 1997 [5], 81. The **European Landscape Convention**, also known as Florence Convention, adopted in 2000, and into force on 1 March 2004, promotes the protection, management and planning of European landscapes and organises European co-operation on landscape issues. It is the first international treaty to be exclusively concerned with all dimensions of European landscape, such as Cornwall and West Devon Mining Landscape, Upper Middle Rhine Valley between Bingen and Koblenz, the Loire Valley between Sully-sur-Loire and Chalonnes, the Aranjuez Cultural Landscape in Spain et al.; ▶historic site); *s* **paisaje** [m] **humanizado histórico (2)** (Paisaje singular creado como resultado de las prácticas tradicionales de gestión del espacio rural del pasado y que frecuentemente está protegido o se considera parte del patrimonio natural y cultural por sus valores estéticos y culturales. Estos paisajes singulares se pueden registrar en la Lista del patrimonio mundial. En octubre del 2000 el Consejo de Europa aprobó el **Convenio Europeo del Paisaje** [Convenio de Florencia] que entró en vigor el 01.03. 2004, según el cual podrían adherirse a él también países no-miembros del consejo. El propósito general de la convención es establecer un instrumento nuevo consagrado exclusivamente a la protección, gestión y ordenación de todos los paisajes de Europa, así como garantizar la cooperación europea en ese campo. Después de unos años la situación evolucionó en el seno de la Convención del patrimonio mundial, para dar un mayor espacio a la protección de paisajes culturales, patrimonio cultural amezado de los siglos pasados como p. ej. en Europa los paisajes mineros de Cornwall y del este de Devon, el valle del Loira, el valle del Alto Rin medio y el paisaje cultural de Aranjuez; ▶sitio histórico); *syn.* paisaje [m] vernáculo histórico; *f* **paysage** [m] **vernaculaire** (Expression popularisée par BRINCKERHOFF JACKSON [un des fondateurs après 1945 de l'enseignement et de la recherche sur les paysages américains] et encore peu connue en France désignant la relation entre le paysage local [fabriqué localement par les habitants] façonné par les pratiques traditionnelles de gestion de l'espace rural avec le paysage habité unique ayant gardé dans la forme de l'habitat et le style architectural, toutes ses caractéristiques jusqu'à nos jours. Ces paysages sont des registres vivants des activités humaines et représentent un patrimoine culturel et historique inappréciable. Ces paysages singuliers et uniques, témoin de l'histoire et de la mémoire, sont souvent inscrits sur la Liste du patrimoine mondial. En juillet 2000 le Conseil de l'Europe a adopté la **Convention paysagère européenne** [Convention de Florence] et depuis quelques années la situation évolue au sein de la Convention du patrimoine mondial afin d'accorder une plus grande place aux paysages culturels, héritage culturel menacé des siècles et millénaires passés p. ex. en Europe, le paysage minier des Cornouailles et de l'ouest du Devon, les paysages viticoles du pourtour méditerranéen, le paysage culturel d'Aranjuez, le Val de Loire entre Sully-sur-Loire et Chalonnes, la vallée

H

du Haut-Rhin moyen ; ▶site historique culturel) ; *syn.* paysage [m] culturel historique ; *g* **historische Kulturlandschaft [f] (2)** (Landschaft oder ein Teil davon, heute oft naturschutzpflegerischer Obhut unterliegende Gebiete, die durch frühere landwirtschaftliche extensive Bewirtschaftungsweisen entstanden sind, z. B. Reste der Lüneburger Heide — durch extensive Heidschnucken-Weidewirtschaft bis ins 18. Jh. offengehaltene *Calluna*-Heidegebiete —, die noch weit bis ins 19. Jh. als eine wüstenhafte Landschaft galt, und deren ästhetische Wertschätzung erst entdeckt wurde, als sie bereits im Verschwinden begriffen war; cf. JES 2004. In D. sind seit 1998 die Ziele des Naturschutzes und der Landschaftspflege auch für **h. K.en** gem. § 2 (1) 13 BNatSchG festgeschrieben. Am 20.10.2000 wurde in Florenz die **Europäische Landschaftskonvention [ELC]** unterzeichnet [Florence Convention], trat am 01.03.2004 in Kraft und fördert den Schutz, das Management und die Entwicklung europäischer Kulturlandschaften sowie die Organisation einer europäischen Zusammenarbeit in diesen Fragen; D. verfügt im Vergleich zu anderen Staaten über ein breites landschaftspolitisches Instrumentarium, wie es im Bundesnaturschutzgesetz, im Bau- und Raumordnungsgesetz oder im Wasserhaushaltsgesetz verankert ist; in Europa sollen u. a. die Bergbaulandschaft von Cornwall und West-Devon, das Obere Mittelrheintal zwischen Bingen und Koblenz, Loiretal zwischen Sully-sur-Loire und Chalonnes, die Kulturlandschaft von Aranjuez in Spanien als kulturelles Landschaftserbe geschützt werden; ▶kulturhistorische Stätte).

historic zoning district [n] [US] *arch. conserv' hist. leg.* ▶historic district [US].

history [n] *geo. hist. landsc.* ▶landscape history.

histosol [n] [US] *constr. pedol.* ▶bog soil; ▶organic soil.

2743 hobby and allotment gardening [n] [UK] *hort. recr.* (All favo[u]rite leisure time gardening activities of amateur gardeners and community gardeners [US]/allotment gardeners [UK]); *syn.* hobby and community gardening [n] [US]; *s* **horticultura [f] y/o jardinería [f] de ocio** (Práctica de la jardinería con fines recreativos en jardines privados o huertos recreativos urbanos); *f* **jardinage [m] amateur** (Pratique du jardinage pendant les loisirs par l'amateur du jardinage) ; *syn.* loisirs [mpl] de jardinage ; *g* **Freizeitgartenbau [m, o. Pl.]** (Alle nicht erwerbsmäßig betriebenen gärtnerischen Tätigkeiten der Hobbygärtner und Kleingartenbesitzer); *syn.* Freizeitgärtnerei [f, o. Pl.], Freizeitgärtnern [n, o. Pl.].

hobby and community gardening [n] [US] *hort. recr.* ▶hobby and allotment gardening [UK].

2744 hoeing operation [n] *constr. hort.* (Tilling of a soil surface with a hoe for weed control or loosening; ▶weed 2; ▶maintenance operation); *s* **operación [m] de binado** (Labrado de la superficie de un suelo con una azada para desherbarlo y mullirlo; ▶escardar, ▶operación de mantenimiento); *syn.* trabajos [mpl] de azadonado; *f* **intervention [f] de binage** (Intervention effectuée sur une aire dans le but d'éliminer les adventices et d'ameublir le sol lors d'une ▶intervention d'entretien ; ▶désherber) ; *g* **Hackgang [m]** (Gärtnerischer Arbeitseinsatz, bei dem eine Fläche zur Bodenlockerung oder Unkrautbekämpfung mit einer Hacke gelockert wird; ▶Unkraut jäten; ▶Pflegegang).

hoggin playing court [n] [UK] *constr.* ▶granular playing court [US].

hoggin playing surface [n] [UK] *constr.* ▶granular playing surface [US].

hoggin surface [n] [UK] *constr.* ▶water-bound surface.

hoggin surface course [n] [UK] *constr.* ▶granular surface course [US].

holding capacity [n] [US] *recr. plan.* ▶accommodation capacity, ▶water-holding capacity.

holdings [npl] *agr. for.* ▶scattered agricultural holdings.

holding site [n] [US] *constr.* ▶site nurserey.

2745 hold-over [n] [US] *for.* (Tree that has escaped logging, wind throw or fire, and currently occupies a dominant position in the stand; SAF 1983, ▶seed bearer); *s* **árbol [m] padre (1)** (Cualquier árbol productor de semilla y, más concretamente, el que se deja en pie, al cortar el monte a matarrasa, para que lo repueble; DB 1985; ▶portagrano); *f* **survivant [m]** (Arbre qui a échappé à la coupe, au vent, au feu, etc. et qui occupe de ce fait une position dominante dans le peuplement qui s'ensuit ; DFM 1975 ; ▶semencier) ; *syn.* rescapé [m] ; *g* **Überhälter [m] (1)** (Einzelner, Baum oder eine Gruppe Bäume, die beim Fällen eines Bestandes oder nach starkem Sturm stehen blieben oder einen Waldbrand überlebten, und somit eine natürliche Verjüngung oder einen Sonneneinstrahlungsschutz ermöglichen; ▶Samenträger, ▶Überhälter 2).

hole [n] *constr.* ▶earth void, ▶fertilizing hole; *conserv. zool.* ▶fish hole; *zool.* ▶nesting hole; *arch. constr.* ▶pipe hole; *constr. hort.* ▶planting hole; *geo. min.* ▶test borehole; *constr.* ▶weep hole.

hole [n]**, drain** *constr.* ▶weep hole.

hole [n] [UK]**, kettle-** *geo.* ▶kettle.

hole [n] [US]**, maintenance** *constr.* ▶manhole.

hole [n] [US]**, trial bore-** *geo. min.* ▶test borehole.

hole [n] [US]**, utility** *constr.* ▶manhole.

2746 hole [n] **in ozone layer** *envir. met.* (▶Ozone layer opening caused by chlorofluorocarbons, CFCs, which break up the ozone into other oxygen compounds); *s* **agujero [m] del ozono** (Parte deteriorada de la ▶capa de ozono de la atmósfera, debido a los efectos de los compuestos fluroclorocarbonados procedentes de los usos humanos como la refrigeración, como agentes propulsores en envases aerosoles y para fabricar espumas sintéticas); *syn.* agujero [m] de la capa de ozono; *f* **trou [m] dans la couche d'ozone** (Zone détruite de la ▶couche d'ozone de l'atmosphère ; les chlorofluorocarbures [CFC] utilisés dans les aérosols, les fluides frigorigènes, les mousses synthétiques et les solvants sont considérés comme étant principalement à l'origine de l'appauvrissement et de la destruction de la couche d'ozone) ; *g* **Ozonloch [n]** (Ungewöhnlich stark zerstörter Bereich in der ▶Ozonschicht der Erdatmosphäre, der um 1980 in der Antarktis und ab 1992 in der Arktis festgestellt wurde. Der Abbau geschieht durch gasförmige Halogenverbindungen, speziell durch vom Menschen zusätzlich in die Atmosphäre gebrachten Fluorchlorkohlenwasserstoffe [FCKW] in Form von Treibgasen in Spraydosen, als Kühlmittel in Kühlschränken und Kältemaschinen sowie als Treibmittel bei der Herstellung von Schaumstoffen).

hole-nesting bird [n] [UK] *zool.* ▶cavity-nesting bird [US].

2747 hole planting [n] [US] *constr. hort.* (SAF 1983, 133; setting a plant in a dug pit or hole with loosened soil; ▶trench planting, ▶slit planting [US]/notch planting [UK], ▶furrow planting); *syn.* pit planting [n] [UK]; *s* **plantación [f] en hoyos** (▶plantación en raja, ▶plantación en zanjas, ▶plantación en surcos); *syn.* plantación [f] hoyo por hoyo; *f* **plantation [f] sur potets** (DFM 1975 ; ▶plantation en fentes, ▶plantation en sillons, ▶plantation en tranchée) ; *syn.* plantation [f] sur trous (DFM 1975) ; *g* **Lochpflanzung [f]** (Setzen von Pflanzen mit lockerer Erde in ein ausgehobenes Loch; ▶Pflanzung in durchgehendem Pflanzgraben, ▶Klemmpflanzung, ▶Riefenpflanzung).

holiday [n] *recr.* ►hiking holiday, ►subsidized holiday.

holiday [n] [UK], **farm** *recr.* ►farm vacation [US]/farmstay holidays [UK].

holiday [n] [UK], **rambling** *recr.* ►hiking holiday.

holiday [n] [UK], **walking** *recr.* ►hiking holiday.

holiday accommodation [n] [UK] *recr.* ►vacation accommodation [US] (1).

holiday accommodation [n] [UK], **rural** *recr.* ►rural vacation accommodation [US].

holiday activity [n] [UK] *recr.* ►vacation activity [US].

holiday area [n] [UK] *plan. recr.* ►tourist area.

holiday cottage [n] [UK] *recr.* ►vacation house [US].

2748 holiday entertainment [n] *recr.* (Provision of organized passive amusements or educationally-orientated performances for vactioners [US]/holidaymakers [UK]; ►physical amusement); *s* **promoción** [f] **recreacional** (Oferta de actividades de ocio o educativas sin fines comerciales para veraneantes o turistas; ►animación); *f* **loisirs** [mpl] **organisés** (±) (Forme de loisirs non lucrative pour laquelle les usagers disposent d'animateurs spécialisés dans les techniques de plein air ; ►animation) ; *g* **durch Pädagogen betreute Erholung [f, o. Pl.]** (Angebot der Freizeitindustrie für eine inhaltliche Gestaltung eines Erholungsaufenthaltes durch dafür ausgebildete Betreuer [= Freizeitpädagogen]; ►Animation).

holidayer [n] [UK] *recr. sociol.* ►vacationist [US].

holiday frequency [n] [UK] *plan. recr. sociol.* ►vacation frequency [US].

holiday home [n] [UK] (1) *recr.* ►vacation accommodation [US] (2)/holiday accommodation [UK].

holiday home [n] [UK] (2) *recr.* ►vacation house [US].

holiday-maker [n] [UK] *recr. sociol.* ►vacationist [US].

holiday-makers [npl] [UK], **accommodation-seeking** *recr. plan.* ►accommodation-seeking vacationers [US].

holiday occupation [n] [UK] *recr.* ►vacation activity [US]/holiday activity [UK].

holiday park [n] [UK] *recr.* ►park-like vacation development [US].

holiday residence [n] [UK] *recr.* ►vacation residence [US].

holidays [npl] [UK] *recr.* ►vacation [US].

holiday traffic [n] *plan. recr. trans.* ►tourist traffic, ►summer holiday traffic.

holiday travellers [npl] [US], **accommodation-seeking** *recr. plan.* ►accommodation-seeking vacationers [US].

holiday village [n] [UK] *plan. recr.* ►resort community [US].

hollow tining [n] *constr.* ►lawn aeration, #6.

home [n] [UK] (1), **holiday** *recr.* ►vacation accommodation [US] (2).

home [n] [UK] (2), **holiday** *recr.* ►vacation house [US].

Home and Neighborhood Improvement Award *hort. plan. urb.* ►"All-America City" [US].

2749 home range [n] (1) *zool.* (Area around the home of an animal, e.g. nest, den, or burrow, that is used for feeding and other daily activities. Commonly, a pair of animals and their offspring share a **h. r.**, though in some species a larger group shares it. The **h. r.** of certain species are characterized by winter-summer animal movements [**h. r.** movements]; cf. LE 1986; ►total range, ►territory; *specific terms* winter **h. r.**, summer **h. r.**); *s* **área** [f] **de actividad** (Zona alrededor de la morada de un animal utilizada por éste para alimentarse y para otras actividades

diarias. Normalmente una pareja de animales comparte un **á. de a.**, aunque en algunas especies la comparten un gran grupo de ellas. A diferencia del ►territorio que es defendido activamente, en el **á. de a.** no se producen interacciones violentas con otros individuos; cf. MARG 1977, 846; ►territorio vital total); *f* **territoire** [m] **d'activité** (±) (Partie de l'aire d'habitation dans laquelle certains individus se déplacent ou qu'ils visitent régulièrement mais qu'ils ne défendent pas activement ; *termes spécifiques* ►domaine vital, ►territoire) ; *g* **Streifgebiet** [n] (Teil des Wohngebietes, in dem sich bestimmte Individuen regelmäßig aufhalten oder das sie regelmäßig aufsuchen, das aber im Gegensatz zum Revier [►Territorium] nicht aktiv verteidigt wird; ►Jahreslebensraum; *UBe* Sommerstreifgebiet, Winterstreifgebiet); *syn.* Aufenthaltsgebiet [n].

2750 home range [n] (2) *zool.* (Area familiar to an animal or a group of the same animal species, which it patrols regularly. Portions of the **h. r.** that are defended constitute the ►territory; ►total range, ►home range 1); *s* **área** [f] **de habitación** (Zona de vivienda de animales residentes en un lugar específico que sirve para la alimentación, reproducción y cría. En muchos casos en el centro de esta área se encuentra otra menor [►territorio] que es defendida ante otros individuos de la misma especie; ►área de actividad, ►territorio vital total); *f* **aire** [f] **d'habitation** (Surface familière à un individu, un couple, une famille ou un groupe d'espèces animales sur laquelle ils se nourrissent, se reproduisent et élèvent leur progéniture ; dans de nombreux cas l'aire d'habitation comprend en son centre le ►territoire défendu contre d'autres individus appartenant le plus souvent à leur espèce ; LAF 1990, 295 et s. ; ►domaine vital, ►territoire d'activité) ; *g* **Wohngebiet** [n] (1) (Gebiet, das von einem seßhaften Einzeltier, einem Paar, einer Familie oder einer Sippe von Tieren zwecks Ernährung, Fortpflanzung und Aufzucht der Jungen regelmäßig benutzt wird. In vielen Fällen liegt im Zentrum dieses Gebietes ein kleineres Gebiet [►Territorium], das gegenüber Eindringlingen gleicher Art verteidigt wird; ►Jahreslebensraum, ►Streifgebiet); *syn.* Wohnbereich [m].

2751 home zone [n] [UK] *recr. urb.* (Section of a street in a residential area specially designated for children's play; the **h. z.** concept, called woonerf, was pioneered in the 1970s in the Netherlands. It is an attempt to strike a balance between vehicular traffic and everyone else who uses the street, the pedestrians, cyclists, business people and residents; motorists are forced to drive with greater care and at lower speeds; many countries support this with legislation allowing the **h. z.s** to enforce a reduced speed limit of 10 miles an hour; **in D.**, traffic law restricts motorists in a "verkehrsberuhigter Bereich" to a maximum speed of 4-10 km/h; in **U.K.**, **h. z.s** do not enjoy similar protection in law); *syn. obs.* playstreet [n]; *s 1* **calle** [f] **de tráfico tranquilizado** (Calle o sección de calle en zona residencial calificada y señalada en la que todos los usuarios —estén motorizados o no— tienen los mismos derechos y se deben respetar mutualmente. Los vehículos deben ir a paso y no se pueden aparcar más que en las zonas marcadas para ello, exceptuando el caso de carga y descarga. El concepto se desarrolló en los años 1970 en los Países Bajos bajo la denominación „Woonerf", fue introducido en D. en los 1980); *s 2* **calle** [f] **para juegos** (Calle o sección de calle en zona residencial calificada y señalada como zona para juegos infantiles, con las correspondientes limitaciones para la circulación de vehículos); *f 1* **zone** [f] **de circulation apaisée** (Terme générique instituant le devoir de prudence accrue des conducteurs à l'égard des usagers vulnérables et désignant l'►aire piétonne, la ►zone de rencontre et la zone 30) ; *f 2* **zone** [f] **de rencontre** (*Catégorie des zones de circulation apaisées* Section ou ensemble de sections de voies en agglomération constituant une zone affectée à la circulation de tous les usagers.

H

Dans cette zone, les piétons sont autorisés à circuler sur la chaussée sans y stationner et bénéficient de la priorité sur les véhicules. La vitesse des véhicules y est limitée à 20 km/h. Toutes les chaussées sont à double sens pour les cyclistes, sauf dispositions différentes prises par l'autorité investie du pouvoir de police ; cf. Décret n° 2008-754 du 30 juillet 2008 ; ce concept avec quelques variantes existe déjà sous cette dénomination en Belgique et en Suisse. Pour autant la réglementation française n'a pas repris intégralement les pratiques de ces deux pays ; CH., la zone de rencontre a été introduite dans la réglementation avec l'ordonnance royale du 28 septembre 2001. Cet espace désigne des secteurs situés dans des quartiers résidentiels ou commerciaux, sur lesquels les piétons peuvent utiliser toute l'aire de circulation pour des activités de jeu, de sport, d'achats, de flâne ou de rencontre. Ils bénéficient de la priorité, peuvent traverser partout, mais ne doivent toutefois pas gêner inutilement les véhicules. Cette restriction dans le code constitue une différence par rapport aux usages aux Pays-Bas qui les premiers ont développé le concept de « Woonerf » dans les années 1970) ; *syn.* zone [f] résidentielle [CDN, L, NL]) ; *f 3* **rue [f] réservée aux jeux d'enfants** (1. B., rue réservée sur toute la largeur de la voie publique pour les jeux, principalement des enfants. Les personnes qui jouent sont considérées comme des piétons; toutefois, seuls les conducteurs des véhicules à moteur habitant dans la rue de même que les véhicules prioritaires, lorsque la nature de leur mission le justifie ainsi que les cyclistes, ont accès aux rues réservées au jeu. Les conducteurs qui circulent dans les rues réservées au jeu doivent le rouler au pas; ils doivent céder le passage aux piétons qui jouent, leur céder la priorité et au besoin s'arrêter. Les cyclistes doivent descendre de leur bicyclette si nécessaire. Les conducteurs ne peuvent pas mettre en danger les piétons qui jouent ni les gêner. Ils doivent en outre redoubler de prudence en présence d'enfants. La rue peut être fermée temporairement pendant un jour, une semaine ou toutes les vacances scolaires, chaque fois pendant les mêmes heures ; cf. Code de la route belge ; **2. D.**, rue fermée à la circulation et pour laquelle les enfants sont autorisés à jouer sur la chaussée et les trottoirs ; du fait que l'interdiction de circuler touche aussi les riverains, l'utilisation de cette forme d'apaisement de la circulation est assez rare) ; *g 1* **verkehrsberuhigter Bereich [m]** (Straße in einem Wohngebiet, die für das Spielen von Kindern bestimmt und gekennzeichnet ist. Das Konzept wurde in den 1970er-Jahren in den Niederlanden mit dem Begriff „Woonerf" entwickelt; die offizielle Einführung erfolgte in D. 1980. Alle Verkehrsteilnehmer, Fußgänger, Radfahrer und Kraftfahrzeuge, sind gleichberechtigt, dürfen sich aber nicht gegenseitg behindern; der Fahrzeugverkehr muss das Schrittempo [nach dt. Rechtsprechung 4-10 km/h] einhalten und das Parken ist außerhalb der dafür gekennzeichneten Flächen nicht zulässig, ausgenommen zum Ein- und Ausladen sowie zum Be- und Entladen; gekennzeichnet ist der **v. B.** mit den rechteckigen blauen Verkehrszeichen 325 und 326 — Anfang und Ende des Geltungsbereiches); *syn.* Begegnungszone [f] [CH], Wohnstraße [f] [A], *ugs. in D.* Spielstraße [f]; *g 2* **Spielstraße [f]** (D., in der Straßenverkehrsordnung wird der Begriff **S.** zwar nicht erwähnt, jedoch wird in der Verwaltungsvorschrift zur Straßenverkehrsordnung [VwV-StVO] mit **S.** eine Straße bezeichnet, die für Fahrzeuge aller Art gesperrt ist; durch das rechteckige Zusatzzeichen Z 1010-10, schwarzes Kind mit Ball auf weißem Grund, wird Kindern erlaubt, auf der Fahrbahn und auf dem Seitenstreifen zu spielen; da die Sperrung der Straße durch das Verkehrszeichen 250, roter Kreis auf weißem Grund, auch die Anlieger betreffen würde, ist die eigentliche **S.** recht selten; cf. wikipedia/Spielstraße).

honey pot [n] *plan. recr.* ▶attraction point.

2752 honorable mention [n] [US]/**honourable mention** [n] [UK] *prof.* (Recognition of competition entries which fail to win the major prizes or monetary awards, but are considered worthy of consideration; ▶final selection [UK] ▶prize); *syn.* runner up [n] [US], award [n] of merit [US]; *s* **mención** [m] (*Resultado* trabajo de concurso que no ha sido ni premiado ni adquirido, pero que se considera de calidad; ▶premio, ▶selección final); *f* **projet** [m] **mentionné** (Prestations des participants sélectionnés à un concours n'ayant pas reçu de ▶prix ou n'ayant pas été primés, mais ayant obtenu une mention particulière ; ▶sélection définitive) ; *syn.* travail [m] mentionné ; *g* **engere Wahl [f, o. Pl.] (2)** (*Ergebnis* Wettbewerbsarbeit, die weder mit einem ▶Preis noch mit einem Ankauf bedacht wurde, sondern eine lobende Anerkennung erhielt; ▶engere Wahl 1); *syn.* lobende Erwähnung [f].

2753 honorarium [n] *prof.* (Voluntary remuneration for professional services rendered without the normal fee); *syn.* token payment [n] [also US]; *s* **honorarios** [mpl] **reconocidos** (Remuneración no acordada cuyo monto depende de la apreciación del comitente de la obra); *f* **honoraire** [m] **de gratification** (Rémunération laissée à l'appréciation du maître d'ouvrage) ; *g* **Anerkennungshonorar** [n] (Nicht vereinbartes Honorar, dessen Höhe im Ermessen des Auftraggebers liegt).

honourable mention [n] [UK] *prof.* ▶honorable mention [US]/honourable mention [UK].

hooked climber [n] *bot.* ▶scandent plant.

hookup charge [n] [US]/**hook-up charge** [n] [UK] *adm. urb.* ▶utility connection charge.

horizon [n] *pedol.* ▶illuvial horizon, ▶organic horizon, ▶organic horizon, ▶soil horizon; *geo. hydr.* ▶spring horizon.

horizon [n], **A** *pedol.* ▶topsoil (1).

horizon [n] [US], **argillic** *pedol.* ▶claypan.

horizon [n], **B** *constr. pedol.* ▶subsoil.

horizon [n], **C** *pedol.* *▶brown earth, *▶exposed stone formation, *▶ranker [UK], *▶rendzina; *▶truncated soil profile.

horizon [n] [US], **illuvial** *pedol.* ▶claypan.

horizon [n], **L** *pedol.* ▶litter layer.

horizon [n], **O** *pedol.* ▶organic horizon.

horizons [npl] *pedol.* ▶exposed soil horizons.

2754 horizontal alignment [n] *plan. trans.* (Delineation of horizontal travelled way [US]/carriage-way [UK] layout determined by geometry of curves and straight segments according to use and design speed requirements; ▶horizontal and vertical alignment, ▶vertical alignment, ▶vertical profile); *s* **alineación [f] horizontal** (Fijación del trazado y dimensionamiento [corte transversal y radios de curvas] de una carretera en el paisaje; ▶alineación horizontal y vertical, ▶alineación vertical, ▶trazado en perfil longitudinal); *syn.* trazado [m] en planta, trazado [m] en perfil transversal; *f* **définition [f] des caractéristiques géométriques principales du tracé** (Dans un avant-projet d'études de routes, détermination du tracé en plan, des profils en long et de ses rayons, des profils en travers-types ; ▶cotation altimétrique et planimétrique d'un tracé, ▶étude de tracé routier, ▶profil en long d'un tracé) ; *syn.* tracé [m] en plan ; *g* **Lagetrassierung [f]** (Festlegung der Lage und Dimensionierung von Verkehrswegen in die Landschaft — Querschnitt und Kurvenradien entsprechend den Entwurfsgeschwindigkeiten; ▶Höhenabwicklung einer Verkehrstrasse, ▶Gradiente einer Trasse, ▶Trassierung in Lage und Höhe); *syn.* Trassierung [f] in Lage, Trassenführung [f] (1).

2755 horizontal and vertical alignment [n] *plan. trans.* (Layout of a road, railroad [US]/railway [UK], or canal as delineated on a plan, profile and typical cross sections showing horizontal stations measured on the centreline and designating proposed elevations or grades at each point; ►transportation corridor selection, ►vertical alignment 1); *s* **alineación [f] horizontal y vertical** (Determinación del trazado de una línea de transporte [carretera, ferrocarril, canal] en corte longitudinal y altitudinal así como en corte transversal; ►gradiente, ►selección de corredor [de transporte]); *syn.* trazado [m] (en planta y corte longitudinal); *f* **étude [f] de tracé routier** (Définition et coordination du tracé général [tracé en plan, profil en long] d'une infrastructure routière ; ►profil en long d'un tracé, ►recherche du tracé optimum) ; *g* **Trassierung [f] in Lage und Höhe** (Festlegung der Linienführung einer Verkehrslinie [Straße, Bahnlinie, Kanal] im Lage- und Höhenplan sowie durch Regelquerschnitte; ►Gradiente einer Trasse, ►Trassenfindung); *syn.* Trassenführung [f] (2).

2756 horizontally-coursed masonry [n] *constr.* (Wall laid with horizontal elements); *s* **mampostería [f] en hiladas horizontales** (Obra de fábrica en la que las juntas de aparejo se encuentran mayormente alineadas horizontalmente); *f* **maçonnerie [f] à appareillage en assises** (Mur de pierres dont l'appareillage laisse une impression d'horizontalité dominante ; on distingue suivant l'épaisseur des assises et la forme des rives en retour : **mur à appareillage en assises régulières, mur à appareillage en assises irrégulières** [genre d'appareillage spécifique aux murs en pierres sèches]) ; *g* **lagerhaftes Mauerwerk [n]** (Mauerwerk mit vorwiegend horizontalem Fugenbild).

2757 horizontal scale [n] *arch. constr. eng. surv.* (Two dimensional drawing measurement ratio to corresponding measurement on a real object or surface; e.g. 1:50 scale indicates one unit on the drawing as representing 50 of the same units on the ground in full scale; ►vertical scale); *s* **escala [f] horizontal** (En un plano o mapa relación entre las longitudes representadas y las reales; ►escala vertical); *syn.* escala [f] de longitudes; *f* **échelle [f] des longueurs** (Rapport entre une longueur dans la réalité et sa transcription sur un plan ou une carte ; ►échelle des hauteurs) ; *g* **Längenmaßstab [m]** (Auf einem Plan oder einer Karte benutztes Verhältnis von dargestellten Längen zu denen in der Wirklichkeit; ►Höhenmaßstab).

2758 horseback riding [n] [US] *recr.* (Movement on a horse, which can be ridden upon for recreational or therapeutic resons; *generic term* riding); *syn.* equestrian riding [n] [US], riding [n] [UK]; *s* **equitación [f]** (1. Arte de montar y manejar bien el caballo. 2. Acción de montar a caballo; CAS 1985); *f* **équitation [f]** (Action et art de monter à cheval) ; *g* **Pferdereiten [n, o. Pl.]** (Fortbewegung auf einem Reitpferd als Freizeitbeschäftigung oder zur Therapie; *OB* Reiten).

2759 horseback riding sport [n] [US] *recr.* (Recreational activity that covers hobby riding on horseback, foxhunting with hounds, thoroughbred horse racing, totting races, polo games, dressage, show and steeple chase jumping); *syn.* equestrian sport [n] [UK], horse riding sport [n] [also UK]; *s* **hípica [f]** (Cualquier tipo de deporte que necesita del caballo para realizarse); *syn.* deporte [m] hípico; *f* **hippisme [m]** (Activités sportives équestres comprenant les courses de chevaux, les courses de trot, la dressure, la chasse à courre, la chasse au renard, la course d'obstacles et la promenade à cheval ; LA 1981, 289) ; *syn.* sport [m] hippique, sport [m] équestre ; *g* **Reitsport [m, o. Pl.]** (...sportarten [pl]; Teil des Pferdesportes, der die Fuchsjagd, das Pferderennen, Trabrennen, Polo, das Dressurreiten, Jagdreiten, Hindernisrennen, Springreiten und das Freizeitreiten umfasst).

horse riding sport [n] [UK] *recr.* ►horseback riding sport [US].

horse riding stable [n] [UK] *recr.* ►riding stable [US].

horse riding trail network [n] [UK] *plan. recr.* ►bridle path network [US].

horticultural company [n] *hort.* ►plant nursery.

2760 horticultural exhibition [n] *hort. urb.* (**1. In U.K.,** national garden shows or horticultural exhibitions were known as **garden festivals** and were conceived as catalysts for the regeneration of large areas of derelict land in Britain's major industrial districts during the 1980s and early 1990s. Only 5 festivals were held: Liverpool Garden Festival [International Garden Festival, 1984], Stoke-on-Trent Garden Festival [1986], Glasgow Garden Festival [1988], Gateshead Garden Festival, Northumbria [1990], Ebbw Vale Garden Festival, Wales [1992]. **2. In U.S.,** such exhibitions often occur in public parks and botanical gardens [Chicago, Brooklyn, Boston, etc.]; horticultural exhibits are included in international expositions. **3. In D.,** exist regional, national or international public showing of horticultural exhibits: h. e.s have been held since the 1920s in large, newly-created permanent urban parks. Since 1951 Federal Garden Shows *'Bundesgartenschauen' [BUGA]* have been held every two years and since 1953 international horticultural exhibitions *[Internationale Gartenbauausstellung, IGA]* have been held every 10 years. **In D.,** h. e.s have the aim of creating new permanent green open spaces for neighbo(u)rhood recreation to improve the overall provision in a city. They have proved to be a major catalyst for development in the cities in which they were created and have had a positive effect on the quality of life, social environment, infrastructure, urban climate as well as the tourism and retail industries. The garden festivals in UK have been less successful in that sustainable after-uses, were not planned from the start. In Liverpool, for example, the festival site has changed hands several times since 1984. Half of the original festival grounds have been used for a residential housing development. The rest of the site, after attempts to create leisure and entertainment facilities now lies empty and derelict awaiting development; in Glasgow the site remained derelict, until it was redeveloped for the Glasgow Science Centre and a media campus; in Ebbw Vale the site is now occupied by a housing development; ►flower show, ►horticultural exhibition area, ►indoor horticultural exhibition); *syn.* garden exhibition [n], garden festival [n] [also UK], garden show [n], horticultural show [n]; *s* **exhibición [f] de horticultura** (Exhibición regional, nacional o internacional de los avances de la horticultura. En Alemania se han realizado este tipo de exposiciones desde los años 1920 en parques existentes o nuevos creados para este fin, que posteriormente fueron conservados; ►exhibición [de horticultura] en pabellones, ►exhibición floral, ►terreno de exposición horticultural); *syn.* exposición [f] horticultural; *f 1* **exposition [f] horticole** (Présentations publiques d'intérêt régional, national ou international des produits, techniques et matériels horticoles, p. ex. hortimat) ; *f 2* **exposition [f] de jardins** (Exposition d'intérêt régional, national ou international souvent réalisée dans un ►parc floral [►jardin d'expositions] utilisant un cadre paysager existant à l'intérieur duquel sont réalisés divers types de jardins, massifs floraux permanents ou temporaires ; D., depuis les années 20, et en particulier depuis les années 50 les **e. de j.** nationales [BUGA depuis 1951] et internationales [IGA depuis 1953] sont de plus en plus intégrées dans des espaces urbains de verdure et de récréation ; ►exposition couverte, ►floralies) ; *g* **Gartenbauausstellung [f]** (Regionale, nationale oder internationale öffentliche Leistungsschau des Gartenbaues, die in D. seit den Zwanzigerjahren des 20. Jhs. immer mehr in große, neu geschaffene, bleibende Park- und Erholungsanlagen eingebunden wird; seit 1951 Bundesgartenschauen [BUGA] und seit 1953 alle 10 Jahre die Internationale Gartenbauausstellung [IGA]. Regionale **G.en** werden Landesgartenschauen [LAGA] genannt. Vorrangiges städtebauliches/grünplanerisches Ziel der **G.en** ist die Gestaltung von Freiräumen und die Schaffung von dauerhaften Grünzonen zur Naherholung und Wohnumfeldverbesserung [Sicherung von Naturressourcen]. **G.en** haben sich als Impulsgeber für eine umfassende Entwicklung in den durchführenden Städten bewährt, mit sehr positiven Auswirkungen auf die Lebensqualität, das soziale Umfeld, die Infrastruktur, das Stadtklima und die wirtschaftliche Entwicklung [u. a. Auswirkungen auf Gastronomie, Tourismus und Einzelhandel]. Vorläufer heutiger Gartenschauen waren private Pflanzenschauen, die von wohlhabenden Bürgern im 19. Jh. initiiert wurden, z. B. vom Sammler und Dahlienliebhaber BREITNER aus Leipzig und von dem Kamelienbetrieb Firma SEIDEL in Dresden. Die erste Pflanzen- und Blumenausstellung in Europa begann 1809 in Belgien mit der Gründung einer Pflanzengesellschaft, und im Jahre 1822 wurde der „Verein zur Förderung des Gartenbaus in den Königlichen Preußischen Staaten" gegründet. In D. sind aktive Gartenschaustädte, die bereits im 19. Jh. **internationale G.en** veranstalteten, Hamburg [1869 mit einer Ausstellung überseeischer Blüten, Pflanzen und Früchte im Alten Elbpark] und Stuttgart [1870 mit der ersten **G.**, bei der am Rande der damaligen Innenstadt der Stadtgarten angelegt wurde]. Die Tradition der Internationalen **G.** haben Hamburg [1953, 1963, 1973] und Stuttgart [1993] im 20. Jh. als IGAs beibehalten; München [1983] und Rostock [2003] haben dieses IGA-Konzept aufgegriffen und als Großereignis für Stadtentwicklung und Freizeitaktivitäten weiterentwickelt; cf. PRE 2004; ▶Blumenschau, ▶Gartenschaugelände, ▶gärtnerische Hallenschau); *syn.* Gartenschau [f].

2761 horticultural exhibition area [n] *hort. urb.* (A public green space in which a ▶horticultural exhibition has been held and later becomes a public park or receives another use; **in D.**, these green spaces always become public parks in order to enlarge the system of public open green spaces for urban recreation and in the long term to improve the climatic conditions; **in U.S.**, h. e.s often occur in existing public parks and botanical gardens [Chicago, Brooklyn, Boston, etc.]); *syn.* horticultural exhibition site [n], garden festival area [n]; *s* **terreno** [m] **de exposición horticultural** (En D. terreno público en·el que se llevan a cabo las ▶exhibiciones de horticultura y que a continuación es utilizado como parque); *f 1* **jardin** [m] **d'expositions** (Aire sur laquelle sont réalisées les ▶expositions de jardins et qui est ensuite utilisée comme parc public par les citadins) ; *f 2* **parc** [m] **floral** (Parc urbain ou périurbain accueillant régulièrement des expositions florales ; ▶exposition horticole) ; *g* **Gartenschaugelände** [n] (In D. öffentlicher Grund, auf dem eine ▶Gartenbauausstellung stattfindet und anschließend zu einem öffentlichen Park wird, um den innerstädtischen Grünflächenbestand zu erweitern, zusätzliche wohnungsnahe Erholungsmöglichkeiten im Freien zu schaffen und langfristig das innerstädtische Klima zu verbessern); *syn.* Gartenbauausstellungsgelände [n].

horticultural exhibition site [n] *hort. urb.* ▶horticultural exhibition area.

horticultural firm [n] *hort.* ▶plant nursery.

2762 horticultural peat [n] *constr. hort.* (Commercially prepared peat; e.g. ▶humified raised bog peat or ▶fibric peat 2 in granular form with mineral additives, which is used in horticultural practice; ▶growing medium, ▶loose peat); *s* **turba** [f] **fertilizante** (Turba preparada industrialmente, como p. ej. ▶turba negra o ▶turba clara granulada, para su aprovechamiento como enmienda de suelos; ▶sustrato vegetal, ▶turba suelta); *f 1* **tourbe** [f] **enrichie** (Tourbe préparée industriellement, p. ex. ▶tourbe noire ou ▶tourbe blonde sous forme de granulés mélangée à des substances nutritives minérales servant à la fabrication d'engrais composés ; ▶mélange terreux) ; *f 2* **tourbe** [f]

horticole (Tourbe généralement composée de sphaignes faiblement décomposées vendue à des fins horticoles ; *termes spécifiques* ▶tourbe blonde, ▶tourbe brune, ▶tourbe enrichie, ▶tourbe en vrac, ▶tourbe fibreuse) ; *syn. anglicismes* mousse [f] de tourbe [CDN], mousse de [f] de sphaignes [CDN] ; *g* **Düngetorf** [m] (Für gartenbauliche Zwecke industriell aufbereiteter Torf, z. B. granulierter ▶Schwarztorf oder ▶Weißtorf 2 mit mineralischen Nährstoffen versetzt; durch Beimengung weiterer Zuschlagstoffe wie z. B. Ton, Sand Rindenhumus wird der Torf zum gärtnerischen Kultursubstrat [▶Erdgemisch] weiterverarbeitet; ▶loser Torf).

horticultural show [n] *hort. urb.* ▶horticultural exhibition.

horticultural show [n]**, indoor** *hort.* ▶indoor horticultural exhibition.

horticulture [n] *hort.* ▶commercial horticulture, ▶ornamental plant horticulture.

hortus [n] **deliciarum** *gard'hist.* ▶pleasure ground, #2.

2763 hose bib [n] *constr.* (Fitting to which a water hose [US]/hosepipe [UK] is attached for irrigation; ▶water supply connection [US]/water supply point [UK]); *s* **racor** [m] **para manguera** (Empalme entre el grifo de agua y la mangera. Si éste tiene boca curva y rosca para adaptarlo a una manguera, se denomina «grifo de mangera»; cf. DACO 1988; ▶acometida de agua); *syn.* empalme [m] para manguera; *f* **raccord** [m] **de tuyau** (Pièce de raccordement à un tuyau d'alimentation ; ▶raccordement à un point d'eau) ; *g* **Schlauchanschluss** [m] (Vorrichtung zum Anbringen eines Bewässerungsschlauches; ▶Wasseranschluss).

hose bib [n] **[UK], underground** *urb.* ▶underground hydrant [US].

hostel [n] **[US]** *recr.* ▶hiking accommodation [US].

hostel [n] **[UK], ramblers** *recr.* ▶hiking accommodation [US].

2764 hostelry [n] *recr.* (Generic term for a facility providing overnight accommodation; e.g. hotel, inn, hostel, ▶alpine hut, ▶hiking accommodation [US]); *s* **albergue** [m] (Lugar de hospedaje sencillo; ▶hostal de senderistas, ▶refugio de alta montaña); *syn.* fonda [f]; *f* **gîte** [m] (Lieu d'hébergement [studio, appartement, maison ou partie de maison] généralement classé par arrêté préfectoral, loué pendant les vacances à une clientèle de passage. C'est une structure indépendante qui n'est pas directement située dans la maison ou l'appartement du propriétaire, la clientèle ayant la possibilité de faire sa propre cuisine, petits déjeuners et repas. Le propriétaire d'un gîte peut choisir de créer et louer son hébergement de façon indépendante ou d'adhérer à un label [Gîtes de France, Clévacances, etc.] ; *terme spécifique* gîte rural ; ▶gîte d'étape, ▶refuge de haute montagne) ; *syn.* auberge [f] ; *f 2* <u>**gîte**</u> [m] <u>**de France**</u> (Logement meublé en général situé dans une commune rurale présentant certaines conditions de confort et d'habitabilité favorables à un séjour de vacances dans un cadre naturel et humain agréable ; le classement des **g. d. F.** relève de la compétence des relais départementaux des gîtes de France ; cf. arr. du 8 janvier 1993) ; *g* **Beherbergungsstätte** [f] (Ort der Beherbergung wie z. B. Hotel, Pension, ▶Alpenhütte, ▶Wanderheim); *syn.* Herberge [f], Übernachtungsheim [n], Übernachtungsstätte [f], Unterkunft [f].

hostile habitat [n] *ecol.* ▶extreme habitat.

host plant [n] *hort. landsc.* ▶woody host plant.

2765 host plant [n] **for bees** *hort. landsc.* (Generic term for any plant providing a food source for bees; *specific term* ▶woody host plant for bees); *syn.* bees forage plant [n], bee plant [n] [also US], nectar plant [n], pollen plant [n]; *s* **planta** [f] **apícola** (Planta con flores abundantes y de larga duración que ofrece gran cantidad de néctar y polen a las abejas; *término específico* ►leñosa apícola); *f* **plante [f] mellifère** (Plante aux fleurs abondantes et à longue floraison offrant aux abeilles nectar et pollen, assurant leur nourriture et la constitution de réserve de miel ; *terme spécifique* ►végétal ligneux mellifère) ; *syn.* plante [f] apicole ; *g* **Bienennährpflanze [f]** (Pflanze mit großer Blüten- zahl und langer Blühdauer, die für Bienen eine umfangreiche Nektar- und Pollennahrungsquelle bietet; *UB* ►Bienennährge- hölz); *syn.* Bienenweidepflanze [f], Trachtpflanze [f].

2766 hot air welding [n] *constr.* (Joining sheets of liner or ►root protection membrane on site by fusion with a hot air gun to prevent moisture penetration; ►waterproofing sheet, ►joint sealing, ►welded joint); *s* **termosoldadura [f]** (Unión, p. ej. de láminas de polivinilo con aire caliente, para impermeabilizar superficies; ►banda de material impermeabilizante, ►junta sol- dada, ►lámina antirraíces, ►sellado de juntas); *syn.* soldadura [f] térmica; *f* **thermosoudure [f]** (Soudure à l'air chaud de lés d'une membrane d'étanchéité en PVC ; ►film [de protection] anti- racines, ►garnissage des lèvres de soudure, ►lé d'étanchéité, ►lèvre de soudure) ; *g* **Verschweißen [n, o. Pl.] der Fügestellen mit Heißluft** (Wasserdichtes Verbinden von ►Dichtungsbahnen, Teichfolien, ►Wurzelschutzfolien etc.; ►Nahtversiegelung, ►Schweißnaht).

2767 hourly billing rate [n] [US] *contr. prof.* (Cost charged per hour for personnel dependent upon qualifications, the inclus- sion of overheads and calculated profit. In D., ►time charges fees are based upon rates governed by the architects'/engineers' con- ditions of engagement and scale of fees [HOAI]; ►daily billing rate, ►hourly wage rate, ►renumeration on a time basis); *syn.* hourly rate [n] [UK]; *s* **tasa [f] de honorarios por hora** (Monto a cobrar por hora dependiendo de la cualificación y la considera- ción de gastos corrientes. En D. los ►honorarios por tiempo empleado se rigen por el reglamento de honorarios para arqui- tectos e ingenieros [HOAI]; ►salario horario, ►tarifa por hora, ►tasa por día); *f 1* **prix [m] horaire** (Prix de revient imputant les dépenses d'un bureau d'études établi pour une mission déter- minée prenant en compte la rémunération du temps passé pour chacun des intervenants y compris les charges sociales, les frais communs, le bénéfice du bureau d'études ; ►honoraire au temps consacré, ►montant journalier, ►prix de revient élémentaire de base [pour main d'œuvre ou matériels] ; ►rémunération horaire) ; *syn.* taux [m] horaire ; *f 2* **prix [m] horaire de vacation** (Rému- nération du temps passé personnellement par l'architecte pour l'exécution de la mission et pour les déplacements correspon- dants. Celle-ci est estimée par référence à un salaire ou un traitement d'un niveau de responsabilités comparable à celui accordé dans l'industrie ou l'administration ; HAC 1989, 446) ; *g* **Stundensatz [m]** (Kalkulierter Geldbetrag je Stunde, abhängig von der Person, vom Gemeinkostenzuschlag und dem bean- spruchten Gewinn. Bei der Vereinbarung eines ►Honorars nach Zeitaufwand gelten die Sätze nach § 6 [2] HOAI 2002. In der HOAI 2009 sind verbindliche Stundensatzvereinbarungen nicht mehr vorgesehen, sie müssen frei vereinbart werden. In F. wird der Stundensatz begrifflich zwischen dem des Angestellten [prix horaire] und dem des Büroinhabers [prix horaire de vacation] unterschieden; ►Stundenlohnsatz 2, ►Stundenvergütung, ►Ta- gessatz); *syn.* Stundenlohnsatz [m] (1), Zeitgrundgebühr [f] [auch A].

hourly rate [n] [UK] *contr. prof.* ►hourly billing rate [US].

hourly rate [n]**, professional fee based on an** *contr. prof.* ►time charges.

hourly wage [n] [US] *constr. contr.* ►hourly wage rate.

2768 hourly wage rate [n] *constr. contr.* (Amount agreed upon for the payment of ►daywork; ►hourly billing rate,

►specifically agreed rate); *syn.* hourly wage [n] [also US], labor cost [n] at an hourly rate [also US], labour cost [n] at a unit rate [also UK]; *s* **tarifa [f] por hora** (Cantidad acordada según la cual se pagan los ►trabajos por salario horario o cada hora de utilización de maquinaria; ►tarifa [de pago] acordada, ►tasa de honorarios por hora); *syn.* remuneración [f] por hora; *f* **prix [m] de revient élémentaire de base (pour main d'œuvre ou maté- riels)** (Prix de revient imputant les dépenses d'une entreprise, établi pour une tâche déterminée prenant en compte les coûts salariaux, l'amortissement, l'entretien et la réparation des véhi- cules, les frais de transport, les frais généraux et la marge béné- ficiaire ; ►prix horaire, ►prix horaire de vacation, ►taux de rémunération, ►travaux exécutés en régie) ; *g* **Stundenlohnsatz [m] (2)** (Vereinbarter Betrag, nach dem ►Stundenlohnarbeiten oder Maschinenstunden abgerechnet werden; bei Architekten- und Ingenieurleistungen spricht man nur vom ►Stundensatz; *OB* ►Verrechnungssatz).

house [n] *arch. urb.* ►courtyard house, ►detached house, ►fac- tory-built house [US], ►individual house, ►patio house, ►row house, ►semidetached house, ►single-family detached house, ►single-family house, ►stepped house; *recr. urb.* ►vacation house [US], ►weekend house [US].

house [n] [US]**, apartment** *arch. urb.* ►multistory building for housing [US].

house [n]**, atrium** *arch.* ►courtyard house.

house [n]**, detached** *arch. urb.* ►single-family detached house.

house [n] [US]**, double** *arch. leg. urb.* ►semidetached house.

house [n]**, out-** *arch.* ►outbuilding.

house [n] [UK]**, pre-assembled** *arch.* ►factory-built house [US].

house [n]**, single** *arch. leg. urb.* ►individual house.

house [n]**, summer-** *gard.* ►gazebo.

house [n]**, terrace** *urb.* ►row house.

house [n]**, terraced** *arch.* ►stepped house.

house [n]**, town** *urb.* ►row house.

house garden [n] *gard. urb.* ►row house garden.

house garden [n] [UK]**, split-level** *arch. constr.* ►terraced townhouse garden [US].

household refuse [n] *envir.* ►household waste.

2769 household waste [n] *envir.* (Discarded, unusable ma- terial from a dwelling; ►waste, ►kitchen waste, ►unauthorized dumped waste); *syn.* domestic refuse [n], domestic waste [n], household refuse [n], garbage [n] [also US]; *s* **residuos [mpl] domésticos** (Todos los restos no aprovechables que se producen en domicilios particulares que deben ser recogidos y eliminados por la empresa pública competente o por una privada encargada de ello; ►basura clandestina, ►residuos sólidos, ►restos de cocina); *syn.* basuras [fpl] sólidas domésticas, desechos [mpl] domésticos [MEX]; *f* **ordures [fpl] ménagères (OM)** (Déchet issu de l'activité domestique des ménages, pris en compte par les collectes usuelles ou séparatives ainsi que les déchets non ménagers collectés dans les mêmes conditions issus des activités professionnelles [déchets produits par les artisans, les com- merçants, bureaux, etc.] assimilables aux ordures des ménages de par leur nature et leur quantité ; ►déchets, ►déchets de cuisine, ►détritus abandonné) ; *g* **Hausmüll [m]** (Alle im Haushalt an- fallenden Abfälle, die von den Entsorgungspflichtigen selbst oder von beauftragten Dritten in genormten, im Entsorgungsgebiet vorgeschriebenen Behältern regelmäßig gesammelt, transportiert und der weiteren Entsorgung zugeführt werden; gem. TA Sied- lungsabfall 1993 und Abfallablagerungsverordnung [AbfAblV]

2001 gibt es auch **hausmüllähnliche Gewerbeabfälle**, die in Gewerbebetrieben, in Geschäften, Dienstleistungsbetrieben, öffentlichen Einrichtungen und Industrie anfallen und nach Art und Menge gemeinsam mit oder wie **H.** entsorgt werden können; ►Abfall 2, ►wilder Müll, ►Küchenabfall); *syn.* häuslicher Abfall [m], häuslicher Müll [m].

2770 household waste collection [n] [US] *envir.* (Gathering of such waste by a municipal refuse department or private waste disposal firm); *syn.* domestic waste collection [n] [UK]; *s* **recogida [f] de residuos sólidos urbanos** (Sistema organizado por una empresa comunal pública o privada para recolectar la basura doméstica en la ciudad); *syn.* recogida [f] de basura (doméstica); *f* **collecte [f] des ordures ménagères** (Ramassage effectué directement sous forme de régie par la commune ou le syndicat de communes, indirectement par une entreprise privée dans le cadre d'un marché de prestation de services [concession]) ; *syn.* ramassage [m] des ordures ménagères ; *g* **Müllsammlung [f]** (Abholen von Hausmüll durch städtischen oder privaten Entsorgungsbetrieb).

house [n] **in a cluster** [US]**, attached single family** *arch. urb.* ►attached dwelling in a cluster.

housing [n] *urb.* ►detached and semidetached housing [US]/ detached and semi-detached housing [UK], ►landscape of proliferated housing, ►low-density housing; *sociol. plan.* ►low-income housing; *urb.* ►multifamily housing, ►poor quality housing, ►slum housing.

housing [n] [UK]**, council** *sociol. plan.* ►low-income housing.

housing [n] [UK]**, publicly assisted** *sociol. plan.* ►low-income housing.

housing areas [npl] *plan.* ►landscape design of housing areas.

housing association [n] *adm. urb.* ►nonprofit housing association [US]/non-profit housing association [UK].

housing block [n] *urb.* ►terraced housing block.

housing density [n]**, reduction in** *urb.* ►wide spacing in a housing development.

housing development [n] *urb.* ►housing subdivision [US], ►new housing development, ►wide spacing in a housing development.

housing estate [n] [UK] *urb.* ►housing subdivision [US].

housing [n] **fit for rehabilitation** *urb.* ►run-down housing fit for rehabilitation.

housing landscape [n] *landsc. urb.* ►landscape planting in built-up areas.

housing site [n]**, multifamily** *arch. urb.* ►residential complex.

housing stock [n] [UK]**, poor quality** *urb.* ►poor quality housing [US].

2771 housing subdivision [n] [US] *urb.* (Group of dwellings in an urban area constructed at the same time on previously undeveloped land, which has been divided into small lots [US]/ small plots [UK]; ►bedroom community [US]/dormitory suburb [UK]); *syn.* housing estate [n] [UK], housing development [n], residential community development [n]; *s* **polígono [m] residencial** (Grupo de edificios de vivienda construidos en el mismo periodo/período en terrenos no utilizados anteriormente; ►ciudad-dormitorio); *f* **ensemble [m] résidentiel** (Groupe important d'habitations collectives présentant un caractère d'unité architecturale souvent situé en zone urbaine ; ►cité dortoir) ; *syn.* ensemble [m] d'habitations, grand ensemble [m], ensemble [m] immobilier ; *g* **Wohnsiedlung [f]** (Siedlung, die ausschließlich

dem Wohnen dient und im selben Zeitraum gebaut wurde; ►Schlafstadt).

housing subdivision [n] [US]**, new** *urb.* ►new housing development.

housing zone [n] [US] *leg. urb.* ►residential land use.

HOV lane [n] *trans.* * ►offside lane.

human activity [n]**, dispersal by** *phyt. zool.* ►anthropogenic dispersal.

2772 human ecology [n] *ecol. sociol.* (Study of interrelationship between humans and the living and nonliving components of the environment, especially the influence of human cultures on biocenosis and biotopes; ►civilization ecology); *s* **ecología [f] humana** (Escuela sociológica que surge en la década de los veinte del siglo XX y se apoya en la obra de los sociólogos de Chicago E. W. BURGESS y R. D. MCKENZIE. La **e. h.** se considera como la ciencia que estudia las relaciones espaciales que mantienen los seres humanos respondiendo en la actuación de un complejo de fuerzas físicas y culturales; cf. DGA 1986; ►ecología de la civilización); *f* **écologie [f] humaine** (Branche de l'écologie étudiant les rapports réciproques entre les êtres humains et le milieu naturel et physique dans lequel ils vivent et se reproduisent, en particulier leur influence sur les biocénoses et les biotopes ; ►écologie des civilisations) ; *g* **Humanökologie [f, o. Pl.]** (Teilbereich der Ökologie, der die Beziehungen und die Wechselwirkungen des Menschen mit seiner belebten und unbelebten Umwelt untersucht, insbesondere auch die Einwirkungen des Menschen auf Biozönosen und Biotope; der Mensch wird nicht länger als Ausnahmespezies betrachtet, der kulturfähig und damit auch außerhalb genetischer Evolution anpassungsfähig ist, und der mehr durch soziale als durch biologische Vorgaben beeinflusst wird. Der Mensch wird somit als eine von vielen Spezies betrachtet, die mit der begrenzten natürlichen Umgebung interagiert. Die Grundlagen dieses Wissenschaftsgebietes wurden in den USA in den 1920er-Jahren besonders durch die Soziologen E. W. BURGESS und R. D. MCKENZIE sowie HOWARD W. ODUM und dem Inder RADHAKAMAL MUKERJEE geschaffen; ►Zivilisationsökologie).

humic location [n] *phyt. zool.* ►mesic site.

2773 humid desulphurization [n] *envir.* (Procedure for the elimination of hydrogen sulphide [H_2S] deriving from gas mixtures by washing out with liquids or by silting up of substances which have oxidizing or neutralizing effects. Most of **h. d.** techniques add calcareous absorbents and create a final product of gipsum which will be dumped [US]/tipped [UK] or processed for the building industry; ►dehydration of sulphur emissions); *s* **desulfuración [f] húmeda** (Proceso de tratamiento de las emisiones para eliminar el azufre contenido en ellas por medio de lavado con líquidos desulfurantes o por tratamiento con sustancias oxidantes o neutralizantes, generalmente absorbentes con cal, por lo que en el proceso se produce yeso que se deposita o se trata para su posterior aprovechamiento en la construcción; ►desulfuración en seco); *f* **désulfuration [f] par voie humide** (Processus de traitement de combustibles gazeux [épuration des gaz de fumée], générateurs d'anhydride sulfureux [H_2S] par utilisation de liquides de lavage ou par le décantage de substances oxydantes ou neutralisantes. La plupart des procédés d'hydro-désulfuration entraîne, par des réactions basées sur le calcium, la formation de plâtre qui, après traitement, est réutilisé par l'industrie du bâtiment ; ►désulfuration par voie sèche) ; *syn.* hydro-désulfuration [f] ; *g* **Nassentschwefelung [f]** (Verfahren zur Entfernung von Schwefelwasserstoff [H_2S] aus Gasgemischen durch Auswaschen mit Flüssigkeiten oder durch Aufschlämmen von Substanzen, die oxidierend oder neutralisierend wirken. Der überwiegende Teil der Nassverfahren erzeugt durch die Zufuhr kalkhaltiger Absorbentien als Endprodukt Gips, der deponiert oder aufbereitet der Bauindustrie zugeführt wird; ▶Trockenentschwefelung).

2774 humification [n] *pedol.* (Microbial breakdown of organic matter in the soil to form humus); *syn.* humus development [n]; *s* **humificación [f]** (Proceso de transformación de la materia orgánica en compuestos húmicos bajo la influencia de los organismos edáficos); *syn.* formación [f] de humus; *f* **humification [f]** (Processus de transformation de la matière organique fraîche en composés humiques sous l'influence des microorganismes du sol) ; *syn.* formation [f] de l'humus ; *g* **Humifizierung [f]** (Vorgang der Umwandlung im Boden von organischer Substanz in Huminstoffe, d. h. in schwach braun bis schwarz gefärbte, postmortale organische Substanzen ohne reproduzierbare chemische Struktur. Es wird in Laugen und Säuren lösliche Fulvosäuren [Fulvinsäuren], in Laugen lösliche und in Säuren ausfallende Huminsäuren und in Laugen und Säuren nicht lösliche Humine unterschieden); *syn.* Huminstoffbildung [f], Humusbildung [f].

2775 humified raised bog peat [n] *pedol.* (Heavily decomposed *Sphagnum*-peat which may contain Alder *[Alnus]* and Birch *[Betula]* in which case it is called 'humified wood-peat', or may be predominantly composed of Bog-Cotton *[Eriophorum]* and is then named 'humified cyperaceous peat'; ▶fibric peat 1); *syn.* sapric peat [n], Michigan peat [n] [also US]; *s* **turba [f] negra** (Turba de esfagno fuertemente descompuesta de la capa inferior de las turberas altas en la que se pueden reconocer restos de leñosas como aliso *[Alnus]* y abedul *[Betula]*, o de erióforo *[Eriophorum]*; ▶turba blanca); *f* **tourbe [f] noire** (Tourbe moyennement à fortement décomposée de la couche inférieure de la tourbière à Sphaignes, de couleur brun-noir laissant apparaître des résidus de végétaux ligneux tels que l'Aulne *[Alnus]* et le Bouleau *[Betula]* ainsi que la Linaigrette *[Eriophorum]* ; ▶tourbe blanche, ▶tourbe fibreuse) ; *syn.* tourbe [f] altérée ; *g* **Schwarztorf [m]** (Mittel bis stark humifizierter, braunschwarz gefärbter Hochmoortorf mit erkennbaren Pflanzenresten von Gehölzen wie Erle *[Alnus]* und Birke *[Betula]* oder Wollgras *[Eriophorum]*; ▶Weißtorf 1); *syn.* stark zersetzter Hochmoortorf [m].

2776 hummock [n] *phyt.* (**1.** Pillow-shaped elevation, usually in ombrotrophic bogs, formed of *Sphagnum* species, additionally with cotton grass *[Eriophorum]* or woodrush *[Luzula]* species; in dryer parts with dwarf shrubs; ▶tussock. **2. hammock** [n] **a]** in southern U.S., a small, fertile, slightly raised area, mound, knoll, or hillock of deep, humus-rich soil which is characterized by hardwood vegetation in a plain or swamp; **b]** an island of dense, tropical undergrowth in the Everglades of Florida; cf. WEB 1993); *syn. for* **2.** hummock [n]; *s* **mamelón [m]** (En turberas, pequeña elevación abombada constituida por esfagnos, según su altura con respecto al nivel de agua de la turbera puede estar poblada por *Eriophorum*, *Scirpus* o por arbustos enanos; ▶macolla); *f* **butte [f]** (*Tourbière* petite élévation bombée constituée de Sphaignes. Suivant sa position par rapport au niveau d'eau propre à la tourbière la butte peut être peuplée par surcroît de touffes de Linaigrettes ou de *Luzula* ou encore d'arbustes nains ; ▶touffe, ▶touradon) ; *syn.* bombement [m] ; *g* **Bult [m]** (Höherer, kissenförmiger, aus Sphagnen gebildeter Buckel im Moor. Je nach seiner Lage zum mooreigenen Wasserspiegel wird er zusätzlich entweder von horstartig wachsenden Wollgräsern oder Simsen oder aber von Zwergsträuchern besiedelt; ▶Horst 1).

2777 hummock-forming peat moss [n] *bot.* (Mostly oligotrophic species of *Sphagnum [Chamaephyta sphagnoïdea]* with unlimited growth; ▶hummock); *s* **esfagno [m] del mamelón** (Especies oligótrofas de Sphagnum *[Chamaephyta sphagnoïdea]* que crecen en medio ácido y forman el ▶mamelón); *syn.* musgo [m] esfagnoide de turbera; *f* **sphaigne [f] de buttes** (Mousse des tourbières, au développement illimité, formant des buttes ; ces espèces de Sphaignes *[Chamaephyta sphagnoïdea]* croissent principalement en milieu acide ; ▶butte) ; *syn.* mousse [f] de buttes ; *g* **Bultmoos [n]** (Bultbildendes Sumpfmoos *[Chamaephyta sphagnoïdea]* mit unbegrenztem Wachstum, hauptsächlich zu den oligotrophen *Sphagnum*-Arten gehörig; ▶Bult).

2778 hummock-hollow complex [n] *phyt.* (Alternating ▶hummocks and pools [▶bog hollow] of a growing raised bog surface); *s* **comunidad [f] de prominencias y cavidades (de la turbera)** (Combinación de ▶mamelones y ▶pocinas de turbera alternantes); *syn.* ciclo [m] «hummock and hollow»; *f* **complexe [m] de buttes et de dépressions** (Alternance de ▶buttes et de dépressions [▶gouille (d'une tourbière)] sur une tourbière naturelle ou régénérée) ; *g* **Bult-Schlenken-Komplex [m]** (Das Nebeneinander von ▶Bulten und ▶Schlenken in natürlichen oder regenerierten Hochmooren).

humose soil [n] [UK] *pedol.* ▶humosic soil [US].

2779 humosic soil [n] [US] *pedol.* (Soil with high ▶humus content); *syn.* humose soil [n] [US], humose soil [n] [UK] (LD 1986, 71); *s* **suelo [m] humífero** (Suelo que tiene gran cantidad de materia orgánica en completa descomposición, con considerable cantidad de suelo mineral y con algunos residuos fibrosos; MEX 1983; ▶humus); *f* **sol [m] humifère** (Sol riche en ▶humus résultant de la décomposition de la matière organique d'origine végétale, caractérisé par sa couleur noire ou très sombre et contenant beaucoup de débris végétaux non décomposés) ; *g* **humoser Boden [m]** (Boden mit hohem Anteil von in Zersetzung befindlichen organischen Stoffen; ▶Humus); *syn.* Humusboden [m], humushaltiger Boden [m].

hump [n] *trans.* ▶speed hump.

2780 humus [n] *pedol.* (Organic material in the process of decomposition of plant and animal tissues in and on the soil, and the soil biomass; an essential basis for supplying nutrient to plants; ▶moder, ▶mor humus, ▶mull, ▶raw humus); *syn.* soil organic matter [n]; *s* **humus [m]** (Materia resultante de la descomposición y síntesis química a partir de los restos orgánicos que recibe el suelo y que se incorporan a él. Es de composición muy compleja; DGA 1986; ▶humus bruto, ▶moder, ▶mor, ▶mull); *f* **humus [m]** (Ensemble des produits d'altération ou en voie d'altération de la matière organique du sol ; DIS 1986 ; ▶humus brut, ▶moder, ▶mor, ▶mull) ; *g* **Humus [m, o. Pl.]** (Gesamtheit der in Umwandlung begriffenen organischen Substanz auf und im Boden; eine wesentliche Grundlage für die Nährstoffversorgung der Pflanzen, zur Verbbesserung der Bodenstruktur und zur Aktivierung des Bodenlebens; ▶Auflagehumus, ▶Moder, ▶Mull, ▶Rohhumus).

humus [n]**, bark** *constr. hort.* ▶shredded bark humus.

humus [n]**, mild** *pedol.* ▶mull.

2781 humus content [n] *pedol. s* **contenido [m] en humus**; *f* **teneur [f] en humus** *syn.* taux [m] humique (PED 1984, 209) ; *g* **Humusgehalt [m].**

humus decomposer [n] *pedol. phyt.* ▶raw humus decomposer.

humus development [n] *pedol.* ▶humification.

2782 humus fertilizer [n] *agr. constr. hort.* (Organic matter for soil improvement; e.g. peat, peat compost, sewage sludge, organic household compost, earth compost, and green manure; such forms of organic matter also increase the amount of organic matter in the soil, the water-holding capacity and availability of water for plants, extend the range between plasic limit and shrinking limit of soil, change soil pH reaction, and enhance the activity of soil organisms; ▶soil amendment [US]/soil ameliorant

[UK], ►soil improvement, ►fertilizer, ►organic fertilizer, ►refuse compost); *s* **abono [m] húmico** (Sustancias orgánicas como abono de turba, compost de turba, cieno de depuradora, compost de basura, compost de tierra o abono verde —clasificadas en D. según DIN 18 915 y denominadas ►materiales de enmienda de suelos— utilizadas para aumentar el contenido de sustancia orgánica en el suelo, mejorar la capacidad de almacenamiento de agua y la disponibilidad de la misma para las plantas, cambiar el pH del suelo y promover la actividad de los microorganismos edáficos; ►compost de residuos, ►enmienda del suelo, ►fertilizante); *f* **amendement [m] humifère** (Apports de produits organiques tels que la tourbe, les boues d'épuration, le compost de tourbe le compost issu de la fermentation de déchets ménagers organiques, le compost végétal, le fumier décomposé et lisier, les engrais verts en vue de l'amélioration des propriétés physiques et chimiques et biologiques du sol ; ►compost de déchets ménagers, ►correction de la terre, ►engrais, ►engrais organique) ; *syn.* amendement [m] organique, amendement [m] humique) ; *g* **Humusdünger [m]** (Organische Stoffe zur ►Bodenverbesserung 2, die geeignet sind, den Gehalt an organischer Substanz zu erhöhen, die Wasserspeicherfähigkeit und die Wasserverfügbarkeit für die Pflanzen zu erhöhen, den Bereich zwischen Ausroll- und Schrumpfgrenze zu erweitern, die Bodenreaktion zu verändern und die Mikroorganismentätigkeit zu fördern. Dazu gehören Dünger auf Torf-, Torfkompost-, Klärschlamm-, ►Müllkompost- [Kompost aus organischen Hausabfällen], Erdkompost- [Kompost aus Laub, Mäh- und Schnittgut, Grasnarbe, Rinde (Rindenhumus) o. Ä.] oder Gründüngungsbasis [Voranbau oder Zwischenbegrünung], die nach der DIN 18 915 als Stoffe zur Bodenverbesserung [►Bodenverbesserungsstoffe 1] bezeichnet werden; ►Dünger, ►organischer Dünger).

humus-forming plant [n]**, raw** *pedol. phyt.* ►raw humus decomposer.

2783 humus type [n] *pedol.* (Group of soil horizons located at or near the surface of the ground, which have formed from organic residues, either separate from or intermixed with mineral material; e.g. ►raw humus, ►moder, ►mull, mostly under aerobic conditions and hydromorphic forms, e.g., ►dy, ►gyttja, ►sphagnum peat, ►sapropel, frequently or totally existing in anaerobic conditions; cf. SST 1997); *s* **tipo [m] de humus** (Diferenciación de los diferentes grupos de humus de los suelos, incluyendo todos los horizontes orgánicos de los mismos. Bajo condiciones aerobias se forman el ►humus bruto, ►moder y ►mull; bajo condiciones anaerobias se forman, el ►dy, ►gyttja y ►sapropel; ►turba de esfagnos); *f* **type [m] d'humus** (Ensemble des horizons humifères ; pour des facteurs écologiques donnés on distingue les types d'humus forestiers, ►humus brut, ►moder, ►mull formés en milieu aéré et les types d'humus formés en station hydromorphe ; ►dy, ►gyttja, ►sapropèle, ►tourbe de Sphaigne) ; *g* **Humusform [f]** (Gesamtheit der Humushorizonte in einem Bildungsmilieu. Man unterscheidet terrestrische Formen, z. B. ►Rohhumus, ►Moder, ►Mull, die vorwiegend unter aeroben Bedingungen und hydromorphe Formen, z. B. ►Hochmoortorf, ►Dy, ►Gyttja, ►Sapropel, die unter zeitweilig bis ständig anaeroben Bedingungen entstehen; modif. nach SS 1979); *syn.* Humustyp [m].

2784 hundred-year flood [n] *hydr.* (Statistical likelihood of a maximum flood in a century, used as the basis in calculating ►channel discharge cross-sections of watercourses); *syn.* 100-year flood [n]; *s* **crecida [f] máxima del siglo** (Altura máxima del nivel de las aguas de un curso de agua que ha sido medida. Se utiliza de base de cálculo para la ►capacidad máxima de desagüe); *f* **crue [f] centennale** (Crue d'ampleur et de fréquence exceptionnelle souvent prise en compte dans le calcul des sections du chenal d'écoulement ou lors de la construction

d'ouvrages routiers en bordure de cours d'eau ; ►section [transversale totale] d'écoulement) ; *syn.* crue [f] du siècle, crue [f] centenaire ; *g* **Jahrhunderthochwasser [n]** (…wasser [pl]. Höchster Hochwasserstand, der für ein Fließgewässer gemessen wurde und als Berechnungsgrundlage für ►Abflussquerschnitte dient).

2785 huntable game [n] *leg. hunt.* (Any species of wildlife for which seasons and bag limits have been prescribed; concessions are normally subject to state laws and regulations; cf. WPG 1976; ►game); *syn.* fair game [n], game species [n], legal game [n]; *s* **especie [f] cinegética** (Especie definida por ley como objeto de ►caza, cuya declaración es responsabilidad de las CC.AA. y que en ningún caso podrá afectar a las especies incluidas en el Listado de Especies en Régimen de Protección Especial o a las prohibidas por la UE; art. 62.1 Ley 42/2007; en España existe también la categoría de «especies objeto de pesca»); *syn.* especie [f] objeto de caza; *f* **espèce [f] chassable** (Animal défini comme tel par la loi, et généralement apprécié pour sa chair, sa fourrure, ses plumes, pour l'amour du sport, ou bien pour son trophée ; DFM 1975. L'anglais « game » inclut aussi l'espèce des poissons ; ►gibier) ; *syn.* espèce [f] cynégétique, gibier [m] chassable, gibier [m] dont la chasse est autorisée ; (LA 1981, 278) ; *g* **jagdbares Tier [n]** (Tierart, die dem Jagdrecht unterliegt; cf. § 2 BJagdG. Hierzu gehören auch Arten, die gem. Artenschutzverordnung nicht gejagt werden dürfen. Nach den zeitlichen Beschränkungen der Jagdausübung wird zwischen ‚ständig jagdbaren Wildarten', ‚zeitweilig jagdbaren Wildarten' und ‚Wildarten ohne Jagdzeit' unterschieden. Der englische Begriff *game* beinhaltet auch Fische; ►Wild); *syn.* jagdbares Wild [n].

hunting [n] *hunt.* ►game hunting; *game'man. hunt.* ►overhunting [US]/over-hunting [UK]; *recr. hunt.* ►recreational hunting.

hunting area [n] *hunt.* ►in an open hunting area.

2786 hunting district [n] *hunt. leg.* (Land which may be hunted upon in accordance with established traditions and hunting laws); *s 1* **terreno [m] cinegético de aprovechamiento común** (En estos terrenos el ejercicio de la caza puede practicarse sin más limitaciones que las generales fijadas en la Ley de Caza y su Reglamento; cf. art. 9, Ley de Caza 1/70, 4 de Abril. NOTA: La ley de caza solo rige en las CC.AA. en las que no se haya aprobado una ley propia); *s 2* **terreno [m] sometido a régimen de caza controlado** (Aquéllos que se constituyan únicamente sobre terrenos cinegéticos de aprovechamiento común, en los cuales la protección, conservación, fomento y aprovechamiento de su riqueza cinegética debieran adaptarse a los planes que con este objeto apruebe el Ministerio de Agricultura; cf. art. 14, Ley de Caza); *s 3* **coto [m] de caza** (Toda superficie continua de terrenos susceptible de aprovechamiento cinegético que haya sido declarada y reconocida como tal, mediante resolución del Servicio de Pesca Continental, Caza y Parques Nacionales; cf. art. 15, Ley de Caza); *f* **domaine [m] de chasse** (Terres sur lesquelles conformément à la législation en vigueur l'exercice de la chasse est autorisé) ; *g* **Jagdbezirk [m]** (Fläche, auf der, nach ihrem Umfang und ihrer Gestalt, die Jagd nach den Erfordernissen der Jagdpflege und Jagdausübung möglich ist; cf. §§ 4 und 5 BJagdG. Die Bayern benutzen im Bayerischen Jagdgesetz abweichend vom BJagdG den Begriff 'Jagdrevier'); *syn.* Jagdrevier [n] [BY u. A].

hunting forest [n] [UK] *for. hist.* ►royal hunting forest.

hunting ground [n] [UK] *hunt.* ►game preserve [US].

2787 hunting laws [npl] *hunt. leg.* (**1.** Objective sense: all laws and regulations related to hunting; **in U.S.,** hunting legislation is done by the states, except for federally-owned land); **2. hunting right** [n] (Subjective sense: individual right given by

the owner of a hunting area to a hunter according to hunting laws; e.g. hunting management, shooting game and taking the meet on a defined area); **s derecho [m] cinegético** (Legislación sobre la caza; en Es. la caza es competencia exclusiva de las Comunidades Autónomas y se rige entre otras por la Ley 37/1966, de 31 de mayo, sobre la creación de reservas nacionales de caza; Ley 1/1970, de 4 de abril; de Caza; Convención de 23 de junio de 1979 sobre la conservación de especies migratorias de animales silvestres, ratificada por Instrumento de 22 de enero de 1985 y diversas leyes a nivel de las CC.AA.); *syn.* legislación [f] de la caza; ***f 1* droit [m] de la chasse** (Ensemble des lois et règlements [regroupés dans le C. rural] régissant les activités de la chasse) ; ***f 2* droit [m] de chasse** (Droit mobilier appartenant au propriétaire d'un domaine de chasse) ; ***f 3* droit [m] de chasser** (Droit personnel que possède les fermiers et métayers sur les terres qu'ils ont en location) ; *syn.* droit [m] cynégétique ; **g Jagdrecht [n, o. Pl.]** (**1.** Alle sich auf die Jagd beziehenden Gesetze und Vorschriften; in D. das Bundesjagdgesetz [BJagdG] und die Jagdgesetze der Länder [objektives **J.**]; durch die Föderalismusreform 2006 haben die Länder das Recht erhalten, vom geltenden BJagdG [Rahmengesetz] weitgehend abzuweichen [konkurrierende Gesetzgebung]. In der CH wird das **J.** durch das Bundesgesetz über die Jagd und zum Schutz wild lebender Säugetiere und Vögel vom 20.06.1986 [JSG] und durch kantonale Jagdgesetze geregelt. **2.** Die Befugnis zur Wildhege, Jagdausübung und die Aneignung jagdbarer Tiere auf einem bestimmten Gebiet im Rahmen der zu **1.** genannten Gesetze und Vorschriften [subjektives **J.**].

2788 hunting management [n] *hunt.* (Measures for hunting grounds concerned with the more important game species and the management of their habitats for the purpose of hunting. These practices do not usually create biological diversity within these communities and habitats as they would do, if the diversity principles of nature conservation are respected; ▶game management, ▶huntable game, ▶integrated species conservation; ▶overprotection [US]/over-protection [UK], ▶wildlife management); **s gestión [f] de la caza (2)** (Medidas de protección y aprovechamiento de las especies cinegéticas y sus hábitats, sin considerar conscientemente la diversidad de especies en el sentido de la protección de la naturaleza; ▶conservación de la fauna silvestre, ▶gestión de la caza 1, ▶protección integral de flora y fauna); **f gestion [f] cynégétique** (Mesures exclusivement destinées à favoriser la protection des espèces remarquables du point de vue cynégétique [▶espèce chassable] ; l'action coordonnée, de la part ou pour le compte de chasseurs, s'exerce sur une partie des espèces sauvages d'un territoire comportant p. ex. l'aménagement du territoire pour favoriser une espèce, la distribution d'aliments pour le grand gibier sauvage et le gibier d'eau [agrainage et affouragement], l'apport de sel, la mise à disposition de cultures destinées au gibier, les lâcher de gibier et surtout le choix raisonné des prélèvements en nombre et en qualité [âge et sexe des animaux] ainsi que des introductions éventuelles [repeuplement] ; pour favoriser le développement de la faune sauvage, les **a**ssociations **c**ommunales de **c**hasse **a**grées [A.C.C.A] doivent mettre au moins 10 % de leurs territoires en réserve de chasse ; tout acte de chasse y est en principe interdit, des mesures complémentaires en faveur de la faune y sont prises. La constitution d'une diversité biologique dans les communautés d'espèces et les habitats selon le principe de la diversité écologique appliqué dans le domaine de la protection de la nature ne constitue pas l'objectif recherché ; ▶conservation et gestion durable de la faune sauvage, ▶conservation intégrée des espèces animales et végétales, ▶gestion de la faune sauvage, ▶gestion et conservation de la faune sauvage et de ses habitats, ▶réglementation excessive de la chasse) ; *syn.* gestion [f] du gibier ; **g jagdliche Hege [f]** (Maß-

nahmen, die einzig auf den Schutz und die Pflege von jagdlich bedeutsamen Arten [▶jagdbare Tiere] oder deren Lebensräume ausgerichtet sind und deshalb i. d. R. nicht der Schaffung einer biologischen Vielfalt in Lebensgemeinschaften und Biotopen im Sinne des ökologischen Vielfaltsprinzips des Naturschutzes dienen; ▶integrierter Artenschutz, ▶Überhege, ▶Wildlife Management, ▶Wildbewirtschaftung); *syn.* Wildhege [f].

hunting right [n] *hunt. leg.* ▶hunting laws.

2789 hunting season [n] *hunt. leg.* (Period of year, established by hunting regulations, in which game may be hunted; ▶huntable game; *opp.* ▶closed season); *syn.* shooting season [n], open season [n]; **s periodo/período [m] de caza** (Temporada del año en la que está permitida la caza que varía según las especies; en el caso de las aves excluye la época de celo, reproducción y crianza, así como el periodo/período durante el cual recorren el trayecto hacia los lugares de cría en el caso de las especies migratorias; art. 62.3 b Ley 42/2007; ▶especie cinegética, ▶veda); *syn.* periodo/período [m] no vedado a la caza; **f période [f] d'ouverture de la chasse** (Temps réglementé dans chaque département durant l'année pendant lequel la chasse est permise [▶espèce chassable] ; ▶période de fermeture de la chasse) ; *syn.* temps [m] d'ouverture de la chasse ; **g Jagdzeit [f]** (Durch das Jagdrecht festgelegte Zeiten innerhalb eines Jahres, in denen die Jagd auf Wild [▶jagdbare Tiere] ausgeübt werden darf; cf. § 22 BJagdG; die einzelnen Bundesländer können Sondervorschriften mit z. T. strengeren Bestimmungen erlassen oder aus besonderen Gründen [Wildseuchenbekämpfung, Vermeidung übermäßiger Wildschäden, für wissenschaftliche Zwecke oder bei Störung des biologischen Gleichgewichts] sogar die Schonzeiten für bestimmte Jagdbezirke aufheben; *opp.* ▶Schonzeit).

husbandry [n]**, livestock** *agr.* ▶livestock industry.

hut [n] *recr.* ▶alpine hut.

hut [n] **[UK], allotment garden** *urb.* ▶community garden shelter [US].

hut [n]**, mountain** *recr.* ▶mountain cabin [US].

hut [n] **[US], trail** *recr.* ▶mountain lodge [US].

2790 hybrid tea [n] *hort. plant.* (Hybrid rose, 80-180cm tall, which produces large flowers, usually one on a long stem. One of the best-known **h. t.s** is the yellow-flowering and very robust 'Peace', bred by Francis Meilland and known in France as 'Mme. A. Meilland', in D. as 'Gloria Dei'. Introduced in 1945, it is the most-planted **h. t.** of all time; **in U.S.**, but not in U.K., there is also a 'grandiflora' class of tall-growing roses, producing large flowers, often in clusters, of a size between the **h. t.** and floribunda; e.g. Queen Elizabeth, Aquarius, Montezuma; cf. YOU 1974; ▶tea rose); *syn.* hybrid tea rose [n]; **s rosa [f] híbrido de té** (Tipo de rosal de altura variable entre 80 y 180 cm, en general uniflora con ramas largas; en los EE.UU., no en GB, también se diferencia una clase de rosas 'grandiflora' de crecimiento alto que producen flores grandes, a menudo en pequeños grupos, y que se encuentra entre la **r. h. de t.** y la floribunda; ▶rosa de té); **f rosier [m] hybride de thé** (Arbuste érigé caractérisé par des tiges robustes atteignant 3 à 1,80 m de haut, produisant des boutons allongés donnant naissance à des grosses fleurs doubles de 20 à 50 pétales, légèrement parfumées, la plupart du temps solitaires au bout d'une tige et par une floraison abondante, sans interruption ; les **r. h. de t.** exigent une bonne protection hivernale et beaucoup d'entretien ; JEAN-BAPTISTE GUILLOT créa en 1867 avec « La France » le premier groupe de roses modernes, appelés hybrides de Thé, à partir du croisement des ▶rosiers Thé ou rosiers à odeur de thé [*Rosa odorata*, rosier Bourbon ou rosier Noisette] et des hybrides remontants [rosier thé *Rosa gallica* ou *Rosa centifolia* ou *Rosa damascena*] ; « Mme A. Meilland » est le plus connu des **r. h. de t.** et la variété qui depuis son obtention

par Meilland en 1945, a été probablement la plus vendue de tous les temps ; il existe aussi le groupe des rosiers Grandiflora qui sont le résultat de croisements entre les hybrides de thé et les rosiers Floribunda [rosiers « Queen Elizabeth », « Aquarius », « Montezuma »] ; rosiers de grande taille, ils portent plusieurs grosses fleurs semi doubles et doubles au bout de chaque tige) ; *syn.* hybride [m] de thé; *g* **Edelrose** [f] (Straff aufrechte Rosenwuchsform, 80-180 m hoch, mit meist einzeln [resp. zu dritt bis fünft] stehenden Blüten an langen Stielen; 1867 schuf JEAN-BAPTISTE GUILLOT mit der Sorte 'La France' den Grundstein für die **E.n** und gleichzeitig den Wendepunkt zwischen ►Historischen Rosen und modernen Rosen, indem er die ►Teerosen [*Rosa odorata*, Bourbon-Rose *(Rosa x borboniana)* und Noisette-Rose *(R. indica noisettiana)*] mit remontierenden Rosen wie *Rosa gallica, R. centifolia* oder *R. damascena* kreuzte; zu den bekanntesten **E.n** zählt die gelblich blühende und sehr robuste 'Gloria Dei' [in F. als „Mme A. Meilland", in englischsprachigen Ländern als „Peace" bekannt], von FRANCIS MEILLAND in Frankreich gezüchtet, die 1945 in D. eingeführt wurde und als die meist gepflanzte **E.** aller Zeiten gilt; in den USA, nicht im UK, gibt es auch die Bezeichnung „Grandiflora-Gruppe" für hoch wachsende Rosen mit großen, oft in Büscheln ausgebildeten Blüten, deren Größe zwischen Teehybride und Floribundarose liegt, z. B. Queen Elizabeth, Aquarius, Montezuma); *syn.* Teehybride [f].

hybrid tea rose [n] *hort. plant.* ►hybrid tea.

2791 hydrant [n] **(1)** *urb.* (Tap/connection point to which a hose can be attached for drawing water from a water main in a public water supply network; ►fire hydrant, ►hydrant 2, ►underground hydrant [US]/underground hose bib [UK]); *s* **boca** [f] **de agua** (Lugar de toma de agua de la red de abastecimiento público. Se diferencia entre la ►boca de agua superficial y la ►boca de agua subterránea; ►boca de incendios); *syn.* hidrante [m], boca [f] de riego e incendios, grifo [m] [PE], pie [m] de agua [YV] (BU 1959); *f* **bouche** [f] **d'eau** (Prise de prélèvement d'eau sur un réseau [public] de distribution d'eau ; on distingue la ►borne d'alimentation en eau et la ►bouche d'alimentation en eau ; VRD 1986, 222 ; ►bouche d'incendie) ; *g* **Hydrant** [m] (Zapfstelle zur Wasserentnahme aus einem [öffentlichen] Versorgungsnetz. Man unterscheidet im Straßenraum zwischen ►Überflurhydranten und ►Unterflurhydranten; ►Hydrant zur Löschwasserentnahme).

2792 hydrant [n] **(2)** *constr. urb.* (Discharge connection to a water main, usually consisting of an upright pipe having one or more hose bibs and controlled by a gate valve; MEA 1985; ►hydrant 1, ►underground hydrant [US]/underground hose bib [UK]); *s* **boca** [f] **de agua superficial** (Tipo de ►boca de agua que sobresale a la superficie y que está equipada con conexiones para mangueras y con válvulas; ►boca de agua subterránea); *syn.* hidrante [m] superficial, boca [f] de riego e incendios superficial; *f* **borne** [f] **d'alimentation en eau** (Prise d'alimentation en saillie hors du sol placée sur un massif en béton en retrait de la bordure de la voirie et utilisée dans la lutte contre l'incendie; *terme spécifique* poteau d'incendie; ►bouche d'alimentation en eau, ►bouche d'eau ; VRD 1986, 222) ; *syn.* poteau [m] d'alimentation en eau ; *g* **Überflurhydrant** [m] (Über Grund stehender ►Hydrant mit Schlauchanschlüssen und Ventilen; *opp.* ►Unterflurhydrant).

2793 hydrant [n] **for street cleaning** *urb.* *s* **boca** [f] **de riego**; *f* **bouche** [f] **d'arrosage** (Prise d'eau destinée au lavage de la chaussée par les Services de la Voirie) ; *g* **Hydrant** [m] **zur Wasserentnahme für die Straßenreinigung**.

2794 hydraulic conductivity [n] *pedol.* (**1.** A soil property that describes the capacity of soil pores to permit water

movement as governed by soil type, porosity and pore configuration in saturated flow. **2.** Combined property of a porous medium and the fluid moving through it in saturated flow, which determines the relationship between the specific discharge and the head gradient causing it; cf. WMO 1974, 0155; ►water permeability); *s* **conductividad** [f] **hidráulica** (WMO 1974, 0155; propiedad combinada de un medio poroso y de un fluído moviéndose a través de él con flujo saturado, que determina la relación llamada Ley de Darcy entre el caudal específico y el gradiente de carga que lo origina; WMO 1974; ►permeabilidad [para el agua]); *f 1* **conductivité** [f] **hydraulique** (PED 1979, 291 ; propriété d'un milieu poreux, combinée a celle du fluide s'écoulant dans ce milieu à l'état de saturation, qui détermine la relation entre le débit spécifique et le gradient hydraulique qui provoque le mouvement ; cf. WMO 1974, 0155 ; ►perméabilité [à l'eau]) ; *f 2* **drainage** [m] **interne** (Désigne l'élimination naturelle de l'excès d'eau par percolation à travers le sol et dépendant de la perméabilité du sol, la saturation de la nappe et la perméabilité des couches sous-jacentes. En pédologie on parle de 'drainage climatique' qui est une valeur théorique dépendant des précipitations, de l'évapotranspiration et des réserves d'eau retenue dans le sol) ; *syn.* drainage [m] naturel ; *g* **Wasserleitfähigkeit** [f] (Fähigkeit eines Bodens oder einer Gesteinsschicht in einer definierten Zeit eine bestimmte Wassermenge in tiefere Schichten sickern [perkolieren] zu lassen; ►Wasserdurchlässigkeit).

2795 hydraulic engineering [n] *constr. eng.* (Science and art of planning, constructing and managing waterworks and all facilities for water supply, irrigation, drainage, flood and waterway control, etc., sometimes employing hydraulic machinery); *s* **ingeniería** [f] **hidráulica** (Ciencia y técnica de planificación, construcción y mantenimiento de medidas e instalaciones de gestión del agua para fines de suministro, regado, drenaje, navegabilidad de cursos de agua y control de crecidas, etc.); *f* **génie** [m] **hydraulique** (Science et techniques qui traitent de l'étude, la planification, l'aménagement, la réalisation et l'entretien des opérations et des ouvrages hydrauliques tels que les systèmes d'arrosage et d'irrigation, d'assainissement et de drainage, la protection contre les marées et les crues, la régulation des cours d'eau et la construction des voies navigables) ; *g* **Wasserbau** [m] (Wissenschaft und Technik, die sich mit der Planung, Durchführung und Unterhaltung wasserbaulicher Maßnahmen und Anlagen wie z. B. Be- und Entwässerungssysteme, Hochwasserschutz, Gewässerregulierung und Wasserstraßenbau befasst).

2796 hydraulic engineering [n] **for agriculture** *agr.* (Measures taken to provide and maintain the water supply in agricultural areas by means of irrigation, drainage, and flood control as well as ►land reclamation 1; ►hydraulic engineering); *s* **ingeniería** [f] **hidráulica para la agricultura** (Medidas para regular duraderamente el régimen hídrico en tierras agrícolas por medio de la irrigación, el drenaje y la protección contra crecidas, y —en los países europeos ribereños del Mar del Norte— para ganar tierras al mar; ►ganancia de tierras 1, ►ingeniería hidráulica); *syn.* obras [fpl] hidráulicas para la agricultura; *f* **génie** [m] **hydraulique agricole** (Mesures de régulation du régime des eaux sur les surfaces agricoles au moyen de l'arrosage, l'assainissement, la protection contre les crues et la ►poldérisation ; ►génie hydraulique) ; *g* **landwirtschaftlicher Wasserbau** [m] (Maßnahmen zur nachhaltigen Regelung des Wasserhaushaltes in landwirtschaftlich genutzten Flächen durch Bewässerung, Entwässerung, Hochwasserschutz sowie zur ►Landgewinnung; ►Wasserbau).

hydraulic excavation [n] *min.* ►dredging (1).

2797 hydraulic filling [n] *constr.* (Technical method of soil disposal in low lying areas by pumping in waterborne

suspensions of solid material); *s* **relleno [m] hidráulico** (Método técnico de relleno de superficies con materia sólida en suspensión por medio de tubería a presión); *syn.* refulado [m]; *f* **remblayage [m] hydraulique** (Méthode technique de remplissage d'un bas-fond par introduction de matériaux fins au moyen de conduites forées [injection sous pression] ; *syn.* remblaiement [m] hydraulique ; *g* **Auflandung [f] (2)** (Technische Bodenablagerungsmethode, bei der Erdmassen mit Druckwasserleitungen in tiefer liegende oder eigens dafür vorbereitete Flächen gespült werden); *syn.* Aufspülung [f].

2798 hydraulic gradient [n] *agr. constr. hydr.* (LAD 1986, 100; run-off made possible by a natural slope or an artificially created elevation of the ground); *s* **gradiente [m] hidráulico** (En un conducto cerrado, pendiente piezométrica; en canales abiertos o cursos de agua, pendiente de la superficie del agua, que permite al agua fluir hacia abajo; cf. WMO 1974); *f 1* **pente [f] d'écoulement** (Possibilité d'écoulement naturel ou d'évacuation artificielle des eaux superficielles par l'aménagement d'une pente générale ou dans certains cas par le relèvement des eaux) ; *f 2* **gradient [m] hydraulique** (Quotient de la différence de charge hydraulique entre deux points d'un milieu poreux saturé, sur une même ligne de courant, par la distance les séparant sur cette ligne de courant) ; *g* **Vorflut [f] (2)** (Möglichkeit des Wassers, mit natürlichem Gefälle oder durch künstliche Hebung abzufließen — „natürliche" oder „künstliche" Vorflut; DIN 4049, 4.01).

hydraulic seeding [n] *constr.* ▶hydroseeding.

hydric forest [n] *geo. phyt.* ▶swamp forest.

hydric soils [npl] *pedol.* ▶hydromorphic soils.

2799 hydrobiology [n] *biol.* (Field of biology which encompasses aquatic plants and animals and their interrelationships with biotic and abiotic environmental factors; h. includes ▶limnology and ▶marine biology); *s* **hidrobiología [f]** (Ciencia que se ocupa de los seres vivos que se desarrollan en medios acuáticos. Se subdivide en ▶limnología y ▶biología marina); *f* **hydrobiologie [f]** (Science qui étudie la vie des organismes végétaux [hydrobotanique] et animaux [hydrozoologie] qui vivent dans l'eau, ainsi que les rapports de ces organismes avec le milieu. L'**h.** comprend la ▶limnologie et la ▶biologie marine) ; *g* **Hydrobiologie [f, o. Pl.]** (Teilgebiet der Biologie, das als Wissenschaft die im Wasser lebenden Pflanzen [Hydrobotanik] und Tiere [Hydrozoologie] und deren Wechselbeziehungen mit den sie umgebenden biotischen und abiotischen Umweltfaktoren untersucht. Die **H.** gliedert sich in ▶Limnologie und ▶Meeresbiologie).

hydrochory [n] *phyt.* ▶water dispersal.

2800 hydroelectric power station [n] *envir.* (Major facility which produces electric power from the potential energy of water; ▶dam); *s* **central [f] hidroeléctrica** (Planta de producción de electricidad sobre la base de la energía potencial del agua; *término específico* minicentral hidroeléctrica [C]; ▶presa); *syn.* planta [f] hidroeléctrica; *f* **centrale [f] hydroélectrique** (Ouvrage hydraulique permettant de disposer de l'énergie potentielle de l'eau des marées, des lacs et des cours d'eau pour la transformer en énergie électrique ; ▶barrage de vallée) ; *syn.* usine [f] hydroélectrique, installation [f] hydroélectrique ; *g* **Wasserkraftwerk [n]** (Kraftwerk, das aus der potenziellen Energie des Wassers elektrischen Strom gewinnt; ▶Talsperre); *syn.* Wasserkraftanlage [f].

2801 hydrogeology [n] *hydr.* (Science dealing with subsurface waters and with related geologic aspects of surface water; ▶hydrology); *syn.* geohydrology [n]; *s* **hidrogeología [f]** (Ciencia que estudia los hechos relacionados con las aguas subterráneas; DGA 1986; ▶hidrología); *f* **hydrogéologie [f]** (Science des eaux souterraines. Branche de la géologie et de

l'▶hydrologie continentale ; elle étudie l'origine, la composition, les propriétés, les modes de gisement, l'extension, le volume et les mouvements des eaux des nappes souterraines ; DG 1984) ; *g* **Hydrogeologie [f, o. Pl.]** (Lehre vom unterirdischen Wasser, das in den Hohlräumen der Gesteine vorhanden ist. **H.** erforscht die Zusammenhänge zwischen Wassereinzugsgebiet, unterirdischen Wasserwegen und den Austrittsstellen; ▶Hydrologie); *syn.* Geohydrologie [f], Quellen- und Grundwasserkunde [f].

2802 hydrological map [n] *hydr.* (Map showing the pattern of circulation and distribution of watercourses and waterbodies, groundwater, groundwater flows, etc. of an area); *s* **mapa [m] hidrológico** (Representación cartográfica que representa los cursos y masas de agua superficiales y subterráneas de una zona determinada); *syn.* mapa [f] hidrográfico; *f* **carte [f] hydrologique** (Représentation cartographique figurant les propriétés physiques, chimiques et la circulation de l'eau pour un espace défini) ; *g* **hydrologische Karte [f]** (Kartografische Darstellung, die das in einem Gebiet ober- und unterirdisch vorkommende Wasser [fließend und stehend], die Grundwasserströme etc. darstellt); *syn.* Gewässerkarte [f].

2803 hydrologic cycle [n] *hydr. met.* (Succession of stages through which water passes from the atmosphere to the earth by ▶evaporation, condensation from the clouds, ▶precipitation 1 and re-evaporation; WMO 1974); *s* **ciclo [m] hidrológico** (Sucesión de etapas que atraviesa el agua al pasar de la atmósfera a la tierra y volver a la atmósfera, ▶evaporación del suelo, mar o aguas continentales, condensación de nubes, ▶precipitación, acumulación en el suelo o en masas de agua y reevaporación; WMO 1974); *syn.* ciclo [m] del agua; *f 1* **cycle [m] de l'eau** (Circulation de l'eau sous ses différents états dans la nature, considérée dans son ensemble — hydrosphère, lithosphère et atmosphère ; ▶évaporation, ▶précipitations ; CILF 1978) ; *f 2* **cycle [m] hydrologique** (Période au début et à la fin de laquelle on se trouve en présence d'un même état de réserve d'eau pour un bassin donné. On peut généralement s'en tenir, avec une bonne approximation, au cycle annuel, d'étiage à étiage ; CILF 1978) ; *g* **Wasserkreislauf [m]** (Natürliche Bewegung des Wassers in seinen unterschiedlichen Aggregatzuständen auf der Erde in seiner Gesamtheit — Hydrosphäre, Lithosphäre und Atmosphäre; ▶Evaporation, ▶Niederschlag).

2804 hydrology [n] *hydr.* (Science dealing with global water, both liquid and solid, its properties, circulation, and distribution on and under the earth's surface and in the atmosphere, including environmental and economic aspects; ▶hydrogeology, ▶limnology, ▶water mangement, ▶water resources conservation); *s* **hidrología [f]** (Ciencia que analiza los fenómenos o hechos concernientes a las aguas y que se divide en las siguientes ramas: hidrología de las aguas continentales, que a su vez comprende la hidrología fluvial [potamología], la hidrología lacustre [limnología] y la glaciología, e hidrología marina. En alemán el término "hidrografía" no es sinónimo de **h.** sino que comprende la parte que se ocupa de la descripción y representación cartográfica cuantitativa de las aguas continentales; cf. DGA 1986 y DINA 1987; ▶gestión de recursos hídricos, ▶gestión del dominio público hidráulico, ▶hidrogeología, ▶limnologia); *syn.* hidrografía [f]; *f* **hydrologie [f]** (Science qui étudie la nature, les propriétés physiques, chimiques et les mouvements des eaux marines et des eaux continentales. Celle-ci a pour thème central l'étude du cycle de l'eau. On distingue l'hydrologie lacustre ou ▶limnologie, de l'▶hydrogéologie et de l'hydrologie fluviale, branche de la potamologie ; DG 1984 ; ▶gestion de la ressource en eau, ▶gestion du cycle de l'eau) ; *g* **Hydrologie [f, o. Pl.]** (1. Lehre von den Erscheinungsformen des Wassers über, auf und unter der Erdoberfläche, die sein Vorkommen, seinen Kreislauf und seine Verteilung [Gewässernetze], unter Einbeziehung

meteorologischer Elemente seine Funktion, seine chemischen und physikalischen Eigenschaften, auch die Wechselwirkungen mit der Umwelt incl. des Menschen, d. h. den ►Wasserhaushalt erforscht. **2.** Sie umfasst vor allem die **Hydrographie** als Beschreibung und quantitative kartografische Darstellung der Binnengewässer [Fließ- und Stillgewässer]. **3.** Weitere Teilgebiete der **H.** sind die **Potamologie [Flusskunde]**, Lehre von den Fließgewässern, die ►**Limnologie [Seenkunde]** und die **Glaziologie [Gletscherkunde]**; die **H.** ist die Grundlage der ►Wasserwirtschaft; ►Hydrogeologie); *syn. zu Ziff. 2* Gewässerkunde [f].

hydro-mesophilic [adj] *ecol.* ►mesic.

hydro-mesophilous [adj] *ecol.* ►mesic.

2805 hydromorphic soils [npl] *pedol.* (Soils that include both those which are permeable and permit groundwater to rise, i.e. ►groundwater soils [US]/ground water soils [UK], and those which are impermeable and prevent surface water from penetrating downward, i.e. ►poorly drained soils. In the German classification system the following ►soil orders may be differentiated according to the type of water [fresh or salt] and its duration in the soil: ►alluvial soil, ►aquepts, ►coastal marsh soil, ►soil affected by impeded water. **In U.S.**, a different soil classification system has been adopted by the Soil Conservation Service of the USDA. In this system the names of hydromorphic soils always include the combination of the three letters 'aqu', e.g. aquent, aqualf, aquox); *syn.* hydric soils [npl]; **s suelo** [m] **semiterrestre** (En la clasificación alemana, ►grupos principales [de suelos] cuya evolución se debe o ha debido al fenómeno de hidromorfia, es decir al exceso de agua. Esto conlleva un déficit de aireación que lleva consigo la reducción de iones con la consecuencia de colores particulares azul verdoso o rojizo. Dependiendo de la duración de la influencia del agua y el tipo de la misma [dulce o salada] se subdividen en las siguientes cinco clases de suelos: ►suelos aluviales [fluvisol], ►suelos de agua estancada, ►suelos de marsch, ►suelo encharcado, ►suelos gley [gleysol], ►suelo semiterrestre de inundación. En EE.UU. según el sistema de clasificación del Servicio de Conservación de Suelos del USDA los nombres de suelo hidromorfos siempre incluyen la combinación de las tres letras 'aqu' como p. ej. aquent, aqualf, aquox); *syn.* suelo [m] hidromorfo no permeable; **f sols** [mpl] **hydromorphes** (►Classe pédologique des sols caractérisés par des phénomènes de réduction ou de ségrégation locale du fer, liés à une saturation temporaire ou permanente des pores par l'eau, provoquant un déficit prolongé en oxygène. Cet engorgement en eau peut être dû à la présence naturelle permanente de la nappe phréatique, à la remontée de la nappe à la suite de la destruction d'un peuplement forestier ou à un mauvais drainage provoquant la formation intermittente d'une nappe perchée. Les sols hydromorphes possèdent 3 sous-classes : sol hydromorphe minéral, sol hydromorphe moyennement organique et sol hydromorphe organique ; dans la classification allemande ces sols sont dénommés « demi-terrestres » et sont divisés en 5 sous-classes pédologiques : 1. ►sol à nappe temporaire perchée, 2. ►sol à gley, 3. sols en lisière de sources et des zones de suintement, 4. ►sol alluvial et 5. ►sol salin à vase marine ; ►sol à nappe phréatique profonde, ►sol engorgé ; cf. PED 1983, 368 et s.) ; **g Grundwasser- und Stauwasserböden** [mpl] (*Klassifikationssystem in D.* ►Bodenabteilung der grund- und stauwasserbeeinflussten Böden, die sich je nach Zeitdauer des Wassereinflusses und der Art des Wassers [Süß- oder Salzwasser] in folgende fünf [Boden]klassen unterscheiden lassen: 1. ►Stauwasserböden, 2. ►Gleye, 3. Böden der Quellwasserbereiche, 4. ►Auenböden und 5. ►Marschböden; ►Grundwasserboden, ►staunasser Boden); *syn.* hydromorphe Böden [mpl].

hydrophilic plant [n] *bot.* ►hygrophyte.

2806 hydrophyte [n] *bot. phyt.* (Greek *hydor* >water< and *phyton* >plant<; perennial aquatic plant other than plankton, whose leaves disappear beneath the water surface in order to survive during the winter; ►floating-leaved plant, ►free-floating plant, ►submerged aquatic plant); *syn.* aquatic plant [n], aquatic [n] [also US]; **s hidrófito** [m] (Todas las plantas acuáticas perennes no pertenecientes al plancton, cuyos órganos invernan bajo el agua; dependiendo de la forma de crecimiento se diferencian: ►planta acuática sumergida con los órganos asimiladores sumergidos, radicante o no radicante, ►hidrófito libremente flotante con órganos asimiladores flotantes, como p. ej. el jacinto de agua *[Eichhornia]*, ►hidatófito radicante con las hojas flotando sobre la superficie del agua, como el nenúfar *[Nymphaea alba]* e hidrófito emergente); *syn.* planta [f] acuática; **f hydrophyte** [m] (Plante aquatique dont les organes végétatifs sont totalement immergés pendant la mauvaise saison ; ►plante flottante et fixée, ►plante flottante et libre, ►plante immergée) ; *syn.* plante [f] aquatique, plante [f] aquatile ; **g Hydrophyt** [m] (**1.** *bot. phyt.* alle nicht zum Plankton zählenden, ausdauernden Wasserpflanzen, deren Erneuerungs-/Überdauerungsknospen während der ungünstigen Jahreszeit am Gewässergrund liegen; cf. BOR 1958. Je nach Lebensweise werden die Wuchsformen **submerse H.en** mit ganz untergetauchten Sprossachsen [►Unterwasserpflanze], wurzelnd oder nicht wurzelnd, ►**Schwimmpflanzen** als frei schwimmende Wasserpflanze, z. B. Wasserhyazinthe *[Eichhornia]*, und **wurzelnde H.en** mit Sprossen auf der Wasseroberfläche [Schwimmblattstauden] und **H.** mit die Wasseroberfläche überragenden Sprossen, z. B. Lotosblume *[Nelumbo nucifera]*, unterschieden. **2.** *bot.* Nach RAUNKIAER [1934] zur ►Lebensformklasse der ►Kryptophyten gehörig, deren Erneuerungsknospen [ständig] unter der Wasseroberfläche liegen; nach HAEUPLER & MUER [2000] Standortsangabe mit weiterer Untergliederung in ►Lebensformen: **therophytischer H.** Wassernuss *[Trapa natans]*, **hemikryptophytischer H.** *[Lemna]*, **kryptophytischer H.** Teichrose *[Nuphar lutea]*, **phanaerophytischer H.** [Mangrove]. **3.** *hort.* Staude der Lebensbereiche „Wasser" und „Wasserrand" mit entsprechenden morphologischen und verbreitungsbiologischen Anpassungen; ►Schwimmpflanze 1, ►Schwimmpflanze 2, ►Unterwasserpflanze); *syn.* Wasserpflanze [f].

hydrophyte [n]**, rooted** *bot. phyt.* ►floating-leaved plant.

2807 hydroseeding [n] *constr.* (Application of seed, mulch, ►soil amendments [US]/soil ameliorants [UK], and nutrients suspended in a water slurry by means of a pressurized spray nozzle to establish vegetation on immature soils, ►entisols [US], rock faces or other poorly accessible areas with little or no topsoil; ►dry seeding, ►mulching material); *syn.* hydraulic seeding [n]; **s riego** [m] **de semillas por emulsión** (Método de siembra en ►suelos minerales brutos por pulverización de semillas, utilizando una mezcla de éstas, ►material de mulching y ►material de enmienda de suelos disuelta en agua; *opp.* ►siembra en seco); *syn.* siembra [f] líquida, siembra [f] húmeda, método [m] de pulverización de semillas; **f procédé** [m] **d'enherbement par projection hydraulique** (Méthode utilisée pour l'ensemencement mécanique de grandes surfaces, de sols naturels ou rocheux, de ►sols peu évolués, d'aires difficilement accessibles ou soumises à l'érosion, grâce à l'emploi de l'hydroseeder et consistant à utiliser un mélange graines, liquide et fixateur ; ►amendement, ►produit de mulching ; *opp.* ►ensemencement à sec) ; *syn.* procédé [m] d'engazonnement par projection hydraulique, procédé [m] d'ensemencement par projection hydraulique ; **g Anspritzverfahren** [n] (Methode zur Begrünung von ►Rohböden, Felsböschungen oder sonstigen schwer zugänglichen Flächen mit geringer oder fehlender Oberbodenschicht durch den

Einsatz von Spritzkanonen oder Mulchgebläsen. Das Saatgut wird mit Wasser als Trägersubstanz, gegebenenfalls unter Beigabe von Dünger und Bodenverbesserungsstoffen, ►Mulchstoffen und Klebern auf die zu sichernden Hangflächen aufgespritzt; diese Ansaatform wurde ursprünglich in den USA in den 1960er-Jahren entwickelt; ►Bodenverbesserungsstoff 1; *opp.* ►Trockensaat); *syn.* Anspritzbegrünung [f], Begrünung [f] durch Hydrosaat, Hydrosaat [f], Nasssaat [f] (cf. DIN 18 918), Nassansaatverfahren [n], *im Fachjargon auch* Hydroseeding [n].

2808 hydrosphere [n] *geo.* (Water layer of the earth, including oceans, rivers and streams, lakes, ice of glaciers and the water vapo[u]r in the atmosphere; ►geosphere, ►lithosphere); *s* **hidrosfera** [f] (Conjunto total de las aguas de la Tierra, que incluye los océanos y mares, las aguas subterráneas, los glaciares, los lagos y ríos, la humedad del suelo y el vapor en la atmósfera; DGA 1986; ►geosfera, ►litosfera); *f* **hydrosphère** [f] (Enveloppe aquatique de la terre comprenant les océans et les mers, les eaux intérieures, la nappe phréatique, les calottes glacières et l'eau contenue dans l'atmosphère ; ►géosphère, ►lithosphère) ; *g* **Hydrosphäre [f]** (Wasserhülle der Erde, die neben den Meeren auch das Grundwasser, alle Binnengewässer, das in Gletschern gebundene Eis und das in der Atmosphäre vorhandene Wasser umfasst; ►Geosphäre, ►Lithosphäre).

hygiene [n] *conserv. envir.* ►environmental hygiene.

hygrophilic plant [n] *bot.* ►hygrophyte.

2809 hygrophilic scrub [n] *phyt.* (Shrub vegetation 1-5m high, usually found at margins of lakes or slow-moving streams on peat or mineral soil, poor to moderately rich in nutrients with a continually high-water table; VIR 1982; ►hygrophyte); *syn.* wetland scrub ; *s* **matorral** [m] **higrófilo** (Formación de leñosas que habitan en suelos de húmedos a saturados, de turba o areno-pantanosos y de pobres a medianamente ricos en nutrientes; ►higrófito); *f* **taillis** [m] **hygrophile** (Végétation arbustive, buissonnante, généralement basse et inextricable localisée sur les sols humides à marécageux, tourbeux ou paratourbeux-sableux ; p. ex. les taillis de l'alliance du *Salicion cinereae* ou le taillis tourbeux de l'alliance de l'*Alnion glutinosae* ; GGV 1979, 335 et s. ; ►hygrophyte) ; *g* **Feuchtgebüsch [n]** (Strauchbestand, z. B. Weiden-Faulbaumgebüsch *[Frangulo-Salicetum cinereae Malc. 1929]*, Gagelgebüsch *[Myricetum gale Jonas 1932]* auf feuchten bis nassen, torfigen oder sandig-anmoorigen Böden; cf. Biotopkartierung SH; ►Hygrophyt).

hygrophilous plant [n] *bot.* ►hygrophyte.

2810 hygrophyte [n] *bot.* (Greek *hygros* >humid< and *phyton* >plant<; **1.** plant growing in or preferring moist habitats. **2.** *hort. plant.* Herbaceous perennial, which grows in the cool, semi-shade of the edge of trees and shrubs, or under shade-giving trees and shrubs or at the edge of water [►helophyte]. **H.s** require fresh to moist soils and often have large leaves, which aid transpiration. They are of particular design significance, when planted to contrast with the linear shaped leaves of plants, which also grow in the same habitat); *syn.* hygrophilous plant [n], hygrophilic plant [n], moisture-loving plant [n] hydrophilic plant; *s* **higrófito** **[m]** (Término de la clasificación de IVERSEN con el que se denomina a las plantas terrestres sin tejidos de aireación que viven en ambientes de atmósfera muy húmeda y reciben del suelo húmedo pero no encharcado, un abundante abastecimiento de agua; DINA 1987); *syn.* planta [f] higrófila, vegetal [m] higrófilo, higrófitas [fpl]; *f* **hygrophyte [m]** (Du grec *hygros* « humide » et *phyton* « plante » ; **1.** plante adaptée aux habitats fortement humides et inondés temporairement. **2.** *hort. plant.* Plante vivace des milieux mi-ombragés et frais des zones boisées, en bordure de massifs ligneux ou des berges des zones humides [►hélophyte], sur sols frais à très humides souvent caractérisée par de

grandes feuilles favorisant la transpiration et d'un grand effet de contraste dans les compositions décoratives) ; *syn.* plante [f] hygrophile ; *g* **Hygrophyt [m]** (Griech. *hygros* >feucht< und *phyton* >Pflanze<: **1.** einen stets feuchten Standort liebende Landpflanze. **2.** *hort. plant.* Staude der Lebensbereiche absonnigkühler Gehölzrand, schattiges Gehölz und Wasserrand [►Helophyt] auf frischen bis nassen Böden mit oft großen, die Transpiration fördernden Blättern. Von besonderer gestalterischer Bedeutung durch effektvolle Kontrastsetzungen zu hier ebenfalls vorhandenen linearen Blattformen [Flächen-Linien-Kontrast bei KARL FOERSTER „Harfe und Pauke"]); *syn.* hygrophile Pflanze [f].

2811 hypertrophication [n] *ecol.* (Excess enrichment of nutrients in a water body; ►eutrophication); *s* **hipertrofización** **[f]** (Sobrealimentación o aporte excesivo de materias alimenticias; ►eutrofización); *syn.* hipertrofia [f]; *f* **hypertrophisation** **[f]** (Enrichissement excessif en substances nutritives ; ►eutrophisation) ; *g* **Hypertrophierung [f]** (Übersteigerte Nährstoffanreicherung; ►Eutrophierung).

hypocotylar tuber [n] *bot.* *∗*►stem tuber.

2812 hypopotamon [n] *limn.* (Biotic community living in area of brackish water; ►flounder zone [UK]); *s* **hipopótamon** **[m]** (Biocenosis de las aguas salobres; ►región salobre); *f* **hypopotamon** [m] (Communauté biologique de la ►zone à éperlan) ; *g* **Hypopotamon [n]** (...potama [pl]; Lebensgemeinschaft der ►Brackwasserregion).

2813 hyporhithron [n] *limn.* (Biotic river community living in the ►grayling zone [UK]); *s* **hipóritron** [m] (Biocenosis de la zona inferior de la región salmonícola de los ríos [►zona de umbra]); *f* **hyporhithron** [m] (Communauté biologique de la ►zone à ombre) ; *g* **Hyporhithron [n]** (...rhithra [pl]; Lebensgemeinschaft der ►Äschenregion).

2814 IBA list [n] *adm. conserv. nat'res.* (**Important Bird Areas List**: Catalogue or inventory of especially important areas for birds, compiled for the EU Commission by the International Council for Bird Preservation [ICBP]; the ICBP was renamed BirdLife International in 1993 [umbrella organization of the associations for the protection of birds]); *syn.* list [n] of Important Bird Areas, list [n] of IBAs; *s* **lista [f] IBA** (**Important Bird Areas list** = lista de Áreas Importantes para las Aves: Inventario de lugares de importancia internacional que es elaborado por la BirdLife International [federación internacional de asociaciones nacionales para la conservación de las aves] por encargo de la Comisión de la UE. **En Es.** están registradas 391 áreas IBA; la organización miembro de BirdLife en España es SEO/BirdLife [Sociedad Española de Ornitología] fundada en 1954. **En AL** solamente en los Andes Tropicales [en Bolivia, Colombia, Ecuador, Perú y Venezuela] hay 455 áreas IBA; cf. www.birdlife.org); *f* **liste [f] LPO Habitat** (*Abrév.* LPO : Ligue pour la protection des oiseaux : Inventaire des sites d'intérêt ornithologique considérable, établi par le Conseil International pour la Protection des Oiseaux [ICBP] sur mission de la commission européenne ; le Conseil International pour la Protection des Oiseaux est devenu, en 1993, *BirdLife International*, la Fédération internationale regroupant les associations de protection des oiseaux qui établie

maintenant la liste des habitats naturels d'intérêt ornithologique) ; *g* **IBA-Liste** [f] („*Important Birds Area*"-Liste[n]: Verzeichnis besonders bedeutsamer Gebiete für den Vogelschutz, im Auftrag der EU-Kommission vom Internationalen Rat für Vogelschutz [ICBP] zusammengestellt; der ICBP ist 1993 in *BirdLife International* [Dachorganisation der Vogelschutzverbände] übergegangen).

ice crystals [npl] *constr. pedol.* ▶formation of ice crystals.

2815 identification [n] *biol.* (Determining the exact specific name of a plant or animal species); *s* **identificación** [f] (Clasificación científica de especies de flora y fauna); *syn.* determinación [f]; *f* **identification** [f] (d'espèces animales et végétales) ; *g* **Bestimmung** [f] **(1)** (Wissenschaftliche Zuordnung einer Tier- oder Pflanzenart).

2816 identification [n] **and mapping** [n] *conserv. conserv'hist.* (Recognizing, classifying, catalog[u]ing and graphic representation of biotopes, monuments and sites, etc.); *s* **inventario** [m] (Identificación, clasificación, catalogación y descripción de biótopos, monumentos histórico-artísticos, etc.); *f* **recensement** [m] (Repérage, identification, description, synthèse cartographique des biotopes, des monuments et des sites, etc.) ; *g* **Erfassung** [f] (Ermittlung, Klassifizierung, Registrierung und Aufzeichnung von Biotopen, Denkmälern/Denkmalen etc.).

illness [n] *envir.* ▶environmental illness.

2817 illumination [n] *constr.* (Lighting by artificial means); *s* **iluminación** [f] **(artificial)** *syn.* alumbrado [m]; *f* **éclairage** [m] (Lumière artificielle) ; *g* **Beleuchtung** [f] (Ausleuchtung mit künstlichem Licht).

2818 illustrative site plan [n] *plan.* (Graphic two dimensional representation of the proposed layout for an area depicting existing and proposed elements and its future organization; ▶site plan); *s* **plano** [m] **de ordenación general** (Un dibujo que representa el concepto general para un área, visualizando bidimensionalmente la conformación futura de la misma; ▶plano de situación); *f* **plan** [m] **de principe** (Plan représentant la distribution des différents éléments décoratifs et utilitaires, expression de la solution d'ensemble — parti général et solutions techniques — préconisée pour l'ensemble des ouvrages ; ▶plan de situation [du terrain]) ; *g* **Gestaltungsplan** [m] **(1)** (Plan, der die gestalterischen und funktionalen Grundzüge darstellt; ▶Lageplan).

2819 illuvial horizon [n] *pedol.* (Soil layer in which material carried from an overlying layer has been precipitated from solution or deposited from suspension; RCG 1982; **in U.S.,** ▶claypan, ▶ortstein [US]/ironpan [UK], ▶podsolization, ▶translocation of clay); *s* **horizonte** [m] **iluvial** (Horizonte del suelo enriquecido con sustancias deslavadas de los superiores, pudiendo llegar a formar capas de gran concentración de la correspondiente sustancia; ▶eluviación de arcilla, ▶horizonte de iluviación, ▶podsolización); *syn.* horizonte [m] de enriquecimiento, horizonte [m] de iluviación (2); *f* **horizon** [m] **illuvial** (Horizon enrichi suite à un apport de matériaux provenant des horizons sus-jacents. Les matériaux accumulés peuvent être de l'argile, du fer, de l'aluminium, de l'humus, des sesquioxydes, des carbonates, des sels solubles ou un mélange de deux ou plusieurs de ces matériaux ; DIS 1986 ; ▶horizon d'accumulation d'argile, ▶podzolisation, ▶processus d'accumulation d'argile) ; *syn.* horizon [m] d'accumulation (DEE 1982, 55), horizon [m] enrichi (LA 1981, 1048) ; *g* **Anreicherungshorizont** [m] (Bodenschicht, in die aus darüberliegenden Schichten/Horizonten ausgewaschene Stoffe eingelagert sind; ▶Ortstein, ▶Podsolierung, ▶Tonverlagerung, ▶Tonband); *syn.* Illuvialhorizont [m].

image [n] *plan. rem'sens.* ▶thermal image.

2820 immature plant [n] *hort. for.* (Young plant propagated by seeds or cuttings, and cultivated until it is large enough for

sale); ▶seedling); *syn.* plantlet [n], liner [n] (in a nursery); *s 1* **plántula** [f] (Plantita recién nacida resultado de reproducción por siembra o por esqueje; ▶plántula de semilla); *syn.* plantilla [f]; *s 2* **postura** [f] *for.* (Arbolillo criado en vivero para repoblar bosques); *f 1* **jeune plant** [m] (Végétal au début de son développement, résultant de semis, marcotte, bouture, éclat, greffe ou tout autre mode de reproduction ou de multiplication, destiné à être proposé à la vente pour plantation directe ou à être replanté à distance en pépinière d'élevage ; N.F. V 12 031 ; ▶semis 2) ; *f 2* **jeune plant** [m] **à recultiver** (Jeune plant issu de semis ou de multiplication végétative, destiné à être remis en culture pour être élevé ; ENA-F 1996, 6) ; *f 3* **jeune touffe** [f] (Végétal provenant d'un jeune plant ayant subit un repiquage en pleine terre, en pot ou en tout autre récipient ; N.F. V 12-031) ; *g* **Jungpflanze** [f] (Durch Aussaat oder Stecklingsvermehrung gezogene Pflanze, die bis zur verkaufsfertigen Ware für den Endverbraucher weiterkultiviert wird; ▶Sämling).

immature soil [n] **[UK]** *pedol.* ▶entisol [US] (1).

immediate runoff [n] *constr. geo.* ▶direct runoff.

2821 immigration [n] *zool.* (▶Migration 2 of one or many individuals into a new territory; ▶invasion 1; *opp.* ▶emigration); *s* **inmigración** [f] (▶Migración 2 de uno o muchos individuos a un territorio; ▶emigración, ▶invasión); *f* **immigration** [f] (Entrée réduite ou massive d'individus dans un autre territoire constituant un déplacement prolongé ou définitif ; ▶invasion, ▶migration 2 ; *opp.* ▶émigration) ; *g* **Immigration** [f] (Wandern einzelner bis vieler Individuen [aus ihrem angestammten Lebensraum] zu einer anderen Population und das dauerhafte Bleiben in deren Gebiet; ▶Invasion, ▶Migration 2; *opp.* ▶Emigration); *syn.* Einwanderung [f].

immobilization [n] *pedol.* ▶nutrient fixation.

impact [n] *conserv. leg. nat'res. plan.* ▶analysis of environmental impact; *ecol.* ▶effects; *conserv. nat'res. plan.* ▶environmental impact; *envir.* ▶gaseous pollution impact; *ecol.* ▶negative impact; *envir.* ▶pollution impact

impact [n]**, adverse** *ecol. envir. recr.* ▶disturbance, #2; ▶environmental stress.

impact [n]**, very adverse** *ecol. envir.* ▶severe disturbance.

2822 impact analysis [n] *landsc. plan.* (Method in landscape planning, as a part of an ▶environmental risk assessment, which investigates the causal and system-related context of ecological compatibility in the proposed layout of land uses); *s* **análisis** [m] **de efectos** (Instrumento de planificación del paisaje en el marco del ▶procedimiento de análisis de riesgo ecológico que consiste en estudiar las interrelaciones causales y sistemarias de los usos del espacio en cuanto a su idoneidad ecológica); *f* **analyse** [f] **des impacts** (Méthodologie utilisée dans la planification des paysages dans le cadre du ▶procédé d'évaluation des incidences environnementales consistant dans l'analyse des relations causales et spécifiques à un système quant à l'opportunité écologique de l'organisation spatiale des usages projetés) ; *g* **Wirkungsanalyse** [f] (Landschaftsplanerisches Instrument im Rahmen des ▶Risikoverfahrens zur Untersuchung von kausalen und systembezogenen Zusammenhängen hinsichtlich der ökologischen Zweckmäßigkeit der räumlichen Anordnung zu planender Nutzungsansprüche); *syn.* ökologische Wirkungsanalyse [f].

2823 impact mitigation [n] *conserv. leg.* (**In Europe**, legal environmental compensation for disturbance of a landscape area to restore or recreate the structural and functional character of the natural habitat and the necessary ecological balance; **in U.S.,** any illegal disturbance of designated wetlands or species protection zone requires the offending party to pay a fine and to restore the disturbed area as part of the EPA oversight process. This may

occur at the local, state or national level; ►environment compensation measure, ►environmental impact statement [EIS], ►intrusion upon the natural environment, ►mitigation measure); *s* **compensación [f] (de un impacto)** (**En Es.**, según la legislación de ►evaluación del impacto ambiental [EIA], obligación de prever medidas para reducir, eliminar o compensar los efectos ambientales negativos significativos de proyectos, obras o actividades sobre el medio ambiente; cf. art. 11 RD 1131/1988. **En D.** la ley de protección de la naturaleza obliga a tomar medidas de **c. de i.** para reestablecer el carácter estructural y funcional de un área en cuanto a su capacidad ecológica y su aspecto palsajístico, que han sido alterados por una obra de construcción o impacto similar; ►intrusión en el entorno natural, ►medida compensatoria, ►medida sustitutiva); *f* **compensation [f] (des conséquences dommageables sur l'environnement)** (Mesure envisagée pour la reconstitution de l'identité structurelle et fonctionnelle ainsi que du potentiel du milieu naturel ayant pour but, nonobstant la suppression ou la réduction de toutes interventions ou atteintes réductibles dommageables à l'environnement lors d'un projet d'aménagement, de compenser les impacts impossibles à supprimer ; cf. décret n° 77-1141 du 12 octobre 1977 et circ. n° 93-73 du 27 septembre 1993 ; ►atteinte au milieu naturel, ►mesure de compensation, ►mesure de substitution, ►procédure d'étude d'impact) ; *g* **Ausgleich [m] (eines Eingriffes)** (Wiederherstellung der von der Gesellschaft nachgefragten funktionalen und strukturellen Identität und des benötigten Leistungsvermögens des Naturhaushaltes eines durch einen ►Eingriff in Natur und Landschaft beeinträchtigten Landschaftsraumes in einer zeitlich angemessenen Frist durch Maßnahmen des Naturschutzes und der Landschaftspflege. Der Ort von Eingriff und Ausgleich sollte möglichst auf Grund des funktionalen Wirkungsgefüges identisch sein; cf. §§ 8 u. 8a-b BNatSchG; Der Begriff des *Ausgleichs* wird im BNatSchG nicht näher erläutert; er lässt sich aber indirekt aus § 8 [2] Satz 4 BNatSchG bestimmen. Danach ist ein Eingriff ausgeglichen, „wenn nach seiner Beendigung keine erhebliche oder nachhaltige Beeinträchtigung des Naturhaushaltes zurückbleibt und das Landschaftsbild landschaftsgerecht wiederhergestellt oder neu gestaltet ist". Eine gleichartige Wiederherstellung des vorherigen Zustandes wird gesetzlich nicht vorausgesetzt und ist im naturwissenschaftlichen Sinne auch nicht möglich. Bei der Beeinträchtigung des Landschaftsbildes wird die Neugestaltung sogar als **A.** ausdrücklich anerkannt. Der Begriff des **A.s** ist daher so zu verstehen, dass die durch den Eingriff beeinträchtigten Funktionen des Naturhaushaltes wiederherzustellen sind; cf. DIFU 1996, 22; in diesem Falle spricht man auch von der „Wiederherstellung der funktionalen Identität in Bezug auf die Leistungen des Naturhaushaltes..."; KW 1982; ►Ausgleichsmaßnahme [des Naturschutzes und der Landschaftspflege], ►Ersatzmaßnahme, ►Umweltverträglichkeitsprüfung); *syn.* Eingriffskompensation [f], Kompensation [f] bei Eingriffen in Natur und Landschaft.

2824 impact mitigation regulation [n] *conserv. leg. plan.* (Legally prescribed directive to implement alleviation of environmental impacts, and to execute ►environmental compensation measures, as required in D. by nature conservation laws and in U.S. by environmental laws; ►compensation payment, ►impact mitigation, ►intrusion upon the natural environment, ►prevention measure [of an intrusion upon the natural environment); *s* **regulación [f] de impactos sobre la naturaleza** (**En D.**, proceso administrativo-legal según el cual se fijan las medidas a tomar para mitigar, corregir o compensar los efectos negativos de una ►intrusión en el entorno natural; cf. § 8 Ley Federal de Protección de la Naturaleza. **En Es.** no existe tal procedimiento, sino que en caso de proyectos de cierta envergadura que impli-

quen riesgos considerables para el medio ambiente, se debe realizar la evaluación del impacto ambiental y tomar las correspondientes medidas para reducir, eliminar o compensar los efectos negativos significativos; cf. art. 1 y Anexos I y II RDL 1302/1986; ►compensación [de un impacto], ►indemnización compensatoria, ►medida correctora, ►medida sustitutiva); *f* **réglementation [f] relative aux atteintes subies par le milieu naturel (±)** (Détermination des ►atteintes au milieu naturel portées par un projet et de leur répercussion sur l'environnement, des mesures de ►compensation [des conséquences dommageables sur l'environnement] ainsi que des ►mesures de substitution, ►mesures de suppression ; cf. art. 2 de la loi n° 76-629 du 10 juillet 1976 relative à la protection de la nature ; ►redevance compensatoire) ; *g* **Eingriffsregelung [f]** (Legaldefinitorische Festlegung der Dauer und Schwere eines ►Eingriffes in Natur und Landschaft und die verfahrensmäßige Behandlung der Eingriffsfolgen resp. Anordnungen, den Eingriff zu vermeiden, auszugleichen, zu ersetzen oder zu mindern. Grundgedanke dieses Instrumentes ist es, dass jeder, der in die Leistungsfähigkeit des Naturhaushaltes oder in das Landschaftsbild erheblich oder nachhaltig eingreift, vermeidbare Beeinträchtigung von Natur und Landschaft zu unterlassen und vermeidbare Beeinträchtigungen von Natur und Landschaft durch entsprechende Maßnahmen des Naturschutzes und der Landschaftspflege zu kompensieren hat. Seit In-Kraft-Treten des Investitionserleichterungs- und Wohnbaulandgesetzes von 1993 ist die naturschutzrechtliche **E.** auf der Ebene des Bauleitplanverfahrens zu berücksichtigen. Die Rechtsfolgen der **E.** sind nach § 8a [1] Satz 1 und [9] BNatSchG durch die Begriffe *Vermeidung, Ausgleich* und *Ersatz* gekennzeichnet. Über diese bereits in der vorhabenbezogenen **E.** des § 8 BNat SchG enthaltenen Begriffe hinaus wird durch § 8a [1] Satz 2 BNatSchG zusätzlich der Begriff *Minderung* eingeführt. Darüber hinaus wird in der Praxis häufig der Begriff *Kompensation* gebraucht; seit 1998 ist die **E.** auch planungsrechtlich durch das BauGB abgesichert. Vermeidung und Ausgleich von Eingriffen unterliegen ausdrücklich der gemeindlichen Abwägungsentscheidung; ►Ausgleichsabgabe, ►Ausgleich [eines Eingriffes], ►Ersatzmaßnahme [des Naturschutzes und der Landschaftspflege], ►Vermeidungsmaßnahme [eines Eingriffes]); *syn.* naturschutzrechtliche Ausgleichsregelung [f] bei Eingriffen.

impact [n] **on the landscape, adverse** *conserv. leg. plan.* ►intrusion upon the natural environment.

impacts [npl] *ecol. land'man.* ►elimination of disturbing impacts.

impact statement [n] *conserv. leg. nat'res. plan.* ►environmental impact statement (EIS).

impact study [n] *envir. phyt.* ►vegetation impact study.

2825 impairment [n] **of landscape** *conserv. land'man.* (Degradation and change that is detrimental to the ►landscape character such that it loses its original unique character; ►detraction from visual quality, ►uncontrolled proliferation of settlements, ►visual disturbance, ►visual quality); *syn.* disfigurement [n] of landscape (VDE 1980); *s* **deterioro [m] del paisaje** (Cambio radical del carácter de un paisaje que conlleva la pérdida de su particularidad original; ►calidad visual, ►desfiguración del paisaje, ►desfiguración del paisaje [rural o urbano], ►desfiguración [visual], ►particularidad del paisaje); *f* **défiguration [f] des sites et des paysages** (Nuisance persistante portant atteinte au caractère originel d'un paysage ; ►altération visuelle du caractère [des sites et des paysages], ►caractère particulier des sites et des paysages ►mitage du paysage, ►nuisance visuelle, ►valeur esthétique) ; *syn.* altération [f] du caractère des sites et des paysages ; *g* **Verunstaltung [f] der Landschaft** (Nachteilige

Veränderung des Charakters einer Landschaft, so dass sie ihrer ursprünglichen ►landschaftlichen Eigenart widerspricht; ►Gestaltqualität, ►visuelle Beeinträchtigung 1, ►visuelle Beeinträchtigung 2, ►Zersiedelung [der Landschaft]); *syn.* Landschaftsverschandelung [f], Landschaftsverunstaltung [f].

impeded drainage [n] *pedol.* ►impeded water.

2826 impeded water [n] *pedol.* (Rainwater which collects near the surface temporarily, as distinguished from ►groundwater [US]/ground water [UK], due to impervious subsoil layers; ►poorly-drained soil, ►soil affected by impeded water, ►waterlogging); *syn.* impeded drainage [n] (FAO 1960); *s* **agua [f] estancada** (Al contrario que la causada por el ►agua subterránea, acumulación temporal de agua de la lluvia en suelos con capas impermeables; ►encharcamiento del suelo, ►suelo de agua estancada, ►suelo encharcado); *syn.* anegamiento [m]; *f* **nappe [f] perchée** (À la différence de l'action d'une ►nappe phréatique permanente, hydromorphie temporaire des horizons de surface gorgés d'eau provoquée par la présence de couches peu perméables en profondeur ; ►processus d'engorgement du sol, ►sol à nappe temporaire perchée, ►sol engorgé) ; *syn.* nappe [f] superficielle, nappe [f] temporaire ; *g* **Staunässe [f]** (Oberflächennahe Wasseranreicherung, die durch schwer wasserdurchlässige Unterbodenschichten entsteht und im Vergleich zum ►Grundwasser nur zeitweise im Jahr vorhanden ist; ►Bodenvernässung, ►staunasser Boden, ►Stauwasserboden); *syn.* Stauwasser [n], Tagwasser [n] [auch A].

2827 impeded water indicator plant [n] *phyt.* (►Indicator plant of water-saturated soils); *syn.* bad drainage indicator plant [n]; *s* **planta [f] indicadora de encharcamiento** (►Planta indicadora de suelos anegados); *f* **plante [f] indicatrice de l'engorgement** (►Plante indicatrice des sols engorgés) ; *g* **Staunässezeiger [m]** (►Zeigerpflanze für staunasse Böden).

2828 imperative reasons [npl] **of overriding public interest** *ecol. leg. plan.* (Reasons, which speak for the execution of a project, because they are essential for a nation's well-being and its society, e.g. public health and safety. The U.S. Constitution is based upon the concept that government is empowered at all levels to protect the public health, safety, and welfare. Environmental Protection Agency regulations are designed to accomplish this goal by protecting habitats, species, and systems deemed vital to ecological viability and human health and welfare. These regulations may be challenged in the courts [local, state, and federal] to argue the merits against broader concerns; e.g. military objectives are often given priority over ecological aims, such as the ruling permitting low frequency sonar communication by the Navy within known whale habitats. All Western cultures retain some authority to override ecological regulations for cultural purposes or political prerogatives in the name of pressing "National interests". In terms of planning, even if an ►environmental impact assessment [EIA] [US]/environmental assessment [UK] comes to the conclusion that a project will have negative results, the implementation of a such project is nevertheless deemed to be absolutely necessary, though it may be contrary to Article 6 [4] of the Habitats Directive, in which compelling reasons are not defined. In addition, if alternative solutions are not possible or existing solutions with regard to the conservation objectives of the directive were to have an even greater environmental impact on the area concerned, the authority responsible is required to implement mitigation measures, in order to ensure that favorable conditions of natural habitats within the bio-geographic region are maintained. I. r. of o. p. i. occur when a project is proven essential within the framework of political and legal requirements based on the protection of basic values [health, public safety, economic and social interest for the state and society]; only public interest as demanded by public or private entities can be weighed up against the merits of conservation objectives of the ►Habitats Directive. If the area concerned contains priority species and habitat types, as defined in Annex I, public interest can only prevail, if fundamental values [health, safety, environment], vital to the lives of citizens, were to be affected); *s* **razones [fpl] imperiosas de interés público de primer orden** (Según la ►Directiva de Hábitats de la UE, si, a pesar de las conclusiones negativas de la evaluación de las repercusiones sobre el lugar y a falta de soluciones alternativas, debiera realizarse un plan o proyecto por **rr. ii. de i. p. de p. o.**, incuidas aquéllas de índole social o económico, el Estado miembro tomará cuantas medidas compensatorias sean necesarias para garantizar que la coherencia global de Natura 2000 quede protegida. En el caso de que el lugar considerado albergue un tipo de hábitat natural y/o una especie prioritaria, únicamente se podrán alegar consideraciones relacionadas con la salud humana y la seguridad pública, o relativas a consecuencias positivas de primordial importancia para el medio ambiente, o bien, previa consulta de la Comisión, otras **rr. ii. de i. p. de p. o.**; cf. art. 6.4 Directiva de Hábitats)); *f* **raisons [fpl] impératives d'intérêt public majeur** (Notion juridique instituée en application de l'article 6, paragraphe 4, de la ►Directive Habitat Faune Flore 92/43/CEE, permettant la réalisation d'un plan ou projet indispensable [santé, sécurité, politiques fondamentales pour l'Etat et pour la société, raisons sociales ou économiques] pour lequel l'évaluation des incidences sur un site Natura 2000 a abouti à des conclusions négatives en l'absence de solutions de remplacement ; dans ce cas, l'État membre prend toute mesure compensatoire nécessaire pour assurer la protection de la cohérence globale de Natura 2000 dans le périmètre de la région biogéographique et informe la Commission des mesures compensatoires adoptées ; lorsque le site concerné est un site abritant un type d'habitat naturel et/ou une espèce prioritaires, seules peuvent être évoquées des considérations liées à la santé de l'homme et à la sécurité publique ou à des conséquences bénéfiques primordiales pour l'environnement) ; *g* **zwingende Gründe [mpl] des überwiegenden öffentlichen Interesses** (Gründe, die für die Durchführung eines Vorhabens, das für die Daseinsvorsorge für Staat und Gesellschaft, Gesundheit und öffentliche Sicherheit unbedingt erforderlich ist, sprechen. In diesem Falle kann entgegen Artikel 6 [4] der ►Fauna-Flora-Habitat-Richtlinie, in dem zwingende Gründe nicht definiert sind, trotz negativer Ergebnisse der Verträglichkeitsprüfung die Durchführung eines unbedingt notwendigen Projektes ermöglicht werden, wenn Alternativlösungen nicht vorhanden sind oder die vorhandenen Lösungen im Hinblick auf die Erhaltungsziele der Richtlinie noch stärkere Umweltbeeinträchtigungen für das betreffende Gebiet zur Folge haben. In diesem Falle hat die zuständige einzelstaatliche Behörde Ausgleichsmaßnahmen zu treffen, die den Beitrag eines Gebietes zur Erhaltung eines günstigen Zustandes natürlicher Lebensräume innerhalb der biogeographischen Region sicherstellen. Z. G. des ü. ö. I. sind dann gegeben, wenn ein Projekt im Rahmen von politischen und gesetzlichen Vorgaben, die auf den Schutz von Grundwerten [Gesundheit, öffentliche Sicherheit, wirtschaftliche und soziale Daseinsvorsorge für Staat und Gesellschaft] abzielen, sich als unerlässlich erweist; allein die von öffentlichen oder privaten Körperschaften geförderten öffentlichen Interessen können gegen die Erhaltungsziele der Richtlinie abgewogen werden. Wenn das betreffende Gebiet Lebensraumtypen nach Anhang I und prioritäre Arten beherbergt, kann das öffentliche Interresse nur dann überwiegen, wenn Grundwerte für das Leben der Bürger [Gesundheit, Sicherheit, Umwelt] tangiert werden).

impermeable stratum [n] *hydr. geo.* ►aquiclude.

imperviousness [n] *constr.* ►watertightness.

2829 implementation [n] (1) *adm. plan.* (Execution of planning proposals after coordination and approval); *s* **implementación** [f] **[AL]** (Ejecución de propuesta de planificación después de su aprobación); *syn.* realización [f]; *f* **imposition** [f] (Capacité à faire adopter un projet par vote ou par procédure d'autorisation ; *contexte* faire adopter un projet, faire prévaloir une idée) ; *g* **Durchsetzung** [f] (Realisierung einer Planung nach einem Abstimmungs- und Genehmigungsprozess; durchsetzen [vb]).

implementation [n] (2) *envir. pol.* ►deficiency in implementation.

2830 implementation [n] **of a plan** *plan.* (Carrying out of a plan or design); *syn.* execution [n] of a plan; *s* **implementación** [f] **de un plan [AL]** *syn.* realización [f] de un plan; *f* **réalisation** [f] **(d'un plan)** ; *g* **Ausführung** [f] **eines Planes** (Realisierung eines Planes oder Entwurfes; ausführen [vb]).

import [n] *envir. limn. pedol.* ►matter import.

import [n]**, nutrient** *limn. pedol.* ►nutrient input.

importance [n] *conserv. ecol.* ►site of community importance.

2831 Important Birds Area [n] *conserv. nat'res.* (*Abbr.* IBA; site, varying in size, on either public or private land, that provides an essential habitat to one or more species of breeding or non-breeding birds. An **IBA** zone is designated by the global ornithological federation, **BirdLife International**, on the basis of internationally-agreed, technical criteria and evaluations of national bird counts, as being a significant bird protection area. Initiated in Europe in the mid-1980s, over 7,000 **IBAs** have been identified in 130 countries; an **IBA** must exist as an actual or potential protected area and has the potential to be managed for bird and general nature conservation. In U.S., the program was introduced in 1995, and is administered by the National Audubon Society. Currently, about 46 states have **IBA** programs, with some 1,500 sites identified. **In U.K.**, 295 **IBAs** have been identified, covering more than 31,000km², or over 12% of the UK. The smallest **IBA**, the rocky Sheep Island in Northern Ireland, is four hectares and is important for breeding seabirds. The largest **IBA**, the Central Highland Hills and Glens in Scotland, is 230,000 hectares and important for golden eagles. **IBA** thresholds are set by regional and national governing organizations. To be listed, a site must meet at least one of the following criteria: **1.** Globally threatened species, included in the IUCN Red List as critically endangered, endangered or vulnerable. **2.** Restricted-range species, i.e. species vulnerable, because they are not widely distributed. **3.** Biome-restricted species, i.e. species vulnerable, because their populations are concentrated in one general habitat type. **4.** Congregations, i.e. species vulnerable, because they congregate together for breeding, feeding, or migration. The criteria are applied and defined at different levels: an **IBA** may be important at the global, continental or regional level as well as at the national or state level); *s* **área** [f] **importante para las aves** (Las IBA —en sus siglas en inglés— son parajes de diferentes tamaños, sean de dominio público o privado, que ofrecen hábitats esenciales a una o más especies de aves, en reproducción o no. Un **a. i. para las a.** es designada por la federación mundial de ornitología «BirdLife International» sobre la base de criterios estandarizados y acordados internacionalmente y evaluaciones de conteos nacionales de aves como área significativa para la protección de las aves. Desde mediados de los 1980 y partiendo de Europa se han identificado hasta la fecha unas 4000 IBAs en Europa y más de 7000 IBAs en un total de 130 países del mundo. Un IBA debe existir como un espacio real o potencialmente protegido que

puede ser gestionado para la protección de aves o de la naturaleza en general. **En Es.** la organización ornitológica SEO/BirdLife implementa el Programa de Conservación de las Áreas Importantes para las Aves. Hasta la fecha se han inventariado 391 IBAs para la conservación de 160 especies, que cubren 16 mill. de ha [32% de la superficie del Estado y hasta el 75% de la Comunidad Autónoma de Extremadura], de las cuales un tercio son de importancia global. Hasta ahora el 40% de estas 391 IBAs están también protegidas como zonas especiales de conservación de la Red Natura 2000. A nivel internacional, SEO/BirdLife está realizando desde 1998 un programa en dos áreas de trabajo: Norte de África y Latinoamérica. En los Andes Tropicales [Bolivia, Colombia, Ecuador, Perú y Venezuela] y en las tierras bajas, costas e islas de esos países, se está llevando a cabo desde 2003 un proyecto de identificación con el resultado de 455 IBAs ya delimitadas que cubren un total de 776 128 km² [17% de la superficie de la región], de las cuales unas 180 tienen algún grado de protección [82% en Venezuela, 45% en Bolivia, 38% en Colombia]. La región de los Andes Tropicales es la más diversa del planeta con el 24% del total de la biodiversidad terrestre y alberga a más de 2700 especies de aves, más de 40 000 especies de plantas vasculares y 878 de anfibios. En ellos se encuentran cuatro de los 34 «hotspots» mundiales —áreas de gran diversidad, endemismo y extremadamente amenazadas. **En EE.UU.** el programa fue introducido en 1995 y es administrado por la National Audubon Society. Actualmente existen programas IBA en 46 estados con unas 1500 áreas identificadas. **En GB** han sido identificadas 295 IBAs que cubren una superficie de más de 31 000 km² [> 12% de la superficie total]. **En F.**, la «Ligue de Protection des Oiseaux» [LPO] es la organización que lleva a cabo este programa. **En D.** la organización responsable es la federación de protección de la naturaleza «*Naturschutzbund Deutschlands*» *[NABU]*; cf. BirdLife 2009; ►zona especial de conservación); *f* **zone** [f] **importante pour la conservation des oiseaux (ZICO)** (Habitat protégé devant assurer la conservation des espèces d'oiseaux menacées reposant sur la directive de l'Union européenne sur la conservation des oiseaux ; l'inventaire des habitats naturels d'intérêt ornithologique est établie par les représentants de *Bird Life International* [Fédération internationale regroupant les associations de protection des oiseaux] ; F., la Ligue de Protection des Oiseaux [LPO] établi la liste des sites ornithologiques majeurs ; chaque année au mois de janvier a lieu sur l'ensemble des zones humides d'Europe [baies, estuaires, zones humides littorales, plaines alluviales, fleuves, plans d'eau, marais, deltas et carrières en eau] un comptage international d'oiseaux d'eaux ; les informations recueillies permettent d'estimer les tailles des populations d'oiseaux d'eau et l'importance relative des sites d'hivernage, d'évaluer leurs tendances d'évolution, et de retenir, sur la base des critères dits de « Ramsar » [pertinence d'au moins 1 des 3 critères suivants: abriter des effectifs importants d'une ou plusieurs espèces d'oiseaux menacés, abriter des espèces à aire de répartition limitée ou liées à des habitats rares, abriter des nombres exceptionnels d'oiseaux migrateurs ou grégaires] les sites prioritaires pour la conservation quant ils abritent des effectifs significatifs ; ►zone spéciale de conservation [ZSC]) ; *g* **IBA-Gebiet** [n] („*Important Birds Area*"; Gebiet, das von dem internationalen ornithologischen Dachverband *Bird Life International* auf der Grundlage fachlicher Kriterien und nationaler Auswertungen von Vogelzählungen als bedeutsam im Sinne eines EU-Vogelschutzgebietes identifiziert und publiziert wird. Mitte der 1980er-Jahre startete in Europa das IBA-Programm und wurde dann weltweit verfolgt, so dass mittlerweile über 4000 Gebiete in Europa und global über 7000 erfasst sind; einige dieser Gebiete wurden von den Bundesländern resp. von Deutschland noch nicht oder nur in Teilflächen der EU als *Special Protection Area [SPA]* — ►besonderes

I

Schutzgebiet — gemeldet; da in der Vergangenheit die Europä-ische Kommission IBA-Publikationen als fachgutachterlichen Beitrag zur Bewertung der SPA-Meldungen durch die Mit-gliedsstaaten herangezogen hat, besitzen gegenwärtig diese Gebiete den Status eines faktischen Vogelschutzgebietes, und zwar solange, bis Deutschland seiner Meldeverpflichtung voll-ständig nachgekommen ist; **in D.** erfolgt die nationale Auswer-tung durch den NABU [Naturschutzbund Deutschlands e. V.]; seit 1967 findet jedes Jahr im Januar die internationale Wasser-vogelzählung, zz. von der Organisation *Wetlands International* koordiniert, statt; sie ist eine wesentliche Informationsgrundlage für die Bestandsermittlung und die räumliche wie zeitliche Ver-änderung von insgesamt 878 Vogelarten sowie auf der Grundlage der Ramsar-Kriterien die Feststellung und Bewertung von prio-ritären Schutzgebieten); *syn.* Vogelparadies [n].

Important Birds Area List [n] *conserv. nat'res.* ►IBA List.

importing [n] [US] *constr.* ►hauling to the site [US].

import [n] **of mineral elements** *envir. limn. pedol.* ►matter import.

imports [npl] *limn. pedol.* ►nutrient imports, ►nutrient input.

impounding [n] *eng. wat'man.* ►coastal impounding.

2832 impounding [n] **of a watercourse** *eng. wat'man.* (Creating a lake or reservoir for the regulation of a watercourse level or for irrigation purposes; ►damming 2, ►dam, ►floodable land, ►impoundment); *syn.* bunding [n] of a body of water [UK]; *s* **embalse [m] de un curso fluvial** (Construcción de embalse en un río para regular crecidas o con fines de riego; ►construcción de presas, ►embalse, ►presa, ►terreno inundable); *f* **endigue-ment [m] d'un bassin fluvial de retenue** (Mesure hydraulique réalisée dans le lit d'un cours d'eau qui a pour objet de relever le plan d'eau d'une rivière en vue de la création d'un bassin de régulation des pointes de crues ou d'une retenue d'arrosage ; ►barrage de vallée, ►construction d'ouvrages de rétention, ►lac de retenue, ►polder d'inondation) ; *g* **Eindeichung [f] eines Gewässeraufstau(e)s** (Wasserbauliche Maßnahmen zur Schaf-fung eines Ausgleichsspeichers an Fließgewässern, die der Regelung von Hochwasserspitzen oder zur Bewässerung dienen; ►Stauanlagenbau, ►Stausee, ►Talsperre, ►Überschwem-mungsbereich); *syn.* Fassung [f] eines Aufstau(e)s, Bau [m] eines Ausgleichsspeichers.

impounding reservoir [n] [UK] *wat'man.* ►impoundment.

2833 impoundment [n] *wat'man.* (Artificial lake for the storage of water, hydroelectric use, flood control or for regulating water levels in watercourses for better navigability; ►dam, ►impounding of a watercourse); *syn.* impounding reservoir [n] [also UK] (DOE 1977, 413); *s* **embalse [m]** (Lago artificial creado al construir una ►presa en un curso fluvial cuyas funcio-nes principales pueden ser almacenar agua, generar energía o regular el nivel de agua del río correspondiente; ►embalse de un curso fluvial); *syn.* pantano [m] (1); *f* **lac [m] de retenue** (*Terme spécifique* lac de barrage hydroélectrique ; ►barrage de vallée, ►endiguement d'un bassin fluvial de retenue) ; *syn.* lac-réservoir [m] (WMO 1984, 350), lac [m] de barrage (DUV 1984, 283), lac [m] d'un barrage-réservoir (PR 1987) ; *g* **Stausee [m]** (Durch Anstauen eines Wasserlaufes mittels eines Staudammes entstan-dener künstlicher See, der zur Trinkwasserversorgung, Elektrizi-tätsgewinnung oder aus wasserwirtschaftlichen Gründen zum Hochwasserschutz oder zur Wasserstandsregulierung für die Schifffahrt angelegt wurde; ►Eindeichung eines Gewässerauf-staues, ►Talsperre); *syn.* Speichersee [m].

2834 impoverishment [n] *conserv. ecol.* (Depletion of structural diversity in a landscape or ecosystem); *syn.* deterio-ration [n]; *s* **empobrecimiento [m]** (Disminución de la diver-sidad estructural del paisaje o de un ecosistema); *f* **banalisation [f]** (Appauvrissement de la diversité structurelle des paysages ou des milieux) ; *syn.* appauvrissement [m] ; *g* **Verarmung [f]** (Abnahme der Strukturvielfalt der Landschaft oder eines Öko-systems).

impregnation [n] [UK]**, pressure** *constr.* ►pressure treat-ment [US].

improvement [n] *agr. pedol.* ►agricultural land improvement; *plan.* ►infrastructure improvement; *plan. urb.* ►neighborhood improvement [US]/neigbourhood improvement [UK], ►park im-provement; *eng. wat'man.* ►riverbed improvement [US]/river-bed improvement [UK]; *constr. pedol.* ►soil improvement; *wat'man.* ►stream improvement; *urb.* ►urban improvement

improvement [n] [US]**, surface** *constr.* ►resurfacing.

improvement [n]**, timber stand** *constr. for.* ►improvement cut, #2.

improvement [n] [US]**, village** *plan. urb.* ►village renewal.

2835 improvement cut [n] [US] *constr. for.* (**1.** Elimination or suppression of less valuable in favo[u]r of more valuable tree growth, typically in mixed uneven-aged forest. In the Common-wealth it includes thinnings, cleanings and liberation fellings; SAF 1983); *syn.* improvement felling [n] [UK]. **2. timber stand improvement** [n] (Loose term comprising all intermediate cut-tings made to improve the composition, constitution, condition and increment of a timber stand. In many countries this includes girdling and poisoning, save when these are done solely to assist regeneration; SAF 1983); *s* **corte [m] de mejora** (Tala de leñosas no deseadas en un rodal denso para favorecer el crecimiento de las especies principales de la población); *f 1* **opération [f] d'amélioration** (*Sensu lato* élimination par coupe ou suppres-sion, des arbres ou végétaux ligneux de moindre valeur immé-diate ou d'avenir, en faveur de ceux de plus haut intérêt. Ces opérations incluent les dégagements, nettoiements et éclaircies ; DFM 1975) ; *syn.* coupe [f] d'amélioration ; *f 2* **coupe [f] de nettoiement retardée [CDN]** (*Sensu stricto* élimination par coupe, annulation ou tout autre moyen, dans un peuplement détérioré ou écrémé qui a dépassé le stade du gaulis, des essences ou sujets de moindre valeur. Quoique très voisin, ce concept est plus étroit que celui auquel correspond l'anglais « improvement felling » ; DFM 1975) ; *g* **Pflegehieb [m]** (Wegschneiden von unerwünschten Gehölzen in einem zu dichten Bestand, damit die wertvolleren, bestandsbildenden Arten [z. B. Zukunftsbäume] sich besser entwickeln können).

improvement felling [n] [UK] *constr. for.* ►improvement cut [US].

2836 improvement [n] **of agrarian structure** *agr. plan.* (Planning of agricultural land use structure in Europe through socio-economic changes, i.e. the reduction in the num-bers of farm units and people employed in agriculture, redistribu-tion of land holdings through ►farmland consolidation, etc.; **in** U.S., the consolidation of farming areas into large land holdings is by private agribusiness); *syn.* improvement [n] of agricultural structure; *s* **mejora [f] de la estructura agraria** (Medidas de mejora del uso del suelo agrícola por medio de cambios socioeco-nómicos como la reducción del número de fincas y de la pobla-ción activa en el campo, redistribución de la tierra por medio de la ►concentración parcelaria); *f* **amélioration [f] des structures rurales** (Amélioration de l'usage des sols ruraux au moyen d'une modification socio-économique des structures de la propriété agricole comme p. ex. la réduction du nombre des exploitations agricoles, la redistribution du parcellaire morcelé, actions fon-cières et ►remembrement, etc.) ; *syn.* amélioration [f] de la struc-ture agricole ; *g* **Agrarstrukturverbesserung [f]** (Verbesserung der landwirtschaftlichen Flächennutzung durch sozioökonomi-

schen Strukturwandel, wie z. B. Schrumpfung der Zahl der Betriebseinheiten und der in der Landwirtschaft Beschäftigten, Arrondierung von Betriebsflächen durch die ►Flurbereinigung etc.); *syn.* Verbesserung [f] der Agrarstruktur.

improvement of agricultural structure [n] *agr. plan.* ►improvement of agrarian structure.

improvement [n] **of a river course** *eng. wat'man.* ►river development.

improvement [n] **of rivers and streams** *leg. wat'man.* ►river engineering measures.

improvement [n] **of tree pits** *constr.* ►revitalization of tree pits.

improvement plan [n] *constr. plan. urb. wat'man.* ►rehabilitation plan.

improvement scheme [n] *constr. plan. urb. wat'man.* ►rehabilitation plan.

2837 impurity [n] **of seed** *agr. constr.* (Presence of foreign matter in seeds); *s* **impurezas [fpl] en semillas**; *f* **impuretés [fpl] des semences** ; *g* **Besatz [m] (1)** (Verunreinigung des Saatgutes).

2838 in a green setting [n] *urb. landsc.* (A spatial context defined by vegetative and physiographic entourage or intentionally planted areas); *s* **en un marco de verde [loc]** (Relativo a un lugar o una ubicación determinada mayormente por vegetación); *f* **dans un cadre [m] de verdure** *syn.* dans un cadre [m] naturel ; *g* **in grüner Umgebung [f]** (Einen Ort oder eine Gegend betreffend, die vorwiegend durch Vegetation geprägt ist).

2839 in an open hunting area [n] *hunt.* (►shooting preserve); *syn.* in the wild [n] [also UK]; *s* **zona [f] de caza libre** (Término de caza que en alemán denomina las zonas no envalladas donde está permitido cazar; ►reserva de caza); *f* **territoire [m] de chasse ouvert** (Dans le langage des chasseurs terme désignant un territoire de chasse non enclos ; ►enclos réserve de chasse) ; *g* **in freier Wildbahn [loc]** (Waidmännischer/weidmännischer Begriff für ein nicht eingegattertes Jagdrevier; *opp.* Gatterrevier: ►Wildgehege).

incineration plant [n] **for toxic/hazardous waste** *envir.* ►incinerator for toxic/hazardous waste.

2840 incinerator [n] **for toxic/hazardous waste** [US] *envir.* (General term for an incinerator, which burns rubbish and other refuse under supervision, the remains of which are then disposed of according to regulations; ►waste incineration plant); *syn.* incineration plant [n] for toxic/hazardous waste; *s* **planta [f] de incineración de residuos industriales peligrosos** (Instalación para quemar residuos tóxicos y peligrosos que cumple con normas técnicas especiales en cuanto a la temperatura de combustión, la filtración de las emisiones y el depósito de los residuos inertes de combustión; ►planta de incineración de residuos sólidos urbanos); *syn.* incineradora [f] de residuos industriales peligrosos; *f 1* **installation [f] d'incinération** (Installation incinérant des déchets industriels spéciaux définis par le décret n° 97-517 du 15 mai 1997 en application de la liste européenne des déchets dangereux, avec ou sans récupération de la chaleur produite, les opérations d'incinération étant déterminées par le contrôle des déchets admis à l'incinération, la conduite du bon déroulement de la combustion, le traitement des émissions gazeuses issues de la combustion des déchets et de la bonne élimination des déchets produits par ce traitement ; *terme spécifique* installation de coincinération [installation d'incinération spécialisée dans l'incinération de déchets industriels par utilisation comme combustible d'appoint dans des installations dont la vocation première n'est pas l'incinération] ; ►usine d'incinération de résidus urbains) ; *f 2* **installation [f] spécialisée d'incinération** (Installation explicitement conçue et réalisée pour

incinérer des déchets industriels spéciaux et dont la première autorisation d'exploiter au titre de la législation sur les installations classées pour la protection de l'environnement a été délivrée pour ce faire) ; *g* **Sondermüllverbrennungsanlage [f]** (Allgemeiner Begriff für Verbrennungsanlage, die überwachungsbedürftige Abfälle und ähnliche Stoffe verbrennt und die Rückstände ordnungsgemäß entsorgt; *cf.* 17. BImSchV vom 23.11.1990/19.08.2003 und durch Verordnung zur Absicherung von Luftqualitätsanforderungen in der Verordnung über Großfeuerungs- und Gasturbinenanlagen und in der Verordnung über die Verbrennung von Abfällen [BGBl I v. 30.01.2009, 129] sowie § 41 KrW-/AbfG; ►Müllverbrennungsanlage).

2841 incised slope [n] *constr.* (Cut slope resulting from excavation on a hillside; ►cut slope); *syn.* exposed cut [n] [also US], exposed face [n] [also US], cut face [n] [also US]; *s* **cara [f] de corte** (►Talud de corte resultante de excavación en una pendiente); *f* **talus [m] de fouille** (Plan incliné formé lors de travaux d'excavation ; ►talus en déblai) ; *syn.* talus [m] d'excavation ; *g* **Einschnittböschung [f]** (Durch Abgrabung in bestehendes Gelände entstandene Böschung; ►Anschnittsböschung)

2842 inclination [n] *constr. geo. plan. trans.* (**1.** *constr. geo. plan. trans.* Deviation from the horizontal, measured as a percentage or ratio 1:n, Rise:Run; ►cross slope [US]/cross fall [UK], ►incline 2, ►slope [US] 2/fall [UK]. **2.** *geo.* Slope of the earth's surface measured against the horizontal); *s* **inclinación [f] (2)** (**1.** *constr. plan. trans.* Desviación de la horizontal, medida en porcentaje o en razón 1:n; ►inclinación del gradiente, ►pendiente, ►pendiente transversal. **2.** *geo.* Ángulo formado entre la superficie de la tierra y la horizontal); *syn.* de *1* declive [m]; *f 1* **déclivité [f]** *constr. plan. trans.* (Inclinaison par rapport au plan horizontal exprimé en % ou dans la proportion de 1:n ; ►déclivité longitudinale, ►pente, ►pente transversale ; *f 2* **inclinaison [f]** *geo.* (Angle formé par la surface du sol avec l'horizontale) ; *g* **Neigung [f] (2)** (**1.** *constr. plan. trans.* Abweichung einer Geraden von einer Bezugsgeraden in % oder 1:n gemessen; ►Gefälle, ►Längsneigung, ►Quergefälle. **2.** *geo.* Neigung der Erdoberfläche gegenüber der Horizontalen); *syn. zu 2.* Inklination [f], Neigungsgrad [m].

incline [n] (1) *geo.* ►slope (1).

2843 incline [n] (2) *trans.* (Longitudinal gradient of a road alignment; ►vertical alignment 1); *s* **inclinación [f] del gradiente** (►Gradiente longitudinal de un trazado de carretera); *f* **déclivité [f] longitudinale** (Inclinaison du ►profil en long d'un tracé d'une route, d'une autoroute, etc.) ; *syn.* inclinaison [f] longitudinale ; *g* **Längsneigung [f]** (Neigung der ►Gradiente einer Trasse).

incline change [n] *constr. trans.* ►gradient change.

inclined [pp] *constr.* ►sloped.

2844 incoming commuter [n] *sociol. trans. urb.* (Person working or studying in one area and travelling from the area of residence; ►commuter pattern; *opp.* ►outgoing commuter); *s* **conmutador, -a [m/f] entrante** (Persona que llega diariamente a trabajar o estudiar a una ciudad; ►movimientos migracionales pendulares; *opp.* ►conmutador, -a saliente); *f* **migrant [m] alternant considéré du point de vue de la collectivité réceptrice (±)** (Personne extérieure à la commune effectuant des ►déplacements domicile-travail et se déplaçant de son domicile vers son lieu de travail ; ►migrant [m] alternant considéré du point de vue de la collectivité émettrice) ; *g* **Einpendler [m]** (Person, die nicht im Gemeindegebiet ihrer Arbeits- resp. Ausbildungsstätte wohnt, sondern aus einer anderen Gemeinde täglich an- und abreist; *cf.* PW 1970; *opp.* ►Auspendler, ►Pendlerbewegungen).

2845 incompatible [adj] *ecol. plan.* (Inconsistent with the natural environment); *s* **incompatible [adj] (con el medio natu-**

ral); *syn.* dañino/a [adj] al medio natural; *f* **incompatible [adj]** (avec la vocation des espaces naturels environnants ; cf. R. 111-14-1) ; *g* **unverträglich [adj]** (So beschaffen, dass es nicht mit den Gegebenheiten harmoniert).

2846 in compliance with the issue of an instruction [loc] *constr.* (According to written directives and orders, issued by the supervising architect/landscape architect or appointed inspector, which must be carried out by the contractor within a specified period; *s* **según las instrucciones del jefe de obra [loc]**; *f 1* **d'après les instructions du chef de chantier [loc]** (oral) ; *f 2* **d'après l'ordre de service [loc]** (écrit) ; *g* **nach Anweisung der Bauleitung [loc]** *syn.* nach Angaben der Bauleitung [loc].

2847 increase [n] *chem. biol.* (Augmentation of nutrient, oxygen content, etc.; ►eutrophication); *syn.* enrichment [n]; *s* **enriquecimiento [m]** (Aumento del contenido de nutrientes, oxígeno, etc.; ►eutrofización); *syn.* acumulación [f]; *f* **enrichissement [m]** (Naturel ou artificiel de la concentration en matières nutritives, en oxygène, etc. ; ►eutrophisation) ; *g* **Anreicherung [f] (2)** (Erhöhung des Gehalts von Nährstoffen, Sauerstoff etc.; ►Eutrophierung).

increase [n]**, population** *plan. sociol.* ►population growth.

2848 increase [n] **in estimated volume** *contr.* (Exceeding of quantities agreed upon contractually in the specifications. A new price is to be negotiated upon request for an excess of more than 10%; *opp.* ►reduction in estimated volume; ►list of bid items and quantities); *s* **aumento [m] de cantidades** (Superación de las cantidades fijadas en el ►resumen de prestaciones. Si el **a.** supera el 10%, hay que fijar un nuevo precio si el contratista lo exige; *opp.* ►reducción de cantidades); *f* **augmentation [f] de la masse des travaux** (Variation en cours d'exécution des travaux de la masse des travaux prévus au marché dans le ►descriptif-quantitatif ; pour une augmentation au-delà de l'augmentation limite [seuil] contractuelle [dans le cas de marchés réglés au prix unitaires, de plus du tiers en plus des quantités portées au détail estimatif] l'entrepreneur a le droit d'être indemnisé du préjudice subi du fait de cette augmentation ; *opp.* ►diminution de la masse des travaux ; CCAG., art. 15 du décret n° 76-87 du 21 janvier 1976) ; *syn.* augmentation [f] des quantités ; *g* **Massenmehrung [f]** (Überschreitung des mit dem ►Leistungsverzeichnis vertraglich vereinbarten Mengenansatzes. Für die über 10 % hinausgehende **M.** ist auf Verlangen ein neuer Preis zu vereinbaren; cf. § 2 VOB Teil B; *opp.* ►Massenminderung); *syn.* Überschreitung [f] des Mengenansatzes.

2849 increase [n] **in fees/charges** *contr. prof.* (Addition monetary requirements due to a number of factors within a contractual agreement); *s* **incremento [m] de honorarios** *syn.* subida [f] de honorario(s); *f* **augmentation [f] des honoraires** (Pour cause d'augmentation de la mission, de la remise en cause du programme ou du calendrier de réalisation, de modification des documents approuvés, etc.) ; *g* **Honorarerhöhung [f]** *syn.* Erhöhung [f] des Honorars.

2850 increase in soil acidity [n] *pedol.* (Lowering of pH value; ►acid soil); *s* **acidificación [f] del suelo** (Reducción del pH del suelo; ►suelo ácido); *f* **acidification [f] du sol** (►sol acide) ; *g* **Bodenversauerung [f]** (Absinken des pH-Wertes; ►saurer Boden).

2851 increase [n] **in stocking density** *agr. syn.* increase [n] in stock ratio; *s* **aumento [m] de densidad en explotación ganadera** (Aumento de número de cabezas en explotación ganadera); *f* **repeuplement [m]** (Augmentation artificielle d'un peuplement animal) ; *g* **Aufstockung [f] eines Tierbestandes** (Erhöhung der Anzahl von Tieren in einem Tierbestand); *syn.* Erhöhung [f] eines Tierbestandes.

increase [n] **in stock ratio** *agr.* ►increase in stocking density.

2852 increase [n] **of staff** *adm. prof.* (Addition of employees); *s* **ampliación [f] de plantilla** *syn.* ampliación [f] de personal; *f* **renforcement [m] (1)** (des effectifs du personnel) ; *g* **Personalaufstockung [f]** (Vergrößerung des Personalbestandes); *syn.* Personalerhöhung [f].

2853 increaser [n] **[US]** *constr.* (MEA 1985; pipe fitting connecting a smaller pipe with a bigger one; *opp.* ►reducing coupling); *syn.* tapered pipe [n] [UK]; *s* **acoplamiento [m] ampliador** (En tubería, tubo con bocas desiguales para conectar a uno mayor; *opp.* ►acoplamiento reductor); *f* **cône [m] d'augmentation** (Pièce accessoire utilisée pour le manchonnage de tubes d'assainissement pour le passage à un diamètre supérieur ; ►cône de réduction) ; *g* **Übergangsstück [n]** (Im Rohrleitungsbau ein Rohr mit Übergang von einer kleineren zur größeren Nennweite; *opp.* ►Reduktionsstück).

2854 increasing [n] **of pH value** *chem. linm. pedol.* (for greater alkalinity); *syn.* raising [n] of pH value; *s* **aumento [m] del pH** *syn.* crecimiento [m] del pH; *f* **augmentation [f] de la valeur du pH** *syn.* augmentation [f] du pH ; *g* **pH-Werterhöhung [f]** *syn.* Erhöhung [f] des pH-Wertes.

2855 incubation period [n] *zool.* (Time required for laying and incubating eggs); *syn.* brooding period [n]; *s* **periodo/período [m] de incubación** (Tiempo requerido para la puesta de huevos y la incubación); *f* **période [f] d'incubation** (Temps nécessaire à la couvée des œufs) ; *g* **Brutzeit [f] (2)** (Zeit, die für das Ausbrüten von Eiern erforderlich ist); *syn.* Bebrütungszeit [f].

independent site inspector [n] **[UK]** *constr. prof.* ►client's project representative [US].

indicator [n] *biol. ecol. envir.* ►environmental stress indicator.

2856 indicator fossil [n] *biol. geo.* (Fossil of flora or fauna which is characteristic of an epoch); *s* **fósil [m] guía** (Especie de flora o fauna característica de una época que se presenta como fósil de forma masiva en una capa geológica no muy espesa); *f* **fossile [m] indicateur** (Espèce végétale ou animale fossilisée dont la découverte sur de grandes étendues mais sur une couche de faible épaisseur est caractéristique d'une époque précise de l'histoire de la terre) ; *g* **Leitfossil [n]** (Für einen bestimmten Zeitabschnitt charakteristische Tier- oder Pflanzenart, die als Versteinerung in möglichst weiter flächenhafter Ausdehnung, aber gering mächtiger Schicht vorkommt).

2857 indicator plant [n] *ecol. phyt.* (Plant which reveals particular environmental conditions; *generic term* ►indicator species; *specific terms* ►acidophilous species, ►aridity indicator plant, ►biological indicator, ►calcareous indicator species, ►impeded water indicator plant, ►light-demanding plant, ►moisture indicator plant, ►nitrogen indicator [species], ►shade plant, ►wetland indicator plant, ►wetness indicator plant); *s* **planta [f] indicadora** (Término genérico para especie vegetal que por su presencia, o también por su ausencia, proporciona información sobre alguna característica del medio o biocenosis en la que se encuentra, como del pH, la humedad, el contenido de nitratos, la contaminación del aire o del agua; cf. DINA 1987; *término genérico* ►especie indicadora; *términos específicos* ►bioindicador, ►especie indicadora de cal, ►especie indicadora de humedad, ►heliófito, ►planta indicadora de acidez, ►planta indicadora de encharcamiento, ►planta indicadora de nitrógeno, ►planta indicadora de sequía, ►planta de sombra); *f* **plante [f] indicatrice** (*Terme générique* ►espèce indicatrice ; *termes spécifiques* ►espèce héliophile, ►espèce indicatrice d'acidité, ►espèce indicatrice de calcaire, ►plante indicatrice de l'engorgement, ►espèce indicatrice d'humidité, ►indicateur biologique, ►plante d'ombre, ►plante indicatrice d'azote, ►plante

indicatrice de forte humidité, ▶plante indicatrice de sécheresse) ;
g **Zeigerpflanze [f]** (Pflanzenart, dessen An- oder Abwesenheit in einem Bestand innerhalb gewisser Grenzen bestimmte biotische oder abiotische Faktoren oder ein Faktorenverhältnis anzeigt, z. B. Bodenreaktion, Feuchtigkeit, Nitratgehalt, Gewässer- oder Luftverschmutzung; *UBe* ▶Bioindikator, ▶Feuchtigkeitszeiger, ▶Kalkzeigerpflanze, ▶Lichtpflanze, ▶Nässezeiger, ▶Säurezeiger, ▶Schattenpflanze, ▶Staunässezeiger, ▶Stickstoffzeiger, ▶Trockenheitszeiger; *OB* ▶Zeigerart); *syn.* Indikatorpflanze [f].

indicator plant [n], bad drainage *phyt.* ▶impeded water indicator plant.

2858 indicator species [n] *ecol. envir.* (Any plant or animal that, by its presence, its frequency or its vigo[u]r, indicates any particular characteristic of the ▶habitat 2; cf. SAF 1983; *specific terms* ▶acidophilous species, ▶aridity indicator plant, ▶biological indicator, ▶calcareous indicator species, ▶light-demanding plant, ▶indicator plant, ▶nitrogen indicator [species], ▶shade plant, ▶site indicator [species], ▶wetland indicator plant, ▶wetness indicator plant); *s* **especie [f] indicadora** (Término genérico para especie vegetal o animal que, por su presencia, su vigor y su abundancia en un ▶biótopo revela determinadas propiedades ecológicas o el grado de contaminación de éste. *Términos específicos* ▶bioindicador, ▶especie indicadora de humedad, ▶especie indicadora de cal, ▶heliófito, ▶planta de sombra, ▶planta indicadora, ▶planta indicadora de acidez, ▶planta indicadora de encharcamiento, ▶planta indicadora de estación, ▶planta indicadora de nitrógeno, ▶planta indicadora de sequía); *syn.* indicadora [f]; *f* **espèce [f] indicatrice** (Espèce végétale ou animale dont la présence ou absence dans un ▶biotope est caractéristique des liens existants entre cette espèce et un facteur ou un ensemble de facteurs biotiques ou abiotiques, [p. ex. acidité du sol, humidité, concentration en nitrates, pollution de l'eau ou atmosphérique] : une espèce particulièrement sensible aux conditions de l'environnement peut être utilisée pour évaluer la qualité des milieux naturels ; *termes spécifiques* ▶espèce héliophile, ▶espèce indicatrice d'acidité, ▶espèce indicatrice de calcaire, ▶espèce indicatrice de la station, ▶espèce indicatrice d'humidité, ▶indicateur biologique, ▶plante d'ombre, ▶plante indicatrice, ▶plante indicatrice d'azote, ▶plante indicatrice de forte humidité, ▶plante indicatrice de sécheresse) ;
g **Zeigerart [f]** (Pflanzen- oder Tierart, dessen An- oder Abwesenheit in einem ▶Biotop innerhalb gewisser Grenzen bestimmte biotische oder abiotische Faktoren oder ein Faktorenverhältnis anzeigt, z. B. Bodenreaktion, Feuchtigkeit, Nitratgehalt, Gewässer- oder Luftverschmutzung; *UBe* ▶Bioindikator, ▶Feuchtigkeitszeiger, ▶Kalkzeiger, ▶Lichtpflanze, ▶Nässezeiger, ▶Säurezeiger, ▶Schattenpflanze, ▶Standortzeiger, ▶Stickstoffzeiger, ▶Trockenheitszeiger, ▶Zeigerpflanze); *syn.* Indikatorart [f], Indikatororganismus [m].

2859 indicator value [n] *phyt.* (Significance of a plant or organism in the indication of specific biotic or abiotic factors; ▶indicator species); *s* **valor [m] indicador** (Capacidad de una planta o un organismo de indicar la presencia de factores bióticos o abióticos específicos; ▶especie indicadora); *f* **valeur [f] indicatrice** (Capacité d'une plante, d'un organisme, d'un groupement à réagir aux variations des facteurs biotiques ou abiotiques et souvent caractérisés par un indice ; les systèmes de bioindication, utilisant la présence/absence des plantes, reposent sur une évaluation du caractère bioindicateur des espèces établie de façon empirique à partir de l'expérience des naturalistes ; ▶espèce indicatrice) ; *g* **Zeigerwert [m, o. Pl.]** (Bedeutung einer Pflanze oder eines Organismus, einen bestimmten biotischen oder abiotischen Faktor oder ein Faktorenverhältnis anzuzeigen; ▶Zeigerart).

2860 indifferent species [n] *phyt.* (That species, according to the ▶degree of fidelity, which occurs as a ▶companion plant 1 and, therefore, is neither a ▶character species nor a ▶differential species); *s* **especie [f] indiferente** (Especie que por su ▶grado de fidelidad se presenta como acompañante en la comunidad vegetal, por lo que no es ni ▶especie característica ni ▶especie diferencial); *f* **espèce [f] indifférente** (Espèce qui en rapport avec le ▶degré de fidélité des espèces dans les peuplements végétaux n'est qu'une ▶espèce compagne ne croissant que plus ou moins abondamment dans plusieurs groupements et ne pouvant être considérée ni comme ▶espèce caractéristique [exclusive] ou comme une ▶espèce différentielle) ; *g* **vage Art [f]** (Art, die hinsichtlich des ▶Treuegrades bei Pflanzengesellschaften als ▶Begleiter 1 vorkommt und somit weder als ▶Kennart noch als ▶Trennart gelten kann).

2861 indigenous plant species [n] *phyt. plant.* (1. Descriptive term applied to a species originating in a natural area before the last glacial epoch. 2. Species native to a particular site and thereby able to thrive under the prevailing ecological conditions; ▶endemic species, ▶species suited to site conditions; *opp.* ▶exotic species); *syn.* autochthonous plant species [n], native plant species [n]; *s* **especie [f] vegetal autóctona** (Especie vegetal natural de un país, no introducida ni naturalizada, sino indígena; cf. DB 1985; *opp.* ▶especie exótica, ▶especie adaptada a la residencia ecológica, ▶especie endémica); *syn.* especie [f] vegetal indígena, especie [f] vegetal nativa; *f* **espèce [f] végétale indigène** (1. qui appartient à un peuplement végétal spontané, indigène en Europe avant l'apparition de l'être humain. 2. Espèce végétale spontanée originaire d'un territoire dans lequel on la rencontre depuis très longtemps, et y vivant conformément au climax et qui croît naturellement sur un milieu en fonction des conditions stationnelles de celui-ci ; l'indigénat d'une espèce peut être constaté par analyse pollinique ; ▶espèce bien adaptée aux conditions de la station, ▶espèce endémique, *opp.* ▶espèce exotique) ; *syn.* espèce [f] végétale autochtone ; *g* **bodenständige Pflanzenart [f]** (1. Zum alten Bestand der Flora zugehörig; *in Europa* vor dem Erscheinen des Menschen an Ort und Stelle heimisch. 2. Art, die auf Grund der Landschaftsfaktoren auf einem bestimmten Standort, immer mit einem strengen Gebietsbezug gemäß der potenziellen natürlichen Vegetation lebt [entstanden ist] und nicht eingeführt wurde; ob eine Pflanzenart bodenständig ist, kann mit Hilfe der Pollenanalyse und des genetischen Bildes, das für die entsprechende geografische Region typisch ist, nachgewiesen werden. In D. wird der synonyme Begriff ‚heimisch' aus naturschutzfachlicher Sicht leider vereinzelt auch für Arten verwendet, die eingeführt oder eingeschleppt wurden und sich etabliert haben. 3. Der Landschaftsarchitekt ALWIN SEIFERT definierte 1939 den Begriff **bodenständig** aus ästhetischer Sicht folgendermaßen: „Bodenständig ist im Garten jede fremde und jede heimische Pflanze, die das volle Maß ihrer Schönheit erreicht und mit ihrer engeren und weiteren Umgebung in künstlerischer und biologischer Harmonie steht"; SEI 1939, 44; ▶endemische Art, ▶standortgerechte Art; *opp.* ▶exotische Art); *syn.* autochthone Pflanzenart [f], biotopeigene Pflanzenart [f], gebietseigene Pflanzenart [f], einheimische Pflanzenart [f], heimische Pflanzenart [f], standortheimische Pflanzenart [f].

2862 indigenous species [n] *phyt. zool.* (Animal or plant species native to a particular region and thereby able to thrive under the prevailing ecological conditions and which was not introduced; ▶endemic species, ▶species suited to site conditions; *opp.* ▶exotic species); *syn.* autochthonous species [n], native species [n], naturally occurring species [n]; *s* **especie [f] indígena** (Especie vegetal o animal natural de un país, no introducida ni naturalizada, sino indígena; cf. DB 1985; *opp.*

►especie exótica; ►especie adaptada a la estación, ►especie endémica); *syn.* especie [f] autóctona, especie [f] nativa; *f* **espèce [f] indigène** (Espèce végétale ou animale qui habite ou croît naturellement dans un pays, qui n'a pas été introduite ; ►espèce bien adaptée aux conditions de la station, ►espèce endémique ; *opp.* ►espèce exotique) ; *syn.* espèce [f] autochtone ; *g* **einheimische Art [f]** (Tier- oder Pflanzenart, die auf Grund der Landschaftsfaktoren in einem bestimmten Gebiet lebt und nicht eingeführt wurde; **e. A.** bedeutet auch, dass sie über die naturraumtypische genetische Vielfalt innerhalb ihrer Art verfügt; ►endemische Art, ►standortgerechte Art; *opp.* ►exotische Art); *syn.* autochthone Art [f], biotopeigene Art [f], gebietsheimische Art [f], heimische Art [f], indigene Art [f], standortheimische Art [f];

2863 individual distance [n] *zool.* (Congenitally established minimum distance toward[s] which individuals of the same species will approach each other without provoking aggression); *s* **distancia [f] entre individuos** (Distancia mínima que necesitan los individuos de cada especie entre sí, que si no es respetada por otros provoca actitudes agresivas); *f* **distance [f] individuelle** (Distance pour laquelle un individu provoque l'agression ou l'évitement chez un autre individu de la même espèce ; LAF 1990, 295) ; *g* **Individualdistanz [f]** (Bei sozial lebenden Distanzarten erblich festgelegter Mindestabstand, bis zu welchem sich artgleiche Individuen einander nähern können, ohne dass Aggressionen ausgelöst werden); *syn.* Individualabstand [m].

2864 individual house [n] *arch. leg. urb.* (Generic term for detached and semi-detached house; ►detached building development); *syn.* single house [n]; *s* **casa [f] individual** (►edificación discontinua); *syn.* villa [f]; *f 1* **bâtiment [m] indépendant** *arch.* (Construction isolée ou séparée d'une autre par un mur mitoyen) ; *syn.* construction [f] isolée ; *f 2* **bâtiment [m] isolé** *leg. urb.* (Construction implantée isolément ; ►implantation des constructions en ordre discontinu, ►marge d'isolement) ; *syn.* habitation [f] isolée ; construction [f] isolée, maison [f] isolée ; *g* **Einzelhaus [n]** (Freistehendes Gebäude, das in keiner Weise mit einem anderen Haus durch gemeinsame Wände verbunden ist und über ein selbständiges Erschließungssystem verfügt; zu den **E.häusern** zählen freistehende Einfamilienhäuser, aber auch Wohnblocks, Hochhäuser und Fabrikhallen unter 50 m Länge, unabhängig von ihrer Höhe und Nutzung, und bauordnungsrechtlich durch die ►offene Bauweise 2 definiert).

2865 individual overnight accommodation [n] *recr.* (Lodging in hotels, inns, or second homes, etc. by a person on vacation [US]/holiday [UK]; ►hostelry); *s* **hospedaje [m] individual** (Alojamiento de turistas individuales en hotel, fonda, pensión o casa particular, en contraposición al hospedaje colectivo; ►albergue); *f* **hébergement [m] individuel** (de vacanciers solitaires dans des hôtels, des pensions, etc. ; ►gîte) ; *g* **Beherbergung [f] von Einzelpersonen** (Unterbringung von einzelnen Personen in Gasthöfen, in privaten Zweitwohnungen etc.; ►Beherbergungsstätte); *syn.* Unterbringung [f] von Einzelpersonen.

2866 individual sewage disposal system [n] [US] *envir.* (A system of sewage treatment tanks and disposal facilities—often a three chambers system [US]—designed for a single building, establishment or lot, not served by a public sewer; DAC 1975); *syn.* miniature sewage treatment plant [n] [UK], private sewage treatment plant [n] [UK], septic system [n] [US], septic tank disposal system [n] [US]; *s* **instalación [f] individual de depuración de agua** (Pequeña planta de tratamiento de aguas de una casa o un grupo de ellas, una granja, etc. situada en el mismo terreno); *f 1* **microstation [f] d'épuration** (Une installation d'épuration biologique à boues activées assurant le traitement de l'ensemble des eaux usées domestiques — eaux vannes et eaux ménagères — d'une seule parcelle bâtie dans les communes

rurales. Ce dispositif est le résultat d'une miniaturisation des stations d'épuration urbaine ; cf. ASS 1987, 150) ; *syn.* installation [f] d'épuration individuelle, système [m] d'épuration individuelle ; *f 2* **fosse [f] septique** (Un dispositif d'épuration individuelle assurant un traitement préalable des eaux usées domestiques); *g* **Kleinkläranlage [f]** (Kläranlage, die die Abwässer eines Hauses, Häuserkomplexes, Einzelhofes, Wanderheimes etc. auf dem Grundstück in Klärgruben je nach Ausstattung nur vorklären [Mehrkammeranlage] oder bis zur biologischen Stufe voll reinigt); *syn.* Grundstückskläranlage [f], Hauskläranlage [f].

2867 individual territory [n] *zool.* (►Home range 1 inhabited alone by a single animal; ►territoriality); *s* **territorio [m] de un individuo** (►Área de actividad de un solo animal o una pareja; ►territorialidad); *f* **territoire [m] exclusif** (►Territoire d'activité habité par une seule espèce animale ; LAF 1990, 888-890 ; ►comportement territorial); *g* **Individualterritorium [n]** (Ein ►Streifgebiet, das von einem einzelnen Tier bewohnt wird; ►Territorialverhalten); *syn.* Individuen-Territorium [n].

2868 individual urn grave [n] *adm. landsc.* (Ground burial of cremation ashes); *s* **tumba [f] individual para urna**; *f* **jardin [m] d'urnes individuelles** ; *g* **Urneneinzelgrabstätte [f]** (Kleine Grabstätte für eine Aschenbeisetzung).

2869 individual woody plant [n] *landsc. plant.* (Single tree or shrub in an area; ►specimen tree/shrub); *s* **leñosa [f] solitaria** (Árbol o arbusto aislado; ►ejemplar, ►espécimen); *f* **arbre/arbuste [m] isolé** (Végétal ligneux en position isolée ; *contexte* arbre/arbuste [à planter] en isolé) ; *syn.* ligneux [m] isolé, *synonymes peu usités et à éviter* plante [f] solitaire, solitaire [m] ; *g* **Solitärgehölz [n] (1)** (Baum oder Strauch in Einzelstellung); *syn.* Einzelgehölz [n].

2870 indoor bathing complex [n] [US] *recr.* (*opp.* ►outdoor bathing complex [US]/open air swimming baths [UK]); *syn.* indoor swimming baths [npl] [UK]; *s* **piscina [f] cubierta** (Instalación de natación con todo el equipamiento necesario; *opp.* ►piscina al aire libre); *syn.* alberca [f] cubierta [MEX], pileta [f] cubierta [RA], instalación [f] de piscinas cubiertas; *f* **piscine [f] couverte** (Piscine aménagée dans un bâtiment y compris toutes les installations annexes nécessaires ; *terme spécifique* piscine sports-loisirs ; *opp.* ►piscine de plein air) ; *g* **Hallenbad [n]** (Eingehaustes Schwimmbad mit allen dazu notwendigen Einrichtungen; *opp.* ►Freibad); *syn.* Hallenschwimmbad [n].

indoor exhibition [n] *hort.* ►indoor horticultural exhibition.

2871 indoor horticultural exhibition [n] *hort.* (►flower show); *syn.* indoor (horticultural) show [n]; *s* **exhibición [f] (de horticultura) en pabellones** (Exposición horticultural dentro de edificios; ►exhibición floral); *f* **exposition [f] horticole sous chapiteau** (Exposition ornementale qui se déroule à l'intérieur de grands bâtiments ; ►floralies) ; *syn.* exposition [f] en intérieur, exposition [f] horticole couverte ; *g* **gärtnerische Hallenschau [f]** (Ausstellung in einem Gebäude, das aus einem großen, hohen Raum besteht und die gärtnerische Leistungsfähigkeit in den unterschiedlichen Produktionsbereichen und im Garten- und Landschaftsbau zeigt, z. B. bei Gartenbauausstellungen oder in Messen; ►Blumenschau).

indoor horticultural show [n] *hort.* ►indoor horticultural exhibition.

indoor pool [n] [UK] *recr.* ►swimming pool hall [US].

indoor show [n] *hort.* ►indoor horticultural exhibition.

indoor steps [npl] *arch. constr.* ►steps.

indoor swimming baths [npl] [UK] *recr.* ►indoor bathing complex [US].

indoor tennis courts [npl] [UK] *recr.* ▶indoor tennis facility [US].

2872 indoor tennis facility [n] [US] *recr.* (IBDD 1981); *syn.* tennis building [n] [US], indoor tennis courts [npl] [UK]; *s* **pabellón** [m] **de tenis** (Instalación para el deporte de tenis, que generalmente consiste en un solo gran espacio dividido en pistas [ES]/canchas [AL] individuales); *f* **court** [m] **de tennis couvert** ; *g* **Tennishalle** [f] (Anlage für den Tennissport, die vorwiegend aus einem einzigen großen Raum besteht); *syn.* Hallentennisanlage [f].

2873 industrial area [n] *urb.* (Generic term; ▶industrial park [US]/industrial estate [UK]); *s* **zona** [f] **industrial** (Término genérico; ▶parque industrial); *f* **zone** [f] **industrielle** (Ensemble de terrains à destination industrielle, à implantation spontanée ou aménagés [viabilisés et comportant des équipements collectifs pour les entreprises] ; ▶parc d'activités, ▶parc industriel) ; *syn.* région [f] industrielle ; *g* **Industriegebiet** [n] **(1)** (allgemeiner Begriff; ▶Industriepark); *syn.* Industriefläche [f].

industrial development [n] *plan. urb.* ▶established industrial development, ▶planned industrial development.

2874 industrial effluent treatment [n] *envir.* (Processing of industrial waste streams to separate toxic chemicals and other properties from common sewerage); *s* **tratamiento** [m] **de vertidos industriales** *syn.* tratamiento [m] de aguas residuales industriales; *f* **traitement** [m] **des effluents industriels** *syn.* traitement [m] des eaux usées (LA 1981, 473), traitement [m] des rejets industriels, traitement [m] des eaux résiduaires industrielles ; *g* **Abwasserbehandlung** [f] **der Industrieabwässer**.

industrial estate [n] [UK] *plan. urb.* ▶industrial park [US], ▶location of commercial facilities and light industries.

2875 industrial gaseous emissions [npl] *envir.* (Exhaust fumes emitted during industrial processes, typically regulated to curb toxicity or air quality degradation); *s* **contaminación** [f] **atmosférica de origen industrial** *syn.* emisión [f] industrial, efluente [m] gaseoso industrial; *f* **pollution** [f] **atmosphérique en provenance des installations industrielles** ; *g* **Industrieabgase** [npl] (Durch industrielle Produktion entstehende Abgase).

2876 industrial green area [n] *landsc. urb.* (Predominantly green open space integrated within industrial areas and designed to diminish the effects of air pollution and to provide opportunities for recreation in work breaks); *s* **espacios** [mpl] **verdes industriales** (Zonas verdes que rodean a emplazamientos industriales y tienen funciones protectoras, estéticas, de ordenamiento urbano y a veces recreativas); *f* **espace** [m] **vert d'accompagnement industriel** (Espaces à prédominance végétale constitués des abords des installations industrielles aménagés pour l'accueil des visiteurs et du personnel [parking et circulations], l'insertion des bâtiments dans le paysage, la protection contre les pollutions et le bruit) ; *syn.* jardin [m] d'usine ; *g* **Industriegrün** [n] (Grünbestimmte Freiräume, die der Eingliederung, Erschließung, Erhaltung von Industrie- und Gewerbeanlagen und dem Schutz vor Immissionen sowie der Erholung in Arbeitspausen dienen); *syn.* Gewerbegrün [n].

2877 industrial landscape [n] *geo. urb.* (Landscape predominantly characterized by industrial development; ▶industrial region 1); *s* **paisaje** [m] **industrial** (Paisaje caracterizado predominantemente por la presencia de industrias; ▶región industrial); *f* **paysage** [m] **industriel** (Paysage caractérisé par la prédominance d'installations industrielles ; ▶région industrielle) ; *g* **Industrielandschaft** [f] (Landschaft, die vorwiegend durch große Produktionsbetriebe mit den dazugehörenden Trassen für die Infrastruktur [Verkehrsinfrastruktur und Umschlaganlagen für Güter sowie Ver- und Entsorgungsinfrastruktur] geprägt ist; ▶Industrieregion).

2878 industrial land use [n] *leg. urb.* (▶Urban land use category covering either light or heavy industry, as shown on a ▶comprehensive plan [US]/Local Development Framework [LDF] [UK]); ▶commercial and light industry area, ▶industrial use zone); *s* **suelo** [m] **industrial** (**Es.**, ▶categoría de suelo urbano en la planificación urbana. **D.**, categoría de uso de suelo en plan general municipal de ordenación que se diferencia en el ▶plan parcial [de ordenación] en ▶zona de actividades comerciales e industriales no contaminantes y en ▶zona de uso industrial); *f* **zone** [f] **à vocation d'activités économiques** (Catégorie de zonage dans un ▶plan local d'urbanisme [PLU]/plan d'occupation des sols [POS]; une différenciation des activités a lieu suivant la vocation choisie, industrielle, commerciale ou artisanale ; ▶zone à vocation économique, ▶zone d'activités industrielles, ▶zone d'aménagement concerté à usage industriel, ▶zone d'entreprises ; **D.**, catégorie de zonage du « *Flächennutzungsplan* » [schéma de cohérence territoriale allemand] définissant les terrains susceptibles d'accueillir des activités commerciales et industrielles ; cette zone au niveau du « *Bebauungsplan* » [PLU/POS allemand] pouvant être affectée en zone soit à vocation commerciale, soit à vocation industrielle ; *g* **gewerbliche Bauflächen** [fpl] (*Abk. im Flächennutzungsplan* G; gemäß § 5 [2] Nr.1 BauGB im FNP nach der allgemeinen Nutzung — gewerblich oder industriell — darzustellende Flächen, die im weiteren Planungsprozess im ▶Bebauungsplan in Gewerbegebiet oder Industriegebiet unterschieden werden; cf. auch § 1 BauNVO; ▶Baufläche, ▶Industriegebiet 2, ▶Gewerbegebiet 1).

2879 industrial park [n] [US] *plan. urb.* (Large tract of land that has been planned, developed and operated as an integrated facility for different industrial uses with emphasis on coordinated circulation, parking, utilities, aesthetics and a high proportion of well-designed open space; cf. IBDD 1981. The concept was created in the **U.S.**; e.g. "Silicon Valley"; **in U.K.**, the first **b. p.s** were Aztec West Business Park [Bristol] and Stockley Park; ▶location of commercial facilities and light industries); *syn.* business park [n] [UK], industrial estate [n] [UK]; *s* **parque** [m] **industrial** (Zona situada a las afueras de centros urbanos en emplazamiento bien comunicado diseñada con arquitectura y paisajismo atractivos para atraer industrias y servicios modernos y centros de investigación de las nuevas tecnologías y así servir de impulso económico para la correspondiente región. El concepto fue creado en los EE.UU. y puesto en práctica en lugares como Silicon Valley en California y se ha extendido a todo el mundo industrializado; ▶zona industrial-comercial); *syn.* parque [f] tecnológico; *f 1* **parc** [m] **d'activités** (Espace aménagé pour accueillir des activités offrant aux établissements qui s'y installent une gamme plus diversifiée d'équipements et de services ; selon les destinations on parle de **parcs d'activités** [faible densité d'activités diversifiées], **parcs industriels, parcs artisanaux, parcs d'affaires, parcs de bureaux, parcs technologiques** [technopôle], etc. ; cf. DUA 1996 ; la traduction littérale « parc industriel » ne prend pas en compte les diverses activités que comporte le terme allemand ; ▶zone d'activités économiques ; *f 2* **parc** [m] **industriel** (**F.**, espace aménagé pour accueillir des activités spécifiquement industrielles) ; *g* **Industriepark** [m] (Von öffentlichen oder privaten Trägern im Außenbereich, im ländlichen Raum, in strukturschwachen oder stillgelegten Industriegebieten erschlossenes und verwaltetes Gelände mit anspruchsvoller Freiraumgestaltung und hohem Freiflächenanteil zur Ansiedlung von Industrie- und Gewerbebetrieben sowie Forschungseinrichtungen, vielfach solche der Spitzentechnologie; das Konzept entstand in den USA, z. B. „Silicon Valley" in Kalifornien; ▶Gewerbegebiet 2).

I

2880 industrial plant [n] *urb.* (Factory for manufacturing or assembling equipment of goods for sale on essentially developed land); **s planta [f] industrial** *syn.* fábrica [f]; *f* **installation [f] industrielle** (Ensemble des bâtiments, installations techniques et terrains d'une entreprise industrielle) ; *syn.* usine [f], établissement [m] industriel (BON 1990, 353) ; **g Industrieanlage [f]** (Gesamtheit der industriellen Produktions- und Forschungsstätten mit den dazugehörenden Verwaltungs- und Sozialbauten, Lagerhallen, technischen Großbauten und die unmittelbar dazugehörende Erschließungsinfrastruktur wie Straßen, Bahngleise etc.).

2881 industrial plant abandonment [n] *econ. envir.* (decommissioning of industrial plants); *syn.* (industrial) plant closing [n] [also US]; **s cierre [m] de planta industrial** (Abandono de plantas industriales, cese del uso de líneas de ferrocarril, etc.); *syn.* abandono [m] de planta industrial; *f* **fermeture [f] d'usines** (Arrêt des activités dans un établissement, cessation définitive de la production d'un établissement industriel) ; *syn.* fermeture [f] d'établissements industriels, fermeture [f] de sites industriels ; **g Stilllegung [f] von Industrieanlagen** (Außerbetriebsetzung von z. B. Industriebetrieben, Eisenbahnstrecken).

2882 industrial plant operation [n] *envir.* (Running of a factory); **s operación [f] de una fábrica** *syn.* explotación [f] de una fábrica; *f* **fonctionnement [m] d'une installation industrielle** ; **g Betrieb [m] einer Industrieanlage.**

2883 industrial plant toxic accident [n] *envir. leg.* (Unexpected disruption in the usual operation of a plant whereby toxic matter is released with present or potential endangerment to human beings and the environment, e.g. nuclear plant accident); **s accidente [m] en planta industrial** (Interrupción imprevista del funcionamiento de una instalación industrial o de generación de energía en el proceso de la cual pueden liberarse sustancias tóxicas o peligrosas que pueden causar grandes daños a la salud de la población y al medio ambiente como ya ha ocurrido en el caso de accidentes en centrales nucleares [Harrysburg, USA o Tschernobyl, Ucrania] o en plantas de producción química [Bophal, India o Seveso, Italia], por mencionar sólo las más graves); *f* **accident [m] majeur** (Événement tel qu'une émission, un incendie ou une explosion de caractère majeur, en relation avec un développement incontrôlé d'une activité industrielle, entraînant un danger grave, immédiat ou différé pour l'homme à l'intérieur ou à l'extérieur de l'établissement, et/ou pour l'environnement, et mettant en jeu une ou plusieurs substances dangereuses ; directive SEVESO I du 24 juin 1982) ; *syn.* panne [f] ; *g* **Störfall [m]** (Abweichung vom bestimmungsgemäßen Betrieb einer Anlage, in der hochgiftige Stoffe frei werden oder frei zu werden drohen und dadurch bereits durch eine einmalige Einwirkung eine Gemeingefahr für Menschen und Umwelt hervorgerufen werden kann; N+R 1980).

2884 industrial region [n] (1) *geo.* (Area, characterized by its industry and manufacturing activity, in which agriculture and traditional crafts have lost their significance; ▶industrial landscape); **s región [f] industrial** (Región caracterizada por la presencia de industria en la que la agricultura y los gremios artesanos han perdido importancia; ▶paisaje industrial); *f* **région [f] industrielle** (Région empreinte par la présence d'équipements industriels et commerciaux, les activités agricoles et de l'artisanat traditionnel ayant perdu leur prépondérance ; ▶paysage industriel) ; **g Industrieregion [f]** (Durch Industrieanlagen resp. produzierendes Gewerbe geprägte Region, in der die Landwirtschaft und das traditionelle Handwerk an Bedeutung verloren hat; ▶Industrielandschaft).

industrial region [n] (2) *adm. plan.* ▶old industrial region.

2885 industrial revitalization [n] [US] *plan. urb.* (Removal of buildings in need of demolition, removal of hazardous waste material, improvement of infrastructure, etc. in order to transform a decaying industrial area into a well-functioning complex fulfilling the demands of modern society; ▶decontamination of soil, ▶recycling of derelict sites); *syn.* redevelopment [n] of industrial areas [UK]; **s saneamiento [m] de ruinas industriales** (Demolición de edificios obsoletos, remoción o limpieza de suelos contaminados y mejora de la infraestructura, etc. con el fin de hacer posible el uso del suelo para otros fines; ▶recuperación de ruinas industriales, ▶saneamiento de suelos); *syn.* rehabilitación [f] de zonas industriales; *f* **rénovation [f] de zones industrielles** (Démolition de bâtiments industriels vétustes, des équipements d'infrastructure, etc. rétablissant la fonction d'une zone industrielle, concourant à la dépollution d'un site pollué par les déchets historiques, etc. ; ▶décontamination d'un sol pollué, ▶réhabilitation des friches industrielles) ; **g Sanierung [f] von Industriegebieten** (Beseitigung abrissreifer Gebäude, Beseitigung von Altlasten, Verbesserung der Infrastruktur etc., um ein Industriegebiet in einen funktionsfähigen, den modernen Anforderungen entsprechenden Zustand zu versetzen; ▶Flächenrecycling, ▶Bodensanierung).

2886 industrial sewage [n] *envir.* (Liquid waste originated by industrial processes; **i. s.** is normally from mixed industrial and residential areas; cf. DES 1991, 311; **in D.,** industrial liquid pollutants not entering water are separately identified; whereas in the English language 'effluent' can also contain other liquid waste products without water); *syn.* industrial wastewater [n]; **s aguas [fpl] residuales industriales** (Aguas contaminadas con residuos procedentes de la producción industrial; el término «efluente» se utiliza —al igual que en inglés— para los vertidos de sólidos o líquidos no necesariamente disueltos en agua); *syn.* vertido [m] industrial; *f* **effluent [m] industriel** (Rejets liquides en provenance des installations industrielles constituant la pollution des eaux douces et des eaux de mer) ; *syn.* eaux [fpl] résiduaires industrielles (C. urb., art. R. 111-12) ; **g Industrieabwasser [n]** (...abwässer [pl]; durch industrielle Produktion, Verarbeitung oder Kühlprozesse entstehendes Abwasser; der britische Begriff *effluent* kann auch andere flüssige, nicht wasserhaltige Abfallprodukte enthalten); *syn.* industrielles Abwasser [n].

2887 industrial site [n] *plan.* (Land area designated for industrial use); **s emplazamiento [m] industrial** *syn.* ubicación [f] industrial; *f* **site [m] industriel** ; **g Industriestandort [m].**

industrial site [n], **abandoned** *plan. urb.* ▶derelict land.

2888 industrial use zone [n] *leg. urb.* (Area or district for industries which are not permitted in other zoned areas; **in U.S.,**—in comparison to Germany—residential and commercial uses are also permitted within industrial use zones); *syn.* area [n] for industrial use, industrial use district [n] [also US]; **s zona [f] de uso industrial** (Área prevista en plan general municipal de ordenación o plan parcial para la ubicación de industrias; en D. una de las dos categorías pertenecientes al tipo de uso de suelo industrial en la que pueden emplazarse actividades molestas o peligrosas, al contrario que en las zonas de actividades comerciales e industriales no contaminantes); *f 1* **zone [f] d'activités industrielles** (Appellé souvent **Zone UE** ; dans le zonage du PLU/POS, zone urbaine destinée à l'accueil d'activités industrielles [bureaux, locaux de fabrication, entrepôts lié à l'activité]) ; *syn.* zone [f] à vocation industrielle ; *f 2* **zone [f] d'aménagement concerté à usage industriel («ZAC industrielle»)** (Zone dans laquelle un collectivité publique ou un établissement public réalise ou fait réaliser une opération d'aménagement et d'équipement de terrains en vue de la réalisation d'une zone industrielle ; le plan d'aménagement de zone [PAZ] est soumis à enquête publique et est incorporé au PLU/POS) ; *g* **Industriegebiet [n] (2)** (Gemäß § 9 [1] BauGB und §§ 1 [2] Nr. 9 und 9 BauNVO nach der besonderen Art seiner baulichen Nutzung darzustellendes Baugebiet, das vorwiegend der Unterbringung von Gewerbebetrieben dient, die im Gewerbegebiet und in allen anderen Baugebieten unzulässig sind).

2889 industrial waste [n] *envir.* (Waste generated by industrial production, which is allowed to be treated, dumped or disposed of only at authorized waste disposal facilities and sites or the sake of public health and convenience; ▶hazardous materials, ▶hazardous waste, ▶industrial plant, ▶waste); *s* **residuos [mpl] industriales** (Término específico de ▶residuos sólidos; residuos generados en la producción industrial o en talleres que —si no se pueden reutilizar o reciclar— deben ser eliminados de acuerdo a la legislación en instalaciones técnicas o depósitos adecuados para ellos, de manera que no produzcan daños a la salud humana o al medio ambiente; ▶planta industrial, ▶residuos industriales peligrosos, ▶residuos tóxicos y peligrosos); *syn.* desechos [mpl] industriales; *f* **déchets [mpl] industriels** (Terme spécifique relatif aux ▶déchets ; les déchets de l'industrie font l'objet d'une prévention ou d'une réduction de la production et de la nocivité, leur élimination ou leur stockage par des moyens appropriés dépend de leur caractère polluant ou dangereux ; suivant la réglementation on distingue les ▶déchets générateurs de nuisances, les ▶déchets dangereux, les déchets industriels spéciaux et les déchets industriels spéciaux ultimes en provenance des ▶installations industrielles) ; *g* **Gewerbemüll [m, o. Pl.]** (UB zu ▶Abfall 2; bei industrieller Produktion oder industriellen Verfahren anfallender Müll, der, solange er nicht ordnungsgemäß wieder oder weiter verwertet werden kann, einer geordneten Beseitigung zur Wahrung des Wohls der Allgemeinheit nur in den dafür zugelassenen Anlagen oder Einrichtungen [Abfallbeseitigungsanlagen] behandelt, gelagert oder abgelagert werden darf; cf. § 27 KrW-/AbfG; ▶besonders überwachungsbedürftiger Abfall, ▶Industrieanlage, ▶umweltgefährdender Abfall); *syn.* gewerblicher Abfall [m], gewerblicher Müll [m], *ugs.* Industrieabfall [m], *ugs.* Industriemüll [m].

industrial waste land [n] *plan. urb.* ▶derelict land.

industrial wastewater [n] *envir.* ▶industrial sewage.

industries [npl] *urb.* ▶relocation of industries.

industry [n] *agr.* ▶fishing industry; *constr. prof.* ▶landscape contracting industry; *recr.* ▶leisure industry; *agr.* ▶livestock industry; *envir.* ▶waste treatment and disposal industry.

industry [n]**, disposal** *envir.* ▶waste treatment and disposal industry.

industry [n]**, landscape** *constr. prof.* ▶landscape contracting industry.

industry [n]**, tourism** *recr. trans.* ▶tourism (1).

inert construction [n] *constr. eng.* ▶conventional engineering.

2890 inert construction material [n] *constr.* (Inorganic material used in building construction); *syn.* building material [n]; *s* **material [m] de construcción (inerte)**; *f* **matériau [m] inerte** *syn.* matériau [m] de construction ; *g* **toter Baustoff [m]** *syn.* Werkstoff [m], Baumaterial [n].

2891 infestation [n] *agr. for. hort. phytopath.* (Massive attack of ▶pests on ▶useful plants species or desirable plants; ▶fungus disease); *syn.* attack [n]; *s* **plaga [f] de depredadores** (Presencia masiva de ▶depredores en cultivos de ▶plantas útiles u ornamentales; ▶micosis); *f* **attaque [f] des déprédateurs** (Apparition massive de ▶déprédateurs sur les ▶plantes utiles ou les plantes d'agrément ; ▶maladie cryptogamique) ; *syn.* invasion [f] massive/généralisée des déprédateurs, attaque [f] parasitaire (LA 1981, 29), agression [f] parasitaire (LA 1981, 401) ; *g* **Schädlingsbefall [m]** (Massives Auftreten von ▶Schädlingen an ▶Nutzpflanzen oder Zierpflanzen; ▶Pilzkrankheit).

2892 infill development [n] *plan. urb.* (CPF-IV 1986, 106; urban planning concept with the aim of slowing down urban sprawl by constructing buildings on unbuilt sites, e.g. within a row of houses or a block, or by developing property in urban areas rather than on the outkirts of a community or city; *specific term* infill housing; **i. d.** focuses on the reuse and repositioning of sub-standard houses or underutilized buildings and sites in order to renew blightes neighbo[u]rhoods; **in U.S.**, 'brownfields' are developed or contaminated sites where state-of-the-art restoration technology is applied to enable redevelopment; ▶odd-lot development [US]/gap-stopping [UK]); *s* **densificación [f] urbana** (≠) (Proceso de concentración urbana iniciado en los años 1980 en las grandes ciudades de algunos países de Europa y en los EE.UU. por el cual se promueve la construcción de viviendas dentro de la trama urbana tanto en los desvanes de las casas existentes como en los solares sin edificar para aprovechar más intensamente el suelo urbano y las infraestructuras técnicas y sociales y así reducir el gasto de suelo y la urbanización caótica; ▶relleno de solares desocupados); *f 1* **construction [f] dans le bâti existant** (Réalisation de projets de construction à l'intérieur d'une périmètre bâti ; ▶reconstitution du front bâti) ; *f 2* **construction [f] nouvelle en zone urbanisée** (Extension de l'urbanisation réalisée en continuité avec les agglomérations et villages existants dans le but de lutter contre le mitage et d'agir pour une gestion économe de l'espace ; ▶politique de la ville compacte) ; *f 3* **politique [f] de la ville compacte** (Politique d'urbanisme menée, à partir des années 1980, par certaines grandes villes européennes pour limiter ou enrayer les effets négatifs du desserrement de la population et des activités [développement des constructions isolées ou lotissements en milieu naturel grandes consommatrices d'espaces] au cours de la génération précédente ; politique favorisant la reconstitution du tissu urbain dans les centres-villes ou les quartiers existants [regroupement de l'habitat, augmentation de la densité d'occupation du sol, ▶reconstitution du front bâti] ; DUA 1996) ; *g* **Nachverdichtung [f]** (Städtebauliches Konzept, bei dem in einem im Zusammenhang bebauten Ortsteil, z. B. durch Aufstockung, Anbau, ▶Schließen von Baulücken oder Wiedernutzbarmachung von Flächen, ein unnötiger Landschaftsverbrauch im Außenbereich für weitere Siedlungsflächen verhindert oder zeitlich verzögert wird resp. verzögert werden soll); *syn.* Bauen [n, o. Pl.] im Bestand, innere Verdichtung [f], Innenentwicklung [f].

2893 infilling [n] *constr.* (**1.** Correcting of irregularities in a surface by filling depressions with a suitable material; BS 3975: part 5. **2.** process of depositing fill on low-lying land, which may be marshy or water areas to create usable land; cf. IBDD 1981; ▶back filling, ▶dumping of earth [US]/tipping of earth [UK], ▶land filling); *syn.* filling [n]; *s* **relleno [m] (2)** (**1.** Corrección de las irregularidades de una superficie donde se encuentran pequeñas hondonadas. **2.** Proceso de depositar material sobre tierras bajas que pueden ser pantanosas o acuáticas para conseguir superficie de tierra aprovechable; ▶relleno del terreno, ▶relleno de zanjas y hoyos, ▶terraplenado); *f* **comblement [m]** (Utilisation en remblai de terres déblayées mises en place pour combler un fossé ; ▶comblement du terrain, ▶exhaussement, ▶mise en remblai, ▶remplissage [d'une tranchée/d'un trou]) ; *syn.* remblayage [m] ; *g* **Auffüllen [n, o. Pl.]** (Aufhöhung von tiefer liegenden Flächen [z. B. Geländemulden] oder von ebenen Flächen [z. B. Auffüllberg] mit geeignetem Boden/Material; ▶Aufschüttung, ▶Geländeauffüllung, ▶Verfüllen); *syn.* Auffüllung [f], Erdauffüllung [f].

infill planting [n] [US] *constr.* ▶replacement planting.

2894 infiltration [n] *pedol.* (**1.** Slow penetration of surface water into the soil. **2.** Flow from a porous medium into a channel, drain, reservoir, or conduit; ▶leachate, ▶percolation, ▶seepage water; *opp.* ▶capillary rise; ▶trench drain infiltration, ▶surface infiltration, ▶swale infiltration); *syn.* seepage [n] (1); *s* **infiltración** [f] (**1.** Entrada de aguas superficiales a los poros y huecos de los suelos, sedimentos, rocas y otros materiales subsuperficiales; DINA 1987. **2.** Flujo de agua de un medio poroso en un canal, dren, embalse o conducto; WMO 1974; ▶agua de percolación, ▶infiltración por hoyos, ▶infiltración superficial, ▶líquidos lixiviados contaminados, ▶percolación, ▶percolación por acequia de infiltración; *opp.* ▶ascenso capilar); *f* **infiltration** [f] (**1.** Processus de pénétration lente de l'eau à travers les pores d'un sol due à l'action combinée de la gravité et des forces de succion. **2.** Écoulement d'eau d'un milieu poreux dans un canal, un tuyau, un drain, une retenue ou un conduit ; WMO 1974) ; ▶eau de gravité, ▶infiltration par fossé drainant, ▶infiltration superficielle, ▶lessiviat, ▶lixiviat, ▶infiltration par noues, ▶percolation ; *opp.* ▶remontée capillaire) ; *g* **Versickerung** [f] (**1.** Langsames Eindringen von Oberflächenwasser in den Boden; die umweltverträgliche Bauweise bei der Entstehung neuer Baugebiete fordert die **V.** zur Verringerung der Abflussraten des Regenwassers in die ▶Vorfluter zur Förderung der Grundwasseranreicherung, zur Entlastung der Regenrückhaltebecken und auch zur Abmilderung der Hochwasserspitzen; ▶Durchsickerung, ▶Sickerwasser 1; *opp.* ▶kapillarer Aufstieg; *UBe* ▶Flächenversickerung, ▶Muldenversickerung, ▶Rigolenversickerung. **2.** Sickerbewegung des Wassers von einem durchlässigen Medium in eine Dränleitung, in ein Becken, in einen Entwässerungsgraben etc.); *syn.* Infiltration [f], Sickerbewegung [f].

infiltration [n]**, drain** *constr.* ▶trench drain infiltration.

2895 infiltration capacity [n] *constr. pedol.* (Capacity of soil to absorb surface water within a specified period of time; ▶infiltration, ▶water permeability); *syn.* permeability [n]; *s* **capacidad** [f] **de infiltración** (Velocidad máxima por unidad de superficie y en ciertas condiciones, a la que el agua puede ser absorbida por el suelo; WMO 1974; ▶infiltración, ▶permeabilidad [para el agua]); *f* **capacité** [f] **d'infiltration** (Capacité d'un sol ou d'un revêtement de sol perméable d'emmagasiner une certaine quantité d'eau par unité de temps ; à cet effet sont effectués des tests de ▶perméabilité, l'▶infiltration des surfaces plantées étant plus importante que celle des surfaces enherbées) ; *g* **Versickerungsvermögen** [n, o. Pl.] (Fähigkeit eines Bodens oder wasserdurchlässigen Oberflächenbelages, eine bestimmte Wassermenge je Zeiteinheit aufnehmen zu können; die ▶Versickerung nach extremen Regenereignissen ist Untersuchungen zufolge bei bepflanzten Flächen besser als bei Rasenflächen; cf. TASPO 1999, Nr. 41, 12; ▶Wasserdurchlässigkeit); *syn.* Versickerungsleistung [f].

2896 infiltration swale [n] *constr. eng.* (**1.** Usually a depression or elongated ditch parallel to roads and paths, which allows surface water run-off to percolate into the soil in built-up areas, generally filled with broken stone and gravel to create cavities in which the water is retained before percolation); **2. ponding area** [n] (Large or small-scale depression for surface runoff infiltration created to control potential flooding and maintain flood storage capacity as well as to reduce the extent of flood damages downstream and to regenerate groundwater capacity. Large **p. a.s** regulate and provide surface storage of storm water runoff, after major rainfall events as well as providing for the evaporation of a portion of the runoff. Ponding design depths maximize the surface area to facility depth ratio, ensuring that the hydrologic loading capacity is not compromised. Ponding

times may be kept low by utilizing appropriate soil mediums: the filter media being selected as a function of the time required to drain the water. In arid regions, where precipitation is minimal and extensive pipe drainage systems impractical, surface drainage is often directed into **p. a.s**, where the percolating water helps in leaching soils of salts and encourages the growth of indigenous desert species after winter rains. A natural **p. a.** on the Arabian Peninsula is known as a **rowdah**; an Arabic word for a shallow depression ranging from a few metres to 20 metres in depth and from a few hundred metres to a couple of kilometres in diameter. **Rowdah** receive runoff water after storms, in addition to wind and water-borne material, and so they are usually densely vegetated. Where deposits are deep, the dominant plant is the shrub 'Sidr' *[Zizyphus nummularia]*; in shallower deposits, species of Acacia are more typical: *'samr' [Umbrella Thorn—Acacia tortilis]* and *'salam', [Acacia ehrenbergiana].* Shrubs and grasses, together with seasonal flowers, present an attractive picture, particularly after autumn and late winter rainfall; ▶drainage swale, ▶trench drain infiltration, ▶swale infiltration); *s* **acequia** [f] **de infiltración** (Zanja artificial para recoger y permitir la infiltración de aguas pluviales en zonas edificadas; ▶canal de drenaje [subterráneo], ▶infiltración por hoyos, ▶percolación por acequia de infiltración); *syn.* acequia [f] de percolación; *f 1* **noue** [f] **d'infiltration** (**1.** Large fossé enherbé peu profond avec un profil à pentes douces aménagé en milieu urbain ou le long des infrastructures routières pour la collecte, l'évacuation et l'infiltration des eaux pluviales de ruissellement provenant des surfaces imperméabilisées ; cette technique permet de ralentir l'évacuation de l'eau, avec un écoulement et un stockage à l'air libre ; l'infiltration s'effectue par le passage des eaux à travers une couche de terre végétale, d'une épaisseur minimale de 20 cm, caractérisée par une bonne conductivité hydraulique ; les noues présentent ainsi l'avantage de piéger et dégrader les polluants au fil de l'écoulement, sans les concentrer. **2.** Ouvrage linéaire, elle a pour spécificité de structurer l'espace ou de s'adapter à la géographie et à l'aménagement du site ; ▶infiltration par noue, ▶fossé de pied de talus, ▶noue de drainage) ; *syn.* fossé [m] d'infiltration ; *f 2* **cuvette** [f] **d'infiltration** (Dépression de faible étendue aménagée dans les zones bâties en vue de la collecte, l'évacuation et la filtration des eaux pluviales de ruissellement provenant des surfaces imperméabilisées ; ▶infiltration par fossé drainant) ; *syn.* bassin [m] d'infiltration ; *g* **Versickerungsmulde** [f] (Flächenhaft vertiefte oder muldenförmige Rinne als Entwässerungseinrichtung in bebauten Gebieten zur Aufnahme des Niederschlagswassers und Ableitung in den Untergrund, i. d. R. mit Schotter- oder Kiespackungen zur Schaffung hohlraumreicher Speicher ausgestattet; ▶Entwässerungsmulde, ▶Muldenversickerung, ▶Rigolenversickerung); *syn.* Fangmulde [f] mit Rasendecke (TWW 1982, 478), Muldenrigole [f], Versickerungsrigole [f].

2897 inflow [n] (1) *wat'man.* (Amount of water flowing into an aquifer, lake, reservoir, etc.; ▶spring discharge); *s* **caudal** [m] **afluente** (WMO 1974, 0580; cantidad de agua que fluye en una capa freática, en un lago o embalse, etc.; ▶descarga de una fuente); *f* **débit** [m] **entrant** (Quantité d'eau arrivant dans une formation aquifère, un lac, un réservoir, etc. ; ▶débit d'une source) ; *g* **Wasserspende** [f] (Diejenige Wassermenge, die in die Grundwasserschichten, in einen See, Stausee etc. fließt; ▶Quellschüttung); *syn.* Wasserzufluss [m].

inflow [n] (2) *limn. pedol.* ▶nutrient inflow, ▶nutrient input.

2898 influencing factor [n] *ecol.* (Significant attribute or phenomenon that affects adjacent or surrounding condition(s); *s* **factor** [m] **de influencia**; *f* **facteur** [m] **d'influence** ; *g* **Einflussfaktor** [m].

2899 informal garden [n] *gard.* (Non-axial garden using an asymmetrical design; ▶landscape garden, *opp.* ▶formal garden); **s jardín** [m] **informal** (*Término no bien definido* en comparación con el ▶jardín geométrico; jardín en el que predomina el diseño asimétrico no formalista; ▶jardín paisajístico; **f jardin** [m] **de style libre** (Par comparaison au ▶jardin régulier, jardin dont la conception est caractérisée par une grande liberté dans le choix des idées ; ▶jardin paysager) ; *syn.* jardin [m] informel ; **g nicht formaler Garten** [m] (*Nicht genau definierter Begriff* im Vergleich zum ▶formalen Garten ein nach freien Gestaltungsideen konzipierter, i. d. R. vorwiegend vegetationsbestimmter Freiraum; ▶Landschaftspark).

informal manner [loc] *gard. plan.* ▶designed in an informal manner.

informal settlement [n] *urb.* ▶slum.

information notice [n] [UK] *leg. plan.* ▶position statement [US].

information system [n] *adm. geo. landsc. plan.* ▶geographic information system (GIS).

information system [n]**, natural resources** *adm. geo. landsc. plan.* ▶landscape information system.

2900 infrared color aerial photograph [n] [US]/**infrared colour aerial photograph** [n] [UK] *envir. plan. rem'sens. surv.* (▶infrared photography); *syn.* false colo(u)r aerial photograph [n]; **s fotografía** [f] **aérea infrarrojo color** (▶fotografía infrarrojo 1); *syn.* foto [f] aérea infrarrojo color, fotografía [f] aérea de colores falsos, foto [f] aérea de colores falsos; **f photographie** [f] **aérienne infrarouge couleur (I.R.C.)** (▶photographie infrarouge) ; *syn.* photographie [f] aérienne fausse-couleur ; **g Infrarotluftbild** [n] **in Farben** (▶Infrarotfotografie); *syn.* Falschfarbenluftbild [n].

2901 infrared color photograph [n] [US]/**infrared colour photograph** [n] [UK] *envir. plan. rem'sens. surv.* (▶infrared photography, ▶remote sensing); *syn.* false colo(u)r photograph [n]; **s fotografía** [f] **infrarrojo color** (Técnica de ▶teledetección que utiliza la composición coloreada del infrarrojo color en la que se aplican a las bandas espectrales verde, roja e infrarrojo próximo, los colores naturales azul, verde y rojo, respectivamente. Es muy útil para detectar cambios en las superficies vegetales; cf. CHUV 1990; ▶fotografía infrarrojo 1); *syn.* foto [f] infrarrojo color, foto [f] de colores falsos, fotografía [f] de colores falsos; **f photographie** [f] **infrarouge couleur** (▶photographie infrarouge, ▶télédétection [de terrain]) ; *syn.* photographie [f] fausse-couleur ; **g Infrarotbild** [n] **in Farben** (▶Fernerkundung, ▶Infrarotfotografie); *syn.* Falschfarbenbild [n].

2902 infrared photograph [n] *envir. plan. rem'sens. surv.* (Image produced by ▶infrared photography; ▶remote sensing); **s fotografía** [f] **infrarrojo (2)** (Imagen creada con ayuda de la ▶fotografía infrarrojo 1); *syn.* foto [f] infrarrojo; **f image** [f] **infrarouge** (Image réalisée au moyen de la ▶photographie infrarouge, ▶télédétection [de terrain]) ; **g Infrarotbild** [n] (Mit Hilfe der ▶Infrarotfotografie hergestelltes Bild, ▶Fernerkundung); *syn.* Infrarotaufnahme [f].

2903 infrared photography [n] *envir. plan. rem'sens.* (Photographic technique based upon the special physical characteristics of infrared radiation which may be used for environmental pollution surveillance; ▶monitoring, ▶infrared color photograph [US]/infrared colour photograph [UK]; ▶remote sensing); **s fotografía** [f] **infrarrojo (1)** (Técnica de registro de datos ampliamente utilizada en la ▶teledetección basada en las características físicas especiales del espectro de radiación infrarrojo que abarca entre 0,7 y 100 micras y se divide en infra-

rrojo cercano [0,7 a 1,3 micras], medio [1,3 a 3,0] y lejano [7,0 a 15,0]. Este último se conoce como infrarrojo térmico. Entre otras aplicaciones está la del ▶monitoreo ambiental; cf. CHUV 1990; ▶fotografía infrarrojo color); **f photographie** [f] **infrarouge** (Technique de photographie en ▶télédétection [de terrain] basée sur les propriétés physiques du rayonnement infrarouge et utilisée dans la surveillance de l'environnement [▶monitorage] ; on distingue les clichés noirs et blancs et la ▶photographie infrarouge couleur) ; **g Infrarotfotografie** [f] (1. Fotografische Aufnahmetechnik, die sich der besonderen physikalischen Eigenschaften infraroter Strahlung bedient und deshalb zur Umweltüberwachung [▶Monitoring] verwendet wird; man unterscheidet zwischen Schwarzweißbildern und ▶Infrarotbildern in Farbe [Falschfarbenbilder]; ▶Fernerkundung). 2. Einzelnes Infrarotlichtbild); *syn. o. V.* Infrarotphotographie [f]; *syn. zu 2.* Infrarotaufnahme [f].

2904 infrastructure [n] *plan.* (Discussion about the definition of this term continues. According to BOESLER **i.** is the furnishing of a space [community, settlement, region] with public facilities provided by public money conducive to increases in productivity as well as social security, and thus in the broadest sense promoting optimum opportunities for competition and living standards for society and its members; technological **i.** [roads, utilities, telecommunication, etc.] and social **i.** [schools, hospitals, libraries, theatres/theaters, etc.] are coordinated in development planning; ▶public infrastructure, ▶recreational infrastructure, ▶transportation infrastructure [US]/traffic infrastructure [UK], ▶vehicular and pedestrian infrastructure); **s infraestructura** [f] (Conjunto de instalaciones públicas o privadas de servicio a la población que se diferencian entre «infraestructura técnica» e «infraestructura social». La primera incluye el conjunto de redes viarias, de comunicación, abastecimiento [agua, electricidad, gas, etc.] y de recogida y tratamiento de aguas residuales. La segunda los centros docentes, religiosos, asistenciales, sanitarios y demás servicios de interés público y social; ▶infraestructura de abastecimiento público, ▶equipamiento de recreo, ▶infraestructura de tráfico y transportes [Es]/infraestructura de tránsito y transportes [AL], ▶superficies viarias); **f 1 infrastructure** [f] (F., équipements de transport, communication, réseaux divers [électricité, assainissement, etc.], les installations de traitement des déchets; **D.,** terme générique désignant les équipements techniques, sociaux, culturels, administratifs et commerciaux ; ▶circulations, ▶équipements collectifs, ▶équipements collectifs de loisirs, ▶infrastructure de transport, ▶voirie) ; **f 2 superstructure** [f] (Équipements scolaires, universitaires, sociaux, culturels, hospitaliers, administratifs, commerciaux) ; **g Infrastruktur** [f] (Gesamtheit der Ausstattung eines organisatorischen oder räumlichen Bereiches [Gemeinde oder Region] mit öffentlichen Einrichtungen aus Mitteln des sozialen Kapitals, die der Sicherung resp. der Erhöhung der Produktivität sowie der sozialen Sicherheit sowie Versorgung kultureller Ansprüche und damit im weitesten Sinne der Schaffung optimaler gesellschaftlicher Wettbewerbschancen und Lebensbedingungen dieses Bereiches und seiner Bewohner dienen; man unterscheidet zwischen technischer und sozialer **I.**; cf. BOESELER; die Diskussion über die Definition dieses Begriffes ist noch nicht abgeschlossen. ▶Freizeitinfrastruktur, ▶Versorgungsinfrastruktur, ▶Verkehrsinfrastruktur ▶Verkehrsflächen); *syn.* Ausstattung [f].

infrastructure [n] [UK]**, traffic** *plan. trans.* ▶transportation infrastructure [US].

2905 infrastructure development [n] *plan. trans. urb.* (Provision of utilities, services and connections to individual plots or development areas, where the road and public transport network is being built; ▶land developed with roads); **s urbani-**

zación [f] infraestructural (Conjunto de medidas constructivas para conectar lotes individuales o nuevas zonas urbanizadas a las redes de transporte y servicio público; ►suelo urbanizado 2); *syn.* desarrollo [m] de infraestructura; *f* **viabilité [f]** (Ensemble des travaux d'aménagement pour un terrain, lotissement ou zone d'activité afin de les raccorder à la voirie et aux réseaux divers ; ►terrain équipé, ►terrain viabilisé) ; *syn.* équipements [mpl] d'infrastructure ; *g* **infrastrukturelle Erschließung [f]** (**1.** Gesamtheit der Maßnahmen, die einzelne Grundstücke oder ganze Gebiete an das öffentliche Verkehrs- und Versorgungsnetz anschließen; ►mit Straßen erschlossenes Bauland; **2.** Ergebnis von 1.).

2906 infrastructure improvement [n] *plan.* *s* **mejora [f] en infraestructura**; *f* **amélioration [f] des équipements d'infrastructure** ; *g* **Infrastrukturverbesserung [f]** *syn.* Verbesserung [f] der Infrastruktur.

infrastructure plan [n] [UK]**, green** *landsc. urb.* ►general plan for urban open spaces [US]/green open space structure plan [UK].

2907 inherited genes [npl] *biol.* (All the units of heredity, composed of DNA and RNA, that determine the particular characteristics of a plant, animal or man; ►gene pool); *syn.* genetic potential [n] (TEE 1980); *s* **patrimonio [m] genético** (Conjunto de características o comportamientos hereditarios de un individuo; ►banco de genes); *f* **patrimoine [m] génétique** (Ensemble des caractères ou comportements héréditaires d'un individu ; ►pool génique populationnel) ; *syn.* patrimoine [m] héréditaire; *g* **Erbgut [n, o. Pl.]** (Gesamtheit der in den Genen eines Individuums festgelegten Fähigkeiten charakteristische Merkmale oder Verhaltensweisen im Zusammenwirken mit der Umwelt zu entwickeln; ►Genpool); *syn.* Erbanlagen [fpl], Erbgefüge [n], Erbmasse [f], Erbsubstanz [f], Genpotential [n], *o. V.* Genpotenzial [n].

2908 inhibiting [ppr] **(plant) growth** *biol. hort.* (Descriptive term for stopping or slowing down plant growth); *syn.* stunting [ppr] (plant growth); *s* **inhibitorio/a del crecimiento (de las plantas) [loc]** (Término descriptivo para agente que frena el desarrollo de las plantas; ►inhibidor del crecimiento de plantas); *syn.* que inhibe el crecimiento (de las plantas) [loc]; *f* **inhibiteur, trice de croissance [loc]** (Qui ralentit ou arrête le processus de croissance des végétaux ; ►substance inhibitrice) ; *g* **wachstumshemmend [ppr/adj]** (So beschaffen, dass das Streckungswachstum von Pflanzen verlangsamt oder gebremst wird; ►Hemmstoff).

inhibitor [n]**, growth** *chem. agr. hort.* ►growth retardant.

initial phase of top drying [n] [UK] *arb. for.* ►initial phase of top-kill [US].

2909 initial phase of top-kill [n] [US] *arb. for.* (**1.** First signs of a dying tree crown [deciduous and conifer tree]. **2.** The final phase, when all the needles have fallen, is termed **black-top** [US&CDN] ; ►dead crown, ►opening up of forest, ►sorrel-top [US&CDN]); *syn.* initial phase of top drying [n] [UK]; *s* **puntiseco [m] incipiente** (Fase inicial de muerte de coníferas o de frondosas que puede deberse a la edad o ser consecuencia de plagas de hongos, de insectos o de la contaminación; ►aclareo 1, ►copa muerta, ►puntiseco incipiente en conífera); *f* **phase [f] initiale du dépérissement terminal** (Premiers signes indicateurs de la mort de la couronne d'un arbre ; ►éclaircie, ►dépressage, ►houppe morte, ►rougissement et brunissement des conifères) ; *g* **beginnende Wipfeldürre [f]** (**1.** Anfangsstadium des Absterbens der Baumwipfel bei Laub- und Nadelgehölzen, das altersbedingt oder als Folge von Pilzbefall, Insektenfraß oder Rauchschaden auftritt. **2.** Bei Koniferen sind erste Anzeichen für eine absterbende Baumkrone meist Rötungen und Verbraunungen

der Nadeln; bei Schadensfortschritt tritt die sog. **Nadelröte** ein, Endstadium ist die völlige **Entnadelung** oder völlige ►Verlichtung; ►abgestorbene Baumkrone; ►Rötungen und Verbraunungen bei Koniferen).

2910 initial pioneers association [n] **of alpine belt** *phyt.* (TGG 1984, 187; plant community which grows slowly on an extreme site above the alpine tree line [►timberline], usually not pioneer vegetation with succession but rather an ►initial plant community which remains unchanged for hundreds of years [►permanent community]; ►pioneer vegetation, ►sere); *syn.* association [n] of scattered pionieer species (TEE 1980, 211); *s* **comunidad [f] inicial del piso alpino** (Comunidad vegetal rala, de crecimiento lento, en hábitats extremos en el ►límite del bosque alpino. Generalmente no es una comunidad pionera con su correspondiente ►serie [de sucesión], sino una ►comunidad inicial que a menudo se mantiene inalterada durante siglos [►communidad permanente]; ►vegetación pionera); *f* **groupement [m] pionnier de l'étage alpin** (Formations végétales de faible recouvrement et de composition inconstante vivant dans des milieux écologiquement très variés, aux conditions extrêmes situées au-dessus de la ►limite des forêts, déterminant une sélection rigoureuse des espèces, tels que les groupements rupicoles, les groupements des sols mobiles, les groupement des combes à neige, les groupements aquatiques et de marais ; certaines associations ne présentent pas de séries évolutives [blocage dû aux conditions climatiques extrêmes de haute altitude], elles sont de fait un ►groupement initial non caractérisé par une ►série de végétation et pouvant constituer un ►groupement permanent ; VCA 1985, 249-256) ; *g* **Vorpostengesellschaft [f]** (Offene, langsam wachsende Pflanzengesellschaft auf äußerst extremen Standorten oberhalb der alpinen ►Waldgrenze. Sie ist meist keine Pioniergesellschaft [►Pionierflur] mit ►Sukzessionsserie, sondern eine ►Initialgesellschaft, die oft jahrhundertelang unverändert bleibt [►Dauergesellschaft]; ELL 1978, 548).

2911 initial plant community [n] *phyt.* (First plant community—initial floristic composition [IFC]—to have evolved spontaneously on land completely cleared of vegetation, on raw soils or on sites with extreme conditions; ►pioneer vegetation, ►initial pioneers association of alpine belt); *s* **comunidad [f] inicial** (Comunidad vegetal que coloniza inicialmente un área nueva no ocupada por otras especies; ►comunidad inicial del piso alpino, ►vegetación pionera); *syn.* comunidad [f] pionera; *f* **groupement [m] initial** (Association végétale s'établissant spontanément sur les coupes à blanc, les défrichements de prairies, les sols bruts et les stations aux conditions extrêmes [arides ou mobiles], p. ex. le ►groupement pionnier de l'étage alpin ; ►végétation pionnière) ; *syn.* association [f] initiale ; *g* **Initialgesellschaft [f]** (Pflanzengesellschaft, die sich spontan auf Kahlschlägen, Wiesenumbrüchen, Rohböden oder Extremstandorten als Erstbesiedlung einstellt; ►Pionierflur; *auf ganz extremen Standorten* ►Vorpostengesellschaft); *syn.* Vorgesellschaft [f].

2912 initial site analysis [n] *plan.* (First assessment of existing site conditions of a planning area; ►site analysis); *s* **análisis [m] del estado inicial** (Primer estudio de la situación de la ubicación de un proyecto; ►estudio del terreno); *f* **analyse [f] de l'état initial** (Analyse des données relatives à un site et son environnement affecté par un aménagement ; ►analyse du site) ; *syn.* analyse [f] du milieu initial, diagnostic [m] et analyse [f] de l'existant, étude [f] des caractéristiques du milieu ; *g* **Bestandsanalyse [f]** (Untersuchung der Gegebenheiten eines Planungsgebietes; ►Standortanalyse); *syn.* Zustandsanalyse [f].

2913 initiating authority [n] *adm. prof.* (Agency, public authority or organization responsible for originating and financing a public awards competition, including the provision of prize or money; ▶awarding of prizes, ▶sponsor of a competition); *syn.* awarding authority [n]; *s* **promotor, -a [m/f] de concurso (1)** (Institución o empresa que convoca un concurso de ideas o de realización y financia y otorga los premios o las menciones; ▶entrega de premios, ▶promotor, -a de un concurso 2); *syn.* convocador, -a [m/f] de concurso; *f* **pouvoir [m] adjudicateur d'un concours** (Maître d'ouvrage public responsable de toutes les opérations relatives au déroulement du concours ; ▶remise des prix ; ▶promoteur du concours) ; *syn.* organisateur [m] du concours ; *g* **auslobende Behörde/Organisation [f]** (Diejenige öffentliche Stelle, die einseitig verspricht, dass sie einen Wettbewerb ausschreibt, durchführt und für die besten Ergebnisse Preise in Geld vergibt und Auszeichnungen und Belobigungen verleiht; ▶Preisverleihung, ▶privater Auslober/private Ausloberin); *syn.* öffentlicher Auslober [m], öffentliche Ausloberin [f].

initiator [n] **of an impact/intrusion** *envir. pol.* ▶causer of an impact/intrusion.

injunction [n] **on alteration** *leg. plan.* ▶preliminary injunction on alteration.

injunction [n], **permanent** *leg. plan.* ▶preliminary injunction on alteration.

2914 injurious insect [n] *agr. for. hort. phytopath.* (Specific term for ▶pest; *generic term* ▶injurious organism); *syn.* harmful insect [n]; *s* **insecto [m] nocivo** (Término específico de ▶depredador; *término genérico* ▶organismo nocivo); *f* **insecte [m] nuisible** (*Terme spécifique pour* ▶dépredateur ; *terme générique* ▶organisme nuisible) ; *syn.* insecte [m] dépredateur ; *g* **Schadinsekt [n]** (UB zu ▶Schädling; OB ▶Schadorganismus).

2915 injurious organism [n] *agr. for. hort. phytopath.* (Pathogenic agent as well as viruses at various stages in their development, which cause considerable damage to plants and plant products; ▶pest, ▶weed 1); *s* **organismo [m] nocivo** (Agente patógeno vegetal o animal que puede causar daños considerables en las plantas o productos resultantes de ellas; ▶depredador, ▶planta advenediza); *syn.* especie [f] nociva, organismo [m] dañino; *f* **organisme [m] nuisible** (LA 1981, 357 ; animal, champignon, bactérie ou virus qui affecte les végétaux ou encore ▶plante adventice qui concurrence les plantes ; LA 1981, 462 ; *impropre* parasite ; ▶dépredateur) ; *syn.* ennemi [m] des cultures (LA 1981, 462) ; *g* **Schadorganismus [m]** (Tierisches und pflanzliches Lebewesen, das erheblichen Schaden an Pflanzen oder Pflanzenerzeugnissen verursachen kann, z. B. Virus in allen Entwicklungsstadien, schädliche Bakterien, parasitische höhere Pflanzen wie Moose, Algen, Flechten, Pilze und Unkräuter; cf. PflSchG 1998; *UBe* ▶Schädling, ▶Unkraut); *syn.* Schaderreger [m], Pflanzenparasit [m], Tierparasit [m].

2916 injurious to health [loc] *envir.* *syn.* detrimental [adj] to health; *s* **insalubre [adj]** *syn.* deletéreo/a [adj], malsano/a [adj]; *f 1* **insalubre [adj]** (État d'une chose malsaine, nuisible pour la santé, p. ex. un logement, un quartier insalubre ; est insalubre tout immeuble, bâti ou non, vacant ou non, dangereux pour la santé des occupants ou des voisins du fait de son état ou de ses conditions d'occupation ; l'insalubrité associe la dégradation du bâti à des effets négatifs sur la santé. Elle s'analyse au cas par cas et après visite des lieux, en se référant notamment à une liste de critères. Parmi ces critères, on peut citer les murs fissurés, l'humidité importante, le terrain instable, l'absence de raccordement aux réseaux d'électricité ou d'eau potable ou encore l'absence de système d'assainissement ; cf. www.var. equipement.gouv.fr) ; *f 2* **nocif, ive [adj]** (Substance qui nuit à la santé, qui, par inhalation, ingestion ou pénétration cutanée peut entraîner des risques de gravité limitée ; Directive du Conseil n° 91/689/CEE du 12 décembre 1991, annexe III) ; *syn.* délétère [adj] ; *g* **gesundheitsschädigend [ppr/adj]** (So beschaffen, dass es der Gesundheit abträglich ist).

2917 injury limit [n] *agr. for. hort.* (Level above which the infestation of a plant, by pest attack or disease, causes damage); *syn.* injury threshold [n]; *s* **límite [m] de resistencia ante plagas** (Grado de intensidad de presencia de organismos nocivos pasado el cual se presentan daños en las plantas o en los cultivos); *f* **valeur [f] limite de constatation du dommage (±)** (Valeur de l'intensité de l'action des organismes nuisibles à partir de laquelle le dommage peut être constaté) ; *g* **Schadgrenze [f]** (Befallsintensität von Schadorganismen, ab der ein Schaden festzustellen ist).

injury threshold [n] *agr. for. hort.* ▶injury limit.

2918 ink-strips [npl] *phyt.* (Blue epilithic algae mat on rock walls formed as ink-like lines that are prolonged with the trickling of water); *s* **rayas [fpl] de tinta (≠)** (Crecimiento en hilera de algas azules sobre paredes rocosas que se prolongan debido al agua de lluvia que gotea); *f* **traînée [f] d'algues bleues** (Traces laissées par le couvert linéaire d'algues bleues le long des parois rocheuses et se prolongeant par un ruissellement lent de fines gouttelettes d'eau) ; *g* **Tintenstriche [mpl]** (Streifenförmiger Blaualgenbewuchs an Felswänden, der sich durch tropfendes Regenwasser verlängert).

2919 inland dune [n] *geo.* (Areas of sand which occur in the interior of continents, predominantly in deserts, compared to coastal ridges, mounds or hills of drifted sand); *s* **duna [f] interior** (En comparación con una duna litoral, montículos de arena existentes en el interior de los continentes, generalmente en desiertos); *f* **dune [f] continentale** (DG 1984 ; par opposition aux dunes littorales, relief de sable édifié à l'intérieur des continents en général dans les régions désertiques) ; *g* **Binnendüne [f]** (Im Vergleich zu den Küstendünen diejenigen Sandanhäufungen, die im trockenen Innern der Kontinente, meist in Wüsten, entstanden sind); *syn.* Inlanddüne [f].

2920 inland waters [npl] *geo. leg. pol.* (General term for bodies of water, such as lakes, rivers, canals, and bays under the sovereignty of state or federal authorities; ▶territorial waters); *s* **aguas [fpl] interiores** (Conjunto de aguas continentales [ríos, lagos, lagunas, etc.] y marítimas interiores [puertos marítimos, bahías, radas, estuarios y desembocaduras de ríos] que pertenecen a la jurisdicción de un Estado; ▶aguas territoriales); *f* **eaux [fpl] intérieures** (Eaux sur lesquelles un État exerce les pouvoirs découlant de sa souveraineté, situées soit à l'intérieur des terres telles que lacs, fleuves, soit sur la frange côtière comme les eaux marines intérieures limitées dans le cadre des ▶eaux territoriales telles que rades ports, estuaires et mers de Waden) ; *g* **Binnengewässer [npl]** (Der Gebietshoheit eines Staates unterstehende Gewässer wie Seen, Flüsse und Kanäle sowie die durch ▶Territorialgewässer begrenzten maritimen Binnengewässer wie z. B. Seehafen, Binnenmeere, Mündungsgebiete, Meeresbuchten und Wattenmeere).

inlet [n] *constr.* ▶catch basin inlet [US], ▶curb inlet [US]/kerbinlet [UK], ▶grate inlet [US], ▶surface inlet

inlet [n] [UK], **combined** *constr.* ▶combined curb and grate [US].

inlet [n] [US], **drain** *constr.* ▶catch basin inlet [US], ▶grate inlet [US].

inlet [n] [US], **drop** *constr.* ▶yard drain [US].

inlet [n] [UK], **gutter** *constr.* ▶catch basin inlet [US], ▶grate inlet [US].

inlet [n] [UK], **kerb-** *constr.* ▶curb inlet [US].

inlet grate [n] [US] *constr.* ▶storm drain grate [US].

inlet structure [n] [US]**, drainage** *constr. eng.* ▶surface inlet [US].

2921 in-line parking [n] [US] *trans. urb.* (Parking of cars, one behind the other, on a strip along the road in the direction of traffic; ▶in-line parking layout [US]/parallel parking layout [UK]); *syn.* parallel parking [n] [UK]; *s* **aparcamiento [m] en hilera** (Ordenación de las plazas de aparcamiento paralelas a la calle; ▶disposición [de aparcamiento] en hilera); *syn.* estacionamiento [m] en hilera [CS], parqueo [m] en hilera [CA y C]; *f* **stationnement [m] longitudinal** (▶rangement longitudinal) ; *syn.* stationnement [m] en file ; *g* **Längsparken [n, o. Pl.]** (Abstellen von parkenden Fahrzeugen in Fahrtrichtung hintereinander; ▶Längsaufstellung).

2922 in-line parking layout [n] [US] *plan. trans.* (Layout for parking cars, one behind the other, in the direction of traffic; ▶in-line parking, ▶ninety-degree parking layout, ▶angle parking layout); *syn.* parallel parking layout [n] [UK]; *s* **disposición [f] (de aparcamiento) en hilera** (Ordenación de las plazas de aparcamiento a lo largo de la calle; ▶aparcamiento en hilera, ▶disposición [de aparcamiento] en diagonal, ▶disposición [de aparcamiento] en perpendicular); *syn.* disposición [f] de estacionamiento en hilera [CS], disposición [f] de parqueo en hilera [C y CA]); *f* **rangement [m] longitudinal** (Disposition des véhicules en stationnement parallèle à l'axe de la chaussée ; ▶stationnement longitudinal, ▶stationnement en épi, ▶stationnement perpendiculaire) ; *syn.* rangement [m] en file, disposition [f] en file, disposition [f] en ligne ; *g* **Längsaufstellung [f]** (Aufstellungsanordnung von parkenden Fahrzeuge in Fahrtrichtung hintereinander; ▶Längsparken, ▶Schrägaufstellung, ▶Senkrechtaufstellung).

in-line parking row [n] [US] *trans.* ▶parking lane.

2923 in multiple rows [loc] *hort.* (e.g. planting of trees; TGG 1984, 189); *s* **de varias hileras** [loc] (Plantación o avenida formada por dos o más filas de árboles u otras plantas); *syn.* con varias hileras [loc]; *f* **à rangées multiples** [loc] *syn.* à rangées nombreuses [loc] ; *g* **mehrreihig** [adj] (Eine Pflanzung betreffend, die aus mehr als zwei Reihen besteht; z. B., Windschutzpflanzung, Baumpflanzung).

2924 inner city [n] *urb.* (General term covering downtown area; ▶core area 1, ▶central business district [CBD], ▶central city environs [US]/city centre precincts [UK]); *syn.* downtown [n] [also US], city center [n] [also US]; *s* **centro [m] urbano** (Parte central de una gran ciudad caracterizada por multiplicidad de usos comerciales, administrativos, en parte residenciales y recreativos; ▶perímetro edificado, ▶zona central, ▶zona residencial céntrica); *syn.* centro [m] de ciudad, casco [m] urbano; *f* **centre-ville [m]** (Zone souvent très commerciale, aux constructions denses dans le centre d'une ville ; ▶zone urbaine 1, ▶quartiers péricentraux, ▶zone urbaine consacrée aux activités économiques) ; *syn.* cité [f], centre [m] urbain ; *g* **Innenstadt [f]** (Innerer Teil von Großstädten, der vorwiegend durch Hauptgeschäftsstraßen und Verwaltungsgebäude gekennzeichnet ist; traditionell wird die **I.** als belebtes Zentrum, in dem man wohnt, arbeitet, einkauft, flaniert, kommuniziert und sich kulturell bildet, verstanden. Tatsächlich hat sich dies in den letzten Jahrzehnten so stark verändert, dass ein Großteil der Bevölkerung in Neubaugebieten am Stadtrand wohnt und in Discountern und Einkaufszentren „auf der grünen Wiese" einkauft und nur Teile seiner Arbeits- und Freizeit in der **I.** verbringt; ▶Innenbereich, ▶innerstädtisches Verdichtungsgebiet, ▶Kerngebiet); *syn.* City [f], Stadtzentrum [n], Zentrum [n] (einer Stadt).

inner-city express highway [n] [US] *trans.* ▶inner-city freeway [US] (1)/inner-city expressway [UK].

inner-city expressway [n] [UK] *trans.* ▶inner-city freeway [US] (1).

2925 inner-city freeway [n] [US] (1) *trans. urb.* (Usually a grade-separated highway in a conurbation with two or more traffic lanes in each direction; ▶inner-city freeway [US] 2/inner-city motorway [UK], ▶freeway [US]/motorway [UK], ▶limited vehicular access, ▶major cross-town artery [US]/through road [UK], ▶major street); *syn.* inner-city express highway [n] [US], inner-city expressway [n] [UK]; *s* **autovía [f] urbana** (Carretera urbana por lo menos con dos carrilles en cada dirección, con restricción de acceso total para tránsito no motorizado y generalmente con cruces con diferentes niveles; ▶arteria [principal] de tráfico, ▶autopista, ▶autopista urbana, ▶restricción de acceso de vehículos, ▶vía de tránsito); *f* **voie [f] express urbaine** (Routes ou sections de routes appartenant au domaine public de l'État, des départements ou des communes, souvent à double voie pour chaque sens de circulation, accessibles seulement en des points aménagés à cet effet ; ▶limitation d'accès pour les véhicules, ▶autoroute, ▶voie de transit, ▶voirie primaire, ▶autoroute urbaine) ; *syn.* autoroute [f] urbaine, voie [f] rapide ; *g* **innerörtliche Schnellverkehrsstraße [f]** (Innerörtliche Kraftfahrzeugstraße mit mindestens zwei Fahrspuren je Richtung, vollständiger ▶Zufahrtsbeschränkung sowie oft mit Knotenpunkten in mehreren Ebenen; ▶Autobahn, ▶Durchgangsstraße, ▶Hauptverkehrsstraße 2, ▶Stadtautobahn); *syn.* Schnellstraße [f].

2926 inner-city freeway [n] [US] (2) *trans.* (Urban road with separate travel lanes [US]/carriage-ways [UK], limited access from adjoining properties and interchanges with various levels; ▶beltway [US], ▶ring road); *syn.* inner-city motorway [n] [UK]; *s* **autopista [f] urbana** (Autopista situada dentro del perímetro de una ciudad; ▶anillo periférico, ▶autopista periférica); *f* **autoroute [f] urbaine** (Voirie urbaine à chaussées séparées, caractérisée par un accès limité à quelques points et un dispositif de croisement avec la voirie secondaire sur différents niveaux ; ▶rocade, ▶rocade autoroutière) ; *g* **Stadtautobahn [f]** (Städtische Kraftfahrzeugstraße mit Richtungsfahrbahnen, Zufahrtsbeschränkungen und ausnahmslos kreuzungsfreien Verknüpfungen; ▶Autobahnring, ▶Ringstraße).

inner-city motorway [n] [UK] *trans.* ▶inner-city freeway [US] (2).

2927 inner city park [n] [US] *urb.* (Public recreation area of varying size with facilities to serve a dense population; **in U.S.**, the 'Public Garden' on the Boston Common in Massachusetts was developed in the late 19th century with 'swan boats' for paddling groups of people around garden ponds in a landscaped setting. Similar specialized public parks in the U.S. feature intensive planting of specimen trees, shrubs and flowering plants; e.g. Dumbarton Oaks, Washington, D.C., Bellingrath Gardens, Mobile, Alabama; Hershey Gardens, Hershey, Pennsylvania; DIL 1987; ▶neighborhood park [US]/distrik park [UK], ▶proximity green space, ▶urban park); *syn.* downtown park [n] [US], public garden [n] [UK]; *s* **jardín [m] público** (Pequeña zona verde en el centro de una ciudad; ▶jardín de barrio, ▶parque urbano, ▶zona verde de vecindario); *syn.* jardín [m] municipal; *f* **jardin [m] public** (Espace vert public urbain de petite dimension prenant souvent l'aspect d'un parc ; ▶espace vert urbain de proximité, ▶parc central, ▶parc de quartier, ▶parc de voisinage, ▶parc urbain) ; *g* **Stadtgarten [m]** (Kleinere öffentliche Parkanlage in der Innenstadt; ▶wohnungsnahe Grünfläche; ▶Stadtpark, ▶Stadtteilpark); *syn.* Bürgergarten [m].

inner suburbs [npl] [UK] *urb.* ▶central city environs [US].

inoculation [n] *for. hort.* ▶mycorrhiza inoculation.

inoculation [n] [US]**, mycorrhizal** *for. hort.* ▶mycorrhiza inoculation.

2928 inorganic fertilizing [n] *agr. constr. for. hort.* (Spreading of industrially manufactured mineral ▶soil amendment [US]/soil ameliorant [UK]; ▶fertilizer, ▶soil conditioner; *opp.* ▶organic fertilizing); *s* **fertilización [f] mineral** (Aplicación de abono de origen industrial; ▶fertilizante, ▶material de enmienda de suelos, ▶«soil conditioner»; *opp.* ▶fertilización natural); *syn.* abonado [m] mineral; *f* **fertilisation [f] minérale** (▶Amendement fabriqué industriellement ; *opp.* ▶fumure organique ; ▶amendement synthétique, ▶engrais) ; *syn.* fumure [f] minérale (LA 1981, 554), amendement [m] minéral ; *g* **anorganische Düngung [f]** (Ausbringen von industriell hergestellten Bodenverbesserungsstoffen; ▶Bodenverbesserungsstoff 1, ▶Bodenverbesserungsstoff 2, ▶Dünger; *opp.* ▶organische Düngung).

input [n] *landsc. plan.* ▶environmental planning input; *limn. pedol.* ▶nutrient input.

input [n]**, pollution** *envir.* ▶pollution impact.

input [n]**, salt** *envir.* ▶salination (1).

input [n] **of mineral elements** *envir. limn. pedol.* ▶matter import.

insect [n]**, harmful** *agr. for. hort. phytopath.* ▶injurious insect.

2929 insecticide [n] *chem. envir.* (Chemical agent used in controlling insects); *s* **insecticida [m]** (Producto para la lucha contra las plagas de insectos); *f* **insecticide [m]** (Produit antiparasitaire à usage agricole utilisé dans la lutte contre les insectes nuisibles aux cultures ou aux produits récoltés et néfastes à l'hygiène ; on distingue les insecticides minéraux, végétaux ou biologiques et parmi les insecticides de synthèse [organochlorés, etc.], les insecticides de contact ou systémiques, spécifiques ou polyvalents) ; *g* **Insektizid [n]** (Mittel zur Bekämpfung von Insekten im Pflanzen-, Vorrats-, Materialschutz und Hygienebereich).

in-situ (cast) concrete [n] [UK] *constr.* ▶poured-in-place concrete [US].

in-situ concrete slab [n] [UK] *constr.* ▶poured-in-place concrete slab [US].

2930 in situ gravel [n] *constr. min.* (Gravel which occurs naturally and is exposed on site); *s* **grava [f] in situ**; *f* **gravier [m] en place** ; *g* **anstehender Kies [m]** (Kies, der vor Ort hervortritt, zu Tage liegt).

in-situ material [n] *constr.* ▶excavation, haulage, loading and transport of in-situ material.

2931 in situ sand [n] *constr. min.* (Sand which has its original occurrence on the site); *s* **arena [f] in situ**; *f* **sable [m] en place** ; *g* **anstehender Sand [m]** (Sand, der vor Ort hervortritt, zutage liegt).

in situ soil [n] *constr. pedol.* ▶site soil.

2932 insolation [n] *met.* (Intensity and quality of the sun's energy when it reaches the surface of the earth after direct exposure to the sun's rays; amount of exposure depends on solar access; ▶duration of insolation, ▶light exposure); *syn.* solar exposure [n] (CPF 1986-III, 129); *s* **insolación [f]** (Cantidad de radiación solar directa incidente por unidad de superficie horizontal, a un determinado nivel durante cierto periodo/período de tiempo; DM 1986; ▶iluminación, ▶período de exposición solar); *syn.* soleamiento [m], solación [f]; *f* **ensoleillement [m]** (1. État d'un lieu ensoleillé et **2.** durée, intensité et qualité du rayonnement solaire à un lieu donné ; ▶insolation ; ▶exposition à la lumière) ; *syn.* rayonnement [m] solaire global (CILF 1978, 113) ; *g* **Besonnung [f]** (Intensität und Qualität der direkten Sonneneinstrahlung an einem Ort — ohne Behinderung durch Wolken und Horizont; ▶Belichtung, ▶Sonnenscheindauer); *syn.* Insolation [f], (direkte) Sonneneinstrahlung [f].

inspection [n] *constr.* ▶construction site inspection, ▶final site inspection; *adm. leg.* ▶open for public inspection [US]; *hort. leg.* ▶phytosanitary inspection; *landsc.* ▶site inspection (2); *prof.* ▶site inspection (3); *adm. leg.* ▶tree inspection.

inspection [n] (1) [UK]**, site** *adm. leg.* ▶building code control [US].

2933 inspection [n] **and approval** [n] **of plants** *constr. contr.* (SPON 1974, 112; certified acceptance of delivered plants according to contract at the construction site); *s* **examen [m] de plantas** (Aceptación certificada de las plantas entregadas a obra para plantación); *f* **vérification [f] des plantes avant plantation** (Acceptation avec ou sans réserve de la provenance et de la qualité des végétaux lors de la livraison sur le chantier) ; *g* **Abnahme [f] der Pflanzen bei Lieferung** (Verbindliche Anerkennung auf der Baustelle von vertragsgerecht gelieferten Pflanzen); *syn.* Pflanzenabnahme [f].

inspection chamber [n] [UK] *constr.* ▶cleanout chamber [US].

2934 inspection [n] **of additional bid documents** [US] *constr. contr.* (Contractors examination of supplementary documents provided by the agency/authority making a request for proposals during bidding [US]/tendering [UK] period; ▶inspection of bid documents [US]/inspection of tender documents [UK]); *syn.* inspection [n] of additional tender documents [UK]; *s* **inspección [f] de documentos adicionales de concurso** (Estudio por parte del licitante dentro del plazo de entrega de ofertas de material informativo adicional no entregado por la administración comitente con los documentos de concurso; ▶inspección de las bases de concurso); *f* **consultation [f] du dossier d'appel d'offres** (Obtention de renseignements complémentaires et prise de connaissance des documents annexes auprès du maître de l'ouvrage pendant le délai de réception des offres/soumissions ; ▶étude du dossier d'appel d'offres) ; *syn.* consultation [f] des documents ; *g* **Einsicht [f] in Ausschreibungsunterlagen** (Prüfende Durchsicht seitens eines Bieters von zusätzlichen Unterlagen bei der ausschreibenden Stelle während der Angebotsfrist; *zum Vergleich* ▶Durchsicht der Ausschreibungsunterlagen).

inspection [n] **of additional tender documents** [UK] *constr. contr.* ▶inspection of additional bid documents [US].

2935 inspection [n] **of bid documents** [US] *constr. contr.* (Contractor's review of tender/bid documents provided by the agency/authority making a request for proposals during bidding period [US]/tendering period [UK]; ▶inspection of additional bid documents [US]/inspection of additional tender documents [UK]); *syn.* inspection [n] of tender documents [UK]; *s* **inspección [f] de las bases de concurso** (Estudio por parte del licitante de los documentos de licitación [textos explicativos, planos, etc.] entregados por la administración comitente a los licitantes para que éstos puedan elaborar una oferta de presupuesto [plica]; ▶inspección de documentos adicionales de concurso); *f* **étude [f] du dossier d'appel d'offres** (Prise de connaissance des différents documents constituant le dossier de consultation des entreprises afin de remettre une offre ; ▶consultation du dossier d'appel d'offres) ; *g* **Durchsicht [f] der Ausschreibungsunterlagen** (Lesen von Ausschreibungstexten, Studium von Plänen und sonstigen Unterlagen sowie Prüfung auf Vollständigkeit seitens eines Bieters, um ein Angebot sicher kalkulieren zu können; *zum Vergleich* ▶Einsicht in Ausschreibungsunterlagen).

inspection [n] **of tender documents** [UK] *constr. contr.* ▶inspection of bid documents [US].

inspection [n] **of the works** [UK] *constr. prof.* ▶supervision of works.

inspection shaft [n] [UK]**, drain(age)** *constr.* ▶drainage cleanout pipe [US].

inspector [n] *constr.* ▶public works inspector.

install [vb] *constr. contr.* ▶furnish and install.

install [vb]**, provide/supply** [vb] **and** *constr. contr.* ▶furnish and install.

2936 install [vb] **a laying course** *constr. s* **preparar** [vb] **el lecho de asiento** *syn.* tender [vb] el lecho de asiento; *f* **établir** [vb] **un lit de pose** ; *g* **Pflasterbett herrichten** [vb].

2937 install [vb] **at a prescribed level** *constr.* (To place at level specified in contract documents; ▶spreading of soil to true contours); *s* **tender** [vb] **al nivel de cota final** (▶relleno de tierra hasta nivel de rasante); *syn.* colocar [vb] al nivel de cota final; *f* **mettre** [vb] **en place aux cotes finies** (Mettre en œuvre les différents matériaux conformément aux cotes indiquées sur le plan ; ▶profilage aux pentes indiquées) ; *g* **höhengerecht einbauen** [vb] (Entsprechend den planerisch vorgegebenen Höhen herstellen; ▶profilgerechter Einbau).

2938 installation depth [n] *constr.* (Thickness of spread fill material, crushed rock, etc.; ▶thickness of a layer); *s* **espesor** [m] **a poner en obra** (p. ej. de capa de tierra vegetal; ▶profundidad de capa); *f 1* **épaisseur** [f] **de mise en œuvre** (Épaisseur relative à une couche de matériaux à mettre en place) ; *syn.* épaisseur [f] de mise en place ; *f 2* **épaisseur** [f] **de pose** (P. ex. pour le lit de pose en sable, en mortier, etc. d'un pavage, d'un revêtement ; ▶épaisseur de couche) ; *g* **Einbaustärke** [f] (Höhe des Auftrages von Schüttgütern oder Erdgemischen; ▶Schichtstärke 2); *syn.* Einbauhöhe [f].

2939 installation [n] **of a sculpture** *arch.* (Placement of a stucture on aspecified site); *s* **instalación** [f] **de una escultura** *syn.* colocación [f] de una escultura; *f* **érection** [f] **d'une sculpture** (Action de mise en place d'œuvre d'art d'un sculpteur, d'une statue ; ériger [vb] une statue) ; *g* **Aufstellung** [f] **einer Plastik** (Platzierung eines Werkes der Bildhauerkunst; eine Plastik aufstellen [vb]); *syn.* Installierung [f] einer Plastik.

installation [n] **of soil** *constr.* ▶placement of soil.

2940 installation [n] **of street furniture** *constr.* (▶piece of street furniture); *syn.* placement [n] of street amenities; *s* **instalación** [f] **de mobiliario urbano** (▶elemento de mobiliario urbano); *f* **implantation** [f] **de mobilier urbain** (▶élément de mobilier de voirie) ; *syn.* mise [f] en place de mobilier urbain ; *g* **Aufstellung** [f] **von Straßenmöblierungselementen** (Hinstellen oder Einbau von ▶Straßenmöblierungselementen).

2941 installation [n] **of utilities** *constr.* (Laying of utilities; *specific terms* laying of underground services, laying of overhead lines, laying of gas and water pipes, laying of district heating pipelines); *syn.* laying [n] of utilites; *s 1* **tendido** [m] **de tuberías o líneas** (Colocación de tuberías de agua, gas, calefacción o líneas eléctricas o de teléfonos bajo la superficie; cf. BU 1959; tender [vd] tuberías o líneas); *syn.* colocación [f] de tuberías o líneas; *s 2* **instalación** [f] **de tuberías o líneas** (Montaje de tuberías de agua, gas, calefacción o líneas eléctricas o de teléfonos en un edificio; cf. BU 1959; instalar [vb] tuberías o líneas); *f* **pose** [f] **des installations de distribution des fluides** (Mise en place des réseaux de distribution d'eau, gaz, électricité, d'eau surchauffée et du réseau de téléphone et de télédistribution) ; *syn.* installation [f] des réseaux divers ; *g* **Verlegen** [n, o. Pl.] **von Versorgungsleitungen** (Einbau/Legen von z. B. Fernwärme-, Gas-, Wasser-, Strom-, Telekommunikationsleitungen. Beim **V.** z. B. einer einzelnen Rohrleitung spricht man auch von **Strangverlegung**; die Begriffe *Verlegen* und *Verlegung* beinhalten die oberirdische und unterirdische Verlegung); *syn.* Leitungsverlegung [f], Verlegung [f] von Versorgungsleitungen; *syn. für das unterirdische Verlegen* Einlegen [n, o. Pl.] von Versorgungsleitungen, Einlegung [f] von Versorgungsleitungen, Leitungseinlegung [f].

2942 instinctual migration [n] *zool.* (Permanent migration of wild animals, induced by a species-specific migration drive, even when the environmental conditions do not urge this behaviour; ▶nomadism); *s* **vagabundismo** [m] (Movimiento de animales de área en área por un instinto de la especie, sin que las condiciones ambientales les obligen a ello; ▶nomadismo); *f* **vagabondage** [m] (Migration permanente qui fait quitter aux animaux sauvages, poussés par un instinct de migration, leur environnement familier sans que les conditions du milieu les y obligent ; forme de la migration exploratoire, le vagabondage de jeunesse étant souvent appelé dispersion post-juvénile ; cf. LAF 1990 ; ▶nomadisme) ; *g* **Nomadentum** [n, o. Pl.] **(1)** (Ständiges Umsiedeln/Umherstreifen von wild lebenden Tieren, jedoch durch einen arteigenen Wandertrieb bedingt, auch wenn die Umweltverhältnisse nicht dazu zwingen; diese Form des **N.s** nennt man auch *Zigeunertum*, *Vagabundentum* oder *Vagabundismus*; SCH 1979; ▶Nomadentum 2); *syn.* Zigeunertum [n], Vagabundentum [n], Vagabundismus [m].

2943 institutional land use [n] *leg. recr. urb.* (Land use category in urban development plans for churches, hospitals, schools, etc.; ▶comprehensive plan [US]/Local Development Plan [UK], ▶urban land use category); *syn.* institutional use [n]; *s* **reserva** [f] **de terreno para equipamiento público** (±) (Prescripción legal de prever superficies para servicios comunitarios como parques y jardines públicos, zonas deportivas y de recreo, centros culturales y docentes públicos y privados, templos, centros asistenciales y sanitarios y demás servicios de interés público y social en ▶plan general municipal de ordenación urbana y en ▶plan parcial [de ordenación]; arts. 12 y 13 RD 1346/1976. En D. ▶categoría de suelo urbano fijada en plan municipal de ordenación reservada para obras de infraestructura técnica y social de gran envergadura como universidades, hospitales, centros de investigación de energías alternativas, puertos, ferias de muestra, centros comerciales, etc., así como para zonas de recreo); *f* **emplacement** [m] **réservé pour ouvrage public, installation d'intérêt général ou espace vert** (F., emplacement réservé des PLU/POS prévus pour les équipements d'infrastructure, de superstructure ou à fonction collective ; ▶zone urbaine 2 ; D., réservations fixées dans le cadre du « Flächennutzungsplan » [▶schéma de cohérence territoriale allemand]) ; *g* **Sonderbaufläche** [f] (Im ▶Flächennutzungsplan dargestellte Bauflächenart, u. a. für Hochschulgebiete, Klinikgebiete, [soweit wird der engl. Begriff abgedeckt]. In D. gehören aber auch Hafengebiete, Gebiete für Einkaufszentren, Messen, Kongresse etc., Gebiete für Anlagen, die der Erforschung, Entwicklung oder Nutzung erneuerbarer Energien dienen sowie Erholungsgebiete [Wochenendhaus-, Ferienhaus- und Campingplatzgebiete] dazu; ▶Baufläche).

institutional use [n] *leg. urb.* ▶institutional land use.

instruction [n] *constr.* ▶in compliance with the issue of an instruction.

instruction [n] **for rectification** [UK] *constr. contr.* ▶rectification directive [US].

instruction [n] **requiring defects to be made good** [UK] *constr. contr.* ▶rectification directive [US].

instruction [n] **to rectify defective work** [UK] *constr. contr.* ▶rectification directive [US].

instruction to the contractor [n] *constr. contr.* ►change order [US].

2944 insulated roof membrane assembly [n] [US] *arch. constr.* (Non-ventilated, single-ply, flat roof with similar sequence of layers to that of a ►warm roof, with the significant difference that the waterproofing layer of the roof is protected by the thermal insulation layer installed on top of it, i.e. the inverse of layers for warm roof construction; as a rule the sequence of layers from bottom to top is as follows: **1.** Load-bearing construction [roof ceiling rafters], **2.** Protection fabric, **3.** Waterproof sealing, **4.** Thermal insulation layer [e.g. pressure resistant, extruded polystyrene hard foam board, resistant to humidity], **5.** Surface protection [e.g. gravel, vegetation layer]; cf. FLA 1983, 36; ►cold roof]; *syn.* inverted roof [n] [UK]; *s* **tejado** [m] **invertido** (Tejado plano de una hoja no ventilado con secuencia similar de capas a la del ►tejado caliente solo que en el **t. i.** la capa impermeabilizante está protegida por la de aislamiento térmico que se coloca por encima, es decir es la inversión de capas con respecto al tejado caliente. En general la secuencia de capas, de abajo a arriba, es la siguiente: **1.** Capa portante [vigas y viguetas del techo], **2.** Tela protectora, **3.** Membrana impermeabilizante de tejado, **4.** Capa de aislamiento térmico, **5.** Protección de superficie [p. ej. gravilla, capa portante de vegetación]; ►tejado frío); *f* **toiture** [f] **inversée** (La toiture plate constituée **1.** du support, **2.** de l'étanchéité, **3.** de l'isolant, **4.** d'une natte de protection et **5.** du lestage. L'étanchéité est placée directement sur le support porteur, l'isolant venant se poser sur celle-ci ; elle réalise une double fonction, étanchéité à l'eau et pare-vapeur ; le support de toiture et l'étanchéité sont protégés des actions thermiques et des rayons solaires par l'isolant et le lestage qui doit pouvoir résister à l'arrachement et au soulèvement de l'isolant par le vent ; le Polystyrène Extrudé [XPS], est le seul isolant utilisé pour la réalisation de toitures inversées ; il ne doit pas absorber l'eau, doit résister à la chaleur et doit résister aux attaques de l'eau superficielle [cycle gel/dégel, acidité de l'eau, etc.] ; les bâtiments où la température intérieure est supérieure à 35 °C ou les chambres froides ne peuvent contenir une toiture inversée sous risque de provoquer une condensation dans l'isolant ; dans les toitures inversées, l'eau qui s'infiltre sous l'isolant entraîne des déperditions calorifiques et doit être rapidement évacuée par une pente de toiture suffisante ; si l'étanchéité est en elle-même tout à fait apte à faire face à des stagnations d'eaux dues à des pentes nulles ou faibles, dans le cas d'une toiture inversée on impose d'exclure toutes stagnations d'eaux et la pente conseillée de la toiture sera > 4 % ; ►toiture froide, ►toiture chaude) ; *g* **Umkehrdach** [n] (Unbelüftetes einschaliges Flachdach mit ähnlicher Schichtenfolge wie beim ►Warmdach, jedoch bei dem die Dachdichtung von der Wärmedämmung geschützt wird [Reihenfolge dieser beiden Schichten umgekehrt zum Warmdachaufbau]; die Schichtenfolge besteht i. d. R. von unten nach oben: **1.** Tragschicht [Dachdecke], **2.** Schutzgewebe, **3.** Abdichtung, **4.** Wärmedämmschicht [z. B. aus druckfestem, wenig Feuchtigkeit aufnehmendem extrudiertem Polystyrol-Hartschaum], **5.** Oberflächenschutz [z. B. Bekiesung, Vegetationstragschicht]; cf. FLA 1983, 36; ►Kaltdach).

insulation [n] *constr.* ►heat insulation 1, ►heat insulation 2.

insulation [n], **thermal insulation** *constr.* ►heat insulation (1).

insurance [n] *contr. prof.* ►obligation to take out professional liability insurance, ►professional liability insurance.

insurance [n] [US], **errors and ommission** *prof.* ►professional liability insurance.

insurance [n], **liability for professional** *prof.* ►obligation to take out professional liability insurance.

intake area [n] *hydr.* ►aquifer recharge area.

integrated environmental protection [n] *conserv. pol.* ►complex environmental protection.

2945 integrated landscape conservation [n] *conserv. land'man.* (Comprehensive measures aimed at combining economic land use with the safeguarding of geographic, ecological, aesthetic, and social viability of large-scale natural landscapes; ►protection of natural area); *s* **protección** [f] **integral del paisaje** (Término específico de la ►protección de áreas naturales. Conjunto de medidas que sirven para preservar paisajes y los procesos ecológicos en un sistema de usos del suelo de orientación económica); *f* **conservation** [f] **intégrée des paysages** (Mesures favorisant la protection de l'espace naturel, la ►sauvegarde des sites naturels et le maintien d'une exploitation du sol respectueuse des équilibres écologiques ou de productions compatibles avec les exigences de la protection de l'environnement) ; *g* **integrierter Landschaftsschutz** [m, o. Pl.] (Maßnahmen des ►Flächenschutzes, die der Sicherung eines Bestandes geogener Erscheinungen und der Aufrechterhaltung ökologischer Prozesse innerhalb der wirtschaftsorientierten Bodennutzung dienen).

2946 integrated nature conservation [n] *conserv. land'man.* (All measures to safeguard the natural environment and to maintain ecological processes within economically-focused land uses; ►integrated landscape conservation, ►nature protection); *s* **protección** [f] **integral de la naturaleza** (En comparación con la ►protección íntegra de la naturaleza, todas las medidas que sirven para asegurar el funcionamiento de la naturaleza y de los procesos ecológicos en el marco del uso económico de suelo; ►protección integral del paisaje); *f* **protection** [f] **intégrée de la nature** (En comparaison avec une conception restrictive de la protection de la nature toutes mesures de sauvegarde des espaces naturels et des paysages, de maintien des équilibres écologiques dans le cadre d'une exploitation et utilisation rationnelle des sols et des ressources ; ►conservation intégrée des paysages, ►gestion conservatoire/conservatrice du milieu naturel, ►sanctuarisation [du milieu naturel]) ; *g* **integrierter Naturschutz** [m, o. Pl.] (Im Vergleich zum konservierenden N. alle Maßnahmen, die der Sicherung von Natur und Landschaft und der Aufrechterhaltung ökologischer Prozesse innerhalb der wirtschaftsorientierten Bodennutzung dienen; ►integrierten Landschaftsschutzes, ►konservierender Naturschutz).

2947 integrated pest control [n] *agr. for. hort. phytopath.* (Financial, ecological and toxicological methods of keeping injurious organisms below the threshold of economic damage, and deliberately giving precedence to natural control methods; ►pest and weed control); *syn.* integrated pest management [n]; *s* **control** [m] **integrado de plagas** (Sistema de control de una población de organismos nocivos en el que, teniendo en consideración el medio y la dinámica de la especie, se emplean todas las técnicas y métodos idóneos, de la forma más compatible posible, y se mantiene la densidad a un nivel tan bajo que no pueda causar perjuicios económicos; DINA 1987; ►control antiplaguicida); *f* **lutte** [f] **intégrée** (Ensemble de méthodes culturales et prophylactiques, biologiques, physiques, chimiques, pour maintenir les dégâts des ennemis des cultures à un niveau économiquement acceptable, tout en satisfaisant les exigences toxicologiques sur les végétaux récoltés ; cf. LAP 1979 ; ►protection des végétaux) ; *g* **integrierter Pflanzenschutz** [m, o. Pl.] (Verfahren, bei denen alle wirtschaftlich, ökologisch und toxikologisch vertretbaren Methoden verwendet werden, um Schadorganismen unter der wirtschaftlichen Schadensschwelle zu

halten, wobei die bewusste Ausnutzung natürlicher Begrenzungs-
faktoren im Vordergrund steht; ▶Pflanzenschutz).

integrated pest management [n] *agr. for. hort. phytopath.* ▶integrated pest control.

integrated planning [n] *plan.* ▶multidisciplinary planning.

2948 integrated species conservation [n] *conserv.
land'man.* (Coordinated measures to protect and promote the
upkeep of animal and plant habitats as vital natural resources in
the face of such activities as hunting, fishing, agriculture, and
forestry; ▶hunting management); *s* **protección** [f] **integral de
flora y fauna** (Conjunto de medidas tomadas para mantener los
hábitats naturales de especies de flora y fauna que se aprovechan
económicamente [caza, pesca, agricultura, silvicultura, etc.];
▶gestión de la caza 2); *f* **conservation** [f] **intégrée des espèces
animales et végétales** (Mesures concernant les activités économiques de chasse, pêche, agriculture et sylviculture visant à
introduire les pratiques et techniques culturales compatibles avec
les exigences de protection des espèces sauvages et de leurs
milieux naturels ; politique de gestion du patrimoine biologique
en relation avec le concept de développement rural intégré ;
▶gestion cynégétique) ; *g* **integrierter Artenschutz** [m, o. Pl.]
(Alle Maßnahmen zur Beeinflussung und Förderung der direkt in
die Tier- und Pflanzenwelt eingreifenden Wirtschaftsformen
[Jagd, Fischerei, Land- und Forstwirtschaft usw.] hinsichtlich der
pfleglichen und nachhaltigen Nutzung der Tier- und Pflanzenwelt
als Naturgut; ▶jagdliche Hege).

2949 integration [n] **of structures into the landscape** *arch. landsc.* (Blending of one or more buildings in
height and siting into the scenery with grading and planting that
is appropriate to the site); *s* **integración** [f] **al paisaje** (Construcción de edificios considerando el paisaje en el que se sitúan, tanto
en cuanto a alturas y volúmenes como por medio de plantaciones
apropiadas a la ubicación); *syn.* adaptación [f] al paisaje; *f* **intégration** [f] **paysagère (au site)** (d'un projet dans son environnement) ; *syn.* insertion [f] (au site), intégration [f] au paysage,
intégration [f] à l'environnement (L. 145-5), incorporation [f]
dans un paysage, incorporation [f] paysagère ; *g* **landschaftliche
Einbindung** [f] (Integration von einem oder mehreren
Bauwerken in baulichen Anlagen in Höhe und Lage, ggf. mit
einer landschaftsgerechten Bepflanzung eingegrünt); *syn.*
landschaftliche Einfügung [f], landschaftliche Eingliederung [f].

intensity [n] *ecol. phyt.* ▶light intensity.

intensity [n]**, rainfall** *constr. eng. hydr.* ▶rate of rainfall.

intensity [n]**, traffic** *trans.* ▶traffic density.

2950 intensity [n] **of agricultural and forest land
use** *agr. for. hort.* (Degree to which measures are undertaken to
promote the yield of one growing season); *syn.* intensity [n] of
crop production; *s* **intensidad** [f] **de explotación del suelo agrosilvícola** (Conjunto de medidas tomadas para fomentar la productividad de los cultivos agrosilvícolas); *f* **intensité** [f] **de l'exploitation du sol** (Ensemble de toutes les facteurs de production mis
en jeu pour obtenir un certain rendement pendant une saison) ;
syn. intensité [f] de l'utilisation du sol ; *g* **Nutzungsintensität** [f]
land- und forstwirtschaftlicher Flächen (Summe aller Aufwendungen zur Förderung der Ertragsleistung während einer Vegetationsperiode); *syn.* Intensität [f] der Bodenbewirtschaftung, land-
und forstwirtschaftliche Nutzungsintensität [f].

intensity [n] **of crop production** *agr. for. hort.* ▶intensity
of agricultural and forest land use.

intensive crop production [n] *agr. hort.* ▶intensive
farming.

2951 intensive farming [n] *agr. hort.* (Cultivation of crops
characterized by the high degree of work involved and a

correspondingly high yield); *syn.* intensive crop production [n];
s **cultivo** [m] **intensivo** (Tipo de agricultura caracterizada por el
empleo de gran cantidad de insumos y de mano de obra con la
consecuente influencia en la productidad, como p. ej. en la
producción de frutas y verduras); *f* **culture** [f] **intensive** (Forme
d'agriculture caractérisée par un important investissement dans le
capital naturel [grandes exploitations ou spécialisation avec
recours massif à l'épandage d'engrais, de produits antiparasitaires, à l'achat de semences de grandes cultures sélectionnées,
etc.], du capital technique [forte mécanisation et équipements
automatisés] et du capital humain [ressources en main d'œuvre
par unité de surface élevées] avec l'objectif d'obtenir de forts
rendements dans le but de maximiser les profits retirés de l'exploitation) ; *g* **Intensivkultur** [f] (Pflanzenbau, der durch hohen
Einsatz an Kapital [hohe Düngegaben, Spritzmittel, Erwerb von
hochleistungsfähigen Pflanzensorten], erhöhten Arbeitsaufwand
je Flächeneinheit und hohen Ertrag gekennzeichnet ist, z. B. die
Produktion von Gemüse, Obst, Zuckerrüben).

2952 intensively grazed pasture [n] *agr. phyt.* (Heavily-
used pasture fertilized regularly by grazing livestock. Pasture
weeds are rare and occur only at 'hot spots' where droppings
have collected. The pasture is mown at the end of the growing
period to clear away weeds and 'hot spots'; ▶hayfield [US]/hay
meadow [UK], ▶rotation pasture [US]); *s* **pastizal** [m] **permanente** (Pasto aprovechado intensamente, a menudo abonado sólo
por el ganado que pasta en él, en el cual se presentan pocas hierbas advenedizas en los lugares de concentración de excrementos.
Al final del periodo/período de crecimiento se suele segar una
vez para eliminar las malas hierbas; ▶prado de siega, ▶pastizal
en rotación); *syn.* pasto [m] permanente; *f* **pâturage** [m] **libre**
(Prairie sur laquelle les animaux disposent en permanence
pendant toute la période de pâturage de la totalité de la surface
fourragère à pâturer ; mode d'exploitation intensif, en général
sans fertilisation complémentaire, avec entretien à la fin de la
période de végétation [fauche des refus, déboisage, désherbage] ;
LA 1981, 821/903 ; ▶pâturage tournant, ▶prairie à pâturage
rationné, ▶prairie à pâturage en paddocks) ; *syn.* pâturage [m]
continu ; *g* **Standweide** [f] (Intensiv genutzte Weide; wird
vielfach nur von dem auf ihr weidenden Vieh gedüngt. Weide-
unkräuter selten, nur an sog. ‚Geilstellen', die durch Kot bedingt
sind. Am Ende der Vegetationsperiode erfolgt oft ein Säube-
rungsschnitt, der Unkräuter und Geilstellen abmäht; ▶Mähweide,
▶Umtriebsweide).

2953 intensive planting [n] (1) *constr.* (Dense planting in a
small area; *opp.* ▶thin planting); *s* **plantación** [f] **densa** *opp.*
▶plantación rala; *f* **plantation** [f] **dense** (*opp.* ▶plantation clairsemée) ; *syn.* plantation [f] intensive ; *g* **dichte Anpflanzung** [f]
(Ergebnis einer geschlossenen Pflanzung; *opp.* ▶lockere Anpflanzung); *syn.* dichte Bepflanzung [f], dichte Pflanzung [f].

intensive planting [n] [UK] (2) *constr.* ▶high maintenance
planting.

intensive recreation area [n] *plan. recr.* ▶concentration
of recreation facilities.

2954 intensive roof planting [n] *constr.* (Establishment of
vegetation on relatively moist roofs which normally have a soil or
growing medium layer of more than 10cm. Depth and moisture
level of this layer determine to what extent more demanding
plants may be cultivated and the degree of maintenance necessary; *opp.* ▶extensive roof planting); *syn.* deep soil roof planting
[n] [also US]; *s* **revestimiento** [m] **vegetal intensivo de tejados**
(Establecimiento de vegetación en tejados sobre capa de sustrato
o tierra de 15 o más cm. Dependiendo del espesor pueden culti-
varse plantas más exigentes en cuanto a la ubicación y al mantenimiento; *opp.* ▶revestimiento vegetal extensivo de tejados); *syn.*

plantación [f] intensiva de tejados; *f* **végétalisation [f] intensive des toitures** (Terrasse-jardin accessible d'un entretien intensif constituée d'un support de culture supérieur à 15 cm permettant l'implantation d'une végétation d'agrément ; *opp.* ►végétalisation extensive des toitures) ; *syn.* aménagement [m] (intensif) de terrasses-jardins, aménagement [m] (intensif) de jardins sur dalles ; *g* **intensive Dachbegrünung [f]** (Bauwerksbegrünung auf Dächern mit einer i. d. R. mehr als 12 cm mächtigen Erdschicht/Substratschicht. Je nach Höhe dieser Schicht und Menge verfügbaren Wassers können anspruchsvolle und aufwendig zu pflegende Pflanzen kultiviert werden; *opp.* ►extensive Dachbegrünung); *syn.* Intensivbegrünung [f] eines Daches.

interactions [npl] **and interrelationships** [npl] *ecol.* ►ecological interactions and interrelationships.

interactions [npl] **and relationships** [npl] *ecol.* ►pattern of interactions and interrelationships.

2955 interactions [npl] **of ecosystems** *ecol.* (Connectivity or pattern of various parts of an ecosystem; e.g. forest, lake, damp meadow, field, within a geographical area; ►ecosystem, ►pattern of ecotopes); *s* **sistema [f] de interacciones en un ecosistema** (Conectividad entre diferentes partes de un ecosistema como p. ej. bosque, lago, pradera húmeda, campo de cultivo; ►ecosistema, ►mosaico de ecotopos); *f* **tissu [m] d'interactions d'un écosystème** (Réseau de différentes parties d'écosystèmes [forêts, lacs, prairies humides, champs, etc.] dans le milieu naturel ; ►écosystème, ►réseau d'écotopes) ; *g* **Ökosystemgefüge [n]** (Verbund oder Muster von verschiedenen Ökosystemteilen, z. B. Wald, See, Feuchtwiese, Acker, in einer Landschaft; ►Ökosystem, ►Ökotopengefüge).

2956 intercepting ditch [n] *constr.* (Channel constructed at top, bottom or across a slope to intercept and control surface water runoff); *syn.* diversion ditch [n]; *s* **contracuneta [f]** (Pequeño canal construido para captar el agua de un talud o una ladera); *syn.* cuneta [f] de guardia, zanja [f] de retención; *f* **fossé [m] de maîtrise du ruissellement** (VRD 1994, partie 2, chap. 1.3.2.1.3, p. 14 ; fossé creusé un peu au-dessus de la crête d'un talus en déblai interceptant les eaux de ruissellement, p. ex. en tête de talus) ; *syn.* fossé [m] de captage du ruissellement, fossé [m] d'arrêt (WIB 1996), fossé [m] de rétention (WIB 1996), fossé [m] de crête, fossé [m] de garde ; *g* **Fanggraben [m]** (Graben, der Tagwasser, z. B. oberhalb des Böschungskopfes, aufnimmt).

interceptor [n] [UK] *constr.* ►silt fence [US].

interceptor pipe [n] [US] *constr.* ►drain pipe.

2957 interchange [n] *trans.* (Junction of traffic routes on different levels as the connecting point of a subordinate road at a freeway [US]/motorway [UK]; ►cloverleaf interchange [US], ►diamond-shaped interchange, ►freeway interchange, ►highway ramp, ►grade separated junction); *syn.* freeway junction [n] [also US], motorway junction [n] [also UK]; *s* **nudo [m] de enlace** (Conexión a diferentes niveles entre una autopista y una carretera; ►cruce de autopistas, ►nudo de enlace con forma de rombo, ►nudo sin cruces a nivel, ►rampa de acceso, ►trébol); *syn.* nudo [m] de empalme, entronque [m] [MEX], crucero [m] [AL]; *f* **diffuseur [m]** (Raccordement entre une autoroute et une route secondaire s'effectuant sur deux niveaux différents ; ►bretelle d'accès, ►échangeur, ►échangeur autoroutier, ►raccordement de type losange, ►trèfle) ; *syn.* point [m] d'accès (au réseau routier) ; *g* **Anschlussstelle [f]** (*Straßenbau* Verkehrsknoten zwischen einer Autobahn und einer untergeordneten Straße, der planfrei geführt wird; ►Auffahrtsrampe, ►Autobahnkreuz, ►Kleeblatt, ►planfreier Knotenpunkt, ►Rautenanschluss).

intercompatibility [n] *landsc. plan.* ►matrix on intercompatibility.

interconnectedness [n] *conserv. ecol.* ►network (1).

2958 interdependency [n] **within an ecosystem** *ecol.* (Functional pattern of any response or adaptation to the environment by plants or animals depending on each other); *syn.* dynamics [npl] of relationships (TGG 1984, 185); *s* **sistema [m] de interacciones e interrelaciones** (Estructura funcional de un ecosistema); *f* **système [m] de relations interdépendantes** (Organisation fonctionnelle entre les différentes composantes d'un écosystème) ; *g* **Wirkungsgefüge [n]** (Funktionelle Organisation in einem Ökosystem); *syn.* Wirkungsbeziehungen [fpl].

2959 interdisciplinary planning [n] *plan.* (Planning carried out jointly by associated professionals in several specialist fields); *s* **planificación [f] interdisciplinaria** (Planeamiento que integra diferentes disciplinas); *syn.* planificación [f] multidisciplinaria, planeamiento [m] interdisciplinario, planeamiento [m] multidisciplinario; *f* **aménagement [m] interdisciplinaire** (Pratique associant plusieurs domaines d'aménagement ; *syn.* aménagement [m] pluridisciplinaire, aménagement [m] multidisciplinaire ; *g* **interdisziplinäre Planung [f]** (Mehrere Fachdisziplinen einbeziehende Planung).

interest [n], **community** *conserv. ecol.* ►natural habitat of community interest.

interest [n], **overriding public** *ecol. leg. plan.* ►imperative reasons of overriding public interest.

interest [n] **in land** *leg.* ►land requirements.

interference [n] *conserv. leg. plan.* ►intrusion upon the natural environment.

interfingering open space pattern/system [n] *landsc. urb.* ►open space pattern/system of peninsular interdigitation.

2960 interim agreed measurement [n] **of completed work** [US]**/works** [UK] *constr. contr.* (Computation of executed work during the construction process, which usually cannot be identified after completion of the whole project; ►agreed measurement of completed works/project); *s* **medición [f] intermedia de los trabajos terminados** (Revisión de obras realizadas durante una fase de construcción que no pueden ser o son difíciles de controlar una vez terminada la obra completa; ►medición acordada de los trabajos terminados); *f* **métré [m] provisoire** (Par comparaison au ►métré général et définitif, métré servant de base à l'attachement en travaux terminés, établi au fur et à mesure du déroulement du chantier en vue de faciliter l'établissement, en fin de travaux, d'un décompte général précis) ; *g* **Zwischenaufmaß [n]** (Feststellung von fertiggestellten Bauleistungen während des Baufortschrittes, die bei der Fertigstellung des Gesamtprojektes nicht oder nur schwer nachvollzogen werden können; cf. § 14 [2] VOB Teil B; ►örtliches Aufmaß).

interim bill [n] *contr.* ►interim invoice.

interim certificate [n] *contr.* ►issue of a sectional completion certificate.

2961 interim invoice [n] *contr.* (Interim project bill submitted for design or construction executed at a specified stage in a contract; ►final account invoice, ►interim payment); *syn.* interim bill [n]; *s* **liquidación [f] intermedia** (Elaboración de la factura para servicios y/o materiales suministrados que son parte de un encargo total; ►liquidación final de obra, ►pago a plazos); *syn.* factura [f] intermedia; *f* **situation [f]** (Document qui fait état de l'avancement des travaux et de leur montant, en vue du paiement d'un acompte ; DIC 1993 ; ►décompte final, ►paiement par acompte) ; *g* **Zwischenrechnung [f]** (Rechnung für erbrachte Leistungen und/oder Lieferungen, die Teil eines Gesamtauftrages sind; ►Abschlagszahlung, ►Schlussabrechnung).

2962 interim payment [n] *constr. contr.* (**1.** Payment for partially-executed construction work; ▶interim invoice); *syn.* progress payment [n] [also US], partial payment [n]; **2. advances [npl] on account [UK]** (According to General Conditions of Contract for Building and Civil Engineering Works [GC/Works/1], a payment which the contractor is entitled to receive during the progress of the execution of the works at monthly intervals; BCD 1990); *s* **pago [m] a plazos (2)** (**1.** Liquidación de parte de los trabajos de construcción; ▶liquidación intermedia); **2. abono [m] a cuenta [Es]** (De acuerdo a la Ley de Contratos de las Administraciones Públicas, pago a buena cuenta que se puede realizar mensualmente tras certificación de la administración competente, sujeto a las rectificaciones y variaciones que se produzcan en la medición final y sin suponer, en forma alguna, aprobación y recepción de las obras que comprenden; art. 145 Ley 13/1995); *f* **paiement [m] par acompte** (Paiement effectué en fonction de l'avancement des travaux ou des stipulations inclues au marché ; ▶situation) ; *syn.* acompte [m] ; *g* **Abschlagszahlung [f] (2)** (Teil einer zu zahlenden Rechnung für erbrachte Bauleistungen; ▶Zwischenrechnung. Im britischen Bauvertragswesen *[General Conditions of Contract for Building and Civil Engineering Works]* gibt es die Möglichkeit, vorher vereinbarte, dem Baufortschritt entsprechende, monatliche Abschlagszahlungen zu gewähren: *advances on account*); *syn.* Akontozahlung [f], Teilzahlung [f].

interim report [n] *plan.* ▶provisional report.

2963 interim storage [n] **of recyclable materials** *constr.* (Pile [US]/heap [UK] of material suitable for reuse from demolition or excavation during construction of a project; *topsoil stockpiling* [n], *stockpile* [vb], *stack* [vb] on site [also UK] (SPON 1974, 88); ▶reuse of building rubble); *s* **almacenamiento [m] provisional de materiales de construcción reciclables** (▶reutilización de material de derribo); *syn.* almacenamiento [m] intermedio de materiales de construcción reutilizables; *f* **mise [f] en dépôt provisoire de matériaux réutilisables** (Mise en dépôt provisoire sur le chantier de matériaux à la suite de démolitions, affouillements ou de démontage en vue d'une réutilisation ultérieure ; *terme spécifique* conservation pour réemploi [lors de la dépose de dalles et autres revêtements] ; ▶réutilisation de matériaux de démolition) ; *syn.* mise [f] en dépôt pour réemploi ; *g* **Zwischenlagerung [f] von verwertbaren Baumaterialien** (Zeitlich begrenzte Lagerung von Baustoffen, die auf einer Baustelle durch Abbruch, Aushub oder Ausbau gewonnen wurden und nach entsprechender Aufbereitung zur Wiederverwendung vorgesehen sind; ▶Wiedereinbau von Abbruch-/Aufbruchmaterial).

2964 interlocking block pavement [n] *constr.* (Wearing surface made of connecting blocks of prefabricated units; ▶interlocking paver); *s* **adoquinado [m] de ensamblaje** (Pavimento de ▶adoquines de ensamblaje); *f* **pavage [m] en pavés autobloquants** (Revêtement de surface constitué de ▶pavés autobloquants) ; *g* **Verbundpflaster [f]** (Belag aus ▶Verbundpflastersteinen); *syn.* Verbundsteinpflaster [n].

interlocking concrete paving unit [n] [AUS] *constr.* ▶interlocking paver [US].

2965 interlocking paver [n] [US] *constr.* (TSS 1988, 840-17; connected paver [US]/paving block [UK] which is designed to distribute shearing, torsion, and vibration forces into the surrounding pavers; ▶concrete paver [US]/concrete paving unit [UK]); *syn.* interlocking paving block [n] [UK] (SPON 1986, 368), interlocking concrete paving unit [n] [also AUS]; *s* **adoquín [m] de ensamblaje** (Tipo de adoquín cuya forma transmite la energía de empuje, torsión y vibración a los adoquines vecinos; ▶piedra de hormigón); *syn.* adoquín [m] de aparejo; *f* **pavé [m]**

autobloquant (Terme générique pour les pavés formant une liaison horizontale ou horizontale et verticale entre les éléments du pavage et assurant la transmission des charges aux pavés voisins ; on distingue les pavés autobloquant à emboîtement ou à emboîtement/épaulement ; ▶pavé en béton) ; *g* **Verbundpflasterstein [m]** (Pflasterstein, der durch eine besondere Formgebung auftretende Schub-, Walk- und Schwingkräfte auf die umgebenden Steine abgibt; ▶Betonpflasterstein 1).

interlocking paving block [n] [UK] *constr.* ▶interlocking paver [US].

2966 intermediate [adj] *hort. plant.* (Term applied to understock with medium vigo[u]r on which cultivars are grafted; ▶vigorous-growing); *s* **de crecimiento medio [loc]** (Término horticultor aplicado p. ej. a patrón sobre el que ha sido injertada una planta; ▶vigoroso/a); *f* **à croissance moyenne [loc]** (▶vigoureux) ; *g* **mittelstark wüchsig [adj]** (Eine Pflanze betreffend, die in ihrer Wachstumvitalität in etwa im Mittel zwischen stark und schwach wüchsigen Pflanzen liegt, z. B. **m. w.e** Veredlungsunterlage; ▶wüchsig); *syn.* mittelstark wachsend [ppr/adj].

2967 intermediate planting [n] *constr.* (Sowing of Fabaceae/Leguminosae, such as Yellow Lupine *[Lupinus lúteus]*, Alexandrine Clover *[Trifolium alexandrinum]*, Brassicaceae/Cruciferae, such as White Mustard *[Sinapis alba]* and Radish *[Raphanus sativus var. oleiformis]* or of Italian rye grass for stabilization, shading or keeping free of weeds on temporarily open ground and topsoil piles [US]/topsoil heaps [UK], as well as for mobilization of nutrients and improvement of soil structure; ▶green manuring, ▶green manure catch crop); *s* **plantación [f] intermedia** (Siembra de fabaceas/leguminosas como el altramuz o lupino *[Lupínus albus, L. lúteus]*, el trébol *[Trifolium alexandrinum]* o brassicaceas/cruciferas como la mostaza blanca *[Sinapis alba]* y el rábano *[Raphanus sativus var. oleiformis]* para fijar temporalmente el suelo desnudo o la tierra vegetal en acopios, evitar el crecimiento de malas hierbas y mejorar la textura; ▶abonado verde, ▶intercalado de cultivos de abonado verde); *f* **ensemencement [m] intermédiaire** (Semis de légumineuses, p. ex. le Lupin jaune *[Lupínus lúteus]*, le Trèfle d'Alexandrie *[Trifolium alexandrinum]* ou de crucifères telles que la Moutarde blanche *[Sinapis alba]* et le Radis cultivé *[Raphanus sativus var. oleiformis]* effectué en protection des terres excédentaires mises en dépôt provisoire, en attente prolongée d'enlèvement ou de réemploi sur un chantier ; ▶apport d'engrais vert, ▶fertilisation par apport d'engrais vert) ; *syn.* enherbement [m] intermédiaire ; *g* **Zwischenbegrünung [f]** (*Landschaftsbau* Einsaat von Leguminosen *[Fabaceae]* wie z. B. Gelbe Lupine *[Lupínus lúteus]*, Alexandriner Klee *[Trifolium alexandrinum]* oder Kreuzblütlern *[Brassicaceae/Cruciferae]* wie Weißer Senf *[Sinapis alba]* und Ölrettich *[Raphanus sativus var. oleiformis]* zur vorübergehenden Festlegung, Beschattung oder Unkrautfreihaltung von offenen Bodenflächen und Oberbodenmieten sowie zur Bodenaufschließung und Verbesserung des Bodengefüges; ▶Gründüngung, ▶Gründüngung als Zwischenkultur).

2968 intermediate storage space [n] *envir.* (TGG 1984, 250; interim storage area for toxic wastes from ▶nuclear plants or very toxic refuse from chemical plants until they are recycled or finally disposed of; ▶final disposal); *s* **almacén [m] provisional** (Lugar de almacenamiento temporal de residuos de ▶instalaciones nucleares o de otras sustancias tóxicas antes de su reproceso o ▶almacenamiento definitivo de residuos); *syn.* depósito [m] provisional (de residuos radiactivos); *f* **entreposage [m]** (Stockage temporaire des déchets radioactifs solides et des effluents radioactifs ; ▶installation nucléaire, ▶stockage définitif) ; *g* **Zwischenlager [n]** (Aufbewahrungsort abgebrannter Brennelemente aus ▶kerntechnischen Anlagen oder sonstiger hochgiftiger Abfälle vor der Wiederaufbereitung oder ▶Endlagerung).

2969 interment capacity [n] *adm. plan.* (Total number of graves according to gross lot size which can be placed in a cemetery); *s* **capacidad [f] neta máxima** (Número total de tumbas que caben en un cementerio); *f* **capacité [f] maximale d'un cimetière (±)** (Surface totale brute des tombes dans un cimetière) ; *g* **Gesamtbelegungsfläche [f]** (Summe der Bruttograbgrößen auf einem Friedhof); *syn.* Gesamtbestattungsfläche [f].

2970 intermittent riprap [n] *constr. eng. wat'man.* (Rock groups separated by vegetation to prevent undercutting of outer banks on slow-flowing watercourses; ▶streambank erosion, ▶riprap); *s* **escollera [f] intermitente (≠)** (En cursos fluviales de flujo lento, montones de rocas partidas colocadas sobre los taludes de las márgenes alternando con tramos de vegetación cuya función es proteger las márgenes contra la erosión lateral; ▶erosión de márgenes de cursos de agua, ▶escollera); *f* **plots [mpl] en enrochements** (Ouvrage discontinu de protection de pied de talus lors du recalibrage du lit suite à l'▶érosion des berges des cours d'eau ; ▶cailloutage des berges) ; *g* **Steinschüttungspakete [npl] (≠)** (Abfolge von geschütteten Steinpackungen auf Uferböschungen zum Schutz gegen Seitenerosion bei langsam fließenden Gewässern; ▶Erosion an Ufern von Fließgewässern, ▶Steinschüttung an Uferböschungen).

2971 intermittent stream [n] *geo. hydr.* (**1.** Stream flowing occasionally at sporadic intervals; ▶intermittent watercourse); **2. runnel** [n] (In arid regions a small runoff channel [US]/run-off channel [UK]); *s* **arroyo [m] intermitente** (Arroyo que fluye en intervalos irregulares dependiende del suministro de agua de deshielo o de lluvia; ▶curso de agua intermitente); *f* **lit [m] d'un ruisseau à écoulement intermittent** (Lit dans lequel l'écoulement des eaux réapparaît régulièrement entre deux interruptions ; ▶cours d'eau intermittent) ; *g* **zeitweilig Wasser führendes Bachbett [n]** (Durch gelegentliche Wasserführung gekennzeichnetes kleines Gewässerbett; *OB* ▶zeitweilig Wasser führendes Fließgewässer); *syn.* intermittierender Bach(lauf) [m], periodisch Wasser führender Bach(lauf) [m].

2972 intermittent watercourse [n] *geo. hydr.* (Watercourse, which only occasionally contains water; ▶intermittent stream); *s* **curso [m] de agua intermitente** (Río o arroyo cuyo caudal varía dependiendo de las precipitaciones, pudiendo llegar a secarse totalmente. Se presentan en estepas, desiertos o sabanas; ▶arroyo intermitente); *f* **cours [m] d'eau intermittent** (Dont le débit s'arrête et reprend par intervalles, telles que certaines sources, torrents ou ruisseaux) ; *g* **zeitweilig Wasser führendes Fließgewässer [n]** (Meist in Steppen, Wüsten oder Savannen vorkommender Gewässerlauf mit zeitlich auftretendem Wasser im selben Gewässerabschnitt. Gemeint ist nicht die petrografisch oder geomorphologisch bedingte abschnittsweise Wasserführung von Fließgewässern); *syn.* intermittierendes Fließgewässer [n], intermittierender Gewässerlauf [m], zeitweilig Wasser führender Gewässerlauf [m].

2973 internal diameter [n] *constr.* (Standardized size of the ▶clearance space of a pipe; ▶clearance width); *syn.* tubing size [n] [also US]; *s* **anchura [m] de paso** (Medida estandarizada del diámetro interior de tubería; ▶medida interior, ▶vano); *syn.* diámetro [m] nominal de paso; BU 1959); *f* **diamètre [m] nominal (d'écoulement)** (Mesure du diamètre intérieur d'un tuyau ; ▶largeur dans-œuvre, ▶mesure dans-œuvre) ; *syn.* diamètre [m] intérieur utile ; *g* **Nennweite [f]** (*Kurzzeichen* DN; Kenngröße für den Innendurchmesser von Rohren; ▶lichtes Maß, ▶lichte Weite); *syn.* Rohrnennweite [f].

2974 internal friction [n] *constr. stat.* (Resistance to sliding between granular soil particles); *s* **fricción [f] interna** (Resistencia que se opone al movimiento relativo de un cuerpo sobre otro, p. ej. entre las partículas del suelo); *syn.* rozamiento [m] interno [Es]; *f* **frottement [m] interne** (Résistance dans le mouvement relatif de deux parties d'un corps, p. ex. entre les particules d'un sol) ; *syn.* friction [f] interne ; *g* **innere Reibung [f]** (Hemmung der relativen Bewegung sich berührender Teile eines Körpers gegeneinander, z. B. eines Erdkörpers).

2975 International Council [n] **for Bird Protection** *conserv. pol.* (*Abbr.* ICBP; international union founded in 1922 to coordinate technical and political endeavo[u]rs for bird protection with continental and national divisions; cf. Inter-American Commission on Nature Protection and Wildlife Preservation which formulated the Convention on Nature Protection and Wildlife Preservation in the Western Hemisphere in 1940); *s* **Consejo [m] Internacional de Protección de Aves (CIPA)** (Asociación internacional fundada en 1922 con el fin de coordinar las medidas prácticas y políticas para proteger a las aves; está organizada en secciones continentales y nacionales); *f* **Conseil [m] International pour la Préservation des Oiseaux (C.I.O.P.)** (Association internationale créée en 1922 afin d'assurer la coordination technique et politique des actions en faveur de la conservation des oiseaux sauvages ; cet organisme possède des sections nationales et continentales) ; *g* **Internationaler Rat [m] für Vogelschutz** (*Abk.* IRV; 1922 gegründete internationale Vereinigung zur Koordinierung der fachlichen und politischen Bemühungen für den Vogelschutz; hat kontinentale und nationale Sektionen).

internship [n] [US] *prof.* ▶practical training.

2976 interpretation [n] **of aerial photographs** *rem' sens. surv.* (▶satellite data analysis); *syn.* aerial photographic interpretation [n]; *s* **interpretación [f] de foto(grafía)s aéreas** (Lectura de fotos aéreas de alta definición y características especiales con el fin de ganar información sobre el estado de la vegetación, para aplicaciones cartográficas o para otros usos; ▶interpretación de imágenes de satélite); *f* **interprétation [f] des photographies aériennes** (Analyse des photographies aériennes [photographies à axe vertical, photographies obliques ou images satellite] présentant une haute résolution et des caractéristiques particulières telles que les images infrarouges, les images en vraies et fausses couleurs servant à l'établissement de cartes topographiques [photogrammétrie] et de cartes thématiques [unités végétales, utilisation du sol, pollution des eaux] ou à des études plus techniques [recherche archéologique] ; *termes spécifiques* photo-identification [repérage des formes et des objets], photo-interprétation [étude des faits observés et non visibles par déduction] ; cf. DUA 1996 ; ▶interprétation des images-satellite ; *syn.* interprétation [f] des images aériennes ; *g* **Luftbildauswertung [f]** (Deutung von Luftbildern [Senkrechtbilder, Schrägbilder oder Satellitenbilder] mit hohem Auflösungsvermögen und speziellen Eigenschaften wie z. B. schwarz-weiß Infrarotbilder, Falschfarbenbilder, die kartografische Darstellungen von Vegetationseinheiten, Flächennutzungsmustern, Temperaturverhältnissen zu bestimmten Tageszeiten, die Bestimmung von Höhenunterschieden, Gewässerverschmutzungen, die Erforschung archäologischer Tatbestände etc. ermöglichen; ▶Satellitenbildauswertung); *syn.* Auswertung [f] von Luftbildern, Luftbildinterpretation [f].

2977 interregional traffic [n] *plan. trans.* (Long-distance traffic crossing several statistically defined regional units; ▶regional planning policy); *s* **tráfico [m] interregional** (Tráfico de larga distancia que atraviesa diferentes regiones tal como se entienden en la ▶ordenación del territorio), *syn.* tránsito [m] interregional [AL], tráfico [m] de larga distancia, tránsito [m] a larga distancia [AL]; *f* **transport [m] interrégional** (Mode de transport par grandes liaisons/voies de communication dans le cadre de l'▶aménagement du territoire permettant la desserte de grandes villes, d'importants équipements touristiques ou de zones d'activités industrielles et dépassant les limites d'une région à l'échelon interrégional) ; *g* **großräumiger Verkehr [m]** (Verkehr über größere Strecken, der mehrere statistische räumliche Einheiten im Sinne der ▶Raumordnung überspannt, z. B. Verbindung von Ballungsräumen); *syn.* überregionaler Verkehr [m].

interrelated land uses [npl] *plan.* ▶pattern of interrelated land uses.

2978 interrelationship [n] *ecol.* (Interdependency, e.g. of living beings and the abiotic environmental factors which affect them); *s* **interrelación [f]** (Interdependencia p. ej. de los seres vivos entre sí y de los factores ambientales abióticos que les influyen); *f* **interaction [f]** (Actions réciproques entre deux facteurs écologiques, entre deux espèces, ou entre un facteur et un organisme ; DEE 1982) ; *syn.* incidence [f] et répercussion [f], interdépendance [f] ; *g* **Wechselwirkung [f]** (Bezeichnung für gegenseitige Beeinflussung und Abhängigkeit, z. B. Lebewesen untereinander und mit ihren abiotischen Umweltfaktoren).

interrelationships [npl] *ecol.* ▶ecological interactions and interrelationships.

interrupt [vb] **a hard-edged pathway** [US] *gard.* ▶soften a pathway's hard edge.

interrupted alignment [n] *urb.* ▶interrupted appearance.

2979 interrupted appearance [n] *urb.* (Aspect of openings; e.g. of voids in a frontage, treeline, etc.); *syn.* gap-tooth appearance [n] (PSP 1987, 197), interrupted alignment [n]; *s* **alineación [f] discontinua** (Impresión incompleta en un frente de fachadas o una fila de árboles por falta de elementos constituyentes); *f* **aspect [m] discontinu** (p. ex. alignement discontinu/lacunaire de façades de bâtiment ou d'une rangée d'arbres) ; *g* **lückiges Bild [n]** (Unvollständiger Gesamteindruck, z. B. Baulücken in einer geschlossenen Häuserfront [lückige Häuserzeile], Fehlstellen in einer Flächenpflanzung [lückige Pflanzung], fehlende Bäume in einer Allee [lückige Baumreihen]).

intersection [n] **[US], grade-separated** *trans.* ▶grade-separated junction.

intersection [n] **[UK], motorway** *trans.* ▶freeway interchange [US].

2980 interspecific [adj] *ecol.* (Relationship—▶competition—between different plant or animal species; ▶intraspecific); *s* **interespecífico/a [adj]** (Relaciones [▶compentencia] entre diferentes especies de flora resp. fauna); ▶intraespecífico/a); *f* **interspécifique [adj]** (qui a trait aux relations [▶compétition] entre des espèces différentes ; ▶intraspécifique) ; *g* **interspezifisch [adj]** (Beziehungen [▶Konkurrenz] zwischen verschiedenen Pflanzen- resp. Tierarten betreffend; ▶intraspezifisch); *syn.* zwischenartlich [adj].

2981 interstate highway [n] **[US]** *adm. leg. trans.* (In US., major highway in a nation-wide freeway system of controlled and limited access highways with grade separated intersections, which is funded on a 90/10%, federal/state basis. In U.K., the correspondent **motorway and trunk road** is a main inter-urban road built and maintained by central government; ▶freeway [US]/motorway [UK], ▶major highway [US] 2/expressway [UK]); *syn.* federal highway [n] [US]); *s 1* **carretera [f] interestatal** (En EE.UU. vía de tránsito interestatal perteneciente al sistema federal de autopistas y autovías, financiado en un 90% por el gobierno federal y el 10% por el correspondiente estado por donde trascurre); *s 2* **carretera [f] estatal** (En Es. ▶autopista, ▶autovía o carretera de conexión entre grandes ciudades, planeada y financiada por la administración central del Estado y que forma parte de la Red de carreteras del Estado. **En GB** autopista o carretera interurbana financiada por el gobierno

central); *s 3* **carretera [f] federal** (D., carretera interurbana o autopista, que —tras la determinación de la demanda— es planeada y financiada por el gobierno federal); *f* **grand itinéraire [m]** (*Code de l'urbanisme — R. 111-5* ; catégorie de routes du réseau routier comprenant les routes à grande circulation [route nationale, ▶autoroute, ou route assimilée à cet itinéraire]) ; *g* **Bundesfernstraße [f]** (D., OB zu Bundesautobahn [▶Autobahn] und Bundesstraße. Die Bedarfsermittlung und Baulast obliegt dem Bund als Eigentümer. Geplant und verwaltet werden die **B.n** von den Ländern oder von den nach Landesrecht zuständigen Verwaltungskörperschaften im Auftrage des Bundes; ▶Schnellverkehrsstraße [f] [CH]).

interstice [n] *constr.* ▶earth void.

2982 interurban traffic [n] *plan. trans.* (Traffic between urban communities or parts thereof in short distances); *s* **tráfico [m] interurbano** (Tránsito entre ciudades; en alemán también entre partes de ellas); *syn.* tránsito [m] interurbano [AL]; *f* **transport [m] interurbain** (Déplacement dans les aires d'emploi des grands centres d'activités) ; *g* **zwischenörtlicher Verkehr [m]** (Verkehr zwischen Orten resp. Ortsteilen im Nahbereich).

interwoven lattice fence [n] *constr.* ▶interwoven wood fence.

2983 interwoven wood fence [n] *constr.* (Fence consisting of panels formed from thin slats of wood woven horizontally between framed uprights. The panels are supported between wooden posts); *syn.* interwoven lattice fence [n], fence [n] of lattice work, woven wood fence [n]; *s* **valla [f] de enrejado de madera** (Cercado que consiste en paneles formados por láminas estrechas de madera entrelazadas horizontalmente entre postes verticales); *f* **clôture [f] à lamelles tressées** (Clôture constituée de panneaux de fines lattes de bois tressées horizontalement constituant souvent des éléments de remplissage fixés sur les poteaux) ; *syn.* clôture [f] à lattes tressées (HEC 1985, 66) ; *g* **Holzflechtzaun [m]** (Zaun aus Feldern mit horizontal geflochtenen, dünnen Holzstreifen, die an Pfosten befestigt werden. Meist werden in Kantholzrahmen eingespannte Fertigteilzaunelemente verwendet, die zwischen Pfosten montiert werden); *syn.* Flechtzaun [m] (als Sichtschutzeinrichtung), Flechtwerkzaun [m], Lamellenzaun [m].

in the wild [n] **[UK]** *hunt. syn.* ▶in an open hunting area.

2984 intraspecific [adj] *ecol.* (Relationship—▶competition—between individuals of the same species; ▶interspecific); *s* **intraespecífico/a [adj]** (Relaciones [▶competencia] entre individuos de una especie de flora resp. fauna; ▶interespecífico/a); *f* **intraspécifique [adj]** (qui a trait aux relations [▶compétition] entre des individus de la même espèce ; ▶interspécifique) ; *g* **intraspezifisch [adj]** (Beziehungen [▶Konkurrenz] zwischen Individuen der gleichen Art betreffend; ▶interspezifisch); *syn.* innerartlich [adj].

2985 introduced species [n] *phyt. zool.* (Species living after ▶introduction 1 in an area where it is not native or established; ▶exotic species, ▶strange species); *syn.* alien species [n]; *s* **especie [f] intrusa** (Especie introducida por ▶suelta [masiva] en un área a donde no pertenecía anteriormente; ▶especie exótica, ▶especie extraña); *syn.* especie [f] alóctona; *f* **espèce [f] introduite** (Espèce non indigène, qui n'est pas originaire ou présente, à l'état naturel, dans un pays particulier, introduite volontairement ou non hors de son aire d'origine ; ▶introduction, ▶espèce étrangère, ▶espèce exotique) ; *syn.* espèce [f] allochtone ; *g* **gebietsfremde Art [f]** (**1.** Art, die in einem Gebiet nach einer ▶Aussetzung lebt, aber nicht [ein]heimisch/nicht bodenständig ist. **2.** Von außerhalb eines bestimmten Biotopes stammende Art; *opp.* autochthone Art); *syn.* allochthone Art [f], arealfremde Art [f], biotopfremde Art [f], landes- und standortfremde Tier- und Pflanzenart (Art. 23 u. 27 NHG) [CH], *z. T. auch* ►exotische Art, ►fremde Art.

2986 introduction [n] **(1)** *zool.* (Release of animals, either caught in their native habitat or bred in captivity, into another natural habitat; ►colonization 1, ►naturalization, ►reintroduction); *s* **suelta** [f] **(masiva)** (Puesta en libertad de individuos de una especie de fauna —procedentes de otras zonas o de cría artificial— en biótopos que corresponden a sus necesidades ecológicas; ►naturalización, ►reintroducción de especies extintas); *f* **introduction** [f] (Mise en liberté d'animaux de capture ou d'élevage dans un biotope dont les conditions stationnelles correspondent à leurs besoins vitaux ; ►naturalisation, ►réintroduction 2) ; *syn.* lâcher [m] dans la nature ; *g* **Aussetzung** [f] (Bewusste Freilassung von Tierindividuen — aus Freifang oder Zucht — in Biotope, die ihren Lebensbedürfnissen entsprechen; ►Einbürgerung, ►Wiederansiedelung); *syn.* Ausbringung [f], Auslassung [f], Aussetzen [n, o. Pl.], Auswilderung [f], Auswildern [n, o. Pl.], Freilassung [f].

introduction [n] **(2)** *hort.* ►new introduction.

2987 introduction [n] **of plants or seeds** *ecol. phyt.* (Dissemination of plant species into an area outside their natural or present range of distribution; ►introduction 1, ►introduced species); *s* **introducción** [f] **de plantas o semillas** (Diseminación de especies de plantas en un areal ajeno; ►suelta [masiva], ►especie intrusa); *f* **introduction** [f] **d'espèces végétales** (Ensemencement volontaire ou involontaire ou transplantation d'espèces végétales hors de leur aire d'origine ; ►introduction, ►espèce introduite) ; *g* **Einbringung** [f] **von Pflanzen** (Aussäen oder Auspflanzen von derzeitig im Betrachtungsraum nicht vorhandenen Pflanzenarten; einbringen [vb]; ►Aussetzung, ►gebietsfremde Art); *syn.* Einbringen [n, o. Pl.] von Pflanzen.

intrusion [n] *conserv. leg. plan.* ►reducing of intrusion.

2988 intrusion [n] **upon the natural environment** *conserv. leg. plan.* (Encroachment of human activities, which have a long-term adverse impact on the landscape, the functioning of an ecosystem and impairment of attractive scenery; ►prevention measure [of an intrusion upon the environment]; ►impact mitigation regulation); *syn.* interference [n], adverse impact [n] on the landscape; *s* **intrusión** [f] **en el entorno natural** (Acto de transformación de un área de tal manera que se ven afectados los factores ecológicos de la misma; ►medida correctora, ►regulación de impactos sobre la naturaleza); *syn.* interferencia [f] humana en el paisaje; *f* **atteinte** [f] **au milieu naturel** (Transformation artificielle d'un système naturel sur lequel les incidences exercées par les différents modes d'occupation du sol sont si importantes qu'elles altèrent à long terme le caractère et la fonction de tout ou parties du milieu concerné, entraînent ainsi une dégradation notable de la résilience de l'écosystème et peuvent remettre en cause les usages qui en sont fait ; porter [vb] atteinte au milieu naturel ; ►mesure de suppression, ►réglementation relative aux atteintes subies par le milieu naturel) ; *syn.* agression [f] paysagère, atteinte [f] à l'environnement ; *g* **Eingriff** [m] **in Natur und Landschaft** (Konkrete, anthropogene Veränderung [keine diffuse wie z. B. Eintrag von Luftschadstoffen] in Natur und Landschaft, die durch das Ausmaß der raumwirksamen Nutzungsansprüche die Gestalt oder Funktion von Natur und Landschaft oder Teile davon und dadurch das betroffene Ökosystem mit seinen Nachbarräumen so erheblich oder nachhaltig beeinträchtigen kann, dass die Funktionsfähigkeit des Naturhaushalts oder das Landschaftsbild für Ansprüche der Gesellschaft erheblich vermindert ist. Gemäß § 1a BauGB sind **E.e in N. u. L.** soweit wie möglich zu vermeiden und bei zu erwartenden **E.en** ist ein Ausgleich nach § 8

BNatSchG vorzunehmen; ►Eingriffsregelung, ►Vermeidungsmaßnahme [eines Eingriffs]); *syn.* Landschaftseingriff [m].

2989 inundated meadow [n] [US] *agr.* (Grassy lowland subject to controlled inundation); *syn.* water meadow [n] [UK]; *s* **pradera** [f] **inundada artificialmente** (Pradera ripícola que se inunda artificialmente para abonarla desviando el agua de crecida o elevando el nivel del agua por medio de una represa); *f* **prairie** [f] **irriguée par inondation provoquée** (Mode d'amendement d'une prairie en plaine alluviale par déviation volontaire des eaux de crue) ; *g* **bewässerte Auenwiese** [f] (Bewirtschaftete Auenwiese, die durch Umleitung des Hochwassers mittels technischer Einrichtungen oder durch Anstau des nährstoffhaltigen Wassers eines Fließgewässers [Stauwässerung] zwecks Düngung überflutet wird [Wiesenwässerung]); *syn.* Wässerwiese [f].

inundation [n] *geo. wat'man.* ►flood.

invade [vb] *bot. hort. plant.* ►overrun.

invader [n] *hort. phyt.* ►invasive species.

2990 invasion [n] **(1)** *zool.* (Irregular, unforeseeable immigration of animals into areas which are not inhabited by members of the same species. Frequently the reason for invasion is an insufficient food supply for the population in its ancestral habitat; ►migration 2); *s* **invasión** [f] (Inmigración irregular, no previsible, de animales a áreas no habitadas hasta el momento por miembros de la misma especie. Generalmente por falta de alimento en su hábitat habitual; ►migración 2); *f* **invasion** [f] (Immigration massive d'une population dans une région nouvelle pour elle et exerçant des effets préjudiciables sur certaines populations autochtones de la région envahie ; DUV 1984, 17 ; ►migration 2) ; *g* **Invasion** [f] (Unregelmäßige, nicht vorher bestimmbare Tierwanderung in Räume, die von Vertretern der Art nicht besiedelt sind. Häufiger Invasionsgrund ist ein unzureichendes Nahrungsangebot für Populationen im angestammten Lebensraum; ►Migration 2).

invasion [n] **(2)** *phyt.* ►heath invasion.

2991 invasive species [n] *hort. phyt. plant.* (**1.** Intrusive plant or animal, often exotic, which threatens to take over an undisturbed area and extirpate native species—often without the direct assistance of people—causing significant changes in species composition, and structure of ecosystem processes; in U.S., e.g. Eulalia *[Microstegium vimineum]*, common reed *[Phragmites australis]*, mile-a-minute *[Polygonum perfoliatum]* or woody species, such as tree-of-heaven *[Ailanthus altissima]*, empress tree *[Paulownia tomentosa]*, golden bamboo *[Phyllostachys aurea]*. **2.** Intrusive weed, which invades disturbed areas, garden, parks or agricultural land); *syn. phyt.* aggressive plant [n], suppressing species [n], invader [n] (LD 1994 [4], 11), invasive [n]; *s* **especie** [f] **invasora** (**1.** Especie animal o vegetal, a menudo exótica, que se desarrolla con gran masividad en biótopos naturales o cuasi-naturales, generalmente sin influencia humana directa, amenazando así a las especies autóctonas, a la composición de especies y así a la estructura del ecosistema. **2.** Malas hierbas o leñosas que se expanden de forma significativa en área de cultivo); *syn.* invasora [f]; *f* **espèce** [f] **envahissante** (**1.** Espèce animale ou végétale, souvent introduite, ayant tendance, souvent sans influence anthropogène, à occuper, parfois jusqu'à leur éviction totale, l'aire naturelle des espèces indigènes ; espèce dont la prolifération est susceptible de provoquer de graves déséquilibres biologiques ; les **e. e.** en Europe centrale sont p. ex. la Verge d'or *[Solidago]*, la Renouée du Japon *[Reytounia japonica]*, le Robinier faux-acacia *[Robinia pseudoacacia]* ou récemment la *Caulerpa taxifolia* en mer Méditerranée. **2.** Plantes adventices ou végétaux ligneux ayant un développement particulièrement important sur les terrains cultivées ou dégénérées) ; *syn.* espèce [f] à forte extension ;

g **sich stark ausbreitende Art [f]** (**1.** Sich über eine Fläche verbreitende, oft exotische Art, die heimische Arten in natürlichen oder naturnahen Biotopen meist ohne direkten Einfluss des Menschen unterdrückt oder verdrängt und somit erhebliche Veränderungen in der Artenzusammensetzung bewirkt und damit die Struktur eines Ökosystems ändert; in Mitteleuropa gehören z. B. folgende Pflanzen dazu: Goldrute *[Solidago canadensis* und *S. gigantea]*, Gewöhnlicher Japanischer Flügelknöterich *[Fallopia japonica var. japonica, syn. Reynoutria japonica]*, Sachalin-Knöterich *[Fallopia sachalinensis]*, Indisches Springkraut *[Impatiens glandulifera]*, Scheinakazie *[Robinia pseudoacacia]* und neuerdings die „Massenalge" *[Caulerpa taxifolia]* im Mittelmeer. **2.** Flexible, anpassungsfähige Pflanze, die in ihrem Herkunftsgebiet durchaus unauffällig ist, sich in neuer Umgebung aber stark ausbreitet, heimische Arten verdrängt und den ganzen organischen Anteil eines Ökosystems verändern kann; weltweit gibt es viele Beispiele für invasive Arten, und angesichts globaler Handelsströme und zunehmender Mobilität wird das Problem größer. Zu den anpassungsfähigen exotischen Arten gehören auch Schädlinge [z. B. Kastanienminiermotte vom Ohrider See in Süd Mazedonien — *Cameraria ohridella*; aus N-Amerika eingeschleppte Platanennetzwanze — *Corythuca ciliata*] und Pflanzenkrankheiten [z. B. der aus N-Amerika stammende Erreger des ▶Ulmensterbens — *Ophiostoma novo-ulmi*]; cf. S+G 2003, H. 3, 5. **3.** Sich über eine kultivierte oder gestörte Fläche signifikant verbreitendes [heimisches] Unkraut oder Gehölz); *syn.* invasive (gebietsfremde) Art [f], wuchernde Art [f].

inventory [n] *adm. envir. leg.* ▶atmospheric emission inventory, ▶damage inventory; *for.* ▶forest stand inventory; *landsc. plan.* ▶landscape inventory; *plan. recr.* ▶recreation resources inventory; *pedol.* ▶soil inventory; *phyt. zool.* ▶species inventory; *adm. envir.* ▶toxic site inventory [US].

inventory [n], emission ▶atmospheric emission inventory.

inventory [n] [US], **soil resource** *geo. pedol.* ▶soil mapping.

inventory [n] [US], **urban wilds** *conserv. phyt. zool.* ▶urban habitat mapping.

2992 inventory [n] **of fauna and flora** *phyt. zool.* (Listing of fauna and flora in an area; ▶sampling 2, ▶vegetation survey); *s* **inventario [m] florístico y faunístico** (El registro de la flora, la fauna o cualquier otro elemento natural en un área determinada; inventariar [vb]; ▶inventario fitosociológico, ▶inventario de los tipos de vegetación); *f 1* **relevé [m] floristique et faunistique** (Technique d'échantillonnage d'un peuplement sur une aire de dimension réduite et choisie en fonction de la nature de la biocénose ; ▶relevé de formations végétales, ▶relevé phytosociologique) ; *syn.* relevé [m] de flore et de faune, inventaire [m] floristique et faunistique, inventaire [m] faune et flore ; *terme utilisé en général pour le règne animal ou les peuplements aquatiques* prélèvement [m] floristique et faunistique ; *f 2* **élaboration [f] d'un inventaire faunistique et floristique** (F., élaboration, p. ex. des inventaires locaux et régionaux du patrimoine floristique et faunistique) ; *syn.* élaboration [f] d'un inventaire des espèces de faune et de flore ; *g* **Inventarisierung [f] der Tier- und Pflanzenarten** (Erfassung und Darstellung in Karten und Tabellen von Flora und Fauna eines definierten Gebietes; ▶Vegetationsaufnahme, ▶großräumige Vegetationsaufnahme); *syn.* Aufnahme [f] der Flora und Fauna, Bestandsaufnahme [f] der Tier- und Pflanzenarten, Bestandsinventur [f] der Tier- und Pflanzenarten.

inversion [n] *met.* ▶atmospheric inversion.

2993 inversion frequencies [npl] *met.* *s* **frecuencia [f] de inversiones**; *f* **fréquence [f] du phénomène d'inversion thermique** ; *g* **Inversionshäufigkeit [f]**.

2994 inversion layer [n] *met.* (Blocking air layer in the atmosphere, which stops vertical air movements, causing an accumulation of dust and haze and concentration of noxious substances in the air layers close to the ground); *s* **capa [f] de inversión** (Capa atmosférica, aproximadamente horizontal, en cuyo seno, y de forma excepcional, la temperatura aumenta con la altura; dentro de ella los movimientos verticales del aire están inhibidos por razones de estabilidad estática. Si la **c. de i.** se sitúa junto al suelo, se denomina «inversión de superficie» y surge como consecuencia del enfriamiento radiativo sufrido por el suelo durante la noche; DM 1986); *f 1* **couche [f] d'inversion thermique** (Couche de l'atmosphère s'opposant au mouvement vertical de l'air entraînant l'augmentation de la concentration des particules de poussière et de l'humidité au-dessus des agglomérations urbaines causant une forte concentration des substances polluantes dans les couches proches du sol) ; *f 2* **surface [f] de subsidence** (Surface horizontale séparant deux masses d'air, la supérieure animée d'un mouvement vertical vers le bas, l'inférieure étant immobile, plus froide et plus humide ; la **s. de s.** est caractérisée par une inversion de température) ; *g* **Inversionsschicht [f]** (Bei austauscharmen Wetterlagen entstehende Sperrschicht in der Atmosphäre, die Vertikalbewegungen abbremsen, wodurch es zu einer Anreicherung von Staub und Dunst und besonders in Ballungsgebieten zu einer Erhöhung der Schadstoffkonzentration in den bodennahen Luftschichten kommt).

2995 inversion weather [n] *met.* (Episode of stagnant weather conditions with reversal of normal meteorological gradient; ▶atmospheric inversion); *syn.* general stagnation situation [n] [also US]; *s* **condiciones [fpl] de inversión térmica** (Situación meteorológica caracterizada por la estagnación de las capas de aire con la consiguiente ▶inversión térmica); *f* **conditions [fpl] d'inversion thermique** (Situation météorologique caractérisée par la stabilité des couches de l'atmosphère et de faibles déplacements d'air et caractéristique du phénomène d'▶inversion thermique) ; *syn.* situation [f] d'inversion ; *g* **Inversionswetterlage [f]** (Wetterlage mit stabiler Luftschichtung und geringer Luftbewegung, die durch eine ▶Inversion gekennzeichnet ist); *syn.* austauscharme Wetterlage [f], stagnierende Wetterlage [f] [auch A].

inverted siphon [n] *eng.* ▶sag pipe.

2996 invert elevation [n] *constr.* (Lowest level of a constructed channel or pipe, quoted with reference to an established benchmark; ▶channel invert, ▶drain invert level); *s* **nivel [m] de fondo** (Cota inferior de una tubería o canal definida en relación a una cota de referencia establecida; ▶invert, ▶lecho de canal de desagüe); *f 1* **cote [f] du fil d'eau d'un caniveau** ; *f 2* **cote [f] de la génératrice inférieure intérieure du tuyau** (▶fil d'eau, ▶génératrice inférieure intérieure du tuyau) ; *f 3* **cote [f] du fond de tranchée** *syn.* cote [f] du fond de fouille ; *g* **Sohlenhöhe [f]** (**1.** Höhe der tiefsten Linie einer Abflussrinne oder der unterste, innere Teil eines Rohres, angegeben in Metern über NN oder in Relation zu einem Vermessungspunkt. **2.** Dito, jedoch Höhe der Sohle eines Grabens; ▶Rinnensohle, ▶Rohrsohle).

investigation [n] *ecol. geo. pedol.* ▶field investigation; *constr.* ▶on-site soil investigation; *plan.* ▶preliminary investigation; *pedol.* ▶soil investigation.

investigation [n], **geotechnical** *constr. eng.* ▶exploration for foundation.

investigation [n] [UK], **prior** *contr.* ▶background work [US]/background research [n] [UK].

investigation area [n] *ecol. pedol.* ▶research area.

investigation [n] **of site conditions** *constr.* ▶precontract investigation of site conditions.

investigative report [n] *plan.* ►final investigative report.

invigoration pruning [n] [US] *arb. constr. hort.* ►rejuvenation pruning.

2997 invitation [n] **for a bid** [US] *constr. contr.* (Solicitation of competitive formal ►bids [US] 1/tenders [UK] or ►price quotations, sometimes from an unpublished and selected list of potential tenderers/bidders; ►bid advertisement [US]/tender notice [UK]); *syn.* call [n] for bids [US]/call [n] for tenders [UK], invitation [n] for a tender [UK]; *s* **invitación** **[f] a oferta** (Solicitud a empresas competitivas de presentar oferta sin que se convoque a concurso; solicitar [vb] ofertas; ►anuncio público de concurso-subasta, ►oferta a la demanda); *f* **lancement** **[m] d'appel d'offres** (Consultation des entreprises dans le cadre d'une procédure d'appel d'offres ; ►devis hors mise en concurrence ; ►avis d'appel d'offres) ; *syn.* lancement [m] de la consultation des entreprises ; *g* **Aufforderung [f] zur Angebotsabgabe** (Bitte an potenzielle Bieter, an einer Ausschreibung teilzunehmen; cf. §§ 17 [2] und 18 [3] VOB Teil A; ►Ausschreibungsankündigung, ►Angebot auf Anfrage); *syn.* Aufforderung [f] zur Einreichung von Angeboten (cf. § 3 VOB Teil A), Angebotseinholung [f], Einholen [n, o. Pl.] von Angeboten.

invitation [n] **for a tender** [UK] *constr. contr.* ►invitation for a bid [US].

invoice [n] *contr.* ►final account invoice, ►final fee invoice, ►interim invoice.

invoice amount [n] *contr.* ►final invoice amount.

involvement [n]**, community** *leg. plan. pol.* ►public participation.

iron [n]**, step** [UK] *constr.* ►step rod [US].

2998 iron pan [n] (1) *pedol.* (Natural subsurface horizon, only a few mm thick, seemingly cemented with iron oxides and hard for roots to penetrate; ►hardpan); *s* **capa [f] ferruginosa** (►Horizonte petrificado de pocos milímetros de profundidad formado por la lixiviación de los óxidos de hierro que es muy difícil de penetrar para las raíces y prácticamente impermeable); *f* **alios [m] ferrigneux** (►Horizon d'accumulation durci constitué de sable cimenté par de la matière organique [alios humique] mais principalement par des oxydes de fer et pouvant constituer dans les podzols des plaines sableuses une couche de faible épaisseur infranchissable pour l'eau et les racines ; DIS 1986, 9 ; ►horizon de fragipan) ; *g* **Eisenschwarte [f]** (Durch Verlagerung von Fe-Ionen entstandener, nur wenige Millimeter mächtiger Verdichtungshorizont, der Wasser stauend und kaum durchwurzelbar ist; ►Verdichtungshorizont).

ironpan [n] [UK] (2) *pedol.* ►ortstein [US].

2999 irregular bond [n] *constr.* (►lay with staggered joints); *syn.* irregular pattern [n]; *s* **opus [m] incertum** (Ensamblaje irregular de los mampuestos en obra de fábrica; ►colocar con juntas alternadas); *syn.* aparejo [m] irregular, trabazón [m] irregular; *f* **appareillage [m] en opus incertum** (Disposition des moellons de parement d'une maçonnerie, de dalles d'une terrasse ou d'un cheminement, de formes irrégulières, dont les angles saillants s'adaptent avec les angles rentrants des pierres déjà posées et présentant un ouvrage à joints incertains ; cet ouvrage ne présente pas de joint aligné, jamais plus de trois joints ne se recoupant ; *ne pas confondre avec* « opus insertum » ; ►poser à joints décalés) ; *syn.* appareillage [m] irrégulier ; *g* **unregelmäßiger Verband [m]** (Zusammenfügung der Steine eines Mauerwerkes [z. B. ►unregelmäßiges Schichtenmauerwerk], eines Pflastersteinbelages [Wildsteinpflaster] oder Plattenbelages [►römischer Verband, ►polygonaler Verband], so dass keine geraden, durchgehenden Fugen entstehen; ►im Verband verlegen); *syn.* Opus incertum [n].

3000 irregular coursed ashlar masonry [n] *arch. constr.* (Natural stone masonry in which the heights of the stones within each course vary; the courses, too, may to a certain extent vary in height; ►broken range work); *syn.* irregular rangework [n]; *s* **mampostería [f] aparejada en hiladas irregulares** (Tipo de mampostería en la que los mampuestos se asientan sobre superficies sensiblemente planas, aunque pueden variar un poco en la altura así como la altura de las hiladas entre sí; ►sillería ordinaria); *f* **maçonnerie [f] de moellons assisés d'inégales hauteurs** (Maçonnerie en pierres naturelles, à moellons assisés, de hauteurs inégales dans le rang et dans l'alternance des lits ; ►maçonnerie à appareillage à l'anglaise) ; *syn.* mur [m] à assises irrégulières réglées ; *g* **unregelmäßiges Schichtenmauerwerk [n]** (Natursteinmauerwerk aus lagerhaften Steinen; die Ansichtsflächen der Steine sind rechtwinklig gearbeitet. Die Schichthöhe darf innerhalb einer Schicht und in den verschiedenen Schichten in mäßigen Grenzen variieren; cf. DIN 1053 Teil 1; ►Wechselmauerwerk).

irregular pattern [n] *constr.* ►irregular bond.

irregular rangework [n] *arch. constr.* ►irregular coursed ashlar masonry.

3001 irrigation [n] *agr. constr.* (Artificial delivery of water to an area of lawn, planting, or growing crops; ►basin check method of irrigation, ►channel invert, ►cross-slope furrow irrigation, ►drip irrigation, ►flood irrigation, ►irrigation by pop-up sprinklers, ►sprinkler irrigation, ►underground irrigation); *s* **irrigación [f]** (Provisión artificial de agua a un cultivo, una plantación, un césped, etc.; ►irrigación por aspersión, ►irrigación por canales, ►irrigación por goteo, ►irrigación subterránea, ►método de irrigación del surco oblicuo a la pendiente, ►método de riego en tablares, ►riego automático, ►riego por inundación); *syn.* riego [m] (1); *f 1* **arrosage [m]** (1) (Action de répandre de l'eau sur un sol ou une culture, caractérisée par la dose, la fréquence, le débit, la durée et l'intensité ; DAV 1984 ; le terme **a.** par son acceptation plus générale est souvent utilisé comme synonyme d'irrigation ; ►arrosage au goutte à goutte, ►arrosage automatique intégré/enterré, ►arrosage par aspersion, ►irrigation par canaux, ►irrigation par submersion, ►irrigation par submersion contrôlée, ►irrigation souterraine, ►méthode d'irrigation par ruissellement avec épis) ; *f 2* **irrigation [f]** (Ensemble des techniques utilisées [captage, transport et distribution] pour apporter au sol l'eau nécessaire à la croissance optimale des plantes ou pour pallier aux effets de la sécheresse en particulier dans les régions ou l'aridité rend toute récolte impossible. On distingue l'irrigation de surface, l'irrigation par contrôle de nappe, l'irrigation par aspersion et la micro-irrigation ou micro-aspersion) ; *g* **Bewässerung [f]** (Alle technischen und organisatorischen Maßnahmen, die erforderlich sind, um einem Stück Land mit einem Regner, Schlauch, mit Gräben oder durch Überflutung Wasser zuzuführen, damit die Pflanzen über das nötige Wasser verfügen können, das für ein gesundes Wachstum und für einen wirtschaftlichen Ertrag erforderlich ist; ►Beregnung, ►Bewässerung durch oberirdisches Fluten, ►Bewässerung mit Kanälen, ►Bewässerung mit Versenkregnern, ►Rieselbewässerung, ►Querfurchenbewässerung, ►Tröpfchenbewässerung, ►Unterflurbewässerung).

irrigation [n]**, controlled flood** *agr.* ►basin check method of irrigation.

irrigation [n] [UK]**, trickle** *agr. hort.* ►drip irrigation.

3002 irrigation [n] **by pop-up sprinklers** *constr.* (Watering with automatic equipment for lawns and shrub borders, golf courses, etc.; ►sprinkler irrigation, ►pop-up sprinkler); *s* **riego [m] automático** (Sistema de riego subsuelo con aspersor escamoteable; ►aspersor escamotable, ►irrigación por aspersión); *syn.* irrigación [f] con aspersor subsuelo escamoteable; *f* **arrosage [m] automatique intégré/enterré** (►arrosage par aspersion, ►arroseur escamotable) ; *syn.* arrosage [m] automatique par arroseur escamotable, arrosage [m] automatique par asperseur escamotable ; *g* **Bewässerung [f] mit Versenkregnern** (Zuführung von Wasser mit automatischer, unter Flur eingebauter Beregnungsanlage; ►Beregnung, ►Versenkregner).

3003 irrigation canal [n] *agr.* *s* **canal [m] de riego** *syn.* acequia [f], canal [m] de irrigación; *f* **canal [m] d'irrigation** ; *g* **Bewässerungskanal [m]**.

irrigation controller [n] *constr.* ►automatic irrigation controller.

3004 irrigation equipment manhole [n] *constr.* (Structure allowing access to irrigation pipes, fittings, valves, and/or wiring for inspection or repair work); *s* **pozo [m] de riego (subterráneo)** (Instalación subterránea para la irrigación y el control de las cantidades de agua necesarias para el mantenimiento y cuidado de una zona verde); *f* **regard [m] d'arrosage** *syn.* boîtier [m] d'arrosage ; *g* **Bewässerungsschacht [m]** (Unterirdische Einrichtung zur Bewässerung und Steuerung der Wassermenge, die zur Pflege und Unterhaltung einer Grünanlage notwendig ist).

3005 irrigation system [n] *agr. constr. hort.* (Infrastructure required to deliver water to plants, lawns or crops); *s* **sistema [m] de irrigación** (Infraestructura necesaria para suministrar agua a las plantas, los céspedes o los cultivos); *syn.* sistema [m] de riego; *f* **procédé [m] d'irrigation** (VRD 1994, partie 4, chap. 8.2.1.2. p. 1) ; *syn.* système [m] d'irrigation ; *g* **Bewässerungssystem [n]** (Rohrsystem, das Grünanlagen und Kulturflächen mit Wasser versorgt und meist von einem Rechner gesteuert wird).

3006 irruptive [n] *zool.* (NAB 1992; invasive bird species without permanent migratory behavio[u]r, which migrate to other areas for food due to excessive population density or climatic extremes. In North America there are birds whose numbers fluctuate widely from winter to winter, such as pine siskins *[Carduelis pinus]* or evening grosbeak *[Coccothraustes vespertinus]*); *s* **ave [f] invasiva** (Especie de ave sin comportamiento migratorio definido que por gran densidad de población en un área o por situaciones climáticas extremas se traslada a otros lugares); *f* **oiseau [m] envahissant** (Espèce d'oiseau sans comportement migratoire affirmé qui, par suite d'un changement des conditions climatiques ou d'une explosion de population, colonise un habitat nouveau à la recherche de nourriture [migration d'implantation]) ; *g* **Invasionsvogel [m]** (Vogelart ohne festes **Zugverhalten**, die wegen zu großer Populationsdichte oder wegen klimatischer Extremverhältnisse in andere Nahrungsräume ausweicht).

island [n] *landsc. urb.* ►green island; *ecol.* ►habitat island; *met. urb.* ►heat island; *constr. urb.* ►planting island.

island [n]**, remnant habitat** *conserv. ecol. land'man.* ►remnant habitat.

island [n]**, rock** *geo.* ►meander lobe.

island [n] **[UK], turnaround with central** *trans. urb.* ►loop turnaround.

islands [npl] *ecol.* ►ecology of habitat islands.

3007 isochrone [n] *recr. trans.* (Line on a map or chart connecting all points of equal travelling or walking time in relation to a point of origin); *s* **isocrona [f]** (Línea en mapa que conecta los puntos que están a igual distancia de un punto de referencia sea a pie o en vehículo); *f* **courbe [f] isochrone** (Courbe représentant sur une carte tous les points écartés d'une durée égale par rapport à un point d'origine) ; *syn.* isochrone [m], ligne [f] isochrone, ligne [f] isochronique ; *g* **Isochrone [f]** (Linie gleicher

Fahr- oder Gehzeit, bezogen auf einen Ausgangsort); *syn.* Zeitgleiche [f].

3008 isochronous plan [n] *recr. trans.* (Plan showing lines which connect a given location with those points which can be reached in an equal amount of time); *s* **mapa [m] isócrono** (Plano que muestra líneas que conectan una ubicación determinada con todos los puntos que se pueden alcanzar en el mismo lapso de tiempo); *f* **carte [f] d'isochrones** (Carte sur laquelle figurent les lignes [isochrones] reliant tous les points écartés d'une durée égale par rapport à un point d'origine ; ces cartes permettent p. ex., d'apprécier l'efficacité d'un ou de divers réseaux de transport dans les agglomérations urbaines) ; *g* **Isochronenplan [m]** (Plan, der alle Punkte, die von einem gegebenen Ort aus innerhalb der gleichen Zeit erreicht werden können, durch Linien darstellt); *syn.* Zeitlinienplan [m].

isolated [pp/adj] (1) *urb.* ►detached.

3009 isolated [pp/adj] (2) *landsc.* (Descriptive term for e.g. a single plant); *syn.* set [pp] apart [also US], solitary [adj]; *s* **aislado/a [pp/adj]** (Término descriptivo para la situación, p. ej. de un árbol, una casa); *syn.* separado/a [pp/adj] (2); *f* **isolé, ée [pp/ adj]** (En position éloignée de tout autre élément sur une surface donnée, p. ex. bâtiment, arbre) ; *g* **frei stehend (2) [ppr/ adj]** (So beschaffen, dass z. B. ein Baum oder ein Gehölzgruppe auf einer Fläche einzeln steht); *syn.* als *Präfix* Einzel...

isolated patches [npl] **of habitat** *ecol.* ►habitat island.

3010 isolated patch habitat [n] *ecol.* (Individual, small-scale, mostly natural habitat which occurs sporadically in an otherwise characteristic landscape and is a modification of the normal surroundings; a man-made habitat, such as an abandoned stone quarry, may also be an **i. p. h.**; ►location 1); *s* **habitat [m] natural aislado** (Pequeño espacio natural atípico para la zona en la que se encuentra, p. ej. una cantera abandonada; ►estación); *f* **habitat [m] relictuel isolé** (Milieu de faible étendue, en général naturel mais totalement atypique pour la région dans laquelle il est situé. Un site anthropogène tel qu'une carrière abandonnée peut être considéré comme un **h. r. i.** ; ►station) ; *syn.* îlot [m] relictuel isolé) ; *g* **Sonderstandort [m]** (Einzelner kleinflächiger, meist natürlicher Lebensraum als Einsprengsel in einer sonst völlig anders gearteten Landschaft, in der er als Abweichung von den „normalen" Gegebenheiten verstanden wird; ein anthropogener Lebensraum wie z. B. ein ‚aufgelassener Steinbruch'kann auch ein **S.** sein; ►Standort 1).

3011 isolation effect [n] *biol.* (Appearance of a serious depletion in inherited genes and pathological symptoms of inbreeding after a long period of isolation of a population or its component groups); *s* **efecto [m] de aislamiento** (Empobrecimiento genético de una población animal o parte de ella por encontrarse aislada); *f* **mécanisme [m] d'isolement** (1. Appauvrissement du patrimoine génétique et apparition de caractères défavorables et maladifs après un isolement prolongé [séparation/ isolement géographique] d'une partie ou de la totalité d'une population animale, p. ex. due à la constitution de barrières physiques sur l'aire de répartition d'une espèce. 2. *Biologie animale* favorise le maintien de la pureté des races [non-dilution du pool génétique] et contribue à la formation de nouvelles espèces [spéciation sympatrique], p. ex. chez les espèces très sédentaires habitants des habitats spécifiques) ; *g* **Isolierungseffekt [m]** (1. Erscheinung schwerwiegender Verarmung der Erbanlagen und krankhafter Inzuchtsymptome nach längerer Isolation von Tier- und Pflanzenpopulationen oder deren Teile durch z. B. massive Zerschneidung von Lebensräumen. 2. *Tierbiologie* der **I.** fördert die Artenreinheit und kann die Bildung neuer Arten oder Unterarten ermöglichen).

3012 isolation joint [n] *constr.* (Continuous joint in foundations or separating structures to prevent cracking [fissures] when the footing is subject to unequal loads; ▶expansion joint); *s* **junta** [f] **de separación** (Separación en una obra de hormigón que se utiliza para aislar completamente una parte de una losa de hormigón o estructura de otra; DACO 1988; ▶junta de dilatación); *f* **joint** [m] **de rupture** (Joint ménagé entre deux bâtiments d'inégale charge ; DTB 1985 ; ▶joint de dilatation) ; *g* **Trennfuge** [f] (Durchlaufende Fuge im Fundament und Baukörper zur Vermeidung von Rissbildungen, wenn auf dem Fundament unterschiedliche Belastungen erwartet werden; auch beim Anstoßen einer freistehenden Mauer an ein Bauwerk wird eine **T.** vorgesehen; ▶Dehnungsfuge); *syn.* Raumfuge [f].

3013 isolation [n] **of trees** *for. landsc.* (Removal of unwanted, woody plant species in a stand of trees, in order to separate and promote the growth of trees; ▶crop tree); *syn.* setting [n] apart of trees [also US]; *s* **corta** [f] **de liberación** (En silvicultura, limpieza de leñosas no deseadas alrededor de ▶árboles de élite para promover el crecimiento de los mismos); *f* **isolement** [m] **(d'arbres)** (Conservation sur pied des ▶arbres d'élite à des fins touristiques ou en vue de la régénération naturelle dans un peuplement ; un arbre isolé est un arbre accompagné de voisins situés à une distance supérieure à leur hauteur ou assez éloignés pour que ses branches basses aient pu se développer à la lumière et sans contact avec les branches voisines ; ▶arbre plus ; *syn.* maintien [m] sur pied ; *f 2* **détourage** [m] (Opération localisée qui consiste à couper les arbres gênant la croissance d'un bel arbre, dit « arbre d'avenir » ; elle permet de valoriser au mieux le potentiel des meilleurs arbres même en nombre limité, les autres arbres et arbustes servant d'accompagnement) ; *g* **Freistellung** [f] **von Bäumen** (Entfernen von unerwünschten Gehölzen in einem Bestand um ▶Zukunftsbäume herum, um diese zu fördern).

3014 isometric drawing [n] *plan.* (Three-dimensional projection in which all of the principle planes are drawn at an angle of 30° or 45° parallel to corresponding established axes and at true dimensions; cf. MEA 1985; ▶perspective drawing); *syn.* isometric projection [n]; *s* **proyección** [f] **isométrica** (Dibujo de un proyecto en tres dimensiones; ▶dibujo en perspectiva); *syn.* dibujo [m] isométrico; *f* **isométrie** [f] (Représentation tridimensionnelle [dans le plan ou dans l'espace] d'un objet par laquelle les hauteurs et longueurs de l'objet sont projetées à l'échelle sur deux axes formant entre eux un angle de 30° ou de 45° ; ▶perspective) ; *syn.* projection [f] isométrique ; *g* **Isometrie** [f] (Dreidimensionale Darstellung eines Objektes, bei der auf zwei schrägen Achsen im Winkel von 30° oder 45° maßstabsgetreu Höhen und Längen aufgetragen werden; ▶Perspektive).

isometric projection [n] *plan.* ▶isometric drawing.

3015 issue [n] **of a certificate of compliance** [US] *contr. leg.* (Written recognition and acceptance of completed work [US]/works [UK], in accordance with the contract specifications; ▶certificate of practical completion, ▶issue of a certificate of final completion, ▶issue of a sectional completion certificate, ▶provisional certificate of completion); *syn.* issue [n] of a certificate of practical completion [UK]; *s* **aceptación** [f] **de obra** (En el mes siguiente al cumplimiento del plazo de garantía; ▶aceptación condicional [de obra], ▶aceptación final de obra, ▶aceptación parcial de obra, ▶acta de recepción/aceptación [de obra]); *syn.* recepción [f] de obra; *f* **réception** [f] (Reconnaissance par le maître d'œuvre qui constate que les ouvrages ou travaux achevés ont été exécutés conformément aux conditions générales du contrat ; ▶procès-verbal de réception, ▶réception avec réserves, ▶réception définitive des travaux, ▶réception partielle, ▶visite de préréception) ; *g* **Abnahme** [f] (*einer An-*

lage, von Bauleistungen verbindliche Anerkennung seitens des Auftraggebers oder seines Bevollmächtigten [Planers], dass ein Bauwerk dem ▶Leistungsvertrag entsprechend hergestellt, Material- und Pflanzenlieferungen sowie Ausstattungsgegenstände vertragsgerecht geliefert wurden; ▶Abnahmeniederschrift, ▶Abnahme unter Vorbehalt, ▶Schlussabnahme, ▶Teilabnahme).

3016 issue [n] **of a certificate of final completion** *constr.* (Planner's written recognition and acceptance of completed work [US]/works [UK], in accordance with contract specifications, after issue of sectional completion certificates, and then taking possession of the project by the client, whereby risks such as 'Act of God', [▶force majeure], war, etc. are simultaneously transferred to the owner; ▶issue of a certificate of compliance [US]/issue of a certificate of practical completion [UK]); *s* **aceptación** [f] **final de obra** (Reconocimiento escrito de que los trabajos se han completado como estaba previsto en el contrato, de manera que la obra pasa a las manos del comitente quien a partir de ese momento es responsable de posibles daños causados por ▶fuerza mayor; ▶aceptación de obra); *syn.* recepción [f] final de obra; *f* **réception** [f] **définitive des travaux** (±) (Terme utilisé couramment pour la réception de l'ensemble des travaux, de fournitures ou de services entre l'entrepreneur et le maître d'ouvrage ou son représentant, ce dernier prenant à ses risques les cas de ▶force majeure ; ▶réception) ; *g* **Schlussabnahme** [f] (Umgangssprachlicher Ausdruck, der den Zeitpunkt der Übernahme eines Bauwerkes vom Auftragnehmer an den Auftraggeber festlegt, — wenn vorher eine oder mehrere Teilabnahmen stattgefunden hatten —, wobei auf letzteren gleichzeitig die Gefahr [▶höhere Gewalt, Krieg etc.] übergeht; *der juristische Begriff heißt* ▶Abnahme); *syn. ugs.* Endabnahme [f].

issue [n] **of a certificate of practical completion** [UK] *contr. leg.* ▶issue of a certificate of compliance [US].

issue [n] **of an instruction** *constr.* ▶in compliance with the issue of an instruction.

3017 issue [n] **of a sectional completion certificate** *contr.* (Certification requested by the contractor for the completion of a self-contained part of the work [US]/works [UK] or such parts which cannot be inspected at a later stage due to their continuous progress; ▶issue of a certificate of compliance [US]/issue of a certificate of practical completion [UK], ▶issue of certificate of final completion); *syn.* interim certificate [n]; *s* **aceptación** [f] **parcial de obra** (Recepción parcial de aquellas partes de una obra que son susceptibles a ser ejecutadas por fases y/o que no pueden ser inspeccionadas en una fase posterior de la obra; ▶aceptación definitiva de obra, ▶aceptación de obra); *syn.* recepción [f] parcial de obra; *f 1* **réception** [f] **partielle** (Réception de tranches de travaux, d'ouvrages ou de parties d'ouvrage effectuée à la demande de l'entrepreneur ou fixée par le marché hors du délai global d'exécution ; ▶réception, ▶réception définitive des travaux) ; *f 2* **visite** [f] **de préréception** (Dans le cadre du délai contractuel, il est souvent procédé à une visite détaillée afin de dresser la liste des reprises ou finitions nécessaires en vue de la réception des travaux ; au plus tard à la date contractuelle pour l'achèvement des travaux, l'entrepreneur devra avoir effectué la totalité des reprises ou finitions de telle façon qu'il puisse présenter les travaux à la réception) ; *g* **Teilabnahme** [f] (Auf Verlangen des Auftragnehmers abzunehmende, in sich geschlossene Teile der Bauleistung oder solche Teile, die durch den Baufortschritt der Prüfung und Feststellung entzogen werden; cf. § 12 [2] VOB Teil B; ▶Abnahme, ▶Schlussabnahme); *syn.* Zwischenabnahme [f].

issuing [n] **a call for entries** [US] *adm. prof. leg.* ▶design competition announcement [US].

3018 item [n] (1) *constr. contr.* (Numbered article of work and materials in contract specifications; ►collective specification item; *specific term* work item); *s* **partida** [f] (Sección numerada en un ►resumen de prestaciones en la que se especifica una unidad de obra; ►partida global; *términos específicos* partida de obra, partida de suministro, partida de servicio); *f* **numéro** [m] **de prix décomposé** (≠) (U.K., D., il définit un élément d'une prestation de travaux, de livraison ou de services dans un ►descriptif-quantitatif ; *opp.* ►poste forfaitaire) ; *g* **Einzelposition** [f] (Numerierter Absatz eines ►Leistungsverzeichnisses, in dem die Bauleistung, Lieferungsleistung oder Dienstleistung im Vergleich zur ►Sammelposition einzeln spezifiziert ist; *UBe* [Einzel]position für Bauarbeiten, [Einzel]position für Lieferungen, [Einzel]position für Dienstleistungen freiberuflich Tätiger); *syn.* Leistungsposition [f].

item [n] (2) *constr. contr.* ►add-on item, ►alternate specification item, ►basic specification item, ►bid item [US]/tender item [UK], ►collective specification item, ►optional specification item, ►separate specification item.

item [n], **lump-sum** *contr.* ►collective specification item.

item [n], **provisional specification** *contr.* ►optional specification item.

item [n] [UK], **tender** *contr.* ►bid item [US].

item [n] [UK], **work** *contr.* *►item (1).

item cost [n] *constr. contr.* ►unit price.

itemised work [n] [UK] *constr. contr.* ►construction segment [US].

J

3019 jetty [n] [US] *wat'man.* (Structure built into a water body to control or to divert a current, protect a harbo[u]r or the like, or a wharf or pier; WEB 1993); *syn.* groyne [n] [UK]; *s* **espigón** [m] (Estructura construida en perpendicular a la orilla del mar o de un río para desviar la corriente del agua y reducir así su influencia en una zona específica); *syn. para el mar* rompeolas [m]; *f* **épis** [m] **de protection** (1. ouvrage de protection du littoral placé perpendiculairement à la côte. 2. Ouvrage de régulation d'un cours d'eau établi suivant un certain angle contre la berge pour fixer la forme de son lit, réduire la largeur du lit majeur ou reconstituer par autocurage les berges détruites par des affouillements ; AEP 1976, 215 ; *termes spécifiques* épi noyé, déflecteur) ; *g* **Buhne** [f] (Senkrecht zur Küste oder quer zur Gewässerfließrichtung verlaufendes Regelungsbauwerk, das die Küste schützt und bei Fließgewässern den Wasserlaufquerschnitt auf die Regelungsbreite einschränkt; nach der Bauweise unterscheidet man **Einwandbuhnen** [mit oder ohne Abstand eingebaute Pfähle aus Stahlbeton oder Holz, manchmal auch zweireihig auf Lücke gesetzt], **geböschte B.en, geböschte B.en mit befahrbarer Dammkrone, Kastenb.en** und **Flachb.en**; cf. TWW 1982).

3020 joint [n] (1) *geo.* (Cleft, crack or fissure intersecting a mass of rock; ►cleft rock); *syn.* diaclase [n], diaclasis [n]; *s* **diaclasa** [f] (Fractura que no supone desplazamiento de los bloques afectados. En las series estratificadas el plano de diaclasación o de ruptura se dispone en forma perpendicular u oblicua respecto de los planos de estratificación; en las rocas masivas, como el granito, dicho plano puede coincidir con una superficie curva. Por su origen cabe diferenciar **dd.** tectónicas, resultado de la actuación de fuerzas tectónicas sobre una masa rocosa, y **dd.** atectónicas, generadas por causas tales como el enfriamiento y solidificación del magma o la desecación de determinadas rocas. Por la apertura que una **d.** presenta ésta puede ser **d.** latente, **d.** cerrada, **d.** abierta y **d.** sellada. Las **dd.** juegan en geomorfología un papel fundamental, puesto que constituyen zonas de debilidad frente a los agentes erosivos; DGA 1986; ►roca fisurada); *f* **diaclase** [f] (Fente étroite dans les roches dures. On distingue les fentes ouvertes ou fentes de retrait et les fentes étroites ; ►roche fissurée) ; *syn.* fissure [f] ; *g* **Kluft** [f] (1. Klüfte [pl]; Riss, der ein Gestein durchzieht; es werden offene [= Spalten] und geschlossene Klüfte unterschieden; ►klüftiges Gestein. 2. **Klüfte** oder **Spalten** entstehen durch Rissbildung auf Grund von thermischen, chemischen oder physikalischen Vorgängen; hingegen sind ►**Fugen** immer anthropogenen Ursprungs und kommen nur in der vom Menschen gestalteten Umwelt als schmale Zwischenräume zwischen Bauteilen und Materialien vor); *syn.* Spalte [f].

3021 joint [n] (2) *constr.* (Juncture of two surfaces; ►asphalt joint, ►butt joint, ►coursing joint, ►expansion joint, ►heading joint, ►isolation joint, ►overlapping joint, ►recessed joint, ►stack bond joint, ►staggered joint, ►turf-filled joint, ►welded joint, ►welded lap joint); *s* **junta** [f] (1. En carpintería, lugar en que se encuentran dos superficies. 2. En albañilería, el mortero de relleno entre dos mampuestos. 3. En obras de hormigón, separación estrecha entre las losas para controlar las grietas y su posible desplazamiento; DACO 1988; *términos específicos* ►junta alternada, ►junta a tope, ►junta cruzada, ►junta de asiento, ►junta de bitumen, ►junta de césped, ►junta de dilatación, ►junta de separación, ►junta de solape, ►junta de superposición, ►junta prensada, ►junta remetida, ►junta soldada); *f* **joint** [m] (Intervalle étroit entre deux éléments dans un ouvrage ; *termes spécifiques* ►joint au bitume, ►joint croisé, ►joint décalé, ►joint de dilatation, ►joint délit, ►joint de rupture, ►joint en creux, ►joint gazonné, ►joint montant, ►joint serré) ; *g* **Fuge** [f] (1. Schmaler Zwischenraum zwischen zwei Bauteilen; eine **F.** kann drei Funktionen erfüllen: die technisch-konstruktive, ökologische und optische oder gestalterische; *UBe* ►Bitumenfuge, ►Dehnungsfuge, ►Kreuzfuge, ►Lagerfuge, Pflasterfuge, ►Pressfuge, ►Rasenfuge, ►Schattenfuge, ►Stoßfuge, ►Trennfuge, ►Überlappung, ►versetzte Fuge. 2. Die **Fuge** ist im Vergleich zum Riss, zur Spalte und zur ►Kluft immer menschengemacht).

joint [n], **contraction** *constr.* ►shrinkage crack.

joint [n], **grass-filled** *constr.* ►turf-filled joint.

3022 joint [n] **filled with liquid mortar** *constr.* *s* **junta** [f] **rellena de mortero líquido**; *f* **joint** [m] **bourré au coulis de mortier** ; *g* **mit flüssigem Mörtel verfüllte Fuge** [f].

joint fitting [n], **pipe** *constr.* ►pipe coupling [US]/pipe junction piece [UK, AUS].

3023 jointing pattern [n] *constr.* (LD 1988 [10], 48; overall layout of joints in masonry or paving; ►jointing pattern plan); *s* **patrón** [m] **de juntas** (Aspecto general de las juntas de una obra de mampostería o de un pavimento; ►plan de aparejo); *f* **calepinage** [m] (Aspect général des joints d'un ouvrage maçonné, d'un dallage ou d'un pavage ; ►plan de calepinage) ; *g* **Fugenbild** [n] (Gesamterscheinungsbild der Fugen eines Mauerwerkes, eines Plattenbelages oder Pflasters; ►Verlegeplan).

3024 jointing pattern plan [n] *constr. plan.* (Setting-out diagram[me], showing the joints and direction of laying pavers or masonry; ►jointing pattern); *s* **plan** [m] **de aparejo** (Dibujo que representa el ►patrón de juntas y la dirección de tendido resp. colocación de las piezas en pavimento o mampostería); *f* **plan** [m] **de calepinage** (Représentation graphique fournissant les détails nécessaires à l'exécution de l'appareillage d'un mur, d'un pavage, dallage, carrelage [dimension des éléments, direction de pose, aspect et disposition des joints ; ►calepinage) ; *g* **Verlege-plan [m]** (Werkzeichnung, die das ►Fugenbild und Verlege-richtungen von Werksteinen in einem Belag oder Mauerstein-verband darstellt).

3025 joint sealing [n] *constr.* (Sealing of welded joints in sheets of waterproofing membranes, ►root protection membranes, etc. using, e.g. liquid PVC; ►welded joint); *s* **sellado [m] de juntas** (Técnica de impermeabilización con polivinilo líquido de ►juntas soldadas de láminas de polivinilo, ►láminas antirraíces, etc.); *f* **garnissage [m] des lèvres de soudure** (lors de la mise en place d'un film P.V.C. antiracines, d'un ►film [de protection] antiracines, membrane d'étanchéité ; ►lèvre de soudure) ; *g* **Nahtversiegelung [f]** (Technik, die ►Schweißnähte von PVC-Folien, ►Wurzelschutzfolien etc. mit einer Flüssigkeit wasserundurchlässig macht, z. B. mit Flüssig-PVC).

journal [n]**, technical** *prof.* ►professional journal.

journey-to-work pattern [n] [UK] *plan. sociol. trans. urb.* ►commuter pattern [US].

jumbo cobblestone [n] [US] *constr.* ►large-sized paving stone [US]/large-sized paving sett [UK].

3026 jump area [n] *recr.* (Generic term for ►high jump installation, ►pole vault facility, ►long jump area); *s* **instalación [f] de salto** (Término genérico para ►instalación de salto de altura, ►instalación de salto de pértiga e ►instalación de salto de fondo); *f* **aire [f] de saut** (Terme générique pour l'►installation de saut en hauteur, l'►aire de saut à la perche, l'►aire de saut en longueur) ; *g* **Sprunganlage [f]** (Sporteinrichtung der Leicht-athletik und des Schwimmsportes; *UBe* ►Hochsprunganlage, ►Stabhochsprunganlage und ►Weitsprunganlage sowie Turmsprunganlage im Schwimmbad mit verschiedenen Plattformen/Sprungbrettern).

jump installation [n] *recr.* ►high jump installation.

3027 junction [n] **(1)** *constr.* (Joining of two or more things; e.g. interface between paving and grassed area, existing and proposed levels; ►meeting finished grade); *syn.* coming [n] together [also US], connection [n] to, meeting [n]; *s* **juntura [f]** (Área de contacto p. ej. entre losas y césped, entre el nivel existente del terreno y el planeado; ►enrase); *syn.* unión [f]; *f* **raccordement [m]** (Par exemple entre un revêtement en dalles et une surface engazonnée, entre le niveau de la plate-forme existante et le niveau futur fini du terrain ; ►raccordement au niveau existant) ; *g* **Anschluss [m]** (1. Verbindung zwischen zwei Bau- oder Gestaltungselementen, z. B. von Platten an den Rasen. 2. Anbindung eines vorhandenen Geländes an eine geplante Geländehöhe. 3. Anbindung einer neu verlegten Leitung an ein bestehendes Netz; ►Höhenanschluss).

junction [n] **(2)** *trans.* ►at-grade junction, ►grade-separated junction.

junction [n] [UK]**, cloverleaf** *trans.* ►cloverleaf interchange [US].

junction [n] [US]**, freeway** *trans.* ►interchange.

junction [n] [UK]**, motorway** *trans.* ►interchange.

junction fitting [n] *constr.* ►pipe coupling [US]/pipe junction piece [UK, AUS].

junction piece [n] [UK]**, pipe** *constr.* ►pipe coupling [US]/pipe junction piece [UK, AUS].

jurisdiction [n] *leg. plan.* ►planning jurisdiction.

jurisdiction [n] [US]**, authority having** *adm. leg.* ►law enforcement agency [US].

3028 karst [n] *geo.* (Area of limestone or dolomite with associated phenomena of sinkholes, underground drainage and cave-systems, resulting from carbonation-solution; DNE 1978; ►karst landscape); *s* **karst [m]** (Término tomado de la región yugoslava de kars, en transcripción alemana, y que se ha generalizado en geomorfología para designar el modelado que caracteriza a medios macizos calcáreos y, por extensión, a otras rocas capaces de ser afectadas por fenómenos de disolución, como es el caso del yeso; DGA 1986; ►paisaje cárstico/kárstico); *f* **karst [m]** (Ensemble des formes liées à l'action particulière des eaux sur les roches solubles calcaires ou salines [gypse] dans laquelle la dissolution joue le rôle principal ; DG 1984 ; le mot « karst » vient initialement du slave « kras » et désignait à l'origine les plateaux calcaires du NW de l'ancienne Yougoslavie [Slovénie] qui est le modèle même du ►paysage karstique) ; *g* **Karst [m]** (Gesamtheit des Formenschatzes, der durch die Wirkung von kohlensäurehaltigem Grund- und Oberflächenwasser in Kalk- und Gipsgesteinen entstanden ist; nach MEL 1975, Bd. 13. Ursprünglich bezeichnete **K.**, ein Wort aus dem Indogermanischen, die bodenarme und steinige alpin-dinarische Landschaft *Kras* nordöstlich von Triest und wurde im 19. Jh. von deutschen Geografen für ähnliche Landschaften eingeführt; ►Karstlandschaft).

3029 karstification [n] *geo.* (**1.** All processes, chemical and physical, which lead to the creation of a bare limestone area with poor vegetation cover); *syn.* karst processes [npl]. **2.** Result of tree felling processes in the Mediterranean which encouraged soil erosion and laid bare calcareous or other rock surfaces is also called **denudation**); *s* **karstificación [f]** (Procesos químicos y físicos de disolución de rocas [yesos, calizas, dolomitas, rocas detríticas con clastos y otras] por acción del agua superficial y subterránea que tienen como resultado la formación de paisajes desnudos; ►karst); *f* **karstification [f]** (Processus chimique de destruction des paysages dans les régions calcaires ou salines [gypse] par dissolution des roches calcaires avec formation d'un modelé karstique spécifique [lapiaz, dolines, pitons, grottes, etc.], dénudation des roches et développement d'une végétation xérophile dans les dépressions formées ; ►karst) ; *g* **Verkarstung [f]** (1. Chemische und physikalische landschaftszerstörende Prozesse [Lösungsverwitterung], die kahle, vegetationsarme Flächen auf Carbonatgestein wie Kalk, Gips oder Dolomit, z. T. auch in Sulfat- und Salzgesteinen, entstehen lassen. 2. Im mediterranen Raum das Ergebnis der großflächigen Waldrodungen, die die Bodenerosion förderten und kahle Kalkfelsen und Felspartien anderer Gesteine freilegten; ►Karst).

3030 karst landscape [n] *geo.* (Topography formed over limestone, dolomite or gypsum by dissolution, and characterized by sinkholes, caves and underground drainage; ►karst); *s* **paisaje [m] cárstico/kárstico** (Paisaje pobre en vegetación formado por lixiviación de rocas calcáreas, dolomitas o yeso y caracterizado por la presencia de dolinas, cuevas y masas de agua subterránea; ►karst); *f* **paysage [m] karstique** (Paysage dont les formes possèdent un modelé karstique formé sur des roches calcaires [dolomie] ou salines [gypse] rongées par les agents chimiques de l'eau avec formation de fentes profondes de dissolution [lapiaz], dépôt de matériaux colluviaux dans les dépressions [dolines] qui sont les formes dominantes dans le karst tempéré, formation de grottes et constitution d'un réseau d'eaux souterraines et de résurgences ; ►karst) ; *g* **Karstlandschaft [f]** (Landschaftstyp, der auf Carbonatgestein wie Kalk, Dolomit oder Gips durch Auswaschung [Lösungsverwitterung] entstanden und durch Geländemulden, Dolinen, Höhlen und unterirdische Gewässer gekennzeichnet ist. Das typische Landschaftsbild wurde meist durch menschliche Eingriffe hervorgerufen und ist durch kahle Felsen, Hänge, die von tiefen Erosionsrinnen zerfurcht sind, durch ausgetrocknete Fließgewässerbetten und stark degradierte Pflanzengemeinschaften geprägt; **K.en** sind ca. auf einem Fünftel der terrestrischen Oberfläche des Globus verbreitet; zu den bedeutendsten Karstgebieten Europas gehören die nackten, zerklüfteten Kalkgebiete der Alpen [insbesondere der Ostalpen] sowie die ausgedehnten Karstgebiete auf dem Balkan. Weitere große **K.en** sind in den südfranzösischen Hochflächen der Causses, in der Schwäbischen und Fränkischen Alb, in den Mittelgebirgen des Schweizer und Französischen Juras, in den unwirtlichen Ebenen Irlands, in den Hügellandschaften Englands, in den „grünen Hügeln" Burgunds und in der Champagne; ►Karst).

karst processes [npl] *geo.* ►karstification.

katabatic wind [n] *geo. met.* ►downslope wind.

3031 keeping [n] **open of riparian land** *leg. recr.* (Obligation of owners of riparian private property to permit free public access along the border of lakes and watercourses); *s* **mantenimiento [m] de zona de servidumbre de acceso a las orillas** (Obligación de propietarios privados de dejar una franja de 5 metros de anchura en las riberas de ríos para uso público; art. 6, Ley 29/1985, de Aguas); *f* **servitude [f] de libre passage sur les berges** (**1.** servitude de libre passage sur les berges des cours d'eau non navigables ni flottables ; décret 59-96 [obligation des riverains, dans la limite d'une largeur de quatre mètres à partir de la rive]. **2.** Servitude de passage à l'usage des pêcheurs le long des cours d'eau et plans d'eau domaniaux ; loi n° 84-512) ; *g* **Uferfreihaltung [f]** (Verpflichtung von Anliegern an Gewässern, Einfriedungen und bauliche Anlagen genügend weit vom Gewässerrand fernzuhalten, u. a. damit das unentgeltliche Betreten des Uferstreifens für jedermann möglich ist; cf. Art. 22 BayNatSchG).

3032 keep [vb] **soil beside an excavation** [US] *constr. syn.* leave [vb] soil lying by side of an excavation [UK] (SPON 1986, 189); *s* **acopiar [vb] tierra vegetal (en obra)** (Proceso de almacenamiento provisional de tierra al borde de una excavación en forma de caballón o artesa); *f* **terre [f] jetée sur le côté** (*Terme spécifique* terre jetée sur un seul côté, en cordon) ; *syn.* terre [f] mise en cavalier, terre [f] mise en berge ; *g* **Boden seitlich lagern [vb]** (Erdmaterial zeitlich befristet neben einer Aushubstelle [eines Grabens, einer Grube] deponieren).

kerb [n] [UK] *constr.* ►curb [US].

kerb [n] [UK]**, battered** *constr.* ►battered curb(stone) [US].

kerb [n] [UK]**, drop** *constr.* ►curb cut [US], ►drop curb [US].

kerb [n] [UK]**, edge** *constr.* ►barrier curb [US].

kerb [n] [UK]**, flush** *constr.* ►drop curb [US].

kerb [n] [UK]**, splayed** *constr. eng.* ►mountable curb [US].

kerb edging [n] [UK] *constr.* ►curbing [US].

kerb height [n] [UK] *constr.* ►curb height [US].

kerb-inlet [n] [UK] *constr.* ►curb inlet [US].

kerb level [n] [UK] *constr.* ►curb level [US].

kerbstone [n] [UK] *constr.* ►curbstone [US].

kerbstone [n] **with gutter** [UK] *constr.* ►curb and gutter [US].

3033 kettle [n] *geo.* (DNE 1978; circular depression in glacial drift, commonly waterfilled or occupied by a bog, caused by the melting of a detached block of stagnant ice, formerly wholly or partly buried, and the subsequent collapse of the overlying drift); **k.s** range in the depth from about 1-10 m, and in diameter to as much as 13 km); *syn.* glacial kettle [n], kettle basin [n], kettle-hole [n] [also UK]; *s 1* **kettle [m]** (Depresión circular en llanura aluvial *sandur* o *de outwash* [acumulación de detritos glaciares en forma de cono aplanado], generalmente rellena de agua u ocupada por turbera baja, causada por el derretimiento posterior del bloque de hielo, antiguamente enterrado total o parcialmente, y por el colapso subsiguiente de la capa de cobertura; cf. DINA 1987); *s 2* **lago [m] glaciar** (*Término regional* **ibón [m]**: Nombre dado en los Pirineos a los múltiples lagos glaciares existentes allí; DGA 1986); *f* **soll [m]** (*Terme allemand* petite dépression intercalée dans un complexe morainique, le plus souvent située entre le front du glacier et la moraine frontale à l'emplacement d'un culot de glace morte recouvert de sédiments et dont la fonte provoque la formation d'une cuvette qui, après colmatage, peut se remplir d'eau ou former un lac tourbeux ; DG 1984 ; *pl. allemand* sölle) ; *syn.* kettle [m], cuvette [f] de glace morte, kettler [m], cuvette [f] de fusion glaciaire [CDN] ; *g* **Toteisloch [n]** (In der Eiszeit entstandene, oft als See oder Moor ausgebildete kessel- oder trichterförmige Hohlform [Erdsenke] in Grund- oder Endmoränengebieten, die durch einen großen abgebrochenen, von Moränenmaterial überdeckten Eisblock, der später abschmolz, entstand); *syn.* Soll [n], (Sölle [pl]), Toteiskessel [m], glaziales Wasserloch [n], Wasser führende Hohlform [f].

kettle basin [n] *geo.* ►kettle.

kettle-hole [n] [UK] *geo.* ►kettle.

key [n] *plan. surv.* ►legend.

kick-about area [n] [UK] *recr.* ►practice game area [US].

kill [n] [north-east US] *geo.* ►brook.

3034 kindergarten [n] *sociol. urb.* (Private or public learning facility enrolling four or more children between two and five years old; ►child care centre [UK], ►day care center [UK], ►day nursery); *syn.* child care center [n] [also US], nursery school [n] (facility for children one to five years old) [also US]; *s* **jardín [m] de infancia** (Centro de atención de niños y niñas en edad preescolar; ►casa-cuna, ►guardería infantil 1, ►guardería infantil pedagógica); *syn.* kinder(garten) [m], círculo [m] infantil [C], guardería [f] infantil (2) [Es]; *f 1* **jardin [m] d'enfants** (Équipement socio(-)pédagogique assurant la garde des enfants de 3 à 6 ans pendant la journée ; ►crèche, ►garderie d'élèves, ►halte garderie, ►garderie d'enfants) ; *f 2* **école [f] maternelle** (Établissement public d'enseignement primaire pour les enfants de 2 à 6 ans) ; *f 3* **halte [f] garderie** (Établissement équipé plus modestement dans lequel les enfants jusqu'à 6 ans sont gardés de manière occasionnelle ; cf. DUA 1996) ; *syn.* garderie [f] d'enfants ; *f 4* **centre [m] de loisirs maternels (CLM)** (Équipements rattachés à un établissement scolaire accueillant les enfants âgés de 2 ans et demi à 6 ans en dehors du temps scolaire. Ils fonctionnent entre 7 h 30 et 19 h dans les locaux des écoles maternelles dont certaines salles ont fait l'objet d'aménagements spécifiques. Ils ont pour objectif la sécurité physique, affective et morale de chaque enfant, développer l'épanouissement des petits à travers le jeu et la découverte ainsi que de favoriser leur socialisation ; ils sont ouverts le matin avant l'école, le soir après la classe, ainsi que les mercredis et vacances scolaires) ; *g* **Kindergarten [m]** (Sozialpädagogische Einrichtung, in der Kinder von 3-6 Jahren resp. bis zum Schulbeginn betreut und in ihrer Entwicklung aus der kindlichen Lebenssituation heraus gefördert werden. Der Begriff **K.** stammt von dem Pädagogen FRIEDRICH FRÖBEL, der 1837 in Bad Blankenburg, Thüringen, eine „Anstalt zur Pflege des schaffenden Beschäftigungsbetriebes und des Selbsttuns der Kindheit und Jugend" gründete, die er ab 1840 **K.** nannte. Dieser Terminus setzte sich dann in Deutschland und später in vielen anderen Ländern durch; cf. MON 2006, 39. Zuvor richtete der franz. Pfarrer JOHANN FRIEDRICH OBERLIN ab 1769 im Elsaß mehrere Mädchenstrick- und Kleinkinderschulen ein *[École à tricoter]*, die Fürstin ZUR LIPPE gründete 1802 die erste deutsche „Kinderverwahranstalt" [►Kinderkrippe], und der wallisische Fabrikant und Sozialreformer ROBERT OWEN schuf 1816 die erste „Infant School"; cf. BROCK 1990, Bd. 11; ►Kinderhort, ►Kindertagesstätte).

kingdom [n] *bot.* ►plant kingdom.

3035 kitchen garden [n] *gard.* (Part of a private garden or a ►school garden in which herbs and vegetables are grown; ►fruit and vegetable garden, ►herb garden, ►cut flower garden); *syn.* vegetable garden [n] [also US]; *s* **huerta [f] (2)** (Parte de ►jardín escolar pedagógico o de jardín de una casa que se utiliza para cultivar yerbas aromáticas y verduras; ►huerto, ►jardín de hierbas finas, ►jardín de flores para cortar); *f* **jardin [m] potager** (Partie d'un jardin d'habitation ou scolaire sur lequel sont cultivés des plantes potagères et condimentaires ; ►jardin scolaire, ►jardin de rapport, ►jardin d'herbes, ►jardin bouquetier) ; *syn.* potager [m] ; *g* **Küchengarten [m]** (Teil eines Haus- oder ►Schulgartens, in dem Küchenkräuter und Gemüse angepflanzt werden; ►Nutzgarten, ►Kräutergarten, ►Schnittblumengarten); *syn.* Gemüsegarten [m].

3036 kitchen waste [n] *envir.* (All waste from a kitchen, organic and inorganic; ►organic kitchen waste; *generic term* ►household waste); *syn.* garbage [n] [also US]; *s* **restos [mpl] de cocina** (*Término genérico* ►residuos domésticos); *f* **déchets [mpl] de cuisine** (Déchets produits lors de la préparation des repas, de la restauration ; *terme générique* ►ordures ménagères) ; *g* **Küchenabfall [m]** (Abfall, der bei der Zubereitung von Speisen entsteht; *OB* ►Hausmüll).

3037 knob [n] *arb. hort.* (ARB 1983; enlarged branch end caused by annual pollarding; ►pollarded tree, ►pollarding); *s* **nudo [m] de desmoche** (Protuberancia en el extremo de una rama causada al desmochar continuamente; ►árbol desmochado, ►desmoche); *f* **tête [f] de chat** (Formation de bourrelets de cicatrisation à l'extrémité d'une branche par suite de la suppression régulière des pousses annuelles ; *terme spécifique* tête de saule ; ►arbre têtard, ►étêtage, ►étêter) ; *syn.* têtard [m] ; *g* **Astkopf [m]** (Durch ständiges ►auf den Kopf setzen entstandenes, verdicktes Astende; ►Kopfbaum).

3038 knoll [n] *geo.* (**1.** Small, low rounded hill); *syn.* hillock [n], mound [n]. **2.** The rounded top of a hill or mountain; GFG 1997); *syn.* knowe [n] [SCOT], know [n] [SCOT]; *s 1* **montículo [m]** (Colina pequeña con cima redondeada); *s 2* **cima [f] redondeada** (de un monte o una duna); *f 1* **monticule [m]** (Légère élévation de terrain) ; *syn.* bosse [f], butte [f] ; *f 2* **mamelon [m]** (Partie supérieure arrondie d'une colline, d'une dune, etc.) ; *g 1* **kleiner Hügel [m]**; *g 2* **Kuppe [f] (1)** (Oberster Teil eines Berges, Dünenhügels etc.).

knowledge [n] *plant.* ►plant knowledge; *geo.* ►regional research and knowledge.

3039 krummholz [n] *phyt.* (Woody plants whose stems or branches are bent over by pressure of snow in winter or through other unfavo[u]rable growing conditions at the ►tree line, and therefore are dwarfed or otherwise distorted; ►fen covered with pine krummholz, ►pine krummholz); *syn.* gnarled wood [n], twisted wood [n] [also US]; *s* **leñosa [f] achaparrada** (Árbol o arbusto que —debido al peso de la nieve o a otras condiciones desfavorables para el crecimiento en el ►límite de los árboles— tiene porte enano y cuyo tronco o cuyas ramas crecen torcidas y nudosas; ►paular con pinos mugo, ►pino mugo); *f* **krummholz [m]** (Végétaux ligneux, dont les parties aériennes ont un port prostré et une forme naine dus à des conditions de développement défavorables [vent, neige] ; ►brousse à Pins rampants, ►pin rampant) ; *g* **Krummholz [n]** (...hölzer [pl]; Holzgewächse, deren Stämme oder Äste durch winterlichen Schneedruck oder durch sonstige ungünstige Wachstumsbedingungen [an der ►Baumgrenze] niederliegen und deshalb zwergig und verbogen wachsen; ►Latsche, ►Latschenfilz); *syn.* Knieholz [n].

3040 krummholz community [n] *phyt.* (Slow-growing and dwarf broadleaf or coniferous vegetation at the ►tree line; ►pinus-dominated krummholz formation); *syn.* elfin forest [n] (SAF 1983), elfinwood [n] (TEE 1980, 492); *s* **matorral [m] de altura** (Comunidad vegetal de árboles de copa densa y porte torcido y de arbustos de porte almohadillado que se presenta en la altitud cuando por el efecto del frío, insolación, viento y la rarefacción del aire es imposible la existencia del bosque; DINA 1987, 587; ►límite de los árboles, ►matorral de pino mugo); *f 1* **formation [f] de krummholz** (Terme générique pour les formations arborées basses, à port rabougri et étalé, à feuilles caduques ou à feuilles persistantes, caractéristiques de l'étage subalpin à la ►limite [supérieure] des arbres et des arbustes [limite des arbres] et des milieux arctiques ; *termes spécifiques* brousse de Pin mugo, ►formation de brousse à *Pinus mugo*, brousse d'Aulne vert) ; *syn.* forêt [f] rabougrie ; *f 2* **forêt [f] de lutins [CDN]** (Forêt d'arbres rabougris généralement recouverts de lichens ou mousses, caractéristique des étages alpins humides de la côte Pacifique ; DFM 1975) ; *g* **Krummholzgesellschaft [f]** (Langsam wachsende und klein bleibende Laub- oder Nadelbaumformation am Rande der Waldgrenze [►Baumgrenze]; *UBe* Legerlengebüsch [= Grünerlengebüsch], ►Legföhrenformation); *syn.* Krummholzgebüsch [n], Krummholzformation [f].

L

3041 label [n] *hort.* (Small identifying piece of plastic, metal, etc. for attachment to a plant); *syn.* tag [n] [also US]; *s* **etiqueta [f]** (Pequeña pieza de identificación de plástico, metal u otro material que se cuelga de las plantas); *f* **étiquette [f]** (Marque en carton, bois ou plastique sur laquelle est portée la désignation des plants et fixée au végétaux, sur la motte ou à un pieu, conformément à la règlementation en vigueur) ; *g 1* **Etikett [n]** (Kleines Namensschild in einem Pflanzentopf, an einem in die Erde gesteckten Pflock oder an einem Gehölz); *g 2* **Markenetikett [n]** (Kleines Hinweisschild für Pflanzen, die in anerkannten Markenbaumschulen, Staudengärtnereien oder Rosenbaumschulen herangezogen werden; auf diesen Etiketten sind z. B. vermerkt: Güteklasse A, Deutsche Markenbaumschule, Betriebs- oder Lizenznummer, Sortenname, Veredlungsunterlage, Anschrift der Baumschule).

3042 labeling [n] *hort.* (*also* **labelling**; marking and attaching of labels to plants in nurseries, demonstration gardens or parks); *syn.* tagging [n] [also US]; *s* **etiquetado [m]** (**1.** Colocación de etiquetas de identificación en jardines botánicos; poner [vb] etiqueta. **2.** Colocación de etiquetas de precio en productos de jardinería); *f* **étiquetage [m]** (Action d'attacher une étiquette sur les végétaux dans les jardins de démonstration, les parcs, sur les produits destinés à la commercialisation tels que les semences et les plants, les engrais, les produits phytosanitaires, les médicaments vétérinaires) ; *g* **Etikettierung** [f] (**1.** Beschilderung von Pflanzen in Schaugärten, Parks oder Gärtnereien. **2.** Preisauszeichnung an gärtnerischen Verkaufsprodukten).

labor [n] [US] *constr.* ▶manual labor [US]/manual labour [UK], ▶provision of labo(u)r/personnel.

laboratory [n] *conserv.* ▶natural area laboratory.

3043 labor costs [npl] [US]/**labour costs** [npl] [UK] *constr. contr.* (Gross wages for workers used as a calculation basis for a quotation or bid [US]/tender [UK]); *s* **costos** [mpl] **de mano de obra** (Costos brutos de los obreros como base de cálculo para oferta); *syn.* costes [mpl] de mano de obra [Es]; *f* **coûts** [mpl] **de main d'œuvre** (Taux horaires moyens de main d'œuvre compris dans les prix unitaires servant à l'établissement des devis) ; *g* **Lohnkosten** [pl] (Bruttoarbeitsentgelt für Arbeiter als Kalkulationsgröße beim Angebot).

labor costs [npl] [US]**, proportion of** *constr. contr.* ▶percentage of labor costs [US]/percentage of labour costs [UK].

3044 labor costs [npl] **and wage rates** [npl] [US] *constr. contr.* (Gross wages for workers and salaries for employees used as a calculation basis for a quotation or bid [US]/tender [UK]); *syn.* labour costs [npl] and wage rates [npl] [UK]; *s* **costos** [mpl] **salariales** (Costos brutos de salarios de empleados y trabajadores como base de cálculo para oferta); *syn.* costes [mpl] salariales [Es]; *f* **dépenses** [fpl] **de main d'œuvre** (Frais occasionnés par les salariés et les ouvriers intervenant dans le calcul des prix d'une offre) ; *syn.* frais [mpl] de main d'œuvre ; *g* **Lohn- und Gehaltskosten** [pl] (Summe des Bruttoarbeitsentgeltes für Arbeiter und Angestellte, die in einem definierten Zeitraum vom Arbeitgeber an Arbeitnehmer gezahlt wird und als Kalkulationsgröße bei Angeboten dient).

laborer [n] [US] *constr. contr.* ▶construction worker.

labor-intensive maintenance [n] [US] *constr. hort.* ▶high maintenance.

labour [n] [UK] *constr. prof.* ▶provision of labo[u]r/personnel.

labour cost [n] **at an unit rate** [UK] *constr. contr.* ▶hourly wage rate.

labour costs [npl] [UK]**, proportion of** *constr. contr.* ▶percentage of labor costs [US]/percentage of labour costs [UK].

labour costs [npl] **and wage rates** [npl] [UK] *constr. contr.* ▶labor costs and wage rates [US].

labour-intensive maintenance [n] [UK] *constr. hort.* ▶high maintenance.

3045 lacustrine community [n] *phyt.* (Herbaceous plant association in or adjacent to freshwater lakes, ponds, pools, etc.; ▶rooted floating-leaf community); *s* **comunidad** [f] **lacustre** (Comunidad herbácea de aguas dulces estancadas; ▶comunidad de hidrófitos radicantes); *f* **groupement** [m] **des eaux calmes** (Formations herbacées des eaux douces libres permanentes ou à plan variable [bras morts des rivières, étangs, lacs, mares, etc.] ; ▶groupement flottant, fixé) ; *syn.* groupement [m] lentique/lacustre, groupement [m] d'eaux stagnantes ; *g* **Stillgewässergesellschaft** [f] (Krautige Pflanzengesellschaft in Süßwasserseen, Teichen, Tümpeln etc. oder in deren unmittelbaren Einflussbereichen; ▶wurzelnde Schwimmblattgesellschaft); *syn.* Stillwassergesellschaft [f].

lacustrine community [n]**, free-floating** *phyt.* ▶free-floating freshwater community.

3046 lacustrine reed zone [n] *phyt.* (Reed zone of a lake or pond with Lake bulrush [Scirpus lacustris], narrow-leaved and common cattail *[Typha angustifolia et latifolia]*, manna-grass

[Glyceria maxima], reed *[Phragmites communis]*, twig-rush *[Cladium mariscus]*, bur-reed *[Sparganium erectum]*, salt-marsh bulrush [Scirpus maritimus], etc.); *syn.* lacustrine zone [n] of emergent vegetation (CWD 1979, 12); *s* **carrizal** [m] **lacustre** (Término genérico para las comunidades vegetales del cinturón de colmatación de lagos *[Phragmition]* con presencia de juncos *[Scirpus lacustris* y *Scirpus maritimus]*, espadañas *[Typha angustifolia et latifolia]*, carrizo *[Phragmites communis]*, etc.); *f* **roselière** [f] **lacustre** (Terme générique désignant les associations d'atterrissement appartenant à l'alliance des peuplements denses d'hélophytes *[Phragmition]* et dominée par les plantes sociales telles que le Jonc des tonneliers *[Scirpus lacustris]*, les Massettes à large feuilles et à feuilles étroites *[Typha angustifolia et latifolia]*, la Glycérie aquatique *[Glyceria maxima]*, le Roseau commun *[Phragmites communis]*, la Marisque *[Cladium mariscus]*, le Rubanier dressé *[Sparganium erectum]*, le Triangle *[Scirpus maritimus]*, etc., pouvant former en marge des lacs et étangs des peuplements unispécifiques ; *terme spécifique* roselière de ceinture) ; *syn.* ceinture [f] de roselière, zone [f] des Roseaux ; *g* **Röhrichtzone** [f] **am Seeufer** (Verlandungsgesellschaft des Verbandes Süßwasserröhrichte *[Phragmition]* an stehenden Gewässern mit Teichbinse *[Scirpus lacustris]*, Schmalblättriger und breitblättriger Rohrkolben *[Typha angustifolia et latifolia]*, Wasserschwaden *[Glyceria maxima]*, Schilf *[Phragmites communis]*, Schneidebinse *[Cladium mariscus]*, Igelkolben *[Sparganium erectum]*, Meerbinse *[Scirpus maritimus]* etc.); *syn.* Seeröhrichtzone [f].

lacustrine zone [n] **of emergent vegetation** *phyt.* ▶lacustrine reed zone.

ladder [n] *wat'man. zool.* ▶fish ladder.

3047 lagg [n] *geo.* (Peripheral wet zone of a ▶raised bog into which surface water drains, collects and comes into contact with the mineral soil water of the surrounding area); *s* **lagg** [m] **periférico** (Zona periférica húmeda de una ▶turbera alta en la que se concentra el agua que fluye de ésta y se junta con la del suelo mineral de los alrededores); *f* **dépression** [f] **périphérique d'une tourbière ombrogène** (Marge très hygrophile d'une ▶tourbière haute dans laquelle les eaux acides en provenance de la tourbière rejoignent les eaux de ruissellement des sols minéraux sur lesquels elle s'adosse [impluvium des tourbières topogènes]) ; *syn.* lagg [m] ; *g* **Lagg** [n] (Nasse Randzone eines ▶Hochmoores, in der sich das vom meist gewölbten Moorkörper abfließende Wasser sammelt und mit dem „Mineralbodenwasser" der Umgebung zusammentrifft; ELL 1987); *syn.* Hochmoorrandsumpf [m].

3048 lagoon [n] *geo.* (**1.** Coastal bay separated from the sea by a ▶barrier-beach of land usually created by ▶longshore drift. **2.** Area of water within the coral reef of an atoll; ▶bodden); *s 1* **albufera** [f] (Laguna litoral separada del mar por un ▶cordón litoral, más o menos extensa, en la que existen una o más bocas, por las que tanto sale agua como entra; cf. DGA 1986; ▶bodden, ▶deriva litoral); *s 2* **lagoon** [m] (Masa de agua dentro de los arrecifes de coral de forma circular, conectada con el mar a través de algún canal; cf. DGA 1986); *f 1* **lagune** [f] (Étendue d'eau salée ou saumâtre isolée de la mer par une ▶flèche littorale ou un cordon littoral, un lido suite à la ▶dérive littorale ; ▶bodden) ; *f 2* **lagon** [m] (Étendue d'eau salée ou saumâtre formée le long des côtes bordées d'un atoll ou de récifs-barrières) ; *g* **Lagune** [f] (**1.** Durch eine ▶Nehrung/einen Lido abgetrennte Meeresbucht, i. d. R. durch ▶Küstenversetzung entstanden; ▶Bodden; **2.** Wasserfläche innerhalb des Korallenriffs eines Atolls); *syn. für 1.* Haff [n].

lagoon [n]**, tailings** *min.* ▶tailings pond.

3049 laid on edge [loc] *constr.* (Placed on the narrow side of a stone); *s* **sentado de canto** [loc] (DACO 1988; mampuesto o ladrillo colocado sobre su borde longitudinal más estrecho); *f* **posé, ée sur chant** [loc] (Face étroite la plus longue placée horizontalement) ; *g* **hochkant versetzt** [pp] (Auf der langen Schmalseite eines Steines verlegt); *syn.* hochkant verlegt [pp/adj], hochkant eingebaut [pp/adj].

3050 lake [n] (1) *geo.* (Large inland body of standing water which has collected in a natural, enclosed hollow or basin. Compared to a ►man-made pond, a lake is characterized by ►thermal stratification; ►bog lake, ►clear-water lake, ►kettle, ►oxbow lake); *s 1* **lago** [m] (Masa de agua cuyo origen se debe a la aparición de una barrera que atraviesa un sistema fluvial o a la formación de una depresión cerrada en el relieve. Se diferencia del ►estanque [artificial] en que en aquél existe una ►estratificación térmica de las masas de agua dulce; cf. DINA 1987; ►kettle, ►lago de aguas claras, ►lago glaciar); *s 2* **laguna** [f] (Acumulación de agua en depresiones del terreno o áreas hundidas, de extensión pequeña, en la que la zona litoral es relativamente grande y la región hipolimnética es pequeña o está ausente; DINA 1987); *f* **lac** [m] (Grand plan d'eau calme, plus ou moins profond, dont les eaux sont retenues par une contre-pente naturelle ou artificielle, à la différence de l'►étang artificiel, les eaux sont animées par des mouvement verticaux et présentent une ►stratification thermique ; cf. DG 1984 ; ►lac d'eau claire, ►soll) ; *g* **See** [m] (Großes Stillgewässer als Wasseransammlung in einer natürlichen, geschlossenen Hohlform. Im Vergleich zum ►Teich ist der See durch eine thermische ►Schichtung 3 gekennzeichnet; clear-water lake ►Klarwassersee, ►Toteisloch).

lake [n] (2) *limn.* ►dystrophic lake, ►mesotrophic lake, ►polytrophic lake.

lake [n], **brown-water** *limn.* ►dystrophic lake.

lake [n], **oligotrophic** *limn.* ►oligotrophic lake or watercourse.

lake [n], **mort-** *geo.* ►oxbow lake.

3051 lake and river water quality map [n] (±) *wat' man.* (Map showing **1.** the degree of pollution of rivers and streams according to their easily degradable substances—heavy metals and salts are not included—and divided into seven ►water quality categories. **2.** The degree of pollution of stagnant waters is classified into four ►trophic level); *s* **mapa** [m] **de calidad de las aguas continentales** (En D., representación gráfica de la clasificación de los cursos de agua en siete clases de contaminación biológica [que no incluye ni metales pesados ni sales] y de los lagos en cuatro ►niveles tróficos. **En Es.** no existe mapa equivalente; ►clase de calidad de las aguas continentales); *f 1* **carte** [f] **départementale d'objectifs de qualité** (F., cartographie généralisée à l'ensemble du territoire national des eaux superficielles destinées à la production d'eau alimentaire, des eaux de baignades, des eaux piscicoles et d'autres usages [loisirs, usage industriel, abreuvement des animaux, etc.] avec application des valeurs guide et des valeurs impératives des paramètres indiqués dans les directives d'objectifs de qualité du Conseil des Communautés Européennes) ; *syn.* carte [f] de qualité des cours d'eau ; *f 2* **carte** [f] **de pollution des cours d'eau** (D., carte établie à l'échelle nationale représentant le degré de pollution des eaux continentales caractérisé d'une part par la teneur en éléments biodégradables — non compris les métaux lourds et les sels minéraux — et comprenant 4 classes de pollution et 3 classes intermédiaires pour les eaux courantes et d'autre part par 4 ►niveaux trophiques pour les eaux stagnantes ; la cartographie des eaux côtières et des embouchures des grands fleuves n'a pas encore été établie ; ►classe de pollution des cours d'eau, ►niveau de qualité des cours d'eau) ; *g* **Gewässergütekarte** [f]

(Kartografische Darstellung der Gewässergüte der Fließgewässer in einem Gebiet nach Farben differenziert darstellt; **D.**, für das gesamte Bundesgebiet gibt es eine Karte, die einmal den Verschmutzungsgrad der **Fließgewässer** hinsichtlich ihrer leicht abbaubaren Stoffe [Schwermetalle und Salze werden nicht berücksichtigt] anhand vier ►Gewässergüteklassen mit drei Übergangsstufen, somit sieben Stufen darstellt, zum anderen **stehende Gewässer** in vier Trophiestufen [►Trophiestufe eines stehenden Gewässers] einteilt; die Biologischen **G.n** werden seit 1975 alle fünf Jahre von der Bund/Länderarbeitsgemeinschaft Wasser [LAWA] veröffentlicht; anhand dieser Karten kann man sehr gut die längerfristige Entwicklung der Gewässergüte verfolgen; Küstengewässer und Mündungsbereiche der großen Flüsse sind noch nicht erfasst).

lakebank [n] *geo. landsc.* ►lakeshore.

3052 lake beach [n] *geo.* (Flat, sandy or pebbly edge of an inland lake; ►beach 1, ►lakeshore); *s* **playa** [f] **de lago** (►orilla de un lago, ►playa); *syn.* playa [f] lacustre; *f* **plage** [f] **d'un lac** (Rive sableuse ou de galets [rocheuse] d'un lac intérieur ; ►rivage lacustre, ►plage) ; *g* **Seeuferstrand** [m] (Flacher, sandiger oder kiesiger [steiniger] Rand eines Binnensees; ►Seeufer, ►Strand).

Lake District [n] *geo. hydr. landsc.* ►lake landscape, #2.

lake edge [n] *geo. landsc.* ►lakeshore.

lakefront [n] *geo. landsc.* ►lakeshore, ►water's edge, #3.

3053 lake landscape [n] *geo. hydr. landsc.* (**1.** Landscape characterized by many lakes); **2. lake district** [n] (Regional landscape charcterized by a system or series of lakes; e.g. Lake District of N.W. England [Cumbria]; ►waterscape); *s* **paisaje** [m] **lacustre** (Paisaje caracterizado por presencia abundante de lagos. En algunos casos es resultado de las glaciaciones p. ej. en el norte y este de Alemania en Brandenburgo, Mecklenburgo y en Holstein, en Finlandia, en el noroeste de Inglaterra [Cumbria] el **distrito de los lagos** o en Norteamérica [Canadá y EE.UU.] la región de los **grandes lagos**; *término genérico* ►paisaje acuático); *syn.* paisaje [m] de lagos; *f 1* **paysage** [m] **lacustre** (Paysage caractérisé par la présence de nombreux lacs, p. ex. le **Lake Distrikt** situé dans le Comté de Cumbria dans le dans le nord-ouest de l'Angleterre, le **delta du Po** sur la côte adriatique, les **Grands Lacs** en Amérique du Nord constituant le groupe de lacs à eau douce le plus étendu au monde ; *terme générique* ►paysage aquatique) ; *f 2* **paysage du plateau** [m] **lacustre** (Paysage typique caractérisé par l'extrême abondance de lacs formés en majorité sur les plaines d'Europe centrale et occidentale au cours des dernières glaciations ; les régions du Mecklemburg et de Pomeranie dans la plaine d'Allemagne du Nord avec ses lacs glaciaires et ses plaines alluviales proglaciaires [plateau des lacs du Mecklemburg, plateau lacustre mecklenbourgeois] ; en France les **lacs du Jura** formés à la fin de l'ère jurassique par les glaciers qui en se retirant entaillèrent le plateau, formant des cuvettes [►soll] ; les **étangs de la Dombes** [plus de mille sur le plateau de la Dombes] sur un plateau d'origine morainique au nord-est de Lyon [►soll], les **étangs de la plaine de Sologne** sont d'origine humaine) ; *syn. de 1* paysage [m] de lacs ; *g 1* **Seenlandschaft** [f] (Landschaft, die durch Seen geprägt ist, z. B. der durch eiszeitliche Vergletscherungen und Zertalung entstandene **Lake Distrikt** im Kumbrischen Bergland in NW-England, der aus Karseen, die von Moränen abgedämmt wurden, besteht; das Podelta an der Adriatischen Küste; die größte zusammenhängende Süßwasserfläche der Erde bilden die fünf im pleistozänen Eiszeitalter entstandenen **Großen Seen** in Nordamerika; in Norddeutschland entstand durch glaziale Schmelzwasserrinnen und durch die aushobelnde Wirkung der Gletscherzungen die seenreiche **Holsteinische Schweiz**); *g 2* **Seenplatte**

L

[f] (Überwiegend durch das Nordische Inlandeis geprägte Tief-landbereiche in Mittel- und Nordeuropa mit vielen Seen, z. B. die **Finnische, Mecklenburgische und Pommersche S**. Sie sind oft eine Abbildung des Kluftnetzes der Gesteine, weil die Eis- und Schmelzwassererosion sich an den vorgegebenen Strukturen im Gestein orientierte; cf. WAG 2001; *OB* ▶Gewässerlandschaft); *syn.* Seengebiet [n].

lake margin [n] *geo. landsc.* ▶lakeshore.

lake [n] **or watercourse** [n] *limn.* ▶oligotrophic lake or watercourse.

3054 lakeshore [n] *geo. landsc. syn.* lakebank [n], lake edge [n], lake margin [n], lakefront [n]; **s orilla [f] de un lago** *syn.* orilla [f] lacustre; *f* **rivage [m] lacustre** *syn.* bord [m] de lac, rive [f] de plan d'eau intérieur ; *g* **Seeufer [n]**.

land [n] *sociol.* ▶abandoned land; *agr. plan.* ▶agricultural production land; *agr. for. hort.* ▶cultivated land; *leg.* ▶riparian land; *plan. urb.* ▶undeveloped land, ▶unserved land.

land [n]**, abandoned farm-** *sociol.* ▶abandoned land.

land [n] **[UK], agricultural** *sociol.* ▶agricultural district [US] (1).

land [n]**, consolidation of agricultural** *adm. agr. leg.* ▶farmland consolidation [US].

land [n]**, dereliction of** *plan. urb.* ∗▶derelict land.

land [n] **[US], disused** *plan. urb.* ▶derelict land.

land [n] **[US], extensive open** *landsc.* ▶open country.

land [n]**, industrial waste** *plan. urb.* ▶derelict land.

land abandonment [n] *agr.* ▶agricultural land abandonment.

3055 land acquisition [n] *adm. plan.* (Purchase of full land ownership; ▶land assembly policy [US]/land acquisition policy [UK]); **s adquisición [f] de terrenos** (Compra de tierras tras firma de contrato ante notario y pago completo del precio acordado; *opp.* enajenación de terrenos/suelo; ▶constitución de patrimonio público de suelo); *f* **acquisition [f] foncière** (L'achat de biens fonciers ; ▶maîtrise foncière) ; *syn.* acquisition [f] de terrains, acquisition [f] de/du foncier ; *g* **Grunderwerb [m]** (Ankauf von Grund und Boden nach Klärung der Finanzierung und Abschluss eines Kaufvertrages mit notarieller Beurkundung, die nach dem BGB immer erforderlich ist; sonstige schriftliche oder mündliche Absprachen haben in D. keinen Bestand; ▶Bodenvorratshaltung).

land acquisition policy [n] **[UK]** *plan. pol.* ▶land assembly policy [US].

3056 land assembly policy [n] **[US]** *plan. pol.* (**1**. Acqui-sition of land [rather than developing] by a planning agency [US]/authority [UK] for planning or development purposes or for future land uses; **in U.S.**, the Fifth Amendment of the US Constitution grants Federal and State Agencies the right of eminent domain to acquire privately held land for public uses [roads, parks, schools, reservoirs, hospitals and other public purposes; *syn.* condemnation [n], taking [n], expropriation [n]. Planning agencies or public building authorities have this right to acquire land provided "just compensation" at market value is awarded to the land owner, with or without the consent of the owner]; **in U.K.**, acquisition of land for public purposes is governed by the Town and Country Planning Acts [cf. TCPA 1990, Part 9]); **2. land banking [n]** (Used in U.S. as a synonym for **l. a. p.**, is used in U.K. for speculative investment schemes in which land is purchased with the intent to sell it again at a profit after its value has risen; ▶land acquisition, ▶land bank, ▶right of first refusal); *syn.* land acquisition policy [n] [UK], land assembly strategy [n] [US], land banking [n] [US]; **s constitución [f] de patrimonio público de suelo** (▶Adquisición de terrenos

por parte de las corporaciones locales para formar el patrimonio de suelo; ▶derecho de tanteo, ▶reserva de suelo); *f* **1 maîtrise [f] foncière** (Mesures d'acquisition de ▶réserves foncières prises par l'État, les collectivités locales ou leurs groupements ayant compétence en matière d'urbanisme, les syndicats mixtes et les établissements publics d'aménagement permettant l'achat de biens en prévision de l'extension d'agglomérations, de l'amé-nagement d'espaces naturels les entourant, de la création de villes nouvelles ou de stations de tourisme, la conservation des milieux naturels ; il en est de même pour la rénovation urbaine et l'amé-nagement des villages ; cf. C. urb., art. L. 221-1 ; ▶acquisition foncière, ▶droit de préemption) ; *syn.* constitution [f] de réserves foncières, politique [f] de réserves foncières, acquisition [f] de réserves foncières ; *f* **2 municipalisation [f] des sols** (*Terme spécifique de* ▶maîtrise foncière réalisée par la commune) ; *g* **Bodenvorratshaltung [f]** (Maßnahmen von Planungsträgern oder Gebietskörperschaften zum Grundstückserwerb, z. B. durch ▶Vorkaufsrecht, zur Schaffung eines Bestandes, um zukünftige Flächennutzungen, Naturschutzmaßnahmen oder sonstige raum-bedeutsame Entwicklungen verwirklichen zu können oder als Flächenpool, um geeignete Grundstücke für ▶Ausgleichsmaß-nahmen [des Naturschutzes und der Landschaftspflege] anbieten zu können; ▶Bodenvorrat, ▶Grunderwerb); *syn.* Bevorratung [f] von Flächen, Bodenvorratspolitik [f], Flächenbevorratung [f], Flächenhaushaltspolitik [f], Flächensicherung [f].

land assembly strategy [n] **[US]** *adm. leg.* ▶land assembly policy [US]/land acquisition policy [UK].

3057 land bank [n] *adm. plan.* (**1**. Land acquired by a public agency for future use and development; **in U.S.**, the term **l. b.** is used permitted only where land is acquired through gift or purchase; not through eminent domain, condemnation or exces-sive regulatory schemes); **2. land banking [n] [US]** (▶land assembly policy [US]/land acquisition policy [UK]); **s reserva [f] de suelo** (Adquisición de suelo por la administración pública con el fin de disponer de espacio para gestionar el desarrollo urbano; ▶constitución del patrimonio público del suelo); *syn.* patrimonio [m] público de suelo [Es]; *f* **réserves [fpl] foncières** (Fonds bâtis ou non bâtis acquis par l'état, les collectivités locales ou les établissements publics d'aménagement afin de pouvoir définir la destination définitive ou les orientations futures de l'occupation des sols et d'effectuer des actions ou des opérations d'aménage-ment concernant l'habitat, les activités économiques, les loisirs et le tourisme, les espaces naturels, etc. ; ▶maîtrise foncière, ▶municipalisation des sols) ; *g* **1 Flächenpool [m]** (**1**. Von Pla-nungsträgern erworbenes Land, um zukünftige Nutzungen oder Entwicklungen ermöglichen zu können; ▶Bodenvorratshaltung. **2**. *conserv. plan. nat'res.* Sammlung von potenziellen ▶Aus-gleichsflächen, um zukünftige ▶Eingriffe in Natur und Land-schaft durch Maßnahmen des Naturschutzes und der Landschafts-pflege [Maßnahmenpool] zu kompensieren); *syn.* Bodenvorrat [m], Flächenreserve(n) [f(pl)]; *syn. zu 2.* Bodenbevorratung [f]; *g* **2 Ökokonto [n]** (Art der Bewirtschaftung des Maßnahmen-pools für Naturschutz und Landschaftspflege. Wie bei einem gewöhnlichen Bankkonto, werden vorab durchgeführte Maß-nahmen auf das Konto „eingebucht" und im Falle des Zurück-greifens auf die Maßnahmen zur Eingriffskompensation „abge-bucht". Die Maßnahmenbevorratung wird oft zunächst auf Kostern der Gemeinde durchgeführt. Die Refinanzierung erfolgt durch den Eingriffsverursacher, der diese Maßnahmen für seine Kompensationsverpflichtung nutzen will. Flächen- und Maß-nahmenpools werden auf Grundlage eines naturschutzfachlichen Gesamtkonzeptes entwickelt und häufig im Landschaftsplan oder Flächennutzungsplan dargestellt; den rechtlichen Rahmen bilden meist die Landesnaturschutzgesetze. In der Praxis erfolgt die Verwendung der Begriffe Flächenpool, Maßnahmenpool, Ökokonto und Ökopool häufig synonym; BFN 2006, 33); *syn.* Ökopool [m]).

land banking [n] *adm. leg.* ►land assembly policy [US]/land acquisition policy [UK].

3058 land banking plan [n] [US] *adm. agr. leg.* (Documentation of the results of the ►farmland consolidation procedure, showing footpath networks, plans for streams and channels together with the ►development mitigation plan [US]/landscape envelope plan [UK], common and public facilities, the existing pattern of land ownership, etc.); *syn.* land consolidation plan [n] [UK]; *s* **plan [m] de concentración parcelaria** (Documentación de los resultados del ►procedimiento de concentración parcelaria que muestra la red de caminos y otras infraestructuras previstas junto con el ►plan paisajístico complementario y el mosaico de propiedad de las tierras); *f* **plan [m] définitif de remembrement** (Plan de redistribution des terres établi à la suit d'une ►procédure de remembrement et désignant un remodelage des infrastructures grâce à la réalisation de travaux « connexes », p. ex. l'établissement de chemins d'exploitation, l'arrachage de haies, l'arasement de talus et le creusement de fossés d'assainissement, etc.) ; *g* **Flurbereinigungsplan [m]** (Darstellung der Ergebnisse des ►Flurbereinigungsverfahrens: Wege- und Gewässerplan, Landschaftspflegerische Begleitplan, Darstellung der gemeinschaftlichen und öffentlichen Anlagen, der alten Grundstücke etc.; § 56 ff FlurbG).

3059 land capability class [n] [US] *agr.* (*U.S. Soil Conservation Service* land classification category based on its suitability or limitations for various uses; this planning term typically includes topography, soils, geographical limitations, potential for flooding, etc.; a **l. c. c.** is the broadest category in the system: class codes I to VIII indicate progressively greater limitations and narrower choices for agriculture; the numbers are used to represent both irrigated and nonirrigated land capability; the second category in this system is the **land capability subclass** with class codes **e** [erosion problems]—►extreme soil erosion], **w** [wetness problems], **s** [root zone limitations], and **c** [climatic limitations]; cf. RCG 1982, 86-87). Each local government may establish its own classification system for land capability in preparing development plans; ►arable land rank [US]/arable land grade [UK], ►Land Capability Classification [US]; **in U.K.**, ►**agricultural land grade [UK]** (Land class broadly defining the inherent value for crop production; ►land use capability classification [UK]); *s* **clase [f] de aptitud del suelo** (Según el Servicio de Clasificación de Suelos de los EE.UU., clase principal en la valoración detallada de los suelos que se realiza según su productividad para los cultivos comunes y para plantas prátícolas, teniendo en cuenta los peligros de erosión y otras restricciones del uso agrícola. La clasificación se subdivide en 8 clases principales [►*land capability classes*] y 4 subclases *[land capability subclasses]*. Las clases principales diferencian los niveles crecientes de restricción del uso y la reducción de la productividad agrícola, también considerando la capacidad de absorción del agua de irrigación; las subclases se distinguen con 4 letras: **E** grado de peligro de erosión [►destrucción del suelo], **W** grado de humedad del suelo, **S** limitaciones para el enraizamiento y **C** limitaciones climáticas; ►clasificación de aptitud de suelos, ►índice de fertilidad de suelos, ►índice de fertilidad de tierras de cultivo); *f* **classe [f] d'aptitude d'un sol (≠)** (U.S., classification des sols du *Soil Conservation Service* : classe dans laquelle sont groupés les sols en fonction de leur possibilités de production agricole en tenant compte de facteurs et de limitations fonctionnels ; la classification distingue **8 classes principales** [►land capability classes] dépendantes de l'importance des facteurs limitants, en ordre croissant des sols ayant peu de limitations jusqu'aux sols pour lesquels les limitations excluent toute possibilité de culture ou de pâturage permanent et **4 sous-classes** de subdivision des classes en fonction de facteurs limitatifs représentés par les lettres **E** pour les risques liés au degré d'érosion [►dévastation des sols], **W** pour le degré d'humidité du sol, **S** pour la capacité à l'enracinement, **C** pour des conditions climatiques défavorable ; **U.S.**, ►classification des aptitudes culturales et pastorales des sols ; la classification FAO de l'aptitude des terres évalue et groupe les terres sur la base de leur aptitude absolue ou relative à un mode d'utilisation particulier ; **F.**, ►classification des sols agricoles ; ►coefficient de fertilité d'un sol, ►indice de fertilité physique, ►indice d'estimation des sols) ; *g* **Bodeneignungsklasse [f] (≠)** (Bewertungsgröße für Böden des *U.S. Soil Conservation Service*: in den USA werden Land und Böden detailliert gemäß ihrer Produktionskraft für normale Feldfrüchte und Grünlandgräser beurteilt — bei gleichzeitiger Berücksichtigung der langfristigen Gefahren der Bodenerosion bis zur ►Bodenverheerung, der Nutzungseinschränkung für die Landwirtschaft oder anderer Nutzungen. Die Bodeneignungseinteilung gliedert sich in **acht Hauptklassen** *[►land capability classes]* und **vier Unterklassen** *[land capability subclasses]*. Die Hauptklassen bezeichnen in Stufen die Zunahme der Nutzungsbeschränkung und der Abnahme landwirtschaftlicher Ertragsfähigkeit, auch unter Berücksichtigung der Wasseraufnahmekapazität bei Beregnung; die Unterklassen sind mit vier Buchstaben gekennzeichnet: **E** Grad der Erosionsgefährdung, **W** Grad der Bodenfeuchtigkeit, **S** Begrenzungen der Durchwurzelbarkeit und **C** für klimatische Einschränkungen; ►Ackerzahl, ►Bodeneignungsbewertung, ►Bodenzahl).

3060 Land Capability Classification [n] [US] *agr.* (System of *U.S. Soil Conservation Service* with detailed grouping of land and soils primarily on the basis of their capability to produce common cultivated crops and pasture plants without deteriorating over a long period; i.e. according to the risks of land damage or the limitations of land use for cultivation and other purposes; **L. C. C.** is subdivided into *capability classes* and *capability subclasses* nationally. A ►land capability class is the broadest category in the system: class codes I to VIII indicate progressively greater limitations and narrower choices for agriculture; the numbers are used to represent both irrigated and nonirrigated land capability; the second category in this system is the **[land] capability subclass** with class codes **e** [erosion problems], **w** [wetness problems], **s** [root zone limitations], and **c** [climatic limitations]. Each local government may establish its own classification system for land capability in preparing development plans; cf. RCG 1982, 86; ►agricultural land grade, ►arable land rank, ►extreme soil erosion, ►soil quality analysis); *s* **clasificación [f] de aptitud de suelos** (En EE.UU., sistema del Servicio de Conservación de Suelos con agrupación detallada de las tierras y los suelos primordialmente sobre la base de su aptitud para los cultivos comunes y para plantas prátícolas sin deteriorarse durante un largo periodo; la **c. de a. de ss.** se subdivide en *clases y subclases de aptitud* a nivel nacional; sin embargo, cada administración local puede establecer su propio sistema de clasificación de aptitud de suelos en los planes de desarrollo; ►clase de aptitud del suelo, ►destrucción del suelo, ►índice de fertilidad de suelos, ►índice de fertilidad de tierras de cultivo, ►valoración del suelo); *syn.* clasificación [f] de tierras (MEX 1983); *f 1* **classification [f] des aptitudes culturales et pastorales des sols (±)** (U.S., élaborée par le *U.S. Soil Conservation Service* la classification des terres et sols est établie **1.** en huit classes qui désignent la limitation progressive des choix d'utilisation des sols à des fins agricoles sur la base de leur capacité à l'irrigation, **2.** en quatre sous-classes de dangers causés par l'érosion du sol, les problèmes d'humidité, la diminution de la

L

zone racinaire et les limites climatiques ; ►classe d'aptitude d'un sol, ►coefficient de fertilité d'un sol, ►dévastation des sols, ►évaluation de la qualité du sol, ►indice de fertilité physique, ►indice d'estimation des sols) ; *f 2* **classification [f] des sols et des paysages agricoles de premier choix et marginaux [CDN]** (Cadre régissant l'inventaire des terres du Canada [ITC] et classant les terres selon leur potentiel agricole au moyen d'un système interprétatif qui évalue l'influence limitative exercée par diverses caractéristiques de sol et de climat sur l'aptitude des terres à la production des grandes cultures communes) ; *f 3* **classification [f] de l'aptitude des terres** (FAO., évaluation et groupement, ou encore processus d'évaluation et de groupement, des terres sur la base de leur aptitude absolue ou relative à un certain mode d'utilisation) ; *g 1* **Bodeneignungsbewertung [f] (±)** (System des *U.S. Soil Classification Service*, nach dem Land und Böden detailliert gemäß ihrer Produktionskraft für normale Feldfrüchte und Grünlandgräser beurteilt werden — bei gleichzeitiger Berücksichtigung der langfristigen Gefahren der Bodenerosion bis zur ►Bodenverheerung, der Nutzungseinschränkung für die Landwirtschaft oder anderer Nutzungen. Die Bodeneignungseinteilung gliedert sich in acht Hauptklassen *[land capability classes]* und vier Unterklassen *[land capability subclasses]*. Die Hauptklassen bezeichnen in Stufen die Zunahme der Nutzungsbeschränkung und die Abnahme landwirtschaftlicher Ertragsfähigkeit, auch unter Berücksichtigung der Wasseraufnahmekapazität bei Beregnung; die Unterklassen sind mit vier Buchstaben gekennzeichnet: **E** Grad der Erosionsgefährdung, **W** Grad der Bodenfeuchtigkeit, **S** Begrenzungen der Durchwurzelbarkeit und **C** für klimatische Einschränkungen; ►Bodenbewertung 2, ►Bodeneignungsklasse); *g 2* **Bodenschätzung [f]** *agr. leg.* (D., Wertermittlung in Form einer Bestandsaufnahme und Feststellung der Ertragsfähigkeit landwirtschaftlicher Kulturböden [Ackerböden und Grünlandböden], die in D. aus fiskalischen Gründen 1934 nach der Reichsbodenschätzung begonnen wurde; Maßstab war hierzu der beste Ackerboden, der Schwarzerdeboden in der Magdeburger Börde mit der Wertzahl 100; für Hangneigungen werden Abschläge berechnet; ►Ackerzahl, ►Bodenzahl); *syn. obs.* Bodenbonitierung [f].

land classification [n] [UK]**, agricultural** *agr.* ►agricultural land grade [UK].

3061 land clearance [n] [US] *agr. land'man.* (Large scale removal of trees, shrubs and hedgerows from areas where intensive agriculture is practiced; ►extensively cleared land [for cultivation], ►hedgerow clearance); *syn.* landscape clearance [n] [UK]; *s* **desmonte [m] total del paisaje** (Eliminación total de árboles, arbustos o setos en un paisaje; ►paisaje desnudo, ►tala de setos interparcelarios); *f* **dégagement [m] du paysage** (Débarrasser une vaste étendue d'un site de toute sa végétation arbustive, taillis, haies et arbres divers ; ►paysage ouvert, ►enlèvement des haies) ; *g* **Ausräumung [f] der Landschaft** (Großflächige Entblößung der Landschaft von jeglichem Baum-, Strauch- und Heckenbewuchs; ausräumen [vb]; ►ausgeräumte Landschaft, ►Schleifen von Feldhecken).

land consolidation [n] [UK] *adm. agr. leg.* ►farmland consolidation [US].

3062 land consolidation act [n] [UK] *adm. agr. leg.* (**In UK.,** an historical process dating back to the enclosure movement of the sixteenth and seventeenth centuries, widescale ►farmland consolidation to adjust the fragmentation of land holdings took place in the late eighteenth to nineteenth centuries with many acts being passed between 1801 and 1862 [in Scotland, 1868]; little legislation [amendments] has taken place since then; **in U.S.,** there is no such comparable act); *syn.* farmland consolidation act [n] [US] (≠); *s* **ley [f] de concentración parcelaria** (En D., ley que regula los procesos de agrupación de parcelas de terreno y la

creación de infraestructuras viarias para mejorar la explotación agraria; ►concentración parcelaria); *f* **loi [f] relative au remembrement des propriétés rurales** (Les diverses dispositions législatives relatives au ►remembrement rural sont regroupées dans le C. rural, livre I l'aménagement et l'équipement de l'espace rural, titre II, chapitre III ; la première loi française relative au réaménagement de la propriété date de 1918, les premières opérations de remembrement rural sur la base d'échanges volontaires de terres sont lancés en 1941 par la loi relative à la réorganisation de la propriété foncière ; le regroupement de parcelles de terrain, afin d'accroître la productivité des exploitations agricoles pour en faire naître de véritables entreprises, s'est affirmé ensuite avec la loi d'orientation agricole du 5 août 1960 ; *g* **Flurbereinigungsgesetz [n]** (**D.,** *Abk.* FlurbG; Gesetz, in dem das Verfahren [►Flurbereinigungsverfahren] zur Verbesserung der Produktions- und Arbeitsbedingungen in der Land- und Forstwirtschaft sowie die Förderung der allgemeinen Landeskultur und der Landentwicklung durch Neuordnung ländlichen Grundbesitzes innerhalb eines festgelegten Gebietes [►Flurbereinigungsgebiet] geregelt wird; ►Flurbereinigung); *syn.* Flurverfassungs-Grundsatzgesetz [n] [A].

land consolidation area [n] [UK] *agr. leg.* ►farmland consolidation area [US].

3063 land consolidation authority [n] *adm. agr. leg.* (In U.S. and U.K., no such governmental power exists; ►farmland consolidation [US]/land consolidation [UK]); *s* **administración [f] de agricultura** (Administración pública responsable del procedimiento de ►concentración parcelaria); *f* **administration [f] compétente pour le remembrement rural** (F., la procédure de ►remembrement rural est dirigée par la commission communale ou intercommunale d'aménagement foncier) ; *g* **Flurbereinigungsbehörde [f]** (Für die ►Flurbereinigung zuständige Behörde); *syn.* Amt [n] für Flurneuordnung und Landentwicklung [BW].

land consolidation plan [n] [UK] *adm. agr. leg.* ►land banking plan [US].

land consolidation procedure [n] [UK] *adm. agr. leg.* ►farmland consolidation procedure [US].

3064 land consolidation resolution [n] [UK] *adm. agr. leg.* (**In Europe,** governmental decree designating a ►farmland consolidation area; **in U.S.,** no such government power exists; ►farmland consolidation [US]/land consolidation [UK]); *s* **decisión [f] (legal) de proceder a la concentración parcelaria** (Decisión oficial de empezar el proceso de ►concentración parcelaria en una zona determinada; ►área de concentración parcelaria); *f* **arrêté [m] de remembrement** (Décision préfectorale clôturant le ►remembrement rural et entraînant le transfert des propriétés ; ►périmètre de remembrement) ; *g* **Flurbereinigungsbeschluss [m]** (**D.,** Feststellen des ►Flurbereinigungsgebietes und Anordnung der ►Flurbereinigung durch die obere Flurbereinigungsbehörde).

3065 land consumption [n] *landsc. plan.* (Transformation of near-natural landscape by a certain amount of land use change [quantity aspect] as well as a depreciation in scenic value [quality aspect], especially as a result of piecemeal division, noise pollution, direct interferences with natural systems, e.g. by land drainage, or removal of natural landscape elements, e.g. land forms, vegetation; ►intrusion upon the natural environment, ►land requirements); *syn.* sequent occupance [n] of landscape (LE 1986, 301), land take [n] [also UK] (TAN 1975, 137); *s* **gasto [m] de paisaje** (Deterioro causado por el cambio de usos y su consiguiente alteración del aspecto del paisaje [= aspecto cuantitativo] y la pérdida de valor del mismo [= aspecto cualitativo], sobre todo debido a su subdivisión, a la contaminación acústica y atmosférica, como consecuencia de ▶intrusión en el entorno natural [p. ej. drenaje] o de la eliminación de componentes naturales del paisaje [modelado, vegetación]; ▶demanda de suelo); *syn.* consumo [m] de paisaje, gasto [m] de suelo; *f* **consommation [f] d'espace** (Transformation de l'espace au sol nécessaire pour l'habitat et les transports, soit par changement d'usage [aspect quantitatif], soit par dépréciation de son caractère [aspect qualitatif] provoquée en particulier par le bruit, la pollution atmosphérique, les ▶atteintes au milieu naturel [p. ex. le drainage] ou la suppression d'éléments naturels du paysage [p. ex. le relief, la végétation] ; ▶besoins fonciers ; *syn.* mitage [m] du paysage ; *g* **Landschaftsverbrauch [m]** (Durch Inanspruchnahme von Land[schafts]fläche für Siedlungs- und Verkehrszwecke [Nutzungsänderungen] bedingte Umformung der naturnahen oder Kulturlandschaft [= quantitativer Aspekt] sowie dadurch bedingte Wertminderung der Landschaft [= qualitativer Aspekt], insbesondere infolge von Zerschneidung, Verlärmung, Schadstoffbelastung, durch direkte ▶Eingriffe in Natur und Landschaft [z. B. Entwässerungen] oder durch Beseitigung natürlicher Landschaftselemente [z. B. Oberflächenformen, Vegetation]; nach den vom Bundesamt für Naturschutz [BfN] zusammengetragenen „Daten zur Natur 1999" wurden in Deutschland täglich ca. 120 ha Fläche für Siedlungs- und Verkehrszwecke verbraucht — korrekterweise müsste man sagen umgewidmet, da die Fläche ja noch vorhanden ist und unsere Kulturlandschaft entscheidend mitprägt; lt. BfN-Pressearchiv vom 28.05.05 waren es 2005 noch ca. 90 ha **L.**; ▶Flächenanspruch); *syn.* Flächeninanspruchnahme [f], Flächenverbrauch [m], Freiflächenverbrauch [m], Landverbrauch [m].

landcover [n] [US] *land'man.* ▶spatial landscape characteristics/resources.

3066 land developed with roads [loc] *urb.* (▶Approved development site with dedicated public rights-of-way and required utilitiy infrastructure); *s* **suelo [m] urbanizado (2)** (▶Suelo edificable que ha sido equipado con infraestructura viaria; ▶suelo urbanizado 1); *f* **terrain [m] (constructible) desservi par la voirie** (Terrain desservi par une voie publique ou privée répondant à l'importance ou à la destination de l'ouvrage envisagé ; ▶terrain viabilisé) ; *g* **mit Straßen erschlossenes Bauland [loc]** (▶Bauland, das in der Rechtswirksamkeit des Bebauungsplanes liegt, d. h. nach öffentlich-rechtlichen Vorschriften baulich nutzbar ist, bei dem die Erschließung durch Herstellung von öffentlichen Straßen, Wegen, Plätzen und Parkflächen entsprechend den Erfordernissen der Bebauung und des Verkehrs abgeschlossen ist; ▶fertiges Bauland).

land disposal site [n] [US] *envir.* ▶dumpsite [US]/tipping site [UK].

3067 land drainage [n] *agr. eng. wat'man.* (Measures taken to lower the water table over relatively large agricultural areas, in order to eliminate waterlogging of the soil; ▶agricultural drainage, ▶berm ditch, ▶drainage channel, ▶drainage ditch, ▶field drainage, ▶French drain, ▶lowering of groundwater table, ▶soil drainage 1); *s* **drenaje [m] a gran escala** (Medidas tomadas para reducir el nivel freático en zonas relativamente extensas de explotación agrícola, a menudo a través de obras de ingeniería hidráulica; ▶canal de avenamiento, ▶drenaje agrícola, ▶drenaje agrícola subterráneo, ▶drenaje del suelo, ▶dren francés, ▶reducción del nivel freático, ▶surco en curva de nivel, ▶zanja de avenamiento); *f* **drainage (agricole) [m] à ciel ouvert** (Mesures ou ouvrages hydrauliques en vue de l'abaissement artificiel de la nappe phréatique sur les grandes étendues agricoles et l'élimination de l'engorgement en eau du sol ; ▶canal de drainage, ▶drainage agricole, ▶drainage agricole souterrain, ▶drainage de sol, ▶fossé de drainage, ▶fossé gradin, ▶rabattement de la nappe phréatique, ▶tranchée drainante) ; *syn.* drainage [m] par

fossés ; *g* **Wasserbaumaßnahme [f] zur Entwässerung landwirtschaftlicher Flächen** (Meistens Baumaßnahmen zur Absenkung des Grundwasserspiegels in einem größeren landwirtschaftlich genutzten Gebiet zur Beseitigung der Bodenvernässung; ▶Bodenentwässerung, ▶Dränung landwirtschaftlicher Flächen, ▶Entwässerungsgraben 1, ▶Entwässerungsgraben 2, ▶Entwässerungsgraben 3, ▶Entwässerungsgraben in einer Böschung, ▶Entwässerung landwirtschaftlicher Flächen, ▶Grundwasserabsenkung).

land exchange [n] *agr. leg.* ▶voluntary land exchange.

3068 land filling [n] *constr.* (Sand, gravel, earth or other materials of any composition whatsoever placed or deposited by human action on low areas; ▶infilling); *syn.* fill [n] [also US]; *s* **relleno [m] del terreno** (Resultado de la deposición de material sobre tierras bajas que pueden ser pantanosas o acuáticas para conseguir superficie de tierra aprovechable; ▶relleno 2); *f* **comblement [m] du terrain** (Remplissage de surfaces situées au-dessus du niveau du terrain naturel [fosse, dépression, cuvette] avec des matériaux appropriés ; ▶comblement ; *terme spécifique* comblement d'une carrière) ; *g* **Geländeauffüllung [f]** (Ergebnis der Aufschüttung von tiefer liegenden Flächen [z. B. Geländemulden, Senken, Steinbrüche oder sonstige Entnahmestellen] mit geeignetem Boden/Material; *UB* Auffüllung eines Steinbruchs; *Vorgang* ▶Auffüllen); *syn.* Geländeaufhöhung [f], Geländeaufschüttung [f].

landfill site [n] [UK] *envir.* ▶sanitary landfill [US] (2).

3069 landfill [n] **to original contour** *min. envir.* (Dumping [US]/tipping [UK] of waste in a previously excavated pit to the original contour; ▶dumpsite [US]/tipping site [UK], ▶high spoil pile [US]/high tip [UK], ▶land filling, ▶partially infilled pit); *s* **vertedero [m] en hondonada (1)** (Forma de depósito de residuos en lugar de préstamo hasta la cota original; ▶depósito de residuos sólidos, ▶depósito sobre el nivel de superficie, ▶relleno del terreno, ▶vertedero en zona de préstamo); *f* **décharge [f] par enfouissement** (Installation de stockage de déchets située dans une dépression naturelle ou artificielle du sol provenant, p. ex. d'une extraction de matériaux ; ▶comblement du terrain, ▶décharge publique, ▶décharge par dépôt, ▶décharge par enfouissement partiel) ; *syn.* décharge [f] pour mise à niveau (GEN 1982-I, 212) ; *g* **Flurkippe [f]** (Deponieform in einer ausgebeuteten Entnahmestelle, bei der das verkippte Material die ehemalige Bodenoberfläche erreicht; ▶Deponie, ▶Unterflurkippe, ▶Hochkippe, ▶Geländeauffüllung).

3070 landform [n] *geo. landsc.* (Single topographic configuration of a landscape); *s* **topografía [f] del terreno** (Configuración topográfica de un área); *f* **morphologie [f] du terrain** (Configuration topographique d'un paysage) ; *g* **Geländeform [f]** (Größenunabhängige, singuläre Einzelausformung in einer Landschaft, z. B. Hügel, Mulde, Hangterrasse); *syn.* Reliefeinzelform [f].

3071 land grab [n] [US] *agr.* (TGG 1984, 25; taking possession of land by settlement and cultivation whereby the land may or may not have been legally-owned; ▶overnight land squatting, ▶shifting cultivation); *syn.* land occupation [n] [UK]; *s* **colonización [f] de tierras yermas** (Toma de tierras desocupadas o sin cultivar; ▶agricultura itinerante, ▶paracaidismo); *syn.* ocupación [f] de tierras sin cultivar (o infraexplotadas); *f* **colonisation [f] sauvage** (Prise de possession, appropriation, usurpation de terres individuelles ou sans propriétaire, peuplement et mise en culture ; ▶agriculture itinérante [sur brûlis], ▶établissement d'un campement sauvage) ; *syn.* occupation [f] des terres ; *g* **Landnahme [f]** (Durch Ansiedlung und Bearbeitung erfolgende Inbesitznahme von herrenlosem oder einem Eigentümer gehörenden Grund und

L

Boden; ▶Bauen von Notbehausungen über Nacht, ▶Wanderfeld-bau).

land grade [n] [UK] *agr.* ▶agricultural land grade.

3072 land-holding agency [n] [US] *adm.* (Municipal, State or federal agency owning real estate; *specific term* city property department [US&CDN]) [PSP 1987, 193], county recorders office [US]: administering the land records); *syn.* land registry office [n] [UK]; *s* **departamento** [m] **de bienes inmuebles** (Administración estatal, autonómica o municipal de bienes inmuebles de propiedad pública); *syn.* administración [f] de bienes raíces; *f* **service** [m] **des transactions immobilières (F.,** désignation variant selon les communes et caractérisant le service municipal responsable de la gestion des biens immobiliers communaux ; *cf.* organigramme de l'administration de la ville de Strasbourg ; **D.,** service communal, régional ou fédéral) ; *syn.* service [f] des mutations foncières [Aix en Provence] ; *g* **Liegenschaftsamt** [n] (Städtische, Landes- oder Bundesbehörde, die alle in ihrem Besitz sich befindlichen Grundstücke [Immobilien] verwaltet).

land holdings [npl] *agr. for. leg. urb.* ▶enlargement of land holdings, ▶re-organisation of land holdings [UK].

land improvement [n] *agr. pedol.* ▶agricultural land improvement.

land information system [n] *adm. geo. landsc. plan.* ▶landscape information system.

3073 landing [n] *constr.* (Horizontal platform between two flights of steps or in a ramp); *syn.* landing tread [n] [also UK], pace [n] [also US], stair landing [n]; *s* **descansillo** [m] (Plataforma horizontal en que desemboca un tramo de escalera para cambiar de dirección o terminar; DACO 1988); *syn.* rellano [m] de escalera, meseta [f] (1); *f* **palier** [m] (Plate-forme dans un escalier en haut et en bas de chaque plan incliné ; *terme spécifique* palier de repos pour le cheminement des personnes handicapées) ; *g* **Podest** [n] (Verebnung zwischen zwei Treppenläufen oder in einer Rampe); *syn. für Treppen* Treppenpodest [n], Treppenabsatz [m].

3074 landing length [n] *constr.* (Stretch/span of landing between two flights of steps); *s* **largo** [m] **de rellano** (Extensión del descansillo entre dos tramos de escalera); *f* **longueur** [f] **de palier** (d'un escalier) ; *g* **Podestlänge** [f] (Die P. einer Treppenanlage wird nach folgender Formel berechnet: P = n x Schrittmaß + 1 Auftritt); *syn.* Podesttiefe [f].

3075 landing pier [n] [US] *recr.* (Structure built out into the water on piles for use as a landing place); *syn.* boating pier [n], landing stage [n] [UK]; *s* **embarcadero** [m] **de botes** (Estructura construida a partir de la orilla sobre el agua donde pueden atracar botes); *syn.* atracadero [m] de botes; *f 1* **appontement** [m] (**1.** Plate-forme en bois ou en métal sur pilotis permettant l'accostage et l'amarrage des bateaux ; *syn.* lieu [m]/aire [f] d'appontement, zone [f] d'appontage. **2.** À distinguer de **ponton** [m] qui est une plate-forme flottante à laquelle peuvent s'amarrer les bateaux ; *termes spécifiques* ponton de pêche, ponton de débarcadère ou d'embarcadère) ; *f 2* **estacade** [f] (Construction légère sur pilotis s'avançant en mer et permettant l'accostage des embarcations ou servant à prolonger un bajoyer pour guider les bateaux à l'entrée d'une écluse ou leur permettre de s'amarrer) ; *g* **Bootssteg** [m] (Vom Ufer ausgehender, über dem Wasser gebaute Konstruktion, an der Boote anlanden können); *syn.* Anlegesteg [m], Bootsanleger [m], Bootsanlegesteg [m], Bootslände [f], Landesteg [m].

landing stage [n] [UK] *recr.* ▶landing pier [US].

landing tread [n] [UK] *constr.* ▶landing.

land management [n] *agr. for.* ▶ecologically-based land management, ▶economic land management.

land management information system [n] *adm. geo. landsc. plan.* ▶landscape information system.

3076 land management measures [npl] *agr. for.* (Steps which retain and enhance the productivity and efficiency of agriculture and forestry; ▶agricultural land improvement, ▶ecologically-based land management); *s* **medidas** [fpl] **de gestión ecológica de zonas agrosilvícolas (≠)** (Todo tipo de medidas que sirven para preservar y aumentar la capacidad de producción de los suelos agrosilvícolas; ▶gestión ecológica de zonas agrosilvícolas, ▶mejoramiento de suelos); *f* **mesures** [fpl] **de gestion des territoires agricoles et forestiers** (Toute mesure ayant pour objectif la sauvegarde, la mise en valeur et le développement des ressources et de la productivité agricole et sylvicole dans le respect de l'équilibre biologique ; ▶amélioration du sol, ▶gestion du terroir) ; *g* **landeskulturelle Maßnahmen** [fpl] (Alle Maßnahmen, die dazu dienen, die land- und forstwirtschaftliche Leistungsfähigkeit der Kulturlandschaft zu erhalten, zu fördern und zu verbessern; ▶Landeskultur, ▶Melioration).

landmark building [n] [US] *conserv'hist.* ▶listed landmark building.

3077 landmark feature [n] *landsc. urb.* (Typical and widely-recognized natural landscape feature or man-made structure which serves as an orientation point and conspicuous guide); *s* **punto** [m] **de orientación significativo** (Elemento urbanístico o paisajístico muy visible a distancia que sirve de orientación); *syn.* punto [m] destacado, punto [m] prominente; *f* **repère** [m] **remarquable** (Élément typique et facilement reconnaissable des sites et des paysages naturels ou artificiels permettant une localisation aisée) ; *g* **markanter Orientierungspunkt** [m] (Ein stark ausgeprägter und weit sichtbares Stadt- oder Landschaftselement, das für die Umgebung typisch ist. Ein **m. O.** in der Landschaft wird auch ‚Landschaftszeichen' und seit einigen Jahren auch ‚Landmarke' [vom Englischen *landmark*] genannt. Ein weithin sichtbarer Punkt an der Küste, der dem Navigator zur Orientierung nützt, heißt auch ‚Landmarke'); *syn.* Blickpunkt [m], Landmarke [f].

landmark tree [n] [US] *conserv. leg.* ▶monarch tree [US].

land needed for development [n] *plan.* ▶open land needed for development.

land occupation [n] [UK] *agr.* ▶land grab [US].

3078 land ownership pattern [n] [US] *agr. plan. urb.* (Graphic representation showing distribution of land parcels in a defined area); *syn.* land ownership structure [n] [UK]; *s* **estructura** [f] **parcelaria** (Estado de distribución y propiedad de las parcelas de una zona específica); *f* **régime** [m] **foncier** (État du morcellement parcellaire d'une commune) ; *syn.* structure [f] parcellaire ; *g* **Grundstücksbestand** [m] (Zustand der Verteilung von Grundstücken und deren Eigentümer in einem definierten Gebiet).

land ownership structure [n] [UK] *agr. plan. urb.* ▶land ownweship pattern [US].

land parcels [npl] *agr. geo.* ▶agricultural land parcels.

land pasture [n] *agr.* ▶common grazing.

land planning [n] *plan.* ▶overall land planning.

land rank [n] [US] *agr.* ▶arable land rank.

3079 land reassignment [n] [UK] *agr. leg.* (**In Europe**, reallocation of newly configured [p]lots to land owners participating in a land consolidation procedure); *s* **reparto** [m] **nuevo de tierras** (Redistribución de las parcelas de tierra entre sus propietarios después de la concentración parcelaria); *f* redistribution [f] des parcelles (et prise de possession des nouvelles parcelles par un propriétaire lors d'une procédure de remembrement) ; **g Landabfindung [f]** (Neue Besitzeinweisung für einen Teilnehmer des Flurbereinigungsverfahrens, der für seine eingebrachten Grundstücke unter Berücksichtigung der für die gemeinschaftlichen und öffentlichen Anlagen erfoderlichen Abzüge mit Land von gleichem Wert abzufinden ist; **L.en** müssen in möglichst großen Grundstücken ausgewiesen werden; cf. § 44 FlurbG).

3080 land reclamation [n] (1) *hydr. eng. land'man. nat'res.* (Reclaiming of areas of ground by ▶aggradation of coastal mudflats and river deltas, damming up of large stretches of marine beaches, draining of wetlands, filling of open pits, or reclaiming of derelict dock land for settlement or agricultural purposes; ▶polder, ▶reclamation of derelict land); *syn.* reclamation [n] of land (1); **s ganancia [f] de tierras (1)** (Conjunto de medidas para aumentar la superficie de la tierra respecto al mar o en zonas húmedas desecadas para su aprovechamiento agrícola o urbano por medio de la ▶agradación artificial de zonas marítimas costeras o lagos interiores, ▶pólder, ▶recultivo de tierras yermas); **f 1 poldérisation [f]** (Mesures prises en vue de gagner des terres sur la mer par alluvionnement de zones des marais littoraux ou des embouchures de fleuves, par endiguement de grandes surfaces marines ou l'assèchement de lacs ; *termes spécifiques* poldérisations industrielles, ▶poldérisation agricole ; ▶mise en valeur de terres incultes [récupérables], ▶polder) ; **f 2 conquête [f] de terres nouvelles** (Réalisé par ▶atterrissement de zones basses sur le littoral [conquête sur la mer] par comblement le long des cours d'eau [bras mort], par utilisation des alluvionnements des fleuves à leur embouchure, par assèchement des lacs, tourbières ou marais dans le but de gagner des terres pour l'agriculture ou l'urbanisation ; ▶mise en valeur de terres incultes [récupérables], ▶polder) ; **g Landgewinnung [f]** (**1.** Gewinnung von Bodenflächen zur Besiedelung oder landwirtschaftlichen Nutzung durch ▶Auflandung 1 von Flächen im Watten- und Flussdeltabereich, durch Abdämmung größerer Meeresgebiete oder durch Trockenlegung von Binnenseen; **2.** Gewinnung landwirtschaftlicher Flächen durch Entwässerung von Mooren und Sümpfen; ▶Landgewinnung durch Trockenlegung; ▶Polder, ▶Ödlandrekultivierung); *syn.* Neulandgewinnung [f], Neugewinnung [f] von Land.

3081 land reclamation [n] (2) *agr. for. land'man.* (Redesign and restoration of land disturbed or destroyed by human activity, especially amelioration measures employed to improve soil [▶agricultural land improvement], predominantly soil material which has been dumped [US]/tipped [UK] during mining operations, or the utilization of dumpsites [US]/tipping sites [UK], ▶gravel extraction sites, etc. according to goals set by regional planning, nature conservation regulations and recreational planning; ▶reclamation of derelict land, ▶restoration 1, ▶revegetation; *specific term* gravel pit reclamation); *syn.* land restoration [n] (1), reclamation [n] of land (2), restoration [n] of land; **s recuperación [f] de tierras** (Recultivo y rehabilitación paisajística y ecológica de áreas degradadas por actividades industriales o de minería a cielo abierto como ▶graveras, etc. y de escombreras y vertederos por medio de medidas de construcción paisajística y de ▶mejoramiento de suelos; ▶recultivo de tierras yermas, ▶renaturalización, ▶revegetalización); *syn.* rehabilitación [f] de tierras; **f remise [f] en état des lieux après exploitation** (Remise en culture et recolonisation des territoires détruits par les activités humaines grâce aux techniques paysagères d'▶amélioration du sol, en particulier des sols recouvrant les haldes, les dépôts d'ordures, les anciennes carrières ou ▶sites d'extraction de gravier, etc. ; dispositions régies par différentes réglementations relatives aux installations classées après cessa-

tion d'activités, aux mines et carrières, etc. ; ▶mise en valeur des terres incultes, ▶reconstitution en l'état naturel, ▶revégétalisation [de zone d'emprunt]) ; *syn.* réaménagement [m], restauration [f], réhabilitation [f] des paysages ; **g Rekultivierung [f]** (Ordnungsgemäße Neugestaltung und Wiedernutzbarmachung eines durch menschliche Eingriffe zerstörten Gebietes durch landschaftsplanerische Maßnahmen der Bodenverbesserung [▶Melioration], besonders bei solchem Bodenmaterial, das als Folge des Bergbaues umgelagert wurde oder bei Nutzbarmachung von Deponien, ▶Kiesentnahmestellen 1 etc. unter Beachtung des öffentlichen Interesses, insbesondere der Ziele und Erfordernisse der Raumordnung und Landesplanung, des Naturschutzes und der Landschaftspflege sowie der Erholung; cf. § 4 [4] BBergG; ▶Ödlandrekultivierung, ▶Renaturierung, ▶Wiederbegrünung); *syn.* Wiedernutzbarmachung [f].

3082 land reclamation [n] (3) *agr.* (Conversion of land covered by its original vegetation, e.g. heath, bog, virgin forest into agricultural or forested areas by ▶clear felling, ▶deep plowing [US]/deep ploughing [UK], drainage, etc., ▶tree clearing and stump removal [operation]); *syn.* reclamation [n] of land (3); **s 1 roturación [f] de tierra virgen** (Arar por primera vez las tierras para cultivarlas después de su ▶roturación o ▶tala a matarrasa; ▶arado en profundo); *syn.* puesta [m] en cultivo de tierra virgen; **s 2 roza-tumba-quema [f]** (Acto de desmonte o derribe de la vegetación, así como su fraccionamiento para su desecación y destrucción por medio del fuego; MEX 1983); **f défrichement [m] et mise [f] en culture/pâture** (PR 1987, 473 ; action de mettre en culture un terrain resté en friche ou rendre propre à la culture une forêt, une lande, etc. par ▶coupe à blanc, ▶labour profond, drainage, etc. ; ▶abattage par extraction de souche) ; *syn.* labour [m] de défrichement ; **g Urbarmachung [f]** (Umwandlung von Land in land- oder forstwirtschaftliche Nutzflächen durch ▶Kahlschlag, ▶Rodung, ▶Tiefpflügen, Entwässerung o. Ä.; derartige Flächen sind meist mit Urvegetation bestockt, z. B. Heide, Moor, Urwald).

3083 land reform [n] *agr. leg.* (Changes in the ownership of agricultural or building land, usually involving the consolidation or redistribution of land holdings; no US equivalent); **s 1 reforma [f] agraria** (Cambio en las relaciones agrarias con el objetivo de mejorar la situación social y económica de los sectores más pobres o infraempleados de la población agraria y aumentar la producción agraria para asegurar la alimentación y el desarrollo de los países del Tercer Mundo. Por estas últimas razones la FAO ha impulsado la **r. a.** en los últimos años; cf. DGA 1986); **s 2 reforma [f] urbana [C, NIC]** (Transformación de las relaciones de tenencia de suelo y viviendas en Cuba, Nicaragua y otros países después de un cambio revolucionario con el fin de solucionar el problema de la vivienda y de la especulación del suelo); **f réforme [f] foncière** (Changement du régime juridique des terrains à usage agricole ou de construction par la socialisation de la propriété foncière ou de sa redistribution ; *terme spécifique* réforme agraire) ; **g Bodenreform [f]** (Veränderung der Rechtsverhältnisse an landwirtschaftlich genutztem Boden resp. am Bauland entweder durch Überführung des Grund und Bodens in Gemeindeeigentum oder durch seine Umverteilung. *UB* Agrarreform); *syn.* Bodenbesitzreform [f].

3084 land register [n] *adm. leg.* (Register containing all legal status and deed information on land maintained by a Land Registry Office; **in U.S.,** land registers are maintained by county governments by an official usually called "the Register/Registrar of Deeds". Deed or land registries are not maintained by cities or state governments; ▶real property identification map [US]/land register [UK]); **s catastro [m]** (Registro legal de todos los datos sobre la propiedad inmobiliaria de un municipio que incluye el plano catastral; ▶catastro de bienes inmuebles); *syn.* registro [m] de la propiedad [Es], libro [m] de propiedad, libro [m] fundiario [AL]; *f* **cadastre [m]** (Ensemble des documents administratifs définissant la propriété foncière et comportant : **1.** Le **registre des états de section** [description de la parcelle]. **2.** Le **plan cadastral** : repérage géographique parcellaire. **3.** La **matrice cadastrale** : détermination de l'impôt foncier ; ►cadastre parcellaire) ; *g* **Grundbuch [n]** (Buch, in dem alle Beurkundungen über die Rechtsverhältnisse an Grundstücken aufgenommen werden, das von einer Behörde, dem ,Grundbuchamt' [beim zuständigen Amtsgericht], geführt wird; cf. § 1 GBO; ►Liegenschaftskataster).

land register [n]**, abandoned** *adm. urb.* ►derelict land register.

land registration map [n] [UK] *adm. surv* ►metes and bounds map [US].

land registry [n] [UK]**, contaminated** *adm. envir.* ►toxic site inventory [US].

land registry map [n] [UK] *adm. surv.* ►area ownership map [US].

land registry office [n] [UK] *adm.* ►land-holding agency [US].

land requirement [n]**, open** *plan.* ►open land needed for development.

3085 land requirements [npl] *plan.* (**1.** Demand for land for particular, designated uses and important planning objectives; ►land consumption); **2. interest** [n] **in land** *leg.* (Entitlement in use of property); *s 1* **demanda [f] de suelo** *plan.* (Necesidad de suelo para un uso específico u objetivo de planeamiento importante); *s 2* **derecho [m] de superficie** *leg.* (Derecho real que permite al titular del mismo [superficiario] edificar en un suelo que no sea de su propiedad, y usar y disfrutar de lo construido, a cambio de un canon durante un tiempo determinado y limitado, de manera que, transcurrido el mismo, todo lo construido pasa a ser propiedad del propietario del suelo en el que se realizó la construcción; DıcUʀʙ 2008; cf. arts. 287-289 RDL 1/1992; ►gasto de suelo); *f 1* **besoins [mpl] fonciers** (Manque d'espaces à bâtir pour la réalisation d'une opération immobilière ou d'importants projets d'infrastructure) ; *f 2* **droit [m] sur un territoire** *leg.* (droit de revendiquer un territoire/un terrain ; ►consommation d'espace ; *syn.* droit [m] au territoire ; *g* **Flächenanspruch [m]** (**1.** *plan.* Bedarf an Flächen/Grundstücken für die Realisierung von Vorhaben oder für Nutzungsansprüche bei raumbedeutsamen Planungen; ►Landschaftsverbrauch; **2.** *leg.* Recht oder Anrecht auf Flächen/Grundstücke); *syn. für 1.* Flächenbeanspruchung [f], Flächenbedarf [m]).

3086 land [n] **reserved for environmental purposes** *leg. urb.* (Area to halt the spread of detrimental effects; i. e., noise or noxious fumes from landfills or industrial operations, reserved according to a legally-binding land-use plan such as a ►zoning map [US]/Proposals Map/Site Allocations Development Plan Document [UK]; ►buffer zone); *s* **zona [f] no edificable** (En D. en los ►planes parciales [de ordenación], zona reservada a medidas de protección contra los efectos negativos de instalaciones industriales o para separar las mismas de otros usos; ►zona de amortiguación); *f* **zone [f] non aedificandi** (±) (Zone sur les ►documents de planification urbaine, les espaces réservés dans le périmètre de laquelle la construction est interdite ; ►zone tampon, ►distance d'isolement ; **D.**, périmètre prescrit par le « Bebauungsplan » [plan local d'urbanisme [PLU]/plan d'occupation des sols allemand] réservées aux installations et mesures en faveur de la protection de l'environnement prévues par la loi relative à la lutte contre les pollutions atmosphériques et les odeurs) ; *g* **freizuhaltende Schutzfläche [f]** (Im ►Bebauungsplan oder durch besondere Vorschrift festgesetzte Fläche für

besondere Anlagen und Vorkehrungen zum Schutze vor schädlichen Umwelteinwirkungen im Sinne des BImSchG oder für Belange von Naturschutz und Landschaftspflege; cf. auch § 9 [1] 24 BauGB; ►Abstands- und Schutzzone 2).

land restoration [n] (**1**) *agr. for. land'man.* ►land reclamation (2).

3087 land restoration [n] (**2**) *conserv'hist. landsc.* (Act or process of accurately returning the form, features, and general or precise character of a piece of land, as it appeared at a particular period of time, by removal of features from other periods in history and ►reconstruction of missing features from the restoration period; LA 1997 [5], 78); *s* **restauración [f] de un paisaje** (Medidas que sirven para restaurar el estado original de un paisaje degradado o destruido o de partes del mismo; restaurar [vb]/regenerar [vb] el paisaje; ►reconstrucción); *syn.* regeneración [f] de un paisaje; *f 1* **remise [f] en état d'un site** (Mesures — souvent ordonnées par l'administration ou un tribunal — ayant pour but de reconstituer la forme, le caractère et les éléments d'un paysage conformément à son aspect antérieur ; ►reconstruction) ; *syn.* rétablissement [m] dans l'état antérieur (PPH 1991, 101), remise [f] dans l'état initial ; *f 2* **réaffectation [f] à l'usage antérieur** (Mesures ayant pour but de rétablir l'usage initial, p. ex. pâture, forme d'exploitation sylvicole, etc.) ; *g* **Wiederherstellung [f] einer Landschaft** (Maßnahmen, die dazu dienen, Form, Charakter und Einzelbestandteile einer Landschaft oder eines Teils davon so wiederherzustellen, wie sie zu einem bestimmten früheren Zeitpunkt gewesen waren; ►Rekonstruktion); *syn.* Restaurierung [f] einer Landschaft, Wiederherstellung [f] einer Landschaft in den ursprünglichen Zustand.

land ripe for development [n] [US] *urb.* ►unzoned land ripe for development [US]/white land [UK].

3088 land rotation [n] *agr.* (Agricultural system whereby cultivation is carried on for a few years and then the land is allowed to rest perhaps for a considerable time, before the scrub or grass which has grown is cleared and the land recultivated. In such areas, farms or settlements on which cultivation takes place are permanent; cf. GGT 1979; ►crop rotation, ►shifting cultivation); *s* **cultivo [m] en rotación** (Sistema agrícola utilizado en la antigüedad que consiste en el cultivo de un campo durante unos años y su abandono durante varios más para que se recupere el suelo: «descanso». Esta forma de uso surgió al hacerse sedentaria la población, en contraste con el cultivo itinerante de la agricultura primitiva; ►agricultura itinerante, ►rotación de cultivos); *f* **assolement [m]** (Forme culturale extensive ancienne pratiquant l'alternance de cultures sur plusieurs soles ; *termes spécifiques* assolement à jachère, assolement à l'anglaise [alternance de céréales, prairies semées et plantes sarclées] ; ►agriculture itinérante [sur brûlis], ►rotation des cultures) ; *g* **Landwechselwirtschaft [f, o. Pl.]** (Extensive Form der Landwirtschaft durch wechselweise Nutzung eines Landstückes als Acker- oder Grünland mit Brachezeiten zur Erholung des Bodens. Diese Nutzungsform entstand bei Dauersiedlung im Gegensatz zum ►Wanderfeldbau; ►Fruchtfolge).

3089 landscape [n] (**1**) *geo.* (Part of the earth's surface characterized by its pictorial aspect [natural scenery, ►visual quality of landscape] or functional aspect [►landscape ecosystem], and consisting of ►interactions of ecosystems or ►pattern of ecotopes; ►cultural landscape, ►natural landscape, ►virgin area); *s* **paisaje [m]** (Porción de la superficie terrestre, provista de límites naturales, donde los componentes naturales [rocas, relieve, clima, aguas, suelos, vegetación, mundo animal] forman un conjunto de interrelación e interdependencia; FGB 1981; ►aspecto escénico del paisaje, ►mosaico de ecótopos,

►naturaleza y paisaje, ►paisaje humanizado, ►paisaje virgen, ►régimen ecológico del paisaje, ►sistema de interacciones de un ecosistema); *f* **paysage [m]** (Partie de la surface terrestre caractérisée par sa structure [conditionné par la perception : ►image paysagère], ses fonctions [aspect physique et biotique : ►écosystème paysager], composée des ►tissus d'interactions d'un écosystème ou d'un ►réseau d'écotopes ; notion affectée d'une très grande diversité de valeurs selon le point de vue et les méthodes des scientifiques [diverses écoles de la science du paysage] traitant du paysage dans ses représentations comme dans ses éléments matériels ; le paysage est d'abord le produit de la pratique, de l'action quotidienne, d'une pratique exercée sur le monde physique entre la simple retouche et l'artefact intégral ; MG 1993 ; ►espaces naturels, ►espaces naturels et paysages, ►espace vierge, ►paysage culturel) ; *g* **Landschaft [f]** (Nach Struktur [►Landschaftsbild] und Funktion [►Landschafts-haushalt] geprägter Ausschnitt der Erdoberfläche, die aus einem ►Ökosystemgefüge oder ►Ökotopengefüge besteht; cf. ANL 1984. Jede L. hat eine Naturgeschichte und eine Nutzungs-geschichte und zeichnet sich deshalb dadurch aus, dass sie relativ beharrlich ist und doch stetem, manchmal sogar einem sprung-haften Wandel unterliegt; deshalb wird unter ►Kulturlandschaft und ►Naturlandschaft 1 unterschieden. Die Geburt des Begriffes L. wird häufig auf den 16.04.1336 datiert, als der Dichter FRANCESCO PETRARCA den Mount Ventoux, ein Bergmassiv am östlichen Rand des Rhônetals in Südfrankreich, bestieg und die Aussicht auf die Landschaft entdeckte. L. ist ein ästhetischer und kultureller Begriff und wird in allen Kulturkreisen unterschied-lich gesehen; L. in Deutschland ist nicht L. in der Schweiz, nicht *landscape* in England und schon gar nicht *paysage* in Frankreich! Immer handelt es sich um Produkte der Phantasie, die je nach Zeitepoche in unterschiedlichsten Bildern ausgedrückt wird: In D. z. B. waren oder sind es Leitvorstellungen der Romantik und Landschaftsmalerei [Vorstellung arkadischer Ideallandschaft als Projektion utopischer Ideale], die Entdeckung des Wilden sowie die Uminterpretation von städtischen Räumen zur „Landschaft". Die einen Betrachter sehen mehr den ästhetischen, die anderen mehr den ökologischen und/oder funktionalen Aspekt, wieder andere sehen L. als politisch-kulturellen Raum — als vom Menschen gestaltete Kulturlandschaft. Durch die heutige Wahr-nehmung der L. bei zunehmender Flexibilisierung der Gesell-schaft und fortschreitender Urbanisierung und den dadurch entstehenden postindustriellen Landschaften, die sich in immer schneller werdenden Zyklen verändern, wird auch der Land-schaftsbegriff weg von der arkadischen, szenischen Utopie allmählich neu gedacht werden. In heutiger Zeit, in der eine auf Zuwachs angelegte Stadtentwicklung mangels Nachfrage immer mehr ins Leere läuft, gewinnt L. für die Entwicklung unserer Stadtumwelt immer mehr an Bedeutung. Gefragt ist vielmehr ein dynamisches **L.sverständnis**, das die Eindimensionalität eines starren **L.sbildes** überwindet; ►Natur und Landschaft).

landscape [n] (2) *agr.* ►agrarian cultural landscape, ►agricul-tural landscape; *conserv. landsc.* ►beauty and amenity of the landscape [US]/beauty and amenity of the countryside [UK]; *ecol. plan. recr.* ►carrying capacity of landscape; *conserv. landsc. plan. recr.* ►characteristic of a landscape; *geo.* ►coastal landscape; *agr.* ►compartmentalization of a landscape; *geo.* ►cultural landscape; *landsc.* ►current state of a landscape; *land'man.* ►hedgerow landscape; *conserv'hist. land'man. leg.* ►historic landscape, ►historic vernacular landscape; *conserv. land'man.* ►impairment of landscape; *geo.* ►industrial land-scape, ►karst landscape, ►lake landscape; *conserv. landsc.* ►natural landscape; *landsc.* ►near-natural landscape, ►park-like landscape; *plan. recr. sociol* ►perception of the landscape; *ecol. landsc.* ►potentials of a landscape; *conserv.* ►protected cultural

heritage landscape; *hydr. landsc.* ►river landscape; *geo.* ►rolling landscape; *conserv.* ►unique natural character of a landscape; *urb.* ►urban landscape; *geo. nat'res.* ►virgin landscape; *landsc. plan. recr.* ►visual quality of landscape.

landscape [n], **aesthetic value of a** *landsc. plan. recr.* ►visual quality of landscape.

landscape [n], **agrarian** *agr. plan.* ►agricultural landscape.

landscape [n], **analysis of the visual appearance of a** *plan. landsc.* ►analysis of the visual appearance of a landscape.

landscape [n], **cultural heritage** *conserv.* ►protected cultural heritage landscape.

landscape [n], **disfigurement of** *conserv. land'man.* ►impairment of landscape.

landscape [n], **excellence of** *conserv. landsc.* ►beauty and amenity of the landscape [US]/beauty and amenity of the countryside [UK].

landscape [n], **fabric of the** *landsc.* ►landscape structure.

landscape [n], **hard** *constr.* ►hardscape.

landscape [n], **heritage** *conserv.* ►historic landscape, ►protected cultural heritage landscape.

landscape [n], **hilly** *geo.* ►rolling landscape.

landscape [n], **historic designed** *conserv'hist. land'man. leg.* ►historic landscape.

landscape [n], **housing** *landsc. urb.* ►landscape planting in built-up areas.

landscape [n], **man-made** *geo.* ►cultural landscape.

landscape [n] [UK], **middle** *plan.* ►transition zone.

landscape [n], **patterning of a** *agr.* ►compartmentali-zation of a landscape.

landscape [n], **physiognomy of a** *landsc. plan. recr.* ►visual quality of landscape.

landscape [n] [US], **reclaimed surface-mined site** *min. landsc.* ►landscape of reclaimed surface-mined land.

landscape [n] [UK], **recultivated mining** *min. landsc.* ►landscape of reclaimed surface-mined land.

landscape [n], **soft** *constr.* ►softscape.

landscape [n], **totally cleared agrarian** *agr. land'man.* ►extensively cleared land (for cultivation).

landscape [n], **undulating** *geo.* ►rolling landscape.

landscape [n], **visual** *landsc. plan. recr.* ►visual quality of landscape.

3090 landscape aesthetics [npl *but usually singular*] *land'man. landsc. recr.* (Basic principle of landscape architecture dealing with the allocation of artistic values to the visual appear-ance of a landscape with regard to composition and visual character; ►perceived environmental quality, ►visual quality); *syn. o.v.* landscape esthetics [npl *but usually singular*]; *s* **estética [f] del paisaje** (Ámbito de la planificación del paisaje y de la conservación de la naturaleza que se ocupa de definir y preservar la ►calidad visual del paisaje; ►calidad vivencial del ambiente); *f* **esthétique [f] du paysage** (Valeur conceptuelle concernant l'architecture du paysage reposant sur l'étude de la sensibilité visuelle du paysage ; ►qualité perceptuelle, ►valeur esthétique) ; *syn.* esthétique [f] paysagère ; *g* **Landschaftsästhetik [f, o. Pl.]** (Wesentliches Element der Landschaftsarchitektur, das sich mit dem visuellen Erlebenswert der Landschaft befasst; ►Erlebnis-qualität, ►Gestaltqualität).

landscape aesthetics management [n] *landsc. plan.* ►visual resource management.

L

3091 landscape analysis [n] *plan.* (Study of various elements of a landscape [both animate and inanimate] within the scope of landscape planning: includes a survey and assessment of their spatial arrangement and interaction as well as their actual use, proposed changes in use and associated measures to be undertaken; ▶analysis of the visual appearance of a landscape); *s* **estudio** [m] **del paisaje** (Estudio de los diferentes elementos [bióticos y abióticos] y de sus interdependencias e interrelaciones, así como de los usos actuales de un paisaje en el marco de la planificación del paisaje, ▶análisis del aspecto escénico del paisaje); *syn.* análisis [m] paisajístico, prospección [f] del paisaje; *f* **analyse** [f] **paysagère** (Phase de recueil des données et évaluation des paysages à l'occasion d'un aménagement ou d'une planification des paysages ; elle comprend aussi bien les éléments biotiques qu'abiotiques et leurs interactions dans l'organisation spatiale, l'occupation des sols existante et projetée ainsi que les mesures afférentes ; *terme spécifique* analyse de l'état initial du paysage ; F., terme également utilisé dans le cadre de la recherche des paysages [appréciation de la valeur esthétique ou écologique des paysages] ; ▶étude sitologique) ; *f 3* **reconnaissance** [f] **du paysage** (Phase d'études dans l'élaboration d'un plan de paysage permettant de mettre en évidence les caractéristiques du paysage, de déterminer les enjeux et leur territorialisation ; Circulaire n° 95-23 du 15 mars 1993) ; *g* **Landschaftsanalyse** [f] (Bestandsaufnahme und Bewertung der natürlichen, naturnahen und durch den Menschen veränderten Landschaftselemente und Prozesse incl. ihrer Wechselwirkungen. Sie umfasst sowohl die abiotischen und biotischen Landschaftselemente und deren Wirkungsgefüge in ihrer räumlichen Anordnung als auch deren reale Nutzung, vorgesehene Nutzungsänderungen und damit verbundene Maßnahmen; ▶Landschaftsbildanalyse).

landscape appeal [n] *recr.* ▶appeal of a landscape.

3092 landscape architect [n] *prof.* (**1.** Person, who as a planner, works with the aim of protecting, improving and maintaining the sustainable development of landscapes or parts thereof, in order to ensure the functioning of ecological systems; moreover, he or she designs open spaces in [densely] populated areas according to aesthetic, functional and environmental considerations; the field of activity is a design-oriented and applied form of engineering on an ecological basis; a **l. a.** executes the work in a position of trust for his or her client. **2.** In U.S., **l. a.** is a protected professional title, which can only be used when a person passes a landscape architectural licensing examination approved by the ▶Council of Landscape Architectural Registration Boards [CLARB]; ▶Council of Architectural Registration Boards [CARB] [US]/Architects Registration Council [UK]. This requires an accredited bachelor's or master's degree, several years of practical work and a national licensing examination, the so-called Landscape Architect Registration Examination [LARE]. Most architects belong to the American Institute of Architects and use the abbreviation AIA or FAIA [fellow of AIA] after their names; many **l. a.s** belong to the American Society of Landscape Architects, and use the abbreviation ASLA or FASLA [fellow of ASLA] after their names. **3.** U.K., professional title which can only be adopted when a person is a member of the Architects Registration Council; the first British designer who adopted the title of **l. a.** was PATRICK GEDDES [1854-1932], trained as a biologist; in 1907 he was advertizing his practice in Scotland as "Landscape Architects, Park and Garden Designers". **4.** D., **l. a.** is a professional title protected by federal law and may only be used by those who are registered in ▶Chamber of Architects of the federal state in which he or she lives; ▶architectural register, ▶Master of Landscape Architecture, ▶project landscape architect); *s* **arquitecto/a** [m/f] **paisajista** (**1.** Profesional de planificación que trabaja con el objetivo de proteger, mejorar y man-

tener el desarrollo sostenible/sustentable de paisajes o partes de los mismos, de manera que se garantice el funcionamiento de los sistemas ecológicos. Además diseña espacios libres en áreas [densamente] pobladas teniendo en cuenta criterios estéticos, funcionales y ambientales, siendo el campo de actividad una forma aplicada de la ingeniería orientada al diseño y fundamentada en la ecología. El/la **a. p.** desarrolla su actividad profesional en una relación de confianza hacia el/la comitente. **2. En EE. UU.** es un título protegido que solo puede ser utilizado si la persona aprueba un examen de licencia de arquitectura paisajista del Consejo de Juntas de Registro de Arquitectos Paisajistas [Council of Landscape Architectural Registration Boards]. Esto exige un título acreditado de Bachelor o Máster, varios años de experiencia profesional y un examen de licencia, llamado Examen de Registro de Arquitecto/a Paisajista. **3. En GB** título profesional que solo puede ser adquirido cuando la persona es miembro del Consejo de Registro de Arquitectos. **4. En Es.** no hay reconocimiento legal de la profesión de paisajista y por lo tanto tampoco del título oficial de **a. p.**; ▶Colegio Oficial de Arquitectos, ▶paisajista diplomado/a, ▶registro de arquitectos); *syn.* paisajista [m/f]; *f* **architecte** [m/f] **paysagiste** (F., titre professionnel habilité au plan international par le bureau des professions de Genève, non protégé en France, qualifiant les hommes de l'art intervenant dans les domaines les plus variés de la création et de planification environnementale en milieu urbain et rural [sites et espaces classés, équipements d'infrastructures, industriels, touristiques, sportifs, espaces verts publics ou privés, etc.] ; profession complémentaire et partenaire de l'écologue, de l'architecte et de l'urbaniste ; D., titre protégé attribué à tout paysagiste membre de l'▶ordre des architectes et inscrit sur le ▶registre des architectes ; HAC 1989, 463 ; ▶paysagiste, ▶paysagiste diplômé) ; *g* **Landschaftsarchitekt/-in** [m/f] (**1.** Jemand, der als Planer/-in, die nachhaltige Sicherung und Verbesserung der Funktionsfähigkeit des Naturhaushalts von Landschaften oder deren Teile betreibt sowie nicht überbaute Flächen im [dicht] besiedelten Bereich nach ästhetischen, funktionalen und ökologischen Gesichtspunkten gestaltet; die Tätigkeit ist eine gestalterisch angewandte Ingenieurwissenschaft auf ökologischer Grundlage; der **L.** übt seine Tätigkeit als Vertrauensperson des Auftraggebers aus. **2.** D., durch Architektengesetze der Bundesländer geschützte Berufsbezeichnung, die nur führen darf, wer in die ▶Architektenliste der ▶Architektenkammer des jeweiligen Bundeslandes eingetragen ist. **3.** In den **USA** ist der Begriff **L.** eine geschützte Berufsbezeichnung, deren Verwendung der Erwerb einer professionellen Lizenz voraussetzt. Diese erfordert jedoch einen akkreditierten Bachelor- oder Masterabschluss, mehrere Jahre praktische Tätigkeit und eine nationale Lizenzprüfung, die so genannte LARE, **L**andscape **A**rchitect **R**egistration **E**xamination; ▶Diplomingenieur/-in für Landschaftsarchitektur); *syn.* Garten- und Landschaftsarchitekt/-in [m/f].

3093 landscape architect [n] **employed by a public agency/authority** *prof.* *s* **arquitecto/a** [m/f] **paisajista funcionario/a** (empleado/a en la administración); *f* **architecte** [m/f] **paysagiste fonctionnaire** (employé, ée dans les services publics) ; *g* **angestellte/-r Garten- und Landschaftsarchitekt/-in** [m/f] **in einer Behörde** (Jemand, der als Planer/-in einem vertraglichen Arbeitsverhältnis mit monatlichen Gehaltsbezügen bei einem öffentlichen Auftraggeber beschäftigt ist).

3094 landscape architect [n] **in charge of a branch office** [US]; *s* **arquitecto/a** [m/f] **paisajista jefe/a de sucursal** (de gabinete de arquitectura); *f* **architecte** [m/f] **paysagiste en charge d'un établissement secondaire** ; *g* **Garten- und Landschaftsarchitekt/-in** [m/f] **in leitender Stellung (2)** *prof.* (Büroleiter/-in eines privaten Zweigstellenbüros).

L

3095 landscape architect [n] **in private practice**
prof. s **arquitecto/a [m/f] paisajista empleado/a** (en gabinete de arquitectura paisajista); *f* **architecte [m/f] paysagiste salarié, ée** (Employé, ée dans un bureau d'études) ; *g* **angestellte/-r Garten- und Landschaftsarchitekt/-in [m/f] in einem Planungsbüro** (Jemand, der als Landschaftsarchitekt/-in in einem vertraglichen Arbeitsverhältnis mit monatlichen Gehaltsbezügen in einem freien Planungsbüro beschäftigt ist).

3096 landscape architect principal [n] *prof.* (Owner or partner of a landscape architectural firm); *s* **arquitecto/a [m/f] paisajista jefe/a** (Profesional con su propio gabinete de arquitectura paisajista); *f* **architecte [m/f] paysagiste indépendant, ante** (Directeur, trice d'un bureau d'études) ; *syn.* paysagiste [m/f] indépendant, ante, paysagiste [m/f] conseil ; *g* **freie/-r Garten- und Landschaftsarchitekt/-in [m/f]** (Jemand, der als Planer/-in freiberuflich/selbständig tätig und Inhaber/-in eines Planungsbüros ist); *syn.* freischaffende/-r Landschaftsarchitekt/-in [m/f].

3097 landscape architecture [n] *landsc. recr.* (A profession and academic discipline that employs principles of art and the physical and social sciences to the processes of environmental planning, design and conservation, which serve to ensure the long-lasting improvement, sustainability and harmony of natural and cultural systems or landscape parts thereof, as well as the design of outdoor spaces with consideration of their aesthetic, functional and ecological aspects. The term has been used first by SCOTT, in his book 'On the Landscape Architecture of the great painters of Italy' [1828], and subsequently by FREDERIC LAW OLMSTED and VAUX for the Central Park in New York [1858]; cf. JEL 1986; ▶general urban green space planning); *syn.* environmental planning [n], design [n] and conservation [n]; *s* **paisajismo [m]** (Es el arte y la ciencia de adaptar el paisaje al uso humano. Arte porque responde a la característica primordial de la expresión artística que es la ordenación de impresiones visuales en formas coherentes y completas. Ciencia porque se vale de conceptos científicos, de la observación y la experimentación para el desarrollo de proyectos y para la comprensión de la relación entre la ciencia y la sociedad. Esta mezcla le da a la profesión su carácter propio, siendo la función del arquitecto paisajista la de solucionar proyectos desde el punto de vista de su adaptación a la naturaleza. En el ámbito de la ciudad se asemeja, si es que no coincide con ella, a la jardinería, y busca compensar la ausencia de espacios naturales con la creación de otros espacios, pseudo-naturales, naturalizados, gracias a la presencia de la vegetación. Fuera de los espacios urbanos, en los paisajes alterados por las grandes obras públicas, el **p.** busca minimizar la alteración, integrar la acción en el paisaje; cf. COFF 1991 y DINA 1987; ▶planificación de zonas verdes y espacios libres urbanos); *syn.* arquitectura [f] paisajista [BOL, COL, EC, RA, tb. Es], arquitectura [f] del paisaje [MEX]; *f* **architecture [f] des paysages (1.** Art et pratiques relatifs à l'organisation paysagère de l'environnement humain ayant pour objectifs la sauvegarde et le développement durables des potentialités du milieu naturel ainsi que l'aménagement des espaces urbains et territoriaux non construits dont la conception est fondée sur la prise en compte d'exigences fonctionnelles, esthétiques et écologiques ; ▶planification urbaine des espaces verts ; **2. D.**, le synonyme « Landespflege » [▶gestion du terroir] est un terme généralement utilisé dans le domaine de l'enseignement et de la recherche qui a tendance aujourd'hui à être remplacé par une terminologie plus reconnue internationalement telle que l'architecture des paysages) ; *syn.* art [m] du paysage, paysagisme [m] ; *g* **Landschaftsarchitektur [f, o. Pl.] (1.** Teilbereich der Umweltplanung und des Umweltschutzes mit Zielen und Maßnahmen, die der nachhaltigen Sicherung und Verbesserung der Funktionsfähigkeit des Naturhaushalts von Landschaften oder Landschaftsteilen dienen sowie die Gestaltung von nicht überbauten Flächen [Freiräume] im [dicht] besiedelten Bereich nach ästhetischen, funktionalen und ökologischen Gesichtspunkten; **L.** befindet sich stets in der kulturellen Auseinandersetzung mit Natur und Zeit auf technischer, planerischer und ästhetischer Ebene. **2.** Traditionellerweise wird der Terminus **L.** für die planerische Gestaltung von Grünanlagen oder vegetationsgeprägten städtischen Plätzen gebraucht. Das Begriffspaar **Naturschutz und Landschaftspflege** kann synonym für den ersten Teil der Definition verwendet werden. Der Begriff **L.** wurde 1828 zuerst von SCOTT in seinem Buch „On the Landscape Architecture of the great painters of Italy" eingeführt und anschließend von FREDERIC LAW OLMSTED and VAUX für den Central Park in New York [1858] weiterverwendet; cf. JEL 1986. Synonym ist der 1903 durch R. MIELKE geprägte Begriff **Landespflege**, der 1942 durch E. MÄDING erneut im Schrifttum erscheint und nach dem Kriege erstmals im Flurbereinigungsgesetz vom 14.07.1953 (BGBl. I 591) verwendet wird. Der Begriff Landespflege, der insbesondere in Hochschulkreisen oft als ungenau, zu wenig verbreitet und unzureichend bezeichnet wird, ist ein vorrangiger Begriff für das Berufsfeld im akademischen Bereich und deckt hier die Arbeitsbereiche Naturschutz und Landschaftspflege sowie die ▶Grünordnung ab; cf. S+G 1999 [7], 444. Dennoch zeichnet sich ab, dass in den Hochschulen immer mehr dazu übergegangen wird, die Studiengänge in die international verständliche Bezeichnung Landschaftsarchitektur [als OB] umzubenennen und je nach Studienschwerpunkt mit dem Zusatz Landschaftsplanung, Umweltplanung etc. zu differenzieren); *syn.* Landespflege [f, o. Pl.].

landscape assessment [n]**, aesthetic** *plan. landsc.* ▶visual landscape assessment.

3098 landscape change [n] *ecol.* (Alteration of the characteristics or appearance of a landscape due to natural evolution or human influence); *syn.* modification [n] of landscape; *s* **modificación [f] del carácter de un paisaje** (Alteración del aspecto de un paisaje por causas naturales o antropógenas); *syn.* cambio [m] del aspecto de un paisaje; *f* **modification [f] de/du caractère d'un paysage** (Altération des éléments ou de l'aspect d'un paysage provoqué par une intervention anthropogène) ; *syn.* changement [m] de l'aspect des lieux (PPH 1991, 99) ; *g* **Veränderung [f] der Landschaft** (Das durch natürlichen oder menschlichen Einfluss im Wesen oder in der Erscheinung Anderswerden einer Landschaft); *syn.* Landschaftsveränderung [f].

3099 landscape character [n] *conserv. landsc. plan. recr.* (Physical nature of landscape determined by various landscape factors or formed by patterns composing it; ▶visual quality of landscape; ▶area of outstanding scenic beauty); *syn.* visual character [n] of a landscape, unique character [n] of a (scenic) landscape, uniqueness [n] of a landscape, countryside character [n] [also UK]; *s* **particularidad [f] del paisaje** (Conjunto de características de un paisaje determinadas por sus componentes que le dan un carácter propio; ▶aspecto escénico del paisaje, ▶paraje natural de gran belleza escénica); *syn.* carácter [m] de un paisaje; *f* **caractère [m] particulier des sites et des paysages** (Paysages présentant un intérêt esthétique particulier, un caractère remarquable, soit par leur unité et leur cohérence, soit par leur richesse particulière en matière de patrimoine, soit comme témoins de mode de vie et d'habitats ou d'activités et de traditions industrielles, artisanales agricoles et forestières ; art. 1 du décret n° 94-283 du 11 avril 1994 ; ▶espaces les plus remarquables du patrimoine naturel et culturel, ▶image paysagère) ; *syn.* intérêt [m] esthétique d'un paysage, spécificité [f] du paysage ; *g* **landschaftliche Eigenart [f] (1.** Summe der wahrgenommenen, unverwechselbaren, individuellen Eigenheiten oder Einmaligkeit einer konkreten Landschaft zu einem be-

stimmten Zeitpunkt. Es gibt keine wissenschaftliche Methode, die die **E.** einer Landschaft allgemeingültig erfassen kann, da sie an die menschliche Psyche gebunden ist. **2.** Das geistige Gebilde, das sich ein Betrachter einer Landschaft aus einer Vielzahl von Einzelelementen oder Gestaltelementen macht; cf. NOHL 1997; ▶landschaftlich besonders schönes Gebiet, ▶Landschaftsbild); *syn.* Charakter [m] des Landschaftsbildes, Eigenart [f] des Landschaftsbildes, Landschaftscharakter [m].

landscape characteristics/resources [npl] *land'man.* ▶spatial landscape characteristics/resources.

landscape clearance [n] [UK] *agr. land'man.* ▶land clearance [US].

landscape component [n] *conserv. land'man. landsc.* ▶landscape feature.

3100 landscape component [n] **for recreation** *plan. recr. syn.* recreational feature [n] in the landscape, landscape feature [n] for recreation; *s* **componente [m] recreativo del paisaje**; *f* **élément [m] du paysage présentant un intérêt récréatif** (Espaces et curiosités naturels à fonction distractive) ; *syn.* élément [m] du paysage à forte valeur récréative, élément [m] paysager à forte valeur récréative ; *g* **erholungswirksames Landschaftselement [n]**.

3101 landscape conservation [n] *conserv. landsc.* (**1.** Measures concerned with coordination and plans for the safeguarding against undesirable changes of landscape resources and the prevention of exploitation, destruction, and neglect); **2. landscape preservation [n] [US]** (Retention of existing landscape conditions, e.g. historic landscapes by comprehensive protective measures to safeguard natural, near-natural or cultural landscapes, or parts thereof, for adherence to the aims of nature conservation and landscape management; ▶bestowed protection status); *syn.* landscape protection [n] [also UK]; *s 1* **protección [f] del paisaje** (Término genérico para conjunto de medidas dirigidas a proteger, preservar y gestionar los paisajes y la naturaleza para su aprovechamiento y disfrute sostenibles; ▶procedimiento de declaración (de espacio natural protegido); *s 2* **preservación [f] del paisaje** (Término específico para conjunto de medidas dirigidas a conservar paisajes naturales, poco alterados o antropógenos, o partes de ellos, en su estado actual con el fin de contribuir a la protección de la naturaleza y a la gestión del paisaje); *syn.* conservación [f] del paisaje; *f* **sauvegarde [f] des paysages** (Ensemble des mesures relatives à la préservation et la conservation des sites naturels ou milieux naturels, semi-naturels ou anthropogènes ; ▶procédure de mise sous protection, ▶inscription) ; *syn.* protection [f] des sites naturels, protection [f] de l'espace naturel ; *g* **Landschaftsschutz [m, o. Pl.]** (**1.** Gesamtheit der Maßnahmen entsprechend den Zielen von Naturschutz und Landschaftspflege, die der Sicherung und Pflege natürlicher, naturnaher oder anthropogener Landschaften oder ihrer Teile dienen; nach dem ▶Bundesnaturschutzgesetz [BNatSchG] können besonders schutzwürdige Landschaften als ▶Landschaftsschutzgebiete ausgewiesen werden; **2.** gem. Europäischer Landschaftskonvention [ELC]: Maßnahmen zur Erhaltung und Pflege der maßgeblichen oder charakteristischen Merkmale einer Landschaft, die durch ihren kulturhistorischen Wert begründet oder durch ihr natürliches Erscheinungsbild und/oder die Tätigkeit des Menschen zurückzuführen sind; ▶Unterschutzstellung).

landscape conservation [n] *conserv. land'man.* ▶integrated landscape conservation.

3102 landscape contracting firm [n] *constr. hort.* (Business firm devoted to ▶landscape construction); *syn.* landscape construction firm [n], landscape contractor's firm [n]; *s* **empresa [f] de construcción paisajística** (Firma que se dedica

a la ▶construcción paisajística 1); *syn.* empresa [f] de construcción de zonas verdes; *f* **entreprise [f] paysagiste** (Établissement dont les activités appartiennent au secteur commercial et agricole de la création et de l'entretien d'espaces verts et de terrains de sport ; ▶création, aménagement et entretien des/ d'espaces verts) ; *syn.* entreprise [f] de jardins (et d'espaces verts), entreprise [f] d'aménagement des parcs et jardins, entreprise [f] de jardins ; *g* **Garten- und Landschaftsbaubetrieb [m]** (Gewerbliches Unternehmen, das der Tätigkeit des ▶Garten-, Landschafts- und Sportplatzbaues nachgeht); *syn.* Ausführungsbetrieb [m] des Garten- und Landschaftsbau(e)s, Ausführungsfirma [f] des Garten- und Landschaftsbau(e)s, Betrieb [m] des Garten- und Landschaftsbau(e)s, Garten-, Landschafts- und Sportplatzbaubetrieb [m], Garten- und Landschaftsbaufirma [f], Garten- und Landschaftsbauunternehmen [n], Landschaftsbaufirma [f], Landschaftsbauunternehmen [n].

3103 landscape contracting industry [n] *constr. prof.* (Separate business dealing with ▶landscape construction and maintenance work); *syn.* landscape industry [n]; *s* **industria [f] de construcción paisajística** (Cuerpo profesional independiente que se dedica comercialmente a los ▶trabajos de paisajismo); *syn.* industria [f] de construcción de parques y jardines; *f* **profession [f] des entrepreneurs paysagistes** (F., profession dont les activités sont définies dans le Code Rural au niveau de ses prestations de services dans le cadre des métiers auxiliaires de l'Agriculture et tombent également sous l'application de l'article 632 du Code du Commerce ; ▶travaux d'espaces verts) ; *syn.* entrepreneurs [mpl] de paysages ; *g* **Garten- und Landschaftsbau [m]** (*Abk.* GaLaBau; selbständiger Berufsstand der gewerblichen Wirtschaft, der Baumaßnahmen der Garten- und Landschaftsarchitektur und Landschaftsplanung zur Gestaltung, Sanierung, Sicherung und Wiederherstellung von vorwiegend vegetationsbestimmten Freianlagen im besiedelten Bereich sowie den Bau von Sportplätzen und ▶landschaftsgärtnerische Arbeiten in der freien Landschaft durchführt. Die Innenraumbegrünung ist teilweise im Produktionsbereich Zierpflanzenbau enthalten).

3104 landscape construction [n] *constr. prof.* (**1.** Term used for professional execution of a new green area such as a garden, park, or sports area and the development, upkeep and maintenance of open spaces in populated areas and in the open countryside; upkeep and maintenance of a scheme after implementation is not included within the English term. **2.** The construction of private gardens is often known as **garden design**, whereby the design component is also provided by the company, which carries out the construction; ▶landscape maintenance, ▶landscape practice, ▶softscape); *syn.* landscaping [n]; *s* **construcción [f] paisajística (1)** (**1.** Término que denomina la actividad profesional de crear nuevos espacios libres [parques y jardines, campos de deportes, cementerios, zonas de juegos infantiles, etc.]; en alemán el término incluye los trabajos de mantenimiento. **2.** La construcción de jardines privados se denomina generalmente **diseño [m] de jardines**; ▶construcción paisajística 2, ▶gestión paisajística de espacios verdes, ▶superficie blanda); *syn.* ingeniería [f] del paisaje (IAP 1983); *f* **création [f], aménagement [m] et entretien [m] des espaces verts** (**1.** L'activité professionnelle dans le domaine de l'aménagement, la mise en valeur et l'entretien pour les particuliers et les collectivités locales des espaces extérieurs en milieux urbains et dans les espaces naturels [parcs et jardins, aménagements urbains, aires de jeux, terrains de sport et parcours sportifs, golfs, aménagements paysagers] ; l'▶entretien des espaces verts n'est pas compris dans l'acceptation du terme anglais *landscape construction* ; *traduction littérale du terme allemand* création et entretien des jardins, paysages et terrains de sport ; ▶éléments vivants d'un espace libre, ▶travaux de paysagisme ; **2.** l'activité

d'aménagement de jardins pour les particuliers est appelée **création de jardin**) ; *g* **Garten-, Landschafts- und Sportplatzbau [m]** (**1.** Bezeichnung für die berufliche Tätigkeit der Neuanlage, des Ausbaus und der Pflege und Unterhaltung von Freiräumen im Siedlungsbereich und in der freien Landschaft; im Englischen wird bei dem Begriff *landscape construction* die ►Pflege und Unterhaltung von Grünflächen nicht mitgedacht. **2.** Die Anlage von Privatgärten wird auch **Gartengestaltung** genannt; ►Landschaftsbau, ►vegetationsbestimmte Freianlage).

3105 landscape construction and maintenance

work [n] *constr.* (Whole range of activities in landscaping and related maintenance work; ►landscape contracting industry); *s* **trabajos [mpl] de paisajismo** (Toda la gama de trabajos de ►industria de construcción paisajística y de mantenimiento de parques, jardines y zonas verdes); *syn.* obras [fpl] de paisajismo; *f* **travaux [mpl] d'espaces verts** (Travaux réalisés par la ►profession des entrepreneurs paysagistes comprenant les fosses des plantations, reprise, mise en place et réglage des terres végétales, formation grosso modo, fertilisants et produits phytosanitaires, façons culturales, consolidation de talus, travaux de plantation et d'engazonnement ; cf. C.C.T.G. fascicule n° 35 : travaux d'espaces verts, d'aires de sport et de loisirs) ; *syn.* travaux [mpl] de paysagisme ; *g* **landschaftsgärtnerische Arbeiten [fpl]** (Arbeiten des ►Garten- und Landschaftsbaues); *syn.* Landschaftsbauarbeiten [fpl].

landscape construction firm [n] *constr. hort.* ►landscape contracting firm.

3106 landscape construction techniques [npl]

constr. (Sum of activities and skills, as well as the use of equipment and machines, necessary for the accomplishment of landscape projects and their satisfactory maintenance according to ecological and economic aspects as well as established objectives; ►landscape construction, ►landscape practice); *s* **técnica [f] de construcción paisajística** (Conjunto de habilidades y actividades manuales y aplicaciones de equipo y maquinaria necesarias para realizar proyecto de construcción paisajística y para su mantenimiento, teniendo en cuenta aspectos y objetivos ecológicos y económicos; ►construcción paisajística 1, ►construcción paisajística 2); *f* **techniques [fpl] de travaux paysagers** (Ensemble des activités et techniques artisanales, emploi des matériels et machines nécessaires à la réalisation et l'entretien des projets d'espaces verts et de paysages dans le souci des exigences écologiques et économiques ; ►création, aménagement et entretien des/d'espaces verts, ►travaux de paysagisme) ; *g* **Technik [f] des Landschaftsbau(e)s** (Gesamtheit der handwerklichen Fertigkeiten und Tätigkeiten sowie der Einsatz von Geräten und Maschinen, die notwendig sind, damit sowohl landschaftsgärtnerische Projekte realisiert als auch erwünschte Pflegeergebnisse nach ökologischen und ökonomischen Gesichtspunkten und Zielsetzungen erzielt werden können; ►Garten-, Landschafts- und Sportplatzbau, ►Landschaftsbau).

3107 landscape contractor [n] *hort. constr.* (Individual

or firm devoted to landscape construction; ►landscape contracting firm); *s* **empresario/a [m/f] paisajista** (Titular de una ►empresa de construcción paisajística); *syn.* empresario [m] de construcción paisajística; *f* **entrepreneur [m] paysagiste** (Chef d'une ►entreprise paysagiste) ; *syn.* entrepreneur [m] de jardins ; *g* **Garten- und Landschaftsbauunternehmer/-in [m/f]** (Inhaber/-in eines ►Garten- und Landschaftsbaubetriebes); *syn.* Landschaftsbauunternehmer/-in [m/f], Unternehmer/-in [m/f] des Garten- und Landschaftsbau(e)s, *Jargon* GaLa-Bauer [m].

landscape contractor's firm [n] *constr. hort.* ►landscape contracting firm.

3108 landscape damage [n] *landsc.* (Impairment of a

landscape patch, as a small part of a landscape matrix, caused by a wide variety of man-made or natural disturbances. Impairments are assessed and recorded according to usage, and are to be seen as a collective term for all those disturbances which must be taken into consideration in landscape planning; ►cultural landscape); *s* **daño [m] al paisaje** (Deterioro directo, indirecto, de origen natural o antropógeno causado sobre un ►paisaje cultural o sobre uno o varios de los componentes estructurales del mismo. Los **dd. al p.** se registran [en D] y se valoran desde la perspectiva de los usos. El término **d. al p.** no define una categoría específica de daños-tipo sino que es un término genérico utilizado en la planificación del paisaje para todo tipo de distorsiones que han de ser consideradas en la planificación y gestión); *f* **altération [f] du paysage** (Nuisances anthropogènes directes et indirectes [*terme spécifique* pollution physique solide (du territoire)], DUV 1984, 249] causées par les déchets, carrières, grandes infrastructures, urbanisation, etc. ou dégradations naturelles [accidents climatiques] portant atteinte au caractère ou à l'intérêt des paysages naturels ou des ►paysages culturels) ; *syn.* pollution [f] paysagère ; *g* **Landschaftsschaden [m]** (**1.** Direkte, indirekte, anthropogene sowie naturbedingte Beeinträchtigungen des geordneten Gefüges einer ►Kulturlandschaft [in den Bereichen Boden, Wasser, Luft, Klima, Lebewelt, Siedlungsraum]. Diese Beeinträchtigungen werden nutzungsbezogen beurteilt und erfasst. Landschaftsschäden bilden keine einheitliche Schadenskategorie, sondern sind als Sammelbegriff für alle Beeinträchtigungen anzusehen, die von der Landschaftsplanung berücksichtigt werden müssen. **2.** Der von KIEMSTEDT [1971] vorgeschlagene Begriff für **Nutzungsschaden** deckt nur die vorwiegend durch ökonomische Wertmaßstäbe erfassbaren Beeinträchtigungen ab).

3109 landscape data bank [n] *adm. landsc.* (Computer-

ized collection of basic ecological information and associated data of other professional disciplines stored for use in environmental planning, design and conservation; ►lanscape information system); *syn.* landscape database [n]; *s* **banco [m] de datos paisajísticos** (Sistema de información computarizado de datos ecológicos y referentes al uso del suelo, registrados desde la perspectiva de la protección y gestión de la naturaleza y el paisaje; ►sistema de información paisajística); *f* **banque [f] de données paysagères** (Ensemble informatisé des informations écologiques relatives aux paysages recensées dans le domaine de la protection de la nature, de la sauvegarde et de la gestion des paysages ; ►système d'information paysager) ; *syn.* banque [f] de données des espaces naturels ; *g* **Landschaftsdatenbank [f]** (Datei, in der flächenhaft erhobene landschaftsökologische Grundlagendaten aus der Sicht von Naturschutz und Landschaftspflege und raumwirksame Daten anderer Fachdisziplinen gespeichert sind; ►Landschaftsinformationssystem).

landscape database [n] *adm. landsc.* ►landscape data bank.

3110 landscape design [n] *land'man. landsc.* (Creative,

aesthetic organization or layout of open space to attain an optimum functional balance of landscape factors and human needs, with emphasis on the visual aspects; ►landscape architecture, ►landscape management, ►landscape planing); *s* **diseño [m] paisajístico** (Organización estética y creativa de los espacios libres para obtener un equilibrio óptimo entre los factores del paisaje y las exigencias humanas, pero que prioriza los aspectos visuales; ►gestión y protección del paisaje, ►paisajismo, ►planificación del paisaje); *f* **création [f] de paysages** (Terme ambivalent désignant toute forme d'organisation créative d'espaces paysagers relevant de la technique et de l'art avec pour objectif la recherche d'un équilibre entre le respect des composantes naturelles et la satisfaction des besoins humains ;

les professionnels établissant les études, les plans et les dossiers liés aux ouvrages paysagers en espaces verts, parcs et jardins, ou terrains de sport et qui en surveille la conformité sont dénommés **concepteur du/en paysage** et **concepteur paysagiste** ; **D.**, terme utilisé souvent pour désigner la création d'ensembles paysagers structurés selon la loi de la bonne forme et de l'aspect visuel ; ►architecture des paysages, ►gestion des milieux naturels, ►planification des paysages) ; *syn.* aménagement [m] de(s) paysages ; *g* **Landschaftsgestaltung [f]** (Sich überschneidender Begriff mit ►Landschaftspflege; umfasst neben der erforderlichen Ausgewogenheit von Nutzungsansprüchen in einem Landschaftsraum hauptsächlich den visuellen Aspekt; ►Landschaftsarchitektur, ►Landschaftsplanung).

3111 landscape design [n] **of housing areas** *plan.* (Planning and execution of open spaces within the sites of specific housing projects according to functional, aesthetic and ecological aspects); *s* **diseño [m] de espacios libres** (Planificación y construcción de zonas verdes en zonas residenciales siguiendo criterios funcionales, estéticos y ecológicos); *f* **aménagement [m] des espaces extérieurs** (Conception et exécution d'espaces verts d'accompagnement en zone résidentielle en tenant compte de critères techniques, écologiques et artistiques) ; *g* **Gestaltung [f] der Außenanlagen** (Planung und Herstellung von Freianlagen im Wohngebiet nach funktionalen, ästhetischen und ökologischen Gesichtspunkten).

3112 landscape development [n] *plan.* (Comprehensive measures in landscape planning to ensure the sustainable use of natural resources, as well as the preservation and improvement of diversity, character and beauty of nature and the use of landscape for recreation; ►landscape architecture, ►landscape design, ►landscape management); *s* **desarrollo [m] del paisaje** (Planificación paisajística con el fin de posibilitar el aprovechamiento sostenible de los recursos naturales y de preservar y aumentar la diversidad, particularidad y belleza de la naturaleza y los paisajes para fines recreativos; ►diseño paisajístico, ►gestión y protección del paisaje, ►paisajismo); *f* **maîtrise [f] de l'évolution d'un paysage** (Orientations, mesures, programmes et planification en faveur d'une utilisation durable des ressources naturelles et de la sauvegarde et l'amélioration des paysages en raison de leur qualités pour l'exercice des activités de loisirs et de détente ; cf. C. urb., L. 123-1 ; ►architecture des paysages, ►création de paysages, ►gestion des milieux naturels) ; *g* **Landschaftsentwicklung [f]** (**1.** *Im allgemeinen Sprachgebrauch* natürliche oder anthropogene Veränderungen einer Landschaft. **2.** *In der Landschaftsplanung* Planung und Maßnahmen des Naturschutzes und der ►Landschaftspflege, die der Gestaltung von Natur und Landschaft und der nachhaltigen Nutzungsfähigkeit der Naturgüter sowie der Erhaltung und Verbesserung der Vielfalt, Eigenart und Schönheit von Natur und Landschaft für die Erholung dienen; ►Landschaftsarchitektur, ►Landschaftsgestaltung).

3113 landscape diversity [n] *land'man. landsc.* (Extent to which a landscape or part thereof contains different physical settings, e.g. forests, forest edges, hedges, rows of trees, groups of trees, copses, solitary trees, lakes and rivers, topography [flat or sloping], fields or grassland); *syn.* landscape heterogeneity [n] (LE 1986, 463); *s* **heterogeneidad [f] del paisaje** (Grado de presencia de elementos estructurales naturales en un paisaje o una sección del mismo, como p. ej. bosque, lindero de bosque, setos, alamedas, bosquetes, ejemplares o ríos y lagos, relieve [activo], campos de cultivo y prados); *f* **diversité [f] des paysages** (Présence d'éléments structurants naturels pris dans leur valeur individuelle ou leur combinaison caractérisant tout ou parties de paysages perçus tels que les lisières de forêt, haies, alignements d'arbres, bosquets, groupes d'arbres, arbres isolés, cours d'eau et eaux calmes, topographie, champs ou prairies composant le support de l'impression paysagère et dégageant une impression d'organisation et une impression paysagère) ; *syn.* diversité [f] paysagère ; *g* **landschaftliche Vielfalt [f]** (Grad der Ausstattung einer Landschaft oder eines Landschaftsteiles mit natürlichen Strukturelementen im Sinne von z. B. Wald, Waldrand, Hecke, Baumreihe, Baumgruppe, Feldgehölz, Einzelbaum oder Gewässer [fließend oder stehend], Relief [eben o. geneigt, Rinnenlage], Acker oder Grünland).

landscaped open space [n] *landsc. urb.* ►green space.

3114 landscape ecologist [n] *prof.* (Scientist who studies ecological systems of a landscape); *s* **ecólogo/a [m/f] del paisaje** (Científico/a que estudia las relaciones ecológicas del paisaje); *f* **écologue [m/f] du paysage** (Terme utilisé dans les pays anglo-saxons pour désigner les scientifiques dont la formation et les activités professionnelles qui touchent à l'écologie, la protection de la nature, l'aménagement du territoire, la biogéographie, la gestion de l'environnement et de la biodiversité, l'architecture du paysage et sont regroupées dans le concept de l'►écologie du paysage) ; *g* **Landschaftsökologe [m]/Landschaftsökologin [f]** (**1.** Wissenschaftler, der die ökologischen Systemzusammenhänge in der Landschaft erforscht. **2.** Landschaftsplaner, der die wissenschaftlichen Erkenntnisse der Landschaftsökologie praxisorientiert im Rahmen von raumbedeutsamen Planungen anwendet); *syn.* Geoökologe [m]/Geoökologin [f].

3115 landscape ecology [n] *ecol.* (Science of ►landscape ecosystems); *s* **ecología [f] del paisaje** (Ciencia de la estructura, el funcionamiento y el desarrollo del paisaje; ►régimen ecológico del paisaje); *f* **écologie [f] du paysage** (Concept apparu vers 1940 et qui dans les années 1970-1980 s'est établi comme discipline scientifique étudiant les effets de l'hétérogénéité et de l'évolution dans le temps et l'espace des composantes biologiques, physiques et sociales de l'écosystème paysage [écopaysage ou unités paysagères] sur les fonctionnements écologiques et les facteurs qui les induits; ►écosystème paysager) ; *g* **Landschaftsökologie [f]** (Wissenschaft von der Struktur, Funktion und Entwicklung der Landschaft; sie untersucht das räumliche, zeitliche und funktionale Wirkungsgefüge zwischen den dort lebenden Pflanzen und Tieren und ihrer abiotischen Umwelt; ►Landschaftshaushalt); *syn.* Geoökologie [f], Landschaftsforschung [f].

3116 landscape ecosystem [n] *ecol.* (Interrelationships between inert and living ►landscape factors, or between artificial elements and their environment, and the dynamic interaction of energy, matter, and information processes; ►natural systems); *syn.* landscape equilibrium [n] (±); *s* **régimen [m] ecológico del paisaje** (Interrelación entre los ►factores del paisaje bióticos y abióticos y los elementos antropógenos que tiene lugar a nivel energético, material e informacional; ►régimen ecológico de la naturaleza); *f* **écosystème [m] paysager [CDN, CH, ± F]** (Approche physiologique dans la conception allemande de la géographie du paysage ; terme désignant le champ d'action des éléments biotiques, abiotiques [complexe naturel] et anthropiques du paysage ainsi que les interactions et la dynamique des substances, des masses, des énergies et des informations avec leur environnement ; ►facteur du paysage, ►système naturel) ; *g* **Landschaftshaushalt [m]** (Wirkungsgefüge aus biotischen und abiotischen natürlichen ►Landschaftsfaktoren sowie vom Menschen geschaffenen Bestandteilen, die untereinander und mit ihrer Umwelt in energetischen, stofflichen und informatorischen Wechselwirkungen stehen; der Begriff ►Naturhaushalt umfasst im Vergleich dazu nur die natürlichen Faktoren).

landscape element [n] *conserv. land'man. landsc.* ►landscape feature.

landscape envelope plan [n] [UK] *leg. landsc. plan.*
▶development mitigation plan [US].

landscape envelope planning [n] [UK] *leg. landsc. plan.*
▶development mitigation planning.

landscape equilibrium [n] *ecol.* ▶landscape ecosystem.

3117 landscape evaluation [n] *plan. landsc. recr.* (Appraisal of a landscape or a particular area to determine its suitability or potential for a proposed land use. The analytical process aims to demonstrate the ecological and visual impact of proposed land development upon the functioning of natural systems, with due regard for environmental planning, design and conservation); *s* **evaluación** [f] **del paisaje** (Según criterios de planificación paisajística, valoración de la idoneidad de un paisaje para los usos previstos en él, teniendo en cuenta en especial los posibles impactos ecológicos y escénicos sobre el régimen de la naturaleza y ante las exigencias para la recreación); *syn.* valoración [f] del paisaje; *f* **évaluation** [f] **paysagère** (Instrument normatif d'analyse des paysages dans le but d'apprécier l'aptitude d'un paysage à divers usages et d'évaluer les impacts écologiques et visuels [sensibilité] sur l'environnement naturel du point de vue de la protection des paysages) ; *syn.* évaluation [f] du paysage ; *g* **Landschaftsbewertung** [f] (Planerisches Instrumentarium zur Eignungsbewertung einer Landschaft oder deren Teile für geplante Nutzungsansprüche mit dem Ziel, die entstehenden ökologischen und visuellen Beeinträchtigungen [Empfindlichkeit] für die Funktionsfähigkeit des Naturhaushalts und für die Belange von Freizeit und Erholung aus der Sicht von Naturschutz und Landschaftspflege darzulegen).

3118 landscape factor [n] *geo. landsc.* (Influential element within the dynamics and relationships of the landscape, e.g. topography, soils and rocks, water, climate, vegetation, wildlife; ▶landscape feature); *s* **factor** [m] **del paisaje** (Aspecto funcional en las interrelaciones de un paisaje, como topografía, suelo, agua, clima, vegetación, fauna; ▶componente del paisaje); *syn.* factor [m] paisajístico; *f* **facteur** [m] **paysager** (Élément fonctionnel dans le système paysage tel que le relief, le sol et les roches, l'eau, le climat, la végétation, la faune sauvage ; ▶élément de paysages) ; *g* **Landschaftsfaktor** [m] (Funktional wirkende Kraft im Wirkungsgefüge Landschaft wie Relief, Boden und Gestein, Wasser, Klima, Vegetation, frei lebende Tierwelt; ▶Landschaftsbestandteil); *syn.* Geofaktor [m].

3119 landscape feature [n] *conserv. land'man. landsc.* (Generic term for an individual unit contributing to the significance of a landscape, which is determined by the visual aspect or natural environment, whether it is of natural or artificial origin. This may be a single element or group of elements in a landscape, e.g. a solitary tree, group of trees, small woodland, hedge, cliff, etc., also colonies of birds, typical for the landscape; due to their age, natural beauty or significance for the visual quality of the landscape; ▶landscape factor, ▶landscape inventory, ▶landscape sector, ▶landscape structure, ▶protected landscape feature); landscape component [n], landscape element [n] (LE 1986, 12), visual element [n] (±); *s* **componente** [m] **del paisaje** (Elemento individual del paisaje de especial importancia para éste, como setos interparcelarios, bosquetes, ejemplares de árboles, rocallas o colonias de aves características; ▶componente protegido del paisaje, ▶estructura del paisaje, ▶factor del paisaje, ▶inventario paisajístico, ▶sector del paisaje); *syn.* elemento [m] del paisaje, elemento [m] paisajístico, elemento [m] natural particularizado; *f* **élément** [m] **de paysages** (Éléments remarquables [▶structure paysagère] des espaces naturels ou paysagers, repère topographique tels que les arbres isolés, un alignement d'arbres, les haies, les haies bocagères, les réseaux de haies,

les rideaux d'arbres, une trame végétale, les murets, les terrasses, et dont la protection est assurée par des prescriptions spéciales ou pouvant être classés comme site [naturel] protégé, ou comme espaces boisés ; ▶élément de paysages identifié, ▶facteur paysager, ▶partie de paysage, ▶relevé des éléments du paysage ; *terme spécifique* repère topographique) ; *syn.* composant [m] d'un milieu naturel, élément [m] d'un milieu naturel ; *g* **Landschaftsbestandteil** [m] (*Baustein der Geosphäre und des Landschaftsraumes* einzelnes, relevantes und räumlich klar definierbares Element oder Elementgruppe einer Landschaft, z. B. Einzelbaum, Baumgruppe, kleines Wäldchen, Feldhecke, Felswand, auch für die Landschaft charakteristische Vogelkolonien; die Zusammensetzung und Anordnung der **L.e** bestimmt die ▶Landschaftsstruktur; **L.e** sind wegen des Alters, der natürlichen Schönheit oder wegen der Bedeutung für das Landschaftsbild oft gesetzlich geschützt; ▶Aufnahme der Landschaftsbestandteile, ▶geschützter Landschaftsbestandteil, ▶Landschaftsfaktor, ▶Landschaftsteil); *syn.* Bestandteil [m] von Natur und Landschaft (cf. § 19 [2] BNatSchG), Landschaftselement [n], Landschaftserscheinung [f].

landscape feature [n] **for recreation** *plan. recr.* ▶landscape component for recreation.

landscape features [npl] *land'man. landsc.* ▶small landscape features.

landscape features [npl]**, natural** *conserv. landsc. recr.* ▶natural site characteristics.

3120 landscape garden [n] *gard'hist.* (**1.** In contrast to formal Renaissance and Baroque gardens, a large-scale, romantic, landscaped area or public open space designed along informal, asymmetrical lines, and characterized by serpentine lakes, groups of trees and expanses of lawns or meadows with sweeping vistas; especially applied to naturalistic eigtheenth- and nineteenth-century English gardens [the English landscape style] and their successors on the European continent and in the United States. The first landscape gardener who used the naturalistic line in large compositions was CHARLES BRIDGEMAN [† 1738]. He popularized the "Ha-ha", a dry ditch with a raised retaining wall, making it possible to expand estate vistas without interruption and yet contain roving livestock; the popular landscape gardener WILLIAM KENT [1685-1748] succeeded BRIDGEMAN and derived new design ideas from his studies of Italian painters and Italian countryside during his visit in Italy; then followed LANCELOT [CAPABILITY] BROWN [1716-1783] and HUMPHRY REPTON [1752-1818]. REPTON was deeply influenced by CAPABILITY BROWN and established the planting of exotics, which became fashionable at that time; ▶garden, ▶park 1, ▶park-like landscape, ▶public park); *syn.* naturalistic park [n] [also US], landscape park [n]; **2.** pastoral park [n] [UK] (TGG 1984, 182; romantic park with varied landscape forms, quaint structures, and large wildflower meadows, which was usually designed in accordance with the gardens of the eighteenth-century English Landscape School; ▶park-like landscape); *s* **jardín** [m] **paisajístico** (Al contrario que los jardines formales del renacimiento o el barroco, el **j. p.** imita el paisaje de la campiña inglesa utilizando formas irregulares y caracterizándose por su riqueza de arbolado y amplias superficies de césped o hierba, grupos de árboles, lagos, ríos con meandros y caminos sinuosos. Para evitar la interrupción visual entre el jardín y paisaje adyacente en sus principios se hizo uso de las vallas o cercas rehundidas [Ha-Ha]. Su origen se encuentra en el movimiento romántico inglés de principios del siglo XVIII que se inspiró sobre todo en los paisajes italianos y la pintura paisajista; ▶jardín, ▶paisaje campestre, ▶parque, ▶parque público); *syn.* parque [m] inglés, jardín [m] inglés, jardín [m] paisaje, jardín [m] romántico; *f 1* **parc** [m] **paysager** (Parc conçu et réalisé pour rappeler un

paysage naturel. Il est pensé comme une succession de scènes naturelles, pittoresques utilisant les éléments visuels tels que pièces d'eau, cascades, torrents, rochers, mouvement de sol, végétation. Le principe général de conservation de l'aspect de la nature entraîne l'utilisation de tracés courbes, souples, les plantations devant accentuer les perspectives, le réseau d'allées devant être aussi effacé que possible ; à l'intérieur du parc on retrouve le mobilier urbain : kiosques à musique, bancs, abris ; parcs paysagers de renom en France : le parc Borely à Marseille, le parc d'Ermenonville, le parc des Buttes-Chaumont à Paris ; ▶paysage champêtre, ▶parc, ▶parc public) ; *f 2* **jardin [m] paysager** (Désigne un jardin privatif de superficie réduite de conception paysagère, caractérisée par l'usage de massifs arborescents et de vivaces à port naturel, de grands espaces engazonnés, d'allées sinueuses, l'emploi d'éléments naturels comme les petits plans d'eau, les enrochements ou rocailles et parfois l'utilisation d'éléments d'architecture en provenance de pays étrangers ; ce terme est de temps à autre employé comme synonyme de ▶parc paysager, les « jardins publics » des grandes villes [p. ex. les jardins du Trocadéro à Paris] étant en général des ▶parcs publics ; ▶jardin) ; *syn.* jardin [m] anglais, jardin [m] paysage ; *g* **Landschaftspark [m]** (**1.** Im Gegensatz zum formalen Garten der Renaissance und des Barocks eine Gartenschöpfung oder Gestaltung von öffentlichen Freiräumen mit freien asymmetrischen Formen, die durch geschwungene Wasserflächen, prägnant und kulissenartig angeordnete Baumbestände, weite Rasen- oder Wiesenflächen und mehrere aufeinander folgende Blickachsen geprägt ist. Das zentrale Thema der **L.idee** war das Sichhinwenden zur Natur. Diese neue Bewegung wurde im Zeitalter der Aufklärung, der Gesellschaftskritik am Absolutismus, von Dichtern und Philosophen ins Leben gerufen, quasi als Beginn der Demokratisierung der Gesellschaft. Der erste Landschaftsgärtner, der in großen Kompositionen die freie Linie wählte, war CHARLES BRIDGEMAN [† 1738]. Er führte den „Aha", einen trockenen Grenzgraben mit einer Stützmauer in den **L.** ein, der es ermöglichte, die weite Landschaft in den Garten mit einzubeziehen und die weidenden Tiere von dem Grundstück fernzuhalten. Auf BRIDGEMAN folgte der populäre Landschaftsgärtner WILLIAM KENT [1685-1748], der auf seiner Italienreise neue Gestaltungsideen beim Studium der italienischen Landschaftsmaler und Landschaften erhielt; dann folgten LANCELOT [CAPABILITY] BROWN [1716-1783] and HUMPHRY REPTON [1752-1818]. REPTON war stark von CAPABILITY BROWN beeinflusst und führte die Verwendung von exotischen Arten, die damals in Mode kamen, ein. In Deutschland waren die herausragendsten Landschaftsparkgestalter nach englischem Vorbild JOHANN FRIEDRICH EYSERBECK [1734-1818] mit dem Park in Wörlitz bei Dessau [seit August 2001 in das Welt[kultur]erbe aufgenommen], HERMANN LUDWIG HEINRICH FÜRST VON PÜCKLER-MUSKAU [1785-1871] mit den Parks in Bad Muskau, Branitz bei Cottbus, Babelsberg etc. und PETER JOSEPH LENNÉ [1789-1866], Generaldirektor der königlichen Gärten in Preußen mit der Pfaueninsel bei Berlin, dem Tiergarten in Berlin, den ▶Volksparks in Magdeburg, Frankfurt a. d. Oder und Dresden sowie FRIEDRICH LUDWIG VON SCKELL [1750-1823] mit dem Park Schönbusch in Hessen, dem Englischen Garten in München [Zitat anlässlich der Eröffnung: „Öffnung des Parks für alle Stände"], dem Landschaftsteil im Park vom Schloss Nymphenburg etc. Den englischen Gartenstil findet man überall auf der Welt im öffentlichen Grün als unterschiedlich geglückten Versuch, ein landschaftliches Schönheitsideal international in organischer Formensprache zu kultivieren: der Rasen, ein immergrünes Nutz- und Schmuckelement als Symbol einer bäuerlichen Weidelandschaft, Sträucher, Einzelbäume und Baumgruppen als Visualisierungen des „Waldes", Teiche und Wasserläufe als Erinnerung an Seen und Flüsse. Die so geschaffene „freie Land-

schaft" symbolisiert politisch eine freie Gesellschaft mit freien selbstbestimmenden Menschen; cf. MIL 2002, 17. **2.** Seit Ende des 20. Jhs. wird der Begriff Landschaftspark auch für umgestaltete Industrielandschaften, die der Freizeit und Erholung dienen, verwendet; ▶Garten, ▶Park, ▶Parklandschaft 2); *syn.* Englischer Garten [m], Landschaftsgarten [m].

3121 landscape gardener [n] *hort. prof.* (In **A.**, **CH**, **D.** and **U.K.**, a professional title for qualified gardeners, who are responsible for the execution and maintenance of gardens and open spaces. **In U.S.**, there is normally no formal apprenticeship for such a professional position); *s* **jardinero [m] paisajista** (≠) (En **A.**, **CH**, **D.** y **GB**, título profesional de jardineros calificados responsables de construir y mantener jardines y zonas verdes; en Es. no existe tal título); *f* **jardinier [m] paysagiste/jardinière [f] paysagiste** (Titre professionnel désignant tout personnel qui intervient dans les travaux de création, d'aménagement et d'entretien des espaces verts, [parcs, jardins], de sols sportifs [terrains de sport football, rugby, golf, tennis, etc.] ou d'espaces naturels à l'aide d'outils manuels ou motorisés et d'engins motorisés. Cet emploi/métier est accessible à partir de formations de niveau V [CAPA et BEPA], de niveau IV [BTA], de niveau III [BTSA] dans le domaine de l'horticulture avec des options telles que « jardins et espaces verts » ou « pépinières et entreprises de jardins », etc. ; l'accès est également possible pour des salariés ayant une expérience dans d'autres secteurs de l'agriculture ; les jardiniers de golf sont formés dans des centres de formation agréés par la Fédération française de golf ; dans le secteur public, l'emploi/métier est accessible par concours sous certaines conditions de recrutement et de niveau de formation ; il travaille principalement dans les entreprises du paysage ou les services espaces verts des collectivités territoriales ; cf. www.anpe.fr. code ROME) ; *syn.* ouvrier [m] du paysage ; *g* **Landschaftsgärtner [m]** (*Gärtner der Fachrichtung Garten- und Landschaftsbau* in **A.**, **CH** und **D.** staatlich anerkannter Ausbildungsberuf und Berufsbezeichnung für gelernte Gärtner/-innen, die die Anlage sowie Pflege und Unterhaltung von Freiräumen im Siedlungsbereich und in der freien Landschaft durchführen).

3122 landscape geographical research [n] *geo.* (Scientific discipline of regional geography concerned with nature and/or the cultural landscape as well as its development); *s* **investigación [f] geográfica del paisaje** (Rama de la geografía que estudia el paisaje natural o cultural y su desarrollo); *f* **analyse [f] géographique dans la recherche paysagère** (Techniques et pratiques dans la science des paysages d'inspiration géographique ayant pour objet d'analyse le paysage ; *syn.* géophysique [f] du paysage ; *g* **geografische Landschaftsforschung [f]** (Die **g. L.** wird als ‚Landschaftsgeografie' bezeichnet und ist die Sammelbezeichnung für jene regionalgeografisch arbeitenden Disziplinen, deren Gegenstand die Natur- und/oder Kulturlandschaft sowie deren Entwicklung ist; cf. WAG 1984); *syn.* Landschaftsgeografie [f], *o. V.* Landschaftsgeographie [f], *o. V.* geographische Landschaftsforschung [f].

landscape grass seeds [n] [UK] *constr.* ▶low-maintenance grass type.

landscape heterogeneity [n] *land'man. landsc.* ▶landscape diversity.

3123 landscape history [n] *geo. hist. landsc.* (Cultural and natural development which has led to the present land forms and settlement patterns, dynamics of relationships and appearance of a landscape, with repercussions for the future); *s* **geografía [f] histórica del paisaje** (Desarrollo natural y cultural que ha llevado a la estructura y al aspecto escénico actuales de un paisaje con sus interrelaciones dinámicas y las repercusiones para el futuro); *f* **histoire [f] des paysages** (Évolution naturelle et

culturelle des paysages jusqu'à leurs formes, structures et aspects actuels et futurs) ; *g* **Landschaftsgeschichte [f]** (Geschichtliche Entwicklung, die zu dem gegenwärtigen Wirkungsgefüge, der derzeitigen Struktur und dem heutigen Erscheinungsbild einer Landschaft geführt hat und darüber hinaus in die Zukunft weiterführt; BUCH 1978).

landscape industry [n] *constr. prof.* ►landscape contracting industry.

3124 landscape information system [n] *adm. geo. landsc. plan.* (Computerized information system specifically designed for nature conservation and landscape management based on a ►landscape data bank, with the objective of attaining more and greater accuracy and dissemination of information for better efficiency in implementing nature conservation and landscape management; ►geographic information system [GIS]); *syn.* land information system [n], natural resources information system [n], land management information system [n], resource data management system [n]; *s* **sistema [m] de información paisajística** (Sistema de datos e información computarizado diseñado en especial para la gestión y conservación de la naturaleza y el paisaje que se basa en el ►banco de datos paisajísticos, y que tiene como fin registrar más exactamente más información para implementar la gestión y conservación con eficacia; es una aplicación de los ►sistemas de información geográfica); *f* **système [m] d'information paysager** (Système informatique comportant une ►banque de données paysagères détaillée, rassemblant les informations cartographiées sur les espaces protégés ; ►système d'information géographique [SIG]) ; *g* **Landschaftsinformationssystem [n]** (*Abk.* LINFOS, LANIS; EDV-gestützte Daten- und Informationsbasis vor allem für den Bereich von Naturschutz und Landschaftspflege auf der Grundlage der ►Łandschaftsdatenbank mit dem Ziel, durch mehr und genauere Daten eine höhere Effizienz in der Durchsetzung der Belange von Naturschutz und Landschaftspflege zu erhalten; ►Geografisches Informationssystem [GIS]).

3125 landscape inventory [n] *landsc. plan.* (Record of landscape features within a study area); *s* **inventario [m] paisajístico** (Registro de los elementos del paisaje en un área a estudiar/planificar); *syn.* registro [m] de componentes del paisaje; *f* **relevé [m] des éléments du paysage** (Inventaire sur l'aire d'étude ou de planification des composantes paysagères, des repères et impacts visuels d'un paysage à partir de relevés sur le terrain ou de photos aériennes en vue d'une analyse et d'un diagnostic) ; *syn.* cartographie [f] des éléments paysagers, inventaire [m] des éléments du paysage ; *g* **Aufnahme [f] der Landschaftsbestandteile** (Erfassung einzelner Elemente oder Elementgruppen einer Landschaft in einem Planungs-/Untersuchungsgebiet).

3126 landscape maintenance [n] *adm. hort. landsc. urb.* (Sum of all public or private measures to control and promote plant growth and to keep areas of vegetation, as well as technical and built facilities, tidy. Measures are aimed to ensure the continuous development and functioning of green space and associated facilities, and include such aspects as public safety, replacement or additions to planting, new seeding or replanting. The main goal is the sustainable functioning of a green area to the highest standard; ►administration and maintenance of public green spaces, ►establishment maintenace, ►maintenance management, ►maintenace work); *s* **gestión [f] paisajística de espacios verdes** (Todas las medidas necesarias para promover el crecimiento de la vegetación y su estado adecuado, para limpiar las áreas verdes y las instalaciones técnicas y constructivas situadas en espacios libres, así como aquéllas necesarias para desarrollar y mantener en funcionamiento los espacios libres y sus instalaciones o para garantizar la seguridad. Según el «Pliego general tipo de condiciones facultativas para el servicio de parques y jardines», este programa incluye las labores de conservación [riegos, siega de céspedes, resiembra de céspedes, recorte y poda con tijeras, abonados, recebados con arena en caminos y paseos, aireación, entrecavado, rastrillado, limpiezas, tratamientos fitosanitarios, conservación del trazado, setos y perfilado de las praderas], labores de reposición y labores de poda; cf. PGT 1987; ►mantenimiento de zonas verdes, ►mantenimiento inicial, ►mantenimiento y rehabilitación de zonas verdes, ►trabajos de mantenimiento); *f* **entretien [m] des espaces verts** (Tous travaux de maintenance d'espaces verts privés, collectifs ou publics, de développement des végétaux, de nettoyage des espaces végétalisés, des équipements techniques et installations diverses, ainsi que les opérations ayant pour objet le développement et la préservation durable des espaces libres et de leurs équipements y compris semis de regarnissage, remplacement des végétaux et livraison des fournitures afférentes ; ►entretien courant, ►gestion des espaces verts, ►maintenance 1, ►travaux d'entretien pendant l'année de garantie) ; *g* **Pflege [f, o. Pl.] und Unterhaltung von Grünflächen** (Summe aller öffentlichen oder privaten Maßnahmen zur Lenkung und Förderung des Pflanzenwachstums und zur Säuberung von Vegetationsbeständen, technischen und baulichen Einrichtungen sowie Maßnahmen, die der Entwicklung und Erhaltung der Funktionsfähigkeit der Freianlagen und deren Einrichtungen sowie der Verkehrssicherheit dienen, incl. aller Ersatzbeschaffungen, Ergänzungen oder Nachsaaten und Nachpflanzungen sowie habitusgerechte Gehölzschnittarbeiten. Ziel ist es, nachhaltig die Funktion der Grünanlage und einen möglichst hohen Qualiltätsstandard vorzuhalten; ►Entwicklungspflege, ►Pflege, ►Pflege und Unterhaltung von öffentlichen Grünflächen, ►Unterhaltung); *syn.* Pflege [f. o. Pl.] und Unterhaltung von Grünanlagen.

3127 landscape management [n] *landsc.* (Comprehensive measures which take care of, preserve and enhance the long-term beneficial use of natural resources as well as the heterogeneity, character, and beauty of the environment; ►development mitigation plan, ►European Landscape Convention, ►general urban green space planning, ►landscape architecture, ►landscape design, ►nature conservation); *s* **gestión [f] y protección [f] del paisaje** (Conjunto de medidas que sirven para asegurar el aprovechamiento sostenido de los recursos naturales y mantener, conservar y mejorar la diversidad, particularidad y belleza de la naturaleza y el paisaje para la recreación; ►conservación de la naturaleza, ►Convenio Europeo del Paisaje, ►diseño paisajístico, ►paisajismo, ►plan paisajístico complementario, ►planificación de zonas verdes y espacios libres urbanos); *f 1* **gestion [f] des milieux naturels** (Terme générique qualifiant l'ensemble des mesures en faveur d'une utilisation durable des espaces, ressources et milieux naturels [sol, eau, air, climat, faune et flore, qualité des paysages] assurant et développant le bien-être et la détente des populations actuelles et futures ; ►architecture des paysages, ►Convention européenne du paysage, ►création de paysages, ►plan de protection et de mise en valeur des paysages, ►planification urbaine des espaces verts, ►protection de la nature, ►schéma directeur paysager ; *termes spécifiques* **1. opération grand site** : mesures de restauration de la qualité paysagère de sites classés confrontés à un problème de fréquentation touristique, bénéficiant au développement local ; **2. opération locale agri-environnementale** : mesures incitant le maintien de l'agriculture dans les zones sensibles, où les agriculteurs exercent une activité de protection des ressources naturelles, de l'espace naturel et des paysages ainsi qu'introduction de pratiques agricoles compatibles avec les exigences de la protection de l'environnement) ; *syn.* entretien [m] de l'espace naturel/des espaces naturels ; *f 2* **gestion [f] des paysages** (►*Convention européenne* *du paysage* actions visant, dans une perspective de développement durable, à entretenir le paysage afin de guider et d'harmoniser les transformations induites par les évolutions sociales, économiques et environnementales ; cf. Art. 1, définitions, Convention de Florence) ; *g* **Landschaftspflege [f, o. Pl.]** (1. Nach dem ►Europäischen Landschaftsübereinkommen die Gesamtheit der Maßnahmen, die eine nachhaltige Entwicklung und den Erhalt der Kulturlandschaften mit dem Ziel gewährleisten, dass Veränderungen, die durch gesellschaftliche, wirtschaftliche und ökologische Prozesse hervorgerufen werden, gesteuert und abgestimmt werden. 2. D., Gesamtheit der Maßnahmen, die dazu dienen, die nachhaltige Nutzungsfähigkeit der Naturgüter Boden, Wasser, Luft, Klima, Tier- und Pflanzenwelt sowie die Vielfalt, Eigenart und Schönheit von Natur und Landschaft in ihrer reichen Gliederung für das Wohlbefinden und für die Erholung des Menschen zu sichern und zu verbessern; ►Landschaftsgestaltung, ►Grünordnung, ►Landschaftsarchitektur, ►Landschaftspflegerischer Begleitplan, ►Naturschutz); *syn.* Landschaftsordnung [f], *z. T.* Heimatschutz [m] [CH].

landscape management [n]**, prerequisites of nature conservation and** *conserv. nat'res.* ►considerations of nature conservation and landscape management.

landscape management [n] [US]**, urban** *adm. hort. landsc. urb.* ►administration and maintenance of public green spaces.

landscape management [n]**, visual** *landsc. plan.* ►visual resource management.

3128 landscape management area [n] *landsc.* (Area not within a ►natural area preserve [US]/nature reserve [UK] or resource conservation area [US] in which important ecological functions of natural systems, being in contact with intensively used areas, should be preserved and enhanced by landscape management program[me]s); *s* **zona [f] de gestión paisajística** (Área no protegida que cumple funciones ecológicas importantes por estar cerca de áreas de uso intenso, por lo que han de preservarse y potenciarse por medio de medidas de gestión paisajística; ►espacio natural protegido 1); *f* **zone [f] de mesures environnementales (≠)** (D., zone située à l'extérieur du périmètre des ►sites classés ou inscrits, dans lesquelles les fonctions écologiques de secteurs naturels au contact avec des zones culturales intensives doivent être, du point de vue de la protection de la nature, sauvegardées et mise en valeur par des mesures de gestion paysagère) ; *g* **Landschaftspflegebereich [m]** (Bereich außerhalb von ►Naturschutzgebieten 2 und Landschaftsschutzgebieten, in dem aus der Sicht von Naturschutz und Landschaftspflege wichtige ökologische Funktionen von naturnahen Bereichen im Kontakt mit intensiv genutzten Flächen erhalten und durch landschaftspflegerische Maßnahmen gesichert werden sollen).

landscape modification [n]**, gradient of** *ecol.* ►degree of landscape modification.

landscape mosaic pattern [n] *landsc.* ►structuring of landscape.

3129 landscape [n] **of proliferated housing** *urb. plan. landsc.* (Disfigured landscape as a result of scattered housing developments in outlying areas; ►uncontrolled proliferation of settlements, ►urban sprawl); *s* **paisaje [m] desfigurado** (Resultado de la expansión desordenada de una población en zona rural; ►desfiguración del paisaje, ►urbanización caótica); *f* **paysage [m] mité** (Résultat de la multiplication des constructions éparses en milieu naturel ou rural et constituant un habitat diffus suite à la rurbanisation et à la constitution de l'habitat périurbain ; ►accroissement désordonné d'une agglomération, ►mitage du paysage) ; *syn.* paysage [m] menacé par l'habitat diffus ;

g **zersiedelte Landschaft [f]** (Ergebnis eines ungeordneten Ausbreitens von Siedlungsflächen im ländlichen Raum; ►ungeordnete Ausbreitung einer Stadt, ►Zersiedelung der Landschaft).

3130 landscape [n] **of reclaimed surface-mined land** *min. landsc. syn.* reclaimed surface-mined site landscape [n] [also US], recultivated mining landscape [n] [also UK]; *s* **paisaje [m] minero recultivado** (Paisaje resultante de medidas de restauración en antigua área de explotación minera a cielo abierto); *f* **paysage [m] réaménagé après l'exploitation minière** *syn.* paysage [m] réhabilité après l'exploitation minière ; *g* **Bergbaufolgelandschaft [f]** (Nach Beendigung des Tagebaus rekultivierte Landschaft).

landscape park [n] *gard'hist.* ►landscape garden.

3131 landscape patrimony [n] *conserv. land'man.* (All existing historical monuments and natural sites, natural and cultural landscapes which represent an aesthetic, historic, scientific or special cultural interest which originates of the collective memory of a country); *syn.* landscape resource heritage [n]; *s* **patrimonio [m] paisajístico** (Conjunto de monumentos y paisajes naturales o antropógenos con valor estético, histórico, científico o cultural especial que conforman la herencia común a proteger de un país); *f* **patrimoine [m] paysager** (Ensemble des monuments ou sites naturels, paysages naturels ou construits qui présentent un intérêt esthétique, historique ou culturel particulier, enraciné dans la mémoire collective d'une nation) ; *g* **Gesamtheit [f] des landschaftlichen Erbes** (Alle Baudenkmäler außerhalb der Städte, natürlichen Landschaften und Kulturlandschaften sowie Parks und Gärten von besonderem ästhetischen, wissenschaftlichen und kulturhistorischen Interesse, die das zu schützende Erbe eines Landes, einer Region ausmachen).

landscape pattern [n]**, ecological** *ecol.* ►ecological distribution of spatial patterns.

3132 landscape plan [n] *landsc.* (Planning documents, usually in the form of reports and drawings, for promoting the long-term preservation and development of natural and cultural landscape, to meet the demands of nature conservation and recreation); *syn.* landscape strategy plan [n] [also UK]; *s* **plan [m] de gestión del paisaje (≠)** (Instrumento de planificación, normalmente en forma de informes, mapas y planes, para promover el uso óptimo del suelo con el fin de conservar y desarrollar a largo plazo el paisaje natural y cultural, considerando en especial los intereses de la conservación de la naturaleza y de la recreación; ►planificación del paisaje, ►plan municipal de zonas verdes y espacios libres, ►plan paisajístico complementario); *f 1* **étude [f] paysagère** (F., étude réalisée dans le cadre d'une planification sectorielle ou d'un aménagement et ne s'appuyant pas sur une réglementation particulière) ; *f 2* **plan [m] de paysage** (F., démarche partenariale de protection des espaces remarquables et de prise en compte globale des paysages promue par la loi paysage de 1993 aboutissant à l'établissement d'un document de référence entre l'État et les collectivités locales transcrivant un projet de devenir du paysage sur un territoire correspondant à une unité paysagère pertinente dont la mise en œuvre aboutit à un programme d'action pouvant prendre la forme d'un ►contrat pour le paysage ; les plans de paysage correspondent aux objectifs de qualité paysagère de la ►Convention européenne du paysage) ; *f 3* **contrat [m] pour le paysage** (Contrat signé entre l'État et une ou plusieurs collectivités territoriales pour la mise en œuvre d'un programme d'actions concrètes, traduisant un projet pour le paysage pouvant être élaboré dans le cadre d'un plan de paysage, d'une charte paysagère, d'une étude paysagère ; pour les communes rurales on parle de contrat paysage rural) ; *syn.* contrat [m] paysage ; *f 4* **charte [f] pour l'environnement** (Document contractuel par lequel une collectivité territoriale [communes ou groupement de communes — *charte municipale pour l'environnement* ou département — *charte départementale pour l'environnement*] s'engage en s'associant avec l'Etat à l'amélioration de l'environnement et de la qualité de la vie sur son territoire ; ce document expose la politique d'environnement menée par la collectivité territoriale ; le 3 octobre 2008, le Conseil d'État a consacré l'opposabilité de la charte de l'environnement à l'égard des citoyens) ; *syn.* charte [f] environnementale ; *f 5* **directive [f] de protection et de mise en valeur des paysages** (Instrument réglementaire [documents d'urbanisme compatibles avec les orientations de la directive] ayant pour objectif la protection et la mise en valeur des éléments caractéristiques constituant la structure d'un territoire remarquable par son intérêt paysager, s'appliquant sur tout ou parties du territoire d'une ou plusieurs communes, établie par l'État en concertation avec les collectivités territoriales ; GPE 1998, fiche n° 6 ; ►planification des paysages, ►plan paysage, ►schéma directeur paysager, ►volet paysager d'un PLU) ; *syn.* directive [f] paysagère ; *f 6* **schéma [m] vert** (F., *terme spécifique pour le département de Seine St Denis* dispositif d'évaluation des besoins, des règles de protection et démarche de développement mis en place en 1997 pour pallier les déficits d'espaces verts liés à l'intense et brutale urbanisation du département) ; *f 7* **schéma [m] de paysage** (≠) (D., document fixant les orientations fondamentales de l'aménagement du territoire d'une commune en matière de protection de la nature conformément à la loi fédérale sur la protection de la nature ; le **s. de p.** définit l'organisation de l'espace en fonction des programmes établit par les administrations supérieures, les documents d'urbanisme devant être compatibles avec ses dispositions) ; *g* **Landschaftsplan [m]** (...pläne [pl]; **1.** Flächen- und grundstücksbezogene Planung gemäß §§ 5 und 6 Bundesnaturschutzgesetz und Landespflegegesetze/Naturschutzgesetze der Bundesländer als Beitrag zur vorbereitenden Bauleitplanung; in **L.plänen** werden zur Verwirklichung von definierten Umweltqualitätszielen die Ziele und Erfordernisse des Naturschutzes und der Landschaftspflege sowie die dazu erforderlichen Maßnahmen für ein Gemeindegebiet flächendeckend in einem Text- und Kartenteil dargestellt und sind somit ein Instrument der ►Landschaftsplanung; in einigen Bundesländern ein eigenständiger, rechtsverbindlicher Fachplan der Landespflege [Berlin, Bremen, HH, NRW], der durch Rechtsverordnung oder Satzung beschlossen wird. In den übrigen Ländern erhält die kommunale Landschaftsplanung erst durch eine inhaltliche Übernahme in die Bauleitplanung Rechtswirkung gegenüber allen Behörden. **2.** Beitrag zu Fachplanungen, z. B. zu Rekultivierungsvorhaben, zum Naturschutz, für Freizeit und Erholung; ►Grünordnungsplan, ►Landschaftspflegerischer Begleitplan).

3133 landscape planner [n] *landsc. plan. prof.* (**1.** Person who works upon ►general plan for urban open spaces [US]/green open space structure plan [UK], ►development mitigation plan [US]/landscape envelope plan [UK], ►environmental impact studies, ►landscape plans or other functional plan [US]/sectoral plan [UK] for nature conservation and landscape management as well as provisions for recreation; ►landscape architect); **2. land scientist** [n] **[UK]** (Landscape planner who puts into practice the scientific findings of landscape ecology within the context of spatial planning); *s* **planificador, -a [m/f] del paisaje** (Profesional que se dedica a poner en la práctica los conocimientos científicos de la ecología del paisaje en el contexto de la planificación territorial; **en D.** entre las tareas de esta profesión se encuentra la elaboración de ►estudios de impacto ambiental, ►planes de gestión del paisaje, ►planes municipales de zonas verdes y espacios libres, ►planes paisajísticos complementarios y de otros trabajos relacionados con la protección de la naturaleza y la recreación; ►arquitecto/a paisajista); *f* **paysagiste [m/f] d'amé-**

nagement (Professionnel du paysage impliqué dans l'élaboration des documents de planification de l'espace telle que les ►études d'impact, ►les études paysagères, les ►plans de paysage, les ►plans de protection et de mise en valeur des paysages, les ►plans paysage, les ►schémas directeurs paysagers ou autres études sectorielles relatives à la protection de la nature, à la mise en valeur des paysages et aux loisirs ; ►architecte paysagiste) ; *syn.* concepteur, trice [m/f] paysagiste ; *g* **Landschaftsplaner/-in [m/f]** (Jemand, der sich mit der Erarbeitung von ►Grünordnungsplänen, ►Landschaftspflegerischen Begleitplänen, ►Landschaftsplänen, ►Umweltverträglichkeitsstudien oder anderen Fachplänen für Naturschutz und Landschaftspflege sowie für die Erholungsvorsorge befasst; er versteht sich nicht nur als Anwalt der bedrohten Natur, sondern als Manager einer besseren Zukunft, indem er die erkannten Probleme partizipatorisch löst, Umweltqualitätsziele definiert und versucht, sie auf Grund gegebener Rahmenbedingungen mit Systemdenken, Sozialkompetenz, geschickter Moderationstechnik und politischem Gespür verständlich zu machen; ►Landschaftsarchitekt/-in).

3134 landscape planning [n] *landsc.* (Planning process for implementation of the objectives of environmental planning, design and conservation in built-up areas and the countryside; ►consultant's report on landscape planning, ►development mitigation plan, ►environmental impact assessment [EIA], ►functional landscape planning [US]/sectoral landscape planning [UK], ►general plan for urban open spaces, ►landscape plan, ►landscape structure plan, ►pertaining to landscape planning, ►regional landscape program [US]/regional landscape programme [UK]); *s* **planificación [f] del paisaje** (Disciplina de planificación con instrumentos legales [en A, D, CH] que tiene como fin la gestión, el diseño y la conservación del medio ambiente y las zonas verdes urbanas y rurales; ►estudio de los efectos ecológicos y paisajísticos, ►evaluación del impacto ambiental, ►plan de gestión del paisaje, ►plan de ordenación de los recursos naturales, ►planificación sectorial del paisaje, ►plan marco de gestión del paisaje y la naturaleza, ►plan municipal de zonas verdes y espacios libres, ►plan paisajístico complementario, ►programa de gestión del paisaje y la naturaleza, ►relacionado/a con la planificación/gestión del paisaje); *syn.* gestión [f] del paisaje; *f* **planification [f] des paysages** (**1. F.**, outils de préservation des paysages naturels institués par différentes législations telles que **1.** la loi du 2 mai 1930 protégeant les sites naturels dont la conservation présente, au point de vue artistique, historique, pittoresque d'un intérêt général. **2.** La loi N° 76-629 du 10 juillet 1976 qui fait entrer les préoccupations de protection de l'environnement dans les opérations d'aménagement au sens large et qui institue ►les procédures d'études d'impact. **3.** La circulaire du 24 juin 1991 qui défini le champ d'application des plans municipaux et départementaux d'environnement et la Lettre-circulaire du 5 janvier 1993 du Ministère de l'Environnement qui transforme ces Plans municipaux en Chartes d'écologie urbaine. **4.** La loi N°93-24 du 8 janvier 1993 sur la protection et mise en valeur du paysage [loi paysage] qui, avec une approche avant tout visuelle, privilégiant les paysages remarquables, institue pour de vastes sites des directives de protection et de mise en valeur des paysages, [charte des Parcs naturels régionaux, zones de protection du patrimoine architectural, urbain et paysager] ainsi qu'au niveau des collectivités territoriales le volet paysager du permis de construire et des documents d'urbanisme [►volet paysager d'un PLU, charte paysagère, plan paysage, contrats de paysage]. **D.**, Terme désignant l'ensemble des instruments à portée réglementaire spécifiques à la République fédérale d'Allemagne dans le domaine de la protection de la nature, définissant aux différents échelons administratifs la politique de protection, de mise

L

en valeur et de développement de la nature et des paysages en milieu urbain et rural ; ►charte pour l'environnement, ►contrat pour le paysage, ►directive de protection et de mise en valeur des paysages, ►étude paysagère, ►expertise paysagère, ►inventaire départemental du patrimoine naturel, ►plan de paysage, ►plan régional d'orientation pour la protection et la mise en valeur des paysages, ►plan vert départemental, ►plan vert régional, ►programme régional de protection et de mise en valeur des paysages, ►schéma de paysage, ►schéma directeur paysager) ; *syn.* planification [f] paysagère ; **2.** ►*Convention européenne du paysage* actions présentant un caractère prospectif particulièrement affirmé visant la mise en valeur, la restauration ou la création de paysages ; cf. Art. 1, définitions, Convention de Florence ; **g Landschaftsplanung [f]** (Raumbezogenes, vorsorgeorientiertes Planungsinstrument mit gesetzlicher Regelung zur Verwirklichung der Ziele und Grundsätze von Naturschutz und Landschaftspflege nach §§ 1 und 2 BNatSchG [Umweltqualitätsziele] im besiedelten und unbesiedelten Raum unter der Prämisse einer nachhaltigen Landnutzung. L. ist ein vorlaufender Beitrag zur räumlichen Gesamtplanung, also Landes-, Regional- und vorbereitenden Bauleitplanung sowie für andere Planungen und bildet ferner die Grundlage und Bewertungsmaßstäbe für die Umweltfolgeabschätzung bei ►Umweltverträglichkeitsprüfungen [UVPs] und Strategischen Umweltprüfungen [SUP-Verfahren]. Für die einzelnen ►Landschaftsfaktoren werden die Leistungs- und Funktionsfähigkeit ermittelt, bewertet und dargestellt. Aus den zu erarbeitenden Zielen für die zukünftige Landschaftsentwicklung eines bestimmten Gebietes werden die Erfordernisse des Naturschutzes und der Landschaftspflege abgeleitet und Maßnahmen des Naturschutzes und der Landschaftspflege formuliert, die von Planungsträgern und Naturschutzbehörden umzusetzen sind. Inhalte und Darstellungen hängen jeweils von der Planungsebene ab: Es gibt die **überörtliche Landschaftsplanung** mit den Instrumenten ►Landschaftsprogramm und ►Landschaftsrahmenplan und die **örtliche Landschaftsplanung** mit den Instrumenten ►Landschaftsplan und ►Grünordnungsplan. Bis in die 1980er-Jahre bestand in **D.** die Illusion, dass man die Landschaft in Richtung bestimmter Leitbilder durch eine querschnittsübergreifende **L.** prägen könnte [Agrarwirtschaft, Erholungsplanung, Flurbereinigung, Verkehrsplanung, Wasserwirtschaft]. Mittlerweile wird in D. durch Landespflegeverbände oder durch Träger der Agenda 21-Prozesse mehr Landschaft verändert als durch das Instrument der Landschaftspläne; **in D.** ist die **L.** im Bundesnaturschutzgesetz [BNatSchG] und in den jeweiligen Ländernaturschutzgesetzen rechtlich verankert. Sie wurde erstmals 1976 mit dem BNatSchG bundesweit eingeführt. In der **Schweiz** hat zumindest bis 2001 die **L.** nicht explizit in die für die Raumplanung relevanten Gesetze Eingang gefunden und ist somit keine institutionalisierte Planung. Dennoch wurden schon seit den 1960er-Jahren in den Kantonen Zürich und Aargau erste Landschaftspläne erarbeitet. In den 1970er-Jahren folgte Basel mit dem „Regionalplan Landschaft Basel" als wegweisende Richtlinie für die örtliche und regionale Planung. Durch das Bundesgesetz über die Raumplanung [RPG] vom 01.01.1980 wurde die **L.** zumindest auf kantonaler Ebene in der Richtplanung [= Bauleitplanung] in materieller Hinsicht als ökologische und Landschaftsnutzungsplanung aufgewertet; cf. LAS 2001. In **Österreich** gibt es im bundesstaatlichen System das gesetzlich verankerte Instrument des Landschaftsplanes nicht, auch keine Rahmengesetzgebung des Bundes für den Naturschutz. Für Raumordnung und Naturschutz liegt die Kompetenz ausschließlich bei den Bundesländern. Besonders schwierig für einen Vergleich und für die Verständigung über die Instrumente sind dabei die z. T. erheblich voneinander abweichenden Bezeichnungen — z. B. Landschaftskonzept, Freiraumkonzept — sowie die unterschiedlichen Bezugsebenen der Landschaftspläne.

Die **L.** hat im kommunalen Bereich einen geringen Stellenwert. Im ländlichen Raum bei Tourismus, Landwirtschaft, Erholungsnutzung, aber auch als Regulativ der Siedlungsentwicklung gewinnt sie immer mehr an Bedeutung; cf. GRI 2001. **Frankreich** verfügt seit dem Landschaftsgesetz von 1993 über ein eingeführtes Instrumentarium zur **L.** Seit dem wurde ein relativ differenziertes Instrumentarium geschaffen, das die französische Praxis in starkem Maße in Form vertraglicher Regelungen zwischen zentralstaatlichen Behörden und den Gemeinden aufgegriffen hat. Dieses Instrumentarium besteht aus folgenden Teilen: **1.** Richtlinien zum Landschaftsschutz und zur Landschaftspflege *[Directives de protection et de mise en valeur des paysages]* mit weiteren Details im *Décret n° 94-283* und *Circulaire n° 94-88.* **2.** Seit 1991 gemeindliche und departementale [einen Regierungsbezirk betreffende] Umweltpläne *[Plans départementaux/municipaux d'environnement]*, die 1993 durch **3.** Umweltchartas *[Chartes pour l'environnement]* ersetzt wurden. **4.** Seit 1993 Landschaftspläne *[Plans de paysage* auf der Grundlage des Landschaftsgesetzes]. **5.** Seit 1993 Landschaftsverträge für ländliche Gemeinden *[Contrats pour le paysage ou Contrats paysage rural]* auf der Grundlage des Landschaftsgesetzes]. **6.** Seit 1993 Landschaftschartas *[Chartes paysagères* auf der Grundlage des Landschaftsgesetzes]; ►Landschaftspflegerischer Begleitplan, ►landschaftsplanerisch, ►landschaftsplanerisches Gutachten, ►Landschaftsprogramm, ►Landschaftsrahmenplan).

landscape planning [n] [UK]**, sectoral** *landsc. leg. plan.* ►functional landscape planning [US].

3135 landscape planning measures [npl] *land'man. landsc.* **s medidas [fpl] de planificación del paisaje; *f* mesures [fpl] de planification des paysages** ; **g landschaftsplanerische Maßnahmen [fpl]**.

3136 landscape planning proposals [npl] *landsc. plan.* (Creative part of a landscape planning process, which includes suggestions for development according to the aims of nature conservation and landscape management; ►mandatory landscape development plan); **s capítulo [m] de desarrollo de un plan general del paisaje** (Parte creativa del proceso de planificación del paisaje que incluye proposiciones de medidas de protección y desarrollo paisajístico de acuerdo con los objetivos de conservación de la naturaleza y de gestión del paisaje; ►plan vinculante de desarrollo paisajístico); *f* **rapport [m] d'orientation du plan de paysage** (Pièce du plan de paysage exposant les objectifs et les orientations poursuivis en ce qui concerne la protection de la nature, la gestion et la mise en valeur des paysages ; ►carte des orientations et principes du plan de paysage) ; *g* **Entwicklungsteil [m] einer Landschaftsplanung** (*Kreativer Teil der Landschaftsplanung* Planungsvorschlag, der Entwicklungs-, Pflege- und Schutzmaßnahmen aus der Sicht von Naturschutz und Landschaftspflege darstellt; ►Entwicklungs- und Festsetzungskarte); *syn.* Planungsteil [m] einer Landschaftsplanung.

landscape planning requirements [npl] *leg. landsc.* ►statutory designation of landscape planning requirements.

3137 landscape planting [n] **in built-up areas** *landsc. urb.* (Vegetation provided for open spaces of settlement areas; ►external spaces); *syn.* housing landscape [n] (GP 2003, Dec. p. 19); **s zonas [fpl] verdes en áreas residenciales** (Espacios libres en zonas edificadas determinados predominantemente por vegetación; ►zona ajardinada); *f* **espaces [mpl] verts d'une zone urbanisée** (Espaces verts privés d'accompagnement d'immeubles dans un ensemble résidentiel, ►espaces extérieurs d'un grand ensemble) ; *syn.* espace [m] vert d'un complexe résidentiel ; *g* **Siedlungsgrün [n, o. Pl.]** (Vegetationsbestimmte ►Außenanlagen in bebauten Flächen).

3138 landscape policy [n] *pol.* (Statement [and course] of public action, primarily in state and local politics, which aims to implement the objectives of nature conservation and landscape management. *COMMENT* there is a difference between a *course of action* and a *statement*. Policy in US parlance is limited to oral or written *statements* or *declarations* not involving activities or program[me]s; ►regional landscape program [US]/regional landscape programme [UK]); *s* **política [f] de preservación del paisaje y la naturaleza** (Acción política pública a nivel estatal y local que tiene como fin apoyar la implementación de los objetivos de la gestión del paisaje y la conservación de la naturaleza; ►plan de ordenación de los recursos naturales, ►programa de gestión del paisaje y la naturaleza); *syn.* política [f] paisajística; *f* **politique [f] du paysage** (Action publique au niveau de l'État et des collectivités territoriales destinée à renforcer la prise en compte du paysage ; ►plan vert régional, ►programme régional de protection et de mise en valeur des paysages) ; *syn.* politique [f] de défense des paysages, politique [f] en faveur des paysages ; *g* **Landschaftspolitik [f,** selten Pl.**]** (Öffentliche Tätigkeit vorwiegend im Rahmen der Staats- und Kommunalpolitik, die auf die Durchsetzung der Ziele von Naturschutz und Landschaftspflege gerichtet ist; ►Landschaftsprogramm).

landscape potential [n] *geo.* ►natural landscape potential.

3139 landscape practice [n] *landsc. obs.* (Sphere of ►landscape planning aimed at the use, preservation and improvement of cultural and naturalistic landscapes in an ecological, human and economic sense, with living materials [plants, plant communities] and inert construction materials [e.g. pavement, soil, wood, stone, masonry, plastics, metal]. 'Landscaping' is usually concerned with living and inert materials but seldom exclusively with the latter; ►landscape construction, ►landscape contracting industry); *s* **construcción [f] paisajística (2)** (Campo de la ►planificación del paisaje que tiene como fin contribuir a preservar y mejorar el paisaje [cuasi] natural y antropógeno en sentido ecológico, humano y económico, utilizando materiales vivos [plantas] y materiales inertes [pavimentos, madera, piedra, mamposterías, plásticos, metales] para reformar artísticamente zonas. En la **c. p.** es usual la aplicación conjunta de materiales vivos e inertes o sólo de materiales vivos, muy excepcionalmente se utilizan sólo materiales inertes; ►construcción paisajística 1, ►industria de construcción paisajística); *f* **travaux [mpl] de paysagisme** (Ensemble des activités réalisées dans le domaine de la ►planification des paysages et des aménagements paysagers ayant pour objectifs la sauvegarde et l'amélioration des potentialités écologiques, humaines et économiques du milieu naturel grâce à l'emploi de matériaux vivants [plantes et communautés de végétaux] ou inertes [terre végétale, bois, roche, matières plastiques, métal, etc.] ; la caractéristique principale des travaux de paysagisme consiste dans l'utilisation de matériaux vivants ou leur combinaison avec des matériaux inertes mais rarement dans l'emploi exclusif de ces derniers ; ►création, aménagement et entretien des espaces verts, ►profession des entrepreneurs paysagistes) ; *syn.* travaux [mpl] paysagers ; *g* **Landschaftsbau [m, o. Pl.]** (Arbeitsgebiet der ►Landschaftsplanung mit der Zielsetzung, durch Verwendung lebender [Pflanzenteile, Pflanzengemeinschaften] und toter Baustoffe [z. B. Boden, Holz, Stein, Kunststoff, Metall] zur Realisierung, Erhaltung und Verbesserung des landschaftsökologischen, humanökologischen und ökonomischen Leistungsvermögens von anthropogenen und naturnahen Objekten und deren Nutzung beizutragen. Für den Landschaftsbau ist typisch, dass entweder mit lebenden und toten Baustoffen gemeinsam oder nur mit lebendem Material, selten aber ausschließlich mit toten Baustoffen gebaut wird; ►Garten- und Landschaftsbau, ►Garten-, Landschafts- und Sportplatzbau).

landscape preservation [n] [US] *conserv. landsc.* ►landscape conservation, #2.

landscape program(me) [n] *landsc. pol.* ►regional landscape program [US]/regional landscape programme [UK].

landscape protection [n] [UK] *conserv. landsc.* ►landscape conservation.

3140 landscape-related [adj] *landsc.* (►landscape 1, ►scenic); *s* **relacionado/a al paisaje [loc]** (►paisaje; ►paisajístico/a, ►pintoresco); *f* **en relation avec le paysage [loc]** (Qui a trait au ►paysage ; ►en terme de paysage, ►paysager, ►paysagique, ►pittoresque) ; *g* **landschaftsgebunden [adj]** (Zur [freien] ►Landschaft gehörig; *Kontext* **l.e** Erholung; ►landschaftlich); *syn.* landschaftsbezogen [adj].

landscape resource heritage [n] *conserv. land'man.* ►landscape patrimony.

landscape resources [npl] *conserv. landsc. recr.* ►natural site characteristics.

landscape resources [npl]**, spatial** *land'man.* ►spatial landscape characteristics/resources.

3141 landscape sector [n] *landsc.* (►landscape feature); *syn.* portion [n] of a landscape (LE 1986, 13); *s* **sector [m] del paisaje** (Sección de un paisaje; ►componente del paisaje); *f* **partie [f] de paysage** (Partie de la nature ou élément de paysages ; ►élément de paysages protégé) ; *syn.* secteur [m] de paysage, fraction [f] de/du paysage ; *g* **Landschaftsteil [m]** (Landschaftlicher Teilraum; § 2 [1] 12 BNatSchG; ►Landschaftsbestandteil).

landscape strategy plan [n] [UK] *landsc.* ►landscape plan.

3142 landscape structure [n] *landsc.* (Landscape fabric with different kinds, sizes, shapes, colo[u]rs, material compositions and spatial arrangment of ►landscape features); *syn.* fabric [n] of the landscape; *s* **estructura [f] del paisaje** (Trama espacial de ordenación de un paisaje o de sus ►componentes [del paisaje] diferenciada por el tipo, tamaño, forma, color, composición espacial material y ordenación de los mismos); *f* **structure [f] paysagère** (Agencement ou combinaison d'éléments végétaux, minéraux, hydrauliques, agricoles ou urbains dont le rôle structurant caractérisé par la taille, la forme, la couleur, la composition, l'agencement spatial des ►éléments de paysage contribue à former un ensemble ou un système cohérent spécifique ; circulaire n° 94-88 du 21 novembre 1994) ; *syn.* structure [f] du paysage ; *g* **Landschaftsstruktur [f]** (Raumgefüge der Landschaft oder eines Landschaftsteiles, das durch Art, Größe, Formenwelt, Farben, stoffliche Zusammensetzung und räumliche Anordnung der ►Landschaftsbestandteile geprägt ist); *syn.* Landschaftsgefüge [n].

3143 landscape structure plan [n] *landsc. plan.* (Broad framework for landscape planning and design on a state and regional level, defining measures to be taken for nature conservation and landscape management; ►regional landscape program [US]/regional landscape programme [UK]); *s* **plan [m] marco de gestión del paisaje y la naturaleza** (En D. con carácter de ley; se desarrolla a nivel regional; ►programa de gestión del paisaje y la naturaleza); *f 1* **plan [m] vert départemental** (F., plan présentant les objectifs et les orientations de la collectivité en matière d'environnement, paysage et espaces verts en général pour une période de 10 ans ; seuls quelques départements de la région parisienne se sont dotés de cet outil de protection de la nature et des paysages ; ►plan vert régional, ►programme régional de protection et de mise en valeur des paysages) ; *syn.* schéma [m] vert départemental ; *f 2* **plan [m] régional d'orientation pour la protection et la mise en valeur des paysages** (≠) (**D.**, rapport d'orientation et document graphique ayant pour objectif la protection et la mise en valeur des paysages à l'échelle du *Landkreis* [département allemand]) ; *f 3* **inventaire [m] départemental du patrimoine naturel** (F., inventaire régulièrement mis à jour, recensant les sites, paysages et milieux naturels ainsi que les mesures de protection de l'environnement existantes et pour lequel est élaboré un rapport d'orientation élaboré par l'État indiquant les mesures prises pour assurer la protection, la gestion et la mise en valeur du patrimoine naturel départemental) ; *g* **Landschaftsrahmenplan [m]** (*Abk.* LRP ; überörtliches Planwerk der Landespflege, entwickelt aus der nächsthöheren Planungsebene des ▶ Landschaftsprogramms, das auf der Ebene der Regionalplanung überörtliche Ziele und Maßnahmen des Naturschutzes und der Landschaftspflege sowie der Erholungsvorsorge für das Gebiet einer Region, eines Regierungsbezirkes oder eines Landkreises flächendeckend darstellt; zu den Inhalten gehören weiterhin die Schaffung eines Biotopverbundes, die Vorbereitung der ▶ Eingriffsregelung, z. B. durch die Schaffung eines ▶ Flächenpools; es werden ▶ Vorranggebiete und ▶ Vorbehaltsgebiete für Natur und Landschaft ausgewiesen; ferner werden aktuelle und mögliche zukünftige Konflikte zu anderen Nutzungsansprüchen anderer Fachplanungen und wie diese vermieden oder vermindert werden können, aufgezeigt; gesetzliche Grundlagen sind das Bundesnaturschutzgesetz und die in den Naturschutzgesetzen der Länder näher ausgeführten Regelungen; für die Erstellung der LRPläne sind je nach Landesrecht die Träger der Regionalplanung oder die oberen Naturschutzbehörden zuständig; der LRP besteht aus einem Text- und Kartenteil; cf. BFN 2006); *syn.* Landschaftsentwicklungsplan [m].

3144 landscape structure planning [n] *landsc.* (Broad landscape planning within a regional context; ▶ landscape structure plan); *s* **planificación [f] marco de gestión del paisaje** (Se desarrolla a nivel regional; ▶ plan marco de gestión del paisaje); *f* **planification [f] régionale des paysages** (≠) (Planification définissant à l'échelle du *Landkreis* [département allemand] la politique de protection, de mise en valeur et de développement de la nature et des paysages en milieu urbain et rural ; ▶ inventaire départemental du patrimoine naturel, ▶ plan régional d'orientation pour la protection et la mise en valeur des paysages, ▶ plan vert départemental) ; *g* **Landschaftsrahmenplanung [f]** (Flächendeckende Landschaftsplanung auf der Ebene der Regionalplanung, jeweils für das Gebiet einer Region, eines Regierungsbezirkes oder eines Landkreises, dargestellt in einem ▶ Landschaftsrahmenplan; gesetzliche Grundlagen sind das Bundesnaturschutzgesetz und die in den Naturschutzgesetzen der Länder näher ausgeführten Regelungen).

3145 landscape survey [n] *plan.* (Inventory of existing landscape conditions in a specific area, identified by on-site inspection or aerial photographs for delineation on a map); *s* **inventariación [f] paisajística** (Registro de datos paisajísticos de un área utilizando fotografías aéreas y/o levantamiento sobre el terreno); *syn.* toma [f] de datos de un paisaje; *f* **inventaire [m] des paysages** (Identification et description des entités paysagères en vue de la définition et caractérisation des types de paysages pour un territoire donné) ; *syn.* inventaire [m] paysager, étude [f] de paysage ; *g* **Bestandsaufnahme [f] einer Landschaft** (Erfassung und Kartierung eines landschaftlichen Zustandes durch Aufnahmen vor Ort und/oder durch Luftbildauswertung); *syn.* Bestandskartierung [f] einer Landschaft, Ist-Aufnahme [f] einer Landschaft.

landscape treatment [n] *landsc.* ▶ planting treatment.

3146 landscape type [n] *landsc. geo.* (LE 1986, 499; typical character of a landscape resulting from relatively homogeneous combinations of landform and landcover according to its ecology, geomorphology, history and socio-economic back-

ground; ▶ landscape character); *s* **tipo [m] de paisaje** (Diferenciación de paisajes en cuanto a su carácter particular determinado por su ecosistema, su estructura, fisionomía, historia y su desarrollo socioeconómico; ▶ particularidad del paisaje); *f* **type [m] de paysage** (Caractères généraux homogènes [écologiques, physiogéographiques, historiques et socio-économiques] permettant l'identification et la normalisation d'un paysage ; ▶ caractère particulier des sites et des paysages) ; *g* **Landschaftstypus [m]** (Landschaftskategorie, die durch ihre ökologische, strukturelle, physiognomische, historische und sozioökonomische Eigenart und Qualität geprägt ist; ▶ landschaftliche Eigenart).

3147 landscape unit [n] *land' man. plan.* (Component used in landscape planning in defining the ▶ ecological distribution of spatial patterns in a landscape with the objective of determining its possible aquirement for use by the community. Term cannot be applied to a scientific classification of individual landscape factors or a combination of them, e.g. soil types or geological mapping, ▶ physiographic province); *s* **unidad [f] paisajística** (En el marco de la ▶ estructuración ecológica del territorio, definición de una unidad de un paisaje con el fin de determinar posibilidades de uso. Este término no se aplica en la clasificación científica de factores específicos del paisaje o de combinaciones de ellos; ▶ tesela); *syn.* unidad [f] de paisaje; *f* **unité [f] paysagère** (Unité homogène formée dans le cadre de la division d'un territoire dans l'étude du paysage pouvant être définie à travers les caractéristiques visuelles, géographiques et culturelles d'un territoire, les éléments naturels d'un paysage ; GEP 1991, 28 ; ▶ district naturel, ▶ division de l'espace en territoires naturels, ▶ secteur physiographique) ; *syn.* unité [f] de paysage, unité [f] paysagique, entité [f] paysagère ; *g* **Landschaftseinheit [f]** (Im Rahmen der ▶ ökologischen Raumgliederung landschaftsplanerisch ausgewiesene Einheit einer Landschaft mit dem Ziel einer möglichen Aneignung und Nutzung durch die Gesellschaft. Mit diesem Begriff wird keine fachwissenschaftliche Klassifikation einzelner Landschaftsfaktoren oder deren Kombinationen, z. B. nach ihrer Entstehung definierte Bodentypen oder geologische Kartierungen verstanden; ▶ naturräumliche Einheit); *syn.* ökologische Raumeinheit [f].

3148 landscape urbanism [n] *landsc. urb.* (Concept largely developed at the University of Pennsylvania during the 1990s after IAN L. MCHARG'S [1921-2001] term there as professor. The theory, which has emerged in the last ten years, argues that landscape, rather than architecture, is more capable of organizing the city and enhancing the urban experience. **L. u.** accepts the horizontality and sprawl of big agglomerations and suburban fringes as a new urban reality and suggests that traditional programs are not suitable for this diffuse urban condition, because their scope is small and limiting. 'Territories' and 'potential' are terms used to define a place's use, instead of 'program'; adaptable 'systems' instead of rigid 'structures' are a better way to organize space. Landscape urbanists argue for a conflation of landscape and building. SIR GEOFFREY JELLICOE [1900-19969] was convinced that landscape rather than architecture is the real organizational force not just of built environment, but of the human spirit as well. Rather than being just a cosmetic, the practice of landscape ought to be bolder and much more assertive in spatial organization of the urban fabric, for example, using systematically planned open space defined by vegetation); *s* **urbanismo [m] paisajista** (≠) (Concepto desarrollado ampliamente por la Universidad de Pensilvania en los 1990 después del periodo de IAN L. MCHARG [1921-2001] como catedrático. La teoría que surgió entonces argumenta que el paisaje, más que la arquitectura, es más idóneo para organizar la ciudad y mejorar la calidad de vida. El **u. p.** acepta la horizontalidad y dispersión de las grandes aglomeraciones y los bordes suburbanos como una nueva realidad y propone que los programas tradicionales no son adecuados para este tipo de desarrollo urbano difuso, porque su margen de acción es pequeño y limitado. «Territorios» y «potenciales» son términos utilizados en vez de «programas» para definir el uso de un lugar; «sistemas» adaptables en vez de «estructuras» rígidas son la mejor manera de organizar el espacio. Urbanistas paisajísticos argumentan en favor de la fusión de paisaje y construcción. Sir Geoffrey Jellicoe [1900-1996] estaba convencido de que el paisaje —más que la arquitectura— es la verdadera fuerza organizacional, no sólo de un medio construido, sino también del espíritu humano. En vez de ser sólo cosmética, la práctica del paisajismo debería ser más audaz y mucho más aseverativa en la organización espacial de la trama urbana, utilizando p. ej. espacios libres planificados sistemáticamente y definidos por vegetación); *f* **urbanisme [m] paysager** (Notion d'urbanisme développée par Ian Mac Harg [1921-2001] à l'université de Pennsylvanie à la fin des années 1990 et déjà apparue en Amérique et en Europe du Nord à la fin des années 1960, sous les appellations de planification écologique et paysagère ; cette notion considère que le paysage plus que l'architecture est le moteur du développement local urbain et d'une nouvelle conception et organisation de la vie urbaine ; cette notion de planification prend acte de l'expansion des grandes agglomérations, des régions urbaines et de leurs franges suburbaines mais considère le champ d'action de la planification urbaine actuelle étroit et limité, ses concepts et son organisation structurelle impropre à solutionner les problèmes de à l'urbanisation diffuse. « Territoires » et « potentiel » sont des termes susceptibles de définir l'utilisation d'un espace plutôt qu'un programme, d'un système adaptable plutôt qu'une structure rigide et d'assurer enfin une meilleure organisation de l'espace. Les urbanistes paysagistes, pour lesquels le paysage est au cœur de toute intervention urbanistique, argumentent en faveur du fusionnement du paysage avec les constructions. Sir Geoffrey Jellicoe [1900-1996], paysagiste anglais, était persuadé que ce sont les espaces verts et non l'architecture qui détiennent la véritable force organisatrice non seulement de l'environnement bâti mais encore comme source de l'esprit humain. Au lieu de considérer les espaces verts comme des accessoires futiles, il apparaît nécessaire que les pratiques paysagistes structurent de manière plus résolue et déterminante la physionomie urbaine, par exemple par l'usage systématique d'espaces libres à dominance végétale ; la notion de *landscape urbanism* désigne à l'heure actuelle les activités de planification et d'aménagement de l'espace ouvert sous influence urbaine ; elle met l'accent sur les « bonnes pratiques » nouvelles de l'aménagement de la ville avec la participation des habitants dans son contexte paysager, historique et géographique dans une perspective de développement soutenable admettant la réversibilité de l'usage des espaces ouverts et les incertitudes du développement économique) ; *g* **grünraumorientierte Stadtplanung** (Stadtplanerisches Konzept, das von Ian MacHarg [1921-2001] an der University of Pennsylvania in den 1990er-Jahren entwickelt wurde, das besagt, dass die Grünplanung besser als die Hochbauarchitektur das Leben in der Stadt fördert. Dieser Planungsansatz akzeptiert die Flächenausdehnung großer Städte und Ballungsräume als Realität, sieht jedoch in den herkömmlichen Planungsansätzen und -strukturen keine adäquaten Lösungen. Eine Stadtplanung hingegen, die vornehmlich ein schlüssiges **System vegetationsbestimmter Freiräume** schafft, definiert Flächen durch ihr Potenzial, z. B. durch anpassungsfähige und veränderbare Nutzungen, statt durch starre Strukturen und Gebautes. Der englische Landschaftsarchitekt Sir Geoffrey Jellicoe (1900-1996) war davon überzeugt, dass Grünräume und nicht Hochbauarchitektur die eigentliche organisatorische Kraft nicht nur für alles Gebaute

besitzen, sondern auch für den menschlichen Geist. Statt die Grünanlagen nur als kosmetisches Beiwerk zu betrachten, sollte die Grünplanung entschiedener und maßgeblich das Stadtbild durch Grünflächensysteme strukturieren).

landscaping [n] *constr. prof.* ►landscape construction.

landscaping [n] [US]**, cemetery** *constr.* ►cemetery gardening.

landscaping [n] [US]**, roadside** *leg. trans. landsc.* ►right-of-way planting [US]/roadside planting [UK].

landscaping [n] [US]**, street** *leg. trans. landsc.* ►right-of-way planting [US]/roadside planting [UK].

3149 landscaping [n] **of built-up areas** *urb. landsc.* (Greening of ►developed areas [US]/settlement areas [UK] with private or public green open spaces, tree-lined avenues [US]/avenues [UK], or with a few trees or clumps of trees; ►low density building development, ►low-density housing, ►planting treatment); *syn.* greening [n] of built-up areas; *s* **tratamiento [m] paisajístico (1)** (Dotación de ►áreas de asentamiento con zonas verdes privadas y públicas y con elementos de vegetación en las calles; ►edificación de baja densidad, ►tratamiento paisajístico 2, ►edificación residencial de baja densidad); *f* **traitement [m] paysager (1)** (Réaménagement d'un ►site urbanisé au moyen d'espaces verts privés ou publics, d'arbres isolés, d'allées ou de groupes d'arbres ; ►habitat diffus, ►habitat dispersé, ►traitement paysager 2, ►urbanisation aérée) ; *syn.* aménagement [m] paysager, paysagement [m] [CDN] ; *g* **Durchgrünung [f]** (Ausstattung von Siedlungsgebieten mit privaten oder öffentlichen Grünflächen sowie mit Einzelbäumen, Baumalleen oder Baumgruppen; ►aufgelockerte Wohnbebauung, ►Eingrünung, ►lockere Bebauung, ►Siedlungsflächen); *syn.* Begrünung [f] von Stadtteilen.

land scientist [n] [UK] *landsc. plan. prof.* ►landscape planner.

land shaping [n] *constr.* ►ground modeling [US]/ground modelling [UK].

3150 landslide [n] *constr. geo.* (**1.** Type of shallow ►mass slippage [US]/mass movement [UK] of rock material, soil, artificial fill, or a combination of these, lubricated by rainwater [US]/rain-water [UK], down a slope, the result of a [tectonic] earthquake, or subsequent to inadequate road construction operations on ►slip planes following contours; ►creeping soil, ►slope failure, ►soil creep 2, ►unstable slope. Types of movement include falls, topples, slides, spreads or flows. If the slope failure in which the shear plane is largely parallel to the slope surface is shallow, the sliding will extend downward and outward along a broadly planar surface. Following types of **l.s** are distinguished: **a] translational slide**, which usually occurs along structural features, such as a bedding plane or between bedrock and weaker, overlying material; **b] rotational slide:** downward movement, in which the surface of rupture is curved concavely upward and the slide movement is roughly rotational about an axis that is parallel to the ground surface and transverse across the slide; there is a downward intermittent movement of rock debris, caused by the gradual removal of material at the foot of the slope; cf. NAUS 2009; *syn.* rotational slip [n], rotational slumping [n], slump [n]; **c] block slide:** overlying material moves as a single, little-deformed mass. **L.s** which have a more fluid character are referred to as **flows**, e.g. **d]** ►debris flow [n]: form of rapid mass movement in which a combination of loose soil, rock, organic matter, air, and water mobilize as a slurry that flows downslope, including <50% fines; cf. NAUS 2009; **e]** ►debris avalanche, **f]** ►debris slide, **g]** ►earthflow, **h]** ►mudflow or **i] creep** (►soil creep 1); *syn.* landslip [n] [UK]; *s* **deslizamiento [m] [Es, CO, EC, YV] (1.** Desplazamiento masivo de una vertiente, o de parte de ella, que se produce a través de una o

varias superficies de rotura bien definidas [▶plano de desliza-miento]. La masa generalmente se desplaza en conjunto, comportándose como una unidad, prácticamente sin deformación interna, en su recorrido. Puede ser causado por procesos tectó-nicos, terremotos o por obras de construcción viaria inadecuadas tras fuertes precipitaciones; cf. DGA 1986 & UGR; ▶caída de un talud, ▶movimientos gravitacionales, ▶plancha de solifluxión, ▶reptación hídrica del suelo, ▶talud inestable 2. **2.** En función de la forma de la superficie de rotura se pueden diferenciar en **deslizamiento traslacional o planar**, cuando la superficie es un plano con una inclinación más o menos constante, y ▶desliza-miento rotacional. Los **dd. tt.** se producen generalmente sobre materiales heterogéneos con superficies de discontinuidad bien definidas. Cuando los movimientos de ladera tienen una super-ficie de rotura con una geometría mixta se denominan **desliza-mientos compuestos**; cf. UGR); *syn.* corrimiento [m] de tierras [Es, RA], deslizamiento [m] de tierras [RA, RCH, YV], derrumbe [m] [PA] (MESU 1977); *f 1* **glissement [m] de terrain** (▶Mouvement de masse de matériaux le long d'un versant [▶plan de faille, ▶plan de glissement] provoqué par un tremble-ment de terre, des secousses tectoniques ou par des pluies violentes suite à une mise en place défectueuse de terres ou d'un cheminement parallèlement à une pente ; ▶boue reptante, ▶glissement de talus, ▶reptation d'un versant ; ▶versant instable en état de rhexistasie) ; *syn.* glissement [m] de versant, glissement [m] en plaque (WIB 1996) ; *f 2* **glissement [m] translationnel** (Déplacement lent et continu d'un grand volume de matériaux meubles sur une pente le long d'un plan de glissement, provoqué par des secousses tectoniques ou par des pluies) ; *g 1* **Hangrutschung [f]** (Durch Erdbeben, tektonische Störungen oder z. B. durch unsachgemäßen Erd- oder Wegebau auf hangparallelen ▶Gleitflächen nach stärkeren oder sehr lang anhaltenden Niederschlägen verursachter flachgründiger ▶Mas-senversatz am Hang; die Größe der **H.en** variieren von wenigen m² bis einigen km²; die großen **H.en** werden **Bergrutsch** ge-nannt; die Bewegungsgeschwindigkeit variiert von wenigen mm/a bis über 80 km/h; www.ilews.de; ▶Gekriech, ▶Hang-gleiten, ▶Hangkriechen, ▶Rutschhang); *g 2* **Translations-rutschung [f]** (Eine besondere Form der ▶Hangrutschung, bei der eine stabile Schicht auf einer unstabilen, gleitfähigen Schicht abrutscht, wenn die Scherspannung einen gewissen Schwellen-wert [Grenzscherspannung oder Grenzschubspannung] über-schreitet resp. die Haftreibung vermindert wird. Gefördert wird die **T.** durch Destabilisierung des Unterhanges, wenn z. B. auf-grund einer Tiefenerosion eines Gewässers am Unterhang der Gegendruck fehlt und Zugrisse entstehen. Diese weiten sich, bis eine Scholle abrutscht. Dann entstehen weitere Zugrissen ober-halb, so dass Scholle für Scholle nachrutschen; weitere Formen der Hangrutschung/des Massenversatzes: ▶Bergrutsch, ▶Erd-schlipf, ▶Geländebruch, ▶Mure, ▶Schuttrutschung, ▶Stein-lawine); *syn.* gravitative Massenbewegung [f], *ugs.* Erdrutsch [m].

3151 landslide hazard [n] *land'man.* **s peligro [m] de deslizamiento de tierras** *syn.* riesgo [m] de corrimiento de tierras; *f* **danger [m] de glissement de terrain** *syn.* risque [m] de glissement de terrain ; *g* **Bodenrutschungsgefahr [f]**.

landslip [n] *geo.* ▶edge of a landslip, ▶landslide.

land swap [n] [US]**, voluntary** *agr. leg.* ▶voluntary land exchange.

land take [n] [UK] *landsc.* ▶land consumption.

land transactions [npl] [UK] *leg. urb.* ▶transfers of real estate [US].

3152 land use [n] (1) *agr. plan. urb.* (**1.** Utilization of land. **2.** Category designated on community development plans); **s uso**

[m] del suelo (**1.** Tipo de aprovechamiento del terreno. **2.** Cate-goría de uso del terreno prevista en plan general municipal de ordenación urbana o en plan parcial [de ordenación]); *f 1* **utili-sation [f] du sol** (*Terme général* affectation du sol à des usages déterminés) ; *f 2* **occupation [f] du sol** (Terme spécifique à l'urbanisme) ; *g* **Landnutzung [f]** (**1.** Nutzung von Grund und Boden. **2.** Im Bauleitplan vorgesehene/ausgewiesene Nutzung für eine Fläche einer Gemeinde); *syn.* Bodennutzung [f], Flächen-nutzung [f], Raumnutzung [f].

land use [n] (2) *agr.* ▶agricultural land use; *leg. urb.* ▶industrial land use, ▶institutional land use; *agr. for. hort.* ▶intensity of agricultural and forest land use; *leg. urb.* ▶mixed land use (1), ▶mixed land use (2), ▶recreational land use, ▶residential land use; *conserv. leg. plan.* ▶prohibition of a land use.

land use [n]**, agricultural and forest** *agr. for. hort.* ▶intensity of agricultural and forest land use.

land-use [n]**, pressure of competitive** *plan.* ▶competing land-use pressure.

land-use area [n] *plan.* ▶designated land-use area.

3153 Land Use Capability Classification [n] [UK] *agr. plan.* (More detailed grading of land developed in U.K. by soil survey to present results of soil surveys in a form suitable for planners, agricultural advisers, farmers and other land users); **s clasificación [f] de aptitud de uso del suelo** (≠) (En GB clasificación detallada de suelos desarrollada por la investigación edáfica para presentar los resultados en una forma más adecuada para la planificación, la asesoría agrícola, la agricultura y otros aprovechamientos del suelo); *f* **classification [f] de l'aptitude culturale des sols** (±) (**U.K.**, Classification détaillée des terres et des sols établie sur la base de relevés pédologiques et mise à la disposition d'aménageurs, de conseillers agricoles, d'agriculteurs ou d'autres utilisateurs) ; *syn. F., B., CH.* classification [f] de l'aptitude culturale des terres, classification [f] de l'aptitude des terres à l'agriculture ; *g* **Bodenbewertung [f] (1)** (**U.K.**, detail-lierte Einteilung von Landflächen und Böden auf Grund von Bodenaufnahmen für Planer, landwirtschaftliche Berater, Land-wirte und andere Bodennutzer).

3154 land use conflict [n] *plan.* (COU 1978, 193; non-compatibility of land uses because they mutually exclude or adversely affect each other when situated together or adjacently); **s conflicto [m] entre usos** (No compatibilidad entre diferentes usos en una misma área por lo que se excluyen o dificultan mutuamente); *f* **conflit [m] d'usages** (Non compatibilité de différentes utilisations qui par suite de leur voisinage ou de leur superposition dans l'espace s'excluent ou se gênent mutuelle-ment) ; *g* **Nutzungskonflikt [m]** (Nichtverträglichkeit von Nut-zungen, da sie sich gegenseitig ausschließen oder infolge räum-licher Überlagerung oder Benachbarung beeinträchtigen).

land use density [n] [US] *leg. urb.* ▶density of development.

3155 land use intensity standards [npl] [US] *urb.* (System of ▶bulk regulations, designed primarily for large scale developments, and based on the physical relationship between specific development factors; IBDD 1981; ▶density of develop-ment, ▶maximum bulk, ▶zoning district category [US]/use class [UK]); **s estándar [m] de intensidad de uso del suelo** (≠) (En EE.UU. sistema de regulación de la ▶densidad de edificación utilizado primordialmente para proyectos de gran escala, basado en las condiciones topográficas del terreno; ▶categoría de zonificación, ▶regulación del volumen edificable, ▶volumen edificable); *f* **règlement [m] d'urbanisme pour les grands ensembles** (≠) (**U.S.**, système de ▶règles relatives à l'implanta-tion et les dimensions des constructions définissant la ▶densité de construction à travers la détermination du type d'occupation ou d'utilisation du sol, de l'implantation et de la dimension des

constructions ; ►conditions de l'occupation du sol, ►nature de l'affectation des sols) ; **g Art [f] und Maß [n] der baulichen Nutzung für große Neubaugebiete (≠) (U.S.**, planungsrechtliche Vorschriften über die Darstellung und Festsetzung der ►Bebauungsdichte durch die Art und den Umfangs baulicher Nutzung in Bauleitplänen in Abhängigkeit der topografischen Gegebenheiten; ►Art der baulichen Nutzung, ►Maß der baulichen Nutzung, ►Vorschriften zur Bestimmung des Maßes der baulichen Nutzung).

3156 land use pattern [n] *plan.* (Result of the spatial arrangement and distribution of individual land-uses within an area; ►pattern of interrelated land uses); *s* **mosaico [m] de usos del suelo** (Resultado de la ordenación y distribución espacial de usos individuales del terreno dentro de una zona específica; ►articulación de usos); *syn.* esquema [m] de articulación de usos, pattern [m] de usos del suelo; *f* **modèle [m] d'affectation des sols** (Arrangement et distribution spatiale des diverses occupations du sol dans le périmètre d'une commune ; ►articulation des usages) ; *g* **Flächennutzungsmuster [n]** (Anordnung der einzelnen Flächennutzungen in einem Gemeindegebiet; ►Nutzungsverflechtung); *syn. als Ergebnis* Flächennutzungsverteilung [f]; *im nutzungsökologischen Sinne* Landnutzungssystem [n].

land-use plan [n], **legally-binding** *adm. leg. urb.* ►zoning map [US]/Proposals Map/Site Allocations Development Plan Document [UK].

land-use pressure [n] *plan.* ►competing land-use pressure.

3157 land-use requirements [npl] *plan.* (Need for existing, proposed or expected use of an area); *s* **demanda [f] de suelo (para un uso);** *f* **exigence [f] d'utilisation d'un sol** (Usage existant, prévu ou prévisible sur une surface de sol) ; *syn.* exigence [f] d'affectation ; *g* **Nutzungsanspruch [m]** (Eine Forderung, ein Verlangen oder Recht von jemandem, eine vorhandene oder geplante Fläche oder Einrichtung zu nutzen oder nutzen zu wollen).

land uses [npl] *plan.* ►pattern of interrelateded land uses, ►segregation of land uses.

land uses [npl], **combination of various** *plan.* ►layering of different uses.

land uses [npl], **designation of** *adm. leg. plan.* ►statutory land use specification.

land uses [npl], **interrelated** *plan.* ►pattern of interrelated land uses.

land uses [npl], **mapping of existing** *plan.* ►survey of existing uses.

land uses [npl] [UK], **overlapping of** *plan.* ►layering of different uses.

land use specification [n] *adm. leg. plan.* ►statutory land use specification.

3158 land use structure [n] *plan.* (Composition of land uses, mostly in three dimensional terms in comparison with the two-dimensional representation of ►land use pattern); *s* **estructura [f] de usos del suelo** (Composición de usos del suelo, generalmente pensada en tres dimensiones a diferencia del ►mosaico de usos del suelo); *f* **structure [f] d'utilisation du sol** (Terme utilisé en général dans une acceptation tridimensionnelle par comparaison avec le ►modèle d'affectation des sols) ; *g* **Flächennutzungsstruktur [f]** (Im Vergleich zum ►Flächennutzungsmuster meist dreidimensional gedacht).

3159 land [n] **zoned for development** *urb.* (Area for future use and settlement with different planning status: ►approved development site, ►undeveloped zoned land, ►unzoned land ripe for development [US]/white land [UK]); *s* **suelo [m] edificable** (Término genérico que incluye los tipos de suelo susceptibles a ser edificados. Se puede encontrar en diferentes fases de planificación; ►suelo edificable no urbanizado, ►suelo urbanizable, ►suelo urbanizado l); *f* **zone [f] constructible** (Terme générique dans la planification urbaine pour les terrains affectés à la construction et soumis aux documents d'urbanisme ; ►terrain non viabilisé, ►terrain viabilisé, ►zone d'urbanisation future) ; *syn.* zone [f] urbanisable ; *g* **Bauland [n, o. Pl.]** (Die der Bauleitplanung unterstehenden Flächen für eine zukünftige Besiedlung mit verschiedenen Stadien des planerischen Entwicklungsfortgangs: ►Bauerwartungsland, ►fertiges Bauland, ►Rohbauland).

lane [n] *trans.* ►acceleration lane, ►bike lane [US], ►deceleration lane lane, ►directional lane, ►fire lane [US], ►offside lane, ►parking lane, ►slow vehicle lane.

lane [n] [US], **allee** *gard. landsc. urb.* ►tree-lined avenue [US].

lane [n] [UK], **bicycle** *trans.* ►bike lane [US].

lane [n], **commuter** *trans.* ►offside lane, #2.

lane [n], **diamond** *trans.* ►offside lane, #2.

lane [n], **high occupancy vehicle** *trans.* *►offside lane.

lane [n] [US], **merger** *trans.* ►acceleration lane.

lane [n], **parallel parking** *trans.* ►parking lane.

lane [n] [US], **paved pull-off** *trans.* ►hard shoulder [US].

lane [n] [UK], **skidding** *for.* ►skidding road [US&CDN].

lane [n], **sunken** *agr. for.* ►sunken path.

lane [n] [AUS, NZ], **transit** *trans.* ►offside lane, #2.

lane [n] [US], **truck** *trans.* ►slow vehicle lane.

langlauf [n] [US] *recr.* ►cross-country skiing.

langlaufer [n] [US] *recr.* ►cross-country skier.

3160 lapidary garden [n] *gard'hist.* (Collection of historic stone artifacts [US]/stone artefacts [UK]); *s* **jardín [m] lapidario** (Jardín en el que se encuentran esculturas que han sido sustituidas por copias en el edificio original de emplazamiento o que provienen de edificios que ya no existen); *syn.* lapidario [m]; *f* **jardin [m] lapidaire** (Jardin dans lequel la pierre, pouvant revêtir les aspects les plus divers, est l'élément dominant de la composition ; cf. DIC 1993) ; *g* **Lapidarium [n]** (Sammlung von originalen Bauskulpturen, die am Bau selbst durch Kopien ersetzt wurden oder von nicht mehr existierenden Bauten stammen [= Spolien]).

3161 large-caliper tree/large-calliper tree [n] *hort. arb.* (Tree with a large diameter trunk; **in U.S.**, measured > 30cm [12 inches] above the ground); *s* **árbol [m] de diámetro de tronco grande;** *f* **gros sujet [m]** (Les différentes catégories d'arbres sont définies par classe de dimension en fonction de la circonférence du tronc mesurée à 1 m du sol pour les feuillus et de la hauteur pour les conifères ; les gros sujets sont des arbres tiges de grande taille, régulièrement transplantés afin de garantir un important chevelu racinaire et une reprise optimale et dont la circonférence du tronc mesurée à 1 m du sol est supérieure à 18 cm pour les feuillus et la hauteur supérieure à 2,5 m pour les résineux ; *syn.* fort sujet [m], sujet [m] de fort diamètre ; *g* **Baum [m] mit großem Stammdurchmesser** (In D. werden Bäume in Baumschulen nur mit Stammumfängen — nicht mit Stammdurchmessern — sortiert und gehandelt).

large groups [npl] *phyt.* ►growing in large groups.

large nursery stock [n] *hort. plant.* ►specimen tree/shrub.

3162 large-scale planning [n] *plan.* (Concept for land use development over a large area); *s* **planificación [f] a gran escala;** *f* **planification [f] sur une grande échelle** ; *g* **großräumige Planung [f].**

L

large scale planning research [n] [US] *plan. pol.*
▶regional research.

large-sedges fen [n] [UK] *phyt.* ▶tall sedge swamp.

3163 large shrub [n] *hort. plant.* (▶Shrub usually taller than
5m; *opp.* ▶small shrub); *syn.* tall shrub [n]; *s* **arbusto [m]
grande** (▶Arbusto ramificado desde la base, generalmente de
más de 5 m de altura; ▶subarbusto; *opp.* ▶pequeño arbusto);
f 1 **grand arbrisseau [m]** (Végétal ligneux de moins de 7 m de
hauteur, ramifié dès la base, et donc dépourvu de tronc ; ▶arbris-
seau, ▶arbuste ; *opp.* ▶petit arbuste) ; *f 2* **grand arbuste [m]**
(Végétal ligneux à tige simple et nu à la base, mais n'atteignant
pas 7 m de hauteur à l'état adulte, mais offrant par ailleurs tous
les caractères d'un arbre ; p. ex. *Acer ginnala, Cercis siliquas-
trum, Lagerstroemia indica*) ; *g* **Großstrauch [m]** (I. d. R. über
5 m hoher ▶Strauch 2; *opp.* ▶Kleinstrauch).

large-sized cobblestone [n] [US] *constr.* *▶large-sized
paving stone [US]/large-sized granite sett [UK].

large-sized concrete paver [n] [US] *constr.* *▶large-sized
paving stone [US]/large-sized concrete paving sett [UK].

large-sized concrete pavestone [n] [US] *constr.*
*▶large-sized paving stone [US]/large-sized concrete paving sett
[UK].

large-sized concrete paving sett [n] [UK] *constr.*
*▶large-sized paving stone [US].

large-sized granite sett [n] [UK] *constr.* *▶large-sized
paving stone [US].

large-sized natural pavestone [n] [US] *constr.* *▶large-
sized paving stone [US]/large-sized granite sett [UK].

large-sized natural paving stone [n] *constr.* *▶large-
sized paving stone [US]/large-sized granite sett [UK].

3164 large-sized paving stone [n] [US] *constr.* (Dressed
block of natural stone or concrete, shaped or selected for the
wearing surface of a stone-block pavement, for edgings and
paving strips; **in U.S.**, rough-cut, rectangular-shaped, 150-
250mm [6 to 10 inches]; cf. MEA 1985; such stones are referred
to as **Belgian block pavers** [approx. 100 x 170 x 250mm] cut
from granite, often set vertically for heavy-duty vehicular paving;
in U.K., usually granite according to BS 435: 1931, length not
less than 127mm, nor more than 254mm; ▶cobblestone [US]/
natural stone sett [UK]); *syn. for natural stones* large-sized
cobblestone [n] [US], jumbo cobblestone [n] [US], large-sized
natural paving stone [n], large-sized natural pavestone [n] [US],
large-sized granite sett [n] [UK]; *syn. for concrete pavers* large-
sized concrete pavestone [n] [US], large-sized concrete paver [n]
[US], large-sized concrete paving sett [n] [UK]; *s* **adoquín [m]
grande** (Adoquín de piedra natural con superficies trabajadas y
de tamaño normado que se utiliza para pavimentos o para
encintados y bordillos; **en EE.UU.** de corte bruto y forma
rectangular, y de 150-250 mm [6-10 pulgadas] de largo; **en GB**
generalmente de granito, de largo mínimo de 127 mm y máximo
de 254 mm; **en D.** con aristas de 120-220-290 mm; los **aa. gg.**
también pueden fabricarse de hormigón; ▶adoquín natural);
f **gros pavé [m]** (14 x 20 cm ; Norme P 98.301 ; ▶pavé en pierre
naturelle) ; *syn.* pavé [m] de gros module (VRD 1994, II, part. 4,
chap. 6.4.2.2.1, p. 9) ; *g* **Großpflasterstein [m]** (**1.** Pflasterstein
aus Granit oder Beton; **2.** nach DIN 18 502 beschriebene
Güteklasse und Größeneinteilung für Naturpflastersteine mit
Kantenlängen von 120-220-290 mm; **G.** aus Granit [Granit-
großpflasterstein] wird ugs. auch *Katzenkopf* genannt; ▶Natur-
steinpflasterstein).

3165 large spoil area [n] [US] *envir. min.* (Generic term
for dump [US]/tip [UK] of waste material; ▶dumpsite [US]/
tipping site [UK], ▶rubble pile [US]/rubble tip [UK] ▶slag pile,

▶spoil bank [US] 2/spoil heap [UK]); *syn.* large tip [n] [UK];
s **escombrera [f] (1)** (Término genérico de depósito de materiales residuales; ▶depósito de residuos sólidos, ▶escorial,
▶montón de escombros, ▶vaciadero de gangas); *f* **amoncellement [m]** (Terme générique pour tout dépôt de matériaux sur
un terrain naturel ; ▶crassier, ▶décharge publique, ▶dépôt de
gravats, ▶dépôt d'ordures, ▶halde, ▶terril) ; *g* **Halde [f] (1)**
(OB zu Aufschüttung auf unverritztem Gelände; ▶Bergehalde,
▶Deponie, ▶Schlackenhalde, ▶Schutthalde 1).

large tip [n] [UK] *envir. min.* ▶large spoil area [US].

3166 late-blooming plant [n] *gard. hort. plant.* (Plant
coming into flower after the usual peak flowering period or
flowering in autumn; ▶autumn-blooming plant); *syn.* late flowerer [n]; *s* **planta [f] de floración tardía** (Planta que florece
después del periodo de floración principal o en otoño; ▶planta de
floración otoñal); *f* **plante [f] à floraison tardive** (Plante dont la
floraison a lieu plus tard que la majorité des individus de l'espèce
ou qui fleurit à l'automne ; ▶plante à floraison automnale) ;
g **Spätblüher [m]** (*Gärtnerische Bezeichnung* Pflanze, die später
als der Hauptflor einer jeweiligen Art oder diejenige, die im
Herbst blüht; ▶Herbstblüher).

late clean-up [n] [US] *constr.* ▶late cut.

3167 late cut [n] *constr.* (Final grass cut of lawn or meadow
in late autumn or very early spring; *specific terms* spring clean-
up, fall clean-up [also US]); *syn.* late clean-up [n] [also US];
s **corte [m] tardío** (Último corte de un césped o de un prado en
otoño o a principios de la primavera); *syn.* corte [m] de hierba
tardizo [AL]; *f* **fauche [f] tardive** (L'action de sectionner les
tiges d'un gazon ou d'une prairie sur pied en automne) ; *syn.*
fauchage [m] tardif ; *g* **Reinigungsschnitt [m]** (Letzter Schnitt
eines Rasens oder einer Wiese — im Spätherbst oder zeitigen
Frühjahr inklusive Müllentfernung).

late flowerer [n] *gard. hort. plant.* ▶late-blooming plant.

3168 late frost [n] *met.* (Frost occurring relatively late in
spring, often causing damage to plants just coming into leaf; *opp.*
▶early frost); *s* **helada [f] blanca de primavera** (Helada que se
presenta una vez comenzada la primavera; *opp.* ▶helada blanca
de otoño); *syn.* helada [f] tardía; *f* **gelée [f] tardive** (Gelée de
printemps dite de rayonnement ; *opp.* ▶gelée hâtive) ; *syn.* gelée
[f] printanière ; *g* **Spätfrost [m]** (Nach Ende des phänologischen
Winters [in Mitteleuropa von Dezember bis Februar] im Frühling
eintretender Frost — kann sich je nach geografischer Lage bis in
den Juni hineinziehen; *opp.* ▶Frühfrost).

lateral [n] [US] *constr.* ▶lateral drain.

3169 lateral drain [n] *constr.* (LAD 1986, 146, 215; drain
which serves to remove excess water from the soil and discharges
the water into the ▶collector drain); *syn.* lateral [n] [also US];
s **dren [m] secundario** (Conducto que sirve para recoger el agua
superflua del suelo y conducirla al ▶colector); *syn.* dren [m]
lateral; *f* **drain [m] adducteur** (Drain d'adduction des eaux ;
▶collecteur) ; *syn.* adducteur [m] ; *g* **Sauger [m]** (Drän, der der
Wasseraufnahme im Boden dient und das Wasser in einen
▶Sammler ableitet); *syn.* Saugdrän [m], *o. V.* Saugdrain [m].

lateral drains [n] *constr.* ▶spacing of lateral drains.

3170 lateral moraine [n] *geo.* (Deposition of scree along the
edges of glaciers; ▶pebbles); *s* **morrena [f] lateral** (Formada por
los materiales que el hielo va acumulando en los bordes del
trayecto; ▶canto rodado); *f* **moraine [f] latérale** (Accumulation
de débris en marge d'un glacier provenant d'éboulements sur les
versants qui le domine ; ▶galet) ; *syn.* moraine [f] riveraine ;
g **Randmoräne [f]** (Schuttwall entlang von Gletschern; ▶Ge-
röll); *syn.* Seitenmoräne [f].

3171 lateral root [n] *bot.* (Root that branches from a parent root); *syn.* side root [n] (RRST 1983); *s* **raíz** [f] **lateral** (Raíz secundaria que se ramifica lateralmente de la principal); *f* **racine** [f] **latérale** (Racine secondaire se détachant de la racine principale) ; *g* **Seitenwurzel** [f] (Von einer Hauptwurzel abzweigende Wurzel).

3172 late wood [n] *arb. bot.* (Dense, small-celled, late-formed part of ►annual ring developed toward the end of a growing season; *opp.* ►early wood); *syn.* summer wood [n]; *s* **leño** [m] **tardío** (La madera más densa con células pequeñas, generada al final de una fase de crecimiento; ►anillo anual; *opp.* ►leño temprano); *syn.* leño [m] de verano; *f* **bois** [m] **final** (Partie d'un ►cerne annuel dont le bois est le plus dense, à cellules plus petites et le dernier formé ; *opp.* ►bois initial) ; *syn.* bois [m] d'été ; *g* **Spätholz** [n] (Das dichtere und letztgebildete Holz eines ►Jahresringes mit englumigen Zellen; MGW 1964; *opp.* ►Frühholz).

3173 lattice [n] *constr.* (Open-work structure of wood, metal, etc. made by crossing laths, rods or bars, and forming a network or open pattern of squares or diamonds, used as a screen, fence or ►climber support for vines or other creeping plants; ►metal grid, ►perimeter fencing, ►wooden lattice); *syn.* trellis [n], grillwork [n] [also US]; *s* **enrejado** [m] (Término genérico para estructura de listones, barras o varas de madera, metal u otros materiales, que en cruzado forman una red y se utiliza como cerca, pantalla o como soporte para trepadoras; ►enrejado de madera, ►enrejado [de pista de juegos de pelota], ►enrejado de soporte de trepadoras, ►reja de hierro); *syn.* reja [f]; *f* **treillis** [m] (Ouvrage en bois ou en métal dont les barreaux ou les lattes s'entrecroisent pour former un maillage ; dans la confection de dalles en béton le **treillis soudé** constitue l'armature métallique de renforcement ; ►grillage pareballon, ►grille, ►support de plantes grimpantes, ►treillage en bois) ; *g* **Gitter** [n] (Bauteil aus gekreuzten Stäben oder Latten; ein **G.** aus Latten wird auch ►**Lattengerüst** genannt; ►Ballfanggitter, ►Rankgerüst, ►Stabgitter); *syn.* Gitterwerk [n].

lattice concrete block [n] [US] *constr.* ►grass paver.

lattice concrete blocks [npl] [US]**, surfacing of** *constr.* ►grid pavement [US]/pavement of grass pavers [UK].

lattice fence [n]**, interwoven** *constr.* ►interwoven wood fence.

lattice work concrete block [n] [US] *constr.* ►grass paver.

lava [n] *constr. geo.* ►processed lava, ►scoria.

3174 law [n] *adm. leg. pol.* (**1.** Law is a system of rules, enforced through a set of institutions, used as an instrument to underpin civil obedience, politics, economics and society. Law serves as the foremost social mediator in relations between people. Law consists of a wide variety of separate disciplines; adopted by a state or federal legislative body); **2. statute** [n] (Another word for a law, it is established by an **act** of the legislature and considered to be the primary authority; an **s.** is a formal written enactment of a legislative authority and declares a policy; term is often used to distinguish law made by legislative bodies from the judicial decisions of common law and regulations issued by government agencies. As opposed to secondary authority, in some countries, an **s.** must be agreed upon by the highest government executive and published as part of a code before it becomes law. In US, statutory law is distinguished from and subordinate to constitutional law: legislatures make the statutes and courts interpret the law [cases]. Most state statutes are published in books referred to as codes; e.g. building code, welfare code. The term **s.** is sometimes used to refer to an international treaty that establishes an institution); **3. act** [n] (An Act is the statement of a law made by a legislative body. **In U.S.**, it is known as an Act of Congress; **in U.K.** and Commonwealth countries, it is referred to as an Act of Parliament, i.e. a statute enacted as primary legislation by a national or sub-national parliament. Before an Act becomes law a Bill is drafted that Parliament proposes to enact. It may be initiated in either House of Parliament or the House of Representatives. After a Bill has been passed, it is presented for royal assent [UK]. Once assent is given, the Bill is an Act, which then becomes a part of statute law; ►city ordinance); *s* **ley** [f] (**1.** Cualquier norma legal escrita aprobada por un órgano legislativo [parlamento o asamblea legislativa] competente durante el ejercicio de su poder legislativo. **En Es.** a nivel estatal las leyes una vez aprobadas por las Cortes son sancionadas por el Rey y entran en vigor generalmente un día después de su publicación en el Boletín Oficial del Estado. Las leyes pueden ser ordinarias o reglamentarias. La **ley ordinaria** es una norma de rango legal que constituye, generalmente, el último escalón en la jerarquía jurídica de las leyes de un Estado, tras la Constitución y las leyes orgánicas u otras equivalentes. Las leyes ordinarias inician su tramitación, bien a iniciativa del propio parlamento, o bien por iniciativa del poder ejecutivo. En algunos sistemas, además, se admite que sea a través de una iniciativa popular. Son también leyes ordinarias las dictadas por los órganos legislativos de los estados federados, territorios o comunidades autónomas que, dentro de un Estado federal, regional o de autonomías, tienen atribuida esta capacidad. La **ley reglamentaria** es una ley secundaria que detalla, precisa y sanciona uno o varios preceptos de la Constitución con el fin de articular los conceptos y medios necesarios para la aplicación constitucional que regulan. **2.** En España y muchos países latinoamericanos la Constitución permite en algunos casos la aprobación de normas jurídicas [decretos] por el poder ejecutivo. El **decreto** [m] **legislativo** es el que el Gobierno puede utilizar para dictar normas en materia delegada por las Cortes, sobre materias que no necesiten ser reguladas por ley orgánica. El **decreto-ley** [m] es la disposición de carácter legislativo que, sin ser sometida al órgano adecuado, se promulga por el poder ejecutivo, en virtud de alguna excepción circunstancial o permanente, previamente determinada; cf. BCN 2009, DJ 2009, DJM 2009; ►ordenanza municipal); *f* **loi** [m] (**1.** Au sens matériel on entend sous le terme de loi toutes règles de droit écrites, formulées de manière générale par un organe étatique compétent dans l'exercice du pouvoir exécutif ou législatif [gouvernement et Parlement] et en France promulgué par le Président de la République et le cas échéant après décision du Conseil constitutionnel ; au sens formel, toutes dispositions émanant de l'organe étatique investit du pouvoir législatif par la constitution [Parlement] et élaborée selon les formes prévues par la constitution ; la plupart des lois formelles sont aussi matérielles ; la loi de finances votée chaque année par le Parlement est une loi formelle car elle n'a aucune incidence directe sur les droits et devoirs des citoyens et n'est opposable qu'aux organes et institutions étatiques ; **2.** ►arrêté municipal) ; *g* **Gesetz** [n] (**1.** **G.** *im materiellen Sinne* **1.a** jede Rechtsnorm, d. h. jede hoheitliche Anordnung, die für eine unbestimmte Vielzahl von Menschen allgemein verbindliche Regelungen enthält. **1.b** *G. im formellen Sinne* jeder Beschluss der zur G.gebung zuständigen Organe, der im verfassungsmäßig vorgesehenen förmlichen G.gebungsverfahren ergeht, ordnungsmäßig ausgefertigt und verkündet ist; die meisten formellen **G.e** sind zugleich materielle G.e. Ein rein **formelles G.** ist z. B. der Haushaltsplan, weil er keine unmittelbaren Rechte und Pflichten für die Staatsbürger erzeugt, sondern nur für die Staatsorgane/Verwaltungsorgane verbindlich ist; RWB 2007; **2.** ►Ortssatzung).

3175 law enforcement agency [n] [US] *adm. leg.* (Department of government with the competency and responsibility of carrying out a specific program[me] to implement enacted legislation); *syn.* Central Authority [n] [England&Wales], authority [n] having jurisdiction; *s* **autoridad [f] competente** (Responsable del cumplimiento de los objetivos [ambientales] previstos por la ley); *syn.* administración [f] competente, administración [f] implementadora [AL]; *f* **autorité [f] exécutive** (Personne ou administration compétente responsable de la mise en œuvre et l'application de la législation) ; *g* **Vollzugsbehörde [f]** (Zuständige Behörde, die für die Vollziehung der Gesetze verantwortlich ist).

3176 lawn [n] *constr.* (**1.** Generic term for a dense, compact plant cover composed of one or more species of grass with roots that form a thick mat within the uppermost soil layer [▶root-zone layer], which is regularly mown. Lawns usually have no agricultural value. According to their use they may also contain leguminous or other herbaceous plants; *specific terms* ▶car park lawn, ▶clipped lawn, ▶crushed aggregate lawn, ▶hard-wearing lawn, ▶low-maintenance gras type, ▶play lawn [US] 1/all-around lawn [UK], ▶playfield lawn [US]/play area lawn [UK], ▶playfield turf [US]/sportsground turf [UK], ▶rolled turf, ▶sod layer [US]/turf layer [UK], ▶shade-tolerant lawn, ▶sunbathing lawn. **2. In U.S.**, *yard* most often is synonymous with the ubiquitous "lawn"); *s* **césped [m]** (Término genérico para capa de vegetación de diferentes especies de gramíneas, densa y bien enraizada con el ▶sustrato de plantación que se corta con regularidad y no se usa con fines agrícolas. Dependiendo del uso previsto puede contener también leguminosas y especies herbáceas; *términos específicos* ▶césped de aparcamiento, ▶césped de bajo mantenimiento, ▶césped de campos de deporte, ▶césped decorativo, ▶césped de juegos, ▶césped de tepes, ▶césped de tepes enrollados, ▶césped de umbría, ▶césped normal, ▶césped para transplante, ▶césped resistente, ▶césped sobre piedra partida); *f* **gazon [m]** (Tapis dense et vert recouvrant le ▶support de culture et constitué de l'association d'un nombre variable de graminées, pouvant aussi comprendre suivant l'utilité quelques herbacées ou légumineuses, dont la variation de proportion, la présence ou l'absence de certaines espèces et variétés déterminent les utilisations en fonction de la nature des terrains et de leur exposition ; *termes spécifiques* ▶gazon adapté à l'ombre, ▶gazon d'agrément et de détente, ▶gazon d'aires de stationnement, ▶gazon d'ornement, gazon de regarnissage, ▶gazon en rouleaux précultivé, ▶gazon pour terrains de jeux, ▶gazon pour terrain de sport, ▶gazon précultivé, ▶gazon rustique, ▶gazon sur grave [concassée], ▶gazon utilitaire) ; *g* **Rasen [m]** (Dichte, fest verwachsene, mit der ▶Vegetationstragschicht durch Wurzeln und Ausläufer verbundene Pflanzendecke aus einer oder mehreren Grasarten. Ein **R.** wird meist regelmäßig gemäht und unterliegt keiner landwirtschaftlichen Nutzung. Entsprechend dem Verwendungszweck und Standort kann sie auch niedrig wachsende Kräuter und Moose enthalten; *UBe* ▶belastbarer Rasen, ▶Fertigrasen, ▶Gebrauchsrasen, ▶Landschaftsrasen, ▶Parkplatzrasen, ▶Rollrasen, ▶Schattenrasen, ▶Schotterrasen, ▶Spielrasen, ▶Sportplatzrasen, ▶Zierrasen).

lawn [n] [UK]**, all-around** *constr.* ▶play lawn [US] (1)/all-around lawn [UK].

lawn [n] [UK]**, amenity** *constr.* ▶play lawn [US] (1)/all-around lawn [UK].

lawn [n]**, edging up of a** *constr. hort.* ▶lawn edge clipping.

lawn [n] [UK]**, heavy-use** *constr.* ▶play lawn [US] (1)/all-around lawn [UK].

lawn [n] [US]**, parking lot** *constr.* ▶car park lawn.

lawn [n] [UK]**, play area** *constr.* ▶playfield lawn [US]/play area lawn [UK].

lawn [n]**, pleasure** *constr.* ▶clipped lawn.

3177 lawn aeration [n] *constr.* (Generic term for perforation of clipped turf to increase entry of air; **in U.K.**, the following specific terms are distinguishable: **1. turf perforation [n]:** making a series of holes through a grass sward into the soil beneath with round, flat, or hollow tines to improve the passage of air, moisture or nutrients; BS 3975: part 5; *syn.* piercing [n]; **2. slitting [n]:** use of a flat-bladed implement to make a series of cuts through a fibrous mat of turf into the soil; BS 3975: part 5; **3. spiking [n]:** perforation of turf with a solid-tined implement; **4. pricking [n]:** the shallow perforation [spiking] of turf with a fork; *syn.* pricking [n] over; **5. springing [n]:** easing of soil compaction, usually under turf, by lightly levering with a fork: BS 3975: part 5; **6. hollow tining [n]:** perforation of turf with a hollow-tined implement to relieve compaction or to penetrate undesirable material; e.g. into heavy clay or fiber [US]/fibre [UK], and to allow the entry of air, water, nutrients, or conditioners; BS 3975: part 5; ▶dethatching); *syn.* turf corning [n] [UK]; *s* **aireación [f] del césped** (Operación para mejorar el suministro de aire en las superficies de césped haciendo —con ayuda de herramientas especiales— muchos pequeños orificios en la capa del suelo ocupada por las raíces; ▶escarificación); *f* **aération [f] de pelouse** (Opération d'entretien des surfaces engazonnées afin d'améliorer la pénétration de l'air et de l'eau sur les surfaces très piétinées. On distingue **1.** le **carrotage** : consistant à percer le gazon de trous [au minimum 200 par m²] d'un diamètre minimum de 1 cm et de 5-8 cm de longueur ; ces « carottes » sont extraites et ramassées ; *syn.* piquage [m], perforage [m]. **2.** Le **tranchage** : procédé utilisant une lame en acier qui coupe les tiges latérales du pâturin et de l'agrostide afin de forcer la croissance latérale de la racine et d'épaissir le gazon ; *syn.* aération [f] à lames. **3.** Le **carrotage et sablage** qui est une intervention après carrotage par laquelle les « petits puits » formés par l'extraction des « carrotes » sont remplis de sable par balayage afin d'assurer une pénétration en profondeur ; ▶scarification, ▶verticutage) ; *g 1* **Belüftung [f] des Rasens** (Versorgung der oberen Bodenschicht mit Luft durch viele kleine, in die Grasnarbe eingebrachte, kleine Löcher; ▶Vertikutieren); *syn.* Aerifizieren [n, o. Pl.], Lüften [n, o. Pl.] des Rasens, Rasenbelüftung [f]; *g 2* **Löchern [n, o. Pl.]** (Ausdünnen des Rasenfilzes, indem in die oberfläch verdichtete Vegetationstragschicht mindestens 200 Löcher je m² von mindestens 5 cm Tiefe und einem Durchmesser von mindestens 1 cm eingebracht werden; cf. DIN 18 919); *g 3* **Schlitzen [n, o. Pl.]** (Fräsen von Schlitzen in eine verdichtete Vegetationstragschicht mit einer Arbeitstiefe von i. d. R. 10 cm; der Schlitzabstand ist abhängig vom Verdichtungsgrad; cf. DIN 18 919); *g 4* **Lüften [n, o. Pl.] mit Hohlstacheln** (Form des Löcherns, bei der die Löcher anschließend mit Sand verfüllt werden); *syn.* Aerifizieren [n, o. Pl.] mit Hohlstacheln.

lawn area [n] *constr. recr.* ▶turf area.

lawn areas [npl] *constr.* ▶establishment of lawn areas.

lawn care [n] [US] *constr.* ▶lawn maintenance.

3178 lawn clippings [npl] *constr. hort.* (Cropped ends of mown grass; ▶clippings); *syn.* grass clippings [npl] [also US]; *s* **hierba [f] de césped cortado** (Conjunto de restos de tallos de ▶hierba segada); *f* **produit [m] de tonte de gazon** (Totalité de l'herbe coupée lors de la tonte ; *hort.* ▶produit de tonte/coupe ; *agr.* ▶produit de fauche ; *g* **Rasenschnittgut [n, o. Pl.]** (Summe der beim Mähen eines Rasens abgeschnittenen Grashalme; ▶Mähgut); *syn.* Rasenmähgut [n, o. Pl.], Grasschnittgut [n, o. Pl.].

3179 lawn edge [n] *constr.* (Border of an area of lawn that is cut straight by an edging tool or bordered by a ►mowing strip or by paved surfaces); *s* **borde** [m] **de césped** (Límite de un césped que puede estar cortado con laya, encintado por piedra de bordillo o por superficie pavimentada; ►borde de corte); *f* **filet** [m] **de pelouse** (Bord d'une aire engazonnée découpé au moyen d'une bêche, délimitée par une bande métallique, une rangée de pavés, une bordurette ou une aire stabilisée en limite d'allée ; en bordure de massif cette limite s'appelle « contre-filet » ; ►bordure de pelouse) ; *g* **Rasenkante** [f] (Begrenzung einer Rasenfläche, z. B. mit dem Spaten gestochen, durch Mähkantensteine [mit Rhizomsperre], Metallband oder befestigte Flächen eingefasst; ►Mähkante).

3180 lawn edge clipping [n] *constr. hort.* (Cutting of grass that has overgrown the edge of a lawn with shears or edger/edge trimmer); *syn.* lawn edging [n], edging up [n] of a lawn, lawn edge cutting [n]; *s* **recorte** [m] **con tijeras del borde del césped** (Corte de la hierba que ha rebosado el borde de un césped con tijeras podadoras; recortar [vb] el borde del césped con tijeras podadoras); *f* **ébarbage** [m] **des bordures** (Exécution d'une bordure de pelouse soignée en limite des aires stabilisées ou des revêtements de sol au moyen d'une tondeuse ou d'un ciseau) ; *syn.* découpage [m] des bordures ; *g* **Nachschneiden** [n, o. Pl.] **der Rasenkante** (Bilden eines ‚gepflegten' Rasenabschlusses zu befestigten Flächen oder nach dem Stechen der Rasenkante mit Schere oder Mäher); *syn.* Schneiden [n, o. Pl.] der Rasenkante.

lawn edge cutting [n] *constr. hort.* ►lawn edge clipping.

3181 lawn edge spade cutting [n] *constr. hort.* (Cutting of grass that has overgrown the edge of a lawn, with a spade); *syn.* lawn edging [n] with a spade; *s* **recorte** [m] **con laya del borde del césped** (Corte de la hierba que ha rebosado el borde de un césped con laya; cortar [vb] con laya el borde del césped); *f 1* **découpe** [f] **de filet** (Exécution d'une bordure de gazon avec la bêche en limite d'allées) ; *syn.* découpage [m] de filet ; *f 2* **découpe** [f] **de contre-filet** (Exécution d'une bordure de gazon avec la bêche en limite de massifs) ; *g* **Stechen** [n, o. Pl.] **der Rasenkante** (Bilden eines Rasenabschlusses zu Beetpflanzungen mit dem Spaten).

lawn edging [n] (1) *constr. hort.* ►lawn edge spade cutting.

lawn edging [n] (2) *constr. hort.* ►lawn edge clipping.

lawn edging with a spade [n] *constr. hort.* ►lawn edge spade cutting.

3182 lawn grass [n] *constr.* (Species of grass that are especially suitable for use in a ►turf sward); *s* **gramínea** [f] **para césped** (Especie de gramínea muy idónea para céspedes; ►tapiz de césped); *f* **graminée** [f] **pour gazon** (Graminée croissant dans les gazons ou présentant un intérêt pour l'installation des pelouses ; ►tapis de graminées) ; *g* **Rasengras** [n] (...gräser [pl]; auf Rasen wachsende oder für Rasen besonders gut geeignete Grasart; ►Rasennarbe).

3183 lawn maintenance [n] *constr. syn.* lawn care [n] [also US]; *s* **mantenimiento** [m] **del césped** *syn.* labores [fpl] de conservación del césped; *f* **entretien** [m] **d'un gazon** *syn.* entretien [m] d'une pelouse ; *g* **Rasenpflege** [f, o. Pl.].

3184 lawn reseeding [n] [US]/**lawn re-seeding** [n] [UK] *constr.* (Restoration of an existing grass sward or turf by reseeding after scarification or chemical destruction; BS 3975: part 5; ►over-seeding [US]/over-sowing [UK]); *s* **resiembra** [f] **de césped (1)** (En trabajos de mantenimiento paisajístico, siembra después de la escarificación; ►resiembra de césped 2); *f* **regarnissage** [m] **de zones mal levées** (Opération de réfection d'une surface engazonnée en mauvais état comprenant des travaux aratoires profonds, un nivellement et un ensemencement ; ►semis de regarnissage) ; *g* **Wiederherstellung** [f] **von Rasenflächen durch Nachsaat** (Landschaftsgärtnerische Arbeiten, die vorhandene [ältere] Rasenflächen nach Aufreißen und Feinplanie der zerstörten Teile durch Wiederansaat erneuern; ►Nachsaat).

3185 lawn seeding [n] *constr.* (►seeding of meadows); *syn.* grassing [n]; *s* **establecimiento** [m] **de césped/hierba con semillas** (►siembra de pradera); *f* **engazonnement** [m] **par semis** (►enherbement) ; *syn.* végétalisation [f] par un gazon ; *g* **Begrünung** [f] **durch Raseneinsaat** (►Wiesenansaat).

3186 lawn substitute plant [n] *gard. plant.* (Low ►ground cover plant which is used instead of grasses due to unfavo[u]rable conditions such as extensive shade, aridity or to avoid intensive maintenance); *s* **planta** [f] **tapizante de sustitución del césped** (Pequeña planta tapizante que se utiliza en ubicaciones con condiciones mesológicas extremas como sequía o umbría excesivas, o para evitar el mantenimiento intenso, ►coberturas); *f* **plante** [f] **de substitution de gazon** (Plante vivace tapissante utilisée dans des situations défavorables en remplacement d'un gazon, p. ex. en situation ombragée, sur emplacement sec ou en raison d'un entretien malaisé ; en région méditerranéenne on utilise du gazon en godet comme le kikouyou, le *pennisetum* ou le *dichondra* ; ►plante couvre-sol) ; *g* **Rasenersatzpflanze** [f] (Niedrige, bodendeckende Pflanze, die an Stelle von Gräsern wegen ungünstiger Standortverhältnisse, z. B. bei zu großem Schatten, zu großer Trockenheit sowie aus pflegetechnischen oder gestalterischen Gründen, verwendet wird; ►Bodendecker).

3187 lawn type [n] *constr.* (Five categories in the ►landscape contracting industry may be distinguished: ►car park lawn, ►clipped lawn, ►low-maintenance grass type, ►playfield lawn [US]/play area lawn [UK], ►play lawn [US] 1/all-around lawn [UK]); *s* **tipo** [m] **de césped** (D., Según la norma DIN 18 917 en ►industria de construcción paisajista se diferencian 5 tipos según su función: ►cesped de bajo mantenimiento, ►césped de juegos, ►césped de aparcamiento, ►césped decorativo y ►césped normal); *f* **type** [m] **de gazon** (Unité de classification des gazons dans la ►profession des entrepreneurs paysagistes ; **F.**, la réglementation mise en application le 1er octobre 1984 a pour but de relever le niveau de qualité des mélanges de semences ; **D.**, conformément à la norme DIN 18 917, on distingue cinq types de gazon : ►gazon d'agrément et de détente, ►gazon d'aires de stationnement, ►gazon d'ornement, ►gazon pour terrains de jeux, ►gazon utilitaire) ; *syn.* catégorie [f] de gazon ; *g* **Rasentyp** [m] (Rasenklassifikationseinheit im ►Garten- und Landschaftsbau gemäß DIN 18 917; es werden sieben Typen unterschieden: ►Zierrasen, ►Gebrauchsrasen, ►Sportrasen, Golfrasen, ►Parkplatzrasen, Rasen für extensive Dachbegrünung und ►Landschaftsrasen).

lawn weed [n] *gard. plant.* ►wildflowers in the lawn.

lawn work [n] *constr.* ►remedial lawn work.

laws [npl]**, building** *leg. urb.* ►zoning and building regulations.

lay-by [n] [UK] *trans.* ►pull-off [US]; ►pullout [US].

3188 layer [vb] *hort.* (Fastening a shoot down in the soil to take root while still attached to the parent plant); *syn.* propagate [vb] by layering; *s* **acodar** [vb] (En jardinería, método de reproducción vegetativa en el que se entierra un vástago en el suelo para que eche raíces sin separarlo de la planta madre; acodo [m]); *f* **marcotter** [vb] (Moyen de reproduction des végétaux consistant dans la séparation de la nouvelle plante de son pied d'origine lorsque celle-ci a émis des racines pouvant satisfaire à son développement ; marcottage [m]) ; *g* **absenken** [vb] (*Gärtnerische Pflanzenvermehrung* oberirdische Pflanzentriebe [Absenker] einer Mutterpflanze so in die Erde stecken, dass sie sich bewurzeln und neue Pflanzen entstehen).

layer [n] *met.* ►air layer; *constr.* ►blinding layer, ►clay sealing layer; *phyt.* ►closely united layer, ►dependent layer; *constr.* ►drainage layer; *min.* ►dumping layer [US]; *constr.* ►earth layer, ►filter layer; *met.* ►ground-level air layer; *arb.* ►growth layer; *constr.* ►hedge brush layer, ►hedge brush layer; *phyt.* ►herb layer; *met.* ►inversion layer; *pedol.* ►litter layer; *phyt.* ►moss layer, ►overlapping layer; *met.* ►ozone layer; *constr.* ►resiliant layer, ►root-zone layer, ►turf root-zone layer, ►seepage layer; *phyt.* ►shrub layer; *constr. hort.* ►sod layer [US]/turf layer [UK]; *constr.* ►stone layer, ►surface layer; *pedol.* ►top bog layer; *phyt.* ►tree layer; *constr.* ►turf root-zone layer; *phyt.* ►vegetation layer; *constr.* ►waterproof layer.

layer [n] [UK]**, bedding** *constr.* ►laying course for a paved stone surface.

layer [n]**, capping** *constr.* ►blinding.

layer [n]**, concrete protection** *constr.* ►protective screed.

layer [n]**, confining** *hydr.* ►aquitard.

layer [n]**, discontinuity** *limn. ocean.* ►thermocline.

layer [n]**, elastic** *constr.* ►resiliant layer.

layer [n] [US]**, filter stopping** *constr.* ►filter layer.

layer [n] [US]**, levelling** *constr.* ►base course (1).

layer [n]**, pavement** *constr.* ►surface layer.

layer [n] [UK]**, tipping** *min.* ►dumping layer [US].

layer [n] [UK]**, turf** *constr. hort.* ►sod layer [US].

layer [n] [UK]**, vapour control** *arch.* ►vapor barrier [US]/vapour barrier [UK].

layer [n]**, weatherproofing** *constr.* ►roof membrane.

3189 layer composition [n] *constr.* (e.g. the sequence of layers for a roof garden); *syn.* layer protocol [n]; *s* **disposición** [f] **de capas** (Orden de las capas p. ej. de un tejado para posibilitar su revestimiento vegetal); *f* **constituants** [mpl] **du procédé** (Disposition et épaisseurs des couches pour des jardins sur dalles) ; *syn.* organisation [f] des couches ; *g* **Schichtenaufbau** [m] **(1)** (Abfolge, z. B. der Schichten eines Dachaufbaues für eine Dachbegrünung); *syn.* Schichtenfolge [f] (FLA 1983, 11, 61).

layering [n] *constr.* ►brush layering, ►hedge layering.

layering [n]**, branch** *constr.* ►brush layering.

layering [n]**, contour brush-** *constr.* ►brush layering.

layering [n]**, propagate by** *hort.* ►layer [vb].

3190 layering [n] **of different uses** *plan.* (cf. PSP 1987, 197; coincidence of several land uses on the same plots; e.g. recreation and agriculture or forestry, nature conservation and recreation; ►multiple use, ►use 1); *syn.* combination [n] of various land uses, overlapping [n] of land uses [also UK], layering [n] of various land uses; *s* **superposición** [f] **de usos** (Presencia de diferentes funciones o usos en una misma área, como agricultura o silvicultura con recreación o protección de la naturaleza y recreación; ►uso, ►uso múltiple); *f* **superposition** [f] **des usages** (sur un même espace, p. ex. le tourisme et l'agriculture, la sylviculture, la protection de la nature et le tourisme ; ►usage, ►usage multiple, ►utilisation) ; *syn.* superposition [f] des affectations des sols ; *g* **Nutzungsüberlagerung** [f] (Zusammentreffen mehrerer Nutzungen auf derselben Fläche, z. B. Erholung und Landwirtschaft oder Forstwirtschaft, Naturschutz und Erholung; ►Mehrfachnutzung, ►Nutzung); *syn.* Überlagerung [f] verschiedener Nutzungen.

layering [n] **of plant communities** *phyt.* ►stratification of plant communities.

layering [n] **of various land uses** *plan.* ►layering of different uses.

layer [n] **of crusher-run aggregate** *constr.* ►gravel subbase [US]/gravel sub-base [UK].

layer [n] **of gravel chippings** [UK] *constr.* ►layer of gravel chips [US].

3191 layer [n] **of gravel chips** [US] *constr.* (Protective layer on ►flat roof; ►gravel-covered roof); *syn.* layer [n] of gravel chippings [UK]; *s* **lecho** [m] **de grava** (Capa de grava sobre ►tejado plano; ►azotea recubierta de grava); *f* **couche** [f] **de gravier** (Recouvrant une ►toiture à pente nulle, p. ex. sur une terrasse inaccessible ; ►toiture recouverte de gravier) ; *syn.* protection [f] de gravier, protection [f] lourde ; *g* **Kiesschüttung** [f] (Kiesschicht auf einem ►Flachdach; ►Kiesdach); *syn.* Kiesschicht [f].

layer [n] **of headers-on-edge** [UK] *constr.* ►rowlock course [US].

3192 layer [n] **of raw humus** *pedol.* (Uppermost layer of only slightly decomposed organic debris on top of nutrient-poor soils; ►needle straw [US]/needle litter [UK], ►raw humus); *s* **capa** [f] **de humus bruto** (Capa superior de suelo terrestre constituida por materia orgánica poco descompuesta; ►capa de barrujo, ►humus bruto); *f* **couche** [f] **d'humus brut** (Couche supérieure de la matière première végétale peu biodégradable à la surface du sol minéral ; ►humus brut, ►litière d'aiguilles) ; *syn.* couche [f] de mor ; *g* **Rohhumusauflage** [f] (Oberste Schicht aus kaum zersetzter organischer Substanz auf terrestrischen Böden; ►Rohhumus, ►Nadelstreu).

layers [npl] *phyt.* ►fusion of layers; *constr.* ►surface layers.

laying [n] *constr.* ►pipe laying.

laying [n] [UK]**, brush** *constr.* ►brush matting with live plant material [US].

laying [n] [US]**, sod** *constr.* ►sod laying and seeding techniques [US]/laying of turves and seeding techniques [UK].

laying course [n] *constr.* ►install a laying course.

3193 laying course [n] **for a paved stone surface** *constr.* (Sand or crushed aggregate layer on which pavers, bricks, etc. are laid; ►install a laying course; *specific terms* ►mortar bed, ►sand bed; ►bedding sand); *syn.* bedding layer [n] [UK] (SPON 1986, 369), setting bed [n] [also US]; *s* **lecho** [m] **de asiento (de adoquinado)** (Capa de arena o mortero sobre la cual se colocan los adoquines, ladrillos, etc.; *términos específicos* ►lecho de arena, ►lecho de mortero; ►arena para adoquinados); *f* **lit** [m] **de pose du pavage** (Lit de sable, en mortier, etc. sur lequel on établit un pavage ou un revêtement de sol ou de pavement ; cf. DTB 1985, 62 ; *termes spécifiques* ►lit de pose en mortier, ►lit de sable ; ►sable de pavage) ; *syn.* couche [f] de pose, *terme peu usité* couchis [m] de pose [B] ; *g* **Pflasterbett** [n] (Sand- oder Splittschicht in der Körnung 0-2 mm, 0-5 mm oder 0-8 mm, auf die Pflastersteine, Klinker etc. verlegt werden. Das **P.** hat die Aufgabe, Höhentoleranzen der Steine und die zulässigen Maßungenauigkeiten der Tragschicht auszugleichen sowie in manchen Fällen auch Wasser abzuleiten; ►Pflasterbett herrichten; *UBe* ►Sandbett, ►Mörtelbett; ►Pflastersand); *syn.* Bettung [f], Bettungsschicht [f].

3194 laying length [n] *constr.* (Length of a construction unit to be put in place; e.g. a pipe or prefabricated concrete element, steel building beam; ►overall length); *syn.* effective length [n]; *s* **largo** [m] **de tendido** (Largo de una unidad de construcción que debe ser instalada en su lugar definitivo, como p. ej. una tubería o un elemento de hormigón prefabricado; ►largo total); *f 1* **longueur** [f] (d'un tuyau, d'un élément préfabriqué en béton, d'une poutre métallique ; ►longueur totale) ; *f 2* **linéaire** [m]

(Longueur réelle mesurée ; *contexte* le linéaire de tranchée) ; *g* **Baulänge [f]** (Länge eines Bauteils, z. B. eines Rohres, eines Betonfertigteiles, Stahlträgers; ▶Gesamtlänge).

3195 laying [n] **of prefabricated stones** *constr.* (e.g. curbstones [US]/kerbstones [UK], pavers [US]/paving units [UK], fitted pieces, etc.; ▶lay with staggered joints); *syn.* laying work [n]; *s* **colocación [f] de sillares** (Disposición de piedras de encintado, adoquines, losas, etc. en lugar definitivo; ▶colocar con juntas alternadas; colocar [vb] sillares, tender [vb] losas); *syn. para pavimentos* tendido [m] de losas); *f* **mise [f] en œuvre de matériaux de construction minéraux** (Mise en place aux dimensions prévues de bordures, de pavés, etc. en pierre naturelle [matériau de construction en pierre naturelle] ou en béton [pierre reconstituée] ; ▶poser à joints décalés ; *terme spécifique* pose de pavés préfabriqués) ; *g* **Einbau [m, o. Pl.] von Werksteinen** (Verlegen von vorgefertigten Bauteilen, z. B. Kantensteine, Pflastersteine, oder Einsetzen von Passstücken etc. aus Naturstein [Naturwerkstein] oder Beton [Betonwerkstein] mit vorgegebenen Abmessungen; ▶im Verband verlegen); *syn.* Verlegen [n, o. Pl.] von Werksteinen, *z. T* Verlegearbeiten [fpl].

laying [n] **of turves and seeding techniques** [npl] [UK] *constr.* ▶sod laying and seeding techniques [US].

laying [n] **of utilities** *constr.* ▶installation of utilities.

laying [n] **work** *constr.* ▶laying of prefabricated stones.

3196 lay [vb] **in mortar** *constr.* (▶mortar bed); *syn.* lay [vb] up in mortar bed [also US] (PSP 1987, 201); *s* **tender [vb] en mortero** (▶lecho de mortero); *syn.* colocar [vb] en mortero; *f* **poser [vb] à bain de mortier** (▶lit de pose en mortier) ; *g* **in Mörtel verlegen [loc]** (Werksteine wie Pflastersteine, Platten oder Kacheln in ein▶Mörtelbett setzen).

3197 layout [n] (1) *plan. prof.* (**1.** Design concept in a two-dimensional form. **2.** Arrangement of text and graphic elements as well as other presentations on a plan, poster, placard, etc.); *s 1* **plan [m] de diseño** (Concepto de diseño representado en un plan bidimensional); *s 2* **compaginación** (Técnica de composición gráfica [textos, imágenes, etc.] para un espacio delimitado [página de papel, poster, página web, etc.]); *syn.* composición [f] (2), layout [m]; *f* **mise [f] en page** (Techniques de composition graphique [textes, images, etc.] représentant un contenu informationnel dans un espace [feuille de papier, page web, etc.] de manière hiérarchique et harmonieuse) ; *syn.* layout [m] ; *g* **Gestaltung [f]** (3) (**1.** Zweidimensionales Entwurfskonzept einer planerischen Idee. **2.** Anordnung von Text und Bildelementen sowie weiterer Darstellungsformen auf einem Plan, Poster, Plakat etc.); *syn.* Layout [n] (2).

3198 layout [n] (2) *plan.* (Spacial arrangement of design components; ▶design 2, ▶parking layout); *s* **concepto [m] estético-funcional** (Organización espacial de los componentes funcionales y de diseño en un plan de proyecto; ▶diseño de un proyecto); *syn.* concepto [m] espacial; *f* **conception [f]** (2) (Ensemble des principes généraux régissant l'affectation et la localisation de l'espace pour un aménagement ; ▶conception 1) ; *g* **Anlage [f]** (1) (Gestalterisch-funktionaler Aspekt eines Planungsentwurfes; ▶Gestaltung 1); *syn.* Layout [n] (3).

layout [n]**, topographic** *arch. constr.* ▶configuration of an area.

layout plan [n] *plan.* ▶site plan, ▶overall plan (1).

3199 lay [vb] **pavers on mortar** [US] *constr. syn.* lay [vb] setts on mortar [UK]; *s* **empedrar [vb] en (lecho de) mortero** *syn.* adoquinar [vb] en (lecho de) mortero; *f* **pose [f] à bain de mortier** (Pose traditionnelle sur une chape fraîche dite « à bain de mortier » ; *syn.* pose [f] sur plein bain de mortier [B] ; *g* **in**

Mörtel pflastern [loc] *syn.* Pflastersteine in Mörtel verlegen [loc], in Mörtel pflästern [loc] [CH].

3200 lay [vb] **pavers on sand** [US] *constr.* (▶bedding sand); *syn.* lay [vb] setts on sand [UK]; *s* **empedrar [vb] en (lecho de) arena** (▶arena para adoquinados); *syn.* adoquinar [vb] en (lecho de) arena; *f* **pose [f] sur lit de sable** (Technique de pose de matériaux lourds [dalles, pavés] sur une couche de sable dont l'épaisseur dépend de l'utilisation du revêtement ; ▶sable de pavage) ; *g* **in Sand pflastern [loc]** (▶Pflastersand); *syn.* Pflastersteine in Sand verlegen [loc], in Sand pflästern [loc] [CH].

lay [vb] **setts on mortar** [UK] *constr.* ▶lay pavers on mortar [US].

lay [vb] **setts on sand** [UK] *constr.* ▶lay pavers on sand [US].

lay [vb] **to falls** [UK] *constr.* ▶lay to slope [US].

3201 lay [vb] **to slope** [US] *constr.* (*Context* to lay a pavement to falls/slope of 1 : 40); *syn.* lay [vb] to falls [UK]; *s* **colocar [vb] adoquinado con pendiente** (*Contexto* tender un pavimento con pendiente de p. ej. 2,5%); *f* **pose [f] des pavés avec une pente de** (2,5 %) ; *g* **ins Gefälle verlegen [vb]** (*Kontext* eine Pflasterung mit 2,5 % Gefälle verlegen).

lay [vb] **up in mortar bed** [US] *constr.* ▶lay in mortar.

3202 lay with a (5cm) blinding layer of sand [loc] *constr.* (SPON 1986, 373); *s* **tender sobre lecho de arena (de 5 cm) [loc]**; *f* **mise en œuvre sur un lit de sable de 5 cm d'épaisseur [loc]** ; *g* **auf ein (5 cm hohes) Sandbett verlegen [loc]** *syn.* in (50 mm) Sand verlegen [loc].

3203 lay [vb] **with staggered joints** *constr.* (Built in a random, rectangular or random, irregular pattern of structural units for pavement or masonry; *specific term for brick work* lay [vb] in stretcher bond/running bond; ▶bond; *opp.* lay with stack bond joints); *s* **colocar [vb] con juntas alternadas** (Disposición de los mampuestos, adoquines o losas de pavimentación siguiendo un ▶aparejo dado de tal manera que no se formen juntas cruzadas; *opp.* colocar con juntas cruzadas); *syn. para pavimentos* tender [vb] con juntas alternadas; *f* **poser [vb] à joints décalés** (Disposition de pierre de taille, de pavés, de dalles, dont le principe d'assemblage ne permet pas la formation de joints croisés ; ▶appareillage ; *opp.* poser à joints croisés) ; *syn.* poser [vb] à joints rompus, appareillage [m] en opus insertum ; *g* **im Verband verlegen [loc]** (Zusammenfügen von Mauersteinen, Pflastersteinen oder Belagsplatten zu einem vorgegebenen Muster dergestalt, dass keine Kreuzfugen entstehen; ▶Verband 2; *opp.* mit Kreuzfuge verlegen).

3204 leachate [n] *envir.* (Leached water from waste dump [US]/waste tip [UK] or sanitary landfill [US] 2/landfill site [UK]); *syn.* seep water [n] [US]; *s* **líquidos [mpl] lixiviados contaminados** (Agua contaminada con sustancias residuales que fluye de un depósito de residuos hacia el subsuelo); *syn.* lixiviados [mpl] contaminados; *f 1* **lessiviat [m]** (Eaux pluviales de ruissellement des routes, de décharges ou autres équipements contenant des substances polluantes et se déversant dans le milieu naturel après traitement ou non ; *terme spécifique* pluviolessiviat [m] routier) ; *f 2* **lixiviat [m]** (Eau d'infiltration dans une installation de stockage, entrée en contact avec des déchets et, suivant la nature de l'installation [décharge existante ou nouvelle installation de stockage] et celle des déchets, récupérée ou non par des équipements de collecte, de stockage et de traitement avant son rejet dans le milieu naturel) ; *g* **Sickerwasser [n]** (1) (...wässer [pl]; belastetes Wasser, das von Straßenflächen, aus Mülldeponien oder Auffüllplätzen in den Untergrund sickert, wenn keine entsprechenden Kläreinrichtungen [Vorbehandlungsanlagen] vorgesehen sind).

3205 leaching [n] **of nutrients** *pedol.* (Removal of nutrients by percolating water to lower soil horizons or into groundwater, whereby the nutrients are then no longer available for uptake by the plant; ▶available to plants, ▶enhancement of soil fertility); *s* **lixiviación [f]** (Remoción o migración de materiales del suelo en solución, pseudodisolución o en suspensión coloidal arrastrados por las aguas de percolación, de manera que no están ▶disponibles para las plantas. La velocidad y la intensidad de la **l.** depende de la textura del suelo, la concentración de nutrientes en el agua edáfica y de la cantidad de ▶agua de percolación, pero sobre todo del tipo, el momento y la intensidad de las medidas de ▶mejora de la fertilidad del suelo y del abonado; cf. DINA 1987); *syn.* levigación [f], lavado [m] edáfico de nutrientes, eluviación [f] de nutrientes; *f* **lixiviation [f]** (Entraînement vers les horizons profonds ou dans la nappe phréatique des éléments solubles par l'▶eau de gravité si bien qu'ils ne sont plus ▶disponibles pour les plantes ; la lixiviation dépend non seulement des caractéristiques du sol, de la concentration des sels solubles et de la quantité d'eau gravitaire mais encore des techniques de ▶mise en valeur du sol par la culture ainsi que des conditions d'apport d'engrais) ; *g* **Nährstoffauswaschung [f]** (Verlagerung von Nährstoffen durch ▶Sickerwasser 2 in tiefere Bodenschichten oder ins Grundwasser, so dass sie nicht mehr ▶pflanzenverfügbar sind. Zeitlicher Ablauf und Ausmaß der **N.** hängt von der Bodenart, Nährstoffkonzentration des Bodenwassers und der Sickerwassermenge, insbesondere aber von Art, Zeitpunkt und Intensität der ▶Bodenpflege und Düngung ab); *syn.* Auslaugung [f] von Nährstoffen, Nährstoffauslaugung [f].

3206 leader [n] *arb. hort.* (Main stem that forms the apex of a tree); *s* **retoño [m] principal** (Tronco principal que forma el ápice de un árbol); *syn.* retoño [m] terminal, brote [m] principal, brote [m] terminal; *f* **flèche [f]** (Axe principal terminal dans le prolongement du tronc, portant après la taille l'ensemble des ramifications) ; *syn.* pousse [f] apicale (DFM 1975) ; *g* **Leittrieb [m]** (Haupt- oder Mitteltrieb eines Baumes oder Stämmlings [nach erfolgtem Schnitt] als durchgehende Verlängerung der Stammachse); *syn.* Mitteltrieb [m], Stammverlängerung [f], Terminale [f], Terminaltrieb [m].

leader [n]**, contractor's project** *constr. prof.* ▶contractor's agent.

leader [n]**, project** *plan.* ▶project manager.

leader drain [n] *agr. constr.* ▶collector drain.

3207 leaf [n] *bot.* (Flat, broad-shaped foliage of seed-bearing plants and ferns. The leaf of a fern is called a 'frond'); *s* **hoja [f]** (Órgano generalmente lateral de los briófitos, pteridófitos y antófitos que brota del tallo o de las ramas y —en general— tiene forma laminar y estructura dorsiventral. La función principal de las **hh.** consiste en la asimilación de los hidratos de carbono, gracias a su abundante contenido de clorofila y su forma, estructura y disposición en el tallo; *término específico para hoja de helecho [pteridófitos]* fronda [f]/fronde [m]; cf. DB 1985); *syn.* nomofilo [m] (DB 1985); *f* **feuille [f]** (Organe fondamental [limbe assimilateur] des végétaux supérieurs [Angiospermes et Gymnospermes] ; chez les Ptéridophytes [fougères] on utilise le terme de « fronde ») ; *g* **Laubblatt [n]** (Flächig ausgebreitetes Blatt der Samenpflanzen und Farne. Blätter der Farne werden **Wedel** genannt).

leafage [n] *bot. hort.* ▶foliage.

leafage [n] **of woody plants** *bot. hort.* ▶foliage of woody plants.

3208 leaf area index [n] *phyt.* (*Abbr.* LAI; Ratio of total leaf area covered by a particular species to the total sample area. Many crops, such as corn, have LAI of about 4, meaning that for every m^2 of ground 4m^2 of leaves lie above it; TEE 1980, 134);

s **índice [m] (de superficie) foliar** (Relación entre la superficie total de hojas de una especie y la superficie de suelo); *f* **indice [m] (de surface) foliaire (LAI)** (Indice défini comme la moitié de la surface foliaire totale d'une espèce déterminée ou d'un couvert végétal par unité de surface au sol ; indice utilisé pour l'évaluation du potentiel agronomique ou du rendement de certaines cultures) ; *g* **Deckungsgrad [m] der Blattfläche(n)** (Prozentzahl der Bodenflächeneinheit, die von der Blattfläche einer bestimmten Pflanzenart überschirmt ist); *syn.* Blattflächenindex [m].

3209 leaf bud [n] *bot.* *s* **yema [f] foliífera** (Yema que produce una ramita hojosa al desarrollarse); *syn.* yema [f] foliar; *f* **bourgeon [m] foliaire** (Excroissance apparaissant à l'aisselle des feuilles ou à l'extrémité des rameaux et donnant naissance à une feuille) ; *g* **Blattknospe [f]** (Knospe, die nur junge Blattanlagen enthält).

leaf burn [n] *bot. hort.* ▶leaf necrosis.

leaf canopy [n] *for. phyt.* ▶tree canopy.

leaf drop [n] [US] *bot. hort.* ▶premature shedding o leaves.

leaf fall [n] [US] *bot.* ▶defoliation.

leaf fall [n] [US]**, premature** *bot. hort.* ▶premature shedding of leaves.

3210 leafing out [n] *bot.* (Production of new leaves and shoots [in spring]; leaf [vb]; ▶coppice shoot, ▶emerging leaves); *syn.* unfolding foliage [n], spring bud break [n] [also US] (TEE 1980, 114), shooting out [n]; *s* **brotación [f]** (Desarrollo de las yemas de las plantas y formación de vástagos o flores; cf. DB 1985; ▶brote de cepa, ▶foliación); *f* **pousse [f]** (Poussée de croissance des bourgeons et des feuilles au printemps ; pousser [vb] ; ▶feuillaison, ▶rejet de taillis) ; *syn.* poussée [f], reprise) ; *syn.* reprise [f] de végétation (au printemps) ; *g* **Austrieb [m]** (Das Hervorbringen von neuen Blättern und Trieben [im Frühling] aus [Überwinterungs]knospen; austreiben [vb], ausschlagen [vb]; ▶Laubaustrieb, ▶Stockaustrieb 1).

3211 leafless [adj] *bot.* (Descriptive term for a plant without leaves); *syn.* bare [adj]; *s* **sin hojas [loc]** *syn.* desnudo/a [pp/adj]; *f* **dénudé, ée [pp/adj]** (Plante dépourvue de feuilles) ; *syn.* sans feuilles [loc] ; *g* **unbelaubt [pp/adj]** (So beschaffen, dass eine Pflanze ohne Blätter ist).

3212 leaf litter [n] *bot. pedol.* (Uppermost layer, the L layer, of organic debris on the ground, i. e., the freshly fallen or slightly decomposed foliate material; ▶needle straw [US]/needle litter [UK], ▶organic horizon); *s* **hojarasca [f] (2)** (Conjunto de hojas que se han caido de los árboles y forman el ▶mantillo; ▶capa de barrujo); *syn.* hojas [fpl] secas; *f* **litière [f] feuillue** (Couche supérieure [sous-horizon L] du sol, résultant de la chute de litière, c.-à-d. les feuilles récemment tombées, ou seulement légèrement décomposées ; ▶horizon organique, ▶litière d'aiguilles) ; *g* **Laubstreu [f]** (Auf der Bodenoberfläche liegender, noch nicht oder kaum zersetzter Bestandesabfall von Laubgehölzen, der die ▶Humusauflage bildet; ▶Nadelstreu); *syn.* Bestandsabfall [m], Laubdecke [f].

leaf litter decay [n] *pedol.* ▶leaf litter decomposition.

3213 leaf litter decomposition [n] *pedol.* (▶decomposition 1); *syn.* leaf litter decay [n]; *s* **descomposición [f] de la hojarasca** (▶descomposición); *f* **décomposition [f] de la litière des feuilles** (La ▶décomposition en matière minérale de la matière organique constituant la litière par les êtres vivants du sol entraîne la formation de l'humus) ; *g* **Laubstreuzersetzung [f]** (Humus bildender, biologischer Abbau von Falllaub durch Bakterien und Kleintierlebewesen im und auf dem Boden; ▶Abbau 1).

3214 leaf mold [n] [US]/**leaf-mould** [n] [UK] *pedol. hort.*
(Garden soil chiefly composed of naturally decaying, decayed or
composted fallen shrub, ▶dead leaves and herbage; ▶compost-
ing, ▶leaf litter); *syn.* leaf soil [n]; *s* **mantillo [m] de hojarasca**
(Enmienda de horticultura originada naturalmente por descomposición de la ▶hojarasca 1 o fabricada por ▶compostaje de la
misma); *f* **terreau [m] de feuilles** (Amendement horticole fabriqué artificiellement par ▶compostage ou élaboré naturellement et
résultant de la décomposition de ▶feuilles mortes, ▶de litière
feuillue, des produits de tonte ; *g* **Lauberde [f]** (Durch natürliche
Zersetzung von ▶abgestorbenem Laub oder durch ▶Komposti-
rung von ▶Laubstreu entstandene gärtnerische Erde).

3215 leaf necrosis [n] *bot. hort. phytopath.* (Death of living
tissue in a leaf; ▶necrosis); *syn.* leaf burn [n]; *s* **necrosis [f] de la
hoja** (Muerte localizada o general de las células de una hoja. Se
distinguen diferentes tipos como p. ej. **n.** de las venas o neural y
n. macular, cf. DB 1985; ▶necrosis); *syn.* necrosis [f] foliar;
f **nécrose [f] foliaire** (Maladie foliaire provoquant la mort des
tissus du limbe et caractérisée par une croissance réduite de
certaines portions du lobe foliaire ; ▶nécrose) ; *g* **Blattnekrose
[f]** (Absterben der Zellen eines Blattbereiches; es gibt Blatt-
randnekrosen, Intercostalnekrosen [*intercostal* >zwischen den
Rippen<] und punktförmige **B.n**; ▶Nekrose).

leaf soil [n] *pedol. hort.* ▶leaf mold [US]/leaf-mould [UK].

3216 leaf surface [n] *bot.* (Outside of an individual leaf);
s **superficie [f] foliar**; *f* **surface [f] foliaire** ; *g* **Blattfläche [f]**
syn. Blattspreite [f].

3217 leaf tendril [n] *bot.* (Modified leaf outgrowth that curls
around anything for support; ▶tendril); *s* **zarzillo [m] foliar**
(Órgano filamentoso enroscado y haptotrópico formado a partir
de la hoja como en muchas leguminosas. Puede corresponder a
toda la hoja, como en la almorta *[Lathyrus aphaca]*, o a parte de
ella, al raquis y algunos foliolos, como en el guisante *[Pisum
sativum]*; ▶zarzillo) ; *f* **vrille [f] foliaire** (Organe filiforme d'une
plante grimpante résultant de la transformation des feuilles, de
section en général circulaire, capable de s'enrouler autour d'un
support par tropisme de contact ; ▶vrille) ; *g* **Blattranke [f]**
(Fadenförmig umgewandelter, kontaktreizbarer, ausschließlich
dem Klettern dienender Teil der Blattspreite, z. B. bei Erbse
[Pisum sativum], verschiedenen Platterbsen *[Lathyrus]* und
Gartenkürbis *[Cucurbita pepo]*; ▶Ranke).

3218 leakage [n] *envir.* (Escape of potable or wastewater from
a pipe which is not watertight into the soil; leak [vb]); *syn.* seepage [n] (2), pipe leakage [n]; *s* **escape [m]** (Derrame de agua
potable o de aguas residuales por rotura o porosidad de tubería);
f **fuite [f]** (Écoulement fin de matériaux liquides ou gazeux au
travers de conduits non étanches) ; *syn.* suintement [m] (2) ;
g **Aussickern [n, o. Pl.]** (Auslaufen von Abwasser oder Trinkwasser ins Erdreich aus undichten Leitungen; aussickern [vb],
ausströmen [vb]); *syn.* Ausströmen [n, o. Pl.].

lean concrete [n] *constr.* ▶lean-mixed concrete.

3219 lean-mixed concrete [n] *constr.* (Concrete with a
relatively low cement content); *syn.* lean concrete [n]; *s* **hormigón [m] magro** (Hormigón con baja proporción de cemento);
syn. hormigón [m] pobre, concreto [m] magro/pobre [C]; *f* **béton
[m] maigre** (Béton peu dosé en ciment [teneur en ciment
inférieure à 200 g de ciment par mètre cube de béton] et en fines,
ce qui en réduit l'ouvrabilité et limite la résistance, employé entre
autre pour des remplissages de cavités, des fondations importantes, certains remblais lorsque des apports de terre ne pourraient en garantir la stabilité) ; *g* **Magerbeton [m]** (Unbewehrter
Beton der Festigkeitsklasse C 8/10 [*bis 2003* B 5] mit einem
geringen Zementanteil).

leasehold [n] *agr.* ▶agricultural leasehold (property).

3220 leasehold garden [n] *urb.* (Garden for which a tenant
pays rent to the lessor for the use of the land and the harvesting of
vegetables or fruit; ▶community garden [US]/allotment garden
[plot] [UK], ▶rented garden [UK], ▶tenant garden); *s* **jardín
[m] arrendado** (Jardín para el cual el arrendador paga una renta
de arrendamiento al arrendatario para poder utilizarlo); ▶huerto
recreativo urbano, ▶jardín alquilado, ▶jardín de inquilinos);
f **jardin [m] affermé** (Jardin dont l'usage est cédé par un bailleur
à un preneur moyennant une rente annuelle) ; ▶jardin collectif,
▶jardin locatif, ▶jardin loué, ▶jardin ouvrier) ; *g* **Pachtgarten
[m]** (Garten, den der Verpächter dem Pächter gegen Entgelt
[Pachtzins] zum Gebrauch, zum Heranziehen und zum Genuss
der Früchte überlässt; ▶Kleingarten, ▶Mietergarten, ▶Miet-
garten).

leasehold property [n] *agr.* ▶agricultural leasehold (property).

leaves [npl], **new** *bot. hort.* ▶emerging leaves.

leave [vb] **soil lying by side of an excavation** [UK]
constr. ▶keep soil beside an excavation [US].

3221 leave [vb] **uncultivated** *agr. syn.* let agricultural land
lie fallow [loc]; *s* **dejar [vb] tierras en barbecho** *syn.* dejar [vb]
tierras yermas; *f* **laisser [vb] en friche** (Abandonner l'exploitation de terres agricoles par suite de manque de main d'œuvre ou
pour non-rentabilité ; l'abandon des activités et de la maîtrise des
sols suite à la *déprise agricole* provoque le retour à la friche) ;
g **brach fallen [vb]** (Kulturflächen unbestellt lassen; *Kontext* die
Ackerflächen fallen brach; Brachfall [m] von [landwirtschaftli-
chen] Flächen [FRE 2003, 42]; im Franz. gibt es den Begriff der
déprise agricole, der das allmähliche Brachfallen landwirtschaftlicher Flächen durch Aufgabe der Bewirtschaftung beinhaltet).

3222 ledge [n] (1) *constr. eng.* (Narrow horizontal shelf cut
into an artificially created, steep slope); *syn.* terrace [n] [also US],
berm(e) [n] [also US]; *s* **berma [f] (1)** (Bancal horizontal
estrecho en taludes muy pendientes); *syn.* lisera [f], bancal [f],
banqueta [f] [Es, MEX] (1), banquina [f] [RA] (2), zarpa [f]
[RCH]; *f* **risberme [f] (1.** Étroite surface horizontale intermédiaire aménagée lors de fouilles profondes ou du remodelage
du sol afin d'interrompre l'érosion sur les talus à forte pente.
2. Plate-forme aménagée en pied, en milieu ou en tête de talus de
déblai ou de remblai pour augmenter sa stabilité et faciliter son
entretien) ; *syn.* cran [m] de pente ; *g* **Berme [f] (1)** (Schmaler,
horizontaler Absatz in steilen [Anschnitt]böschungen, der einmal
zur Stabilisierung und zum anderen zur Erleichterung der Pflege
und Unterhaltung der Böschungsflächen dient).

3223 ledge [n] (2) *geo.* (Shelf-like projection on the side of a
rock or mountain); *s* **plataforma [f] saliente** (Escalón en forma
de terraza en una pared rocosa); *syn.* cama [f] de roca; *f* **replat
[m]** (Partie plate [terrasse] en épaulement sur une paroi rocheuse,
un versant d'une vallée glaciaire) ; *g* **Felsterrasse [f]** (Terrassen-
förmiger Absatz in einer Felswand).

lee side [n] *met.* ▶leeward side.

3224 leeward side [n] *met.* (Sheltered side, away from the
wind; *opp.* ▶windward side); *syn.* lee side [n]; *s* **sotavento [m]**
(*Generalmente se utiliza sin artículo*; lado protegido del viento;
opp. ▶barlovento); *f* **côté [m] sous le vent** (1. Côté le plus
éloigné du vent sur un bateau ; 2. *pour une île* côte protégée des
vents dominants, en général la côte occidentale ; ▶face au vent) ;
g **Lee [f** *auch* **n]** (*meist ohne Artikel gebräuchlich*; dem Wind
abgekehrte Seite; *opp.* ▶Luv).

3225 leftover area [n] *plan. urb.* (Small, often irregular
shaped piece of land, remaining after site planning and usually
treated as a green area); *syn.* residual parcel [n] of land, outlot [n]
[also US] (UDH 1986, 34), leftover land [n] [also US] (TGG

1984, 185), leftover landform [n] [also US] (VIA 1980, 63); *s* **espacio [m] residual** (Superficie sobrante en la planificación de una zona que se «aprovecha» como zona verde en miniatura); *syn.* parcela [f] sobrante; *f* **espace [m] résiduel** (Surface restante lors d'un aménagement, ce qui reste entre les édifices, affectée souvent à un usage de mini-espace vert) ; *g* **Restzwickel [m]** (Bei Planungen im besiedelten Bereich entstehende Restfläche, die oft für eine Minigrünfläche übrigbleibt); *syn.* Restfläche [f].

leftover land [n] [US] *plan. urb.* ►leftover area.

leftover landform [n] [US] *plan. urb.* ►leftover area.

legal announcement [n] [US] *adm. leg. plan.* ►public announcement.

3226 legal basis [n] *leg.* (Legislative enactments and court decisions which together constitute the fundamental legal principle for all lawful acts; **in U.S.**, constitutional provisions, legislative enactments, and judicial decisions that provide the basis for all lawful acts and program[m]s); *s* **base [f] legal** (Base legislativa o sentencia de tribunal de un acto legal, como la Constitución, las leyes estatales, de las CC.AA., ordenanzas o normas municipales); *f* **fondement [m] légal** (Source ou base d'un acte légal régissant le droit objectif en vigueur pour tous les citoyens, p. ex. la constitution [**D.**, la Loi Fondamentale], les lois, les décrets, les arrêtés municipaux) ; *syn.* base [f] légale ; *g* **Rechtsgrundlage [f]** (Quelle oder Fundstelle eines Rechtssatzes des für alle geltenden objektiven Rechts, z. B. Grundgesetz, Bundes- und Landesgesetze, Rechtsverordnungen, städtische Satzungen).

legal designation [n] *leg. plan.* ►statutory designation.

legal game [n] *leg. hunt.* ►huntable game.

3227 legally-binding [ppr] *adm. leg.* (Descriptive term for the fulfillment/fulfilment of formal conditions required in order for a law to come into effect after its approval by an administrative body, e.g. a public notice; **in U.S.**, the term is applied to development plans only in a specialized zoning process for 'planned unit developments', 'special use permits', 'conditional use permits', and the like); *syn.* legally effective [adj], binding in law [loc], with legally-binding effect [loc]; *s* **obligatorio [adj]** (Término descriptivo para el hecho de que un plan ha cumplido todas las condiciones formales para entrar en vigor después del procedimiento administrativo correspondiente, como p. ej. en el caso de un ►plan parcial [de ordenación] aprobado definitivamente en el momento de la publicación); *syn.* legal [adj] y válido [adj], con carácter jurídicamente vinculante [loc] (BECHER 1999); *f* **légal, ale [adj] et valide [adj]** (Reconnaissance formelle de la validité et de l'entrée en application d'un acte juridique suite à une procédure administrative, p. ex. après la publication de l'approbation du ►plan local d'urbanisme [PLU]/plan d'occupation des sols [POS] rendu public, celle-ci ne devenant exécutoire qu'après les mesures de publicités) ; *syn.* exécutoire [adj] ; *g* **rechtsverbindlich [adj]** (Kennzeichnung für die Erfüllung der formellen Voraussetzungen für den Eintritt der Rechtswirkungen nach Abschluss eines verwaltungsrechtlichen Verfahrens; z. B. mit der öffentlichen Bekanntmachung des als Satzung beschlossenen ►Bebauungsplanes wird dieser rechtsverbindlich); *syn.* rechtswirksam [adj].

legally-binding land-use plan [n] *adm. leg. urb.* *►zoning map [US]/Proposals Map/Site Allocations Development Plan Document [UK].

legally effective [adj] *adm. leg.* ►legally-binding.

3228 legally protected habitat [n] *conserv. ecol. leg.* (Habitat in which threatened plants and animals live and are protected; the cornerstone of European nature conservation policy is the ►Habitats Directive [more formally known as **Council Directive 92/43/EEC on the Conservation of natural habitats and of wild fauna and flora**]. The Directive lays down measures for the preservation, protection and improvement of the quality of important, rare, and threatened natural habitats and specific species of plant and animals. It aims to protect some 220 habitats and approximately 1000 species listed in the directive's Annexes, which are considered to be of European interest, following criteria given in the directive, requiring the restoration or maintenance of natural habitats and species of 'Community interest' at favourable conservation status. Each member state must design and implement the necessary measures to achieve favourable conservation status for the habitats and species within those sites. Measures include consideration of the sites in local and regional land use plans, special assessments of the impacts of certain activities on the ►conservation status 2 of designated habitat types and species within the site. The State is required to monitor the conservation status of protected habitats. The directive led to the setting up of a network of ►**Special Areas of Conservation**, which together with the existing **Special Protection Areas** form a network of protected sites across the European Union called ►Natura 2000. **In UK** strengthened habitat regulations came into force in 2008, transposing the European Habitats Directive and providing further safeguards for European Protected Species. The new laws require stronger consideration for animals and affect trade and possession of protected species; they are enforceable through amendments to the Habitats Regulations and the introduction of new Offshore Marine Conservation Regulations); *s* **hábitat [m] protegido legalmente** (Biótopo en el que moran especies amenazadas de flora y fauna que están bajo protección. La base de la política de protección de la naturaleza de la UE es la ►Directiva de Hábitats. Esta directiva define medidas de preservación, protección y mejora de la calidad de los hábitats naturales importantes, raros y amenazados, y de especies específicas de plantas y animales. Tiene como objetivo proteger unas 1000 especies y alrededor de 220 hábitats que son considerados de interés europeo y deben ser rehabilitados y conservados en un ►estado de conservación favorable y protegidos declarándolos ►zonas especiales de conservación. Los Estados miembros están obligados a supervisar este estado de conservación. **En Es.** el capítulo III de la ley 42/2007, así como las correspondientes leyes autonómicas, define las medidas a tomar para cumplir con los objetivos de la Directiva de Hábitats en este respecto. En los anexos I, II y III están listados los hábitats de interés comunitario, las especies de interés comunitario para cuya conservación es necesario designar zonas especiales de conservación y los criterios de selección de los lugares que pueden clasificarse como lugares de importancia comunitaria y designarse zonas especiales de conservación; cf. Directiva 92/43/CEE de Hábitats, Ley 42/ 2007); *f* **biotope [m] protégé par arrêté préfectoral (±)** (Protection des milieux naturels régie par le Code de l'environnement ; art. L. 411-1 et L. 411-2 ; art. R. 411-15 à R. 411-17 ; partie du territoire constituée par des formations naturelles peu exploitées, où l'exercice des activités humaines est réglementé soit pour préserver les biotopes nécessaires à la survie d'espèces animales ou végétales protégées, soit pour protéger l'équilibre biologique de certains milieux ; la protection est édicté par un arrêté préfectoral de protection de biotope [APB ou dans le langage courant un **arrêté de biotope**] après avis de la commission départementale des sites, perspectives et paysages et de la chambre régionale d'agriculture, et le cas échéant de l'office national des forêts [si le territoire bénéficie du régime forestier]. Aucune enquête publique ou consultation préalable des propriétaires et des communes concernées n'est prévue par les textes. En pratique, cette consultation est presque toujours effectuée pour assurer une meilleure efficacité dans l'application de cet outil réglementaire. Cette procédure est rapide à mettre en place, si elle ne rencontre pas d'opposition manifeste. Elle peut concerner des sites de petite surface. Cet arrêté délimite le périmètre géographique concerné et fixe les mesures tendant à favoriser la conservation des biotopes : interdiction ou réglementation des activités susceptibles de porter une atteinte effective au milieu ; en 2003, 516 arrêtés de protection de biotope ont été pris et couvrent 0,5 % du territoire national ; DIREN Picardie, 2006 ; ▶Directive Habitat Faune Flore [DHFF], ▶zone spéciale de conservation [ZSC]) ; **g gesetzlich geschützter Biotop [m]** (*auch* gesetzlich geschütztes Biotop [n]; Lebensraum, in dem Pflanzen und Tiere leben, die in ihrem Bestand gefährdet und deshalb zu schützen sind; Gebiete von europäischer Bedeutung sind nach den Maßgaben des Art. 4 Abs. 4 der ▶Fauna-Flora-Habitat-Richtlinie 92/43/EWG [FFH-RL] zu schützen. In dieser Schutzverordnung werden aber besondere Schutzzweck, die dementsprechenden Erhaltungsziele sowie die dafür erforderlichen Gebote, Verbote und Gebietsbegrenzungen unter Berücksichtigung der Einwirkungen von außen festgelegt. Das Ziel ist, ca. 220 Biotoptypen und annähernd 1000 Arten, die europaweit bedeutend und in den Anhängen der FFH-RL aufgelistet sind, nach vorgegebenen Kriterien so wiederherzustellen und zu pflegen, dass sie einen „günstigen ▶Erhaltungszustand" erreichen. Die Staaten sind verpflichtet diesen Erhaltungszustand zu überwachen; **in D.** sind in § 30 BNatSchG sowie in den jeweiligen Paragrafen der Landesnaturschutzgesetze die Listen der **g. g. B.**e aufgeführt und festgeschrieben, dass diese Lebensräume gegen jede Maßnahme, die zur Zerstörung, Beschädigung oder erheblichen/nachhaltigen Beeinträchtigung des charakteristischen Zustandes führen kann, zu schützen sind; eine vom Gesetz abhängige verkürzte Bezeichnung [nach der Artikel-/Paragrafen-Nr. der jeweiligen Landesnaturschutzgesetze] wird in den einzelnen Bundesländern verwandt, z. B. 13d-Biotop in Bayern, 24a/24b-Biotop in Niedersachsen, 28a-Biotop in Nordrhein Westfalen etc.; ▶besonderes Schutzgebiet).

legal notice [n] [US] *adm. leg. plan.* ▶public announcement.

3229 legal obligation [n] *leg. plan.* (**1.** Formal and binding acknowledgment of a liability to carry out a specific action. **2.** *leg. plan.* Statutory requirement to comply with publicly-announced plans, e.g. ▶community development plans); *syn.* binding force [n]; **s vinculación [f] legal** (Carácter obligatorio de planes [p. ej. ▶planes urbanísticos] después de su aprobación y al entrar en vigencia); *syn.* obligatoriedad [f] legal; **f opposabilité [f] (au tiers)** (Caractère obligatoire et exécutoire d'un plan après avoir été rendu public ou approuvé, p. ex. des ▶documents de planification urbaine) ; **g Verbindlichkeit [f]** (Rechtswirksamkeit von Plänen durch Verkündung oder öffentliche Bekanntmachung, z. B. von ▶Bauleitplänen).

legal protection [n] *conserv. leg.* ▶place under legal protection.

3230 legend [n] *plan. surv.* (Explanation of drawing symbols on plans or maps); *syn.* key [n]; **s leyenda [f]** (Explicación de símbolos en planos o mapas); **f légende [f]** (Texte explicatif des signes ou éléments graphiques divers figurant sur un plan ou une carte) ; **g Legende [f]** (Zeichenerklärung auf Plänen oder Karten).

legislation [n] [UK], **"right to roam"** *leg. recr.* ▶right of public access.

3231 legislation [n] **on environmental protection** *envir. leg. pol.* (Laws for ▶environmental protection, and sanctions for prohibited conduct; ▶polluter-pays principe, ▶precautionary principe, ▶public responsability principle); **s legislación [f] ambiental** (Conjunto de normas legales que regulan la ▶protección ambiental que son actualizadas de tiempo en tiempo y en los países de la Unión Europea cada vez están más determinadas por la legislación ambiental comunitaria; ▶principio de causalida, ▶princpio de cooperación, ▶principio de prevención [de

deterioros ambientales]); *syn.* derecho [m] de protección de la naturaleza y del medio ambiente; **f droit [m] (de la protection) de l'environnement** (Branche du droit administratif comportant l'ensemble des lois et réglementations nationales et internationales relatives à la ▶protection de l'environnement ; droit en constante évolution et largement influencé par la législation européenne ; droit basé sur [1] le ▶principe de précaution, le ▶principe d'action préventive et de correction, [2] le ▶principe pollueur-payeur, [3] le ▶principe de responsabilité publique [de l'État et des collectivités publiques]) ; **g Umweltrecht [n]** (Gesamtheit der nationalen und internationalen Rechtsnormen, die regeln, welches Verhalten der Gesetzgeber zum Schutze der Umwelt [▶Umweltschutz] gebietet oder verbietet und welche Sanktionen für verbotswidriges Verhalten drohen. Das **U.** im heutigen Sinne, das erst seit den 1970er-Jahren praktiziert wird, basiert auf drei Prinzipien: **1.** ▶Vorsorgeprinzip — Umweltschäden gar nicht erst entstehen zu lassen, **2.** ▶Verursacherprinzip — wer verschmutzt, ist für die Sanierung verantwortlich und **3.** ▶Gemeinlastprinzip — Einbeziehung der Privaten bei Entscheidungen der öffentlichen Hand. In Europa werden die einschlägigen nationalen Rechtsnormen ständig weiterentwickelt und immer mehr durch das Umweltrecht der Europäischen Union und das Völkerrecht beeinflusst).

legislation [n] **upon nature conservation** *conserv. leg.* ▶nature conservation enactments.

3232 leisure [n] *plan. recr.* (Period of time in which a person is free to engage in self-chosen activities, ranging from sleeping, relaxation, eating, personal hygiene to such other leisure pursuits as physical activities. The term **l.** is often confused with recreation, which is only one aspect of leisure; ▶recreation); *syn.* free time [n], spare time [n]; **s 1 tiempo [m] libre** (Totalidad del tiempo no determinado por el trabajo. Se diferencia entre el t. **l.** dedicado al ▶descanso y la restitución física [actividades domésticas y necesidades biológicas de regeneración] y el **t. l.** efectivo en el que las personas pueden elegir libremente el tipo de actividad a realizar); *syn.* ocio [m]; **s 2 tiempo [m] libre y recreación [f]** (Pareja de términos frecuentemente utilizada para denominar todo tipo de actividades de recreo activo o pasivo); *syn.* tiempo [m] libre y recreo [m], tiempo [m] libre y esparcimiento [m]; **f 1 temps [m] libre** (La totalité du temps libre en dehors du temps consacré au travail. On distingue le temps de reproduction [activités domestiques et besoins physiologiques] et le temps libre effectif, disponible en dehors de toute occupation de la vie courante et de ses contraintes et effectivement utilisé pour les activités de loisirs ; *opp.* temps contraint) ; *syn.* loisirs [mpl]. **f 2 loisirs [mpl] et récréation [f]** (Termes souvent utilisés ensemble pour désigner toutes les formes d'activités de loisirs actifs et passifs ; ▶détente) ; *syn.* loisirs [mpl] et tourisme [m] ; **g 1 Freizeit [f]** (Die gesamte Zeit, die der Mensch außerhalb seiner vor allem fremdbestimmten Erwerbsarbeit für sich zur Verfügung hat. Sie lässt sich einteilen in ‚reproduktive Zeit' [zum Schlafen, Ausruhen, Essen und zur Körperpflege] und in ‚verhaltensbeliebige Zeit', der sog. effektiven Freizeit; MEL 1973, Bd. 9); **g 2 Freizeit [f] und Erholung [f]** (Das Begriffspaar ist der OB zu aktiver und passiver Erholungsform und wird in der Planungspraxis häufig synonym verwendet. ▶Erholung 1 ist jedoch nur ein Aspekt der Freizeit! Im Englischen wird das Begriffspaar *leisure and recreation* nicht verwendet, da in *leisure* Erholung stets mitgedacht wird).

leisure activities [npl] *plan. recr.* ▶area for concentrated leisure activities, ▶promotion of leisure activities, ▶study of leisure activities.

3233 leisure activity pattern [n] *plan. recr.* (**1.** Manner in which a society chooses to spend free time. **2.** *In planning* a statistical trend or a spatial map indicating user actions; ▶leisure behavior [US]/leisure behaviour [UK]); **s pauta [f] de actividades de tiempo libre** (**1.** Manera en que p. ej. una sociedad industrial utiliza primordialmente su tiempo libre. **2.** En la planificación gráfico o mapa geográfico en el que se representan estadísticas o tendencias de actividades de tiempo libre; ▶comportamiento de tiempo libre); *f* **organisation [f] des loisirs** (La manière avec laquelle la société organise une politique et sociale des loisirs ; ▶comportement de loisirs) ; *g* **Freizeitgestaltung [f] (1)** (**1.** Art und Weise wie z. B. die Industriegesellschaft ihre Freizeit gestaltet. **2.** Statistische Darstellung oder geografische Karte, auf der Trends der Freizeitaktivitäten dargestellt sind; ▶Freizeitverhalten).

leisure avocation [n] [US] *plan. recr.* ▶leisure pursuit.

3234 leisure-based [adj/pp] *plan. recr.* (Descriptive term for recreation requiring leisure time); **s relacionado/a al recreo [loc]**; *f* **relatif, ive aux loisirs [loc]** ; *g* **erholungsbezogen [adj]**.

3235 leisure behavior [n] [US] *plan. recr.* (Manner of using leisure; ▶leisure activity pattern, ▶organized use of leisure time); *syn. o.v.* leisure behaviour [n] [UK]; **s comportamiento [m] de tiempo libre** (Forma de aprovechar el tiempo libre en una sociedad; ▶pauta de actividades de tiempo libre, ▶uso [individual] del tiempo libre); *f* **comportement [m] de loisirs** (La manière avec laquelle la société exploite un large éventail d'activités de loisirs ; ▶organisation des loisirs, ▶pratique d'activités de loisirs) ; *syn.* comportement [m] ludique ; *g* **Freizeitverhalten [n, o. Pl.]** (Art und Weise, wie eine Gesellschaft das weit gefächerte Freizeittätigkeitsspektrum nützt; ▶Freizeitgestaltung 1, ▶Freizeitgestaltung 2).

leisure behaviour [n] [UK] *plan. recr.* ▶leisure behavior [US].

3236 leisure budget [n] *plan. recr.* (Available amount of spare time in which self-chosen activities can be pursued); **s cantidad [f] de tiempo libre disponible** *syn.* tiempo [m] libre disponible; *f* **temps [m] libre disponible** (Temps libre effectif pendant lequel sont pratiquées des activités de loisirs) ; *syn.* budget [m] temps libre ; *g* **Freizeitbudget [n]** (Effektiv freie Zeit, in der Freizeitaktivitäten ausgeübt werden können); *syn.* Freizeitmenge [f].

3237 leisure center [n] [US]/**leisure centre** [n] [UK] *plan. recr.* (Easily accessible, well-provided complex on the periphery of a heavily-populated area; ▶amusement park, ▶concentration of recreation facilities); **s centro [m] de recreo** (Complejo fácilmente accesible y bien equipado con instalaciones de juegos, deporte y ocio, situado en la periferia de una aglomeración; ▶nucleo recreativo, ▶parque de atracciones); *f 1* **centre [m] ludique** (Infrastructure à la périphérie des agglomérations urbaines qui offre aux citadins pendant la semaine et le week-end, pour la pratique d'activités récréatives des équipements de jeux, de sports et culturels ; ▶base de plein air et de loisirs, ▶parc d'attraction) ; *syn.* parc [m] multi-activités ; *f 2* **centre [m] de sports et de loisirs** (Équipements collectifs de loisirs couverts ou en plein-air proposant diverses activités de sports et de jeux ainsi que les équipements de loisirs passifs ou contemplatifs ; *terme spécifique* centre de loisirs rattaché à l'école [CLAE]) ; *g* **Freizeitzentrum [n]** (Gut erschlossener, großräumiger Erholungsbereich in der Randzone eines Verdichtungsraumes zur Freizeitgestaltung mit verschiedenen Einrichtungen für Spiel, Sport und Unterhaltung; ▶Erholungsschwerpunkt, ▶Vergnügungspark).

3238 leisure harbor [n] [US]/**leisure harbour** [n] [UK] *recr.* (Harbour/harbor for private boat use; generic term for ▶sailboat harbor [US]/sailing-harbour [UK]; ▶yachting habor [US]/yachting harbour [UK]; **s puerto [m] de recreo** (Término genérico para ▶puerto de veleros y ▶puerto de yates); *f* **port [m]**

de plaisance (Infrastructure de tourisme située en bord de mer ou sur une voie navigable intérieure consacrée au stationnement d'en général plusieurs centaines de bateaux de plaisance à voile et à moteur. Il accueille des bateaux résidant à l'année ou des bateaux visiteurs en offrant des services divers aux plaisanciers ; *termes spécifiques* ▶port de navires de plaisance, ▶port de voiliers de plaisance) ; *g* **Freizeithafen [m]** (OB zu ▶Segelboothafen und ▶Yachthafen).

3239 leisure industry [n] *recr.* (Branch of industry serving the ways in which people desire to use their free time); **s industria [f] del ocio** (Rama de la economía que se dedica a producir bienes y ofrecer servicios para actividades de tiempo libre; *término específico* industria turística); *f* **industrie [f] des loisirs** (Industrie offrant aux usagers des services divers pour les activités pendant les temps libres) ; *g* **Freizeitbranche [f]** (Wirtschaftszweig, der mit Touristikunternehmen, Freizeitparks und sonstigen Dienstleistungen auf die Bedürfnisse der Menschen in ihrer Freizeit eingeht; *UB* Touristikbranche); *syn.* weiße Industrie [f].

leisure [n] **in the countryside** [UK] *recr.* ▶countryside recreation.

leisure [n] **in the open** *recr.* ▶outdoor recreation.

leisure motorist [n] [UK] *recr.* ▶auto tourist [US].

3240 leisure-oriented society [n] *recr. sociol.* (Society now learning how to use an ever-increasing amount of free time); **s civilización [f] del ocio** (Sociedad que necesita aprender a utilizar adecuadamente el creciente tiempo libre); *f* **société [f] de loisirs** (Société confrontée aux problèmes liés à l'augmentation constante du temps de loisirs) ; *g* **Freizeitgesellschaft [f]** (Gesellschaft, die mit der Herausforderung einer ständigen Zunahme an Freizeit und deren sinnvolle und gesundheitsfördernde Gestaltung konfrontiert ist).

3241 leisure park [n] *plan. recr.* (▶Recreation area 1 with play, amusement, and sports facilities within a densely populated area, easily reached by public transport, often used on weekends; no U.S. equivalent; ▶amusement park, ▶leisure center [US]/leisure centre [UK], ▶park 1, ▶recreation complex, ▶recreation space); *syn.* recreation park [n]; **s parque [m] recreativo regional** (Nucleo recreativo en una aglomeración dotado de instalaciones de juego, deporte y entretenimiento, asequible por medio de transporte público; ▶área de recreo, ▶centro de recreo, ▶complejo recreacional, ▶parque, ▶parque de atracciones, ▶zona de uso priorizado para el recreo); *f* **parc [m] de loisirs** (▶Espace touristique aménagé comportant des moyens de loisirs de type récréatif, culturel et/ou sportif et pouvant offrir en sus un terrain de camping et de caravanage et/ou des habitations légères de loisirs [H.L.L.] ; ▶aire de loisirs, ▶aire de récréation, ▶centre de sports et de loisirs, ▶centre ludique, ▶parc, ▶parc d'attraction) ; *g* **Freizeitpark [m]** (Der Nah- und Wochenenderholung dienender, mit Spiel-, Sport- und Vergnügungseinrichtungen ausgestatteter und durch öffentliche Verkehrsmittel erreichbarer Erholungsschwerpunkt innerhalb dichtbesiedelter Gebiete; ▶Erholungsfläche, ▶Erholungsgebiet, ▶Freizeiteinrichtung 1, ▶Freizeitzentrum, ▶Park, ▶Vergnügungspark); *syn. im Ruhrgebiet* Revierpark [m].

leisure perspective [loc]**, from a** *plan. recr.* ▶leisure-related.

3242 leisure provisions [npl] *plan. recr.* (Existing facilities for free time pursuits; ▶recreation opportunities); **s oferta [f] de equipamiento de tiempo libre** (Posibilidades e instalaciones existentes para aprovechamiento del tiempo libre; ▶oferta recreativa); *f* **offre [f] en équipements de loisirs** (Totalité des équipements collectifs utilisés pour les loisirs ; ▶offre touristique) ; *g* **Freizeitangebot [n]** (Vorhandene Möglichkeiten und Einrichtungen, die in der Freizeit genutzt werden können; ►Erholungsangebot); *syn.* Angebot [n] von Freizeiteinrichtungen.

3243 leisure pursuit [n] *plan. recr.* (Generic term for all activities which can be pursued in spare time); *syn.* leisure avocation [n] [also US], free time activity [n] [also US], leisure time activity [n] [also US]; *s* **actividad [f] de tiempo libre** (Término genérico para todo tipo de actividades a las que se puede dedicar una persona en sus horas libres; *término específico* actividad lúdica); *syn.* actividad [f] de ocio; *f* **activité [f] de loisirs** (Terme générique pour toutes les activités corporelles ou de l'esprit auxquelles on se livre librement pendant le temps libre effectif ; *terme spécifique* activités ludiques campagnardes) ; *syn.* activité [f] récréative, pratique [f] ludique ; *g* **Freizeitaktivität [f]** (Betätigung, der in der verhaltensbeliebigen Freizeit nachgegangen werden kann); *syn.* Freizeitbeschäftigung [f], Freizeitbetätigung [f], Freizeittätigkeit [f].

3244 leisure-related [adj/pp] *plan. recr. syn.* from a leisure perspective [loc], recreational [adj]; *s* **relacionado/a con el tiempo libre [loc]** *syn.* relativo/a al ocio [loc]; *f* **du point de vue récréatif [loc]** *syn.* du point de vue des loisirs [loc] ; *g* **freizeitbezogen [adj]** (die Freizeit betreffend).

leisure time [n] *plan. recr.* ►organized use of leisure time.

leisure time activity [n] [US] *plan. recr.* ►leisure pursuit.

3245 leisure value [n] *plan. recr.* (Usefulness as a leisure resource); *s* **valor [m] recreativo** (Potencial de una ciudad o un paisaje para cumplir funciones recreativas y turísticas); *f* **valeur [f] récréative** (Valeur esthétique et d'agrément, qualité de l'environnement, diversité des milieux en tant qu'expression des potentialités d'un site pour la fréquentation touristique) ; *syn.* valeur [f] ludique ; *g* **Freizeitwert [m]** (Wert z. B. einer Stadt oder einer Landschaft hinsichtlich der Erholungs- und Erlebniswirkung für den Betroffenen oder der Wert resp. die Sinnhaftigkeit, die eine Freizeitbeschäftigung für die Entspannung hat).

length [n] *constr.* ►grip length, ►landing length, ►laying length, ►overall length, ►tread length.

length [n], **bond** *arch. constr.* ►grip length.

length [n], **effective** *constr.* ►laying length.

length [n] **of a flight of steps** *arch. constr.* ►total run of a stairway [US].

3246 lens-shaped raised bog [n] *geo. syn.* concentric domed bog [n] (WET 1993, 376); *s* **turbera [f] abombada** (►Turbera alta que, debido al crecimiento en altura, tiene forma redondeada y que ha perdido el contacto con la capa freática); *f* **tourbière [f] à forme de lentille** (Tourbière formée quand l'accumulation de tourbe est si importante que la végétation perd le contact avec la nappe phréatique et n'est plus alimentée que par les précipitations atmosphériques ; la tourbière prend alors une forme bombée) ; *syn.* tourbière [f] (à surface) bombée, tourbière [f] (à surface) bosselée en coupole ; *g* **urglasförmiges Hochmoor [n]** (►Hochmoor, das auf Grund ständigen Höhenwachstums rund gewölbt ist und keinen Kontakt mehr zum Grundwasser hat).

3247 lenticel [n] *arb.* (Small pore, usually eliptical, in the bark of woody plants, serving for the exchange of gases between the air and stem tissues); *s* **lenticela [f]** (En la peridermis de las plantas leñosas protuberancia visible a simple vista y con una abertura de forma lenticular que, reemplazando a los estomas de la desaparecida epidermis, utiliza la planta para el cambio de gases; DB 1985); *f* **lenticelle [f]** (Souvent une ouverture lenticulaire, constituée par des cellules de liège faiblement subérifiées qui laissent entre elles des méats. Sert aux échanges de gaz à travers l'épiderme des plantes ligneuses ; cf. MGW 1964) ; *g* **Lentizelle [f]** (Meist linsenförmige, aus locker geordneten

Zellen bestehende, leicht verkorkte Öffnung auf den Korkhäuten der Holzpflanzen, die dem Gasaustausch dienen); *syn.* Korkwarze [f].

3248 lepidopterology [n] *zool.* (Science of lepidoptera, e.g. butterflies and moths); *s* **lepidopterología [f]** (Ciencia que estudia el orden de los lepidópteros, es decir las mariposas); *f* **lépidoptérologie [f]** (Science étudiant le développement et la répartition des papillons) ; *syn.* étude [f] des lépidoptères ; *g* **Schmetterlingskunde [f]** (Wissenschaft von der Entwicklung und Verbreitung der Schmetterlinge); *syn.* Lepidopterologie [f].

lessen [vb] **a slope** *constr.* ►flatten a slope.

lessening [n] **of water resources** [US] *wat'man.* ►diminution of water resources.

let agricultural land lie fallow [loc] *agr.* ►leave uncultivated.

letter [n] **of appointment** [UK] *prof.* ►design contract [US].

letting [n] **of contract** [UK] *constr. contr.* ►awarding of contract [US].

letting [n] **to subcontractors** [UK] *contr.* ►contract subletting.

levee [n] (1) *agr.* ►dirt bank [US]/dyke [UK].

3249 levee [n] [US] (2) *eng. wat'man.* (Elongated mound to prevent flooding; generic term covering ►coastal dike, ►foredike, and ►river embankment; ►summer dike); *syn.* flood berm [n], flood bank [n] [UK]; *s* **dique [m] de contención** (Dique para proteger contra inundaciones; término genérico para ►dique anterior, ►dique costero, ►dique de verano y ►dique fluvial); *syn.* dique [m] de abrigo; *f* **digue [f]** (*Terme générique* Ouvrage linéaire en terre ou enrochement exécuté pour protéger un village, des habitations, une plaine alluviale contre les crues ; *termes spécifiques* ►digue avancée, ►digue côtière, ►digue d'été, ►digue fluviale) ; *syn.* levée [f] de retenue, digue [f] de protection, talus [m] de protection ; *g* **Deich [m] (2)** (Geschüttetes Schutzbauwerk gegen Überschwemmungen an der Küste und entlang von Fließgewässern; *UBe* ►Flussdeich, ►Küstendeich, ►Sommerdeich, ►Vordeich); *syn.* Hochwasserdeich [m], Hochwasserdamm [m], Hochwasserschutzdeich [m], Schutzdamm [m], Schutzdeich [m].

levee [n] [US], **river** *wat'man.* ►river embankment.

level [n] *envir.* ►ambient noise level; *constr.* ►bottom level, ►curb level [US]/kerb level [UK], ►drain invert level; *plan. land'man.* ►environmental sensitivity level; *constr.* ►existing ground level, ►finished floor level; *hydr.* ►fluctuating water level; *constr.* ►install at a prescribed level; *wat'man.* ►lowest low-water level, ►low-water level, ►mean-water level; *hydr.* ►measured water level; *envir.* ►noise decibel level; *plan.* ►planning level; *limn. wat'man.* ►saprobic level; *constr. envir. leg.* ►sound pressure level; *agr. game'man.* ►stocking level; *biol. ecol. envir.* ►stress level; *constr. stat.* ►subsoil level [US]; *arch. constr.* ►top level; *ecol.* ►trophic level; *constr. hydr.* ►water level.

level [n], **datum** *constr.* ►datum.

level [n] [UK], **decibel** *constr. envir. leg.* ►sound pressure level.

level [n] [UK], **excavation to formation** *constr.* ►subbase grade preparation [US].

level [n] [UK], **existing** *constr. plan.* ►existing ground level.

level [n] [UK], **finished** *constr. plan.* ►finished elevation [US].

level [n] **(FFL)** [US], **finish floor** *constr.* ►finished floor level.

level [n], **floor** *constr.* ►finished floor level.

level [n] [UK], **formation** *constr.* ▶subbase grade [US], ▶subsoil level [US].

level [n], **groundwater** *hydr.* ▶watertable.

level [n], **high-water** *wat'man.* ▶average high-water level/mark.

level [n], **maximum** *envir.* ▶maximum value.

level [n], **noise pollution** *envir.* ▶noise decibel level.

level [n] [UK], **normal water** *leg. wat'man.* ▶mean-water level.

level [n], **oscillating water** *constr.* ▶fluctuating water level.

level [n], **peak high-water** *wat'man.* ▶record high-water level/mark.

level [n] [UK], **proposed** *constr. plan.* ▶finished elevation [US].

level [n] [US], **stable soil** *constr. eng.* ▶consolidated subgrade.

level [n], **tree canopy** *phyt.* ▶tree layer.

3250 level [vb] **a laying course** *constr.* (Make even a bed of sand, mortar, poured concrete, etc.); *syn.* screed [vb] a laying course [also US]; *s* **enmaestrar** [vb] **lecho de asiento** (Aplanar un lecho de asiento de arena, mortero u hormigón con un listón de madera o metal; enmaestrado [m] de lecho de asiento); *syn.* nivelar [vb] lecho de asiento; *f* **niveler** [vb] **un lit de pose à la règle** (Établissement d'un lit de pose de sable, de mortier, d'une chape de béton d'épaisseur constante au moyen d'une règle en bois ou métal tirée sur gabarits) ; *g* **Pflasterbett mit einer Latte herrichten** [vb] (Ein Pflasterbett aus Sand, Mörtel, Beton etc. mit einer Holz- oder Metalllatte höhengerecht einebnen).

3251 level difference [n] *constr.* (Difference in height between two elevations); *syn.* change [n] of level; *s* **desnivel** [m] **de altura** *syn.* diferencia [f] de altura; *f* **différence** [f] **de hauteur** *syn.* dénivellement [m] (GEN 1982-II, 84) ; *g* **Höhenunterschied** [m] (2) (Differenz zwischen zwei Höhenangaben).

level ground [n] *landsc.* ▶flat terrain.

leveling [n] *constr.* ▶grading and leveling [US]/grading and levelling [UK].

leveling layer [n] [US] *constr.* ▶base course (1).

levelling [n] [UK] *constr.* ▶grading and leveling [US]/grading and levelling [UK].

levelling course [n] [UK], **manhole** *constr.* ▶manhole adjustment ring.

levelling layer [n] [UK] *constr.* ▶base course (1).

levelling plan [n] [UK] *constr. plan.* ▶grading plan [US].

level [n] **of air pollution** *envir. leg.* ▶acceptable level of air pollution, ▶threshold level of air pollution.

level [n] **of a road/path** [UK], **formation** *constr.* ▶subbase grade of a road/path [US].

level [n] **of artificialization** *ecol.* ▶degree of landscape modification.

level [n] **of human-caused landscape modification** *ecol.* ▶degree of landscape modification.

level [n] **of planning** *leg. plan.* ▶higher administrative level of planning.

level [n] **of pollution** [UK], **basic** *envir.* ▶background level of pollution [US].

3252 level [n] **of recreation use** *plan. recr.* (Degree of use intensity by recreation users); *s* **frecuentación** [f] **turística** (Grado de intensidad de uso por recreacionistas); *syn.* intensidad

[f] de uso recreativo, grado [m] de frecuentación turística; *f* **degré** [m] **de fréquentation touristique** (Résultat d'une étude de fréquentation d'un site basée sur des approches quantitative, qualitative et comportementale, s'appuyant sur la collecte de données de terrain adaptées à la spécificité du site) ; *g* **Erholungsnutzung** [f] (1) (Grad der Nutzungsintensität durch Erholungsuchende).

level stake [n] [UK] *constr. surv.* ▶grade stake [US].

ley [n] [UK] *agr.* ▶temporary grassland [US].

L horizon [n] *pedol.* ▶litter layer.

liability [n] *envir. leg.* ▶causer liability.

liability [n] **for professional insurance** *contr. prof.* ▶obligation to take out professional liability insurance.

liability insurance [n] *contr. prof.* ▶obligation to take out professional liability insurance, ▶professional liability insurance.

liability period [n] *contr. leg.* ▶defects liability period.

3253 liana [n] *bot.* (True lianas are climbing woody plants which grow towards the light using various supports. They are characteristic of tropical virgin forests. Only a few of them grow in cold temperate zones; e.g. Clematis, Ivy *[Hedera]*, Honeysuckle *[Lonicera]* and Virginia creeper *[Parthenocissus]*; ▶climbing plant); *s* **bejuco** [m] (Voz caribe para planta trepadora, voluble o no, generalmente de largos tallos sarmentosos, que suele encaramarse a las copas de los árboles en busca de la luz; en las zonas tropicales y subtropicales existen muchas especies como los bejucos de agua *[Vitis esp.]*, el bejuco legítimo *[Bignonia unguiscati]*, el bejuco loco *[Cissus sicyoïdes]* y muchos otros; en las zonas templadas existen pocas especies como *Clematis*, yedra *[Hedera]*, madreselva *[Lonicera]*, vid silvestre *[Parthenocissus]*; el término «liana» es un galicismo; cf. DB 1985, *término genérico* ▶planta trepadora); *syn.* liana [f]; *f* **liane** [f] (Plante terrestre ligneuse dont les tiges très longues ne sont pas rigide et s'appuient sur divers supports, en particulier d'autres plantes, p. ex. *Clematis, Hedera, Lonicera* et *Parthenocissus*) ; *g* **Liane** [f] (*Phanerophyta scandentia* in der Literatur nicht immer einheitlich abgegrenzter Begriff; zu den eigentlichen **L.n** zählen Holzgewächse, die sich unter Zuhilfenahme von Stützpunkten zum Lichte durcharbeiten und besonders charakteristisch für den Tropenurwald sind. In der gemäßigt-kalten Zone gibt es nur wenige Gattungen: *Clematis*, Efeu *[Hedera]*, Geißblatt *[Lonicera]*, Jungfernrebe/Wilder Wein *[Parthenocissus]*; cf. BOR 1958, LB 1978, 199; *OB* ▶Kletterpflanze).

life [n] *plan. sociol.* ▶quality of life.

life community [n] *biol. ecol.* ▶biocenosis.

3254 life form [n] *biol.* (Term used to describe a type of plant, in a classification system, which is characterized by its mode of adaptation to the environment. According to RAUNKIAER [1934] the following terms describe distinct life forms: ▶chamaephyte, ▶geophyte, ▶hemicryptophyte, ▶phanerophyte, ▶therophyte. The term 'growth form' is nearly identical with **l. f.** The first means pure morphological phenomena, the latter implies significantly developed adaptations to the environment; ▶habit); *syn.* growth form [n] (WES 1985, 373); *s* **forma** [f] **biológica** (Categoría dentro de la cual se incluyen los vegetales, de posición sistemática cualquiera, que concuerdan fundamentalmente en su estructura morfológico-biológica y de un modo especial en los caracteres relacionados con la adaptación al ambiente ecológico. La clasificación mas usada es la de RAUNKIAER que comprende, como divisiones de primer orden las formas de ▶caméfito, ▶fanerófito, ▶geófito, ▶hémicriptófito y ▶terófito; DB 1985. El término «forma de crecimiento» es casi idéntico al de **f. b.** El primero incluye sólo aspectos morfológicos, mientras que el segundo incluye también las adaptaciones al medio más características; ►hábito de crecimiento); *syn.* forma [f] biótipo, tipo [m] biológico, biótipo [m]; *f 1* **type [m] biologique** (Physionomie que prend un organisme et qui résulte de tous les processus biologiques dus aussi bien à son comportement intrinsèque qu'à l'action des facteurs du milieu. On distingue notamment selon RAUNKIAER, les types suivants : ►chaméphyte, ►géophyte, ►hémicryptophyte, ►phanérophyte, ►thérophyte ; DEE 1982) ; *syn.* forme [f] biologique ; *f 2* **forme [f] naturelle** (Le mode de développement des branches, leur orientation vont donner à l'arbre sa silhouette, son port spécifique ; ►habitus) ; *g 1* **Lebensform [f]** (**1.** Bezeichnung für Organisationstypen von Organismen, die an bestimmte Bedingungen ihrer Umgebung durch gleichartige Anpassungserscheinungen gekennzeichnet sind. **2.** *bot.* OB für Pflanzen, die gleichartige ökologische Anpassungsformen an ihre Umwelt aufweisen; für die heimische Flora und viele andere Floren in entsprechenden Klimaten, hat der Temperaturwechsel zwischen Sommer- und Wintermonaten zu einer Reihe besonderer Anpassungsstrategien geführt. Die **L.n** beschreiben den Gesamtbau von Pflanzen und umfassen Gruppen nicht näher verwandter Pflanzen, die auf Grund ähnlicher Lebensweise entsprechend deutlich ausgeprägte, gleichartige Anpassungserscheinungen an den Lebenshaushalt und die Umwelt aufweisen. Kennzeichnende Merkmale der Lebensform sind: **a.** Art und Form jahreszeitlicher Anpassung, **b.** Dauerhaftigkeit der oberirdischen Organe; **c.** Lage und Form der Überdauerungsorgane; **d.** Lage der Erneuerungsknospen zur Erdoberfläche; **e.** Lebensdauer; **f.** Vermehrungsstrategie und **g.** Wuchsform oder Wuchstyp. Nach RAUNKIAER [1934] werden je nach Lage/Höhe der Erneuerungs- oder Überdauerungsknospen in Bezug zum Erdboden als Anpassung an die ungünstige Jahreszeit fünf **Haupt-L.en** unterschieden: ►Chamaephyt, ►Geophyt, ►Hemikryptophyt, ►Phanerophyt, ►Therophyt); *syn. zu 2.* Vegetationsform [f], Überwinterungsform [f]; *g 2* **Wuchsform [f]** *bot.* (Der Begriff **W.** ist mit Lebensform nicht ganz identisch. Ersterer wird rein morphologisch, d. h. bauplanorientiert [und damit genetisch festgelegt] verstanden, während letzterer deutlich ausgeprägte, standörtlich veränderliche Anpassungen an Umweltbedingungen beinhaltet; **W.en** sind z. B. **halbparasitisch, lianenartig wachsend [scandent], schaftig wachsend [scapos], kriechend wachsend [repent], horstig wachsend, vollparasitisch, W.** kann in Kombination mit Lebensform gebraucht werden: so ist die Waldrebe *[Clematis vitalba]* ein lianenartig wachsender ►Phanerophyt, der Gemeine Hopfen *[Humulus lupulus]* ein lianenartig wachsender ►Hemikryptophyt und das Klebkraut *[Galium aparine]* ein lianenartig wachsender ►Therophyt; ►Habitus); *syn. zu 3.* Lebensstrategie [f].

3255 **life form spectrum** [n] *phyt.* (Proportional range of ►life forms, which determines the composition of a plant community); *s* **espectro [m] biotípico** (Proporción de cada una de las ►formas biológicas que determinan la composición de una comunidad vegetal); *f* **spectre [m] biologique** (Proportion de l'espèce occupée par chaque ►type biologique dans un groupement végétal) ; *g* **Lebensformspektrum [n]** (Anteil der ►Lebensformen, die den Aufbau einer Pflanzengesellschaft bestimmen).

3256 **life pyramid** [n] *ecol.* (Term describing the quantitative relationship between nutrients, ►biomass or living organisms at various ►trophic levels within the ►food chains and ►food webs of an ecosystem. The pyramidal shape depicts the diminishing degree of exploitation of the nutrients or biomass produced as well as the decline in the number of individual organisms from the ►producer level up to the highest level of consumers); *s* **pirámide [f] ecológica** (Representacion gráfica de la estructura trófica de un ecosistema, es decir de la distribución de la ►biomasa en los diferentes ►niveles tróficos y ►cadenas

tróficas. La base la constituyen los ►productores [primer nivel trófico] sobre los que se encuentran los consumidores primarios [segundo nivel trófico] y los secundarios [tercer nivel trófico]. La **p. e.** muestra así mismo la reducción de la eficiencia de transmisión de energía de un nivel a otro y la consiguiente disminución de la producción de materia orgánica por unidad de tiempo, alcanzando la producción entre el 10 y 20% del nivel inmediato inferior; cf. DINA 1987; ►red trófica); *f* **pyramide [f] écologique** (Diagramme représentant l'importance des divers ►niveaux trophiques d'une biocénose sur lequel sont indiqués les niveaux de production de ►biomasse formée par les ►producteurs et les différents consommateurs, la productivité ou le rendement de chaque niveau successif diminuant avec le nombre d'individus de la ►chaîne trophique ou du ►réseau trophique); *g* **Nahrungspyramide [f]** (Darstellung der quantitativen Beziehungen der von ►Produzenten gebildeten Nährstoffe, ►Biomasse oder der davon lebenden Organismen auf verschiedenen Konsumentenstufen in den ►Nahrungsketten und ►Nahrungsnetzen eines Ökosystems, wodurch der abnehmende Ausnutzungsgrad der erzeugten Nährstoffe oder Biomasse sowie die Abnahme der Individuenzahlen von der Produzenten- zur höchsten Konsumentenebene ersichtlich wird; ANL 1984; ►Trophiestufe).

3257 **lifting** [n] *hort.* (Digging-up of plants with care in nurseries, done by hand-tool or machine, e.g. a plant lifter); *s* **desarraigado [m]** (En vivero, desenterrar con cuidado plantas para transportarlas al lugar de plantación. Se puede realizar manualmente o con máquinas de transplante; desenterrar [vb]; HDS 1987); *syn.* desenterrado [m]; *f* **arrachage [m]** (En pépinière opération de préparation des jeunes plants effectuée manuellement ou à l'aide d'une arracheuse en vue de leur transplantation) ; *g* **Roden [n, o. Pl.] von Gehölzen** (**1.** Das Herausnehmen in Baumschulen von lieferfähigen Gehölzen ohne Ballen; **2.** das Herausnehmen von Pflanzen mit Ballen mit Handspaten oder Verpflanzungsmaschine für den Versand wird i. d. R. nicht als **R.**, sondern als **Ausgraben von Ballenpflanzen/Ballenware** bezeichnet; **3.** ►Rodung); *syn. zu 2.* Ausgraben [n, o. Pl.] von Gehölzen.

3258 **lifting** [n] **of plants by frost** *constr. hort.* (Ground-raising of plants that are not planted deep enough in the autumn; ►frost-heave); *syn.* frost-heave [n] of plants [also US]; *s* **levantamiento [m] de plantas por helada** (Elevación de las plantas causada por congelación del suelo que se presenta si éstas no se han plantado con suficiente profundidad; ►levantamiento por congelación); *f* **déchaussement [m] de plantes par le gel** (Tendance que peuvent avoir les jeunes végétaux à sortir de terre et se déraciner lorsque la motte n'a pas été suffisamment enfouie à l'automne et que le sol gonfle sous l'action du gel ; ►foisonnement par le gel) ; *g* **Hochfrieren [n, o. Pl.] von Pflanzen** (Durch Frosteinwirkung bewirktes Herausdrücken aus dem Boden von im Herbst zu flach eingepflanzter Gehölze oder Stauden; ►Frosthebung).

lift-pruning [n] [US] *constr. for. hort.* ►crown lifting.

3259 **light availability** [n] *for. phyt.* (for plants; ►relative light availability); *syn.* light conditions [npl]; *s* **condiciones [fpl] de iluminación** (Luz disponible para las plantas; ►luminosidad relativa); *f* **conditions [fpl] lumineuses** (Données relatives à la lumière pour le développement des végétaux ; ►luminosité relative) ; *syn.* conditions [fpl] de lumière ; *g* **Lichtverhältnisse [npl]** (Das Licht betreffende Gegebenheiten, z. B. für Pflanzen; ►relativer Lichtgenuss).

light conditions [n] *phyt.* ►light availability.

light demander [n] *for. phyt.* ►light-demanding woody species.

3260 light-demanding plant [n] *for. phyt.* (Plant that needs great amounts of light for optimum growth; ▶light-demanding woody species; *opp.* ▶shade plant); *syn.* heliophilous species [n]; *s* **helíofito [m]** (Planta que necesita ubicaciones soleadas o muy iluminadas para crecer de forma óptima; ▶especie leñosa de solana; *opp.* ▶planta de sombra); *syn.* planta [f] de solana, planta [f] heliófila, planta [f] de sol; *f* **espèce [f] héliophile** (▶essence de lumière ; *opp.* ▶plante d'ombre) ; *syn.* plante [f] de lumière, héliophyte [m], *for.* essence [f] de lumière (1) ; *g* **Lichtpflanze [f]** (Pflanze, die einen sonnigen oder sehr hellen Standort für optimales Wachstum benötigt; *for.* ▶Lichtholzart; *opp.* ▶Schattenpflanze); *syn.* Heliophyt [m], Sonnenpflanze [f].

3261 light-demanding woody species [n] *for. phyt.* (Plant that is intolerant of shade; *opp.* ▶shade bearer); *syn.* heliophilous woody species [n], *for.* light demander [n] (SAF 1983); *s* **especie [f] leñosa de solana** (Árbol o arbusto que necesita mucha luz para crecer; *opp.* ▶leñosa esciófila); *syn.* leñosa [f] heliófila, leñosa [f] de solana; *f* **essence [f] de lumière** (Espèce ligneuse nécessitant une exposition en pleine lumière et qui ne supporte pas l'ombre ; DFM 1975, 281 ; *opp.* ▶essence d'ombre) ; *g* **Lichtholzart [f]** (Gehölzart mit relativ großer Lichtbedürftigkeit für optimales Wachstum; *opp.* ▶Schattholzart).

3262 light exposure [n] *bot. met.* (Subjection to artificial light or sun light; *specific term* sun exposure; ▶illumination, ▶insolation, ▶strong light exposure); *s* **iluminación [f]** (Incidencia de radiación de la zona visible del espectro entre 0,40 y 0,78 micras sobre un cuerpo; DINA 1987; en meteorología siempre se refiere a la debida a la radiación solar, pero en el caso de que ésta no contribuya directamente se habla de **i.** difusa; cf. DM 1986; ▶iluminación [artificial], ▶insolación); *f* **exposition [f] à la lumière** (Produit de l'intensité lumineuse d'une source lumineuse naturelle ou artificielle et de la durée d'exposition, avec laquelle une surface est éclairée ; ▶éclairage, ▶ensoleillement) ; *syn.* éclairement [m] ; *g* **Belichtung [f]** (Einwirkung von künstlichem oder Sonnenlicht; ▶Beleuchtung, ▶Besonnung).

light exposure [n]**, high level of** *bot.* ▶strong light exposure.

light exposure [n]**, low level of** *bot.* ▶poor light exposure.

light exposure [n]**, weak** *bot.* ▶poor light exposure.

light industries [npl] *plan. urb.* ▶allocation of commercial facilities and light industries, ▶location of commercial facilities and light industries.

light industry zone [n] *leg. urb.* ▶commercial and light industry area.

3263 light intensity [n] *ecol. phyt.* (Measurable amount of the strength of natural or artificial light); *s* **intensidad [f] de luz** (Medida de la potencia de la luz natural o artificial); *syn.* intensidad [f] de iluminación; *f* **intensité [f] lumineuse** (Mesure exprimant le quotient du flux lumineux quittant une source lumineuse naturelle ou artificielle et se propageant dans un élément d'angle solide, par cet élément d'angle solide ; *unité* candela) ; *g* **Lichtintensität [f]** (Maß für die Stärke der natürlichen oder künstlichen Lichteinwirkung).

light plane [n] [US] *leg. urb.* ▶bulk plane [US].

light shrub [n] *hort.* ▶young shrub transplant.

light soil [n] [US] *agr. hort.* ▶easily workable soil.

light well [n] *arch.* ▶basement light well.

light well grid [n] [UK] *constr.* ▶window well grate [US].

3264 lignicolous species [n] *phyt. zool.* (Species living in wood is called xylobiontic); *syn.* species [n] living on wood; *s* **especie [f] lignícola** (Especie que vive o se desarrolla sobre la madera, como los hongos parásitos de la madera; cf. DB 1985); *syn.* lignícola [f], especie [f] xilógena; *f* **espèce [f] lignicole** (Espèce qui vit principalement dans le bois [certains insectes] ou qui pousse sur le bois [certains champignons]) ; *syn.* espèce [f] épixyle ; *g* **auf Holz lebende Art [loc]** (Eine Art, die Holz bewohnt, wird **xylobionte Art** genannt); *syn.* lignikole Art [f].

lignify [vb] *bot.* ▶become woody.

3265 limb breakage [n] *arb. hort.* (Breakage of tree branches); *s* **rotura [f] de rama** (Quebradura de ramas de árboles); *f 1* **cassure [f] de branche** *syn.* effondrement [m] de branche ; *f 2* **arrachement [m] de la branche** (La branche casse lors de coupures mal exécutées, elle est entrainée par son poids avant d'être complètement sectionnée, peut pivoter autour de son point d'attache et générer des plaies importantes au tronc ; cf. IDF 1988, 117) ; *g* **Astbruch [m]** (Abbrechen oder Zerbrechen eines Astes).

lime [vb] *agr. for. hort.* ▶liming.

3266 lime-avoiding species [n] *phyt.* (Plant which will not grow on limestone, e.g. ▶peat garden plant [UK]; *syn.* calcifuge species [n], calciphobe species [n]; *s* **especie [f] calcífuga** (Especie vegetal que no tolera el calcio, por lo que no crece en suelo cálcicos, sino en suelos ácidos; ▶planta de tierra turbosa); *f* **espèce [f] calcifuge** (Végétal se développant mal sur les sols calcaires ou espèce acidophile marquant une préférence pour les sols acides ; ▶plante de terre de bruyère) ; *g* **kalkfliehende Art [f]** (Pflanzenart, die durch hoch konzentrierte Kalklösungen geschädigt oder in ihrer Entwicklung gehemmt wird oder Säure liebende Art, die nur auf saurem Boden wächst; ▶Moorbeetpflanze); *syn.* Kalkflieher [m], Kalkmeider [m].

3267 lime content [n] *limn. pedol.* (calcium carbonate content); *s* **contenido [m] de calcio** (Cantidad de carbonato cálcico [CO_3Ca] en un suelo o en un curso o un cuerpo de agua); *f* **teneur [f] en calcaire** (Teneur en Ca dans le sol, dans un cours d'eau ou dans un plan d'eau stagnante) ; *g* **Kalkgehalt [m]** (Calciumcarbonat [$CaCO_3$] im Boden oder in einem Gewässer; Böden mit einem hohen Kalkgehalt sind meist sehr basen- und nährstoffreich, zeigen eine neutrale Bodenreaktion und sind gut gepuffert. Durch eine vom Calcium verursachte Ausflockung der Bodenkolloide werden Krümelung, Wasserführung, Durchlüftung und Temperatur für Pflanzen positiv beeinflusst. Stark kalkhaltige Böden werden durch den Bodentyp der Rendzina oder Pararendzina repräsentiert).

lime-loving species [n] *phyt.* ▶calcicolous species.

lime marl [n] *pedol.* ▶marly limestone.

3268 limestone [n] *geo. pedol.* (▶Sedimentary rock consisting mainly [sometimes defined as 80%, by others 50%] of calcium carbonate, $CaCO_3$; DNE 1978; ▶marl, ▶marly limestone); *s* **caliza [f]** (▶Roca sedimentaria con contenido entre 70 y 100% de carbonato cálcico [CO_3Ca]; ▶marga, ▶marga caliza); *syn.* roca [f] calcárea; *f* **roche [f] calcaire** (▶Roche sédimentaire essentiellement formée de carbonate de calcium, $CaCO_3$; ▶marne, ▶marne calcaire) ; *g* **Kalkgestein [n]** (▶Sedimentgestein mit 70-100 % $CaCO_3$-Gehalt; ▶Mergel, ▶Kalkmergel); *syn.* Kalkstein [m], Karbonatgestein [n].

3269 limestone grassland [n] *phyt.* (Area with open or closed, species-rich, grassed plant community on calcium-rich, dry and warm soils; occurring naturally on rocky ground, secondarily in extensively grazed areas; VIR 1982; ▶nutrient-poor grassland, ▶semidry grassland [US]/semi-dry grassland [UK]; *opp.* ▶acidic grassland); *syn.* limestone prairie [n] [also

US], chalk grassland [n] [also UK]; **s pastizal [m] calcícola seco** (Comunidad de ▶prados oligótrofos rica en especies sobre suelo pobre en nutrientes pero rico en calcio. Dependiendo del grado de sequía de la ubicación se diferencian los pastizales calcícolas secos y los ▶pastizales semi-secos; *opp.* ▶pastizal [oligótrofo] silicícola); *syn.* pastizal [m] oligótrofo calcícola; *f* **pelouse [f] calcicole maigre** (▶Pelouse oligotrophe naturelle rase, à végétation ouverte, riche en espèces herbacées basses, sur sol calcaire rocailleux, très pauvre en matières organiques, appartenant à la classe du *Festuco-Brometea* et constituant l'ordre des *Brometalia* ; GGV 1979, 302 ; dans les massifs préalpins les pelouses calcicoles constituent l'ordre des *Sesleretalia variae* et se répartissent en trois alliances d'écologie différente : pelouses maigres calcicoles du *Seslerion variae* et de l'*Elynion,* pelouses mésophiles calcicoles du *Caricion ferrugineae* ; VCA 1985, 256 ; ▶pelouse héliophile dense ; *opp.* ▶pelouse silicicole) ; *syn.* pelouse [f] oligotrophe crayeuse, pelouse [f] oligotrophe calcicole, pelouse [f] calcaire ; **g Kalkmagerrasen [m]** (Artenreicher ▶Magerrasen [z. B. Klasse der *Festuco-Brometea*] auf nährstoffarmen, kalkhaltigen Böden. Je nach Wärmebedürfnis und Trockenheitsgrad der Böden werden **Kalktrockenrasen** und **Kalkhalbtrockenrasen** unterschieden; ▶Halbtrockenrasen; *opp.* ▶Silikatmagerrasen); *syn.* Kalktrockenrasen [m], basiphiler Magerrasen [m] (OBER 1978, 86).

3270 limestone groove vegetation [n] [US] *phyt.* (Association of plants growing in chemically and mechanically-eroded limestone furrows filled with very fine soil between flat-topped ridges); *syn.* clint vegetation [n] [UK] (DNE 1978); **s vegetación [f] de lapiaces** (Formación vegetal que crece en superficie rocosa de origen kárstico [lapiaz] con relieve irregular, microcanales, oquedades, alveolos, etc. de diferentes tamaños rellenos de tierra; cf. DINA 1987); *f* **association [f] des lapiaz/lapiés/lapiez** (Végétaux poussant dans les fentes de dissolution parallèles, rainures ou cannelures sans profondeur colmatées par la terre, souvent envahies par la garrigue dans les paysages karstiques) ; *syn.* végétation [f] des lapiaz/lapiés/lapiez; **g Schrattenvegetation [f]** (*Karstgebiet mit dickbankigen Kalken* Pflanzenwuchs in mit Feinerde ausgefüllten, chemisch [Lösungsverwitterung] und mechanisch ausgespülten Gesteinsrinnen, z. B. im Kalkgestein); *syn.* Karrenfeldvegetation [f], Karrenvegetation [f].

limestone prairie [n] [US] *phyt.* ▶limestone grassland.

3271 liming [n] *agr. for. hort.* (Spreading of lime on a soil surface in order to increase its pH value; lime [vb]); **s abonado [m] cálcico** (Aporte de cal a un suelo para incrementar el pH; abonar [vb] con cal); *f* **chaulage [m]** (Apport d'amendements calciques ou calcomagnésiens à un sol dont il est nécessaire de réduire l'acidité ; LA 1981, 281) ; **g Kalkung [f]** (Ausbringen von Kalk auf eine Bodenfläche, um den pH-Wert zu erhöhen; kalken [vb]); *syn.* Kalkdüngung [f].

limit [n] *adm. envir. leg.* ▶adulteration limit; *phyt. zool.* ▶distribution limit; *agr. for. hort.* ▶injury limit; *trans.* ▶road limit.

limit [n]**, range** *phyt. zool.* ▶distribution limit.

limit [n]**, tree** *for. phyt.* ▶tree line.

limitation period [n] *contr. leg.* ▶period of limitation.

limited-access highway [n] [US] *trans.* ▶freeway [US]/motorway [UK].

3272 limited design competition [n] *adm. plan. prof.* (Competition in which a limited number of planning/design firms are invited to participate); **s concurso [m] de ideas restringido** (Concurso en el que sólo se pide participar a un grupo pequeño de gabinetes de planificación); *f* **appel [m] d'offres restreint avec concours (d'idées)** (Mise en compétition d'un nombre restreint de bureaux d'études pour un concours d'idées) ; **g beschränkte Ausschreibung [f] (1)** (Ausschreibung, bei dem nur ein kleiner Teilnehmerkreis von Planungsbüros für einen Ideen- oder Bauwettbewerb aufgefordert wird; beschränkt ausschreiben [vb]).

3273 limited recreation use [n] *plan. recr.* (Recreation use of an area with low visitor frequency); **s uso [m] recreativo extensivo** (Bajo grado de aprovechamiento de una zona, una instalación, etc. para usos recreativos); *syn.* frecuentación [f] turística extensiva; *f* **fréquentation [f] touristique diffuse** (Par opposition à la surfréquentation, activité [p. ex. tourisme sportif de nature] générant un faible flux récréatif ou touristique ; le ▶tourisme vert est une forme de tourisme diffus) ; *syn.* tourisme [m] doux ; **g extensive Erholungsnutzung [f]** (Nutzung, die durch geringes Aufkommen von Erholungsuchenden gekennzeichnet ist).

3274 limited vehicular access [n] *trans.* (Restricted access to a road under public law aimed at giving priority to through traffic by allowing vehicles to enter the road only at certain points); **s restricción [f] de acceso de vehículos** (Limitación administrativa de acceso de vehículos a una calle/carretera para dar prioridad al tráfico de paso, por la que los vehículos solo pueden entrar por determinados accesos); *f* **limitation [f] d'accès pour les véhicules** (Limitation administrative des droits des riverains à pouvoir accéder directement à la voie publique et les obligeant à emprunter des points d'accés définis afin de ne pas gêner la circulation de transit) ; **g Zufahrtsbeschränkung [f]** (Eine öffentlich-rechtliche Beschränkung der Rechte von Anliegern oder anderen Personen auf Zufahrt zu einer Straße, bei der zum Zwecke der Bevorzugung des durchgehenden Verkehrs die Zufahrt nur von bestimmten Zufahrtsstellen her gestattet ist).

limit [n] **of soil** *constr. eng. pedol.* ▶liquid limit of soil, ▶plastic limit (of soil).

3275 limnetic [adj] *limn.* (living in fresh water); *syn.* limnic [adj]; **s dulciacuícolo/a [adj]** (Referente a especie que vive en agua dulce; se distinguen los términos léntico [que vive en aguas quietas] y lótico [que vive en aguas corrientes]); *syn.* límnico/a [adj]; *f 1* **dulçaquicole [adj]** (Vivant en eaux douces, on distingue les termes lentique [vivant dans les eaux calmes] et lotique [vivant dans les eaux courantes]) ; *f 2* **limnique [adj]** (Terme plus spécifique ; vivant à la surface des milieux lacustres ou des eaux douces dormantes ; DG 1984, 21) ; **g im Süßwasser lebend [loc]** (Pflanzen und Tiere betreffend, die in Fließ- und Stillgewässern leben; nach den unterschiedlichen Strömungsverhältnissen leben in schnell fließenden Gewässern **lotische Arten**, in langsam fließenden Gewässerabschnitten **lentische Arten**; in Stillgewässern **lakustrische Arten**); *syn.* limnisch [adj].

3276 limnetic zone [n] *limn.* (Open-water region of a lake, especially in areas too deep to support rooted aquatic plants; RCG 1982; ▶benthic zone); **s zona [f] limnética** (Zona de aguas continentales profundas. Aplícase también a los organismos planctónicos de las aguas dulces separadas del litoral en sentido equivalente a pelágico, usado cuando se trata del plancton marino; DB 1985; ▶zona del bentos); *f* **zone [f] limnétique** (Eau libre d'un lac, colonisée essentiellement par du phytoplancton [Diatomées, Chlorophycées, Cyanophycées] parfois si dense qu'il forme des « fleurs d'eau », par du bactérioplancton et par un zooplancton extrêmement diversifié ; DUV 1984 ; ▶zone benthique) ; **g Limnopelagial [n]** (Freies Wasser der Binnengewässer von der Oberfläche bis zur größten Tiefe. Gliederung bei Binnengewässern in oberes [duchlichtete trophogene Zone] und unteres **L.** [tropholytische Zone]; ▶Benthal).

limnic [adj] *limn.* ▶limnetic.

3277 limnology [n] *limn.* (Part of hydrobiology concerned with biological, chemical, geographical, and physical features of lakes and watercourses); *s* **limnología [f]** (Ciencia que estudia las aguas continentales y sus biocenosis; es parte de la hidrología); *f* **limnologie [f]** (Science qui étudie les eaux continentales aussi bien du point de vue biologique, écologique que physique ; DEE 1982) ; *syn.* hydrologie [f] lacustre ; *g* **Limnologie [f, o. Pl.]** (Teilgebiet der Hydrobiologie, das sich mit den Binnengewässern [Süßgewässern] und deren Lebewelt befasst).

line [n] *plan.* ►branch rail line [US]/branch railway line [UK]; *envir.* ►cable television line, ►electric power line; *arb. hort.* ►drip line; *plan.* ►form line; *surv.* ►front lot line [US]; *constr. envir.* ►high tension power line; *leg. surv.* ►lot line; *leg. urb.* ►mandatory building line; *constr. eng.* ►no cut-no fill line; *constr. envir.* ►overhead power line, ►overhead telephone line; *trans. urb.* ►public supply utility line; *surv.* ►rear lot line [US]/rear plot boundary [UK]; *leg. urb.* ►setback line; *constr.* ►shadow line; *geo. wat'man.* ►shoreline [US]/shore line [UK]; *surv.* ►side lot line; *urb.* ►skyline; *geo.* ►snow line; *constr.* ►subsurface drainage line; *for. geo. phyt.* ►timber line, ►tree line; *envir.* ►underground power line; *constr.* ►water supply line; *trans.* ►yellow edge line [US].

line [n], **building** *leg. urb.* ►mandatory building line, ►setback line.

line [n] [US], **building restriction** *leg. urb.* ►setback line.

line [n], **collector drainage** *agr. constr.* ►collector drain.

line [n], **drainage** *constr.* ►subsurface drainage line.

line [n] [US], **electric power transmission** *constr. envir.* ►high tension power line.

line [n] [US], **front setback** *leg. urb.* ►setback line.

line [n] [US], **high voltage transmission** *constr. envir.* ►high tension power line.

line [n], **main drainage** *agr. constr.* ►collector drain.

line [n] [UK], **sight** *trans.* ►sight triangle.

line [n], **spring** *geo. hydr.* ►spring horizon.

line [n], **subsurface drain** *constr.* ►subsurface drainage line.

line [n], **supply** *constr.* ►feed pipe.

line [n], **utility** *trans. urb.* ►public supply utility line.

line [n], **valley** *geo.* ►thalweg.

line [n] [US], **water distribution** *constr.* ►water main.

line [n], **yard** *leg. urb.* ►setback line.

line [n] **and levels** [npl] *constr.* ►graded to proper line and levels.

linear cutting [n] [UK] *constr.* ►linear excavation [US].

3278 linear excavation [n] [US] *constr.* (Ground surface removal in a trapezoidal profile for, e.g. road construction, railroad construction [US]/railway construction [UK], etc.; *specific term* railway cutting [UK]; ►cut 1, ►cut shelf); *syn.* cut area [n] [US], linear cutting [n] [UK]; *s* **corte [m] en el terreno [Es]** (Desmonte generalmente en perfil trapezoidal para construcción de vía de transporte; *término específico para vía de ferrocarril* **trinchera [f]**, *termino específico para vía fluvial* **paso [m]**; ►desmonte, ►perfil de desmonte [en una pendiente]); *syn.* incisión [f] en el terreno [AL]; *f* **excavation [f] linéaire** (Profil d'extraction d'un terrassement en général de forme trapézoïdale avec formation d'un talus effectué pour la réalisation de voies de communication terrestres ; ►déblaiement, ►profil de déblai) ; *syn.* fouille [f] en déblai ; *g* **Einschnitt [m]** (Meist trapezförmiges Bodenabtragsprofil in vorhandener Geländeoberfläche für Verkehrsanlagen; ►Abtrag 1, ►Anschnitt); *syn.* Geländeeinschnitt [m].

linear feature [n] *geo.* ►major linear feature.

3279 linear open space pattern/system [n] *landsc. urb.* (Urban open space system along rivers or in narrow valleys, whereby ►green space corridors are located parallel to watercourses and the geomorphology of the area; *specific term* stream valley corridor [US]); *syn.* open space corridor [n] [also US]; *s* **trama [f] verde en forma de banda** (Sistema de espacios libres urbanos a lo largo de ríos y arroyos o en valles estrechos, en el que los ►corredores verdes van paralelos a la morfología natural de la zona); *syn.* red [f] de espacios libres en forma de banda, sistema [m] de espacios libres en forma de banda; *f* **trame [f] linéaire d'espaces libres** (Système de disposition des espaces libres le long d'un axe ou parallèle à certaines données géomorphologiques en milieu urbain tels que des ►coupures vertes le long d'un fleuve, d'un vallon, etc.) ; *g* **bandförmiges Freiraumsystem [n]** (Freiraumanordnung besonders bei Städten an Flüssen oder in engen Tälern, wo ►Grünzüge parallel den Fließgewässern oder geomorphologischen Gegebenheiten folgen); *syn.* linienförmiges Freiraumsystem [n].

3280 linear section [n] *constr. eng. plan.* (Planning term, e.g. for a part of a road project, a stretch of a watercourse); *s* **tramo [m]** (Término que se utiliza para denominar una sección de proyecto de construcción de obra vial); *f* **tronçon [m]** (Terme qualifiant une partie du tracé d'un projet de route, de restauration d'un cours d'eau, etc.) ; *g* **Teilstrecke [f]** (*Planungsbegriff* Abschnitt eines linearen Projektes, z. B. einer projektierten Verkehrsstrasse, eines zu sanierenden Gewässerlaufes); *syn.* Teilabschnitt [m].

line-out [vb] [UK] *hort.* ►plant trees or shrubs in nursery rows.

liner [n] *constr.* ►waterproofing membrane; *for. hort.* ►whip; *arb. hort.* ►strongly branched liner [US]/one-year-old feathered tree [UK].

liner [n], **forest tree** *for. hort.* ►forest tree seedling.

liner [n] **(in a nursery)** *hort. for.* ►immature plant.

lining [n] *wat'man.* ►channel lining.

lining [n] [US], **dumped stone** *wat'man.* ►channel lining.

link [n] *urb. plan.* ►pedestrian link.

linkage [n] *conserv. ecol.* ►becoming interlocked; *trans. urb.* ►connection (1), ►traffic linkage.

lip [n] *constr.* ►step lip.

3281 liquidated damages [npl] *constr. contr.* (Contractually pre-determined sum of money which must be paid by a contractor/consultant if the contract deadline is exceeded, causing significant adverse effects; *opp.* ►bonus payment); *s* **multa [f] por retraso** (Penalización acordada por contrato que debe ser pagada por el contratista/consultor en caso de superar la fecha límite de entrega. **En Es.** para los contratos de las administraciones públicas las cantidades están determinadas en el art. 96 de la ley 13/1995; ►gratificación por entrega adelantada); *syn.* penalidad [f]; *f* **pénalité [f] pour retard** (Somme fixée dans le marché **1.** en cas de retard sur le planning d'exécution des travaux ou sur la production d'un document dans le cas de marché de travaux, **2.** en cas de dépassement du délai contractuel d'exécution des prestations dans le cas d'un marché de maîtrise d'œuvre ; ►prime d'avance) ; *syn.* pénalité [f] de retard (CCM 1984, 13) ; *g* **Vertragsstrafe [f]** (Eine vertraglich festgelegte Geldsumme, die bei Überschreitung von Vertragsfristen, wenn dadurch erhebliche Nachteile entstehen, seitens des Auftragnehmers bezahlt werden muss; cf. § 12 VOB Teil A; eine Höchstbegrenzung bis maximal 10 % der Bausumme ist erforderlich, da sie sonst rechtlich unwirksam ist; *opp.* ►Beschleunigungsvergütung); *syn.* Konventionalstrafe [f].

3282 liquid cement [n] *constr.* (Thin mortar; e.g. used for pouring into joints); *s* **mezcla [f] líquida cemento-agua** (Mortero líquido utilizado p. ej. para rellenar juntas); *syn.* suspensión [f] de cemento, derretido [m] de cemento [C]; *f* **barbotine [f] de ciment** (Mélange pâteux mais fluide de ciment et d'eau pour remplir des joints) ; *syn.* coulis [m] de ciment ; *g* **Zementschlempe [f]** (Flüssiger Zementmörtel, z. B. zum Ausgießen von Fugen).

3283 liquid limit [n] **of soil** *constr. eng. pedol.* (Limit of ►consistency between plastic and fluid states for soils with more than 10% clay content; ►plastic limit [of soil]); *s* **límite [m] de liquidez (del suelo)** (Límite de consistencia de suelos cohesivos [>10% de arcillas] entre el estado plástico y líquido; ►consistencia, ►límite de plasticidad [del suelo]); *f* **limite [f] de liquidité** (Limite de la ►consistance d'un sol plastique [> 10 % d'argile] définie par la quantité d'eau contenue dans celui-ci quand un sillon de dimensions bien définies se referme sur 1 cm en 25 coups donnés par un appareil standardisé ; DIS 1986, 126 ; ►limite de plasticité) ; *g* **Fließgrenze [f]** (Konsistenzgrenze bei bindigen Böden mit > 10 % Tonanteil zwischen plastischem und flüssigem Zustand; ►Ausrollgrenze, ►Konsistenz).

3284 liquid manure [n] *agr.* (Mixture of water and animal urine decomposed by microorganisms, together with small quantities of excrement and bedding straw, used for fertilizing agricultural land; ►semiliquid manure [US]/semi-liquid manure [UK]); *s* **purín [m]** (Mezcla de agua y orina animal descompuesta por microorganismos que, junto con pequeñas cantidades de excrementos animales y paja de camada, se utiliza para fertilizar tierras de cultivo; ►licuame); *syn.* estiércol [m] líquido; *f* **purin [m]** (Fraction liquide qui s'écoule du fumier mis en tas et constitué par les urines des animaux, les eaux de pluie, pouvant contenir des éléments fibreux [paille] et qui constitue un bon engrais ; LA 1981 ; ►lisier) ; *g* **Jauche [f]** (**1.** Durch Mikroorganismen zersetztes Gemisch von Wasser und Tierharn, das mit geringen Bestandteilen von Kot und Einstreu vermischt ist, in Jauchegruben aufgefangen wird, und zur Düngung landwirtschaftlicher Flächen dient. **2.** Inhalt von Klär- und Jauchegruben von Wohnhäusern, die nicht an die Kanalisation angeschlossen sind. **3.** Mit kaltem Wasser und Pflanzenteilen zur Gärung angesetzte Brühe, z. B. Brennnesseljauche, die in der Landwirtschaft und im Gartenbau zum biologischen Pflanzenschutz verwendet wird; ►Gülle).

3285 liquid manuring [n] *agr.* (Fertilizing with fermented urine from livestock, often with a little excrement); *s* **abonado [m] de purín**; *f* **purinage [m]** (Action d'épandre du purin) ; *g* **Jauchedüngung [f]** (Düngung mit vergorenem Harn der Stalltiere, oft mit Kot vermischt); *syn.* *Südwestdeutschland und CH* Gülledüngung [f].

liquid mortar [n] *constr.* ►joint filled with liquid mortar.

list [n] *conserv. zool.* ►Blue List [US]; *adm. conserv. nat'res.* ►IBA List (Important Bird Areas List); *landsc. plant.* ►plant list; *constr. contr.* ►price list, ►punch list [US]/snagging list [UK]; *conserv. phyt. zool.* ►Red List; *prof.* ►short list; *constr. contr.* ►unit price list.

List [n] [US]**, Endangered Species** *conserv. phyt. zool.* ►Red List, #2.

List [n]**, Important Birds Area** *conserv. nat'res.* ►IBA List.

list [n] [UK]**, snagging** *constr. contr.* ►punch list [US]/ snagging list [UK].

list [n]**, species** *landsc. plant.* ►plant list.

list [n]**, undesirable plant** *landsc.* ►list of undesirable plants.

listed building [n] [UK] *conserv'hist.* ►listed landmark building [US].

3286 listed landmark building [n] [US] *conserv'hist.* (**U.S.**, this is a building or structure listed as a landmark building, or a contributing building located in a legally-designated historic district by local or state governments, and subject to regulatory limitations on the owner's right to change, add to, move, or demolish the building. **U.S.**, federally-designated historic buildings or landmarks are kept by the Keeper of the National Register in the National Park Service, acting under delegated authority from the Secretary of the Interior. Moreover, National Register properties impose no regulatory restrictions on private property owners, who may refuse listing if they wish. Once listed, the property owner may lose some tax benefits by failing to follow the Secretary's standards. Under locally-designated historic district regulations, non-contributing buildings are equally subject to local regulations as are contributing buildings. **U.K.**, building or structure of historic or outstanding architectural interest protected by law and entered into a list. The Planning (Listed Buildings and Conservation Areas) Act 1990 consolidates enactments relating to buildings and areas of special architectural or historic interest. The Secretary of State compiles lists of such buildings or approves lists compiled by the Historic Buildings and Monuments Commission; ►Registry of Historic Landmarks [US]/list of historic buildings of special architectural or historic interest [UK]; ►historic preservation measures [US]/measures for conservation of historic monuments [and sites] [UK]); *syn.* listed building [n] [UK]; *s* **edificio [m] declarado monumento** (Edificio u otra construcción protegida por ley y registrada en el ►Registro General de Bienes de Interés Cultural [Es]; cf. Ley 16/1985; ►gestión del Patrimonio Histórico Español [Es]/gestión de monumentos histórico-artísticos. **En EE.UU.**, edificio o estructura registrada como monumento o parte de un distrito histórico declarado legalmente como tal por el gobierno local o del estado, que conlleva limitaciones de los derechos de propiedad en cuanto a realizar cambios, añadir construcciones o demolirlo. Los **ee. dd. m.** declarados por el Gobierno Federal son registrados por el Archivero del Registro Nacional del Servicio Nacional de Parques, que actúa bajo la autoridad delegada del Ministro del Interior. Sin embargo, el registro de monumentos a nivel federal no implica restricciones sobre los propietarios privados que pueden retrazar el registro de sus propiedades. Una vez registrados, los propietarios pueden perder algunas ventajas fiscales si no acotan los estándares del Ministerio. **En GB**, estructura o edificio de interés histórico o arquitectónico excepcional, protegido por ley y registrado en una lista. La ley de planificación de 1990 regula la actuación sobre estos edificios o áreas de interés arquitectónico o histórico especial. El Ministro del Interior elabora las listas y aprueba las elaboradas por la Comisión de Edificios y Monumentos Históricos); *syn.* edificio [m] protegido; *f 1* **bâtiment [m] classé ou inscrit au titre des monuments historiques** (Bâtiment ancien ou exceptionnel régit par la loi sur les monuments historiques et inscrit sur la ►liste générale des monuments classés ou à l'►inventaire supplémentaire des monuments historiques ; ►gestion du patrimoine monumental) ; *syn.* objet [m] immobilier classé ; *f 2* **bâtiment [m] protégé** (Elément, édifice patrimonial repéré au titre de l'article L 123-1 7° du code de l'urbanisme, faisant l'objet de mesures de protection spécifiques parce qu'il possède une qualité architecturale remarquable, ou constitue un témoignage de la formation et de l'histoire de la ville ou d'un quartier, ou assure par sa volumétrie un repère particulier dans le paysage urbain, ou appartient à une séquence architecturale remarquable par son homogénéité ; les bâtiments protégés font l'objet d'une présentation par fiche dans le rapport de présentation du PLU, dans le guide du patrimoine bâti ; il s'agit en général d'un patrimoine urbain dont une partie peut être jugée de qualité « modeste » au

L

regard des critères classiques de l'histoire de l'architecture et qui, de ce fait, ne peut bénéficier d'une protection au titre des monuments historiques ; néanmoins, il reste constitutif et représentatif de l'histoire de l'agglomération et de son développement et mérite à ce titre d'être protégé) ; *syn.* bâtiment [m] protégé identifié, bâtiment [m] distingué ; *f 3* **bâtiment [m] désigné** (Elément patrimonial repéré au titre de l'article L 123-3-1° du code de l'urbanisme permettant de désigner dans les zones agricoles, les bâtiments agricoles qui, en raison de leur intérêt architectural ou patrimonial, peuvent faire l'objet d'un changement de destination, dés lors que ce changement de destination ne compromet pas l'exploitation agricole ; PLU Palaiseau) ; *g* **denkmalgeschütztes Gebäude [n]** (Gebäude, das unter Denkmalschutz steht und im Denkmalbuch oder in einer Denkmalliste eingetragen ist; ►Denkmalbuch, ►Denkmalpflege).

3287 list [n] **of bid items and quantities** [US] *constr. contr.* (Compilation of clearly defined items of work [US]/works [UK] and services to be provided for a construction project as the basis for a ►submission of bids [US]/submission of tenders [UK]; ►bid documents [US]/tender[ing] documents [UK], ►calculation of quantities, project requirements for bidding [US]/project requirements for tendering [UK], ►shedule of quantities); *syn.* schedule [n] of tender items [UK]; *s* **resumen [m] de prestaciones** (Lista de todos los trabajos a realizar durante una obra como base para la ►entrega de ofertas; ►cálculo de cantidades, ►documentos de contratos de las administraciones públicas, ►lista de cantidades, ►pliego de condiciones [generales] de concurso); *f* **descriptif [m] quantitatif** (Pièce écrite [faisant partie du cahier des prescriptions spéciales dans le dossier de consultation des entrepreneurs] constituant une liste de décomposition élémentaire et de désignation des travaux avec mention des unités, des quantités servant de base à la ►remise des offres/soumissions ou d'un devis [quantitatif – estimatif] ; F., le devis quantitatif – estimatif [*syn.* détail estimatif] est souvent constitué de deux pièces : le devis descriptif et le cadre du bordereau des prix ; ►dossier d'appel d'offres, ►dossier de consultation, ►établissement du quantitatif, ►pièces contractuelles constitutives du marché) ; *syn.* détail [m] quantitatif ; *g* **Leistungsverzeichnis [n]** (Auflistung von eindeutig und erschöpfend beschriebenen Teilleistungen einer Bauaufgabe als Grundlage für die ►Angebotsabgabe; ►Ausschreibungsunterlagen, ►Massenaufstellung 2, ►Massenberechnung, ►Verdingungsunterlagen); *syn.* Übernahmebedingungen [fpl] [CH].

list [n] **of defects** [US] *constr. contr.* ►punch list [US]/ snagging list [UK].

3288 list [n] **of expenses** *contr.* *s* **especificación [f] de costos/costes [Es]**; *f* **liste [f] des frais** ; *g* **Kostenaufstellung [f]** (**1.** Ergebnis der Auflistung des Finanzaufwandes für ein Vorhaben; **2.** Auflistung bezahlter Rechnungen/finanzieller Aufwendungen/Ausgaben); *syn. zu 2.* Aufstellung [f] bezahlter Kosten.

list [n] **of historic buildings, monuments and sites** [UK] *adm. leg. conserv'hist.* ►Registry of Historic Landmarks [US].

list [n] **of historic buildings of special architectural or historic interest** [UK] *adm. leg. conserv'hist.* ►Registry of Historic Landmarks [US].

List [n] **of IBAs** *adm. conserv. nat'res.* ►IBA List.

List [n] **of Important Bird Areas** *adm. conserv. nat'res.* ►IBA List.

list [n] **of professional services** *contr. prof.* ►scope of professional services.

3289 list [n] **of species** *phyt. zool.* (Species compilation as a result of a floristic or faunal inventory; ►species inventory); *s* **lista [f] de especies** (Compilación de especies como resultado de un ►inventario de especies); *f* **liste [f] des espèces** (Liste établie pour un ►inventaire des espèces lors d'analyses biocénotiques) ; *g* **Artenliste [f]** (Bei vegetations- oder tierökologischen Aufnahmen erstellte Zusammenstellung aller in einem definierten Gebiet gefundenen Arten; ►Arteninventar).

3290 list [n] **of undesirable plants** *landsc.* (Schedule of intrusive or incompatible plants which should not be planted in order to avoid a conflict with the particular planting scheme); *syn.* undesirable plant list [n]; *s* **lista [f] de plantas no utilizables** (Lista incluida en algunos planes del paisaje o planes de zonas verdes para evitar el uso de plantas no deseadas por ser intrusivas, repetitivas o posibles causas de peligro para la población, como plantas con frutos venenosos, etc.); *f* **liste [f] des végétaux indésirables** (Végétaux ligneux incompatibles avec les objectifs d'aménagement fixés et dont l'utilisation est explicitement déconseillée sur une liste, p. ex. dans le cadre de prescriptions d'urbanisme, des plantations d'accompagnement, des aires de jeux, etc.) ; *syn.* liste [f] des végétaux impropres à la plantation ; *g* **Gehölznegativliste [f]** (Zur Verwirklichung bestimmter Bepflanzungsabsichten das Aufführen von nicht zu pflanzenden Gehölzarten in den Festsetzungen der Grünordnungspläne, ggf. Bebauungspläne oder Landschaftspläne).

lithosol [n] [US] *pedol.* ►skeletal soil.

3291 lithosphere [n] *geo.* (Part of the ►geosphere rock mantle or solid portion of the Earth); *s* **litosfera [f]** (Parte de la ►geosfera: la capa rocosa o porción sólida de la corteza terrestre); *f* **lithosphère [f]** (Partie solide de l'écorce terrestre; ►géosphère) ; *g* **Lithosphäre [f, o. Pl.]** (*Teil der* ►Geosphäre Gesteinshülle der Erde).

litigation [n] **brought by a non-profit organization** [UK] *conserv. leg.* ►non-profit organization suit [US].

3292 litter [n] (1) *pedol. phyt.* (Organic debris on the ground, consisting of freshly fallen or only slightly decomposed plant material, mainly foliate but also bark fragments, twigs, flowers, fruits etc; ►leaf litter, ►litter layer, ►needle straw [US]/needle litter [UK]); *syn.* organic detritus [n]; *s* **litter [m]** (Residuos orgánicos frescos; ►capa de barrujo, ►capa de litter, ►hojarasca 2); *f* **débris [mpl] végétaux** (Déchets organiques frais recouvrant le sol constitués par les feuilles, les rameaux morts accumulés sur le sol et constituant la source essentielle de la matière organique fraîche ; ►couverture morte, ►litière d'aiguilles, ►litière feuillue) ; *g* **Streu [f]** (Bestandsabfall der Vegetation, der auf dem Boden liegt; ►Laubstreu, ►Nadelstreu, ►Streuschicht).

litter [n] (2) *phyt.* ►driftline litter; *for. pedol.* ►forest floor litter; *envir.* ►scattered rubbish [US]/litter [UK].

litter [n], **conifer** *pedol.* ►neddle straw [US]/needle litter [US].

litter [n], **floodmark** *phyt.* ►driftline litter.

litter [n], **forest** *for. pedol.* ►forest floor litter.

litter [n] [UK], **needle** *pedol.* ►needle straw [US].

litter bin [n] [UK] *envir.* ►trash can [US] (2).

litter bin support [n] **with sackholder** [UK] *envir.* ►trash can support with bag liner [US].

litter breakdown [n] *biol. chem.* ►litter decomposition.

3293 litter decomposition [n] *biol. chem.* (Molecular breakdown of ►litter 1 resulting in physical/chemical separation into its elements or simpler constituents); *syn.* litter breakdown [n]; *s* **descomposición [f] de litter** (Degradación molecular y descomposición físico-química del ►litter); *f* **décomposition [f] de la litière** (Décomposition de la matière organique fraîche

composant la litière [▶débris végétaux] par des processus bio-chimiques, mécaniques et enzymatiques ; PED 1983, 27) ; *syn.* altération [f] de la litière, dégradation [f] de la litière ; *g* **Streuabbau** [m] (Molekulare Zersetzung und physikalisch-chemische Veränderung [Vereinfachung] der ▶Streu).

3294 litter layer [n] *pedol. (Abbr.* L-layer; soil horizon primarily comprising non-decomposed litter; ▶forest floor litter, ▶mor humus, ▶organic horizon); *syn.* L horizon [n]; *s* **capa [f] de litter** (Capa superior del suelo constituida fundamentalmente por restos vegetales, recientemente caídos y ligeramente descompuestos, en los que todavía es reconocible la estructura original: hojas, acículas, ramillas, corteza, etc.; DINA 1987; ▶litter del bosque, ▶mantillo, ▶moor); *f* **couverture [f] morte** (Horizon organique peu décomposé, constitué par l'accumulation sur le sol des débris végétaux provenant des strates sus-jacentes et qui contribuent à la formation de l'humus ; ▶horizon organique Ao, ▶litière forestière, ▶mor) ; *syn.* horizon A [m], litière [f] ; *g* **Streuschicht [f]** (Oberster Bodenhorizont, der aus weitgehend unzersetztem Bestandsabfall besteht; ▶Auflagehumus, ▶Humusauflage, ▶Waldbodenstreu); *syn.* L(itter)-Horizont [m], O₁-Horizont [m], Streudecke [f], *for.* Streuauflage [f].

litter meadow [n] [UK] *agr. land'man. phyt.* ▶straw meadow [US].

littoral [n] [UK] *geo. limn. ocean.* ▶littoral zone [US].

3295 littoral vegetation [n] *phyt.* (Existing plant growth along the shores of the sea and lakes; whilst in English and German the term ▶riparian vegetation is applied only to vegetation growing on the edge of flowing water, lakes and ponds, the term, 'littoral' in English and German refers to habitat of the shores of the sea and inland waters; 'littoral' does not apply to the banks of flowing water in France, yet, in francophone Canada, it does [littoral fluvial]; ▶driftline community, ▶exposed mud vegetation, lakeshore vegetation); *s* **vegetación [f] del litoral y de las riberas de las aguas continentales** (Conjunto de especies de plantas que crecen en las orillas del mar y de las aguas continentales. En inglés y en alemán, el término «littoral/litoral» se aplica tanto para el mar como para las aguas continentales, en francés y castellano sólo se utiliza para la zona de las orillas del mar; ▶vegetación de bordes de estanques, ▶vegetación de las acumulaciones de restos orgánicos, ▶vegetación halófila de los cúmulos de residuos orgánicos, ▶vegetación ripícola); *f* **végétation [f] du littoral et des rives des cours d'eau** (Espèces végétales qui poussent en bordure des côtes maritimes et sur les berges des lacs, des étangs et des cours d'eau ; ▶flore d'ourlet halonitrophile, ▶groupement des grèves d'étangs, ▶végétation des grèves, ▶végétation ripicole ; *terme spécifique* végétation des rives/rivages lacustres) ; *g* **Vegetation [f, o. Pl.] entlang von Küsten und Ufern** (1. Vorhandener Bewuchs an der Küste und an Ufern von Fließ- und Stillgewässern; *UBe* ▶Spülsaumflur, ▶Teichuferflur; Seeufervegetation. 2. Während im Deutschen sich **Ufervegetation** [▶Uferbewuchs] nur auf Vegetation am Rande von fließenden und stehenden Gewässern bezieht, beinhaltet der Terminus *litoral/littoral* im Deutschen und Englischen den Lebensbereich des Meeres und der Binnengewässer; in Frankreich wird *littoral* nicht für die Ufer der Fließgewässer verwendet, jedoch im frankophonen Kanada *[littoral fluvial]*); *syn.* Litoralvegetation [f].

3296 littoral zonation [n] *phyt.* (Shoreline sequence of plant communities, usually ribbon-like and spreading from water to land; e.g. along sea and lake shores; ▶riparian zonation); *s* **cliserie [f] litoral** (Zonación de la vegetación en las orillas de ríos y lagos; ▶zonación de la vegetación ribereña); *f* **zonation [f] de la végétation littorale** (Succession par bandes ou en ceintures des groupements végétaux en bordure des cours d'eau et plans d'eau ; *termes spécifiques* zonation de la végétation lacustre, zonation de la végétation des rives, zonation de la végétation des grèves alluviales ; ▶zonation de la végétation riveraine des cours d'eau) ; *g* **Vegetationszonierung [f] an Ufern** (Meist bandförmige Abfolge von Pflanzengesellschaften an Ufern von Fließ- und Stillgewässern; ▶Vegetationszonierung an Fließgewässern).

3297 littoral zone [n] [US] *geo. limn. ocean.* (Term pertaining to the ▶benthic zone or depth zone between high and low water, or to the organisms of that environment. Multiple zonation of plant communities in freshwater bodies: tall sedge swamp, reed swamp, rooted floating plants, pondweed meadows, and subaqueous Stonewort swards. The belt of submerged growing plants is called 'sublittoral'. 'Eulittoral' is a riverine zone of emergent vegetation—in freshwater habitats the zone of wave action or oscillating water table and at the coast an area affected by tides; ▶shallow water area); *syn.* littoral [n] [UK]; *s 1* **litoral [m]** (En la costa del mar, espacio que se extiende entre los niveles de pleamar y bajamar. Se subdivide en: zona supralitoral, z. litoral media, z. infralitoral, z. sublitoral. Según algunos autores también pertenece al **l.** la zona de salpicadura; cf. DINA 1987); *syn.* estrán [m], *leg. Ley de Costas* ribera [f] del mar; *s 2* **zona [f] litoral** (Faja lateral de los cauces de ríos y de los lagos situada entre el nivel de aguas de crecida y el límite inferior de la vegetación radicante; es una zona de alta producción primaria, en la que algunas plantas como la enea *[Typha]* baten récords de productividad; alberga muchos animales [crustáceos, gusanos, moluscos, larvas de insectos y, en las agus mediterráneas, muchos peces que no son aptos para comer plancton] y sirve de zona de alimento para muchas ranas, insectos y aves acuáticas; cf. MARG 1981; ▶zona de aguas poco profundas); *f 1* **zone [f] benthique littorale** (▶Zone de faible profondeur correspondant à la partie périphérique de plans d'eau douce stagnante ou courante caractérisée par une végétation enracinée sur le fond ; de la rive vers le large, la **z. b. l.** présente les ceintures de colonisation suivantes : la prairie à grandes laîches [magnocariçaie], la roselière [phragmitaie], la scirpaie, la nupharaie, la zone à Potamots [potamaie] et la zone des plantes entièrement submergées [charaie]) ; *f 2* **zone [f] marine littorale** (Dans la série littorale marine indépendamment des grandes différences de distribution des peuplements floristiques et faunistiques rencontrés en fonction de la nature des marées et des substrats [meuble ou compact] ou de l'exposition du trait de côte on distingue les zonations suivantes : ceinture supralittorale, ceinture médiolitorale, ceinture infralittorale [herbiers marins]) ; *f 3* **littoral [m]** (1. Domaine géomorphologique compris, au sens strict, entre les plus hautes et les plus basses mers, mais en fait étendu à l'espace influencé par les forces marines agissant au contact du continent et qui comprend : la ▶côte et le ▶rivage 1, l'▶estran et l'avant-côte ; DG 1984. 2. Ensemble des sites naturels en bordure de mer dont la situation particulière confère une qualité exceptionnelle très estimée et qui mérite d'être protégée) ; *g* **Litoral [n]** (Uferbereich der Gewässer zwischen dem äußersten, vom Wasser bedeckten Gewässerrand und der untersten Grenze des Pflanzenwuchses. Bei Süßgewässern ist das **L.** gegliedert in Großseggengürtel, Röhrichtgürtel, Schwimmblattpflanzengürtel, Laichkrautgürtel und submerse Armleuchteralgenrasen am Gewässergrund. Die Zone der untergetaucht wachsenden Pflanzen wird oft als **Sublitoral** bezeichnet. Als **Eulitoral** gilt bei Süßgewässern der Bereich des Wellenschlages resp. der jahreszeitlichen Wasserspiegelschwankungen [Wasserwechselzone], am Meer der Bereich der Gezeitenzone; ▶Flachwasserzone).

3298 live brushlayer barrier [n] *constr.* (Protective ▶biotechnical construction technique to stop undercutting and enlargement of streambank erosion or to control the median level

of water flow); *syn.* brush jetty [n] [also US] (WRP 1974, 127); *s* **barrera [f] (viva) de ramajes** (▶Construcción de bioingeniería protectora para frenar la erosión de las márgenes de cursos de agua); *f* **traverse [f] de branchage à rejets** (▶Technique de génie biologique dans certains aménagements hydrauliques par la mise en place d'une barrière perpendiculaire au cours des eaux constituée de boutures de saule tenue en place par un enrochement ou recouverte de terre en vue de la stabilisation du profil en long du cours d'eau, la protection contre les affouillements) ; *syn.* enrochement [m] et bouturage sur berge (WIB 1996), enrochement [m] naturel végétalisable ; *g* **Buschbautraverse [f]** (▶Ingenieurbiologische Bauweise, bei der zur Beruhigung und Verlandung von Ausuferungen und Kolken oder zur Festlegung des Bereiches um die Mittelwasser-Linie an Fließgewässern quer zur Fließrichtung Buschlagen aus 100-150 cm langen, lebenden Weidenzweigen und Astwerk in die Ufer in Gräben eingebaut und mit groben Steinen und Boden abgedeckt werden).

live brush mattress [n] [US] *constr.* ▶brush mat.

live [n] **deciduous leaves** *bot. hort.* ▶foliage of woody plants.

3299 live picket [n] [US] *constr.* (*Bioengineering* a long, thin stem cutting [1.5 to 2.5m in height] used in steep slope planting of willows and poplars); *syn.* set [n] [UK]; *s* **vareta [f] viva** (*En bioingeniería* ▶esqueje leñoso recto y poco ramificado de 1,5-2,5 m de altura, utilizado para plantar en pendientes, p. ej. de sauces o chopos); *f* **plançon [m]** (Longue et forte ▶bouture de rameau aoûté, [1-2 m de hauteur], p. ex. d'un Saule ou Peuplier, pour la multiplication des espèces ligneuses utilisées dans les techniques de génie biologique) ; *syn.* plantard [m] ; *g* **Setzstange [f]** (Lebender Pflanzenteil eines Gehölzes in Form eines 1,5 bis 2,5 m langen, geraden, wenig verzweigten Leittriebes [▶Steckholz] für ingenieurbiologische Baumaßnahmen; cf. DIN 18 918).

live plant material [n] [US] *constr.* ▶brush matting with live plant material [US]/brush laying [UK].

3300 live stake [n] *constr.* (Cutting of a green woody plant in the form of a straight branch > 3cm thick and at least 50cm long for use in ▶bioengineering construction); *s* **estaca [f] viva** (Parte de tallo recto de leñosa de >3 cm de grosor y >50 cm de largo utilizada en medidas de ▶construcción biotécnica); *f* **pieu [m] en bois vivant** (Partie de végétaux ligneux constituée d'une branche droite, d'un diamètre minimum de 3 cm et d'une longueur d'environ 50 cm, utilisée pour les travaux utilisant les ▶techniques de génie biologique) ; *g* **lebender Pflock [m]** (Pflanzenteil eines Gehölzes in Form eines geraden Triebes von > 3 cm Dicke und mindestens 50 cm Länge für ▶ingenieurbiologische Baumaßnahmen; cf. DIN 18 918).

3301 livestock [n] *agr.* (Domestic farm animals); *s* **cabezas [fpl] de ganado** (Conjunto de animales domésticos en una granja); *syn.* capital [m] vivo; *f* **cheptel [m] vif** (Ensemble des troupeaux de bétail d'une exploitation agricole, d'une région) ; *g* **Viehbestand [m]** (Vorhandene Menge an Vieh eines landwirtschaftlichen Betriebes); *syn.* Viehstand [m] [CH].

livestock farming [n] *agr.* ▶livestock industry.

livestock husbandry [n] *agr.* ▶livestock industry.

3302 livestock industry [n] *agr.* (Branch of agricultural land use for the production of meat, milk, eggs, wool, etc. as well as cattle breeding); *syn.* livestock husbandry [n], livestock farming [n]; *s* **ganadería [f] industrial** (Rama del aprovechamiento agrícola del suelo que se dedica a producir carnes, leche, huevos, lana, etc. y a criar ganado); *syn.* producción [f] ganadera, producción [f] pecuaria; *f* **industrie [f] liée à l'élevage** (Branche de l'industrie de l'agro-alimentaire spécialisée dans l'élevage du bétail, l'abattage et la découpe, la charcuterie, la laiterie, la fromagerie, la production d'aliments du bétail, etc.) ; *g* **Vieh-**

wirtschaft [f] (Zweig der landwirtschaftlichen Bodennutzung zur Produktion von Fleisch, Milch, Eiern, Wolle, Häuten, Haaren etc. sowie zur Viehzucht); *syn.* Viehhaltung [f].

3303 living area [n] *arch. gard.* (Space in a house or apartment [US]/flat [UK], which may also include a part of the garden, a courtyard or balcony/terrace and serves as room for sitting, receptions or other activities of the home; ▶courtyard living area); *s* **área [f] de convivencia** (±) (Zona de la vivienda dedicada a la convivencia familiar, en la que se incluye el jardín, el balcón o patio; ▶patio habitación); *f* **domaine [m] d'habitation** (±) (Espace dans un appartement dédié à la vie familiale commune et pouvant parfois inclure le jardin, la cour intérieure ou le balcon ; ▶cour d'habitation intérieure) ; *syn.* espace [m] convivial ; *g* **Wohnbereich [m] (1)** (Das Wohnzimmer einer Wohnung und ggf. der Bereich des Gartens, des Innenhofes oder des Balkons, der dem Wohnen dient; ▶Wohnhof).

3304 living conditions [npl] *ecol. plan. sociol.* (Circumstances under which man, animals and plants actually exist on earth); *syn. also* survival conditions [npl] (±); *s* **condiciones [fpl] de vida** (Circunstancias bajo las cuales viven concretamente los seres humanos, la flora y la fauna en la tierra); *f* **conditions [fpl] de vie** (Données climatiques, sociales, etc. sous lesquelles les êtres vivants se développent) ; *syn.* conditions [fpl] d'existence ; *g* **Lebensbedingungen [fpl]** (Umstände, unter denen Menschen, Tiere und Pflanzen faktisch in dieser Welt leben).

3305 living construction unit [n] *constr.* (Live element used in ▶biotechnical construction techniques, e.g, ▶sod [US] 1/turf [UK], ▶seed mat or living ▶fascine, ▶living plant material; ▶sod layer [US]/turf layer [UK]; ▶wattlework/wattle-work [UK]); *s* **elemento [m] de construcción vivo** (Cualquier tipo de ▶material vivo que se utiliza en la ▶construcción biotécnica, como ▶tepes, ▶estera de semillas, ▶fajina viva o ▶varaseto de ramas; ▶césped para trasplante); *f* **assemblage [m] de matériaux vivants** (Végétaux ou parties de plantes utilisés dans les ▶techniques de génie biologique des sols, tels que les ▶plaques de gazons, le ▶gazon précultivé, ▶natte préensemencée, ▶fascine ou ▶claie de branches à rejets ; ▶matériau vivant) ; *g* **lebender Bauteil [m]** (Für ▶ingenieurbiologische Bauweisen verwendete Teile wie ▶Rasensode, ▶Saatmatte, lebende ▶Faschine oder ausschlagfähige ▶Flechtwerke/Geflechte; cf. DIN 18 918; ▶Fertigrasen; *OB* ▶lebender Baustoff).

living fence [n] [US] *constr. hort.* ▶clipped hedge.

3306 living [ger] **in the soil** *ecol.* (▶species of soil fauna); *s* **hipogeo/a [adj]** (Referente a animal que vive en el suelo; ▶animal edáfico); *f* **endogé, ée [adj]** (Vivant dans le sol ; ▶pédobionte) ; *g* **im Boden lebend [loc]** (▶tierisches Bodenlebewesen); *syn.* endogäisch [adj].

living [ger] **on wood, species** *phyt. zool.* ▶lignicolous species.

3307 living plant material [n] *constr.* (Generic term for plants or parts of plants, such as seed, cuttings, branches, and twigs as well as ▶living construction units used in bioengineering and ▶biotechnical construction technique; ▶fascine); *syn.* animate material [n] [also US]; *s* **material [m] vivo** (Plantas o parte ellas, como semillas, esquejes, ramas y ramillas así como material de construcción vivo que se utilizan en la ▶construcción biotécnica; ▶elemento de construcción vivo, ▶fajina); *syn.* material [m] natural; *f* **matériau [m] vivant** (Végétaux ou parties de plantes utilisés dans les ▶techniques de génie biologique utilisée dans les techniques paysagères de stabilisation des sols, tels que les semences, enherbements, boutures, branches à rejets, branchages, ramilles ; ▶assemblage de matériaux vivants, ▶fascine) ; *g* **lebender Baustoff [m]** (Für ▶ingenieurbiologische Bauweisen verwendete Pflanzen oder deren Teile wie Saatgut,

Steckholz, Äste und Zweige sowie ▶lebende Bauteile; cf. DIN 18 918; ▶Faschine).

3308 living space [n] *phyt. zool. leg.* (Comprises the total space on earth in which an individual living being or species of organism exists and has room for growth; ▶habitat 2, ▶home range 2, ▶location 1, ▶range 1, ▶total range); *syn.* habitat range [n] [also US]; *s* **dominio** [m] **vital** (Incluye todo el espacio de la tierra en el cual existe y puede crecer un individuo o una especie de organismos; ▶área de distribución, ▶área de habitación, ▶biótopo, ▶estación, ▶territorio vital total); *syn.* espacio [m] vital; *f 1* **domaine** [m] **d'existence** (Grande unité spatiale naturelle sur laquelle vit un individu ou une espèce d'individus ; LAF 1990, 295) ; *syn.* espace [m] vital ; *f 2* **habitat** [m] **naturel** (Zone terrestre ou aquatique se distinguant par ses caractéristiques géographiques, abiotiques, qu'elle soit entièrement naturelle ou semi-naturelle ; Art. 1, Directive 92/43/CEE ; ▶aire d'habitation, ▶aire de répartition, ▶biotope, ▶domaine vital, ▶habitat naturel, ▶station) ; *syn.* domaine [m] naturel ; *g 1* **Lebensraum** [m] (**1.** Umfasst den gesamten Raum auf der Erdoberfläche mit allen abiotischen und biotischen Lebensbedingungen, in dem sich das Dasein eines Individuums oder einer Organismenart abspielt; **2.** wird oft im Sinne von ▶Biotop, d. h. als Lebenstätte einer Lebensgemeinschaft oder im Sinne von ▶Habitat, d. h. als Lebensstätte der Populationen einer Art, verwendet; *g* **natürlicher Lebensraum** [m] *leg.* (Durch geografische, abiotische und biotische Merkmale gekennzeichnetes völlig natürliches oder naturnahes terrestrisches oder aquatisches Gebiet; Art. 1 Fauna-Flora-Habitat-Richtlinie 92/43/EWG; auf Grund der abiotischen Merkmale, der Vegationsstrukturen und –typen sowie der Artenzusammensetzung lassen sich Lebensräume in verschiedene ▶Lebensraumtypen abgrenzen ; ▶Areal, ▶Jahreslebensraum, ▶Standort 1, ▶Wohngebiet 1).

living space [n] **[UK], garden** *gard.* ▶outdoor living space [US].

llano [n] *geo. phyt.* ▶steppe.

load [n] *geo. hydr.* ▶bed load; *eng.* ▶design load; *envir.* ▶effluent pollution load, ▶environmental pollution load, ▶existing environmental pollution load, ▶pollution load; *recr. plan.* ▶recreation design load, ▶recreation load; *hydr.* ▶sediment load; *eng.* ▶service load; *arb. constr.* ▶snow load; *eng. stat.* ▶structural load(ings); *envir. plan. trans.* ▶traffic load; *hydr.* ▶wash load; *arb. constr. stat.* ▶wind load.

load [n]**, background** *envir.* ▶background level of pollution [US].

load [n]**, debris** *geo. hydr.* ▶bed load.

load [n]**, thermal** *envir. limn.* ▶thermal pollution of watercourses.

3309 load-bearing capacity [n] *eng.* (Maximum load that ground or structures can support before subsiding or failing; cf. RCG 1982; ▶load-bearing construction, ▶soil-bearing capacity); *syn.* design loading [n]; *s* **capacidad** [f] **de carga** (Peso máximo que pueden sostener estructuras portantes como puentes, tejados, etc.; ▶capacidad de soporte del suelo, ▶estructura portante); *f* **capacité** [f] **de charge (d'exploitation)** (Charge maximale que peut supporter un pont, une toiture, une dalle armée, etc. ; ▶forme support, ▶portance d'un sol) ; *syn.* possibilité [f] de charge d'exploitation, capacité [f] limite, capacité [f] porteuse ; *g* **statische Belastbarkeit** [f] (Maximale Last, die von Bauwerken oder deren Teile, z. B. Brücken, Dachflächen, Tiefgaragendecken, Fundamente, aufgenommen werden kann, ohne dass sie sich setzen, Bauschäden oder andere beeinträchtigende Veränderungen entstehen; ▶Tragfähigkeit eines Bodens, ▶Tragkonstruktion).

3310 load-bearing construction [n] *constr.* (Structural members; e.g. beams for a roof; ▶load-bearing capacity); *syn.* load-bearing structure [n]; *s* **estructura** [f] **portante** (Partes de una construcción responsables de soportar la carga, p. ej. vigas para un tejado; ▶capacidad de carga); *syn.* estructura [f] soportante; *f* **forme** [f] **support** (Structure qui supporte un élément de construction, p. ex. un toit ; ▶capacité de charge [d'exploitation]) ; *syn.* structure [f] portante ; *g* **Tragkonstruktion** [f] (Tragende Baukonstruktion, z. B. für ein Dach; ▶statische Belastbarkeit).

3311 load-bearing soil [n] *constr.* (Soil that is capable of supporting loads without the occurrence of subsidence; ▶soil bearing capacity); *s* **suelo** [m] **portante** (Suelo que puede soportar cargas sin que ocurran hundimientos; ▶capacidad de soporte del suelo); *syn.* suelo [m] apto para construcciones, suelo [m] resistente; *f* **sol** [m] **porteur** (Bonne aptitude comme couche de fondation ; ▶portance d'un sol) ; *g* **tragfähiger Boden** [m] (Als Baugrund zur Aufnahme von Lasten geeigneter Boden; ▶Tragfähigkeit eines Bodens).

load-bearing structure [n] *constr.* ▶load-bearing construction.

load-bearing test [n] *constr.* ▶plate load-bearing test.

loading [n] *constr.* ▶excavation, haulage, loading and transport of in-situ material.

3312 loading [n] **of soil** *constr.* (Loading a transporting vehicle with soil); *s* **cargado** [m] **de tierra** (Acción de cargar con tierra un vehículo de transporte; cargar [vb] tierra); *f* **chargement** [m] **de terre** (Action de remplir un véhicule de terre) ; *g* **Laden** [n, o. Pl.] **von Bodenmassen** (Ein Transportmittel mit Boden beladen).

loadings [pl] *eng. stat.* ▶structural load(ings).

3313 loam [n] *pedol.* (Soil mixture of sand, silt and clay particles; ▶alluvial silt, ▶glacial loam, ▶soil textural class); *s* **limo** [m] (1) (▶*Clase textural de suelo* mezcla de arcilla, limo y arena que, dependiendo del contenido de cada fracción, se diferencia en limo arenoso, limo limoso y limo arcilloso; ▶harina glaciar, ▶limo aluvial); *f* **limon** [m] (Matériau meuble [alluvion fluviatile ou dépôt éolien] de fraction granulométrie comprise entre 2 et 50 µm ; on distingue les limons fins [2-20 µm] et les limons grossiers [20-50 µm] ; ▶classe de texture de sol, ▶limon alluvial, ▶limon glaciaire décarbonaté) ; *g* **Lehm** [m] (▶Bodenart aus einem Korngrößenfraktionengemisch der Hauptbodenarten Ton, Schluff und Sand; je nach Korngrößenverteilung unterscheidet man sandigen L. [sL], schluffigen L. [uL] und tonigen L. [tL]; ▶Auenlehm, ▶Geschiebelehm).

3314 loamy [adj] (Consisting of or like loam); *s* **limoso/a** [adj]; *f* **limoneux, euse** [adj] ; *g* **lehmig** [adj] *pedol.* (Aus Lehm bestehend oder Lehm enthaltend).

lobby [n] *pol. sociol.* ▶pressure group.

lobe [n] *geo.* ▶meander lobe.

local administrative authority/agency [n] *adm. leg. pol.* ▶local/regional administrative authority/agency.

local agency [n] [US] *adm.* ▶local authority.

3315 local authority [n] *adm.* (**In U.S.**, **l. a.** has a distinct and separate meaning from "authority" as used in other countries. In U.S., "authority" refers to a separate and independent federal state, or local public corporation with broad authority to build, own, and operate government businesses providing a public service: water, sewerage, gas, electricity, etc. Authority governing board members are appointed, not elected, possess authority to levy charges or fees, borrow money [issue bonds], etc. and may not be voted out of office; ▶local/regional administrative authority/agency; **in U.K., l. a.** is a term generally used to describe a statutory corporation with elected members, charged with a range of functions, such as the provision of services, over a limited geographical area. A **l. a.** is also the local planning authority with the power to acquire appropriate land and develop it, either alone or in authorized association. The two-tier system of planning and development control, comprising shire counties and metropolitan counties with their respective district councils, is governed by the Town and Country Planning Act 1990); *syn.* local agency [n] [also US], municipal authority [n], local government department [n]; *s* **autoridad [f] local** (**1. En Es.** conjunto de organismos públicos de una ciudad o municipio, cuyas competencias y los servicios que están obligados a prestar a la población están regulados en los arts. 25 y 26 de la Ley 7/1985, de 2 de abril, Reguladora de las Bases del Régimen Local, y dado el caso en la legislación de la correspondiente Comunidad Autónoma. **2.** El término también se puede emplear para denominar a los/las representantes elegidos para gobernar una ciudad o un municipio: alcalde/intendente [CS]/presidente municipal [MEX] y al consejo; ►entidad local [territorial]); *syn.* autoridad [f] municipal, administración [f] local/municipal; *f* **autorité [f] territoriale** (Assemblée délibérante ou exécutif local, distincte de l'administration de l'État, assurant des services publics et responsables de l'exécution des tâches en leur compétence ; depuis les lois de décentralisation de 1982-1983 la France connaît trois niveaux de structures locales : les communes, départements et régions, l'État ayant transféré au profit des collectivités territoriales certaines compétences et les ressources correspondantes ; ce terme est souvent utilisé dans le langage courant pour désigner le dernier niveau de pouvoir celui des communes ; ►collectivité territoriale ; *syn.* autorité [f] locale, administration [f] locale ; *g* **örtliche Behörde [f]** (Stelle, die als Organ einer selbständigen Gemeinde Aufgaben der öffentlichen Verwaltung, also Dienstleistungen für die Daseinsvorsorge, wahrnimmt; in der EU-Terminologie „Dienstleistungen von allgemeinem [resp. allgemeinem wirtschaftlichem] Interesse — DA[W]I". Die **ö. B.** entsteht durch organisationsrechtliche Vorschriften und erledigt die Aufgaben, die ihr gemäß Zuständigkeitsordnung nach sachlichen, instanziellen und räumlichen Kriterien obliegt; ►Gebietskörperschaft).

Local Authority Building Control [n] [UK] *adm. leg.* ►building code control [US].

Local Authority Building Control Department [n] [UK] *adm. leg.* ►Building Permit Office [US].

local authority land [n] [US] *plan.* ►community land [US]/ local authority land [UK].

local authority nursery [n] *adm. hort.* ►municipal nursery.

3316 local authority project [n] *plan.* (Project executed by municipal agencies; e.g. a city or town); *s* **proyecto [m] de obras públicas municipal/comunal** *syn.* proyecto [m] de construcción municipal/comunal; *f* **opération [f] immobilière communale** ; *g* **städtisches Bauvorhaben [n]** (Seitens einer Stadtverwaltung durchgeführtes Bauprojekt); *syn.* kommunales Bauvorhaben [n], kommunales Bauprojekt [n], städtisches Bauprojekt [n].

3317 local authority reorganization [n] *adm.* (Planning and execution of measures to adjust the existing system of local and regional administration); *s* **reforma [f] del régimen local y reestructuración [f] de las entidades locales** (Medidas para adaptar las estructuras administrativas heredadas del pasado a las necesidades actuales, que incluye la redefinición de fronteras municipales); *f* **réforme [f] administrative et territoriale** (Mesures prises pour adapter les structures administratives aux impératifs d'une société moderne ; **F.**, elle constitue la décen-

tralisation administrative entreprise principalement par les lois de 1982 à 1984 devant favoriser le transfert du pouvoir d'administration de l'État vers les collectivités locales ; cf. DUA 1996, 222) ; *g* **Gebiets- und Verwaltungsreform [f]** (Planungen und Maßnahmen zur Anpassung von gewachsenen Verwaltungsstrukturen, die meist noch aus vorindustriellen Gesellschaftsordnungen stammen, an die Erfordernisse der heutigen Zeit, indem größere Verwaltungseinheiten auf der Ebene der Gemeinden, Landkreise resp. Kreise, Stadtkreise und kreisfreien Städte gebildet werden; **in D.** setzte die **G. und V.** gegen Ende des 19. Jhs. ein und wurde Ende der 1960er bis zum Ende der 1970er-Jahre abgeschlossen. Nach der deutschen Wiedervereinigung wurde die Gemeindereform auch in den östlichen Bundesländern begonnen); *syn.* kommunale Neugliederung [f], Gemeindereform [f].

3318 local community area [n] *adm. pol. urb.* (Entire land of a community); *s* **territorio [m] municipal** *syn.* perímetro [m] municipal/comunal, territorio [m] comunal; *f* **territoire [m] communal** (Le terme allemand ancien « Gemarkung » désigne encore aujourd'hui dans certaines régions la totalité du territoire communal) ; *g* **Gemeindegebiet [n]** (Gesamtes Gebiet, das zu einer Gemeinde gehört); *syn.* Gemarkung [f] (1).

Local Development Framework (LDF) [n] [UK] *leg. urb.* ►comprehensive plan [US].

local distributor road [n] [UK] *plan. trans. urb.* ►local feeder road [US].

local district [n] [UK] *adm. agr. urb.* ►section of a community area [US].

local express train [n] [UK] *trans.* ►rapid transit vehicle [US].

3319 local feeder road [n] [US] *plan. trans. urb.* (Small public road which connects ►collector roads with ►arterial road; ►major street); *syn.* local distributor road [n] [UK]; *s* **arteria [f] (secundaria) de tráfico** (Calle que recoge el tráfico de las ►calles de distribución y lo conduce a las ►arterias [principales] de tráfico; ►carretera principal); *f 1* **voirie [f] secondaire** (Catégorie de trafic : ensemble des voies [voirie] sur lesquelles s'effectuent la circulation entre quartiers ou autres unités de voisinage ainsi que la circulation principale interne à un quartier assurant les services propres à l'habitation et au stationnement ; BON 1990, 358 ; assure la collecte de la circulation des ►voies de distribution et le raccordement à la ►voirie primaire ; ►route à grande circulation) ; *f 2* **voie [f] secondaire** (Terme utilisé pour désigner une voie unique de la voirie secondaire) ; *syn. pour les ingénieurs de la circulation* artère [f], voie [f] artérielle ; *g* **Verkehrsstraße [f]** (Innerörtliche, öffentliche Straße, die den Verkehr der ►Sammelstraßen aufnimmt und zur ►Hauptverkehrsstraße 1 führt; ►Hauptverkehrsstraße 2).

local government department [n] *adm.* ►local authority.

3320 locality [n] *phyt. zool.* (Defined area where a plant or animal species is found; ►location 1); *s* **localización [f]** (Término geográfico que denomina al lugar ocupado por una especie dentro del biótopo ocupado por la biocenosis de la cual aquella especie forma parte y depende de sus necesidades ecológicas y de la distribución de las estaciones dentro del biótopo. Especies de ecología similar coinciden en las mismas ►estaciones y se dice que pertenecen a un mismo tipo de localización; DB 1985); *f* **localité [f]** (Unité géographique qui correspond à des coordonnées — longitude et latitude — bien précises et porte souvent un nom, dans laquelle se trouve une espèce ou un ensemble d'espèces ; à ne pas confondre avec la ►station qui est une unité écologique ; DEE 1982, 56) ; *g* **Fundort [m]** (Geografische Örtlichkeit des Vorkommens einer Art resp. in der eine Art gefunden wurde; im floristischen Sinne syn. mit *Wuchsort*; nicht zu verwechseln mit dem ökologischen Begriff ►Standort 1); *syn. nur phyt.* Wuchsort [m], Wuchsstelle [f].

3321 local migrant [n] *zool.* (Bird species which flies in flocks [US]/swarms [UK] within a short range after brooding; ►migratory bird, ►partial migrant); *s* **ave [f] de dispersión posgenerativa** (Aves que después de la reproducción se pasean, generalmente en bandadas, por un amplio territorio alrededor de sus nidos; ►ave migradora, ►ave migradora parcial); *f* **oiseau [m] erratique** (Oiseau non nicheur qui en dehors des périodes de reproduction effectue des déplacements souvent considérables sans but défini ou vers des zones productrices ou climatiquement plus clémentes ; ces déplacements [irruption] dans une aire géographique inhabituelle sont souvent effectués en bandes et présentent parfois des caractères d'invasion, ces déplacements ne doivent pas être confondus avec les ►migrations 2 ; ►oiseau migrateur, ►oiseau migrateur partiel) ; *g* **Strichvogel [m]** (Vogelart, die nach der Brutzeit meist schwarmweise in weitem Umkreis umherschweift; ►Teilzieher, ►Zugvogel).

local park [n] [UK] *urb.* ►proximity green space.

3322 local planning agency/office [n] [US] *adm. urb.* (Local agency/authority chiefly concerned with urban or town planning); *syn.* department [n] of planning and development [CDN], local planning authority [n] [UK]; *s* **departamento [m] de urbanismo** (Autoridad o agencia local que se ocupa mayormente de la planificación urbana); *syn.* dirección [f] de urbanismo, dirección [f] del área de urbanismo, departamento [m] de planificación urbana; *f* **service [m] d'urbanisme municipal** (Administration communale responsable de l'urbanisme réglementaire) ; *syn.* agence [f] (municipale) d'urbanisme ; *g* **Stadtplanungsamt [n]** (Kommunale Behörde, die vornehmlich die städtebauliche Planung betreibt).

local planning authority [n] [UK] *adm. urb.* ►local planning agency/office [US].

local postal and telecommunications services [n] *trans. urb.* ►local post services.

3323 local post services [npl] *trans. urb.* (Transferral of goods, communications, payments and monetary transactions within a populated area [community, town or city]; many private companies now compete to provide these postal services since the privatization of state-run post offices during the 1990s; ►local telephone traffic); *syn.* local postal and telecommunications services [npl]; *s* **servicio [m] de correos local** (Conjunto de medidas y actividades necesarias para la transmisión de bienes y mensajes, incluyendo a veces transacciones monetarias y pagos, dentro de una ciudad o municipio. **En Es.** el **s. de c.** no incluye el servicio de telefonía que está privatizado); *syn.* servicio [m] postal local, tráfico [m] postal local; *f* **traffic [m] postal local** (Collecte, acheminement et distribution des objets de correspondance, de communication, de messagerie, de presse [traffic courrier] ainsi que les transactions commerciales et financières, formalités fiscales ou sociales effectuées dans des établissements postaux à l'échelle d'un village, d'une ville, d'une commune) ; *syn.* service [m] postal local ; *g* **örtlicher Postverkehr [m]** (Gesamtheit der Übermittlungen von Gegenständen/Waren und Nachrichten sowie die Abwicklung des Zahlungs- und Geldverkehrs innerhalb eines Siedlungsgebietes durch die Deutsche Post AG mit ihren drei Gesellschaften Deutsche Post, DHL und Postbank; im Vergleich zu D. und A. gehört in der CH auch heute noch der Personentransport mit dem Postauto zum **ö. P.**; ►Ortsverkehr, Ziff. 2); *syn.* örtliches Post- und Fernmeldewesen [n, o. Pl.].

3324 local public transportation system [n] [US] *trans.* (Mass transportation within conurbations, served by express trains, rapid transit trains [US], underground trains [UK],

trams [UK], buses and trolley-buses; ►public transportation vehicle, ►transportation system); *syn.* mass transit [n] [US] (TGG 1984, 68), local transport system [n] [UK]; *s* **transporte [m] público urbano** (Sistema de transporte masivo en las ciudades que utiliza diferentes tipos de vehículos como tren de cercanías, metro, autobus, tranvía, trolebus, taxis, etc. [►medio de transporte colectivo]; ►red de transportes); *f* **transports [mpl] suburbains** (Transport collectif de masse dans les grandes agglomérations utilisant différents moyens de transport tels que le train, autocar, bus, métro, tramway, taxi ; ►moyens de transport collectif, ►moyens de transport suburbain, ►réseau de transport) ; *g* **öffentlicher Personennahverkehr [m]** (*Abk.* ÖPNV; durch Personenbeförderung verursachter Massenverkehr in Ballungsgebieten, der mit Nahverkehrszügen, Stadtbahnen, U-Bahnen, Omnibussen, Taxen etc. [►öffentliches Verkehrsmittel] bewältigt wird; ►Verkehrssystem).

3325 local recreation [n] *recr.* (Recreation within the vicinity of residential areas; ►neighbo[u]rhood recreation, ►short-stay recreation); *s* **recreo [m] local** (Recreo dentro de la zona de residencia; ►puente, ►recreo en el vecindario/barrio, ►vacaciones cortas); *f* **loisirs [mpl] de proximité** (Accessibilité proche des zones d'habitation ; ►loisirs à proximité de quartiers résidentiels, ►tourisme de passage) ; *g* **Naherholung [f, o. Pl.]** (Erholung in der näheren Umgebung der Siedlungsgebiete; ►Kurzzeiterholung, ►wohnungsnahe Erholung).

3326 local recreational traffic [n] *recr. trans.* (Vehicular movement between urban agglomerations and places of recreation in the vicinity); *syn.* day-trips [npl]; *s* **tráfico [m] de recreo local** (Circulación de vehículos entre aglomeraciones urbanas y zonas de recreo vecinas); *f* **circulation [f] de loisirs de proximité (±)** (Circulation de véhicules générée par les utilisateurs des zones de plein air et de loisirs à proximité des grandes agglomérations) ; *g* **Naherholungsverkehr [m]** (Verkehr, der durch Erholungsuchende in der näheren Umgebung von Ballungsräumen entsteht).

3327 local recreation area [n] *plan. recr.* (Landscape of scenic beauty in close vicinity of towns, cities, or agglomerations, which is predominantly used for relaxation after a working day and weekend recreation, and where additional overnight accommodation is usually not provided; ►local recreation, ►short-stay recreation); *syn.* day-trip recreation area [n] [UK], suburban recreation area [n]; *s* **zona [f] de recreo local** (Terreno cercano a la ciudad [situado a 50 km o menos de ciudad o aglomeración urbana] aprovechable para actividades de recreo de un día o menos de duración, generalmente sin oferta de alojamiento; ►puente, ►recreo local, ►vacaciones cortas); *syn.* área [f] de recreación diaria, zona [f] suburbana de recreo; *f* **base [f] périurbaine de plein air et de loisirs** (Espace libre, animé, situé entre 10 et 60 km d'une agglomération importante ouvert à l'ensemble de la population réunissant des équipements qui offrent à ses usagers les éléments permettant la détente et la pratique d'activités sportives, culturelles de plein-air dans un site naturel ; circulaire du 20 janvier 1964 ; ►loisirs de proximité, ►tourisme de passage) ; *syn.* espace [m] de fin de semaine ; *g* **Naherholungsgebiet [n]** (Landschaftlich reizvolles Gebiet in der näheren Umgebung von Städten und Ballungsräumen bis 50 km entfernt, das vorwiegend für die Feierabend- und Wochenenderholung genutzt wird und i. d. R. keine zusätzlichen Übernachtungsmöglichkeiten vorhält; ►Kurzzeiterholung, ►Naherholung); *syn.* Gebiet [n] für die Naherholung, Naherholungsbereich [m], stadtnahes Erholungsgebiet [n], Tageserholungsgebiet [n].

3328 local/regional administrative authority/agency [n] *adm. leg. pol.* (Public body with jurisdiction over a legally established area; e.g. a city, portion of a state, or a multistate region [US]; ►local authority); *s* **entidad** [f] **local (territorial)** (En Es., las **ee. ll. tt.** están constituidas por los municipios, las provincias, las islas en los archipiélagos balear y canario, las entidades de ámbito territorial inferior al municipal instituidas o reconocidas por las Comunidades Autónomas; las comarcas u otras entidades que agrupen varios municipios, instituidas por las Comunidades Autónomas conforme a lo establecido en sus correspondientes estatutos de autonomía; las áreas metropolitanas y las mancomunidades de municipios; art. 1, Reglamento de Organización, Funcionamiento y Régimen Jurídico de las Entidades Locales, RD 2568/1986; ►autoridad local); *syn.* corporación [f] local; *f* **collectivité** [f] **territoriale** (Structure administrative de droit public dotée de la personnalité morale et juridique dont la souveraineté confiée par le législateur s'exerce sur une partie limitée du territoire national — circonscription administrative ; ►autorité territoriale) ; *syn. obs. depuis 2003* collectivité [f] locale ; *g* **Gebietskörperschaft** [f] (Körperschaft des öffentlichen Rechts, deren Hoheit einen räumlich begrenzten Teil eines Staates erfasst, z. B. Gemeinde, Landkreis; im weiteren Sinne auch Land und Bund; ►örtliche Behörde).

3329 local street [n] **with access only for residents** *trans. urb.* (Minor access road designed to provide vehicular access to abutting property and to discourage through traffic; ►cul-de-sac, ►loop street 1, ►residential street, ►residential traffic, ►service road); *s* **calle** [f] **de vecindario (≠)** (Calle secundaria diseñada para permitir el acceso a la correspondiente zona residencial y no para atraer tráfico de paso; *términos específicos* ►calle residencial; ►calle semicircular, ►fondo de saco, ►tráfico de residentes, ►vía de servicio); *f* **rue** [f] **riveraine** (Rue réservée à la ►circulation riveraine ; *termes spécifiques* ►boucle de desserte, ►contre-allée de desserte, ►rue résidentielle, ►voie sans issue) ; *syn.* voie [f] de desserte ; *g* **Anliegerstraße** [f] (Straße, die dem ►Anliegerverkehr 2 dient. *UBe* ►Einhang, ►Nebenfahrbahn, ►Sackgasse, ►Wohnstraße).

local telephone traffic [n] *trans. urb.* ►local traffic, #2.

3330 local topography [n] *geo.* (Natural configuration of a small area of ground/countryside; ►landform); *s* **microrrelieve** [m] (Configuración natural del relieve de un área reducida; ►topografía del terreno); *f* **modelé** [m] **de terrain** (Configuration naturelle du relief d'une unité de paysage de faible étendue ; ►morphologie du terrain) ; *g* **Geländemodellierung** [f] (Zustand der natürlichen Ausgestaltung des Bodenreliefs eines kleinräumigen Landschaftsraumes; ►Geländeform; *constr.* ►Bodenmodellierung).

3331 local traffic [n] *trans. urb.* (**1.** Vehicular traffic within a developed area [community, town or city]; ►local post services); **2. local telephone traffic** [n] (Entire telecommunication services within a developed area [community, town or city]); *s* **tráfico** [m] **local** (**1.** Tránsito de vehículos dentro de un área urbana [ciudad, pueblo]; ►servicio de correos local. **2.** En D. el término también se utiliza para denominar el conjunto de tráfico de correos y telecomunicaciones dentro de una ciudad); *f* **trafic** [m] **local** (**1.** Déplacement dont l'origine et la destination sont intérieures à l'aire d'une agglomération. **2. D.**, terme aussi utilisé pour désigner l'ensemble du trafic des Postes et Télécommunications à l'intérieur d'une agglomération ; *termes spécifiques* ►trafic postal local, trafic téléphonique local) ; *g* **Ortsverkehr** [m] (**1.** Straßenverkehr innerhalb eines Siedlungsgebietes. **2.** Im Deutschen wird je nach Zusammenhang unter Ortsverkehr auch der Post- und Telekommunikationsverkehr als Gesamtheit der Übermittlungen von Gegenständen und Nachrichten verstanden: ►örtlicher Postverkehr und **örtlicher Telefonverkehr**); *syn.* örtlicher Verkehr [m].

local transport system [n] [UK] *trans.* ►local public transportation system [US].

local visual amenities [npl] *urb.* ►preservation of local visual amenities.

3332 location [n] (1) *ecol.* (Living place of a species or of a community, including all environmental factors [►ecological factors], except competition, that influence them. The exact place of occurrence is called ►locality or station; ►habitat 1, ►habitat 2, ►isolated patch habitat); *syn.* locality [n] [also US], station [n]; *s* **estación** [f] (Totalidad de los puntos en que se repiten los factores físicos, químicos y bióticos [►factores ecológicos], exceptuando las relaciones de competencia, que constituyen el medio de una masa vegetal; el medio físico donde crecen y se desarrollan se llama ►biótopo; no se debe confundir con el término geográfico ►localización; cf. DINA 1987; ►habitación, ►habitat natural aislado); *syn.* residencia [f] ecológica; *syn. parcial* hábitat [m]; *f* **station** [f] (Espace défini par des conditions écologiques [►facteurs écologiques] homogènes où se développe une communauté biologique homogène [►biotope], en particulier un type de groupement végétal ou animal bien défini ; cf. DEE 1982 ; ►habitat 1, ►habitat relictuel isolé, localité) ; *g* **Standort** [m] (1) (Autökologischer Begriff für die Gesamtheit der am Wohn- oder Wuchsort eines Organismus auf diesen einwirkenden Umweltfaktoren/►Ökofaktoren; **S.** darf nicht mit dem geografischen Begriff ►Fundort verwechselt werden; der synökologische Begriff ist ►Biotop; ►Habitat, ►Sonderstandort); *syn. phyt.* Pflanzenstandort [m], Wuchsort [m].

location [n] (2) *agr. urb.* ►hillside location; *plan.* ►marginal location; *phyt. plant.* ►semishaded location [US], ►shady location; *plan.* ►site 1; *plant.* ►sunny location.

location [n], **humic** *phyt. zool.* ►mesic site.

3333 location in a basin [n] *geo.* (Topographical situation in a circular or oval valley surrounded by steep hillsides; lie [vb] in a basin; ►depression); *syn.* bowl-shaped situation [n], bowl-shaped basin [n]; *s* **localización** [f] **en valle cerrado** (Situación de un asentamiento en un valle rodeado de montañas por todos los lados. Un valle cerrado es más profundo que una ►depresión); *f* **situation** [f] **en cuvette** (Dépression profonde entourée sans discontinuité par des hauteurs ; ►dépression) ; *syn.* situation [f] dépressionnaire ; *g* **Kessellage** [f] (Situation eines Ortes, der rings von Höhenrücken umgeben ist [Talkessel. Ein Kessel ist tiefer als eine ►Geländemulde).

3334 location [n] **of commercial facilities and light industries** *plan. urb.* (Area in which several commercial establishments have been established. Formerly known under the term **industrial estate**, it is now often referred to as a **business park**; ►industrial park); *s* **zona** [f] **industrial-comercial** (Área de servicios e industria ligera; ►parque industrial); *f* **zone** [f] **d'activités économiques** (Résultat de la localisation des activités ; selon l'usage prévu, on parle de ►zone industrielle, zone industrielle verticale, zone artisanale ; cf. DUA 1996 ; ►parc d'activités, ►parc industriel) ; *syn.* zone [f] d'activités diverses et de petites industries ; *g* **Gewerbegebiet** [n] (2) (*Ergebnis* Ort, an dem sich mehrere Gewerbebetriebe niedergelassen haben. In neuerer Zeit angelegte **G.e** werden häufig **Gewerbepark** genannt; ►Industriepark); *syn.* Gewerbeansiedlung [f], Gewerbepark [m].

3335 lockable bike rack [n] [US] *trans. urb. syn.* lockable cycle rack [n] [UK]; *s* **apoyo** [m] **para bicicleta con seguro**; *f* **support** [m] **de bicyclettes avec étrier antivol** ; *g* **Fahrradständer** [m] **mit Verschlussbügel** *syn.* abschließbarer Fahrradständer [m], Fahrradständer [m] mit Anschließbügel.

lockable cycle rack [n] [UK] *trans. urb.* ►lockable bike rack [US].

3336 loess [n] *pedol.* (Fine yellowish or greyish-brown loamy earth with a high percentage of fine sand and silt fractions, carbonate of lime, and easily soluble alkaline salts. It is generally considered to be mainly material transported by wind during and after the Glacial Period. It is, in effect, wind-sorted morainic material, covering extensive areas in North America, especially in the Mississippi basin, north central Europe, Russia, and Asia, especially eastern China; cf. GGT 1979; ►black earth, ►loose sedimentary deposit); *s* **loes(s)** [m] (Depósitos de limo originados por la deposición de las partículas transportadas por las tormentas de polvo durante miles de años. Es de color amarillento, delez-nable y carece de estratificación. El polvo procede del material detrítico de las morrenas y de las llanuras de inundación de las glaciaciones pleistocénicas. Los suelos derivados contienen del 50-70% de limos no consolidados, con buen drenaje. Son suelos ricos [►chernosem] muy aptos para el cultivo de cereales, que se presentan en Ucrania, la pampa argentina, el norte de China y las grandes llanuras de América del Norte; DINA 1987; ►sedimento no consolidado); *f* **lœss** [m] (Formation limoneuse, d'origine éolienne, qui s'inscrit, en Europe, dans un cycle morpho-génétique de climat froid et aride dont les modalités varient en fonction des conditions climatiques régionales. Le **l.** est surtout constitué de quartz de micas, de feldspaths, de carbonate de calcium. Ces fractions très fines libèrent par altération des éléments échangeables ce qui rend ce matériau très fertile pour l'agriculture. Le **l.** recouvrant le nord de la France, la Moyenne Belgique et l'Allemagne date des glaciations de Riss et Würm et constitue donc un dépôt périglaciaire ; DIS 1986 ; ►chernozem, ►sédiment détritique) ; *g* **Löss** [m] (Aus dem Allemannischen *lösch* >locker< stammend; meist karbonathaltiges, gelblich gefärbtes ►Lockersediment mit einem ausgeprägten Korngrö-ßenmaximum zwischen 10 und 60 µm Durchmesser [ca. 60 %], das vom Wind aus glazialen Schotter- und Sanderflächen, den Hochflutabsätzen der Urstromtäler und den periglazialen Frost-schutzzonen ausgeweht und in andere Regionen abgelagert wurde; aus dem **L.** haben sich fruchtbarste Böden, z. B. die ►Schwarzerde entwickelt; cf. SS 1979. Große **L.flächen** gibt es in Nordfrankreich, Mittelbelgien, Deutschland, in den USA vor-nehmlich im Mississippibecken, in der Ukraine und in großen Teilen Nord- und Ostchinas).

loessleem [n] [UK] *pedol.* ►loess loam [US].

3337 loess loam [n] [US] *pedol.* (GGT 1979, DG 1984, 274; postglacial transformation of ►loess caused by the leaching of carbonates); *syn.* loessleem [n] [UK] (DNE 1978, 182); *s* **limo** [m] **de loes(s)** (Transformación posglacial del ►loes causada por la lixiviación de los carbonatos); *f* **limon** [m] **lœssique** (►Lœss récemment décalcifié par lessivage superficiel [décarbonatation] ; PED 1983, 319) ; *syn.* terre [f] à brique (DG 1984, 274) ; *g* **Löss-lehm** [m] (Nacheiszeitlich umgewandelter ►Löss durch Auswa-schung der Carbonate [*ugs.* Karbonate]).

3338 loess plain [n] *geo.* (Extensive areas of land originating during the cold periods of the Pleistocene period which in North China comprise aeolean deposits in layers up to 400m thick. In the prairies of Illinois, Missouri, Nebraska and Arkansas several thousands of square kilometers of land are charcterized by a loess horizon up to 40m in thickness); *s* **llanura** [f] **de loes(s)** (Grande extensión de tierra en donde se ha acumulado el loes en parte con gran profundidad, p. ej. en el norte de China donde llega hasta los 400 m, o las praderas de Illinois, Missouri, Nebraska y Kansas donde se extiende por miles de km cuadrados, aunque con una profundidad mucho menor de 30-40 m; cf. DINA 1987); *f* **plaine** [f] **lœssique** (DUV 1984, 259 ; dépôts périglaciaires d'origine éolienne pouvant atteindre dans le nord de la Chine 400 m et dans la Prairie nord-américaine de l'Illinois, du Missouri, du Nebraska et de l'Arkansas sur des milliers de km² jusqu'à 40 m d'épais-seur ; en Europe, l'épaisseur maximale des couches de lœss varient entre 10 et 40 m d'épaisseur) ; *g* **Lössebene** [f] (Vorzeit-lich während der Kaltzeiten des Pleistozäns entstandene riesige Landfläche, die z. B. in N-China bis zu 400 m und in den Prärien Illinois, Missouri, Nebraska und Arkansas über tausende von Quadratkilometern äolische Ablagerungen bis zu 40 m Mächtig-keit aufweist. In Europa gibt es 10 m bis maximal 40 m mächtige Lössschichten).

3339 logging [n] *for.* (Felling and extraction of timber, partic-ularly as logs; ►timber extraction); *s* **explotación** [f] **forestal de tala y arrastre** (El término inglés «logging» incluye la tala y la extracción de troncos por arrastre; ►arrastre); *f* **abattage** [m] **et débusquage** [m] (Opérations permettant de couper et de tirer les bois du lieu d'abattage au layon ou à une piste de ►débardage. Ce travail peut se faire avec un cheval, un petit tracteur [débus-queuse], ou à l'aide de treuils ; l'anglais « logging » ne corres-pond qu'à l'exploitation du bois en forêt et plus spécialement des bois ronds et des grumes ; DFM 1975, 3562) ; *g* **Fällen** [n, o. Pl.] **und Rücken** [n, o. Pl.] (Holzeinschlag und [Zwischen]trans-port der Baumstämme vom Fällort zum Lkw-Verladeplatz; ►Rücken).

logging off [n] [AUS] *constr. for.* ►saw into pieces.

London smog [n] *envir.* ►smog.

long distance footpath [n] [UK] *plan. recr.* ►national scenic trail [US].

longeval [adj] *hort. plant.* ►long-lived.

3340 longevity [n] *hort. plant.* (Capability of plants to live for a long time); *s* **longevidad** [f] (Capacidad de las plantas de vivir durante muchos años); *f* **longévité** [f] (Capacité des végétaux à vivre plusieurs années) ; *syn.* longue durée [f] de vie ; *g* **Lang-lebigkeit** [f] (2) (Vermögen von Pflanzen, Jahrzehnte bis Jahr-hunderte zu leben); *syn.* lange Lebensdauer [f].

long falls [n, *only pl.*] [UK] *constr.* ►longitudinal gradient.

long grass cutting [n] *agr. constr.* ►hay cutting.

longitudinal fall [n] [UK] *constr.* ►longitudinal gradient.

3341 longitudinal gradient [n] *constr.* (Linear slope mea-sured along the road or path centreline; ►cross slope [US]/cross fall [UK], ►crowing of paved surfaces); *syn.* longitudinal fall [n] [also UK], long falls [n, *only pl*] [UK]; *s* **pendiente** [f] **longitu-dinal** (Inclinación, p. ej. de un camino en dirección longitudinal; ►bombeo de superficies, ►pendiente transversal); *f* **pente** [f] **longitudinale** (Inclinaison d'une surface prise dans son sens longitudinal, généralement mesurée en pourcentage de la verti-cale sur l'horizontale ; ►dévers, ►pente transversale, ►profil en fome de toit) ; *syn.* gradient [m] longitudinal ; *g* **Längsgefälle** [n] (Neigung der Mittellinie eines Verkehrsweges gegen die Horizontale, gemessen in Grad oder in Prozenten zur Länge; ►Quergefälle, ►Dachgefälle).

3342 longitudinal section [n] *constr.* (Lengthwise section shown on a drawing; ►cross-section); *s* **sección** [f] **longitudinal** (Corte longitudinal de una construcción mostrado en un dibujo; ►sección transversal); *syn.* corte [m] longitudinal, perfil [m] longitudinal; *f* **coupe** [f] **longitudinale** (Section dans un terrain représentant dans le sens longitudinal la disposition de couches ou les différentes côtes d'un projet ; dans le domaine de la voirie on utilise de préférence le terme de ►profil en long ; ►coupe en travers, ►profil en dévers) ; *syn.* profil [m] longitudinal ; *g* **Längsschnitt** [m] (Schnitt in Längsrichtung eines Körpers oder eines Projektes; im Straßenbau wird meistens von einem **Längsprofil** gesprochen; ►Querschnitt); *syn.* Längsprofil [n].

L

longitudinal valley profile [n] *geo.* ▶thalweg.

3343 long jump area [n] *recr.* (Runway track of up to 40m in length to a take-off board and jumping pit filled with sand; *generic term* ▶jump area); *s* **instalación [f] de salto de fondo** (Instalación deportiva con rampa de hasta 40 m para carrerilla hasta plataforma de salto y fosa de aterrizaje llena de arena; *término genérico* ▶instalación de salto); *f* **aire [f] de saut en longueur** (Installation sportive pour le saut en longueur constituée d'une piste d'élan de 40 m maximum, d'une planche d'appel et d'une fosse remplie de sable ; *terme générique* ▶aire de saut) ; *g* **Weitsprunganlage [f]** (Sportanlage für den Weitsprung, die aus einem bis zu 40 m langen Anlauf, einem Absprungbalken und einer mit Sand gefüllten Sprunggrube besteht; *OB* ▶Sprunganlage).

long-lasting disturbance [n] *ecol. envir.* ▶persistent environmental disturbance.

3344 long-lasting storage [n] *envir.* (Accumulation over a long period, e.g. of noxious substances in plant and animal tissues; ▶bioaccumulation); *s* **almacenamiento [m] a largo plazo** (Acumulación de sustancias nocivas en tejidos vegetales y animales; ▶bioacumulación); *f 1* **accumulation [f] de longue durée** (Stockage pendant une longue durée de substance nocives, p. ex. dans les tissus des organismes végétaux ou animaux ; ▶bioaccumulation) ; *f 2* **rétention [f] de longue durée** (Immobilisation prolongée par les sols de substances toxiques) ; *g* **Langzeitspeicherung [f]** (Anreicherung von Schadstoffen über einen langen Zeitraum, z. B. in pflanzlichen oder tierischen Geweben; ▶Bioakkumulation).

3345 long-lived [adj] *hort. plant.* (Capable of having a long life; *context* long-lived ▶perennial 1); *syn.* longeval [adj]; *s* **longevo/a [adj]** (▶perenne); *f* **à/de longue durée de vie [loc]** (▶pérenne) ; *syn.* longévif, ve [adj] ; *g* **langlebig [adj]** (eine lange Lebenszeit besitzend; viele Jahre lebend; ▶ausdauernd).

3346 long-range transportation plan [n] [US] *trans. pol.* (Plan comprising the present situation, based on traffic studies, and the future development of traffic facilities of a community or city; ▶traffic planning); *syn.* long-range transport plan [n] [UK]; *s* **plan [m] general de tráfico y transportes [Es]/plan [m] general de tránsito y transportes [AL]** (Plan que describe la situación actual en su momento y planifica el desarrollo futuro del tráfico en una ciudad; ▶planificación de tráfico y transportes [Es]/planificación de tránsito y transportes [AL]); *f 1* **plan [m] de circulation** (Plan décrivant la situation actuelle de la circulation dans une commune ou agglomération urbaine et définissant les actions futures permettant d'accroître l'efficacité de la voirie existante en fonction des divers scénarios de développement ; *termes spécifiques* schéma directeur de pistes cyclables, de randonnée pédestre, de randonnée automobile ; ▶planification des transports) ; *syn.* schéma [m] directeur de circulation ; *f 2* **plan [m] de déplacement urbain (PDU)** (F., plan dont les agglomérations françaises de plus de 100,000 habitants doivent se doter, visant, selon la loi du 30 décembre 1996 [Loi sur l'air et l'utilisation rationnelle de l'énergie], à assurer un « équilibre durable entre les besoins en facilité d'accès d'une part, et la protection de l'environnement et de la santé d'autre part » ; les orientations fixées par la loi sont **a]** la diminution du trafic automobile, **b]** le développement des transports collectifs et des modes économes et les moins polluants [marche à pied et vélo], **c]** l'aménagement et la gestion du réseau principal de voirie en l'affectant aux différents modes de transport, **d]** l'organisation du stationnement sur le domaine public, sur voirie et souterrain, **e]** la réduction de l'impact sur la circulation et sur l'environnement du transport et de la livraison des marchandises, **f]** l'encouragement pour les entreprises et les collectivités

publiques à favoriser le transport de leur personnel par l'utilisation des transports en commun et du ▶covoiturage) ; *g* **Generalverkehrsplan [m]** (*Abk.* GVP; Plan, der die gesamte gegenwärtige Situation und zukünftige langfristige Entwicklung des Verkehrs in einer Gemeinde, einer Stadt oder für ein ganzes Bundesland darlegt; es werden die Grundlinien der Verkehrspolitik in einer Zusammenschau mit der Gesellschafts-, Wirtschafts-, Umwelt- und Raumordnungspolitik für den Geltungsbereich beschrieben, festgelegt und die wichtigsten Verkehrsprojekte benannt; dabei werden die vier Fachkonzepte Straßenbau, Öffentlicher Personenverkehr, Güterverkehr und Luftverkehr behandelt; die Öffentlichkeit und Träger öffentlicher Belange werden dabei angehört; ▶Verkehrsplanung).

long-range transport plan [n] [UK] *trans. pol.* ▶long-range transportation plan [US].

3347 longshore drift [n] *geo.* (Drift of material along a beach as a result of waves breaking at an angle. A breaker [swash] sweeps material obliquely up the beach, the backwash drags some down again at right angles, causing a net movement along the beach; DNE 1978); *syn.* coastal drift [n]; *s* **deriva [f] litoral** (Flujo de material paralelo a la costa originado por el oleaje que puede conducir a la deriva de playas; cf. MOPU 1985); *syn.* flujo [m] de material por el litoral; *f* **dérive [f] littorale** (Charriage de matériaux sur le bas de plage provoqué par les houles obliques qui engendrent des courants de dérive littorale) ; *g* **Küstenversetzung [f]** (Verlagerung von Küstensedimenten längs der Küste durch das schräge Auflaufen der Brandung, den rückflutenden Sog und die Küstenströmungen; GEO 1977); *syn.* Küstenversatz [m], Strandversetzung [f].

3348 long-stay recreation [n] *plan. recr.* (Relaxing holiday of several weeks duration, statistically > 5 days, which has a lasting recuperative effect upon one's health; ▶short-stay recreation, ▶vacation); *syn.* long-term recreation [n]; *s* **vacaciones [fpl] largas** (Vaciones de descanso de varias semanas de duración —para la estadística de > 5 días— que tienen efectos positivos físicos y [p]síquicos ; ▶vacaciones, *opp.* ▶vacaciones cortas); *syn.* recreo [m] a largo plazo; *f* **tourisme [m] de séjour** (Séjour touristique de plusieurs semaines ; ▶vacances ; *opp.* ▶tourisme de passage) ; *syn.* tourisme [m] de longue durée ; *g* **Langzeiterholung [f, o. Pl.]** (Mehrwöchige Erholung [statistisch > 5 Tage], die der nachhaltigen Regeneration der Kräfte dient; ▶Urlaub; *opp.* ▶Kurzzeiterholung); *syn.* Urlaubserholung [f, o. Pl.].

3349 long-term benefits [npl] *landsc. plan. urb.* (Long-lasting effectiveness of measures for ameliorating the environment; *generic term* long-term effect; ▶amenity benefits of a forest, ▶social benefits of open areas, ▶social benefits of the forest); *s* **efecto [m] a largo plazo** (Efecto de larga duración de medidas para mejorar el medio ambiente; ▶efectos beneficiosos del bosque, ▶efectos psicológicos del bosque, ▶función beneficiosa [de zonas verdes]); *f* **effet [m] à long terme** (Action sur une longue durée de mesures d'amélioration de l'environnement ; ▶fonction sanitaire et sociale de la forêt, ▶fonction sanitaire et sociale des espaces verts, ▶rôle de la forêt sur la santé corporelle et l'équilibre psychique) ; *syn.* effet [m] sur le long temps ; *g* **Langzeitwirkung [f]** (Sich über einen langen Zeitraum erstreckende Wirksamkeit, z. B. von Maßnahmen zur Verbesserung der Umwelt; ▶umweltpsychologische Wohlfahrtswirkungen, ▶Wohlfahrtswirkungen des Waldes, ▶Wohlfahrtswirkungen von Grünanlagen).

long-term recreation [n] *plan. recr.* ▶long-stay recreation.

loop [n] *ecol.* ▶feedback.

loop [n]**, microbial** *ecol.* ▶organic turnover.

loop highway [n] [US] *trans.* ▶bypass [US]/by-pass [UK].

loop road [n] *urb. trans.* ▶loop street (1).

3350 loop street [n] **(1)** *urb. trans.* (▶Local street with access only for residents which has its own ingress and egress at two points on the same ▶collector road; IBDD 1981); *syn.* loop road [n]; *s* **calle [f] semicircular** (≠) (▶Calle de vecindario con forma de bucle que empieza y termina en la misma ▶calle de distribución); *f* **boucle [f] de desserte** (Système de desserte dans un lotissement caractérisé par le décrochement de forme semicirculaire [boucle] d'une ▶rue riveraine sur la voie principale ou sur une ▶voie de distribution) ; *g* **Einhang [m]** (Erschließungsart in Wohngebieten, bei dem eine ▶Anliegerstraße von einer ▶Wohnsammelstraße bogenförmig abzweigt und an anderer Stelle wieder in diese einmündet); *syn.* Rucksackerschließung [f], Schleife [f], Schleifenerschließung [f], Straßenschleife [f].

loop street [n] [US] **(2)** *urb. trans.* ▶bypass [US]/by-pass [UK].

3351 loop trail [n] **(1)** *recr.* (Prepared circular route created for ▶cross-country skiing); *s* **pista [f] de esquí de fondo** (Pista preparada para ▶esquiadores de fondo); ▶esqui de fondo); *syn.* pista [f] de esquí escandinavo; *f* **piste [f] de ski de fond** (Piste tracée sur la neige et banalisée pour les ▶fondeurs ; ▶ski de fond ; *syn.* piste [f] tracée) ; *g* **Loipe [f]** (Präparierte Spur für ▶Skilangläufer/-innen ; ▶Skilanglauf).

loop trail [n] [US] **(2)** *plan. recr.* ▶circular path [US]/circular pathway [UK].

3352 loop turnaround [n] *trans. urb.* (Vehicle area at the end of a ▶cul-de-sac which has a large, usually planted, central island to enable trams or buses to turn; ▶turnaround); *syn.* turnaround [n] with open centre [UK], turnaround [n] with central island [UK]; *s* **lazo [m] de (re)vuelta** (≠) (Ampliación de la calzada al final de un ▶fondo de saco con isla central grande alrededor de la cual los vehículos pueden dar la vuelta; *término genérico* ▶revuelta); *syn.* lazo [m] de giro (≠); *f* **boucle [f] de retournement** (Voirie à l'extrémité d'une ▶voie sans issue constituée par un îlot central de verdure aménagé pour le retournement des tramways ou des omnibus ; *terme générique* ▶dispositif de retournement) ; *g* **Wendeschleife [f]** (Verkehrsfläche am Ende einer ▶Sackgasse mit einer großen Mittelinsel, meist als Verkehrsgrünfläche gestaltet, zum Wenden von Straßenbahnen und Omnibussen; *OB* ▶Wendeanlage).

3353 loosening [n] **(of soil)** *pedol.* (*The opposite of* ▶soil compaction an increase in the volume of soil caused by burrowing animals, root penetration or frost as well as cultivation of the soil); *s* **esponjamiento [m] del suelo** (Proceso contrario a la ▶compactación del suelo que ocurre por aumento del volumen de los poros gracias a la acción de la fauna edáfica, por presión de las raíces o por formación de lentejas de hielo); *syn.* ahuecamiento [m] del suelo, mullimiento [m] del suelo; *f* **ameublissement [m]** (Processus naturel ou pratique culturale opposé à celui du ▶compactage du sol favorisant l'aération, la fragmentation et la perméabilité d'un sol ; ▶tassement du sol) ; *g* **Lockerung [f]** (Der ▶Bodenverdichtung entgegengesetzter Vorgang, der durch die Vergrößerung des vorhandenen Volumens durch grabende Tiere, Wurzeldruck oder Eislinsenbildung sowie durch Bodenbearbeitung entsteht).

3354 loose peat [n] *constr. hort.* (Fibrous uncompacted material used as a soil ameliorant; ▶peat ball); *s* **turba [f] suelta** (Material fibroso no compactado utilizado para enmienda de suelos; ▶fardo de turba); *f* **tourbe [f] en vrac** (Produit industriel d'extraction utilisé pour l'amendement des sols et conditionné sans emballage ; ▶balle de tourbe) ; *g* **loser Torf [m]** (Aus gestochenen Torfsoden industriell hergesteller Torfmull [pH-Wert ca. 3,5] in verschiedenen Fraktionen, für Topfpflanzenkulturen 0-7 mm, 7-20 mm, für ▶Moorbeete 20-40 mm und

> 40 mm und zur Senkung des pH-Wertes als Bodenverbesserungsmittel [*Handelsname* ▶Düngetorf] verwendet; *opp.* ▶Torfballen).

3355 loose rock dam [n] *constr. wat'man.* (Check dam of stones constructed at right angles, or a stone layer in erosion gullies or stream beds to prevent vertical erosion; ▶brush and rock dam, ▶ground sill); *s* **solera [f] de piedras sueltas** (Muro o capa de piedras construido transversalmente en cursos de agua o cárcavas de erosión para evitar mayor socavación; ▶solera de fondo, ▶solera de piedras y ramas vivas); *f* **seuil [m] en enrochements** (Ouvrage de confortement du fond du lit d'un cours d'eau constitué par un mur en pierres transversal ou de roches pour dissiper l'énergie des eaux et régulariser le cours d'une rivière ou protéger les fossés du creusement par érosion ; ▶seuil en enrochements et branchages ; ▶seuil transversal d'un cours d'eau) ; *g* **Steinschwelle [f]** (Aus Steinen aufgesetzte Quermauer oder Steinlage in Erosionsgräben oder in Wasserläufen zur Verhinderung einer weiteren Eintiefung; ▶Sohlenschwelle, ▶Steinschwelle mit Ast- oder Zweigbesatz).

3356 loose sedimentary deposit [n] *geo.* (**1.** Unconsolidated weathered material, that has accumulated in layers by water, ice and/or wind action; e.g. ▶loess and other unconsolidated sedimentary rock, ▶loose sedimentary rock, **2.** sediment formed by the accumulation of detritus and transported by water or ice to the place of deposition); *syn.* unconsolidated sedimentary deposit [n], detrital sediment [n]; *s* **sedimento [m] no consolidado** (Productos de meteorización transportados por el agua, el aire o el viento y sedimentados en un determinado lugar, como ▶loes y otras ▶rocas sedimentarias no consolidadas); *f* **sédiment [m] détritique** (Produits de la destruction des roches provenant de processus d'érosion fluviatile [alluvions], glaciaire [moraines] et éolienne [▶lœss] ; ▶roche sédimentaire meuble) ; *syn.* accumulation [f] détritique ; *g* **Lockersediment [n]** (Durch Wasser, Eis und/oder Windeinfluss transportierte und in Schichten abgelagerte Verwitterungsprodukte, z. B. ▶Löss und andere ▶Lockergesteine).

3357 loose sedimentary rock [n] *geo.* (Loose unconsolidated fragmental material transported and deposited by wind, water, or ice, chemically precipitated from solution, or secreted by organisms, and formed in layers; e.g. blocks of stone, boulders/coarse rock debris, gravel, sand, loess, silt, mud, clay, marl, loam; ▶loose sedimentary deposits); *s* **roca [f] sedimentaria no consolidada** (Material rocoso no consolidado transportado por el aire, el agua o el hielo y depositado en otro lugar, en capas ordenadas según la granulometría: bloques, cantos rodados o piedras partidas, grava, arena, loes[s], limo, légamo/lodo, arcilla, margas; ▶sedimento no consolidado); *syn.* roca [f] exógena no consolidada; *f* **roche [f] sédimentaire meuble** (Roche provenant de la désagrégation d'une autre roche [érosion mécanique, altération chimique] et constituée de fragments divers et classée selon la taille des débris p. ex. alluvions sédimentaires : argile, lœss, limons, marnes, sables, poudingues, brèches, grès etc. ▶sédiment détritique) ; *syn.* roche [f] détritique ; *g* **Lockergestein [n]** (Unverfestigtes Sedimentgestein nach Korngrößen geordnet: Blöcke, Geröll/Schotter, Kies, Sand, Löss, Schluff, Schlick/Schlamm, Ton, Mergel und Lehm; ▶Lockersediment).

3358 loose soil [n] *agr. hort.* (**1.** Soil loosened by soil cultivation equipment. **2.** *constr.* Soil extracted during earthworks by digging, scraping, etc.; ▶loose soil material); *s* **suelo [m] mullido** (Suelo descompactado tras operación de desfonde; ▶tierra suelta); *f* **sol [m] ameubli** (**1.** Sol décompacté au moyen d'outils. **2.** *constr.* État du sol provoqué par son extraction lors de travaux d'excavation ; ▶terre foisonnée) ; *g* **gelockerter Boden [m]** (**1.** Durch Bodenbearbeitungsgeräte gelockerter Boden.

2. *constr.* Im Erdbau durch Lösen gelockerter Boden: ►lose Bodenmasse); *syn.* aufgelockerter Boden [m].

3359 loose soil material [n] *constr.* (Disturbed or handled soil; ►bulking, ►loose soil); *s* **tierra** [f] **suelta** (En terracerías, tierra desplazada por los trabajos de movimiento de tierras que tiene más volumen que en estado natural; ►estado descompactado del suelo, ►suelo mullido); *syn.* suelo [m] suelto, material [m] suelto); *f* **terre** [f] **foisonnée** (Sol déplacé lors de travaux de terrassement et ayant augmenté de volume ; cet accroissement de volume varie en fonction de la nature du sol et est proportionnel au volume en place ; cette proportionnalité est caractérisée par un coefficient appelé *coefficient de foisonnement* — pour le sable il est de 1,1, pour l'argile de 1,33 ; ►foisonnement 2, ►sol ameubli) ; *syn.* sol [m] foisonné) ; *g* **lose Bodenmasse** [f] (Im Erdbau durch Lösen gelockerter Boden, der beim Transport und Wiedereinbau ein größeres Volumen hat; ►Bodenauflockerung 1, ►gelockerter Boden).

lopped tree [n] *agr.* ►fodder tree.

3360 lopping [n] *arb. constr.* (Drastic cutting back of main branches without consideration for the natural shape of a tree; BS 3975: Part 5; ►branch stub [US]/branch stump [UK], ►topping 3); *syn.* stubbing [n] [also US] (ARB 1983, 387); *s* **trasmocho** [m] (Poda drástica realizada en ejemplares adultos, eliminando el ramaje prácticamente hasta el tronco del árbol, cuyo resultado es que quedan ►tocones de ramas; cf. PODA 1994; ►descopado); *f 1* **rapprochement** [m] (Taille radicale et exceptionnelle [procédé de sauvetage et non d'entretien] des branches charpentières environ au tiers de leur longueur ; ►chicot 1, ►éhoupage) ; *f 2* **ravalement** [m] (Une opération encore plus radicale de rabattage des branches qui sont sectionnées à même le tronc ; cf. IDF 1988, 179) ; *g* **Stummelschnitt** [m] (Unsachgemäßer Baumschnitt, bei dem Reste eines Astes [►Aststummel] am Stamm stehen bleiben; ►Kronenkappung).

Los Angeles smog [n] *envir.* ►smog.

3361 loss [n] *constr. hort.* (Failure of vegetation establishment by seeding or planting; ►bare patch); *s* **marra** [f] (En una siembra o plantación, pérdida de plantas que no enraizan y se mueren; ►calva); *f* **perte** [f] (lors d'un ensemencement ou d'une plantation, lorsque les semences ou les végétaux ne se développent pas et meurent ; ►pelade, ►trou) ; *g* **Ausfall** [m] (Verlust eines Vegetationsbestandes oder eines Teiles [Teilausfall] davon, z. B. einer Aussaat, wenn sie z. B. nicht aufläuft resp. Verlust einer Bepflanzung, die nicht anwächst und abstirbt; ►Fehlstelle).

3362 loss [n] **of nutrients** *pedol. syn.* nutrient loss [n]; *s* **pérdida** [f] **de nutrientes**; *f* **perte** [f] **de substances nutritives** *syn.* perte [f] d'éléments fertilisants ; *g* **Nährstoffverlust** [m].

3363 lost gradient [n] *trans.* (Rise in height of a traffic route which is subsequently cancelled out by a fall in relation to the overall uphill gradient); *s* **gradiente** [m] **perdido** (Subida en un trazado de carretera que es contrarrestada seguidamente por una bajada de manera que en total no llega a una altura superior); *f* **déclivité** [f] **perdue** (±) (Tracé routier sur lequel une hauteur atteinte peut être abandonnée lorsqu'intervient sur le parcours général de la montée un secteur descendant important) ; *g* **verlorene Steigung** [f] (Steigung einer Verkehrstrasse, bei der durch eine Gefällstrecke in der Gesamtsteigungsrichtung eine einmal gewonnene Höhe wieder aufgegeben wird).

lot [n] *urb.* ►building plot; *leg. urb.* ►neighboring lot [US]/neighbouring plot [UK]; *trans. urb.* ►parking lot; *recr.* ►sandlot [US], ►tot lot [US]/toddlers play area [UK]; *urb.* ►vacant lot.

lot [n] [US]**, abutting** *leg. urb.* ►neighboring lot [US]/neighbouring plot [UK].

lot [n] [US]**, adjoining** *leg. urb.* ►neighboring lot [US]/neighbouring plot [UK].

lot [n] [US]**, play** *recr.* ►tot lot [US]/toddlers play area [UK].

lot area [n] *urb.* ►plot area.

lot coverage [n] [US] *leg. urb.* ►built portion of a plot/lot.

3364 lot depth [n] [US] *leg. surv. urb.* (Length or width of a plot/lot measured from the road to the rear boundary. To avoid the use of the misleading term *plot length* in planning and building regulations, the terms **l. d.** together with ►plot width are used); *syn.* plot depth [n] [UK]; *s* **profundidad** [f] **de solar** (Longitud de un solar medido desde el frente de la calle hasta el límite trasero del predio; ►anchura de solar); *syn.* profundidad [f] de lote de terreno/predio; *f* **profondeur** [f] **d'une parcelle** (±) (Pour une parcelle ou une unité foncière de forme quadrilatère, distance entre la limite de fond de parcelle et l'alignement ou la limite d'emprise des voies publiques ; ►largeur d'une parcelle) ; *g* **Grundstückstiefe** [f] (Länge oder Breite eines Grundstückes, die ausgehend von der Straßenseite bis zur Hintergrenze gemessen wird. Um den missverständlichen Begriff der *Grundstückslänge* im Planungs- und Baurecht zu vermeiden, spricht man von der straßenseitigen ►Grundstücksbreite und der **G.**).

lotic species [n] *phyt. zool.* ►riverine species.

3365 lot line [n] [US] *leg. surv.* (Line of record bounding a lot, which divides one lot from another lot, or from a public or private street or any other public space; IBDD 1981, 120; ►front lot line [US]/front plot boundary [UK], ►rear lot line [US]/rear plot boundary [UK], ►side lot line [US]/side plot boundary [UK]); *syn.* plot boundary [n] [UK]; *s* **delimitación** [f] **de solar** (Línea que delimita un trozo de terreno descrito como solar en título de propiedad; cf. DACO 1988; ►límite delantero de solar, ►límite lateral de solar, ►límite trasero de solar); *syn.* delimitación [f] de lote de terreno/predio; *f 1* **limite** [f] **parcellaire** *leg. surv.* (Limite entre deux unités foncières contiguës) ; *syn.* limite [f] de parcelle, ligne [f] séparative d'un fonds, limite [f] (séparative) de propriété, limite [f] de terrain ; *f 2* **limite** [f] **séparative** *leg. urb.* (*Urbanisme* ensemble des limites parcellaires matérialisées ou non par des bornes de repère, à la cote des sols existants, délimitant la surface d'une propriété ; il existe deux types de limites séparatives : les ►limites séparatives latérales et les ►limites séparatives de fond de parcelle ; ►limite séparative avec l'espace public) ; *g* **Grundstücksgrenze** [f] (Im umgangssprachlichen Sinne OB zu ►vorderer Grundstücksgrenze, ►hinterer Grundstücksgrenze und ►seitlicher Grundstücksgrenze; im Sinne des Liegenschaftskatasters spricht man nur von **Flurstücksgrenze**); *syn. adm. leg.* Flurstücksgrenze [f].

lot shape [n] *surv.* ►plot shape.

lot survey [n] [US] *surv.* ►boundary survey.

lot width [n] *leg. surv. urb.* ►plot width.

3366 low bog [n] *pedol. phyt.* (**1.** A peat-accumulating subhydrous area or freshwater swamp, i.e. a bog where the vegetation is influenced by groundwater, supporting marshlike vegetation, as opposed to a ►raised bog. **2.** Open land with peaty soil covered with coarse grasses, sedges, Sphagnum—wetter than a heath; ►fen 1, ►peatland [US]/moor [UK]); *syn.* minerotrophic peatland moor [n], fen [n] (2), low moor [n]; *s* **turbera** [f] **baja (2)** (►Turbera de origen infraacuático que se forma en terrenos que tienen un nivel freático muy alto, ocupando las depresiones y las partes bajas del relieve o aquellos lugares en los que el agua aflore a la superficie, su agua puede ser caliza y rica en elementos. Se compone principalmente de ciperáceas a las que se añaden, más o menos, gramíneas, herbétum de numerosas familias, y lignétum, que puede llegar hasta el arborétum, así como abundante muscinétum. En ella faltan generalmente las especies del género Sphagnum y plantas carnívoras, y de micorriza. Está relacionada con suelos de gley; cf. DB 1985 y DINA 1987; ▶turbera alta, ▶turbera baja 1); *syn.* turbera [f] infraacuática, prado [m] turboso, turbera [f] límnica, turbera [f] tropógena, turbera [f] cálcica, fen [m]; *f* **tourbière [f] basse** (▶Tourbière en formation correspondant aux premiers stades de la colonisation du milieu aquatique ou à la reconstitution de la tourbière après exploitation, [étape transitoire dans la formation d'une tourbière alcaline ou acide] ; GGV 1979, 243 ; ▶tourbière haute, ▶tourbière plate) ; *g* **Niedermoor [n]** (Subhydrisch, d. h. durch Grundwasser beeinflusstes Moor, im Gegensatz zum ▶Hochmoor; ▶Flachmoor, ▶Moor 1); *syn.* topogenes Moor [n].

3367 low-cost grassing [n] *constr.* (Creation of grassed areas requiring low maintenance at less construction cost; ▶wildflower meadow); *syn.* cost-effective grassing [n]; *s* **establecimiento [m] de césped a bajo costo** (Creación de césped o superficie de hierba que necesita poco mantenimiento; ▶pradera silvestre); *f* **enherbement [m] à entretien réduit** (Création de surfaces engazonnées dont les coûts de création et d'entretien sont faibles ; ▶prairie extensive ensemencée) ; *g* **kostengünstige Begrünung [f] durch Raseneinsaat** (Schaffung von Rasenflächen mit geringen Herstellungskosten, die wenig Pflege und Unterhaltung benötigen; ▶naturnahe Wiese).

3368 low-density building development [n] *arch. urb.* (Building groups which are characterized by a high percentage of private or public green open space; ▶low-density housing, ▶wide spacing in a housing development); *s* **edificación [f] de baja densidad** (Área edificada caracterizada por una gran presencia de zonas verdes públicas y privadas; ▶descongestión urbana, ▶edificación residencial de baja densidad y con ajardinamiento); *syn.* urbanización [f] de baja densidad; *f* **urbanisation [f] aérée** (Occupation du sol en ordre discontinu, de densité réduite ; ▶aération de l'habitat, ▶habitat diffus, ▶habitat dispersé) ; *syn.* habitat [m] aéré, urbanisation [f] lâche, tissu [m] aéré ; *g* **lockere Bebauung [f]** (Durch einen hohen Anteil an privaten oder öffentlichen Grünflächen gekennzeichnete Bebauung; ▶Auflockerung einer Bebauung, ▶aufgelockerte Wohnbebauung).

3369 low-density housing [n] *urb.* (Widely-spaced residential development usually characterized by a high proportion of planted ▶open spaces; ▶landscaping of built-up areas; ▶low-density building development); *s 1* **edificación [f] residencial de baja densidad** (Término genérico para tipo de edificación de viviendas caracterizada por su baja densidad y su gran cantidad de ▶espacios libres, generalmente con vegetación; ▶edificación de baja densidad, ▶tratamiento paisajístico 1); *s 2* **urbanización [f]** (En Es. término específico para zonas residenciales en ámbito rural y generalmente con carácter turístico o recreativo); *f 1* **habitat [m] dispersé** (Zone d'habitations caractérisée par une forte proportion d'▶espaces libres en général plantés ; ▶traitement paysager, ▶urbanisation aérée) ; *f 2* **habitat [m] diffus** (Résidences principales ou secondaires, éparses ou en lotissement implantées par hasard dans les espaces ruraux et naturels et responsable du ▶mitage du paysage) ; *g* **aufgelockerte Wohnbebauung [f]** (Mit Wohngebäuden bebauter Bereich, der meist durch einen hohen Prozentsatz an vegetationsbestimmten ▶Freiräumen geprägt ist; ▶Durchgrünung; ▶lockere Bebauung).

3370 low-density residential area [n] *urb. landsc.* (Spread-out dwelling area with a high proportion of private and public landscaped open spaces; ▶garden suburb, ▶landscaping of built-up areas); *s* **barrio [m] residencial ajardinado** (Zona de la ciudad con gran porcentaje de zonas verdes públicas y privadas; ▶suburbio ajardinado, ▶tratamiento paisajístico 1); *f* **quartier [m] paysager** (Quartier résidentiel qui possède un fort pourcentage d'espaces verts publics et privés ; ▶traitement

paysager 1, ▶ville verte) ; *syn.* lotissement [m] paysager ; *g* **durchgrüntes Wohngebiet [n]** (Wohngebiet, das einen hohen Anteil an privaten oder öffentlichen Grünflächen aufweist; ▶Durchgrünung, ▶Gartenstadt 2).

3371 lowering [n] **of groundwater table** *constr. wat' man.* (Artificial lowering of a water table caused by extraction of groundwater by pumping or open-pit excavation; ▶natural groundwater recession); *syn.* drawdown [n] of groundwater table [also US]; *s* **reducción [f] del nivel freático** (Descenso artificial de la capa freática por exceso de aprovechamiento; ▶descenso natural del agua subterránea; reducir [vb] el nivel de las aguas subterráneas, rebajar [vb] el nivel freático); *syn.* descenso [m] (artificial) de la capa freática, rebajamiento [m] del nivel freático, *f* **rabattement [m] de la nappe phréatique** (Abaissement local artificiel de la nappe causé, p. ex. par pompage, par une exploitation minière à ciel ouvert ; ▶abaissement naturel de la nappe phréatique) ; *syn.* rabattement [m] de la nappe aquifère (PED 1983, 392) ; *g* **Grundwasserabsenkung [f]** (Durch Eingriff lokal erfolgte Senkung des Grundwasserspiegels; ▶Grundwasserabsinken); *syn.* Senkung [f] des Grundwasserspiegels.

3372 lowering [n] **of pH value** *chem. limn. pedol.* (for greater acidity); *syn.* reducing [n] of pH value; *s* **reducción [f] del pH**; *f* **abaissement [m] du pH** ; *g* **pH-Wertabsenkung [f]** *syn.* Absenkung [f] des pH-Wertes.

3373 lowering [n] **of watercourses** *wat'man.* (Artificial deepening of riverbed [US]/river-bed [UK] for the purpose of flood alleviation; ▶dredging 1, ▶dredging 2, ▶dredging of navigable channels); *s* **dragado [m] de cursos de agua** (Profundización del cauce de ríos para permitir el desagüe más rápido en el caso de avenidas. No es un método muy adecuado ni ecológico por la perturbación de la fauna del fondo del río y porque puede conllevar la inundación de zonas río abajo; ▶dragado de lodo/fango, ▶dragado en profundo, ▶dragado subacuático); *f* **approfondissement [m] (du lit) d'un cours d'eau** (Mesures ayant pour but d'accélérer l'écoulement des crues ; ▶approfondissement du chenal, ▶désenvasement, ▶extraction de matériaux alluvionnaires en nappe) ; *g* **Eintiefung [f] von Gewässerläufen** (Ausbaggerung der Gewässersohle zur schnelleren Hochwasserabführung oder zur ▶Eintiefung von Gewässerfahrrinnen; einen Gewässerlauf tiefer legen [vb]; ▶Nassbaggerung, ▶Schlammräumung).

lowering [n] **of water resources** *wat'man.* ▶diminution of water resources.

3374 lower reaches [npl] *geo.* (Part of the stream channel in the lower drainage basin; ▶headstream, ▶middle reaches; *opp.* ▶upper reaches); *s* **tramo [m] inferior (de un curso de agua)** (Parte del cauce en la zona inferior de la cuenca de drenaje; ▶cabecera [de río], ▶tramo medio [de un curso de agua]; *opp.* tramo superior [de un curso de agua]); *f* **cours [m] inférieur** (Partie du lit d'un cours d'eau dans la région la moins élevée du bassin versant ; ▶cours moyen, ▶cours supérieur près de la source ; *opp.* ▶cours supérieur) ; *g* **Unterlauf [m]** (Unterer Teil eines Fließgewässers; ▶Mittellauf, ▶Quelllauf; *opp.* ▶Oberlauf).

lower riparian alluvial plain [n] [US] *phyt.* ▶regularly flooded alluvial plain.

lower riparian woodland [n] *phyt.* ▶regularly flooded riparian woodland.

lower riverine alluvial plain [n] *phyt.* ▶regularly flooded alluvial plain.

lower riverine woodland [n] *phyt.* ▶regularly flooded riparian woodland.

3375 lower river terrace [n] *geo.* (Lowest ►gravel river terrace, which is usually created by frequent periods of accumulation and erosion during the last pleistocene in central European valleys, and which is seldom reached by highwater); *s* terraza [f] fluvial baja (La más inferior de las ►terrazas [fluviales] de aluviones, creada por cambios frecuentes entre periodos/períodos de acumulación y de erosión durante el último pleistoceno en Europa Central); *syn.* terraza [f] fluvial inferior; *f* terrasse [f] (fluviatile) inférieure (►Terrasse alluviale inférieure taillée dans la partie plus ancienne de la plaine d'inondation, en général jamais atteinte par les crues, constituée par les alluvions accumulées et érodées pendant la dernière période de glaciation du Pléistocène) ; *g* Niederterrasse [f] (Die unterste ►Schotterterrasse, oft durch mehrfachen Wechsel von Akkumulation und Erosion während der letzten pleistozänen Kaltzeit in [mitteleuropäischen] Tälern entstanden, die vom Hochwasser meist nicht erreicht wird).

3376 lowest bidder [n] [US] *contr.* (One who offers the lowest price for proposed construction project work); *syn.* lowest tenderer [n] [UK]; *s* licitante [m] a la baja (Empresa que ha entregado la oferta más barata en un concurso-subasta o en una oferta a la demanda); *f* concurrent [m] le moins-disant (Entreprise ayant souscrit le devis ou le prix le plus bas lors d'une procédure d'appel d'offres) ; *g* billigster Bieter [m]/billigste Bieterin [f] (Firma, die bei einer Ausschreibung oder Preisanfrage das niedrigste Angebot abgegeben hat); *syn.* Bieter/-in [m/f] des billigsten Angebotes, Mindestfordernde/-r [f/m].

3377 lowest low-water level [n] *wat'man.* (Abbr. LLW; lowest water level of an inland waterbody which has ever occurred in a defined period of time; ►low-water level); *s* nivel [m] de estiaje mínimo registrado (Altura mínima del nivel del agua de una masa de agua continental en un periodo/período determinado; ►estiaje); *f* niveau [m] d'étiage absolu (Niveau de l'eau d'un cours correspondant au débit journalier le plus faible de la période de référence ; ►étiage) ; *g* niedrigster Niedrigwasserstand [m] (...stände [pl]; Abk. NNW; der niedrigste ►Niedrigwasser[stand] eines Binnengewässers in einem bestimmten Zeitraum).

lowest tenderer [n] [UK] *contr.* ►lowest bidder [US].

3378 low-income housing [n] *sociol. plan.* (Dwellings with tax abatement which are subsidized by state or local government [in England formerly 'councils'] for persons and families with low incomes); *syn.* publicly assisted housing [n] [UK], council housing [n] [UK]; *s* construcción [f] de viviendas de protección oficial [Es] (Actividad constructiva subvencionada por el estado o las CC.AA. para hacer accesible la vivienda para personas o familias de ingresos reducidos); *syn.* construcción [f] de viviendas sociales, construcción [f] de viviendas de utilidad pública; *f* habitat [m] social (Logements construits avec l'aide de l'État et des collectivités territoriales, destinés aux personnes et familles à faibles revenus ; *termes spécifiques* habitat ouvrier, habitation à loyer modéré [HLM] ; selon le régime et le régime de financement on distingue plusieurs types de logements qualifiés d'HLM ; cf. DUA 1996, 393) ; *syn.* logement [m] social ; *g* sozialer Wohnungsbau [m] (Von Bund, Ländern und Gemeinden mit öffentlichen Mitteln geförderter und steuerlich begünstigter Wohnungsbau für Personen und Familien mit geringem Einkommen).

3379 lowland [n] (1) *geo.* (General term referring to low or level tract of land along a watercourse; a bottom; ►lowland 2); *s* tierras [fpl] bajas (1) (Término genérico referente a zonas planas a la altura del nivel del mar o a lo largo de ríos [►llanura aluvial]; *término específico* llanura costera; ►tierras bajas 2); *f* basse plaine [f] (1) (Terme générique pour désigner une surface

continentale étendue, plane et peu élevée en bordure des cours d'eau ou du littoral ; ►basse plaine 2) ; *g* Niederung [f] (Allgemein sprachliche Bezeichnung für nahe dem Meeresspiegel oder an Flüssen [Flussniederung] gelegenes ebenes Gebiet, meist mit Schwemmlandböden; bei ausreichenden Niederschlägen sind **N.en** oft von Altwassern, Sümpfen, Mooren und Nassböden geprägt; ►Tiefebene).

3380 lowland [n, *often used in the plural*] (2) *geo.* (Vague term with no precise meaning: extensive low-lying and relatively level land, especially near the coast and extended plains along big rivers; generally referring to land below 180m above sea level; cf. DNE 1978; ►high plateau, ►plain); *s* tierras [fpl] bajas (2) (*Término no bien definido para* superficie continental plana situada entre el nivel del mar y una altitud no superior a los 200 m; ►llanura; *opp.* ►meseta); *f 1* basse plaine [f, *très souvent utilisé au pluriel*] (2) (Terme générique désignant une surface continentale étendue située en bordure des océans et dont l'altitude est inférieure à 200 m au-dessus du niveau de la mer [ainsi que deltas et estuaires] ; ►plaine ; *opp.* ►haute plaine ; *syn.* basses terres [fpl] [CDN]; *f 2* basse vallée [f, *très souvent utilisé au pluriel*] (Vaste étendue alluviale inondable, submergée périodiquement, souvent caractérisée par une grande diversité d'un patrimoine naturel remarquable mais menacé) ; *f 3* (grande) plaine [f] (Vaste étendue de terres sans relief, parcourue par des cours d'eau lents à la forme sinueuse, de nature sédimentaire souvent ouverte sur un océan p. ex. la grande plaine d'Allemagne du Nord, la plaine de Sibérie occidentale, d'Amazonie, la grande plaine du Gange) ; *g* Tiefebene [f] (1. Ausgedehnte Landoberfläche in Küstennähe nahe dem Meeresspiegelniveau mit geringen Höhenunterschieden; ca. < 200 m ü. NN; meist im Mündungsgebiet großer Flüsse, z. B. die Norddeutsche **T.**, die Westsibirische **T.** zwischen Ural und Jenissei, das Amazonasbecken, die Gangestiefebene. 2. Die **T.** ist ein nicht genau von **Tiefland** unterschiedener Begriff; im Vergleich zur **T.** weist das Tiefland in sich größerflächig verbreitete Hügel- und/oder Plattenlandschaften auf, wodurch das gesamte Gebiet reliefiert ist; ►Ebene; *opp.* ►Hochebene); *syn.* ± Tiefland [n].

3381 lowland forest [n] *for. phyt.* (►lowland 2); *s* bosque [m] de tierras bajas (Bosque que crece en ►tierras bajas 2); *f* forêt [f] de plaine (►basse plaine 2) ; *syn.* forêt [f] de basses terres [aussi CDN]; *g* Tieflagenwald [m] (Wald, der in der ►Tiefebene wächst; ELL 1967).

low level [n] **of light exposure** *bot.* ►poor light exposure.

3382 low maintenance area [n] *constr. hort.* (*opp.* high maintenance area); *s* área [f] de poco mantenimiento (*opp.* área de mucho mantenimiento); *f* espace [m] d'entretien limité (*opp.* surface de prestige) ; *syn.* espace [m] peu entretenu ; *g* Fläche [f] mit geringem Pflegeaufwand (*opp.* Fläche mit hohem Pflegeaufwand).

3383 low-maintenance grass type [n] *constr.* (Blend of grasses sown in the open landscape, e.g. on roadsides or riverbanks and areas not usually used for agriculture, which are drought-resistant, provide protection against erosion, and require low maintenance; seed mixtures that produce a dwarf grass and flowers of uniform nature are referred to as **landscape grass seeds** [UK], **conservation grass mix** [US], **meadow mix** [US]; ►lawn type); *s* césped [f] de bajo mantenimiento (►Tipo de césped utilizado por la construcción paisajística para espacios verdes secundarios, como redes viarias, que es resistente a la sequía, sirve para proteger contra la erosión y necesita poco mantenimiento); *f* gazon [m] utilitaire (►Type de gazon souvent préparé à la demande de l'utilisateur et utilisé en vue du recouvrement des talus, des berges, des bords de routes et autoroutes, des pistes de ski, etc. ; LEU 1987, 197) ; *syn.* gazon [m] de fixation, gazon [m] de végétalisation ; *g* **Landschaftsrasen [m]** (Im Landschaftsbau für die freie Landschaft verwendeter ▶Rasentyp, der in der Regel keiner landwirtschaftlichen Nutzung unterliegt, widerstandsfähig gegen Trockenheit ist, hohen Erosionsschutz bietet und geringe Pflegeansprüche benötigt; je nach Standort werden verschiedene Regelsaatgutmischungen [RSM 7.1] im Handel angeboten: z. B. Standardmischung ohne Kräuter [RSM 7.1.1] und mit Kräutern [RSM 7.1.2], Mischungen für Trockenlagen [RSM 7.2] und Feuchtlagen [RSM 7.3] sowie Halbschattenflächen [RSM 7.4]; cf. DIN 18 917 und www.f-l-l.de).

3384 low maintenance planting [n] *constr.* (Establishment of vegetative cover requiring minimal upkeep; ▶extensive roof planting); *syn.* extensive planting [n] [UK]; *s* **revestimiento [m] vegetal extensivo** (Establecimiento de una capa de vegetación generalmente sobre un sustrato de poco espesor que requiera pocos cuidados; ▶revestimiento vegetal extensivo de tejados); *syn.* plantación [f] extensiva; *f* **végétalisation [f] extensive** (Réalisation d'une couverture végétale sur de faibles épaisseurs de substrat, ca. 4-15 cm. Cette végétalisation est caractérisée par l'utilisation d'espèces — mousses, plantes grasses, plantes herbacées, graminées, vivaces — nécessitant un entretien minimum ; ▶végétalisation extensive des toitures) ; *g* **extensive Begrünung [f]** (Erstellung einer anspruchslosen Bepflanzung oder Vegetationsdecke, die nur einer geringen Pflege bedarf; ▶extensive Dachbegrünung).

low maintenance roof planting [n] *constr.* ▶extensive roof planting.

3385 low-maintenance vegetation [n] *constr. hort. landsc.* (Plants that require minimal maintenance in planted areas); *s* **vegetación [f] de poco mantenimiento**; *f* **couverture [f] végétale d'entretien réduit** ; *g* **pflegeextensiver Bewuchs [m, o. Pl.]** (Pflanzenbestand, der einen geringen Pflegeaufwand benötigt); *syn.* pflegeextensiver Pflanzenbestand [m].

low moor [n] *pedol. phyt.* ▶low bog.

low-moor peat [n] [US] *pedol.* ▶minerotrophic peat.

3386 low-pitch roof [n] *arch. constr.* (Roof with a pitch less than 5°; ▶pitched roof); *s* **tejado [m] plano con pendiente maxima de 5°** (▶tejado inclinado) ; *f* **toiture-terrasse [f] plate** (▶toiture-terrasse rampente) ; *g* **Flachdach [n] (2)** (Dach mit sehr geringer Neigung: 1-5°; ▶geneigtes Dach); *syn.* Gefälledach [n].

low productivity [n] *agr.* ▶pasture of low productivity.

low roof wall [n] [US] *arch.* ▶roof parapet.

3387 low sedge swamp [n] *phyt.* (Plant formation primarily composed of low sedges *[Carex]*, rushes *[Juncus]*, bulrushes *[Scirpus]* or cotton grass *[Eriophorum]* on previous fen woodlands [US]/carrs [UK] at the edge of sediment-filled ponds or lakes, on fens or marshes; ▶sedge swamp, ▶tall sedge swamp); *syn.* small-sedges fen [n] [also UK] (ELL 1988, 573); *s* **formación [f] de pequeños cárices** (Vegetación perteneciente a la alianza *Scheuchzerio-Caricetea*, compuesta sobre todo de cárices pequeños, juncos, scirpus, etc., que se presenta en antiguas ubicaciones de bosques turbícolas, en zonas de colmatación de aguas estancadas, en turberas bajas o humedales; ▶ciénaga de grandes cárices, ▶formación de cárices); *syn.* vegetación [f] de pequeños cárices; *f 1* **cariçaie [f] à petites Laiches/Laîches** (Peuplement herbacé bas et dense, composé de petites Laiches/Laîches *[Carex]*, Joncs *[Juncus]*, Scirpes *[Scirpus]* ou Linaigrettes *[Eriophorum]*, vivant sur les anciennes stations des taillis tourbeux en zone de terrassement des eaux calmes, sur les tourbières basses ou autres sols tourbeux ; appartient à la classe du *Scheuchzerio-Caricetea* ; ▶cariçaie, ▶magnocariçaie, ▶prairies à Laiches/Laîches) ; *f 2* **prairie [f] à petites Laiches/Laîches**

(Localisation du peuplement) ; *g* **Kleinseggenried [n]** (...e [pl]; Pflanzenformation, die zum *Scheuchzerio-Caricetea* gehört, vorwiegend aus niedrigen Seggen *[Carex]*, Binsen *[Juncus]*, Simsen *[Scirpus]* oder Wollgräsern *[Eriophorum]* zusammengesetzt ist und auf ehemaligen Bruchwaldstandorten, in Verlandungszonen an Stillgewässern, auf Niedermooren oder moorähnlichen Standorten anzutreffen ist; auf kalkreichen Standorten sind Kalk-Kleinseggensümpfe *[Tofieldietalia]* anzutreffen; cf. ELL 1978; ▶Großseggenried, ▶Seggenried); *syn.* Kleinseggenrasen [m] (KNA 1971), Kleinsimsenried [n].

low solubility, of [loc] *chem. pedol.* ▶poorly soluble.

3388 low wall [n] *arch. constr.* *s* **murete [m]** (Pequeño muro de poca altura); *f* **muret [m]** (Petit mur, de faible hauteur) ; *g* **Mäuerchen [n]** (Kleine, niedrige Mauer).

3389 low-waste technology [n] *envir.* (Production method of producing a minimum of waste and reducing the harmfulness of residues; ▶clean technology); *s* **proceso [m] productivo pobre en residuos** (Proceso industrial hipo-generador de desechos; ▶tecnología limpia); *f 1* **technologie [f] faiblement génératrice de déchets** (Au sens littéral du mot allemand, technologie assurant la fabrication de produits ayant une incidence moindre sur l'environnement ; ▶technologie non génératrice de déchets, ▶technologie propre) ; *syn.* technologie [f] faiblement productrice de déchets, procédé [m] industriel faiblement générateur de déchets [et de pollution] ; *f 2* **technologie [f] propre (2)** (F., procédé innovant permettant le recyclage de l'eau, des polluants dans les industries consommatrices de matières premières, ou techniques n'engendrant peu ou pas de déchets ou permettant une valorisation maximale par réemploi l'entreprise ; ▶technologie propre 1 ; lois n° 75-633 du 15 juillet 1975 et n° 76-663 du 19 juillet 1976) ; *g* **abfallarmes Verfahren [n]** (Fertigungsmethode, Produkte so herzustellen, dass möglichst wenig Abfallstoffe entstehen und die Schädlichkeit der Rückstände vermindert wird; im Franz. wird **a. V.** *und* ▶abfallfreies Verfahren mit dem OB *technologie propre* übersetzt).

3390 low-water level [n] *wat'man.* (Abbr. LWL; lowest level of a body of water during a certain period of time. In general, also that level which is remarkably below that of the mean level; ▶high-water [level] of watercourses, ▶lowest low-water level); *s* **estiaje [m]** (1. Nivel que alcanzan las aguas, en condiciones de caudal mínimo; WMO 1974; *opp.* ▶avenida 1. 2. Caudal mínimo de un curso de agua, estero, lago o laguna y de la época en que se produce; DGA 1986; ▶nivel de estiaje mínimo registrado); *f 1* **niveau [m] d'étiage** (Niveau de la surface libre d'un cours d'eau dans la période pendant laquelle les débits sont très bas et issus en général des nappes souterraines en voie d'épuisement ; DG 1984 ; ▶crue, ▶niveau d'étiage absolu) ; *syn.* niveau [m] des basses eaux (ARI 1986, 46) ; *f 2* **étiage [m]** (Période de l'année pendant laquelle le débit journalier d'un cours d'eau est le plus faible ; *opp.* crue) ; *g* **Niedrigwasser(stand) [n/(m)]** (...wasser [pl], ...stände [pl]; Abk. NW; niedrigster Wasserstand eines Gewässers in einem bestimmten Zeitraum. Allgemein auch ein Wasserstand, der erheblich unter dem mittleren Wasserstand liegt; ▶niedrigster Niedrigwasserstand; *opp.* ▶Hochwasser).

3391 L-shaped retaining wall unit [n] *constr.* (Supporting wall unit usually made of precast concrete utilizing cantilever structural principles to achieve resistance stability); *s* **elemento [m] de hormigón en L** (Pieza prefabricada en forma de «L» para construir muros de retención); *f* **élément [m] en béton armé à semelle** (Éléments préfabriqués en L, en béton armé pour le soutènement des talus ou la délimitation de différents niveaux de plantation) ; *syn.* élément [m] pour mur de soutènement ; *g* **Stehtraverse [f]** (Vorgefertigter Betonwinkelstein zum Bau von Stützmauern); *syn.* Betonstützwinkel [m] in L-Form, L-Stein [m], Mauerscheibe [f].

3392 lumber connection [n] [US] *constr.* (Fabricated connections of pieces of wood, used for example in pergola construction); *syn.* timber connection [n] [UK]; *s* **ensamblaje** [m] **de piezas de madera** (Unión de elementos de madera prefabricados para construir p. ej. una pérgola o un tejado); *f* **assemblage** [m] **de pièces de bois** (En particulier dans la construction d'un toit ou d'une pergola) ; *g* **Holzverbindung** [f] (Zusammenfügung von Holzelementen, z. B. bei Dachkonstruktionen und beim Pergolenbau; es gibt die traditionelle **zimmermannsmäßige H.**, die **mechanische H. mit metallischen Verbindungsmitteln** wie Nägel, Bolzen, Dübel, Schrauben, Klammern und Lochblechen/Lochplatten sowie **verleimte H.en**, z. B. Leimbinder, und bedingt auch eine Kombination der Verfahren).

lump sum [n] [UK] *constr. contr.* ▶lump-sum price.

3393 lump-sum contract [n] *constr. contr.* (Contractually-binding agreement between the owner and the contractor which stipulates the final cost of a project to be executed); *s* **contrato** [m] **global** (Contrato entre el comitente y el contratista sobre el precio final de un proyecto); *f* **marché** [m] **au forfait** (Contrat, marché passé entre le maître d'ouvrage et l'entrepreneur fixant par avance pour un ensemble de prestations un prix invariable) ; *syn.* adjudication [f] forfaitaire ; *g* **Pauschalvertrag** [m] (Vertrag zwischen Bauherrn und Unternehmer über den verbindlichen Endpreis eines auszuführenden Projektes).

3394 lump-sum fee [n] *contr. prof.* (Payment for professional services in the form of an overall fee, the amount of which is fixed by prior agreement between the planner and his client; ▶retainer fee); *s* **honorarios [mpl] globales** (Pago de los servicios profesionales de arquitecto/a en forma de suma global acordada por anticipado entre el/la comitente y el/la profesional; ▶adelanto de honorarios); *f* **honoraire** [m] **au forfait** (Rémunération au forfait des activités professionnelles librement convenue entre le maître d'ouvrage et le concepteur s'appliquant lorsque mission et programme sont parfaitement définis et connus ; HAC 1989 ; ▶avance d'honoraires) ; *syn.* forfait [m] d'honoraire, rémunération [f] forfaitaire, forfait [m] de rémunération) ; *g* **Pauschalhonorar** [n] (Endgültige Honorarsumme, die zwischen Planer und Bauherrn auf Grund festgelegter Leistungen vereinbart wird; ▶Honorarvorauszahlung); *syn.* Honorarpauschale [f], Pauschalvergütung [f], Pauschalentgelt [n] [auch A] (HRLA o. J., p. 8).

lump-sum item [n] [US] *contr.* ▶collective specification item.

3395 lump-sum price [n] [US] *constr. contr.* (Sum proposed for a collective item, not described in detail, as part of a bid [US]/tender [UK]); *syn.* lump sum [n] [UK]; *s* **precio** [m] **global** (Precio indicado en oferta para una partida grande no descrita en detalle); *f* **prix** [m] **forfaitaire** (Prix fixé dans une offre dans le cadre d'un marché au forfait, appliqué à tout ou partie du marché [ensemble de prestations] quelles que soient les quantités et limité aux cas ou l'exécution des prestations ne présente pas de problèmes techniques ; *contexte* rémunération à prix forfaitaire) ; *g* **Pauschalpreis** [m] (Im Angebot aufgeführter und im Rahmen der Auftragserteilung vertraglich festgelegter Preis für eine genau beschriebene, umfangreiche Leistung, wobei die Ausführungsmengen jedoch nicht näher quantitativ detailliert werden); *syn.* Festpreis [m].

3396 maar [n] *geo.* (Low-relief, broad crater caused by the explosion of a volcano without lava flow. In the Eifel region of Western Germany the term is used for lakes which have been formed in the craters, though they are sometimes dry; cf. GGT 1979); *s* **maar** [m] (Término específico para lago en cráter volcánico en el macizo de Eifel en Alemania); *f* **maar** [m] (*de l'allemand* cratère d'explosion, en général de forme circulaire et à bords verticaux, occupé par un lac) ; *syn.* lac [m] de cratère ; *g* **Maar** [n] (Maare [pl]; durch rein explosiven Vulkanismus ohne Förderung von Lava oder Lockergesteinen entstandene, kraterartige Hohlform im anstehenden Gestein. **M.e** sind in der Eifel mit Wasser gefüllt [Kratersee], können aber auch verlanden oder trocken sein).

3397 macadam [n] [US] *constr.* (Paving with layers of uniformly-graded and compacted, coarse aggregate. Voids are filled with a finer aggregate, and the surface is sealed with asphalt; *specific terms* waterbound macadam, asphalt bound macadam, mixed macadam, spread macadam); *syn.* coated macadam surface [n] [UK], tar macadam surface [n], *abbr.* tarmac [n]; *s* **pavimento** [m] **de macadam** (Revestimiento bituminoso utilizado para pavimentar caminos que consiste en capas múltiples de grava de grosor uniforme, compactadas con grava fina en los intersticios y selladas con brea o asfalto; *términos específicos* macadam [m] mezcla, macadam [m] esparcido y macadam [m] a penetración); *syn.* firme [m] de macadam; *f* **revêtement** [m] **en macadam** (Couche de chaussée constituée par une couche de base constituée de pierres concassées [empierrement] de granulométrie serrée fortement cylindrée dont l'agrégation [stabilisation] est effectuée à l'eau [macadam à l'eau] par apport de sable légèrement argileux et très mouillé ou par un liant bitumineux [tarmacadam] ; cf. GEN 1982-II ; *contexte* allée empierrée, allée en macadam) ; *syn.* empierrement [m] à l'eau, couche [f] de chaussée en macadam goudronné ; *g* **Makadamdecke** [f] (Bituminöse Decke im Wegebau, die vor allem aus grobkörnigem Gestein [Geröll, Splitt] besteht, das mit Teer oder Asphalt gebunden wird. Es werden Mischmakadam-, Streumakadam- und Tränkmakadamdecken unterschieden. Der Begriff erinnert an den schottischen Straßenbauer JOHN LOUDON MACADAM [1755-1836] und bezeichnete ursprünglich sandeingeschlämmte Schotterdecken).

macadam surface [n] [UK], **coated** *constr.* ▶macadam [US].

macadam surface [n], **tar** *constr.* ▶macadam [US].

3398 macchia [n] *geo. phyt.* (Secondary vegetation growing in place of the original broad-leaved evergreen sclerophyllous forests in the form of low, evergreen scrub [1-5m high] on acid soils of hilly and low-lying coastal ranges of the Mediterranean; ▶garrigue); *s* **maquis** [m] (Voz originaria de Córcega utilizada para monte bajo mediterráneo, principalmente perennifolio y arbustivo, en general cerrado y más o menos exuberante, pudiendo alcanzar hasta dos y tres metros. En el resto de la región mediterránea este tipo de vegetación se encuentra en medios análogos, es decir, sobre suelos silíceos [sialíticos y aún oxihúmicos] y en estaciones frescas por su situación respecto al régimen hidrográfico o en la cliserie altitudinal; DB 1985; ▶garriga); *syn.* maquia [f], maqui [m]; *f* **maquis** [m] (Formation végétale xérique secondaire, broussailleuse dense, riches en plantes épineuses, couvrant les terrains siliceux en milieu méditerranéen et issue de la destruction ou de la dégradation anthropogène de la forêt primitive de chêne verts ou de la subéraie ; VOC 1979, 162 ; on distingue le maquis haut [fourré très dense avec deux espèces principales l'arbousier et la bruyère arborescente] et le maquis bas [peuplement plus ouvert caractérisé par la callune et la bruyère à balai] ; ►garrigue) ; *g* **Macchia** [f] (Sekundärvegetation für ursprüngliche Hartlaubwälder in Form einer niedrigen, immergrünen Gebüschformation [1-5 m hoch] auf sauren Böden feuchter, küstennaher Hügel- und niederen Gebirgslagen des Mittelmeergebietes; ►Garrigue); *syn.* Macchie [f].

3399 machinery stock [n] *constr.* (Total stock of equipment **1.** in a works yard or at a place of business, or **2.** on a construction job site, typically demarcated with a secure fence); *syn.* machinery [n], site plant [n] [UK], yard equipment [n] [US]; *s* **equipo** [m] **de máquinas** (Conjunto de máquinas y equipamiento de construcción perteneciente a una empresa constructora o utilizado en una obra); *f* **parc** [m] **des matériels** (de chantier/ d'emprise) ; *g* **Maschinenpark** [m] (Gesamtheit aller Maschinen und Geräte **1.** eines Betriebes/Werkhofes oder **2.** auf einer Baustelle).

3400 machine usage [n] *constr.* (Use of machines for specific construction work); *s* **uso** [m] **de maquinaria** (Empleo de máquinas en trabajos de construcción); *syn.* utilización [f] de maquinaria; *f* **mise** [f] **en œuvre d'engins de chantier** (Utilisation de matériels mécaniques pour des travaux définis sur un chantier) ; *g* **Maschineneinsatz** [m] (Das Verwenden von Maschinen für bestimmte Bauarbeiten); *syn.* Einsatz [m] von Maschinen.

3401 machine work [n] *constr.* (Work which is executed with machines; ►manual labor [US]/manual labour [UK]); *s* **trabajo** [m] **a máquina** (►trabajo hecho a mano); *syn.* operación [f] mecánica; *f* **travail** [m] **mécanique** (Travail exécuté au moyen d'une machine ; ►travail exécuté manuellement) ; *g* **Maschinenarbeit** [f] (Mit Maschinen ausgeführte Arbeit ; ►Handarbeit).

3402 macroclimate [n] *met.* (Climate of large areas monitored by weather stations, which measure data at more than 20 km intervals; ►microclimate, ►topoclimate); *s* **macroclima** [m] (Clima de una extensa región geográfica, de un continente globalmente considerado e incluso de todo el globo terráqueo, DM 1986; ►mesoclima, ►microclima); *f* **macroclimat** [m] (Climat influant sur une grande région géographique les stations de mesure étant éloignées entre elles de plus de 20 km ; la représentation cartographique est au 1 : 10 000 000 ou plus réduite ; ►climat local, ►microclimat) ; *g* **Makroklima** [n] (…ta [pl], *fachsprachlich* …mate [pl]; Klima größerer Gebiete, das durch Messwerte der 20 km von einander entfernten Messstellen erfasst wird. *Zum Unterschied* ►Geländeklima, ►Mikroklima); *syn.* Großklima [n].

3403 macrophanerophyte [n] *bot.* (Plant belonging to the life form category of trees; after RAUNKIAER); *s* **macrofanérito** [m] (En la nomenclatura de BRAUN-BLANQUET Y PAVILLARD, síntesis de las tres categorías biotípicas de megafanerófitos, mesofanerófitos y microfanerófitos de RAUNKIAER, o sea, fanerófitos de 2 m de talla en adelante; DB 1985); *f* **macrophanérophyte** [m] (Classification des végétaux vasculaires selon RAUNKIAER désignant une plante vivace ligneuse dont la taille est supérieure à 5 m — arbre et liane) ; *g* **Makrophanerophyt** [m] (Pflanze, die nach RAUNKIAER der Lebensformklasse der Bäume angehört); *syn.* Baum [m] (1).

magazine [n]**, professional** *prof.* ►professional journal.

main bough [n] [UK] *arb.* ►scaffold branch.

main branch [n] *arb.* ►scaffold branch.

3404 main collector drain [n] *constr.* (Largest drain pipe of a drainage system, which gathers water from smaller drains in an area; ►collector drain); *s* **colector** [m] **principal** (Dren principal de un sistema de drenaje o de alcantarillado; ►colector); *f* **collecteur** [m] **principal** (Canalisation principale d'un réseau d'assainissement constituée de tuyaux enterrés de tracé rectiligne et de pente constante entre les ouvrages de visite ; ►collecteur) ; *g* **Hauptsammler** [m] (Größter Leitungsstrang eines Entwässerungssystems, der die Abwässer aus Nebensammlern aufnimmt und zur Kläranlage führt; ►Sammler).

3405 main dike [n] [US] *wat'man.* (Embankment serving principally to control flooding and for drainage of high water in the interests of water management); *syn.* main dyke [n] [UK]; *s* **dique-guía** [m] (Dique que debe servir principalmente para dirigir las aguas de crecida y de esa manera evitar inundaciones); *f* **digue** [f] **de canalisation des crues** (Digue servant principalement à l'écoulement dirigé des hautes eaux et à la protection contre les inondations) ; *g* **Leitdamm** [m] (Damm, der hauptsächlich dazu dient, ein Hochwasser im Sinne der Wasserwirtschaft ordnungsgemäß abfließen zu lassen).

main drainage line [n] *agr. constr.* ►collector drain.

main dyke [n] [UK] *wat'man.* ►main dike [US].

main limb [n] [US] *arb.* ►scaffold branch.

3406 maintaining [n] **of plant equipment and site installations** *constr.* (Keeping construction machinery, equipment, and site facilities in fit condition for continued use on a construction site); *s* **disponibilidad** [f] **de maquinaria y materiales de obra** (≠) (Obligación del contratista de mantener los equipos de obra en buen estado y los materiales necesarios a disposición para poder ejecutar la obra contratada); *f* **mise** [f] **à disposition des matériels de chantier** (de matériaux, des matériels de chantier ou de divers équipements en vue de réaliser des travaux) ; *g* **Vorhalten** [n, o. Pl.] (Zeitlich befristetes Bereitstellen von Stoffen, Bauteilen und Maschinen in ausreichendem Umfang, um eine Bauaufgabe durchführen zu können).

3407 maintenance [n] (1) *arch. constr. urb.* (Upkeep of, e.g. green areas, buildings, etc.; ►building maintenance, ►preservation of historic buildings, ►maintenance work); *s* **mantenimiento** [m] (Todas las medidas necesarias para mantener el estado y la funcionalidad de edificios, zonas verdes, etc.; ►mantenimiento de edificios, ►preservación de edificios históricos, ►trabajos de mantenimiento); *f* **conservation** [f] (Dans le sens de l'exécution de travaux garantissant le maintien dans le même état de propreté et esthétique des espaces verts, des plantations, de fonctionnement de bâtiments, d'édifices, etc. ; ►entretien immobilier, ►entretien des monuments historiques classés, ►entretien courant) ; *g* **Erhaltung** [f, o. Pl.] (Alle Maßnahmen, die den Bestand und die Funktionsfähigkeit einer Grünanlage, eines Bauwerkes bewahrt; ►Instandhaltung von Gebäuden, ►Pflege, ►Pflege von historischen Gebäuden).

maintenance [n] (2) *constr.* ►establishment maintenance; *constr. hort.* ►garden maintenance [US]; *agr. land'man. landsc.* ►hedgerow maintenance; *constr. hort.* ►high maintenance; *adm. hort. landsc. urb.* ►landscape maintenance; *constr.* ►lawn maintenance; *constr. for.* ►pathway maintenance; *contr.* ►project development maintenance [US]; *leg. wat'man.* ►river maintenance; *constr.* ►scope of maintenance, ►separate contract maintenance; *arb. constr.* ►tree maintenance.

maintenance [n] [US]**, follow-up** *constr.* ►separate contract maintenance.

maintenance [n]**, heavy** *constr. hort.* ►high maintenance.

maintenance [n]**, labo(u)r-intensive** *constr. hort.* ►high maintenance.

maintenance [n] [UK]**, pre-handover** *contr.* ▶project development maintenance [US].

maintenance [n] [US]**, trail** *constr. for.* ▶pathway maintenance.

3408 maintenance [n] **and repair costs** *constr.* (Monetary expenses which are necessary or spent for maintenance and repair work; ▶maintenance costs, ▶maintenance work); *s* **coste(s) [m(pl)] de mantenimiento y rehabilitación [Es]** (▶costo[s] de mantenimiento, ▶trabajos de mantenimiento); *syn.* costo(s) [m(pl)] de mantenimiento y rehabilitación [AL]; *f* **coûts [mpl] d'entretien (2)** (Dépenses monétaires nécessaires à la réalisation des travaux d'entretien ; ▶coûts d'entretien 1, ▶entretien courant) ; *syn.* dépenses [fpl] d'entretien ; *g* **Unterhaltungskosten [pl]** (Zur Durchführung einer Unterhaltungsmaßnahme notwendige oder ausgegebene Finanzmittel; ▶Pflege, ▶Pflegekosten).

3409 maintenance [n] **and repair work** [n] *conserv' hist. constr.* (Upkeep and putting back into working order of built structures and open spaces, including the care of vegetation to ensure both proper growth and public safety as well as the regular maintenance of hardscape areas; ▶maintenance management, ▶establishment maintenance, ▶maintenance work); *syn.* upkeep [n] and renovation [n]; *s* **trabajos [mpl] de mantenimiento y rehabilitación** (Todas las operaciones artesanales que sirven para el ▶mantenimiento y rehabilitación de zonas verdes y para garantizar la seguridad de las construcciones y de los componentes vegetativos de espacios libres; ▶mantenimiento inicial, ▶trabajos de mantenimiento); *f* **travaux [mpl] d'entretien** (Terme qui englobe tous les travaux relatifs à la ▶maintenance et la sécurité des bâtiments et espaces libres ; dans le domaine de la création des parcs et jardins ces travaux comprennent le remplacement des éléments usagés et les travaux de petites réparations ; ▶travaux d'entretien pendant l'année de garantie, ▶entretien courant) ; *g* **Unterhaltungsarbeiten [fpl]** (Alle Arbeiten, die der ▶Unterhaltung und Verkehrssicherheit von Bauwerken und Freianlagen dienen. Für den Garten- und Landschaftsbau sind Leistungen an Vegetationsflächen, die zur Entwicklung und zur Erhaltung der Funktionsfähigkeit der Vegetation erforderlich sind in DIN 18 320 Ziff. 3.8, DIN 18 919 Ziff. 2 näher beschrieben; ▶Entwicklungspflege, ▶Pflege).

3410 maintenance condition [n] *constr.* (Quality and appearance of a plantation, lawn, etc. at a certain point in time); *s* **estado [m] de conservación** (Calidad y aspecto estético de una plantación, un césped, etc.); *f* **état [m] d'entretien** (Qualité et aspect extérieur d'une plantation, d'un gazon à une période déterminée) ; *g* **Pflegezustand [m]** (Beschaffenheit und Erscheinungsbild einer Pflanzung, eines Rasens etc. zu einem bestimmten Betrachtungszeitpunkt).

3411 maintenance costs [npl] *constr.* (Monetary expenses which are necessary or spent for maintenance work; ▶maintenance work, ▶maintenance and repair costs); *syn.* running costs [npl]; *s* **costo(s) [m(pl)] de mantenimiento** (Gastos financieros causados por los ▶trabajos de mantenimiento de una obra; ▶coste[s] de mantenimiento y rehabilitación); *syn.* costo(s) [m(pl)] de conservación, coste(s) [m(pl)] de mantenimiento/conservación [Es]; *f* **coûts [mpl] d'entretien (1)** (Moyens financiers nécessaires à la réalisation de travaux d'entretien ; ▶entretien courant, ▶coûts d'entretien 2) ; *g* **Pflegekosten [pl]** (Zur Durchführung einer Pflegemaßnahme notwendige oder ausgegebene Finanzmittel; ▶Pflege, ▶Unterhaltungskosten).

maintenance hole [n] [US] *constr.* ▶manhole.

3412 maintenance management [n] *adm. conserv'hist. constr. urb.* (All measures necessary for planning and scheduling the development, upkeep, functioning, and restoration of struc-

tures, paved areas, planting and open spaces, as well as for public safety; ▶building maintenance, ▶maintenance and repair work, ▶maintenace work); *s 1* **mantenimiento [m] de zonas verdes** (Todas las medidas que sirven para mantener y desarrollar la funcionalidad de las instalaciones y zonas verdes y para prevenir posibles accidentes causados por árboles; ▶trabajos de mantenimiento de zonas verdes, ▶trabajos de mantenimiento); *s 2* **mantenimiento [m] y rehabilitación [f] de zonas verdes** (Todas las medidas de planificación y prácticas que sirven para mantener y desarrollar la funcionalidad de las zonas verdes incluídas sus instalaciones y para garantizar la seguridad ante posibles accidentes causados por árboles; en comparación con los ▶trabajos de mantenimiento, incluye también medidas de restauración, sustitución o resiembra; las operaciones puramente artesanales se denominan ▶trabajos de mantenimiento y rehabilitación; ▶mantenimiento de edificios); *f* **maintenance [f]** (Conservation et développement des plantations et engazonnements, conservation en bon état de fonctionnement des ouvrages, voiries, équipements et matériels techniques divers ; le terme allemand, par comparaison à l'▶entretien courant comprend le remplacement des éléments morts, usagés ou inutilisables, les travaux de petites réparations, les reprises et réfections diverses ; ▶travaux d'entretien, ▶entretien immobilier) ; *g* **Unterhaltung [f, o. Pl.]** (Alle planerischen und handwerklichen Maßnahmen, die der Entwicklung und Erhaltung der Funktionsfähigkeit von [Frei]anlagen und deren Einrichtungen sowie der Verkehrssicherheit dienen. Im Deutschen beinhaltet der Begriff **U.** im Vergleich zur ▶Pflege auch Ersatzbeschaffungen, Reparaturarbeiten und notwendige Ergänzungen; reine handwerkliche Tätigkeiten, die der Unterhaltung dienen, werden ▶**Unterhaltungsarbeiten** genannt; ▶Instandhaltung von Gebäuden).

3413 maintenance obligation [n] *leg. trans. wat'man.* (Legal responsibility to maintain water bodies, roads, parks, etc.); *syn.* maintenance responsibility [n]; *s* **obligación [f] de mantenimiento** (Responsabilidad legal de los organismos del Estado, los gobiernos autónomos [Es]/de estados federados, de departamentos o de administraciones municipales fijada en diferentes leyes de mantener en buen estado ríos y canales, vías de comunicación, parques y jardines públicos, etc.); *syn.* responsabilidad [f] de mantenimiento; *f* **obligation [f] d'entretien (±)** (Responsabilité réglementaire incombant à une personne physique ou morale de veiller à l'entretien de la voirie, des cours d'eaux, des parcs et jardins, etc. ; *contexte* les routes départementales sont **à la charge** des départements) ; *g* **Unterhaltungslast [f]** (Durch diverse Gesetze bestehende Verpflichtung der Gebietskörperschaften, Eigentümer oder sonstigen in den gesetzlichen Regelungen definierten juristischen Personen zur Unterhaltung von Gewässern, Straßen, Kanälen, Parkanlagen etc.; cf. auch § 29 WHG und Straßengesetze der Länder); *syn.* Unterhaltungspflicht [f] (§ 51 WG-BW).

3414 maintenance [n] **of green spaces** *adm. landsc. urb.* (Routine day-to-day or week-to-week operations involved in the upkeep of parks, gardens and other areas of public green space; ▶administration and maintenance of public green spaces, ▶maintenance management, ▶maintenance work); *s* **trabajos [mpl] de mantenimiento de zonas verdes** (Conjunto de trabajos rutinarios necesarios para mantener los espacios verdes en un estado adecuado para su uso; ▶gestión paisajística de espacios verdes, ▶mantenimiento y rehabilitación zonas verdes, ▶trabajos de mantenimiento); *syn.* mantenimiento [m] de zonas verdes; *f* **entretien [m] des espaces verts** (Interventions régulières qui maintiennent la pérennité des végétaux et des parties minérales ou construites des espaces verts publics ou privés de superficies importantes ; **D.**, terme ne comprenant pas les réparations de revêtements des sols, d'ouvrages ou le remplacement de végétaux ; le terme « entretien des jardins » ne concerne que l'entretien des jardins particuliers de petite et moyenne surface ; ▶gestion des espaces verts publics, ▶maintenance ; *terme générique* ▶entretien courant) ; *syn.* entretien [m] des parcs et jardins ; *g* **Grünflächenpflege [f, o. Pl.]** (Maßnahmen, die der reinen Pflege von Grünanlagen ohne Ersatzbeschaffungen oder Ergänzungen dienen — im Vergleich zum Begriff der ▶Unterhaltung, der auch Reparaturen und Ersatzbeschaffungen beinhaltet; **G.** wird i. d. R. bei öffentlichen Anlagen angewandt, bei Privatgärten spricht man meist von Gartenpflege oder gärtnerischer Pflege in/von Privatgärten; ▶Pflege und Unterhaltung von öffentlichen Grünflächen; *OB* ▶Pflege).

maintenance [n] **of public green spaces** *adm. hort. landsc. urb.* ▶administration and maintenance of public green spaces.

3415 maintenance operation [n] *constr. hort.* (Maintenance action/undertaking as a means of controlling and encouraging the growth of planting and lawn areas. In comparison with ▶maintenance work, the term does not include the replacement of failed plants or the ▶overseeding of grassed areas; ▶hoeing operation); *s* **operación [f] de mantenimiento** (Medida específica de cuidado de la vegetación, como poda, corte de césped, ▶resiembra de césped 2, escarda, etc.; ▶operación de binado); *f* **intervention [f] d'entretien** (Travaux effectués sur les peuplements végétaux existants afin de diriger et favoriser le développement des plantations et des aires engazonnées. Par comparaison avec les ▶travaux d'entretien, ce terme n'inclue pas les plantations de complément et de remplacement ou les ▶semis de regarnissage ; ▶intervention de binage) ; *g* **Pflegegang [m]** (Bestimmter Arbeitseinsatz in Vegetationsbeständen zur Lenkung und Förderung des Wachstums von Pflanzungen und Rasenflächen [▶Nachsaat], der in Ausschreibungen oft zeitlich oder durch einen definierten Qualitätsstandard vorgegeben ist; ▶Hackgang); *syn.* Pflegedurchgang [m], Pflegeeingriff [m].

3416 maintenance pathway [n] *constr.* (Route providing access for the maintenance of public open spaces or for landscape plantings); *s* **camino [m] de mantenimiento** (En parques o jardines o en plantaciones paisajísticas, vía de acceso para realizar trabajos de mantenimiento); *syn.* vía [f] de mantenimiento; *f 1* **chemin [m] d'entretien** (*Terme spécifique* caractérise un élément du réseau) ; *f 2* **circulations [fpl] d'entretien** (*Terme générique* réseau de cheminements d'accès permettant l'entretien des espaces verts ou des plantations paysagères) ; *g* **Pflegeweg [m]** (Weg oder Pfad, der zur Pflege von Grünanlagen oder landschaftlichen Pflanzungen dient).

maintenance program [n] [US] *constr. hort.* ▶plant maintenance program.

maintenance responsibility [n] *leg. trans. wat'man.* ▶maintenance obligation.

3417 maintenance work [n] *constr. hort.* (Measures for the upkeep of green areas, technical equipment, building facilities and auxilliary structures, both private and public, including encouragement and control of plant growth, mowing, weeding, cleaning of vegetation areas, etc.; ▶administration and maintenance of public green spaces, ▶establisment maintenance, ▶landscape construction and maintenance work, ▶maintenace and repair work, ▶maintenance of green spaces, ▶project development maintenance, ▶separate contract maintenance); *syn.* maintenance [n], upkeep [n]; *s* **trabajos [mpl] de mantenimiento** (Conjunto de trabajos necesarios para promover el crecimiento de la vegetación y su estado adecuado, para limpiar las áreas verdes, las instalaciones técnicas y constructivas situadas en espacios libres. En comparación con los ▶trabajos de mantenimiento y rehabilitación, aquéllos no incluyen ningún tipo

de medidas de restauración, sustitución o resiembra; *término genérico* ▶gestión paisajística de espacios verdes; *términos específicos* ▶mantenimiento inicial, ▶operación de mantenimiento, ▶trabajos de mantenimiento de zonas verdes, ▶trabajos de mantenimiento pre-entrega, ▶trabajos regulares de mantenimiento); *syn.* mantenimiento [m], labores [fpl] de mantenimiento, labores [fpl] de conservación, trabajos [mpl] de conservación); *f* **entretien [m] courant** (Tous travaux nécessaires à la conservation et au développement normal des plantations [entretien du sol, tuteurage et haubanage, tailles et élagages, arrosage, fertilisation, protection phytosanitaire] ou des semis [tonte, découpage des bordures, aération des pelouses, roulage] au maintien en bon état de propreté, esthétique des espaces verts en général, des voiries diverses, des sols sportifs de plein air, des ouvrages, des réseaux et équipements divers ; ▶entretien des espaces verts, ▶entretien [d'espaces verts] jusqu'à la réception, ▶gestion des espaces verts publics, ▶intervention d'entretien, ▶travaux d'entretien courant, ▶travaux d'entretien pendant la période de garantie) ; *g* **Pflege [f, o. Pl.]** (Die Summe aller Maßnahmen zur nachhaltigen Lenkung und Förderung der Funktion eines Pflanzenbestandes als Ganzes sowie einer einzelnen Pflanze in einem Bestand und die Säuberung von Vegetationsbeständen, technischen und baulichen Einrichtungen [in öffentlichen und privaten Freianlagen] sowie die Erhaltung deren Gebrauchsfähigkeit. Im Vergleich zu ▶Unterhaltungsarbeiten beinhaltet **P.** weder Ersatzbeschaffungen noch Ergänzungen oder Nachsaaten; ▶Pflege und Unterhaltung von öffentlichen Grünflächen; *UBe* ▶Entwicklungspflege, ▶Erhaltungspflege, ▶Fertigstellungspflege, Gartenpflege, gärtnerische Pflege, ▶Grünflächenpflege, Parkpflege, Rasenpflege, Staudenpflege; ▶Pflegegang); *syn.* Bestandspflege [f, o. Pl.], Pflegearbeiten [fpl], Pflegeleistungen [fpl]).

main thoroughfare [n] *trans.* ▶major street.

main water supply line [n] [UK] *constr.* ▶water main.

3418 major cross-town artery [n] [US] *trans.* (Roadway carrying high volumes of traffic and passing through a city or town from one edge to the other; the capacity of a **t. r./m. c.-t. a.** depends among other things upon the number of directional lanes and traffic lights and, most important, upon the length of the road sections, which contain at-grade or grade-separated junctions; **in U.S., cross-town streets** are those which encircle the central business district); *syn.* troughway [n], *o.v.* thruway [n] [US], through road [n] [UK], through traffic road [n] [UK]; *s* **vía [f] de tránsito** (Calle, carretera o autovía con gran densidad de tráfico que atraviesa una población de un extremo a otro, cuya capacidad depende de la cantidad de carriles unidireccionales, de la frecuencia de semáforos y de la longitud de las secciones sin cruces a nivel); *syn.* arteria [f] de tránsito, calle [f] de tránsito; *f* **voie [f] de transit** (Voirie assurant la circulation générale d'une entrée à la sortie d'une agglomération) ; *syn.* rue [f] pour le trafic de transit ; *g* **Durchgangsstraße [f]** (Hauptverkehrsstraße, die einen Ort vom Ortseingang bis zum Ortsausgang durchquert. Die Leistungsfähigkeit einer **D.** hängt u. a. von der Anzahl der Richtungsfahrbahnen und Lichtsignalanlagen und davon ab, wie lang die Abschnitte mit plangleichen und planfreien Knotenpunkten/Kreuzungen sind); *syn.* Straße [f] für den Durchgangsverkehr.

3419 major dump [n] *min. landsc.* (Landscape form created by dumping [US]/tipping [UK], which is approximately 100m high and > 1,000m wide. This 3rd generation of tipping/dumping comes after conical and 'whale back' [table mountain] forms, and has grown in size over several decades with the addition of ▶mining spoil); *s* **vaciadero [m] gigante** (Gran escombrera de unos 100 m de altura y > 1000 m de anchura utilizada durante varios decenios para depositar ▶zafras); *f* **grand ouvrage [m] paysager (±)** (Amoncellement important d'environ 100 m. de hauteur et 1000 m de largeur caractérisant la mise en dépôt, p. ex.

de ►stériles, d'ordures pendant plusieurs décennies) ; **g Landschaftsbauwerk [n]** (Ca. 100 m hohe und > 1000 m breite Großhalde [dritte Haldengeneration nach Spitzkegel- und Tafelberghalde], die mehrere Jahrzehnte mit ►Berge 3 beschickt und während des Schüttens oder anschließend für Freizeit- und Erholungsnutzung gestaltet und ausgestattet wird/wurde).

major grading [n] [UK] *constr.* ►rough grading [US].

major group [n] [UK] *pedol.* ►soil order [US].

3420 major highway [n] [US] (1) *trans.* (**In U.S.**, a major roadway that carries high volumes of traffic; the breakdown in descending order is: **Interstates** (90/10 federal/state financing), **US-numbered highways** (50/50 federal/state financing); **state roads** (no federal, state funding only) and **local roads and streets** (local financing only); **in U.K.**, a major road, usually a **motorway** or dual carriage-way; ►interstate highway [US], ►major highway 2); *syn.* trunk highway [n] [US], arterial highway [n] [US], major road [n] [UK], trunk road [n] [UK] (2); **s vía [f] interurbana** (Autopista o carretera de tránsito rápido que une las principales ciudades de un país o países vecinos entre sí; ►autovía, ►carretera estatal, ►carretera federal, ►carretera interestatal); *syn.* carretera [f] principal; ƒ**infrastructure [f] routière à grande distance** (Infrastructure routière et autoroutière à grande vitesse et grande capacité reliant les régions ou les états entre eux ; terme générique pour les autoroutes européennes, autoroutes, routes nationales ; ►grand itinéraire, ►voie express) ; **g Fernstraße [f]** (Straße, die Länder eines Staates oder Staaten miteinander verbindet; OB zu Europastraße, Autobahn und Bundesstraße; ►Bundesfernstraße, ►Schnellverkehrsstraße).

3421 major highway [n] [US] (2) *trans.* (Major road for high-speed traffic with at least two lanes in both directions. According to the traffic situation the travelled ways [US]/carriageways [UK] may be grade-separated or at grade); *syn.* expressway [n] [UK]; **s autovía [f]** (En España, carretera interprovincial de características similares a las ►autopistas, pero en las que no hay que pagar peaje); ƒ**voie [f] express** (Routes ou sections de routes appartenant au domaine public de l'État, des départements ou des communes, accessibles seulement en des points aménagés à cet effet et qui peuvent être interdites à certaines catégories d'usagers et de véhicules) ; *syn.* voie [f] rapide, route [f] à grand débit [CH] ; **g Schnellverkehrsstraße [f]** (Mindestens zweibahnige [vierspurige] Hauptverkehrsstraße mit je nach Verkehrssituation kreuzungsfreien oder plangleichen Knotenpunkten); *syn.* Schnellstraße [f], Hochleistungsstraße [f] [auch CH].

3422 major linear feature [n] *geo.* (Linear geomorphological form which characterizes a landscape; e.g. river valley, mountain ridge); **s motivo [m] lineal predominante** (Forma geomorfológica lineal que caracteriza un paisaje, como p. ej. valle de río, cresta de montaña); ƒ**ligne [f] de force d'un paysage** (Forme géomorphologique linéaire [horizontale, verticale ou radiale], axes à fort impact visuel déterminant la structuration d'un paysage telle qu'une vallée fluviale, une ligne de crête, etc.) ; **g Leitlinie [f] einer Landschaft** (Lineare geomorphologische Reliefform, die das Hauptstrukturmerkmal einer Landschaft ist, z. B. ein Flusstal, Höhenrücken).

3423 major natural event [n] *leg.* (Catastrophic natural event which can not be prevented by man; e.g. earthquake, cyclonic disturbance, volcanic eruption, heavy stormtide, tidal wave; ►natural hazard, ►force majeure); **s catástrofe [f] natural** (Suceso catastrófico natural como erupción volcánica, terremoto, huracán o marea viva, que no es evitable por los seres humanos, aunque la magnitud de los daños causados sí puede ser limitada con medidas preventivas, como de calidad de construcción en zonas de temblores de tierra, evacuación masiva de posibles damnificados por huracanes o avenidas, etc.; ►riesgo

natural, ►fuerza mayor); *syn.* calamidad [f] natural; ƒ*1* **catastrophe [f] naturelle** (Événement ou aléas géologiques exceptionnel [séismes, éruptions volcaniques, mouvements de terrains, etc.] événements ou aléas climatiques [précipitations cycloniques, inondations, etc.] ; ƒ*2* **risque [m] naturel majeur** (Événement naturel de caractère exceptionnel [faible fréquence] tel que tremblement de terre, ouragan, éruption volcanique, violente marée de tempête pouvant potentiellement provoquer de nombreuses victimes et/ou des dommages importants pour les biens et/ou l'environnement et nécessitant donc une forte mobilisation dans l'organisation des secours ; ►risque naturel, ►force majeure) ; **g schweres Naturereignis [n]** (Außergewöhnliches Ereignis in der Natur, gegen das der Mensch machtlos ist, z. B. Erdbeben, verheerender Wirbelsturm, Vulkanausbruch, schwere Sturmflut; ►natürliche Gefahr, ►höhere Gewalt).

major road [n] *trans.* ►arterial road.

major road [n] [UK] *trans.* ►major highway [US] (1).

major soil group [n] [UK] *pedol.* ►soil order [US].

3424 major street [n] *trans.* (Important traffic road in cities and suburban areas); *syn.* main thoroughfare [n]; **s arteria [f] (principal) de tráfico** (Calles centrales en la ciudad para la circulación de vehículos que enlazan puntos neurológicos del interior de la misma o con líneas de enlace exterior; cf. DGA 1986); *syn.* calle [f] de tráfico; ƒ*1* **voirie [f] primaire** (*Catégorie de trafic* ensemble des voies [voirie] de trafic normal à important utilisées pour la circulation générale à l'intérieur d'une agglomération et caractérisée par des largeurs de chaussée) ; ƒ*2* **voie [f] primaire** (Terme utilisé pour désigner une voie unique de la voirie primaire) ; **g Hauptverkehrsstraße [f] (2)** (Für den Verkehr wichtige Straße innerhalb von Ortschaften).

3425 make a finished grade flush with [loc] [US] *constr.* (e.g. existing levels of a road or structure to meet a finished grade; ►defined elevation, ►meeting finished grade); *syn.* marry finish grade with [loc] [UK] (SPON 1986, 153), match finished grade with [loc] [US]; **s equiparar [vb] alturas** (p. ej. en relación a un edificio existente, en la modelación de un terreno; ►cota de nivel, ►enrase); ƒ**se raccorder au niveau fini [loc]** (Travaux réalisés conformément au niveau indiqué aux plans ou au niveau final de la voirie ou des bâtiments existants ; ►raccordement au niveau existant, ►cote de raccordement) ; *syn.* se raccorder à la cote altimétrique [loc] ; **g höhengerecht anschließen [vb]** (Entsprechend den planerisch vorgegebenen Höhen, z. B. an ein Bauwerk eine Erdmodellierung anarbeiten oder einen neu zu verlegenden Weg an einen vorhandenen anpassen; ►Anschlusshöhe, ►Höhenanschluss); *syn.* höhengerecht anpassen [vb].

make available for public inspection [loc] [UK] *adm. leg.* ►open for public inspection [US].

3426 making available for development [loc] *urb* *syn.* making [n] ready for development; **s urbanización [f] de terrenos** (Construcción de infraestructura en terrenos que van a ser edificados); ƒ**viabilisation [f]** (Action de viabiliser ; processus de mise en viabilité d'un terrain, c.-à-d. les travaux de raccordement aux réseaux divers [voirie, eau, gaz, électricité, télécommunication, réseau d'assainissement] à exécuter avant toute opération immobilière) ; **g Baureifmachung [f]** (Die völlige Erschließung eines Geländes für eine Bebauung); *syn.* Baulanderschließung [f].

3427 making building land available [loc] *plan.* **s puesta [f] a disposición de terrenos** *syn.* cesión [f] de terrenos; ƒ**cession [f] de terrains** (céder [vb] un terrain) ; **g Bereitstellung [f] von Grundstücken.**

making flush with finished grade [loc] [US] *constr.* ►meeting finished grade [US].

making good of defects [loc] [UK] *constr. contr.* ▶remedying defects.

making ready for development [loc] *urb.* ▶making available for development.

mallee scrub [n] [AUS] *phyt.* ▶brushy area [US], #2.

managed [pp] **(landscape)** *ecol.* ▶near-natural.

management [n] *agr.* ▶agricultural management; *conserv. game'man.* ▶balanced game management; *plan. recr.* ▶countryside management; *agr. for. nat'res. obs.* ▶ecologically-based land management; *agr. for.* ▶economic land management; *for.* ▶forest management (1), ▶forest management (2); *conserv. game'man.* ▶game management; *conserv. land'man.* ▶habitat management; *hunt.* ▶hunting management; *landsc.* ▶landscape management; *adm. conserv'hist. constr. urb.* ▶maintenance management; *nat'res.* ▶natural resources management; *ecol. urb.* ▶near-natural storm water runoff management; *plan.* ▶project management; *plan. recr.* ▶recreational area management; *agr. for.* ▶soil management; *ecol. econ. nat'res.* ▶sustainable management; *for. hort. land'man.* ▶vegetation management; *landsc. plan.* ▶visual resource management; *envir. urb.* ▶waste management; *adm. hydr. nat'res. wat'man.* ▶water management; *conserv. nat'res. wat'man.* ▶water quality management; *adm. leg. wat'man.* ▶water resources management; *conserv.* ▶wildlife management.

management [n]**, air quality** *envir. leg. nat'res.* ▶air pollution control.

management [n]**, chemical pest** *agr. for. hort. phytopath.* ▶chemical pest control, ▶chemical plant disease prevention and pest control.

management [n]**, compost** *envir.* ▶compost recycling.

management [n]**, contractor's site** *contr. constr.* ▶supervision by the person-in-charge.

management [n]**, grassland** *agr.* ▶grassland farming.

management [n] [UK]**, heritage** *adm. conserv'hist.* ▶historic preservation measures [US]/measures for conservation of historic monuments [and sites] [UK].

management [n]**, integrated pest** *agr. for. hort. phytopath.* ▶integrated pest control.

management [n]**, landscape aesthetics** *landsc. plan.* ▶visual resource management.

management [n]**, nature conservation and landscape** *conserv. nat'res.* ▶considerations of nature conservation and landscape management.

management [n]**, pest** *agr. for. hort. phytopath.* ▶pest and weed control.

management [n]**, silvicultural** *for.* ▶forest management (2).

management [n]**, tree** *arb. constr.* ▶tree maintenance.

management [n] [US]**, urban landscape** *adm. hort. landsc. urb.* ▶administration and maintenance of public green spaces.

management [n]**, visual landscape** *landsc. plan.* ▶visual resource management.

management [n] [UK]**, wildflower meadows** *constr. landsc.* ▶wildflower meadow gardening.

management information system [n]**, land** *adm. geo. landsc. plan.* ▶landscape information system.

management measures [npl] *agr. for.* ▶land management measures.

3428 management [n] **of fish population** *conserv. leg. zool.* (Legal requirement to preserve a stock of fish in relation to the size and conditions of a body of water; ▶protected fish habitat area, ▶protected spawning area); *s* **cuidados** [mpl] **ictiogénicos** (Exigencias legales para el aprovechamiento acuícola de terrenos acotados sobre la base de un plan técnico para proteger y fomentar la riqueza y el equilibrio de especies; ▶zona de freza protegida, ▶zona vedada a la pesca); *f* **aménagement** [m] **piscicole et halieutique** (Ensemble des interventions à effectuer pour régler l'exploitation d'un cours d'eau, d'un plan d'eau, de façon à en obtenir le rendement le plus avantageux pour le mode d'exploitation défini et retenu ; AEP 1976 ; ▶réserve de pêche, ▶réserve statique, ▶réserve d'élevage) ; *g* **Fischhege** [**f, o. Pl.**] (Nach dem Fischereirecht bestehende Verpflichtung für Fischereiberechtigte einen der Größe und Beschaffenheit von Gewässern entsprechenden Fischbestand zu erhalten; ▶Fischschonbezirk, ▶Laichschonbezirk).

management [n] **of green spaces, near-natural** *adm. for. hort. landsc. urb.* ▶urban forestry.

3429 management [n] **of lakes and streams** *land' man. landsc.* (*Specific term* river management; ▶river maintenance); *s* **gestión** [**f**] **de ríos y lagos** (≠) (En comparación con el ▶mantenimiento de cursos de agua 1 en alemán este término incluye también el mantenimiento y desarrollo de aguas estancadas y se realiza partiendo de una visión compleja y espacial, considerando toda la cuenca o gran parte de ella); *f* **gestion** [**f**] **des eaux intérieures** (Terme générique caractérisant la conservation et l'utilisation durable des ressources en eau pris sous l'aspect global de la totalité ou parties de leurs bassins hydrographiques ; ▶entretien des cours d'eau) ; *g* **Gewässerpflege** [**f, o. Pl.**] (Im Vergleich zur ▶Gewässerunterhaltung ein allgemeiner Begriff, der auch die Unterhaltung und Entwicklung der Stillgewässer umfasst und meist eine ganzheitliche und räumliche Betrachtungsweise vertritt, die das ganze oder weite Teile des Einzugsgebietes mit einbezieht).

management [n] **of migratory species of wild animals** *conserv.* ▶effective management of migratory species of wild animals.

3430 management [n] **of wastewater and sewage treatment** *envir. plan.* (Generic term encompassing all technical, economic, chemical, biological, legal, and organizational aspects of sewage and wastewater treatment); *s* **sistema** [**m**] **de tratamiento de aguas residuales** (Término que incluye todos los aspectos técnicos, económicos, químicos, biológicos, legales y administrativos concernientes al manejo de aguas residuales); *f* **assainissement** [**m**] (Terme désignant l'ensemble des actions techniques, scientifiques, financières, législatives, réglementaires et administratives ayant pour objet d'assurer l'évacuation des eaux usées et pluviales ainsi que leur rejet dans des exutoires naturels sous des modes compatibles avec les exigences de la santé publique et de l'environnement) ; *g* **Abwasserwesen** [**n, o. Pl.**] (Sämtliche technischen, wirtschaftlichen, chemischen, biologischen, rechtlichen und organisatorischen Tätigkeiten zur Erreichung einer optimalen Abwasserbehandlung).

3431 management plan [n] (1) *adm. conserv'hist. recr.* (Plan which lays down development goals and all of the measures necessary for the care, further development and long-term upkeep of nature conservation and recreation areas, nature parks, public green open spaces, historic monuments, etc. as well as the necessary deadlines, financial resources and other needs; ▶establishment maintenance); *s* **plan** [**m**] **de gestión** (Plan que contiene el conjunto de las medidas de mantenimiento, desarrollo y de conservación a largo plazo de la capacidad funcional de espacios naturales protegidos, zonas de esparcimiento, parques naturales,

M

parques y jardines públicos, etc.; ▶mantenimiento inicial); *f 1* **plan** [m] **de gestion** (Plan définissant les mesures et programmes nécessaires au développement et la préservation durable des sites classés, espaces de loisirs, parcs naturels, espaces verts publics, les sites du patrimoine historique et culturel, etc. ; ▶travaux d'entretien pendant la période de garantie) ; *f 2* **plan** [m] **de gestion écologique** (Plan spécifique aux réserves naturelles établit le plus souvent pour cinq ans dont les mesures visent à la conservation ou l'enrichissement de leur patrimoine naturel et plus généralement la préservation durable de la biodiversité ; circulaire n° 95-47 du 28 mars 1995) ; *g* **Pflege- und Entwicklungsplan** [m] (Plan, der Pflege- und Entwicklungsziele und die Gesamtheit der Maßnahmen zur Pflege, Weiterentwicklung und langfristigen Erhaltung der Funktionsfähigkeit von Naturschutzgebieten, Erholungsgebieten, Naturparken, Garten- und Parkanlagen mit ihren Vegetationsbildern, kulturhistorischen Anlagen etc. beschreibt und die dafür notwendigen Termine, Finanzmittel und sonstigen Erfordernisse benennt; ▶Entwicklungspflege Ziffer 2).

management plan [n] (2) *conserv. ecol.* ▶Habitats Directive Management Plan; *leg. wat'man.* ▶water management plan.

manager [n] *plan.* ▶project manager.

manager [n] [UK], **contractor's construction** *constr. prof.* ▶contractor's agent.

manager [n], **supervision by the site** *contr. constr.* ▶supervision by the person-in-charge.

3432 mandatory building line [n] *leg. urb.* (Imaginary line parallel to the front property line, which is required to be built upon; e.g. in order to obtain a desired straight line of buildings, or to preserve a minimum amount of open space between the street and the nearby building to separate street traffic from residences thereby minimizing the likelihood of accidents; a building line may be established for a subdivision, by restrictive covenants in deeds or leases, by building codes, or by zoning ordinances; **in U.S.**, this **m. b. l.** is contained in special zoning regulations that require a particular building massing arrangement, such as in a historical district, a so-called, "New Urbanism" development, or a local zoning law that seeks to control the façade placement for reasons of architectural continuity, or other aesthetic or functional purposes; **in U.K.**, a **b. l.** is less mandatory and is contained in Local Development Framework documents such as the Supplementary Planning Document, where it may be described in design guides pertaining to the form, layout, character and quality of a new development for those preparing planning applications; ▶setback line); *s* **línea** [f] **de fachada** (En urbanismo, línea que limita la zona en la que efectivamente se puede construir con el fin de conseguir una trama regular de la calle o plaza donde se construye. Puede estar más al interior que la ▶línea de la calle, ▶línea de retranqueo. **En Es.** la **l. de f.** se puede definir en los Estudios de Detalle. **En EE.UU.** se puede fijar en regulaciones especiales de zonificación para conseguir efectos estéticos o funcionales especiales o para proteger conjuntos históricos. **En F. y D.** se define en el ▶plan parcial [de ordenación]); *syn.* retranqueo [m] frontal; *f* **ligne** [f] **d'implantation** (Règle d'urbanisme réglementant les conditions d'occupation du sol dans le ▶plan local d'urbanisme [PLU]/plan d'occupation des sols [POS] ; l'implantation des constructions nouvelles peut être réglementée par l'obligation de positionner leur façade bâtie le long d'une ligne inscrite dans le plan de zonage afin p. ex. d'obtenir une forme ou un volume architectural déterminé [place, alignement des constructions, etc.], d'obtenir un front urbain plus harmonieux au fur et à mesure des nouvelles constructions [maintien de perspective] ; cette procédure doit être clairement distinguée de la procédure d'implantation de la construction sur la limite de l'alignement, dans le prolongement des constructions existantes ou sur la ▶limite de constructibilité ; la marge de reculement peut être fixée en bordure des voies de circulation ou par rapport aux limites séparatives ; ▶polygone d'implantation) ; *syn.* ligne [f] d'implantation obligatoire ; *g* **Baulinie** [f] (**D.**, im ▶Bebauungsplan durch Zeichnung und roter Farbe festgesetzte Linie, auf der im Vergleich zur ▶Baugrenze gebaut werden muss, um eine einheitlich ausgerichtete Anordnung der Bauten, z. B. eine geschlossene Begrenzung eines Straßenraumes oder Platzes, zu erreichen; cf. § 23 [2] BauNVO; ein Vor- oder Zurücktreten von Gebäudeteilen kann in geringfügigem Ausmaß zugelassen werden. In verbindlichen Bauleitplänen, die vor In-Kraft-Treten des BBauG 1960 rechtskräftig wurden, wurden straßenseitige Gebäudefronten durch Fluchtlinien festgesetzt); *syn. obs.* Bauflucht [f], Baufluchtlinie [f], Fluchtlinie [f].

3433 mandatory landscape development plan [n] *landsc. leg.* (Part of a legally-binding ▶landscape plan which sets the most important tasks to be fulfilled in the development of a landscape and stipulates areas for conservation, development and management measures to be undertaken by public law and adhered to by property owners); *s* **plan** [m] **vinculante de desarrollo paisajístico** (≠) (**D.** parte del ▶plan de gestión del paisaje en la que se informa sobre las tareas principales de desarrollo urbanístico y sobre las zonas protegidas así como las medidas de desarrollo, mantenimiento e infraestructura a realizar en aquéllas, que son vinculantes para personas jurídicas públicas y dentro de lo razonable para los propietarios o usuarios de predios); *f* **carte** [f] **des orientations et principes du plan de paysage** (≠) (**D.**, document graphique constituant une des pièces du ▶schéma de paysage identifiant pour le périmètre d'application du plan les orientations et principes fondamentaux de développement des paysages et localisant les espaces dans lesquels sont fixées les mesures de protection, de développement, de gestion ou d'entretien opposables à toute personne publique et, dans les limites du raisonnable, aux propriétaires de fonds ; ▶charte pour l'environnement, ▶contrat pour le paysage, ▶plan de paysage) ; *g* **Entwicklungs- und Festsetzungskarte** [f] (Teil des ▶Landschaftsplanes, der über das Schwergewicht der im Planungsgebiet zu erfüllenden Aufgaben der Landschaftsentwicklung behördenverbindlich Auskunft gibt und Schutzausweisungen für Flächen und Entwicklungs-, Pflege und Erschließungsmaßnahmen häufig verbindlich für juristische Personen des öffentlichen Rechts und im Rahmen des Zumutbaren auch für Grundstückseigentümer oder -besitzer festsetzt; in manchen Bundesländern hat diese Karte nur empfehlenden Charakter).

3434 man/day calculation [n] *contr.* (Calculation of a rate which is appropriate for the payment of work per man/worker for one day); *s* **cálculo** [m] **de hombre/día** (Cuantificación de la remuneración apropiada por el trabajo de una persona por día basándose en el salario y los gastos de seguridad social, seguro de accidentes, etc.); *f* **calcul** [m] **des prix à taux journalier** (Détermination d'un coût unitaire journalier, d'un taux de rémunération calculé à partir du salaire, des charges et frais divers afférents au personnels, à la main d'œuvre) ; *g* **Mann-pro-Tag-Kalkulation** [f] (Berechnung eines Verrechnungssatzes für den Einsatz eines Mitarbeiters je Tag).

3435 man-developed [pp] *ecol.* (Descriptive term applied to the highest ▶degree of landscape modification landscape modification: ecosystems created artificially by cultural intervention, the existence of which is dependent upon the control and supply of energy by human efforts. The system breaks down, if the human influence ceases or it may become subject to change by succession; ▶near-natural; *opp.* ▶virgin); *syn.* unnatural [adj], anthropogenic [adj] (2); developed [pp] by man; *s* **antropógeno/a**

[adj] (2) (Término descriptivo para el mayor ►grado de modificación [de ecosistemas y paisajes]: ecosistemas biológicos o técnicos creados artificialmente, cuya existencia depende del control humano y el abastecimiento artificial de energía; ►antrópico/a; *opp.* ►virgen; ►poco alterado); *f* **fortement artificialisé, ée [loc]** (Terme utilisé pour caractériser un écosystème au ►degré d'artificialisation le plus élevé dont l'existence dépend des apports externes d'énergie et de l'action régulatrice de l'homme ; les systèmes se désagrègent lorsque cesse l'influence humaine ou se transforment successivement en une autre communauté biologique tels que les systèmes d'agriculture intensive, urbains et industriels ; ►anthropogène 1 ; *opp.* ►d'artificialisation nulle, ►faiblement artificialisé) ; *syn.* artificiel, elle [adj], anthropogène [adj] (2) ; *g* **naturfern [adj]** (Den geringsten ►Natürlichkeitsgrad betreffend: vom Menschen geschaffene Bio- oder Techno-Ökosysteme, deren Existenz von der Steuerung und Energiezufuhr durch den Menschen abhängig ist. Hierzu gehören z. B. intensiv gepflegte Park-, Sportplatz- und Gartenrasen sowie Pflanzungen von Zierstauden, Wechselflor, Ziersträuchern, aber auch intensiv kultivierte landwirtschaftliche Flächen. Hört der menschliche Einfluss auf, so zerfällt das Ökosystem resp. die Lebensgemeinschaft und verändert sich durch ►Sukzession; ►anthropogen; *opp.* ►natürlich 2, ►naturnah).

maneuver space [n] **[US]** *trans. plan.* ►parking maneuver space [US]/parking manoeuvre space [UK].

3436 mangrove stand [n] *geo. phyt.* (Plant community composed almost entirely of evergreen sclerophyllous broad-leaved trees and shrubs with either stilt roots or pneumatophores, occurring in the tidal range of the tropical and subtropical zones; ELL 1967); *syn.* mangrove wood [n]; *s* **manglar** [m] (Comunidad de composición florística simple, cuya altura es en general de 3-5 m. Se encuentra en las orillas bajas y fangosas de las costas tropicales y subtropicales de los océanos Atlántico y Pacífico y es característico de esteros, desembocaduras de ríos y otros lugares cercanos, en donde el suelo es de origen aluvial y se inunda periódicamente por aguas salobres de oleaje fuerte. Está compuesto casi únicamente de árboles y arbustos latifoliados esclerófilos semperviventes, con raíces fúlcreas o con neumatóforos; MEX 1983 y UNE 1973); *syn.* bosque [m] de manglares, bosque [m] (tropical) de mangles; *f* **mangrove [f]** (Peuplement arboré ou arborescent littoral composé d'une végétation d'arbres et d'arbustes feuillus sclérophylles sempervirents avec des racines-échasses ou des pneumatophores, localisé dans la zone intertidale du littoral plat et souvent sablonneux des régions tropicales et subtropicales ; ce peuplement n'existe seulement que dans la zone intertidale souvent sablonneuse des régions tropicales et subtropicales ; UNE 1973) ; *g* **Mangrove [f]** (**1.** Durch hohe, den Schlick festhaltende Stelzwurzeln [Pneumatophoren] charakterisierte, vorwiegend immergrüne Hartlaubvegetationsformation im Gezeitenbereich der flachen, oft sandigen Küstenzonen der Tropen und Subtropen; nach dem Artenreichtum der **M.n** werden a] floristisch reiche „östliche M.n" des Indopazifischen Ozeans und b] artenärmere „westliche M.n" des Atlantischen Ozeans unterschieden; **2.** OB für verschiedene verholzende Salzpflanzen [z. B. *Avicennia germinans, Rhizóphora mangle*], die sich an das Leben im Gezeitenbereich tropischer Küstenregionen angepasst haben und dort geschlossene Bestände bilden); *syn.* Mangrove(n)wald [m], Gezeitenwald [m].

mangrove wood [n] *geo. phyt.* ►mangrove stand.

3437 manhole [n] *constr.* (Vertical access shaft from the surface of the ground to a sewer or underground utilities, usually at a junction for connection, cleaning, inspection and repairs; typically, it is made of precast concrete, brick or concrete masonry units; **m.s** subdivide underground channels and pipes in sections, mostly at connection points, or where direction of utility lines,

pipe diameters or gradients change; according to their functions, it consists of the following components [from bottom to top]: ►cleanout chamber, manhole base, ►drywell sump, intermediate ring, ►riser unit [US]/chamber section [UK], sleeve unit with ogee joints, shaft, manhole adjustment ring, ►cover slab on top and ►step rods [US]/step irons [UK], if necessary; minimum internal dimensions of rectangular **m.s** are 1 200x750mm; circular for sewers minimum diameter 1050mm; minimum opening size of any **m.** is 600x 600mm; ►access eye, ►irrigation equipment manhole); *syn.* utility hole [n] [also US], maintenance hole [n] [also US]; *s* **pozo** [m] **de registro (2)** (Construcción vertical en el subsuelo, de sección circular o cuadrada, con diferentes funciones de control de las conducciones subterráneas, que está provista de una tapa de diámetro suficiente para el paso de una persona que tenga que visitarlas; ►fondo de pozo, ►placa de cobertura de pozo de registro, ►peldaño de hierro, ►pozo de registro 1, ►pozo de riego [subterráneo], ►registro, ►sumidero); *f* **regard** [m] **visitable** (Ouvrage généralement vertical de section circulaire ou carrée, constitué suivant sa fonction du bas vers le haut : d'un ►socle de regard, d'une cheminée pouvant être munie de ►poses pied en acier et constituée de viroles et de piédroits, d'une hotte et selon la nature du regard d'un dispositif de fermeture avec ►rehausse sous cadre ; *termes spécifiques* ►regard d'arrosage, ►regard de curage, ►regard de visite, ►puisard) ; *g* **Einstiegsschacht** [m] (...schächte [pl]; enger, meist senkrecht-röhrenförmiger oder rechteckiger Raum in der Verkehrsfläche über Abwasser- oder Versorgungsleitungen, aus Betonfertigteilen oder gemauert, ggf. mit ►Steigeisen versehen, in dem Wartungsarbeiten, Kontroll- oder Anschlussarbeiten durchgeführt werden. **E.schächte** unterteilen Rohrleitungen in mehrere Haltungen, meist an Zusammenführungen von Leitungen, bei Änderung der Richtung, Rohrdurchmesser, des Leitungsgefälles und der Höhendifferenzen; ►Bewässerungsschacht, Kabelschacht, ►Kontrollschacht 1, ►Kontrollschacht 2, ►Schachtabdeckplatte mit Einstiegsöffnung, ►Sickerschacht); *syn.* Einsteigschacht [m], besteigbarer Kontrollschacht [m], besteigbarer Revisionsschacht [m], besteigbarer Wartungsschacht [m].

manhole adjusting ring [n] *constr.* ►manhole adjustment ring.

manhole adjustment course [n] **[UK]** *constr.* ►manhole adjustment ring.

3438 manhole adjustment ring [n] *constr.* (Precast, flat concrete circular segment or mortar layer installed upon the uppermost ►riser unit [US]/chamber section [UK] or ►cone 1 to bring the level of the cover slab to the same grade as the surrounding surface area); *syn.* manhole adjusting ring [n], manhole adjustment course [n] [also UK], manhole levelling course [n] [also UK]; *s* **anillo** [m] **de ajuste de registro** (Anillo de hormigón que se coloca sobre el ►anillo de pozo superior o sobre el ahusamiento de pozo de registro para nivelar la cobertura del pozo de registro con la superficie del terreno); *f 1* **rehausse** [f] **de regard** (Élément terminal en béton de la cheminée d'un regard placé sur la dernière ►virole ou sur la ►hotte de regard afin de permettre la mise à niveau du dispositif de fermeture) ; *syn.* rallonge [f] de cône ; *f 2* **gouttière** [f] **de mise à niveau** (Élément en fonte, en forme de gouttière, du dispositif de fermeture d'un regard, posé sur l'élément terminal de la cheminée, pouvant être rempli par un laitier d'asphalte compacté, sur lequel repose le cadre et le tampon, le niveau variant en fonction de la quantité de produit mis en œuvre) ; *g* **Schachtauflagering** [m] (Flacher Betonring zum Aufsetzen auf den obersten ►Schachtring oder ►Schachthals zum Angleichen der Schachtabdeckung an die Geländeoberflächenhöhe); *syn.* Auflagering [m].

3439 manhole base [n] *constr.* (Bottom of a manhole shaft, either a precast element or cast in situ); *syn.* manhole floor [n];

M

s **fondo [m] de pozo** (Parte inferior de un pozo que puede ser prefabricada o construida con hormigón in situ); *f* **socle [m] de regard** (Élément de fond de regard pouvant être constitué par un radier en béton comportant une cunette assurant la continuité hydraulique ou par un élément préfabriqué et pouvant constituer une fosse de dessablement) ; *syn.* plage [f] de regard de visite ; *g* **Schachtboden [m]** (Unterster Teil eines Schachtes, entweder als vorgefertigtes Element oder in Ortbeton hergestellt); *syn.* Bodenstück [n] eines Schachtes.

manhole brick [n] *constr.* ▶concentric manhole brick.

3440 manhole cover [n] *constr.* (Sealing device [ductile and cast iron, galvanized steel or precast concrete] of a ▶manhole, comprising the cover itself and a frame; many paving projects use recess tray covers in which block pavers or setts are inserted into the frame to match with the design of the surrounding paving; **m. c.s** are classified according to their load-bearing capacity, heavy duty [H.D.] or light duty [L.D.]. *NOTE* not to be confused with [manhole] cover slab [typically reinforced concrete] which is installed to carry the cover); *syn.* chamber cover [n] [UK] (SPON 1986, 170s), manhole frame [n] and cover [n] [also UK], manhole lid [n], shaft lid [n]; *s* **tapa [f] de pozo de registro** (Pieza de cobertura de ▶pozo de registro 2 clasificada según su capacidad de carga); *f* **tampon [m]** (Dispositif de fermeture d'un ▶regard visitable de forme circulaire ou carrée qui, selon les classes de résistance aux charges roulantes, répond en F. à la norme N.F. P 98 312 [classes B-E] ; en D. à la norme E DIN 1229 [classes A-F]) ; *syn.* tampon [m] d'obturation ; *g* **Schachtdeckel [m]** (Verschlusselement eines ▶Einsteigschachtes; abhängig von der Belastbarkeit klassifiziert gemäß E DIN 1229 nach Einbaustelle in Klassen A-F: Prüfkräfte von 15 kN bis 900 kN); *syn.* Schachtabdeckung [f]; *für Kanalschacht* Kanaldeckel [m].

manhole floor [n] *constr.* ▶manhole base.

manhole frame [n] **and cover** [n] [UK] *constr.* ▶manhole cover.

manhole levelling course [n] [UK] *constr.* ▶manhole adjustment ring.

manhole lid [n] *constr.* ▶manhole cover.

man-made [pp] *ecol. land'man.* ▶anthropogenic (1).

3441 man-made humic soil [n] *pedol.* (Humous soil made by repeated cultivation, in Europe over centuries, into more ▶friable soil; ▶man-made soil, ▶topsoil 1); *syn.* garden soil [n] [also US]; *s* **suelo [m] hortisol** (▶Suelo friable resultante de lustros a siglos de aprovechamiento horticultural; *término genérico* ▶cultosol; ▶tierra vegetal); *syn.* hortisol [sm]; *f* **sol [m] horticole** (Sol ayant été cultivé pendant des décennies ou des siècles à des fins horticoles et constituant un ▶sol à structure grumeleuse ; *terme générique* ▶sol de culture ; ▶horizon de surface, ▶sol superficiel, ▶terre végétale) ; *syn.* terre [f] horticole ; *g* **Gartenboden [m]** (Durch jahrzehnte- bis jahrhundertelange gartenbauliche Kultur entstandener ▶krümeliger Boden; *OB* ▶anthropogener Boden; ▶Oberboden); *syn.* Hortisol [m].

man-made landscape [n] *geo.* ▶cultural landscape.

3442 man-made pond [n] *constr.* (Small artificial body of shallow water with controllable water supply and drain outlet. In contrast to a ▶lake 1, water and nutrients are homogeneously mixed; ▶fish pond, ▶fire suppression pond [US]/fire pond [UK], ▶natural pond); *syn.* artificial pond [n]; *s* **estanque [m] (artificial)** (Pequeño cuerpo de agua construido con suministro controlable de agua y salida de drenaje. Al contrario que en un ▶lago, los nutrientes están repartidos homogéneamente. El término **charco [m]** denomina una pequeña superficie cubierta de agua temporalmente, sea de origen natural o artificial; ▶estanque natural, estanque ornamental, ▶estanque de reserva, ▶estanque

de vivero [de peces]); *f* **étang [m] artificiel** (Petit plan d'eau calme, à fond peu profond, à niveau d'eau variable provenant de l'activité humaine comme p. ex. l'▶étang piscicole, l'étang de jardin, le ▶point d'eau de défense contre l'incendie. Il se distingue du ▶lac par l'homogénéité de ses eaux et des substances nutritives qu'elles contiennent ; ▶étang naturel) ; *g* **Teich [m]** (Vom Menschen geschaffenes, kleines, seichtes, stehendes Gewässer mit regulierbarem Zu- und Abfluss wie z. B. ▶Fischteich, Gartenteich, ▶Feuerlöschteich, Dorfteich. Der **T.** unterscheidet sich vom ▶See dadurch, dass beim Teich die Nährstoffe im Wasser homogen durchmischt sind. Der Begriff **Tümpel [m]**, eine kleine, flache, periodisch austrocknende Wasseransammlung, ist künstlich oder natürlich entstanden, mit oder ohne Zu- und Ablauf; ▶Weiher).

3443 man-made soil [n] *agr. constr. hort. pedol.* (Soils with thick topsoil that has been amended by human intervention or disturbed soil more than 40cm thick; cf. Soil Survey of England and Wales cit. in CURTIS et al. 1976; ▶soil order [US]/major soil group [UK], ▶plaggen soil, ▶man-made humic soil); *syn.* altered soil [n] [also US], anthropic soil [n] (MAR 1964); *s* **cultosol [m]** (Suelo alterado en la estructura del perfil por su uso para fines de cultivo. En la clasificación alemana [▶grupos principales (de suelos)] **c.** es un orden con tres clases de suelos: a] cultosoles terrestres con ▶suelo de tipo «plaggen», ▶suelo hortisol y regosuelo, b] cultosoles hidromorfos y c] turberas antropógenas); *f 1* **sol [m] de culture** (Sol utilisé pour les cultures maraîchères, l'agriculture ou la sylviculture dont le profil a été totalement bouleversé par l'action de l'homme ; *syn.* sol [m] anthropogène ; *f 2* **sols [mpl] anthropogènes** (▶Classe pédologique utilisée dans la classification des sols allemande comprenant trois sous-classes a] les sols dits terrestres [▶sol de type « plaggen », ▶sol horticole], b] les sols semi-terrestres [ou sols hydromorphes de culture] et c] les sols hydromorphes organiques [tourbes] recouverts d'une couche de sable [culture sur sable sur les tourbières basses] ou à mélange de sable [tourbe sableuse]) ; *g* **anthropogener Boden [m]** (1. Gärtnerisch, landwirtschaftlich oder forstwirtschaftlich genutzter Boden, bei dem der ursprüngliche Bodentyp oder das gesamte Bodenprofil von Menschenhand völlig verändert oder geformt wurde. 2. In der deutschen Bodensystematik ist der a. B. eine ▶Bodenabteilung mit drei Bodenklassen: a] terrestrische Kultosole mit ▶Plaggenesch, Hortisol [▶Gartenboden] und Rigosol, b] hydromorphe Kultosole und c] anthropogene Moore mit ca. 15 cm Sand bedeckt [Sanddeckkultur bei der Niedermoorkultivierung] und Sand gemischt [Sandgemischte Moore]; cf. SS 1979, 353 ff); *syn.* künstlicher Boden [m], *pedol.* Kultosol [m].

manoeuvre space [n] [UK], **parking** *trans. plan.* ▶parking maneuver space [US].

3444 manual excavation [n] *constr.* (Digging by hand with a shovel, spade or pick; ▶excavate by hand); *s* **excavación [f] manual** (▶excavar manualmente); *f* **excavation [f] exécutée à la main** (Exécution d'une tranché au moyen du pic et de la pelle ; ▶exécuter une excavation à la main) ; *syn.* excavation [f] exécutée manuellement, fouille [f] à la main ; *g* **Handschachtung [f]** (Ausheben eines Grabens/Loches mit Schaufel, Spaten und Pickel; ▶in Handschacht ausführen).

3445 manual labor [n] [US]/**manual labour** [n] [UK] *constr.* (Work done by hand, in contrast to ▶machine work; ▶excavate by hand); *syn.* manual work [n] [UK], hand work [n] [also US]; *s* **trabajo [m] hecho a mano** (▶excavar manualmente; *opp.* ▶trabajo a máquina); *syn.* trabajo [m] manual; *f* **travail [m] exécuté manuellement** (▶exécuter une excavation à la main, manuellement ; *opp.* ▶travail mécanique) ; *syn.* travail [m] manuel, travail [m] effectué manuellement ; *g* **Handarbeit**

[f, o. Pl.] (Körperliche, mit der Hand ausgeführte Arbeit, im Gegensatz zur ▶Maschinenarbeit; ▶in Handschacht ausführen).

manual work [n] [UK] *constr.* ▶manual labor [US]/manual labour [UK].

manufactured home [n] [US] *recr. urb.* ▶mobile home.

manure [n] *agr.* ▶liquid manure; *agr. constr. for. hort.* ▶organic fertilizer, ▶semiliquid manure [US]/semi-liquid manure [UK].

manure [n], **barn** *agr.* ▶straw dung.

manure [n], **stable** *agr.* ▶straw dung.

manure crop [n] *agr. constr. hort.* ▶green manure crop.

manure plant [n] *agr. constr. hort.* ▶green manure plant.

manuring [n] *agr. constr. for. hort.* ▶fertilizing, ▶green manuring; *agr.* ▶liquid manuring.

map [n] *adm. surv.* ▶area ownership map; *plan. recr.* ▶avalanche location map; *plan.* ▶base map (1), ▶base map (2); *met.* ▶climatic map; *geo.* ▶contour map (1); *hydr.* ▶hydrological map; *wat'man.* ▶lake and river water quality map; *adm. surv.* ▶metes and bounds map [US]/land registration map [UK]; *plan.* ▶noise contour map; *rem'sens. surv.* ▶orthophotographic map; *plan.* ▶ownership map; *plan. sociol.* ▶population density map; *adm. surv.* ▶real property identification map [US]; *plan.* ▶reference map; *geo. plan.* ▶relief map; *plan.* ▶slope analysis map; *pedol.* ▶soil map; *phyt.* ▶vegetation map; *leg. urb.* ▶zoning map [US]/Proposals Map/Site Allocations Development Plan Document [UK].

Map [n] [UK], **adoption of a Proposals** *leg. urb.* ▶adoption of a zoning map [US].

map [n], **avalanche hazard** *plan. recr.* ▶avalanche location map.

map [n] [US], **cadastral** *adm. surv.* ▶metes and bounds map [US]/land registration map [UK].

map [n] [UK], **land registration** *adm. surv* ▶metes and bounds map [US]/land registration map [UK].

map [n] [UK], **land registry** *adm. surv.* ▶area ownership map.

map [n] [US], **noise exposure** *plan.* ▶noise contour map.

Map [n] [UK], **Proposals** *leg. urb.* ▶zoning map [US].

3446 map [n] **of avalanche courses** *plan.* (In France a graphic presentation of likely avalanches in a defined area which categorizes three avalanche risks: **1. white zone:** no risk presumed; ▶avalanche location map; **2. red zone:** very high avalanche risk sometimes with enormous snowmasses; **3. blue zone:** occurrence of small avalanches; *s* **plan** [m] **de zonas expuestas a aludes** (F., plan que contiene mapas con tres tipos de zonas de riesgo de aludes: **1. Zona blanca:** presumiblemente sin riesgo. **2. Zona roja:** muy peligrosa. **3. Zona azul:** expuesta a aludes de menor intensidad— y un informe; ▶mapa de inventario de aludes); *f* **plan** [m] **de zones exposées aux avalanches** (Plan comprenant des documents cartographiques à l'échelle de 1 : 2000 à 1 : 5000 qui indiquent les limites de trois zones : **1.** La **zone blanche** : présumée sans risques. **2.** La **zone rouge** estimée très dangereuse, où les avalanches sont particulièrement redoutables. **3.** La **zone bleue** : exposée à des avalanches de moindre intensité et un rapport de présentation ; circ. n° 74-201 du 05 décembre 1974 relative au plan des zones exposées aux avalanches [cf. CPCU] ; ▶carte d'inventaire d'avalanches) ; *g* **Lawinenrisikokarte** [f] (≠) (F., eine kartografische Darstellung im Maßstab 1:2000 bis 1:5000 von drei Lawinenwahrscheinlichkeiten in einem bestimmten Gebiet: „weiße Zone": kein Lawinenrisiko; „rote Zone": sehr hohes Lawinenrisiko mit z. T. gewaltigen Schneemassen; „blaue Zone": Auftreten kleinerer Lawinen; ▶Lawinengefahrenkarte).

map [n] **of trees** *adm. landsc. surv.* ▶cadastral map of trees.

3447 mapping [n] *landsc. pedol. phyt. zool.* (Precise documentation in cartographical form of information gathered and evaluated after a ▶field survey 1, e.g. ▶contours, land uses and such landscape elements as trees, soil types, vegetational units and fauna; ▶habitat mapping, ▶identification and mapping, ▶mapping of grassland vegetation, ▶site mapping, ▶soil mapping, ▶topographic mapping, ▶tree mapping, ▶urban habitat mapping); *s* **inventario** [m] **cartográfico** (Representación cartográfica de diferentes tipos de informaciones, como ▶curvas de nivel, usos, componentes del paisaje, árboles, tipos de suelo, unidades de vegetación, etc., sobre la base del ▶trabajo de campo y la correspondiente evaluación de los datos; hacer [vb] un inventario cartográfico, inventariar [vb]; ▶cartografía de árboles, ▶cartografía de suelos, ▶cartografía topográfica, ▶inventario, ▶inventario cartográfico de biótopos, ▶inventario de biótopos urbanos, ▶inventario de suelos [de una ubicación], ▶inventario de vegetación herbácea); *f* **cartographie** [f] (Mise en forme cartographique des ▶études préliminaires de terrain ou relevés effectués préalablement sur le terrain, p. ex. les éléments caractéristiques d'un paysage, arbres, groupe de sols, unités végétales, groupements faunistiques, occupation des sols, des données topographiques [p. ex. les ▶courbes de niveaux] ; ▶cartographie de formations herbacées, ▶cartographie des arbres, ▶cartographie des biotopes, ▶cartographie des biotopes urbaines, ▶cartographie des sols, ▶cartographie de[s] station[s]) ; *g* **Kartierung** [f] (Grafische Darstellung eines generalisierten Abbildes räumlicher Gegebenheiten auf Grund von vorausgegangenen ▶Feldaufnahmen und Auswertungen von Luftbildern; die Zusammentragung der topographischen Gegebenheiten, z. B. ▶Höhenlinien, Nutzungen, Landschaftsbestandteile, Bäume, Bodentypen, Vegetationseinheiten, Faunenbestände ergeben Karten, Pläne, ein Atlas oder ein andere Dokumentationsform; ▶Baumkartierung, ▶Biotopkartierung, ▶Bodenkartierung, ▶Grünlandkartierung, ▶Stadtbiotopkartierung, ▶Standortkartierung).

mapping [n], **biotope** *conserv. phyt. zool.* ▶habitat mapping.

mapping [n] **of existing land uses** *plan.* ▶survey of existing uses.

3448 mapping [n] **of forest functions** *for. nat'res. envir. recr.* (Survey and recording of forest areas according to the predominant ▶forest functions of component parts and their graphic representation in a plan, e.g. as a wood-producing forest, protective forest—water protection, erosion control, climatic amelioration, avalanche protection, pollution control, etc.—and recreation woodland; ▶mapping); *s* **inventario** [m] **de funciones del bosque** (Registro e inventariado de áreas de bosque según las funciones [▶función del bosque] predominantes del conjunto o de sus partes, como p. ej. bosque de explotación forestal, bosque protector, bosque recreativo; ▶inventario cartográfico; *término específico en España* monte [m] de utilidad pública); *f* **inventaire** [m] **des fonctions de la forêt** (±) (Recensement et synthèse cartographique de terrains boisés réalisés à partir de la ▶cartographie des fonctions prédominantes de la forêt telles que les forêts de protection — protection des eaux, protection contre l'érosion, amélioration climatique, protection contre les avalanches, protection contre la pollution — et la forêt récréative ou touristique ; ▶fonction de la forêt) ; *g* **Waldfunktionskartierung** [f] (Erfassung und grafische Darstellung von Waldflächen nach den vorherrschenden ▶Waldfunktionen der einzelnen Teile und deren Darstellung, z. B. als Produktionswald, Schutzwald — Wasserschutz, Erosionsschutz, Klimaschutz, Lawinenschutz, Immissionsschutz etc. — und Erholungswald; *OB* ▶Kartierung).

3449 mapping [n] **of grassland vegetation** *phyt.* *s* **inventario** [m] **de vegetación herbácea** *syn.* muestreo [m] de vegetación pratícola; *f* **cartographie [f] de formations herbacées** ; *g* **Grünlandkartierung [f]** (Grafische Darstellung der Grünlandflächen eines bestimmten Gebietes mit ihren Vegetationseinheiten).

mapping [n] of urban wild sites [US] *conserv. phyt. zool.* ►urban habitat mapping.

marginal agricultural site [n] *agr. (for.)* ►marginal soil.

marginal land [n] *agr. (for.)* ►marginal soil.

3450 marginal location [n] *plan.* (Situation at the edge of cities, industrial areas, mountains, forests, etc.); *s* **ubicación [f] periférica** (Área o superficie situada al borde de una ciudad, aglomeración, zona industrial, zona montañosa, espacio natural, etc.); *syn.* ubicación [f] limítrofe; *f* **situation [f] périphérique** (Abords, p. ex. d'une ville, de centres industriels, de massifs montagneux, etc.) ; *syn.* situation [f] marginale ; *g* **Randlage [f]** (Fläche oder Gebiet am Rande, z. B. von Städten, Industriezentren, Metropolregionen, von Gebirgen, Schutzgebieten, Wäldern etc.).

marginal moraine [n] *geo. obs.* ►terminal moraine.

3451 marginal protection area [n] *conserv. landsc.* (Protective zone to prevent ►intrusion upon the natural environment around the periphery of, e.g. a nature preserve, in France and Spain, but not usually part of the protected area in other countries); *syn.* perimeter protection zone [n]; *s 1* **Zona [f] Periférica de Protección [Es]** (Área limítrofe de un espacio natural protegido que se puede establecer para evitar ►intrusiones en el entorno natural, impactos ecológicos o paisajísticos procedentes del exterior. Cuando proceda, en la propia norma de creación, se establecerán las limitaciones necesarias; art. 37, Ley 42/2007); *s 2* **Área [f] de Influencia Socioeconómica [Es]** (Área que se puede establecer con el fin de contribuir al mantenimiento de los ►espacios naturales protegidos y favorecer el desarrollo socioeconómico de las poblaciones locales de forma compatible con los objetivos de conservación del espacio. Debe estar integrada, al menos, por el conjunto de términos municipales donde se encuentre ubicado el espacio natural de que se trate y su ►zona periférica de protección; art. 38, Ley 42/2007); *f 1* **périmètre [m] de protection** (Zone instituée par l'autorité administrative ayant pour objet la protection de l'environnement d'un ►site classé ou inscrit, dans un périmètre souvent assez large autour du site ou de la réserve proprement dit. Dans ce périmètre, des prescriptions sont établies pour les constructions et les grands travaux soumis à autorisation préalable ; depuis la loi n° 83-8 du 7 janvier 1983 cette disposition tend a être remplacée par les Z.P.P.A.U.P. ; *f 2* **zone [f] de protection** (Protection des abords d'une réserve naturelle réglementant toute intervention susceptible de dégrader la réserve et de porter ►atteinte au milieu naturel, établie par décret en Conseil d'État ; C. rural, art. L. 242-18) ; *g* **geschützte Randzone [f]** (*Frankreich und Spanien* äußerer Bereich eines NSG, LSG oder eines flächenhaften Kulturdenkmals, das nicht Bestandteil des Schutzgebietes ist, jedoch mit förmlichem Verfahren festgesetzt wird, um ►Eingriffe in Natur und Landschaft oder in das Ensemble zu vermeiden. Bauvorhaben in diesem Bereich sind beim zuständigen Ministerium genehmigungspflichtig).

3452 marginal soil [n] *agr. (for.)* (Agricultural land or soil, which is no longer cultivated because of its poor potential yield; ►arable land rank [US]/arable land grade [UK]; ►grassland yield index); *syn.* marginal land [n], marginal agricultural site [n]; *s* **suelo [m] marginal** (Suelo pobre en la zona agrícola marginal con ►índice de fertilidad de tierras de cultivo o ►índice de fertilidad de suelos [de prados] inferior a 25); *f* **terre [f] marginale** (Zone agricole défavorisée en raison de leur faible valeur culturale et qui n'est plus cultivée. Le problème de la mise en valeur des terres incultes est particulièrement préoccupant en France, et notamment en zones de montagne, où les conséquences de l'abandon des terres peuvent devenir désastreuses — avalanches, incendies, etc. ; LA 1981, 1094 ; ►coefficient de valeur agricole des prairies, ►indice d'estimation des sols) ; *syn.* terre [f] inculte ; *g* **Grenzertragsboden [m]** (Auf Grund seiner geringen wirtschaftlichen Ertragsfähigkeit nicht mehr bewirtschafteter landwirtschaftlicher Boden, da durch einen zu hohen Arbeitsaufwand die Erträge so niedrig sind, dass die Produktionskosten nachhaltig nicht gedeckt werden können; alle Flächen mit einer ►Ackerzahl oder ►Grünlandzahl unter 25); *syn.* Grenzertragsstandort [m], Grenzertragsfläche [f], Grenzboden [m].

marginal trees [npl] of a forest *for.* ►edge trees.

3453 marina [n] *recr.* (Italien origin; from latin *mare* >sea<; **1.** seaside promenade or esplanade. **2.** Pleasure haven providing secure moorings for sailing boots, yachts and motorboats; ►marina village, ►sailboat harbor [US]/sailing-harbour [UK], ►yachting habor [US]/yachting harbour [UK]); *s* **marina [f] [C]** (Puerto deportivo o centro de deportes acuáticos; en Es. este término no es usual; ►pueblo de marina, ►puerto de veleros, ►puerto de yates); *f* **marina [f]** (*De l'italien* ensemble touristique aménagé au bord de l'eau sur le littoral maritime ou les grands plans d'eaux, comportant un port de plaisance et dont les infrastructures offrent de nombreux services et activités de loisirs aux plaisanciers de passage ; la **m.** peut être un véritable ensemble balnéaire littoral, port de plaisance directement intégré dans un vaste complexe immobilier avec équipements de service, d'animation et de loisirs intégrés aux installations individuelles ou collectives [hôtels, restaurants, clubs, commerces, etc.] ayant vocation à héberger des touristes et des plaisanciers et dont la capacité d'accueil atteint souvent plusieurs milliers de lits ; les marinas sont nombreuses sur le littoral languedocien ; Port-Camargue avec ses 4436 places à flot réparties sur 70 hectares de plan d'eau est le plus grand port de plaisance d'Europe ; ►port [de voiliers] de plaisance, ►port de navires de plaisance, ►village marina) ; *g* **Marina [f]** (Italienisch, zu lat. *mare* >Meer<; **1.** moderner Siedlungstyp in unmittelbarer Meeresnähe und an großen Inlandsgewässern, der durch Fremdenverkehr und Freizeitaktivitäten entstanden und mit Hotels, Klubhäusern, Restaurants Boots- und Yachthafen sowie anderen Freizeiteinrichtungen ausgestattet ist; die **M.** mit dem größten europäischen Freizeithafen ist in Port-Camargue in Südfrankreich mit 4436 Liegeplätzen auf 70 ha Wasserfläche; ►Segelboothafen, ►Yachthafen. **2.** Das Siedlungsgebiet wird auch ►**Marinadorf** genannt).

3454 marina village [n] *urb. recr.* (Small community with a boat basin for safe anchorage in close vicinity to the sea); *syn.* marina community [n]; *s* **pueblo [m] marina** (Zona residencial o asentamiento turístico con puerto y otras instalaciones de recreo); *f* **village [m] marina (±)** (Petite enclave balnéaire fermée, aménagée au bord de l'eau, en général d'accès réglementée par terre et par mer intégrant port de plaisance, habitations et équipements de service et de loisirs) ; *g* **Marinadorf [n]** (Siedlungsgebiet mit Hotels, Klubhäusern, Restaurants, Boots- und Yachthafen sowie anderen Freizeiteinrichtungen in unmittelbarer Meeresnähe).

3455 marine biology [n] *bot.* (Science of living organisms in the sea); *s* **biología [f] marina** (Ciencia que estudia la vida del mar); *f* **biologie [f] marine** (Branche de l'hydrologie se consacrant à l'étude de la vie, du comportement, de la répartition et de la physiologie de la faune et de la flore marine ainsi que des interactions dans les chaînes trophiques) ; *g* **Meeresbiologie [f, o. Pl.]** (Wissenschaftszweig der Hydrobiologie, der sich mit dem Leben, Verhalten, der Verbreitung und der Physiologie meeresbewohnender Pflanzen und Tiere und deren Wechselbeziehungen befasst).

marine crest vegetation [n] *phyt.* ►glasswort vegetation.

Marine Environment [n] **and the Coastal Regions of the Mediterranean, Convention for the Protection of the** *conserv. land'man. leg.* ►Barcelona Convention.

marine forest [n] [US]**, protective** *conserv. land'man.* ►protective coastal woodland.

3456 marine pollution [n] *envir.* (Direct or indirect introduction of material by man into the sea, as a result of which negative effects occur, such as damage to flora and fauna or to the whole marine ecosystem; ►dumping at sea); *syn.* pollution [n] of the sea; *s* **contaminación** [f] **marina** (Vertido directo o indirecto de materiales extraños en los mares que pueden producir efectos nocivos sobre la vida de los organismos marinos y el equilibrio de los ecosistemas marinos; ►descarga de residuos en alta mar); *syn.* polución [f] marina, contaminación [f] del mar/de los océanos, polución [f] del mar/de los océanos; *f* **pollution** [f] **marine** (Résultat du rejet ou déversement dans la mer de matières nocives ou de déchets tels que principalement les rejets industriels et urbains à partir de la côte, les ►immersions des déchets en mer qu'ils soient industriels et radioactifs, le déversement opérationnel de résidus d'hydrocarbures, les déversements accidentels d'hydrocarbures responsables de la dégradation de la flore et de la faune sous-marine, des écosystèmes marins ou de la qualité des eaux de baignade) ; *syn.* pollution [f] des mers, pollution [f] des eaux de mer ; *g* **Meeresverschmutzung** [f] (Unmittelbare oder mittelbare Zuführung von Stoffen durch den Menschen in das Meer, aus der sich abträgliche Wirkungen wie eine Schädigung der Tier- und Pflanzenwelt oder des ganzen marinen Ökosystems ergeben kann; die **M.** erfolgt **1.** durch Schadstoffeintrag über die Luft resp. den Regen [saurer Regen, Stickoxide], **2.** durch Schadstoffeintrag über die Flüsse, **3.** durch ►Verklappung und **4.** durch den Schiffsverkehr wie z. B. bei Tankerunfällen oder durch Ablassen von Bilgenölen).

3457 maritime casualty [n] *envir. leg.* (Collision of ships, stranding or other accident of navigation, or other occurrence on board a ship or external to it, resulting in material damage or imminent threat of material damage to a ship or cargo; ICOP 1992; ►MARPOL Convention, ►shore oil pollution, ►oil pollution accident; ►oil tanker accident); *s* **accidente** [m] **marítimo** (Choque entre buques, varado de buque o cualquier otro tipo de accidente que puede conllevar daños en el barco o un derrame de los productos almacenados en él; ►accidente con derrame de hidrocarburos, ►accidente de petrolero, ►Convenio de MARPOL, ►marea negra); *f* **accident** [m] **en mer** (Abordage, échouement ou autre incident de navigation ou autre événement survenu à bord ou à l'extérieur d'un navire qui aurait pour conséquence soit des dommages matériels, soit une menace immédiate dont pourrait être victime un navire ou sa cargaison dans le cas d'un accident entraînant ou pouvant entraîner une pollution par les hydrocarbures ; ICOP 1992 ; ►accident entraînant une pollution par les hydrocarbures, ►Convention de Marpol, ►accident de navire pétrolier, ►marée noire) ; *syn.* accident [m] maritime ; *g* **Seeunfall** [m] (Schiffzusammenstoß, Stranden oder ein anderer nautischer Vorfall, ein sonstiges Ereignis an Bord oder außerhalb eines Schiffes, wodurch Sachschaden an Schiff oder Ladung entsteht oder unmittelbar zu entstehen droht; IÜÖ 1992; ►MARPOL-Abkommen, ►Ölpest, ►Ölverschmutzungsunfall, ►Tankerunfall).

3458 market gardening [n] *hort.* (Usually small-scale production of flowers, fruit, and vegetables, often for local consumption; ►commercial horticulture, ►plant nursery); *syn.* truck gardening [n] [also US]; *s* **jardinería** [f] (Empresa de horticultura de pequeña escala generalmente orientada al mercado local; ►em-

presa de horticultura, ►horticultura comercial); *f* **jardinerie** [f] (**1.** Établissement commercial, souvent de dimension importante, offrant tout ce qui concerne le jardin, principalement pour les résidences situées à proximité des grandes agglomérations ; il comporte généralement une surface d'exposition-vente, une pépinière, un parc de stationnement ; ►entreprise horticole, ►horticulture. **2.** Lorsque la part de la propre production de végétaux est importante on parle souvent d'une **pépinière jardinerie**) ; *g* **Betrieb** [m, o. Pl.] **einer Endverkaufsgärtnerei** (Produktion von unterschiedlichen Kulturpflanzen für den ortsnahen Bedarf, teilweise mit Verkauf von zugekauftem, vorkultiviertem Pflanzenmaterial, Pflanzenvermietung, evtl. mit Binderei und Pflege kleiner Gärten; eine Gärtnerei, die über 30 % der zu verkaufenden Waren zukauft, ist eine **Handelsgärtnerei**; ►Erwerbsgartenbau, ►Gartenbaubetrieb); *syn.* Betreiben/Führen [n, o. Pl.] einer Endverkaufsgärtnerei.

3459 marking [n] *constr.* (Determination by pegs, colo[u]r, lines, boards or labels, etc.; e.g. paved or painted lines delineating individual parking spaces in a parking lot [US]/car park [UK]); *s* **señalización** [f] (Marcación por medio de jalones, colores, líneas o señales indicadoras, p. ej. **s.** de plazas individuales de aparcamiento); *f* **marquage** [m] (Action de délimiter un espace au moyen de piquets, potelets, couleurs, lignes, panneaux, p. ex. la bande de guidage le long d'une route, l'emplacement de stationnement) ; *g* **Markierung** [f] (Kennzeichnung durch Pflöcke, Farbgebung, Linien, Schilder etc.; z. B. **M.** von Stellplätzen auf Parkflächen).

3460 marking [n] **of hiking trails** *recr. for.* (Identifying trails with signs or markers); *s* **señalización** [f] **de pasillos verdes** (Marcación de caminos, pistas de esquí de fondo, etc. por medio de pictogramas o carteles); *f* **balisage** [m] (Signalisation des chemins de randonnée au moyen de couleurs, signes, panneaux, pictogrammes, etc. ; *termes spécifiques relatifs au mobilier signalétique* panneau d'information, borne de sentier, borne thématique) ; *g* **Markierung** [f] **von Wanderwegen** (Kennzeichnung von Wegen, Loipen, Pisten etc. durch Schilder, Piktogramme etc.); *syn.* Kennzeichnung [f] von Wanderwegen, Wanderwegekennzeichnung [f], Wanderwegemarkierung [f].

3461 marl [n] *geo. pedol.* (**1.** Unconsolidated earthy deposits primarily composed of calcium carbonate and clay with 10-70% $CaCO_3$ content; an old term; used as a fertilizer on soils deficient in lime; ►limestone; ►till [US]/boulder-clay [UK], ►marly limestone, ►sedimentary rock); **2. marlstone** [n] (Indurated or hardened rock of similar composition as marl, i.e. an earthy or impure argillaceous limestone); *syn. to 2.* marlite [n]; *s* **marga** [f] (►Roca sedimentaria constituida por una mezcla de ►caliza [CO_3Ca: carbonato cálcico] y arcillas; ►till, ►marga caliza); *syn.* marna [f] [RA]; *f* **marne** [f] (►Roche sédimentaire clastique meuble, composite argilo-carbonaté contenant 35 à 65 % d'argile souvent illitique et 35 à 65 % de $CaCO_3$ [►roche calcaire] ; DIS 1986 ; ►argile à blocaux, ►marne calcaire) ; *g* **Mergel** [m] (**1.** ►Sedimentgestein, vorwiegend aus ►Kalkgestein und Ton zusammengesetzt, mit 10-70 % $CaCO_3$-Gehalt; der eigentliche **M.** verfügt über 35 % Kalk [Calcit] und 65 % Ton; ►Kalkmergel enthält bis zu 65 % Calcit und 35 % Ton; **Tonmergel** hingegen nur 25 % Calcit und 75 % Ton; **Gips-M**. besteht zu einem großen Teil aus Gips; ►Geschiebemergel. **2.** Verfestigte **M.** sind z. B. **Mergelschiefer**).

marl [n]**, lime** *pedol.* ►marly limestone.

marlite [n] *geo. pedol.* ►marl.

marlstone [n] *geo. pedol.* ►marl.

3462 marly limestone [n] *pedol.* (Soil developed from deposited, stratified or sedimentary rock, containing up to 65% calcite and 35% clay); *syn.* lime marl [n]; *s* **marga** [f] **caliza**

(Suelo muy rico en cal, desarrollado a partir de rocas sedimentarias, y que contiene hasta un 65% de calcita y un 35% de arcilla); *f* **marne [f] calcaire** (Roche sédimentaire argilo-carbonatée meuble contenant jusqu'à 65 % de calcite [spath] et 35 % d'argile) ; *syn.* calcaire [m] marneux (PED 1983, 6) ; *g* **Kalkmergel [m]** (Sehr kalkreicher Boden, der sich auf Ablagerungs-, Schicht- oder Sedimentgesteinen entwickelt hat und bis zu 65 % Calcit [Kalkspat] und 35 % Ton enthält).

3463 MARPOL Convention [n] *envir. leg.* (Common name for the International Convention for the Prevention of Pollution from Ships, which was adopted under the auspices of the International Maritime Organization [IMO] in 1973 and became operational in 1983. The Convention contains five annexes: I. pollution by oil, II. pollution from noxious liquid substances, III. pollution from hazardous waste in packaged form, IV. sewage disposal, and V. pollution by dumping of refuse, which includes all kinds of victual, domestic or operational waste; cf. DED 1993; ►marine pollution); *s* **Convenio [m] de MARPOL** (Acuerdo internacional para la prevención de la ►contaminación marina provocada por vertidos desde buques y aeronaves, firmado por primera vez el 15 de febrero de 1972, en España ratificado por Instrumento de 19 de febrero de 1973 y Protocolo de enmienda de 2 de marzo de 1983 al que España se adhirió por Instrumento de 15 de junio de 1989. El Convenio tiene cinco anexos: **1.** contaminación por hidrocarburos, **2.** contaminación por líquidos tóxicos, **3.** contaminación por residuos tóxicos embalados, **4.** flujo de efluente y **5.** vertido de residuos sólidos de cualquier tipo); *f* **Convention [f] de Marpol** (Convention relative à la ►pollution marine par les hydrocarbures, entrée en vigueur le 2 octobre 1983 et instituant le principe d'une interdiction générale avec certaines exceptions du déballastage à la mer du mélange eau-hydrocarbures contenu dans les citernes des navires pétroliers ; elle précise les règles de prévention de la pollution par les ordures des navires et définit les conditions de sécurité et de contrôle dans le transport maritime pétrolier) ; *g* **MARPOL-Abkommen [n]** (Internationales Übereinkommen vom 02.11. 1973 zur Verhütungung der Meeresverschmutzung durch Schiffe und Protokoll vom 17.02.1978 zu diesem Übereinkommen. Das Abkommen wurde in London am 04.03.1974 und das dazugehörige Protokoll am 17.02.1978 von der Bundesrepublik Deutschland unterzeichnet. Beide Texte traten für die BRD am 02.10.1983 in Kraft [BGBl. 1983 II 632]. Seit 1995 gibt es eine Neufassung der amtlichen deutschen Übersetzung zum Übereinkommen von 1973 und Protokoll von 1978 in der seit 15.06.1995 gültigen Fassung [BGBl. 1996 II 399]. Das Abkommen enthält fünf Anhänge über **I.** Regeln zur Verhütung der Verschmutzung durch Öle; **II.** Regeln zur Überwachung der Verschmutzung durch als Massengut beförderte, schädliche flüssige Stoffe; **III.** Regeln zur Verhütung der ►Meeresverschmutzung durch Schadstoffe, die auf See in verpackter Form befördert werden; **IV.** Regeln zur Verhütung der Verschmutzung durch Schiffsabwässer und **V.** Regeln zur Verhütung der Verschmutzung durch Schiffsmüll).

marry finish grade with [loc] [UK] *constr.* ►make a finished grade flush with [US].

marrying with existing levels [loc] [UK] *constr.* ►meeting finished grade [US].

3464 marsh [n] *geo. phyt.* (A periodically wet or continually flooded area where the surface is not deeply submerged; covered dominantly with sedges, cattails, rushes or other hydrophytic plants, essentially without the formation of peat; *specific terms* ►coastal marshland, ►cordgrass marsh, freshwater marsh, saltwater marsh [►salt marsh], ►tall forb reed marsh [US]/tall forb reed swamp [UK]); *s* **marisma [f]** (Terreno un poco por debajo del nivel del mar que se inunda con las aguas de la creciente marina y de los ríos, o de ambas a la vez; DGA 1986; *término específico* marjal [m]; ►marisma de espartina, ►marisma marítima, ►marisma salina, ►pantano megafórbico); *f* **marais [m]** (Terres basses inondées en permanence et envahie par la végétation en zone littorale par les eaux de mer [marais littoral], dans les estuaires par les eaux douces ou saumâtres [marais d'estuaire], en bordure des lacs [marais lacustre] ; est également une étendue d'eau stagnante peu profonde due à un affleurement permanent de la nappe phréatique [marais continental] ; ►marais maritime) ; *syn.* molière [f], marais [m] mouillé ; *g* **Sumpf(land) [m/(n)]** (**1.** Ein periodisch nasses oder wiederholt flach überflutetes Gebiet, das hauptsächlich mit Seggen, Rohrkolben, Binsen und anderen Nässe liebenden Pflanzen bewachsen ist und keine Torfbildung aufweist. **2.** ►Marsch); *syn.* Sumpfgebiet [n] (1).

marsh [n]**, coastal** *geo.* ►coastal marshland.

marsh [n]**, sea-** *geo. phyt.* ►salt marsh.

marsh [n] [US]**, tidal** *geo.* ►coastal marshland, ►salt marsh.

marsh plant [n] *bot.* ►helophyte.

marsh soil [n] *pedol.* ►coastal marsh soil.

3465 masonry [n] *arch. constr.* (Art of shaping, arranging, and uniting stone, brick, building blocks, etc., to form walls and other parts of a building; DAC 1975. Also construction composed of shaped or molded units; *specific terms* brickwork, **m.** without mortar is ►dry masonry, according to how the stone is dressed or installed there is ►broken range work, ►coursed ashlar masonry, and ►coursed dressed ashlar masonry, ►double-sided masonry wall, ►horizontally-coursed masonry, ►irregular coursed ashlar masonry, ►natural stone masonry; masonry with especially large natural stones is a ►cyclopean wall; the following are distinguished according to the position in the ground: ►masonry foundation wall and ►above-ground masonry; a ►retaining wall, the inner side of which is constructed against undisturbed subgrade is ►one-sided masonry; one-tier masonry [US]/one-leaf masonry [UK], *further specific terms* ►polygonal masonry, ►random rubble ashlar masonry, ►regular coursed masonry of natural stones, ►rubble masonry with pinned joints, ►split-face dry masonry, ►squared rubble masonry, ►two-tier masonry [US]/two-wythe masonry [UK], ►veneer masonry, ►brick veneer); *syn.* masonry construction [n], stonework [n]; *s* **mampostería [f]** (Término genérico para las construcciones de muros o paredes con o sin mortero; ►fábrica de sillarejo, ►fábrica de sillarejo en hiladas, ►mampostería aparejada en hiladas irregulares, ►mampostería aparejada en hiladas regulares, ►mampostería ciclópea, ►mampostería compacta, ►mampostería con paramento de cara vista bilateral, ►mampostería con paramento de ladrillo, ►mampostería de cara vista [unilateral], ►mampostería de dos capas, ►mampostería de piedras de corte natural, ►mampostería de piedras naturales, ►mampostería enchapada, ►mampostería en hiladas, ►mampostería en hiladas horizontales, ►mampostería enripiada, ►mampostería en seco, ►mampostería ordinaria [de piedra partida], ►mampostería poligonal, ►sillería almohadillada, ►sillería aparejada con mampuestos naturales, ►sillería de fábrica, ►sillería ordinaria, ►sillería ordinaria con mampuestos de labra tosca); *syn.* fábrica [f] de mampostería, sillería [f], albañilería [f]; *f* **maçonnerie [f]** (Ouvrage réalisé par la juxtaposition de matériaux solides : pierres, briques ou agglomérés, liés ou non entre eux, et formant un ensemble stable, de forme et de dimensions déterminées ; MAÇ 1981, 5 ; *termes spécifiques* ►maçonnerie à appareillage à l'anglaise, ►maçonnerie à appareillage cyclopéen, ►maçonnerie à appareillage en assises, ►maçonnerie à appareillage polygonal, ►maçonnerie à deux parements, ►maçonnerie à double paroi, ►maçonnerie à parement, ►maçonnerie à parement en briques,

▶maçonnerie à parois simple, ▶maçonnerie à sec, ▶maçonnerie de moellons assisés à parement bossagé, ▶maçonnerie de moellons assisés d'égales hauteurs, ▶maçonnerie de moellons assisés finement équarris, ▶maçonnerie de moellons de taille rugueuse, ▶maçonnerie de moellons assisés d'inégales hauteurs, ▶maçonnerie en élévation, ▶maçonnerie en moellons à opus quadratum, ▶maçonnerie en moellons assisés, ▶maçonnerie en moellons équarris irrégulier, ▶maçonnerie en moellons pleins, ▶maçonnerie en moellons pleins, grossièrement équarris, ▶maçonnerie en moellons pleins, grossièrement équarris, à joints rocaillés, ▶maçonnerie en pierres bossagées, ▶maçonnerie en pierres naturelles, ▶maçonnerie paysagère) ; *g* **Mauerwerk [n]** (Baukörper als Gefüge aus künstlichen oder natürlichen Steinen, mit oder ohne Mörtel errichtet, „gemauert"; ein **M.** ohne Mörtel ist ein ▶**Trockenmauerwerk**; je nach Art und Weise der Bearbeitung der Mauersteine gibt es das **Backsteinmauerwerk**, ▶**Bruchsteinmauerwerk**, **Feldsteinmauerwerk**, **Findlingsmauerwerk**, ▶**Quadermauerwerk**, **Werksteinmauerwerk**; ein **M.** aus besonders großen Natursteinen heißt ▶**Zyklopenmauerwerk**; je nach Lage des **M.es** unterscheidet man zwischen **Fundamentmauerwerk** [▶Grundmauer] und ▶**aufgehendes Mauerwerk**; eine ▶**Stützmauer**, die mit einer Seite gegen den gewachsenen Boden stößt, ist ein ▶**einhäuptiges Mauerwerk**; *weitere UBe* ▶aufgehendes Mauerwerk, ▶ausgezwicktes Bruchsteinmauerwerk, ▶Bossenwerk, ▶einhäuptiges Mauerwerk, ▶einschaliges Mauerwerk, ▶hammerrechtes Schichtenmauerwerk, ▶hammerrechtes unregelmäßiges Bruchsteinmauerwerk, ▶lagerhaftes Mauerwerk, ▶Natursteinmauerwerk, ▶Polygonalmauerwerk, ▶Quadermauerwerk, ▶Raumauerwerk, ▶regelmäßiges Schichtenmauerwerk, ▶Schichtenmauerwerk, ▶Schichtenmauerwerk mit bossierten Steinen, ▶unregelmäßiges Schichtenmauerwerk, ▶Verblendmauerwerk, ▶Wechselmauerwerk, ▶Wechselmauerwerk mit bossierten Steinen, ▶Ziegelverblendmauerwerk, ▶zweihäuptiges Mauerwerk, ▶zweischaliges Mauerwerk).

masonry [n], above-grade masonry *constr.* ▶above-ground masonry.

masonry [n], bossage *arch. constr.* ▶rusticated masonry.

masonry [n], coursed dressed ashlar *arch. constr.* ▶coursed dressed ashlar masonry.

masonry [n] [UK], one-leaf *constr.* ▶one-tier masonry [US].

masonry [n], quarry faced *arch. constr.* ▶rusticated masonry.

masonry [n], random embossed ashlar *arch. constr.* ▶random rough-tooled ashlar masonry.

masonry [n], range *arch. constr.* ▶coursed ashlar masonry.

masonry [n], rubble *constr.* ▶squared rubble masonry.

masonry [n] [UK], single-wythe *constr.* ▶one-tier masonry [US].

masonry [n] [UK], two-leaf *constr.* ▶two-tier masonry [US].

masonry [n] [UK], two-wythe *constr.* ▶two-tier masonry [US].

3466 masonry bond [n] *arch. constr.* (Arrangement of masonry units created by a setting pattern which laps units one over the other; ▶header bond pattern, ▶stretcher bond pattern, ▶masonry); *s* **aparejo [m] de mampuestos** (En albañilería, diferentes posiciones en las que se asientan las piedras o ladrillos en una fábrica, formando hiladas yuxtapuestas [▶aparejo a tizón y ▶aparejo a soga], en prevención de que las llagas o juntas verticales no caigan una sobre otra; cf. DACO 1988; ▶mampostería; *término específico* aparejo de ladrillos); *syn.* trabazón [m] de mampuestos; *f* **appareillage [m] d'une maçonnerie** (Assemblage des pierres de taille ou de moellons, disposition des briques [▶appareillage en boutisses, ▶appareillage à panneresses] dans une ▶maçonnerie pour former un dessin d'appareil mis en relief par le type de joints ; *termes spécifiques* appareillage à l'anglaise, appareillage à la française) ; *syn.* appareil [m] d'un mur ; *g* **Mauersteinverband [m]** (...verbände [pl]; Art und Weise der Zusammenfügung von Mauersteinen zu einem ▶Mauerwerk; man unterscheidet **M.verbände** für Ein-Steinwände [▶Binderverband, Blockverband und Kreuzverband] und solche für ½-Steinwände [▶Läuferverband und viele Zierverbände, die vielfach nach den Regionen, in denen sie hauptsächlich vorkommen, bezeichnet werden, z. B. Flämischer Verband, Gotischer Verband, Holländischer Verband, Märkischer Verband/Wendischer Verband]; *UB* Klinkerverband, Ziegelverband, Zierverband); *syn.* Verband [m] im aufgehenden Mauerwerk, Mauerwerksverband [m], Mauerverband [m].

masonry construction [n] *constr.* ▶masonry.

3467 masonry foundation wall [n] *constr. eng.* (Part of a wall foundation, made of stones or nowadays of concrete, which serves to transmit the load of the structure to the subbase [US]/sub-base [UK], and which is usually below ground level; ▶above-ground masonry); *s* **muro [m] de cimentación** (Muro de piedra o de hormigón sobre el subsuelo que soporta la construcción transmitiendo su peso al subsuelo; ▶muro en elevación); *syn.* muro [m] de fundación; *f* **mur [m] de cave** (MAÇ 1981, 325 ; mur de soutènement adossé au terrain, en pierre ou en béton constituant le support de l'ouvrage sur lequel reposent les ▶murs en élévation) ; *syn.* mur [m] de fondation ; *g* **Grundmauer [f]** (In der Erde liegende, aus festen Steinen oder heute meist aus Beton bestehende Mauer, die die Last des Bauwerkes auf den Untergrund überträgt; ▶aufgehendes Mauerwerk); *syn.* Fundamentmauer [f], Fundationsmauer [f].

masonry wall [n] *constr.* ▶double-sided masonry wall.

masonry wall [n], rough faced *constr.* ▶split-face dry masonry.

masonry wall base [n] *arch. constr.* ▶bottom of wall base.

masonry [n] with pinned joints *constr.* ▶rubble masonry with pinned joints.

3468 mass death [n] of fish *envir. zool.* (Large scale death of fish as a result of water pollution or disease. In the 1990s on the eastern seaboard of the United States, the mass death of fish was caused by *Pfiesteria piscicida*, which was able to spread due to the massive and uncontrolled introduction of nutritive substances from pig and poultry farming; *syn.* massive fish mortality [n], mass die-offs [npl] of fish; *s* **mortandad [f] acusada de peces** (Muerte masiva de peces causada por la contaminación, por contenido muy reducido de oxígeno del agua o por epidemia); *syn.* muerte [f] masiva de peces; *f* **mortalité [f] piscicole importante** (Mortalité de poissons provoquée par une pollution massive aiguë, une dystrophisation des cours d'eau ou en période de sècheresse à une rupture d'écoulement et la formation d'assecs) ; *g* **Fischsterben [n] (2)** (Vor allem durch Sauerstoffmangel auf Grund starker Gewässerbelastung, z. B. durch Verschmutzung, Vergiftung durch eingeschwemmte Chemikalien oder Infektionskrankheit verursachtes massenhaftes Sterben von Fischen. An der Ostküste der USA wurde in den 1990er-Jahren das **F.** durch *Pfiesteria piscicida* verursacht, da sich diese Krankheit bei massivem und unkontrolliertem Nährstoffeintrag durch Abwassereinleitungen großer Schweine- und Geflügelbetriebe schnell ausbreiten konnte).

mass die-offs [npl] of fish *envir. zool.* ▶mass death of fish.

massed planting [n] *constr. hort.* ▶mass planting.

mass effect [n] *zool.* ▶crowding effect.

mass erosion [n] [UK] *geo.* ▶mass slippage [US].

M

massif [n] *geo.* ▶mountain range.

3469 massive bedrock [n] *constr.* (*Soil classification for earthworks*) solid rock with few fissures or little weathering, which has a high structural resistance and internal mineral bonding; e.g. unweathered compacted argillaceous slate, layers of nagelfluh conglomerates); *s* **subsuelo** [m] **de roca** (En terracerías, clase de subsuelo de roca poco fisurada o meteorizada con gran resistencia estructural y cohesión mineral interna; también se denominan así capas compactas de esquisto arcilloso o de escoria de minería); *f* **sol** [m] **rocheux** (Classe de sols [classe R] dans la classification des matériaux de terrassement, issus de roches sédimentaires, magmatiques et métamorphiques caractérisée par la nature pétrographique des roches rencontrées ainsi que son état et ses caractéristiques mécaniques ; cf. Norme N.F. P 11 300) ; *syn.* matériaux [mpl] rocheux ; *g* **schwer lösbarer Fels** [m] (*Bodenklasse im Erdbau nach DIN 18 300* nur wenig klüftige oder verwitterte Felsarten mit hoher Gefügefestigkeit und innerem, mineralisch gebundenem Zusammenhalt, auch unverwitterte, fest gelagerte Tonschiefer, Nagelfluhschichten, Hüttenschlackenhalden); *syn. obs.* schwerer Fels [m].

massive fish mortality [n] *envir. zool.* ▶mass death of fish.

mass movement [n] [UK] *geo.* ▶mass slippage [US].

3470 mass planting [n] *constr. hort.* (Profuse grouping of a single species or variety); *syn.* massed planting [n]; *s* **plantación** [f] **masiva** (Plantación en superficie grande con plantas de una sola especie); *f* **plantation** [f] **en grand massif** (Plantation d'un grand nombre d'espèces ou de variétés végétales sur une grande surface) ; *g* **Massenpflanzung** [f] (Großflächige Pflanzung mit einer Art); *syn.* flächenhafte Pflanzung [f].

3471 mass slippage [n] [US] *geo.* (Generic term for all forms of downslope movement of rock or soil material; ▶debris slide, ▶debris avalanche, ▶earthflow, ▶ground settlement, ▶landslide, ▶mud flow, ▶rock creep, ▶rockfall/rock-fall [UK], ▶rockslide [US]/rock-slide [UK], ▶soil creep 1, ▶soil creep 2, ▶solifluction, ▶subsidence, ▶talus creep); *syn.* mass wasting [n] (BIS 1982, 10), mass erosion [n] (BIS 1982, 10), mass movement [n] [UK]; *s* **movimiento** [m] **gravitacional** (Tipo de desplazamiento de rocas o formaciones superficiales cuya única causa de movimiento es la gravedad. En ellos el material transportado es el único que interviene, sin que pueda definirse un agente geológico de transporte [hielo, viento, etc.]. Pueden ser **mm. gg. lentos** [▶reptación hídrica del suelo, ▶solifluxión, ▶reptación de detritus de talud, ▶reptación de rocas, ▶selifluxión, ▶asentamiemto y ▶subsidencia] y **mm. gg. rápidos** [▶alud de tierras, ▶colada de barro, ▶avalancha de rocas, ▶deslizamiento de rocas, ▶corrimiento de tierras, ▶deslizamiento planar, ▶deslizamiento de detritos, ▶caída de rocas y desplome]; cf. DINA 1987); *syn.* movimientos [mpl] en masa, movimientos [mpl] de derrubios, coluvionamientos [mpl], procesos [mpl] de vertiente; *f* **mouvement** [m] **de masse** (Terme générique qualifiant toutes les formes de déplacement des roches et des sols ; ▶avalanche de pierres, ▶chute de pierres, ▶coulée de terre, ▶éboulement continu, ▶gélifluxion, ▶glissement d'éboulis, ▶glissement de montagne, ▶glissement de terrain, ▶lave torrentielle, ▶reptation d'éboulis, ▶reptation d'un versant, ▶rupture de terrain, ▶solifluxion, ▶subsidence, ▶tassement) ; *syn.* déplacement [m] de masse, (LA 1981, 477 et s.) ; *g* **Massenversatz** [m, o. Pl.] (1. OB zu flach- oder tiefgründiges Abgleiten von Bodenmassen am Hang oder an Erdbauwerken. 2. Form der Umlagerung von Bodenkörpern in der Landschaft. Man kann vier Formen des **M.es** unterscheiden: **a.** langsam fließende Formen wie ▶Hangkriechen [soil creep 2], ▶Bodenfließen [soil creep 1], ▶Schuttgleitung [talus creep], ▶Versatzbewegung von Einzelblöcken [rock creep] und ▶Solifluktion [solifluction], **b.** schnell fließende Formen wie ▶Erdschlipf [earthflow], ▶Mure [mud flow], ▶Steinlawine [debris avalanche], **c.** ▶Bergrutsch [rockslide], ▶Hangrutschung [landslide], ▶Geländebruch [shear failure of embankment], ▶Schuttrutschung [debris slide], ▶Steinschlag [rockfall] und **d.** ▶Senkung [ground settlement]); *syn.* Bodenrutschung [f], Massenbewegung [f].

mass sport [n] [US] *recr.* ▶participatory sport [US].

3472 mass tourism [n] *recr. trans.* (Tourist activities in which a large proportion of the population participates due to rising incomes, fewer working hours, longer vacation [US]/holidays [UK] and a diverse wide-ranging transportation network; ▶tourism 1); *s* **turismo** [m] **de masas** (Viajes turísticos organizados por touroperadores con oferta de programa común para muchas personas; resultado del crecimiento de los ingresos y del aumento de las vacaciones de la gran mayoría de la población en los países industrializados; ▶turismo); *f* **tourisme** [m] **de masse** (Activité de détente pratiquée par de larges couches de la population consécutive à l'amélioration du pouvoir d'achat, la réduction du temps de travail, l'accroissement du nombre et de la qualité des équipements touristiques et des infrastructures de transport ; ▶tourisme) ; *g* **Massentourismus** [m, o. Pl.] (Reiseverkehr, an dem ein Großteil der Bevölkerung auf Grund steigender Einkünfte, sinkender Arbeitszeit resp. längeren Urlaubs und einer dicht ausgebauten und vielfältigen Verkehrsinfrastruktur teilnimmt; ▶Fremdenverkehr).

mass transit [n] [US] *trans.* ▶local public transportation system [US]/local transport system [UK].

mass transit vehicle [n] [US] *trans.* ▶public transportation vehicle.

mass wasting [n] [UK] *geo.* ▶mass slippage [US].

mast [n]**, mobile phone** *urb.* ▶mobile phone antenna.

3473 Master [n] **of Architecture (M.ARCH)** [US] *prof.* (In U.S., a graduate degree given at universities with programs accredited by the U.S. Department of Education and by the National Architectural Accrediting Board [NAAB], representing the American Institute of Architects [AIA], Association of Collegiate Schools of Architecture [ACSA], National Council of Architectural Registration Boards [NCARB]); *syn.* Master of Arts (MA) [UK]; *s 1* **arquitecto/a** [m/f] **licenciado/a** (±) (Es. la licenciatura es el título académico para personas que se han graduado en una universidad después de realizar cinco años de estudios; dentro de las facultades universitarias de arquitectura se puede realizar la especialización en urbanismo y en paisajismo que se terminan con el título de Master en Urbanismo o ▶Master en Paisajismo); *s 2* **ingeniero/a** [m/f] **diplomado/a en arquitectura** (Título académico que reciben las personas que han cursado estudios de arquitectura en una Universidad Politécnica); *f 1* **Master** [m] **d'Architecture** (F., titre académique ; les études d'architecture s'inscrivent désormais dans le schéma européen d'harmonisation des cursus d'enseignement supérieur et sont structurées en trois cycles sur la base du L M D (licence, master, doctorat) ; un premier cycle d'études d'une durée de 3 ans mène au diplôme d'études en architecture conférant le **grade de licence** ; un deuxième cycle d'études d'une durée de 2 ans mène au diplôme d'État d'architecte conférant le **grade de master** ; celui-ci permet d'entrer dans un cabinet d'architecte ou d'exercer en tant qu'architecte-conseil toutefois ce diplôme ne permet ni d'ouvrir un cabinet ni de signer de contrats de maîtrise d'œuvre ; la capacité à exercer la maîtrise d'œuvre requiert une année de formation théorique et un stage obligatoire ; un troisième cycle d'études d'une durée de 3 ans mène au doctorat d'architecture conférant le **grade de docteur** ; l'architecte porte alors le titre DPLG [diplômé par le gouvernement], DESA pour le diplôme de l'Ecole spéciale d'architecture à Paris [école privée reconnue par l'Etat] et DENSAIS pour l'Ecole nationale supérieure des arts et industries de Strasbourg [école publique placée sous l'égide du Ministère de l'Éducation] ; *f 2* **ingénieur [m] diplômé en architecture (±)** (**D.**, Grade académique obsolète délivré à la fin des études d'architecture [formation en 8 semestres minimum, stage pratique de 6 mois] dans les universités techniques) ; *g* **Diplomingenieur [m] für Architektur/Master [m] of Science in Architektur** (*Abk. in D.* Dipl.-Ing., *in A. auch* DI; akademischer Grad, den [technische] Universitäten nach Abschluss eines mindestens achtsemestrigen wissenschaftlich-theoretischen Studiums und eines sechsmonatigen Praktikums verleihen. Seit der „Bologna-Erklärung" haben die Europäischen Bildungsminister eine einheitliche europaweite Umstellung der Studienabschlüsse auf **Bachelor** und **Master** bis 2010 festgeschrieben. Zur Qualifizierung gibt es Aufbau-, Ergänzungs- und Zusatzstudiengänge, die direkt im Anschluss an einen ersten 8-semestrigen, berufsqualifizierenden Abschluss *Bachelor of Science* oder *Bachelor of Arts* in einem Studiengang Architektur oder konsekutiv resp. als Weiterbildungsstudiengang für Diplomingenieure postgraduell nach einer Phase der Berufspraxis absolviert werden können und mit dem akademischen Grad eines **Master of Science [M.Sc.]** in unterschiedlichen Sparten abschließen; z. B. **M. Sc. in Architektur, M. Sc. in Stadtplanung, M. Sc. in Europäischer Urbanistik, M. Sc. in Real Estate Management, M. Sc. in Bauen im Bestand**; die Verleihung des universitären **Dipl.-Ing.-Titels** wird bis 2010 aufgegeben).

Master [n] of Arts (MA) [UK] *prof.* ▶Master of Architecture (M.ARCH) [US].

Master [n] of Arts in Landscape Design (MALD) **[UK]** *obs. prof.* ▶Master of Landscape Architecture (MLA).

Master [n] of Fine Arts in Landscape Architecture (MFLA) [US] *prof.* ▶Master of Landscape Architecture (MLA).

3474 Master [n] of Landscape Architecture (MLA) *prof.* (**In U.S.**, graduate university degree accredited by the U.S. Department of Education and by the Council of Landscape Architectural Registration Boards [CLARB] of the American Society of Landscape Architects [ASLA]; many graduate schools offer landscape architecture at a graduate level leading to master degrees [MS, MA] as well as PhD programs in landscape architecture; **in U.K.**, the following universities offer accredited masters courses leading to an MA in Landscape Architecture; e.g. Birmingham City University; Edinburgh College of Art; Leeds Metropolitan University; University of Greenwich; University of Sheffield. Former degree titles were known as Master of Arts in Landscape Design [MALD], Master of Philosophy in Landscape Design [MPhilLD], Diploma in Landscape Design [DipLD]; ▶Bachelor of Landscape Architectur [BLA]; **in D.**, since 2005, several universities offer an international masters course; e.g. the International Master's Programme in Landscape Architecture [known as the IMLA programme for short] at three higher education institutions: Nürtingen-Geislingen University, Baden-Württemberg; Weihenstephan University of Applied Sciences, Bavaria, both in Germany, in association with the University of Applied Sciences, Rapperswil/St. Gallen, Switzerland; **in A.** and **D.**, academic graduate qualification in landscape architecture/landscape planning); *syn.* Master [n] of Fine Arts in Landscape Architecture (MFLA) [US], Master [n] of Landscape Design (MLD) [UK], Diploma [n] in Landscape Architecture (DipLA) [UK]; *s 1* **paisajista [m/f] diplomado/a** (En Es. título no homologado que reciben las personas que han terminado su carrera en una institución académica privada, ya que no existe oficialmente la carrera universitaria de paisajismo); *s 2* **Master [m] en Paisajismo** (Título concedido a personas que han realizado estudios de

posgrado en esta especialidad en una Facultad de Arquitectura); *s 3* **ingeniero/a [m/f] de planificación del paisaje y de espacios libres** (En A. y D. título académico superior concedido después de estudios universitarios de cuatro años [como mínimo] y seis meses de prácticas; ▶ingeniero/a técnico/a de planificación del paisaje y de espacios libres); *f 1* **Master [m] en architecture des paysages** (**US.**, grade universitaire délivré par le Ministère de l'Education, par la commission d'admission de l'ordre national des architectes paysagistes [CLARB] et l'union des architectes paysagistes [ASLA]) ; *f 2* **paysagiste [m/f] D.P.L.G.** (**F.**, les paysagistes diplômés par le gouvernement [D.P.L.G.] sortent de l'École Nationale Supérieure du Paysage à Versailles, des Écoles Nationales Supérieures d'Architecture et de Paysage de Bordeaux et de Lille ; les études s'effectuent en quatre ans ; les écoles sont accessibles, par voie de concours commun, à tout étudiant titulaire d'un diplôme de niveau Baccalauréat + 2 et n'offre chaque année qu'une centaine de places ; certains concours ou appels d'offres sont réservés seulement aux paysagistes D.P.L.G. tout comme l'accès à certains postes administratifs ; ▶paysagiste, ▶ingénieur paysagiste) ; *syn.* paysagiste [m/f] diplomé, ée ; master [m] architecte paysagiste [B] ; *f 3* **paysagiste [m/f] diplômé, ée** (**F.**, les **p. d.** sortent des diverses écoles supérieures spécialisées dans l'art et les techniques des jardins et du paysage auxquels sont délivrés les titres correspondants, p. ex. Ingénieur des techniques horticoles et paysagères [École nationale supérieure des techniques horticoles et paysagères], paysagiste diplômé en architecture des jardins [École supérieure d'architecture des jardins] ; durée des études 3 à 4 ans) ; *g 1* **Diplomingenieur/-in [m/f] für Landschaftsarchitektur** (*Abk. in D.* Dipl.-Ing., *in A. auch* DI; akademischer Grad, den [technische] Universitäten nach Abschluss eines mindestens achtsemestrigen wissenschaftlich-theoretischen Studiums und eines sechsmonatigen Praktikums verliehen haben. In D. verleiht nur noch die Technische Universität Dresden den Titel *Diplomingenieur Landschaftsarchitektur* nach einem 10-semestrigen Studium, dem ein 12-wöchiges Vorpraktikum vorausgehen und bei dem ein 12-wöchiges Zwischenpraktikum absolviert werden muss. Demnächst wird auch in Dresden auf den Bachelor- und Masterstudiengang umgestellt; die alten Abschlussbezeichnungen waren: TU Berlin *Dipl.-Ing. für Landschaftsplanung*, TU Dresden s. o., Universität Hannover *Dipl.-Ing. für Landschafts- und Freiraumplanung*, TU München, Freising-Weihenstephan *Dipl.-Ing. für Landschaftsarchitektur und Landschaftsplanung*, GH Kassel [Gesamthochschule mit integriertem Studiengang] *Dipl.-Ing. für Landschaftsplanung* und an der Universität Rostock *Dipl.-Ing. für Landeskultur und Umweltschutz*); *syn. obs.* Diplomingenieur/-in [m/f] für Landespflege; *g 2* **Master [m] of Science Landscape Architecture** (**1.** 1999 hatten mit der „Bologna-Erklärung" die Europäischen Bildungsminister eine einheitliche europaweite Umstellung der Studienabschlüsse auf **Bachelor** und **Master** bis 2010 festgeschrieben. Deshalb wird die Verleihung des universitären **Dipl.-Ing.-Titels** bis 2010 aufgegeben. **2. In D.** werden an vier Universitäten, an denen der Masterstudiengang in Landschaftsarchitektur konsekutiv von dem vorangehenden 6-8-semestrigen Bachelorstudium — je nach Bundesland: in Bayern acht Semester — studiert werden kann, z. T. verschiedene Bezeichnungen des akademischen Grades *Master of Science* verliehen: TU Berlin *Master of Science Landscape Architecture* und *Master of Science Environmental Planning*, Universität Hannover *Master of Science Landscape Architecture* und *Master of Science Umweltplanung*, Universität Kassel *Master of Science Landschaftsarchitektur und Landschaftsplanung*, TU München *Master of Science Landschaftsarchitektur*; je nach technischer Vertiefungsrichtung können z. B. an der FH Osnabrück zu dem Bachelorstudiengang nach einem konsekutiven Studium folgende akademische Grade verliehen

M

werden: *Master of Engineering Landschaftsarchitektur und Regionalentwicklung, Master of Science Bodennutzung und Bodenschutz, Master of Engineering Management im Landschaftsbau* und *Master of Science International Facility Management*; ▶Bachelor of Engineering [Landschaftsarchitektur]/Bachelor of Science [Landschaftsarchitektur]).

Master [n] of Landscape Design (MLD) [UK] *prof.* ▶Master of Landscape Architecture (MLA).

Master [n] of Philosophy in Landscape Design (MPhilLD) [UK] *obs. prof.* ▶Master of Landscape Architecture (MLA).

3475 master plan [n] (1) *plan.* (Written and graphic document for the development of a large area, often divided into several building phases; **in U.S.**, mostly for institutional use, e.g. hospital, university, military bases); *syn.* master plan [n] for a project [UK]; *s* **plan [m] maestro** (Documento escrito o gráfico para el desarrollo de un ámbito de la administración o de un gran área; p. ej. residuos sólidos urbanos, hospital, universidad, cinturón verde); *syn.* master plan [m] [AL]; *f* **plan [m] d'ensemble d'un projet** (Document graphique et écrit conforme à l'exécution représentant l'ensemble d'un aménagement, d'une ou plusieurs phases d'un aménagement) ; *g* **Gesamtplan [m] eines Projektes** (Grafische und textliche Gesamtdarstellung einer abgeschlossenen Planung oder von mehreren Planungsabschnitten eines Vorhabens).

3476 master plan [n] (2) *plan.* (Long-range plan showing proposed ultimate development of a defined area, including graphic and written segments; ▶general plan, ▶master planing 1); *syn.* development plan [n], *in common usage* general plan [n]; *s* **plan [m] de desarrollo** (Plan en el que están representados los objetivos de la ▶planificación del desarrollo; ▶plan director); *f* **plan [m] de développement** (Plan représentant les objectifs déterminés par la ▶planification spatiale sous forme de cartes et d'un rapport de présentation ; ▶schéma directeur sectoriel) ; *g* **Entwicklungsplan [m]** (Plan, der die Ziele der ▶Entwicklungsplanung in Form von Plänen und eines Erläuterungsberichtes darstellt; ▶Leitplan).

master plan [n] **for a project** [UK] *plan.* ▶master plan (1).

master plan [n] **for play areas** [US] *plan. recr. urb.* ▶overall master plan for play areas.

master plan [n] **for sports fields** [US] *plan. recr. urb.* ▶overall masterplan for sports and physical activities [US].

master plan [n] **for sports pitches** [UK] *plan. recr. urb.* ▶overall masterplan for sports and physical activities [US].

3477 master planning [n] (1) *adm. plan.* (*General term for the* formulation, coordination and implementation of objectives and policies for development of defined areas and facilities on different governmental levels: state, regional, city, and local communities); *s* **planificación [f] del desarrollo** (*Término general para* la formulación de la política de desarrollo que incluye tanto el nivel general de ▶ordenación del territorio como el ▶planeamiento urbanístico); *f* **planification [f] spatiale** (Action visant à fixer, pour un territoire donné, les objectifs de developpement et de localisation harmonieuse des hommes, de leurs activités, des équipements et des moyens de communication ; elle peut s'exercer à différentes échelles : aménagement du territoire [territoire national], planification régionale [région, massif, bande littorale], urbanisme [quartier, ville, agglomération], composition urbaine [îlot ou groupe de bâtiments] ; DUA 1996 ; ▶planification urbaine) ; *g* **Entwicklungsplanung [f]** (Planung zur Koordinierung und Durchführung möglichst aller öffentlicher Planungsaufgaben auf verschiedenen Planungsebenen. D., in der Landesplanung als Landesentwicklungsplanung, in der Regio-

nalplanung in Form von Regionalplänen, regionalen Raumordnungsplänen, Kreis- oder Gebietsentwicklungsplänen, in der gemeindlichen Planung in Form von Stadt- oder Gemeindeentwicklungsplanung resp. der ▶Bauleitplanung).

3478 master planning [n] (2) *plan.* (Generic term in widespread use, which describes the overall planning process with some emphasis on the documentary/map and product of the process itself—predominantly for preparing a document for institutional or large-scale area development; **m. p.** is often aimed at the considered and controlled development of urban areas including town and city centers [US]/centres [UK], local neighbo(u)rhoods and parks and contains individual design elements such as special buildings, schools, housing, tertiary education institutions, parks and recreational facilities, green and open spaces, transport systems and water courses as well as infrastructure systems; **m. p.** comprises: environmental assessment, site and landscape analysis, market research, stakeholder consultation, cost management, specialist fields, consultation and approval procedures with authorities); *s* **planificación [f] maestra** (Proceso de preparación de un gran proyecto público, como universidad, obra de infraestructura, etc.); *f* **étude [f] d'un grand projet** (Aménagement en matière de grands projets tels qu'un complexe universitaire, une station d'épuration, etc.) ; *g* **Gesamtplanung [f] für eine öffentliche Einrichtung** (Planerische Tätigkeit für ein umfangreiches öffentliches Großprojekt, z. B. eine Universität, Kläranlage).

master planning [n] **for play areas** [US] *plan. recr. urb.* ▶overall master planning for play areas.

master planning [n] **for sports and physical activities** [US] *plan. recr. urb.* ▶overall master planning for sports and physical activities.

mat [n] *constr.* ▶brush mat, ▶filter mat, ▶floating sphagnum mat, ▶planting mat, ▶seed mat, ▶substrate mat, ▶vegetative mat.

mat [n], **precultivated** *constr.* ▶planting mat.

mat [n], **pregrown** *constr.* ▶planting mat.

mat [n], **preplanted** *constr.* ▶planting mat.

material cycle [n] *ecol.* ▶biogeochemical cycle.

mating ritual [n] *zool.* ▶courtship.

3479 mating territory [n] *zool.* (Area exclusively or primarily used for the reproduction of a wildlife species; ▶breeding territory); *s* **territorio [m] de reproducción** (Área necesaria para la reproducción normal de una pareja de animales salvajes; ▶territorio de cría); *f* **territoire [m] de reproduction** (Aire exclusive de reproduction de la faune sauvage défendue sélectivement de grandeur variable selon les espèces sur laquelle ont lieu tout ou partie des activités d'un couple, p. ex. parades, accouplement, gestation, nidification/constitution d'abris, recherche de nourriture, élevage des jeunes ; ▶territoire de nidification) ; *g* **Fortpflanzungsrevier [n]** (Gebiet, das ausschließlich oder vorwiegend der Fortpflanzung eines Paares dient [Balz, Paarung und Brutpflege]; ▶Brutrevier).

mat [n] **of floating aquatics** [US] *phyt.* ▶floating meadow.

3480 matrix [n] **on ecological intercompatibility** *landsc. plan.* (DEN 1971, 144; assessment check-list of environmental concerns which demonstrates the influences of certain demands upon spatial uses or functions or upon ecologically important changes in the quality of landscape factors, as a means of ascertaining or quantifying the interrelationships and environmental compatibility); *syn.* environmental compatibility matrix [n]; *s* **mátrix [f] de interdependencias ecológicas** (Instrumento de planificación que muestra gráficamente los efectos de usos del territorio sobre otras funciones antropógenas del espacio y los

M

cambios de importancia ecológica previsibles de los factores del paisaje con el fin de determinar y calcular las interrelaciones y la compatibilidad con el medio de los usos planeados); *f* **matrice [f] des interdépendances écologiques** (Instrument de planification permettant de mettre en évidence les impacts de certains usages sur d'autres, sur les fonctions écologiques ou les changements importants du caractère des facteurs paysagers afin de déterminer ou d'évaluer les interdépendances et la compatibilité des usages avec le milieu) ; *g* **ökologische Verflechtungsmatrix [f]** (...matrizes *auch* ...matrices [pl]; planerisches Instrument zum Aufzeigen der Einwirkungen von Nutzungen auf andere Raumnutzungen, Raumfunktionen oder ökologisch bedeutsame Eigenschaftsveränderungen von Landschaftsfaktoren zur Ermittlung/Abschätzung der Wirkungszusammenhänge resp. der Umweltverträglichkeit); *syn.* ökologische Konfliktmatrix [f].

3481 matrix [n] **on intercompatibility** *landsc. plan.* (DEN 1971, 144; assessment chart for testing each land use against all others, in order to determine their compatibility or incompatibility with natural resources); *s* **mátrix [f] de interdependencias** (Instrumento de planificación que muestra gráficamente diferentes usos del territorio y la superposición entre ellos en relación con su compatibilidad o incompatibilidad con los recursos naturales); *f* **matrice [f] d'interdépendances** (Matrice représentant les différents usages et la superposition d'usages, en relation avec leur compatibilité ou non-compatibilité avec la gestion durable des ressources naturelles) ; *g* **Verflechtungsmatrix [f]** (...matrizes *auch* ...matrices [pl]; Matrix, die unterschiedliche Nutzungen und Nutzungsüberlagerungen in ihrer Verträglichkeit oder Unverträglichkeit mit den natürlichen Ressourcen darstellt).

matted dwarf shrub carpet [n] [US] *phyt.* ▶creeping dwarf shrub carpet [US]/creeping dwarf-shrub thicket [UK].

matted dwarf-shrub thicket [n] [UK] *phyt.* ▶creeping dwarf shrub carpet [US]/creeping dwarf-shrub thicket [UK].

matter [n] *envir.* ▶diffused particulate matter, ▶particulate matter (1); *hydr.* ▶suspended particulate matter.

matter [n], **soil organic** *pedol.* ▶humus.

3482 matter import [n] *envir. limn. pedol.* (Fine material conveyed by wind or water and deposited in a certain area, e.g. ▶particulate matter 1, mineral substances, chemical components; *specific terms* ▶nutrient input, ▶dust precipitation); *syn.* import [n] of mineral elements, input [n] of mineral elements; *s* **aporte [m] de sustancias** (▶Partículas sólidas, minerales, complejos químicos, etc. transportados por el viento o el agua, que se depositan en un lugar determinado; *término específico* aporte atmosférico de sustancias; ▶aporte de nutrientes, ▶deposición de polvo, ▶polvo atmosférico); *f* **apport [m] d'éléments minéraux** (▶Poussière, substances minérales, composés chimiques, etc. apportés par le vent ou la pluie dans une région définie ; ▶apport de substances nutritives, ▶émission particulaire) ; *syn.* apport [m] de minéraux extérieurs ; *g* **Stoffeintrag [m]** (Durch die Luft oder durch Oberflächenwasser in ein bestimmtes Gebiet transportierte ▶Stäube, Mineralstoffe, chemische Verbindungen etc. Durch elektrische Entladungen [Blitze] entstehen z. B. in der Atmosphäre nitrose Gase, die, umgewandelt in Nitrat, den Boden jedes Jahr mit einigen Kilogramm Stickstoff anreichern; cf. PRO 2001; *UBe* atmosphärischer Stoffeintrag, ▶Nährstoffeintrag, ▶Staubeintrag).

mattress [n], **willow** *constr.* ▶brush mat.

3483 mature tree [n] *arb. hort.* (Tree with ultimate growth; ▶large-caliper tree/large-calliper tree, ▶large shrub); *s* **árbol [m] maduro** (Árbol de varias décadas de edad; ▶árbol de diámetro grande, ▶arbusto grande); *syn.* árbol [m] grande; *f* **arbre [m] mature** (Arbre arrivé à maturité, en sylviculture arbre âgé de

60 ans et d'un diamètre d'environ 20 cm, comme fruitier arbre ayant déjà fructifié au moins trois fois ; ▶grand arbrisseau, ▶grand arbuste, ▶arbre à tronc de fort diamètre) ; *syn.* arbre [m] adulte ; *g* **Großgehölz [n]** (*Nicht klar definierter Begriff* **1.** meist ein Baum, der in einer Baumschule kultiviert und mehrere Jahrzehnte alt ist; ▶Baum mit großem Stammdurchmesser; **2.** ein ausgewachsener ▶Großstrauch); *syn.* Altgehölz [n].

3484 maximum bulk [n] [US] *leg. urb.* (Three-dimensional space within which a structure is permitted to be built on a lot, and which is defined by maximum height regulations, yard setbacks, and sky exposure plane regulations; IBDD 1981; ▶cubic content ratio [of a building] [US], ▶density of development, ▶floor area ratio [FAR] [US]/floor space index [UK], ▶building coverage [US]/plot ratio [UK], ▶permitted number of stories [US]/permitted number of storeys [UK], ▶zoning district category [US]/use class [UK]); *s* **volumen [m] edificable** (En EE.UU., espacio tridimensional dentro del cual está permitido construir una estructura en un solar y que se define a través de las regulaciones de altura máxima, retranqueo y regulación del plano de exposición al cielo. En Es. no existe regulación estatal sobre el v. e., ésta existe sólo a nivel de las CC.AA. En D. norma constructiva que describe la relación entre la superficie o el volumen construidos y la superficie del solar a construir. Se fija en la planificación urbana por medio del ▶número de plantas completas, el ▶coeficiente de ocupación del suelo, ▶coeficiente de edificación y la ▶edificabilidad; ▶categoría de zonificación, ▶densidad de edificación); *syn.* volumen [m] máximo de edificación); *f* **conditions [fpl] de l'occupation du sol (et de l'espace)** (**1.** U.S., caractéristiques de l'espace tridimensionnel sur lequel peut être érigée une construction ; le volume extérieur est défini par les règles régissant la hauteur des constructions, la ▶limite de constructibilité, le ▶polygone d'implantation et la ▶bande constructible ainsi que par les prescriptions fixant les règles d'ensoleillement des façades. **2.** F., règles d'urbanisme ou d'architecture relatives à la desserte des terrains [accès, desserte, alimentation en eau potable, assainissement], à l'implantation et la dimension des constructions [▶coefficient d'emprise au sol, ▶coefficient d'occupation des sols, ▶coefficient d'utilisation du sol], l'implantation des constructions par rapport aux voies, aux limites séparatives, aux constructions les unes par rapport aux autres, aspect extérieur et hauteur des bâtiments, ▶nombre des niveaux de la construction] et aux espaces libres [aires de stationnement, espaces verts et plantations] applicables aux constructions comprises dans une zone déterminée et précisées dans le règlement du PLU/POS, etc. ; cf. C. urb., art. R. 123 – 21-22 ; ▶nature de l'affectation des sols, ▶densité de construction) ; *g* **Maß [n] der baulichen Nutzung** (**1.** Im Bebauungsplan festgesetzte planungsrechtliche Kennzeichnung des Umfangs der baulichen Nutzung als Verhältnis von überbauter Fläche [GRZ], Geschossfläche [GFZ] oder Baumasse [BMZ] zur Fläche des Baugrundstückes; in den Bebauungsplänen wird dieses Maß als Höchstmaß, das unterschritten werden darf, festgelegt; cf. §§ 5 und 9 BauGB in Verbindung mit den §§ 16 bis 21 a BauNVO. In Bebauungsplänen durch die ▶Geschossflächenzahl, ▶Grundflächenzahl, ▶Baumassenzahl, Festsetzung der Gebäudetiefe, Höhe baulicher Anlagen, ▶Zahl der Vollgeschosse und Anrechenbarkeit von Stellplätzen oder Garagen dargestellt und festgesetzt; ▶Art der baulichen Nutzung, ▶Bebauungsdichte. **2.** In den USA wird das **M. der b. N.** durch einen dreidimensionalen Raum, innerhalb dessen ein Gebäude oder eine bauliche Anlage auf einem Grundstück gebaut werden darf, bestimmt. Das äußere Bauvolumen wird durch Vorschriften über Gebäudehöhen, Abstände zu den Nachbargrundstücken [Baugrenzen und Baulinien] und ausreichende Belichtung und Besonnung [Abstandsflächen] bestimmt); *syn.* Nutzungsmaß [n] [CH] (RBU 1999, 131), Nutzungsziffer [f] [CH] (§§ 254-259 PBG).

maximum discharge [n] *wat'man.* ►flood discharge, ►peak discharge.

maximum flood [n] *wat'man.* ►probable maximum flood (PMF).

maximum level [n] *envir.* ►maximum value.

3485 maximum value [n] *envir.* (Stipulated or measured top value, indicating the maximum amount of a substance or its concentration; ►threshold value); *syn.* maximum level [n]; *s* **contenido [m] máximo** (Valor fijado o medido de la cantidad o concentración máxima de una sustancia o grupo de sustancias que se tolera en un medio; ►valor límite); *f* **valeur [f] maximale** (Valeur fixée ou mesurée indiquant la quantité maximale ou le taux de concentration maximal d'une substance dans un milieu ; ►valeur limite) ; *g* **Höchstwert [m]** (Festgesetzter oder gemessener Wert, der die Höchstmenge eines Stoffes oder der Konzentration von Stoffkombinationen in einem Medium anzeigt; ►Grenzwert).

3486 meadow [n] (1) *agr.* (Plant formation composed of grasses —predominantly species of the grass family *[Gramíneae]* rather than species of the sedge family *[Cyperáceae]*—and herbs without or with very few woody species; there are various agricultural uses such as productive **m.s** ►hayfield [US]/hay meadow [UK], ►traditional orchard meadow [UK], ►fertilized meadows and ►nutrient-poor grassland. A **m.** used for grazing is known as a ►pasture); *s* **pradera [f]** (Formación herbácea compuesta predominantemente de diferentes especies de gramíneas *[Gramíneae]* y de hierbas con pocas especies arbustivas; se diferencian las praderas cultivadas (►prado de siega, ►prado de huerta frutal], los ►prados jugosos y los pobres [►prado oligótrofo]. Las **pp.** utilizadas para pastoreo se llaman ►pastizales. En España es costumbre llamar **p.** al prado sembrado, para distinguirlo así del que salió del monte rozado; cf. DINA 1987); *syn.* prado [m] de diente (DB 1985); *f* **prairie [f]** (Formation végétale à dominante herbacée — en majorité graminées, plantes annuelles et quelques cypéracées — fermée et dense, à caractère principalement hygrophile et mésophile, dont le recouvrement des espèces ligneuses ne dépasse pas 30 % ; on distingue les prairies ou pelouses alpines, les prairies humides et prairies sèches [►pelouse oligotrophe] ainsi que les prairies secondaires dites 'permanentes' [►prairie (mésophile) de fauche, ►pâturage tournant, ►pâturage 2, ►prairie complantée d'arbres fruitiers]) ; *g* **Wiese [f]** (Aus Gräsern — vor allem aus Süßgräsern *[Gramíneae]*, seltener aus Riedgräsern *[Cyperáceae]* — und Kräutern zusammengesetzte, gehölzfreie oder -arme Pflanzenformation; es werden landwirtschaftliche Nutzwiesen [►Mähweide, ►Streuobstwiese], ►Fettwiesen und Magerwiesen [►Magerrasen] unterschieden; die zur Beweidung genutzte **W.** wird ►**Weide** genannt).

meadow [n] (2) *phyt.* ►damp meadow, ►development of a meadow, ►dry meadow, ►fertilized meadow, ►floating meadow; *agr.* ►fodder meadow, ►hay meadow, ►inundated meadow; *phyt.* ►natural meadow; *agr. hort.* ►once-a-year mown meadow; *geo.* ►riparian meadow; *phyt.* ►rush meadow, ►salt meadow; *agr. conserv.* ►traditional orchard meadow [UK]; *agr. phyt.* ►wet meadow; *gard. landsc.* ►wildflower meadow; *phyt.* ►xerothermous meadow.

meadow [n] [US], **alpine** *phyt.* ►alpine grassland.

meadow [n] [UK], **annual cut of a** *agr. constr. hort.* ►single mowing [US].

meadow [n], **fodder** *agr.* ►hay meadow (2).

meadow [n] [UK], **hay** *agr. (phyt.)* ►hayfield [US].

meadow [n] [UK], **litter** *agr. land'man. phyt.* ►straw meadow [US].

meadow [n] [UK], **moist soil** *phyt.* ►damp meadow.

meadow [n] [UK], **once-a-year cut** *agr. hort.* ►once-a-year mown meadow [US].

meadow [n] [US], **open orchard** *agr. conserv.* ►traditional orchard meadow [UK].

meadow [n], **subaqueous** *phyt.* ►chara vegetation.

meadow [n] [US], **successional** *gard. landsc.* ►wildflower meadow.

meadow [n] [UK], **water** *agr.* ►inundated meadow [US].

meadow cut several times per annum [n] [UK] *agr. hort.* ►meadow mown several times per year [US].

3487 meadow cutting [n] **from inside to outside** *ecol. land'man.* (Maintenance of ecologically valuable meadow land, such as traditional hay meadows and wetlands, which allows animals to escape in front of machines, often carried out in the context of nature protection with the aim of safeguarding the habitat of meadow-breeding birds; an alternative method is to cut from left to right, but under no circumstances should mowing take place from the outside to the inside in a circular manner, thus trapping animals in the centre. This protective method is used together with other maintenance methods according to the circumstances, e.g. a cut at reduced speed [maximum. 6 to 7 km/h], with a minimum mown height of 5-7cm or leaving an uncut strip as a refuge area at least 2m wide with each cut parallel to the longitudinal mowing direction as well as the practice of a late cut); *s* **prado [m] de corte de dentro para afuera** (Método de siega utilizado para mantener prados de valor ecológico, como prados húmedos o prados de heno tradicionales, que permite a los animales escapar por delante de las máquinas. Se aplica en el contexto de la conservación de la naturaleza para proteger los hábitats de aves que nidifican en el suelo. Otro método alternativo es cortar por bandas, pero en ningún caso se debe segar de afuera hacia dentro en círculos concéntricos que atrapan a los animales en el centro. Este método se utiliza junto con otros como cortar a velocidad reducida [máximo 6-7 km/h], con una altura mínima de corte de 5-7 cm o dejar sin segar una banda paralela a la dirección de corte de 2 m de ancho como área de refugio o realizar la siega tardía); *f* **fauche [f] centrifuge** (Technique de fauche partant du centre de la parcelle pour aller vers l'extérieur pour permettre aux animaux [oiseaux, mammifères, insectes] et en particulier aux nicheurs au sol de fuir vers les bordures et d'y trouver refuge ; au contraire, une fauche centripète classique, partant des bords de la parcelle pour aller vers le centre, emprisonne les animaux au centre, les mettant en danger ; méthode conservatoire souvent contractuelle utilisée sur les prairies naturelles fourragères riche du point de vue faunistique ; celle-ci peut être réalisée conjointement avec d'autre mesures telles qu'une vitesse de fauche moins élevée [vitesse au pas avec un max. de 6-7 km/h], une bande herbeuse refuge [largeur min. 2] parallèle au plus grand axe fauché, la fauche tardive) ; *syn.* fenaison [f] centrifuge ; *g* **Mahd [f] von innen nach außen** (Pflege von ökologisch wertvollen Wiesenflächen, z. B. Feuchtwiesen und traditionelle Heuwiesen, oft im Rahmen des Vertragsnaturschutzes zur Sicherung des Lebensraumes von Wiesenbrütern; alternativ kann streifenweise von links nach rechts und keinesfalls darf einkreisend von außen nach innen gemäht werden, um Tieren die Flucht zu ermöglichen. Diese schonende Methode wird zeitgleich mit anderen Pflegemaßnahmen und -gesichtspunkten durchgeführt, z. B. eine Mahd mit verringerter Geschwindigkeit [Schritttempo bis max. 6 bis 7 km/h] unter Berücksichtigung einer Mindestschnitthöhe von 5-7 cm und das Belassen eines ungemähten, mindestens 2 m breiten

Streifens bei jeder Mahd parallel zur Längsachse als Rückzugsgebiet sowie eine späte Mahd).

3488 meadow flower [n] *bot.* (Herbaceous blooming plant growing in grassland according to soil and moisture conditions; ►wildflowers in the lawn, ►natural meadow, ►wildflower); *s* **flor** [f] **de prado** (Planta herbácea que crece en los prados y florece con flores vistosas; ►flor silvestre de césped, ►hierba silvestre, ►prado natural); *syn.* flor [f] de pradera; *f* **fleur** [f] **des prés** (Plante herbacée vivace et de floraison remarquable ; ►plantes herbacées des pelouses, ►pelouse, ►prairie climacique, ►herbe sauvage) ; *g* **Wiesenblume** [f] (In Wiesen je nach Feuchtigkeit und Bodeneigenschaften wachsende krautige Pflanze mit auffälligen Blüten, ►Rasenkräuter, ►natürliche Wiese, ►Wildkraut).

meadow in a forest [n] *for. recr.* ►open meadow in a forest.

3489 meadow [n] **mown several times per year** [US] *agr. hort.* (►be mown once a year [US]/be cut once a year [UK], ►cut a few times per year); *syn.* meadow [n] cut several times per annum [UK]; *s* **pradera** [f] **de siega múltiple** (Prado que se siega varias veces al año; ►segado/a anualmente, ►segado/a varias veces al año); *f* **prairie** [f] **fauchée plusieurs fois par an** (►fauché, ée plusieurs fois par an, ►fauché, ée une fois par an) ; *g* **mehrschürige Wiese** [f] (Wiese, die mehrmals im Jahr gemäht wird; ►einschürig, ►mehrschürig); *syn.* mehrmals gemähte Wiese [f], mehrmals geschnittene Wiese [f].

3490 meadow strip [n] *constr.* (Narrow band of grass, usually along roads, which is usually cut 3 times per year or according to weather conditions and the required degree of maintenance; ►grass strip 1, ►turf strip 1); *s* **banda** [f] **de herbáceas** (►Franja de hierba estrecha que se corta como máximo tres veces al año; ►franja de césped); *syn.* franja [f] de herbáceas; *f* **accotement** [m] **enherbé** (Espace linéaire aménagé le long des chaussées sur lequel sont effectuées au maximum trois tontes annuelles ; ►bande engazonnée, ►accotement engazonné) ; *g* **Wiesenstreifen** [m] (Schmaler, meist straßenbegleitender ►Grasstreifen, der je nach Pflegeaufwand maximal 3-mal pro Jahr geschnitten wird; ►Rasenstreifen 1).

3491 meadow valley [n] *landsc.* (Deforested or natural high altitude valley characterized by a grass cover); *s* **valle** [m] **de praderas** (Valle sin bosque o de alta montaña caracterizado por la abundancia de prados); *f* **vallée** [f] **à végétation herbacée** (Vallée après un déboisement ou vallée de haute montagne où prédominent les prés et les pelouses continues) ; *syn.* vallée [f] herbeuse ; *g* **Wiesental** [n] (Entwaldetes oder Hochgebirgstal, das durch Wiesen geprägt ist).

3492 meander [n] *geo.* (Winding bed of a river or stream caused by the gradual erosion of the ►bluff 1 and the ensuing accumulation of material along the ►slip bank [US]/slip-off slope [UK]; ►cutting a meander, ►meander neck); *s* **meandro** [m] (Cada una de las sinuosidades o curvas que dibuja un río. Los **mm.** presentan en planta un sector cóncavo, en el que la erosión es muy activa que conduce a una orilla abrupta [►orilla cóncava], y uno convexo, en el que domina la acumulación aluvial, dando lugar a una orilla de pendiente suave [►orilla convexa]. *Términos específicos* meandro abandonado, meandro libre, meandro encajado, ►cuello de meandro, ►separación de un meandro; cf. DGA; cf. DGA 1986); *f* **méandre** [m] (Sinuosité régulière décrite par le lit ordinaire d'un cours d'eau au chenal bien calibré mais dissymétrique. La berge de la ►rive concave sapée par le courant est abrupte ; la ►rive convexe où se déposent momentanément des alluvions est en pente douce. Les méandres sont généralement disposés en séries ou forment un train de méandres ; DG 1984 ; ►étranglement d'un méandre, ►scindement d'un méandre) ; *g* **Mäander** [m] (Mäander [pl];

aus dem Griech. *maiandros* >Fluss *Menderes* im westl. Anatolien<; durch Abtragung am ►Prallhang und anschließender Aufschüttung am ►Gleithang entstandene Abfolge von Windungen eines Fließgewässers; mäandern [vb], mäandrieren [vb]; ►Kappen eines Mäanders, ►Mäandersporn); *syn. bei Flüssen* Flussschleifen [fpl], Flussschlingen [fpl].

meander core [n] *geo.* ►meander lobe.

3493 meandering river [n] *geo.* (►meander, ►meander neck); *syn.* winding river [n], meandering watercourse [n] (±); *s* **curso** [m] **(fluvial) divagante** (►meandro, ►cuello de meandro), *syn.* curso [m] divagante de un río; *f* **cours** [m] **d'eau à méandre libre** (Cours d'eau dans une plaine alluviale caractérisé par une forte sinuosité [méandres libres]. Par migration vers l'aval, un méandre peut en rejoindre un autre et créer ainsi un bras mort de forme arquée, les recoupements successifs tendent à calibrer la vallée aux dimensions de « trains » de méandres ; ►méandre, ►étranglement du/d'un méandre) ; *syn.* cours [m] d'eau à méandre divaguant (DG 1984, 287) ; *g* **natürlich mäandrierender Flusslauf** [m] (Gewässerlauf, der in der Ebene durch eine Vielzahl aufeinanderfolgender Flussschlingen, die durch keine menschlichen Eingriffe verändert wurden, und durch viele Altarme gekennzeichnet ist; ►Mäander, ►Mäanderhals).

meandering watercourse [n] *geo.* ►meandering river.

3494 meander lobe [n] *geo.* (**1.** The elevated, tongue-shaped area of land enclosed within a stream ►meander; ►meander neck. **2.** The central hill encircled or nearly enclosed by a stream meander is a **meander core**); *syn. to 1.* meander tongue [n]; *syn. to 2.* cutoff spur [n], rock island [n]; *s* **lóbulo** [m] **de meandro** (Trozo de tierra elevado y con forma de lengua que está rodeado por un ►meandro; ►cuello de meandro); *f* **lobe** [m] **de méandre** (Espace compris dans la boucle formée par un ►méandre qui lors l'abandon du chenal par recoupement [déconection par l'amont] au niveau de l'►étranglement du/d'un méandre forme une butte appelée terrasse ou talus polygénique) ; *g* **Mäandersporn** [m] (Von einer Fließgewässerschlinge umflossenes Landstück, das beim Durchbruch des ►Mäanderhalses zum **Umlaufberg** wird).

3495 meander neck [n] *geo.* (Narrow strip of land between the two limbs of a ►meander, that connects a ►meander lobe with the mainland; GFG 1997); *s* **cuello** [m] **de meandro** (Parte más estrecha de tierra entre los dos brazos de un ►meandro que conecta al ►lóbulo de meandro con la vega adyacente); *f* **étranglement** [m] **du/d'un méandre** (Partie la plus étroite de la boucle [►lobe de méandre] formée par un ►méandre entre le chenal amont et le chenal aval, appelée racine ou pédoncule si la boucle est très accentuée) ; *g* **Mäanderhals** [m] (Engste Stelle eines ►Mäanders, an dem sich die bergseitige und talseitige Laufstrecke des Fließgewässers nähern; ►Mäandersporn).

meander tongue [n] *geo.* ►meander lobe.

3496 mean-water level [n] *leg. wat'man.* (Arithmetic average of the water level in rivers or lakes during the past 20 years. Where there have been no accurate observations of the water level, the average water level may be determined by the limit of plant growth on the edges); *syn.* normal water level [n] [UK]; *s* **nivel** [m] **medio (del agua)** (Media aritmética de los niveles de aguas de los últimos 20 años); *f* **niveau** [m] **des moyennes eaux** (Moyenne arithmétique des niveaux d'eau moyens journaliers pendant les 20 dernières années ; le **n. d. m. e.** se détermine aussi par la limite du couvert végétal) ; *syn.* niveau [m] d'eau moyen (WMO 1974) ; *g* **Mittelwasserstand** [m] (Arithmetisches Mittel der Wasserstände der letzten 20 Jahre. Fehlen Pegelbeobachtungen, so bestimmt sich der **M.** nach der Grenze des Pflanzenwuchses entlang des Ufers; cf. Wassergesetze der Bundesländer).

M

measure [n] *envir. leg. plan.* ►alleviation measure; *conserv. urb.* ►conservation measure; *conserv. leg.* ►environmental compensation measure, ►mitigation measure; *constr.* ►pace measure; *envir. leg. plan.* ►prevention measure (of an intrusion upon the natural environment), ►preventive measure; *plan. pol. sociol.* ►preventive measures; *constr. envir. landsc.* ►remedial measure.

measure [n]**, compensation** *conserv. leg.* ►environmental compensation measure.

3497 measured surface area [n] *surv.* (Numerical extent of two-dimensional space which is measured, e.g. in m², acres, hectares. In the context of a garden, city or country, however, the word 'area' alone is used); *s* **superficie** [f] **medida** (≠) (Tamaño de un terreno medido en m², km²; *contexto* la **s. m.** en un jardín, un terreno, una ciudad, etc.); *f* **superficie** [f] (Nombre caractérisant l'étendue d'une surface exprimée en m², a, ha, km², etc. ; *contexte* la **s.** p. ex. d'un pays, d'un terrain, mais par contre la surface d'une forme géométrique) ; *syn. suivant le contexte* surface [f] ; *g* **Flächeninhalt** [m] (Größe einer Fläche, gemessen z. B. in m², a, ha, km²; Kontext der **F.** einer geometrischen Figur, aber die Fläche eines Gartens, einer Stadt oder eines Landes); *syn. je nach Kontext* Fläche [f].

3498 measured water level [n] *hydr.* (Measured height of a water surface in relation to a fixed level gauge, as measured at a specific time; ►water level); *s* **nivel** [m] **del agua (1)** (Altura de la superficie del agua en relación con una cota de nivel definida; ►nivel del agua 2); *f* **niveau** [m] **d'eau à l'échelle des eaux** (Hauteur du niveau d'eau mesuré à partir d'un niveau de référence, p. ex. fixé à l'échelle fluviométrique ; ►niveau de l'eau 2) ; *g* **Wasserstandshöhe** [f] (Höhe des Wasserspiegels bezogen auf einen Bezugshorizont, z. B. Pegel; ►Wasserspiegel); *syn.* Pegelhöhe [f], Pegelstand [m].

measurement [n]**, earthworks** *constr. contr.* ►real earthworks measurement.

measurement [n] [UK]**, method of** *constr. contr.* ►method of site survey [US].

3499 measurement(s) [n(pl)] *constr.* (**1.** Determination of dimensions, area, volume, etc. **2.** Amount determined by measuring); *syn.* measure [n]; *s* **medida** [f] *syn.* dimensión [f]; *f* **mesure** [f] (Unité de détermination des quantités ou des grandeurs des prestations de travaux) ; *syn.* dimension [f] ; *g* **Maß** [n] (Maße [pl]; **1.** Einheit, mit der die Größe oder Menge von Bauleistungen gemessen wird; **2.** Gemessenes).

measurement [n] **of completed project** *constr. contr.* ►measurement of completed work [US]/works [UK].

3500 measurement [n] **of completed work** [US]/ **measurement** [n] **of completed works** [UK] *constr. contr.* (Computation of executed construction during final inspection, as the basis for issuing a certificate of completion for payment; ►agreed measurement of completed work [US]/works [UK], ►interim agreed measurement of completed work [US]/works [UK]); *syn.* measurement [n] of completed project; *s* **medición** [f] **de los trabajos terminados** (Cómputo de las tareas ejecutadas durante la inspección final de obra como base para la liquidación de cuentas final de un proyecto; ►medición acordada de los trabajos terminados, ►medición intermedia de los trabajos terminados); *syn.* medida [f] de los trabajos terminados; *f 1* **attachement** [m] (Document établi contradictoirement entre le maître d'œuvre et l'entreprise à partir des relevés, fait sur le chantier, des travaux exécutés en vue de l'établissement des situations provisoires ; ►métré provisoire) ; *f 2* **métré** [m] **définitif** (Détermination précise des quantités de travaux terminées en vue de l'établissement d'un ►décompte final, ►métré

général et définitif) ; *g* **Aufmaß** [n] **(1)** (Das Ausmessen fertiger Bauteile, Bauleistungen etc. im Rahmen der Abnahme und als Grundlage zur Abrechnung; ►örtliches Aufmaß, ►Zwischenaufmaß); *syn.* Aufmessung [f] [auch A].

measurement [n] **of completed works** [UK]**, agreed** *constr. contr.* ►agreed measurement of completed work [US].

measures [npl] **against noise** *envir. plan.* ►precautionary measures against noise.

measures [npl] **for conservation of historic monuments (and sites)** [UK] *adm. conserv'hist.* ►historic preservation measures [US].

3501 mechanical pest control [n] *agr. hort. phytopath.* (Capture of pests, e.g. collection in ditches, lime rings, baited traps, repellence by means of fences and nets, or their deterrence, e.g. by scarecrows); *s* **lucha** [f] **mecánica contra parásitos** (Por captura con medios mecánicos, como trampas, anillos, etc., por protección [con vallas o redes] o por disuación p. ej. con espantapájaros, de los animales nocivos); *syn.* control [m] mecánico de organismos nocivos; *f* **procédé** [m] **mécanique de lutte contre les ennemis des cultures** (Capture [p. ex. le ramassage, les fossés de capture, les pièges], interposition d'obstacles [p. ex. au moyen de clôtures, de filets] et effarouchement [p. ex. au moyen d'épouvantail] des déprédateurs) ; *g* **mechanische Schädlingsbekämpfung** [f] (Abfangen [z. B. durch Absammeln, Anlegen von Fanggräben, Fang- und Leimringen, Aufstellen von beköderten Fallen], Abwehr [z. B. durch Zäune, Netze] und Abschreckung [z. B. durch Vogelscheuchen] der Schädlinge).

3502 mechanical sowing [n] *agr. constr. hort.* (Seed sowing with a sowing machine; ►hydroseeding); *syn.* broadcasting [n] with a (e.g. cyclone) seeder [also US]; *s* **siembra** [f] **mecánica** (►riego de semillas por emulsión); *f* **semis** [m] **effectué à la machine** (Semis réalisé p. ex. au moyen d'une machine à engazonner, semis à l'engazonnement par projection hydraulique ; *terme spécifique* semis au semoir ; ►procédé d'enherbement par projection hydraulique) ; *syn.* semis [m] mécanique ; *g* **Maschinensaat** [f] (Mit einer Sämaschine erfolgende Aussaat; ►Anspritzverfahren); *syn.* maschinelle Saat [f].

median [n] [US] *trans.* ►median strip [US].

3503 median strip [n] [US] *trans.* (Barrier located along the center line [US]/centre line [UK] of a road, often comprising curbstones [US]/kerbstones [UK] and/or a guardrail [US]/crash barrier [UK], which separates traffic flowing in opposite directions; such strips may vary in width and contain planting, according to the situation [landscaped median]; ►freeway median strip [US]/motorway central reservation [UK]; *generic term* ►separator); *syn.* central reservation [n] [UK], median [n] [US]; *s* **franja** [f] **divisoria central** (Banda de vegetación que separa las calzadas; ►franja divisoria central de autopista; *término genérico* ►banda de separación); *syn.* banda [f] central de separación, bandejón [m] [RCH]; *f* **terre-plein** [m] **central** (*Abrév.* T.P.C. ; séparation matérielle des deux sens de la circulation, dans le cas où les tracés en plan et les profils en long des deux chaussées ne sont pas indépendants ; la bande médiane supporte entre autre les plantations et les dispositifs de sécurité ; ICTA 1985, 18 ; ►terre-plein central d'une autoroute ; *terme générique* ►bande de séparation) ; *syn.* berme [f] centrale (DTB 1985) ; *g* **Mittelstreifen** [m] (In Straßenmitte mit Markierungslinien und oft mit Leitplanken angelegter Streifen, der die entgegengesetzten Fahrtrichtungen trennt, je nach Breite auch als Grünstreifen ausgestattet; ►Autobahnmittelstreifen; *OB* ►Trennstreifen).

3504 medicinal herb [n] *bot. plant.* (Plant with curative properties either planted or collected from natural areas, and used in medicinal products); *s* **planta** [f] **medicinal** (Planta silvestre o cultivada que tiene propiedades curativas); *syn.* hierba [f] medicinal, yerba [f] medicinal, planta [f] oficinal; *f* **plante [f] médici-nale** (Plante cultivée ou ramassée dans la nature pour ses pro-priétés curatives et à partir de laquelle sont obtenus des médica-ments) ; *g* **Heilkraut [n]** (Pflanze, die wegen ihres Gehaltes an heilenden Wirkstoffen für medizinische Zwecke kultiviert oder in der freien Natur gesammelt wird); *syn.* Arzneipflanze [f], Heil-pflanze [f], offizinelle Pflanze [f].

3505 medicinal herb garden [n] *gard.* (As early as the Middle Ages physicians and apothecaries commonly maintained this type of garden to supplement ▶medicinal herbs and costly, imported drugs; cf. JEL 1986, 252, 363); *syn.* physic garden [n] [also UK]; *s* **jardín [m] medicinal** (Variante del jardín de hierbas finas en el que se cultivan sobre todo ▶plantas medicinales); *syn.* jardín [m] de plantas medicinales; *f* **jardin [m] médicinal** (Jardin caractérisé par la culture des ▶plantes médicinales) ; *syn.* jardin [m] d'apothicaire, jardin [m] de simples ; *g* **Heilpflanzengarten [m]** (Variante eines Kräutergartens, in dem vorwiegend ▶Heil-kräuter und Giftpflanzen kultiviert werden. Im Mittelalter und später diente er vorwiegend für die Ausbildung der Medizinstu-denten, um Arzneipflanzen kennenzulernen. Die ersten **H.gärten** wurden in Italien in botanischen Gärten als ‚offizinelle Abtei-lungen' in Pisa [1543], Padua [1545], Bologna [1567] angelegt und in D. entstand in Leipzig 1542 der erste **H.** als ‚Hortus medicus'); *syn.* Apothekergarten [m], Heilkräutergarten [m], Hortus [m] medicus.

medicine [n] *conserv.* ▶environmental medicine.

medieval garden [n] *arch. hist.* ▶enclosed medieval garden.

3506 medium gravel [n] *constr.* (**In U.S.**, according to NCSS, diameters of all gravel are from 12.5 to 25.4mm; ▶gravel 1, ▶soil textural class); *s* **grava [f] media** (Fracción de las partículas del suelo cuyo grosor comprende entre 6,3 y 20 mm de diámetro; ▶clase textural, ▶grava); *f* **gravier [m] moyen** (F., en terme de granulométrie, matériau sédimentaire meuble, grossier dont la classe granulaire comprend les grains entre 6,3 et 20 mm ; ▶classe de texture de sol, ▶gravier) ; *g* **Mittelkies [m]** (**D.**, ▶Bodenart mit einem Durchmesser von > 6,3 bis 20 mm; *Abk.* mG; cf. DIN 4022 und 4023; ▶Kies).

3507 medium root [n] *arb.* (Root with a diameter from 20 to 50mm; ▶fibrous root, ▶structural root); *syn.* coarse root [n]; *s* **raíz [f] secundaria** (Raíz con un espesor de 5-50 mm entre la ▶raíz fibrosa y la ▶raíz primaria); *f* **racine [f] secondaire** (Caractérise une racine de diamètre compris entre 20 et 50 mm dont la taille est comprise entre celle de la ▶radicelle et de la ▶racine maîtresse ; les **r. s.** assurent l'approvisionnement en eau et en substances nutritives, la constitution de réserves ainsi que la stabilité des végétaux) ; *g* **Grobwurzel [f]** (Bezeichnung einer Wurzel mit einem Durchmesser von 20-50 mm; Größe zwischen Schwachwurzel und ▶Starkwurzel; **G.n** dienen dem Wasser- und Nährstofftransport, der Speicherung von Reservestoffen sowie der Verankerung des Baumes; cf. QBB 1992; ▶Faserwurzel).

3508 medium sand [n] *constr.* (According to BS mineral particles with diameter from 0.2 to 0.6mm [UK], according to NCSS diameter from 0.25 to 0.5mm [US]; ▶soil textural class); *s* **arena [f] media** (▶Clase textural de suelo con un diámetro de > 0,2 a 0,6 mm); *f* **sable [m] moyennement grossier** (F., en terme de granulométrie, matériau meuble formé de quartz [grains de sable] dont les dimensions sont comprises entre 0,2 à 0,5 mm ; cf. DIS 1986, 192 ; ▶classe de texture des sols) ; *g* **Mittelsand [m]** (**D.**, ▶Bodenart mit einem Durchmesser von > 0,2 bis 0,6 mm; *Abk.* mS; cf. DIN 4022 und 4023).

3509 medium silt [n] *constr.* (**In U.S.**, according to NCSS: diameter from > 0.005 to 0.05mm; ▶soil textural class, ▶silt 1); *s* **limo [m] medio** (▶Clase textural de suelo con un diámetro de > 0,006 a 0,02 mm; ▶limo 2); *f* **limon [m] moyen** (F., n'existent

que les fractions granulométriques : limon fin et limon grossier ; **D.**, fraction granulométrique dans la ▶classe de texture des sols dont les dimensions sont comprises entre 0,006 et 0,02 mm ; ▶limon) ; *g* **Mittelschluff [m]** (**D.**, ▶Bodenart mit einem Durch-messer von > 0,006 bis 0,02 mm; *Abk.* mU; cf. DIN 4022 und 4023; ▶Schluff).

meeting [n] *constr.* ▶junction (1).

3510 meeting finished grade [loc] [US] *constr.* (e.g. height at which paving connects with finished grade [US]/finished level [UK]; ▶junction 1, ▶make a finished grade flush with [US]/marry finish grade with [UK]); *syn.* making flush with finished grade [loc] [US], marrying with existing levels [loc] [UK] (SPON 1986, 156); *s* **enrase [m]** (Igualado de la altura con materiales de construcción o con elementos constructivos al nivel existente en obra; ▶equiparar alturas, ▶juntura); *f* **raccorde-ment [m] au niveau existant** (Raccordements divers sur ou-vrages émergents ou existants ; ▶raccordement, ▶se raccorder au niveau fini) ; *g* **Höhenanschluss [m]** (Anpassen mit Baustof-fen oder Konstruktionselementen an eine vorhandene Höhe auf der Baustelle; ▶Anschluss, ▶höhengerecht anschließen); *syn.* Höhenanpassung [f].

3511 mellow soil [n] *hort.* (Very soft, very friable, porous loamy soil for optimum plant cultivation; ▶soil mellowness); *s* **suelo [m] con sazón** (Suelo muy friable de calidad óptima para el cultivo de plantas; ▶sazón); *f* **sol [m] à forte stabilité structu-rale** (Sol cultural, caractérisé par une structure durable de ses agrégats, une bonne aération et une forte capacité de rétention pour l'eau et présentant par conséquent des conditions optimales pour la culture ; ▶bon état de structure grumeleuse) ; *syn.* sol [m] à forte stabilité de structure, sol [m] en bon état structural ; *g* **garer Boden [m]** (Durch ständige Bodenbearbeitung und Bodenverbesserung erzielter Kulturboden, der auf Grund seiner Beständigkeit der Krümelstruktur, guten Durchlüftung und seines hohen Wasserhaltevermögens ein optimales Pflanzenwachstum erwarten lässt; ▶Bodengare).

member [n] **of operatives** [UK] *adm. constr. contr.* ▶staff person.

member state [n]**, contracting** *contr. pol.* ▶party to a contract.

membrane [n] *constr.* ▶thermal stress on roof membrane.

membrane [n]**, plastic** *constr.* ▶plastic sheet.

membrane [n] [US]**, root repellant** *constr.* ▶root protec-tion membrane.

membrane [n] [US]**, waterproof roofing** *constr.* ▶roof membrane.

3512 memorial [n] *adm. hist.* (**1.** Structure/building designed to commemorate or preserve the memory of an individual or historical event. **2.** Periodic observance of a past event; e.g. a war, battlefield. **3.** Tree or other plant, plaque or other cultural resource [US]/cultural asset [UK] of historic commemorative significance); *s* **monumento [m] conmemorativo** (Monumento que tiene la función de preservar la memoria de una personalidad o un acontecimiento histórico importante); *f* **mémorial [m]** (Monument commémoratif funéraire, tel le mausolée, l'obé-lisque, etc.) ; *g* **Gedenkstätte [f] (1)** (**1.** Monument zum An-denken an eine Persönlichkeit oder ein historisches Ereignis, z. B. ein Mausoleum, eine Gedenkmauer, ein Obelisk. **2.** Gedenk-zeichen, das dazu dient, den Opfern von Krieg und Gewaltherr-schaft in besonderer Weise zu gedenken und für zukünftige Generationen die Erinnerung daran wach zu halten, welche schrecklichen Folgen Krieg und Gewaltherrschaft haben); *syn. zu* 2. Mahnmal [n].

M

3513 memorial grave [n] *adm. conserv'hist.* (Burial place memorializing a notable person; ►burial site, ►memorial, ►memorial site) *syn.* memorial tomb [n]; *s* **tumba [f] conmemorativa** (►Tumba con monumento para recordar a una personalidad; ►lugar conmemorativo, ►monumento conmemorativo); *f* **tombeau [m] commémoratif** (Lieu de sépulture commémoré par un monument funéraire ; ►lieu commémoratif, ►mémorial, ►tombe) ; *syn.* sépulture [f] monumentale, tombeau [m] monumental ; *g* **Ehrengrabstätte [f]** (Besondere ►Grabstätte mit Ehrenmal zur Erinnerung an eine bedeutende Persönlichkeit des öffentlichen Lebens; ►Gedenkstätte 1, ►Gedenkstätte 2); *syn.* Ehrengrab [n].

3514 memorial site [n] *adm. hist.* (Place that serves to memorialize a person or historic event; this term also includes places of "commemorative" importance); *s* **lugar [m] conmemorativo** (Lugar que tiene la función de preservar la memoria de una personalidad o un acontecimiento histórico importante); *f* **lieu [m] commémoratif** (Lieu d'évocation de la mémoire d'une personnalité ou d'un événement historique sur lequel est érigé un mémorial) ; *g* **Gedenkstätte [f] (2)** (Ort, an dem zum Andenken an eine Persönlichkeit oder an ein historisches Ereignis ein Denkmal errichtet wurde).

merger [n] [US] *trans.* ►acceleration lane.

merger lane [n] [US] *trans.* ►acceleration lane.

3515 merit award [n] *prof.* (Intermediate category of award given to competition entries which receive monetary awards, but are placed between the top prize winners and ►honorable mention [US]/honourable mention [UK]; **in U.S.**, winning entries are usually given monetary awards for first, second and third places, and certificates for honorable mentions; ►design competition announcement [US]/design competition invitation [UK], ►prize); *syn.* recommended scheme [n] [also UK]; *s* **mención [f] honorífica** (Categoría de premio dada en concurso para participantes que no han logrado los primeros premios; ►anuncio de un concurso [de ideas], ►mención, ►premio); *f* **travaux [mpl] primés** (Récompense pour les travaux de concours non désignés comme lauréat ; ►appel public à la concurrence [pour un concours], ►prix, ►projet mentionné) ; *g* **Ankauf [m]** (*Ideenwettbewerb* Anerkennung von Wettbewerbsarbeiten, die nicht unter den ersten Preisträgern sind; ►Auslobung, ►Preis, ►engere Wahl 2).

3516 mesic [adj] *ecol.* (TEE 1980, 72; descriptive term for a soil temperature regime applied to environment with medium moisture and moderate temperature; living organisms in this environment are called 'mesophytics' or 'mesophilics'); *syn.* mesophytic [adj] (TEE 1980, 140), hydro-mesophilous [adj], hydro-mesophilic [adj]; *s* **mesófilo/a [adj]** (Referente a organismos que prefieren un grado de humedad media. Aquéllos que crecen en medios muy húmedos o encharcados son **higrófilos**); *f* **mésophile [adj]** (Caractérise les organismes ou communautés biologiques préférant se développer dans des conditions d'humidité ; caractère intermédiaire entre hygrophile et xérophile. Les plantes qui vivent dans ces conditions sont des « mésophytes » ; DEE 1982) ; *syn.* hydromésophile [adj] (TSF 1985) ; *g* **mesophil [adj]** (Organismen betreffend, die mittlere Feuchtigkeitsverhältnisse und gemäßigte Temperaturen bevorzugen; Organismen, die feuchtere oder nasse Standorte lieben, sind **hygrophile** Arten); *syn.* feuchtigkeitsliebend [ppr/adj].

mesic habitat [n] *ecol.* ►wetland habitat.

3517 mesic site [n] *phyt. zool.* (TEE 1980, 72; place characterized by a relatively high level of moisture in the soil; ►wetland, ►wetland habitat, ►wetland of international significance); *syn.* humic location [n]; *s* **estación [f] mesofítica** (Lugar caracterizado por suelo con contenido de agua relativamente alto;

►biótopo humedo, ►zona húmeda, ►zona húmeda de importancia internacional); *syn.* residencia [f] ecológica mesofítica; *f* **station [f] humide** (Station caractérisée par une forte teneur en eau du sol ; ►biotope humide, ►zone humide, ►zone humide d'importance internationale) ; *syn.* site [m] humide ; *g* **feuchter Standort [m]** (Standort, der durch einen relativ hohen Wassergehalt im Boden gekennzeichnet ist; ►Feuchtbiotop, ►Feuchtgebiet, ►Feuchtgebiet internationaler Bedeutung).

mesophil species [n] *phyt.* ►moisture indicator plant.

mesophilic species [n] *phyt.* ►moisture indicator plant.

mesophyte [n] *phyt.* ►moisture indicator plant.

mesophytic [adj] *ecol.* ►mesic.

3518 mesotrophic [adj] *ecol.* (Descriptive term applied to an intermediate state of eutrophication with moderate nutrient capacity: between ►eutrophic and ►oligitrophic); *s* **mesótrofo/a [adj]** (Término descriptivo para indicar contenido moderado de nutrientes; ►oligótrofo, ►eutrófico); *syn.* mesotrófico/a [adj]; *f* **mésotrophe [adj]** (Terme désignant un milieu moyennement riche en éléments nutritifs, modérément acide et permettant une activité biologique moyenne ; TSF 1985 ; situé entre le stade ►eutrophe et ►oligotrophe) ; *g* **mesotroph [adj]** (**1.** Ein mittleres Nährstoffangebot in einem Lebensraum betreffend: zwischen ►eutroph und ►oligotroph liegend. **2.** *Gewässergüte bei Stillgewässern* so beschaffen, dass eine mäßige Nährstoffbelastung und mäßige Algenproduktion vorliegt, zeitweise Algenblüten möglich sind, eine mittlere Sichttiefe und eine geringe Sauerstoffzehrung im Tiefenwasser und in der ►Sprungschicht anzutreffen ist).

3519 mesotrophic lake [n] *limn.* (Lake with moderate nutrient capacity, medium plankton production, and a visible depth of over 2 m, which is saturated in deep water at the end of the ►stagnation phase with 30-70% oxygen; ►trophic level); *s* **lago [m] mesótrofo** (Lago con capacidad nutritiva baja, producción moderada de plancton y una visibilidad de más de 2 m, cuyas capas inferiores de agua están saturadas de oxígeno en un 30-70% al final del periodo/período de ►estagnación; ►nivel trófico); *syn.* lago [m] mesotrófico; *f* **lac [m] mésotrophe** (État transitoire dans l'évolution d'un lac du stade oligotrophe au stade eutrophe caractérisé par une faible concentration en éléments nutritifs, une abondance de plancton moyenne, une transparence supérieure à 2 m et dont les eaux profondes à la fin de la période de ►stagnation ont une concentration en oxygène comprise entre 30 et 70 % ; ►niveau trophique) ; *g* **mesotropher See [m]** (See mit geringem Nährstoffangebot, mäßiger Planktonproduktion und Sichttiefen von über 2 m, der im Tiefenwasser am Ende der ►Stagnation zu 30-70 % mit Sauerstoff gesättigt ist; ►Trophiestufe, ►Trophiestufe eines stehenden Gewässers).

mesotrophic peatland [n] *pedol.* ►transition bog.

3520 mesotrophy [n] *limn. pedol.* (An intermediate state of eutrophication; ►eutrophy, ►oligotrophy, ►trophic state); *s* **mesotrofia [f]** (Estado intermedio en el proceso de ►eutrofia; ►estado trófico, ►oligotrofia); *f* **mésotrophie [f]** (Teneur moyenne en substances nutritives ; ►eutrophie, ►oligotrophie, ►degré trophique) ; *g* **Mesotrophie [f]** (**1.** Mittlere Versorgung von Böden und Gewässern mit Nährstoffen. **2.** *limn. Klassifizierung der Gewässergüte* Grad einer mäßigen Nährstoffbelastung mit mäßiger Algenproduktion, zeitweise Algenblüten möglich, mit einer mittleren Sichttiefe und einer geringen Sauerstoffzehrung im Tiefenwasser und in der ►Sprungschicht; ►Trophie, ►Eutrophie, ►Oligotrophie).

3521 metal anchor support [n] *constr.* (Lower part of, e.g. a wooden pergola post to avoid accelerated decay; *specific terms* pin **a. s.**, plate **a. s.**, double plate **a. s.**, bracket **a. s.**; ►pergola post anchor); *s* **soporte [m] metálico** (Parte inferior, p. ej.

de un poste de pérgola, construida de metal para evitar el deterioro prematuro causado por efectos de la intemperie; ►pie de pérgola); *f* **platine-support [f] en métal** (Pièce en métal destinée à supporter, p. ex. le pilier d'une pergola, d'un jeu pour enfants ; ►pied de pergola) ; *g* **Metallfuß [m]** (Unterer Teil eines Bauteiles, z. B. eines Holzpergolapfostens, der aus Gründen der Witterungsbeständigkeit aus Metall gefertigt ist; UBe sind z. B. **M.** mit Dorn, Metallbandfuß, U-förmiger **M.**; ►Pergolenfuß); *syn.* Pfostenschuh [m].

3522 metal fence [n] *constr.* (Enclosure constructed of sectional steel standards with panels of steel bars, steel mesh, or corrugated wire; ►metal grid); *syn.* cyclone fence [n] [*trademark also US*]; *s* **valla [f] de metal** (Cerca contruida de paneles con marcos de hierro perfilado rellenos de barras de metal, malla de acero o alambre ondulado; ►reja de hierro); *f* **clôture [f] métallique** (Clôture réalisée par panneaux monolithiques [nappes et mailles standard] en acier galvanisé et constituée d'une entretoise [ronde, carrée, torsadée] et d'une barre porteuse [profilé plat] fixées entre deux poteaux ; *termes spécifiques* clôture serrurerie, clôture grillagée, clôture à fil de fer ; ►grille métallique) ; *g* **Stahlzaun [m]** (Zaun, dessen Zaunfelder aus Profileisen gebildet und mit Gitterstäben, Baustahlgewebematten oder Wellengitter ausgefüllt sind; ►Stabgitter).

3523 metal grid [n] *constr.* (Grill[e] composed of steel or wrought iron bars of square or round section for gates, fences, balustrades, etc.; ►lattice); *syn.* grill(e) [n]; *s* **reja [f] de hierro** (Rejilla de barras perfiladas de hierro forjado o acero que se utiliza para verjas, puertas, barandillas, etc.; ►reja); *f* **grille [f] métallique** (Assemblage de barreaux métalliques de section ronde ou rectangulaire établissant une séparation [clôture serrurerie] ou fermant une ouverture [portail] ; ►treillis) ; *g* **Stabgitter [n]** (Gitter aus Profilstäben in Stahl oder Schmiedeeisen für Zäune, Tore, Geländer etc.; ►Gitter).

3524 metal grid fence [n] *constr.* *s* **valla [f] de enrejado de metal**; *f* **clôture [f] métallique à caillebotis** ; *g* **Metallgitterzaun [m]**.

metallophyte [n] *phyt.* ►heavy metal-tolerant plant.

metal mesh [n] *constr.* ►chain link metal mesh, ►galvanized chain link metal mesh, ►plastic-coated chain link metal mesh.

3525 metal sheet [n] *constr.* (Metal formed into thin, flat sheets from 0.1 to 1.5mm thick); *syn.* sheet metal [n]; *s* **hoja [f] de metal** (Pieza ancha y fina de metal de un grosor de 0,1-1,5 mm); *f* **feuille [f] de métal** (Produit plat, enroulé ou non, de section transversale pleine rectangulaire, homogène et flexible, en métal tel que l'aluminium, le cuivre, l'étain, etc., de 0,1 à 1,5 mm d'épaisseur) ; *g* **Metallfolie [f]** (Flächiges, in sich homogenes und flexibles Gebilde aus Metall, z. B. Alluminium-F. oder Zinnf. [Staniol] mit Dicken von 0,1-1,5 mm).

metal-tolerant [adj] *phyt.* ►heavy metal-tolerant.

3526 metapotamon [n] *limn.* (Life community of the ►bream zone [UK]); *s* **metapótamon [m]** (Biocenosis de la ►zona de madrilla); *f* **métapotamon [m]** (Communauté animale de la ►zone à brème) ; *g* **Metapotamon [n]** (*Fließgewässer* Lebensgemeinschaft der ►Brachsenregion).

3527 metes and bounds map [n] [US] *adm. surv.* (In U.S., a land map showing property boundaries by directions in degrees and minutes and by distances in meters or feet from a known point of reference; IBDD 1981; in U.K., generic term for a large scale map or plan [from 1:500 to 1:5,000] showing property boundaries and the distribution of land ownership; ►area ownership map); *syn.* cadastral map [n] [US], land registration map [n] [UK]; *s* **mapa [m] catastral** (Mapa que representa los límites parcelarios, formado por compilación de ►planos catastrales; DGA 1986; en D. además de los límites de

las parcelas y de las relaciones de propiedad en el **m. c.** se indica además el valor de las parcelas agrícolas a partir de la clasificación de aptitud de suelos; en los últimos tiempos en los que cada vez se utilizan más los sistemas cartográficos y de recopilación de datos automatizados, la tendencia en D. es concentrar todos los datos y mapas catastrales de diferentes tipos en lo que se viene a denominar «mapa automatizado de bienes inmuebles»); *f* **plan [m] cadastral** (Plan à grande échelle [M 1:500 à 1:5000] institué par la loi du 15 septembre 1807 créant un cadastre parcellaire général sur lequel figurent toutes les parcelles de propriétés et leur numérotation ; depuis 2008 le cadastre numérique [dit PCI ou Plan Cadastral Informatisé], géré par la Direction Générale des Impôts [DGI], est partiellement accessible en ligne ; **plan terrier :** plan des propriétés seigneuriales ayant précédé le plan cadastral dans certaines régions ; ►parcellaire rural) ; *syn.* plan [m] parcellaire ; *g* **Katasterkarte [f]** (*OB* kartenmäßige Darstellung der Besitzgrenzen und Eigentumsverhältnisse in Form einer ►Flurkarte [M 1:500 bis 1:5000] und der Schätzungskarte, die auf der Grundlage der Flurkarte die Ergebnisse der ►Bodenschätzung enthält. Bei den Entwicklungsarbeiten für die Grundstücksdatenbank hat man für den darstellenden Teil des Liegenschaftskatasters den neuen Begriff **Liegenschaftskarte** als Bezeichnung für die Automatisierte Liegenschaftskarte [ALK] geprägt. Dieser Begriff setzt sich immer mehr als Ersatz für die Kataster- [und Flur-]Karte durch).

metes and bounds survey [n] [US] *surv.* ►boundary survey.

method [n] **of measurement** [UK] *constr. contr.* ►method of site survey [US].

3528 method [n] **of site survey** [US] *constr. contr. syn.* method [n] of measurement [UK]; *s* **método [m] de medición** (Técnica utilizada para medir un terreno antes de la planificación o la construcción); *f* **mode [m] d'établissement des métrés** (Méthode permettant la prise en compte des prestations de travaux par des mesures sur site en vue de l'établissement des attachements et des situation, de l'exécution des plans de recollement) ; *g* **Aufmaßverfahren [n]** (**1.** Methode zur Erfassung von vertraglich vereinbarten Bauleistungen nach Fertigstellung für die Abrechnung von Bauleistungen und Lieferungen oder zur Erstellung von Angeboten. **2.** Methode zur Vermessung eines Grundstückes vor Planungs- oder Baubeginn).

metropolitan area [n] *adm. plan. urb.* ►metropolitan region.

3529 metropolitan area planning [n] *plan.* (Complete planning covering several local political jurisdictions; ►regional planning); *syn.* metropolitan region planning [n]; *s* **planificación [f] regional metropolitana (≠)** (Ordenamiento del territorio en toda un área metropolitana, independientemente de las fronteras administrativas; ►planificación regional 1); *f* **planification [f] régionale d'aires métropolitaines** (Planification couvrant le territoire d'une aire métropolitaine ; ►aménagement du territoire au niveau régional) ; *g* **Regionalplanung [f] für Ballungsräume** (►Regionalplanung).

3530 Metropolitan Council [n] **of Governments (COG)** [US] *adm. plan.* (In U.S., an advisory organization of elected representatives from local jurisdictions, authorized by State law to prepare and recommend a coordinated sub-regional plan and, in many States, a Council of Governments or regional planning agency; in U.K., a voluntary association of local planning authorities to coordinate planning policies; in D., an association of local planning authorities usually required by law whose aim is to prepare and implement a coordinated subregional plan); *syn.* Standing Conference [n] of local planning authorities [UK]; *s* **mancomunidad [f] (de municipios)** (En Es., organismo creado por municipios vecinos voluntaria u obligatoriamente para el desarrollo de sus competencias urbanísticas; cf. art. 7 y 9, RD 3288/1978; **en EE.UU.**, organización asesora de representantes elegidos en municipios, autorizada por ley de estado para preparar y recomendar un plan subregional y —en muchos estados— consejo de gobiernos o agencia de planificación regional; **en GB**, asociación voluntaria de autoridades municipales de planificación para coordinar las políticas de planeamiento); *syn.* agrupación [f] de municipios; *f 1* **Établissement [m] Public de Coopération Intercommunale [EPCI]** (Un EPCI est une structure publique administrative regroupant des communes et dotée de la personnalité morale et de l'autonomie financière ayant choisi de développer des « projets communs de développement au sein de périmètres de solidarité » comme par exemple les transports en commun, les travaux publics ; on distingue a] les EPCI à fiscalité propre tels que les **communautés de communes** regroupant les communes de moins de 50 000 habitants, les **communautés d'agglomérations** regroupant les communes entre 50 000 et 500 000 habitants les **communautés urbaines** regroupant les communes de plus de 500 000 habitants, b] les EPCI sans fiscalité propre tels que les syndicats intercommunaux à vocation unique [SIVU] — syndicats intercommunaux d'alimentation en eau potable syndicats intercommunaux d'électrification, etc. — ainsi que les syndicats intercommunaux à vocation multiple [SIVOM] ; comme tous les établissements publics, l'EPCI est régi par le principe de spécialité, il ne peut intervenir que dans le champs des compétences qui lui ont été transférées [principe de spécialité fonctionnelle] et à l'intérieur de son périmètre [principe de spécialité territoriale] ; il n'a donc pas la compétence générale d'une commune ; depuis la loi sur la solidarité et le renouvellement urbain du 13 décembre 2000 les EPCI peuvent adopter des ►schémas de cohérence territoriale [SCOT] ; *f 2* **communauté [f] d'agglomération** (Établissement public de coopération intercommunale à caractère administratif créé dans un périmètre d'urbanisation nouvelle et institué par la loi dite « Chevènement » du 12 juillet 1999 ; il remplace la **communauté [f] d'agglomération nouvelle** [CAN] et le **syndicat [m] d'agglomération nouvelle** [SAN] institués par la loi dite « Rocard » du 13 juillet 1983 pour organiser les villes nouvelles et se substituant eux-mêmes à la structure précédente du **syndicat [m] communautaire d'aménagement** [SCA] ; cette association intercommunale exerce les compétences des communes en matière de programmation et d'investissement dans les domaines du développement économique, de l'urbanisme, du logement, des transports [schéma directeur des transports], de la politique de la ville, des réseaux divers et de la création de voies nouvelles ; Code général des collectivités territoriales, art. L 5333-1 ; cf. agglomérations nouvelles) ; *g* **Nachbarschaftsverband [m]** (Planungsgemeinschaft in Ballungsräumen zwischen einer Großstadt und deren Nachbarschaftsgemeinden mit dem vordringlichen Ziel, für das gesamte Verbandsgebiet eine abgestimmte Flächennutzungs- und Landschaftsplanung zu erarbeiten und für ihre Verwirklichung einzutreten; Nachbarschaftsverbände existieren in A. nicht); *syn.* Kommunalverband [m], Regionale Planungsgemeinschaft [f], Planungsverband [m], Stadtverband [m], Regionalverband [m].

metropolitan park [n] *recr. urb.* ►urban park (2).

3531 metropolitan region [n] *adm. plan. urb.* (**1.** Region within the immediate sphere of influence of a large city, including the city and its suburbs. **2.** Area where economic and social life is influenced by a metropolis and where boundaries are roughly defined by the commuting limits to the center city; IBDD 1981. **3. In U.S., m. r.s** are often linked by common economic, transportation, and cultural factors to create extended **m. r.s** such as the Northeast region from Maine to Virginia, the transportation corridor between Illinois and Pennsylvania, or the West coastal region from San Francisco to San Diego. Large Metropolitan planning districts may include a number of adjacent Metro Regions; ►conurbation, ►Metropolitan Council of Governments [US]/Standing Conference of local planning authorities [UK]); *syn.* metropolitan area [n]; *s* **área [f] metropolitana** (**1.** Término utilizado por primera vez por la Oficina del Censo de los EE.UU. en 1910 para referirse al territorio en el que aparece una ciudad de más de 200 000 habitantes que no se incluye en el área de influencia de ninguna otra gran ciudad. El concepto ha ido modificándose, pero los criterios en los que se apoyan la mayoría de las definiciones son: presencia de una ciudad central con tamaño determinado y gran complejidad funcional, movimientos migratorios pendulares de los trabajadores desde los núcleos satélites, que exista contigüidad y un sistema de transportes y comunicaciones muy desarrollado que garantice las relaciones entre la ciudad central y su área de influencia; cf. DGA 1986. **2.** Unidad de planificación que incluye una ciudad grande y los municipios conlindantes con los que tiene relaciones funcionales; ►mancomunidad [de municipios]); ►aglomeración urbana, ►región urbana); *syn.* región [f] metropolitana, zona [f] metropolitana; *f 1* **aire [f] métropolitaine** (*Aménagement du territoire* grandes agglomérations ou groupes d'agglomérations pour lesquelles ont été élaborés dans les années 1970 des schémas directeurs d'aménagement devant assurer la maîtrise de la croissance urbaine, favoriser l'amélioration du cadre de vie et réaliser les équipements indispensables à la création d'activités nouvelles ; ►communauté d'agglomération, ►Établissement Public de Coopération intercommunale, ►région urbaine, ►zone de concentration urbaine) ; *f 2* **métropole [f] régionale** (Capitale d'un espace régional constituant un carrefour de communication et offrant en sus des fonctions politiques et religieuses des anciennes capitales de province, des activités commerciales diversifiées, des équipements administratifs, sociaux, culturels, de loisirs et sportifs ; la politique de réanimation de la province par la promotion des métropoles régionales a été mise en œuvre dans le cadre de l'aménagement du territoire à travers les ►métropoles d'équilibre ; DUA 1996 ; ►aire urbaine, ►aire métropolitaine) ; *f 3* **métropole [f] d'équilibre** (*Aménagement du territoire* ville ou agglomération importante susceptible de constituer un pôle de développement capable d'attirer des activités et les habitants, échappant ainsi à l'attraction d'une région urbaine, destinée dans les années 60 et 70 à faire contrepoids à la capitale ; créées en 1963 dans le cadre du cinquième plan, huit grandes agglomérations ont ainsi bénéficié d'importants investissements : Lyon, Marseille, Lille-Roubaix-Tourcoing, Nantes-St Nazaire, Strasbourg, Toulouse, Bordeaux, Metz-Nancy-Thionville ; en 1970 la liste s'est élargie aux villes assimilées aux métropoles d'équilibre Clermont-Ferrand, Dijon, Nice, Rennes, Rouen et inclus désormais les villes de Caen, Limoges et Montpellier ; aujourd'hui l'agglomération lyonnaise inclue les villes de Saint-Étienne et Grenoble ; DUA 1996) ; **CH.**, les métropoles suisses sont des pôles économiques majeurs qui se distinguent par la productivité du travail, la spécialisation des emplois et surtout leur rayonnement mondial ; on compte 5 métropoles : Zurich, Bâle, Berne, la métropole lémanique [Genève-Lausanne] et la métropole tessinoise) ; *g* **Ballungsraum [m]** (**1.** Durch besonders starke Konzentration von Wohnungen und Arbeitsstätten dicht besiedelte Region und das dazu gehörende Hinterland; in D. hat GERHARD ISENBERG (1902-1982) den **B.** als eine zusammenhängende Fläche definiert, auf der mehr als eine halbe Million Menschen bei einer Bevölkerungsdichte von ca. 1000 EW je km² leben; dies kann durch eine einzelne Stadt oder durch mehrere Städte erreicht werden. Die größten **B.räume** in D. sind das Rhein-Ruhr- und Rhein-Main-Gebiet, Berlin, Hamburg, München, Stuttgart und Dresden-Leipzig-Chemnitz. Die Ministerkonferenz für Raumordnung [MKRO] hatte 1995 mit ihrem Beschluss zum raumordnungspolitischen Handlungsrahmen die europäische Bedeutung der **B.räume** unterstrichen und sieben **Metropol-regionen** ausgewiesen, die als Motoren der gesellschaftlichen, wirtschaftlichen, sozialen und kulturellen Entwicklung mit guter Erreichbarkeit auf europäischer und internationaler Ebene und weiter Ausstrahlung auf das Umland wirken: Rhein-Ruhr [10,2 Mio. EW], Frankfurt/Rhein-Main [5,3], Hamburg [4,3], Berlin-Brandenburg [4,2], Sachsendreieck [Halle-Leipzig-Dresden-Chemnitz-Zwickau; 3,2], München [2,6] und Stuttgart [2,5]; 2005 kamen noch vier kleinere hinzu: Hannover-Braunschweig-Göttingen [3,9 Mio. EW], Nürnberg [2,5], Rhein-Neckar [2,4] und Bremen-Oldenburg mit über 2,3 Mio. EW. Seit der Ausweisung der „kleineren" Europäischen Metropolregionen 2005 ist jede deutsche Stadt über 400 000 EW Kernstadt einer **M.**; somit gibt es in D. 41 **B.räume** um kleinere Kernstädte; in der **Schweiz** gibt es fünf **M.** — auch **Metropolitanregion** genannt, z. T. grenzübergreifend: Zürich [1,6 Mio. EW], Genf-Lausanne [1,2], Basel [0,8], Bern [0,7] und Südtessin [0,5]; in **Österreich** ist es die Ostregion um Wien [2 Mio.]; die vier größten europäischen **M.** sind Moskau [14,5 Mio. EW], London [12,5], Istanbul [12] und Île-de-France im Pariser Becken [11]; cf. Wikipedia. 2. Städtische **B.räume** sind eine typische Erscheinung des 20. Jhs. Dieses Phänomen erkannte zuerst der amerikanische Historiker und Soziologe LEWIS MUMFORD, der 1938 den Terminus **Megalopolis** einführte. Der franz. Geograph JEAN GOTTMANN untersuchte dann 1961 in den USA die ausgedehnte Verstädterungszone zwischen Boston und Washington, kurz *Boswash* genannt. Weitere **M.polen** sind die „M. der Großen Seen" und „Sansan" in Kalifornien; in der übrigen Welt z. B. Peking, Schanghai-Nanking, Tokio-Kōbe, Mumbay, Kalkutta, Delhi, Melbourne, Sydney, Kairo, Johannesburg, Buenos Aires, São Paulo, Rio de Janeiro und die Stadt Mexiko; ▶Stadtregion, ▶Verdichtungsraum); *syn.* Ballungsgebiet [n], Ballungszentrum [n], Metropolregion [f], Metropolitanregion [f] [auch CH]; *g 2* **Großraum [m]** (*Großräumige Planungseinheit* Großstadt mit umliegenden Gemeinden; ▶Nachbarschaftsverband).

metropolitan region planning [n] *plan.* ▶metropolitan area planning.

Michigan peat [n] [US] *pedol.* ▶humified raised bog peat.

microbial loop [n] *ecol.* ▶organic turnover.

3532 microclimate [n] *met.* (Climatic conditions of a small area resulting from local differences in elevation, exposure, or vegetation cover; ▶site climate, ▶topoclimate); *s* **microclima [m]** (Conjunto de condiciones climáticas que caracterizan una zona limitada, con frecuencia de dimensiones muy reducidas, y a la que se hace referencia al estudiar no sólo pequeños organismos sino también diversos hábitats o medios; DINA 1987; *términos específicos* microclima del suelo, microclima de cultivo, microclima del bosque, ▶microclima de estación; ▶mesoclima); *f* **microclimat [m]** (Ensemble des conditions atmosphériques [température, vent, humidité du sol, etc.] qui caractérisent un espace homogène de faible étendue et souvent dépendantes des facteurs topographiques biotiques ou abiotiques locaux ; LA 1981 ; ▶climat local, ▶climat stationnel, ▶mésoclimat) ; *g* **Mikroklima [n]** (…ta [pl]; *fachsprachlich* …mate [pl]; in Bodennähe durch wechselnde Art und Feuchte des Bodens, durch geringfügige Neigungen der Bodenoberfläche und durch die Art und Höhe der Vegetationsbedeckung bedingte, vom Mesoklima abweichende meteorologische Einzelerscheinungen; cf. GEI 1961; ▶Geländeklima, ▶Standortklima); *syn.* Kleinklima [n], Lokalklima [n], Ortsklima [n].

3533 microorganism [n] *biol. pedol.* (Minute unicellular ▶soil organisms, mostly only a few thousandths of a millimeter in size; e.g. algae, bacteria, fungae, protozoa, and viruses);

s **microorganismo [m]** (Planta o animal sumamente pequeño, invisible a simple vista; ▶edafon); *f* **micro-organisme [m]** (Organisme unicellulaire vivant dans le sol. Les **m-o.s** jouent un rôle fondamental dans la fertilité du sol. Ils sont plusieurs milliards par gramme de sol ; LA 1981 ; ▶édaphon) ; *syn.* microbe [m] ; *g* **Mikroorganismus [m]** (Kleinster, meist nur wenige tausendstel Millimeter großer, einzelliger Organismus der ▶Bodenlebewelt); *syn.* Kleinlebewesen [n].

middle landscape [n] [UK] *plan.* ▶transition zone.

3534 middle reaches [npl] *geo.* (Midway stretches of a stream or river; ▶headwater stream, ▶lower reaches, ▶upper reaches); *s* **tramo [m] medio (de un curso de agua)** (Parte del cauce en la zona media de la cuenca de drenaje; ▶cabecera [de río], ▶tramo inferior [de un curso de agua], ▶tramo superior [de un curso de agua]); *f* **cours [m] moyen** (▶cours inférieur, ▶cours supérieur, ▶cours supérieur près de la source) ; *g* **Mittellauf [m]** (Mittlerer Teil eines Fließgewässers; ▶Oberlauf, ▶Quelllauf, ▶Unterlauf).

3535 middle river terrace [n] *geo.* (Medial ▶gravel river terrace in river valleys which has been formed by a continual process of accumulation and erosion during the last Pleistocene ice age; ▶lower river terrace, ▶upper river terrace); *s* **terraza [f] fluvial media** (▶Terraza [fluvial] de aluviones entremezclados con lechos de arenas y limos en valle fluvial que se formó en la última glaciación del pleistoceno gracias a procesos de acumulación [sedimentación] y erosión; cf. DGA 1986; ▶terraza fluvial alta, ▶terraza fluvial baja); *f* **terrasse [f] (fluviatile) moyenne** (Étendue intermédiaire de la ▶terrasse alluviale taillée dans une partie plus ancienne [▶terrasse (fluviatile) inférieure] du lit d'un cours d'eau abandonné, constituée par les alluvions accumulées et érodées pendant la dernière période de glaciation du Pléistocène ; ▶terrasse [fluviatile] supérieure) ; *g* **Mittelterrasse [f]** (Mittlere ▶Schotterterrasse in Flusstälern, die durch mehrfachen Wechsel von Akkumulation und Erosion während der letzten pleistozänen Kaltzeit entstanden ist. Die **M.** liegt über der ▶Niederterrasse; ▶Hochterrasse).

3536 mid-rise dwellings development [n] *arch. urb.* (Series of linear apartment buildings containing from three to seven stories and situated perpendicular to a road; ▶block development); *s* **edificación [f] en hilera** (Tipo de ordenación de edificios de viviendas de varios pisos en perpendicular a la calle. En comparación con la ▶edificación en bloque cerrado tienen la desventaja de que el ruido de la calle afecta también a los espacios libres intermedios. Sin embargo, como los patios interiores de los bloques son pequeños, la **e. en h.** tiene la ventaja de permitir mejor iluminación y aeración de todos los espacios de habitación); *f* **construction [f] en ligne** (Implantation des constructions en ordre continu le long de la voirie ; ▶construction en îlot) ; *g* **Zeilenbebauung [f] (±)** (Besondere Bauweise, bei der zeilenartig im Allgemeinen mehrgeschossige Wohngebäude senkrecht zur Straße angeordnet werden oder sind. Im Vergleich zur ▶Blockbebauung, bei der das Innere vom Straßenlärm abgeschirmt ist, dringt bei der **Z.** besonders der Verkehrslärm ungehindert in die Freiräume zwischen den Gebäudereihen); *syn.* Zeilenbauweise [f].

migrant [n] *zool.* ▶local migrant, ▶partial migrant.

migrant [n]**, passage** *zool.* ▶bird of passage.

migrant bird [n] *zool.* ▶migratory bird.

3537 migration [n] (1) *sociol.* (Process of spatial movements of people or large parts of a population from one area to settle in another for economic, political, social, climatological, demographic, religious or other reasons; *context* emigration from a town to outskirt areas; other forms of spatial mobility, such as travel and commuting do not count as **m.**, even though the latter

M

is sometimes viewed as a circulatory form of migration; cf. BROCK 1994, vol. 23, 569; **m**. distinguishes the following: **exodus** [emigration from an area], **internal migration** [migration within a country]: the **migration gain** in one country leads to the **migration loss** in the other: cf. brain drain; ►depopulation); *s* **migración [f] (1)** (Proceso de desplazamiento duradero de personas o grupos de población de una región a otra para establecerse en ella, sea en el mismo país [**migración interior**] o en el extranjero [**migración internacional**], aunque ésta sólo supone el 3% del total mundial. Para medir la importancia de la **m**. se utilizan diferentes índices, siendo muy común el del **saldo migratorio**, o sea la diferencia entre la emigración y la ►inmigración para un territorio y un periodo/período de tiempo dados. Las causas de la **m**. pueden ser, entre otras, económicas, políticas, sociales, ambientales, étnicas, religiosas, pudiendo ser libre o espontánea, dirigida o forzada. La modalidad fundamental de la **m**. interior fue históricamente el ►éxodo rural que se produjo sobre todo a partir de la revolución industrial y que perdura en los países en desarrollo. Otra modalidad muy común en países en desarrollo es la **migración estacional**, es decir, el desplazamiento de trabajadores/as rurales a los lugares de cosecha; este tipo de **m**. prevalece puntualmente en los países industrializados, aunque generalmente se trata de trabajadores extranjeros quienes realizan las labores agrícolas. En América Latina las principales causas de la **m. i**. son económicas, siendo los EE.UU. el principal país de destino, pero también en creciente grado los países latinoamericanos más desarrollados [Argentina, Chile, México] así como Europa, y políticas, siendo actualmente Colombia el país con más población desplazada del mundo —afuera y dentro de sus fronteras—, debido a las luchas internas [guerrilla, paramilitares] y a la expulsión forzada de los campesinos de sus tierras por grandes terratenientes. La **migración pendular** o alternante, o sea el desplazamiento de una persona entre su lugar de residencia y su lugar de trabajo, no se considera como **m**. propiamente dicha, sino que es una modalidad de la **migración habitual**, que es el desplazamiento repetitivo y de corta duración que no supone un cambio de residencia de las personas que lo realizan; cf. DGA 1986; ►desertización, ►despoblamiento); *syn*. emigración [f]; *f* **migration [f] (1)** (Déplacement d'une personne ou de populations qui quittent leur lieu de naissance ou de résidence pour un autre lieu. La migration est le plus souvent une **migration interne**, c'est le cas des personnes se mouvant dans leur pays d'origine, p. ex. qui quittent la campagne pour aller s'installer en ville. La **migration externe** [migration internationale] est rare et ne touche que 3 % de la population mondiale ; pour le territoire d'origine on parle de **perte migratoire** pour le territoire de destination du **gain migratoire** et d'**excédent migratoire** si le nombre des entrées est supérieur au nombre des départs. On peut distinguer les différentes formes de migration selon les motivations [économiques, familiales, politiques, religieuses] ou selon les statuts légaux des personnes concernées. Le terme **migrant** est généralement compris comme toute personne vivant de façon temporaire ou permanente dans un État dont elle n'est pas ressortissante. Pour le pays de départ, le migrant est un **émigrant**, et pour celui d'arrivée, un **immigrant**. F., les démographes appellent « **immigrés** » toute personne née étrangère à l'étranger. Ils excluent des immigrés les personnes nées à l'étranger de parents français, comme les enfants d'expatriés. En 1999, la France comptait 7,3 % d'immigrés et 5,6 % d'étrangers. Une partie des immigrés sont en effet devenus français après leur arrivée. La part des immigrés est stable dans la population française, l'arrivée de nouveaux immigrés étant compensée par les départs et la mortalité. Le **solde migratoire** — différence entre le **solde migratoire intérieur** et le **solde migratoire extérieur** — est d'environ 100 000 personnes par an [en 2004]. Il contribue pour près d'un tiers à la croissance démographique, les

deux autres tiers venant du solde naturel ; cf. site de la cité des sciences et de l'industrie. La **migration circulaire** est une forme de migration favorisant les allers-retours [retour durable]. Elle incite les migrants à investir dans le pays de départ et qui reste, du coup, le principal milieu de vie et permet d'éviter les drames liés à la migration clandestine ; ►dépeuplement) ; *g* **Migration [f] (1)** (Prozesse räumlicher Bewegungen von Leuten oder großer Bevölkerungsteile aus einem Gebiet, um sich woanders aus ökonomischen, politischen, sozialen, klimatologischen, demografischen, religiösen oder anderen Gründen niederzulassen; *Kontext* Abwanderung aus einer Stadt ins Umland; andere Formen räumlicher Mobilität wie Reisen und Pendeln zählen nicht zur **M**., auch wenn Letzteres zuweilen als Form der **zirkularen Wanderung** angesehen wird; cf. BROCK 1994, Bd. 23, 569; man unterscheidet **Außenwanderung** [Abwanderung aus einem Gebiet] und **Binnenwanderung** [Wanderungsbewegungen innerhalb eines Staates von Land zu Land: der **Wanderungsgewinn** des einen Landes führt zum **Wanderungsverlust** des anderen; cf. ALG 2004, 71]; ►Entvölkerung); *syn*. Abwanderung [f], Wanderung [f].

3538 migration [n] **(2)** *zool*. (Generic term for re-locating of one or many individuals of an animal population including ►emigration, ►immigration, ►instictual migration, ►invasion 1, ►nomadism, ►permigration, ►return migration); *s* **migración [f] (2)** (Término genérico para los diferentes cambios de lugar de estancia de los individuos de una población faunística como ►emigración, ►inmigración, ►migración transeunte e ►invasión; ►migración de contrapasa, ►nomadismo, ►vagabundismo); *f* **migration [f] (2)** (Déplacement géographique d'un animal ou d'un groupe d'animaux ; ►émigration, ►immigration, ►invasion, ►migration de déplacement, ►migration de retour, ►nomadisme) ; *g* **Migration [f] (2)** (Form der Ortsveränderung einzelner bis vieler Individuen einer Tierpopulation wie ►Emigration, ►Immigration, ►Invasion und ►Permigration; ►Nomadentum 1, ►Nomadentum 2, ►Rückzug von Zugvögeln).

migration [n]**, capillary** *pedol*. ►capillary rise.

migration [n] **from rural areas** *agr. sociol*. ►rural exodus.

3539 migratory bird [n] *zool*. (Bird species, which migrates annually to its specific breeding ground or ►overwintering area as opposed to a ►permanent resident. If all the individuals in a population migrate to the hibernation area, they are referred to as true migratory birds, if only part of the species migrate they are termed ►**partial migrants**; ►breeding range, ►local migrant, ►overwintering (migratory) bird); *syn*. migrant bird [n]; *s* **ave [f] migradora** (Ave que —a diferencia de las ►aves sedentarias— se traslada dos veces al año a grandes distancias como respuesta al cambio de estaciones, en otoño buscando ambientes más cálidos a su ►área de invernada y en primavera a su cuartel de verano donde se reproduce [►área de cría]. Cuando sólo una parte de los individuos de una especie migra o todos los individuos de una especie se trasladan pero a corta distancia, se habla de ►**aves migradoras parciales**; ►ave de dispersión posgenerativa, ►ave invernante); *syn*. ave [f] migratoria; *f* **oiseau [m] migrateur** (Oiseau qui, à la différence des ►oiseaux sédentaires, effectue des voyages réguliers, cycliques entre les régions vouées à la reproduction au printemps et aux aires de repos l'hiver ; lorsque l'espèce hiverne totalement en dehors de l'►aire de nidification, on parle de visiteurs d'été et si une partie de l'espèce quitte en hiver le nord de son aire de distribution, on parle alors d'►oiseaux migrateurs partiels ; ►oiseau erratique, ►oiseau hivernant, ►quartiers d'hiver) ; *syn*. migrateur [m] ; *g* **Zugvogel [m]** (Vogel, der im Unterschied zum ►Standvogel jedes Jahr zu seinem artspezifischen Brutgebiet [►Brutareal] resp. ►Winterquartier 2 zieht. Wenn alle Individuen einer Art in ihr Winterquartier ziehen, spricht man von „echten Zugvögeln", zieht nur ein Teil dieser Art, werden diese Individuen ▶Teilzieher genannt; ▶Strichvogel, ▶Überwinterer).

migratory path [n] *game'man. hunt.* ▶game trail.

3540 migratory species [n] *conserv. zool.* (Entire population or any geographically separate part of the population of any species or lower taxon of wild animals, a significant proportion of whose members cyclically and predictably cross one or more national jurisdictional boundaries; Art. I Convention on the Conservation of Migratory Species of Wild Animals: ▶Bonn Convention, ▶Convention on the Conservation of African-Eurasian Migratory Waterbirds [AEWA]); **s especie [f] migratoria** (El conjunto de la población o todas las poblaciones parciales separadas geográficamente de todas las especies o todos los taxones de animales silvestres, de cuyos cuales una parte importante cruza cíclicamente y de forma previsible uno o varios límites de jurisdicción nacional; art. I, Convención de 23 de junio de 1979 sobre conservación de especies migratorias de animales silvestres; ▶Convención de Bonn, ▶Acuerdo para la Conservación de las Aves Acuáticas Migratorias de África y Eurasia [AEWA]); *syn.* especie [f] migradora; *f* **espèce [f] migratrice** (Ensemble de la population ou toute partie séparée géographiquement de la population de toute espèce ou de tout taxon inférieur d'animaux sauvages, dont une fraction importante franchit cycliquement et de façon prévisible une ou plusieurs des limites de juridiction nationale ; cf. art. I, Convention sur la conservation des espèces migratrices appartenant à la faune sauvage : ▶Accord sur la Conservation des Oiseaux d'Eau Migrateurs d'Afrique-Eurasie [AEWA], ▶Convention de Bonn) ; *g* **wandernde (wild lebende) Tierart [f]** (Gesamtpopulation oder eine geographisch abgegrenzte Teilpopulation jeder Art oder jedes niedrigeren Taxon wild lebender Tiere, von denen ein bedeutender Anteil zyklisch und vorhersehbar eine oder mehrere nationale Zuständigkeitsgrenzen überquert; Art. 1 ▶Bonner Konvention, ▶Abkommen zur Erhaltung der afrikanisch-eurasischen wandernden Wasservögel [AEWA]).

migratory species [n] **of wild animals** *conserv.* ▶effective management of migratory species of wild animals.

Migratory Species [n] **of Wild Animals, Convention on the Conservation of** *conserv.* ▶Bonn Convention.

migratory trail [n] *game'man. hunt.* ▶game trail.

migratory waterbirds [npl] *conserv. ecol. pol.* ▶Convention on the Conservation of African-Eurasian Migratory Waterbirds [AEWA].

mild humus [n] *pedol.* ▶mull.

milieu [n], **urban** *urb.* ▶urban environment.

mine dump [n] [US] *min.* ▶spoil bank [US] (2)/spoil heap [UK].

3541 mineral aggregate [n] *constr.* (Well-graded mixture of several inert materials such as sand, gravel, slag, or crushed stone); **s material [m] pétreo** (Mezcla de materiales minerales de diferentes tamaños, como arena, grava, piedra partida, sin material ligante); *syn.* agregado [m] pétreo, agregado [m] mineral, árido [m] pétreo; *f* **grave [f]** (Mélange de matériaux naturels ou non, en une seule ou deux fractions à granulométrie étalée [les granulométries types des graves sont 0/14 mm, 0/20 mm, 0/32 mm et 0/63 mm] continue [cailloux, graviers et sable], non traité ; *termes spécifiques* grave concassée, grave traitée [mélange de grave et d'un liant tels que grave-bitume, grave-cendre, grave-ciment]) ; *g* **Mineralbeton [m]** (Mischung ohne Bindemittel aus kornabgestuften mineralischen Stoffen, z. B. Sand, Kies, Schotter, in den Körnungen 0/16, 0/32, 0/45 mm); *syn.* Mineralgemisch [n].

mineral compounds [npl] *chem. pedol.* ▶mineral elements.

3542 mineral cycle [n] *chem. pedol.* (Permanent process of releasing nutrients through mineralization to make them available again for plant growth; ▶mineralization, ▶nutrient cycling); **s ciclo [m] de (los elementos) minerales** (Proceso continuo por el cual se liberan los nutrientes del suelo para suministrar a las plantas por medio de la ▶mineralización; ▶ciclo de nutrientes); *f* **cycle [m] biogéochimique des éléments minéraux** (Retour annuel à la surface du sol, au sein de la litière, d'une grande partie des éléments prélevés en profondeur par les racines [les substances nutritives libérées par ▶minéralisation — réserves mobilisables — deviennent utilisables par les plantes], l'apport régulier compensant les pertes par entraînement ; PED 1983 ; ▶cycle biologique des éléments nutritifs) ; *g* **Mineralstoffkreislauf [m]** (Ständiger Vorgang, bei dem durch ▶Mineralisierung freigesetzte Nährstoffe dem Pflanzenwachstum wieder zur Verfügung stehen; ▶Nährstoffkreislauf).

3543 mineral elements [npl] *chem. pedol. syn.* mineral substances [npl], mineral compounds [npl]; **s nutrientes [mpl] anorgánicos** *syn.* sustancia [f] mineral, nutriente [m] mineral, elementos [mpl] minerales; *f* **éléments [mpl] minéraux** *syn.* nutriments [mpl] minéraux (DUV 1984, 165) ; *g* **Mineralstoffe [mpl]** *syn. agr. for. hort.* mineralische Nährstoffe [mpl].

mineral elements [npl], **import/input of** *envir. limn. pedol.* ▶matter import.

mineral exploitation [n] *min.* ▶mineral working.

3544 mineral extraction [n] *min.* (*Generic term* process of obtaining valuable minerals, clay, sand, gravel, or rock, etc., using various types of earth moving equipment; ▶borrow pit, ▶dredging 1, ▶dry excavation, ▶extraction site, ▶gravel extraction, ▶mineral working, ▶open-pit mining [US]/open-cast mining [UK], ▶quarrying, ▶sand extraction, ▶tape site, ▶permit for mineral extraction [US]/planning permission for mineral extraction [UK]); *syn.* mineral excavation [n]; **s 1 explotación [f] superficial de recursos minerales** (Extracción superficial de minerales; ▶dragado subacuático, ▶explotación minera, ▶extracción de grava, ▶extracción en seco [de rocas y arenas], ▶minería a cielo abierto, ▶punto de alumbramiento); **s 2 extracción [f] de minerales superficiales o rocas industriales** (Término genérico; *términos específicos* ▶extracción de arenas, ▶extracción de grava, ▶explotación minera, ▶minería en cantera; ▶zona de extracción, ▶zona de préstamo); *f* **extraction [f] des substances minérales** (Enlèvement à ciel ouvert des ressources minérales ; ▶exploitation des substances minérales, ▶exploitation [minière/de carrière] à ciel ouvert ; *termes spécifiques* ▶extraction de gravier, ▶extraction de matériaux alluvionnaires en nappe, ▶extraction de matériaux alluvionnaires hors nappe, ▶extraction de pierres ; ▶extraction de sable, ▶lieu d'emprunt, ▶point de prélèvement, ▶site d'extraction de matériaux) ; *syn.* extraction [f] minière ; *g* **Materialentnahme [f]** (Oberirdische Entnahme von Steinen und Erden; ▶Abbau 3, ▶Entnahmestelle 2, ▶Entnahmestelle 3, ▶Gesteinsabbau, ▶Kiesabbau, ▶Nassbaggerung, ▶Sandabbau, ▶Seitenentnahme, ▶Tagebau, ▶Trockenabbau); *syn.* Gewinnung [f] von Steinen und Erden, Bodenabbau [m], Bodengewinnung [f], Materialgewinnung [f] [CH].

mineral extraction [n] [UK], **planning permission for** *adm. leg. min.* ▶permit for mineral extraction [US].

3545 mineral extraction [n] **by an excavator** *min.* (Removal of gravel, sand, clay, etc. with the help of heavy equipment; ▶excavation 1); **s extracción [f] mineral con excavadora** (Término genérico para el proceso de ganancia de minerales, arenas, arcillas o grava; extraer [vb]; ▶excavación de tierras o rocas); *f* **extraction [f] réalisée avec une pelle** (Terme générique pour l'action d'enlèvement de matériaux tels que gravier, de sable, etc. au moyen d'une pelle, d'un bulldozer, etc. ; ►fouille) ; *g* **Ausbaggerung [f] (2)** (Gewinnung von Lockergesteinen mit Hilfe eines Baggers, z. B. Kies, Sand, Ton, Schlick/Schlamm; ►Abgrabung).

3546 mineral fertilization [n] *agr. constr. for. hort.* (Application of ►mineral fertilizer to the soil); *s* **fertilización [f] mineral** (Aplicación de ►fertilizante mineral); *syn.* abonado [m] mineral; *f* **fertilisation [f] avec des engrais minéraux** (Apport d'►engrais minéral) ; *g* **Mineraldüngung [f]** (Ausbringen von ►Mineraldünger).

3547 mineral fertilizer [n] *agr. constr. for. hort.* (Chemically or mechanically manufactured inorganic material added as a soil amendment to supply elements necessary for plant growth; ►chemical fertilizer; *opp.* ►organic fertilizer); *s* **fertilizante [m] mineral** (►fertilizante [mineral] artificial; *opp.* ►fertilizante orgánico); *syn.* abono [m] mineral, abono [m] anorgánico); *f* **engrais [m] minéral** (Matière inorganique obtenue à partir de sels minéraux naturels ou artificiels dont la fonction principale est d'apporter aux plantes cultivées des éléments directement utiles à leur nutrition ; ►engrais artificiel, ►engrais organique) ; *g* **Mineraldünger [m]** (Chemisch oder mechanisch durch Gesteinszerkleinerung hergestellter Dünger; ►Kunstdünger; *opp.* ►organischer Dünger); *syn.* anorganischer Dünger [m].

3548 mineralization [n] *pedol.* (Conversion of an element from an organic form to an inorganic state as a result of microbial decomposition; RCG 1982; ►nutrient mobilization); *s* **mineralización [f]** (Liberación de la fracción de materia mineral contenida en los compuestos orgánicos por efecto de la actividad de los microorganismos del suelo; FAO 1960; ►movilización de nutrientes); *f* **minéralisation [f]** (Transformation lente de l'humus en matière minérale sous l'influence des micro-organismes du sol ; DIS 1986 ; ►libération de/des substances nutritives) ; *g* **Mineralisierung [f]** (Vorgang der Zersetzung organischer Verbindungen durch Mikroorganismen in anorganische Verbindungen, bei dem z. T. Mineralstoffe freigesetzt werden; SS 1979; ►Mobilisierung von Nährstoffen).

3549 mineral resources [npl] *geo. min.* (Geological formations of potential economic value to man, when ►deposits are exploited by deep or surface mining; ►development of mineral resources; ►natural resources 1); *s* **recursos [mpl] minerales** (Materias primas inertes con valor económico que se encuentran en la corteza terrestre y pueden ser extraídas por minería subterránea o a cielo abierto; ►ganancia de recursos geológicos, ►recursos naturales, ►yacimiento); *syn.* recursos [mpl] geológicos, riquezas [fpl] del subsuelo; *f* **richesses [fpl] naturelles du sous-sol** (Formations géologiques de grande valeur économique pour l'homme, sous forme de ►gisements de minéraux et de fossiles exploitables à ciel ouvert ou en sous-sol ; ►recherche et exploitation des ressources minérales, ►ressources fondamentales naturelles) ; *syn.* ressources [fpl] minières, ressources [fpl] en minéraux ; *g* **Bodenschätze [mpl]** (In ►Lagerstätten vorhandene, für den Menschen wertvolle Rohstoffe, die im Untertage- oder Tagebau abgebaut werden können; ►Erschließung von Bodenschätzen; ►natürliche Grundgüter).

mineral resources [npl], **exploitation of** *min.* ►mineral working.

3550 mineral soil [n] *pedol.* (Soil consisting predominantly of, and having its properties determined predominantly by, mineral matter usually containing less than 20 percent organic matter; RCG 1982; ►organic soil); *s* **suelo [m] mineral** (Suelo terrestre, semiterrestre o subacuático constituido mayormente por material anorgánico; *opp.* ►suelo orgánico); *f* **sol [m] minéral** (Sol terrestre, semiterrestre ou subaquatique constitué en grande

partie d'éléments minéraux ; *opp.* ►sol organique) ; *g* **Mineralboden [m]** (Terrestrischer, semiterrestrischer oder subhydrischer Boden, der überwiegend aus anorganischen Bestandteilen besteht; *opp.* ►organischer Boden).

mineral substances [npl] *chem. pedol.* ►mineral elements.

3551 mineral working [n] *min.* (Generic term for exploitation of mineral resources by ►open-pit mining [US]/open-cast mining [UK] or ►deep mining; ►dredging 1, ►dry excavation, ►excavation 1, ►mineral extraction, ►mining; ►quarry); *syn.* exploitation [n] of mineral resources, mineral exploitation [n], mineral extraction [n]; *s* **explotación [f] minera** (Término genérico para ►dragado subacuático, ►excavación de tierras o rocas, ►extracción de minerales superficiales o rocas industriales, ►extracción en seco [de rocas y arenas], ►minería, ►minería a cielo abierto, ►mineria en galeria; ►cantera); *syn.* explotación [f] de yacimientos y recursos geológicos; *f* **exploitation [f] des substances minérales** (Terme générique pour l'exploitation de gisements à ciel ouvert et l'exploitation souterraine ; ►carrière de roches massives, ►exploitation [minière/de carrière] à ciel ouvert, ►exploitation minière, ►exploitation minière souterraine, ►extraction de matériaux alluvionnaires en nappe, ►extraction de matériaux alluvionnaires hors nappe, ►extraction des substances minérales, ►fouille) ; *syn.* extraction [f] des ressources minérales ; *g* **Abbau [m, o. Pl.] (3)** (OB zu ►Abgrabung, ►Materialentnahme, ►Nassbaggerung, ►Tagebau, ►Trockenabbau; und ►Untertagebau; ►Bergbau, ►Steinbruch); *syn.* Gewinnung [f] von Steinen, Erden und anderen Bodenschätzen; § 9 [1] 17 BauGB).

3552 mineral working site [n] *min. plan.* (Location of mineral working activities; ►extraction site); *syn.* pits [npl] and quarries [npl]; *s* **área [f] de explotación minera (mina o cantera)** (►zona de extracción); *syn.* perímetro [m] de explotación minera (mina o cantera), área [f] minera (mina o cantera) [AL]; *f* **emprise [f] d'extraction** (Périmètre de la surface d'extraction des substances minérales ou fossiles ; ►site d'extraction de matériaux) ; *syn.* surface [f] d'extraction ; *g* **Abbaufläche [f]** (Fläche, die zur Gewinnung von Steinen, Erden oder anderen Bodenschätzen bestimmt ist; ►Entnahmestelle 2); *syn.* Entnahmefläche [f], Abbaubereich [m], Rohstoffgewinnungsfläche [f].

3553 minerotrophic peat [n] *pedol.* (VIR 1982, 345; peat developing in a ►low bog); *syn.* low-moor peat [n] [also US] (SAF 1983); *s* **turba [f] minerótrofa** (Turba formada en ►turbera baja); *syn.* turba [f] topógena; *f* **tourbe [f] topogène** (Tourbe formée dans les ►tourbières basses) ; *syn.* tourbe [f] à Hypnacées ; *g* **Niedermoortorf [m] (1)** (Im ►Niedermoor entstandener Torf); *syn.* topogener Torf [m].

minerotrophic peatland [n] *geo.* ►fen (1).

minerotrophic peatland moor [n] *pedol. phyt.* ►low bog.

miniature fruit tree [n] [US] (1) *hort.* ►dwarf fruit tree [US].

3554 miniature fruit tree [n] [US] (2) *hort.* (Cultivar of a fruit tree grafted on a seedling or standardized understock, and patented in U.S.; ►pyramidal dwarf fruit tree [US]/pyramidal fruit bush [UK], ►dwarf fruit tree [US]/short standard [UK]); *s* **frutal [m] enano** (*Fruticultura* especie frutal de forma libre con altura de tronco de unos 60 cm, el siguiente menor en tamaño es el ►arbusto de tipo «Spindel», el siguiente mayor el ►árbol de pie bajo); *f* **buisson [m] (d'arbres fruitiers)** (*Culture fruitière* arbre de forme libre [non palissée], commercialisé jusqu'à six ans, présentant des charpentières régulièrement réparties autour du tronc, la plus basse étant située à environ 30 à 45 cm audessus de la greffe ; ►arbre courte-tige, ►arbre mini-tige, ►quenouille [d'arbres fruitiers]) ; *g* **Buschbaum [m]** (*Obstbau* auf Typenunterlage oder Sämling veredelte Obstbaumform mit einer Stammhöhe von 60 cm. Die nächst niedrige Obstbaumform ist der ▶Spindelbusch, die nächst höhere der ▶Niederstamm); *syn.* Busch [m].

3555 miniature garden [n] *gard.* (Garden occupying an extremely small area; e.g. rock garden, plant container or trough on roofs, terraces or balcony balustrades. They may be up to 100 m² in size but have no direct contact with natural ground; ▶window box, ▶balustrade planter, ▶planter serving as a handrail, ▶container garden [US]/trough garden [UK]); *syn.* postagestamp yard [n] [also US] (NAB 1992, 22); *s* **jardín [m] miniatura** (Jardín de superficie muy reducida como en rocalla en miniatura, jardinera en terrazas, balcones o azoteas; ▶jardinera en barandilla, ▶jardín miniatura en jardinera, ▶jardinera para balcón, ▶jardinera para pretil); *syn.* jardincito [m]; *f* **jardin [m] miniature** (Jardin établi sur une surface extrêmement réduite, p. ex. une petite rocaille, bacs de plantation sur toitures-terrasses, terrasses ou balustrades ; le **j. m.** peut atteindre au maximum 100 m² mais n'est jamais en contact avec le sol naturel ; ▶jardinière pour balcon, ▶jardinière pour balustrade, ▶jardinière garde-corps, ▶jardin en forme de jauge) ; *g* **Miniaturgarten [m]** (Garten auf kleinster Fläche, z. B. kleiner Steingarten, Pflanzenkübel, Betontröge auf Dächern, Terrassen oder als Balkonbrüstung — bis ca. 100 m² groß, jedoch ohne Anschluss an gewachsenen Boden; ▶Balkonkasten, ▶Pflanzentrog als Mauerbrüstung, ▶Pflanzentrog an einem Geländer, ▶Troggarten).

3556 miniature rose [n] *hort. plant.* (A naturally dwarf rose growing to a height of about 30cm); *s* **rosal [m] miniatura** (Rosal de unos 30 cm de altura); *f* **rosier [m] miniature** (Rosier dont la taille naturelle ne dépasse pas 30 cm) ; *syn.* rosier [m] nain ; *g* **Zwergrose [f]** (Rosenwuchsform mit einer Wuchshöhe um 30 cm, die vermutlich auf *Rosa chinensis* 'Minima', auch *Rosa semperflorens* 'Minima' genannt, zurückgehen soll); *syn.* Miniaturrose [f], *im Blumenhandel* Kussröschen [n]).

miniature sewage treatment plant [n] [UK] *envir.* ▶individual sewage disposal system [US].

3557 minibus [n] *trans.* (Small pickup bus or taxi operated by a company, which picks up and transports various passengers along an unscheduled route); *syn.* minicab [n] [UK]; *s* **colectivo [m] [AL]** (En países con sistemas de transporte público poco eficiente, tipo de servicio de taxi en automóvil o bus pequeño utilizado para transportar a personas a lo largo de rutas determinadas, pero sin paradas fijas, y por el que se paga una tarifa fija por persona dependiendo del largo del trayecto); *f* **taxi [m] collectif** (Voiture ou minibus mis en service par une entreprise de taxi transportant plusieurs personnes sur un itinéraire donné et pouvant supporter de légères modifications de parcours) ; *syn. en Afrique* taxi-brousse [m] ; *g* **Sammeltaxi [n]** (Von Taxen- oder Mietwagenunternehmen als PKW oder Kleinbus vorgehaltenes Fahrzeug, bei dem der Fahrgast damit rechnen muss, mit anderen, ihm fremden Personen zu fahren und dementsprechend gewisse Umwege in Kauf zu nehmen).

minicab [n] [UK] *trans.* ▶minibus.

3558 minimal area [n] *phyt.* (Smallest area within which species of a community are sustainably represented. The **m. a.** may be determined by a ▶species area curve. The resulting sample quadrat, based on the concept of **m. a.**, is called relevé); *s* **área [f] mínima** (Superficie mínima que necesita una comunidad vegetal para su desarrollo normal. Al **á. m.** le corresponde un número mínimo de especies, **á. m.** y número mínimo de especies son distintos para cada asociación y deben determinarse empíricamente; BB 1979; ▶curva de especies y área); *syn.* espacio [m] mínimo; *f* **aire [f] minimale** (La plus petite surface nécessaire pour identifier la plupart des espèces d'un groupement homogène ; ▶courbe d'aire) ; *syn.* espace [m] minimum (BB 1928) ; *g* **Minimalfläche [f]** (Mindestfläche, die eine Pflanzengesell-

schaft zu ihrer normalen Entwicklung benötigt. Der **M.** entspricht eine Mindestartenzahl. **M.** und Mindestartenzahl sind für jede Assoziation verschieden und müssen empirisch festgestellt werden; ▶Arealkurve); *syn.* Minimumareal [n], Minimalareal [n], Minimalraum [m].

3559 minimum building spacing [n] *leg. urb.* (Lowest allowable distance between buildings or side yard requirements to allow enough daylight and fresh air in rooms of the buildings, enough green open spaces, and enough access space for emergency vehicles, maintain privacy, prevent spread of fire, etc.; ▶building setbacks); *syn.* minimum yard requirement [n] [US], minimum set-back requirements [npl] [UK]; *s* **retranqueo [m] legal entre edificios** (Distancia mínima entre edificios y hasta el límite del predio vecino fijada para asegurar suficiente iluminación, ventilación y [en países fríos] insolación de los edificios, el acceso de servicios de emergencia y eventualmente proveer de espacios libres para usos recreativos; ▶distancia legal de construcción); *f* **gabarit [m] de prospect** (Moyen de définition d'une surface théorique de bâtiments sur la base du ▶prospect de la hauteur de bâtiment, ensoleillement ; cf. R. 111-16/17/18/19 ; ▶espacement minimal entre les constructions) ; *g* **Mindestabstand [m] von Gebäuden** (Der **M. von G.** untereinander und von den Nachbargrenzen ist in den Landesbauordnungen unterschiedlich geregelt. Er dient zur Sicherung der Belichtung, Belüftung und Besonnung der zum dauernden Aufenthalt von Menschen bestimmten Räumen, zur Sicherung der nötigen Freiflächen für die Nutzung der Grundstücke, für die Zugänglichkeit von Notdiensten etc.; ▶Abstandsflächen).

minimum set-back requirements [npl] [UK] *leg. urb.* ▶minimum building spacing.

minimum sight triangle [n] *trans.* ▶sight triangle.

minimum yard [npl] [US] *leg. urb.* ▶building setbacks.

minimum yard requirement [n] [US] *leg. urb.* ▶minimum building spacing.

3560 mining [n] *min.* (Removal of mineral resources; *specific terms* ▶deep mining, ▶deep pit mining, ▶open pit mining [US]/open-cast mining [UK]); *s* **minería [f]** (Término genérico para la actividad de ganancia de recursos minerales; *términos específicos* ▶explotación a cielo abierto profunda, ▶minería en galería, ▶minería a cielo abierto); *f* **exploitation [f] minière** (Industrie primaire qui se consacre à la prospection, l'exploration et l'exploitation dans le sous-sol des roches et des minéraux solides ayant une valeur économique ; ▶exploitation [minière/de carrière] à ciel ouvert, ▶exploitation minière à ciel ouvert, ▶exploitation minière souterraine) ; *syn.* exploitation [f] de mines ; *g* **Bergbau [m]** (Wirtschaftszweig, der sich mit der Suche nach Lagerstätten [Prospektion], deren sorgfältige Erforschung [Exploration] zur Feststellung der abbauwürdigen Mengen und der Gewinnung von Bodenschätzen befasst; *UBe* ▶Tagebau, ▶Tieftagebau, ▶Untertagebau).

mining [n] [UK], **open-cast** *min.* ▶open-pit mining [US].

mining [n] [US], **surface** *min.* ▶open-pit mining [US].

mining [n] [US], **underground** *min.* ▶deep mining.

3561 mining area [n] *min. plan.* (Generic term covering both deep mining area and ▶open-pit mining areas [US]/open-cast mining area [UK]); *s* **cuenca [f] minera** (Término genérico que se usa para zonas mineras de ambos tipos de minería: subterránea y a cielo abierto; ▶zona de minería a cielo abierto); *syn.* región [f] minera, zona [f] minera; *f* **bassin [m] minier** (Vaste gisement minier sur lequel l'extraction est effectuée à ciel ouvert ou en sous-sol ; *terme spécifique* ▶région de mines à ciel ouvert) ; *syn.* région [f] minière ; *g* **Bergbaugebiet [n]** (OB zu ▶Tagebaugebiet und Untertagebaugebiet).

M

mining area [n] [UK]**, open-cast** *min.* ▶open-pit mining area [US].

mining gob [n] [US]**, disposal of** *min.* ▶disposal of mining spoil.

mining gob [n] [US]**, underground disposal of** *min.* ▶underground disposal of mining spoil.

mining landscape [n] [UK]**, recultivated** *min. landsc.* ▶landscape of reclaimed surface-mined land.

mining operation [n] *min.* ▶derelict land resulting from mining operation.

mining permit [n] [US] *adm. leg. min.* ▶permit for mineral extraction [US]/planning permission for mineral extraction [UK].

3562 mining spoil [n] *min.* (Waste rock resulting from mining operations; *specific terms* liquefied and solid ▶tailings slurry; ▶disposal of mining spoil, ▶overburden 2, ▶underground disposal of mining spoil); *syn.* gob [n] [also US]; *s* **zafras [fpl]** (Material restante de la minería de carbón y metales; ▶cascote, ▶depositar zafras, ▶lodo de minería, ▶relleno de minas [subterráneas] con zafras); *syn.* roca [f] estéril, escombros [mpl] (2), estériles [fpl]; *f* **stérile [m]** (Roche restante produit de l'exploitation minière et inutilisable comme minéral ; terme très souvemt utilisé au pluriel ; ▶matériaux de recouvrement, ▶stérile de sable lavé, ▶stockage de stériles, ▶stockage souterrain de stériles) ; *g* **Berge [pl] (3)** (Taubes Gestein, das bei der Gewinnung von Kohle und Erzen und nach deren Aufbereitung als Abfall übrig bleibt; ▶Abraum, ▶Bergeversatz, ▶Bergeverbringung, ▶Waschberge); *syn.* Bergematerial [n].

mining subsidence [n] *min.* ▶damage due to mining subsidence.

minor grading [n] *constr.* ▶fine grading.

3563 minutes [npl] **of bid opening** [US] *adm. constr. contr.* (Record of the opening of bids [US]/tenders [UK] on a specific date; ▶certificate of practical completion); *syn.* submissions report [n] [UK], minutes [npl] of submission date [UK]; *s* **acta [f] de apertura de pliegos** (▶acta de recepción/aceptación [de obra]); *syn.* acta [f] de apertura de proposiciones; *f* **procès-verbal [m] des opérations d'ouverture** (Acte dressé par la commission d'adjudication ou d'appel d'offres relatant les opérations et décisions prises lors de la séance d'ouverture des plis ; ▶procès verbal de réception) ; *g* **Niederschrift [f] über den Eröffnungstermin** (Protokoll über den Eröffnungstermin; die unabdingbaren Inhalte der Niederschrift sind in § 22 Nr. 4-6 u. § 23 Nr. 4 VOB Teil A dargelegt; ▶Abnahmeniederschrift); *syn.* Submissionsniederschrift [f].

minutes [npl] **of submission date** [UK] *adm. constr. contr.* ▶minutes of bid opening [US].

mire [n] *geo.* ▶peatland [US].

mire [n]**, shallow** *geo. min.* ▶shallow bog.

3564 mitigation measure [n] *conserv. leg. urb.* (Compensating action to offset degradation of existing ecological conditions at a current project site with state-of-the-art techniques; ▶alleviation measure, ▶compensation area 1, ▶compensation payment [UK], ▶environmental compensation measure, ▶impact mitigation, ▶impact mitigation regulation, ▶prevention measure [of an intrusion upon the natural environment], ▶reducing of intrusion); *s* **medida [f] compensatoria** (Medidas de restauración o de efecto contrario al de la acción emprendida que se deben implementar para compensar la degradación de un área en la que se realiza una obra y su subsiguiente actividad, cuando no son posibles medidas correctoras para atenuar o suprimir los efectos ambientales negativos de las mismas; cf. Art. 11 Rd 1131/1988; ▶área de compensación, ▶compensación [de un

impacto], ▶indemnización compensatoria, ▶medida correctora, ▶medida mitigante, ▶medida sustitutiva; ▶reducción de impacto, ▶regulación de impactos sobre la naturaleza); *syn.* medida [f] contrarrestante; *f* **mesure [f] compensatoire** (F., mesures prises dans le cadre d'une étude d'impact en vue de compenser les effets dommageables non réductibles d'un projet sur l'environnement lorsqu'un impact résiduel significatif subsiste, cf. C. envir. L 122-1 à L 122-3 ; il ne faut pas confondre ces mesures avec celles qui sont définies au titre du décret du 20 décembre 2001, ces dernières étant traitées dans les études d'incidences sur les sites Natura 2000 ; D., pour tout projet portant atteinte au milieu naturel que ce soit en milieu rural ou en milieu urbain ; ▶compensation [des conséquences dommageables] sur l'environnement, ▶mesure de réduction, ▶mesure de substitution, ▶mesure de suppression, ▶réduction des atteintes à l'environnement, ▶redevance compensatoire, ▶réglementation relative aux atteintes subies par le milieu naturel, ▶surface compensatoire) ; *syn.* mesure [f] pour compenser les impacts impossibles à supprimer, mesure [f] de compensation ; *g* **Ausgleichsmaßnahme [f] (des Naturschutzes und der Landschaftspflege)** (Maßnahmen des Naturschutzes und der Landschaftspflege, mit denen erhebliche Beeinträchtigungen der Umwelt vermieden, vermindert oder soweit möglich ausgeglichen werden, die ein Planungsträger vor, während oder nach einem Eingriff in Natur und Landschaft auf Grund eines nach öffentlichem Recht vorgesehenen Fachplanes oder ▶Landschaftspflegerischen Begleitplanes vornehmen soll; cf. § 8 [4] BNatSchG und § 6 [3] Nr. 3 UVPG und Art. 5 [3] RL 85/337/EWG; auf Grund § 200 a i. V. m. § 5 [2 a] und § 9 [1] BauGB können **A.n** räumlich getrennt, planextern und zeitlich entkoppelt vom Eingriff vorgesehen werden [zu letzterem cf. § 135 a (2) Satz 2 BauGB], vorausgesetzt, sie sind mit einer geordneten städtebaulichen Entwicklung, mit den Zielen der Raumordnung und denen von Naturschutz und Landschaftspflege vereinbar. Es kommen nur solche Flächen in Betracht, die aufwertungsbedürftig und -fähig sind. Diese Voraussetzung erfüllen sie, wenn sie in einen Zustand versetzt werden können, der sich im Vergleich mit dem früheren als ökologisch höherwertig einstufen lässt; cf. BAUR 1999, 484; ▶Ausgleichsabgabe, ▶Ausgleichsfläche, ▶Ausgleich [eines Eingriffes], ▶Eingriffsminderung, ▶Eingriffsregelung, ▶Ersatzmaßnahme [des Naturschutzes und der Landschaftspflege], ▶Vermeidungsmaßnahme [eines Eingriffes], ▶Verminderungsmaßnahme [eines Eingriffes]); *syn.* Kompensationsmaßnahme [f] (des Naturschutzes und der Landschaftspflege).

mitigation regulation [n] *conserv. leg. plan.* ▶impact mitigation regulation.

mix [n] [US]**, conservation grass** *constr.* ▶low-maintenance grass type.

mix [n] [US]**, meadow** *constr.* ▶low-maintenance grass type.

mix [n] [US]**, planting** *constr. hort.* ▶growing medium.

mix [n]**, topsoil** *constr. hort.* ▶growing medium.

3565 mixed border [n] *gard. hort.* (Elongated flower bed with a mixture of different annuals and/or perennials; ▶border 1); *s* **arriate [m] de flores** (Macizo de flores alargado con mezcla de flores anuales y/o vivaces; ▶arriate/a); *syn.* arriata [f] de flores; *f* **mixed-border [m]** (Plate-bande étroite, souvent adossée à un mur ou une haie dans un parc ou un jardin, garnie d'un mélange de plantes vivaces, de plantes herbacées, d'annuelles, de bulbes et d'arbustes ; ▶bordure végétale, plate-bande) ; *syn.* bordure [f] herbacée ; *g* **Blumenrabatte [f]** (Beet in privater oder öffentlicher Grünanlage, mit Sommerblumen und/oder prächtig blühenden Stauden bepflanzt; *OB* ▶Rabatte).

3566 mixed crop production [n] *agr. for.* (Various kinds of crops cultivated on the same portion of land simultaneously;

►monoculture, ►pure stand, ►single crop system); *syn.* mixed farming [n]; *s* **policultivo** [m] (Cultivo simultáneo de varias especies vegetales en una misma superficie; ►cultivo monoespecífico, ►monocultivo, ►rodal monoespecífico); *syn.* cultivo [m] mixto; *f* **polyculture** [f] (Culture simultanée de différentes espèces végétales sur une même exploitation ou dans une même région ; cf. LA 1981 ; ►boisement monospécifique, ►culture monospécifique, ►monoculture) ; *g* **Mischkultur** [f] (Gleichzeitiger Anbau mehrerer Pflanzenarten auf derselben Fläche in der Land- und Forstwirtschaft; ►Monokultur, ►Reinbestand, ►Reinkultur).

3567 mixed effluent [n] *envir.* (In a ►combined sewer system discharged sullage, ►sewage 1, and ►precipitation water); *s* **aguas [fpl] residuales mixtas** (Agua de abastecimiento después de haber sido empleada para diversos usos. Suele ser una combinación de residuos líquidos y materias en suspensión de tipo doméstico, municipal e industrial, y además del agua superficial, subterránea y de lluvia que pueda presentarse; WMO 1974; ►aguas residuales 2, ►aguas de precipitación, ►sistema de alcantarillado unitario); *f* **eaux [fpl] usées et eaux pluviales** (►Eaux de précipitations et ►eaux usées évacuées par un ►réseau [d'assainissement] unitaire) ; *g* **Mischwasser** [n] (...wässer [pl]; das im ►Mischverfahren abgeführte ►Schmutzwasser und ►Niederschlagswasser).

mixed farming [n] *agr. for.* ►mixed crop production.

3568 mixed forest [n] *for.* (Forest composed of several deciduous tree species or of deciduous and coniferous trees; *opp.* ►pure stand); *s* **bosque [m] mixto** (Monte compuesto de especies caducifolias y acicufolias; *opp.* ►rodal monoespecífico); *f 1* **forêt [f] mélangée** (Un peuplement ou une forêt composé de deux ou plusieurs essences soit feuillues, soit résineuses ; *opp.* ►boisement monospécifique) ; *f 2* **forêt [f] mixte** (Forêt composée de feuillus et de résineux) ; *g* **Mischwald [m]** (Wald, der aus mehreren Laubholzarten oder aus Laub- und Nadelbäumen besteht; *opp.* ►Reinbestand); *syn.* Mischbestand [m].

3569 mixed land use [n] (1) *leg. urb.* (►Urban land use category permitting several types of land use in the same area, as shown on a ►comprehensive plan [US]/Local Development framework [LDF] [UK]; ►central business district, ►mixed use development, ►rural district); *s* **suelo [m] urbano de uso mixto (±)** (D. categoría de uso de suelo en los ►planes generales municipales de ordenación urbana en la cual se permiten diversos tipos de uso como comercial, residencial, etc. Según la combinación de usos en el ►plan parcial [de ordenación] se diferencia en ►núcleo de población rural, ►zona central y ►zona urbana mixta. **En Es.** no existe esta ►categoría de suelo urbano; ►categoría de suelo urbano); *f* **zone [f] d'activités mixtes (±)** (Secteur d'occupation des sols à vocations diverses [habitat, emplois, équipements] ; **F.**, les documents graphiques du ►schéma de cohérence territoriale [SCOT] ne comportent plus de carte de destination générale des sols ; certains schémas directeurs avaient fait de la carte de destination générale des sols l'élément pivot du dispositif sans définir d'orientations ; leur système de zonage les rapprochait du POS et certains interféraient par leur précision cartographique avec le domaine de compétence des POS ; la loi SRU [relative à la **s**olidarité et au **r**enouvellement **u**rbains] a voulu corriger cette dérive en donnant aux SCOT la dimension prospective et fédératrice dont certains schémas directeurs étaient dépourvus ; la représentation des lignes forces du projet se trouve ainsi renforcée dans sa dimension stratégique et globalisante ; institut d'aménagement et d'urbanisme de la région d'Île-de-France ; loi n°2000-1208 du 13 décembre 2000 relative à la solidarité et au renouvellement urbains — sept 2002 ; **D.**, catégorie de division du territoire fixée par le « *Flächennutzungsplan* » [►schéma de cohérence territoriale (SCOT) allemand]

englobant approximativement les zones d'agglomération continue avec implantation d'activités de bureau et de commerce, les zones réservées aux activités commerciales ; à l'échelon du « Bebauungsplan » [plan local d'urbanisme allemand] cette catégorie de zonage correspond selon l'occupation du sol définie à la ►zone de noyaux villageois, la ►zone à urbaniser à vocation d'activités mixtes ou bien la ►zone urbaine consacrée aux activités économiques ; ►zone urbaine 2) ; *g* **gemischte Baufläche [f]** (Im ►Flächennutzungsplan für die Bebauung vorgesehene Fläche nach der allgemeinen Art ihrer baulichen Nutzung dargestellte Bauflächenart; *Abk.* M; im Bebauungsplan wird sie dann nach der besonderen Art ihrer baulichen Nutzung als ►Dorfgebiet, ►Mischgebiet oder ►Kerngebiet ausgewiesen; ►Baufläche).

3570 mixed land use [n] (2) *urb.* (Residential area with commercial facilities; ►mixed use development, ►layering of differents uses); *s* **uso [m] mixto del suelo** (Barrio residencial en el que también hay talleres y pequeñas industrias; ►superposición de usos, ►zona urbana mixta); *f* **zone [f] d'activités mêlées à l'habitat existant** (Zone résidentielle à l'intérieur de laquelle sont implantées des constructions, dépôts ou entrepôts liés à une activité commerciale ou artisanale ; ►superposition des usages, ►zone à urbaniser à vocation d'activités mixtes) ; *syn.* occupation [f] des sols mixte ; *g* **Gemengelage [f] (1)** (Mit Gewerbebetrieben durchmischtes Wohngebiet; ►Mischgebiet; ►Nutzungsüberlagerung); *syn. o. V.* Gemenglage [f].

3571 mixed use development [n] *leg. urb.* (Urban ►zoning district [US]/development zone [UK] comprising housing as well as commercial facilities and other uses which do not conflict with its residential use); *s* **zona [f] urbana mixta** (En D., categoría en ►plan parcial [de ordenación] para ►sector urbano con funciones residenciales, comerciales e industriales que no sean molestas para la primera; **en Es.** no existe esta categoría; ►suelo urbano de uso mixto); *f* **zone [f] à urbaniser à vocation d'activités mixtes (±)** (Catégorie de zonage d'un document de planification urbaine ; *PLU/POS* la dénomination du secteur est variable selon les communes ; ►secteur urbain, ►zone d'activités mixtes, ►zone urbaine) ; *syn. liste non exhaustive* il n'existe pas de réglementation concernant l'appellation précise de cette zone si bien qu'il est possible d'avoir autant de synonymes que de plans locaux d'urbanisme : zone [f] urbaine d'extension de l'agglomération avec une occupation du sol discontinue, zone [f] urbaine d'habitat mixte composé de collectifs bas, individuels, de services et d'activités, zone [f] urbaine de fonction résidentielle avec installation de locaux d'activités non génératrices de nuisances — intégration habitat-emploi-loisirs, zone [f] urbaine de quartiers mixtes à forte présence d'activités économiques, zone [f] urbaine de quartiers mixtes à dominante habitat, zone [f] urbaine de tissu diversifié, secteur [m] multifonctionnel à urbaniser, zone [f] d'activités économiques mixtes [B] ; *g* **Mischgebiet [n]** (In der Bauleitplanung ein ►Baugebiet, das dem Wohnen und der Unterbringung von Gewerbebetrieben, die das Wohnen nicht wesentlich stören, dient; cf. § 6 BauNVO; ►gemischte Baufläche).

3572 mixing [n] **of air masses** *met.* (Intermingling of atmospheric air caused by quick-changing air movements; ►air exchange; mix [vb]); *s* **mezcla [f] de masas de aire** (Entremezclado de las masas de aire a pequeña escala provocado por el movimiento rápido del aire; ►ventilación 1); *f* **échange [m] atmosphérique** (Brassage de l'air atmosphérique provoqué par des mouvements changeants rapides sur des zones réduites; ►brassage de l'air) ; *g* **Durchmischung [f] der Luft** (Durchmischung der atmosphärischen Luft durch kleinräumige, schnell wechselnde Luftbewegungen; Luft durchmischen [vb]; ►Luftaustausch).

M

mixture [n] *agr. constr. hort.* ►grass-seed mixture, ►seed mixture; *constr.* ►sports grass mixture, ►standard grass-seed mixture.

mixture [n]**, leaf mold/sand** *constr. hort.* ►soil improvement by working in a leaf mold/sand mixture [US]/soil improvement by working in a leaf-mould/sand mixture [UK].

mixture [n] **[US], planting** *constr. hort.* ►growing medium.

mixture [n]**, topsoil** *constr. hort.* ►growing medium.

mixture [n] **[US], turf-seed** *constr. hort.* ►grass-seed mixture.

3573 mobile home [n] *recr. urb.* (Mobile accommodation, often transported from one place to another while on holiday, **in Europe**, sometimes parked and used as a second home for vacations; **in U.S.**, a **m. h.** is mostly used as a year-round residence, often located in a mobile home park; ►recreation vehicle [RV] [US]/motor home [UK]); *syn.* motor home [n] [also US], manufactured home [n] [also US]; *s* **casa** [f] **móvil** (Vehículo motor aparcado y utilizado normalmente como segunda vivienda para las vacaciones; en EE.UU. una **c. m.** se usa frecuentemente como residencia habitual y se estaciona con muchas otras en los llamados «parques de casas móviles»; ►autocaravana); *f* **résidence** [f] **mobile** (Logement à utilisation temporaire pouvant être transporté ou tracté d'un lieu à un autre, en général non assimilé à une caravane à cause de sa mobilité limitée et dont le stationnement est réglementé selon la nature des terrains utilisés ; cf. CPEN Protection de la nature § 39 ; ►camping-car) ; *syn.* mobile home [m] ; *g* **Mobilheim** [n] (Umbauter Einfamilienwohnraum der zum mehrmaligen Transport von Standort zu Standort geeignet ist und in Europa i. d. R. nicht ständig bewohnt wird, sondern vorwiegend der Erholungsnutzung dient. In den USA werden **M.e**, wenn sie einmal in einem *mobile home park* ihren Platz gefunden haben, meist nicht mehr bewegt; ►Wohnmobil).

3574 mobile phone antenna [n] *urb.* (antennas [pl], antennae [pl]; device mounted on a tall mast to receive or to broadcast electromagnetic waves for the transmission of telecommunication data by mobile [cell] phones, e-mail, internet. **In D.**, development mitigation plans have to be prepared for the approval of mobile phone masts, in order to evaluate and minimize their impact on nature and landscape); *syn.* mobile phone mast [n]; *s* **antena** [f] **de telefonía móvil** (Instalación en poste alto o sobre edificio alto para la recepción y el envío de las ondas electromagnéticas que transmiten los datos de redes de telecomunicación como teléfonos celulares, correo electrónico, internet); *syn.* antena [f] de telecomunicación; *f* **pylône** [m] **de téléphone mobile** (Équipement de radiotéléphonie mobile fixé sur un mât et utilisant les ondes hertziennes pour transmettre la voix humaine ; afin de favoriser une meilleure insertion des équipements de télécommunication dans le paysage, le projet d'implantation fait l'objet d'une étude préalable visant la prise en compte des paysages et de ses structures géographiques [vallées, plaines, forêts, littoral, montagne, bourgs et villages, villes, etc.] sur la base de l'inventaire des paysages ; la hiérarchisation [espaces banals, espaces sous surveillance, entités paysagères, espaces règlementés] correspond au degré de sensibilité des lieux désignés, à leur fragilité, face à l'implantation d'un nouvel équipement radiotéléphonique et à l'attention qu'il convient d'y apporter lors de l'insertion envisagée ; à chaque catégorie d'espaces correspond un ensemble de conditions, à défaut desquelles le projet d'implantation pourra être rejeté ; cf. charte nationale de recommandations environnementales entre l'État et les opérateurs de radiotéléphonie mobile du 12 juillet 1999) ; *g* **Mobilfunkantenne** [f] (Auf Masten erhöht angebrachte Vorrichtung zum Empfang oder zur Ausstrahlung elektromagnetischer Wellen für die Datenübertragung telekommunikativer Systeme, wie z. B.

Handy, E-Mail, Internet. In D. müssen zur Genehmigung von Mobilfunkmasten Landschaftspflegerische Begleitpläne erstellt werden, um den Eingriff in Natur und Landschaft zu bewerten und so gering wie möglich zu halten).

mobile phone mast [n] *urb.* ►mobile phone antenna.

3575 mobilization [n] *pedol.* (Making available of soil nutrients through the action of plant growth; ►available to plants, ►nutrient mobilization); *s* **movilización** [f] **de nutrientes** (Puesta a disposición de los nutrientes en suelos brutos a través de plantas [pioneras]; ►disponible para las plantas, ►movilización de nutrientes); *f* **désagrégation** [f] (Action chimique des végétaux [pionniers] dont les substances qu'ils excrètent corrodent ou dissolvent les roches-mères en libérant ainsi des substances nutritives ; ►disponible pour la plante, ►libération de/des substances nutritives) ; *g* **Aufschließen** [n, o. Pl.] (Lösen und Verfügbarmachung von Nährstoffen in [Roh]böden durch [Pionier]pflanzen; ►pflanzenverfügbar, ►Mobilisierung von Nährstoffen).

mobilization [n] **of nitrate** *pedol.* ►volatilization of nitrogen.

3576 modal-split [n] *plan. trans.* (Distribution of traffic between various forms of transport or means of transport, including bicycles. **In Europe**, modern ecological traffic planning has the objective to promote the portion of environment-friendly transportation vehicles [public transportation vehicles, bicycles and walking] in order to slow down the increase in traffic of private motor vehicles); *s* **modalsplit** [m] (Reparto del tráfico entre los diferentes medios de transporte, incluida la bicicleta. En muchos países de Europa la planificación moderna del tráfico y transporte busca promover el uso de sistemas de transporte favorables al medio ambiente, como el transporte público y la bicicleta, para así reducir el crecimiento del tráfico de vehículos de motor privados y con ello la contaminación atmosférica y acústica así como el gasto de energía no renovable que favorece el cambio climático); *f* **séparation** [f] **des modes de déplacement** (Répartition du trafic entre les différentes voies de communication ou moyens de transport ; dans une politique en faveur du développement des déplacements doux la voirie est aménagée de telle sorte que piétons et cyclistes [modes doux] disposent de leur propre espace de circulation séparé, sur un axe appelé « axe fort sécurisé ») ; *g* **Verkehrsmittelwahl** [f] (Aufteilung des Verkehrs auf verschiedene Verkehrssysteme resp. Transportmittel einschließlich des Fahrrades. Die moderne ökologische Verkehrspolitik hat das Ziel, den Anteil der umweltschonenden Verkehrsmittel incl. das Zu-Fuß-Gehen zu fördern, um das weitere Anwachsen des Autoverkehrs zu verlangsamen); *syn.* Modalsplit [m].

model [n] *envir.* ►emission distribution model.

model [n]**, emission dispersal** *envir.* ►emission distribution model.

3577 model garden [n] *gard. hort.* (Garden intended to give visitors, e.g. at garden shows, ideas about design, construction tips and hints about uses of gardens); *s* **jardín** [m] **modelo** (Jardín p. ej. en exposición de horticultura que tiene la función de mostrar al público diferentes técnicas constructivas, ideas de diseño y posibilidades de uso); *f* **jardin** [m] **spécimen** (Désigne lors de floralies une surface conçue pour procurer aux visiteurs des idées nouvelles en matière de conception, d'utilisation et de réalisation des jardins) ; *syn.* jardin [m] modelé ; *g* **Mustergarten** [m] (Garten, z. B. in einer Gartenschau, der den Besuchern gestalterische und technische Anregungen sowie Nutzungshinweise geben soll).

3578 moder [n] *pedol.* (**1. In Europe**, an uppermost loose layer of forest humus, transitional between mull and mor, with partially decomposed organic matter, containing mostly all horizons [O_l-O_f-O_h], together with a distinct A_h horizon. **M.** occurs usually in deciduous and coniferous forests with only low herb cover and on relatively nutrient-poor parent rock or under cool, humid climatic conditions. **2. In U.S. & CDN,** the equivalent term is **duff mull** and is similar to ►mor humus in that it generally features an accumulation of partially to well-humified organic materials resting on the mineral soil. It is similar to **mull** in that it is zoologically active; a **d. m.** usually has four horizons: Oi[L], Oe[F], Oa[H], and A; sometimes differentiated into the following groups: Mormoder, Leptomoder, Mullmoder, Lignomoder, Hydromoder, and Saprimoder; cf. SST 1997; ►humus type); *syn.* duff mull [n] [US&CDN]; *s* **moder [m]** (Humus que se forma en suelos ácidos o secos donde se desmenuzan primero los formadores de humus por acción de los antrópodos y luego se descomponen algo, principalmente por acción de hongos. La materia orgánica se mezcla poco con la mineral; BB 1979, 413; ►moor, ►tipo de humus); *syn.* moderhumus [m]; *f* **moder [m]** (Humus forestier de milieu aéré peu évolué à incorporation moyenne de matière organique peu transformée, à rapport C/N compris entre 15 et 25 ; humus forestier non submergé par une nappe perchée, à engorgement nul ou temporaire, présentant une couche holorganique en surface, différenciée en horizon A_h de diffusion par l'intermédiaire d'un horizon OAh ; DUCHAUFOUR ET DELECOUR cit. in DIS 1986 ; ►mor, ►type d'humus) ; *g* **Moder [m]** (Oberste, lockere Schicht auf terrestrischen Böden, mit meist allen Auflagehorizonten [O_l-O_f-O_h-Horizont], die teilweise zersetzt sind und deutlich ausgeprägtem A-Horizont. **M.** bildet sich vor allem unter krautarmen Laub- oder Nadelwäldern auf relativ nährstoffarmen Gesteinen oder unter kühlfeuchten Klimaverhältnissen; SS 1979; ►Auflagehumus, ►Humusform).

3579 modernization [n] (1) *arch. urb.* (Measures taken to increase the use value of a structure to meet current needs, including ►rehabilitation 1, ►remodeling [US]/remodelling [UK], and ►repair; ►building rehabilatation); *s* **modernización [f]** (Conjunto de obras de ►rehabilitación de edificios cuyo fin es incrementar el valor del objeto, incluyendo las obras de ►reparación de edificios y reestructuración necesarias para ello; ►rehabilitación de edificios, ►remodelación); *f* **modernisation [f]** (1) (Opération de remise à neuf pour rehausser le niveau de confort d'un appartement ou d'un immeuble, sans modification importante du gros œuvre y compris les travaux de ►remise en état et réparations provoqués par cette mesure ; ►réaménagement, ►réfection, ►réhabilitation, ►rénovation immobilière, ►restauration immobilière) ; *g* **Modernisierung [f]** (1. Bauliche Maßnahmen zur nachhaltigen Erhöhung des Gebrauchswertes eines Objektes incl. der durch diese Maßnahmen verursachten ►Instandsetzungen und Umbauarbeiten. 2. Anpassung und Weiterentwicklung des Wohn-, Arbeitsplatz- und Dienstleistungsumfeldes sowie der vegetationsbestimmten Freiräume einer Stadt an neuzeitliche Erfordernisse und Entwicklungen; ►Gebäudesanierung, ►Sanierung, ►Umbau).

3580 modernization [n] (2) *envir.* (Addition to or redesign of technical structures or machines for adaptation to modern standards); *s* **reequipamiento [m]** (Ampliación o reconstrucción de instalaciones técnicas o máquinas para que cumplan estándares modernos); *f* **modernisation [f]** (2) (Complément ou transformation d'équipements techniques ou de machines afin de les rendre conforme aux conditions standard) ; *g* **Nachrüsten [n, o. Pl.]** (Ergänzung oder Umbau von technischen Anlagen oder Maschinen, damit diese dem aktuellen Verfahrensstand entsprechen).

modification [n]**, gradient of landscape** *ecol.* ►degree of landscape modification.

modification [n]**, level of human-caused landscape** *ecol.* ►degree of landscape modification.

modification [n] **of a plan** *plan.* ►plan revision.

modification [n] **of landscape** *ecol.* ►landscape change.

modified organism [n] *biol. ecol. leg.* ►genetically modified organism (GMO).

module paver [n] [US] *constr.* ►precast paving block.

moist soil meadow [n] [UK] *phyt.* ►damp meadow.

moisture [n] *hort.* ►soil moisture

moisture capacity [n] *hort.* ►field capacity.

3581 moisture indicator plant [n] *phyt.* (Plant which grows on relatively moist, well-aerated soil, also known as 'mesophyte' or 'mesophil[ic]' species; ►indicator plant, ►mesic); *syn.* mesophyte [n], mesophil(ic) species [n]; *s* **especie [f] indicadora de humedad** (Planta que se considera indicadora de estaciones con alta humedad. Las plantas que crecen en suelos semihúmedos, bien aireados se llaman también especies ►mesófilas, ►planta indicadora); *syn.* mesófito [m] (DB 1985); *f* **espèce [f] indicatrice d'humidité** (Plante dont la présence est caractéristique d'un sol à forte humidité, dont la fréquence varie de façon significative avec le facteur d'humidité du sol ; les plantes vivant dans un milieu humide sont des « mésophytes » ; ►mésophile, ►plante indicatrice) ; *g* **Feuchtigkeitszeiger [m]** (Pflanze, die als Indikator für einen Standort mit höherer Bodenfeuchtigkeit gilt. Pflanzen, die auf mäßig feuchten, gut durchlüfteten Böden wachsen, werden auch **Mesophyten** oder **mesophile Arten** genannt; hinsichtlich der Wasserregime im Boden werden in der Skala 1-10 folgende Feuchtezahlen unterschieden: **1.** Starktrockniszeiger, **3.** Trockniszeiger, **5.** Frischezeiger, **7.** Feuchtezeiger, **9.** Nässezeiger, **10.** Wechselwasserzeiger; ►mesophil, ►Zeigerpflanze).

moisture-loving plant [n] *bot.* ►hygrophyte.

3582 mole drainage [n] *agr. constr.* (Underground drainage in heavy soils as well as peat in which a mole plow [US]/mole plough [UK] is pulled through the ground by a tractor or cable winch; the device consists of a vertical or slightly inclined cutting blade and a mole with an attached cone, which presses the soil to the sides thus creating a tube-shaped, unlined drain); *s* **drenaje [m] topo** (Evacuación subterránea de las aguas en suelos pesados por medio de cavidades de 50 a 150 mm en el subsuelo creadas por medio de un arado topo); *f* **drainage [m] par charrue-taupe** (Évacuation souterraine des eaux au moyen d'un système de drainage composé de galeries tubulaires forées pour permettre l'écoulement de l'eau dans le sol ; s'applique également à l'assainissement du sol par ce procédé) ; *g* **Maulwurfdränung [f]** (Unterirdische Wasserabführung in einem Dränsystem, das im standfesten Lehm- oder Tonböden sowie im gewachsenen Torf mit einem Maulwurfpflug gezogen wurde; das Gerät, das von einem Traktor oder Seilzug gezogen wird, besteht aus einem senkrechten oder leicht geneigten schneidenden Schwert [Messer] mit Presskörper [Maulwurf] und angehängtem Ziehkegel/ Presskegel, der die Dränröhre [Hohlgang ohne Rohr] bildet); *syn.* rohrlose Dränung [f], Erddränung [f].

3583 molluscicide [n] *chem. envir. agr. hort.* (Chemical agent for destroying snails; ►plant protection agent); *s* **molusquicida [m]** (►Plaguicida empleado para combatir a los caracoles); *f* **mollusquicide [m]** (►Pesticide, ►produit antiparasitaire [à usage agricole] qui détruit les gastéropodes pulmonés nuisibles aux cultures [limaces par exemple], utilisable en jardins à base de métaldéhyde) ; *syn.* hélicide [m], substance [f] active hélicide, limaticide [m] ; *g* **Molluskizid [n]** (Zur Bekämpfung von Schnecken eingesetztes ►Pflanzenschutzmittel).

3584 monarch tree [n] [US] *conserv. leg.* (In U.S., city tree ordinances have designated the category of a protected monarch tree, where a tree is of historical significance, has high aesthetic

M

value or unique character, is a rare or unusual species or will be the focal point of a project. Usually **m. t.s** must have a life expectancy of more than 15 years and are selected, if their diameter equals or exceeds a specific diameter [normally more than 60 cm]. A **m. t.** can only be removed with city council authorization. A stand of **m. t.s** may also be recognized, if there is a group of trees, in which a specified minimum number of individuals have attributes complying with the criteria for a single **m. t**. **M. t.s** are listed by the American Forestry Association in the National Register of Big Trees in order to protect the giants of the plant world; ►National Register of Natural Areas, ►natural landmark [US]); *syn.* champion tree [n] [US], landmark tree [n] [US]; *s* **árbol [m] monumento** (Ejemplar arbóreo que por su singularidad notoria, rareza o belleza merece ser objeto de una protección especial; art. 33, Ley 42/2007; ►monumento natural, ►registro de monumentos naturales. **En EE.UU.**, los **«árboles monarcas»** están registrados en la Lista Nacional de Grandes Árboles de la Asociación Americana de Silvicultura con el fin de proteger los ejemplares gigantes del reino vegetal); *f 1* **arbre [m] classé comme monument naturel** (F., arbre ayant un caractère remarquable et classé en vertu de la loi du 2 mai 1930 en raison de l'intérêt général que présente sa conservation ou sa préservation) ; *f 2* **arbre [m] remarquable d'intérêt national** (Arbre figurant sur une liste établie à l'échelon régional et national des arbres rares et exceptionnels par leur âge, leurs dimensions [hauteur de l'arbre ou circonférence du tronc], leurs formes parfois étranges [arbres sculptés, cépées originales, têtards, émondes, alignement], leur passé [évoque un événement historique] ou encore leur légende [à l'origine d'une croyance, de pratiques religieuses]. En 2002 l'association A.R.B.R.E.S. et l'Office National des Forêts ont signé une convention de partenariat où ils s'engagent dans une démarche de préservation des arbres remarquables. Un label « Arbre remarquable de France » est attribué aux communes qui, possédant un arbre exceptionnel, signent un accord de partenariat et s'engagent à entretenir, sauvegarder et mettre en valeur l'arbre considéré comme patrimoine naturel et culturel ainsi que de mettre en place un panneau de présentation de l'arbre ; ►liste des monuments naturels et des sites, ►monument naturel) ; *f 3* **arbre [m] patrimonial** (*Terme spécifique* arbres anciens isolés ou en alignements souvent protégés, caractéristiques du paysage de certaines régions, p. ex. les ►arbres têtards, les arbres âgés dans les grands parcs urbains) ; *g* **Baum-Naturdenkmal [n]** (Baum, der auf Grund seiner Größe, Seltenheit, Eigenart oder Schönheit rechtsverbindlich als Einzelschöpfung festgesetzt ist; gesetzliche Grundlagen sind § 17 BNatSchG und Naturschutzgesetze der Bundesländer]. **B.e** sind bei der unteren Naturschutzbehörde im ►Naturdenkmalbuch eingetragen ►Naturdenkmal).

3585 monastery garden [n] *gard'hist.* (Formal garden, which was established in medieval times by monks outside of the monastery for cultivation of medicinal plants, culinary herbs, vegetables, and cut flowers; ►cloister garden); *s* **jardín [f] monástico** (En la Edad Media, jardín de forma regular situado en el interior del recinto de un monasterio, en el que se cultivaban plantas medicinales, condimenticias, verduras y flores; ►jardín de claustro); *f* **jardin [m] monastique** (Jardin de forme régulière aménagé au moyen-âge par les moines à la périphérie du monastère et utilisé pour la culture des fleurs, des plantes médicinales, condimentaires et potagères ; ►jardin de cloître, ►jardin de curé) ; *g* **Klostergarten [m]** (Formaler Garten, der im Mittelalter von den Mönchen am Rande des Klosters angelegt wurde und der Anzucht von Heilpflanzen, Gewürzkräutern, Gemüsen und Schnittblumen diente; ►Kreuzgarten).

3586 monitoring [n] *envir.* (Process of measurement and observations of the characteristics of a population or an ecosys-

tem, as well as uninhabited areas, to form the basis for determining conditions of the environment and changes to it; ►air pollution monitoring, ►biomonitoring, ►technical monitoring); *s* **monitoreo [m] ambiental** (Seguimiento de cambios ambientales y ecológicos debidos a la industria y a otras actividades humanas; ►biomonitoring, ►evaluación de la calidad del aire, ►inspección técnica); *f* **monitorage [m]** (Technique de mesure et de surveillance utilisée pour définir l'état actuel et l'évolution de l'environnement ; ►surveillance biologique, ►surveillance de la qualité de l'air, ►surveillance technique) ; *syn.* surveillance [f] environnementale ; *g* **Monitoring [n, o. Pl.]** (Prozess der langfristigen, regelmäßig wiederholten, zielorientierten Messung resp. Erfassung und Beobachtung sowie Dokumentation an Raum-Zeit-Serien von Merkmalen in Populationen oder Ökosystemen sowie in unbelebten Bereichen, die geeignet sind, repräsentative Aussagen über den Zustand der Umwelt im Allgemeinen und von Natur und Landschaft im Besonderen und der auftretenden Veränderungen zu treffen; je nach Fragestellung werden z. B. ►Biomonitoring, technisches Monitoring [►technische Überwachung], ►Überwachung der Luftverunreinigung, Waldmonitoring unterschieden); *syn.* Umweltmonitoring [n], Umweltüberwachung [f], Umweltbeobachtung [f]).

monitoring network/system [n] *envir.* ►air quality monitoring network/system.

3587 monoculture [n] *agr. for.* (Repetitive raising of the same crop on the same piece of land over many years whether it be annual or perennial crops in agriculture or forestry; ►mixed crop production, ►pure stand, ►single crop system); *s* **monocultivo [m]** (Cultivo intensivo de una sola especie [de animales o plantas] en un territorio dado; DINA 1987; ►cultivo monoespecífico, ►policultivo, ►rodal monoespecífico); *f* **monoculture [f]** (Le fait d'élever ou cultiver des peuplements monospécifiques annuels ou pluriannuels ; ►boisement monospécifique, ►culture monospécifique, ►polyculture) ; *syn.* culture [f] monospécifique ; *g* **Monokultur [f]** (In der Land- oder Forstwirtschaft langjährig wiederholter, alleiniger Anbau einer ein- oder mehrjährigen Pflanzenart auf derselben Fläche; ►Mischkultur, ►Reinbestand, ►Reinkultur).

3588 monodominant [adj] *phyt. plant. zool.* (VIR 1982, 322. 1. Descriptive term applied to a vegetative cover of only one species. 2. *zool.* Consisting of one species); *syn.* monospecific [adj]; *s* **monoespecífico/a [adj]** (1. Superficie de vegetación que está poblada por una sola especie. 2. Consistente de una sola especie); *f* **unispécifique [adj]** (1. Plantation, peuplement végétale composé d'une seule espèce. 2. *zool.* Relatif à une seule espèce) ; *syn.* monospécifique [adj] ; *g* **aus einer Art bestehend [loc]** (1. Eine Pflanzung, eine Vegetationsfläche betreffend, die nur mit einer Art bepflanzt wurde oder auf der sich nur eine Art angesiedelt hat. 2. *zool.* Bestand, der nur aus einer Art besteht); *syn.* einartig [adj].

monospecific [adj] *phyt. zool.* ►monodominant.

montane belt [n] *geo. phyt.* ►montane zone.

3589 montane plant [n] *bot.* (Plant which grows predominantly in mountainous regions; ►high mountain plant); *s* **planta [f] montana** (Planta de montaña de mediana altitud; ►orófito); *f* **plante [f] montagnarde** (Espèce végétale qui se développe dans l'étage montagnard ; ►orophyte) ; *g* **Mittelgebirgspflanze [f]** (Pflanze, die vornehmlich in Mittelgebirgen beheimatet ist; ►Gebirgspflanze); *syn.* montane Pflanze [f].

3590 montane zone [n] *geo. phyt.* (TEE 1980, 494; ►altitudinal belt between ►colline zone and ►subalpine zone; ►vegetation altitudinal zone); *syn.* montane belt [n]; *s* **piso [m] montano** (►Piso de vegetación que, partiendo del nivel del mar, sigue al ►piso colino, a partir de los 500 m y llega a los 900 m, donde empieza el ▶piso subalpino; DGA 1986; ▶zonación altitudinal); *f* **étage [m] montagnard** (▶Étage de végétation de 800 à 1700 m environ situé entre l'▶étage collinéen et l'▶étage subalpin correspondant à l'étage forestier par excellence ; on peut distinguer l'étage montagnard supérieur — en général formé de conifères — et l'étage montagnard inférieur — à base d'essence feuillues ; cf. DG 1984 ; ▶étagement de la végétation en altitude) ; *g* **montane Stufe [f]** (Hinsichtlich der ▶Höhenstufung der Bereich zwischen ▶kolliner Stufe und ▶subalpiner Stufe; die Höhenangaben variieren, je nach Quelle und in Abhängigkeit von der Höhe über NN und geografischer Lage, zwischen 300-1500 m in den Mittelgebirgen und zwischen 700-1700 m in den Alpen; ▶Vegetationsstufe); *syn.* Bergstufe [f].

3591 monument [n] *conserv'hist.* (Man-made object of historical, cultural, architectural, or aesthetic interest, typically a building or structure, erected or maintained in memory of the dead or to preserve the remembrance of a person, event, or action; generic term for ▶burial monument, ▶buried cultural monument, ▶cultural monument, ▶historic structure, ▶listed landmark building, ▶National Monument [US], ▶natural landmark [US] and ▶monarch tree); *syn.* protected resource [n]; *s 1* **monumento [m]** (Bien inmueble que constituye realización arquitectónica o de ingeniería, u obra de escultura colosal siempre que tenga interés histórico, artístico, científico o social; cf. art. 15 Ley 16/1985; ▶árbol monumento, ▶bién de interés cultural, ▶edificio declarado monumento, ▶edificio histórico-artístico, ▶monumento, ▶monumento funerario, ▶monumento natural, ▶restos arqueológicos cubiertos); *s 2* **Bien [m] de Interés Cultural [Es]** (Término jurídico para todos los bienes muebles o inmuebles integrantes del ▶Patrimonio Histórico Español; entre los bienes inmuebles se encuentran los «Monumentos, Jardines, Conjuntos y Sitios Históricos y Zonas Arqueológicas»; cf. arts. 1 y 14 Ley 16/1985); *f* **monument [m]** (Terme générique pour un ▶terrain qui renferme des stations ou des gisements préhistoriques ; ▶arbre classé comme monument naturel, ▶arbre remarquable d'intérêt national, ▶bâtiment classé ou inscrit au titre des monuments historiques, ▶monument funéraire, ▶monument historique 1, ▶monument historique 2 et ▶monument naturel, ▶monument naturel étendu) ; *g* **Denkmal [n]** (...mäler [pl], in A. nur so, sonst auch ...male [pl]; OB zu ▶Baudenkmal, ▶Baum-Naturdenkmal, ▶Bodendenkmal, ▶Grabdenkmal, ▶Kulturdenkmal und ▶Naturdenkmal; ▶denkmalgeschütztes Gebäude).

monument [n] [UK]**, natural** *conserv. leg.* ▶natural landmark [US].

monument [n]**, soil-covered cultural** *conserv'hist. land' man.* ▶buried cultural monument.

Monuments Commission [n] **for England [UK], Historic Buildings and** *adm. conserv'hist.* ▶historic district commission [US].

moor [n] [UK] *phyt.* ▶heather moor; *geo.* ▶peatland [US].

moorland [n] *geo.* ▶peatland [US].

3592 moorpan [n] *pedol.* (Dark, very hard ▶illuvial horizon on mostly humid sites; e.g. bog margins or moors covered by *Erica* or *Calluna* vegetation; ▶podzol, ▶ortstein [US]/ironpan (2) [UK]); *s* **capa [f] de ortstein de humus** (▶Horizonte iluvial muy oscuro y muy duro que se forma en los ▶podsoles en zonas húmedas, p. ej. bordes de turberas cubiertas de brezos; ▶ortstein); *f* **alios [m] humique** (Matériau peu perméable et dur d'un ▶horizon illuvial formé de sable cimenté par des revêtements humifères constituant un horizon d'accumulation dans le ▶podzol humique hydromorphe dans les plaines sableuses en bordure de tourbière ou dans les landes à éricacées ; ▶alios durci) ; *g* **Humusortstein [m]** (Bei ▶Podsolen dunkler, sehr harter

▶Anreicherungshorizont an meist feuchten Standorten, z. B. an Moorrändern und unter *Erica*-Heiden; ▶Ortstein).

3593 moraine [n] *geo.* (Accumulation of unsorted and unstratified masses of clay and stone drift carried and deposited by a glacier; etymologically a French term which was used by Alpine peasants in the 18th century for any heap of earth and stony debris; *specific terms* ▶ground moraine, ▶lateral moraine, ▶terminal moraine); *s* **morrena [f]** (Formación producida por la acumulación de materiales arrastrados y deformados por la acción directa del hielo y de los glaciares. Usualmente están compuestas de materiales heterogéneos, de nula a moderada consolidación, poco o parcialmente estratificados, con mezclas de arcillas, limos, arenas, gravas y materiales gruesos; DINA 1987; ▶morrena de fondo, ▶morrena frontal, ▶morrena lateral); *f* **moraine [f]** (Matériaux hétérométriques transportés et déposés par un glacier ; ▶moraine de fond, ▶moraine frontale, ▶moraine latérale) ; *g* **Moräne [f]** (Gesteinsschicht, die ein Gletscher verfrachtet und beim Abschmelzen ablagert; der Begriff stammt ursprünglich aus dem Französischen, da Bergbauern der Alpen im 18. Jh. jedwede Anhäufung von Erde und Gesteinsschutt als *moraine* bezeichneten; *UBe* ▶Endmoräne, ▶Grundmoräne, ▶Randmoräne).

moraine [n]**, end** *geo.* ▶terminal moraine.

moraine [n]**, frontal** *geo.* ▶terminal moraine.

moraine [n]**, marginal** *geo. obs.* ▶terminal moraine.

3594 more frequent mowing [n] *agr. constr. hort.* (ELL 1988, 575; several cuts of a meadow annually; ▶cut a few times per year); *s* **siega [f] múltiple** (*Contexto* ▶prado de siega múltiple; ▶segado/a varias veces al año); *f* **fauche [f] exécutée plusieurs fois par an** (▶fauché, ée plusieurs fois par an) ; *g* **mehrmalige Mahd [f]** (mehrmalig mähen [vb], mehrmals mähen [vb]; ▶mehrschürig); *syn.* mehrmaliger Schnitt [m].

3595 mor humus [n] *pedol.* (Humus type [O_l-O_f-O_h horizons] lying on the surface of a mineral soil and unmixed with it; ▶forest floor litter, ▶litter layer, ▶organic horizon, ▶raw humus); *s* **moor [m]** (Humus de suelos terrestres poco descompuesto y sin ligar con la parte mineral del suelo. En él, el complejo adsorbente está muy lejos de la saturación, y la reacción es fuertemente ácida. Es típico de la taiga y de las landas; cf. DGA 1986; ▶capa de litter, ▶humus bruto, ▶litter del bosque, ▶mantillo); *syn.* humus [m] de superficie; *f* **mor [m]** (Humus forestier de milieu aéré, peu évolué, superposé ou à incorporation faible de matière organique peu transformée, à rapport C/N supérieur à 25 ; humus forestier non submergé à engorgement nul ou temporaire par une nappe perchée présentant une couche holorganique en surface différenciée en horizon O plus ou moins épais à humus très acide, pauvre en déjections animales, surmontant un horizon A_h de diffusion généralement mince ; DUCHAUFOUR et DELECOUR cit. in DIS 1986 ; ▶couverture morte, ▶horizon organique [Ao], ▶humus brut, ▶litière forestière) ; *syn.* humus [m] de couverture, litière [f] humifiée, litière [f] structurée, humus [m] peu actif ; *g* **Auflagehumus [m]** (Oberste Schicht auf terrestrischen Böden, die aus noch nicht [völlig] zersetzter organischer Substanz [O_l-O_f-O_h-Horizonte] besteht; ▶Humusauflage, ▶Rohhumus, ▶Streuschicht, ▶Waldbodenstreu).

morphogenesis [n]**, root** *bot.* ▶root growth.

3596 mortar [n] *constr.* (Sand/cement/lime/water mixture used in masonry construction, paving and plastering; ▶joint filled with liquid mortar); *s* **mortero [m]** (Cualquier material pastoso con consistencia suficiente que sirva para cubrir muros, tabiques y techos de edificios y para rellenar juntas en muros y pavimentos. Antiguamente se utilizaba un **m.** a base de cal, arena, fibra y agua. Hoy en día se emplea la mezcla del cemento Portland con arena y agua; cf. DACO 1988; ▶junta rellena de mortero líqui-

M

do); *f* **mortier [m]** (Mélange de sable [généralement siliceux] et de liant hydraulique [chaux ou ciment] gâché avec une certaine quantité d'eau, utilisé pour la mise en œuvre de matériaux entrant dans la constitution de maçonnerie ; ►joint bourré au coulis de mortier) ; *g* **Mörtel [m]** (Aus dem Lateinischen *mortarium*; Gemisch aus Sand, Wasser und Zement, Kalk oder Gips, das als Bindemittel für den Mauerbau, für Platten- oder Pflasterbeläge sowie zum Verputzen verwendet wird [Putzmörtel]; ►mit flüssigem Mörtel verfüllte Fuge).

3597 mortar bed [n] *constr.* (SPON 1986, 360; layer of mortar upon the ►subbase [US]/sub-base [UK] for laying setts, bricks, pavers, etc.); *syn.* mortar setting bed [n] [also US] (TSS 1988, 910-3); *s* **lecho [m] de mortero** (Capa de mortero sobre la ►capa portante sobre la que se tienden los adoquines, los ladrillos clinker, las losas, etc.); *syn.* asiento [m] de mortero; *f* **lit [m] de pose en mortier** (Couche de mortier située au-dessus de la ►couche de fondation sur laquelle est effectuée la pose de pavés, dalles, klinkers, etc. ; poser [vb] à bain de mortier) ; *syn.* mortier [m] de pose, bain [m] de mortier ; *g* **Mörtelbett [n]** (Mörtelschicht über der ►Tragschicht zur Verlegung von Pflastersteinen, Klinkern, Platten etc.); *syn.* Mörtelbettung [f].

mortar setting bed [n] [US] *constr.* ►mortar bed.

mortlake [n] *geo. obs.* ►oxbow lake.

3598 mosaic block paving [n] [US] *constr. eng.* (Paved surface of ►mosaic pavers [US]/mosaic paving setts [UK] made of natural stone or concrete); *syn.* mosaic sett paving [n] [UK]; *s* **pavimento [m] de mosaico** (Superficie pavimentada con ►adoquín mosaico); *syn.* adoquinado [m] mosaico, empedrado [m] mosaico; *f* **pavage [m] en pavés mosaïques** (Revêtement réalisé avec des ►pavés mosaïques en pierre naturelle ou en béton) ; *g* **Mosaikpflaster [n]** (Belag mit ►Mosaikpflastersteinen aus Naturstein oder Beton verlegt).

3599 mosaic paver [n] [US] (≠) *constr.* (In D., according to DIN 18 502 such pavers have standard dimensions of 40-60mm and are quarried; in U.K. and U.S., such dimensions are not usual; small sized paving stones vary from square to rectangular and have standard sizes of approximately 100mm [4 inches]; **m. p.s** are often manufactured as pre-cast concrete pavers with stone or ceramic patterns pressed into the surface to create a unique or repetitive pattern; the term may also refer to tiles. In addition ready-made paving patterns are manufactured by glueing setts together and installation by machines to create a paver pattern [rosette, fish scale, or running bond]; pre-cast units range from 300, 400, and 600mm; ►small paving stone); *syn.* mosaic paving sett [n] [UK] (≠); *s* **adoquín [m] mosaico** (En Es. adoquín de hormigón o cerámica de 40-50 mm de largo y ancho; en piedra natural de cara vista con longitud entre 100-120 mm, ancho entre 80-100 mm y altura de 80 mm, que corresponde al ►adoquín pequeño; cf. PGT 1987, 118; **en D.** según la norma DIN 18 502 el **a. m.** tiene caras vistas entre 40-60 mm de largo); *f* **pavé [m] mosaïque en pierre naturelle** (Pavé de dimensions comprises entre 70 et 100 mm et dont les tolérances dimensionnelles et de taille sont définies par les normes N.F. B 10 513 et B 10 508 ; ►petit pavé) ; *g* **Mosaikpflasterstein [m]** (1. D., nach DIN 18 502 beschriebene Güteklassen und Größeneinteilungen für Naturpflastersteine mit Kantenlängen von 40-60 mm. **2.** Die Betonindustrie stellt auch Betonmosaikpflastersteine her).

mosaic paving sett [n] [UK] *constr.* ►mosaic paver [US].

mosaic sett paving [n] [UK] *constr. eng.* ►mosaic block paving [US].

3600 moss [n] **(1)** *bot.* (Small cryptogamous plant belonging to the two subdivisions of **liverworts** *[Marchantiophytina = Hepaticae]* und **mosses** *[Bryophytina]*; to the latter belong the two classes of **sphagnum mosses** *[Sphagnopsida]* and **mosses**

[Bryopsida = Musci]; ►hummock-forming peat moss); *syn.* bryophyte [n]; *s* **briófito [m]** (*Bryophyta sensu latu* constituye una división del reino vegetal de plantas autótrofas no vasculares con unas 24 000 especies que comprende: **1.** las **hepáticas** *[Marchantiophytina = Hepaticae]*, **2.** los **musgos** *[Bryophytina sensu stricto]* con tres subclases: esfagnales *[Sphagnopsida]*, andreales *[Andreaeopsida]* y briales *[Bryopsida]* y **3.** las **antóceras** *[Anthocerophytina]*; cf. DB 1985, LB 2008 y WIKI 1-2009; ►esfago del mamelón); *f* **bryophyte [m]** (Groupe de bryophytes comprenant les embranchements de plantes archégoniates non vascularisées terrestres qui ne possèdent ni racines ni vrai système vasculaire lignifié ; on compte environ 24 000 espèces, regroupant **1.** les **hépatiques** *[Marchantiophytina = Hepaticae]*, **2.** les **bryophytes** au sens large, divisés en trois classes les bryopsidées — **mousses vraies** — *[Bryopsida = Musci]*, les **sphaignes** *[Sphagnopsida]*, les **andréales** *[Andreaeopsida]*, et **3.** les **anthocérotes** *[Anthocerotophyta]* ; ►sphaigne de buttes) ; *syn.* bryophyte [m] ; *g* **Moos [n] (2)** (Autotrophe Sporenpflanze, von der es ca. 24 000 Arten gibt, die stammesgeschichtlich in drei Unterabteilungen gegliedert sind: **1. Lebermoose** *[Marchantiophytina = Hepaticae]* mit den beiden Klassen der thallosen und foliosen Lebermoose, **2. Laubmoose im weiteren Sinne** *[Bryophytina = Musci]* mit den drei Klassen der **Torfmoose** *[Sphagnopsida]*, **Klaffmoose** *[Andreaeopsida]* und **Lebermoose im weiteren Sinne** *[Bryopsida]* und **3. Hornmoose** *[Anthocerophytina]*; cf. LB 2008; ►Bultmoos); *syn.* Moospflanze [f], Bryophyt [m].

moss [n] **(2)** *geo.* ►peatland [US].

3601 moss floor [n] *phyt.* (TEE 1980, 292; extensive, very low vegetation cover predominantly of mosses; ►moss layer); *syn.* moss ground cover [n]; *s* **tapiz [m] muscinal** (Capa de vegetación extensa de muy poca altura, constituida predominantemente de musgos; ►estrato muscinal); *f* **pelouse [f] rase à mousses** (Surface végétale très basse principalement recouverte de mousse ; ►strate muscinale) ; *syn.* tapis [m] muscinal ; *g* **Moosrasen [m]** (Eine größere, sehr niedrige Vegetationsschicht, die vornehmlich aus Moosen besteht; ►Moosschicht).

moss ground cover [n] *phyt.* ►moss floor.

3602 moss layer [n] *phyt.* (Cryptogamous plants growing in dense clusters close to the ground surface and forming the ground layer; ►moss floor, ►stratification of plant communities); *s* **estrato [m] muscinal** (Capa inferior en la ►estratificación de la vegetación constituida principalmente por musgos; ►tapiz muscinal); *f* **strate [f] muscinale** (Dans la ►stratification de la végétation, strate inférieure essentiellement composée de bryophytes, lichens et champignons ; ►pelouse rase à mousses) ; *syn.* strate [f] cryptogamique ; *g* **Moosschicht [f]** (In der ►Vegetationsschichtung die unterste, vorwiegend aus Moosen bestehende Schicht; ►Moosrasen); *syn.* Bodenschicht [f].

3603 moss rose [n] *hort. plant.* (Pink-flowered form of the cultivated centifolia rose [Rosa centifolia], also known as the *Provence rose*, up to 2 meters high with fine hairs on both sides of the leaves, dense, moss-like hairs or bristles at the top of the peduncel and ovary and often strongly plumed sepals); *s* **rosa [f] musgosa** (Cultivar de la rosa centifolia de hasta 2 m de altura de flores rosas y hojas peludas por ambas caras y sépalos muy pinnados, de manera que tiene un aspecto musgoso); *syn.* rosa [f] musqueta; *f* **rosier [m] mousseux** (Variété sélectionnée à partir de la rose cent-feuilles [Rosa centifolia muscosa], atteignant 2 m de hauteur, aux fleurs rose pur, aux grandes fleurs pleines, délicieusement parfumées, dont les feuilles sont velues sur les deux faces leurs donnant l'aspect de la mousse) ; *g* **Moosrose [f]** (Rosa blühende, bis 2 m hohe Zuchtform der Zentifolie *[Rosa centifolia]*, auch *Provencerose* genannt, mit beiderseits behaarten, schlaffen Blättern, dichten, moosartigen Drüsenhaaren oder -borsten oben am Blütenstiel und Fruchtknoten sowie oft stark gefiederten Kelchblättern).

motor home [n] [US] *recr. urb.* ▶mobile home.

motorway [n] [UK] *trans.* ▶freeway [US], ▶major highway [US] (1).

motorway [n] **and trunk road** [n] [UK] *adm. leg. trans.* ▶interstate highway [US].

motorway central reservation [n] [UK] *trans.* ▶freeway median strip [US].

motorway corridor [n] [UK] *plan. trans.* ▶freeway corridor [US].

motorway intersection [n] [UK] *trans.* ▶freeway interchange [US].

motorway junction [n] [UK] *trans.* ▶interchange.

motorway service area [n] [UK] *trans.* ▶service area with all facilities.

motorway services [n] [UK] *trans.* ▶service area with all facilities.

mound [n] (1) *geo.* ▶knoll.

3604 mound [n] (2) *constr.* (Raised mass of earth, soil, stones, or other compacted material in a heap or pile; ▶earth mound, ▶noise attenuation mound); *s* **montículo** [m] **(artificial)** (▶terraplén); *syn.* elevación [f] (artificial); *f* **monticule** [m] **artificiel** (Élévation de terre, de rochers ou autres matériaux résultat de travaux de modelage du terrain ; ▶merlon) ; *syn.* butte [f] artificielle ; *g* **künstlicher Hügel** [m] (Ergebnis einer modellierten Erdaufschüttung; ▶Erdwall).

mound [n] [UK], **acoustic screen** *constr. envir.* ▶noise attenuation mound.

mound [n] [UK], **noise screen** *constr. envir.* ▶noise attenuation mound.

mound [n], **residue** *min.* ▶spoil bank [US] (2)/spoil heap [UK].

mound [n], **slate** *min.* ▶spoil bank [US] (2)/spoil heap [UK].

3605 mound planting [n] *constr. for. hort.* (**1.** Setting out young trees on a small heap of soil, so as to promote aeration and free drainage, particularly on wet sites; SAF 1983; ▶ridge planting. **2. In U.S.**, setting out young trees on dry ground with the roots spread over a cone of soil at the bottom of a dug hole; SAF 1983; ▶hole planting); *s* **plantación** [f] **en montículos** (Plantación de leñosas sobre montículos planos para mejorar el drenaje y así el suministro de aire de las raíces, se emplea sobre todo en ubicaciones muy húmedas; ▶plantación en caballones, ▶plantación en hoyos); *f* **plantation** [f] **sur butte** (Plantation de jeunes plants, ou semis direct, sur de petites buttes de terre, continues ou discontinues, de façon à améliorer l'aération et le drainage, notamment en milieu trop humide, et à réduire la concurrence herbacée ; DFM 1975 ; ▶plantation sur bourrelet, ▶plantation sur potets) ; *g* **Hügelpflanzung** [f] (Pflanzung von Gehölzen auf flachen Erdhügeln zur besseren Luftversorgung der Wurzeln, besonders auf nassen Standorten; ▶Lochpflanzung, ▶Kammpflanzung).

3606 mountable curb [n] [US] *constr. eng.* (Rim of stone or concrete below curb [US]/kerb [UK] height permitting access from a road to a driveway or other paved areas with a cross section slope between 1:1 and 1:3; barrier curb [US]/edge kerb [UK], ▶curb cut [US]/drop kerb [UK], ▶drop curb [US]/flush kerb [UK]; *syn.* flared curb [n] [US], splayed kerb [n] [UK]; *s* **bordillo** [m] **accesible** (▶Bordillo de acera rebajado o biselado de 1:1 a 1:3 para hacer posible el paso de vehículos; ▶bordillo hundido, ▶piedra de bordillo enterrado); *syn.* bordillo [m]

sobrepasable, cordón [m] accesible, cordón [m] sobrepasable [ambos CS]; *f* **bordure** [f] **basse franchissable** (Type A1 et A2, cf. norme N.F. P 98 302 ; ▶bordure de trottoir abaissée permettant à partir du trottoir public l'accès des véhicules sur une parcelle de terrain ; ▶bordure arasée d'épaulement, ▶zone d'abaissement de bordure de trottoir ; *g* **Flachbord** [m] (▶Hochbord, der zum Überfahren an Grundstückseinfahrten stark abgesenkt oder 1:1 bis 1:3 abgeschrägt ist; ▶Bordsteinabsenkung, ▶Tiefbordstein); *syn.* Kante [f] aus Flachbordsteinen, Schrägbord [m].

3607 mountain cabin [n] [US] *recr.* (**1.** Originally, a small, one-story lodging facility for short stays in the mountains; such a facility is either privately owned or rented in state parks and national forests; ▶alpine hut. **2.** Today also greater backcountry houses with 2-3 stories are called **m. c.**); *syn.* mountain hut [n]; *s* **cabaña** [f] **de montaña** (**1.** En EE.UU., originalmente pequeño albergue de un piso para alojamiento de corto plazo en zonas montañosas, que puede ser de propiedad privada o pública de los parques estatales o de los bosques nacionales, en cuyo caso se alquila; ▶refugio de alta montaña. **2.** Hoy en día en los EE.UU. se utiliza también el término «mountain cabin» para chalets de hasta 2-3 pisos construidos en lugares apartados); *f* **chalet** [m] **d'alpage** (Élément majeur du patrimoine montagnard, protégé et mis en valeur dans le cadre de la loi n° 94-112 du 9 février 1994 ; ▶refuge de haute montagne) ; *g* **Gebirgshütte** [f] (▶1. U.S., kleine Übernachtungseinrichtung für kurze Aufenthalte im Gebirge; *cabins* sind im Privatbesitz oder können in *State Parks* oder *National Forests* gemietet werden. **2.** Heute werden auch mehrgeschossige Ferien- und Beherbergungsstätten in abgelegener Lage als *mountain cabin* bezeichnet; ▶Alpenhütte); *syn.* Berghütte [f].

3608 mountain forest [n] *phyt. geo.* (▶protective forest 1, ▶protective forest 2); *s* **bosque** [m] **de montaña** (▶bosque protector 1, ▶bosque protector 2); *syn.* bosque [m] montano; *f* **forêt** [f] **montagnarde** (Forêt située en région montagneuse dont la constitution dépend de l'exposition des versants et qui colonise les milieux d'altitude des étages montagnard et subalpin ; elle peut être déclarée comme ▶espace boisé protégé en vue de la protection contre l'érosion, les chutes de pierres, etc. ; ▶forêt de protection) ; *syn.* forêt [f] de montagne ; *g* **Bergwald** [m] (Wald in Berggebieten, der bis zur Baumgrenze reicht und meist wegen des Erosionsschutzes zum ▶Bannwald 2 erklärt wurde; ▶Schutzwald).

mountain hut [n] *recr.* ▶mountain cabin [US].

3609 mountain lodge [n] [US] *recr.* (Hut built by ski or hiking [US]/rambling [UK] clubs in mountainous areas, with basic cooking and sleeping facilities; a smaller structure with only sleeping facilities is called in U.S. a ▶mountain cabin; ▶alpine hut); *syn.* mountain refuge [n] [UK], mountain shelter [n] [UK], trail hut [n] [US]; *s* **refugio** [m] **de montaña** (Cabaña situada en alta montaña, equipada sencillamente, y que pertenece generalmente a una asociación de montañistas, en la que puede pernoctar cualquier persona que lo necesite; ▶cabaña de montaña, ▶refugio de alta montaña); *f* **refuge** [m] **de montagne** (Équipement touristique construit en montagne par les associations de randonnée ou d'alpinisme, dont l'équipement de base est constitué d'un emplacement pour dormir et d'un coin cuisine ; *termes spécifiques* refuge de haute montagne, refuge de la grande traversée des Alpes ; ▶chalet d'alpage, ▶refuge de haute montagne) ; *g* **Schutzhütte** [f] (1) (Im Gebirge durch einen Verein erbaute Hütte, die für Bergwanderer, Kletterer und Tourenskiläufer mindestens mit Matratzenlager und Kochmöglichkeit ausgerüstet ist; ▶Alpenhütte, ▶Gebirgshütte).

mountain plant [n] *bot.* ▶high mountain plant.

M

3610 mountain range [n] *geo.* (**1.** Mountainous mass or group of connected heights, whether isolated or forming part of a larger mountain system [e.g. the Rocky Mountains] or of a mountain chain [e.g. the Adirondack Mountains in the north of New York State. **2.** Region characterized by mountains; in this case the mountain term is usually used in the plural]; *syn.* massif [n], mountains [npl]; *s* **montaña** [f] (Conjunto de elevaciones de terreno que están separadas entre sí por valles; el término **m.** también denomina a una sola elevación de terreno, en cuyo caso el sinónimo es **monte** [m]; en el caso de la **cordillera** [f] se trata de una serie de montañas entrelazadas entre sí, aunque en geología este término se utiliza sólo para las montañas creadas por plegamiento; cf. DINA 1987); *syn.* sierra [f], cadena [f] montañosa; *f 1* **massif** [m] **montagneux** *geo.* (Groupe de montagnes séparées par des vallées) ; *f 2* **espace** [m] **montagnard** *recr.* (Terminologie de l'aménagement du territoire ou du tourisme) ; *g* **Gebirge** [n] (**1.** Gruppe hoher, markanter Berge, die durch Täler und Hochflächen gegliedert sind; geomorphologisch werden **Mittel-** und **Hochgebirge** nach der Höhe und Hangneigung unterschieden; **Kamm-, Ketten-, Kuppen-** und **Plateau-Gebirge** werden nach der Form und **Vulkan-** und **tektonische Gebirge** nach ihrer Orogenese benannt. **2.** *min.* Das anstehende Gestein resp. der Gesteinsverband, der eine Lagerstätte oder einen Grubenbau umgibt).

mountain refuge [n] [UK] *recr.* ►mountain lodge [US].

mountain region [n] *geo.* ►high mountain region.

mountains [npl] *geo.* ►mountain range.

mountain shelter [n] [UK] *recr.* ►mountain lodge [US].

3611 mountain wind [n] *met.* (Wind down the valley at the beginning of sunset; ►downslope wind, ►valley wind); *s* **brisa** [f] **de montaña** (►Viento catabático o descendente que sopla por la noche y en las primeras horas bajando por las laderas montañosas desde la cima hasta el valle debido al enfriamiento de la superficie causado por la radiación nocturna; cf. DM 1986; *opp.* ►viento de valle); *f* **brise** [f] **de montagne** (Vent catabatique soufflant la nuit et aux premières heures du jour, le long des pentes des montagnes vers les vallées ou les plaines [brise descendante] ; CILF 1978 ; *opp.* ►brise de vallée ; ►vent catabatique) ; *syn.* vent [m] de montagne ; *g* **Bergwind** [m] (Wind, der mit Beginn der Abenddämmerung talwärts weht; ►Hangabwind; *opp.* ►Talwind).

3612 mouth [n] *geo.* (Place where a stream or river enters a larger body of water; ►estuary, ►delta); *s* **desembocadura** [f] (Zona de confluencia de un curso de agua en otro más grande o en el mar; ►delta, ►estuario); *f 1* **embouchure** [f] **d'un cours d'eau** (Extrémité aval d'un cours d'eau se jetant dans la mer ou dans un lac ; ►delta, ►estuaire) ; *f 2* **confluent** [m] (Zone où se joignent deux cours d'eau) ; *g* **Mündung** [f] (Stelle, an dem ein Wasserlauf in ein anderes Gewässer fließt; ►Ästuar, ►Delta); *syn.* Mündungsgebiet [n].

3613 movable outdoor furniture [n] *arch. constr. urb.* (Those items which are not fastened or anchored to the ground; *specific term* movable chairs and tables); *s* **mobiliario** [m] **urbano móvil** (Elemento de mobiliario urbano que no está sujetado o fijado en el suelo); *f* **mobilier** [m] **mobile** (Élément de mobilier urbain ou de jardin posé sur le sol) ; *g* **lose Möblierung** [f] (Nicht im Boden verankerte Möblierungselemente).

moving [n] *constr.* ►transplanting (2).

3614 mow [vb] *agr. constr. hort.* (To cut grass by machine or with a scythe; **in U.K.**, a lawn is mown and a meadow is cut; ►first mowing, ►grass cutting, ►mowing of meadows [US]/cutting of meadows [UK]; *s 1* **segar** [vb] *agr.* (Corte de la hierba para cosecha agraria; ►siega); *s 2* **cortar** [vb] **la hierba** *hort.*

(►Corte del césped en parques y jardines como medida de mantenimiento; ►primer corte); *f* **tondre** [vb] (un gazon d'agrément ; ►étrépage, ►fauche, ►première tonte, ►tonte du gazon) ; *g* **mähen** [vb] (Schneiden von Gras mit einem Mäher, Freischneider oder einer Sense; ►Mahd, ►Rasenschnitt, ►erster Schnitt); *syn.* Gras schneiden [vb].

mowing [n] *agr. constr. hort.* ►first mowing, ►grass cutting, ►more frequent mowing, ►single mowing.

mowing edge [n] *constr.* ►mowing strip.

mowing margin [n] *constr.* ►mowing strip.

3615 mowing [n] **of meadows** [US] *agr. constr. hort.* (Harvesting of fodder meadows or cutting of wildflower meadows in urban green spaces; ►mow); *syn.* cutting [n] of meadows [UK]; *s* **siega** [f] (Cosecha de la hierba para alimentación del ganado; ►cortar la hierba, ►segar); *f 1* **fauche** [f] (À l'origine, coupe d'un pré, d'une prairie au moyen d'une faux ; le terme est aussi utilisé pour la coupe mécanique ; ►tondre) ; *syn.* fauchaison [f], fauchage [m], fenaison [f] ; *f 2* **étrépage** [m] (1) (Coupe de touffes d'herbes pour le recouvrement du silage de betteraves fourragères) ; *g* **Mahd** [f] (...en [pl]; Ernteschnitt bei Futterwiesen oder Blumenwiesenschnitt in öffentlichen Grünanlagen; ►mähen); *syn.* Mähen [n, o. Pl.], Wiesenschnitt [m].

3616 mowing strip [n] *constr.* (Paved edging to an area of lawn, comprising, for example, concrete or brick pavers laid at the same level as the grass, in order to facilitate the cutting of the grass edge during mowing); *syn.* mowing margin [n], mowing edge [n]; *s* **borde** [m] **de corte** (Borde alrededor del césped, generalmente pavimentado, para facilitar el corte de la hierba); *f* **bordure** [f] **de pelouse** (Bordure en béton ou alignement de pavés ou de briques disposé à hauteur d'un gazon pour faciliter les opérations de coupe en bordure des zones engazonnées) ; *syn.* chenette [f] de pavés bordant une pelouse ; *g* **Mähkante** [f] (Befestigtes, mit dem Rasen plangleiches Werksteinband aus Betonwerkstein, Naturstein, oder Klinker am Rande von Rasenflächen zur Erleichterung der Kantenpflege beim Rasenschnitt).

3617 mowing tolerance [n] *constr. hort.* (Relative capability of lawn grass species to survive cutting); *s* **resistencia** [f] **a la siega** (Característica de ciertas especies herbáceas de césped de regenerarse después de múltiples cortes); *f* **adaptation** [f] **aux tontes répétées** (Grâce à leurs rhizomes, fort pouvoir de régénération des graminées pour gazon, leurs permettant de résister à des tontes renouvelées et en particulier aux tontes très rases) ; *syn.* résistance [f] aux tontes ; *g* **Schnittfestigkeit** [f] (Eigenschaft von Rasengräsern mehrmalige Mähgänge zu vertragen oder zu überleben); *syn.* Schnittverträglichkeit [f].

mown [pp] **once a year** [US] *agr. constr. landsc.* ►be mown once a year.

muck [n] *pedol.* ►anmoor.

muck [n] **and plant material** [n] *constr. land'man. wat'man.* ►excavated muck and plant material.

3618 mud [n] *limn. pedol.* (Wet organic sediment in standing water; ►exposed mud, ►gyttja, ►harbor mud [US]/harbour mud [UK], ►organic mud, ►sapropel); *s* **lodo** [m] **lacustre** (Depósito de materias en suspensión y de sustancia orgánica muerta en aguas quietas; ►fango expuesto, ►gyttja, ►légamo portuario, ►sapropel, ►suelo subacuático no turboso); *syn.* fango [m] lacustre; *f* **vase** [f] **lacustre** (Sédiment fin constitué de minéraux argileux et de matières organiques en suspension, formé par mise en flocons dans les eaux calmes ; DG 1984 ; ►boue de dragage portuaire, ►boue organique, ►gyttja, ►sapropèle, ►vase exondée) ; *g* **Schlamm** [m] (1) (Ablagerungen von Schwebstoffen und organischem Bestandsabfall im Stillgewässer; ►Gyttja, ►Hafenschlick, ►Mudde, ►offener Schlammboden, ►Sapropel).

mud [n]**, tidal** *geo. ocean. pedol.* ▶ sediment (2).

3619 muddying [n] **of soil** *pedol.* (Propensity of a silt-rich soil with a clay content of 10-15% to lose its structural stability caused by the effects of overwatering or heavy rainfall; ▶ silting up of soil pores); *s* **riesgo** [m] **de embarramiento del suelo** (Disposición de suelos ricos en arcilla [10-15%] de perder la estabilidad estructural [grumosidad] al encharcarse; ▶ embarramiento del suelo); *syn.* riesgo [m] de enfangamiento del suelo; *f* **mouillabilité** [f] (Capacité que possèdent les agrégats à résister à l'éclatement [dégradation de la stabilité structurale] lors de la pénétration de l'eau dans le sol ; ▶ colmatage des vides d'un sol) ; *syn.* risque [m] de colmatage, possibilité [f] de colmatage ; *g* **Verschlämmbarkeit** [f] (Disposition eines schluffreichen Bodens mit einem Tonanteil von 10-15 % [z. B. Lössböden, Marschen] durch erhöhte Zufuhr von Wasser in seiner Gefügestabilität zerstört zu werden; ▶ Verschlämmung).

mudflat [n] *geo. phyt.* ▶ glasswort mudflat.

3620 mud flow [n] *geo.* (**1.** Rapid flowage type of ▶ mass slippage [US]/mass movement [UK] in the mountains involving the sudden downslope flow of a saturated mass of cohesive soil and abundant coarse-grained materials such as rocks, tree trunks; a **m. f.** contains at least 50 percent sand-, silt-, and clay-sized particles); **2. debris flow** [n] (Form of rapid mass movement in which a combination of loose soil, rock, organic matter, air, and water mobilize as a slurry that flows downslope. Debris flows include <50% fines; cf. NAUS 2009); *syn.* mudslide [n] (LE 1986, 85); *s 1* **colada** [f] **de barro** (Uno de los tipos de solifluxión propio de vertientes arcillosas que han alcanzado el límite de plasticidad. Este estado permite la movilización de la masa fangosa que se desplaza pudiendo transportar cantos y bloques de gran tamaño; DGA 1986; ▶ movimiento gravitacional); *syn.* corriente [f] de barro; *s 2* **corriente** [f] **terrosa** (Tipo de ▶ movimiento gravitacional rápido, aunque más lento que la ▶ colada de barro, que no está confinado a cauces y tiene un contenido de agua inferior y se presenta comúnmente en regiones húmedas; DINA 1987); *f* **lave** [f] **torrentielle** (▶ *Mouvement de masse* dépôt boueux hétérométrique laissé, en fin de crue, par un écoulement torrentiel, dans un chenal d'écoulement ou sur un cône de déjection, fortement chargée de matériaux hétérogènes : blocs, troncs d'arbres. L'anglais « mudflow » désigne plutôt une coulée de ▶ terre fluente sans intervention et d'un ruissellement concentré ; DG 1984) ; *syn.* coulée [f] de débris ; *g* **Mure** [f] (Form des ▶ Massenversatzes im Gebirge nach Starkregen oder plötzlich einsetzender Schneeschmelze als Schlammstrom, der stark mit Gesteinsschutt durchmischt ist); *syn.* Murgang [m], Schlammstrom [m].

mudslide [n] *geo.* ▶ mud flow.

3621 mud trap [n] (1) *constr.* (Device in a sewer or drain system which retains mud sediment, sand or gravel. Dry matter is separated in a ▶ silt box [US]/sediment bucket [UK], liquid mud in a ▶ sediment basin [US]/silt trap [UK]); *s* **colector** [m] **de fango** (1) (Dispositivo para recoger sustancias sólidas en conducciones de aguas residuales o de drenaje. El fango seco se recoge en ▶ cubos de fango, el fango fluido en ▶ arquetas de recogida con filtro); *f* **dispositif** [m] **de rétention des boues** (Élément de canalisations d'égouts ou de drainage destiné à retenir les déchets transportés [substances en suspension, sable ou gravier] par les eaux de ruissellement ; ▶ panier [ramasse-boue], ▶ radier de décantation) ; *g* **Schlammfang** [m] (1) (Vorrichtung in Abwasser- oder Dränleitungen zum Abfangen von Schweb-/Sinkstoffen, Sand oder Kies. Trockenschlamm wird im ▶ Schlammeimer, Nassschlamm im ▶ Sumpf 1 abgeschieden).

3622 mud trap [n] (2) *constr.* (Device for retaining mud before a ▶ grease separator); *s* **colector** [m] **de fango** (2) (Dispo-
sitivo colocado en conducciones antes del ▶ separador de grasa); *f* **débourbeur** [m] (Dispositif en aval d'un ▶ bac séparateur à graisses) ; *g* **Schlammfang** [m] (2) (In einer Rohrleitung einem ▶ Fettabscheider vorgelagerte Vorrichtung zur Abscheidung von Feinteilen).

mud vegetation [n] *phyt.* ▶ exposed mud vegetation.

3623 mulch [n] *agr. constr. hort.* (**1.** Natural or artificial layer of loose material used as a top dressing on soil in order to provide insolation, conserve soil moisture, aid in soil stabilization or to deter weeds; thus providing microclimatic conditions suitable for germination and growth; organic materials, dead leaves, ▶ bark chips, wood chippings, etc. are commonly used; less common is the use of volcanic ash, scoria, coarse sand and gravel as **m.** in dry-lands agriculture; cf. ESS 2008. **2.** In desert landscaping, planted areas are commonly covered with stone mulch; often referred to as gravel, the mineral mulch is clean, washed free of loam, sand, clay, all organic material and substances, screened to 10-25mm in size, and installed to a thickness of approx. 50mm. The mulch hinders evaporation from irrigated soil, is used to hide drip irrigation lines, keeps down weeds and gives a pleasing appearance to the soil surface under shrubs or a desert look to xeriscaping. In **D.**, since 2000, mineral mulch such as fine aggregate [US]/stone chippings [UK] of granite, limestone, porphyry, crushed bricks, etc. are used for mulching of perennial planting for low-cost weed control in a layers of 50-150mm, screened to 8-16mm in size; ▶ bark mulch, ▶ hydroseeding, ▶ mulching, ▶ mulching material, ▶ soil mellowness); *s* **mulch** [m] (Cubierta superficial del suelo, orgánica o inorgánica, p. ej. de cortes de hierba, ▶ astillas de corteza, turba o productos sintéticos, que tiene efecto protector contra la erosión, la transpiración o contra las malas hierbas y de mejoramiento de la ▶ sazón del suelo; en el ▶ riego de semillas por emulsión, capa de paja fijada con bitumen o medios sintéticos utilizada para asegurar taludes contra la erosión; ▶ material de mulching, ▶ mulch de corteza triturada; ▶ mulching); *f* **mulch** [m] (**1.** Couverture de débris végétaux [herbes, paille, ▶ mulch d'écorce fragmentée, ▶ copeaux d'écorce, tourbe ou mousses organosynthétiques], afin de conserver un ▶ bon état de structure grumeleuse, de protéger le sol de l'érosion et du dessèchement et d'éviter la prolifération des mauvaises herbes. **2.** Mélange de paille et bitume ou de produits synthétiques utilisé dans le ▶ procédé d'enherbement par projection hydraulique pour la protection des talus ; ▶ mulching, ▶ produit de mulching) ; *g* **Mulch** [m] (...mulche [pl]; **1.** Bodenbedeckung aus zerkleinertem Pflanzenmaterial, z. B. Grasschnittgut, Häckselstroh, ▶ Rindenmulch, ▶ Rindenschrot, Torf oder synthetisch-organische Schaumstoffe zur Verhinderung der Bodenoberflächenverschlämmung und damit zur Förderung der ▶ Bodengare und Humusversorgung, zum Erosionsschutz, Verdunstungsschutz oder zur Unkrautbekämpfung. **2.** Beim ▶ Anspritzverfahren eine mit Bitumen oder synthetischen Mitteln verklebte Strohschicht zur Erosionssicherung an Böschungen. **3.** Der Begriff **M.** wird auch für eine Bodenbedeckung aus mineralischen Schüttstoffen [Splittmulch oder Schottermulch] in Staudenbeeten mit Wärme liebenden Pflanzen, z. B. bei Steppenpflanzungen oder in pflegeextensiven Straßenbegleitgrünflächen, benutzt; ▶ Mulchen, ▶ Mulchstoff).

3624 mulching [n] *agr. constr. hort.* (Covering of soil with ▶ mulch; a particular form of **m.** in organic farming is **surface composting**, whereby semi-decomposed material, e.g. straw, animal manure and kitchen waste is spread over an area such as a field, thereby allowing the material to decay into the soil); *s* **mulching** [m] (Cubrir el suelo con ▶ mulch); *syn.* acolchamiento [m] de suelos (HDS 1987); *f* **mulching** [m] (**1.** Technique horticole consistant à recouvrir le sol avec du ▶ mulch. **2.** Une forme particulière du mulching est le **compostage de/en surface**

M

qui consiste à épandre fumier, paille ou résidus de culture sur une surface agricole ou un jardin potager ; ces matières organiques seront décomposées sur place par l'activité bactérienne et incorporées dans le sol par l'activité des vers de terre) ; *syn.* paillage [m] (IDF 1988, 292) ; *g 1* **Mulchen [n, o. Pl.]** (Aus dem Englischen *mulch*; Bedeckung des Bodens mit ►Mulch; mulchen [vb]; *g 2* **Flächenkompostierung [f]** (Eine besondere Form des Mulchens im biologischen Landbau, bei der angerotteter halbreifer Kompost, Stroh, Mist und Kulturreste zur Verrottung auf den zu bewirtschaftenden Boden aufgetragen werden); *syn. zu 1.* Abdeckung [f] mit Mulch.

3625 mulching material [n] *agr. constr. hort.* (Mineral or organic material for covering the soil surface); *s* **material [m] de mulching** (Sustancias orgánicas o anorgánicas utilizadas para cubrir el suelo y protegerlo contra la evapotranspiración, el embarramiento y las plantas advenedizas); *f* **produit [m] de mulching** (Matériau utilisé pour le mulching) ; *g* **Mulchstoff [m]** (Mineralisches oder organisches Material zur Abdeckung des Bodens zum Schutz gegen Verdunstung, Verschlämmung, Erosion und zur Unkrautbekämpfung); *syn.* Mulchmaterial [n].

mulching [n] **with compost** *constr. hort.* ►dressing with compost.

mulga scrub [n] [AUS] *phyt.* ►brushy area, #2.

3626 mull [n] *pedol.* (►Humus type without a permanent ►organic horizon composed almost exclusively of brown-grey to black fine humus mixed with clay minerals. **M.** occurs in soils with favo[u]rable hydrological and air conditions, and a relatively high nutrient content, in which litter decomposes rapidly, e.g. beneath steppe vegetation and herb-rich deciduous forests; **m.** is usually also present on meadows and arable land); *syn.* mild humus [n]; *s* **mull [m]** (►Tipo de humus totalmente descompuesto con ausencia casi absoluta de materia orgánica bruta, integrado a la parte mineral del suelo, y una textura grumosa. Su complejo adsorbente está saturado y tiene una reacción básica, salvo casos excepcionales que presentan una leve acidez; DGA 1986; ►mantillo); *f* **mull [m]** (►Type d'humus forestier de milieu aéré et à forte activité biologique, incorporé au milieu minéral [►horizon argilo-humique] à rapport C/N compris entre 10 et 15, selon le type de milieu ; humus forestier non submergé à engorgement nul ou temporaire par une nappe perchée avec absence ou quasi-absence de couche holoorganique car seul un mince horizon O1 plus ou moins discontinu peut être présent à certains moments de l'année ; l'horizon A$_h$ est très caractéristique ; DUCHAUFOUR ET DELECOUR cit. in DIS 1986) ; *syn.* humus [m] doux ; *g* **Mull [m]** (►Humusform, der eine ständige ►Humusauflage völlig fehlt, die fast ausschließlich aus braungrauem bis schwarzem, mit Tonmineralen innig verbundenem Feinhumus besteht. **M.** bildet sich in Böden mit günstigen Wasser- und Luftverhältnissen und relativ hohem Nährstoffgehalt, in denen die Streu rasch abgebaut wird, z. B. unter Steppenvegetation und unter krautreichen Laubwäldern; auch bei vielen Wiesen- und Ackerstandorten liegt in der Regel **M.** vor; SS 1979).

mull [n] [US&CDN], **duff** *pedol.* ►moder.

mullet zone [n] *limn.* ►flounder zone [UK].

3627 multicourse construction [n] *constr.* (**1**. Sequence of layers forming a ►pavement structure. **2.** Term used in roof planting, ►multistrata planting system); *s* **estructura [f] multicapa** (**1.** Secuencia de capas que forman carreteras y caminos; ►estructura del cuerpo de carreteras y caminos. **2.** *Revestimiento vegetal de tejados* ►estructura de plantación multicapa); *f* **assise [f] multicouches** (**1.** de chaussée ; ►structure de la voirie. **2.** *Toitures-terrasses* ►structure pluricouche) ; *g* **mehrschichtiger Aufbau [m, o. Pl.]** (**1**) (**1.** Abfolge von Schichten beim

Wege- und Straßenbau; ►Wegeaufbau. **2.** *Dachbegrünung* ►Mehrschichtenaufbau); *syn.* mehrschichtige Bauweise [f].

3628 multideck car park [n] [US] *trans. urb.* (Open-sided parking structure; ►car park 1, ►parking garage, ►underground parking garage [US]/underground car park [UK]); *syn.* decked car park [n] [US], multistory car park [n] [US]/multi-storey car park [n] [UK]); *s* **edificio [m] de aparcamiento (1)** (Garaje de varias plantas situadas bajo o sobre nivel, en cuyo caso están abiertas lateralmente; ►aparcamiento de residentes, ►edificio de aparcamiento 2, ►garaje subterráneo); *syn.* edificio [m] de estacionamiento [AL] (1); *f* **parc [m] de stationnement à étages** (Parc de stationnement souterrain ou aérien comportant plusieurs niveaux [Arrêté du 27 mai 1992] reliés entre eux par des rampes et dont l'accès des piétons est assuré par des ascenseurs ou des escaliers ; *termes spécifiques* ►garage souterrain, ►parc de stationnement couvert, parking aérien, ►parking souterrain résidentiel) ; *syn.* parc [m] en élévation, parking [m] à étages ; *g* **Parkhaus [n] (1)** (Mehrgeschossiges Garagenhaus mit oberirdischen, seitlich offenen und/oder unterirdischen Parkdecks; ein **P.** mit sehr vielen Stockwerken wird **Parkhochhaus** genannt; ►Anwohnertiefgarage, ►Parkhaus 2, ►Parkplatz, ►Tiefgarage); *syn.* Parkgarage [f].

3629 multidisciplinary planning [n] *plan.* (Coordination of various ►functional planning [US]/sectoral planning disciplines [UK] according to criteria specified by a master plan, regional plan or plans based upon ecological principles); *syn.* integrated planning [n]; *s* **planificación [f] integrada** (Es una planificación de las actividades humanas que incluye, además de los factores técnicos, sociales y económicos implicados, los criterios ambientales, con el fin de minimizar el impacto negativo o efecto adverso de la actuación sobre el medio; cf. DINA 1987; ►planificación sectorial); *f* **planification [f] intégrée** (Planification dont l'objectif est la coordination de ►planifications sectorielles impliquant la prise en compte de l'interdépendance de différents critères et objectifs initialement conçus de façon isolée) ; *syn.* aménagement [m] intégré ; *g* **Querschnittsplanung [f]** (Koordinierung von Fachplanungen nach den Zielkriterien anderer, ressortübergreifender Programmplanungen wie z. B. des Raumordnungsprogramms oder nach ökologisch-gestalterischen Gesichtspunkten; ►Fachplanung); *syn.* querschnittsorientierte Planung [f].

multifamily dwelling (unit) [n] *arch. urb.* ►multi-family housing.

3630 multifamily housing [n] *arch. urb.* (Building containing more than two dwelling units; ►multistory building for housing [US]/multistorey building for housing [UK]); *syn.* multifamily dwelling (unit) [n]; *s* **edificio [m] multifamiliar** (Edificio residencial con tres o más viviendas; ►bloque de pisos, ►casa de pisos); *syn.* casa [f] multifamiliar; *f 1* **habitat [m] collectif (1)** (*Terme* (*Terme générique* habitation comprenant plus de trois appartements ; ►habitat collectif 2) ; *f 2* **habitat [m] intermédiaire** (*Terme spécifique, type d'habitat* ensemble de logements collectifs locatifs ou privatifs ne dépassant pas R+3 dont l'accès est assuré par une cage d'escalier commune ou un passage couvert ; habitat très apprécié des allemands et des anglais ; *syn.* habitat [m] semi-collectif, logement [m] intermédiaire, habitat [m] multilogement [CDN], habitation [f] à plusieurs logements [CH]) ; *f 3* **habitat [m] individuel groupé** (*Terme spécifique, type d'habitat* maisons individuelles jumelées, accolées ou superposées disposant chacune d'une entrée particulière et ne comportant qu'un seul logement ; ►bloc d'habitation) ; *f 4* **habitat [m] collectif individualisé** (*Terme spécifique, type d'habitat* immeuble collectif avec accès individuels, jardins privatifs ou grandes terrasses) ; *g 1* **Mehrfamilienhaus [n]** (**1.** Freistehendes Gebäude mit mehreren abgeschlossenen Wohnungen — als Miet- oder Eigentumswohnungen —, das über ein gemeinsames Treppenhaus erschlossen ist; die häufigste Form ist das Zweifamilienhaus mit zwei Wohneinheiten übereinander, typisch als Vorstadthaus und Siedlungshaus im ländlichen Raum); *g 2* **Mehrwohnungshaus [n]** (Haus mit 2-3 Etagen, bei dem in jedem Vollgeschoss zwei [bis 3] Wohnungen sind; man spricht dann von einem *Zweispänner* oder *Dreispänner*; ▶Geschosswohnungsbau); *syn.* Mehrfamilienwohnhaus [n].

multifamily housing site [n] [US] *arch. urb.* ▶residential complex.

multiple henhouse [n] [US] *agr.* ▶battery of cages.

multiple-purpose forest land [n] [US] *for. recr.* ▶multiple-use forest.

multiple rows [loc] *hort.* ▶in multiple rows.

multiple-stemmed tree [n] [US] *arb.* ▶multistem tree [US]/multistemmed tree [UK].

3631 multiple use [n] *plan.* (Several uses of buildings or land; e.g. developed land, such as a unified developent; i.e. commercial, residential, business use as permitted in a planned unit development; also use of forests for timber production, recreational functions and nature conservation purposes; ▶layering of different uses); *s* **uso [m] múltiple** (Aprovechamiento de edificios o zonas verdes en la ciudad y en el campo para diferentes usos al mismo tiempo; ▶superposición de usos); *syn.* uso [m] variado; *f* **usage [m] multiple** (Utilisation variée d'équipements dans les espaces libres urbains ou dans les espaces naturels ; *contexte* à usage multiple ; ▶superposition des usages) ; *syn.* utilisation [f] multiple ; *g* **Mehrfachnutzung [f]** (Unterschiedliche Nutzungen von Gebäuden oder Flächen; bei Waldflächen z. B. für die Holzproduktion, Erholungsnutzung und für die Belange des Naturschutzes; ▶Nutzungsüberlagerung).

3632 multiple-use forest [n] *for. recr.* (Forest which is administered for several purposes; e.g. sustained yield of forest products, water supply catchment, wildlife habitat, outdoor recreation, game management, erosion control, etc.); *syn.* multiple-purpose forest land [n] [also US]; *s* **bosque [m] de usos múltiples** (Monte gestionado de tal manera que permita la realización de diversas funciones como producción de madera, protección de especies de flora y fauna, protección contra la erosion, caza, recreación de la población, etc.); *f* **forêt [f] à objectifs intégrés** (Forêt créée et aménagée en vue d'atteindre simultanément plusieurs objectifs, p. ex. la production de bois, de gibier, la conservation de la faune sauvage, la détente et les loisirs, la lutte contre l'érosion, etc. ; cf. DFM 1975) ; *syn.* forêt [f] à usage multiple ; *g* **Wald [m] mit Mehrzwecknutzung** (Wald, der forstlich so begründet und gepflegt wird, dass er zwei oder mehrere Funktionen erfüllen kann, z. B. Holzproduktion, jagdliche Wildhege, Erholungsnutzung, Naturschutz, Erosionsschutz, Immissionsschutz etc.).

multiplication [n] *hort.* ▶reproduction.

3633 multipurpose hall [n] *recr.* (Large room or building used for several activities; ▶multiple use); *s* **pabellón [m] multiusos** (Gran edificio construido para cumplir diferentes funciones; ▶uso múltiple); *f* **halle [f] à usages multiples** (Édifice conçu pour une utilisation multiple, des activités diverses ; ▶usage multiple) ; *g* **Mehrzweckhalle [f]** (Gebäude, das für unterschiedliche Nutzungen eingerichtet ist; ▶Mehrfachnutzung).

multistemmed tree [n] [UK] *arb.* ▶multistem tree [US].

3634 multistem tree [n] [US] *arb. syn.* multiple-stemmed tree [n] [US], multistemmed tree [n] [UK]; *s* **porte [m] en cepa** (Árbol de tronco ramificado desde la base); *syn.* árbol [m] de porte en cepa; *f* **cépée [f] (forte branchue)** (Forme d'arbres d'alignement et d'ornement cultivés et formés en cépée, présentant un ensemble de tiges partant au ras du sol d'une même souche ; cf. N.F. V 12 051) ; *syn.* bouquet [m] d'arbres, arbre [m] en cépée ; *g* **mehrstämmiger Baum [m]** (Baum, der aus mehreren Stämmen besteht, sei es, dass mehrere Bäume in ein Pflanzloch zusammengepflanzt wurden oder durch Auf-den-Stock-Setzen entstanden; ein Baum gilt als mehrstämmig, wenn die Stämme bis 0,5 m über der Bodenoberfläche entstanden sind).

multistorey [adj] [UK] *arch. urb.* ▶multistory [US].

multistorey car park [n] [UK] *trans. urb.* ▶multideck car park [US].

multistorey residential building [n] [UK] *arch. urb.* ▶multistory building for housing [US].

3635 multistory [adj] [US] *arch. urb. syn.* multistorey [adj] [UK]; *s* **de varias plantas [loc]**; *f* **à plusieurs étages [loc]** ; *g* **mehrgeschossig [adj]** (Ein Gebäude mit mehr als drei Stockwerken betreffend); *syn.* vielgeschossig [adj].

3636 multistory building [n] for housing [US]/multistorey building [n] for housing [UK] *arch. urb.* (Freestanding, large residential building containing several or many apartments/flats, in which the apartments/flats are arranged above each other on various floors, as compared to the ▶single-family house and ▶multi-family housing; buildings 35 meters or greater in height are known as **high-rise**; ▶high-rise building, ▶high-rise residential building); *syn.* multistorey residential building [n] [UK], apartment house [n] [US], apartment block [n] [UK], block [n] of flats [UK]; *s 1* **casa [f] de pisos** (Edificio de más de tres plantas generalmente construido junto a otros a lo largo de una calle; ▶casa unifamiliar, ▶edificio multifamiliar; ▶edificio alto, ▶torre de pisos); *s 2* **bloque [m] de pisos [Es]** (Edificio de viviendas grande y separado); *syn.* bloque [m] de viviendas [AL]; *f 1* **habitat [m] collectif (2)** (Bâtiment résidentiel comprenant plusieurs logements construits sur plusieurs étages ; ▶habitat collectif 1, ▶habitat collectif individualisé, ▶habitat individuel groupé, ▶habitat intermédiaire, ▶habitat résidentiel collectif, ▶maison individuelle ; ▶immeuble de grande hauteur, ▶tour d'habitations) ; *syn.* immeuble [m] collectif d'habitation ; *f 2* **bloc [m] d'habitation** (Bâtiment isolé d'habitation comportant plusieurs logements et dont la juxtaposition avec d'autres forme un ensemble résidentiel) ; *g* **Geschosswohnungsbau [m]** (Bebauungstyp resp. freistehendes Vielwohnungshaus, bei dem im Vergleich zum ▶Einfamilienhaus und ▶Mehrfamilienhaus die Wohnungen in mehreren Stockwerken übereinander liegen; ▶Wohnhochhaus, ▶Hochhaus); *syn.* Vielwohnungshaus [n].

multistory car park [n] [US] *trans. urb.* ▶multideck car park [US]/multi-storey car park [UK].

3637 multistrata planting system [n] *constr.* (▶Roof planting method where the root-zone layer is composed of several soil layers; ▶single-stratum planting system); *s* **estructura [f] de plantación multicapa** (▶Técnica de revestimiento vegetal de tejados o azoteas en la que la capa portante de la vegetación consiste de varios estratos; *opp.* ▶estructura de plantación monocapa); *f* **structure [f] pluricouche** (▶Procédé de végétalisation des toitures pour lequel le support végétal est constitué de plusieurs couches [couche drainante, couche filtrante, support végétal ; *opp.* ▶structure monocouche) ; *g* **Mehrschichtenaufbau [m]** (...bauten [pl]; ▶Begrünungssystem für Dachflächen, bei dem die Vegetationstragschicht aus mehreren Schichten besteht; *opp.* ▶Einschichtaufbau); *syn.* mehrschichtiger Aufbau [m] (2).

municipal authority [n] *adm.* ▶local authority.

municipal hygiene department [n] *adm. envir.* ▶waste management department.

M

municipality [n] [UK] *adm. pol.* ►community [US].

3638 municipal nursery [n] *adm. hort.* (Plant propagation facility operated by a city parks department); *syn.* local authority nursery [n]; *s* **vivero** [m] **municipal** (Explotación horticultora para la producción de plantas operada por el departamento de parques y jardines de una ciudad o municipio); *syn.* jardinería [f] municipal; *f* **exploitation** [f] **horticole municipale** (Dans le cadre des services municipaux d'espaces verts, entreprise horticole municipale ; *termes spécifiques* fleuriste municipal, pépinière municipale) ; *syn.* établissement [m] horticole municipal [CH] ; *g* **Stadtgärtnerei** [f] (Von einer Gemeindeverwaltung betriebener Gartenbaubetrieb, der vorrangig die Anzucht von Einjährigen zur Frühjahrs-, Sommer- und Herbstbepflanzung für die stadteigenen Grünanlagen und Pflanzenkübel betreibt und Pflanzen für die Dekoration städtischer Gebäude für öffentliche Anlässe, Empfänge und Ausstellungen anzieht, vorhält und pflegt).

municipal sludge [n] [US] *envir.* ►sewage sludge.

municipal woodland [n] *urb.* ►town forest [UK].

3639 mushroom picker [n] *recr.* (Person who collects mushrooms in the wild); *s* **colector, -a** [m/f] **de setas**; *f* **chercheur** [m] **de champignons** (Personne, qui ramasse des champignons) ; *syn.* amateur [m] de champignons, amateur [m] « mycophage » (LA 1981, 264), ramasseur [m] de champignons, chasseur [m] de champignons, cueilleur [m]/cueilleuse [f] de champignons (PPH 1991, 269) ; *g* **Pilzsammler/-in** [m/f] (Jemand, der in der freien Natur Pilze sammelt).

mustache planting [n] [US] *constr.* ►foundation planting.

mutual agreement [loc] *leg.* ►by mutual agreement.

3640 mycorrhiza [n] *bot. pedol.* (mycorrhizae [pl], ...rhizas [pl]; obligate interaction—mutualism—of fungal associations with roots of higher plants. In an **ectotrophic mycorrhiza** [e.g. heath, pine trees] the fungal mycelium [mycorrhizal hyphae] covers the outside of the roots and, in an **endotrophic mycorrhiza** [e.g. orchids], the fungal mycelium filaments grow inside the cells of the root cortex); *syn.* mycorrhizal fungi [npl]; *s* **micorriza** [f] (Asociación simbiótica entre las hifas de algunos hongos y las raíces de plantas superiores. En las **micorrizas endotróficas/endótrofas** [en tejo, brezos, diversas orquídeas] las hifas fúngicas no sólo penetran en la capa cortical de la raíz, sino que se alojan en el interior de sus células, y en parte son digeridas por la planta hospedante que se beneficia de sus albuminoides. En las **micorrizas ectotróficas/ectótrofas** [p. ej. en hayas, abedules, avellanos, coníferas] el hongo forma un manto de hifas que envuelve a las raicillas, mientras que el micelo se extiende entre las células del córtex. La asociación micorrícica permite un intercambio de nutrientes y metabolitos entre los simbiontes que beneficia el crecimiento de ambos resp. la resistencia ante patógenos de las plantas; cf. DB 1985 y DINA 1987); *f* **mycorhize** [f] (Micro-organismes du groupe des champignons formant par symbiose une association avec les racines de plantes supérieures. La plupart des champignons mycorhiziens ne pénètrent pas profondément dans les racines [**ectomycorhizes**, p. ex. chez la Bruyère, le Pin]. Par opposition, les **endomycorhizes** résultent de la présence de certains champignons infestant des plantes herbacées ou ligneuses et dont les filaments [hyphes fongiques] pénètrent profondément dans les tissus des racines, p. ex. les orchidées, le blé ; cf. LA 1981) ; *g* **Mykorrhiza** [f] (...zen [pl]; symbiontisches resp. durch wechselseitigen Parasitismus gekennzeichnetes Zusammenleben der Wurzeln vieler Landpflanzen mit einem Geflecht wuchsfördernder Pilze. Die Pilzhyphen wachsen entweder zwischen den Zellen der Wurzelrinde [= **ektotrophe M.**] oder dringen in das Innere der Rindenzellen ein [= **endotrophe M.**]).

mycorrhiza [n]**, ectotrophic** *bot. pedol.* ►mycorrhiza.

mycorrhiza [n]**, endotrophic** *bot. pedol.* ►mycorrhiza.

3641 mycorrhiza inoculation [n] *for. hort.* (Necessary mycorrhizal hyphae are added to the soil in the form of forest litter compost or in suspension; e.g. for oak, spruce, pine, larch, to increase the success rate of planting on habitats limited to raw soils); *syn.* mycorrhizal inoculation [n] [also US] (ARB 1983, 155); *s* **inyección** [f] **de micorriza** (Método utilizado en ubicaciones extremas para fomentar el crecimiento de determinadas especies arbóreas, como roble, picea, pino o alerce, de añadir al suelo por medio de tierra de compost de hojarasca o por suspensión los filamentos de micorriza necesarios); *f* **inoculation** [f] **de mycorhize** (Introduction de mycélium de champignons mycorhiziens [sur jeune semis avant leur mise en place ou injonction de compost de litière forestière] en vue de l'augmentation de la croissance de certaines espèces d'arbres telles que le Chêne, l'Épicéa, le Pin, le Mélèze sur des sols pauvres) ; *syn.* mycorhization [f] ; *g* **Mykorrhizaimpfung** [f] (Zur Erhöhung des Anwuchserfolges werden auf extremen Standorten bei bestimmten Baumarten, z. B. Eiche, Fichte, Kiefer, Lärche, dem Bodenmaterial mit Komposterde aus Waldstreu oder durch Suspension die notwendigen Pilzfäden [Hyphen] beigemengt).

mycorrhizal fungi [npl] *bot. pedol.* ►mycorrhiza.

mycorrhizal hyphae [npl] *bot. pedol.* ►mycorrhiza.

N

name [loc] *hort.* ►true to name.

3642 nano-phanerophyte [n] *bot.* (Plant belonging to the ►life form category of shrubs; after RAUNKIAER); *s* **nanofanerófito** [m] (Subdivisión del término fanerófito perteneciente a la clasificación de ►formas biológicas de RAUNKIAER, que incluye los fanerófitos de menos de 2 m de altura, como p. ej. muchos brezos, tomillos y jaras; DINA 1987); *f* **nanophanérophyte** [m] (D'après la classification des végétaux vasculaires selon RAUNKIAER désigne une plante vivace ligneuse dont la taille est plus petite que 5 m — arbrisseau ; ►type biologique) ; *g* **Nanophanerophyt** [m] (Nach RAUNKIAER die Pflanze, die der Lebensformklasse der Sträucher angehört; ►Lebensform); *syn.* Strauch [m] (1).

3643 narrow ecological range [n] *ecol.* (Relatively small area of distribution, limited by ecological factors, within which a species can live due to its ►ecological tolerance; *opp.* ►wide ecological range); *s* **valencia** [f] **ecológica estrecha** (Tolerancia ambiental reducida de una especie, por lo que su área de distribución es inferior a la media; *opp.* ►valencia ecológica amplia; ►tolerancia ecológica); *syn.* amplitud [f] ecológica estrecha; *f* **faible valence** [f] **écologique** (Terme qui caractérise **1.** la possibilité réduite pour une espèce végétale ou animale à coloniser des milieux différents, **2.** le milieu limité qu'une espèce, en raison des facteurs stationnels et de sa ►plasticité écologique, est en mesure de coloniser ; une espèce très spécialisée, dont la niche est étroite, autrement dit qui possède une faible valence écologique est une **espèce sténoèce** ; *opp.* ►valence écologique large) ; *g* **enge ökologische Amplitude** [f] (Kleiner Bereich, in dem eine Art zwischen zwei Werten eines Standort-/Umweltfaktors auf Grund ihrer ►ökologischen Potenz leben kann; ein

Lebewesen, das nur in einer **e.n ö.n A.** leben kann, ist eine **stenöke Art**; *opp.* ▶breite ökologische Amplitude).

3644 narrow plant terrace [n] *constr.* (Flat area of small width on planted slope to retard erosion); *s* **berma [f] de plantación** (En plantaciones de taludes pequeño escalón que tiene también funciones antierosivas); *syn.* lisera [f] de plantación, banquina [f] de plantación [RA], zarpa [f] de plantación [RCH]; *f* **redan [m]** (Espace plat étroit le long de plantations sur talus, servant également de protection contre l'érosion) ; *syn.* redent [m] ; *g* **schmale Pflanzterrasse [f]** (Bei Pflanzungen an Böschungen entstehender schmaler Absatz, der auch dem Erosionsschutz dient).

natatorium [n] *recr.* ▶swimming pool (1).

National Building Specification [n] **(N.B.S.)** [UK] *constr. eng.* ▶Uniform Construction Specifications [US].

3645 National Environmental Policy Act [n] [US] *conserv. leg. nat'res. plan.* (*Abbr.* NEPA; **in U.S.**, this 1969 federal law requires first an ▶environmental impact assessment [EIA] [US]/environmental assessment [UK] for all development proposals and then, if they will have a major effect, an ▶environmental impact statement [EIS]; **in U.K.**, legislation is based upon the European Community Environmental Assessment Directive, 85/337/EU, which was incorporated into the 1988 Regulations. Legislation makes two fundamental points. First that the best environmental policy should aim to prevent the creation of pollution or nuisances at source rather than subsequently trying to counteract their effects. Secondly that it is necessary to take all of the effects on the environment into account at the earliest possible stage in all technical planning and decision-making, when it is easiest to deal with potentially damaging effects. (LD 6/94, p. 38); *syn.* Town and Country Planning (Assessment of Environmental Effects) Regulations 1988 [UK]; *s 1* **Real Decreto [m] Legislativo de Evaluación de Impacto Ambiental [Es]** (Decreto 1302/1986, de 28 de junio, con carácter de normativa básica que es desarrollado por el Reglamento para la Ejecución del RD Legislativo de evaluación de impacto ambiental, RD 1131/1988, de 30 de septiembre, que es directamente aplicable a la Administración del Estado y a las de las Comunidades Autónomas que carezcan de competencia legislativa en materia de medio ambiente, así como con carácter supletorio, a aquéllas que la tengan atribuida en sus respectivos Estatutos de Autonomía; cf. Preámbulo de RD 1131/1988); *s 2* **Ley [f] sobre Evaluación de los Efectos de Determinados Planes y Programas sobre el Medio Ambiente [Es]** (Ley promulgada para incorporar al derecho interno del Estado Español la Directiva EC/42/CE del Parlamento Europeo y del Consejo, de 27 de junio de 2001, que tiene como objetivo introducir la evaluación ambiental de planes y programas como un instrumento de prevención que permita integrar los aspectos ambientales en la toma de decisiones sobre planes y programas públicos, tanto en el ámbito de la Administración General del Estado, como en el ámbito autonómico. La también llamada ▶evaluación del impacto ambiental estratégica lleva consigo la elaboración de informes de sostenibilidad ambiental como parte integrante de los planes y programas, que corresponden a la declaración de impacto ambiental de las EIA basadas en el RD 1302/1986 y su reglamento; cf. Ley 9/2006, de 28 de abril); *f* **loi [f] relative aux études d'impact** (≠) (**F.**, l'étude préalable à toute autorisation ou décision d'approbation permettant d'apprécier les incidences de travaux et projets d'aménagement importants sur le milieu naturel, afin de limiter les conséquences dommageables pour l'environnement résulte de l'article 2 de la **loi relative à la protection de la nature** du 10 juillet 1976, modifié par la loi du 30 décembre 1996 ; les différentes dispositions d'application sont contenues dans différents décrets et circulaires [12 octobre 1977, 25 février 1993, circulaire du 11 mars 1996] ; ▶procédure d'étude d'impact) ; *g* **Umweltverträglichkeitsprüfungs-Gesetz [n]** (*Abk.* UVPG; **1.** Gesetz vom 12.02.1990, das die Durchführung der ▶Umweltverträglichkeitsprüfung regelt. Dabei wird sichergestellt, dass bei bestimmten öffentlichen und privaten Vorhaben alle erheblichen Auswirkungen auf die Umwelt frühzeitig umfassend ermittelt, beschrieben und in einem förmlichen Teilverfahren geprüft und zusammenfassend bewertet werden und dass das Ergebnis der UVP so früh wie möglich bei allen behördlichen Entscheidungen über die Zulässigkeit berücksichtigt wird; cf. § 1 UVPG. Mit dem UVP-Gesetz wurde die UVP-Richtlinie der EU in nationales Recht umgesetzt. **2.** Durch Gesetz vom 25.06.2005 [BGBl. I 1746] wurde das Instrument der **Strategischen Umweltprüfung [SUP]** in das **U.** eingeführt. Diese Prüfung stellt eine UVP für Pläne und Programme dar, wenn diese von den Verwaltungsbehörden auf Grund bundesrechtlicher Vorschriften erstellt werden müssen. Dies gilt für wesentliche verwaltungs- und wirtschaftsrechtliche Pläne und Programme, z. B. für Raumordnung, Flächennutzungspläne, Bebauungspläne, Wasserwirtschaft, Bundesverkehrswegeplanung, Industrie und Energie); *syn.* UVP-Gesetz [n].

3646 National Historic Preservation Act [n] **1966, [as amended]** [US] *conserv'hist. leg.* (**In U.S.**, "the" principal national historic preservation law; establishes a "National Register of Historic Places" listing buildings, sites, districts, and objects of architectural, historic, or archeological significance of national, state, local, or tribal importance. Administered in partnership with state, local, and tribal governments. Listing is largely honorific and does not restrict private owners, who may decline to be listed. However, federally financed or subsidized proje cts or those requiring a federal license or permit which may have an "adverse effect" or impact on listed or eligible properties are subject to a non-binding, special environmental review and comment process by the federal Advisory Council on Historic Preservation. The owners of listed or eligible commercial properties may receive income tax benefits or occasionally grants for completing approved preservation projects. Administered by the National Park Service, US Department of the Interior and state historic preservation offices in each state. This law should not be confused with similar state state and local laws listing, regulating and otherwise protecting such resources from inapporopriate change, demolition, or removal; **in U.K.**, controls and powers over listed buildings, conservation areas, ancient monuments and archaeological areas are governed by the Ancient Monuments and Archaeological Areas Act 1979 and the Planning [Listed Buildings and Conservation Areas] Act 1990; ▶protection of historic landmarks); *syn.* Planning [Listed Buildings and Conservation Areas] Act [n] 1990 [UK]; *s* **Ley [f] del Patrimonio Histórico Español [Es]** (Según esta ley, integran el Patrimonio Histórico Español los inmuebles y objetos muebles de interés artístico, histórico, paleontológico, arqueológico, etnográfico, científico o técnico, el patrimonio documental y bibliográfico, los yacimientos y las zonas arqueológicas, así como los sitios naturales, jardines y parques que tengan valor histórico, artístico o antropológico. La administración del PHE se realiza en cooperación con las comunidades autónomas y los municipios, siendo competencia de la Administración del Estado garantizar la conservación del PHE, promover el enriquecimiento del mismo y fomentar y tutelar el acceso de todos los ciudadanos a los bienes comprendidos en él. Así mismo son compentencia estatal la protección de dichos bienes frente a la exportación ilícita y la expoliación, la difusión internacional del conocimiento de los bienes integrantes del PHE, la recuperación de tales bienes cuando hubiesen sido ilícitamente exportados y el intercambio de información cultural, técnica y científica con los demás Estados y

N

con los Organismos internacionales; arts. 1 y 2, Ley 16/1985, de 25 de junio; cf. BOE núm. 296, de 11 de diciembre de 1985; ►protección del Patrimonio Histórico Español); *f* **loi [f] sur les monuments historiques** (Aux termes de la loi sont classés comme monuments historiques en totalité ou en partie les immeubles dont la conservation présente un intérêt public au point de vue de l'histoire ou de l'art ; sont compris parmi les immeubles susceptibles d'être classés : **1.°**les monuments mégalithiques, les ►terrains qui renferment des stations ou gisements préhistoriques, **2.°**les immeubles dont le classement est nécessaire pour isoler, dégager ou assainir un immeuble classé ou proposé pour le classement, **3.°**d'une façon générale, les immeubles nus ou bâtis situés dans le champ de visibilité d'un immeuble classé ou proposé pour le classement ; article 1 loi du 31 décembre 1913 ►protection des monuments historiques) ; *g* **Denkmalschutzgesetz [n]** (**D.**, wegen der Kulturhoheit der Länder vornehmlich in Ländergesetzen geregelter Schutz von historisch bedeutenden Objekten, an deren Erhaltung aus wissenschaftlichen, künstlerischen oder geschichtlichen Gründen ein öffentliches Interesse besteht; ►Denkmalschutz); *syn.* Gesetz [n] zum Schutze der Kulturdenkmale.

National Joint Consultative Council [n] (NJCC) for Building Code of Procedure for Single or Two Stage Tendering [UK] *contr. leg.* ►General Conditions of Contract for Construction (AIA Document A201) [US].

3647 National Monument [n] [US] *conserv. land'man.* (Areas or features of historic, prehistoric, natural or scientific interest which are preserved unimpaired. They have been set aside from the public domain administered by the Bureau of Land Management, U.S. Department of the Interior. Legal authority for their establishment is contained in the American Antiquities Act of 1906. Their jurisdiction was unified under the National Park Service in 1933; similar areas have been established by state legislative acts); *s* **Monumento [m] Nacional (≠)** (En EE.UU., áreas o elementos individuales de valor histórico, prehistórico, natural o científico que se conservan intactos y que se encuentran en terrenos en propiedad del Estado, administrados por la Oficina de Gestión del Territorio [*Bureau of Land Management*] del Ministerio del Interior, y son declarados como **m. n.** por esta institución. La base legal para su protección reside en la Ley de Antigüedades Americanas de 1906 y su jurisdicción fue unificada bajo el Servicio Nacional de Parques en 1933. A nivel de los estados existen áreas similares establecidas sobre la base de leyes de los mismos); *f* **monument [m] historique et naturel national (≠) (U.S.**, site ou monument dont la préservation présente un intérêt national du point de vue historique, préhistorique, écologique ou scientifique. Le monument est inscrit sur un inventaire géré par le Bureau of Land Management, sous la tutelle du Ministère de l'Intérieur. Ces monuments sont régis par la loi « American Antiquities Act » du 8 juin 1906 ; *g* **Nationales Kultur- und Naturdenkmal [n] (≠)** (In USA eine Fläche oder Einzelschöpfung der Natur, die im Besitz der Bundesregierung [*Federal Government of the United States*] ist und von ihr wegen ihrer historischen, prähistorischen, ökologischen oder wissenschaftlichen Bedeutung zum Schutzgebiet oder Schutzobjekt ausgewiesen oder erklärt wird, und zu erhalten ist. *National Monuments* können — im Gegensatz zu Nationalparks — ohne Zustimmung des Kongresses vom Präsidenten direkt zum Schutzgut deklariert werden. Der Großteil der *National Monuments* werden vom *National Park Service*, zahlreiche Schutzgebiete auch von anderen Behörden, z. B. dem *Bureau of Land Management* [Staatliches Liegenschaftsamt], die alle dem Innenministerium unterstehen, verwaltet. Die gesetzliche Grundlage ist das *American Antiquities Act von 1906*. Da in D. die Kulturhoheit bei

den Bundesländern liegt, gibt es keine so genannten ‚nationalen' Denkmale).

3648 national or state surveillance system [n] for air pollution *envir.* (Nation or State-wide series of monitoring devices designed to check for compliance with air qualtiy standards; ►surveillance); *s* **sistema [m] español de información, vigilancia y prevención de la contaminación atmosférica [Es]** (Red de medición de contaminantes del aire repartida por todo un país. En Es. debido a la competencia compartida sobre medio ambiente entre el Gobierno central y las CC.AA., la antigua «red nacional de vigilancia y previsión de la contaminación atmosférica» ha sido sustituída por un sistema descentralizado, en el cual las CC.AA. son responsables de controlar la calidad del aire y de poner a disposición de la Administración General del Estado la información, que coordina el sistema; cf. art. 27 Ley 34/2007, de calidad del aire y protección de la atmósfera; ►vigilancia ambiental); *f 1* **réseau [m] de surveillance** (Réseau de stations de mesure créé en vertu du décret du 13 mai 1974 en vue de contrôler la qualité de l'air en particulier dans les sites où la pollution est présumée la plus forte ; ►inventaire des émissions des substances polluantes, ►surveillance) ; *syn.* réseau [m] d'alerte et de surveillance ; *f 2* **dispositif [m] de surveillance** (Organisme régional agréé, mis en place à partir de 1997 pour les agglomérations de plus de 250 000 habitants et jusqu'en 2000 pour l'ensemble du territoire conformément à la loi n° 96-1236 du 30 décembre 1996 sur l'air et l'utilisation rationnelle de l'énergie auquel est confié ►surveillance de la qualité de l'air et de ses effets sur la santé et sur l'environnement) ; *g* **Überwachungsnetz [n]** (Landesweit eingerichtete Messstationen, die eine kontinuierliche oder punktuelle Kontrolle von Auflagen des Bundes-Immissionsschutzgesetzes [BImSchG] ermöglichen; ►Überwachung).

3649 national park [n] *conserv.* (Large, outstandingly beautiful area of natural landscape or near-natural cultural landscape, which is protected by sometimes strict and all-encompassing nature conservation laws. The aim is to [a] protect the intrinsic character of the park and the ecological integrity of one or more ecosystems with their plants, animals, geological and geomorphological features of special interest for today and future generations, [b] exclude exploitation or occupation inimical to the purposes of designation of the area and [c] provide a foundation for spiritual, scientific, educational, recreational and visitor opporutunities, all of which must be environmentally and culturally compatible; certain areas of the park may be open to the public for recreation; **in U.S.**, the term '**n. p.**' originated with legal establishment of *Yellowstone National Park* in Wyoming in 1872. In 1916 President Woodrow Wilson signed the National Park Service Organic Act, which states that the fundamental purpose of the parks "is to conserve the scenery and the natural and historic objects and the wildlife therein and to provide for the enjoyment of the same in such manner and by such means as will leave them unimpaired for the enjoyment of future generations". **In Europe**, Sweden was the first country to initiate a national parks programme, having enacted its nature conservation law in 1909, which created the first nine **n. p.s** in 1910. The most famous parks are the *Sarek*, *Abisko* and *Stora Sjöfallet N. P.* in northwest Lapland. In some European and non-European countries recreation areas are designated as national parks; their functions are similar to those of the German ►nature park. In some other countries protection is not always strict and all-encompassing); *s* **parque [m] nacional** (Áreas naturales, poco transformadas por la explotación u ocupación humana que, en razón a la belleza de sus paisajes, la representatividad de sus ecosistemas o la singularidad de su flora, de su fauna o de sus formaciones geomorfológicas, poseen unos valores ecológicos, estéticos, educativos y científicos cuya conservación merece una atención preferente y se considere de interés general de la Nación con la atribución al Estado de su gestión. **En Es.** existen 14 **p. n.**, siendo los más antigüos el Parque Nacional de la Montaña de Covadonga [22.07.1918] y el de Ordesa [16.08.1918], ampliado a **P. N.** de Ordesa y Monte Perdido en 1982. Los otros son: Teide [1954], Caldera de Taburiente [1954], Aigüestortes i Estany de Sant Maurici [1955, ampliado 1996], Doñana [1969], Tablas de Daimiel [1973], Timanfaya [1974], Garajonay [1981], el **P. N.** Marítimo-Terrestre del Archipiélago de Cabrera [1991], Picos de Europa [1995], Sierra Nevada [1999] y el **P. N.** Marítimo-Terrestre de las Islas Atlánticas de Galicia [2002]. Los **pp. nn.** son declarados por ley de las Cortes Generales y son gestionados conjuntamente por la Administración General del Estado y la o las CC.AA. en las que se encuentren situados, por medio de un Patronato en el que están representadas las administraciones públicas y aquellas instituciones, asociaciones y organizaciones relacionadas con el Parque, o cuyos fines concuerden con los principios inspiradores de esta ley; arts. 13.1, 22 y 23 bis Ley 4/89 de Conservación de los Espacios Naturales y de la Flora y Fauna Silvestre; ▶reserva natural, ▶zona central); *f* **parc** [m] **national** (Territoire pour lequel la conservation de la faune, de la flore, du sol, du sous-sol, de l'atmosphère, des eaux et en général d'un milieu naturel présente un intérêt spécial et qu'il importe de protéger contre tout effet de dégradation naturelle ou de le soustraire à toute intervention artificielle susceptible d'en altérer l'aspect, la composition et l'évolution ; à cet effet peuvent être soumis à un régime particulier ou à interdictions certaines activités à l'intérieur du parc telles que la chasse et la pêche, les activités industrielles et commerciales, l'exécution des travaux publics et privés, l'extraction des matériaux concessibles ou non, l'utilisation des eaux, la circulation du public, toute action susceptible de nuire au développement naturel de la faune et de la flore et plus généralement, d'altérer le caractère du **p. n.** ; cf. C. rural, art. L. 241-1, 241-3 ; La France compte en 2007 neuf **parcs nationaux** qui couvrent des domaines terrestres et maritimes variés et représentent près de 8 % du territoire français (49 147 km²) : « La Vanoise », un parc de haute montagne en Savoie, créé en 1963 ; « Port-Cros », un parc sous-marin et insulaire du Var, créé en 1963 ; « Pyrénées », créé en 1967 ; « Cévennes », créé en 1970 ; « Écrins », un parc des Hautes-Alpes et d'Isère, créé en 1973 ; « Mercantour » en Alpes-Maritimes et en Alpes-de-Haute-Provence, créé en 1979 ; « Guadeloupe », créé en 1989 et fait partie de la réserve mondiale de la biosphère ; le parc amazonien de Guyane et le parc national de la Réunion créés en 2007 ; le parc national des Calanques [environ 12 500 ha terrestres et 82 000 ha marins] dans la région de Marseille est en projet pour 2010 ; ▶zone centrale) ; *g* **Nationalpark** [m] (Schutzgebietskategorie nach dem Bundesnaturschutzgesetz [BNatSchG] und eine IUCN-Managementkategorie: Rechtsverbindlich festgesetzte, großräumige, durch ihre besondere Eigenart und Schönheit hervorragende Naturlandschaft oder naturnahe Kulturlandschaft, die strengen Schutzbestimmungen im Sinne von Vollnaturschutzgebieten unterworfen ist und in Teilen, soweit es der Schutzzweck erlaubt, dem Erholungsverkehr zur Verfügung stehen kann. Es ist ein Gebiet, das vornehmlich der Erhaltung eines möglichst artenreichen heimischen Pflanzen- und Tierbestandes, geologischer und morphologischer Besonderheiten von speziellem Interesse dient und das durch menschliche Nutzung oder Inanspruchnahme in der Substanz nicht oder wenig beeinflusst wird/werden darf; cf. § 14 BNatSchG. Nachdem in den **USA**, in Wyoming 1872 der weltweit erste **N.**, der *Yellowstone National Park* mit Naturschutzstatus eingerichtet wurde, übernahm für Europa 1910 Schweden die Vorreiterrolle und gründete mit dem 1909 in Kraft getretenen Naturschutzgesetz die ersten neun Nationalparke. Die bekanntesten sind der *Sarek-*,

Abisko- und *Stora-Sjöfallet-***N.** in Nordwest-Lappland. Initiiert wurde dies durch einen Vortrag des deutschen Professors HUGO CONWENTZ 1904 vor der Schwedischen Gesellschaft für Anthropologie und Geographie, in dem er über die Bedrohung natürlicher Landschaften sprach; cf. SKAN 2002. **In D.** gibt es 15 **N.e**; seit 1970 den **N.** *Bayerischer Wald* [243 km²], seit 1978 den **N.** *Berchtesgaden* [218 km²], 1985 **N.** *Schleswig-Holsteinisches Wattenmeer* [4410 km²], 1986 **N.** *Niedersächsisches* Wattenmeer [2780 km²], 1990 **N.** *Hamburgisches Wattenmeer* [117 km²], 1990 **N.** *Vorpommersche Boddenlandschaft* [805 km²], 1990 **N.** *Müritz* [Mecklenburger Seenplatte, 322 km²], 1990 **N.** *Sächsische Schweiz* [93 km²], 1990 **N.** *Hochharz* [89 km²], 1990 **N.** *Jasmund* [Rügen; 30 km²], 1994 **N.** *Harz* [158 km²], 1995 **N.** *Unteres Odertal* [105 km²], 1997 **N.** *Hainich* [Muschelkalkhochfläche am Westrand des Thüringer Beckens; 76 km²], und 2003 **N.** *Eifel* [107 km²] und **N.** *Kellerwald-Edersee* [Hessen; 57 km²]. **In A.** gibt es die **N.e** *Hohe Tauern, Oberösterreichische Kalkalpen* und die *Nockberge* in Kärnten; in der Schweiz den *Schweizerischen* **N.** im Engadin. In einigen europäischen und außereuropäischen Ländern werden als **N.e** Erholungsgebiete bezeichnet, die in ihren Funktionen den deutschen ▶Naturparken ähnlich sind. 1961 wurde erstmalig eine „Weltliste" für **N.e** und gleichwertige Schutzgebiete mit einer Mindestgröße von 1000 ha aufgestellt, die 1967 resp. 1971 fortgeschrieben wurde. 1994 verabschiedete die Generalversammlung der IUCN in Buenos Aires eine weiterentwickelte, modifizierte **N.**-Definition mit folgenden drei Zonen: a] eingriffsfreie ‚strenge Naturzonen', sog. **Kernzonen** *[strict natural areas]*; b] Naturzonen mit Managementmaßnahmen, sog. **Entwicklungs- oder Pflegezonen** *[managed natural areas]* und c] **Wildniszonen** nach amerikanischem Vorbild *[*▶*wilderness areas]*. Auf internationaler Ebene gelten folgende gültigen Merkmale der Schutzkategorie »N.«: [a] **N.e** dienen vorrangig dem großflächigen Schutz der natürlichen [ungestörten] Dynamik von Ökosystemen sowie ihrer charakteristischen Biotope und Lebensgemeinschaften [genetische Ressourcen]; [b] wirtschaftsbestimmte Nutzungen oder sonstige Inanspruchnahmen sind auszuschließen resp. innerhalb einer Übergangsfrist zu beenden; [c] **N.e** sind grundsätzlich für Besucher unter bestimmten Bedingungen [Erbauung, Bildung/Umwelterziehung, „sanfte" Erholung] zugänglich. Der Zutritt muss mit den Schutzzielen in Einklang stehen [Zonierung, Besucherlenkung]; [d] das primäre Schutzziel der Erhaltung des natürlichen Zustandes und die „nutzungsfreie" natürliche Entwicklung des Gebietes ist auf mindestens 75 % der Schutzfläche umzusetzen; cf. N+L 1999 [6], 266 ff; ▶Kernzone).

3650 National Register [n] **of Natural Areas** [US] *conserv. leg.* (List of ▶natural landmarks [US]/natural monuments [UK]. **In U.S.,** 1. the National Register of Natural Areas is maintained by the National Park Service, Department of the Interior. 2. The Register of Rare Native Plant Sites lists special habitats compiled for protection purposes by *The Nature Conservancy* and by state natural heritage programs; **in U.K.,** compiled by the *Nature Conservancy Council*); *syn.* register [n] of natural monuments [UK]; *s* **registro** [m] **de monumentos naturales** (Inventario llevado por las administraciones competentes de los monumentos naturales de la región correspondiente; ▶monumento natural); *syn.* lista [f] de monumentos naturales; *f* **liste** [f] **des monuments naturels et des sites** (Il est établi dans chaque département par la commission départementale des sites, perspectives et paysages une liste des ▶monuments naturels et des sites inscrits dont la conservation ou la préservation présentent, du point de vue artistique, historique, scientifique, légendaire ou pittoresque, un intérêt général. L'inscription sur la liste est prononcée par arrêté du ministre des affaires culturelles ; cf. art. 4 du loi du 02 mai 1930) ; *syn.* inventaire [m] des monuments naturels

et des sites ; *g* **Naturdenkmalbuch [n]** (Bei den zuständigen Naturschutzbehörden geführte Liste [und Karten] über ►Naturdenkmale; in BW wird das **N.** von der unteren Naturschutzbehörde geführt).

3651 national scenic trail [n] [US] *plan. recr.* (Hiking trail or path for rambling, which traverses a whole county, region or states; e.g. Pennine Way [UK], Appalachian Trail [US], Pacific Crest Trail [US]); *syn.* long distance footpath [n] [UK]; *s* **pasillo [m] verde de largo recorrido** (Vía de senderismo que recorre todo un país o varios países, como p. ej. el Camino de Santiago, ruta de peregrinos de la Edad Media, que lleva desde Alemania, Suiza e Italia cruzando Francia y España hasta la ciudad gallega de Santiago de Compostela); *syn.* sendero [m] de largo recorrido; *f* **sentier [m] de grande randonnée** (Sentier balisé traversant un ou plusieurs États ; *terme spécifique* les chemins de Saint-Jacques, itinéraire traditionnel culturel traversant l'Allemagne, la Suisse, la France et l'Espagne pour atteindre Saint-Jacques-de-Compostelle) ; *syn.* chemin [m] de grande randonnée, itinéraire [m] de grande randonnée ; *g* **Fernwanderweg [m]** (Wanderweg, der durch ein ganzes Land oder mehrere Staaten führt; *UB* Jakobsweg, z. B. Pilgerweg durch Deutschland, Schweiz, Frankreich und Spanien nach Santiago de Compostela).

national surveillance system [n] for air pollution *envir.* ►national or state surveillance system for air pollution.

national tourism [n] *econ. plan. recr.* ►promotion of national tourism.

3652 native plants and wildlife corridor [n] *ecol.* (Strip of land with habitat for indigenous flora and fauna, e.g. hedge, watercourse, forest mantle, forest and hedge seam, inner city green space corridor); *s* **corredor [m] vital** (Hábitat alargado de flora y fauna silvestre, p. ej. seto, curso de agua, lindero de un bosque o pasillo verde dentro de una ciudad); *f 1* **espace [m] vital linéaire** (Espace de développement de la faune et la flore indigène de forme étroite tels que les haies, cours d'eaux, manteaux forestiers, lisières forestières et de haies, ceintures vertes urbaines) ; *syn.* habitat [m] linéaire ; *f 2* **corridor [m] biologique** (Axe de déplacement et de migration des espèces animales et la dispersion des espèces végétales les corridors biologiques forment un réseau permettant le déplacement, le refuge, et la nourriture de la faune sauvage [oiseaux, insectes, mammifères] entre différents habitats sur un territoire ; les corridors biologiques sont nécessaires à leur survie, et jouent de nombreux rôles : 1. couloir de dispersion pour certaines espèces ou de migration vers des lieux de nourrissage ou de reproduction, 2. habitat où les espèces effectuent l'ensemble de leur cycle biologique, 3. refuge, 4. habitat-source constituant un réservoir d'individus colonisateurs, 5. brassage génétique des espèces ; un réseau de corridors biologiques s'oppose efficacement à la fragmentation ou l'►isolement de biotopes et assure un continuum vert pour la faune et la flore indispensable à la survie des espèces) ; *syn.* corridor [m] écologique ; *g* **linienförmiger Lebensraum [m]** (Langer, schmaler Lebensraum für heimische Pflanzen und Tiere, z. B. Hecke, Fließgewässer, Waldmantel, Wald- und Heckensaum, innerstädtischer Grünzug).

native plant species [n] *phyt.* ►indigenous plant species.

native species [n] *phyt. zool.* ►indigenous species.

native vertebrates [npl] *conserv.* ►wildlife.

3653 Natura 2000 [loc] *conserv. ecol.* (Term used for an EU wide network of nature protection areas established under the 1992 Habitats Directive. It is the centrepiece of EU nature and biodiversity policy and a major contribution to preserving biological diversity according to the 1992 Rio de Janeiro Convention. All member states of the EU have pledged to safeguard the natural heritage of Europe and ensure the long-term survival of Europe's most valuable and threatened species and habitats. The network is comprised of ►Special Areas of Conservation [SAC] designated by Member States under the Habitats Directive, and also incorporates ►Special Protection Areas [SPAs] which are designated under the 1979 ►Birds Directive. Natura 2000 is not a system of strict nature reserves where all human activities are excluded. Although the network includes nature reserves, most of the land is likely to continue to be privately owned and the emphasis will be on ensuring that future management is sustainable, both ecologically and economically. Natura 2000 applies to Birds Sites, Habitats Sites, which are divided into biogeographical regions and to the marine environment; ►Natura 2000 Protection Area Network, ►preliminary assessment of programme and project impacts in terms of Natura 2000); *s* **Natura 2000 [loc]** (Denominación utilizada para la red de áreas de protección de la naturaleza de toda la Unión Europea, establecida según la ►Directiva de Hábitats de 1992. Es el instrumento central de la política de protección de la naturaleza y la biodiversidad de la UE, y una contribución central para preservar la diversidad biológica de acuerdo a la ►Convención de Biodiversidad de Río de Janeiro de 1992. Todos los Estados miembros de la UE han prometido preservar el patrimonio natural de Europa y asegurar la supervivencia de las especies y los hábitats más valiosos y amenazados. La red está formada por Zonas Especiales de Conservación designadas por los Estados miembros según la Directiva de Hábitats y también incorpora Zonas de Protección Especial [ZOPA], designadas según la ►Directiva de las Aves de 1979. Natura 2000 no es un sistema de reservas naturales estrictas, donde todas las actividades humanas están excluidas. Aunque incluye a éstas, la mayor parte de las tierras puede seguir siendo de propiedad privada y el énfasis está en asegurar que la gestión futura sea sostenible, tanto ecológica como económicamente; ►elaboración de informe de sostenibilidad ambiental de planes y proyectos sobre un hábitat de la red Natura 2000, ►Red Ecológica Europea Natura 2000); *f* **Natura 2000 [loc]** (Terme désignant l'ensemble des sites naturels identifiés pour la rareté ou la fragilité des espèces sauvages, animales ou végétales et de leurs habitats et constituant le ►réseau Natura 2000 ; ce réseau a pour objectif de contribuer à préserver la diversité biologique sur le territoire de l'Union européenne en assurant le maintien ou le rétablissement dans un état de conservation favorable des habitats naturels et des habitats d'espèces de la flore et de la faune sauvage d'intérêt communautaire en application des directives européennes « Oiseaux » de 1979 [►Directive Oiseaux] et « Habitats » de 1992 [►Directive Habitat Faune Flore] ; depuis le 30 avril 2007, le réseau français de sites Natura 2000 comprend 1307 sites d'intérêts communautaires [SIC] et 367 zones de protection spéciales [ZPS] ; ►pré-diagnostic de l'étude [d'évaluation] d'incidence Natura 2000) ; *g* **Natura 2000 [loc]** (Bezeichnung für das zusammenhängende Netz besonderer Schutzgebiete zur Erhaltung europäisch bedeutsamer Lebensräume sowie seltener Tier- und Pflanzenarten in der Europäischen Union. Dieses Netzwerk ist daher ein wesentlicher Beitrag zur Umsetzung des „Übereinkommens über die Biologische Vielfalt", das 1992 anlässlich der Weltkonferenz über Umwelt und Entwicklung in Rio de Janeiro unterzeichnet wurde. Alle Mitgliedstaaten der Europäischen Union — so auch die Bundesrepublik Deutschland — haben sich verpflichtet, durch den Aufbau des ►Natura-2000-Schutzgebietssystems das Naturerbe Europas zu sichern. Rechtliche Grundlage dieses grenzüberschreitenden Naturschutznetzwerkes bilden die ►Fauna-Flora-Habitat-Richtlinie — [FFH-Richtlinie] und die ►Vogelschutzrichtlinie der Europäischen Union. Nach den Vorgaben dieser beiden Richtlinien benennt und schützt jeder Mitgliedstaat Gebiete, die für die Erhaltung seltener Tier- und Pflanzenarten

sowie typischer oder einzigartiger Lebensräume von europäischer Bedeutung wichtig sind; ▶FFH-Vorprüfung).

3654 Natura 2000 Protection Area Network [n] *conserv. leg. pol.* (Coherent Europe-wide system of significant natural and near-natural habitat areas with populations of endangered species flora and fauna, designated for the protection of biodiversity. The system, based upon the ▶Habitats Directive and ▶Birds Directive was established to maintain the European Cultural Heritage for coming generations. Special Protection Areas (SPAs) of European significance have been selected in various ▶biogeographical regions and linked under the auspices of Natura 2000 in three phases; **Phase 1**: Compilation of the national account; for each species for which at least one SPA has been selected, there is an individual species account. These accounts summarise population status and size, distribution, and population structure and trends; **Phase 2:** Determination of an area with community significance; summary of reasons why the particular suite of SPAs was selected for that species; application of selection guidelines; **Phase 3:** Securement of the NATURA 2000 protection area by the year 2004, after which the individual member state is responsible for monitoring and obliged to submit comprehensive reports to the European Commission every six years); *syn.* system [n] of Protection Areas NATURA 2000; **s Red [f] Ecológica Europea Natura 2000** (Sistema europeo de hábitats naturales y cuasinaturales significativos con poblaciones de especies amenazadas de flora y fauna, designados para proteger la biodiversidad. El sistema, basado en la ▶Directiva de las Aves y la ▶Directiva de Hábitats, fue establecido para preservar el patrimonio natural para las generaciones futuras. Las Zonas Especiales de Protección [ZPE] de significado europeo han sido seleccionadas en diversas ▶regiones biogeográficas y articuladas en la Red Natura 2000 en tres fases. **Fase 1:** Compilación de la cuenta nacional; para cada especie para la cual ha sido seleccionada una ZPE existe una cuenta individual de especie. Estas cuentas resumen el estado y tamaño, la distribución y estructura y las tendencias de la población. **Fase 2:** Determinación de un área de importancia comunitaria; sumario de razones por las que el área particular fue elegida para esa especie; aplicación de las guías de selección. **Fase 3:** Protección legal del área de protección Natura 2000 antes de finales de 2004, después de lo cual el Estado miembro individual es responsable del monitoreo y está obligado a entregar informes exhaustivos a la Comisión Europea cada seis años); *syn.* sistema [m] de áreas protegidas Natura 2000; **f réseau [m] Natura 2000** (Réseau écologique européen de zones spéciales de conservation [ZSC] formé par des sites abritant des types d'habitats naturels figurant à l'Annexe 1 et des habitats des espèces figurant à l'Annexe 1 ; il doit assurer le maintien ou, le cas échéant, le rétablissement, dans un état de conservation favorable, des types d'habitats naturels et des habitats d'espèces concernés dans leur aire de répartition naturelle ; il vise à préserver les espèces et les habitats menacés et/ou remarquables sur le territoire européen, tout en tenant compte des exigences économiques, sociales, culturelles et régionales, dans une logique de développement durable et d'arrêt de l'appauvrissement de la biodiversité d'ici 2010 ; le réseau Natura 2000 est constitué de deux types de zones naturelles, à savoir les ▶Zones Spéciales de Conservation [ZSC] issues de la ▶Directive Habitat Faune Flore de 1992 et les ▶Zones de Protection Spéciale [ZPS] issues de la ▶directive Oiseaux de 1979 ; ces deux directives ont été transcrites en droit français par l'ordonnance du 11 avril 2001 ; la mise en place du réseau Natura 2000 s'est effectué en trois étapes : **1.** réalisation de l'inventaire des habitats et espèces concernées et proposition d'une liste nationale de sites susceptibles d'être reconnus d'importance communautaire, **2.** en accord avec chacun des états membres, établissement par la Commission de la liste des sites d'importance communautaire au sein de chacune des ▶régions biogéographiques européennes à partir des listes nationales, **3.** désignation officielle des sites retenus comme Zone Spéciale de Conservation [ZSC] ; s'ensuit l'établissement des mesures de conservation notamment des mesures de protection réglementaires ou des mesures de gestion contractuelle ; les États membres assurent la surveillance de l'état de conservation des espèces et des habitats et transmettent tous les six ans un compte rendu à la Commission) ; **g Natura-2000-Schutzgebietssystem [n]** (Kohärentes Netz von europaweit bedeutsamen Schutzgebieten natürlicher und naturnaher Lebensräume mit Vorkommen gefährdeter Tier- und Pflanzenarten zur Erhaltung der biologischen Vielfalt. Das System basiert auf der ▶Fauna-Flora-Habitat-Richtlinie und ▶Vogelschutzrichtlinie und wird eingerichtet, um das europäische Naturerbe für kommende Generationen zu bewahren. Hierfür werden ausgewählte Lebensräume von europäischer Bedeutung aus verschiedenen ▶biogeografischen Regionen miteinander verknüpft); das Netz Natura 2000 wird in drei Phasen erstellt: **Phase 1:** Erstellung der nationalen Meldeliste; **Phase 2:** Bestimmung der Gebiete von gemeinschaftlicher Bedeutung; **Phase 3:** Sicherung der Natura-2000-Gebiete: Die ausgewählten Gebiete von gemeinschaftlicher Bedeutung mussten in der dritten Phase von den Mitgliedstaaten mit den nationalen Möglichkeiten schnellstmöglich, spätestens jedoch bis zum Jahr 2004, dauerhaft gesichert werden. Nach Abschluss der dritten Phase sind die Mitgliedstaaten für die Zustandsüberwachung [Monitoring] der Natura-2000-Gebiete verantwortlich und berichten der Europäischen Kommission umfassend in einem Turnus von sechs Jahren); *syn.* Schutzgebietssystem [n] Natura 2000, Netz [n] Natura 2000; Schutzgebietsnetz [n] Natura 2000.

3655 natural [adj] *ecol.* (In general use belonging to ▶nature; ▶near-natural); *syn.* natural [adj] in character; **s natural [adj]** (**1.** En uso general, relacionado con la ▶naturaleza. **2.** Sin o con muy poca influencia antropógena; ▶poco alterado); **f naturel, elle [adj]** (**1.** *Langage courant* appartenant à la nature. **2.** Qui n'est pas ou peu influencé, transformé par les activités humaines ; ▶faiblement artificialisé, ▶nature) ; **g natürlich [adj]** (1) (**1.** *Allgemein sprachlich* der ▶Natur zugehörig. **2.** Ohne menschlichen resp. mit sehr geringem menschlichen Einfluss belassen; ▶naturnah); *syn.* naturhaft [adj]; *syn. zu 2.* naturbelassen [adj].

natural and cultural heritage [n] *conserv. hist. land'man.* ▶conservation of the natural and cultural heritage.

3656 natural angle [n] **of repose** *constr.* (Angle of an uncompacted fill slope at rest); **s ángulo [m] de reposo** (Ángulo que se establece naturalmente en un talud sin compactar); *syn.* ángulo [m] natural de talud; **f angle [m] de talus naturel** (Angle que fait naturellement avec l'horizontale la partie latérale d'un tas de matériaux mis en remblai ; DIR 1977) ; *syn.* angle [m] d'éboulement (DTB 1985), angle [m] de stabilité naturelle ; **g natürlicher Böschungswinkel [m]** (Neigungswinkel eines geschütteten Bodens, der sich ohne menschlichen Einfluss einstellt).

3657 natural area laboratory [n] *conserv.* (Area containing many typical or unusual faunistic or floristic types, associations, or other biotic phenomena, or characteristic of outstanding geologic, pedologic, or aquatic features or processes, which has been established and maintained for the primary purpose of research and education on the ecology, successional trends and other aspects of the natural environment. The general public may be excluded or restricted where necessary to protect studies or preserve this area; WPG 1976; ▶forest research natural area [US]/forest nature reserve [UK]; *syn.* natural research area [n] [also US]; **s reserva [f] científica** (Espacios naturales protegidos, que por su riqueza de flora y fauna son especialmente aptos para

N

la formación y el estudio científicos y se utilizan como «laboratorios de campo»; ►reserva forestal integral); *f 1* **laboratoire [m] naturel pour l'étude scientifique** (≠) (US., D., CDN., dénomination caractérisant les sites inscrits, les parcs nationaux et autres espaces protégés présentant des qualités remarquables du point de vue scientifique et sur lesquels sont effectués des programmes de surveillance et d'observations scientifiques) ; *f 2* **réserve [f] intégrale** (F., zone créée dans un parc national **1.** dans un but scientifique afin d'observer les processus d'évolution des biocénoses sur différents biotopes ou **2.** dans un but de plus grande protection de certains éléments de la faune ou de la flore dans une partie déterminée d'un parc national ; pour une meilleure protection de certaines zones d'un parc naturel régional sont utilisées les mesures de classement traditionnelles telles que, la création de réserve naturelle ou d'espace naturel sensible) ; *f 3* **réserve [f] biologique intégrale** (Espace protégé réservés aux scientifiques et aux forestiers qui inventorient et analyse la flore et la faune afin de comparer les zones exploitées avec celles qui évoluent naturellement [forêt de Fontainebleau] ; ►réserve biologique domaniale, ►réserve biologique forestière) ; *f 4* **réserve [f] naturelle intégrale** (Nomenclature UICN dans la classification internationale des aires protégées désignant les espaces terrestres ou marins comportant des écosystèmes pour lesquels le but de protection est de maintenir des processus naturels non perturbés, afin de disposer d'exemples écologiquement représentatifs d'un milieu naturel particulier pour les besoins de la recherche scientifique et/ou de la surveillance continue de l'environnement) ; *syn.* réserve [f] scientifique ; *g* **Freilandlaboratorium [n]** (Bezeichnung für Naturschutzgebiete, Nationalparke oder ähnliche Landschaftsbereiche, die sich wegen ihrer artenreichen Fauna und Flora zu wissenschaftlicher Ausbildung und Forschung besonders eignen; ►Naturwaldreservat).

3658 natural area preserve [n] [US] *conserv. leg.* (Clearly-defined area or parts of a landscape in which nature is either completely protected in its entirety or partially protected where individual components are concerned. An area is so designated in the public interest or because of scientific, ecological, historical or cultural reasons or, **in Europe**, due to its landscape beauty or special characteristics. In **n. a. p.s** the status quo is to be preserved or allowed to evolve naturally. **In U.S.**, these areas are set aside by public or private agencies, especially by *The Nature Conservancy* [founded 1951] temporarily until turned over to a government agency; **in U.K.**, the *Nature Conservancy Council* and local authorities were given powers to enter into agreements with land-owners for establishing ►nature reserves, to compulsorily purchase land, to establish or manage reserves, and to make bye-laws for the protection of these areas; ►Area of Outstanding Natural Beauty [UK], ►natural landmark [US]/natural monument [UK], ►strict nature reserve, ►Wild and Scenic River, ►Wilderness Area); *syn.* nature reserve [n] [UK]; *s* **espacio [m] natural protegido (1)** (En Es. término genérico que incluye todas las categorías de protección legal. En EE.UU., estas áreas son designadas por agencias públicas y privadas, en especial por *The Nature Conservancy* temporalmente hasta que son cedidas a una agencia gubernamental. En UK, el Consejo de Conservación de la Naturaleza [*Nature Conservancy Council*] y sus autoridades locales recibieron competencias para llegar a acuerdos con propietarios de terrenos, establecer espacios protegidos, enajenar terrenos forzosamente, gestionar espacios protegidos y crear reglamentos para proteger estas áreas. **En D.** categoría máxima de protección de espacios naturales, menos estricta que la categoría española de ►reserva natural; ►Área de Gran Belleza Natural, ►monumento natural, ►tramo de río protegido, ►reserva natural [científica]); *f* **site [m] classé** (En F. sites et monuments naturels protégés au titre de l'art. 4 de la loi du 2 mai 1930, [relève depuis le 20 février 2004 du code du patrimoine] dont la conservation ou la préservation présente au point de vue artistique, historique, scientifique, légendaire ou pittoresque un intérêt remarquable ; le classement assure une protection contraignante des sites naturels de grande qualité ; **en D.** cette catégorie constitue la protection la plus contraignante des espaces naturels ; ►réserve naturelle régionale, ►réserve naturelle nationale) ; *g* **Naturschutzgebiet [n] (2)** (In seinen Grenzen rechtsverbindlich festgesetzter Landschaftsraum in der freien Landschaft oder im Siedlungsbereich, in dem die Natur in ihrer Ganzheit [= ►Vollnaturschutzgebiet] oder in Teilen [= **Teilnaturschutzgebiet**] uneingeschränkt geschützt ist. Dies geschieht im öffentlichen Interesse **1.** zur Erhaltung von Lebensgemeinschaften oder Biotopen bestimmter wild lebender Tier- und Pflanzenarten, **2.** aus wissenschaftlichen, naturgeschichtlichen oder landeskundlichen Gründen oder **3.** wegen ihrer Seltenheit, besonderen Eigenart oder hervorragenden Schönheit. Im **N.** ist ein besonderer Schutz von Natur und Landschaft in ihrer Ganzheit oder in einzelnen Teilen erforderlich; cf. § 13 BNatSchG. Die Regierungspräsidien/Bezirksregierungen in den Bundesländern erlassen Verordnungen, die u. a. die Erklärung zum Schutzgebiet, den Schutzgegenstand, Schutzzweck und Verbote beinhalten; ►Auenschutzgebiet, ►Naturdenkmal, ►Naturreservat; ►Landschaftsschutzgebiet. **In den USA** werden viele **N.e** vom Naturschutzbeirat, *The Nature Conservancy*, einer gemeinnützigen Naturschutzorganisation, initiiert und solange betreut, bis sie in die Verwaltung einer Behörde übergeht. **Im UK** können das *Nature Conservancy Council* und örtliche Behörden mit Grundstückseigentümern Verträge über die Einrichtung von Naturschutzgebieten abschließen, Land enteignen, die zu schützenden Gebiete bewirtschaften und Schutzsatzungen erlassen).

3659 natural background radiation [n] *envir.* (Existing lower atmosphere radiation emanating from cosmic and terrestrial sources); *s* **radiación [f] natural (mínima)** (Radiación existente en la atmósfera inferior procedente de fuentes cósmicas o terrestres); *f* **rayonnement [m] naturel** (Rayonnement provenant des étoiles et du soleil [rayons cosmiques], des éléments radioactifs contenus dans la croûte terrestre tels les éléments lourds comme l'uranium 238 et 235 ou des éléments légers comme le potassium 40, de l'irradiation exogène [gaz radon] sous-produit de la dégradation de l'uranium qui se trouve dans la croûte terrestre et notamment dans les roches primaires [granit], ainsi que de la radioactivité interne des êtres vivants) ; *syn.* radiations [fpl] naturelles ; *g* **natürliche Grundbelastung [f]** (In der unteren Atmosphäre natürlich vorhandene Grundstrahlung, die sich aus kosmischer und Erdstrahlung zusammensetzt); *syn.* natürliche Grundstrahlung [f], natürliche Strahlenexposition [f].

natural character [n] **of a landscape** *conserv.* ►unique natural character of a landscape.

3660 natural colonization [n] *phyt.* (Invasion of a bare site by vegetation, usually developing from wind-blown seeds; ►barochory, ►seed rain, ►spontaneous plant, ►weed tree, ►woody wildling); *syn.* spontaneous colonization [n]; *s* **colonización [f] natural de «espacio vacío»** (Resultado de la colonización de zonas vacías por la diseminación anemócora de diásporas de especies pioneras; ►barocoria, ►leñosa adventicia, ►nuevo vuelo, ►vegetación espontánea); *syn.* colonización [f] espontánea; *f* **colonisation [f] spontanée** (Résultat de la dissémination d'espèces végétales par ►flux de semences et ►barochorie, étendant leur territoire au dépens d'espèces moins compétitives ; ►végétation arbustive adventice, ►végétation spontanée) ; *g* **spontane Besiedelung [f]** (Pflanzenbestand als Ergebnis der Pflanzenbesiedelung meist durch ►Anflug 1 und ►Aufschlag, ►Fremdaufwuchs, ►Wildwuchs 2 von Verbreitungseinheiten [wie Früchte, Samen, Spross-/Wurzelabschnitte, Brutknospen], der sich ohne gärtnerisches Zutun auf offenen, meist vegetationsfreien Flächen einstellt, z. T. unerwünscht ist, z. T. in dynamischen Bepflanzungskonzepten toleriert oder bereichernd integriert werden kann, z. B. durch Zuwanderer; ►Wildling 3); *syn.* Aufwuchs [m] (1), *o. V.* spontane Besiedlung [f], spontane Vegetation [f] (1), Spontanvegetation [f].

3661 natural colonization [n] **by seed rain** *phyt. for.* (Invasion of a bare site by vegetation, usually developing from wind-blown seeds; ►natural colonization, ►natural regeneration (1), ►seed rain, ►weed tree, ►woody wildling); *s* **colonización [f] natural de «espacio vacío» por nuevo vuelo** (Resultado de la diseminación anemócora de diásporas de especies pioneras por ►nuevo vuelo; ►colonización natural de «espacio vacío», ►leñosa adventicia, ►regeneración natural); *f* **semis [m] naturel par flux de semences** (Résultat du ►flux de semences ; ►colonisation spontanée, ►régénération naturelle, ►végétation arbustive adventice) ; *syn.* ensemencement [m] naturel ; *g* **Anflug [m, o. Pl.] (1)** (Ergebnis des ►Anfluges 1; ►Fremdaufwuchs, ►Naturverjüngung, ►spontane Besiedelung, ►Wildling 3).

3662 natural danger zone [n] *leg.* (Hazardous area characterized by usually unforseeable and unpreventable natural events, such as seismic zones, steep mountain slopes, river valleys, etc.); *s* **zona [f] expuesta a grandes riesgos naturales** (Generalmente imprevisibles o no localizables exactamente como terremotos, lavinas, inundaciones, etc.); *f* **zone [f] exposée aux risques naturels prévisibles** (Territoire exposé à des risques naturels [inondations, séismes, feux de forêts, avalanches, etc.] pouvant gravement mettre en danger les personnes, les biens et l'environnement. Le Code de l'Environnement impose aux vendeurs et aux bailleurs de tout bien immobilier [bâti ou non bâti] situé dans le périmètre d'un plan de prévention des risques [PPR] ou/et d'une zone de sismicité une obligation d'information. Elle s'applique également pour les biens immobiliers situés dans une zone exposée aux risques délimitée par un plan de prévention des risques naturels prévisibles [PPRN] ; un état des risques naturels, doit être annexé à tout type de contrat de location écrit, à la réservation pour une vente en l'état futur d'achèvement, à la promesse de vente ou à l'acte réalisant ou constatant la vente de ce bien immobilier) ; *g* **natürliche Gefahrenzone [f]** (Durch in der Regel unabweisbare und nicht genau vorhersehbare oder lokalisierbare Naturereignisse geprägte Gebiete wie z. B. Erdbebenregionen, steile Lawinenhänge, Überschwemmungsgebiete, Waldgebiete auf Sandböden etc.).

3663 natural drainage way [n] [US] *geo.* (Natural, small and vegetated gully or ravine formed by surface runoff); *syn.* natural hillside drainage channel [n] [UK]; *s* **canal [m] de drenaje natural** (Pequeño canal de desagüe formado por la escorrentía en una vertiente); *f* **ravineau [m]** (Petite ravine en voie d'évolution et non stabilisée sur un versant de matériaux meubles provoquée par les eaux de ruissellement) ; *g* **Siepen [m]** (Durch Erosion entstandene, kleine, natürliche Abflussrinne/Talrinne oder Ausmuldung in einem Hang); *syn.* Klinge [f] [BW], Seifen [m], Siefen [m], Seipen [m].

natural ecosystems [npl], **effective functioning of** *ecol.* ►natural potential.

3664 natural environment [n] *conserv. ecol. nat'res.* (Environs neither created nor significantly influenced, nor used by man; ►degree of landscape modification); *s* **medio [m] natural** (Conjunto de condiciones físicas, químicas y biológicas que rodean a un individuo. El calificativo «natural» presupone que estas condiciones externas [casi] no han sido variadas por el hombre; DINA 1987; ►grado de modificación [de ecosistema y paisajes]); *f* **milieu [m] naturel** (**1.** Espace n'ayant pas ou peu subi l'influence de l'homme ; ►degré d'artificialisation ; **2.** terme général utilisé par les géographes pour désigner les divers types d'écosystèmes terrestres ; **3.** pour les biologistes ensemble d'éléments [eau, sol, relief, air, climat, faune et flore] qui agissent directement ou indirectement sur tout ou partie des organismes qui l'habitent et par leurs caractéristiques biologiques et physiques assurent le maintien de la diversité biologique animale et végétale ; des perturbations naturelles ou anthropiques peuvent rompre cet équilibre et entraîner la régression ou la disparition du milieu naturel) ; *syn.* environnement [m] naturel ; *g* **natürliche Umwelt [f]** (Das nicht vom Menschen, aber meistens von ihm Beeinflusste, Gemachte, Genutzte: die Umweltfaktoren Gestein, Boden, Wasser, Luft, Vegetation und Tierwelt sowie die von ihnen gebildeten Einheiten, wie Landschaftsräume in ihrer Naturausstattung, die gesamte Erdoberfläche [Geosphäre] und schließlich der Kosmos, von dem das Leben der Erde in seinem Energiehaushalt abhängig ist; cf. BUCH 1978, 2; § 2 BNatSchG nennt die Elemente der **n.n** U. wie Boden, Gewässer, Luft, Klima, Vegetation, wild lebende Tier- und Pflanzenarten sowie historische Kulturlandschaften und deren Teile, die den Zielen des Naturschutzes und der Landschaftspflege unterliegen; ►Natürlichkeitsgrad, ►Umwelt).

3665 natural erosion [n] *geo.* (Generic term for processes of earth-sculpture by such agents as running water, ice, wind, waves that involve transport of material, not including static weathering nor mass movement through gravity. This term includes physical erosion [corrasion] and chemical erosion [solution, carbonation]. It is commonly but incorrectly used synonymously with the *generic term* ►denudation; DNE 1978; the gradual erosion of land used by man which does not greatly exceed natural erosion is called 'normal erosion'; SOIL CONSERV. SOC. AM 1970. **In D.**, the term 'Erosion' is used primarily for linear [point] erosion. The concept is widened by forming compound nouns such as 'Winderosion' [►wind erosion], 'Flächenerosion' [►sheet erosion], etc.; the English language distinguishes between naturally occurring erosion processes and those caused by humans, e.g. soil erosion; ►gully erosion, ►soil erosion, ►stream erosion); *syn.* geologic erosion [n] [also US]; *s* **erosión [f]** (Degradación y progresiva destrucción del relieve como consecuencia de la actuación de una serie de agentes y procesos. Los agentes de la erosión más importantes son el agua, el hielo, el viento, las variaciones térmicas, los organismos vivos y el hombre. Ellos permiten una diferenciación entre erosión hídrica [**e.** pluvial, **e.** fluvial, e. marina], **e.** glaciar, ►erosión eólica, **e.** biológica y **e.** antrópica; cf. DGA 1986; ►denudación, ►erosión del suelo, ►erosión en cárcavas 1, ►erosión en cárcavas 2, ►erosión laminar); *f* **érosion [f]** (Ensemble des phénomènes extérieurs à l'écorce terrestre — ou phénomènes exogènes — qui contribuent à modifier les formes créées par les phénomènes endogènes — tectonique et volcanisme ; DG 1984 ; ►érosion continentale, ►érosion des sols, ►érosion éolienne 2, ►érosion pelliculaire, ►ravinement 1, ►ravinement 2) ; *g* **Erosion [f]** (Im deutschen Sprachraum primär die ausfurchende und den Boden forttragende Tätigkeit linear fließenden Wassers. Durch zusammengesetzte Begriffe wie ►Windabtragung, ►Flächenerosion etc. werden erweiterte Begriffsinhalte möglich. Im englischen Sprachraum wird *erosion* umfassender gedacht [Windschliff, thermische Lösungsvorgänge, Brandungstätigkeit etc.], enthält jedoch weder die örtliche Verwitterung noch Bodenrutschungen. Man unterscheidet im Englischen grundsätzlich zwischen natürlichen Erosionsvorgängen *natural erosion* und solchen, die durch den Menschen verursacht werden *soil erosion*; cf. DNE 1978; ►Abtragung, ►Bodenerosion, ►Grabenerosion 1, Grabenerosion 2).

natural event [n] *leg.* ►major natural event.

natural feature [n] *conserv.* ►outstanding natural feature.

natural feature [n], **unique** *conserv.* ►outstanding natural feature.

3666 natural features [npl] *landsc.* (Organic and inorganic attributes of the landscape which have been hardly influenced or not at all influenced by human activities, in contrast to artificial features); *s* **rasgos** [mpl] **naturales (del paisaje)** (Conjunto de elementos bióticos y abióticos de un paisaje poco influenciados por los usos humanos, en contraste con los elementos creados por la sociedad); *syn.* características [fpl] naturales (del paisaje); *f* **composantes** [fpl] **naturelles (du paysage)** (Éléments biotiques et abiotiques du milieu naturel par opposition aux composantes culturelles) ; *g* **Naturausstattung** [f] (Vom Menschen nicht oder kaum beeinflusste anorganische und organische Gegebenheiten in der Landschaft im Gegensatz zur Kulturausstattung).

natural features [npl], **uniqueness of** [US] *conserv.* ►unique character of natural features.

3667 natural forest [n] *for. land'man.* (Forest which is virtually uninfluenced by human activity, and may well be a ►second-growth [forest] [US]/secondary forest [UK]; SAF 1983; ►virgin forest); *s* **bosque** [m] **natural** (Monte originado sin influencia humana y que se mantiene sin ella, aunque puede ser un ►bosque secundario; ►bosque primario); *f* **forêt** [f] **naturelle** (n'ayant pas subi l'influence des activités humaines ; ►forêt de substitution, ►forêt vierge) ; *g* **Naturwald** [m] (Ohne Einfluss des Menschen entstandener und so belassener Wald; ►Sekundärwald, ►Urwald).

3668 natural forest community [n] *phyt.* (Naturally-occurring ►virgin forest or near-natural woodland community, as opposed to a plant community which originates from a planted forest, and which is a substitute for the ►natural forest. Substitute forest communities are characterized by non-indigenous species and by an unstable equilibrium; ►forest biocoenosis); *s* **comunidad** [f] **forestal** (Comunidad vegetal del ►bosque natural o poco transformado por el hombre, al contrario que la comunidad del bosque artificial, por la cual se entiende la del ►bosque secundario; ►biocenosis forestal); *f* **groupement** [m] **sylvatique** (Désigne l'unité sociologique correspondant à la ►forêt vierge ou à la ►forêt naturelle par opposition à un boisement artificiel qui est caractérisé par la présence d'espèces non indigènes et par un équilibre instable ; ►biocénose forestière) ; *g* **Waldgesellschaft** [f] (Pflanzengesellschaft des natürlichen [►Urwaldes] oder naturnahen Waldes [►Naturwald] im Gegensatz zur **Forstgesellschaft** [TÜXEN 1950], worunter eine anthropogene Ersatzgesellschaft des natürlichen Waldes verstanden wird. Forstgesellschaften zeichnen sich durch standortfremde Baumarten und ein labiles Gleichgewicht aus; ►Waldbiozönose).

natural forest reserve [n] [UK] *conserv. leg. nat'res.* ►forest research natural area [US].

3669 natural garden stone flag [n] *constr.* (Concrete or natural stone unit for surfacing in private gardens; ►flagstone, ►footpath paving slab); *syn.* garden paving slab [n] [UK]; *s* **losa** [f] **de jardín (de piedra natural o artificial)** (Losa no normada de hormigón o de piedra natural que se utiliza para pavimentar caminos en jardines; ►losa de pavimentación, ►losa de piedra natural, ►losa para pasos); *f* **dalle** [f] **de jardin** (Élément non normalisé en béton ou en pierre naturelle utilisé pour les circulations dans un jardin ; ►dalle d'un pas japonais, ►dalle en pierre naturelle pour voirie de surface, ►dalle pour circulation piétonne) ; *g* **Gartenplatte** [f] (Platte aus Beton oder Naturstein für den Wegebau oder zur Verlegung als Schrittplatte für die Gartengestaltung; ►Gehwegplatte, ►Natursteingehwegplatte).

3670 natural grassland [n] *agr.* (Areas of indigenous grasses left unused or unsuitable for cultivation; ►permanent grassland); *s* **prado** [m] **natural** (Pradera que crece en lugares no indicados para el cultivo; ►pradera permanente); *syn.* pradera [f] natural, pradería [f] natural (DINA 1987, 748); *f* **prairie** [f] **naturelle** (Surface enherbée, non assolée, qui n'a été ni labourée ni ensemencée et dont la flore, complexe, est composée d'espèces issues de la végétation locale ; DAV 1984 ; ►prairie permanente) ; *g* **natürliches Grünland** [n] (Landwirtschaftliche Nutzfläche auf Standorten, die für Ackerbau ungeeignet sind; ►Dauergrünland); *syn.* absolutes Grünland [n].

3671 natural ground surface [n] *constr. plan.* (►below ground level); *syn.* natural terrain [n], existing ground surface [n]; *s* **superficie** [f] **natural del terreno** (►bajo el nivel del suelo); *syn.* superficie [f] del suelo; *f* **terrain** [m] **naturel** (Étendue naturelle considérée du point de vue de son relief ; ►sous la surface du sol, ►sous sol fini futur) ; *g* **natürliches Gelände** [n, o. Pl.] (Vom Menschen unveränderte Geländeoberfläche; ►unter Flur); *syn.* Bodenoberfläche [f], Geländeoberfläche [f].

3672 natural groundwater recession [n] *hydr.* (Natural drop in groundwater level; *s* **descenso** [m] **natural del agua subterránea** (WMO 1974); *syn.* recesión [f] natural del nivel freático; *f* **abaissement** [m] **naturel de la nappe phréatique** (Nappe en régime de tarissement ; PED 1983, 392) ; *g* **Grundwasserabsinken** [n, o. Pl.] (Natürliche Senkung der Grundwasseroberfläche); *syn.* Absinken [n, o. Pl.] des Grundwassers, Grundwassersenkung [f], Grundwasserrückgang [m], Rückgang [m] des Grundwassers.

3673 natural habitat [n] **of community interest** *conserv. ecol.* (Natural habitats, which are in danger of disappearance in their natural range or have a small natural range due to their decline or are endangered because of their intrinsically restricted area or display typical characteristics of one or more of the five following biogeographical regions: Alpine, Atlantic, Continental, Macaronesian and Mediterranean; such habitat types are listed or may be listed in Annex I, Council Directive 92/43/EEC); *s* **hábitat** [m] **natural de interés comunitario** (Según la Directiva de hábitats de la UE, hábitat que se encuentra amenazado de desaparición en su área de distribución natural, o bien presenta un área de distribución natural reducida a causa de su regresión o debido a su área intrínsecamente restringida, o bien constituye ejemplo representativo de características típicas de una o de varias de las cinco regiones biogeográficas siguientes: alpina, atlántica, continental, macaronesia y mediterránea; este tipo de hábitats figura o podrá figurar en el Anexo I, Directiva 92/43/CEE); *f* **type** [m] **d'habitats naturels d'intérêt communautaire** (Habitat qui est soit en danger de disparition dans son aire de répartition naturelle ou a une aire de répartition naturelle réduite par suite de sa régression ou en raison de son aire intrinsèquement restreinte ou constitue un exemple remarquable de caractéristiques propres à l'une ou à plusieurs des cinq régions biogéographiques suivantes : alpine, atlantique, continentale, macaronésienne et méditerranéenne ; ce type d'habitat figure ou est susceptible de figurer à l'annexe I, Directive 92/43/CEE) ; *g* **natürlicher Lebensraum** [m] **von gemeinschaftlichem Interesse** (Lebensraum, der im Bereich seines natürlichen Vorkommens vom Verschwinden bedroht ist oder infolge seines Rückgangs oder auf Grund seines an sich schon begrenzten Vorkommens ein geringes natürliches Verbreitungsgebiet hat oder typische Merkmale einer oder mehrerer der folgenden fünf biogeographischen Regionen aufweist: alpine, atlantische, kontinentale, makaronesische und mediterrane; diese Lebensraumtypen sind in Anhang 1, RL 92/43/EWG aufgeführt).

natural habitats [npl] *conserv. leg.* ►conservation of natural habitats and of wild fauna and flora, ►conservation regulations of natural habitats etc. 1994 [UK], ►Convention on the Conservation of European Wildlife and Natural Habitats, ▶Habitats Directive.

3674 natural habitat type [n] *conserv. ecol.* (Natural habitat of community interest within the European Union. Characterized by specific site conditions and the century-long influence of human beings, **n. h. t.** are classified according to a FFH Code and included under Annex I of the European Directive 92/43/EEC dated 21.05.1992. Guidance on the interpretation of habitat types is given in the 'Interpretation Manual of European Union Habitats'. This manual is endorsed by the appointed committee according to the article 20 ["Habitat committee"] and published by the European commission. **n. h. t.** include, for example, oligotrophic standing water bodies, dystrophic lakes, near-natural raised bogs, lime-rich low bogs, siliceous rock and its rock cleft vegetation, alder and ash forests and regularly flooded riparian woodland *[Alno-Padion, Alnion incanae, Salicion albae]* along watercourses, coniferous forests on acidic soils, woodruff-beech forests *[Asperulo-Fagetum]*, dry heathland, limestone pioneer grassland; ▶Habitats Directive, ▶priority natural habitat type); *s* **tipo** [m] **de hábitat natural** (Espacio natural de interés comunitario que cumple un papel preciso en el ciclo de vida de una especie de flora o fauna. Todos los **tt. de hh. nn.** están registrados en el anexo I de la ▶Directiva de Hábitats que actualmente contiene 198 tipos de **hh. nn.** europeos e incluye 65 prioritarios [hábitats en peligro de desaparición y de los cuales el área de distribución natural está incluida mayormente en el territorio de la UE]. La clasificación de los hábitats está basada en la tipología jerárquica de los hábitats europeos, elaborada en primer lugar en el marco del proyecto CORINE Biótopos y reelaborada por el Comité de Hábitats, nombrado según el art. 20, para la versión del anexo I de la Directiva 92/43/EEC aprobada por el Consejo Europeo y publicada en mayo de 1992 como Manual de Interpretación de Hábitats de la UE que es actualizado periódicamente. Esta clasificación está dividida en nueve grupos principales [**1.** hábitats costeros y vegetaciones halofíticas, **2.** dunas marítimas y continentales, **3.** hábitats de agua dulce, **4.** brezales y matorrales de zona templada, **5.** matorrales esclerófilos, **6.** formaciones herbosas naturales y seminaturales, **7.** turberas altas, turberas bajas y áreas pantanosas, **8.** hábitats rocosos y cuevas, y **9.** bosques], que a su vez se subdividen en hábitats específicos, según los medios en los que se encuentran; ▶tipo de hábitat natural prioritario); *syn.* hábitat [m] natural tipo; *f* **type** [m] **d'habitat naturel** (Espace concret remplissant un rôle précis dans le cycle de vie d'une espèce de la faune ou de la flore ; les types d'habitats naturels d'intérêt communautaire sont mentionnés dans l'annexe I la directive « Habitats » qui énumère actuellement 198 types d'habitats naturels européens, y compris 65 prioritaires [types d'habitats en danger de disparition et pour lesquels leur aire de répartition naturelle est majoritairement incluse dans le territoire de l'Union européenne]. La classification des habitats est basée au départ sur la typologie hiérarchique des habitats européens, réalisée par le projet CORINE biotopes, cette typologie étant la seule alors existante à l'échelle européenne ; sur la base de cette typologie a été réalisé, par le Professeur Albert Noirfalise et après nombreuses discussions avec les experts nationaux la version de l'Annexe I approuvée par le Conseil et publiée dans le Journal Officiel en mai 1992 ; depuis 1994 existe le manuel d'interprétation des types d'habitats de l'union européenne approuvé par le Comité Habitats qui a été révisé dans la Version EUR12, elle même actualisée par la version EUR15 ; la plupart des types d'habitats de l'Annexe I sont qualifiés par des termes biogéographiques indiquant que leur distribution principale se trouve dans une certaine région biogéographique ; cette indication n'exclue pas la possibilité de trouver les mêmes types d'habitats dans d'autres régions biogéogra-

phiques ; de plus, ces habitats extra zonaux ont une valeur extraordinaire en termes scientifiques et pour la conservation ; il est donc nécessaire de garder une certaine souplesse d'interprétation par les experts, notamment dans les régions où les types d'habitats sont très fragmentaires et influencés par les activités humaines ; des révisions périodiques de ce manuel sont attendues, de façon à l'adapter aux progrès scientifiques sur la connaissance des habitats ; à titre d'exemple on peut mentionner les types d'habitats suivants : eaux oligotrophes très peu minéralisées des plaines sablonneuses *[Littorelletalia uniflora]*, lacs et mares dystrophes naturels, tourbières hautes actives, tourbières basses alcalines, pentes rocheuses siliceuses avec végétation chasmophytique, forêts alluviales à *Alnus glutinosa* et *Fraxinus excelsior [Alno-Padion, Alnion incanae, Salicion albae]*, forêts acidophiles à Picea des étages montagnard à alpin *[Vaccinio-Piceetea]*, hêtraies du *Asperulo-Fagetum*, landes sèches européennes, pelouses rupicoles calcaires ; cf. Manuel d'interprétation des types d'habitats de l'union européenne, EUR15 octobre 1999, Commission européenne DG Environnement ; ▶type d'habitats naturels prioritaire) ; *g 1* **Lebensraumtyp** [m] (Lebensraum von gemeinschaftlichem Interesse für die Europäische Union. Er ist geprägt durch Standortbedingungen und jahrhundertelanges Einwirken des Menschen und wird durch einen FFH-Code klassifiziert und im Anhang I der Richtlinie 92/43/EWG des Rates vom 21.05.1992 aufgeführt. Im „Interpretationshandbuch der Lebensräume der Europäischen Union" wird eine Orientierungshilfe für die Interpretation der Typen natürlicher Lebensräume gegeben. Dieses Handbuch wird durch den nach Artikel 20 eingesetzten Ausschuss [„Habitat-Ausschuss"] befürwortet und durch die Europäische Kommission veröffentlicht. **L.en** sind z. B. oligotrophe Stillgewässer, dystrophe Seen, naturnahe lebende Hochmoore, kalkreiche Niedermoore, Silikatfelsen und ihre Felsspaltenvegetation, Erlen- und Eschenwälder und Weichholzauenwälder *[Alno-Padion, Alnion incanae, Salicion albae]* an Fließgewässern, bodensaure Nadelwälder, Waldmeister-Buchenwälder *[Asperulo-Fagetum]*, trockene Heiden, Kalk-Pionierrasen; ▶Fauna-Flora-Habitat-Richtlinie; *g 2* **prioritärer Lebensraumtyp** [m] (▶prioritärer natürlicher Lebensraumtyp); *g 3* **Lebensraumtyp** [m] **von gemeinschaftlichem Interesse** (**Europa**, natürlicher Lebensraum, der **1.** im Bereich seines natürlichen Vorkommens vom Verschwinden bedroht ist, **2.** infolge seines Rückganges oder auf Grund seines an sich schon berenzten Vorkommens ein geringes natürliches Verbreitungsgebiet hat oder **3.** typische Merkmale einer oder mehrerer der folgenden fünf biogeografischen Regionen hat: alpin, atlantisch, kontinental, makronesisch oder mediterran).

3675 natural hazard [n] *leg.* (Natural feature, such as an escarpment, cliff, waterfall, etc. which may endanger man; ▶force majeure, ▶major natural event); *s* **riesgo** [m] **natural** (Fenómeno natural generalmente no evitable y no predecible por el hombre como terremotos, inundaciones, tormentas, lavinas; ▶catástrofe natural, ▶fuerza mayor); *f* **risque** [m] **naturel** (Phénomène naturel en général non exactement prévisible ou localisable tels que les tremblements de terre, inondations, avalanches, incendies de forêts représentant un danger pour la population ; ▶catastrophe naturelle, ▶force majeure, ▶risque naturel majeur) ; *g* **natürliche Gefahr** [f] (**1.** Natürliche Gegebenheit wie plötzliche Hangkante, Wasserfall etc., die eine Gefahr für den Menschen darstellen können. **2.** Durch den Menschen in der Regel unabweisbares und nicht genau vorhersehbares oder lokalisierbares Naturereignis wie z. B. Erdbeben, Überschwemmung, Sturm, Lawine, Waldbrand; ▶höhere Gewalt, ▶schweres Naturereignis).

3676 natural heritage [n] *conserv. land'man. nat'res.* (**1.** Biological, physical, geological or physiographical formations

N

[natural features], and precisely-delineated areas which constitute the habitat of threatened species of animals and plants, and natural sites or precisely delineated areas which have an outstanding value from the point of view of science, conservation or natural beauty; cf. Paris Convention of 1972. **2.** Existing natural environment, natural resources, natural habitats, species of flora and fauna as well as biological equilibrium and biodiversity of a defined area); *syn.* natural legacy [n] (TGG 1984, 188), natural patrimony [n]; *s* **patrimonio [m] natural** (**1.** Formaciones biológicas, físicas, geológicas o fisiográficas [formaciones naturales] y zonas delimitadas estrictamente que constituyen el hábitat de especies de flora y fauna amenazadas, que tienen un valor excepcional desde el punto de vista científico, de conservación o de belleza natural; cf. Convención de París, de 23 de noviembre de 1972, para la Protección del Patrimonio Mundial, Cultural y Natural. **2.** Conjunto de espacios, recursos y hábitats naturales, especies animales y vegetales así como la diversidad y el equilibrio dinámico existente entre ellos que forma parte de la herencia de una sociedad para las generaciones futuras); *f* **patrimoine [m] naturel** (**1.** Formations biologiques, physiques, géologiques ou physiographiques [monuments naturels] et les zones strictement délimitées constituant l'habitat d'espèces animales et végétales menacées, qui ont une valeur exceptionnelle du point de vue de la science, de la conservation ou de la beauté naturelle ; cf. Convention de Paris de 1972 ; en France l'État établi un inventaire départemental du **p. n.** qui recense les sites, paysages et milieux naturels ainsi que les mesures de protection et les moyens de gestion et de mise en valeur qui s'y rapportent. **2.** Ensemble des espaces, ressources et milieux naturels, sites et paysages, espèces animales et végétales ainsi que la diversité et les équilibres biologiques auxquels ils participent ; ▶monument naturel) ; *g* **Naturerbe [n]** (**1.** Ästhetisch oder wissenschaftlich bedeutsame biologische, physikalische, geologische oder physiografische Erscheinungsformen [Naturgebilde], genau abgegrenzte Gebiete, die den Lebensraum für bedrohte Pflanzen- und Tierarten bilden sowie Naturstätten oder genau abgegrenzte Naturgebiete, die aus wissenschaftlichen Gründen oder ihrer Erhaltung oder natürlichen Schönheit wegen von großem Wert sind; cf. Pariser Übereinkommen von 1972. **2.** Alle vorhandenen Bodenschätze, bestehenden Natur- und Lebensräume für Flora und Fauna, die Biodiversität und das biologische Gleichgewicht eines Betrachtungsraumes als natürliches Erbe für zukünftige Generationen); *syn.* Gesamtheit [f] der natürlichen Grundgüter.

natural hillside drainage channel [n] [UK] *geo.* ▶natural drainage way [US].

natural [adj] **in character** *ecol.* ▶natural.

3677 naturalistic [adj] (in accordance with, or in imitation of nature); *s* **naturalista** [adj] (Concepto, objeto o espacio que reproduce fielmente la naturaleza); *f* **naturaliste** [adj] (Concept [regard, approche], objet ou espace [paysage, parc] qui reproduit fidèlement la nature) ; *g* **naturalistisch** [adj] (So beschaffen, dass ein gestalteter Gegenstand oder Park der empirisch fassbaren Wirklichkeit von Natur oder vom Natürlichem entspricht: der Natur getreu nachempfunden).

3678 naturalistic garden [n] *gard.* (**1.** A wild **g.** intended to initiate a diverse biocenosis with curvilinear lines, tree masses and lawns, water and soft land shaping or **2.** an "imitation of nature **g.**" in the creation of a "romantic landscape", or **3.** a **g.** where there may be no particular restriction upon the type of use. In contrast to a formal or ornamental **g.**, the **n. g.** is initially characterized by indigenous plants, followed by a natural succession and containing built elements which can be changed or removed at any time; ▶landscape garden); *syn.* natural-like garden [n]; *s* **jardín [m] silvestre** (Espacio de terreno asilvestrado que se utiliza como jardín o diseñado con la

intención de crear un algo natural, o de imitar a la naturaleza, es decir en el que crecen plantas autóctonas y encuentran cobijo pájaros y otros animales menores y en el que generalmente no se aplican pesticidas; ▶jardín paisajístico); *f* **jardin [m] naturel** (LA 1981, 652 ; espace végétal aménagé dont la conception est déterminée en fonction des besoins de l'homme, p. ex. en vue de provoquer le développement de biocénoses ou d'imiter le caractère de paysages romantiques, de permettre l'appropriation non limitative de l'espace par différents groupes d'utilisateurs. Par opposition au ▶jardin régulier ou ordonnancé, l'utilisation de végétaux indigènes pour lesquels il est laissé libre cours à la succession naturelle ainsi que l'emploi de constructions transformables à tout instant entraînent une diminution importante des travaux d'entretien; ▶jardin paysager, ▶parc paysager) ; *g* **naturnaher Garten [m]** (Vegetationsbestimmter Freiraum, z. B. als [Naturschutz]garten zur Initiierung einer vielfältigen Biozönose oder als Naturimitationsgarten zum Anschauen von romantischen Landschaftsbildern oder als Garten zur Aneignung unterschiedlicher Nutzergruppen ohne besondere Nutzungseinschränkungen. Im Gegensatz zu den Repräsentations-, Schmuck- oder Ziergärten durch bodenständige Vegetationsformen, natürliche Sukzession und durch jederzeit veränderbare Baukonstruktionen gekennzeichnet; ▶Landschaftspark); *syn.* Naturgarten [m], *auch* Ökogarten [m].

naturalistic park [n] *gard'hist.* ▶landscape garden.

3679 naturalization [n] *ecol. zool.* (Release of animal species in an area outside their natural or present range of distribution; ▶anthropogenic alteration of the genetic fauna pool, ▶degree of naturalization, ▶naturalized species, ▶reintroduction); *s* **naturalización [f]** (Proceso por el cual, una especie animal adquiere las condiciones necesarias para vivir y perpetuarse en un país distinto del que procede; cf. DINA 1987; ▶adulteración de la fauna endémica, ▶especie naturalizada, ▶grado de naturalización, ▶reintroducción de especies); *f* **naturalisation [f]** (Introduction par l'homme d'espèces animales ou végétales ayant pu se reproduire et retourner à l'état sauvage pour prendre leur place dans les groupements de la faune et de la flore des pays dans lesquels elles n'étaient auparavant pas indigènes ; ▶anthropisation de la composition de la faune, ▶degré de la naturalisation, ▶espèce naturalisée, ▶espèce subspontanée, ▶réintroduction 2) ; *g* **Einbürgerung [f]** (Aussetzen von Tierarten durch den Menschen in Gebiete, in denen sie vorher nicht heimisch waren und deren dauerhafte Ansiedlung; im weiteren Sinne sind es auch eingeschleppte Arten oder solche, die sich ohne Mitwirkung des Menschen angesiedelt haben [Neozoen]; ▶neuheimische Art, ▶Faunenverfälschung, ▶Grad der Einbürgerung, ▶Wiederansiedelung); *syn.* Aussetzen [n, o. Pl.] oder Ansiedeln [n, o. Pl.] gebietsfremder Tiere in der freien Natur (§ 21 BNatSchG), Ansiedeln [n, o. Pl.] landes- und standortfremder Tierarten [CH] (Art. 23 NHG), Naturalisation [f].

3680 naturalized species [n] *phyt. zool.* (**1.** That species, according to the ▶degree of naturalization, first introduced by man, which has, however, proved itself competitive in the almost natural plant community, and which will remain even after a cessation of human intervention; ▶epecophyte, ▶feral species, ▶neophyte; **2.** animal species introduced by man, which had not been previously native in that area; ▶naturalization); *s* **especie [f] naturalizada** (**1.** Planta que, no siendo oriunda de un país, medra en él y se propaga como si fuese autóctona; cf. DB 1985; *términos específicos* ▶epecófito, ▶especie asilvestrada 1, ▶neófito. **2.** Especie animal introducida por el hombre en un área en la que anteriormente no era nativa; ▶grado de naturalización, ▶naturalización); *syn.* especie [f] aclimatada, especie [f] asilvestrada (2); *f 1* **espèce [f] naturalisée** (**1.** Selon son ▶degré de naturalisation, espèce végétale introduite et susceptible de se maintenir de façon permanente à l'état sauvage ou de se reproduire spontanément et s'est fait une place parmi les groupements végétaux locaux. **2.** Espèce animale introduite dans un territoire sur lequel elle n'existait pas initialement ; DEE 1982 ; *termes spécifiques* ►espèce végétale acclimatée, ►plante néophyte ; ►espèce retournée à l'état sauvage) ; *f 2* **espèce [f] subspontanée** (Espèce végétale naturalisée se comportant presque exactement comme des espèces indigènes, p. ex. le robinier ; ►naturalisation) ; *syn.* espèce [f] spontanée étrangère ; *g* **neuheimische Art [f]** (**1.** Nach dem ►Grad der Einbürgerung diejenige Pflanzenart, die zwar erst im Gefolge des Menschen eingewandert ist, sich aber in naturnahen Pflanzengesellschaften als konkurrenzfähig erwiesen hat und auch bei Aufhören des menschlichen Einflusses erhalten bliebe; *UBe* ►Neophyt, ►kulturabhängige Art; ►verwilderte Art. **2.** Vom Menschen in ein Gebiet ausgesetzte oder ausgewilderte Tierart, in dem sie vorher nicht heimisch war; ►Einbürgerung); *syn.* eingebürgerte Art [f].

3681 natural landmark [n] [US] *conserv. leg.* (An individual manifestation of nature essentially untouched by man which is protected by law for scientific, historical or cultural reasons, or due to its rarity, peculiar character or beauty; a natural or cultural feature of outstanding or unique value, representative of aesthetic qualities or cultural significance. Criteria for selection of a natural landmark are: 1. The area should contain one or more features of outstanding significance [e.g. spectacular waterfalls, caves, craters, fossil beds, sand dunes and marine features, along with unique or representative fauna and flora]. Associated cultural features may include cave dwellings, cliff-top forts, archaeological sites, or natural sites, which have heritage significance to indigenous peoples. 2. The area should be large enough to protect the integrity of the feature and its immediately related surroundings; ►National register of Natural Areas [US]/register of natural monuments [UK]. **In UK.**, natural monuments are designated as **Sites of Special Scientific Interest** and, as representative of British habitats, deemed the best examples of wildlife habitats, geological features and landforms in Great Britain. Their designation is based on established criteria including naturalness, diversity, size, fragility and rarity; the aim of an SSSI is to maintain the existing diversity of animals and plants. There are about 6500 **SSSIs**, ranging in size from 62,000 ha to 1 ha, totalling approximately 1.6 million hectares or 9% [1999] of the land area of the United Kingdom); *syn.* Site [n] of Special Scientific Interest (S.S.S.I.) [UK], natural monument [n] [UK]; *s 1* **monumento [m] natural** (Son espacios o elementos de la naturaleza constituidos básicamente por formaciones de notoria singularidad, rareza o belleza, que merecen ser objeto de protección especial, incluyendo árboles singulares y monumentales, formaciones geológicas, yacimientos paleontológicos y mineralógicos demás elementos de la gea que reúnan un interés especial por la singularidad o importancia de sus valores científicos, culturales o paisajísticos; art. 33 Ley 42/2007; ►registro de monumentos naturales); *s 2* **área-monumento [m] natural** (≠) (En D. categoría de protección a nivel de estado federado para zona de hasta 5 ha protegida por su interés ecológico, científico, histórico o cultural y por su singularidad y belleza); *f 1* **monument [m] naturel** (**F.**, les **m. n.** dont la conservation ou la préservation présente au point de vue artistique, historique, scientifique, légendaire ou pittoresque un intérêt général sont, au même titre que les sites naturels, protégés au titre de l'art. 4 de la loi du 2 mai 1930 [relative à la protection des monuments naturels et des sites de caractère artistique, historique, scientifique, légendaire ou pittoresque], et peuvent être inscrits à l'inventaire des sites, être classés ou être répertoriés sur la ►liste des monuments naturels et des sites ; **D.**, objet naturel remarquable [site ponctuel], présentant un intérêt général du point de vue scienti-

fique, historique, culturel ou pour sa beauté et son caractère, protégé par l'art. 17 de la loi fédérale et les lois des *Länder* relative à la protection de la nature) ; *f 2* **monument [m] naturel étendu** (≠) (**F.**, la protection des monuments et sites naturels est assurée au titre du code de l'environnement aux articles L. 341-1 et suivants [codification de la loi de 1930] ; à l'origine, la prévalence de l'approche esthétique a conduit à privilégier la prise en compte de sites naturels ponctuels en raison de certaines de leurs particularités esthétiques ou historiques, le monument naturel étant le pendant du monument historique ; avec le temps une protection élargie à des ensembles plus importants et la notion de « **site étendu** » a été ainsi consacrée par la jurisprudence ; cette notion de site étendu peut englober des espaces très vastes comme des grands sites naturels de plusieurs milliers d'hectares ; dans cette acceptation ces sites protégés se rapprochent du concept de protection allemand du « *Naturschutzgebiet* » ; **D.**, monument naturel étendu [jusqu'à 5 ha], dont la sauvegarde est d'intérêt général pour sa beauté et son caractère remarquable ou du point de vue écologique, scientifique, historique culturel ou patrimonial protégé par les lois des *Länder* relative à la protection de la nature ; ►monument naturel) ; *g 1* **Naturdenkmal [n]** (...mäler [pl] ; in A. nur so, sonst auch ...male [pl]; Schutzgebietskategorie nach dem Bundesnaturschutzgesetz [BNatSchG] und IUCN-Managementkategorie: aus wissenschaftlichen, naturgeschichtlichen oder landeskundlichen Gründen oder wegen ihrer Seltenheit, Eigenart oder Schönheit rechtsverbindlich festgesetzte Einzelschöpfung der Natur incl., wenn nötig, der für den Schutz des **N.s** notwendigen Umgebung [§ 17 BNatSchG und Naturschutzgesetze der Bundesländer]. **N.e** werden i. d. R. durch die untere Naturschutzbehörde der Gemeinde oder des Kreises per Rechtsverordnung ausgewiesen und sind bei dieser Behörde im ►Naturdenkmalbuch eingetragen. Ein **N.** stellt als Schutzgebietskategorie im Vergleich zum ►geschützten Landschaftsbestandteil ideelle Zwecke in den Vordergrund); *g 2* **flächenhaftes Naturdenkmal [n]** (...mäler [pl]; in A. nur so; auch ...male [pl]; in D. durch Rechtsverordnung der Länder unter Schutz gestellte, bis 5 ha große Fläche, deren Erhaltung wegen ihrer hervorragenden Schönheit oder Eigenart oder ihrer ökologischen, wissenschaftlichen, geschichtlichen, volks- oder heimatkundlichen Bedeutung im öffentlichen Interesse liegt; cf. Naturschutzgesetze der Länder; ►Naturdenkmal).

3682 natural landscape [n] *conserv. landsc.* (Area or ecosystem which has been hardly to strongly influenced from little to much by man; ►degree of landscape modification, ►natural landscape unit); *syn.* countryside [n] and nature [n]; *s* **naturaleza [f] y paisaje [m]** (Par de términos no claramente definidos utilizados en alemán para denominar áreas naturales o ecosistemas poco influenciados por la humanidad; ►grado de modificación [de ecosistemas y paisajes], ►unidad de espacio natural); *f* **espaces [mpl] naturels et paysages [mpl]** (Art. 1 de la loi n° 76-629 du 10 juillet 1976 relative à la protection de la nature ; cf. PPH 1991, 268 ; avec la loi n° 95-101 du 2 février 1995 les espaces, ressources et milieux naturels, les sites et les paysages, les espèces animales et végétales, la diversité et les équilibres biologiques auxquels ils participent, font partie du patrimoine commun de la Nation ; ►degré d'artificialisation, ►espace naturel) ; *syn.* sites [mpl], paysages [mpl] et milieux [mpl] naturels ; *g* **Natur [f, o. Pl.] und Landschaft [f, o. Pl.]** (*Kein fest definierter Begriff* Bezeichnung für vom Menschen kaum bis stark beeinflusste Landschaftsräume oder Ökosysteme; ►Natürlichkeitsgrad, ►Naturraum).

natural landscape [n] **area** *geo. landsc.* ►natural landscape unit.

natural landscape features [npl] *conserv. landsc. recr.* ►natural site characteristics.

N

3683 natural landscape potential [n] *geo.* (Physiognomy of a part of the earth's surface which has not been influenced by continuous human interference, but is the result of dynamic relationships within natural systems); *s* **fisionomía [f] natural de un territorio (≠)** (D., término tradicional utilizado para denominar el conjunto de las dotaciones físicas y geográficas de una parte de la superficie de la tierra que no ha sido influenciada por la interferencia humana sino que es el resultado de las interacciones dinámicas de los sistemas naturales que a largo plazo están sometidos a las leyes de la naturaleza); *f* **physionomie [f] naturelle d'un territoire** (Terme définissant l'ensemble d'un territoire ayant peu subi les influences anthropiques et caractérisé par une spécificité naturelle, fruit de la dynamique et des interactions des forces du monde vivant) ; *syn.* physionomie [f] paysagère d'un territoire ; *g* **Landesnatur [f, o. Pl.]** (Gesamtheit der physisch-geografischen Ausstattung eines Teils der Erdoberfläche, die nicht durch ständige neue Eingriffe des Menschen geschaffen oder gestaltet wurde, sondern das Wirkungsgefüge aus Erscheinungen und Kräften der anorganischen und vitalen Welt, das auf lange Sicht der Naturgesetzlichkeit unterliegt); *syn.* natürliches Potential/Potenzial [n] einer Landschaft.

3684 natural landscape unit [n] *geo. landsc.* (Space defined by its natural ►landscape factors; ►landscape unit; ►physiographic division, ►spatial unit of a landscape); *syn.* natural landscape area [n]; *s* **unidad [f] de espacio natural** (Espacio natural diferenciado según factores paisajísticos naturales; ►factor del paisaje, ►tipología del paisaje, ►unidad espacial de paisaje, ►unidad paisajística); *f* **espace [m] naturel** (Étendue terrestre [►espace paysager, espace géographique] constituée par un ensemble d'éléments physiques biotiques et abiotiques de l'espace géographique [►facteurs paysagers naturels] ; l'**e. n.** constitue dans l'écologie des paysages la représentation abstraite d'une entité paysagère réelle dépourvue de ses caractères anthropogènes ; MG 1993 ; ►typologie physiographique des paysages, ►unité paysagère) ; *syn.* milieu [m] naturel ; *g* **Naturraum [m]** (...räume [pl]; allgemeine Bezeichnung für einen Teil der Erdoberfläche [►Landschaftsraum], der mit biotischen und abiotischen ►Landschaftsfaktoren ausgestattet und somit abgrenzbar ist. Der **N.** ist in der Landschaftsökologie eine gedankliche Abstraktion eines realen Landschaftsausschnittes, d. h. Landschaft abzüglich des anthropogen überformten Anteils; eine Weiterentwicklung dieses Ansatzes ist die ►naturräumliche Gliederung).

natural legacy [n] *conserv. land'man. nat'res.* ►natural heritage.

natural-like garden [n] *gard.* ►naturalistic garden.

natural looking [ger] *ecol.* ►near-natural.

naturally occurring species [n] *phyt. zool.* ►indigenous species.

3685 naturally-sprayed area [n] *geo.* (Area of rocky coastline or in proximity of waterfall wetted by spray; ►salt spray); *s* **zona [f] de salpicadura** (Área del litoral rocoso o próxima a cascadas que es humedecida por el agua que salpica; ►salpicadura de sal); *f* **zone [f] des embruns** (Zone de brouillard mouillant à proximité des falaises ou des chutes d'eau ; ►embruns salés) ; *g* **Gischtzone [f]** (Durch ständig spritzendes Wasser und Sprühnebel geprägte Zone an felsigen Küsten oder Wasserfällen; ►Salzgischt).

3686 natural meadow [n] *phyt.* (Grassland with herbaceous plant community confined to locations having varying moisture and soil conditions, and composed primarily of grasses and perennials, more rarely of sedges. The following meadows occur according to location: ►damp meadow, ►dry meadow, ►nutrient-poor grassland, ►salt meadow, ►wet meadow, ►xerothermous meadow); *s* **prado [m] natural** (Comunidad herbácea atada a ubicaciones específicas y compuesta principalmente por hierbas y perennes, menos frecuentemente por juncos. Dependiendo de la ubicación se diferencian ►pastizales secos o ►pastizales xerotérmicos, ►prados higrófilos, ►prados húmedos, ►prados oligótrofos, ►prados salinos); *f 1* **pelouse [f]** (Formation végétale à dominante herbacée fermée, essentiellement constituée de graminées, plantes annuelles et cypéracées xérophiles, sans ou à très faible recouvrement d'espèces ligneuses ; ►pelouse aride, ►pelouse oligotrophe, ►pelouse xérophile et thermophile) ; *f 2* **prairie [f] climacique** (Formation végétale à dominante herbacée fermée des milieux naturels, essentiellement constituée de graminées, plantes annuelles et cypéracées mésophiles et hygrophiles sans ou à très faible recouvrement d'espèces ligneuses ; **F.**, le terme « prairie naturelle » est improprement employé en agriculture pour désigner une formation herbacée anthropique de fauche et de pâture, sans labour et semis, parfois amendée et irriguée ; ►prairie mésophile, ►prairie hygrophile, ►pré salé) ; *g* **natürliche Wiese [f]** (Gehölzfreie oder -arme, vorwiegend aus Süßgräsern und Stauden, seltener aus Sauergräsern gebildete Pflanzengesellschaft, die je nach Feuchtigkeit und Bodenverhältnisse eine bestimmte Artenzusammensetzung hat. Je nach Standort werden z. B. ►Feuchtwiesen, ►Magerrasen, ►Nasswiesen, ►Salzwiesen, ►Trockenrasen oder ►Xerothermrasen unterschieden).

natural monument [n] [UK] *conserv. leg.* ►natural landmark [US].

natural patrimony [n] *conserv. land'man. nat'res.* ►natural heritage.

natural pavestone [n]**, large-sized** *constr.* ►large-sized paving stone [US]/large-sized natural paving stone [UK].

natural paving stone [n]**, large-sized** *constr.* ►large-sized paving stone [US].

3687 natural phenomenon [n] *conserv.* (Occurrence in nature that is apparent to the senses, and that can be scientifically described or appraised); *s* **fenómeno [m] natural** (Proceso que tiene lugar en la naturaleza, puede ser registrado por los sentidos y explicado científicamente); *f* **phénomène [m] naturel** (qui se manifeste à la conscience, est perçu par les sens, pouvant être décrit ou expliqué scientifiquement) ; *g* **Naturerscheinung [f]** (Das sich den Sinnen in der Natur Zeigende, welches wissenschaftlich beschrieben oder erklärt werden kann).

3688 natural pond [n] *limn.* (Naturally-occurring, small water body; ►man-made pond); *s* **estanque [m] natural** (Pequeño cuerpo de agua quieta de origen natural, resultante de acumulación de agua en depresiones del terreno, que puede estar poblado de vegetación ripícola e hidrófitos; ►estanque [artificial]); *syn.* lagunilla [f]; *f* **étang [m] naturel** (Petit plan d'eau fermé, calme, résultant de l'accumulation d'eau dans des dépressions de terrain ou dans des zones d'effondrement à fond peu profond [moyenne inférieur à 5 m], pouvant être colonisé sur toute son étendue par la flore littorale ; ►étang artificiel) ; *g* **Weiher [m]** (Natürlich entstandenes, kleines, flaches Stillgewässer, oft mit Ufervegetation und Wasserpflanzen bestockt; ein sehr großer Weiher wird **Flachsee** genannt; ►Teich).

3689 natural potential [n] *ecol.* (Capacity of natural [eco]systems to function in the face of certain material, structural and competitive human demands; ►potentials of a landscape); *syn.* effective functioning [n] of natural (eco)systems; *s* **potencial [m] natural** (Capacidad de funcionamiento de los sistemas naturales en cuanto a demandas humanas de materiales, estructuras y exigencias competitivas hacia ellas; ►potencial de un paisaje); *syn.* capacidad [f] de rendimiento de los sistemas naturales; *f* **potentialité [f] du milieu naturel** (Mise en évidence des aptitudes des différents éléments du milieu à divers usages potentiels du sol ; ▸potentiel écologique du paysage) ; *syn.* efficacité [f] écologique (DUV 1984, 87), efficacité [f] d'un écosystème, capacité [f] d'accueil du milieu naturel ; *g* **Naturpotenzial [n]** (**1.** Leistungs- und Funktionsvermögen des Naturhaushaltes hinsichtlich der Stoffe, Strukturen und Abhängigkeiten auf bestimmte Nutzungsansprüche bezogen; **2.** Vermögen eines Naturhaushaltes auf Grund des komplexen Wirkungsgefüges seiner Faktoren naturgegebene physikalische, chemische und biologische Prozesse ablaufen zu lassen; die Leistungs- und Funktionsfähigkeit des Naturhaushaltes zu erhalten, gehört zu den im Bundesnaturschutzgesetz formulierten Zielen des Naturschutzes und der Landschaftspflege; ▸Landschaftspotenzial); *syn.* Leistungs- und Funktionsfähigkeit [f] des Naturhaushaltes, *o. V.* Naturpotential [n].

3690 natural process [n] *ecol.* (Progressive continuation or development of natural phenomena marked by a series of gradual changes); *s* **proceso [m] natural** (Desarrollo de un proceso natural caracterizado por una serie de cambios graduales); *f* **processus [m] naturel** (Évolution, développement d'un phénomène naturel caractérisé par des changements incessants ou une suite continuelle de phases) ; *g* **Ablauf [m] des Naturgeschehens** (Verlauf eines natürlichen Ereignisses, der durch bestimmte Phasen des Entstehens und der Veränderung gekennzeichnet ist).

3691 natural range [n] *phyt. zool.* (Geographical limits within which an organism occurs naturally); *s* **área [f] natural de distribución** (Territorio colonizado naturalmente por una especie o una comunidad); *f* **aire [f] naturelle d'une espèce** (Territoire colonisé naturellement par une espèce) ; *g* **natürliches Areal [n]** (Das von einer Tier- oder Pflanzenart [in historischer Zeit, ggf. rezent] auf natürliche Weise besiedelte Gebiet); *syn.* natürliches Verbreitungsgebiet [n].

3692 natural regeneration [n] (1) *for. phyt.* (Spontaneous establishment of vegetation by self-sown seed or by vegetative means—regrowth—from an existing plant community; ▸barochory, ▸seed rain); *syn.* spontaneous regeneration [n], natural reproduction [n]; *s* **regeneración [f] natural** (Reproducción de un bosque por vía natural. La **r. n.** tiene como consecuencia una mayor estabilidad ecológica debido a la diversidad genética resultante de la selección natural. Según expertos, este tipo de bosque es más resistente ante tormentas, sequía, parásitos e incendios forestales, pero también produce menos madera y de peor calidad. Desde los años 1990 la **r. n.** es, junto a la combinación de especies arbóreas adecuadas, parte integrante de conceptos de calidad de bosques de organizaciones internacionales orientadas a la silvicultura sostenible, como el FSC [Forestry Stewardship Council] y el sistema de certificación europeo para la silicultura sostenible PEPC [Certificación Paneuropea de Bosques]; ▸barocoria, ▸brinzal natural, ▸nuevo vuelo); *syn.* regeneración [f] espontánea, rejuvenecimiento [m] natural/espontáneo; *f* **régénération [f] naturelle** (Le renouvellement naturel d'un peuplement végétal par voie de semences ou renouvellement artificiel : le peuplement provenant de la régénération naturelle est dit « peuplement de régénération naturelle » ; ▸barochorie, ▸flux de semences, ▸recolonisation barochore) ; *syn.* repeuplement [m] naturel, régénération [f] spontanée ; *g* **Naturverjüngung [f]** (1) (Begründung eines Bestandes durch Selbstansamung oder vegetative Vermehrung von einem Altbestand aus. Die **N.** führt zu einer größeren ökologischen Stabilität dank der in natürlicher Selektion entstandenen genetischen Vielfalt; Fachleute gehen von einer geringeren Anfälligkeit solcher Wälder gegen Sturm, Dürre, Schädlinge oder Waldbrände aus, aber auch von einer deutlich geringeren und minderwertigen Holzproduktion. Der Vorrang für **N.** ist seit den 1990er-Jahren ebenso wie die Mischung geeigneter Waldbaumarten wichtiger Bestandteil internationaler, auf ökologische Nachhaltigkeit aus-

gerichteter Waldqualitätskonzepte wie des internationalen FSC [Forestry Stewardship Council] und des europäischen PEFC [**P**aneuropäische **F**orst**c**ertifizierung] Zertifizierungssystems für nachhaltige Waldbewirtschaftung]; ▸Anflug 2, ▸Aufschlag); *syn.* natürliche Verjüngung [f], *z. T.* Selbstansamung [f], *z. T.* Selbstaussaat [f].

3693 natural regeneration [n] (2) *for. phyt.* (Young tree crop or other plants established by self-sown seed or vegetative reproduction; ▸natural colonization by seed rain); *syn.* natural reproduction [n], volunteer growth [n] [also US]; *s* **bosquecillo [m] de renuevo** (Resultado de la reproducción vegetativa o la ▸colonización natural de «espacio vacío» por nuevo vuelo); *f* **peuplement [m] de régénération naturelle** (Jeune peuplement issu d'un renouvellement naturel par voie de semences ou d'un renouvellement artificiel ; ▸semis naturel) ; *g* **Naturverjüngung [f]** (2) (*Ergebnis* junger Bestand, der durch Selbstansamung oder vegetative Vermehrung entstanden ist; ▸Anflug 1); *syn.* natürliche Verjüngung [f].

natural reproduction [n] *for. phyt.* ▸natural regeneration (1), ▸natural regeneration (2).

natural research area [n] [US] *conserv.* ▸natural area laboratory.

natural resource characteristics [npl] *conserv. landsc. recr.* ▸natural site characteristics.

3694 natural resources [npl] (1) *ecol. plan.* (Material or organisms in the natural environment, available for/used by the human race. The term encompasses renewable resources, such as plants, water, air, wildlife, soils, etc., and non-renewable resources, such as coal, oil, natural gas, and mineral ores; ▸natural environment, ▸natural heritage, ▸raw material 1); *s* **recursos [mpl] naturales** (Conjunto de materiales y organismos del medio ambiente natural que están a disposición de la humanidad como el agua, el aire, tierras para bosques, prados y cultivos, los animales y las plantas silvestres, recursos minerales, etc.; ▸materias primas, ▸medio natural, ▸patrimonio natural); *f* **ressources [fpl] fondamentales naturelles** (Matériaux, éléments et organismes de la nature utilisés par l'homme tels : les sols agricoles et forestiers, l'eau, l'air, les matières premières minérales ainsi que la flore et la faune ; ▸matières premières, ▸milieu récepteur, ▸patrimoine naturel) ; *g 1* **natürliche Grundgüter [npl]** (**1.** Für die allgemeine Nutzung der in der Natur vorhandenen Stoffe und Organismen durch den Menschen. Man unterscheidet erneuerbare **n. G.** wie Wasser, Luft, land- und forstwirtschaftlich nutzbarer Boden, Pflanzen und Tiere sowie die erneuerbaren Energiequellen und nicht erneuerbare **n. G.** wie fossile Brennstoffe und mineralische Rohstoffe. **2.** Bestandteile des Naturhaushaltes, die als wertvoll betrachtet werden und deshalb im Sinne eines Schutzgutes durch das Umwelt- resp. Naturschutzrecht geschützt werden; in Artikel 20a des Grundgesetzes wurde der Schutz der „natürlichen Lebensgrundlagen" zum Staatsziel erklärt; § 2 BNatSchG nennt die Naturgüter Boden, Gewässer, Luft, Klima, Vegetation, wild lebende Tier- und Pflanzenarten sowie historische Kulturlandschaften und deren Teile, die dem Naturschutz und der Landschaftspflege unterliegen; *syn., aber sprachlich unschön und vermeidbar* natürliche Hilfsquellen; ▸Naturerbe, ▸natürliche Umwelt, ▸Rohstoff); *syn.* Naturgut [n], natürliche Grundlagen [fpl], natürliche Lebensgrundlagen [fpl], natürliche Ressourcen [fpl]; *g 2* **biologische Ressourcen [fpl]** (Genetische Ressourcen, Organismen oder deren Teile, Populationen oder andere biotische Bestandteile von Ökosystemen, die einen tatsächlichen oder potenziellen Wert für den Menschen haben; cf. BFN 2006).

natural resources [npl] (2) *ecol. nat'res. sociol.* ▸consumption of natural resources; *pol. nat'res.* ▸preservation of natural resources; *ecol. land'man.* ►utilization of natural resources.

natural resources information system [n] *adm. geo. landsc. plan.* ►landscape information system.

3695 natural resources management [n] *nat'res.* (Planning and measures to accomplish the integrated utilization of ►natural resources 1, especially air, water, soil, minerals, forest, fish and wildlife for the greatest long-term benefit of society; ►sustainable development); *s* **gestión [f] de los recursos naturales** (Conjunto de medidas para el aprovechamiento sostenible de los recursos naturales renovables; ►desarrollo sostenible, ►recursos naturales); *syn.* planificación [f] de los recursos naturales; *f* **gestion [f] des ressources naturelles** (Sauvegarde et développement de la capacité productrice du milieu naturel ainsi que de l'aptitude à l'utilisation des ►ressources fondamentales naturelles ; ►développement durable) ; *syn.* aménagement [m] des ressources naturelles ; *g* **Bewirtschaftung [f] der Naturgüter** (Planung und Summe der Maßnahmen, die zur Erhaltung und Entwicklung der Funktionsfähigkeit des Naturhaushalts, der Sicherung und Regeneration erneuerbarer und der nachhaltigen Nutzung und Nutzungsfähigkeit nicht erneuerbarer Ressourcen dienen; ►nachhaltige Entwicklung, ►natürliche Grundgüter); *syn.* Ressourcenmanagement [n], Ressourcenplanung [f] (±).

3696 natural rise [n] **of groundwater level** *hydr.* (*opp.* ►natural groundwater recession]; *s* **crecimiento [m] natural del nivel freático** *opp.* ►descenso natural del agua subterránea; *syn.* ascensión [f] natural del nivel freático; *f* **remontée [f] de la nappe phréatique** (Mouvement ascendant naturel du niveau de la nappe phréatique ; *opp.* ►abaissement naturel de la nappe phréatique) ; *g* **Grundwasseranstieg [m]** (Natürliches Ansteigen der Grundwasseroberfläche; *opp.* ►Grundwasserabsinken).

3697 natural river engineering measures [npl] *conserv. landsc. wat'man.* (Bioengineering methods pertaining to rivers or streams with special regard to their ecological context; ►restoration 1); *syn.* bioengineering [n] of rivers and streams [also US], restoration [n] of watercourses; *s* **renaturalización [f] de cursos de agua** (Medidas de ingeniería hidráulica que tienen en especial consideración los aspectos ecológicos entre los cursos de agua y el suelo, la vegetación y la fauna. Generalmente se trata la restauración del curso ondulante original y de la plantación de las márgenes con vegetación autóctona de los ríos antiguamente canalizados; ►renaturalización); *syn.* renaturalización [f] de ríos y arroyos; *f* **renaturation [f] des cours d'eau** (Action paysagère de reconstitution d'ensemble et de revitalisation d'un cours d'eau, en particulier par l'utilisation de méthodes du génie biologique afin de procéder à un retour à l'état naturel [ou proche de la naturalité], tels que la reconnection de bras morts, le reméandrage de cours d'eau rectifiés, etc. ; travaux concernant les aménagements hydrauliques traditionnels, les cours d'eau à l'abandon ; à ne pas confondre avec la **restauration d'un cours d'eau** qui est une intervention visant à retrouver un état de référence initial généralement lié à des objectifs d'usage particuliers [restauration d'un paysage, d'une capacité d'écoulement « vieux fonds — vieux bords », etc.]. La restauration est souvent motivée par l'absence prolongée d'entretien d'un milieu dont le fonctionnement est donc « altéré » au regard de l'état antérieur régulièrement entretenu pour tel ou tel objectif d'usage [cas classique des rivières aménagées au fil de l'eau par divers seuils ou moulins progressivement abandonnés] ; cf. glossaire www. eau-adour-garonne.fr ; ►renaturation) ; *syn.*, renaturalisation [f] des cours d'eau [CDN], réhabilitation [f] paysagère des cours d'eau [CDN], restauration [f] en l'état naturel d'un cours d'eau ; *g* **Gewässerrenaturierung [f]** (Ausbau von Fließgewässern unter besonderer Berücksichtigung der ökologischen Zusammenhänge [Wasser, Boden, Vegetation, Fauna]. Meist ist der Rückbau von

technisch ausgebauten Fließgewässern gemeint; ►Renaturierung); *syn.* naturnaher Gewässerausbau [m, o. Pl.].

natural scenery [n] *landsc. plan. recr.* ►visual quality of landscape.

3698 natural seeding [n] **of conifers** *for.* (Spontaneous growth of conifers caused by ►seed rain or ►animal dispersal of seeds); *s* **siembra [m] natural de coníferas** (Crecimiento espontáneo de coníferas causado por ►nuevo vuelo, ►zoocoria o por la proximidad de árboles portagrano); *f* **ensemencement [m] naturel des conifères** (Régénération d'un peuplement de conifères par dissémination naturelle des graines soit par le vent, les oiseaux, les mammifères, la gravité ou l'eau de ruissellement à partir d'arbres proches d'une surface à régénérer ou des arbres semenciers dispersés sur cette surface ; ►flux de semences) ; *g* **natürliche Versamung [f] von Koniferen** (Durch ►Anflug 2, Verschleppung von Samen oder durch in der Nähe stehende Samenbäume entstandener Nadelholzbestand; ►Verschleppung durch Tiere); *syn.* natürliche Bestockung [f] mit Koniferen.

3699 natural silviculture [n] *for.* (Type of silviculture intended to achieve ►ecological equilibrium in ►forest biocoenoses and sustainability of ►forest functions. Opposite sense 'man-made silviculture' as the plantation of trees for economic forestry, known as 'orchard silviculture'; ►forestry 1); *s* **silvicultura [f] natural** (Está basada en el sistema de monte alto de selección por pies o grupos de árboles, y da al bosque una estructura que trata de imitar la de los bosques vírgenes. La flora herbácea y arbustiva es diversa. Cuando la regeneración se efectúa de forma natural, se contribuye a mantener la competencia interespecífica e intraespecífica, que es favorable para la conservación de líneas dendrológicas de excelente calidad genética. La **s. n.** es perfectamente compatible con la conservación de la ►biocenosis forestal; cf. DINA 1987; ►función del bosque, ►silvicultura artificial); *syn.* selvicultura [f] natural; *f* **sylviculture [f] écologique** (Forme de sylviculture utilisant les techniques favorisant l'équilibre dynamique des ►biocénoses forestières et un développement durable des différentes ►fonctions de la forêt, en opposition à un régime sylvicole favorisant la création d'une forêt artificielle ; ►sylviculture) ; *syn.* éco(-)sylviculture [f] ; *g* **naturnaher Waldbau [m]** (Form des Waldbaues mit dem Ziel ein dynamisches Gleichgewicht der ►Waldbiozönosen und eine Nachhaltigkeit der ►Waldfunktionen zu erreichen. Das Gegenteil wäre in diesem Sinne der ‚künstliche Waldbau' als industrielle Holzplantage; ►Forstwirtschaft).

3700 natural site characteristics [npl] *conserv. landsc. recr.* (Natural elements of a landscape or parts thereof which were not altered by human activities or which may not be changed remarkably in the future); *syn.* natural resource characteristics [npl], natural landscape features [npl], landscape resources [npl]; *s* **componentes [mpl] paisajísticos naturales** (Todos los elementos naturales de un paisaje o parte de él que [casi] no han sido alterados por el hombre); *f* **composantes [fpl] paysagères** (Éléments naturels constitutifs d'un paysage peu soumis à l'influence de l'homme et qui le resteront à long terme) ; *g* **natürliche Ausstattung [f]** (Alle natürlichen Elemente einer Landschaft oder deren Teile, die vom Menschen kaum verändert wurden oder langzeitig kaum wirksam verändert werden); *syn.* natürliche Gegebenheiten [fpl] (einer Landschaft).

3701 natural slope [n] (1) *geo. constr.* (**1.** An inclined plane, such as a sloping hillside, caused by geomorphological processes, e.g. soil erosion and uplift. **2.** The inclination of a pile of granular material as determined by the angle of repose, i.e. the angle of a stable slope according to the friction, cohesion and shape of the particles; the maximum angle of slope is a coefficient of the internal friction of granular materials; in cohesive soils, the angle of a steep slope is more stable than that of less cohesive material. Terminology applied in earthwork construction such as the filling of an earth embankment); *s* **talud [f] natural** (Pendiente lateral de un terraplén no estabilizado; el grado de inclinación corresponde al ángulo de rozamiento del material utilizado; por su capacidad de adhesión en suelos cohesivos la pendiente puede tener mayor inclinación que en suelos sueltos); *f* **talus [m] naturel** (**1.** Plan incliné stable formé naturellement par gravitation par un matériau pulvérulent mis en tas [sable, graviers, blocs rocheux] ou en géomorphologie versant formé par un matériau naturel stable [p. ex. éboulis] ; l'angle de talus naturel formé constitue une caractéristique de la nature [forces de frottement] et de la géométrie [cohésion] des particules du matériau. **2.** Le terme **talus** est en général utilisé pour décrire un plan incliné de nature anthropogène, en géomorphologie pour désigner un élément de relief, versant structurel raide, p. ex. talus d'éboulis, talus intercontinental) ; *g* **natürliche Böschung [f]** (**1.** Eine durch geomorphologische Vorgänge wie Erosion und Bodenhebung entstandene geneigte Seite eines kleinen Hanges oder durch Schüttung entstandene unbefestigte Böschung eines Erdkörpers; die Neigung entspricht dem Winkel der inneren Reibung [Reibungswinkel] des Bodenmaterials; bei bindigen Böden hält die Böschung wegen der zusätzlich wirkenden Kohäsion in einem steileren Winkel als bei rolligem Material. **2.** Unter **Böschung** wird im Deutschen eine meist anthropogen entstandene, schräg abfallende Seitenfläche eines Erdbauwerkes und in der Geomorphologie ein kleiner, kurzer Hang verstanden).

3702 natural slope [n] (2) *geo. landsc.* (Natural ground incline; ►gradient 1, ►natural ground surface); *s* **declive [m] natural** (Grado de inclinación de la ►superficie natural del terreno; ►pendiente en subida); *syn.* pendiente [f] natural; *f* **pente [f] naturelle** (Inclinaison du versant d'un ►terrain naturel considérée dans le sens de la montée ou de la descente, mesurée comme différence de hauteur par rapport à une ligne horizontale de référence, p. ex. 1 m pour 100 m = 1 % ou respectivement 1 :100 ; à ne pas confondre avec le dénivelé qui est la différence de hauteur entre deux points évaluée en mètres ; ►déclivité en rampe) ; *syn.* déclivité [f] (d'un terrain) ; *g* **natürliches Gefälle [n]** (Grad der Neigung einer durch den Menschen nicht veränderten Geländeoberfläche, gemessen als Höhenunterschied auf einer horizontalen Referenzstrecke, z. B. 1 m auf 100 m = 1 % Gefälle resp. 1:100; ►Gelände 1, ►Steigung).

3703 natural soil fertility [n] *ecol. pedol.* (Innate productive capacity of natural soil to support plant growth; ►organic farming, ►soil fertility); *s* **fertilidad [f] natural del suelo** (Estado del suelo en condiciones naturales que en las zonas templadas en estado natural estaría cubierto de bosque. El ►cultivo alternativo favorece la **f. n. del s.** gracias a medidas adecuadas de cuidado del suelo, promoción del humus y renuncia a utilizar fertilizantes artificiales y pesticidas químicos; ►fertilidad del suelo); *f* **fertilité [f] naturelle du sol** (État du sol sous des conditions naturelles ; celles-ci se rencontrent le plus souvent sous une couverture forestière conforme aux conditions stationnelles. C'est là que se développe le type d'humus naturel représentant le critère important de détermination de la ►fertilité du sol naturelle ; ►culture biologique) ; *g* **natürliche Bodenfruchtbarkeit [f]** (Im Sinne des § 2 [1] Nr. 3 u. 4 BNatSchG derjenige edaphische Zustand, der sich unter natürlichen Bedingungen einstellt. Die am ehesten natürlichen oder naturnahen Bedingungen ergeben sich im Boden unter natürlicher [standortgerechter] Waldbedeckung. Hier stellt sich die ‚natürliche Humusform' ein, die eine wesentliche Kenngröße der **n.n B.** darstellt; BML 1981; der ►alternative Landbau fördert durch geeignete Bodenpflegemaßnahmen, geordnete Humuswirtschaft und den weitestge-

henden Verzicht auf synthetische Düngemittel und Agrarchemikalien die **n. B.**; ►Bodenfruchtbarkeit).

3704 natural stone [n] *constr.* (Stone originating from natural deposits used in construction; ►cut stone #2; *opp.* artificial stone; ►queen closer of natural stone); *s* **piedra [f] natural** (Piedra procedente de cantera que se utiliza en construcción paisajística o en mampostería; ►sillar; *opp.* piedra artificial; ►pichulín de piedra natural); *f* **pierre [f] naturelle** (Pierre provenant d'un gisement naturel et utilisée comme matériau ; ►pierre de taille ; *opp.* pierre artificielle ; ►barrette en pierre naturelle) ; *g* **Naturstein [m]** (Aus natürlichem Vorkommen gewonnener Stein, der technisch verwendet wird; ►Naturwerkstein; *opp.* Kunststein; ►Natursteinriemchen).

3705 natural stone deposit [n] *geo. min.* (Place where natural stone occurs); *s* **yacimiento [m] de recursos petrográficos** (Lugar en el que se presenta piedra natural que puede ser explotada); *f* **gisement [m] de pierres naturelles** (Lieu sur lequel la pierre naturelle est ou peut être exploitée) ; *g* **Natursteinvorkommen [n]** (Geografischer Ort, an dem Naturstein ansteht).

3706 natural stone flag [n] *constr.* (Slab for paving purposes or ►stepping stone cut from natural stone for ►sidewalk [US]/pavement [UK] and garden paths and terraces; ►natural garden stone flag); *syn.* precut paving stone [n], flagstone [n], flag [n] [UK, north], *pl. also* flags [US]; *s* **losa [f] de piedra natural** (Placa de piedra natural cortada para pavimentar ►aceras y caminos; ►losa de jardín [de piedra natural o artificial], ►losa para pasos); *f* **dalle [f] en pierre naturelle pour voirie de surface** (Élément non normalisé en pierre naturelle utilisé pour les ►trottoirs, les terrasses et les circulations dans un jardin [►dalle de pas japonais] ; ►dalle de jardin) ; *syn.* dalle [f] en pierre naturelle pour sols extérieurs, dalle [f] en/de pierre naturelle de jardin, dalle [f] en pierre naturelle pour sentier ; *g* **Natursteingehwegplatte [f]** (Natursteinplatte für ►Bürgersteige, Gehwege und Terrassenbeläge, z. B. aus Granit, Kalkstein, Porphyr, Sandstein; ►Gartenplatte, ►Schrittplatte); *syn.* Natursteinplatte [f] für Gartenwege.

3707 natural stone masonry [n] *arch. constr.* (Wall constructed of natural stones with or without mortar; ►dry masonry, ►rubble work, ►rusticated masonry, ►squared rubble masonry; *syn.* natural stone wall [n]; *s* **mampostería [f] de piedras naturales** (Muro construido de piedras naturales, con o sin mortero; ►fábrica de sillarejo, ►mampostería en seco, ►mampostería ordinaria de piedras sin labra); *syn.* sillería [f] de mampuestos, muro [m] de piedra natural, fábrica [f] de mampostería de piedra natural, obra [f] de fábrica de piedra natural; *f* **maçonnerie [f] en pierres naturelles** (Ouvrage réalisé avec des éléments de pierre naturels posés à sec ou sur un lit de mortier ; ►maçonnerie à sec, ►maçonnerie en pierres bossagées, ►maçonnerie en moellons bruts, ►maçonnerie en moellons pleins, grossièrement équarris) ; *g* **Natursteinmauerwerk [n]** (Mauer aus natürlichen Steinen, mit oder ohne Mörtel errichtet; ►Bosenwerk, ►Bruchsteinmauerwerk, ►Bruchsteinmauerwerk aus unbehauenen Steinen, ►Trockenmauerwerk); *syn.* Natursteinmauer [f].

natural stone pavement [n] [US] *constr.* ►natural stone paving.

3708 natural stone paving [n] *constr.* (paved natural stone surface); *syn.* natural stone pavement [n] [also US]; *s* **adoquinado [m] de piedras naturales**; *f* **pavage [m] en pierre naturelle** ; *g* **Natursteinpflaster [n]** (Belag aus Natursteinpflastersteinen).

natural stone sett [n] [UK] *constr.* ►cobblestone [US].

3709 natural stone slab [n] *constr.* (Tooled or untooled, flat construction unit of natural stone used for paving, wall copings or veneering); *s* **losa [f] de piedra (natural)** (Pieza plana de piedra natural utilizada para pavimentos, cobertura de muros o paramentos); *f* **dalle [f] en pierre naturelle** (Élément de pierre naturelle brut ou de taille utilisé pour le dallage, couronnement ou habillage de mur) ; *g* **Natursteinplatte [f]** (Behauenes oder unbehauenes, flaches Bauteil aus Naturstein für Beläge, Mauerabdeckungen oder Wandverkleidungen).

3710 natural stone steps [npl] *constr. syn.* flight [n] of natural stone steps; *s* **escalera [f] rústica en piedra** *syn.* escalera [f] de piedra natural; *f* **escalier [m] rustique** ; *g* **Natursteintreppe [f]**.

natural stone wall [n] *arch. constr.* ►natural stone masonry.

3711 natural systems [npl] *ecol.* (Organization characterized by interactions and interrelationships of living organisms and inanimate natural objects, which have a reciprocal effect upon each other and their environment in terms of energy, matter, and information; ►ecological interactions and interrelationship, ►landscape ecosystem); *s* **régimen [m] ecológico de la naturaleza** (Estructura de las relaciones e interacciones de los organismos vivos y los elementos abióticos de la naturaleza entre sí y de sus influencias recíprocas, que son de carácter energético, material e informativo; ►sistema de interrelaciones e interacciones ecológicas, ►régimen ecológico del paisaje); *f* **système [m] naturel** (Ensemble interactif des organismes vivants et des éléments naturels inertes entretenant un équilibre biologique [énergie, matière et information] avec leur environnement ; ►écosystème paysager, ►tissus d'interactions et de rapports interspécifique écologiques) ; *syn.* écosystème [m] naturel ; *g* **Naturhaushalt [m]** (Wirkungsgefüge aus Lebewesen und unbelebten natürlichen Bestandteilen, die untereinander und mit ihrer Umwelt in energetischen, stofflichen und informatorischen Wechselwirkungen stehen; der Begriff ►Landschaftshaushalt umfasst im Vergleich dazu sowohl die natürlichen Faktoren als auch die anthropogenen Bestandteile; ►ökologisches Gefüge).

natural systems [npl]**, effective functioning of** *ecol.* ►natural potential.

3712 natural terrace [n] **on a slope** *geo.* (Natural shelf on an incline; ►cultivated terrace, ►ledge 2); *s* **terraza [f] en ladera** (Rellano ancho con inclinación horizontal o menor que la de la ladera; ►berma, ►campo en bancal); *f* **replat [m]** (Large étendue horizontale ou de pente plus faible que l'inclinaison générale d'un versant ; ►risberme, ►terrasse de culture) ; *g* **Hangterrasse [f]** (Breiter Absatz in einem Hang, der horizontal oder flacher geneigt ist als der Gesamthang; ►Berme 1, ►Terrassenfeld); *syn.* Hangstufe [f].

natural terrain [n] *constr. plan.* ►natural ground surface.

3713 natural waterbody [n] *geo. landsc.* *s* **aguas [fpl] continentales en estado natural** (*Término específico* lago o río en estado natural); *f* **eaux [fpl] continentales naturelles** ; *g* **natürliches Gewässer [n]**.

3714 nature [n] *ecol.* (In ecology, the entire sphere of animate and inanimate manifestations which have not been created by man; in botany, the genetic make-up of an organism. Interaction of the 'nature' and 'nurture' of an organism affects the characteristics or traits of an individual; ►experiencing nature); *s* **naturaleza [f]** (Conjunto de factores bióticos y abióticos no creados por el ser humano; ►vivencia de la naturaleza); *f* **nature [f]** (Ensemble des éléments biotiques et abiotiques [réalité physique] existant sans l'intervention de l'homme ; ►perception de la nature) ; *g 1* **Natur [f, o. Pl.]** (1. Vielfältig definierter Begriff; im Naturschutz wird darunter i. d. R. die Gesamtheit der nicht vom Menschen erschaffenen, belebten und unbelebten Erscheinungen; cf. ANL 1984; ►Naturerlebnis); *g 2* **unberührte Natur [f]** (Vom Menschen nicht oder kaum beeinflusste Natur); *syn.* ursprüngliche Natur [f].

nature [n]**, countryside and** *conserv. landsc.* ►natural landscape.

Nature Conservancy Council [n] [UK] *conserv. leg.* ►The Nature Conservancy [US].

3715 nature conservation [n] *conserv. leg.* (Comprehensive measures for the protection and perpetuation of species of wild fauna and flora, their natural communities and fundamentals of life as well as for the safeguarding of landscapes and their component parts under natural conditions; ►contract-based nature conservation [UK], ►animal species conservation, ►contract-based nature conservation [UK], ►environmental protection, ►integrated nature conservation, ►related to nature conservation, ►species conservation, ►valuable for nature conservation, ►wildlife conservation); *s* **protección [f] de la naturaleza** (Medidas tomadas para impedir que las intervenciones humanas causen daños en los elementos bióticos y abióticos del medio ambiente; DINA 1987; ►custodia del territorio, ►digno/a de protección, ►protección ambiental, ►protección de especies [de flora y fauna], ►protección de especies de fauna en peligro, ►protección de la fauna salvaje, ►protección integral de la naturaleza, ►relacionado/a con la protección de la naturaleza); *syn.* conservación [f] de la naturaleza, preservación [f] de la naturaleza; *f 1* **protection [f] de la nature** (Ensemble de mesures requises pour préserver, maintenir ou rétablir les espaces, ressources et milieux naturels, les sites et les paysages, les populations d'espèces de faune et de flore sauvage, la diversité et les équilibres biologiques dans un état favorable caractérisé par la totalité des influences pouvant affecter à long terme leur répartition naturelle, leur structure et leurs fonctions ainsi que la survie à long terme des espèces typiques sur la base des principes de précaution, d'action préventive et de correction, du pollueur-payeur et de participations des citoyens ; loi 95-101 du 2 février 1995 relative au renforcement de la protection de l'environnement et cf. dir. n° 92-43 CEE du 21 mai 1992 concernant la conservation des habitats naturels ainsi que de la faune et de la flore sauvages ; ►digne d'être protégé, ée, ►gestion contractuelle de la nature, ►protection de la faune sauvage, ►protection de l'environnement, ►protection des espèces animales, ►protection des espèces [animales et végétales], ►protection des espèces de faune et de flore sauvage, ►protection intégrée de la nature, ►relatif, ve à la protection de la nature ; à ne pas confondre avec la « conservation intégrée » qui consiste à traiter les constructions et les ensembles anciens) ; *f 2* **préservation [f] de la nature** (Maintien dans les conditions actuelles des territoires n'ayant pas encore subis l'influence de l'homme ; *terminologie* les termes préservation et protection sont souvent confondus et utilisés pour désigner un même concept alors que des différences existent) ; *g* **Naturschutz [m, o. Pl.]** (1. Gesamtheit der Maßnahmen zur Erhaltung und Förderung von Pflanzen und Tieren wild lebender Arten, ihrer Lebensgemeinschaften und natürlichen Lebensgrundlagen sowie zur Sicherung von Landschaften oder Landschaftsteilen unter natürlichen Bedingungen; ANL 1984; die Belange des **N.es** werden auf kommunaler Ebene im Rahmen der Flächennutzungsplanung durch das Instrument des Landschaftsplanes und bei Bebauungsplänen i. d. R. mit dem ►Grünordnungsplan vertreten und festgesetzt. Den **N.** regelt das ►Bundesnaturschutzgesetz als Rahmengesetz mit z. T. sehr detaillierten Vorgaben, das durch den Landesgesetzgeber weiter ausgefüllt werden muss. **2.** Moderner **N.** versteht sich als Teil einer nachhaltigen Entwicklung; er ist mehr als bloßer Arten- und Biotopschutz und umfasst den Schutz aller Naturgüter, die landschaftliche Vielfalt, Eigenart und Schönheit der Kulturland-

N

schaften sowie die Schaffung der Vorausetzung für das Natur-erleben des Menschen. **3**. Im allgemeinen Sprachgebrauch auch Kurzbezeichnung für ‚Naturschutz und Landschaftspflege‘. Schon Jahrzehnte vor der Einführung des Begriffes '**N.**' im Jahre 1888 und vor der Institutionalisierung der Naturschutzbewegung im Jahre 1904 [Gründung des Bund Heimatschutzes] gab es vielfältige Aktivitäten eines gezielten ▶Artenschutzes: Bereits um 1850 wurde im Deutschen das Wort „Artenschutz" gebraucht; cf. N+L 2002 [7], 321; **N.** ist eine kulturelle Dimension, d. h. das Ergebnis gesellschaftlicher Prozesse, das z. B. in der Gesetzge-bung durch die Fortschreibung der Naturschutz- und anderer Planungsgesetze seinen Niederschlag findet. Seit der Konferenz für Umwelt und Entwicklung in Rio de Janeiro 1992 wird auch die Geschlechtergerechtigkeit [Gender Mainstream] im Bereich des Naturschutzes gefordert und seit dem Anfang dieses Jhs. beginnen Bestrebungen, geschlechtersensible und geschlechter-gerechte Perspektiven in die **N.politik, -verwaltung, -forschung** und **-praxis** zu integrieren; ▶Faunenschutz, ▶integrierter Naturschutz, ▶naturschutzfachlich, ▶naturschutzwürdig, ▶Tier-artenschutz, ▶Vertragsnaturschutz, ▶Umweltschutz).

nature conservation [n]**, legislation upon** *conserv. leg.*
▶nature conservation enactments.

nature conservation [n] **and landscape manage-ment** [n]**, prerequisites of** *conserv. nat'res.* ▶considera-tions of nature conservation and landscape management.

nature conservation body [n] **[UK]** *adm. conserv.*
▶conservation/recreation agency [US].

3716 nature conservation enactment [n] *conserv. leg.*
(Law which came into force to protect the natural environment and which will be continuously updated according to the con-sciousness of society; **in U.S.**, there is not one single federal law but a large number of federal and state nature conservation enactments such as the *National Park Service Act [1916], Migra-tory Bird Conservation Act [1929], Fish and Wildlife Act [1956], Wilderness Act [1964], Wild and Scenic Rivers Act [1968]*, ▶Endangered Species Act [E.S.A.] [US] [1973], National Forest Management Act [1976], *Fish and Wildlife Conservation Act [1980]*, ▶*National Environmental Policy Act* as amended [NEPA], *Clean Air Act [1955], Federal Water Pollution Control Act [Clean Water Act, 1972], Estuaries and Clean Waters Act [2000]*, and similar state laws which govern nature conservation; most states, as sovereign units of government, have significant legislation either implementing or supplementing federal laws; **in U.K.**, there are several nature conservation acts, such as *The Protection of Animals [Scotland] Act 1912, The National Parks and Access to the Countryside Act 1949, The Countryside Act 1968, The* ▶*Conservation of Wild Creatures and Wild Plants Act 1975 [amended 1978 and 1982], The Endangered Species [Im-port and Export] Act 1976, Wild* ▶*Birds Directive 79/409/EEC, The Wildlife and Countryside Act 1981 [amended by the Coun-tryside and Rights of Way (CRoW) Act 2000 and the Nature Conservation (Scotland) Act 2004]*, and Environmental Protec-tion Act [1990]; **in D.**, there is a basic federal frame law *[Bundes-naturschutzgesetz]* which is more specific in state laws); *s 1* **Ley [f] del Patrimonio Natural y de la Biodiversidad [Es]** (**1**. La ley 42/2007, de 13 de diciembre, viene a sustituir y modernizar la ley 4/1989 de conservación los espacios naturales y la flora y fauna silvestres. Esta ley establece el régimen jurídico básico de la conservación, uso sostenible, mejora y restauración del patrimorio natural y la biodiversidad española, como parte del deber de conservar y del objetivo de garantizar los derechos de las personas a un medio ambiente adecuado para su bienestar, salud y desarrollo. Igualmente recoge normas y recomendaciones internacionales, del Consejo de Europa y de las Naciones Unidas,

así como aquéllas dictadas por la Comisión de las Comunidades Europeas y dispone la creación de mecanismos de coordinación y cooperación entre la Administración General del Estado y las Comunidades autónomas como la Comisión Estatal para el Patri-monio Natural y la Biodiversidad. La ley contiene siete títulos que articulan los objetivos e instrumentos previstos: Título preliminar contiene los objetivos, principios básicos, definiciones y los instrumentos básicos; Título I Instrumentos para el conoci-miento y la planificación del patrimonio natural y de la biodiver-sidad; Título II Catalogación, conservación y restauración de hábitats y espacios del patrimonio natural; Título III Conserva-ción de la biodiversidad; Título IV Uso sostenible del patrimonio natural y de la biodiversidad; Título V Fomento del conoci-miento, la conservación y restauración del patrimonio natural y la biodiversidad; Título VI De las infracciones y sanciones; cf. Ley 42/2007. **2. En Es**, esta ley tiene carácter de legislación básica del Estado sobre protección del medio ambiente, de conformidad con la Constitución, salvo las disposiciones del art. 69 sobre comercio exterior, y la disposición adicional sexta, que constituye competencia exclusiva del Estado en materia de relaciones internacionales. Por lo demás la competencia en este ámbito recae en las CC.AA., por lo cual le corresponde el desarrollo legislativo y la ejecución dentro de su territorio de la legislación básica del Estado; cf. Disposición final segunda, Ley 42/2007); *s 2* **ley [f] de protección de la naturaleza** (Ley aprobada con el fin de proteger la naturaleza y que se va actualizando según las exigencias de protección y el desarrollo de la conciencia de la sociedad en este respecto. **En EE.UU.** no hay una única ley federal, sino una gran cantidad de leyes de conservación de la naturaleza federales y de los estados como la Ley del Servicio Nacional de Parques *[National Park Service Act 1916]*, la Ley de Conservación de las Aves Migratorias *[Migratory Bird Conser-vation Act 1929]*, la Ley de la Fauna Piscícola y Animales Salvajes *[Fish and Wildlife Act 1956]*, la Ley de Zonas Vírgenes *[Wilderness Act 1964]*, la Ley de Protección de Ríos *[Wild and Scenic Rivers Act 1968]*, la ▶Ley de Protección de Especies *[Endangered Species Act 1973]*, la Ley de Gestión de Bosques Nacionales *[National Forest Management Act 1976]*, la Ley de Conservación de la Fauna Piscícola y Animales Salvajes *[Fish and Wildlife Conservation Act 1980]* y leyes similares a nivel de los estados que —en general— como unidades soberanas de gobierno tienen legislación significativa bien para implementar bien para complementar la legislación federal. **En GB**, hay varias leyes de conservación de la naturaleza como la Ley de Protección de Animales de Escocia *[Protection of Animals Act (Scotland) 1912]*, la Ley de Parques Nacionales y Acceso al Campo *[National Parks and Access to the Countryside Act 1949]*, la Ley del Campo *[Countryside Act 1968]*, la Ley de Protección de Especies *[Conservation of Wild Creatures and Wild Plants Act 1975,* reformada en 1978 y 1982], la Ley de Especies en Peligro (Importación y Exportación) *[Endangered Species (Import and Export) Act 1976]*, la Ley de Animales Salvajes y del Campo *[Wildlife and Countryside Act 1981]*, reformada por la Ley del Campo y Servidumbre de Paso *[Countryside and Rights of Way Act 2000]* y la Ley de Conservación de la Naturaleza de Escocia *[Nature Conservation (Scotland) Act 2004])* y la Ley de Protec-ción del Medio Ambiente *[Environmental Protection Act 1990]*. **En D.** rige la ▶Ley Federal de Conservación de la Naturaleza y de Gestión del Paisaje que constituye una ley marco que debe ser desarrollada por leyes de los estados federados); *f* **loi [f] pour la protection de la nature** (Loi promulguée en faveur de la protection de la nature et contribuant par la participation des citoyens et élever la conscience de la population en faveur de la protection des sites, des paysages et des milieux naturels ; **F.**, la ▶loi relative à la protection de la nature [loi n° 76-629 du 10 juillet 1976] porte sur les études d'impact, la protection de la

N

faune et de la flore, la protection de l'animal et les réserves naturelles ; ►code de l'environnement, ►loi relative au renforcement de la protection de l'environnement ; **D.**, ►loi fédérale relative à la protection de la nature du 20 décembre 1976 concrétisée par les lois de protection de la nature des Länder) ; *g* **Naturschutzgesetz [n]** (1. Gesetz, das zum Schutze der Natur erlassen wurde und der Entwicklung des gesellschaftlichen Bewusstseins für Naturschutz und Landschaftspflege entsprechend fortgeschrieben wird; in D. gibt es das ►Bundesnaturschutzgesetz [BNatSchG — überwiegend ein Rahmengesetz], das durch die Naturschutzgesetze der Länder weiter konkretisiert wird. Die neue **N.gebung** von Bund und Ländern hat mit ihrer ökologischen Ausrichtung etwa ab Mitte der 1970er-Jahre das gesamte fachliche Aufgabenfeld von Landschaftsentwicklung und Landschaftsbau vor allem durch die gesetzlich eingeführte ►Landschaftsplanung und ►Grünordnungsplanung sowie die Ausgleichsregelung bei Eingriffen in Natur und Landschaft [►Eingriffsregelung] wesentlich beeinflusst. Später seit den 1990er-Jahren werden die Belange von Naturschutz und Landschaftspflege weiterhin durch die per Gesetz vorgeschriebenen Umweltverträglichkeitsprüfungen für bestimmte Projekte und Maßnahmen berücksichtigt. **2. Im U.K.** regeln die Belange des Naturschutzes z. B. der *National Parks and Access to the Countryside Act, Wildlife and Countryside Act, Environmental Protection Act*).

3717 nature conservation enactments [npl] *conserv. leg.* (Generic term for federal, state laws, and local ordinances, which are implemented by regulations, and international ►conventions on nature conservation which are adopted by agreements between nations); *syn.* legislation [n] upon nature conservation; *s* **derecho [m] de protección de la naturaleza** (Todo tipo de normas legales [leyes, decretos, órdenes, instrumentos y protocolos] relacionados con la conservación de la naturaleza a nivel regional, estatal, comunitario o internacional; ►convención para la conservación de la naturaleza); *syn.* derecho [m] sobre la conservación de la naturaleza; *f* **législation [f] pour la protection de la nature** (Définie par les lois, décrets, arrêtés et circulaires nationaux, les règles et directives du Conseil de la CEE ainsi que les conventions ou traités internationaux ; ►convention [internationale] pour la protection de la nature) ; *g* **Naturschutzrecht [n, o. Pl.]** (Gesamtheit der Rechtsnormen, die den Naturschutz in Gesetzen, Verordnungen, Verwaltungsvorschriften des Bundes und der Länder sowie auf Grund von Verordnungen der Kommission der Europäischen Gemeinschaften, der Richtlinien des Rates der Europäischen Gemeinschaften und ►Naturschutzkonventionen regeln).

nature conservation [n] **on a contract basis** *agr. conserv. pol.* ►contract-based nature conservation.

3718 nature conservation ordinance [n] *conserv. leg.* (Legal enactment by a local jurisdiction for conservation of natural areas administered by a governing authority/board/commission); *s* **instrumento [m] de declaración de espacio natural protegido** (En Es. estas medidas legislativas son compentencia de las CC.AA., con la excepción de la declaración de ►Parque Nacional, que se realiza por ley de las Cortes Generales, después de previo acuerdo favorable de la asamblea legislativa de la o las CC.AA. en cuyo territorio se encuentre situado; arts. 21 y 22); *syn.* ordenanza [f] de protección de espacios naturales; *f* **ordonnance [f] relative à la protection de la nature (±)** (F., ces mesures sont généralement prises par arrêté ministériel ou préfectoral [région, département] ou par décision du conseil municipal ; **D.**, dispositions ou mesures en faveur de la protection de la nature imposées par une commune, un « Landkreis » ou l'administration chargée de la protection de la nature) ; *g* **Naturschutzverordnung [f]** (Rechtsverordnung, die seitens einer

Gemeinde, eines Landkreises oder einer Naturschutzbehörde zur Sicherung von Naturschutzobjekten erlassen wurde).

3719 nature conservation organization [n] *conserv. leg. pol.* (Generic term for private organization which is concerned with the planning, guidance, and supervision of nature conservation projects, e.g. **in U.S.**, National Audubon Society [N.A.S.], National Wildlife Federation [NWF], The Conservation Foundation [C.F.], and other national and state organizations with similar powers and objectives; in **U.K.**: **1.** Royal Society for Nature Conservation [R.S.N.C.], **2.** Royal Society for the Protection of Birds [R.S.P.B.], **3.** Council for the Protection of Rural England [C.P.R.E.], **4.** Council for Environmental Conservation [CoEnCo], **5.** Conservation Society [C.S.], etc.; ►non-profit organization suit); *s* **asociación [f] para la protección de la naturaleza** (Organización privada dedicada a la defensa de la naturaleza. En España existen organizaciones tradicionales como la Real Sociedad Española de Historia Natural, fundada en 1871 o la Sociedad Española de Ornitología [SEO] y organizaciones ecologistas modernas como la Asociación Española para la Defensa de la Naturaleza [ADENA] fundada en 1968, la Asociación Española para la Ordenación del Territorio y el Medio Ambiente [AEORMA] en 1970 [disuelta en 1975]. Posteriormente se fueron creando muchas organizaciones regionales como el Grupo d'Ornitologia Balear [GOB], Grupo Ornitológico Gallego [GOG] en 1973, la Asociación de Estudios y Protección de la Naturaleza [AEPDEN] en 1976 y muchas más fundadas en la fase de transición política y posteriormente. En 1977 se creó la Coordinadora para la Defensa de las Aves y sus Hábitats [CODA] y poco después la Federación de Amigos de la Tierra. Greenpeace España fue fundada en 1984. La red Ecologistas en Acción es una confederación de hoy en día más de 300 grupos ecologistas y forma parte del movimiento de ecologismo social; cf. DINA 1987, www.ecologistasenaccion.org; ►derecho a pleito de organizaciones ecologistas y de protección de la naturaleza); *syn.* asociación [f] naturista, asociación [f] conservacionista; *f* **association [f] de protection de l'environnement** (Association de droit privé agréée exerçant une activité pour la protection de la nature, l'amélioration du cadre de vie, la protection de l'eau, de l'air des sols, des sites et des paysages, l'urbanisme, la lutte contre les pollutions et les nuisances ; les **a. de p. de l'e.** contribuent à la défense de l'environnement soit en participant directement à l'action des organismes publics, soit en faisant valoir en justice les intérêts qu'elles ont pour but de protéger [►action en justice des associations] ; loi n° 95-101 du 2 février 1995) ; *syn.* association [f] de défense de l'environnement et des paysages, association [f] pour la protection de la nature ; *g* **Naturschutzverband [m]** (...verbände [pl]; private, parteipolitisch neutrale, nicht auf Gewinn ausgerichtete Vereinigung, die gemeinnützige Ziele verfolgt, z. B. die Förderung des Natur-, Umwelt- und Landschaftsschutzes; cf. § 29 [2] BNatSchG; **in D.** z. B. Naturschutzbund Deutschlands e.V. [NABU], Deutscher Naturschutzring e.V. [DNR], Bund für Natur- und Umweltschutz Deutschlands [BUND]. Der Satzungszweck solcher Vereine wird insbesondere durch Stellungnahmen zu Planungen und Vorhaben von Behörden und anderen Institutionen verwirklicht, die Natur, Umwelt und Landschaft beeinflussen können, sowie durch Vorschläge zu rechtlichen und organisatorischen Fragen auf den einschlägigen Gebieten; ►Verbandsklage); *syn.* Umweltverband [m].

nature-like [adj] *ecol.* ►near-natural.

3720 nature [n] **of the terrain** *geo.* (State of a land area according to its vegetation cover, incline, soil conditions, and ground relief); *syn.* character [n] of the ground; *s* **características [fpl] naturales del terreno** (Estado de una superficie o una zona en cuanto a su cobertura vegetal, las características del suelo o el relieve); *f* **nature [f] du terrain** (Conditions locales caractérisant l'état d'un terrain telles que la couverture végétale, la disposition topographique, la nature des sols) ; *syn.* caractéristiques [fpl] du terrain) ; *g* **Geländebeschaffenheit [f]** (Zustand einer Fläche oder eines Gebietes hinsichtlich der Reliefgegebenheiten, der Bodeneigenschaften und der Vegetationsbedeckung).

3721 nature protection [n] *conserv.* (Program[me] and measures in individual cases of ►nature conservation, whereby the area remains unaffected by ►intrusion upon the natural environment for a certain period of time, in order that especially endangered or rare species can develop); *s* **protección [f] íntegra de la naturaleza** (En la ►conservación de la naturaleza, método de protección para casos especiales o fases de prohibir cualquier tipo de ►intrusión en el entorno natural y evitar toda molestia antropógena, con el fin de permitir que se recuperen ciertas especies muy raras o que se encuentran en alto peligro de extinción); *syn.* protección [f] absoluta de la naturaleza; *f 1* **gestion [f] conservatoire/conservatrice du milieu naturel** (Méthode en matière de ►protection de la nature visant à soustraire un territoire délimité pendant une certaine période à toute intervention humaine susceptible de nuire au développement naturel des espèces de faune et de flore sauvages rares ou en voie de disparition) ; *syn.* conservation [f] de la nature, gestion [f] conservatoire des espaces naturels ; *f 2* **sanctuarisation [f] (du milieu naturel)** (Notion restrictive de « protection » de la nature, qui exclue l'homme et ses activités et tendrait à s'opposer à la notion d'un usage raisonné des espaces à protéger et à l'approche du développement durable qui intègre les questions écologiques, économiques, sociales et politiques. Terme à connotation négative lorsqu'il est utilisé par les « contestataires agro-sylvo-cynégétiques » dénonçant les mesures de protection fortes de conservation des espaces naturels particulièrement fragiles ou des écosystèmes exeptionnels à haute valeur écologique ; *g* **konservierender Naturschutz [m, o. Pl.]** (Methode/Verfahren des ►Naturschutzes in Einzelfällen oder für bestimmte Zeitperioden ein Gebiet mit gezielten Pflegemaßnahmen so zu erhalten, dass stark gefährdete oder seltene Arten sich entwickeln können oder ein bestimmtes Landschaftsbild erhalten bleibt); *syn.* strenger Naturschutz [m], absoluter Naturschutz [m], restriktiver Naturschutz [m].

nature protection [n] **and wildlife preservation in the western hemisphere** *conserv. pol.* ►Convention on nature protection and wildlife preservation in the western hemisphere.

3722 nature reserve [n] *conserv. land'man.* (Generic term for specific natural areas usually under the protection of a non-profit or governmental agency; e.g. ►biosphere reserve, ►bird sanctuary, ►game reserve, ►natural area preserve [US]/nature reserve [UK], ►protected spawning area, ►strict nature reserve, ►wetland of international significance. **In U.S.**, the National Park Service and equivalent state park and recreation agencies have designated "natural landmarks", essentially untouched by man, which contain unique botanical communities); *s* **reserva [f] natural (1)** (**En Es.** categoría de protección: espacio natural cuya creación tiene como finalidad la protección de ecosistemas, comunidades o elementos biológicos que por su rareza, fragilidad, importancia o singularidad merecen una valoración especial. En las **rr. nn.** estará limitada la explotación de recursos, salvo en aquellos casos en los que esta explotación sea compatible con la conservación de los valores que se pretenden proteger. Con carácter general estará prohibida la recolección de material biológico o geológico, salvo en aquellos casos que por razones de investigación, conservación o educativas se permita la misma, previa pertinente autorización administrativa; art. 31 Ley 42/2007. **En EE.UU.**, el Servicio Nacional de Parques y las insti-

tuciones equivalentes a nivel de estado pueden declarar como monumento natural espacios naturales cuasi vírgenes que albergan comunidades de plantas únicas. **En D.** término genérico que denomina todo tipo de ►espacio natural protegido 1, sea estatal o privado, como ►parques nacionales, ►refugios de fauna, ►reservas de la biosfera, ►reservas regionales de caza, ►reservas naturales [científicas], ►santuario de aves, ►zonas de freza protegidas, ►zonas húmedas de importancia internacional); *f 1* **réserve [f] naturelle nationale (RNN)** (F., parties du territoire d'une ou plusieurs communes dont la conservation de la faune, de la flore, du sol, des eaux, des gisements de minéraux et de fossiles en particulier et du milieu naturel en général présente un intérêt remarquable et qu'il convient de soustraire à toute intervention susceptible de le dégrader [C. rural, art. L. 242-1] en particulier en vue de la préservation d'espèces animales ou végétales ou d'habitats en voie de disparition, la reconstitution de populations animales ou végétales ou de leurs habitats, la préservation de biotopes et de formations géologiques, géomorphologiques ou spéléologiques remarquables, la préservation ou la constitution d'étapes sur les grandes voies de migration de la faune sauvage, la préservation des sites présentant un intérêt particulier pour l'étude de l'évolution de la vie et des premières activités humaines conformément aux dispositions de la loi du 10 juillet 1976 et du code rural [art. L. 242-1 et suivants] ; on distingue **1.** la **réserve [f] naturelle nationale « intégrale »** [RNN « intégrale »] [zone centrale d'un parc national] sur le territoire de laquelle la chasse, la pêche, l'exploitation agricole, forestière ou minière, toute action modifiant l'aspect du terrain ou de la végétation et tout acte de nature à nuire à la faune et la flore sont strictement interdits, la circulation, le camping et les recherches scientifiques n'est admise qu'avec une permission des autorités. **2.** la **réserve [f] naturelle nationale « dirigée »** [RNN « dirigée »], exclusivement réservée à la surveillance et l'étude scientifique ; ►réserve de chasse, ►réserve de chasse et de faune sauvage, ►réserve de la biosphère, ►réserve d'élevage, ►réserve nationale de chasse et de faune sauvage, ►réserve naturelle intégrale, ►réserve ornithologique, ►site classé, ►zone de protection spéciale, ►zone humide d'importance internationale) ; *f 2* **réserve [f] naturelle régionale (RNR)** (**1.** Propriété privée présentant un intérêt particulier pour la faune, la flore, le patrimoine géologique ou paléontologique ou d'une manière générale, pour la protection des milieux naturels dont le statut est défini par la loi relative à la démocratie de proximité du 27 février 2002 ; une réserve naturelle régionale est créée à l'initiative du Conseil Régional ou à la demande des propriétaires concernés ; **2.** anciennement **réserve [f] naturelle volontaire** ; propriété privée pouvant être agréée par le ministre chargé de la protection de la nature afin de protéger les espèces de la flore et de la faune sauvage présentant un intérêt particulier sur le plan scientifique et écologique ; cf. art. 23 loi n° 76-629 du 10 juillet 1976 et PPH 1991, 274) ; *syn.* réserve [f] naturelle de statut libre ; *g* **Naturreservat [n]** (≠) (OB zu staatlich oder privat eingerichtetes Schutzgebiet unterschiedlicher Art, vor allem im englischen, französischen und spanischen Sprachraum. Im deutschen Naturschutzrecht gibt es den Begriff des N.s nicht. ►Biosphärenreservat, ►Feuchtgebiet internationaler Bedeutung, ►Laichschonbezirk, ►Naturschutzgebiet 2, ►Vogelschutzgebiet, ►Vollnaturschutzgebiet, ►Wildschutzgebiet).

3723 nature study center [n] [US]/**nature study centre** [n] [UK] *recr.* (Place, often with a building, usually located in a park or in the open countryside, which contains botanical, zoological, geological or mineralogical exhibits and interpretations, and which serves as an information or educational center [US]/centre [UK] for the public); *s* **museo [m] de información sobre la naturaleza** (Edificio situado en

N

zona natural en el que se presentan informaciones sobre procesos naturales, los ecosistemas naturales, la flora y fauna asi como en donde se realizan actividades educativas de divulgación sobre la naturaleza); *syn.* escuela [f] de naturaleza; *f* **centre [m] permanent d'initiation à l'environnement (C.P.I.E.)** (Équipement pédagogique expérimental dont le but est de développer une initiation à la relation homme/nature. L'accueille aussi bien les enseignants que les jeunes, les groupes que le grand public. Le centre d'hébergement muni d'une centaine de lits avec salles d'études, ateliers, bibliothèques est aussi un outil d'animation du milieu rural) ; *g* **Naturkundehaus [n]** (In der Regel außerhalb von Siedlungsgebieten gelegenes Haus mit Sammlungen von botanischen, zoologischen, geologischen oder mineralogischen Ausstellungsstücken, das der Öffentlichkeit zugänglich ist und Bildungs- und Lehrzwecken dient).

nature study path [n] [UK] *recr.* ▶nature trail [US].

3724 nature trail [n] [US] *recr.* (Trail or path provided with explanatory signs for educational purposes, which provides circulation through a nature study area containing features of interest connected with, e.g. botany, geology, dendrology, ecology, ornithology or history. *Specific term* ▶forest study trail); *syn.* nature study path [n] [UK]; *s* **itinerario [m] de la naturaleza** (Itinerario diseñado como apoyo para la educación ambiental y concebido para poner en relieve en su transcurso una serie de procesos y elementos naturales dificilmente observables de otro modo. Los **ii. de la n.** son, asimismo, infraestructuras que permiten un rápido conocimiento en visitas cortas a Reservas y Parques Naturales; cf. PARRA 1984; ▶itinerario forestal); *syn.* senda [f] ecológica, sendero [m] pedagógico, sendero [m] natural educativo; *f* **sentier [m] de découverte** (Itinéraire signalisé aménagé et conçu pour répondre à l'intérêt des utilisateurs de découverte de la nature ; il présente les particularités d'un site [géologiques, botaniques, zoologiques, ornithologiques, historiques, artistiques, régionales, etc.] tout en respectant l'environnement dans lequel il s'insère ; très souvent il s'agit de sentiers permettant l'ouverture au public de zones protégées [milieux dunettes, humides, etc.] ; cf. DUV 1984, 288 ; ▶sentier de découverte de la forêt) ; *syn.* sentier [m] découverte, sentier [m] pour la découverte de la nature, sentier [m] d'initiation ; *g* **Lehrpfad [m]** (Zu Lehr- und Anschauungszwecken angelegter, beschilderter Weg, der an naturkundlichen, z. B. dendrologischen, geologischen, ornithologischen, historischen oder landeskundlichen Sehenswürdigkeiten resp. Ausstellungsstücken vorbeiführt oder der über Umweltveränderungen Auskunft gibt. *OB zu* Baumkronenpfad, Bodenlehrpfad, geologischer Lehrpfad, Pflanzenlehrpfad, Vogelschutzlehrpfad, ▶Waldlehrpfad, Wildlehrpfad etc.); *syn.* Naturlehrpfad [m].

navigability [n] *trans.* ▶canalization of a waterbody for navigability.

3725 navigable waterway [n] *trans.* (River or canal used for transport by boat); *s* **hidrovía [f]** *syn.* curso [m] de agua navegable; *f* **voie [f] navigable (ou flottable)** (Cours d'eau et lacs, rivières canalisés, canaux y compris leurs dépendances inscrits sur la nomenclature des voies navigables ou flottables) ; *g* **Wasserstraße [f]** (Fluss oder Kanal als Verkehrsweg für Schiffe).

3726 near-natural [adj] *ecol.* (Almost exclusively composed of indigenous spontaneous species, hardly affected by human influences; ▶virgin); *syn. according to the* ▶*degree of landscape modification* managed [pp] (landscape), nature-like [adj], natural looking [ger]; *s* **poco alterado [pp]** (Para diferenciar los ecosistemas según su ▶grado de modificación [de ecosistemas y paisajes]: Se trata de aquéllos compuestos casi exclusivamente de especies autóctonas y que han sufrido pocas alteraciones por la influencia humana; ▶virgen); *f* **faiblement artificialisé, ée [loc]** (Différenciation du gradient d'artificialisation d'un écosystème qui s'est développé spontanément sous une faible pression de l'activité humaine et est composé presque exclusivement d'espèces indigènes, p. ex. dans les systèmes d'agriculture et de sylviculture extensive ; ▶degré d'artificialisation, ▶d'artificialisation nulle) ; *syn.* d'artificialisation faible [loc] ; *g* **naturnah [adj]** (*Zur Unterscheidung von Ökosystemen nach* ▶*Natürlichkeitsgraden* fast ausschließlich aus einheimischen, standorteigenen Arten spontan entstanden und mit geringem menschlichen Einfluss nicht wesentlich verändert; cf. BUCH 1978-II; ▶natürlich 2).

3727 near-natural landscape [n] *landsc.* (Landscape in a naturalistic state which has been created and used by man, whose vegetational cover closely matches the ▶potential natural vegetation over a large proportion of the area and in rare cases is identical to it within a limited area); *s* **paisaje [m] cuasi-natural** (▶vegetación natural potencial); *f* **paysage [m] faiblement artificialisé** (Paysage humanisé dont le couvert végétal, comme indicateur de l'intensité de l'utilisation des sols, se rapproche sur de grandes parties du territoire de la ▶végétation naturelle potentielle et est identique à celle-ci dans les cas les plus rares) ; *g* **naturnahe Kulturlandschaft [f]** (Eine vom Menschen gestaltete und genutzte Landschaft, deren Pflanzendecke als Indikator der Nutzungsintensität auf großen Flächenanteilen der ▶potentiellen natürlichen Vegetation nahe steht und in seltenen Fällen auf begrenzten Flächen mit ihr identisch ist; cf. BUCH 1978).

near-natural management [n] **of green spaces** *adm. for. hort. landsc. urb.* ▶urban forestry.

3728 near-natural storm water runoff management [n] *ecol. urb.* (Comprehensive measures undertaken for the collection and gradual infiltration of rainfall in order to replenish ground water resources, instead of draining the runoff to a piped system. Systems include infiltration in trench drains or basins, soakaways or large, buried rainwater tanks; ▶de-sealing, ▶receiving stream); *s* **gestión [f] natural de las aguas pluviales** (Conjunto de medidas de planificación, organización y técnicas tomadas para almacenar las aguas pluviales con el fin de aumentar su infiltración en el suelo y así alimentar los acuíferos in situ, en vez de dejarlas fluir superficial y directamente a los ▶cursos receptores a través del sistema de alcantarillado. Estos sistemas de g. n. de las a. p. incluyen entre otras medidas la ▶desimpermeabilización del suelo [cubiertas porosas] y las cisternas de infiltración construidas bajo patios o calles); *f* **gestion [f] naturelle des eaux de pluie** (Pratiques nouvelles de développement de systèmes visant, en déconnectant les eaux pluviales des eaux d'assainissement, à ralentir, stocker ou infiltrer les eaux pluviales le plus en amont possible en évitant de les rejeter directement dans un ▶émissaire ou ▶milieu récepteur ; parmi ces techniques on peut citer la ▶reperméabilisation des sols [revêtement poreux], les citernes d'infiltration et de drainage, les puits d'infiltration, les noues et fossés engazonnées, les toitures-terrasses [végétalisées] régulées, les chaussées à structure réservoir, les places et parkings inondables, les bassins naturels de stockage ; afin d'optimiser la gestion des eaux de pluie, il est recommandé d'associer entre elles les différentes techiques) ; *g* **naturnahe Regenwasserbewirtschaftung [f]** (Sämtliche Planungen, organisatorischen und technischen Maßnahmen, die dazu dienen, anfallendes Regenwasser nicht gleich in den nächsten Abwassersammler oder zum nächsten ▶Vorfluter abzuleiten, sondern die anfallenden Niederschlagsmengen ortsnah zu sammeln und sukzessive versickern zu lassen, damit sie dem Grundwasser wieder zugeführt werden können. Dies kann durch Versickerungssysteme wie Muldenrigolen, Versickerungsbecken oder -mulden, Rohr- und Schachtversickeungsanlagen, durch große im Boden eingebaute Regentanksysteme oder durch Straßen, deren

Unterbau als Wasserspeicher ausgebildet ist, erfolgen; ►Entsiegelung des Bodens); *syn.* naturnahes Regenwassermanagement [n].

3729 near threatened species [n] *conserv. phyt. zool.*
(*Faunistic and floristic* ►*status of endangerment* species with a growing rarity of population by occurrence of small numbers in a region, or those living in small numbers on the geographic periphery of their range, which is eligible for listing as a ►critically endangered species; ►threatened species, ►Red List); *syn. obs.* candidate species [n]; *s 1* **especie** [f] **casi amenazada [NT]** (Según las categorías de la ►lista roja de la UICN, un taxón está casi amenazado [Near Threatened] cuando ha sido evaluado según los criterios de la lista roja y no satisface actualmente los criterios para las categorías en peligro crítico, en peligro o vulnerable, pero está próximo a satisfacerlos, o posiblemente los satisfaga, en el futuro cercano; cf. UICN 2001; ►categoría de amenaza de extinción de especies); *s 2* **especie** [f] **potencialmente en peligro (≠)** (D., ►categoría de amenaza de extinción de especies de flora y fauna: Especie que en una zona se presenta en poblaciones pequeñas y con pocos ejemplares o que vive en poblaciones pequeñas al borde de su área; ►especie amenazada, ►libro rojo, ►especie en peligro de extinción); *f* **espèce** [f] **dont une régression s'est manifestée sans qu'il soit possible de définir dans quelle mesure** (►*Catégorie de régression des populations et de distribution des espèces* espèce dont le peuplement est de petite taille ou localisé sur un territoire restreint ou vivant à la limite géographique de son aire de répartition naturelle ; à rapprocher de la catégorie de régression des espèces menacées établie par l'UICN : « espèce quasi menacée » ; ►espèce amenée par sa régression à un niveau critique des effectifs, ►espèce dont la population n'a pas sensiblement diminuée, mais dont les effectifs sont faibles, donc en danger latent, ►liste des espèces menacées, ►Livre rouge [des espèces menacées]) ; *g* **potentiell gefährdete Art** [f] (*Faunistische und floristische* ►*Gefährdungskategorie* Art, die in einem Gebiet nur wenige und kleine Vorkommen aufweist oder eine Art, die in kleinen Populationen am Rande ihres Areals lebt; in D. können diese Arten der Kategorie G [Gefährdung anzunehmen, aber Status unbekannt] zugeordnet werden, im IUCN-Klassifizierungssystem *Near Threatened [NT]*; ►bedrohte Art, ►Rote Liste, ►vom Aussterben bedrohte Art); *syn. IUCN-Klassifikation* gering gefährdete Art [f]; *o. V.* potenziell gefährdete Art [f].

3730 necrosis [n] *bot.* (Irreversible damage to cells of plants, especially leaves, which begins at their edges or in the case of needles, at the tips; ►leaf necrosis); *s* **necrosis** [f] (Muerte localizada o general de los tejidos de una planta o animal, a menudo caracterizada por su color marrón o negruzco; DINA 1987; ►necrosis de la hoja); *f* **nécrose** [f] (En phytopathologie, altération visible d'ordre biochimique très fréquente sur les différents organes d'un végétal [feuilles, rameaux, tiges, racines, etc.] qui se produit dans les tissus après la mort des cellules, due à des agents très divers [bactérie, champignons, pollutions, etc. ; cf. LA 1981 ; ►nécrose foliaire) ; *g* **Nekrose** [f] (Irreversible Zellschädigungen und –zerstörungen an Pflanzenteilen, besonders an Blättern, bei denen vom Blattrand her, bei Nadeln an der Spitze beginnend, das Gewebe abstirbt; ►Blattnekrose).

nectar plant [n] *hort. landsc.* ►host plant for bees.

3731 needle [n] *bot. hort.* (Leaf of a conifer; ►foliage of woody plants); *s* **aguja** [f] (Hoja de las coníferas; ►follaje 2); *f* **aiguille** [f] (Feuille caractéristique des conifères dont le pétiole est très court et le limbe étroit et vernissé est souvent terminé en pointe; ►frondaison) ; *g* **Nadel** [f] (Blatt einer Konifere; die Gesamtheit der Nadeln eines Nadelbaumes heißt **Benadelung**; ►Belaubung).

3732 needle-leaved deciduous forest [n] *for. phyt.*
(UNE 1973, 22; forest consisting entirely or mainly of, e.g. larch [Siberian *Larix*]); *s* **bosque** [m] **aciculifoliado deciduo** (Población vegetal en la que predomina el alerce de Siberia *[Larix sibirica]*; UNE 1973, 77); *f* **forêt** [f] **de résineux caducifoliée** (Forêt essentiellement constituée ou à dominance, p. ex. de Mélèzes de Sibérie *[Larix sibirica]* ; dans l'étage subalpin, avec le Mélèze d'Europe *[Larix decidua]*, on parle de forêt de Mélèzes caducifoliée ; *terme spécifique* mélézein [m]) ; *syn.* forêt [f] de résineux décidus, forêt [f] coniférienne caducifoliée (UNE 1973, 49) ; *g* **sommergrüner Nadelwald** [m] (Pflanzenformation, in der z. B. die sibirische Lärche *[Larix sibirica]* vorherrscht; in den Lärchenwäldern der Alpen die Europäische Lärche *[Larix decidua]*).

3733 needle-leaved species [n] *bot. for.* (Generic term for gymnospermous tree and shrub with needle-shaped leaves, e.g. Yew *[Taxus]*, Spruce *[Picea]*, Pine *[Pinus]*, Larch *[Larix]*, Arbor-vitae *[Thuja]*, Mammoth tree *[Sequoiadendron giganteum]*, Fir *[Abies]*, Juniper *[Juniperus]*, Cypress *[Cupressus]*; ►conifer, ►stocking with conifers); *s* **especie** [f] **acicufolia** (Término genérico para árboles y arbustos gimnoespermos con hojas en forma de aguja, como p. ej. el tejo *[Taxus]*, la picea *[Picea]*, el pino *[Pinus]*, el alerce *[Larix]*, la tuya *[Thuja]*, la secuoya gigante *[Sequoiadendron giganteum]*, el abeto *[Abies]*, el enebro *[Juniperus]*, el ciprés *[Cupressus]*; ►especie conífera, ►repoblación de coníferas); *syn.* especie [f] de aguja; *f* **espèce** [f] **à aiguilles** [m] *for.* (Terme générique peu utilisé qualifiant un arbre provenant d'essences forestières conifères à feuillage en aiguilles opposition aux arbres feuillus ; ►conifère, ►enrésinement [naturel ou artificiel]) ; *syn.* bois [m] résineux, résineux [m] ; *g* **Nadelgehölz** [n] (...gehölze [pl]; OB zu nadelblättriger Nacktsamer wie z. B. Eibe *[Taxus]*, Fichte *[Picea]*, Kiefer *[Pinus]*, Lärche *[Larix]*, Lebensbaum *[Thuja]*, Mammutbaum *[Sequoiadendron giganteum]*, Tanne *[Abies]*, Wacholder *[Juniperus]*, Zypresse *[Cupressus]*; ►Bestockung mit Nadelgehölzen); *syn.* ►Konifere [f], *for.* Nadelholz [n] (...hölzer [pl]).

needle litter [n] [UK] *pedol.* ►needle straw [US].

3734 needle straw [n] [US] *pedol.* (Layer of shed needles of ►conifers forming a ►layer of raw humus; *specific term* pine straw [US]: dead pine needles that are frequently used for mulch; RCG 1982; ►leaf litter, ►litter layer); *syn.* conifer litter [n] (TEE 1980, 96), needle litter [n] [UK]; *s* **capa** [f] **de barrujo** (►Capa de humus bruto en bosques de ►coníferas; ►capa de litter, ►hojarasca 2); *syn.* capa [f] de förna (KUB 1953); *f* **litière** [f] **d'aiguilles** (►Couche d'humus brut constituée d'aiguilles de ►conifères ; ►couverture morte, ►litière feuillue) ; *syn.* litière [f] de conifères ; *g* **Nadelstreu** [f] (Schicht abgeworfener Nadeln der ►Konifere als ►Rohhumusauflage; ►Laubstreu, ►Streuschicht).

negative effect [n] *ecol.* ►negative impact.

3735 negative impact [n] *ecol.* (Disturbance which disrupts the natural balance); *syn.* negative effect [n]; *s* **impacto** [m] **significativo/negativo/no favorable** *syn.* incidencia [f] significativa/negativa/no favorable, efecto [m] significativo/negativo/no favorable; *f* **incidence** [f] **négative** (Modification qualitative, quantitative ou fonctionnelle de l'environnement provoquée par un projet ou un plan qui engendre une perturbation d'un système par rapport à l'état initial, cette perturbation n'ayant que des effets minimes sur la structure du système qui n'en sera pas considérablement modifiée) ; *syn.* séquelle [f], effet [m] négatif, effet [m] nocif (LA 1981, 473) ; *g* **negative Auswirkung** [f] (**1.** Beeinträchtigende Wirkung von Substanzen, natürlichen Ereignissen oder menschlichen Handlungen und Maßnahmen auf den Naturhaushalt oder Teilen davon sowie auf ein bestehendes

N

System oder dessen Teile. **2.** Das Ergebnis von 1.); *syn.* negativer Effekt [m], negative Wirkung [f].

negotiation [n] *contr.* ►bid negotiation [US]; *contr. prof.* ►fee negotiation.

negotiation [n] [UK]**, tender** *contr.* ►bid negotiation [US].

3736 negotiation phase [n] [US] *contr. prof.* (►Work phase [US]/work stage [UK] of a planner's services for appointment of a contractor, which follows the bidding [US]/tendering [UK] procedure; **in U.K.**, tender action and contract preparation are included in one work stage [work stage HJ]; ►Guidance for Clients on Fees [UK]); *syn.* contract preparation [n] [UK]; *s* **fase [f] de negociación (≠)** (►Fase de trabajo en el marco de los servicios profesionales de arquitectos/as [paisajistas] consistente en el estudio comparativo de ofertas para la posterior adjudicación de la ejecución de la obra; ►baremo de honorarios para arquitectos [paisajistas] e ingenieros, ►reglamento de honorarios [D]); *f* **assistance-marché [f] de travaux (A.M.T.)** (Conformément à la ►rémunération des prestations de maîtrise d'œuvre [exercée pour le compte de maîtres d'ouvrage publics], mission élémentaire normalisée désignant l'assistance au maître d'ouvrage dans le choix des entreprises à consulter, mise au point de l'offre retenue et assistance pour l'attribution du marché y compris la mise au point matérielle des projets de documents contractuels ; ►étape de la mission de maîtrise d'œuvre) ; *syn.* étape [f] du choix des entrepreneurs ; *g* **Mitwirkung [f] bei der Vergabe** (Mitwirkung bei der Vergabe von Aufträgen ist eine ►Leistungsphase im Rahmen der Abwicklung von Bauaufträgen durch den Planer gemäß der ►Honorarordnung für Architekten und Ingenieure — HOAI).

3737 neighborhood [n] [US]**/neighbourhood** [n] [UK] *urb.* (Particular urban residential area in which social contact between residents is characterized by permanent personal relationships. Modern ►urban planning seeks to reactivate such neighbo[u]rhood contacts by implementing planning and architectural schemes in large cities where the social structure has been lost); *s* **vecindario [m]** (Colectivo de residentes en un barrio o sector de un asentamiento, dentro del cual existen relaciones sociales duraderas. Por medio de medidas urbanísticas y arquitectónicas el ►urbanismo moderno busca favorecer la reactivación de estas relaciones con el fin de recrear el tejido social deteriorado en las grandes ciudades); *f* **voisinage [m]** (Ensemble des voisins dans un quartier, un ensemble résidentiel, etc. caractérisé par des relations sociales bonnes et durables. L'►urbanisme moderne s'efforce par des mesures favorisant l'amélioration du cadre de vie de faire revivre et de reconstituer le tissu social détérioré) ; *g* **Nachbarschaft [f]** (Gesamtheit der Nachbarn in einem Wohn- und Siedlungsgebiet, in dem die sozialen Kontakte der dort lebenden Menschen durch besonders ausgeprägte persönliche und dauerhafte Beziehungen bestimmt sind. Der moderne ►Städtebau versucht dieses in Großstädten verloren gegangene Sozialgefüge über stadtplanerische und architektonische Maßnahmen zu reaktivieren).

neighborhood [n] [US]**, poverty** *urb.* ►slum.

neighborhood conservation area [n] *leg. urb.* ►neighborhood improvement area [US]/neighbourhood improvement area [UK].

neighborhood green space [n] *urb.* ►proximity green space.

3738 neighborhood improvement [n] [US]**/neighbourhood improvement** [n] [UK] *plan. urb.* (Planning and implementation of measures to improve the quality of life of a ►dwelling environs in need of rehabilitation); *syn.* neighborhood revitalization [n] [US]; *s* **mejoramiento [m] del hábitat (residencial)** (Planificación e implementación de medidas para

mejorar la calidad del ►hábitat [residencial] en barrios deteriorados); *f* **amélioration [f] de l'environnement de l'habitat** (Planification et mesures en faveur de la revalorisation d'un ►environnement de l'habitat dégradé) ; *syn.* amélioration [f] du cadre de vie ; *g* **Wohnumfeldverbesserung [f]** (Planungen und Maßnahmen zur Aufwertung eines sanierungsbedürftigen ►Wohnumfeldes).

3739 neighborhood improvement area [n] [US]**/neighbourhood improvement area** [n] [UK] *leg. urb.* (Area which is selected because of its defects and officially designated by the authorities for renewal measures; **in U.K.**, an area which is selected for comprehensive treatment by development, redevelopment or improvement within a prescribed period is known as an **action area**; cf. OPL 1996, 82; ►poor quality housing [US]/poor quality housing stock [UK], ►run-down housing fit for rehabilitation, ►center city redevelopment [US]/ urban regeneration [UK]); *syn.* neighborhood conservation area [n] [also US], urban renewal area [n]; *s* **área [f] de rehabilitación integrada [Es]** (Zona declarada oficialmente por la autoridad competente con el fin de coordinar las actuaciones de las Administraciónes Públicas y el fomento de la iniciativa privada, dirigidas a rehabilitar de forma integrada los conjuntos urbanos y áreas rurales de interés arquitectónico, histórico-artístico, cultural, ambiental o social. La rehabilitación urbana incluye actuaciones de adecuación constructiva o funcional de las viviendas o edificios dedicados a la vivienda y así mismo adecuaciones del equipamiento comunitario primario, como los espacios libres, las infraestructuras y las dotaciones; cf. arts. 1 y 41 RD 2328/1983, de 28 de julio, sobre Protección a la Rehabilitación del Patrimonio Residencial y Urbano; ►edificación antigua con mal trazado, ►edificación antigua en mal estado, ►rehabilitación urbana); *f 1* **secteur [m] sauvegardé** (Périmètre fixé par décision administrative après enquête publique à l'intérieur duquel sont effectués des travaux de conservation, de restauration et de mise en valeur du patrimoine immobilier ; cf. lois n° 62-903 et n° 85-729 sur la restauration immobilière et C. urb., art. L. 313-1 à 313-3 ; ►restauration urbaine, ►rénovation urbaine) ; *f 2* **zone [f] de rénovation urbaine (Z.R.U.)** (Zone d'aménagement concerté [ZAC] fixée par décision administrative après enquête publique dans laquelle est effectuée une opération de ►rénovation urbaine ; opération réglementée par le décret n° 58-1469 du 31 décembre 1958 et supprimée par la loi n° 85-729 du 18 juillet 1985 ; ►habitat ancien, ►habitat vétuste) ; *f 3* **zone [f] de restauration immobilière** (Périmètre fixé par décision administrative après enquête publique à l'intérieur duquel est réalisée une opération de ►restauration urbaine ; dans la **z. r. u.** sont prises des dispositions spécifiques relatives au maintien et à la création d'activités et d'emplois, à l'aménagement urbain et à l'habitat, à la vie associative ; cf. C. urb., art. L. 313-4 à 313-4-3) ; *f 4* **zone [f] franche urbaine** (Zone créée dans des quartiers de plus de 10.000 habitants particulièrement défavorisés des zones urbaines sensibles, caractérisées par la présence de grands ensembles ou de quartiers d'habitat dégradé et par un déséquilibre entre l'habitat et l'emploi ; cf. C. urb., art. L. 510-1, L. 631-10) ; *f 5* **zone [f] de redynamisation urbaine** (Zone urbaine sensible, caractérisée par la présence de grands ensembles ou de quartiers d'habitat dégradé et par un déséquilibre entre l'habitat et l'emploi, confrontées à des difficultés particulières au vu de leur situation dans l'agglomération et de leurs caractéristiques commerciales ; cf. C. urb., art. L. 325-1) ; *g* **Sanierungsgebiet [n]** (Durch ein Verwaltungsverfahren förmlich festgelegtes Gebiet innerhalb einer Gemeinde, das städtebauliche Missstände aufweist [►Altbebauung 1 und ►Altbebauung 2], deren Beseitigung durch Sanierungsmaßnahmen erforderlich ist; cf. § 3 StBauFG; ►Stadtsanierung); *syn.* Stadtteil [m] mit Erneuerungsbedarf [NW].

3740 neighborhood park [n] [US] *recr. urb.* (Area providing outdoor recreation opportunities within walking distance of residents; ▶urban park); *syn.* district park [n] [UK]; *s* **jardín [m] de barrio** (▶Parque urbano de tamaño medio cuya función principal es servir para la recreación de la población del barrio en el que se encuentra. Cuenta con equipamiento para actividades de ocio para las diferentes edades de la población, como áreas de juegos y zonas tranquilas para el recreo y dencanso de la población del vecindario); *f 1* **parc [m] de voisinage** (Parc de faible étendue, quelques ha., accessible dans un rayon de 5 mm de marche ; DUA 1996, 540) ; *f 2* **parc [m] de quartier** (*Terme général désignant un* parc urbain d'environ 10 ha, composé de pelouses, boisements, aires de jeux ou installations légères de loisirs, dont l'emplacement, le type d'aménagement et l'aire d'influence répond à la satisfaction des besoins d'un quartier ; ▶parc central, ▶parc urbain) ; *syn.* jardin [m] public de quartier ; *g* **Stadtteilpark [m]** (Mittelgroße, meist über 10 ha große Parkanlage, die eine zentrale Versorgungsfunktion für einen Vorort oder Ortsteil einer Großstadt hat. Sie ist für stundenweise allgemeine Freizeitaktivitäten mit Spielangeboten für alle Altersgruppen, mit Einrichtungen für den vereinsungebundenen Freizeitsport und mit ruhigen Grünbereichen für die Anwohner zur Erholung und Entspannung ausgestattet; ▶Stadtpark).

3741 neighborhood recreation [n] [US]/**neighbourhood recreation** [n] [UK] *plan. recr. urb.* (Recreational activity within walking distance of the home; ▶neighborhood park, ▶proximity green space); *s* **recreo [m] en el vecindario/ barrio** (Actividad recreativa que se puede realizar en lugar accesible a pie desde la vivienda; ▶jardín de barrio, ▶zona verde de vecindario); *f* **loisirs [mpl] à proximité de quartiers résidentiels** (Forme de loisirs et de détente proche du lieu d'habitation et pouvant être atteint à pied ; ▶espace vert urbain de proximité, ▶parc de voisinage) ; *syn.* loisirs [mpl] proche des zones d'habitats ; *g* **wohnungsnahe Erholung [f, o. Pl.]** (Erholung, die in fußläufiger Nähe der Wohnung möglich ist; ▶Stadtteilpark, ▶wohnungsnahe Grünfläche).

3742 neighborhood redevelopment [n] [US] *urb.* (Comprehensive urban planning measures introduced for the sustainable improvement of living conditions of an urban population within a part of the city. The term **n. r.** and center city redevelopment [US]/town centre redevelopment [UK] are basically synonymous, because usually only parts of a city are redeveloped; ▶center city redevelopment [US]/urban regeneration [UK]); *syn.* neighborhood revitalization [n] [US], district redevelopment [n] [UK]; *s* **rehabilitación [f] de barrios** (Conjunto de actividades urbanísticas realizadas con el fin de mejorar las condiciones de habitación y vida de la población de un barrio. En principio los términos **r. de b.** y ▶rehabilitación urbana son sinónimos, ya que generalmente la rehabilitación se lleva a cabo en los barrios residenciales); *syn.* renovación [f] de barrios; *f 1* **réhabilitation [f] des quartiers** (Opération programmée visant à l'amélioration de l'habitat et le maintien ou le développement des services de voisinage ; loi n° 91-662 du 13 juillet 1991 ; ▶rénovation urbaine, ▶restauration urbaine) ; *f 2* **réaménagement [m] des quartiers** (Opération ayant pour objet, au-delà de l'amélioration de l'habitat, de faire revivre les quartiers existants, centraux comme périphériques, prévenir leur dégradation, valoriser leur identité, leur redonner une fonction économique et sociale à l'intérieur de l'agglomération, mettre en valeur leur patrimoine architectural et urbain ; restructuration urbaine des quartiers existants ; cf. projet de quartier, circ. n° 84-51 du 27 juillet 1984 ; ▶rénovation urbaine, ▶restauration urbaine) ; *g* **Stadtteilsanierung [f]** (Gesamtheit aller städtebaulichen Maßnahmen, die der nachhaltigen Verbesserung der Wohn- und Lebensbedingungen städtischer Bevölkerungsgruppen in einem Stadtteil dienen. Der Begriff **S.** und ▶Stadtsanierung sind im Prinzip syn. Da i. d. R. nur in Stadtteilen saniert wird, spricht man besser von **S.**).

neighborhood revitalization [n] [US] *plan. urb.* ▶neighborhood improvement [US]/neighbourhood improvement [UK].

3743 neighboring lot [n] [US] *leg. urb.* *syn.* abutting lot [n] [US], adjacent property [n], adjoining lot [n] [US], neighbouring plot [n] [UK]; *s* **predio [m] vecino** *syn.* predio [m] colindante, predio [m] aledaño, propiedad [f] colindante, propiedad [f] aledaña, propiedad [f] vecina; *f* **fonds [m] voisin** (Immeuble bâti ou non bâti immédiatement contigüe à la parcelle voisine) ; *g* **Nachbargrundstück [n]** (Grundstück, das [unmittelbar] an ein anderes angrenzt); *syn.* benachbartes Grundstück [n].

neighbourhood [n] [UK] *urb.* ▶neighborhood [US].

3744 neighbourhood rights [npl] [UK] *leg. urb.* (Generic term for regulations which protect the rights of neigho[u]rs under local public building laws); *s* **derechos [mpl] vecinales** (En derecho civil las regulaciones que limitan los derechos de uso de los predios por parte de sus propietarios en interés de los propietarios de los predios vecinos); *f* **droit [m] de voisinage [CH, B]** (**1.** Dispositions réglementaires limitant l'exercice du droit de propriété en matière immobilière [servitudes d'utilité publique] ; *textes réglementaires* Code de l'urbanisme, de la construction et de l'habitation, etc. **2.** En droit civil, dispositions réglementaires limitant l'exercice du droit de propriété en matière de voisinage [servitudes administratives civiles de voisinage] ; *textes réglementaires* Code civil, arrêté préfectoral, municipal, etc. ; **3. F.,** terme peu utilisé en France ; il est plutôt traité du règlement des **troubles de voisinage**, qui sont définis comme tout inconvénient ou toute gêne de toutes natures que ce soit [bruits, odeurs, fumées, émanation de gaz, poussières, animaux, servitude de passage, empiètements, diminution de l'ensoleillement, de la lumière et de la vue, plantations à proximité des habitations, ruissellement, etc.] causés à un voisin et constituant un abus de droit qui se résout par des interdictions ou des obligations prononcées par un juge notamment par la voie de la procédure d'injonction et par la condamnation à des dommages-intérêts ; les troubles de voisinage sont régis par les règlements administratifs locaux et les usages constants et reconnus, et, à défaut, par différents articles du Code Civil, du Code de la santé publique, du Code de l'urbanisme, du Code pénal, etc. ; en ce qui concerne les distances à respecter en matière de plantation par rapport à la propriété voisine c'est l'article 67 du Code civil qui est déterminant, pour la diminution de l'ensoleillement de la lumière et de la vue, ce sont les articles 678 et suivants du Code civil) ; *g* **Nachbarrecht [n]** (**1.** Nachbarschützende Vorschriften, die die Rechtsstellung des Nachbarn im öffentlichen Baurecht regeln; *gesetzliche Grundlagen* BauNVO, BauGB, LBO etc. **2.** Im Zivilrecht diejenigen Vorschriften, die das Verfügungsrecht des Eigentümers über sein Grundstück im Interesse benachbarter Grundstückseigentümer beschränken; *gesetzliche Grundlage* Bürgerliches Gesetzbuch [BGB] und Nachbarechtsgesetze der Bundesländer); *syn. zu 1.* Baunachbarrecht [n], Anrainerrecht [n] [auch A].

neighbouring plot [n] [UK] *leg. urb.* ▶neighboring lot [US].

3745 neophyte [n] *bot. phyt.* (Plant species which has entered the flora of a region within the 'historic' time scale; ▶naturalized species); *s* **neófito [m]** (Según la clasificación de THELLUNG, planta naturalizada que, desarrollándose en estaciones favorables, no intervenidas por el hombre, podría pasar por indígena de no conocerse la historia de su expansión; DB 1985; ▶especie naturalizada); *f* **plante [f] néophyte** (D'après la méthode de classification d'immigration floristique, espèce végétale dont l'introduction, l'installation et la propagation ont eu lieu en des temps historiques ; ►espèce naturalisée ; *syn.* néophyte [m] ; *g* **Neophyt [m]** (Nach floristischen Klassifizierungsgesichtspunkten hinsichtlich der Einwanderungszeit von Arten in ein [Floren]gebiet diejenige Art, deren Einwanderung und Verbreitung erst in „historischer" Zeit erfolgte. In D. sind bis zum Jahre 2000 mindestens 417 gebietsfremde Pflanzenarten dauerhaft eingebürgert; cf. R+U 2001 [4], 49. Bedeutende, stark lebensraumbesetzende **N.en** in Zentraleuropa sind z. B. die Herkulesstaude *[Heracleum mantegazzianum]*, der Japanische Staudenknöterich *[Fallopia japonica var. japonica,* syn. *Reynoutria japonica]*, Goldrutenarten *[Solidago spec.]*. Während sich der tatsächliche Schaden in Deutschland derzeit noch nicht in Euro und Cent beziffern lässt, beläuft sich einer Studie zufolge diese Summe in den USA auf fast 100 Mrd. €; *opp.* Archaeophyt [m]; ►neuheimische Art); *syn.* Neuadventive [f], zugewanderte Pflanzenart [f].

nestbox [n] [US] *zool.* ►nesting box.

3746 nesting [n] *zool.* (Brooding and hatching of eggs in a nest or nesting box and the rearing of young birds); *s* **anidar [vb]** (Acción de puesta e incubación de huevos en un nido, una cavidad o un nidal); *f* **nicher [vb]** (Action de couver les œufs et élever les jeunes oiseaux dans un nid, une cavité, un nichoir) ; *g* **Nisten [n, o. Pl.]** (Ausbrüten der Eier auf einem Nest, in einer Nisthöhle oder Nistkasten und Großziehen der Jungvögel).

3747 nesting box [n] *zool.* (Wooden or wood-concrete box made for ►cavity-nesting birds); *syn.* bird-nesting box [n], birdhouse [n] [also US], nestbox [n] [also US], bird box [n] [also US]; *s* **nidal [m]** (Caja de madera u hormigón y madera hecha para ►ave que anida en cavidades); *syn.* cajón [m] para anidar, caseta [f] para anidar; *f* **nichoir [m]** (Abri artificiel en bois ou en fibragglo construit pour les ►oiseaux nichant dans des cavités) ; *g* **Nistkasten [m]** (...kästen [pl]; aus Holz oder Holzbeton gefertigter Kasten für höhlenbrütende Vögel; ►Höhlenbrüter).

nesting habitat [n] *zool.* ►breeding site.

3748 nesting hole [n] *zool.* (Hole in a tree or wall in which eggs are laid and hatched); *s* **cueva [f] para anidar** (Cavidad en árbol o en muro en la que anida una pareja de aves para reproducirse); *f* **cavité [f] de ponte** (Trou, cavité naturelle ou creusée, dans les arbres creux, une paroi rocheuse, le sable des falaises ou des berges de cours d'eau ; *terme spécifique* cavité rupestre de nidification) ; *syn.* terrier [m] ; *g* **Nisthöhle [f]** (Loch in einem Baum oder einer Wand zum Brüten).

nesting site [n] *zool.* ►breeding site.

net [n] [UK], **drainage** *geo.* ►drainage pattern.

3749 net floor area [n] *arch.* (Occupied space in a building, which does not include hallways, elevator shafts, stairways, toilets, and wall thicknesses. This area is used to determine rental space and fire code requirements; cf. MEA 1985, 330); *s* **superficie [f] útil de un edificio** (Área neta de vivienda, oficina o almacén en un edificio); *f* **surface [f] utile** (Dans le cas général, somme des surfaces de plancher des locaux abritant les activités principales, des locaux annexes ou d'assistance et des locaux techniques ; dans le cas de l'habitation, la **s. u.** est la somme de la surface habitable et de la surface des annexes [pièces principales, pièces de service, circulations intérieures]) ; *g* **Nutzfläche [f]** (Anteil der Wohn- und Lagerflächen in einem Gebäude, der der Nutzung entsprechend der Zweckbestimmung dient; bei Wohnbauten die Summe der Nettogrundflächen [Innenmaße] aller abgeschlossenen Räume; Wände, Kamine, Schächte, Flure, Treppenhäuser und Funktionsflächen wie Heizungsraum und technische Betriebsräume werden nicht dazugerechnet. Bei Gewerbebauten alle dem Betriebe dienenden Flächen incl. der Grundrissflächen von Treppen, Liftanlagen und mobilen Bauteilen).

3750 network [n] (1) *conserv. ecol.* (Result of joining isolated units to form, e.g. ►habitat network, hedgerow network [►hedgerow landscape]; ►becoming interlocked, ►native plants and wildlife corridor, ►Natura 2000 Protection Area Network ►process of habitat fragmentation); *syn.* interconnected system [n], interconnectedness [n]; *s 1* **reticulación [f]** (Medidas para posibilitar la interconexión de diferentes hábitats de especies y crear redes p. ej. de biótopos, setos interparcelarios [►paisaje de bocage], etc.; ►aislamiento de biotopos, ►articular); *s 2* **red [f] de biótopos** (Interconexión de biótopos por medio de corredores verdes; ►corredor vital, ►Red Ecológica Europea Natura 2000, ►reticulación de biótopos); *f 1* **formation [f] d'un réseau** (Stratégie de gestion restauratoire ou conservatoire de l'environnement pour toutes les espèces que menace la fragmentation de leur habitat grâce au rétablissement ou la restauration d'un enchaînement, d'un réseau de ►corridors biologiques reliant divers biotopes entre eux, p. ex. les haies dans un paysage de ►bocage ; ceux-ci sont à l'heure actuelle indispensable à la survie des espèces ; ►constituer un réseau, ►constitution d'un réseau d'habitats naturels, ►isolement de biotopes) ; *syn.* constitution [f] d'un tissu, trame [f] écologique, maillage [m] écologique [B] ; *f 2* **réseau [m]** (Ensemble de liaisons constitué, p. ex. par des stations aux conditions écologiques homogènes ; ►constitution d'un réseau d'habitats naturels, ►corridor biologique, ►réseau de biotopes, ►réseau Natura 2000) ; *g* **Vernetzung [f]** (Verknüpfung oder netzartige Verbindung von z. B. gleichartigen oder ähnlichen Biotopen oder Biotopstrukturen zu einem Biotopverbund, Hecken zu einem Heckenverbund [►Heckenlandschaft]; ►vernetzen; ►Biotopvernetzung, ►linienförmiger Lebensraum, ►Natura-2000-Schutzgebietssystem, ►Verinselung von Biotopen); *syn.* Verbund [m].

network [n] (2) *recr.* ►hiking network, ►pathway network; *plan. trans. urb.* ►public utiliy network; *trans.* ►road network.

network [n], **circulation** *plan. urb.* ►circulation system, ►transportation system.

network [n], **footpath** *recr.* ►pathway network.

network [n] [UK], **horse riding trail** *plan. recr.* ►bridle path network [UK].

network [n], **open space** *landsc. urb.* ►open space system.

network [n], **services** *plan. trans. urb.* ►public utiliy network.

network [n], **transportation** *plan. trans.* ►transportation system.

network analysis [n] *constr. plan.* ►critical path method.

3751 network plan [n] *constr. plan.* (In C.P.M. [►Critical Path Method] terminology a graphic representation of activities showing their temporal interrelationships; e.g. for a major project); *syn.* progress schedule [n]; *s* **plan [m] reticular** (En la terminología del «Critical Path Method», gráfico en el que se representa el desarrollo temporal de las fases de trabajo, p. ej. de una gran obra de construcción; ►método del «camino crítico»); *f* **projet [m] de réseau** (Document graphique représentant le planning des divers travaux, p. ex. pour un grand chantier ; ►analyse de chemin critique) ; *g* **Netzplan [m]** (Grafischer Plan zur zeitlichen Planung von komplexen Arbeitsabläufen, z. B. für eine Großbaustelle; ►Netzplantechnik).

network planning technique [n] *constr. plan.* ►critical path method.

3752 new foliage [n] **of woody plants** *bot. hort.* (►emerging leaves, ►leafing out, ►spring foliage); *s* **foliación [f] joven** (►brotación, ►foliación, ►foliación prevernal); *f* **jeune frondaison [f]** (►feuillaison, ►frondaison printanière, ►pousse) ; *g* **frische Belaubung [f, o. Pl.]** (►Laubaustrieb im

Frühjahr resp. nach einer Periode der Vegetationsruhe; ▶Austrieb, ▶Frühjahrsbelaubung, ▶Laubaustrieb); *syn.* frisches Grün [n, o. Pl.].

new growth capability [n] [US] *for. hort.* ▶capacity of making new shoots.

3753 new housing development [n] *urb.* (Group of dwellings under construction or just completed on previously undeveloped land); *syn.* new housing subdivision [n] [also US]; *s* **zona [f] de nueva construcción** *syn.* barrio [m] nuevo; *f* **quartier [m] nouveau** (Zone autrefois non bâtie sur laquelle viennent d'être réalisées des constructions nouvelles) ; *g* **Neubaugebiet [n]** (Vormals unbebauter Teil einer Gemarkung, der vor kurzem resp. vor wenigen Jahren fertiggestellt wurde oder gerade mit Wohnhäusern bebaut wird; *opp.* Bestandsgebiet); *syn.* Neubausiedlung [f].

new housing subdivision [n] [US] *urb.* ▶new housing development.

3754 new introduction [n] *hort.* (Latest result of ▶plant breeding which is new to the public or will be put on the market as a novel plant); *syn.* novelty [n]; *s* **nueva variedad [f] (de planta)** (Resultado de trabajos de ▶fitogenética); *f* **variété [f] nouvelle** (Création récente d'une variété par la ▶sélection végétale) ; *g* **Neuheit [f]** (Ergebnis der ▶Pflanzenzüchtung, das neu auf dem Markt ist oder kommen soll).

new leaves [npl] *bot. hort.* ▶emerging leaves.

3755 new project [n] *constr.* (In landscape construction a projected design or the recent execution of a plan; e.g. a green open space, square, etc.); *s 1* **proyecto [m] de construcción de espacio libre** (Objeto a construir, como un parque, una plaza, etc.); *s 2* **instalación [f] nueva** (Objeto ya construido); *f 1* **réalisation [f] d'un équipement neuf** (Projet d'aménagement, p. ex. d'un espace vert, d'une place) ; *syn.* installation [f] neuve ; *f 2* **équipement [m] neuf** (Équipement réalisé) ; *g* **Neuanlage [f]** (1. Im Garten-, Landschafts- und Sportplatzbau neu zu errichtendes oder neu herzustellendes Objekt, z. B. Grünfläche, Platz. 2. Fertig gestellte N.).

niche [n] *ecol.* ▶ecological niche.

3756 nidificating avifauna [n] *zool.* (Nest-building birds); *syn.* nidifying birds [npl]; *s* **avifauna [f] nidificante**; *f* **avifaune [f] nidificatrice** (Oiseaux nicheurs) ; *g* **nestbauende Vogelwelt [f]**.

nidifying birds [npl] *zool.* ▶nidificating avifauna.

3757 night [n] **of ground radiation** *met.* (Cloudless night, which may result in frost [▶radiation frost], after warm air, close to the Earth's surface, cools down due to radiation during a period of cold; cf. WASHINGTON POST May 11, 1998); *s* **noche [f] radiativa** (Noche despejada de nubes y en la que las capas cercanas al suelo pierden calor por radiación, de manera que en la época fría se pueden presentar heladas [▶helada de irradiación]); *syn.* noche [f] de radiación; *f* **nuit [f] en situation radiative** (Nuit à ciel clair ou peu nuageux avec vents faibles ou nuls pendant laquelle les risques de gelée [▶gelée de rayonnement] près du sol sont importants lorsque la température de l'atmosphère est basse [entre 1° et 5 °C]) ; *syn.* nuit [f] claire favorisant le rayonnement ; *g* **Strahlungsnacht [f]** (Nacht, die durch einen wolkenlosen Himmel gekennzeichnet ist und in der die bodennahe Luftschicht durch Ausstrahlung Wärme verliert, so dass während der kalten Jahreszeit mit Frost gerechnet werden muss [▶Strahlungsfrost]).

3758 ninety-degree parking layout [n] *plan. trans.* (Parking at 90° to traffic flow; ▶in-line parking layout [US]/parallel parking layout [UK], ▶angle-parking layout, ▶perpendicular parking); *syn.* perpendicular parking layout [n] [also US] (HALAC 1988-II, 71), nose-to-kerb parking [n] [also UK],

90-degree parking layout [n], right angle parking layout [n]; *s* **disposición [f] (de aparcamiento) en perpendicular** (Orden de colocación de los vehículos en ángulo de 90° a la dirección del tráfico; ▶disposición [de aparcamiento] en hilera, ▶disposición [de aparcamiento] en diagonal, ▶aparcamiento en perpendicular); *syn.* aparcamiento [m] en posición perpendicular, estacionamiento [m] en perpendicular [CS], parqueo [m] en perpendicular [C, CA]); *f* **rangement [m] perpendiculaire** (Disposition des véhicules placés perpendiculairement à l'axe de la chaussée ; ▶rangement longitudinal, ▶stationnement en bataille, ▶stationnement en épi) ; *syn.* rangement [m] en bataille (VRD 1986, 176), rangement [m] à 90°, disposition [f] en bataille ; *g* **Senkrechtaufstellung [f]** (Anordnung der ruhenden Fahrzeuge in einem Winkel von 90° zur Fahrtrichtung; ▶Längsaufstellung, ▶Schrägaufstellung, ▶Senkrechtparken).

3759 ninety-degree parking row [n] *trans.* (Parking strip for ▶ninety-degree parking layout; ▶row of parking); *syn.* perpendicular parking row [n] [US] (HALAC 2, 71), 90° parking row [n]; *s* **fila [f] de estacionamiento en perpendicular** (▶Banda de aparcamiento para ▶disposición [de aparcamiento] en perpendicular en relación a la calle); *syn.* fila [f] de aparcamiento en perpendicular; *f* **bande [f] pour le stationnement perpendiculaire** (▶bande de stationnement [longitudinal], ▶rangement perpendiculaire) ; *syn.* parking [m] perpendiculaire, bande [f] de stationnement à 90°, bande pour le stationnement en bataille ; *g* **Senkrechtparkstreifen [m]** (▶Parkstreifen für Fahrzeuge zur ▶Senkrechtaufstellung).

3760 nitric oxide [n] *chem. envir.* (Toxic nitric monoxide [NO], produced by combustion engines at high temeratures, is converted by the process of oxidation within a short period of time into reddish, strongly smelling and very toxic nitric dioxide [NO_2]); *syn.* nitrogen oxide [n]; *s* **(di)óxido [m] de nitrógeno** (Compuesto de nitrógeno y oxígeno; el monóxido de nitrógeno [NO] tóxico que se produce en motores de combustión se oxida a altas temperaturas para formar el NO_2, sustancia fuertemente tóxica y maloliente); *f* **oxydes [mpl] d'azote** (Composés résultant de la combinaison de l'azote avec l'oxygène, produits principalement par la combustion des combustibles fossiles solides, liquides ou gazeux et jouant un rôle primordial dans les processus photochimiques au sein de la troposphère ; le monoxyde d'azote [NO] gaz toxique émis se transforme par oxydation en un gaz très toxique et fortement odorant, de couleur rouge-brun, le dioxyde d'azote NO_2) ; *syn.* peroxydes [mpl] d'azote [NO_x] ; *g* **Stickoxid [n]** (Verbindung des Stickstoffs mit Sauerstoff; das bei Verbrennungsmotoren mit hohen Temperaturen entstehende giftige Stickstoffmonoxid [NO] verwandelt sich innerhalb kurzer Zeit durch Oxidation in das rotbraune, unangenehm riechende, stark toxische Stickstoffdioxid [NO_2]); *syn.* Stickstoffoxid [n].

3761 nitrification [n] *chem. pedol.* (Oxidation of ammonium to nitrite and the further oxidation of nitrite to nitrate ions, which are important to the soil and plants, because the ions are easily leached); ▶denitrification); *s* **nitrificación [f]** (Oxidación del ión amonio [NH_4^+] para formar nitritos [NO_2] que son importantes para las plantas, ya que son de fácil lixiviación; *opp.* ▶desnitrificación); *f* **nitrification [f]** (Oxydation biologique, dans le sol, qui transforme des dérivés ammoniacaux du sol [ions ammonium [NH_4^+]] en ions nitriques [NO_3^-] en deux phases successives ; la première, sous l'action de micro-organismes des genres *Nitrosomonas* [bactérie nitreuse] ou *Nitrosococcus* l'azote ammoniacal est transformé en azote nitreux — c'est la nitrosation — et la seconde, celle des nitrobacter qui transforme l'azote nitreux en azote nitrique — c'est la nitratation ; DIS 1986, 143 ; *opp.* ▶dénitrification) ; *g* **Nitrifizierung [f]** (Für Böden und Pflanzen bedeutungsvolle Oxidation von Ammonium- [NH_4^+] zu Nitrit-

N

[NO₂]Ionen, da diese leicht der Auswaschung unterliegen; SS 1979; *opp.* ▶Denitrifizierung).

3762 nitrogen content [n] *pedol.* *s* **contenido [m] de nitrógeno**; *f* **teneur [f] en azote** ; *g* **Stickstoffgehalt [m].**

3763 nitrogen enrichment [n] *pedol.* (Increase of nitrogen concentration in the soil); *s* **enriquecimiento [m] de nitrógeno** (Aumento de la concentración de nitrógeno en el suelo); *f* **enrichissement [m] en azote** (Augmentation de la teneur en azote dans le sol) ; *g* **Stickstoffanreicherung [f]** (Zunahme des Stickstoffgehaltes im Boden).

3764 nitrogen fixation [n] *biol. pedol.* (Conversion of atmospheric nitrogen gas into organic ammonium; nitrogen-fixing [adj]); *s* **fijación [f] biológica del nitrógeno** (Conversión de nitrógeno atmosférico en amonio orgánico por medio de microorganismos del suelo [p. ej. *Rhizobium*] o de las aguas dulces y los océanos [p. ej. cianobacterias]; cf. DINA 1987); *f* **fixation [f] d'azote (moléculaire)** (Dans la fixation de l'azote atmosphérique on distingue une fixation non symbiotique par les micro-organismes photosynthétiques anaérobies [cyanophycées] et les diverses bactéries aérobies du genre azotobacter ainsi qu'une fixation symbiotique réalisée par les bactéries des nodules *[rhizobium]* chez les légumineuses et par les champignons à filaments mycéliens très ramifiés) ; *g* **Stickstoffbindung [f]** (Biogene Bindung von Luftstickstoff durch nichtsymbiontische Organismen [einige blaugrüne Algen und verschiedene Bakterien] und symbiontische N₂-Bindung durch Knöllchenbakterien und Strahlenpilze; stickstoffbindend [ppr/adj], *o. V.* Stickstoff bindend [ppr/adj]); *syn.* Stickstoffoxidierung [f].

3765 nitrogen indicator (species) [n] *phyt.* (▶Indicator plant for nitrogen-rich, ▶mellow soil); *s* **planta [f] indicadora de nitrógeno** (▶Planta indicadora para suelos ricos en nitrógeno; ▶suelo con sazón); *f* **plante [f] indicatrice d'azote** (▶Plante indicatrice des sols riches en azote et à forte stabilité structurale ; ▶sol à forte stabilité structurale) ; *g* **Stickstoffzeiger [m]** (▶Zeigerpflanze für stickstoffreichen, ▶garen Boden).

nitrogen oxide [n] *chem. envir.* ▶nitric oxide.

3766 nitrogen-rich [adj] *limn. pedol.* *syn.* rich in nitrogen [loc]; *s* **rico/a en nitrógeno [loc]** *syn.* rico/a en nitratos [loc]; *f* **riche en azote [loc]** *syn.* riche en nitrates [loc] ; *g* **stickstoffreich [adj].**

3767 nitrophilous species [n] *phyt.* (Plant occurring primarily in, or limited to, nitrogen-rich soil; ▶nitrogen indicator [species]); *s* **especie [f] nitrófila** (Planta que crece principalmente sobre suelos ricos en nitrógeno; ▶planta indicadora de nitrógeno); *f* **espèce [f] nitrophile** (En milieu terrestre, espèce croissant de préférence sur les sols naturels riches en azote ou les sols modifiés par l'apport d'engrais azotés comme le fumier, l'épandage, les pollutions, etc. ; ▶plante indicatrice d'azote) ; *syn.* espèce [f] nitratophile (DEE 1982) ; *g* **nitrophile Art [f]** (Pflanze, deren Verbreitung vorzugsweise auf stickstoffreichen Böden anzutreffen oder beschränkt ist; ▶Stickstoffzeiger); *syn.* Nitratpflanze [f], Nitrophyt [m], stickstoffliebende Art [f], *o.V.* Stickstoff liebende Art [f].

nival belt [n] *geo. phyt.* ▶nival zone.

3768 nival zone [n] *geo. phyt.* (TEE 1980, 487; highest ▶altitudinal belt of mountains in which the average annual rate of snow precipitation on a level area is higher than the rate of melting); *syn.* nival belt [n]; *s* **piso [m] nival** (Piso de vegetación que se halla por encima del límite inferior del de las nieves perpetuas; DGA 1986; ▶zonación altitudinal); *f* **étage [m] nival** (Étage de végétation au-dessus de 3000 m constituant la limite supérieure de la végétation en haute montagne et au contact des neiges pérennes ; ▶étagement de la végétation) ; *g* **nivale Stufe [f]** (Hinsichtlich der ▶Höhenstufung der oberste Bereich im vorwiegend schneebedeckten Hochgebirge über 3000 m Höhe, in dem die Schneeniederschlagsraten auf ebener Fläche im Durchschnitt der Jahre höher als die Abschmelzraten sind; in dieser Höhe liegt auch das Nährgebiet der Gletscher); *syn.* Firnschneestufe [f].

'no build' option [n] [US] *plan.* ▶'no project' option.

3769 nocturnal roost [n] *game'man. zool.* (Place for sleeping birds); *s* **lugar [m] de reposo nocturno** (Sitio en el que normalmente duermen pájaros); *f* **site [m] de repos des oiseaux** (Pour le repos nocturne) ; *syn.* lieu [m] de repos des oiseaux ; *g* **Schlafplatz [m] für Vögel** (Platz, auf dem Vögel nachts schlafen).

3770 no cut-no fill line [n] *constr. eng.* (Line on a contour map which has to be determined by earth calculation, either marking the boundary between earth that has to be moved and earth that is left in place, or marking the boundary between ▶cut and fill); *s* **línea [f] de excavación** (En mapa o plano de curvas de nivel, para el cálculo de terracerías línea de referencia que marca la separación entre las zonas de desmonte y las de relleno; ▶desmonte y terraplén); *f* **tracé [m] du projet** (Ligne représentée sur les profils en long et en travers faisant apparaître les déblais et les remblais et permettant d'évaluer la cubature des terres ; ▶déblai et remblai) ; *g* **Nulllinie [f]** (Zur Erdmassenberechnung im Höhenschichtlinienplan zu suchende Linie, die Ab- und Auftragsflächen zum unveränderten Gelände begrenzt; ggf. bildet sie die Trennlinie zwischen ▶Auftrag und Abtrag).

nodule [n] *bot.* ▶root nodule.

3771 nodule bacteria [npl] *bot. pedol.* (Symbiotic nitrogen-fixing soil bacteria forming nodules on the roots of certain plants, e.g. *Leguminosae*; ▶nitrogen fixation); *s* **bacterios [mpl] nodulares** (Organismos bacterianos simbióticos que forman nódulos en las raíces de algunas plantas, p. ej. leguminosas, que enriquecen el suelo de nitrógeno; ▶fijación biológica del nitrógeno); *syn.* bacterios [mpl] de las nudosidades, bacterios [mpl] radícicolas; *f* **bactéries [fpl] des nodules** (Bactéries symbiotiques [rhizobiums] responsables de la formation de nodosités chez les légumineuses, certains angiospermes et gymnospermes et permettant à ceux-ci de fixer l'azote atmosphérique et de synthétiser les substances azotées ; ▶fixation d'azote [moléculaire]) ; *syn.* bactéries [fpl] des nodosités ; *g* **Knöllchenbakterien [npl]** (Symbiontische, Luftstickstoff bindende Bakterien [Rhizobien], die N₂ im Boden anreichern; ▶Stickstoffbindung).

3772 no-fines concrete [n] *constr.* (Composite material with Portland cement, no-fines aggregate of nearly the same size grains and water; ▶porous concrete); *s* **hormigón [m] de un sólo grano** (Material compuesto de cemento Portland, agregados del mismo tamaño y agua; ▶hormigón poroso); *f* **béton [m] caverneux** (Béton léger constitué de granulats de forme et de granulométrie identiques collés ensemble à l'aide d'une pâte de ciment ; ▶béton poreux ; DTB 1985) ; *syn.* béton [m] sans éléments fins ; *g* **Einkornbeton [m]** (Betonstruktur aus Steinzuschlagstoffen der gleichen Größe und Form, ohne Feinanteile; ▶wasserdurchlässiger Beton).

3773 noise [n] *envir.* (Random din or loud persistant incoherent sound or electromagnetic radiation, which is socially and medically undesirable. Sound with its physical properties is measurable by sound pressure levels, noise is perceived by the receiver subjectively; ▶noise pollution); *s* **ruido [m]** (Sonido fuerte y persistente, que implica molestia, riesgo o daño para las personas, las actividades que ellas realicen o para el medio ambiente. Así el término **r.** contiene una valoración negativa de un fenómeno físico neutro. Mientras que la intensidad física del sonido se puede medir objetivamente, el **r.** es percibido subjetivamente por cada individuo. La unidad de medición del sonido es el decibelio [dB(A)]; ▶contaminación acústica); *f* **bruit [m]** (Toute sensation auditive désagréable ou gênante, tout phénomène acoustique produisant cette sensation, tout son ayant un caractère aléatoire qui n'a pas de composantes définies. L'unité de mesure utilisée est le décibel [dB]. Le **b.** est la principale source de nuisance. Sous-produit de la civilisation technologique et urbaine, le **b.** est l'un des facteurs les plus perturbateurs de la vie moderne ; cf. Association française de normalisation, CPEN, section I. ; ▶nuisance sonore) ; *g* **Lärm [m, o. Pl.]** (Als störend empfundener Schall; der Begriff **L.** enthält somit eine negative Wertung eines physikalisch neutralen Phänomens. Während **Schall** eine objektiv messbare Größe ist [dB(A)], nimmt **L.** jeder Mensch subjektiv unterschiedlich wahr; ▶Lärmbelastung).

noise [n] *envir. plan.* ▶consultant's report on noise, ▶precautionary measures against noise, ▶traffic noise.

noise [n] [US]**, background** *envir.* ▶ambient noise level.

3774 noise abatement [n] *envir.* (Policy, program[me]s, measures and activities to avoid or to reduce noise; ▶noise control); *syn.* noise reduction [n]; *s* **lucha [f] contra el ruido** (Conjunto de medidas para prevenir, vigilar y reducir la contaminación acústica, para evitar y reducir los daños que de ésta pueden derivarse para la salud humana, los bienes o el medio ambiente; art. 1 Ley 37/2003, del Ruido; ▶protección contra el ruido); *syn.* lucha [f] contra la contaminación acústica; *f* **lutte [f] contre le bruit** (Toutes mesures et dispositions règlementaires générales et d'urbanisme mises en œuvre pour prévenir, réduire, protéger, contrôler et surveiller les nuisances sonores ; la législation française de lutte contre le bruit est régit par la loi n° 92-1444 du 31 décembre 1992 texte général venant renforcer les mesures existantes et complétées par les directives de la Communauté européenne ; depuis septembre 2000 la loi a été intégrée au Code de l'environnement, Livre V : prévention des pollutions, des risques et des nuisances, Titre VII : prévention des nuisances sonores Chapitre Ier : lutte contre le bruit ; ▶protection contre le bruit) ; *g* **Lärmbekämpfung [f, o. Pl.]** (Planerische und bauliche Maßnahmen, die der Vermeidung oder Reduzierung von Lärmbelästigungen dienen; ▶Lärmschutz).

noise abatement zone [n] [UK] *leg. plan.* ▶noise zone [US].

noise attenuating berm [n] *constr. envir.* ▶noise attenuation mound.

noise attenuating earth structure [n] *constr. envir.* ▶noise attenuation mound.

3775 noise attenuation forest [n] *for. landsc.* (Type of protective forest which absorbs environmental pollution and is characterized by dense, multilayered vegetation; ▶protective forest absorbing air pollution. Mature forests without ground cover or shrub layer have only a minor effect in reducing noise); *s* **bosque [m] protector contra ruidos** (Tipo de ▶bosque protector contra inmisiones caracterizado por su estructura vertical densa y regular, ya que fustales sin sotobosque no tienen apenas efecto reductor); *f* **forêt [f] de protection contre le bruit** (Catégorie des ▶forêts de protection contre la pollution atmosphérique constituée par un peuplement homogène dense pluristrate [arbres de haut jet, arbres intermédiaires et arbustes]) ; *g* **Lärmschutzwald [m]** (Form des ▶Immissionsschutzwaldes, der sich durch einen dichten, vertikalen, möglichst gleichmäßig aufgebauten Bestand auszeichnet; Hallenbestände [▶Hallenwald] ohne Bodenvegetation und Strauchschicht bewirken nur eine geringe Lärmminderung).

3776 noise attenuation mound [n] *constr. envir.* (Earth berm, usually planted, to prevent dispersion of disturbing sounds of road traffic or other noise generators; ▶planted noise attenu-

ation structure); *syn.* noise bund [n] [also UK], noise screen mound [n] [also UK], acoustic screen mound [n] [also UK], noise attenuating earth structure [n], noise attenuating berm [n]; *s* **terraplén [m] antirruidos** (Terraplén con pendiente lo más escarpada posible en el lado de la fuente de ruidos. Una forma especial del **t. a.** es el ▶muro de jardineras antirruidos); *syn.* terraplén [m] de protección contra ruidos; *f* **merlon [m] antibruit** (Remblai longitudinal dont la plus forte pente est située côté de l'émission ; ▶merlon en éléments préfabriqués) ; *g* **Lärmschutzwall [m]** (Erdwall mit zur Lärmquelle möglichst steil angeböschter Seite. Eine besondere Form des **L.s** ist der ▶Steilwall).

3777 noise attenuation planting [n] *landsc.* (Planting of rows of woody plants usually in layers of three different heights to reduce harmful effects of noise pollution. This type of noise control will be perceived psychologically; in terms of dB[A] it is not measurable); *s* **plantación [f] protectora antirruidos** (Plantación de leñosas, normalmente de tres hileras, para reducir las inmisiones de ruido. Estas plantaciones no reducen realmente las inmisiones acústicas, pero tienen efectos [p]sicológicos positivos en las personas afectadas); *syn.* plantación [f] protectora contra ruidos; *f* **plantation [f] contre le bruit** (Bande boisée à trois strates, plantée sur plusieurs lignes en protection contre les nuisances sonores ; dispositif de nature principalement psychologique n'ayant un intérêt technique véritable qu'en relation avec la réalisation d'un merlon ; *syn.* écran [m] végétal acoustique ; *g* **Lärmschutzpflanzung [f]** (Dreistufig aufgebauter, mehrreihiger Gehölzstreifen zur Abwehr schädlicher Geräuschimmissionen. Diese Lärmschutzeinrichtung wird von der Bevölkerung primär psychologisch wahrgenommen — die Lärmreduzierung ist in dB[A] kaum messbar).

noise attenuation structure [n] *constr. trans.* ▶planted noise attenuation structure.

noise baffle [n] [UK] *constr. envir.* ▶noise barrier wall.

3778 noise barrier wall [n] *constr. envir.* (Most frequently used noise protection technique along traffic routes due to its relatively low space requirement. A distinction is made between deflecting, absorptive and high-absorption barriers; ▶noise screening facility); *syn.* acoustic screen wall [n] [also UK], noise baffle [n] [also UK] (TAN 1975, 91); *s* **pantalla [f] antirruidos** (Instalación antirruidos con poca necesidad de espacio que se utiliza muy a menudo a lo largo de carreteras y autopistas. Se diferencia entre pantallas reflectantes, absorventes y muy absorventes); *syn.* pantalla [f] de protección contra ruidos; *f* **écran [m] acoustique** (Protection acoustique constituée en général d'éléments préfabriqués, implantée en bordure des infrastructures de transport terrestres ; on distingue selon la nature des matériaux utilisés les écrans absorbants et les écrans réfléchissants ; *termes spécifiques* mur antibruit, écran antibruit ; ▶dispositif antibruit) ; *g* **Lärmschutzwand [f]** (Entlang von Verkehrswegen auf Grund ihres geringen Platzbedarfes am häufigsten verwendete ▶Lärmschutzeinrichtung. Man unterscheidet zwischen reflektierenden, absorbierenden und hochabsorbierenden Wänden); *syn.* Schallschutzwand [f].

noise bund [n] [UK] *constr. envir.* ▶noise attenuation mound.

3779 noise contour map [n] *plan.* (Graphic presentation showing ▶noise pollution in cities, or portions thereof, or along roads or around individual noise sources, based upon measurements of noise dispersion or sound pressure levels; ▶noise level contouring); *syn.* noise exposure map [n] [also US]; *s* **mapa [m] de ruido** (Representación gráfica de la ▶contaminación acústica de ciudades, barrios o zonas específicas de éstas, o alrededor de fuentes dominantes de ruido, basada en mediciones de difusión del ruido o de niveles de presión acústica. Según art. 15 de la ley

N

de ruidos, estos mapas contienen además los índices acústicos previstos en cada una de las áreas afectadas, los valores límite y objetivos de calidad acústica aplicable a dichas áreas, así como el número estimado de personas, de viviendas, de colegios y de hospitales expuestos a la contaminación acústica en cada área acústica; cf. Ley 37/2003, del Ruido; ►difusión del ruido); *syn.* mapa [m] de contaminación acústica; *f 1* **plan [m] de gêne sonore** (Plan délimitant trois zones selon les indices psophiques [zone I, 96 ; zone II, 96 et 89 ; zone III, 89 et 84 étendue à 89 et 78] et servant à définir les riverains d'un aérodrome pouvant bénéficier d'une aide couverte par la taxe pour l'atténuation des nuisances phoniques payée par les exploitants d'aéronefs) ; *f 2* **plan [m] d'exposition au bruit (PEB)** (Plan fixant les conditions d'utilisation des sols situés autour des aérodromes civils et militaires et délimitant trois zones de bruit fort A et B, et modéré, C selon la gêne due au bruit ; plan soumis à enquête publique, les documents d'urbanisme devant être compatible avec celui-ci) ; *f 3* **carte [f] d'émission sonore** (Représentation cartographique des ►nuisances sonores en milieu urbain, en bordure des infrastructures de transport terrestres, au voisinage d'installations fixes, établie en fonction des mesurages de niveaux sonores ou des niveaux sonores prévisionnels calculés ; ►propagation du bruit) ; *syn.* carte [f] de bruit ; *g* **Lärmkarte [f]** (Darstellung der Lärmsituation in Städten, Stadtteilen, entlang von Straßen oder um einzelne Schallquellen herum auf Grund einer Ausbreitungsrechnung oder mittels Messungen der Schalldruckpegel; ►Lärmausbreitung, ►Lärmbelastung).

3780 noise control [n] *envir.* (All devices and measures taken for the prevention of harmful noise pollution. There are two distinct forms: **1.** 'active' or 'technical' noise control [= primary protection measures to reduce noise at the source], and **2.** 'passive' noise control [= secondary protection measures to prevent noise dispersion, e.g. by walls, mounds); *syn.* sound control [n] [also US]; *s* **protección [f] contra el ruido** (Todo tipo de instalaciones y medidas destinadas a reducir y evitar emisones e inmisiones acústicas nocivas. Existe la «p. pasiva c. el r.» [medidas contra la dispersión del ruido o sea la inmisión] y la «p. activa c. el r.» [medidas para evitar o reducir el ruido en su origen]. **En Es.** las medidas de **p. c. el r.** se fijan **planes de acción en materia de contaminación acústica** que tienen como objetivo a] afrontar globalmente las cuestiones concernientes a la contaminación acústica en la o las correspondientes áreas acústicas, b] determinar las acciones prioritarias a realizar en caso de superación de los valores límite de emisión o inmisión o de incumplimiento de los objetivos de calidad acústica y c] proteger las zonas tranquilas en las aglomeraciones y en el campo abierto contra el aumento de la contaminación acústica; cf. art. 23 Ley 37/2003, del Ruido); *f* **protection [f] contre le bruit** (Toutes mesures et dispositions mises en œuvre pour prévenir, réduire, protéger les nuisances sonores ; il existe divers moyens de lutter contre le bruit, p. ex. **1.** dès la conception du tracé ou de l'aménagement prévu [reculement, éloignement déblais, remblais, **2.** par des actions à la source par l'isolation phonique des sources de production de bruit [*mesures actives contre le bruit*, limitations sonores des moteurs automobiles ou des engins de chantier, régulation de la circulation — fluidité, vitesse, poids lourds], **3.** par des actions sur la propagation [*mesures passives contre le bruit*, murs antibruit, écrans réfléchissants ou absorbants, couverture des voies construites en tranchée, buttes de terre] et **4.** par des actions à la réception telles que l'insonorisation des bâtiments avec isolation acoustique de façade, double vitrage, etc.) ; *g* **Lärmschutz [m, o. Pl.]** (Alle gesetzlich verordneten Vorrichtungen und Maßnahmen zur Abwendung von schädlichen Geräuschimmissionen. Man spricht vom ‚aktiven' oder ‚technischen **L.'** [= primärer Schutz in Form von Maßnahmen zur Reduzierung des Lärms an der Quelle] und ‚passiven **L.'** [= sekundärer Schutz in Form von Maßnahmen gegen Lärmverbreitung]. In D. gelten für den **L.** insbesondere die Vorschriften des BImSchG. Nähere Bestimmungen enthalten vor allem die TA Lärm i. d. F. vom 26.08.1998, die Verkehrslärmschutzverordnung vom 12.06.1990 [16. BImSchV], die Sportanlagenlärmschutzverordnung vom 18.07.1991 [18. BImSchV], die Baumaschinenlärmverordnung vom 10.11.1986 und die Rasenmäherlärmverordnung i. d. F. vom 13.07.1992. In A. wird der **L.** auf Bundesebene in verschiedenen Gesetzen wie Gewerbeordnung, Arbeitnehmerschutzgesetz, Bundesstraßengesetz, Kraftfahrgesetz und durch verstreute Regelungen in Ländergesetzen geregelt. In der CH. wird der **L.** im Umweltschutzgesetz vom 07.10.1983 sowie in der **L.**-VO vom 15.12.1986 geregelt); *syn.* Schallschutz [m, o. Pl.]).

3781 noise corridor [n] *envir. plan.* (Area along a linear source of noise, e.g. transportation route, the width of which may be calculated according to the measurement of sound in dB[A]); *s* **banda [f] de ruido** (Se mide a lo largo de fuentes de ruido lineales, como carreteras o calles, en dB[A]); *syn.* banda [f] de contaminación acústica; *f* **corridor [m] d'étude du bruit** (Zone le long d'infrastructures terrestres dont la largeur doit être pertinente par rapport aux méthodes de calcul de prévision des niveaux sonores [en particulier par la prise en compte des effets des variations météorologiques, vent et température] ; *terme utilisé en F.* zone d'étude de bruit) ; *syn.* couloir [m] d'étude de bruit ; *g* **Lärmband [n]** (Bereich entlang linearer Lärmquellen, z. B. Verkehrswege, der durch die Schallausbreitung, gemessen in dB[A], eingrenzbar ist und anhand von Ausbreitungsberechnungen ermittelt werden kann; cf. auch DIN 18 005 Schallschutz im Städtebau).

3782 noise decibel level [n] *envir.* (Scale of decibels expressed logarithmically and used to measure harmful ►sound pressure levels in dB[A]; ►ambient noise level); *syn.* noise pollution level [n]; *s* **nivel [m] de ruidos** (Escala logarítmica de decibelios —dB[A]— para medir el ►nivel de intensidad del ruido; ►nivel de ruidos ambientales); *syn.* nivel [m] de contaminación acústica; *f* **niveau [m] sonore** (**1.** Échelle de mesure du niveau d'égale sensation du son établie sur une base logarithmique et exprimée en décibels [dB(A)] ; **2.** le ►niveau de pression acoustique mesuré en milieu extérieur et exprimé en dB[A] est dénommé ►niveau sonore ambient) ; *syn.* niveau [m] sonore des bruits émis ; *g* **Lärmpegel [m]** (**1.** Logarithmisch aufgebaute dB-Skala zur Messung des Schalldruckpegels in dB[A]. **2.** An einem bestimmten Ort außerhalb von Gebäuden gemessener ►Schalldruckpegel in dB[A] ist der ►Umgebungslärmpegel); *syn. zu 2.* Lärmpegel [m] außerhalb von Gebäuden).

noise dispersion [n] *envir.* ►traffic noise dispersion.

3783 noise exposure [n] *envir.* (Disturbing effect of ►noise pollution upon life-forms); *s* **exposición [f] al ruido** (Efecto pernicioso de la ►contaminación acústica sobre los seres vivos); *f* **exposition [f] au bruit** (Effet de la gène provoquée par le bruit sur les êtres vivants ; ►nuisance sonore) ; *g* **Lärmeinwirkung [f]** (Belastung von Lebewesen durch störende Geräuschimmissionen; ►Lärmbelastung).

noise exposure map [n] [US] *plan.* ►noise contour map.

noise forecast [n] *plan.* *►consultant's report on noise.

noise level [n] *envir.* ►ambient noise level.

3784 noise level contouring [n] *envir.* (Graphic presentation of noise pollution on a ►noise contour map); *s* **difusión [f] del ruido** (Representación gráfica de bandas de propagación de la contaminación acústica en una zona; ►mapa de ruido); *f* **propagation [f] du bruit** (Représentation graphique ou cartographique des niveaux sonores en milieu extérieur [►carte d'émission sonore] calculés selon une étude acoustique pour un territoire exposé à des nuisances sonores qui résultent des bruits de circulation routière et ferroviaire, du bruit industriel, ainsi que des bruits d'origine aéronautique ; ►plan de gêne sonore, ►plan d'exposition au bruit) ; *g* **Lärmausbreitung [f]** (Zeichnerische Darstellung der Schallbelastung eines Gebietes in einer ►Lärmkarte).

3785 noise level record [n] [US] *plan.* (Systematic collection of data pertaining to noise and its spatial distribution in a regional or supraregional [US]/supra-regional [UK] context); context; in **U.S.**, the term **noise level register** implies that the data or information gathered has special legal status or consequences; *syn.* noise level register [n] [UK]; *s* **registro [m] de contaminación acústica** (Colección sistemática de datos sobre el ruido y su dispersión espacial a nivel regional o supraregional); *f 1* **recensement [m] et classement [m] des infrastructures de transports terrestres** (F., détermination de cinq catégories d'infrastructures de transports terrestres en fonction de niveaux sonores de référence diurnes et nocturnes ainsi que la largeur maximale correspondante des secteurs affectés par le bruit, situés au voisinage de l'infrastructure ; art. 3 du décret n° 95-21 du 9 janvier 1995) ; *f 2* **registre [m] des nuisances sonores (≠)** (D., collecte systématique de données relatives aux nuisances sonores et leur distribution spatiale à l'échelle régionale ou interrégionale) ; *g* **Lärmkataster [m,** *in A nur so* **— oder n]** (Systematische Datenzusammenstellung zur räumlichen Lärmsituation im regionalen oder überregionalen Bereich).

noise level register [n] [UK] *plan.* ►noise level record [US].

3786 noise nuisance [n] *envir.* (Disturbance of human comfort and hearing due to an unacceptably high level of noise; ►noise pollution); *s* **molestia [f] causada por el ruido** (Nivel de ruido que produce molestias físicas, [p]síquicas o sociales; ►contaminación acústica); *syn.* molestia [f] causada por la contaminación acústica; *f* **gêne [f] sonore** (Nuisance ressentie variant en fonction des caractéristiques physiques du bruit et de nombreux facteurs individuels [physiologiques, psychologiques, sociologiques et contextuels]. Les exigences réglementaires pour réduire le bruit des transports et améliorer la qualité acoustique des logements sont essentiellement fondées sur la quantification des niveaux acoustiques. La qualité sonore [design sonore] correspond à une demande de confort et de santé qui tend à être prise en compte par la règlementation ; ►nuisance sonore) ; *g* **Lärmbelästigung [f]** (Das Wohlbefinden störende Wirkungen von Geräuschbelastungen; ►Lärmbelastung); *syn.* Lärmbeeinträchtigung [f].

3787 noise pollution [n] *envir.* (Noise emission, in an area under study, which impairs physical, mental and social wellbeing; ►noise nuisance); *s* **contaminación [f] acústica** (Presencia en el ambiente de ruidos o vibraciones, cualquiera que sea el emisor acústico que los origine, que impliquen molestia, riesgo o daño para las personas, para el desarrollo de sus actividades o para los bienes de cualquier naturaleza, o que causen efectos significativos sobre el medio ambiente; art. 3d, Ley 37/2003, del Ruido; ►molestia causada por el ruido); *f 1* **nuisance [f] sonore** (Pollution par un son de fréquence trop importante qui crée une gêne et est néfaste pour la santé et la qualité de la vie des individus qui la subissent ; l'oreille humaine perçoit les sons à partir de 0 dB [seuil d'audibilité] et jusqu'à 120 dB [seuil de douleur]) ; *syn.* nuisance [f] acoustique, pollution [f] sonore ; *f 2* **bruit [m] d'ambiance** (Ensemble des bruits perçus dans un espace donné) ; *syn.* émissions [fpl] sonores ; *f 3* **nuisances [fpl] sonores des infrastructures terrestres** (Gêne de l'usager des riverains provoquée par les voies routières et lignes ferroviaires ; ►gêne sonore) ; *syn.* bruit [m] des infrastructures de transports terrestres ; *g* **Lärmbelastung [f]** (Umfang derjenigen Geräuschim-

missionen in einem Betrachtungsgebiet, die das körperliche, seelische und soziale Wohlbefinden beeinträchtigen; der Lärm wird in dB[A] gemessen; in der Literatur wird die akustische Schmerzgrenze für den Menschen mit 120 dB[A] angegeben, ab 85 dB ist das menschliche Gehör gefährdet; ►Lärmbelästigung); *syn.* Lärmimmission [f].

noise pollution level [n] *envir.* ►noise decibel level.

3788 noise protection area [n] *leg. plan.* (Area of land in the vicinity of airports, or other landing fields with comparable noise effect, where planning restrictions on residential development are necessary to protect the population against the noise of aircraft; ►noise zone); *s 1* **zona [f] de servidumbre acústica (En Es.,** sectores del territorio delimitados en las ►zonas de ruido, en los que las inmisiones podrán superar los objetivos de calidad acústica aplicables a las correspondientes áreas acústicas y donde se podrán establecer restricciones para determinados usos del suelo, actividades, instalaciones o edificaciones, con la finalidad de, al menos, cumplir los valores límites de inmisión establecidos para aquéllos; art. 3p Ley 37/2003, del Ruido); *s 2* **zona [f] de edificación limitada (En D, UK, EE.UU.,** zona en los alrededores de aeropuertos o aeródromos con contaminación acústica comparable en la cual existen limitaciones de usos —restricción del uso residencial— para proteger a la población del ruido causado por los aviones); *f* **zone [f] de bruit** (Les zones de bruit autour d'un aéroport sont définies par un document graphique d'urbanisme **le plan d'exposition au bruit [PEB]** qui détermine la nature et la densité de l'urbanisme possible en fonction du bruit prévisible à 10-15 ans en définissant un périmètre représenté par des courbes de bruit à l'intérieur duquel s'appliquent des restrictions à l'urbanisme ; le code de l'urbanisme fixe quatre zones de bruit sur la base de l'indice Lden [*Level day evening night* — indice exprimé en décibels [dB] qui caractérise le niveau d'exposition total au bruit des avions sur l'ensemble d'une année] : **zone A :** zone de bruit fort comprise à l'intérieur de la courbe d'indice 70 ; **zone B :** zone de bruit fort comprise entre la courbe d'indice 70 et la courbe d'indice choisie entre 62 et 65, **zone C :** zone de bruit comprise entre la limite de la zone B et la courbe d'indice choisie entre 57 et 55, **zone D :** zone comprise entre la limite de la zone C et la courbe d'indice 50 ; dans chacune des zones, des limitations du droit de construire sont prescrites, en application des dispositions de l'article L. 147-5 du Code de l'Urbanisme ; les constructions autorisées dans les zones de bruit doivent néanmoins satisfaire aux prescriptions d'isolation acoustique, en application des dispositions de l'article L. 147-6 du Code de l'Urbanisme ; le certificat d'urbanisme doit signaler l'existence de la zone de bruit et l'obligation de respecter les règles d'isolation acoustique. De plus, le contrat de location d'un immeuble situé dans une des zones doit comporter une clause claire et lisible précisant la zone du PEB dans laquelle l'immeuble est situé ; cf. Charte de qualité de l'environnement — Aéroport Toulouse-Blagnac) ; ►zone à émergence réglementée, ►zone d'ambiance sonore modérée ►zone de bruit) ; *g* **Lärmschutzgebiet [n]** (In der Umgebung von Flughäfen und sonstigen Flugplätzen mit vergleichbaren Lärmauswirkungen festgesetztes Gebiet, in dem Planungsbeschränkungen für die Siedlungsentwicklung zum Schutze der Bevölkerung vor Fluglärm erforderlich sind; ►Lärmschutzzone); *syn.* Lärmschutzbereich [m].

noise reduction [n] *envir.* ►noise abatement.

noise reduction zone [n] [US] *leg. plan.* ►noise zone [US].

3789 noise screening facility [n] *envir.* (Generic term for installations which reduce harmful effects of noise pollution; *specific terms* ►noise attenuation planting, ►noise attenuation mound, ►noise barrier wall, ►planted noise attenuation struc-

N

ture); *syn.* sound barrier [n]; *s 1* **instalación [f] antirruidos** (Término genérico para instalaciones que reducen la difusión del ruido en una dirección determinada: ▶muro de jardineras antirruidos, ▶pantalla antirruidos, ▶plantación protectora antirruidos, ▶terraplén antirruidos); *s 2* **aislamiento [m] acústico** (Medidas constructivas o técnicas para reducir la inmisión en un lugar o la emisión de ruido de un aparato); *f* **dispositif [m] antibruit** (*Terme générique* Dispositif de réduction du bruit, de protection phonique prévu pour la limitation du bruit [absorption acoustique ou réflexion] des infrastructures routières ; *termes spécifiques* écran réfléchissant, ▶merlon antibruit, ▶écran acoustique, ▶merlon en éléments préfabriqués, ▶plantation contre le bruit) ; *syn.* protections [fpl] acoustiques, écran [m] antibruit ; *g* **Lärmschutzeinrichtung [f]** (Vorrichtung zur Abwendung von Geräuschimmissionen; *UBe* ▶Lärmschutzpflanzung, ▶Lärmschutzwall, ▶Lärmschutzwand, ▶Steilwall); *syn.* Schallschutzeinrichtung [f].

noise screen mound [n] [UK] *constr. envir.* ▶noise attenuation mound.

3790 noise source [n] *envir.* (Location or cause of noise); *s* **emisor [m] acústico** (Cualquier actividad, infraestructura, equipo, maquinaria o comportamiento que genere contaminación acústica; art. 3e Ley 37/2003, del Ruido); *syn.* fuente [f] de ruido, foco [m] de ruido; *f* **source [f] de bruit** (Source de propagation d'un phénomène vibratoire plus ou moins intense ; les principales sources de bruit sont : le voisinage [bruits intérieurs au bâtiment], les activités commerciales artisanales, industrielles ou de chantier, la circulation routière [bruit lié à la propulsion, au roulement et au déplacement d'air], les transports ferroviaires, le trafic aérien, les activités ludiques et sportives) ; *g* **Lärmquelle [f]** (Ort oder Ursache des Lärms).

3791 noise zone [n] [US] *leg. plan.* (Part of a ▶noise protection area designated by a local agency/local authority which is classified in three zones with decreasing noise intensity [decibel levels]); *syn.* noise abatement zone [n] [UK], noise reduction zone [n] [US]; *s* **zona [f] de ruido** (En D. parte de una ▶zona de edificación limitada que está dividida en tres según la intensidad disminuyente de ruido: zona A > 75 dB[A], zona B > 67 dB[A], zona C > 62 dB[A]); *f 1* **périmètre [m] de zone de bruit** (F., périmètre d'une zone de conditions d'utilisation des sols situées autour des différentes catégories d'aérodromes, interdisant ou limitant la construction d'immeubles d'habitation et/ou l'implantation d'équipements publics [prescriptions d'urbanisme] ; ces ▶zones de bruit sont délimitées aujourd'hui sur la base de valeurs l'indice Lden [anciennement indice psophique, indice défini à partir du niveau sonore maximal exprimé en PNDB (niveau instantané PNLmax) perçu lors du passage de chaque avion ; cet indice représente le cumul énergétique des bruits maximum sur une journée moyenne, chaque mouvement de nuit étant compté dix fois] ; **D.,** ▶zone de bruit divisée en trois zones de niveaux sonores décroissants ; Zone A > 75 dB[A], Zone B > 67 dB[A], Zone C > 62 dB[A]) ; *syn.* courbe [f] limitant la zone de bruit ; *f 2* **zone [f] d'ambiance sonore modérée** (**F.,** zone à vocation résidentielle, éloignée de toute source de bruit gênante où le bruit ambiant préexistant avant travaux est globalement inférieur à 65 dB[A] de jour et 60 dB[A] de nuit et à préserver par des exigences sur les limites de bruit plus importantes que sur les autres zones ; circulaire n° 97-110 du 12 décembre 1997) ; *f 3* **zone [f] à émergence réglementée** (**F.,** zone dans lesquelles les émissions sonores ne doivent pas engendrer d'émergence supérieure à des valeurs variant selon le bruit ambiant de la zone et la période jour ou nuit ; cf. § 37 CPEN) ; *g* **Lärmschutzzone [f]** (Bereich, der nach dem Fluglärmgesetz vom 30.03.1971 in der Umgebung von Flughäfen festgesetzt wird; ein ▶Lärmschutzgebiet wird in drei Zonen

abnehmender Lärmintensität eingeteilt: Zone A > 75 dB[A], Zone B > 67 dB [A], Zone C > 62 dB[A]; die **L.n** sollen alle fünf Jahre auf ihre Gültigkeit überprüft werden); *syn.* Lärmschutzbereich [m].

3792 nomadic bird [n] *zool.* (Vagabond bird species which changes its breeding place regularly between different areas, because of rhythmic alterations in the food supply [e.g. spruce seeds], probably determined endogenely—as a form of ▶adaptation; e.g. Common Crossbill *[Loxia curvrostra]*); *s* **ave [f] nomádica** (Especie de ave que —debido a cambios periódicos de la oferta alimenticia y probablemente debido a factores endógenos [como ▶adaptación (evolutiva)]— cambia regularmente de lugar de reproducción); *f* **oiseau [m] explorateur** (Espèce d'oiseau procédant à une migration calculée de déplacement afin de localiser de nouvelles sources de nourriture et changeant donc régulièrement d'aire de nidification mais capable de retourner dans sa région d'origine, [comme résultat d'une ▶adaptation héréditaire], p. ex. le Bec croisé des Sapins *[Loxia curvrostra]*) ; *g* **Nomadenvogel [m]** (OB zu Vagabundenvogel. Art, die auf Grund rhythmischer Veränderungen im Nahrungsangebot [z. B. Fichtensamen] wahrscheinlich endogen bedingt [als ▶Anpassung 2] regelmäßig einen Brutortswechsel zwischen verschiedenen Gebieten vornimmt, wie z. B. der Fichtenkreuzschnabel *[Loxia curvrostra]*).

3793 nomadism [n] *zool.* (Movement of animals from area to area without observable periodicity or pattern); *s* **nomadismo [m]** (Movimiento de animales de área en área sin periodicidad o pauta observable o aparente; DINA 1987); *f* **nomadisme [m]** (Déplacement d'animaux d'un endroit à un autre sans périodicité ou ordre déterminés) ; *g* **Nomadentum [n, o. Pl.] (2)** (Ständiges Umsiedeln/Umherstreifen von wild lebenden Tieren); *syn.* Nomadismus [m], permanente Translokation [f].

non-accessed land [n] [UK] *plan. urb.* ▶unserved land [US].

non-cohesive soil [n] [UK] *pedol.* ▶granular soil [US].

3794 noncompactible substrate [n] [US]/**non-compactible substrate** [n] [UK] *constr.* (Mineral material with such ▶particle-size distribution that, after compaction, enough ▶pore volume is available for drainage and plant growth); *s* **sustrato [m] no compactable** (Material que, debido a la mezcla en el tamaño de las partículas, tiene suficiente ▶porosidad [total] aun cuando se ha asentado; ▶granulometría); *f* **substrat [m] non compactable** (Matériaux qui en raison de son ▶spectre granulométrique garde après compactage une porosité suffisante pour assurer le drainage du sol et la croissance des végétaux ; ▶porosité totale) ; *syn.* substrat [m] de structure stable ; *g* **verdichtungsfreies Substrat [n]** (Material, das auf Grund seines Korngefüges nach einer Verdichtung genügend Luftporenvolumen behält; ▶Kornverteilung, ▶Porenvolumen); *syn.* gerüststabiles Substrat [n].

nondecaying [adj] [US]/**non-decaying** [adj] [UK] *biol.* ▶nonrotting [US]/non-rotting [UK].

nonfishing season [n] [US] *conserv. game'man. hunt. leg.* ▶closed season.

non-for-profit organization [n] [US] *leg. pol. sociol.* ▶nonprofit organization.

3795 nongame species [n] [US]/**non-game species** [n] [UK] *leg. hunt.* (WPG 1976, 233; species of animal which, while not game according to the hunting law, e.g. magpies, jays, may be hunted as long as nature conservation laws are not contravened); *s* **especie [f] no cinegética** (Especie de fauna que no está en el catálogo de especies objeto de caza pero que puede ser objeto de caza si no está protegida); *f* **espèce [f] dont la chasse**

est **prohibée** (Espèces animales non soumises à la législation de la chasse et qui peuvent être chassées lorsqu'elles ne sont pas protégées par la législation de la protection de la nature) ; *syn.* espèce [f] non chassable, espèce [f] protégée dans le cadre du droit de la chasse ; *g* **nicht jagdbares Tier [n]** (Tierart, die nicht dem Jagdrecht unterliegt, z. B. Elster, Eichelhäher, die, wenn sie nicht unter Naturschutz steht, dennoch gejagt werden darf).

nonhunting season [n] [US] *conserv. game 'man. hunt. leg.*
►closed season.

3796 nonindigenous species [n] [US]/**non-indige-nous species** [n] [UK] *phyt. zool.* (Species which is not enduring or adaptable to conditions in a particular location; ►introduced species); *syn.* nonnative species [n] [US]/non-native species [n] [UK]; *s* **especie [f] no adaptada a un hábitat (específico)** (Especie que no es natural de una ubicación o no crece bien en ella; ►especie intrusa); *f* **espèce [f] non adaptée au milieu** (ne pouvant s'établir durablement sur une station ; ►espèce introduite) ; *g* **standortfremde Art [f]** (Art, die für einen bestimmten Standort nicht bodenständig oder nicht geeignet ist; ►gebietsfremde Art).

nonnative species [n] [US]/**non-native species** [n] [UK] *phyt. zool.* ►nonindigenous species [US]/non-indigenous species [UK].

nonplastic soil [n] [US] *pedol.* ►granular soil.

3797 nonpoint source pollution [n] [US]/**non-point source pollution** [n] [UK] *envir.* (Dispersed pollution from a variety of sources, e.g, runoff from roads, urban and agricultural areas and, importantly, from animal feeding operations which adversely affect groundwater, air, soil, and sensitive ecosystems; *opp.* ►point source pollution); *s* **contaminación [f] ubicua** (Polución ambiental dispersa procedente de muchas fuentes no bien localizables; *opp.* ►contaminación focal); *f* **pollution [f] superficielle** (Détérioration de l'environnement par des sources de pollution mobiles ou non exactement localisées telle que la circulation routière, ou encore par des pollutions localisées mais ayant un champ d'action très étendu telle que la pollution agricole des sources et de la nappe phréatique par les nitrates ; *opp.* ►pollution ponctuelle) ; *syn.* pollution [f] par des sources mobiles ; *g* **flächenhafte Verschmutzung [f]** (Kontaminierung der Umwelt durch mobile oder nicht genau lokalisierbare Verschmutzungsquellen, z. B. Straßenverkehr oder lokalisierbare Verschmutzungen, z. B. durch unsachgemäße landwirtschaftliche Bodennutzung; *opp.* ►punktuelle Verschmutzung).

3798 nonpreventable disturbance [n] [US]/**non-preventable disturbance** [n] [UK] *ecol. envir.* *s* **perturbación [f] inevitable** *syn.* daño [m] inevitable; *f* **nuisance [f] inévitable** ; *g* **unvermeidbare Beeinträchtigung [f]**.

3799 nonprofit housing association [n] [US]/**nonprofit housing association** [n] [UK] *adm. urb.* (Private or public enterprise acting in the public interest by building and managing low-income residential accommodations and associated communal facilities such as kindergarten, children's play grounds, etc.; ►low-income housing); *s* **empresa [f] de construcción de viviendas de protección oficial [Es]** (Organismo público o privado que se dedica a la construcción y administración de viviendas para la población con ingresos bajos y por ello tiene un estatus fiscal privilegiado; ►construcción de viviendas de protección oficial); *syn.* empresa [f] de construcción de viviendas sociales; *f* **société [f] anonyme d'habitations à loyer modéré (HLM)** (Organisme privé ayant pour objet de réaliser gérer et entretenir en vue de la location des opérations d'aménagement et de construction ainsi que des hébergements de loisirs à vocation sociale pour le compte d'une collectivité locale ;

►habitat social ; **F.**, parmi les organismes publics et privés d'habitation à loyer modéré on distingue suivant leurs statuts : les offices publics d'aménagement et de construction [OPAC], les offices publics d'habitations à loyer modéré [OPHLM], les sociétés anonymes d'habitations à loyer modéré, les sociétés anonymes coopératives de production d'habitations à loyer modéré, les sociétés anonymes de crédit immobilier [SACI], les fondations d'habitations à loyer modéré) ; *g* **gemeinnütziges Wohnungsunternehmen [n]** (Dem Gemeinwohl verpflichtetes Privatunternehmen oder Unternehmen der öffentlichen Hand, das die Errichtung, Bewirtschaftung und Verwaltung von Wohnungen, deren Gemeinbedarfseinrichtungen und Wohnfolgeeinrichtungen, z. B. Kindergarten, Spielplätze etc., übernimmt; ►sozialer Wohnungsbau); *syn.* gemeinnützige Wohnungsbaugesellschaft [f].

3800 nonprofit organization [n] [US]/**non-profit organization** [n] [UK] *leg. pol. sociol.* (Private or public association organized for educational, religious or charitable purposes without financial gain; e.g. nature conservation association; ►nature conservation organization, ►non-profit organization suit [US]/litigation brought by a non-profit organization [UK]); *syn.* non-for-profit organization [n] [also US]; *s 1* **asociación [f] sin fines de lucro** (Organización privada cuyo objetivo es promover actividades en diferentes campos, como la ciencia, la educación, el medio ambiente o sociales y culturales sin perseguir obtener ganancias de ellas. Pueden estar reconocidas oficialmente o no; ►asociación para la protección de la naturaleza, ►derecho a pleito de organizaciones ecologistas y de protección de la naturaleza, ►entidad de utilidad pública); *syn.* organización [f]/entidad [f] sin fines de lucro; *s 2* **entidad [f] de utilidad pública** (Empresa pública u organización privada cuyos objetivos son de interés público por lo que reciben un trato especial por parte del Estado, como liberación de impuestos, subvenciones, tras ser reconocidas oficialmente; ►asociación sin fines de lucro); *syn.* asociación [f]/organización [f] de utilidad pública, entidad [f] de bien público [RA]; *f 1* **association [f] à but non lucratif** (Association régie par loi du 10 juillet 1901 ; groupement d'individus s'étant fixé l'objectif de défendre et promouvoir des intérêts spécifiques dans les domaines économique, social, culturel et sportif ; ►association reconnue d'utilité publique) ; *f 2* **association [f] reconnue d'utilité publique** (Organisme privé ou public poursuivant des objectifs d'intérêt général [soumis à un contrôle administratif conformément à la loi du 10 juillet 1901] ayant fait l'objet d'un agrément, qui lui confère la personnalité juridique et certaines prérogatives dans le domaine de la protection de la nature et de l'environnement [art. 40 de la loi du 10 juillet 1976], de l'urbanisme [loi du 31 décembre 1962] ; ►action en justice des associations, ►association à but non lucratif, ►association de protection de l'environnement) ; *syn.* association [f] agréée ; *g* **gemeinnütziger Verband [m]** (Private, parteipolitisch neutrale, nicht auf Gewinn ausgerichtete Vereinigung, die gemeinnützige Ziele verfolgt, z. B. die Förderung des Natur-, Umwelt- und Landschaftsschutzes; ►Naturschutzverband, ►Verbandsklage).

non-profit organization [n], **litigation brought by a** [UK] *conserv. leg.* ►non-profit organization suit [US].

3801 non-profit organization suit [n] [US] *conserv. leg.* (Legal possibility to bring an action against nature conservation and other government agencies when false decisions have been made or defects have arisen. **In Europe**, nature conservation organizations have the right to institute legal proceedings—bring suit [US]—against government authorities/agencies to reverse decisions which are harmful to the environment, despite prior participation in public hearings on the planning process; **in U.S.**, nature conservation organizations may have 'standing' [i.e. right of an individual or organization to go to Court to enforce legal obligations] to intervene in judicial actions); *syn.* litigation [n] brought by a non-profit organization [UK]; *s* **derecho [m] a pleito de organizaciones ecologistas y de protección de la naturaleza** (D., en algunos estados federados existe el derecho de las organizaciones ecologistas reconocidas de presentar demandas ante juez contra proyectos del gobierno que tienen impacto negativo sobre el ambiente); *f* **action [f] en justice des associations** (Les associations agréées de défense et de protection de l'environnement, au-delà de leur participation à l'élaboration des documents d'urbanisme et à la gestion des organismes publics, peuvent se porter partie civile et prendre l'initiative d'engager des poursuites pénales devant les tribunaux répressifs pour les faits constituant des infractions aux différentes réglementations protégeant l'environnement ; C. rural, L. 252-3/4) ; *g* **Verbandsklage [f]** (In Fragen des Naturschutzes und der Landschaftspflege über Anhörungen und Mitwirkungsrechte der Naturschutzverbände hinausgehende, rechtliche Möglichkeit, Klage vor Verwaltungsgerichten gegen Naturschutz- und andere Behörden wegen Mängeln, Fehlentscheidungen u. Ä. anzustrengen, als Vollzug des Naturschutzrechts [§ 29 BNatSchG]; wurde in D. zuerst 1979 im Bremer Naturschutzgesetz eingeführt); *syn.* Verbandsbeschwerde [f] [CH].

nonputrefying [adj] [US]/**non-putrefying** [adj] [UK] *biol.* ►nonrotting [US]/non-rotting [UK].

3802 nonrenewable power source [n] [US]/**non-renewable power source** [n] [UK] *envir.* (Origin of fossil energy, e.g. petroleum and natural gas; ►renewable power source); *s* **fuente [f] de energía no renovable** (Son aquéllas de origen fósil, como petróleo, gas o uranio; ►fuente de energía renovable); *f* **source [f] d'énergie non renouvelable** (Source d'énergie qui se dégrade dans le processus de production et d'utilisation par l'homme, qui par opposition aux ►sources d'énergie renouvelable, ne se régénère pas, ne se renouvelle pas ou qui se régénère ou se renouvelle pas assez rapidement pour être considérée comme inépuisable à l'échelle des temps ; les principales sources d'énergies non-renouvelables sont l'énergie fossile [charbon, gaz naturel et pétrole] et l'énergie nucléaire) ; *g* **nicht erneuerbare Energiequelle [f]** (Energiequelle, die sich durch die menschliche Nutzung im Vergleich zu erneuerbaren Energien reduziert und sich nicht in absehbarer Zeit erneuern kann; zu den **n. e.n E.n** gehören alle fossilen Brennstoffe wie Kohle, Erdgas und Erdöl sowie die Kernenergie; ►erneuerbare Energiequelle).

3803 nonrotting [adj] [US]/**non-rotting** [adj] [UK] *biol.* (Descriptive term applied to natural resistance to decay; ►weather-resistant; *opp.* ►biodegradable); *syn.* nonputrefying [adj] [US]/non-putrefying [adj] [UK], nondecaying [adj] [US]/non-decaying [adj] [UK], deterioration resistant [adj]; *s* **no degradable [adj]** (Término descriptivo aplicado a las sustancias orgánicas/materiales que resisten procesos de degradación; ►resistente a la intemperie; *opp.* ►biodegradable); *syn.* no pudrible [adj]; *f* **imputrescible [adj]** (Caractérise un corps ou des matériaux qui résistent au divers processus de dégradation ou de décomposition ; ►résistant, ante aux intempéries ; *opp.* ►biodégradable) ; *g* **verrottungsfest [adj]** (So beschaffen, dass organische Stoffe/Materialien dem Zersetzungsprozess standhalten; ►witterungsbeständig; *opp.* ►abbaubar); *syn.* nicht faulend [ppr/adj], nicht vermodernd [ppr/adj], nicht verrottend [ppr/adj], sich nicht zersetzend [ppr/adj].

3804 nonselective herbicide [n] [US]/**non-selective herbicide** [n] [UK] *agr. chem. for. hort.* (Chemical weed killer destroying any plant growth; ►selective herbicide, ►root herbicide); *s* **herbicida [m] total** (Herbicida que impide el creci-

miento de todas las plantas en el lugar donde se aplica; ►herbicida de translocación, ►herbicida parcial); *f* **herbicide [m] total** (Herbicide qui est susceptible de détruire toute plante adventice en empêchant le développement de la végétation en terrains non cultivés avec des persistances d'action variables ; cf. LAP 1979 ; ►herbicide racinaire, ►herbicide sélectif) ; *syn.* désherbant [m] total, herbicide [m] non sélectif ; *g* **Totalherbizid [n]** (Chemisches Unkrautvernichtungsmittel, das jeglichen Pflanzenwuchs vernichtet; ►Teilherbizid, ► Wurzelherbizid).

nonskid [adj] [US]/**non-skid** [adj] [UK] *constr.* ►nonslip [US]/non-slip [UK].

3805 nonskid property [n] [US]/**non-skid property** [n] [UK] *constr.* (Attribute of a surface resistant to skidding); *syn.* resistance [n] to skidding; *s* **propiedad [f] antideslizante** (Característica, p. ej. de pavimentos de suelos, de dificultar o hacer prácticamente imposible que uno se resbale sobre ellos); *syn.* propiedad [f] antiderrapante; *f* **non-glissance [f]** (Caractéristique des revêtements, p. ex. de la voirie piétonne limitant ou empêchant les risques de chute par sol mouillé) ; *g* **Rutschsicherheit [f]** (Beschaffenheit von Belagsmaterialien, die ein Ausrutschen erschweren oder verhindern).

3806 nonslip [adj] [US]/**non-slip** [adj] [UK] *constr. syn.* nonskid [adj] [US]/non-skid [adj] [UK], nonslippery [adj] [US]/non-slippery [adj] [UK], nonslippy [adj] [US]/nonslippy [adj] [US]/non-slippy [adj] [UK], skidfree [adj] [US]/skid-free [adj] [UK]; *s* **antideslizante [adj]** *syn.* antiderrapante [adj], no resbaladizo/a [loc]; *f* **non glissant, ante [loc]** *syn.* antidérapant, ante [adj] ; *g* **rutschsicher [adj].**

nonslippery [adj] [US]/**non-slippery** [adj] [UK] *constr.* ►nonslip [US]/non-slip [UK].

nonslippy [adj] [US]/**non-slippy** [adj] [UK] *constr.* ►nonslip [US]/non-slip [UK].

nonventilated roof [n] [US]/**non-ventilated roof** [n] [UK] *arch. constr.* ►warm roof.

3807 noosphere [n] *ecol.* (►biosphere which has undergone change due to the conscious cultural intervention); *s* **noosfera [f]** (Parte de la ►biosfera más directamente influenciada por el hombre; cf. DINA 1987); *f* **noosphère [f]** (DUV 1984, 345 ; résultat de la transformation de la ►biosphère par l'intelligence humaine) ; *syn.* technosphère [f] ; *g* **Noosphäre [f]** (Die durch bewusste menschliche Tätigkeit veränderte ►Biosphäre).

3808 'no project' option [n] *plan.* (Opportunity of abandoning a proposed project after an investigation, e.g. an ►environmental impact statement [EIS], which predicts adverse environmental effects); *syn.* 'no build' option [n] [also US]; *s* **opción [f] «no hacer nada»** (En el contexto de la evaluación de impacto ambiental en la planificación, p. ej. de una carretera, alternativa posible de no construir para evitar los efectos negativos de la misma sobre el medio); *syn.* alternativa [f] «no hacer nada», opción [f] cero, alternativa [f] cero; *f* **variante [f] 0** (Situation de référence par rapport à laquelle est étudié dans une étude d'impact un projet d'infrastructure dans le cas ou celui-ci ne serait pas réalisé) ; *syn.* option [f] pas de projet ; *g* **Nullalternative [f]** (Bei der Untersuchung von zwei Vorschlägen zur Durchführung eines beabsichtigten Vorhabens die Möglichkeit, das Projekt nicht zu realisieren, z. B. wegen seiner Umweltunverträglichkeit); *syn.* bei mehreren Vorschlägen Nullvariante [f].

normal water level [n] *leg. wat'man.* ►mean-water level.

north arrow [n] [US] *plan.* ►north point.

3809 north-facing slope [n] *geo. met.* (Hill slope aspect, which faces N. or N.E., and so receives minimum light and warmth in the northern hemisphere; ►ubac); *s* **ladera [f] norte**

(En el hemisferio norte, ladera de montaña orientada al norte o al noreste a la que apenas llegan los rayos de sol, por lo que es más oscura y fría; ►ladera umbría); *syn.* vertiente [f] norte, vertiente [f] septentrional; *f* **versant [m] nord** (Versant de montagne exposé au nord ou au nord-est et recevant dans l'hémisphère nord le minimum de lumière et de chaleur ; ►ubac) ; *g* **Nordhang [m]** (Hang eines Berges, der nach Norden oder NO exponiert ist und somit auf der Nordhalbkugel ein Minimum an Wärme und Licht erhält; ►Schattenseite).

north indicator [n] [US] *plan.* ►north point.

3810 north orientation [n] (North-facing aspect or direction of a slope or facade of a building); *s* **orientación [f] norte** (Ladera o edificio con cara al norte); *syn.* exposición [f] norte; *f* **orientation [f] vers le nord** *syn.* exposition [f] au nord ; *g* **Nordlage [f]** *geo. met.* (Gen Norden gerichteter Hang, gen Norden gerichtete Gebäudeseite etc.).

3811 north point [n] *plan.* (Indicator of compass direction on plans or maps); *syn.* north arrow [n] [also US], north indicator [n] [also US]; *s* **norte [m]** (Signo o flecha en mapa o plan que indica el norte); *f* **nord [m]** (Signe [flèche] apposé sur un plan et déterminant son orientation géographique) ; *g* **Nordpfeil [m]** (Himmelsrichtungsanzeiger auf Plänen).

nose-to-kerb parking [n] [UK] *plan. trans.* ►ninety-degree parking layout.

3812 nosing [n] *arch. constr.* (**1.** Overhang of the tread of an upper step in relation to the lower tread. **2.** Rounded or chamfered tread extension beyond the riser plane); *s* **mampirlán [m]** (Parte de la huella que sobresale a la contrahuella; DACO 1988); *syn.* mamperlán [m]; *f* **débordement [m]** (Dépassement de la dalle de surface du giron d'un escalier par rapport à la contremarche, p. ex. de 2 à 3 cm ; GEN 1982-II, 90) ; *syn.* débord [m] du nez de marche, dépassement [m] (GEN 1982-II, 90) ; *g* **Unterschneidung [f]** (Im Treppenbau das Maß des Überstandes der Auftrittsfläche zur unteren Stufe).

nostalgic rose [n] *hort.* ►old-fashioned rose.

notch planting [n] [UK] *constr. for.* ►slit planting [US].

notice [n] [UK]**, emergency preservation** *conserv. leg. plan.* ►temporary restraining order [US].

novelty [n] *hort.* ►new introduction.

3813 noxious fungus [n] *agr. for. hort. phytopath.* (Pathogenic disease on plants; *generic term* ►injurious organism; ►fungal disease); *s* **hongo [m] nocivo** (Agente patógeno en plantas; *término genérico* ►organismo nocivo; ►micosis); *f* **champignon [m] parasite** (Agent pathogène des végétaux [80 % des maladies des plantes cultivées] ; *terme générique* ►organisme nuisible ; ►maladie cryptogamique) ; *g* **Schadpilz [m]** (Pflanzlicher Krankheitserreger; *OB* ►Schadorganismus; ►Pilzkrankheit).

3814 noxious substance [n] *envir.* (Chemical substances with a negative or destructive effect upon humans, animals, plants or whole ►ecosystems; ►air pollutant, ►ecotoxicology); *syn.* pollutant [n], toxic agent [n], contaminant [n]; *s* **sustancia [f] nociva** (Sustancia química que causa daños en los seres humanos, la fauna, la flora o sobre todo los ►ecosistemas y sobre los materiales; ►contaminante de la atmósfera, ►ecotoxicología); *syn.* sustancia [f] tóxica/contaminante; *f* **substance [f] génératrice de nuisances** (Toute substance de nature à avoir des conséquences préjudiciables, un effet nocif sur le sol, la faune, la flore, à dégrader les sites et les paysages, à polluer l'air ou les eaux, à engendrer des bruits ou des odeurs, et d'une façon générale, à porter atteinte à la santé de l'homme et à l'environnement ; loi n° 75-633 du 15 juillet 1975 ; ►écosystème, ►écotoxicologie, ►polluant atmosphérique) ; *syn.* substance [f] nocive, substance [f] nuisible, produit [m] nocif, substance [f] toxique (±), substance [f] contaminante, substance [f] polluante, agent [m] polluant ; *g* **Schadstoff [m]** (Auf Mensch, Tier, Pflanze oder auf ganze ►Ökosysteme sowie auf Materialien negativ oder zerstörend einwirkende chemische Substanz; ►Luftschadstoff, ►Ökotoxikologie).

noxious substances [npl] *envir.* ►concentration of noxious substances, ►emission of noxious substances.

3815 nuclear fuel reprocessing plant [n] *envir.* (Facility for recycling of re-usable materials, especially uranium and plutonium, which are retrieved by mechanical and chemical processes from spent fuel rods; highly radioactive waste is separated and stored in leak-proof containers in special depositories); *s* **planta [f] de reciclaje de residuos radiactivos** (Instalación para recuperar uranio y plutonio de los residuos de centrales nucleares. Este método de procesamiento es muy controvertido porque tiene como resultado la ganancia de plutonio enriquecido que sirve para producir bombas atómicas); *syn.* planta [f] de proceso de residuos radiactivos, instalación [f] de regeneración de combustibles nucleares irradiados; *f* **station [f] de traitement des déchets radioactifs** (Installation de retraitement de combustibles usés, l'uranium étant recyclé dans la fabrication de combustible neuf, le plutonium sous forme de combustible dans les réacteurs classiques ou le réacteur à neutrons rapides ; les déchets radioactifs à haute activité et à vie longue sont conditionnés, enrobés [p. ex. vitrification] pour l'entreposage et le stockage définitif) ; *syn.* centre [m] de retraitement des déchets radioactifs, centrale [f] de traitement des déchets radioactifs, usine [f] de traitement des déchets radioactifs ; *g* **Wiederaufbereitungsanlage [f] für Kernbrennstoffe** (*Abk.* WWA; Anlage, in der verwertbare Stoffe, insbesondere Uran [U] und Plutonium [Pu], durch mechanische und chemische Zerlegung verbrauchter Brennelemente zurückgewonnen werden. Dabei werden die hochradioaktiven Abfälle [Spaltprodukte] abgetrennt und so eingeschlossen, z. B. durch Einglasung, dass sie für ein Endlager geeignet sind); *syn.* Aufbereitungsanlage [f] für Kernbrennstoffe, Kernbrennstoffwiederaufbereitungsanlage [f].

3816 nuclear plant [n] *envir.* (Generic term covering installations concerned with the cycle of nuclear fuel and the supply and disposal service of a nuclear power station, e.g. the extraction and concentration of uranium, the production of fuel elements, the ►reprocessing of used nuclear fuel and the ►final disposal of radioactive waste); *s* **instalación [f] nuclear** (Término genérico para todo tipo de instalaciones concernientes al ciclo de los combustibles nucleares, como de enriquecimiento del uranio, de producción de elementos de combustión, de ►reproceso de material nuclear y de ►almacenamiento definitivo de los residuos radiactivos); *syn.* planta [f] nuclear; *f* **installation [f] nucléaire** (Terme générique pour les installations mettant en œuvre des techniques nucléaires telles que les centrales nucléaires de base [réacteurs de recherche, centres d'études nucléaires, usines de préparation de fabrication ou de transformation des substances radioactives, installations de stockage, dépôt et utilisation de substances radioactives], les centrales nucléaires de puissance, les centrales électronucléaires, les stations de traitement de déchets radioactifs, les installations d'entreposage ou de stockage de déchets radioactifs ; ►retraitement, ►stockage définitif) ; *g* **kerntechnische Anlage [f]** (Anlage, die dem Brennstoffkreislauf und den damit zusammenhängenden Verfahrensstufen der ►Ver- und Entsorgung von Kernkraftwerken, wie Gewinnung und Anreicherung des Urans, Herstellung der Brennelemente, ►Wiederaufbereitung der abgebrannten Kernbrennstoffe und ►Endlagerung des radioaktiven Abfalls dient).

N

3817 nuclear power opposition group [n] [US] *pol. syn.* anti-nuclear action group [UK]; *s* **grupo** [m] **ecologista antinuclear** *syn.* grupo [m] ciudadano antinuclear; *f* **association [f] de défense contre une centrale nucléaire** ; *g* **Bürgerinitiative [f] gegen Atomkraftwerke**.

3818 nuclear power station [n] *envir.* (►nuclear plant); *s* **central [f] nuclear** (►instalación nuclear); *f* **réacteur** [m] **de puissance électronucléaire** (Installation qui produit de l'énergie en entretenant dans le combustible qui constitue son cœur des réactions en chaîne de fission nucléaire initiées, modérées et contrôlées ; ►installation nucléaire) ; *syn.* centrale [f] atomique, centrale [f] nucléaire, réacteur [m] électronucléaire ; *g* **Kernkraftwerk** [n] (*Abk.* KKW; Wärmekraftwerk zur Gewinnung elektrischer Energie durch kontrollierte Spaltung von Atomkernen; ►kerntechnische Anlage); *syn.* Atomkraftwerk [n], Kernkraftanlage [f].

nuisance [n]**, odorous** *envir.* ►odorous annoyance.

number [n] **of individuals** *phyt. zool.* ►species abundance.

number [n] **of stories** [US]/**number of storeys** [UK] *leg. urb.* ►permitted number of stories [US]/permitted number of storeys [UK].

3819 nurse crop [n] *agr. for. land'man.* (Thinly-sown companion plant, providing protection to the main crop, which is subsequently harvested to make way for the development of the long-term plant community; ►nurse plant); *s* **plantación [f] de plantas nodrizas** (Planta acompañante sembrada ralamente que es segada una vez que ha cumplido la labor de favorecer la introducción de otras plantas hasta conseguir la comunidad vegetal deseada; ►planta nodriza); *f* **peuplement** [m] **d'espèces abri** (Ensemble d'espèces abri qui seront enlevées lorsque sera atteint l'objectif de réintroduction des associations végétales vulnérables ; ►espèce abri) ; *g* **Ammenpflanzenbestand** [m] (Summe der als vorläufige Begleiter eingebrachten ►Ammenpflanzen, die später herausgenommen werden oder bei Wiesenflächen nach der Mahd wegen der Schnittunverträglichkeit nicht wiederkommen, so dass sich die angestrebte Pflanzengesellschaft oder der zu erntende Bestand entwickeln kann).

3820 nurse crop [n] **of woody plants** *for. landsc. phyt.* (Nurse trees and shrubs introduced to foster another, more important crop during its youth; cf. SAF 1983); *s* **cubierta [f] de tutela** (Cultura protectora formada por especies arbóreas de crecimiento rápido, fuerte y heliófilas, que sirven de protección [sombra] para otras especies de crecimiento más lento y más sensibles en su fase juvenil); *f* **culture-abri [f]** (Culture d'arbres [arbre-abri], d'arbrisseaux ou d'autres végétaux destinée à protéger de l'insolation, du froid, du vent les jeunes plantes d'une plantation, et qui permet à des espèces tolérantes à l'ombre et de succession tardive de se développer ; le résultat obtenu est la **plantation-abri** ; *g* **Schirmbestand [m]** (Anspruchslose, raschwüchsige Lichtbaumarten, die als vorübergehender Schirm zum Schutz später einzubringender empfindlicher Baumarten gepflanzt werden, oder Schirmbäume, die durch einen Schirmhieb im Rahmen eines waldbaulichen Verjüngungsverfahrens als aufgelockerter Bestand verbleiben, um mit ihren Kronen die nachwachsende Baumgeneration vorwiegend vor Frost und Hitze zu schützen).

3821 nurse grass [n] *plant.* (Fast and vigorous type of grass, often included in lawn mixtures for quick turf establishment; HLT 1971; *generic term* ►nurse plant); *s* **gramínea [f] nodriza** (*Término genérico* ►planta nodriza); *f* **graminée [f] abri** (*Terme générique* ►espèce abri) ; *g* **Ammengras [n]** (*OB* ►Ammenpflanze).

3822 nurse plant [n] *plant.* (Fast-growing plant species providing shelter for the main planting during its vulnerable early stages of development; ►nurse grass, ►pioneer species); *s* **planta [f] nodriza** (Planta de crecimiento rápido que da protección y permite crecer a otras en fase juvenil; *término específico* ►gramínea nodriza; ►planta pionera); *f* **espèce [f] abri** (Végétaux à croissance rapide constitués de graminées ou de légumineuses cultivés de manière intercalaire [simultanément] sous la protection desquels des espèces délicates peuvent surmonter les stades de vulnérabilité de leur développement ; ils présentent en outre l'avantage de réduire l'érosion, de réduire les pertes d'humidité et de faire concurrence aux adventices ; *terme spécifique* ►graminée abri ; ►espèce pionnière) ; *syn.* plante [f] abri ; *g* **Ammenpflanze [f]** (Schnellwüchsige, schattierende, ggf. Humus und Nährstoffe liefernde Pflanzenart, in deren Schutz empfindliche Pflanzen ihr Jungpflanzenstadium — häufig auf Extremstandorten — durchstehen können; *UB* ►Ammengras; ►Pionierart).

nursery [n] *sociol. urb.* ►day nursery; *for. hort.* ►field nursery, ►forest tree nursery; *adm. hort.* ►municipal nursery; *hort.* ►perennial nursery, ►plant nursery; *constr. hort.* ►site nursery; *hort.* ►sod nursery, ►tree nursery.

nursery [n]**, local authority** *adm. hort.* ►municipal nursery.

nursery [n] [US]**, perennials test** *hort.* ►perennials test garden.

nursery [n]**, temporary site** *adm. hort.* ►site nursery.

nursery [n] [UK]**, turf** *adm. hort.* ►sod nursery [US].

nursery-bred stock [n] [US] *hort. plant.* ►nursery stock.

3823 nursery field [n] *agr. hort.* (Outdoor area of plant cultivation as opposed to greenhouse or cold frame cultivation; ►field nursery, ►forest tree nursery); *s* **campo [m] de cultivo al aire libre** (Zonas de cultivo de plantas en el campo, al contrario que en invernadero. En alemán se habla de **c. de c. al a. l.** cuando se trata de trabajo de investigación y no de la agricultura normal); *f* **pleine terre [f]** (Surface de culture intensive en plein air par opposition à la culture sous serre ou la culture hâtée) ; *syn.* plein champ [m], de/en plein air [m], au champ [m], de/en plein champ [m] ; *g* **Freiland [n, o. Pl.]** (1. Nicht durch Gebäude oder Überdachungen ständig abgedeckte, pflanzenbaulich genutzte Kulturflächen oder gärtnerische Schauanlage; im Gegensatz zu Gewächshaus- und Frühbeetflächen. 2. *Legaldefinition im Pflanzenschutzgesetz* die nicht durch Gebäude oder Überdachung ständig abgedeckten Flächen, unabhängig von ihrer Beschaffenheit oder Nutzung; dazu gehören auch Verkehrsflächen jeglicher Art wie Gleisanlagen, Straßen-, Wege-, Hof- und Betriebsflächen sowie sonstige durch Tiefbaumaßnahmen veränderte Landflächen; § 2 Ziff. 15 PflSchG); *syn.* Freilandfläche [f], Freigelände [n] (außerhalb von Verkaufsgewächshäusern).

nursery-grown stock [n] *hort. plant.* ►nursery stock.

nursery-grown stock [n] [UK]**, advanced** *hort. plant.* ►specimen tree/shrub.

nurseryman [n]**, tree** *arb.* ►tree nursery specialist.

nursery row [n] *hort.* ►transplant in nursery row.

nursery school [n] [US] *sociol. urb.* ►kindergarten.

nursery specialist [n] *arb.* ►tree nursery specialist.

3824 nursery stock [n] *hort.* (►quality of nursery stock, ►quality standards for nursery stock); *syn.* nursery-bred stock [n] [also US] (TGG 1984, 177), nursery-grown stock [n]; *s* **planta [f] de vivero** (Planta cultivada en vivero para su posterior venta; ►calidad de plantas [de vivero], ►estándar de calidad de plantas de vivero; *término específico* leñosa de vivero); *f* **fourniture [f] de pépinière** (Végétaux cultivés en pépinières à aux fins de vente ; ►norme de qualité pour végétaux de pépinière, ►qualité morphologique) ; *syn.* végétaux [mpl] en provenance d'une pépinière, plantes [fpl] de pépinière ; *g* **Baumschulware [f]** (Im Baumschulbetrieb zum Verkauf kultivierte Pflanze[n]; ►Gütebestimmungen für Baumschulpflanzen, ►Pflanzenqualität); *syn.* Baumschulpflanze(n) [f(pl)], Baumschulerzeugnis [n].

nursery stock [n]**, large** *hort. plant.* ►specimen tree/shrub.

nursery stock [n]**, specimen** *hort. plant.* ►specimen tree/shrub.

nursery trade [n] [US] *hort.* ►commercial horticulture.

nursery woman [n]**, tree** *arb.* ►tree nursery specialist.

nutrient [n] *bot. pedol.* ►plant nutrient.

3825 nutrient absorption [n] *limn. pedol.* (Uptake of nutrients from a growing medium by plants, dependent upon the species of plant and the nutrient content of the medium); *s* **absorción [f] de nutrientes** (Proceso de extracción de nutrientes por las plantas del medio en el que crecen); *syn.* absorción [f] de elementos minerales; *f* **prélèvement [m] de substances nutritives** (Processus d'assimilation des végétaux qui puisent pendant la période de végétation active les éléments minéraux et organiques dans leur milieu de développement ; est très dépendant de l'espèce végétale ainsi que de la teneur en éléments nutritifs du milieu de vie) ; *g* **Nährstoffentzug [m, o. Pl.]** (Entnahme von Nährstoffen aus einem Wachstumsmedium durch Pflanzen; abhängig von der Pflanzenart und dem Nährstoffgehalt des Mediums).

nutrient assimilation [n] [US] *pedol.* ►nutrient uptake.

nutrient capital [n] [US] *pedol. chem.* ►nutrient pool.

3826 nutrient content [n] *limn. pedol. chem.* (Amount of nutrient which exists in the soil, determined by chemical analysis. The 'nutrient status' is a neutral statement defining the condition of the nutrient supply in an environmental medium; ►nutrient supply); *s* **contenido [m] de nutrientes** (Cantidad de nutrientes que se encuentran en un medio. Se determinan por análisis químico; ►suministro de nutrientes); *syn.* contenido [m] de elementos minerales; *f* **teneur [f] en substances nutritives** (Quantité d'éléments nutritifs [réserve assimilable] observée sur un support nutritif ; ►nutrition minérale [des plantes]) ; *syn.* teneur [f] en éléments nutritifs ; *g* **Nährstoffgehalt [m]** (In einem Umweltmedium vorhandene Menge an Nährstoffen, die durch chemische Analysen bestimmt werden kann; ►Nährstoffversorgung).

3827 nutrient cycle [n] *ecol.* (Circulation of life-giving elements and inorganic components within an ecosystem; ►mineral cycle); *syn.* nutrient cycling [n]; *s* **ciclo [m] de nutrientes** (Ciclo biogeoquímico de los elementos esenciales para la vida de los organismos de un ecosistema; ►ciclo de [los elementos] minerales); *syn.* ciclo [m] de los bioelementos; *f* **cycle [m] biologique des éléments nutritifs** (Mouvement en circuit fermé des éléments minéraux et des composés anorganiques au sein d'un écosystème [passage successif de la graine à la plante, sa destruction et décomposition par les micro-organismes du sol] ; DIS 1986, 61 ; ►cycle biogéochimique des éléments minéraux) ; *g* **Nährstoffkreislauf [m]** (Umlauf der unmittelbar lebenswichtigen Elemente und anorganischen Verbindungen im Ökosystem oder in Teilen davon; ►Mineralstoffkreislauf); *syn.* Nährstoffzyklus [m].

nutrient cycling [n] *ecol.* ►nutrient cycle.

3828 nutrient deficiency [n] *limn. pedol.* (Lack of required nutrients in an environmental medium); *s* **deficiencia [f] de nutrientes** (Carencia de los elementos minerales necesitados por las plantas en su medio de crecimiento); *syn.* falta [f] de nutrientes; *f* **carence [f] en substances nutritives** (État déficitaire de certains éléments biogènes nécessaires dans un support nutritif) ; *syn.* déficience [f] en éléments nutritifs ; *g* **Nährstoff-**

mangel [m, o. Pl.] (Fehlen oder ungenügendes Vorhandensein von benötigten Nährstoffen in einem Umweltmedium).

nutrient demands [npl] *pedol.* ►nutrient requirements.

3829 nutrient enrichment [n] *limn. pedol.* (Increase in nutrient content; ►eutrophication); *s* **enriquecimiento [m] de nutrientes** (Incremento de nutrientes en el suelo o el agua; ►eutrofización); *syn.* almacenamiento [m] de nutrientes; *f* **enrichissement [m] en substances nutritives** (Augmentation de la concentration en éléments nutritifs ; ►eutrophisation) ; *g* **Nährstoffanreicherung [f]** (Zunahme des Nährstoffgehaltes in einem Umweltmedium; ►Eutrophierung).

3830 nutrient fixation [n] *pedol.* (**1.** Transformation of a nutrient from an easily soluble form to a less soluble one, thus limiting its availability to a plant. **2.** Adsorption of nutrients in soils, e.g. in the form of salts or an exchangeable solution on the surface of organic and inorganic soil particles, not easily exchangeable in clay minerals; *opp.* ►nutrient mobilization); *syn.* nutrient immobilization [n], nutrient retention [n]; *s* **fijación [f] de nutrientes** (**1.** Transición de un nutriente en estado fácilmente a estado poco soluble, de manera que pasa a no estar disponible para las plantas. **2.** Fijación de nutrientes en el suelo, p. ej. en forma de sales o en forma difícilmente intercambiable en la superficie de minerales de arcilla; *opp.* ►movilización de nutrientes); *syn.* adsorbción [f] de nutrientes, inmovilización [f] de nutrientes; *f* **fixation [f] des substances nutritives** (**1.** Processus par lequel les s, els nutritifs essentiels à la croissance des plantes passent d'une forme soluble ou assimilable à une forme peu soluble et non échangeable ; DIS 1986, 88 ; ►libération de/des substances nutritives. **2.** Fixation d'éléments nutritifs sur des particules colloïdales du sol, p. ex. sous forme de sels échangeables à la surface d'un complexe adsorbant organique ou anorganique ou sous forme difficilement échangeable dans les minéraux argileux) ; *syn.* adsorption [f] des substances nutritives, immobilisation [f] ; *g* **Nährstofffestlegung [f]** (**1.** Übergang eines Nährstoffes von einer leicht in eine schwer lösliche und damit nicht pflanzenverfügbare Form. **2.** Adsorbtion von Nährstoffen in Böden, z. B. in Form von Salzen resp. austauschbar an der Oberfläche von organischen oder anorganischen Sorbenten, in schwer austauschbarer Form in Tonmineralen; *opp.* ►Mobilisierung von Nährstoffen); *syn.* Fixierung [f] von Nährstoffen, Immobilisierung [f] von Nährstoffen, Nährstoffimmobilisierung [f], Nährstoffbindung [f], Nährstofffixierung [f].

nutrient immobilization [n] *pedol.* ►nutrient fixation.

nutrient import(s) [n(pl)] *limn. pedol.* ►nutrient input.

nutrient inflow [n] *limn. pedol.* ►nutrient input.

3831 nutrient input [n] *limn. pedol.* (TEE 1980, 266; adding of nutrients into an area; e.g. by wind, water or fertilizing; ►matter import, ►nutrient transport); *syn.* nutrient import(s) [n(pl)] (TEE 1980, 264), nutrient inflow [n]; *s* **aporte [m] de nutrientes** (Entrada de nutrientes en un área determinada provenientes de otras zonas por medios de diseminación como el viento y el agua; *término genérico* ►aporte de sustancias; ►carga de nutrientes); *syn.* aportación [f] de nutrientes, entrada [f] de nutrientes; *f* **apport [m] de substances nutritives** (Transport d'éléments nutritifs dans une aire donnée, sous l'action du vent, de l'eau, sous forme d'engrais ; ►apport d'éléments minéraux, ►charge en substances nutritives ; *syn.* approvisionnement [m] de substances nutritives ; *g* **Nährstoffeintrag [m]** (Transport von Nährstoffen in ein Gebiet, z. B. durch Wind [Nährstoffeinwehung], [Grund]wasser oder Düngergaben; ►Nährstofffracht; *OB* ►Stoffeintrag); *syn.* Eintrag [m] von Nährstoffen, Nährstoffzufuhr [f].

nutrient loss [n] *pedol.* ►loss of nutrients.

3832 nutrient mobilization [n] *pedol.* (Transformation of a nutrient from a not easily soluble form to a relatively easy one, thereby making it freely available to plants; the **release of nutrients** by microbial degradation of organic matter is also known as ▶mineralization; cf. SS 1979; ▶nutrient fixation); *syn.* nutrient release [n]; *s* **movilización [f] de nutrientes** (Transición de un nutriente de un estado poco a uno fácilmente soluble, es decir disponible para las plantas. La liberación de nutrientes en la descomposición microbiana de la sustancia orgánica se denomina ▶mineralización; ▶fijación de nutrientes); *syn.* liberación [f] de nutrientes; *f* **libération [f] de/des substances nutritives** (Processus par lequel les sels nutritifs essentiels à la croissance des plantes passent d'une forme peu soluble et non échangeable à une forme soluble ou assimilable ; la libération de substance nutritives par la décomposition de la substance organique due à l'activité microbienne est la ▶minéralisation ; ▶fixation des substances nutritives) ; *g* **Mobilisierung [f] von Nährstoffen** (Übergang eines Nährstoffes von einer schwer in eine relativ leicht lösliche, d. h. leicht pflanzenverfügbare Form; die **Freisetzung von Nährstoffen** bei mikrobiellem Abbau der organischen Substanz heißt auch ▶Mineralisierung; cf. SS 1979; *opp.* ▶Nährstofffestlegung); *syn.* Nährstofffreisetzung [f], Nährstoffmobilisierung [f].

nutrient needs [npl] *pedol.* ▶nutrient requirements.

3833 nutrient pool [n] *pedol. chem.* (Total sum of soil nutrients which are available to plants in a short, medium or long-term period; *syn.* nutrient capital [n] [also US], nutrient vaults [npl] [TEE 1980, 292], reserve [n] of nutrients; *s* **reserva [f] de nutrientes** (Conjunto de elementos minerales disponibles en el suelo a corto, medio y largo plazo); *syn.* reserva [f] de elementos minerales; *f* **réserves [fpl] en substances nutritives** (PED 1979, 366 ; somme des éléments nutritifs disponibles à court, moyen et long terme dans le sol pour la nutrition des plantes ; ces réserves sont de formes assimilables [immédiatement disponible], mobilisables [progressivement assimilable par altération ou minéralisation] ou stables [très lentement altérable]) ; *g* **Nährstoffvorrat [m]** (Summe der Nährstoffe, die je nach Betrachtungsweise kurz-, mittel- oder langfristig im Boden für Pflanzen verfügbar sind); *syn.* Nährstoffreserven [fpl].

nutrient-poor [adj] *limn. pedol.* ▶oligotrophic.

3834 nutrient-poor grassland [n] *phyt.* (Plant community growing in nutrient-deficient locations, especially on dunes or in mountainous and alpine regions with low vegetation consisting primarily of grass and herbaceous species; *specific terms* ▶acidic grassland, ▶dry meadow, ▶limestone grassland, ▶pasture of low productivity, ▶semidry grassland [US]/semi-dry grassland [UK], ▶xerthermous meadow); *s* **prado [m] oligótrofo** (Término genérico para comunidades de pastizales secos en suelos pobres en nutrientes, en especial sobre dunas y en zonas montañosas y alpinas, cuya vegetación consiste mayormente de especies herbáceas; *términos específicos* ▶pastizal calcícola seco, ▶pastizal [oligótrofo] silicícola, ▶pastizal seco, ▶pastizal semiseco, ▶pastizal xerotérmico, ▶pasto pobre); *syn.* prado [m] pobre, herbazal [m] oligótrofo; *f* **pelouse [f] oligotrophe** (Formation herbacée naturelle rase dominée par les graminées et les herbes basses, constituant une association végétale se développant sur un milieu pauvre en éléments nutritifs comme les dunes, les pentes rocailleuses et les étages montagnards et alpins ; ▶prairie pâturée maigre, ▶pelouse aride, ▶pelouse calcicole maigre, ▶pelouse héliophile dense, ▶pelouse silicicole, ▶pelouse xérophile et thermophile) ; *syn.* pelouse [f] pauvre, pelouse [f] maigre, pelouse [f] à faible recouvrement ; *g* **Magerrasen [m]** (Pflanzengesellschaft auf nährstoffarmen Standorten, besonders auf Dünen oder im montanen und alpinen Bereich mit niedriger Bodenvegetation aus überwiegend Gras- und Krautarten; *UBe* ▶Halbtrockenrasen, ▶Kalkmagerrasen, ▶Magerweide, ▶Silikatmagerrasen, ▶Trockenrasen, ▶Xerothermrasen); *syn.* Magerwiese [f].

nutrient release [n] *pedol.* ▶nutrient mobilization.

3835 nutrient requirements [npl] *pedol.* (Essential nourishment needed by organisms to sustain life and to foster growth); *syn.* nutrient needs [npl], nutrient demands [npl]; *s* **demanda [f] de nutrientes** (Necesidad de sustancias nutritivas para el crecimiento y el mantenimiento de la vida de organismos vivos); *syn.* necesidades [fpl] nutritivas, exigencias [fpl] nutritivas; *f* **besoins [mpl] en substances nutritives** (Quantité d'un élément nutritif donné nécessaire à la vie d'une plante) ; *syn.* besoin [m] alimentaire des végétaux (LA 1981, 152) ; *g* **Nährstoffbedarf [m]** (...bedarfe [pl]; Notwendigkeit des Vorhandenseins von Stoffen für den Aufbau und die Erhaltung von Organismen); *syn.* Nährstoffanspruch [m].

nutrient retention [n] *pedol.* ▶nutrient fixation.

3836 nutrient-rich [adj] *limn. pedol.* (Descriptive term applied to soils and water bodies containing nutrients in sufficient quantities for plant growth; ▶eutrophic; *opp.* nutrient-poor [▶oligotrophic]); *s* **rico/a [adj] en nutrientes** (Referente a suelo o masa de agua con alto contenido de nutrientes; ▶eutrófico; *opp.* pobre en nutrientes [▶oligótrofo]); *f* **riche en substances nutritives [loc]** (Concerne le sol ou un système aquatique disposant d'une teneur en éléments nutritifs satisfaisante ; ▶eutrophe ; *opp.* Pauvre en éléments nutritifs [▶oligotrophe]) ; *g* **nährstoffreich [adj]** (Gewässer oder Böden betreffend, die Nährstoffe in genügender Menge für das Pflanzenwachstum enthalten; ▶eutroph; *opp.* nährstoffarm [▶oligotroph]).

3837 nutrient storage capacity [n] *pedol.* (Capacity of a soil to retain nutrients); *s* **capacidad [f] de retención de nutrientes** (Capacidad de un suelo de almacenar nutrientes); *f* **fertilité [f] minérale du sol** (Capacité d'un sol à fournir à la plante l'ensemble des substances nutritives [absorbées sous forme minérale et stockées sous forme organique] qui lui sont nécessaires ; PED 1984, 87) ; *syn.* pouvoir [m] de rétention en substances nutritives ; *g* **Nährstoffspeicherfähigkeit [f]** (Eigenschaft des Bodens, Nährstoffe für Pflanzen vorhalten zu können, ohne dass sie durch Sickerwasser in tiefere Bodenschichten ausgewaschen werden).

3838 nutrient stress [n] *limn. pedol.* (Strain from oversupply of nutrients in a limnetic or terrestrial ecosystem or parts thereof, which may lead to the impairment or destruction of its ability to function; ▶eutrophication); *s* **exceso [m] de nutrientes** (Aporte excesivo de nutrientes a un ecosistema límnico o terrestre o a partes de él, de manera que se deteriora su capacidad funcional; ▶eutrofización); *f* **charge [f] en éléments nutritifs** (Apport trop important de matière minérale ou organique dans un écosystème limnique ou terrestre, si bien que son fonctionnement en est fortement perturbé ou même totalement bouleversé ; ▶eutrophisation) ; *syn.* apport [m] excessif en substances nutritives, excès [m] en substances nutritives ; *g* **Nährstoffbelastung [f]** (Zu hohe Zufuhr von Nährstoffen in ein limnisches oder terrestrisches Ökosystem oder Teilen davon, so dass seine Funktionsfähigkeit gestört oder nicht mehr gegeben ist; ▶Eutrophierung).

3839 nutrient supply [n] *pedol. chem.* (Availability of nutrients in the soil for plant uptake; ▶nutrient content, ▶supply of sustenance); *s* **suministro [m] de nutrientes** (Cantidad de nutrientes disponible para las plantas; ▶avituallamiento de la vegetación, ▶contenido de nutrientes); *f* **nutrition [f] minérale (des plantes)** (Ensemble des processus d'assimilation et d'utilisation des aliments qui ont lieu chez les végétaux vivants ; ▶alimentation, ▶teneur en substances nutritives) ; *syn.* alimentation [f] des plantes en éléments nutritifs ; *g* **Nährstoffversorgung**

[f] (Vorhandensein und Zuführung von Nährstoffen und die Möglichkeit für Pflanzen diese aufschließen zu können; ▶Nährstoffgehalt, ▶Versorgung der Pflanzen).

3840 nutrient transfer [n] *pedol.* (TEE 1980, 297; transport of nutrients down a soil profile through the action of percolating water; ▶leaching of nutrients, ▶illuvial horizon); *s* **transporte [m] de nutrientes** (Flujo de nutrientes hacia el interior del suelo con el agua de ▶lixiviación; ▶horizonte de iluviación); *f* **transfert [m] (descendant) des substances nutritives** (Déplacement descendant sous l'action de la gravité des éléments nutritifs dissous d'une couche supérieure à l'horizon d'accumulation ; ▶lixiviation, ▶horizon illuvial) ; *g* **Nährstoffverlagerung [f]** (Durch Sickerwasser bedingter Transport von Nährstoffen im Bodenprofil nach unten; ▶Nährstoffauswaschung, ▶Anreicherungshorizont).

3841 nutrient transport [n] *envir. limn. pedol.* (Amount of nutrients transported in a medium; ▶nutrient input); *s* **carga [f] de nutrientes** (Cantidad de nutrientes contenidos en un medio en flujo; ▶aporte de nutrientes); *f* **charge [f] en substances nutritives** (Diffusion et transport d'éléments nutritifs dans un support ; ▶apport de substances nutritives) ; *g* **Nährstofffracht [f]** (In einem Medium transportierte Menge von Nährstoffen; ▶Nährstoffeintrag).

3842 nutrient uptake [n] *pedol.* (TEE 1980, 266; all chemical and physical processes in a living organism dedicated to nutrient absorption); *syn.* nutrient assimilation [n] [also US]; *s* **asimilación [f] de nutrientes** (Procesos físicos y químicos en los organismos vivos para absorber nutrientes); *syn.* asimilación [f] de elementos minerales; *f* **absorption [f] des substances nutritives** (Ensemble des processus physico-chimiques de pénétration des éléments minéraux par les organes aériens ou souterrains des organismes vivants) ; *g* **Nährstoffaufnahme [f]** (Gesamtheit der chemischen und physikalischen Vorgänge, die in einem lebenden Organismus vorgehen, um Nährstoffe aufzunehmen); *syn.* Aufnahme [f] von Nährstoffen.

nutrient vaults [npl] *pedol. chem.* ▶nutrient pool.

nutrition [n] *bot. hort.* ▶plant nutrition.

3843 nutritional resources [npl] *conserv. zool.* (TEE 1980, 121; available food for consumption by humans and animals; ▶natural resources 1); *syn.* nutritional source (TEE 1980, 121); *s* **base [f] alimentaria** (Disponibidad de alimentos para animales y seres humanos; ▶recursos naturales); *syn.* base [f] nutritiva, recursos [mpl] alimentarios/nutritivos; *f* **ressources [fpl] alimentaires** (Exploitées dans la recherche et la sélection de la nourriture par les hommes et les animaux ; ▶ressources fondamentales naturelles) ; *g* **Nahrungsgrundlage [f]** (Basis der Nahrungsbeschaffung für Mensch und Tier; ▶natürliche Grundgüter); *syn.* Nahrungsressource [f].

nutritional source [n] *conserv. zool.* ▶nutritional resources.

nylon anchorage net [n] [US] *constr.* ▶root anchoring fabric [US].

oak copse [n] [UK] *phyt.* ▶oak scrub [US].

3844 oak scrub [n] [US] *phyt.* (**1.** Impenetrable plant community, predominantly of various low branched oak species; **o. s.s** occur in Florida, where they are mixed/occur with some pine

species, e.g. thickets of gnarled sand pines *[Pinus clausa]* and the Longleaf pine; oaks in scrub are evergreen: Myrtle oak *[Quercus myrtifolia]*; sand live oak *[Quercus germinate]*; trees are interspersed with shrubs, e.g. Florida rosemary *[Ceratiola ericoides]* and ground cover includes sparse herbs, lichens [*Cladonia* spp.], and bare sand; 40-60% of species are endemic; intense fires occur every 15-100 years. In **U.K.** wildlife-rich upland heath and bogs have a diverse structure and species composition, comprising dry and wet heath with bryophytes, wet flushes and pools. Bracken, scrub and oak woodland may also occur with scattered clumps of trees, especially pine, as well as gorse and birch/oak scrub; lack of grazing leads to invasion by scrub. **2. In D.**, oak scrubland, primarily in western Schleswig-Holstein, consisting mostly of oaks, hazels and black alders, often as dwarf vegetation, with scattered junipers); *syn.* oak copse [n] [UK], oak scrubland [n]; *s* **matorral [m] rastrero de roble, avellana y arraclán** (Matorral típico de la costa atlántica de Dinamarca y del norte de Alemania con vegetación generalmente rastrera y ejemplares individuales de enebro); *f* **couvert [m] buissonneux de chênes** (Peuplement ligneux constitué principalement de Chênes, Noisetiers, Nerpruns souvent de port rabougri, parfois accompagné de Genévrier) ; *syn.* couvert [m] végétal buissonneux de chênes ; *g* **Kratt [n]** (...e [pl]; **1.** nordd. für Eichengestrüpp, vor allem im westlichen Schleswig-Holstein, das im Wesentlichen aus Eiche, Hasel und Faulbaum besteht, oft im Krüppelwuchs, vereinzelt mit Wacholder. **2. Eichengestrüppe** gibt es auch z. B. in Florida, die eine undurchdringliche Pflanzengesellschaft aus unterschiedlichen Eichenarten bilden, durchsetzt mit Sträuchern und wenigen Kiefernarten); *syn.* Eichengestrüpp [n].

oak scrubland [n] *phyt.* ▶oak scrub [US].

O&D traffic [n] [US] *trans.* ▶origin and destination traffic.

3845 objection [n] *leg.* (Protest of private persons, agencies or authorities which are affected by planning proposals; ▶objections and supporting representations, ▶public partizipation); *s* **objeciones [fpl]** (En el marco de la ▶participación pública en la planificación urbana o en procedimiento de aprobación de un plan, muestra de desacuerdo por parte de una persona, empresa o institución con un plan propuesto; ▶alegaciones, comentarios y proposiciones); *f* **observations [fpl]** (Recueil des observations du public sur un registre lors de l'enquête publique ; ▶observations 2, ▶participation du public) ; *g* **Einwendung [f]** (Äußerungen von Planungsbetroffenen gegen ein Projekt im Rahmen der ▶Beteiligung der Öffentlichkeit 2 bei der Bauleitplanung, beim Planfeststellungsverfahren etc.; cf. § 73 VwVfG; ▶Anregungen [und Bedenken]); *syn.* Einsprachen [fpl] und Begehren [npl] [CH] (gem. Art. 12 NHG).

3846 objections [npl] and supporting representations [npl] *leg. urb.* (Comments voiced by citizens or their representatives affected by a planning proposal at a ▶public hearing [US]/public inquiry [UK]; ▶non-profit organization suit [US]/litigation brought by a non-profit organization [UK], ▶public participation); *s* **alegaciones [fpl], comentarios [mpl] y proposiciones [fpl]** (Opiniones y propuestas de la ciudadanía en el marco de la ▶participación pública en la planificación urbana; ▶audiencia pública, ▶derecho a pleito de organizaciones ecologistas y de protección de la naturazela); *f* **avis [mpl] et observations [fpl]** (F., **1.** du président d'une association agréée lorsqu'il en fait la demande lors de l'élaboration d'un PLU/POS. **2.** Consignation des observations de toute personne physique ou morale, de tout association sur un projet d'intérêt public dans le cadre de l'▶enquête publique ; **3.** des services et personnes publiques lors de l'élaboration d'un plan local d'urbanisme [PLU]/plan d'occupation des sols [POS] ; ▶action en justice des associations, intervention des associations ▶participation du public ; **D.**, dans le cadre de la ►participation du public lors de l'élaboration des plans d'urbanisme) ; *syn.* appréciations [fpl], suggestions [fpl] et contre-propositions [fpl] ; *g* **Anregungen [fpl] (und Bedenken [npl])** (Äußerungen von Planungsbetroffenen im Rahmen der Bürgerbeteiligung bei der Bauleitplanung während der Auslegungsfrist; cf. § 3 BauGB i. d. F. von 1986-1996; 1998 wurde der planungsrechtliche Begriff ‚Bedenken' in dem Wortpaar ‚Bedenken und Anregungen' abgeschafft; es gibt seitdem nur noch ‚Anregungen' [§ 3 (2) BauGB 1988]; seit der Neufassung des BauGB v. 2004 wird der Begriff durch **Stellungnahme** ersetzt; ►Beteiligung der Öffentlichkeit 2, ►öffentliche Anhörung, ►Verbandsklage); *syn.* Stellungnahme [f], Einsprache [f] [CH], Einwendungen [fpl] [A].

objective [n]**, planning** *plan.* ►planning goal.

obligation [n] *contr.* ►binding obligation to a bid [US]/binding obligation to a tender [UK], ►guarantee fulfillment obligation [US]/guarantee fulfilment obligation [UK]; *leg.* ►legal obligation, ►maintenance obligation, ►private owner's obligation, ►social obligation of private property.

obligation [n] **of occupier to make land or premises safe for persons or vehicles** *leg.* ►public responsibility.

obligation [n] **to preserve existing plants** [UK] *conserv. leg. urb.* ►plant preservation requirement [US].

3847 obligation [n] **to take out professional liability insurance** *contr. prof.* (Requirement of a contract holder to obtain and maintain professional indemnity insurance; ►professional liability insurance); *syn.* liability [n] for professional insurance; *s* **obligación [f] de tener seguro profesional** (Responsabilidad de contratista de ser titular de un ►seguro de responsabilidad civil que cubra todas sus actividades profesionales); *f* **obligation [f] de contracter une assurance professionnelle** (Responsabilité civile de l'entrepreneur de couvrir ses activités par une assurance et être titulaire d'une Police Responsabilité Civile Chef d'Entreprise, d'une Police Responsabilité Décennale ; ►assurance responsabilité civile professionnelle) ; *g* **Versicherungspflicht [f]** (Pflicht des Auftragnehmers zum Abschluss und zur Aufrechterhaltung eines Haftpflichtversicherungsvertrages — Berufshaftpflicht; ►Haftpflichtversicherung für Planer).

3848 oblique aerial photograph [n] *envir. plan. rem' sens. surv.* (Photograph of a part of the Earth's surface taken at an angle, usually from an aircraft); *s* **fotografía [f] aérea oblicua** (Tomas fotográficas de la superficie terrestre tomadas con ángulo generalmente desde aviones); *f* **prise [f] de vue aérienne oblique** (Clichés aériens de parties de la surface terrestre à axe oblique) ; *g* **Luftschrägbild [n]** (Fotografische Aufnahme eines Teils der Erdoberfläche, die seitlich, meist aus einem Flugzeug aufgenommen wird).

3849 oblique photograph [n] *envir. plan. rem'sens. surv.* (Photograph taken at an angle from an airplane [US]/aeroplane [UK]; ►vertical aerial photograph); *s* **fotografía [f] oblicua** (Fotografía aérea tomada en ángulo desde un avión; ►fotografía aérea vertical); *f* **photographie [f] oblique** (Photographie prise par avion ayant un angle de prise oblique ; ►photographie verticale) ; *g* **Schrägaufnahme [f]** (Fotografie, die in einem schrägen Winkel aus einem Flugzeug gemacht wird; ►Senkrechtluftbild).

observation [n]**, protective aerie** *conserv. game'man.* ►protective eyrie observation.

observation boardwalk [n] [UK] *recr. zool.* ►wetland boardwalk [US].

3850 observation [n] **of artistic quality** (≠) *constr. prof.* (**In D.**, this term is used for the regard paid during supervision to the quality and aesthetic execution of the work according to the intent of the designer; the duties therefore include special attention to workmanship, use and quality of materials as specified, so that the final effect of the executed work fulfills the aesthetic demands of both designer and client); *syn.* supervision [n] of artistc quality (≠); *s* **supervisión [f] de la calidad artística** (≠) (En D., término utilizado para denominar el cuidado que debe tener la dirección de obra de controlar la calidad técnica y estética de los trabajos realizados de acuerdo con la intención del diseñador o diseñadora; las tareas incluyen por tanto supervisar el buen oficio, el uso y la calidad de los materiales tal como fueron especificados, de manera que el objeto [edifico, zona verde, plaza pública] cumpla con las exigencias estéticas tanto del diseñador o diseñadora como de la propiedad); *f 1* **direction [f] de chantier responsable de la qualité artistique des travaux** (≠) (**D.**, personne missionnée par le maître d'ouvrage et chargée de contrôler et de surveiller les entreprises lors des différentes étapes de l'exécution des travaux afin de remettre des prestations correspondant au projet artistique du concepteur tant dans le choix des matériaux que dans la qualité de leur mise en œuvre) ; *f 2* **maîtrise [f] d'œuvre nécessaire à l'intégration de l'œuvre artistique dans l'ouvrage** (**F.**, mission de conception et d'assistance pour les réalisations artistiques au titre du **1 % artistique** : mesure s'appliquant à toutes les opérations dont la maîtrise d'ouvrage est assurée par l'État ou ses Établissements Publics, les collectivités territoriales ou leurs groupements qui consiste à consacrer, à l'occasion de la construction ou de l'extension de bâtiments ou d'espaces publics ainsi que leur réhabilitation dans le cas d'un changement d'affectation ou d'usage, une somme représentant un pour cent du coût des constructions publiques et permettant la réalisation d'une ou plusieurs œuvres d'art spécialement conçues pour le lieu ou le bâtiment considéré ; cf. Décrets du 29 avril 2002 et du 4 février 2005, circulaire d'application du 16 août 2006) ; *g* **künstlerische Bauoberleitung** (Bei sehr großen Bauvorhaben, bei dem mehrere Planer Teilbereiche eigenständig bearbeiten, z. B. bei Bundes- oder Internationalen Gartenbauausstellungen, die Überwachung der ausgeführten Bauleistungen der einzelnen Gewerke in Übereinstimmung mit den Entwürfen der Planer hinsichtlich der beabsichtigten Gestaltung, der Wahl der Baustoffe und Konstruktionen und der handwerklichen Qualität, damit das Objekt als Ganzes den künstlerischen Ansprüchen des hauptverantwortlichen Planers und Bauherrn entspricht); *syn.* künstlerische Oberleitung [f].

3851 obtaining a mutual agreement on planning proposals [loc] *plan.* (Reaching a common understanding on planning proposals); *syn.* planning coordination [n], coordination [n] of a planning project; *s* **coordinación [f] sobre un plan** (Discusión de proposiciones de planificación para llegar a un acuerdo); *f* **coordination [f] d'un projet** (Discussion de propositions d'aménagement et recherche en commun d'une solution) ; *g* **Abstimmung [f] einer Planung** (Erörterung von Planungsvorschlägen und Finden einer gemeinsamen Lösung); *syn.* Planungsabstimmung [f], Planungskoordination [f].

occasionally flooded riparian woodland [n] *phyt.* ►riparian upland woodland.

occlusion [n] [UK] *arb. bot.* ►callusing.

3852 occupancy permit [n] [US] *adm. leg.* (Occupancy certificate to use a completed building after examination by a ►Building Permit Office [US]/Building Surveyors Department [UK], and the determination that the project meets all requirements of applicable ordinances and regulations); *syn.* buildings

regs sign off [n] [UK]; *s* **recepción [f] oficial de obra** (Revisión de una obra finalizada por parte de la autoridad competente antes de emitir el permiso de ocupació) ►autoridad de inspección urbanística; *syn.* aceptación [f] oficial de obra; *f* **mise [f] en conformité** (Contrôle d'un ouvrage, d'une installation conformément aux règles de construction par les autorités compétentes et délivrance d'un certificat de conformité ; ►bureau de contrôle) ; *g* **baurechtliche Abnahme [f]** (Prüfung eines Bauwerkes nach Fertigstellung durch die Baubehörde, ob entsprechend der Genehmigung und den geltenden öffentlich-rechtlichen Vorschriften gebaut wurde; *UBe* baurechtliche Schlussabnahme, Rohbauabnahme; ►Baurechtsbehörde).

3853 occupied area [n] *constr.* (Space taken up, e.g. by construction work, buildings, development, travelled way [US]/ carriage-way [UK]); *s* **superficie [f] ocupada** (Espacio cubierto, p. ej. por una obra, un edificio, un proyecto de urbanización, etc.); *f* **emprise [f]** (Surface occupée par p. ex. des travaux, un bâtiment, un aménagement, un chantier, une voie routière, etc. *Contexte* **e.** de chantier, **e.** d'un bâtiment) ; *g* **in Anspruch genommene Fläche [f] (±)** (Fläche, auf der gebaut wird oder ein Eingriff erfolgt).

3854 occurrence [n] *biol.* (Existence of plant or animal species in a particular area); *s* **presencia [f]** (Existencia de una especie de flora o fauna en un área específica); *syn.* existencia [f]; *f* **présence [f]** (Existence d'une espèce végétale ou animale sur une aire donnée) ; *g* **Vorkommen [n]** (Das Vorhandensein einer Pflanzen- oder Tierart in einem bestimmten Gebiet).

ocean beach [n] *geo.* ►shore.

3855 odd-lot development [n] [US] *urb.* (Construction of residential buildings to close gaps between existing buildings; ►infill development); *syn.* gap-stopping [n] [UK]; *s* **relleno [m] de solares desocupados** (Construcción de viviendas en solares vacíos existentes en la trama urbana; ►densificación urbana); *f* **reconstitution [f] du front bâti** (Construction d'un terrain non bâti le long d'un alignement de constructions ; ►construction dans la bâti existant, ►construction nouvelle en zone urbanisée, ►politique de la ville compacte) ; *syn.* reconstitution [f] de la continuité bâtie ; *g* **Schließen [n, o. Pl.] von Baulücken** (Bebauung von nicht bebauten Grundstücken innerhalb einer Häuserzeile/Blockseite; Baulücken schließen [vb]; ►Nachverdichtung); *syn.* Auffüllen [n, o. Pl.] von Baulücken, Baulückenauffüllung [f], Baulückenschließung [f], Lückenwohnungsbau [m]).

odor [n] [US] *envir.* ►emission of odo(u)r.

3856 odorous annoyance [n] *envir.* (Offensive scent caused by foul air or noxious smelling gaseous emissions); *syn.* odorous nuisance [n]; *s* **molestia [f] pública por olores** (Malestar causado en la población por emisiones de gases malolientes); *syn.* malos olores [mpl], olores [mpl] molestos; *f* **incommodité [f] causée par les odeurs** (Gêne provoquée par l'air vicié ou l'émission de gaz malodorants) ; *g* **Geruchsbelästigung [f]** (Störender Einfluss durch schlechte Luft oder übelriechende Gasimmissionen).

odorous nuisance [n] *envir.* ►odorous annoyance.

odor trap [n] [US] *constr.* ►water seal.

odour [n] [UK] *envir.* ►emission of odo(u)r.

odour trap [n] [UK] *constr.* ►water seal.

office [n] *adm. leg.* ►Building Permit Office [US]; *adm. urb.* ►local planning agency/office [US]; *adm. conserv'hist.* ►State Historic Preservation Office (SHPO) [US]; *eng. stat.* ►structural engineering office.

Office [n] [US]**, District Surveyor's** *adm.* ►Building Permit Office [US].

office [n] [UK]**, commissioned planning** *contr. prof.* ►contract holder [US].

office [n] [US]**, county recorders** *adm.* *►land-holding agency [US].

office [n] [UK]**, land registry** *adm.* ►land-holding agency [US].

office buildings [n]**, vacancy at** *sociol. urb.* *►vacancy.

office employee [n] *adm. contr.* ►office worker.

officer [n] [UK]**, building control** *constr. prof.* ►site supervisor representing an authority.

3857 office tower [n] *arch. urb.* (►high-rise building, ►tower building); *s* **torre [f] de oficinas** (Edificio de oficinas muy alto y aislado; ►edificio alto, ►torre de viviendas); *f* **tour [f] (à usage) de bureaux** (NEU 1983, 299 ; ►immeuble de grande hauteur, ►tour à noyau central) ; *g* **Bürohochhaus [n]** (►Hochhaus, ►Punkthaus).

3858 office worker [n] *adm. contr.* (Collective term is 'administrative staff' or in U.S. also 'administrative personnel'; *specific term* 'clerical personnel' covering employees who keep accounts and records, do filing, copying, etc.); *syn.* office employee [n]; *s* **administrativo/a [m/f]** (Funcionario/a o empleado/a de la administración pública con tareas administrativas); *f* **personnel [m] administratif** (Fonctionnaire ou salarié de la fonction publique) ; *g* **Arbeitskraft [f] in der Verwaltung** (In der [öffentlichen] Verwaltung beschäftigter Arbeiter, Angestellter oder Beamter); *syn.* Behördenbeschäftigte/-r [f/m], Behördenmitarbeiter/-in [m/f], Beschäftigte/-r [f/m] bei der Behörde, Beschäftigte/-r [f/m] in der Verwaltung.

3859 official [n] **of a competition jury** *prof.* (Local government adviser of a competition jury who is particularly acquainted with the local situation and the aims of the competition—usually a representative of the initiating authority; ►professional member of a competition jury [US]/assessor of a competition jury [UK]); *s* **miembro [m] de jurado de concurso** (Asesor, -a de las autoridades locales en un jurado de concurso que tiene buenos conocimientos de la situación del lugar y de los fines del concurso y generalmente representa al gobierno que inició el mismo; ►miembro profesional de jurado); *f* **membre [m] du jury** (Personnalité compétente justifiant d'une parfaite connaissance du contenu de la mission et de la nature des prestations à fournir dans le cadre du concours avec voie délibérative, en général représentants de l'administration publique ; ►personnalité qualifiée membre du jury) ; *g* **Sachpreisrichter [m]** (Mitglied eines Preisgerichtes, das mit den örtlichen Verhältnissen und der Wettbewerbsaufgabe besonders vertraut ist — meist Vertreter der öffentlichen Verwaltung; ►Fachpreisrichter).

3860 official phytosanitary certificate [n] *agr. for. hort. leg.* (Document issued by the pest control agency of the export country, which certifies that the exported plants, parts thereof, or plant products, are free of diseases and pests; ►phytosanitary inspection); *s* **certificado [m] fitosanitario de plantas** (Documento extendido por la autoridad de un país exportador que garantiza que las plantas, partes de ellas o productos vegetales están libres de plagas; ►control fitosanitario); *f* **certificat [m] phytosanitaire** (Document délivré par les services phytosanitaires étrangers et contrôlé par le Service de la protection des végétaux attestant la qualité, le bon état sanitaire [absence d'organismes nuisibles] et la conformité des végétaux aux normes exigées ; ►contrôle phytosanitaire) ; *g* **Pflanzengesundheitszeugnis [n]** (Vom amtlichen Pflanzenschutz des Exportlandes ausgestelltes Zeugnis, das beurkundet, dass die ausgeführten Pflanzen, Pflanzenteile oder pflanzlichen Erzeugnisse frei

von Krankheiten und Schädlingen sind; ▶Pflanzenbeschau); *syn.* phytosanitäres Zertifikat [n].

offsetting open space [n] [US] *plan.* ▶open space compensation area.

3861 offshore [adj] [US]/**off-shore** [adj] [UK] *geo.* (Located at sea, an indefinite distance from the shore); *s* **costa afuera** [adj] (Término descriptivo para algo localizado en el mar a distancia indefinida de la costa); *syn.* off-shore [adj]; *f* **à proximité du littoral** [loc] *syn.* proche du rivage [loc] ; *g* **küstennah** [adj] *syn. für die Seeseite* der Küste vorgelagert [pp], wasserwärts [adj] der Küstenlinie.

3862 offshore drilling [n] [US]/**off-shore drilling** [n] [UK] *min.* (Boring holes for exploitation of petroleum and natural gas at sea some distance from the coast line); *s* **perforación** [f] **costa afuera** (Técnica de explotación de petróleo o gas natural en plataformas situadas a cierta distancia de la costa); *syn.* perforación [f] off-shore; *f* **forage** [m] **en mer** (Technique d'exploration et d'exploitation des ressources en hydrocarbures et en gaz du fond de la mer à partir de plates-formes offshore) ; *syn.* forage [m] offshore ; *g* **Offshore-Bohrung** [f] (Im Meer, außerhalb der Küstenzone auf Bohrinseln vorgenommene Bohrung nach Erdöl oder Erdgas).

3863 offside lane [n] *trans.* (**1.** Lane for parked vehicles, part of or adjacent to a road, which may be used for parking or temporarily for flowing traffic according to the time of day; *not to be confused with a* ▶parking lane. **2. In U.S.**, in addition, **high occupancy vehicle [HOV] lanes** are in operation in many cities; HOV lanes are being provided by allowing high occupancy vehicles to drive on a separated lane during commuter hours when the freeway is congested ; **in U.K.**, the Highways Agency's feasibility study of 2004 on high occupancy vehicle lanes concluded that introduction of an HOV lane would be feasible at each of the four locations being investigated. The preferred option is to introduce the HOV lane as an 'open access' [i.e. not cordoned-off] peak period off-side running lane); *syn. to* **2.** commuter lane [n], diamond lane [n], transit lane [n] [AUS, NZ]; *s* **carril** [m] **auxiliar** (Vía adicional al borde de carretera que, dependiendo de la hora del día o de otros criterios, puede utilizarse como carril adicional o como aparcamiento; ▶carril de aparcamiento. En muchos países europeos existen carriles exclusivos para autobuses y taxis; en EE.UU. carriles exclusivos para vehículos de alta ocupación); *syn.* trocha [f] auxiliar [RA]; brocha [f] auxiliar [MEX] (BU 1959); *f 1* **couloir** [m] **réservé ouvert au stationnement temporaire** (±) (Couloir en bordure de chaussée, affecté en alternance aux bus et au stationnement résidentiel pendant la période nocturne lorsque les transports publics ne fonctionnent pas à un rythme cadencé ; *à ne pas confondre avec un* ▶parc à voitures longitudinal) ; *f 2* **couloir** [m] **réservé** (*Terme générique* voie de circulation exclusivement empruntée par une catégorie de véhicules, p. ex. les transports en commun en site propre ; *termes spécifiques* **couloir bus**, **couloir cyclable/cyclistes** ou **voie verte** qui est une route exclusivement réservée à la circulation des véhicules non motorisés, des piétons et des cavaliers) ; *syn.* voie [f] réservée ; *f 3* **couloir** [m] **mixte** (*Terme spécifique* voie de circulation empruntée par deux catégorie d'utilisateurs, p. ex. les bus et les cyclistes [couloir mixte bus-vélo] ou les tramways et les cyclistes [couloir mixte tramway-vélo]) ; *syn.* voie [f] mixte ; *f 4* **voie** [f] **réservée aux véhicules à occupation multiple (VOM)** (U.S., CDN., voirie qui, à la différence des transports en commun en site propre, est affectée aux « véhicules à forte occupation » [*High Occupancy Vehicle Lanes*] sur les autoroutes et les artères principales et dont l'accès est restreint et qui sont réservées, durant les heures de pointe ou autres, aux véhicules à occupation multiple, notamment le covoiturage, le covoiturage par fourgonnette et l'autobus ; www.tc.gc.ca) ;

syn. voie [f] réservée aux véhicules multioccupants ; *g 1* **Mehrzweckfahrstreifen** [m] (±) (**1.** Am Rande der Fahrbahn liegende Spur, die je nach Tageszeit oder sonstigen Kriterien dem ruhenden oder fließenden Verkehr dienen kann; ▶Parkspur; der engl. Begriff *offside lane [Fahrstreifen außerhalb der Hauptfahrstreifen]* beinhaltet auch alle Abbiegespuren); *syn.* Haltestreifen [m]; *g 2* **Sonderfahrstreifen** [m] (OB für Fahrstreifen, die durch spezielle Verkehrszeichen [Zeichen 245] und Fahrbahnmarkierungen extra ausgewiesen sind und Omnibussen des Linienverkehrs vorbehalten sind. Durch zusätzliche Schilder wie „Taxi frei" oder „Radfahrer frei" dürfen diese Verkehrsteilnehmer auch diese **S.** nutzen. In den USA gibt es **Sonderfahrstreifen für Fahrgemeinschaften**, sog. *High Occupancy Vehicle Lanes [car pool lanes]*, nur für Fahrzeuge, die zwei oder mehr Personen befördern, um somit den Verkehrsfluss speziell im Berufsverkehr zu fördern); *syn.* Fahrgemeinschaftsspur [f].

3864 off-site disposal [n] **of surplus material** *adm. constr. eng.* (Getting rid of surplus excavated material by loading, transport and orderly dumping [US]/tipping [UK] at another site for ▶final storage of surplus material); *s* **disposición** [f] **de excedentes de terracerías** (▶almacenamiento definitivo de residuos); *f* **enlèvement** [m] **des excédents** (Enlèvement, chargement et transport des déblais et matériaux excédentaires pour stockage sur le lieu de dépôt définitif ou évacuation aux décharges ; ▶stockage définitif) ; *g* **Massenverbringung** [f] (Fördern, Laden, Transportieren und profilgerechtes Schütten von überschüssigen Aushubmassen [Überschussmassen] an einem Bestimmungsort zwecks ▶Endlagerung); *syn.* Verbringen [n, o. Pl.] von Aushub, Verbringen [n, o. Pl.] von Erdmassen, Verbringung [f] von Aushub, Verbringung [f] von Erdmassen.

3865 ogee slope [n] *constr.* (A double curve, resembling an S-shape for bank stabilization to resemble a parabolic surface at the top and bottom of a slope; ▶concave slope, ▶rounding the top of an embankment/slope); *s* **talud** [m] **con forma de S** (▶pie de talud cóncavo, ▶redondeado de talud); *f* **talus** [m] **modelé en forme de S** (▶adoucissement de talus, ▶arrondi de pied de talus, ▶doucine) ; *g* **S-förmig ausgerundete Böschung** [f] (▶ausgerundeter Böschungsfuß, ▶Böschungsausrundung).

O horizon [n] *pedol.* ▶organic horizon.

oil [n] *envir.* ▶waste oil.

3866 oil film [n] *envir.* (Layer of oil floating upon the sea, which has been caused by a tanker collision or the discharge of used oil; ▶shore oil pollution); *syn.* oil slick [n] [also US] (GE 1977); *s* **mancha** [f] **de petróleo** (Capa de petróleo flotante sobre la superficie del mar, causada por accidente de petrolero o por vaciado de los tanques de buques; ▶marea negra); *f* **couche** [f] **de pétrole** (déversée accidentellement ou délibérément et qui flotte sur la mer ; ▶marée noire) ; *syn.* couche [f] d'hydrocarbures ; *g* **Ölteppich** [m] (Auf dem Meer schwimmende Ölschicht, sog. Slicks, die durch Tankerhavarie oder Ablassen von Ölrückständen entsteht; ▶Ölpest).

3867 oil pollution [n] *envir.* (Generic term for illegal defiling of the soil, a waterbody or the sea by discharge of hydrocarbons/petroleum or their derivatives, which causes the deterioration of flora and fauna, groundwater or the whole ecosystem and impedes the permissible use of natural resources; ▶shore oil pollution); *s* **contaminación** [f] **de hidrocarburos** (Descarga accidental o ilegal de petróleo o derivados en el suelo, las aguas o el mar que causa daños en el medio ambiente, la flora y la fauna, los ecosistemas, contamina las aguas superficiales y subterráneas e impide el uso legítimo de los recursos naturales; ▶marea negra); *syn.* polución [f] de hidrocarburos; *f* **pollution** [f] **par les hydrocarbures** (Déversement accidentel ou délibéré d'hydrocarbures dans le sol, les cours d'eau ou dans la mer provoquant la

contamination du sol, la pollution de la nappe phréatique, la dégradation de la flore et la perturbation de la faune ou portant atteinte à un écosystème tout entier ; ▶marée noire) ; *g* **Ölverschmutzung [f]** (Unmittelbares oder mittelbares Ablassen von Erdöl oder der aus ihm gewonnenen Produkte in den Boden, in ein Gewässer oder ins Meer, das eine Schädigung der Tier- und Pflanzenwelt, des Grundwassers oder eines ganzen Ökosystems zur Folge hat und somit die gefahrfreie Nutzung der natürlichen Ressourcen behindert; ▶Ölpest).

3868 oil pollution accident [n] *envir.* (Tanker collision, road accident or pipeline explosion which causes emptying or leakage of petroleum into the sea, rivers or canals or into the soil; ▶maritime casualty); *syn.* oil pollution casualty [n], oil spill [n] [also US]; *s* **accidente [m] con derrame de hidrocarburos** (Vertido de hidrocarburos en el mar, cursos de agua o en el suelo por accidente de petrolero, de camión de transporte de hidrocarburos o por rotura o explosión de oleoducto o gaseoducto; ▶accidente marítimo); *f* **accident [m] entraînant une pollution par les hydrocarbures** (Déversement accidentel d'hydrocarbures dans le sol, les cours d'eau, canaux ou la mer provoqué par une avarie d'un navire pétrolier, d'un accident d'un camion-citerne, d'une rupture d'un oléoduc ; ▶accident en mer) ; *g* **Ölverschmutzungsunfall [m]** (Durch Tankerhavarie, Tankwagenunfall oder Pipelineexplosion verursachtes Auslaufen von Erdöl in das Meer, in Flüsse, Kanäle oder in den Boden; ▶Seeunfall).

oil pollution casuality [n] *envir.* ▶oil pollution accident.

3869 oil separator [n] *envir.* (Device for the removal of oil from wastewater); *s* **separador [m] de hidrocarburos** (Dispositivo para retirar aceites usados de las aguas residuales); *f* **séparateur [m] à hydrocarbures** (Dispositif destiné à la rétention des huiles usées contenues dans les eaux usées) ; *g* **Ölabscheider [m]** (Anlage zum Abtrennen von Öl aus Abwasser).

oil slick [n] [US] *envir.* ▶oil film.

oil spill [n] [US] *envir.* ▶oil pollution accident.

3870 oil tanker accident [n] *envir.* (Accident due to poor crew training, lack of safety standards, poor ship maintenance, or the old age of a ship; ▶oil pollution); *s* **accidente [m] de petrolero** (Accidente que tiene como resultado el derrame de petróleo o sus derivados, causado generalmente por deficiencia en los estándares de seguridad, mal estado de los buques o carencia de capacitación de la tripulación; ▶marea negra); *syn.* accidente [m] de barco cisterna; *f* **accident [m] de navire pétrolier** (Accident généralement provoqué par la formation déficiente de l'équipage, l'insuffisance des équipements et des normes de sécurité ou encore le vieillissement du navire ; ▶marée noire) ; *g* **Tankerunfall [m]** (Unfall eines Tankschiffes auf Grund mangelnder Sicherheitsstandards, schlechter Ausbildung der Besatzung oder Überalterung des Schiffes; ▶Ölpest); *syn.* Tankerunglück [n].

3871 oil tanker security [n] *envir. leg.* (Precautionary safety measures taken to prevent accidents with oil tankers at sea; ▶MARPOL Convention); *s* **seguridad [f] de petroleros** (Medidas de prevención de accidentes de petroleros para evitar la contaminación de los océanos y de las costas; ▶Convenio de MARPOL); *f* **sécurité [f] des navires pétroliers** (Mesures de prévention de la pollution due aux accidents occasionnels par les navires pétroliers ; ▶Convention de Marpol) ; *g* **Tankersicherheit [f]** (Vorkehrungen zur Verbesserung der Verkehrssicherheit von Tankschiffen auf den Meeren; ▶MARPOL-Abkommen).

3872 old-fashioned rose [n] *hort.* (Rose, which was commercially available before 1867 and, due to its long tradition, has a special attraction for rose lovers and collectors. It usually has double, intensely fragrant flowers and blossoms only once a year. **O.-f. r.s** are available as ground cover, shrub or climbing roses. Old roses are occasionally more disease prone than modern varieties. So-called **Romantic** or **Nostalgic Roses** have now been bred, which are more robust, but resemble historic roses in appearance and perfume. Assorted varieties named **English Roses** usually comprise of old and nostalgic roses); *syn.* historic rose [n]; *s* **rosal [m] antiguo** (Rosa que ya se cultivaba y comercializaba en 1867. Por su larga tradición es muy apreciada por los amantes y los coleccionistas de las rosas. Generalmente florece una vez al año y sus flores son de perfume intenso. Se dan en forma rastrera, arbustiva y trepadora. Los **rr. aa.** son más susceptibles a las enfermedades que las variedades modernas, por lo cual se han desarrollado las llamadas **rosas románticas** o **nostálgicas** que son más robustas, pero se parecen en su óptica y su olor a las rosas antiguas); *f* **rosier [m] ancien** (Rosiers introduits dans le commerce avant 1867 ; on compte parmi eux en Europe 1. les **rosiers galliques**, rosiers de taille moyenne dont les fleurs apparaissent en bouquet, 2. les **rosiers de Damas** originaire du Moyen-Orient qui ont donné naissance aux rosiers à fleurs pour fleurs coupées, encore utilisés en parfumerie au Maroc, 3. les **rosiers de Portland** [1790] issus du croisement entre le rosier gallique et le rosier de Damas et ayant donné naissance au rosiers remontants, 4. les **rosiers cent feuilles** ayant donné naissance aux **rosiers mousseux** [1696] ; les roses anciennes sont très à la mode et apprécié des amateurs de roses ; la floraison est courte mais intense ; l'hybridation des variétés anciennes a donné naissance aux **rosiers romantiques** plus résistants aux maladies ; les **rosiers anglais** associent le charme romantique des roses anciennes avec la grande gamme de couleurs les parfums intensifs et la floraison abondante des variétés contemporaines) ; *g* **Historische Rose [f]** (Rose, die bereits vor 1867 im Handel war. Wegen ihrer langen Tradition hat sie einen besonderen Reiz für Liebhaber und Sammler. Sie hat meist gefüllte, intensiv duftende Blüten und blüht jährlich i. d. R. nur einmal. Es gibt sie als Bodendecker, Strauch- oder Kletterrose. Alte Rosen sind zum Teil krankheitsanfälliger als moderne Züchtungen. Daher gibt es inzwischen so genannte **Romantische** oder **Nostalgische Rosen**, die robuster sind, aber in der Optik und im Duft die historischen Rosen nachahmen. Sortimente mit so genannten **Englischen Rosen** bestehen meist aus Historischen und Nostalgischen Rosen); *syn.* Alte Rose [f].

3873 old-field succession [n] *agr. phyt.* (Secondary succession on abandoned cropland; TEE 1980, 204); *s* **sucesión [f] secundaria** (Serie evolutiva de la vegetación en campo de cultivo abandonado); *f* **succession [f] sur friche agricole** (Série évolutive transitoire anthropique sur les terres anciennement cultivées) ; *syn.* succession [f] sur friche résultant de l'abandon des cultures ; *g* **Sukzession [f] auf Ackerbrachland** (Vegetationsentwicklung auf nicht mehr bestellten Ackerböden).

3874 old-growth forest [n] [US] *for.* (Dense woodland of old trees devoid of undergrowth; some ancient forests in North America have been likened to **Cathedral forests**; ▶forest 1); *s* **oquedal [m]** (▶Monte arbóreo sin sotobosque, ya por haberse talado éste, ya por tratarse de un tipo natural de sotobosque exclusivamente herbáceo, p. ej. hayedo de tipo de *Anemone nemorosa*; DB 1985); *f* **forêt [f] à voûte** (Peuplement forestier adulte caractérisé par un couvert dense et dépourvu de strate arbustive) ; *g* **Hallenwald [m]** (▶Wald mit dichtem Laubdach im hohen Reifestadium ohne Gehölzunterwuchs).

3875 old industrial region [n] *adm. plan.* (An area where outmoded industrial plants are upgraded or converted with funds allocated by the European Community; **in U.S.**, an area of outmoded or disused industrial plants; ▶regional funds of the European Community); *s* **zona [f] industrial en declive** (Cate-

goría de promoción del ▶Fondo Regional de las CC.EE. que tiene como fin apoyar la reestructuración de la economía en regiones antiguamente muy industrializadas que se encuentran en crisis debido a la reducción de la producción industrial y/o a la racionalización de puestos de trabajo); *f* **région [f] industrielle en déclin (±)** (Catégorie bénéficiant de l'attribution de l'aide prévue par le ▶fonds européen de développement régional dont les dotations ont pour objet de financer les restructurations dans les régions industrielles en reconversion) ; *g* **altindustrialisierte Region [f]** (Förderkategorie des ▶europäischen Fonds für regionale Entwicklung für Räume der Europäischen Gemeinschaft, die den Zweck verfolgt, in Gebieten mit alten Industrien durch Einsatz von Finanzmitteln Strukturveränderungen herbeizuführen).

old orchard [n] [UK] *agr. conserv.* ▶traditional orchard [UK].

old orchard grassland [n] *agr. conserv.* ▶traditional orchard meadow [UK].

3876 old part [n] **of a city/town/village** *urb.* (A long existing area of town or village; also a municipal section with old and often sub-standard construction, which may be improved by rehabilitation; ▶historic core, ▶poor quality housing [US]/poor quality housing stock [UK], ▶rehabilitation 1); *s* **zona [f] de edificación antigua** (En F. y D., término utilizado para barrio o pueblo construido antes de la 2ª guerra mundial; ▶centro histórico, ▶edificación antigua con mal trazado, ▶rehabilitación); *syn.* barrio [m] antiguo; *f* **vieux quartiers [mpl]** (F., village ou zone urbaine construit avant 1945 ; ▶habitat ancien, ▶réhabilitation, ▶rénovation immobilière, ▶restauration immobilière, ▶vieille ville) ; *g* **Altbaugebiet [n]** (D., ein bis ca. 1960 gebautes Dorf- oder Stadtviertel; ▶Altbebauung 1, ▶Altstadt, ▶Sanierung).

old quarry [n] *min. landsc.* ▶abandoned quarry.

3877 old-style garden [n] *gard'hist.* (Garden with the aim of demonstrating and preserving the old tradition of horticulture and garden design); *s* **jardín [m] clásico** (Jardín en el que se muestran la horticultura y el diseño de jardines tradicionales); *f* **jardin [m] conservatoire** (Jardin ayant pour vocation le maintien et la sauvegarde des traditions horticoles et de l'art des jardins ; DIC 1993) ; *g* **Garten [m] zur Pflege alter Kulturmethoden** (Garten, der mit dem Ziel angelegt wurde, dem Besucher alte gärtnerische Kulturweisen, alte Kultursorten und Beispiele historischer Gartengestaltung zu zeigen).

old town [n] *urb.* ▶historic core.

3878 oligotrophic [adj] *limn. pedol.* (Descriptive term applied to soil or waterbodies containing low nutrient concentrations; ▶oligotrophic lake or watercourse; ▶trophic state, *opp.* ▶eutrophic); *syn.* nutrient-poor [adj]; *s* **oligótrofo/a [adj]** (Término descriptivo para indicar el contenido bajo de nutrientes en una masa de agua o un suelo; ▶aguas oligótroficas, ▶estado trófico; *opp.* ▶éutrofo/a); *syn.* oligotrófico/a [adj], pobre [adj] en nutrientes; *f* **oligotrophe [adj]** (Qualifie un milieu très pauvre en substances nutritives. Les organismes qui peuvent se développer dans ce type de milieu sont dit « oligotrophiques » ; DEE 1982 ; ▶degré trophique, ▶eaux oligotrophes ; *opp.* ▶eutrophe) ; *syn.* pauvre en éléments nutritifs [loc] ; *g* **oligotroph [adj]** (1. Ein Umweltmedium [See, Fließgewässer oder Boden] betreffend, das durch Nährstoffarmut gekennzeichnet ist. 2. *Gewässergüte bei Stillgewässern* so beschaffen, dass eine geringe Nährstoffbelastung, geringe Algenproduktion, hohe Sichttiefe und ganzjährig ein hoher Sauerstoffsättigungsgrad bis zum Seegrund vorliegt; ▶oligotrophes Gewässer; ▶Trophie, *opp.* ▶eutroph); *syn.* nährstoffarm [adj].

oligotrophic condition [n] *limn. pedol.* ▶oligotrophy.

3879 oligotrophic lake [n] **or watercourse** [n] *limn.* (Waters poor in nutrients, near-saturated with dissolved oxygen,

and supporting very little organic production; ▶trophic level of a waterbody); *s* **aguas [fpl] oligótrofas** (Aguas corrientes o estancadas con bajo contenido de nutrientes, casi saturada de oxígeno disuelto y con baja producción de plancton; ▶nivel trófico de aguas estancadas); *syn.* aguas [fpl] oligotróficas; *f* **eaux [fpl] oligotrophes** (Eaux contenant peu de matières nutritives dissoutes et ayant par conséquent un faible taux de minéralisation ; ▶niveau trophique des eaux douces) ; *g* **oligotrophes Gewässer [n]** (Klares, nährstoffarmes Gewässer mit geringer Planktonproduktion, in dem organische Verbindungen durch Mikroorganismen kaum zersetzt werden. In oligotrophen Seen sind am Ende der Stagnationsperiode die tiefen Wasserschichten noch mit über 70 % Sauerstoff gesättigt; ▶Trophiestufe eines stehenden Gewässers).

3880 oligotrophy [n] *limn. pedol.* (Quality or state of soils or bodies of water poorly supplied with nutrients; ▶trophic level, ▶trophic state; *opp.* ▶eutrophy); *syn.* oligotrophic condition [n]; *s* **oligotrofia [f]** (▶Estado trófico de una masa de agua o un suelo caracterizado por su bajo contenido de nutrientes; ▶nivel trófico; *opp.* ▶eutrofia); *f* **oligotrophie [f]** (Qualité ou état d'un sol ou d'un milieu aquatique très pauvre en substances nutritives ; ▶degré trophique, ▶niveau trophique ; *opp.* ▶Eutrophie) ; *g* **Oligotrophie [f]** (1. Geringe Versorgung von Böden oder Gewässern mit Nährstoffen. 2. *limn. Klassifizierung der Gewässergüte* Gewässer mit einer geringen Nährstoffbelastung, geringen Algenproduktion, einer hohen Sichttiefe und mit einem ganzjährig hohen Sauerstoffsättigungsgrad bis zum Seegrund; ▶Trophie, ▶Trophiestufe; *opp.* ▶Eutrophie); *syn.* Nahrungsarmut [f], Nährstoffarmut [f].

ombrogenous peatland [n] *geo. phyt.* ▶raised bog.

ombrophilous bog [n] *geo. phyt.* ▶raised bog.

ombrophilous cloud forest [n]**, tropical** *for. phyt.* ▶cloud forest.

ombrophilous forest [n] [UK] *geo. phyt.* ▶rain forest [US]/rain-forest [n] [UK].

ombrophilous peatland [n] *geo. phyt.* ▶raised bog.

ombrotrophic bog [n] *geo. phyt.* ▶raised bog.

ombrotrophic peatland [n] *geo. phyt.* ▶raised bog.

omnipresent [adj] *phyt. zool.* ▶ubiquitous.

on a time basis [n] *contr. prof.* ▶payment on a time basis, ▶remuneration on a time basis.

once-a-year cut meadow [n] [UK] *agr. hort.* ▶once-a-year mown meadow [US].

3881 once-a-year mown meadow [n] [US] *agr. hort.* (Meadow on an annual cutting schedule; ▶be mown once a year [US]/be cut once a year [UK]); *syn.* once-a-year cut meadow [n] [UK]; *s* **pradera [f] de siega anual** (Prado que se siega una vez al año; ▶segado/a anualmente); *f* **prairie [f] fauchée une fois par an** (▶fauché une fois par an) ; *g* **einschürige Wiese [f]** (Wiese, die einmal im Jahr gemäht wird; ▶einschürig); *syn.* einmal gemähte Wiese [f], einmal geschnittene Wiese [f].

one-crop farming [n] *agr.* ▶single crop system.

one-layer planting system [n] *constr.* ▶single-stratum planting system.

one-leaf masonry [n] [UK] *constr.* ▶one-tier masonry [US].

3882 one-sided masonry [n] *constr.* (Wall with a fair face on only one side where the masonry facing and backing are so bonded as to exert a common action under load; cf. UBC 1979); *syn.* faced wall [n] [also US] (UBC 1979); *s* **mampostería [f] de cara vista (unilateral)** (Muro con una cara visible); *syn.* mampostería [f] con paramento de cara vista unilateral; *f* **maçonnerie [f] à un parement** (Type de mur avec une face extérieure visible constituée de moellons, la face intérieure étant en béton banché, enduite ou brut de décoffrage et sur laquelle s'appuie le talus) ; *syn.* maçonnerie [f] à une face vue ; *g* **einhäuptiges Mauerwerk [n]** (Mauer mit nur einer Ansichtsfläche).

one spit digging [n] [UK] *hort.* ▶single digging.

3883 one-tier masonry [n] [US] *constr.* (Masonry with bonding running through the entire thickness of the wall; the units at the front of the wall are connected to those at the back in a unified bond. Such masonry is thus load-bearing as a whole; ▶two-tier masonry [US]/two-leaf masonry [UK]); *syn.* solid wall [n] [US], one-leaf masonry [n] [UK], single-wythe masonry [n] [UK]; *s* **mampostería [f] compacta** (Obra de fábrica en la que los mampuestos están aparejados a través de todo el muro, de manera que éste es estáticamente portante; ▶mampostería de dos capas; *f* **maçonnerie [f] à paroi simple** (Ouvrage réalisé avec des moellons de la largeur du mur et dont l'appareillage est sensiblement le même sur les deux faces vues ; ▶maçonnerie à double paroi) ; *g* **einschaliges Mauerwerk [n]** (Mauerwerk, bei dem der Mauersteinverband durch die gesamte Mauerdicke greift resp. die Vormauersteine mit der Hintermauerung zu einem einheitlichen Verband vermauert sind. Dadurch ist dieses **M.** im Gesamtquerschnitt statisch tragfähig; ▶zweischaliges Mauerwerk).

one-year-old feathered tree [n] [UK] *arb. hort.* ▶strongly branched liner [US].

3884 one year's growth [n] *bot. hort.* (Non-lignified, herbaceous shoot of perennials or woody plants, which has originated from buds created in the previous year); *syn.* annual shoot [n]; *s* **guía [f] anual** (Retoño herbáceo, no lignificado de vivaz o leñosa resultante de yemas del año anterior); *syn.* vástago [m] anual; *f* **pousse [f] annuelle** (Résultat de l'accroissement pendant une unité de végétation de la tige chez les végétaux pérennes par allongement des entre-nœuds à partir d'un bourgeon formé l'année précédente) ; *g* **Jahrestrieb [m]** (Unverholzter, krautiger Spross, der aus im Vorjahr angelegten Knospen ausdauernder Pflanzen entsteht); *syn.* Innovationstrieb [m] (MEL 1973, Bd. 12).

on proof of verified delivery notes [loc] [UK] *constr. contr.* ▶according to certified delivery [US].

on-site soil [n] *constr.* ▶volume of on-site soil.

3885 on-site soil investigation [n] *constr.* (Field check of soil by e.g. hand proving, determination of pH value, colo[u]r and odo[u]r, for planting operations; ▶soil investigation); *syn.* field check [n] of soil [also US]; *s* **análisis [m] del suelo in situ** (Investigación de campo para fines de construcción paisajística; p. ej. por medición del pH, prueba manual de textura, etc.; ▶prospección de suelos); *syn.* estudio [m] del suelo in situ; *f* **analyse [f] du sol sur site** (Procédés en général empiriques permettant de déterminer les qualités et les défauts d'un sol, tels que l'analyse mécanique du sol, l'étude des horizons culturaux, la détermination de la valeur du pH, de la richesse du sol en azote, acide phosphorique, potasse, l'observation des plantes spontanées, le prélèvement d'échantillons pour une analyse en laboratoire, etc. ; ▶prospection du sol) ; *syn.* analyse [f] de terre ; *g* **Bodenbeurteilung [f]** (Felduntersuchung für vegetationstechnische Zwecke, z. B. durch Fingerprobe, pH-Wert Messung, nach Farbe und Geruch etc.; ▶Bodenuntersuchung).

on-the-job training [n] *prof.* ▶continuing education [US].

open air recreation [n] *recr.* ▶outdoor recreation.

open air swimming baths [npl] [UK] *recr.* ▶outdoor bathing complex [US].

3886 open bidding [n] [US] *adm. constr. contr.* (Public anouncement in the daily press or professional periodicals for invitation of firms to participate in a ▶bidding [US]/tendering [UK]; ▶selective bidding); *syn.* compulsory competitive tendering [n] [UK], open tendering [n] [UK]; *s* **concurso-subasta [m] público** (Invitación a empresas a participar en ▶concurso-subasta de obras o servicios anunciado en la prensa o en revistas profesionales; ▶concurso-subasta restringido); *f 1* **appel [m] d'offres ouvert** (Mode de passation de marchés d'▶appel d'offres par mise en concurrence de toute entreprises désirant remettre une offre avec avis d'appel public à la concurrence) ; *f 2* **adjudication [f] ouverte** (Mode de passation de marchés d'▶adjudication par mise en concurrence de toute entreprises désirant remettre une offre avec avis d'appel public à la concurrence, la commission d'adjudication éliminant les candidats n'ayant pas qualité ou capacité pour présenter une offre ; ▶adjudication restreinte, ▶appel d'offres restreint) ; *g* **öffentliche Ausschreibung [f]** (In der Presse oder in Fachzeitschriften bekanntgemachte Aufforderung an Unternehmer zur Teilnahme an einer ▶Ausschreibung für Bauleistungen oder Lieferungen; öffentlich ausschreiben [vb]; ▶beschränkte Ausschreibung 2); *syn.* offenes Verfahren [n].

open-cast mining [n] [UK] *min.* ▶open-pit mining [US].

open-cast mining area [n] [UK] *min.* ▶open-pit mining area [US].

open-cast workings [npl] [UK] *min.* ▶open-pit mining [US].

3887 open country [n] *landsc.* (**1.** A post-forest landscape in an European area with a social system that directly controls land-use practices; LE 1986, 597; **2. in U.K.**, this term refers to land, which in a colloquial sense is simply not in use; it also refers to land designated for public access as defined under the 1949 National Parks and the 1968 Countryside Acts; used in the 2000 Countryside Right of Way Act to describe 'areas of mountain, moor, heath and down', **o. c.** is land which the public is allowed to access for rambling and other outdoor pursuits; ▶farmstead, ▶fields, ▶rural land 1); *syn.* extensive open land [n] [also US] (TGG 1984, 247), open countryside [n], open field [n], open terrain [n]; *s* **campiña [f] (±)** (Extensión considerable de tierra llana cultivada; **D.**, término utilizado en la ley de protección de la naturaleza para referirse a paisajes caracterizados por grandes áreas de campos de cultivo y prados, con poca superficie de bosque y ▶granjas dispersas, sin influencia visible de pueblos y ciudades; en gran parte de Europa este término puede utilizarse como sinónimo de ▶paisaje rural 2; ▶campo abierto, ▶zona rural); *f 1* **campagne [f] (2)** (▶Espace rural ni montagneux ni très forestier de champs ouverts et parsemés de ▶fermes, par opposition à la ville ; type de paysage encore appelé « champagne » ou « paysage ouvert », caractérisé par l'absence de haies ou de clôture ; LA 1981) ; *syn.* paysage [m] découvert, espace [m] naturel ; *f 2* **paysage [m] d'openfield** (Paysage né de l'activité agricole dans certaines régions de l'Europe occidentale médiévale, qui a laissé de larges paysages découverts à villages groupés ; MG 1993 ; ▶openfield) ; *g 1* **freie Landschaft [f]** (Durch große, unbewaldete Flächen, meist Äcker, Grünland und Heiden, geprägte offene Kulturlandschaft außerhalb der Städte und Dörfer; der Begriff **f. L.** beinhaltet im Vergleich zur ▶Flur 2 auch eine Landschaft mit Gewässern, Wäldern, Felsgebieten und einzeln verstreuten ▶Gehöften; in weiten Teilen Europas kann dieser Begriff synonym mit Agrarlandschaft gebraucht werden; ▶ländlicher Raum); *syn.* freie Natur [f] (±), freies Feld [n], *conserv. hunt.* freie Wildbahn [f] [CH] ; *g 2* **Offenland [n]** (UB zu freie Landschaft; OB für alle unbewaldeten Biotoptypen in der ▶freien Landschaft; ▶Ödland, Ziff. 2).

open countryside [n] *landsc.* ▶open country.

3888 open crown [n] *arb.* (►opening up tree crown); *s* **copa [f] hueca** (►aclareo de la copa); *f* **houppe [f] au port aéré** (►éclaircie de cime) ; *syn.* houppe [f] au port léger ; *g* **lichte Baumkrone [f]** (Krone mit einer lockeren Beastung und einem lichtdurchlässigen Blattwerk, z. B. Gleditschie *[Gleditsia triacanthos]*; ►Verlichtung 2).

open cut technique [n] *eng.* ►cut and cover excavation [US]/cut and cover tunnelling [UK].

3889 open cut tunnel [n] *trans.* (Road tunnel constructed in an open excavation, the roof of which is provided with a continuous slit to allow removal of exhaust fumes and for noise suppression); *s* **túnel [m] construido a cielo abierto y cubierto** (Túnel de carretera construido a zanja abierta cuya cubierta tiene una franja abierta para la evacuación de los gases de escape); *f* **tunnel [m] réalisé en fouille ouverte** (Tunnel routier creusé en fouille ouverte et dont le toit possède une large ouverture longitudinale afin de permettre l'évacuation des gaz d'échappement et d'assurer la protection contre le bruit) ; *syn.* tunnel [m] réalisé en fouille talutée ; *g* **Tunnel [m] in Schlitzlage** (In offener Baugrube gebauter Straßentunnel, dessen Dach zur Abführung der Abgase und zum Lärmschutz in Form eines Schlitzes offen bleibt).

3890 open design competition [n] *adm. plan. prof.* (Public anouncement in the daily press or professional periodicals for invitation of planners to participate in a design competition or final design competition [US]/realization competition [UK]; ►design competition announcement [US]/design competition invitation [UK]); *s* **concurso [m] de ideas público** (Invitación a gabinetes de planificación a participar en concurso de ideas o de realización anunciado en la prensa o en revistas profesionales; ►anuncio de un concurso [de ideas]); *syn.* concurso [m] de ideas abierto; *f* **appel [m] d'offres avec concours** (Mode de passation de marchés de maîtrise d'œuvre par les maîtres d'ouvrage publics s'appliquant au concours [concours d'idées ou de conception-construction] d'architecture et d'ingénierie avec appel public à la concurrence par insertion dans une publication habilitée à recevoir des annonces légales ; décret n° 93-1269 du 29 novembre 1993 ; ►appel public à la concurrence [pour un concours]) ; *g* **öffentliche Ausschreibung [f] für einen Wettbewerb** (In der Presse oder in Fachzeitschriften bekanntgemachte Aufforderung an Planer zur Teilnahme an einem Ideen- oder Bauwettbewerb; öffentlich ausschreiben [vb]; ►Auslobung).

open drain [n] *agr. for.* ►drainage channel.

open field [n] *landsc.* ►open country.

3891 open [vb] **for public inspection** [US] *adm. leg.* (►public announcement, ►publication of planning proposals); *syn.* make [vb] available for public inspection [UK]; *s* **someter [vb] a información pública** (Procedimiento de presentar al público en general los planes de ordenación urbana, proyectos de urbanización, etc. después de su aprobación inicial. En Es. la ley sobre régimen del suelo y valoraciones como legislación básica determina en el art. 6 que «la legislación urbanística garantizará la participación pública en los procesos de planeamiento y gestión, así como el derecho a la información [...] de los particulares», mientras que la ley del suelo como legislación supletoria fija que el plazo es de un mes, si no está regulado de otra manera a nivel de las CC.AA.; cf. Ley 6/1998 y art. 41 RD 1346/1976; ►publicación de un plan, ►sumisión a información pública); *f* **soumettre [vb] à enquête publique** (►enquête publique, ►mise à la disposition du public d'un projet de plan ; ►publication) ; *g* **öffentlich auslegen [vb]** (Einen Plan, der von der Verwaltung erarbeitet wurde, ortsüblich für jedermann zugänglich in der Planauflage zur Einsichtnahme hinlegen; ►Aus-

legung eines Planentwurfes, ►öffentliche Bekanntmachung); *syn.* öffentlich auflegen [vb] [A].

3892 opening [n] **of bids** [US] *constr. contr.* (Opening of submitted bids [US]/tenders [UK] at a precisely defined date; ►bid opening date [US]/submission date [UK]); *syn.* bid opening [n] [US], opening [n] of tenders [UK]; *s* **apertura [f] de ofertas** (Apertura de las sumisiones a concurso en un momento predeterminado; ►fecha de apertura de pliegos); *syn.* apertura [f] de plicas; *f* **ouverture [f] des plis** (Ouverture des enveloppes contenant les offres ou les soumissions ; ►date d'ouverture des plis) ; *syn.* ouverture [f] des offres ; *g* **Angebotseröffnung [f]** (1. Öffnung und Verlesung der im Rahmen einer öffentlichen oder beschränkten Ausschreibung über Bauleistungen [VOB-Verfahren] eingereichten Angebote zu einem vorher genau festgelegten Zeitpunkt. 2. Bei der Öffnung der Angebote für Lieferleistungen, die der Verdingungsordnung für Leistungen [VOL-Verfahren] unterliegen, dürfen wegen der Vertraulichkeit keine Bieter anwesend sein; ►Eröffnungstermin); *syn. obs.* Submission [f].

opening [n] **of tenders** [UK] *constr. contr.* ►opening of bids [US].

3893 opening up [n] **of forest canopy** *for.* (Considerable reduction of canopy density, e g., by lopping, felling or poisoning selected trees, or naturally through pests, disease, drought, mortality, etc.; SAF 1983); *s* **aclareo [m] (1)** (Tratamiento selvícola que se realiza en los estratos de latizal y fustal de las masas forestales, consistente en la corta de ciertos pies con el objeto de acelerar el crecimiento diametral de los árboles restantes, o para seleccionar de entre la masa los de mejor forma y vigor, sin alterar ni romper la permanencia de la cubierta vegetal; DINA 1987; *términos específicos* aclareo sucesivo, aclareo por entresaca); *f 1* **éclaircie [f]** (Réduction de la densité d'un peuplement forestier non arrivé à maturité en vue d'améliorer la croissance et la forme des arbres restants ; TSF 1985) ; *syn.* ouverture [f] du couvert (DFM 1975) ; *f 2* **dépressage [m]** (Éclaircie de jeunes semis ou rejets en densité trop forte en vue d'améliorer la croissance individuelle des plants) ; *g* **Verlichtung [f] (1)** (Auflockerung eines Waldbestandes durch forstwirtschaftliche Maßnahmen oder durch Absterben relativ vieler Bäume nach Kalamitäten, Trockenheit oder durch Sturm; cf. WT 1980).

3894 opening up [n] **of tree crowns** *arb. for.* (Reduction of leaf growth and die-back of a large number of small branches which may be a symptom of disease or mere senescence); *s* **aclarado [m] de la copa** (Reducción de la densidad de la copa de un árbol por enfermedad o por senescencia que se caracteriza por un menor crecimiento de las hojas y la muerte de muchas ramillas resp. por la caída de las agujas en las coníferas); *f* **éclaircie [f] de cime** (Réduction de la densité du feuillage de la cime d'un arbre Réduction de la densité du feuillage de la cime d'un arbre provoquée par un ralentissement de croissance ou par la mort de nombreux rameaux ou des feuilles ou la perte des aiguilles chez les conifères ; chez le pin le degré d'éclaircissement se constate par l'observation des classes d'âge des aiguilles ; pour les peuplements d'épicéa l'état sanitaire est déterminé selon un protocole d'observation [classes de dépérissement] qui prend en compte la mort ancienne, la mort récente, les fenêtres dans le houppier, la présence de rameaux et/ou branches sèches, la perte foliaire, la coloration anormale des aiguilles ; cf. Protocole d'observation du Département de la Santé des Forêts [DSF] 2006, Centre Régional de la Propriété Forestière [CRPF] Midi-Pyrénées) ; *syn.* éclaircissement [m] de la cime ; *g* **Verlichtung [f] (2)** (Krankheitsbild eines Baumes, das durch vermindertes Blattwachstum und Absterben von vielen Feinästen resp. durch Entnadelung bei Koniferen entsteht. Bei Kiefern wird der Grad der **V.** an der Zahl der bestehenden Nadeljahrgänge sichtbar; bei Fichten gilt folgende Typisierung: Top-Dying, Sub-Top-Dying oder Fensterverlichtung, Lamettabehang und die fortschreitende Entnadelung von innen nach außen; cf. HAN 1990, 162-166).

3895 open land [n] (±) *leg. urb.* (**In Europe**, undeveloped open areas generally beyond the confines of a zoning map [US]/ Proposals Map [UK] in rural outskirts; **in U.S. and U.K.**, this is not a zoning term; as long as an area with building potential is not a national park or protected water catchment area, the right of ownership allows, in principle, that anything can be built; in comparison with Europe, **open land** does not have to be protected as a cultural landscape, but is regarded as space for free development and use; ►open country, ►built-up area 2; *opp.* ►core area 1); *syn.* rural outskirts [npl] [US] (TGG 1984, 83), green-field site [n] [UK], non-zoned land [n] (±); *obs.* white land [n] [UK], undeveloped peripheral area [n] [UK], undeveloped outskirts area [n] (FBC 1993, 43), greenfields [npl] [UK]; ***s 1* suelo** [m] **no urbanizable** (1. Es., categoría de la clasificación del suelo dentro del perímetro urbano, que implica que éste ha de ser preservado del proceso de desarrollo urbano y para el que prevé establecer, en su caso, medidas de protección del territorio y del paisaje; cf. art. 11 [3] Ley del Suelo y art. 18 RD 2159/78. **2.** Es., zona no edificable que se encuentra afuera de la zona de vigencia de un plan parcial o de los ►núcleos de población agrupada; *opp.* ►perímetro edificado. **3. En EE.UU.** no existe esta categoría); ***s 2* suelo** [m] **rústico** (Es., categoría para suelo que se encuentra afuera del perímetro urbano; ►campiña); ***f 1* zone** [f] **non urbanisable** (1. F., PLU/POS zone peu ou non équipée, zone non touchée par l'urbanisation qu'il importe souvent de protéger en raison de la valeur agricole ou forestière des terres, de la richesse du sol ou du sous-sol et de son importance pour l'environnement ; celle-ci n'est en général pas comprises dans le périmètre d'agglomération et rassemble deux catégories, **a.** les ►zones agricoles, **b.** les ►zones naturelles et forestières : **2. D.**, zone située à l'extérieur du périmètre fixé par un « Bebauungsplan » — PLU/POS allemand — [cf. § 30 BauGB] ou comprise à l'extérieur des ►secteurs agglomérés ; cf. § 35 BauGB ; **F.**, correspondrait de par sa nature aux ►espaces non urbanisés ; ►campagne 2, ►espace naturel ; *opp.* ►zone urbaine 1) ; ***f 2* zone** [f] **agricole** — zone **ZA** — (*Documents d'urbanisme* [ancienne zone NC des POS], secteurs à protéger en raison du potentiel agronomique, biologique ou économique des terres agricoles) ; ***f 3* zone** [f] **naturelle et forestière** — zone **ZN** (*Documents d'urbanisme* [ancienne zone ND des POS], secteurs à protéger notamment en raison de la qualité des sites, des milieux naturels, des paysages et de leur intérêt, notamment du point de vue esthétique, historique ou écologique) ; ***g* Außenbereich** [m] (D., außerhalb des räumlichen Geltungsbereichs eines Bebauungsplanes i. S. v. § 30 BauGB und außerhalb der ►im Zusammenhang bebauten Ortsteile gelegene Gebiete; cf. § 35 BauGB; ►freie Landschaft; *opp.* ►Innenbereich; **in USA und im U.K.** gibt es im Planungsrecht keine Kategorie des **A.s** wie im dt. Planungsrecht: Solange eine potenzielle Baufläche nicht als Nationalpark oder Trinkwassereinzugsgebiet geschützt ist, erlaubt das Recht des Eigentums, dass im Prinzip alles bebaut werden kann; das Land außerhalb von Siedlungen wird im Vergleich zu Europa nicht als die zu bewahrende, von den Vorfahren ererbte Kulturlandschaft, sondern als Raum zur freien Entfaltung und Nutzung gesehen).

open land [n] [US], **extensive** *landsc.* ►open country.

3896 open land [n] **needed for development** *plan.* (Undeveloped area of land needed for a construction project); *syn.* open land requirement [n]; ***s* requerimiento** [m] **de suelo rústico** (Superficie de terreno necesitada para realizar un proyecto de construcción en una zona no urbanizada); ***f* besoins**

[mpl] **fonciers en zone rurale** (±) (Terres non bâties en vue d'une extension urbaine prévisible ou autres projets d'aménagement) ; ***g* Landschaftsbedarf [m, o. Pl.]** (Erfordernis von Flächen einer unbebauten Landschaft oder deren Teile für die Verwirklichung von [Bau]vorhaben).

open land requirement [n] *plan.* ►open land needed for development.

3897 open meadow [n] **in a forest** *for. recr.* (TEE 1980, 214; moist grass and herb-covered area in a ►forest clearing); ***s* prado** [m] **forestal** (Formación herbácea en ►claro [del bosque]); ***f 1* pré** [m] **de clairière** (Végétation herbacée recouvrant le sol d'une clairière) ; *syn.* clairière [f] pastorale, préclairière [m] ; ***f 2* prairie** [f] **forestière** (Strate herbacée régulièrement entretenue de certaines Hêtraies, Chênaies ou de peuplements de Châtaigniers à régime de futaie claire régulière dans l'étage montagnard et sub-montagnard dans lesquelles est encore pratiquée la fauche ou le ramassage des châtaignes ; ►clairière, ►troué forestière) ; *syn.* pré [m] sous forêt ; ***g* Waldwiese [f]** (Wiesenfläche in einer ►Waldlichtung).

open orchard grassland [n] [US] *agr. conserv.* ►traditional orchard meadow [UK].

open orchard meadow [n] [US] *agr. conserv.* ►traditional orchard meadow [UK].

3898 open-pit mining [n] [US] *min.* (Type of mining in which the overburden is removed from the material being mined and is dumped back after mining; term may also specifically refer to an area from which the overburden has been removed and has not been refilled; cf. GSM 1974; ►deep pit mining; *specific terms* strip mining [US], opencut mining [US], contour strip mining, hillside strip mining; *opp.* ►deep mining); *syn.* open-cast mining [n] [UK], open-cast workings [npl] [UK], surface mining [n] [US]; ***s* minería** [f] **a cielo abierto** (Tipo de minería en la que la capa de sobrante se remueve totalmente para extraer el mineral. Es frecuente en la extracción de minerales de hierro, cobre, zinc y otros metales. Altera totalmente la topografía local y puede causar otros numerosos impactos ambientales; DINA 1987; *términos específicos* explotación por bandas, explotación de gravera, explotación arenera, explotación de placeres; ►explotación a cielo abierto profunda; *opp.* ►minería subterránea); *syn.* explotación [f] a cielo abierto; ***f* exploitation** [f] **(minière/de carrière) à ciel ouvert** (Extraction effectuée à l'air libre de substances minérales ou fossiles sur des gîtes existants à la surface de la terre, sur lesquels les matériaux de recouvrement sont enlevés, mis en dépôt et éventuellement remis en place selon les impératifs économiques ou écologiques ; ►exploitation minière à ciel ouvert ; *opp.* ►exploitation minière souterraine) ; ***g* Tagebau [m]** (Gewinnung von nutzbaren Mineralen oder Gesteinen aus oberflächennahen Lagerstätten von der Erdoberfläche aus, indem die Deckschichten unter vertretbaren Kosten abgetragen, seitlich gelagert und später wieder verfüllt werden können. Es gibt auch Tagebaugebiete, in denen die Deckschichten nicht wieder unter landschaftsgestalterischen Gesichtspunkten zurückgeführt werden; ►Tieftagebau; *opp.* ►Untertagebau).

3899 open-pit mining area [n] [US] *min. plan.* (►Mining area, in which mineral deposits, close to the surface, are extracted above ground); *syn.* open-cast mining area [n] [UK], surface mining area [n]; ***s* zona** [f] **de minería a cielo abierto** (►Cuenca minera en la que los yacimientos están a poca profundidad por lo que se extraen a cielo abierto); ***f* région** [f] **de mines à ciel ouvert** (►Bassin minier sur lequel les gîtes de substances minérales et fossiles sont exploités à ciel ouvert) ; ***g* Tagebaugebiet [n]** (►Bergbaugebiet, in dem oberflächennahe Lagerstätten über Tage abgebaut werden).

O

3900 open recycling system [n] *envir.* (Human method of reusing materials produced from natural resources and returned for remaking of the same kind of product; ►recycling); *s* **ciclo** [m] **abierto de materias** (Sistema de reutilización de materiales elaborados a partir de recursos naturales para producir productos del mismo tipo; ►reciclado); *f* **cycle** [m] **de la matière ouvert** (Réemploi de matériaux élaborés à partir des ressources naturelles en vue de la production du même type de produits ; ►recyclage) ; *g* **offener Stoffkreislauf** [m] (System der Wiederverwendung von aus natürlichen Ressourcen hergestellten Materialien für die Produktion desselben Produkttyps; ►Recycling).

open season [n] *hunt. leg.* ►hunting season.

3901 open space [n] *plan.* (Area or plot of ground predominantly free of buildings in an urban or rural region, which is sometimes protected from development by government action to provide for outdoor recreation, or, in Europe, to preserve prime agricultural land, wooded areas, exceptional views, land and water features, and to channel urban growth; GE 1977; ►common open space [US]/community open space [UK], ►green space, ►open space system, ►outdoor space, ►private open space, ►public open space, ►use of open space); *syn.* open terrain [n]; *s* **espacio** [m] **libre** (Término utilizado en planificación para todas las superficies urbanas no edificadas, con o sin vegetación, que pueden ser utilizadas para el esparcimiento y recreo, pero que también pueden cumplir funciones ecológicas si son suficientemente grandes o parte de una ►trama verde; el término inglés «open terrain» puede traducirse si el espacio se encuentra en la ciudad como «campa silvestre», si se encuetra fuera de ella como «paisaje rural»; ►espacio libre colectivo, ►espacio libre privado, ►espacio libre público, ►exteriores, ►uso de espacios libres, ►zona verde); *syn.* espacio [m] abierto (DINA 1987); *f* **espace** [m] **libre** (Étendue non bâtie, espace minéral ou espace planté en milieu urbain à usage récréatif public, collectif ou privatif ; les ►espaces libres extérieurs lorsqu'ils s'incèrent dans ►trame verte contribuent à développer l'orientation spatiale et la lisibilité du tissus urbain, la continuité et l'identité ; ils permettent de faire naître une culture urbaine grâce à la réalisation des besoins et aspirations ainsi que l'accaparation de l'espace par leurs utilisateurs ; afin de lutter contre la tendance à l'imperméabilisation le règlement du PLU oblige à réaliser des espaces libres dans le cadre de construction nouvelles ; ►espaces libres collectifs, ►espace libre privatif, ►espace libre public, ►espace vert, ►utilisation des espaces libres) ; *syn.* espace [m] extérieur, espace [m] ouvert ; *g* **Freiraum** [m] (*Planungswissenschaftlicher Begriff* alle nicht überdachten, vegetations- oder nicht vegetationsbestimmten Flächen in einem Siedlungsgebiet, die für Freizeit und Erholung nutzbar sind, aber auch stadthygienischen Erfordernissen dienen, wenn sie genügend groß oder Bestandteil eines ►Freiraumsystems sind. Der **F.** verleiht den Städten Orientierung und Lesbarkeit, Kontinuität und Identität, d. h. er lässt durch die Ansprüche und Inbesitznahme der Bürger immer mehr spezifische Stadtkultur entstehen. **F.** als Synthese von vielfältigen Formen urbaner Kultur und Raum dient den Begegnungen von Menschen, der Befriedigung der Neugierde füreinander und dem Alleinsein im öffentlichen Raum. Der US-Begriff *open terrain* kann im Stadtgebiet auch mit ‚freier Wiesenfläche' und außerhalb von im Zusammenhang bebauter Ortsteile mit ‚freier Landschaft' übersetzt werden; ►Außenraum, ►Freiraumnutzung, ►Grünfläche, ►kollektiv nutzbarer Freiraum, ►öffentlich nutzbarer Freiraum, ►privat nutzbarer Freiraum); *syn.* Freifläche [f].

open space [n] [UK], **community** *plan.* ►common open space [US].

open space [n], **green** *landsc. urb.* ►green space.

open space [n], **landscaped** *landsc. urb.* ►green space.

3902 open space compensation area [n] *plan.* (Land located outside of a conurbation with a semi-natural and varied landscape structure or composition which is intended to improve the urban climate, to compensate for lack of urban open space, and to provide needed outdoor recreation facilities. The effectiveness of such areas is disputed); *syn.* offsetting open space [n] [also US]; *s* **area** [f] **de compensación (ecológica)** (*En D. término de política ecológica y regional muy discutido* zona localizada en la vecindad de aglomeraciones con una estructura del paisaje variada y casi natural, a la que se adjudica la función central de servir para mejorar el clima [local] y para fines recreativos); *f* **zone** [f] **naturelle d'équilibre (Z.N.E.)** (F., instrument de planification ayant pour but de créer un équilibre entre les grandes agglomérations urbaines et les territoires périphériques non urbanisés au moyen de [1] mesures de protection [espaces naturels, limitation du développement urbain] et de [2] programmes de mise en valeur [agriculture, loisirs, etc.]) ; *g* **Ausgleichsraum** [m] (*Ökologisch und raumordnerisch strittiger Begriff* den Ballungsgebieten funktional zugeordnete Regenerationsgebiete mit naturnaher oder abwechslungsreicher Landschaftsstruktur, die der stark belasteten Stadtlandschaft z. B. zur Verbesserung des Klimas, der Biotopstruktur, auch des stadtnahen Freizeitangebotes dient).

open space corridor [n] [US] *landsc. urb.* ►linear open space pattern/system.

open space design [n] *landsc. recr. urb.* ►open space planning.

open space needs [npl] *plan. recr.* ►requirements for open spaces.

open space network [n] *landsc. urb.* ►open space system.

open space pattern/system [n] *landsc. urb.* ►combined open space pattern/system, ►concentric open space pattern/system, ►linear open space pattern/system, ►radial open space pattern/system.

open space pattern/system [n], **interfingering** *landsc. urb.* ►open space pattern/system of peninsular interdigitation.

3903 open space pattern/system [n] **of peninsular interdigitation** *landsc. urb.* (Tree-like [dendritic] open space pattern with a series of interlocked linear open spaces [= green fingers] in built-up areas); *syn.* interfingering open space pattern/system [n] (LE 1986, 408); *s* **trama** [f] **verde en franjas (≠)** (Sistema de espacios libres urbanos en forma de franjas paralelas entre las áreas edificadas); *syn.* red [f] de espacios libres en franjas, sistema [m] de espacios libres en franjas; *f* **trame** [f] **d'espaces libres en bandes** (Système de disposition des espaces libres perpendiculairement à la pente d'un versant urbanisé) ; *g* **kammförmiges Freiraumsystem** [n] (In Siedlungsbereichen senkrecht zum Hang konzipiertes F.).

3904 open space planning [n] *landsc. recr. urb.* (Planning tool for the design of recreation areas which are not built upon, as well as for a green open space system fulfilling urban micro-climatic requirements within the scope of ►general urban green space planning; ►landscape architecture); *syn.* open space design [n]; *s* **planificación** [f] **de espacios libres** (Conjunto de instrumentos de planificación y diseño de espacios no construídos para el esparcimiento y recreo de la población así como para crear un sistema de zonas verdes que sirva para asumir funciones microclimáticas de la ciudad; el término ►paisajismo se puede utilizar como sinónimo si en la planificación el diseño es lo primordial; ►planificación de zonas verdes y espacios libres urbanos); *f 1* **aménagement** [m] **des espaces libres** (Instrument de conception permettant l'organisation des divers espaces extérieurs urbains devant assumer les fonctions de circulation, d'équilibre écologique, de détente et de loisirs ; ▶architecture des paysages, ▶planification urbaine des espaces verts) ; *syn.* conception [f] des espaces urbains ; *f 2* **mission [f] d'étude générale de planification et de programmation du paysagiste** (Prestations intellectuelles relatives aux domaines d'études suivants : **1.** études d'environnement, d'impact, de programmation et d'aménagement, **2.** schémas de cohérence territoriale/schémas directeurs d'aménagement et d'urbanisme, plan local d'urbanisme/plan d'occupation des sols, chartes et plans de paysage, plans verts, **3.** zones d'aménagement concerté, zones industrielles et artisanales, **4.** reboisements, remembrement, protection et la mise en valeur des espaces naturels, **5.** études de tracés des grandes infrastructures [routes, voies ferrées, rivières et canaux, lignes électriques, etc.]) ; *g 1* **Freiraumplanung [f]** (Planerisches Instrumentarium zur Gestaltung von nicht überbauten Flächen für Freizeit und Erholung, Arten- und Biotopschutz sowie von Grünflächensystemen für stadthygienische Erfordernisse im Rahmen der Grünordnung unter besonderer Berücksichtigung soziologischer Gesichtspunkte. Freiräume müssen nicht immer vegetationsbestimmt sein. **F.** ist im Vergleich zur ,Grünflächenplanung' und ,Grünplanung' der umfassendere Begriff, da Freiräume nicht immer vegetationsbestimmt sein müssen; ▶Landschaftsarchitektur kann als Synonym benutzt werden, wenn das Formal-Gestalterische bei der Planung im Vordergrund steht; ▶Grünordnung); *g 2* **Flächenplanung [f]** (OB für alle Planungen, die im Vergleich zu Einzelobjekten [Objektplanung] große Flächen und Gebiete umfassen: **1. bauleitplanerische Leistungen** [Bauleitplanung] für die Planarten Flächennutzungsplan und Bebauungsplan und **2. landschaftsplanerische Leistungen** [Landschaftsplanung] für Pläne wie Landschaftspläne, Grünordnungspläne, Landschaftsrahmenpläne und Landschaftspflegerische Begleitpläne zu Vorhaben, die den Naturhaushalt, das Landschaftsbild oder den Zugang zur freien Natur beeinträchtigen können, Pflege und Entwicklungspläne sowie sonstige landschaftsplanerische Leistungen; cf. §§ 17 und 22 HOAI 2009).

3905 open space policy [n] [US] *pol. landsc. urb.* (Strategy followed by governments, legislatures and environmental organizations aimed at preserving or creating green spaces within urban areas. **In U.S.**, a statement on use or reservation of mostly undeveloped land with potential value for specific uses, and to channel urban growth; ▶general urban green space planning); *syn.* green space policy [n] [UK]; *s* **política [f] de zonas verdes** (Conjunto de medidas de la administración pública realizadas para cumplir los fines de la ▶planificación de zonas verdes y espacios libres); *f* **politique [f] d'espaces verts** (Mesures et actions des gouvernements, parlements, collectivités territoriales, organisations de protection de la nature en faveur de la sauvegarde, de la création et de la gestion des espace verts urbains ; ▶planification urbaine des espaces verts) ; *g* **Grünflächenpolitik [f] (1)** (Auf die Durchsetzung der Ziele der ▶Grünordnung gerichtetes Handeln von Regierungen, Parlamenten, Umweltorganisationen etc., um in den Siedlungsgebieten ein bestehendes Grünflächensystem auszubauen, weiterzuentwickeln sowie langfristig zu sichern und damit die Lebens- und Wohnqualität zu verbessern); *syn.* Grünpolitik [f].

open space ring [n] *landsc. urb.* ▶green open space ring.

3906 open space standard [n] *urb. recr.* (Quantitative expression of an open green space planning objective, in amount of land to be provided as open space: often expressed as a ▶guideline [value] or design standard in terms of area per resident; ▶determination of requirements); *s* **estándar [m] de zonas verdes** (Valor cuantitativo de superficie de terreno por

habitante necesaria para cubrir la demanda de espacios libres; ▶determinación de la demanda, ▶valor estándar); *f* **valeur [f] des besoins en espaces libres** (Valeur estimative [▶détermination de la demande] formulant des exigences en matière d'espaces verts, aires de jeux, etc. et considérée comme ▶valeur guide pouvant être prescrite lors de la délivrance d'un permis de construire ou comme valeur indicative dans la définition d'objectifs pour les schémas directeurs) ; *g* **Bedarfszahl [f] für Freiflächen** (Quantitative Bemessung [▶Bedarfsermittlung] von Grundflächen für Freiräume, meist als überschlägiger ▶Richtwert für Leitplanungen); *syn.* Bedarfswert [m] für Freiflächen.

3907 open space system [n] *landsc. urb.* (Spatial interconnection of green areas and other ▶open spaces, e.g. ▶green belts, ▶green finger connections, ▶green space corridor, ▶parks 1, ▶permanent community garden area [US]/permanent allotment site [UK], ▶tree-lined avenues [US]/avenues [UK], ▶urban squares, cemeteries and sports grounds in urban areas. These areas contribute to the planned structure of a city and are included in the design of zoning districts for the creation of recreational facilities close to residential neighbo[u]rhoods and for climatic, ecological, circulation, and cultural purposes; *specific terms* ▶combined open space pattern/system, ▶concentric open space pattern/system, green grid [UK] [GP 2003, Dec. 03, 7], ▶linear open space pattern/system, ▶open space pattern/system of peninsular interdigitation, park system, ▶radial open space pattern/system, system of green areas); *syn.* open space network [n]; *s* **trama [f] verde** (Interconexión espacial de ▶zonas verdes y otros ▶espacios libres en zonas urbanas como ▶avenidas, ▶conexiónes verdes, ▶cinturones verdes y ▶corredores verdes, ▶parques y jardines, ▶plazas, ▶zonas permanentes de huertos recreativos urbanos, cementerios y pistas de deportes. Contribuyen a la estructura planificada de una ciudad y son incluídos en el diseño de los barrios con el fin de crear equipamiento de recreo en o cerca de las zonas residenciales y con fines ecológicos, climáticos y culturales; *términos específicos* ▶trama verde combinada, ▶trama verde concéntrica, ▶trama verde en forma de banda, ▶trama verde en franjas, ▶trama verde radial); red [f] de espacios libres, sistema [m] de espacios libres; *f* **trame [f] verte** (Organisation spatiale hiérarchisée des espaces verts urbains, espaces naturels plantés et autres ▶espaces libres dans la composition urbaine, reliés entre eux par des cheminements bordés d'arbres pour les piétons et les cyclistes, tels que p. ex. les ▶allées d'arbres, ▶ceintures vertes, ▶coulées vertes, ▶coupures vertes, ▶lotissement de jardins [ouvriers], ▶parcs, ▶squares, cimetières et terrains de sport dont leur rôle comme espaces de détente et de loisirs, favorise l'action microclimatique, la vie sociale et culturelle et constitue un élément essentiel de la qualité de la vie en zone urbaine ; cf. DUA 1996, 325 ; *termes spécifiques* ▶trame concentrique d'espaces libres, ▶trame d'espaces libres en bandes, ▶trame linéaire d'espaces libres, ▶trame radiale d'espaces libres, ▶trame verte composée) ; *syn.* trame [f] d'espaces libres, trame [f] de verdure ; *g* **Freiraumsystem [n]** (Räumlicher Verbund von Grünflächen und sonstigen ▶Freiräumen wie z. B. ▶Grüngürtel, ▶Grünverbindungen, ▶Grünzüge, ▶Stadtplätze, ▶Parks, ▶Alleen, ▶Dauerkleingartenanlagen, Friedhöfe und Sportplätze im Siedlungsbereich zur räumlichen Ordnung, städtebaulichen Gliederung und Gestaltung des immer unübersichtlich werdenden Siedlungsgemenges und zur Schaffung von wohnungsnahen Freizeit- und Erholungseinrichtungen und für klimaökologische, verkehrliche und kulturelle Zwecke sowie für Belange des Arten- und Biotopschutzes [Biotopverbundsystem]; *UBe* ▶bandförmiges Freiraumsystem, ▶kammförmiges Freiraumsystem, ▶kombiniertes Freiraumsystem, ▶radiales Freiraumsystem, ▶ringförmiges Freiraumsys-

O

tem); *syn.* Freiraumordnung [f]; *im engeren Sinne der Vegetationsausstattung* Grünflächensystem [n], Grünraumsystem [n].

open space system [n]**, combined** *landsc. urb.* ►combined open space pattern/system.

open space system [n]**, concentric** *landsc. urb.* ►concentric open space pattern/system.

open space system [n] **forming a radial pattern** *landsc. urb.* ►radial open space pattern/system.

open space system [n] **of peninsular interdigitation** *landsc. urb.* ►open space pattern/system of peninsular interdigitation.

3908 open stairway [n] [US] *arch. constr.* (Free-standing flight of steps without lateral walls; ►cheek wall); *syn.* flight [n] of steps without cheek walls [UK]; *s* **escalera** [f] **abierta** (Tramo de escalera sin ►zancas ni paredes laterales); *f* **escalier** [m] **sans limon** (►limon) ; *g* **wangenlose Treppe** [f] (**1.** Im Garten- und Landschaftsbau eine Treppe ohne seitliche Begrenzung; der seitliche Erdanschluss kann über die vordere oder an die hintere Auftrittsflächenbegrenzung geführt werden. **2.** Im Hochbau werden die Stufen nicht in eine Wange eingebaut, sondern durch Stufenabstandhalter, die in eine Wand eingelassen werden und durch Sprossen des Geländers gehalten, wodurch die Treppe optisch eine besondere Leichtigkeit erhält; ►Treppenwange).

3909 open tank treatment [n] [US&CDN] *constr.* (Type of wood preservation treatment in which timber is immersed in a preservative solution, normally at atmospheric temperature, for a period of one to many hours, depending on the species, size and actual moisture content; ►pressure treatment [US]/pressure impregnation [UK]); *syn.* steeping treatment [n] [UK]; *s* **tratamiento** [m] **de la madera por sumersión** (Método de impregnación de la madera por el que se introduce ésta en un baño durante una a varias horas dependiendo del tipo de madera, su grosor y su humedad; ►impregnación a presión); *f* **traitement** [m] **en cuve ouverte** (Traitement de protection du bois consistant à l'immerger dans un bain de préservation sous la pression atmosphérique. Cette forme de traitement doit tenir compte de l'espèce de bois, de son épaisseur et de son humidité ; ►imprégnation à cœur en autoclave) ; *syn.* traitement [m] à cuve ouverte ; *g* **Tränkung** [f] (Imprägnierung von Hölzern durch ein- bis mehrstündiges Tauchbad, abhängig von der Holzart, seiner Dicke und seines Feuchtigkeitsgehaltes; ►Kesseldruckimprägnierung).

open tendering [n] [UK] *adm. constr. contr.* ►open bidding [US].

open terrain [n] *landsc.* ►open country; *plan.* ►open space.

operating agency [n] [US] *adm.* ►operating authority

3910 operating authority [n] *adm.* (Specific agency/authority serving as a branch of a government or as an autonomous administration, e.g. of a local community, which is responsible for a defined field of activities, e.g. forest, parks, nature conservation, roads, lakes and rivers administration); *syn.* branch authority [n] [US], competent authority [n] [UK], operating agency [n] [US]; *s* **autoridad** [f] **competente** (Departamento de la administración de un Estado, una comunidad autónoma, provincia, departamento o un municipio que es responsable de una o varias funciones en la administración pública como p. ej. la administración de montes, protección de la naturaleza, de aguas, de obras públicas o de zonas verdes urbanas); *f* **autorité** [f] **compétente** (Service de l'administration d'État, régionale ou locale, office public responsable de la gestion d'un domaine défini de l'administration publique, p. ex. les espaces verts, les services techniques, l'équipement, la gestion de l'eau, etc.) ; *syn.* service [m] compétent ; *g* **Fachbehörde** [f] (Stelle, die als Organ des Staates oder eines selbständigen Verwaltungsträgers, z. B.

einer Gemeinde, für einen bestimmten Bereich Aufgaben der öffentlichen Verwaltung wahrnimmt, z. B. Forst-, Grünflächen-, Naturschutz-, Straßenbau-, Wasserverwaltung).

operation [n] *constr. hort.* ►hoeing operation; *envir.* ►industrial plant operation, ►maintenance operation.

operations [npl] *constr.* ►planting operations, ►preparatory operations.

opinion poll [n] *plan. sociol.* ►poll.

opinion survey [n] *sociol.* ►representative poll.

opportunities [npl] *plan. recr.* ►recreation opportunities.

opposition group [n] *pol.* ►highway opposition group [US], ►nuclear power opposition group [US].

3911 optical guidance [n] *landsc. trans.* (HALAC 1988-II, 12; leading the gaze of an observer in a certain direction; e.g. by rows of trees along the roadside); *syn.* visual orientation [n]; *s* **orientación** [f] **visual** (Dirigir la atención de un/a observador, -a en una dirección deseada, p. ej. a lo largo de las carreteras por medio de hilera de árboles o hitos con reflejo); *f* **guide** [m] **visuel** (Orientation de la vue dans une direction ou vers un point précis provoqué, p. ex. par un alignement d'arbres le long d'une route) ; *g* **optische Führung** [f] (Leitung eines Betrachters in eine bestimmte Richtung, z. B. durch Baumreihen an Straßenrändern).

3912 optimizing [n] *plan.* (Attempt to produce the best possible result by comparing alternatives); *s* **optimización** [f] *syn.* optimación [f]; *f* **optimisation** [f] ; *g* **Optimierung** [f] (Versuch der Erzielung bestmöglicher Ergebnisse durch vergleichende Betrachtung von Alternativen oder Varianten).

option [n] *plan.* ►'no project' option [US].

option [n] **(BPEO), best practicable environmental** *envir.* ►state-of-the-art (2).

option [n]**, 'no build'** *plan.* ►'no project' option [US].

3913 optional specification item [n] *contr.* (►Bid item [US]/tender item [UK] which can be either a contractor's option or an owner's option for inclusion in the contract; an **o. s. i.** may be specified as an ►add-on item. The costs of this item, in contrast to the ►alternate specification item, are included in the total costs of a tender/bid); *syn.* provisional (specification) item [n]; *s* **partida** [f] **discrecional** (En el resumen de prestaciones ►partida de oferta independiente que no se fija al comienzo de la obra ya que no se sabe si se habrá de realizar y en qué alcance. Puede especificarse paralelamente como ►partida [de obra] adicional. Al contrario que en el caso de la ►partida alzada, los costes [Es]/costos [AL] de la misma están incluidos en la oferta total); *f* **numéro** [m] **de prix provisoire** (COMPU 1993, art. 78, 80 ; prestation exceptionnelle prévue dans le descriptif-quantitatif [►article] ou dans le Bordereau des Prix Unitaires [B.P.U.] lorsqu'au moment de l'appel d'offres il n'est pas encore possible de déterminer si et dans quelle mesure une prestation pourra être exécutée et de conclure le marché à prix définitifs. Les coûts correspondant à cette prestation font partie intégrante de la soumission ; COMPU 1993, art. 78, 80 ; ►numéro de prix pour plus-value, ►numéro de prix pour une variante) ; *g* **Bedarfsposition** [f] (Im Leistungsverzeichnis gekennzeichnete, eigenständige Leistung [►Position], bei der zum Zeitpunkt der Fertigstellung der Ausschreibungsunterlagen noch nicht feststeht, ob und ggf. in welchem Umfang sie tatsächlich ausgeführt wird. Eine **B.** kann zugleich als ►Zulageposition ausgeschrieben werden. Die Kosten dieser Position sind im Vergleich zur ►Alternativposition im zu wertenden Gesamtangebot enthalten); *syn.* Eventualposition [f].

opus [n] **quadratum** *arch. constr.* ►ashlar masonry.

3914 oral agreement [n] *constr. contr. prof.* (Mutual understanding between two or more parties; *specific terms* construction commission, planning commission; ►contract 1); *s* **contratación [f] verbal** (Acuerdo hablado entre dos o más partes para adjudicar un trabajo profesional; ►contrato); *f* **contrat [m] oral** (Passation orale d'un marché de travaux, d'études ; ►contrat) ; *g* **mündlicher Auftrag [m]** (Die mündliche Übertragung von Aufgaben an einen Auftragnehmer; ►Vertrag).

3915 orchard [n] *hort.* (Piece of land which is predominantly used for fruit production; ►open orchard meadow, ►traditional orchard [UK]); *syn.* fruit tree grove [n] [also US]; *s* **huerto [m] frutal** (Jardín que se aprovecha sobre todo para la producción de frutas; ►plantación dispersa de árboles frutales, ►prado de huerta frutal); *syn.* huerta [f] frutal, *poético* vergel [m]; *f* **verger [m]** (Parcelle plantée d'arbres fruitiers et, par extension, ensemble des plantations d'arbres fruitiers d'un pays ; ►peuplement dispersé d'arbres fruitiers, ►prairie complantée d'arbres fruitiers) ; *syn.* jardin [m] fruitier ; *g* **Obstgarten [m]** (Garten, der vorwiegend der Obstproduktion dient; ►Streuobstbestand, ►Streuobstwiese).

orchard [n] [UK], **old** *agr. conserv.* ►traditional orchard [UK].

orchard grassland [n], **open** *agr. conserv.* ►open orchard meadow [UK].

orchard meadow [n] [UK], **open** *agr. conserv.* ►traditional orchard meadow.

3916 order [n] (1) *phyt.* (Phytosociological unit superior to an ►alliance. An o. is composed of a group of alliances, which—in turn—is composed of a group of ►plant associations. The compound suffix *-etalia* is added to the root of the name of one of the most important associations; e.g. *fag-etum* to *fag-etalia*); *s* **orden [m]** (En la sistemática de las asociaciones de BRAUN-BLANQUET, la unidad inmediatamente superior a la ►alianza e inferior a la clase; la denominación se forma añadiendo la terminación *-etalia* a la raíz del nombre de una de las ►asociaciones o alianzas importantes pertenecientes a él: p. ej. *fag-ion* se convierte en *fag-etalia*; cf. BB 1979); *f* **ordre [m]** (En systématique phytosociologique, unité supérieure de végétation qui regroupe les ►alliances ayant des affinités floristico-sociologiques manifestées surtout par la possession d'espèces caractéristiques propres à l'ordre et dont la caractérisation du niveau hiérarchique dans la classification est définie par la terminaison *-etalia* ajoutée au nom générique d'une espèce dominante ou caractéristique, p. ex. *Fag-etalia* ; les ordres sont eux-mêmes regroupés en classes ; cf. BB 1928 ; ►association [végétale]) ; *g* **Ordnung [f]** (Die nächste dem ►Verband 1 übergeordnete Vegetationseinheit. **O.en** werden durch Anfügung der Endung *-etalia* an den Wortstamm einer der wichtigsten zugehörigen ►Assoziationen oder Verbände bezeichnet: z. B. *fag-ion* wird zu *fag-etalia*).

3917 order [n] [US] (2) *leg.* (Written instruction by local, state or federal government authorizing e.g. implementation of laws by local ordinances and regulations; **in U.S** a legislative enactment by a local authority, often implementing state laws, to approve development plans, zoning maps, etc., or to undertake regulatory, spending and other program[m]s; ►condemnation order [US]/compulsory purchase order (COP) [UK], ►enabling act [US]/enabling statute [UK], ►temporary restraining order [US]/emergency preservation notice [UK]); *syn.* statutory instrument [n] [UK]; *s* **reglamento [m] jurídico** (Cualquier orden de una autoridad pública basada en su función legal para cuya implementación se legislan ►decretos; ►declaración de régimen de protección preventiva, ►orden judicial de expropiación); *f* **ordonnance [f] légale** (Décision écrite du domaine de la loi, pour mettre en vigueur, en exécution, en application les mesures émanant d'une autorité administrative ; ►classement d'office, ►décret d'appli-

cation, ►ordonnance d'expropriation) ; *g* **Rechtsverordnung [f]** (Jede Anordnung zur Regelung von Fällen, die auf Grund gesetzlicher Ermächtigung von Behörden getroffen wird. Für die Ausführung gibt es ►Durchführungsverordnungen; ►einstweilige Sicherstellung, ►Enteignungsanordnung).

ordinance [n] *adm. leg. urb* ►aesthetic ordinance [US], ►city ordinance [US]; *conserv. leg.* ►nature conservation ordinance.

organic detritus [n] *pedol. phyt.* ►litter (1).

3918 organic farmer [n] *agr.* (Farmer who manages his land according to ►organic farming methods); *s* **agricultor, -a [m/f] biológico/a** (Productor, -a agrícola que cultiva la tierra con métodos de ►cultivo alternativo); *syn.* agricultor, -a [m/f] orgánico/a; *f* **agrobiologiste [m]** (Agriculteur pratiquant ►culture biologique) ; *syn.* agriculteur [m] biologiste (DUV 1984, 267) ; *g* **Ökobauer [m]** (Landwirt, der ►alternativen Landbau betreibt).

3919 organic farming [n] *agr.* (Unconventional agricultural forms usually involving soil amendment by humus-producing materials and animal manures, with rejection or limited use of chemical fertilizers and pesticides as a common practice; ►traditional farming); *syn.* organic gardening [n] [also US], bio-intensive gardening [n] [also US]; *s* **cultivo [m] alternativo** (Término genérico para diferentes formas de producción agropecuaria que tienen en común la proscripción de abonos y biocidas químicos con el fin de evitar en la mayor medida posible los efectos negativos sobre la naturaleza y de producir alimentos sanos; ►cultivo tradicional); *syn.* agricultura [f] biológica, agricultura [f] alternativa, agricultura [f] orgánica, cultivo [f] biológico, cultivo [m] orgánico; *f* **culture [f] biologique** (Terme générique pour différentes méthodes culturales proscrivant l'usage de produits chimiques — pesticides et engrais — dont l'objectif est au moyen d'opérations très différenciées de bouleverser le moins possible l'équilibre naturel des sols ; ►agriculture traditionnelle) ; *syn.* agriculture [f] biologique (LA 1981, 33) ; *g* **alternativer Landbau [m, o. Pl.]** (OB zu Form der Landbewirtschaftung mit erheblicher Differenzierung der Anwendung bestimmter Hilfsmittel und Präparate und der Bewirtschaftungsmethoden, denen der prinzipielle Verzicht oder die sehr eingeschränkte Anwendung aller chemisch hergestellten Produktionsmittel für Düngung und Schädlingsbekämpfung gemeinsam ist, um zu vermeiden, dass diese Mittel sich ungünstig auf die Umwelt auswirken oder zu Rückständen in den Agrarerzeugnissen führen können; cf. EWG-Verordnung Nr. 2092/91 des Rates vom 24.06.1991 über den ökologischen Landbau und die entsprechende Kennzeichnung der landwirtschaftlichen Erzeugnisse und Lebensmittel; ►konventioneller Landbau); *syn.* biologische Landwirtschaft [f, o. Pl.], naturnahe Landwirtschaft [f, o. Pl.] [auch CH], Ökolandbau [m, o. Pl.], ökologischer Landbau [m, o. Pl.]; *z. T.* alternativer Gartenbau [m, o. Pl.], biologischer Gartenbau [m, o. Pl.], ökologischer Pflanzenbau [m, o. Pl.].

3920 organic fertilizer [n] *agr. constr. for. hort.* (Fertilizer with a high percentage of degradable biological substances; ►fertilizer; *opp.* mineral fertilizer); *syn.* manure [n]; *s* **fertilizante [m] orgánico** (Abono con un gran porcentaje de sustancias biológicas degradables; ►fertilizante; *opp.* ►fertilizante mineral); *syn.* abono [m] orgánico; *f* **engrais [m] organique** (►Engrais constitué d'un fort pourcentage d'éléments décomposables biologiquement, p. ex. fumier, matières fécales, résidus végétaux, boues d'épuration, purin, lisier, compost de déchets ménagers ou plantes d'engrais vert; *opp.* ►engrais minéral) ; *g* **organischer Dünger [m]** (Dünger mit einem großen Anteil biologisch umsetzbarer Bestandteile und einer Vielzahl an Nährstoffen, die für das Pflanzenwachstum wichtig sind; Nährstoffe liegen organisch gebunden vor und müssen durch Mikroorganismen mineralisiert werden, damit sie von den Pflanzen aufgenommen werden können; ►Dünger; *opp.* ►Mineraldünger).

3921 organic fertilizing [n] *agr. constr. for. hort.* (Enrichment of soil, usually with animal manure [dung] or other ►organic fertilizer; ►fertilizer, ►green manuring, ►liquid manure, ►semiliquid manure [US]/semi-liquid manure [UK]); *s* **fertilización [f] natural** (Abonado con ►fertilizante orgánico; ►abonado verde, ►fertilizante, ►licuame, ►purín); *syn.* fertilización [f] orgánica; *f* **fumure [f] organique** (Entretien ou amélioration de la fertilité du sol par apport d'►engrais organiques ; LA 1981, 554 ; ►engrais, ►fertilisation par [apport d']engrais vert, ►lisier, ►purin) ; *syn.* fertilisation [f] organique ; *g* **organische Düngung [f]** (Ausbringen von ►organischem Dünger; ►Dünger, ►Gründüngung, ►Gülle, ►Jauche).

organic gardening [n] [US] *agr.* ►organic farming.

3922 organic horizon [n] *pedol.* (Top layer above mineral soil dominated by fresh or partly decomposed organic matter which is comprised of various sub-horizons: O = L [litter layer], O_1 = F [fermentation layer] and O_2 = H [humus layer]; cf. RCG 1982; ►forest floor litter, ►mor humus, ►litter layer); *syn.* O horizon [n]; *s* **mantillo [m]** (Capa superior del suelo, formada en gran parte por la descomposición de materias orgánicas: O_1 = hojarasca no descompuesta, O_f = capa de descomposición, O_h = capa de materiales húmicos; ►capa de litter, ►litter del bosque, ►moor); *f* **horizon [m] organique (Ao)** (Couche à la surface du sol exclusivement organique constituée de débris végétaux incomplètement décomposés, comprenant les sous-horizons de la litière, la couche de fermentation et la couche humifiée ; PED 1983 ; ►couverture morte, ►litière forestière, ►mor) ; *syn.* couche [f] organique ; *g* **Humusauflage [f]** (Oberste Schicht aus nicht oder wenig zersetzter organischer Substanz auf terrestrischen Böden, die aus dem O_l-[Litter Horizont], O_f-[Fermentationshorizont] und dem O_h-Horizont [Humifizierungshorizont] besteht; SS 1979; ►Auflagehumus, ►Streuschicht, ►Waldbodenstreu); *syn.* Ektohumus [m].

3923 organic kitchen waste [n] *envir* (Animal and vegetable matter resulting from preparation, cooking, and disposing of food in a kitchen; *generic term* ►kitchen waste); *s* **restos [mpl] orgánicos de cocina** (Restos de materia animal o vegetal resultantes de la elaboración de comidas; *término genérico* ►restos de cocina); *f* **déchets [mpl] organiques de cuisine** (*Terme générique* ►déchets de cuisine) ; *g* **organischer Küchenabfall [m]** (Pflanzliche und tierische Reste, die bei der Speisenzubereitung in der Küche anfallen; *OB* ►Küchenabfall).

3924 organic mud [n] *pedol.* (►Subhydric soils such as ►dy, ►gyttja, and ►sapropel which occur at the bottom of lakes and ponds and have a significantly high content of organic material); *s* **suelo [m] subacuático no turboso** (Término genérico para los suelos subacuáticos como ►dy, ►gyttja, ►sapropel y protopedon que se encuentran bajo aguas estancadas y contienen cantidades importantes de sustancia orgánica; cf. KUB 1959, 97c; ►suelo subacuático); *f* **boue [f] organique** (Terme générique pour les sols d'apport peu évolués sur des matériaux récents, humus submergés localisés sur le fond des eaux calmes ou stagnantes comme le ►dy, le ►gyttja et le ►sapropèle ; DUV 1984, 193 ; ►sol subaquatique) ; *syn.* vase [f] de marais ; *g* **Mudde [f]** (Subhydrischer Boden am Grunde nicht fließender Gewässer mit einem deutlich erkennbaren Gehalt an organischer Substanz wie ►Dy, ►Gyttja und ►Sapropel; cf. SS 1979 u. DIN 18 196; ►Unterwasserboden).

3925 organic soil [n] *pedol.* (Soil that contains a high percentage—greater than 20 or 30%—of organic matter throughout the solum—[upper soil profile]; cf. RCG 1982; ►organic mud, ►peat soil 2; *opp.* ►mineral soil); *syn.* histosol [n] [also US];

s **suelo [m] orgánico** (Suelo que contiene un alto porcentaje —mayor de 20-30%— de materia orgánica; ►suelo subacuático no turboso, ►suelo turboso; *opp.* ►suelo mineral); *f* **sol [m] organique** (Sol caractérisé par une importante quantité de matériaux organiques — > 30 % — dans l'horizon A et B ; ►boue organique, ►tourbe ; *opp.* ►sol minéral) ; *syn.* histosol [m] ; *g* **organischer Boden [m]** (Boden mit > 30 % organischer Substanz im A- und B-Horizont; ►Moor 2, ►Mudde; *opp.* ►Mineralboden).

3926 organic turnover [n] *chem. ecol.* (TEE 1980, 125; process of decay of organic material, e.g. litter. The period for complete decomposition is called turnover time; *specific term* **microbial loop** [n]: in aquatic environments a trophic system of reintroduction of dissolved organic carbon [DOC] to the food web through the incorporation into bacteria. Bacteria are consumed mostly by protists such as flagellates and ciliates. These protists, in turn, are consumed by larger aquatic organisms (for example small crustaceans like copepods); *s* **tasa [f] de renovación orgánica** (Velocidad de renovación de la materia orgánica de un determinado nivel trófico; este proceso implica la descomposición de materia orgánica existente y su sustitución por nueva materia; cf. DINA 1987); *syn.* turnover [m] de materia orgánica; *f* **décomposition [f] de la matière organique fraîche** (Processus de transformation [humification, minéralisation] de la matière organique fraîche pour donner naissance à l'humus et aux composés minéraux solubles ou gazeux ; le rythme de décomposition complète [minéralisation primaire et secondaire] de la matière organique est dénommé turnover [renouvellement] ; *terme spécifique* **boucle [f] microbienne :** système presque clos comprenant les bactéries hétérotrophes, les cyanobactéries et les algues eucaryotes de petite taille ainsi qu'une variété de protistes amiboïdes, flagellés et ciliés des réseaux nutritionnels aquatiques ; la biomasse bactérienne entretenue par les protozoaires planctoniques est consommée par les protistes, dont en retour les produits d'excrétion sont consommés par les organismes zooplanctoniques) ; *syn.* décomposition [f] des litières ; *g* **Stoffumsatz [m]** (Vorgang der Umwandlung von organischer Substanz, z. B. Fallaub; die Dauer des Abbaues wird Umsatzzeit genannt; *UB* mikrobieller Stoffumsatz).

organism [n] *biol. ecol. leg.* ►genetically modified organism (GMO).

organism [n] *agr. for. hort. phytopath.* ►injurious organism.

organization [n] *prof.* ►professional organization.

3927 organized use [n] **of leisure time** *plan. recr.* (Orderly manner in which an individual chooses to spend free time); *s* **uso [m] (individual) del tiempo libre** (Forma en la que cada persona aprovecha su tiempo libre); *syn.* aprovechamiento [m] (individual) del tiempo libre; *f* **pratique [f] d'activités de loisirs** (Manière dont une personne occupe ses loisirs sur le plan individuel) ; *syn.* pratique [f] d'activités récréatives ; *g* **Freizeitgestaltung [f] (2)** (Art und Weise wie jemand seine Freizeit verbringt).

organ [n] **with adhesive disks, grasping** *bot.* ►grasping branch with adhesive disks.

3928 orientation [n] *met.* (Facing of a slope, building or plants towards the points of the compass or solar radiation path); *syn.* aspect [n], exposure [n]; *s* **orientación [f]** (Exposición de una ladera, un edificio o de las plantas respecto al sol, respecto a los puntos cardinales); *syn.* exposición [f]; *f* **exposition [f]** (*Termes spécifiques* exposition au nord, exposition ensoleillée) ; *syn.* orientation [f] ; *g* **Himmelslage [f]** (Neigung eines Hanges, Lage eines Bauwerkes oder die von Pflanzenbeständen zur Himmelsrichtung).

orientation [n] **to sun** *gard.* ►solar orientation.

3929 origin and destination traffic [n] *trans.* (►destination traffic, ►origin traffic); *syn.* O&D traffic [n] [also US]; *s* **tráfico** [m] **de origen y de destino** (►tráfico de destino, ►tráfico de origen); *f* **circulation** [f] **de provenance et de destination** (►circulation de destination, ►circulation de provenance); *g* **Quell- und Zielverkehr** [m] (►Quellverkehr, ►Zielverkehr).

3930 origination area [n] [US] *recr. urb.* (Area from which people travel to their workplaces or to use public facilities; ►bedroom community [US]/dormitory suburb [UK], ►recreation area 1, ►recreation origination area); *syn.* catchment area [n] [UK]; *in the vernacular* stockbroker belt [n], commuter belt [n]; *s* **área** [f] **de influencia** (Zona alrededor de una ciudad, una instalación, una zona de recreo, etc. desde la cual la gente se traslada para utilizar un servicio o gozar de una oferta; cf. DGA 1986; ►ciudad-dormitorio, ►zona de origen de recreacionistas, ►zona de uso priorizado para el recreo); *f* **zone** [f] **d'attraction** (Terme générique définissant l'espace avoisinant les villes ou les équipements en provenance duquel les utilisateurs se déplacent pour se rendre sur leur lieu de travail ou pour accéder aux équipements collectifs; ►cité-dortoir, ►espace émetteur de flux touristique, ►espace touristique); *syn.* zone [f] d'influence, espace [m] émetteur; *g* **Einzugsgebiet** [n] (Gebiet im Umland einer Stadt, einer Infrastruktureinrichtung etc., aus dem die Menschen kommen, um angebotene Arbeitsplätze, Gemeinbedarfseinrichtungen oder sonstige Einrichtungen aufzusuchen; ►Erholungsgebiet, ►Erholungsquellgebiet, ►Schlafstadt); *syn.* Einzugsbereich [m].

3931 origin traffic [n] *trans.* (Proportion of traffic which has its source in a particular area; *opp.* ►destination traffic); *syn.* "O" traffic [n] [US]; *s* **tráfico** [m] **de origen** (Parte del tráfico que es originado en una zona específica; *opp.* ►tráfico de destino); *f* **circulation** [f] **de provenance** (Partie du trafic qui s'écoule à partir d'un lieu donné; *opp.* ►circulation de destination); *syn.* déplacements [mpl] en provenance d'un lieu, trafic [m] de provenance, trafic [m] d'origine; *g* **Quellverkehr** [m] (Teil des Verkehrs, der aus einem bestimmten Gebiet herausfließt; *opp.* ►Zielverkehr).

3932 ornamental garden [n] *gard.* (Generic term for a garden, which is enjoyed passively for its location and attractive appearance, in contrast to a ►fruit and vegetable garden; *specific terms* ►estate garden [US]/villa garden [UK], ►front yard [US]/front garden [UK] [partially], ►private garden); *syn.* flower garden [n]; *s* **jardín** [m] **ornamental** (Jardín que, al contrario que el huerto, sirve para representar, admirar y para la contemplación y el descanso; *términos específicos* ►jardín privado, ►jardín de villa, ►jardín delantero); *f* **jardin** [m] **d'agrément** (Terme générique pour les jardins d'habitation, ►jardins de banlieue, ►jardin de façade, ►jardin privé; par opposition au ►jardin de rapport, espace d'agrément et décoratif favorisant la contemplation et la détente; ►jardin particulier); *syn.* jardin [m] d'ornement, *peu utilisé* jardin [m] ornemental; *g* **Ziergarten** [m] (Garten, der im Vergleich zum ►Nutzgarten der Repräsentation, dem Anschauen bzw. der passiven Erholung dient; *UBe* ►Hausgarten, ►Villengarten, *z. T.* ►Vorgarten); *syn.* Schmuckgarten [m].

3933 ornamental grass [n] *gard. hort. plant.* (Grass with a specially attractive appearance, due to its size, brightly colo[u]red or wide blades, its autumn colo[u]rs, flowers or seed heads, which become lignified in the autumn and, if left standing during the winter, are also attractive; ►perennial with lignified seed stalks; *specific terms* ►arching ornamental grass, ►specimen grass); *s* **hierba** [f] **ornamental** (Gramínea especialmente decorativa por su tamaño, el colorido de sus hojas, flores o infru[c]tes-

cencias; ►perenne con infructescencia invernal; *términos específicos* ►gramínea solitaria, ►hierba ornamental llorona); *syn.* gramínea [f] ornamental; *f* **graminée** [f] **ornementale** (Graminée à forte valeur décorative de par sa taille [graminée géante] sa couleur et l'aspect de ses feuilles [larges et retombantes], sa couleur automnale, le caractère de ses chaumes en particulier pour les ►vivaces à infrutescence [lignifiée] pérenne qui ne sont pas coupé à la fin de l'automne; *termes spécifiques* ►graminée [en position] isolée, ►graminée ornementale à longues feuilles et hampes retombantes); *g* **Ziergras** [n] (Gras mit besonderem Schmuckwert durch Größe [Riesengras], Buntlaubigkeit, Breitblättrigkeit, Herbstfärbung, Blüten und Fruchtstände und auch als ►Wintersteher, wenn sie nicht im Spätherbst zurückgeschnitten werden; *UBe* ►Solitärgras, ►Ziergras mit überhängenden Blättern und Blütenständen); *syn.* Schmuckgras [n].

3934 ornamental pattern [n] *constr.* (Decorative paving patterns created when laying brick or stone pavers and ►concrete pavers [US]/concrete paving units [UK]; *specific term* ►herringbone pattern; ►paving pattern); *s* **aparejo** [m] **ornamental** (Término genérico para diferentes ►patrones de aparejo de ladrillos clinker o ►piedras de hormigón; *término específico* ►aparejo en espina); *f* **appareillage** [m] **ornemental** (Terme générique utilisation de différents ►calepinage et association de divers matériaux [pierre, terre cuite, ►pavé en béton] de dimensions et couleurs variées ou emploi de matériaux nobles [pavés mosaïques, dalles en pierre naturelle] avec appareillages recherchés; *termes spécifiques* ►appareillage en chevron, appareillage en épi, appareillage en opus spicatum); *g* **Zierverband** [m] (Dekoratives Verlegemuster von Klinkern oder ►Betonpflastersteinen 1; *UB* ►Fischgrätenmuster; ►Verlegemuster).

3935 ornamental perennial [n] *hort. plant.* (Herbaceous plant, cultivated or selected for its attractive flowers, colo[u]rs, seedheads and leaves, for use in garden beds, borders and park adornment; *specific term* ►decorative foliage plant); *syn.* showy perennial [n] [also US]; *s* **vivaz** [f] **ornamental** (Planta herbácea cultivada por sus flores, sus colores, sus hojas o sus infru[c]tescencias llamativas para decorar parques y jardines; *término específico* ►vivaz de hojas decorativas); *syn.* perenne [f] ornamental, planta [f] vivaz ornamental; *f* **plante** [f] **vivace d'ornement** (Espèce vivace qui a amélioré, par multiplication, ses caractéristiques telles que la richesse, la variété des tons et l'aspect décoratif du feuillage et des fleurs ainsi que développé une période de floraison plus longue; *terme spécifique* ►plante vivace à feuilles ornementales); *syn.* plante [f] vivace ornementale; *g* **Schmuckstaude** [f] (*Terminus für das Wuchsverhalten einer Staude hinsichtlich ihrer ästhetischen Wirkung und gärtnerischen Verwendung* mehrjährige, krautige Pflanze, oft durch Züchtung oder Auslese weiterentwickelt, die wegen ihrer Blütenformen, Farben, Fruchtstände, Blätter oder wegen ihrer besonderen Schönheit in Garten- und Parkanlagenrabatten gepflanzt wird; *UB* ►Blattschmuckstaude); *syn.* Zierstaude [f], *z. T.* Prachtstaude [f].

3936 ornamental plant [n] *hort.* (Plant species grown in gardens or kept indoors or in greenhouses, cultivated because of its attractive flowers, fruits, leaves or scent; *specific terms* pot plant, indoor plant, house plant, flowering plant, decorative foliage plant, etc.); *s* **planta** [f] **ornamental** (Especie vegetal cultivada para fines decorativos, sea en jardines o en interiores); *f* **plante** [f] **ornementale** (Plante cultivée spécialement pour la valeur décorative de ses fleurs, fruits, feuilles ou pour son odeur parfumée et utilisée en appartement, jardin d'hiver et en extérieur; *termes spécifiques* plante en pot, plante d'intérieur, plante d'appartement, plante d'orangerie, plante de serre); *g* **Zierpflanze** [f] (Kulturpflanze, die wegen ihrer Blüten, Früchte, Blätter, ihres Duftes oder Habitus' in Gewächshäusern, in

Räumen oder im Garten verwendet wird. Oft wird sie züchterisch weiterentwickelt; *UBe* Topfpflanze, Zierpflanze für Innenraumbegrünung, Zimmerpflanze).

3937 ornamental plant horticulture [n] *hort.* (Often specialized branch of the greenhouse and nursery trades [▶commercial horticulture], which cultivate and sell ornamental plants and blooms; e.g. pot plants, bedding plants, cut flowers, etc.); *s* **horticultura [f] ornamental** (Rama de la ▶horticultura comercial que se dedica al cultivo de plantas ornamentales como plantas para jardineras, plantas de macizos, flores para cortar); *f* **horticulture [f] d'ornement** (Branche de l'▶horticulture spécialisée dans la culture des plantes ornementales, p. ex. les plantes en pot, les plantes de massif, les plantes coupées) ; *syn.* horticulture [f] ornementale ; *g* **Zierpflanzenbau [m]** (Oft spezialisierter Zweig des ▶Erwerbsgartenbaues, der die Kultivierung von Zierpflanzen, z. B. Topfpflanzen, Beetpflanzen, Schnittblumen, betreibt).

3938 ornamental planting [n] *gard. hort.* (Eye-catching, colo[u]rful plants, usually summer annuals or perennials in parks); *s* **plantación [f] ornamental** (Plantación realizada con fines decorativas utilizando plantas llamativas, generalmente de flores de estación o perennes); *f* **plantation [f] ornementale** (Plantation pour effet décoratif de plantes vivaces ou de à floraison estivale et colorée) ; *g* **Schaupflanzung [f]** (Effektvolle, meist farbenprächtige Wechselflor- oder Staudenpflanzung).

ornamental rose bush [n] [UK] *hort. plant.* ▶cultivated shrub rose.

3939 ornamental shrub [n] *hort. plant.* (Deciduous or evergreen shrub, which is planted in gardens and parks because of its attractive appearance, e.g. leaves, flowers, bark or fruits, and has often been bred or selected for its special attributes; ▶ornemental woody plant); *syn.* partially flowering shrub [n]; *s* **arbusto [m] ornamental** (Arbusto perennifolio o caducifolio que se planta en parques y jardines por la belleza de sus ramajes, sus hojas, flores o frutos, a menudo cultivado o seleccionado por sus atributos especiales; ▶leñosa ornamental); *f* **arbuste [m] d'ornement** (Arbuste à feuilles caduques ou persistantes cultivé pour la beauté de sa ramification, ses feuilles, ses fleurs, son écorce, ses fruits, planté dans les parcs et jardins ; ▶végétaux d'ornement) ; *g* **Zierstrauch [m]** (Laub abwerfender oder immergrüner Strauch, der wegen der Schönheit seiner Verzweigung, seiner Blätter, Blüten, Rinde, Früchte kultiviert, oft auch durch Züchtung oder Auslese entstanden, in Gärten und Parks gepflanzt wird; ▶Ziergehölz).

3940 ornamental tree [n] *hort. gard. plant.* (Attractive deciduous or evergreen tree, which is planted in parks and gardens because of its beauty, the shape of its branches, its leaves, flowers, bark or fruits and has often been bred or specially selected for these attributes; ▶ornemental woody plant); *s* **árbol [m] ornamental** (Árbol perennifolio o caducifolio que se planta en parques y jardines por la belleza de sus ramajes, sus hojas, flores o frutos, a menudo cultivado o seleccionado por sus atributos especiales; ▶leñosa ornamental); *f* **arbre [m] d'ornement** (Arbre à feuilles caduques ou persistantes cultivé pour la beauté de sa ramification, ses feuilles, ses fleurs, son écorce, ses fruits, planté dans les parcs et jardins ; ▶végétaux d'ornement) ; *syn.* arbre [m] d'agrément ; *g* **Zierbaum [m]** (Laub abwerfender oder immergrüner Baum, der wegen der Schönheit seiner Verzweigung, seiner Blätter, Blüten, Rinde, Früchte kultiviert, oft auch durch Züchtung oder Auslese entstanden, in Parks und Gärten gepflanzt wird; ▶Ziergehölz).

3941 ornamental woody plant [n] *hort. plant.* (Generic term covering ▶ornamental tree and ▶ornamental shrub); *syn.* flowering tree [n] or shrub [n]; *s* **leñosa [f] ornamental** (Término

genérico para ▶árbol ornamental y ▶arbusto ornamental); *f* **plante [f] ligneuse d'ornement** (Terme générique pour l'▶arbre d'ornement et l'▶arbuste d'ornement) ; *syn. au pluriel* ligneux [mpl] d'ornement, végétaux [mpl] ligneux d'ornement ; *g* **Ziergehölz [n]** (OB zu ▶Zierbaum und ▶Zierstrauch).

3942 ornithological station [n] *conserv. zool.* (Ornithological institute primarily concerned with banding of birds and the study of bird migration, their age, size of habitat and threats to their existence, as distinguished from a ▶bird protection research center [US]/bird protection research centre [UK]); *s* **estación [f] ornitológica** (Centro de investigación sobre las aves que se dedica sobre todo a estudiar los comportamientos migratorios, la edad, el tamaño de los hábitats y las causas que las amenazan; ▶instituto de ornitología aplicada); *f* **station [f] ornithologique** (Centre dont les activités principales, à la différence de l'▶institut d'ornithologie appliquée, consistent dans le baguage des oiseaux en vue de l'étude de la migration, de l'âge des individus, de la taille des habitats et des menaces qui pèsent sur les populations) ; *g* **Vogelwarte [f]** (Institut für wissenschaftliche Vogelkunde, das sich zum Vergleich mit der ▶Vogelschutzwarte vornehmlich mit der Vogelberingung zum Studium des Vogelzuges, des Alters, der Biotopgröße und der Bestandssituatoin/-bedrohung beschäftigt).

3943 ornithologist [n] *zool.* (Scientist or specialist in ▶ornithology. An amateur is called a 'birder' [US] or 'bird watcher'); *s* **ornitólogo/a [m/f]** (Cientifico/a que se dedica a la ▶ornitología); *syn.* ornitología [m/f]; *f* **ornithologiste [m/f]** (Scientifique ou spécialiste de l'▶ornithologie ; *terme spécifique* ornithologiste amateur) ; *syn.* ornithologue [m] ; *g* **Ornithologe [m]/...login [f]** (Wissenschaftler auf dem Gebiet der ▶Ornithologie); *syn.* Vogelkundler/-in [m/f].

3944 ornithology [n] *zool.* (Branch of zoology dealing with the study of birds); *s* **ornitología [f]** (Parte de la zoología que estudia las aves); *f* **ornithologie [f]** (Branche de la zoologie qui traite des caractères morphologiques et physiologiques des diverses espèces d'oiseaux, leur forme de vie, comportement, habitat et distribution ainsi que leur classification suivant les relations de parenté) ; *g* **Ornithologie [f]** (Wissenschaft, die sich v. a. mit den morphologischen und physiologischen Gegebenheiten der verschiedenen Vogelarten, deren Lebensweise resp. Verhalten sowie den verwandtschaftlichen Zusammenhängen in Bezug auf die Systematik befasst; MEL 1976, Bd. 17); *syn.* Vogelkunde [f].

3945 orphan contaminated site [n] *envir.* (Area with hidden soil contamination where the causer of the pollution cannot be identified and made responsible for remediation. In U.S, the Superfund Law spreads the cost on all producers of such pollution ▶hazardous old dumpsite); *s* **emplazamiento [m] contaminado huérfano (≠)** (Superficie contaminada de la que se desconoce o no se puede hacer responsable al causante de la contaminación; ▶emplazamiento contaminado); *f* **site [m] pollué orphelin** (Terrain pollué, soit par stockage, enfouissement de produits toxiques ou contamination du sol caractérisé par l'insolvabilité de l'exploitant et du propriétaire du terrain ou par l'absence de responsable et dont la réhabilitation est confiée à l'ADEME en vertu de la circulaire du 09 janvier 1989 ; ▶dépôt ancien) ; *syn.* site [m] contaminé orphelin ; *g* **herrenloser Altstandort [m]** (Stillgelegte Deponie oder von Altablagerungen geprägte Fläche, für die der Verursacher der Verseuchung infolge Eigentumverzichtes nicht mehr haftbar gemacht werden kann; ▶Altstandort); *syn.* derelinquierter Altstandort [m].

orphan land [n] [US] *min.* ▶derelict land resulting from mining operation.

orterde [n] [UK] *pedol.* ▶orthod [US].

3946 orthod [n] [US] *pedol.* (Brown-black, illuvial horizon rich in humus, Fe and Al which has been created by ►podzolization, and is less compacted than ►ortstein. In U.S. soil classification system the 'Orthod' belongs to the soil order of Spodosols); *syn.* orterde [n] [UK]; *s* **orterde** [m] (En el proceso de ►podsolización, horizonte de iluviación enriquecido con humus, hierro y aluminio, pero que está menos endurecido que la capa de ►orstein); *syn.* capa [f] endurecida; *f* **alios** [m] **faiblement durci** (Horizon d'accumulation brun foncé formé lors de la ►podzolisation et constitué de sable cimenté par de la matière organique et des oxydes d'alumine et localement de fer ; ►alios durci) ; *g* **Orterde** [f] (Im Rahmen der ►Podsolierung entstandener braunschwarzer, humus- und Fe- oder Al-haltiger Anreicherungshorizont, der nicht so verfestigt ist wie der ►Ortstein).

3947 orthophotograph [n] *envir. plan. rem'sens. surv.* (Photographic copy, prepared from an aerial photograph, in which the displacements of images due to tilt and relief have been removed; RCG 1982; ►remote sensing); *s* **ortofoto** [f] (Fotografía aérea o de satélite corregida de las distorsiones del relieve; cf. CHUV 1990; ►teledetección); *syn.* ortofotografía [f]; *f* **orthophoto** [f] (Photographie aérienne verticale dont les influences de l'inclinaison de prise de vue et du relief du terrain ont été traitées [travail de redressement] pour la rendre parfaitement superposable à une carte photographique ; HAC 1989 ; ►télédétection) ; *syn.* orthophotographie [f] ; *g* **Orthophoto** [n] (Entzerrtes Senkrechtluftbild; ►Fernerkundung); *syn. o. V.* Orthofoto [n].

3948 orthophotographic map [n] *envir. plan. rem'sens. surv.* (Assemblage of several rectified aerial photographs forming an overall picture); *s* **mapa** [m] **de ortofotografías** (Ensamblaje de diversas fotografías aéreas o de satélite corregidas de las distorsiones del relieve para generar una imagen de conjunto); *f* **orthophotoplan** [m] (Assemblage de photographies aériennes, après restitution orthophotographique éliminant les déformations dues à l'inclinaison de l'axe de prise de vue sur la verticale et au relief du terrain et formant un document graphique d'ensemble ; cf. HAC 1989, 567) ; *syn.* photoplan [m] ; *g* **Orthofotokarte** [f] (Zusammenfassung mehrerer Luftbilder, ihre geometrische Entzerrung und Wiedergabe zu einem Gesamtbild); *syn. o. V.* Orthophotokarte [f].

3949 ortstein [n] [US] *pedol.* (Indurated layer in the B horizon of ►podzols in which the cementing material consists of illuviated sesquioxides—mostly iron—and organic matter; RCG 1982; ►duripan, ►fragipan, ►hardpan, ►iron pan 1, ►moorpan, ►orthod [US]/orterde [UK]); *syn.* ironpan [n] [UK] (2.); *s* **ortstein** [m] (Capa endurecida de consistencia casi pétrea que se desarrolla en el horizonte B de determinados ►podsoles por acumulación de óxidos de aluminio y hierro y de humus; ►capa de ortstein de humus, ►capa ferruginosa, ►duripan, ►fragipan, ►horizonte petrificado, ►orterde); *syn.* hardpan [m]; *f* **alios** [m] **durci** (Mot désignant l'horizon B d'accumulation de certains ►podzols dû à la cimentation des grains de sable ou de limon par des oxydes de fer et de manganèse ; ►alios faiblement durci, ►alios humique, ►alios ferrugineux, ►horizon d'accumulation durci, ►horizon de fragipan) ; *syn.* alios [m] ferrugineux ; *g* **Ortstein** [m] (Durch verlagerte Humusteilchen, Al- und Fe-Oxide stark verfestigte, sandige, braunschwarze humusreiche Bodenschicht im B$_h$-Illuvialhorizont, z. B. bei ►Podsolen; **O.** ist fast wasserundurchlässig und für Wurzeln kaum durchdringbar, weshalb sie in trockenen Zeiten nicht an tiefer gelegenes Wasser gelangen; über **O.** wachsen deshalb flachwurzelnde und trockenheitsresistente Pflanzen; weniger verfestigt ist die ►Orterde; ►Eisenschwarte, ►Humusortstein, ►Verdichtungshorizont).

oscillating water level [n] *hydr.* ►fluctuating water level.

3950 osier twig [n] *hort. constr.* (Thin, leafless willow branch used for binding or in making of fascines); *s* **mimbre** [m] (Rama fina de sauce sin hojas que se utiliza como material para atar las fajinas); *f* **brin** [m] **d'osier** (Tige de Saule *[Salix viminalis, S. amygdalina, S. purpurea]* utilisé en vannerie, en horticulture comme lien ou pour le fascinage) ; *g* **Weidenrute** [f] (Vom Laub befreiter, dünner Weidenzweig, der als Bindegut oder zum Faschinenbau verwendet wird); *syn.* Weidengerte [f].

'O' traffic [n] [US] *trans.* ►origin traffic.

3951 outbuilding [n] *arch.* (**1.** Building which has a secondary or accessary function in relation to another on the same property). **2. outhouse** [n] *arch. constr.* (Outbuilding used as a toilet, most often without plumbing, and located temporarily on crowded or construction sites); *syn. for 2.* outdoor toilet [n]; *s* **edificio** [m] **accesorio** (Construcción de función subordinada a un inmueble principal situada en la misma parcela que éste); *syn.* anejo [m], dependencia [f]; *f* **bâtiment** [m] **annexe** (Construction de nature et de destination distinctes mais dépendante d'un bien immeuble principal sur une même parcelle) ; *syn.* annexe [f], bâtiment [m] auxiliaire ; *g* **Nebengebäude** [n] (Gebäude mit untergeordneter Funktion gegenüber einem anderen Gebäude auf demselben Grundstück).

3952 outcrop [n] *geo.* (Area of rock exposed above the ground surface; ►exposed soil horizons); *s* **afloramiento** [m] **natural** (Lugar sobre el terreno en el que se puede ver la estratificación de las rocas o del material meteorizado; ►perfil aflorado); *syn.* corte [m] de prospección natural; *f* **affleurement** [m] (Lieu sur le terrain donnant un aperçu de la stratification des roches ; ►profil pédologique tranché) ; *g* **Aufschluss** [m] (Stelle im Gelände, die Einblick in die Lagerung der Gesteine und des verwitterten Materials zulässt; ►Bodenaufschluss); *syn.* Ausbiss [m], Ausstreichende [n], Ausstrich [m].

outdoor area [n] *constr. urb.* ►external space.

3953 outdoor bathing complex [n] [US] *recr.* (*opp.* ►indoor bathing complex [US]/indoor swimming baths [UK]); *syn.* open air swimming baths [npl] [UK]; *s* **piscina** [f] **al aire libre** (*opp.* ►piscina cubierta); *syn.* piscina [f] descubierta, pileta [f] descubierta [CS], alberca [f] descubierta [MEX]; *f* **piscine** [f] **de plein air** (Conception anglo-saxonne d'une aire de loisirs constituée par des aires de baignade, des zones de repos enherbées ombragées et divers équipement légers d'accompagnement [buvette, aires de jeux, etc.] ; *opp.* ►piscine couverte) ; *syn.* piscine [f] parc, piscine-jardin [f] ; *g* **Freibad** [n] (Öffentliche, meist kostenpflichtige Badeanstalt im Freien; es gibt zwei Formen: **1.** ein im Freien angelegtes Schwimmbad mit Becken von 25 m oder 50 m Länge, oft mit Sprungturm und Planschbecken für Kinder ausgestattet; **2.** ein abgegrenzter Bereich eines fließenden oder Stillgewässers; *opp.* ►Hallenbad).

3954 outdoor furniture [n] *arch. constr. urb.* (**1** Generic term covering equipment in parks, streets, squares, etc., usually seats, tables, lamps, waste bins, etc. *Specific terms* garden furniture, ►movable outdoor furniture, park furniture, ►street furniture); *syn.* amenity facility [n], site furniture [n]; *s* **mobiliario** [m] **urbano** (**1.** Término genérico para todo tipo de equipamiento de las calles, plazas, parques, etc. como p. ej. bancos, mesas, farolas, papeleras. *Términos específicos* ►mobiliario de calles, mobiliario de jardín, mobiliario de parques, etc.; **2. mobiliario** [m] **urbano de zonas verdes:** todo tipo de equipamiento de espacios libres públicos —exceptuando las calles— como p. ej. bancos, mesas, aparatos para juegos, farolas, papeleras); *f* **mobilier** [m] **urbain (d'espaces verts)** (Équipements urbains, de parcs et jardins publics, regroupant le mobilier de repos [bancs, sièges, tables], de propreté [poubelles, corbeilles], de jeux pour enfants, d'abris, d'éclairage public [réverbères, candélabres], d'information de communication et d'orientation, de circulation [barrières, bornes, range-vélos], de protection des arbres [grilles, tuteurs, corsets] ; *termes spécifiques* mobilier de jardin, ▶mobilier de voirie, ▶mobilier mobile) ; *g* **Möblierung [f] für Außenanlagen** (Ausstattung resp. Gesamtheit aller Gegenstände, die zur zweckdienlichen Nutzung von Garten- und Parkanlagen, auch vorübergehend, aufgestellt oder fest installiert sind, insbesondere Sitzgelegenheiten, Tische, Papierkörbe, Spielgeräte, Wartehäuschen, Schaltschränke, Beleuchtungsmasten, Sperrketten, Pfosten, Poller etc.; *UBe* Gartenmöblierung, ▶lose Möblierung, Parkmöblierung, ▶Straßenmöblierung; die Gegenstände heißen *Gartenmöbel, Parkmöbel, Straßenmöbel*); *syn.* Außenanlagenmöblierung [f], Ausstattungsgegenstand/-stände [m/pl] für Außenanlagen, Ausstattungselement(e) [n(pl)] für Außenanlagen, Freiraumausstattungselemente [npl].

3955 outdoor living space [n] [US] *gard.* (Intensively-designed living area around a residential building; ▶courtyard living area); *syn.* garden living space [n] [UK]; *s* **jardín [m] habitación** (Jardín diseñado con el fin de ser utilizado como parte de la vivienda; ▶patio habitación); *f* **jardin [m] d'habitation** (Jardin conçu pour faire partie intégrante de l'habitation ; ▶cour d'habitation intérieure) ; *g* **Wohngarten [m]** (Intensiv gestalteter Garten, der dem Wohnen dient; ▶Wohnhof).

3956 outdoor recreation [n] *recr.* (Generic term for all recreation pursuits in the open air; ▶contryside recreation, ▶outdoor urban recreation, ▶water recreation); *syn.* open air recreation [n], leisure [n] in the open, out-of-doors recreation [n] [also UK]); *s* **recreo [m] al aire libre** (Término genérico para todo tipo de actividades de recreación al aire libre, tanto sea en el campo como en los espacios libres en la ciudad; ▶esparcimiento asociado al agua, ▶recreo en espacios libres urbanos, ▶recreo en la naturaleza); *syn.* esparcimiento [m] al aire libre, recreación [f] al aire libre); *f* **loisirs [mpl] de plein air** (Terme générique comprenant la détente et la récréation en milieu rural et dans les espaces verts urbains ou périurbains ; ▶agrotourisme, ▶détente en espaces libres urbains, ▶loisirs à caractère spécifique aquatique, ▶loisirs de nature, ▶tourisme vert) ; *syn.* vacances [fpl] de plein air, récréation [f] de plein air ; *g* **Erholung [f, o. Pl.] im Freien** (Allgemeiner Begriff für Erholung in freier Natur und in städtischen Freiräumen; ▶Erholung in der freien Landschaft, ▶Freiraumerholung, ▶gewässerbezogene Erholung); *syn.* Erholung [f, o. Pl.] im Grünen (±).

outdoor recreation [n] **in exurban areas** [US] *recr. urb.* ▶outdoor recreation in outlying areas.

3957 outdoor recreation [n] **in outlying areas** *recr. urb.* (Recreation in or beyond of the periphery or fringe of a city or metropolitan area); *syn.* outdoor recreation [n] in exurban areas [also US]; *s* **recreación [f] en la periferia urbana** (Actividades de recreo en zonas cercanas a la ciudad); *f* **loisirs [mpl] périurbains** (Activités de détente situées à la périphérie des centres urbains) ; *g* **Stadtranderholung [f, o. Pl.]** (Erholung im stadtnahen Bereich; sich am Stadtrand erholen [vb]).

outdoor's man [n] [US] *recr. sociol.* ▶recreation user.

3958 outdoor space [n] *plan.* (TGG 1984, 368; landscaped area outside of buildings, in contrast to indoor space; ▶open space); *s* **exteriores [mpl]** (Término utilizado para los espacios fuera de la vivienda en contraposición al del espacio construido «interior»; ▶espacio libre); *f* **espace [m] libre extérieur** (par opposition à l'espace bâti ; ▶espace libre) ; *g* **Außenraum [m]** (Instrumenteller Arbeitsbegriff als Gegensatz zum umbauten Raum — Innenraum; **A.** wird als offener, ungehinderter Bewegungsraum unter freiem Himmel mit öffentlichen, halböffent-

lichen und privaten Zonen verstanden und empfunden; ▶Freiraum).

outdoor steps [npl] *arch. constr.* ▶steps.

outdoor toilet [n] *arch. constr.* ▶outbuilding, #2.

3959 outdoor urban recreation [n] *plan. recr.* (Recreation in urban spaces, such as squares, residential streets, in parks, on waterfront promenades and beaches, in allotment gardens and other green spaces. The concept of ▶outdoor recreation also includes recreation outside populated areas; e.g. ▶countryside recreation, ▶water recreation); *s* **recreo [m] en espacios libres urbanos** (Actividades de recreación realizadas en espacios abiertos de la ciudad. El término ▶recreo al aire libre se refiere tanto a la recreación en zonas rurales como en urbanas; ▶esparcimiento asociado al agua, ▶recreo en la naturaleza); *syn.* recreación [f] en espacios libres urbanos, esparcimiento [m] en espacios libres urbanos; *f* **détente [f] en espaces libres urbains** (Activité de détente exercée sur les espaces non occupés par les constructions [espace (de) détente], lieux privilégiés pour la rencontre et la détente, tels que places, rues, espaces verts, parcs urbains, berges de cours d'eau, jardins familiaux, etc. ; ▶agrotourisme, ▶loisirs à caractère spécifique aquatique, ▶loisirs [de] nature, ▶loisirs de plein air, ▶tourisme vert) ; *syn.* détente [f] en espaces ouverts urbains ; *g* **Freiraumerholung [f, o. Pl.]** (Erholung in städtischen Freiräumen, z. B. auf Plätzen, Wohnstraßen, in Parkanlagen, auf Uferpromenaden und -stränden an Gewässern, in Dauerkleingärten und sonstigen Grünflächen. Der Begriff ▶Erholung im Freien wird im Allgemeinen weiter gedacht und beinhaltet auch die Erholung außerhalb von Siedlungsgebieten, z. B. ▶Erholung in der freien Landschaft, Erholung in freier Natur, ▶gewässerbezogene Erholung).

3960 outer bark [n] *arb. bot.* (Outer skin of a woody trunk, outside the secondary, vascular cambium layer, which is dead tissue protecting the living tissue [phloem]. The outer zone is penetrated by cork layers [periderms] formed from cork cambia [phelloderms], and is sometimes called 'rhytidome'. There are three distinct outer bark forms: "scaling bark", [e.g. *Betula, Prunus*]; "chipping bark", [e.g. *Quercus, Platanus, Pinus*], and "fibrous bark", [e.g. *Clematis, Lonicera, Vitis*]; ▶bark); *syn.* rhytidome [n]; *s* **ritidoma [m]** (En los troncos, ramas y raíces de los árboles y arbustos, conjunto de tejidos muertos que los recubren, situados al exterior de los estratos de súber, formados por el felógeno, y generalmente rugosos y resquebrajados. Son raros los árboles que, como el haya, tienen el mismo felógeno durante toda su vida, de modo que el súber formado es también constantemente el mismo, lo cual da a los troncos y ramas de dichos árboles una tremenda lisura, y el **r.** falta en ellos. En los otros árboles, dicho felógeno cesa más pronto o más tarde en su actuación, y surgen otros, más hacia el interior, que son los que, aislando masas de tejidos vivos y que tienen que morir indefectiblemente, forman el **r.** Cuando el **r.** se desprende en placas o láminas [plátano de sombra] se le llama «r. escamoso», cuando lo hace en forma de anillos regulares [abedul] «r. anillado o anular»; cf. DB 1985; ▶corteza); *syn.* corteza [f] externa; *f* **écorce [f]** (Tissus cellulaires externes morts de l'▶écorce, exfoliés par la formation en profondeur de couches successives de liège ; on distingue une exfoliation en plaques [**rhytidome écailleux** du Platane, du Pin], en anneaux [**rhytidome annulaire** du Cerisier, du Bouleau], en manchons dilacérés [**rhytidome fibreux** de la Vigne, de la Clématite, du Chèvrefeuille] ; cf. MVV 1977, 237) ; *syn.* rhytidome [m] ; *g* **Borke [f]** (Durch Korkschichten abgetrenntes Gewebe der ▶Rinde. Die **B.** dient dem Schutz von Bast, Kambium und Holz. Man unterscheidet **Ringelborke** [z. B. Betula, *Prunus*], **Schuppenborke** [z. B. Eiche *(Quercus)*, Platane *(Platanus)*, Kiefer *(Pinus)*] und

Streifenborke [z. B. *Clematis*, Geißblatt *(Lonicera)*, Weinrebe *(Vitis)*]); *syn.* äußere Rinde [f], Außenrinde [f].

3961 outfall [n] **(of sewage)** *envir.* (Point of discharge of ▶wastewater from a pipe; ▶drop structure, ▶water polluting firm); *syn.* outlet point [n] [also US], point [n] of discharge (WET 1993, 563); *s* **punto** [m] **de vertido** (Lugar en el que se introducen las ▶aguas residuales 2 en el curso receptor; ▶acometida a colector principal, ▶causante de vertidos directos); *syn.* punto [m] de descarga; *f* **point** [m] **de rejet (des effluents)** (Lieu [▶ouvrage de rejet] dans lequel se déversent les ▶effluents urbains ; ▶auteur d'un déversement direct [des effluents]) ; *g* **Einleitungsstelle** [f] (Ort, an dem ▶Abwässer eingeleitet werden; ▶Direkteinleiter, ▶Einlaufbauwerk 1); *syn.* Abwassereinleitungsstelle [f].

3962 outfall pipe [n] *constr.* (Pipe ending in a ▶drop structure); *s* **tubería** [f] **de salida** (Conducción que termina en una ▶acometida a colector principal); *syn.* emisario [m]; *f* **tuyau** [m] **d'arrivée** (de l'effluent en tête d'▶ouvrage de rejet) ; *g* **Einlaufrohr** [n] (Rohr, das an einem ▶Einlaufbauwerk 1 endet).

outfall structure [n] [US] *constr. envir.* ▶drop structure.

3963 outflow [n] *constr.* (Pipe discharge of water into a ▶receiving stream); *syn.* discharge [n]; *s* **caudal** [m] **efluente (de tubería)** (Volumen de agua que se vierte al ▶curso receptor a través de una tubería); *syn.* efluente [m] (de tubería); *f* **débit** [m] **sortant** (Volume d'eau s'écoulant à la sortie d'un tuyau, d'une canalisation, d'un bassin, d'un cours d'eau, d'une formation aquifère ; ▶émissaire, ▶milieu récepteur) ; *g* **Abfluss** [m] **(1)** (Wasserabfluss aus einem Rohr in einen ▶Vorfluter).

outflow [n] **of permeable layers** *hydr. geo.* ▶seepage spring.

3964 outgoing commuter [n] *sociol. trans. urb.* (Person travelling to work or study away from a place of residence; ▶commuter, *opp.* ▶incoming commuter); *s* **conmutador, -a** [m/f] **saliente** (Persona que se desplaza a trabajar a la periferia de la ciudad o a otra ciudad; *opp.* ▶conmutador, -a entrante; ▶conmutador, -a); *f* **personne** [f] **effectuant des migrations alternantes considérée du point de vue de la collectivité réceptrice (±)** (Personne se déplaçant quotidiennement entre son domicile et son lieu de travail vers une autre commune ; ▶migrant alternant ; *opp.* ▶migrant alternant considéré du point de vue de la collectivité émettrice) ; *syn.* migrant [m] extra-urbain (±) ; *g* **Auspendler** [m] (Person, die sich täglich zur Arbeit oder zur Ausbildung von ihrer Wohngemeinde in eine andere Gemeinde begibt; cf. PW 1970; ▶Pendler; *opp.* ▶Einpendler).

outhouse [n] *arch. constr.* ▶outbuilding, #2.

outlet ditch [n] *constr. eng. wat'man.* ▶receiving stream.

outlet pipe [n]**, drain** *constr.* ▶drain outfall.

outlet point [n] *envir.* ▶outfall (of sewage).

outlet stream [n] *constr. eng. wat'man.* ▶receiving stream.

3965 outline [n] *plan.* (Defined basic concept for a design); *s* **esquema** [m] **básico de diseño** (Concepto preliminar de distribución espacial de todos los componentes de un espacio a planificar); *f* **schéma** [m] **de principe** (Distribution spatiale de tous les composants d'un jardin lors de l'étude préliminaire); *g* **Konzept** [n] **einer Planung** (Gedankliches und funktionales Grundgerüst einer Planung); *syn.* Anlage [f] einer Planung, Grundidee [f] einer Planung.

outline and sketch scheme proposals [npl] [UK] **(work stage C&D)** *contr. plan.* ▶schematic design phase [US].

outline planning application [n] [UK] *adm. plan.* ▶application for preliminary building permit [US].

3966 outline specifications [npl] *constr. contr.* (Skeletal descriptive list of numbered items covering design requirements for a construction project, in order to make a preliminary cost estimate for the client/owner); *s* **resumen** [m] **de especificaciones técnicas** (Breve descripción de las especificaciones técnicas y generales desarrolladas en la fase de concepción del proyecto); *f 1* **description** [f] **sommaire des ouvrages au stade de l'avant-projet** (Présentation de la solution d'ensemble indiquant l'implantation, la répartition et les caractéristiques fonctionnelles des ouvrages ; une des parties du mémoire d'avant-projet sommaire) ; *f 2* **description** [f] **sommaire des ouvrages au stade du projet** (Présentation de la solution d'ensemble des principaux composants des ouvrages dans le cadre du mémoire de l'avant-projet détaillé ; dans le cas où l'appel d'offres est lancé sur avant-projet ou exceptionnellement sur avant-projet sommaire on parle souvent de « spécifications techniques sommaires » ; cf. HAC 1989, 33-34 ; **D.**, le **d. s. d. o.** ne figure qu'au stade de l'avant-projet détaillé) ; *g* **verkürzte Leistungsbeschreibung** [f] (**F.**, Stichwortartige Rohfassung/Kurzfassung eines Leistungsverzeichnisses im Entwurfsstadium einer Planung zur überschlägigen Kostenermittlung. In F. wird jeweils zum Vorentwurf und Entwurf eine **v. L.** gemacht).

outlot [n] [US] *plan. urb.* ▶leftover area.

out-of-doors recreation [n] [UK] *recr.* ▶outdoor recreation.

outskirts area [n]**, undeveloped** *urb.* ▶open land.

outskirts [npl] **of a city/town** *urb.* ▶urban fringe.

3967 outstanding natural feature [n] *conserv.* (Unique or rare ▶natural phenomenon of outstanding natural beauty, often protected as a ▶natural landmark [US]/natural monument [UK]; e.g. ▶monarch tree—under conservation laws); *syn.* unique natural feature [n]; *s* **elemento** [m] **natural singular** (▶Fenómeno natural geomorfológico o biótico de gran belleza, particularidad o escasez que a menudo se protege como ▶árbol monumento, ▶monumento natural); *f* **objet** [m] **naturel remarquable** (Terme utilisé dans le cadre de la protection de l'espace naturel et désignant un ▶phénomène naturel géomorphologique ou floristique présentant une importance particulière et un caractère pittoresque, en général classé comme ▶monument naturel au titre de l'art. 4 de la loi du 2 mai 1930 relative à la protection des monuments naturels et des sites de caractère artistique, historique, scientifique, légendaire ou pittoresque ; ▶arbre classé comme monument naturel, ▶arbre remarquable d'intérêt national) ; *syn.* espace [m] ponctuel naturel remarquable ; *g* **Einzelschöpfung** [f] **der Natur** (Im deutschen Naturschutzrecht eine geomorphologische oder pflanzliche ▶Naturerscheinung, die von hervorragender Schönheit und Eigenart oder Seltenheit ist und aus wissenschaftlichen, naturgeschichtlichen oder landeskundlichen Gründen oft als ▶Naturdenkmal festgesetzt wird; cf. § 17 BNatSchG und Naturschutzgesetze der Bundesländer; ▶Baum-Naturdenkmal).

outstanding scenic beauty [n] *landsc. recr.* ▶area of outstanding scenic beauty.

3968 overall area planning [n] *plan.* (Process of preparing a planning document for area development; ▶general planning, ▶functional planning [US]/sectoral planning [UK]); *syn.* general development planning [n]; *s* **planificación** [f] **general** (Término no claramente definido que incluye todos los tipos de planes que afectan a una región o una ciudad; ▶planificación directriz, ▶planificación sectorial); *f* **planification** [f] **globale** (Planification englobant l'ensemble des ▶planifications sectorielles pour une zone déterminée ; ▶planification référencielle) ; **g Gesamtplanung [f]** (Kein gesetzlich verankerter Begriff; i. d. R. erfasst er alle ►Fachplanungen für einen bestimmten Raum, sei es auf lokaler, regionaler oder Landesebene; ►Leitplanung).

3969 overall costs [npl] *contr.* (Total project expenditure); **s costo [m] global** (Inversión financiera total de un proyecto que incluye los costos de terreno, infraestructura, edificación, espacios libres, maquinaria y trabajos complementarios); *syn.* coste [m] global [Es], costo [m] total, coste [m] total [Es]; *f* **coût [m] global** (Terme générique désignant le volant financier d'une opération comprenant les coûts suivants : terrain, voirie et réseaux divers, bâtiment, espaces verts, travaux complémentaires) ; **g Gesamtkosten [pl]** (Gesamter finanzieller Aufwand für ein Vorhaben; nach DIN 276 Teil 2 folgende sieben Kostengruppen: 1. Kosten des Baugrundstücks, 2. Kosten der Erschließung, 3. Kosten des Bauwerks, 4. Kosten des Gerätes, 5. Kosten der Außenanlagen, 6. Kosten für zusätzliche Maßnahmen und 7. Baunebenkosten); *syn.* Gesamtkostenaufwand [m], *ugs.* Herstellungskosten [pl]).

3970 overall fixed costs [npl] *contr. prof.* (A figure contractually agreed upon by the designer and owner for the overall costs of a project, including construction costs, professional fees, land investigation costs, etc.); *syn.* total fixed costs [npl]; **s costo [mpl] total acordado** (Suma total a pagar por el comitente, acordada entre éste y el arquitecto, que incluye los costos de construcción, los honorarios y los costos adicionales); *f* **coût [m] d'objectif définitif** (F., somme du coût prévisionnel des ouvrages et du forfait de rémunération [honoraires] fixée contractuellement entre le ►maître d'ouvrage et le concepteur pour un projet d'aménagement) ; **g vertraglich festgelegte Gesamtkosten [pl]** (Eine zwischen Planer und Bauherrn für ein Bauvorhaben vertraglich festgelegte Summe, bestehend aus Baukosten, Honorar- und Nebenkosten).

3971 overall land planning [n] *plan.* (Process of planning large land areas on a local, regional or county [and state] level); **s planificación [f] general integral** (Esbozo de planificación de una gran área); *syn.* planeamiento [m] general integral; *f* **planification [f] spatiale globale** (Organisation d'activités englobant un espace géographique important à l'échelon local ou régional) ; **g räumliche Gesamtplanung [f]** (Planerische Tätigkeit für einen Landschaftsraum auf lokaler, regionaler oder Landesebene).

3972 overall length [n] *constr.* (Linear measurement, e.g. of pipe, wall); **s largo [m] total** (Medida de la longitud de una conducción, un muro, etc.); *f* **longueur [f] totale** ; **g Gesamtlänge [f]** (Maß der räumlichen Ausdehnung in einer Richtung, z. B. einer Leitung, einer Mauer, eines Deiches).

3973 overall master plan [n] **for play areas** [US] *plan. recr. urb.* (Design and description as a result of the process of overall planning of children's playgrounds [in a community]; ►overall master planning for play areas [US]/play spaces strategic planning [UK]); *syn.* strategic plan for play spaces [n] [UK]; **s plan [m] de desarrollo de áreas de juegos y recreo infantiles** (≠) (Resultado de la ►planificación integrada de áreas de juegos y recreo infantiles para un área determinada); *f* **schéma [m] directeur d'aires de jeux** (≠) (D., plan établi dans le cadre de la ►planification d'aires de jeux) ; **g Spielflächenleitplan [m]** (Ergebnis der ►Spielflächenleitplanung in einem Siedlungsgebiet; er ist Orientierungsrahmen und Informationsgrundlage für die wohnumfeldorientierte Stadtplanung und gewährleistet eine fundierte Abwägung und Entscheidung in der städtebaulichen Planung).

3974 overall master plan [n] **for sports and physical activities** [US] *plan. recr. urb.* (Result of the process of ►overall master planning for sports and physical activities

[US]/strategic planning for sports and physical activities [UK]); *syn.* city plan [n] for sports fields [UK], city plan [n] for sports pitches [UK], master plan [n] for sports fields [US]; **s plan [m] de desarrollo de equipamientos deportivos** (Resultado del ►programa de gestión de equipamientos deportivos); *f* **programme [m] d'équipements sportifs** (Résultat de la politique de ►programmation des équipements sportifs) ; **g Sportstättenentwicklungsplan [m]** (Ergebnis der ►Sportstättenentwicklungsplanung).

3975 overall master planning [n] **for play areas** [US] *plan. recr. urb.* (Survey and analysis of play areas and assessment of requirements according to nationally or internationally established guidelines with the objective of overall coverage and safe-guarding of play areas within the scope of urban land-use and town planning); *syn.* play spaces strategic planning [n] [UK], strategic planning [n] of play areas [UK]; **s planificación [f] integrada de áreas de juegos infantiles** (≠) (Registro de las instalaciones existentes y análisis de la demanda de superficies dedicadas a juegos infantiles. En D se realiza sobre la base de las directrices de la Sociedad Olímpica Alemana y la norma DIN 18 034 con el fin de asignar y asegurar suelo para esta función en el marco del desarrollo y del planeamiento urbanístico); *f* **planification [f] d'aires de jeux** (Étude de l'état existant et analyse des besoins en aires de jeux pour différents secteurs d'une commune, établie sur la base de normes, directives, valeurs guides et caractéristiques nationales ou internationales dans le but de déterminer l'affectation, la fixation des emplacements, et permettre la sauvegarde des aires de jeux en fonction des besoins des habitants, dans le cadre du développement urbain et de l'urbanisme réglementaire) ; **g Spielflächenleitplanung [f]** (Erfassung des Bestandes und des Fehlbedarfes von Spielflächen, gemessen an den Richtzahlen der Deutschen Olympischen Gesellschaft [DOG] und der DIN 18 034 in allen Teilplanungsbereichen einer Gemeinde mit dem Ziel der bedarfsdeckenden Ausweisung, Zuordnung und nachhaltigen Sicherung von Spielflächen, Spielerlebnis- und Erfahrungsräumen im Rahmen der städtebaulichen Entwicklung/Bauleitplanung, um den Interessen von Kindern und Jugendlichen strukturell und langfristig zu entsprechen Die **S.** ist ein kontinuierlicher Prozess und eine Daueraufgabe der Gemeinden); *syn.* Spielleitplanung [f] [RP]).

3976 overall master planning [n] **for sports and physical activities** [US] *plan. recr. urb.* (Survey, analysis and assessment of community requirements for different sports facilities, based upon certain targets or guidelines. Long-term aim is comprehensive coverage, integration and retention of sports facilities within the scope of ►community development planning [US]/urban land-use planning [UK]); *syn.* strategic planning [n] for sports and physical activities [UK]; **s programa [m] de gestión de equipamientos deportivos** (Análisis de la demanda de equipamiento de deportes de la población y planificación y construcción de las correspondientes áreas e instalaciones, sobre la base de directivas específicas según cada país o región); *f* **programmation [f] des équipements sportifs** (Inventaire communal des équipements sportifs existants et analyse du niveau d'adéquation du parc en fonction d'éléments de référence tel que la surface souhaitable par habitant, de règles fédérales, d'objectifs quantitatifs, qualitatifs et fonctionnels ou des attentes de la population) ; **g Sportstättenentwicklungsplanung [f]** (Erfassung des Bestandes und des Fehlbedarfes von Sportstätten in einer Gemeinde gemessen an den Zielwerten für den Sportstättenbedarf, z. B. nach den „Richtlinien für die Schaffung von Erholungs-, Spiel- und Sportanlagen" der Deutschen Olympischen Gesellschaft [DOG], der Richtlinien und Empfehlungen des Internationalen Arbeitskreises Sport- und Freizeiteinrichtungen [IAKS], der Internationalen Akademie für Bäder-, Sport- und Freizeitbau [IAB] oder des Sportförderungsgesetzes mit dem langfristigen Ziel der bedarfsdeckenden Ausweisung, Zuordnung und Sicherung von Sportstätten im Rahmen der städtebaulichen Entwicklung/▶Bauleitplanung); *syn.* Sportstättenleitplanung [f].

3977 overall plan [n] (1) *plan.* (Plan showing total site of a project in relation to its surroundings; ▶overall site plan, ▶site plan); *syn.* layout plan [n]; *s* **plano [m] general** (Plano de vista general de localización de un proyecto; ▶plan general [de ubicación], ▶plano de situación); *f* **plan [m] général** (Plan replaçant le site d'un projet dans son environnement en général topographique, urbain, etc. et comportant différentes données afférentes à l'opération ; ▶plan d'ensemble, ▶plan de situation [du terrain]) ; *syn.* plan [m] d'ensemble ; *g* **Übersichtsplan [m]** (Plan, der die Lage eines Projektes im örtlichen Zusammenhang zeigt; ▶Gesamtlageplan, ▶Lageplan).

overall plan [n] (2) *constr. plan.* ▶overall site plan.

3978 overall project planning [n] *plan.* *s* **planificación [f] general de un proyecto** *syn.* planeamiento [m] general de un proyecto; *f* **étude [f] globale d'aménagement d'un projet** ; *g* **Gesamtplanung [f] eines Projektes**.

3979 overall site plan [n] *constr. plan.* (Graphic presentation of a complete area); *syn.* overall plan [n] (2); *s* **plano [m] general (de ubicación)** *syn.* plano [m] de conjunto; *f* **plan [m] d'ensemble** (Représentation graphique donnant la situation générale d'un terrain déterminé) ; *g* **Gesamtlageplan [m]** (Grafische Darstellung einer vollständigen örtlichen Situation einer bestimmten Fläche oder eines definierten Gebietes).

3980 over-browsing damage [n] *agr. for.* (Wounds caused to plants by gnawing game; *specific term* ▶scaling damage); *s* **recomido [m]** (Heridas causadas en las plantas por mordiscos de especies de caza; *término específico* ▶daño por descortezamiento); *f* **abroutissement [m]** (Dégâts provoqués par les cerfs, chevreuils et daims lorsqu'ils broutent les pousses terminales et latérales des plantes ou semis des essences ligneuses ; TSF 1985 ; *terme spécifique* ▶dégâts d'écorçage) ; *syn.* dégâts [mpl] d'abroutissement ; *g* **Wildverbiss [m]** (Durch Wild verursachte Fraßstellen an Pflanzen oder Pflanzenteilen; *UB* ▶Schälschaden) ; *syn.* Verbissschaden [m].

3981 overburden [n] (1) *geo. min.* (Surface geological strata that are clearly differentiated from the underlying [workable] mineral strata; ▶overburden 2); *syn.* overlying strata [npl]; *s* **montera [f] [Es, YZ]** (MESU 1977; capa mineral no aprovechable que se encuentra sobre la de mineral explotable; ▶cascote); *syn.* capa [f] de desperdicio [MEX] (BU 1959); espesor [m] del suelo [EC], sobrecarga [f] [CO, MEX, RA, RCH], sobrecarga [f] de tierra [Es, YZ tb.] (todos MESU 1977); terreno [m] de recubrimiento; *f* **formation [f] superficielle** (▶Matériaux de recouvrement de la roche en place ou des gisements) ; *g* **Deckgebirge [n]** (Gesteinskomplex, der sich vom darunterliegenden Grundgebirge oder von auszubeutenden Bodenschätzen in seiner Struktur deutlich unterscheidet; ▶Abraum).

3982 overburden [n] (2) *min.* (Surface strata which have to be removed in the course of overcast mining of mineral deposits; ▶mining spoil, ▶overburden 1); *s* **cascote [m]** (▶Montera que se retira para la explotación del mineral subyacente en minería a cielo abierto; ▶zafras); *syn.* primera tierra [f], escombro [m], capa [f] de sobrante; *f* **matériaux [mpl] de recouvrement** (Couche géologique ou pédologique de recouvrement sus-jacente aux ressources minérales sujettes à l'exploitation minière ; ▶formation superficielle, ▶stérile) ; *syn.* matériaux [mpl] de découverte ; *g* **Abraum [m, o. Pl.]** (▶Deckgebirge über Lagerstätten, das durch den Tagebau entfernt wird; ▶Berge 3).

overburden dump [n] [US] *min.* ▶overburden pile.

3983 overburden dumping site [n] *min.* (Area for deposit of removed overburden from surface mining [US]/opencast mining [UK]; ▶overburden pile, ▶spoil bank [US] 2/spoil heap [UK]); *s* **depósito [m] de cascote** (Area para depositar los materiales no aprovechables retirados para llegar hasta el mineral a explotar; ▶montón de zafras, ▶vaciadero de gangas); *syn.* depósito [m] de escombro, depósito [m] de primera tierra; *f* **décharge [f] de matériaux de recouvrement** (Aire de dépôt des matériaux de recouvrement enlevés dans une exploitation à ciel ouvert ; ▶dépôt de matériaux d'extraction, ▶terril) ; *g* **Abraumdeponie [f]** (Ablagerungsstelle zum Verkippen des im Tagebau abgetragenen Deckgebirges; ▶Abraumhalde, ▶Bergehalde).

3984 overburden pile [n] *min.* (Dumping pile [US]/dumping heap [UK] of overburden removed in the course of surface mining [US]/open-cast mining [UK]; ▶overburden dumping site); *syn.* overburden dump [n] [also US]; *s* **montón [m] de zafras** (Amontonamiento del material no aprovechable en minería a cielo abierto; ▶depósito de cascote); *f* **dépôt [m] de matériaux d'extraction** (Accumulation d'éléments stériles, matériaux rocheux ou terreux provenant du décapage de surface des mines à ciel ouvert ou sortis de terre pour accéder aux veines de minerais lors de la prospection minière ; ▶décharge de matériaux de recouvrement) ; *syn.* halde [f] de mort-terrain [CDN] ; *g* **Abraumhalde** (Aufschüttung von im Tagebau abgetragenen Deckgebirge; *OB* ▶Abraumdeponie).

overexploitation [n] *agr. for. land'man. min. nat'res.* ▶overuse [US]/over-use [UK], ▶wasteful exploitation.

overfall [n] *eng. wat'man.* ▶spillway.

3985 over-fertilization [n] *agr. for. hort.* (Input/use of too much fertilizer per unit area; ▶eutrophication); *s* **abonado [m] excesivo** (Aporte de más fertilizante del necesitado para el cultivo por unidad de superficie; ▶eutrofización); *syn.* fertilización [f] excesiva; *f* **fertilisation [f] excessive** (Apport excessif de fertilisants par unité de surface ; ▶eutrophisation) ; *syn.* suramendement [m] ; *g* **Überdüngung [f]** (Ausbringen von zuviel Dünger je Flächeneinheit; ▶Eutrophierung).

3986 overfilling [n] **of root zone** *arb. constr.* (Raising of soil level on existing ground surface over roots, which prevents adequate penetration of air and rainwater; ▶groundfill around a tree); *s* **recubrimiento [m] de la zona de raíces** (▶Relleno con tierra [alrededor del tronco] de árboles viejos que impide la penetración suficiente aire y agua de lluvia en el suelo); *f* **recouvrement [m] de la zone racinaire** (Mise en œuvre de terre sur l'emprise des racines autour d'un arbre ; ▶enfouissement du tronc) ; *g* **Überfüllung [f] des Wurzelbereiches** (Bodenauffüllung auf vorhandene Bodenoberfläche im Wurzelbereich meist älterer Bäume; ▶Stammeinschüttung); *syn.* Auffüllen [n, o. Pl.] des Wurzelbereichs mit Boden, Bodenauftrag [m] im Wurzelbereich.

3987 overfishing [n] *agr.* (Excessive catches of fish which threaten the sustainable reproductive capacity of fish stock); *s* **sobrepesca [f]** (Pesca por encima del rendimiento máximo sostenido; DINA 1987, 714); *syn.* overfishing [m]; *f* **surpêche [f]** (Capture de quantités excessives de poissons menaçant certaines espèces lorsqu'elles dépassent le taux de renouvellement des populations) ; *syn.* pêche [f] excessive ; *g* **Überfischung [f]** (Fangen von zu hohen Mengen, so dass die nachhaltige Reproduzierbarkeit des Fischbestandes gefährdet ist; überfischen [vb]).

overflow car park [n] *plan.* ▶overflow parking area.

3988 overflow parking area [n] *plan.* (Informal, additional car parking space, often unpaved and available for peak periods; e.g. garden festivals, trade fairs, sports stadiums); *syn.* overflow car park [n]; *s* **aparcamiento [m] de reserva** (Zona de

O

aparcamiento adicional, generalmente no equipada, reservada para ocasiones excepcionales en las que se necesita mucho más espacio para este fin); *syn.* parking [m] de reserva, parqueo [m] de reserva [AL], estacionamiento [m] de reserva [CS]; *f* **parking [m] de délestage** (Terme spécifique pour toute aire utilisée exceptionnellement pour le stationnement des véhicules pour résorber les pointes de fréquentations, p. ex. terrain d'exposition, de sport ; en zone rurale le stationnement s'effectue souvent au moyen d'un « parking sur prairie drainée ») ; *syn.* parc [m] de stationnement de secours ; *g* **Ausweichparkplatz [m]** (Zusätzlich verfügbare, weniger aufwendig oder gar nicht ausgebaute Parkfläche für Spitzenbedarfszeiten, z. B. bei Gartenschauen, Messen, Sportveranstaltungen); *syn.* Bedarfsparkplatz [m].

3989 overflow pipe [n] *constr.* (Outlet from a water basin or an underground irrigation reservoir [e.g. roof garden] used to prevent flooding or to keep a constant water level); *syn.* riser [n] [also US]; *s* **tubo [m] de sobrante** (Tubo de desagüe de un estanque o un almacén subterráneo de riego [p. ej. en una azotea ajardinada] que sirve para mantener el nivel de agua deseado y prevenir inundaciones); *syn.* tubo [m] de rebosamiento (DACO 1988); *f* **bonde [f] de surverse** (Tuyau ou ouvrage hydraulique placé au point le plus bas servant à la formation d'une réserve d'eau, d'évacuer le trop plein d'eau par le fond ou en surface, de vidanger un bassin ou la nappe d'arrosage stockée dans la couche drainante pour les jardins sur dalles) ; *syn.* bonde [f] de trop-plein, bonde [f] de déversement, bonde [f] à surverse (GEN 1982-II, 141), moine [m] (AEP 1976, 322) ; *g* **Anstaurohr [n]** (Rohr, das dem Wasseranstau in einem Wasserbecken oder einem unterirdischen Bewässerungshorizont, z. B. im Dachgarten, dient); *syn.* Anstauelement [n], Mönch [m], Standrohr [n], Überlauf [m], Überlauf(stand)rohr [n].

3990 overgrazing [n] *agr. game´man.* (Grazing so heavy and prolonged that it impairs future forage production and causes damage to plants or soil, or both; RCG 1982; ▶overstocking); *s* **sobrepastoreo [m]** (Sobreexplotación de una zona de pastos debido a la ▶sobrecarga de ganado); *syn.* pastoreo [m] abusivo; *f* **surpâturage [m]** (Pâturage trop fréquent ou population trop importante [▶surcharge pastorale] si bien que l'herbe ne peut pas se reconstituer entre deux exploitations) ; *syn.* surexploitation [f] pastorale, surfréquentation [f] des pâturages, pâturage [m] excessif, surpâture [f] ; *g* **Überbeweidung [f]** (Durch zu hohen Tierbesatz [▶Überbesatz] bedingtes übermäßiges Abweiden, so dass die Regeneration der Pflanzendecke nicht gewährleistet ist; Bodenerosion ist meist die Folge).

overgrow [vb] *bot. hort. plant.* ▶overrun.

3991 overgrowing [n] *phyt.* (**1.** Displacement or invasion of cultivated plants by wild plants, weeds and bushes due to the lack of maintenance. **2.** The result of this process is called **overgrowth**; *specific terms* ▶spontaneous colonization by scrub, ▶weed growth; ▶feral species); *syn.* overgrowth [n]; *s* **invasión [f] de advenedizas** (**1.** Desplazamiento de las plantas cultivadas por plantas silvestres, malas hierbas y arbustos, en un jardín o un cultivo por falta de cuidado. **2.** El resultado de este proceso es un **estado [m] de abandono**; *términos específicos* ▶crecimiento de malas hierbas, ▶invasión de leñosas; ▶especie asilvestrada 1); *f 1* **envahissement [m] par les plantes sauvages** (Processus de développement des espèces adventices, broussailles et arbustes et d'invasion progressive des prairies abandonnées ; le résultat est l'▶embroussaillement ; *termes spécifiques* ▶envahissement par les adventices, ▶envahissement par les espèces buissonnantes ; ▶espèce retournée à l'état sauvage) ; *syn.* envahissement [m] par les espèces végétales indésirables, envahissement [m] par les mauvaises herbes/les espèces sauvages ; *f 2* **réapparition [f] des espèces sauvages** (Implantation des espèces de friches lors de soins négligés ou l'abandon de la culture) ; *g* **Verwilderung [f]**

(2) (**1.** Vorgang der Überwucherung von Kulturpflanzen durch Wildkraut und Gehölzaufwuchs, ausgelöst durch mangelnde Pflege. **2.** Das Ergebnis der Überwucherung durch krautige Pflanzen ist die ▶Verkrautung oder Verunkrautung, durch Gehölze die ▶Verbuschung; ▶verwilderte Art).

overgrowth [n] *phyt.* ▶overgrowing.

3992 overhang [n] *constr.* (Projection of a roof, beam, etc., beyond the construction beneath); *s* **alero [m]** (Parte del tejado que sobresale del muro; DACO 1988; *término específico* alero [m] corrido); *f* **débord [m]** (Partie de construction qui n'est pas à l'aplomb de l'élément sous-jacent, p. ex. d'un mur sur le soubassement) ; *g* **Überstand [m]** (Herausragen oder Vorspringen eines Daches, Balkons etc. aus der darunterliegenden Konstruktion).

overhanging [adj] *hort. plant.* ▶arching.

3993 overhanging growth [n] *constr. hort. leg.* (Parts of plants growing beyond a property line or into right-of-way space); *s* **ramaje [m] sobresaliente** (≠) (Partes de árboles que sobresalen los límites de un predio); *f* **empiétement [m] de branches** (Parties de branches d'un arbre ou d'un arbuste dépassant la ligne séparative et qui surplombent le fonds du voisin ou l'emprise de la voirie) ; *syn.* extension [f] de branches sur le fonds voisin ; *g* **Überwuchs [m]** (*Fachsprachlich* ...wüchse [pl], *sonst ohne Pl.*; über die Grundstücksgrenze hinaus- oder in den Verkehrsraum hineinwachsende Pflanzenteile; cf. N+L 1980 [6], 256 u. N+R 1982, 38); *syn.* Überhang [m].

3994 overhead [n] *contr.* (General business expenses necessarily incurred in the operation of a planning office that cannot be charged against one particular project; e.g. management and administration costs, office rent, operational costs, expenditure for advertising, corporate presentation, legal advice as well as taxation. These expenses, known simply as 'overhead' are added as a supplement when calculating a planning fee according to a company determined formula or as a percentage of the contract sum, usually between 6-8%); *s* **costos [mpl] generales de empresa** (Aquellos gastos de una empresa que no se pueden liquidar en un proyecto específico); *syn.* costes [mpl] generales de empresa [Es]; *f* **frais [mpl] généraux (d'entreprise)** (Frais fixes ou, du moins, non proportionnels aux quantités produites, au chiffre d'affaires réalisé, et qui se répartissent sur l'ensemble de l'exploitation tels que les achats consommables [fournitures de bureau, produits d'entretien, petit outillage, énergie et fluides, etc.], les prestations de services techniques [frais de sites, logistique, maintenance, sécurité, transports et voyages, informatique, etc.], les prestations intellectuelles [assurances, communication, juridique, publicité, recherche, environnement, etc.] les investissements, les impôts, taxes et cotisations ; PR 1987 ; l'ensemble des frais non stratégiques d'une entreprise peuvent représenter jusqu'à 30 % du chiffre d'affaires ; pour établir le prix de vente d'un produit ou d'un service les entreprises utilisent souvent un barème [coefficient multiplicateur du prix de revient] dans lequel la prise en compte des frais généraux entre en général dans 6 à 8 % du prix de revient) ; *g* **allgemeine Geschäftskosten [pl]** (Kosten, die zum Betrieb eines Unternehmens oder eines Planungsbüros notwendig sind, sich aber nicht einem Produkt direkt zuordnen lassen, z. B. Kosten der Unternehmensleitung oder –verwaltung, Büromieten, Betriebskosten, Kosten des Bauhofes, Werbung und Unternehmenspräsentation, Rechtsberatung, Steuern. In der Vollkostenrechnung werden die **a.n G.** mit Hilfe eines Verrechnungsschlüssels den einzelnen Produkten oder abzurechnenden Leistungen als Zuschlagssatz, der im Allgemeinen zwischen 6-8 % der Auftragssumme liegt, zugerechnet); *syn.* Gemeinkosten [pl].

3995 overhead expenses [npl] *contr. prof.* (Expenditures incurred during the performance of a planning contract by a

planner or a third-party, which are not directly covered by the planning fee; e.g. postal and telecommunication charges, reproduction costs, travel and accommodation expenses, costs for a construction site office, etc. These costs are often reimbursed on a lump-sum basis or expressed as a percentage of net fees instead of the reimbursement of individual invoices); *syn.* reimbursable expenses [npl] [also US] (AIA documents B727&B141/CM); *s* **gastos [mpl] corrientes** (Costes generales que se presentan al realizar un proyecto que no están considerados en los honorarios); *f* **frais [mpl] particuliers** (Frais engagés par l'architecte ou pour des travaux confiés à des entreprises extérieures, occasionnés pendant l'exécution du contrat par les prestations imprévues accomplies en sus de la mission confiée et qui ne sont donc pas couverts par les honoraires) ; *g* **Nebenkosten [pl] (eines Planungsauftrages)** (Bei der Ausführung eines Auftrages entstehende Auslagen, die durch Leistungen Dritter oder eigene Leistungen verursacht werden und nicht durch das Honorar abgedeckt sind, z. B. Gebühren für Post- und Telekommunikationsleistungen, Kosten für Vervielfältigungen, Fahrtkosten, Kosten für ein Baustellenbüro; diese Kosten werden i. d. R. statt auf Einzelnachweis pauschal mit einem Prozentsatz des Nettohonorars abgerechnet).

overhead power cable [n] *constr. envir.* ▶overhead power line.

3996 overhead power line [n] *envir.* (Cable above ground for the supply of electricity or telecommunications; ▶high tension power line, ▶electric power line, ▶underground power line; ▶overhead telephone line); *syn.* overhead power cable [n]; *s* **línea [f] aérea** (Conducción aérea de distribución de electricidad, telecomunicación, etc.; ▶cable eléctrico subterráneo, ▶conducción electrica, ▶línea de alta tensión, ▶línea telefónica aérea); *f* **ligne [f] aérienne électrique** (Réseau de distribution d'énergie électrique et de télécommunication aérien ; ▶ligne à haute tension, ▶ligne électrique, ▶ligne électrique souterraine, ▶ligne téléphonique aérienne) ; *g* **Freileitung [f] (1.** Oberirdisch geführte Stromversorgungs-, Fernsprech- oder Telegrafenleitung. **2.** *UB* ▶oberirdische Telefonleitung; ▶Hochspannungsleitung, ▶Stromleitung, ▶unterirdisches Stromkabel; *syn. für die freie Landschaft* Überlandleitung [f].

3997 overhead shelter [n] *arch.* (Protective structure; *specific term* shade structure [LD 1988 [9], 61]; ▶roofing); *s* **techado [m]** (Estructura protectora contra la intemperie o la lluvia; ▶construcción de tejado); *f* **couverture [f]** (Ensemble des matériaux imperméables et de leurs supports destinés à mettre le bâtiment à l'abri des intempéries ; ▶mise hors d'eau, ▶recouvrement) ; *syn.* toiture [f] ; *g* **Überdachung [f] (1)** (Zustand des Überdachtseins, das Dach selbst).

3998 overhead sign structure [n] *trans.* (Signpost spanning the travelled way [US]/carriage-way [UK] of a motorway containing directions for each lane); *syn.* span structure [n] for signage; *s* **puente [m] de señalización (≠)** (Estructura metálica que atraviesa autopistas o autovías en la que están montados carteles que indican los destinos); *f* **portique [m] de signalisation** (Dispositif de signalisation enjambant la voirie d'autoroutes et de voies express et apportant une information pour chaque voie) ; *g* **Schilderbrücke [f]** (Eine Schnellstraße oder Autobahn überspannende Beschilderung, die für jede Fahrbahn Verkehrsschilder vorsieht).

3999 overhead telephone line [n] *constr. envir.* (Telephone cable above ground, mounted on masts); *s* **línea [f] telefónica aérea** (Cable de transmisión de teléfono que se apoya en postes); *f* **ligne [f] téléphonique aérienne** (Câble téléphonique supporté par des poteaux) ; *g* **oberirdische Telefonleitung [f]** (Auf Masten geführte Telefonleitung).

overhead timber [n] [US] *constr.* ▶hanging beam.

4000 overhunting [n] [US]/**over-hunting** [n] [UK] *game' man. hunt.* (Over-catching of the total population of a game species, of a gender, or of a particular age group of game; *opp.* ▶overprotection [US]/over-protection [UK]); *s* **sobreexplotación [f] de la caza** (Exceso de caza de ejemplares de una población, un sexo o un determinado grupo de edad de una especie salvaje; *opp.* ▶sobreprotección de la caza); *syn.* sobreexplotación [f] de especies cinegéticas; *f* **chasse [f] excessive (±)** (Abattage d'un très grand nombre d'animaux d'une population, d'individus d'espèces sauvages de même sexe ou de même classe d'âge ; *opp.* ▶réglementation excessive de la chasse) ; *g* **Überjagung [f]** (Zu hoher Abschuss von Tieren eines Gesamtbestandes, eines Geschlechts oder bestimmter Altersgruppen einer Wildart; *opp.* ▶Überhege).

4001 overland flow [n] *geo. hydr.* (Proportion of total rainfall which is not intercepted by vegetation and does not infiltrate the soil, but which runs over the ground surface as ▶sheetwash; DNE 1978; ▶sheet erosion, ▶sheet flow); *s* **escorrentía [f] superficial (2)** (Flujo de agua de las precipitaciones en el terreno sin cauce definido, que no infiltra en el suelo o no se ve interceptada por la vegetación. La pendiente, la permeabilidad del suelo, la densidad de la vegetación y el tipo de uso del suelo son los factores que más influencia tienen sobre el porcentaje de e. s.; ▶arrastre laminar, ▶erosión laminar, ▶escorrentía laminar); *syn.* escorrentía [f] de superficie; *f* **écoulement [m] superficiel (2)** (Circulation des eaux pluviales le long du trajet des cours d'eau ; ▶ablation, ▶érosion laminaire, ▶érosion pélliculaire, ▶ruissellement en nappes) ; *syn.* circulation [f] des eaux, ruissellement [m] concentré ; *g* **Oberflächenwasserabfluss [m] (2)** (Großräumiger Regenabfluss auf der Erdoberfläche; Hangneigung, Wasserdurchlässigkeit des Bodens, Dichte des Vegetationsbestandes und die Art der Bodennutzung beeinflussen die Intensität des Abflusses am meisten; ▶Abspülung, ▶Flächenabfluss 2, ▶Flächenerosion); *syn.* Flächenabfluss [m] (1).

4002 overlap [n] *constr.* (Joint laid with; e.g. 10cm overlap); *s* **superposición [m] de los bordes** (*Contexto* superponer en 10 cm los bordes de junta de soldadura de lámina de tejado); *f* **recouvrement [m] (des lés)** (Joints/jointoiement effectué par recouvrement des lés de 10 cm) ; *g* **Überlappung [f] (1)** (Überdeckung eines Bereiches dergestalt, dass ein Teil des einen sich auf dem anderen befindet; *Kontext* Naht einer Kunststoffbahn mit 10 cm Überlappung verlegen).

4003 overlapping joint [n] *constr.* (Junction between two sheets of material in which one sheet overlays the adjacent one by a constant, usually specified, amount); *s* **junta [f] de superposición** (Partes de dos láminas de material que se superponen en un ancho constante, p. ej. de 5-10 cm); *syn.* superposición [f] de capas; *f* **zone [f] de recouvrement (des lés)** (Bande étroite [5 à 10 cm] correspondant au recouvrement de deux lés d'étanchéité) ; *syn.* bande [f] de recouvrement ; *g* **Überlappung [f] (2)** (Bereich, in dem sich z. B. zwei Kunststoffbahnen überdecken); *syn.* Nahtüberdeckung [f].

4004 overlapping layer [n] *phyt.* (Plant community layer in which the individual sub-layers are associated with two or more superior layers; ▶fusion of layers, ▶stratification of plant communities); *s* **estrato [m] transgresivo** (En la estructura vertical de una comunidad [forestal], uno de los subtipos de ▶estratos relacionados en el que aparecen dos juntos, como es el caso del estrato muscinal y el estrato arbustivo de algunas comunidades de turbera alta; cf. BB 1979; ▶estratificación de la vegetación); *f* **strate [f] transgressive** (Association végétale dans laquelle les strates inférieures se trouvent en relation avec deux ou plusieurs

strates supérieures ; ▶solidarité écologique des strates, ▶stratification de la végétation) ; *g* **übergreifende Schicht [f]** (Unter dem Gesichtspunkt der ▶Schichtbindung einer Pflanzengesellschaft vorkommende Schicht, bei der die einzelnen Unterschichten mit zwei oder mehreren Oberschichten verbunden sind; ▶Vegetationsschichtung).

overlapping [n] of land uses [UK] *plan.* ▶layering of different uses.

overlook [n] [US], scenic *recr.* ▶view (1).

overlying strata [npl] *geo. min.* ▶overburden (1).

overnight accommodation [n] *recr.* ▶group overnight accommodation, ▶individual overnight accommodation, ▶rural overnight accommodation.

4005 overnight land squatting [n] (≠) *urb.* (Overnight settlement of people in a ▶shanty town without having a right or title; the results are "squatter camps" which become permanent; ▶land grab [US]/land occupation [UK]); *syn.* squatter camping [n] [ZA]; *s* **paracaidismo [m]** (Fenómeno de ocupación nocturna de tierras urbanas para la construcción de viviendas primitivas que se da en México en los alredededores de México D.F. por el hecho de que la ley mexicana lo permite en terrenos públicos si se construyen las «casas» de un día para otro; ▶barrio de chabolas, ▶colonización de tierras yermas); *f* **établissement [m] d'un campement sauvage (≠)** (À Mexico ou dans d'autres villes de pays sous-développés, prise de possession sauvage d'un terrain pendant la nuit à des fins de campement par une population de sans-abris à la périphérie de grandes agglomérations et ne pouvant, selon la législation locale en vigueur, être détruit ultérieurement s'il est implanté sur un terrain public ; ▶bidonville, ▶colonisation sauvage) ; *g* **Bauen [n, o. Pl.] von Notbehausungen über Nacht (≠)** (In Mexiko gibt es den sog. ‚Paracaidismo' [*wörtlich übersetzt* Fallschirmspringen]. Darunter versteht man das plötzliche Auftauchen von obdachlosen Menschen in Randbereichen großer Städte, um während der Nacht ein Stück Land in Besitz zu nehmen, worauf eine Hütte gebaut wird, die nach geltendem Recht nicht abgerissen werden darf, wenn sie auf öffentlichem Grund steht; ▶Barackenstadt, ▶Landnahme).

4006 overprotection [n] [US]/**over-protection** [n] [UK] *hunt.* (Aspect of hunting which leads to an extremely high ▶game density—mostly even-hoofed game—and which is no longer adapted to the landscape and the cultural situation and, therefore to a certain extent, causes great ▶game damage; ▶game thinning; *opp.* ▶overhunting [US]/over-hunting [UK]); *s* **sobreprotección [f] de la caza** (Medidas de cuidado de la caza que llevan a la sobrepoblación de especies cinegéticas, lo cual puede conllevar graves ▶daños causados por la caza en la naturaleza; ▶caza selectiva, ▶densidad de población cinegética; *opp.* ▶sobreexplotación de la caza); *syn.* sobreprotección [f] de la fauna cinegética); *f* **réglementation [f] excessive de la chasse (±)** (Mesures cynégétiques favorisant une protection extrême des espèces chassables occasionnant des dommages aux activités humaines et qui se trouve en décalage avec les pratiques culturelles et la gestion des équilibres biologiques ; ▶chasse régulatrice, ▶dégâts causés par le gibier, ▶densité de la population cinégétique ; *opp.* ▶chasse excessive) ; *syn.* surchasse [f], surprotection [f] de la chasse ; *g* **Überhege [f]** (Jagdbetriebliche Maßnahmen, die eine zu große ▶Wilddichte [meist bei Schalenwild] aufkommen lassen, die den landschaftlichen und landeskulturellen Verhältnissen nicht mehr angepasst ist und deshalb z. T. große ▶Wildschäden verursachen; ▶Hegeabschuss; *opp.* ▶Überjagung).

overriding public interest [n] *ecol. leg. plan.* ▶imperative reasons of overriding public interest.

4007 overrun [vb] *bot. hort. plant.* (To invade other plants; to encroach on other plants, so that these are suppressed; ▶invasive species, ▶spread 2, ▶vigorous); *syn.* overgrow [vb], invade [vb]; *s* **invadir [vb]** (Proceso de algunas plantas de ▶extenderse y sofocar a otras; ▶de crecimiento vigoroso, ▶especie invasora); *f* **envahir [vb]** (Très forte extension d'une espèce qui élimine les végétaux voisins ; ▶à croissance rapide, ▶espèce envahissante, ▶s'étendre) ; *g* **überwachsen [vb]** (Sich stark ausbreiten und andere Pflanzen unterdrücken; ▶sich ausbreiten, ▶sich stark ausbreitende Art, ▶stark wüchsig); *syn.* sich übermäßig stark ausbreiten [vb].

4008 overseas tourism [n] *recr. trans.* *s* **turismo [m] intercontinental**; *f* **tourisme [m] intercontinental** ; *g* **Ferntourismus [m]** (interkontinentaler Reiseverkehr).

4009 over-seeding [n] [US] *constr.* (Repair of thin areas of sward or turf by lightly cultivating the surface and sowing additional seed; BS 3975: part 5; ▶seeding); *syn.* reseeding [n] [US]/re-seeding [n] [UK], over-sowing [n] [UK]; *s* **resiembra [f] de césped (2)** (Repetición de la ▶siembra 1 donde la primera no tuvo éxito); *f 1* **semis [m] de regarnissage** (Effectué manuellement ou à la machine sur des pelades, des parties dégradées d'un gazon en place ou après certaines opérations d'entretien, [aération, scarification, etc.]) ; *syn.* semis [m] de renforcement (de gazon) ; *f 2* **réensemencement [m]** (Reprise des parties malvenues dans le cadre de la garantie ; ▶semis 1) ; *g* **Nachsaat [f]** (Nochmaliges Aussäen, wenn Saatgut nicht aufgelaufen ist oder zur Schließung eines lückigen Gras- oder Kräuterbewuchses; nachsäen [vb]; ▶Ansaat 1).

over-sowing [n] [UK] *constr.* ▶over-seeding [US].

overspill town [n] [UK] *urb.* ▶satellite city.

4010 overstocking [n] *agr. ecol.* (Excessive number of animals on a given area; ▶overgrazing; *s* **sobrecarga [f] de ganado** (Número excesivo de cabezas de ganado en un área determinada; ▶sobrepastoreo); *f* **surcharge [f] pastorale** (Population trop importante d'animaux par unité de surface ; ▶surpâturage) ; *syn.* charge [f] animale excessive ; *g* **Überbesatz [m]** (Zu hohe Populationsdichte von Nutztieren je Flächeneinheit; ▶Überbeweidung).

overtime charges [n] *contr.* ▶overtime payment.

4011 overtime payment [n] *contr.* (Remuneration for time spent working after hours); *syn.* overtime charges [npl]; *s* **pago [m] de horas extras** *syn.* pago [m] de horas extraordinarias; *f* **rémunération [f] des heures supplémentaires** ; *g* **Überstundenvergütung [f]** (Bezahlung, die bei Ableistung von Überstunden gewährt wird).

4012 overuse [n] [US]/**over-use** [n] [UK] *agr. ecol. for. recr.* (General term; ▶overfishing, ▶overgrazing, ▶overhunting [US]/over-hunting [UK]); *syn. agr. for* overexploitation [n]; *s* **sobreexplotación [f]** (Agotar o menoscabar una fuente de riqueza, sacando de ella mayor provecho que el debido; ▶sobreexplotación de la caza, ▶sobrepastoreo, ▶sobrepesca); *f 1* **surexploitation [f]** *agr. ecol. for.* (Exploitation excessive d'un écosystème ou d'un de ses éléments provoquant sa régression, sa dégradation ou sa destruction ; ▶chasse excessive, ▶surpâturage, ▶surpêche) ; *f 2* **surfréquentation [f]** *recr.* (Pression excessive des visiteurs, des touristes ou des usagers supérieure à la capacité de charge d'un milieu naturel et entraîne des dégâts notires) ; *g* **Übernutzung [f]** (Zu hohe Nutzung einer Fläche oder eines Ökosystems, die eine Gefährdung, Schädigung oder Zerstörung des vorhandenen Bestandes verursachen; ▶Überbeweidung, ▶Überfischung, ▶Überjagung); *syn.* Übernutzen [n, o. Pl.].

4013 overwintering [n] (1) *bot.* (Survival of plants through the winter period, whereby plant growth comes to a standstill or is hardly noticeable due to low temperatures; ▶dormancy); *syn.* wintering [n]; *s* **invernación** [f] (2) (Fase de descanso de la vegetación en invierno en la que —debido a las bajas temperaturas— las plantas o sus diásporas apenas crecen; ▶dormancia); *syn.* hibernación [f] (1); *f* **hivernation** [f] (1) (Transformations dans l'organisme des végétaux pour survivre aux périodes froides ; ▶dormance ; *g* **Überwinterung** [f] (2) (Verbringen des Winters in einem Zustand, bei dem auf Grund niedriger Temperaturen Wachstumsvorgänge bei Pflanzen oder deren Diasporen stillstehen oder kaum merklich vorangehen; ▶Vegetationsruhe).

4014 overwintering [n] (2) *zool.* (Spending winter in an overwintering area; ▶hibernation area/region, ▶winter sleep); *s* **invernación** [f] (3) (Paso del invierno en ▶área de invernación; ▶hibernación 2); *f* **hivernation** [f] (2) (Passer l'hiver dans certains biotopes plus propices ; les oiseaux migrateurs sur l'▶aire d'hivernage sont dénommés visiteurs d'hiver ; ▶sommeil hibernal, ▶territoire d'hibernation ; *terme spécifique pour les oiseaux migrateurs* **hivernage** [ces oiseaux sont dénommés hivernants]) ; *g* **Überwinterung** [f] (3) (**1.** Verbringen der kalten Jahreszeit durch jahresperiodische Wanderungen in ein wärmeres Gebiet; Zugvögel im ▶Überwinterungsgebiet werden „Wintergäste" genannt. **2.** Insekten können durch unterschiedliche Entwicklungsstadien, z. B. durch Eier, Larven und Puppen oder durch zusätzliche Bildung von Gefrierschutzproteinen, überwintern; ▶Winterschlaf).

4015 overwintering area [n] *zool.* (Region to the south of a breeding range in which migratory birds spend the winter; ▶hibernation area/region); *s* **área** [f] **de invernada** (DINA 1987; zona en que las aves migratorias pasan el invierno; ▶área de invernación); *syn.* zona [f] de invernada; *f* **quartiers** [mpl] **d'hiver** (Aire située au sud de l'aire de reproduction, sur laquelle les oiseaux migrateurs passent l'hiver ; ▶aire d'hivernage, ▶territoire d'hibernation) ; *g* **Winterquartier** [n] (2) (Südlich vom Brutgebiet gelegenes Gebiet, in dem Zugvögel den Winter verbringen; ▶Überwinterungsgebiet).

4016 overwintering (migratory) bird [n] *zool.* (Species of bird, which stays over during the winter in a climatically favorable region and breeds in the far north; ▶stopover bird); *syn.* wintering bird [n]; *s* **ave** [f] **invernante** (Especie de ave que permanece durante el invierno en una región climáticamente favorable y cría en el extremo norte en el hemisferio norte resp. en el hemisferio sur en el extremo sur; ▶ave en etapa migratoria); *f* **oiseau** [m] **hivernant** (Espèce venant dans nos régions pour profiter du climat océanique ou méditerranéen en hiver, et repartant pour nicher plus au nord ou à l'est de l'Europe ; ▶oiseau en étape migratoire) ; *syn.* hivernant [m] ; *g* **Überwinterer** [m] (Vogelart, die in einer klimatisch günstigen südlichen Region überwintert und in der wärmeren Jahreszeit weiter nördlich oder östlich brütet; ▶Rastvogel).

4017 overwintering part [n] *bot.* (Generic term covering bud, bulb, corm, rhizome, tuber, seed, etc.). Above ground **o. p.s** of trees, shrubs and many perennials [▶hemicryptophytes] are distiguished from underground **o. p.s** of e.g. ▶geophytes and perennial rhizomes; *syn.* perennating part [n] (TEE 1980, 6); *s* **hibernáculo** [m] (DB 1985; en su sentido prístino, linneano, yema, bulbo, etc., destinado a proteger a un vástago rudimentario durante el invierno; hoy el concepto de **h.** se limita a las producciones gemarias que, desprendiéndose de la planta madre al presentarse condiciones mesológicas desfavorables, son capaces de sobrevivirla y perpetuarla asexualmente; DB 1985; ▶hemicriptófito, ▶geófito); *syn.* órgano [m] hibernante; *f* **organe** [m] **d'hivernation** (Terme générique désignant les organes assurant

pendant la mauvaise saison la pérennance des végétaux tels que bourgeons, bulbes, rhizomes, tubercules, semences, etc. on distingue les **organes d'hivernation superficiels** chez les arbres, arbustes et nombreuses vivaces [▶hémicryptophytes] et les **organes d'hivernation souterrains** comme chez les ▶géophytes ou les vivaces rhizomateuses) ; *g* **Überwinterungsorgan** [n] (OB zu Überwinterungsknospe, Zwiebel, Rhizom, Knolle, Samen etc.; es werden **oberirdische Ü.e** — bei Bäumen, Sträuchern und vielen Stauden [▶Hemikryptophyten] — und **unterirdische Ü.e** — z. B. bei ▶Geophyten und rhizombildenden Stauden — unterschieden).

owner [n] [UK] *contr.* ▶developer [US] (3).

owner's duty to consent [loc] *leg.* ▶private owner's obligation.

4018 ownership map [n] *plan.* (Part of a larger plan used to indicate the distribution and ownership of land holdings during a planning process, not usually a cadastral map; **in U.S.**, property maps do not typically show or indicate land use or buildings, yield, or any information beyond ownership; ▶real property identification map [US]/land register [UK]); *s* **mapa** [m] **de registro de la propiedad** (Mapa en el que están registrados la distribución, propiedad de y los derechos de usufructo sobre los terrenos; ▶catastro de bienes immuebles); *f* **plan** [m] **parcellaire du régime de la propriété foncière** (≠) (Plan graphique établi selon les besoins lors d'études d'urbanisme sur lequel figurent divers renseignements tels que les limites d'emprise et les numéros de parcelle, la liste des propriétaires, locataires, fermiers ou métayers ainsi que celle de tout ayant-droits, les servitudes conventionnelles ou légales, etc. ; ▶cadastre parcellaire) ; *g* **Besitzstandskarte** [f] (**1.** Kartenausschnitt [Inselkarte] für einen jeweiligen Planungsfall mit Angaben über Flurstücksnummern, im Grundbuch eingetragene Eigentümer, Pächter oder sonstige Nutzungsberechtigte. **2.** *urb.* Von der Umlegungsstelle gefertigte Karte der Grundstücke eines Umlegungsgebietes, die mindestens die bisherige Lage und Form der Grundstücke des Umlegungsgebietes und die auf ihnen befindlichen Gebäude mit Angabe der Eigentümer ausweist; ▶Liegenschaftskataster); *syn.* Besitzstandsplan [m].

4019 ownership pattern [n] *adm. sociol.* (Structure of land holdings); *s* **régimen** [m] **de propiedad**; *f* **régime** [m] **de propriété** ; *g* **Besitzverhältnisse** [npl] (Verteilung des Grundbesitzes in einem definierten Gebiet).

own-rooted [adj] [UK] *hort.* ▶self-rooted [US].

4020 own root plant [n] *hort.* (Plant with its own roots, rather than understock roots; ▶self-rooted); *s* **planta** [f] **de raíces propias** (Planta cultivada de semillas o esquejes con raíces propias en vez de sobre la de un patrón; ▶de raíces propias); *f* **plante** [f] **franche** (Végétaux cultivés obtenus par semis, bouturage ou marcottage se développant par ses propres racines au lieu de celles du porte-greffe ; ▶franc, franche de pied) ; *syn.* plante [f] franche de pied, plante [f] de franc pied ; *g* **wurzelechte Pflanze** [f] (Pflanze mit eigenen Wurzeln [aus Sämling oder Steckling gezogen], statt mit Wurzeln einer Unterlage; ▶wurzelecht).

4021 oxbow lake [n] *geo.* (Water body in the former meandering arm or bed of a river, which has become completely separated from the main channel); *syn.* cut-off [n], *obs.* mortlake [n]; *s* **brazo** [m] **muerto** (Laguna originada por un meandro o por un brazo de río aislado); *syn.* laguneta [f]; *f* **bras** [m] **mort** (Boucle formée par incision d'un méandre lors d'une période de creusement actif d'un cours d'eau — bras situé au-dessus du lit actuel et sans relation avec celui-ci) ; *g* **Altwasser** [n] (1) (…wasser [pl]; natürliches Stillgewässer, das durch eine total abgetrennte Flussschleife entstand); *syn.* Altarm [m], Altlauf [m].

oxide [n] *chem. envir.* ►nitric oxide.

oxide [n], **nitrogen** *chem. envir.* ►nitric oxide.

4022 oxygenation [n] *envir. limn.* (Natural or artificial addition of oxygen; e.g. to water; ►oxygen supply; oxidize [vb]); *s* **oxigenación** [f] **(de aguas)** (Aporte natural o artificial de oxígeno al agua; ►suministro de oxígeno); *f* **oxygénation** [f] **(de l'eau)** (Processus naturel ou artificiel d'apport, p. ex. dans les eaux douces ; ►apport d'oxygène) ; *g* **Sauerstoffzuführung** [f] (Natürliche oder künstliche Zufuhr von Sauerstoff, z. B. in Gewässern, ►Sauerstoffversorgung).

4023 oxygen balance [n] **in water bodies** *limn.* (Relationship between the production and consumption of oxygen in water within a certain period of time; ►saprobic level); *s* **balance** [m] **de oxígeno** (Relación entre la producción y el consumo de oxígeno en el agua en un período dado de tiempo; ►saprobiedad); *f* **bilan** [m] **d'oxygène de l'eau** (Relation entre la production et la consommation d'oxygène dans les eaux pendant un temps donné ; ►saprobie) ; *g* **Sauerstoffhaushalt** [m] (Wirkungsgefüge/Bilanz zwischen Sauerstoffproduktion und -verbrauch [Atmung] in einem Gewässer in einem bestimmten Zeitraum; ►Saprobie).

4024 oxygen consumption [n] *limn. pedol.* (Assimilation of oxygen by microorganisms, etc. in waterbody); *s* **consumo** [m] **de oxígeno** (Asimilación de oxígeno, p. ej. por microorganismos en un cuerpo de agua); *f* **consommation** [f] **d'oxygène** (p. ex. par les micro-organismes dans un milieu aquatique) ; *g* **Sauerstoffzehrung** [f] (Verbrauch von Sauerstoff in einem Gewässer durch Mikroorganismen etc.); *syn.* Sauerstoffverbrauch [m].

4025 oxygen content [n] *limn. pedol.* (Amount of oxygen present in an environmental medium; e.g. in water, soil); *s* **contenido** [m] **en oxígeno** (Cantidad de oxígeno presente en un medio, como p. ej. el agua o el suelo); *f* **teneur** [f] **en oxygène** (Quantité d'oxygène contenue dans un milieu) ; *g* **Sauerstoffgehalt** [m] (Menge an Sauerstoff, die in einem Umweltmedium vorhanden ist, z. B. in einem Gewässer, Boden etc.).

oxygen demand [n] *limn.* ►biochemical oxygen demand.

4026 oxygen supply [n] *envir. limn. pedol.* (Availability/presence of oxygen in a body of water or in soil; ►oxygenation); *s* **suministro** [m] **de oxígeno** (Disponibilidad de oxígeno en un cuerpo de agua o en el suelo; ►oxigenación [de aguas]); *syn.* aporte [m] de oxígeno; *f* **apport** [m] **d'oxygène** (p. ex. dans les eaux douces ; ►oxygénation [de l'eau]) ; *syn.* oxygénation [f] ; *g* **Sauerstoffversorgung** [f] (Vorhandensein von Sauerstoff, z. B. in Gewässern, im Boden, ►Sauerstoffzuführung).

4027 ozone layer [n] *met.* (Stratum of a form of oxygen produced by electrical discharges in the atmosphere, 20-30 km above the Earth's surface, which screens out harmful ultra-violet rays and prevents skin cancer, eye disease, climatic change, etc.; ►hole in ozone layer); *syn.* ozone shield [n]; *s* **capa** [f] **de ozono** (Capa de la atmósfera localizada en la estratosfera entre 20-30 km de altura que evita el acceso a la tierra de parte de los rayos ultravioleta de onda corta a los cuales los organismos no están adaptados; ►agujero del ozono); *syn.* ozonosfera [f], pantalla [f] de ozono; *f* **couche** [f] **d'ozone** (Mince strate située à 20 km d'altitude de la surface terrestre qui protège la terre des rayons UV de courte longueur d'onde d'origine solaire et empêche ainsi l'apparition de cancers de la peau, de maladies des yeux et les changements météorologiques et climatiques ; ►trou d'ozone) ; *syn.* ozonosphère [f] (DUV 1984, 329) ; *g* **Ozonschicht** [f] (In der Stratosphäre, in 20-30 km Höhe liegende Schicht der Erdatmosphäre, in der sich durch das Ultraviolettlicht der Sonne besonders viel Ozon bildet. Diese Schicht schützt die Erdoberfläche vor übermäßig vieler kurzwelliger UV-Strahlung.

Durch den vom Menschen verursachten beschleunigten Abbau der **O.** [►Ozonloch] durch die Produktion von Fluor-Chlor-Kohlenwasserstoffen [FCKW] und anderer Gase, gelangt mehr UV-Licht bis zur Erdoberfläche, besonders über der Antarktis. Organismen können sich an die veränderten Strahlungsbedingungen nicht so schnell anpassen, weshalb vermehrt an Pflanzen Verbrennungen, beim Menschen Hautkrebs und Augenkrankheiten [Grauer Star] auftreten werden).

ozone layer [n] *envir. met.* ►hole in ozone layer.

ozone shield [n] *met.* ►ozone layer.

P

pace [n] [US] *constr.* ►landing.

4028 pace measure [n] *constr.* (Length of average human stride used in calculating step length and spacing of stepping stones); *s* **dimensión** [f] **en pasos** (Longitud media de un paso humano para calcular la altura de las contrahuellas y la distancia entre losas para pasos); *syn.* medida [f] de pasos; *f* **mesure** [f] **d'un pas** (Longueur moyenne d'un pas en vue de définir la hauteur d'une marche ou la distance entre deux dalles, p. ex. pour un pas japonais) ; *g* **Schrittmaß** [n] (Länge eines durchschnittlichen Schrittes für die Berechnung des Treppenstufensteigungsverhältnisses und der Schrittplattenabstände).

4029 packed fascine-work [n] *constr.* (WRP 1974; bioengineering term for the protection of slopes and banks prone to erosion whereby bundles of brushwood are laid crosswise in layers 20-30cm thick); *syn.* brush packing [n] [also US]; *s* **relleno** [m] **de fajinas** (Método de construcción biotécnica en el que se colocan capas de desbrozo de 20-30 cm en cruzado sobre los pies de terraplenes o pendientes para protegerlos contra la erosión); *f* **ouvrage** [m] **de stabilisation par amas de fascines, de terre et de pierres** (Technique de confortement de pied de talus ou d'affouillement de berge) ; *g* **Packfaschinat** [n] (Ingenieurbiologische Bauweise, bei der die 20-30 cm dicke, kreuzweise zueinander verlegte Schichten von Faschinenreisig in angerissene Hang- oder Böschungsfüße oder in größere Uferabrisse eingebracht werden; cf. DIN 19 657); *syn.* Raupackung [f], *f. S.* Rauhpackung [f].

pad [n], **adhesive** *bot.* ►adhesive disc/disk.

paddling pool [n] [UK] *recr.* ►toddlers wading pool [US].

pads [npl] [US] *bot.* ►self-clinging vine with adhesive pads [US].

pale fence [n] *constr.* ►picket fence.

paling [n] [UK] *constr.* ►picket fence.

paling fence [n] [US] *constr.* ►picket fence.

4030 palisade [n] *constr.* (Round-section stake or post which is embedded close to others in the ground, or in a concrete foundation, to form a barrier or screening fence); *syn.* timber pile [n] (RIW 1984, 153); *s* **estaca** [f] **de empalizada** (Poste redondo de madera, hormigón/concreto o plástico reciclado que se fija muy estrechamente junto a otros postes en el suelo o en un fundamento de hormigón para formar una barrera o una valla de protección visual); *syn.* estacada [f], estaca [f] de valla; *f* **palissade** [f] **(1)** (Clôture constituée d'une rangée de rondins plantés serrés dans le sol ou scellés dans une fondation en béton) ; *g* **Palisade** [f] (Runder Pfahl aus Holz, Beton oder rezykliertem Kunststoff,

der mit anderen Pfählen dicht nebeneinander in den Boden oder in Betonfundamente eingelassen wird).

4031 palisade construction [n] *constr.* (Bioengineering method using palisades to obstruct water-free V-shaped channels by installing thin stem cuttings or ►live pickets thereby slowing down the velocity flow of surface water runoff, and allowing the deposition of sediment. The palisade is secured by a horizontally-laid live picket of appropriate thickness across the upper third of the downstream side of the palisade); *s* **construcción [f] de empalizada** (Método de construcción biotécnica usado para obstruir canales secos de forma de V instalando ►varetas vivas o ►estacas muertas para reducir la velocidad de flujo de escorrentía superficial y con ello permitir la sedimentación. La empalizada se asegura con un palo redondo de grosor adecuado que se coloca transversalmente en el tercio superior del lado orientado canal abajo); *f* **mise [f] en œuvre de palissade (de réduction du ruissellement)** (Technique de génie biologique utilisant une palissade transversale constituée de ►plançons ou de ►pieux en bois mort plantés dans un fossé sans plafond afin de réguler le débit solide, de réduire la vitesse et de dissiper l'énergie des eaux de surface ; la palissade est maintenue transversalement dans son tiers supérieur par un rondin) ; *g* **Palisadenbau [m]** (...bauten [pl]; ingenieurbiologische Baumaßnahme zur Sperrung von [nicht Wasser führenden] V-Runsen durch ►Setzstangen oder ►tote Pflöcke zur Abbremsung der Fließgeschwindigkeit des Oberflächenwassers und zur Ablagerung mitgeführter Bodenteile. Die Palisade wird durch ein angemessen dickes Rundholz, das auf der Talseite im oberen Drittel quer gelegt wird, gesichert; cf. DIN 18 918).

4032 palisade fence [n] *constr.* (Free-standing wall or screen constructed by installing a row of round stakes close together); *s* **valla [f] de empalizada [f]** (Cerca o panel construido con estacas unidas entre sí sin dejar hueco); *syn.* estacada [f]; *f* **clôture [f] jointive en rondins** (Pieux plantés côte à côte pour former un mur de protection) ; *syn.* palissade [f] en/de rondins ; *g* **Palisadenzaun [m]** (Aus dicht nebeneinander gesetzten Rundpfählen gebildete Wand. Bauweise, die schon bei den Römern zum Bau der Befestigungsanlage des Limes verwendet wurde); *syn.* Palisadenwand [f].

4033 palmette espalier [n] *hort.* (►Espaliered fruit tree which has one vertical trunk with parallel or V-shaped laterals like the ribs of a fan; ►double-U espalier); *syn.* fan-trained fruit tree [n]; *s* **palmeta [f] regular (de ramas horizontales/oblicuas)** (En fruticultura, ►frutal en espaldera con un tronco vertical con la horquilla a unos 40 cm del suelo y con ramificación horizontal o en forma de V; ►palmeta en U doble); *f* **palmette [f] à branches horizontales/obliques** (Forme palissée d'un arbre fruitier de forme plate composé d'un axe vertical et d'une charpente formée d'une ou plusieurs séries de deux rameaux latéraux dont les plus bas sont taillés à 20-30 cm du sol ; on distingue différentes formes palissées, des étages de branches charpentières horizontales [palmettes à branches horizontales ou palmette Cossonnet], obliques [palmette oblique] ou en forme de U [palmette en U simple ou en U double] ; cf. DAV 1984, 348 ; ►palmette Verrier, ►arbre fruitier taillé en espalier ; cf. DAV 1984, 348 ; ADT 1988, 43-53) ; *g* **Spalier [n] mit waagerechten/schrägen Ästen** (*Obstbau* ►Formobstbaum mit einer Stammhöhe von 40 cm und möglichst gegenständiger, horizontaler oder V-förmiger Astverzweigung; ►Verrierpalmette).

4034 paludification [n] *pedol. phyt.* (WET 1993, 372s; process of permanent ►waterlogging of soils which encourages ►swamp vegetation); *s* **empantanamiento [m]** (Proceso de ►encharcamiento de suelos con la consecuencia de la introducción de ►helophytia); *f* **paludification [f]** (Processus d'engorgement durable des sols et évolution vers un milieu à ►végétation

des marais ; ►processus d'engorgement du sol) ; *g* **Versumpfung [f]** (Prozess der Dauervernässung von Böden mit der sich dann einstellenden ►Sumpfvegetation; ►Bodenvernässung).

4035 palustrine area [n] *geo.* (**1.** All nontidal wetlands dominated by trees, shrubs, persistent emergents, emergent mosses or lichens. **2.** All such wetlands that occur in tidal areas where salinity due to ocean-derived salts is below 0.5%; ►marsh, ►swamp 1, ►swamp forest); *syn.* palustrine region [n]; *s* **zona [f] pantanosa** (Terreno cenagoso ocupado por una capa de agua estancada, generalmente superficial, e invadido por vegetación acuática. Se origina en suelos impermeables y sin pendientes o en depresiones con un nivel alto de la capa freática; MEX 1983; ►bosque pantanoso, ►marisma, ►pantano 2); *f 1* **région [f] marécageuse** (Zone périodiquement inondée et recouverte par une végétation herbacée hydrophile, arbustive ou arborescente ; ►forêt humide, ►forêt marécageuse, ►forêt tourbeuse, ►marécage) ; *f 2* **région [f] palustre** (Zone inondée en permanence en milieu littoral, lacustre ou continental ; ►marais ; *syn.* zone [f] de marais, milieu [m] palustre ; *g* **Sumpfgebiet [n] (2)** (Ständig von Wasser durchtränkter, zeitweilig oder während des ganzen Jahres unter Wasser stehender, großflächiger Bereich im Litoral, in Tiefebenen, in abflusslosen Plateaus oder Senken mit undurchlässigem Untergrund, mit angepassten Pflanzengesellschaften bestockt; ►Sumpf 2, ►Sumpf[land], ►Sumpfwald).

palustrine region [n] *geo.* ►palustrine area.

pampa [n] *geo. phyt.* ►steppe.

4036 pan [n] *pedol.* (Impermeable, compacted or indurated subsoil layer causing bad drainage by impeding the percolation of water; ►hardpan, ►ortstein); *s* **capa [f] de cementación** (KUB 1953, 304; horizonte impermeable del suelo que causa estancamiento de agua temporal o permanentemente; ►horizonte petrificado, ►ortstein); *f* **substratum [m] de la nappe perchée** (Partie supérieure de la couche imperméable responsable de la nappe perchée temporaire ou permanente ; ►alios durci, ►horizon de fragipan) ; *g* **Stauwassersohle [f]** (Verdichtete, wasserundurchlässige Schicht, die temporäres oder dauerndes Stauwasser erzeugt; ►Ortstein, ►Verdichtungshorizont); *syn.* Staunässesohle [f].

pan [n] **[UK], plough** *agr. pedol.* ►plow pan [US].

pan [n] **[US], tillage** *agr. pedol.* ►plow pan [US]/plough pan [UK].

panel construction [n] *urb.* ►prefabricated concrete panel construction.

panel fence [n]**, close board** *constr.* ►visual screen with vertical boards.

4037 panning [n] *pedol.* (Compacting of a layer of soil due to the action of water, mechanical pressure by normal tillage operations or chemical change; ►hardpan); *s* **endurecimiento [m] de un horizonte** (Generación de un ►horizonte petrificado por presión del arado o por lixiviación de sustancias en el suelo); *f* **induration [f] d'un horizon** (L'endurcissement dans les sols par concrétionnement et cimentation, p. ex. par le fer [►horizon d'accumulation durci]) ; *g* **Entstehung [f] eines Verdichtungshorizontes** (Verdichtung durch mechanischen Druck [Bodenbearbeitung] oder durch Verlagerung chemischer Substanzen; ►Verdichtungshorizont).

4038 panoramic view [n] *recr.* (An all-encompassing view from an elevated point; ►road with panoramic view); *s* **vista [f] panorámica** (Vista en todas las direcciones desde un lugar elevado; ►ruta panorámica); *f* **vue [f] panoramique** (Vue qui, à partir d'un point haut, embrasse un champ très large d'au moins 180° et permet de découvrir un vaste paysage ; ►route de vision panoramique) ; *g* **Panoramablick [m]** (Rundumaussicht von

P

einem erhöhten Punkt aus; ▶Panoramastraße); *syn.* Panorama [n], Rundblick [m].

4039 parabolic dune [n] *geo.* (Sand dune of crescentic outline with concave windward side and steeper convex leeward side; **p. d.s** are often covered with sparse vegetation, and are often found along coasts where strong onshore winds are supplied with abundant sand; cf. GFG 1997; ▶barkhan), *syn.* upsiloidal dune [n]; *s* **duna** [f] **parabólica** (Duna con planta en arco de círculo, con frente abrupto convexo en cuya dirección avanza la duna; DGA 1986; ▶barkhana); *f* **dune** [f] **parabolique** (Dune dissymétrique en forme de fer à cheval, à concavité au vent ; ▶barkhane) ; *g* **Parabeldüne** [f] (Dünenform mit konkaver Luv- und konvexer Leeseite; bei dieser in semiariden Gebieten und an Küsten mit starken Seewinden, die viel Sand vor sich hertreiben, weit verbreiteten Dünenform hinken die langgezogenen Sichelenden dem rascher wandernden Dünenmittelkörper hinterher; oft sehr locker mit Vegetation bedeckt; ▶Sicheldüne); *syn.* Bogendüne [f], Paraboldüne [f] (WAG 2001).

parallel parking [n] [UK] *trans. urb.* ▶in-line parking [US].

parallel parking layout [n] [UK] *plan. trans.* ▶in-line parking layout [US].

4040 parameter [n] *plan.* (Constant limiting factor which indicates maximum extent or scope, and is taken into account in planning or in giving expert opinions); *s* **parámetro** [m] (Elemento constante o cambiante utilizado como base de información para representar un conjunto de características); *f* **paramètre** [m] (Élément constant ou quantité variable caractéristique d'un ensemble de questions prise en compte lors de la définition d'un projet ou d'une expertise) ; *g* **Parameter** [m] (Charakteristische Konstante oder veränderliche Größe, die bei Planungsüberlegungen oder Gutachten als Informationseinheit berücksichtigt wird).

4041 paramo [n] *geo. phyt.* (Sparse, evergreen, treeless, tropical ▶alpine zone with permanent rainy and misty climatic conditions, which is characterized by grasses as well as rosette plants and, in South America, especially by stem-forming plants of the genus Espeletia and, in Africa, by treelike [US]/tree-like [UK] groundsel/ragwort *[Senecio]*, lupin[e]s *[Lupinus]*, lobelias *[Lobelia aberdarica]*, and everlasting flower *[Helichrysum formosissimum]*); *s 1* **páramo** [m] **(de montaña tropical)** (Término castellano que se utiliza para denominar en los Andes húmedos [desde Venezuela hasta el extremo norte del Perú] las elevadas altiplanicies del ▶área cacuminal intertropical por encima de los bosques tropicales, caracterizadas por un tipo especial de vegetación; son vicariantes de las «punas» de los Andes secos [del Perú a la Argentina y Chile]. Aunque los **pp.** representan un medio húmedo por sus frecuentes lluvias y nieves, su vegetación tiene características xerofíticas, debido a la altitud y a la acidez. La vegetación dominante es abierta y se compone de géneros comunes con Europa [como *Poa, Hordeum, Anemone, Draba, Alchemilla, Lupinus, Viola, Gentiana, Senecia*], pero en gran parte también de géneros especiales de saxifragáceas *[Escallonia]*, tropeoláceas *[Tropaeolum]*, leguminosas *[Krameria]*, entre muchos otros; a diferencia de en las punas faltan las cactáceas; cf. DB 1985); *s 2* **páramo** [m] (Llanura alta, rasa y desabrigada frente a los vientos, cuya vegetación muestra adaptaciones xéricas por la fuerte radiación, sequedad ambiental, etc. Está representada en el centro de España por los restos de la planicie sedimentaria correspondientes a los estratos calizos miocenos más resistentes a la erosión; DINA 1987); *syn.* alcarria [f]; *f* **páramo** [m] (Terme désignant l'▶étage alpin en climat tropical caractérisé par une formation végétale herbacée ouverte, sempervirente constituée en Amérique du Sud par les espèces du genre *Espeletia* et en Afrique par les espèces arbustives des Séneçons *[Senecio]*, Lupins *[Lupinus]*, Lobelias *[Lobelia aber-*

darica], Immortelles *[Helichrysum formosissimum]*) ; *g* **Páramo** [m] (Bezeichnung für eine lichte, immergrüne, baumlose, tropische ▶alpine Stufe mit dauernd regnerischen und nebligen Klimaverhältnissen, die neben Gräsern durch Rosettenpflanzen — in Südamerika speziell durch stammbildende Pflanzen der Gattung *Espeletia* und in Afrika durch baumförmige Greiskrautarten *[Senecio]*, Lupinen *[Lupinus]*, Lobelien *[Lobelia aberdarica]*, Strohblumen *[Helichrysum formosissimum]* — gekennzeichnet ist).

parapet [n] *arch.* ▶roof parapet.

4042 pararendzina [n] *pedol.* (In European soil taxonomy, A_h-C soil corresponding to the ▶ranker composed of sandy and loamy marl, rich in quartz and silica; in U.S. classification, a suborder of inceptisols [with shallow A horizon] or mollisols [with thick A horizon]); *s* **pararendsina** [f] (▶Rendsina formada sobre rocas dolomíticas, calcáreas o silícicas ricas en calcio, con microesqueleto predominantemente silícico o de silicatos; según la clasificación norteamericana un suborden de los inceptisoles [con horizonte A plano] o de los mollisoles [con horizonte A grueso]; ▶ranker); *syn.* rendsina [f] mixta; *f* **pararendzine** [f] (▶Rendzine développée sur matériau dolomitique ou calcaire gréseux dont l'altération libère un important squelette sableux, ce qui diminue la proportion des complexes humiques stabilisés par les carbonates dans l'horizon A_1 ; DIS 1986 ; ▶ranker) ; *g* **Pararendzina** [f] (Dem ▶Ranker 1 entsprechender AC-Boden aus quarz- und silikatreichem Sand- und Lehmmergel. Die **P.** entwickelte sich aus Löss, Geschiebemergel, carbonathaltigen Schottern, Sanden und Sandstein, durch Humusakkumulation und mäßige Carbonatverarmung. Im US-Bodenklassifizierungssystem werden sie bei gering mächtigem A-Horizont unter den *Inceptisolen*, bei mächtigerem A-Horizont unter den *Mollisolen* geführt; SS 1979).

parcel [n] (1) *urb.* ▶building plot.

4043 parcel [n] (2) *adm. leg. surv.* (Piece of property registered under a numbered ownership, which may have various uses; ▶building plot, ▶plot 1, ▶land register, ▶real property identification map [US]/land register [UK], ▶real property parcel [US]/plot [UK] 2, ▶section of a community area [US]/local district [UK]); *s* **parcela** [f] **de propiedad** (Porción de tierra registrada en el ▶catastro de bienes inmuebles con un número específico. Puede estar dedicada a diferentes usos; ▶catastro, ▶finage 2, ▶parcela, ▶solar); *syn.* lote [m] de propiedad; *f* **parcelle** [f] **(de propriété)** (1. Portion de terre d'un seul propriétaire, désignée par un numéro ou une lettre et constituant l'unité d'un ▶cadastre ; plusieurs parcelles forment le ▶parcellaire ; une ou plusieurs parcelles constituent le ▶fonds de terre au sens du cadastre. 2. En sylviculture, unité de gestion dont la surface oscille entre 10 et 15 ha ; cf. LA 1981, 812 ; ▶cadastre parcellaire, ▶finage, ▶parcelle [d'exploitation], ▶terrain à bâtir) ; *g* **Flurstück** [n] (Teil der Erdoberfläche, der von einem geschlossenen Linienzug begrenzt und im ▶Liegenschaftskataster unter einer besonderen Nummer [Flurstücksnummer] geführt wird. Ein **F.** kann mehrere Nutzungsarten enthalten. Mehrere **F.e** ergeben eine ▶Gemarkung 2. **F.e** bilden einzeln oder zu mehreren zusammengefasst ein ▶Grundstück im Sinne des ▶Grundbuches; ▶Baugrundstück, ▶Parzelle).

parcel [n] [US]**, real estate** *surv.* ▶real property parcel [US]/plot [UK] (2).

parcel [n] [US]**, vacant** *urb.* ▶vacant lot.

parch blight [n] *agr. bot. for. hort.* ▶frost-desiccation.

4044 parching [n] *pedol.* (Deep drying out of soil in arid and semiarid areas [US]/semi-arid areas [UK]); *s* **aridecimiento** [m] (Desecación completa de los suelos en regiones áridas y semiáridas); *f* **dessiccation** [f] (1. Perte d'eau dans les couches

profondes des sols sous les climats arides et semi-arides. **2.** Élimination de l'humidité contenue dans un corps gazeux ou solide) ; *g* **Austrocknung [f] (3)** (Tiefes Austrocknen des Bodens in ariden und semiariden Gebieten; austrocknen [vb]).

parcours(e) [n] [US] *recr.* ▶fitness trail.

4045 parent material [n] *pedol.* (In situ geological stratum from which the site soil develops; ▶bedrock [US]/bed-rock [UK], ▶solid rock); *syn.* parent rock [n]; *s* **roca [f] madre** (Roca de la que proceden los materiales de un suelo, de un sedimento, de una roca, etc. Se aplica también a aquellas rocas en las que se forman los hidrocarburos naturales; DINA 1987; ▶roca in situ, ▶roca dura); *syn.* material [m] parental, sustrato [m] geológico; *f* **roche-mère [f]** (Matériau à partir duquel ou dans lequel se forme le sol ; dans le profil, l'horizon R désigne la roche mère non altérée et l'horizon C la roche mère altérée ; les roches mères sont massives [granites, calcaires cristallins], consolidées [grès, schistes] ou meubles [alluvions, lœss, marnes, sables, etc.] ; DG 1984 ; ▶roche en place, ▶roche massive) ; *syn.* matériau [m] minéral, substrat [m] géologique, substratum [m] géologique, matériau [m] originel, matériau [m] parental ; *g* **Ausgangsgestein [n]** (An der Erdoberfläche anstehendes Gestein oder Lockersediment, auf dem durch Verwitterung der Bodenbildungsprozess stattfindet; ▶anstehendes Gestein, ▶Festgestein); *syn.* Ausgangsmaterial [n], Muttergestein [n].

parent plant [n] [UK] *hort.* ▶stock plant [US].

parent rock [n] *pedol.* ▶parent material.

4046 Paris Convention [n] **of 1972** *conserv'hist. leg. pol.* (Convention concerning the protection of world cultural and natural heritage; adopted 16.11.1972); *s* **Convención [f] de París de 1972** (Convención sobre la protección del patrimonio mundial, cultural y natural; aprobada 16 de noviembre de 1972); *f* **Convention [f] de Paris de 1972** (Convention concernant la protection du patrimoine mondial culturel et naturel ; adoptée le 16.11.1972) ; *g* **Pariser Übereinkommen [n] von 1972** (Übereinkommen zum Schutze des Kultur- und Naturerbes der Welt vom 16.11.1972, veröff. im BGBl. 1977 II 215 ff, mit dem Ziel, Teile des Kultur- und Naturerbes von außergewöhnlicher Bedeutung, die Bestandteil des Welterbes der Menschheit sind, zu schützen und zu erhalten); *syn.* Welterbekonvention [f], Welterbeübereinkommen [n].

4047 Paris Convention [n] **of 2003** *conserv'hist. leg. pol.* (**Convention for the Safeguarding of the Intangible Cultural Heritage**, adopted at Paris, 17. October 2003; "intangible cultural heritage" means the practices, representations, expressions, knowledge, skills—as well as the instruments, objects, artefacts and cultural spaces associated therewith—that communities, groups and, in some cases, individuals recognize as part of their cultural heritage. This intangible cultural heritage, transmitted from generation to generation, is constantly recreated by communities and groups in response to their environment, their interaction with nature and their history, and provides them with a sense of identity and continuity, thus promoting respect for cultural diversity and human creativity; art. 2); *s* **Convención [f] de París de 2003** (**Convención de la UNESCO para la salvaguardia del patrimonio cultural inmaterial**, que fue acordada en París el 17 de octubre de 2003. Se entiende por «patrimonio cultural inmaterial» los usos, representaciones, expresiones, conocimientos y técnicas —junto con los instrumentos, objetos, artefactos y espacios culturales que les son inherentes— que las comunidades, los grupos y en algunos casos los individuos reconozcan como parte integrante de su patrimonio cultural. Este patrimonio cultural inmaterial, que se transmite de generación en generación, es recreado constantemente por las comunidades y grupos en función de su entorno, su interacción con la naturaleza

y su historia, infundiéndoles un sentimiento de identidad y continuidad y contribuyendo así a promover el respeto de la diversidad cultural y la creatividad humana; art. 2); *f* **Convention [f] de Paris de 2003** (**Convention pour la sauvegarde du patrimoine culturel immatériel**, adoptée le 17.10.2003 ; on entend par « patrimoine culturel immatériel » les pratiques, représentations, expressions, connaissances et savoir-faire — ainsi que les instruments, objets, artefacts et espaces culturels qui leur sont associés — que les communautés, les groupes et, le cas échéant, les individus reconnaissent comme faisant partie de leur patrimoine culturel ; art. 2) ; *g* **Pariser Übereinkommen [n] von 2003** (**UNESCO-Übereinkommen zur Bewahrung des immateriellen Kulturerbes** vom 17.10.2003. Unter immateriellem Kulturerbe sind Praktiken, Darbietungen, Ausdrucksformen, Kenntnisse und Fähigkeiten sowie die damit verbundenen Instrumente, Objekte, Artefakte und Kulturräume zu verstehen, die Gemeinschaften, Gruppen und gegebenenfalls Individuen als Bestandteil ihres Kulturerbes ansehen. Dieses immaterielle Kulturerbe, das von einer Generation an die nächste weitergegeben wird, wird von Gemeinschaften und Gruppen in Auseinandersetzung mit ihrer Umwelt, ihrer Interaktion mit der Natur und ihrer Geschichte fortwährend neu geschaffen und vermittelt ihnen ein Gefühl von Identität und Kontinuität; cf. Art. 2).

parish [n] [UK] *adm. agr. urb.* ▶section of a community area [US].

4048 park [n] **(1)** *landsc. recr. urb.* (**1. In U.S.,** a natural or man-made area with trees and lawns/meadows, which is preserved and managed for its scenic and recreational benefits; **in Europe,** man-made area of open space characterized by tree cover and extensive areas of lawn or meadow, which has been designated for preservation of natural features and/or for public or private recreational use. **2. In U.S.,** **park** has also the meaning of ▶town forest [UK] ; ▶landscape garden, #2, ▶leisure park, ▶neighborhood park [US]/district park [UK], ▶public park, ▶regional park [US]/country park [UK] (2), ▶urban park); *s* **parque [m]** (Espacio verde grande situado en el interior de una ciudad, caracterizado por la presencia de cantidad considerable de árboles grandes y de superficies de césped o hierba; dependiendo de su carácter específico puede contener plantaciones de vivaces, instalaciones de agua como fuentes, estanques, etc. y edificios; ▶jardín de barrio, ▶jardín paisajístico, ▶parque natural, ▶parque público, ▶parque recreativo regional, ▶parque urbano); *f* **parc [m]** (Espaces libres aménagés, caractérisés par un peuplement arborescent d'âge adulte, par des aires engazonnées ou des prairies naturelles, parcourus par des chemins de promenade et souvent agrémentés de plantations de plantes vivaces, de pièces d'eau et de bâtiments divers [fabriques] ; l'origine de ces espaces remonte en France au XVIème siècle avec l'aménagement de réserves de chasse autour des grands demeures seigneuriales et c'est au XVIIIème avec les jardins classiques en France et principalement au XIXème siècle avec les jardins paysagers et romantiques en provenance d'Angleterre que le parc, comme réaction à la croissance anarchique des villes, la dégradation des conditions de vie de la population et la pollution engendrées par la révolution industrielle, entraîne en Europe une véritable métamorphose des espaces extérieurs urbains ; ▶parc central, ▶parc de loisirs, ▶parc de quartier, ▶parc de voisinage, ▶parc naturel, ▶parc naturel régional, ▶parc naturel urbain, ▶parc paysager, ▶parc public, ▶parc urbain) ; *g* **Park [m]** (Parks [pl], *seltener* Parke [pl], *CH* meist Pärke [pl]; meist im Siedlungsbereich liegender Freiraum, der durch einen starken Baumbestand geprägt und mit weiten Rasen oder Wiesenflächen ausgestattet ist; weitere Elemente sind häufig Staudenpflanzungen, Wasseranlagen und Gebäude. Eine genaue Definition ist angesichts der vielfältigen Erscheinungsformen heutiger Parkanlagen äußerst

P

schwierig, ganz abgesehen davon, dass in den USA der Begriff ‚Park' weiter gefasst ist. In Europa entstand der Begriff aus den klassischen Landschaftsgärten des 18. und 19. Jhs., die damals zuerst in England geschaffen und als eine „scheinbar heile Welt Arkadiens, als das verlorene Paradies" in einer zunehmend vom Menschen manipulierten und naturentfremdeten Welt angesehen wurden; cf. WOR 1976. Während früher der Park ein bürgerlicher Spazierpark war, ist zumindest heute der öffentliche P. ein Ort gesellschaftlicher Freiheit, der die Aneignung dieses Grünraumes als emanzipatorischen Prozess ermöglicht; ▶Freizeitpark, ▶Landschaftspark, ▶Naturpark, ▶Stadtpark, ▶Stadtteilpark, ▶Volkspark); *syn.* Parkanlage [f].

park [n] (2) *recr.* ▶amusement park; *trans. urb.* ▶car park (1); *trans. urb.* ▶communal car park; *recr.* ▶country park (1), ▶game park (2), ▶health resort park; *plan. urb.* ▶industrial park [US]/industrial estate [UK]; *urb.* ▶inner city park [US]/public garden [UK]; *trans. urb.* ▶multideck car park; *conserv.* ▶national park; *landsc. recr. urb.* ▶riverside park, ▶suburban park.

park [n] [UK], **business** *plan. urb.* ▶industrial park [US]/industrial estate [UK].

park [n] [UK], **country** *conserv. land'man. recr.* ▶regional park [US].

park [n] [US], **decked car** *trans. urb.* ▶multideck car park [US]/multi-storey car park [UK].

park [n] [UK], **district** *recr. urb.* ▶neighborhood park [US].

park [n] [US], **dog** *landsc. urb.* ▶pet exercise area [US]/dogs' loo [n] [UK].

park [n] [US], **downtown** *urb.* ▶inner city park [US]/public garden [UK].

park [n] (1), **game** *conserv. game'man. hunt. leg. obs.* ▶game reserve.

park [n] [UK], **holiday** *recr.* ▶park-like vacation development [US].

park [n], **landscape** *gard'hist.* ▶landscape garden.

park [n] [UK], **local** *urb.* ▶proximity green space.

park [n], **metropolitan** *recr. urb.* ▶urban park (2).

park [n] [UK], **multi-storey car** *trans. urb.* ▶multideck car park [US].

park [n] [US], **multi-story car** *trans. urb.* ▶multideck car park [US].

park [n], **naturalistic** *gard'hist.* ▶landscape garden.

park [n], **overflow car** *plan.* ▶overflow parking area.

park [n] [UK], **pastoral** *gard'hist.* ▶landscape garden, #2.

park [n], **pleasure** *recr.* ▶amusement park.

park [n], **recreation** *plan. recr.* ▶leisure park.

park [n] [US], **stream valley** *landsc. recr. urb.* ▶riverside park.

park [n] [UK], **underground car** *urb.* ▶underground parking garage [US].

park [n], **zoological** *recr.* ▶zoological garden.

parkade [n] [West CDN] *trans. urb.* ▶multideck car park [US].

4049 park-and-ride system [n] *trans. urb.* (Procedure developed in the U.S. to relieve the congestion of car traffic in inner cities. The system, devised especially for ▶ommuters, provides for the parking of cars in fringe parking lots or structures on the periphery of a city from which point the commuter may continue his journey using public transport; ▶car pool); *s* **sistema** [m] «**park and ride**» (Denominación anglosajona de un sistema para evitar la congestión del interior de las grandes ciudades por los vehículos privados. Consiste en ofertar, fuera del núcleo urbano en cuestión, aparcamientos bien comunicados con el centro por medio de transporte público; cf. PARRA 1984; ▶comunidad de conmutadores, ▶conmutador, -a); *f* **Parc** [m] **Relais (P+R)** (Modèle de circulation mis au point aux États-Unis afin de décongestionner les grandes agglomérations en aménageant à la périphérie urbaine des grands parcs de stationnement à proximité d'une station de transports collectifs sur lesquels les automobilistes laissent leurs véhicules et se dirigent vers le centre ville au moyen des transports en commun ; cette dénomination a été retenue pour que ses initiales coïncident avec celles du concept britannique de « Park and Ride » ou « P R » d'Europe du Nord ; ▶covoiturage, ▶migrant alternant) ; *syn.* (système de) park and ride [m], Parc [m] de Stationnement Régional (PSR), Parc [m] d'Intérêt Régional (PIR), parc [m] de liaison, parc [m] de dissuasion, parc [m] de rabattement ; *g* **Park-and-Ride(-System)** [n] (In den USA entwickeltes Verfahren zur Entlastung der Innenstädte vom Kraftfahrzeugverkehr. Insbesondere Berufspendler sollen ihre Fahrzeuge auf Parkplätzen am Stadtrand abstellen und von dort aus ihre Fahrt [in die Innenstadt] mit öffentlichen Verkehrsmitteln fortsetzen; ▶Fahrgemeinschaft; ▶Pendler).

4050 park bench [n] *constr. landsc.* (Bench installed in a public or private park); *s* **banco** [m] **de parque** (Banco de asiento instalado en un parque público o privado); *f* **banc** [m] **de parc** (Banc placé dans un parc public ou privé) ; *g* **Parkbank** [f] (In einem öffentlichen oder privaten Park aufgestellte Bank).

park conservancy programme [n] [UK] *conserv'hist. gard.* ▶park management program [US]/park management programme [UK].

4051 park department [n] [US]/**parks department** [n] [UK] *adm. urb.* (Public authority/agency responsible for recreational green open spaces, and sometimes for street trees and planting of other public facilities; in U.S., the highway department is often responsible for all road right-of-way work); *syn.* parks and cemetery department [n] [UK], parks and recreation department [n] [US]; *s* **administración** [f] **de parques y jardines** (Autoridad pública responsable de la planificación y el mantenimiento de zonas verdes y de la vegetación existente en la ciudad. Dependiendo de su gama de responsabilidades tienen diferentes denominaciones); *f* **service** [m] **des espaces verts** (En F. les services municipaux ont des dénominations les plus différentes : service [m] des espaces verts, service [m] d'espace vert et de l'environnement, service [m] d'écologie et d'espaces verts ; [GEN 1982-III] ont opté pour la dénomination : service [m] municipal d'espaces verts) ; *syn.* service [m] de jardins, service [mpl] des parcs et promenades [aussi CH] ; *g* **Grünflächenamt** [n] (Je nach Aufgabenzuordnung unterscheiden sich die Bezeichnungen der Ämter, die für die Grünflächenverwaltung zuständig sind: Gartenamt [n], Gartenbauamt [n], Garten- und Forstamt [n], Garten- und Friedhofsamt [n], Amt [n] für Grünflächen und Umweltschutz, Fachamt [n] für Stadtgrün und Erholung etc. Seit Ende der 1990er-Jahre werden Voraussetzungen geschaffen, die **G.ämter** zu betriebswirtschaftlich orientierten Betriebsbehörden zu machen).

parked traffic [n] *plan. trans.* ▶stationary traffic.

parked vehicle [n] *trans.* ▶space covered by a parked vehicle.

4052 parked vehicles [npl] *trans.* (Vehicles parked for short or long periods of time along roadsides or in parking lots or areas [US]/car parks [UK]; ▶stationary traffic); *syn.* stored automobiles [npl] [also US]; *s* **vehículos** [mpl] **aparcados** [Es] (Vehículos parados al borde de la calle o en otras superficies de aparcamientos durante periodos/períodos más o menos largos; ▶tráfico quieto); *syn.* vehículos [mpl] estacionados; *f* **véhicules**

[mpl] en stationnement (Véhicules laissés sur une aire de stationnement pendant une durée indéterminée [courte ou longue] le long de la voirie publique ou sur un parc de stationnement ; ►circulation intermittente) ; *g* **parkender Verkehr [m]** (Kurz- oder langfristig abgestellte Fahrzeuge innerhalb oder außerhalb des Straßenraumes; ►ruhender Verkehr).

4053 park improvement [n] *plan. urb.* (Amelioration, re-design, or enlargement of an existing park; e.g. layout of paths); *s* **revalorización [f] de un parque** (Mejora, reestructuración o ampliación de un parque existente); *f* **développement [m] d'un parc** (Amélioration de la qualité d'un parc existant : construction de chemins, d'allées, plantations nouvelles, etc.) ; *g* **Ausbau [m, o. Pl.] eines Parkes** (Verbesserung, Umbau oder Erweiterung einer bestehenden Parkanlage).

parking [n] [UK]**, nose-to-kerb** *plan. trans.* ►ninety-degree parking layout.

parking [n] [US]**, residents' underground** *urb.* ►underground parking garage [US]/underground car park [UK].

4054 parking access [n] *trans.* (Way to reach a car park or parking lot [US]; ►parking access road); *s* **carril [m] de acceso a estacionamiento** (Vía para acceder a aparcamiento); ►calle de acceso a estacionamiento); *f* **accès [m] à l'aire de stationnement** (Possibilité d'accéder à une aire de stationnement ; ►voie d'accès à l'aire de stationnement) ; *g* **Parkplatzzufahrt [f]** (**1.** Fahrmöglichkeit zu einem Parkplatz. **2.** Fahrweg, auf dem man zum Parkplatz gelangt; ►Parkplatzzufahrtsstraße).

4055 parking access road [n] *trans.* (Drive leading to a car park); *syn.* car park approach [n]; *s* **calle [m] de acceso a estacionamiento** (Calle que lleva hasta un aparcamiento); *f* **voie [f] d'accès à l'aire de stationnement** (Voie de desserte qui permet d'accéder à une aire de stationnement) ; *g* **Parkplatzzufahrtsstraße [f]** (Fahrweg, der zu einem Parkplatz führt).

parking area [n] *trans. urb.* ►car park (1); *plan.* ►overflow parking area; *trans. urb.* ►parking lot [US].

4056 parking bay [n] *trans. urb.* (HUL 1978, 234; parking module consisting of one or two rows of parking and the access aisle from which motor vehicle enter and leave the spaces; IBDD 1981; ►parking lane); *s* **área [f] de aparcamiento** (Pequeña superficie de estacionamiento con una o dos filas de plazas que tiene un acceso de entrada y otro de salida a la calle; no se debe confundir con el ►carril de aparcamiento); *syn.* área [f] de estacionamiento; *f* **baie [f] de stationnement** (Parc de stationnement réduit constitué de une ou deux rangées de stationnement possédant une desserte sur la voirie principale ; ne pas confondre avec le ►parc à voitures longitudinal ; ne pas confondre avec voie de stationnement) ; *syn.* créneau [m] de stationnement) ; *g* **Parkbucht [f]** (Kleine Parkfläche mit einer oder zwei Parkstandreihen, die eine Zufahrt und Ausfahrt zur Straße hat; nicht zu verwechseln mit ►Parkspur); *syn.* Parktasche [f].

parking bay [n] [UK] *trans. urb.* ►parking space.

4057 parking deck [n] *trans. urb.* (Area of car parking on the roof of a single-stor[e]y building or within a ►multideck car park [US]/multi-storey car park [UK]; ►covered parking deck); *s 1* **cubierta [f] de aparcamiento** (Zona de aparcamiento en el tejado de un edificio de una planta o de un ►edificio de aparcamiento 1); *syn.* cubierta [f] de estacionamiento; *s 2* **piso [m] de aparcamiento** (Planta en un edificio de aparcamiento); *syn.* piso [m] de estacionamiento; *f* **plate-forme [f] de stationnement** (Aire de stationnement constituant un niveau, située sur une terrasse d'un immeuble ou dans un parc de stationnement couvert ; ►parc de stationnement couvert) ; *syn.* plate-forme [f] de parking, dalle-parking [f], étage [m] de parking ; *g* **Parkdeck [n]** (Stockwerk als Parkplatzfläche auf einem eingeschossigen Bauwerk oder in einem ►Parkhaus 1); *syn.* Parkplatte [f], Parkpalette [f].

4058 parking garage [n] *trans. urb.* (Aboveground multi-level indoor parking building; ►multideck car park [US]/multi-storey car park [UK]); *syn.* parkade [n] [West CDN]; *s* **edificio [m] de aparcamiento (2)** (Garaje de varias plantas situadas sobre nivel y constructivamente cerradas lateralmente, al contrario que en el ►edificio de aparcamiento 1, ►garaje subterráneo, ►aparcamiento de residentes); *syn.* edificio [m] de estacionamiento [AL] (2); *f* **parc [m] de stationnement couvert** (Terme générique désignant les parcs installés dans un bâtiment réservé ou non à ce seul usage se trouvant dans un immeuble bâti en superstructure ou en infrastructure ; ►garage souterrain, ►parc de stationnement à étages, ►parking souterrain résidentiel) ; *syn.* parking [m] couvert *(par opposition au parking ouvert)*, parking [m] à différents niveaux ; *g* **Parkhaus [n] (2)** (Mehrgeschossiges, seitlich baulich geschlossenes Garagenhaus mit oberirdischen und/oder unterirdischen Parkdecks; ►Anwohnertiefgarage, ►Parkhaus 1, ►Tiefgarage); *syn.* Parkgarage [f].

4059 parking lane [n] *trans.* (Area reserved for parking parallel and immediately adjacent to a lane of traffic which enables cars to park in a longitudinal parking row; ►in-line parking layout [US]/parallel parking layout [UK]); *syn.* in-line parking row [n] [also US], parallel parking lane [n]; *s* **carril [m] de aparcamiento** (Zona de aparcamiento al borde de la calle de colocación en hilera; no se debe confundir con el ►carril auxiliar; ►disposición [de aparcamiento] en hilera); *f* **parc [m] à voitures longitudinal** (Emplacement de stationnement situé le long de la voirie dans le cas du ►rangement longitudinal des véhicules) ; *syn.* voie [f] de stationnement, bande [f] de stationnement (NEU 1983, 365) ; *g* **Parkspur [f]** (Parkfläche zur ►Längsaufstellung im Straßenraum unmittelbar neben den Fahrspuren; *nicht zu verwechseln mit* Standspur [►Standstreifen]); *syn.* Längsparkstreifen [m].

parking lane [n]**, parallel** *trans.* ►parking lane.

4060 parking layout [n] *plan. trans.* (Arrangement of parking spaces within a car park; *specific terms* ►angle parking layout, ►in-line parking layout [US]/parallel parking layout [UK], ►ninety-degree parking layout); *s* **disposición [f] de plazas de aparcamiento** (Orden de las plazas individuales para los vehícuos en un estacionamiento; *términos específicos* ►disposición [de aparcamiento] en diagonal, ►disposición [de aparcamiento] en hilera, ►disposición [de aparcamiento] en perpendicular); *syn.* orden [f] de lotes de aparcamiento; *f* **disposition [f] des emplacements de stationnement** (Disposition des véhicules sur une aire de stationnement. On parle du ►rangement longitudinal, du ►stationnement en épi, du ►rangement perpendiculaire) ; *g* **Anordnung [f] der Parkstände** (Anordnung der Stellflächen auf einem Parkplatz; *UBe* ►Längsaufstellung, ►Schrägaufstellung, ►Senkrechtaufstellung); *syn.* Parkstandanordnung [f], Aufstellungsrichtung [f] der Fahrzeuge auf Parkständen.

parking layout [n] [US]**, echelon** *plan. trans.* ►angle-parking layout.

parking layout [n] [UK]**, parallel** *plan. trans.* ►in-line parking layout [US].

parking layout [n] [US]**, perpendicular** *plan. trans.* ►ninety-degree parking layout.

parking layout [n]**, right angle** *plan. trans.* ►ninety-degree parking layout.

4061 parking lot [n] [US] *trans. urb.* (An off-street area, usually surfaced and improved, for the temporary storage of motor vehicles, usually of smaller size than a ►car park 1); *syn.* parking area [n] [UK] (2); *s* **aparcamiento [m] (2)** (Superficie pública o privada generalmente asfaltada, situada fuera de la calle, que se

utiliza para estacionar vehículos; el término inglés «parking lot» se utiliza para aparcamientos de menor tamaño que el «car park»; ▶aparcamiento 1); *syn.* estacionamiento [m] (3) [AL], aparcadero [m] (3) [CA], parqueo [m] (3) [CAR]; *f* **parc [m] de stationnement hors de la voie publique** (Aire stabilisée de stationnement aménagée pour plusieurs véhicules hors de la voirie publique, par opposition au stationnement sur rue ; ▶parc de stationnement) ; *g* **Parkplatz [m] (2)** (Befestigte Fläche, die außerhalb des Straßenraumes dem Abstellen von Fahrzeugen dient; ▶Parkplatz 1).

parking lot lawn [n] [US] *constr.* ▶car park lawn.

4062 parking maneuver space [n] [US]/**parking manoeuvre space** [n] [UK] *trans. plan.* (Movement space for ingressing and egressing vehicles to and from parking; ▶row of parking); *s* **espacio [m] de maniobra** (Superficie junto a ▶banda de aparcamiento necesaria para que los vehículos puedan aparcar y desaparcar); *f* **bande [f] de manœuvre** (Espace nécessaire à la circulation permettant l'accès d'un véhicule à une ▶bande de stationnement [longitudinal]) ; *syn.* bande [f] d'accès, bande [f] de circulation, zone [f] de dégagement ; *g* **Manövrierstreifen [m]** (Für das Ein- und Ausparken in ▶Parkstreifen notwendige Fläche).

parking place [n] [US] *trans. urb.* ▶parking space.

parking row [n] *trans.* ▶row of parking, ▶ninety-degree parking row.

parking row [n] [US]**, in-line** *trans.* ▶parking lane.

parking row [n]**, perpendicular** *trans.* ▶ninety-degree parking row.

4063 parking space [n] *trans. urb.* (**1.** Term used for a specific area delineated for the parking of an individual motor vehicle comprising the ▶space covered by a parked vehicle and the required clearance to adjacent spaces including space for door opening and demarcation lines. P. s.s are usually identified by linear road markings, by a different surface material or by paving bands. **2.** German planning terminology distinguishes between public and private parking spaces, whereby a permit-holder parking stall in a public traffic area can be used exclusively for parking a private vehicle with a parking permit; in new development the owner is required to provide car parking space within his own property; ▶car park 1, ▶carport [US]/car port [UK], ▶covered parking space, ▶communal car park, ▶parking lot [US]); *syn.* parking place [n] [US], parking bay [n] [UK] (HUL 1978, 234), parking stall [n] [US] (IBDD 1981: 25, 182); *s* **aparcamiento [m] (3)** (**1.** Término genérico para un lugar para dejar un vehículo aparcado/estacionado/parqueado; *término de planificación* plaza de aparcamiento/estacionamiento. **2.** Área necesaria para aparcar un vehículo que se calcula sumando la ▶superficie ocupada por un vehículo a los espacios intermedios entre vehículos y, en su caso, a las bandas de separación o protección; ▶aparcamiento 1, ▶aparcamiento 2, ▶aparcamiento de comunidad de vecinos, ▶garaje sin puertas); *syn.* aparcadero [m] (2) [CA], estacionamiento [m] (2) [AL], parqueo [m] (2) [CAR]; *f* **emplacement [m] de stationnement** (Installation propre à assurer le stationnement d'un véhicule ; *terme spécifique* parc à voiture ; ▶abri de voiture, ▶abri de voiture fermé, ▶parc de stationnement, ▶parc de stationnement collectif, ▶parc de stationnement hors de la voie publique, ▶surface occupée par un véhicule) ; *syn.* place [f] de parking, place [f] de stationnement, parc [m] de stationnement ; aire [f] de parcage [CH, B] ; *g* **Parkplatz [m] (3)** (**1.** Die zum Aufstellen eines einzelnen Fahrzeuges bestimmte und markierte Fläche, die sich aus der ▶Stellfläche und den erforderlichen Zwischenräumen zu benachbarten Parkständen resp. Schutz- oder Trennstreifen zusammensetzt. Die einzelnen Parkstände können durch Markierungen,

unterschiedliche Beläge oder durch Pflastersteinstreifen gegeneinander abgegrenzt werden. **2.** *Allgemeinsprachlicher Begriff* einzelner Platz zum Parken eines Fahrzeuges; *planerischer Begriff für den öffentlichen Raum* **Parkstand [m]**; *für private Flächen oder mit Berechtigungsschein privat nutzbarer Platz im Straßenraum* **Stellplatz [m]**; ▶Carport, ▶Einstellplatz, ▶Gemeinschaftsstellplatz, ▶Parkplatz 1, ▶Parkplatz 2); *syn.* Abstellplatz [m].

4064 parking space requirement [n] *plan. trans.* (Needed space for parking in a defined area); *s* **demanda [f] de aparcamientos** (Superficie necesitada para estacionar vehículos en una zona dada); *syn.* necesidad [f] de aparcamientos; *f* **besoins [mpl] en matière de stationnement** (Surface des aires de stationnement nécessaires dans une zone urbaine en fonction des exigences de desserte des différentes occupations des sols [habitation collective, habitation individuelle, locaux commerciaux, bâtiments publics, etc.]) ; *g* **Parkraumbedarf [m]** (...bedarfe [pl]; benötigte Flächen für den ruhenden Verkehr in einem Untersuchungsgebiet; der Bedarf ist abhängig von verschiedenen Nachfragegruppen wie Anwohner, Besucher, Berufstätige, Inder-Ausbildung-Stehende, Einkaufende, Lieferanten und Gewerbetreibende; für die Ermittlung des Bedarfes dient das Regelwerk *Empfehlungen für Anlagen des ruhenden Verkehrs [EAR 2005]* der Forschungsgesellschaft für Straßen- und Verkehrswesen).

parking spaces [n] *plan. trans.* ▶total parking spaces.

4065 parkland [n] *geo. phyt.* (Term in vegetation geography describing an edaphic or topographically influenced mosaic of large groups of trees or woodland islands in an open grass-covered landscape; e.g. palm tree savanna[h], termite savanna[h], floodplain savanna[h], Entre Ríos, Argentina, Pantanal do Mato Grosso, Brazil; ▶flood savanna[h], ▶savanna[h]); *syn.* park savanna(h) [n]; *s* **sabana [f] arbolada** (Término fitogeográfico que determina un paisaje de gramíneas y cuperáceas, salpicado con bosquetes o grupos de árboles, p. ej. sabana de palmas, **s.** de acacias altas, **s.** de termitas, ▶savana de inundación como la de Entre Ríos en la Argentina y el Pantanal en el Mato Grosso del Brasil; ▶sabana); *f* **savane-parc [f]** (Terme biogéographique désignant une formation végétale herbacée ouverte [▶savane] qui, sous l'influence des facteurs édaphiques ou des conditions topographiques, est parsemée d'une mosaïque de bouquets et d'îlots d'arbres, p. ex. la savane palmier, la savane termite, la ▶savane-galerie inondable comme l'Entre Rios en Argentine ou le Pantanal du Mato Grosso au Brésil ; ▶savane inondable ; DG 1984, 407) ; *g* **Parklandschaft [f] (1)** (Vegetationsgeografische Bezeichnung für ein edaphisch resp. reliefbedingtes Mosaik von größeren Baumgruppen oder Waldinseln in einer offenen Graslandschaft; z. B. Palmsavanne, Termitensavanne, ▶Überschwemmungssavanne, Entre Ríos in Argentinien sowie das Große Pantanal do Mato Grosso in Brasilien; ▶Savanne).

4066 park-like cemetery [n] *landsc. urb.* (Open landscape cemetery, which was first created in the U.S., when America's cities began establishing large, park-like cemeteries outside their boundaries. Mount Auburn Cemetery in Cambridge, Massachusetts, was established in 1831 and is said to be "America's first garden cemetery." It is 70 ha [175 acres], and features rugged terrain and floral landscaping, which at the time was unusual in both Europe and the US. It included carriage-paths and ornamental plantings as well as public structures organized around vistas and focal points. The style and public access concept heavily influenced the design elements of Central Park in New York some 26 years later in 1857. The concept emphasized the cemetery as a place for "the living," and well as a place for "the departed". This concept was introduced to Germany with the construction of the Hamburg-Ohlsdorf Cemetery in 1877. The park-like cemetery has less space for burial and crypts than

geometrically laid out cemeteries, and may have horizontal tablets in place of vertical gravestones; ►forest cemetery, ►landscape garden); *s* **cementerio [m] paisajístico** (Cementerio diseñado al estilo del ►jardín paisajístico, que se realizó primero en los EE.UU. y fue introducido en Alemania en 1877 con la construcción del cementerio de Hamburg-Ohlsdorf. Este tipo de cementerio tiene menos capacidad que el cementerio geométrico clásico; ►cementerio boscoso); *f* **cimetière [m] paysager** (Par opposition à la conception française très dense et d'aspect minéral, cimetière dont la conception s'inspire du ►jardin paysager ; forme de cimetière réalisée pour la première fois aux USA et introduite en D. en 1877 avec le cimetière de Hamburg-Ohlsdorf ; pour des contraintes géologiques et des durées de rotation comparables le **c. p.** a une capacité plus faible et une surface quatre à cinq fois plus importante que le cimetière traditionnel de forme géométrique ; ►cimetière en ambiance forestière, ►cimetière vert, ►parc paysager) ; *syn.* cimetière parc [m], parc [m] funéraire (DIC 1993) ; *g* **Parkfriedhof [m]** (Dem Landschaftsgarten nachempfundener Friedhof, der in den USA zuerst realisiert wurde und mit dem Friedhof in Hamburg-Ohlsdorf 1877 in D. eingeführt wurde. Der **P.** hat eine geringere Bestattungsausnutzung als geometrisch ausgelegte Friedhofsformen; ►Landschaftspark, ►Waldfriedhof).

4067 park-like garden [n] *gard.* (Large private garden with park-like design which recognizes the owner's prominence); *s* **jardín [m] parque** (Jardín particular de grandes dimensiones diseñado al estilo de un parque); *f* **jardin [m] d'apparat** (Grand jardin conçu sous forme de parc et déployant la magnificence de son propriétaire ; cf. DIC 1993) ; *syn.* jardin [m] parc, jardin [m] seigneurial ; *g* **parkartiger Garten [m]** (Großer Garten, der wie ein Park gestaltet ist und den Wohlstand seines Eigentümers repräsentiert); *syn.* Parkgarten [m], *hist.* Herrengarten [m], Herrschaftsgarten [m].

4068 park-like landscape [n] *landsc.* (Area of managed landscape which conveys the impression of a ►park 1 due to the frequent transition from groups of trees, to meadows, fields, lakes, ponds, and woodland; ►landscape garden); *s* **paisaje [m] campestre** (Paisaje que —debido al cambio continuo entre grupos de árboles, prados, tierras de cultivo, lagunas y bosquetes— da la impresión de ser un ►parque; ►jardín paisajístico); *f* **paysage [m] champêtre** (Paysage culturel riche en composantes naturelles telles qu'arbres isolés, groupes d'arbres, prés, champs, cours et plans d'eau, bosquet et forêts dont l'alternance et les formes de transition procurent au paysage aspect de ►parc ; terme d'inspiration rurale patrimoniale ; ►jardin paysager, ►parc paysager) ; *g* **Parklandschaft [f] (2)** (Kulturlandschaft, die durch ständigen Wechsel von Gehölzgruppen, Wiesen, Äckern, Wasser und Waldstücken optisch den Eindruck eines ►Parks vermittelt; ►Landschaftspark).

4069 park-like vacation development [n] [US] *recr.* (Attractively situated, extensive tourist settlement of houses and apartments with wide range of recreational facilities); *syn.* holiday park [n] [UK]; *s* **parque [m] de recreo** (Colonia de recreo [casas y viviendas] emplazada en paisaje atractivo y con gran oferta de actividades de tiempo libre); *f 1* **parc [m] de vacances (±)** (Lotissement de tourisme [résidence et appartements de vacances] dans un site naturel remarquable offrant une gamme étendue d'activités diversifiées de plein air et de loisirs de qualité) ; *f 2* **parc [m] résidentiel de loisirs** (F., terrain aménagé pour l'accueil d'au moins 35 habitations légères de loisirs, souvent installé sur le territoire d'un village de vacances) ; *g* **Ferienpark [m]** (Landschaftlich reizvoll gelegene, großzügige Fremdenverkehrssiedlung [Ferienhäuser und -wohnungen] mit reichhaltigem Erholungsangebot an Freizeitbetätigung und Unterhaltung).

4070 park management program [n] [US]/**park management programme** [n] [UK] *conserv'hist. gard.* (Guidelines, plans and work program[me]s, including instructions for long-term maintenance and a rough estimate of the costs, which have been developed for the lasting care and management of a park or in Europe, of an historic garden, based upon a detailed survey and analysis of the existing situation; in U.S., management plans are prepared by local, state as well as national agencies); *syn.* park conservancy programme [n] [also UK]; *s* **plan [m] de conservación de parque patrimonial** (Guía de preservación y programa de trabajo de conservación de un parque perteneciente al patrimonio); *f* **programme [m] de gestion et d'entretien d'un parc** (Expertise réalisée sur l'état de conservation d'un parc ou d'un jardin historique et de ses diverses composantes, comportant une analyse historique, la définition des objectifs, le programme de travaux d'entretien à long terme, y compris une évaluation sommaire du coût des travaux ; un programme de gestion et d'entretien a également pour but de présenter les idées maîtresses ayant été à l'origine de la conception et ayant participé au développement d'un jardin historique) ; *syn.* plan [m] de gestion et de conservation d'un parc ; *g* **Parkpflegewerk [n]** (Ein auf Anlagenforschung beruhendes Bestandsgutachten, Leitbild, Planwerk und Arbeitsprogramm einschließlich langfristiger Pflegeanweisungen und einer überschlägigen Kostenschätzung für die nachhaltige Pflege, Unterhaltung und Entwicklung eines Parks oder eines Gartendenkmals. Ein **P.** soll auch die Entstehung und Entwicklung einer historischen Anlage nachvollziehen und deren kulturhistorische Einordnung und Bedeutung aufzeigen. Für kleinere, weniger bedeutende Grünanlagen reicht i. d. R. ein einfacher Erläuterungsbericht, in dem die gestalterische Grundidee, das Bepflanzungs- und Nutzungskonzept sowie die gärtnerischen Entwicklungsziele dargelegt sind).

4071 park rose [n] *hort. plant.* (Bushy rose planted in a park; ►shrub rose); *s* **rosal [m] de parque (≠)** (En la clasificación de rosas, un tipo de aplicación que gracias a la fitogenética mantiene su carácter de rosa silvestre y se planta en parques como ►rosal arbustivo); *f* **rosier [m] des parcs (et des jardins)** (►Rosier arbustif planté dans un parc ; ►rosier buisson) ; *g* **Parkrose [f]** (In der Rosenklassifizierung ein Verwendungstyp, der durch Züchtungen seinen Wildrosencharakter noch behalten hat und in Parks als ►Strauchrose gepflanzt wird).

parks and cemetery department [n] [UK] *adm. urb.* ►park department [US]/parks department [UK].

parks and recreation department [n] [US] *adm. urb.* ►park department [US]/parks department [UK].

park savanna(h) [n] *geo. phyt.* ►parkland.

4072 park shelter [n] *recr.* (Usually an open structure in a forest or the open landscape which serves as temporary refuge in bad weather; ►canopy 1, ►pedestrian shelter, ►picnic shelter); *s* **techado [f] contra la lluvia** (Estructura generalmente abierta, situada en bosque o en el campo, prevista para servir de refugio temporal a excursionistas cuando llueve; ►marquesina 1, ►marquesina 2, ►techado para picnics); *f* **hutte [f]** (Abri sommaire de structure légère ouverte, en général en bois, utilisé temporairement comme protection contre les intempéries en forêt ou en pleine campagne ; ►auvent, ►abri de pique-nique, ►abri couvert) ; *g* **Schutzhütte [f] (2)** (Meist offene Hütte im Wald oder in der freien Landschaft zum kurzfristigen Schutz vor Wetterunbilden; ►Picknickhütte, ►Schutzdach 1, ►Schutzdach 2).

4073 parkway [n] [US] *trans. recr.* (Limited access road in scenic environments, where trucks and commercial development are not permitted. **In U.S.**, this term is sometimes used for a local scenic road); *syn.* scenic route [n] [UK]; *s* **autovía [f] panorá-**

mica (≠) (En EE.UU. carretera de varios carriles que transcurre por zona de bellos paisajes y por la que no pueden circular vehículos pesados. En algunos casos el término «parkway» se utiliza para tramo de carretera local que atraviesa una zona paisajísticamente atractiva); *f* **route-parc [f]** (±) (**U.S.**, route à plusieurs voies aménagée dans un cadre paysager exceptionnel [spectacle paysager], sur laquelle la circulation des poids lourds est interdite et sur les bords de laquelle le développement de zones industrielles ou commerciales n'est pas autorisée) ; *syn.* route [f] paysagère (±) ; *g* **Schnellverkehrsstraße [f] in landschaftlich schöner Umgebung** (≠) (In den USA in landschaftlich besonders schönen Gebieten mit vielen Ausblicken in die Landschaft angelegte, mehrspurige, für Schwerlastverkehr gesperrte Autobahn, in deren visuellen Einflussbereich keine Gewerbe- und Industriegebiete erschlossen werden dürfen. In den USA wird *parkway* manchmal auch für eine örtliche, landschaftlich besonders schöne Strecke benutzt; es gibt aber auch lange Landstraßen, z. B. den *Blue Ridge Parkway*, der mit über 755 km die südlichen Appalachen von Waynesboro in Virginia bis zum *Great Smoky Mountains National Park* in North Carolina durchquert).

4074 parterre [n] *gard'hist.* (Level area of garden with ornately patterned flower beds, and an absence of trees and tall shrubs; created during the French Baroque period as an expression of the subjugation and regulation of nature. In the 1760 edition of the "Théorie et pratique du jardinage" from A. J. DEZALLIER D'ARGENVILLE, a variety of parterre patterns are mentioned: according to their functions, e.g. *p. d'orangerie*; according to their main elements, e.g. *p. d'eau, p. fleuriste, p. de gazon*; according to their structuring, ornamentation and design features, e.g. *p. en broderie* [▶embroidered parterre], *p. de compartiment, p. de pièces coupées*; or according to their origin, *p. à l'Anglaise, boulingrin*. Richer parterre designs are laid out in patterns using agraffe, acanthus, tendril, twig or leaf forms); *s* **parterre [m]** (Área plana de un jardín con carácter representativo formada por muchos macizos de flores ornamentales, sin leñosas grandes. El origen se encuentra en el jardín francés del Barroco como muestra de la naturaleza dominada y bien organizada. En la «Théorie et Pratique du Jardinage» de A. J. DEZALLIER D'ARGENVILLE, editada en 1760, se presentan gran variedad de **pp.** clasificados según sus funciones, como el *p. d'orangerie* por sus elementos principales, p. ej. el *p. d'eau, p. fleuriste, p. de gazon*; por sus elementos estructurales, de ornamentación y de diseño, p. ej. *p. en broderie* [▶parterre bordado], *p. de compartiment, p. de pièces coupées* o por su origen *p. à l'Anglaise, boulingrin*. Formas más ricas de **p.** están dotadas de mosaicos de agrafes, acanthus, ramas u hojas); *f* **parterre [m]** (Principal élément de décoration des jardins classiques français du XVII et XVIIIème siècle (appelé jardin baroque par les allemands) ; surface décorative plane composée de nombreux massifs floraux d'ornement, de surfaces de gazon entourés de Buis *[Buxus]*, de Santoline, de plantes condimentaires, de gravier ou autres matériaux de couleur constituant un dessin de forme homogène [compartiment, broderie]. L'ouvrage « Théorie et pratique du jardinage », dans son édition de 1790, A. J. DEZALLIER D'ARGENVILLE présente un recueil de dessins de parterres classés selon leur fonction, p. ex. le *p. d'orangerie*, selon les principaux éléments de composition utilisés, p. ex. le *p. d'eau, p. fleuriste, p. de gazon*, selon leur organisation, les ornements et éléments de décoration, p. ex. *p. en broderie* [▶parterre de broderie], *p. de compartiment, p. de pièces coupées* ou selon leur origine, p. ex. le *p. à l'Anglaise, le boulingrin*) ; *g* **Parterre [n] (2)** (Ebene Gartenanlage repräsentativ-zierenden Charakters mit [vielen] ornamentalen Blumenbeeten ohne hohe Gehölze, entstanden zur Zeit des französischen Barocks als Ausdruck einer bezwungenen, übersichtlich geordneten Natur. In der *„Théorie et pratique du jardinage"* von A. J.

DEZALLIER D'ARGENVILLE, Auflage von 1760, werden eine Fülle von **P.**-Mustern nach ihren Funktionen, z. B. *parterre d'orangerie*, nach den Hauptelementen, z. B. *parterre d'eau, p. fleuriste, p. de gazon*, nach ihrer Gliederung, Ornamentierung und Ausstattung, z. B. *parterre en broderie* [▶Parterre mit starker Ornamentierung], *p. de compartiment, p. de pièces coupées* oder nach ihrer Herkunft — *parterre à l'Anglaise, boulingrin* — bezeichnet. Reichere Parterreformen sind mit Agraffen-, Palmetten-, Zweig-, Ranken-, Akanthus- oder Blattwerkmustern ausgestattet; *cf.* GDG 1965, 137 f); *syn.* Gartenparterre [n], Parterreanlage [f], Luststück [n] (GDG 1965, 153).

4075 parterre [n] **with formal flower borders** *gard'hist.* (Type of level, ornamental garden created during the French baroque era with a very decorative character comprising flower beds of low growing plants; ▶parterre); *s* **parterre [m] de flores** (Tipo de ▶parterre en jardín ornamental llano creado en Francia durante el barroco que consiste en macizos de flores de poca altura y sin leñosas); *f* **parterre [m] de pièces coupées pour les fleurs** (Type de parterre aménagé dans les jardins du XVIIème siècle pour recevoir les fleurs à longues tiges ; DIC 1993 ; ▶parterre) ; *syn.* parterre [m] fleuriste ; *g* **Parterre [n] mit formal gestalteten Blumenbeeten** (Zur Zeit des französischen Barocks entstandene ebene Gartenanlage repräsentativ-zierenden Charakters mit ornamentalen Blumenbeeten ohne hohe Gehölze; ▶Parterre 2).

4076 partially infilled pit [n] *min. envir.* (Extraction site backfilled below the original contour; ▶dumpsite [US]/tipping site [UK], ▶landfill to original contour, ▶refilling, ▶high spoil pile [US]/high tip [UK]); *s* **vertedero [m] en zona de préstamo** (▶Depósito de residuos sólidos que aprovecha el hueco dejado por la extracción de minerales rellenándolo hasta el nivel de la superficie, ▶depósito sobre el nivel de superficie, ▶relleno de explotaciones a cielo abierto, ▶vertedero en hondonada 1); *syn.* vertedero [m] en hondonada (2); *f* **décharge [f] par enfouissement partiel** (±) (▶Décharge publique localisée dans une aire d'extraction de matériaux dans laquelle les déchets n'atteignent pas le niveau supérieur de l'excavation ; ▶décharge par dépôt, ▶décharge par enfouissement, ▶remblayage d'une excavation) ; *g* **Unterflurkippe [f]** (▶Deponie in einer Entnahmestelle, bei der das verkippte Material die ursprüngliche Bodenoberfläche nicht erreicht; ▶Flurkippe, ▶Innenverkippung, ▶Hochkippe).

4077 partial migrant [n] *zool.* (Bird species of which the northern population usually leaves the northern part of their range during the winter; ▶bird of passage, ▶local migrant, ▶migratory bird); *s* **ave [f] migradora parcial** (Especie de ▶ave migradora de la cual sólo una parte de su población migra o si migran todos los individuos de ella, lo hacen sólo a corta distancia; ▶ave de dispersión posgenerativa, ▶ave de paso); *syn.* migrador [m] parcial; *f* **oiseau [m] migrateur partiel** (Espèce d'oiseau pour laquelle la plupart des individus quittent leur aire de distribution pour l'hivernage ; ▶oiseau de passage, ▶oiseau erratique, ▶oiseau migrateur) ; *g* **Teilzieher [m]** (Vogelart, von der nur ein Teil der Individuen [der meist nördlichen Populationen] während des Winters das brutzeitliche Aufenthaltsgebiet verlässt; ▶Durchzügler, ▶Strichvogel, ▶Zugvogel).

partial payment [n] *constr. contr.* ▶interim payment.

4078 partial redesign [n] *arch. plan.* (Alteration of a part of a project; ▶design 2, ▶redesign 1, ▶remodeling [US]/remodelling [UK], ▶replanning); *s* **reconfiguración [f] parcial** (Modificación del diseño de partes de un objeto existente; ▶diseño [de un proyecto], ▶reconfiguración, ▶remodelación, ▶replanificación); *f 1* **remodelage [m]** (Transformation de l'aspect d'un ensemble urbain, de parcs, jardins, etc. ; opération partielle par comparaison avec ▶rénovation ; remodeler [vb], refaçonner

[vb] ; ►conception, ►composition 1) ; *syn.* remaniement [m] ; *f 2* **recomposition [f] (1)** (Réorganisation de l'arrangement des éléments de conception d'un aménagement, d'un site, d'une espace ; *comme verbe* refaçonner ; ►réaménagement, ►refonte d'un projet) ; *syn.* transformation [f], réorganisation [f], restructuration [f] ; *g* **Umgestaltung [f] (1)** (Veränderung der Gestaltung eines bestehenden Objektes; im Vergleich zur ►Neugestaltung werden i. d. R. nur Teile verändert; ►Gestaltung, ►Umbau, ►Umplanung).

partial shade [n] [US] *phyt. plant.* ►semi-shade.

participant [n] *prof.* ►competition entrant

participation [n] *adm. leg.* ►statutory provision for participation.

4079 participation conditions [npl] *prof.* (**1.** Prerequisites for entering any competition of any kind. **2.** Prerequisites and qualifications for a ►design competition 1 or ►final design competition [US]/realization competition [UK]; ►bidding requirements [US]/preconditions for tenderers [UK]); *s* **condiciones [fpl] de participación en concurso** (**1.** Prerequisitos para participar en un concurso de cualquier tipo. **2.** Prerequisitos y restricciones para participar en ►concurso de ideas o en ►concurso de realización; ►condiciones de participación en concurso-subasta); *f* **conditions [fpl] de participation (2)** (Modalités requises pour la participation à un ►concours d'idée ou à un ►concours de conception-construction ; ►conditions de participation 1) ; *g* **Teilnahmebedingungen [fpl] (2)** (**1.** Voraussetzungen, die für die Teilnahme an jedwedem Wettbewerb erfüllt sein müssen. **2.** Voraussetzungen, die für die Teilnahme an einem ►Ideenwettbewerb oder ►Realisierungswettbewerb erfüllt sein müssen; ►Teilnahmebedingungen 1).

4080 participatory sport [n] [US] *recr.* (**1.** Sport activity engaged in by many people; e.g. soccer, swimming, skiing); *syn.* mass sport [n] [US], sport [n] enjoyed by the masses [UK]; **2. spectator sport** [n] (Team sport watched by large numbers of people); *s 1* **deporte [m] masivo** (Actividad deportiva en la que participan activamente muchas personas, como p. ej. fútbol, natación, béisbol, etc.); *syn.* deporte [m] de masas; *s 2* **deporte [m] para espectadores** (Tipo de deporte al que acude una gran cantidad de público, como fútbol en Suramérica y España o béisbol en Centroamérica y el Caribe); *f* **sport [m] de masse** (Sport pratiqué par de larges couches de la population) ; *g* **Massensport [m, o. Pl.]** (Sport, der von großen Teilen der Bevölkerung betrieben wird, z. B. Fußball, Schwimmen, Skilaufen).

particle fall out [n] *envir.* ►particulate deposition.

particles [npl], **dust** *envir.* ►particulate matter (1).

4081 particle-size class [n] *constr. pedol.* (Term used in soil taxonomy to characterize the grain-size composition of a whole soil—excluding organic matter and salts more soluble than gypsum; classification system of the National Cooperative Soil Survey and the U.S. Department of Agriculture, Natural Resources Conservation Service; cf. RCG 1982 and KST 1996, 589; ►graded sediment group [US]/particle size group [UK], ►soil textural class); *s* **clase [f] granulométrica** (En la definición de las propiedades físico-químicas de los suelos, agrupación de varios ►grupos granulométricos en una ►clase textural); *syn.* clase [f] de granulación; *f* **classe [f] granulométrique** (Dans la définition des propriétés physico-chimiques des sols, regroupement de plusieurs ►fractions granulométriques pour former une ►classe de texture des sols, de matériau) ; *syn.* classe [f] de granulation ; *g* **Korngrößenbereich [m]** (Zusammenfassung mehrerer ►Korngruppen zu einer ►Bodenart; DIN 18 196); *syn. pedol.* Körnungsklasse [f].

4082 particle-size distribution [n] *constr.* (Amount of various soil separates in a soil sample, usually expressed as weight percentages; ►graded sediment group [US]/particle size group [UK]); *syn.* grain-size distribution [n] *s* **granulometría [f]** (Distribución según sus medidas de los componentes sólidos del suelo que se expresa en porcentaje de peso y proporciona información sobre las características del mismo en cuanto a su friabilidad, capacidad de carga, permeabilidad, resistencia a las heladas, etc.; ►grupo granulométrico); *syn.* distribución [f] granulométrica, gradación [f] de partículas; *f 1* **granularité [f]** *constr.* (Ensemble des caractéristiques d'un granulat mesurées par l'analyse granulométrique ; *contexte* grave de granularité 0/20 ; DIR 1977) ; *f 2* **spectre [m] granulométrique** *pedol.* (La composition granulométrique des sols peut être définie par une analyse pour quantifier pondéralement les particules minérales élémentaires cristallines groupées en classes granulométriques ; PED 1979, 229 et s. ; ►fraction granulométrique) ; *syn.* granulométrie [f], composition [f] granulométrique ; *g* **Kornverteilung [f]** (Zusammensetzung der einzelnen ►Korngruppen eines Bodens oder Zuschlagstoffes, die Aufschluss über die plastischen Eigenschaften [Bearbeitbarkeit], die Verdichtbarkeit, Belastbarkeit, Wasserdurchlässigkeit, Frostempfindlichkeit etc. gibt; cf. LEHR 1981); *syn.* Korngrößenverteilung [f], Korngrößenzusammensetzung [f], Körnung [f], Körnungsaufbau [m].

4083 particle-size distribution curve [n] *constr. pedol.* (Graphic representation of the composition of soil or aggregate sizes, in the form of a flat, steep or undulating curve, based on the amount and grading of sediment particles); *s* **curva [f] de granulometría** (Representación gráfica de la granulometría de un suelo o un árido en forma de curva que según el porcentaje de cada grupo granulométrico puede ser llana, empinada u ondulada); *syn.* curva [f] granulométrica; *f* **courbe [f] granulométrique** (**1.** Représentation graphique sous forme de courbe [en ordonnée les pourcentages des fractions, en abscisse logarithmique le diamètre des particules] des résultats de l'analyse granulométrique et précisant la composition granulométrique des éléments minéraux d'un sol ; la c. g. peut être plate, inclinée ou ondulée en fonction du nombre et de la présence de palier. **2.** Composition granulométrique et proportion [dosage] des composants solides des bétons [granulats et liants]) ; *g* **Kornverteilungskurve [f]** (**1.** Grafische Darstellung des Körnungsaufbaus eines Bodens oder Zuschlagstoffes in Form einer Kurve, die je nach Anzahl und Abgestuftheit der Korngruppen flach, steil oder wellenförmig verlaufen kann. **2.** Kornzusammensetzungen der Zuschläge für die Herstellung von Beton, die durch Siebversuche ermittelt werden, heißen **Sieblinien**); *syn.* Körnungskurve [f], Körnungssummenkurve [f], Siebkurve [f], Sieblinie [f].

particle-size group [n] [UK] *constr. (pedol.)* ►graded sediment group [US].

4084 particulate air pollution [n] *envir.* (High concentration of particles in the air, which are harmful to the health of organisms or can have a lasting effect on the environment); *s* **contaminación [f] de partículas sólidas** (Fuerte concentración de polvo en el aire que tiene efectos nocivos sobre la salud de los organismos y sobre el medio ambiente); *syn.* contaminación [f] de polvo, carga [f] de partículas sólidas; *f* **charge [f] en poussières (fines)** (Forte concentration en poussières dans l'atmosphère ayant des effets nocifs sur la santé des organismes ou présentant un risque de dégradation de l'environnement) ; *syn.* empoussièrement [m] ; *g* **Staubbelastung [f]** (Hohe Staubkonzentrationen in der Luft, die die Gesundheit von Organismen mindern oder nachhaltig beeinträchtigen).

4085 particulate deposition [n] *envir.* (Settling from rain of fine solid particles [air pollutants] found in air or emissions of dust, smoke, mist, fumes, or smog; ►acid rain); *syn.* particle fall out [n]; *s* **deposición [f] seca** (Sedimentación de partículas contaminantes por gravedad o por inhalación o absorción en la superficie; DINA 1987, 95; ►lluvia ácida); *f* **déposition [f] atmosphérique sèche** (Dépôt de particules toxiques sur les végétaux. Sous l'action de la pluie, du brouillard, de la rosée celles-ci se transforment en une solution acide concentrée bloquant le métabolisme et détruisant les tissus foliaires, écorce, etc.; ►pluie acide); *g* **trockene Deposition [f]** (Ablagerung [von Luftschadstoffen] auf Oberflächen durch Sedimentation oder Einlagerung. Diese **D.en** auf Blättern, Nadeln oder Rinden können bei Regen, Nebel oder Tau in Lösung gehen und konzentrierte Säuren bilden, die zu direkten Gewebeschäden führen; ►saurer Regen); *syn.* Staubablagerung [f].

4086 particulate matter [n] (1) *envir.* (**1.** Separate fine solid particles, such as ash, dust particles, fume, aerosols, pollen, spores, grits, that are suspended in the air or the inert dust particles are released from the combustion process in exhaust gases; the solids often provide extended surfaces due to their irregularities and therefore other pollutants can be carried along; ►air pollutant, ►diffused particulate matter; *syn.* particulates [npl]; **2. dust particles [npl]** (Fine substances in the air. Particle sizes are measured in microns [μ]: 1μ = 0.001mm. A convenient division is to use the terms suspended particles up to 20μ; dust from 29μ to 75μ and grit for larger particles. Fly ash particles are 3 to 80μ; TAN 1975); *s 1* **partículas [fpl] sólidas (1.** Componentes sólidos, como cenizas, partículas de polvo, aerosoles, polen y esporas, que están suspendidos en el aire o son emitidos como partículas de polvo inertes en la combustión incompleta de sustancias de carbón. Estos gases de escape contienen a menudo grandes porcentajes de sustancias nocivas que se adhieren a la superficie irregular de las **pp. ss.** y pueden así ser transportadas a grandes distancias. En la atmósfera, el contenido natural de **pp. ss.** está formado por silicio, aluminio, calcio y sus cloruros. La mayor parte tiene su origen en zonas desérticas, en Europa p. ej. en el Sahara y el Sahel; ►contaminante de la atmósfera, ►partículas en suspensión); *s 2* **polvo [m] atmosférico** (Aerosol sólido compuesto de partículas inferiores a 100 micras [μ]: 1 μ = 0.001 mm que se encuentra suspendido en el aire. Según el tamaño de las partículas se habla de **p.** finísimo, **p.** fino y **p.** grueso; cf. DICE 1975); *f* **poussière [f]** (►Particules en suspension dans l'atmosphère d'un diamètre compris entre 0,01 et 50 μm, la concentration des particules émises à ne pas dépasser est réglementée par des valeurs limites ; ►polluant atmosphérique) ; *g* **Staub [m]** (Stäube [pl]) **1.** Mengenbegriff für kleinste, feste, in der Luft schwebende Partikelchen von unterschiedlicher Form, Struktur und Dichte im Größenbereich von ca. 0,01-50 μm. Der natürliche Anteil in der Atmosphäre besteht aus Silicium, Aluminium, Calcium und deren Chloriden. Hinzu kommen organische Bestandteile wie Pollen, Sporen und Bakterien. In Abgasen aus Verbrennungsvorgängen sind oft hohe Schadstoffanteile, die sich auf den unregelmäßig geformten Oberflächen der **Staubkörner** festsetzen und somit weit verbreitet werden können; in den Städten finden sich mehr Rußteilchen, auf dem Lande mehr Pflanzensporen, an der See durch die Gischt Salzkristalle. Der meiste in der Atmosphäre transportierte Staub stammt aus Wüstengebieten, z. B. aus der südlichen Sahara und dem Sahel, der größtenteils über den Atlantik und teilweise auch nach Europa geweht wird. Er düngt z. B. mit seiner Phosphatfracht den Regenwald im Amazonasgebiet; cf. STZ 2009; ►Luftschadstoff, ►Schwebstaub); *g 2* **Staubpartikel [n]** (Einzelnes Staubteilchen).

particulate matter [n] (2) *hydr.* ►suspended particulate matter.

particulates [npl] *envir.* ►particulate matter (1).

particulates [npl], suspended *hydr.* ►suspended particulate matter.

part shade [n] [US] *phyt. plant.* ►semi-shade.

4087 part-time agricultural business [n] *agr.* (Agricultural business operated as a side-line, because it is not profitable enough to support a family and forces the owner to pursue a full-time job in a non-agricultural occupation); *syn.* agricultural side-line [n] [also UK]; *s* **granja [f] agrícola marginal** (Pequeña granja que no produce lo suficiente para mantener a una familia, de manera que el amo [= agricultor a tiempo parcial] está obligado a ganarse la vida en otro trabajo); *f* **entreprise [f] agricole à temps partiel** (DUV 1984, 14 ; unité de production pour laquelle l'activité agricole n'est pas la principale activité, l'exploitant exerçant une activité artisanale ou salariée qui constitue son revenu principal) ; *g* **landwirtschaftlicher Nebenerwerbsbetrieb [m]** (Nebenberuflich bewirtschafteter Landwirtschaftsbetrieb, der keine Lebensgrundlage für eine Familie zu erbringen vermag und den Besitzer [= Nebenerwerbslandwirt] zwingt, einem nicht landwirtschaftlichen Hauptberuf nachzugehen; MEL, Bd. 17); *syn.* (landwirtschaftliche) Nebenerwerbsstelle [f].

party [n], contracting *contr. pol.* ►party to a contract.

4088 party [n] responsible for maintenance *leg.* (Person or authority/agency obligated [US]/obliged [UK] by law to provide maintenance and make money available for maintenance and repair work; ►maintenance obligation); *s* **parte [f] responsable del mantenimiento** (Persona o autoridad/agencia pública obligada a realizar los trabajos de mantenimiento de una instalación o infraestructura y de poner a disposición los fondos necesarios para ellos; ►obligación de mantenimiento); *syn.* responsable [m] del mantenimiento; *f* **partie [f] responsable de l'entretien** (Personne physique ou morale de droit public ou privé responsable légalement de l'exécution et des dépenses relatives aux travaux d'entretien ; ►obligation d'entretien) ; *syn.* partie [f] chargée de l'entretien ; *g* **Unterhaltungspflichtige/-r [f/m]** (Der per Gesetz Zuständige, der für die Planung, Durchführung und Finanzierung von Unterhaltungsarbeiten verpflichtet ist; ►Unterhaltungslast).

4089 party [n] to a contract *contr.* (Person, group, public authority/agency, etc., engaged in a contract); *syn.* contracting party [n], *syn. pol.* contracting member state [n]; *s* **parte [f] contratante** (Persona, grupo, empresa o autoridad/agencia pública que firma o ha firmado un contrato con otra parte); *syn.* contratante [m], contraparte [f], parte [f] firmante; *f* **partie [f] contractante** (Personne civile ou morale, établissement public, collectivité territoriale ou autorité territoriale contractant ou ayant passé un marché avec un tiers ; CCM 1991, 1) ; *syn.* service [m] contractant, collectivité [f] contractante, établissement [m] public contractant ; *g* **Vertragspartei [f] (1.** Person, Gruppe, öffentliche Hand o. Ä., die mit [einem] anderen einen Vertrag schließt resp. geschlossen hat. **2.** *pol.* Staat, der mit [einem] anderen einen Vertrag oder ein Abkommen schließt oder geschlossen hat); *syn.* Vertragspartner [m], Vertragsgegner [m]; *syn. pol.* Vertragsstaat [m].

party wall [n] [US] *constr. leg.* ►common boundary wall.

pass [n] *zool.* ►fish pass.

passage migrant [n] *zool.* ►bird of passage.

passage visitor [n] *zool.* ►bird of passage.

passageway [n] [US], pedestrian *urb. plan.* ►pedestrian link.

4090 passive recreation [n] *recr.* (Non-physical forms of pleasurable activity, e.g. reading, listening to radio, watching television); *s* **recreación [f] pasiva** (Recuperación de la salud y de la capacidad de rendimiento por medio de actividades de esparcimiento sin actividad física, como leyendo, escuchando música, viendo la televisión, etc.); *f* **loisirs [mpl] passifs** (Forme

P

passive, non physique des activités de loisirs [lecture, radio, télévision] permettant la régénération de l'équilibre corporel) ; *syn.* récréation [f] passive ; *g* **passive Erholung [f, o. Pl.]** (Zurückgewinnung von Gesundheit und Leistungsfähigkeit, die sich fast ausschließlich auf z. B. Lesen, Radiohören, Fernsehen beschränkt).

pastoral park [n] [UK] *gard'hist.* ▶landscape garden, #2.

4091 pasture [n] *agr.* (Usually fenced, man-made grassland used for grazing. The opposite is a natural pasture without seeding, mowing or fertilizing; ▶abandoned pasture, ▶alpine pasture, ▶fertilized pasture, ▶forest pasture, ▶intensively grazed pasture, ▶permanent pasture, ▶pasture of low productivity, ▶rotation pasture [US], ▶sheep pasture); *s 1* **pastizal** [m] (Terreno utilizado para pastar/apacentar al ganado; *términos específicos* pasto común, pasto de rastrojo, pasto de invierno, pasto de verano; ▶pastizal de alta montaña, ▶pastizal en rotación, ▶pastizal forestal, ▶pastizal jugoso, ▶pastizal permanente, ▶pasto de ovejas, ▶pasto pobre); *syn.* pasto [m], saltus [m]; *s 2* **dehesa [f]** (Tierra acotada destinada a pastos con dominio particular, aunque también puede ser de dominio de los pueblos [= dominio de propios]. Según el uso al que se destinen reciben diferentes denominaciones [**d.** arbitrada, **d.** boyal, **d.** de yeguas o de potros, **d.** carnicera]. Pueden además ser de puro pasto o de uso alternante de pasto y cultivo. En el área mediterránea la **d.** consiste en un bosque aclarado [bosque hueco] con utilización de los espacios libres para pastos o cultivos y cuyas especies arbóreas suelen ser la encina, el rebollo o el fresno; DINA 1987); *f* **pâturage [m] (2)** (Prairie souvent réensemencée — permanente, temporaire ou artificielle —, en général clôturée sur laquelle les animaux vont paître ; ▶alpage, ▶herbage, ▶herbage permanent, ▶pâturage à ovins, ▶pâturage en forêt, ▶pâturage libre, ▶pâturage permanent, ▶prairie à pâturage rationné, ▶prairie pâturée maigre) ; *syn.* pâture [f], pacage [m] (2); *g* **Weide [f]** (Zur Beweidung meist eingezäunte Grasflur oder angelegtes Grünland. Das Gegenteil wäre die Naturweide — ohne Aussaat, Pflege und Düngung; ▶Alm, ▶Dauerweideland, ▶Fettweide, ▶Magerweide, ▶Schafweide, ▶Standweide, ▶Umtriebsweide, ▶Waldweide); *syn.* Weideland [n].

pasture [n]**, land** *agr.* ▶common grazing.

pasture [n] [US]**, high mountain** *agr.* ▶alpine pasture.

pasture [n]**, reverted** *agr.* ▶abandoned pasture.

pasture [n]**, wood** *agr. for.* ▶forest pasture.

4092 pasture farming [n] *agr.* (Type of agriculture where the use of pasture for grazing animals is predominant. The opposite term is 'stable farming'; ▶grassland farming); *s* **pasticultura [f]** (Tipo de agricultura en la que predomina la función ganadera de los pastos; al contrario que la ganadería de establo; ▶praticultura); *syn.* explotación [f] de pastos; *f* **élevage [m] pastoral** (Forme de l'élevage en plein air [élevage à l'herbe] fondée sur l'exploitation des prairies naturelles pour le pâturage, par *opp.* à « l'élevage à l'étable » ; *termes spécifiques* transhumance, estivage ; ▶exploitation des prairies) ; *syn.* agriculture [f] herbagère ; *g* **Weidewirtschaft [f]** (Form der Landwirtschaft, bei der die Pflege und Nutzung der Weideflächen für Weidevieh vorrangig betrieben wird; im Ggs. zur Stallviehhaltung; ▶Grünlandwirtschaft).

4093 pasture [n] **of low productivity** *agr.* (Nutrient-poor grassland characterized by permanent grazing or hay cutting without fertilizing; ▶common grazing); *s* **pasto [m] pobre** (Pastizal pobre en nutrientes debido al aprovechamiento como tal o como prado de siega durante muchos años sin fertilizar. Fitosociológicamente se presenta un ▶prado oligótrofo; ▶pastizal malo); *f* **prairie [f] pâturée maigre** (Prairie naturelle de très faible productivité, avec une alimentation en eau faible, fortement

appauvrie suite à une pâture intensive prolongée sans apport de fertilisants, p. ex. les pâturages maigres calcaires ou les pâturages maigres acides à *Nardus stricta* ; phytosociologiquement évolue vers une ▶pelouse oligotrophe [p. ex. alliances du *Mesobromion, Nardo-Galion*] ; ▶parcours) ; *syn.* prairie [f] pacagée médiocre, pâture [f] maigre, prairie [f] pâturée sèche, pacage [m] maigre ; *g* **Magerweide [f]** (Durch jahrelange Beweidung oder Heumahd ohne Düngung entstandene Weide mit gewisser Nährstoffarmut. Pflanzensoziologisch stellt sich dann ein ▶Magerrasen [z. B. *Mesobromion, Nardo-Galion*] ein; ▶Hutung).

4094 pasture weeds vegetation [n] *phyt.* (Plant community on extensively-used pasture land with thistles [*Cirsium, Carduus, Carlina*, etc.], docks, sorrels *[Rumex]*, stiff grasses [*Carex, Nardus stricta, Juncus*, etc.], as well as woody weeds, e.g. sloe *[Prunus spinosa]* and juniper *[Juniperus]*; *specific terms* ▶vegetation of animal rest areas, ▶weed community); *s* **comunidad [f] arvense de pastizales** (Comunidad de plantas advenedizas sobre pastizales de explotación extensiva que se caracteriza por especies de cardos [*Cirsium, Carduus, Carlina*, etc.], especies de *[Rumex]*, hierbas duras [*Carex, Nardus stricta, Juncus*, etc.] y también algunos arbustos como el endrino *[Prunus spinosa]* y el enebro *[Juniperus]*; *término específico* ▶vegetación de lugares de reposo; *término genérico* ▶communidad de malas hierbas); *syn.* comunidad [f] advenediza de pastizales; *f* **végétation [f] des prairies pacagées négligées** (Groupement d'adventices des prairies pacagées mésophiles négligées [pacage extensif] formé d'espèces coriaces comme les Patiences *[Rumex]* ou piquantes comme les Chardons [*Cirsium, Carduus, Carlina*, etc.], de graminées [*Carex, Nardus stricta, Juncus*, etc.] et de broussailles épineuses comme p. ex. l'Épine noire *[Prunus spinosa]* ou le Genévrier *[Juniperus]* ; *terme spécifique* ▶végétation de reposoir ; *terme générique* ▶flore adventice de cultures et de prairies ; ▶groupement anthropique) ; *syn.* flore [f] d'adventices des prairies pacagées ; *g* **Weidenkrautflur [f]** (Unkrautgesellschaft auf extensiv bewirtschafteten Weideflächen mit Distelarten [*Cirsium, Carduus, Carlina* etc.], Ampferarten *[Rumex]*, Hartgräsern [*Carex, Nardus stricta, Juncus* etc.] und auch holzigen ‚Weideunkräutern' wie z. B. Schlehe *[Prunus spinosa]* oder Wacholder *[Juniperus]*; *UB* ▶Lägerflur; *OB* ▶Unkrautflur).

pasture woodland [n] *agr. for.* ▶grazed woodland.

pasturing [n] *agr.* ▶grazing.

patch [n] *ecol.* ▶environmental resource patch; *hunt.* ▶food patch.

path [n] *plan. recr.* ▶bridle path [US]/bridle way [UK], ▶circular path [US]/circular pathway [UK]; *landsc. urb.* ▶foot path; *constr.* ▶stepping stone path, ▶subbase grade of a road/path [US]/formation level [n] of a road/path [UK]; *agr. for.* ▶sunken path.

path [n] [UK]**, bicycle** *trans.* ▶bikeway [US].

path [n] [UK]**, cycle** *trans.* ▶bikeway [US].

path [n] [UK]**, fire** *constr. urb.* ▶fire lane [US].

path [n] [UK]**, forest** *for.* ▶timber road [US].

path [n] [UK]**, hiking foot-** *recr.* ▶hiking trail [US].

path [n] [UK]**, long distance foot-** *plan. recr.* ▶national scenic trail [US].

path [n]**, migratory** *game'man. hunt.* ▶game trail.

path [n] [UK]**, nature study** *recr.* ▶nature trail [US].

path [n] [UK]**, rambling foot-** *recr.* ▶hiking trail [US]/hiking footpath [UK].

path [n] [US]**, towing/towing-path** [n] [UK] *recr. trans.* ▶towpath [US].

P

path [n] [US]**, walking** *recr.* ▶hiking trail [US]/hiking footpath [UK].

4095 pathside strip [n] [US] *agr. constr. ecol.* (Roadside; ▶road limit); *syn.* wayside [n] [UK]; *s* **borde** [m] **de camino** (Banda herbácea a lo largo de un camino; ▶borde de carretera); *f* **lisière** [f] **d'un sentier** (Bande enherbée le long d'un chemin ; ▶accotement, ▶bord de route) ; *syn.* bord [m] de chemin, bordure [f] de chemin ; *g* **Rain** [m] **(2)** (i. S. v. Wegrand; ▶Straßenrand); *syn.* Wegrain [m], Wegrand [m].

4096 path surfacing [n] *constr.* (*generic term* ▶surface layers; ▶paved stone surface; ▶road surfacing, ▶surface layer); *s* **pavimento** [m] **de caminos** (*Término genérico* ▶revestimiento de superficies; ▶adoquinado, ▶pavimento 1; ▶pavimento de carreteras/calles); *syn.* pavimentación [f] de caminos; *f* **revêtement** [m] **de cheminement** (*Terme générique* ▶revêtement de surface assurant la circulation des promeneurs ; ▶pavage, ▶revêtement de sol, ▶revêtement d'une rue/route) ; *syn.* revêtement [m] de sol pour cheminement ; *g* **Wegebelag** [m] (*OB* ▶Belag 2; ▶Decke, ▶Pflaster, ▶Straßenbelag); *syn.* Wegebefestigung [f], Wegedecke [f].

pathway [n] *landsc. urb.* ▶footpath; *plan. trans.* ▶alignment of a pathway; *gard. constr.* ▶stepped pathway.

pathway [n] [UK]**, circular** *plan. recr.* ▶circular path [US].

4097 pathway excavation [n] *constr.* (Dug out bed for construction of paths; ▶construction of hard surfaces, ▶subbase grade of a road/path [US]/formation level of a road/path [UK]); *syn.* footpath excavation [n]; *s* **vaciado** [m] **para carretera** (Excavación necesaria como parte de los ▶trabajos de construcción viaria; ▶lecho de carretera/camino); *f* **décaissement** [m] **des circulations** (Dans le cas d'une voie construite dans un déblai, volume évidé pour recevoir l'ensemble des couches de chaussée ; ▶travaux de voirie, ▶arase de terrassement) ; *g* **Wegekoffer** [m] (Die für die Aufnahme des ▶Wegebaues notwendige Ausschachtung; bei Straßen spricht man von **Straßenkoffer**; ▶Koffersohle).

4098 pathway maintenance [n] *constr. for.* (Upkeep to ensure the functioning of paths and public safety in woodlands, parks, open green spaces and recreation areas); *syn.* trail maintenance [n] [also US]; *s* **mantenimiento** [m] **de caminos** (Todas las medidas necesarias para conservar la funcionalidad y la seguridad de los caminos en parques, jardines, instalaciones de recreo y zonas verdes públicas); *f* **entretien** [m] **des voies** (Toutes mesures servant à garantir l'état, la fonction et la sécurité des circulations dans les forêts, les parcs, les espaces verts et autres aires de loisirs) ; *g* **Wegeunterhaltung** [f, o. Pl.] (Alle Maßnahmen, die der Erhaltung der Funktionsfähigkeit und der Verkehrssicherheit von Wegen im Wald, in Park- und [öffentlichen] Grünanlagen sowie in sonstigen Freizeitanlagen dienen).

4099 pathway network [n] *recr.* (System of pathways for pedestrians and hikers; ▶bikeway network, ▶bridle path network [US]/horse riding trail network [UK], ▶hiking network); *syn.* footpath network [n] (COU 1978, 144); *s* **red** [f] **de caminos** (Sistema de vías para peatones y senderistas; ▶red de pasillos verdes, ▶red de pistas para bici[cletas], ▶red de senderos ecuestres); *f* **réseau** [m] **de sentiers de promenade et de randonnée pédestre** (Système de divers itinéraires pour les promeneurs et les randonneurs ; *terme spécifique* réseau de parcours et de circuits récréatifs/thématiques ; ▶réseau de randonnée cyclotouristique, ▶réseau [de randonnée] équestre, ▶réseau de randonnée pédestre) ; *g* **Wegenetz** [n] (Ein System von Wegen für Fußgänger, Wanderer, Radfahrer und gesondert für Reiter; *UBe* ▶Radwegenetz, ▶Reitwegenetz, ▶Wanderwegenetz).

4100 pathway network plan [n] *plan.* (Plan showing an existing or proposed pathway system); *s* **plan** [m] **de circulación**

(Plano de la red viaria existente o planificada de una zona); *f 1* **plan** [m] **de sentiers pédestres** (Document reproduisant les liaisons pédestres existantes ou proposées dans un territoire donné) ; *syn.* plan [m] d'itinéraires pédestres ; *f 2* **plan** [m] **départemental de promenade et de randonnée pédestre** (Document établi au niveau départemental) ; *g* **Wegeplan** [m] (Plan, der ein Wegenetz in einer Anlage oder einem Gebiet darstellt).

patio [n] *arch.* ▶courtyard.

4101 patio house [n] *arch.* (**1.** House type built in the Spanish colonies of Latin America, whereby the rooms are arranged around a rectangularly-shaped courtyard; originating in Andalusia where they were built by the Moors early in the 8th century. **2.** Dwelling with an enclosed outdoor area onto which the living room windows are oriented and/or access is provided; ▶courtyard house); *syn.* patio house dwelling [n] [also US]; *s* **casa** [f] **de patio (2)** (**1.** Tipo de casa colonial en América Latina en la que las habitaciones rodean a un patio rectangular. Originalmente fue creada por los árabes e introducida por ellos en España, sobre todo en Andalucía. **2.** Edificio de viviendas construido alrededor de un gran patio hacia el cual se orientan aquéllas; ▶casa de patio 1); *f* **maison** [f] **patio** (**1.** Type d'habitation de l'époque coloniale espagnole en Amérique latine dans laquelle les pièces sont disposées autour d'une cour intérieure [patio]. **2.** Maison de forme rectangulaire comprenant une cour intérieure à ciel ouvert vers laquelle sont orientées les ouvertures des pièces d'habitation ; ▶maison atrium) ; *syn.* maison [f] à patio ; *g* **Patiohaus** [n] (**1.** Kolonialspanische Hausform in Lateinamerika, bei der die Zimmer bei einem rechteckigen Hausgrundriss um den Innenhof [Patio] angeordnet sind; stammt aus dem arabischen Raum. **2.** Mit einem Innenhof/Lichthof versehenes Wohnhaus, dessen Wohnräume sich zum Innenhof öffnen; ▶Gartenhofhaus).

patio house dwelling [n] [US] *arch.* ▶patio house.

patrimony [n] *conserv. land'man. nat'res.* ▶forest patrimony, ▶landscape patrimony.

patrimony [n]**, cultural** *conserv.'hist.* ▶cultural heritage.

patrimony [n]**, natural** *conserv. land'man. nat'res.* ▶natural heritage.

pattern [n] *constr.* ▶basket-weave pattern; *arb. hort.* ▶branching pattern; *landsc. urb.* ▶combined open space pattern/system; *sociol. trans. urb.* ▶commuter pattern; *landsc. urb.* ▶concentric open space pattern/system; *geo.* ▶drainage pattern; *constr.* ▶header bond pattern, ▶herringbone pattern, ▶jointing pattern; *agr. plan. urb.* ▶land ownership pattern; *plan.* ▶land use pattern; *plan. recr.* ▶leisure activity pattern; *landsc. urb.* ▶linear open space pattern/system; *constr.* ▶ornamental pattern; *adm. sociol.* ▶ownership pattern; *constr.* ▶pavement pattern, ▶paving in curved pattern, ▶paving pattern, ▶plan of a paving pattern; *landsc. urb.* ▶radial open space pattern/system; *constr.* ▶slab paving pattern; *pedol.* ▶soil distribution pattern; *biol. ecol.* ▶species pattern; *constr.* ▶stretcher bond pattern; *constr. hort.* ▶tree planting pattern.

pattern [n]**, activity** *plan. recr.* ▶leisure activity pattern.

pattern [n]**, dissemination** *phyt.* ▶dispersion (1).

pattern [n]**, drainage piping** *constr.* ▶subsurface drainage system.

pattern [n]**, ecological landscape** *ecol. plan.* ▶ecological distribution of spatial patterns.

pattern [n] [UK]**, fan** *constr.* ▶fish scale paving [US].

pattern [n] [US]**, heading bond** *arch. constr.* ▶header bond pattern.

pattern [n]**, interfingering open space** *landsc. urb.* ▶open space pattern/system of peninsular interdigitation.

pattern [n]**, irregular** *constr.* ▶irregular bond.

pattern [n] [UK]**, journey-to-work** *sociol. trans. urb.* ▶commuter pattern [US].

pattern [n]**, landscape mosaic** *landsc.* ▶structuring of landscape.

pattern [n]**, open space system forming a radial** *landsc. urb.* ▶radial open space pattern/system.

pattern [n]**, paving** *constr.* ▶pavement pattern.

pattern [n]**, planting** *constr. hort.* ▶planting grid.

pattern [n]**, random irregular** *constr.* ▶random irregular bond.

pattern [n]**, random rectangular** *constr.* ▶random rectangular bond.

pattern [n]**, random-jointed rectangular paving** *constr.* ▶random rectangular bond.

pattern [n]**, running bond** *arch. constr.* ▶stretcher bond pattern.

pattern [n]**, segment arc** *constr.* ▶paving in curved pattern.

pattern [n]**, soil** *pedol.* ▶soil distribution pattern.

pattern [n]**, square** *constr. hort.* ▶square planting grid.

4102 patterned concrete paving slab [n] *constr.* (Decorative slab with precast covering. *Specific term* concrete slab with imitation stone); *s* **losa** [f] **de hormigón con dibujo** (Placa de hormigón con superficie simulando emsamblaje de adoquines pequeños); *f* **dalle pavé(e)** [f] (Dalle en béton dont l'aspect de surface présente un assemblage de pavés) ; *g* **Betonplatte** [f] **mit integriertem Belagsmuster** (Serienmäßig vorgefertigte Betonplatte, deren Oberfläche die Struktur kleiner Pflastersteine darstellt).

patterning [n] **of a landscape** *agr.* ▶compartmentalization of a landscape.

4103 pattern [n] **of ecotopes** *ecol.* (Distribution and linkage of several, smallest homogeneous units at the spatial scale of a landscape); *syn.* pattern [n] of tesserae [also US] (LE 1986, 600); *s* **mosaico** [m] **de ecotopos** (Mezcla de pequeñas unidades homogéneas a la escala espacial de un paisaje); *syn.* «pattern» [m] de ecotopos; *f* **réseau** [m] **d'écotopes** (Ensemble d'unités spatiales homogènes réduites dans un paysage) ; *g* **Ökotopengefüge** [n] (Verbund oder Muster von mehreren kleinsten räumlichen Einheiten einer Landschaft).

4104 pattern [n] **of interactions and relationships** *ecol.* (Natural organization of ecological components; ▶interdependency within an ecosystem); *s* **relaciones** [fpl] **de interdependencia** (Interrelación compleja entre los elementos de un ecosistema; ▶sistema de interacciones e interrelaciones); *syn.* relaciones [fpl] estructurales; *f* **relations** [fpl] **structurelles** (entre les éléments d'un écosystème ; ▶système de relations interdépendantes) ; *syn.* système [m] d'interactions, relations [f] d'interdépendances ; *g* **Beziehungsgefüge** [n] (Vielfältig strukturierte Organisation der Verflechtung von Beziehungen und Abhängigkeiten in einem Ökosystem; ▶Wirkungsgefüge); *syn.* Beziehungsstruktur [f].

4105 pattern [n] **of interrelated land uses** *plan.* (Interconnection of different land uses; ▶land use pattern); *s* **articulación** [f] **de usos** (Interconexión entre diferentes usos del suelo; ▶mosaico de usos del suelo); *f* **articulation** [f] **des usages** (dans l'organisation de l'occupation des sols, p. ex. de zones d'activités et de zones de calmes ; ▶modèle d'affectation des sols) ; *g* **Nutzungsverflechtung** [f] (Verflechtung von Flä-chennutzungen; ▶Flächennutzungsmuster); *syn.* Anordnung [f] von Flächennutzungen, Verflechtung [f] von Nutzungen.

pattern [n] **of tesserae** [US] *ecol.* ▶pattern of ecotopes.

pattern plan [n] *constr. plan.* ▶jointing pattern plan.

4106 pave [vb] *constr.* (To cover a surface with pavers, bricks, etc.; ▶paved stone surface); *s* **empedrar** [vb] (Cubrir una superficie con adoquines; ▶adoquinado); *syn.* adoquinar [vb]; *f* **paver** [vb] (Action de poser un revêtement en pavés ; ▶pavage 1) ; *g* **pflastern** [vb] (In eine Fläche ein ▶Pflaster einbauen); *syn.* pflästern [vb] [CH].

4107 paved area [n] *constr. urb.* (Surface paved, e.g. with asphalt, in situ concrete, pavers or synthetic surfacing; ▶hardscape; *opp.* ▶unpaved area); *syn.* hard surface [n], pavement area [n] [also US], hardscape area [n] [also US] (LA 10/89, 128); *s* **superficie** [f] **pavimentada** (Área revestida con algún tipo de pavimento, p. ej. de asfalto, hormigón vertido in situ, adoquines o losas; ▶componentes duros de espacios libres; *opp.* ▶superficie no pavimentada); *syn.* área [f] pavimentada; *f* **surface** [f] **stabilisée** (*Terme spécifique* sol dallé ; ▶éléments inertes d'un espace libre ; *opp.* ▶surface semi-stabilisée) ; *syn.* sol [m] stabilisé, aire [f] en sol stabilisé ; *g* **befestigte Fläche** [f] (Bodenoberfläche, die z. B. mit Asphalt, Ortbeton, Pflastersteinen oder einem Kunststoffbelag befestigt ist; ▶Technisch-Bauliches einer Freianlage; *opp.* ▶unbefestigte Fläche); *syn.* Hartfläche [f] [CH], harte Oberfläche [f] (UG 2004, 53ff).

paved embankment [n] *eng.* ▶bank revetment.

paved pull-off lane [n] [US] *trans.* ▶hard shoulder.

4108 paved rubble drop chute [n] [US] *constr.* (Depressed channel composed of vertically packed rocks of various heights to allow for occasional drainage on steep slopes); *syn.* rock-lined ditch [n], rough bed channel [n] [UK]; *s* **canal** [m] **de desagüe rugoso** (≠) (Sistema de drenaje de taludes muy pendientes en forma de hondonada construida de montones de piedras irregulares colocadas en montones de diferentes alturas para permitir la evacuación del agua); *f* **fossé** [m] **rugueux** (Fossé d'assainissement en forme de noue dans le fond duquel sont disposées sous forme d'un épi serré des pierres brutes de différentes hauteurs afin de réduire la vitesse de l'eau sur les talus de forte pente) ; *syn.* caniveau [m] frein ; *g* **Raubettrinne** [f] (Muldenartig ausgeformte Entwässerungsrinne aus aufrecht, in dichter Packung in unregelmäßiger Höhe aufgestellten Bruchsteinen zur zeitweiligen Wasserableitung an steilen Böschungen).

paved shoulder [n] [US] *trans.* ▶hard shoulder.

4109 paved stone surface [n] *constr.* (Pavement of roads, trails, courts or squares with natural stone or concrete pavers, cobblestones [US], pebbles, wood blocks, etc.; generic term for ▶block paving [US]/sett paving [UK], ▶brick paving, ▶grass-filled modular paving [US]/grass setts paving [UK], ▶interlocking block pavement, ▶mosaic block paving [US]/mosaic sett paving [UK], ▶natural stone paving, ▶pavement of concrete pavers, ▶paving 1, ▶random cobblestone paving [US]/random sett paving [UK], ▶small block paving [US]/small sett paving [UK], ▶wood paving); *syn.* stone pavement [n] [also US]; *s* **adoquinado** [m] (1) (Término genérico para diversos tipos de pavimento de piedras naturales o artificiales, como ▶adoquinado de ensamblaje, ▶adoquinado de piedra de hormigón, ▶adoquinado de piedras naturales, ▶adoquinado pequeño, ▶adoquinado sin orden, ▶pavimento de adoquines con césped, ▶pavimento de adoquines grandes, ▶pavimento de ladrillo clinker, ▶pavimento de madera, ▶pavimento de mosaico; ▶adoquinado 2); *syn.* pavimento [m] de adoquín, empedrado [m] (1); *f* **pavage** [m] (1) (Terme générique englobant les revêtements d'aires de circulation constitués de pavés naturels ou artificiels tels que ▶pavage désordonné en pierre, ▶pavage en bois, ▶pavage engazonné,

►pavage en gros pavés, ►pavage en pavés autobloquants, ►pavage en pavés en béton, ►pavage en pavés mosaïques, ►pavage en petits pavés, ►pavage en pierre naturelle, ►pavage en terre cuite ; ►pavage 2) ; *syn.* pavement [m], revêtement [m] en pavés, dallage [m] en pavés ; *g* **Pflaster [n]** (Straßen-, Wege- oder Platzbefestigung aus aneinander gesetzten Natur- oder Betonsteinwürfeln, Kieseln oder Holzpflasterelementen; *UBe* ►Betonsteinpflaster, ►Großsteinpflaster, ►Kleinpflaster, ►Klinkerpflaster, ►Mosaikpflaster, ►Natursteinpflaster, ►Rasenpflaster, ►Verbundpflaster und ►Wildsteinpflaster, ►Holzpflaster; *MERKE*, Plaster bedeutet Belag; *Pflasterbelag* und *Pflasterdecke* sind tautologische Begriffe! Pflastersteinbelag und Pflastersteindecke sind logisch richtige Formen; ►Pflasterung 2).

pavement [n] *constr.* ►concrete pavement, ►grid pavement [US]/pavement of grass pavers [UK], ►interlocking block pavement.

pavement [n] [UK] *trans. urb.* ►sidewalk [US].

pavement [n], **bituminous** *constr.* ►black-top [US&CDN] (1), *►surface layers.

pavement [n] [US], **bituminous concrete** *constr.* *►surface layers.

pavement [n], **block** *constr.* ►interlocking block pavement.

pavement [n] [UK], **concrete block** *constr.* ►pavement of concrete pavers [US].

pavement [n] [US], **natural stone** *constr.* ►natural stone paving.

pavement [n], **rigid** *constr.* ►concrete pavement.

pavement [n], **stone** *constr.* ►paved stone surface.

pavement area [n] [US] *constr. urb.* ►paved area.

4110 pavement construction courses [npl] *constr.* (Total thickness of construction upon the ►subgrade [US] 2/subgrade [UK] comprising the following layers: ►frost resistant subbase, ►subbase [US]/sub-base [UK], ►base course 1 and ►surface layer; ►pavement structure, ►subbase grade [US]/formation level [UK]); *s* **composición [f] estratificada de superficies revestidas** (Conjunto de capas de pavimento de vías o plazas: ►capa de protección contra las heladas, ►capa portante, a veces ►capa de enrase y ►firme; ►estructura del cuerpo de carreteras y caminos, ►subbase, ►subrasante, ►subsuelo 1); *f* **éléments [mpl] constructifs de la voirie** (►Structure de la voirie: constituée par la ►couche antigel, la ►couche de fondation, éventuellement la ►couche de base et la ►couche de roulement ; ►couche de forme, ►partie supérieure des terrassements, ►fond de forme) ; *syn.* couches [fpl] de chaussée, couches [fpl] de base et de surface ; *g* **Oberbau [m]** (*Wegeaufbau* Straßen-, Wege oder Platzbefestigung, die aus einer ►Frostschutzschicht, ►Tragschicht, evtl. einer ►Ausgleichsschicht und einer ►Decke besteht; ►Untergrund, ►Unterbau, ►Erdplanum, ►Wegeaufbau).

4111 pavement frost damage [n] *constr. eng.* (Surface fracturing of asphalt or concrete caused by the weight of moving vehicles during thawing period: damage is thus not caused directly by frost; ►building frost damage, ►frost heave); *s* **dislocamiento [m] causado por helada** (Término específico de ►daños causados por heladas, consistente en rotura del pavimento de carreteras en época de deshielo causada por el peso de los vehículos sobre el pavimento afectado por las heladas; ►levantamiento por congelación); *f* **gonflement [m] lors du dégel** (*Certains* ►*dommages causés par le gel* détérioration des revêtements routiers ou des cheminements sous l'action de la circulation des véhicules pendant la période du dégel ; ►foisonnement par le gel) ; *g* **Frostaufbruch [m]** (UB zu ►Frostschaden [an Bauwerken]: Aufbrechen des Deckenaufbaues im Straßen-

und Wegebau während der Tauperiode bei Belastung durch Fahrzeuge. Es sind keine unmittelbaren Frostschäden; ►Frosthebung).

pavement layer [n] *constr.* ►surface layer.

4112 pavement [n] **of concrete pavers** [US] *constr.* (Surface made of ►concrete pavers [US]/concrete paving units [UK]); *syn.* concrete block pavement [n] [UK]; *s* **adoquinado [m] de piedra de hormigón** (Pavimento de ►piedra de hormigón); *f* **pavage [m] en pavés en béton** (Surface recouverte de ►pavés en béton préfabriqués) ; *g* **Betonsteinpflaster [n]** (Belag aus ►Betonpflastersteinen 1).

pavement [n] **of grass pavers** [UK] *constr.* ►grid pavement [US].

4113 pavement pattern [n] *constr.* (Arrangement of a paved stone surface; e.g. ►paving in curved pattern, ►random cobblestone paving [US]/random sett paving [UK], ►paving in running bond); *syn.* paving pattern [n] (LD 1988 [10], 48); *s* **aparejo [m] de adoquinado** (Pattern de ordenación de los adoquines en una superficie, p. ej. ►adoquinado sin orden, ►pavimento en arco, ►pavimento en aparejo a soga; ►patrón de aparejo); *f* **calepinage [m] du pavage** (Forme d'exécution d'un revêtement en pavés, p. ex. ►pavage en arc de cercle, en losange ou damier ou ►pavage en bandes ; ►pavage désordonné en pierre, ►calepinage) ; *syn.* pavement [m] ; *g* **Pflasterung [f] (1)** (Art und Weise [Muster] wie ein Belag aus Pflastersteinen [►Pflaster] ausgeführt wird, z. B. ►Reihenpflasterung, ►Segmentpflasterung; ►Wildsteinpflaster; ►Verlegemuster); *syn.* Pflästerung [f] [CH].

pavement planting [n] *constr.* ►paving joint vegetation.

4114 pavement structure [n] *constr. eng.* (MEA 1985, 359; overall depth of ►pavement construction courses with successive layers of specified materials installed on existing or improved ►subgrade [US] 1/sub-grade [UK]; ►multicourse construction); *syn.* pavement system [n] [also US] (MEA 1985, 495), paving structure [n] [also UK]; *s* **estructura [f] del cuerpo de carreteras y caminos** (Secuencia de capas [de abajo a arriba] para estabilizar carreteras y caminos: ►subbase o ►subsuelo y capa[s] superior[es]; ►composición estratificada de superficies revestidas, ►estructura multicapa); *syn.* estructura [f] de suelos revestidos; *f* **structure [f] de la voirie** (Éléments constitutifs de chemins, rues et routes présentant du bas vers le haut: ►couche de forme ou ►partie supérieure des terrassements et les ►éléments constitutifs de la voirie ; ►assise multicouche) ; *syn.* disposition [f] des couches (d'une chaussée), organisation [f] des couches (d'une chaussée), éléments [mpl] constructifs de chaussée, structure [f] d'une chaussée ; *g* **Wegeaufbau [m]** (Schichtenfolge zur Befestigung von Wegen und Straßen von unten nach oben: ►Unterbau oder ►Untergrund und ►Oberbau; ►mehrschichtiger Aufbau); *syn.* Schichtenaufbau [m] von Wegen, *für Straßen* Straßenaufbau [m].

pavement system [n] [US] *constr.* ►pavement structure.

4115 paver [n] [US] **(1)** *constr.* (One who lays pavements, or whose occupation is to pave); *syn.* pavier [n] [UK], pavior [n] [UK], paviour [n] [UK]; *s* **empedrador [m]** (Nombre profesional de persona que coloca el pavimento); *syn.* adoquinador [m]; *f* **paveur [m]** (*Profession* spécialiste dans la pose de revêtements en pavés) ; *syn.* poseur [m] de pavés ; *g* **Pflasterer [m]** (Berufsbezeichnung für jemanden, der einen Pflastersteinbelag herstellt); *syn.* Steinsetzer/-in [m/f], Pflästerer/-in [m/f] [CH].

paver [n] [US] **(2)** *constr.* ►concrete paver [US]/concrete paving unit [UK], ►grass paver, ►interlocking paver [US]/interlocking paving block, ►mosaic paver [US]/mosaic paving sett [UK], ►paving stone [US]/paving sett [UK], ►standard non-interlocking concrete paver, ▶tinted concrete paver [US]/coloured paving block [UK].

paver [n] **[US], brick unit** *constr.* ▶paver brick [US]/ clinkerbrick for paving [UK].

paver [n] **[US], concrete unit** *constr.* ▶concrete paver [US]/ concrete paving stone [UK].

paver [n] **[US], grass-crete** *constr.* ▶grass paver.

paver [n] **[US], large-sized concrete** *constr.* *▶large-sized paving stone [US]/large-sized concrete paving sett [UK].

paver [n] **[US], module** *constr.* ▶precast paving block [US]/precast concrete paving slab [UK].

paver [n] **[US], precast concrete** *constr.* ▶precast paving block [US]/precast concrete paving slab [UK].

paver [n] **[US], small stone** *constr.* ▶small paving stone [US]/small paving sett [UK].

paver [n] **[US], turf** *constr.* ▶grass paver.

4116 paver brick [n] **[US]** *constr.* (DIL 1987; usually rectangular-shaped brick of clay material with or without admixtures which is hard-baked [sintered] for use in ▶brick paving); *syn.* brick paving unit [n] [US], brick unit paver [US], clinker brick [n] for paving [UK]; *s* **ladrillo** [m] **clinker de pavimentación** (Ladrillo generalmente rectangular de menor grosor que los ladrillos normales, de arcilla con o sin aditivos, que es cocido hasta la vitrificación; ▶pavimento de ladrillo clinker); *syn.* clinker [m] de pavimentación, ladrillo [m] vitrificado de pavimentación, ladrillo [m] holandés de pavimentación, ladrillo [m] pavimentador, ladrillo [m] para pavimento/firme (BU 1959); *f 1* **brique** [f] **de pavage** (Brique très dure de forme en général rectangulaire fabriquée par frittage de l'argile et dont la porosité ne doit pas dépasser 4 % ; cf. DTB 1985 ; ▶pavage en terre cuite) ; *syn.* pavé [m] en terre cuite ; *f 2* **pavé** [m] **en grès étiré** (constitué d'un mélange à base de terre cuite) ; *syn.* pavé [m] en clinker ; *g* **Straßenbauklinker** [m] (Vorwiegend rechteckig geformter Klinker aus tonigem Material mit oder ohne Zusatzstoffen, bis zur Sinterung gebrannt, zur Herstellung einer Klinkerpflasterung verwendet; cf. DIN 18 503: 2003-12 und DIN EN 1344; ▶Klinkerpflaster); *syn.* Pflasterklinker [m], Pflasterziegel [m].

paver [n] **with exposed aggregate** **[US]** *constr.* ▶exposed aggregate paving block.

paver [n] **with exposed crushed basalt** **[US]** *constr.* ▶concrete paver with exposed crushed basalt [US]/exposed basalt aggregate paving block [UK].

paver [n] **with protective coating** **[US]** *constr.* ▶concrete paver with protective coating [US]/concrete paving block with hard-wearing surface layer [UK].

pavestone [n] **[US], large-sized concrete** *constr.* ▶large-sized paving stone [US]/large-sized concrete paving sett [UK].

pavestone [n] **[US], large-sized natural** *constr.* ▶large-sized paving stone [US]/large-sized natural paving stone.

pavier [n] **[UK]** *constr.* ▶paver [US] (1).

4117 paving [n] **(1)** *constr.* (Process of laying paving stone [US]/paving setts [UK]); *s* **adoquinado** [m] **(2)** (Proceso de colocar adoquines siguiendo un patrón); *syn.* empedrado [m] (2); *f* **pavage** [m] **(2)** (Action de poser des pavés) ; *syn.* pose [f] d'un pavage ; *g* **Pflasterung** [f] **(2)** (Vorgang des fluchtgerechten und höhengleichen Setzens von Pflastersteinen sowie das Verlegen der Steine im vorgeschriebenen Muster/ Verband); *syn.* Pflasterverlegung [f].

paving [n] **(2)** *constr.* ▶block paving [US]/sett paving [UK], ▶cobble paving, ▶grass-filled modular paving [US]/grass setts paving [UK], ▶mosaic block paving, ▶natural stone paving,

▶random cobblestone paving [US], ▶wood disk paving [US]/timber disk paving [UK], ▶wood paving.

paving [n] **[UK], baked clay** *constr.* ▶brick paving.

paving [n]**, cobblestone** *constr.* ▶cobble paving.

paving [n] **[US], flagstone** *constr.* *▶surface layers, #8.

paving [n] **[UK], grass setts** *constr.* ▶grass-filled modular paving [US].

paving [n] **[UK], mosaic sett** *constr. eng.* ▶mosaic block paving [US].

paving [n] **[UK], random sett** *constr.* ▶random cobblestone paving [US].

paving [n] **[UK], round wood** *constr.* ▶wood disk paving [US]/timber disk paving [UK].

paving [n] **[UK], sett** *constr.* ▶block paving [US].

paving [n] **[UK], small sett** *constr.* ▶small stone paving [US].

paving [n] **[UK], timber disk** *constr.* ▶wood disk paving [US].

paving [n] **[UK], timber sett** *constr.* ▶wood block paving.

paving [n] **[UK], wood sett** *constr.* ▶wood block paving.

4118 paving alignment [n] *constr.* (Direction of pavers or rows of paving slabs); *s* **dirección** [f] **de colocación/tendido** (Sentido en el que se colocan los adoquines o las losas); *f* **direction** [f] **de pose** (Sens de mise en œuvre de pavés, de bandes de dalles) ; *g* **Verlegerichtung** [f] (Ausrichtung von Pflasterstein- oder Plattenreihen).

paving block [n] *constr.* ▶exposed aggregate paving block, ▶precast paving block [US]/precast concrete paving slab [UK].

paving block [n] **[UK], coloured** *constr.* ▶tinted concrete paver [US].

paving block [n] **[UK], exposed basalt aggregate** *constr.* ▶concrete paver with exposed crushed basalt [US].

paving block [n] **[US], grass-filled** *constr.* ▶grass paver.

paving block [n] **[UK], interlocking** *constr.* ▶interlocking paver [US].

paving block [n]**, standard concrete** *constr.* ▶standard non-interlocking concrete paver [US]/standard concrete paving block [UK].

paving block [n] **with hard-wearing surface layer** **[UK]** *constr.* ▶concrete paver with protective coating [US].

paving block [n] **without exposed aggregate** *constr.* ▶standard concrete paving block without exposed aggregate.

4119 paving [n] **in a forward direction** *constr.* (System of paving whereby the workman lays setts on a prepared sand bed while positioned on the finished surface); *s* **adoquinado** [m] **hacia adelante** (Sistema de pavimentar en el que el trabajador se posa sobre la superficie ejecutada); *f* **pavage** [m] **face à l'avancement** (Mode de pose d'un pavage pour laquelle le poseur travaille sur le pavage exécuté et pose les pavés sur un lit de pose déjà préparé) ; *g* **Pflasterung** [f] **vorwärts** (Arbeitsvorgang, bei dem sich der Ausführende auf dem fertig gepflasterten Belag bewegt und die weiteren Pflastersteine oder Platten auf das vorbereitete, vor ihm liegende Sandbett verlegt).

4120 paving [n] **in a reverse direction** *constr.* (System of paving whereby the workman operates from a position on the sand bed and installs sand under each sett or slab); *s* **adoquinado** [m] **hacia atrás** (Sistema de pavimentar en el que el trabajador opera desde el lecho de asiento); *f* **pavage** [m] **exécuté devant le poseur** (Mode de pose d'un pavage pour laquelle le poseur travaille le dos tourné à la direction de pose, le lit de pose devant

P

être renouvelé pour chaque pavé ou chaque dalle) ; *g* **Pflasterung [f] rückwärts** (Arbeitsvorgang, bei dem der Pflasterer den fertig verlegten Belag vor sich hat, sich auf dem Sandbett befindet und für jeden Stein oder für jede Platte das Sandbett gesondert herrichtet).

4121 paving [n] **in curved pattern** *constr.* (SPON 1986, 361; ▶pavement pattern, ▶paving in running bond); *syn.* segment arc pattern [n] [also UK] (LD 10/88, 49); *s* **pavimento [m] en arco** (Pavimento en hileras curvas de adoquines; ▶aparejo de adoquinado, ▶pavimento en aparejo a soga); *f* **pavage [m] en arc de cercle** (▶calepinage du pavage, ▶pavage en bandes) ; *syn.* pavage [m] à la parisienne, pavage [m] avec pose en arceau ; *g* **Segmentpflasterung [f]** (▶Pflasterung 1 aus bogenförmigen Pflasterzeilen; ▶Reihenpflasterung); *syn.* Bogenpflasterung [f].

4122 paving [n] **in running bond** *constr.* (Paving in which bricks or stones are laid lengthwise with alternate joints); *s* **pavimento [m] en aparejo a soga** (Revestimiento de suelos en el cual los ladrillos o adoquines son tendidos a lo largo con juntas alternantes); *f* **pavage [m] en bandes** (Revêtement des sols en pavés disposés selon un motif de bandes parallèles) ; *syn.* revêtement [m] en bandes ; *g* **Reihenpflasterung [f]** (In parallelen Reihen angeordnete Pflastersteine).

4123 paving joint vegetation [n] *constr.* (Plants which have been sown or planted in deliberately-wide joints between pavers, or which have become established there spontaneously; ▶soil pocket, ▶vegetation of wall joints, rock crevices or paving joints); *syn.* pavement planting [n] (DIL 1987); *s* **vegetación [f] de las juntas de adoquinado** (Vegetación espontánea o sembrada que crece entre los intersticios de los adoquines; ▶junta de plantación, ▶vegetación de rendijas); *f* **végétation [f] d'interstices de pavage** (Végétation semée [joint planté] ou poussant naturellement dans les interstices de pavés ; ▶niche de plantation, ▶végétation saxicole et muricole) ; *syn.* végétation [f] de joints de pavage ; *g* **Pflastergrün [n]** (Gesamtheit der in breiten Pflasterfugen eingesäten, gepflanzten oder sich spontan ansiedelnden Pflanzen; ▶Fugenbewuchs, ▶Pflanzfuge).

4124 paving [n] **of a square/plaza** *constr.* (▶Surfacing 3 of a square/plaza); *s* **pavimento [m] de plaza** (Tipo y patrón de firme de una plaza; ▶estabilización de plazas y caminos); *f* **revêtement [m] d'une place** (Type d'aménagement de sols pour une place ; ▶consolidation) ; *g* **Platzbefestigung [f]** (Art der Ausbildung der Oberfläche/Decke eines Platzes; ▶Befestigung 2).

4125 paving pattern [n] *constr.* (Arrangement of pavers on pathways and squares; ▶bond, ▶pavement pattern, ▶plan of a paving pattern, ▶slab paving pattern); *s* **patrón [m] de aparejo** (Arreglo de los adoquines, losas, ladrillos clinker, etc. en caminos y plazas; ▶aparejo, ▶aparejo de adoquinado, ▶aparejo de losas, ▶representación gráfica del patrón de aparejo); *f* **calepinage [m]** (Disposition des pavés, des dalles, briques utilisés en extérieur en tant que revêtement de surface [voies, places, parkings, cours] ; ▶appareillage, ▶appareillage de dalles, ▶calepinage du pavage, ▶principe de calepinage) ; *syn.* assemblage [m] des matériaux ; *g* **Verlegemuster [n]** (Anordnung von Pflastersteinen, Gehwegplatten, Klinkern etc. auf Straßen-, Wege- und Platzflächen oder Gartenterrassen; ▶Pflasterung 1, ▶Plattenverband, ▶Verband 2, ▶zeichnerische Darstellung eines Verlegemusters).

paving pattern [n]**, random-jointed rectangular** *constr.* ▶random rectangular bond.

paving sett [n] [UK] *constr.* ▶paving stone [US].

paving sett [n] [UK]**, large-sized concrete** *constr.* *▶large-sized paving stone [US].

paving sett [n] [UK]**, mosaic** *constr.* ▶mosaic paver [US].

paving sett [n] [UK]**, small** *constr.* ▶small paving stone [US].

4126 paving slab [n] *constr.* (Flat construction block for surfacing; ▶footpath paving slab, ▶natural garden stone flag, ▶natural stone slab, ▶patterned concrete paving slab, ▶precast paving block [US]/precast concrete paving slab [UK]); *s* **losa [f]** (Elemento plano de pavimentación; ▶losa de hormigón, ▶losa de hormigón con dibujo, ▶losa de jardín [de piedra natural o artificial], ▶losa de pavimentación, ▶losa de piedra natural); *f* **dalle [f]** (Plaque pour revêtement de sol ; ▶dalle de jardin, ▶dalle en pierre naturelle, ▶dalle pavé[e], ▶dalle pour circulation piétonne, ▶dalle préfabriquée en béton pour voirie de surface) ; *g* **Platte [f]** (Ebenes, gleichmäßig dickes, flächiges Bauelement zur Befestigung einer Bodenoberfläche; ▶Betonplatte mit integriertem Belagsmuster, ▶Gehwegplatte aus Beton, ▶Gartenplatte, ▶Gehwegplatte, ▶Natursteinplatte).

paving slab [n] [UK]**, garden** *constr.* ▶natural garden stone flag.

paving slab [n] [UK]**, precast concrete** *constr.* ▶precast paving block [US].

4127 paving stone [n] [US] *constr.* (Prefabricated concrete unit or natural stone: ▶cobblestone [US]/natural stone sett [UK], ▶cobblestone [US]/random sett [UK], ▶concrete paver [US]/concrete paving unit [UK], ▶interlocking paver [US]/interlocking paving block [UK], ▶large-sized paving stone [US]/large-sized natural stone sett [UK], ▶mosaic paver [US]/mosaic paving sett [UK], ▶small paving stone [US]/small paving sett [UK]); *syn.* paver [n] [US], paving sett [n] [UK], paving unit [n] [US]; *s* **adoquín [m]** (Pieza prefabricada de hormigón o piedra natural utilizada para pavimentos; *términos específicos* ▶adoquín de ensamblaje, ▶adoquín grande, ▶adoquín mosaico, ▶adoquín natural, ▶adoquín pequeño, ▶ladrillo clinker de pavimentación, ▶piedra de hormigón); *f* **pavé [m]** (Matériau normalisé en pierre ou en béton utilisé comme revêtement des circulations ; *termes spécifiques* ▶brique de pavage, ▶gros pavé, ▶pavé autobloquant, ▶pavé en béton, ▶pavé en grès étiré, ▶pavé en pierre naturelle, ▶pavé mosaïque en pierre naturelle, ▶petit pavé) ; *g* **Pflasterstein [m]** (Natur- oder Betonwerkstein zum Befestigen einer Verkehrsfläche; *UBe* ▶Betonpflasterstein 1, ▶Großpflasterstein, ▶Kleinpflasterstein, ▶Mosaikpflasterstein, ▶Natursteinpflasterstein, ▶Straßenbauklinker, ▶Verbundpflasterstein).

paving stone [n]**, large-sized natural** *constr.* *▶large-sized paving stone [US].

paving stone [n]**, precut** *constr.* ▶natural stone flag.

4128 paving strip [n] *constr.* (Single row or several rows of pavers; e.g. for pathway edging); *s* **bordillo [m] de adoquines** (Una o varias filas de adoquines p. ej. para encintado de caminos o de piezas jardineras); *f* **bordure [f] en pavés** (Alignement d'une ou plusieurs rangées de pavés constituant, p. ex. la bordure d'un chemin pour piétons) ; *syn.* ceinture [f] en pavés ; *g* **Pflastergurt [m]** (Ein- oder mehrzeiliger Streifen aus Pflastersteinen, z. B. als Wegeeinfassung; ein einreihiger **P.** wird auch **Pflasterzeile** genannt); *syn.* Pflasterstreifen [m].

paving structure [n] [UK] *constr. eng.* ▶pavement structure.

paving unit [n] [US] *constr.* ▶paving stone [US]/paving sett [UK].

paving unit [n] [US]**, brick** *constr.* ▶paver brick [US]/clinkerbrick for paving [UK].

paving unit [n] [UK]**, concrete** *constr.* ▶concrete paver [US].

paving unit [n] [AUS]**, interlocking concrete** *constr.* ▶interlocking paver [US].

pavior [n] [UK] *constr.* ▶paver [US] (1).

paviour [n] [UK] *constr.* ▶paver [US] (1).

payment [n] *contr. prof.* ▶bonus payment, ▶compensation payment, ▶interim payment; *contr.* ▶overtime payment, ▶remuneration; *constr. contr.* ▶retention of payment.

payment [n], **partial** *constr. contr.* ▶interim payment.

payment [n] [US], **progress** *constr. contr.* ▶interim payment.

payment [n] [US], **service connection** *adm. urb.* ▶utility connection charge.

payment [n] [US], **token** *prof.* ▶honorarium.

payment [n] **of professional charges/fees** *contr. prof.* ▶remuneration of professional fees.

4129 payment [n] **on a time basis** *contr.* (Listing of hours worked by labo[u]r in construction, machine work or landscape maintenance; ▶daywork); *s* **liquidación** [f] **de cuentas sobre la base del tiempo empleado** (Enlistado de las horas trabajadas por la mano de obra en trabajos de construcción, con maquinaria o en mantenimiento de zonas verdes; ▶trabajos por salario horario); *f* **règlement** [m] **au temps passé** (Établissement du décompte des heures passées pour des prestations de travaux, main d'œuvre ou de matériels dans le cas d'un marché en régie ; ▶travaux exécutés en régie) ; *g* **Abrechnung** [f] **nach Zeit(aufwand)** (Aufstellung einer [Schluss]rechnung über geleistete Stunden für Bauleistungen, Maschinen- oder gärtnerische Pflegearbeiten; ▶Stundenlohnarbeiten).

4130 pea gravel [n] *constr.* (OEH 1990, 242; washed round gravel, 2-8mm in particle size, used as wearing course when mixed in a stabilized surface [US]/hoggin surface [UK] of garden and park walks); *syn.* pea shingle [n] [UK], shingle [n] [UK]; *s* **garbancillo** [m] (Tipo de gravilla redondeada y lavada, de granulometría entre 2 y 8 mm, que se utiliza para firmes de revestimiento compactado en caminos de parques y jardines); *f* **gravillon** [m] (Gravier lavé, rond, de granulométrie comprise entre 2 et 8 mm, utilisé comme revêtement des allées stabilisées aux liants hydrauliques dans les parcs et jardins) ; *syn.* mignonnette [f] ; *g* **Gartenriesel** [m, o. Pl.] (Gewaschener, runder Kies, i. d. R. mit einer Körnung 2-8 mm als Deckschicht für wassergebundene Garten- oder Parkwege).

4131 peak blooming [n] *hort.* (Maximum flowering period of one or several species); *s* **floración** [f] **principal** (Época del año en la que florecen al máximo una o varias especies); *f* **période** [f] **de floraison principale** (Époque de l'année pendant laquelle la plupart des fleurs d'une espèce ou d'une composition fleurissent) ; *g* **Hauptblüte** [f] (Zeit, in der die meisten Blüten einer Art oder Artenkombination blühen).

4132 peak discharge [n] *eng. hydr. wat'man.* (Top quantity of water flowing out of pipe or watercourse in a given period of time); *syn.* maximum discharge [n] (WMO 1974); *s 1* **caudal** [m] **de punta** (Valor máximo del caudal de un río para un periodo/período determinado); *syn.* caudal [m] máximo instantáneo; *s 2* **caudal** [m] **máximo de salida de agua** (Cantidad máxima de agua que puede salir de una tubería); *f* **débit** [m] **de pointe** (Volume d'eau maximum franchissant une section transversale d'une canalisation, du lit d'un fleuve, mesué lors d'un évènement) ; *syn.* débit [m] extrême, débit [m] maximal ; *g* **Abflussspitze** [f] (1) (Höchstmaß an Wasseraufkommen in Rohrleitungen oder im Flussbett, das abfließt); *syn.* Abflussextrem [n].

peak high-water level/mark [n] *wat'man.* ▶record high-water level/mark.

4133 peak runoff [n] *constr.* (Top quantity of rainwater [US]/rain-water [UK] flowing from a land surface); *s* **escorrentía** [f] **máxima** (1) (Cantidad máxima de aguas pluviales que fluye

en superficialmente en una zona); *f* **ruissellement** [m] **maximal** (Écoulement rapide correspondant au maximum des eaux de pluie ou de fusion nivale à la surface des versants) ; *syn.* écoulement [m] de pointe ; *g* **Abflussspitze** [f] (2) (Höchstmaß an Regenwasseraufkommen auf der Landoberfläche, das abfließt); *syn.* Spitzenabfluss [m].

4134 peak visitor use [n] *plan. recr.* (Extremely high numbers of visitors to a recreation area bringing the greatest pressure to the area; ▶visitor number, ▶visitor pressure); *s* **cantidad** [f] **máxima de visitantes** (Número extremadamente alto de personas visitando una zona o un evento; ▶afluencia de visitantes, ▶presión turística); *syn.* cantidad [f] máxima de usarios; *f* **pointe** [f] **de fréquentation (touristique)** (Nombre très important de visiteurs dans un même, lieu à un moment donné souvent pris en compte dans la valeur de fréquentation quotidienne maximale [jour de pointe] ; ▶affluence touristique, ▶pression exercée par les visiteurs) ; *g* **Spitzenbesucheraufkommen** [n] (Extrem hohe Besucheranzahl je Zeiteinheit und Fläche oder je Veranstaltung; ▶Besucheraufkommen, ▶Besucherdruck).

pea shingle [n] [UK] *constr.* ▶pea gravel.

4135 peat [n] *conserv. pedol.* (Hydromorphous, non-terrestrial ▶humus type in ▶low bogs, ▶transition bogs, ▶raised bogs; ▶growing on peat; *specific terms* ▶fibric peat 1, ▶fibric peat 2, ▶humified raised bog peat, ▶minerotrophic peat, ▶sedge peat [US]/fen peat [UK], ▶sphagnum peat, transition bog peat); *s* **turba** [f] (Materia carbonácea, blanda, parda, más o menos oscura, constituida por restos vegetales variados en diversos grados de descomposición. La **t.** se forma en el seno de las aguas, en los trampales y tremedales, con poco oxígeno. En gran parte procede de la descomposición de musgos del género *Sphagnum*; ▶tipo de humus de la ▶turbera baja, de la ▶turbera de transición y de la ▶turbera alta; cf. DB 1985; ▶turfófilo/a; *términos específicos* ▶turba blanca, ▶turba clara, ▶turba de esfagnos, ▶turba minerótrofa, ▶turba negra, ▶turba para jardines); *f* **tourbe** [f] (1) (▶Type d'humus hydromorphe produit de la décomposition et de l'humidification lente de certains végétaux dans un milieu mal aéré et saturé en eau dans les ▶tourbières basses, les ▶tourbières de transition et les ▶tourbières hautes ; ▶turficole ; *termes spécifiques* ▶tourbe de Sphaigne, ▶tourbe blanche, ▶tourbe blonde, ▶tourbe brune, ▶tourbe fibreuse, tourbe intermédiaire, ▶tourbe noire, ▶tourbe topogène) ; *g* **Torf** [m] (Hydromorphe, nicht terrestrische ▶Humusform im ▶Niedermoor 1, ▶Übergangsmoor und ▶Hochmoor; ▶auf Torf wachsend; *UBe* ▶Niedermoortorf 1, ▶Niedermoortorf 2, Übergangsmoortorf, ▶Hochmoortorf; ▶Schwarztorf, ▶Weißtorf 1, ▶Weißtorf 2).

peat [n] [UK], **fen** *constr.* ▶sedge peat [US].

peat [n] [US], **low-moor** *pedol.* ▶minerotrophic peat.

peat [n] [US], **Michigan** *pedol.* ▶humified raised bog peat.

peat [n], **raised bog** *pedol.* ▶shagnum peat.

peat [n], **sapric** *pedol.* ▶humified raised bog peat.

4136 peat ball [n] *constr. hort.* (▶loose peat); *s* **fardo** [m] **de turba** (Cubo bien prensado de turba seca; *opp.* ▶turba suelta); *syn.* paca [f] de turba; *f* **balle** [f] **de tourbe** (Tourbe déchiquetée puis comprimée en balle ; ▶tourbe en vrac) ; *g* **Torfballen** [m] (Fest gepresster Kubus aus trockenem Torf; *opp.* ▶loser Torf).

4137 peat bed [n] (1) *constr. hort.* (Part of a garden intended for plants which require a bog soil; ▶peat garden plant [UK]); *s* **macizo** [m] **de tierra turbosa** (Arriate para plantas de la familia de las Ericáceas que requieren ese tipo de tierra para su crecimiento óptimo; ▶planta de tierra turbosa); *f* **plate-bande** [f] **à plantes de terre de bruyère** (Plate-bande conçue principalement pur la plantation de plantes de la famille des Ericacées ;

P

►plante de terre de bruyère) ; *g* **Moorbeet [n]** (Ein für die Bepflanzung von ►Moorbeetpflanzen angelegtes Beet).

peat bed [n] (2) *geo.* ►peat mass.

peat bog [n] *geo.* ►peat mass.

4138 peat-cut area [n] *min.* (Peat bog excavated by hand or machine; ►peat cutting [US]/peat-cutting [UK], ►worked bog); *s* **área [f] de extracción de turba** (►extracción de turba, ►turbera explotada) ; *f* **fosse [f] de tourbage** (Aire de dimension réduite dans une tourbière exploitée manuellement ou machinellement ; ►exploitation industrielle de la tourbe, ►exploitation de la tourbe, ►tourbière entièrement exploitée) ; *syn.* fosse [f] d'extraction de la tourbe, fosse [f] d'exploitation, creuse [f] [CH] ; *g* **Torfstich [m]** (Mit Hand oder maschinell abgetorfte kleinere Moorfläche; ►abgetorftes Moor, ►Torfabbau).

4139 peat cutting [n] [US]/**peat-cutting** [n] [UK] *min.* (VIR 1982, 341) (Extraction of peat by manual or mechanical methods; ►peat-cut area, ►worked bog); *syn.* peat-winning [n] [also UK]; *s* **extracción [f] de turba** (Ganancia de turba manual o industrial en ►área de extracción de turba; ►turbera explotada) ; *f* **exploitation [f] de la tourbe** (Extraction manuelle ou industrielle de la tourbe ; ►fosse de tourbage ; *termes spécifiques* **1.** exploitation manuelle de la tourbe [coupe manuelle de blocs [ou briques] de tourbe extraits dans une tranchée ou une fosse au moyen d'une bêche et déposés en petites meules sur les terre-pleins latéraux pour le séchage ; *syn.* exploitation [f] traditionnelle de la tourbe ; **2.** exploitation industrielle de la tourbe [récolte par découpage mécanique de mottes ou par broyage et aspiration après assèchement de la tourbière] ; ►tourbière entièrement exploitée) ; *g* **Torfabbau [m]** (Gewinnung von Torf durch manuellen ►Torfstich oder industrielle Abbaumethoden; ►abgetorftes Moor); *syn.* Torfgewinnung [f].

4140 peat formation [n] *geo.* (Partial decomposition of vegetable matter resulting over time under waterlogged [anaerobic] conditions in peatland [US]/moor [UK]; ►raised bog or ►low bog; cf. DNE 1978; ►peat-forming species); *s* **turbificación [f]** (Descomposición parcial de partes de plantas bajo condiciones anaerobias en ►turbera; ►especie turbígena) ; *f* **turbification [f]** (Processus de décomposition lente et incomplète de la matière organique fraîche dans les milieux saturés d'eau en permanence ; ►tourbière, ►espèce formatrice de tourbe ; PED 1983, 394) ; *syn.* activité [f] turfigène, formation [f] de la tourbe ; *g* **Torfbildung [f]** (Teilweise Zersetzung von Pflanzenteilen unter anaeroben, semiterrestrischen Bedingungen in Mooren [►Hochmoor, ►Niedermoor]; ►Torf bildende Art).

4141 peat-forming species [n] *phyt.* *s* **especie [f] turbígena** *syn.* especie [f] formadora de turba; *f* **espèce [f] formatrice de tourbe** ; *g* **Torf bildende Art [f]** *o. V.* torfbildende Art [f].

4142 peat garden plant [n] [UK] *hort.* (Shrub which thrives upon an acid, organic soil such as peat, ►bog soil or ►leafmold [US]/leaf-mould [UK]; e.g. many rhododendron species, Dwarf Birch *[Betula nana]*, Scotch Heather *[Calluna vulgaris]*, Creeping Snowberry *[Gaultheria procumbens]*, etc.); *syn.* ericaceous plant [n] (±) (2); *s* **planta [f] de tierra turbosa** (Término horticultor para arbusto que crece preferentemente en suelos orgánicos ácidos de turba, ►suelo de turba y ►mantillo de hojarasca, como p. ej. diferentes especies de rododendron, el abedul enano *[Betula nana]* y la brecina *[Calluna]*); *syn.* planta [f] ericácea (±); *f* **plante [f] de terre de bruyère** (Végétal ligneux se développant de préférence dans un substrat acide, riche en substances organiques composé de tourbe, de ►terre de bruyère, de ►terreau de feuilles, p. ex. diverses espèces de Rhododendron, la Gaulthérie *[Gaultheria procumbens]*, le Camélia *[Camellia]*, le Skimmia *[Skimmia japonica]*) ; *g* **Moorbeetpflanze [f]** (*Gärtnerischer Begriff* Gehölz, das zum guten Gedei-

hen einen sauren, sehr humosen Boden aus Torf, ►Moorerde und ►Lauberde bevorzugt, z. B. diverse Rhododendronarten, Zwergbirke *[Betula nana]*, Besenheide *[Calluna]*, Rote Teppichbeere *[Gaultheria procumbens]*. Nach neueren Forschungen wachsen Rhododendronarten auf allen Bodenarten, selbst auf reinen Sandstandorten oder auf schluffigen Böden, wenn z. B. eine Auflageschicht aus Streu von Rhododendronblättern, Kiefernnadeln oder Eichenlaub oder auch Rindenhumus [nicht Rindenmulch!] vorhanden ist; cf. TASPO 1999, Nr. 22, 8).

4143 peatland [n] [US] *geo.* (WET 1993, 32; extensive area of ground overlaid with peat or acid peaty soil, usually more or less wet. In popular usage the word 'moor' is restricted to European moors, in which heather is often the prevailing plant; but similar, phytogeographical areas occur elsewhere. The English word moor cannot be used as an ecological term in the German sense of an area covered with deep ►peat: ►Hochmoor; WEBSTER cit. in GGT 1979; ►fen 1, ►raised bog, ►transition bog); *syn.* moorland [n], mire [n], moor [n] [UK], moss [n]; *s* **turbera [f]** (Terreno anegado y esponjoso cubierto por una capa superficial de vegetación pobre, constituida básicamente por especies del género *Sphagnum* y otras acidófilas, cuyas partes muertas se acumulan y estratifican en los fondos, en condiciones anaerobias, dando lugar a la formación de la ►turba. Son propias de climas fríos. Según su origen se diferencian: ►turbera alta, ►turbera baja 1 y ►turbera de transición; cf. DGA 1986); *f* **tourbière [f]** (Terme désignant un paysage ou une portion de celui-ci, caractérisé par des zones mouillées en permanence, pauvre en espèces animales et riche en groupements végétaux vivant sur une strate de ►tourbe parfois importante. Suivant leurs modes de formation, on distingue : ►tourbière haute, la ►tourbière basse, la ►tourbière de transition) ; *syn.* paysage [m] de tourbières ; *g* **Moor [n] (1)** (Bezeichnung für eine Landschaft oder deren Teile, die durch dauernd feuchte, tierarme Bereiche mit artenarmen Pflanzengesellschaften auf einer oft mächtigen Torfschicht gekennzeichnet ist. Nach ihrer Entstehung unterscheidet man ►Flachmoor, ►Hochmoor und ►Übergangsmoor; ►Torf); *syn.* Torfmoor [n].

peatland [n]**, shallow** *geo. min.* ►shallow bog.

4144 peat mass [n] *geo.* (Whole body of bog or peat soil which has developed under acid conditions); *syn.* peat bog [n], peat bed [n] (2); *s* **corpus [m] de turbera abombada** (Conjunto de la masa de turba que se ha desarrollado bajo condiciones ácidas, sobre todo a partir de especies del género *Shagnum*); *syn.* mamelón [m] de turba, corpus [m] de turbera mamelonada; *f* **corps [m] de la tourbière** (Ensemble de la tourbe constituant la tourbière bombée) ; *g* **Moorkörper [m]** (Gesamtheit der urglasförmigen Torfmasse eines Moores, das sich unter sauren Bedingungen vorwiegend aus Torfmoosen [*Sphagnum*-Arten] gebildet hat).

4145 peat moss [n] *phyt.* (Generic term for mosses of the *Sphagnum* genus, which are the major component of peat; ►hummock-forming peat moss); *syn.* Sphagnum [n]; *s* **esfagno [m]** (Término genérico para musgos del género *Sphagnum*; ►esfagno del mamelón); *syn.* musgo [m] esfagnófilo, musgo [m] del género *Sphagnum*; *f* **mousse [f] de tourbière** (Terme générique pour les mousses de l'espèce des Sphaignes *[Sphagnum]*, participant principalement à la formation de la tourbe ; ►sphaigne de buttes) ; *syn.* Sphaigne [f] ; *g* **Torfmoos [n]** (OB zu Moos der Gattung *Sphagnum,* das massenmäßig an der Torfbildung beteiligt ist; ►Bultmoos); *syn.* Sphagnum [n].

peat moss hollow community [n]**, emerged** *phyt.* ►floating sphagnum mat.

4146 peat soil [n] (1) *pedol.* (Generic term for an hydromorphic, organic soil the peat horizon of which is more than

30cm deep and contains 50% organic matter, which is only slightly decomposed; cf. RCG 1982; ►anmoor, ►bog soil); *s* **suelo [m] turboso** (Término genérico para suelos orgánicos, hidromorfos cuyo horizonte de turba tiene 30 o más cm de profundidad y gran cantidad [> 30%] de materia orgánica poco descompuesta; ►anmoor, ►suelo de turba); *syn.* histosol [m]; *f* **tourbe [f] (2)** (Terme générique pour les sols hydromorphes dont la teneur en matière organique dépasse 30 % dans les horizons superficiels, la couche organique atteignant plus de 40 cm ; ►anmoor, ►sol hydromorphe organique, ►terre de bruyère, ►tourbière) ; *g* **Moor [n] (2)** (Voll hydromorpher, organischer Boden mit über 30 cm mächtigem Torfhorizont —häufig mehrere Meter mächtig —, der mindestens 30 % organische Substanz enthält; die geografische Bezeichnung *Moor* [►Moor 1] wird auch für die Böden dieser Landschaft verwendet; cf. SS 1979, 351; ►Anmoor, ►Moorerde).

4147 peat soil [n] (2) *pedol.* (ELL 1988, 573; *imprecise term for* organic, acid soil in low bogs or arising after agriculture has improved the soil of a bog, usually drained and containing minerals. Peat overlain with mud is known as peat marsh on the North Sea coastline; ►bog soil); *syn.* peaty soil [n]; *s* **suelo [m] de turba** (*Término no claramente definido* suelo orgánico ácido de turbera baja o generado de turbera por medidas de mejora de suelos agrícolas en turbera, generalmente drenado y mineralizado; ►suelo de turba); *f 1* **sol [m] hydromorphe organique (2)** (Sous-classe des sols hydromorphes caractérisée par des sols avec une matière organique de type tourbe et une hydromorphie totale et permanente ; DIS 1986, 111) ; *syn.* sol [m] organique tourbeux [classification F.A.O.], tourbe [f] ; *f 2* **sol [m] à tourbe (±) (D.,** terme imprécis désignant les sols organiques acides sur tourbières basses entropies ou sur tourbières après mesures agricoles de mise en valeur, en général drainés et minéralisés ; ►sol hydromorphe organique) ; *syn.* sol [m] tourbeux ; *g* **Torfboden [m]** (*Kein fest definierter Begriff* saurer organischer Boden auf Niedermooren oder durch landwirtschaftliche Meliorationen in Mooren entstanden, meist entwässert und mineralisiert; an der Nordseeküste werden überschlickte Torflagen als **Torfmarsch** bezeichnet; cf. ELL 1978, 751; SS 1979, 349; ►Moorerde).

peat-winning [n] [UK] *min.* ►peat cutting [US]/peat-cutting [UK].

peaty soil [n] *pedol.* ►peat soil (2).

4148 pebbles [npl] *geo.* (Round stone particles larger than a granule and smaller than a cobble with diameters between 4-64mm; **p.** have been rounded by abrasion during transport and deposited by water; ►gravel 1); *s* **canto [m] rodado** (Piedra redondeada por el efecto del agua y la frotación, que es transportada por ésta a lo largo de arroyos y ríos y se deposita cuando cesa la fuerza de la corriente; ►grava); *syn.* guijarro [m]; *f* **galet [m]** (Fragment de roche dure usé et poli par le frottement, que la mer dépose sur le rivage ou qu'on trouve dans le lit des cours d'eau ; cf. PR 1987 ; ►gravier) ; *syn.* caillou [m] roulé ; *g* **Geröll [n]** (Gerölle [pl]; rundliche Gesteinsbruchstücke, die durch Wellenschlag oder fließendes Wasser geformt, transportiert und abgelagert werden; eine Anhäufung von Geröllen nennt man ►Kies oder Schotter).

pedaler's crossing [n] [US] *trans.* ►bike crossing [US]/cycle crossing [UK].

4149 pedestal [n] *constr.* (Supporting base for a sculpture, column or for fixing wooden bench slats); *s* **pedestal [m]** (Base de soporte para una escultura o una columna); *f 1* **socle [m]** (Désigne la pierre inférieure de soubassement d'une colonne, d'un piédestal) ; *f 2* **piédestal [m]** (Désigne d'une manière générale le support d'une statue, d'un vase) ; *g* **Sockel [m]** (Konsole/Unter-

konstruktion zum Aufstellen einer Plastik, Säule, eines Brunnens, zum Befestigen von Banklatten etc.); *syn.* Unterkonstruktion [f].

pedestrian and vehicular right-of-passage [n] *leg. urb.* ►vehicular and pedestrian easement.

4150 pedestrian circulation [n] *urb. trans.* *syn.* pedestrian traffic [n]; *s* **tráfico [m] peatonal** *syn.* circulación [f] peatonal; *f* **circulation [f] piétonnière** ; *g* **Fußgängerverkehr [m, o. Pl.]**.

4151 pedestrian circulation plan [n] *urb.* *s* **plan [m] de vías de peatones** ; *f* **plan [m] de cheminements piétonniers** (Plan souvent mis en œuvre dans le cadre de l'élaboration du plan de déplacement urbain [PDU]) ; *syn.* plan [m] de cheminement piéton, schéma [m] directeur de cheminements piétonniers ; *g* **Fußgängerverkehrsplan [m]**.

4152 pedestrian guardrail [n] [US]/**pedestrian guard-rail** [n] [UK] *urb. trans.* (Barrier to direct or contain safe pedestrian movement and access; **p. g.** may be a one- or two-rod guardrail. A low guardrail to keep pedestrians away from lawn or plantings is called 'trip rail' [UK]; ►guardrail [US] 1/crash barrier [UK]); *syn.* safety barrier [n]; *s* **barrera [f] de protección** (Obstáculo a lo largo de la acera en lugares peligrosos o ante salidas de escuelas, etc. para canalizar o proteger a los peatones; ►quitamiedo); *syn.* baranda [f] de protección; *f* **barrière [f] de protection** (Mobilier urbain de protection de personnes et de réglementation de la circulation des véhicules placé le long de la voirie, et en général à proximité des passages pour piétons situés près des carrefours dangereux ou des établissements scolaires. On distingue les barrières à une lisse — supérieure — et à deux lisses — supérieure et médiane ; *termes spécifiques* barrière de canalisation des piétons ; ►glissière de sécurité ; cf. BON 1990, 169) ; *g* **Abschrankung [f]** (Leithindernis entlang eines Bordsteines, meist neben Fußgängerüberwegen, an gefährlichen Straßenkreuzungen, an Ausgängen von Schulen, Kindergärten, Spielplätzen etc.; man unterscheidet einholmige und zweiholmige **A.en**; ►Leitplanke); *syn.* Absperrgeländer [n].

pedestrian infrastructure [n] *trans. urb.* ►vehicular and pedestrian infrastructure.

4153 pedestrianization [n] *urb.* (**1.** Conversion of a road to pedestrian use, often planted and provided with street furniture and amenities. **2.** Removal of vehicular traffic to create a ►pedestrian zone or mall [US]; ►traffic calming, ►traffic rerouting [US]/traffic redirection [UK]); *s* **peatonalización [f]** (Cierre a la circulación de vehículos y la subsiguiente reestructuración y reconstrucción de vías en zonas urbanas para convertirlas en superficies multifuncionales para el uso predominante de los peatones; ►apaciguamiento del tráfico [Es]/apaciguamiento del tránsito [AL], ►restricción del tráfico de paso [Es]/restricción del tránsito de paso [AL], ►zona peatonal); *syn.* creación [f] de zonas peatonales; *f* **création [f] de zone piétonne** (Transformation de rues en espaces multifonctionnels réservés à la circulation piétonnière [►zone piétonne] et caractérisés par l'absence de trottoirs, la plantation d'arbres d'alignement, l'installation de bacs et de divers mobiliers urbains ; ►circulation apaisée, ►restriction de la circulation automobile) ; *syn.* création [f] d'espaces piétons, piétonnisation [f] ; *g* **Schaffung [f] von Fußgängerzonen** (Umgestaltung und Umbau von Verkehrsstraßen im Innenstadtbereich zu multifunktionalen fußgängerfreundlichen Flächen, meist ohne Bordsteine, ausgestattet oft mit Bäumen, Pflanzbeeten und Straßenmobiliar; ►Fußgängerzone, ►Herausnahme des Verkehrs, ►Verkehrsberuhigung); *syn.* Einrichtung [f] von Fußgängerzonen.

4154 pedestrian link [n] *urb. plan.* (Corridor, free of wheeled vehicles, connecting blocks of flats, city suburbs or

P

districts; ▶pedestrian thoroughfare); *syn.* pedestrian passageway [n] [also US]; *s* **conexión [f] para peatones** (Corredor libre de vehículos motores que conecta barrios de una ciudad entre sí; ▶conexión peatonal principal); *f* **liaison [f] piétonnière** (Voie strictement réservée aux piétons et reliant les bâtiments d'un ensemble résidentiel, différents quartiers entre eux ; ▶liaison piétonne principale) ; *syn.* liaison [f] piétonne, desserte [f] piétonne, cheminement [m] piétonnier ; *g* **Fußgängerverbindung [f]** (Für Fußgänger vorgesehene motorverkehrsfreie Verbindung von Wohnblocks, Quartieren oder Stadtteilen; ▶Hauptfußgängerverbindung).

pedestrian passageway [n] [US] *urb. plan.* ▶pedestrian link.

pedestrian precinct [n] *urb.* ▶pedestrian zone.

4155 pedestrian shelter [n] *recr. urb.* (Protective roof structure, which may be open-sided, one- or three-sided; e.g. at public transit stops or on hiking trails; ▶park shelter, ▶public transit shelter; *generic term* ▶shelter); *s* **marquesina [f] (2)** (Estructura a modo de tejado que se proyecta desde un muro hacia el exterior, p. ej. para proteger la entrada de un edificio; ▶garita de espera, ▶techado contra la lluvia; *término genérico* ▶refugio); *syn.* tejabana [f]; *f* **abri [m] couvert** (Équipement de protection contre les intempéries, p. ex. à un arrêt de transport en commun, en forêt ; ▶abri, ▶abribus, ▶hutte ; *terme générique* ▶mobilier de protection contre les intempéries) ; *terme spécifique* marquise) ; *g* **Schutzdach [n] (2)** (Schutzeinrichtung zum Unterstellen, z. B. an Haltestellen öffentlicher Verkehrsmittel, Wanderwegen; ▶Schutzhütte 2, ▶Wartehalle, *OB* ▶Witterungsschutzeinrichtung); *syn.* Unterstand [m], Unterstehdach [n].

4156 pedestrian thoroughfare [n] *urb. plan.* (Important, open-ended passageway for foot traffic); *s* **conexión [f] peatonal principal** (Vía entre dos lugares que es muy frecuentada por peatones); *f* **liaison [f] piétonne principale** (Voie principale de desserte piétonne) ; *g* **Hauptfußgängerverbindung [f]** (Von Fußgängern viel frequentierter Weg, der zwei Orte verbindet); *syn.* wichtige Fußgängerverbindung [f].

pedestrian traffic [n] *urb. trans.* ▶pedestrian circulation.

4157 pedestrian zone [n] *urb.* (In Europe, town center [US]/centre [UK] area closed completely and permanently, or during a specified period, to all wheeled vehicles; **in U.S.**, any urban area closed to vehicular traffic); *syn.* pedestrian precinct [n]; *s* **zona [f] peatonal** (Espacio de circulación dentro de la ciudad con prioridad para peatones); *f* **zone [f] piétonne** (Espace de circulation à priorité piétonne pour lequel un arrêté municipal définit les règles de circulation [ouverture à certaines catégories de véhicules] et de comportement des différents usagers [riverains] ; *termes spécifiques* rue piéton[ne], zone piétonne commerçante, plateau piétonnier, rue promenade, itinéraire piétonnier protégé, quartier piétonnier) ; *syn.* zone [f] piétonnière, espace [m] piéton ; *g* **Fußgängerzone [f]** (Innerstädtische Verkehrsfläche, die für den Fahrverkehr aller Art durch Verbot ganz oder zeitweilig gesperrt und dem Fußgängerverkehr vorbehalten ist).

pedogenesis [n] *pedol.* ▶soil formation.

4158 pedogenetic [adj] *pedol.* (Pertaining to or having to do with ▶soil formation); *syn.* pedogenic [adj] (RCG 1982); *s* **edafogénico/a [adj]** (Relativo a la ▶edafogénesis); *f* **pédogénétique [adj]** (Qui a trait au développement d'un sol sur un substrat minéral ; ▶pédogénèse) ; *g* **bodenbildend [ppr/adj]** (▶Bodenbildung).

pedogenic [adj] *pedol.* ▶pedogenetic.

pedology [n] *pedol.* ▶soil science.

ped stability [n] [US] *pedol.* ▶structural stability of soil.

4159 pelagic zone [n] *limn. ocean.* (Open sea, away from the shore, especially as distinguished from coastal waters; ▶benthic zone); *s* **zona [f] pelágica** (Zona de mar abierto, alejada de la costa; ▶zona del bentos); *syn.* piélago [m]; *f* **zone [f] pélagique** (Eau libre de la mer, colonisée par le plancton, organismes végétaux et animaux microscopiques, flottant passivement dans l'eau. L'ensemble des végétaux ou animaux marins — espèces pélagiques — ne vivant pas en contact avec le fond, est le « pélagos » ; ▶zone benthique) ; *g* **Pelagial [n]** (Freies Wasser der Meere und Binnengewässer von der Oberfläche bis zur größten Tiefe. Gliederung bei Binnengewässern [Limnopelagial] in oberes [durchlichtete trophogene Zone] und unteres **P.** [tropholytische Zone]; beim Meer [Halipelagial] in Epipelagial und Bathypelagial; ▶Benthal).

peninsular interdigitation [n] *landsc. urb.* ▶open space system of peninsular interdigitation.

4160 perceived environmental quality [n] *plan. recr. sociol.* (Personal impression of the environment according to cultural traditions and values shared by the observer and his contemporaries; ▶visual quality); *syn.* value [n] of the environmental experience, perceived scenic quality [n], subjective quality [n] of the environmental experience; *s* **calidad [f] vivencial del ambiente** (Impresión personal del entorno que depende de las tradiciones culturales y los valores compartidos por el/la observador, -a y sus contemporáneos; ▶calidad visual); *f* **qualité [f] perceptuelle** (Représentation subjective de l'environnement en rapport avec les normes et les valeurs édifiées par la tradition culturelle qu'une personne partage avec les membres d'un groupe ou d'une couche sociale ; ▶valeur esthétique) ; *g* **Erlebnisqualität [f]** (Subjektive Abbilder der Umwelt, auf Grund kultureller Tradition durch Normen und Werthaltungen geprägt, die die erlebenden Personen mit den Mitgliedern ihrer Bezugsgruppen und ihrer gesellschaftlichen Schicht teilen; ▶Gestaltqualität); *syn.* Erlebniswert [m].

perceived scenic quality [n] *plan. recr.* sociol ▶perceived environmental quality.

percentage basis [n] *contr. prof.* ▶remuneration on a percentage basis.

4161 percentage [n] **of labor costs** [US]/**percentage** [n] **of labour costs** [UK] *constr. contr.* (Proportion of wages in a calculation of total costs); *syn.* proportion [n] of labo[u]r costs; *s* **proporción [f] de costos de mano de obra** *syn.* porcentaje [m] de costos de mano de obra; *f* **pourcentage [m] de main d'œuvre** (Coûts intervenant dans le prix global des travaux) ; *syn.* part [f] de la main d'œuvre ; *g* **Lohnkostenanteil [m]** (Anteil der Arbeitskosten bei der Kalkulation von Gesamtkosten).

percentage [n] **of vegetative cover** *phyt.* ▶degree of total vegetative cover.

4162 perception [n] **of the landscape** *plan. recr. sociol.* (Ability to have inner feeling/enjoyment and emotional impressions of landscape); *s* **percepción [f] de un paisaje** (Capacidad subjetiva de las personas de gozar de la belleza y tener una impresión emocional de un paraje); *f* **perception [f] du paysage** (Capacité à ressentir les impressions que dégage un paysage) ; *g* **Erlebbarkeit [f] der Landschaft** (Innerliches Empfinden und gefühlsmäßiges Erleben einer Landschaft durch Betrachten, Hören und Wandern/Autofahren etc.).

perch [n] *conserv. zool.* ▶roost.

4163 perched boulder [n] *geo.* (Large ▶erratic block or a block transported downhill by a ▶landslide and lying in an unstable position on a hillside); *s* **pedrejón [m] errático** (▶Bloque errático o roca grande que ha sido transportada o descubierta por un movimiento gravitacional y que yace en una posición inestable; ▶corrimiento de tierras); *f* **bloc [m] perché (≠)** (▶Bloc erratique ou bloc rocheux entraîné et dégagé lors d'un ▶glissement de terrain) ; *g* **Wanderblock [m]** (Auf instabiler Böschung durch ▶Hangrutschung freigelegter und durch Schwerkraft hangabwärts verlagerter ▶Findling 2 oder Felsblock).

4164 percolation [n] *pedol.* (Downward movement of water through pores, joints or crevices within a mass of soil or rock; cf. DNE 1978; the percolation rate or ▶hydraulic conductivity is the numerical rate at which **p.** occurs; e.g. cm^3 · cm^{-2} · sec^{-1} or litre/liter per second [l/s]; *opp.* ▶capillar rise; ▶infiltration, ▶seepage water); *s* **percolación [f]** (Flujo del agua a través de los poros y grietas hacia las capas inferiores del suelo o de las rocas por efecto de la gravedad; *opp.* ▶ascenso capilar; ▶agua de percolación, ▶conductividad hidráulica, ▶infiltración); *f* **percolation [f]** (Écoulement par les fissures, les galeries d'animaux ou de racines de l'eau de précipitation ou d'irrigation à travers le sol sous le simple effet de la gravité [▶drainage interne] ; la percolation des eaux alimente la nappe phréatique ; *opp.* ▶remontée capillaire ; ▶conductivité hydraulique, ▶eau de gravité, ▶infiltration ; DAV 1984) ; *g* **Durchsickerung [f]** (Abwärtsbewegung des Wassers durch Poren oder Spalten im Boden oder Fels bis zum Grund- oder Stauwasser. Das Ausmaß der Wasserbewegung ist u. A. abhängig von der Durchlässigkeit oder ▶Wasserleitfähigkeit des Bodens; die perkolierende Wassermenge wird in cm^3 · cm^{-2} · sec^{-1} gemessen; *opp.* ▶kapillarer Aufstieg; ▶Versickerung, ▶Sickerwasser 2); *syn.* Perkolation [f].

perennating part [n] *bot.* ▶overwintering part.

4165 perennial [n] **(1)** *bot. hort. plant.* (**1.** Generic term for herbaceous plant having a life cycle of more than two years, with above-ground dieback in winter, and persistent roots and storage organs that produce new annual shoots in spring; *specific terms* ▶accent perennial, ▶carpet-forming perennial, ▶climbing perennial, ▶cushion plant, ▶geophyte, ▶ground cover perennial, ▶hemicryptophyte, ▶ornamental perennial, ▶randomly planted perennial [UK], ▶shade-tolerant perennial, ▶specimen perennial, ▶tall perennial, ▶wild perennial. **2.** Any perennial herb excluding grasses, rushes and sedges is called **perennial forb**; ▶tall forb); *syn.* perennial herb [n] (TEE 1980, 47); *s* **planta [f] vivaz** (Vegetal que vive tres o más años. Aunque en principio las leñosas también son perennes, el término se utiliza mayormente para denominar a aquellas plantas herbáceas con órganos subterráneos persistentes, como rizomas, tubérculos, bulbos, etc.; cf. DB 1985; *términos específicos* ▶geófito, ▶hemicriptófito, ▶hemicriptófito trepador, ▶megaforbia, ▶perenne silvestre, ▶planta pulviniforme, ▶planta vivaz cespitosa, ▶planta vivaz esciófila, ▶planta vivaz tapizante, ▶vivaz alta, ▶vivaz de acentuación, ▶vivaz ornamental, ▶vivaz solitaria); *syn.* planta [f] perenne, vivaz [f], perenne [f]; *f* **plante [f] vivace** (Plante herbacée [qui ne fait pas de bois], dont la souche vit plusieurs années [pérenne], qui disparaît généralement en hiver [période de repos] pour réapparaitre au printemps ; *termes spécifiques* ▶géophyte, ▶hémicryptophyte, ▶mégaforbe, ▶plante grimpante vivace, ▶plante vivace couvre-sol, ▶plante vivace des emplacements ombragées, ▶plante vivace dominante, ▶plante vivace d'ornement, ▶plante [vivace] en coussin[et], ▶plante vivace gazonnante, ▶plante vivace haute, ▶plante vivace isolée, ▶plante vivace sauvage) ; *syn.* plante [f] pérenne ; *g* **Staude [f]** (Mehrjährig lebende [perennierende], krautige, wiederholt blühende und fruchtende, winterharte höhere Pflanze, die im oberirdischen Teil [fast] nicht verholzt und am Ende ihrer Vegetationsperiode teilweise oder vollständig abstirbt und ihre Assimilate einzieht; OB für folgende ▶Lebensformen: ▶Kryptophyt, ▶Hemikryptophyt, hapaxanther Hemikryptophyt [kurzlebige Staude], ▶Bienne [Zweijährige Pflanze] und krautiger

▶Chamaephyt, z. T. auch ▶Epiphyt; zu **S.n** zählen im gärtnerischen Sinne auch mehrjährige Ziergräser/Staudengräser, Farne, Sumpf- und Wasserpflanzen sowie verholzende Kräuter und Halbsträucher; **in A.** wird im allgemeinen Sprachgebrauch unter **S.** ein Gehölz [Strauch] verstanden: *ethymologisch* mhd. *stūde*, ahd. *stūda* Staude, Strauch, Busch. *MERKE, nicht zu verwechseln* mit perennierende Pflanze; *UBe* ▶bodendeckende Staude, ▶Geophyt, ▶Hochstaude 2, ▶Klimmstaude, ▶Leitstaude, ▶Polsterpflanze, ▶rasenbildende Staude, remontierende Staude, ▶Schattenstaude, ▶Schmuckstaude, ▶Solitärstaude, ▶Streupflanze, ▶Wildstaude; *syn.* ausdauernde krautige Pflanze [f], perenne krautige Pflanze [f], perennierende krautige Pflanze [f], Mehrjahresblume [auch A], winterharte Blumen, Gräser und Farne [pl] [auch A].

4166 perennial [adj] **(2)** *bot.* (Descriptive term applied to plants which live more than two years, such as herbaceous species, semi-shrubs and woody plants; in comparison to annual and biennial; ▶long-lived); *syn.* root hardy [adj] [also US]; *s* **perenne [adj]** (Referente a las plantas que —a diferencia de las anuales o bianuales— viven varios años; ▶longevo/a); *syn.* vivaz [adj]; *f* **pérenne [adj]** (Plante herbacée [plante vivace], arbuste et autres ligneux qui vivent plusieurs années, par opposition aux plantes annuelles ou bisannuelles ; ▶à/de longue durée de vie) ; *syn.* vivace [adj] ; *g* **ausdauernd [ppr/adj]** (Krautige Pflanzen [Stauden], Halbsträucher und Gehölze betreffend, die mehrere Jahre leben; im Vergleich zu einjährig und zweijährig; ▶langlebig); *syn.* perennierend [ppr/adj].

perennial [n]**, dense-mat-forming** *hort. plant.* ▶carpet-forming perennial.

perennial [n]**, focal point** *hort. plant.* ▶accent perennial.

perennial [n] **[US], showy** *hort. plant.* ▶ornamental perennial.

perennial [n]**, theme** *hort. plant.* ▶accent perennial.

perennial [n] **[US], wildflower** *hort. plant.* ▶wild perennial.

4167 perennial border [n] *gard. hort.* (Elongated planting bed, usually of mixed perennials; ▶border 1); *s* **macizo [m] de vivaces** (Platabanda plantada mayormente o sólo con vivaces; ▶arriate/a); *f* **massif [m] de plantes vivaces** (Ensemble uniforme, association florale ornementale composée principalement ou exclusivement d'un groupement d'espèces vivaces regroupées en un même espace ; ▶bordure végétale) ; *g* **Staudenrabatte [f]** (Vorwiegend oder nur mit Stauden bepflanztes Beet; *OB* ▶Rabatte); *syn.* Staudenfläche [f], Staudenbeet [n].

perennial forb [n] *bot. hort. plant.* ▶perennial (1).

4168 perennial forb community [n] *phyt.* (Perennial herbaceous plant association at the edge of forests, along driftlines, in ruderal or clear cutting areas [US]/clear felling areas [UK] as well as weed communities on cultivated land; ▶tall forb community); *syn.* perennial herb community [n]; *s* **vegetación [f] de herbáceas perennes** (Formación vegetal de herbáceas en orlas de bosques, de cúmulos de residuos orgánicos, en áreas ruderales o claros de tala de bosques así como formaciones de «malas hierbas» en cultivos; ▶formación megafórbica); *f* **végétation [f] herbacée pérenne** (Groupements herbacés pérennes des marges forestières, berges, coupes, friches, ainsi que la végétation adventice des cultures ; ▶mégaphorbiaie) ; *g* **Staudenflur [f]** (Ausdauernde Krautflur an Waldsäumen, Spülsäumen, auf Ruderal- und Kahlschlagflächen sowie mehrjährige Unkrautbestände in Pflanzenkulturen; ▶Hochstaudenflur).

4169 perennial [n] **for cutting** *hort.* (Herbaceous ▶perennial plant useful for cut flowers); *s* **vivaz [f] para cortar** (▶Planta vivaz que dura mucho después del corte); *f* **plante [f] vivace à couper** (▶Plante vivace donnant des fleurs à couper utilisée pour

P

sa bonne tenue en vase) ; *syn.* plante [f] vivace pour fleurs à couper ; *g* **Schnittstaude [f]** (▶Staude, die sich besonders gut als Schnittblume eignet).

perennial gardener [n] *hort. prof.* ▶perennial grower, #3.

4170 perennial grower [n] *hort. prof.* (**1.** Gardener specialized in breeding or cultivation of perennials); **2. perennial nurseryman** [n] (Owner of a perennial nursery); **3. perennial gardener** [n] (Gardener specialized in planting and maintaining perennial beds and borders); *s* **jardinero/a [m/f] especializado/a en vivaces** ; *f 1* **pépiniériste [m] spécialiste en plantes vivaces** (Horticulteur spécialisé dans la production, la culture, le développement et la vente de plantes vivaces) ; *syn.* pépiniériste [m] spécialisé dans la culture des plantes vivaces ; *f 2* **jardinier [m] spécialiste en plantes vivaces** (Professionnel spécialisé dans la plantation et l'entretien des plantes vivaces) ; *g* **Staudengärtner/-in [m/f]** (*Gärtner der Fachrichtung Staudengärtnerei* in A., CH und D. staatlich anerkannter Ausbildungsberuf und Berufsbezeichnung für gelernte Gärtner/-innen, die Stauden heranziehen, kultivieren oder Staudenpflanzungen fachkundig anlegen und pflegen).

perennial herb [n] *bot. hort. plant.* ▶perennial (1).

perennial herb border [n] *phyt.* ▶herbaceous edge [US]/herbaceous seam [UK].

perennial herb community [n] *phyt.* ▶perennial forb community.

4171 perennial nursery [n] *hort.* (Place where perennials are bred and cultivated for sale); *s* **vivero [m] de vivaces** (Vivero especializado en el cultivo de plantas perennes para su venta); *f* **pépinière [f] spécialisée en plantes vivaces** (Pépinière de culture pour plantes vivaces) ; *g* **Staudengärtnerei [f]** (Anzuchtbetrieb für Stauden und deren Vertrieb); *syn.* Staudenbetrieb [m].

perennial nurseryman [n] *hort. prof.* ▶perennial grower.

4172 perennial plant breeder [n] *hort.* (Gardener who propagates new varieties of herbaceous perennials by selective breeding, cross breeding [e.g. of two species] or mutation breeding; *generic term* ▶plant breeder); *syn.* perennial plant hybridizer [n] [also US] (HORT I, 1989, 30); *s* **cultivador, -a [m/f] de vivaces/perennes** (Jardinero/a que se dedica a seleccionar y cultivar nuevas variedades de vivaces; *término genérico* ▶fitogeneticista); *f* **sélectionneur [m] de plantes vivaces** (Jardinier utilisant les différentes techniques de la sélection végétale telles que les sélections généalogique et massale, l'hybridation ou la mutation afin d'assurer la création de variétés nouvelles ou le maintien des caractéristiques propres de variétés de plantes vivaces ; *terme générique* ▶sélectionneur de plantes) ; *g* **Staudenzüchter/-in [m/f]** (Gärtner, der durch Auslesezüchtung, Kombinationszüchtung [z. B. Kreuzung] oder Mutationszüchtung neue Staudensorten heranzieht; *OB* ▶Pflanzenzüchter).

4173 perennial plant breeding [n] *hort.* (Creation of new varieties of perennials by hybridization, selection and mutation, in order to produce special forms of growth or colour as well as pest-resistant species; ▶perennial plant breeder, ▶plant breeding); *s* **cultivo [m] de vivaces/perennes** (Creación de nuevas variedades de vivaces por cruce, selección o mutación artificial para conseguir formas de crecimiento y colores especiales y variedades resistentes a los parásitos; ▶cultivador, -a de vivaces/perennes, ▶fitogenética); *f* **sélection [f] de plantes vivaces** (Conservation de variétés de plantes vivaces en vue du maintien des caractéristiques spécifiques des variétés obtenues ou création de variétés de plantes vivaces nouvelles afin d'obtenir des formes de croissance et des couleurs particulières et d'améliorer la résistance aux déprédateurs en ayant recours aux modifications phylogénétiques ; ▶sélection végétale, ▶sélectionneur de plantes vivaces) ; *g* **Staudenzüchtung [f]** (Schaffung neuer Staudensorten durch Kreuzung, Selektion und künstlich erzeugte Mutation, um besondere Wuchsformen und Farben sowie schädlingsresistente Sorten hervorzubringen; ▶Pflanzenzüchtung, ▶Staudenzüchter).

perennial plant hybridizer [n] [US] *hort.* ▶perennial plant breeder.

perennials [npl] *hort.* ▶testing of perennials.

4174 perennials test garden [n] *hort.* (Permanently-used area of land on which perennial plants are tested and evaluated according to their reactions to environmental conditions and their suitability for cultivation. Such areas are termed 'breeding gardens' in plant breeding nurseries; ▶perennial plant breeding, ▶testing of perennials); *syn.* perennials trial garden [n] [also UK] (LD 1994 [4], 9), perennials test nursery [n] [also US]; *s* **jardín [m] experimental de vivaces/perennes** (Jardín que sirve a jardineros y botánicos para estudiar las plantas vivaces y experimentar diferentes métodos de cultivo y de composición; ▶cultivo de vivaces/perennes, ▶evaluación comparativa de vivaces/perennes); *f* **jardin [m] d'essais comparatifs des plantes vivaces** (Espace de présentation en vue de l'évaluation comparative des plantes vivaces ; ▶évaluation comparative des plantes vivaces/pérennes, ▶sélection des plantes vivaces) ; *syn.* champ [m] d'essais comparatifs des plantes vivaces ; *g* **Staudensichtungsgarten [m]** (Ein dauernd genutztes Bodenareal, auf dem nach wissenschaftlichen Methoden ausdauernde Freilandpflanzen hinsichtlich ihres Verhaltens zur Umwelt, ihrer Verwendbarkeit und Kultur untersucht und bonitiert werden; in Pflanzenzuchtbetrieben werden diese Flächen **Zuchtgärten** genannt; **in D.** war KARL FOERSTER ein unermüdlicher Vorkämpfer für die ▶Staudensichtung. **S.gärten** entstanden planmäßig erst nach dem 2. Weltkrieg, weil die Zunahme der Staudenverwendung eine zentrale Sichtung und Auswertung notwendig machte. 1948 gründete RICHARD HANSEN den **S. Weihenstephan**. 1952 wurde die „Arbeitsgemeinschaft Selektion und Züchtung der Blütenstauden" gegründet. Aus dieser vom Bundesminister für Ernährung, Landwirtschaft und Forsten geförderten Einrichtung ging später die Arbeitsgemeinschaft Staudensichtung der Sondergruppe Stauden im Zentralverband Gartenbau [ZVG], dem heutigen Bund deutscher Staudengärtner [BdS], hervor. In über ganz D. verteilten **S.gärten** werden nach einem einheitlichen und abgestimmten Bewertungsverfahren jährlich Bonitierungslisten und Blütenkalender geführt und in der zentralen Auswertungsstelle Weihenstephan erfasst und ausgewertet; ▶Staudenzüchtung).

perennials test nursery [n] [US] *hort.* ▶perennials test garden.

perennials trial garden [n] [UK] *hort.* ▶perennials test garden.

4175 perennial [n] **with lignified seed stalks** *hort. phyt.* (Herbaceous plant, the stem of which becomes wood-like and remains in place with seed heads, at least until spring and longer; ▶fog [UK] 1); *s* **perenne [f] con infructescencia invernal** (Planta herbácea cuyo tallo seco se mantiene con sus semillas por lo menos hasta la primavera siguiente; ▶gramínea resistente a heladas); *f* **vivace [f] à infrutescence (lignifiée) pérenne (±)** (Plante vivace dont les parties aériennes lignifiées portant l'infrutescence perdurent jusqu'au printemps et gardent un aspect décoratif en hiver ; ▶graminée vivace résistante au gel) ; *g* **Wintersteher [m]** (Staude mit verholztem oberirdischen Teil, deren Fruchtstände mindestens bis zum Frühling stehen bleiben; ▶winterständiges Gras).

4176 perforated drain pipe [n] **with soil separator** *constr.* (e.g. drain pipe with fibrous material or geotextiles; ▶wrapping with soil separator); *s* **dren [m] enfundado** (Dren cubierto de material protector, como fibra de coco; ►enfundado de dren); *syn.* tubo [m] de drenaje enfundado; *f* **drain** [m] **enrobé** (Drain utilisé pour les réseaux de drainage agricole et de génie civil enterrés et enrobé de fibre de coco, d'un géotextile ou de polypropylène pour éviter toute pollution de l'eau récoltée ; ►enrobage d'un drain) ; *g* **ummanteltes Dränrohr [n]** (Dränrohr, das z. B. mit Kokosstricken, Filtervlies [Geotextil, Polypropylen] oder einem Maschen-Filtermaterial umwickelt ist, um abschlämmbare Bodenteile vom abzuführenden Wasser zu filtern; ►Dränrohrummantelung).

4177 performance bond [n] *contr.* (Guarantee deposited with the employer/owner in the form of a certificate, until the contractor/consultant has satisfactorily completed his contractual obligations; in U.S., alternatives to the 'bond' exist, e.g. a lender may give a mortgage on property subject to improvement); *s* **garantía** [f] **de cumplimiento de contrato** (Cantidad en metálico, en valores públicos o en valores privados avalados por el Estado de un porcentaje entre 2 y 4% del presupuesto del proyecto que el contratista debe depositar para garantizar su solvencia para cumplir el contrato. La garantía también puede presentarse a través de un aval prestado, en la forma y condiciones reglamentarias, por algunos de los Bancos, Cajas de Ahorros, Cooperativas de Crédito y Sociedades de Garantía Recíproca autorizados para operar en España o por contrato de seguro de caución celebrado en la forma y condiciones que reglamentariamente se establezcan, con entidad aseguradora autorizada para operar en el ramo de caución; cf. arts. 36 y 37 Ley 13/1995; ►condiciones generales de contratos de las administraciones públicas); *f 1* **garantie** [f] **de bonne fin des travaux** (Garantie exigée de la part des entreprises sous forme d'une caution bancaire fournie à la signature du marché garantissant l'exécution des engagements de l'entrepreneur ; celui-ci est tenu, à partir de la date d'effet de la réception, à l'obligation de parfait achèvement ; pendant la durée de la garantie de parfait achèvement l'entrepreneur est tenu d'effectuer la réparation de tous les désordres signalés par le maître d'oùvrage ; cf. CCM 1984, 105) ; *f 2* **assurance** [f] **de bonne fin des travaux** (**U.K.**, contrat d'assurance souscrit par le maître de l'ouvrage pour un montant qui peut atteindre celui de marché) ; *g* **Vertragserfüllungsbürgschaft** [f] (Bei öffentlichen Aufträgen eine Sicherheit durch eine Bürgschaft, die beim Auftraggeber in Form einer Urkunde, ausgestellt durch ein Kreditinstitut oder Kreditversicherer, sofern das Kreditinstitut oder der Kreditversicherer in der Europäischen Gemeinschaft oder in einem Staat der Vertragsparteien des Abkommens über den Europäischen Wirtschaftsraum oder in einem Staat der Vertragsparteien des WTO-Übereinkommens über das öffentliche Beschäftigungswesen zugelassen ist, solange hinterlegt wird, bis der Auftragnehmer seine vertraglichen Leistungen erfüllt hat; cf. § 17 Nr. 2 im Ergänzungsband 1996 zur VOB Teil B. Eine **V.** muss der Auftragnehmer beibringen, um die vertragsgemäße Ausführung der Leistung incl. Abrechnung, Mängelansprüche und Schaden[s]ersatz, die Zahlung von Vertragsstrafen und die Erstattung von Überzahlungen incl. Zinsen sicherzustellen. Nach der Abnahme wird diese Bürgschaft zur Bürgschaft für Mängelansprüche).

Performance Standard Codes [npl] [US] *constr. contr.* ►standards for construction [US]/Codes of Practice (C.P.) [UK].

4178 pergola [n] *constr.* (Overhead structure of posts or columns, beams and joists for covering a walk or sitting place. Used for the structuring of an open space and as a support for climbing plants. Instead of a system of joists on beams, a pergola roof may also be constructed with connecting panels in the shape of a grid); *s* **pérgola** [f] (Estructura abierta al cielo, generalmente de maderos, que sostienen plantas trepadoras); *f* **pergola** [f] (Sorte de tonnelle comportant des poteaux ou colonnes et des poutrelles à claire-voie formant toiture, qui sert de support à des plantes grimpantes) ; *g* **Pergola** [f] (Gerüst aus Pfosten oder Pfeilern, Pfetten und Auflagehölzern, das der offenen Überdeckung eines Ganges oder Sitzplatzes dient — zur Gliederung eines Freiraumes und als Rankhilfe für Kletterpflanzen; statt Pfetten und Auflagehölzer gibt es auch die Konstruktion der Kassettenausbildung).

4179 pergola post anchor [n] *constr.* (Post-to-footing connection of a pergola); *s* **pie** [m] **de pérgola** (Conexión entre el poste de una pérgola y su fundamento); *f* **pied** [m] **de pergola** (Pièce métallique assurant la fixation d'un montant de pergola avec la fondation) ; *g* **Pergolenfuß** [m] (Verbindung zwischen Pergolenpfosten und Fundament; bei Holzkonstruktionen i. d. R. aus einem Metallschuh gefertigt).

perimeter block development [n] *urb.* ►block development.

4180 perimeter fencing [n] *constr. recr.* (Enclosure around ball-playing areas; ►baseball backstop [US]); *syn.* ball stop fencing [n] [UK]; *s* **enrejado** [m] **(de pista de juegos de pelota)** (Valla alta para evitar que la pelota se salga de la zona de juego; ►baranda trasera [de campo de béisbol]); *f* **grillage-pareballon** [m] (Clôture grillagée entourant la totalité ou une partie d'une aire de jeux de balles ; ►filet d'arrêt de base-ball) ; *g* **Ballfanggitter** [n] (Hohe Zaunanlage um einen Ballspielplatz oder Teilen davon; ►Baseball-Fangzaun).

perimeter protection zone [n] *conserv. landsc.* ►marginal protection area.

4181 perimeter wall [n] *arch.* (LD 1991 [11], 5; wall around a building or a plot of land); *syn.* enclosing wall [n]; *s* **muro** [m] **de cercamiento** (Muro alrededor de un edificio o lote de terreno); *f* **mur** [m] **de clôture** (MAÇ 1981 ; mur servant à clore une propriété ou un espace réservé) ; *syn.* mur [m] d'enceinte non couverte (MAÇ 1981, 315, 320) ; *g* **Umfassungsmauer** [f] (Mauer, die ein Gebäude, ein Grundstück oder Teile davon nach außen abschließt); *syn.* Außenmauer [f].

period [n] *limn.* ►autumn circulation period; *constr. contr.* ►bidding period, ►binding period, ►construction period, ►contract period; *contr. leg.* ►defects liability period; *hort. plant.* ►fall-flowering period [US]/autumnal flowering season [UK], ►flowering period; *zool.* ►incubation period; *constr. for. hort.* ►planting period; *zool.* ►spawning period; *limn.* ►spring circulation period; *hort. plant.* ►spring flowering period, ►summer-flowering period; *leg. envir.* ►transition period.

period [n], **blooming** *hort. plant.* ►flowering period.

period [n], **brooding** *zool.* ►incubation period.

period [n], **contract awarding** *contr.* ►period for acceptance of a bid [US].

period [n] [US], **guarantee** *contr.* ►defects liability period.

period [n], **liability** *contr. leg.* ►defects liability period.

period [n], **limitation** *contr. leg.* ►period of limitation.

period [n], **planning** *plan.* ►planning phase.

period [n] [UK], **tender** *contr.* ►bidding period [US].

4182 period [n] **for acceptance of a bid** [US] *contr.* (Fixed period from ►bid opening date [US]/submission date [UK] to the ►awarding of contract [US]/letting of contract [UK] during which the contractor is bound to his bid/tender price; the contract award deadline is the end of the period; ►acceptance of a bid [US] 1, ►acceptance of a bid [US] 2, ►binding period); *syn.* acceptance period [n] of a bid [US], contract awarding period [n]; *s* **periodo/período** [m] **de remate** (Periodo fijo entre la ►fecha de apertura de ofertas y la ►adjudicación del remate en el cual el licitante está obligado a mantener su oferta; ►límite

P

de validez de ofertas, ▶remate); *f* **délai [m] de validité des offres** (Période fixée par le cadre d'acte d'engagement entre la remise des offres et l'▶attribution du marché [sur appel d'offres/sur adjudication] pour laquelle le candidat/soumissionnaire est lié à son offre/sa soumission ; ▶date d'ouverture des plis, ▶notification [de l'attribution] du marché à l'entreprise retenue, ▶période de validité des offres) ; *syn.* limite [f] de validité des offres ; *g* **Zuschlagsfrist [f]** (Festgelegter Zeitraum vom ▶Eröffnungstermin bis zur ▶Zuschlagserteilung, in dem der Bieter an sein Angebot gebunden ist; cf. § 19 VOB Teil A; ▶Bindefrist, ▶Zuschlag).

period [n] **for acceptance of a tender** [UK] *contr.* ▶period for acceptance of a bid [US].

4183 periodicity [n] *phyt.* (Recurrent change in the condition of individual plants or plant communities at intervals during a year); *s* **periodicidad [f]** (Conjunto de variaciones cuantitativas y cualitativas de las comunidades e individuos animales y vegetales a lo largo del día y de las estaciones; DINA 1987; *términos específicos* **p.** dial, **p.** estacional); *f* **périodicité [f]** (Alternance de phases de développement ou d'aspect d'un organisme ou d'une communauté biologique au cours d'un cycle complet. La **p.** peut se rapporter à un cycle annuel — phase vernale, estivale, automnale, hivernale par exemple — ou un cycle pluriannuel, p. ex. la phase de développement d'une futaie ; DEE 1982) ; *g* **Periodizität [f]** (**1.** In regelmäßigen Abständen sich wiederholende Zustandsänderung innerhalb eines Jahres bei Einzelpflanzen oder Pflanzengesellschaften; insbesondere bei mehrjährigen Pflanzen auftretende Jahresabschnitte im Wechsel von Wachstums-, Laubfall- und Ruhephase, i. d. R. abhängig von genetischer Veranlagung und klimatischen Bedingungen; **2.** Wechsel der Aspekte in bestimmten Pflanzengesellschaften während eines Jahres oder eine mehrjährige **P.**, die z. B. die Entwicklungsphasen eines Hochwaldes charakterisiert); *syn.* Periodik [f].

periodicity [n]**, diurnal** *biol.* ▶circadian rhythm.

4184 period [n] **of limitation** *contr. leg.* (Legal or contract-prescribed deadline after which claims are no longer valid; **in U.S.**, there are time 'statutes of limitation'); *syn.* limitation period [n], period [n] of prescription; *s* **plazo [m] de prescripción** (Periodo/período fijado por ley o contrato en el que expira un derecho); *f* **délai [m] de prescription** (Date déterminée par une loi ou fixée par contrat à l'expiration de laquelle il n'est plus possible d'engager une action en responsabilité civile contre toute personne physique ou morale) ; *g* **Verjährungsfrist [f]** (Per Gesetz oder Vertrag bestimmter Zeitraum, nach dem die Durchsetzbarkeit eines Anspruches verloren geht).

period [n] **of prescription** *contr. leg.* ▶period of limitation.

4185 period [n] **of stay** *recr. zool.* (Time span for remaining at a place of, e.g. migrant birds, recreational users); *s* **duración [f] de estancia** (**1.** Lapso de tiempo en el que permanecen las personas buscando recreo en algún lugar. **2.** Lapso de tiempo en el que permanecen animales en un área, p. ej. las aves migratorias); *syn.* periodo/período [m] de estancia; *f* **durée [f] de séjour** (**1.** Laps de temps pendant lequel les estivants séjournent dans un hébergement. **2.** Temps passé par les oiseux migrateurs sur une aire de repos lors d'une halte migratoire [printanière automnale] ; *terme spécifique* durée de séjour au nid) ; *syn.* période [f] de séjour ; *g* **Aufenthaltsdauer [f]** (**1.** Zeitspanne, in der Erholung Suchende verweilen. **2.** Zeitspanne, in der Tiere in einem Gebiet verweilen).

4186 peripheral building development [n] *urb.* (Built-up area, e.g. on the edge of a park or the boundary with open fields; ▶edge of a settlement, ▶urban fringe); *syn.* peripheral built area [n]; *s* **edificación [f] periférica** (Construcción al borde de un parque o de una población; ▶borde de ciudad, ▶franja

rururbana); *syn.* edificación [f] limítrofe; *f* **constructions [fpl] périphériques** (Constructions situées en bordure d'un parc ou constituant les derniers éléments d'un secteur urbanisé à la limite du milieu naturel ; ▶limite du site construit) ; *g* **Randbebauung [f]** (Bebauung z. B. am Rande eines Parkes oder als Abschluss einer Siedlungsfläche zur freien Landschaft; ▶Siedlungsrand, ▶Übergangsbereich einer Stadt zur offenen Kulturlandschaft).

peripheral built area [n] *urb.* ▶peripheral building development.

peripheral distributor road [n] [UK] *plan. trans* ▶peripheral highway.

4187 peripheral highway [n] [US] *plan. trans.* (Term used in road planning for a major traffic artery which touches part of a city tangentially; ▶bypass [US]/by-pass [UK], ▶beltway [US]); *syn.* peripheral distributor road [n] [UK], peripheral road [n] [UK], peripheral route [n] [also SCOT]; *s* **carretera [f] periférica** (En alemán el término «tangente» se utiliza en la planificación viaria para denominar una arteria de tráfico/tránsito principal que bordea tangencialmente una ciudad; ▶carretera de circunvalación, ▶autopista periférica); *f* **périphérique [m]** (Terme utilisé dans l'aménagement des routes et désignant une route à grande circulation dont le tracé est tangent à la périphérie d'une agglomération ; à Paris, malgré sa forme de rocade complète, on utilise le terme de périphérique ou boulevard périphérique ; ▶rocade autoroutière ; *terme spécifique* autoroute périphérique ; ▶route de contournement) ; *syn.* boulevard [m] périphérique, route [f] périphérique ; *g* **Tangente [f]** (Straßenplanerische Bezeichnung für die Verlaufrichtung einer Hauptverkehrsstraße, die Städte oder Stadtteile an ihrer Peripherie berührt/„tangiert"; ▶Umgehungsstraße, ▶Autobahnring).

peripheral planting [n] *constr. hort.* ▶boundary planting.

peripheral road [n] [UK] *plan. trans* ▶peripheral highway.

peripheral route [n] [SCOT] *plan. trans* ▶peripheral highway.

peripheral zone [n] **of a city** *urb.* ▶city periphery.

permanent allotment site [n] [UK] *leg. urb.* ▶permanent community garden area [US].

4188 permanent campground [n] [US] *recr.* (Several rentable camping spaces for the use of travel trailers [US]/caravans [UK], tents or mobil homes for an extended period; ▶campground [US]/campsite [UK] 2; in D. for more than two months); *syn.* permanent caravan site [n] [UK], static caravan ground [n] [UK]; *s* **camping [m] permanente** (Camping en el cual los campistas alquilan parcelas para largo tiempo, utilizándolo como base de veraneo o como segunda residencia para los fines de semana; generalmente no se mezclan los dos tipos de camping en una sola instalación; ▶camping 1); *syn.* camping [m] duradero; *f* **terrain [m] de camping permanent** (Terrain aménagé sujet à autorisation sur lequel sont reçus de façon habituelle plus de 20 campeurs sous tentes, soit plus de 6 tentes ou caravanes à la fois ; **D.**, la période d'exploitation minimale est de 2 mois ; ▶terrain de camping) ; *g* **Dauercampingplatz [m]** (Mehrere verpachtete Parzellen zum Aufstellen von Wohnwagen [Caravans], Steilwandzelten oder Mobilheimen für mindestens zwei Monate; ▶Campingplatz).

permanent caravan site [n] [UK] *recr.* ▶permanent campground [US].

4189 permanent community [n] *phyt.* (Plant community which has not yet reached the end point, the ▶climax, and yet, from whatsoever cause, remains unchanged for a long time and maintains its community relationships; e.g. ▶rock crevice plant community, alpine grasslands characterized by *Polytrichum*, riparian woodlands of big rivers); *s* **comunidad [f] permanente**

(Es una comunidad vegetal que por cualquier motivo no ha alcanzado el estado final climáticamente posible [▶climax] o que no lo puede alcanzar, pero que se mantiene durante mucho tiempo y conserva su característica sociológica, como es el caso de las ▶comunidades fisurícolas; cf. BB 1979); *syn.* comunidad [f] duradera; *f* groupement [m] permanent (Association végétale capable de se maintenir longtemps et de conserver ses caractéristiques sociologiques sans toutefois avoir atteint son état de ▶climax ; p. ex. les ▶associations rupicoles, pelouses alpines à *Polytrichum*, forêts alluviales) ; *g* **Dauergesellschaft [f]** (Pflanzengesellschaft, die ohne ihren klimatisch möglichen Endzustand erreicht zu haben, sich sehr lange behauptet und ihre soziologische Eigenart bewahrt, wie z. B. ▶Felsspaltengesellschaften, *Polytrichum*-Rasen alpiner Schneeböden, Auenwälder im Grundwasserbereich längs der großen Flüsse; ▶Klimax).

4190 permanent community garden area [n] [± US] *(leg.) urb.* (Privately used open space used for ▶community gardens [US]/allotment garden [plots] [UK], composed of many individual plots with communal facilities, whose long-term use as such is ensured either through planning designation or leasing of garden plots; such plots are used not only for the non-commercial prodution of fruits, flowers and vegetables for own consumption, but also for recreation; **in UK., allotment gardens** are characterized by up to several hundred parcels of land [from perhaps 200 to 400 m²] 'allotted' to individual families, who are members of the allotment association for a small fee. Each garden is cultivated individually, whereas community gardens may be looked after collectively. The association leases the land, which may be owned publicly or privately. A parcel usually has a shed for tools and shelter, but it cannot be used residentially. Allotment gardeners usually have to abide with local rules, such as regular hedge cutting or stipulations about how much paved area is allowed on the plot); *syn.* permanent allotment site [n] [UK]; *s* **zona [f] permanente de huertos recreativos urbanos** (En D, A, GB y CH zona urbana utilizada a largo plazo como ubicación de ▶huertos recreativos urbanos); *syn.* zona [f] permanente de huertos familiares urbanos; *f* **lotissement [m] de jardins (ouvriers)** (Zone de ▶jardins ouvriers regroupant jusqu'à 400 lots, réglementée par un document de planification urbaine et pour laquelle ont été établis des contrats de bail de longue durée) ; *syn.* lotissement [m] de jardins familiaux ; *g* **Dauerkleingartenanlage [f]** (Privat nutzbarer öffentlicher Freiraum aus mehreren Einzelgartenparzellen mit gemeinschaftlichen Einrichtungen [Wege, Spielflächen, Vereinshaus], der durch Ausweisung in der Bauleitplanung und durch langfristige Pachtverträge gesichert ist. Eine solche Anlage dient der nicht erwerbsmäßigen gärtnerischen Nutzung, insbesondere zur Gewinnung von Gartenbauerzeugnissen für den Eigenbedarf, und der Erholung; cf. § 1 [1] Satz 1 BKleingG; ▶Kleingarten); *syn.*, nicht *leg.* Dauerkolonie [f], Kleingartenanlage [f], Kleingartenkolonie [f], *obs.* Schrebergartenkolonie [f].

permanent cropping [n] *agr. hort.* ▶permanent cultivation.

4191 permanent cultivation [n] *agr. for. hort. syn.* permanent cropping; *s* **cultivo [m] permanente**; *f* **culture [f] continue** (Pratique agricole qui consiste à cultiver une parcelle chaque année sans laisser d'année de jachère, ou à cultiver la même espèce végétale sur la même parcelle d'une année sur l'autre sans rotation des cultures) ; *g* **Dauerkultur [f]** (**1.** Das Kultivieren von mehrjährigen Nutzpflanzen, die wiederkehrende Erträge liefern, z. B. Obstanlagen, Spargel, Hopfen, Rosen, Baumschule; nach EU-Beihilferecht müssen die Kulturen mindestens fünf Jahre auf derselben Fläche betrieben werden; cf. Art. 2 der EG-Verordnung Nr. 795/2004; *opp.* Wechselkultur. **2.** Fläche mit Nutzpflanzen, die mehrjährige Erträge liefern).

4192 permanent feature [n] *landsc. recr.* (Longlasting facility or structure; ▶permanent site); *s* **instalación [m] permanente** (▶terreno permanente de recreo); *f* **installation [f] permanente** (Équipement, construction installé et utilisé à demeure à un endroit pré-définie ; ▶aménagement permanent) ; *g* **Dauereinrichtung [f]** (Auf Dauer installierte bauliche Anlage oder sonstiger Ausstattungsgegenstand; ▶Daueranlage).

4193 permanent formwork [n] *constr.* (Formwork which is not removed after the concrete has set); *s* **encofrado [m] perdido** (Encofrado que no se quita una vez que el hormigón se ha endurecido); *syn.* encofrado [m] permanente; *f* **coffrage [m] perdu** (Coffrage qui est laissé dans le sol après la prise du béton) ; *g* **verlorene Schalung [f]** (Schalung, die nach dem Aushärten des Betons meist im Boden verbleibt).

4194 permanent grassland [n] *agr.* (Grassland used continually for grazing or to provide litter/hay; ▶abandoned land, ▶hayfield [US]/hay meadow [UK], ▶natural grassland, ▶permanent pasture); *s* **pradera [f] permanente** (Prados utilizados permanentemente como pastizales o para producir heno o paja. Los prados abandonados por razones económicas o sociales [▶tierra abandonada] no se consideran **pp. pp.**; ▶pastizal permanente, ▶prado de siega, ▶prado natural); *f 1* **prairie [f] permanente** (Une surface engazonnée, de durée en principe illimitée, donc non assolée, qui n'a été ni labourée, ni ensemencée [▶friche sociale] et dont la flore, complexe, est composée d'espèces issues de la végétation herbacée locale ; LA 1981, 903 ; ▶herbage permanent, ▶pâturage permanent, ▶pâturage tournant, ▶prairie naturelle) ; *f 2* **surface [f] toujours en herbe (S.T.H.)** (Part statistique de la superficie agricole utilisée : S.A.U. ; LA 1981, 1071) ; *g* **Dauergrünland [n]** (Alle Graslandflächen, die ohne Unterbrechung durch andere Kulturen zur Futter- oder Streugewinnung oder zur Abweidung bestimmt sind. **Absolutes Grünland** sind Flächen, die für die Nutzung als Ackerland ungeeignet sind; Wiesen, die aus sozialen, wirtschaftlichen oder anderen Gründen nicht mehr genutzt werden [▶Sozialbrache], zählen nicht zum Dauergrünland; ▶Dauerweideland, ▶Mähweide, ▶natürliches Grünland); *syn.* Dauergrünlandfläche [f].

permanent injunction [n] *leg. plan.* *▶preliminary injunction on alteration.

4195 permanently installed bench [n] *gard. landsc.* (Bench attached to the ground; bolted down or with legs set in concrete); *s* **banco [m] fijo** (Banco en parque, jardín o la calle sujetado al suelo por medio de tornillos o con las patas enterradas en hormigón); *f* **banc [m] scellé** (Banc dont les pieds sont fixés au sol par l'intermédiaire d'une pièce de serrurerie ou enterrés dans le sol par un socle en béton) ; *g* **fest verankerte Sitzbank [f]**.

4196 permanent pasture [n] *agr.* (▶Permanent grassland used for ▶pasture farming); *s* **pastizal [m] permanente** (▶Pradera permanente que se utiliza en ▶pasticultura); *syn.* pasto [m] permanente; *f 1* **pâturage [m] permanent** (▶Prairie permanente de qualité moyenne, qui peu recevoir des soins d'entretien, mais dont la productivité est, malgré tout, insuffisante pour l'engraissement des bovins destinés à la boucherie ou pour l'alimentation des fortes laitières ; LA 1981) ; *f 2* **herbage [m] permanent** (▶Prairie permanente de qualité supérieure, apte à l'embouche et à l'alimentation des fortes laitières ; LA 1981 ; ▶élevage pastoral) ; *g* **Dauerweideland [n]** (▶Dauergrünland, das der ▶Weidewirtschaft dient).

permanent quadrat [n] *phyt.* ▶permanent sample plot.

4197 permanent resident [n] *zool.* (Bird which occupies a site for the whole of the year as opposed to a ▶local migrant or ▶migratory bird); *syn.* resident bird [n] [also US], sedentary bird

[n] (HH 1981); *s* **ave** [f] **sedentaria** (Ave que —al contrario que las ►aves de dispersión posgenerativa y las ►aves migradoras— pasan toda su vida en un lugar o cambian de territorio sólo con desplazamientos cortos. El término se refiere únicamente a las aves que se mueven sólo en cortas distancias [máximo 100 km] y nunca fuera del área de cría donde están presentes todos los meses del año; cf. DINA 1987); *syn.* ave [f] residente; *f* **oiseau** [m] **sédentaire** (Hôte régulier ou vu chaque année, qui ne migre pas et ne quitte pas son territoire pendant la mauvaise saison ; ►oiseau erratique, ►oiseau migrateur) ; *syn.* sédentaire [m], oiseau [m] résident permanent, résident [m] permanent ; *g* **Standvogel** [m] (Vogel, der im Unterschied zum ►Strichvogel und ►Zugvogel während des ganzen Jahres in der Nähe seines Nistplatzes bleibt und keine größeren Ortsveränderungen vornimmt).

4198 permanent sample plot [n] *phyt.* (Square area laid down on a plant association to estimate the number of plants enclosed and to determine the character of successional changes; ►sample plot); *syn.* permanent quadrat [n] (TEE 1980, 211); *s* **cuadrado** [m] **permanente** (►Superficie de muestreo delimitada para tomar muestras de vegetación en investigaciones fitosociológicas a largo plazo); *f* **carré** [m] **permanent d'échantillonnage** (►Aire d'échantillonnage floristique pour des relevés phytosociologiques de longue durée) ; *syn.* placette [f] permanente ; *g* **Dauerquadrat** [n] (Abgesteckte ►floristische Probefläche für langjährige pflanzensoziologische Untersuchungen).

4199 permanent site [n] *landsc. recr.* (Permanently established green space or recreation area; ►permanent feature); *syn.* static site [n] [also US]; *s* **terreno** [m] **permanente (de recreo)** (►instalación permanente); *f* **aménagement** [m] **permanent** (Unité technique fixe sur laquelle se déroulent une ou plusieurs activités de récréation ou de loisirs ; ►installation permanente) ; *g* **Daueranlage** [f] (Auf Dauer angelegte Grün- oder Erholungsfläche; ►Dauereinrichtung).

permanent state forest [n] [NZ] *for. leg.* ►reserved forest.

permanent structure [n] [US]**, supportive** *leg. urb.* ►accessory structure.

permeability [n] *constr. pedol.* ►infiltration capacity; *pedol.* ►water permeability.

permeability [n] **of soil, drainage** *pedol.* ►water permeability.

permeable soil [n] *constr. pedol.* ►well-drained soil.

4200 permigration [n] *zool.* (Movement of individual or many animals of a population, which do not settle but continue their ►migration 2; ►bird of passage); *s* **migración** [f] **transeunte** (Cambios de lugar de estancia de individuos o de grupos de una población sin que lleven al asentamiento definitivo en la zona, sino que después de un tiempo continúan su ►migración 2; ►ave de paso); *f* **migration** [f] **de déplacement** (Déplacement saisonnier, périodique et temporaire pour lequel des individus ou un ensemble d'individus d'une population sont de passage sur un territoire et continuent leur progression vers un autre milieu géographique ; ►migration 2, ►oiseau de passage) ; *syn.* migration [f] de transit ; *g* **Permigration** [f] (Wanderung von einzelnen oder vielen Individuen einer Population, bei der es nicht zur Ansiedlung kommt, sondern bei der die Tiere weiterziehen; ►Durchzügler, ►Migration 2); *syn.* Durchwanderung [f].

4201 permissible built area [n] **of a lot/plot** *leg. urb.* *s* **superficie** [f] **edificable;** *f* **surface** [f] **constructible** (Terrain propre à la construction sur une parcelle située sur une zone à bâtir. Elle se calcule en appliquant à la surface du terrain le ►coefficient d'occupation des sols [COS] défini dans le règlement du plan local d'urbanisme [PLU]/plan d'occupation des sols [POS]. Il s'agit d'une surface maximum théorique, car d'autres

règles du PLU (prospect, hauteur, emprise, etc.) peuvent s'opposer à la construction effective de la totalité de la surface autorisée sur chaque parcelle) ; *syn.* espace [m] constructible ; *g* **bebaubare Grundstücksfläche** [f] (Der im Bebauungsplan durch Baugrenzen und Baulinien gekennzeichnete Teil eines Grundstückes).

permissible floor area [n] [ZA] *leg. urb.* ►floor area ratio (FAR) [US]/floor space index [UK].

permission [n] [UK] *adm. leg.* ►permit [US].

4202 permit [n] [US] *adm. leg.* (Generic term for action by a public authority/agency, after establishment of the legality of a project according to a set procedure to authorize planning or measures for specified project development; *specific terms* ►building permit [US]/building permission [UK], ►conditional development permit [US]/conditonal planning permission [UK], ►demolition permit [US]/demolition permission [UK], ►permit for mineral extraction [US]/planning permission for mineral extraction [UK], zoning permit [US]); *syn.* permission [n] [UK]; *s* **permiso** [m] **definitivo** (Acto administrativo de una autoridad en el que se comprueba la legalidad de un proyecto o se aprueba la puesta en marcha y el funcionamiento de una instalación industrial siguiendo un procedimiento prefijado; ►licencia de construcción, ►permiso con condiciones, ►permiso de derribo, ►permiso de explotación); *syn.* licencia [f] definitiva; *f* **autorisation** [f] (*Terme générique* acte en droit public, par lequel est exercé dans une procédure administrative le contrôle de la légalité d'un projet d'occupation ou d'utilisation du sol. Certaines autorisations administratives conditionnent la délivrance d'un ►permis de construire, d'autres valent permis de construire, quelques unes sont parallèles à l'octroi du permis de construire ; *contexte* soumis à autorisation, demande d'autorisation ; ►autorisation accordée sous réserve de l'observation de prescriptions spéciales, ►autorisation d'exploitation, ►autorisation d'extraction, ►permis de démolir) ; *g* **Genehmigung** [f] (Im öffentlichen Recht ein Verwaltungsakt, bei dem die Gesetzmäßigkeit eines Vorhabens in einem geordneten Verfahren geprüft und festgestellt wird und somit im Voraus eine Zustimmung zur Vornahme bestimmter Handlungen erteilt wird; z. B. **G.** eines Bauantrages durch die zuständige Baurechtsbehörde. **G.** [cf. z. B. §§ 6, 10, 19, 22, 109, 144 BauGB; § 58 BauO-BW; § 6 BImSchG] wird anders als im Zivilrecht je nach Gesetz synonym oder teilsynonym mit **Bewilligung** [cf. § 8 WHG] oder **Erlaubnis** [cf. § 7, 7a WHG; § 21 FischG-BW] benutzt; ►Abbaugenehmigung, ►Abbruchgenehmigung, ►Baugenehmigung, ►Genehmigung mit Auflagen).

permit application [n] [US] *adm. leg.* ►building and site plan permit application [US]/detailed planning application [UK].

4203 permit application documents [npl] [US] *adm. leg. urb.* (Documents completed by an architect and owner of the property which has to be submitted to a ►Building Permit Office [US]/Building Surveyors Department [UK]; ►building and site plan permit application [US]/detailed planning application [UK], ►building permit [US]/building permission [UK]); *syn.* planning application documents [npl] [UK]; *s* **documentos** [mpl] **de solicitud de licencia de construcción** (►comisión de urbanismo, ►licencia de construcción, ►solicitud de licencia de construcción); *f* **dossier** [m] **(de demande) de permis de construire** (CPEN, Feuillets 107, p. 8346 ; document fourni par l'architecte et transmis au ►bureau de contrôle permettant de contrôler le respect des règles d'urbanisme en vue de la délivrance du ►permis de construire ; ►service de l'application du droit des sols) ; *syn.* projet [m] de permis de construire ; *g* **Bauvorlagen** [fpl] (Vom Architekten anzufertigende Unterlagen wie z. B. Lagepläne, Ansichten, Bauzeichnungen, Baubeschreibung, statische Nachweise, die der ▶Baurechtsbehörde zur ▶Bauge-nehmigung eines beabsichtigten Vorhabens vorzulegen sind; ▶Bauantrag).

4204 permit [n] **for mineral extraction** [US] *adm. leg. min.* (Authorization to operate a mineral extraction site; e.g. gravel pit; ▶extraction right); *syn.* mining permit [n] [also US], planning permission [n] for mineral extraction [UK]; *s* **permiso [m] de explotación** (Autorización para explotar recursos minerales superficiales; ▶derechos extractivos); *syn.* autorización [f] de explotación; *f 1* **autorisation [f] d'exploitation** (Acte de concession ou permis délivré par l'administration publique aux fins de l'exploitation de substances minérales ; ▶droit d'exploitation) ; *syn.* autorisation [f] d'exploiter, autorisation [f] d'ouverture de carrière, permis [m] d'exploitation [pour les gîtes géothermiques] ; *f 2* **autorisation [f] d'extraction** (La demande d'autorisation d'exploitation d'une carrière située sur le domaine public de l'État doit être précédée d'une autorisation d'extraction) ; *syn.* autorisation [f] d'extraire ; *g* **Abbaugenehmigung [f]** (Behördliche Genehmigung zur Betreibung von Materialentnahmestellen, z. B. Kiesgruben; bergrechtlich spricht man auch von ▶Gewinnungsberechtigung); *syn.* Entnahmegenehmigung [f], Gewinnungsbewilligung [f] [A].

4205 permitted number [n] **of stories** [US]/**permitted number of storeys** [UK] *leg. urb.* (Number of building floors laid down by a legally-binding land-use plan, such as a ▶zoning map [US]/Proposals Map/Site Allocations Development Plan Document [UK]; included within this figure are: **1.** floors with a clear height of more than 1.80m below the roof gutter. **2.** Basements [US]/cellar storeys [UK] with walls on average more than 1.40m above grade. **3.** Garage stories which project more than 2m on average above the established final grade; ▶stories [US]/storeys [UK]); *s* **número [m] de plantas completas** (En planes parciales, cantidad de ▶plantas completas de un edificio; en D. se deben considerar: **1.** plantas con altura libre de más de 1,80 m bajo borde superior del canalón, **2.** plantas de sótano con una media superior a 1,40 m de altura y **3.** plantas de garaje que se elevan 2 m o más sobre el nivel del terreno); *syn.* número [m] de pisos completos; *f* **nombre [m] des niveaux de la construction** (F., nombre maximum de niveaux d'un bâtiment fixé par le permis de construire lorsque celui-ci ne comporte qu'un seul bâtiment ou le maximum des nombres de niveaux dans le cas d'une construction de plusieurs bâtiments ; D., ▶nombre de niveaux d'une construction fixé pour chaque zone ou partie de zone d'affectation dans le « Bebauungsplan » [plan local d'urbanisme allemand] ; dans ce nombre sont compris les niveaux des combles d'une hauteur libre de 1,80 m, les sous-sols de 1,40 m de hauteur, les garages dont la hauteur atteint en moyenne 2 m au-dessus du terrain) ; *g* **Zahl [f] der Vollgeschosse** (*Abk.* Z; im ▶Bebauungsplan festgesetzte Anzahl [ohne Dezimalstelle] der ▶Vollgeschosse eines Gebäudes; die Z. der V. kann als Höchstgrenze [z. B. II], als Mindest- und Höchstgrenze [z. B. II-IV] oder als zwingend [z. B. III] festgesetzt werden; Auf die **Z. der V.** sind anzurechnen: **1.** Geschosse mit einer lichten Höhe von mehr als 1,80 m unterhalb der Traufoberkante, **2.** Kellergeschosse, die im Mittel mehr als 1,40 m und **3.** Garagengeschosse, die im Mittel mehr als 2 m über die festgelegte Geländeoberfläche hinausragen — ansonsten cf. Ausführungen unter ▶Vollgeschoss; cf. § 20 BauNVO); *syn. obs.* Geschosszahl [f].

4206 perpendicular parking [n] *trans.* (Parking of cars at a 90° angle to the direction of travel; ▶ninety-degree parking layout); *s* **aparcamiento [m] en perpendicular** (▶disposición [de aparcamientos] en perpendicular); *f* **stationnement [m] en bataille** (Disposition de l'aire de stationnement perpendiculairement à l'axe de circulation ; ▶rangement perpendiculaire) ; *syn.*

rangement [m] à 90° ; *g* **Senkrechtparken [n, o. Pl.]** (Parken von Fahrzeugen im Winkel von 90° zur Fahrtrichtung; ▶Senkrechtaufstellung).

perpendicular parking layout [n] [US] *plan. trans.* ▶ninety-degree parking layout.

perpendicular parking row [n] [US] *trans.* ▶ninety-degree parking row.

perpetual flowering rose [n] *hort.* ▶remontant rose.

persistence [n] *ecol.* ▶ecological stability.

4207 persistent environmental disturbance [n] *ecol. envir.* (Enduring negative effects of man-made influences upon an environmental system or object without the capacity for restoration of its destroyed or lost functions or its repair or regeneration, even after a period of time; e.g. flora, fauna, buildings or terrain); *syn.* long-lasting disturbance [n]; *s* **perturbación [f] persistente** (Efectos ambientales negativos continuos sobre un sistema o un objeto, cuya capacidad de regeneración se ve deteriorada o se pierde totalmente, y no se recupera incluso después de pasado un periodo/período de tiempo considerable); *syn.* daño [m] persistente, daño [m] duradero; *f* **nuisance [f] persistante** (Effets négatifs affectant durablement la santé, la qualité, les caractéristiques, la fonction, etc. des personnes, des objets et de l'environnement) ; *syn.* nuisance [f] durable ; *g* **nachhaltige Beeinträchtigung [f]** (Die fortwährende negative Auswirkung von Umwelteinflüssen auf ein System oder Objekt, dessen erheblich gestörte oder verlorengegangene Funktionszusammenhänge oder Eigenschaften auch nach einem Zeitraum der Regenerationsmöglichkeit oder durch Reparatur nicht wiederhergestellt werden können); *syn.* anhaltende Beeinträchtigung [f], anhaltende Schädigung [f], dauerhafte Beeinträchtigung [f], dauerhafte Schädigung [f], nachhaltige Schädigung [f].

person [n] *sociol.* ▶handicapped person; *adm. constr. contr.* ▶staff person.

person [n], **disabled** *sociol.* ▶handicapped person.

4208 person [n] **affected by planning measures/ proposals** *plan. pol.* (Someone who is influenced by a planning project materially and immaterially); *s* **afectado/a [m/f] por un plan o proyecto** *syn.* ciudadano/a [m/f] afectado/a por un plan o proyecto; *f* **personne [f] concernée par un projet d'aménagement** (Individu affecté directement ou indirectement par un processus de planification et d'aménagement) ; *syn.* personne [f] touchée par un aménagement) ; *g* **Planungsbetroffene/-r [f/m]** (Jemand, auf den sich ein Planungsvorhaben unmittelbar oder mittelbar in materieller oder immaterieller Hinsicht auswirkt).

personnel [n] *constr. prof.* ▶provision of labo(u)r/personnel.

personnel [n] [US], **administrative** *adm. contr.* ∗▶office worker.

person [n] **responsible for direct discharge** [UK] *envir. leg.* ▶water polluting firm [US].

4209 perspective drawing [n] *plan.* (Graphic presentation which gives a three-dimensional picture; ▶isometric drawing); *s* **dibujo [m] en perspectiva** (Representación gráfica de un objeto que da una impresión tridimensional; ▶proyección isométrica); *f* **perspective [f]** (Représentation graphique des objets permettant de retranscrire la perception visuelle de l'espace ; ▶isométrie) ; *g* **Perspektive [f]** (Zeichnung oder Abbildung, die einen Eindruck des Räumlichen vermittelt; ▶Isometrie).

perspective view [n] *plan. recr.* ▶prospect.

4210 pertaining to landscape planning [loc] *plan.* *s* **relacionado/a con la planificación/gestión del paisaje [loc]** *syn.* relativo/a a la planificación/gestión del paisaje [loc]; *f* **relatif à l'aménagement des paysages [loc]** ; *syn.* relatif à la gestion des paysages [loc], du point de vue de l'aménagement des paysages [loc], qui a trait à l'aménagement des paysages [loc] ; *g* **landschaftsplanerisch [adj]** (die Landschaftsplanung oder die landespflegerische Fachplanung betreffend).

pervious surface [n] [US] *constr.* ►water-bound surface.

4211 pest [n] *agr. for. hort. phytopath.* (►*Pest control* animal species which impairs the development of a ►useful plant species by gnawing, sucking, or by the production of noxious substances; ►injurious insect, ►pest control, ►plant pest; *generic term* ►injurious organism; *opp.* ►beneficial species); *s* **depredador [m]** (►*Control antiplaguicida* animal que degrada o destruye alguna cosa considerada útil por el hombre, como las ►plantas útiles; ►depredador [de plantas de cultivo], ►insecto nocivo, ►lucha contra parásitos; *término genérico* ►organismo nocivo; *opp.* ►animal beneficioso); *syn.* especie [f] nociva, especie [f] perjudicial; *f* **déprédateur [m]** (►*Protection des végétaux* animal qui commet des dégâts importants sur une ►plante utile ou sur des denrées, le plus souvent dans le but de se nourrir ; ►déprédateur des cultures, ►insecte nuisible, ►lutte contre les parasites ; *terme générique* ►organisme nuisible ; *opp.* ►auxiliaire) ; *syn.* ravageur [m] ; *g* **Schädling [m]** (1. ►*Pflanzenschutz* Tierart, die durch ihre Fraß- oder Saugtätigkeit sowie durch ihre Entwicklung von Schadstoffen die Entwicklung von ►Nutzpflanzen beeinträchtigt; ►Pflanzenschädling, ►Schadinsekt, ►Schädlingsbekämpfung; *OB* ►Schadorganismus; *opp.* ►Nützling. **2.** Tierart, die dem Menschen lästig ist — z. B. Stallfliegen, Stechmücken, Köcherfliegen — werden **Lästlinge** genannt; cf. N+L 2000, 165); *syn.* Schadtier [n].

4212 pest and weed control [n] *agr. for. hort. phytopath.* (Generic term for biological, chemical, physical and biotechnical methods for the prevention and control of diseases, and pests which affect the development of plants. Integrated pest control includes the introduction of beneficial insects, plants or viruses to prevent the spreading of pests or diseases; the German term *Pflanzenschutz* also includes weed control; ►beneficial species, ►biological pest control, ►chemical pest control, ►injurious organism, ►integrated pest control, ►mechanical pest control, ►pest, ►pest control); *syn.* plant pest and disease control [n], pest management [n] (ARB 1983); *s* **control [m] antiplaguicida** (Conjunto de métodos utilizados para la prevención y lucha contra enfermedades y plagas; ►animal beneficioso, ►control biológico, ►control integrado de plagas, ►depredador, ►lucha antiplaguicida química, ►lucha contra parásitos, ►lucha mecánica contra plagas, ►organismo nocivo); *syn.* lucha [f] antiplaguicida; *f* **protection [f] des végétaux** (Abrév. P.V. ; ensemble des techniques utilisées pour la lutte contre les ennemis et les maladies des plantes et des produits récoltés ; ►auxiliaire, ►déprédateur, ►lutte biologique, ►lutte chimique, ►lutte contre les parasites, ►lutte intégrée, ►organisme nuisible, ►procédé mécanique de lutte contre les ennemis des cultures) ; *syn. agr.* protection [f] des cultures, lutte [f] contre les parasites (LA 1981, 32), protection [f] phytosanitaire ; *g* **Pflanzenschutz [m, o. Pl.]** (OB zu biologische, chemische, physikalische, biotechnische oder integrierte Maßnahme zur Verhütung und Bekämpfung von Krankheits-, Schädlings- und Unkrautbefall, der die Entwicklung von Pflanzenbeständen stört. Zum P. gehören auch die Verwendung und der Schutz von Tieren, Pflanzen und Viren, durch die das Auftreten oder die Verbreitung von Schadorganismen oder Krankheiten verhütet oder bekämpft werden kann sowie Maßnahmen zum Vorratsschutz. Während in der Pflanzenproduktion der P. nach wie vor zur Gesunderhaltung der Bestände einen festen Platz einnimmt und in vielen Kulturen des Gartenbaues sowie der Land- und Forstwirtschaft den Ertrag für die Produzenten sichert, spielt er im Gegensatz hierzu bei der Pflege und Unterhaltung öffentlicher Grünflächen seit den 1960er-Jahren eine immer

geringere Rolle. Stadtökologische und umweltideologische Ansichten sowie Veränderungen insbesondere im Naturschutz-, Pflanzenschutz- und Wasserrecht bewirkten eine Reduzierung von **P.mitteln**. Gleichzeitig wurden vermehrt alternative **P.verfahren** gefordert und entwickelt. Ob diese Haltung bei zunehmender Globalisierung und den damit eingeschleppten Schaderregern beibehalten werden kann, wird die Zukunft zeigen. Hinzu kommt, dass der Pflanzenbestand in Siedlungsgebieten in jahrzehntelangen Standzeiten mit erhöhten Stressfaktoren, die die Vitalität nachhaltig beeinträchtigen, zu kämpfen hat und eine entsprechend lange Pflege und Unterhaltung benötigt; ►biologischer Pflanzenschutz, ►chemischer Pflanzenschutz, ►integrierter Pflanzenschutz, ►mechanische Schädlingsbekämpfung, ►Nützling, ►Schädling, ►Schädlingsbekämpfung, ►Schadorganismus).

4213 pest control [n] *agr. for. hort. phytopath.* (Measures to combat insects and other animals which are harmful to plants and their products; ►biological pest control, ►biotechnical pest control, ►chemical pest control, ►ecological pest control, ►integrated pest control, ►mechanical pest control, ►pest and weed control, ►physical pest control, ►use of beneficial animal species); *s* **lucha [f] contra parásitos** (Medidas para combatir animales que dañan a las plantas de cultivo; ►aprovechamiento de animales beneficiosos, ►control antiplaguicida); *syn.* lucha [f] contra las plagas de parásitos; *f* **lutte [f] contre les parasites** (LA 1981, 32 ; mesures de lutte contre les espèces animales qui occasionnent des dommages aux plantes et leurs produits par l'usage de produits antiparasitaires ; expression se recoupant avec celle de ►protection des végétaux ; ►utilisation d'auxiliaires) ; *syn.* lutte [f] antiparasitaire, traitement [m] antiparasitaire ; *g* **Schädlingsbekämpfung [f]** (Maßnahmen zur Bekämpfung von Tierarten, die schädigend auf Pflanzen und deren Produkte wirken; überschneidender Begriff mit ►Pflanzenschutz; ►mechanische Schädlingsbekämpfung, ►Nützlingseinsatz).

4214 pest control act [n] *agr. for. hort. leg. phytopath.* (**In U.S.**, pest control is governed by the Federal Food, Drug and Cosmetic Act [1938 as amended], Federal Insecticide, Fungicide and Rodenticide Act [FIFRA] [enacted in 1947], Federal Food Quality Protection Act [1996 as amended]; **in U.K.**, pest control is governed by the Control of Pesticide Regulations, 1986 [SI 1986 No. 1510]; **in D.**, federal law with regulations and directives for approval of products which afford protection of growing plants from pests and diseases [pest control], and products for protection of harvested foodstuffs from pests and diseases [protection of stored food products], as well as to avert possible harm to man or animals from application of ►plant protection agent; ►pest and weed control); *s* **Ley [f] de sanidad vegetal [Es]** (En España, la primera ley aprobada para regular la protección fitosanitaria fue la Ley de 21 de mayo de 1908, de Plagas del Campo, seguida por la Ley de 20 de diciembre de 1952, de Defensa de los Montes contra Plagas Forestales. Desde aquellas fechas hasta la aprobación de la actual **l. de s. v.** en noviembre de 2002, se aprobaron un sinnúmero de decretos y órdenes que regulan aspectos específicos de la protección fitosanitaria. Esta nueva ley tiene carácter de normativa básica, a excepción de los arts. 10, 11 y 12, relacionados con el comercio exterior y sanidad exterior, que son vinculantes por ser esos ámbitos compentencia exclusiva del Estado; cf. Ley 43/2002, de 20 de noviembre; ►control antiplaguicida, ►plaguicida); *f* **loi [f] sur la protection des végétaux** (Loi du 25.03.1941 instituant la création du « service de la protection des végétaux » [organisme faisant suite à l'inspection phyto-pathologique de la production horticole créée par le décret du 1er mai 1911] dont la mission est la défense contre les ennemis des cultures, l'interdiction de l'importation, la détention et le transport de parasites réputés dangereux, l'obligation de lutter contre les grands fléaux des cultures, le contrôle sanitaire des pépinières, des semences et des plants ainsi que l'inspection phytosanitaire contrôle de tous les végétaux à l'importation et à l'exportation, l'organisation du service de la protection des végétaux ; ce service a subsisté jusqu'à aujourd'hui avec des rattachements variables au sein de l'administration centrale du ministère de l'Agriculture. Depuis le 21 février 2005 [loi n° 2005-153] la France a approuvé la convention internationale pour la protection des végétaux, telle qu'elle résulte des amendements adoptés à Rome par la 29ème session de la conférence de l'organisation des Nations Unies pour l'alimentation et l'agriculture. La recherche et la mise au point de substances nouvelles sont contrôlées par la Commission d'étude de la toxicité des ►produits antiparasitaires [à usage agricole] 2, les essais devant fournir des résultats sur la toxicité à l'égard de la flore et de la faune, de la rémanence de ces substances et de leurs métabolites sur le sol, les végétaux et l'eau ; LA 1981, 920 ; ►protection des végétaux) ; *g* **Pflanzenschutzgesetz [n]** (Abk. PflSchG; D. Bundesgesetz zum Schutze der Kulturpflanzen vom 14.05.1998 [BGBl. I Nr. 28, 971, zuletzt geändert am 05.03.2008, BGBl I Nr. 8, S. 284] mit Vorschriften über u. A. die Anwendung von und den Verkehr mit Pflanzenbehandlungsmitteln, um Pflanzen von Schadorganismen und Krankheiten [►Pflanzenschutz] und Pflanzenerzeugnisse vor Schadorganismen zu schützen [*Vorratsschutz* Verhinderung eines Befalles von Vorräten durch Schädlinge resp. Bekämpfung bei Schädlingsbefall] und Schäden abzuwenden, die bei der Anwendung von ►Pflanzenschutzmitteln oder anderen Maßnahmen des Pflanzenschutzes, insbesondere für die Gesundheit von Mensch und Tier entstehen können. Ferner hilft es, Rechtsakte der Europäischen Gemeinschaft im Bereich des Pflanzenschutzrechts durchzuführen; cf. Richtlinie 91/414/EWG über das Inverkehrbringen von Pflanzenschutzmitteln, Zubereitungsrichtlinie 1999/45/EG und die Verordnung [EWG] 2092/91 des Rates über den ökologischen Landbau und die entsprechende Kennzeichnung der landwirtschaftlichen Erzeugnisse und Lebensmittel. Das **P.** schreibt allgemein vor, dass Pflanzenschutz nur nach guter fachlicher Praxis erfolgen darf und dass dabei die Grundsätze des integrierten Pflanzenschutzes und der Schutz des Grundwassers zu berücksichtigen sind).

pest-eating animal species [n] *agr. for. hort. phytopath.*
►beneficial species.

4215 pesticide [n] *agr. chem. envir. for. hort. phytopath.* (Any substance, that prevents, destroys, repells or mitigates fungal diseases or animal pests. Often used as a generic term covering acaricide, bactericide, ►fungicide, ►herbicide, ►insecticide, molluscicide, nematicide, rodenticide; ►biocide, ►plant protection agent; **in U.S.**, by regulation, Environmental Protection Agency [EPA] has defined **p.** to exclude microorganisms [viruses, bacteria, etc.]; cf. Antimicrobial Regulatory Technical Correction Act [ARTCA] of 1998; **in D.**, herbicide is generally not included under **p.**); *s* **pesticida** [m] (**1.** Cualquier sustancia que sirve para evitar, combatir o mitigar plagas causadas por organismos animales o vegetales en las plantas de cultivo o en los productos agrícolas. En castellano, se utiliza frecuentemente como sinónimo de ►plaguicida y de ►biocida. Entre ellos se pueden distinguir los acaricidas, algicidas, bactericidas, fasciolicidas, ►fungicidas, ►herbicidas, ►insecticidas, molusquicidas, nematicidas y rodenticidas. **2.** En alemán, los **pp.** sólo incluyen las sustancias que combaten plagas causadas por organismos animales, es decir que los insecticidas no se consideran **pp.**); *f* **pesticide** [m] (Substance destinée à détruire les organismes animaux ou végétaux nuisibles aux cultures. *On peut distinguer les catégories principales* : les ►insecticides, les ►herbicides, les ►fongicides, les bactéricides, les acaricides, les ►limaticides, les nématicides et les rodenticides [contre les rongeurs] ; ►produit

antiparasitaire [à usage agricole]) ; *g* **Pestizid** [n] (...e [pl]; aus dem Angelsächsischen *pests* >Schädlinge<; **1.** Im engeren Sinne vornehmlich gegen tierische Schädlinge eingesetztes Bekämpfungsmittel. **2.** Im weiteren Sinne OB zu chemischer Stoff, der im Pflanzenschutz und Vorratsschutz tierische und pflanzliche Organismen abtöten soll; im Deutschen nicht eindeutig definierter Begriff, weshalb ►Herbizide mal zu den **P.en** und mal nicht zu ihnen gerechnet werden. *MERKE*, es ist deshalb besser, den Terminus ►Pflanzenschutzmittel zu verwenden; ►Biozid, ►Fungizid, ►Insektizid).

4216 pesticide application [n] *agr. envir. for. hort. phytopath.* (Utilization of a protective substance as a pest control for animals and plants; ►pesticide); *syn.* pesticide use [n]; *s* **uso** [m] **de pesticidas/plaguicidas** (►pesticida); *syn.* aplicación [f] de pesticidas/plaguicidas; *f* **emploi** [m] **de pesticides** (►pesticide) ; *syn.* application [f] de pesticides ; *g* **Pestizideinsatz** [m] (Anwendung von ►Pestiziden); *syn.* Begiftung [f].

4217 pesticide residues [npl] *chem. envir.* (Leftovers or biodegraded products which remain in plant cells or in the soil after ►pesticide application); *s* **residuos** [mpl] **de pesticidas** (Restos o productos de descomposición de ►pesticidas que quedan en los órganos de las plantas o en el suelo); *syn.* residuos [mpl] de plaguicidas; *f* **résidus** [mpl] **de pesticides** (Restes ou produits de biodégradation stockés dans les cellules des végétaux ou dans le sol après un épandage de pesticides ; ►pesticide) ; *syn.* dépôt [m] rémanent ; *g* **Pestizidrückstände** [mpl] (Reste resp. Abbauprodukte, die nach einer Anwendung von ►Pestiziden in den Pflanzenzellen oder im Boden zurückbleiben).

pesticide use [n] *agr. envir. for. hort. phytopath.* ►pesticide application.

pest management [n] *agr. for. hort. phytopath.* ►pest and weed control.

pest management [n]**, chemical** *agr. for. hort. phytopath.* ►chemical pest control, ►chemical plant disease prevention and pest control.

pest management [n]**, integrated** *agr. for. hort. phytopath.* ►integrated pest control.

4218 pet cemetery [n] *adm. urb.* (Place where pet animals are buried); *s* **cementerio** [m] **de animales** (Lugar de entierro de animales caseros, sobre todo de perros, existentes en países anglosajones y en Alemania); *f* **cimetière** [m] **pour animaux** (Enclos communal destiné à l'inhumation d'animaux de compagnie et en particuliers des chiens ; l'enterrement des animaux domestiques est régi par l'article L.226-1 et L226-2 du Code rural. Les animaux domestiques sont admis à la crémation par arrêté du 4 mai 1992 relatif aux centres d'incinération de cadavres d'animaux de compagnie ; ils peuvent donc faire l'objet d'une incinération collective ou individuelle si le propriétaire de l'animal le demande et les cendres pourront lui être remises ; *termes afférents* monument funéraire animalier, crématorium animalier) ; *syn.* cimetière [m] animalier ; *g* **Tierfriedhof** [m] (Zur Bestattung von Haustieren, vor allem Hunden, angelegtes, eingefriedetes Grundstück).

4219 pet exercise area [n] [US] *landsc. urb.* (Specially provided space in some public parks where dogs can relieve themselves); *syn.* dog park [n] [US], dogs' loo [n] [UK], poop-scoop area [n] [UK]; *s* **área** [f] **sanitaria para perros** (≠) (En algunas zonas verdes, área en la que los perros pueden hacer sus necesidades); *syn.* «pipi-can» [m] [Catalunya]; *f* **équipement** [m] **sanitaire pour chien** (Équipement fixe destiné à l'élimination des déjections canines, installé dans certains espaces publics en bordure de trottoir, sur les espaces piétonniers ou aux abords des parcs et jardins, tels qu'un enclos pour chien, trottoir promenade pour chien, « vespachien » ; LEU 1987, 302) ; *syn.* édicule [m]

P

d'aisance pour chien, sanitaire [m] pour chien, sani-canin [m] ; *g* **Hundeklo** [n] (Spezielle Einrichtung in manchen öffentlichen Grünanlagen, in der sich Hunde lösen können); *syn.* Hundeauslauffläche [f].

petrol trap [n] [UK] *envir.* ▶gasoline trap [US].

petrophyte [n] *phyt.* ▶rock plant.

4220 petting zoo [n] *recr.* (Small animal park in which mostly domesticated animals can be observed and also stroked; ▶children's farm); *s* **zoo(lógico)** [m] **para niños/as** (Pequeño parque zoológico en el que se pueden observar y también acariciar animales domesticados; ▶granja juvenil); *f* **zoo** [m] **pour enfants** (≠) (Petit parc animalier dans lequel les animaux, en général domestiques, peuvent être observés et aussi caressés ; ▶ferme pour enfants) ; *syn.* jardin [m] zoologique pour enfants ; *g* **Streichelzoo** [m] (Tiergarten oder -gehege, in dem Tiere, meist Haustiere, beobachtet und auch gestreichelt werden können; ▶Jugendfarm).

4221 phanerophyte [n] *bot.* (Term for a plant ▶life form, principally trees and shrubs, but not ▶dwarf shrubs, according to the RAUNKIAER classification; ▶surface plant); *s* **fanerófito** [m] (División de primer orden en la clasificación biotípica y simorfial de RAUNKIAER: conjunto de ▶formas biológicas vegetales en que las yemas de reemplazo se elevan en el aire a más de 25 cm del suelo; cf. DB 1985; ▶caméfito, ▶mata); *f* **phanérophyte** [m] (Classification des végétaux vasculaires selon RAUNKIAER désignant le ▶type biologique de végétaux dont les bourgeons hivernaux sont à plus de 50 cm du sol tels arbres, arbustes, arbrisseaux et lianes ligneuses ; ▶arbuste nain, ▶chaméphyte) ; *g* **Phanerophyt** [m] (Nach RAUNKIAER ▶Lebensform der Pflanzen, vorwiegend Bäume und Sträucher außer ▶Zwergsträucher und Halbsträucher [▶Chamaephyt], deren Überdauerungs-/ Erneuerungsknospen [Tropenpflanzen bilden im Allgemeinen keine Knospen!] höher als 50 cm über dem Boden liegen, weshalb sie der Kälte besonders ausgesetzt sind. Anpassungseinrichtungen: Blattfall, Knospenschutz, besondere Blattstrukturen bei ▶wintergrünen und ▶immergrünen Pflanzen [Hartblatt, Nadelblatt, Rollblatt]; weitere Untergliederung in **Makro-Ph.** [Bäume], **Nano-Ph.** [Sträucher] und **Pseudo-Ph.** [Scheinsträucher, oberirdisches Sprosssystem meist nur zweijährig, dann absterbend, sich ständig erneuernd, z. B. bei Brombeere und Himbeere *(Rubus)*]); *syn.* Luftpflanze [f].

4222 phase [n] (1) *phyt.* (Stage of development within a plant community in which small, floristic and ecological modifications occur. The phases of an ▶plant association are referred to as initial, optimal, retrogressional or terminal; ▶displacement of species (composition or balance); *s* **fase** [f] (Pequeña ▶modificación de la composición de especies en la sucesión de comunidades vegetales. Se diferencian las fases inicial, óptima y final de una ▶asociación; cf. BB 1979); *f 1* **stade** [m] (Légère ▶évolution de la constitution en espèces d'une population dans la succession de groupements végétaux depuis le sol nu jusqu'à la végétation climacique dans une série évolutive qui, du stade pionnier ou stade initial, passe par des stades intermédiaires pour atteindre le stade final ; ▶association [végétale]) ; *f 2* **phase** [f] (Période comprise entre chaque stade de développement) ; *g* **Phase** [f] (Kleine ▶Artenverschiebung in der Sukzessionsfolge von Pflanzengesellschaften. Man unterscheidet Initial-, Optimal- und Schlussphase einer ▶Assoziation).

phase [n] (2) *contr. prof.* ▶bidding and negotiation phase [US]/ tender action and contract preparation [UK], ▶construction document phase [US]/detailed design phase [UK], *constr.* ▶construction phase; *contr. prof.* ▶negotiation phase [US]/contract preparation [UK]; *plan.* ▶planning phase; *contr. plan.* ▶schematic design phase [US]/outline and sketch scheme proposals

[UK]; *limn.* ▶stagnation phase, ▶summer stagnation phase; *landsc. plan. prof.* ▶survey phase; *limn.* ▶winter stagnation phase; *contr. prof.* ▶work phase.

phase [n] [US]**, conceptual design** *contr. plan.* ▶schematic design phase [US].

phase [n] [US]**, design development** *contr. prof.* ▶final design stage.

phase [n] [UK]**, detailed design** *contr. prof.* ▶construction document phase [US].

phase [n] [US]**, soil** *pedol.* ▶soil series [US], #2.

4223 phenology [n] *phyt. zool.* (Study of seasonal biological phenomena in flora and fauna and their correlation with changes in weather; in the case of plants, for example, the observation and documentation of emerging leaves, flowering, seeding, shedding of foliage; in animals, the arrival and departure of migratory birds, the duration of hibernation, the start of the mating season, etc.; *specific terms* phytophenology, zoophenology); *s* **fenología** [f] (Estudio de los fenómenos biológicos acomodados a cierto ritmo periódico, como la brotación, la florescencia, la maduración de los frutos, etc.; *términos específicos* fitofenología, zoofenología; DB 1985); *f* **phénologie** [f] (Science qui étude les phases de développement des organismes — dites **phases phénologiques** ou **phénophase** — en relation avec les variations des conditions saisonnières, essentiellement climatiques. Ces phases sont limitées par des points de repère définis avec précision, les **stades phénologiques** ou **phénostades**. Par exemple pour une plante, la germination, la montaison, la floraison, la dissémination des graines, etc. sont des stades phénologiques séparés par des phénophases ; DEE 1982 ; *termes spécifiques* phytophénologie, zoophénologie) ; *g* **Phänologie** [f] (Wissenschaft von den jahresperiodischen Erscheinungen an Pflanzen und Tieren in Abhängigkeit vom Witterungsverlauf. Bei Pflanzen werden z. B. Blattaustrieb, Blühen, Fruchten, Laubfall, bei Tieren Ankunft und Wegzug der Zugvögel, Dauer des Winterschlafes, Beginn der Paarung etc. beobachtet und dokumentiert; *UBe* Phytophänologie und Zoophänologie).

phenomenon [n] *conserv.* ▶natural phenomenon.

pH factor [n] *chem. limn. pedol.* ▶pH value.

4224 phloem [n] *arb.* (Principal food-conducting tissue of vascular plants. In the stems of most gymnosperms and eudicots [dicotyledons] the secondary phloem is separated from the secondary ▶xylem by the ▶cambium from which it is derived; MGW 1964); *s* **floema** [m] (Tejido que sirve para el transporte de la savia elaborada en las plantas vasculares. Puede distinguirse entre un **f. primario** [liber duro], dispuesto en haces liberoleñosos cuando no hay crecimiento de espesor, y un **f. secundario** [**liber secundario o blando**], generado por el ▶cambium hacia el exterior, en el que no se distingue el producido un año de otro y del que sólo es funcional el producido en el último. El de las coníferas es más sencillo que el de las eudicotiledóneas, que está formado por tubos cribosos, células anejas, células parenquimáticas, fibras liberianas y radios floemáticos; cf. DINA 1987; ▶xilema); *syn.* líber [m], corteza [f] secundaria; *f* **liber** [m] (Principal tissu servant au transport des substances élaborées chez les plantes vasculaires. Dans les tiges de la plupart des Gymnospermes et des Eudicotylédones [Dicotylédones vraies] le liber est séparé du bois par le ▶cambium qui le produit ; MGW 1964 ; ▶xylème) ; *syn.* phloème [m] secondaire ; *g* **Bastteil** [m] (Hauptgewebe für den Nährstofftransport in Gefäßpflanzen. In den Stämmen der meisten Gymnospermen und Eudikotyledonen ist der **B.** durch das ▶Kambium vom sekundären ▶Xylem getrennt und wird, wie das Xylem, vom Kambium gebildet; nach MGW 1964); *syn.* Bast [m], Innenrinde [f], Phloem [n].

phone antenna [n] *urb.* ▶mobile phone antenna.

phone mast [n]**, mobile** *urb.* ►mobile phone antenna.

4225 photogrammetry [n] *rem'sens. surv.* (Science of preparing charts and maps from aerial photographs using stereoscopic equipment and methods for a controlled mosaic to which have been added a reference grid, scale, and other pertinent data; cf. RCG 1982; ►aerial photogrammetry, ►remote sensing, ►topographic mapping); *s* **fotogrametría** [f] (Procedimiento técnico de elaboracion de mapas a partir de datos suministrados por fotos aéreas o teledetección utilizando equipos y métodos estereoscópicos; cf. DINA 1987; ►cartografía topográfica, ►fotogrametría aérea, ►teledetección); *f* **photogrammétrie** [f] (Procédé technique d'exploitation des informations livrées par les photographies aériennes dans un réseau de points géodésiques réalisé par redressement et stéréorestitution graphique ou numérique [photo-identification] permettant d'établir des cartes topographiques ; ►photogrammétrie aérienne, ►phototopographie, ►télédétection [de terrain]) ; *g* **Photogrammetrie** [f] (Verfahrenstechnik der Einpassung von Fernerkundungsinformationen in ein geodätisches Netz [Kartenwerk] durch Entzerrung, Bildpunktanpassung etc. und die grafische oder numerische Auswertung [Luftbildauswertung]; ►Erdbildmessung, ►Luftbildmessung, ►Fernerkundung); *syn. o. V.* Fotogrammetrie [f].

photograph [n] *envir. plan. rem'sens. surv.* ►infrared photograph, ►infrared colo(u)r aerial photograph [US], ►oblique photograph, ►oblique aerial photograph.

photograph [n]**, false colo(u)r aerial** *envir. plan. rem' sens. surv.* ►infrared color aerial photograph [US].

photography [n] *envir. plan. rem'sens. surv.* ►infrared photography.

4226 photosynthesis [n] *bot. chem.* (Process in green plants in which complex organic compounds [glucosis $C_6H_{12}O_6$] are synthesized from water [H_2O] and carbon dioxide [CO_2] and oxygen [O_2] is produced using energy absorbed from sunlight by chlorophyll. The reaction takes place in chloroplasts in green plants and in chromatophores in blue-green algae; cf. TOB 1983); *s* **fotosíntesis** [f] (Proceso biológico que tiene lugar en los organismos fotosintetizadores [cianobacterias, fitoplancton y vegetales] que contienen clorofila gracias al cual se transforma la energía solar en energía química utilizable por los seres vivos); *f* **photosynthèse** [f] (Ensemble des processus chimiques par lesquels les organismes chlorophylliens [végétaux supérieurs, algues, algues bleues et bactéries phototropes] sous l'influence de l'énergie solaire transforment le gaz carbonique de l'air [CO_2] et l'eau [H_2O] en sucres [Glucose — $C_6H_{12}O_6$] et oxygène [O_2] ; la réaction a lieu par l'intermédiaire des pigments de chlorophylle contenus dans les chloroplastes et chez les algues bleues par les pigments contenus dans les chromatophores) ; *g* **Photosynthese** [f] (Durch den Einfluss des Sonnenlichtes verursachte chemische Umwandlung in chlorophyllhaltigen Organismen [höhere Pflanzen, Algen, Blaualgen und fototrophe Bakterien] von Kohlendioxid der Luft [CO_2] und Wasser [H_2O] zu Zucker [Glucose — $C_6H_{12}O_6$] und Sauerstoff [O_2]. Die Reaktion findet in Chlorophyllkörnern, den Chloroplasten, und bei Blaualgen in Chromatophoren statt).

phreatic water [n] *hydr. wat'man.* ►groundwater [US]/ground water [UK].

phrygana [n] [Greece] *phyt.* ►garide.

4227 pH value [n] *chem. limn. pedol.* (Abbr. for *potentia hydrogenii*; figure on a numerical scale devised to measure the concentration of hydrogen in solution and therefore an indication of the solution's acidity or alkalinity. The pH scale ranges from 0.0 [pure acid] to 7.0 [neutral] and 14.0 [pure alkaline]. Soils tested at or below 4.5 are very acid, those of 6.5 to 7.0 are effectively neutral, and soils over 7.0 are alkaline; ►increasing of

pH value, ►lowering of pH value); *syn.* pH factor [n]; *s* **pH** [m] (Escala química numérica que mide el grado de acidez o de alcalinidad de una solución. Cuantitativamente, es el logaritmo negativo de la concentración activa de iones hidrógeno; DINA 1987; ►aumento del pH, ►reducción del pH); *syn.* valor [m] de pH; *f* **valeur** [f] **(du) pH** (Abréviation du terme « potentiel hydrogène » ; cologarithme décimal de la concentration d'une solution en ions H^+ permettant de désigner le caractère acide neutre ou basique d'une solution ; l'échelle du pH varie de 1 [très acide] à 14 [très basique] en passant par 7 [neutre] ; le pH des sols varie de 3,5 à 6,5 pour les sols acides et de 7,5 à 9,5 pour les sols basiques ; ►augmentation de la valeur du pH, ►abaissement du pH) ; *g* **pH-Wert** [m] (Abk. für *potentia hydrogenii*; Maßzahl für die in Lösungen enthaltene Wasserstoffionenkonzentration [Hydroniumionenaktivität] und damit für den sauren oder basischen Charakter einer Lösung; die pH-Skala reicht von 0,0 [extrem sauer] über 7 [neutral] bis 14,0 [extrem basisch]; Böden mit einem pH-Wert von 4,5 oder weniger sind sehr sauer, solche von 6,5 bis 7,0 liegen quasi im neutralen, > 7,0 im basischen Bereich; ►pH-Werterhöhung, ►pH-Wertabsenkung).

pH value [n]**, raising of** *chem. limn. pedol.* ►increasing of pH value.

pH value [n]**, reducing of** *chem. limn. pedol.* ►lowering of pH value.

pH value [n] **of soil** *chem. pedol.* ►soil reaction.

4228 physical amusement [n] *recr.* (Bodily activities of vacationists at resorts with participation by specially trained staff, which are made available as part of a travel package by an agency/tour operator); *s* **animación** [f] (Oferta de actividades lúdicas, recreativas y deportivas en el contexto de vacaciones programadas y ofertadas por touroperadores turísticos para cuya orientación éstos contratan a los llamados «animadores»); *f* **animation** [f] (Offre commerciale proposée par une agence de voyage [tour-opérateur] mettant à la disposition des vacanciers pendant leur séjour des animateurs responsables de l'organisation d'activités diverses) ; *g* **Animation** [f] (Kommerzielles Angebot hinsichtlich der inhaltlichen Gestaltung eines Ferienaufenthaltes durch einen entsprechenden Reiseunternehmer [Touroperator] mit dafür bereitgestellten Animateuren).

4229 physical benefits [npl] **of building materials** *arch.* (Physical merit, value or advantage of construction materials, e.g. of a roof garden for insulation, weatherproofing, thermal equalization, etc., and also reciprocal values of roof garden fabric for insolation and weatherproofing; ►roof planting); *s* **valor** [m] **higrotérmico** (Ventaja medible de un tipo de material de construcción, p. ej. de un ►revestimiento vegetal de tejado, para contrarrestar los extremos térmicos en un edificio, proteger las membranas de tejados planos contra las inclemencias del tiempo, etc.); *f* **qualités** [fpl] **hygrothermiques** (Caractéristiques d'un isolant thermique ou p. ex. d'une ►végétalisation des toitures à limiter la déperdition de chaleur par transmission à travers la toiture à réduire les pointes de chaleur à l'intérieur du bâtiment, à protéger l'étanchéité contre les effets naturels, etc.) ; *g* **bauphysikalischer Wert** [m] (**1.** Wert, der die physikalische Eigenschaft eines Baustoffes oder Baukonstruktion beziffert, z. B. den Durchgang von Wärme [Wärmedurchlasskoeffizent], Schall, Feuchtigkeit [Diffusion und Sorption] und Luft. **2.** Bewertung z. B. einer ►Dachbegrünung hinsichtlich ihrer positiven Wirkung auf den Ausgleich von Wärmeextremen im Gebäude [Wärmeschutz], auf die Schonung der Dachhautmaterialien etc.).

4230 physical pest control [n] *agr. constr. land'man. phytopath.* (Use of physical agents such as temperature, humidity, electric shock, radioactivity, to destroy pests or inhibit their growth and reproduction, e.g. sterilizing garden soil with steam,

P

flame-throwing to remove locusts, burning-off weeds; acoustic signals as well as periodic flooding of the soil to deter rodents); *s* **lucha [f] física contra parásitos** (Aplicación de frío, calor, radiaciones para conservar alimentos en la industria agroalimenticia, de vapor de agua o de presión en horticultura para desinfectar la tierra de semilleros o de macizos, empleo de lanzallamas contra plaguas de langosta migratoria, quema de las «malas hierbas» al borde de cultivos, señales acústicas o encharcamiento de suelos para espantar a roedores); *syn.* control [m] físico contra parásitos; *f* **procédé [m] physique de lutte contre les ennemis des cultures** (Utilisation du froid, de la chaleur ou du rayonnement dans la conservation des aliments dans l'industrie agro-alimentaire, de la vapeur d'eau sous pression en horticulture pour la désinfection [superficielle ou profonde] de mélanges terreux pour semis, repiquages, cultures en pots, l'emploi de lance-flammes contre les criquets migrateurs ou pour le désherbage des circulations, de procédés acoustiques ou la submersion du sol contre les rongeurs) ; *g* **physikalische Schädlingsbekämpfung [f]** (Kälte-, Hitze- oder Strahlenanwendung zur Lebensmittelkonservierung, das Dämpfen gärtnerischer Anzuchterden oder von Pflanzbeeten, der Einsatz von Flammenwerfern gegen Wanderheuschrecken, das Abflammen von ‚Unkraut' auf Wegen, akustische Signale sowie das zeitweise Überfluten des Bodens gegen Massenauftreten von Nagetieren).

4231 physical planning [n] *plan.* (Strategy to determine feasible methods of implementing regional planning aims); *syn.* spatial planning [n]; *s* **planificación [f] territorial** (Planeamiento con el fin de alcanzar los objetivos de la ordenación del territorio. La diferenciación entre los términos alemanes *«Raumordnung»* y *«Raumplanung»* no se hace en castellano, refieriéndose el primero más a los objetivos y directrices, o sea al desarrollo de las líneas principales de planificación, mientras que el segundo implica el desarrollo de planes y de medidas para su implementación); *syn.* planificación [f] física [C] (1); *f* **planification [f] territoriale** (Planification dont l'objectif est de contrôler, concrétiser et préparer les plans et programmes de mise en œuvre des orientations de la politique d'aménagement du territoire) ; *g* **Raumplanung [f]** (Planung, die dazu dient, Raumordnungsziele hinsichtlich ihrer Operationalität zu prüfen und zu konkretisieren, Aktionspläne aufzustellen und ihre Durchführung vorzubereiten).

4232 physical properties [npl] **of soils** *constr. pedol.* (▶ soil mechanics); *s* **características [fpl] físicas del suelo** (▶ mecánica de suelos); *f* **propriétés [fpl] physiques du sol** (Les propriétés physiques d'un sol dépendent de ses trois éléments constituants, les grains solides, l'eau et l'air [ou du gaz] ; l'assemblage des grains solides forme le squelette du sol ; lorsque l'eau remplit tous les vides, il n'y a pas d'air et le sol est dit saturé, le **degré de saturation en eau** est le rapport du volume effectivement occupé par l'eau au volume total des vides. Dans le cas contraire, l'eau se dépose par attraction capillaire [▶ capillarité] en un film plus ou moins épais autour des grains solides. Entre les grains, l'eau et l'air circulent dans les vides [interstices]. On appelle indice des vides le rapport du volume des vides [remplis ou non d'eau] au volume des pleins. La porosité est le rapport du volume des vides au volume total. Lorsque le sol est chargé, l'indice des vides diminue. Dans ce cas le sol se tasse. Les tassements doivent être prévus avant la construction sinon ils risquent d'amener la ruine surtout s'ils ne se passent pas de façon homogène. Si des points tassent plus que d'autres, on dit qu'il y a des **tassements différentiels** qui peuvent amener des ruptures des radiers et plus généralement des structures ; www.planete-tp.com ; ▶ mécanique des sols) ; *g* **bodenphysikalische Eigenschaften [fpl]** (Charakteristika eines Bodens wie z. B. Festigkeit, Formänderungsverhalten, Korngefüge [Korngröße, Kornform,

Kornbindung, räumliche Anordnung], Wassergehalt und -speicherung, Zersetzungsgrad etc., die für die Gewinnung, den Transport, die Belastung des Fahrplanums oder die Bauwerksgründung bedeutsam sind; ▶ Bodenmechanik).

physic garden [n] [UK] *gard.* ▶ medicinal herb garden.

physiognomy [n] **of a landscape** *landsc. plan. recr.* ▶ visual quality of landscape.

4233 physiographic division [n] *geo.* (Designation of separate homogeneous provinces having a similar set of biophysical characteristics and processes due to effects of climate, geology, and which results in patterns of soils and broad-scale plant communities; the defined area is a ▶ physiographic province; ▶ ecological distribution of spatial patterns, ▶ natural landscape potential); *s* **tipología [f] del paisaje** (División del paisaje de una región en unidades homogéneas [▶ teselas] basándose en sus características fitosociológicas; ▶ estructuración ecológica del territorio, ▶ fisionomía natural de un territorio); *f* **typologie [f] physiographique des paysages** (Découpage d'une région en ▶ secteurs physiographiques homogènes en utilisant des éléments caractéristiques [p. ex. la lithologie, les groupes de sols, le relief, le microclimat] et susceptibles par synthèse et définition d'unités paysagiques d'être utilisés pour différents milieux naturels comme base méthodologique de l'aménagement du territoire et des paysages ; cf. PED 1979, 424 ; ▶ division de l'espace en territoires naturels, ▶ physionomie naturelle d'un territoire) ; *syn.* division [f] physique du territoire, division [f] physiogéographique des paysages ; *g* **naturräumliche Gliederung [f]** (Ausscheiden von homogenen Raumeinheiten in der Landschaft unter Zugrundelegung der ▶ Landesnatur [Naturausstattung wie z. B. geologische Schichtung, Bodentyp, Relief, Mikroklima]; die G. der Landschaft in unterschiedliche Naturräume dient z. B. als Grundlage für Raumordnung und Landschaftsplanung; ▶ naturräumliche Einheit, ▶ ökologische Raumgliederung); *syn.* Naturraumgliederung [f], Naturpotentialgliederung [f], *o. V.* Naturpotenzialgliederung [f], Naturraumpotentialgliederung [f], *o. V.* Naturraumpotenzialgliederung [f], Landschaftsgliederung [f].

4234 physiographic province [n] *geo. ecol.* (Homogeneous spatial unit which is distinguished by its different geologic and vegetative characteristics from adjacent units; ▶ natural landscape unit, ▶ physiographic division); *syn.* physiographic region [n] (NRM 1996); *s* **tesela [f]** (Unidad elemental de la corología o fitogeografía. Se trata de un territorio o superficie geográfica, de mayor o menor extensión, homogéneo ecológicamente. Lo que quiere decir que solo posee un único tipo de vegetación potencial y por consiguiente una sola secuencia de comunidades de sustitución; DINA 1987; ▶ tipología del paisaje, ▶ unidad de espacio natural); *f 1* **secteur [m] physiographique** (Unité spatiale homogène d'un espace naturel définie par des critères géomorphologiques, topologiques et chorologiques ; ▶ espace naturel, ▶ typologie physiographique des paysages) ; *syn.* unité [f] d'espace naturel ; *f 2* **district [m] naturel** (*Terme spécifique* Entité géographique, définie en 1977 par le Centre Ornithologique Rhône-Alpes [CORA], caractérisant le découpage biogéographique de la région Rhône-Alpes en terroirs ; ceux-ci sont caractérisés par une bonne homogénéité physique [climat, substrat, géologie] et biologique [fondée essentiellement sur l'organisation de la végétation], le relief [lignes de crêtes et fonds de vallée] constituant les frontières essentielles de la plupart de ces districts. La région Rhône-Alpes compte 60 districts naturels, notion utilisée pour la cartographie de la présence/absence de certains Mammifères ; DIREN Rhône-Alpes) ; *g* **naturräumliche Einheit [f]** (Nach fachwissenschaftlichen [z. B. pedologischen, geologischen, phytosoziologischen] Kriterien definierte homogene Raumeinheit in einer Landschaft; ▶ Naturraum, ▶ naturräumliche Gliederung); *syn.* landschaftsräumliche Grundeinheit [f].

physiographic region [n] *geo. ecol.* ▶physiographic province.

phytocenosis [n] [US] *phyt.* ▶phytocoenosis.

4235 phytocoenosis [n] *phyt.* (Distinctive group of vegetal species living under the same general environmental conditions in the same space. The **p.** represents the vegetative part of an ecosystem; ▶biocoenosis, ▶plant association); *syn.* phytocenosis [n] [also US]; *s* **fitocenosis** [f] (1) (Voz preferida por la mayoría de los geobotanistas en Suiza, y más o menos empleada también en otros países, para expresar la unidad más general de la colectividad vegetal. Encierra la idea, no sólo de cohabitación en un medio, sino de una cierta relación objetiva de las plantas entre sí. Si se admite esta relación, es lógico preferir el término de fitocenosis. Los que niegan o creen dudosa tal relación, prefieren el de sinecia; DB 1985; ▶asociación, ▶biocenosis); *f* **phytocénose** [f] (Communauté végétale rassemblant en un lieu déterminé et à un moment donné diverses populations végétales vivant en équilibre dynamique ; La **p.** correspond à une portion homogène du tapis végétal qui représente généralement un groupement végétal ; DEE 1982, 69 ; ▶association [végétale], ▶biocénose) ; *g* **Phytozönose** [f] (Pflanzengesellschaft; Lebensgemeinschaft als Vergesellschaftung von Pflanzen, die durch gegenseitige Beeinflussung und Abhängigkeit in Wechselbeziehungen stehen [dynamisches Gleichgewicht]. Die **P.** stellt den pflanzlichen Anteil eines Ökosystems dar; ▶Assoziation, ▶Biozönose).

phytogeography [n] *geo. phyt.* ▶plant geography.

4236 phytomass [n] *bot. ecol.* ([Dry] weight of plant material existing at any given time. There is a distinction between **p.** above and **p** below ground; *generic term* ▶biomass); *syn.* standing crop biomass [n] (TEE 1980, 238); *s* **fitomasa** [f] (Desde el punto de vista ecológico, la **f.** es la materia total de las plantas presentes en un ecosistema determinado. Se expresa generalmente en unidad de peso seco por unidad de superficie; DINA 1987; *término genérico* ▶biomasa); *syn.* biomasa [f] vegetal; *f* **phytomasse** [f] (La ▶biomasse végétale constituée par les végétaux d'une phytocénose ; on distingue la **p.** aérienne ou épigée et la **p.** souterraine ou hypogée) ; *g* **Phytomasse** [f] (Gewicht einer zu einem bestimmten Zeitpunkt vorhandenen Pflanze oder eines Pflanzenbestandes je Flächen- oder Volumeneinheit als Frisch- oder Trockengewicht. Es wird zwischen oberirdischer und unterirdischer **P.** unterschieden; *OB* ▶Biomasse).

phytome [n] *ecol.* ▶biome.

phytophenology [n] *phyt.* ▶phenology.

phytosanitary certificate [n] *agr. for. hort. leg.* ▶official phytosanitary certificate.

4237 phytosanitary inspection [n] *agr. hort. leg.* (Inspection of imported plants legally required by health authorities, to prevent the spreading of plant diseases or pests, as well as injurious organisms of stored products; ▶official phytosanitary certificate); *s* **control** [m] **fitosanitario** (Inspección de plantas importadas para evitar la propagación de plagas o enfermedades de plantas; ▶certificado fitosanitario de plantas); *f* **contrôle** [m] **phytosanitaire** (F., contrôle exécuté par le Service de la protection des végétaux pour éviter l'introduction ou d'arrêter la dissémination d'organismes dangereux pour les cultures et les végétaux, et de limiter les dégâts provoqués par ceux-ci ; LA 1981, 357 ; ▶certificat phytosanitaire) ; *syn.* inspection [f] phytosanitaire ; *g* **Pflanzenbeschau** [f] (Gesetzlich vorgeschriebene Prüfung des Gesundheitszustandes von eingeführten Pflanzen zur Verhütung der Verschleppung von Pflanzenkrankheiten sowie Pflanzen- und Vorratsschädlingen; ▶Pflanzengesundheitszeugnis).

phytosociologic [adj] *geo. phyt.* ▶phytosociological.

4238 phytosociological [adj] *geo. phyt.* (**1.** Related to vegetation science [▶phytosociology]. **2.** Pertinent to or involved in floristic interrelations); *syn.* phytosociologic [adj]; *s* **fitosociológico/a** [adj] (Relativo a la ▶fitosociología, o sea al estudio de las comunidades vegetales); *f* **phytosociologique** [adj] (relatif à la ▶phytosociologie) ; *g* **vegetationskundlich** [adj] (Die Vegetationskunde [▶Pflanzensoziologie] betreffend).

4239 phytosociological classification [n] *phyt.* (Plant communities are classified in a taxonomic hierarchy according to their floristic composition and uniform physiognomy. The lowest rank is an ▶plant association, followed by an ▶alliance, ▶order 1, and ▶class 1. The higher the rank the wider is the ▶ecological amplitude of an unit); *s* **clasificación** [f] **fitosociológica** (Las comunidades vegetales se clasifican en una jerarquía taxonómica de acuerdo a su composición florística y su fisionomía uniforme. El nivel más inferior es la ▶asociación, seguida por la ▶alianza, el ▶orden y la ▶clase. Cuanto más alto el nivel jerárquico, mayor es la ▶amplitud ecológica de una unidad); *f* **systématique** [f] **phytosociologique** (Classification des groupements végétaux selon l'homogénéité physionomique de la station observée et leur composition floristique ; par ordre décroissant dans la hiérarchie on distingue la ▶classe, l'▶ordre, l'▶alliance, l'▶association [végétale], la sous-association, la variante et le faciès, la ▶valence écologique augmentant avec le niveau hiérarchique du groupement dans la classification) ; *syn.* classification [f] phytosociologique ; *g* **Klassifizierung** [f] **der Vegetationseinheiten** (Vegetationseinheiten werden entsprechend ihrer floristischen Zusammensetzung und ihres einheitlichen physiognomischen Erscheinungsbildes taxonomisch hierarchisch gegliedert. Die unterste Einheit ist die ▶Assoziation, gefolgt vom ▶Verband 1, von der ▶Ordnung und der ▶Klasse. Je höher die Hierarchiestufe, desto breiter ist die ▶ökologische Amplitude einer Einheit).

4240 phytosociologist [n] *phyt. prof.* (Scientist who studies and practices vegetation ecology); *s* **fitosociólogo/a** [m/f] (Científico/a que estudia las comunidades vegetales); *f* **phytosociologue** [m] (Scientifique effectuant des travaux de recherche sur les associations végétales) ; *g* **Pflanzensoziologe** [m]/**Pflanzensoziologin** [f] (Jemand, der sich wissenschaftlich mit der Erforschung der Pflanzengesellschaften befasst); *syn.* Vegetationskundler/-in [m/f].

4241 phytosociology [n] *phyt.* (Science of interrelationships between plant species [floristic composition] and their environmental conditions within plant communities; ▶association, ▶succession, ▶synchorology, ▶syndynamics); *syn.* plant sociology [n], vegetation ecology [n], vegetation science [n]; *s* **fitosociología** [f] (Término preferido por las escuelas de Upsala y de Zurich-Montpellier para expresar la parte de la geobotánica que estudia las comunidades vegetales y sus relaciones con el medio, englobando los dos conceptos de sinecología y ▶sinecología; cf. DB 1985; ▶asociación, ▶sincorología, ▶sindinamismo, ▶sucesión); *f* **phytosociologie** [f] (Science qui étudie la façon dont sont groupées les plantes dans la nature ; elle se préoccupe de la définition et de la mise en évidence d'▶associations [végétales], de leur classification [syntaxonomie], de leur écologie [▶synécologie], de leur dynamique [▶syndynamique], de leur répartition géographique [▶synchorologie] et de leurs potentialités ; DEE 1982, 70 ; ▶succession) ; *syn.* phytocénotique [f] ; *g* **Pflanzensoziologie** [f] (Wissenschaft von dem Beziehungsgefüge der Pflanzen untereinander [floristische Zusammensetzung] und ihren Umweltbedingungen in den jeweiligen Pflanzengesellschaften; sie beschäftigt sich mit der **1.** Gesellschaftssystematik durch Bestimmung von Pflanzengesellschaften anhand von

Vergleichen der Pflanzenlisten vieler Standorte [Syntaxonomie] und **2.** deren Einordnung anhand von standorttypischen biotischen und abiotischen Umweltfaktoren [►Synökologie], **3.** der standortbezogenen Wuchsdynamik und Gesellschaftsentwicklung [►Syndynamik] sowie mit den Gesellschaftsfolgen am Standort [►Sukzession], **4.** der *geografischen Verbreitung von Gesell*schaften [►Synchorologie] und deren Aussagefähigkeiten auf die Standortsituationen und **5.** der Gesellschaftsgeschichte [Synchronologie], d. h. der Erforschung der erdgeschichtlichen Wanderungen der Pflanzengesellschaften; cf. ÖKO 1983; ►Assoziation); *syn.* Phytosoziologie [f] Vegetationskunde [f].

picket [n] *constr.* ►stake (2).

4242 picket fence [n] *constr.* (Wooden fence built of vertical wooden pales fixed at [close] regular intervals to arris rails, secured between posts. The pickets may be sawn or planed from hardwood or softwood; of rectangular [or in D. also half-rounded] section; picket tops may be pointed, flat or rounded; ►close-boarded fence, ►decorative panel fence); *syn.* lath fence [n], pale fence [n] [also UK], vertical slatted fence [n], paling [n] [UK], paling fence [n] [also US]; *s 1* **valla** [f] **de listones**; *syn.* valla [f] de estacas; *s 2* **valla** [f] **de listones verticales** (Cerca en la que los listones están unidos verticalmente a intervalos regulares a las traviesas horizontales; ►valla de tablas perfiladas 1, ►valla de tablones); *f 1* **clôture** [f] **à lattes** (Constituée à partir de lattes de bois [ou échalas] fixées sur les lisses souvent assemblées sous forme de panneaux de lattes verticales ou croisées ; ►clôture en planches jointives, ►clôture [en bois] non jointive à lattes verticales) ; *f 2* **clôture** [f] **(en bois) non jointive à lattes verticales** (Clôture dont les lattes en bois éclatées sont fixées sur un barreau avec un espacement ; cf. GEN 1982-II, 155) ; ►clôture de planches ajourée ; *syn.* clôture [f] non jointive « type palissade », clôture [f] ajourée à lattes verticales ; *g* **Lattenzaun** [m] (Aus runden, halbrunden, vierkantigen Holzleisten oder schmalen Brettern mit unterschiedlichen Kopfausprägungen gefertigte Einfriedigung; ►Bretterzaun; *UB* ►Profilbretterzaun); *syn.* Staketenzaun [m], Senkrechtlattenzaun [m].

picking up [n] **litter and weeding** [n] **on planted areas** [UK] *constr.* ►cleaning up planted areas [US].

4243 picnic [n] *recr.* (A daytrip with a packed meal eaten outdoors); *s* **picnic** [m] *syn.* merienda [f] campestre; *f* **pique-nique** [m] (Consommation des aliments emportés au cours d'une sortie journalière) ; *g* **Picknick** [n] (...e und ...s [pl]; Verzehr von mitgebrachten Speisen im Freien).

picnic area [n] *recr.* ►picnic site, #2

4244 picnic shelter [n] *recr.* (Usually a roofed open structure with table-bench furniture for food consumption by family or group; ►park shelter); *s* **techado** [m] **para picnics** (Construcción ligera en zona de excursiones, generalmente abierta por los lados, equipada con mesas y bancos para que la gente se siente a comer o descansar; ►techado contra la lluvia); *f* **abri** [m] **de pique-nique** (Abri en général ouvert, équipé d'une table et de bancs permettant aux promeneurs ou randonneurs de prendre un repas ; ►hutte) ; *g* **Picknickhütte** [f] (Meist offene ►Schutzhütte mit Tisch-Bank-Einrichtung zum Verzehr von mitgebrachten Speisen).

4245 picnic site [n] *recr.* (**1.** Informal facility in the open for social interaction and food consumption); *syn.* picnic spot [n]; **2. picnic area** [n] (Planned outdoor facility consisting of one to several tables, benches and rubbish bins, often also equipped with grill and shelter to which packed food is brought and consumed); *syn.* picnic unit [n]; *s 1* **merendero** [m] (En Es., venta en la que la gente puede consumir su comida traída y sólo ordenar bebidas al servicio); *s 2* **área** [f] **de picnic** (Instalación en zona de recreación equipada con mesas y bancos, de manera que la gente

pueda sentarse para su picnic); *f* **aire** [f] **de pique-nique** (Aire de plein-air aménagée en forêt, en bordure de routes, etc. et équipée de tables et de bancs, souvent d'un grill, permettant de prendre un repas) ; *g* **Picknickplatz** [m] (Einrichtung im Freien, oft mit Feuerstelle, zum Verzehr von mitgebrachten Speisen).

picnic spot [n] *recr.* ►picnic site, #1.

picnic unit [n] *recr.* ►picnic site, #2.

picture [n] **of a town, visual** *urb.* ►townscape.

piece [n] **[UK, AUS], pipe junction** *constr.* ►pipe coupling [US].

piece made-to-measure [loc] **[UK]** *constr.* ►fitted piece [US].

4246 piece [n] **of street furniture** *arch. urb.* (Permanently fixed or mobile feature of functional or aesthetic character installed in street spaces or public squares, e.g. bench, kiosk, shelter, telephone booth, trash receptacle; ►movable outdoor furniture); *syn.* streetscape element [n] [also US] (PSP 1987, 226), street amenity [n] [also US] (PSP 1987, 196), street design element [n]; *s* **elemento** [m] **de mobiliario urbano** (Cualquier tipo de pieza fija o móvil de equipamiento instalada en las calles o plazas de la ciudad sea con fines funcionales o estéticos, como bancos, quioscos/kioscos, cabinas de teléfonos, refugios de espera, papeleras, etc.; ►mobiliario urbano móvil); *f* **élément** [m] **de mobilier de voirie** (Ensemble des équipements utilitaires ou décoratifs, mobiles [►mobilier mobile] ou fixes rencontrés sur les espaces d'agrément collectifs ou publics urbains concourant à l'amélioration de la vie ; *termes spécifiques* mobilier de jeux, de circulation, de repos, de protection, de décoration, d'information, de signalisation, etc.) ; *syn.* mobilier [m] urbain ; *g* **Straßen-möblierungselement** [n] (Jedes ortsgebundene oder bewegliche, temporäre oder Dauerausstattungsstück von funktionalem oder ästhetischem Charakter, das zur zweckdienlichen Nutzung von Straßen oder öffentlichen Plätzen aufgestellt oder angebracht ist; ►Möblierung); *syn.* Straßenmöbel [n], Ausstattungselement [n] für Straßen und Plätze; *Gesamtheit der Elemente* Straßenmobiliar [n].

piece of work [n], **execution of** *constr. contr.* ►execution of construction items.

piedmont [n] *geo.* ►foothills.

4247 pier [n] **(1)** *arch. constr.* (Structure extending over water used as a landing place for ships or as an amusement area; e.g. Brighton Pier [UK]; generic term covering ►landing pier [US]/landing stage [UK], ►fishing pier); *s* **embarcadero** [m] (Término genérico; ►embarcadero de botes, ►pontón de pesca); *syn.* atracadero [m]; *f* **embarcadère** [m] (*Terme générique* structure en bordure de mer ou en rive de cours d'eau ou de plan d'eau [quai, jetée - ouvrages de maçonnerie -, ►appontement — ouvrage en bois ou en métal] le long de laquelle un bateau peut accoster afin d'embarquer ou de débarquer des personnes ou du matériel ; ce terme est aussi utilisé pour désigner l'emplacement réservé pour le départ et le retour des promenades touristiques en barques traditionnelles mais aussi, aux canoës, aux pédalos, aux house-boats ; *termes spécifiques* ponton, ►ponton de pêche) ; *g* **Ufersteg** [m] (Direkte Zugangsmöglichkeit an ein Gewässer, z. B. in Form eines Holzdecks, oder Anlegeeinrichtung am Rande eines Gewässers; *OB* für ►Angelsteg, ►Bootssteg); *syn.* Gewässersteg [m].

4248 pier [n] **(2)** *arch. constr.* (Intermediate support/pillar for the adjacent ends of two spans of a bridge); *s* **pila** [f] **de puente** (Pilar de apoyo para dos extremos contiguos de un tramo de un puente); *syn.* machón [m] [RCH], pilastra [f] [CA] (ambos BUKSCH 1959); *f* **pile** [f] **de pont** (**1.** Appui intermédiaire en général vertical qui supporte un pont ou sur lequel viennent porter les arches d'un pont ; l'appui d'extrémité enterré sur tout ou partie de sa hauteur est dénommé « pile culée » ; dans les cours d'eau par nécessité hydrodynamique afin de faciliter l'écoulement de l'eau de part et d'autre de la pile ou à éviter ou limiter les contre-courants qui se forment en arrière de celle-ci, la pile a une forme de plan pointu, demi rond ou ogival : **l'avant-bec** en face amont, **l'arrière-bec** en face aval. **2.** Le terme **pilier** ayant l'acceptation plus générale d'un support vertical cylindrique, carré ou rectangulaire, dans une construction) ; *syn.* pilier [m] de pont (±) ; *g* **Brückenpfeiler** [m] (Mittlere Stütze von langen Brücken; in Fließgewässern sind diese Pfeiler strömungsgünstig ausgebildet und heißen **Strompfeiler**).

4249 pierced wall [n] *arch. constr.* (Freestanding ornamental wall containing voids within or between units); *s* **muro** [m] **calado** (Muro libre con función decorativa que contiene huecos entre los ladrillos o mampuestos); *f* **claustra** [f] (Mur de clôture ou de fermeture de baies constitué en général par un mur bahut sur lequel sont maçonnés des éléments en pierre ou en terre cuite ajourés) ; *syn.* mur [m] ajouré en briques ; *g* **Mauer** [f] **aus durchbrochenen Steinen** (Frei stehendes Ziermauerwerk aus auf Lücke gesetzten Mauerziegeln oder aus Tonhohl- oder Betonziersteinen).

piercing [n] *constr.* ▶lawn aeration, #1.

pigweed vegetation [n] *phyt.* ▶root-crop weed community.

pile [n] *hort.* ▶compost pile; *envir.* ▶high spoil pile [US]; *min.* ▶overburden pile; *envir.* ▶rubble pile [US]/rubble tip [UK]; *min.* ▶sand pile [US]/sand tip [UK], ▶silt pile [US]/silt heap [UK], ▶slag pile [US]/slag heap [UK]; *constr.* ▶soil pile [US]/soil heap [UK], ▶topsoil pile [US]/topsoil heap [UK].

pile [n] [US], **spoil** *min.* ▶spoil bank [US] (2)/spoil heap [UK].

pile [n], **timber** *constr.* ▶palisade.

pile dumping [n] [US], **elongated** *constr.* ▶front end dumping.

pillar foundation [n] [US] *constr.* ▶post foundation [US].

4250 pilot study [n] *ecol. landsc. plan.* (Empirical investigation used as a practical guide for the collection of data and criteria necessary for a demonstration research project in landscape management or a related discipline); *syn.* demonstration study [n]; *s* **estudio** [m] **piloto** (Investigación empírica aplicada como guía práctica para recolectar datos o criterios necesarios para un proyecto de demostración en gestión paisajística o en disciplina vecina); *f* **étude** [f] **pilote** (Étude empirique utilisée comme participation méthodologique à la détermination de divers facteurs, critères, échelles, p. ex. dans le domaine de la préservation du patrimoine naturel et des paysages) ; *g* **Modelluntersuchung** [f] (Bei der planungs- und anwendungsbezogenen Disziplin der Landespflege eine Untersuchung als methodischer Beitrag zum Herausarbeiten der für Forschungsvorhaben erforderlichen Daten, Maßstäbe und Kriterien); *syn.* Modellstudie [f], Pilotprojekt [n], Pilotstudie [f].

4251 pine krummholz [n] *bot.* (Shrublike species of *Pinus mugo*, which occurs in Central and Eastern Europe as far east as the Balkan and Carpathian Mountains at an altitude of up to 2,600m; ▶fen covered with pine krummholz); *s* **pino** [m] **mugo** (Forma arbustiva del pino de montaña *[Pinus mugo]* que crece hasta una altitud de 2600 m en Europa Central y Meridional hasta los Balcanes; ▶paular con pinos mugo); *f* **pin** [m] **rampant** (Formes arbustives [2 à 3 m de haut] du Pin mugo (*Pinus mugo*) et spontanées de l'étage subalpin jusqu'à une altitude de 2600 m dans les Alpes orientales, les Balkans et les Carpates, aux branches rampantes et extrémités redressées (*Pinus m. var. mughus*) ou dressées (*Pinus m. var. pumilio* ; ▶brousse à Pins rampants) ; *syn.* pin [m] mugo, pin [m] rabougri ; *g* **Latsche** [f]

(Strauchartige Form der Bergkiefer *Pinus mugo,* die in Mittel- und Osteuropa bis zum Balkan und in den Karpaten bis in 2600 m Höhe vorkommt: *P. m. var. mughus, P. m. var. pumilio*; ▶Latschenfils); *syn.* Krummholzkiefer [f], Legföhre [f].

4252 pinewood [n] *phyt.* *s* pinar [m]; *f* forêt [f] de pins *syn. appellation gascogne* pignade [f], *appellation provençale* pinède [f] ; *g* Kiefernwald [m].

pinned joints [npl] *constr.* ▶rubble masonry with pinned joints.

pinus-dominated elfinwood [n] [US] *phyt.* ▶pinus-dominated krummholz formation.

4253 pinus-dominated krummholz formation [n] *phyt.* (▶Krummholz community, primarily of twisted pines on the periphery of the ▶tree line; ▶coniferous evergreen thicket/shrubland, ▶timberline); *syn.* pinus-dominated elfinwood [n] [also US]; *s* **matorral** [m] **de pino mugo** (Comunidad vegetal de pinos de montaña *[Mugo-Ericetum]* que crece en el ▶límite de los árboles; cf. BB 1979; ▶limite del bosque, ▶matorral achaparrado de coníferas, ▶matorral de altura); *f* **formation** [f] **de brousse à** *Pinus mugo* (▶Formation de krummholz, groupement climacique *[Mugeto-Rhododendretum hirsuti* BRAUN-BLANQUET] croissant sur sols calcaires en pente, escarpements, couloirs d'avalanche, formant ordinairement la ▶limite supérieure de la végétation arborescente ; VCA 1985, 216 ; ▶brousse sempervirente de résineux, ▶limite des forêts) ; *syn.* brousse [f] à *Pinus mugo,* formation [f] de pin rabougri ; *g* **Legföhrenformation** [f] (Vegetationsdecke am Rande der ▶Waldgrenze/▶Baumgrenze, die vorwiegend aus Krummholzkiefern besteht; ▶Krummholzgesellschaft, ▶immergrünes Nadelgebüsch); *syn.* Bergkieferngebüsche [npl], Latschengebüsche [npl], Legföhrengebüsche [npl].

4254 pioneer crop [n] **of woody plants** *for. landsc. phyt.* (Crop of trees and shrubs [pioneers] which has invaded bare sites and colonized them, and is later supplanted by succession species; ▶nurse crop of woody plants); *s* **rodal** [m] **pionero** (Fase pionera de la vegetación de árboles en el desarrollo natural de ésta; ▶cubierta de tutela); *f* **groupement** [m] **préforestier** (Stade initial de développement de la végétation arborescente lors de la colonisation d'un site dans le cadre de la succession naturelle ; ▶culture-abri) ; *syn.* peuplement [m] (forestier) pionnier, peuplement [m] (forestier) colonisateur ; *g* **Vorwald** [m] (Pionierphase der Baumvegetation in der natürlichen Vegetationsentwicklung; ▶Schirmbestand); *syn.* Pionierbestand [m] eines Waldes.

pioneer plant [n] *phyt.* ▶pioneer species, ▶woody pioneer plant/species.

4255 pioneer plant [n] **on entisol** [US] *phyt.* (▶pioneer species); *syn.* pioneer plant [n] on immature soil [UK]; *s* **pionera** [f] **de suelo bruto** (▶especie pionera); *syn.* planta [f] pionera de suelo bruto; *f* **plante** [f] **pionnière de sols bruts** (▶Espèce pionnière liée aux sols ou substrats bruts, secs comme les rocailles ou humides comme les rives de cours d'eau) ; *g* **Rohbodenbesiedler** [m] (▶Pionierart auf Rohböden).

pioneer plant [n] **on immature soil** [UK] *phyt.* ▶pioneer plant on entisol [US].

4256 pioneer population [n] *phyt. zool.* (Total number of an animal or plant species which has been introduced to colonize an area, or small numbers of immigrated species just beginning to spread naturally); *s* **población** [f] **pionera** (Conjunto de especies de flora y fauna que han sido introducidos en una zona para recolonizarla o pequeña cantidad de especies inmigrantes que comienzan a dispersarse naturalmente); *f* **population** [f] **pionnière** (**1.** Ensemble des individus d'une espèce animale ou végétale introduite pour coloniser une station, **2.** peuplement réduit d'espèces en voie de coloniser un espace défini lors de leur dispersion naturelle) ; *syn.* population [f] colonisatrice ; *g* **Pionierpopulation [f]** (Gesamtbestand einer Tier- oder Pflanzenart, die zwecks Ansiedlung in einem Gebiet ausgesetzt wurde resp. kleine Bestände [= Vorposten] der frisch auf dem Wege der natürlichen Ausbreitung in ein Gebiet eingewanderten Arten).

pioneers association [n] **of alpine belt** *phyt.* ▶initial pioneers association of alpine belt.

4257 pioneer species [n] *phyt.* (Plant which initially colonizes an unfavo[u]rable site in terms of soil and microclimate, and, by improving the soil enables the subsequent growth of more demanding plants; ▶nurse plant; ▶pioneer plant on entisol [US]/pioneer plant on immature soil [UK]); *syn.* pioneer plant [n]; *s* **especie [f] pionera** (Especie que coloniza inicialmente un área nueva no ocupada por otras especies. Son especies eurioicas, con gran capacidad de multiplicación y esperanza de vida muy pequeña [oportunistas]; producen diásporas de pequeño tamaño, muy fáciles de transportar, y en gran número, lo que les proporciona gran capacidad de dispersión [cinetófila]; son las primeras en ocupar espacios vacíos [pioneras]. En las etapas más avanzadas de la sucesión son desplazadas por especies especializadas en condiciones estables, más longevas y eficientes, con las que no pueden competir; DINA 1987; ▶pionera de suelos brutos, ▶planta nodriza); *syn.* especie [f] fugitiva, especie [f] oportunista, especie [f] pródiga, especie [f] apocrática, especie [f] cinetófila, especie [f] fugaz; *f* **espèce [f] pionnière** (Espèce végétale colonisant, pendant la phase initiale de développement d'une communauté biologique, les stations aux conditions édaphiques et climatiques défavorables [sols nus, abandonnés, perturbés, privés de végétation] ; les **e. p.** favorisent la formation d'un substrat sur lequel se développeront ensuite les espèces plus exigeantes qui les supplanteront définitivement ; parmi les espèces pionnières on dénomme « **opportunistes** » les espèces qui profitent de circonstances favorables [absence de compétition, richesse en éléments nutritifs, etc.] pour se développer en très forte abondance ; ces espèces peu compétitives envahissent les terrains vagues, les jachères et les décharges publiques ; ▶espèce abri, ▶plante pionnière de sols bruts) ; *syn.* plante [f] colonisatrice ; *g 1* **Pionierart [f]** (Pflanze, die sich auf ungünstigen Standorten hinsichtlich der Boden- und Lokalklimabedingungen zuerst ansiedelt, die Bodenbildung fördert und anspruchsvolleren Pflanzen später das Wachstum ermöglicht und von diesen verdrängt wird; unter den **P.en** gibt es auch **Opportunisten**, die die biotischen und abiotischen Vorteile [kein Konkurrenzkampf, nährstoffreicher Boden etc.] meist künstlicher Nischen, z. B. Brachen, Deponien, nutzen und besiedeln); *g 2* **R-Stratege [m]** (Von J. P. GRIME 1979 mit seinem Strategiekonzept für Pionierpflanzen eingeführter Begriff, der eine reproduktionsstarke [r = reproduction/Vermehrung] krautige Pflanze beschreibt: ein Strategietyp mit hoher generativer Reproduktionsrate durch Selbstsaat, als Individuum kurzlebig, nicht ortsfest, mit geringer Konkurrenzkraft, auf Offenflächen mit schneller Entwicklung bis zum Blühstadium, geht mit Verdichtung der Vegetation durch konkurrenzstarke langlebige **C-Strategen** [c = competition/Wettbewerb] zurück; cf. JPG 1979; ▶Ammenpflanze, ▶Rohbodenbesiedler); *syn.* Erstbesiedler [m], Pionierpflanze [f], R-Stratege [m].

pioneer species [npl], **association of scattered** *phyt.* ▶initial pioneers association of alpine belt.

pioneer species [n], **woody** *landsc. phyt.* ▶woody pioneer plant/species.

4258 pioneer vegetation [n] *phyt.* (Plant community which has first colonized an extreme or unvegetated site before a permanent community has established itself by way of ▶succes-

sion; ▶initial pioneers association of alpine belt, ▶initial plant community); *syn.* primary (plant) growth [n]; *s* **vegetación [f] pionera** (Comunidad vegetal que ocupa los «espacios vacíos», con tasas de multiplicación muy elevadas de las especies que la constituyen y adaptadas a condiciones cambiantes en el tiempo y en el espacio; MARG 1977; ▶comunidad inicial, ▶comunidad inicial del piso alpino, ▶sucesión); *f* **végétation [f] pionnière** (Peuplement végétal colonisant les sols nus ou les stations soumis à des conditions extrêmes [p. ex. aride, mobile] avant que ne se développe progressivement au cours de la ▶succession un groupement stable ; ▶groupement initial, ▶groupement pionnier de l'étage alpin) ; *syn.* peuplement [m] pionnier, groupement [m] pionnier, végétation [f] initiale, végétation [f] colonisatrice ; *g* **Pionierflur [f]** (Pflanzengesellschaft, die sich auf extremen oder unbewachsenen Standorten zuerst ansiedelt, bevor sich eine entsprechende Dauergesellschaft durch ▶Sukzession einstellt; ▶Initialgesellschaft, ▶Vorpostengesellschaft); *syn.* Pioniergesellschaft [f], Pioniervegetation [f].

pipe [n] *agr. constr.* ▶bell pipe [US]/socket pipe [UK], ▶clay pipe, ▶drainage cleanout pipe [US]/drain(age) inspection shaft [UK], ▶drain pipe, ▶earth-covered drainage cleanout pipe [US]/earth-covered drain(age) inspection shaft [UK]; *envir. urb.* ▶effluent discharge pipe; *constr.* ▶feed pipe, ▶outfall pipe, ▶overflow pipe, ▶sag pipe, ▶vitrified clay pipe.

pipe [n] [US], **cleanout** *constr.* ▶drainage cleanout pipe [US]/drain(age) inspection shaft [UK].

pipe [n] [US], **drain cleanout** *constr.* ▶drainage cleanout pipe [US]/drain(age) inspection shaft [UK].

pipe [n], **drain outlet** *constr.* ▶drain outfall.

pipe [n] [US], **earth-covered drain cleanout** *constr.* ▶earth-covered drainage cleanout pipe [US]/earth-covered drain(age) inspection shaft [UK].

pipe [n] [UK], **glazed stoneware** *constr.* ▶vitrified clay pipe [US].

pipe [n] [US], **interceptor** *constr.* ▶drain pipe.

pipe [n], **sewage discharge** *envir. urb.* ▶effluent discharge pipe.

pipe [n] [UK], **sewage disposal** *envir. urb.* ▶effluent discharge pipe.

pipe [n] [UK], **socket** *constr.* ▶bell pipe [US].

pipe [n] [UK], **tapered** *constr.* ▶increaser [US].

4259 pipe branching [n] *constr.* (Pipe fitting for a side connection, e.g. ▶pipe elbow, ▶pipe fitting, ▶saddle fitting, tees or other connectors used in assembling pipes); *s* **bifurcación [f] de tubería** (Pieza especial para tuberías de alimentación o de desagüe; ▶accesorio con brida, ▶codo, ▶tubo bifurcado); *f* **raccord [m] de tuyau** (Terme générique ; *termes spécifiques* ▶coude, ▶pièce d'embranchement, ▶raccord de picage à plaquette) ; *g* **Rohrverzweigung [f]** (Formstück in Abwasser- oder Wasserleitungen zur Sammlung oder Verteilung des Wassers aus unterschiedlichen resp. in unterschiedliche Richtungen; ▶Abzweig, ▶Bogen, ▶Sattelstück).

4260 pipe casing [n] *constr.* (Concrete covering of a drainage pipe to increase its strength and load-bearing capacity; ▶wrapping of perforated drain pipe with soil separator); *s* **envoltura [f] de tubería** (Reforzamiento de tubería para darle más capacidad de carga; ▶enfundado de dren); *syn.* camisa [f] de tubería; *f 1* **revêtement [m] extérieur d'un tuyau** (±) (Renforcement de tuyaux d'assainissement au moyen d'un manche en béton afin d'en augmenter la résistance mécanique) ; *f 2* **protection [f] extérieure d'un tuyau** (Revêtement extérieur pour protéger le tuyau contre l'action du sol et du milieu environnant) ; *f 3* **enrobage [m] d'une canalisation** (Partie du remblaiement en sable constitué du lit de pose, de l'assise, du remblai de protection latérale, et du remblai de protection supérieur ; C.C.T.G. fascicule n° 70 ; ►enrobage d'un drain) ; **g Rohrummantelung [f]** (Verstärkung der Entwässerungsrohre mit Beton zur Tragfähigkeitserhöhung; ►Dränrohrummantelung).

4261 pipe connection [n] *constr.* (Fitting which is used to join together drain, irrigation, water or sewer pipes, e.g. with a groove, sleeve or thread; ►pipe branching); **s conexión [f] de tuberías** (Manera de conectar las tuberías de desagüe o irrigación entre sí; ►bifurcación de tubería); *syn.* unión [f] de tuberías, unión [f] de tubos; *f* **jonction [f] de tuyau** (Processus ; le résultat est le ►raccord de tuyau) ; *syn.* emboîtement [m] d'un tuyau, assemblage [m] d'un tuyau ; **g Rohrverbindung [f]** (Art des Zusammenfügens von Ent- oder Bewässerungsrohren, z. B. mit Falz, Muffe resp. Verschraubung, Klammerverbindung; ►Rohrverzweigung).

4262 pipe coupling [n] **[US]** *constr.* (Unit for joining sections of piping; ►fitted piece [US]/piece made-to-measure [UK], ►pipe fitting); *syn.* pipe joint fitting [n], junction fitting [n], pipe junction piece [n] [UK, AUS]; **s pieza [f] de juntura** (Unidad para juntar dos secciones de tubería; ►pieza de empalme, ►tubo bifurcado); *syn.* tubería [f] de empalme; *f* **pièce [f] de raccord** (Pour canalisation ; ►pièce ajustée, ►pièce d'embranchement) ; *syn.* raccord [m] (de tuyauterie) ; **g Rohrverbindungsstück [n]** (Verbindungs- oder Anschlussstück für Rohrleitungen; ►Abzweig, ►Passstück); *syn.* Rohranschlussstück [n], Fitting [n].

4263 pipe culvert [n] *constr. trans.* (►Culvert with a segmental, fully circular, or elliptical arch cross-section; ►arch culvert, ►culvert); **s alcantarilla [f] tubular** (Paso transversal de sección circular o elíptica bajo vía de tránsito para permitir el flujo de pequeño curso de agua; ►alcantarilla, ►alcantarilla abovedada); *syn.* atarjea [f] de caño, puentecillo [m], tajea [f] [C], pontón [m] [Es, YV], conducto [m] pluvial [RA]; *f* **passage [m] busé circulaire** (Passage de section circulaire pour l'écoulement de cours d'eau peu importants, p. ex. pour une traversée de route ; ►passage busé, ►passage vouté) ; **g runder Durchlass [m]** (Queröffnung bei Verkehrstrassen für den Durchfluss von kleinen Wasserläufen; ►Durchlass, ►Gewölbedurchlass); *syn.* Rohrdurchlass [m].

4264 pipe discharge cross-section [n] *constr.* (Inside cross-sectional area of a drainage pipe measured to determine the maximum discharge capacity); **s sección [m] transversal de descarga** (Diámetro de tuberías de desagüe medido para calcular la capacidad máxima de caudal); *f* **section [f] diamètre intérieur** (Section définissant le débit d'écoulement d'eau admissible dans les canalisations pour l'assainissement) ; **g Abflussquerschnitt [m] (2)** (Durchflussquerschnitt bei Entwässerungsrohren); *syn.* Durchflussprofil [n], Durchflussquerschnitt [m].

pipe [vb] **ditches or watercourses** *wat'man.* ►culverting.

4265 piped stream [n] **[US]** *wat'man.* (**1.** Culvert carrying water under a road, railway or embankment. **2.** Conduit carrying a stream underground; ►culverting); *syn.* culverted stream [n] [UK]; **s arroyo [m] entubado** (Pequeño curso de agua conducido bajo tierra por tubería; ►entubado de arroyos); *f* **rivière [f] busée** (Ruisseau à conduit souterrain/à passage busé ; ►busage) ; *syn.* ruisseau [m] busé ; **g verdolter Bach [m]** (Ein unterirdisch geführter Bach ; ►Verdolung); *syn.* verrohrter Bach [m], eingedolter Bach [m].

4266 pipe elbow [n] *constr.* (Pipe fitting for providing a sharp change of direction in a pipeline; a 90° elbow is also called an **ell** [US]); **s codo [m]** (Pequeño trozo de tubería que sirve para conectar dos alineaciones rectas de la conducción; DACO 1988; *términos específicos* codo de 90°, codo de 180°, tubo en L, tubo

en T); *f* **coude [m]** (Accessoire de canalisation d'assainissement ou d'alimentation en eau de 15° à 90°) ; **g Bogen [m]** (Formstück für Entwässerungs- oder Bewässerungsleitungen in 15°-, 30°-, 45°- und 90°-Ausführung).

4267 pipe fitting [n] *constr.* (Standardized connecting fixture for pipelines; e.g. 45° [pipe] fitting, ►T-jointed pipe fitting, etc.; ►pipe branching, ►saddle fitting); **s tubo [m] bifurcado** (Pieza de tubería para conductos de aguas; ►accesorio con brida, ►bifurcación de tubería, ►tubo bifurcado a 90°); *f 1* **pièce [f] d'embranchement [m]** (Terme générique pour un dispositif de raccordement à l'égout ; ►raccord de picage à plaquette, ►raccord de tuyau, ►té de raccordment à 90°) ; *syn.* raccord [m] de branchement ; *f 2* **culotte [f] (de branchement)** (Dispositif de raccordement intercalé sur le réseau réservé pour des branchements de diamètre inférieur à 400 mm) ; **g Abzweig [m]** (Formstück in Abwasser- oder Wasserleitungen zur Sammlung oder Verteilung der Wassermengen aus unterschiedlichen resp. in unterschiedliche Richtungen; ►Abzweig 90°, ►Sattelstück, ►Rohrverzweigung).

4268 pipe gradient [n] *constr.* (Slope of a length of pipe, usually expressed in percentage or degrees); **s gradiente [m] de tubería** (Grados de inclinación de una tubería, expresados en porcentaje); *syn.* pendiente [m] de tubería, inclinación [f] de tubería; *f* **pente [f] de la conduite/du collecteur** (Exprimée en %, avec une pente minimale de 0,5 % ; *terme spécifique* pente [f] des drains ; GEN 1982-II, 30) ; **g Leitungsgefälle [n]** (Gefälle eines Leitungsstranges mit mindestens 0,5 %).

4269 pipe hole [n] *arch. constr.* (Opening for pipe, e.g. in ceilings, roofs, concrete walls, or slabs; ►blockout); **s paso [m] de tuberías y conducciones** (Apertura a través de paredes, techos, muros de hormigón o losas para dar paso a conducciones o tuberías; ►roza); *f* **passage [m] de conduites** (p. ex. à travers un mur ou une dalle en béton ; ►réservation) ; **g Durchführung [f] von Rohrleitungen** (Öffnung, z. B. in Betonwänden oder -decken; ►Aussparung); *syn.* Rohrleitungsdurchführung [f].

pipe [n] **into the sea, discharge** *envir.* ►underwater effluent discharge pipe into the sea.

pipe junction piece [n], *constr.* ►pipe coupling [US]/pipe junction piece [UK, AUS].

4270 pipe laying [n] *constr.* (Process of installing pipes); **s tendido [m] de tuberías** (Proceso de instalación de tuberías); *syn.* colocación [f] de tuberías; *f* **mise [f] en place de tuyaux** (Pose des tuyaux comprenant bardage si nécessaire, examen des éléments de canalisation avant la pose, coupe des tuyaux, réalisation du lit de pose, mise en place des canalisations en tranchées ou en apparent ; *syn.* pose [f] de tuyaux ; **g Rohrverlegung [f]** (Unterirdischer Einbau oder oberirdisches Legen von Rohrleitungen).

pipe leakage [n] *envir.* ►leakage.

4271 pipeline [n] *envir.* (Long-distance pipe for conveying natural gas, oil, potable water, etc.; *specific terms* gas pipeline, oil pipeline); **s «pipeline» [m]** (Conducto destinado al transporte de petróleo o gas a larga distancia; *términos específicos* oleoducto, gaseoducto); *f* **pipeline [m]** (Tuyau d'assez grand diamètre, servant au transport à grande distance de carburants liquides, gaz naturel, eau potable, etc. ; *termes spécifiques* oléoduc, gazoduc) ; *syn. o. v.* pipe-line [m] ; **g Pipeline [f]** (Eine über größere Strecken verlegte Rohrleitung zum Transport von Erdgas, Öl, Trinkwasser etc.; *UBe* Gaspipeline, Ölpipeline).

4272 pipe sleeve [n] *constr.* (Extra pipe insert laid after excavation of utility/service trench, or cast in a concrete wall or floor, for later passage of a pipe; MEA 1985); **s manguito [m]** (Tubo subterráneo que se utiliza para la instalación posterior de cables de electricidad o teléfono en parques y jardines); *f 1* **four-**

P

reau [m] (Tuyau placé **1.** dans les bâtiments permettant le passage de câbles ou de conduites à travers les murs, **2.** en fond de tranchées permettant la protection contre les racines, d'éviter la démolition des ouvrages [passage sous chaussée] lors de travaux de réfection ou une pose ultérieure ; *contexte* fourreau en attente) ; *f 2* **gaine [f]** (Conduit placé dans les murs, en plafond ou en plancher permettant le passage et la protection des câbles des réseaux téléphoniques et électriques, des conduites d'eau et de gaz) ; *syn.* conduit [m] ; *g* **Leerrohr [n]** (Rohr, das in Mauern, Decken oder nach dem Ausheben von Leitungsgräben zusätzlich zu den erforderlichen Rohren eingelegt wird, um bei späterem Bedarf, z. B. zum Verlegen von Strom- oder Telekommunikationskabeln, kostenaufwendige Meißel- oder Aushubarbeiten zu vermeiden).

pipe tee [n] [US] *constr.* ▶T-jointed pipe fitting.

pipe [n] with soil separator *constr.* ▶perforated drain pipe with soil separator.

piping [n] of ditches or watercourses *wat'man.* ▶culverting.

piping pattern [n] [US], drainage *constr.* ▶subsurface drainage system.

pit [n] *min.* ▶borrow pit, ▶clay pit; *landsc. recr.* ▶flooded gravel pit; *min. envir.* ▶partially infilled pit; *pedol.* ▶sample pit; *min.* ▶sand and gravel pit; *recr.* ▶sand pit; *constr. pedol.* ▶test pit [US]/trial pit [UK]; *constr. hort.* ▶tree pit; *min.* ▶wet gravel pit.

pit [n], excavate a planting *constr. hort.* ▶excavate a planting hole.

pit [n], flooded borrow *landsc. recr.* ▶flooded gravel pit.

pit [n], gravel *landsc. recr.* ▶flooded gravel pit; *min.* ▶gravel extraction site; *landsc. recr.* ▶wet gravel pit.

pit [n] [US], plant *constr. hort.* ▶planting hole.

pit [n], planting *constr. hort.* ▶planting hole.

pit [n], sand borrow *min.* ▶sand extraction site.

pit [n], seepage *constr.* ▶drywell sump [US]/soakaway [UK].

pit [n] [UK], trial *constr. pedol.* ▶test pit [US].

pitch [n] [US] *constr.* ▶slope [US] (2)/fall [UK].

pitch [n] [UK], all-weather *constr. recr.* ▶all-weather court [US].

pitch [n] [UK], change of *constr.* ▶gradient change.

pitch [n] [UK], grass playing *constr. recr.* ▶grass sports field.

pitch [n] [UK], hard *constr. recr.* ▶hard court [US].

pitch [n] [UK], sports *recr.* ▶sports ground.

pitch [n] [UK], turfed *recr.* ▶play lawn [US] (2).

4273 pitched roof [n] *arch.* (A roof having one or more surfaces with a pitch/slope greater than 10%); *s* **tejado [m] inclinado** (Tejado con una o más aguas de inclinación superior al 10%; *términos específicos* **t.** a un agua, **t.** a dos aguas, **t.** a cuatro aguas, **t.** en V, **t.** con dos faldones; *syn.* tejado [m] en pendiente; *f 1* **toiture-terrasse [f] rampante** (Pente comprise entre 5 % et 15 %) ; *f 2* **toiture [f] inclinée** (Pente supérieure à 15 %) ; *g* **geneigtes Dach [n]** (Neigung unter 15°) *syn.* flach geneigtes Dach [n].

pit edging [n] [US] *constr.* ▶tree pit edging.

4274 pit gravel [n] *constr.* (Gravel which is used directly from its extraction source without sieving to size and other preparation); *s* **grava [f] de cantera** (BU 1959; grava utilizada directamente después de extraída sin ser clasificada por tamaños u otra preparación); *syn.* grava [f] de gravera, grava [f] de mina; *f* **gravier [m] tout-venant** (Gravier non criblé mis en place

directement après son extraction d'une carrière) ; *syn.* grave [f] tout-venant ; *g* **Wandkies [m]** (Kies, der aus einer Kiesentnahmestelle ohne vorherige Sortierung und Aufbereitung eingebaut wird); *syn.* Grubenkies [m].

pit planting [n] [UK] *constr. hort.* ▶hole planting [US].

pits [npl] and quarries [npl] *min. plan.* ▶mineral working site.

place [n] *prof.* ▶prize; *recr.* ▶rest place [US]/rest spot [UK] (1), ▶stopping place [US]/rest spot [UK] (3)

place [n], burial *adm.* ▶burial site.

place [n] [UK], central *plan.* ▶population growth center [US].

place [n] [US], parking *trans. urb.* ▶parking space.

place [n], pleasuring *gard'hist.* ▶pleasure ground.

place [n], winter resting *zool.* ▶hibernation site.

4275 placement [n] of soil *constr.* (Transport of excavated soil, fill material, etc. to the work site and grading to desired contours; ▶furnish and install, ▶topsoil spreading); *syn.* installation [n] of soil; *s* **relleno [m] de tierra (3)** (Colocación de la tierra vegetal o no para dar forma a una superficie; ▶extensión de tierra vegetal, ▶proveer e instalar); *f* **remblaiement [m] de/en terre** (Transport de terre et sa mise en place [dans une tranchée, autour d'une fondation, etc.] selon la forme souhaitée ; ▶fourniture et mise en place, ▶mise en place de la terre végétale) ; *syn.* mise [f] en œuvre de la terre (végétale) ; *g* **Einbau [m, o. Pl.] von Boden** (*Oberboden/Füllboden* Entladen des gelösten Bodens am Einbauort und Herstellen des gewünschten Erdkörpers; LEHR 1981, 107; ▶Andecken von Oberboden, ▶liefern und einbauen); *syn.* Bodeneinbau [m].

placement [n] of street amenities *constr.* ▶installation of street furniture.

place [n] of public worship [UK] *leg. urb.* ▶church [US].

4276 place [vb] under legal protection *conserv. leg.* (To ensure the preservation of areas or individual objects such that their outstanding characteristics or uniqueness is preserved for the future; ▶bestowed protection status); *s* **declarar [vb] protegido** (Llevar a cabo un ▶procedimiento de declaración de protección de un área o un objeto debido a su interés especial para la preservación de la naturaleza o del patrimonio cultural); *f* **classer [vb]** (Moyen d'assurer, avec le plus de rigueur, la protection de sites naturels de grande qualité, d'espaces naturels sensibles, la conservation d'espaces boisés et des espaces verts, etc., pris à l'initiative de l'administration ou de la commission compétente et faisant l'objet d'une réglementation et d'une procédure de classement spécifiques ; ▶procédure de classement, ▶procédure d'inscription) ; *g* **unter Schutz stellen [vb]** (Flächen oder Einzelobjekte insbesondere aus Gründen des Denkmalschutzes oder des Naturschutzes durch ein förmliches Verfahren so sichern, dass die zu schützenden Flächen und Objekte in ihrer Besonderheit oder Einmaligkeit für die Zukunft erhalten werden können; ▶Unterschutzstellung).

plaggen epipedon [n] [US] *pedol.* ▶plaggen soil.

4277 plaggen soil [n] *pedol.* (Anthropogenic surface layer, produced by continuous manuring since medieval times. Sods cut from the heath and used for bedding livestock were spread on fields being cultivated; ▶heath sod); *syn.* plaggen epipedon [n] [also US]; *s* **suelo [m] de tipo «plaggen»** (Suelo antropógeno causado por el aporte de materia orgánica por medio de ▶tepes de turba); *syn.* plaggenboden [m]; *f* **sol [m] de type « plaggen »** (Sol anthropique, transformé par l'apport au cours des siècles de matière organique, les ▶plaques tourbeuses et provoquant la formation d'un sol très noir, caractérisé par des strates parallèles ;

cf. PED 1983, 149) ; *g* **Plaggenesch [m]** (Anthropogener Boden, der durch jahrhundertelangen Auftrag von ▶Plaggen entstanden ist); *syn.* Plaggenboden [m], geplaggter Boden [m].

4278 plain [n] *geo.* (Extensive land surface with very little difference in elevation. ▶Lowland [up to ca. 200 m] and ▶high plains are differentiated from each other according to their height above sea level; ▶alluvial plain, ▶loess plaine, ▶plateau 1); *s 1* **llanura [f]** (Terreno de topografía plana o casi plana; DINA 1987; ▶altiplano, ▶llanura de loes[s], ▶meseta, ▶tierras bajas); *syn.* planicie [f], planada [m]; *s 2* **llano [m]** (Llanura de superficie apreciable; DINA 1987); *syn.* llanada [f]; *f* **plaine [f]** (Terme du langage courant pour désigner une surface continentale étendue, plane et peu élevée en bordure des cours d'eau ou d'un littoral. Une plaine de niveau de base est à pente très faible, moins de 1 %, et se raccorde insensiblement au niveau de l'océan ou au niveau d'un lac ou d'une mer intérieure ; DG 1984, 350 ; on distingue la ▶basse plaine et la ▶haute plaine ; ▶basse vallée, ▶plaine alluviale, ▶plateau, ▶plaine lœssique) ; *g* **Ebene [f]** (**1.** Ausgedehnte Landoberfläche mit sehr geringen Höhenunterschieden. **2.** OB; nach der Höhenlage über dem Meeresspiegel wird zwischen ▶Tiefebene [bis ca. 200 m ü. NN] und ▶Hochebene, hinsichtlich geogener Prozesse nach *Aufschüttungsebene* und *Abtragsebene* unterschieden; ▶Lössebene); *syn.* Flachland [n].

4279 plan [n] (1) *plan.* (Graphic result of planning; ▶approved plan [US]/approved scheme [UK], ▶final plan, ▶final plan for approval, ▶preliminary planning proposals); *s* **plan [m]** (Resultado gráfico del planeamiento; ▶ante-proyecto de un plan de ordenación, ▶plan aprobado definitivamente, ▶plan aprobado provisionalmente, ▶plan definitivo); *f* **plan [m]** (Résultat graphique d'un projet d'aménagement ou d'un document d'urbanisme ; ▶plan approuvé, ▶plan approuvé par arrêté préfectoral, ▶projet de plan approuvé, ▶projet de plan arrêté) ; *g* **Planfassung [f]** (Dargelegtes Ergebnis einer Planung; ▶endgültige Planfassung, ▶genehmigte Planfassung, ▶genehmigungsfähige Planfassung, ▶vorläufige Planfassung).

plan [n] (2) *contr. plan.* ▶as-built plan; *plan. trans.* ▶bikeway plan [US]/bikeway scheme [UK]; *leg.* ▶clean air plan; *leg. urb.* ▶community development plan [US]/urban development plan [UK], ▶comprehensive plan [US]/Local Development Framework (LDF) [UK]; *constr. contr.* ▶delivered plan [US]/handover plan [UK]; *leg. landsc.* ▶development mitigation plan [US]/landscape envelope plan [UK]; *plan.* ▶enlargement of a plan, ▶floor plan, ▶functional plan [US]/sectoral plan [UK], ▶general plan; *constr.* ▶grading plan [US]/levelling plan [UK]; *conserv. ecol.* ▶habitats directive management plan; *plan.* ▶illustrative site plan, ▶implementation of a plan; *recr. trans.* ▶isochronous plan; *constr.* ▶jointing pattern plan; *adm. agr. leg.* ▶land banking plan; *landsc.* ▶landscape plan; *landsc. plan.* ▶landscape structure plan; *plan.* ▶layout plan; *pol. trans.* ▶long-range transportation plan [US]; *adm. conserv'hist. recr.* ▶management plan (1); *landsc. leg.* ▶mandatory landscape developement plan; *plan.* ▶master plan (1), ▶master plan (2); *constr. plan.* ▶network plan; *plan.* ▶overall master plan for sports and physical activities [US]/strategic plan for sports and physical activities [UK], ▶overall plan (1), ▶overall site plan, ▶pathway network plan; *urb.* ▶pedestrian circulation plan; *plan.* ▶planting plan, ▶project plan, ▶reduction of a plan, ▶regional plan; *constr. plan. urb. wat'man.* ▶rehabilitation plan; *constr.* ▶reinforcing steel plan [US]/reinforcement plan [UK]; *plan.* ▶revision of a plan, ▶site plan; *constr. surv.* ▶staking-out plan; *leg. plan.* ▶state development plan [US]/regional policy/strategy plan [UK]; *plan.* ▶subregional plan, ▶survey plan (1); *surv.* ▶survey plan (2); *adm. constr. landsc.* ▶tree survey plan; *urb.* ▶Urban

Design Framework Plan; *trans. urb.* ▶utilities plan; *leg. wat'man.* ▶water management plan; *min. envir.* ▶working plan.

plan [n] [UK], **allotment garden development** *leg. urb.* ▶community garden development plan [US].

plan [n] [UK], **allotment garden subject** *leg. urb.* ▶community garden development plan [US].

plan [n], **amendment of a** *leg. plan.* ▶plan revision.

plan [n] [UK], **area action** *leg. urb.* ▶zoning map [US]/Proposals Map/Site Allocations Development Plan Document [UK].

plan [n], **circulation** *urb.* ▶pedestrian circulation plan.

plan [n], **crown** *arb. phyt.* ▶crown cover.

plan [n], **execution of a** *plan.* ▶implementation of a plan.

plan [n] [US], **existing use** *plan.* ▶survey of existing uses.

plan [n] [UK], **green infrastructure** *landsc. urb.* ▶general plan for urban open spaces [US]/green open space structure plan [UK].

plan [n] [UK], **green open space structure** *landsc. urb.* ▶general plan for urban open spaces [US].

plan [n] [UK], **green space strategy** *landsc. urb.* ▶general plan for urban open spaces [US].

plan [n] [UK], **handover** *constr. contr.* ▶delivered plan [US].

plan [n], **improvement** *constr. plan. urb. wat'man.* ▶rehabilitation plan.

plan [n] [UK], **land consolidation** *adm. agr. leg.* ▶land banking plan [US].

plan [n] [UK], **landscape envelope** *leg. landsc. plan.* ▶development mitigation plan [US].

plan [n] [UK], **landscape strategy** *landsc.* ▶landscape plan.

plan [n], **layout** *plan.* ▶site plan, ▶overall plan (1).

plan [n], **legally-binding land-use** *adm. leg. urb.* *∗*▶zoning map [US]/Proposals Map/Site Allocations Development Plan Document [UK].

plan [n] [UK], **levelling** *constr. plan.* ▶grading plan [US].

plan [n] [UK], **long-range transport** *pol. trans.* ▶long-range transportation plan.

plan [n], **management** *conserv. ecol.* ▶Habitats Directive Management Plan; *leg. wat'man.* ▶water management plan.

plan [n], **modification of a** *plan.* ▶plan revision.

plan [n], **pattern** *constr. plan.* ▶jointing pattern plan.

plan [n] [D], **preparatory land-use** *leg. urb.* ▶comprehensive plan [US]/Local Development Framework (LDF) [UK].

plan [n] [UK], **regional policy** *leg. plan.* ▶state development plan [US].

plan [n] [US], **regional strategic** *plan.* ▶regional plan.

plan [n] [UK], **regional strategy** *leg. plan.* ▶state development plan [US].

plan [n] [UK], **reinforcement** *constr.* ▶reinforcing steel plan [US].

plan [n], **revision of a** *plan.* ▶plan revision.

plan [n], **revitalization** *constr. plan. urb. wat'man.* ▶rehabilitation plan.

plan [n] [UK], **sectoral** *plan.* ▶functional plan [US].

plan [n] [UK], **setting-out** *constr. surv.* ▶staking-out plan.

plan [n], **statutory regional** *plan.* ▶regional plan report.

P

plan [n] **[UK], unitary development** *leg. urb. obs.* ▶comprehensive plan [US]/Local Development Framework (LDF) [UK].

plan [n] **[UK], urban development** *leg. urb.* ▶community development plan [US].

plan alteration [n] *plan.* ▶plan revision.

planar belt [n] *geo. phyt.* ▶planar zone.

4280 plan area [n] **of a zoning map** [US] *leg. urb.* (Applicable land area, e.g. as designated on a ▶zoning map [US]/ Proposals Map/Site Allocations Development Plan Document [UK]); *s* **área** [f] **de vigencia de un plan parcial** (Parte de un municipio para el que se aprobó un ▶plan parcial [de ordenación]); *f* **périmètre** [m] **d'un plan local d'urbanisme (PLU)/ plan d'occupation des sols (POS)** (Territoire à l'intérieur duquel le plan rendu public et approuvé est opposable aux tiers ; cf. C. urb., art. R. 122-2 ; *contexte* territoire à l'intérieur du périmètre d'un PLU/POS) ; *g* **Geltungsbereich** [m] **eines Bebauungsplanes** (Der Teil einer Gemarkung, für den ein ▶Bebauungsplan rechtsverbindlich verabschiedet wurde); *syn.* Bebauungsplangebiet [n].

4281 planar zone [n] *geo. phyt.* (▶Altitudinal belt covering flat lowlands; ▶zonation of vegetation); *syn.* planar belt [n]; *s* **piso** [m] **planar** (En la ▶zonación altitudinal el piso de vegetación que se desarrolla en las llanuras; ▶cliserie); *f* **étage** [m] **de plaine** (En plaine étage de végétation de faible variation altitudinale ; ▶étagement de la végétation, ▶zonation de végétation) ; *syn.* étage [m] planitaire (DEE 1982, 40) ; *g* **planare Stufe** [f] (Hinsichtlich der ▶Höhenstufung der Bereich des Flachlandes, der Tiefebene; ▶Vegetationszonierung).

plan [n] **determined by Secretary of State** [UK] *adm. leg.* ▶approved plan for major projects [US].

plan [n] **for approval** *leg. urb. trans.* ▶final plan for approval.

plan [n] **for play spaces** [UK], **strategic** *plan. recr. urb.* ▶overall master plan for play areas [US].

plan [n] **for sports and physical activities** [UK], **strategic** *plan. recr. urb.* ▶overall masterplanning for sports and physical activities [US].

plan [n] **for sports fields** [UK], **city** *plan. recr. urb.* ▶overall master plan for sports and physical activities [US].

plan [n] **for sports fields** [US], **master** *plan. recr. urb.* ▶overall master plan for sports and physical activities [US]/strategic plan for sports and physical activities [UK].

plan [n] **for sports pitches** [UK], **city** *plan. recr. urb.* ▶overall master plan for sports and physical activities [US].

plan [n] **for urban open spaces** [US] *plan. urb.* ▶general plan for urban open spaces [US]/green open space structure plan [UK].

4282 planking [n] **and strutting** [n] *constr.* (Temporary sheet piling supporting the soil at both sides or one side of an excavation, and constructed of timber planking, steel sheet piling or ▶shotcrete [US]/guncrete [UK]; *specific term* ▶trench bracing; ▶riverbank stabilization [US]/river-bank stabilization [UK], ▶torrent control); *s* **encofrado** [m] **y apuntalado** [m] (**1.** Métodos de sujetar los bordes de canales o zanjas de construcción por medio de tablones de madera, paredes tablestacadas o ▶gunita; *término específico* ▶arriostramiento de una zanja. **2.** Protección de orillas y contra aludes; ▶barrera contra aludes, ▶corrección de torrentes; ▶fijación de orillas. **3.** *Construcción biotécnica* ▶consolidación de taludes); *f 1* **blindage** [m] (Étaiement de parois de fouilles larges ou de tranchées pour éviter les éboulements en terrain instable au moyen de planches ou madriers bois, planches métalliques, palfeuilles ou ▶béton

projeté ; *termes spécifiques* ▶étaiement d'une fouille ; *f 2* **soutènement** [m] **de(s) berges** (**D.**, travaux réalisés dans la recherche de la stabilisation et le renforcement des berges ou des rives d'un cours d'eau afin d'éviter l'érosion ; ▶confortement des berges, ▶stabilisation d'un torrent) ; *g* **Verbau** [m] **(2)** (**1.** Sicherung von Gräben- und Baugrubenwänden, z. B. durch Holzbohlen [z. B. Hamburger Verbau], Verbaukorb, Spundwand oder ▶Spritzbeton; *UBe* ▶Grabenverbau, Grubenverbau. **2.** Befestigung von Ufern [Uferverbau]: ▶Uferbefestigung; ▶Wildbachverbauung; *Schutz vor Lawinen* ▶Lawinenverbauung. **3.** *Ingenieurbiologische Bauweisen* ▶Ingenieurbiologie, Hangverbau [▶Hangsicherung]).

planned establishment [n] **of industries** *plan. urb.* ▶planned industrial development.

4283 planned industrial development [n] *plan. urb.* (Areas reserved for industry during the planning process; ▶relocation of industries); *syn.* planned establishment [n] of industries; *s* **gestión** [f] **del desarrollo industrial** (Asignación de suelo para la instalación de nuevas industrias; ▶relocalización de industrias); *f* **implantation** [f] **d'activités industrielles** (Projet d'aménagement visant la création ou l'extension d'activités industrielles dans des zones aménagées) ; *syn.* développement [m] industriel, accueil [m] d'activités industrielles ; *g* **geplante Industrieansiedlung** [f] (Planerische Absicht, dass sich ein Industriebetrieb auf einem Gelände oder in einem Gebiet niederlässt; ▶Aussiedlung von Industrie- oder Gewerbebetrieben).

4284 planner [n] *prof.* (Someone who plans; ▶designer, ▶landscape planner, ▶planning 1, ▶planning 2, ▶planning consultant [US]/town planning consultant [UK], ▶preliminary project planner [US]/scheme designer [UK], ▶regional planner, ▶site designer, ▶urban planner [US]/town planner [UK]); *s* **planificador, -a** [m/f] (Profesional que se dedica a la planificación regional, urbanística o paisajística; ▶diseñador, -a, ▶planificación, ▶planificador, -a del paisaje, ▶planificador, -a regional, ▶proyectar, ▶urbanista, ▶urbanista consultor, -a); *f 1* **aménageur** [m] (Personne qui réalise un aménagement ; ▶concepteur, conceptrice, ▶concepteur-maître d'œuvre, ▶concepteur, trice [paysagiste] d'opération, ▶études, ▶paisagiste d'aménagement, ▶planification, ▶planificateur de l'aménagement régional, ▶urbaniste, ▶urbaniste libéral) ; *syn.* planificateur [m] ; *f 2* **projeteur** [m] (Personne réalisant la conception d'un objet, d'un ouvrage selon des règles esthétiques de la technique et de l'art) ; *f 3* **maître** [m] **d'œuvre** (Personne physique ou morale publique ou privée qui, pour sa compétence technique, est missionné [mission de ▶maîtrise d'œuvre] par le ▶maître d'ouvrage ou par la personne responsable du marché, dans les conditions de délais, de qualité, de performances, de sécurité et de coût définies dans un contrat, de concevoir, diriger, coordonner et contrôler de l'exécution de travaux, la réalisation d'un ouvrage ainsi que de proposer leur réception et leur règlement ; cf. Article 7 de la loi n° 85-704 du 12 juillet 1985 et Décret n° 93-1268 du 29 novembre 1993) ; *g* **Planer/-in** [m/f] (**1.** Jemand, der etwas plant. **2.** Jemand, der die relevanten Aspekte für seine Planungsaufgabe und die möglichen Konsequenzen der praktischen Funktionalität und ästhetischen Wahrnehmung einer Lösung oder von Lösungsvarianten in seine Überlegungen einbezieht und das Ergebnis rational nachvollziehbar macht; ▶Gestalter, ▶Landschaftsplaner/-in, ▶Objektplaner/-in, ▶Planung 1, ▶Planung 2, ▶Stadtplaner 2).

planner [n] **[US], city** *prof.* ▶urban planner [US]/town planner [UK].

planner [n] **[US], community** *prof.* ▶urban planner [US]/ town planner [UK].

planner [n], **state** *prof.* ▶regional planner.

planner [n] [UK], **town** *prof.* ►urban planner [US].

4285 planning [n] (1) *plan. prof.* (*Sensu lato* processes involved in the creation of solutions and goal-oriented concepts particularly with regard to the future); *s* **planificación** [f] (*Sensu lato* proceso de pensar soluciones a largo plazo); *syn.* planeamiento [m]; *f* **planification** [f] (Réflexion anticipatrice d'un individu, d'une institution, d'une collectivité territoriale afin de résoudre les problèmes d'organisation et de développement économique ou spatial sur la base d'objectifs prospectifs et comportant les actions et les étapes de réalisation pour les atteindre) ; *g* **Planung** [f] (1) (*Sensu lato* gedankliche Vorwegnahme eines Planenden für zukünftiges, zielgerichtetes Handeln).

4286 planning [n] (2) *plan. prof.* (*Sensu stricto* general term for activities of a ►planner at local, urban, regional, and state levels of land use development; ►project planning); *s* **proyectar** [vb] (Actividad profesional de un/a ►planificador, -a; ►planificación de un proyecto); *syn.* planificar [vb]; *f* **études** [fpl] (Activité d'un ►aménageur dans le domaine urbain, régional ou territorial ; ►mission opérationnelle) ; *g* **Planung** [f] (2) (*Sensu stricto* Tätigkeit eines/einer ►Planers/-in im objekt-, stadt-, regional- oder landesplanerischen und landschaftsplanerischen Bereich; ►Objektplanung).

planning [n] (3) *agr. plan.* ►agricultural planning; *plan.* ►change in planning; *adm. leg. urb.* ►community development planning; *recr. urb.* ►community garden development planning; *leg. landsc. plan.* ►development mitigation planning; *conserv. nat' res. plan.* ►ecological planning; *landsc. leg. plan.* ►functional landscape planning [US]/sectoral landscape planning [UK]; *plan.* ►functional planning [US]/sectoral planning [UK]; ►general planning [US]/strategic planning [UK]; *landsc. urb.* ►general urban green space planning; *plan.* ►guideline in planning; *leg. plan.* ►higher administrative level of planning; *plan.* ►interdisciplinary planning; *landsc.* ►landscape planning, ►landscape structure planning; *plan.* ►large-scale planning; *adm. plan.* ►master planning (1), ►master planning (2); *plan.* ►metropolitan area planning, ►multidisciplinary planning; *landsc. recr. urb.* ►open space planning; *plan.* ►overall area planning, ►overall land planning; *plan. recr. urb.* ►overall master planning for sports and physical activities [US]/strategic planning for sports and physical activities [UK]; *plan.* ►overall project planning, ►physical planning; *agr. plan.* ►preliminary agrarian structure planning; *plan. prof.* ►project planning; *landsc. plan. pol. recr.* ►recreation area planning, ►recreation planning; *plan.* ►regional planning, ►replanning; *plan. trans.* ►road planning; *plan.* ►rural area planning [US]/countryside planning [UK]; *landsc. plan. pol. recr.* ►rural recreation planning [US]/countryside recreation planning [UK]; *plan. prof.* ►site planning; *plan.* ►State regional planning [US]/regional planning [UK] (2), ►target planning; *adm. plan. pol. trans.* ►traffic planning; *plan.* ►transfrontier regional planning; *landsc. plan. pol. recr.* ►urban area recreation planning; *urb.* ►urban planning; *plan. urb.* ►village redevelopment planning; *adm. hydr. wat'man.* ►water management planning.

planning [n], **agrarian structure** *agr. plan.* ►preliminary agrarian structure planning.

planning [n] [UK], **allotment garden development** *recr. urb.* ►community garden development planning [US].

planning [n], **area** *plan.* ►metropolitan area planning, ►overall area planning; *landsc. plan. pol. recr.* ►recreation area planning; *plan.* ►rural area planning [US]/countryside planning [UK].

planning [n] [US], **city** *urb.* ►urban planning.

planning [n] [US], **community** *urb.* ►urban planning.

planning [n] [UK], **countryside** *plan.* ►rural area planning [US].

planning [n] [UK], **countryside recreation** *landsc. plan. pol. recr.* ►rural recreation planning [US].

planning [n], **detail** *plan.* ►detailed design.

planning [n], **general development** *plan.* ►overall area planning.

planning [n], **green space** *landsc. urb.* ►general urban green space planning.

planning [n], **highway** *plan. trans.* ►road planning, #2.

planning [n], **integrated** *plan.* ►multidisciplinary planning.

planning [n] [UK], **landscape envelope** *urb.* ►development mitigation planning.

planning [n], **metropolitan region** *plan. trans.* ►metropolitan area planning.

planning [n] [UK], **play spaces strategic** *plan. recr. urb.* ►overall master planning for play areas [US].

planning [n] [UK], **sectoral** *plan.* ►functional planning [US].

planning [n] [UK], **sectoral landscape** *landsc. leg. plan.* ►functional landscape planning [US].

planning [n], **spatial** *plan.* ►physical planning.

planning [n] [UK], **strategic** *plan.* ►general planning [US].

planning [n], **structure** *landsc.* ►landscape structure planning; *agr. plan.* ►preliminary agrarian structure planning.

planning [n] [UK], **town and country** *adm. leg. urb.* ►community development planning [US].

planning [n], **transboundary** *plan.* ►transfrontier regional planning.

planning [n], **transnational** *plan.* ►transfrontier regional planning.

planning [n], **transportation** *adm. plan. pol. trans.* ►traffic planning.

planning [n], **urban development** *adm. leg. urb.* ►community development planning [US]/urban land-use planning [UK].

planning [n] [UK], **urban land-use** *urb.* ►community development planning [US].

planning aim [n] *plan.* ►planning goal.

planning alteration [n] *leg. plan.* ►change in planning.

planning amendment [n] *leg. plan.* ►change in planning.

4287 planning [n] **and design** [n] **of the environment** *conserv. nat'res. plan.* (Imprecise term, which sometimes corresponds to the terms ►landscape design and ►landscape management according to the context in which it is used. **P. a. d. o. t. e.** is also used in ►urban planning, if the design covers functional, ecological and aesthetic aspects; ►design 2, ►landscape planning, ►planning 2); *syn.* environmental design [n]; *s* **planificación** [f] **y diseño** [m] **ambiental** (Término no claramente definido que dependiendo del contexto puede tener contenidos similares al ►diseño paisajístico y a la ►gestión y protección del paisaje; ►diseño [de un proyecto], ►planificación, ►planificación del paisaje, ►urbanismo); *f* **aménagement** [m] **de l'environnement** (Terme peu précis, utilisé suivant le contexte souvent dans le sens de la ►création de paysages et de la ►gestion des milieux naturels [►gestion des paysages] mais aussi en ►urbanisme pour tout ce qui a trait à l'aménagement des composantes environnementales du milieu bâti selon des critères esthétiques, fonctionnels et écologiques ; ►conception 1, ►planification, ►planification des paysages) ; *g* **Umweltgestaltung** [f] (Kein fest definierter Begriff; je nach Kontext gibt es Begriffsüberschneidungen mit ►Landschaftsgestaltung und ►Landschaftspflege, aber auch im städtebaulichen Zusammenhang wird der Begriff **U.** benutzt, wenn eine Gestaltung nach funktionalen, ökologischen und formal-ästhetischen Gesichtspunkten [visueller Aspekt] gemeint ist; ►Gestaltung, ►Landschaftsplanung, ►Planung, ►Städtebau); *syn.* Gestaltung [f] der Umwelt.

planning application [n] [UK], **detailed** *adm. leg.* ►building and site plan permit application [US].

planning application [n] [UK], **outline** *adm. plan.* ►application for preliminary building permit [US].

planning application documents [n] [UK] *adm. leg. urb.* ►permit application documents [US].

4288 planning approach [n] *plan.* (Intellectual/conceptual viewpoint from which to proceed in solving a planning problem); *s* **enfoque** [m] **de planificación** (Visión conceptual/intelectual desde la cual se parte para desarrollar soluciones en la planificación); *f* **démarche** [f] **du parti d'aménagement** (Représentation conceptuelle clairement définie guidant l'aménageur au long des différentes phases de l'étude d'aménagement, de la planification) ; *syn.* approche [f] du parti d'aménagement ; *g* **Planungsansatz** [m] (Gedanklicher Hintergrund, mit dem eine Planung erfolgen soll oder durchgeführt wurde).

4289 planning approval procedure [n] *adm. leg.* (Course of actions by an authority to verify compliance with the preconditions for a planning proposal; persons concerned and participating authorities/agencies are given an opportunity to raise objections, usually at a public hearing, and the agency/authority then issues a written approval. **In U.S.**, there are a number of approval procedures in connection with zoning, subdivision regulation, creation of historic and other forms of overlay districts. The planning staff reviews an application, certifies compliance, issues permits, etc. This is nominally known as 'staff review and approval'; ►permit [US]/permission [UK]); *s* **procedimiento** [m] **de aprobación o de autorización** (Conjunto de acciones realizadas por la administración competente sobre la base de la correspondiente legislación para aprobar planes o autorizar construcciones urbanas e industriales; ►permiso definitivo); *syn.* tramitación [f] de aprobación o de autorización; *f* **procédure** [f] **d'autorisation administrative** (Ensemble des formalités régissant le contrôle des conditions préalables, la préparation et le décret d'un acte administratif [= ►autorisation], la procédure permettant la consultation des personnes publiques associées à l'élaboration et la participation du public par l'enquête publique); *g* **Genehmigungsverfahren** [n] (Nach außen wirkende Tätigkeit der Behörden, die auf die Prüfung der Voraussetzungen für die Gesetzmäßigkeit eines Vorhabens, die Vorbereitung und den Erlass eines Verwaltungsaktes [= ►Genehmigung] gerichtet ist. Das Verfahren endet entweder mit einer Zustimmung, d. h. dass das Vorhaben realisiert werden darf, oder mit einer Ablehnung. Dabei muss gewährleistet sein, dass im Rahmen einer Öffentlichkeitsbeteiligung die Betroffenen und die beteiligten Behörden Einwendungen geltend machen können; cf. z. B. §§ 6 u. 10 BauGB, §§ 49 ff BauO-BW, § 10 BImSchG. Zur Genehmigung größerer staatlicher Planungen, z. B. Verkehrswegeplanungen, Planungen von Kraftwerken, müssen ►Planfeststellungsverfahren durchgeführt werden); *syn.* Bewilligungsverfahren [n] (§ 9 WHG).

4290 planning area [n] (1) *plan.* (Portion of land which requires an overall planning solution; ►planning district [US]/sub-area [UK], ►research area, ►study area); *s* **área** [f] **de planificación** (1. Región que exige planificación para su desarrollo armonioso. 2. D., término utilizado en urbanismo y ordenación del territorio para denominar la zona sobre la que se prevé elaborar o se está elaborando un plan; ►área de estudio, ►área

tipificada, ►zona de estudio); *syn.* zona [f] de planificación/planeamiento; *f* **aire** [f] **d'aménagement** (Territoire déterminé concerné par une planification, un projet d'aménagement ; ►unité territoriale, ►zone d'étude 1, ►zone d'étude 2) ; *syn.* périmètre [m] d'aménagement ; *g* **Planungsraum** [m] (**1.** Definiertes Gebiet, das einer Gesamtplanung unterliegt. **2.** Definiertes Gebiet, das stadtplanerisch oder raumordnerisch bearbeitet wird; ►Gebietseinheit, ►Untersuchungsgebiet, ►Bearbeitungsgebiet); *syn.* Planungsgebiet [n].

planning area [n] (2) *plan.* ►study area.

planning atlas [n], **regional** *plan.* ►regional map atlas [US].

planning authority [n] [UK], **local** *adm. urb.* ►local planning agency/office [US].

4291 planning basis [n] *plan.* (Relevant information to form the foundation for planning of a particular project; e.g. existing planning solutions and their possible consequences, funding, current regulations and enactments, etc.; ►planning data); *s* **requerimientos** [mpl] **legales de la planificación** (Planes territoriales o sectoriales, normas y usos existentes son vinculantes a la hora de redactar planes a niveles inferiores; ►bases técnicas de la planificación); *f* **contraintes** [fpl] **d'une opération** (Tous les éléments communiqués par le maître de l'ouvrage et par les autorités intéressées nécessaires à la définition des ouvrages tels que les programme et budget de l'opération, les contraintes du site, les informations administratives relatives aux servitudes ou prescriptions d'urbanisme ainsi que les informations juridiques [titre de propriété, mitoyenneté, etc.] ; ►données d'une l'opération [d'aménagement]) ; *g* **Planungsvorgabe** [f] (Alle für ein Planungsvorhaben vorgegebenenen oder zu berücksichtigenden Ideen/Informationen und relevante Bestandsdaten wie z. B. Nutzungsgegebenheiten, bestehende [übergeordnete] Planungen und deren mögliche Entwicklung, Förderungsmöglichkeiten durch öffentliche Mittel, bestehende Festsetzungen und Vorschriften etc.; ►Planungsgrundlage).

planning blight [n] [UK] *plan.* ►property devaluation caused by planning [US].

planning body [n] [US] *adm. plan.* ►public planning body.

planning comments [npl] *urb. plan.* ►submission of planning comments.

planning commission [n] [UK] *contr. prof.* ►planning contract.

4292 planning conditions [npl] (1) *adm. leg. plan.* (Requirements which must be fulfilled for issuance of a ►permit; ►conditional development permit [US]/conditonal planning permission [UK]); *s* **prescripción** [f] **urbanística** (Condición fijada por la administración competente al dar un ►permiso definitivo para la realización de un proyecto urbanístico; ►permiso con condiciones); *f* **prescription** [f] (Conditions législatives et réglementaires fixées lors de la délivrance d'une ►autorisation dans un cahier de charges ; ►autorisation accordée sous réserve de l'observation de prescriptions spéciales) ; *g* **Auflage** [f] (Zu erfüllende Bedingung, um eine ►Genehmigung zu erteilen/zu erhalten; ►Genehmigung mit Auflagen).

4293 planning conditions [npl] (2) *adm. plan.* (Stipulations imposed by authorities/agencies on a project to remedy defects or violations of existing law and statutes, as well as requirements which have to be fulfilled during the implementation of a planning project); *s* **requerimientos** [mpl] **de la administración** (Condiciones impuestas por las autoridades para corregir defectos o violaciones de la ley en el contexto de un proyecto que han de ser cumplidas al realizarlo); *f 1* **prescription** [f] **administrative d'aménagement** (Règles élaborées par l'autorité administrative et **1.** imposées en vue de remédier aux infractions à la législation existante sur l'aménagement et l'urbanisme ou **2.** devant être prises en compte lors de la réalisation de projets d'aménagement) ; *f 2* **prescriptions [fpl] d'aménagement et d'urbanisme** (F., prescriptions nationales ou prescriptions particulières à certaines parties du territoire [zone de bruit des aéroports, zones de protection et mise en valeur de la montagne, de protection et d'aménagement du littoral, etc.] fixées en application des lois d'aménagement et d'urbanisme) ; *g* **Planungsauflagen [fpl]** (Behördlich auferlegte Vorschriften für ein Einzelvorhaben zur Abhilfe von festgestellten Mängeln [Verstoß gegen geltende gesetzliche Vorschriften] bei Planungstätigkeiten oder gesetzliche Verpflichtungen/Vorgaben, die bei der Durchführung einer Planung erfüllt werden müssen).

4294 planning consultant [n] [US] *prof.* (Self-employed urban planner [US]/self-employed town planner [UK]; ►urban planner); *syn.* town planning consultant [n] [UK]; *s* **urbanista [m/f] consultor, -a** (Profesional de planificación resp. arquitecto/a independiente que ofrece sus servicios de planificación y asesoría en el mercado; ►urbanista); *f 1* **urbaniste [m] libéral** (Aménageur travaillant à son compte au sein d'un structure privée et réalisant des missions d'urbanisme ; ►urbaniste) ; *syn.* consultant [m] urbaniste ; *f 2* **architecte [m] conseil de l'équipement** (Architecte libéral désigné par le directeur départemental de l'équipement pour assister ce dernier dans la préparation et l'élaboration des documents d'urbanisme, dans la programmation de logements sociaux et dans le domaine de constructions publiques ; C. urb., art. A. 614-1 à A 614-4) ; *g* **freischaffender Stadtplaner [m]/freischaffende Stadtplanerin** (Jemand, der als Selbständiger Stadtplanung/►Städtebau betreibt; *OB* ►Stadtplaner).

4295 planning contract [n] *contr. prof.* (Oral or written agreement to perform professional planning services, which may be in the form of a statement or a contract); *syn.* planning commission [n] [also UK], contract [n] for planning services; *s* **contratación [f] de un/a planificador, -a** (Adjudicación de un contrato a un/a profesional para realizar un proyecto de planificación o trabajos de inventariación de datos paisajísticos); *syn.* encargo [m] de un/a planificador, -a; *f 1* **mission [f] d'études** (passé avec un concepteur ; missionner [vb] un concepteur) ; *syn.* commande [f] (PPH 1991, 559) ; *f 2* **mission [f] de maîtrise d'œuvre** (Prestations d'ingénierie et d'architecture confiées à un concepteur [personne ou groupement de personnes de droit privé] pour le compte de maîtres d'ouvrage publics suivant les étapes normalisées de la maîtrise d'œuvre ; cf. décret n° 93-1268 du 29 novembre 1993) ; *g* **Planungsauftrag [m]** (Zur Erledigung mündlich oder schriftlich übertragene Planungsarbeiten oder Beauftragung von Erfassungen oder Erhebungen für bestimmte Fragestellungen); *syn.* Auftrag [m] an den Planer.

4296 planning contribution [n] *plan.* (Provision of an idea or professional service by someone involved in a planning project); *s* **contribución [f] al planeamiento** (Aporte de un servicio profesional en el contexto de un proyecto de planificación); *syn.* contribución [f] a la planificación; *f* **contribution [f] au projet d'aménagement** (Services, prestations apportés à la définition, la réalisation d'un projet d'aménagement) ; *g* **Planungsbeitrag [m]** (Planerische Mitwirkung oder Leistung, mit der sich jemand bei einem Planungsvorhaben beteiligt).

planning coordination [n] [US]/**planning co-ordination** [n] [UK] *plan.* ►obtaining a mutual agreement on planning proposals.

4297 planning costs [npl] *prof.* (Professional fees and overhead charged for the services of a planner; in public authorities/agencies these costs are wages and salaries as well as related overhead); *s* **costos [mpl] de planificación** (Honorarios profesionales y gastos corrientes causados por un proyecto de planificación); *syn.* costes [mpl] de planificación [Es]; *f* **coût [m] de l'étude** (Honoraires et frais particuliers provenant de la rémunération des missions d'étude ; pour les études réalisées par les services de l'administration le **c. de l'é.** est constitué par les salaires et les frais généraux) ; *g* **Planungskosten [pl]** (Honorar- und Nebenkosten, die durch die Beauftragung eines Planers entstehen. Bei der behördlichen Planung sind es die Gehalts- und anfallenden allgemeinen Geschäftskosten).

4298 planning data [npl] *plan.* (Facts contained in site map, site analysis, resource information, data collection, database, development suitability map, etc., which are necessary for the planning of a project; ►analysis of planning data, ►base map 2, ►planning basis); *syn.* base data [npl]; *s* **bases [fpl] técnicas de la planificación** (Conjunto de datos, generalmente en forma de planos catastrales, mapas temáticos, recopilación de datos, etc. que son necesarios para desarrollar un proyecto de planificación; ►análisis de datos de planificación, ►plan básico, ►requerimientos legales de la planificación); *syn.* materiales [mpl] básicos de planificación; *f* **données [fpl] d'une opération (d'aménagement)** (Documents remis par la maître d'ouvrage ou obtenus auprès des services administratifs et techniques lors des investigations préalables pour une mission d'études, tels que les informations techniques [►fonds de plan, plan de situation, plan de nivellement, levé topographique, données cadastrales, constructions, etc.], les servitudes de droit privé, etc. ; ►analyse des données de planification, ►contraintes d'une opération) ; *syn.* plus généralement données [fpl] d'une étude de planification ; *g* **Planungsgrundlage [f]** (Grundstückspläne, thematische Einzelkarten, Datenerhebungen, Budgetvorgaben etc., die für die Planung eines Vorhabens notwendig sind; ►Auswertung von Planungsunterlagen, ►Planungsvorgabe, ►Planunterlage); *syn.* Planungsunterlage [f].

planning data [npl], **evaluation** [n] of *plan.* ►analysis of planning data.

4299 planning development control [n] *adm.* (Exercise of approval [US]/permission [UK] of planning and execution of a project; ►building code control [US]/Local Authority Building Control [UK]); *s* **inspección [f] urbanística** (Función de control de realización de los planes aprobados por parte de una administración; ►inspección de la construcción); *syn.* permiso [m] y control [m] de obras; *f* **tutelle [f]** (Caractérise l'action de contrôle de l'ensemble des mesures dont dispose une administration pour faire respecter les Plans et Programmes approuvés ; ►contrôle technique des ouvrages) ; *g* **Aufsicht [f, o. Pl.] einer Behörde** (Ausübung der Überwachungsfunktion bei der Realisierung von genehmigten Plänen eines Bauprojektes; ►Bauaufsicht).

4300 planning district [n] [US] *plan.* (Demarcated part, zone or geographic section of a municipality, county or state for planning purposes; ►sub-area [n] [UK]; *s* **área [f] tipificada** (Sección definida según criterios específicos de una región más amplia. En alemán también denominación de 38 regiones en la que está dividida la [antigua] RFA para efectos de la ordenación del territorio); *f* **unité [f] territoriale** (Aire délimitée en fonction de critères particuliers) ; *g* **Gebietseinheit [f]** (Nach ausgewählten Merkmalen abgegrenzter Bereich eines größeren Raumes resp. genau definierte regionalplanerische Einheit); *syn.* Planungsregion [f].

Planning Documents [n] [UK], **Supplementary** *leg. urb.* *►comprehensive plan [US]/Local Development Framework (LDF) [UK].

planning [n] **for play areas** *plan. recr. urb.* ►overall master planning for play areas [US]/play spaces strategic planning [UK].

P

planning [n] for sports and physical activities *plan. recr. urb.* ▶overall master planning for sports and physical activities [US]/strategic planning for sports and physical activities [UK].

4301 planning [n] for the public welfare *adm. plan. sociol.* (Establishment and maintenance of fair and just planning policies and decisions in the long term to provide public health, safety and general welfare; ▶preventive measures); *s* **planificación [f] en beneficio público** (Definición e implementación a largo plazo por parte de las administraciones públicas de políticas de planeamiento de servicios públicos, como salud, infraestructuras, seguridad, etc. a un precio razonable para así garantizar el bienestar en general. En alemán el término denomina la obligación constitucional del Estado de garantizar las necesidades mínimas de la población e implica la puesta a disposición de la infraestructura y los servicios correspondientes; ▶prevención); *f* **solidarité [f] nationale** (Mise en œuvre et maintien d'une politique de justice sociale grâce à une politique prévoyante de l'administration afin d'éviter les conflits sociaux ; ▶prévention) ; *g* **Daseinsvorsorge [f, selten Pl.]** (Stetige, dauerhafte und möglichst preisgünstige Vorhaltung von öffentlichen Dienstleistungen zur Herstellung und Aufrechterhaltung sozialer Gerechtigkeit, um soziale Konflikte durch vorausschauende Planung der Politik und Verwaltung zu vermeiden; d. h. die Bereitstellung notwendiger oder nützlicher Leistungen für die Gesellschaft, was kommunale Einrichtungen für Bildung, Soziales, Gesundheit, Kultur und Freizeitgestaltung wie auch elementare Leistungen der Energie- und Wasserversorgung, Wohnungsbau, Abwasser- und Müllentsorgung, Straßenwesen und Personennahverkehr, aber auch Telekommunikation, Kreditwesen, Rettungsdienste und anderes mehr betrifft; *OB* ▶Vorsorge).

4302 planning goal [n] *plan.* (**1.** Objective toward(s) which a planner strives to obtain a desired result. **2.** Objective in regional or state planning whereby conurbations are located in relation to each other so that they can fulfill their functional, economic, social, cultural and leisure time requirements with due concern for environmental protection and nature conservation); *syn.* planning purpose [n], planning objective [n], planning aim [n]; *s* **objetivo [m] de la planificación** (**1.** Meta concreta que se desea alcanzar a través de un proyecto de planificación. **2.** En la ordenación del territorio, el desarrollo deseado para las regiones de manera que con su estructura espacial puedan cumplir sus funciones económicas, sociales, culturales e infraestructurales, así como garantizar la protección ambiental y de la naturaleza); *syn.* fin [m] del planeamiento, objetivo [m] del planeamiento, fin [m] de la planificación; *f* **parti [m] d'aménagement** (**1.** Objectif retenu parmi plusieurs guidant la démarche d'un planificateur. **2.** Objectifs de la planification spatiale visant à favoriser un développement et une localisation harmonieuse des activités, des équipements et des espaces ouverts dans la structure spatiale. C'est ainsi que les grandes agglomérations peuvent remplir leur fonction dans les domaines économiques, sociaux, culturels, la protection de la nature, la détente et les loisirs ; ces objectifs concernent les zones de concentration urbaine et les espaces périurbains, les zones rurales, les zones de faiblesse structurelle, les zones de protection de l'espace naturel, etc.) ; *g* **Planungsziel [n]** (**1.** Angestrebtes Ergebnis einer Planung. **2.** In der Regional- oder Landesplanung die angestrebte Entwicklung, dass Gebiete in ihrer räumlichen Struktur so zugeordnet werden, dass sie ihre funktionalen, wirtschaftlichen, sozialen und kulturellen Aufgaben erfüllen können und dass durch Ausweisung von ausreichenden Freiräumen unterschiedliche ökologische Funktionen und die ortsnahe Erholung gesichert sind; es wird z. B. zwischen Entwicklungszielen für Freiräume, für Verdichtungsräume und deren Randzonen, für den ländlichen Raum, für Räume mit Struktur-

schwächen, für den Umweltschutz, Schutz von Natur und Landschaft, Denkmalschutz unterschieden); *syn.* z. T. Entwicklungsziel [n].

4303 planning goal fulfillment [n] [US]/planning goal fulfilment [n] [UK] *prof.* (Result of planning work which is successfully achieved after completing the ▶scope of professional services); *s* **meta [f] de planificación** (Resultado deseado una vez realizados todos los trabajos nombrados en la ▶descripción de servicios de planificación); *syn.* fin [m] de planificación, objeto [m] de planificación; *f* **objectif [m] de la mission** (Résultat de l'activité planificatrice après achèvement de la mission ; ▶forme et étendue de la mission) ; *g* **Leistungsziel [n]** (Beabsichtigtes Ergebnis planerischer Arbeit, das nach Erfüllung des ▶Leistungsbildes entsteht).

planning input [n] *landsc. plan.* ▶environmental planning input.

4304 planning jurisdiction [n] *leg. plan.* (Legal administrative authority over planning and regulations; *syn.* planning sovereignty [n] [also UK]); *s* **competencia [f] (exclusiva) de planificación** (Derecho de las administraciones locales, territoriales, etc. para aprobar sus propios planes de ordenación urbana. **En Es.** la ordenación del territorio y el urbanismo son —con excepción del País Vasco, donde las competencias están compartidas por la Comunidad Autónoma misma y los Órganos Forales de los Territorios Históricos [provincias]— c. e. de las CC.AA. que a su vez pueden delegar competencias a los municipios; En D. la planificación urbana es, por derecho constitucional, c. e. de los municipios); *syn.* jurisdicción [f] de planificación; *f* **compétence [f] en matière de planification** (Souveraineté, droit reconnu ou attribué à un organisme, une administration pour exercer les pouvoirs en matière de planification régionale ou urbaine tels que l'aménagement de l'espace [établissement des plans d'urbanisme] et le développement économique, la protection de l'environnement, la voirie et les transports urbains, etc.) ; *g* **Planungshoheit [f]** (Nur durch Rechtsvorgaben eingeschränkte Souveränität über regionale oder städtische Planung, d. h. z. B. durch Flächennutzungspläne und Bebauungspläne die Gestaltung des Gemeindegebietes zu bestimmen und zu leiten; cf. Art. 28 GG u. §§ 1 [3] u. 2 [1] BauGB); *syn.* Planungskompetenz [f].

4305 planning laws [npl] *leg. plan.* (Cumulative body of public enactments in the fields of ▶regional planning policy, ▶State regional planning [US], ▶community development planning [US]/urban land-use planning [UK] and ▶functional planning [US]/sectoral planning [UK]; ▶regional planning); *s* **derecho [m] de planificación** (Conjunto de leyes y normas públicas a través de las cuales se regula la ▶ordenación del territorio, la ▶planificación regional 1, el ▶planeamiento urbanístico y la ▶planificación sectorial; ▶planificación regional 2); *syn.* legislación [f] de planificación; *f* **législation [f] relative à la planification** (Ensemble des lois et de ses applications en matière d'▶aménagement du territoire au niveau national et régional [▶aménagement du territoire au niveau régional], de ▶planification régionale, de ▶planification urbaine et relatives aux ▶planifications sectorielles) ; *g* **Planungsrecht [n]** (Gesamtheit aller öffentlich-rechtlichen Gesetze und Vorschriften, durch die die ▶Raumordnung, die ▶Landesplanung, die ▶Regionalplanung, die ▶Bauleitplanung und die ▶Fachplanung nach Inhalt, Rechtswirkung und Verfahren geregelt werden).

4306 planning level [n] *plan.* (Public administrative level at which planning is carried out; e.g. by a federal, state, regional, county, community or local agency/department); *s* **nivel [m] de planificación** (Nivel administrativo en el cual se elaboran planes, p.ej en España: Estado, Comunidad Autónoma, provincia o municipio); *f* **niveau [m] de planification** (Niveau hiérarchique de l'administration publique responsable de l'élaboration des plans, p. ex. l'État, la région, le département, la commune) ; *g* **Planungsebene [f]** (Ebene der öffentlichen Verwaltung, auf der Planung durchgeführt wird, z. B. in D: Bund, Land, Regierungsbezirk/Bezirksregierung, Region, Kreis, Gemeinde, Fachbehörde); *syn.* Planungsstufe [f].

Planning [Listed Buildings and Conservation Areas] Act [n] 1990 [UK] *conserv'hist. leg.* ►National Historic Preservation Act [US].

planning measures [npl] *land'man. landsc.* ►landscape planning measures; *plan. pol.* ►person affected by planning measures/proposals.

4307 planning methodology [n] *plan.* (Discipline dealing with a logical approach to planning processes); *s* **metodología [f] de planificación** (Ciencia que trata del enfoque metodológico y de los instrumentos de la planificación); *f* **méthodologie [f] de planification** (Science consacrée à l'étude des différentes approches ou démarches pour aborder les processus de planification) ; *g* **Planungsmethodik [f]** (Wissenschaft von der Vorgehensweise bei Planungen).

planning objective [n] *plan.* ►planning goal.

planning office [n] [UK], commissioned *contr. prof.* ►contract holder [US].

planning [n] on a higher statutory level [UK] *leg. plan.* ►higher administrative level of planning.

planning period [n] *plan.* ►planning phase.

planning permission [n] [UK] *adm. leg.* ►building permit [US].

planning permission [n] [UK], conditional *adm. leg.* ►conditional development permit [US].

planning permission [n] for mineral extraction [UK] *adm. leg. min.* ►permit for mineral extraction [US].

4308 planning phase [n] *plan.* (Period of time in which the planning of a project, or part of it, takes place; ►work phase [US]/work stage [UK]); *syn.* planning stage [n], planning period [n]; *s* **fase [f] de planificación** (Periodo/período que dura el planeamiento de un proyecto; ►fase de trabajo); *syn.* fase [f] de redacción de un proyecto técnico; *f* **phase [f] d'études** (Période pendant laquelle ont lieu les études d'un projet d'aménagement qui précède la phase de réalisation; ►étape de la mission de maîtrise d'œuvre) ; *g* **Planungsphase [f]** (1. Zeitraum, in dem die Planung eines Projektes abläuft. 2. Zeitlicher Teilabschnitt einer Planung; ►Leistungsphase).

planning policy [n] *leg. plan. pol.* ►regional planning policy.

planning powers [npl] [UK], public authority or agency with *adm. plan.* ►public planning body [US].

4309 planning practice [n] *plan. prof.* (Professional activities involved in describing and delineating land use development, as conceived by authorities/agencies/private owners, and based on experience in the planning field); *s* **experiencia [f] en planificación** (Acumulación de conocimientos prácticos debido a la experiencia profesional en planificación); *syn.* práctica [f] profesional en planificación; *f* **expérience [f] en matière d'aménagement** (Expérience acquise pendant les activités professionnelles avec la réalisation d'études d'aménagement) ; *g* **Planungspraxis [f]** (...praxen [pl]; durch Berufstätigkeit erworbene Erfahrung in der Planung von Projekten).

4310 planning procedure [n] (1) *plan.* (Course of step-by-step action in a ►planning process); *s* **esquema [m] de actuación en una planificación** (Conjunto de pasos previstos durante la planificación de un proyecto o plan que suelen representarse

gráficamente; ►proceso de planificación); *f* **planning [m]** (Déroulement des phases d'un projet, d'une étude ; ►processus de planification) ; *syn.* programme [m] de phasage, calendrier [m] de phasage ; *g* **Ablaufschema [n] einer Planung** (...schemas [pl], ...schemata [pl]; festgelegte und meist grafisch dargestellte Ablaufschritte einer Planung; ►Planungsablauf).

planning procedure [n] (2) *leg. plan.* ►regional planning procedure.

4311 planning process [n] *plan.* (Sequence of logical planning steps); *syn.* design process [n]; *s* **proceso [m] de planificación** (Secuencia lógica de las diferentes fases en procedimiento de planificación de un proyecto o plan); *syn.* proceso [m] de planeamiento; *f* **processus [m] de planification** (Suite logique des différentes phases d'une procédure d'aménagement) ; *g* **Planungsablauf [m]** (Abfolge/Verlauf der einzelnen Planungsschritte); *syn.* Planungsprozess [m].

planning program [n] [US] *plan.* ►regional planning program [US]/regional planning programme [UK].

planning programme [n] [UK], regional *plan.* ►regional planning program [US]/regional planning programme [UK].

4312 planning project [n] *plan.* (►project 1); *s* **proyecto [m] de planificación** (►proyecto 1); *f* **projet [m] d'aménagement** (►projet) ; *g* **Planungsvorhaben [n]** (►Projekt 1); *syn.* Planungsprojekt [n].

planning project [n], coordination of a *plan.* ►obtaining a mutual agreement on planning proposals.

planning proposal [n] *leg. urb. trans.* ►preliminary planning proposal.

planning proposal [n], draft *leg. urb. trans.* ►preliminary planning proposal.

planning proposals [npl] *landsc. plan.* ►landscape planning proposals, ►obtaining a mutual agreement on planning proposals.

planning purpose [n] *plan.* ►planning goal.

planning-related [adj] *plan. pol.* ►regional planning-related.

planning report [n] *plan.* ►final planning report.

planning requirements [npl] *plan.* ►fulfillment of planning requirements [US]/fulfilment of planning requirements [UK].

4313 planning services [npl] *plan. prof.* (Performance of work provided by a planner for the fulfillment/fulfilment of planning tasks; ►basic professional services, ►scope of professional services, ►special professional services; *generic term* ►professional services); *s* **servicios [mpl] de planificación** (Servicios de un/a profesional o un equipo para realizar una tarea de planificación, si están incluidos en la ►descripción de servicios de planificación se clasifican en ►servicios profesionales básicos y ►servicios complementarios de planificación; *término genérico* ►servicios profesionales); *f 1* **prestation [f] de maîtrise d'œuvre** (Terme général caractérisant les services rendus par une ou plusieurs personnes dans la réalisation de la mission ; *f 2* **mission [f] normalisée de maîtrise d'œuvre** (Prestations réalisées dans le cadre d'études de conception comprenant des éléments normalisés ; ►forme et étendue de la mission, ►prestation complémentaire, ►prestations élémentaires ; *terme générique* ►prestations [intellectuelles]) ; *g* **Planungsleistung [f]** (Leistung eines oder mehrerer Planer zur Erfüllung einer Planungsaufgabe; soweit diese **P.en** in ►Leistungsbildern erfasst sind, gliedern sie sich in ►Grundleistungen und ►besondere Leistungen 2; *OB* ►Leistung).

planning services [npl] [UK], additional *contr. prof.* ►special professional services.

planning services [npl]**, contract for** *contr. prof.* ▶planning contract.

planning sovereignty [n] [UK] *leg. plan.* ▶planning jurisdiction.

planning stage [n] *plan.* ▶planning phase.

4314 planning status [n] *plan. prof.* (Stage of project progress or completion); *s* **avance [m] de un proyecto** (Grado de adelanto de un proyecto); *syn.* estado [m] de planificación de un proyecto; *f* **état [m] d'avancement des études** (HAC 1989, 34 ; d'un projet) ; *syn.* situation [f] des études ; *g* **Planungsstand [m]** (Erreichter Fortschritt/Konkretisierungsgrad eines Projektes); *syn.* Stand [m] der Planung.

4315 planning task [n] *plan.* (Specified piece of work which has to be undertaken); *s* **cometido [m] del planeamiento** (Tarea específica a realizar o problema a solucionar por medio de una planificación) *syn.* cometido [m] de la planificación; *f 1* **mission [f] d'aménagement** (Tâche de planification, d'aménagement dévolue à une ou un groupe de personnes ; *syn.* mission [f] de planification ; *syn.* mission [f] d'études ; *f 2* **mission [f] de maîtrise d'œuvre** (Tâche relative à l'établissement d'un projet d'aménagement et au contrôle de son exécution) ; *g* **Planungsaufgabe [f]** (Eine dargelegte Problemstellung oder eine beschriebene Aufgabe, die planerisch zu lösen ist).

4316 planning work [n] *plan.* (Planning activities); *s* **trabajo [m] de redacción de planes** *syn.* trabajos [mpl] (preparatorios) del planeamiento, trabajos [mpl] (preparatorios) de la planificación; *f* **études [fpl]** (Activités prospectives planificatrices) ; *g* **Planungsarbeit [f]** (Planerische Tätigkeit).

Planning Zone [n] [UK]**, Simplified** *leg. urb.* *▶comprehensive plan [US]/Local Development Framework (LDF) [UK].

plan notation [n] [UK] *plan.* ▶plan symbol [US].

plan notation regulations [npl] [UK] *leg. plan.* ▶plan symbol conventions [US].

4317 plan [n] **of a paving pattern** *constr.* (Drawing showing the paving layout of a specific bond of pavers; ▶jointing pattern plan); *s* **representación [f] gráfica del patrón de aparejo** (Dibujo del ensamblaje previsto para pavimento de losas, adoquines o clinker; ▶plan de aparejo); *f* **principe [m] de calepinage** (Représentation graphique d'appareillages pour des revêtements de surface [pavage, dallage, revêtement en clinker] ; ▶plan de calepinage) ; *syn.* dessin [m] d'appareillage ; *g* **zeichnerische Darstellung [f] eines Verlegemusters** (Werkzeichnung von Verbänden für Platten-, Pflasterstein- oder Klinkerbelag; ▶Verlegeplan).

plan report [n] *plan.* ▶regional plan report.

4318 plan revision [n] *plan.* (Process of up-dating, redesign, alteration or modification of a plan; the result is a revised version, amended version or up-dated version of a plan; ▶revised version); *syn.* plan alteration [n], modification [n] of a plan, revision [n] of a plan; *syn. leg. plan.* amendment [n] of a plan; *s 1* **revisión [f] de un plan** (Proceso de actualización de un plan por adopción de nuevos criterios respecto de la estructura general y orgánica del territorio o de la clasificación del suelo, motivada por la elección de un modelo territorial distinto o por la aparición de circunstancias sobrevenidas, de carácter demográfico o económico, que incidan sustancialmente sobre la ordenación, o por agotamiento de la capacidad del plan; art. 154.3 RD 2159/1978; ▶versión revisada); *s 2* **modificación [f] de un plan** (Alteración de algunos elementos que constituyen un plan, sin adoptar nuevos criterios en la estructura general y orgánica del territorio, aunque puede llevar consigo cambios aislados en la clasificación o calificación del suelo; cf. art. 154 RD 2159/1978); *f 1* **révision [f] du plan** (*Révision d'un document d'urbanisme*

procédure analogue à celle de l'élaboration d'un plan, le plan révisé n'entrant en vigueur que lorsqu'il est approuvé après enquête publique, l'ancien plan continuant de s'appliquer ; DUA 1996 ; ▶version révisée) ; *f 2* **modification [f] du plan** (1. Changements apportés au contenu ou à la forme du dessin d'un plan, d'un document graphique ou d'un projet d'aménagement. 2. Procédure abrégée, qui s'opère sans la consultation des personnes publiques associées à l'élaboration du plan ; DUA 1996) ; *g* **Überarbeitung [f] eines Planes** (Umgestaltung, Verbesserung oder Abänderung eines Planes/Planungsergebnisses; ▶überarbeitete Fassung); *syn.* Planänderung [f], Planüberarbeitung [f].

4319 plan symbol [n] [US] *plan.* (Figure used graphically to represent a land use or feature on a plan, which is shown in a chart or legend; ▶plan symbol conventions [US]/plan notation regulations [UK]); *syn.* plan notation [n] [UK]; *s* **símbolo [m] de leyenda** (▶reglamento de símbolos de planificación); *syn.* símbolo [m] de un plan; *f* **symbole [m] normalisé** (Symbole graphique en noir et blanc ou de couleur utilisé dans une légende de plan, p. ex. pour la représentation des différentes servitudes d'utilité publique ou les servitudes d'urbanisme édictées par le PLU/POS ; ▶arrêtés réglementaires concernant les légendes affectant les servitudes d'un PLU/POS) ; *syn.* symbole [m] graphique ; *g* **Planzeichen [n]** (Darstellungssymbol in schwarzweißer oder farbiger Ausfertigung, um ein Planwerk allgemein verständlich und leicht lesbar zu machen; ▶Planzeichenverordnung); *syn.* Plansignatur [f].

4320 plan symbol conventions [US] *leg. plan.* (Customary or required symbols shown in a plan legend for graphic uniformity in the preparation and presentation of development plans); *syn.* plan notation [n] regulations [UK]; *s* **reglamento [m] de símbolos de planificación** (D., norma legal que rige para los planes de ordenación urbana que incluye la normativa de redacción de planes y su representación gráfica); *f* **arrêtés [mpl] réglementaires concernant les légendes affectant les servitudes d'un PLU/POS** (Arrêtés précisant la représentation des différents servitudes d'urbanisme [C. urb., art. 123-1] et d'utilité publique [C. urb., art. 126-1] pouvant figurer sur les documents graphiques du plan local d'urbanisme [PLU]/plan d'occupation des sols [POS]) ; *g* **Planzeichenverordnung [f]** (*D. Abk.* PlanzV; Gesetzliche Verordnung über die Ausarbeitung der Bauleitpläne und die Darstellung des Planinhaltes; cf. PlanzV 90).

plant [n] (1) *phyt.* ▶adventitious plant; *bot. phyt.* ▶alpine plant; *phyt.* ▶aridity indicator plant; *hort. plant.* ▶autumn-blooming plant; *hort.* ▶balled and potted plant [US]; *for. hort.* ▶bare-rooted plant; *hort.* ▶bedding plant; *bot.* ▶biennial plant; *landsc. plant.* ▶bird refuge plant; *bot. hort.* ▶bulbous plant; *hort. plant.* ▶carpet-forming ground cover plant; *plant.* ▶characteristic plant; *bot. hort.* ▶climbing plant; *phyt.* ▶companion plant (1); *agr. hort.* ▶companion plant (2); *hort.* ▶container-grown plant; *gard. hort.* ▶container plant; *hort. plant.* ▶creeping ground cover plant; *agr. for. hort.* ▶cultivated plant (1); *hort. phyt.* ▶cultivated plant (2); *bot. hort. plant.* ▶cushion plant; *envir.* ▶decontamination plant; *hort. plant.* ▶decorative foliage plant; *bot. hort.* ▶deep-rooting plant; *phyt.* ▶driftline plant; *gard. hort. plant.* ▶dry wall plant; *phyt.* ▶emergent aquatic plant (1); *bot.* ▶emergent aquatic plant (2); *bot.* ▶ericaceous plant (1); *hort.* ▶field-grown plant; *bot. phyt.* ▶floating-leaved plant; *hort. plant.* ▶flowering woody plant; *ecol. landsc. plant.* ▶food plant; *gard. plant.* ▶fragrant plant; *bot. phyt.* ▶free-floating water plant; *bot.* ▶germinated young plant; *agr. constr. hort.* ▶green manure plant; *hort. plant.* ▶ground cover plant; *phyt.* ▶heavy metal-tolerant plant; *hort.* ▶hedge plant; *bot.* ▶high mountain plant; *hort. landsc.* ▶host plant for bees; *hort. for.* ▶immature plant; *phyt.* ▶impeded water indicator plant; *ecol. phyt.* ▶indicator plant; *landsc. plant.* ▶individual woody plant; *gard. hort. plant.* ▶late-blooming plant; *gard. plant.* ▶lawn substitute plant; *for. phyt.* ▶light-demanding plant; *phyt.* ▶moisture indicator plant; *bot.* ▶montane plant; *plant.* ▶nurse plant; *hort.* ▶ornamental plant, ▶ornamental woody plant, ▶own root plant; *bot.* ▶peat garden plant [UK]; *phyt.* ▶pioneer plant on entisol [US]/pioneer plant on immature soil [UK]; *phyt.* ▶rock plant; *constr. hort.* ▶root-balled plant; *bot. hort. plant.* ▶rosette plant; *phyt.* ▶ruderal plant; *bot.* ▶scandent plant, ▶shade plant; *hort. plant.* ▶shade-tolerant woody plant; *bot. hort.* ▶shallow rooting plant; *hort. plant.* ▶specimen plant; *gard. hort. plant.* ▶spring-blooming plant; *hort.* ▶stock plant [US]/parent plant [UK]; *phyt.* ▶submerged aquatic plant; *gard. hort. plant.* ▶summer-blooming plant; *bot.* ▶surface plant; *bot. hort.* ▶tap-rooted plant, ▶tuberous-rooted plant, ▶tussock plant; *hort. plant.* ▶wall-climbing plant [US]/house wall climber [UK]; *phyt.* ▶wetness indicator plant; *hort.* ▶widely-spaced plant; *plant.* ▶wild plant; *hort. landsc.* ▶woody host plant, ▶woody host plant for bees, *landsc. plant.* ▶woody host plant for birds; *phyt.* ▶zinc-tolerant plant.

plant [n], **accent** *hort. plant.* ▶specimen plant.

plant [n], **agressive** *hort. phyt.* ▶invasive species.

plant [n], **annual** *bot. hort. phyt. plant.* ▶annual.

plant [n], **aquatic** *bot. phyt.* ▶hydrophyte, ▶emergent aquatic plant (1), ▶emergent aquatic plant (2), ▶submerged aquatic plant.

plant [n], **aromatic** *gard. plant* ▶fragrant plant.

plant [n], **bad drainage indicator** *phyt.* ▶impeded water indicator plant.

plant [n] [UK], **balled container** *hort.* ▶balled and potted plant [US].

plant [n] [US], **bee** *hort. landsc.* ▶host plant for bees.

plant [n], **bees forage** *hort. landsc.* ▶host plant for bees.

plant [n], **bird forage** *landsc. plant.* ▶woody host plant for birds.

plant [n] [US], **broadleaf woody** *bot.* ▶broad-leaved woody species.

plant [n] [US], **bunch** *bot. hort.* ▶tussock plant.

plant [n], **carpet** *hort. plant.* *∗*▶carpet-forming groundcover plant.

plant [n], **chasmophytic** *phyt.* ▶rock plant.

plant [n] [US], **deciduous woody** *bot.* ▶broad-leaved woody species.

plant [n] [US], **deep-rooted** *bot. hort.* ▶deep-rooting plant.

plant [n], **electric power** *envir.* ▶power plant.

plant [n], **ephemeral** *phyt.* ▶casual species.

plant [n], **epilithic** *phyt.* ▶rock plant.

plant [n], **ericaceous** *bot.* ▶peat garden plant [UK].

plant [n] [US], **fall-blooming** *hort.* ▶autumn-blooming plant.

plant [n] [UK], **hedging** *hort.* ▶hedge plant.

plant [n], **hydrophilic** *bot.* ▶hygrophyte.

plant [n], **hygrophilic** *bot.* ▶hygrophyte.

plant [n], **hygrophilous** *bot.* ▶hygrophyte.

plant [n], **marsh** *bot.* ▶helophyte.

plant [n], **moisture-loving** *bot.* ▶hygrophyte.

plant [n], **nectar** *hort. landsc.* ▶host plant for bees.

plant [n] [UK], **parent** *hort.* ▶stock plant [US].

plant [n], **pioneer** *phyt.* ▶pioneer species, ▶woody pioneer plant/species.

plant [n], **pollen** *hort. landsc.* ▶host plant for bees.

plant [n], **raw humus-forming** *pedol. phyt.* ▶raw humus decomposer.

plant [n], **recently introduced** *phyt.* ▶adventitious plant.

plant [n] [UK], **reed** *phyt.* ▶emergent aquatic plant (1).

plant [n], **representative** *plant.* ▶characteristic plant.

plant [n], **rock cleft** *phyt.* ▶rock plant.

plant [n], **rock crevice** *phyt.* ▶rock plant.

plant [n], **rooted water** *bot. phyt.* ▶floating-leaved plant.

plant [n], **rubble** *phyt.* ▶ruderal plant.

plant [n], **rupicolous** *phyt.* ▶rock plant.

plant [n], **sciophilous** *bot.* ▶shade plant.

plant [n] [US], **shallow-rooted** *bot. hort.* ▶shallow rooting plant.

plant [n], **spring-flowering** *gard. hort. plant.* ▶spring-blooming plant.

plant [n] [US], **submergent** *phyt.* ▶submerged aquatic plant.

plant [n], **summer-blooming** *gard. hort. plant.* ▶summer-flowering plant.

plant [n] [US], **swamp** *bot.* ▶helophyte.

plant [n], **tendril-climbing** *bot.* ▶tendril climber.

plant [n], **woody** *bot. hort.* ▶woody species (1).

plant [n], **woody bee** *hort. landsc.* ▶woody host plant for bees.

plant [n], **woody bee forage** *hort. landsc.* ▶woody host plant for bees.

plant [n], **woody food** *hort. landsc.* ▶woody host plant for birds.

plant [n], **woody nectar** *hort. landsc.* ▶woody host plant for bees.

plant [n], **woody pollen** *hort. landsc.* ▶woody host plant for bees.

plant [n], **xerophile** *bot.* ▶xerophyte.

plant [n], **xerophilous** *bot.* ▶xerophyte.

plant [n], **zinc-enduring** *phyt.* ▶zinc-tolerant plant.

plant [n] (2) *urb.* central heating plant [US]/district heating plant [UK]; *wat'man.* ▶drinking water extraction and treatment plant; *urb.* industrial plant; *envir.* ▶nuclear plant, ▶power plant, ▶refuse compost production plant, ▶sewage treatment plant, ▶treatment plant, ▶waste incineration plant, ▶waste recycling plant, ▶waste separation plant, ▶water treatment plant, ▶wind energy plant.

plant [n] [UK], **district heating** *urb.* ▶central heating plant [US].

plant [n], **electric power** *envir.* ▶power plant.

plant [n], **fuel reprocessing** *envir.* ▶nuclear fuel reprocessing plant.

plant [n] [UK], **miniature sewage treatment** *envir.* ▶individual sewage disposal system [US].

plant [n], **power distribution** *envir. plan.* ▶power distribution facility.

plant [n] [UK], **private sewage treatment** *envir.* ▶individual sewage disposal system [US].

plant [n], **sewage disposal** *envir.* ▶sewage treatment plant.

plant [n] [US], **waste compost production** *envir.* ▶refuse compost production plant.

P

plant [n] [US]**, waste reprocessing** *envir.* ►waste recycling plant.

plant [n]**, water extraction and treatment** *envir. wat'man.* ►drinking water extraction and treatment plant.

plant [n]**, water purification** *envir. wat'man.* ►water treatment plant.

plant abandonment [n] *econ. envir.* ►industrial plant abandonment.

4321 plant association [n] *phyt.* (In phytosociology plant community having a definite floristic composition, uniform physiognomy and occurring in uniform habitat conditions, representing the lowest rank in the taxonomic hierarchy. An association is defined according to its species composition, especially with regard to its own ►character species and ►differential species; ►biome, ►phytosociological classification, ►phytocoenosis); *syn.* plant community [n] (1); *s* **asociación** [f] (En sentido geobotánico, una colectividad vegetal de composición florística determinada, unidad de condiciones estacionales y unidad fisiognómica: es la unidad fundamental de la sinecología [Congreso de Bruselas 1910]. La expresión «**comunidad vegetal**» se puede aplicar a una **a.** bien definida por su combinación característica de especies, lo mismo que a un tipo de vegetación débilmente diferenciado o cuyo valor fitosociológico no se puede precisar con exactitud. La expresión «►**fitocenosis**» es empleada por la mayoría de los geobotanistas suizos para denominar la unidad más general de la colectividad vegetal y encierra la idea, no sólo de cohabitación en un medio, sino de una cierta relación objetiva de las plantas entre sí. Los que niegan o creen dudosa tal relación, prefieren el término de ►sinecia; cf. DB 1985; ►bioma, ►clasificación fitosociológica ►especie característica, ►especie diferencial; *syn.* comunidad [f] vegetal (1), fitocenosis [f] (2) (±); *f* **association** [f] (**végétale**) (Unité phytosociologique fondamentale caractérisée par un ensemble floristique spécifique, stable et en équilibre avec le milieu ambiant, possédant une physionomie et des caractéristiques stationnelles homogènes, représentant dans la hiérarchie taxonomique l'échelon inférieur. La dénomination des associations est définie par l'ajout du suffixe -*etum* au nom générique ou spécifique d'une espèce dominante. Les ►espéces caractéristiques [exclusives] et les ►espèces différentielles peuvent aider à différencier les associations entre elles ; ►biome, ►phytocénose, ►systématique phytosociologique) ; *syn.* groupement [m] végétal ; *g* **Assoziation** [f] (Pflanzensoziologische Einheit von bestimmter floristischer Zusammensetzung, einheitlichen Standortbedingungen und einheitlicher Physiognomie, die in der systematischen Rangordnung die unterste darstellt; durch Anfügung der Endung -*etum* an den Gattungs- oder Artnamen soziologisch wichtiger Arten nomenklatorisch gekennzeichnet; durch ►Kennarten und ►Trennarten von anderen Assoziationen unterschieden; ►Biom, ►Klassifizierung der Vegetationseinheiten, ►Phytozönose); *syn.* Pflanzenassoziation [f], Pflanzengemeinschaft [f], (Pflanzen)gesellschaft [f].

4322 plant association individual [n] *phyt.* (Seperate group of associated plants being studied within a plant community); *s* **individuo** [m] **de asociación** (Poblaciones vegetales homogéneas concretas que responden al concepto abstracto que tenemos formado de una asociación determinada; DB 1985); *syn.* representante [m] de asociación; *f* **individus** [mpl] **d'association** (Peuplement végétal déterminé d'une association végétale étudiée) ; *syn.* peuplement [m] végétal isolé ; *g* **Assoziationsindividuum** [n] (Vegetationsbestand einer bestimmten [untersuchten] Pflanzengesellschaft); *syn.* Einzelbestand [m] einer Pflanzengesellschaft.

4323 plant association [n] **of pavement joints** *phyt.* (Community of certain plant species, which appear primarily in the joints between paving stones and slabs of footpaths, squares or roads. This community belongs, e.g. in central Europe to the alliance of *Polygonion avicularis* and to the association of *Bryo-Saginetum procumbentis*; ►paving joint vegetation, ►vegetation of wall joints); *s* **comunidad** [f] **de rendijas de pavimentos** (Comunidad vegetal de una constitución florística específica que se presenta mayormente en las juntas de superficies pavimentadas. Pertenece a la alianza *Polygonion avicularis* y a la asociación *Bryo-Saginetum procumbentis*; ►vegetación de rendijas, ►vegetación de las juntas de adoquinado); *f* **groupement** [m] **des interstices des pavés** (Association végétale vivant principalement dans les fentes des pavages, dallages ou dans les fissures des revêtements de diverses circulations, appartenant à l'alliance de *Polygonion avicularis* et à l'association de *Bryo-Saginetum procumbentis* ; ►végétation d'interstices de pavage, ►végétation saxicole et muricole) ; *syn.* association [f] des interstices de(s) pavés ; *g* **Pflasterritzengesellschaft** [f] (Pflanzengesellschaft von bestimmter floristischer Zusammensetzung, die vorwiegend in Fugen von gepflasterten oder plattierten Wege-, Platz- oder Straßenbelägen vorkommt. Sie gehört z. B. in Mitteleuropa zur Assoziation der Mastkraut-Trittgesellschaft *[Bryo-Saginetum procumbentis]* im Verband der Vogelknöterich-Trittgesellschaften *[Polygonion avicularis]*; ►Fugenbewuchs, ►Pflastergrün); *syn.* Pflasterfugengesellschaft [f].

4324 plantation [n] (1) *constr. for. hort.* (**1.** *Result of a planting operation* area occupied by planted vegetation; ►planting 1, ►area of planting, ►intensive planting 1. **2.** Large-scale planting, especially of trees); *s* **plantación** [f] (1) (*Resultado de trabajos de plantación* superficie cubierta de vegetación plantada; ►plantación 2, ►plantación densa, ►plantío); *f 1* **plantation** [f] (1) (Résultat de l'opération de plantation ; ►aire de plantation, ►aire plantée, ►plantation 2, ►plantation dense) ; *f 2* **plant** [m] (Ensemble de végétaux de même espèce planté dans un même terrain) ; *g* **Pflanzung** [f] (1) (*Ergebnis eines Pflanzvorgangs* mit Pflanzen angepflanzte Fläche; ►Bepflanzungsfläche, ►dichte Anpflanzung, ►Pflanzung 2); *syn.* Anpflanzung [f] (1), Bepflanzung [f], Neuanpflanzung [f].

plantation [n] (2) *landsc.* ►wind-penetrable plantation.

4325 plantation [n] **appropriate to the site** *landsc. plant.* (Planted vegetation which has growing requirements corresponding to local site conditions [►habitats conditions]. The appropriate planting may deviate considerably from the indigenous plants to be found there, depending upon the degree of anthropogenic influences; ►indigenous plant species); *s* **plantación** [f] **adecuada a la ubicación** (Plantación de especies cuyas necesidades coinciden con las ►condiciones mesológicas del lugar; ►especie vegetal autóctona); *f* **plantation** [f] **appropriée au site** (Plantation d'espèces végétales adaptées aux conditions locales d'un site [►conditions de la station] ; les espèces appropriées peuvent considérablement différer des ►espèces végétales indigènes en fonction de l'intensité des influences anthropogènes sur le site considéré) ; *g* **standortgerechte Pflanzung** [f] (Pflanzung von Arten, Unterarten oder Sorten, deren Ansprüche den ►Standortbedingungen 1 und den pflanzenspezifischen Eigenarten hinsichtlich der Nutzer- und Anliegerinteressen, z. B. Blatt- und Fruchtfall, Duft, Giftigkeit, entsprechen. Je nach Veränderung des Pflanzstandortes durch anthropogene Einflüsse, kann die standortgerechte Pflanzung sehr stark von der Pflanzung mit gebietseigenen/autochthonen Arten abweichen, z. B. Stadtstandort, Deponie; ►bodenständige Pflanzenart).

plant bed [n] *constr. hort.* ►planting bed.

4326 plant bowl [n] *hort.* (Round, shallow, dish-shaped ►plant container); *s* **cuenco** [m] **de plantas** (►Recipiente de plantas plano y redondo); *syn.* maceta [f] plana, tiesto [m] plano;

P

f 1 **coupe [f] à plantes** (Petit ▶contenant à plantes de forme évasée) ; *syn.* coupe [f] pour plantation ; *f 2* **vasque [f] à plantes** (Coupe large) ; *g* **Pflanzenschale [f]** (**1.** Bepflanzte Schale. **2.** *Flache Form eines* ▶*Pflanzengefäßes* Pflanzschale [f]; cf. Erläuterung bei ▶Pflanzengefäß).

4327 plant boxing [n] *arb. constr.* (Assembling of [wooden] planks around a root ball for tree moving or shipment; the planks are reinforced with exterior bracing; ▶transplantation of semi-mature trees; ▶balled and platformed [US]); *s* **encajonamiento [m] del cepellón** (Montaje de tablones [de madera] alrededor del cepellón para ▶transplante de árboles grandes; ▶embalado en malla metálica y enpaletado [del cepellón]); *f* **mise [f] en bac** (Confection d'un bac en bois pour la protection des racines des gros végétaux afin de les transplanter ; ▶à motte grillagée livrée sur palette, ▶transplantation de gros végétaux) ; *g* **Einkübeln [n, o. Pl.]** (Montage eines Holzkastens um einen eingegrabenen Wurzelballen bei ▶Großbaumverpflanzungen; ▶balliert und palettiert); *syn.* Eindauben [n, o. Pl.], Einkisten [n, o. Pl.], in Kisten verpacken [vb].

4328 plant breed [n] *agr. for. hort.* (Result of plant breeding: new ▶cultivar; TGG 1984, 180; ▶plant breeding); *s* **planta [f] seleccionada** (Variedad de planta creada por selección con el fin de mejorar cada vez más su capacidad de adaptación al medio; ▶cultivar, ▶fitogenética); *syn.* cultivar [m] nuevo; *f* **plante [f] sélectionnée** (Variété crée par sélection afin d'être de mieux en mieux adaptée aux conditions environnementales ; ▶sélection végétale, ▶cultivar) ; *g* **gezüchtetes Pflanzengut [n]** (Durch Züchtung neu geschaffene Sorte einer [lange in Kultur befindlichen] Kulturpflanze; ▶Kultivar, ▶Pflanzenzüchtung).

4329 plant breeder [n] *agr. for. hort.* (Person who produces new cultivars by cross-breeding, selection and artificially-induced mutations or who recultivates old cultivars for propagation or who reintroduces an old cultivar by re-crossing older varieties; ▶plant breeding, ▶perennial plant breeder, ▶rose breeder); *s* **fitogeneticista [m/f]** (Persona que se dedica a producir nuevas plantas de cultivo por medio de cruces, selección y mutaciones provocadas artificialmente y a propagar antiguas plantas de cultivo cultivándolas o reproduciéndolas por medio de cruces retroactivos; ▶fitogenética, ▶cultivador, -a de vivaces/perennes, ▶cultivador, -a de rosas); *syn.* criador, -a [m/f] de plantas; *f* **sélectionneur [m] de plantes** (Spécialiste qui **1.** sélectionne des nouvelles variétés de culture au moyen de méthodes de croisement, de sélection et de mutation artificielle et **2.** plante les espèces anciennes pour la multiplication ou rétabli par croisement les espèces anciennes de plantes cultivées ; ▶sélection végétale, ▶sélectionneur de plantes vivaces, ▶sélectionneur, euse de roses) ; *syn.* phytogénéticien [m] ; *g* **Pflanzenzüchter/-in [m/f]** (Person, die neue Kulturpflanzensorten durch Kreuzung, Selektion und künstlich erzeugte Mutationen hervorbringt resp. alte Kulturpflanzen zur Vermehrung wieder anpflanzt oder durch Rückkreuzung wiederherstellt; ▶Pflanzenzüchtung, ▶Staudenzüchter, ▶Rosenzüchter).

4330 plant breeding [n] *agr. for. hort.* (**1.** Development of new cultivars which are adapted to specific site requirements or to changeable cultivation methods and human needs, as well as the conservation of old cultivars. In applied genetics, **p. b.** causes the hereditary transformation of cultivated plants, by creating new combinations of genetic make-up and the production of mutations. **2.** Creation of new crop varieties **[new plant breeding, ▶new introduction]**, which are tolerant of specific site conditions [e.g. aridity and salinity] or better adapted to changes in cultivation methods and human requirements [e.g. productivity increases, disease resistance, improved fertilizer use, improved product quality] as well as the preservation of existing crop varieties **[cultivation of a plant under conservation]**. In partic-

ular, **p. b.** as a form of applied genetics, is concerned with the hereditary modification of crops, mainly through new combinations of hereditary characteristics and the introduction of mutations as well as the additional implantation of genes [gene technology]; mankind has manipulated genomes of cultivated crops for approximately the past 10,000 years; ▶genetically modified organism, ▶plant breed); *s* **fitogenética [f]** (Rama de la genética aplicada que se dedica a crear ▶nuevas variedades de plantas para el cultivo que se adapten mejor a las condiciones mesológicas y sean más productivas o —en el caso de las plantas ornamentales— cumplan mejor las funciones estéticas y decorativas a las que se destinan y a preservar las variedades de cultivo existentes hasta la fecha. En las últimas décadas, la **f.** —que ya tradicionalmente modificaba las plantas genéticamente por técnicas de mutagénesis o de fusión celular y empleaba, entre otras, técnicas de fertilización «in vitro»— ha comenzado a manipular las plantas insertando en ellas genes de otros organismos para proporcionarles características que no tienen naturalmente. Esta tendencia es muy criticada mundialmente por los peligros que conlleva para la salud humana y para los organismos vivos [contaminación genética], así como para los ecosistemas naturales; ▶organismo modificado genéticamente); *f* **sélection [f] végétale** (**1.** Adaptation des plantes aux besoins de l'homme et permettant l'amélioration de la productivité, l'adaptation des plantes au milieu ainsi que l'amélioration qualitative des produits. **2.** Ensemble de méthodes qui ont pour objet l'ajustement génétique [amélioration génétique végétale] des plantes ; la **s. v.** comprend trois activités distinctes : la **sélection créatrice** [ou amélioratrice/améliorante], qui assure la création de ▶variétés nouvelles [création variétale], la **sélection conservatrice**, qui veille au maintien des caractéristiques spécifiques des variétés obtenues et le **génie génétique** qui, par modification du matériel génétique, confère à un organisme végétal une caractéristique ou une propriété nouvelle ; ce processus donne naissance à un ▶organisme génétiquement modifié [OGM] ; LA 1981, 1023 ; ▶plante sélectionnée) ; *g* **Pflanzenzüchtung [f]** (**1.** Anpassung von Pflanzen an die Erfordernisse des Menschen. **2.** Die Schaffung neuer Kulturpflanzensorten **[Neuzüchtung, ▶Neuheit]**, die den besonderen Standortverhältnissen [z. B. Toleranz gg. Trocken- und Salzstress] oder den veränderlichen Anbaumethoden und Ansprüchen des Menschen [z. B. Ertragssteigerung, Krankheitsresistenz, verbesserte Düngerverwertung, verbesserte Produktqualität] angepasst sind sowie die Erhaltung bisheriger Kultursorten **[Erhaltungszüchtung]**. Insbesondere ist **P.** vorwiegend als angewandte Genetik die erbliche Veränderung von Kulturpflanzen, vor allem durch Neukombination von Erbmerkmalen und Auslösung von Mutationen sowie durch zusätzliche Einpflanzung von Genen **[Gentechnik, ▶gentechnisch veränderter Organismus]**; der Mensch verändert seit ca. 10.000 Jahren das Erbgut der Kulturpflanzen; ▶gezüchtetes Pflanzengut); *syn.* Pflanzenzucht [f].

4331 plant care product [n] *agr. chem. for. hort.* (Generic term for ▶plant protection agent, ▶growth regulator and plant enhancement product; ▶pesticide); *s* **productos [mpl] fitosanitarios** (Término genérico para ▶plaguicidas, ▶sustancias reguladoras del crecimiento y sustancias de limpieza o conservación; en España según orden de 2 de agosto de 1976 se incluyen en esta categoría también los productos destinados a influir en los procesos biológicos, con excepción de los nutrientes; ▶pesticida); *f* **produit [m] de traitement phytosanitaire** (Terme générique pour les ▶produits antiparasitaires [à usage agricole], ▶régulateurs de croissance et produits d'entretien ; ▶pesticide) ; *syn.* produit [m] phytosanitaire, produit [m] phytopharmaceutique ; *g* **Pflanzenbehandlungsmittel [n]** (OB zu ▶Pflanzenschutzmittel, Pflanzenstärkungsmittel, ►Wachstumsregler und Pflegemittel; ►Pestizid).

plant communities [npl]**, layering of** *phyt.* ►stratification of plant communities.

plant communities [npl] [US]**, reestablishment of** *phyt.* ►reintroduction.

plant community [n] (1) *phyt.* ►plant association.

4332 plant community [n] (2) *phyt.* (Designation for a plant community growing on a specific site. Generic term for, e.g. ►aquatic plant community, ►clearance herb formation, ►driftline community, ►dwarf-shrub plant community, ►fire plant community, ►forb vegetation, ►forest seam formation, ►glasswort vegetation, ►heavy metal-tolerant vegetation, ►initial plant community, ►perennial forb community, ►pioneer vegetation, ►plant community in silting up ponds or lakes, ►rock crevice plant community, ►scree vegetation, ►spring vegetation, ►successional plant community, ►tall forb community, ►tree-fall gap community, ►vegetation of animal rest area, ►zinc tolerant plant community; ►composition of a plant community, ►range of a plant community, ►structural characteristics of a plant community); *s* **comunidad [f] vegetal (2)** (Designación para una sinecia que crece en un lugar específico. Término genérico para ►comunidad acuática, ►comunidad de [la zona de] colmatación, ►comunidad de pedregales, ►comunidad herbácea del lindero del bosque, ►comunidad fisurícola, ►comunidad inicial, ►comunidad parietal, ►comunidad pirófita, ►comunidad sucesora, ►formación megafórbica, ►matorral bajo, ►vegetación calaminar, ►vegetación de bosques talados, ►vegetación de claros, ►vegetación de herbáceas perennes, ►vegetación de las acumulaciones de restos orgánicos, ►vegetación de lugares de reposo, ►vegetación del metalofita, ►vegetación de salicor, ►vegetación fontinal, ►vegetación halófila de los cúmulos de residuos orgánicos, ►vegetación herbácea, ►vegetación pionera; ►área de una comunidad/asociación, ►composición de una comunidad vegetal, ►estructura de una comunidad); *syn.* vegetación [f], formación [f] vegetal; *f* **végétation [f] (1)** (*Terme générique* dans une communauté végétale, l'ensemble des plantes qui vivent sur un territoire déterminé ; ►association des fissures de rocher, ►association des zones d'atterrissement, ►association pariétale, ►association pyrophyte, ►association rupicole, ►association successive, ►flore de clairière, ►flore de coupe à blanc[-étoc], ►flore des éboulis, ►flore des grèves [alluviales], ►flore d'ourlet halonitrophile, ►groupement calaminaire, ►groupement d'arbrisseaux nains, ►groupement des sols contaminés par les métaux lourds, ►groupement d'hydrophytes, ►groupement d'ourlet préforestier à mégaphorbes, ►groupement initial, ►mégaphorbiaie, ►végétation herbacée, ►peuplement à salicornes, ►végétation de reposoir, ►végétation fontinale, ►végétation herbacée pérenne, ►végétation pionnière ; ►aire de répartition d'une association, ►organisation d'un groupement végétal, ►organisation des groupements végétaux) ; *syn.* groupement [m] végétal, flore [f] ; *g* **Flur [f] (3)** (Bezeichnung für eine Pflanzengesellschaft, die auf einem bestimmten Standort vorkommt. *OB zu z. B. folgenden Gesellschaften* Felsspaltenflur [►Felsspaltengesellschaft], ►Galmeiflur, ►Gesteinsschuttflur, ►Hochstaudenflur, ►Kahlschlagflur, ►Krautflur, ►Lägerflur, ►Lichtungsflur, ►Pionierflur, ►Quellerflur, ►Quellflur, ►Schwermetallflur, ►Spülsaumflur, ►Staudenflur, ►Waldsaumflur).

plant community [n]**, ecotonal** *phyt.* ►ecotonal association.

plant community [n]**, seral** *phyt.* ►successional plant community.

plant community [n]**, structure of a** *phyt.* ►structural characteristics of a plant community.

plant community [n]**, terrestrialization** *phyt.* ►plant community in silting up ponds or lakes.

4333 plant community [n] **in silting up ponds or lakes** *phyt.* (Lakeside plant association—littoral helophytes—growing in deposited silt and organic debris; ►sediment accumulation zone); *syn.* terrestrialization plant community [n], plant community [n] of hydrosere succession (cf. ELL 1988, 276); *s* **comunidad [f] de (la zona de) colmatación** (Comunidad vegetal de la ►zona de colmatación de lagos y lagunas; cf. BB 1979); *f* **association [f] des zones d'atterrissement** (Association des stations peu profondes, des ►zones d'atterrissement des eaux calmes ou stagnantes ; *terme spécifique* groupement des grèves alluviales) ; *g* **Verlandungsgesellschaft [f]** (Pflanzengesellschaft der ►Verlandungszone, die in stehenden oder langsam fließenden Gewässern durch Festhalten des jährlichen Bestandesabfalles, der Sand- und Schlickzufuhr sowie der angelagerten Schwebstoffe und durch Vorrücken des Bestandes in das offene Gewässer dieses ständig flacher und kleiner macht bis es [das Stillgewässer] zugewachsen ist).

plant community [n] **of hydrosere succession** *phyt.* ►plant community in silting up ponds or lakes.

4334 plant community [n] **of trampled areas** *phyt.* (Plant community which has adapted to continuous trampling, belonging to the Great plantain Class—*Plantaginetea majoris Tx. et Prsg. 1950*); *s* **comunidad [f] pisoteada** (BB 1979, 478; comunidades de vegetales de la clase *Plantaginetea majoris*, resistentes al pisoteo, que encuentran su desarrollo óptimo en los caminos y demás lugares pisoteados con frecuencia moderada; cf. BB 1979); *f* **groupement [m] des lieux piétinés** (Végétation basse — moins de 15 cm — des sols très piétinés et généralement secs, appartenant à la classe du *Plantaginétea majoris Tx. et Prsg. 1950* ; *terme spécifique* végétation des lieux ensoleillés et piétinés et des pelouses urbaines) ; *g* **Trittpflanzengesellschaft [f]** (An extreme Standortbedingungen durch ständiges Betreten und an starke Bodenverdichtung angepasste Pflanzengesellschaft, z. B. die Klasse der *Plantaginetea majoris Tx. et Prsg. 1950*); *syn.* Trittgesellschaft [f], Trittflur [f].

4335 plant container [n] *gard. hort.* (Generic term for separate plant receptacle, such as ►balustrade planter, ►flower planter, ►plant bowl, ►planter, ►planter serving as a handrail, ►window box; ►miniature garden); *syn.* plant receptacle [n]; *s* **recipiente [m] de plantas** (Término genérico para todo tipo de macetas de plantas como ►cuenco de plantas, ►jardinera, ►jardinera en barandilla, ►jardinera para balcón, ►jardinera para pretil, ►tiesto de flores; ►jardín miniatura); *f* **contenant [m] à plantes** (Terme générique englobant les termes ►bac à fleurs, ►bac de plantation, ►coupe à plantes, ►jardinière garde-corps, ►jardinière pour balcon, ►jardinière pour balustrade, ►vasque à plantes ; ►jardin miniature) ; *g 1* **Pflanzengefäß [n]** (1. OB zu ►Balkonkasten, ►Blumenkübel, ►Pflanzentrog an einem Geländer, ►Pflanzenkübel, ►Pflanzenschale, ►Pflanzentrog als Mauerbrüstung; ►Miniaturgarten); *syn.* Pflanzencontainer [m]; *g 2* **Pflanzengefäß [n]** (Leerer Behälter *zum* Bepflanzen; OB für Pflanzkübel, Pflanzschale, Pflanztrog; *MERKE*, der oft gebrauchte Terminus **P.** für einen mit Pflanzen bestückten Behälter ist ein Fehlname, da das Präfix *Pflanz* immer die Tätigkeit des Einpflanzens oder Verpflanzens, nie den Inhalt bezeichnet).

4336 plant cover [n] *phyt.* (Total vegetation in a given area; ►herbaceous vegetation 2, ►forest cover); *syn.* vegetative cover [n], vegetal cover [n] [also US]; *s* **cobertura [f] vegetal** (Conjunto de plantas que crecen en una superficie dada; *términos*

P

específicos ►cobertura forestal, ►cobertura herbácea); *syn.* cubierta [f] vegetal; *f* **couverture [f] végétale** (Ensemble de la végétation sur une surface donnée ; *termes spécifiques* couverture arborescente, ►couverture forestière, ►couverture herbacée) ; *syn.* couvert [m] végétal ; *g 1* **Bewuchs [m, o. Pl.]** *hort. phyt.* (1. Gesamtheit eines Pflanzenbestandes auf einer Fläche; *UBe* ►Waldbedeckung, ►Krautbewuchs); *syn.* Pflanzenbedeckung [f], Pflanzendecke [f], Vegetation [f], Vegetationsbedeckung [f], Vegetationsdecke [f]; *g 2* **bewachsen [vb/pp]** *hort.* (Aus gärtnerischer Sicht das Bestocktsein einer Fläche mit unerwünschtem Aufwuchs; *Kontext* die Fläche ist mit Unkraut bewachsen); *syn.* bedeckt [pp].

4337 plant disease [n] *constr. hort.* (Unhealthy physical condition of plants caused by injurious bacterium, fungus, virus or insufficient nutrition; ►injurious organism); *s* **enfermedad [f] de plantas** (Debilitamiento de la salud de las plantas causado por ►organismo nocivo o por falta de nutrientes); *f* **maladie [f] des plantes** (Affaiblissement des plantes provoqué par des ►organismes nuisibles [champignons et bactéries parasites, mycoplasmes, virus] ou une insuffisance nutritionnelle ; *terme spécifique* maladie cryptogamique) ; *g* **Pflanzenkrankheit [f]** (Durch ►Schadorganismen oder Mangelernährung verursachte Schwächung von Pflanzen).

plant ecological survey [n] *phyt.* ►vegetation survey.

4338 plant ecologist [n] *phyt.* (Scientist who studies interrelationships between plants and their environment); *s* **ecólogo/a [m/f] vegetal** (Científico/a que estudia las interrelaciones de las plantas con el medio que las rodea); *syn.* fitoecólogo/a [m/f]; *f* **phyto-écologue [m]** (Scientifique qui étudie les relations entre les plantes et leur environnement) ; *g* **Pflanzenökologe/Pflanzenökologin [m/f]** (Wissenschaftler, der sich mit den Wechselbeziehungen zwischen Pflanzen und ihrer Umwelt befasst).

4339 plant ecology [n] *phyt.* (Science dealing with interrelationships between plants and their environment; ►ecology); *syn.* vegetation ecology [n]; *s* **ecología [f] vegetal** (Estudio de la influencia del medio orgánico e inorgánico sobre las plantas; ►ecología); *syn.* fitoecología [f]; *f* **écologie [f] végétale** (Science qui étude les relations entre les espèces végétales et le milieu ; ►écologie) ; *syn.* phyto-écologie [f] ; *g* **Pflanzenökologie [f]** (Wissenschaft von den Wechselbeziehungen zwischen Pflanzen und ihrer Umwelt; ►Ökologie).

4340 planted bed [n] *gard. hort.* (Separate area already planted; ►border 2, ►flower bed, ►planting bed); *s* **macizo [m] de plantas** (Área delimitada ya plantada); ►bancal para plantación, ►macizo de flores, ►platabanda); *syn.* bancal [m] de plantas; *f* **massif [m] de plantation** (1. Ensemble uniforme de plantes fleuries et/ou d'arbustes regroupés sur un même espace ; 2. l'aire plantée ; ►massif à planter ►massif floral [d'annuelles], ►plate-bande, ►plate-bande florale/fleurie) ; *g* **Pflanzenbeet [n]** (Mit Pflanzen bestücktes Beet; ►Blumenbeet, ►längliches Pflanzenbeet, ►Pflanzbeet).

4341 planted façade [n] *gard. urb.* (*also* planted facade; building wall covered with ►climbing plants, window boxes or espaliered woody plants; *specific term* façade with window boxes; ►façade planting); *syn.* vine-covered façade [n] [also US], green wall [n], vegetation [n] on building facades; *s* **fachada [f] vegetalizada** (Frente de edificio recubierto de ►plantas trepadoras; ►plantación de fachadas); *syn.* fachada [f] vegetal, fachada [f] tapizada de trepadoras; *f* **façade [f] tapissée (de)** (Façade recouverte ou agrémentée au moyen de plantes grimpantes, de plantes palissées ; ►végétalisation des façades) ; *syn.* façade [f] recouverte de plantes grimpantes, façade [f] végétalisée, jardin [m] vertical ; *g* **begrünte Fassade [f]** (Mit ►Kletterpflanzen, Blumenkästen oder Spalierbäumen bewachsene/gestaltete Haus-

fassade; *UB* mit Blumenkästen gestaltete/geschmückte Fassade; ►Fassadenbegrünung).

4342 planted flat roof [n] *constr. urb.* (►roof garden, ►roof planting); *s* **tejado [m] plano plantado** (►azotea ajardinada, ►revestimiento vegetal de tejados); *f* **toiture-terrasse [f] végétalisée/plantée** (►jardin sur toiture-terrasse, ►végétalisation des toitures) ; *g* **begrüntes Flachdach [n]** (►Dachgarten, ►Dachbegrünung).

4343 planted noise attenuation structure [n] *constr. trans.* (Free-standing protective barrier constructed of steel/earth, concrete/earth, or timber/earth crib systems; ►noise attenuation mound); *s* **muro [m] de jardineras antirruidos** (Muro libre construido de jardineras de metal, hormigón o madera cuya función principal es proteger contra ruidos; ►terraplén antirruidos); *f* **merlon [m] en éléments préfabriqués (±)** (Écran de protection contre les vues ou les émissions, autoportant libre, végétalisable sur ses deux faces dont la structure est constituée par un assemblage ou empilement d'éléments en métal, en béton, en bois remplis de terre végétale ; ►merlon antibruit) ; *g* **Steilwall [m]** (Freistehende Emissions- und Sichtschutzwand aus pyramidenförmigen Stahl-Erde-, Beton-Erde- oder Holz-Erdekonstruktionen/Systemen; ►Lärmschutzwall); *syn.* Lärmschutzsteilwall [m], Pflanzwall [m], Pflanzmauer [f], terrassenförmige Lärmschutzwand [f].

4344 planted roof [n] *constr.* (►roof planting); *s* **tejado [m] verde** (►revestimiento vegetal de tejados); *syn.* tejado [m] ajardinado; *f* **toiture [f] plantée** (►végétalisation des toitures, ►aménagement de terrasse-jardin) ; *syn.* toiture [f] végétalisée, toit [m] végétalisé ; *g* **begrüntes Dach [n]** (►Dachbegrünung); *syn.* begrünte, unterbaute Fläche [f], Gründach [n].

plant equipment [n] **and site installations** [npl] *constr.* ►maintaining of plant equipment and site installations.

4345 planter [n] *gard. hort.* (►Plant container with growing medium and plants, which is placed on paved surfaces, or constructed as an integral extension of a structure; ►balustrade planter, ►flower planter, ►plant bowl, ►window box); *syn.* plant trough [n] [also UK], planting box [n] [also US]; *s* **jardinera [f]** (Recipiente grande para plantas que se coloca p. ej. en zonas asfaltadas para su decoración; *término genérico* ►recipiente de plantas; ►cuenco de plantas, ►jardinera para balcón, ►jardinera para pretil, ►módulo de jardinera, ►tiesto de flores); *syn.* macetón [m]; *f* **bac [m] de plantation** (Gros contenant la plante et sur support végétal, placé sur les surfaces imperméabilisées ; *terme générique* ►contenant à plantes ; ►bac à fleurs, ►coupe à plantes, ►jardinière pour balcon, ►jardinière pour balustrade, ►module de plantation, ►vasque à plantes) ; *g 1* **Pflanzkübel [m]** (Großes Gefäß, das der Aufnahme von Pflanzen und Boden dient und auf versiegelten Flächen aufgestellt wird; ►Balkonkasten, ►Blumenkübel, ►Pflanzenschale, ►Pflanzkübelelement, ►Planzentrog als Mauerbrüstung; *OB* ►Pflanzgefäß); *syn.* Pflanzcontainer [m], Pflanztrog [m], Plantainer [m]; *g 2* **Pflanzenkübel [m]** (Mit Pflanzen bestücktes Behältnis; *cf.* Erläuterung bei ►Pflanzengefäß); *syn.* Pflanzencontainer [m], Pflanzentrog [m], Plantainer [m].

planter [n] **[US], confined** *constr.* ►planter with a raised edge.

planter [n] **[US], raised** *constr.* ►raised planting bed.

planter element [n] *gard. landsc.* ►planter unit.

planter module [n] *gard. landsc.* ►planter unit.

4346 planter seat [n] *gard. landsc.* (Sitting place formed by or attached to a planter wall); *s* **banco [m] adosado a jardinera**; *f* **banc [m] intégré à une jardinière** *syn.* banc [m] périphérique à jardinière incorporée ; *g* **an einem Pflanzentrog befestigte Sitzbank [f]**.

4347 planter [n] **serving as a handrail** *gard.* (Raised planter in place of a handrail along a flight of steps); *s* **jardinera [f] en barandilla**; *f* **jardinière [f] garde-corps** ; *g* **Pflanzentrog [m] an einem Geländer** *syn.* Geländerpflanzentrog [m].

4348 planter unit [n] *gard. landsc.* (Prefabricated mass-produced element which can be installed with several other units to form a desired pattern and planted with permanent or seasonal plant displays); *syn.* planter element [n], planter module [n]; *s* **módulo [m] de jardinera** (Elemento prefabricado que junto con varias unidades sirve para formar jardineras de diferentes formas); *f* **module [m] de plantation** (Contenant préfabriqué produit en série dont plusieurs éléments peuvent être juxtaposés selon un ordonnancement particulier et constituer une plantation permanente ou saisonnière) ; *syn.* élément [m] modulaire (de plantation), bac [m] juxtaposable (de plantation), élément [m] juxtaposable (de plantation) ; *g* **Pflanzkübelelement** [n] (Meist Industriell vorgefertigtes Serienprodukt, das auch zu mehreren Stücken passgenau in einem beliebigen Verband aufgestellt werden kann und für Dauer- oder Wechselbepflanzungen vorgesehen ist).

4349 planter [n] **with a raised edge** *constr.* (Planting bed [in a paved area] with a raised edging); *syn.* confined planter [n] [US] (MET 1985, 38); *s* **macizo [m] de plantas con encintado elevado** (Superficie plantada [en calle o plaza] separada de las pavimentadas por bordes elevados); *f* **massif [m] de plantes avec bordure surélevée** (Surface plantée séparée [des zones de circulation automobile ou piétonnes] par une bordure surélevée) ; *syn.* massif [m] surélevé ; *g* **eingefasstes Pflanzenbeet** [n] (Pflanzfläche, die durch eine erhöhte Einfassung begrenzt ist, z. B. im Straßen- und Platzbereich).

plant expert [n] *bot. hort.* ▶plantsman.

4350 plant formation [n] *phyt.* (Ecological classification of vegetative cover according to its physiognomy; ▶pinus-dominated krummholz formation, ▶plant association, ▶rain forest [US]/rain-forest [UK], ▶raised bog, ▶savanna[h]); *s* **formación [f] vegetal** (Una cohabitación botánica individualizada por la forma biológica que en ella domina; p. ej. bosque, prado, estepa, etc.; DB 1985; ▶asociación, ▶matorral de pino mugo, ▶pluviisilva, ▶sabana, ▶turbera alta); *f* **formation [f] végétale** (Classification physionomique des peuplements végétaux correspondant à un type de milieu, p. ex. ▶forêt ombrophile, lande, prairie, ▶savane, ▶tourbière haute, etc., sans vouloir ou pouvoir faire appel à un stade quelconque de la hiérarchie phytosociologique; ▶formation de brousse à *Pinus mugo* ; ▶association [végétale]) ; *g* **Pflanzenformation** [f] (Physiognomisch-ökologisch klassifizierter Vegetationsbestand, der einheitliche resp. gleichartige äußere Merkmale aufweist und für einen Lebensraum charakteristisch ist, z. B. Halbstrauchgebüsch, ▶Hochmoor, ▶Legföhrenformation, ▶Regenwald, ▶Savanne; ▶Assoziation).

plant [n] **for shade** *bot.* ▶shade plant.

plant [n] **for toxic/hazardous waste, incineration** *envir.* ▶incinerator for toxic/hazardous waste.

4351 plant frost injury [n] *agr. arb. for. hort.* (Death, damage or deformation of cultivated plants as a result of [severe] frost; ▶frost crack); *s* **daños [mpl] causados por heladas** (Destrucción, deterioro o alteración causadas en las plantas de cultivo o en construcciones por la acción de las heladas; ▶grieta por congelación); *f* **dommages [mpl] causés par le gel** (Destruction, détérioration ou modification provoquées par le gel sur les végétaux ou les constructions ; ▶gélivure) ; *syn.* dégâts [mpl] causés par le gel ; *g* **Frostschaden [m] an Pflanzen** (Zerstörungen, Beschädigungen oder Veränderungen an Kulturpflanzen

als Folge von Frosteinwirkungen; ▶Frostriss); *syn. arb. hort.* Frostschädigung [f] an Pflanzen.

4352 plant geography [n] *geo. phyt.* (Science of the spatial and chronological distribution of the earth's plants with respect to both the individual species and the plant communities. A branch of **p. g.** is ▶phytosociology or vegetation science which is, to a large extent, called ▶synecology; ▶distribution of plants, ▶geobotany, ▶vegetation geography); *syn.* phytogeography [n]; *s* **fitogeografía [f]** (En acepción lata, se ha empleado en el sentido de ▶geobotánica antes de que se creara esta última voz. En sentido estricto, es la parte de la geobotánica que estudia la localización, en la superficie terrestre, de la vida vegetal; cf. DB 1985; ▶biogeografía, ▶distribución de especies, ▶fitosociología, ▶sinecología); *syn.* geografía [f] botánica, corología [f] vegetal; *f* **phytogéographie [f]** (Science étudiant la ▶répartition des espèces et des formations végétales à la surface du globe ; la ▶phytosociologie et la ▶synécologie constituent des branches de la **p.** ; nouvellement, terme parfois remplacé par les termes de ▶géographie botanique et de ▶biogéographie) ; *g* **Pflanzengeografie [f]** (Wissenschaft über die räumliche und zeitliche Verteilung der Pflanzen auf der Erde, sowohl hinsichtlich der einzelnen Arten als auch der Pflanzengesellschaften. Ein Teilgebiet der **P.** ist die ▶Pflanzensoziologie [auch Vegetationskunde genannt] als weitgehend botanische ▶Synökologie. In neuerer Zeit wird **P.** durch die Begriffe ▶Geobotanik [für die botanische Forschungsrichtung] bzw. ▶Vegetationsgeographie [für die geografische Forschungsrichtung] ersetzt; ▶Pflanzenverbreitung); *syn.* Biogeografie [f] der Pflanzen, *o. V.* Pflanzengeographie [f], Phytogeografie [f], *o. V.* Phytogeographie [f].

4353 plant growth [n] (1) *bot.* (**1.** Process of plant development. **2.** All plants growing in a defined area; ▶habit, ▶slow-growing, ▶fast-growing); *s 1* **crecimiento [m] de las plantas** (▶hábito de crecimiento, ▶de crecimiento lento, ▶de crecimiento rápido); *s 2* **cobertura [f] de plantas** (Conjunto de plantas que crecen en un lugar); *f 1* **croissance [f] des végétaux** (Processus de développement de la plante) ; *f 2* **végétation [f]** (2) (Ensemble des végétaux croissant en un lieu donné) ; ▶habitus, ▶à croissance lente, ▶à croissance rapide) ; *g* **Pflanzenwuchs [m]** (**1.** *fachsprachlich* ...wüchse [pl], *sonst ohne Pl.*; das Wachsen der Pflanzen. **2.** Gesamtheit der an einem bestimmten Ort wachsenden Pflanzen; ▶Habitus, ▶langsam wüchsig, ▶schnell wüchsig).

plant growth [n] (2) *bot. ecol.* ▶deterioration of plant growth; *biol. hort.* ▶inhibiting (plant growth).

plant growth [n], **stunting** *biol. hort.* ▶inhibiting (plant growth).

4354 plant [vb] **in a semi-shaded location** [US]/ **plant** [vb] **in a semi-shaded position** [UK] *hort.* *s* **plantar** [vb] **en semisombra** *syn.* colocar [vb] en semisombra; *f* **planter** [vb] **en exposition mi-ombragée** *syn.* planter [vb] à mi-ombre ; *g* **Pflanzen in Halbschattenlage setzen** [vb].

4355 planting [n] (1) *constr. for. hort.* (Process of planting; plant [vb]); *syn.* planting process [n]; *s* **plantación [f]** (2) (Proceso de plantar; plantar [vb]); *f* **plantation [f]** (2) (Processus de plantation ; planter [vb]) ; *g* **Pflanzung [f]** (2) (**1.** Vorgang des Pflanzens; pflanzen [vb]; **2.** *Ergebnis* ▶Pflanzung 1); *syn.* Anpflanzen [n, o. Pl.], Anpflanzung [f] (2), Bepflanzen [n, o. Pl.], Bepflanzung [f], Pflanzen [n, o. Pl.], Pflanzvorgang [m].

planting [n] (2) *landsc. trans.* ▶anti-dazzle planting; *gard. plant.* ▶courtyard planting; *constr. hort.* ▶boundary planting, ▶climber planting; *gard. plant.* ▶courtyard planting; *agr. conserv.* ▶ditch corridor tree planting; *constr. landsc.* ▶dune planting; *for.* ▶enrichment planting; *constr.* ▶extensive roof planting; *gard. urb.* ▶façade planting; *constr.* ▶foundation planting, ▶furrow planting; *hort. plant.* ▶grave planting; *constr.* ▶grid planting of reed stems, ▶high maintenance planting, ▶hole planting [US]/pit planting [UK], ▶intensive planting 1, ▶intensive roof planting, ▶intermediate planting, ▶low maintenance planting, ▶mass planting; *constr. for. hort.* ▶mound planting; *landsc.* ▶noise attenuation planting; *gard. hort.* ▶ornamental planting; *landsc. urb.* ▶pre-construction planting; *agr. for. land' man. landsc.* ▶protective planting; *hort.* ▶pruning at planting; *constr.* ▶reed plug planting, ▶replacement planting, ▶ridge planting; *leg. trans. landsc.* ▶right-of-way planting [US]/roadside planting [UK]; *arch. constr. urb.* ▶roof planting; *agr. constr. for. hort.* ▶row planting; *landsc. urb.* ▶schoolground planting [US]/school-ground planting [UK]; *gard. landsc.* ▶screen planting (2); *plant.* ▶shade-tolerant planting; *constr. for.* ▶slit planting [US]/notch planting [UK]; *landsc.* ▶street planting; *gard. landsc.* ▶structure planting; *constr.* ▶thin planting; *constr. plant.* ▶three-level planting [US]; *constr. landsc.* ▶tree and shrub planting; *constr. hort.* ▶trench planting; *landsc.* ▶windbreak planting.

planting [n]**, angle** *constr. for.* ▶T-notching.

planting [n] [US]**, antiglare** *landsc. trans.* ▶anti-dazzle planting.

planting [n] [US]**, base** *constr.* ▶foundation planting.

planting [n]**, concealment** *gard. landsc.* ▶screen planting (2).

planting [n] [US]**, deep soil roof** *constr.* ▶intensive roof planting.

planting [n] [US]**, dispersed fruit tree** *agr. conserv.* ▶traditional orchard [UK].

planting [n]**, divisional** *gard. landsc.* ▶structure planting.

planting [n] [US]**, dry habitat roof** *constr.* ▶extensive roof planting.

planting [n] [UK]**, extensive** *constr.* ▶low maintenance planting.

planting [n]**, functional roof** *constr.* ▶extensive roof planting.

planting [n]**, gravesite** *hort. plant.* ▶grave planting.

planting [n] [US]**, infill** *constr.* ▶replacement planting.

planting [n] [UK]**, intensive** *constr.* ▶high maintenance planting.

planting [n]**, low maintenance roof** *constr.* ▶extensive roof planting.

planting [n]**, massed** *constr. hort.* ▶mass planting.

planting [n] [US]**, mustache** *constr.* ▶foundation planting.

planting [n] [UK]**, notch** *constr. for.* ▶slit planting [US].

planting [n]**, pavement** *constr.* ▶paving joint vegetation.

planting [n] [UK]**, pit** *constr. hort.* ▶hole planting [US].

planting [n]**, peripheral** *constr. hort.* ▶boundary planting.

planting [n]**, preparation for** *constr. hort.* ▶planting preparation.

planting [n]**, reed rhizome and shoot clumps** *constr. hort.* ▶planting of reed rhizome clumps.

planting [n] [US]**, repair** *for.* ▶replacement planting.

planting [n] [UK]**, roadside** *leg. trans. landsc.* ▶right-of-way planting [US].

planting [n] [US]**, rooftop** *arch. constr. urb.* ▶roof planting.

planting [n]**, screen** *agr. for. land'man. landsc.* ▶protective planting, ▶screen planting (2).

planting [n] [UK]**, shade** *plant.* ▶shade-tolerant planting.

planting [n] [US]**, shallow soil roof** *constr.* ▶extensive roof planting.

planting [n]**, side** *constr. for.* ▶ridge planting.

planting [n]**, step** *constr. for.* ▶ridge planting, #2.

planting [n] [UK]**, three-layered** *constr. plant.* ▶three-level planting [US].

planting [n]**, woody species** *constr. landsc.* ▶tree and shrub planting.

4356 planting area [n] *constr. hort.* (**1.** Area intended for planting. **2.** Area prepared for planting; ▶area of planting); *s* **superficie** [f] **a plantar** (**1.** Área prevista para ser plantada. **2.** Área preparada para plantar; ▶plantío; *syn.* superficie [f] de plantación; *f* **surface** [f] **de plantation** (**1.** Surface devant être plantée. **2.** Surface préparée en vue de la plantation ; ▶aire de plantation, ▶aire plantée) ; *syn.* aire [f] de plantation ; *g* **Pflanzfläche** [f] (**1.** Fläche, die bepflanzt werden soll. **2.** Für eine Bepflanzung vorbereitete Fläche; ▶Bepflanzungsfläche).

4357 planting [n] **at 8m on centers** [US]/**planting** [n] **at 8m centres** [UK] *constr. hort.* (▶planting distance); *s* **espaciamiento** [m] **a 8 m entre centros de plantío** (▶distanciamiento de plantación); *f* **planter à intervalles de** [loc] (p. ex. planter à intervalles de 8 m ; ▶distance de plantation) ; *g* **im Abstand von … pflanzen** [loc] (z. B. im Abstand von 8 m pflanzen; ▶Pflanzabstand).

planting basin [n] [US] *constr. hort.* ▶planting saucer [US]/watering hollow [UK].

4358 planting bed [n] *constr. hort.* (Place prepared or to be prepared for planting bulbs, herbaceous plants, groundcover plants or shrubs; ▶planted bed; *specific terms* ▶planter with a raised edge, ▶raised planting bed); *syn.* plant bed [n]; *s* **bancal** [m] **para plantación** (Bancal preparado para ser plantado con bulbos, plantas herbáceas, coberturas o arbustos; *términos específicos* ▶bancal macizo de plantas, ▶macizo de plantación elevado, ▶macizo de plantas con encintado elevado); *syn.* bancal [m] de plantación; *f* **massif** [m] **à planter** (Aire préparée ou à préparer en vue d'une plantation future ; ▶massif de plantation, ▶massif de plantes avec bordure surélevée, ▶plate-bande surelevée) ; *g* **Pflanzbeet** [n] (Zum Pflanzen vorzubereitendes oder vorbereitetes Beet; ▶Pflanzenbeet, ▶eingefasstes Pflanzenbeet, ▶Hochbeet).

planting bed [n]**, raised** *constr.* ▶raised planting bed.

planting box [n] [US] *gard. hort.* ▶planter.

planting composition [n]**, seasonally attractive** *phyt. plant.* ▶seasonally changing visual dominance.

planting corridor [n] *constr. hort.* ▶planting trench.

4359 planting depth [n] *constr. hort. s* **profundidad** [f] **de plantación**; *f* **profondeur** [f] **de plantation** (Dimensionnement en profondeur d'un trou ou d'une fosse en vue de la plantation des plantes bulbeuses et tubéreuses, des arbres et des arbustes ; la profondeur de plantation varie selon que les végétaux sont livrés en racines nues, en mottes ou en conteneur ; pour les végétaux à racines nues la profondeur doit être suffisante pour contenir les racines sans les comprimer, le collet au niveau du sol, pour les rosiers avec le point de greffe légèrement au-dessus du niveau du sol ; comme règle générale, les plantes bulbeuses et tubéreuses doivent être recouvertes d'une épaisseur de terre égale au double de la grosseur du bulbe ; pour les végétaux à petit développement, arbustes, rosiers, végétaux grimpants, végétaux de terre de bruyère profondeur comprise entre 40 et 70 cm ; pour les sujets à grand développement, arbres feuillus et conifères profondeur comprise entre 80 et 100 cm) ; *syn.* profondeur [f] du trou de plantation ; *g* **Pflanztiefe** [f] (Maß, bis zu dem eine

P

Mulde, ein Loch oder eine Grube ausgehoben werden muss, um eine Knolle, Zwiebel, krautige Pflanze, einen Strauch oder Baum artgerecht zu pflanzen).

4360 planting design [n] *landsc. plant.* (Selection of various plants on the basis of their ecological growing requirements to create a desired composition); *syn.* use [n] of plant material; **s utilización [f] de plantas** (Selección de diversas especies de plantas de acuerdo con sus condiciones de crecimiento y el efecto a alcanzar); *syn.* aplicación [f] de plantas; *f* **choix [m] végétal** (Sélection des végétaux en fonction des caractéristiques écologiques requises pour créer la composition désirée) ; *syn.* utilisation [f] des végétaux, emploi [m] des végétaux, choix [m] des végétaux (LA 1981, 653) ; *g* **Pflanzenverwendung [f]** (Planerische Auswahl von Pflanzenarten entsprechend ihrer Standortansprüche und für bestimmte Nutzungen und Funktionen unter Berücksichtigung der Gestaltungsziele und Verdeutlichung der Gesetze sinnlicher Wahrnehmung); *syn.* Verwendung [f] von Pflanzen.

4361 planting distance [n] *constr. hort.* (Distance between two plants in planting operations); *syn.* plant spacing [n] [also US]; **s distanciamiento [m] de plantación** (Espacio intermedio entre plantas o entre éstas y líneas de avenamiento; RA 1970, 63); *syn.* distancia [f] de plantación, espaciamiento [m]; *f 1* **espacement [m]** (Distance qui sépare les plantes cultivées sur une même ligne et espace laissé entre les lignes de cultures ; LA 1981, 479) ; *f 2* **distance [f] de plantation** (Droits et obligations d'un propriétaire relatifs à la plantation de végétaux en bordure de fonds privés, entre deux fonds privés ou sur une propriété privée en bordure du domaine public ; cf. C. civil, art. 671 et 672 ; VRD 1994, Tome III, 5.2.3., p. 6) ; *g* **Pflanzabstand [m]** (Entfernung zwischen zwei Pflanzen bei Pflanzarbeiten); *syn.* Pflanzweite [f].

planting for shade [n] [UK] *plant.* ►shade-tolerant planting.

4362 planting [n] **for traffic control** [US] *landsc.* (Roadside planting which serves for the visual guidance of traffic); *syn.* planting [n] for traffic guidance [UK]; **s plantación-guía [f] (del tráfico)** (Plantación a lo largo de carreteras que sirve de orientación al tráfico rodado); *f* **plantation [f] directionnelle** (Bande boisée plantée en bordure des routes élément d'orientation visuelle pour le trafic) ; *syn.* plantation [f] guide ; *g* **Leitpflanzung [f]** (Gehölzpflanzung an Straßen, die der optischen Führung des Verkehrs dient).

planting [n] **for traffic guidance** [UK] *landsc.* ►planting for traffic control [US].

4363 planting furrow [n] *constr. hort.* (**1.** In bioengineering, a trench 30cm wide and 30cm deep for planting small shrubs at an angle of 15° on a slope to control erosion. **2.** Shallow slit for lining out stock; e.g. in a nursery); **s surco [m] de plantación** (**1.** *En bioingenería* una zanja de 30 cm de ancho sobre el terreno en pendiente de unos 15° para plantar leñosas; ►plantación en surcos. **2.** Zanjas de poca profundidad donde se colocan las plántulas para ser criadas, p. ej. en vivero de plantas); *f* **rigole [f] de plantation** (Tranchée de largeur de 30 cm sur des terrains en pente pour la plantation des végétaux ligneux ; ►plantation en sillons) ; *syn.* tranchée [f] de plantation (1) ; *g* **Pflanzriefe [f]** (**1.** *Ingenieurbiologie* 30 cm breiter und tiefer ►Pflanzgraben am Hang, in dem Junggehölze im Winkel von ca. 15° gepflanzt werden. **2.** Flache Riefe zum Aufschulen von Jungpflanzen in der Baumschule; ►Riefenpflanzung).

4364 planting grid [n] *constr. hort.* (Rectangular or triangular pattern of plants in a plantation of several rows; *specific terms* ►triangular planting grid and ►square planting grid); *syn.* planting pattern [n]; **s esquema [m] de plantación (1)** (Ordenación regular de las plantas en plantaciones de varias hileras; término

genérico de ►esquema de plantación triangular y ►esquema de plantación rectangular); *f* **disposition [f] des végétaux** (Arrangement régulier de végétaux dans des plantations de plusieurs rangées ; ►disposition en quinconce, ►disposition rectangulaire) ; *g* **Pflanzenverband [m]** (Regelmäßige [rechteckige oder dreieckige] Pflanzenanordnung bei mehrreihigen Pflanzungen; OB zu ►Dreiecksverband und ►Vierecksverband; während des Pflanzens spricht man auch vom **Pflanzverband**).

planting grid [n]**, quincunx** *constr. hort.* ►triangular planting grid.

planting grid [n]**, tree** *constr. hort.* ►tree planting pattern.

4365 planting hole [n] *constr. hort.* (Excavated pit for a plant; ►bathtub effect, ►hole planting [US]/pit planting [UK]; ►tree pit; ►excavate a planting hole); *syn.* planting pit [n], plant pit [n] [also US] (HLT 1971); **s hoyo [m] de plantación** (Fosa excavada para plantar plantas; en el caso de árboles o arbustos debe tener una anchura y profundidad mínima de 1,3 veces [de ancho y alto] que las del cepellón o sistema radical; cf. PGT 1987; ►efecto de encharcado, ►foso de plantación, ►plantación en hoyos; ►apertura de hoyo [de plantación]); *syn.* hoyadura [f] de plantación; *f* **trou [m] de plantation** (Fosse effectuée en vue de la plantation des végétaux ; l'excavation réalisée pour la plantation d'arbres est dénommée ►fosse de plantation ; ►effet fosse de plantation, ►plantation sur potets ; ►creuser le trou de plantation) ; *g* **Pflanzloch [n]** (...löcher [pl]; zum Pflanzen einer Pflanze mit Spaten oder Maschine ausgehobenes Loch, das gem. DIN 18 916 mindestens in einer Breite auszuheben ist, die dem 1,5-fachen Durchmesser des Wurzelwerkes oder Ballens entspricht, damit sich das Wurzelwerk gut entwickeln kann; Verfestigungen an den Wänden oder der Sohle des **P.es** müssen zur Vermeidung des ►Blumentopfeffektes aufgelockert werden; ein großes **P.**, z. B. zur Pflanzung eines Baumes oder Großstrauches, wird auch **Pflanzgrube** genannt; ►Lochpflanzung, ►Pflanzgrube; ►Pflanzloch ausheben).

planting hole [n] [US]**, dig a** *constr. hort.* ►excavate a planting hole.

planting [n] **in built-up areas** *landsc. urb.* ►landscape planting in built-up areas.

4366 planting island [n] *constr. urb.* (Planted or grassed linear strip in a parking area between ►rows of parking; ►vegetated strip); **s banda [f] verde** (Franja plantada o sembrada de hierba entre las ►bandas de aparcamiento; ►banda verde central); *f* **îlot [m] de plantation** (Plantation délimitée par les aires de stationnement, par la voirie sur les giratoires, etc. ; ►bande de stationnement [longitudinal], ►bande verte d'accompagnement) ; *syn.* îlot [m] planté ; *g* **Pflanzenstreifen [m]** (Von ►Parkstreifen eingeschlossene, bepflanzte oder mit Rasen bestockte Fläche; ►Grünstreifen).

4367 planting mat [n] *constr.* (►Substrate mat made of foam interspersed with seeds for use in rooftop **gardening**); *syn.* planting slab [n] [also UK], preplanted mat [n], precultivated mat [n], pregrown mat [n]; **s estera [f] sintética sembrada** (►Estera de sustrato sintético con semillas incorporadas que se utiliza para plantar tejados); *f* **natte [f] synthétique préensemencée** (utilisée pour la végétalisation extensive de toitures-terrasses ; ►natte de substrat synthétique) ; *g* **Begrünungsmatte [f]** (Zur Dachbegrünung verwendete ►Substratmatte aus Schaumstoff mit vorgefertigter Einsaat); *syn.* Fertigbegrünungsmatte [f], vorkultivierte Substratmatte [f].

4368 planting material [n] *constr. hort. for.* (Species or cultivars as well as parts thereof to be planted, such as layers, tubers, cuttings, bulbs and corms; *generic term* ►plant material); *syn.* planting stock [n] [also US]; **s material [m] a plantar** (Especies, cultivares o partes de plantas [bulbos, esquejes, estolones, tubérculos] previstas para plantar; *término genérico* ▶matrial vegetal); *syn.* material [m] de plantación; *f* **végétaux [mpl] à planter** (Espèces ou cultivars à planter ou partie de végétaux propres à la production de nouveaux sujets tels que jeunes plants, marcotte, bouture ; *terme générique* ▶végétaux) ; *g* **Pflanzmaterial [n, o. Pl.]** (Zum Auspflanzen vorgesehene Arten oder Kulturformen sowie deren Teile [z. B. Absenker, Knollen, Stecklinge, Zwiebeln]; *OB* ▶Pflanzenmaterial); *syn.* Pflanzgut [n, o. Pl.], Pflanzware [f].

planting method [n] *constr.* ▶roof planting method.

4369 planting method [n] **for buildings** *constr.* (Landscape construction technique of establishment of vegetation on buildings; e.g. ▶roof planting, ▶façade planting); *s* **técnica [f] de revestimiento vegetal de edificios** (Término específico para la vegetalización de fachadas de edificios; ▶plantación de fachadas, ▶revestimiento vegetal de tejados); *f* **procédé [m] de végétalisation des bâtiments** (Technique paysagère de ▶végétalisation des toitures, ▶végétalisation des façades) ; *syn.* procédé [m] de verdissement des bâtiments ; *g* **Begrünungssystem [n] für Gebäude** (Bauverfahren für eine ▶Dachbegrünung oder ▶Fassadenbegrünung an Gebäuden).

4370 planting method [n] **for extreme sites** *constr.* (Landscape construction technique for the establishment of vegetative cover on hostile sites; e.g. on roofs, steep slopes, exposed façades, rock faces, xeric sites; ▶roof garden system, ▶bioengineering); *s* **técnica [f] de revestimiento vegetal para ubicaciones extremas** (Término genérico para técnicas de construcción paisajística con cuya ayuda se posibilita el crecimiento de vegetación en ubicaciones extremas como tejados, taludes, fachadas, etc.; ▶sistema de plantación de tejados, ▶bioingeniería); *f* **procédé [m] de végétalisation en milieux extrêmes** (Techniques particulières permettant le développement d'un couvert végétal sur des supports artificiels en milieux extrêmes [toitures, façades] ou sur des stations aux conditions extrêmes [talus, falaises, espaces dégradés en altitude] ; ▶procédé d'aménagement de jardins sur dalles, ▶génie biologique) ; *syn.* procédé [m] de végétalisation de sites en conditions extremes ; *g* **Begrünungssystem [n] für extreme Standorte** (Technische Konstruktionsmethode, mit deren Hilfe Pflanzenwuchs auf extremen Standorten, z. B. auf Dächern, steilen Böschungen, an Hausfassaden, Felswänden, möglich wird; ▶Dachgartensystem, ▶Ingenieurbiologie); *syn.* Begrünungsaufbau [m] für extreme Standorte.

planting mix(ture) [n] [US] *constr. hort.* ▶growing medium.

4371 planting [n] **of a streambank** *landsc.* (Seeding with grasses and herbs as well as planting of perennials or woody species; ▶planting of riparian vegetation); *s* **vegetalización [f] de riberas** (Siembra de gramíneas y herbáceas así como plantación de leñosas o vivaces en taludes de orillas; ▶plantación de riberas); *f* **végétalisation [f] des berges** (Ensemencement de graminées et de plantes herbacées, la plantation de végétaux ligneux ou de plantes vivaces sur un talus de berge ; ▶plantation des berges) ; *g* **Uferbegrünung [f]** (Aussaat von Gräsern und Kräutern sowie Bepflanzung von Gehölzen oder Stauden an einer Böschung eines Gewässerrandes; ▶Uferbepflanzung).

planting [n] **of conifers** *for.* ▶stocking with conifers.

4372 planting [n] **of hedgerows and woodland patches** *agr. land'man.* (Planting of trees, shrubs and hedges of indigenous species in an agricultural area to increase ecological diversity and improve the visual quality of the landscape); *s* **plantación [f] de setos y manchas boscosas en parajes rurales** (Plantación de especies arbóreas y arbustivas autóctonas o indígenas en setos o grupos con el fin de aumentar la diversidad ecológica y mejorar la calidad visual de zonas de cultivo);

f **plantation [f] des ligneux champêtres** (Espèces végétales ligneuses rustiques et adaptées aux conditions du milieu, plantées sous forme de haie, groupes d'arbres, peuplement de berges, etc. ; elles concourent l'accroissement de la diversité écologique et à la mise en valeur des paysages en milieu rural) ; *g* **Flurholzanbau [m, o. Pl.]** (Anbau von bodenständigen und standortgerechten Gehölzen [Flurgehölzen] in Form von Hecken, Baumgruppen, Uferpflanzungen etc. in der Agrarlandschaft zur Erhöhung der ökologischen Diversität und zur Belebung des Landschaftsbildes).

4373 planting [n] **of reed rhizome clumps** *constr.* (RIW 1984, 220; reed planting technique using emergent aquatic plants ['reeds']; e.g. for bank protection; ▶reed clump); *syn.* reed rhizome and shoot clumps planting [n] (RIW 1984, 220); *s* **plantación [f] de rizomas de carrizo** (Método de plantación en el que se utilizan ▶tepes de carrizo); *f* **plantation [f] de mottes de Roseaux** (Technique de plantation pour la protection des berges, utilisant des ▶touffes de Roseaux découpées à la bêche) ; *g* **Röhrichtballenpflanzung [f]** (Pflanzmethode, die ausgestochene Röhrichtballen verwendet, z. B. zur Befestigung von Gewässerrändern; ▶Schilfsode).

planting [n] **of reed stems** *constr.* ▶grid planting of reed stems.

4374 planting [n] **of riparian vegetation** *constr. landsc.* (Setting out of plants along a watercourse); *s* **plantación [f] de riberas** (Vegetalización de bordes de cursos de agua con leñosas o vivaces); *f* **plantation [f] des berges** (Végétalisation au moyen de végétaux ligneux ou de plantes vivaces) ; *g* **Uferbepflanzung [f]** (Begrünung von Ufern mit Gehölzen oder Stauden).

4375 planting operations [npl] *constr.* (Work process for planting; **in U.S.**, such national standards do not exist, but there are guidelines issued by some chapters of the ASLA, dealing with planting operations and guidelines issued by college and university departments of agriculture/horticulture; **in U.K.**, the standards for planting of individual trees, shrubs, hedges, climbers, herbaceous plants and bulbs are specified in BS 4428: 1969; ▶plant protection agent); *s* **trabajos [mpl] de plantación** (En construcción paisajística conjunto de actividades necesarias para garantizar buenos resultados a la hora de plantar; ▶plaguicida); *f* **plantations [fpl]** (Notion de planter des végétaux comprenant **1.** les travaux principaux comprenant le déchargement, la mise en jauge éventuelle, la répartition sur le chantier, l'ouverture du trou de plantation en terre végétale, préalablement préparé, la taille des branches, l'habillage des racines, le pralinage éventuel des racines, la mise en place du végétal, le comblement en terre végétale, le tassement, la façon de cuvette, le premier arrosage, **2.** les travaux annexes comprenant le tuteurage et le haubanage, **3.** l'entretien comprenant la taille, les façons sur sols divers, la fertilisation, le désherbage et le traitement [▶produit antiparasitaire (à usage agricole)] ; SDP 1998) ; *syn.* travaux [mpl] de plantation ; *g* **Pflanzarbeiten [fpl]** (Umgang mit Pflanzen und dem Pflanzvorgang bei Maßnahmen des Garten- und Landschaftsbaues; Art und Umfang der **P.** richten sich nach dem Zeitpunkt der Pflanzung, den Ansprüchen der einzelnen Arten sowie den Standortverhältnissen. Anforderungen an das Pflanzenmaterial bei Anlieferung, Anforderungen an Stoffe für Pflanzarbeiten [z. B. Pfähle, Mulchstoffe, ▶Pflanzenschutzmittel], Pflanzverfahren [Bodenvorbereitung, Pflanzzeit, Pflanzschnitt, Pflanzvorgang], Fertigstellung und Fertigstellungspflege sind in dem Regelwerk DIN 18 916 „Vegetationstechnik im Landschaftsbau, Pflanzen und Pflanzarbeiten" zur Qualitätssicherung der Ausführung und in den „Empfehlungen für Baumpflanzungen, Teil 2: Standortvorbereitungen für Neupflanzungen; Pflanzgruben und Wurzelraumerweiterung, Bauweisen und

Substrate", Hrsg.: Forschungsgesellschaft Landschaftsentwick-
lung, Landschaftsbau e. V. [FLL 2004], beschrieben; ergänzende
Vorschriften, speziell für Pflanzungen im Straßenbau, finden sich
in den „Zusätzlichen Technischen Vertragsbedingungen und
Richtlinien für Landschaftsbauarbeiten im Straßenbau" cf.
ZTVLa-StB 1999); *syn.* Pflanzleistungen [fpl].

planting pattern [n] *constr. hort.* ►planting grid, ►tree
planting pattern.

4376 planting period [n] *constr. for. hort.* (Season in which
a plant may best be planted or transplanted); *s* **periodo/período
[m] de plantación** (Época del año adecuada para plantar o transplantar una especie); *f 1* **période [f] de plantation** (La saison la
plus propice à la transplantation des végétaux) ; *f 2* **époque [f]
de plantation** (Date à laquelle est prévue la plantation par le
descriptif) ; *g* **Pflanzzeit [f]** (Zeitraum, der sich am besten zum
Verpflanzen einer Art eignet. Im Regelfall sind Laub abwerfende
Gehölze in der Wachstumsruhe zu pflanzen; Immergrüne mit
Ballen können gem. DIN 18 916 ganzjährig gepflanzt werden,
mit Ausnahme während der Zeit des Austriebes. Pflanzen in
Töpfen oder Containern können ebenfalls ganzjährig gepflanzt
werden. Gehölze ohne Ballen dürfen bei Frost nicht gepflanzt
werden. Die ZTVLa-StB schreibt vor, dass Frühjahrspflanzungen
bis zum 30. April abgeschlossen sein müssen; Stauden sollten je
nach Klimazone spätestens 4-5 Wochen vor Frostbeginn gepflanzt werden, damit die Wurzeln etwas einwachsen können).

planting pit [n] *constr. hort.* ►planting hole.

planting pit [n], **excavate a** *constr. hort.* ►excavate a
planting hole.

4377 planting plan [n] *plan. syn. s* **esquema [m] de plantación (2)** *syn.* plan [m] de plantación; *f* **plan [m] de plantation** ;
g **Bepflanzungsplan [m]** *syn.* Pflanzplan [m].

4378 planting preparation [n] *constr. hort.* (Making a site
ready for planting with such measures as soil improvement, pruning at planting time, root puddling, etc. for successful planting
and rooting); *syn.* preparation [n] for planting; *s* **preparación [f]
de la plantación** (Todo tipo de medidas, como la preparación del
suelo con enmiendas, la poda de plantación, el pralinaje de raíces,
etc. que sirven para garantizar la plantación de acuerdo a las
normas profesionales y asegurar que arraiguen y crezcan las plantas); *f* **travaux [mpl] de préparation des végétaux** (Travaux
principaux effectués avant la plantation favorisant la bonne
reprise des végétaux et comprenant : la mise en jauge, la préparation de la terre végétale, l'ouverture du trou de plantation, le
trempage des racines dans une solution fungicide, la taille des
branches, l'habillage des racines, occasionnellement le pralinage ; VRD 1994, 4/8.2.1.1, p. 2) ; *syn.* préparation [f] de la
plantation ; *g* **Pflanzvorbereitung [f]** (Alle Maßnahmen wie
z. B. Bodenlockerung, Bodenverbesserung, Pflanzschnitt, ggf.
Tauchen der Wurzeln in Lehmschlämme etc., die dazu dienen,
Pflanzen fachgerecht zu setzen und ihnen ein sicheres Anwachsen zu ermöglichen).

planting process [n] *constr. for. hort.* ►planting (1).

4379 planting prototype [n] [US] *constr. plant.* (Typical
planting layout for a section of a whole scheme: **1.** Planting of
woody species: a planting plan that indicates multi-rows of plants
to be repeated in blocks for afforestation or linear shelter belts.
2. Perennials and seasonal planting plan: showing species,
subspecies and varieties, as a list with percentages and/or a
graphic representation of the quantities to be distributed within a
specific area to be planted); *syn.* prototypical planting scheme [n]
[UK]; *s* **esquema [m] de plantación (3)** (**1.** Plan tipo de mezcla
de especies para repoblaciones o plantaciones protectoras que se
repiten por filas. **2.** *En plantación de perennes o flores de estación* plan de plantación que indica en una lista los porcentajes de

especies, subespecies o variedades, o representa en gráfico el
reparto de cantidades para una superficie a plantar); *f* **schéma
[m] de plantation** (Plan type définissant l'implantation répétitive
des végétaux sur une section réduite d'une plantation à plusieurs
rangées, telles qu'une plantation linéaire de protection, une
plantation d'accompagnement sur aménagements routiers, etc.) ;
g **Pflanzschema [n]** (...schemata [pl]); **1.** *Gehölzpflanzung* Pflanzplan als sich wiederholender Teilabschnitt einer mehrreihigen
Pflanzung für Aufforstungen oder lineare Schutzpflanzungen.
2. *Stauden- und Wechselflor[be]pflanzung* Bepflanzungsplan, der
Arten, Unterarten und Sorten als Liste mit prozentualen Anteilen
und/oder als grafische Darstellung deren Mengenverteilung für
eine zu bepflanzende Fläche festlegt).

planting requirements [npl] [US]**, directive on** *leg.
urb.* ►directive on planting requirements [US].

planting requirements [npl] [UK]**, enforcement notice on** *leg. urb.* ►directive on planting requirements [US].

4380 planting saucer [n] [US] *constr. hort.* (Saucer-like
depression in soil placed around the trunk of a newly planted tree,
for the purpose of water containment; RCG 1982); *syn.* watering
depression [n], planting basin [n] [US], watering hollow [n]
[UK]; *s* **alcorque [m] de riego** (Hueco circular en la superficie,
con centro en la planta, formando un caballón horizontal alrededor de unos 25 cm de altura, que permita el almacenamiento de
agua; PGT 1987); *f* **cuvette [f] d'arrosage** (Bourrelet de terre
disposé autour du collet des végétaux ligneux nouvellement
plantés pour recevoir les eaux d'arrosage) ; *g* **Gießmulde [f]**
(Bewässerungshilfe für neu gepflanzte Gehölze in Form eines
erhöhten Erdrandes, der die Baumscheibe umgibt; die ZTVLa-
StB fordern für alle Gehölze Gießmulden, die mindestens die
Größe des Pflanzloches haben).

planting scheme [n] [UK]**, prototypical** *constr. plant.*
►planting prototype [US].

planting screen [n] *gard. landsc.* ►screen planting (2).

4381 planting setback [n] [UK] *leg. urb.* (Distance between trees and tall shrubs and the boundary line of a neighboring lot [US]/neighbouring plot [UK] or intersection. This limit
may be governed by local civil laws [UK]; **in U.S.**, planting—
except in historic districts—is not regulated between private
property owners or adjoining neighbors. It is only required from
street intersections on both public and private property to
preserve ►sight triangles of drivers of motor vehicles waiting in
the intersections); *s* **retranqueo [m] de plantación** (Distancia
entre árboles y arbustos grandes y la línea divisoria del terreno
con el del vecino); *f 1* **distance [f] de plantation par rapport à
la limite séparative** (Éloignement d'un arbre, d'un arbuste,
d'une haie par rapport à la limite séparative des deux propriétés) ;
syn. recul [m] de plantations par rapport aux limites de parcelles ;
f 2 **recul [m] de plantation** (À défaut de règlements ou d'usages
particuliers [règlement de PLU, etc.] la distance fixée par les
articles 671 à 673 du Code Civil [les plantations dont la hauteur
dépasse 2 mètres doivent être au moins distant de 2 mètres de la
limite parcellaire, d'au moins 0,50 mètres pour les autres
plantations] ; cette règlementation constitue la **servitude de recul
de plantation**) ; *g* **Grenzabstand [m] von Gehölzen** (Abstand
von Bäumen und Sträuchern zu Nachbargrundstücken).

planting [n] **shoots of emergent plants** *constr.* ►reed
plug planting.

planting slab [n] [UK] *constr.* ►planting mat.

4382 planting soil [n] *constr. hort.* (**1.** Soil prepared in a
planting bed, improved for use by adding amendments, fertilizer,
etc. **2.** Soil which has been manufactured for bulk use or
packaged in sacks; ►growing medium); *s* **tierra [f] para**

plantaciones (**1.** Tierra de bancal de plantación mejorada con enmiendas, fertilizantes, etc. en la que se planta. **2.** Tierra que ha sido fabricada para usos a granel o está empaquetada en sacos; ►sustrato vegetal); *f* **terre [f] végétale améliorée** (Préparation d'un sol, d'un massif, d'une plate-bande **1.** par amendement et injonction de produits fertilisant, **2.** par apport de substrat de culture préparé en usine et épandu en vrac ou conditionné en sacs ; ►mélange terreux) ; *syn.* sol [m] amélioré, sol [m] amendé ; *g* **Pflanzerde [f]** (**1.** Durch Bodenverbesserungsstoffe, Dünger etc. vorbereitete Erde eines Pflanzbeetes, in die gepflanzt werden soll; **2.** wie vor, jedoch in Erdwerken als Schüttgut produziert oder in Säcken abgepackt; ►Erdgemisch).

planting stock [n] *constr. hort.* ►size of planting stock.

planting stock [n] [US] *constr. hort.* ►planting material.

4383 planting strip [n] *constr.* (Long, narrow piece of ground to be planted; ►grass strip 1, ►vegetated strip); *s* **banda [f] de plantación** (Superficie lineal que se desea plantar; ►banda verde central, ►franja de hierba); *f* **bande [f] plantée** (Espace linéaire planté ou à planter [enherbé et planté d'arbres] en bordure des aires de stationnement, en limite de parcelle le long des voies de circulation ou entre les voies de circulation et contribuant à la hiérarchisation de la voirie ; ►bande engazonnée, ►bande verte d'accompagnement) ; *syn.* bande [f] de plantation ; *g* **Pflanzstreifen [m]** (Zu bepflanzende lineare Fläche; ►Grünstreifen, ►Grasstreifen).

4384 planting structure [n] *constr. landsc.* (Plant composition e.g. of a shelter belt, ►protective planting, or bedding perennials); *s* **estructura [f] de plantación protectora** (Estructura vertical y composición de ►plantación protectora); *f* **structure [f] d'une plantation** (Aspect d'une plantation caractérisé par la palette des végétaux utilisés [hauteur, largeur, forme particulière] ainsi que leur répartiton sur la surface plantée ; ►plantation de protection) ; *g* **Aufbau [m, o. Pl.] einer Pflanzung (1)** (Höhenmäßige Gestaltung durch Auswahl der Pflanzenarten, z. B. für ein Staudenbeet, eine ►Schutzpflanzung).

planting system [n] *constr.* ►multistrata planting system, ►single-stratum planting system.

planting system [n]**, one-layer** *constr.* ►single-stratum planting system.

4385 planting technique [n] *constr.* (Landscape construction method of establishing plant growth on unvegetated areas; e.g. ►hydroseeding, ►landscape construction and maintenance work, ►plugging, ►turfing 2); *syn.* vegetating technique [n]; *s* **técnica [f] de plantación** (Método de construcción paisajística para implantar superficies de vegetación; ►establecimiento de césped con tepes, ►plantación de césped con tepes o con macollas, ►riego de semillas por emulsión, ►trabajos de paisajismo); *f* **techniques [fpl] de végétalisation (±)** (Techniques de création d'espaces de végétation [plantations, engazonnement] ; ►engazonnement par placage, ►engazonnement par bouturage, ►procédé d'enherbement par projection hydraulique, ►travaux d'espaces verts) ; *g* **Vegetationstechnik [f]** (Bauweise zur Herstellung von Vegetationsflächen im Landschaftsbau; cf. DIN 18 915; ►Anspritzverfahren, ►Herstellung von Rasenflächen durch Fertigrasen, ►Herstellung von Rasenflächen durch Auspflanzen von Rasen- oder Grasteilen, ►landschaftsgärtnerische Arbeiten).

4386 planting treatment [n] *landsc.* (Planting to ameliorate the visual impact of buildings, industries, etc.; ►boundary planting, ►integration of structures into the landscape, ►landscaping of built up areas); *syn.* screening [n] with plants, greening [n] [also UK], landscape treatment [n]; *s* **tratamiento [m] paisajístico (2)** (Plantación en los alrededores de una construcción para mejorar su aspecto visual y reducir la desfiguración del paisaje; ►integración al paisaje, ►plantación alrededor de edificio o solar, ►tratamiento paisajístico 1); *f* **traitement [m] paysager (2)** (Intégration d'immeubles, de zones industrielles, de décharges et bâtiments divers dans le paysage ; ►intégration paysagère [au site], ►plantation d'une clôture végétale, ►traitement paysager 1) ; *syn.* aménagement [m] d'écrans de verdure ; *g* **Eingrünung [f]** (Integrierung von Gebäuden, Industriegebieten, Deponien und sonstigen baulichen Anlagen in das städtische Umfeld oder in die Landschaft durch Bepflanzung; ►Durchgrünung, ►landschaftliche Einbindung, ►Umpflanzung).

4387 planting trench [n] *constr. hort.* (Excavated longitudinal rooting space, instead of individual tree pits, for placing of prepared soil and planting of hedgerows or tree avenues); *syn.* planting corridor [n] (MET 1985: 5, 40); *s* **zanja [f] de plantación** (Hondonada alargada para plantar setos o hileras de árboles); *f* **tranchée [f] de plantation (2)** (Fouille étroite et longue effectuée en vue de la plantation d'une haie ou d'une rangée d'arbres) ; *g* **Pflanzgraben [m]** (Schmale, lange ausgehobene Vertiefung, z. B. für Heckenpflanzungen oder Baumalleen [zur Verbindung einzelner Baumstandorte], um speziell für Bäume ein großes strukturstabiles Bodensubstratvolumen für eine ausreichende Wurzelentwicklung einzubauen — statt in kleine Einzelgruben); *syn.* Wurzelgraben [m].

4388 plant kingdom [n] *bot.* (Entire realm of plant organisms; ►floristic kingdom); *s* **reino [m] vegetal** (Jerarquía fitogeográfica superior: holártico, paleotropical, neotropical, del Cabo, australiano, antártico, oceánico; DINA 1987; ►reino floral); *f* **règne [m] végétal** (Ensemble des végétaux sauvages en milieu terrestre et marin ; ►empire floristique) ; *g* **Pflanzenreich [n]** (Gesamtheit der pflanzlichen Organismen; ►Florenreich); *syn.* Pflanzenwelt [f].

4389 plant knowledge [n] *plant.* (Understanding of and experience with plants); *s* **conocimiento [m] de las plantas** (Saber científico y práctico sobre las plantas); *f* **connaissance [f] des végétaux** (Savoir théorique et pratique sur les végétaux) ; *g* **Pflanzenkenntnis [f]** (Sach- und Erfahrungswissen über Pflanzen).

plantlet [n] *hort. for.* ►immature plant.

plantlet [n]**, germinated** *bot.* ►germinated young plant.

4390 plant list [n] *landsc. plant.* (Planned palette of species and cultivars for a planting scheme; ►list of undesirable plants); *syn.* species list [n], plant schedule [n] [also UK]; *s* **lista [f] de especies a plantar** (►lista de plantas no utilizables); *syn.* lista [f] de plantas; *f* **liste [f] des végétaux** (Détail des végétaux [arbres, arbustes, plantes vivaces] choisis pour la plantation dans le cadre d'un aménagement ou recommandés aux autorités locales et aux particuliers lors de plantations ou pour des aménagements particuliers ; ►liste des végétaux indésirables) ; *g* **Pflanzenliste [f]** (Schriftliche Zusammenstellung von Pflanzen für eine Bepflanzungsmaßnahme; ►Gehölznegativliste); *syn.* Pflanzenzusammenstellung [f].

plant list [n]**, undesirable** *landsc.* ►list of undesirable plants.

4391 plant maintenance program [n] [US] *constr. hort.* (Formulation of a specified sequence for working operations to encourage long-lasting growth conditions of healthy vegetation); *syn.* programme [n] of after care [UK]; *s* **programa [m] de mantenimiento de la vegetación (≠)** (Concepto y secuencia de trabajos para mantener y promover la salud de la vegetación de un jardín, zona verde, etc.); *f* **programme [m] de travaux d'entretien paysager** (Établissement d'un planning d'interventions d'entretien en vue de favoriser et sauvegarder à long terme la croissance d'une végétation saine) ; *g* **gärtnerisches Pflegeprogramm [n]** (Konzeption und festgelegte Abfolge von

P

Arbeitsgängen zur Förderung und langfristigen Erhaltung eines gesunden Vegetationsbestandes).

4392 plant material [n] *constr. hort. for.* (**1.** Species or cultivars, cultivated in nurseries. **2.** Species or cultivars, used in planting design by planners. **3.** Plant parts suitable for the propagation of new plants. Those plants chosen for planting are known collectively as ►planting material; ►brush matting with live plant material [US]/brush laying [UK], ►dead plant material, ►excavated muck and plant material, ►living plant material); *s* **material** [m] **vegetal** (**1.** Especies o cultivares criados en viveros. **2.** Especies o cultivares que se utilizan en el paisajismo. **3.** Partes de plantas que sirven para producir nuevas plantas; ►material a plantar, ►material vivo); *f 1* **végétaux** [mpl] (**1.** Espèces ou cultivars à planter cultivés en pépinière. **2.** Espèces ou cultivars utilisés par l'aménageur pour une plantation. **3.** Partie d'une plante propre à la production de nouveaux sujets ; ►végétaux à planter, ►matériau vivant) ; *f 2* **matériel** [m] **végétal** (Terme utilisé dans la sélection génétique, la production et multiplication des végétaux désignant une ou plusieurs espèces botaniques et, le cas échéant, un groupe de variétés très typé à l'intérieur d'une même espèce botanique ; cf. www.projetgiea.fr) ; *g* **Pflanzenmaterial** [n, o. Pl.] (**1.** Arten oder Kulturformen, die in Gärtnereien herangezogen werden. **2.** Arten oder Kulturformen, mit denen Planer gestalten. **3.** Für die Produktion neuer Pflanzen geeignete Pflanzenteile. Zur Pflanzung vorgesehene Pflanzen oder deren Teile nennt man ►Pflanzmaterial; ►lebender Baustoff).

4393 plant nursery [n] *hort.* (**1.** Place where vegetables, flowers, fruits, trees, shrubs, and perennials are grown for sale, **2.** Nursery firm/grower which cultivates plants; ►commercial horticulture, ►field nursery, ►forest tree nursery, ►market gardening, ►perennial nursery, ►tree nursery); *syn.* horticultural firm/company [n], grower [n]; *s* **empresa** [f] **de horticultura** (Centro de producción de plantas ornamentales, frutales, leñosas, vivaces, flores y semillas; una ►jardinería es una **e. de h.** pequeña; ►horticultura comercial, ►vivero [de árboles], ►vivero de vivaces, ►vivero forestal, ►vivero forestal volante); *syn.* jardinería [f]; *f* **entreprise** [f] **horticole** (Entreprise spécialisée dans la culture intensive et la production de plantes dans le domaine des cultures maraîchères, de la floriculture, de l'arboriculture fruitière et forestière, de la culture de la vigne, de la culture des graines et des plantes d'ornement ; ►horticulture, ►jardinerie, ►pépinière, ►pépinière forestière, ►pépinière spécialisée en plantes vivaces, ►pépinière volante) ; *syn.* établissement [m] horticole ; *g* **Gartenbaubetrieb** [m] (Wirtschaftlicher Betrieb zum intensiven Anbau und zur Erzeugung von pflanzlichen Produkten im Bereich des [Blumen- und] Zierpflanzenbaus, der ►Baumschulen, des Gemüsebaus, der ►Staudengärtnerei und des Obstbaus; ►Betrieb einer Endverkaufsgärtnerei, ►Erwerbsgartenbau, ►Forstbaumschule, ►temporäre Forstbaumschule); *syn.* Gärtnerei [f].

4394 plant nutrient [n] *bot. pedol.* (Chemical element or compound in a certain amount or quantity ratio to be absorbed by autotrophic plants for their growth; ►fertilizer); *s* **nutriente** [m] **(de plantas)** (Elemento o componente químico que necesitan las plantas autótrofas para su crecimiento y que es absorbido por ellas en cantidad y relación específica; ►fertilizante); *f* **élément** [m] **nutritif** (Élément chimique absorbé en une quantité définie par les végétaux autotrophes afin d'assurer leur développement ; ►engrais) ; *syn.* nutriment [m] ; *g* **Pflanzennährstoff** [m] (Chemisches Element oder Verbindung, das/die von autotrophen Pflanzen für deren Wachstum in bestimmter Menge oder in einem bestimmten Mengenverhältnis aufgenommen wird; ►Dünger).

4395 plant nutrition [n] *bot. hort.* (Generic term for the sum of processes by which a plant or soil takes in, absorbs and assimilates nutrients in any form or by any method; cf. BS 3975: part 5); *syn.* feeding [n]; *s* **nutrición** [f] **de (las) plantas** (Proceso de absorción y asimilación de nutrientes por las plantas); *f* **nutrition** [f] **des plantes** (Processus d'assimilation et d'utilisation [nutrition organique et minérale] des substances nutritives par les plantes après prélèvement dans le sol) ; *g* **Pflanzenernährung** [f] (Versorgung der Pflanzen mit organischen und mineralischen Stoffen entsprechend ihrem quantitativen und qualitativen Nahrungsbedarf. Durch die Arbeiten des Chemikers JUSTUS VON LIEBIG [1803-1873] wurde die moderne **P.** und Agrikulturchemie begründet).

plant [n] **on immature soil** [UK], **pioneer** *phyt.* ►pioneer plant on entisol [US].

plant operation [n] *envir.* ►industrial plant operation.

plant part [n] *bot.* ►aerial plant part, ►subterranean plant part.

plant part [n]**, underground** *bot.* ►subterranean plant part.

plant parts [npl] *bot.* ►developing of plant parts.

4396 plant pest [n] *agr. for. hort.* (Insect or other animal organism which harms plants, sometimes despite human intervention; *specific terms* forestry pest, garden pest, fruit-tree pest, vine pest; *generic term* ►injurious organism); *s* **depredador** [m] **(de plantas de cultivo)** (Animal que degrada o destruye alguna planta considerada útil por el hombre; DINA 1987; *término genérico* ►organismo nocivo); *f* **déprédateur** [m] **des cultures** (Animal qui commet des dégâts sur une plante utile ou sur une denrée, principalement dans le but de se nourrir ; ne confonde pas avec ►prédateur ; *terme générique* ►organisme nuisible) ; *syn.* ravageur [m] des cultures ; *g* **Pflanzenschädling** [m] (Tierischer Organismus, der entgegen menschlicher Bestrebungen Nutzpflanzen schädigt. *UBe* Forstschädling, Gartenschädling, Obstbaumschädling, Rebschädling; *OB* ►Schadorganismus); *syn.* Pflanzenparasit [m].

plant pest and disease control [n] *agr. for. hort.* *phytopath.* ►pest and weed control.

plant pit [n] [US] *constr. hort.* ►planting hole.

4397 plant preservation requirement [n] [US] *conserv. leg. urb.* (**In Europe**, legal requirement to preserve certain existing vegetation, such as trees, shrubs, hedges or special vegetation on land to be developed; cf. TCPA 1971 part IV, section 59-60; **in U.S.**, such a requirement is typically expressed by regulations or property deed; ►directive on planting requirements [US]/enforcement notice on planting requirements [UK]); *syn.* obligation [n] to preserve existing plants [UK]; *s* **prescripción** [f] **de preservar la vegetación** (De acuerdo con la legislación urbanística y de gestión del paisaje, prohibición de destruir o alterar elementos naturales del paisaje protegidos o por proteger tal como árboles, arbustos, setos, eriales, lagunas, etc.; ►prescripción de plantar); *f* **obligation** [f] **de préservation de la végétation existante** (Prescription d'urbanisme à respecter stipulant par classement la conservation des espaces boisés et des plantations existantes dans le cadre des documents d'urbanisme ou des plans d'urbanisme ; ►obligation de réaliser des plantations) ; *g* **Pflanzbindung** [f] (Nach Maßgabe gesetzlicher Bestimmungen, die die Bauleitplanung oder Landschaftsplanung regeln, Verbot der Zerstörung oder Veränderung geschützter oder zu schützender Landschaftsbestandteile — z. B. Bepflanzungen, Bäume, Sträucher, Hecken, Brachflächen; cf. § 9 [1] 25b BauGB; ►Pflanzgebot); *syn.* Bindung [f] für Bepflanzungen.

4398 plant protection agent [n] *chem. envir. agr. for. hort.* (Chemical or biological product intended to prevent, destroy, repel or mitigate plant pests or pathogenes of diseases, or to kill weeds; ►biocide, ►fungicide, ►herbicide, ►insecticide, ►molluscicide, ►pesticide, ►plant care product); *s* **plaguicida [f]** (Compuestos químicos utilizados en el control y la destrucción de las plagas y enfermedades de las plantas; según el tipo de plaga que se combata se diferencian en acaricidas, algicidas, bactericidas, fasciolicidas, ►fungicidas, ►herbicidas, ►insecticidas, ►molusquicidas, nematicidas y rodenticidas; *término genérico* ►producto fitosanitario; los términos ►pesticida y ►biocida se utilizan en castellano como sinónimos); *f* **produit [m] antiparasitaire (à usage agricole)** (Substance chimique, mélangée ou non à d'autres substances, utilisée contre les ennemis des cultures ou préparation utilisée pour la lutte contre les ennemis des cultures et des produits récoltés : acaricides, bactéricides, corvicides, ►fongicides, ►herbicides, ►insecticides, ►mollusquicides, nématicides, rodenticides [contre les rongeurs] ; ►biocide ; *terme générique* ►produit de traitement phytosanitaire ; ►pesticide) ; *syn.* produit [m] de protection des plantes, produit [m] agropharmaceutique ; *g* **Pflanzenschutzmittel [n]** (Chemisches oder biologisches Mittel, das dazu bestimmt ist, Pflanzenschädlinge oder Erreger von Krankheiten bei Pflanzen oder Pflanzenerzeugnissen zu vernichten oder fernzuhalten, Konkurrenzwirkungen von Unkraut zu verhindern sowie Lebensvorgänge in Pflanzen zu beeinflussen, ohne ihrer Ernährung zu dienen; legaldefinitorisch werden auch Stoffe dazugerechnet, die das Wachstum von Pflanzen hemmen oder verhindern; cf. § 2 PflSchG. Je nach dem Zweck der Bekämpfung unterscheidet man: Akarizid, Bakterizid, ►Fungizid, ►Herbizid, ►Insektizid, ►Molluskizid, Nematizid, Rodentizid; ►Biozid; *UB* Schädlingsbekämpfungsmittel; *OB* ►Pflanzenbehandlungsmittel; ►Pestizid).

plant receptacle [n] *gard. hort.* ►plant container.

4399 plant requiring protection [n] *phyt.* (Plant which grows under the shelter of larger plants more resistant to inclement weather); *s* **planta [f] necesitada de abrigo para crecer (≠)** (Planta que sólo crece si se encuentra protegida de las inclemencias del tiempo por otras especies más resistentes); *f* **plante [f] qui demande un abri** (Plante protégée des intempéries qui nécessite l'abri de végétaux plus résistants pour assurer son développement, ou qui ne peuvent se développer qu'à l'abri des plantes pionnières colonisatrices) ; *syn.* plante [f] qui requiert un abri, plante [f] qui nécessite un abri ; *g* **Pflanzenschützling [m]** (Pflanze, die im Schutze größerer und gegen die Unbilden des Klimas resistentere Pflanzen aufwächst).

plant schedule [n] [UK] *landsc. plant.* ►plant list.

plant shipment [n] *constr. hort.* ►shipping of plants.

4400 plant size [n] *hort.* (Measurement expressed in height, spread, trunk circonference, trunk caliper, shipping weight or number of canes, etc.; ►size of planting stock); *s* **tamaño [m] de planta** (Dimensiones utilizadas para categorizar plantas ante su venta, que pueden incluir la altura, anchura, circunferencia del tronco, diámetro del tronco, número de brotes, etc.; ►tamaño del material de plantación); *f* **taille [f] des végétaux** (Caractéristiques dimensionnelles des végétaux de pépinières classés selon un échelonnement [classe de hauteur ou de diamètre] qui est fonction **1.** de la longueur de leurs branches exprimées en cm pour les jeunes plants, **2.** du diamètre moyen en cm ou du nombre de branches et de leur longueur pour les jeunes touffes ; cf. NF V 12-037 ; ►taille des végétaux) ; *syn.* hauteur [f] en mètres (pour plantations d'alignement) ; *g* **Pflanzengröße [f]** (Verkaufs- oder Endgröße, angegeben in Höhe, Breite, Stammumfang, Stammdurchmesser oder Anzahl der Triebe etc.; ►Pflanzgröße).

4401 plantsman [n] *bot. hort.* (Individual who has amassed considerable knowledge about plant materials through identification, growth characteristics, experimentation, habitat needs,

pest control, etc.; *specific terms* ►dendrologist, wildflower expert); *syn.* plant expert [n]; *s* **experto/a [m/f] en plantas** (Persona que tiene grandes conocimientos sobre las plantas; *término específico* ►dendrólogo); *f* **spécialiste [m] des plantes** (Expert ayant acquis une très bonne connaissance du monde végétal, la détermination des espèces, leur forme de croissance, leurs exigences stationnelles, la lutte antiparasitaire, etc. ; ►dendrologue) ; *syn.* botaniste [m] expérimenté, botaniste [m] averti ; *g* **Pflanzenfachmann [m]** (...fachleute [pl]; Experte, der über überdurchschnittliche, umfassende Pflanzenkenntnisse hinsichtlich der Bestimmung von Arten, Wuchsformen, über Standortansprüche, Schädlingsbekämpfung etc. verfügt und entsprechende Erfahrung hat; *UBe* ►Dendrologe, Wildpflanzenkenner); *syn.* Pflanzenkenner/-in [m/f], Pflanzenfachfrau [f], Pflanzenspezialist/-in [m/f].

4402 plant sociability [n] *phyt. plant.* (Spatial relationship of individual plants of a species, with regard to how they form groups. The following scale is used in describing the clustering: 1 = growing singly, 2 = forming small but dense clumps; grouped or tufted, 3 = in troops, small patches, or cushions, 4 = in small colonies, in extensive patches, or forming carpets, 5 = in great colonies or large groups [pure stand]); *syn.* gregariousness [n]; *s* **sociabilidad [f]** (Forma de agrupación de los individuos de una especie: 1 = individuos aislados [o vástagos o troncos aislados], 2 = creciendo en pequeños grupos, 3 = creciendo en grupos mayores [pequeños rodales o almohadillas], 4 = creciendo en pequeñas colonias o en rodales o tapices extensos, 5 = población continua; BB 1979, 40); *f* **sociabilité [f]** (Concerne la manière dont sont disposés les uns par rapport aux autres, les individus [ou les pousses] d'une même espèce, à l'intérieur d'une population donnée. On peut distinguer cinq dispositions principales : 1 = espèces croissant isolément ; dispersées, isolées, 2 = espèces croissant en groupes/en petites touffes, 3 = espèces croissant en troupes/en touffes moyennes, espacées, 4 = espèces croissant en petites colonies ; espèces croissant en larges touffes discontinues, 5 = espèces croissant en peuplements serrés et continus) ; *g* **Häufungsweise [f]** (Vegetationskundliches Strukturmerkmal in Pflanzengesellschaften als Art des Individuenzusammenschlusses resp. der Gruppierung oberirdischer Sprosse einer Art. Die Häufung der einzelnen Arten wird nach BRAUN-BLANQUET mit einer fünfgliedrigen Gesellligkeitsskala beschrieben: 1 = einzeln wachsend [Einzelsprosse, Einzelstämme], 2 = gruppen- oder horstweise wachsend, 3 = truppweise wachsend [kleine Flecken oder Polster], 4 = in kleinen Kolonien wachsend oder ausgedehnte Flecken oder Teppiche bildend, 5 = große Herden bildend); *syn.* Soziabilität [f], Gesellligkeit [f].

plant sociology [n] *phyt.* ►phytosociology.

plant spacing [n] [US] *constr. hort.* ►planting distance.

4403 plant species [n] *bot. phyt.* (For specific terms refer to species; ►indigenous plant species, ►useful plant species); *s* **especie [f] vegetal** (Para términos específicos véase especie; ►especie vegetal autóctona); *f* **espèce [f] végétale** (Pour les termes spécifiques se réfère au terme espèce ; ►espèce végétale indigène) ; *g* **Pflanzenart [f]** (UBe siehe bei Art; ►bodenständige Pflanzenart).

plant species [n], **autochthonous** *phyt.* ►indigenous plant species.

plant species [n], **economic** *agr. for. hort.* ►useful plant species.

plant species [n], **native** *phyt.* ►indigenous plant species.

plant species stock [n] *phyt.* ►existing area vegetation.

plant terrace [n] *constr.* ►narrow plant terrace.

plant tissue [n] *arb.* ►underprovided plant tissue.

P

4404 plant tolerance [n] *agr. for. hort.* (Characteristics of plants which are not damaged or stunted in growth by application of, e.g. insecticides, herbicides, or of plants which accept unfavo[u]rable light conditions or exposure); *s* **no dañino/a [adj] para las plantas** (Característica de plantas de no reducir el crecimiento ante condiciones poco favorable como p. ej. falta de luz, aplicación de insecticidas o herbicidas, etc.); *f* **tolérance [f] des végétaux** (Capacité des végétaux de supporter dans certaines limites sans dommages et réduction de croissance l'action d'un facteur écologique, de produits antiparasitaires etc. ; cf. DEE 1982) ; *g* **Pflanzenverträglichkeit [f]** (Beschaffenheit von Pflanzen, die z. B. auf ungünstige Lichtverhältnisse oder auf Anwendungen von Insektiziden, Herbiziden, Streusalz etc. ohne Schäden oder mit keinen oder nur geringen Wachstumseinbußen resp. Ertragsausfällen reagieren; *UB* Leistungstoleranz); *syn.* Toleranz [f] von Pflanzen.

4405 plant [vb] **trees or shrubs in nursery rows** *hort.* (To line out seedlings of woody species in a nursery; lining out [n]; ►transplanting 1, ►transplant in nursery row, ►widely-spaced plant); *syn.* line-out [vb] [also UK]; *s* **plantar [vb] leñosas en hileras** (En vivero, plantación de plántulas de leñosas en alineación para su ulterior cultivo; ►planta de gran espaciamiento, ►trasplante 1, ►trasplantar en hilera de vivero); *f* **planter [vb] des végétaux ligneux sur des lignes de cultures** (Transplanter les jeunes plants dans une pépinière ; ►contre-planter, ►plante à grand écartement, ►repiquer, ►transplantation 3) ; *g* **aufschulen [vb]** (**1.** In einer Baumschule Jungpflanzen in Reihen zur weiteren Kultivierung auspflanzen. **2.** Auspflanzen von Pflanzen, die auf der Baustelle angeliefert werden und bis Ende der Pflanzzeit nicht gepflanzt werden können, in einem art- und größenentsprechenden Abstand; cf. Ziff. 4.3.4 DIN 18 916; ►Pflanze aus extra weitem Stand, ►Verpflanzung 1, ►verschulen).

plant trough [n] [UK] *gard. hort.* ►planter.

4406 plan view [n] *plan.* (Two-dimensional graphic layout on a horizontal plane; *specific terms* ►floor plan, ►site plan, ►top view); *s* **planta [f]** (Dibujo o plano que muestra una sección horizontal de un edificio o proyecto paisajístico, en el que se señala la cimentación, disposición de espacios, muros, tabiques, puertas y ventanas; cf. DACO 1988; ►planta de piso, ►plano de situación, ►vista de planta); *syn.* plano [m] de planta, plano [m] horizontal, dibujo [m] en planta; *f 1°* **dessin [m] en plan** (En technique de dessin, projection en plan d'un objet ; ►dessin en plan d'un étage/niveau, ►plan de situation [du terrain], ►vue en plan) ; *f 2°* **plan [m] de masse** (**1.** *Permis de construire* plan qui présente l'emplacement d'un projet de construction par rapport à son voisinage immédiat, en indiquant les limites et l'orientation du terrain, l'implantation et la hauteur de la construction, le tracé des voies de desserte et des raccordements ; **2.** les services du cadastre délivrent un plan de masse sur lequel figurent les limites des parcelles et l'implantation des constructions et de la voirie) ; *g* **Grundriss [m]** (Im technischen Zeichnen die Projektion eines Gegenstandes auf eine Ebene; ►Draufsicht, ►Geschossgrundriss, ►Lageplan).

4407 plastic-coated [adj] *constr.* *s* **plastificado [adj]**; *f* **plastifié, ée [adj]** ; *g* **kunststoffbeschichtet [pp/adj]** *syn.* kunststoffummantelt [adj].

4408 plastic-coated chain link metal mesh [n] *constr.* *s* **malla [f] metálica recubierta de plástico**; *f* **grillage [m] plastifié** ; *g* **kunststoffummantelter Maschendraht [m]**.

4409 plastic limit [n] **(of soil)** *constr. eng. pedol.* (Limit of consistency for colloidal soils [US]/cohesive soils [UK] [> 10% clay content] between semi-compaction and plasticity; ►liquid limit of soil); *s* **límite [m] de plasticidad (del suelo)** (Límite de

consistencia de suelos arcillosos [> 10% de arcilla] entre estado semicompacto y plástico; ►límite de liquidez [del suelo]); *f* **limite [f] de plasticité** (DIS 1986, 25 ; limite de la consistance d'un sol plastique [> 10 % d'argile] définie par la quantité d'eau contenue dans celui-ci quand on peut le rouler à la main pour en faire de petits boudins ; DIS 1986, 25 ; ►limite de liquidité) ; *g* **Ausrollgrenze [f]** (Konsistenzgrenze bei bindigen Böden [> 10 % Tonanteil] zwischen halbfestem und plastischem Zustand; ►Fließgrenze).

plastic membrane [n] *constr.* ►plastic sheet.

4410 plastic sheet [n] *constr.* (Broad, homogenous and flexible piece of plastic material, with a thickness of 0.5 to 2mm); ►waterproofing membrane, ►waterproofing sheet); *syn.* plastic membrane [n]; *s* **membrana [f] de plástico** (Hoja ancha flexible de plástico de un grosor de 0,5-2 mm; ►banda de material impermeabilizante, ►membrana de impermeabilización); *f* **film [m] plastique** (Membrane monocouche plastique en PVC non régénéré, en élastomères vulcanises, inattaquable par les racines, ou feuille bicouche en PVC plastifiée armée ou bitume élastomère armé, d'épaisseur comprise entre 0,5 et 2 mm, pour les étanchéités de toitures-terrasses, bassins, réservoirs, etc. ; ►lé d'étanchéité, ►membrane d'étanchéité) ; *syn.* feuille [f] plastique, liner [m], film [m] polyane, polyane [m] ; *g* **Kunststofffolie [f]** (Flächiges, in sich homogenes und flexibles Gebilde aus z. B. Polyäthylen [PE], Polyvinylchlorid [PVC], Polypropylen etc. mit Dicken von 0,5-2 mm; ►Dichtungsbahn, ►Dichtungsfolie).

plastic soil [n] [US] *pedol.* ►colloidal soil.

4411 plateau [n] **(1)** *geo.* (**1.** Extensive area of land, more than 200m above sea level, without any great differences in height; used in contrast to ►lowland); **2. high plains [npl]** (Specifically the *Great Plains* in USA, 1,000 to 2,000m in altitude, west of the prairies); *syn.* high-veld [n] [ZA]; *s 1* **meseta [f] (2)** (Superficie llana o ligeramente inclinada en una determinada dirección cortada por valles y situada a una cierta altitud con respecto al nivel del mar; DGA 1986; *opp.* ►tierras bajas 2); *s 2* **altiplano [m]** (Extensa ►meseta 2 situada a gran altitud, como el **a.** en los Andes entre Bolivia y el Perú que supera los 3000 m.s.n.m.); *syn.* altiplanicie [f]; *f 1* **haute plaine [f]** (Vaste étendue à faible différence d'altitude située à plus de 200 m au-dessus du niveau de la mer ; *opp.* ►basse plaine) ; *f 2* **plateau [m]** (Vaste étendue de terrain haute entourée de vallées profondes ou de vallées à versants raides ; DG 1984 ; un plateau calcaire est une « causse ») ; *g* **Hochebene [f]** (Ausgedehnte Landoberfläche mit sehr geringen Höhenunterschieden, höher als 200 m über dem Meeresspiegel; cf. MEL 1973, Bd. 7, 373; *opp.* ►Tiefebene); *syn.* Hochlandebene [f].

plateau [n] **(2)** *landsc.* ►spoil plateau [US].

plateau [n] [UK], **tip** *landsc.* ►spoil plateau [US].

4412 plate load-bearing test [n] *constr.* (Method used during earthworks to assess the elasticity, plasticity, and weight-bearing capacity of a soil); *s* **ensayo [m] de carga con placa** (Método utilizado en terracerías para determinar el comportamiento elástico y plástico de los suelos); *syn.* prueba [f] de carga con placa; *f 1* **essai [m] à la plaque** (Méthode de définition de la densité d'un sol ou d'un matériau compte tenu de sa composition et de sa teneur en eau [mesure de la portance d'un sol sous revêtement rigide], effectuée par chargement à l'aide d'une plaque de 60 à 80 cm de côté ; VRD 1986, 56) ; *syn.* essai [m] de plaque, essai [m] de Westergaard ; *f 2* **essai [m] à la dynaplaque** (Essai effectué par chute d'une masse sur une plaque d'appui reposant sur le sol par l'intermédiaire de ressorts) ; *g* **Plattendruckversuch [m]** (Im Erdbau angewandte Methode zur Bestimmung des elastischen und plastischen Verhaltens von Böden, um deren Druckfestigkeit und Tragfähigkeit zu bestimmen; die Durchführung ist in der DIN 18 134 *Baugrund; Versuche und Versuchsgeräte — Plattendruckversuch* geregelt); *syn.* Lastplattenversuch [m], Lastplattendruckversuch [m].

platformed [pp/adj] *hort.* ▶balled and platformed [US].

play [n] *recr.* ▶active play.

playable space [n] *recr.* ▶playground.

4413 play activity [n] *recr.* (▶recreation activity, ▶leisure pursuit); *s* **actividad [f] lúdica** (▶actividad de recreo, ▶actividad de tiempo libre); *f* **activité [f] ludique** (▶activité de loisirs, ▶activité de récréation) ; *g* **spielerische Aktivität [f]** (▶Erholungsaktivität, ▶Freizeitaktivität); *syn.* Spielaktivität [f].

play apparatus [n] [US] *recr.* ▶play equipment.

play apparatus [n] [US]**, playground with** *recr.* ▶playground with play equipment.

play area [n] *recr.* ▶forest play area, ▶free play area, ▶playground.

play area [n] [US]**, unstructured** *recr.* ▶free play area.

play area lawn [n] [UK] *constr.* ▶playfield lawn [US].

play areas [npl] [US] *plan. recr. urb.* ▶overall master planning for play areas [US]/play spaces strategic planning [UK].

play areas [npl] [UK]**, strategic planning of** *plan. recr. urb.* ▶overall master planning for play areas [US].

4414 play equipment [n] *recr.* (Facility required for a particular game, usually permanently installed, e.g. swings, climbing frame, slide, seesaw or mobile devices such as stilts, balls, Frisbees, sand buckets. The latter are collectively known as **toys**; ▶playground with play equipment); *syn.* play apparatus [n] [also US]; *s* **aparato [m] para juegos** (En área de juegos infantiles, término genérico para elementos de equipamiento de madera, plástico, metal u hormigón que permiten a niños, niñas y adolescentes realizar los movimientos elementales como saltar, deslizarse, balancearse, trepar, suspenderse, etc.; ▶área de juegos equipada); *f 1* **mobilier [m] de jeux** (Terme général éléments en bois, matières plastiques, métalliques, béton reconstituant les mouvements élémentaires des enfants et adolescents tels que courir, sauter, glisser, se balancer, grimper, se suspendre, se cacher, etc. ; ▶aire de jeux équipée) ; *f 2* **équipements [mpl] d'aires collectives de jeux** (Terme juridique, réglementation technique ; décret n° 94-699 du 10 août 1994) ; *syn.* équipements [mpl] de jeux à usage collectif pour enfants [N.F. S 54 201], équipements [mpl] d'aires de jeux [N.F. S 54 202] ; *g* **Spielgerät [n]** (Für ein bestimmtes Spiel erforderlicher Ausrüstungsgegenstand, z. B. fest installiert wie Schaukel, Klettergerüst, Rutsche, Wippe oder bewegliche Geräte wie Stelzen, Ball, Frisbeescheibe, Sandförmchen. Kleine bewegliche Geräte werden unter dem Sammelbegriff **Spielzeug** zusammengefasst; eine Fläche, die mit **S.en** ausgestattet ist, wird ▶Gerätespielplatz genannt).

4415 playfield lawn [n] [US] *constr.* (▶Lawn type consisting of a mixture of grasses used in landscape construction for playgrounds, sun-bathing areas, private gardens, etc., which tolerate heavy use during the whole year [▶trampling] and require an average to high level of maintenance); *syn.* play area lawn [n] [UK]; *s* **césped [m] de juegos** (▶Tipo de césped utilizado en la construcción paisajística para áreas de juegos y recreo, campas para reposo, jardines privados, etc. por su capacidad de carga durante todo el año [▶pisoteo] y por su exigencia de mantenimiento entre media y alta); *f* **gazon [m] pour terrains de jeux** (▶Type de gazon très rustique utilisé pour les espaces verts collectifs, les terrains d'entraînement, les jardins privatifs faisant l'objet d'une fréquentation importante ; ▶piétinement) ; *syn.* gazon [m] pour terrains d'entraînement et plaines de jeux ; *g* **Spielrasen [m]** (Strapazierfähiger Rasen für Spielplätze,

Liegewiesen, Hausgärten etc., der eine ganzjährige hohe Beanspruchung [▶Trittbelastung] verträgt und je nach Nutzungsintensität einen mittleren bis hohen Pflegeaufwand benötigt; nach DIN 18 917 dem ▶Rasentyp Gebrauchsrasen zugehörig).

4416 playfield turf [n] [US] *constr.* (Heavy-use lawn for sports activities, which typically requires special sub-grade and base preparation for loading and drainage functions); *syn.* sports ground turf [n] [UK]; *s* **césped [m] de campos de deporte** (Cubierta de césped muy resistente al pisoteo); *f* **gazon [m] pour terrain de sport** (Couverture végétale constituée d'un mélange spécial de graminées, adapté aux différentes disciplines sportives et présentant un très bon comportement au piétinement) ; *syn.* gazon [m] sport, pelouse [f] de sport ; *g* **Sportrasen [m]** (**1.** Rasendecke einer Sportfläche im Freien, die so strapazierfähig ist, dass sie regelmäßig bespielbar ist; cf. DIN 18 035 Teil 4. **2.** Im Garten- und Landschaftsbau gemäß DIN 18 917 verwendeter ▶Rasentyp, der als Grassamenmischung für Neuanlagen [Regelsaatgutmischung RSM 3.1] oder für Regenerationsarbeiten als RSM 3.2 im Handel erhältlich ist); *syn. zu 1.* Sportplatzrasen [m].

4417 playground [n] *recr.* (**1.** Area of land used for and usually having special facilities and equipment for children's play. **2.** Generic term for all types of play areas: ▶adventure playground, ▶children's playground, ▶forest play area, ▶playground with play equipment, ▶playground with water feature, ▶practice game area [US]/kickabout area [UK], ▶supervised children's playground, ▶teenage playground [US]/teenage hangout [UK], ▶tot lot [US]/toddlers play area [UK]. **3.** Area not specifically equipped, but suitable for spontaneous or informal play activities is called a **playable space**); *syn.* play space [n] (TGG 1984, 197), play area [n]; *s* **área [f] de juegos y recreo** (**1.** Terreno reservado para juegos infantiles y equipado generalmente con instalaciones y aparatos de juego. **2.** Término genérico para ▶área de juegos con agua, ▶área de juegos educativos, ▶área de juegos equipada, ▶área de juegos infantiles, ▶área de juegos para niños y niñas pequeñas, ▶parque infantil de juegos de aventura, ▶pista de juegos de pelota, ▶terreno forestal de juegos infantiles, ▶terreno para juegos juveniles); *f 1* **terrain [m] de jeux** (Terme générique désignant toutes les catégories d'aires de jeux telles que : ▶aire de jeux d'eau, ▶aire de jeux en forêt, ▶aire de jeux équipée, ▶aire de jeux [pour enfants], ▶aire de jeux pour petits enfants, ▶espace multisports, ▶terrain d'aventure, ▶terrain de jeux éducatifs, ▶terrain de jeux pour adolescents) ; *syn.* espace [m] de jeux, aire [f] de jeux ; *f 2* **aire [f] collective de jeux** (Terme juridique ; désigne toute zone, y compris dans un parc aquatique ou parc d'attraction, spécialement aménagée et équipée pour être utilisée, de façon collective, par des enfants à des fins de jeux ; décret n° 96-1136 du 18 décembre 1996) ; *g* **Spielplatz [m]** (**1.** Gelände, das zum Spielen für Kinder dient und i. d. R. mit verschiedenen Spielgeräten und je nach Hauptnutzergruppe mit sonstigen Einrichtungen ausgestattet ist. **2.** OB zu Spielplatzart wie ▶Abenteuerspielplatz, ▶betreuter Kinderspielplatz, ▶Bolzplatz, ▶Gerätespielplatz, ▶Kinderspielplatz, ▶Spielplatz für Jugendliche, ▶Spielplatz für Mutter und Kind, ▶Waldspielplatz, ▶Wasserspielplatz); *syn.* Spielfläche [f].

4418 playground [n] **for children, 5 to 8 years old** *urb. recr.* *s* **área [f] de juegos para niños y niñas de 5 a 8 años**; *f* **plaine [f] de jeux libres pour des enfants de 5-8 ans** ; *g* **Kinderspielplatz [m] für 5-8-Jährige**.

4419 playground [n] **with basic play equipment** *recr.* (especially in rural areas); *s* **jardín [m] infantil con equipamiento básico** (Área de juegos infantiles ubicada generalmente en zonas rurales y caracterizada por su equipamiento sencillo); *f* **plaine [f] de jeux sommaire** (Aire de jeux en général

P

en milieu rural caractérisée par un équipement léger) ; *g* **Spielplatz [m] mit einfacher Ausstattung** (meist im ländlichen Raum).

playground [n] **with play apparatus** [US] *recr.* ▶playground with play equipment.

4420 playground [n] **with play equipment** *recr. syn.* playground [n] with play apparatus [also US]; *s* **área [f] de juegos equipada** (Zona de juegos infantiles equipada con aparatos como toboganes, columpios, estructuras metálicas para trepar, etc.); *f* **aire [f] de jeux équipée** (Aire collective de jeux pourvue de toboggan, tourniquets, d'équipements de balancement, etc.) ; *g* **Gerätespielplatz [m]** (Mit Rutschen, Schaukeln, Klettergerüsten und sonstigen Spielgeräten ausgestatteter Kinderspielplatz); *syn.* Spielplatz [m] mit Geräten.

4421 playground [n] **with water feature** *recr.* (Children's play area with hand water pump[s], occasionally equipped with devices for raising water to a higher level or for scooping); *s* **área [f] de juegos con agua** (Parque infantil en el que el agua es el principal medio de juego y hay instalaciones de bombeo, etc. para moverla); *f* **aire [f] de jeux d'eau** (Aire de superficie restreinte et de faible profondeur permettant aux enfants de patauger ou de s'asperger ; *terme spécifique* pataugeoire) ; *syn.* plan [m] de jeux d'eau ; *g* **Wasserspielplatz [m]** (Spielplatz für Kinder mit Wasserpumpe[n], Wasserstauanlage und manchmal auch mit Matschanlage und Wasserschöpfgeräten ausgestattet).

4422 playing field [n] *recr.* (Area or facilities suitable for recreational pastimes or sport; e.g. soccer field [US]/football pitch [UK], baseball field, football field [US]; ▶small playing field, ▶sports area); *s* **campo [m] de juegos** (Terreno adecuado para juegos y actividades deportivas informales; ▶campo de deportes pequeño, ▶suelo para usos deportivos); *f* **terrain [m] de jeu** (Espace et installation appropriés pour les activités sportives et de loisirs ; ▶sols sportifs, ▶terrain de jeux réduit) ; *g* **Spielfeld [n]** (Für spielerisch-sportliche Freizeitbetätigung geeignete Fläche und Anlage; ▶Kleinspielfeld, ▶Sportflächen).

playing field [n], **turf** *constr. recr.* ▶grass sports field.

playing surface [n] [UK], **hoggin** *constr.* ▶granular playing surface [US].

playing surface [n] [US], **water-bound** *constr.* ▶granular playing surface [US]/hoggin playing surface [UK].

4423 play lawn [n] [US] (1) *constr.* (▶Lawn type for general public or private recreational use); *syn.* all-around lawn [n] [UK], heavy-use lawn [n], amenity lawn [n] [UK]; *s* **césped [m] normal** (▶Tipo de césped utilizado en construcción paisajística para zonas verdes públicas y privadas de mucha frecuentación); *f* **gazon [m] d'agrément et de détente** (▶Type de gazon de composition choisie pour une utilisation dans les espaces verts publics et privés ; **D.**, mélange conforme à la norme DIN 18 917) ; *syn.* gazon [m] rustique ; *g* **Gebrauchsrasen [m]** (▶Rasentyp des Garten- und Landschaftsbaues gem. DIN 18 917 mit einer Grassamenmischung, die für nutzbare öffentliche oder private Grünflächen und für Spielrasen geeignet ist. Durch entsprechende Regelsaatgutmischungen [RSM 2.1 bis RSM 2.4] können für unterschiedliche Nutzungsansprüche und Nutzungsintensität die entsprechenden Samenmischungen im Handel bezogen werden. Für den Erhalt der Gebrauchseigenschaften ist eine mittlere bis hohe Pflege notwendig).

4424 play lawn [n] [US] (2) *recr.* (Area with closely-clipped sward used for games; ▶turf area); *syn.* turfed pitch [n] [UK] (SEE 1977, 159); *s* **campo [m] de césped** (Superficie sembrada con gramíneas y utilizada para juegos; ▶superficie de césped); *f* **aire [f] de jeu engazonnée** (Aire aménagée pour des jeux sur le gazon ; ▶pelouse) ; *g* **Rasenplatz [m]** (Für Spiele oder sportliche Tätigkeiten angelegte grasbewachsene Fläche; ▶Rasenfläche).

play lot [n] [US] *recr.* ▶tot lot [US]/toddlers play area [UK].

playscape [n]**, sand** *recr.* ▶sand playground.

4425 play section [n] *recr.* (Portion of a children's playground for specific play functions); *s* **sección [f] de tipo de juego (≠)** (Parte de un área de juegos y recreo con una función específica); *f* **secteur [m] de jeux** (Espace réservé à un fonction ludique dans le périmètre d'une aire de jeux) ; *syn.* secteur [m] d'activités ludiques ; *g* **Spielbereich [m]** (Teilfläche für einzelne Spielfunktionen innerhalb eines Spielplatzes oder einer Spielfläche; Ziff. 3.6 DIN 18 034); *syn.* Spielteilbereich [m].

play space [n] *recr.* ▶playground.

play spaces strategic planning [n] [UK] *plan. recr. urb.* ▶overall master planning for play areas [US].

playstreet [n] *recr. urb.* ▶home zone [UK].

plaza [n] *constr.* ▶paving of a square/plaza; *urb.* ▶urban square.

pleasance [n] *gard'hist.* ▶pleasure ground, #2.

pleasaunce [n] *gard'hist.* ▶pleasure ground, #2.

4426 pleasure driving [n] *recr.* (Car driving for enjoyment; ▶auto tourist [US]/pleasure motorist [UK]); *s* **excursión [f] en automóvil** (Actividad de viajar en automóvil con fines recreativos o turísticos; ▶excursionista motorizado/a); *f 1* **randonnée [f] en voiture** (Découverte en voiture d'un pays, de son patrimoine culturel et environnemental) ; *f 2* **randonnée [f] motorisée** (Activité motorisée [4x4, quad, moto], très contestée par les naturalistes, qui est pratiquée individuellement, sur un circuit aménagé ou en pleine nature, dans le cadre de randonnées accompagnées ou organisées ; activité régie par la loi n° 91-2 du 3 janvier 1991 relative à la circulation des véhicules); *g* **Autowandern [n, o. Pl.]** (Mit dem Auto spazieren fahren; ▶Autowanderer).

4427 pleasure ground [n] *gard'hist.* (**1.** Old-fashioned term used for an ornamentally designed garden with lawns, flower beds, paths, and sitting areas, used predominantly for recreation and enjoyment. Such a garden was already known in antiquity. After the 16th century this type of garden, as distinguished from the fruit and vegetable garden, became wide spread in Europe and its size was in accordance with its importance. **In U.K.**, the term was often used for gardens on promenades laid out for the enjoyment of holidaymakers at coastal resorts); *syn.* pleasuring place [n], hortus [n] deliciarum; **2. pleasaunce** [n] (**a)** Small medieval garden ["garden of love"] enclosed within the ramparts of a castle, primarily used for entertaining ladies; DIL 1987. **b)** Pavilion within an estate garden; DIL 1987]); *syn.* pleasance [n]; *s* **jardín [m] de las delicias** (En la edad media, jardín especialmente diseñado para el recreo de damas); *f* **jardin [m] de plaisance** (**1.** Jardin d'agrément ou partie de jardin aménagé pour le plaisir de la promenade, du repos contemplatif ou de la fête. **2.** Au moyen-âge, petit jardin, enclos par un mur sur lequel est adossé une banquette engazonnée, planté d'arbres fruitiers et de massif de fleurs, souvent agrémenté d'un bassin ou d'une fontaine, espace de conversation courtoise avec les dames ; la littérature du bas moyen-âge le présentera comme le théâtre des scènes d'amour (Roman de la Rose) ; cf. GDG 1962]) ; *syn.* jardin [m] des plaisirs, jardin [m] des amours ; *g 1* **Lustgarten [m]** (Mit Rasen, Blumenbeeten und Wegen angelegter Garten oder Gartenteil, der durch seine Gestaltung ausschließlich der Zerstreuung und Erholung dient. Reine L.gärten kannte man bereits in der Antike. Der **L.** wird vom gartenbaulich bewirtschafteten Nutzgarten unterschieden. Seit dem 16. Jh. ist der **L.** weit verbreitet und wird i. d. R. je nach Größe und Bedeutung vom Nutzgarten räumlich getrennt); *g 2* **Plaisance [f]** (Mittelalterlicher Lustgarten, der der ritterlich-höfischen Gesellschaft als Sinnbild des irdischen Paradieses galt, ein Garten, der auf der einen Seite mit einer Wiesenfläche, die auf der Süd- und West-seite mit Obst-, Schmuck- und Schattengehölzen sowie mit Rasenbänken vor den Mauern und oft mit einer Brunnenanlage ausgestattet ist, und auf der anderen Seite Kräuter- und Blumen-beete aufweist, oft unter den Fenstern der Frauenwohnungen einer Burganlage angelegt. Schon die Römer unterschieden den *hortus*, den einfachen Garten aller Schichten, der vorwiegend ein **L.** war, vom *viridarium*, dem Baum- und **L.** der Wohlhabenden; cf. GOT 1926-I, 191-196; GDG 1962, 42-56); *syn.* Hortus [m] deliciarum.

pleasure lawn [n] *constr.* ►clipped lawn.

pleasure motorist [n] [UK] *recr.* ►auto tourist [US].

pleasure park [n] *recr.* ►amusement park.

pleasuring place [n] *gard'hist.* ►pleasure ground.

plinth [n] **of a wall** *constr.* ►wall base course.

4428 plot [n] **(1)** *surv. obs.* (Parcel of land with a single land use in Europe, such as grassland, arable field, fallow land, deciduous wood, mixed forest, etc.; **in U.S.**, any defined and usually small piece of ground); **s parcela** [f] (Trozo de terreno con un único uso como cultivo, bosque, prado, erial, etc.; ►parcela propiedad); **f parcelle** [f] **(d'exploitation)** (En agri-culture ou en sylviculture terme désignant une unité présentant un seul mode d'utilisation du sol tel que p. ex. un pâturage, une friche, une forêt de feuillus, une forêt mixte et présentant un aspect homogène ; ►parcelle [de propriété]) ; **g Parzelle** [f] (Frühere Bezeichnung für ►Flurstück — fachsprachlich heute obsolet; eine **P.** bestand aus einer oder mehreren Nutzungsarten, z. B. Grünland, Brachland, Laubwald, Mischwald; heute noch ugs. verwendet).

plot [n] [UK] **(2)** *surv.* ►real property parcel [US].

plot [n]**, burial** *adm.* ►burial site.

plot [n] [UK]**, empty** *urb.* ►vacant lot.

plot [n] [UK]**, neighbouring** *leg. urb.* ►neighboring lot [US].

4429 plot area [n] *urb. syn.* lot area [n]; **s superficie** [f] **de solar** *syn.* superficie [f] de terreno, superficie [f] de lote, super-ficie [f] de parcela; **f superficie** [f] **d'un fonds** ; **g Grund-stücksfläche** [f].

plot boundary [n] [UK] *leg. surv.* ►lot line [US].

plot depth [n] [UK] *surv.* ►lot depth [US].

plot ratio [n] [UK] *leg. urb.* ►building coverage [US].

4430 plot shape [n] *surv. syn.* configuration [n] of a plot/lot; **s forma** [f] **de solar** *syn.* forma [f] de lote de terreno, forma [f] de lote de predio, configuración [f] de solar, configuración [f] de lote de terreno, configuración [f] de lote de predio; **f 1 configu-ration** [f] **parcellaire** (Terme générique, terme qualifiant la superficie, la forme, la situation, l'accessibilité d'une parcelle) ; **f 2 découpe** [f] **parcellaire** (Disposition des unités parcellaires traduisant leur forme vue en plan ; selon l'orientation des struc-tures ou du tissu parcellaires on parle de découpage « en lanière », découpage rectangulaire ou orthogonal) ; *syn.* décou-page [m] parcellaire ; **g Grundstückszuschnitt** [m] (Die äußere, senkrecht projizierte Form eines Grundstückes).

plot survey [n] *surv.* ►boundary survey.

4431 plot width [n] *leg. surv. urb.* (►lot depth); *syn.* lot width [n]; **s anchura** [f] **de solar** (Dimensión de una parcela de terreno medida paralelamente a la calle o al espacio público; ►profundidad de solar); **f largeur** [f] **d'une parcelle** (Largeur d'un terrain mesurée parallèlement à l'alignement de la voirie de l'espace public ; ►profondeur d'une parcelle) ; **g Grundstücks-breite** [f] (Im Planungs- und Baurecht die straßenseitige Abmessung eines Grundstückes; ►Grundstückstiefe).

ploughed land [n] **with fruit trees** [UK] *agr. conserv.* ►plowed land with fruit trees [US].

ploughing [n] [UK] *agr. for.* ►plowing [US].

ploughing-up [n] **of grassland** [UK] *agr.* ►plowing up of grassland [US]/ploughing-up of grassland [UK].

plough pan [n] [UK] *agr. pedol.* ►plow pan [US].

4432 plowed land [n] **with fruit trees** [US]/**ploughed land** [n] **with fruit trees** [UK] *agr. conserv.* (**In Europe**, arable land planted with fruit trees and other crops between the rows; ►traditional orchard [UK]); **s campo** [m] **cultivado con frutales intercalados** (En Europa Central y Occidental, tierras de cultivo plantadas con frutales entre los cuales se cultiva la tierra; ►plantación dispersa de árboles frutales); **f champ** [m] **com-planté d'arbres fruitiers** (►Peuplement dispersé d'arbres fruitiers sur une surface agricole) ; **g Baumacker** [m] (Flächen-hafte Anpflanzung von Hochstammobstbäumen auf ackerbaulich oder gärtnerisch genutzten Flächen; ►Streuobstbestand); *syn.* Streuobstacker [m].

4433 plowing [n] [US]/**ploughing** [UK] *agr. for.* (Turning up the earth, especially before sowing, changing of the crop, after tree clearing, etc.; ►deep plowing [US]/deep ploughing [UK], ►plowing up of grassland [US]/ploughing-up of grassland [UK]); **s roturado** [m] **de la tierra** (Arado de la tierra, especial-mente antes de sembrar, de la rotación de cultivos o después de tala a matarrasa; roturar [vb] la tierra; ►arado en profundo, ►roturación de prados); **f retournement** [m] **du sol** (Pratique agricole de labour, nécessaire à l'enfouissement des résidus de la récolte précédente, d'un engrais vert, du fumier ou des engrais minéraux, etc. ou pratique sylvicole après une coupe à blanc ; *terme spécifique*, ►retournement de prairie ; ►labour profond) ; *syn.* défoncement [m] du sol ; **g Bodenumbruch** [m] (Um-pflügen des Bodens vor Ansaat, bei Fruchtwechsel, nach einem Kahlschlag etc.; ►Grünlandumbruch, ►Tiefpflügen); *syn.* Pflügen [n, o. Pl.], Umpflügen [n, o. Pl.].

4434 plowing up [n] **of grassland** [US]/**ploughing-up** [n] **of grassland** [UK] *agr.* (**1.** Conversion of grassland into arable field. **2.** Digging up of existing grassland and resowing with more productive fodder grasses); *syn.* turning-over [n] of grassland; **s roturación** [f] **de prados** (**1.** Conversión de prados en campos de cultivo. **2.** Arado de praderas existentes para re-sembrar con gramíneas forrajeras más productivas); **f retourne-ment** [m] **de prairie** (**1.** Transformation d'une prairie en champ. **2.** Réensemencement pour régénération ou reconstitution) ; **g Grünlandumbruch** [m] (**1.** Umwandlung von Grünland in Ackerland. **2.** Umpflügen einer vorhandenen Wiese und Neu-einsaat von ertragsreicheren Futtergräsern).

4435 plow pan [n] [US] *agr. pedol.* (Compacted soil horizon at the depth of tillage, caused by ►plowing [US]/ploughing [UK] in wet soil; *generic term* pressure or induced pan); *syn.* plough pan [n] [UK], plowsole [n] [US], tillage pan [n] [US] (RCG 1982); **s piso** [m] **de arado** (Horizonte edáfico compactado formado al arar la tierra en época de mucha humedad; ►roturado de la tierra); **f semelle** [f] **de labour** (Zone de structure compacte dans un sol agricole défavorable à la pénétration des racines et constitué par le fond du sillon de labour tassé et lissé sous l'influence du passage répété du tracteur et de la charrue toujours à la même profondeur ; LA 1981, 1027 ; ►retournement du sol) ; **g Pflugsohle** [f] (Durch Pflügen/►Bodenumbruch bei zu nasser Witterung entstandener verdichteter Horizont, meist in der Unter-bodenschicht).

plowsole [n] [US] *agr. pedol.* ►plow pan [US]/plough pan [UK].

4436 plugging [n] *constr.* (Vegetative establishment of turf-grasses by planting small plugs [approximately 5 cm] containing the top growth, roots, rhizomes or stolons of the turfgrass; method often employed in semiarid [US]/semi-arid [UK] and arid regions; cf. HLT 1971; ▶grass plugging, ▶stolonizing); *s* **plantación [f] de césped con tepes o con macollas** (Método de establecimiento de superficies herbáceas empleado en regiones áridas o semiáridas; ▶establecimiento de césped con estolones, ▶establecimiento de césped por tallos rizomatosos); *syn.* plantación [f] de césped con panes o con motas [C]; *f* **engazonnement [m] par bouturage** (Méthode de multiplication végétative par éclatage de touffes, par plantation de fragments de rhizomes des graminées en particulier dans les régions arides et semi-arides ; ▶engazonnement par jet de stolons de graminées, ▶plantation d'un tapis de graminées) ; *g* **Herstellung [f] von Rasenflächen durch Auspflanzen von Rasen- oder Grasteilen** (Bedecken einer Bodenfläche z. B. mit geteilten Rasensoden, Rhizomen oder deren Teile, geteilten Grashorsten — besonders in semiariden Gebieten; ▶Begrünung mit Grasstolonen; ▶Pflanzen einer Grasnarbe).

4437 poaching [n] *leg. hunt.* (Catching game or fish illegally by pursuing, trapping or misappropriation or possessing, damaging or destroying of an object which is subject to hunting or fishing laws; *specific terms* ▶fish poaching and ▶game poaching); *s* **caza [f] y pesca [f] furtivas** (Infracción de la legislación de caza y pesca en cuanto al hecho de cazar o pescar en lugar o tiempo no autorizado o en el método utilizado para ello; *términos específicos* ▶caza furtiva, ▶pesca furtiva); *f* **braconnage [m]** Infractions à la législation de la chasse et de la pêche telles que l'approche, la poursuite, la capture et l'enlèvement d'espèces de gibier ou de poissons ainsi que la modification et la destruction illicite de biens régis par la législation de la chasse et de la pêche ; *termes spécifiques* ▶braconnage de gibier, ▶braconnage de poissons) ; *g* **Wilderei [f]** (Verletzung fremden Jagd- oder Fischereirechtes durch Nachstellen, Fangen oder Aneignen von ▶Wild oder Fischen oder die Aneignung, Beschädigung oder Zerstörung einer Sache, die dem Jagdrecht resp. Fischereirecht unterliegt; § 293 StGB; *UBe* ▶Fischwilderei und ▶Jagdwilderei).

4438 podzol [n] *pedol.* (*FAO-System* soils with illuvial accumulations of organic matter and compounds of aluminum and usually iron. These soils are formed in acid, mainly coarse-textured materials in cool-temperate to temperate, humid climates under coniferous or mixed coniferous and deciduous forest and characterized particularly by a highly-leached, whitish gray A2 horizon, where iron oxide, alumina and organic matter have been removed from the A horizon and deposited in the B horizon; cf. RCG 1982; ▶podzolization); *syn.* podzolic soil [n], spodosol [n] [also US]; *s* **podsol [m]** (Tipo de suelos terrestres formados sobre rocas pobres en Ca y Mg, como areniscas o granito, muy evolucionados con perfil O_l-O_f-A_h-A_e-B_h-B_s-C, en los que la evolución está condicionada por la presencia de un ▶humus bruto [mor] muy ácido y de descomposición lenta. El horizonte A_e de máxima eluviación de las sustancias coloidales es de color entre blancuzco y ceniza, debido a la abundancia de granos de cuarzo y limo residuales, mientras que los horizontes B de iluviación contienen acumulaciones amorfas de materia orgánica y concreciones de hierro, manganeso y a veces de aluminio. Es propio de zonas frías y húmedas como el bosque boreal de coníferas; ▶podsolización); *f* **podzol [m]** (Sol très évolué, à profil ABC, dont l'évolution est conditionnée par la présence d'un ▶humus brut [▶mor] très acide â décomposition lente. Le mor est très épais [10 à 20 cm] et très acide. L'horizon E est décoloré et cendreux. L'horizon B contient, à l'état amorphe, de la matière organique, de l'aluminium et du fer. Ce profil se rencontre sous climat humide et sous végétation produisant un humus acide ; ▶podzolisation) ; *g* **Podsol [m]** (...e [pl]; terrestrischer Boden mit O_l-O_f-A_h-A_e-B_h-B_s-C-Profil in kühlgemäßigten humiden Klimabereichen auf Ca- und Mg-armen, durchlässigen Gesteinen, z. B. Sandstein, Granit; mit einer Vegetation bewachsen, die geringe Nährstoffansprüche hat; ▶Podsolierung); *syn. obs.* Bleicherde [f], Bleichsand [m].

4439 podzolic [adj] *pedol.* (Descriptive term applied to a soil which has characteristics of a ▶podzol); *s* **podsólico/a [adj]** (Término descriptivo de un suelo con características similares a los ▶podsoles); *f* **podzolisé, ée [pp/adj]** (Caractéristique d'un sol possédant les propriétés d'un ▶podzol, c.-à-d. d'un sol dont l'évolution est surtout conditionnée par le climat et affecte des matériaux d'origine variée ; PED 1984, 162) ; *g* **podsolig [adj]** (Einen Boden betreffend, der Eigenschaften eines ▶Podsols hat).

podzolic soil [n] *pedol.* ▶podzol.

4440 podzolization [n] *pedol.* (Process of soil formation caused by downwards shifting of Al and Fe ions together with organic materials in the soil, which results in the genesis of ▶podzols and podzolic soils, and is mostly associated with the occurrence of certain plant communities and a strongly acidified soil; ▶illuvial horizon); *s* **podsolización [f]** (Proceso físico-químico de formación de suelos ▶podsol o podsólicos en los que en condiciones de clima húmedo, vegetación acidificante, formación de humus ácido, generalmente de tipo mor, a veces moder y escasa actividad biológica, los compuestos orgánicos son ácidos y complejantes. La acidez y la actividad complejante provocan la meteorización de los minerales, con liberación de sus componentes. La complejación posibilita la translocación de hierro, aluminio y humus [queluviación] de la parte superior del suelo y su acumulación en horizontes inferiores del mismo; cf. EDAFO 1994; ▶horizonte iluvial); *f* **podzolisation [f]** (Processus physico-chimique évolutif sur des roches-mères filtrantes relativement pauvres en minéraux altérables, caractéristique des régions à climat humide ou très humide, tropical et boréal. Il est caractérisé par la formation et la subsistance de composés organiques qui sont mobilisés et s'immobilisent à une certaine profondeur, formant ainsi un horizon d'accumulation. Al et Fe, qui sont insolubles dans les conditions du milieu et peuvent être complexés et entraînés et s'accumuler avec les composés organiques ; DIS 1986 ; ▶horizon illuvial, ▶podzol) ; *g* **Podsolierung [f]** (Prozess der Bodenbildung, die durch abwärts gerichtete Umlagerungen von Al- und Fe-Ionen zusammen mit organischen Stoffen einen ▶Podsol oder podsolige Böden entstehen lässt und meist mit dem Auftreten bestimmter Pflanzengesellschaften und starker Bodenversauerung verbunden ist; ▶Anreicherungshorizont).

point [n] *plan. recr.* ▶attraction point; *surv.* ▶datum point; *envir.* ▶emission point; *plan. recr.* ▶focal point; *plan. pol.* ▶growth point; *plan.* ▶north point.

point [n], **central growth** *plan. pol.* ▶growth point.

point [n] [US], **outlet** *envir.* ▶outfall (of sewage).

point [n], **security anchorage** *constr. leg.* ▶protection from falling.

point [n] [UK], **stopping** *recr.* ▶stopping place [US].

point [n], **vantage** *recr.* ▶viewpoint [US].

pointed [pp/adj] *constr.* ∗▶tooling of stone [US]/dressing of stone [UK], #4.

4441 pointing [n] *constr.* (Finishing of mortar joints between courses of bricks or stones in masonry or pavement; ▶grouting); *s* **rejuntado [m]** (Operación de acabado de las juntas en mampostería o en pavimentos; ▶rejuntado con mortero líquido); *syn.* relleno [m] de juntas; *f* **jointoiement [m]** (Action de combler et finir les joints d'un parement [maçonnerie], sol, etc. ; jointoyer [vb] ; ►jointoiement avec une barbotine) ; *g* **Verfugen [n, o. Pl.]** (Verfüllen von Fugen im Mauerwerk, Plattenbelag etc.; verfugen [vb]; ►Verfugen mit flüssigem Mörtel); *syn.* Verfugung [f].

point of discharge [n] *envir.* ►outfall (of sewage).

4442 point source pollution [n] *envir.* (Concentrated pollution of waterbodies from drains and gullies, due to industrial effluents, sewage disposal, etc.; *opp.* ►nonpoint source pollution [US]/non-point source pollution); *s* **contaminación [f] focal** (Polución de cursos y masas de agua que tiene su origen en vertidos de instalaciones fijas, generalmente industriales; *opp.* ►contaminación ubicua); *syn.* contaminación [f] puntual; *f* **pollution [f] ponctuelle** (Dégradation de la qualité des eaux provoquée par le rejet d'effluents industriels à partir d'un émissaire ; *opp.* ►pollution superficielle ; PPC 1989, 112) ; *g* **punktuelle Verschmutzung [f]** (Verunreinigung von Gewässern durch punktuelle Einleitungen von ortsfesten Industrieanlagen, Abwassereinleitern, sonstigen festen Einrichtungen, Anlagen oder Ausrüstungen; *opp.* ►flächenhafte Verschmutzung); *syn.* Verschmutzung [f] durch ortsfeste Quellen.

4443 polder [n] *land'man.* (Tract of land near, at or below sea level reclaimed from the sea by dyking and draining; empolder [vb]; ►impounding of a watercourse, ►land reclamation 1, ►floodable land); *s* **polder [m] (1.** Tierra de marisma ganada al mar mediante construcción de diques; ►embalse de un curso fluvial, ►ganancia de tierras 1; **2.** polder inundable: ►terreno inundable); *f* **polder [m]** (Terre gagnée sur la mer par endiguement et située au niveau ou au-dessous du niveau de celle-ci ; par extension, terrain très humide protégé des inondations marines ou fluviales par des digues ; ►endiguement d'un bassin fluvial de retenue, ►polder d'inondation, ►poldérisation) ; *g* **Polder [m]** (**1.** Dem Meer abgewonnenes und durch Deiche geschütztes Marschland; einpoldern [vb]; ►Eindeichung eines Gewässeraufstau[e]s, ►Landgewinnung. **2.** *Überschwemmungspolder* ►Überschwemmungsbereich); *syn. zu 1.* Koog [m].

4444 pole [n] *for.* (Young tree of generally 20-25cm d. b. h. in a dense forest stand; ►pole-stage forest); *s* **fustal [m] (2)** (Árbol joven en una masa arbórea generalmente de 20 o más cm de diámetro; ►fustal 2); *f* **perche [f]** (Jeune arbre de diamètre moyen compris entre 10 et 15 cm dans un jeune peuplement de futaie régulière ; ►perchis) ; *g* **Stange [f]** (Baum mit einem mittleren Stammdurchmesser bis zu 20 cm in einem mittelalten Waldbestand; ►Stangenholz).

4445 pole-stage forest [n] *for.* (Developmental stage of a forest following the sapling stage; ►sapling stage forest); *s* **fustal [m] (3)** (Conjunto de pies arbóreos en la fase de desarrollo de un bosque posterior al ►latizal); *f* **perchis [m] (1.** jeune peuplement de futaie régulière dont les tiges ont un diamètre moyen de l'ordre de 10 à 15 cm, et qui est donc justiciables d'éclaircies. **2.** Stade de développement d'une futaie régulière, éventuellement divisée en bas-perchis et haut-perchis, à qui succède le ►gaulis et précède la futaie ; TSF 1985, 133) ; *g* **Stangenholz [n]** (Natürliche Entwicklungsstufe eines mittelalten Hochwaldes bis zum Erreichen einer mittleren Stammstärke von 20 cm; folgt dem ►Gertenholz).

4446 pole vault facility [n] *recr.* (Sports equipment for pole vaulting; *generic term* ►jump area); *s* **instalación [f] de salto de pértiga** (*Término genérico* ►instalación de salto); *f* **aire [f] de saut à la perche** (Installation sportive pour le saut à la perche ; *terme spécifique* ►aire de saut) ; *g* **Stabhochsprunganlage [f]** (Sporteinrichtung der Leichtathletik für dem Stabhochsprung; *OB* ►Sprunganlage).

policemen [n], **sleeping** *trans.* ►speed hump.

policy [n] *conserv. pol.* ►environmental policy; *landsc. pol. urb.* ►greenbelt policy; *plan. pol.* ►land assembly policy [US]/land acquisition policy [UK]; *pol.* ►landscape policy; *landsc. pol. urb.* ►open space policy; *leg. plan. pol.* ►regional planning policy.

policy [n] [UK], **green space** *landsc. pol. urb.* ►open space policy [US].

policy [n] [UK], **land acquisition** *plan. pol.* ►land assembly policy [US].

policy plan [n] [UK], **regional** *leg. plan.* ►state development plan [US].

polished [pp/adj] *constr.* *►tooling of stone [US]/dressing of stone [UK], #7.

4447 poll [n] *plan. sociol.* (Random canvas of the opinions of a large number of people with regard to a certain topic; ►questionnaire, ►representative poll); *syn.* opinion poll [n]; *s* **encuesta [f]** (Método de investigación sociológica por el que se entrevista a un porcentaje representativo de la población o de determinado grupo de personas para conocer la opinión sobre temas específicos; ►cuestionario, ►encuesta representativa); *syn.* sondeo [m] de opinión; *f* **enquête [f]** (Mode d'investigation sociologique, psychologique, économique, politologique, qui procède par interrogation, en vue de la connaissance d'une population ; DUA 1996 ; ►questionnaire, ►sondage d'opinion) ; *syn.* sondage [m] ; *g* **Umfrage [f]** (Befragung einer größeren Anzahl von Personen, um deren Meinung zu bestimmten Sachverhalten herauszufinden; ►Fragebogen, ►Repräsentativbefragung); *syn.* Meinungsumfrage [f], Befragung [f].

poll [n], **opinion** *plan. sociol.* ►poll.

4448 pollarded tree [n] *agr. conserv.* (Tree with the top branches regularly cut back to the trunk producing a crown of many small dense shoots. This was originally done as a means of managing timber production from a tree. **In Europe**, this method is often used now for preservation of the cultural landscape; ►fodder tree, ►green pruning, ►pollarded willow, ►pollarding); *s* **árbol [m] desmochado** (Árbol de gran capacidad de regeneración, como sauces, chopos, carpes, que se poda regularmente hasta el tronco para aprovechar sus ramas jóvenes como leña o material flexible para cestería. Hoy en día los **áa. dd.** tienen cada vez más importancia para la preservación de la naturaleza, ya que son lugares de retirada de especies de fauna en peligro de extinción [►lista roja]; ►árbol de forraje, ►desmoche, ►poda en verde, ►sauce de desmoche); *syn.* árbol [m] de desmoche; *f* **arbre [m] têtard** (Arbre caractérisé par une bonne capacité de régénération comme le Saule, le Peuplier, le Charme, etc., dont la cime sera supprimée après la plantation et dont les branches seront régulièrement coupées toujours à la même hauteur en vue de la production de bois de chauffage ou de bois de tressage ; les **a. t.** sont à l'heure actuelle d'une importance particulière dans le domaine de la protection de la nature car ils constituent souvent les derniers refuges de nombreuses espèces protégées ; ►arbre d'émonde, ►élagage en vert, ►étêtage, ►saule têtard) ; *syn.* arbre [m] en tête de chat, têtard [m] ; *g* **Kopfbaum [m]** (Baum mit besonders gutem Regenerationsvermögen wie Weide, Pappel, Hainbuche etc., der nach der Pflanzung in Höhe des Kronenansatzes gekappt wird und dessen Äste meist im Abstand von mehreren Jahren immer in derselben Höhe geschnitten werden, um Brennholz oder flexibles Binde- und Flechtmaterial zu gewinnen. Heute haben Kopfbäume eine zunehmende Bedeutung für den Naturschutz, weil sie oft letzte Rückzugsorte für viele Arten der Roten Liste sind; ►auf den Kopf setzen, ►Kopfweide, ►Schneitelbaum, ►Schneiteln).

4449 pollarded willow [n] *agr. conserv.* (Willow [e.g. osier—Salix viminalis] with a thickened top of trunk, caused by regular pruning of annual shoots; *generic term* ▶pollarded tree); *s* **sauce [m] de desmoche** (Sauce [p. ej. mimbre: *Salix viminalis*] con extremo del tronco engruesado por la poda regular de los brotes anuales; *término genérico* ▶árbol desmochado); *f* **saule [m] têtard** (Saule [p. ex. l'osier blanc Salix viminalis] dont le sommet du tronc forme une ou plusieurs têtes volumineuses provoquée au rabattage régulier aux mêmes endroits des rejets [formation de têtes de saule] ; *terme générique* ▶arbre têtard) ; *g* **Kopfweide [f]** (Weide [z. B. Korbweide — Salix viminalis] mit verdicktem Stammende, das durch ständigen, regelmäßigen Schnitt der Jahrestriebe entsteht; *OB* ▶Kopfbaum).

4450 pollarding [n] *arb. hort.* (Means of tree management involving regularly cutting back the crown of a tree to produce a close head of shoots; pollard [vb]; ▶pollarded tree, ▶knob, ▶coppice 2); *syn.* topping [n] (2), pollard pruning [n]; *s* **desmoche [m]** (Poda severa de árboles para provocar el crecimiento de ramas jóvenes, como en el caso de los sauces, cuyo resultado en un ▶árbol desmochado; desmochar [vb]; ▶nudo de desmoche, ▶poda hasta el tocón); *syn.* desmochadura [f]; *f* **étêtage [m]** (Supprimer la tête d'un arbre, le sommet d'une pousse afin de provoquer sa ramification ; étêter [vb] ; ▶arbre têtard, ▶recéper, ▶tête de chat) ; *syn.* étêtement [m] ; *g* **auf den Kopf setzen [vb]** (Baumschnitt, z. B. bei Weiden; das Ergebnis ist ein ▶Kopfbaum; ▶Astkopf; ▶auf den Stock setzen).

pollard pruning [n] *arb. hort.* ▶pollarding.

4451 pollard system [n] *agr. for.* (Ancient system of cutting pollard shoots for emergency animal fodder or for firewood, with due provision for the replacement of exhausted or defective pollards in order to provide continued forage for domestic animals; ▶green pruning); *s* **tratamiento [m] de montes de trasmocho** (Aprovechamiento de árboles por desmoche o descabezamiento; ▶poda en verde); *f* **traitement [m] en taillis sur têtards** (Régime sylvicole ancien par lequel est appliqué un émondage régulier, sans toucher à la cime, sur un peuplement forestier de taillis en vue de la production de bois de chauffe ou la récolte du feuillage pour la litière ; DFM 1975, 273 ; ▶élagage en vert) ; *g* **Schneitelbetrieb [m, o. Pl.]** (Oft niederwaldartige Entnahme seitlicher Äste bei Laubbäumen unter Belassung des Wipfels zur Gewinnung von Laub und Feinreisig als Futter in Notzeiten und zur Brennholzgewinnung; ▶Schneiteln); *syn.* Kopfholzbetrieb [m, o. Pl.], Kopfholzwirtschaft [f, o. Pl.].

pollen plant [n] *hort. landsc.* ▶host plant for bees.

pollutant [n] *envir.* ▶air pollutant, ▶noxious substance.

pollutants [npl] *envir.* ▶residual pollutants.

polluted air space [n] [US] *envir.* ▶contaminated airspace [US].

4452 polluter [n] *envir. leg.* (Legal entity, commercial firm or natural person responsible for harmful or objectionable emissions; ▶polluter-pays principle); *s* **emitente [m]** (Persona jurídica o física que es responsable de una emisión; ▶principio de causalidad); *f* **émetteur [m]** (Personne juridique ou naturelle responsable d'un rejet; ▶principe pollueur-payeur) ; *syn.* pollueur [m] ; *g* **Emittent [m]** (Juristische oder natürliche Person, die für Emissionen verantwortlich ist; ▶Verursacherprinzip).

4453 polluter-pays principle [n] *envir. pol.* (Political principle of making the polluter of air, soil, or water pay, because of his responsibility for environmental damages, or in order to prevent damages, since subsequent remedy is often nearly impossible from the technical point of view. This principle also covers the ▶effluent discharge fee for the disposal of liquid waste into watercourses, the solid waste fees and also standards for production processes and products; the alternative is the ▶public responsibility principle); *syn.* principle [n] of causal responsibility, "polluter should pay" principle [n], principle [n] of

causation; *s* **principio [m] de causalidad** (Principio según el cual se imputan los costos de prevención o restauración al causante de contaminación o de un impacto ambiental; ▶principio de cooperación; *syn.* principio [m] «el que contamina paga»; *f* **principe [m] pollueur-payeur** (Principe adopté en 1972 par l'OCDE et figurant parmi les principes fondamentaux de l'Acte unique européen en 1987, selon lequel les frais résultant des mesures de prévention, de réduction et de lutte contre la pollution doivent être supportées par le pollueur ; C. rural, art. L. 200-1 ; cette « internalisation » des coûts de l'environnement reçoit son application sous forme de redevances et de taxes diverses, les fonds collectés étant utilisés pour la mise en place de dispositifs d'épuration, l'application de systèmes préventifs ou la réalisation d'aménagements compensateurs ; l'alternative au **p. p. p.** est le ▶principe de coopération ; ▶principe de responsabilité publique [de l'État et des collectivités publiques]) ; *syn.* principe [m] de la causalité ; *g* **Verursacherprinzip [n]** (Politischer Grundsatz, der in D. 1976 in das Umweltprogramm der Bundesregierung aufgenommen wurde, nach der der Verursacher von Umweltschädigungen als Verantwortlicher die finanziellen Folgen tragen oder Schäden erst gar nicht entstehen lassen soll, da eine nachträgliche Beseitigung sehr oft technisch undurchführbar ist. Hierzu gehören auch Abgaberegelungen wie z. B. Abwasserabgaben für das Einleiten ungeklärter Abwässer in ▶Vorfluter, Müllgebühren, aber auch Verfahrens- und Produktnormen; die Alternative ist das ▶Gemeinlastprinzip).

"polluter should pay" principle [n] *envir. pol.* ▶polluter-pays principle.

polluting firm [n] *envir. leg.* ▶water polluting firm [US]/ person responsible for direct discharge [UK].

4454 pollution [n] (1) *envir.* (Contamination of air, water or land by polluting substances; ▶environmental nuisance); *s* **contaminación [f]** (▶Impacto ambiental negativo sobre el aire, agua o suelo causado por sustancia nociva); *syn.* polución [f]; *f* **concentration [f] de substances polluantes** (dans l'air, dans l'eau; ▶nuisance d'environnement) ; *syn.* niveau [m] de substances polluantes ; *g* **Schadstoffbelastung [f]** (Belastung eines Umweltmediums — Luft, Wasser, Boden — durch Schadstoffe; ▶Umweltbelastung).

pollution [n] (2) *envir.* ▶acceptable level of air pollution, ▶air pollution 1; *agr. envir.* ▶agricultural pollution; *envir.* ▶damage caused by environmental pollution, ▶effect of environmental pollution, ▶effluent pollution; *envir. leg.* ▶environmental pollution; *envir. wat'man.* groundwater pollution [US]/ground water pollution [UK]; *envir.* ▶marine pollution, ▶national or state surveillance system for air pollution, ▶noise pollution, nonpoint source pollution [US]/non-point source pollution [UK], ▶oil pollution, ▶particulate air pollution, ▶point source pollution, ▶protection against pollution; *for. landsc.* ▶protective forest against air pollution; *envir.* ▶radiation pollution; *hort.* resistance to air pollution; *envir. wat'man.* ▶remediation of groundwater pollution; *envir. hydr.* ▶risk of groundwater pollution; *envir.* ▶river pollution; *ecol. envir. limn.* ▶salt pollution (2); *envir.* ▶shore oil pollution; *envir. limn.* ▶thermal pollution of watercourses; *envir. leg.* ▶threshold level of air pollution, ▶transboundary air pollution; *envir.* ▶water pollution.

pollution [n], **atmospheric** *envir.* ▶air pollution (1).

pollution [n], **background level of environmental** *envir.* ▶existing environmental pollution load.

pollution [n], **remedial measures for groundwater** *envir. wat'man.* ▶remediation of groundwater pollution.

pollution [n], **salt** *limn. pedol.* ▶salination (2).

pollution [n], **transfrontier air** *envir. leg.* ▶transboundary air pollution.

pollution accident [n] *envir.* ▶oil pollution accident.

Pollution Act [n] [UK]**, Control of** *envir. leg.* ▶Federal Clean Air Act [US].

pollution casuality [n]**, oil** *envir.* ▶oil pollution accident.

4455 pollution control [n] (1) *envir. leg.* (Administrative mechanisms and comprehensive management measures designed to protect human beings, animals, plants and inert objects against pollution and prevent its continued occurrence; **in U.K.**, the administrative control is effected by legislation, e.g. Alkali Act 1906, Clean Air Act 1956 and 1968, Control of Pollution Act 1974, Health and Safety at Work Act 1974, and Environmental Protection Act 1990, and its enforcement and implementation through statutory bodies such as the Waste Disposal Authority for a region; the National Rivers Authority, Pollution Inspectorate, and the Health and Safety Executive; cf. DES 1991; ▶pollution impact); *s* **protección** [f] **contra inmisiones** (Conjunto de medidas legales que tienen como objetivo proteger a las personas, la fauna y la flora y los objetos de los efectos negativos de la contaminación atmosférica [▶inmisión] y de la contaminación acústica); *f* **lutte** [f] **contre les émissions** (Moyens juridiques divers qui tendent à surveiller, éviter, empêcher, réduire ou supprimer les pollutions atmosphériques, préserver la qualité de l'air, réduire les émissions sonores et vibratoires, économiser et utiliser rationnellement l'énergie en vue de protéger l'environnement dans son ensemble ainsi que la santé des personnes ; cf. loi relative à la lutte contre le bruit n° 92-1444 du 31 décembre 1992, dir. n° 96/62/CE du 27 septembre 1996 et loi n° 96-1236 du 30 décembre 1996 sur l'air et l'utilisation rationnelle de l'énergie ; ▶émission) ; *g* **Immissionsschutz [m, o. Pl.]** (Zur Luftreinhaltung und Lärmbekämpfung gesetzlich vorgeschriebene Gesamtheit der Maßnahmen, die dazu dienen, Menschen, Tiere, Pflanzen und nicht belebte Sachgüter vor schädlichen anthropogenen Umwelteinwirkungen [▶Immission] zu schützen und dem Entstehen dieser vorzubeugen; cf. BImSchG).

pollution control [n] (2) *envir. leg. nat'res.* ▶air pollution control, ▶air quality control region.

pollution control district [n] *envir. leg.* ▶air quality control region.

pollution control registering [n] *ecol. envir.* ▶biomonitoring.

pollution dispersal [n] [UK] *envir.* ▶pollution dispersion [US].

4456 pollution dispersion [n] [US] *envir.* (Spread by ▶noxious substances); *syn.* pollution dispersal [n] [UK]; *s* **dispersión** [f] **de contaminantes** (▶sustancia nociva/tóxica/contaminante); *f* **dissémination** [f] **des matières polluantes** (▶substance génératrice de nuisances) ; *g* **Ausbreitung [f] von Schadstoffen** (▶Schadstoff).

pollution dome [n] *envir. met.* ▶air pollution dome.

pollution fee [n] *adm. envir. leg.* ▶air pollution fee.

4457 pollution impact [n] *envir.* (Intake of air pollution, noise, vibrations, light, heat radiation, and other discharges from adjacent or distant property, which adversely affect human beings, animals, plants, and inert objects; the term 'immission' is not used in the English language!; ▶emission, ▶gaseous pollution impact); *syn.* pollution input [n]; *s* **inmisión** [f] (Contaminación atmosférica, sonora, de vibraciones, radiaciones, calor, etc. que incide en una zona determinada y a los seres vivos y objetos que la habitan; ▶emisión, ▶inmisión gaseosa); *f* **immission** [f] (F., terme considéré du point de vue de la zone réceptrice et utilisé pour définir les concentrations ambiantes de substances et particules polluantes dans l'atmosphère, c'est-à-dire en unité de masse par unité de volume d'air ; **en France** l'utilisation de ce terme est réduite aux polluants atmosphériques et n'inclue pas comme en Allemagne ou en Suisse les « immissions immatérielles » telles que le bruit, les vibrations, la lumière, la chaleur et le rayonnement affectant les différents éléments d'un milieu [hommes, animaux, plantes, air, eau, sols, bâtiments, etc.] et provoquant une gêne ou une nuisance ; ▶émission, ▶rejet à l'atmosphère de polluants gazeux) ; *syn.* rejet [m] à l'atmosphère, rejet [m] polluant dans l'atmosphère ; *g* **Immission** [f] (Auf Menschen, Tiere, Pflanzen, den Boden, das Wasser, die Atmosphäre sowie Kultur- und sonstige Sachgüter einwirkende Luftverunreinigung, Geräusche, Erschütterungen, Licht, Wärme, Strahlen und ähnliche Erscheinungen; cf. § 3 [2] BImSchG; im Deutschen wird zwischen ▶Emission, Transmission und **I.** unterschieden; im Englischen gibt es das Wort **I.** nicht; im Französischen in Frankreich reduziert sich **I.** nur auf Luftverschmutzungen, im Französischen in der Schweiz gilt derselbe Begriffsumfang wie im Deutschen; ▶Gasimmission).

pollution input [n] *envir.* ▶pollution impact.

pollution level [n]**, noise** *envir.* ▶noise decibel level.

4458 pollution level tolerance [n] **of a water body** *envir.* (Level of water pollution of a river, lake or sea by waste water or poisonous substances, which can be tolerated without destroying the ▶self-cleansing capacity); *s* **capacidad** [f] **asimilativa de las aguas continentales** (Nivel de carga contaminante [aguas residuales, sustancias nocivas] que puede ser tolerada por un curso o cuerpo de agua sin superar su ▶capacidad de autodepuración); *f* **capacité** [f] **de charge des eaux** (Quantité d'effluents ou de matières polluées pouvant être assimilée par un cours d'eau sans en altérer durablement la ▶capacité d'autoépuration) ; *g* **Belastbarkeit** [f] **eines Gewässers** (Menge an Abwässern, Giftstoffen oder sonstigen Verunreinigungen in einem Gewässer, die die ▶Selbstreinigungskraft nicht dauerhaft beeinträchtigen).

4459 pollution load [n] *envir.* (Measurable quantity of polluted substances carried in a watercourse; ▶effluent pollution load, ▶existing environmental pollution load); *s* **carga** [f] **de contaminantes** (Cantidad medible de sustancias contaminantes que es transportada por un curso de agua; ▶carga de aguas residuales, ▶nivel de contaminación de fondo); *f* **charge** [f] **de pollution (2)** (Quantité de matières polluantes en suspension transportée par un cours d'eau ; LA 1981, 474; ▶charge de pollution 1, ▶niveau de pollution dans l'air ambiant) ; *syn.* charge [f] polluante ; *g* **Schmutzfracht** [f] (Die von einem Fließgewässer mitgeführte Menge an verschmutzenden Bestandteilen [Schadstoffen] pro Zeiteinheit; ▶Abwasserlast, ▶Immissionsvorlast).

pollution monitoring [n] *envir. leg.* ▶air pollution monitoring.

pollution [n] **of the sea** *envir.* ▶marine pollution.

pollution-resistant [adj] *hort. plant.* ▶air pollution-resistant.

4460 pollution source [n] *envir.* (Origin of a pollutant substance which emits into an environmental medium; ▶polluter); *syn.* emission source [n]; *s* **foco** [m] **emisor de contaminantes** (Instalación industrial que emite sustancias contaminantes; ▶emitente); *f* **source** [f] **d'émission** (Installations industrielles ou d'utilité publique susceptibles de causer une pollution ; ▶émetteur) ; *syn.* point [m] de rejet d'émission ; *g* **Emissionsquelle** [f] (Ort, an dem Stoffe austreten, die die Umwelt verunreinigen oder von dem sonstige Beeinträchtigungen ausgehen, z. B. Lärm, störende Lichtreize, Gestank; ▶Emittent).

pollution-tolerant [adj] *hort. plant* ▶air pollution tolerant.

pollution zone [n] [US]**, air** *envir.* ▶contaminated air space.

4461 polyantha rose [n] *hort. plant.* (Generally small, shrubby garden rose with repeat-flowering habit which originated by crossing *Rosa chinensis* with *Rosa multiflora* [from Japan], as well as with *Rosa polyantha* [from Japan, Korea]. The hybridization with ►hybrid teas produced polyantha hybrids; ►bedding rose, ►floribunda rose); *syn.* cluster rose [n] [also US]; *s* **rosa [f] polyantha** (Rosal arbustivo, de pequeña talla y refloreciente, resultado del cruce de *Rosa chinensis* y *Rosa multiflora* [del Japón] con *Rosa polyantha* [del Japón, Corea]. Por medio del cruce con ►rosas híbridos de té se crearon los híbridos de polyantha; ►rosa de macizo, ►rosa floribunda); *f* **rosier [m] polyantha** (Rosier issu du croisement du rosier Bengale *[Rosa chinensis]* et du rosier multiflora *[Rosa multiflora]* — Japon — ainsi que de *Rosa polyantha* — Japon, Corée — généralement nain, à floraison abondante continue pendant toute la saison avec de petites fleurs réunies en bouquet ; de l'hybridation avec des ►rosiers hybrides de thé sont issus les rosiers polyantha hybrides ; ►rosier buisson à fleurs groupées, ►rosier floribunda) ; *syn.* rosier [m] à fleurs groupées ; *g* **Polyantharose [f]** (Aus *Rosa chinensis* und *Rosa multiflora* [aus Japan] sowie *Rosa polyantha* [aus Japan, Korea] gekreuzte, meist niedrige, buschige, viel blühende Gartenrose. Durch Kreuzung mit Teehybriden entstanden Polyanthahybriden; ►Beetrose, ►Edelrose, ►Floribundarose).

4462 polygonal masonry [n] *arch. constr.* (►Masonry construction which is constructed of stones having smooth, irregular, polygonal faces with a minimum of three sides and angles; ►cyclopean wall); *s* **mampostería [f] poligonal** (Obra de fábrica de mampuestos labrados con caras irregulares, poligonales con un mínimo de tres lados resp. ángulos; ►mampostería, ►mampostería ciclópea); *f* **maçonnerie [f] à appareillage polygonal** (►Maçonnerie de pierres à appareillage irrégulier en opus de gros moellons irréguliers à angles vif ; *terme spécifique* maçonnerie à appareillage polygonal à décrochements ; ►maçonnerie à appareillage cyclopéen) ; *g* **Polygonalmauerwerk [n]** (►Mauerwerk, dessen Ansichtsfläche aus unregelmäßig vieleckigen Steinen besteht; ►Zyklopenmauerwerk).

4463 polytechnic institute [n] [US] *prof.* (**In U.S.**, technical university with graduate schools, formerly state agricultural college, offering four years curricula with a practical bias, now expanded to a university with graduate schools); *syn.* state college [n] [US], university [n] of applied sciences [UK]; *s* **Escuela [f] Superior** (Institución de educación superior en la que se cursan estudios de orientación práctica de 3 ó 4 años de duración, a diferencia de la universidad con orientación científico-teórica); *f 1* **Institut [m] Universitaire Technique (I.U.T.)** (F., établissement public d'enseignement supérieur dispensant une formation théorique technique et préparant aux diverses professions de l'industrie) ; *f 2* **université [f] technique spécialisée (≠)** (D., établissement public d'enseignement supérieur dispensant une formation orientée vers la pratique professionnelle d'une durée de quatre ans) ; *g* **Fachhochschule [f]** (Hochschule, an der ein 3-4-jähriges, praxisorientiertes, berufsqualifizierendes Studium absolviert wird).

4464 polytrophic lake [n] *limn.* (Lake with a very high continually available nutrient content; deep water lacking oxygen in summer and occasionally infiltrated by hydrogen sulphide; surface water occasionally strongly saturated with oxygen; very low depth of visibility; ►trophic level of a waterbody); *s* **lago [m] polítrofo** (Lago con gran contenido de nutrientes siempre disponibles, cuyas capas profundas pueden carecer de oxígeno en el verano por lo que se puede generar azufre por descomposición anaerobia, cuyas capas superficiales están temporalmente saturadas de oxígeno, en el que hay una gran producción de plancton y cuya visibilidad es muy reducida; ►nivel trófico de aguas estancadas); *f* **lac [m] polytrophe** (Lac caractérisé par une très forte concentration en éléments nutritifs mobilisables en permanence,

une pénurie d'oxygène à partir de l'été et une production occasionnelle d'hydrogène sulfuré dans les eaux profondes, une saturation en oxygène occasionnelle des eaux de surface, une forte turbidité, un développement de masse du phytoplancton ; ►niveau trophique des eaux douces) ; *g* **polytropher See [m]** (**1.** See mit schr hohem, stets frei verfügbarem Nährstoffangebot; Tiefenwasser schon im Sommer sauerstofffrei mit zeitweiser Schwefelwasserstoffentwicklung; Oberflächenwasser zeitweise stark mit Sauerstoff übersättigt; Sichttiefe sehr gering; Massenentwicklung von Phytoplankton. **2.** *Gewässergüte* See mit einer übermäßig hohen Nährstoffbelastung, massiven Algenproduktion — oft mit Blaualgendominanz, mit einer Sichttiefe von nur wenigen Zentimetern und ganzjährig übermäßig hoher Sauerstoffzehrung, die den größten Teil des Wasserkörpers umfasst; ►Trophiestufe eines stehenden Gewässers).

pond [n] *constr.* ►fire suppression pond [US]; *agr.* ►fish pond; *constr.* ►man-made pond; *limn.* ►natural pond; *min.* ►tailings pond.

pond [n], **artificial** *constr.* ►man-made pond.

pond [n] [UK], **farm** *agr.* ►fish pond.

pond [n] [UK], **fire** *constr.* ►fire suppression pond [US].

pond [n] [UK], **slurry** *min.* ►tailings pond.

pond [n], **storm water detention** *hydr. wat'man.* ►storm water detention basin.

pond [n] [US], **storm water management** *hydr. wat'man.* ►storm water detention basin.

4465 pond bank [n] *land'man.* (Sloping edge of a pond); *s* **borde [m] de estanque** *syn.* orilla [f] de estanque; *f* **grève [f] d'étang** *syn.* rive [f] d'étang ; *g* **Teichufer [n]** (Randbereich eines Teiches).

4466 pond farming [n] *agr.* (Breeding and production of edible fish in specially constructed ponds); *s* **piscicultura [f] de estanques** (Cría y producción de peces para la alimentación en estanques construidos para este fin); *f* **pisciculture [f] en étangs** (Exploitation d'étangs ou de plans d'eau artificiels en vue de l'élevage et de la production de poissons ; AEP 1976, 339 ; LA 1981, 97) ; *g* **Teichwirtschaft [f]** (Bewirtschaftung von Teichen zur Zucht und Produktion von Speisefischen).

4467 pony stable [n] *recr.* *s* **picadero [m] de ponys**; *f* **centre [m] équestre pour poneys** ; *g* **Ponyreiterhof [m]** *syn.* Reiterhof [m] für Ponys.

4468 pool [n] **(1)** *constr.* (Ornamental basin in a garden/park or in an urban open space; *specific terms* ►garden pool, ►precast pool; ol); *syn.* basin [n] (2); *s* **estanque [m]** (Instalación con funciones ornamentales en jardín o espacio libre urbano; ►estanque ornamental, ►estanque prefabricado); *f* **pièce [f] d'eau** (Terme générique pour les plans d'eau artificiels allant du très petit bassin jusqu'à l'étang artificiel ; ►bassin ornemental, ►bassin préfabriqué) ; *syn.* bassin [m] ; *g* **Wasserbecken [n]** (►Zierbecken im Garten/Park oder in städtischen Freiräumen; ►Fertigbecken).

pool [n] **(2)** *zool.* ►anthropogenic alteration of the genetic fauna pool; *trans.* ►car pool; *recr.* ►diving pool; *urb. gard.* ►fountain pool; *constr. gard.* ►garden pool; *biol.* ►gene pool; *geo. phyt.* ►heathland pool; *pedol. chem.* ►nutrient pool; *recr.* ►public swimming pool; *geo. hydr.* ►spring-fed pool; *recr.* ►swimming pool (1), ►swimming pool (2), ►toddlers wading pool [US], ►wading pool [US], ►wave pool.

pool [n], **cycling** *ecol.* ►ecological cycle.

pool [n], **fauna** *zool.* ►anthropogenic alteration of the genetic fauna pool.

pool [n], **flood control** *wat'man.* ►floodable land.

pool [n] [UK], **indoor** *recr.* ▶swimming pool hall [US].

pool [n] [UK], **paddling** *recr.* ▶toddlers wading pool [US].

pool [n], **resurgence** *geo. hydr.* ▶spring-fed pool.

pool [n] [US], **stilling** *wat'man.* ▶stilling basin.

pool [n] **for non-swimmers** [UK] *recr.* ▶wading pool [US].

pool spring [n] *geo. hydr.* ▶vauclusian spring.

poop-scoop area [n] [UK] *landsc. urb.* ▶pet exercise area [US].

4469 poor light exposure [n] *bot. syn.* weak light exposure [n], low level [n] of light exposure; *s* **iluminación [f] baja** *syn.* iluminación [f] leve, iluminación [f] suave; *f* **faible exposition [f] à la lumière** ; *g* **geringe Belichtung [f]** *syn.* schwache Belichtung [f].

4470 poor load-bearing soil [n] *constr.* *s* **suelo [m] no/ poco portante** *syn.* suelo [m] no apto para construcciones; *f* **sol [m] peu porteur** ; *g* **wenig tragfähiger Boden [m]**.

4471 poorly drained soil [n] *agr. hort.* (Soil characterized by ▶impeded water; ▶soil affected by impeded water, ▶waterlogged soil [US]/water-logged soil [UK]); *syn.* drainage-poor soil [n]; *s* **suelo [m] encharcado** (Suelo con mal drenaje que se caracteriza por la presencia frecuente de ▶agua estancada; ▶suelo de agua estancada, ▶suelo saturado); *f* **sol [m] engorgé** (Sol caractérisé par une humidité excessive temporaire ou permanente dans lequel ont lieu des phénomènes de réduction ou de ségrégation du fer ; ▶nappe perchée, ▶sol à nappe temporaire perchée, ▶sol saturé par l'eau) ; *g* **staunasser Boden [m]** (Boden, der durch Stauwasser stark geprägt ist; ▶Staunässe, ▶Stauwasserboden, ▶wassergesättigter Boden).

poorly-provided recreation facility [n] *recr.* ▶simply-provided recreation facility.

4472 poorly soluble [adj] *chem. pedol.* (Characteristic applied to soil nutrients of difficult solubility in water; *opp.* ▶easily soluble); *syn.* of low solubility [loc]; *s* **poco soluble [loc]** (Característica aplicada a sustancias, p. ej. nutrientes, que no se disuelven bien en agua; *opp.* ▶de gran solubilidad); *f* **difficilement soluble [loc]** (Dont les caractéristiques ne permettent pas de dissolution rapide, p. ex. les substances nutritives dans le sol ; *opp.* ▶facilement soluble) ; *syn.* peu soluble [loc], de faible solubilité [loc] ; *g* **schwer löslich [adj]** (Substanzen betreffend, die, sich in Flüssigkeit schwer auflösen, z. B. Nährstoffe im Boden; *opp.* ▶leicht löslich).

4473 poor quality housing [n] [US] *urb.* (Old building or groups of buildings, often with an out-of-date or uneconomic floor plan and inadequate sanitary facilities; such buildings are usually condemned for occupancy; ▶old part of a city/town/ village); *syn.* poor quality housing stock [n] [UK]; *s* **edificación [f] antigua con mal trazado** (Edificio o conjunto de edificios construidos antes de la segunda guerra mundial con planta anticuada y servicios sanitarios actualmente inadecuados; ▶zona de edificación antigua); *f* **habitat [m] ancien** (Tout immeuble ou lotissement dont la division parcellaire et l'équipement ne sont plus adaptés aux règles d'urbanisme en vigueur; ▶vieux quartiers) ; *g* **Altbebauung [f] (1)** (Alte Gebäude oder Gebäudekomplexe mit meist schlechtem, unwirtschaftlichem Grundflächenzuschnitt und mangelhaften sanitären Einrichtungen; ▶Altbaugebiet); *syn.* z. T. alte Bausubstanz [f].

poor quality housing stock [n] [UK] *urb.* ▶poor quality housing [US].

4474 populated area [n] *plan. urb.* (*Term used in urban planning* area developed for habitation and served by urban infrastructure with access to roads and required utilities; ▶open land [US]/greenfield site [UK]; ▶built-up area 1, ▶built-up area 2; *opp.* ▶unpopulated area); *s* **zona [f] edificada (2)** (Término utilizado en el marco de la planificación regional y urbana; *opp.* ▶suelo no urbanizable); *syn.* zona [f] urbanizada; *f* **zone [f] urbanisée** (Terme utilisé en aménagement du territoire ; ▶zone naturelle ; *opp.* ▶zone non urbanisée ; *syn.* espace [m] urbanisé ; *g* **besiedelter Bereich [m] (2)** (*Begriff der Stadt- und Regionalplanung* bewohntes Gebiet, das mit allen nötigen Infrastruktureinrichtungen ausgestattet und an das vorhandene Straßennetz angeschlossen ist; ▶Außenbereich [cf. § 35 BauGB]; *opp.* ▶unbesiedelter Bereich).

4475 population [n] **(1)** *ecol. leg.* (Number of all individuals of one species inhabiting the same area in a given period and generally sharing a common genetic continuity; ▶existing area vegetation); *s* **población [f] (2)** (Conjunto de individuos de una misma especie —y por tanto genéticamente afines— que habitan un área determinada en un momento dado; DINA 1987; ▶población 1); *f* **population [f]** (Ensemble des individus appartenant à la même espèce vivant généralement dans des conditions de milieu homogènes, donc dans une même communauté biologique, à un moment donné. Par extension, dans le règne animal, le terme de population peut désigner l'ensemble des individus d'une même espèce dans une région donnée ; DEE 1982 ; ▶peuplement végétal) ; *g* **Population [f]** (**1.** Gesamtheit der Individuen einer Art, die zur gleichen Zeit einen bestimmten, zusammenhängenden Lebensraumabschnitt [Areal] bewohnen und im Allgemeinen auf Grund ihrer Entstehungsprozesse miteinander verbunden sind, eine Fortpflanzungsgemeinschaft bilden und somit durch mehrere Generationen genetische Kontinuität zeigen. **2.** *leg.* Eine biologisch oder geografisch abgegrenzte Zahl von Individuen; cf. Art. 2 Verordnung EG Nr. 338/97 des Rates v. 9.12.1996; ▶Bestand).

population [n] **(2)** *zool.* ▶animal population; *biol.* ▶decrease in population; *zool.* ▶fish population; *game'man.* ▶game population; *conserv. leg.* ▶management of fish population; *phyt. zool.* ▶pioneer population; *zool.* ▶re-establishment of an animal population; *phyt. zool.* ▶residual population; *game'man.* ▶restocking of game population; *phyt. zool.* ▶restocking of population; *conserv. zool.* ▶roosting population.

population [n], **remnant** *phyt. zool.* ▶residual population.

population decline [n] *sociol.* ▶depopulation.

4476 population density [n] **(1)** *plan. sociol. urb.* (**1.** Number of persons per unit area in a specified region. **2.** Closeness of human habitation; ▶density of development); *syn.* settlement density [n]; *s* **densidad [f] de población (1)** (Cantidad de habitantes por superficie dada; ▶densidad de edificación); *syn.* densidad [f] de poblamiento; *f* **densité [f] de (la) population** (Nombre moyen d'habitants par unité de surface [en général par km^2] sur une aire donnée ; ▶densité de construction) ; *g* **Bevölkerungsdichte [f] (1)** (Anzahl der Einwohner je Flächeneinheit [i. d. R. je km^2] in einem bestimmten Gebiet; ▶Bebauungsdichte); *syn.* Besiedelungsdichte [f] (1).

4477 population density [n] **(2)** *biol.* (Number of organisms per unit area); *s* **densidad [f] de población (2)** (Cantidad de organismos por unidad de superficie); *f* **intensité [f] de peuplement** (Nombre d'individus par unité de surface) ; *g* **Besiedelungsdichte [f] (2)** (Anzahl der Organismen je Flächeneinheit).

4478 population density [n] **(3)** *phyt. zool.* (Quantitative term denoting the closeness of individuals of a particular species in its habitat; ▶species abundance, ▶species density, ▶population dynamics); *s* **densidad [f] de población (3)** (Número de individuos de un grupo definido presentes en una unidad territorial determinada. En el caso de las plantas, se relaciona con la distancia media entre individuos adyacentes; DINA 1987;

P

▶abundancia [de una especie]), ▶densidad de especies, ▶dinámica de poblaciones); *f* densité **[f] des individus** (Nombre d'individus d'une espèce par unité de surface ou de volume sur un territoire donné ; ▶abondance, ▶densité des espèces, ▶dynamique des populations) ; *syn.* densité [f] de population ; *g* **Individuendichte [f]** (Mengenbezeichnung für das Auftreten einer Art in einem Lebensraum; ÖKO 1983; ▶Abundanz, ▶Artendichte, ▶Populationsdynamik); *syn.* absolute Individuenabundanz [f], Bevölkerungsdichte [f] (2), Populationsdichte [f], Siedlungsdichte [f].

4479 population density map [n] *plan. sociol.* (Graphic representation by different sized dots to show varying concentrations of inhabitants); *s* **mapa [m] de distribución demográfica** (Mapa gráfico en el cual se indica con puntos de diferentes tamaños o en cantidades diferentes la población de secciones espaciales dadas); *syn.* mapa [m] de distribución de la población; *f* **carte [f] de densité de population** (Représentation graphique de la densité de la population dans une zone définie au moyen de points dont la grosseur dépend du nombre d'habitants) ; *g* **Einwohnerpunktekarte [f]** (Kartenmäßige Darstellung der Bevölkerungsdichte in einem Untersuchungsraum durch Punkte, bei denen die Punkte je nach Größe durch Einwohnerzahlen definiert sind).

4480 population distribution [n] *plan. sociol. urb.* *s* **distribución [f] de la población** *syn.* distribución [f] demográfica; *f* **répartition [f] de la population** *syn.* répartition [f] démographique ; *g* **Bevölkerungsverteilung [f]**.

4481 population dynamics [npl] *ecol.* (Periodic fluctuation of the density [abundance] of individuals in a population, dependent upon abiotic factors;—e.g. climatic—and biotic factors; e.g. food availability, enemies. Some ecologists, who see this fluctuation only in relation to population density, will say that **p. d.** describes the changes in the structure of a population; e.g. oscillations in the composition of sexually active and inactive male and female, as well as parasitized or ill, individuals; ▶population fluctuation); *s* **dinámica [f] de poblaciones** (Estudio de los cambios de densidad de poblaciones vegetales y animales debidos a factores bióticos o abióticos; cf. MARG 1977; ▶fluctuación de poblaciones); *f* **dynamique [f] des populations** (Étude de la structure et de l'évolution d'une population végétale ou animale [abondance] en relation avec les facteurs abiotiques [climat extrême] et biotiques [p. ex. manque de nourriture, présence de prédateurs] du milieu ; DEE 1982, 32 ; ▶variation de la densité de population) ; *g* **Populationsdynamik [f]** (Zeitliche Schwankungen in der Individuendichte [Abundanz] einer Population in Abhängigkeit von abiotischen [z. B. Witterungseinflüssen] und biotischen Umwelteinflüssen [z. B. Nahrungsangebot, Feinde]. Von manchen Ökologen wird diese allein auf die Populationsdichte resp. den Massenwechsel bezogene Betrachtungsweise als **Abundanzdynamik** bezeichnet, während sich dann die **P.** als Beschreibung der Veränderungen in der Struktur einer Population versteht — z. B. Schwankungen in der Zusammensetzung einer Population bezüglich des Anteils an männlichen und weiblichen geschlechtlich aktiven und inaktiven Individuen, an parasitierten resp. erkrankten Individuen; MEL 1977, Bd. 19; ▶Populationsbewegung).

4482 population ecology [n] *ecol.* (Branch of ecology dealing with the interrelationships within a plant or animal population; ▶autecology, ▶synecology); *syn.* demecology [n]; *s* **ecología [f] de poblaciones** (Rama de la ecología que estudia las relaciones entre y dentro de las poblaciones de animales y plantas; ▶autoecología, ▶sinecología); *syn.* demografía [f] (DINA 1987); *f* **démécologie [f]** (Branche de l'écologie étudiant le comportement des animaux ou des végétaux à l'intérieur d'une population ; ▶autécologie, ▶synécologie) ; *g* **Demökologie [f]**

(Teilgebiet der Ökologie, das sich mit den Beziehungen der Tiere oder Pflanzen innerhalb einer Population befasst; ▶Autökologie, ▶Synökologie); *syn.* Populationsökologie [f].

4483 population fluctuation [n] *zool.* (Change in a given population density of a animal species during the course of one generation [– intracyclical **f.**] or from one generation to another [= intercyclical **f.**] also known as 'population change'; ▶fluctuation of species, ▶population dynamics); *syn.* abundance dynamics [npl], population movement [n]; *s* **fluctuación [f] de poblaciones** (Variación del número de individuos que forman una población, como consecuencia de la intervención en su dinámica de diversos factores de mortalidad. Las fluctuaciones pueden ser regulares [el número de individuos se repite de forma más o menos periódica] o irregulares; DINA 1987; ▶dinámica de poblaciones, ▶fluctuación); *f* **variation [f] de la densité de population** (Oscillation de la densité de population d'une espèce au cours d'une génération [fluctuation saisonnière] ou de générations en générations ; ▶dynamique des populations, ▶fluctuation) ; *syn.* mouvement [m] de population ; *g* **Populationsbewegung [f]** (Veränderung der Bevölkerungsdichte einer Art im Laufe einer Generation [= intrazyklische **P.**] oder von Generation zu Generation [= interzyklische **P.**], auch **Massenwechsel** genannt; ÖKO 1983; ▶Fluktuation, ▶Populationsdynamik); *syn.* Abundanzdynamik [f], Massenwechsel [m].

4484 population growth [n] *plan. sociol. syn.* population increase [n]; *s* **incremento [m] demográfico** *syn.* incremento [m] de la población, crecimiento [m] demográfico; *f* **croissance [f] de la population** (Variation de la population [naturelle et migratoire] dans le temps pour un territoire donné) *syn.* croissance [f] démographique ; *g* **Bevölkerungswachstum [n]** (Zunahme der Weltbevölkerung oder der Einwohnerzahl in einem definierten Gebiet); *syn.* Bevölkerungszuwachs [m].

4485 population growth center [n] [± US] *plan.* (In U.S., a city or community with the most viable mix of facilities for habitation, business and industry, and some service to suburban settlements; in U.K. and D., according to the central place theory a hierarchy of settlements—Regional Centre, Sub-Regional Centre, Key Inland Town, Sub-Urban Town, Key Settlement—that will provide the most economical mix of facilities throughout the entire settlement pattern; cf. COU 1978, 102-115; ▶growth point, ▶state development plan [US]); *syn.* central place [n] [UK]; *s* **lugar [m] central** (En la teoría del **l. c.** CHRISTALER concibe la ciudad por su función específica de ser un centro abastecedor de bienes y servicios a la población residente en la ciudad y en su área de influencia. Se utiliza el término de **l. c.**, ya que para llevar a cabo esta función, la ciudad debe localizarse en el centro de su área de influencia. Los **ll. cc.** de mayor rango ofrecen una mayor gama de productos, disponen de más establecimientos comerciales y de servicios y poseen mayor población y áreas de influencia, así mismo su número es inversamente proporcional al rango del **l. c.**; cf. DGA 1986. En D., población con funciones importantes para sus alrededores o su región de influencia, determinada en ▶plan de desarrollo territorial, los **ll. cc.** están clasificados jerárquicamente según un sistema de grados de especialización de equipamientos y la diferente frecuencia de utilización en centros superiores, medios, inferiores y pequeños; ▶polo de desarrollo); *f 1* **pôle [m] urbain** (*Terme statistique* Centre urbain concentrant population et activités ; selon l'Institut National de la Statistique et des Études Économiques [INSEE], une unité urbaine offrant au moins 5000 emplois et qui n'est pas située dans la couronne périurbaine d'un autre pôle urbain) ; *f 2* **bassin [m] de vie** (Classification administrative de l'INSEE ; territoire présentant une cohérence géographique, sociale, culturelle et économique, exprimant des besoins homogènes en matière d'activités et de services. La délimitation d'un **b. de v.** correspond à des zones d'activités homogènes reposant sur des besoins locaux et structurés à partir du flux migratoire quotidien de la population et de la capacité d'attraction des équipements et services publics et privés [transport, enseignement, santé, action sociale]) ; *f 3* **lieu [m] central (≠)** (D. *aménagement du territoire* ville ou commune possédant une fonction centrale et remplissant les conditions d'octroi d'une aide privilégiée orientée vers la concentration des activités de production et des équipements publics collectifs d'accueil de la population des zones périphériques ou du réseau reliant les centres entre eux [treillage] ; la hiérarchisation des différents niveaux de centralité dans l'espace géographique est définie dans le ▶schéma régional d'aménagement et de développement du territoire des Länder qui distingue les centres de niveau supérieur, intermédiaire, inférieur ainsi que les petits centres ; cf. MG 1993 ; *F. aménagement du territoire* ▶pôle de développement) ; *g* **zentraler Ort [m]** (D., Stadt oder Gemeinde mit zentraler Funktion der Daseinsvorsorge für das Umland oder für den Verflechtungsbereich [Bündelung von Einrichtungen für die überörtliche Versorgung eines Bereiches mit Gütern und Dienstleistungen]; z. O.e sind entsprechend des abgestuften Spezialisierungsgrades der zentralen Einrichtungen und der unterschiedlichen Häufigkeit der Inanspruchnahme hierarchisch in Ober-, Mittel-, Unter- und Kleinzentren gegliedert und im ▶Landesentwicklungsplan ausgewiesen; cf. § 2 [2] Ziff. 3 ROG von 2008; ▶Wachstumspol; *syn.* Gemeinde [f] mit zentralörtlicher Bedeutung, Entwicklungsschwerpunkt [m].

population increase [n] *plan. sociol.* ▶population growth.

population movement [n] *zool.* ▶population fluctuation.

4486 population pressure [n] *ecol.* (Stress placed by all individuals in a ▶population 1 upon a particular habitat); *s* **presión [f] de población** (Conjunto de los efectos causados por todos los individuos de una ▶población 2 en un determinado hábitat); *f* **pression [f] de population sur le milieu** (Ensemble des facteurs démographiques exercés sur un milieu, p. ex. la croissance démographique accentue la pression foncière qui entraîne une dégradation du milieu morphologique et biologique ; ▶population ; *syn.* pression [f] démographique sur le milieu ; *g* **Populationsdruck [m]** (Gesamtheit der Einwirkungen aller Individuen einer ▶Population auf einen Lebensraum).

population pyramid [n] [US] *plan. sociol.* ▶demographic pyramid.

4487 pop-up sprinkler [n] *constr.* (Irrigation fitting, installed flush with the soil on top of a riser from an underground pipe; when in use, the ▶sprinkler is pushed above the soil by water pressure; ▶irrigation by pop-up sprinklers); *s* **aspersor [m] escamoteable** (▶Aspersor introducido en el suelo y conectado a conexión de agua subterránea; ▶riego automático); *f* **arroseur [m] escamotable** (Tuyère, turbine ou asperseur à battant incurporés dans le corps de l'arroseur et raccordés à un tuyau enterré, se soulevant de quelques cm au-dessus du sol sous l'effet de la pression ; ▶arrosage automatique intégré/enterré, ▶arroseur) ; *syn.* arroseur [m] intégré, arroseur [m] enterré ; *g* **Versenkregner [m]** (Bündig mit Oberkante Boden in einem Gehäuse eingebauter, an eine Unterflurwasserzuleitung angeschlossener ▶Regner, der im Betriebsfalle mit Wasserdruck wenige Zentimeter aus dem Boden hinausgeschoben wird; ▶Bewässerung mit Versenkregnern); *syn.* Unterflurregner [m].

pop-up sprinklers [npl] *constr.* ▶irrigation by pop-up sprinklers.

4488 porch [n] *arch.* (Open vestibule of a house); *s* **porche [m]** (Espacio cubierto que en algunas casas precede a la entrada; DPD 2005); *syn.* zaguán [m], portal [m] [C]; *f* **porche [m]** (Construc-

---

tion en saillie qui abrite la porte d'entrée d'un édifice) ; *g* **Laube [f]** (Offene Vorhalle eines Hauses).

pore space [n] *pedol.* ▶pore volume.

4489 pore volume [n] *pedol.* (**1.** Capacity of the air-filled voids in dry soil, dependent on particle-size distribution, aggregate sizes, content of organic matter and soil formation. **2.** Degree to which the total volume of a soil is permeated with pores or voids, which is generally expressed as a percentage of the whole volume unoccupied by solid particles; *specific terms* air porosity, capillary porosity; *syn.* (total) porosity [n], pore space [n] (TGG 1984, 105); *s* **porosidad [f] (total)** (Volumen total de los poros del suelo. Depende de la textura y la estructura del mismo y se expresa por la relación entre el volumen ocupado por gases y líquidos y el volumen total del suelo; DINA 1987); *f* **porosité [f] totale** (Volume des espaces laissés libres dans les particules, entre les particules et les unités structurales d'un sol en pourcentage du volume total d'un matériau donné ; selon la dimension des vides on parle de microporosité [ou porosité capillaire] et de macroporosité ; selon l'assemblage des vides on parle de porosité ouverte ou de porosité close, de porosité texturale ou de porosité structurale [d'interstices, lacunaire, fissurale] ; PED 1979, 240 ; l'espace poral peut être classé selon la taille [macrovides > 75 μm, mésovides 35-75 μm, microvides 5-35 μm, ultramicrovides < 5 μm], la forme, l'arrangement ou la morphologie ; *termes spécifiques* microporosité [porosité capillaire], macroporosité [porosité non capillaire, *syn.* volume lacunaire] ; cf. DIS 1986, 168 ; cf. PED 1979, 240) ; *syn.* volume [m] de l'espace poral ; *g* **Porenvolumen [n]** (**1.** Gesamtvolumen der Hohlräume im Boden, abhängig von Körnung und Kornform, vom Gehalt an organischer Substanz und von der Bodenentwicklung. Es werden Porengrößenbereiche nach den Porendurchmessern und der Wasserspannung [cm Wassersäule resp. pF-Wert] unterschieden wie z. B. weite Grobporen [> 50 μm], enge Grobporen [50-10 μm], Mittelporen [10-0,2 μm] und Feinporen [< 0,2 μm]. Das **P.** wird auch als **Porosität** bezeichnet und als Bruch angegeben, z. B. 0,45 oder 45 %; cf. SS 1979, 139 ff. **2.** Die makroskopische Porenklassifizierung kann auch mit den Parametern erfolgen: stark grobporös, stark mittelporös, stark feinporös, mittel gobporös, mittel mittelporös, mittel feinporös und schwach grobporös, schwach mittelporös. schwach feinporös sowie mit makroskopisch keine Poren vorhanden; der Anteil am Porenvolumen ist bei stark > 5 %, mittel 2-5 % und schwach < 2 %; der überwiegende Porendurchmesser bei grob > 2 mm, mittel 1-2 mm und fein < 1 mm; cf. BLU 1996); *syn.* Porosität [f], Porenanteil [m].

4490 porosity [n] *constr.* (Properties of a material charcterized by the proportion of the volume of voids to the total volume of the material; ▶air space ratio, ▶pore volume); *s* **porosidad [f]** (Porcentaje entre el volumen de huecos de un material y su volumen total; DACO 1988; ▶contenido de aire, ▶porosidad total); *f* **porosité [f]** (Capacité d'un corps/matériau à présenter des vides le rendant perméable à l'eau et à l'air ; ▶capacité en air, ▶porosité totale) ; *g* **Porosität [f] (2)** (Eigenschaft eines Körpers/Baustoffes durch Poren für Luft und Wasser durchlässig zu sein; ▶Luftgehalt, ▶Porenvolumen).

porosity [n]**, air** *pedol.* *▶pore volume.

porosity [n]**, capillary** *pedol.* *▶pore volume.

4491 porous concrete [n] *constr.* (Composite material with Portland cement, no-fines aggregate of different size grains and water; ▶no-fines concrete); *s* **hormigón [m] poroso** (Hormigón sobre la base de materiales porosos; ▶hormigón de un solo grano); *syn.* concreto [m] poroso [AL]; *f* **béton [m] poreux** (Béton à base d'agrégats poreux ; ▶béton caverneux) ; *g* **was-**

serdurchlässiger Beton [m] (Beton aus grobkörnigen Zuschlagstoffen ohne Feinanteile; ▶Einkornbeton).

porous windbreak [n] *landsc.* ▶wind-penetrable plantation.

4492 porte cochère [n] *arch.* (**1.** Roofed structure over a driveway at the entrance to a building; *syn.* carriage porch [n]. **2.** A doorway large enough to let a vehicle pass from the street to an interior parking area/courtyard); *syn.* roofed vehicular entrance [n]; *s 1* **marquesina [f] de portal** (Estructura con tejado sobre la llegada de vehículos a un edificio); *syn.* techado [m] de entrada de vehículos; *s 2* **puerta [f] cochera** (Entrada ancha de la calle a un patio interior que permite el paso de vehículos); *f 1* **porche [m] d'entrée de/du/au garage** (Auvent, appenti ou préau édifié en saillie d'un bâtiment pour abriter un véhicule devant l'entrée du garage) ; *f 2* **porte-cochère [f]** (*Anciennement: coche* « grande voiture tirée par des chevaux »; baie en forme de grande porte dans la façade d'un bâtiment [maison bourgeoise, hôtel] permettant le passage d'un véhicule [autrefois le cocher pour sa voiture] pour accéder à une cour arrière, une aire intérieure de stationnement, une maison de fond de cour et pouvant être fermée par un portail à deux battants) ; *g 1* **überdachte Gebäudevorfahrt** (Mit einem Dach versehener Eingangsbereich für eine Autovorfahrt); *g 2* **Gebäudedurchfahrt [f]** (Torförmiger Durchlass durch eine Gebäudefront, um mit Fahrzeugen in das Innere einer Blockrandbebauung oder eines Innenhofes zu gelangen).

portion [n] **of a landscape** *landsc.* ▶landscape sector.

portrait [n] **of a profession** *prof.* ▶professional profile.

4493 position statement [n] [US] *leg. plan.* (Declaration of one party to inform another or others of its intentions and invite comments, without trying to come to a mutual agreement; context to serve an **i. n.** information; provided as information; ▶by mutual agreement); *syn.* information notice [n] [UK]; *s* **consultar a (alguien) [loc]** (A diferencia de las decisiones ▶por mutuo acuerdo, **c. a (a.)** no implica necesariamente que se llegue a un acuerdo); *f* **communication pour avis [loc]** (Consultation de toute personne qualifiée lors de l'élaboration d'un plan sans obtenir nécessairement son accord. Contexte le plan est communiqué pour avis ; ▶en accord avec) ; *syn.* consultation [f] pour avis ; *g* **im Benehmen mit [loc]** (*Juristen- und Verwaltungssprache* sich mit jemandem ins Benehmen setzen: mit einer Dienststelle wegen eines Verwaltungsaktes Verbindung aufnehmen, damit diese Gelegenheit zu einer Stellungnahme mit dem Ziel der Verständigung und Kenntnisnahme hat, jedoch ohne Bindung an ein Einverständnis [Einvernehmen]; unterbleibt die nach einer Rechtsvorschrift erforderliche Mitwirkung einer anderen Behörde, wurde sie also nicht ins Benehmen gesetzt, so ist der Verwaltungsakt deswegen nicht nichtig, aber fehlerhaft; cf. § 44 III Nr. 4 VwVerfG; ▶im Einvernehmen mit); *syn.* zur Vernehmlassung geben [loc] [CH].

4494 post [n] (**1**) *arch. constr.* (Round or square-shaped, vertically-positioned structural member to support a horizontal beam; e.g. of a pergola, or as a part of a fence; ▶fence post); *s* **poste [m]** (Pieza alargada redonda o cuadrada, de madera, metal u hormigón utilizada como elemento portante en construcción, p. ej. en pérgola, valla, etc.; ▶poste de valla); *f* **poteau [m]** (Élément en bois en métal ou en béton utilisé comme support de pergola, de clôture etc. ; ▶poteau de clôture); *g* **Pfosten [m]** (Senkrechtes, rundes oder kantiges Holz, Metall- oder Betonelement als tragendes Bauteil, z. B. für eine Pergola, einen Zaun; ▶Zaunpfahl).

post [n] (**2**) *urb. trans.* ▶barrier post, ▶car park post, ▶hinged post, ▶removable post, ▶retractable post, ▶road edge guide post.

post [n], **collapsible** *urb. trans.* ▶hinged post.

post [n], **guide** *trans.* ▶road edge guide post.

postage-stamp yard [n] [US] *gard.* ▶miniature garden.

postal and telecommunications services [n], **local** *trans. urb.* ▶local post services.

post and rail fence [n] *constr. agr.* ▶wooden rail fence.

post-completion advisory services [npl] [UK] *prof.* ▶concluding project review.

4495 post-emergent herbicide application [n] *agr. for. hort. envir.* (Treatment with an herbicide during the growing season; *opp.* ▶pre-emergent herbicide application); *s* **aplicación [f] de herbicidas después de la foliación** (Tratamiento con herbicida durante el periodo/período de crecimiento; *opp.* ▶aplicación de herbicidas antes de la foliación); *f 1* **traitement [m] herbicide de post-levée** (Pour les plantes en général, emploi d'un herbicide en période de croissance active de la végétation ; le traitement herbicide de post-levé précoce associe un herbicide de pré-levée et un herbicide de post-levée permettant un apport avant la levée de la culture mais après celle des plantes adventices ; *opp.* ▶traitement herbicide de pré-levée) ; *f 2* **traitement [m] herbicide de post-semis** (Traitement effectué aussitôt après le semis ; *opp.* traitement herbicide de pré-semis) ; *g* **Herbizideinsatz [m] während der Wachstumsperiode** (▶Herbizideinsatz vor Austrieb); *syn.* Herbizidanwendung [f] während der Wachstumsperiode.

4496 post foundation [n] [US] *constr.* (Constructed column that transfers structural loads to bearing soils or bed rock; cf. MEA 1985); *syn.* pillar foundation [n] [also US], spot foundation [n] [UK]; *s* **cimentación [f] puntual** (Columna de fundamentación que transfiere la carga estructural al subsuelo portante o a la roca); *f* **fondation [f] sur pieux** (Fondation profonde effectuée lorsque les caractéristiques mécaniques de la couche d'assise sont insuffisantes ou la couche d'assise n'est plus directement accessible ; on distingue les pieux à tube battu pilonnés ou moulés, les pieux forés ; COB 1984, 50) ; *syn.* fondation [f] sur puits ; *g* **Punktfundament [n]** (Einzelfundamentpfeiler mit quadratischer, rechteckiger oder runder Grundfläche auf tragfähigem Boden zur Unterstützung von Stahlbetonplatten oder -balken. Ein Fundamentpfeiler mit ausgespartem Hohlraum zum späteren Einbau einer Stütze wird **Köcherfundament** genannt); *syn.* Punktgründung [f].

4497 postgraduate course [n] *prof.* (Study after graduation from high school, college or university); *s* **estudios [mpl] de posgrado** *syn.* carrera [f] de posgrado; *f* **études [fpl] de spécialiation** ; *g* **Aufbaustudium [n]** (Studium, das nach einer vorausgegangenen [wissenschaftlichen] Ausbildung weitere Kenntnisse vermittelt).

potable water [n] [US] *wat.'man.* ▶drinking water (1).

4498 potamal [n] *limn.* (Section of a river covering the ▶barbel zone [UK], ▶bream zone [UK], and ▶flounder zone [UK]; ▶river zone—European terms; ▶potamon); *s* **región [f] ciprinícola (≠)** (Zona inferior de los ríos que comprende la ▶zona de barbo y la ▶zona de madrilla; el término «potamal» se utiliza poco en castellano y se aplica a todo el curso del río, no sólo al tramo inferior del mismo; ▶región salobre, ▶zona biológica de un río, ▶potamium); *f* **potamal [m]** (Dans la classification des cours d'eau basée sur la composition des communautés d'invertébrés vivant sur et dans le fond des rivières, zone des cours d'eau correspondant à la région cyprinicole de la ▶zone à barbeau et de la ▶zone à brème, ▶zone à éperlan, ▶zones écologiques des cours d'eau ; ▶potamon) ; *syn.* région [f] cyprinicole ; *g* **Potamal [n]** (Flussabschnitt, der die ▶Barbenregion, ▶Brachsenregion und ▶Brackwasserregion umfasst; ▶Flussregion, ▶Potamon); *syn.* Cyprinidenregion [f].

4499 potamon [n] *limn.* (Life community of the ▶potamal; ▶epirhithron, ▶rhithron); *s* **potamium** [m] (Biocenosis de los cursos fluviales; en Europa Central se aplica al tramo inferior de los ríos [▶región ciprinícola y ▶región salobre]; ▶epiritrón, ▶ritrón); *f* **potamon** [m] (Terme de zonation écologique des eaux courantes, caractérisant le cours moyen et inférieur à courant faible dans lequel vit la communauté des invertébrés vivant dans la région cyprinicole ; ▶épirhithron, ▶potamal, ▶rhithron) ; *g* **Potamon** [n] (Lebensgemeinschaft des ▶Potamals; ▶Epirhithron, ▶Rhithron).

pot-binding effect [n] [UK] *constr. arb.* ▶bathtub effect [US].

potential [n] *geo.* ▶erosion potential, ▶natural landscape potential; *ecol.* ▶natural potential; *landsc. recr.* ▶recreation potential.

potential [n]**, genetic** *biol.* ▶inherited genes.

potential [n]**, landscape** *geo.* ▶natural landscape potential.

4500 potential groundwater yield [n] *hydr. min. nat' res.* (Possible groundwater yield); *s* **potencial** [m] **hídrico de agua subterránea** (Riqueza de agua subterránea en una zona dada); *f* **potentiel** [m] **hydrique d'un aquifère** (Richesse en eau souterraine espérée et caractérisée par le débit de la nappe) ; *syn.* potentiel [m] hydrique de la nappe ; *g* **Grundwasserhöffigkeit** [f] (*Bergmannssprachlich* Verheißung reicher Grundwasserausbeute resp. eines nutzbaren ergiebigen -vorkommens, das in Wassermenge je Zeiteinheit gemessen wird).

4501 potential natural vegetation [n] *phyt.* (Combination of species that would eventually come into existence under the prevailing environmental conditions of today if humans no longer exerted any influence and if the plant succession had time to reach its final stage); *s* **vegetación** [f] **natural potencial** (Estado final de la vegetación que se establecería en una zona si cesara la influencia humana); *f* **végétation** [f] **naturelle potentielle** (Concept hypothétique et imaginaire décrivant l'état naturel théorique de la végétation affranchie de l'influence de l'homme, en équilibre avec le climat et le sol, sans prise en compte des changements climatiques qui se seraient produits au cours de la longue période requise pour l'évolution naturelle de la végétation. Ce concept est néanmoins de grande valeur car il indique le potentiel biologique de chaque parcelle de territoire ainsi que les tendances de l'évolution constante de la végétation sauvage ; VOC 1979, 58, comptes rendus bibliographiques, http://id.erudit.org/iderudit/020536ar) ; *g* **potentielle natürliche Vegetation** [f] (Rein gedanklich vorzustellende, höchst entwickelte Vegetation — in Mitteleuropa i. d. R. der Wald —, die sich auf einem Wuchsort den gegenwärtigen Standortbedingungen entsprechend einstellen würde, wenn der Einfluss des Menschen aufhörte); *syn. o. V.* potenzielle natürliche Vegetation [f].

4502 potential range [n] *phyt. zool.* (Likely range in which a particular animal or plant species would occur under favo[u]rable conditions; often the previously occupied part of a range of a species; ▶historic range); *s* **área** [f] **virtual (de distribución)** (Territorio que ocuparía una especie de flora o fauna en condiciones favorables; a menudo corresponde con el ▶área histórica); *f* **aire** [f] **potentielle d'une espèce** (Territoire sur lequel une espèce n'a pu être localisée mais pouvant constituer son aire de répartition) ; *g* **potentielles Areal** [n] (Gebiet, von dem angenommen wird, dass eine bestimmte Tier- oder Pflanzenart, die in diesem Areal fehlt, dort ebenfalls vorkommen könnte; oft handelt es sich hierbei auch um das ehemalige [historische] Teilareal einer Art; ▶historisches Areal); *syn. o. V.* potenzielles Areal [n], potentielles Verbreitungsgebiet [n], *o. V.* potenzielles Verbreitungsgebiet [n].

4503 potentials [npl] **of a landscape** *ecol. landsc.* (Possible functional uses and values of natural systems; ▶natural potential); *s* **potencial** [m] **de un paisaje** (Conjunto de recursos naturales y características de un paisaje así como su capacidad funcional; ▶potencial natural); *f* **potentiel** [m] **écologique du paysage** (Complexe d'éléments physico-chimiques et biologiques et d'interactions participant à la dynamique d'un paysage ; ▶potentialité du milieu naturel); *g* **Landschaftspotenzial** [n] (Funktionsvermögen der natürlichen Ressourcen und sonstigen Gegebenheiten einer Landschaft; ▶Naturpotenzial); *syn. o. V.* Landschaftspotential [n].

4504 potential yield [n] *agr. for.* (Possible capacity of an agricultural or forest ecosystem to produce a certain yield from a specified area over a stated period; ▶biomass, ▶soil fertility); *syn.* yield capabilities [npl]; *s* **capacidad** [f] **de rendimiento** (Productividad potencial de ecosistemas agrícolas y silvícolas; ▶biomasa, ▶fertilidad del suelo); *f 1* **productivité** [f] **(1)** (Capacité de production de ▶biomasse des écosystèmes agraires et forestiers dans un milieu donné pour des conditions optimales de culture pendant un temps déterminé ; ▶fertilité du sol) ; *syn.* rendement [m] maximal ; *f 2* **rendement** [m] (Mesure de ce produit par rapport à ce qui a été investi en travail, capital, temps, etc. ou la mesure égale au poids, au volume ou, même, au nombre d'organes végétaux par unité de surface ; cf. DFM 1975 et LA 1981) ; *g* **Ertragsfähigkeit** [f] (Potenzial von Agrar- und Forstökosystemen oder deren Teile, Erträge an ▶Biomasse hervorzubringen; ▶Bodenfruchtbarkeit).

4505 poured-in-place concrete [n] [US] *constr. syn.* cast-in-place concrete [n] [US], in-situ (cast) concrete [n] [UK], site-cast concrete [n] [US]; *s* **hormigón** [m] **vertido «in situ»** (Hormigón fresco preparado directamente y aplicado en obra); *f* **béton** [m] **coulé sur place** (Technique de bétonnage consistant à apporter le béton liquide sur le chantier et à le verser dans les coffrages, des moules ou banches appropriés ; pendant sa mise en œuvre on utilise des aiguilles vibrantes dans le béton pour faire remonter en surface les bulles d'air qui se sont formées pendant son malaxage, son transport et sa mise en œuvre ; la vibration a aussi pour effet une répartition homogène du béton dans le coffrage, autour des armatures et sur les parties visibles ultérieurement) ; *syn.* béton [m] coulé en place ; *g* **Ortbeton** [m] (Auf der Baustelle als Frischbeton hergestellter und verarbeiteter Beton).

4506 poured-in-place concrete slab [n] [US] *constr. syn.* in-situ concrete slab [n] [UK]; *s* **losa** [f] **(de hormigón/ cemento) hecha a pie de obra**; *f* **dalle** [f] **en ciment coulée sur place** (Revêtement de sol ou de terrasse) ; *syn.* chape [f] de béton ; *g* **Betonplatte** [f] **in Ortbeton** (Auf der Baustelle hergestellte Betonplatte, z. B. für einen Gehwegbelag).

poverty neighborhood [n] [US]/**poverty neighbourhood** [n] [UK] *urb.* ▶slum.

power [n]**, absorbing** *bot.* ▶root absorbing power.

power cable [n]**, overhead** *envir.* ▶overhead power line.

4507 power distribution facility [n] *envir. plan.* (Subsidiary station of an electricity generation, transmission and distribution system where voltage is transformed from high to low or the reverse using transformers); *syn.* power distribution plant [n], electrical substation [n] [also US]; *s* **central** [f] **transformadora** (Planta de distribución de electricidad en la que se transforma el voltaje de la electricidad de alta a baja tensión o al revés); *syn.* substación [f] de transformación; *f* **unité** [f] **de distribution électrique** *syn.* installation [f] de distribution électrique ; *g* **Umspannwerk** [n] (Teil eines elektrischen Versorgungsnetzes, das die Stromspannung mit Hilfe von Transformatoren von zwei unterschiedlichen Spannungsebenen oder

Spannungsnetzen verbindet; ein kleines **U.** wird **Umspann-station**, **Transformatorenstation** oder kurz **Trafostation** genannt); *syn.* Stromverteilungsanlage [f].

power distribution plant [n] *plan.* ▶power distribution facility.

power line [n] *envir.* ▶electric power line, ▶high tension power line, ▶overhead power line, ▶underground power line.

4508 power plant [n] *envir.* (Generic term for installations concerned with the production of electricity, e.g. ▶nuclear power station, geo-thermic power station, tidal **p.** station, ▶solar power station, heating power station, and ▶hydroelectric power station); *syn.* power station [n], electric power plant [n]; *s* **central [f] eléctrica** (*Término génerico* instalación donde se produce electricidad; *términos específicos* ▶central nuclear, central térmica, ▶central hidroeléctrica, ▶central solar); *syn.* central [f] de energía; *f* **centrale [f] électrique** (Terme générique pour une installation électrique de production d'énergie telle que la **c.** nucléaire [▶réacteur de puissance électronucléaire], la **c.** thermique, la **c.** géothermique, la **c.** marémotrice, la ▶centrale solaire et la ▶centrale hydroélectrique) ; *syn.* usine [f] électrique, centrale [f] de puissance ; *g* **Kraftwerk [n]** (Anlage der elektrischen Energiegewinnung wie ▶Kernkraftwerk, geothermisches **K.**, Gezeiten**k.**, ▶Sonnenkraftwerk, Wärme**k.** und ▶Wasserkraftwerk); *syn.* Stromerzeugungsanlage [f].

power plant [n]**, electric** *envir.* ▶power plant.

4509 power source [n] *envir.* (Origin or natural resource for producing usable energy. The most important energy sources of today are coal, petroleum, gas, biofuels, sunlight, wind, and water as well as enriched uranium; ▶nonrenewable power source [US]/non-renewable power source [UK], ▶renewable power source); *syn.* energy source [n]; *s* **fuente [f] de energía** (Recurso natural o fuente natural para producir energía. Las más aprovechadas actualmente son el carbón [hulla y lignito], el petróleo, el gas natural, el agua y el uranio enriquecido, y en creciente medida las energías renovables; ▶fuente de energía removable, ▶fuente de energía no renovable); *f 1* **source [f] d'énergie** (Les plus importantes sources d'énergie actuelles sont le charbon, le pétrole, le gaz et l'eau ainsi que l'uranium enrichi ; on distingue les formes d'énergie renouvelables [solaire, éolienne, hydraulique, géothermique, biomasse] et les formes non renouvelables des énergie fossiles [combustibles minéraux solides, hydrocarbures] et de l'énergie nucléaire ; ▶source d'énergie renouvelable, ▶source d'énergie non renouvelable) ; *syn.* source [f] énergétique) ; *f 2* **ressource [f] énergétique** (Éléments naturels qui possèdent une énergie primaire ou à partir desquels, par transformation, il est possible de produire une énergie secondaire) ; *f 3* **énergie [f] finale** (Énergie mise à la disposition du consommateur [énergie mécanique, électrique, thermique et rayonnante] ; cette énergie transformée à partir de ses propres équipements est dénommée **énergie utile**) ; *g 1* **Energiequelle [f]** (Quelle, die Energie durch Umwandlung aus einem Energieträger zur Verfügung stellt; umgangssprachlich werden **E.**, **Energieträger** und **Energie** synonym verwendet, was wissenschaftlich nicht korrekt ist); *g 2* **Energieträger [m]** (Mengenmäßig messbarer Rohstoff, welcher Primärenergie enthält oder aus dem Sekundärenergie gewonnen und übertragen wird; es werden **regenerative** [Sonne, Wind, Wasser, Biomasse und Erdwärme], **fossile** [Kohle, Mineralöl, Erdgas] sowie **nukleare** Formen [angereichertes Uran und Plutonium] unterschieden); *g 3* **Endenergie [f]** (die vom Verbraucher genutzte Sekundärenergie; ▶erneuerbare Energiequelle, ▶nicht erneuerbare Energiequelle).

power source [n]**, alternative** *envir.* ▶renewable power source.

power source [n]**, sustainable** *envir.* ▶renewable power source.

power station [n] *envir.* ▶hydroelectric power station, ▶nuclear power station, ▶power plant, ▶solar power station.

power supply [n] *plan.* ▶electric power supply.

power transmission line [n] [US]**, electric** *constr. envir.* ▶high tension power line.

practical completion [n] *constr. contr.* ▶certificate of practical completion.

practical completion [n] [UK] *constr.* ▶project completion [US].

practical completion [n] [UK]**, situation of** *constr. contr.* ▶condition for final acceptance [US].

4510 practical training [n] *prof.* (Practice period of a student in advance of or during formal studies in order to obtain on-the-job experience in a future professional field); *syn.* internship [n] [also US]; *s* **prácticas [fpl] profesionales** (Periodo/período de formación práctica durante o al final de la carrera para adquirir experiencia en la profesión futura); *f* **stage [m] (de formation)** (Période de formation pratique, préalable ou effectuée pendant les études, imposée à certaines professions) ; *g* **Praktikum [n]** (...ka [pl]; vor oder während der theoretischen Ausbildung abzuleistende praktische Tätigkeit).

practice [n] *constr.* ▶best practice; *prof.* ▶landscape architect in private practice; *landsc. obs.* ▶landscape practice; *plan. prof.* ▶planning practice; *prof.* ▶professional practice;

Practice [n] [UK]**, Codes of** *constr. contr.* ▶standards for construction [US].

practice [n]**, tree maintenance** *arb. constr.* ▶tree maintenance.

4511 practice game area [n] [US] *recr.* (Recreation facility, usually in an urban area, provided for informal ball games, which may be enclosed with high ▶perimeter fencing; ▶free play area); *syn.* kick-about area [n] [UK]; *s* **pista [f] de juegos de pelota** (Zona de recreo de pequeñas dimensiones para juegos informales de pelota rodeada por una valla alta, generalmente en zona urbana; ▶área de juegos libres, ▶enrejado [de pista de juegos de pelota]); *syn.* cancha [f] de juegos de pelota; *f* **espace [m] multisports** (Équipement de loisirs, principalement réservé aux jeux de ballons et à la libre évolution ; de taille réduite plus ou moins normalisé, p. ex. 24 x 12 m ou 20 x 40 m, sur un sol sablé ou stabilisé et entouré d'un haut ▶grillage-pareballon en zone urbaine ; ▶aire de jeux libres) ; *syn.* plateau [m] de jeux libres, aire [f] d'évolution combinée ; *g* **Bolzplatz [m]** (Vornehmlich für Fußballspiele vorgesehene Freizeiteinrichtung von ca. 20 x 40 m Größe im Siedlungsbereich, die i. d. R. mit hohen ▶Ballfanggittern umgeben ist; ▶Tummelplatz); *syn.* Wetzplatz [m] [BW].

prairie [n] *geo. phyt.* ▶steppe, #2.

prairie [n] [US]**, limestone** *phyt.* ▶limestone grassland.

prairie [n] [US]**, tall grass** *phyt.* ▶tall grass steppe.

4512 prairie gardening [n] [US] *constr. landsc.* (Installation and management of dry grassland areas: one cut per year; ▶dry meadow); *s* **jardinería [f] de las praderas norteamericanas (±)** (Cuidado de zonas de los ▶pastizales secos de la región de las Grandes Praderas en los EE.UU.; ▶cuidado de praderas silvestres, ▶praderas silvestres); *f* **aménagement [m] et entretien [m] des pelouses sèches de la Prairie** (U.S., travaux horticoles nécessaires au développement et maintien de ▶pelouses arides et plantation réalisées avec des plantes vivaces de la Prairie américaine ; ▶entretien de prairies extensives ensemencée) ; *g* **Anlage [f], Pflege [f, o. Pl.] und Unterhaltung [f, o.**

Pl.] **naturnaher Prärie-Trockenrasen** (≠) (Alle gärtnerischen Maßnahmen, die zur Anlage und Erhaltung prärieartiger Wiesen und Pflanzungen mit amerikanischen Präriestauden notwendig sind; ►Pflege naturnaher Wiesen, ►Trockenrasen).

prairie-woodland edge [n] [US] *ecol.* ►field-woodland edge [US].

pre-assembled house [n] [UK] *arch.* ►factory-built house [US].

precast concrete element [n] *constr.* ►precast concrete unit.

precast concrete paving slab [n] [UK] *constr.* ►precast paving block [US].

4513 precast concrete tree vault [n] [US] *constr.* (TGG 1984, 191; concrete well with wide lateral openings admitting irrigation and drainage and allowing roots to grow out into the surrounding soil. The open center [US]/centre [UK] of the removable concrete lid and air space between lid and soil mixture permit air circulation and prevent soil compaction; ►tree well); *syn.* precast concrete tree well [n] [UK]; *s* **jardinera [f] de hormigón para árbol** (≠) (Gran tiesto prefabricado de hormigón de unos 3 m Ø y 1,50-2 m de altura, con agujeros laterales para permitir la aereación, el riego y el crecimiento lateral de las raíces. El fondo de hormigón y el espacio libre entre éste y el suelo permiten la circulación del aire y evitan la compactación del suelo; ►murete de protección de árbol); *f* **anneau [m] en béton de protection d'arbre** (Anneau préfabriqué en béton [Ø 3 m, hauteur 1,5-2 m] comportant d'importantes réservations latérales pour les dispositifs de drainage et d'arrosage et afin d'assurer un bon enracinement dans le sol avoisinant. Le couvercle en béton ainsi que l'espace entre le couvercle et le mélange terreux assurent une bonne aération et conservent le sol meuble ; ►muret de protection d'arbre) ; *g* **Betonbaumschacht [m]** (Vorgefertigter ‚Betonblumentopf' [Ø 3 m; Höhe 1,50-2 m] mit großen seitlichen Durchbrüchen für Be- und Entwässerungseinrichtungen und für seitliches Wurzelwachstum in den angrenzenden Boden. Der Betondeckel sowie der Luftraum zwischen Deckel und Bodengemisch dienen der Belüftung sowie der Erhaltung eines lockeren Bodens; ►Baumschacht).

precast concrete tree well [n] [UK] *constr.* ►precast concrete tree vault [US].

4514 precast concrete unit [n] *constr.* *syn.* prefab(ricated) concrete compound unit [n] [also UK], precast concrete element [n]; *s* **pieza [f] prefabricada de hormigón/concreto [AL]** (Elemento de hormigón manufacturado industrialmente, como p. ej. losas de pavimento, adoquines de hormigón, elementos de fachadas, etc.); *f* **élément [m] en béton préfabriqué** (Produit de construction coulé séparément et durci avant son emploi sur le chantier, p. ex. dalles, bordures, élément de mur à semelle, panneaux mur) ; *syn.* béton [m] manufacturé (DTB 1985) ; *g* **Betonfertigteil [n]** (Betonelement, das industriell [in Serienherstellung] vorgefertigt auf die Baustelle geliefert und eingebaut wird, z. B. Gehwegplatten, Mauerscheiben, Pflastersteine, Fassadenelemente); *syn.* Betonfertigbauteil [n], Fertigbetonteil [n].

4515 precast garage [n] *constr.* (Vehicle storage structure that is made in one piece; *NOTE* a prefab[ricated] garage is made in several sections and erected on site); *s* **garage [m] prefabricado** (Estructura prefabricada para guardar vehículos, consistente en varias piezas y eregida «in situ»); *f* **garage [m] préfabriqué** ; *g* **Fertiggarage [f]** (Aus einem Stück vorgefertigte Garage).

4516 precast paving block [n] [US] *constr.* (Industrially prefabricated concrete unit for sidewalks [US]/pavements [UK] and footpaths); *syn.* module paver [n] [US], precast concrete paving slab [n] [UK], precast concrete paver [n] [US]; *s* **losa [f]**

de hormigón (Pieza de hormigón de fabricación industrial utilizada para pavimentar aceras y caminos); *syn.* losa [f] de cemento prefabricada, losa [f] de pavimentación prefabricada; *f* **dalle [f] préfabriquée en béton pour voirie de surface** (Terme générique pour les matériaux normalisés fabriqués industriellement utilisés comme revêtement de voirie et de sols extérieurs) ; *syn.* dalle [f] béton pour sols extérieurs, dalle [f] béton pour sentier, dalle [f] béton pour cheminement ; *g* **Betongehwegplatte [f]** (OB für eine im Straßen-, Garten- und Landschaftsbau zur Befestigung von Bürgersteigen, Gartenwegen und Sitzterrassen industriell gefertigte Betonplatte, die hinsichtlich Größe und Biegezugfestigkeit genormt ist); *syn.* Betongehwegplatte [f], Bürgersteigplatte [f], Gartenplatte aus Beton, Gehwegplatte [f] aus Beton.

4517 precast pool [n] *constr.* (Prefab[ricated] pool for ornamental ponds; e.g. fiberglass garden pool); *s* **estanque [m] prefabricado** (Pequeño estanque para decoración de jardines de materiales como fibra de vidrio o polietileno, que se pueden adquirir en diferentes tamaños y profundidades); *f* **bassin [m] préfabriqué** (Petit bassin d'agrément dans un jardin ou sur une terrasse, en général en résine armée de fibres de verre ou en polyéthylène, de formes multiples [zones de différentes profondeurs] pour une capacité pouvant atteindre 3000 litres) ; *g* **Fertigbecken [n]** (Vorgefertigtes Zierbecken aus Glasfieber, Polyäthylen o. Ä. für kleine Wasseranlagen in Hausgärten, lieferbar in unterschiedlichen Größen, mit unterschiedlichen Wassertiefen).

precaution(ary) measures [npl] *plan. pol. sociol.* ►preventive measures.

4518 precautionary measures [npl] **against noise** *envir. plan.* (Preventive measures for noise reduction or attenuation; ►noise control); *s* **medidas [fpl] de prevención del ruido** (►protección contra el ruido); *f* **prévention [f] contre le bruit** (Prescriptions, dispositions réglementaires relatives aux objets, dispositifs et activités destinées réduire les nuisances sonores par limitation à la source du bruit ; ►protection contre le bruit) ; *g* **Lärmvorsorge [f, selten Pl.]** (Vorbeugende Maßnahmen und Vorkehrungen zur Lärmminderung; ►Lärmschutz).

4519 precautionary principle [n] *pol. sociol.* (Political postulate aimed at providing guaranteed protection against unexpected natural phenomena or uncontrolled technical developments and their negative effects upon human health, animals and plants, as well as upon the overall environment); *s* **principio [m] de prevención (de deterioros ambientales)** (Postulado que rige la política ambiental en muchos países y que implica que las actividades que tengan influencia sobre el medio deben tratar de evitar de antemano posibles daños sobre el medio ambiente); *syn.* principio [m] de precaución; *f 1* **principe [m] de précaution** (Principe selon lequel l'absence de certitudes, compte tenu des connaissances scientifiques et techniques du moment, ne doit pas retarder l'adoption de mesures effectives et proportionnés visant à prévenir un risque de dommages graves et irréversibles à l'environnement à un coût économiquement acceptable ; C. rural, art. L. 200-1) ; *f 2* **principe [m] d'action préventive et de correction** (Principe selon lequel une action préventive des atteintes à l'environnement est effectuée, par priorité à la source, en utilisant les meilleures techniques disponibles à un coût économiquement acceptable ; C. rural, art. L. 200-1) ; *g* **Vorsorgeprinzip [n]** (Politischer Grundsatz oder politisches Postulat des garantierten Schutzes vor überraschenden Naturereignissen oder ungezügelten technischen Entwicklungen und deren negativen Wirkungsketten auf die Gesundheit von Mensch, Tier und Pflanze sowie auf die Umwelt als Ganzes; zum **V.** gehören auch der Grundsatz Vorkehrungen zu treffen, um Pflanzen und Tiere nicht erst das Gefährdungsstadium erreichen zu lassen).

P

4520 precipice [n] *geo.* (Vertical, high steep rock face; ▶coastal cliff, ▶scarp slope); *s* **precipicio** [m] (Escarpadura de mucha altura y pendiente; ▶costa acantilada, ▶frente de cuesta); *syn.* despeñadero [m]; *f* **précipice** [m] (Versant rocheux très raide, presque perpendiculaire ; ▶côte à falaises, ▶versant anaclinal) ; *syn.* à pic [m] ; *g* **Abhang** [m] (Sehr steile, oft fast senkrechte Felspartie; ▶Steilküste, ▶Stufenhang); *syn.* Steilhang [m].

4521 precipitation [n] (1) *met.* (Generic term for rain, hail, mist droplets, hoar-frost, snow, dew; ▶acid rain, ▶amount of precipitation, ▶heavy rainfall); *s* **precipitación** [f] (Hidrometeoro formado por un conjunto de partículas acuosas, líquidas o sólidas, cristalizadas o amorfas, que caen procedentes de una nube o grupo de nubes y llegan hasta el suelo; DM 1986; término genérico para lluvia, llovizna [*términos locales* sirimiri [m] (EUS), orballo [m] (GA)]; rocío, niebla, escarcha, granizo, nieve; ▶lluvia ácida, ▶pluviosidad, ▶lluvia torrencial); *f* **précipitations** [fpl] (Toutes formes d'eau tombant du ciel sur la surface de la Terre ; on distingue deux types de précipitations : les précipitations stratiformes qui sont d'une faible intensité mais qui durent longtemps et les précipitations convectives qui sont d'une forte intensité et peuvent être accompagnées d'orage ou de grêle mais qui ne durent pas longtemps. Les précipitations peuvent tomber sous plusieurs formes suivant la température de l'air : forme liquide [pluie, bruine], forme verglacée [pluie et bruine verglaçante], forme solide [neige, grésil et grêle], forme déposée ou occulte [rosée, gelé blanche, givre] et forme particulière [▶pluie acide, pluie de sable] ; ▶pluie torrentielle, ▶quantité de précipitations) ; *g* **Niederschlag** [m] (Wasser, das aus Wasserdampf inclusive seiner Verunreinigungen, aus der Atmosphäre in flüssiger [Regen, Sprühregen, nässender Nebel] oder fester Form [Eiskörner, Graupel, Hagel, Schnee, Schneegriesel] auf die Erde fällt, oder sich als **abgesetzter N.** in Form nässenden Nebels, Taus und Reifs an Vegetation und inerten Objekten absetzt; ▶saurer Regen, ▶Niederschlagsmenge, ▶Starkregen).

precipitation [n] (2) *envir. limn. pedol.* ▶dust precipitation.

precipitation [n]**, acid** *envir.* ▶acid rain.

precipitation [n]**, range of** *met. wat'man.* ▶precipitation regime.

4522 precipitation regime [n] *met. wat'man.* (DM 1986; distribution of varying amounts of precipitation in a given period of time and in a defined area; e.g. in a month, year); *syn.* range [n] of precipitation; *s* **régimen** [m] **de las precipitaciones** (Carácter de la distribución estacional de la precipitación en un determinado lugar. Depende de la influencia de los procesos atmosféricos, la estructuración morfológica de la superficie y de las influencias del calentamiento desigual de las aguas costeras y de la superficie terrestre; cf. DM 1986 y SILV 1979, 36; *términos específicos* régimen pluvial/pluviométrico, régimen nival); *f* **régime** [m] **des pluies** (Caractère de la répartition des hauteurs de précipitations, p. ex. entre les divers mois de l'année en un point quelconque ; CILF 1978) ; *syn.* régime [m] pluvial, régime [m] des précipitations ; *g* **Niederschlagsverteilung** [f] (Mengenangaben von Niederschlägen in einem bestimmten Gebiet in Zeiteinheiten, z. B. Monat, Jahr); *syn.* Niederschlagsregime [n].

4523 precipitation water [n] *met. wat'man.* (Generic term covering rainwater [US]/rain-water [UK], water of melted snow and hail); *s* **agua** [f] **de precipitación** (Término genérico para agua de lluvia y agua de deshielo de nieve o granizo); *f* **eau** [f] **de précipitations** (*Terme spécifiques* eau pluviale, eau en provenance de la fonte de la neige ou de la grêle) ; *g* **Niederschlagswasser** [n] (OB zu Regenwasser, Wasser aus geschmolzenem Schnee und Hagel).

4524 precocial animal [n] *zool.* (Young born in a developed state with feathers and open eyes; e.g. hens, ducks, wingless birds, hares, and hoofed animals. Precocial birds already have open eyes, when they hatch out, well-developed plumage, and functioning legs; *opp.* ▶altricial animal); *syn.* precocial nestling [n]; *s* **animal** [m] **nidífugo** (Al contrario que el ▶animal nidícolo, uno que nace en un estado muy desarrollado de manera que puede abandonar relativamente pronto su lugar de nacimiento resp. su nido, como p. ej. pollos, patos, liebres y especies de ungulados); *f* **animal** [m] **nidifuge** (Par opposition aux ▶animaux nidicoles, jeunes se développant sans bénéficier de l'aide de leurs parents et quittant rapidement le nid et l'aire de nidification, p. ex. les gallinacés, canards, limicoles, lièvres et ongulés ; les oiseaux nidifuges possèdent à l'éclosion, des yeux ouverts, un plumage bien développé et des pattes aptes à la marche ; LAF 1990, 828) ; *g* **Nestflüchter** [m] (Im Gegensatz zum ▶Nesthocker im weit entwickeltem Zustand geborenes Tier, das schnell den Geburtsort resp. das Nest verlässt; z. B. Hühner-, Enten- und Laufvögel, Hase und Huftiere. **N.** unter den Vögeln haben beim Schlüpfen bereits geöffnete Augen, ein gut ausgebildetes Nestlingskleid und funktionsfähige Beine).

precocial nestling [n] *zool.* ▶precocial animal.

preconditions [npl] **for tenderers** [UK] *contr.* ▶bidding requirements [US].

4525 pre-construction planting [n] *landsc. urb.* (Advance planting prior to rehabilitation of abandoned commercial or industrial sites as well as to future urban developments before building, using young plants, adapted to site soil and climatic conditions, as the landscape architectural approach to immediately improving the appearance of the sites); *s* **plantación** [f] **anticipada** (El término francés «préverdissement» denomina un planteamiento paisajístico de plantar las áreas no utilizadas o antiguas zonas industriales antes de su reutilización para fines urbanos); *f* **préverdissement** [m] (Approche paysagère ayant pour objectif de végétaliser les friches industrielles ou les zones à urbaniser grâce à des mesures appropriées de plantation de jeunes plants indigènes effectuées avant le début des travaux) ; *g* **Begrünung** [f] **vor Baubeginn** (≠) (Grünplanerischer Ansatz, bei dem z. B. Industriebracheflächen oder zukünftige Baugebiete durch gezielte Bepflanzungsmaßnahmen vor Beginn der Bauarbeiten mit jungen, [einheimischen] Pflanzen, besonders Bäumen, eingegrünt werden); *syn.* vorgezogene Begrünung [f].

4526 precontract investigation [n] **of site conditions** *constr.* (Prior check for determining, e.g. existing utilities, underground construction, and previous damages for which the contractor is not responsible, before beginning of the work [US]/works [UK]); *s* **inspección** [f] **del solar** (Estudio del terreno de obra antes del comienzo de la misma para comprobar la existencia de conducciones o instalaciones, o de daños que no son responsabilidad del contratista); *f* **relevé** [m] **de l'état des lieux** (Établissement par l'expert d'un constat de l'état des lieux avant le début des travaux) ; *syn.* dresser [vb] l'état des lieux ; *g* **Bestandsaufnahme** [f] **vor Baubeginn** (Feststellung von vorhandenen Leitungen, Einrichtungen und evtl. Mängeln, die der Auftragnehmer nicht zu vertreten hat, auf einem Baugelände vor Beginn der Bauarbeiten).

precultivated mat [n] *constr.* ▶planting mat.

precut paving stone [n] *constr.* ▶natural stone flag.

4527 predator [n] *zool.* (Animal that lives by preying on other animals; ▶predatory bird); *s* **depredador** [m] (Animal que persigue a otros animales a los que atrapa y mata para obtener alimento; cf. DINA 1987; ▶ave rapaz); *syn.* predador [m]; *f* **prédateur** [m] (Un animal qui se nourrit de proies, c.-à-d. de matière organique animale ; ▶oiseau de proie) ; *g* **Prädator** [m]

(Tierart [Säugetier, Vogel, Wirbellose], die zur eigenen Ernährung andere Tiere erbeutet; ▶Raubvogel); *syn.* Beutegreifer [m] (ohne Wirbellose), Räuber [m], Fressfeind [m].

4528 predatory bird [n] *zool.* (Predacious bird that preys on other animals for food; e.g. ▶raptor); *syn.* bird [n] of prey; *s* **ave [f] rapaz** (Orden de aves carnívoras que en Europa tiene cuatro familias: dos de ▶rapaces diurnas [accipítridos y falcónidos] y dos de rapaces nocturnas [titónidos y estrígidos]); *syn.* ave [f] depredadora, ave [f] de presa, ave [f] de rapiña, rapaz [f] (rapaces [fpl]); *f* **oiseau [m] de proie** (*Terme spécifique* terme peu utilisé par les spécialistes pour dénommer les ▶rapaces diurnes et nocturnes, mais souvent employé par les amateurs de la pratique ancestrale de la chasse au vol et désignant les oiseaux carnivores qui s'alimentent d'animaux qu'ils capturent vivants) ; *g* **Raubvogel [m]** (Fachsprachlich nicht mehr verwendeter OB zu ▶Greifvogel und Eule).

4529 predatory price [n] [US] *constr. contr.* (Sum which is bid [US]/tendered [UK] below market price to counter competition); *syn.* dumping price [n] [UK]; *s* **precio [m] dumping** (Bien o servicio profesional ofrecido a un precio mucho más bajo que el normal); *f* **prix [m] bradé** (Prix d'une marchandise ou d'une prestation inférieur aux prix de revient des concurrents) ; *syn.* prix [m] cassé ; *g* **Dumpingpreis [m]** (Geldbetrag, der deutlich unter dem handelsüblichen Preis für eine Ware oder Leistung angeboten wird).

4530 pre-emergent herbicide application [n] *agr. for. hort. envir.* (Treatment with an herbicide before the growing season; *opp.* ▶post-emergent herbicide application); *s* **aplicación [f] de herbicidas antes de la foliación** (Empleo de herbicida sobre la zona de cultivo antes del comienzo del periodo/período de crecimiento; *opp.* ▶aplicación de herbicidas después de la foliacíon); *f 1* **traitement [m] herbicide de pré-levée** (Pour les plantes en général, emploi d'un herbicide sur une surface plantée après la préparation du sol et avant le début de la période de végétation ; *opp.* ▶traitement herbicide de post-levée) ; *f 2* **traitement [m] herbicide de pré-semis** (Application d'un herbicide effectuée avant le semis de la culture ; *opp.* ▶traitement herbicide de post-semis) ; *g* **Herbizideinsatz [m] vor Austrieb** (Ausbringen eines Herbizidmittels auf eine Vegetationsfläche vor Wachstumsbeginn. Die Ausbringung in flüssigem Zustand wird **Herbizidspritzung** genannt; *opp.* ▶Herbizideinsatz während der Wachstumsperiode); *syn.* Herbizidanwendung [f] vor Austrieb, Herbizidbehandlung [f] vor Austrieb.

pre-emption [n] [UK]**, right of** *adm. leg.* ▶right of first refusal [US].

preemptive option [n] [US] *adm. leg.* ▶right of first refusal [US].

pre-emptive right [n] [UK] *adm. leg.* ▶right of first refusal [US].

prefab [n] [US] *arch.* ▶factory-built house [US].

prefab(ricated) concrete compound unit [n] [UK] *constr.* ▶precast concrete unit.

4531 prefabricated concrete panel construction [n] *urb.* (Industrial building method developed in the 1960s in the former GDR and Eastern Europe, which allowed the less expensive construction of housing with prefabricated concrete elements; ceilings and walls as large as the rooms with already inserted windows and doors were so manufactured that they only need to be assembled on site to construct a building. Under the leadership of LE CORBUSIER, the Charter of Athens [1933] propagated the ideal of a new town [The functional city] whereby residential buildings are of uniform design and loosely arranged—with the advantage of large green spaces between them—and to prevent social segregation; with a growing urban population, the need arose for such large settlements to be constructed at low cost and within a short period of time, meaning that production had to be standardized; with post-World War II housing shortages and Communist ideology, planners in Communist countries were required to provide large quantities of affordable housing and to cut costs by employing uniform housing designs, thus fostering the ideals of a collective. In U.K., prefabricated housing after World War II was looked upon as a short-term solution; such houses were known as **pre-fabs.**; housing subdivision [US]/housing estate [UK] constructed with pre-fabricated panels; ▶development of prefabricated panel buildings; *syn.* precast concrete panel construction, prefabricated construction with concrete slabs; *s* **construcción [f] prefabricada de placas de hormigón/concreto** (Forma de construcción industrial de viviendas característica de los antiguos países socialistas y de Cuba que tenía como fin construir viviendas de bajo precio con ayuda de piezas prefabricadas; ▶polígono residencial de edificios prefabricados con grandes placas); *f* **construction [f] en plaques de béton préfabriquées** (Forme de construction caractéristique de l'urbanisme des dirigeants des anciens pays socialistes ; l'objectif dans la ville était de supprimer toute ségrégation sociale par l'habitat ; guidé par une démarche de rentabilité des coûts par l'industrialisation et la standardisation, ce type de construction fut appliqué à partir des années 60 en RDA selon le concept de la « ville compacte » une politique de construction de grands immeubles à partir de plaques de béton préfabriquées [Plattenbau], érigés selon un type de construction standard et construits la plupart du temps à la périphérie des centres urbains ; ▶grand ensemble construit en plaques de béton préfabriqué) ; *g* **Plattenbauweise [f]** (In den 1960er-Jahren in der ehemaligen DDR und in Osteuropa entwickeltes industrielles Bauverfahren, bei der preisgünstiger Wohnungsbau durch vorgefertigte Stahlbetonelemente ermöglicht wurde; raumgroße Deckenplatten und Wandscheiben mit eingesetzten Fenstern und Türen wurden so gefertigt, dass sie auf der Baustelle zusammengesetzt werden konnten. Auf der Grundlage der Charta von Athen [1933], bei der unter Federführung von LE CORBUSIER eine neue Stadtplanung [Die funktionale Stadt] mit aufgelockerter und gleichförmiger Bauweise zur Vermeidung von sozialer Segregation propagiert wurde, ergab sich bei zunehmenden Einwohnerzahlen in den Städten die Notwendigkeit, Großsiedlungen in kostengünstiger und bauzeitenverringernder Bauweise mit Elementen standardisierter Serienproduktion zu bauen; ▶Plattenbausiedlung); *syn.* Großtafelbauweise [f].

4532 preferential species [n] *phyt.* (Species abundantly present in several communities but predominantly or with better vitality in one certain community; ▶degree of fidelity); *s* **especie [f] preferente** (Especie más o menos abundante en varias comunidades, pero con preferencia por una determinada; BB 1979; ▶grado de fidelidad); *f* **espèce [f] préférante** (Espèce existant plus ou moins abondamment dans plusieurs groupements et préférant cependant un groupement déterminé ; ▶degré de fidélité ; BB 1928) ; *g* **holde Art [f]** (Art, die hinsichtlich des ▶Treuegrades bei Pflanzengesellschaften in mehreren Gesellschaften reichlich vorkommt, aber vorwiegend oder mit starker Vitalität in einer bestimmten Gesellschaft anzutreffen ist).

pregrown mat [n] *constr.* ▶planting mat.

pre-handover maintenance [n] [UK] *contr.* ▶project development maintenance [US].

4533 preliminary agrarian structure planning [n] *agr. plan.* (Preliminary planning for the development of consolidation program[me]s, which involves the preparation of proposals for ▶improvement of agrarian structure, in accordance with the objectives of state and regional land use policy);

s **planificación [f] estructural agraria** (Fase anterior a la ▶ concentración parcelaria en la que se plantean metas y proposiciones para la mejora de la estructura económico-agraria, considerando los fines de la planificación regional y del territorio; ▶ mejora de la estructura agraria); *f 1* **étude [f] préopérationnelle d'aménagement agricole (≠)** (D., planification rurale phase de planification préliminaire préalable au remembrement rural définissant les orientations d'▶ amélioration des structures rurales tout en tenant compte des directives de l'aménagement du territoire et des orientations régionales) ; *f 2* **plan [m] d'aménagement rural (PAR)** (F., le PAR a pour objectif de définir les perspectives souhaitables de développement des activités socio-économiques et de localisation des équipements, à contribuer à la préservation de l'espace naturel sur les territoires à vocation rurale ; décret n° 70-487 du 8 juin 1970) ; *g* **agrarstrukturelle Vorplanung [f]** (Die der Flurbereinigung vorausgehende Planung zur Entwicklung von Zielvorstellungen und Vorschlägen für die Verbesserung der Agrarstruktur, unter Einbeziehung der Ziele der Raumordnung und Landesplanung; ▶ Agrarstrukturverbesserung).

4534 preliminary assessment [n] **of programme and project impacts in terms of Natura 2000** *conserv. leg. plan.* (Under the EC ▶ Habitats Directive—*Council Directive 92/43/EEC*—on the conservation of natural habitats and of wild fauna and flora, the first phase of an ▶ environmental impact assessment [EIA] [US]/environmental assessment [UK] of a construction program[me] or project to establish whether a full assessment is required. The network of protected areas is known as Natura 2000); *s* **elaboración [f] de informe de sostenibilidad ambiental de planes y proyectos sobre un hábitat de la red Natura 2000** (Según el art. 6.3 de la ▶ Directiva de Hábitats cualquier plan o proyecto que pueda afectar de forma apreciable a una zona de la Red Natura 2000 debe ser sometido a una adecuada evaluación de sus repercusiones en el lugar. En Es. los instrumentos para implementar esta norma son la ley 9/2006, sobre evaluación de los efectos de determinados planes y programas en el medio ambiente, así como el RD legislativo 1302/1986, de evaluación de impacto ambiental [modificado por la ley 9/2006] y su reglamento RD 1131/1988 que regulan la evaluación de proyectos); *f* **pré-diagnostic [m] de l'étude (d'évaluation) d'incidence Natura 2000** (Première phase de l'évaluation des incidences d'un PPTOA ; elle a pour but d'établir la nécessité d'une étude d'évaluation d'incidences selon que les programmes ou les projets de travaux, d'ouvrage ou d'aménagement [PPTOA], non directement liés ou nécessaires à la gestion du site, sont susceptibles d'affecter le site Natura 2000 [Zones de Protection Spéciale — ZPS — et Zones Spéciales de Conservation — ZSC] de manière significative, individuellement, ou en conjugaison avec d'autres plans et projets ; il fournit une description du programme ou projet par rapport au site ou au réseau Natura 2000 ainsi qu'une analyse des effets notables [temporaires, permanents, directs et indirects, cumulatifs, etc.] sur l'état de conservation ; si le projet se poursuit et si des effets notables sont suspectés, le diagnostic dans l'étude d'incidence précisera les mesures de suppression ou de réduction de ces effets et les dépenses correspondantes ; si des effets persistent après ces mesures, le pétitionnaire devra justifier les raisons d'intérêt public et exposer les mesures compensatoires, conformément à la Directive 92/43/CEE du Conseil des Communautés Européennes du 21 mai 1992 concernant la conservation des habitats naturels ainsi que de la faune et de la flore sauvages et au Code de l'environnement art. L 414-4 ; ▶ Directive Habitat Faune Flore [DHFF]) ; *syn.* évaluation [f] préalable ; *g* **FFH-Vorprüfung [f]** (Erste Phase bei einer Flora-Fauna-Habitat-Verträglichkeitsprüfung, in der vor Zulassung oder Durchführung eines Planes oder Projektes festgestellt wird, ob dieser Plan oder dieses Vor-

haben geeignet ist, die Erhaltungsziele eines Gebiet von gemeinschaftlicher Bedeutung [Natura-2000-Gebiet] oder ein Europäisches Vogelschutzgebiet erheblich zu beeinträchtigen. Bei Vorliegen von Beinträchtigungen ist eine vertiefte FFH-Verträglichkeitsprüfung erforderlich; diese Prüfung ist in der ▶ Fauna-Flora-Habitat-Richtlinie 92/43/EWG gefordert und im § 34 BNatSchG geregelt.

4535 preliminary cost estimate [n] *contr.* (Rough calculation of total project costs, used as the provisional basis for financial considerations; ▶ calculation of costs); *syn.* rough cost estimate [n]; *s* **estimación [f] aproximada de costos** (Cálculo a grosso modo de los costes [Es]/costos [AL] totales de un proyecto como base para planificar el financiamiento del mismo; ▶ cálculo de costos); *syn.* estimación [f] aproximada de costes [Es], presupuesto [m] de orientación, presupuesto [m] aproximado (de costos/costes [Es]); *f* **estimation [f] sommaire du coût des travaux** (Estimation des dépenses de premier établissement établie pendant la phase d'Avant-Projet Sommaire ; HAC 1989, 33 ; ▶ estimation financière des travaux) ; *syn. pour l'architecte* estimation [f] globale indicative ; *g* **Kostenschätzung [f]** (Überschlägige Ermittlung der Gesamtkosten eines Projektes im Rahmen der Vorplanung als vorläufige Grundlage für Finanzierungsüberlegungen nach DIN 276 oder nach dem wohnungsrechtlichen Berechnungsrecht; ▶ Kostenermittlung); *syn.* Kostenvoranschlag [m].

4536 preliminary design [n] *plan. prof.* (Preparatory drawing for a project, sometimes accompanied by alternative solutions, after ▶ schematic design phase [US]/outline and sketch scheme proposals [UK] [work stage C&D], ▶ final project design); *s* **anteproyecto [m]** (Dibujo preparatorio para un proyecto de un jardín, parque o plan, a veces con varias alternativas; ▶ anteproyecto detallado, ▶ fase de anteproyecto); *f* **avant-projet [m] sommaire (A.P.S.)** (1. Parti général d'aménagement retenu par le maître d'ouvrage et comprenant une notice descriptive et explicative sommaire, un dossier de solution d'ensemble [plan masse, plans de principe, croquis, esquisses, notes de calcul, etc.], une évaluation sommaire globale des dépenses. 2. Élément normalisé de l'étape de l'▶ avant-projet dans le cas d'une mission normalisée de maîtrise d'œuvre ; ▶ avant-projet détaillé [APD]) ; *g* **Vorentwurf [m]** (1. Zeichnerische Darstellung und Bewertung eines ersten Planungskonzeptes, evtl. mit Variantenlösungen und erläuterndem Text. 2. A., Bezeichnung der Leistungsphase ▶ Vorplanung gem. § 22 lit. a HRLA; ▶ Entwurf).

4537 preliminary injunction [n] **on alteration** *leg. plan.* (Legal ruling [after the decision has been taken to prepare a zoning map/binding land-use plan] to prevent additions or alterations to private property merely to increase its value or which would inhibit planned public use, prior to adjudication at a hearing; ▶ temporary restraining order [US]/emergency preservation notice [UK]; a **permanent injunction [n]** is a denial of property owner's application); *s* **prohibición [f] provisional de alteración** (Ordenanza para una zona en fase de planificación de no realizar cambios que puedan dificultar la implementación del plan previsto o que causen aumentos de valor de los predios o los edificios situados en ellos; ▶ anotación preventiva, ▶ declaración de régimen de protección preventiva); *f* **sursis [m] à statuer** (Décision administrative par laquelle l'administration se refuse à se prononcer immédiatement, de façon positive ou négative, sur des demandes de travaux, constructions ou installations de nature à compromettre ou à rendre plus onéreuse l'exécution d'un plan en préparation ; cf. C. urb., art. L. 111-7 à 11 ; ▶ classement d'office) ; *g* **Veränderungssperre [f]** (Von einer Gemeinde beschlossene Satzung für Planungsgebiete mit dem Ziel, dass auf deren Flächen wesentlich wertsteigernde oder die Durchführung von geplanten Vorhaben erheblich erschwerende Veränderungen nicht vorgenommen werden dürfen; cf. § 14 ff BauGB, § 36 a WHG, § 9 a FStrG, § 34 FlurbG. Die **V.** ergeht als Satzung resp. Rechtsverordnung und wird durch Ablehnung eines Bauantrages durchgesetzt und kann wie jede hoheitliche Belastung im Rechtsstaat gerichtlich im Wege der Normenkontrolle überprüft werden. Die **V.** ist an das Plankonzept gebunden, das zum Zeitpunk ihres Erlasses verfolgt wurde, d. h. ein Austausch der Planungsabsichten führt zu ihrer Unwirksamkeit; eine weniger einschneidende oder vorläufige Maßnahme ist die Zurückstellung von Baugesuchen; ►einstweilige Sicherstellung); *syn.* Bausperre [f] [A].

4538 preliminary investigation [n] *plan.* (Assessment of all aspects of a project, which may include a topographic survey, before formulating the best planning solution); *s* **estudio** [m] **preliminar** (Investigación inicial en el marco de una planificación para buscar la solución óptima a la tarea que se plantea); *syn.* investigación [f] preliminar; *f 1* **étude** [f] **préalable** (Étude destinée à faire avancer la réflexion sur des aspects d'aménagement, d'éclairer des choix, d'envisager des scénarios dans le cadre de l'établissement de documents d'urbanisme) ; *syn.* étude [f] préliminaire ; *f 2* **étude** [f] **préopérationnelle** (Précède les études opérationnelles et destinée à préciser les interventions et l'ordre des opérations, définir les périmètres opérationnels, mettre au point le programme etc. dans le cadre d'une opération d'aménagement) ; *g* **Voruntersuchung** [f] (Untersuchung eines Bearbeitungsgebietes vor Planungsbeginn, um eine optimale Lösung hinsichtlich der Planungsaufgabe zu finden).

4539 preliminary planning proposal [n] *leg. urb. trans.* (Graphic and written presentation of proposed land development for an area, incorporating comments and directives of various ►public and semipublic authorities [US]/public and semi-public authorities [UK] for consideration by the public. **In U.K.**, a **p. p. p.** for a Proposals Map or public enquiry is called 'deposit draft'); *syn.* draft planning proposal [n], deposit draft [n] [also UK]; *s* **ante-proyecto** [m] **de plan de ordenación** (Representación gráfica de los usos del suelo previstos en el perímetro municipal o de partes de él teniendo en cuenta las proposiciones de las ►autoridades e instituciones públicas y semipúblicas); *f* **projet** [m] **de plan arrêté** (Forme d'un document de planification urbaine soumis à la délibération des organes chargés d'arrêter le plan **a]** pour le SCOT ayant au préalable fait l'objet d'un débat pour définir les orientations du projet d'aménagement et de développement durable ou PADD avant ou **b]** après avis des ►personnes publiques [associées à l'élaboration du plan] avant d'être mis à la disposition du public [schéma directeur] ou **c]** pour le PLU ou le POS avant d'être soumis à l'enquête publique) ; *g* **vorläufige Planfassung** [f] (Erläuterungsbericht und zeichnerische Darstellung der zweckmäßigen und städtebaulich erwünschten Flächennutzung eines Gemeindegebiets oder Teilen davon unter Berücksichtigung der Anregungen und Hinweise der ►Träger öffentlicher Belange).

4540 preliminary project planner [n] [US] *plan. prof.* (Planner who is charged with the preliminary planning phase); *syn.* scheme designer [n] [UK]; *s* **planificador, -a** [m/f] **de ante-proyecto** (≠) (Profesional encargado de la primera fase de planificación); *f* **concepteur-maître d'œuvre** [m] (Titulaire d'un marché d'étude portant sur une mission de première catégorie) ; *g* **der/die mit der Vorplanung beauftragte Planer/-in** [loc] (Jemand, der die Vorplanung erstellt).

premature leaf fall [n] [US] *bot. hort.* ►premature shedding of leaves.

4541 premature shedding [n] **of leaves** *bot. hort.* (Early defoliation caused by poor growing conditions or disease); *syn.* premature leaf fall [n] [also US], leaf drop [n] [also US];

s **defoliación** [f] **prematura** (Caída prematura de las hojas debida a cambios bruscos del ambiente [heladas tardías] o a plagas de insectos o a hongos patógenos; cf. DB 1985); *f* **chute** [f] **prématurée des feuilles** (Chute provoquée par des conditions climatiques défavorables ou par des insectes défoliateurs ou des agents ou microorganismes patogènes) ; *g* **frühzeitiger Laubfall** [m, o. Pl.] (Durch schlechte Standortbedingungen oder durch Krankheit bereits im [Spät]sommer verursachtes Abfallen der Blätter von Bäumen und Sträuchern); *syn.* prämaturer Laubfall [m, o. Pl.], vorzeitiger Laubfall [m, o. Pl.].

premixed concrete [n] [US] *constr.* ►ready-mixed concrete.

4542 preparation [n] **(1)** *contr. prof.* (Development of ►total construction cost estimate [US]/final construction cost estimate [n] [UK], ►list of bid items and quantities [US]/schedule of tender items [UK], ►specifications, ►price list, fee invoice); *s* **elaboración** [f] (Proceso de desarrollo de un ►presupuesto de gastos, del ►resumen de prestaciones, de una ►lista de precios, de honorarios, etc.); *f* **établissement** [m] (Processus d'►estimation du coût prévisionnel des travaux, d'un ►devis, ►descriptif quantitatif, des honoraires, d'une ►série des prix) ; *g* **Aufstellung** [f] (Vorgang der Erarbeitung eines ►Kostenanschlages, ►Leistungsverzeichnisses, ►Preisverzeichnisses, einer Honorarrechnung); *syn.* Erstellung [f].

preparation [n] **(2)** *constr. contr.* ►work preparation [US]/works preparation [UK].

preparation [n] **for planting** *constr. hort.* ►planting preparation.

4543 preparation [n] **of a community development plan** [US]/**preparation** [n] **of an urban development plan** [UK] *adm. plan.* (Detailed development of community planning documents, such as a ►community development plan [US]/urban development plan [UK]); *s* **elaboración** [f] **de un plan** (Desarrollo detallado de los documentos necesarios para un ►plan general municipal de ordenación urbana o un ►plan parcial [de ordenación]); *f* **établissement** [m] **des documents d'urbanisme** (Procédure formalisée en vue de l'élaboration des ►documents de planification urbaine, en conformité aux articles L122-1 à L122-19 pour les Schémas de cohérence territoriale, aux articles L123-1 à L123-20 pour les Plans locaux d'urbanisme et aux articles L124-1 à L124-4 pour les Cartes communales) ; *syn.* élaboration [f] des documents d'urbanisme ; *g* **Aufstellung** [f] **eines Bauleitplanes** (Förmliches Verfahren zur Erarbeitung und Inkraftsetzung eines ►Bauleitplanes entsprechend den gesetzlichen Vorschriften; einen Bauleitplan aufstellen [vb]); *syn.* Erstellung [f] eines Bauleitplanes, Erarbeitung [f] eines Bauleitplanes.

4544 preparation [n] **of a plan** *adm. leg.* (Formal procedure of drawing up a plan, making revisions and additions to it and finally obtaining official approval; e.g. for ►community development planning [US]/urban land-use planning [UK], ►State regional planning [US]/regional planning [UK] 2, ►functional planning [US]/sectoral planning [UK]); *s* **procedimiento** [m] **de redacción y aprobación de un plan** (Procedimiento formalizado para elaborar, reformar, ampliar y legalizar planes a nivel del ►planeamiento urbanístico, de la ►planificación regional 2 y de la ►planificación sectorial según las normas legales vigentes; ►aprobación de proyectos públicos); *f* **élaboration** [f] **d'un plan** (Procédure [décentralisée depuis le 1er octobre 1983 pour les plans d'occupation des sols] comprenant l'établissement du projet, l'association des personnes publiques et des associations, l'►enquête publique et l'approbation des plans relatifs à la ►planification urbaine, la ►planification régionale et les ►planifications sectorielles) ; *syn.* mise [f] en route d'un plan, établissement [m] d'un plan ; *g* **Planaufstellung**

[f] (Förmliches Verfahren zur Anfertigung, Änderung, Ergänzung und Inkraftsetzung von Plänen der ▶Bauleitplanung, der ▶Landesplanung und der ▶Fachplanung entsprechend den gesetzlichen Vorschriften; ▶Planfeststellung).

preparation [n] **of flower beds** *constr. hort.* ▶construction of flower beds.

4545 preparation [n] **of site facilities** *constr.* *s* **preparación [f] de las instalaciones de obra**; *f* **préparation [f] du chantier** (Prestations comprenant l'amenée, l'installation générale, la préparation et la signalisation temporaire du chantier ; *contexte* installation et repliement du chantier) ; *g* **Baustelleneinrichtung [f] (2)** (Vorbereitung einer Baustelle; *Kontext im Leistungsverzeichnis* Einrichten und Räumen der Baustelle); *syn.* Einrichten [n, o. Pl.] einer Baustelle, Einrichtung [f] einer Baustelle.

4546 preparatory operations [npl] *constr.* (Work [US]/works [UK] carried out at the beginning of a project by a contractor to be paid in a lump sum or measured amount; e.g. protection of existing vegetation, topsoil, or utilities and measures for protection against erosion, as well as the erection of site facilities, as long as these are not considered ▶additional work and services; cf. BS 4428:1969, 27; ▶site clearance, ▶work preparation [US]/works preparation [UK]); *s* **trabajos [mpl] preliminares** (Aquellas actividades que han de realizarse al comienzo de una obra y que se abonan con un monto global o por medida, como protección de la vegetación existente, protección contra la erosión así como el montaje de las instalaciones de obra, siempre que no se consideren ▶servicios adicionales; ▶preparación del terreno, ▶preparación de las obras); *f* **travaux [mpl] preliminaries** (Travaux exécutés au forfait ou sur métré par l'entrepreneur tels que : aménagement du chantier et installations provisoires [p. ex. clôture provisoire du chantier], reconnaissance des sols [p. ex. sondages], préparation et amélioration du sol [p. ex. ▶nettoyage du terrain, mesures de protection de la végétation à conserver sur place, décapage et stockage de la terre végétale à réemployer], implantation des ouvrages, le cas échéant fouilles et terrassements [p. ex. protection des talus] ; ▶préparation du chantier, ▶sujétions d'exécution) ; *syn.* travaux [mpl] préparatoires ; *g* **Vorarbeiten [fpl] (2)** (Durch Pauschale oder Aufmaß zu erfassende Leistungen des Auftragnehmers, z. B. Sicherung der vorhandenen Vegetation, des Oberbodens, der Versorgungsleitungen und Maßnahmen, die dem Erosionsschutz dienen, auch das Einrichten der Baustelle, wenn diese nicht als ▶Nebenleistungen anzusehen sind; ▶Baufeldfreimachung, ▶Bauvorbereitung).

4547 prepare [vb] **an expert opinion** *prof.* *s* **elaborar [vb] un dictamen**; *f* **procéder [vb] à une expertise** *syn.* préparer [vb] une expertise, établir [vb] une expertise ; *g* **Gutachten erstellen [vb]** (Erarbeiten eines Gutachtens).

4548 prepare [vb] **a recommendation for contract award** *constr.* (To elaborate a proposal for awarding a contract by evaluating bids [US]/tenders [UK] in order to find the tenderer offering the best value for the price bidder [US]/best value for money [UK]); *s* **preparar [vb] una proposición de adjudicación** (Examen de las ofertas utilizando criterios específicos para seleccionar a la mejor para proponerla para el remate); *f* **préparer [vb] le jugement d'appel d'offres** (Examen des offres suivant une hiérarchisation établie préalablement par des critères d'analyse et de jugement conduisant à dégager l'offre la plus intéressante pour la proposition du marché) ; *g* **Vergabevorschlag vorbereiten [vb]** (Notwendige Auswertung und Gewichtung der Angebote vornehmen, um den Bieter mit dem annehmbarsten Angebot herauszufinden).

preplanted mat [n] *constr.* ▶planting mat.

prequalification bidding notice [n] [US] *contr.* ▶public prequalification bidding notice [US]/public prequalification tendering notice [UK].

4549 prequalification document [n] *prof.* (Requested brochure submitted by a planning and design firm, containing a written and delineated record of the firm's expertise and experience, to be kept on file by a government agency and prequalify them for selection to perform future work when funds become available; ▶unsolicited prequalification document); *s* **documentación [f] para la admisión previa** (Documentos a presentar a la correspondiente autoridad que le permita apreciar las obras de análoga naturaleza y las de mayor o menor cuantía ejecutadas por la empresa, o en su defecto, si ésta reúne suficientes condiciones técnicas y económicas para poder ejecutarlas, mediante presentación de los certificados oportunos; ▶documentación de referencia [de empresa paisajista]); *f* **justifications [fpl] de candidature** (Dans le cadre d'un avis d'appel public à la concurrence, documents à produire à une administration quant aux qualités et capacités d'un bureau d'études candidat sous forme d'un mémoire justificatif, d'une liste de références, de certificats relatifs à des travaux similaires ; ▶brochure de prospection) ; *g* **Bewerbungsunterlagen [fpl] zur Vorauswahl** (Angeforderte Unterlagen, meist zur Vorlage an eine Behörde, die die Qualifikation und Erfahrung eines Planungsbüros in Form von Beschreibungen und Plänen abgewickelter Projekte darstellen. Anhand dieser Unterlagen wird zu gegebener Zeit die Auswahl für die Beauftragung eines Büros getroffen; ▶Bewerbungsunterlagen eines Planungsbüros).

prequalification tendering notice [n] [UK], **public** *contr.* ▶public prequalification bidding notice [US].

prerequisites [npl] **of nature conservation and landscape management** *conserv. nat'res.* ▶considerations of nature conservation and landscape management.

prescribed level [n] *constr.* ▶install at a prescribed level.

prescription [n], **period of** *contr. leg.* ▶period of limitation.

4550 presence [n] *phyt.* (The more or less persistent occurrence of a species in all the stands of a certain plant community; ▶degrees of presence, ▶determination of presence, ▶syndynamics); *s* **presencia [f]** (Existencia de una especie en la composición de una sinecia; DB 1985; ▶determinación de la presencia, ▶grado de presencia, ▶sindinamismo); *f* **présence [f]** (Existence ou absence des individus d'une espèce dans les peuplements étudiés d'une association déterminée ; BB 1928 ; ▶coefficient de présence, ▶détermination de la presence [d'une espèce], ▶syndynamique) ; *g* **Gesellschaftsstetigkeit [f]** (Das ständige oder weniger ständige Vorhandensein einer Art in den untersuchten Einzelbeständen einer bestimmten Pflanzengesellschaft; ▶Stetigkeitsgrad, ▶Stetigkeitsbestimmung, ▶Syndynamik); *syn.* Präsenz [f], Stetigkeit [f].

presentation [n] *plan.* ▶characteristic presentation.

presentation [n] **of awards** *prof.* ▶awarding of prizes.

preservation [n] *conserv. landsc.* landscape conservation, #2; *chem. constr.* ▶wood preservation.

Preservation Act [n] *conserv'hist. leg.* ▶National Historic Preservation Act 1966 [US].

preservation [n] **and management** [n] **of landscape character** *landsc. recr.* ▶preservation and management of scenic quality.

4551 preservation [n] **and management** [n] **of scenic quality** *landsc. recr. syn.* preservation [n] and management [n] of landscape character; *s* **preservación [f] de la belleza escénica del paisaje**; *f* **préservation [f] de l'état et de**

l'aspect des paysages (p. ex. les sites et territoires classés ne peuvent pas être modifiés dans leur état ou leur aspect ; art. 12 loi n° 76-1174 et art. 23 loi n° 76-629 relative à la protection de la nature) ; *g* **Erhaltung [f, o. Pl.] und Pflege [f, o. Pl.] des Landschaftsbildes.**

preservation measures [npl] *adm. conserv'hist.* ►historic preservation measures [US]/measures for conservation of historic monuments (and sites) [UK].

preservation notice [n] **[UK], emergency** *conserv. leg. plan.* ►temporary restraining order [US].

4552 preservation [n] **of building groups** *conserv'hist. urb.* (Preservation of groups of buildings having architecturally historic interest; ►historic district [US]/conservation area [UK], ►preservation of historic landmarks [US]/conservation of historic monuments [UK], ►preservation of historic structures); *s* **protección [f] de conjuntos históricos [Es]** (Por medio de la declaración como ►bien de interés cultural [Es] en el marco de la ►protección del Patrimonio Histórico Español [P.H.E.] [Es]. Un ►Conjunto Histórico [Es] es la agrupación de bienes inmuebles que forman una unidad de asentamiento, continua o dispersa, condicionada por una estructura física representativa de la evolución de una comunidad humana por ser testimonio de su cultura o constituir un valor de uso y disfrute para la colectividad. También es **C. H.** cualquier núcleo individualizado de inmuebles comprendidos en una unidad superior de población que reúna esas mismas características y pueda ser claramente delimitado; art. 15 [3] Ley 16/1985; ►conservación de monumentos histó-rico-artísticos); *syn.* conservación [f] de conjuntos históricos; *f* **préservation [f] d'un ensemble de constructions** (Relatif aux constructions présentant un intérêt historique, architectural, patrimonial, tels des quartiers anciens, des ensembles d'habitat urbain ou rural, du patrimoine préindustriel ou industriel, etc. En F. leur protection s'effectue dans le cadre des zones de protection du patrimoine architectural et urbain ; ►garde et conservation des monuments historiques, ►protection des monuments historiques, ►secteur sauvegardé, ►zone de protection du patrimoine architectural, urbain et paysager) ; *g* **Ensembleschutz [m, o. Pl.]** (*Planerjargon* Erhaltung einer Gebäudegruppe oder eines ganzen Straßenzuges im Rahmen des ►Denkmalschutzes; ►Baudenkmalpflege, ►flächenhaftes Kulturdenkmal); *syn. leg.* Schutz [m] einer Sachgesamtheit; *syn. für größere Gebiete* Schutz [m] einer Gesamtanlage.

4553 preservation [n] **of cultural resources** *conserv' hist.* (Protection of tangible and intangible, culturally important heritage against destruction, deterioration, or exportation, e.g. monuments, works of art, historic sites, etc. In American practice this term is gaining acceptance as also referring to the intangible ►cultural heritage: folk arts and folklore, costume, dialect, musical arts, etc. in addition to the physical remnants of our culture, such as buildings, structures, districts and objects and other artifacts; ►cultural resource [US]/cultural asset [UK], ►preservation of historic landmarks [US]/conservation of historic monuments [UK]); *s* **protección [f] del patrimonio histórico (español)** (Conservación de todos los bienes muebles e inmuebles que forman parte del ►patrimonio histórico español, incluyendo las medidas necesarias para evitar su exportación ilícita y su expoliación; ►bien de interés cultural, ►patrimonio cultural, ►Patrimonio Histórico Español); *syn.* protección [f] de los bienes de interés cultural, protección [f] del patrimonio cultural; *f* **protection [f] du patrimoine culturel national** (Protection contre la destruction, la dégradation, la vente et l'exportation des ►biens culturels tels que les monuments historiques, les monuments naturels et les sites, le patrimoine architectural les œuvres et objets d'art, le patrimoine ethnologique, etc. ; ►patri-

moine culturel) ; *syn.* protection [f] du patrimoine historique et esthétique de la France ; *g* **Kulturgüterschutz [m, o. Pl.]** (Erhaltung und konservatorische Pflege kulturell, wissenschaftlich und historisch bedeutender Objekte, z. B. Bau- und Kunstdenkmäler, historische Stätten etc., gegen Zerstörung, Verfall, Verkauf ins Ausland. In D. ist **K.** Angelegenheit der Länder; cf. auch Gesetz zum Schutz deutschen Kulturgutes gegen Abwanderung v. 08.07.1999; BGBl. I Nr. 42 v. 11.08. 1999, p. 1754 ff; ►Kulturerbe, ►Kulturgut).

4554 preservation [n] **of existing trees** *arb. constr. landsc. urb.* (►protection of trees, ►tree maintenance, ►tree ordinance [US]/tree preservation order [UK], ►tree surgery); *s* **mantenimiento [m] de árboles (1)** (►cirugía arbórea, ►mantenimiento de árboles 2, ►protección de árboles [en obras], ►ordenanza de protección de árboles); *syn.* conservación [f] de árboles; *f* **conservation [f] des arbres existants** (Végétaux remarquables à conserver sur place, p. ex. sur un chantier ; ►chirurgie arboricole, ►protection des arbres, ►réglementation concernant la protection des arbres, ►travaux d'entretien des arbres) ; *syn.* conservation [f] des végétaux existants ; *g* **Erhaltung [f, o. Pl.] von Bäumen** (►Baumchirurgie, ►Baumpflege, ►Baumschutz, ►Baumschutzverordnung); *syn.* Baumerhaltung [f], Erhaltung [f] des vorhandenen Baumbestandes.

4555 preservation [n] **of existing vegetation** *constr. leg.* (Safeguarding of vegetative cover on a construction site for protection in place or transplantation to another site; ►salvaging of plants, ►transplanting 1, ►transplanting 2); *s* **protección [f] de la vegetación existente** (Medidas para preservar árboles u otras plantas en ubicación de obra que pudieran ser dañadas durante las operaciones de construcción; ►transplante 1, ►transplante 2, ►transplante de especies vegetales); *f 1* **respect [m] des plantations existantes** (Mesure visant la protection de boisements linéaires, de haies ou de plantations d'alignement pendant la durée d'opérations de remembrement) ; *f 2* **protection [f] des végétaux existants (à préserver)** (Mesures conservatoires des arbres et végétaux divers pendant la durée du chantier et pouvant être détériorés ou détruits lors des travaux ; *contexte descriptif* protection des végétaux à conserver ; ►déplacement de végétaux, ►transplantation 3, ►récupération et transplantation des végétaux à conserver) ; *syn.* protection [f] des végétaux à conserver ; *g* **Schutz [m, o. Pl.] von Bäumen, Pflanzenbeständen und Vegetationsflächen** (Maßnahmen zur Erhaltung vorhandener Bäume und sonstiger Pflanzen- oder Vegetationsbestände, die durch den Baustellenbetrieb in ihrem Bestand gefährdet oder zerstört werden können; ►Umsetzen von Pflanzen[arten], ►Verpflanzung 1, ►Verpflanzung 2); *syn.* Schutz [m] von Vegetationsbeständen.

Preservation Officer [n] *conserv'hist.* ►State Historic Preservation Officer (SHPO) [US].

4556 preservation [n] **of floodplain zone** **[US]/preservation** [n] **of flood-plain zone** **[UK]** *leg. wat'man.* (Construction measures authorized by law to reduce the velocity of floodwaters for protection of people living in lower reaches of a watercourse); *s* **preservación [f] de la llanura de inundación** (Protección legal de las zonas inundables de un curso de agua para permitir la acumulación del agua de crecida y evitar así inundaciones en zonas pobladas aguas abajo); *f* **conservation [f] des champs d'inondation** (Mesures législatives et réglementaires de défense contre les inondations prises pour éviter de faire obstacle à l'écoulement des eaux et éviter de restreindre d'une manière nuisible les champs d'inondation ; cf. loi n° 82-600 du 13 juillet 1982, art. 5-5-1) ; *g* **Sicherung [f] des Überschwemmungsgebietes** (Gesetzliche Maßnahmen zur Schaffung von Überschwemmungsbereichen, die zum Schutze der Unterlieger

P

jederzeit einen ungehinderten, verlangsamten Hochwasserabfluss gewährleisten).

4557 preservation [n] **of historic buildings** *conserv' hist.* (Planning and management measures for the protection of historic structures; ▶preservation of historic landmarks [US]/conservation of historic monuments [UK]); *s* **preservación [f] de edificios históricos** (Planificación y gestión de medidas de mantenimiento y rehabilitación de edificios pertenecientes al patrimonio cultural; ▶protección del patrimonio histórico español); *f* **entretien [m] des monuments historiques classés** (Études et travaux relatifs à la conservation des monuments historiques ; ▶protection des monuments historiques ; *syn.* entretien [m] des édifices classés ; *g* **Pflege [f, o. Pl.] von historischen Gebäuden** (Alle planerischen und organisatorischen Maßnahmen zur Erhaltung von Gebäuden im Sinne des ▶Denkmalschutzes).

4558 preservation [n] **of historic landmarks** [US] *adm. conserv'hist. leg.* (**In U.S.**, federal involvement in preservation has been shaped by several major pieces of legislation, beginning with the Antiquities Act of 1906 which offers protection to prehistoric and historic sites located on federal properties. A national policy of preserving historic resources of national significance for public use and inspiration was established by the Historic Sites Act of 1935. The National Historic Preservation Act of 1966 [as amended] calls for the preservation of cultural properties [defined as buildings, sites, structures, districts, and objects] of State and local as well as national significance. Federal participation and action in preservation activities was emphasized by Executive Order No. 11593, signed May 13, 1971, which has now been incorporated by Congressional action into the national law itself as §110 of the National Historic Preservation Act; more effective historic landmark preservation program[me]s fall under the jurisdiction of state and local governments; **in U.K.**, there is the Ancient Monument and Archaeological Areas Act 1979; an ancient monument will enjoy protected status if it is included in the schedule maintained under the powers granted by the 1979 Act; **in D.**, guaranteed protection of archeological, architectural, and other ▶cultural monuments and artifacts by federal and, particularly, state laws. The federal government has no comprehensive authority for historic preservation, only to establish a framework and guidelines. Individual states and communities identify their own historic monuments; ▶historic district [US]/conservation area [UK], ▶historic preservation measures [US]/measures for conservation of historic monuments [and sites] [UK]); *syn.* conservation [n] of historic monuments [UK]; *s* **protección [f] del Patrimonio Histórico Español (P.H.E.)** [Es] (Conjunto de medidas legislativas existentes para proteger los bienes muebles e inmuebles de interés cultural; en España la **p. del P.H.E.** se basa en la Ley 16/1985, de 25 de junio, del Patrimonio Histórico Español y el Real Decreto 111/1986, de 10 de enero, de desarrollo parcial de la Ley 16/1985; ▶Conjunto Histórico [Es], ▶gestión del Patrimonio Histórico Español [Es], ▶monumento); *syn.* protección [m] de monumentos histórico-artísticos; *f* **protection [f] des monuments historiques** (En F., la législation relative à la **p. d. m. h.** repose sur un grand nombre de lois, notamment la loi du 31 décembre 1913 modifiée sur les monuments historiques, loi du 2 mai 1930 modifiée relative à la protection des monuments naturels et des sites de caractère artistique, historique, scientifique, légendaire ou pittoresque, loi du 27 septembre 1941 modifiée portant réglementation des fouilles archéologiques, loi du 7 juillet 1983 modifiée instituant dans l'urbanisme les secteurs sauvegardés et les zones de protection du patrimoine architectural urbain et paysager ; *g* gestion du patrimoine monumental, ▶monuments historiques, ▶secteur sauvegardé, ▶zone de protection du patrimoine architectural, urbain et paysager [Z.P.P.A.U.P.]. **En D.**, protection des terrains renfermant des stations ou des gisements préhistoriques, du patrimoine immobilier historique et artistique réglementée par les lois des Länder et dans une moindre mesure du Bund) ; *syn.* sauvegarde [f] du patrimoine et des sites ; *g* **Denkmalschutz [m, o. Pl.]** (**D.**, durch Bundes- und vor allem durch Landesgesetze sichergestellter Schutz von Boden-, Bau- und sonstigen Kulturdenkmälern mit der Möglichkeit des behördlichen Eingriffs in das Privateigentum. Der Bund hat keine umfassende Gesetzgebungskompetenz beim **D**. *Gesetzliche Grundlagen* Gesetz zur Berücksichtigung des Denkmalschutzes im Bundesrecht vom 01.07.1980 und die Denkmalschutzgesetze der Bundesländer. **In A.** fällt der **D.** in die Kompetenz des Bundes und ist im **D.gesetz** von 1923 geregelt. In der Schweiz obliegt der **D.** dem Bund und den Kantonen; *gesetzliche Grundlagen* NHG; Bundesbeschluss betreffend die Förderung der Denkmalpflege von 1958, kantonale Denkmal- und Heimatschutzgesetze; ▶Denkmalpflege, ▶flächenhaftes Kulturdenkmal, ▶Kulturdenkmal).

4559 preservation [n] **of historic structures** *conserv' hist.* (Safeguarding of historic monuments and sites; *specific term* ▶historic preservation measures [US]/measures for conservation of historic monuments [and sites] [UK]); *s* **conservación [f] de monumentos histórico-artísticos** (▶gestión del Patrimonio Histórico Español); *syn.* preservación [f] de monumentos histórico-artísticos; *f* **garde [f] et conservation des monuments historiques** (▶gestion du patrimoine monumental) ; *syn.* préservation [f]/entretien [m] des monuments historiques ; *g* **Baudenkmalpflege [f, o. Pl.]** (▶Denkmalpflege bei Gebäuden und sonstigen baulichen Anlagen).

4560 preservation [n] **of historic structures and sites** *conserv. leg. urb.* (City or county ordinance [US]/city bye-law [UK] prohibiting demolition, re-modelling or alterations to individual buildings, roads, squares, bridges, etc., in order to preserve the character of an old structure, townscape, or countryside; ▶preservation of building groups of special architectural or historic interest); *s* **conservación [f] de edificios** (Posibilidad de prohibir la transformación o el derribo u obligar a rehabilitar edificios por su valor de conjunto o por su importancia estética o social por medio del ▶plan parcial [de ordenación] o de ordenanza municipal específica; ▶protección de conjuntos históricos [Es]); *f* **préservation [f] des immeubles** (Disposition réglementaire dans un PLU/POS ou les ▶secteurs sauvegardés soumettant à autorisation préalable les travaux de démolition, de changement d'occupation, de transformation ou de modification de l'aspect d'un ensemble d'immeubles pouvant affecter le caractère esthétique, artistique, architectural de sites urbains ou paysagers ; cf. C. urb., art. L. 123-2°, 313-1 ; ▶préservation d'un ensemble de constructions) ; *syn.* sauvegarde [f] des immeubles ; *g* **Erhaltung [f, o. Pl.] baulicher Anlagen** (Zur Erhaltung der Eigenart von bestimmten Teilen eines Gemeindegebietes kann ein Gebiet innerhalb eines ▶Bebauungsplanes oder per Erhaltungssatzung auf Grund der städtebaulichen Eigenart oder zur Erhaltung der Zusammensetzung der Wohnbevölkerung so festgesetzt werden, dass Nutzungsänderungen, Umbau oder Abbruch einer behördlichen Genehmigung bedürfen — zum Schutze des Ortsbildes, der Stadtgestalt, des Landschaftsbildes oder aus städtebaulichen und künstlerischen Gründen; cf. § 172 BauGB; bei Vorliegen entsprechender rechtlicher Voraussetzungen können städtebauliche Gebote gem. § 176 [1] Nr. 2 BauGB [Baugebot] und § 177 BauGB [Modernisierungs- und Instandsetzungsgebot] erlassen werden; ▶Ensembleschutz).

4561 preservation [n] **of local visual amenities** *urb.* (Measures taken for the preservation of the beauty and special characteristic features of a town or village, or rural areas;

▶aesthetic ordinance [US]/building development bye-laws [UK], ▶community appearance); *s* **protección [f] del aspecto escénico urbano** (Conjunto de medidas realizadas para preservar la belleza o las características típicas de un pueblo o una ciudad; ▶ordenanza municipal de protección de la calidad visual urbana, ▶fisionomía del hábitat residencial); *syn.* protección [f] de la calidad visual urbana; *f* **préservation [f] du caractère de l'agglomération** (Mesures en faveur du maintien et de la conservation et mise en valeur de la nature, les caractéristiques et l'intérêt paysagers, architecturaux et culturels d'une agglomération, d'un site urbain ; ▶physionomie de l'habitat, ▶règle [d'urbanisme] concernant l'aspect extérieur des constructions) ; *g* **Ortsbildpflege [f, o. Pl.]** (Gesamtheit der Maßnahmen zur Erhaltung der Schönheit und Eigenart eines gewachsenen Ortscharakters; ▶Gestalt des Wohnumfeldes, ▶Ortsbildsatzung); *syn.* Pflege [f, o. Pl.] des Ortsbildes.

4562 preservation [n] **of natural resources** *pol. nat' res.* (▶natural resources management); *s* **preservación [f] de los recursos naturales** (▶gestión de recursos naturales); *f* **préservation [f] des ressources naturelles** (▶gestion des ressources naturelles) ; *g* **Ressourcensicherung [f]** (▶Bewirtschaftung der Naturgüter).

4563 preservation [n] **of the surrounding area** *conserv'hist. leg.* (Protection of the area adjacent to a cultural monument, historic buildings, etc., where this is necessary to preserve the character, beauty, or cultural identity of the monument; ▶historic district [US]/conservation area [UK], ▶preservation of building groups); *s* **protección [f] de zonas periféricas** (Protección del entorno de un espacio natural o cultural protegido si es necesario para preservar el carácter, la belleza o la identidad cultural de aquél; ▶Conjunto Histórico [Es], ▶protección de conjuntos históricos [Es]); *f* **protection [f] des abords** (Monuments historiques autorisation préalable pour toute construction nouvelle ou modification de nature à affecter l'aspect d'un immeuble situé dans le champ de visibilité d'un monument classé ou inscrit à l'inventaire des monuments historiques ; ▶préservation d'un ensemble de constructions, ▶secteur sauvegardé, ▶zone de protection du patrimoine architectural, urbain et paysager [Z.P.P.A.U.P.]) ; *g* **Umgebungsschutz [m, o. Pl.]** (Schutz des Umfeldes von Kultur-, Bau- oder Bodendenkmälern, soweit dies für die Erhaltung der Eigenart und Schönheit [kulturgeschichtlich gewachsene Identität] des Denkmals von erheblicher Bedeutung ist; cf. § 2 [1] 13 BNatSchG und §§ 2 [3] und 15 [3] DSchG-BW; ▶Ensembleschutz, ▶flächenhaftes Kulturdenkmal); *syn.* Schutz [m] der Umgebung eines Kulturdenkmals.

preserve [n] [US] *hunt.* ▶game preserve [US]/hunting ground [UK]; *conserv. leg.* ▶natural area preserve; *conserv. hunt.* ▶shooting preserve [US]/shooting reserve [UK].

4564 pressing [n] **of seeds** *constr. hort.* (Pushing down of seeds into the soil, using appropriate equipment; e.g. a lightweight roller; cf. SPON 1974, 119; ▶rolling); *s* **pase [m] de rodillo (1)** (Acción de hundir las semillas sembradas al voleo en el suelo utilizando una herramienta adecuada, como un rodillo; ▶pase de rodillo 2); *f* **plombage [m] après semis** (Action de tasser la terre après un semis au moyen d'un rouleau ou d'une batte pour éviter que ne se forme des poches d'air autour des graines enfouies ; contexte plombage d'une terre, plombage au rouleau ; ▶roulage) ; *g* **Andrücken [n, o. Pl.] des Saatgutes** (Herstellen des Erdschlusses mit ausgesäten Samenkörnern, z. B. mit einer Gitterwalze oder einem anderen geeigneten Gerät; DIN 18 917, 5.2.4; ▶Walzen); *syn.* Andrücken [n, o. Pl.] von Saatflächen, Andrücken [n, o. Pl.] von Saatgut.

pressure [n] *plan.* ▶competing land-use pressure; *urb.* ▶development pressure [US]/settlement pressure [UK]; *constr. eng.* ▶earth pressure; *ecol.* ▶population pressure; *plan. recr.* ▶visitor pressure.

pressure [n]**, competition** *ecol.* ▶pressure of competition.

pressure [n]**, competitive land-use** *plan.* ▶competing land-use pressure.

pressure [n]**, grazing** *agr.* ▶grazing stress.

pressure [n]**, recreation** *recr.* ▶recreation load.

pressure [n]**, soil** *constr. eng.* ▶earth pressure.

pressure [n] [US]**, uplift wind** *met. constr.* ▶wind suction.

pressure [n]**, wind** *arb. constr. stat.* ▶wind load.

4565 pressure group [n] *pol. sociol.* (Organized association of persons with the aim of promoting common interests of their members; a special group of **p. g.s** are ▶nature conservation organizations, which want to influence also the legislation; ▶nonprofit organization [US]/non-profit organization [UK]); *syn.* lobby [n]; *s* **grupo [m] de presión** (Organización o conjunto de individuos que colaboran entre sí para hacer presión ante la política y así alcanzar metas específicas; entre ellos están las ▶asociaciones para la protección de la naturaleza; ▶asociación sin fines de lucro); *syn.* grupo [m] de lobby; *f* **association [f] locale d'usagers** (Groupement de personnes organisées dans le but de défendre les intérêts communs de leurs adhérents et de les faire valoir auprès d'autres groupements ou de l'État ; ▶association à but non lucratif, ▶association de protection de l'environnement) ; *g* **Interessenverband [m]** (Zusammenschluss von Personen[gruppen] mit dem Ziel, in organisierter Form die gemeinsamen Interessen ihrer Mitglieder in der Öffentlichkeit zu vertreten und gegenüber anderen Gruppen oder dem Staat/-Land/der Gemeinde durchzusetzen. Eine besondere Gruppe von Interessenverbänden sind die ▶Naturschutzverbände, die auch auf die Gesetzgebung Einfluss nehmen wollen; ▶gemeinnütziger Verband).

pressure impregnation [n] [UK] *constr.* ▶pressure treatment [US].

4566 pressure [n] **of competition** *ecol.* (1. Extent of competition. 2. Stress caused by heavy use; ▶competition); *syn.* competition pressure [n], competition stress [n] (TEE 1980, 83); *s* **presión [f] de competencia** (1. Grado de ▶competencia a la que está sometida una especie. 2. Estrés producido por una gran cantidad de organismos compitiendo por el mismo espacio); *syn.* intensidad [f] de competencia; *f* **intensité [f] de la compétition** (1. Degré d'activité des organismes végétaux dans leur lutte pour l'occupation de l'espace. 2. Pression produite par ces mêmes organismes ; ▶compétition) ; *syn.* intensité [f] de la concurrence vitale ; *g* **Konkurrenzdruck [m, o. Pl.]** (1. Ausmaß der ▶Konkurrenz. 2. Durch starke Konkurrenz entstehender Zwang, in einem Raum oder auf einer Fläche zu bestehen, um den Bestand zu sichern oder verdrängt zu werden).

pressure [n] **of competitive land-use** *plan.* ▶competing land-use pressure.

4567 pressure treatment [n] [US] *constr.* (Process of impregnating lumber or other wood products with various chemicals for preservation and fire retardance; ▶wood preservation); *syn.* pressure impregnation [n] [UK]; *s* **impregnación [f] a presión** (Proceso de impregnar madera para usos al aire libre con productos químicos para protegerla y retardar la ignición en caso de incedio; ▶protección de la madera); *f* **imprégnation [f] à cœur en autoclave** (Technique industrielle de traitement du bois au moyen de produits fongicides et anticryptogamiques ; ▶protection du bois) ; *g* **Kesseldruckimprägnierung [f]** (Industrielles Verfahren zur Imprägnierung von Holz für Außenanlagen; ▶Holzschutz).

P

4568 prestressed concrete [n] *constr.* (Type of reinforced concrete which is strengthened by high tensile steel cable or rods, pretensioned or posttensioned); *syn.* pretensioned concrete [n]; **s hormigón** [m] **pretensado** (Tipo de hormigón armado en el que la armadura es reforzada con cables o varas de acero pretensado o postensado; este tipo de hormigón se utiliza para obras de ingeniería civil y para elementos de construcción de gran porte que deben aguantar grandes cargas); *syn.* hormigón [m] comprimido [AL], hormigón [m] preforzado [Mex]; **ƒ béton** [m] **précontraint** (Béton qui a subi une technique de précontrainte pratiquée sur le site en génie civil et pour des éléments de construction de grande portée, pouvant être mise en œuvre par pré ou post-tension de câbles d'acier et maintenant dans le béton un état de compression); **g Spannbeton** [m] (Variante des Stahlbetons, bei der die Bewehrung aus vorgespannten Stahleinlagen, den so genannten Spanngliedern, besteht; da Beton nur geringe Zugspannungen aufnehmen kann, aber hohe Druckspannungen, wird die Druckspannung durch die Verkehrslast ganz oder teilweise abgebaut und Verformungen/Durchbiegungen treten durch die Steifheit nur reduziert auf; Anwendung bei Brücken und großen Deckenflächen).

4569 pretreatment facility [n] *envir.* (Installation designed, e.g. for the purification of seepage water from refuse dumps [US]/refuse tips [UK], before it is discharged into a ▶receiving stream); **s planta** [f] **de pretratamiento** (Instalación destinada para la primera fase de tratamiento p. ej. de aguas de infiltración de depósito de residuos); **ƒ installation** [f] **de prétraitement** (Installation de traitement primaire permettant l'élimination des matières en suspension, p. ex. décanteur d'une station d'épuration, bassin de rétention des eaux d'infiltration d'une décharge avant leur rejet dans le réseau public d'assainissement dans le ▶milieu récepteur); **g Vorbehandlungsanlage** [f] (Anlage, die der Klärung von z. B. Deponiesickerwässern dient, bevor sie in einen ▶Vorfluter gelangen).

4570 prevailing price [n] [US] *constr. contr.* (Amount of money usually paid for a particular construction project/work in accordance with accepted practice in the region in which the project is to be implemented); *syn.* local price [n] [UK]; **s precio** [m] **usual** (Precio que se paga normalmente para un trabajo específico de construcción en una región dada); **ƒ prix** [m] **courant** (Valeur usuelle rétribuée pour un matériau ou une prestation de travaux dans la région dans laquelle est réalisé le projet); **g ortsüblicher Preis** [m] (Geldbetrag, der allgemein für einen Baustoff oder eine bestimmte [Bau]leistung in der Region, in der ein Vorhaben durchgeführt werden soll, bezahlt wird).

4571 prevailing wind direction [n] *met.* (Direction from which the wind usually blows); **s dirección** [f] **del viento dominante** *syn.* dirección [f] principal del viento; **ƒ direction** [f] **principale des vents** ; **g Hauptwindrichtung** [f].

4572 preventable disturbance [n] *ecol. envir.* **s perturbación** [f] **evitable** *syn.* daño [m] evitable; **ƒ nuisances** [fpl] **évitables** ; **g vermeidbare Beeinträchtigung** [f].

prevention [n], **flood** *wat'man.* ▶flood control.

4573 prevention measure [n] **(of an intrusion upon the natural environment)** *envir. leg. plan.* (Action to be taken to prevent the significantly disadvantageous, environmental impact of a project; ▶alleviation measure, ▶mitigation measure); **s medida** [f] **correctora** (En la ▶evaluación del impacto ambiental [EIA], medida determinada para evitar daños o consecuencias negativas previsibles del proyecto sobre el medio o la población; ▶medida compensatoria, ▶medida mitigante); *syn.* medida [f] preventiva; **ƒ mesure** [f] **de suppression** (Mesures visant à supprimer les impacts réductibles d'un projet sur l'environnement ; ▶mesure compensatoire, ▶mesure de réduc-

tion) ; **g Vermeidungsmaßnahme** [f] **(eines Eingriffs)** (Maßnahme, die der Vermeidung von bedeutenden nachteiligen Auswirkungen eines Projektes auf die Umwelt dient, d. h., dass einzelne Beeinträchtigungen von Natur und Landschaft gänzlich unterbleiben; ▶Ausgleichsmaßnahme [des Naturschutzes und der Landschaftspflege], ▶Verminderungsmaßnahme [eines Eingriffs]).

4574 preventive measure [n] *envir. leg.* (Measure taken to prevent or minimize environmental damage); **s medida** [f] **preventiva** (En la protección ambiental medida específica tomada para prevenir o minimizar daños en el medio ambiente); **ƒ mesures** [fpl] **de sauvegarde** (Mesures prise pour prévenir ou limiter le dommage par pollution) ; *syn. conserv.* mesure [f] conservatoire (PPH 1991, 275), mesure [f] de protection (PPH 1991, 276) ; **g Schutzmaßnahme** [f] (Handlung oder Regelung mit dem Ziel, Umweltschäden zu verhindern oder einzuschränken).

4575 preventive measures [npl] *plan. pol. sociol.* (Safeguards taken, e.g. for environmental protection, which prevent the later development of possible or additional environmental deterioration; ▶environmental precautions, ▶planning for the public welfare); *syn.* precaution(ary) measures [npl]; **s prevención** [f] (Conjunto de medidas p. ej. en la protección ambiental que sirven para evitar daños ambientales; ▶planificación en beneficio público, ▶prevención ambiental); *syn.* medidas [fpl] preventivas [f]; **ƒ prévention** [f] (Ensemble des mesures prises, p. ex. dans le cadre de la protection de l'environnement en vue d'éviter à la source les atteintes à l'environnement ; la loi n° 95-101 du 2 février 1995 institue entre autres les principes de précaution, d'action préventive et de correction ; ▶prévention des atteintes à l'environnement, ▶solidarité nationale) ; **g Vorsorge** [f, selten Pl.] (Alle Maßnahmen, die dazu dienen, öffentliche Dienstleistungen für jedermann dauerhaft und möglichst preisgünstig vorzuhalten, Unannehmlichkeiten, Belastungen oder vorhersehbare Schäden oder Zerstörungen zu vermeiden oder auf ein nach dem Stand der Technik erreichbares Mindestmaß zu beschränken; z. B. im Umweltschutz solche Maßnahmen, die das Entstehen möglicher späterer oder zusätzlicher Umweltbelastungen grundsätzlich vermeiden oder stark einschränken; *OB* ▶Daseinsvorsorge, ▶Umweltvorsorge).

price [n] *contr.* ▶bidder offering best value for the price [US]/tenderer offering the best value for the money [UK], ▶bid price [US]/tender sum [UK], ▶contract price, ▶excessive price, ▶lump-sum price [US]/lump sum [UK], ▶predatory price [US]/dumping price [UK], ▶prevailing price [US]/local price [UK], ▶reasonable price, ▶unit price.

price [n] [UK], **dumping** *constr. contr.* ▶predatory price [US].

price [n], **reduced** *constr. contr.* ▶price discount.

4576 priced bidding documents [npl] [US] *constr. contr.* (Set of ▶bid documents [US]/tender[ing] documents [UK], which include prices or rates [unit prices for unit price contracts or lump-sum price for lump-sum contracts] quoted by a bidder [US]/tenderer [UK] in his offer for work to be performed or materials delivered. The documents also contain ▶general conditions of contract for project execution, ▶additional contractual conditions, sometimes ▶special conditions of [a] contract, ▶technical specifications recording unit prices or lump-sum price, ▶bidder's affidavit [US]/tenderer's confirmation [UK], etc., which have to be submitted to the authority/agency before the ▶date for submission of bids [US]/date for submission of tenders [UK]; ▶project requirements for bidding [US]/project requirements for tendering [UK], ▶Standard Form of Building Contract [UK]); *syn.* priced tendering documents [npl] [UK]; **s expediente** [m] **de concurso-subasta** (Al **e. de c.** le corres-

ponden las condiciones generales de contrato, las condiciones adicionales de contrato, si ha el caso, las ►condiciones especiales de contrato y las ►especificaciones técnicas; ►declaración del licitante, ►documentos de contratos de las administraciones públicas, ►fecha tope de entrega de ofertas, ►pliego de cláusulas administrativas generales para la contratación de obras, ►pliego de condiciones [generales] de concurso, ►pliego de prescripciones administrativas particulares, ►pliego de prescripciones técnicas particulares); *f* dossier [m] des pièces constitutives du marché (Les pièces constitutives du marché remises par l'entrepreneur sont les suivantes : l'►acte d'engagement, l'ensemble des documents techniques [►description des ouvrages, ►spécifications techniques détaillées], le Cahier des Clauses Administratives Particulières [CCAP], le Cahier des Clauses Techniques Particulières [CCTP], l'état des prix forfaitaires, le bordereau des prix unitaires, le détail estimatif, les décompositions des prix forfaitaires ; ►cahier des clauses administratives générales [des marchés publics] [C.C.A.G.], ►cahier des clauses administratives particulières [C.C.A.P.], ►date limite de réception des offres, ►dossier d'appel d'offres, ►pièces contractuelles constitutives du marché) ; *g* Angebotsunterlagen [fpl] (Alle ►Verdingungsunterlagen mit Preisangaben [Einzelpreise beim Einheitspreisvertrag oder Pauschalpreis beim Pauschalvertrag], zu denen der Bieter die Leistungserbringung anbietet. Zu den **A.** gehören die ►allgemeinen Vertragsbedingungen für die Ausführung von Bauleistungen, die ►zusätzlichen Vertragsbedingungen, gegebenenfalls die ►besonderen Vertragsbedingungen, die mit Einheitspreisen ausgefüllte ►Leistungsbeschreibung 2 oder ein Pauschalpreis, ►Bietererklärung etc., die bei der ausschreibenden Stelle bis zum ►Abgabetermin der Angebote einzureichen sind; ►Ausschreibungsunterlagen).

priced bills [npl] **of quantities** [UK] *contr.* ►total construction cost estimate [US].

4577 price determination [n] *constr. contr.* (Calculation of unit prices for submission of bids [US]/tenders [UK]); *s* cálculo [m] de precios; *f* établissement [m] des prix (Calcul des prix unitaires pour une remise d'offre) ; *g* Preisermittlung [f] (Kalkulation von Einheitspreisen zur Angebotsabgabe); *syn.* Preiskalkulation [f].

4578 price discount [n] *constr. contr.* (**1.** Reduction of a submitted quotation or bid, granted usually in the form of a percentage; **2. in U.K.** codes of procedure for tendering e.g. those prepared by the National Joint Consultative Committee for Building, which foresees the negotiation of a **reduced price [n]**, usually a percentage, if the tender exceeds the estimated cost. Such a procedure is applied for regularly occurring works such as maintenance, where prices may be adjusted in an upward or downward direction; ►unit price reduction); *s* rebaja [f] de precio (Reducción del precio respecto a la oferta, ►rebaja de precio unitario); *f* remise [f] (Réduction sur le montant d'une offre consentie à un maître d'ouvrage ; ►moins-value) ; *g 1* Preisnachlass [m] (Reduzierung des angebotenen [Gesamt]-preises durch Gewährung eines Rabattes, meist in Form eines prozentualen Abzugs; Preisnachlass gewähren [vb]; ►Abschlag); *g 2* Abgebot [n] (Rabatt in Form eines prozentualen Wertes, den der Auftragnehmer auf Preise, die vom Auftraggeber bei einer Ausschreibung vorgegeben werden, im Rahmen des Auf- und Abgebotsverfahrens nach § 6 [2] VOB Teil A gewährt; dieses Verfahren wird bei regelmäßig wiederkehrenden Unterhaltungsarbeiten angewandt); *syn. zu 2.* Rabatt [m].

priced tendering documents [npl] [UK] *constr. contr.* ►priced bidding documents [US].

price inquiry [n] *constr. contr.* *►price quotation on request.

4579 price list [n] *constr. contr.* (Price list compiled on the basis of bids [US]/tenders [UK] used by a public authority or agency, which contains ►unit prices of previous tenders/bids as a basis for cost estimates and contract award negotiations; ►unit-price list); *s* lista [f] de precios (Listado realizado por la administración de los ►precios unitarios de los últimos concursos-subasta que sirve de base para calcular el coste [Es]/costo [AL] y para la negociación de la oferta; ►listado de precios unitarios); *f 1* série [f] des prix de l'administration publique (±) (**D.**, liste des prix établie par l'administration publique ayant rassemblé les ►prix unitaires des derniers appels d'offres et servant de base à l'établissement de l'estimation sommaire du coût des travaux ou aux négociations avec les entreprises contractantes ; ►bordereau des prix unitaires [B.P.U.]) ; *f 2* série [f] des prix (**F.**, ouvrage de référence mis à la disposition des maîtres d'ouvrage, des entrepreneurs, des experts, etc. ; cette estimation des prix unitaires établie par le service économique de l'Académie d'Architecture est de nature à faciliter en premier lieu pour les maîtres d'ouvrage la définition de la soumission et l'appréhension de l'offre des entrepreneurs concernant les différents éléments d'ouvrages dans les bâtiments ; SDP 1998) ; *syn. obs.* Série Centrale des Prix des Travaux de Bâtiment ; *g* Preisverzeichnis [n] (Preiszusammenstellung der öffentlichen Verwaltung, die ►Einheitspreise der letzten Ausschreibungen beinhaltet und als Grundlage für Kostenschätzungen und Verdingungsverhandlungen dient; ►Verzeichnis der Einheitspreise).

4580 price quotation [n] **on request** *constr. contr.* (Direct submittal of a price, obtained by making a **price inquiry** to selected contractors without a formal bidding [US]/tendering [UK] process; ►invitation for a bid [US]/invitation for a tender [UK]); *s* oferta [f] a la demanda (Oferta para la que no se convoca un concurso, sino que se solicita a empresas escogidas; ►invitación a oferta); *f* devis [m] hors mise en concurrence (Devis établi sans procéder à une mise en concurrence des entreprises ; le marché fera donc l'objet d'un devis à la demande adressé à des entreprises que le client aura choisies ; ►lancement d'appel d'offres) ; *syn.* devis [m] sans appel d'offres, devis [m] sans dossier de consultation ; *g* Angebot [n] auf Anfrage (Angebot, das nicht einer Ausschreibung unterliegt und gezielt von ausgesuchten Firmen abverlangt wird, oft auch als **Preisanfrage** genutzt; ►Aufforderung zur Angebotsabgabe).

4581 price reduction [n] *constr. contr.* (Lowering of a contractually-fixed bid due to a deficiency in performance or quality on the part of the contractor; ►unit price reduction); *s* reducción [f] de precio (Posibilidad de la administración comitente de reducir el precio fijado en el contrato de una obra por deficiencias de calidad bajo responsabilidad del licitante; ►rebaja de precio unitario); *syn.* minoración [f] del precio; *f* réfaction [f] sur les prix (Réduction sur le prix des matériaux ou produits non conformes aux spécifications du marché ; ►moins-value) ; *g* Preisminderung [f] (Herabsetzung des vertraglichen Einheitspreises wegen eines vom Bieter zu vertretenden Sach- oder Qualitätsmangels oder geminderten Qualitätsstandards; ►Abschlag); *syn.* Minderung [f] des Preises.

price-revision clause [n] *constr. contr.* ►escalation clause.

pricing [n] *trans. pol.* ►road pricing.

pricking [n] *constr.* ►lawn aeration, #4.

pricking [n] **over** *constr.* ►lawn aeration, #4.

4582 prickle [n] *bot.* (Short woody pointed outgrowth, often recurved, arising from the epidermal tissue of a stem, leaf, or some species of fruit; to be distinguished from ►thorn; prickly [adj]); *s* acúleo [m] (Formación rígida y punzante, puramente epidérmica, distinta de la ►espina. El rosal, la zarzamora y el tojo, p. ej., tienen acúleos; cf. DB 1985); *syn.* aguijón [m]; *f* aiguil-

P

lon [m] (Piquant d'origine épidermique ; ne pas confondre avec ►épine) ; **g Stachel [m]** (Spitze, starre, oft hakenförmig gebogene Ausstülpung/Emergenz, die ausschließlich vom Rindengewebe gebildet wird, z. B. bei *Rosa*; nicht zu verwechseln mit ►Dorn).

primordial forest [n] *phyt.* ►virgin forest.

4583 primary colonization [n] *phyt.* (Beginning of vegetative growth on bare ground; ►initial plant community); **s primera colonización [f]** (Comienzo del desarrollo de comunidades vegetales sobre suelo desnudo; cf. BB 1979; ►comunidad inicial); *syn.* colonización [f] de ambientes vacíos, colonización [f] de «espacio vacío»; *f* **colonisation [f] primaire** (Début du développement de végétaux sur un sol nu ; ►groupement initial) ; **g Erstbesiedlung [f]** (Beginn der Vegetationsentwicklung auf vegetationslosem Boden; ►Initialgesellschaft).

primary consumer [n] *ecol.* * ►consumer.

primary dune [n] *geo. phyt.* ►embryo dune.

primary (plant) growth [n] *phyt.* ►pioneer vegetation.

4584 primary production [n] *ecol.* (Total quantity of autotrophic organisms [green plants] created primarily from organic matter [►biomass] within a given period of time and area; ►production 1, ►secondary production); **s 1 producción [f] primaria bruta** (Cantidad de energía química de enlace, fijada en forma de materia vegetal [►biomasa], creada creada por fotosíntesis en una zona y un tiempo definidos); **s 2 producción [f] primaria neta** (Cantidad de energía a disposicion de los consumidores después de restar a la producción primaria bruta la necesaria para la propia respiración de los productores; cf. DINA 1987; ►producción, ►producción secundaria); *f* **productivité [f] primaire** (La vitesse avec laquelle l'énergie est emmagasinée par l'activité photosynthétique des producteurs [plantes vertes], sous forme de matières organiques pouvant constituer une augmentation de la ►biomasse végétale et être utilisées comme aliments par les consommateurs ; DEE 1982 ; ►production, ►productivité secondaire) ; *syn.* production [f] primaire ; **g Primärproduktion [f]** (Gesamtmenge der innerhalb eines definierten Zeitraumes und innerhalb eines definierten Gebietes resp. Volumens von autotrophen Organismen [Pflanzen] aus überwiegend anorganischen Stoffen gebildeten pflanzlichen ►Biomasse; ►Produktion, ►Sekundärproduktion).

primeval [adj] *ecol.* ►virgin.

primeval forest [n] *phyt.* ►virgin forest.

primitive recreation [n] [US] *recr.* ►recreation in the natural environment.

primordial forest [n] *phyt.* ►virgin forest.

principal [n] *prof.* ►landscape architect principal.

principle [n] *envir. pol.* ►polluter-pays principle, ►precautionary principle, ►public responsibility principle.

principle [n]**, "polluter should pay"** *envir. pol.* ►polluter-pays principle.

principle [n]**, sustainability** *agr. ecol. for.* ►principle of sustained yield.

principle [n] **of causal responsibility** *envir. pol.* ►polluter-pays principle.

principle [n] **of causation** *envir. pol.* ►polluter-pays principle.

principle [n] **of persisting supply** [US] *agr. ecol. for.* ►principle of sustained yield.

4585 principle [n] **of sustained yield** *agr. ecol. for.* (Capacity of a living organism or natural system to sustain itself after being used or incurring losses, without substantial depletion of its productive capability; e.g. yield that a forest can produce continually at a given intensity of management; cf. SAF 1983; ►sustainable development, ►sustainable use); *syn.* sustainability principle [n], principle [n] of persisting supply [also US]; **s sostenibilidad [f]** (Capacidad de un sistema vivo de mantener su productividad sin agotarse a pesar de la explotación continuada de sus recursos; ►aprovechamiento sostenible, ►desarrollo sostenible); *syn.* sustentabilidad [f] [CS], principio [m] de rendimiento constante y aprovechamiento sostenible, principio [m] de rendimiento sostenible; *f* **principe [m] d'exploitation durable** (Capacité d'un organisme ou d'un système vivant à utiliser et compenser les pertes en ressources [exploitation des ressources] tout en produisant continuellement le même rendement ; ►développement durable, ►usage durable) ; *syn.* exploitation [f] soutenable, exploitation [f] viable, principe [m] du rendement soutenu [aussi CDN] ; **g Nachhaltigkeit [f, o. Pl.] der Nutzung** (Fähigkeit eines lebenden Systems bei Nutzung und Ausgleich der Verluste dauerhaft gleiche Leistungen zu erbringen, ohne sich zu erschöpfen; cf. ANL 1984; der Begriff *Nachhaltigkeit* stammt aus der Forstwirtschaft und wurde von HANS KARL VON CARLOWITZ [1645-1714] in seiner 1713 erschienenen *Sylvicultura oeconomica oder haußwirthliche Nachricht und naturmäßige Anweisung zur wilden Baum-Zucht* zum ersten Mal erwähnt. Er bezeichnet damit die Bewirtschaftungsweise eines Waldes, bei welcher lediglich so viel Holz entnommen werden darf wie nachwachsen kann; ►nachhaltige Entwicklung, ►nachhaltige Nutzung); *syn.* Prinzip [n] der Nachhaltigkeit.

prior investigation [n] *contr.* ►background work [US]/ background research [n] [UK].

4586 priority agricultural area [n] (±) *agr. plan.* (Area of land for which agricultural use takes precedence over all other uses when land use patterns are being restructured); **s área [f] priorizada para uso agrícola** (D. categoría de planificación regional); *f* **zone [f] à vocation rurale** (Zones à économie rurale dominante dans lesquelles l'affectation des sols est exclusivement réservée à l'usage agricole et ou le maintien de l'agriculture prévaut sur tout autre forme d'activité) ; **g landwirtschaftliche Vorrangfläche [f]** (Fläche, die auf Grund ihrer besonderen Eignung für die Landwirtschaft im Falle von Nutzungsumordnungen Vorrang vor allen anderen Nutzungen eingeräumt wird; gesetzliche Grundlage ist § 8 [7] 1 ROG).

4587 priority area [n] *plan.* (Area of varying size which is particularly suitable for one or several functions of a larger area and, therefore, is to be protected and developed according to ►regional planning policies. **P. a.** include those for drinking water catchment, for recreation and leisure, for nature conservation and agricultural uses; ►raw material protection zone); **s área [f] de uso priorizado** (Zona que —debido a su vocación específica o a otros factores— se considera apta para cumplir una o varias funciones para toda una región, y que se define como tal en los planes de ►ordenación del territorio. En D. hay, p. ej. ►zonas de aprovechamiento mineral, zonas para la ganancia de aguas potables, zonas para la recreación al aire libre, la conservación de la naturaleza, la producción agrícola, etc.); *f* **zone [f] prioritaire** (Zones d'étendue variable qui en fonction de leurs caractéristiques et potentialités naturelles, économiques, humaines ou culturelles ont la vocation d'assurer une ou plusieurs fonctions de l'espace dans le cadre de l'►aménagement du territoire ; on peut ainsi distinguer des zones à vocation agricole, récréative, etc. ; ►zone d'exploitation et de réaménagement coordonnés de carrières) ; **g 1 Vorranggebiet [n]** (Gebiet unterschiedlicher Ausdehnung, das auf Grund besonderer Eignung oder auf Grund anderer Zuerkennung eine oder mehrere Funktionen für den Gesamtraum übernehmen kann und dementsprechend durch die Raumordnungspolitik gesichert und entwickelt werden muss; BRÖSSE 1975, zit. in BUCH 1978-80, Bd. 3, 20. Legale Grundlage ist das Raumordnungsgesetz, in dem **V.e** als Darstellungskategorie für „bestimmte raumbedeutsame Funktionen oder Nutzungen vorgesehen sind und andere raumbedeutsame Nutzungen in diesem Gebiet ausschließen, soweit diese mit den vorrangigen Funktionen oder Nutzungen nicht vereinbar sind" [§ 8 (7) 1 ROG], und in Programmen und Plänen der Landesplanung und Regionalplanung bezeichnet werden. Es gibt Vorrangflächen für die unterschiedlichsten Nutzungen, z. B. für die Trinkwassergewinnung, für Natur und Landschaft, für Freizeit und Erholung, für landwirtschaftliche Nutzung, Windenergieanlagen, ▶Rohstoffsicherungsflächen, **Vorrangstandorte** für großindustrielle Anlagen etc.; im Unterschied zu Vorbehaltsgebieten lassen **V.e** beispielsweise für Natur und Landschaft nur Nutzungern zu, die sich den Zielen des Naturschutzes und der Landschaftspflege unterordnen, z. B. Schutzgebiete oder Schutzobjekte mit hohem Schutzstatus wie Naturschutzgebiet, Natura-2000-Gebiet, Nationalpark oder gesetzlich geschützte Biotope); **g 2 Vorbehaltsgebiet [n]** (Im Raumordnungsgesetz verankerte Darstellungskategorie für Gebiete, in denen bestimmten raumbedeutsamen Funktionen oder Nutzungen bei der Abwägung mit konkurrierenden raumbedeutsamen Nutzungen ein besonderes Gewicht beizumessen ist [§ 8 (7) 2 ROG]; somit stellt das Vorbehaltsgebiet eine Ergänzung zum Vorranggebiet dar und ist für den Fall einzurichten, wenn die Zielsetzung eines Gebietes noch nicht endgültig feststeht; somit dürfen Nutzungen nur realisiert werden, wenn sie der Verwirklichung der Ziele, die mit den Vorbehaltsdarstellungen verbunden sind, nicht entgegenstehen; es gibt V.e z. B. für Natur und Landschaft, Erholung, Landwirtschaft und Wasserwirtschaft; als **V.e** für Natur und Landschaft kommen grundsätzlich Bereiche in Frage, auf denen sich die Ziele des Naturschutzes und der Landschaftspflege zusammen mit anderen Nutzungen verwirklichen lassen, z. B. Landschaftsschutzgebiete, Naturparke und Biosphärenreservate; ferner eignen sich Flächen, die eine hohe Bedeutung für den Naturhaushalt oder das Landschaftsbild haben oder die sich zum Aufbau eines Biotopverbundes eignen; dargestellt werden diese Flächen in Landschaftsplänen; cf. BFN 2006); ▶Raumordnung); *syn.* Vorrangfläche [f].

4588 priority natural habitat type [n] *conserv. ecol.* (Natural habitat type in danger of disappearence, and for the conservation of which the Community has particular responsibility in view of the proportion of their natural range; these priority natural habitat types are indicated by an asterisk [*] in Annex I, Council Directive 92/43/EEC); **s tipo [m] de hábitat natural prioritario** (Hábitat natural amenazado de desaparición cuya conservación supone una especial responsabilidad para la Comunidad habida cuenta de la importancia de la proporción de su área de distribución natural; estos tipos de hábitats naturales prioritarios se señalan con un asterisco [*] en el Anexo I, Directiva 92/43/CEE); **f type [m] d'habitats naturels prioritaire** (Habitat naturel en danger de disparition et pour la conservation desquels la Communauté porte une responsabilité particulière, compte tenu de l'importance de la part de leur aire de répartition naturelle ; ce type d'habitats naturels prioritaire est indiqué par un astérisque [*] à l'annexe I, Directive 92/43/CEE) ; **g prioritärer natürlicher Lebensraumtyp [m]** (Vom Verschwinden bedrohter Lebensraumtyp, für dessen Erhaltung der europäischen Gemeinschaft auf Grund der natürlichen Ausdehnung dieses Lebensraumtyps besondere Verantwortung zukommt; die prioritären natürlichen Lebensraumtypen sind in Anhang 1 der FFH-Richtlinie 92/43/EWG mit einem Sternchen [*] gekennzeichnet).

4589 pristine [adj] *nat'res. conserv.* (Descriptive term applied to a place or object untouched or unaltered by human activity); **s primigenio/a [adj]** (Término descriptivo aplicado a los lugares que se encuentran en su estado original, es decir que no han sido modificados por la acción humana); *syn.* prístino/a [adj]; **f vierge [adj]** (Caractère d'un territoire, d'un paysage, d'une formation exempt de toute influence humaine et donc non altéré) ; **g ursprünglich [adj]** (Einen Ort, ein Objekt betreffend, das durch den Menschen unberührt resp. nicht unmittelbar verändert wurde).

pristine forest [n] *phyt.* ▶virgin forest.

4590 private contribution [n] *econ.* (Monetary payment by a private person or non-profit organization; ▶public grant); *syn.* private grant [n] [also US]; **s aportación [f] privada** (Contribución monetaria a un proyecto o a una actividad realizada por persona privada o por una organización sin fines de lucro o una fundación; ▶contribución financiera pública); *syn.* contribución [f] financiera privada; **f contribution [f] financière du privé** (▶contribution financière des pouvoirs publics) ; **g privater finanzieller Beitrag [m]** (Von Privaten oder gemeinnützigen Gesellschaften gewährter finanzieller Betrag; ▶Zuschuss der öffentlichen Hand).

4591 private garden [n] *gard.* (**1.** Enclosed and ornamentally planted piece of land located near a dwelling; ▶backyard garden [US]/back garden [UK], ▶front yard [US]/front garden [UK], ▶kitchen garden. **2. In U.S.,** yard [n] typically refers to the area around the building defined by the property line, and is most often synonymous with the ubiquitous "lawn"; it can also have a special function such as, service yard, dog run yard, play yard, etc.); *syn.* residential garden [n] [also US], rear garden [n] [also UK], yard [n] [US]; **s jardín [m] privado** (Espacio libre cercado plantado con plantas ornamentales y situado alrededor de una casa particular; ▶huerta 2, ▶jardín delantero, ▶jardín trasero); **f 1 jardin [m] particulier** (Terme générique pour les différentes catégories de jardins privés de petite et moyenne surface, liés aux résidences principales urbaines ou suburbaines, tels que cour-jardins, jardins de façade, jardins de banlieue, etc. ; cf. ATJ 1981 ; ▶jardin arrière, ▶jardin de façade, ▶jardin potager) ; **f 2 jardin [m] privatif** (Jardin d'usage privatif par opposition au jardin public, englobant le jardin privé, le jardin locatif d'habitation, le jardin ouvrier) ; *syn.* jardin [m] individuel ; **f 3 jardin [m] privé** (Espace privé d'agrément, de surface réduite, attenant à l'habitation) ; *syn.* jardin [m] d'habitation ; **g Hausgarten [m]** (Eingefriedetes und gärtnerisch gestaltetes Grundstück, das an einem resp. um ein Wohnhaus angelegt ist; ▶Hintergarten, ▶Küchengarten, ▶Vorgarten); *syn.* Privatgarten [m].

private grant [n] [US] *econ.* ▶private contribution.

private green area [n] *leg. urb.* ▶private green space.

4592 private green space [n] *leg. urb.* (▶private garden); *syn.* private green area [n]; **s espacio [m] verde privado** (Zona verde de propiedad privada; ▶jardín privado); **f espace [m] vert privé** (Espace utilisé par les particuliers qui en sont propriétaires ou en ont la jouissance ; ▶jardin particulier) ; **g private Grünfläche [f]** (Grünfläche in Privatbesitz; ▶Hausgarten); *syn.* private Grünanlage [f], privates Grün [n, o. Pl.].

4593 private open space [n] *plan.* (Open land restricted to owner or tenant; *specific terms* ▶private garden, ▶tenant garden); **s espacio [m] libre privado** (*Términos específicos* ▶huerto recreativo urbano, ▶jardín de inquilinos, ▶jardín privado); **f espace [m] libre privatif** (Espace réservé à la jouissance des locataires ou des copropriétaires ; ▶jardin collectif, ▶jardin locatif, ▶jardin particulier, ▶jardin privatif) ; **g privat nutzbarer Freiraum [m]** (Vegetations- oder nicht vegetationsbestimmte Freianlage, die ausschließlich privat genutzt wird; *UBe* ▶Hausgarten, ▶Kleingarten, ▶Mietergarten).

P

4594 private owner's obligation [n] *leg.* (Requirement for owners and land users to respect legally imposed measures or rights for public use of a portion of their property; **in U.K.**, exists a "wayleave" which is a right of way over, under or through land for such things as a pipeline, an electric transmission line, or for carrying goods across the land. Many statutory authorities, such as electricity boards, may apply to the appropriate minister for a compulsory wayleave over land where the owner refuses his consent. A wayleave is a kind of ▶easement); *syn.* owner's duty [n] to consent; *s* **obligación [f] de tolerar servidumbres** (Exigencia hacia los propietarios o usuarios de terrenos privados de tolerar todas las medidas resultantes de las regulaciones en vigor o de carácter de interés público, como obras de interés público en sus terrenos, servidumbre de paso, obligación de plantar, etc.; ▶servidumbre inmobiliaria); *f* **obligation [f] à tolérer les servitudes d'utilité publique** (Obligation du propriétaire ou des ayants droits de terrain à tolérer toutes mesures résultant de réglementations en vigueur ou à caractère d'intérêt public, p. ex. le droit de passage, l'utilisation de réseaux d'alimentation, l'obligation à planter ; ▶servitude) ; *g* **Duldungspflicht [f]** (Pflicht der Eigentümer oder Nutzungsberechtigten von Grundflächen Maßnahmen auf Grund oder im Rahmen von erlassenen Rechtsvorschriften oder die Durchführung von Maßnahmen, z. B. gemäß §§ 41, 126, 179 und 209 BauGB, wie z. B. die Verlegung von Versorgungsleitungen, Abbruch von baulichen Anlagen oder Maßnahmen gemäß § 30 WHG, zu dulden; cf. auch § 10 [1] BNatSchG und Landeswassergesetze, ▶Grunddienstbarkeit); *syn.* Pflicht [f] zur Duldung, Verpflichtung [f] zur Duldung.

private practice [n] *prof.* ▶landscape architect in private practice.

private property [n] *leg. pol. sociol.* ▶social obligation of private property.

private sewage treatment plant [n] [UK] *envir.* ▶individual sewage disposal system [US].

4595 private transport [n] *trans.* (Movement of individuals by any means of personal conveyance, especially by private motor vehicles; *opp.* ▶local public transportation system [US]/ local transport system [UK], ▶public transit [US]/public transport [UK]); *s* **transporte [m] individual** (Desplazamiento de personas a pie o con cualquier tipo de medio de transporte individual, generalmente con vehículos motorizados; *opp.* ▶transporte público, ▶transporte público urbano); *syn.* transporte [m] privado; *f* **transport [m] individuel** (Déplacement d'une personne à pied ou avec un véhicule privé par opposition aux transports collectifs ; ▶transports suburbains, ▶transports suburbains) ; *g* **Individualverkehr [m]** (Fortbewegung einzelner Personen zu Fuß oder mit Privatfahrzeugen im Vergleich zum öffentlichen Verkehr; ▶öffentlicher Personennahverkehr, ▶öffentlicher Verkehr).

4596 prize [n] *prof.* (Category of award given to winner[s] in a design or planning competition; ▶honorable mention [US]/ honourable mention [UK], ▶merit award, ▶design competition announcement [US]/design competition invitation [UK], ▶short list); *syn.* place [n]; *s* **premio [m]** (Categoría superior de recompensa monetaria que recibe el ganador de un concurso de ideas o de realización y es otorgada sobre la base de la decisión del jurado entre los trabajos clasificados para la selección final; ▶anuncio de un concurso [de ideas], ▶lista selectiva, ▶mención, ▶mención honorífica); *f* **prix [m]** (Récompense financière [indemnité] attribuée par l'organisateur d'un concours aux lauréats [concours d'idée, de conception-construction] qui auront été sélectionnés sur proposition du jury pour la dernière phase du concours ; ▶appel public à la concurrence [pour un concours],

▶projet mentionné, ▶sélection préalable, ▶travaux primés, ▶prononciation de l'attribution d'un marché) ; *g* **Preis [m]** (Vom Auslober bereitgestellte Belohnung in Form eines Geldbetrages für die ersten Sieger eines [Ideen- oder Realisierungs]wettbewerbes, die auf der Grundlage der Rangfolge der in die ▶engere Wahl 2 genommenen Arbeiten vom Preisgericht zuerkannt wird; ▶Ankauf, ▶Auslobung, ▶engere Wahl 3).

4597 prizewinner [n] *prof.* (Someone who has won a prize in a design competition, or someone officially awarded with a prize for outstanding performance or achievement); *s* **premiado/a [m/f]** (Persona o equipo que ha salido ganador de un concurso de ideas o de realización o que ha recibido un premio oficial por una acción ejemplar); *syn.* titular [m/f] de premio, ganador, -a [m/f] de premio, lauredado/a [m/f]; *f* **lauréat [m]** (Participant à un concours [concours d'idée, de conception-construction] qui aura été sélectionné ou primé sur proposition du jury) ; *g* **Preisträger/-in [m/f]** (Jemand, der bei einem [Ideen- oder Realisierungs]wettbewerb einen Preis gewonnen hat oder jemand, dem für eine besondere Leistung ein offizieller Preis zuerkannt wurde); *syn.* Gewinner/-in [m/f] eines Preises.

4598 probable maximum flood [n] **(PMF)** *wat'man.* (TWW 1982, 313; greatest flood that may be expected, usually expressed in years, such as in a "50-year flood"; ▶hundred-year flood); *s* **crecida [f] máxima probable** (Máxima crecida que cabe esperar teniendo en cuenta todos los factores condicionantes de situación, meteorología, hidrología, geología, suelos, etc.; WMO 1974; ▶crecida máxima del siglo); *f* **crue [f] maximale probable** (Désigne la plus grande crue qui puisse survenir compte tenu de tous les facteurs conditionnels, géographiques, météorologiques, hydrologiques et géologiques ; WMO 1974 ; ▶crue centennale) ; *g* **Maximalhochwasser [n]** (…wasser [pl]; das wahrscheinlich höchste Hochwasser; ▶Jahrhunderthochwasser).

procedure [n] *adm. leg.* ▶administrative procedure; *adm. contr.* ▶bidding procedure [US]/tendering procedure [UK]; *constr. contr. prof.* ▶contract awarding procedure; *plan. trans.* ▶corridor approval procedure [US]; *plan.* ▶evaluation procedure; *adm. agr. leg.* ▶farmland consolidation procedure [US]/land consolidation procedure; *adm. leg.* ▶planning approval procedure; *plan.* ▶planning procedure 1; *leg. plan.* ▶regional planning procedure.

procedure [n] [UK]**, approval for route selection** *plan. trans.* ▶corridor approval procedure [US].

procedure [n] [US]**, route approval** *plan. trans.* ▶corridor approval procedure [US].

procedure [n]**, work** *plan.* ▶sequence of operations.

procedure [n] **for route selection** [UK]**, approval** *plan. trans.* ▶corridor approval procedure [US].

process [n] *adm. leg.* ▶design and location approval process [US]/determination process of a plan [UK]; *constr.* ▶excavation process; *ecol.* ▶natural process; *plan.* ▶planning process.

process [n]**, building** *constr. eng.* ▶construction (1).

process [n]**, design** *plan.* ▶planning process.

process [n]**, planting** *constr. for. hort.* ▶planting (1).

processed aggregate surface [n] [UK] *constr.* ▶waterbound surface.

processed lava [n] *constr. geo.* ▶scoria.

4599 process [n] **of habitat fragmentation** *ecol.* (Dynamic alteration of a natural system over a long period caused by natural processes, such as sea floods, movements of the Earth's crust, separation due to fluctuations in climate or by man-made processes, such as different land uses. These processes may result in the isolation of habitats of flora and fauna causing ▶remnant habitats or ▶habitat islands; ▶ecology of habitat islands); *syn.* fragmentation [n] into isolated habitats; *s* **aislamiento [m] de biótopos** (Proceso de alteraciones del paisaje a través de un largo periodo/período de tiempo con la consecuencia de la separación de biótopos de su contexto natural, tanto por procesos naturales como inundaciones, movimientos de la corteza terrestre, cambios climáticos [glaciaciones] o por influencia antropógena sobre todo a nivel de usos del suelo. Estos procesos pueden llevar a la reducción de grandes espacios vitales a ▶biótopos relícticos o ▶hábitats residuales; ▶ecología de biótopos aislados); *syn.* fragmentación [f] de hábitats; *f* **isolement [m] des biotopes** (Processus dynamique de variation de la répartition des systèmes écologiques évoluant sur des périodes très longues et provoqué par des phénomènes naturels tels que les mouvements de la croûte terrestre, les transgressions marines, des changements climatiques ou des actions anthropogènes dans l'utilisation du sol ; ces processus peuvent provoquer le morcellement en îlots de certains biotopes [▶biotope relictuel, ▶habitat relictuel] et favoriser l'endémisme ; les populations ainsi isolées peuvent donner naissance soit à des écotypes [différenciation physiologique] soit à des races géographiques [variation génétique] ; ▶écologie des biocénoses relictuelles) ; *syn.* insularisation [f] des biotopes, disjonction [f] de biotopes, fragmentation [f] de biotopes, fractionnement [m] de biotopes ; *g* **Verinselung [f] von Biotopen** (Dynamischer Prozess landschaftsökologischer Systemveränderung über einen längeren Zeitraum, bedingt durch natürliche Vorgänge wie z. B. durch Meeresüberflutung, Krustenbewegungen, Isolierung durch Klimaschwankungen oder durch anthropogene Vorgänge wie z. B. durch unterschiedlichste Formen der Bodennutzung. Diese Prozesse können zusammenhängende Lebensräume der Tier- und Pflanzenwelt zu ▶Biotopresten oder ▶Habitatinseln verkümmern lassen. Diese Fragmentierung führt einerseits zu einer Verkleinerung von Populationen auf Resthabitaten und andererseits zur räumlichen Isolation der Populationen. Bei seltenen und bedrohten Arten bewirken sowohl geringe Habitatgrößen als auch räumliche Isolation markante Veränderungen der Populationsstruktur und einen Rückgang der genetischen Vielfalt, welche das langfristige Überleben dieser Arten gefährden können; ▶Inselökologie); *syn.* Fragmentierung [f] von Biotopen, Habitatfragmentierung [f].

4600 proctor compaction [n] *constr. eng.* (Maximum possible consolidation of a soil with an optimum moisture content, according to the specified Proctor compaction test. The test establishes the density-moisture relationship of a soil); *s* **densidad [f] Proctor** (Consolidación máxima posible de un suelo con contenido óptimo de humedad. Se mide por medio de test de Proctor que establece la relación entre la densidad y la humedad del suelo); *f* **compacité [f] Proctor** (Valeur de la densité sèche obtenue lors du compactage d'un granulat lors d'un essai Proctor pour une teneur en eau donnée ; valeur exprimée en pourcentage de vides exprimé en volume ; la valeur maximale de la densité sèche constitue l'optimum Proctor) ; *g* **Proktordichte [f]** (Das größte, durch den Proktorversuch erreichbare Trockenraumgewicht eines zu verdichtenden Bodens mit dem dazugehörenden ‚optimalen Wassergehalt').

4601 producer [n] *ecol.* (Organism that can use radiant energy to synthesize organic substances from inorganic materials; RCG 1982; ▶consumer); *s* **productor [m]** (Organismo autótrofo, macro- o microscópico, capaz de fijar la energía solar mediante la fotosíntesis y de sintetizar materia orgánica a partir de compuestos inorgánicos. La materia así producida es fuente de alimento para los restantes organismos: los ▶consumidores o heterótrofos; DINA 1987); *f* **producteur [m]** (Un organisme, macro- ou microscopique, transformant l'énergie photique en

énergie chimique potentielle s'accumulant dans des composés organiques [glucides, protides, lipides] élaborés à partir de matières minérales fournis par le milieu extérieur abiotique ; DUV 1984 ; ▶consommateur) ; *syn.* organisme [m] producteur ; *g* **Produzent [m]** (Organismus [Pflanze], der mittels der Photosynthese chemosynthetisch organische Substanz [Glucosen, Proteine, Fette] aus anorganischen Stoffen erzeugt; ▶Konsument).

4602 production [n] (1) *ecol.* (Total amount of organic matter produced within a given period of time and area [▶biomass], ▶primary production and ▶secondary production may be differentiated; ▶productivity 1); *s* **producción [f]** (Cantidad total de materia orgánica [▶biomasa] creada en una zona y un tiempo definidos. Se diferencia entre ▶producción primaria y ▶producción secundaria; ▶productividad); *f* **production [f]** (Quantité de matière vivante élaborée par unité de temps, de surface ou de volume [▶biomasse] ; **en F.**, on distingue la production brute et la production nette, la ▶productivité primaire et la ▶productivité secondaire ; ▶productivité 2) ; *g* **Produktion [f]** (Gesamtmenge der innerhalb eines definierten Zeitraumes und innerhalb eines definierten Gebietes resp. Volumens gebildeten organischen Substanz [▶Biomasse]. Es wird zwischen ▶Primärproduktion und ▶Sekundärproduktion unterschieden; ▶Produktivität).

production [n] (2) *met.* ▶dew production; *agr. for.* ▶mixed crop production.

production [n], **crop** *agr. for. hort.* ▶crop husbandry; *agr. for.* ▶mixed crop production.

production [n] [US], **curtailment of agricultural** *agr. pol.* ▶agricultural reduction program [US]/extensification of agricultural production [UK].

production [n], **de-intensification of agricultural** *agr. pol.* ▶agricultural reduction program [US]/extensification of agricultural production [UK].

production [n], **extensification of agricultural** *agr. pol.* ▶agricultural reduction program [US].

production [n], **intensity of crop** *agr. for. hort.* ▶intensity of agricultural and forest land use.

production [n], **intensive crop** *agr. hort.* ▶intensive farming.

production drawing [n] *constr. plan.* ▶shop drawing.

4603 production forest [n] *for.* (Extensive area managed primarily for the production of timber and other forest products, or maintained as woodland for such indirect benefits as protection of catchment areas or recreation. *NOTE* connotes a larger area than a wood [▶woods (US)]; cf. SAF 1983; ▶forest 1, ▶natural forest); *syn.* cultivated forest [n]; *s* **bosque [m] de explotación forestal** (Monte dedicado a la producción de madera o a otros fines productivos; ▶monte; *opp.* ▶bosque natural); *syn.* monte [m] cultivado; *f* **forêt [f] (de production)** (Zone affectée à la production de bois d'œuvre, et d'autres produits forestiers, ou que l'on maintient boisée pour en tirer des avantages divers tels que la protection des bassins-versants, bassins de réception, la récréation, etc. ; DFM 1975. En sens de l'▶inventaire forestier national [en F], toute surface d'au moins quatre hectares, d'une largeur moyenne en cime d'au moins 25 mètres, où l'état boisé est acquis ; TSF 1985 ; ▶forêt; *opp.* ▶forêt naturelle) ; *syn.* forêt [f] soumise au régime forestier ; *g* **Forst [m]** (*Im deutschen Sprachgebrauch* ein nach forstwirtschaftlichen Grundsätzen bewirtschafteter ▶Wald 1. Der engl. Begriff *forest* umfasst grundsätzlich großflächige Bestände und den Urwald; *opp.* ▶Naturwald); *syn.* Wirtschaftswald [m].

production land [n] *agr. plan.* ▶agricultural production land.

production land [n] [UK]**, forestry** *for.* ▶forest products area [US].

4604 productivity [n] (1) *ecol.* (**1.** General term for all aspects considered in the studies of organic matter and energy production by organisms. **2.** Rate at which organic matter is stored in organisms; ▶production 1); *s* **productividad** [f] (Velocidad de producción de materia orgánica creada por un organismo o grupo de ellos en un tiempo determinado; se diferencia entre la productividad primaria y la secundaria; ▶producción); *f* **productivité** [f] (2) (Quantité de matière vivante produite dans un temps donné indépendamment de la biomasse initiale ou finale ; on peut distinguer la productivité primaire, qui est la quantité de matière organique synthétisée dans les organismes, la « productivité nette » qui est l'augmentation de la biomasse pérenne au bout d'un temps déterminé ; VOC 1979 ; ▶production) ; *g* **Produktivität** [f] (**1.** Allgemeiner Ausdruck für alle Aspekte, die beim Studium der Produktion von Stoff und Energie bei Organismen berücksichtigt werden; cf. ÖKO 1983. **2.** Produktionsrate organischer Substanz, die in Organismen angelagert werden; ▶Produktion).

productivity [n] (2) *agr.* ▶pasture of low productivity; *agr. ecol. for.* ▶site productivity; *agr. hort. pedol.* ▶soil productivity.

4605 professional body [n] *prof.* (Collective term for the group of individuals practising the same profession and registered and/or qualified in a corresponding institute or association); *s* **cuerpo** [m] **profesional** (Conjunto de personas que ejercen la misma profesión o tienen una formación profesional reconocida); *f* **ordre** [m] **professionnel** (Terme désignant l'ensemble des personnes exerçant la même activité professionnelle et possédant une référence professionnelle) ; *syn. contexte* la profession [f] ; *g* **Berufsstand** [m] (Gesamtheit aller, die innerhalb einer sozialen Ordnung den gleichen Beruf ausüben und über eine bestimmte Qualifikation verfügen).

professional charges [npl] *prof.* ▶professional fee (2).

professional contract [n] *prof.* ▶design contract [US]/letter of appointment [UK].

4606 professional contribution [n] *plan.* (Input of a particular professional discipline, e.g. to the planning and design of a project); *s* **contribución** [f] **profesional** (Aportación de una disciplina profesional específica a la planificación o el diseño de un proyecto); *f* **contribution** [f] **spécialisée** (Concours apporté par une discipline dans l'établissement de programmes, dans la réalisation d'un projet d'aménagement) ; *g* **Fachbeitrag** [m] (**1.** Leistungsanteil einer Fachdisziplin zu einer Planung oder zu einem Gesamtprojekt. **2.** Fachliche Stellungnahme zu einer Planung oder zu einem bestimmten Fachgebiet).

professional development [n] **(CPD)** [UK]**, continuing** *prof.* ▶continuing education [US].

4607 professional education [n] *prof. syn.* professional training [n]; *s* **formación** [f] **profesional** *syn.* capacitación [f] profesional; *f* **formation** [f] **professionnelle** ; *g* **Berufsausbildung** [f].

professional fee [n] (1) *contr. prof.* ▶Architektenhonorar.

4608 professional fee [n] (2) *contr. prof.* (Remuneration for professional services; ▶honorarium, ▶remuneration of professional fees; *specific terms* ▶architect's fee, ▶lump-sum fee); *syn.* professional charges [npl]; *s* **honorarios** [mpl] **profesionales** (Remuneración por los servicios profesionales; ▶honorarios reconocidos, ▶pago de honorarios; *términos específicos* ▶honorarios de arquitecto, ▶honorarios globales); *f* **honoraire** [m] (Rémunération d'un concepteur pour une ▶mission de maîtrise d'œuvre confiée par un maître d'ouvrage ; ▶honoraire de gratification, ▶rémunération des honoraires ; *termes spécifiques*

▶honoraire au forfait, ▶honoraire d'architecte) ; *g* **Honorar** [n] (Entgelt für die in Auftrag gegebenen Leistungen an einen Planer/eine Planerin; ▶Anerkennungshonorar, ▶Honorarvergütung; *UBe* ▶Architektenhonorar, ▶Pauschalhonorar); *syn.* Honorierung [f].

professional fee [n] **based on an hourly rate** *contr. prof.* ▶time charges.

professional garden preservationist [n] *conserv'hist.* ▶historic garden expert [US]/garden conservator [n] [UK].

professional insurance [n]**, liability for** *contr. prof.* ▶obligation to take out professional liability insurance.

4609 professional journal [n] *prof.* (Periodical, which covers topics of a particular specialist field or profession); *syn.* technical journal [n], professional magazine [n]; *s* **revista** [f] **profesional** (Publicación periódica especializada en temas de una especialidad o una profesión académica, científica o técnica; *términos específicos* revista científica, revista técnica); *syn.* revista [f] especializada; *f* **revue** [f] **professionnelle** (Publication paraissant périodiquement, généralement illustrée et traitant d'un domaine ou d'un secteur d'activités professionnelles déterminé) ; *syn.* journal [m] professionnel, revue [f] spécialisée, périodique [m] professionnel, magazine [m] professionnel ; *g* **Fachzeitschrift** [f] (Regelmäßig erscheinende Zeitschrift, in der aktuelle Themen eines bestimmten Fachgebietes oder Berufszweiges abgehandelt werden); *syn.* Fachzeitung [f].

4610 professional liability insurance [n] *prof.* (Insurance to indemnify injury, damage or loss to others in connection with activities of a professional person; ▶obligation to take out professional liability insurance); *syn.* errors and ommission insurance [n] [also US]; *s* **seguro** [m] **de responsabilidad civil** (Seguro que cubre los riesgos en relación con actividades profesionales; ▶obligación de tener seguro profesional); *f* **assurance** [f] **responsabilité civile professionnelle** (Couverture contractée par un entrepreneur individuel ou un architecte contre les risques aux biens ou aux personnes consécutifs à son activité professionnelle ; les architectes peuvent assurer les dommages causés aux ouvrages consécutifs à des plans défectueux ou une direction des travaux insuffisante, les dommages causés par perte ou troubles de jouissance subis par le client [responsabilité biennale, décennale et de bon fonctionnement] ; les membres d'un ordre professionnel ont l'obligation d'adresser chaque année au Conseil régional de l'Ordre dont il relève une attestation de leur organisme assureur établissant qu'il sont couverts pour l'année en cours [le défaut de production de cette attestation, comme le défaut d'assurance, est passible de sanctions disciplinaires] ; ▶obligation de contracter une assurance professionnelle) ; *g* **Haftpflichtversicherung** [f] **für Planer** (Versicherung, die Architekten und Ingenieuren Schutz bei Schadenersatzansprüchen Dritter, die bei Planungs- und Bauüberwachungsfehlern entstehen, gewährt; i. d. R. werden Personen- und Sachschäden ersetzt. Zum Schutz der Bauherren schreibt die Berufsordnung der Architektenkammer freiberuflich tätigen Kammermitgliedern vor, eine ausreichende Berufshaftpflichtversicherung abzuschließen; dasselbe gilt für Auftragnehmer der öffentlichen Hand; ▶Versicherungspflicht); *syn.* Berufshaftpflichtversicherung [f].

professional magazine [n] *prof.* ▶professional journal.

4611 professional member [n] **of a competition jury** [US] *prof.* (Qualified expert who is competent to make judgements concerning particular disciplines such as regional, town and country planning, landscape planning, open space planning, building design, interior design, or design of construction elements; ▶competition jury, ▶official of a competition jury); *syn.* assessor [n] of a competition jury [UK]; *s* **miembro**

[m] **profesional de jurado** (Especialista en un ámbito profesional [p. ej. urbanismo, planificación del paisaje, arquitectura o paisajismo] nombrado para decidir sobre los premios en un concurso de ideas o de realización; ▶jurado, ▶miembro de jurado de concurso); *f* **personnalité [f] qualifiée membre du jury** (Personnalité membre d'un ▶jury compétente dans la matière qui fait l'objet d'un concours ; il s'agit de spécialistes, maîtres d'œuvre ou hommes de l'art dans les domaines de la planification regionale, l'urbanisme, l'aménagement des paysages, l'architecture, l'architecture d'intérieur ; souvent avec voie consultative ; ▶membre du jury) ; *g* **Fachpreisrichter [m]** (Qualifizierter Fachmann eines ▶Preisgerichtes, der je nach Wettbewerbsaufgabe das Fachgebiet Regionalplanung, städtebauliche Planung, Landschaftsplanung, Freiraumplanung, Bauwerksplanung, Innenraumplanung oder Elementplanung abdeckt; ▶Sachpreisrichter).

4612 professional organization [n] *prof.* (Membership association of persons practicing the same profession, which is formed with the objective of setting and upholding minimum standards of work, and representing the collective interests of its members; ▶Council of Landscape Architectural Registration Boards [US]/Architects Registration Council [UK]); *s* **organización [f] profesional** (Asociación de personas de la misma profesión creada con el objetivo de fijar y mantener normas mínimas de trabajo y de representar los intereses colectivos de sus miembros; ▶Colegio Oficial de Arquitectos); *f* **organisme [m] professionnel** (Association regroupant à titre volontaire des professionnels. Ses activités consistent dans la défense et la représentation de ses adhérents dans les domaines juridiques, fiscaux, sociaux commerciaux ; *pour les entrepreneurs de jardins* U.N.S.E.P.R.F., C.N.I.H., I.T.I.H. ; *pour les architectes paysagistes* F.F.P. — Fédération française du paysage ; ▶ordre des architectes ; ▶Union nationale des syndicats français d'architectes) ; *syn.* organisation [f] professionnelle ; *g* **Berufsorganisation [f]** (Vorwiegend auf freiwilliger Basis gebildete Vereinigung mit dem Ziel, gemeinsame berufliche, wirtschaftliche und auch kulturelle Interessen der Mitglieder zu wahren und nach außen hin zu vertreten; ▶Architektenkammer); *syn.* berufsständische Vertretung [f], Berufsverband [m].

4613 professional practice [n] *prof.* (Field of professional involvement and activities, licensed by professional bodies [UK] or states [US]; **in U.S.**, "professional" is limited in this context to activities that are licensed by state or local governments: i.e. doctors lawyers, engineers, barbers, cosmotologists, etc. whose activities are closely related to public health and safety, as a legal/constitutional matter, under state law); *s* **práctica [f] profesional** (Ámbito de actividades profesionales en una rama específica; algunas profesiones —como arquitectura, medicina, derecho, artesanado— están organizadas en organizaciones profesionales resp. gremiales); *f* **pratique [f] d'une profession** (Activité continue régulière dans un domaine professionnel précis ; les membres des professions libérales [p. ex. architectes médecins, avocats, juristes] sont organisés en ordres professionnels) ; *g* **Berufsausübung [f]** (Regelmäßige Tätigkeit in einem beruflichen Arbeitsfeld; bestimmte Berufe [z. B. Architekten, Ärzte, Handwerker, Juristen] sind in berufsständischen Vereinigungen organisiert).

4614 professional profile [n] *prof.* (Range or scope of services of a particular profession with all its duties and fields of engagement); *syn.* portrait [n] of a profession; *s* **esquema [m] descriptivo profesional**; *f* **profil [m] professionnel** (**1.** Définition, activités, compétences, conditions de travail et qualifications requises caractérisant un métier, une profession. **2.** Caractère, aptitudes, valeurs, motivations, compétences, permettant le choix d'une profession ou d'une formation pour un chercheur d'emploi) ; *syn.* profil [m] de la profession ; *g* **Berufsbild [n]** (Umfang der Aufgabenbereiche und Tätigkeiten eines Berufsstandes).

professional responsibilities [npl] *contr. prof.* ▶fulfillment of professional responsibilities [US]/fulfilment of professional responsibilities [UK].

4615 professional services [npl] *contr. prof.* (Tasks carried out by a trained planner, consultant or expert according to an agreed scope of work; ▶basic professional services, ▶description of professional services, ▶planning services, ▶scope of professional services; ▶special professional services, ▶supplemental professional services); *s* **servicios [mpl] profesionales** (Prestaciones realizadas por un/a arquitecto/a o ingeniero/a. Mientras los **ss. pp.** no estén definidos en listas de trabajo, se dividen en ▶servicios profesionales básicos y ▶servicios complementarios de planificación. *Términos específicos* servicios de arquitectura, servicios de asesoría, servicios de peritaje, servicios de ingeniería y ▶servicios de planificación; ▶descripción de servicios de planificación, ▶servicios suplementarios); *f* **prestations [fpl] (intellectuelles)** (Service intellectuel apporté à un donneur d'ordre par un concepteur, un ingénieur dans le cadre de contrats dans les domaines les plus divers [assistance, conseil, recherche, études, maîtrise d'œuvre, expertise, etc.] ; dans une ▶mission normalisée de maîtrise d'œuvre on distingue les ▶prestations élémentaires et les ▶prestations complémentaires ; ▶prestations supplémentaires ; *termes spécifiques* **p.** d'architecture, **p.** de maîtrise d'œuvre, **p.** d'ingénierie, **p.** d'études ; ▶forme et étendue de la mission, ▶mission normalisée, ▶prestation de maîtrise d'œuvre) ; *g* **Leistung [f]** (Nach Art und Umfang durch einen Vertrag näher bestimmte Tätigkeiten/Dienstleistungen eines Planers, Beraters oder Gutachters; soweit z. B. eine Planerl. in ▶Leistungsbildern erfasst ist, gliedert sie sich in ▶Grundleistungen und ▶besondere Leistungen 2; *UBe* Architektenl., Beratungsl., Gutachterl., Ingenieurl., Planerl.; ▶Planungsleistung, ▶zusätzliche Leistung 2).

professional services [npl], **additional** *contr. prof.* ▶supplemental professional services.

4616 professional services [npl] **of an architect** *prof.* (Tasks carried out by an architect according to an agreed scope of planning work); *s* **servicios [mpl] profesionales de arquitecto/a** (Prestaciones de un/a arquitecto/a en el marco de un contrato de planificación); *f* **prestation [f] d'architecte** (Mission d'ingénierie ou d'architecture effectuée par un architecte) ; *g* **Architektenleistung [f]** (Von einem Architekten/einer Architektin erbrachte planerische Leistung eines vereinbarten Leistungsbildes).

professional training [n] *prof.* ▶professional education.

profile [n] *prof.* ▶professional profile; *pedol.* ▶soil profile; *pedol.* ▶truncated soil profile; *plan. trans.* ▶vertical profile.

profile [n] [US], **gradient** *trans.* ▶vertical alignment.

profile [n], **longitudinal valley** *geo.* ▶thalweg.

4617 profundal zone [n] *limn.* (Deep, floor area in freshwater lakes with mud deposits, including all those living organisms, which can not perform photosynthesis due to the lack of light; ▶benthic zone, ▶littoral zone [US]/littoral [UK]); *s* **zona [f] profunda de un lago** (Zona del fondo de lagos de agua dulce con depósito de fango, incluyendo los organismos que habitan en él, en la que —debido a la falta de luz— los organismos fotoautótrofos no pueden realizar la fotosíntesis; ▶litoral, ▶zona del bentos); *f* **zone [f] benthique profonde** (Zone profonde du fond d'une étendue d'eau douce recouverte de boue ou de vase, dépourvue de plantes aquatiques et habitée par les organismes photo-autotrophes de la vase [pelon], la photosynthèse n'étant plus possible ; ▶zone benthique littorale, ▶zone benthique); *g* **Profundal [n]** (Tiefere Bodenregion von Süßwasserseen mit Schlammablagerungen einschließlich der Gesamtheit der Lebewesen, in der wegen Lichtmangels positive Photoautotrophe nicht

mehr assimilieren können. Das **P.** liegt unterhalb des ►Litorals; ►Benthal).

program(me) [n] *agr. pol.* ►agricultural reduction program [US]/extensification of agricultural production [UK]; *conserv. pol.* ►environmental program [US]/environmental programme [UK]; *constr. hort.* ►plant maintenance program [US]; *landsc. pol.* ►regional landscape program [n] [US]/regional landscape programme [n] [UK].

programme [n] **and project impacts in terms of Natura 2000** *conserv. leg. plan.* ►preliminary assessment of programme and project impacts in terms of Natura 2000.

programme [n] **of after care** [UK] *constr. hort.* ►plant maintenance program [US].

4618 progress chart [n] *constr. plan.* (Graph showing the various operations in a construction project, such as excavating, foundations, work by each trade, with planned start and finish dates in the form of horizontal bars depicting progress by filling in the bars; ►construction program [US]/construction programme [UK]); *s* **diagrama** [m] **de etapas de construcción** (Diagrama o plan en el que está representado el transcurso de un proyecto; ►calendario de ejecución de obras); *f 1* **planning** [m] **d'exécution des travaux** (Diagramme déterminant les dates et durées d'intervention des corps d'état et servant de base à la coordination des travaux) ; *syn.* calendrier [m] d'exécution, calendrier [m] de travaux et de fournitures (HAC 1989, 423) ; *f 2* **planning** [m] **d'avancement des travaux** (Planning établi par le maître d'œuvre rendant compte de l'avancement des travaux et permettant le cas échéant d'appliquer des pénalités éventuelles en cas de retard sur le ►planning d'exécution des travaux ; ►programme d'une opération) ; *syn.* diagramme [m] d'avancement des travaux ; *g* **Bauzeitenplan** [m] (Balkendiagramm, das den geplanten Bauablauf von der Ausschreibung bis zur Abnahme und den aktuellen Stand des Baufortschrittes anzeigt; er dient der Koordination und Kontrolle aller erforderlichen planerischen Tätigkeiten und Gewerke, um den geplanten Fertigstellungstermin eines Bauprojektes einzuhalten; ►Bauablaufplan; *OB* Zeitplan); *syn.* Projektmanager [m], Baukalender [m].

4619 progress [n] **of work** *contr.* (Advancement and sectoral completion of a defined scope of work); *s* **avance** [m] **del trabajo** *syn.* progreso [m] del trabajo; *f* **avancement** [m] **des travaux** *syn.* progression [f] des travaux ; *g* **Arbeitsfortschritt** [m] (Voranbringen und abschnittsweise Erledigung eines definierten Arbeitsumfanges).

progress payment [n] [US] *constr. contr.* ►interim payment.

4620 progress report [n] *constr. plan.* (Written or oral statement of ongoing status of a planning or construction project, and of research or investigation; ►final planning report, ►explanatory report, ►provisional report); *s* **avance** [m] **del estado de los trabajos** (Informe oral o escrito sobre el progreso de un proyecto de planificación o de construcción; ►informe final de una planificación, ►informe descriptivo, ►informe provisional); *f* **rapport** [m] **d'avancement** (d'un projet, d'une étude ; ►rapport définitif d'une étude, ►rapport de synthèse, ►rapport de présentation, ►rapport provisoire [d'avancement]) ; *g* **Bericht** [m] **über den Stand der Arbeiten/Planung** (Schriftliche oder mündliche Darlegung über die derzeitige Situation/den Fortschritt der Bauarbeiten/einer Planung; ►Abschlussbericht einer Planung, ►Erläuterungsbericht, ►vorläufiger Bericht); *syn.* Zwischenbericht [m].

progress schedule [n] *constr. plan.* ►network plan.

progress schedule [n]**, construction** *constr. plan.* ►construction program [US]/construction programme [UK].

4621 prohibition [n] **of a land use** *conserv. leg. plan.* (Denial of planning permission for use of land or property by planning law or other regulations. Development permission is often refused on the grounds of non-compatibilty with the needs of nature conservation; ►setting-aside of arable land [UK], ►use restriction); *s* **prohibición** [f] **del uso** (Negación del permiso de uso del suelo o de un terreno o del aprovechamiento de los recursos naturales sobre la base de leyes u otras regulaciones, generalmente por razones de protección de la naturaleza o del medio rural; ►abandono de tierras, ►restricción del uso); *f 1* **interdiction** [f] **de tout mode d'occupation ou d'utilisation des sols** (F., prescription prévue dans le règlement de zone d'un PLU/POS interdisant toute opération soumise à réglementations, p. ex. lotissements, installations classées, parcs d'attraction, aires de jeux et de sports ouvertes au public, aires de stationnement ouvertes au public, affouillements et exhaussements, terrains de camping ou de camping et de caravaning ; disposition souvent prise pour préserver la qualité d'un site ; ►enfrichement, ►gel environnemental, ►restriction d'usage) ; *f 2* **interdiction** [f] **d'utilisation des sols** (±) (D., Mesures permanentes ou temporaires, fixées par décret ou par arrêté selon la nature de la zone considérée, qui pour des raisons d'utilité publique, visent à interdire des activités agricoles, forestières et pastorales, l'extraction de matériaux, l'utilisation des eaux, dans le cadre de la protection des espaces naturels et des paysages, de la préservation des espèces animales et végétales et de leurs habitats) ; *syn.* interdiction [f] d'usage du sol ; *g* **Nutzungsverbot** [n] (Untersagung der Nutzungsmöglichkeit von Grund und Boden durch Gesetze oder sonstige Vorschriften [cf. § 32 BauGB u. § 3 b BNatSchG], z. B. **N.** durch Naturschutzverordnung oder Anordnungen der für Naturschutz und Landschaftspflege zuständigen Behörden, den Abbau von Bodenbestandteilen zur Verwirklichung der Ziele des Naturschutzes und der Landschaftspflege zu untersagen. Dabei werden die Eigentümerbefugnisse bei der Bodennutzung vor dem Hintergrund der Sozialbindung des Eigentums unter ausgewogener Abwägung der Interessen des Grundstückseigentümers und der öffentlichen Belange des Naturschutzes definiert; ►Flächenstilllegung, ►Nutzungsbeschränkung).

4622 project [n] (1) *plan. prof.* (Work undertaken by a planner; ►project planning, ►planning project); *s* **proyecto** [m] (1) (Tarea de planificación o de un proyecto [de construcción], ►proyecto de planificación); *f* **projet** [m] (Caractérisant un aménagement ; ►mission opérationnelle, ►projet d'aménagement) ; *g* **Projekt** [n] (1) (Gegenstand einer Planung; ►Objektplanung, ►Planungsvorhaben); *syn.* Vorhaben [n].

4623 project [n] (2) *plan. ecol.* (Generic term in European Nature Conservation for a plan, program[me] and measure undergoing studies for the feasibility of their implementation within or outside a Natura 2000 area; the investigation is carried out to establish whether a **p.** would significantly impact, either individually or jointly, a ►site of community importance or an European Bird Sanctuary. Plans or projects that deal directly with the administration of sites of community importance or the European bird sanctuaries are excepted from the process; cf. Article 6, paragraph 3, Habitats Directive; ►special area of conservation [SAC]); *s* **proyecto** [m] (2) (Término genérico de la política europea de protección de la naturaleza para un plan o proyecto que pueda afectar de forma apreciable a una ►zona especial de conservación o a una ►zona especial de conservación para las aves según la Directiva Europea de las Aves, ya sea individualmente o en combinación con otros planes y proyectos, se debe someter a una adecuada evaluación de sus repercusiones en el lugar teniendo en cuenta los objetivos de conservación de dicho lugar; cf. Art. 6.3 de la Directiva de Hábitats; ►lugar de

importancia comunitaria); *f* **programme [m] et projet de travaux, d'ouvrage ou d'aménagement (PPTOA)** (Terme générique désignant tout programme et projet situé à l'intérieur ou à l'extérieur d'un site Natura 2000 soumis à étude ou notice d'impact ou document d'incidences « loi sur l'eau », qui individuellement ou en conjugaison avec d'autres plans et projets est susceptibles d'affecter de façon notable un ou plusieurs sites d'intérêt communautaire ou une ▶zone spéciale de conservation [ZSC] et qui est soumis à une évaluation d'incidences [cf. circulaire DNP/SDEN N°2004-1 du 5 oct. 2004 et Art. R. 214-34 2 du code de l'environnement] ; sont exclus les plans directement liés ou nécessaires à la gestion des ▶sites d'importance communautaire [SIC] ou des zones de protection spéciale [réf. Art. 6 § 3 de la directive « Habitats »]) ; *g* **Projekt [n] (2)** (*Europäischer Naturschutz* Oberbegriff für ein konkret zu prüfendes Vorhaben, für einen zu prüfenden Plan, zu prüfende Programme und Maßnahmen innerhalb oder außerhalb eines Natura-2000-Gebietes, soweit sie, einzeln oder im Zusammenwirken mit anderen Projekten oder Plänen geeignet sind, ein ▶Gebiet von gemeinschaftlicher Bedeutung oder ein Europäisches Vogelschutzgebiet erheblich zu beeinträchtigen. Ausgenommen sind Pläne oder Projekte, die unmittelbar der Verwaltung der Gebiete von gemeinschaftlicher Bedeutung oder der Europäischen Vogelschutzgebiete dienen; cf. Art. 6 Abs. 3 FFH-RL; ▶besonderes Schutzgebiet).

project [n] (3) *constr. plan.* ▶construction project, ▶contractor's supervision of the project, ▶federal government construction project, ▶local authhority project, ▶new project, ▶planning project, ▶public construction project, ▶state government construction project; *constr. contr.* ▶suspension of project [US]/ suspension of construction works [UK],

project [n], agreed measurement of completed *constr. contr.* ▶agreed measurement of completed work [US]/ agreed measurement of completed works [UK].

project [n], building *constr. plan.* ▶construction project.

project [n], coordination of a planning *plan.* ▶obtaining a mutual agreement on planning proposals.

4624 project brief [n] *plan. prof.* (Client's or sponsor's general expression of needs to be fulfilled by a planner or competing planners. *Specific terms* competition brief, consultant's brief, design brief, planning statement [also US]; ▶planning contract); *syn.* statement [n] of the project [also US]; *s* **avance [m] de programa de necesidades** (Descripción general de las funciones a satisfacer en un concurso o un trabajo de planificación; ▶contratación de un/a planificador, -a); *f* **objet [m] de la mission** (Données préalables qui définissent le cadre de réalisation d'une prestation dans le cadre d'un concours, de la conception et de la maîtrise d'œuvre de projets d'espaces extérieurs ainsi que de la planification concernant la protection et la mise en valeur des paysages ; ▶mission d'études, ▶mission de maîtrise d'œuvre) ; *syn.* contenu [m] de la mission ; *g* **Aufgabenstellung [f] (1)** (Vorgabe eines generellen Aufgabenrahmens zu einem Wettbewerb, einer Objekt- oder Landschaftsplanung etc., die für alle Bewerber im gleichen Sinne verstanden werden können; ▶Planungsauftrag); *syn.* Aufgabenbeschreibung [f].

4625 project completion [n] [US] *constr.* (▶site clearance after project completion); *syn.* completion [n] of work [US]/completion [n] of works [UK], practical completion [n] [UK]; *s* **terminación [f] de obras** (▶despeje del lugar de la obra); *syn.* conclusión [f] de obras; *f* **achèvement [m] des travaux** (▶repliement du chantier) ; *g* **Fertigstellung [f] der Bauarbeiten** (▶Räumung der Baustelle nach Fertigstellung); *syn.* Fertigstellung [f] der Bauleistungen.

4626 project control [n] *plan.* (Complete range of services, usually carried out by the Client/Owner or an appointed representative, which are necessary during the realization of a project: e.g. organization according to set targets, coordination, guidance and supervision of execution processes and, in the case of large-scale projects, special attention to economic and technical aspects, deadlines as well as administrative and legal matters; tasks include a] clarification of the scope of work and preparation and coordination of the program for the construction of the entire project; b] setting up and monitoring of organization, time schedules and payment plans necessary for a project and its participants, including the monitoring of costs according to a predetermined cost framework; c] coordination and control of all project participants, with the exception of the contractor; d] clarification of the need to engage additional planners and other experts; e] assistance in the participation and care of those affected by the project; f] clarification of conflicting aims; g] ongoing duty of reporting to the client/▶project manager on the progress of the project and providing information in good time about the necessity of meeting outstanding decisions; h] coordination and control of budget expenditures, the handling of financial assistance programs and approval procedures); *syn.* project controlling [n]; *s* **control [m] de calidad de proyecto** (Conjunto de servicios de dirección, generalmente por parte del propietario de obra o una persona encargada por él, necesarios para llevar a cabo con éxito un proyecto: coordinación, gestión y control de todos los procesos —en especial en proyectos grandes— teniendo en cuenta todos los aspectos técnicos y económicos, fechas de entrega, así como asuntos administrativos y legales. Los servicios incluyen: a] aclaración del alcance de los trabajos y preparación y coordinación del programa para la realización del proyecto completo; b] establecer y monitorear los planes de organización, calendarios y de pagos necesarios para el proyecto y sus participantes, incluyendo el monitoreo de los costes [Es]/costos [AL] de acuerdo con el marco de costes [Es]/costos [AL] predeterminado; c] coordinación y control de todos los participantes en el proyecto, exceptuando al contratista; d] aclaración de la necesidad de contratar a planificadores/as adicionales o a otros/otras expertos/as; e] preparación y asesoría en la participación de la población afectada por el proyecto; f] aclaración de conflictos; g] obligación continua de informar a la propiedad de obra/al ▶project manager sobre el avance del proyecto y de proveer información a tiempo en el caso de necesidad de tomar decisiones pendientes; h] coordinación y control de los gastos presupuestarios, manejo de programas de subvenciones financieras y del procedimiento de aprobación del proyecto); *f 1* **pilotage [m] d'un projet** (Ensemble des missions incombant au maître de l'ouvrage ou au maître d'ouvrage délégué nécessaires à la réussite d'un projet et comprenant, avec un but précis, l'organisation, la coordination, la conduite et le contrôle du déroulement de toutes les activités, et particulièrement dans le cadre de grands projets, sur les plans économique, opérationnel, technique, administratif et juridique ; la mission comporte a] l'analyse et la définition des objectifs ainsi que l'établissement et la coordination du programme de planning général du projet ; b] la mise au point et le contrôle du respect du planning d'organisation du chantier, du calendrier des travaux et de l'échéancier financier ainsi que le respect de l'enveloppe financière de l'opération ; c] la coordination et le contrôle de tous les intervenants à l'exception des entreprises exécutantes ; d] définition d'un programme permettant le missionnement d'aménageurs ou de professionnels pour la réalisation du projet ; e] la préparation et le suivi de la participation des personnes susceptibles d'être touchées par le projet ; f] gestion des conflits d'objectifs ; g] l'information régulière du maître d'ouvrage ou de

P

son représentant ainsi que l'information ponctuelle sur les éléments nécessaires à la prise de décisions éventuelles ; h] la coordination et le contrôle de procédures financières, de demande de subventions, d'autorisations administratives) ; *syn.* conduite [f] d'un projet, *terme à éviter* controling [m] de projet ; *f 2* **mission [f] ordonnancement, pilotage, coordination [mission OPC]** (*Gestion de la construction* Mission de ▶maître d'œuvre instaurée par la loi MOP n° 85-704 du 12 juillet 1985 et comprenant l'organisation **pour l'ordonnancement**, la préparation et la surveillance de la planification du chantier, c.-à-d. l'inventaire et la mise à jour des contraintes techniques et administratives ou formalités conditionnant les travaux de conception et de réalisation du programme, l'établissement du planning général pour l'opération, la sélection des entreprises, la planification des études et des tâches administratives préliminaires, l'élaboration du planning contractuel des entreprises, la mise au point du calendrier détaillé d'exécution, l'édition d'un échéancier lié à l'achèvement des constructions, **pour le pilotage,** des interventions et les actions assurant le bon déroulement du chantier en fonction de la planification, c.-à-d. l'examen des pièces contractuelles, le contrôle du respect des dates butoirs pour l'établissement des documents de conception du projet, le suivi des démarches administratives pour l'obtention de toutes les autorisations concernées par le projet, l'organisation des réunions de coordination « études », le contrôle du respect du planning, la mise au point du planning de rattrapage, tous diagrammes et documents permettant de maîtriser les travaux modificatifs, les présentations et choix de matériaux, les intempéries, les avancements et retards, les mises au point diverses, l'organisation des visites de contrôle pour parachèvement et obtention de la conformité ; **pour la coordination,** l'organisation des relations entre entreprises et entre les différents intervenants, c.-à-d. la diffusion des informations et documents à tous les intervenants concernés, la proposition d'une procédure claire pour faciliter la communication des informations et circulation des documents, l'examen de l'organisation générale du chantier pour les installations, les alimentations, les accès, circulations, la mise au point des relations interentreprises [prorata, gravois, nettoyage], l'établissement systématique d'un compte-rendu de chantier, l'information du maître d'ouvrage et de l'architecte de tous problèmes pouvant générer du retard, ou des conflits, la participation aux opérations de réception ; www.batimexpert.com, glossaire) ; *g* **Projektsteuerung [f]** (Gesamtheit der Leistungen, die i. d. R. seitens eines Bauherrn oder eines von ihm Beauftragten bei der Realisierung von Projekten erfüllt werden müssen: zielgerichtete Organisation, Koordination, Lenkung und Überwachung aller Geschehensabläufe — speziell bei größeren Projekten — in wirtschaftlicher, terminlicher und technischer sowie verwaltungsverfahrensmäßiger und rechtlicher Hinsicht. Zu den Leistungen gehören z. B. a] Klärung der Aufgabenstellung und Erstellung und Koordinierung des Programms für den Ablauf des Gesamtprojektes; b] Aufstellen und Überwachung von Organisations-, Termin- und Zahlungsplänen, die für ein Projekt und Projektbeteiligte nötig sind und Überwachung des vorgegebenen Kostenrahmens; c] Koordinierung und Kontrolle aller Projektbeteiligten, mit Ausnahme der ausführenden Firmen; d] Klärung der Voraussetzung für den Einsatz von Planern und anderen an der Planung fachlich Beteiligten; e] Vorbereitung und Betreuung der Beteiligung von Projektbetroffenen; f] Klärung von Zielkonflikten; g] laufende Berichtspflicht gegenüber dem Auftraggeber/▶Projektleiter/-in 2 über den Stand des Projektes und rechtzeitige Information über die Notwendigkeit, noch ausstehende Entscheidungen/Beschlüsse herbeizuführen; h] Koordinierung und Kontrolle der Bearbeitung von Finanzierungs-, Förderungs- und Genehmigungsverfahren; cf. § 31 HOAI 2002; die **P.** beginnt mit der Einrichtung einer Kostenstelle); *syn.* Projektcontrolling [n].

4627 project description [n] *contr.* (General explanation of a project and its *in situ* condition; ▶technical specifications, ▶design and build program [US]/design and build programme [UK]); *s* **memoria [f] descriptiva (de una obra)** (Representación de la tarea de una obra y un ▶resumen de prestaciones ordenado según las prestaciones individuales; ▶especificaciones técnicas, ▶memoria descriptiva de obra con programa de trabajo); *f* **descriptions [fpl] des ouvrages** (Fixation de l'étendue de la mission, définition de la forme de la mission, définition du contenu de la mission, ▶description des ouvrages, ▶dossier de consultation avec programme) ; *syn.* descriptif [m] ; *g* **Baubeschreibung [f]** (*Garten- und Landschaftsbau* detaillierte Darstellung der Bauaufgabe und der örtlichen Verhältnisse resp. Besonderheiten als Teil der Unterlagen zur ▶Leistungsbeschreibung 2, ▶Leistungsbeschreibung mit Leistungsprogramm; sie wird damit zum Bestandteil des ▶Leistungsvertrages).

project designer [n] *prof.* ▶site designer.

4628 project development maintenance [n] [US] *contr.* (Maintenance of landscape construction work until issuance of the certificate of compliance [US]/certificate of practical completion [UK]; ▶establishment maintenance); *syn.* pre-handover maintenance [n] [UK]; *s* **trabajos [mpl] de mantenimiento pre-entrega (±)** (Todos los servicios de mantenimiento necesarios para asegurar el crecimiento adecuado de la vegetación hasta la emisión del acta de aceptación de obra. Según la norma alemana DIN 18 1916 generalmente el plazo de mantenimiento se prolonga hasta el final del primer ▶periodo/período de crecimiento, ya que normalmente es entonces cuando se puede tener la seguridad de que las plantas han arraigado; en lugares de clima duro o en caso de sequía inusual puede llegar hasta los tres años. En las superficies de césped la resiembra de ▶calvas también es parte de los **tt. de m. pre-e.**; ▶mantenimiento inicial); *syn.* mantenimiento [m] previo a la entrega de la obra (±); *f* **entretien [m] (d'espaces verts) jusqu'à la réception des travaux** (Travaux de parachèvement effectués après la réalisation des ouvrages, la mise en place des végétaux et des engazonnements pendant la ▶période de végétation afin d'assurer la consistance des aménagements paysagers, des aires de sports et de loisirs [tailles, regarnissage des zones de ▶pelades, etc.] pendant la période s'écoulant jusqu'à la réception des travaux ; ▶travaux d'entretien pendant la période de garantie) ; *g* **Fertigstellungspflege [f, o. Pl.]** (Alle Leistungen, die zur Erzielung eines abnahmefähigen Zustandes von landschaftsgärtnerischen Arbeiten erforderlich sind, um eine gesicherte Weiterentwicklung einer hergestellten Anlage resp. erines Biotopes zu ermöglichen [cf. DIN 18 1916: 2000-8: Sicherheit des Anwuchserfolges. Der Pflegezeitraum ist i. d. R. bis zum Ende der ersten ▶Vegetationsperiode nach Fertigstellung, da meistens erst danach die Sicherheit über den Anwuchserfolg besteht; er kann aber auch, z. B. durch trockene Vegetationsperioden oder in klimatisch ungünstigen Lagen, bis zu drei Jahren dauern; bei Rasenflächen gehört auch die Nachsaat von ▶Fehlstellen noch zur **F.**; ▶Entwicklungspflege); *syn.* Anwuchspflege [f, o. Pl.].

4629 project director [n] *plan.* (*Depending on the management hierarchy and organizational structure of a project,* the Client/Owner himself or, in the case of a local authority, public agency or similar company, an appointed, competent representative with special responsibility for the organization and execution of a complex [large-scale] project. His/her tasks include: a] determination of requirements, clarification of the scope of work and the establishment of planning objectives as well as the extent of the spatial program and installation components for the financing [planning of investment funds] of the entire project; b] timely achievement of the necessary decisions on the project's economic deadlines and objectives;

c] decisions on the appointment of planners and, if necessary, a project controller, as well as technical experts and the associated contract award negotiations and awards; d] decisions on awarding contracts to firms and monitoring compliance with relevant contract conditions of award and contract regulations; e] ►project control, if this task is not awarded to a specially designated controller; f] presentation of the project to governing committees and the public, as well as attendance and organization of meetings with the users and public hearings [►public participation]; g] control and acceptance of the services and work of all project stakeholders; these tasks are the responsibility of the Client and cannot be delegated to a third party in comparison with those tasks of the project controller); **s gerente [m/f] de proyecto** (Dependiendo de la estructura de jerarquía y de organización del proyecto, el cliente o la propiedad misma o, en el caso de una autoridad local, agencia pública o similar, un/a representante muy competente designado/a con especial responsabilidad para organizar y ejecutar un proyecto [grande] complejo. Sus tareas son: a] determinar los requisitos, aclarar el alcance de los trabajos y establecer los objetivos de planificación, así como la envergadura del programa espacial y de los componentes de la instalación para la planificación financiera de todo el proyecto; b] garantizar la toma de decisiones oportuna sobre los objetivos y topes financieros del proyecto; c] decidir sobre la contratación de planificadores/as y, si fuese necesario, de un responsable de control de calidad así como de expertos/as técnicos/as, incluyendo las correspondientes negociaciones y adjudicaciones de contratos; d] decidir sobre la adjudicación de contratos a empresas y monitorear el cumplimiento de las condiciones de adjudicación de contrato; e] ►control de calidad de proyecto, si esta tarea no ha sido asignada a un especialista; f] presentación del proyecto al comité responsable y al público, así como atender y organizar reuniones con futuros/as usuarios/as y audiencias públicas [►participación pública en la planificación]; g] control y aceptación de los servicios y trabajos de todos los participantes en el proyecto; estas tareas son responsabilidad de la propiedad de obra y no se pueden delegar a terceros, en comparación con las tareas del responsable de control de calidad); *syn.* director, -a [m/f] de proyecto, jefe/a [m/f] de proyecto; *f 1* **directeur [m] de travaux** (Personne assurant des missions de pilotage fonctionnel sur un ensemble de projets, en s'appuyant sur une ou plusieurs équipes placées [chefs de projet] sous sa responsabilité. Il planifie et manage la réalisation de chantiers dans leur intégralité depuis la phase de négociation préalable à la signature des contrats jusqu'à la livraison finale. Il détermine les moyens humains nécessaires au déploiement des projets dans les différentes phases d'études [études préalables et études de programme] et de réalisation [marchés de travaux et marchés de ►maîtrise d'œuvre] et est responsable du suivi budgétaire des projets ainsi que des délais ; selon la complexité du projet on parle alors de **directeur de programme** ; ►pilotage d'un projet) ; *syn.* directeur [m] de chantier, responsable [m] de la maîtrise d'œuvre d'exécution, directeur [m] de construction ; *f 2* **chef [m] de projet** (Personne chargée de l'avancement d'un projet dès la phase de l'étude, responsable de la détermination du temps nécessaire à la réalisation du projet, du budget à lui affecter ainsi que de la composition et de l'organisation de l'équipe de développement ; il coordonne le travail des différents intervenants, adapte le planning en fonction de l'avancement ; son objectif est de terminer le projet dans les délais fixés sans dépassement de budget) ; *f 3* **directeur [m] d'investissement** (*Terme spécifique* dans les marchés publics, représentant du ►maître d'ouvrage dont l'action est de fixer les impératifs financiers au stade de la définition des ouvrages) ; *g* **Projektleiter/-in [m/f] (1)** (*Je nach hierarchischer oder nichthierarchischer Leitungsstruktur und Form der Projektorganisation* Bauherr resp. bei Behör-

den oder behördenähnlichen Gesellschaften der mit besonderer Kompetenz und Verantwortung ausgestattete Vertreter des Bauherrn, dem für die Organisation und Abwicklung eines komplexen [Groß]projektes u. a. folgende Aufgaben obliegen: a] Bedarfsfeststellung, Klärung der Aufgabenstellung und die Festlegung der Planungsziele sowie des Umfangs des Raum- und Ausstattungs-/Einrichtungsprogramms für die Finanzplanung [Planung der Investitionsmittel] für das Gesamtprojekt; b] rechtzeitige Herbeiführung der erforderlichen Entscheidungen über die wirtschaftlichen und terminlichen Projektziele; c] Entscheidung über den Einsatz von Planern, ggf. eines Projektsteuerers, und anderen fachlich Beteiligten sowie die dazugehörenden Vertragsverhandlungen und –abschlüsse; d] Entscheidung über Vergaben an Firmen und Überwachung der Einhaltung einschlägiger Vergabevorschriften und Vertragsbedingungen; e] ►Projektsteuerung, wenn diese Aufgabe nicht an einen extra beauftragten Steuerer vergeben wurde; f] Projektdarstellung in übergeordneten Gremien und in der Öffentlichkeit sowie Wahrnehmung und Organisation von Terminen zu Nutzer- und Anliegeranhörungen [►Bürgerbeteiligung]; g] Kontrolle und Abnahme der Leistungen aller Projektbeteiligten; die Aufgaben eines **P.** sind Bauherrenleistungen und im Vergleich zu denen eines Projektsteuerers nicht an Dritte delegierbar); *syn.* Leiter/-in [m/f] eines Projektes.

4630 project documentation [n] *prof.* (Systematic compilation of all relevant working drawings, ►as-built plans, inventory of all facilities, maintenance instructions, etc.); *s* **documentación [f] de un proyecto** (Recopilación sistemática de representaciones gráficas y resultados de un proyecto, elaboración de planes de resultado, inventarios e instrucciones de mantenimiento, etc.; ►plano de realización); *f* **dossier [m] des ouvrages exécutés** (Dossier constitué et remis en fin d'exécution des travaux au maître d'ouvrage contenant la collection des notices de fonctionnement des ouvrages ainsi que des plans d'ensemble et de détails conformes à l'exécution, les pièces contractuelles utiles à l'exploitation et à l'entretien des ouvrages ; HAC 1989, 41 ; ►plan de récolement 1) ; *syn.* dossier [m] documentaire et des ouvrages exécutés ; *g* **Dokumentation [f]** (Systematisches Zusammenstellen der zeichnerischen Darstellungen und rechnerischen Ergebnisse eines Projektes, Erstellen von ►Bestandsplänen 1, Inventarverzeichnissen und Wartungsanweisungen etc.; cf. Leistungsphase 9 des § 15 HOAI 2002 resp. §§ 33, 38, 42 u. 46 HOAI 2009).

projection [n], **isometric** *plan.* ►isometric drawing.

4631 project landscape architect [n] *prof.* (Chief landscape architect in charge of a project and accountable to the principal or supervisor); *s* **arquitecto/a [m/f] paisajista jefe/a de obra** (Profesional que dirige un proyecto y responde personalmente de los resultados); *f* **architecte [m/f] paysagiste maître d'œuvre** (Personne physique ou morale chargée par le ►maître d'ouvrage de la conception des ouvrages, du contrôle de l'exécution des travaux, de leur réception et de leur règlement) ; *g* **verantwortlicher Garten- und Landschaftsarchitekt/-in [m/f]** (Jemand, der als Landschaftsarchitekt/-in für ein bestimmtes Projekt voll verantwortlich ist und persönlich haftet).

project leader [n] *plan.* ►project manager.

project leader [n], **contractor's** *constr. prof.* ►contractor's agent.

4632 project management [n] *prof.* (Entire spectrum of actions carried out by an owner in the organization and execution of a complex, large-scale project, including all necessary planning, management and control measures as well as the coordination of project participants according to pre-determined objectives and current state-of-the-art technology. **P. m.** also comprises the integration within the planning procedures, (and, where possible, during the process of construction) of those, who will be involved or affected by the project. Among the basic tasks of **P. m.** are a] the designation of those primarily responsible for the project, b] the establishment of an organization plan, construction schedule or network plan, c] the drawing up of a cost control system, d] the laying down of a system for submitting reports, which is mandatory for all parties, e] the establishment of methods for drafting resolutions and decision-making processes as well as on target initiation of committee decisions necessary for awarding contracts to project participants [planning and construction firms]. The ▶project manager is usually appointed for the **p. m.**; ▶project director); *s* **gestión [f] de proyecto** (Espectro completo de actividades realizadas por una propiedad de obra para organizar y llevar a cabo un proyecto complejo de gran escala, incluyendo todas las medidas de planificación, gestión y control así como la coordinación de los participantes en el proyecto de acuerdo con los objetivos predeterminados y la mejor tecnología de construcción asequible. La **g. de p.** también incluye la integración en procesos de planificación y —donde fuera posible durante el proceso de construcción— de aquellas personas que están involucradas o se vean afectadas por el proyecto. Las tareas básicas de la **g. de p.** son: a] designar a las personas responsables del proyecto; b] elaborar un plan de organización, calendario de construcción o un organigrama; c] establecer un sistema de control de costes [Es]/costos [AL]; d] determinar un sistema de presentación de informes obligatorio para todas las partes participantes; e] establecer métodos para la elaboración de propuestas de resolución/decisión así como para la toma de decisiones oportuna de los gremios de resolución competentes para la contratación de los participantes en el proyecto [gabinetes de planificación y empresas de construcción]. Generalmente es el ▶gerente de proyecto quien es nombrado para llevar a cabo la **g. del p.**; ▶project manager); *f 1* **maîtrise [f] d'ouvrage** (Ensemble des missions incombant au maître de l'ouvrage ou au maître d'ouvrage délégué [assistance à (la) maîtrise d'ouvrage] comprenant : **1.** à l'échelon technique [urbanisme, architecture, infrastructure, environnement, paysages] les études à mener en vue de la formalisation d'un programme [études prospectives sur l'évaluation des besoins, études de faisabilité et d'impact], et du suivi de sa réalisation [examen des choix techniques selon lesquels l'ouvrage sera étudié en phase conception, préparation du choix du maître d'œuvre, signature et gestion du contrat de maîtrise d'œuvre, approbation des avants-projets, passation des marchés, et suivi de leur bonne exécution, suivi des travaux et opérations de réception], suivi de l'intégration des principes de developpement durable à toutes les étapes de la conception et de la réalisation, **2.** au niveau administratif et juridique la mise en place de méthodes et outils permettant l'établissement d'un cadre référentiel [programme, coût, délais, qualité, performances], la mesure de l'avancement des travaux et des écarts avec le prévisionnel, le montage des consultations, la gestion des procédures, des contrats et des contentieux, **3.** du point de vue financier, à établir le montage financier [études économiques, cadrage financier et estimation de l'enveloppe prévisionnelle de l'opération], à gérer les budgets, liquider les dépenses, simuler la gestion de l'exploitation ; cf. Sept. 2005, missions d'assistance à décideur et maître d'ouvrage, www.urbanisme.equipement. gouv.fr ; ▶coordinateur de travaux, ▶directeur de travaux) ; *f 2* **gestion [f] de projet** (Démarche du maître d'ouvrage visant à structurer, assurer et optimiser le bon déroulement d'un projet suffisamment complexe pour devoir être planifiée dans le temps, être budgétée [étude préalable des coûts et avantages ou revenus attendus en contrepartie, des sources de financement, étude des risques operationnels et financiers et des impacts divers, etc.], faire intervenir de nombreuses parties prenantes [organisations qui identifient ▶maîtrise d'œuvre et ▶maîtrise d'ouvrage], responsabiliser le

chef de projet ou le directeur de projet, mettre en place un comité de pilotage, suivre des enjeux opérationnels et financiers importants ; www.techno-science.net, glossaire) ; *syn.* conduite [f] de projet ; *g* **Projektmanagement [n]** (Ganzheitliche Führungsaufgabe des Bauherren für die Organisation und Abwicklung eines komplexen [Groß]projektes, die alle notwendigen Planungen, die Steuerung und Kontrolle sowie den aufeinander abgestimmten Einsatz aller Projektbeteiligten im Sinne der vorgegebenen Ziele nach den allgemein anerkannten Regeln der Baukunst umfasst sowie die Integrierung der Planungsbetroffenen in den Planungs- und wo möglich in den Bauprozess. Zu den Grundvoraussetzungen eines **P.s** gehört a] die Benennung des Hauptverantwortlichen, b] die Aufstellung eines Organisationsplanes, Bauzeiten- resp. Netzplanes, c] die Erstellung eines Kostenkontrollsystems, d] die Festsetzung eines für alle Beteiligten verbindlichen Berichtswesens, e] das Erstellen von Entscheidungs-/Beschlussvorlagen und das rechtzeitige Herbeiführen der nötigen Entscheidungen der zuständigen Beschlussgremien zur Beauftragung der Projektbeteiligten [Planungsbüros und Baufirmen]. Mit dem **P.** wird i. d. R. der/die ▶Projektleiter/-in 1 beauftragt; ▶Projektleiter/-in 2).

4633 project manager [n] *plan.* (Person working in an office entrusted with the management of a project by its client/owner, responsible for the technical, organizational and economic aspects of a project; ▶project director); *syn.* project leader [n]; *s* **project manager [m/f]** (En gabinete de planificación persona responsable ante la propiedad de obra y encargado de la planificación y de todos los aspectos [organizativos, económicos, técnicos] de la realización oportuna de un proyecto de acuerdo a la descripción de servicios de planificación; ▶gerente de proyecto); *f* **coordinateur [m] de travaux** (Service technique public ou concepteur privé [bureau d'études] apportant au maître de l'ouvrage son assistance technique et au stade de la réalisation. Sous la responsabilité du ▶directeur de travaux, il assure la maîtrise d'œuvre de la phase travaux d'un projet ; il est le garant du bon déroulement des travaux [délais, coûts, relations contractuelles et administratives] et assure les missions de direction de l'exécution des contrats de travaux [DET] et d'assistance aux opérations de réception [AOR] pour l'ensemble d'un chantier) ; *g* **Projektleiter/-in [m/f] (2)** (In einem Planungsbüro der Treuhänder des Bauherrn und Verantwortliche für die planerische, organisatorische, wirtschaftliche, terminliche und technische Abwicklung eines Projektes gem. Leistungsbild nach § 15 [2] 1-9 HOAI 2002 resp. §§ 33, 38, 42 u. 46 HOAI 2009; ▶Projektleiter/-in 1); *syn.* Leiter/-in [m/f] eines Projektes.

4634 project plan [n] *prof.* (Graphic presentation of a project proposal; ▶project planning); *s* **plan [m] de proyecto** (Representación gráfica del resultado de la ▶planificación de un proyecto [de construcción]); *f* **plan [m] de projet** (Représentation graphique du projet d'une ▶mission opérationnelle) ; *g* **Objektplan [m]** (Grafische Darstellung als Ergebnis der ▶Objektplanung).

project planner [n] *plan. prof.* ▶preliminary project planner [US]/scheme designer [UK].

4635 project planning [n] *plan. prof.* (Process of design and preparation of construction drawings and specifications for a project; ▶overall project planning); *s* **planificación [f] de un proyecto (de construcción)** (Proceso de diseño y preparación de proyecto de construcción, incluidas todas las especificaciones necesarias para realizar la obra; ▶planificación general de un proyecto); *f* **mission [f] opérationnelle d'ingénierie et d'architecture** (Mission constituée de prestations intellectuelles de la compétence de prestataires d'ingénierie et d'architecture relative à la création d'ouvrages avec ou sans maîtrise d'œuvre [mission complète ou mission partielle] ; *terme spécifique* urbanisme opérationnel ; ▶étude globale d'aménagement d'un projet) ; *syn.*

mission [f] de création, étude [f] opérationnelle, planification [f] operationnelle ; *g* **Objektplanung [f]** (Entwurfs- und baureife Planung eines Bauvorhabens, z. B. eines Gebäudes, einer öffentlichen oder privaten Grünanlage, eines Parks, öffentlichen oder privaten Platzes; ▶Gesamtplanung eines Projektes; *UB* ▶Objektplanung für Freianlagen).

4636 project program [n] [US]/project programme [n] [UK] *plan.* (Detailed statement of the client's or sponsor's requirements to be provided for execution of the ▶project brief, which covers the various areas and facilities to be developed); *s* **programa [m] detallado de necesidades** (Descripción detallada de las exigencias del cliente que se debe elaborar para realizar el ▶avance de programa de necesidades y que cubre las diferentes áreas y el equipamiento del proyecto); *f* **programme [m] d'équipement** (Données d'une classification détaillée, présentées sous la forme d'un programme d'aménagement; ▶objet de la mission) ; *g* **Aufgabenstellung [f] (2)** (Vorgabe der detaillierten Gliederung eines generellen Aufgabenrahmens [▶Aufgabenstellung 1] in Form eines Raum-/Einrichtungsprogrammes).

project representative [n] *constr. prof.* ▶client's project representative [US]/clerk of the works [UK].

4637 project requirements [npl] for bidding [US] *constr. contr. prof.* (All documents and data such as specifications or design and build programs/programmes, contract conditions and contractual clauses, plans and working drawings provided by the contracting agency/client to prospective bidders [US]/tenderers [UK], as a basis for calculation and submission of ▶bids [US] 1/tenders [UK] by construction firms; ▶priced bidding documents [US]/priced tendering documents [UK]); *syn.* project requirements [npl] for tendering [UK]; *s* **pliego [m] de condiciones (generales) de concurso** (Especificaciones, condiciones generales de contrato, planos, etc. entregados por el promotor a los licitantes que sirven de base para presentar una oferta a concurso; ▶expediente de concurso-subasta); *syn.* expediente [m] de condiciones (generales) de concurso; *f 1* **dossier [m] de consultation** (Marché sur appel d'offres ouvert) ; *f 2* **dossier [m] d'appel d'offres** (*Marché sur appel d'offres restreint, marché négocié* dossier constitué de pièces administratives et techniques permettant aux entreprises/concepteurs de présenter une offre et comprenant : le cadre d'acte d'engagement, l'ensemble des documents techniques relatif au projet de conception tels que les spécifications techniques détaillées et les plans d'exécution des ouvrages, le cahier des prescriptions spéciales telles que le cadre du bordereau des prix, le cadre du détail estimatif, le ▶cahier des clauses administratives générales [des marchés publics] [C.C.A.G.], le ▶cahier des clauses administratives particulières [C.C.A.P.], le ▶cahier des clauses techniques générales — documents types [C.C.T.G.], le cahier des clauses techniques particulières [C.C.T.P.] ; *termes spécifiques* dossier de consultation des entreprises, dossier de consultation des concepteurs ; ▶dossier des pièces constitutives du marché) ; *syn.* documents [mpl] d'appel d'offres [CDN] ; *g* **Ausschreibungsunterlagen [fpl]** (Alle Unterlagen wie z. B. Ausschreibungstexte oder Leistungsprogramme, ▶besondere Vertragsbedingungen, Pläne und Ausführungszeichnungen, die von der ausschreibenden Stelle an Bieter als Grundlage zur Ausarbeitung und Abgabe eines Angebotes zur Verfügung gestellt werden; ▶Allgemeine Technische Vertragsbedingungen für Bauleistungen, ▶Angebotsunterlagen).

project requirements [npl] for tendering [UK] *constr. contr. prof.* ▶project requirements for bidding [US].

project review [n] [US] *prof.* ▶concluding project review.

project section [n] *constr.* ▶construction project section.

proliferated housing [n] *urb. plan. landsc.* ▶landscape of proliferated housing.

proliferation [n] of settlements *plan. urb.* ▶uncontrolled proliferation of settlements.

4638 promenade [n] (1) *urb.* (Wide, linear, public pedestrian pathway for taking walks, strolling, often in spas or holiday resorts, which is usually planted with standard street trees [US]/avenue trees [UK] or ▶seasonal flower beds. In U.S., may also be a 'boardwalk' at the seaside); *s* **paseo [m]** (Calle peatonal amplia con árboles de alineación y a veces con ▶macizos de flores estacionales que cumple funciones recreativas. En las ciudades mediterráneas son frecuentes además la venta de productos y la realización de actividades culturales informales); *syn.* rambla [f] [Catalunya]; *f* **voie [f] promenade** (Large espace linéaire piétonnier en général agrémenté d'arbres d'alignement ou de ▶plate-bandes [florales] saisonnières invitant à la promenade) ; voie [f] piétonne ; *g* **Promenade [f]** (Breiter, linearer Fußgängerbereich zum Spazierengehen oder Flanieren, oft mit Alleebäumen oder ▶Wechselbeeten ausgestattet, besonders in Kur- und Ferienorten; *tautologische Fehlbezeichnung* Fußgängerpromenade).

promenade [n] (2) *recr.* ▶beach promenade; *urb.* ▶waterfront promenade.

promenade [n], coastal *recr.* ▶beach promenade.

4639 prominent urn burial site [n] *adm. landsc.* (Special grave for cremation ashes container which is situated in a prominent area of the cemetery and gives the possibility for the burial of several deceased persons, in comparison to an urn in a row there is a longer ▶use period of grave [US]/rest period of grave [UK]); *s* **tumba [f] conmemorativa para urna** (Tumba especial para depositar varias urnas después de la cremación, situada en un lugar central del cementerio y para la cual —en comparación con las tumbas para urnas en hilera— el ▶periodo/período de descanso de tumbas es más prolongado); *f* **jardin [m] d'urnes familial** (Emplacement funéraire aménagé pour le dépôt des urnes cinéraires, en général dans un caveau comportant plusieurs niches funéraires ; pour ce type de sépulture il est prévu une ▶durée d'inhumation beaucoup plus importante) ; *g* **Urnensondergrabstätte [f]** (Grabstätte für eine Aschenbeisetzung in bevorzugter Lage mit der Möglichkeit zur Bestattung mehrerer Verstorbener; im Vergleich zum Urnenreihengrab besteht für diesen Grabtyp eine wesentlich länger bemessene Ruhezeit; ▶Ruhefrist); *syn.* Urnensondergrab [n], Urnensondergrabstelle [f], Urnenwahlgrab [n], Urnenwahlgrabstelle [f], Wahlaschenstätte [f].

promotion [n] of foreign tourism *econ. plan. recr.* ▶promotion of national tourism, #2.

4640 promotion [n] of leisure activities *plan. recr.* (▶promotion of foreign tourism, ▶promotion of national tourism); *s* **promoción [f] recreativa** (▶promoción turística); *f* **promotion [f] d'activités de loisirs et de récréation** (▶promotion du tourisme) ; *g* **Entwicklung [f] von Freizeit und Erholung** (▶Fremdenverkehrsförderung).

4641 promotion [n] of national tourism *econ. plan. recr.* (1. In-country fostering of tourism on an international, national, regional and local level by the travel industry and sometimes a public authority/agency); **2. promotion [n] of foreign tourism** (In-country fostering of tourism from abroad); *s* **promoción [f] turística** (Todo tipo de medidas a nivel internacional, nacional, regional o local para fomentar la infraestructura turística y crear una oferta atractiva así como para dar a conocer las atracciones turísticas de una zona y atraer a turistas a ella); *f* **promotion [f] du tourisme** (Mesures et actions aux niveaux international, national, régional et local engagées en faveur du développement touristique comme facteur de produc-

P

tion par la réalisation d'infrastructures et d'équipements [hébergement, accueil et restauration, équipements sportifs et ludiques, infrastructures d'accès, etc.] avec l'objectif de la création d'emplois et de la conquête de nouvelles sources de recettes) ; *syn.* mise [f] en valeur touristique ; *g* **Fremdenverkehrsförderung [f]** (Maßnahmen und Handlungen auf internationaler, nationaler, regionaler und örtlicher Ebene, die dazu dienen, den Fremdenverkehr als Produktionsfaktor z. B. mit touristisch attraktiven Infrastrukturen und Einrichtungen [Beherbergungs- und Verpflegungsangebote, Sport- und andere Freizeiteinrichtungen, Erschließungswege etc.] so auszubauen, dass zusätzliche Arbeitsplätze geschaffen und weitere Einnahmequellen erschlossen werden); *syn.* Förderung [f] des Fremdenverkehrs, Tourismusförderung [f].

4642 proof [n] **of executed work** *constr. contr.* (Presentation of documents verifying execution of work [US]/works [UK]; e.g. by ▶issue of a sectional completion certificate, ▶daywork sheet[s], ▶weigh bill[s]); *s* **verificación [f] de trabajos cumplidos** (Presentación de los documentos que justifican la ejecución de los trabajos, p. ej. por medio de ▶aceptación parcial de obra, ▶lista de trabajo por horas, ▶certificado de peso, etc.); *syn.* prueba [f] de trabajos cumplidos; *f* **justification [f] des prestations réalisées** (Présentation par l'entrepreneur de pièces justificatives telles que constat contradictoire, ▶réception partielle, ▶rapport de chantier journalier, ▶bon de pesage) ; *g* **Nachweis [m] der erbrachten Leistung** (Darlegung, dass ausgeführte [Bau]leistungen oder Lieferungen tatsächlich erbracht wurden, z. B. durch ▶Teilabnahme[n], ▶Stundenlohnzettel, ▶Wiegekarte[n]).

proof of verified delivery [n] [UK]**, on** *constr. contr.* ▶according to certified delivery [US].

prop [n] *arb. hort.* ▶tree prop.

propagate [vb] **by layering** *hort.* ▶layer [vb].

propagated [pp] **from seed** *for. hort.* ▶seed-propagated (stock).

propagation [n] *hort.* ▶reproduction.

4643 propagation [n] **by cuttings** *hort.* (Vegetative reproduction of a plant by ▶cutting, ▶dormant cutting; *generic term* ▶reproduction); *s* **reproducción [f] por esquejes** (Método de propagación vegetativa por medio de ▶esquejes o ▶esquejes leñosos; *término genérico* ▶multiplicación); *f* **bouturage [m]** (Méthode de multiplication asexuée des végétaux consistant à prélever sur un pied-mère en vue de son enracinement un fragment de végétal [rameau herbacé, rameau semi-aoûté, rameau aoûté ou ligneux, racine, etc.] en vue de reconstituer une plante entière ; *termes spécifiques* bouturage en sec, avec des rameaux qui ont perdu leurs feuilles à l'automne et qui possèdent des bourgeons au repos, bouturage en vert utilisé en été avec des rameaux feuillés ligneux ou herbacés ; ▶bouture, ▶bouture de rameau aoûté, ▶bouture de rameau semi-aoûté ; *termes génériques* ▶multiplication végétative, ▶reproduction) ; *g* **Stecklingsvermehrung [f]** (Vegetative Vermehrung einer Art oder Sorte durch ▶Steckling oder ▶Steckholz; *OB* ▶Vermehrung); *syn.* Vermehrung [f] durch Stecklinge.

4644 propagation [n] **by root suckers** *bot. hort.* (Natural or artificial reproduction of plants using shoots of their roots); *s* **propagación [f] por brotes de raíz** (Reproducción natural o artificial por medio de vástagos de la raíz); *f* **drageonnage [m]** (Reproduction naturelle ou artificielle par les bourgeons adventifs sur les racines) ; *g* **Vermehrung [f] durch Wurzelschösslinge** (Natürliche oder künstliche Vermehrung durch Wurzelaustriebe).

propellant [n] *envir.* ▶gas propellant.

proper line [n] **and levels** [npl] *constr.* ▶graded to proper line and levels.

property [n]**, adjacent** *leg. urb.* ▶neighboring lot [US]/neighbouring plot [UK].

property [n]**, leasehold** *agr.* ▶agricultural leasehold (property).

property [n]**, private** *leg. pol. sociol.* ▶social obligation of private property.

property [n] [US]**, unimproved** *plan. urb.* ▶undeveloped land.

4645 property boundary wall [n] *arch. constr.* (Wall constructed along a property boundary; ▶common boundary wall); *s* **muro [m] de deslinde** (Muro que encierra un área o define una linde; DACO 1988; ▶muro medianero); *syn.* muro [m] de linde; *f* **mur [m] séparatif de propriété** (Mur implanté entre deux terrains contigus et, à la différence du ▶mur mitoyen, appartenent à un propriétaire) ; *g* **Grundstücksmauer [f]** (Mauer, die an die Grenze eines Grundstückes gebaut wurde; ▶Grenzmauer).

4646 property devaluation [n] **caused by planning** [US] *plan.* (Decline in property value caused by redesignation of land use due to statutory rezoning. **In U.K.,** the resulting uncertainty about the market value or full development potential of a property usually entitles an owner to compensation; **in U.S.,** p. d. is not compensable; courts have made that very clear); *syn.* planning blight [n] [UK]; *s* **daños [mpl] y perjuicios [mpl] de planificación** (Reducción del valor de una propiedad inmobiliaria causada a propietario particular por alteración del planeamiento. En Es. el derecho a indemnización existe si la modificación o revisión del planeamiento se produce a] antes de transcurrir los plazos previstos para la ejecución del aprovechamiento urbanístico posible según la categoría de ordenación a la que pertenecía el terreno, b] si existiera para el terreno licencia de construcción en vigor, sin que haya sido comenzada la obra, c] si la edificación ya se hubiese iniciado y la Administración modificara o revocara la licencia, d] si las ordenaciones impusieran vinculaciones o limitaciones singulares en orden a la conservación de edificios, más allá de los deberes legalmente establecidos, o que lleven consigo una restricción del aprovechamiento urbanístico del suelo que no pueda ser objeto de distribución equitativa entre los interesados así como e] en todo caso son indemnizables los gastos producidos por los deberes inherentes al proceso urbanizador que resulten inservibles por el cambio de planificación; cf. arts. 41-44 Ley 6/1998); *f* **préjudice [m] causé par des documents d'urbanisme** (Il s'agit en général de la modification ou de l'abrogation de prescriptions réglementant l'utilisation du sol [droit de construire en relation avec le COS, sursis à statuer, droit de préemption], de la réservation d'un fond bâti ou non pour une installation d'intérêt général dans le cadre d'un ▶plan local d'urbanisme [PLU]/plan d'occupation des sols [POS] entraînant une perte de valeur ou de jouissance des biens du propriétaire et entraînant le droit à une indemnité ou l'obligation d'acquisition d'un terrain par la collectivité) ; *g* **Planungsschaden [m]** (Auf planerische Neufestsetzungen, insbesondere auf Aufhebung oder Änderung der zulässigen baulichen oder sonstigen Nutzung im Bebauungsplan beruhende Wertminderung an Immobilien und die Unsicherheit der Eigentümer hinsichtlich der Wertbeständigkeit ihres Eigentums, weil sich der Verkehrswert oder die Ausnutzbarkeit eines Grundstückes verringert oder die Verfügungsgewalt darüber ändert. Der Eigentümer hat dann einen begründeten Anspruch auf Entschädigung oder Ausgleich für diesen Schaden); *syn.* Vermögensnachteile [mpl] durch Planungsabsichten der Behörden.

property developer [n] [UK] *plan.* ▶developer (2).

property parcel [n] *surv.* ►real property parcel [US]/plot [UK].

4647 property partition [n] *agr. urb.* (Division of property into separate lots/plots or parcels; ►reorganisation of land holdings [UK]); *s* **parcelación** [f] (División de propiedades de suelo en diferentes parcelas; ►ordenación del suelo); *f* **division** [f] **parcellaire** (Morcellement d'un ensemble immobilier à bâtir, dans le cadre d'un remembrement ; ►réorganisation foncière) ; *syn.* parcellement [m] ; *g* **Grundstücksteilung** [f] (Im Sinne der ►Bodenordnung Aufteilung und Neuzuordnung von einzelnen Liegenschaften im ländlichen Raum oder in einem Baugebiet; grundbuchrechtlich werden Grundstücke ‚geteilt', Flurstücke hingegen ‚zerlegt'); *syn.* Grundstücksaufteilung [f], *obs. ugs.* Parzellierung [f].

property register [n] **[US], abandoned property** *adm. urb.* ►derelict land register.

property survey [n] *surv.* ►boundary survey.

4648 proportion [n] **of an overall area** *plan.* (Percentage of entire area); *s* **proporción** [f] **de superficie** (Parte de una superficie total); *syn.* proporción [f] en relación al área total; *f* **proportion** [f] **de la surface** (Rapport entre une partie et son tout, p. ex. la proportion de la surface agricole par rapport à la surface totale) ; *syn.* proportion [f] de l'espace ; *g* **Flächenanteil** [m] (Teil einer Fläche im Gesamtgebiet).

proportion [n] **of labo(u)r costs** *constr. contr.* ►percentage of labor costs [US]/percentage of labour costs [UK].

proposal [n] *constr. contr.* ►addendum proposal.

proposals [npl] *adm. contr.* ►agency requesting proposals; *landsc. plan.* ►landscape planning proposals.

Proposals Map [n] **[UK]** *leg. urb.* ►zoning map [US].

Proposals Map [n] **[UK], adoption of a** *leg. urb.* ►adoption of a zoning map [US].

4649 proposed extraction site [n] *min. plan.* (Location precisely-defined by permit for exploitation of rock, soil or other mineral resources); *s* **zona** [f] **de explotación localizada** (Lugar claramente definido por permiso donde pueden ser extraídas rocas, tierras u otro recurso mineral); *f* **site** [m] **d'exploitation** (Dont les limites d'exploitation des substances minérales ou fossiles sont précisément définies [par bornage]) ; *syn.* site [m] d'extraction, étendue [f] du titre minier ; *g* **Abbaugebiet** [n] (Definitiv für die Gewinnung von Steinen, Erden oder anderen Bodenschätzen festgelegtes Gebiet).

proposed level [n] **[UK]** *constr. plan.* ►finished elevation [US]/finished level [UK].

propping [n] *arb. hort.* ►tree prop, #2.

prop root [n] *arb. bot.* ►stilt root.

4650 prospect [n] *plan. recr.* (Panoramic view within the field of vision of an observer from a particular standpoint; ►visual landscape assessment); *syn.* viewshed [n] [also US], perspective view [n]; *s* **cuenca** [f] **visual** (Campo visible total desde un punto determinado; ►evaluación del aspecto escénico del paisaje); *syn.* campo [m] de visión, ángulo [m] de visión; *f* **champ** [m] **de vision** (Dans le cadre de l'►évaluation de l'image paysagère, désigne l'angle de vision d'un observateur à partir d'un point d'observation) ; *syn.* champ [m] visuel, champ [m] de vue ; *g* **Sichtfeld** [n] (Im Rahmen der ►Landschaftsbildbewertung das gesamte Blickfeld eines Betrachters von einem Standpunkt aus); *syn.* Blickfeld [n].

prospective drilling [n] *geo. min.* ►test boring.

4651 prostrate shrub [n] *bot.* (*Chamaephyta velantia* woody plant with branches growing horizontally along the ground); *syn.* trailing shrub [n]; *s* **arbusto** [m] **en espaldera**

(Según la clasificación de las formas vitales de RAUNKIAER tipo de caméfito *[Chamaephyta velantia]* leñosa rastrera de los climas fríos que forma frecuentemente tapices que cubren el suelo; su tronco crece horizontalmente al ras del suelo, aprovechando así el calor del mismo. Puede ser de follaje persistente o caduco; cf. BB 1979, 147); *syn.* arbusto [m] rastrero; *f* **arbuste** [m] **en espalier** (*Chamaephyta velantia* végétal ligneux nain dont les tiges et les feuilles sont plaquées sur le sol, p. ex. le Saule nain alpin ; GOR 1985, 204 ; se rencontre souvent sous les conditions écologiques sévères des étages alpin et nival) ; *g* **Spalierstrauch** [m] (*Chamaephyta velantia* Holzgewächs, dessen Stämme flach auf dem Boden liegen und sich horizontal ausbreiten); *syn.* Teppichstrauch [m], Kriechstrauch [m].

4652 protected aquifer recharge forest [n] *for. wat'man.* (Wooded area which must be protected because of its capacity to regenerate groundwater; *generic term* ►protective forest 1); *s* **bosque** [m] **protector de área de recarga de acuíferos** (Bosque a proteger por su capacidad de promover la recarga de acuíferos; *término genérico* ►bosque protector 1); *f* **forêt** [f] **protectrice de la nappe aquifère** (Forêt à protéger en raison de sa capacité à renouveler la nappe phréatique ; *terme générique* ►forêt de protection) ; *g* **Speicherschutzwald** [m] (Wald, der wegen seiner Grundwasserneubildungskapazität in seinem Bestand gesichert bleiben muss; *OB* ►Schutzwald).

protected archaeological area [n] **for future digging** **[UK]** *conserv'hist. leg.* ►archaeological probability area [US]/protected archaeological area for future digging [UK].

4653 protected area [n] *conserv. leg.* (Territory demarcated for its significant conservation value, which has been given legal status or contractually agreed upon and in which necessary measures are undertaken to guarantee the preservation or restoration of natural habitats and/or populations of species belonging to that area; generic term for ►Area of Outstanding Natural Beauty [UK], ►biogenetic reserve, ►bird sanctuary, ►biosphere reserve, ►European Diploma Area, ►European Wilderness Reserve, ►forest research natural area [US]/forest nature reserve [UK], ►game reserve, ►national park, ►Natura 2000 Protection Area Network, ►natural area preserve [US]/nature reserve [UK], ►nature reserve, ►protected landscape feature, ►special area of conservation [SAC], ►wetland of international significance, ►wilderness area [US], etc.); *s* **espacio** [m] **natural protegido** **(2)** (Territorio dotado de algún tipo de protección, establecida por la autoridad competente, en orden a preservar determinados valores naturales mediante ciertas limitaciones de acceso y uso; DINA 1987; término genérico para ►área de diploma europeo, ►componente protegido del paisaje, ►espacio natural protegido 1, ►paisaje protegido, ►parque nacional, ►Red Ecológica Europea Natura 2000, ►refugio de fauna [Es], ►reserva de la biosfera, ►reserva forestal integral, ►reserva regional de caza [Es], ►reserva natural, ►reserva natural biogenética, ►reserva natural europea, ►santuario de aves, ►zona especial de conservación, ►zona húmeda de importancia internacional; cf. Ley 42/2007, de 13 de diciembre, del Patrimonio Natural y de la Biodiversidad); *syn.* área [f] protegida; *f* **zone** [f] **protégée** (Terme générique désignant les sites, les paysages, les milieux naturels protégés au titre de diverses réglementations telles que les ►éléments de paysages identifiés, les ►espaces naturels sensibles [ENS], les espaces verts boisés ou non, les ►parcs nationaux, le ►réseau Natura 2000, les ►réserves biogénétiques, les ►réserves biologiques domaniales, les ►réserves biologiques forestières, les ►réserves de chasse, les ►réserves de chasse et de faune sauvage, les ►réserves de la biosphère, les ►réserves européennes, les ►réserves naturelles nationales [RNN], les ►réserves naturelles régionales [RNR], les ►sites classés, les ►sites inscrits, les ►zones d'un Diplôme européen d'espaces

P

protégés, les ▶zones de protection du patrimoine architectural, urbain et paysager, les ▶zones de protection spéciales [ZPS], les ▶zones humides d'importance internationale, les ▶zones spéciales de conservation [ZSC] ; cf. art. 1 Directive du Conseil n° 92/43 du 21 mai [CEE], Code de l'environnement, Code rural, Code du patrimoine) ; *syn.* périmètres [mpl] protégés ; *g* **Schutzgebiet [n]** (Durch eine Rechts- oder Verwaltungsvorschrift und/oder eine vertragliche Vereinbarung als ein von gemeinschaftlicher Bedeutung ausgewiesenes Gebiet, in dem die Maßnahmen, die zur Wahrung oder Wiederherstellung eines günstigen Erhaltungszustandes der natürlichen Lebensräume und/oder Populationen der Arten, für die das Gebiet bestimmt ist, erforderlich sind, durchgeführt werden; Art. 1 RL 92/43/EWG; weitere Schutzziele sind die Vielfalt, Eigenart oder Schönheit des Landschaftsbildes sowie die besondere Bedeutung für die Erholung; die Regierungspräsidien/Bezirksregierungen oder andere zuständige Stellen der einzelnen Bundesländer erlassen Verordnungen, die u. a. die Erklärung zum Schutzgebiet, den Schutzgegenstand und Schutzzweck, Verbote und Erlaubnisvorbehalte resp. zulässige Handlungen beinhalten; OB zu ▶besonderes Schutzgebiet [FFH-Gebiet], ▶biogenetisches Reservat 1, ▶Biosphärenreservat, ▶Europadiplomgebiet, ▶Europareservat, ▶Feuchtgebiet internationaler Bedeutung, ▶geschützter Landschaftsbestandteil, ▶Landschaftsschutzgebiet, ▶Nationalpark, ▶Naturreservat, ▶Naturschutzgebiet 2, ▶Naturwaldreservat, EU-▶Vogelschutzgebiet, ▶Wildschutzgebiet).

4654 protected cultural heritage landscape [n] *conserv.* (Area recommended for protection by the IUCN's Commission on National Parks and Protection Areas wthin the context of establishing biospere reserves. Such a landscape is protected for continuance of certain agricultural practices or as pastureland, for limited harvesting of natural resources, which is consistent with their perpetuation, support of the local economy, and preservation of their genetic potential as well as their special aesthetic value. **In U.S.**, the term is taken increasingly to refer to the totality of an important scenic area, or even an important popular landscape if, through cooperative efforts such as those of *The Nature Conservancy*, the area is protected for ultimate transfer to a state or federal agency; e.g. the National Park Service, U.S. Department of the Interior. A number of 'heritage landscapes' have also been designated as protected federal areas); *s* **paisaje [m] cultural protegido** (Area recomendada a proteger por la Comisión de Parques Nacionales y Areas Protegidas de la UICN en el contexto de la declaración de ▶reservas de la biosfera. Estos paisajes, resultantes del uso agrícola o ganadero, han de ser protegidos para preservar el potencial genético y por su valor estético especial); *f* **patrimoine [m] paysager protégé (±)** (Catégorie de protection proposée par la Commission des parcs nationaux et des aires protégées de l'UICN [Union internationale pour la conservation de la nature] dans le cadre de la désignation de ▶Réserves de la Biosphère, ayant pour objectif la protection de paysages bénéficiant d'une protection juridique appropriée et présentant un intérêt pour la présence de certaines pratiques agricoles et de pâturage les caractérisant, la sauvegarde du patrimoine génétique et en particulier pour leur valeur esthétique) ; *g* **geschützte Kulturlandschaft [f]** (Von der *Commission on National Parks and Protected Areas* der IUCN *[International Union for Conservation of Nature and Natural Resources]* vorgeschlagene Schutzgebietskategorie im Rahmen der Ausweisung von ▶Biosphärenreservaten, die sich auf Landschaften bezieht, die durch bestimmte land- und weidewirtschaftliche Nutzungen entstanden sind und auf Grund der Erhaltung des genetischen Potenzials und ihres besonders ästhetischen Wertes als Landschaftsschutzgebiete erhalten werden sollen).

4655 protected fish habitat area [n] *conserv. leg. zool.* (Contained section of water in which angling and fishing are prohibited according to fishing laws, because of its importance for reproduction of fish and fish migration—▶fish pass, ▶fish ladder—or for the hibernation of fish; ▶hibernation site, ▶protected spawning area); *s* **zona [f] vedada a la pesca** (Tramo de río o lago protegido contra la pesca por su importancia para la reproducción de los peces, la migración [▶paso (natural) de peces, ▶escala de peces] o para la hibernación de los mismos [▶hibernáculo]; ▶zona de freza protegida); *syn.* zona [f] de veda a la pesca; *f* **réserve [f] de pêche** (Portions de cours d'eau du domaine public fluvial et des eaux non domaniales jouant un rôle essentiel à certaines étapes de la vie du poisson telles que la croissance [zones d'alimentation, de réserves de nourriture] la reproduction [frayère], la migration [▶itinéraire de migration des poissons, ▶échelle à poissons] ou les périodes de repos [▶hibernacle] dont les limites sont fixées réglementairement et sur lesquels l'interdiction de pêche ou de capture est absolue en tout temps ; en F. on distingue les réserves permanentes de pêche et les réserves temporaires de pêche, les réserves de pêche nationales et les réserves de pêche d'A.A.P.P. [association agréée de pêche et de pisciculture] ; GPE 1998, fiche 26 ; ▶réserve d'élevage, ▶réserve statique) ; *g* **Fischschonbezirk [m]** (Durch das Fischereirecht bestimmter, abgegrenzter Abschnitt eines Fließ- oder Stillgewässers, in dem Angeln und Fischfang untersagt ist, weil er für die Fortpflanzung der Fische, die Fischwanderung [▶Fischwechsel, ▶Fischleiter] oder für die Winterruhe [▶Winterlager] der Fische wichtig ist; ▶Laichschonbezirk); *syn.* Fischschongebiet [n].

protected forest habitat [n] *conserv. leg. nat'res.* ▶protected habitat forest.

protected habitat [n] *conserv. ecol. leg.* ▶legally protected habitat.

4656 protected habitat forest [n] *conserv. leg. nat'res.* (Protection category for ecologically-valuable land, which is not adequate for a statutory designation because of its small size; ▶forest research natural area [US]/forest nature reserve [UK]); *syn.* protected forest habitat [n]; *s* **biótopo [f] forestal protegido (±)** (D. categoría de protección para superficies forestales ecológicamente muy valiosas, pero que por su tamaño reducido no son adecuadas para ser declaradas como ▶reserva forestal integral); *f* **biotope [m] forestier protégé (±)** (F., les espaces forestiers, de superficie en général limitée peuvent être classés en « **zone naturelle d'intérêt écologique, faunistique et floristique [ZNIEFF] de type** I » ; cf. circ. n° 91-71 du 14 mai 1991 ; **D.**, catégorie de protection des ensembles forestiers de faible étendue qui ne sont pas appropriés à être protégé dans le cadre de la législation sur les « Naturwaldreservate » [▶réserve biologique forestière] ; ▶réserve biologique domaniale) ; *g* **Biotopschutzwald [m]** (Schutzkategorie für ökologisch sehr wertvolle Waldflächen, die dem Schutz und der Erhaltung von seltenen Waldgesellschaften sowie von Lebensräumen seltener wild wachsender Pflanzen und wild lebender Tiere dienen und die sich auf Grund ihrer geringen Flächenausdehnung nicht für eine förmliche Ausweisung als ▶Naturwaldreservat eignen).

4657 protected landscape feature [n] *leg. conserv.* (Individual unit protected by local or state law against defacement, destruction or removal, e.g. trees, small lakes, cliffs, etc.; *generic term* ▶landscape feature); *s* **componente [m] protegido del paisaje** (En D. categoría de protección en la ley federal de protección de la naturaleza y en las de los estados federados; *término genérico* ▶componente del paisaje); *f* **élément [m] de paysage identifié (F.,** Les sites naturels ponctuels, ▶éléments de paysages dignes d'une protection particulière [arbres isolés, alignement d'arbres, haies, haies bocagères, réseaux de haies, rideaux d'arbres, trame végétale, murets, terrasses, etc.] identifiés dans les PLU/POS peuvent être protégés au titre de prescriptions spéciales ou comme espaces boisés ; cf. C. urb., art. L. 123-1-7°, L. 130-1, L. 442-2 ; tous travaux ayant pour effet de détruire un élément de paysage identifié par un plan local d'urbanisme [PLU]/plan d'occupation des sols [POS] en application du 7° de l'article L. 123-1 et non soumis à un régime d'autorisation doivent faire l'objet d'une autorisation préalable ; il en est de même, dans une commune non dotée d'un plan local d'urbanisme, des travaux non soumis à un régime d'autorisation préalable et ayant pour effet de détruire un élément de paysage à protéger et à mettre en valeur, identifié par une délibération du conseil municipal, prise après enquête publique ; **D.**, [traduction littérale : élément de paysage protégé] élément ou site naturel ponctuel remarquable, menacé de dégradation ou de destruction, protégé dans le cadre du § 18 de la loi fédérale et des lois des Länder relative à la protection de la nature, tels une rangée d'arbres, groupe d'arbres, arbres isolés, eaux calmes, falaise etc.) ; *syn.* élément [m] de paysage à protéger et à mettre en valeur identifié ; *g* **geschützter Landschaftsbestandteil [m]** (**D.**, durch § 18 BNatSchG und durch die Naturschutzgesetze der Bundesländer rechtsverbindlich festgesetzter Teil von Natur und Landschaft wie Baumreihe, Baumgruppe, Einzelbaum, kleines Stillgewässer, Kliff etc., der die Leistungsfähigkeit des Naturhaushaltes sichert und zur Belebung, Gliederung oder Pflege des Orts- und Landschaftsbildes dient; *OB* ▶Landschaftsbestandteil); *syn.* geschützter Landschaftsteil [m] [A].

protected resource [n] *conserv'hist.* ▶monument.

4658 protected spawning area [n] *conserv. zool. leg.* (Stretch of a river or lake protected under fishing rights to ensure the undisturbed reproduction of fish during spawning and breeding periods; ▶protected fish habitat area); *s* **zona [f] de freza protegida** (Tramo de río o lago protegido en épocas de freza y cría para permitir la reproducción de la fauna piscícola; ▶zona vedada a la pesca); *f 1* **réserve [f] d'élevage** (Tronçon de cours d'eau dont les limites sont fixées réglementairement produisant davantage de poissons que dans des conditions naturelles et dans lequel sont enlevés et répartis dans les secteurs ouverts à la pêche tous les poissons de taille légale de capture et remplacés par des jeunes poissons en provenance de ruisseaux de grossissement ; AEP 1976, 121) ; *f 2* **réserve [f] statique** (Cours d'eau constituant un refuge occasionnel ou permanent de poissons chassés des parcours de pêche voisins par une quête trop intense ou de carnassiers sédentaires et dont l'exploitation est ouverte par roulement et correspond à la mise en réserve d'un tronçon de cours d'eau de même valeur et non contigu; ▶réserve permanente de pêche ; AEP 1976, 120) ; ▶réserve de pêche ; AEP 1976, 120) ; *g* **Laichschonbezirk [m]** (Durch das Fischereirecht festgelegter Abschnitt eines Fließ- oder Stillgewässers zum Schutze der ungestörten Fortpflanzung von Fischen während der Laich- und Aufzuchtzeit; ▶Fischschonbezirk); *syn.* Laichschutzgebiet [n].

4659 protected species [n] *conserv. phyt. zool.* (Generic term covering those species of wild animals and plants which are placed under protection. These species are included in the Conservation of Wild Creatures and Wild Plants Act and ▶Endangered Species Act [E.S.A.] [US]/Conservation of Wild Creatures and Wild Plants Act 1975 [UK], including ▶threatened species and ▶critically endangered species; ▶endangered species, ▶Red List); *s* **especie [f] protegida** (Especie de flora o fauna silvestre acogida a norma de protección por estar amenazadas de extinción o por su alto valor científico o ecológico; ▶Catálogo Español de Especies Amenazadas, ▶especie amenazada, ▶especie en peligro [EN], ▶especie en peligro crítico

[CR], ▶ley de protección de especies, ▶lista roja); *f* **espèce [f] protégée** (Les espèces animales non domestiques ou végétales non cultivées sont à protéger d'après l'article 4 de la loi relative à la protection de la nature — un décret en Conseil d'État ou un Arrêté ministériel [p. ex. celui du 20 janvier 1982] en fixe la liste limitative ; ▶arrêté réglementant la protection de la faune et de la flore sauvages ; ▶espèce en danger [EN], ▶espèce en danger critique d'extinction [CR], ▶espèce menacée, ▶liste des espèces menacées, ▶Livre rouge [des espèces menacées]) ; *g 1* **geschützte Art [f]** (Gem. § 22 BNatSchG bestimmte wild wachsende Pflanzen- und frei lebende Tierart, die unter besonderen Schutz zu stellen ist. Diese Arten sind in der ▶Artenschutzverordnung des Bundes [BArtSchV] aufgeführt; ▶Rote Liste, ▶stark gefährdete Art, ▶vom Aussterben bedrohte Art, *OB* ▶bedrohte Art); *g 2* **besonders geschützte Art [f]** (Art, die unter den zu schützenden wild lebenden Tier- und Pflanzenarten einen besonderen resp. einen strengen Schutz genießt; in § 1 der Bundesartenschutzverordnung [BArtSchV] wird auf die Anlage 1 verwiesen, in der in zwei Spalten die beiden Schutzkategorien, *besonders geschützte Art* und *streng geschützte Art*, gekennzeichnet sind).

protection [n] *trans.* ▶antiglare protection; *constr.* ▶benchmark protection; *conserv. leg.* ▶bird protection; *game'man.* ▶browsing and debarking protection; *envir. pol.* ▶climate protection; *conserv. land'man.* ▶coastal protection; *conserv. pol.* ▶complex environmental protection; ▶conservation of forests; *leg.* ▶domestic animal protection; *constr. landsc.* ▶dune protection; *conserv. pol.* ▶environmental protection; *agr. hort. landsc.* ▶frost protection; *constr. leg.* ▶heat protection; *conserv. pol.* ▶legislation on environmental protection; *conserv. leg.* ▶place under legal protection; *conserv.* ▶nature protection; *hunt.* ▶overprotection [US]/over-protection [UK]; *phyt.* ▶plant requiring protection; *envir.* ▶radiation protection; *arb.* ▶root zone protection; *conserv. pol.* ▶technical protection of the environment; *arb. constr.* ▶trunk protection; *wat'man.* ▶watercourse bed protection; *conserv. leg. zool.* ▶wildlife conservation.

protection [n], **animal** *leg.* ▶domestic animal protection.

protection [n], **bank** *constr. landsc.* ▶bank erosion control.

protection [n] [US], **drip line** *arb.* ▶root zone protection.

protection [n], **erosion** *constr. land'man.* ▶erosion control.

protection [n], **flood** *wat'man.* ▶flood control.

protection [n], **forest** *landsc. conserv.* ▶conservation of forests.

protection [n], **integrated environmental** *conserv. pol.* ▶complex environmental protection.

protection [n] [UK], **landscape** *conserv. landsc.* ▶landscape conservation.

protection [n], **shoreline** *conserv. land'man.* ▶coastal protection.

protection [n], **tree** *arb. constr.* ▶protection of trees.

4660 protection [n] **against pollution** *envir.* (Measures to reduce pollution of air and water, and contamination of soil for environmental protection; ▶pollution control 1); *s* **protección [f] contra la contaminación** (Medidas para reducir la polución de las aguas, la atmósfera, el suelo; ▶protección contra inmisiones); *syn.* control [m] de la contaminación/polución; *f* **protection [f] contre la pollution** (Mesures visant à garantir la qualité et de l'eau ou de prévenir et limiter la pollution atmosphérique et la contamination du sol ; ▶lutte contre les émissions) ; *g* **Reinhaltung [f]** (Maßnahmen und Handlungen des Umweltschutzes, die dazu dienen, Boden, Gewässer oder Luft sauber zu halten; ▶Immissionsschutz).

P

4661 protection [n] **against transpiration** *arb. constr.* (Measures or protective installation to reduce plant ►transpiration, e.g. by spraying with ►anti-transpirants; ►trunk wrapping, ►wrapping with burlap [US]/wrapping with hessian strips [UK], ►wrapping with straw ropes); *s* **protección** [f] **contra la transpiración** (Medida o dispositivo para reducir la ►transpiración de las plantas, p. ej. injectándoles un ►antitranspirante, con ►envoltura de paja, ►envoltura del tronco o ►saco de protección del tronco); *f* **protection** [f] **contre la transpiration** (Mesures de réduction de la ►transpiration des plantes, p. ex. ►paillage de tronc, ►tampon de protection en jute, ►tampon protecteur de tronc, utilisation d'►antitranspirants) ; *g* **Verdunstungsschutz** [m, o. Pl.] (Maßnahme oder Einrichtung, die die ►Transpiration bei Pflanzen reduziert, z. B. Spritzen von ►Verdunstungsschutzmitteln, Verwendung einer ►Stammumwicklung, ►Strohbandage, eines ►Lehm-Juteverbandes).

protection area [n]**, special** *conserv. ecol.* ►special area of conservation (SAC).

Protection Area Network [n] *conserv. leg. pol.* ►Natura 2000 Protection Area Network.

protection forest [n] *for. land'man* ►avalanche protection forest, ►dust and aerosol protection forest.

4662 protection [n] **from avalanches** *envir.* (Measures and facilities of ►avalanche control techniques to prevent the formation of avalanches or snowslides); *s* **protección** [f] **contra aludes** (Conjunto de medidas para evitar aludes o para controlar su trayectoria; ►barrera contra aludes); *syn.* protección [f] contra avalanchas; *f* **protection** [f] **contre les avalanches** (Ensemble des mesures et installations de prévention contre la formation d'avalanches ou des ►techniques de défense contre les avalanches) ; *g* **Lawinenschutz** [m, o. Pl.] (Alle Vorrichtungen und Maßnahmen zur Verhinderung von Lawinenabgängen oder Maßnahmen der ►Lawinenverbauung).

4663 protection [n] **from cold** *landsc. phyt.* (Measures to reduce the amount of nocturnal radiation or frost damage; *specific term* ►frost protection); *s* **protección** [f] **contra la irradiación nocturna** (Medidas para reducir la cantidad de irradiación nocturna o de los efectos de heladas sobre las plantas); *f* **protection** [f] **contre le froid** (Mesures permettant d'éviter ou d'atténuer les dangers des gelées soit en limitant les pertes de chaleur par rayonnement, soit en les compensant par apport de chaleur ; LA 1981, 565 ; la protection des plantes contre le froid peut atténuer l'effet d'un fort gel et estomper les fortes différences de températures ; les plantes en pleine terre peuvent être protégées par un tapis de feuille, du compost, un paillis, des branches de sapins. Les plantes en bacs ou en pots seront protégées en isolant les pots et emballant les plantes avec une toile ou un film, les troncs des arbres fruitiers au moyen d'un paillasson ou d'un plastique à bulles, etc. ; *terme spécifique* ►protection contre le gel) ; *g* **Kälteschutz** [m, o. Pl.] (Gegebenheiten oder Maßnahmen zur Verminderung der Wirkung nächtlicher Ausstrahlung oder der Frosteinwirkung; *UB* ►Frostschutz).

4664 protection [n] **from falling** *constr. leg.* (**1.** Legally prescribed protective barrier, e.g. parapet, railings or balustrade required on all unenclosed floors or openings, and for open and glazed sides of landings and ramps, balconies or porches which are more than 90-110cm above grade; railings may also be constructed for pedestrian safety in outdoor situations; *specific terms* handrail, railing to prevent falling; ►pedestrian guardrail [US]/pedestrian guard-rail [UK]. **In U.K., guarding design** [n] is covered in the Building Regulations and BS 6180: 1995, Code of practice for protective barriers); **2. fall arrest equipment** [n] (*Roof planting* safety equipment in form of stainless steel nets,

highly durable plastic fabric or stainless steel grid elements which are installed on anchorage points or a system of rails for the fastening of harnesses; the nets and fabric sheets are weighted down with gravel or soil); *syn. for #2* security anchorage point [n]; *s* **protección** [f] **contra el riesgo de caídas** (**1.** Instalación protectora en cualquier parte abierta de un edificio, como balcones, huecos, escaleras y rampas, etc., que esté a más de 1 metro sobre nivel para evitar la caida de personas; *término específico* barandilla [f] de seguridad ►barrera de protección. **2.** Instalación de suelos adecuados para favorecer que las personas no tropiecen, resbalen o se dificulte la movilidad; cf. Art. 12.1. Exigencia básica SU 1, CTE, RD 314/2006. **3.** *Revestimiento vegetal de tejados* equipamiento de seguridad en forma de redes de acero inoxidable, fibra plástica de gran duración o elementos reticulados de acero inoxidable que son instalados en puntos de anclaje o en un sistema de raíles para fijar arreos. Las redes y hojas de tejido se sujetan en su sitio por medio de grava o tierra); *f* **protection** [f] **contre les (risques de) chutes de hauteur** (**1.** Ouvrages ou installations soumis à des règles de sécurité et protégeant les personnes contre une chute fortuite ou involontaire dans le vide lorsque la hauteur de chute excède 1 m, p. ex. balustrade, garde-corps, main courante, acrotère, ►barrière de protection. **2.** *Sur toitures végétalisées* ouvrages ou installations de protection antichute tels que garde-corps fixes, points d'ancrage permanents [crochets d'arrimage] pour fixation d'une longe pour harnais avec enrouleur automatique et absorbeur d'énergie, filets d'arrimage) ; *g* **Absturzsicherung** [f] (**1.** Gesetzlich vorgeschriebene Vorrichtung, die verhindert, dass Personen von Bauwerksteilen, die höher als 1 m sind, hinunterfallen, z. B. Brüstung, Geländer, Handlauf, Attika. **2.** *Dachbegrünung* flächige Sicherheits- oder Sicherungseinrichtungen in Form von Edelstahlnetzen, hochfestem Kunststoffgewebe oder Edelstahlrasterelementen ohne Dachdurchdringungen auf Flachdächern, die mit Anschlagpunkten oder einem Schienensystem zum Befestigen der Sicherungsleinen des Pflegepersonals versehen und mit Kies oder Bodensubstrat als Auflast beschwert sind; diese **A.** wird auch **Sekurant** genannt; ►Abschrankung).

4665 protection [n] **from light** *phyt.* (For prevention of exposure to direct sunlight; e.g. for ►shade plants); *syn.* sunshield [n], sunshade [n]; *s* **protección** [f] **contra la luz** (p. ej. de ►plantas de umbría); *f* **protection** [f] **contre la lumière** (►plante d'ombre) ; *syn.* couvert [m] contre la lumière ; *g* **Lichtschutz** [m, o. Pl.] (Gegebenheiten oder Erfordernisse zur Verhinderung des direkten Lichteinfalls, z. B. bei ►Schattenpflanzen).

4666 protection [n] **of an unexcavated area** *conserv' hist. leg.* (Safeguarding of archaeologically-important sites to prevent their disturbance of assumed ►culturel resources [US]/cultural assets [UK] of historical or prehistoric interest for future excavation); *s* **protección** [f] **de zonas arqueológicas** (Medidas destinadas a evitar la destrucción de zonas arqueológicas. En España la ley prevé la declaración de las **zz. aa.** como ►bienes de interés cultural [Es] y la redacción de Planes Especiales de Protección de área afectada; cf. art. 20 Ley 16/1985, del Patrimonio Histórico Español); *f* **protection** [f] **des sites archéologiques** (Dispositions réglementaires protégeant contre toutes atteintes les terrains pouvant intéresser la préhistoire, l'histoire, l'art ou l'archéologie et les ►biens culturels) ; *g* **Grabungsschutz** [m, o. Pl.] (Nutzungsschutz für archäologisch wertvolle Gebiete, die für wissenschaftliche Ausgrabungen von jeder anderen Nutzung freigehalten werden müssen, um im Boden verborgene ►Kulturgüter nicht zu zerstören).

4667 protection [n] **of conservation areas** *conserv. pol.* (Designation of certain community, state or federal areas, containing exceptional scenery and wildlife for nature conservation, such as ►areas of outstanding natural beauty [UK], ►bird sanctuary, ►forest research natural area [US]/forest nature reserve [UK], ►special areas of conservation [SAC], ►biosphere reserve, ►national park, ►natural area preserve [US]/nature reserve [UK], ►natural landmark [US]/site of spezial scientific interest [UK], ►wildlife sanctuary; ►protection of natural areas); *s* **protección [f] de espacios** (La designación de zonas protegidas [como parques, reservas naturales, áreas marinas protegidas, etc.] con el fin de asegurar la protección de especies por medio de la conservación de las biocenosis y sus espacios vitales, es uno de los instrumentos más utilizados internacionalmente. **En Es.** tendrán la consideración de espacios naturales protegidos aquellos espacios del territorio nacional, incluidas las aguas continentales, y las aguas marítimas bajo soberanía o jurisdicción nacional, incluidas la zona económica exclusiva y la plataforma continental, que cumplan al menos uno de los requisitos siguientes y sean declarados como tales: a] Contener sistemas o elementos naturales representativos, singulares, frágiles, amenazados o de especial interés ecológico, científico, paisajístico, geológico o educativo. b] Estar dedicados especialmente a la protección y el mantenimiento de la diversidad biológica, de la geodiversidad y de los recursos naturales y culturales asociados; art. 27, Ley 42/2007; ►espacio natural protegido, ►monumento natural, ►paisaje protegido, ►parque nacional, ►protección de áreas naturales, ►reserva de la biosfera, ►reserva forestal integral, ►reserva natural, ►santuario de aves, ►santuario de fauna salvaje, ►zona de especial protección para las aves, ►zona especial de conservación); *f* **protection [f] de zones** (Mesures prises dans le domaine de la protection de la nature en faveur de la conservation des espèces par la sauvegarde des communautés vivantes au travers de zones de protection comme p. ex. les Z.N.I.E.F., Z.I.C.O., Z.S.C., Z.R.S., Natura 2000 ; ►sauvegarde des sites naturels) ; *syn.* protection [f] de sites/territoires ; *g* **Gebietsschutz [m, o. Pl.]** (Gesamtheit der Maßnahmen zur Erhaltung der Leistungs- und Funktionsfähigkeit des Naturhaushaltes sowie der Regenerationsfähigkeit und nachhaltigen Nutzungsfähigkeit der Naturgüter und der daraus abzuleitenden Nutzungsbeschränkungen; die Tier- und Pflanzenwelt sowie ihre Lebensstätten und Lebensräume sind zu erhalten und zu entwickeln; weitere Schutzziele sind die Vielfalt, Eigenart und Schönheit sowie der Erholungswert von Natur und Landschaft; zu den Instrumenten des **G.es** gehört die Ausweisung von ►Schutzgebieten, z. B. ►besonderes Schutzgebiet [FFH-Gebiet], ►Biosphärenreservat, ►Landschaftsschutzgebiet, ►Nationalpark, ►Naturdenkmal, ►Naturschutzgebiet 2, ►Naturwaldreservat, ►Vogelschutzgebiet; gesetzliche Grundlagen sind das Bundesnaturschutzgesetz und die Naturschutzgesetze der Länder; die einschlägigen Rechtsverordnungen enthalten auch Bestimmungen zur Pflege und Entwicklung; ►Schutz von Natur und Landschaft); *syn.* Flächenschutz [m].

4668 protection [n] **of local/national heritage** [CH] *arch. conserv'hist. land'man.* (Actions and measures to protect and care for the cultural and natural assets of Swiss landscapes, e.g. the cultivated landscape, parks and gardens, towns or villages or parts thereof, individual buildings, historic sites or monuments, instruments or technical installations, dialects, folk art or local customs; no UK/U.S. specific equivalent; ►landscape management); *s* **protección [f] del patrimonio natural y cultural [CH]** (En CH todas las medidas para proteger y cuidar los componentes naturales y culturales del paisaje, incluyendo paisaje agrícola, parques y jardines, pueblos y ciudades, o parte de ellas, monumentos y sitios históricos, instalaciones técnicas, dialectos, artes populares y costumbres locales; ►gestión y protección del paisaje); *f* **conservation [f] du patrimoine historique et artistique national [CH]** (Toutes mesures de conser-

vation et d'entretien du patrimoine monumental, culturel et paysager ; font objet de la protection les sites et paysages, les parcs et jardins, tout ou partie de villes et villages, les bâtiments isolés ou les ensembles, les villes ou monuments historiques, les objets mobiliers, les outils et installations techniques, les dialectes, les œuvres d'art populaire ou les traditions populaires ; ►gestion des milieux naturels) ; *g* **Heimatschutz [m, o. Pl.] [CH]** (Alle Handlungen und Maßnahmen, die das landschaftliche Kulturgut und die Landesnatur in der Schweiz schützen und pflegen. Schutzgegenstände sind z. B. Kulturlandschaften, Parks und Gärten, Städte resp. Ortschaften oder Teile davon, Einzelbauwerke, historische Stätten oder Denkmäler, Werkzeuge, Geräte oder technische Anlagen, Dialekte, Volkskunst oder Volksbräuche; ►Landschaftspflege).

4669 protection [n] **of natural areas** *conserv. land'man.* (Generic term for the protection of integral parts of the landscape such as Areas of Outstanding Natural Beauty [UK], habitats, etc.; ►protected area, ►protection of conservation areas); *s* **protección [f] de áreas naturales** (Término genérico para la ►protección de espacios, protección de biótopos, de paisajes naturales y partes integrales del paisaje; ►espacio natural protegido); *f* **sauvegarde [f] des sites naturels** (DUV 1984, 285 ; mesures de protection ou amélioration des milieux naturels intéressants pour obtenir la densité du capital biologique au plus haute niveau ; ►protection des zones, ►zone protégée) ; *g* **Schutz [m, o. Pl.] von Natur und Landschaft** (Maßnahmen des Naturschutzes und der Landschaftspflege, die Teile von Natur und Landschaft zur Erhaltung der Leistungs- und Funktionsfähigkeit des Naturhaushaltes sowie der Regenerationsfähigkeit und nachhaltigen Nutzungsfähigkeit der Naturgüter durch ►Schutzgebiete unter Schutz stellen; weitere Schutzziele sind die Vielfalt, Eigenart und Schönheit sowie der Erholungswert von Natur und Landschaft; gesetzliche Grundlagen sind das Bundesnaturschutzgesetz und die Naturschutzgesetze der Länder; ►Gebietsschutz).

4670 protection [n] **of spring water** *wat'man.* (Improvement and preservation of the purity of ►spring water for its sustainable use as potable water); *s* **protección [f] de aguas de manantial** (Conjunto de medidas para proteger y preservar la pureza del ►agua de manantial y su aprovechamiento como recurso hídrico); *f* **protection [f] des eaux de source** (Ensemble des mesures de préservation de la qualité des ►eaux de source en vue de leur utilisation pour l'alimentation humaine ; *terme plus générique* protection des eaux minérales et des eaux de source ; compte tenu que l'instauration de périmètres de protection autour des captages et plus spécialement des aires d'alimentation des sources d'eau n'offre qu'une protection limitée, certains experts préconisent la création de « parcs naturels hydrogéologiques », vastes espaces de terres non cultivées mais entretenues, dont la fonction essentielle serait de préserver les nappes d'eau ayant une qualité irréprochable et sur lesquels toute activité polluante serait proscrite) ; *syn.* protection [f] des eaux vives ; *g* **Quellwasserschutz [m, o. Pl.]** (Gesamtheit der Maßnahmen zur Förderung und Erhaltung der Reinheit des ►Quellwassers und seiner dauerhaften Nutzung als Trinkwasser).

protection of the environment [n] *conserv. pol.* ►technical protection of the environment.

Protection [n] **of the Marine Environment and the Coastal Regions of the Mediterranean, Convention for the** *conserv. land'man. leg.* ►Barcelona Convention.

4671 protection [n] **of trees** *arb. constr.* (Safeguarding of trees on construction sites; ►additional technical contract conditions for tree work, ►boarding up of tree trunks, ►protective fencing around trees, ►preservation of existing trees);

s **protección [f] de árboles (en obras)** (Medidas de protección mientras se realizan obras; ►valla protectora de árboles, ►enrejado de tablones, ►mantenimiento de árboles 1); *f* **protection [f] des arbres** (Mesures générales de protection des arbres pendant la durée des chantiers de construction ; ►barrière de protection d'arbres, ►gaine en planches, ►conservation des arbres existants) ; *g* **Baumschutz [m, o. Pl.]** (Maßnahmen des Garten- und Landschaftsbaues bei Bauvorhaben, die jegliche Beschädigung und Wachstumsbeeinträchtigung an vorhandenen Bäumen verhindern. Bei Baumaßnahmen ist zwischen Schadensbegrenzung und Baumschutz zu unterscheiden! In D. wird der **B.** durch die DIN 18 920, die Richtlinien für die Anlage von Straßen — Teil: Landschaftsplanung [RAS-LP 4] und die ►Zusätzlichen Technischen Vertragsbedingungen — Baumpflege geregelt; ►Baumschutzzaun, ►Bohlenummantelung, ►Erhaltung von Bäumen).

protection [n] of wildlife *conserv. leg. zool.* ►wildlife conservation.

protection status [n] *adm. conserv. conserv'hist. leg.* ►bestowed protection status.

protection structure [n], coastal *conserv. eng.* ►sea defenses [US]/sea defences [UK].

4672 protection suitability [n] *conserv. leg. plan.* (Condition of a spatial unit, a landscape feature, or a heritage site worthy of protection according to nature conservation or historic monument conservation laws, based upon conservation criteria; ►worthy of protection); *s* **idoneidad [f] para la protección** (Características de un espacio natural o de un lugar histórico que le recomiendan para ser declarado como protegido; ►a proteger); *f* **aptitude [f] à la protection** (Appréciation portée sur un espace paysager, un élément de paysage ou un site d'intérêt culturel et historique qui retiennent une attention particulière en raison de leurs valeurs écologique, biologique, paysagère, historique, esthétique et dépend de critères de préservation en fonction desquels un classement peut être proposé ou prononcé au titre de la protection de la nature et des monuments historiques ; ►qui mérite d'être conservé ; *syn.* disposition [f] à la mise sous protection ; *g* **Schutzwürdigkeit [f]** (Betrachtung eines Landschaftsraumes, Landschaftsbestandteiles oder einer kulturhistorisch bedeutsamen Stätte hinsichtlich einer Unterschutzstellung auf Grund von Schutzkriterien — insbesondere im Sinne des Natur- oder Denkmalschutzes; ►schützenswert).

protection technology [n] *envir.* ►environmental protection technology.

protection zone [n] *plan.* ►raw material protection zone.

protection zone [n], perimeter *conserv. landsc.* ►marginal protection area.

protective aerie observation [n] *conserv. game'man.* ►protective eyrie observation.

protective belt [n] *landsc.* ►shelterbelt (1).

4673 protective coastal woodland [n] *conserv. land'man.* (Forest with form, size and location to make it a useful tool in supporting a system of protective measures for the seacoast against tidal actions); *syn.* protective marine forest [n] [also US]; *s* **bosque [m] protector del litoral** (Población forestal que por sus características y su localización en el litoral puede servir para proteger la costa contra los efectos destructores de mareas vivas); *f* **forêt [f] de protection côtière** (Peuplement forestier, dont la composition et la situation permettent dans le cadre d'un système de protection intégré de compléter la protection de la bande côtière contre l'activité de la houle et du vent lors des grandes marées [envaissement des eaux et des sables]) ; *syn.* forêt [f] de protection du littoral ; *g* **Küstenschutzwald [m]** (Waldbestand, der durch seine Beschaffenheit und Lage geeignet ist, im Rahmen eines geschlossenen Schutzsystems die Sicherung der Meeresküste gegen Brandungstätigkeiten bei Sturmfluten zu unterstützen).

4674 protective covering [n] *hort.* (Overspread material shielding sensitive plants against the effects of winter or solar radiation; ►mulching); *s* **acollado [m]** (Capa de tierra al pie de las plantas para protegerlas contra heladas, viento o radiación solar; cf. RA 1970, 70; ►mulching); *f* **paillis [m] (de protection)** (Couche mince de paille, de feuilles, de tourbe ou d'autre matières organiques disposée à la surface du sol en vue de sa protection contre le froid ou l'insolation ; ►mulching) ; *g* **Abdeckung [f] von Pflanzen** (Bedeckung als Kälte- oder Sonnenschutz für witterungsempfindliche Pflanzen; ►Mulchen).

4675 protective eyrie observation [n] *conserv. game'man.* (Watching a nest of endangered birds of prey to prevent disturbance caused by egg collectors, bird keepers, photographers or tourists); *syn. o.v.* protective aerie observation [n]; *s* **protección [f] de nidos de aves rapaces** (Observación de nido de especies rapaces en peligro contra las perturbaciones causadas por colectores de huevos, fotógrafos, turistas, etc.); *f* **protection [f] des nids** (Mesures de protection du lieu de nidification des oiseaux de proies ou des grands oiseaux contre la destruction ou l'enlèvement des œufs, la capture ou le désairage des jeunes oiseaux, les prises de vue et de son, les activités touristiques) ; *g* **Horstbewachung [f]** (Schutzmaßnahme für Nistplätze von seltenen Greif- und Großvögeln gegen Störungen durch Eiersammler, Vogelhalter, Fotografen und Touristen).

4676 protective fencing [n] around trees *constr.* (Temporary fencing during construction period, erected preferably to enclose the perimeter of the branch spread of existing trees to be retained; ►boarding up of tree trunks); *s* **valla [f] protectora de árboles** (Protección temporal de uno o varios árboles mientras se realizan obras de construcción que incluye la zona del suelo cubierta por la copa; ►enrejado de tablones); *f* **barrière [f] de protection d'arbres** (Équipement mis en place pour la protection d'un ou d'un groupe d'arbres existants pendant la durée d'un chantier ; ►gaine en planches) ; *g* **Baumschutzzaun [m]** (Absperrvorrichtung zum Schutze eines Baumes oder einer Baumgruppe im Kronentraufbereich während der Bauzeit, damit der Wurzelbereich weder durch Verdichtung noch durch Ablagerung von Baumaterialien und Abstellen von Geräten und Maschinen belastet wird; ►Bohlenummantelung).

4677 protective forest [n] (1) *for. landsc.* (Generic term for ►protective forest for soil conservation purposes, ►avalanche protection forest, ►dust and aerosol protection forest, ►noise attenuation forest, ►protective forest absorbing air pollution, ►protective forest against air pollution, ►protected aquifer recharge forest, ►protected habitat forest); *s* **bosque [m] protector (1)** (Término genérico para bosques con diferentes funciones protectoras, p. ej. ►bosque protector del suelo, ►biótopo forestal protegido, ►bosque protector contra inmisiones, ►bosque protector contra ruidos, ►bosque protector contra aludes, ►bosque protector de áreas de recarga de acuíferos, ►bosque protector contra inmisiones gaseosas, ►bosque protector contra la inmisión de polvo y aerosoles); *f* **forêt [f] de protection** (Terme générique pour la forêt classée pour cause d'utilité publique [statut défini dans le code forestier, aux articles L. 411-1 et suivants] et dont la conservation est, soit pour des raisons écologiques, soit pour le bien-être de la population, reconnue nécessaire au maintien des terres en montagne et sur les pentes, assurant la protection contre l'érosion, les glissements de terrains, les chutes de pierres, le maintien des berges de torrent, contre les avalanches, et les envaissements des eaux et des sables, la pollution ; ►biotope forestier protégé, ►forêt de défense contre les avalanches, ►forêt de protection contre la pollution atmosphérique, ►forêt de protection contre le bruit, ►forêt de protection contre l'érosion, ►forêt de protection contre les polluants gazeux, ►forêt de protection contre les poussières, ►forêt protectrice de la nappe aquifère) ; *g* **Schutzwald** [m] (OB zu ►Biotopschutzwald, ►Bodenschutzwald, ►Immissionsschutzwald, ►Lärmschutzwald, ►Lawinenschutzwald, ►Speicherschutzwald, Schutzwald gegen schädliche Umwelteinwirkungen, ►Schutzwald gegen gasförmige Immissionen, ►Staubschutzwald und Aerosolschutzwald).

4678 protective forest [n] (2) *for. land'man.* (Wooded area to protect against avalanches, landslides, and soil erosion in mountain areas; ►social benefits of the forest, ►protective forest 1, ►avalanche protection forest); *s* **bosque** [m] **protector** (2) (En alemán, término especial del **b. p.** en los Alpes que, además de sus ►efectos beneficiosos del bosque, cumple funciones de protección contra la escorrentía, la erosión o la generación de aludes; ►bosque protector 1, ►bosque protector contra aludes); *f* **espace** [m] **boisé protégé** (±) (D., zone forestière qui au-delà de la ►fonction sanitaire et sociale de la forêt contribue comme ►forêt de protection dans les pays alpins aussi bien à la restauration des terrains, la protection contre les avalanches, les glissements de terrain, la chute de pierres, etc. qu'à assurer les ressources en eau ou protéger les sources thermales ; statut de protection comparable à celui décrit dans l'article L 411-1 du Code forestier pour les forêts dont la conservation est reconnue nécessaire au maintien des terres sur les montagnes et sur les pentes, à la défense contre les avalanches, les érosions et les envahissements des eaux et des sables ; ►forêt de défense contre les avalanches) ; *g* **Bannwald** [m] (2) (Wald, der neben den allgemeinen Wohlfahrtswirkungen zur Abwehr bestimmter Gefahren von Menschen, menschlichen Siedlungen und Anlagen oder kultiviertem Boden sowie in Alpenländern als ►Schutzwald gegen Lawinen, Erosionen, Erdabrutschungen, Steinschlag, Bergsturz, Hochwasser, Wind oder ähnlichen Gefahren, aber auch der Sicherung eines Wasservorkommens oder zum Schutz von Heilquellen dient; cf. § 27 ForstG; ►Lawinenschutzwald, ►Wohlfahrtswirkung des Waldes).

4679 protective forest [n] **absorbing air pollution** *for. landsc.* (Forest which reduces the harmful impact of emissions, especially noise, dust, aerosols, gases or radiation, and thus protects residential, work and recreation areas, agricultural land, forests, etc.; *specific terms* ►dust and aerosol protection forest, ►noise attenuation forest, ►protective forest against air pollution); *s* **bosque** [m] **protector contra inmisiones** (*Términos específicos* ►bosque protector contra inmisiones gaseosas, ►bosque protector contra la inmisión de polvo y aerosoles, ►bosque protector contra ruidos); *f* **forêt** [f] **de protection contre la pollution atmosphérique** (Forêt dont la fonction est de protéger les zones urbaines, de loisirs et de tourisme, certains espaces agricoles et forestier, constructions et monuments contre l'action des polluants atmosphériques ; *termes spécifiques* ►forêt de protection contre les polluants gazeux, ►forêt de protection contre les poussières, ►forêt de protection contre le bruit) ; *g* **Immissionsschutzwald** [m] (Wald, der schädliche oder belästigende Einwirkungen, insbesondere durch Lärm, Staub, Aerosole, Gase oder Strahlen hervorgerufen, mindert und damit Wohn-, Arbeits- und Erholungsbereiche, land- und forstwirtschaftliche Nutzflächen sowie andere schutzbedürftige Objekte vor nachteiligen Wirkungen dieser Immissionen schützt; *UBe* ►Schutzwald gegen gasförmige Immissionen, ►Staubschutzwald und Aerosolschutzwald, ►Lärmschutzwald).

4680 protective forest [n] **against air pollution** *for. landsc.* (Type of ►protective forest absorbing air pollution; the species of trees planted tolerate the existence of strongly plant-

hostile air pollution; ►dust and aerosol protection forest); *s* **bosque** [m] **protector contra inmisiones gaseosas** (Tipo de ►bosque protector contra inmisiones; ►bosque protector contra la inmisión de polvo y aerosoles); *f* **forêt** [f] **de protection contre les polluants gazeux** (Type de ►forêt de protection contre la pollution atmosphérique ; le choix des espèces ligneuses la constituant dépend exclusivement de leur tolérance à la nature et la concentration de la pollution rencontrée ; ►forêt de protection contre les poussières) ; *g* **Schutzwald** [m] **gegen gasförmige Immissionen** (Form des ►Immissionsschutzwaldes; Auswahl und Anbau der Gehölzarten richtet sich nach der Immissionsverträglichkeit der Pflanzen; ►Staubschutzwald und Aerosolschutzwald).

4681 protective forest [n] **for soil conservation purposes** *conserv. land'man.* (Area of woodland cover with the primary function of preventing soil erosion; ►rill erosion, ►gully erosion, ►sheet erosion, ►mass slippage [US]/mass movement [UK], ►avalanche protection forest, ►protected aquifer recharge forest, ►forest function); *s* **bosque** [m] **protector del suelo** (Bosque que tiene la función de proteger el suelo sobre el que crece y los vecindarios contra las consecuencias de la ►erosión en regueros, la ►erosión en cárcavas 1, la ►erosión laminar, contra aludes y movimientos gravitacionales; ►función del bosque, ►movimiento gravitacional; *términos específicos* ►bosque protector contra aludes, ►bosque protector de área de recarga de acuíferos); *f* **forêt** [f] **de protection contre l'érosion** (Dans le sens originel relatif à la protection du sol ; forêt devant protéger le milieu naturel et les zones limitrophes contre l'érosion provoquée par les agents climatiques tels que la pluie [►érosion pelliculaire, ►ravinement 1, ►ruissellement en filets], le vent, la neige, par l'appauvrissement du sol, les chutes de pierres ou les ►glissements de terrain [►mouvement de masse] ; ►fonction de la forêt ; *termes spécifiques* ►forêt de défense contre les avalanches, ►forêt protectrice de la nappe aquifère) ; *g* **Bodenschutzwald** [m] (Wald, der seinen Standort sowie benachbarte Flächen vor den Auswirkungen von ►Rillenerosion, ►Grabenerosion 1 und ►Flächenerosion, Lawinenabgängen, Aushagerung, Steinschlag und Bodenrutschungen [►Massenversatz] schützen soll; hierzu zählen besonders rutschgefährdete Hänge, felsige oder flachgründige Steilhänge, zur Verkarstung neigende Standorte und Flugsandböden; cf. § 30 LWaldG-BW; ►Waldfunktion; *UBe* ►Lawinenschutzwald, ►Speicherschutzwald).

4682 protective forest [n] **for water resources** *nat'res. wat'man.* (Forest which helps to maintain the purity and recharge of groundwater as well as bodies of water, rivers and streams and also maintains a continual ►inflow 1; ►forest function; *generic term* ►soil and water conservation district [US]/water conservation area [UK]); *s* **bosque** [m] **protector de acuífero** (Bosque que ayuda a mantener la pureza de los acuíferos y de las aguas superficiales y estabiliza el ►caudal afluente contribuyendo así a recargar los acuíferos; ►función del bosque; *término genérico* ►perímetro de protección de acuífero); *f* **forêt** [f] **de protection des ressources en eau** (±) (Forêt protégée pour sa fonction hydrologique prépondérante favorisant l'approvisionnement et la filtration des eaux des nappes aquifères, la régulation du régime des cours d'eaux ; ►débit entrant, ►fonction de la forêt ; *terme générique* ►périmètre de protection d'un captage d'eau potable) ; *g* **Wasserschutzwald** [m] (Wald, der der Reinhaltung des Grundwassers sowie stehender und fließender Oberflächengewässer dient und die Stetigkeit der ►Wasserspende verbessert; ►Waldfunktion; OB ►Wasserschutzgebiet).

4683 protective hazardous facility perimeter [n] *leg. urb.* (Outer boundary around a dangerous facility demarcated for health and security of the population and environment in ►community development plans [US]/urban development plan

P

[UK]); *s* **perímetro [m] de protección** (≠) (En D., en ►plan general municipal de ordenación urbana superficie de uso restringido o dedicada a medidas de protección contra influencias ambientales negativas de usos vecinos); *f* **périmètre [m] de servitude d'utilité publique autour des installations classées pour la protection de l'environnement** (Périmètre délimité autour d'une installation classée susceptible de créer par danger d'explosion ou d'émanation des produits nocifs, des risques très importants par la santé ou la sécurité de populations voisines et pour l'environnement ; cf. art. 7-1 loi n° 76-663 relative aux installations classées pour la protection de l'environnement ; le projet définissant les servitudes et le périmètre est soumis à enquête publique, les servitudes sont ensuite annexées au ►plan local d'urbanisme de la commune) ; *g* **Fläche [f] für Nutzungsbeschränkungen oder für Vorkehrungen zum Schutze gegen schädliche Umwelteinwirkungen** (Im ►Flächennutzungsplan gem. § 5 [2] 6 und Nr. 15.6 der Anlage zur PlanzV darzustellende Fläche, auf der die Nutzung eingeschränkt werden kann resp. auf der Vorkehrungen zum Schutze gegen schädliche Umwelteinwirkungen im Sinne des Bundes-Immissionsschutzgesetzes [BImSchG] getroffen werden können).

protective marine forest [n] [US] *conserv. land'man.* ►protective coastal woodland.

4684 protective planting [n] *agr. for. land'man. landsc.* (Generic term for a linear planting of trees or shrubs designed to provide ►erosion control, ►pollution control 1, ►noise control, ►visual screening, ►wind protection or to slow down the drying out of soils as well as for channeling cold air drainage; ►shelterbelt 1, ►windbreak, ►windbreak planting); *syn.* screen planting [n] (1); *s* **plantación [f] protectora** (Término genérico para plantaciones de leñosas que protegen contra la erosión, las inmisiones, el ruido, el viento, etc. y sirven para reducir la desecación del suelo; ►cortina rompevientos, ►franja protectora, ►protección contra el ruido, ►protección contra el viento, ►protección contra inmisiones, ►protección contra la erosión, ►protección visual); *syn.* pantalla [f] protectora; *f* **plantation [f] de protection** (Terme générique désignant une plantation d'une à plusieurs lignes d'arbres et d'arbustes pouvant satisfaire à un des objectifs suivants : ►lutte contre l'érosion, ►lutte contre les émissions, ►protection contre le bruit, ►protection contre les vues, ►protection contre le vent ; la **p. d. p.** agit en retardant le dessèchement des sols, la canalisation des courants d'air froid, etc. ; ►plantation brise-vent, ►rideau forestier) ; *syn.* plantation [f] protectrice, écran [m] végétal, plantation [f] écran, écran [m] de verdure, bande [f] boisée de protection (LA 1981, 598) ; *g* **Schutzpflanzung [f]** (OB zu ein bis mehrreihige Pflanzung von Bäumen und Sträuchern, die dem ►Erosionsschutz, ►Immissionsschutz, ►Lärmschutz, ►Sichtschutz, ►Windschutz dient oder die Austrocknung von Böden verlangsamt, die Abwehr und seitliche Ableitung von Kaltluftströmen bewirkt etc.; ►Schutzstreifen, ►Windschutzpflanzung); *syn.* Abpflanzung [f].

4685 protective screed [n] *constr.* (Layer of mortar or concrete on roof ►waterproofing membranes which are vulnerable to damage, e.g. during topsoiling and planting; ►protective sheet); *syn.* concrete protective slab [n] (OSM 1999, 163) [also US], concrete protection layer [n] (OSM 1999); *s* **capa [f] protectora de mortero** (Capa fina de mortero o cemento sobre ►membrana de impermeabilización para protegerla de posibles daños; ►capa protectora); *f* **chape [f] de protection** (Couche de mortier mise en place, p. ex. en protection lourde sur un revêtement d'étanchéité ; ►film protecteur et isolateur, ►membrane d'étanchéité) ; *g* **Schutzestrich [m]** (Schützende Mörtelschicht, z. B. auf verletzbaren Dachdichtungsbahnen/►Dichtungsfolien etc.; ►Trenn- und Schutzlage).

4686 protective sheet [n] *constr.* (Protection mat, used in ►roof planting and laid to protect the ►root protection membrane or the ►waterproof layer of a roof against chemical or mechanical damage; ►protective screed); *s* **capa [f] protectora** (Se utiliza en ►revestimiento vegetal de tejados y sirve de protección mecánica y química de la ►lámina antirraíces y de la ►capa de impermeabilización. Se coloca directamente encima de la cubierta del tejado; ►capa protectora de mortero); *f* **film [m] protecteur et isolateur** (►*Végétalisation des toitures* matériau assurant la protection physique et chimique du ►film [de protection] antiracines ou du ►revêtement d'étchantéité sur une toiture ; ►chape de protection) ; *syn.* couche [f] de protection et d'isolation ; *g* **Trenn- und Schutzlage [f]** (►*Dachbegrünung* Kunststoffschicht, die dem chemischen und mechanischen Schutz der ►Wurzelschutzfolie oder der ►Dichtungsschicht über der Dachkonstruktion dient; verrottungsfeste, chemisch und biologisch neutrale **T.- u. S.n** grenzen chemisch unverträgliche Schichten auf dem Dach voneinander ab und schließen das Ausschlagen von PVC-Weichmachern aus; ►Schutzestrich); *syn.* Trenn- und Schutzvlies [n], Trennschicht [f].

4687 protective tree barrier [n] *urb.* (Metal framework installed to prevent damage to a tree); *s* **barra [f] de protección de árbol** (Tubo de metal arqueado que sirve para evitar daños mecánicos a los árboles de las calles al evitar que choquen o aparquen automóviles sobre el alcorque de los mismos); *f* **arceau [m] de protection** (Installation en métal pour la protection des arbres ou des plantations d'accompagnement contre les dommages mécaniques) ; *g* **Baumschutzbügel [m]** (Stabile Metallvorrichtung zum Schutze von Straßenbäumen gegen mechanische Schäden).

prototype [n] [US] *constr. plant.* ►planting prototype.

prototypical planting scheme [n] [UK] *constr. plant.* ►planting prototype [US].

provide [vb] **and install** [vb] *constr. contr.* ►furnish and install.

province [n] *geo. ecol.* ►physiographic division.

4688 provisional certificate [n] **of completion** *contr.* (Interim certificate which is issued with instructions to correct any existing defects or omissions); *s* **aceptación [f] condicional (de obra)** (Aceptación provisional de obra de construcción, a pesar de la existencia de defectos que han de ser corregidos); *syn.* recepción [f] condicional (de obra); *f* **réception [f] avec réserves** (Réception prononcée lorsque certaines épreuves doivent être exécutées après une durée déterminée de service des ouvrages ou à une certaine période de l'année, lorsque certaines prestations prévues au marché n'ont pas été exécutées, lorsqu'existent certaines imperfections ou malfaçons) ; *g* **Abnahme [f] unter Vorbehalt** (Anerkennung von Bauleistungen mit festgestellten Mängeln, die in der Abnahmeniederschrift aufzunehmen sind und nachgebessert werden müssen; cf. § 12 VOB Teil B; das Werk wird mit der Mängelvorbehaltsbemerkung im Wesentlichen als vertragsgerecht vollendet gebilligt und die Mängelansprüche werden aus dem Bürgerlichen Gesetzbuch [BGB] geltend gemacht).

4689 provisional certificate [n] **of completion with reduction in payment for contract item** *contr.* (Issue of certificate with optional instructions not to correct existing defects, which are thus not completely remunerated, or which must be corrected within a specified period. If defects are not corrected, a reduction in the remuneration will be agreed upon; ►price reduction); *s* **aceptación [f] de obra con reducción de precio** (Aceptación de una obra con defectos que deben de corregirse dentro de un determinado plazo o que se aceptan, en cuyo caso se acuerda una ►reducción del precio); *syn.* recepción [f] de obra con reducción de precio; *f* **réception [f] avec réfac-**

tion (sur les prix) (Procédure adoptée lorsque certains ouvrages ou parties d'ouvrages sont imparfaits tout en étant aptes à l'usage et n'étant pas de nature à porter atteinte à la sécurité. Le responsable du marché peut renoncer à ordonner la réfection de l'ouvrage jugé défectueux et proposer à l'entrepreneur une ▶réfaction sur les prix) ; *g* **Abnahme [f] unter Vorbehalt bei Minderung der Vergütung** (Abnahme mit Feststellung von Mängeln, die entweder nicht nachgebessert werden müssen und deshalb nicht voll vergütet werden oder in einer Frist nachgebessert werden müssen. Die Mängelbeseitigung ist grundsätzlich anzustreben. Werden die Mängel nicht beseitigt, wird eine Minderung der Vergütung vereinbart; ▶Preisminderung).

4690 provisional report [n] *plan.* (Written or oral statement of preliminary project planning status or research findings); *syn.* stopgap report [n] [also US], interim report [n]; *s* **informe [m] provisional** (Informe escrito que describe el estado actual de un proceso de planificación o de un proyecto); *f* **rapport [m] provisoire (d'avancement)** (Rapport fourni dans un délai fixé contractuellement après la notification d'un marché présentant l'état d'avancement des études et constituant la base de la rédaction du rapport définitif ; le **r. p.** peut être précédé de pré-rapports d'avancement) ; *g* **vorläufiger Bericht [m]** (Bericht, der den aktuellen Sachstand einer Planung oder Untersuchung wiedergibt).

provisional (specification) item [n] *contr.* ▶optional specification item.

4691 provision [n] **of access for the public** *plan. trans. urb.* (**1.** Making an area accessible; i.e. to connect it to the road and public transport system; open up [vb] to the public; ▶traffic linkage. **2.** Constructing or improving of travelled ways [US]/carriage ways [UK] and access ways for entering and using of an area); *s* **planificación [f] viaria** (Desarrollo de sistema de calles y caminos dentro de la zona de actuación y de enlace al sistema de comunicación viaria existente en el marco de un proyecto de actuación urbanística; ▶urbanización viaria); *f* **viabilisation [f]** (Rendre en état de viabilité [action de viabiliser] un terrain, une zone à urbaniser ou en voie d'urbanisation en effectuant les travaux de desserte, de raccordement aux différents réseaux [voirie, eau, gaz, électricité, télécommunication, assainissement] ; DUA 1996 ▶desserte) ; *g* **Erschließung [f] für den Verkehr** (Planerisches Vorhaben mit dem Ziel, einzelne Grundstücke oder ein ganzes Gebiet an ein bestehendes Verkehrsnetz anzubinden und im Gebiet selbst Straßen und Wege anzulegen; ▶Verkehrserschließung).

provision [n] **of basic facilities and major grading** [UK] *constr.* ▶provision of basic facilities and rough grading [US].

4692 provision [n] **of basic facilities and rough grading** [US] *constr.* (Construction of roadways, installation of utilities and rough grading to proposed contours required, for example, by an horticultural exhibition, industrial park, office park); *syn.* provision [n] of basic facilities and major grading [UK]; *s* **construcción [f] de infraestructura básica y nivelado [m] bruto** (Primera fase de construcción en proyecto paisajístico como base para construcción ulterior); *f* **équipements [mpl] et travaux de base** (Construction de cheminements, mise en place ou mise à disposition de conduites d'eau et exécution du nivellement grosso modo constituant p. ex. la contribution de l'organisateur d'une exposition florale dans le cadre des réalisations paysagères) ; *g* **Grundausbau [m, o. Pl.]** (Anlage von Wegen, Straßen, Einbau oder Vorhalten von Ver- und Entsorgungsleitungen und Erstellen des Grobplanums, z. B. für den Bau eines landschaftsgärtnerischen Ausstellungsbeitrages bei Gartenschauen, für ein Industrie- oder Bürogelände).

4693 provision [n] **of data and aims** *plan.* (Furnishing of data and goals for a project by an agency/authority or client); *s* **puesta [f] a disposición de datos y objectivos** (p. ej. por parte de administración comitente o cliente); *f* **mise [f] à disposition des informations et des documents** (par l'administration ou par le client) ; *g* **Bereitstellung [f] von Daten und Zielen** (Das Zurverfügungstellen von wesentlichen Grundlagen, Ausgangsdaten und Zielvorstellungen an den Planer zur Projektbearbeitung, z. B. durch Behörden oder sonstige Auftraggeber).

4694 provision [n] **of green spaces** *landsc. urb.* (Amount of soft landscape to serve public needs in an urban area; ▶open space, ▶landscaping of built-up areas); *s* **dotación [f] de espacios verdes** (Cantidad de áreas verdes de uso público en una zona urbana. En la jerga de planificación en D. se utiliza despectivamente el término «Grüngarnierung» = decoración verde, cuando los espacios verdes existentes tienen vegetación pobre o no están equipados adecuadamente para permitir su aprovechamiento por la población; ▶espacio libre, ▶tratamiento paisajístico 1); *f* **équipement [m] en espaces verts** (Ensemble des ▶espaces libres végétalisés en milieu urbain, par opposition aux espaces minéraux ; **D.**, terme technique à connotation négative exprimant dans le jargon des aménageurs un déficit dans l'équipement en espaces verts de certaines zones urbaines ou une insuffisance dans l'utilisation de ces mêmes espaces ; ▶traitement paysager 1) ; *g* **Grünausstattung [f]** (Umfang der vegetationsbestimmten ▶Freiräume in einem Siedlungsgebiet. Im Planerjargon gibt es auch den negativ besetzten Begriff **Grüngarnierung** für eine unzureichende Ausstattung mit Vegetation und unzureichenden Nutzungsmöglichkeiten von Grünflächen; ▶Durchgrünung).

4695 provision [n] **of labo(u)r/personnel** *constr. prof.* (Coordinated decision on required types and deployment of labo[u]r/personnel); *s* **utilización [f] de mano de obra** (Cálculo concerniente a la cantidad y el tipo de mano de obra necesaria para una obra); *f* **intervention [f] des personnels** (Décision concernant le nombre de personnels à faire intervenir, p. ex. sur un chantier ; faire [vb] intervenir un personnel) ; *g* **Einsatz [m] von Arbeitskräften/Personal** (Organisationsentscheidung über die einzusetzende Anzahl von Arbeitskräften; *syn.* Arbeitskräfteeinsatz [m], Beschäftigung [f] von Arbeitskräften/Personal, Einsetzen [n, o. Pl.] von Arbeitskräften, Personaleinsatz [m].

4696 provision [n] **of open spaces** *pol. urb.* (Generic term for making open spaces available for public and private use); *s* **dotación [f] de espacios libres** (Previsión, mantenimiento y cuidado de zonas verdes públicas y privadas); *f* **dotation [f] en espaces libres** (Politique de création de maintien, de développement et d'entretien d'espaces verts publics et privés pour satisfaire les besoins des citadins) ; *g* **Freiraumversorgung [f, o. Pl.]** (Bereitstellung und nachhaltige Pflege und Unterhaltung von öffentlichen und privaten Grünflächen).

4697 provision [n] **of services** *econ. trans. wat'man.* (Developing public infrastructure such as water and sewer systems, public transport, electrical power, etc., for the population; ▶public utility network); *s* **abastecimiento [m] de servicios** (Puesta a disposición de servicios como agua, electricidad, transporte público para la población; ▶red de abastecimiento público); *syn.* suministro [m] de servicios; *f 1* **approvisionnement [m]** (Fourniture de biens de consommation nécessaires à la satisfaction des besoins de la population ; ▶réseaux [d'alimentation]) ; *f 2* **couverture [f]** (Espace ou temps pris en compte par les ressources naturelles [couverture alimentaire] ou les services et les moyens de transport et de télécommunication, p. ex. la couverture par le réseau de transport en commun, la couverture énergétique) ; *g* **Versorgung [f]** (Im Rahmen der öffentlichen

Daseinsvorsorge das Bereitstellen von Energie, Trinkwasser, öffentlichem Nahverkehr etc. für die Bevölkerung; ▶Versorgungsnetz).

4698 provisions [npl] **for recreation** *plan. recr.* (Development of recreation areas with consideration for preserving and managing the natural attributes of landscapes or parts thereof due to their variety, character or natural beauty, as a pre-condition for recreational use in order to achieve society's far-sighted objectives); *syn.* recreational provision [n]; *s* **previsión** [f] **de posibilidades de recreo** (Responsabilidad de las instituciones del Estado de gestionar los espacios naturales adecuados para su utilización para el recreo al aire libre de manera que éstos no se deterioren o sean destruídos y así asegurar su aptitud recreativa a largo plazo); *f* **politique** [f] **de prévoyance concernant les loisirs et le tourisme** (Principes directeurs, actions et mesures visant à assurer et promouvoir une politique de tourisme durable sur le plan économique, social, territorial, environnemental et culturel qui garantisse la fréquentation, l'attrait et un bon état de conservation des sites naturels et historiques ainsi que la valorisation de leurs potentiels paysagers, afin de permettre leur transmission aux générations futures) ; *g* **Erholungsvorsorge** [f] (Gesellschaftlicher Auftrag durch Planung und Maßnahmen, das Naturpotenzial in [Stadt]landschaften oder Teilen davon in ihrer Vielfalt, Eigenart und Schönheit als Voraussetzung für die Erholung nachhaltig zu schützen, zu pflegen und zu entwickeln).

4699 proximity green space [n] *urb.* (Relatively small, public or semi-public, open space serving local residents as a green area within close walking distance of the home; ▶neighborhood park [US]/district park [UK]); *syn.* neighborhood green space [n] [US]; *s* **zona** [f] **verde de vecindario** (Espacio verde público o semipúblico situado dentro de un barrio residencial, de manera que la población de éste puede acceder a él a pie; ▶jardín de barrio); *f* **espace** [m] **vert urbain de proximité** (Espace vert public rapidement accessible à pied à partir des îlots habités ; ▶parc de voisinage, ▶parc de quartier) ; *g* **wohnungsnahe Grünfläche** [f] (Öffentliche, auch halböffentliche Grünfläche, die in kurzer, fußläufiger Entfernung zum Wohngebiet liegt; ▶Stadtteilpark); *syn.* Wohnumfeldgrün [n], wohnungsnahes Grün [n], quartierbezogene Grünfläche [f], stadtteilbezogene Grünfläche [n].

pruning [n] *arb. hort.* ▶branch pruning; *arb. constr.* ▶crown pruning; *hort.* ▶fruit tree pruning; *agr.* ▶green pruning; *arb. constr. hort.* ▶rejuvenation pruning; *constr. hort.* ▶root pruning (1); *constr.* ▶root pruning (2); *arb. constr.* ▶routine crown pruning; *hort.* ▶severe pruning, ▶tip pruning; *constr. hort.* ▶tree and shrub pruning; *hort.* ▶tree pruning;

pruning [n] [US], **drop** *constr. for. hort.* ▶crown lifting.

pruning [n] [US], **drop crotch** *arb. constr.* ▶drop-crotching.

pruning [n] [UK], **dry** *arb. hort.* ▶dead wooding [US] (2).

pruning [n], **formative** *hort.* ▶training of young trees.

pruning [n] [US], **invigoration** *arb. constr. hort.* ▶rejuvenation pruning.

pruning [n] [US], **lift-** *constr. for. hort.* ▶crown lifting.

pruning [n], **pollard** *arb. hort.* ▶pollarding.

pruning [n] [US], **severe crown** *arb. constr.* ▶drop-crotching.

pruning [n], **summer** *agr.* ▶green pruning.

4700 pruning [n] **at planting** *hort.* (ARB 1983, 407; cutting back of upper plant parts and damaged roots during planting operations; ▶root pruning 1); *s* **poda** [f] **de plantación** (Poda de las partes aéreas de las plantas o de raíces dañadas al realizar la plantación; ▶poda de raíces); *f 1* **taille** [f] **de plantation** (*Terme générique* taille des parties aériennes des végétaux,

raccourcissement des racines endommagées effectuée lors des travaux de plantation ; ▶rafraîchissement des racines) ; *f 2* **habillage** [m] **des parties aériennes** (*Terme spécifique* taille des branches avant plantation) ; *g* **Pflanzschnitt** [m] (Rückschnitt der oberirdischen Pflanzenteile und beschädigten Wurzeln bei Pflanzarbeiten mit dem Ziel, das Gleichgewicht zwischen Wurzelwerk und zu versorgendem oberirdischem Teil wiederherzustellen, da beim Roden in der Baumschule ein Teil der Wurzeln verloren geht; DIN 18 916 schreibt vor, dass bei Heistern, Stammbüschen, Halb- und Hochstämmen die natürliche Wuchsform durch den Schnitt zu erhalten ist [habitusgerechter Schnitt]; die ZTVLa-StB 1999 geben für verschiedene Gehölzarten genaue Hinweise wie sie zu schneiden sind. Bei Stauden entfällt üblicherweise der P., da sie i. d. R. in Töpfen geliefert werden; bei ballenlos gelieferten oder umzupflanzenden Stauden werden sehr lange und beschädigte Wurzeln gekürzt; Arten mit fleischigen Wurzeln werden nicht gekürzt; ▶Wurzelschnitt).

pruning [n] **for public safety** *arb. constr.* ▶crown pruning for public safety.

4701 prunings [npl] *agr. constr. hort.* (Generic term for cut stems or parts thereof; ▶clippings, ▶lawn clippings); *syn.* trimmings [npl]; *s* **material** [m] **de poda o siega** (Término genérico para restos de tallos cortados o partes de los mismos; ▶hierba segada, ▶hierba de césped cortado); *f* **produit** [m] **de coupe** (*Terme générique* désignant toute partie végétative des plantes soumise à l'action de taille ou de coupe ; ▶produit de fauche, ▶produit de tonte/de coupe, ▶produit de tonte de gazon) ; *g* **Schnittgut** [n, o. Pl.] (Abgeschnittene, vegetative Sprossachsen oder Teile davon, z. B. Gehölzschnittgut, Grasschnittgut, ▶Rasenschnittgut, Staudenschnittgut; *UB* ▶Mähgut); *syn.* Grünschnitt [m].

4702 pruning wound [n] *hort.* (Branch or stem removal scar); *s* **herida** [f] **de poda**; *f* **plaie** [f] **de taille** (Blessure provoquée lors de la taille) ; *g* **Schnittwunde** [f] (Durch einen Schnitt entstandene Verletzung).

4703 psammophilous [adj] *phyt. zool. syn.* arenicolous [adj], sand-dwelling [adj]; *s* **psamófilo** [adj] (Calificativo ecológico de las plantas y sinecias que requieren suelos arenosos; DB 1985); *syn.* arenícola [adj]; *f* **psammophile** [adj] (Vivant sur ou dans le sable) ; *syn.* arénicole [adj] ; *g* **auf Sand lebend** [loc] *syn.* Sand liebend [ppr/adj], Sand bewohnend [ppr/adj].

pseudogley [n] [UK] *pedol.* ▶aqualfs [US].

public access [n] *leg. recr.* ▶right of public access.

4704 public access [n] **to rural land** *leg. recr.* (Admittance to or through private uncultivated rural land for recreational use; by easement in the U.S.; ▶right of public access); *s* **acceso** [m] **público al campo** (Posibilidad para toda la población de pasear con fines recreativos por caminos privados, al borde de los campos de cultivo, por zonas yermas y otras no utilizadas para la agricultura; en D. regulada por el art. 27 de la Ley Federal de Protección de la Naturaleza; ▶servidumbre de tránsito); *f 1* **ouverture** [f] **au public de l'espace naturel** (F., l'ouverture au public des milieux naturels est régit par le Code rural, les documents d'urbanisme et la législation relative à la protection de la nature ; *termes spécifiques* ouverture au public des espaces naturels protégés, ouverture au public des milieux naturels remarquables, agricoles et forestiers ; *syn.* ouverture [f] au public des sites naturels, accès [m] du public aux espaces naturels ; *f 2* **accès** [m] **au public de l'espace rural** (D., l'article 27 de la loi fédérale sur la protection de la nature autorise, pour les activités de détente, la circulation des personnes sur les chemins ruraux, les lisières des champs, les talus, les jachères et autres espaces agricoles non cultivés. La législation des Länder en précise les différentes modalités ; ▶servitude de passage des

piétons) ; *syn.* accès [m] (du public) aux espaces/biens/territoires ruraux ; **g Betreten [n, o.Pl.] der Flur (D.**, gemäß § 27 BNatSchG ist das Betreten von privaten Wegen, Feldrainen, Böschungen, Öd- und Brachflächen und anderen landwirtschaftlich nicht genutzten Flächen zum Zwecke der Erholung gestattet. In den Ländergesetzen werden die Einzelheiten [z. B. Einschränkungen] näher festgelegt; ▸Betretungsrecht); *syn.* Betreten [n, o. Pl.] der freien Landschaft (§ 37 NatSchG-BW).

public agency [n] [US] *adm.* ▸contracting public agency [US], ▸public authority.

public agency review [n] [US]**, coordinated** *adm. leg.* ▸statutory consultation with public agencies [US].

public and civic site [n] [US] *leg. urb.* ▸site for public facilities.

4705 public announcement [n] *adm. leg. plan.* (Official notice published in local newspapers [US]/formal statement in official gazettes [UK] or internet, announcing required public hearing on planning or regulatory matters, as well as adoption of an approved plan [US]/a statutory plan [UK]; ▸give legal notice [vb]); *syn.* public notice [n], legal notice [n] [also US], legal announcement [n] [also US]; *s 1* **anuncio [m] preceptivo** (Aviso oficial de la aprobación inicial o provisional de un plan y de su sumisión a información pública que debe realizarse en el boletín oficial correspondiente y en uno de los diarios de mayor difusión de la provincia correspondiente; art. 27 Texto refundido de la Ley sobre Régimen del Suelo y Ordenación Urbana, RD 1346/1976; ▸anunciar públicamente); *s 2* **publicación [f] de un plan** (Los acuerdos de aprobación definitiva de todos los instrumentos de planeamiento se publicarán en el Boletín Oficial correspondiente. En el caso de la aprobación de planes parciales deben ser notificados personalmente todos los propietarios; arts. 134, 139, 140 y 141 RD 2159/1978); *f 1* **publication [f]** (Le PLU/POS approuvé est rendu public par des actes tels que : **1.** la mention dans les journaux régionaux ou locaux, **2.** l'affichage à la mairie, **3.** la publication par voie d'affiches, **4.** la publication au recueil des actes administratifs, etc., et est tenu à la disposition du public pour être consulté dans la mairie de la commune intéressée ou en tout autre lieu dûment mentionné ; ▸rendre public) ; *f 2* **promulgation [f] (de textes législatifs)** (Acte attestant officiellement l'existence d'une loi et la rendant exécutoire) ; **g öffentliche Bekanntmachung [f]** (Verwaltungsanordnung, die den als Satzung beschlossenen und von der höheren Verwaltungsbehörde genehmigten Bebauungsplan ortsüblich bekannt macht und für jedermann zur Einsicht bereithält; im Rahmen der Bürgerbeteiligung werden in einem früheren Stadium auch Entwürfe bekannt gemacht oder können bekannt gemacht werden; ▸öffentlich bekannt machen [vb]); *syn.* Kundmachung [f] [CH].

4706 publication [n] **of planning proposals** *adm. leg.* (Legally defined requirement of planning authorities to make proposals available for inspection by the general public; publish [vb] planning proposals; ▸community development plan [US]/ urban development plan [UK], ▸objections and supporting representations, ▸open for public inspection [US]); *s* **sumisión [f] a información pública** (Exigencia legal de posibilitar a la población de informarse sobre un plan, programa o proyecto aprobado inicialmente. En Es., deben ser presentados durante un mes para dar oportunidad a la población de presentar sus ▸alegaciones, comentarios y proposiciones, a no ser que la **s. a i. p.** esté regulada a nivel de CC.AA.; cf. art. 41 [1] Ley del Suelo; ▸plan urbanístico, ▸someter a información pública); *f 1* **mise [f] à la disposition du public** (Procédure réglementaire permettant d'informer le public et à toute personne physique ou morale, à tout groupement de consigner ses ▸avis et observations, ses suggestions sur des projets d'intérêt public, p. ex. lors de l'élaboration des ▸documents de planification urbaine, etc. ; cf. loi n° 83-630 du 12 juillet 1983 ; ▸soumettre à enquête publique) ; *f 2* **enquête [f] publique** (**1.** Procédure réglementaire ayant pour objet d'informer le public [habitants, associations, acteurs économiques ou simple citoyen] et de recueillir ses appréciations, suggestions et contre-propositions sur un projet de règlement ou d'aménagement préparé et présenté par une collectivité publique ou privée ou par l'État. **2.** La réalisation d'ouvrages ou de travaux, exécutés par des personnes publiques ou privées, doit être précédée d'une enquête publique lorsqu'en raison de leur nature, de leur consistance ou du caractère des zones concernées, ces opérations sont susceptibles d'affecter l'environnement [Art. L 123-1, Code de l'Environnement]) ; **g Auslegung [f] eines Planentwurfes** (Rechtlich vorgeschriebene öffentliche Gewährung der Einsichtnahme in einen Plan [z. B. ▸Bauleitplan, Fernstraßenplan] für jedermann. Während dieser Zeit haben die Bürgerinnen und Bürger Gelegenheit Stellungnahmen/▸Anregungen [und Bedenken] während der Auslegungsfrist vorzubringen; cf. § 3 [2] BauGB; auslegen [vb]; ▸öffentlich auslegen [vb]); *syn.* öffentliche Auslegung [f], öffentliche Auflage [f] [A], *obs.* Offenlage [f].

4707 public authority [n] *adm.* (Any administrative department/agency of local, state or federal government; ▸public or semipublic authority [US]/public or semi-public authority [UK]); *syn.* public agency [n] [also US]; *s* **autoridad [f]** (Término que designa a toda persona jurídica del derecho público [gobiernos centrales, autónomos, diputaciones y municipios] que actúan como administradoras de los bienes públicos o como empresarios públicos; ▸autoridades e instituciones públicas y semipúblicas); *syn.* administración [f] pública; *f* **pouvoirs [mpl] publics** (Terme désignant la personne juridique de droit public de l'État et des collectivités territoriales, administrateurs des biens publics, donneur d'ordre ou autorité contractante ; ▸personnes publiques [associées à l'élaboration du plan]) ; *syn.* administration [f] publique ; **g öffentliche Hand [f]** (Bezeichnung für die Gesamtheit der juristischen Personen des öffentlichen Rechts [Bund, Länder und Gemeinden] als Verwalter des öffentlichen Vermögens, die als Auftraggeber oder „Unternehmer" [z. B. als Regiebetriebe] auftreten; ▸Träger öffentlicher Belange).

public authority [n] **or agency** [n] **with planning powers** [UK] *adm. plan.* ▸public planning body [US].

4708 public bath(s) [n(pl)] *urb.* *s* **baños [mpl] públicos** (Equipamiento público destinado a la higiene corporal); *f* **bains [mpl] publics** (Établissements de bains-douches implantés au cours du XIXème siècle dans la plupart des villes, permettant à tous d'accéder à l'hygiène corporelle ; avec la généralisation des bains privés dans les appartements individuels après la Deuxième Guerre Mondiale, on assistera à leur fermeture) ; *syn.* bains [mpl] populaires ; **g Badeanstalt [f] (1)** (Öffentliches Bad zur Körperpflege als es in den Wohnungen noch keine Badezimmer gab; als in der Mitte des 20. Jhs. Frei- und Hallenbäder errichtet wurden, wurden die vormals Volksbäder genannten **B.en** geschlossen); *syn. obs.* Volksbad [n].

public cleansing and waste disposal [n] [UK] *envir.* ▸waste treatment and disposal [US]/public cleansing and waste disposal [UK].

public client [n] *adm. contr.* ▸contracting public agency [US].

4709 public construction project [n] *plan.* (Generic term covering a municipal, State or federal project); *s* **proyecto [m] de obras públicas**; *f* **opération [f] immobilière du secteur public** (Projet immobilier [production de logements] de l'État et des collectivités territoriales) ; **g öffentliches Bauvorhaben [n]** (OB zu Bauprojekt der öffentlichen Hand).

P

public contract manager [n] [US] *constr. prof.* ▶site supervisor representing an authority.

public enquiry [n] [UK] *adm. leg. plan.* ▶public hearing [US].

4710 public facility [n] (1) *urb.* (**1.** Public building and utilities shown on a ▶comprehensive plan [US]/local development framework [LDF] [UK], as well as on a ▶zoning map [US]/Proposals Map/Site Allocations Development Plan Document [UK]; ▶public facility 2. **2.** Installation which serves to provide the population of an area with services as part of the ▶public infrastructure); *syn.* community facility [n] [also US]; *s 1* **instalación [f] de equipamiento comunitario** (Construcción existente para usos de interés general [educativos, sociales, religiosos, deportivos, etc.] o para la cual se reserva suelo en el ▶plan general municipal de ordenación urbana o en ▶plan parcial [de ordenación]; ▶instalación de infraestructura); *s 2* **instalación [f] técnica de abastecimiento** (Infraestructura de suministro de agua, luz, gas, etc. que es parte de la ▶infraestructura de abastecimiento público); *f 1* **installation [f] d'intérêt général** (**1.** Terme d'urbanisme qualifiant un équipement ou une installation d'intérêt public et dont l'affectation est réglementée par les documents d'urbanisme. **2.** Équipement prévu pour la satisfaction des besoins matériels et immatériels de la population résidente et des entreprises d'un espace déterminé ; C. urb., art. L. 123-1, R123-1 ; ▶équipements collectifs) ; *f 2* **équipement [m] collectif** (Bâtiments et installations qui permettent d'assurer à une population les services collectifs dont elle a besoin dans les domaines sociaux, médicaux culturels, de transport, de loisirs ; dans le ▶plan local d'urbanisme [PLU]/plan d'occupation des sols [POS] certaines zones peuvent être affectées à la fonction spécifique d'espaces réservés aux *équipements collectifs*) ; *f 3* **équipement [m] urbain d'intérêt public** (Installation à caractère social, culturel, sportif, récréatif, etc., contribuant à la satisfaction des besoins de la collectivité urbaine) ; *g 1* **Gemeinbedarfseinrichtung [f]** (Der Allgemeinheit dienende bauliche Anlage oder Einrichtung, die nach Art und Flächenbedarf im Flächennutzungsplan gem. § 5 [2] 2 BauGB und als Fläche für den Gemeinbedarf im ▶Bebauungsplan gem. § 9 [1] 5 BauGB festzusetzen ist); *syn.* Einrichtung [f] für den Gemeinbedarf, Einrichtung [f] des Gemeinbedarfs, Gemeinbedarfsanlage [f]; *g 2* **Versorgungseinrichtung [f]** (Einrichtung, die die Versorgung der Bevölkerung eines Raumes mit materiellen oder immateriellen Gütern vorsieht — Teil der ▶Versorgungsinfrastruktur; ▶Infrastruktureinrichtung).

4711 public facility [n] (2) *plan.* (Generic term covering public buildings, facilities and utilities as basic element in the infrastructure of a society that ensures and promotes productivity increases of the region it serves, as well as cultural and social security of the population; ▶infrastructure, ▶planning for the public welfare, ▶communal facility); *s* **instalación [f] de infraestructura** (▶infraestructura, ▶planificación en beneficio público, ▶instalación colectiva); *syn.* equipamiento [m] de infraestructura; *f 1* **équipement [m] d'infrastructure** (Réseaux et aménagements au sol ou en sous-sol [voiries et stationnement, transports et communications, eaux et canalisations, énergie, espaces collectifs aménagés [parcs, jardins, cimetières, terrains de sport] ; ▶infrastructure, ▶installation collective, ▶solidarité nationale) ; *f 2* **équipement [m] de superstructure** (bâtiments à usage collectif [administratifs, éducatifs, sanitaires commerciaux, culturels, sportifs, etc.] ; ▶superstructure ; *terme spécifique* services publics) ; *g* **Infrastruktureinrichtung [f]** (Jede öffentliche Anlage, die der Sicherung resp. der Erhöhung der Produktivität in einem Raum sowie der sozialen Sicherheit der Bevölkerung dieses Bereiches dient; ▶Infrastruktur, ▶Daseinsvorsorge, ▶Gemeinschaftsanlage 1); *syn.* Infrastrukturanlage [f].

4712 public funds [npl] *adm. econ.* (Monetary resources in a public budget, which are expended by a ▶public authority); *s* **fondos [mpl] públicos** (Recursos monetarios procedentes del presupuesto de una ▶autoridad); *f* **fonds [mpl] publics** (Mis à disposition par les ▶pouvoirs publics dans le cadre du budget) ; *g* **öffentliche Mittel [npl]** (Gelder aus einem öffentlichen Haushalt, das die ▶öffentliche Hand bereitstellt, nachdem sie von den entsprechenden parlamentarischen Gremien beschlossen/bewilligt wurden); *syn.* öffentliche Gelder [npl].

public garden [n] [UK] *urb.* ▶inner city park [US].

4713 public grant [n] *econ.* (Financial allocation from the public purse which has not to be payed back. The generic term is **subsidy** and includes e.g. public grant and tax credits); *s* **contribución [f] financiera pública** (Aporte financiero de una autoridad pública a un organismo privado); *syn.* ayuda [f] financiera pública; *f* **contribution [f] financière des pouvoirs publics** (Allocation financière attribuée par les pouvoirs publics pour le financement de mesures permettant p. ex. des acquisitions foncières et immobilières, la construction de logements sociaux, etc.) ; *g* **Zuschuss [m] der öffentlichen Hand** (Von der öffentlichen Hand gewährter Betrag, um die Finanzierung von privaten Bauprojekten, Forschungs- und Entwicklungsvorhaben etc. zu fördern. Dieser Zuschuss muss nicht zurückbezahlt werden); *syn.* nicht rückzahlbarer öffentlicher Zuschuss [m], finanzieller Beitrag [m] der öffentlichen Hand.

public green area [n] *leg. urb.* ▶public green space.

4714 public green space [n] *leg. urb.* (Open space for public use, such as park, playground, trail, other recreational area, cemetery, right-of-way planting [US]/roadside planting [UK], etc.; ▶green space, ▶public use; *generic term* public area); *syn.* public green area [n]; *s* **parques [mpl] y jardines [mpl] públicos** (Conjunto de ▶zonas verdes de ▶uso público, como parques, jardines, áreas de juego y recreo de niños, etc.); *f* **espace [m] vert public** (En zone urbaine et périurbaine les milieux non bâtis ouverts au public et appartenant aux collectivités territoriales tels que jardins, parcs et forêts, promenades, places et squares, espaces verts liés aux infrastructures routières, ferroviaires et navigables, zones récréatives de loisirs et de détente, terrains de sport, cimetières, espaces naturels, etc. ; ▶espace vert, ▶usage public) ; *g* **öffentliche Grünfläche [f]** (▶Grünfläche incl. der darin befindlichen Wege und Plätze, Anpflanzungen und Einrichtungen, die der Gemeinde, dem Land oder dem Bund gehört und dem ▶Gemeingebrauch dient; die Summe aller öffentlichen Grünflächen ist das **öffentliche Grün**); *syn.* öffentliche Grünanlage [f].

4715 public hearing [n] [US] *adm. leg. plan.* (Statutory possibility for persons affected by planning measures/proposals making comments, objections and suggestions to community development plans and functional/sectoral plans; **in U.S.**, the term has various meanings. At one extreme, **p. h.** is a generic term describing the formal or informal opportunities guaranteed to all citizens to comment publicly regarding any activity or action by an official body in accordance with state and federal Constitutions giving all citizens the right to petition their governments for a "redress of grievances". More narrowly, a **p. h.** may be required evidentiary or quasi-judicial procedure and similar to the formal "public inquiry" in British planning legislation. Both types of open meeting require public notice of time, place, etc.; **in U.K.**, statutory requirement for the public to be given opportunity to voice their comments on planning proposals; according to U.K. planning law, established by the Planning and Compulsory Purchase Act 2004, Statements of Community Involvement are required to be produced by Local Authorities to explain to the public how they will be involved in the preparation of Local Development Documents. They are intended to set out the standards to be met for community involvement, based upon the minimum requirements laid down by the regulations and planning policy statements. Consultation with the public begins at the earliest stages of each document's development so that communities are given the fullest opportunity to participate in planning; ▶design and location approval process [US]/determination process of a plan [UK], ▶public participation, ▶submission of planning comments); *syn.* public inquiry [n] [UK], *o.v.* public enquiry [n] [UK]; **s audiencia [f] pública** (Reunión abierta al público en general a través de la cual se implementa el derecho legal del público y de las personas interesadas [físicas y jurídicas] de presentar alegaciones y sugerencias en el marco de la planificación urbana y ambiental; en Es. el derecho básico a la participación está legislado a nivel estatal a través de la ley 6/1998, sobre el régimen del suelo y valoraciones y de la ley 27/2006 [...de participación pública...en materia de medio ambiente]; la regulación específica de los procedimientos es competencia de las CC.AA.; cf. art. 6 Ley 6/1998, Titulo III Ley 27/2006; ▶entrega de comentarios o proposiciones, ▶participación pública en la planificación, ▶procedimiento de aprobación de proyectos públicos); *f 1* **enquête publique** (F., 1. Instrument légal de participation ayant pour but d'informer le public et de recueillir ses appréciations, suggestions et contre-propositions [cf. loi du 12 juillet 1983]. 2. *Intervention des associations* les associations locales d'usagers agréés peuvent être consultées sur demande pour l'élaboration d'un PLU/POS — C. urb., art. R. 123-4. Les délégués de groupement représentatifs peuvent être consultés sur demande pour l'élaboration des schémas de cohérence territoriale [SCOT]/schémas directeur d'aménagement et d'urbanisme [SDAU] — C. urb., art. R. 122-9 ; ▶participation du public, ▶présentation d'un rapport, ▶procédure d'instruction de grands projets publics) ; *f 2* **consultation [f] du public** (Instrument de participation par lequel les citoyens sont associés dans un processus de prise de décision ; chacun peut participer à la « consultation du public » en prenant connaissance d'un document et faire part de ses observations sur un registre ; la consultation se déroule dans les préfectures, les sous-préfectures, au siège des administrations responsables du projet, du plan, du programme, etc. ; cf. p. ex. directive cadre sur l'eau, consultation du public sur le programme opérationnel « compétitivité » et son évaluation environnementale ; consultation du public sur Internet sur les mesures à prendre pour améliorer l'environnement urbain en Europe ; *contexte* document soumis à consultation du public) ; *g* **öffentliche Anhörung [f]** (Gesetzlich verankerte Gelegenheit für Planungsbetroffene zur Äußerung und zur Erörterung von Bauleitplänen [§ 3 BauGB: Beteiligung der Öffentlichkeit] und Fachplanungen, z. B. gemäß § 18 FStrG; ▶Abgabe einer Stellungnahme, ▶Beteiligung der Öffentlichkeit, ▶Planfeststellungsverfahren); *syn.* Beteiligung [f] der Öffentlichkeit (1), Öffentlichkeitsbeteiligung [f]).

public hygiene department [n] *adm. envir.* ▶waste management department.

4716 public infrastructure [n] *plan.* (Entire range of installations which serve to provide the population of an area with goods and services; ▶infrastructure, ▶public facility 2); **s infraestructura [f] de abastecimiento público** (Conjunto de equipamientos e instalaciones necesarias para abastecer a la población de un área de bienes y servicios; *término específico* servicios [mpl] subterráneos; ▶infraestructura, ▶instalación técnica de abastecimiento, ▶instalación de equipamiento comunitario); *syn.* infraestructura [f] pública; *f* **équipements [mpl] collectifs** (Ensemble des équipements [▶infrastructure et ▶superstructure] assurant la satisfaction des besoins de la population résidente et des entreprises en biens matériels et immatériels ; C. urb., art. L.

123-1, R 123-1 ; ▶équipement collectif, ▶installation d'intérêt général) ; *syn.* infrastructure [f] publique [CDN] ; *g* **Versorgungsinfrastruktur [f]** (Gesamtheit der Einrichtungen, die die Versorgung der Bevölkerung eines Raumes mit materiellen und immateriellen Gütern gewährleistet; ▶Infrastruktur, ▶Versorgungseinrichtung).

public inspection [n] *adm. leg.* ▶open for public inspection [US]/make available for public inspection [UK].

public inspection [n] [UK], **make available for** *adm. leg.* ▶open for public inspection [US].

public interest [n] *ecol. leg. plan.* ▶imperative reasons of overriding public interest.

public involvement [n] [US] *leg. plan. pol.* ▶public participation.

publicly assisted housing [n] [UK] *sociol. plan.* ▶low-income housing.

public notice [n] *adm. leg. plan.* ▶public announcement.

4717 public open space [n] *plan. sociol.* (Publicly-used ▶open space; all exterior urban spaces—with or without vegetation—which are elementary components of social infrastructure of a city or town which are usable for communication and recreation, and therefore an essential part for the public welfare. **P. o. s.** becomes more and more a place for closer contact with culture and nature; **in U.S.**, emphasis is on aesthetic and recreation objectives in addition to social purposes); **s espacio [m] libre público** (Área no edificada en zonas urbana, determinada o no por vegetación, cuya función principal es servir para el esparcimiento y recreo de la población; ▶espacio libre); *f* **espace [m] libre public** (Espace extérieur ouvert en zone urbaine réservé aux activités de loisirs et de détente ; ▶espace libre) ; *syn.* espace [m] public ; *g* **öffentlich nutzbarer Freiraum [m]** (Alle nicht überdachten, vegetations- oder nicht vegetationsbestimmten Flächen als elementarer Bestandteil der sozialen Infrastruktur einer Stadt, die für Freizeit und Erholung nutzbar sind — und somit wesentliche Aufgaben der Daseinsvorsorge übernehmen. Öffentlich nutzbarer Freiraum wird immer mehr zum Ort der Wiederannäherung von Kultur und Natur, in dem sich städtische Individuen unterschiedlichster Herkunft gesellschaftlich einbetten können; dieser Begriff klärt nicht den eigentumsrechtlichen Status vor Ort, da oft nicht ablesbar; insofern ist die Verwendung des Terminus **öffentlicher Freiraum** in diesem Zusammenhang etwas ungenau; ▶Freiraum); *syn.* öffentlich zugänglicher Freiraum [m], öffentlicher Freiraum [m] *z. T.*

4718 public or semipublic authority [n] [US]/**public or semi-public authority** [n] [UK] *adm. plan.* (Body responsible for public works, e.g. an associated company in a city's administration, which participates in the planning or execution of the project); *syn.* public or semi-public authority [n] [UK]; **s autoridades [fpl] e instituciones [fpl] públicas y semipúblicas** (Órganos de la administración pública o instituciones semipúblicas que son consultadas en la planificación); *f* **personnes [fpl] publiques (associées à l'élaboration du plan)** (Personnes, services, commissions, associations, intéressés par un projet ou associées à l'élaboration d'un plan) ; *g* **Träger [m] öffentlicher Belange** (Träger [pl]; Behörde, öffentliches Unternehmen/öffentliche Beteiligungsgesellschaft einer Stadtverwaltung und/oder gesellschaftliche Institution, die an einem Planungsvorhaben beteiligt werden oder nach gesetzlichen Vorschriften zu beteiligen sind); *syn.* öffentliche Stelle [f].

4719 public park [n] *gard'hist. recr. urb.* (**1.** Extensive urban park with the appearance of a 'natural' landscape and in Europe created after the Industrial Revolution for the rising population of industrial workers and was provided with facilities to meet the demands of the new industrial cities; e.g. Englischer Garten in Munich, Regent's Park in London, Stadtpark and Volkspark in Hamburg, Central Park in New York City, designed in 1868 by Calvert Vaux and Frederick Law Olmsted, founder of the landscape architectural profession in the U.S., who had proposed state and national park systems in 1865; F. L. Olmsted wrote a pamphlet "Public Parks and the Enlargement of Towns" in 1870; **in Europe**, **p. p.s** can be divided into three main types: 1. Open spaces made accessible as public promenades, usually by royal or aristocratic grace and favo[u]r. **2.** The *Volksgarten*, a peculiarly German development; the first, and most influential statement of the idea of a *Volksgarten* is in C. C. L. Hirschfeld's *"Theorie der Gartenkunst"* [5 vol. Leipzig 1779-1785]; cf. JEL 1986, 459; **3.** The **p. p.** proper is a dedicated planted green space set aside for a wide variety of public recreation uses including strolling paths, concert facilities, large recreation fields, etc.; a **p. p.** belongs to the public as of right; ▸park 1, ▸urban park); **s parque [m] público** (Parque generalmente de tipo inglés y de bastante extensión creado después de la revolución industrial en los países del centro de Europa y en Gran Bretaña, p. ej. Regentspark en Londres, Stadtpark y Volkspark en Hamburgo, Tiergarten en Berlín, Englischer Garten en Munich, Casa de Campo en Madrid; ▸parque, ▸parque urbano); **f parc [m] public** (Avec la révolution industrielle pendant la seconde moitié du XIXème siècle verra s'accroître un nombre important des jardins et parcs publics, grand îlot de nature à trame paysagère dans un milieu urbain, lieu de promenade publique, p. ex. le jardin anglais à Munich, le Central Park à New York, le parc de Cristal Palace à Londres, les jardins du Trocadéro à Paris, le parc de la Tête d'Or à Lyon; LEU 1987 ; ▸parc, ▸parc central, ▸parc urbain) ; *syn.* jardin [m] public ; **g Volkspark [m]** (Meist landschaftlich gestaltete, großflächige Parkanlage, in Europa nach Beginn der Industrialisierung entstanden, z. B. der Englische Garten in München, Winterhuder und Harburger Stadtpark in Hamburg, Altonaer Volkspark, Tiergarten in Berlin, Central Park in New York, Jardins du Trocadéro in Paris, Parc de la Tête d'Or in Lyon; die erste einflussreiche Abhandlung über das Konzept des **V.s** erfolgte durch C. C. L. Hirschfeld in seiner „Theorie der Gartenkunst" [5 Bde. Leipzig 1779-1785]. 1874 veröffentlichte GRÄFIN DOHNA-PONINSKI unter dem Pseudonym „Arminius" das Buch „Die Großstädte in ihrer Wohnungsnoth und die Grundlagen einer tiefgreifenden Abhilfe" und bereitete damit intellektuell die Anlage von **V.s** vor, um die durch die Industrialisierung bedingten sozialen und gesundheitlichen Probleme zu mindern. Im Gegensatz zu den bürgerlichen Parks und Anlagen des 19. Jhs., die in erster Linie dem Spaziergang im Freien dienten, sollten die **V.anlagen** für alle Bevölkerungsschichten auf unterschiedliche Weise für Spiel, Sport, Lagern, Plantschen, Reiten, ferner für Musikgenuss mit Tanz, Betrachten von Kunst, Anschauen von schönen Blumenbeeten nutzbar sein; ▸Park, ▸Stadtpark); *syn.* Volksgarten [m].

4720 public participation [n] *leg. plan. pol.* (Involvement of the public as individuals or organized groups in decisions taken as part of the planning process, sometimes on the basis of legal provisions in the planning legislation; ▸public hearing [US]/public inquiry [UK], ▸objections and supporting representations); *syn.* citizen participation [n] [also US], community involvement [n], public involvement [n] [also US]; **s participación [f] pública en la planificación** (▸Audiencia pública, ▸alegaciones, comentarios y proposiciones); *syn.* cogestión [f] de la planificación, intervención [f] ciudadana en los procesos de planificación y en la gestión del planeamiento; **f participation [f] du public** (Information et discussion publique des orientations générales au cours de la préparation des documents de planification ou du choix de projets d'aménagement d'intérêt public ;

pendant la procédure d'▸enquête publique toute personne physique ou morale, toute association a la possibilité d'intervenir et de consigner ses appréciations, suggestions et contre-propositions ; ▸consultation du public, ▸avis et observations) ; *syn.* concertation [f] avec le public, participation [f] de la population ; **g Beteiligung [f] der Öffentlichkeit (2)** [Gesetzlich verankerte öffentliche Darlegung und Diskussion der allgemeinen Ziele und Zwecke von Raum beanspruchenden Planungen während eines Planungsablaufes. Den Bürgern und Bürgerinnen ist nach einem förmlich vorgeschriebenen Verfahren die Gelegenheit zur Äußerung und Erörterung [▸öffentliche Anhörung] zu geben, z. B. gem. § 3 BauGB, § 18 FStrG; ▸Anregungen [und Bedenken]); *syn.* Beteiligung [f] der Bürger, Bürgerbeteiligung [f], Partizipation [f] der Bürger, Öffentlichkeitsbeteiligung [f] (§ 59 BNatSchG).

4721 public planning body [n] [US] *adm. plan.* (Public agency/authority which has responsibility for planning projects; ▸client); *syn.* public authority [n] or agency [n] with planning powers [UK]; **s entidad [f] administrativa responsable de la planificación** (Ministerio, consejería o entidad local responsable de llevar a cabo el planeamiento territorial o de proyectos; ▸propiedad de obra); **f établissement [m] public d'aménagement (EPA)** (Organisme d'État menant une activité d'intérêt public à caractère industriel et commercial dont l'objet social, l'organisation statutaire et les modalités de fonctionnement sont régis par le Code de l'urbanisme. Il est compétent pour réaliser pour son compte, celui de l'État, d'une collectivité locale ou d'un autre établissement public, ou faire réaliser toutes interventions foncières et opérations d'aménagement prévues par ce Code ; DUA 1996, 329 ; ▸maître d'ouvrage) ; **g Planungsträger [m]** (Eine Behörde, Gebietskörperschaft oder eine Stelle, die in eigener Verantwortung für ihren Zuständigkeitsbereich Planungen durchführt; ▸Bauherr/in).

4722 public prequalification bidding notice [n] [US] *contr.* (Public announcement inviting especially qualified construction firms to submit evidence of their qualifications to perform certain work [US]/works [UK] and take part in a ▸selective bidding [US]/selective tendering [UK]); *syn.* public prequalification tendering notice [n] [UK]; **s aviso [m] de candidaturas a concurso público** (Anuncio público para que aquellos gabinetes especialmente cualificados se presenten a ▸concurso-subasta restringido); **f appel [m] public de candidatures pour un appel d'offres** (Recherche sélective de candidatures, au moyen d'une publicité préalable identique à celle de l'appel d'offres ouvert d'entreprises qualifiées possédant les références suffisantes en vue de la réalisation d'ouvrages. Procédure précédant l'▸appel d'offres restreint ; cf. CCM 1984) ; **g öffentlicher Teilnahmewettbewerb [m]** (Öffentlicher Wettkampf zwischen voneinander unabhängigen Unternehmen/Firmen, zu bestimmten Aufgabenstellungen die erforderliche Qualifikation und Eignung darzulegen. Die ausschreibende Stelle führt anschließend nach Auswertung aller eingegangener Unterlagen mit einem reduzierten Bewerberkreis besonders qualifizierter Firmen eine ▸beschränkte Ausschreibung 2 durch; cf. § 3 [3] 2 VOB Teil A).

public prequalification tendering notice [n] [UK] *contr.* ▸public prequalification bidding notice [US].

4723 public responsibility [n] *leg.* (Legal accountability and duty of a public or private land owner to act in such a way that bodily harm, property damage, or violation of other rights is not caused; in D., the standard of due care owed by every citizen to do no harm to the person or property of others, intentionally or unintentionally, upon pain of making or giving restitution for the resulting damage, regardless of whether the right is established by statute, as in Germany, or by common law in U.K. and U.S.;

▶public security); *syn.* obligation/duty [n] of occupier to make land or premises safe for persons or vehicles; **s responsabilidad [f] civil (de garantizar la seguridad pública)** (Según el derecho civil obligación legal de evitar daños contra la vida o la salud de las personas, contra bienes y de no perjudicar cualquier otro derecho sea intencional o involuntariamente por parte de toda autoridad, institución, empresa o persona física o jurídica que produzca algún tipo de fuente de peligro, como p. ej. zanjas en la calle, obras en tejados, árboles con peligro de rotura de ramas, etc.; ▶seguridad pública); *f* **garantie [f] de la sécurité publique** (Responsabilité de la personne publique assurant la protection des personnes et des biens ; le maire peut p. ex. prescrire par arrêté le soutènement, l'élagage ou l'abattage de tout arbre menaçant de s'effondrer, de compromettre la ▶sécurité publique ou n'offrant pas les garanties de solidité nécessaires au maintien de la sécurité publique) ; *g* **Verkehrssicherungspflicht [f]** (Der allgemeinen Haftungsregelung nach § 832 [1] BGB unterliegende Rechtspflicht, das Leben, den Körper, die Gesundheit, das Eigentum oder ein sonstiges Recht eines anderen nicht widerrechtlich zu verletzen. „Der **V.** ist genügt, wenn die nach dem jeweiligen Stand der Erfahrungen und Technik als geeignet und genügend erscheinende Sicherungen getroffen sind, also den Gefahren vorbeugend Rechnung getragen wird, die nach Einsicht eines besonnenen, verständigen und gewissenhaften Menschen erkennbar sind". Weiter heißt es im Urteil in Bezug auf Straßenbäume: „Eine schuldhafte Verletzung der **V.** liegt in solchen Fällen vor, wenn Anzeichen verkannt oder übersehen worden sind, die nach der Erfahrung auf eine weitere Gefahr durch den Baum hinweisen"; Urteil BGH v. 21.01.1965 — III ZR 217/63 [Koblenz], *in* NJW 1965, H. 18, 815. Die **V.** beruht auf dem Gedanken, dass jeder, der Gefahrenquellen schafft, die notwendigen Vorkehrungen zum Schutze Dritter treffen muss. Die **V.** ist nicht darauf angelegt, jeden Unfall auszuschließen. Es sind diejenigen Sicherheitsvorkehrungen zu treffen, die nach den Sicherheitserwartungen des jeweiligen Verkehrs im Rahmen des wirtschaftlich Zumutbaren geeignet sind, Gefahren von Dritten tunlichst abzuwehren; ▶Verkehrssicherheit).

4724 public responsibility principle [n] *envir. pol.* (Policy holding that an activity detrimental to the public interest and well-being is corrected by the perpetrator or with public funds, e.g. environmental pollution; ▶polluter-pays principle); **s principio [m] de cooperación** (Principio según el cual la sociedad en su totalidad o parte de ella tiene que responsabilizarse de los daños del medio ambiente, en los casos en que el principio de causalidad no funciona; *opp.* ▶principio de causalidad); *f* **principe [m] de responsabilité publique (de l'État et des collectivités publiques)** (F., principe selon lequel la collectivité publique peut être rendue responsable **1.** de dommages causés par la pollution provenant de particuliers ou d'entreprises lorsque celle-ci n'a pas usé des pouvoirs que lui donne la loi pour la prévenir ou la réprimer [inaction ou carence des autorités administratives]. **2.** Des dégâts causés par des espèces animales protégées sur les cultures ; **D.,** principe selon lequel la résorption des dommages causés par la pollution est prise en charge par la collectivité suite à la défaillance ou de la disparition du responsable de cette pollution ; *opp.* ▶principe pollueur-payeur) ; *g* **Gemeinlastprinzip [n, o. Pl.]** (Politische Einstellung, bei der sichergestellt ist, dass die Folgen z. B. von Umweltschädigungen und deren Beseitigung durch die Allgemeinheit getragen werden, wenn der Verursacher nicht festgestellt werden kann; *opp.* ▶Verursacherprinzip); *syn.* Kooperationsprinzip [n].

4725 public right-of-way [n] *adm. leg. trans.* (Legal term covering designation and use of public streets, walks, and squares; *not to be confused with* the right-of-way in 'traffic regulations'; the parcel of land on which a **p. r. of w.** exists may

be privately owned, but another party or the general public has the legal right to cross the land. Legal term for **p. r. of w.** is expressed in a deed as an easement. **2. In U.K.,** a **p. r. of w.** is determined by law; a network of trails, footpaths, bridleways, byways and roads over 190,000 km in length extends throughout Great Britain. **3.** In urban planning, a strip of land granted and reserved for a rail line, highway, or other transportation facility is known as a right-of-way [*abbr.* ROW]; ▶vehicular and pedestrian easement); **s derecho [m] vial** (Conjunto de normas legales que regulan la construcción, el funcionamiento y el uso de carreteras, calles, caminos y plazas públicas, que no se debe confundir con el código de circulación; ▶servidumbre de acceso); *f* **code [m] de la voirie routière** (Ensemble des prescriptions ayant pour objet les servitudes foncières en bordure de la voie publique ; ▶servitude de passage) ; *g* **Straßen- und Wegerecht [n]** (**1.** Gesamtheit der Vorschriften, die die Rechtsverhältnisse an öffentlichen Straßen, Wegen und Plätzen regeln; *nicht zu verwechseln mit dem* ‚Straßenverkehrsrecht'; ▶Geh- und Fahrrecht. **2. Im U.K.** gibt es die *Public Rights of Way*, ein durch verschiedene Gesetze festgeschriebenes Wegerecht über 190 000 km Gesamtlänge, das sich als feingliedriges Netz von Wanderwegen über ganz Großbritannien erstreckt).

public safety [n] *arb. constr.* ▶crown pruning for public safety.

4726 public security [n] *leg.* (Safety of users and protection of materials on a building site, in traffic, on private property, etc. by provision of security measures to prevent accidents or damage; ▶public responsibility, ▶road safety); **s seguridad [f] pública** (Seguridad para público y bienes en instalaciones, obras, en el espacio público para evitar accidentes o daños a terceros, por medio de medidas de seguridad, como vallas, señalizaciones, cuidado de los árboles para que no causen daños por rotura de ramas, etc.; ▶seguridad del tráfico, ▶responsabilidad civil [de garantizar la seguridad pública]); *f* **sécurité [f] publique** (Absence de risques pour les usagers et les biens à l'intérieur d'un bâtiment, sur un chantier et sur les circulations ainsi que les mesures de préventions des accidents, des désordres et des dommages dans ces domaines ; ▶sécurité routière, ▶garantie de la sécurité publique) ; *syn.* sécurité [f] des personnes et des biens ; *g* **Verkehrssicherheit [f]** (Sicherheit für Nutzer und Einrichtungen oder Ausstattungsgegenstände in einer Anlage, auf einer Baustelle, im Verkehr etc. durch getroffene Sicherungsvorkehrungen zur Vermeidung von vorhersehbaren Unfällen oder Sachbeschädigungen; ▶Straßenverkehrssicherheit, ▶Verkehrssicherungspflicht).

4727 public space [n] *sociol. urb.* (An open urban area, freely accessible to everybody, which forms a structural element in a city [squares, parks, roads], and is a characteristic part of a system of urban spaces. The design of **p. s.s** and the multifunctionality of their use [predominantly as meeting places] often determine the image of a city and are of significant economic importance as a location factor. There is no uniform understanding of the concept of a **p. s.s** The term **p. s.s** is strongly influenced in theory and practice by a background of social-interaction and various interests; urban planners often refer to **p. s.** as being enclosed: open space planners refer to it as being green, e.g. a park; the operators and managers of shopping malls see their properties as parts of public space, because the demarcation between private and public is not always detectable for customers; cf. BBR 2003, 125 ss); **s espacio [m] público** (Todos los espacios de la ciudad, accesibles sin restricciones para toda la población, que como elementos de articulación [plazas, parques, calles] en su totalidad hacen visible y experimentable a la ciudad misma. El aspecto de los espacios públicos y de la vida que trascurre en ellos determinan con su multifuncionalidad de usos en gran medida la imagen de la ciudad y tienen un significado económico considerable como factor de atractivo. No existe una comprensión única sobre este término. Tanto en la bibliografía como en la práctica está influenciado por el contexto de actuación resp. la perspectiva desde la cual se analizan. Así en el urbanismo por **e. p.** se entienden principalmente espacios situados en áreas edificadas, mientras que la planificación de espacios libres se refiere más a parques y zonas verdes); *f* **espace [m] public** (Espace de vie collective, de pratiques sociales et culturelles, accessible à tous, constitué d'éléments structurant [places, rues, parcs, etc.] dont la lisibilité et l'imagibilité permettent de les intégrer dans l'ensemble cohérent du système de l'espace urbain ; l'aspect physique et la multifonctionnalité des usages [lieu de rencontre], considérés comme facteurs de qualité, ont au cours des dernières décennies accentué l'importance économiques de ces espaces ; il n'existe pas de large consensus dans l'appréhension du concept d'espaces publics ; dans la littérature comme dans la pratique il est fortement empreint du contexte des actions envisagées et des intérêts qui sous-tendent les analyses ; les urbanistes le considèrent plutôt comme l'espace entourant le bâti tandis que les paysagistes y voient les parcs et espaces verts, quant aux gérants de centres commerciaux, un espace semi-public, la différence entre le « privé » et le « public » n'étant pas toujours perceptible par la clientèle) ; *g* **öffentlicher Raum [m]** (Für jedermann frei zugänglicher Raum als Gliederungselement in der Stadt (Plätze, Parks, Straßen), das als Zusammenhang oder „System" den Stadtraum lesbar und erlebbar macht. Die Gestalt der öffentlichen Räume und das in ihnen stattfindende Leben prägen mit ihrer Multifunktionalität der Nutzung [vorwiegend Begegnungsfunktion] wesentlich das Bild der Stadt und haben eine nicht unwesentliche ökonomische Bedeutung als Standortfaktor. Es gibt kein einheitliches Verständnis über diesen Begriff. Er ist in Literatur und Praxis stark vom Handlungshintergrund resp. Analyseinteresse geprägt: So verstehen Stadtplaner oft eher den umbauten Raum, während Freiraumplaner Parks und Grünräume meinen; Betreiber und Manager privater Einkaufszentren sehen ihre Objekte als Teile des öffentlichen Raumes, da für Kunden die Abgrenzung zwischen privat und öffentlich nicht immer wahrnehmbar ist; cf. BBR 2003, 125ff).

4728 public spirited attitude [n] *pol. sociol.* (Social, intellectual and moral approach to program[me]s for public benefit); *s* **utilidad [f] pública** (Estatus legal especial para organizaciones e instituciones que desarrollan actividades sociales, culturales, científicas o benéficas de interés público y cumplen ciertos requisitos fijados por ley); *syn.* estatus [m] sin fines de lucro, bien [m] público [RA]; *f* **statut [m] à but non lucratif** (Activité statutaire exercée dans les domaines matériel, intellectuel ou moral sans but lucratif) ; *g* **Gemeinnützigkeit [f]** (Dem Gemeinwohl verpflichtete Verhaltensweise, die sich auf eine profitlose Förderung auf materiellem, geistigem oder sittlichem Gebiet richtet).

public square [n] *urb.* ▶urban square.

4729 public supply utility line [n] *trans. urb.* (Generic term for conductor of, e.g. gas, water, electricity, and telecommunication. In English, the generic term **public utility line** covers also sewage disposal pipes; ▶public utility network); *s* **conducción [f] de abastecimiento** (Conducto de transporte de agua, gas, electricidad, telecomunicación hasta el punto de consumo; ▶red de abastecimiento público); *syn.* conducto [m] de suministro); *f* **conduite [f] d'alimentation (2)** (Tout tuyau transportant l'eau ou le gaz jusqu'à un point d'utilisation ; cf. DTB 1985 ; ▶réseaux [d'alimentation]) ; *g* **Versorgungsleitung [f] (2)** (Leitung, in der z. B. Fernwärme, Gas, Wasser, Strom, Telekommunikation bis zur Gebrauchsstelle transportiert wird; ▶Versorgungsnetz); *syn.* Sparte [f].

public swimming baths [npl] [UK] *recr.* ▶public swimming pool.

4730 public swimming pool [n] *recr. syn.* public swimming baths [npl] [also UK]; *s* **piscina [f] pública** (Instalación pública de natación con los correspondientes servicios); *f* **piscine [f] publique** (Bassin de natation couvert ou non et installations attenantes apparus à partir du XIXème siècle pour la pratique de la natation) ; *g* **Badeanstalt [f] (2)** (Öffentliches Schwimmbad als Hallenbad oder im Freien; *syn.* öffentliches Schwimmbad [n].

4731 public transit [n] [US] *trans.* (Movement of people by means of public conveyance, such as bus, tram, railway, taxicab; ▶local public transportation system [US]/local transport system [UK]; *opp.* ▶private transport); *syn.* public transport [n] [UK], public transportation [n]; *s* **transporte [m] público** (Servicio de traslado de personas en vehículos públicos como autobuses, trenes, metros, taxis, etc.; ▶transporte público urbano; *opp.* ▶transporte individual); *f* **transports [mpl] publics** (Service routier ou ferroviaire régulier de transport de voyageurs soumis à la réglementation générale des transports tels que les trains, autocars, bus, métro, tramways ; *opp.* ▶transport individuel ; ▶transports suburbains) ; *syn.* transports [mpl] collectifs ; *g* **öffentlicher Verkehr [m]** (Fortbewegung von Personen mit öffentlichen Verkehrsmitteln wie z. B. Bus, Straßenbahn, Eisenbahn, Taxi; ▶öffentlicher Personennahverkehr; *opp.* ▶Individualverkehr).

4732 public transit shelter [n] [US] *trans.* (Roofed facility for weather protection at bus, tram or taxi stops; ▶pedestrian shelter, ▶shelter); *syn.* public transport shelter [n] [UK]; *s* **garita [f] de espera** (▶Marquesina 2 en parada de autobús o tranvía para proteger a los viajeros contra la lluvia; *término genérico* ▶refugio); *f 1* **abri [m]** (Terme générique pour une construction rudimentaire ouverte [▶mobilier de protection contre les intempéries] destinée à protéger le voyageur aux arrêts de train, d'autobus, de taxi, etc. ; ▶abri couvert) ; *f 2* **abribus [m]** (Terme spécifique pour un abri à l'arrêt d'autobus) ; *syn.* aubette [f] [B, Ouest de la F.] ; *g* **Wartehalle [f]** (Offene ▶Witterungsschutzeinrichtung zum Warten an Bus-, Straßenbahnhaltestellen oder Taxiständen; ▶Schutzdach; *UB* Buswartehalle); *syn.* Wartehäuschen [n].

public transit vehicle [n] [US] *trans.* ▶public transportation vehicle.

public transport [n] [UK] *trans.* ▶public transit [US].

public transportation [n] *trans.* ▶public transit [US].

public transportation system [n] *trans.* ▶local public transportation system [US]/local transport system [UK].

4733 public transportation vehicle [n] *trans. urb.* (Generic term covering rapid transit vehicles [US]/high-speed railways [UK] [railroad car (US)/wagon (UK)], underground train systems, trams, buses and trolley-buses for rapid transit within conurbations as well as taxicabs [US]/taxis [UK]; the system is called ▶local public transportation system [US]/local transport system [UK]); *syn.* mass transit vehicle [n] [also US], public transit vehicle [n], mode [n] of public transport [UK]; *s* **medio [m] de transporte colectivo** (Término genérico para cualquier tipo de sistema de transporte público utilizado en las ciudades para transportar a grandes cantidades de gente en trayectos relativamente cortos, como tren de cercanías, metro, tranvía, tranvía subterráneo, autobús y trolebús, y entre las ciudades como los ferrocarriles y autobuses; ▶tranporte público urbano); *syn.* transporte [m] colectivo, (medio [m] de) transporte [m] público, medio [m] de transporte público (urbano); *f 1* **moyen [m] de transport collectif** (Terme générique pour les moyens de transport publics ou privés, bus, car, métro, taxi, train, tramway, avion, bateau ; *terme spécifique* transport scolaire ; ▶transports suburbains) ; *syn.* transports [mpl] collectifs, moyen [m] de transport en commun, transports [mpl] publics, transports [mpl] en commun ; *f* **2 moyens [mpl] de transport suburbain** (Terme générique recouvrant les différents modes de transport de voyageurs sur de courtes distances tels que les chemins de fer urbains, RER [réseau **e**xpress **r**égional — réseau ferroviaire de transport en commun desservant Paris et son agglomération], métro, tramway, les transports en commun routiers, bus, trolleybus) ; *g* **öffentliches Verkehrsmittel [n]** (Beförderungsmittel wie Schnellbahn [S-Bahn], Untergrundbahn [U-Bahn], Straßenbahn [Strab oder Tram], Unterpflasterstraßenbahn [U-Strab], Omnibus [Bus], Oberleitungsomnibus [Obus] und Taxi, das Personen über kürzere Entfernungen in einem Ballungsgebiet transportiert; das System ist der ▶öffentliche Personennahverkehr. Zu den **ö. V.n** gehören auch die Fernbahnen, so sie nicht privatisiert sind); *syn. für eine Region* Nahverkehrsmittel [n].

public transport shelter [n] [UK] *trans.* ▶public transit shelter [US].

4734 public use [n] *leg.* (Legal term covering the free use of public areas and facilities without special permission; ▶communal use); *s* **uso [m] público** (Término jurídico que denomina el derecho de aprovechamiento de las instalaciones de ▶equipamiento comunitario por parte de cualquier ciudadano/a sin ningún permiso especial); *f* **usage [m] public** (Terme juridique désignant la libre utilisation par la collectivité des établissements ou équipements publics ; ▶services collectifs) ; *g* **Gemeingebrauch [m]** (Juristischer Begriff, der die freie Benutzung öffentlicher Einrichtungen oder Anlagen für jeden Bürger ohne besondere Erlaubnis bezeichnet; ▶Gemeinbedarf).

public use site [n] [US] *leg. urb.* ▶site for public facilities.

4735 public utility network [n] *plan. trans. urb.* (Systems for provision of, e.g. gas, potable water, electricity, telecommunication, and mass transit. *Specific terms* underground service, overhead wires, underground utility; HOU 1984, 115; ▶feed pipe, ▶public supply utility line, ▶public utility services); *syn.* services network [n]; *s* **red [m] de abastecimiento público** (Conjunto de ▶conducciones de abastecimiento y servicios de transporte público colectivo de una determinada zona; ▶servicios de infraestructura técnica); *f* **réseaux [mpl] (d'alimentation)** (Ensemble de fonctions, de services et d'objets techniques généralement essentiels à la vie urbaine, tels que les canalisations [gaz, eau, assainissement], les lignes électriques et de télécommunication, [▶conduite d'alimentation] les moyens de transport en commun structurés en réseau ; DUA 1996 ; ▶services d'alimentation, d'assainissement et d'évacuation des déchets) ; *syn.* réseaux [mpl] divers ; *g* **Versorgungsnetz [n]** (Für den Einzugsbereich eines Gebietes geplante oder vorgehaltene ▶Versorgungsleitungen sowie die Einrichtungen des öffentlichen Personennahverkehrs; ▶Ver- und Entsorgung).

4736 public utility services [npl] *urb.* (Provision of potable water supply, electricity, gas, telecommunication, etc., and disposal of sewage, waste water and solid waste; ▶public utility network); *s* **servicios [mpl] de infraestructura técnica** (Conjunto de actividades para suministrar una zona con bienes y servicios de infraestructura como agua, electricidad, gas, teléfono, etc. y de recogida de basuras y evacuación de aguas residuales; ▶red de abastecimiento público); *f* **services [mpl] d'alimentation, d'assainissement et d'évacuation des déchets** (Ensemble des missions relatives à la mise en œuvre et l'entretien des réseaux et des installations d'alimentation [eau potable, électricité, gaz, télécommunications], d'assainissement et d'évacuation des déchets, y compris tout équipement d'infrastructure correspondant ; celles-ci sont réalisées par les services techniques

des collectivités territoriales ; ▶réseaux [d'alimentation]) ; *g* **Ver- und Entsorgung [f]** (Gesamtheit der Versorgung eines Gebietes mit Trinkwasser, Elektrizität, Gas, Telekommunikationseinrichtungen und die Beseitigung von Abfallstoffen und Abwässern mit den dafür nötigen Infrastruktureinrichtungen; ▶Versorgungsnetz).

4737 public water supply [n] *plan. urb.* (TGG 1984, 139; all measures and facilities which serve to distribute potable water to the public, agriculture, and industry; systems of water lines are under the control of local water authorities/water control boards; **in U.K.**, Water Boards deal with the local distribution of water); *syn.* drinking water distribution [n]; *s* **suministro [m] de agua (potable)** (Todas las medidas e instalaciones necesarias para alumbrar, explotar y tratar los recursos hídricos para suministrar agua a la población, a la industria y a la agricultura); *syn.* abastecimiento [m] de agua potable; *f* **alimentation [f] en eau potable** (Toutes mesures et installations satisfaisant l'alimentation de la population en eau potable ainsi que les autres activités humaines) ; *syn.* approvisionnement [m] en eau, gestion [f] des services publics de l'eau ; *g* **öffentliche Wasserversorgung [f]** (Alle Maßnahmen und Einrichtungen als staatliche Daseinsvorsorge, die der Erschließung, Gewinnung und Aufbereitung dienen sowie die Verteilung von Trinkwasser in der durch die Trinkwasserverordnung vorgeschriebenen Qualität in ausreichender Menge und mit dem notwendigen Leitungsdruck an die Verbraucher sicherstellen. In einzelnen Bereichen, in denen kein öffentliches Leitungsnetz besteht, gibt es die **Eigen-** oder **Einzeltrinkwasserversorgung**); *syn.* öffentliche Trinkwasserversorgung [f]).

4738 public works [npl] *constr.* (Construction projects for public benefit; e.g. schools, water supply, power supply, road and highway construction, docks especially when owned and financed by government; ▶building construction, ▶civil engineering works, ▶ground civil engineering); *s* **obras [fpl] públicas** (Proyectos de construcción de infraestructura financiados por la administración pública; ▶edificación, ▶ingeniería civil, ▶trabajos de obras públicas); *f* **travaux [mpl] des chantiers publics** (▶bâtiment ▶travaux de voirie et de réseaux divers, ▶travaux publics) ; *g* **öffentliche Bauarbeiten [fpl]** (Bauarbeiten des ▶Hochbaus und ▶Tiefbaus, die von der öffentlichen Hand finanziert werden; ▶Tiefbauarbeiten); *syn.* öffentliche Baumaßnahme [f]).

4739 public works department [n] *adm.* (Municipal office responsible for roadworks, earthworks, hydraulic engineering and sewage disposal); *syn.* civil engineering department [n] [also UK], Highways and Transportation Services [npl] [also US], Department [n] of Public Works [also CDN]; *s* **departamento [m] de obras públicas** (Administración municipal responsable de la construcción de infraestructuras como carreteras, obras hidráulicas o de saneamiento); *syn.* administración [f] de obras públicas; *f* **département [m] des travaux publics (±)** (Administration communale généralement responsable des travaux de voirie, de terrassement, des eaux et d'assainissement ; **F.**, ces compétences font parties des services techniques responsables des travaux de maintenance de la voirie, de l'assainissement, des espaces verts et des bâtiments communaux et qui sont répartis dans des départements et directions distincts, leur dénomination et organisation variant d'une commune à l'autre ; p. ex. direction de l'eau et de l'assainissement, direction de la propreté et de l'entretien de la voirie, service voirie-assainissement, service des équipements urbains, service bâtiments, service circulation et stationnement) ; *syn.* direction [f] des travaux publics (±), service [m] de la voirie et de l'assainissement ; *g* **Tiefbauamt [n]** (Kommunales Bauamt, das i. d. R. für die Arbeiten des Verkehrswege- und Brückenbaus, des Erd- und

P

Grundbaus sowie für den Wasserbau und für die Abwasserbeseitigung zuständig ist. Letztere Aufgabe kann auch durch einen Eigenbetrieb *Stadtentwässerung* durchgeführt werden).

4740 public works inspector [n] *constr.* (Representative of a [public] client responsible for supervision and controlling of construction or maintenance work; on large-scale projects in cooperation with the ►site supervisor representing an authority; tasks include: a] the monitoring of the execution in compliance with the building permit, the working drawings and the specifications as well as with state-of-the-art technology and relevant regulations; b] records in the site log; c] issue of notice of default to the contractor; d] checking of all invoices for the release of payments; e] cost control and evaluation of contractor services in comparison with the contract prices; f] monitoring of budget expenditures with monthly reports; g] acceptance of work and deliveries, if necessary, in co-operation with the responsible planning or engineering office supervisor and the issue of a certificate of practical completion; h] acceptance of the construction or maintenance services; i] measurement of the completed works with the contractor and if necessary, with the responsible planning or engineering office supervisor; j] determination of defects and supervision of their removal; k] schedule of the guarantee periods); *s* **inspector** [m] **de obra pública** (Representante de un comitente [público] de obra cuya función es vigilar el desarrollo correcto de las obras respecto al proyecto contratado, en proyectos de gran envergadura en cooperación con el ►interventor, -a en pie de obra. Las tareas son: a] monitorear la ejecución de acuerdo con el permiso de construcción, los planos de trabajo y las especificaciones así como que se ejecute con la mejor tecnología de construcción asequible y de acuerdo a las regulaciones pertinentes; b] rellenar el libro de obra; c] avisar al contratista en el caso de falta de cumplimiento por parte de empresa subcontratada; d] revisar todas las facturas para liberar los pagos; e] controlar los costes [Es]/costos [AL] y evaluar los servicios del contratista en comparación con los precios del contrato; f] monitorear el flujo de pagos y elaborar informes mensuales; g] aceptar los trabajos y los suministros, si fuese necesario en cooperación con el arquitecto o ingeniero director de obra responsable del gabinete de planificación, y elaborar el acta de recepción/aceptación de obra; h] aceptar los servicios de construcción o mantenimiento; i] medición de los trabajos completados junto con el contratista y —si fuese necesario— con el arquitecto o ingeniero director de obra responsable del gabinete de planificación; j] determinar los defectos y supervisar su eliminación; k] elaborar listado de los plazos de garantía); *f* **personne** [f] **responsable des travaux (et de la maintenance)** (Représentant légal du maître d'ouvrage ou personne physique mandaté par le maître d'ouvrage [public] responsable de a] du suivi de chantier, de la bonne conduite des travaux et du contrôle de l'application des règles de construction [normes, règles de calcul, méthodes de construction, etc.], b] l'établissement d'un journal de chantier, c] l'analyse et la mise à jour du planning de travaux, le contrôle de l'état d'avancement des travaux, le respect des délais d'exécution, d] le contrôle les situations ou décomptes mensuels, la mise au point des acomptes et des avances forfaitaires, du règlement des travaux, e] l'actualisation et la révision des prix, ainsi que le décompte général et définitif, f] la gestion financière journalière du chantier en fonction des moyens alloués, g] la vérification et la réception des prestations des entreprises, h] la réalisation de la réception des travaux, i] la coordination des prestations des entreprises et établissement des métrés et attachements, j] la constatation des vices et malfaçons, l'établissement de la liste des réserves et le contrôle de la levée des réserves, k] l'établissement de la déclaration d'achèvement des travaux et du dossier des ouvrages exécutés, des la définition

des délais de garanties ; ►conducteur d'opération, ►représentant du maître d'ouvrage public) ; *syn.* contremaître [m] ; *g* **Bauaufseher** [m] (Vertreter eines [öffentlichen] Bauherrn zur Beaufsichtigung und Kontrolle der Bau- oder Pflegeleistungen, bei größeren Vorhaben in Zusammenarbeit mit dem ►Bauleiter/-in 3; zu den Aufgaben gehört u. a. a] die Überwachung der Ausführung auf Übereinstimmung mit der Baugenehmigung, den Ausführungsplänen und Leistungsbeschreibungen sowie mit den allgemein anerkannten Regeln der Technik und einschlägigen Vorschriften; b] Führen eines Bautagebuches; c] Inverzugsetzen der ausführenden Firmen; d] Prüfung aller Rechnungen für die Freigabe der Zahlungen; e] Kostenkontrolle durch Überprüfung der Leistungsabrechnung der bauausführenden Firmen im Vergleich zu den Vertragspreisen; f] Überwachung des Abflusses der zur Verfügung stehenden Finanzmittel mit monatlicher Berichtspflicht; g] Abnahme von Leistungen und Lieferungen ggf. unter Mitwirkung des verantwortlichen Bauleiters des Planungs- oder Ingenieurbüros unter Fertigung einer Abnahmeniederschrift; h] Abnahme der Bau- resp. Pflegeleistungen; i] gemeinsames Aufmaß mit den bauausführenden Unternehmen und ggf. des verantwortlichen Bauleiters des Planungs- oder Ingenieurbüros; j] Feststellen von Mängeln und Überwachen der Beseitigung der festgestellten Mängel; k] Auflistung der Gewährleistungsfristen; der **B.** ist von der Ausbildung her i. d. R. ein Meister oder Techniker).

4741 pull-off [n] [US] *trans.* (Area adjacent to a public road outside the built-up area, which permits a vehicle to park without disturbing traffic; ►rest place [US]/rest spot [UK] 1); *syn.* lay-by [n] [UK]; *s* **área** [f] **de parada sin servicios** (Zona de aparcamiento al borde de autopista o carretera, generalmente equipada con instalaciones básicas como bancos, papeleras, etc.; ►lugar de descanso); *f 1* **aire** [f] **d'arrêt** (Aire aménagée le long des routes, dépourvue d'installations sanitaires, en général équipée de tables de pique-nique permettant l'arrêt et le stationnement des véhicules, et destinée au repos et à l'agrément des usagers ; ►aire de repos 2) ; *f 2* **aire** [f] **de repos (1)** (Aire aménagée le long des autoroutes ou des voies à grande circulation équipée d'installations sanitaires, poubelles et de tables de pique-nique, permettant l'arrêt et le stationnement et destinée au repos et à l'agrément des usagers) ; *g* **unbewirtschaftete Rastanlage** [f] (Anlage ohne sanitäre und Gaststätteneinrichtungen für den ruhenden Verkehr an öffentlichen Straßen außerhalb geschlossener Ortschaften; ►Rastplatz).

pull-off lane [n] [US]**, paved** *trans.* ►hard shoulder.

pull-off strip [n] [US] *trans.* ►pullout [US]/lay-by [n] [UK].

pull-off track [n] [US] *trans.* ►railroad siding [US]/works siding [UK].

4742 pullout [n] [US] *trans.* (Stopping space for vehicles adjacent to traffic lanes; ►public transit shelter [US]; *specific term* ►bus pullout [US]/bus lay-by [UK]); *syn.* lay-by [n] [UK], pull-off strip [n] [US]; *s* **apartadero** [m] (Área al borde de la calzada reservada para parada de vehículos; *término específico* ►apartadero de autobús); *f* **créneau** [m] **d'arrêt** (Espace parallèle à une voie de circulation réservé à l'arrêt temporaire ou le stationnement limité de véhicules ; *terme spécifique* ►créneau d'arrêt de bus) ; *g* **Haltebucht** [f] (Verkehrsfläche neben der Fahrbahn zum Halten von Fahrzeugen; *UB* ►Bushaltebucht).

pump [n] *constr.* ►recirculating pump, ►submersible pump.

4743 punch list [n] [US] *constr. contr.* (Record of ►defects to be removed by the contractor); *syn.* list [n] of defects [US], snagging list [n] [UK]; *s* **lista** [f] **de defectos** (Registro de los ►defectos [de obra] que deben ser eliminados por el contratista); *syn.* lista [f] de deficiencias, lista [f] de reclamaciones; *f 1* **liste** [f] **des imperfections ou malfaçons** (Lors des opérations préalables

P

à la réception des travaux, constatation établie par le maître d'œuvre d'►imperfections ou de ►malfaçons et établissement dans le cadre d'un procès-verbal de la liste des imperfections ou malfaçons à réparer par l'entrepreneur ; ►vice de construction) ; *syn.* mémoire [m] de réclamation, liste [f] des reprises ou finitions ; *f 2* **liste [f] des réserves** (Liste établie lorsque la réception des travaux est prononcée avec réserves par la personne responsable du marché) ; *g* **Mängelliste [f]** (Auflistung der ►Mängel, die durch den Auftragnehmer zu beseitigen sind).

purchase order [n] [UK]**, compulsory** *adm. leg.* ►condemnation order [US]/compulsory purchase order [n] (CPO) [UK].

purchase proceedings [npl] [UK]**, compulsory** *adm. leg.* ►condemnation proceedings [US].

pure forest [n] *for. phyt.* ►pure stand.

4744 pure stand [n] *for.* (Forest, crop or stand, composed principally of at least 90% of the same tree species; ►monoculture, ►mixed crop production, ►mixed forest, ►single crop system); *syn. for.* pure forest [n]; *s* **rodal [m] monoespecífico (1)** (Bosque o cobertura vegetal constituida mayormente por plantas de una sola especie; ►bosque mixto, ►cultivo monoespecífico, ►monocultura, ►policultivo); *syn. for.* bosque [m] monoespecífico; *f* **boisement [m] monospécifique** (Peuplement forestier ou végétal composé principalement d'une espèce constituant 90 % de celui-ci ; il s'agit en général de parcelles qui ont été plantées, les forêts naturelles monospécifiques relativement stables n'existant que sites aux conditions stationnelles limitantes, p. ex. pour les hêtraies, les facteurs édaphiques [complexe climacique de la forêt de Fontainebleau] ou la situation topographique en exposition froide sur les pentes de l'étage montagnard entre 650 et 1100 m environ, les forêts de pins sur les sols sableux, les forêts résineuses naturelles de l'étage subalpin [épicéa, mélèze] ; ►culture monospécifique, ►forêt mélangée, ►forêt mixte, ►monoculture, ►polyculture) ; *syn.* peuplement [m] pur ; *g* **Reinbestand [m]** (Waldbestand aus überwiegend einer Art bestehend, mindestens 90 %; bei natürlichen Gegebenheiten gibt es einen **R.** nur unter extremen Standortbedingungen, z. B. Kieferwälder auf [Dünen]sand, subalpine Fichtenwälder, alpine Lärchenwälder; ►Monokultur, ►Mischkultur, ►Mischwald, ►Reinkultur).

purification [n] *limn.* ►biological self-purification; *envir.* ►flue gas purification.

purification [n]**, water** *envir. wat'man.* ►water treatment.

4745 purified sewage [n] *envir.* (Liquid from treated sewage); *s* **efluente [m] de planta depuradora** (Caudal de aguas residuales vertidas por una planta de tratamiento); *f* **effluent [m] pur** (Eaux traitées rejetées par une station d'épuration) ; *g* **gereinigtes Abwasser [n]** (…wässer [pl]; Produkt eines Klärwerkes).

pursuit [n] *plan. recr.* ►leisure pursuit.

pursuit [n]**, recreation** *plan. recr.* ►recreation activity.

4746 putting green [n] *recr.* (Closely clipped, fine grass areas around the 'hole' on a golf course); *s* **green [m]** (Áreas alrededor de los hoyos en campos de golf en las que el césped es muy denso y se mantiene muy corto); *f* **green [m]** (Surface engazonnée soigneusement entretenue et tondue très ras située sur un parcours de golf et comportant en son centre le trou) ; *g* **Grün [n, o. Pl.]** (Besonders gepflegtes und kurz gehaltenes Rasenstück am Ende einer Golfspielbahn, in das das Loch eingeschnitten ist).

pyramid [n] *plan. sociol.* ►demographic pyramid; *ecol.* ►life pyramid.

pyramid [n] [US]**, population** *plan. sociol.* ►demographic pyramid.

4747 pyramidal dwarf fruit tree [n] [US] *hort.* (Small, trained and grafted fruit tree with a stem height of 40-50cm; ►miniature fruit tree [US] 2, ►pyramidal fruit tree); *syn.* pyramidal fruit bush [n] [UK]; *s* **arbusto [m] de tipo «Spindel»** (PODA 1994, 85; especie arbustiva de porte piramidal; ►frutal enano, ►frutal de porte piramidal); *f* **quenouille [f] (d'arbres fruitiers)** (Forme fruitière en volume dans laquelle les branches charpentières partent presque toutes de la base du tronc et sont arrêtées lorsqu'elles atteignent la hauteur de l'axe principal [2,50 m] ; LA 1981 ; ►buisson [d'arbres fruitiers], ►fuseau [d'arbres fruitiers]) ; *g* **Spindelbusch [m]** (Auf Typenunterlage veredelte kleine Obstbaumform mit einer Stammhöhe von 40-50 cm; ►Buschbaum, ►Spindel).

pyramidal fruit bush [n] [UK] *hort.* ►pyramidal dwarf fruit tree [US].

4748 pyramidal fruit tree [n] *arb. hort.* (Trained, coneshaped fruit tree with a main central stem and secondary lateral branches, the strength and length of which diminish towards the apex of the tree; ►cordon 1); *s* **frutal [m] de porte piramidal** (Forma podada de la copa de un frutal con eje central muy definido y ramas laterales secundarias cuya longitud y grosor disminuyen con la altura; ►cordón); *f* **fuseau [m] (d'arbres fruitiers)** (Forme fruitière non palissée, constituée par une tige principale verticale [flèche], sur laquelle s'insèrent des branches charpentières permanentes, dont la longueur décroît de la base vers le sommet ; ►cordon ; cf. LA 1981) ; *g* **Spindel [f]** (Geschnittene Kronenform eines Obstbaumes mit betonter Mittelachse und untergeordneten Seitensprossen, deren Stärke und Länge von unten nach oben abnehmen. Ein mit kurzen Fruchtsprossen garnierter ►Schnurbaum wird auch **Superspindel** genannt).

Q

quaking bog [n] *phyt.* ►floating sphagnum mat.

qualification [n] *constr. contr.* ►references for qualification; *prof.* ►request for qualification [US].

quality [n] *envir. leg.* ►air quality; *sociol.* ►dwelling quality; *envir. pol.* ►environmental quality (1); *plan. recr.* ►perceived environmental quality, ►perceived environmental quality, ►visual quality; *landsc. plan. recr.* ►visual quality of landscape.

quality [n]**, aesthetic** *recr. plan.* ►visual quality.

quality [n]**, observation of artistic** *constr. prof.* ►observation of artistic quality.

quality [n]**, recreation** *plan. recr.* ►recreation value.

quality [n]**, scenic** *landsc. plan. recr.* ►visual quality of landscape.

quality [n]**, visual resource** *landsc. plan. recr.* ►visual quality of landscape.

4749 quality class [n] [US] *constr. eng.* (High standard of condition and suitability in classification of quality standards for construction material); *syn.* quality grade [n] [UK]; *s* **clase [f] de calidad** (Clasificación de materiales de construcción según estándares de calidad); *syn.* tipo [m] de calidad de materiales de construcción; *f* **classe [f] de qualité** (Classement des matériaux d'après leurs caractéristiques qualitatives) ; *g* **Güteklasse [f]**

(Einteilung der Baustoffe nach bestimmten Normen und Qualitätsmerkmalen).

4750 quality control [n] *constr.* (Procedure for checking compliance with standards and requirements laid down contractually for construction, building materials, or services); *s* **control** [m] **de calidad** (Procedimiento de revisión de los materiales o piezas prefabricadas utilizadas en obra o de la construcción misma por parte del comitente); *f* **vérification** [f] **qualitative des matériaux et produits** (Le Maître d'œuvre peut faire procéder à tous les contrôles de qualité de matériaux ou d'ouvrages qu'il jugera nécessaire, par prélèvements d'échantillons et essais de laboratoires et vérifier que ceux-ci et les prestations exécutées sont conformes aux spécifications des documents contractuels et aux prescriptions des normes françaises homologuées ; C. marchés publ., art. 24) ; *syn.* essai [m] de contrôle ; *g* **Kontrollprüfung** [f] (Prüfung durch den Auftraggeber oder seines Bevollmächtigten, ob die vertraglich festgelegte Beschaffenheit der gelieferten resp. eingebauten Baustoffe, Fertigbauteile oder der fertigen Leistungen dem Istzustand entspricht).

quality grade [n] [UK] *constr. eng.* ▶quality class [US].

quality management [n] *conserv. nat'res. wat'man.* ▶water quality management.

quality management [n], **air** *envir. leg. nat'res.* ▶air pollution control.

4751 quality [n] **of life** *plan. sociol.* (Complex term without a generally accepted definition. Improvement of the **q. of l.** as a political aim includes freedom and security, solidarity and political participation, justice in distribution and sustainability for future generations, in humanization of working conditions, decontamination of the environment and food, creation of equal education and promotion opportunities, provision and maintenance of amenities in the physical and social sense, etc.); *s* **calidad** [f] **de vida** (Término complejo suscito a definición política según qué aspectos de la vida se consideren prioritarios. En general, en la **c. de v.** se incluyen la seguridad del puesto de trabajo, los servicios sociales, la igualdad de derechos y de posibilidades de educación para todos y un medio ambiente sano. En el sentido material, por **c. de v.** se entiende el gozo de posibilidades de consumo de bienes y servicios); *f* **qualité** [f] **de la vie** (Terme d'acceptation générale floue et hétérogène utilisé dans le langage politique pour définir un grand nombre de valeurs et d'objectifs tels que la liberté, la sécurité, la solidarité et la participation, la justice dans le partage des richesses pour les générations futures, l'humanisation des conditions de travail, la lutte contre la pollution de l'environnement, l'égalité devant la formation et la promotion professionnelle, l'amélioration du cadre de vie, etc. ; l'aspect matériel de la **q. de la v.** peut être défini par la notion de 'standing de vie', orientée vers le surpassement de la société de pénurie grâce à l'approvisionnement des ménages en produits de consommation et prestations de service) ; *g* **Lebensqualität** [f] (Lehenübersetzung des englischen Begriffs *quality of life* ohne allgemein anerkannte Definition. Die Steigerung der **L.** als politisches Ziel wird in der Berücksichtigung übergreifender Werte und Ziele wie Freiheit und Sicherheit, Solidarität und politische Mitbestimmung, Verteilungsgerechtigkeit und Vorsorge für zukünftige Generationen sowie die Humanisierung der Arbeitswelt, der Entgiftung der Umwelt und Nahrungsmittel, Schaffung gleicher Bildungs- und Aufstiegschancen etc. gesehen. Materielle Aspekte der **L.** umfasst der Begriff des **Lebensstandards**, orientiert am Leitbild der Überwindung der Mangelgesellschaft durch die Versorgung der Haushalte mit Ver- und Gebrauchsgütern sowie mit Dienstleistungen).

4752 quality [n] **of nursery stock** *constr. hort.* (Condition of delivered nursery stock with respect to external, quality characteristics such as health status, height, trunk caliper, the strength of the trunk/stem, development of the crown according to habit [arrangement of branches specific to the species] number of shoots and mature shoots, compliance with the requirements for root balls and root mass, as well as varietal purity; internal, quality characteristics, which cannot be ascertained at the time of handing-over include the adaptation and resistance to stress such as drought, frost and disease; **in U.K.**, the basis for checking the **q. of n. s.** is contained in BS 3936: *Specification for Nursery Stock* and BS 3975: *Glossary for Landscape Work*. The European Nurserystock Association [ENA] contains European technical and quality standards for hardy nursery stock, including roses, fruit trees, and herbaceous perennials [November 1996]; ▶quality standards for perennials; *s* **calidad** [f] **de plantas (de vivero)** (**1.** Estado de material de vivero suministrado a obra en cuanto a sus **características externas** como altura, diámetro del tronco, estado no dañado del tronco, porte de la copa correspondiente a la especie, número de retoños [maduros], cumplimiento de las exigencias para el cepellón y la masa de raíces así como certificado de variedad; **características internas** que no pueden ser constatadas al aceptar el material como la capacidad de adaptación y resistencia contra situaciones extremas como sequía, heladas y plagas. Como base para el control de la calidad de leñosas sirven los ▶estándares de calidad de plantas de vivero. **2.** Para valorar la calidad de perennes se utiliza la ▶clasificación de calidades de plantas vivaces); *f 1* **qualité** [f] **des végétaux (de pépinières)** (Spécificités des végétaux dont les conditions d'élevage doivent répondre à des critères définis tels le mode, la zone et la conduite de culture, les caractéristiques dimensionnelles, la qualité marchande des végétaux, etc. conformément aux normes françaises AFNOR de décembre 1990 décrivant les spécifications générales ou particulières applicables 1. aux jeunes plants et des jeunes touffes de pépinières fruitières et ornementales, 2. aux jeunes plants d'arbres fruitiers, 3. aux jeunes plants et jeunes touffes d'arbres et d'arbustes d'ornement à feuilles caduques ou persistantes, 4. aux arbres et plantes de pépinières fruitières et ornementales, 5. aux arbres fruitiers, 6. aux rosiers, 7. aux conifères d'ornement, 8. aux arbres d'alignement et d'ornement, 9. aux plantes grimpantes et sarmenteuses, 10. aux plantes dites de terre de Bruyère) ; *f 2* **qualité** [f] **phytosanitaire** (Qualité des végétaux et produits végétaux quant à la présence ou l'absence d'organismes nuisibles) ; *syn.* état [m] phytosanitaire ; *f 3* **qualité** [f] **morphologique** (Identification et appréciation des caractères morphologiques des produits végétaux [racines, feuilles, fleurs, fruits, etc.], en conformité avec les normes régionales/nationales ou selon les critères de sécurité alimentaire) ; *syn.* état [m] morphologique ; *g* **Pflanzenqualität** [f] (**1.** Beschaffenheit gelieferter Baumschulware hinsichtlich **äußerer Qualitätsmerkmale** wie Gesundheitszustand, Höhe, Stammdurchmesser, Unversehrtheit des Stammes, habitusgerechter Kronenaufbau [artspezifische Astgarnierung], Triebzahl und ausgereifte Triebe, Einhaltung der Anforderungen an Ballen und Wurzelmasse und Sortenreinheit; **innere Qualitätsmerkmale**, die bei der Abnahme nicht festgestellt werden können, wären die Anpassungs- und Widerstandsfähigkeit gegen Belastungen wie Trockenheit, Frost und Krankheiten; Grundlagen zur Überprüfung der Gehölzqualität sind die ▶Gütebestimmungen für Baumschulpflanzen und die FLL-Gütebestimmungen für Baumschulpflanzen sowie DIN 18916 „Vegetationstechnik im Landschaftsbau; Pflanzen und Pflanzarbeiten, Beschaffenheit von Pflanzen, Pflanzverfahren". **2.** Zur Beurteilung der Qualität von Stauden gelten die ▶Gütebestimmungen für Stauden, die bei der Forschungsgesellschaft Landschaftsentwicklung Landschaftsbau

Q

e. V. [FLL] seit 1988 [letzte Version 2004] herausgegeben werden); *syn.* Beschaffenheit [f] von Pflanzen.

quality |n| **of rivers and lakes** *wat'man.* ▶water quality of rivers and lakes.

quality |n| **of the environmental experience, subjective** *plan. recr. sociol* ▶perceived environmental quality.

quality standard [n] *adm. envir. pol.* ▶air quality standard.

quality standards [npl] *envir. pol.* ▶environmental quality standards, ▶water quality standards.

4753 quality standards |npl| **for nursery stock** *hort.* (cf. BS 3936, e.g. trueness to variety, good root system, health of species and variety, corresponding shape, height, width, number of leaders, circumference of trunk, etc., selected according to sizes; *cf. also* American Standard for Nursery Stock, ANSI 1986, sponsored and published by the American Association of Nurserymen, Inc.; ▶true to name); *s* **estándar [m] de calidad de plantas de vivero** (Conjunto de estándares de calidad que deben cumplir las plantas de vivero para su venta como buen sistema radical, vigor correspondiente a la especie y variedad, forma, número de ramas, perímetro de tronco, etc.; ▶de variedad certificada); *f* **norme [f] de qualité pour végétaux de pépinière** (Spécifications générales et particulières pour les produits de pépinières — arbres et plantes de pépinières fruitières et ornementales — devant être certifiées par le pépiniériste telles que l'authenticité variétale [genre, espèce, variété ou cultivar], la qualité du développement radiculaire avec formation d'un chevelu abondant, le parfait état sanitaire, la forme, les caractéristiques d'aspect, les caractéristiques dimensionnelles, etc. ; cf. normes N.F. V 12 051, 12 052, 12 054, 12 055, 12 057 ; ▶en conformité spécifique et variétale) ; *g* **Gütebestimmungen [fpl] für Baumschulpflanzen** (Für den Verkauf von Gehölzen geforderte Qualitätsmerkmale wie Sortenechtheit [▶arten- und sortenecht], gute Bewurzelung, Gesundheit der Art und Sorte, entsprechende Wuchsform, nach Höhe, Breite, Triebzahl, Stammumfang etc. sortierte Größen. Die **G.** entsprechen dem Handelsbrauch sowie der gewerblichen Verkehrssitte und sind als anerkannte Regel der Technik zu werten; cf. auch DIN 18 916: „Vegetationstechnik im Landschaftsbau; Pflanzen und Pflanzarbeiten, Beschaffenheit von Pflanzen, Pflanzverfahren"; in D. gibt es 1999 zunächst in elf Baumschulen zusätzlich für Alleebäume das Prüfsiegel der Centralen Marketingorganisation der Agrarwirtschaft [CMA]. Mit diesem CMA-Etikett wird versichert, dass die gelieferten Bäume mit minutiös dokumentierter Historie einzeln durch neutrale Gutachter auf Übereinstimmung mit den FLL-Gütebestimmungen für Baumschulpflanzen kontrolliert wurden und aus einer kontrollierten Pflanzenproduktion stammen — Feststellungen, die eigentlich in jeder normalen Ausschreibung gefordert werden und der oben genannten DIN 18 916 entsprechen. Bayern beginnt ab dem Jahr 2000 mit dem Verkauf von zertifizierten autochthonen Gehölzen, initiiert durch die Erzeugergemeinschaft für autochthone Baumschulerzeugnisse [EAB]).

4754 quality standards |npl| **for perennials** *hort.* (Required standards, e.g. trueness to variety, good root system, health of species and cultivar; ▶true to name); *s* **clasificación [f] de calidades de plantas vivaces** (Conjunto de estándares de calidad que deben cumplir las perennes de vivero para su venta como ser variedad auténtica, buen sistema radical, vigor correspondiente a la especie y variedad, etc.; ▶de variedad certificada); *f* **norme [f] de qualité pour les plantes vivaces** (Spécifications pour les produits de pépinières de plantes vivaces devant être certifiées par le pépiniériste telles que l'authenticité variétale, la qualité du développement radiculaire avec un chevelu apparent sur les parois de la motte au dépotage, le parfait état sanitaire,

etc. ; la norme est élaborée dans une commission de normalisation aux travaux de laquelle participent des représentants des producteurs de l'horticulture et des pépinières, de l'administration et d'organismes publics ; ▶en conformité spécifique et variétale) ; *g* **Gütebestimmungen [fpl] für Stauden** (Für den Verkauf von Stauden geforderte Qualitätsmerkmale wie z. B. Sortenechtheit, gute Bewurzelung, Gesundheit der Art und Sorte sowie die Mindestopfgrößen für die unterschiedlichen Staudengruppen, wie z. B. Polsterstauden, niedrige bis halbhohe Stauden, halbhohe bis hohe Stauden, Gräser, Farne, Sumpf- und Wasserpflanzen. Eine Besonderheit stellen die ▶Repositionspflanzen sowie Pflanzen und Pflanzenteile zur extensiven Dachbegrünung dar. Letztere müssen entsprechend den Anforderungen der FLL-Dachbegrünungsrichtlinien in speziellem Substrat mit hohem mineralischen Anteil in maximal 5 cm hohen Töpfen mit einem Mindestinhalt von 50 cm³ kultiviert werden. Das FLL-Regelwerk „Gütebestimmungen für Stauden" erschien 1988 und wird in einer Arbeitsgruppe, bestehend aus Vertretern der Produzenten, Planer und Verwender seit 1988 beraten und fortgeschrieben. Die **G.** entsprechen somit dem Handelsbrauch sowie der gewerblichen Verkehrssitte und sind als anerkannte Regel der Technik zu werten; cf. auch DIN 18 916: „Vegetationstechnik im Landschaftsbau; Pflanzen und Pflanzarbeiten, Beschaffenheit von Pflanzen, Pflanzverfahren"; ▶arten- und sortenecht).

quantities [npl] *contr.* ▶bill of quantities; *constr. plan.* ▶estimate of quantities; ▶list of bid items and quantities [US]/ schedule of tender items [UK]; *constr. eng.* ▶rough estimate of earthworks quantities.

quantities [npl] [US]**, calculation of contract bid** *constr. plan.* ▶calculation of quantities.

quantities [npl]**, earthworks** *constr. eng.* ▶rough estimate of earthworks quantities.

quantities [npl] [UK]**, priced bills of** *contr.* ▶total construction cost estimate [US].

4755 quantity [n] **(of an item)** *constr. contr.* (Calculated amount of a specified item in a contract specification clause; ▶increase in estimated volume, ▶reduction in estimated volume); *syn.* amount [n]; *s* **cantidad [f]** (Monto de materiales asignados a la construcción en el marco del resumen de prestaciones; ▶aumento de cantidades, ▶reducción de cantidades); *f* **quantité [f]** (des prestations prévues dans le descriptif-quantitatif ; ▶augmentation de la masse des travaux, ▶diminution de la masse des travaux) ; *g* **Menge [f]** (Ermittelter Umfang einer ausgeschriebenen Leistung im Leistungsverzeichnis; ▶Massenmehrung, ▶Massenminderung).

quantity take-off [n] *contr.* ▶calculation of quantities.

4756 quarry [n] *min.* (Place where stone is extracted for construction purposes by drilling, cutting or blasting; ▶abandoned quarry); *s* **cantera [f]** (Lugar de donde se extraen piedras u otros materiales pétreos [mármol] para la construcción; ▶cantera abandonada); *f* **carrière [f] de roches massives** (Lieu d'exploitation à ciel ouvert sur lequel des substances minérales utiles sont extraites par percement, sciage ou dynamitage ; ▶carrière désaffectée) ; *g* **Steinbruch [m]** (Entnahmestelle, in der durch Bohren, Sägen oder Sprengen nutzbares Gestein abgebaut wird; ▶stillgelegter Steinbruch).

quarry [n]**, old** [US] *min. landsc.* ▶abandoned quarry.

4757 quarry-faced [pp/adj] *constr.* (Descriptive term applied to untooled quarried stone; *∗*▶tooling of stone [US]/ dressing of stone [UK], #2&9); *syn.* rockfaced [pp/adj], splitfaced [pp/adj]; *s* **sin labrar [vb]** (Característica de piedras naturales que llegan sin labrar de la cantera; tipo[s] de ▶labra de piedra natural o artificial, n° 2 y 9); *f* **brut, brute d'extraction [loc]** (Caractéristique des blocs de pierres naturelles obtenus par

Q

clivage et non traités en carrière ▶mode de taille des pierres, n° 2 et 9) ; *syn.* brut, brute de carrière [loc] ; *g* **bruchrau [adj]** (*f. S.* bruchrauh; Eigenschaft von Natursteinen, die unbearbeitet aus dem Steinbruch kommen; ▶Steinbearbeitung, Nr. 2 und 9).

quarry-faced masonry [n] *arch. constr.* ▶rusticated masonry.

4758 quarry flagstone [n] *constr.* (Any natural stone slab cut roughly without any special finishes for paving purposes; **q. f.s** may be large or small in size); *syn.* raw stone flag [n], roughly-hewn flag [n]; *s* **losa [f] sin labrar** (Placa de piedra sin elaborar tal como sale de la cantera); *f* **dalle [f] clivée brute** (Élément de pierre naturelle brute de taille, brute de carrière) ; *syn.* dalle [f] tranchée ; *g* **unbehauene Natursteinplatte [f]** (Spaltraue, unbearbeitete Platte aus dem Steinbruch).

4759 quarrying [n] *min.* (Extraction of rock, soil, or mineral from the ground; ▶mineral extraction); *s* **minería [f] en cantera** (▶extracción de minerales superficiales o rocas industriales); *syn.* explotación [f] en cantera, explotación [f] de áridos (y arenas); *f* **extraction [f] de pierres** (Exploitation de matériaux dans une carrière ; ▶extraction des substances minérales) ; *g* **Gesteinsabbau [m]** (▶Materialentnahme); *syn.* Gewinnung [f] nutzbaren Gesteins, Gesteinsgewinnung [f].

quarrying [n] **of gravel** *min.* ▶gravel extraction.

quarrying [n] **of sand** *min.* ▶sand extraction.

4760 quarry rock face [n] *min.* (Cut vertical face in a piedmont or mountain area, using explosives or sawing to extract usable stone blocks); *s* **frente [m] de explotación** (Pared de una cantera abierto a explotación por medio de explosivos o de sierras para extraer bloques de piedra; cf. DINA 1987, 805); *f* **front [m] de taille d'une carrière** (Face d'un massif rocheux mise à nu par explosion ou sciage pour l'extraction de roches dans une carrière) ; *g* **Abbruchwand [f] eines Steinbruch(e)s** (Durch Sprengen und Sägen freigelegte Wand eines Bergmassivs, aus dem nutzbares Gestein abgebaut wird); *syn.* Steinbruchabbauwand [f].

4761 queen closer [n] *arch. constr.* (DAC 1975; brick of normal face dimensions which has been cut in half along its length; a narrow facing brick of natural or artificial stone used on veneering walls; ▶veneering, ▶veneer masonry, ▶queen closer of natural stone); *syn.* soap [n]; *s* **ladrillo [m] pichulín** (Ladrillo cuya cara mayor mide aproximadamente los dos tercios de la cara de un ladrillo normal utilizado para ▶paramento de muros; cf. DACO 1988; ▶enchapado, ▶mampostería enchapada, ▶pichulín de piedra natural); *syn.* pichulín [m]; *f* **barrette [f]** (Plaque de ▶parement étroite utilisée pour les revêtements verticaux, en pierre naturelle ou en matériau préfabriqué ; ▶barrette en pierre naturelle, ▶maçonnerie à parement) ; *g* **Riemchen [n]** (Schmaler Vormauerstein aus Natur- oder Kunststein zur ▶Verblendung von Mauern; ▶Natursteinriemchen, ▶Verblendmauerwerk).

4762 queen closer [n] **of natural stone** *arch. constr.* (Thin piece of natural stone, 5-8cm wide, for masonry veneer; ▶veneering); *s* **pichulín [m] de piedra natural** (Pieza delgada de piedra natural que se utiliza para el chapado de muros; ▶enchapado); *f* **barrette [f] en pierre naturelle** (Pierre naturelle de faible épaisseur utilisée pour le ▶parement de mur en parpaings ou en béton) ; *g* **Natursteinriemchen [n]** (Schmaler Vormauerstein aus Naturstein, 5-8 cm breit, zur ▶Verblendung von Mauern).

4763 questionnaire [n] *plan. sociol.* (Prepared series of questions submitted to a number of persons in order to obtain data for a survey or report); *s* **cuestionario [m]** (Lista de preguntas con la que se hacen entrevistas [representativas] para obtener datos específicos); *f* **questionnaire [m]** (Formulaire où sont inscrites des questions méthodiquement posées en vue d'une enquête) ; *g* **Fragebogen [m]** (Vordruck, der eine Reihe von Fragen eines bestimmten Themenkomplexes für eine Umfrage enthält).

quick-drying cement [n] *constr.* ▶rapid-hardening cement.

4764 quicksand [n] *geo. pedol.* (Bed of extremely loose and supersaturated sand with upward flowing water and consisting of smooth rounded grains with little tendency to mutual adherence, easily yielding to pressure and thus readily swallowing up any heavy object resting on it. **Q.** is frequent on some coasts, rivers or near river mouths); *s* **arena [f] movediza** (En ciertas costas planas y desembocaduras de ríos, acumulación de arena suelta muy saturada y sin ningún contenido de arcilla, formada por granos suavemente redondeados con muy poca capacidad de adherencia mutua. Debido a la influencia de la presión hidroestática la **a. m.** fluye y se traga cualquier objeto pesado que se pose sobre ella); *f* **sables [mpl] mouvants** (Dans les zones de courants maritimes du littoral bas, sable saturé en eau, exempt de particules argileuses, soumis à un mouvement permanent qui cesse lors d'une diminution de la teneur en eau ; les objets déposés sur les **s. m.** sont entourés par les particules de sables et s'enfoncent inexorablement) ; *g* **Treibsand [m]** (In der Meeresströmung leicht beweglicher, tonfreier Sand an Flachküsten, der bei vorhandenem Wassergehalt Fließbewegungen durchmacht, die Bewegungen aber mit abnehmendem Wassergehalt einstellt. Auf dem **T.** liegende Objekte werden von Sandpartikeln umlagert und sinken ein); *syn.* Triebsand [m].

quickset hedge [n] [UK] *geo. land'man.* ▶elevated hedgerow [US].

quiet area [n] *plan. recr.* ▶rest area.

quincunx planting grid [n] *constr.* ▶triangular planting grid.

quotation [n] *constr. contr.* ▶price quotation.

R

race [n], **ecological** *ecol.* ▶ecotype.

radial header [n] [UK] *constr.* ▶concentric manhole brick.

4765 radial open space pattern/system [n] *landsc. urb.* (Open space system in urban areas, whereby mostly wedge-shaped green corridors extend in a radial pattern from the periphery into the city center [US]/city centre [UK]); *syn.* open space system [n] forming a radial pattern (TGG 1984, 273); *s* **trama [f] verde radial** (Sistema de espacios libres urbanos en el que las conexiones verdes se internan como cuñas hasta el centro de la ciudad); *syn.* sistema [m] radial de espacios libres, red [f] radial de espacios libres; *f* **trame [f] radiale d'espaces libres** (Système de disposition des espaces libres urbains qui de la périphérie rejoignent sous forme de pointe le centre de la cité) ; *syn.* trame [f] verte en doigt-de-gant (RO 1970, 476), liaisons [fpl] radiales vertes, coulées [fpl] vertes radiales) ; *g* **radiales Freiraumsystem [n]** (Freiraumsystem im Siedlungsbereich, bei dem meist keilförmige Grünzüge oder Grünverbindungen von der Stadtperipherie strahlenförmig bis in die Stadtmitte reichen); *syn.* radialförmiges Freiraumsystem [n] (RICH 1981), sternförmiges Freiraumsystem [n].

radial pattern [n], **open space system forming a** *landsc. urb.* ▶radial open space pattern/system.

4766 radiant heat [n] *met.* (TGG 1984, 153; heat radiation from the Earth's surface; i.e. radiation of heat stored in the ground. There is also the counterradiation of heat from the

atmosphere; ►reradiation of heat); *s* **radiación [f] térmica** (Radiación del calor acumulado en el suelo hacia la atmósfera; ►reradiación del calor); *f* **radiation [f] thermique** (Rayonnement lumineux ou obscur réfléchi à la surface du sol [coefficient d'albedo] ainsi que le rayonnement diffus en provenance de l'atmosphère ; ►rayonnement thermique des surfaces) ; *syn.* rayonnement [m] thermique (DUV 1984, 56) ; *g* **Wärmestrahlung [f]** (*Strahlung an der Erdoberfläche* Abstrahlung der im Boden gespeicherten Wärme und die atmosphärische Gegenstrahlung; ►Wärmerückstrahlung).

radiation [n] *envir.* ►electromagnetic radiation, ►natural background radiation; *met.* ►night of ground radiation, ►solar radiation, ►terrestrial radiation.

4767 radiation frost [n] *met.* (**1.** Principally, frost occurring during calm, cloudless nights with excessive outgoing radiation, which affects the air layer close to the ground; **r. f.s** are produced locally. **2.** In contrast to **r. f.** the coldness of **advective frost** is caused when cold air from another region moves into an area and winds remain relatively strong. Frosts are frequently classified as either **advective** or **radiative**, depending on the atmospheric conditions under which they occur: both types often cause frost damage to crops through their interaction: whereby incoming cold air masses lower the temperature to a dangerous level, which can lead to the subsequent death of sensitive, cultivated plants; cf. GEI 1961, 534; ►night of ground radiation); *s* **helada [f] de irradiación** (**1.** Helada que tiene sobre todo efecto en las capas de aire cercanas al suelo y que se presenta en las noches de invierno despejadas y sin viento, debido a que la superficie de la tierra pierde calor por causa de la irradiación. **2.** El frío de la **helada negra o helada de advección** se origina —al contrario que el de la **h. de i.**— por el transporte de aire frío por medio del viento. Ambos tipos de heladas causan —frecuentemente en interacción— daños a las plantas de cultivo: la llegada del aire frío reduce la temperatura hasta niveles peligrosos, la consiguiente irradiación de calor puede llevar hasta la muerte de las planta sensibles; ►noche radiativa); *f* **gelée [f] de rayonnement** (Formation de gelée blanche provoquée par la perte de chaleur par rayonnement terrestre nocturne dans la couche d'air proche du sol pendant les nuits calmes et claires [nuit de gelée blanche/nuit de gelée de rayonnement], fréquente surtout au printemps et en automne ; les dégâts causés sur les végétaux interviennent par apport d'air froid [**gelée d'advection** encore appelée **gelée de plein vent**] qui, associé au rayonnement terrestre, peut provoquer la mort des plantes ; la gelée d'advection qui se produit plutôt en hiver prend le nom de *gelée noire* lorsque la végétation que le vent endommage, gelée intérieurement, prend un aspect noirci, alors que la gelée de rayonnement, elle, est grande productrice de *gelée blanche* ; ►nuit en situation radiative) ; *g* **Strahlungsfrost [m]** (**1.** Vorwiegend für die bodennahe Luftschicht wirksamer Frost während einer windstillen und wolkenlosen Nacht, in der die Erdoberfläche infolge großer Abstrahlung Wärme verliert. **2.** Die Kälte des **Advektivfrostes** entsteht im Unterschied zum **S.** durch den Antransport von Kaltluft während einer großräumigen Wetterlage und ist deshalb mit Wind verknüpft. Beide Frostarten verursachen oft durch ihr Zusammenwirken Schäden an Kulturpflanzen: der Antransport der Kaltluftmasse senkt das Temperaturniveau bis zum Gefahrenbereich ab, die nachfolgende Ausstrahlung kann bis zum Tod empfindlicher Pflanzen führen; cf. GEI 1961, 534; ►Strahlungsnacht).

4768 radiation pollution [n] *envir.* (Environmental threat caused by natural or artificial radiation, and especially radiation from radioactive material); *s* **contaminación [f] de radiación** (Carga ambiental de radiación natural y artificial, sobre todo radiactiva); *f* **charge [f] d'irradiation** (Dangers et inconvénients pour l'environnement présentés par l'émission de rayonnements

naturels ou artificiels [rayons ionisants]) ; *syn.* charge [f] irradiante ; *g* **Strahlenbelastung [f]** (Belastung der Umwelt mit natürlicher oder künstlicher, vor allem radioaktiver [ionisierende Strahlen] und elektromagnetischer Strahlung; im engeren Sinne ist meist die ionisierende **S.** gemeint).

4769 radiation protection [n] *envir.* (Facilities and measures to protect human beings from the harmful effects of radiation, especially in the fields of medicine and the use of nuclear energy); *s* **protección [f] contra la radiación** (Conjunto de medidas técnicas y legales para evitar daños causados por radiaciones ionizantes en la población); *f* **radioprotection [f]** (Mesures de protection sanitaires de la population contre les rayonnements ionisants, en particulier dans le domaine de la médecine et contre les effluents gazeux des centrales nucléaires) ; *syn.* protection [f] contre les radiations, protection [f] contre les rayons ionisants ; *g* **Strahlenschutz [m]** (Einrichtungen und Maßnahmen zum Schutze des Menschen vor schädlichen Wirkungen ionisierender Strahlung, vor allem in der Medizin und Technik sowie bei der Nutzung der Kernenergie).

radiative frost [n] *met.* * ►radiation frost.

radii [npl] *trans. plan.* ►transition radii.

4770 radioactive fallout [n] *envir.* (Particles descending from radioactive clouds through the atmosphere following a nuclear explosion or accident; cf. WEB 1993); *s* **precipitación [f] radiactiva** *syn.* fallout [m]; *f* **retombée [f] radioactive** (Dépôt à la surface de la Terre de particules [poussières] radioactives invisibles et légères, rejetées dans l'atmosphère à la suite d'explosions nucléaires ou d'émissions de déchets par les installations nucléaires) ; *g* **radioaktive Deposition [f]** (Radioaktiver Niederschlag in Form von Aerosolen [winzige Staubteilchen — v. a. Spaltprodukte von Explosionsprodukten, welche auch in Regen und Schnee gelöst sind], die bei Kernwaffenexplosionen oder bei schweren Betriebsunfällen in Kernkraftwerken [z. B. Tschernobyl 1986] in die freie Atmosphäre gelangt sind); *syn.* Fallout [m], radioaktiver Niederschlag [m].

4771 radioactive waste [n] *envir.* (Non-usable refuse composed of radionuclides caused by disintegration of atomic nuclei; ►disposal of radioactive waste, ►final disposal site for radioactive waste); *s* **residuos [mpl] radiactivos/nucleares** (Materiales que contienen radionuclidos, por lo que son contaminantes, y para los que no está previsto ningún uso; ►cementerio nuclear, ►disposición de residuos radi[o]activos/nucleares); *syn.* desechos [mpl] nucleares/radiactivos/radioactivos *(v.o.)*, residuos [mpl] radioactivos *(v.o.)*; *f* **déchets [mpl] radioactifs** (Déchets **1.** dont l'activité massique est >2 microcuries/kg et **2.** dont l'activité totale est > 0,1 microcurie pour les éléments du groupe I, 1 microcurie pour le groupe II A, 10 microcuries pour le groupe II B et 100 microcuries pour le groupe III, ces groupes définissant la radiotoxicité plus ou moins élevée des radioéléments conformément à la réglementation définissant les principes généraux de protection contre les rayonnements ionisants et fixant les normes de protection des travailleurs contre ces dangers ; cf. CPEN, avis du 6 juin 1970 ; ►centre de stockage de déchets radioactifs, ►élimination des déchets radioactifs) ; *syn.* rejets [mpl] radioactifs ; *g* **radioaktiver Abfall [m]** (Materialien, die Radionuklide enthalten, hierdurch kontaminiert sind und für die kein Verwendungszweck vorgesehen ist, wenn die Werte der spezifischen Aktivität der Anlage III Teil A Nr. 1 und die Freigrenzen der Anlage IV Tabelle IV Spalte 4 der Strahlenschutzverordnung überschritten werden; AtAV 1998; ►Endlager für radioaktive Stoffe, ►Entsorgung des Atommülls); *syn.* Atommüll [m].

4772 radius pavers [npl] [US] *constr.* (Specially manufactured row of interlocking pavers to follow curves in pathways); *syn.* radius setts [npl] [UK] (SPON 1986, 373); *s* **set [m]**

R

de piedras de pavimento para curvas (Conjunto de piezas de pavimento interconectadas fabricadas especialmente para caminos en curva); *f* **assortiment [m] de pavés pour courbes** (Rangée de pavés autobloquants spécialement prévue pour un pavement de forme sinueuse) ; *g* **Kurvensatz [m]** (...sätze [pl]; besonders gefertigte Verbundpflastersteinreihe für runde Wege); *syn.* Kurvenkeilsteine [mpl], Kurvensteine [mpl].

radius setts [npl] [UK] *constr.* ▶radius pavers [US].

4773 ragstone [n] [US] *constr.* (Building stone quarried in thin blocks); *syn.* flat undressed stone [n] [UK]; *s* **laja [f]** (Término aplicado a las piezas irregulares de piedra que se utilizan para enlosar paseos y terrazas; DACO 1988); *syn.* losa [f] de piedra; *f* **barrette [f] brute** (VRD 1994, partie 4, chap. 6.4.2.3, 16) ; *syn.* barrette [f] éclatée quernée (VRD 1994, partie 4, chap. 6.4.2.3, 16) ; *g* **flacher Bruchstein [m]** (Im Steinbruch gebrochener flacher Naturstein, der zum Mauerbau verwendet wird).

4774 ragwork [n] *constr.* (Crude ▶masonry construction, laid in a random pattern of thin-bedded, undressed stone [like flagging]; most commonly set horizontally; DAC 1975); *s* **mampostería [f] de lajas** (Muro de piedras alargadas [de pizarra] sin labrar, colocadas en horizontal en hiladas de 2-3 cm de espesor; ▶mampostería); *f* **maçonnerie [f] en pierres plates, grossièrement équarries, appareillées à l'horizontale** (▶Maçonnerie en pierres plates, longues et fines, dont la mise en œuvre est réalisée par assises réglées horizontalement avec une épaisseur des lits et des joints ne dépassant guère 2 cm d'épaisseur ; ces pierres sont en général des chutes de dalles ou des barrettes d'ardoise) ; *g* **Bruchsteinmauerwerk [n] aus dünnem, plattigem Material (≠)** (▶Mauerwerk aus langen, dünnen, plattigen, bruchrauen [Schiefer]gesteinsabfällen, in allen Lagerfugenbereichen horizontal, lagerhaft verlegt. Die Schichtstärken sind ca. 2-3 cm dick).

rail [n] *constr.* ▶arris rail.

railing [n] **to prevent falling** *constr. leg.* ▶protection from falling, #2.

railroad lines [n] [US] *leg. trans.* ▶railroad right-of-way [US]/railway right-of-way [npl] [UK].

4775 railroad right-of-way [n] [US] *leg. trans.* (**1.** Acquired strip of land for railroads [US]/railways [UK] for locomotive-drawn trains or other wheeled vehicles; *syn.* railway right-of-way [n] [UK]. **2.** Linear cut areas and embankments for railroads [US]/railways [UK] are **railroad lines [npl] [US]/railway lines [npl] [UK]**; ▶railroad siding [US]/works siding [UK]); *syn.* railroad tracks [npl] [US], trackage [n] [US]; *s* **instalaciones [m] ferroviaras** (Conjunto de áreas por donde transcurren vías de ferrocarril incluida su correspondiente construcción de soporte; ▶vía muerta); *f* **installations [fpl] ferroviaires** (Ensemble de l'emprise occupée par les voies ferrées ; ▶voie de garage) ; *syn.* en partie voie [f] ferrée ; *g* **Gleisanlage [f]** (**1.** Flächen und Bahndämme, die dem schienengebundenen örtlichen und überörtlichen Verkehr dienen. **2.** Die Gesamtheit von Schienen, Schwellen und Gleis[schotter]bett wird **Gleiskörper** genannt; ▶Abstellgleis); *syn.* Bahnanlage [f].

4776 railroad siding [n] [US] *trans.* (Dead-end railroad track to provide a storage place for temporarily idle cars; ▶railroad right-of-way [US]/railway right-of-way [UK]); *syn.* pull-off track [n] [US], railway siding [n] [UK], sidetrack [n] [US], sidetrack [n] [UK], works siding [n] [UK]; *s* **vía [f] muerta** (Vía de ferrocarril sin salida que sirve para estacionar trenes provisionalmente o en desuso; ▶instalaciones ferroviaras); *f* **voie [f] de garage** (Voie ferrée de service, en général à proximité d'une gare, affectée au manœuvres et **1.** utilisée en cul de sac pour l'immobilisation des wagons de chemin de fer et terminée par un

heurtoir ou **2.** parallèle au réseau principal et reliée à la voie principale par des aiguillages afin d'assurer le garage des trains lents ; ▶installations ferroviaires) ; *g* **Abstellgleis [n]** (Toter Schienenstrang zum Abstellen von Eisenbahnwaggons; ▶Gleisanlage).

railroad tracks [n] [US] *leg. trans.* ▶railroad right-of-way [US]/railway right-of-way [UK].

railway lines [n] [UK] *leg. trans.* ▶railroad right-of-way [US]/railway right-of-way [UK], #2.

railway right-of-way [n] [UK] *leg. trans.* ▶railroad right-of-way [US].

railway siding [n] [UK] *trans.* ▶railroad siding [US].

rain [n] *envir.* ▶acid rain; *met.* ▶heavy rainfall; *phyt. for.* ▶seed rain;

raindrop erosion [n] [UK] *geo. pedol.* ▶splash erosion.

rainfall [n] *met.* ▶heavy rainfall; *constr. eng. hydr.* ▶rate of rainfall.

rainfall intensity [n] *constr. eng. hydr.* ▶rate of rainfall.

4777 rain forest [n] [US]/**rain-forest** [n] [UK] *geo. phyt.* (Primarily evergreen natural forest located in permanently wet areas of the tropics [tropical rainforest], the subtropics [subtropical rainforest] and other frost-free regions [temperate rainforest]); *syn.* ombrophilous forest [n]; *s* **pluviisilva [f]** (Bosque siempreverde, cerrado, de muchos pisos y con abundancia de epífitos, lianas, etc., con gran variedad de árboles, que se presenta en zonas tropicales con calor y humedad constantes, sobre suelos zonales y secundarios; DINA 1987; *términos específicos* pluviilignosa, pluviifruticeta); *syn.* bosque [m] pluvial tropical, bosque [m] de lluvia, bosque [m] ombrófilo, selva [f] lluviosa; *f* **forêt [f] ombrophile** (Forêt naturelle en général dense, sempervirente dans les régions très pluvieuses de la zone tropicale [forêt tropicale humide ou forêt équatoriale] des régions subtropicales [forêt de feuillus sempervirents] et des pays tempérés [forêt ombrophile, forêt valdivienne dans l'hémisphère sud] ; DG 1984, 458) ; *syn.* pluvisilve [f] ; *g* **Regenwald [m]** (Ein vorwiegend immergrüner natürlicher Wald in ganzjährig feuchten Gebieten der Tropen [tropischer Regenwald], der Subtropen [subtropischer Regenwald] und der frostfreien Außertropen [temperierter Regenwald]. In der Übergangszone mit hohen Niederschlägen und einer geringen Anzahl arider Monate hin zum Dornbuschwald in der Savanne mit geringen Jahresniederschlägen befindet sich der feuchte Monsunwald. Wenn er durch immergrünes Unterholz und eine Laub abwerfende Baumschicht charakterisiert ist, wird er je nach Feuchteregime als **halbimmergrüner R.**, regengrüner **Monsunwald** oder feuchter **Laub abwerfender Monsunwald** bezeichnet; cf. WAG 1984 et ÖTS 1984, 115 f).

4778 rainshadow [n] *met.* (Area situated on the lee side of a range of mountains or hills, which has relatively low average rainfall); *s* **vertiente [f] protegida de la lluvia** (Zona de baja pluviosidad situada en el lado seco de una cadena de montañas); *f* **zone [f] à l'abri de la pluie** (Versant situé sur le côté abrité du vent d'une montagne ou d'une chaîne de montagnes et recevant peu de précipitations) ; *syn.* zone [f] protégée de la pluie, versant [m] protégé de la pluie ; *g* **Regenschattenseite [f]** (Gebiet, das im Lee eines Berges oder Gebirgszuges wenig Regen erhält).

4779 rainwater [n] (**1**) *met. wat'man.* (**1.** ▶Precipitation 1 in liquid form; ▶direct runoff, ▶heavy rainfall, ▶precipitation water); **2. storm water** [n] *constr. urb.* (Rainwater, which runs off roads, roofs and other paved areas without percolation and is collected by means of surface drainage measures such as ditches, infiltration swales, or percolates in planted areas, lawns and meadows, etc., or is discharged into a system of underground pipes); *s 1* **agua [f] de lluvia** *met. wat'man.* (▶Precipitación en forma

líquida; ►agua de precipitación, ►lluvia torrencial); *syn.* agua [f] pluvial; *s 2* aguas [fpl] de lluvia *constr. urb.* (Aguas de precipitación en calles, tejados o en otras superficies pavimentadas, que se infiltran en el suelo por medio de zanjas, hoyos de infiltración o en superficies de césped o hierba y si no tienen que ser evacuadas por la canalización; ►escorrentía superficial 1); *syn.* aguas [fpl] pluviales; *f 1* eau [f] de pluie *met. wat'man.* (►Précipitations atmosphériques sous forme liquide ; ►eau de précipitations, ►pluie torrentielle) ; *f 2* eaux [fpl] pluviales *constr. urb.* (Eaux issues du ruissellement sur les surfaces imperméabilisées [voiries, toitures, terrasses, parkings, etc.] et qui rejoignent après traitement le milieu naturel ; ►ruissellement) ; *g 1* Regenwasser [n, o. Pl.] (1) *met. wat'man.* (►Niederschlag in flüssiger Form; ►Niederschlagswasser, ►Starkregen); *g 2* Regenwasser [n, o. Pl.] (2) *constr. urb.* (Das nicht versickernde Niederschlagswasser auf Straßen, Dächern und sonstigen befestigten Flächen, das oberirdisch in Gräben, Versickerungsmulden, auf Rasen- und Wiesenflächen etc. versickert oder durch die Kanalisation abgeführt wird; ►Oberflächenwasserabfluss 1).

rainwater [n] (2) *constr. wat'man.* ►excess rainwater; *envir. urb. wat'man.* ►retention of rainwater.

4780 rainwater interceptor basin [n] *wat'man.* (Reservoir designed to trap mud, silt, etc. in surface runoff after a heavy storm, before the water enters a sewage treatment plant. It is then usually discharged directly into a ►receiving stream; ►flood control reservoir, ►storm water detention basin); *s* embalse [m] de sobrecarga (En estación depuradora de aguas residuales, embalse interceptor para acoger las aguas de precipitaciones fuertes; ►curso receptor, ►embalse de retención, ►embalse de regulación de aguas pluviales); *f* déversoir [m] d'orage (Ouvrage permettant d'évacuer dans un ►émissaire ou dans un ►milieu récepteur la partie excédentaire du débit des eaux pluviales par rapport au débit maximal admissible par la station d'eau pluviale ; ►bassin de rétention des crues, ►bassin de retenue d'eaux pluviales) ; *g* Regenüberlaufbecken [m] (Einer Kläranlage vorgelagertes ober- oder unterirdisches Becken, das bei Starkregenanfall die Menge des Abwassers, die sie nicht mehr verarbeiten kann, auffängt und die Schmutz- und Abschwemmstoffe des Oberflächenwassers trennt; das Wasser wird anschließend meist direkt in einen ►Vorfluter geleitet oder bei Trockenwetter in die Kläranlage zur Reinigung gepumpt; ►Hochwasserrückhaltebecken, ►Regenrückhaltebecken).

rain-water run-off [n] [UK], **heavy** *constr. plan.* ►heavy storm runoff [US].

rain-water sewer [n] [UK] *urb. wat'man.* ►storm sewer [US].

4781 raised bog [n] *geo. phyt.* (Concentric or excentric domed accumulation of peat, mostly Sphagnum moss, in a shallow basin. The vegetation is only influenced by rainwater, never by groundwater; ►bog cultivation, ►bog growth, ►lens-shaped raised bog; *opp.* ►fen 1, ►low bog); *syn.* ombrotrophic peatland/bog, ombrogenous peatland, ombrophilous peatland/bog (KUL 1949); *s* turbera [f] alta (Paisaje caracterizado y vegetación determinada por el encharcamiento del suelo que está constituida por la asociación de musgos del género Sphagnum, a la que se unen musgos de otros géneros, herbétum y lignétum, pudiendo éste llegar hasta el arborétum [Betula o Pinus silvestris en Europa nórdica]. La t. a. es de origen supraacuático, su agua es ácida, sin cal y pobre en elementos nutritivos; cf. DB 1985; ►crecimiento de turbera, ►cultivo de turbera, ►turbera abombada, ►turbera baja 2; *opp.* ►turbera baja 1); *syn.* turbera [f] de esfagnos, turbera [f] ombrotrófica, turbera [f] oligotrófica, turbera [f] supraacuática, turbera [f] ácida, bolsón [m] de turba [EC], bolsillo [m] de turba [YV]; *f* tourbière [f] haute (Terme générique caracté-

risant les paysages marécageux généralement en forme de lentille s'élevant au-dessus du plan d'eau, dont l'économie en eau est indépendante de la nappe phréatique, dont l'évapotranspiration et le drainage externe sont inférieurs à la pluviosité [pluviosité excédentaire]. On les qualifie encore de « marais rouge » car les Sphaignes qui constituent l'essentiel de la végétation ont généralement des teintes vives rougeâtres ou brunâtres ; *termes spécifiques* tourbière oligotrophe [Tourbière de pH inférieur à 5 et dont la strate muscinale est formée principalement de Sphaignes], tourbière ombrogène [Tourbière à pluviométrie excédentaire sur substrat sans calcaire et essentiellement oligotrophe ou sur substrat calcaire mais établie sur une tourbière topogène alcaline], tourbière topogène oligotrophe [Tourbière à pluviosité déficitaire et alimentation par les eaux de ruissellement d'un impluvium à sols acides], tourbière de pente formée au niveau d'une source, tourbière adossée s'appuyant sur les marges sèches d'un marais ; GGV 1979, 240-262 ; DEE 1982, 90 ; ►épaississement de la tourbière, ►mise en culture de tourbières, ►tourbière à forme de lentille, ►tourbière basse ; *opp.* ►tourbière plate) ; *syn.* tourbière [f] bombée, tourbière [f] supra-aquatique, haute tourbière [f], haut marais [m], tourbière [f] ombrotrophe [aussi CDN]), tourbière [f] à Sphaignes ; *g* Hochmoor [n] (Meist deutlich aufgewölbter Landschaftsbereich, der außerhalb des Grundwassereinflusses nur auf die Niederschlagsversorgung angewiesen ist; hauptsächlich auf Hochmoortorfmoosen aufgebaut und extrem nährstoffarm; H.e sind im Laufe der Entwicklung über den Grundwasserspiegel emporgewachsen oder haben sich in niederschlagsreichen Klimagebieten direkt aus Rohhumusauflagen von Podsolen entwickelt; viele H.gebiete wurden nach Trockenlegung mit unterschiedlichen Kulturtechniken landwirtschaftlich genutzt; ►Moorkultur, ►Moorwachstum, ►Niedermoor, ►urglasförmiges Hochmoor; *opp.* ►Flachmoor); *syn.* ombrogenes Moor [n], Regenmoor [n], Regenwassermoor [n], *in BY auch* Filz [m]).

raised bog [n], **ancient lake** *geo. phyt.* ►raised bog on a silted-up lake.

4782 raised bog [n] **on a silted-up lake** *geo. phyt.* (►Raised bog developed over a long period of time from a very lime-deficient [dystrophic], stagnant body of water containing a high amount of organic substances); *syn.* ancient lake raised bog [n] (ELL 1988, 328); *s* turbera [f] lacustre (≠) (En climas templados-fríos y relativamente secos, como en Polonia o el este de Alemania, ►turbera alta con asociaciones de musgos de turbera como *Sphagnetum magellanici* y *Sphagnetum fusci* generada por colmatación de aguas estancadas distrofas y muy pobres en calcio); *f* tourbière [f] lacustre (►Tourbière haute formée par le colmatage progressif d'eaux dormantes ; *termes spécifiques* tourbière neutro-alcaline [Tourbière formée par le colmatage d'une nappe d'eau, d'un lac — colonisation par les hydrophytes et les hélophytes et envahissement progressif — dans le fond des grandes vallées ; GGV 1979, 249], tourbière lacustre acide [Tourbière haute formée par le colmatage progressif d'eaux dormantes acides et riches en matières organiques — dystrophes — dans la région du Mark Brandebourg ou en Pologne, avec colonisation par les associations de Sphaignes — *Sphagnetum magellanici, Sphagnetum fusci*]) ; *g* Verlandungshochmoor [n] (In einem kühlgemäßigten und relativ trockenen Klima, z. B. in der Mark Brandenburg oder in Polen, aus sehr kalkarmem und humusreichem [dystrophem] Stillgewässer entstandenes ►Hochmoor mit Torfmoosgesellschaften wie z. B. *Sphagnetum magellanici, Sphagnetum fusci*; cf. ELL 1978, 440).

raised bog peat [n] *min. pedol.* ►humified raised bog peat, ►sphagnum peat.

raised planter [n] [US] *constr.* ►raised planting bed.

4783 raised planting bed [n] *constr.* (Planting area with an elevated stone or wood/railroad tie [US]/sleeper [UK] edging); *syn.* raised planter [n] [also US] (TGG 1984, 178, 194); *s* **bancal [m] de plantación elevada** (Área de plantación elevada encintada con murete de unos 30-50 cm de altura); *f* **plate-bande [f] surélevée** (Surface plantée entourée par un muret de pierres, en bois/travée de bois) ; *g* **Hochbeet [n]** (Durch Stein- oder Holz-/Eisenbahnschwellenmauer eingefasstes Pflanzenbeet).

raised soil level [n] **near a tree trunk** *arb. constr.* ▶ groundfill around a tree.

raised trail [n] *recr.* ▶ boardwalk.

raised walk [n] *recr.* ▶ boardwalk.

raising [n] **of pH value** *chem. limn. pedol.* ▶ increasing of pH value.

raising [n] **the canopy/crown/head** [US] *constr. for. hort.* ▶ crown lifting.

raising [n] **the temperature** *envir. limn.* ▶ heating up.

4784 rake [vb] *constr. hort.* (1. To clean an open area with an implement having a pole and crossbar toothed like a comb, for loosening up and smoothing the soil surface. 2. To collect leaves, twigs, stones, etc. with a rake); *s* **rastrillar** [vb] (1. Limpiar o igualar un área no plantada con un rastrillo. 2. Limpiar un área de hojarasca, desbrozos, piedrecillas, etc. con un rastrillo); *f* **ratisser** [vb] (Nettoyer ou niveler la surface du sol, enlever les feuilles, les pierres, etc. au moyen d'un râteau) ; *g* **abharken** [vb] (1. Mit einer Harke [einem Rechen] eine Fläche einebnen oder säubern. 2. Laub, Reisig, Steine etc. mit einer Harke [einem Rechen] entfernen); *syn.* abrechen [vb] [südd.].

rambler rose [n] *hort.* ▶ climbing rose.

ramblers' hostel [n] [UK] *recr.* ▶ hiking accommodation [US].

ramblers' route [n] [UK] *recr.* ▶ hiking trail [US]/hiking footpath [UK].

rambling area [n] [UK] *plan. recr.* ▶ hiking area.

rambling footpath [n] [UK] *recr.* ▶ hiking trail [US]/hiking footpath [UK].

rambling holiday [n] [UK] *recr.* ▶ hiking holiday.

rambling pathway [n] [UK] *recr.* ▶ hiking trail [US]/hiking footpath [UK].

rambling rose [n] *hort.* ▶ climbing rose.

4785 ramp [n] (1) *constr. trans.* (Incline that bridges a difference in elevation; ▶ highway ramp [US]/slip road [UK]); *s* **rampa** [f] (Plano inclinado que conecta dos superficies de diferente elevación; ▶ rampa de acceso); *f* **rampe** [f] (2) (Plan incliné permettant de relier deux niveaux différents ; *terme spécifique* rampe d'accès ; ▶ bretelle d'accès) ; *g* **Rampe** [f] (Schiefe Ebene zur Überwindung von Höhenunterschieden oder unterschiedlich hoch gelegener Flächen; ▶ Auffahrtsrampe).

ramp [n] (2) *trans.* ▶ cloverleaf ramp; *constr.* ▶ step ramp; *wat'man.* ▶ stream ramp.

ramp [n] [UK], **speed check** *trans.* ▶ speed hump.

ramp [n], **stepped** *trans.* ▶ stepped pathway.

rampart [n] [UK], **stream-bed** *wat'man.* ▶ stream ramp.

4786 ramp step [n] *constr.* (Step within a steep ramp, which has a long and sloping tread; ▶ step ramp); *s* **escalón** [m] **de rampa** (▶ rampa escalonada); *f* **marche** [f] **d'une gradine** (Giron et contremarche sur une ▶ gradine) ; *g* **Rampenstufe** [f] (Stufe in einer steilen Rampe; ▶ Stufenrampe).

4787 Ramsar Convention [n] *conserv. leg.* (Convention on wetlands of international importance especially waterfowl habi-

tats; in Iran 1971; ▶ wetland of international significance); *s* **Convención** [f] **de Ramsar** (Convenio de 2 de febrero de 1971 relativo a humedales de importancia internacional, especialmente como hábitat de aves acuáticas. Hasta la fecha se han adherido 159 países, en los que hay 1.867 humedales protegidos con una superficie total de 183,7 millones de hectáreas. España se adhirió por Instrumento de 18 de marzo de 1982. **En Es.** hay 63 zonas en el catálogo MAR [UICN] con una superficie total de 282.228 ha —entre ellas Doñana, Tablas de Daimiel, Salinas del Cabo de Gata, S'Albufera de Mallorca, Albufera de Valencia, Delta del Ebro, Ria de Mundaka-Guernika, Mar Menor, Marismas de Santoña, Bahía de Cádiz y Parque Nacional de Aiguestortes i Estany de Sant Maurici— y además existe un Inventario Nacional de Zonas Húmedas. **En AL**, el país con más lugares es México con 113 con una superficie total de 8.156.746 ha, seguido de Argentina con 18 [5.315.376 ha], Perú 13 [6.784.042 ha], Ecuador 13 [201.126 ha], Costa Rica 11 [510.050 ha], Brasil 9 [6.441.086 ha], Chile 9 [159.154 ha], Bolivia 8 [6.518.073 ha], Nicaragua 8 [405.691 ha], Guatemala 7 [628.592 ha], Cuba 6 [1.188.411 ha], Paraguay 6 [785.970 ha], Honduras 6 [223.320 ha], Colombia 5 [458.525 ha], Panamá 4 [159.903 ha], El Salvador 3 [125.769 ha], Uruguay 2 [424.904 ha] y República Dominicana 1 [20.000 ha]; cf. RAM 2009; ▶ zona húmeda de importancia internacional); *f* **Convention** [f] **de Ramsar** (La Convention du 2 février 1971 signée à Ramsar, en Iran, est un traité intergouvernemental relatif aux zones humides d'importance internationale ayant pour objet de d'assurer la qualité de l'eau et la production vivrière et de préserver les fonctions écologiques fondamentales des zones humides en tant que régulateur du régime des eaux et en tant qu'habitats d'une flore et d'une faune caractéristiques et particulièrement des oiseaux d'eau [diversité biologique] ; la Convention a, actuellement, 159 Parties contractantes qui ont inscrit 1854 zones humides, pour une superficie totale d'environ 182 millions d'hectares ; sur la Liste de Ramsar des zones humides d'importance internationale figurent pour la France : 1. la Camargue, 2. les Étangs de la Champagne humide, 3. les Étangs de la Petite Woëvre, 4. les Marais du Cotentin et du Bessin, la Baie des Veys, 5. le Golfe du Morbihan, 6. La Brenne, 7. les Rives du Lac Léman, 8. l'Étang de Biguglia, 9. le Lac de Grand-Lieu, 10. les Basses Vallées Angevines, 11. La Petite Camargue, 12. la Baie de la Somme, 13. le Bassin du Drugeon, 14. les Étangs du Lindre, 15. la forêt du Romersberg et zones voisines, 16. le Lac du Bourget — Marais de Chautagne, 17. le Marais du Fier d'Ars, 18. les étangs littoraux de la Narbonnaise, 19. les Mares temporaires de Tre Padule de Suartone, 20. le Rhin supérieur/Oberrhein, 21. l'Estuaire du fleuve Sinnamary, 22. l'Étang de Palo, 23. les Étang des Salines, 24. l'Étang d'Urbino, 25. les Étangs palavasiens, 26. l'Impluvium d'Evian, 27. le Lagon de Moorea, 28. le marais audomarois, 29. les Étangs de Villepey, 30. la Réserve Naturelle Nationale des Terres Australes Françaises, 31. les Salines d'Hyères ; ▶ zone humide d'importance internationale) ; *g* **Ramsar-Konvention** [f] (In Ramsar [Iran] verabschiedetes internationales „Übereinkommen über Feuchtgebiete, insbesondere als Lebensraum für Wasser- und Watvögel, von internationaler Bedeutung", das von der Bundesrepublik Deutschland am 02.02. 1971 unterzeichnet wurde. Das Abkommen ist nach seinem Artikel 10 [2] für die Bundesrepublik Deutschland am 25.06. 1976 in Kraft getreten [BGBl. II 1265, geändert durch das Protokoll vom 03.12.1982 [BGBl. 1990 II 1670] und zuletzt geändert auf der außerordentlichen Konferenz der Vertragsparteien in Regina, Saskatchewan, Canada [v. 28.05.-03.06. 1987], für D. in Kraft getreten am 01.05.1994 [BGBl. 1995 II 218. Die anfänglich 17 benannten Feuchtgebiete sind mittlerweile auf über 20 angestiegen: 1. Wattenmeer, Elbe-Weser-Dreieck, 2. Wattenmeer, Jadebusen u. westliche Wesermündung, 3. Wattenmeer, Ostfriesisches Wattenmeer mit Dollart, 4. Niederelbe, 5. Elbaue zw. Schnackenburg und Lauenburg, 6. Dümmer, 7. Diepholzer Moorniederung, 8. Steinhuder Meer, 9. Rhein zw. Eltville und Bingen, 10. westlicher Bodensee, Teilgebiete Wollmatinger Ried — Giehrenmoos, Hegnebucht des Gnadensees u. Mindelsee bei Radolfzell, 11. Donauauen und Donaumoos im Reg.-Bez. Schwaben, 12. Lech-Donau-Winkel, 13. Ismaninger Speichersee mit Fischteichen, 14. Ammersee, 15. Starnberger See, 16. Chiemsee, 17. Unterer Inn zw. Haiming u. Neuhaus, 18. Rieselfelder Münster, 19. Unterer Niederrhein, 20. Weserstaustufe Schlüsselburg; für das UK trat das Abkommen am 05.05.1976 [BGBl. II 1265], für die USA am 18.12.1986 [BGBl. 1991 II 507], für F. am 01.10.1986 [BGBl. 1991 II 507] und für Spanien am 04.09. 1982 [BGBl. II 767], geändert am 27.05.1987 [BGBl. 1990 II 1670] in Kraft; ▶Feuchtgebiet internationaler Bedeutung).

4788 random cobblestone paving [n] [US] *constr.* (Laying of irregularly shaped, usually rectangular natural stones which are relatively uniform in size without a geometric ▶paving pattern in all directions [random pattern], so that diagonal lines are not created; ▶paving stone [US]/paving sett [UK]); *syn.* random sett paving [n] [UK] (LD 1988 [10], 48); *s* **adoquinado [m] sin orden** (Pavimento de ▶adoquines más o menos rectangulares y de tamaños similares que se tiende sin aparejo geométrico en todas las direcciones, de manera que no crean líneas diagonales; ▶patrón de aparejo); *f* **pavage [m] désordonné en pierre** (Revêtement de sol en pavés irréguliers [de forme et de pose aléatoire] en pierre naturelle, de forme rectangulaire, posés sans ▶calepinage géométrique, dans toutes les directions ; ▶pavé) ; *syn.* pavage [m] à calepinage aléatoire, pavage [m] en pierre à calepinage désordonné, pavage [m] à calepinage irrégulier, pavage [m] à calepinage « pose sauvage » ; *g* **Wildsteinpflaster [n]** (Belag aus unregelmäßigen, vorwiegend rechteckigen Natursteinen, die [in der Größe ziemlich einheitlich sortiert sind und] ohne geometrisches ▶Verlegemuster kreuz und quer ohne Schräglinien angeordnet sind; ▶Pflasterstein); *syn.* Wildpflaster [n].

random embossed ashlar masonry [n] *arch. constr.* ▶random rough-tooled ashlar masonry.

4789 random irregular bond [n] *constr.* (Pattern created by polygonal stones in masonry, as well as by irregular slabs in paved surfaces); *syn.* random irregular pattern [n]; *s* **aparejo [m] poligonal** (Tipo de ensamblaje de las losas de pavimento o de los mampuestos en obra de fábrica); *f* **appareillage [m] polygonal** (Disposition des dalles ou des pierres irrégulières taillées les unes en fonction des autres souvent caractérisée par la quasi absence de joints) ; *g* **polygonaler Verband [m]** (Anordnung von vieleckigen Platten auf Belagsflächen oder von vieleckigen Steinen im Mauerwerk).

random irregular pattern [n] *constr.* ▶random irregular bond.

random-jointed rectangular paving pattern [n] *constr.* ▶random rectangular bond.

4790 randomly planted perennial [n] [UK] *hort. plant.* (*Term used to describe the growth habit of a perennial with regard to its aesthetic effect and used in garden planting* species of freestanding plant planted individually or in clusters at various intervals in a lower-growing carpet of other non competitive plants or in a gravelled area, scattered in such a way that it accentuates the planting and its form or character is shown off to its advantage. Equal spacing of the plants produces a grid planting effect. Term is usually applied to perennials whose attractive form, though not pronounced enough for specimen planting, is enhanced when planted in random groups with a background of less conspicuous plants, e.g. Japanese sedge *[Carex morrowii* 'Variegata'*]*, Leatherleaf sedge *[Carex buchananii]*; as well as ▶geophytes which only flower for short time before they wither and disappear, e.g. Foxtail lily *[Eremurus]* or Allium *[Allium christophii]*, as well as other perennials which die back on the surface and have a long period of dormancy [e.g. *Oenothera macrocarpa, Gypsophila paniculata]*; *s* **perenne [f] plantada al azar (±)** (Término utilizado en GB en la plantación de jardines para denominar el comportamiento de crecimiento de una planta vivaz en relación a su efecto estético; ▶geófito); *f* **espèce [f] pérenne à répartition aléatoire (±)** (*Terme désignant le comportement de croissance d'une plante vivace en regard de son effet esthétique et de son utilisation dans la composition des jardins* espèce herbacée plantée isolée ou en groupe à distances inégales dans un massif de plantes vivaces ou en couverture végétale basse dans une surface empierrée de galet ou de rocaille ; la répartition des espèces plantées se fait de manière aléatoire mais par leur aspect et leur caractère celles-ci provoquent un effet remarquable ; il s'agit en général de végétaux qui par leur forme caractéristique méritent d'être mis en évidence mais, en raison de leur port et/ou de leur taille, ne peuvent prendre une position isolée, p. ex. la laîche japonaise *[Carex morrowii* 'Variegata'*]*, la laîche rouge *[Carex buchananii]* ; ainsi que les ▶géophytes à effet de courte durée, comme l'Aiguille de Cléopatre *[Eremurus]* ou les ails rustiques *[Allium christophii]*, ainsi que diverses plantes vivaces à long repos végétatif [telles que *Oenothera macrocarpa, Gypsophila paniculata]* ces dernières, utilisées en tapis, laissant après la floraison une surface dénudée) ; *syn.* vivace [f] à disposition aléatoire ; *g* **Streupflanze [f]** (*Terminus für das Wuchsverhalten einer Staude hinsichtlich ihrer ästhetischen Wirkung und gärtnerischen Verwendung* in einer Staudenpflanzung frei stehende Pflanzenart oder -sorte, die zu mehreren in einen verträglichen [konkurrenzschwachen], deutlich niedrigeren Pflanzenteppich oder eine Geröll-/Schotterfläche in spannungsvoll ungleichen Abständen eingestreut wird, damit sie durch ihre Form und ihren Charakter besonders wirkt. [Gleiche Abstände führen zur Rasterpflanzung]. Gilt vorwiegend für solche Stauden, deren wirkungsvoller Formcharakter hervorgehoben werden soll, für die aber auf Grund nicht ausreichender Größe und/oder Substanz eine Solitärstellung gewöhnlich nicht in Frage kommt, z. B. Japan-Segge *[Carex morrowii* 'Variegata'*]*, Fuchsrote Segge *[Carex buchananii]*; hierzu gehören auch nur kurze Zeit wirksame, bald einziehende ▶Geophyten, z. B. Steppenkerze *[Eremurus]* oder Sternkugellauch *[Allium christophii]*, sowie andere oberirdisch absterbende Stauden mit langer Vegetationsruhe [wie *Oenothera macrocarpa, Gypsophila paniculata]*, die bei flächendeckender Verwendung nach der Blühphase Kahlstellen erzeugen).

4791 random-range dressed-faced ashlar [n] *arch. constr.* *s* **sillería [f] ordinaria con mampuestos de cara vista pulida** *syn.* mampostería [f] ordinaria con mampuestos de cara vista pulida; *f* **maçonnerie [f] à appareillage à l'anglaise avec parement dressé (±)** ; *g* **Wechselmauerwerk [n] aus Steinen mit ebenen Ansichtsflächen**.

4792 random rectangular bond [n] *constr.* (Pattern created by rectangular slabs of different sizes, and laid without stack bond joints; ▶paving pattern); *syn.* random rectangular pattern [n], random-jointed rectangular paving pattern [n]; *s* **aparejo [m] romano** (En pavimentos, aparejo sin juntas cruzadas con losas rectangulares de diferentes tamaños; ▶patrón de aparejo); *f 1* **appareillage [m] en opus quadratum** (Dans un dallage disposition irrégulière de dalles de forme rectangulaire et de grandeurs diverses sans formation de joints croisés ; souvent en maçonnerie, les boutisses et les panneresses alternent régulièrement dans chaque assise ; *contexte* dallage en opus quadra-

R

tum/à l'anglaise ; ►calepinage) ; *syn.* appareillage [m] à l'anglaise ; *f 2* **appareillage [m] en opus romain** (Dans un dallage disposition irrégulière de dalles rectangulaires, carrées ou trapézoïdales et de grandeurs diverses sans formation de joints croisés ; *contexte* dallage en opus romain/à l'antique) ; *syn.* appareillage [m] à l'antique ; *g* **römischer Verband [m]** (Anordnung von unterschiedlich großen, viereckigen Platten bei Boden- oder Wegebelägen; die Schwierigkeit der Verlegung besteht u. a. darin, dass nie mehr als zwei gleich große Platten nebeneinander liegen, Fugen nicht zu lang sein und keine Kreuzfugen auftreten dürfen; ► Verlegemuster).

random rectangular pattern [n] *constr.* ►random rectangular bond.

4793 random rough-tooled ashlar masonry [n] *arch. constr.* (Ashlar masonry laid in horizontal courses of different heights with rough-tooled quarry stones; ►broken range work, ►bossage [US]/rustication [UK]); *syn.* random embossed ashlar masonry [n]; *s* **sillería [f] ordinaria con mampuestos de labra tosca** (Tipo de fábrica en la que los mampuestos toscamente labrados se colocan en capas de diferentes alturas; ►labra tosca, ►sillería ordinaria); *syn.* mampostería [f] ordinaria con mampuestos de labra tosca; *f* **maçonnerie [f] à appareillage à l'anglaise en pierres bossagées** (Avec utilisation de pierres équarries traitées en taille bossagée ; ►bossage, ►maçonnerie à appareillage à l'anglaise) ; *syn.* maçonnerie [f] à appareillage à l'anglaise en moellons de bossage ; *g* **Wechselmauerwerk [n] mit bossierten Steinen** (► Wechselmauerwerk, ►Bossierung).

4794 random rubble ashlar masonry [n] *constr.* (Masonry in which stones are set without continuous joints and may have a large repeated pattern); *syn.* random rubble range ashlar [n], random rubble range work [n]; *s* **sillería [f] ordinaria** (Tipo de fábrica de mampuestos de piedra natural que no constituyen aparejo alguno y están colocados con juntas estrechas); *f* **maçonnerie [f] en moellons équarris irrégulier** (Mur en moellons dont les quatre arêtes de parement sont taillées à l'équerre, la face vue étant généralement traitée — moellon de bossage ou de face adoucie ; cf. VRD 1994, partie 4, chap. 6.4.2.3) ; *syn.* maçonnerie [f] en moellons pleins, grossièrement équarris ; *g* **hammerrechtes unregelmäßiges Bruchsteinmauerwerk [n]** (Natursteinmauerwerk aus zugeschlagenen Steinen mit bossierten Ansichtsflächen, fugeneng verlegt und mit in den Schichten variierenden Steinhöhen).

random rubble range ashlar [n] *constr.* ►random rubble ashlar masonry.

random rubble range work [n] *constr.* ►random rubble ashlar masonry.

random sett [n] [UK] *constr.* ►cobblestone [US].

random sett paving [n] [UK] *constr.* ►random cobblestone paving [US].

random work [n] [US] *constr.* ►broken range work.

4795 range [n] (1) *phyt. zool.* (*Generic term* the geographical limits within which a taxon occurs; *specific terms* ►breeding range, ►disjunctive range, ►historic range, ►home range (1), ►home range (2), ►natural range, ►potential range, ►range of a plant community, ►real range, ►total range; ►biotope, ►habitat, ►minimal area); *s* **área [f] de distribución** (Término genérico para el espacio en el que puede existir una especie o una comunidad; ►área de actividad, ►área de habitación, ►área mínima, ►área natural de distribución, ►biótopo, ►hábitat, ►territorio vital total; *términos específicos* ►área de cría, ►área de una comunidad/asociación, ►área disyunta, ►área histórica, ►área virtual [de distribución]); *f* **aire [f] de répartition** (Ensemble des localités pouvant être occupées par les individus d'une espèce en fonction de leurs besoins vitaux ; ►aire d'habi-

tation, ►aire minimale, ►biotope, ►domaine vital, ►habitat 1 ; *termes spécifiques* ►aire de nidification, ►aire de répartition d'une association, ►aire disjointe, ►aire historique d'une espèce, ►aire naturelle d'une espèce, ►aire potentielle d'une espèce, ►aire réelle d'une espèce) ; *syn.* aire [f] de dispersion ; *g* **Areal [n]** (Raum, der von Individuen einer Art entsprechend ihrer Lebensansprüche bewohnt werden kann; ANL 1984; ►Biotop, ►Habitat, ►Jahreslebensraum, ►Minimalfläche, ►Streifgebiet, ►Wohngebiet 1; *UBe* ►Brutareal, ►disjunktes Areal, ►Gesellschaftsareal, ►historisches Areal, ►natürliches Areal, ►potentielles Areal, ►reales Areal); *syn.* Verbreitungsgebiet [n].

range [n] (2) *geo.* ►coastal range, ►mountain range; *ecol.* ►narrow ecological range, ►wide ecological range.

range [n]**, ecological** *ecol.* ►ecological amplitude; ►narrow ecological range, ►wide ecological range.

range [n] **[US], habitat** *phyt. zool. leg.* ►living space.

ranged rubble [n] *constr.* ►squared rubble masonry.

range limit [n] *phyt. zool.* ►distribution limit.

range masonry [n] *arch. constr.* ►coursed ashlar masonry.

4796 range [n] **of a plant community** *phyt.* (Area covered by all separate growing places of a plant association; ►range 1); *s* **área [f] de una comunidad/asociación** (Superficie total de todos los lugares donde crece una comunidad vegetal; ►área de distribución); *f* **aire [f] de répartition d'une association** (Aire renfermant toutes les localités sur lesquelles se développe un groupement végétal ; ►aire de répartition) ; *g* **Gesellschaftsareal [n]** (Gebiet, das von sämtlichen Einzelvorkommnissen einer Pflanzengesellschaft eingenommen wird; ►Areal).

range [n] **of precipitation** *met. wat'man.* ►precipitation regime.

range [n] **of species cover** *phyt.* ►degree of species cover.

rangework [n] *arch. constr.* ►coursed ashlar masonry.

4797 ranker [n] [UK] *pedol.* (Soil with AC profile lacking in carbonates. The A horizon occurs directly upon the parent rock and is often full of gravel and stones when the rock is solid. The upper part of the C horizon is often split mechanically by frost; no US equivalent); *s* **ranker [m]** (Suelo pobre en caliza, cuyo horizonte de humus descansa inmediatamente sobre el material de partida que consiste, generalmente, en rocas silícicas o de silicatos pobres en calcio; KUB 1953, 198); *f* **ranker [m]** (Sol à horizon humifère bien développé, très acide, homogène, formant une limite nette avec l'horizon C, avec fragments de roche-mère) ; *g* **Ranker [m]** (1) (Terrestrischer Boden mit A-C-Profil aus karbonatfreiem Festgestein. Der A-Horizont liegt unmittelbar dem Gestein auf und ist bei festen Gesteinen oft kies- und steinreich. Der obere Teil des C-Horizontes ist oft durch Frostsprengung mechanisch zerteilt; SS 1979).

4798 rapid-hardening cement [n] *constr.* (High quality early-strength cement); *syn.* rapid-setting cement [n], quick-drying cement [n]; *s* **cemento [m] rápido** (Cemento de alta calidad que se endurece muy rápidamente); *syn.* supercemento [m]; *f 1* **ciment [m] prompt** (Ciment à prise rapide) ; *f 2* **accélérateur [m] de prise** (Produit qui, ajouté à un mortier ou en béton, en accélère la prise ; DTB 1985) ; *g* **Schnellbinder [m]** (**1.** Schnell abbindender Zement. **2.** Zuschlagstoff, der das Abbinden von Mörtel oder Beton beschleunigt); *syn.* Schnellhärter [m].

rapid-setting cement [n] *constr.* ►rapid-hardening cement.

4799 rapid transit vehicle [n] [US] *trans.* (Transportation rolling stock using a separate right-of-way to carry people within a conurbation; ►public transportation vehicle, ►local public transportation system [US]/local transport system [UK]); *syn.* local express train [n] [UK]; *s* **medio [m] ferrocarril de transporte público urbano** (Medio de transporte público urbano sobre la base de una vía de ferrocarril propia, sea un tren convencional, metro, tranvía u otro sistema; ►medio de transporte colectivo, ►tranporte público urbano); *f* **moyens [mpl] de transport suburbain ferroviaire** (Terme générique recouvrant les différents modes de transport ferroviaires de voyageurs sur de courtes distances tels que les chemins de fer urbains RER [réseau express régional — réseau ferroviaire de transport en commun desservant Paris et son agglomération], métro, tramway ; ►moyen de transport collectif, ►moyens de transport suburbain, ►transports suburbains) ; *g* **schienengebundenes Nahverkehrsmittel [n]** (Beförderungsmittel wie Schnellbahn [S-Bahn], Untergrundbahn [U-Bahn], Straßenbahn [Strab], Unterpflasterstraßenbahn [U-Strab], das Personen in Ballungsgebieten transportiert; ►öffentliches Verkehrsmittel, ►öffentlicher Personennahverkehr).

4800 raptor [n] *zool. hunt.* (►Predatory bird; e.g. falcon, hawk, eagle, or owl, which has feet with sharp talons or claws adapted for seizing prey, and a hooked beak for tearing flesh); *s* **rapaz [f] diurna** (Término genérico para las ►aves rapaces de las familias de rapaces de América Neotropical [catártidos], de Europa las de los azores [accipítridos] y los halcones [falcónidos], así como los *Sagittariidae* que son una familia de rapaces que habita las estapas africanas; cf. DINA 1987); *syn.* ave [f] rapaz diurna; *f* **rapace [m]** (Terme générique pour les groupes d'►oiseaux de proie, adaptés à un régime carnivore et à la chasse, appartenant aux familles des rapaces du nouveau monde **1.** des rapaces qui chassent le jour [diurnes] : vautour noir, condor *[Cathartidae]*, vautour, aigle, buse, épervier, milan, busard, serpentaire *[Accipitridae]*, faucon *[Falconidae]* et grand serpentaire *[Sagittariidae]*, balbuzard *[Pandionidae]*, **2.** des rapaces qui chassent la nuit [nocturnes] : chouette, hibou *[Strigidae]*, chouette effraie et phodile *[Tytonidae]* ; le rapace se caractérise par son bec crochu, acéré et tranchant, muni d'une membrane appelée cire ; les tarses [jambes] sont partiellement ou entièrement recouverts de plumes ; les serres [doigts] sont au nombre de 4 [3 à l'avant, 1 à l'arrière], pourvus d'ongles arqués, rétractiles ; à l'exception des vautours, nécrophages, les rapaces sont prédateurs d'un grand nombre de vertébrés [petits mammifères, reptiles]) ; *syn.* oiseau [m] rapace ; *g* **Greifvogel [m]** (Tagaktiver ►Raubvogel, der zur Ordnung der *Falciniformes* mit den Familien der Neuweltgeier *[Cathartidae]*, Habichtartigen *[Accipitridae]* — vor allem Adler, Bussard, Milan, Habicht, Sperber, Geier und Weihe —, Falken *[Falconidae]* oder der in der afrikanischen Steppe heimischen Sekretäre *[Sagittariidae]* gehört); *syn.* Greif [m], Taggreif [m].

rareness [n] *conserv.* ►rarity.

4801 rarity [n] *conserv.* (Criterion for the statutory protection of rare animal or plant species); *syn.* rareness [n]; *s* **escasez [f]** (En D. criterio legal de protección de especie de flora y fauna. En Es. la **e.** no es un criterio legal, sino que se incluye en el criterio **singularidad**); *f 1* **rareté [f]** (Critère de protection des espèces de faune et de flore utilisé pour caractériser les espèces dont les populations sont de petite taille et qui, bien qu'elles ne soient pas actuellement en danger ou vulnérables, risquent de le devenir ; ces espèces sont localisées dans des aires géographiques restreintes ou éparpillées sur une plus vaste superficie) ; *f 2* **raréfaction [f]** (Critère de protection des espèces de faune et de flore ; état de diminution du nombre des individus d'espèces ou de groupement d'espèces) ; *g* **Seltenheit [f]** (Kriterium für einen besonderen gesetzlichen Schutz von Tier- oder Pflanzenarten sowie bestimmten Lebensräumen. In Spanien wird **S.** in den Gesetzen nicht erwähnt, sondern in dem Begriff *singularidad* mitgedacht).

rate [n] *prof.* ►daily billing rate [US]/daily rate [UK]; *hydr.* ►flow rate; *constr. wat'man.* ►runoff rate; *contr. prof.* ►hourly billing rate [US]/hourly rate [UK]; *constr. contr.* ►hourly wage rate; *constr. wat'man.* ►runoff rate; *constr. hort.* ►sowing rate; *contr. prof.* ►specifically agreed rate.

rate [n] [US]**, labour cost at an hourly** *contr. prof.* ►hourly wage rate.

rate [n] [UK]**, labour cost at a unit** *contr. prof.* ►hourly wage rate.

rate [n]**, professional fee based on an hourly** *contr. prof.* ►time charges.

4802 rate [n] **of rainfall** *constr. eng. hydr.* (Amount of rainfall in mm per hour in one particular area used in making drainage calculations. In U.S., most large cities, the Weather Bureau, and the Department of Agriculture have records and other data available concerning the intensity, duration and frequency of rainfall); *syn.* rainfall intensity [n]; *s* **intensidad [f] de lluvia** (Precipitación máxima por unidad de tiempo y espacio que puede presentarse en un lugar determinado); *syn.* intensidad [f] de precipitación; *f* **débit [m] d'eaux pluviales** (Quantité maximale d'eaux pluviales à assainir par unité de temps pour une surface donnée compte tenu de sa pente et de son coefficient de ruissellement) ; *g* **Regenspende [f]** (Niederschlagsmenge in mm pro Zeit- und Flächeneinheit, die an einem Ort zur Entwässerung anfallen kann; Auskunft über die örtliche **R.** erteilen in Deutschland die zuständigen Tiefbauämter).

ratio [n] *pedol.* ►air space ratio; *leg. urb.* ►cubic content ratio (of a building [US]); ►floor area ratio (FAR) [US]/floor space index [UK]; *arch. constr.* ►riser/tread ratio; *arb. bot.* ►root-crown ratio; *constr.* ►slope ratio [US].

ratio [n]**, increase in stock** *agr.* ►increase in stocking density.

ratio [n] [UK]**, plot** *leg. urb.* ►building coverage.

ratio [n]**, root-top** *arb. bot.* ►root-crown ratio.

rat run traffic [n] [UK] *trans.* ►shortcut traffic [US]/short-cut traffic [UK].

4803 ravine [n] *geo.* (**1.** Long, narrow, steep-sided valley, larger than a gully and smaller than a canyon, and usually worn down by running water; ►watercourse bed erosion); **2. gorge** [n] (**a.** Ravine with steep rocky walls, smaller than a canyon. **b.** Narrow steep-walled canyon or a particularly narrow steep-walled part of a canyon; cf. WEB 1993; ►gulch [US]); *s* **desfiladero [m]** (Paso estrecho entre montañas con vertientes abruptas; ►barranco 2, ►erosion del lecho); *syn.* garganta [f], quebrada [f] [AL]; *f* **gorge [f]** (Vallée étroite et profonde, aux versants rocheux escarpés ; VOG 1979 ; ►érosion du lit d'un cours d'eau, ►gorge de raccordement) ; *syn.* vallée en gorge, canyon [m] ; *g* **Schlucht [f]** (Im Gebirge ein breiter, tiefer Einschnitt im [felsigen] Gelände mit nur wenig abgeschrägten Wänden; die Vertiefung erfolgt durch Strudellochbildung und ►Sohlenerosion; ►Klamm).

4804 ravine forest [n] *phyt.* (Stand of trees which grows mostly in open ►ubacs or ravines; e.g. an Ash-Maple ravine forest *[Fraxino-Aceretum]*); *s* **bosque [m] de umbría** (Bosque de *Fraxino-Aceretum* que crece en laderas de montaña de orientación norte, en desfiladeros húmedos; ►ladera umbría); *f* **forêt [f] de gorge** (Ensemble hétérogène d'associations de l'étage montagnard [p. ex. *Tilio-Fagetum, Aceri-Tilietum, Aceri-Fraxinetum,* etc.] se développant dans des situations écologiques où l'humidité atmosphérique est élevée comme les vallons encaissés, les ravins, les ►ubacs dans lesquels la présence de brume est très fréquente ; VCA 1985, 181) ; *syn.* forêt [f] de ravin ; *g* **Schluchtwald [m]** (An feuchten, freien Schatthängen [►Schattenseite]

R

oder in Schluchten wachsender Wald, z. B. ein Eschen-Ahorn-Schatthangwald *[Fraxino-Aceretum]*); *syn.* Schatthangwald [m], Gesteinsblockwald [m], Kleebwald [m] (OBER 1992-IV, 187).

4805 raw humus [n] *pedol.* (Any appreciable accumulation of slightly decomposed organic matter, especially on top of very nutrient-poor and coarse soils. The existing vegetation produces very slowly decomposable and nutrient-poor litter; ▶layer of raw humus, ▶mor humus, ▶litter layer); *s* **humus** [m] **bruto** (Acumulación de materia orgánica poco descompuesta o sin descomponer sobre suelo terrestre. Se presenta en especial en suelos pobres bajo la capa de vegetación que proporciona un mantillo pobre en nutrientes y de descomposición difícil; ▶capa de humus bruto, ▶capa de litter, ▶moor); *syn.* humus [m] ácido; *f* **humus** [m] **brut** (Accumulation épaisse de matériaux organiques à la surface du sol minéral pauvre en substances nutritives, formée pour la plus grande partie de débris végétaux encore structurés, donc incomplètement décomposés des humus forestiers ; comprend les deux sous-horizons de la litière et de fermentation de l'horizon organique ; cf. PED 1983, 27 ; ▶couche d'humus brut, ▶couverture morte, ▶mor) ; *syn.* mor [m] (sous les forêts de résineux) ; *g* **Rohhumus** [m, o. Pl.] (Oberste [bis 30 cm mächtige] Schicht auf terrestrischen Böden aus kaum zersetzter organischer Substanz, besonders bei extrem nährstoffarmen und grobkörnigen Böden unter einer Vegetationsdecke, die schwer abbaubare und nährstoffarme Streu liefert; cf. SS 1979; ▶Auflagehumus, ▶Rohhumusauflage, ▶Streuschicht); *syn.* Mör [n].

4806 raw humus decomposer [n] *pedol. phyt.* (Plant, the leaves [▶litter 1] of which break down very slowly; e.g. heather *[Calluna vulgaris]*, Norway spruce *[Picea abies]*; ▶raw humus, ▶needle straw [US]/needle litter [UK]); *syn.* raw humus-forming plant [n]; *s* **planta** [f] **productora de humus bruto** (Planta cuyas hojas o agujas se descomponen muy lentamente como la brecina *[Calluna vulgaris]* o el abeto rojo *[Picea abies]*; ▶capa de barrujo, ▶humus bruto, ▶litter); *f* **plante** [f] **à feuilles à décomposition lente** (Espèce à forte capacité de conservation des nutriments, à forte robustesse foliaire ayant pour effet une décomposition lente des litières avec formation de ▶débris végétaux, p. ex. la Callune fausse-bruyère *[Calluna vulgaris]*, l'Épicéa commun *[Picea abies]* ; ▶humus brut, ▶litière d'aiguilles) ; *syn.* plante [f] donnant une litière acidifiante ; *g* **Rohhumusbildner** [m] (Pflanze, deren Blätter [▶Streu] sehr langsam abgebaut werden, z. B. Besenheide *[Calluna vulgaris]*, Rotfichte *[Picea abies]*; ▶Rohhumus, ▶Nadelstreu).

raw humus-forming plant [n] *pedol. phyt.* ▶raw humus decomposer.

4807 raw material [n] (1) *nat'res. plan.* (▶Mineral resources and other natural substances used industrially within a particular area. Reproducible and non-reproducible/non-renewable **r. m.s** are differentiated; ▶deposit, ▶natural resources 1); *s* **materias** [fpl] **primas** (▶Recursos minerales u otras sustancias naturales utilizadas industrialmente en un área específica. Se diferencia entre las **mm. pp.** renovables y no renovables; ▶recursos naturales, ▶yacimiento); *f* **matières** [fpl] **premières** (▶Richesses naturelles du sous-sol et autres produits dans espace naturel donné ; on distingue les matières renouvelables et les matières non renouvelables ; ▶gisement de minéraux et de fossiles, ▶ressources fondamentales naturelles, ▶richesses naturelles du sous-sol) ; *g* **Rohstoff** [m] (Wirtschaftlich nutzbarer Stoff aus der Natur. Man unterscheidet reproduzierbare/nachwachsende Rohstoffe und nicht reproduzierbare/nicht erneuerbare Rohstoffe; als Synonym wird auch **natürliche Ressource** verwendet, die die in der Urproduktion Land- und Forstwirtschaft, Fischerei und im Bergbau gewonnenen Güter bezeichnet; ▶Bodenschätze, ▶Lagerstätte, ▶natürliche Grundgüter).

raw material [n] (2) *envir.* ▶recovery of raw material.

4808 raw material protection zone [n] *plan.* (▶*Regional planning policy* area given priority in regional planning for the protection of raw materials close to the surface and their controlled extraction, with the intention of prohibiting use of the area for other developments should these obstruct future extraction); *s* **zona** [f] **de aprovechamiento mineral** (±) (**En D.** categoría de ▶ordenación del territorio para priorizar el aprovechamiento de recursos minerales superficiales ante otros posibles usos de un territorio específico. **En Es.** no existe tal categoría); *f 1* **schéma** [m] **départemental des carrières** (F., les conditions générales d'implantation des carrières sont définies par le **s. d. d. c.** en prenant en compte l'inventaire des ressources et l'évaluation des besoins, la protection des sites et des milieux naturels sensibles, la nécessité d'une gestion équilibrée de l'espace ; celui-ci prend en compte les dispositions des réglementations antérieures **1. zone d'exploitation et de réaménagement coordonnés de carrières** [zone sur laquelle une coordination d'ensemble de l'exploitation des carrières et de la remise en état du sol est nécessaire pour éviter la dégradation du milieu environnant et permettre le réaménagement des terrains après exploitation ; art. 109 C. *min.*] ; **2. zone spéciale de recherche et d'exploitation** [zone sur laquelle peuvent être délivrés des autorisations de recherche ou des permis exclusifs en vue de satisfaire les besoins des consommateurs, l'intérêt économique de la nation ou de la région ; art. 109-1 C. *min.*]) ; *f 2* **zone** [f] **prioritaire de recherche et d'exploitation des ressources minérales** [D] (D., zone prioritaire fixée dans le cadre de l'▶aménagement du territoire délimitant les gisements superficiels et déterminant le mode d'exploitation dans le but d'empêcher tout usage ou toute planification pouvant avoir un impact négatif sur l'exploitation future ; *g* **Rohstoffsicherungsfläche** [f] (▶*Raumordnung* planerisch festgesetzte Vorrangfläche zur Sicherung oberflächennaher Rohstoffe und zur Ordnung des Rohstoffabbaues mit dem Ziel, diese Flächen für andere Planungen und Nutzungen zu sperren, wenn diese den Abbau zukünftig verhindern); *syn.* Rohstoffsicherungsgebiet [n].

4809 raw sewage [n] *envir. syn.* untreated sewage [n]; *s* **aguas** [fpl] **residuales (sin depurar)**; *f* **effluent** [m] **non épuré** *syn.* eaux [fpl] usées non épurées, effluent [m] non traité ; *g* **ungeklärtes Abwasser** [n].

raw soil [n], **subaqueous** *pedol.* ▶subhydric soil.

raw stone flag [n] *constr.* ▶quarry flagstone.

reaches [npl] *geo.* ▶lower reaches, ▶middle reaches, ▶upper reaches.

4810 ready [adj] **for checking** *contr. prof.* (Available for auditing in an easy format to determine correctness of items, remuneration rates and unit prices, either in a bid [US]/tender [UK] at the initiation of a project, or in an invoice for completed work; *noun* verifiability [n]); *s* **revisable** [adj] (Obligación de presentar ofertas o facturas ordenadas según partidas y tasas de tarifas de precios o precios unitarios de manera que se puedan controlar sin exigir demasiado trabajo, sea para una oferta, sea para al presentar la factura de trabajos terminados); *syn.* controlable [adj]; *f* **contrôlable** [adj] (Qualité nécessaire à l'appréciation de la convenance des prix des offres ou à la vérification des factures supposant une définition précise et détaillée des éléments d'ouvrages ou de prestations ainsi que les paramètres déterminant les prix unitaires dans le descriptif et le bordereau des prix unitaires) ; *g* **prüfbar** [adj] (1. Angebote oder Rechnungen betreffend, die nach Leistungspositionen und Vergütungssätzen oder Einheitspreisen so nachvollziehbar aufgeschlüsselt und aufbereitet sind, dass den Informations- und Kontrollinteressen des Auftraggebers entsprechend eine leicht nachvollziehbare Überprüfung der rechnerischen und rechtlichen Richtigkeit möglich ist; cf. Grundsatzurteil des BGH v. 18.06. 1998 – VII ZR 189/97 mit Folgeurteil des BGH v. 08.10.1998 – VII ZR 296/97. **2.** Zur **Prüfbarkeit [f]** der Honorarschlussrechnungen von Architekten und Ingenieuren hat der Bundesgerichtshof objektive Mindestanforderungen aufgestellt, wie Angaben zur Honorarzone, der das Objekt angehört, zu den anrechenbaren Kosten, zu dem berechneten Tafelwert des anwendbaren Honorarsatzes und zu den Prozentsätzen des jeweiligen Leistungsbildes, um die sachliche und rechnerische Prüfung des Honorars zu ermöglichen; cf. § 8 [1] HOAI 2002/§ 15 HOAI 2009 und BGH-Urteil v. 27.11. 2003 — VII ZR 288/02. Um Zahlungen zu beschleunigen, wurde § 16 [3] VOB Teil B in der Fassung von 2006 so geändert, dass sich der Auftraggeber inhaltlich mit der Schlussrechnung auseinandersetzen muss: „Werden Einwendungen gegen die Prüfbarkeit unter Angabe der Gründe hierfür nicht spätestens innerhalb von zwei Monaten nach Zugang der Schlussrechnung erhoben, so kann der Aufftraggeber sich nicht mehr auf die fehlende Prüfbarkeit berufen"; *syn. zu 1.* prüffähig [adj]; *syn. zu 2.* Prüffähigkeit [f].

4811 ready [adj] **for development** *urb.* (Descriptive term for a fully serviced site with access and utilities); *s* **urbanizado/a [adj]** (Terreno equipado con toda la infraestructura necesaria para ser utilizado como zona de edificación); *f* **entièrement viabilisé, ée [loc]** (Terrain desservi par la voirie et les réseaux divers apte à être construit) ; *g* **baureif [adj]** (Ein Grundstück betreffend, das mit Verkehrsanschluss sowie Ver- und Entsorgungseinrichtungen voll erschlossen und zur Bebauung freigegeben ist oder werden kann); *syn.* erschlossen [adj], bebaubar [adj].

ready for development [loc], **making** *urb.* ▶making available for development.

4812 ready-mixed concrete [n] *constr.* (Concrete slurry manufactured for delivery in a truck mixer from a central mixing plant to a purchaser); *syn.* premixed concrete [n] [also US] (UBC 1979, 32), transit-mixed concrete [n] [also US] (MEA 1985, 407); *s* **hormigón [m] amasado durante el transporte** (DACO 1988; mezcla de hormigón preparada en fábrica y elaborada en camión para entregar en obra); *syn.* hormigón [m] preamasado (DACO 1988), hormigón [m] mezclado en central, hormigón [m] preparado, hormigón [m] premezclado, hormigón [m] fabricado, hormigón [m] elaborado, hormigón [m] a domicilio [también RA] (BU 1959); *f* **béton [m] prêt à l'emploi (B.P.E.)** (Béton livré sur le lieu d'emploi avec la consistance requise dans des récipients munis d'un dispositif d'agitation ; DTB 1985) ; *g* **Transportbeton [m]** (Im Betonwerk nach Gewicht zugemessener, im Werk oder im Transportfahrzeug gemischter, zur Baustelle transportierter und einbaufertig ausgelieferter Beton; LEHR 1981); *syn. als Firmenbezeichnung* Fertigbeton [m].

reafforestation [n] *for.* ▶reforestation.

4813 real earthworks measurement [n] *constr. contr.* (Actual volumes of earth-moving done in order to establish the amount of payment due to the contractor; ▶earthworks calculation); *s* **cómputo [m] final del movimiento de tierras** (Volumen real de tierras movidas que se debe calcular para la facuración de los trabajos; ▶cálculo del movimiento de tierras) *syn.* cómputo [m] final de terracerías; *f* **calcul [m] des cubatures (en déblai et/ou remblai)** (Calcul de la quantité des terres remuées aussi bien en déblai qu'en remblai effectué pour la facturation ; ▶évaluation des cubatures) ; *g* **Erdmassenabrechnung [f]** (Ermittlung der tatsächlich bewegten Erdmassen, um die erbrachten Bauleistungen zur Vergütung in Rechnung zu stellen; ▶Erdmassenberechnung); *syn.* Erdmengenberechnung [f].

real estate development [n] *urb.* ▶subdivision [US].

real estate parcel [n] [US] *surv.* ▶real property parcel [US]/plot [UK] (2).

realignment [n] *landsc.* ▶channel realignment.

realization competition [n] [UK] *prof.* ▶final design competition [US].

reallocation of farmland [n] [US] *adm. agr. leg.* ▶farmland consolidation [US].

4814 real property identification map [n] [US] *adm. surv.* (Complete, continually up-dated list—register of property—and map showing all parcels/plots of land and buildings, their location, owners, type of use as well as all the most important informations about the yield; ▶area ownership map, ▶ownership map); **in U.S.**, property maps do not typically show or indicate land use or buildings, yield, or any information beyond ownership; *syn.* deed register [n] [US], land register [n] [UK]; *s* **catastro [m] de bienes inmuebles** (Registro legal de todos los datos sobre la propiedad inmobiliaria de un municipio que incluye el ▶plano catastral; ▶mapa de registro de la propiedad); *syn.* catastro [m] de bienes raíces [RCH]; *f* **cadastre [m] parcellaire** (Institué par la loi du 15 septembre 1807 créant un cadastre parcellaire, ensemble des documents, en général régulièrement mis à jour qui, dans chaque commune, définissent la propriété foncière et servent à la répartition de l'impôt foncier, constitué du ▶plan parcellaire du régime de la propriété foncière, du registre des états de section et de la matrice cadastrale [numéro, nature de l'usage, classe, contenance de chaque parcelle] ; LA 1981 ; ▶parcellaire rural) ; *g* **Liegenschaftskataster [m,** *in A nur so* — **oder n]** (Vollständiges und ständig fortgeschriebenes amtliches Verzeichnis [Liegenschaftsbuch] sämtlicher Flurstücke einer Gebietseinheit mit den Liegenschaftskarten sowie den vermessungstechnischen Unterlagen; es beschreibt die Bodenflächen und Gebäude/baulichen Anlagen, Lage, führt Eigentümer, Erbbauberechtigte mit den im Grundbuch geführten Unterscheidungsmerkmalen, Eigentumsanteilen und Nutzungsart sowie alle wichtigen Angaben über die Nutzung auf, dient der Sicherung des Grundeigentums, dem Grundstücksverkehr, der Ordnung von Grund und Boden und ist Grundlage für flächenbezogene Informationssysteme; ▶Besitzstandskarte, ▶Flurkarte).

4815 real property parcel [n] [US] *leg. surv.* (Local real estate unit of one or more lots [US]/plots [UK] or tract of land; ▶parcel 2); *syn.* plot [n] [UK] (2), real estate parcel [n] [US]; *s* **propiedad [f]** (Unidad de terreno perteneciente a una persona física o jurídica incluyendo todos los componentes fijos sobre el suelo, sobre todo edificios; ▶parcela de propiedad); *syn.* propiedad [f] inmueble/raíz/real/territorial, inmueble [m], bien [m] inmueble/raíz, heredad [f], predio [m]; *f 1* **fonds [m] de terre** (Bien immobilier constitué par une ou plusieurs parcelles de sol que l'on exploite ; *terme spécifique* fonds privé ; ▶parcelles [de propriété] 1) ; *syn.* propriété [f], immeuble [m] (nu ou bâti) ; *f 2* **terrain [m]** (Portion de terre d'un seul propriétaire constituée d'une ou plusieurs parcelles) ; *g* **Grundstück [n]** (**1.** *Zivilrechtlich im Sinne der Grundbuchordnung* örtliche und wirtschaftliche Einheit von einem oder mehreren Flurstücken eines Eigentümers mit allen wesentlichen Bestandteilen, d. h. mit dem Grund und Boden fest verbundene Sachen, insbesondere Gebäude. **2.** In seinen Grenzen durch Vermessung geometrisch festgelegtes Stück Land; oft wird auch der Begriff ‚Gelände' synonym verwendet. *Kontext* ein ‚Gelände' [Baugrundstück] verkaufen; *auch* ‚Grund und Boden' verkaufen; ▶Flurstück. **3.** *Steuerrechtlich* das nicht steuerbefreite inländische **G.** i. S. des Bewertungsgesetzes); *syn.* Gelände [n, o. Pl.].

4816 real range [n] *phyt. zool.* (Actual area colonized by an animal or plant species today); *s* **área [f] efectiva** (Espacio real de distribución de una especie o de una cornunidad); *f* **aire [f] réelle d'une espèce** (Ensemble des localités étant aujourd'hui occupées par une espèce végétale ou animale) ; *g* **reales Areal [n]** (Das von einer Tier- oder Pflanzenart heute tatsächlich besiedelte Gebiet); *syn.* reales Verbreitungsgebiet [n].

4817 reapplication [n] of waste [US]/re-application [n] of waste *envir. nat'res. constr.* (Reintroduction of reusable waste into the economic cycle, as a means of waste reduction by physical, chemical, thermal, or biological reprocessing; e.g. the use of waste paper in paper production or noise protection barriers manufactured from recycled plastic; ▶recycling, ▶waste exchange service [UK]. In construction operations, e.g. excavation materials from roads is reused after reprocessing for road construction, rather than hauled away to a dumpsite [US]/ tipping site [UK]; ▶reuse of waste); *s* **valorización [f] de residuos** (En el marco de la ▶gestión de residuos sólidos [urbanos], todo procedimiento que permita el aprovechamiento de los recursos contenidos en los residuos sin poner en peligro la salud humana y sin utilizar métodos que puedan causar perjuicios al medio ambiente; art. 3 k] Ley 10/1998. Esto supone p. ej. la reutilización de ciertas materias [plásticos, metales, cartones] tras su reprocesamiento para otros fines, como p. ej. producción de barreras antiruido a partir de los plásticos; ▶bolsa de gestión de residuos tóxicos y peligrosos, ▶material [de construcción] de reciclado, ▶reciclado, ▶reutilización de residuos); *syn.* recuperación [f] de residuos; *f* **valorisation [f] des déchets** (Opération consistant à réutiliser des déchets récupérés pour un usage différent de celui prévu à l'origine [régénération] ou à réintroduire ces déchets dans un autre cycle de production afin d'obtenir des matériaux réutilisables ou de l'énergie ; p. ex. fabrication de poudrette pour terrain de sport ou construction de mur anti-bruit avec les pneumatiques usagés ; ▶bourse de déchets [industriels], ▶gestion des déchets, ▶industrie de la récupération, ▶matériau de recyclage, ▶récupération, ▶recyclage, ▶réemploi des déchets) ; *syn.* réutilisation [f] des déchets ; *g* **Abfallverwertung [f]** (Im Rahmen der ▶Abfallwirtschaft die Rückführung von wiederverwertbaren Abfallstoffen [Rückständen] in einen neuen Verwendungsbereich als Mittel zur Abfallreduzierung nach dem Einsatz eines physikalischen, chemischen, thermischen oder biologischen Aufbereitungsverfahrens, z. B. Einsatz von Altpapier in der Papierproduktion, die Herstellung von Lärmschutzwänden aus Kunststoffverbunden. Auf dem Bausektor werden z. B. Aushubmassen des Straßen- und Wegebaus nach einer Zwischenbehandlung wieder als Straßenbaustoffe verwendet, statt auf Müll- oder Schuttdeponien gefahren zu werden: Wiederverwendung von aufbereitetem Straßenaufbruch; ▶Abfallbörse, ▶Recycling, ▶Weiterverwendung von Abfallstoffen); *syn.* Weiterverwertung [f] von Abfallstoffen, Wiederverwertung [f] von Abfallstoffen, Müllverwertung [f], Wirtschaftsgutverwendung [f].

4818 rear elevation [n] *plan.* (Graphic representation of the back side of a structure or object; *generic term* ▶elevation); *s* **alzado [m] posterior** (Representación gráfica de la parte trasera de un edificio o una estructura; *término genérico* ▶alzado); *syn.* fachada [f] trasera; *f* **élévation [f], vue arrière** (Représentation graphique de la partie arrière d'un objet ; *terme generique* ▶élévation) ; *g* **Rückansicht [f]** (Zeichnerische Darstellung eines Objektes von der Rückseite; *OB* ▶Ansicht).

rear garden [n] [UK] *gard.* ▶private garden.

4819 rear lot line [n] [US] *surv. syn.* rear plot boundary [n] [UK]; *s* **límite [m] trasero de solar** *syn.* límite [m] trasero de un lote de terreno/predio; *f* **limite [f] (séparative) de fonds de parcelle** (Limite en fonds de parcelle qui sépare deux propriétés contigües, sans avoir de contact avec les voies ou les emprises

publiques) ; *syn.* limite [f] parcellaire arrière ; *g* **hintere Grundstücksgrenze [f]** *syn. adm. leg.* hintere Flurstücksgrenze [f].

rear plot boundary [n] [UK] *surv.* ▶rear lot line [US].

4820 rearrangement [n] of lots [US] *leg. urb.* (**1.** ▶Reorganization of land holdings or subdivisions in a plan area as designated on a legally-binding plan for the redistribution of developed or undeveloped [p]lots of land such that their shape and size comply with future land development or ▶rezoning [US]/rescheduling [UK]. **2. In US parlance**, rearrangement of lots would mean simply the revision or rearrangement of boundaries between adjoining landowners by voluntary or court order; ▶farmland consolidation [US]/land consolidation [UK]); *syn.* reorganisation [n] of plot boundaries [UK], subdivision [n] [US] (1); *s* **reordenación [f] de parcelas** (Con el fin de facilitar el aprovechamiento de las mismas de acuerdo con el ▶plan parcial [de ordenación]; ▶concentración parcelaria [Es], ▶ordenación del suelo, ▶redefinición de uso[s]); *f* **remembrement [m] parcellaire** (Redistribution des parcelles bâties ou non bâties en vue de la restructuration urbaine des grands ensembles et quartiers d'habitat dégradé ou d'une meilleure utilisation des droits à construire prévus par les documents d'urbanisme ; C. urb., art. L. 322 ; ▶déclassement, ▶remembrement rural, ▶réorganisation foncière) ; *syn.* remembrement de parcelles, remembrement [m] foncier ; *g* **Umlegung [f]** (Bodenordnende Maßnahme im Geltungsbereich eines ▶Bebauungsplanes mit dem Ziel, bebaute oder unbebaute Grundstücke so neu aufzuteilen, dass nach Lage, Form und Größe für die zukünftige Nutzung zweckmäßig gestaltete Grundstücke entstehen; cf. § 45 BauGB; ▶Bodenordnung, ▶Flurbereinigung, ▶Umwidmung); *syn.* Neuordnung [f] von Grundstücken.

4821 reasonable price [n] *constr. contr.* (Submitted amount of money which is, according to experience, adequate for a specified material or construction work; ▶predatory price [US]/ dumping price [UK]; ▶excessive price); *s* **precio [m] razonable** (En ofertas, precio ofrecido para un determinado material o servicio que por experiencia se considera adecuado; ▶precio dumping, ▶precio excesivo); *f* **prix [m] raisonnable** (Lors de la remise de la soumission ; ▶prix bradé, ▶prix excessif) ; *g* **angemessener Preis [m]** (Der Erfahrung entsprechend bemessener und auskömmlicher Preis für eine bestimmte Ware oder Leistung bei Angebotsabgaben; ein Preis ist auskömmlich, wenn kein grobes Abweichen vom **a. P.** „sofort ins Auge fällt". Als Anhaltspunkt dient der Abstand zum Nächstbietenden. Bei einer 15-prozentigen Abweichung muss der Auslober nachfragen; cf. Beschluss der Vergabekammer Sachsen vom 26.07.2001 [1/SVK/73-01]; ▶Dumpingpreis, ▶überhöhter Preis); *syn.* auskömmlicher Preis [m].

reasons [npl] of overriding public interest *ecol. leg. plan.* ▶imperative reasons of overriding public interest.

4822 receiving stream [n] *constr. eng. wat'man.* (RCG 1982, 30, 132; open channel or stream serving to collect surface water run-off or groundwater); *syn.* carrier channel [n] (LAD 1986, 146), outlet ditch [n], outlet stream [n], outlet channel [n] (LAD 1986, 214ss), receiving water [n]; *s* **curso [m] receptor** (Curso de agua que recibe aguas de otros cursos o de un emisario artificial como puede ser una planta de depuración); *f 1* **émissaire [m]** (Déversoir naturel, milieu récepteur d'écoulement [fossé, cours d'eau, lacs ou étangs, torrent dans ou sous un glacier] ou ouvrage de grande dimension, disposés en bout de collecteur recueillant les eaux d'assainissement ou de drainage, dans lesquel transitent les eaux usées vers une station d'épuration ; *terme spécifique* émissaire de drainage) ; *f 2* **milieu [m] récepteur** (Tout cours d'eau naturel dans lequel on déverse les eaux épurées ou non, p. ex. mer, fleuve, rivière, fossé, lac, étang, marais, etc.) ;

g **Vorfluter [m]** (Wasserlauf oder offener Graben zur Aufnahme und Weiterleitung der Abflussspende einer zu entwässernden Fläche).

4823 recent colonization [n] *phyt. zool.* (Invasion and settlement of an area by a species whose members had not previously populated the region; ▶colonization 2, ▶recolonization, ▶reintroduction); *s* **écesis [f]** *phyt.* (Invasión o establecimiento de un vegetal en un área determinada; DINA 1987; ▶colonización natural, ▶recolonización natural, ▶reintroducción de especies extintas); *syn.* colonización [f] reciente (de un área por especies nuevas); *f* **apparition [f] d'une espèce nouvelle** (Établissement d'une espèce sur une aire qu'elle n'avait préalablement jamais occupé ; ▶colonisation naturelle, ▶recolonisation naturelle, ▶réintroduction 2) ; *syn.* colonisation [f] récente ; *g* **Neubesiedlung [f]** (Ansiedlung einer Art in einem Raum, der zuvor nie von Angehörigen dieser Art besiedelt worden war; ▶Besiedlung 1, ▶Wiederbesiedelung, ▶Wiederansiedlung).

recently introduced plant [n] *phyt.* ▶adventitious plant.

4824 recessed joint [n] *constr.* (In masonry, a joint in which the mortar is pressed back by pointing, about 6mm from the wall face); *s* **junta [f] remetida** (En albanilería, junta de mortero que queda retrasada respecto al paramento; DACO 1988); *syn.* junta [f] rehundida; *f* **joint [m] (en) creux** (DTB 1985 ; intervalle entre deux éléments d'une maçonnerie non rempli de mortier ; on distingue les joints creux carrés, à gorge, à la baleine ; *assemblage avec joint creux* pose d'éléments en bois en métal en pierre entre lesquels on laisse une rainure ; lorsque la rainure du joint est comblée par une masse de remplissage [masticage] et lissée, il s'agit d'un **joint creux chanfreiné**) ; *syn.* joint [m] en retrait, joint [m] dégagé, joint [m] d'ombre [CH] ; *g* **Schattenfuge [f] (1)** (Fuge im Mauerwerk, die nicht bündig mit Mörtel verfüllt ist).

recession [n] *hydr.* ▶natural groundwater recession.

recharge [n] *hydr.* ▶aquifer recharge area; *hydr. wat'man.* ▶groundwater recharge; *for. wat'man.* ▶protected aquifer recharge forest.

4825 recharged groundwater storage [n] *hydr.* (Water artificially recharged into an ▶aquifer or a groundwater reservoir for future exploitation; WMO 1974; ▶groundwater recharge); *s* **almacenamiento [m] de agua subterránea** (Recarga artificial de un ▶acuífero para su posterior aprovechamiento; ▶recarga de acuíferos); *f* **recharge [f] d'eau souterraine** (Opération consistant à réapprovisionner artificiellement en eau une ▶formation aquifère ou un réservoir souterrain en vue d'une exploitation future de la réserve ; WMO 1974 ; ▶alimentation d'une nappe phréatique) ; *g* **Grundwasserspeicher [m] (2)** (Künstlich in ▶Grundwasserleiter oder Grundwasserbecken eingebrachtes Wasser für spätere Nutzungen; ▶Grundwasseranreicherung).

4826 recirculating pump [n] *constr.* (Pump which reutilizes the water in pools for water features such as fountains); *s* **bomba [f] de circulación** (Bomba instalada para hacer circular el agua en juegos de agua o fuentes); *f* **pompe [f] à révolution** (Pompe installée pour la réutilisation et la circulation de l'eau en circuit fermé) ; *g* **Umwälzpumpe [f]** (Pumpe zur Wiederverwendung des Wassers in Wasseranlagen).

reclaimed surface-mined land [n] *min. landsc.* ▶landscape of reclaimed surface-mined land.

reclaimed surface-mined site landscape [n] [US] *min. landsc.* ▶landscape of reclaimed surface-mined land.

reclamation [n] *agr.* ▶desert reclamation; *land'man. landsc.* ▶gravel pit reclamation; *hydr. eng. land'man. nat'res.* ▶land reclamation (1); *agr. for. land'man.* ▶land reclamation (2); *agr.* ▶land reclamation (3); *land'man. landsc.* ▶spoil reclamation [US]/tip reclamation [UK]; *envir. plan.* ▶toxic site reclamation.

reclamation [n] [UK]**, tip** *land'man. landsc.* ▶spoil reclamation [US].

4827 reclamation [n] **after subsidence** *eng. landsc. min.* (Movement of earth to reconstruct the original ground level of an ▶area subject to mining subsidence); *s* **relleno [m] para rectificar hundimiento** (▶zona de subsidencia [Es, CO]); *syn.* relleno [m] para rectificar asentamiento; *f* **remise [f] à niveau des zones de subsidence minière** (Remise au niveau du terrain naturel de zone d'affaissement minier dans le cadre de travaux de terrassement ; ▶zone d'effondrement minier, ▶zone de subsidence minière) ; *g* **Senkungsausgleich [m]** (Wiederherstellen der ursprünglichen Geländehöhe oder von profilgerechten Flächen durch Erdbaumaßnahmen in ▶Bergsenkungsgebieten).

4828 reclamation [n] **of derelict land** *land'man.* (Restoration for cultivation of land which has been destroyed by man, using land management measures and soil amelioration; ▶agricultural land improvement, ▶derelict land, ▶land reclamation 2); *s* **recultivo [m] de tierras yermas** (Restauración de tierras para usos agrosilvícolas que han sido destruidas por usos humanos, utilizando métodos de gestión del paisaje y de mejoramiento de suelos; ▶mejoramiento de suelos, ▶recuperación de tierras, ▶ruina industrial); *f* **mise [f] en valeur de terres incultes (récupérables)** (Récupération et mise en valeur de friches agricoles, pastorales ou forestières ; *contexte* récupération et mise en valeur des terres incultes récupérables ; *terme spécifique* remise en culture de terrains incultes ; ▶amélioration du sol, ▶friche trielle, ▶remise en état des lieux après exploitation) ; *g* **Ödlandrekultivierung [f]** (Wiedernutzbarmachung eines durch menschliche Eingriffe zerstörten Gebietes durch landschaftsplanerische Maßnahmen der Bodenverbesserung; ▶Industriebrache, ▶Melioration, ▶Rekultivierung).

reclamation [n] **of hazardous waste sites** *envir. plan.* ▶toxic site reclamation.

reclamation [n] **of land** *agr. for. land'man.* ▶land reclamation (1), ▶land reclamation (2), ▶land reclamation (3).

reclamation [n] **of used resources** *envir. nat'res.* ▶recycling.

4829 reclamation [n] **of wetlands** *agr. land'man.* (Reclaiming of bogs, marshes, and swamps by drainage schemes; ▶bog cultivation, ▶drying out 2); *s* **ganancia [f] de tierras (2)** (Ganancia de nuevas tierras agrícolas por desecación de turberas o pantanos; ▶cultivo de turbera, ▶desaguado); *f* **poldérisation [f] agricole** (Terres conquises par assèchement des marais, des lacs ou des tourbières ; ▶polder industriel [DG 1984, 148], ▶assèchement, ▶mise en culture de tourbières) ; *syn.* conquête [f] du sol ; *g* **Landgewinnung [f] durch Trockenlegung** (Gewinnung von landwirtschaftlichen Bodenflächen durch Entwässerung von Mooren und Sümpfen; ▶Moorkultur, ▶Trockenlegung).

4830 recolonization [n] *phyt. zool.* (Natural return of a species to an area which it had not populated for a long period of time; ▶reintroduction, ▶reoccurrence of a species); *s* **recolonización [f] natural** (Reaparición de una especie en una zona en la que no había existido durante mucho tiempo; ▶reintroducción de especies extintas, ▶reaparición de una especie); *f* **recolonisation [f] naturelle** (Réapparition d'une espèce dans une aire qu'elle occupait antérieurement : processus d'évolution naturelle des populations de flore et de faune sauvages dans une zone touchée par une catastrophe naturelle [inondation, incendie, etc.] ou par l'arrêt de l'action humaine [carrière, tourbière exploitée, etc.]; ▶réintroduction 2, ▶résurgence d'une espèce) ; *syn.* reconquête [f] d'une espèce, repeuplement [m] naturel ; *g* **Wiederbe-**

R

siedelung **[f]** (Natürliche Rückkehr einer Art in einen Raum, in dem sie lange Zeit nicht mehr gelebt hatte; ►Wiederansiedelung, ►Wiederauftauchen einer Art); *syn.* Rekolonisation [f], *o. V.* Wiederbesiedlung [f].

4831 recommendation [n] for contract award *contr.* (After analysis and evaluation of bids [US]/tenders [UK], the proposal for award to the lowest tenderer/bidder or to the tenderer offering the best value for the price bid [US]/best value for money [UK]; ►prepare a recommendation for contract award); **s proposición [f] de adjudicación** (Propuesta de adjudicar el remate al licitante con la oferta más barata o técnicamente mejor después de evaluar todas las ofertas; ►preparar una proposición de adjudicación); *f* **proposition [f] d'attribution du marché** (Après dépouillement, examen et jugement des offres, désignation du candidat le moins disant ou choix de l'offre la plus intéressante ; ►préparer le jugement d'appel d'offres) ; *g* **Vergabevorschlag [m]** (Nach Prüfung und Auswertung der Angebote die Empfehlung, dem billigsten oder demjenigen Bieter mit dem annehmbarsten/wirtschaftlichsten Angebot den Zuschlag zu erteilen; ►Vergabevorschlag vorbereiten).

recommended scheme [n] [UK] *prof.* ►merit award.

4832 recomposition [n] *arch. plan.* (Modification of an arrangement of elements; ►composition 1, ►partial redesign, ►plan revision, ►reconfiguration of an area/terrain, ►replanning); **s recomposición [f]** (Modificación de un arreglo de elementos, p. ej. de un plan [►modificación de un plan], de una zona verde, de una exposición; ►composición, ►reconfiguración parcial, ►reconfiguración de un área, ►replanificación); *f 1* **requalification [f] (d'un espace public)** (Action de traitement et de recomposition d'un axe, d'une place, d'un carrefour, etc. ; celle-ci s'accompagne le plus souvent d'un meilleur partage entre les modes d'utilisation, de déplacements, etc.) ; *f 2* **recomposition [f] (2)** (Redéfinition d'un espace, d'un site, d'un paysage, d'un territoire à partir d'une nouvelle organisation de son tissu [arrangement des éléments constitutifs structurants] ; *contexte* la recomposition sociale du territoire, la recomposition de l'espace rural, territoire/espace en recomposition ; ►modification du plan, ►recomposition 1, ►remodelage, ►remodelage d'un site) ; *syn.* transformation [f], réorganisation [f], restructuration [f] ; *g* **Umgestaltung [f] (2)** (Neuordnung der gestalterischen Elemente, z. B. einer Planung [►Überarbeitung eines Planes], einer Grünanlage, einer Ausstellung; ►Umgestaltung 1, ►Umgestaltung eines Geländes, ►Umplanung); *syn.* Neuordnung [f] (von Gestaltungselementen).

reconditioning [n] of rivers and streams *wat'man.* ►river restoration.

4833 reconfiguration [n] of an area/terrain *arch. plan.* (Changes to the topographic layout, technical facilities and/or planting of an area); *syn.* remodeling [n] of a site [US]/remodelling [n] of a site [UK]; **s reconfiguración [f] de un área** (Transformación de la modelación del terreno, del equipamiento técnico y/o de la vegetación); *f* **remodelage [m] d'un site** (Transformation **1.** du modelé d'un terrain, **2.** de l'organisation technique [équipements, mobilier urbain, revêtements] paysagère d'un espace) ; *g* **Umgestaltung [f] eines Geländes** (Veränderung der Bodenmodellierung, technischen Ausstattung und/oder Bepflanzung einer Fläche); *syn.* Geländeumgestaltung [f].

4834 reconstruction [n] *conserv'hist.* (Act or process of rebuilding or restoring a structure or building complex, as well as replicating the form, features, and detailing of a non-surviving site, landscape, building, structure, or object for the purpose of recreating its appearance at a specific period of time and in its historic location); **s reconstrucción [f]** (Conjunto de medidas que sirven para reconstruir o rehabilitar exactamente en forma, carác-

ter y partes constituyentes de un edificio o partes del mismo tal y como era en el pasado); *f* **reconstitution [f]** (Reconstruction d'un ou d'une partie de bâtiment, d'un édifice ou d'un ensemble d'édifice disparu ou très endommagé dans son état initial, sur la base de documents écrits et/ou iconographiques) ; *g* **Rekonstruktion [f]** (Maßnahmen, die dazu dienen, Form, Charakter und Einzelbestandteile eines Gebäudes, Bauwerkes, Gartens oder eines Teils davon detailgenau so wiederherzustellen, wie sie zu einem bestimmten früheren Zeitpunkt gewesen waren); *syn.* Wiederherstellung [f] in den ursprünglichen Zustand.

reconstruction [n], urban *urb.* ►urban redevelopment.

record [n] *constr.* ►dumping record [US]/tipping record [UK], *plan.* ►noise level record [US].

4835 record high-water level/mark [n] *wat'man.* (Highest level of watercourse ever documented; ►flood peak, ►hundred-year flood); *syn.* peak high-water level/mark [n]; **s crecida [f] máxima posible** (WMO 1974; nivel de agua más alto que ha sido registrado históricamente en un curso de agua; ►crecida máxima del siglo, ►punta de crecida); *f* **crue [f] maximale** (Plus forte crue qui puisse survenir en un lieu donné ou enregistré sur l'échelle de crue, compte tenu des précipitations et de l'écoulement maximum ; ►pointe de crue, ►crue centennale) ; *syn.* plus forte crue [f], niveau [m] maximal des hautes eaux ; *g* **höchstes Hochwasser [n]** (...wasser [pl]; *Abk.* hHW; höchster erreichter Pegelstand eines Hochwassers; ►Hochwasserspitze, ►Jahrhunderthochwasser); *syn.* höchster Hochwasserstand [m].

4836 recoverability [n] *conserv.* (Possibility of restoring damaged or degraded ecosystems, habitats, etc. to their former sustainable state and function by measures of ►landscape management; ►restoration 1); *syn.* feasibility [n] of recreating ecosystems/habitats; **s capacidad [f] de restaurar/reestablecer ecosistemas o biótopos** (Posibilidad de restaurar el estado original sostenible de ecosistemas o biótopos degradados o destruidos por medio de medidas de ►gestión y protección del paisaje; ►renaturalización); *f* **capacité [f] de restauration** (Propriété des écosystèmes dégradés à retrouver leur intégrité perdue grâce à l'emploi de techniques de ►gestion des milieux naturels ; ►renaturation ; *syn.* capacité [f] de reconstitution ; *g* **Wiederherstellbarkeit [f]** (Möglichkeit, durch Maßnahmen der ►Landschaftspflege zerstörte oder degradierte Ökosysteme, Biotope etc. in ihrer vorherigen Art und Funktion so wiederherzustellen, dass sie dauerhaft bestehen können; ►Renaturierung).

4837 recoverability [n] from trampling *constr. phyt.* (Resistance of soil and vegetation to treading on lawns and sports areas, as well as by man and beast on ground cover of forest and grassland communities; ►trampling); **s resistencia [f] al pisoteo** (Capacidad de resistencia de césped de juegos o deportes o de la vegetación natural de comunidades de bosques o de pradera ante el efecto de pisadas; ►pisoteo); *f* **capacité [f] au piétinement** (Résistance de gazons sportifs, d'agrément et de détente, de la végétation au sol des forêts naturelles ou des prairies ; ►piétinement) ; *syn.* résistance [f] au piétinement ; *g* **Belastbarkeit [f] durch Tritt** (Widerstandsfähigkeit von Spiel- oder Sportrasen, von Bodenvegetation natürlicher Wald- oder Wiesengesellschaften gegen Störungen durch Tritt, so dass die dauerhafte Funktion dieser Flächen nicht gemindert wird; ►Trittbelastung); *syn.* Trittverträglichkeit [f], Trittfestigkeit [f].

recoverable materials [npl], reuse of *envir.* ►reuse waste.

recovery [n], resource *envir. nat'res.* ►recycling.

recovery [n] (of natural habitat) *landsc. land'man. conserv.* ►restoration (1).

4838 recovery [n] **of raw material** *envir.* (Salvage of economically usable raw material during waste treatment and recycling, reusing or reclaiming of demolition waste; ▶reuse of on-site construction material); *s* **recuperación** [f] **de materias primas** (Separación de materias reutilizables en el proceso de tratamiento de residuos; ▶reutilización de materiales de construcción); *f* **récupération** [f] **des matières premières** (Séparation des matériaux réutilisables à des fins de réemploi, de réutilisation ou de recyclage, p. ex. recyclage des enrobés routiers et des assises de chaussées lors de la réfection de voies routières ; ▶reprise des matériaux stockés) ; *g* **Rohstoffrückgewinnung** [f] (Zurückgewinnen von wirtschaftlich nutzbaren, sog. Sekundärrohstoffen bei der Behandlung von Abfallstoffen, z. B. bei der Abfallentsorgung; beim Bauschutt, Straßenaufbruch oder bei Steinbruchabfällen handelt es sich um Sekundärrohstoffe, die als Straßen- und Wegebaumaterial nach entsprechender Aufbereitung wieder verwendet werden; die rechtlichen Vorgaben sind im Kreislaufwirtschafts- und Abfallgesetz [KrW-/AbfG] und den entsprechenden Verordnungen verankert; ▶Wiedereinbau von zwischengelagerten Baustoffen); *syn.* Rohstoffrecycling [n], Rezyklierung [f] von Rohstoffen, Rohstoffrezyklierung [f].

recreating [n] **ecosystems/habitats, feasibility of** *conserv.* ▶recoverability.

4839 recreation [n] *recr.* (Refreshment of normal mental, physical and spiritual capabilities which were reduced by daily work; ▶active recreation, ▶countryside recreation, ▶daily recreation, ▶forest recreation, ▶local recreation, ▶long-stay recreation, ▶neighborhood recreation [US]/neighbourhood recreation [UK], ▶outdoor recreation, ▶outdoor recreation in outlying areas, ▶outdoor urban recreation, ▶passive recreation, ▶recreation in the natural environment, ▶recreational weekend [US]/ weekend break [UK], ▶recuperation, ▶short-stay recreation, ▶water recreation); *s* **descanso** [m] (*Sensu stricto* recuperación física y [p]síquica después del trabajo; ▶descanso de fin de semana, ▶puente, ▶recreación diaria, ▶recreación en la periferia urbana, ▶recreación pasiva, ▶recreo, ▶recreo al aire libre, ▶recreo en el bosque, ▶recreo en el vecindario/barrio, ▶recreo en espacios libres urbanos, ▶recreo en la naturaleza, ▶recreo extensivo, ▶recreo local, ▶regeneración, ▶vacaciones cortas, ▶vacaciones largas); *syn.* reposo [m], reproducción [f] física y (p)síquica; *f* **détente** [f] **(1)** (En liaison avec la régénération des capacités intellectuelles, mentales et corporelles nécessaires aux activités professionnelles; ▶détente en espaces libres urbains, ▶détente et récréation quotidienne, ▶esparcimiento asociado al agua, ▶loisirs à caractère spécifique aquatique, ▶loisirs actifs, ▶loisirs à proximité de quartiers résidentiels, ▶loisirs [de] nature, ▶loisirs de proximité, ▶loisirs de plein air, ▶loisirs de week-end, ▶loisirs en milieu naturel, ▶loisirs et récréation en forêt, ▶loisirs passifs, ▶loisirs périurbains, ▶régénération, ▶tourisme de passage, ▶tourisme de séjour) ; *syn.* délassement [m] physique et intellectuel ; *g* **Erholung** [f, o. Pl.] **(1)** (*Sensu stricto* in der Polarität zur meist fremdbestimmten Arbeit die Wiederherstellung der normalen geistigen, seelischen und körperlichen Leistungsfähigkeit; ▶aktive Erholung, ▶Erholung (2), ▶Erholung im Freien, ▶Erholung im Wald, ▶Erholung in der freien Landschaft, ▶Freiraumerholung, ▶gewässerbezogene Erholung, ▶Kurzzeiterholung, ▶Langzeiterholung, ▶Naherholung, ▶naturnahe Erholung, ▶passive Erholung, ▶Stadtranderholung, ▶Tageserholung, ▶Wochenenderholung, ▶wohnungsnahe Erholung); *syn.* Regeneration [f], Reproduktion [f].

recreation [n] [US], **boondocks** *recr.* ▶recreation in the natural environment.

recreation [n], **extensive** *recr.* ▶recreation in the natural environment.

recreation [n] **in exurban areas** [US], **outdoor** *recr. urb.* ▶outdoor recreation in outlying areas.

recreation [n], **long-term** *plan. recr.* ▶long-stay recreation.

recreation [n] [UK], **neighbourhood** *plan. recr. urb.* ▶neighborhood recreation [US].

recreation [n], **open air** *recr.* ▶outdoor recreation.

recreation [n] [UK], **out-of-doors** *recr.* ▶outdoor recreation.

recreation [n] [US], **primitive** *recr.* ▶recreation in the natural environment.

recreation [n], **short-term** *recr.* ▶short-stay recreation.

recreation [n] [UK], **weekend** *recr.* ▶recreational weekend [US].

recreation [n] [US], **wilderness** *recr.* ▶recreation in the natural environment.

4840 recreation activity [n] *recr.* (Generic term covering all types of recreation pursuits for individuals or groups; ▶vacation activity [US]/holiday activity [UK]); *syn.* recreation pursuit [n]; *s* **actividad** [f] **de recreo** (Término genérico para cualquier tipo de actividad de tiempo libre realizada individualmente o en grupo con fines recreacionales; ▶actividad de vacaciones); *syn.* actividad [f] de esparcimiento; *f* **activité** [f] **de récréation** (*Terme spécifique* activité de plein air ; ▶pratique vacancière) ; *syn.* pratique [f] d'activités récréatives ; *g* **Erholungsaktivität** [f] (OB für jegliche Betätigung in der Freizeit, die der Erholung dient, sei es alleine oder in Gruppen; ▶Urlaubsbetätigung).

recreation agency [n] *adm. conserv.* ▶conservation/-recreation agency [US]/nature conservation body [UK].

recreational [adj] *plan. recr.* ▶leisure-related.

4841 recreational area management [n] *plan. recr.* (**1.** Organization, development and maintenance of ▶recreation areas 1, recreation centers [US]/recreation centres [UK], and nature parks by one managing body, e.g. an agency/authority, registered club, company, local union. **2.** Sum of measures involving the long-term planning and maintenance of areas of land used primarily for recreation purposes, in order to ensure their lasting function as recreation areas; ▶countryside management [UK]); *syn.* recreation area management [n]; *s* **gestión** [f] **recreacional (1)** (**1.** Planificación, desarrollo y mantenimiento de parques naturales con fines recreativos por parte de la autoridad o institución responsable. **2.** Conjunto de medidas de planificación y mantenimiento de espacios naturales utilizados primordialmente para el recreo con el fin de preservarlos de manera que puedan cumplir a largo plazo su función recreacional; ▶gestión de zonas rurales con fines recreativos); *syn.* gestión [f] de zonas con vocación recreativa, gestión [f] de zonas turísticas; *f 1* **gestion** [f] **des équipements de loisirs** (Création, développement et entretien des ▶espaces touristiques, des bases de plein air et de loisirs et des parcs naturels par des acteurs tels que l'État, les collectivités territoriales et locales, les associations ou les opérateurs privés ; ▶gestion des espaces récréatifs) ; *f 2* **gestion** [f] **des sites touristiques** (Mise en œuvre par les collectivités locales d'actions de réorganisation de certains ▶espaces touristiques ayant pour objectif la maîtrise du développement touristique dans le respect des identités culturelles et des milieux naturels fortement sollicités par les visiteurs afin d'assurer à long terme le fonctionnement des sites et la pratique touristique, notamment par la gestion des flux, la modification du zonage des espaces en fonction de leur sensibilité, l'acquisition de réserves immobilières et foncières, en favorisant l'aménagement d'hébergements banalisés, etc. ; cf. TP 1997, 161 et s.) ; *g* **Pflege [f, o. Pl.] und Unterhaltung [f, o. Pl.] von Erholungsgebieten**

(**1.** Entwicklung und Unterhaltung von Erholungsgebieten, Erholungsschwerpunkten und Naturparken zur Freizeitnutzung durch einen Träger, z. B. Behörde, eingetragener Verein, Kapitalgesellschaft, Zweckverband. **2.** Sicherung, Entwicklung sowie Pflege und Unterhaltung von bestimmten ►Erholungsgebieten [innerhalb von Städten oder in der freien Landschaft] mit dem Ziel, durch gezielte Maßnahmen eine langfristige Erholungsnutzung zu ermöglichen, z. B. Lenkung von Besucherströmen oder Vermeidung sich gegenseitig störender Erholungsansprüche oder Vermeidung des Nutzungskonfliktes Erholung und Naturschutz. **3.** Zielorientierte planerische Maßnahmen zur Sicherung, Pflege und Entwicklung von außerstädtischen Bereichen, in denen der Erholung eine besondere Bedeutung zukommt, um Konflikte durch konkurrierende Nutzungsansprüche so abzubauen, dass eine nachhaltige Erholungsnutzung durch Sicherung oder Verbesserung des Landschaftsbildes gegeben ist; ►Pflege und Erhaltung der freien Landschaft zum Zwecke der Erholung).

4842 recreational behavior [n] [US]/**recreational behaviour** [n] [UK] *plan. recr.* (Conduct of recreation users during their leisure time, which is revealed in their spectrum of recreational activities); *s* **comportamiento** [m] **recreacional** (Gama de actividades de recreo desarrolladas por una persona durante su tiempo libre); *f* **comportement** [m] **récréatif** (Éventail des activités de loisirs pratiquées par les usagers ; *terme spécifique* comportement vacancier) ; *g* **Erholungsverhalten** [n] (Art und Weise, wie Erholungsuchende sich während ihrer Freizeit verhalten); *syn.* Freizeitverhalten [n].

4843 recreational demand [n] *plan. recr.* (Popular requirements for an available supply of recreation opportunities expressed in terms of resources or facilities); *syn.* recreation need [n], recreational requirement(s) [n(pl)]; *s* **demanda** [f] **recreacional** (Exigencia de diferentes grupos de usuarios de tener a disposición instalaciones y posibilidades de recreo de acuerdo a sus necesidades); *syn.* demanda [f] vacacional; *f* **besoin** [m] **en activités de loisirs** (Nécessité de l'existence d'équipements ou d'aménagements satisfaisant différents groupes d'usagers) ; *syn.* besoin [m] récréationnel, demande [f] touristique ; *g* **Erholungsbedarf** [m] (Bedarf und Nachfrage nach Erholung und der daraus erwachsenden Notwendigkeit, dass Erholungsmöglichkeiten für unterschiedliche Nutzergruppen vorhanden sind); *syn.* Erholungsanspruch [m], Erholungsnachfrage [f].

recreational feature [n] **in the landscape** *plan. recr.* ►landscape component for recreation.

4844 recreational hunting [n] *recr. hunt.* (DED 1993, 127; hunting, which has evolved into a leisure pursuit either at home or whilst on holiday in other countries); *s* **caza** [f] **deportiva** (Actividad cinegética practicada en tiempo libre y con fines deportivos o recreativos); *f* **chasse** [f] **amateur** (Activité cynégétique pratiquée pendant les temps de loisirs) ; *syn.* chasse [f] récréative ; *g* **Freizeitjagd** [f] (Jagdliche Aktivitäten, die in der Freizeit im eigenen Lande oder in fremden Ländern ausgeübt werden).

4845 recreational infrastructure [n] *plan. recr.* (Essential elements of a recreational complex or system of facilities, such as sports grounds, play areas or ornamental parks, which are either primarily or exclusively used for recreational activities; ►active recreation); *s* **equipamiento** [m] **de recreo (1)** (Instalaciones de deportes y juegos o edificios o partes de tales que tienen como función principal servir para realizar actividades de tiempo libre; ►recreo); *syn.* infraestructura [f] de recreo, infraestructura [f] de esparcimiento, infraestructura [f] de tiempo libre; *f* **équipements** [mpl] **collectifs de loisirs** (Espaces et ouvrages éducatifs et sportifs de base utilisés exclusivement ou partiellement pour les activités de loisirs ; ►loisirs actifs) ; *syn.*

infrastructure [f] de loisirs, équipement [m] d'accueil récréatif ; *g* **Freizeitinfrastruktur** [f] (Gesamtheit der in einem organisatorischen oder geografischen Bereich an unterschiedlichen Standorten für eine Vielzahl von Nutzern bereitstehenden, der Freizeit dienenden Sport- und Spielanlagen sowie Bauwerke oder Teile davon, um hauptsächlich oder ausschließlich Freizeitaktivitäten zu ermöglichen; ►aktive Erholung).

4846 recreational land use [n] *leg. urb.* (Land use category, as shown on ►community development plans [US]/urban development plan [UK]); *s* **reserva** [f] **de suelo para fines recreativos** (Categoría de clasificación de suelo en el planeamiento urbanístico; en Es. los ►planes parciales [de ordenación] deben prever **r. de s.** para parques y jardines, zonas deportivas públicas de recreo y expansión en proporción adecuada a las necesidades colectivas, como mínimo un 10 por ciento de la superficie ordenada, independientemente de las superficies destinadas en el ►plan general municipal de ordenación urbana a espacios libres o zonas verdes para parques públicos; art. 13b RD 1346/1976 Texto refundido, art. 45c RD 2159/1978); *f* **zone** [f] **affectée aux activités sportives et de loisirs** (Catégorie d'occupation du sol dans les ►documents de planification urbaine ; sa dénomination diffère suivant les communes ; *syn.* zone [f] d'activités récréatives, zone [f] destinée à être aménagée pour des activités de loisirs de plein air et des activités sportives, zone [f] aménagée en vue d'activités de loisirs et de sport de plein air, zone [f] réservée aux activités de sports et de loisirs, zone [f] d'urbanisation future à vocation d'activités sportives, de loisirs et touristiques, etc.) ; *g* **Fläche** [f] **für Freizeit und Erholung** (In ►Bauleitplänen dargestellte Nutzungskategorie).

4847 recreational precedence [n] *plan. recr.* (Priority of a landscape unit with intrinsic quality for a specific recreational use, which has primacy over other uses; ►priority area); *s* **vocación** [f] **recreativa preferente** (≠) (Alta aptitud recreativa de una zona por sus características específicas de manera que se prioriza ese uso sobre otros posibles; ►área de uso priorizado); *syn.* capacidad [f] intrínseca preferente para usos recreativos (≠), idoneidad [f] preferente para usos recreativos (≠); *f* **vocation** [f] **préférentielle pour des activités de loisirs** (Aptitude aux activités de loisirs des structures spatiales d'un territoire ; ►zone prioritaire) ; *g* **vorrangige Erholungseignung** [f] (Tauglichkeit der räumlichen Voraussetzungen für den Nutzungsanspruch Erholung vor anderen Nutzungsansprüchen im Sinne einer Vorrangnutzung; ►Vorranggebiet).

recreational provision [n] *plan. recr.* ►provisions for recreation.

recreational requirement(s) [n(pl)] *plan. recr.* ►recreational demand.

4848 recreational suitability [n] *plan. recr.* (Usability of a landscape, according to qualitative criteria, e.g. ease of accessibility for special groups or all of the population; ►accessibility, ►recreation capacity); *s* **aptitud** [f] **recreativa** (Potencial de un paisaje para acoger actividades recreativas por su cualidades estéticas, su buena ►accesibilidad, etc.; ►cabida recreativa); *syn.* potencialidad [f] recreativa de un paisaje, vocación [f] recreativa, capacidad [f] intrínseca para usos recreativos, idoneidad [f] para usos recreativos; *f* **vocation** [f] **touristique** (Capacité d'utilisation d'un paysage d'après des critères qualitatifs, p. ex. ►accessibilité pour toutes ou certaines catégories de population ; *terme spécifique* aptitude des campagnes en matière d'accueil touristique ; ►capacité récréative) ; *syn.* touristicité [f], aptitude [f] aux activités de loisirs, aptitude [f] touristique, aptitude [f] à des fins récréatives ; *g* **Erholungseignung** [f] (Tauglichkeit einer Landschaft für die Zwecke der Freizeit und Erholung nach qualitativen Kriterien, z. B. hinsichtlich der leichten ►Erreichbarkeit für alle oder einzelne Bevölkerungsgruppen, Ausbaumöglichkeit für ein bestimmtes Aktivitätsspektrum; ▶Erholungskapazität); *syn.* Erholungstauglichkeit [f].

4849 recreational traffic [n] *plan. recr. trans.* (Movement of recreation travellers on travelled ways [US]/carriage-ways [UK]; e.g. ▶local recreational traffic, ▶tourism 1, ▶tourist traffic, ▶weekend traffic); *s* **tráfico** [m] **recreacional** (Tráfico causado por personas en búsqueda de lugares de actividades de recreo; ▶tráfico de fin de semana, ▶tráfico de recreo local, ▶tráfico turístico, ▶turismo); *f* **trafic** [m] **touristique** (Circulation générée par les excursionnistes, les vacanciers lors de leurs déplacements ; ▶circulation de loisirs de proximité, ▶tourisme, ▶trafic de fin de semaine, ▶trafic des vacances) ; *g* **Erholungsverkehr** [m] (Verkehr, der durch Erholungsuchende verursacht wird; ▶Fremdenverkehr, ▶Naherholungsverkehr, ▶Urlaubsreiseverkehr, ▶Wochenendverkehr).

4850 recreational use [n] *plan. recr.* (▶area intended for general recreational use; ▶heavy recreation use, ▶level of recreation use, ▶limited recreation use, ▶reduction in recreation use); *syn. also* recreation use [n]; *s* **uso** [m] **recreativo** (▶frecuentación turística, ▶perturbación del uso recreativo, ▶uso recreativo extensivo, ▶uso recreativo intensivo, ▶zona de recreo extensivo); *syn.* uso [m] turístico; *f* **fréquentation** [f] **touristique** (*Terme spécifique* fréquentation récréative des espaces boisés ; (▶degré de fréquentation touristique, ▶espace de récréation diffuse, ▶fréquentation touristique dense, ▶fréquentation touristique diffuse, ▶perturbation de la fréquentation touristique) ; *syn.* fréquentation [f] ludique, fréquentation [f] de loisirs et de tourisme, fréquentation [f] de loisirs et de vacances ; *g* **Erholungsnutzung** [f] **(2)** (▶Beeinträchtigung der Nutzung für Freizeit und Erholung, ▶Erholungsnutzung 1, ▶extensive Erholungsnutzung, ▶Gebiet zur extensiven Erholung, ▶intensive Erholungsnutzung); *syn.* Freizeitnutzung [f].

4851 recreational weekend [n] [US] *plan. recr.* (Relaxation at the end of the working week, often away from home [**in D.**, usually up to 50 km], and involving an overnight stay); *syn.* weekend break [n] [UK], weekend recreation [n] [UK]; *s* **descanso** [m] **de fin de semana** (Actividades recreativas al aire libre realizadas para la recuperación física y [p]síquica después de la semana de trabajo que se llevan a cabo a menudo fuera de la casa, a veces en forma de una corta excursión. En D. se recorren generalmente distancias de hasta 50 km); *syn.* recreación [f] de fin de semana; *f* **loisirs** [mpl] **de week-end** (Activité récréatrices et de détente de courte durée pendant la fin de semaine effectuée souvent en plein-air [jusqu'à une distance de 50 km] avec, en general, une nuitée passée à l'extérieur du lieu d'habitation) ; *g* **Wochenenderholung** [f, o. Pl.] (Am Wochenende ausgeübte Kurzzeiterholung, die der Entspannung dient und häufig außerhalb des Wohnortes [meist bis zu 50 km entfernt] stattfindet, oft mit einer Übernachtung verbunden).

4852 recreation area [n] **(1)** *recr.* (Generic term for a tract of land predominantly designated for recreational use; ▶recreation space, ▶tourist area, ▶local recreation area); *s* **zona** [f] **de uso priorizado para el recreo** (≠) (Espacio natural de alta idoneidad para usos recreativos o turísticos; ▶área de recreo, ▶zona de recreo local, ▶zona de veraneo); *f* **espace** [m] **touristique** (Espace dont l'affectation principale est le tourisme ; les territoires ludiques majeurs sont la mer et la montagne ; ▶aire de récréation, ▶base périurbaine de plein air et de loisirs, ▶zone de tourisme) ; *syn.* espace [m] de loisirs ; *g* **Erholungsgebiet** [n] (Vorrangig der Erholung dienender Landschaftsraum, der verschiedene Erholungseinrichtungen wie z. B. Campingplätze, Rad- und Reitwegenetze, Bauernhöfe zur Feriennutzung, Museen, Wassersportmöglichkeiten anbietet; ▶Erholungsfläche, ▶Ferien-

erholungsgebiet, ▶Naherholungsgebiet); *syn.* Erholungsraum [m].

recreation area [n] **(2)** *plan. recr.* ▶countryside recreation area, ▶day-use recreation area [US] (1), ▶day-use recreation area [US] (2)/day-trip recreation area [UK], ▶local recreation area; *landsc. plan. pol. recr.* ▶recreation area planning; *recr.* ▶rural recreation area.

recreation area [n] [UK]**, day-trip** *plan. recr.* ▶day-use recreation area [US] (2).

recreation area [n]**, intensive** *plan. recr.* ▶concentration of recreation facilities.

recreation area [n]**, suburban** *plan. recr.* ▶local recreation area.

recreation area [n] **for day-trippers** [UK] *recr.* ▶day-use recreation area [US] (1).

recreation area management [n] *plan. recr.* ▶recreational area management.

4853 recreation area planning [n] *landsc. plan. pol. recr.* (Development of policies, strategies and measures to make an area attractive for recreation users); *s* **planificación** [f] **de áreas turísticas y de recreo** (Planificación y aplicación de medidas para aumentar el atractivo de una zona para el turismo o para usos recreativos); *syn.* gestión [f] de áreas turísticas/de recreo; *f* **étude** [f] **d'un schéma de développement des loisirs** (F., étude et organisation de l'aménagement des loisirs élaborées afin d'améliorer l'offre de loisirs et de mettre en valeur un espace particulier ; *résultat* schéma régional de développement des loisirs) ; *g* **Erholungsplanung** [f] **für ein Gebiet** (Konzeptionelle Erarbeitung von Strategien, Vorschlägen und Maßnahmen, die ein Gebiet [lokal oder regional] für die Erholung attraktiv machen); *syn.* gebietsbezogene Erholungsplanung [f].

recreation benefits [npl] *recr.* ▶having recreation benefits.

4854 recreation capacity [n] *plan. recr.* (Capability of an area to accommodate planned visitor uses without detracting from its ▶recreation value; ▶recreational suitability); *s* **cabida** [f] **recreativa** (Capacidad de un determinado territorio natural o artificial para acoger o absorber actividades recreativas; ▶aptitud recreativa, ▶valor recreativo); *syn.* capacidad [f] recreativa; *f* **capacité** [f] **récréative** (Valeur permettant de déterminer l'aptitude des espaces récréatifs en matière d'accueil touristique ; la capacité d'hébergement touristique constitue un critère décisif de la capacité récréative d'un site ; ▶valeur récréative, ▶vocation touristique) ; *g* **Erholungskapazität** [f] (Leistung einer Landschaft hinsichtlich ihrer Eignung zur Aufnahme von Besuchermengen und ihres Angebotes an Übernachtungsmöglichkeiten; ▶Erholungseignung, ▶Erholungswert).

recreation center [n] [US]/**recreation centre** [n] [UK] *plan. recr.* ▶concentration of recreation facilities.

4855 recreation complex [n] *plan. recr.* (Area containing a variety of resources and facilities providing for different types of recreation; ▶recreational infrastructure); *s* **complejo** [m] **recreacional** (Área que contiene diferentes instalaciones de deporte, juegos y recreo sea en pabellones o al aire libre; ▶equipamiento de recreo); *f* **aire** [f] **de loisirs** (Espace aménagé comportant divers équipements récréatifs tels que des jeux pour enfants, une aire de pique-nique, un parcours de santé, un plateau multisport, un barbecue, etc. ; ▶équipements collectifs de loisirs) ; *g* **Freizeiteinrichtung** [f] **(1)** (Bereich mit diversen Sport- und Spielanlagen in Hallen und im Freien sowie Anlagen, die der passiven/kontemplativen Erholung dienen; ▶Freizeitinfrastruktur).

4856 recreation design load [n] *recr. plan.* (Desired maximum number of people during a specific time period for

R

which a recreation area is planned and developed; ►competing land-use pressure, ►visitor pressure); *s* **presión [f] recreacional máxima prevista** (Cantidad máxima de personas por unidad de tiempo para la cual se planea y desarrolla un área de recreo; ►presión turística; *término genérico* ►presión por competencia entre usos); *f* **pression [f] touristique prévisionnelle** (Charge probable que peut supporter un site dans le cadre d'un projet d'aménagement touristique ; ►pression exercée par les visiteurs, ►pression exercée par l'occupation du sol) ; *g* **geplanter Erholungsdruck [m]** (Bei der Erholungsplanung die zu Grunde gelegte verträgliche Belastung eines Gebietes durch Erholungssuchende; ►Besucherdruck; *OB* ►Nutzungsdruck).

4857 recreation destination area [n] *recr.* (Travel target area with attractive recreational opportunities; ►destination traffic); *s* **zona [f] de destino de recreacionistas** (Paraje natural que acoge a personas en búsqueda de esparcimiento y recreo; ►tráfico de destino); *f 1* **bassin [m] touristique** (Territoire comprenant les éléments paysagers et les prestations nécessaires à la constitution d'une offre touristique complète ; cf. LTV 1996, 48 ; ►circulation de destination) ; *syn.* pays [m] récepteur de touristes, espace [m] récepteur de touristes ; *f 2* **destination [f] touristique** (Distribution de la demande vacancière vers différentes catégories spatiales [mer, montagne, campagne, ville, circuit] ; LTF 1984, 56) ; *g* **Erholungszielgebiet [n] (1)** (Landschaftsraum, der Erholungsuchende aufnimmt; ►Zielverkehr).

recreation facilities [npl] *plan. recr.* ►concentration of urban recreation facilities.

4858 recreation facility [n] *plan. recr.* (Installation and equipment provided for recreation; ►simply-provided recreation facility, ►well-provided recreation facility); *s* **equipamiento [m] de recreo (2)** (Conjunto de instalaciones utilizables con fines recreativos existentes en una zona específica; ►instalación de recreo extensivo, ►instalación de recreo intensivo); *syn.* instalaciones [fpl] de recreo, facilidades [fpl] para el recreo; *f* **équipement [m] (social) de loisirs (et de vacances)** (►aménagement touristique léger, ►équipement de loisirs lourd) ; *syn.* équipement [m] touristique ; *g* **Erholungseinrichtung [f]** (Gesamtheit der [landschaftsbezogenen] Ausstattung, die für die Erholung in einem definierten Bereich bereit steht; ►Erholungseinrichtung mit niedrigem Ausstattungsgrad, ►Erholungseinrichtung mit hohem Ausstattungsgrad); *syn.* Freizeiteinrichtung [f] (2).

recreation facility [n], **poorly-provided** *recr.* ►simply-provided recreation facility.

4859 recreation forest [n] *for. plan. recr.* (Woodland not managed solely for economic purposes but also for the health of visitors and their enjoyment of nature. Its attraction mainly lies in the ease of accessibility, natural features and the provision of recreational facilities; ►town forest [UK]); *syn.* amenity woodland [n] [UK], recreational forest [n]; *s* **monte [m] recreativo** (Término utilizado para denominar a bosques que no se aprovechan exclusivamente para fines económicos, sino que también sirven para el esparcimiento de la población; ►bosque urbano); *syn.* bosque [m] explotado con fines recreativos [C]; *f* **forêt [f] récréative** (Forêt d'agrément ouverte au public à des fins touristiques et récréatives ; ►forêt urbaine) ; *syn.* forêt [f] touristique, forêt [f] d'agrément ; *g* **Erholungswald [m]** (Nicht rein wirtschaftlich genutzten Wald in verdichteten Siedlungsräumen, in der Nähe von Städten und größeren Siedlungen, Heilbädern, Kur- und Erholungsorten, der der Gesundheit, Freude, Abwechslung und dem Naturgenuss seiner Besucher dient. Seine Anziehungskraft beruht im Wesentlichen auf der guten Erreichbarkeit, der besonderen Naturausstattung sowie dem Vorhandensein von Erholungseinrichtungen. Die forstliche Bewirtschaftung

nimmt besondere Rücksicht auf die Waldfunktion Erholung; ►Stadtwald); *syn.* Parkwald [m].

4860 recreation function [n] *plan. recr.* (Role of a forest or scenic landscape in providing for leisure pursuits; ►recreation capacity, ►recreational suitability); *s* **función [f] recreativa** (Vocación o clasificación de un espacio, p. ej. de un bosque, de zonas naturales, para servir fines recreativos; ►aptitud recreativa, ►cabida recreativa); *f* **fonction [f] récréative** (Rôle ou vocation que peuvent jouer les espaces boisés, les ensembles paysagers pour favoriser la détente et les loisirs ; ►capacité récréative, ►vocation touristique) ; *g* **Erholungsfunktion [f]** (Leistung oder Rolle, z. B. eines Waldes, eines Landschaftsraumes, Erholung zu ermöglichen; ►Erholungseignung, ►Erholungskapazität).

4861 recreation [n] **in the natural environment** *recr.* (**1.** Stay or activities in the countryside, such as lodging, hiking and climbing in an area with opportunities for solitude and remoteness, where the imprint of man's work is substantially unnoticeable, e.g. in a wilderness area; cf. WPG 1976, 231. **2.** The term **extensive recreation** [n] describes activities that are usually dispersed over a large [natural] area and require few or no facilities; ►eco-tourism, ►farm vacation [US]/farmstay holidays [UK]); *syn.* recreation [n] in wild places, boondocks recreation [n] [also US], wilderness recreation [n] (±), primitive recreation [n] [also US], extensive recreation [n]; *s* **recreo [m] extensivo** (Uso de los espacios naturales para actividades recreativas, que no se concentra en un lugar determinado ni requiere infraestructura y servicios: excursionismo, paseo, etc.; DINA 1987; ►agroturismo, ►turismo ecológico); *syn.* recreación [f] extensiva, esparcimiento [m] extensivo; *f* **loisirs [mpl] en milieu naturel** (Accueil et séjour des touristes — principalement promeneurs, randonneurs — dans des lieux où l'influence de l'environnement industriel est relativement peu importante et permettant de procurer calme et tranquillité ; ►vacances à la ferme, ►écotourisme) ; *g* **naturnahe Erholung [f, o. Pl.]** (Wiederherstellung der geistigen, seelischen und körperlichen Leistungsfähigkeit durch einen Aufenthalt und besonders durch Bewegung [vorwiegend Spazierengehen, Wandern, Bergsteigen] in einem Gebiet, in dem [u. a.] der optische und akustische Einfluss der technischen Umwelt relativ gering ist, d. h. ‚Lärmfreiheit' und ‚Abgeschiedenheit' empfunden werden; ►Ferien auf dem Bauernhof, ►Ökotourismus).

recreation [n] **in wild places** *recr.* ►recreation in the natural environment.

4862 recreation load [n] *recr. plan.* (Pressure from the demands placed upon an area by those seeking recreation; ►visitor pressure; ►recreation design load; *generic term* ►competing land-use pressure); *syn.* recreation pressure [n]; *s* **presión [f] recreacional** (Sobrecarga de una zona por la presencia masiva de excursionistas; ►presión recreacional máxima prevista, ►presión turística; *término genérico* ►presión por competencia entre usos); *f 1* **pression [f] touristique** (exercée sur un site touristique par les visiteurs ; ►pression exercée par les visiteurs ; ►pression touristique prévisionnelle) ; *f 2* **pression [f] récréative** (En général permanente exercée sur les espaces péri-urbains par les flux récréatifs de masse en provenance des agglomérations proches ; *terme générique* ►pression exercée par les usages du sol) ; *g* **Erholungsdruck [m]** (Bei der Erholungsplanung die voraussichtliche, geplante oder tatsächliche Belastung eines Gebietes durch Erholungsuchende; ►Besucherdruck; ►geplanter Erholungsdruck; *OB* ►Nutzungsdruck).

recreation need [n] *plan. recr.* ►recreational demand.

4863 recreation opportunities [npl] *plan. recr.* (Natural possibilities or man-made range of facilities which make recreation possible); *s* **oferta [f] recreativa** (Condiciones naturales

R

propicias para desarrollar actividades de recreo e instalaciones creadas para ese fin); *syn.* oportunidades [fpl] de recreo; *f 1* **offre [f] touristique** (Éléments naturels existants ou équipements autour desquels s'organisent les loisirs) ; *syn.* support [m] d'activités de loisirs, support [m] d'activités touristiques ; *f 2* **offre [f] patrimoniale** (Ensemble des éléments naturels et culturels [p. ex. sites, monuments, musées, jardins historiques, villages de caractère, espaces et curiosités naturels, etc.] existants ; cf. TP 1997, 5) ; *g* **Erholungsangebot [n]** (Natürlich vorhandene Möglichkeit und künstlich geschaffene Einrichtungen, die Erholung ermöglichen; die Gesamtheit des für den Erholungsuchenden verfügbaren natürlichen und kulturellen Erbes heißt im Französischen *offre patrimoniale*).

4864 recreation origination area [n] *plan. recr.* (Source of recreation-seeking people who leave their neighbo[u]rhood for another place, due to a lack of recreational opportunities at home; ▶recreation destination area, ▶recreation resort community); *s* **zona [f] de origen de recreacionistas** (*opp.* ▶zona de destino de recreacionistas, ▶lugar receptor de recreacionistas); *f* **espace [m] émetteur de flux touristiques** (Territoire sous-équipé en aménagements récréatifs originaire des flux de visiteurs ; *opp.* ▶bassin touristique, ▶collectivité réceptrice) ; *syn.* collectivité [f] émettrice, pays [m] émetteur de touristes ; *g* **Erholungsquellgebiet [n]** (Gemeinde, von der Erholungsverkehr ausgeht, die hinsichtlich ihres freizeitrelevanten Infrastrukturangebotes unterversorgt ist, weshalb Erholungsuchende in andere Gebiete ausweichen; ▶Erholungszielgebiet 1, ▶Erholungszielgebiet 2).

recreation park [n] *plan. recr.* ▶leisure park.

4865 recreation planning [n] *landsc. plan. pol. recr.* (Development of policies, strategies and measures for the provision of recreation areas and facilities; ▶urban area recreation planning); *s* **gestión [f] recreacional (2)** (Desarrollo de políticas, estrategias y medidas para prever una región con zonas y equipamiento de recreo; ▶planificación de zonas urbanas de recreo); *syn.* planificación [f] recreacional; *f* **aménagement [m] touristique** (Organisation de l'espace, des équipements et de l'animation dans le secteur des loisirs ; *terme législatif* aménagement [m] des espaces à vocation touristique ; ▶planification du tourisme urbain) ; *syn.* planification [f] du tourisme ; *g* **Erholungsplanung [f]** (Konzeptionelle Erarbeitung von Strategien und Maßnahmen, die für ein Gebiet Flächen, Einrichtungen und Ausstattungen für die Erholung vorhält; ▶Erholungsplanung im städtischen Raum).

recreation planning [n] [UK]**, countryside** *landsc. plan. pol. recr.* ▶rural recreation planning [US].

4866 recreation policies [npl] *plan. recr.* (Official strategies and measures adopted to make land available for provision of recreational facilities and ▶public access to rural land; ▶right of public access); *s* **política [f] de promoción del turismo rural y del recreo (≠)** (Conjunto de medidas legislativas y prácticas dirigidas a posibilitar el uso recreativo de los espacios naturales, como p. ej. la ▶servidumbre de tránsito, ▶acceso público al campo); *f* **mesures [fpl] en faveur d'une politique de prévoyance concernant les loisirs et le tourisme** (Ensemble des actions adoptées par les pouvoirs publics en vue de l'affectation de parcelles à des fins de loisirs et de détente en milieux rural et urbain et rendant possible l'▶ouverture au public de l'espace naturel ; ▶servitude de passage des piétons) ; *g* **Maßnahmen [fpl] für die Erholungsvorsorge** (Gesamtheit der von politischen Kräften getragenen Maßnahmen, die der Bereitstellung von Grundstücken zum Zwecke der Erholung dient und in der freien Landschaft das ▶Betreten der Flur ermöglicht; cf. §§ 27 u. 28 BNatSchG, § 2 [2] 4 ROG von 2008, für den Siedlungsbereich § 9 [1] BauGB; ▶Betretungsrecht).

4867 recreation potential [n] *landsc. recr.* (Capacity of land and water areas and associated developments which provide existing and future opportunities for outdoor recreation; ▶recreational suitability, ▶recreation capacity); *syn.* recreation resources [npl]; *s* **potencial [m] recreativo** (Aptitud actual o futura de un paraje para acoger actividades de recreo; ▶aptitud recreativa, ▶cabida recreativa); *f* **potentiel [m] récréatif** (Ensemble des composantes dominantes d'un cadre paysager, expression de sa capacité à être ou devenir un support d'activités récréatives ; ▶vocation touristique, ▶capacité récréative ; *syn.* potentiel [m] ludique ; *g* **Erholungspotenzial [n]** (Derzeitige und zukünftige Eignung von Landschaftsräumen oder Teilen davon für Freizeit und Erholung; ▶Erholungseignung, ▶Erholungskapazität); *syn. o. V.* Erholungspotential [n].

recreation pressure [n] *recr. plan.* ▶recreation load.

recreation pursuit [n] *plan. recr.* ▶recreation activity.

recreation quality [n] *plan. recr.* ▶recreation value.

4868 recreation resort community [n] *recr.* (Town or city offering a high level of recreation opportunities; ▶recreation destination area, ▶recreation origination area); *s* **lugar [m] receptor de recreacionistas** (Municipio que acoge a personas en búsqueda de esparcimiento y recreo resp. que posee ▶equipamiento de recreo atractivo para personas procedentes de otros lugares; ▶zona de origen de recreacionistas, ▶zona de destino de recreacionistas); *f* **collectivité [f] réceptrice** (Commune accueillant les visiteurs ; ▶espace émetteur de flux touristiques, ▶bassin touristique) ; *g* **Erholungszielgebiet [n] (2)** (Gemeinde, die Erholungsuchende aufnimmt resp. die erforderliche ▶Freizeitinfrastruktur vorhält; ▶Erholungsquellgebiet, ▶Erholungszielgebiet 1).

recreation resources [npl] *landsc. recr.* ▶recreation potential.

4869 recreation resources inventory [n] *plan. recr.* (Compilation of data and graphic representation detailing the distribution and types of recreation facilities and landscape features within a specific area); *s* **mapa-inventario [m] de recursos recreativos** (Compilación de datos y representación gráfica de facilidades para usos recreativos y turísticos así como de elementos paisajísticos de una región determinada); *f* **inventaire [m] touristique et récréatif** (Relevé et représentation cartographique des équipements récréatifs existants et des éléments paysagers présentant un intérêt récréatif) ; *g* **Erholungskataster [m,** *in A nur so* — oder **n]** (Bestandsaufnahme aller Erholungseinrichtungen und erholungswirksamen Landschaftselemente in Form einer genauen kartografischen Darstellung).

recreation seeker [n] [UK] *recr. sociol.* ▶recreation user.

4870 recreation space [n] *plan. recr.* (An open or wooded area affording active or passive leisure-time activities; ▶proximity green space, ▶recreation complex); *s* **área [f] de recreo** (Superficie de pequeña extensión con algún equipamiento de recreo; ▶complejo recreacional, ▶zona verde de vecindario); *f* **aire [f] de récréation** (Espace réduit propice à la détente ou aux activités de découverte ; ▶espace vert urbain de proximité ; ▶aire de loisirs) ; *syn.* unité [f] ludique, centre [m] ludique ; *g* **Erholungsfläche [f]** (Freifläche, die vorherrschend aktiven oder kontemplativen Freizeitaktivitäten dient; im Vergleich zum Erholungsgebiet meist kleinere Flächen; ▶wohnungsnahe Grünfläche, ▶Freizeiteinrichtung 1); *syn.* Erholungsanlage [f], Freizeitanlage [f].

4871 recreation time category [n] *plan. recr.* (Generic term for short- or long-term periods of recreation, such as ▶local recreation, day-trips, weekend stays or vacations [US]/holidays [UK]; ▶long-stay recreation, ▶recreational weekend [US]/

R

weekend recreation [UK], ►recreation type, ►short-stay recreation); *s* **categoría [f] recreativa** (Diferenciación de tipos de recreo según la duración como ►vacaciones cortas, ►vacaciones largas, ►recreo local, ►descanso de fin de semana, etc.; ►tipo de recreo); *f* **catégorie [f] d'activités touristique et de loisirs** (Terme générique englobant le ►tourisme de passage, ►tourisme de séjour, ►loisirs de proximité, ►loisirs de week-end, ►type d'activités de détente) ; *g* **Erholungsform [f]** (OB zu ►Kurzzeiterholung und ►Langzeiterholung, ►Naherholung, ►Wochenenderholung, Ferienerholung; ►Erholungstyp).

4872 recreation type [n] *plan. recr.* (Form of recreational activities, such as active [rock climbing, hiking, canoeing] or passive [e.g. birdwatching, reading]; ►active recreation, ►leisure pursuit, ►passive recreation, ►recreation activity, ►recreation time category, ►type of leisure); *s* **tipo [m] de recreo** (Clase de actividad de recreo de tipo pasivo [p. ej. observación de la naturaleza, lectura] o activo [montañismo, excursionismo, piragüismo]; ►actividad de recreo, ►actividad de tiempo libre, ►categoría recreativa, ►recreación pasiva, ►recreo, ►tipo de actividad de tiempo libre); *f* **type [m] d'activités touristiques et de loisirs** (Formes d'activités de détente telles que le délassement [découverte de la nature, lecture] ou les ►loisirs actifs [p. ex. alpinisme, excursion, rafting] ; ►activité de loisirs, ►activité de récréation, ►catégorie de loisirs, ►loisirs passifs, ►type d'activités de loisirs) ; *syn.* forme [f] d'activité de récréation, type [m] d'activités de détente ; *g* **Erholungstyp [m]** (Form der Erholungsaktivität, z. B. ruhige Erholung [Naturbeobachtung, Lesen] oder aktivitätsbetonte ►Erholung [z. B. Bergsteigen, Wandern, Wildwasserfahren]; ►aktive Erholung, ►Erholungsaktivität, ►Erholungsform, ►Freizeitaktivität, ►Freizeittyp, ►passive Erholung).

recreation use [n] *recr* ►area intended for general recreational use; *plan. recr.* ►heavy recreation use, ►level of recreation use, ►limited recreation use, ►recreational use, ►reduction in recreation use.

4873 recreation user [n] *recr. sociol.* (Individual who seeks regeneration; e.g. in a green open space or in the countryside; ►vacationist [US]/holidayer [UK]); *syn.* recreation visitor [n] [also US], recreation seeker [n] [also UK], outdoor's man [n] [also US]; *s* **recreacionista [m/f]** (Persona que busca regenerar sus capacidades físicas o [p]síquicas, p. ej. realizando actividades al aire libre; ►vacacionista); *f* **séjournant, ante [m/f]** (Usager d'un espace récréatif visiteur d'un équipement de loisirs ; ►vacancier) ; *syn.* utilisateur [m] de loisirs, consommateur [m] de loisirs ; *g* **Erholungsuchende/-r [f/m]** (Individuum, das sich z. B. in einer Grünanlage oder in der freien Landschaft erholen möchte; ►Urlauber/-in); *syn. o. V.* Erholung Suchende/-r [f/m].

4874 recreation value [n] *plan. recr.* (Scenic, inspirational, aesthetic and educational qualities which contribute to visitor enjoyment, and are enhanced by absence of air, water and noise pollution); *syn.* recreation quality [n]; *s* **valor [m] recreativo** (Cualidades escénicas, estéticas, educativas y ausencia de ruidos e inmisiones que hacen posible el gozo de una zona natural); *f* **valeur [f] récréative** (Qualité des composantes paysagères, esthétiques artistiques, historiques et culturelles ainsi que l'absence de pollutions [air, eau, bruit] contribuant à l'intérêt porté à un espace de loisirs) ; *syn.* valeur [f] d'agrément ; *g* **Erholungswert [m]** (Landschaftliche, ästhetische, informative und geistig belebende Gegebenheiten eines Erholungsraumes, abhängig von der natürlichen oder künstlichen Ausstattung, die das Erleben durch optische Reize oder Aktivitätsangebote steigert. Voraussetzung für einen hohen **E.** ist das Fehlen von beeinträchtigenden Luft-, Lärm- und Geruchsimmissionen); *syn.* Erholungsqualität [f].

4875 recreation vehicle (RV) [n] [US] *recr.* (Generic term for a motorized vehicle up to 7.5 t or more, used for living purposes, whether a modified small bus, van or a small truck with special equipment or a bus equipped for holidaying or weekend trips; **in U.K., R.V.** may mean a dune buggy, or an ATV [All Terrain Vehicle]; *specific terms* ►camper van [UK], living van [n] [US], ►mobile home, motor caravan [n], motor home [n] [UK]); *s* **autocaravana [f]** (Término genérico para cualquier vehículo equipado para viajar y vivir en él; ►camioneta de camping, ►casa móvil); *syn.* vehículo-vivienda [m], vehículo [m] de camping; *f* **camping-car [m]** (*Terme générique* véhicule motorisé aménagé en habitation, de la taille d'un minibus ou d'un camion léger d'un tonnage inférieur à 7,5 t spécialement équipé pour voyager pendant les vacances ou les fins de semaine ; ►résidence mobile) ; *syn.* car camping [m] ; *g* **Wohnmobil [n]** (Selbstfahrender Wohnwagen, sei es ein umgebauter Kleinbus [►Campingbus] oder Kleinlastwagen mit entsprechenden Spezialaufbauten oder ein zum Ferien- oder Wochenendwohnen eingerichteter Bus bis 7,5 t oder darüber. Je nach Ausstattungsgrad werden die kleineren Busse auch ,Reisemobil' genannt; ►Mobilheim); *syn.* Reisemobil [n], Motorcaravan [m], selbstfahrender Wohnwagen [m].

recreation visitor [n] [US] *recr. sociol.* ►recreation user.

recreation zone [n] [US]**, day-use** *plan. recr.* ►day-use recreation area [US] (2)/day-trip recreation area [UK].

4876 recruitment [n] **of a forest** *for.* (Any supplementation of a ►standing forest crop; i.e. the successive aspect of natural or anthropogenic ►regeneration 1; cf. SAF 1983; ►natural regeneration 1, ►natural regeneration 2); *s* **bosque [m] de renuevo** (Resultado de la ►regeneración natural o artificial o con plántulas de renuevo; ►bosquecillo de renuevo, ►población forestal); *f 1* **peuplement [m] de régénération (naturelle/artificielle)** (Issu d'une ►régénération naturelle par voie de semences ou artificielle effectuée à partir de semis manuels ou mécaniques, de ►plantules ou de plantations ; ►peuplement de régénération naturelle, ►peuplement forestier 2, ►régénération [forestière]) ; *f 2* **recrû [m]** (Jeune peuplement obtenu par voie végétative, p. ex. d'un taillis ; DFM 1975, 238) ; *g* **verjüngter Waldbestand [m]** (Ergebnis der natürlichen oder anthropogenen ►Verjüngung aus Keimlingen oder Forstjungpflanzen; ►Naturverjüngung 1, ►Naturverjüngung 2, ►Waldbestand 2).

rectangular bond [n] *constr.* ►random rectangular bond.

rectangular pattern [n]**, random** *constr.* ►random rectangular bond.

rectangular paving pattern [n]**, random-jointed** *constr.* ►random rectangular bond.

rectification [n] [UK]**, instruction for** *constr. contr.* ►rectification directive [US].

4877 rectification directive [n] [US] *constr. contr.* (Written order/instruction to the contractor requiring the remedying of ►defects during the ►defects liability period); *syn.* directive [n] to rectify [US], instruction [n] for rectification [UK], instruction [n] to rectify defective work [UK], instruction [n] requiring defects to be made good [UK]; *s* **reclamación [f] de defectos** (Demanda escrita a la empresa constructora para que elimine los ►defectos [de obra] en el ►plazo de garantía); *syn.* demanda [f] de reparación; *f* **mise [f] en demeure de parfait achèvement des travaux** (Notification écrite à l'entrepreneur par le responsable du marché de se conformer aux dispositions du marché et d'y satisfaire dans un délai déterminé ; cf. décret n° 76-88 du 21 janvier 1976 ; ►délai de garantie, ►imperfection, ►malfaçon) ; *g* **Mängelrüge [f]** (Schriftliche Aufforderung an den Auftragnehmer, ►Mängel im Rahmen der ►Gewährleistungsfrist zu beseitigen; cf. § 13 [5] VOB Teil B).

recultivated mining landscape [n] [UK] *min. landsc.*
▶landscape of reclaimed surface-mined land.

4878 recuperation [n] *recr.* (Recovery of mental, physical and spiritual fitness over an extended period of more than three weeks, or after a period of illness, often during a stay in a convalescent home; ▶health resort recuperation, ▶recreation); *s* **regeneración** [f] (*Sensu lato* reestablecimiento de la capacidad física, [p]síquica e intelectual durante un periodo/período largo de tres semanas como mínimo; ▶cura de reposo, ▶descanso); *f* **régénération** [f] (En liaison avec le renouvellement des capacités intellectuelles, mentales et corporelles pendant une période d'au moins trois semaines ; ▶cure de repos, ▶détente) ; *g* **Erholung** [f, o. Pl.] (2) (*Sensu lato* Wiederherstellung der geistigen, seelischen und körperlichen Leistungsfähigkeit während eines längeren Zeitraumes von mindestens drei Wochen; ▶Erholung 1, ▶Kurerholung).

4879 recurrent clutch [n] **of eggs** *zool.* (Repetition of the production of eggs by a brooding bird after premature loss of all or some of the eggs laid); *s* **puesta** [f] **suplementaria** (En aves en fase de reproducción repetición de la puesta de huevos en el caso de pérdida de los huevos de la primera); *f* **ponte** [f] **supplétive** (±) (Nouvelle ponte effectuée par les oiseaux ayant perdu la totalité ou une partie de la couvée) ; *g* **Nachgelege** [n] (Die erneute Eiablage brütender Vögel bei frühzeitigem Verlust der Eier oder Teilen davon).

recurrent flowering rose [n] [US] *hort.* ▶remontant rose.

recyclable material [n] *envir.* ▶reusable material.

recyclable materials [npl] *constr.* ▶interim storage of recyclable materials.

4880 recycled material [n] *constr. envir.* (Building material, which is re-used or reclaimed after processing; e.g. material re-used in road construction, which has been obtained from recycled building rubble or broken-up road surfaces); *s* **material** [m] **(de construcción) reciclado** (Tipo de material de construcción, como escombros de edificios o de carreteras, que se prepara para su reutilización en la construcción vial); *f* **matériau** [m] **de recyclage** (Matériau de construction récupéré et réintroduit, après transformation, dans le cycle économique pour un usage analogue) ; *g* **Recyclingmaterial** [n] (Baustoff, der nach einem Aufbereitungsverfahren weiter- oder wiederverwertet wird, z. B. aufbereiteter Straßenaufbruch und Bauschutt zur Verwendung im Straßenbau; *syn.* recycelter Baustoff [m], Recyclingbaustoff [m], Rezyklat [n], rezyklierter Baustoff [m], weiterverwerteter Baustoff [m].

recycled material [n], **reuse of** *envir.* ▶reapplication of used materials.

4881 recycling [n] *envir. nat'res.* (Process by which new materials are created from used products. Their reintroduction into the production chain serves in the manufacture of economically-useful goods and conserves finite resources; ▶reapplication of used materials, ▶reusable material, ▶reuse of waste, ▶reuse of waste); *syn.* reclamation [n] of used resources, resource recovery [n]; *s* **reciclado** [m] (Reintroducción de materiales recuperados al ciclo de producción como sustitutos de materias primas para así ahorrar recursos minerales y evitar o reducir residuos; ▶material reutilizable, ▶reutilización de residuos, ▶valorización de residuos [Es]); *syn.* reciclaje [m]; *f 1* **recyclage** [m] (*Terme générique* réintroduction d'un matériau récupéré dans le cycle de production comme substitut à la matière première ; *terme spécifique* recyclage des polluants atmosphériques : réutilisation d'une substance réglementée récupérée, au moyen d'opérations telles que filtrage, séchage, distillation et traitement chimique, afin de restituer à la substance des caractéristiques opérationnelles déterminées ; règlement du Conseil

n° 3093/94/CE ; ▶réemploi des déchets, ▶valorisation des déchets, ▶matériau réutilisable) ; *f 2* **récupération** [f] (Tri et collecte sélective de certains déchets ménagers ou industriels en vue de leur réemploi ou de réintroduction sans transformation dans le cycle industriel) ; *g* **Recycling** [n, o. Pl.] (Alle Maßnahmen, die darauf abzielen, aus Abfällen Stoffe zu gewinnen, die erneut einem Produktionsprozess zugeführt werden können und der Herstellung volkswirtschaftlich nützlicher Güter dienen; dazu gehört nicht die energetische Verwertung. **R.** ist in der Abfallwirtschaft nach der Vermeidung von unerwünschten Rückständen die sinnvollste Möglichkeit, Rohstoffe und Energie zu sparen und Mülldeponien und Müllverbrennungsanlagen zu entlasten; ▶Abfallverwertung, ▶Weiterverwendung von Abfallstoffen, ▶Wertstoff); *syn.* Rezyklierung [f], Wiederverwendungskreislauf [m], stoffliche Wiederverwertung [f], Wiedergewinnung [f] von Wertstoffen, Wertstoffwiedergewinnung [f] und Verwertung [f].

recycling [n], **site** *plan. urb.* ▶recycling of derelict sites.

4882 recycling economy [n] *agr. envir. for. hort.* (Measures promoting the natural ▶biogeochemical cycle and saving energy; in a broader sense, this term includes all strategies to avoid or decrease the amount of waste and undesirable residues, as well as ▶recycling and greater longevity of natural materials and energy in the overall **e.**; ▶natural resources management, ▶compost recycling); *s* **economía** [f] **de reciclaje** (**1.** Todas las medidas que favorecen el ▶ciclo biogeoquímico natural y el ahorro de energía. **2.** En sentido más amplio este término incluye todas las estrategias para evitar o reducir la cantidad de residuos, así como el ▶reciclado y la promoción de una larga vida de la materia y energía en el sistema económico; ▶flujo de energía, ▶gestión de los recursos naturales, ▶gestión de residuos aptos para compostaje); *f* **gestion** [f] **des cycles naturels** (**1.** Ensemble des actions qui tendent à promouvoir la gestion équilibrée des cycles biogéochimiques et des ▶flux d'énergie ainsi qu'à favoriser les économies d'énergie ; **2.** Au sens large, ensemble des stratégies favorisant l'économie et l'utilisation rationnelle de l'énergie, le développement des énergies renouvelables, la prévention et la réduction de la production de déchets, la valorisation et le ▶recyclage des déchets ; ▶cycle biogéochimique, ▶flux d'énergie, ▶gestion des ressources naturelles, ▶industrie du compost) ; *g* **Kreislaufwirtschaft** [f, o. Pl.] (**1.** Alle Maßnahmen, die dazu dienen, den natürlichen ▶Stoffkreislauf und die Energieeinsparung zu fördern. **2.** Im weiteren Sinne alle Strategien der Vermeidung und Verminderung von Abfällen oder unerwünschten Rückständen sowie das ▶Recycling und die Förderung der Langlebigkeit von Materie und Energie im Wirtschaftssystem; cf. KrW-/AbfG; ▶Bewirtschaftung der Naturgüter, ▶Kompostwirtschaft).

recycling [n] **of brownfields** *plan. urb.* ▶recycling of derelict sites.

4883 recycling [n] **of derelict sites** *plan. urb.* (Redevelopment of old, often contaminated and abandoned industrial areas for more productive land use and prevention of urban sprawl; ▶conversion 2, ▶industrial revitalization [US]/redevelopment of industrial areas [UK]); *syn.* recycling [n] of brownfields, site recycling [n]; *s* **recuperación** [f] **de ruinas industriales** (Saneamiento de superficies industriales abandonadas, construidas o no, que han sido utilizadas para actividades industriales y están degradadas de manera que un nuevo uso no es posible más que tras un profundo trabajo de recuperación. Los fines de la **r. de rr. ii.** son la reutilización del suelo para otros usos urbanísticos y así evitar el consumo de suelo, la renovación urbanística, etc.; cf. Estudio-Inventario de Ruinas Industriales en CA Vasca 1998; ▶conversión, ▶saneamiento de ruinas industriales); *syn.* reciclaje [m] de suelo industrial, reutilización [f] de

R

antiguos emplazamientos industriales, recuperación [f] de suelos; *f* **réhabilitation [f] des friches industrielles** (Politique de traitement des friches industrielles soit par le réemploi industriel ou la requalification des sites après dépollution des sols à des fins agricoles, de réserves foncières pour des activités diversifiées ultérieures avec l'objectif de limiter la consommation spatiale ; cf. DUA 1996, 367 ; ►conversion, ►rénovation de zones Industrielles) ; *g* **Flächenrecycling [n, o. Pl.]** (Sanierung von alten, oft belasteten, brachliegenden oder untergenutzten Industrie- und Gewerbeflächen zwecks Wiederverwendung für städtebauliche Nutzungen mit dem Ziel, den Landschaftsverbrauch einzuschränken, geschlossene Siedlungsflächen zu erhalten und besser auszunutzen, die Siedlungsstruktur zu verbessern und eine bessere Auslastung der Infrastruktureinrichtungen zu erreichen. Die Wiedernutzung ist in der Regel abhängig von der Finanzierbarkeit der erforderlichen Altlastensanierung. Die Kosten der Sanierung können in vielen Fällen durch die Inwertsetzung des Grundstückes gedeckt werden; ►Umnutzung, ►Sanierung von Industriegebieten); *syn.* Brachflächenrecycling [n], Brachflächenrevitalisierung [f], Neunutzung [f] von Industriebrachen, Wiedernutzung [f] von Industriebrachen, Wiederverwendung [f] brachliegender Flächen.

recycling system [n] *envir.* ►closed recycling system, ►open recycling system.

red blaes [n] [SCOT] *min.* ►red shale.

Red Data Book [n] *conserv. phyt. zool.* ►Red List.

4884 redesign [n] (1) *gard. urb.* (Revision in style and appearance or refurbishment of an existing object or area; ►partial redesign); *syn.* reshaping [n]; *s* **reconfiguración [f]** (Transformación total del diseño de un objeto o un área; ►reconfiguración parcial); *f* **rénovation [f]** (Transformation totale de la forme et des équipements d'un objet existant [bâtiment, espace libre] et impliquant la plupart du temps sa démolition ; ►recomposition 1, ►remodelage) ; *syn.* reconfiguration [f], refonte [f] ; *g* **Neugestaltung [f]** (Völlig neue Formgebung oder Ausstattung eines bestehenden Objektes oder einer Fläche; ►Umgestaltung 1); *syn.* Neuplanung [f].

redesign [n] (2) *urb. trans.* ►road redesign.

redevelopment [n] *urb.* ►center city redevelopment [US]/ urban regeneration [UK], ►comprehensive redevelopment, ►neighborhood redevelopment [US], ►urban redevelopment.

redevelopment [n] **of courtyards** *constr. urb.* ►courtyard rehabilitation scheme.

redevelopment [n] **of industrial areas** [UK] *plan. urb.* ►industrial revitalization [US].

redevelopment planning [n] *plan. urb.* ►village redevelopment planning.

redirection [n] [UK]**, traffic** *trans. urb.* ►traffic rerouting [US].

4885 redirection [n] **of a watercourse** *wat'man.* ([Temporary] rerouting of the course of an existing stream or river or parts thereof through another area); *s* **trasvase [m]** (Medida muy criticada por ecologistas y científicos de redistribuir el caudal de un río hacia otra vertiente con el fin de alimentar zonas que no tienen suficiente agua); *f* **déplacement [m] d'un cours d'eau** (*Réaménagement de cours d'eau* aménagement hydraulique par lequel est créé un nouveau tracé temporaire ou définitif) ; *g* **Umleitung [f] eines Fließgewässers** ([Temporäres] Führen eines bestehenden Fließgewässers oder Teile davon durch ein anderes Gebiet).

4886 Red List [n] *conserv. phyt. zool.* (**1.** Registers of endangered species of plants and animals requiring protection, which are compiled by the Species Survival Commission of the International Union for Conservation of Nature [IUCN], now called World Conservation Union. Presently the following categories are distinguished: 1. Extinct [EX]: ►extinct species, 2. Extinct in the Wild [EW]: ►extinct in the wild species, 3. Critically Endangered [CR]: ►critically endangered species, 4. Endangered [EN]: ►cndangcred species, 5. Vulnerable [VU]: ►vulnerable species, 6. Near Threatened [NT]: ►near threatened species, 7. Least Concern [LC], 8. Data Deficient [DD], 9. Not Evaluated [NE]; cf. IUCN Red List Categories and Criteria, Version 3.1, 2001; ►Endangered Species Act [E.S.A.] [US]/ Conservation of Wild Creatures and Wild Plants Act 1975 [UK], ►Blue List [US]). **2. Endangered Species List [n] [US] (In U.S.**, under the Endangered Species Act, the Fish and Wildlife Service lists the status of endangered species of plants and animals periodically in Federal Register notices, according to the following general categories: 1. endangered, 2. threatened, 3. candidate. Many states also have laws protecting rare species; ►status of endangerment); *syn. obs.* Red Data Book [n]; *s 1* **Catálogo [m] Español de Especies Amenazadas** (Medida de carácter administrativo de registrar aquellas especies, subespecies o poblaciones de la flora y fauna silvestres que requieran medidas específicas de protección. En Es. según la legislación básica del Estado, registro establecido en el seno del Listado de Especies Silvestres en Régimen de Protección Especial [art. 53, Ley 42/ 2007], que incluirá, cuando exista información técnica o científica que así lo aconseje, los taxones o poblaciones de la biodiversidad amenazada, incluyéndolos en alguna de las categorías siguientes: a] en peligro de extinción: taxones o poblaciones cuya supervivencia es poco probable si los factores causantes de su actual situación siguen actuando, b] vulnerable: taxones o poblaciones que corren el riesgo de pasar a la categoría anterior en un futuro inmediato si los factores adversos que actúan sobre ellos no son corregidos; las CC.AA. pueden establecer en sus respectivos ámbitos territoriales otras categorías específicas y, en su caso, incrementar el grado de protección de las especies del **C. E. de EE. AA.** en sus catálogos autonómicos, incluyéndolas en una categoría superior de amenaza; art. 55, Ley 42/2007; ►categoría de amenaza de extinción de especies, ►ley de protección de especies); *s 2* **lista [f] roja** (Registro de taxones que necesitan protección, que son recopilados por la Comisión para la Supervivencia de Especies de la Unión Internacional de Conservación de la Naturaleza [UICN], hoy en día denominada Unión Mundial para la Naturaleza. Actualmente se distinguen las siguientes categorías: 1. extinto [EX]: ►especie extinta, 2. extinto en estado silvestre [EW]: ►especie extinta en estado silvestre, 3. en peligro crítico [CR]: ►especie en peligro crítico, 4. en peligro [EP]: ►especie en peligro, 5. vulnerable [VU]: ►especie vulnerable, 6. casi amenazado [NT]: ►especie casi amenazada, 7. preocupación menor [LC], 8. datos insuficientes [DD]; cf. UICN 2001; ►Lista Azul [US]); *syn.* libro [m] rojo; *f 1* **Livre [m] rouge (des espèces menacées)** (Ouvrage scientifique s'inspirant des critères définis par l'Union Internationale pour la Conservation de la Nature [UICN] concernant les espèces rares ou menacées à l'échelle du territoire national n'ayant pas de rôle réglementaire. Il identifie sur la base de critères objectifs et mondialement admis, les espèces particulièrement menacées d'un territoire, précise les principaux facteurs de menaces qui pèsent sur chacune de ces espèces menacées, synthétise les caractéristiques les plus significatives de ces espèces [aires de répartition, milieux d'apparition, exigences écologiques] et est un outil de référence pour apprécier l'état de santé des espèces au niveau national. ; les critères de l'UICN distinguent neuf catégories de classement des espèces ou sous-espèces : **1. éteint, 2. éteint à l'état sauvage, 3. en danger critique d'extinction, 4. en danger, 5. vulnérable, 6. quasi menacé, 7. préoccupation mineure, 8. données insuffisantes** et **9. non évalué** ; la classification d'une espèce ou d'une sous-espèce dans l'une des trois catégories d'espèces menacées d'extinction [en danger critique d'extinction, en danger ou vulnérable] s'effectue par le biais d'une série de cinq critères quantitatifs basés sur différents facteurs biologiques associés au risque d'extinction : taux de déclin, population totale, zone d'occurrence, zone d'occupation, degré de peuplement et fragmentation de la répartition. **Le Livre rouge de la flore menacée de France**, tome publié en 1995 par le Muséum national d'histoire naturelle et le Ministère de l'Environnement, traite de 486 espèces prioritaires, c'est-à-dire celles qui sont **1.** soit strictement endémiques du territoire national métropolitain, **2.** soit subendémiques, **3.** soit présentes en France et dans au moins deux autres pays, mais qui sont rares ou menacées sur l'ensemble de leur aire ou seulement en France. Un second tome, portant sur les autres espèces menacées, est en cours de rédaction, sous la coordination du Conservatoire botanique national de Porquerolles. En France, des livres rouges ont également été publiés, toujours inspirés des critères définis par l'UICN tels que le livre rouge de la faune nationale (1995), livre rouge des espèces végétales rares et menacées (1995), livre rouge des espèces végétales rares et menacées de la région Provence-Alpes-Côte d'Azur (1994) ; *f 2* **liste [f] rouge** (La liste rouge de l'Union Internationale pour la Conservation de la nature constitue l'inventaire mondial le plus complet de l'état de conservation global des espèces végétales et animales. Elle s'appuie sur une série de critères précis pour évaluer le risque d'extinction de nombreuses espèces et sous-espèces. La première liste rouge des espèces menacées dans le monde a été publiée en 1966. Depuis, elle fait l'objet de mises à jour régulières en fonction de l'évolution de la situation de ces espèces. F., depuis 1979, à la demande du ministère chargé de l'environnement, le Muséum national d'Histoire naturelle collecte et concentre des données sur la faune et la flore de France métropolitaine et des Collectivités d'outre-mer. Ces données servent essentiellement à évaluer la qualité de notre Patrimoine Naturel et d'en déterminer l'état de conservation. En 2007, le Comité français de l'UICN et le Muséum national d'Histoire naturelle ont lancé la réalisation de la Liste rouge des espèces menacées en France. Les premiers volets de la liste rouge nationale concernent les reptiles et amphibiens [mars 2008], les oiseaux nicheurs [décembre 2008] et les mammifères [février 2009] ; cf. www.uicn.fr) ; *f 3* **liste [f] des espèces menacées** (La première loi sur la protection de l'environnement en France, (loi n° 76-629 du 10 juillet 1976) reprise par les articles L 211-1 et 2 du Code Rural et les articles L 411-1 et 2 du Code de l'Environ instaure la préservation des espèces animales non domestiques ou végétales non cultivées lorsqu'un intérêt scientifique particulier ou les nécessités de la préservation du patrimoine biologique national justifient leur conservation ainsi que l'établissement de listes nationales et régionales des espèces protégées Il fallut attendre l'arrêté de 1982 pour voir publier une première liste de plantes à protéger au niveau national qui fixe les interdictions qui y sont liées. Ces textes instaurent actuellement trois régimes de protection : **1. protection intégrale :** destruction, coupe, mutilation, arrachage, cueillette et enlèvement, colportage, mise en vente, vente ou achat de tout ou partie des spécimens sauvages des espèces de la liste ; **2. protection partielle :** espaces rares faisant l'objet d'une certaine utilisation ; **3. réglementation préfectorale temporaire** pour les espèces rares dont l'exploitation peut devenir préoccupante pendant certaines périodes de l'année. En 1986 fut établie la première liste pour une région [Corse]. Puis en 1989 est prise une liste d'espèces susceptibles d'être réglementées par arrêté préfectoral au niveau de chaque département. Les premières listes nationales d'espèces menacées protégées par arrêtés ministériels concernent **les espèces végétales** (20 janvier 1982, modifié en 1995), **les mammifères** (17 avril 1981), **les mammifères marins** (27 juillet 1995), **les oiseaux** (17 avril 1981), **les reptiles et les amphibiens** (24 avril 1979), **les poissons** (8 décembre 1988), **les crustacés** (21 juillet 1983), **les insectes** (22 juillet 1993), **les tortues marines** (17 juillet 1991), **les mollusques** (7 octobre 1992) ; F., on distingue cinq catégories de régression des populations de vertébrés : **1.** ►espèces disparues, **2.** ►espèces amenées par leur régression à un niveau critiques des effectifs, **3.** ►espèces affectées d'une régression forte et continue et qui a déjà disparue de nombreuses régions, **4.** ►espèces dont la population n'a pas sensiblement diminuée, **5.** ►espèces dont une régression s'est manifestée sans qu'il soit possible de définir dans quelle mesure ainsi que trois critères correspondant à une certaine sensibilité biologique : **a]** espèces remarquables, **b]** espèces endémiques, **c]** espèces migratrices ; ►arrêté réglementant la protection de la faune et de la flore sauvages, ►catégorie de régression des populations et de distribution des espèces, ►Liste bleue [US]) ; *f 4* **inventaire [m] local/régional du patrimoine faunistique et floristique** (L'élaboration de l'**i. l./r. du p. f. et f.** décidé par l'État a été lancé en 1982 et est élaboré sous la responsabilité scientifique du muséum national d'histoire naturelle et communiqué pour information aux communes lors de l'élaboration d'un plan local d'urbanisme [PLU]/plan d'occupation des sols [POS] ; cf. art. 23 loi Paysages n° 93-24 du 8 janvier 1993 ; il s'agit aujourd'hui de l'inventaire des ►zones naturelles d'intérêt écologique, faunistique et floristique [ZNIEFF] dont la mise à jour de la cartographie est en cours de lancement) ; *g* **Rote Liste [f]** (Verzeichnisse gefährdeter Pflanzen- und Tierarten, die besonderen Schutz benötigen. **R. L.n** treffen Aussagen zum vermuteten oder tatsächlich festgestellten Rückgang der Arten. Dabei kann der Begriff ‚gefährdet' sehr unterschiedlich verstanden werden: Viele Arten, die in einem Land gefährdet sind, kommen in anderen Ländern häufiger vor, d. h., dass die in nationalen und regionalen Listen aufgenommenen Arten deshalb nicht *per se* in ihrem Bestand bedroht sein müssen. Die international von der IUCN empfohlenen Gefährdungskategorien sind folgende: **1.** Extinct [EX] — ausgestorben: ►ausgestorbene Art, **2.** Extinct in the Wild [EW] — in freier Wildbahn ausgestorben: ►verschollene Art, **3.** Critically Endangered [CR] — vom Aussterben bedroht: ►vom Aussterben bedrohte Art, **4.** Endangered [EN] — stark gefährdet: ►stark gefährdete Art, **5.** Vulnerable [VU] — gefährdet: ►gefährdete Art, **6.** Near Threatened [NT] — gering gefährdt, **7.** Least Concern [LC] — nicht gefährdt, **8.** Data Deficient [DD] — keine ausreichenden Daten und **9.** Not Evaluated [NE] — nicht bewertet; cf. IUCN Red List Categories and Criteria, Version 3.1, 2001. In D. werden nachstehende Gefährdungskriterien mit folgenden Symbolen unterschieden: **0** ausgestorben oder verschollen, **1** vom Aussterben bedroht, **2** stark gefährdet, **3** gefährdt, **G** potentiell gefährdet [Gefährdung anzunehmen, aber Status unbekannt], **R** sehr seltene Arten resp. Arten mit geografischer Restriktion, **V** Arten der Vorwarnliste und **D** Daten defizitär; cf. ROT 1998; Mitte der 1970er-Jahre wurde in D, A und CH. begonnen, nach dem Vorbild der internationalen *Red Lists* **R. L.en** der gefährdeten Pflanzen- und Tierarten zu erstellen; in D werden vom Bundesamt für Naturschutz die „R. L. gefährdeter Tiere Deutschlands" und die „R. L. gefährdeter Pflanzen Deutschlands" erarbeitet; cf. ROT 1998; für die Bundesländer können von den zuständigen Stellen entsprechende **R. L.n** aufgestellt werden; auf Ortsebene können z. B. untere Naturschutzbehörden oder lokale Naturschutzverbände diese Listen erstellen; ►Artenschutzverordnung, ►Blaue Liste [US], ►Gefährdungskategorie).

4887 red shale [n] *min.* (►Coal mine shale [US]/colliery shale [UK] which is smoldering as a result of spontaneous combustion; in U.S., a pile of **r. s.** is called 'red dog' after it has burned; ►coal mine spoil [US]/colliery spoil [UK]); *syn.* red blaes [n] [also SCOT]; *s* escoria [f] (Zafras quemadas por combustión espontánea; ►zafras de carbón, ►zafras de esquisto); *f* schistes [mpl] brûlés (Stériles sur un terril ayant brûlé sous l'action d'un feu couvant incontrôlé ; phénomène très lent et général, pas encore totalement élucidé, dû à la présence de charbon résiduel combustible [5 à 15 % dans les schistes déposés en terrils] et de pyrite qui, en présence d'eau et d'oxygène, dégage de la chaleur et amorce ainsi le processus de combustion ; 80 % des terrils brûlent ou ont brûlé [*résultat* terril brûlé]. La combustion, quoique se produisant dans toute la masse du terril, est plus active au sommet et se manifeste par l'échappement de vapeur d'eau et de gaz sulfureux. Les températures sont très variables : au sommet, de 40° à 100 °C dans les premiers cm, 200° à 700 °C à 90 cm de profondeur, 1300° à 2000 °C dans les grandes profondeurs. La combustion est un phénomène très lent, certains terrils brûlent pendant un siècle ; la combustion provoque un tassement du terril, entraîne le rougissement des schistes et la formation d'agglomérats durs et cohérents. Dès lors, le terril devient plus stable et donc plus facilement colonisable par la végétation thermophile qui se déplace avec la zone de combustion active ; ►houille stérile, ►schistes stériles) ; *g* ausgebrannte Berge [pl] (Beim Kohleabbau angefallenes, auf Halde geschüttetes taubes Gestein, das durch Schwelbrand sich unkontrolliert entzündet; ►Berge 1, ►Berge 2).

red-top [n] [US&CDN] *arb.* ►dead crown.

reduced price [n] *constr. contr.* ►price discount.

reducer [n] (1) *biol. ecol.* ►decomposer (1).

reducer [n] (2) *constr.* ►reducing coupling.

4888 reducing coupling [n] *constr.* (Reduction fitting to connect a large diameter pipe with a small one; *specific term* reducing tee [US]; *opp.* increaser [US]/tapered pipe [UK]); *syn.* reducer [n] (2) (DAC 1975), reducing pipe fitting [n] (MEA 1985); *s* acoplamiento [m] reductor (En tuberías, un tubo cuyas bocas son desiguales, que sirve para cambiar de diámetro grande a más pequeño; *opp.* ►acoplamiento ampliador); *f* cône [m] de réduction (Élément de branchement permettant le raccordement sur un tuyau de canalisation de diamètre nominal inférieur ; *opp.* ►cône d'augmentation) ; *g* Reduktionsstück [n] (Im Rohrleitungsbau ein Rohr mit Übergang von größerer zur kleineren Nennweite; *opp.* ►Übergangsstück).

4889 reducing elbow [n] *constr.* (Angled pipe unit connecting two pipes of different diameters where the pipe changes direction; ►increaser [US]/tapered pipe [UK], ►reducing coupling); *syn.* taper elbow [n]; *s* codo [m] (Pequeño trozo curvado de tubería que sirve para conectar dos alineaciones rectas de la conducción; DACO 1988; ►acoplamiento reductor, ►acoplamiento ampliador); *f* coude [m] de diminution/augmentation (Pièce d'un angle maximal de 67° 30' intercalée sur le réseau et permettant le branchement sur des collecteurs de diamètre inférieur ou égal ; ►cône de réduction, ►cône d'augmentation) ; *syn.* coude [m] de réduction ; *g* Übergangsbogen [m] (1) (Gebogenes Rohrleitungsstück zur Verbindung unterschiedlicher Durchmesser; ►Reduktionsstück, ►Übergangsstück).

reducing [n] **of a crown** [UK] *arb. constr.* ►drop-crotching.

4890 reducing [n] **of intrusion** *conserv. leg. plan.* (Legal requirement to minimize long-term adverse impacts on the environment by execution of ►mitigating measures. **In U.S.**, no legal equivalent regarding landscapes as such, but this covers measures to deter encroachment; ►impact mitigation, ►impact mitigation regulation, ►prevention measure [of an intrusion upon the environment]); *s* reducción [f] de impacto (En D., obligación resultante del § 8 [1] de la Ley Federal de Protección de la Naturaleza de reducir al máximo los efectos negativos de ►intrusiones en el entorno natural causadas por obras de construcción por medio de

►medidas compensatorias o ►medidas correctoras; si éstas no son posibles se recurre a medidas sustitutivas; ►compensación [de un impacto], ►regulación de impactos sobre la naturaleza); *syn.* mitigación [f] de impacto; *f* réduction [f] des atteintes à l'environnement (Obligation résultant de l'art. 2 de la loi du 10 juillet 1976 visant dans le cadre d'une étude d'impact à décrire les mesures pour supprimer ou réduire les impacts réductibles et compenser les conséquences dommageables pour l'environnement et impossibles à supprimer ; cf. circ. 93-73 du 27 septembre 1993 ; ►compensation [des conséquences dommageables sur l'environnement], ►mesure de suppression, ►mesure compensatoire, ►réglementation relative aux atteintes subies par le milieu naturel) ; *syn.* réduction [f] des conséquences dommageables à/sur l'environnement, réduction [f] des impacts ; *g* Eingriffsminderung [f] (Verpflichtung, dass Eingriffe in Natur und Landschaft mit ihren zu erwartenden Beeinträchtigungen der Leistungsfähigkeit des Naturhaushaltes oder des Landschaftsbildes zu mindern sind. Ist dies nicht möglich, müssen Ausgleichs- oder Ersatzmaßnahmen vorgenommen werden; cf. § 8 a [1] BNatSchG; ►Ausgleich [eines Eingriffs], ►Ausgleichsmaßnahme [des Naturschutzes und der Landschaftspflege], ►Eingriffsregelung, ►Vermeidungsmaßnahme [eines Eingriffs]); *syn.* Minderung [f] eines Eingriffs in Natur und Landschaft, Teilausgleich [m] eines Eingriffs.

reducing [n] **of pH value** *chem. limn. pedol.* ►lowering of pH value.

reducing pipe fitting [n] *constr.* ►reducing coupling.

reduction [n] *contr.* ►cost reduction; *arb. constr.* ►crown reduction; *constr. contr.* ►price reduction; *plan. trans.* ►traffic reduction; *constr. contr.* ►unit price reduction; *envir. sociol.* ►waste reduction.

reduction [n]**, agricultural yield** *agr. pol.* ►agricultural reduction program [US]/extensification of agricultural production [UK].

reduction [n]**, noise** *envir.* ►noise abatement.

reduction [n] **in charges** *contr. prof.* ►reduction in fees/charges.

4891 reduction [n] **in estimated volume** *contr.* (Lessening of quantities which fall short of the amount agreed upon contractually in the ►list of bid items and quantities [US]/schedule of tender items [UK]. A new price is to be negotiated upon request for a reduction of more than 10%; *opp.* ►increase in estimated volume); *s* reducción [f] de cantidades (Disminución de las cantidades fijadas en el ►resumen de prestaciones. Si la **r. de cc.** supera el 10%, hay que fijar un nuevo precio si el contratista lo exige; *opp.* ►aumento de cantidades); *f* diminution [f] de la masse des travaux (Modification en cours d'exécution des travaux de la masse des travaux prévus au marché dans le ►descriptif-quantitatif ; si une **d. de la m. d. t.** est supérieure à la diminution limite contractuelle [dans le cas de marchés réglés au prix unitaires, plus du quart en moins des quantités portées au détail estimatif] l'entrepreneur a droit à être indemnisé en fin de compte du préjudice qu'il a éventuellement subi ; CCAG, art. 16 du décret n° 76-87 du 21 janvier 1976 ; *opp.* ►augmentation de la masse des travaux) ; *syn.* diminution [f] dans la masse des travaux ; *g* Massenminderung [f] (Unterschreitung des mit dem ►Leistungsverzeichnis vertraglich vereinbarten Mengenansatzes. Für die über 10 % hinausgehende **M.** ist auf Verlangen ein neuer Preis zu vereinbaren; cf. § 2 VOB Teil B; *opp.* ►Massenmehrung); *syn.* Minderung [f] des Leistungsumfanges, Unterschreitung [f] des Mengenansatzes.

4892 reduction [n] **in fees/charges** *contr. prof. syn.* cut [n] in fees/charges; *s* reducción [f] de honorarios; *f* réduction

R

[f] de l'honoraire ; **g Honorarminderung [f]** (Herabsetzung des Honorars); *syn.* Minderung [f] des Honorars.

reduction [n] in housing density *urb.* ►wide spacing in a housing development.

4893 reduction [n] in recreation use *recr.* (Resulting from decrease in recreation use value); **s perturbación [f] del uso recreativo** (Efectos negativos sobre el valor de una zona para la recreación); **ƒ perturbation [f] de la fréquentation touristique** *syn.* conséquences [fpl] dommageables sur la fréquentation touristique ; **g Beeinträchtigung [f] der Nutzung für Freizeit und Erholung** (Negative Auswirkungen auf die Nutzung für Freizeit und Erholung); *syn.* Beeinträchtigung [f] des Freizeit- und Erholungswertes, Nutzungsbeeinträchtigung [f] für Freizeit und Erholung.

reduction [n] in species diversity *phyt. zool.* ►decline in number of species.

reduction [n] in visual quality *landsc.* ►detraction from visual quality.

4894 reduction [n] of a plan *plan.* (Modification of a plan by reducing the scale; *opp.* ►enlargement of a plan); **s reducción [f] de un plano** (Modificación de un plano reduciendo la escala; *opp.* ►ampliación de un plano); **ƒ réduction [f] de plan** (Modification de l'échelle d'un plan à une échelle plus petite ; *opp.* ►agrandissement de plan) ; **g Planverkleinerung [f]** (Änderung eines Planes durch Verkleinern des Maßstabes; *opp.* ►Planvergrößerung).

reduction [n] of permissible building volume [UK] *leg. urb.* ►downzoning [US]/reduction of permissible building volume [UK].

reduction program [n] [US] *agr. pol.* ►agricultural reduction program [US]/extensification of agricultural production [UK].

reed-bank zone [n] of rivers or streams [UK] *phyt.* ►riverine reed belt.

reed bed [n] *phyt.* ►reed swamp.

reed belt [n] *phyt.* ►reed swamp.

4895 reed canary grass swamp [n] [US] *phyt.* (River edge, ditch, or wetland, dominated mainly by the reed community of reed canary grass [US]/reed-grass [UK] *[Phalaris arundinacea]*); *syn.* reed strip [n] [UK] (ELL 1988, 248); **s carrizal [m] de hierba cinta** (Comunidad vegetal en la que predominan carrizos de la especie *Phalaris arundinacea* que crece en orillas de ríos, en acequias o en prados húmedos); **ƒ phalaridaie [f]** (Groupement végétal dense d'hélophytes [roselières] dans lequel domine le l'Alpiste *[Phalaris arundinacea]* localisé en marge des fleuves et petites rivières, des cours d'eau calmes, des fossés de prairies, très rarement des étangs, des mares et des prairies marécageuses et inondables ; appartient à l'association du Phalaridetum arundinaceae ; *terme spécifique* phalaridaie fluviale) ; **g Rohrglanzgrasröhricht [n]** (Aus Rohrglanzgras *[Phálaris arundinácea]* bestehende Pflanzengesellschaft an Ufern, Gräben, in Feuchtwiesen, die zur Gesellschaft des *Phalaridetum arundinaceae* gehört).

4896 reed clump [n] *constr.* (Approximately 30 x 30cm square piece of reed vegetation cut out of the roots and used for transplanting; ►planting of reed rhizome clumps); *syn.* reed sod [n]; **s tepe [m] de carrizo** (Trozo de 30 x 30 cm de raíces de carrizo utilizado para transplantar; ►plantación de rizomas de carrizo); **ƒ touffe [f] de Roseaux** (Partie du système rhizomateux [ca. 30 x 30 cm] d'un peuplement de Roseau, arraché à la bêche en vue de la transplantation sur les bords des eaux et les lieux humides ; ►plantation de mottes de Roseaux) ; *syn.* motte [f] de Roseaux ; **g Schilfsode [f]** (Der Verpflanzung dienendes, ca. 30 x 30 cm großes ausgestochenes Stück aus dem Wurzelbereich eines Schilfröhrichts; ►Röhrichtballenpflanzung).

4897 reed cutting [n] *constr.* (Mowing of an area with reed growth); **s corte [m] de carrizo** *syn.* guadañado [m] de carrizo; **ƒ faucardage [m] de la roselière** (LA 1981, 502 ; méthode d'entretien des milieux humides par fauche des peuplements de la roselière ; la période propice au faucardage commence à partir de la fin août, lorsque la floraison et la reproduction des espèces végétales et animales est achevée) ; **g Röhrichtmahd [f]** (Schnitt von Röhrichtbeständen als Maßnahme zur Erhaltung und Entwicklung emerser Gewässerrandvegetation von Ende August bis in das zeitige Frühjahr; die Abfuhr des Mähgutes sorgt für eine Reduzierung des Nährstoffeintrages).

reedmace swamp [n] [UK] *phyt.* ►cattail swamp [US]/reedmace swamp [UK].

reed marsh [n] [US] *phyt.* ►tall forb reed marsh [US]/tall forb reed swamp [UK].

reed plant [n] [UK] *phyt.* ►emergent aquatic plant (1).

reed plugging [n] *constr.* ►reed plug planting.

4898 reed plug planting [n] *constr.* (Riverbank or lakeside planting in shallow water using 80-120cm long, rootless reed shoot cuttings from newly developing swamps); *syn.* reed plugging [n], planting [n] shoots of emergent plants (WET 1993, 612); **s plantación [f] de tallos de carrizo** (Plantación de orillas por medio de tallos de fragmitea sin raíz de unos 80-120 cm); **ƒ plantation [f] de(s) chaumes de Roseau** (Plantation de berges constituée par des chaumes dépourvus de racines, de 80 à 120 cm de longueur, pris sur un jeune peuplement et enterrés dans une profondeur d'eau de 40 cm) ; **g Schilfhalmpflanzung [f]** (Uferbepflanzung in knöcheltiefem Wasser aus 80-120 cm langen, wurzellosen Halmstecklingen aus jungem Entwicklungsbestand); *syn.* Halmpflanzung [f], Halmstecklingsbesatz [m].

reed rhizome and shoot clumps planting [n] *constr. hort.* ►planting of reed rhizome clumps.

4899 reed roll [n] *constr.* (*Bioengineering method* reed and swamp grass sods and clumps bound together in a wire netting to form a roll and installed on the banks of standing or slow-flowing water, in order that a permanent reed-zone can develop to stabilize the shoreline; ►reed swamp); **s fajina [f] de carrizo** (*Método de construcción biotécnica* tepes de carrizo u otra planta similar se atan con alambres formando rollos para el transporte y se colocan en la orilla de aguas estancadas o lentas para que se desarrolle un cinturón de carrizo permanente que proteja y estabilice la línea de orilla; ►carrizal de fragmitea); **ƒ fascine [f] d'hélophytes** (*Génie biologique* fagot serré constitué de mottes enracinées d'hélophytes, fixées par un treillis métallique pour former un rouleau qui sera mis en œuvre le long des berges de cours d'eau ou au bord des rives d'eaux calmes afin de constituer et développer une roselière de ceinture ; ►phalaridaie fluviale, ►roselière, ►roselière à Roseaux) ; *syn.* fagot [m] d'hélophytes, fagot [m] de Roseau ; **g Röhrichtwalze [f]** (*Ingenieurbiologie* Rolle aus Ballen und Soden, die von Sumpfgräsern durchwurzelt und mit Maschendraht zusammengebunden sind; **R.en** werden an Stillgewässerufern oder langsam fließenden Gewässern eingebaut, damit sich ein dauerhafter Röhrichtgürtel entwickeln kann; ►Röhricht); *syn.* Röhrichtfaschine [f].

reed sod [n] *constr.* ►reed clump.

reed stems [npl] *constr.* ►grid planting of reed stems.

reed strip [n] [UK] *phyt.* ►reed canary grass swamp [US].

4900 reed swamp [n] *phyt.* (**1.** Swampy areas near lakes, in stagnant or slightly running rivers and streams 0.2-3m deep and in parts of fens, covered by tall clonal perennial ►helophytes, predominantly Reed [US&UK]/Roseau [CDN-Quebec] *[Phrag-*

R

mites]; cf. VIR 1982. **2.** Vegetation of tall clonal perennial helophytes [Phragmitetalia], mostly poor in species, often monodominant; ►bulrush swamp [US]/club-rush swamp [UK], ►cattail swamp [US]/reedmace swamp [UK], ►emergent aquatic plant 1, ►emergent aquatic plant 2, ►reed canary-grass swamp [US]/reed strip [UK], ►riverine reed belt, ►saw sedge swamp, ►tall sedge swamp); *syn. to 1.* reed bed [n], reed belt [n]; *syn. to 2.* reed swamp formation [n], reed swamp vegetation [n]; *s* **carrizal [f] de fragmitea** (**1.** Lugar donde crecen en grupos grandes helófitos en zonas de colmatación de lagos o en orillas de ríos de circulación lenta. **2.** Comunidad vegetal de grandes ►helófitos *[Phragmitetalia]*, generalmente pobre en especies, a menudo monoespecífica; compuesta principalmente por carrizos *[Phragmites]* que crece sobre suelos fangosos de ríos y lagos entre éutrofos y mesótrofos; *términos específicos* ►carrizal de hierba cinta, ►carrizal fluvial, ►ciénaga de grandes cárices, ►espadañal, ►juncar de scirpus; ►marisma de ciperáceas, ►planta acuática emergente, ►planta semiacuática); *f 1* **roselière [f]** (Lieu où croissent les groupements à grands hélophytes du *Phragmitetalia*, principalement constitués de Roseaux *[Phragmites]* ; *f 2* **roselière [f] à Roseaux** (Peuplement de grands ►hélophytes du *Phragmitetalia* formant de vastes colonies généralement unispécifiques mêlées de plantes plus basses localisés en marge des étangs et mares en voie d'atterrissement, bras morts et des rives des cours d'eaux calmes ; ►amphiphyte, ►cladiaie, ►magnocariçaie, ►phalaridaie, ►phalaridaie fluviale, ►plante des roselières, ►roselière de rive de cours d'eau, ►scirpaie, ►typhaie) ; *syn.* marais [m] à Roseaux, marais [m] roselier ; *f 3* **phragmitaie [f]** (Peuplement dense de hauts ►hélophytes, formant la végétation spontanée de la marge des cours d'eau calmes, étangs et mares oligotrophes, sur alluvions minérales, principalement constitué de Roseaux *[Phragmites]* appartenant aux peuplements des roselières à grands hélophytes de l'alliance du *Phragmition* ; ►marais à Marisques) ; *g* **Röhricht [n]** (**1.** Fläche, auf der vorwiegend hohe perennierende ►Sumpfpflanzen, wie Schilf *[Phragmites]* im Boden verlandender Seen oder am Ufer langsam fließender Gewässer wachsen; ►Großseggenried, ►Röhrichtpflanze, ►Überwasserpflanze); *syn.* Schilfröhricht [n], Röhrichtzone [f]. **2.** Pflanzenformation der **Großröhrichte** und **Großseggensümpfe** [Ordnung der *Phragmitetalia*] auf schlammigen Böden eutropher [bis mesotropher] Gewässer, die je nach Standort, Trophiegrad und Wassertiefe in Mitteleuropa z. B. aus Schilf *[Phragmites]*, Großseggen *[Carex]*, Teichsimse *[Scirpus]*, Rohrglanzgras *[Phalaris]*, Rohrkolben *[Thypha]* und Schwaden *[Glyceria]* bestehen; *UBe* ►Flussröhricht, ►Rohrglanzgrasröhricht, ►Rohrkolbenröhricht, ►Schwertriedröhricht, ►Simsenröhricht); *syn. zu 2.* Röhrichtbestand [m].

reed swamp [n] [UK]**, tall forb** *phyt.* ►tall forb reed marsh [US].

reed swamp [n] [UK]**, tall herb** *phyt.* ►tall forb reed marsh [US]/tall forb reed swamp [UK].

reed swamp formation [n] *phyt.* ►reed swamp.

reed swamp vegetation [n] *phyt.* ►reed swamp.

4901 re-establishment [n] **of an animal population** *zool. conserv.* (Measures to reintroduce extinct animals into the wild or to increase the population of endangered animals; ►reintroduction); *s* **restauración [f] de poblaciones animales** (Medidas para reintroducir en una zona especies extintas o en peligro de extinción; ►reintroducción de especies extintas); *f* **restauration [f] des populations** (Opérations visant à reconstituer les populations d'espèces animales disparues ou menacées de disparition dans une aire donnée par réintroduction de ces animaux à partir de souches élevées en captivité ou capturées à

l'état adulte ; LR 1983, 8 ; ►réintroduction 2) ; *syn.* reconstitution [f] des populations animales (PPH 1991, 272) ; *g* **Wiederherstellung [f] von Tierpopulationen** (Maßnahmen, die dazu dienen, in einem Gebiet ausgestorbene Tiere durch Auswilderung wiederanzusiedeln oder eine vom Aussterben bedrohte Tierpopulation aufzustocken; ►Wiederansiedlung).

reestablishment [n] **of plant communities** [US]**/ re-establishment** [n] **of plant communities** [UK] *phyt.* ►reintroduction.

4902 reference map [n] *plan.* (Topographical map or thematic plan illustrating specific information used for reference in planning); *s* **material [m] cartográfico** (Mapa topográfico o temático que presenta información utilizable como referencia en la planificación); *f* **document [m] cartographique** (Carte topographique ou carte thématique utilisé comme document de travail dans un projet d'aménagement ou une étude) ; *g* **Kartenunterlage [f]** (Topografische oder Themenkarte, die als Arbeitsunterlage für eine Planung dient).

4903 references [npl] **for qualification** *constr. contr.* (Presentation of evidence of the capacity of, e.g. a firm to participate in a public prequalification bidding procedure [US]/public prequalification tendering procedure [UK]; ►public prequalification bidding notice [US]/public prequalification tendering notice [UK]); *s* **acreditación [f] de capacidad y solvencia** (Presentación de documentos justificativos de la capacidad de la empresa o de escritura de constitución en el Registro Mercantil; ►aviso de candidaturas a concurso público); *syn.* referencias [fpl] de calificación; *f* **références [fpl] de qualification** (►appel public de candidatures pour un appel d'offres) ; *syn.* justification [f] quant aux qualités et capacités du candidat ; *g* **Leistungsnachweis [m]** (Darlegung der Leistungsfähigkeit, z. B. einer Firma beim ►öffentlichen Teilnahmewettbewerb); *syn.* Qualifikationsnachweis [m], Nachweis [m] der Leistungsfähigkeit.

4904 refilling [n] *min.* (Dumping back of waste or spoil and overburden into an excavated open cast/pit mine or quarry); *s* **relleno [m] de explotaciones a cielo abierto** (Operación de depositar zafras o residuos en una mina a cielo abierto una vez terminada su explotación); *f* **remblayage [m] d'une excavation** (Opération réalisée sur des carrières à ciel ouvert en fin d'exploitation lors de la remise en état ; cf. circ. n° 96-52 du 2 juillet 1996) ; *g* **Innenverkippung [f]** (Verfüllung von ausgebeuteten Tagebauflächen).

4905 reforestation [n] *for.* (Replanting of ►forest tree seedlings after ►clear-cutting system [US]/clear felling system [UK] felling; ►afforestation); *syn.* reafforestation [n]; *s* **reforestación [f]** (Operación de reestablecer el bosque plantando ►posturas en un área después de ►explotación a matarrasa; ►forestación); *syn.* repoblación [f] forestal; *f* **reforestation [f]** (Reconstitution d'un peuplement forestier après une coupe à blanc ; ►exploitation forestière par coupes rases, ►boisement, ►plant forestier, ►reboisement) ; *g* **Wiederaufforstung [f]** (Wiederanpflanzung von ►Forstpflanzen bei ►Kahlschlagswirtschaft; ►Aufforstung); *syn.* Anbau [m, o. Pl.], Wiederbewaldung [f].

refuge [n] [US] *conserv. leg.* ►wildlife sanctuary [US].

refuge [n] [US]**, game** *conserv. game'man. hunt. leg.* ►game reserve.

refuge [n] [UK]**, mountain** *recr.* ►mountain lodge [US].

4906 refuge area [n] *ecol.* (Natural area of survival for endangered species of animals, plants and their biocoenosis due to its favo[u]rable location, e.g. climatic, natural seclusion— islands such as Galapagos—or due to the lack of cultivation; ►remnant habitat, ►habitat island, ►ecology of habitat islands);

s **refugio [m] ecológico (de flora y fauna)** (Lugar donde permanecen condiciones anteriores a un cambio y se encuentran, por tanto, manifestaciones de la flora y fauna propias de tales condiciones; DINA 1987; ►biótopo relicto, ►ecología de biótopos aislados, ►hábitat residual); *syn.* área [f] de supervivencia (MARG 1977, 249), área [f] de refugio (de flora y fauna); *f* **territoire [m] refuge** (Aire qui, **1.** pour des raisons climatiques favorables [îles des Galapagos] ou l'inexistence d'activités humaines, constitue une zone de survie pour les espèces animales et végétales menacées ou en voie d'extinction ainsi que leur habitat. **2.** Aire restreinte sur laquelle, pour des phénomènes consécutifs à d'importantes variations climatiques [glaciation], certaines espèces ont été rejetées ; ►biotope relictuel, ►écologie des biocénoses relictuelles, ►habitat relictuel) ; *syn.* aire [f] de retrait ; *g* **Rückzugsgebiet [n]** (Gebiet, das durch seine begünstigte, z. B. durch Klima, natürliche Abgeschlossenheit [Inseln wie die Galapagos] oder durch Nichtbewirtschaftung eine geeignete Überlebensregion für gefährdete Tier- und Pflanzenarten und deren Biozönosen ist; ►Biotoprest, ►Habitatinsel, ►Inselökologie); *syn.* Refugialgebiet [n], Refugium [n].

refuse [n] *envir. min.* ►dumping of refuse [US]/tipping of refuse [UK]; *envir.* ►urban refuse; *envir. leg.* ►waste.

refuse [n]**, domestic** *envir.* ►household waste.

refuse [n]**, household** *envir.* ►household waste.

refuse [n] [UK]**, tipping of** *envir. min.* ►dumping of refuse [US].

4907 refuse compost [s, no pl.] *envir. pedol.* (►Compost produced with 15-40% organic and nutrient content, as well as toxic matter, e.g. heavy metals; ►waste); *syn.* waste compost [n]; *s* **compost [m] de residuos** (►Compost producido a partir de ►residuos sólidos urbanos con 15-40% de materia orgánica, que puede contener también sustancias contaminantes, como p. ej. metales pesados); *f* **compost [m] de déchets ménagers** (Produit obtenu par le broyage-compostage de résidus urbains et devant répondre à la norme U-44.051 s'il est p. ex. utilisé comme amendement organique en culture ; reçoivent la dénomination « compost urbain » les produits qui présentent en moyenne les caractéristiques suivantes : criblage 90 %, teneur en carbone > à 5 % sur matière sèche, teneur en azote > 0,3 % sur matière sèche, valeur du rapport C/N comprise entre 10 et 25 ; circ. du 22 avril 1966 ; ►compost, ►déchets) ; *syn.* compost [m] de résidus ménagers, compost [m] d'origine domestique, compost [m] urbain ; *g* **Müllkompost [m]** (Aus ►Abfall 2 gewonnener ►Kompost mit 15-40 % organischem Stoffanteil, Nähr- und Schadstoffen, z. B. Schwermetallen).

4908 refuse composting [n] *envir.* (Method of recycling household and yard waste by decomposing mainly organic matter containing microorganisms, to give a friable, soil-like and nutrient-rich growing medium; ►composting); *s* **compostaje [m] de residuos orgánicos domésticos** (Proceso de reciclaje de residuos de jardines y domésticos mayormente orgánicos por descomposición bacteriana en condiciones aeróbicas. Como producto final se consigue una sustancia grumosa, terrosa y rica en nutrientes llamada **tierra [f] de compost**; ►compostaje); *syn.* composteo [m] de residuos orgánicos domésticos); *f* **compostage [m] de déchets ménagers** (Procédé d'élimination de déchets qui après broyage subissent une fermentation qui les transforme en un terreau riche en substances nutritives ; ►compostage) ; *syn.* compostage [m] des résidus urbains, compostage [m] des ordures ménagères ; *g* **Müllkompostierung [f]** (Verfahren der Hausmüll- und Gartenabfallverwertung, bei dem vorwiegend organische Stoffe unter Sauerstoffeinfluss durch Bakterien zersetzt werden und als Endprodukt eine krümelige, erdige, nährstoffreiche Substanz [Komposterde] entsteht; ►Kompostierung).

4909 refuse compost production plant [n] *envir.* (Commercial enterprise which produces refuse compost; ►refuse composting); *syn.* waste compost production plant [n] [also US]; *s* **planta [f] de compostaje de residuos orgánicos** (Empresa que se dedica al ►compostaje de residuos orgánicos domésticos); *syn.* planta [f] de composteo de residuos orgánicos; *f* **installation [f] de compostage des ordures ménagères** (Entreprise exploitant et commercialisant les produits provenant du ►compostage de déchets ménagers) ; *syn.* unité [f] de compostage des ordures ménagères ; *g* **Müllkompostwerk [n]** (Gewerblicher Betrieb, der die ►Müllkompostierung besorgt); *syn.* Müllkompostierungsanlage [f].

refuse disposal [n] *envir.* ►waste disposal.

4910 refuse dump [n] [US] *envir.* (Disposal site for domestic waste; ►dumping of refuse [US]/tipping of refuse [UK], ►dumpsite [US]/tipping site [UK], ►hazardous waste disposal site, ►large spoil area [US]/large tip [UK], ►sanitary landfill [US] 2/landfill site [UK], ►unauthorized dumpsite [US]/fly tipping site [UK], ►unmanaged dumpsite [US]/unmanaged tipping site [UK]); *syn.* trash dump [n] [US], refuse tip [n] [UK]; *s* **depósito [m] de residuos sólidos urbanos** (Lugar de deposición de materiales residuales, sean de origen doméstico o industrial, pero que no sean tóxicos ni peligrosos. Estos se depositan en ►depósitos de residuos peligrosos; *término específico* depósito de residuos domésticos; ►descarga de residuos, ►depósitos de residuos sólidos, ►escombrera 1, ►vertedero controlado, ►vertedero clandestino, ►vertedero incontrolado); *f 1* **décharge [f] de déchets ménagers** (Installation existante avant la loi du 13 juillet 1992 dite de classe 2 recevant des déchets ménagers ; ►amoncellement, ►centre de stockage de déchets spéciaux ultimes et stabilisés, ►centre de stockage des déchets ultimes, ►décharge de déchets industriels [spéciaux], ►décharge contrôlée, ►décharge [de déchets] sauvage, ►décharge non contrôlée, ►décharge publique, ►dépôt de déchets) ; *syn.* décharge [f] d'ordures ménagères, dépôt [m] d'ordures ménagères ; *f 2* **installation [f] de stockage des déchets ménagers et assimilés** (Installation nouvelle instaurée par la loi du 13 juillet 1992 dite de classe II recevant des déchets ménagers et assimilés) ; *syn.* centre [m] de stockage de classe 2, dépôt [m] de déchets ménagers et assimilés ; *g* **Abfalldeponie [f]** (Fläche zur zeitlich unbegrenzten Ablagerung von Hausmüll und sonstigen nicht gefährlichen Abfallstoffen; die Ablagerung von gefährlichen Abfällen erfolgt auf einer ►Sonderabfalldeponie; *UB* Hausmülldeponie [f]; ►Ablagerung von Abfällen, ►Deponie, ►geordnete Deponie, ►Halde 1, ►ungeordnete Deponie, ►wilde Deponie); *syn.* Mülldeponie [f], Müllkippe [f], Müllabladeplatz [m], Müllablagerung [f], Müllablagerungsfläche [f].

refuse-landfill site [n] *envir.* ►sanitary landfill [US] (2)/landfill site [UK].

refuse tip [n] [UK] *envir.* ►refuse dump [US].

4911 regenerated woody growth [n] *landsc.* (Young woody growth after ►clearance of unwanted spontaneous woody vegetation); *s* **regeneración [f] de leñosas** (Nuevo crecimiento de ramas después del ►desbroce); *f* **repousse [f] arbustive** (après une mesure de ►débroussaillage) ; *syn.* végétation arbustive adventice) ; *g* **Aufwuchs [m] von Gehölzen** (*Fachsprachlich* ...wüchse [pl], *sonst ohne Pl.*; junger Gehölzbestand nach einer ►Entkusselung[smaßnahme]; ►Fremdaufwuchs); *syn.* Gehölzaufwuchs [m].

4912 regeneration [n] **(1)** *for.* (Process of renewal of a tree crop, whether by natural or artificial means. The natural and artificial regeneration result in **restocking** of the area concerned; SAF 1983; ►natural regeneration 1, ►natural regeneration 2, ►recruitment of a forest, ►regeneration cutting); *s 1* **regenera-**

R

ción [f] natural o artificial (Renuevo de un bosque de forma natural o artificial; ▶corta de regeneración, ▶poda de rejuvenecimiento, ▶regeneración natural); *s 2* población [f] de renuevo (Vegetación resultante de la regeneración natural o artificial; ▶bosque de renuevo); *f 1* régénération [f] (forestière) (Renouvellement naturel d'un peuplement forestier par voie de semences, ou renouvellement artificiel par semis ou par plantations ; DFM 1975 ; ▶coupe de régénération [forestière], ▶peuplement de régénération [naturelle/artificielle], ▶régénération naturelle) ; *f 2* rajeunissement [m] de peuplement (Renouvellement par voie végétative, p. ex. d'un taillis ; ▶recru, ▶taille de rajeunissement) ; *g* Verjüngung [f] (1. Summe der natürlichen Ereignisse und waldbaulichen Maßnahmen zur Erneuerung eines Waldbestandes. 2. Population der Verjüngung wie z. B. Sämlinge, Lohden oder Stangenholz. 3. ▶Verjüngungshieb; ▶Naturverjüngung, ▶verjüngter Waldbestand, ▶Verjüngungsschnitt); *syn.* Waldverjüngung [f].

regeneration [n] (2) *conserv.* ▶bog regeneration; *conserv. landsc. land'man.* ▶restoration (1); ▶area of regeneration.

regeneration [n] [US], **downtown** *urb.* ▶center city redevelopment [US]/urban regeneration [UK].

regeneration [n], **spontaneous** *for. phyt.* ▶natural regeneration (1).

regeneration [n] [UK], **urban** *urb.* ▶center city redevelopment [US].

4913 regeneration cutting [n] [US] *for.* (Removal of trees in the stand intended to assist regeneration already present or to make regeneration possible; cf. SAF 1983); *syn.* regeneration felling [n] [UK], reproduction cutting [n]; *s* corta [f] de regeneración (Medida forestal de renuevo de un bosque, como la remoción de árboles individuales, para promover la regeneración natural y el desarrollo de la estratificación y de la estructura natural de edades); *f* coupe [f] de régénération (forestière) (Coupe de récolte par laquelle, en fonction des perspectives, prévisions et objectifs de la gestion forestière, on enlève les gros bois et qui est destinée à provoquer la régénération ou favoriser la croissance des jeunes semis. La coupe peut être progressive dans un peuplement arrivé à maturité ou définitive [coupe rase]. La coupe avec protection de la régénération et des sols, la coupe progressive d'ensemencement, la coupe avec réserve de semenciers et la coupe par bandes sont des coupes de régénération) ; *g* Verjüngungshieb [m] (Waldbauliche Maßnahme wie z. B. Plenterung, Saum-, Schirm- oder Kahlschlag, die das Altersgefüge oder die Schichtungsentwicklung eines Waldbestandes dem Bestockungs- und Betriebsziel anpasst); *syn.* Verjüngungsschlag [m].

regeneration felling [n] [UK] *for.* ▶regeneration cutting [US].

regeneration mix [n] *constr.* *▶sports grass mixture.

regeneration of city cores [n] *urb.* ▶center city redevelopment [US]/urban regeneration [UK].

4914 regenerative capacity [n] *biol. ecol.* (Potential of plants, animals or ecosystems, to regrow or replace lost or injured tissue, or to re-establish a greatly impaired functional system); *s* capacidad [f] de regeneración (Potencial de plantas, animales, ecosistemas, etc. de reemplazar tejido herido o perdido o de reestablecer la capacidad funcional); *f* pouvoir [m] de régénération (Capacité que possèdent les végétaux, les animaux, les écosystèmes, etc. à reconstituer les tissus détériorés ou détruits ou de rétablir l'unité structurelle des divers facteurs du milieu) ; *g* Regenerationskraft [f] (Vermögen von Organismen, ihr verletztes resp. zerstörtes Gewebe neu zu bilden oder bei Ökosystemen die Fähigkeit, ein stark beeinträchtigtes Struktur-, Bezie-

hungs- oder Funktionsgefüge nach Beendigung der Störung wiederherzustellen).

regime [n] *hydr.* ▶groundwater regime; *met. wat'man.* ▶precipitation regime.

region [n] *envir. leg.* ▶air quality control region; *geo.* ▶arid region; *geo. phyt. zool.* ▶biogeographical region; *phyt.* ▶brackish water area/region; *agr. for. hort.* ▶crop-growing region; *zool.* ▶hibernation area/region; *geo.* ▶high mountain region, ▶industrial region (1); *adm. plan.* ▶old industrial region; *adm. plan. urb.* ▶metropolitan region, ▶urban region.

region [n], **alpine** *geo.* ▶high mountain region.

region [n], **benthic** *limn. ocean.* ▶benthic zone.

region [n], **coastal** *geo. recr.* ▶coastal area.

region [n], **palustrine** *geo.* ▶palustrine area.

region [n], **physiographic** *geo. ecol.* ▶physiographic province.

region [n], **river** *limn.* ▶river zone.

region [n] [US], **semiarid** *geo.* ▶semidesert [US]/semi-desert [UK].

region [n] [US], **tourist** *plan. recr.* ▶tourist area.

regional administrative authority/agency [n] *adm. leg. pol.* ▶local/regional administrative authority/agency.

4915 regional funds [n] **of the European Community (EC)** *adm. plan.* (Budget item of the E.C. for promoting the regional economic structure in underdeveloped areas; ▶old industrial region); *s* fondo [m] regional de las CC.EE. (F.E.D.E.R.) (Título financiero de la Comunidad Europea para promocionar económicamente regiones de la comunidad menos desarrolladas o necesitadas de reestructuración; ▶zona industrial en declive); *f* fonds [m] européen de développement régional (F.E.D.E.R.) (Fond structurel de la Communauté européenne affecté à l'aide au développement des structures régionales dans les zones peu développées des pays de la Communauté ; ▶région industrielle en déclin) ; *g* Europäischer Fonds [m] für regionale Entwicklung (*Abk.* EFRE; Finanztitel der Europäischen Gemeinschaft zur Förderung der Regionalstruktur in weniger entwickelten Gebieten der Gemeinschaft durch Unterstützung mittelständischer Unternehmen zur Schaffung von dauerhaften Arbeitsplätzen, zur Durchführung von wichtigen Infrastrukturprojekten und sonstigen technischen Maßnahmen; ▶altindustrialisierte Region).

regional green buffer [n] *landsc. urb.* ▶regional green corridor.

4916 regional green corridor [n] *landsc. urb.* (1. Large-scale, linear green space in urban agglomerations, which may have intermittent agricultural and wooded areas, and is intended to prevent further agglomeration of the urban areas; ▶green belt [US]/green belt [UK]); *syn.* regional green buffer [n]; *s* corredor [m] verde regional (Conjunto de espacios libres amplios con forma de banda existentes en grandes aglomeraciones urbanas incluidas áreas de cultivo y bosques, que tienen como fin evitar el que se unan las superficies construidas de municipios vecinos; ▶cinturón verde); *f* coupure [f] verte périurbaine (Grands espaces naturels de discontinuité souvent de forme linéaire tels que les espaces agricoles sans caractéristiques paysagère ou environnementale, sylvicoles ou de forêt, les cours d'eau et petites rivières dont la fonction est de constituer une zone de discontinuité entre les zones d'urbanisation et de canaliser la croissance urbaine de différentes communes en zone périphérique des grandes agglomérations urbaines ; ces coupures d'urbanisation peuvent recevoir en France des équipements privés ou publics à usage sportif ou récréatif adaptés ; ces espaces

R

sur lesquels l'urbanisation est interdite ont un rôle et un impact prédominant sur le paysage de l'agglomération urbaine ; *terme spécifique* zone rurale de discontinuité ; ►ceinture verte) ; *syn.* coupure [f] verte régionale ; *g* **regionaler Grünzug [m]** (**D.**, großräumige, bandförmige Freiräume in großen Städteagglomerationen, die neben einigen öffentlichen Grünflächen zum großen Teil aus land- und forstwirtschaftlichen Nutzflächen bestehen. Derartige **G.züge** sollen das Zusammenwachsen der Bauflächen einzelner Gemeinden oder Gemeindeteile verhindern und bilden somit eine grüne Schneise [**Grünzäsur**] zwischen den dicht bebauten Flächen; ►Grüngürtel); *syn.* regionale Grünzäsur [f].

4917 regional landscape program [n] [US]/**regional landscape programme** [n] [UK] *landsc. pol.* (Report outlining regional requirements and measures necessary to realize the objectives of ►nature conservation and landscape management—**in D.**, at a State level according to nature conservation laws and 'Raumordnungsgesetz'); *s 1* **plan [m] de ordenación de los recursos naturales** (En Es., instrumento específico para la delimitación, tipificación, integración en red y determinación de su relación con el resto del territorio de los sistemas que integran el patrimonio y los recursos naturales de un determinado ámbito espacial, con independencia de otros instrumentos que pueda establecer la legislación autonómica. Estos planes tienen como objetivo: a] identificar y georeferenciar los espacios y los elementos significativos del patrimonio natural de un territorio y, en particular, los incluidos en el Inventario del Patrimonio Natural y la Biodiversidad, los valores que los caracterizan y su relación e integración con el resto del territorio; b] definir y señalar el estado de conservación de los componentes del patrimonio natural, biodiversidad y geodiversidad y de los procesos ecológicos y geológicos en el ámbito territorial de que se trate; c] identificar la capacidad e intensidad de uso del patrimonio natural y la biodiversidad y geodiversidad y determinar las alternativas de gestión y las limitaciones que deban establecerse a la vista del estado de conservación; d] formular los criterios orientadores de las políticas sectoriales y ordenadores de las actividades económicas y sociales, públicas y privadas, para que sean compatibles con las exigencias contenidas en la presente ley; e] señalar los regímenes de protección que procedan para los diferentes espacios, ecosistemas y recursos naturales presentes en su ámbito territorial de aplicación, al objeto de mantener, mejorar o restaurar los ecosistemas, su funcionalidad y conectividad; f] prever y promover la aplicación de medidas de conservación y restauración de los recursos naturales y de los componentes de la biodiversidad y geodiversidad que lo precisen y g] contribuir al establecimiento y la consolidación de redes ecológicas compuestas por espacios de alto valor natural, que permitan los movimientos y la dispersión de las poblaciones de especies de la flora y la fauna y el mantenimiento de los flujos que garanticen la funcionalidad de los ecosistemas; cf. art. 16 y 17 Ley 42/2007); *s 2* **programa [m] de gestión del paisaje y la naturaleza** (En D., documento programático con carácter de ley, en el cual se describen las exigencias y medidas necesarias para alcanzar los objetivos de ►protección de la naturaleza y de ►planificación del paisaje, así como para mejorar las condiciones de vida de la población en el territorio correspondiente. Los **pp. de g. del p.** se desarrollan a nivel de estado federado y se concretizan a nivel regional y local en los correspondientes ►planes de gestión del paisaje y la naturaleza); *f 1* **plan [m] vert régional** (Plan paru en 1995, spécifique à la région Ile de France, servant de cadre pour les politiques de protection et de mise en valeur des espaces naturels et de loisirs de la région ; ►plan vert départemental) ; *f 2* **programme [m] régional de protection et de mise en valeur des paysages** (±) (**D.**, document fixant les orientations fondamentales de l'amé-

nagement du territoire d'une région *[Bundesland]* en matière de ►protection de la nature conformément à la loi fédérale sur la protection de la nature) ; *g* **Landschaftsprogramm [n]** (*Abk.* LaPro; **D.**, flächendeckende Darstellung der Erfordernisse und Maßnahmen zur Verwirklichung der überörtlichen Ziele von ►Naturschutz und Landschaftspflege für das Gebiet eines Bundeslandes gemäß § 5 [1] BNatSchG und den entsprechenden Regelungen in den Landesnaturschutzgesetzen; das LaPro formuliert die landesweit geltenden Leitlinien, Leitbilder und Ziele für den Artenschutz, Biotopschutz, Bodenschutz, Gewässerschutz, Klimaschutz und Landschaftsschutz sowie für die Erholungsvorsorge und setzt somit den Rahmen für die nachgeordneten Planungsebenen. Mit der Ausweisung von ►Vorranggebieten und ►Vorbehaltsgebieten für Natur und Landschaft sowie mit der Darstellung potenzieller Schutzgebiete schafft das LaPro die konzeptionellen Grundlagen für einen landesweiten Biotopverbund. Auf Grund der Hoheitlichen Gewalt der Bundesländer weisen die landesrechtlichen Regelungen Unterschiede bezüglich der Verbindlichkeit und Integration der LaPro in die Landesraumordnungsprogramme auf. Zuständig für die Erstellung der LaPro mit Text- und Kartenteil sind i. d. R. die obersten Naturschutzbehörden; bei dem Planungsverfahren müssen Öffentlichkeit und ►Träger öffentlicher Belange beteiligt werden); *syn.* Landschaftsrahmenprogramm [n] [BW u. BY], Landespflegeprogramm [n] [RP].

4918 regionally-significant [adj] *plan.* (Spatially- or structurally-important for planning development of a particular region/area); *s* **espacialmente significativo/a [adj]** (Referente a política, proyecto o medida que tiene influencia en el uso de espacio); *f* **qui exerce une influence sur l'espace [loc]** (Mesures d'aménagement conduisant à une modification de la structure du foncier ou influençant le développement d'une zone) ; *syn.* à forte incidence sur l'espace [loc] ; *g* **raumbedeutsam [adj]** (Planungen oder Maßnahmen betreffend, die Grund und Boden in Anspruch nehmen oder die räumliche Entwicklung oder Funktion eines Gebietes beeinflussen; cf. § 3 [1] 6 ROG von 2008); *syn.* raumbeanspruchend [ppr/adj], raumbeeinflussend [ppr/adj], raumrelevant [adj], raumwirksam [adj].

4919 regional map atlas [n] [US] *plan.* (Regional inventory and analysis maps collection of maps illustrating the existing conditions and development potential of an area undergoing planning); *syn.* regional planning atlas [n]; *s* **atlas [m] regional de planificación** (Conjunto de mapas de inventario y análisis regional que representan las condiciones existentes y los potenciales de un área); *f* **atlas [m] régional** (Représentation cartographique présentant au moyen de cartes topographiques et thématiques les différents aspects de l'espace d'une région) ; *g* **Planungsatlas [m]** (...atlanten [pl] u. ...atlasse [pl]; Kartenwerk, das an Hand von topografischen und thematischen Karten die Gegebenheiten und Entwicklungsmöglichkeiten eines Planungsraumes darlegt).

regional park [n] [US] *conserv. land'man. recr.* (Large, publicly-owned area which has been uniformly developed and looked after due to its special suitability for recreation, and which is a designated recreational area according to the principles and aims laid down by a regional plan. **In U.S.**, an extensive area developed for different recreational uses and administered by a regional park authority/board; **in U.K.**, the equivalent concept of **country p.s** also offer considerable potential for practical research into discovering systems and techniques for reconciling wildlife interests with public use; cf. WED 1979, 230; **countryside recreation areas** are recognized by the Countryside Commission as eligible for financial assistance under the Countryside Act 1968; ►park 1); *syn.* country park [n] [UK] (2); *syn. for a larger scale* regional park [n] [UK]; *s* **parque [m]**

R

natural (Paraje natural, a veces con inclusión de elementos de paisaje antropógenos y asentamientos humanos autóctonos, preservado y accesible al público. En D. tipo de espacio natural protegido dedicado en especial al esparcimiento y descanso de la población; cf. DINA 1987; ▶parque); *f 1* **parc [m] naturel** (Territoire relativement étendu, qui présente un ou plusieurs écosystèmes, généralement peu ou pas transformés par l'exploitation et l'occupation humaine, sur lequel les espèces végétales et animales offrent un intérêt spécial du point de vue scientifique et récréatif, dans lequel ont été prises des mesures pour y empêcher l'exploitation ou l'occupation et pour y faire respecter les entités écologiques, géomorphologiques ou esthétiques ayant justifié sa création, à des fins récréatives, éducatives ou culturelles ; Union internationale pour la conservation de la nature d'après DUA 1996, 547 ; *f 2* **parc [m] naturel régional** (F., territoire à l'équilibre fragile, au patrimoine naturel et culturel riche et menacé, faisant l'objet d'un projet de développement, fondé sur la préservation et la valorisation du patrimoine et caractérisé par des actions expérimentales en faveur de l'aménagement du territoire, de la gestion des milieux naturels et des paysages, du développement économique, social, culturel et de la qualité de la vie, de l'accueil, l'éducation et de l'information du public pour les activités de loisirs et de tourisme ; cf. C. rural, art. L. 244-1) ; les parcs naturels régionaux constituent un cadre privilégié des actions menées par les collectivités publiques en faveur de la préservation des paysages et du patrimoine naturel et culturel ; CPEN Art. L 333-1) ; *f 3* **parc [m] naturel urbain** (*Abrév.* PNU ; **F.**, espace naturel situé dans un milieu urbain présentant un caractère remarquable et qu'il importe de protéger contre toute atteinte naturelle ou artificielle pouvant l'altérer et de promouvoir auprès du public ; le périmètre du ▶parc peut inclure une zone périphérique urbanisée, destinée à assurer la cohérence de la protection et de la valorisation du milieu naturel, qui peut être soumise au respect de prescriptions architecturales particulières ; à l'intérieur des espaces protégés, peuvent être soumises à un régime particulier ou, le cas échéant, interdites les activités susceptibles d'altérer le caractère du parc ; les modalités de protection, d'aménagement et de mise en valeur du PNU font l'objet d'une charte ; CPEN Article L. 335-1 et 2) ; *g* **Naturpark [m]** (...parks [pl], seltener ...parke [pl], *schweiz. meist* ...pärke [pl]; einheitlich zu entwickelndes und zu pflegendes großräumiges Gebiet [überwiegend Landschaftsschutzgebiet oder Naturschutzgebiet], das sich wegen seiner landschaftlichen Voraussetzungen für die Erholung besonders eignet und nach den Grundsätzen und Zielen der Raumordnung und Landesplanung für die Erholung oder den Fremdenverkehr vorgesehen ist; § 16 BNatSchG; ▶Park).

4921 regional plan [n] *plan.* (Plan and written program[me] devised for the implementation of ▶regional planning policy; e.g. in U.K. for a county; **in U.S.**, e.g. the Regional Plan of New York–New Jersey–Connecticut comprises 31 counties; cf. NEUE LANDSCHAFT 4/99, 216; and **in D.**, in regions of states); *s* **plan [m] regional** (Plan y programa para la ▶ordenación del territorio para una región o partes de ella; en Es. es competencia exclusiva de las CC.AA. la elaboración de los ▶planes directores territoriales de coordinación [de una comarca]); *f 1* **directive [f] territoriale d'aménagement (DTA)** (F., outil juridique permettant à une collectivité [aire métropolitaine, département, aire géographique], sur certaines parties du territoire national présentant des enjeux particulièrement importants en matière d'aménagement, de développement, de protection et de mise en valeur, de formuler et de fixer **1.** les orientations fondamentales de l'État en matière d'aménagement et d'équilibre entre les perspectives de développement, de protection et de mise en valeur des territoires, **2.** les principaux objectifs de l'État en

matière de localisation des grandes infrastructures de transport et des grands équipements et de préservation des espaces naturels, des sites et des paysages ; les DTA ont été créées par la loi « Pasqua » n° 95-115 du 4 février 1995 d'orientation pour l'aménagement et le développement du territoire, modifiée par la loi « Voynet » n° 99-533 du 25 juin 1999 d'orientation pour l'aménagement et le développement durable du territoire puis par la loi n° 2000-1208 du 13 décembre 2000 relative à la solidarité et au renouvellement urbains ; elles peuvent s'appliquer par exemple dans le cadre du schéma régional d'aménagement et de développement, sur terre ou sur le domaine public maritime ; c'est à la fois un document d'aménagement du territoire et un document d'urbanisme, élaboré sous la responsabilité de l'État en association avec les collectivités territoriales et les groupements de communes concernés, puis approuvé par décret en Conseil d'État ; les ▶schémas de cohérence territoriale [SCOT] et les schémas de secteur doivent être compatibles avec les DTA ; en l'absence de ces schémas, et seulement dans ce cas, les DTA sont également opposables aux PLU, aux cartes communales et aux documents en tenant lieu) ; *f 2* **schéma [m] régional d'aménagement du territoire** (UK., US. et D., programme et documents graphiques fixant, les orientations fondamentales de l'▶aménagement du territoire sur le territoire d'une métropole régionale) ; *g* **Regionalplan [m]** (Planwerk der Regionalplanung, das für ein Teilgebiet eines Bundeslandes die Grundsätze der ▶Raumordnung sowie die im Landesraumordungsprogramm formulierten Ziele näher ausführt und Maßnahmen zu deren Umsetzung nennt; **R.pläne** sind die Schnittstelle zwischen der übergeordneten ▶Landesplanung und der örtlichen ▶Bauleitplanung. Ferner sind sonstige städtebauliche Planungen, die von den Gemeinden beschlossen wurden, zu berücksichtigen. Je nach Bundesland werden die Regionalpläne von Regionalverbänden, von Planungsbehörden der Kreise resp. Regierungsbezirke oder von der Landesplanungsbehörde in einem Maßstab zwischen 1:50 000 und 1:100 000 für eine Dauer von 10-15 Jahren erstellt. In den **R.** werden Erfordernisse und Maßnahmen des Naturschutzes und der Landschaftspflege des ▶Landschaftsrahmenplans unter Abwägung mit anderen raumbedeutsamen Planungen und Maßnahmen aufgenommen; gesetzliche Grundlage ist das Raumordnungsgestz [ROG]; cf. BFN 2006); *syn.* Gebietsentwicklungsplan [m], regionaler Raumordnungsplan [m].

regional plan [n]**, statutory** *plan.* ▶regional plan report.

4922 regional planner [n] *prof.* (**1.** Someone who is involved in large-scale ▶regional planning for parts of a state); **2. state regional planner** [n] (Someone who is involved in large-scale regional planning on a state level]); *s* **planificador, -a [m/f] regional** (Profesional dedicado/a a elaborar planes y programas de ordenación del territorio a escala regional; ▶planificación regional 1); *f* **planificateur, trice [m/f] de l'aménagement régional** (DG 1984, 393 ; personne qui élabore des plans et recommandent des politiques d'aménagement régional ; ▶planification régionale) ; *syn.* aménageur [m] spécialiste de l'aménagement régional, aménageur [m] en planification régionale, planificateur [m] régional [CDN] ; *g* **Regionalplaner/-in [m/f]** (Jemand, der sich mit ▶Regionalplanung befasst); *syn.* Landesplaner/-in [m/f].

4923 regional planning [n] *plan.* (**In Europe**, comprehensive planning for socio-economic, cultural and environmental requirements in regional areas according to laws and governmental policies; **in U.S.**, comprehensive planning based on socioeconomic, cultural, and environmental conditions according to governmental policies. In comparison to D. there are no equivalent legal fundamentals in U.S. Regional planning problems are resolved by initiatives, mainly by businessmen, on the basis of communal or state policies; public funds are available to a small extent while private persons, associations and foundations are

R

financing the main part; according to American understanding, „planning by higher hierarchy" is not desired; ►regional planning policy, ►State regional planning); *s* **planificación [f] regional (1)** (1. Término utilizado para denominar medidas tendentes a la ordenación de las actividades económicas y su distribución espacial con el fin de paliar las desigualdades regionales existentes. En Europa comenzaron a desarrollarse en algunos países después de la 2ª Guerra Mundial, incrementándose con posterioridad, cuando se hicieron patentes los efectos polarizadores generados por el crecimiento económico y la aparición de extensas áreas deprimidas. Pese a notables diferencias entre países, la mayoría de las actuaciones se ha dirigido a potenciar la formación de polos de desarrollo en regiones atrasadas y promover la descongestión de las grandes aglomeraciones urbano-industriales; cf. DGA 1986. **2. En D.** ►ordenación del territorio en parte de estado federado; **en F.** ordenación territorial de zonas atrasadas; **en EE.UU.** no existe legislación básica equivalente a la europea. Los problemas de planificación regional son resueltos por iniciativas, generalmente de empresarios, sobre la base de las políticas comunales o estatales. En pequeña medida existen fondos públicos para este fin, mientras que la mayor parte de los gastos son financiados por personas privadas, asociaciones y fundaciones. Según el entendimiento norteamericano, la «planificación desde arriba» no es deseable; ►planificación regional); *f* **aménagement [m] du territoire au niveau régional (F.,** ►aménagement du territoire à l'échelle d'une région, de régions littorales limitrophes ou de régions concernées par un massif de montagne ; **D.,** ►planification régionale au niveau d'un *Land*) ; *syn.* planification [f] régionale ; *g* **Regionalplanung [f] (1. D.,** ►Landesplanung für einen Teil oder Teile [Regionen] eines Bundeslandes, bei der die Grundsätze der ►Raumordnung sowie die im Landesraumordnungsprogramm enthaltenen Ziele in Regionalplänen dargestellt werden. Je nach Bundesland wird die **R.** von Regionalverbänden, von Planungsbehörden der Kreise resp. Regierungsbezirke oder von der Landesplanungsbehörde durchgeführt. Gesetzliche Grundlage ist das Raumordnungsgesetz des Bundes als Rahmengesetz. **2. A., R.** wird auf Grundlage der jeweiligen österreichischen Landesgesetze mit Plänen und Programmen für das gesamte Bundesland [Landesplanung] oder für Teilgebiete [Regionalplanung] von der Landesverwaltung durchgeführt. Die **R.** wird in manchen Bundesländern auch in Zusammenarbeit mit den Gemeinden und/oder Regionalverbänden durchgeführt. **3. CH.,** die Verantwortung für die **R.** liegt bei den Kantonen. **4.** In den **USA** gibt es kein rechtliches Instrumentarium, das der deutschen **R.** vergleichbar wäre; regionalplanerische Fragestellungen werden dort von Initiativen, vorwiegend von Geschäftsleuten, auf der Grundlage kommunaler/staatlicher Politik gelöst; öffentliche Gelder fließen nur zu einem sehr geringen Prozentsatz, während Private, Verbände und Stiftungen den Hauptanteil tragen; „Planung von oben" ist vom amerikanischen Grundverständnis her nicht erwünscht: Wenn Betroffene ihre Lebensqualität beeinträchtigt sehen, suchen sie selber eine Lösung; ►Raumordnung).

4924 regional planning policy [n] *leg. plan. pol.* (High-level, politically-established planning doctrine, the aim of which is to improve coordination between the planning efforts of the Federal Government on Federal lands and of State and local governments on non-Federal lands and to organize, manage and develop the sum of all human and natural forces within a large area or its parts in a coordinated manner; ►multidisciplinary planning, ►physical planning, ►State regional planning [US]); *s* **ordenación [f] del territorio** (Es la expresión espacial de la política económica, social, cultural y ecológica de toda la sociedad, siendo a la vez una disciplina científica, una técnica administrativa y una política concebida como un enfoque interdis-

ciplinario y global cuyo objetivo es un desarrollo equilibrado de las regiones y la organización física del espacio según un concepto rector; cf. DINA 1987; ►planificación integrada, ►planificación regional, ►planificación territorial); *syn.* planificación [f] física (2) [C], ordenamiento [m] territorial/del territorio [AL]; *f* **aménagement [m] du territoire** (Orientations politiques fondamentales d'aménagement de l'espace national dans un but d'une meilleure utilisation et d'un développement harmonieux, rationnel et humain de l'espace en ville, à la campagne, en montagne et sur le littoral en s'attachant, par une action de synthèse et d'arbitrage, à répondre aux aspirations des citoyens dans ses désirs de justice sociale, de qualité de la vie et de protection de l'environnement, tout en tenant compte des disparités et déséquilibres régionaux ; ►planification intégrée, ►planification régionale, ►planification territoriale) ; *g* **Raumordnung [f] (1.** Ein politisch bestimmtes räumliches Leitprinzip oberhalb der Gemeindeebene [Rahmenplanung], das alle sozialen und wirtschaftlichen Nutzungsansprüche der Gesellschaft mit den natürlichen Grundlagen und ökologischen Funktionen eines Gesamtraumes [große Gebietseinheit] und seiner Teilräume entsprechend einer nachhaltigen Tragfähigkeit in Einklang zu bringen versucht, um eine dauerhafte, großräumig ausgewogene Ordnung, Gestaltung und Entfaltung zu ermöglichen [nachhaltige Raumentwicklung]; cf. §§ 1 und 2 ROG 2008. Seit 1999 verständigten sich europäische Mitgliedstaaten und Kommissionen mit dem Europäischen Raumentwicklungsprogramm [EUREK] auf gemeinsame räumliche Ziele für die zukünftige Raumentwicklung in der EU. Durch die Zusammenführung der wichtigsten wirtschaftlichen, demografischen und ökologischen Tendenzen in der EU soll die Koordination der nationalen Raumordnungspolitiken erleichtert werden; ►Landesplanung, ►Querschnittsplanung, ►Raumplanung). **2.** Gesetzliche Grundlage der **R.** ist das Raumordnungsgesetz [ROG] des Bundes als Rahmengesetz und die Landesplanungsgesetze. Seit der Neufassung des ROG umfasst **R.** als OB die Raumplanung des Bundes [cf. §§ 17-25 ROG 2008] sowie der Länder und Regionen [cf. §§ 8-16 ROG 2008]. Planerisches Mittel zur Erarbeitung, Aufstellung und Durchsetzung der raumordnerischen Ziele ist die ►Raumplanung. Die politischen Ziele auf Bundesebene werden alle vier Jahre u. a. im Raumordnungsbericht der Bundesregierung auf Grund der räumlichen Gegebenheiten und der Analyse der Entwicklungstendenzen im Bundesgebiet formuliert und entsprechende Maßnahmen zur gesamträumlichen Entwicklung dargelegt. Die **R.** der Länder wird über die ►Landesplanung verwirklicht. **3.** Die Gesamtheit der Verfahren und Handlungen von Parlamenten, Behörden, sonstiger Institutionen und Organisationen, die öffentliche Belange der **R.** durch Entscheidungen regeln, wird **Raumordnungspolitik** genannt).

4925 regional planning procedure [n] *leg. plan.* (Legally-prescribed courses of action to make certain that development projects and measures are consistent with regional requirements, and, at the same time, to determine how spatially-significant projects and measures can be harmonized with other aspects of ►regional planning policy; ►design and location approval process [US]/determination process of a plan [UK], ►environmental impact assessment [US]); *s* **procedimiento [m] de ordenación del territorio** (Procedimiento formal en el cual se constata si los proyectos planificados están de acuerdo con las directrices de ►ordenación del territorio y con las exigencias del desarrollo regional; ►evaluación del impacto ambiental, ►procedimeineto de aprobación de proyectos públicos); *f 1* **procédure [f] de mise en conformité avec l'aménagement du territoire (≠) (D.,** procédure réglementaire engagée dans le cadre de projets ayant un impact significatif sur l'environnement [►procédure d'instruction de grands projets publics] et ayant pour but de vérifier et

R

d'évaluer la conformité et la compatibilité des aménagements et les installations soumis à autorisation ou à approbation administrative avec les objectifs, principes et dispositions diverses de l'aménagement du territoire et de protection des paysages. Procédure de mise en conformité [compatibilité] d'opérations, d'aménagements ou d'installations d'utilité publique avec les objectifs de l'►aménagement du territoire ; **F.**, les objectifs de l'aménagement du territoire sont pris en compte lors de la procédure d'obtention du permis de construire, de l'étude d'impact ainsi que des autorisations administratives spécifiques au projet ; ►procédure d'étude d'impact) ; *syn.* procédure [f] d'aménagement du territoire [CH] ; *f 2* **procédure [f] de mise en compatibilité des documents d'urbanisme** (Mise en compatibilité de leur élaboration des documents d'urbanisme avec les orientations des documents de planification territoriale de la hiérarchie supérieure) ; *g* **Raumordnungsverfahren [n]** (Dem ►Planfeststellungsverfahren vorgeschaltetes förmliches Verfahren, bei dem festgestellt wird, ob überörtlich raumbedeutsame Planungen und Maßnahmen [z. B. Fernstraßen, Kraftwerke, Bahntrassen, Flugplätze und deren Erweiterungen] mit den Zielen und Erfordernissen der Raumordnung übereinstimmen. Gleichzeitig wird vorgeschlagen, wie raumbedeutsame Planungen und Maßnahmen unter den Gesichtspunkten der ►Raumordnung mit anderen Planungs- und Vorhabenträgern abgestimmt werden können; zum **R.** gehört auch eine raumordnerische ►Umweltverträglichkeitsprüfung; gesetzliche Grundlagen sind § 15 ROG 2008 und Landesplanungsgesetze der Länder. Das **R.** hat keine unmittelbare Rechtswirkung nach außen und ist verwaltungsgerichtlich nicht anfechtbar).

4926 regional planning program [n] [US]/**regional planning programme** [n] [UK] *plan.* (Presentation of sequential actions for implementation of ►regional planning policy and ►State regional planning [US]); *s* **programa [m] de ordenación del territorio** (Descripción de los objetivos de ►ordenación del territorio y ►planificación regional); *f* **programme [m] d'aménagement du territoire (±)** (Présentation des orientations d'organisation du territoire ; ►aménagement du territoire, ►planification régionale) ; *g* **Raumordnungsprogramm [n]** (Darstellung der Ziele der ►Raumordnung und ►Landesplanung).

4927 regional planning-related [adj] *plan. pol.* *s* **relativo/a a la planificación territorial** [loc] *syn.* relativo/a a la planificación física [loc] ; *f* **relatif,ive à l'aménagement du territoire** [loc] ; *g* **raumplanerisch** [adj] (die Raumplanung betreffend).

4928 regional plan report [n] *plan.* (Document analyzing and delineating the aims of ►regional planning policy and ►State regional planning [US]); *syn.* statutory regional plan [n]; *s 1* **Plan [m] Director Territorial de Coordinación (1)** (En Es., plan de ordenación territorial a nivel de las Comunidades Autónomas [CC.AA.] que establece, de conformidad con la planificación económica y social y las exigencias del desarrollo regional, las directrices para la ordenación del territorio, y el modelo territorial en que han de coordinarse los planes y normas a que afecte. A raíz de la sentencia del Tribunal Constitucional sobre la Ley sobre Régimen de Suelo y Ordenación Urbana [RD 1/1992, de 26 de junio], el Estado Español se vió obligado a derogar la mayor parte de las regulaciones contenidas en ella, ya que la política urbanística y de ordenación territorial es competencia exclusiva de las CC.AA. Por ello a nivel estatal ya no existe el antiguamente llamado «plan nacional de ordenación» y la ordenación territorial se reduce a dos niveles: **P. D. T. de C.** y ►Plan General Municipal de Ordenación Urbana); *s 2* **plan [m] de ordenación del territorio** (D., plan y memoria descriptiva en la que se representan y describen los objetivos de la ►ordenación

del territorio y la ►planificación regional, es decir el aprovechamiento ordenado del territorio, considerando en especial las exigencias de la agricultura, la silvicultura, del dominio público hidráulico, de la industria, el tráfico y transporte, los asentamientos, la protección de los bienes culturales y la recreación); *f 1* **schéma [m] national d'aménagement et de développement du territoire (F.**, plan proposant une organisation du territoire fondée sur la notion de bassins de vie, organisés en pays, et de réseaux de ville ; à cet effet sont élaborées des directives d'aménagement fixant les orientations fondamentales de l'État en matière d'aménagement et d'équilibre entre les perspectives de développement, de protection et de mise en valeur des territoires ainsi que les principaux objectifs de l'État en matière de préservation des espaces naturels, des sites et des paysages ; cf. loi n° 95-115 du 4 février 1995 ; *anciennement* schéma national de l'aménagement du territoire ; *aménagement du territoire au niveau d'une région* ►schéma régional d'aménagement et de développement du territoire ; ►directive territoriale d'aménagement, ►schéma régional d'aménagement du territoire ; *f 2* **schéma [m] d'aménagement du territoire** (**D.**, plan fixant les orientations de l'►aménagement du territoire, les actions prévisionnelles et les principes d'organisation de l'espace à moyen terme pour une vaste région ; il détermine les grandes lignes directives de développement de la structure spatiale dans les domaines de l'habitat [pôles urbains, pôles et axes de développement], de l'exploitation des ressources naturelles, de l'environnement, du tourisme, des transports, etc.) ; *g* **Raumordnungsplan [m]** (Zusammenfassender und übergeordneter Plan und Erläuterungsbericht, die die Ziele der ►Raumordnung für größere Gebietseinheiten mittelfristig darstellen. Festgelegt werden im **R.** vor allem Raumstruktur respektive Siedlungsstruktur [Raumkategorien, Zentrale Orte, besondere Gemeindefunktionen wie Entwicklungsschwerpunkte und Entlastungsorte, Siedlungsentwicklungen und Achsen], die anzustrebende Freiraumstruktur [größenmäßig übergreifende Freiräume und Freiraumschutz, Nutzungen im Freiraum sowie deren Funktionen, — z. B. Naturschutz und Landschaftspflege, Freizeit und Erholung — sowie Standorte für die vorsorgende Sicherung], die geordnete Aufsuchung und Gewinnung von standortgebundenen Rohstoffen, Sanierung und Entwicklung von Raumfunktionen und zu sichernde Standorte und Trassen für die Infrastruktur [Verkehrsinfrastruktur und Umschlaganlagen für Güter, Ver- und Entsorgungsinfrastruktur]. Gesetzliche Grundlage bilden das Raumordnungsgesetz des Bundes als Rahmengesetz und die Landesplanungsgesetze); *syn.* raumordnerischer Leitplan [m].

regional policy plan [n] [UK] *leg. plan.* ►state development plan [US].

4929 regional research [n] *plan. pol.* (Interdisciplinary branch of research and analysis of large-scale natural, demographic, economic, social and political structures within a region as a basis for political decision-making processes); *syn.* large scale planning research [n] [also US]; *s* **investigación [f] territorial** (Rama de la ciencia de carácter multidisciplinario que estudia las estructuras naturales, demográficas, económicas, sociales, administrativas y políticas de grandes regiones como base para los procesos de toma de decisiones políticas); *syn.* investigación [f] regional; *f* **recherche [f] en matière d'aménagement spatial** (Branche interdisciplinaire de la recherche scientifique ayant pour objet la description et l'analyse du développement des structures naturelles, démographiques, économiques, sociales et politiques de grandes régions comme instrument de base dans les processus politiques d'aménagement et de décision) ; *g* **Raumforschung [f]** (Interdisziplinäre Forschungsrichtung zur Beschreibung und Entwicklungsanalyse der natürlichen, demografischen, ökonomischen, sozialen und politischen Strukturen größerer Regionen als wichtige Grundlage politischer Planungs- und Entscheidungsprozesse; MEL 1977, Bd. 19); *syn.* Regionalforschung [f].

4930 regional research [n] **and knowledge** [n] *geo.* (Research into and information gained about a certain area or region which takes into account all of its determining factors, principally its geographical characteristics, its natural resources, and its historic(al) economic and social development); *s* **conocimiento [m] sobre la cultura y la naturaleza de una región/ un país** *syn.* civilización [f] de un país; *f* **recherche [f] sur la culture régionale/locale** (Connaissance d'un lieu, d'une région donnés prenant en compte la totalité des facteurs [naturels, historiques, économiques et sociaux] agissant sur son organisation spatiale) ; *syn.* recherche [f] sur les terroirs ; *g* **Landeskunde [f, o. Pl.]** (Die Erforschung und Kenntnis eines bestimmten Raumes oder Gebietes in der Gesamtheit seiner gestaltenden Faktoren, v. a. der natürlichen Gegebenheiten und der historischen Entwicklungen, der Geo- und Soziofaktoren; MEL, Bd. 14, 1975).

regional strategic plan [n] [US] *leg. plan.* ▸regional plan.

regional strategy plan [n] [UK] *leg. plan.* ▸state development plan [US].

4931 regional survey [n] *plan.* (Inventory of existing land conditions for the planning of highly significant projects in a region; ▸site analysis); *s* **inventariación [f] regional** (En el marco de la ordenación del territorio, ▸estudio del terreno y levantamiento de datos para un gran proyecto de construcción); *f* **recueil [m] de données à l' échelle de la planefication régionale** (*Études d'aménagement du territoire* reconnaissance préalable à une importante opération d'aménagement ; ▸analyse du site) ; *syn.* inventaire [m] des données à l' échelle de la planefication régionale ; *g* **regionalplanerische Bestandsaufnahme [f]** (*Raumordnung, Regionalplanung* Zustandserfassung und Kartierung der relevanten Landschaftsdaten für die Planung eines großen Bauvorhabens; ▸Standortanalyse); *syn.* Regionalplanerische Ist-Aufnahme [f].

4932 regional traffic [n] *plan. trans.* (Traffic in a region or traffic which cosses regional boundaries); *s* **tráfico [m] regional** (Tránsito dentro de una región o entre distancias cortas); *f* **transport [m] régional** (Déplacement d'usagers au sens de l'aménagement du territoire et défini au niveau d'une région [schéma régional de transport], p. ex. entre une métropole régionale et une zone de montagne, ou sur des liaisons régionales à longues distances) ; *syn.* transport [m] collectif régional ; *g* **regionaler Verkehr [m]** (Verkehr innerhalb einer Region oder mit geringen Fahrweiten auch regionsüberschreitend).

region [n] **of a city/town** *urb.* ▸surrounding region of a city/town.

register [n] *prof.* ▸architectural register; *adm. urb.* ▸derelict land register; *adm. leg.* ▸land register; *conserv. leg.* ▸National Register of Natural Areas.

register [n] [US]**, abandoned land** *adm. urb.* ▸derelict land register.

register [n] [US]**, abandoned property** *adm. urb.* ▸derelict land register.

register [n] [US]**, deed** *adm. surv.* ▸real property identification map [US]/land register [UK].

register [n] [UK]**, noise level** *plan.* ▸noise level record [US].

registering [n]**, pollution control** *ecol. envir.* ▸biomonitoring.

register [n] **of architects** *prof.* ▸architectural register.

Register [n] **of Natural Areas** *conserv. leg.* ▸National Register of Natural Areas [US].

register [n] **of natural monuments** [UK] *conserv. leg.* ▸National Register of Natural Areas [US].

registry [n] [UK]**, contaminated land** *adm. envir.* ▸toxic site inventory [US].

4933 Registry [n] **of Historic Landmarks** [US] *adm. leg. conserv'hist.* (**In U.S.**, any official list or registry of historic buildings, sites, districts, structures or objects significant in American history, architecture, archeology, and culture [▸cultural monument], which are maintained by national, state or local governments by means of grant aid, tax subsidies, regulation, or recognition, whether kept in the form of a national, state or local registry, book, a list or lists, or some other form. The National Trust for Historic Preservation is a privately-funded organization that also acquires and preserves historic buildings and sites; **in U.K.**, there are listed buildings, scheduled ancient monuments, designated areas of archaeological importance, and conservation areas; ▸National Register of Natural Areas [US]/register of natural monuments [UK], ▸State Historic Preservation Office [US]); *syn.* list [n] of historic buildings of special architectural or historic interest [UK], list [n] of historic buildings, monuments and sites [UK]; *s 1* **Registro [m] General de Bienes de Interés Cultural [Es]** (Anotación e inscripción de los actos que afecten a la identificación y localización de los bienes integrantes del Patrimonio Histórico Español declarados de interés cultural. Está adscrito a la Dirección General de Bellas Artes y Archivos del Ministerio de Cultura; art. 21 RD 111/1986; ▸monumento, ▸organismo responsable del Patrimonio Histórico Español, ▸registro de monumentos naturales); *syn. obs.* Inventario [m] del Patrimonio Histórico-Artístico y Arqueológico de España; *s 2* **Inventario [m] General de Bienes Muebles [Es]** (Término específico. Incluye bienes muebles integrantes del P.H.E. no declarados de interés cultural, que tengan singular relevancia por su notable valor histórico, arqueológico, artístico, científico, técnico o cultural; art. 24 RD 111/1986); *syn. obs.* Inventario [m] del Tesoro Artístico Nacional; *f 1* **liste [f] générale des monuments classés** (Les immeubles dont la conservation présente un intérêt d'histoire ou d'art et classés comme ▸monument historique 1 sont inscrit sur la **l. g. d. m. c.** à l'initiative du ministre chargé de la culture) ; *f 2* **inventaire [m] supplémentaire des monuments historiques** (Les immeubles présentant un intérêt d'histoire ou d'art suffisant pour en rendre désirable la préservation mais qui ne justifient pas de classement immédiat peuvent être inscrit sur l'**i. s. d. m. h.**.) ; *f 3* **inventaire [m] général des monuments et richesses artistiques de la France** (Recueil scientifique informatisé des richesses monumentales immobilières et richesses d'art mobilières actuellement sans valeur juridique rattaché à la direction patrimoine du ministère de la culture ; ▸Conservation régionale des monuments historiques, ▸liste des monuments naturels et des sites) ; *g* **Denkmalbuch [n]** (Bei den höheren Denkmalschutzbehörden [= Regierungspräsidien] geführtes Buch, in das ▸Kulturdenkmäler von besonderer Bedeutung zu ihrem zusätzlichen Schutz eingetragen werden; cf. §§ 12 u. 14 DSchG-BW. Kulturdenkmäler, die keinen oder noch keinen zusätzlichen Schutz durch das **D.** genießen, werden in [Kulturdenkmal]listen bei den unteren Denkmalschutzbehörden und bei den ▸Landesdenkmalämtern geführt; ▸Naturdenkmalbuch).

regular centers [npl] [US]/**regular centres** [npl] [UK] *constr.* ▸regular spacing.

4934 regular coursed masonry [n] **of natural stones** *arch. constr.* (Natural stone wall with mortar: the exposed surfaces of the primarily stratified stones are laid at right angles to one another with head and coursing joints, approximately 12cm deep. The height of the course and the height of stones within the course may be changed at random; ►ashlar masonry); *syn.* coursed ashlar [n] [also US] (DAC 1975); **s sillería [f] aparejada con mampuestos naturales** (Tipo de mampostería cn la que los mampuestos están unidos con mortero y las caras visibles ordenadas en ángulos rectos con juntas de asiento y a tope de unos 12 cm de profundidad. La altura de las hiladas puede variar aleatoriamente; ►sillería de fábrica); *f* **maçonnerie [f] de moellons assisés finement équarris** (Maçonnerie en pierres naturelles à moellons assisés sur mortier, assemblés de telle manière que les joints montants et filants de la face de parement possèdent une profondeur minimale de 12 cm ; les lits de moellons assisés pouvant être d'inégales hauteurs ; ►maçonnerie en moellons à opus quadratum) ; *syn.* maçonnerie [f] de moellons assisés têtués ; *g* **hammerrechtes Schichtenmauerwerk [n]** (Natursteinmauerwerk, lagerhaft mit Mörtel, bei dem die Ansichtsflächensteine in rechtwinklig zueinander stehenden 12 cm tief bearbeiteten Stoß- und Lagerfugenflächen liegen. Die Schichthöhen dürfen, auch in der Schicht, beliebig wechseln; DIN 1053; ►Quadermauerwerk).

4935 regularly flooded alluvial plain [n] *phyt.* (Area within the floodplain of a watercourse which is regularly inundated during high water and mainly supports fast-growing tree species, such as alder *[Alnus]*, poplar *[Populus]* and willow *[Salix]*; ►upper riparian alluvial plain); *syn.* lower riparian/riverine alluvial plain [n] [also US] (WRP 1974); **s vega [f] aluvial regularmente inundada** (Sección de la zona de inundación de cursos de agua que es inundada con regularidad durante la época de más afluencia de aguas y en la que crecen sobre todo árboles y arbustos de madera blanda, como alisos *[Alnus]*, álamos *[Populus]* y sauces *[Salix]*; ►vega aluvial ocasionalmente inundada); *f* **plaine [f] alluviale à bois tendres** (*Terme peu utilisé* partie du lit majeur d'un cours régulièrement inondée au moment des crues et occupée principalement par des espèces de bois tendre comme l'Aulne *[Alnus]*, le Peuplier *[Populus]*, le Saule *[Salix]* et qui constituent les saulaies riveraines ou les aulnaies plus ou moins tourbeuses ; ►plaine alluviale de forêt à bois durs) ; *syn.* plaine [f] alluviale régulièrement inondée, zone [f] alluviale à forêt de bois tendres ; *g* **Weichholzaue [f]** (Bereich innerhalb eines Überschwemmungsgebietes von Fließgewässern, der bei Hochwasser regelmäßig überflutet wird und hauptsächlich mit Weichholzarten wie Erle *[Alnus]*, Pappel *[Populus]*, Weide *[Salix]* bestockt ist; ►Hartholzaue).

4936 regularly flooded riparian woodland [n] *phyt.* (Periodically-inundated woodland with fast-growing trees of having soft and short-lasting timber; e.g. alder *[Alnus]*, poplar *[Populus]*, willow *[Salix]*; ►riparian woodland); *syn.* lower riparian/riverine woodland [n]; **s bosque [m] ripícola regularmente inundado** (►Bosque ripícola que es inundado con periodicidad por lo que está constituido por especies de crecimiento rápido como los sauces *[Salix]*, alisos *[Alnus]* y álamos *[Populus]*); *syn.* bosque [m] de vega (regularmente inundado), bosque [m] ribereño (regularmente inundado), bosque [m] de ribera (regularmente inundado), bosque [m] de ribera alta (BB 1979, 592); *f 1* **forêt [f] alluviale à bois tendres** (Forêt riveraine sur alluvions des marges des eaux courantes sujettée à des inondations périodiques, constituée du Saule *[Salix]*, de l'Aulne *[Alnus]* et du Peuplier *[Populus]* appartenant à l'alliance du Salicion albae ; ►forêt alluviale) ; *syn.* forêt [f] à bois tendres, ripisilve [f], forêt [f] riveraine (inondable) à bois tendres, forêt [f] ripariale ; *f 2* **ripisilve [f] basse** (Terme spécifique de la forêt alluviale à bois tendres : groupement riparial linéaire dont la couverture arbustive est principalement constituée de *Salix*

incana et *Salix purpurea* ; ►forêt galerie) ; *syn.* saulaie [f] ripariale ; *g* **Weichholzauenwald [m]** (Auenwald, der periodisch überflutet wird und deshalb mit schnell wachsenden Gehölzen bestockt ist, deren Holz weich und wenig haltbar ist, wie z. B. Weide *[Salix]*, Erle *[Alnus]* und Pappel *[Populus]*; ►Auenwald); *syn. o. V.* Weichholzauwald.

4937 regular spacing [n] *constr. syn.* even spacing [n], regular centers [US]/regular centres [UK]; **s espaciamiento [m] regular** (*Contexto* plantar con **e. r.**); *f* **intervalle [m] régulier** (Espacement régulier) ; *g* regelmäßiger Abstand [m].

regulating [n] of topsoil *constr.* ►fine grading.

regulation [n] *leg. urb.* ►building setback regulation; *conserv. leg.* ►impact mitigation regulation; *adm. leg.* ►transitional regulation; *conserv. plan.* ►use regulation.

regulation [n], auto- *ecol.* ►self-regulation (1), ►self-regulation (2).

4938 regulation [n] of watercourses *hydr.* (Balancing of water level fluctuations in watercourses by ►river engineering measures or water control structures; ►stream improvement); **s regulación [f] del curso de ríos y arroyos** (Medidas de ►ingeniería hidráulica en arroyos y torrentes o de instalaciones técnicas para equilibrar las variaciones del nivel del agua; ►ingeniería hidráulica de ríos y arroyos); *f* **régulation [f] des cours d'eau** (►Travaux d'aménagement des cours d'eau [recalibrage, reprofilage, rectification] ou tout ouvrage technique permettant un équilibrage des variations du niveau des eaux sur les cours d'eau ; ►travaux d'aménagement d'une rivière/d'un ruisseau) ; *syn.* correction [f] d'un cours d'eau *g* **Wasserlaufregelung [f]** (Ausgleich von Wasserstandsschwankungen der Fließgewässer durch ►Gewässerausbau oder technische Einrichtungen; ►Bachausbau).

4939 regulations [npl] *adm. leg.* (In U.S., detailed requirements to implement a federal or state law or local ►order [US] 2/statutory instrument [UK]; in U.K., written prescription of the legislative body; i.e. an Act of Parliament); **s reglamentación [f]** (Término genérico para regulación legal de cualquier tipo; ►reglamento juridico); *f* **réglementation [f]** (Terme générique regroupant prescriptions, règlements, décrets, arrêtés, directives, ►ordonnances légales et toutes règles établies sur une base législative et assurant leur mise en œuvre ; les règlements permettent l'exécution des prérogatives communales) ; *g* **Bestimmung [f] (1)** (OB zu Satzung, ►Rechtsverordnung, Vorschrift, Anordnung, Verfügung; Satzungen werden im Rahmen der kommunalen [oder sonstigen] Selbstverwaltung erlassen; Rechtsverordnungen beruhen demgegenüber auf einem Gesetz, das sie ausfüllen. **B.** hat aber auch eine zivilrechtliche Bedeutung, z. B. als Vertragsbestimmung oder als Leistungsbestimmung durch einen Dritten).

Regulations [npl] 1988, Town and Country Planning (Assessment of Environmental Effects) [UK] *conserv. leg. nat'res. plan.* ►National Environmental Policy Act [US].

regulator [n] *chem. agr. for. hort.* ►growth regulator.

4940 rehabilitation [n] (1) *constr. urb.* (Act or process of making possible a compatible use for a property through repair, alterations, and additions while preserving those portions or features that convey its historical, cultural, or architectural values; LA 1997 [5], 78; *specific term* ►building rehabilitation; ►center city redevelopment, ►environmental damage, ►modernization 1, ►repair, ►run-down housing fit for rehabilitation); **s rehabilitación [f]** (Conjunto de medidas de construcción para mejorar la calidad de edificaciones o instalaciones [p. ej. parques y jardines] que están en mal estado y necesitan ser adecuadas a

R

necesidades modernas o de ser preservadas en buenas condiciones como parte del ►patrimonio cultural, arquitectónico o histórico; en el caso de la rehabilitación de edificios de viviendas, ésta se realiza tanto a nivel estructural como funcional, equipándolas con infraestructuras modernas y adaptándolas a las necesidades actuales de los usuarios; ►daño ambiental, ►edificación antigua en mal estado, ►mantenimiento de edificios, ►modernización, ►rehabilitación urbana, ►reparación de edificios); *f 1* **réhabilitation [f]** (Remise en état en conservant les caractéristiques majeures, l'aspect initial de bâtiments, de jardins ; ►entretien immobilier, ►modernisation 1) ; *f 2* **réfection [f]** (p. ex. de la voirie consistant dans la démolition et construction des corps de chaussée et des trottoirs existants, dépose et pose de bordures et caniveaux, construction d'ouvrages de collecte et d'évacuation des eaux pluviales) ; *f 3* **rénovation [f] immobilière** (Opération d'urbanisme de remise à neuf d'un immeuble impliquant la destruction complète du bâtiment existant et sa reconstruction ; ►habitat vétuste, ►remise en état, ►rénovation urbaine, ►restauration urbaine) ; *f 4* **réparation [f] de dommages causés à l'environnement** *envir.* (Mesures préventives ou de réparation appropriée des dommages, directs ou indirects, causés au milieu naturel [restauration ou remplacement par des éléments naturels identiques, similaires ou équivalents, soit sur le lieu de l'incident, soit, si besoin est, sur un site alternatif], aux espèces et habitats naturels protégés au niveau communautaire par la directive « oiseaux » de 1979 et par la directive « habitats » de 1992, [remise de l'environnement en l'état antérieur au dommage] ainsi que la contamination, directe ou indirecte, des sols qui entraîne un risque important pour la santé humaine ; ►dommage à l'environnement ; *syn.* réparation des dommages environnementaux/écologiques ; *g* **Sanierung [f]** (**1.** Alle Maßnahmen, die erforderlich sind, wenn die ►Instandhaltung von Gebäuden, Anlagen und Einrichtungen längere Zeit vernachlässigt wurde, um moderne funktionale, wirtschaftliche, ökologische und ästhetische Nutzeransprüche zu befriedigen. **2.** *Sanierung eines Einzelgebäudes* Beseitigung mangelhafter sanitärer Einrichtungen, des schlechten, unwirtschaftlichen Grundrisszuschnitts etc. und Anpassung an neue Nutzeransprüche; ►Modernisierung; *syn.* Bausanierung [f], Gebäudesanierung [f], Modernisierung [f] eines Gebäudes. **3.** Abriss und Wiederaufbau eines oder mehrerer Gebäude; ►Altbebauung 2, ►Instandsetzung von Gebäuden, ►Stadtsanierung. **4.** *envir. wat'man.* Jede Maßnahme, um einen ►Umweltschaden nach Maßgabe der fachtechnischen Vorschriften zu sanieren); *syn. zu 1.-3.* Grundüberholung [f], z. T. Renovierung [f], Umbau [m] und Verbesserung [f].

rehabilitation [n] (2) *landsc. land'man. conserv.* ►restoration (1).

4941 rehabilitation [n] (3) *zool.* (Restorative measures to prepare endangered animals for survival in the wild after being raised in captivity); *s* **readaptación [f]** (Medidas que sirven para preparar a animales [en peligro de extinción] que se han críado en captura a sobrevivir en su medio natural); *f* **réhabilitation [f]** (Opération visant à préparer à la survie en liberté les animaux élevés en captivité et à les installer sur une aire de réhabilitation ; LR 1983, 8) ; *g* **Wiedereingliederung [f]** (Maßnahmen, die dazu dienen, [bedrohte] Tiere, die in Gefangenschaft groß geworden sind, so auf das Leben in freier Wildbahn vorzubereiten, dass sie überleben können).

4942 rehabilitation plan [n] *constr. plan. urb. wat'man.* (Drawing up of a concept containing measures for the improvement of unsatisfactory, existing conditions, such as the demolition or renovation of derelict buildings and the construction of new ones); *syn.* improvement plan [n], improvement scheme [n], revitalization plan [n]; *s 1* **programa [m] de rehabilitación integrada** (Conjunto de planteamientos y medidas para llevar a

cabo la recuperación urbana en una zona declarada como área de rehabilitación integrada o en partes de ella); *s 2* **guía [f] técnica de saneamiento** (Documento técnico orientador para diseñar medidas de recuperación de suelos, aguas contaminadas, etc.); *f 1* **parti [m] d'un projet de rénovation** (Conception fixant les objectifs et les mesures pour la réalisation d'une opération de réhabilitation) ; *f 2* **parti [m] d'un projet de réhabilitation** (Conception fixant les objectifs et les mesures pour la réalisation d'une opération de réhabilitation) ; *g* **Sanierungskonzept [n]** (Planerische Ausarbeitung des Ablaufes durchzuführender Maßnahmen, die veraltete, unzumutbare bauliche Zustände durch Abriss und Neubau, Umbau oder Modernisierung beseitigen oder ändern).

4943 rehydration [n] *land'man. pedol.* (Measures to moisten a dehydrated soil or area to the extent that a wetland may be re-established); *s* **rehumectación [f]** (Medidas orientadas a proporcionar agua al suelo de una zona desecada para restaurar un humedal); *f* **réengorgement [m]** (Réalimentation artificielle d'un sol ou d'un espace après une période de dessèchement afin de reconstituer une zone humide) ; *g* **Wiedervernässung [f] einer Fläche** (Maßnahmen, die dazu dienen, einen ausgetrockneten Boden oder ein trocken gefallenes Gebiet mit so viel Wasser zu versorgen, dass wieder eine Feuchtfläche entsteht).

reimbursable expenses [npl] [US] *contr. prof.* ►overhead expenses.

4944 reinforced turf [n] *constr.* (Lawn sown in ►grass pavers, or other subsurface structural matrix to help withstand wear and tear of heavy use or vehicular loading); *s* **césped [m] armado** (Césped sembrado en revestimiento de ►adoquines pavicésped para ayudar a soportar peso y desgaste); *f* **gazon [m] armé** (Gazon stabilisé par des ►dalles-gazon ou un treillage métallique) ; *g* **armierter Rasen [m]** (Rasen, der durch ►Rasengittersteine, Draht- oder Kunststoffgeflecht befestigt ist).

reinforcement plan [n] [UK] *constr.* ►reinforcing steel plan [US].

4945 reinforcing steel plan [n] [US] *constr.* (Drawing indicating arrangement, amount, diameters, form and position of reinforcing rods or mats, types of steel, etc.); *syn.* reinforcement plan [n] [UK]; *s* **plan [m] de armadura** (Dibujo que indica el aparejo, la cantidad, el grosor, la forma y el lugar donde colocar las varrillas de acero así como el tipo de acero a utilizar); *f* **plan [m] de ferraillage** (Dessin définissant la position, le nombre, le diamètre, la longueur, la qualité des aciers utilisés) ; *syn.* dessin [m] de ferraillage, dessin [m] d'armature ; *g* **Bewehrungsplan [m]** (Zeichnung, die u. a. Anordnung, Anzahl, Durchmesser, Form und Lage der einzelnen Bewehrungsstäbe sowie Stahlsorten angibt); *syn.* Armierungsplan [m].

reinstallation [n] **of stockpiled materials** *constr.* ►reuse of on-site construction material.

reinstallment [n] **of a habitat** *landsc. conserv.* ►habitat restoration.

4946 reinstatement [n] *constr.* (Removal of temporary installations, planting beds, ►demonstration gardens 1, etc., which are not intended as permanent features of a park, e.g. after the end of a ►horticultural exhibition, and restoration to their original condition; ►recyling); *s* **vuelta [f] al estado original** (Desmontaje de ►jardines tipo, macizos de flores y demás instalaciones especiales en un parque al terminar una ►exhibición de horticultura; ►reciclado); *syn.* reconstitución [f] al estado original; *f* **remise [f] en état initial** (Démontage des ►jardins de démonstration, des équipements et installations temporaires à la fin d'une ►exposition de jardins ; ►exposition horticole ; ►recyclage) ; *g 1* **Rückbau [m]** (...bauten [pl]; **1.** Entfernen von temporären baulichen Anlagen, Rabatten, ►Schaugärten etc., die nach einer ►Gartenbauausstellung nicht mehr Bestand der Grün-anlage sein sollen); *g 2* **selektiver Rückbau [m]** (Möglichst zerstörungsfreie Demontage mit dem Ziel, komplette, noch in-takte ganze Bauelemente und -steine erneut zu verwenden, z. B. beim Abriss alter Gebäude werden Fenster, Türen und Beschläge sowie Schlösser oder Dielen und Dachziegel ausgebaut, aufge-arbeitet und wiederverwendet; dadurch werden Entsorgungs-kosten gespart oder Einnahmen durch Verkauf von wieder ver-wertbaren Bauteilen erzielt; ►Recycling).

reinstatement costs [npl] of habitats/ecosystems *adm. plan. ecol.* ►replacement costs of habitats/ecosystems.

4947 reintroduction [n] *game'man. phyt. zool.* (Successful ►introduction 1 of animals or plants in a region in which they formerly lived; ►recolonization); *syn.* relocation [n] (TGG 1984, 213); *syn. phyt.* reestablishment [n] of plant communities [US]/ re-establishment [n] of plant communities [UK]; *s* **reintroduc-ción [f] de especies extintas** (►Suelta exitosa de especies de fauna o plantación de especies de flora en una zona en la que habían existido anteriormente; ►recolonización natural); *f* **réintroduction [f] (2)** (►Introduction couronnée de succès d'une espèce végétale ou animale qui a déjà été observée dans une zone sous la forme d'une population présente à l'état naturel et viable dans les temps historiques mais qui a décliné ou en a disparu à la suite d'une intervention humaine ou d'une catastrophe naturelle ; ►recolonisation naturelle) ; *syn.* opération [f] de réintroduction ; *g* **Wiederansiedelung [f]** (Erfolgreiche ►Aussetzung von Pflanzen oder Tieren in ein Gebiet, in dem sie früher schon mal gelebt haben; ►Wiederbesiedelung); *syn.* Wiedereinbürgerung [f], Reintegration [f], *zool.* Rückbürgerung [f], *o. V.* Wiederansiedlung [f].

rejuvenation cut [n] *arb. constr. hort.* ►rejuvenation pruning.

4948 rejuvenation [n] of trees *arb. constr.* (Severe prun-ing of mature trees with the aim of stimulating new, vigorous growth; ►rejuvenation pruning); *s* **rejuvenecimiento [m] de árboles** (Poda severa en árboles viejos para estimular un creci-miento más vigoroso; ►poda de rejuvenecimiento); *f* **rajeunis-sement [m] d'un arbre** (Taille sévère des branches d'un arbre adulte ; ►taille de rajeunissement) ; *g* **Baumverjüngung [f]** (Kräftiger Rückschnitt bei älteren Bäumen zur Erzielung eines kräftigen Austriebes resp. eines Kronenneuaufbaues; ►Ver-jüngungsschnitt).

4949 rejuvenation pruning [n] *arb. constr. hort.* (Pruning of trees or shrubs which show no symptoms other than extreme lack of vigo[u]r; they may be pruned in a "kill or cure" operation; ARB 1983; ►rejuvenation of trees); *syn.* invigoration pruning [n] [also US], rejuvenation cut [n]; *s* **poda [f] de rejuvenecimiento** (Fuerte poda de árboles o arbustos viejos sin síntomas de enfermedad para que rebroten y así dar nueva forma a la copa; ►rejuvenecimiento de árboles); *f* **taille [f] de rajeunissement** (Suppression des branches ou des gros rameaux d'une haie ou d'un arbre vieillissant [rabattage] dans le but de provoquer le développement de nouvelles pousses vigoureuses ; une **taille de reformation** est ensuite effectuée pour reconstituer la charpente et rééquilibrer la couronne de l'arbre ; ►rajeunissement d'un arbre) ; *syn.* coupe [f] de rajeunissement ; *g* **Verjüngungsschnitt [m]** (Starker Rückschnitt bei älteren Bäumen oder Sträuchern, um einen kräftigen Austrieb oder Neuaufbau der Krone zu erzielen; nach wenigen Jahren werden Astbüschel in der Krone noch mal nachgeschnitten, damit sich eine habitusgerechte Astverzweigung entwickeln kann. Diesen Schnitt nennt man im Französischen *taille de reformation*; ►Baumverjüngung).

4950 related to nature conservation [loc] *conserv.* (Actions and measures taken in connection with nature conser-vation and landscape management); *s* **relacionado/a [pp] con la** protección de la naturaleza (Se dice de acciones o medidas tomadas en el contexto de la protección de la naturaleza y la gestión del paisaje); *f* **relatif, ive à la protection de la nature [loc]** (Se dit d'actions ou de mesures visant la protection de la nature et des paysages) ; *g* **naturschutzfachlich [adj]** (Hand-lungen und Maßnahmen betreffend, die den Zielen des Natur-schutzes und der Landschaftspflege dienen).

4951 related to the environment [loc] *conserv.* *s* **rela-cionado/a con el medio ambiente [loc]** *syn.* relativo [adj] al medio ambiente; *f* **relatif, ive à l'environnement [loc]** *syn.* concernant [part] l'environnement ; *g* **umweltbezogen [adj]** (die Umwelt betreffend).

relationships [npl] *ecol.* ►pattern of interactions and relation-ships.

relative coverage [n] *phyt.* ►total estimate.

4952 relative light availability [n] *for. phyt.* (Amount of light available in a certain location in relation to the intensity of the total amount of light at the same time as measured in the open air; ►light intensity); *s* **luminosidad [f] relativa** (Relación entre la intensidad luminosa en la estación y la ►intensidad de la luz total al aire libre al mismo tiempo); *f* **luminosité [f] relative** (Rapport entre l' ►intensité lumineuse de la station et l'intensité totale à l'air libre) ; *g* **relativer Lichtgenuss [m]** (Verhältnis der Lichtstärke am Standort zur Intensität des gleichzeitigen Gesamt-lichtes im Freien, d. h., das den Pflanzen zur Verfügung stehende Licht; ►Lichtintensität).

4953 relaxation [n] *recr.* (Calming down of physical, mental or psychic stress); *s* **relajación [f] (física y mental)** (Liberación del cansancio físico y mental y del estrés; relajarse [vb/refl]); *syn.* relax [m]; *f* **délassement [m]** (Libération de la tension physique, intellectuelle et psychique) ; *syn.* détente [f] (2), relaxation [f] ; *g* **Entspannung [f]** (Befreiung von körperlicher, geistiger und seelischer Anspannung; *Kontext* E. suchen, E. finden).

relaxation area [n] *recr.* ►sunbathing area.

relaxation [n] of building regulations [UK] *adm. leg. urb.* ►variance [US].

release [n], nutrient *pedol.* ►nutrient mobilization.

relevé [n] *phyt.* ►sampling (2).

relict [n] *phyt. zool.* ►glacial relict.

4954 relict area [n] *phyt. zool.* (Remaining space occupied by a species with a once much greater geographic range; ►habitat island, ►process of habitat fragmentation, ►remnant habitat island); *s* **área [f] relíctica** (Área ocupada por un taxón super-viviente de una flora primitiva ya desaparecida en la zona correspondiente. Suele ser de tamaño reducido. También se da el caso de formaciones relícticas completas, que constituyen restos de floras anteriores desaparecidas por cambios climáticos u otras causas, como p. ej., las laurisilvas de los Canutos de Cádiz o los encinares con laurel de la Cornisa Cantábrica, restos ambos de una flora antigua; cf. DINA 1987; ►aislamiento de biótopos, ►biótopo relíctico, ►hábitat residual); *f* **aire [f] relictuelle** (Reste d'une aire d'extension importante d'une espèce ; ►bio-tope relictuel, ►habitat relictuel, ►isolement de biotopes) ; *g* **Reliktareal [n]** (Rest eines ursprünglich großen Verbreitungs-gebietes einer Art; ►Biotoprest, ►Habitatinsel, ►Verinselung von Biotopen).

4955 relict species [n] *phyt. zool.* (Remnant species from a former development period, when it was more widely distri-buted); *s* **especie [f] relíctica** (Especie vegetal o animal supervi-viente de una población que fue abundante y que en la actualidad se encuentra aislada y lejana de otros representantes de esa población; DINA 1987, 831); *syn.* relicto [m]; *f* **espèce [f]**

relictuelle (Espèce résultant du stade évolutif précédant le peuplement actuel) ; *syn.* organisme [m] relique (DEE 1982), espèce [f] relicte (DEE 1982) ; *g* **Reliktart** [f] (Art, die als Entwicklungsstand einer Evolutionsrichtung übriggeblieben ist).

4956 relief [n] *arch.* (Sculptural projection of figures and forms from a flat surface; there are three kinds: low or bas relief, half and high relief); *s* **relieve** [m] **(2)** (Término genérico para obra plástica no exenta; *términos específicos* bajorrelieve, mediorrelieve, altorrelieve); *f* **relief** [m] **(2)** (En sculpture, motif se détachant en saillie sur un fond ; suivant l'importance de la saillie obtenue, on distingue le bas-relief, le demi-relief et le haut-relief) ; *g* **Relief** [n] **(2)** (In der Bildhauerkunst an einen Hintergrund gebundene plastische Darstellung. Je nachdem wie hoch die Komposition herausgearbeitet wurde, unterscheidet man Flach- oder Basrelief, Halb- und Hochrelief).

relief [n] *envir.* ▶environmental relief; *geo.* ▶ground relief; *geo. ecol.* ▶terrain relief.

4957 relief map [n] *geo. plan.* (Graphic representation depicting the configuration of an area of the Earth's surface by any method; e.g. contours, form lines, hachures, shading, or tinting); *s* **mapa** [m] **de relieve** (Mapa en el que están representadas las formas topográficas del terreno); *f 1* **carte** [f] **du relief** (Carte spéciale sur laquelle l'altimétrie ou les formes de terrains sont représentées au moyen de lignes de niveau, de hachures, d'ombre ou de couleurs) ; *f 2* **carte** [f] **en relief** (Représentation topographique à trois dimensions : les échelles planimétriques et altimétriques sont à des échelles différentes ; DG 1984, 63) ; *g* **Reliefkarte** [f] (Kartentyp, der die Höhenverhältnisse und Geländeformen der Erdoberfläche durch Höhenschichtlinien, Schraffen, Höhenstrukturlinien, Schatten- oder Farbwerte darstellt).

4958 relief road [n] *trans. urb.* (Additionally-constructed road or existing road which is widened to reduce traffic congestion by providing an alternative route for through traffic; ▶bypass [US]/by-pass [UK]); *s* **calle** [f] **de desvío** (Calle construida o ampliada especialmente para acoger tránsito adicional; ▶carretera de circunvalación); *syn.* carretera [f] de desvío, desvío [m]; *f 1* **route** [f] **de délestage** (Voie nouvelle ou existante permettant de décongestionner une voie routière ; ▶route de contournement) ; *f 2* **itinéraire** [m] **de délestage** (Parcours routier proposé pour décongestionner une région ou des itinéraires très encombrés par la circulation) ; *syn.* itinéraire-bis [m] ; *g* **Entlastungsstraße** [f] (1. Zusätzlich gebaute oder so ausgebaute vorhandene Straße, die Verkehrsverdichtungen reduzieren und Verkehr der Umgebung aufnehmen kann; ▶Umgehungsstraße. 2. Straße, die in Hauptreisezeiten zusätzlichen Verkehr aufnehmen kann).

4959 relocated farmstead [n] [US] *agr.* (Removed farmstead [US]/farm holding [UK] from the center [US]/centre [UK] of a village to the openfield [US]/open countryside [UK]; ▶farmstead resettlement [US]); *syn.* resited farm holding [n] [UK]; *s* **granja** [f] **relocalizada** (Nueva construcción de una granja a las afueras de un pueblo en el marco de la política de ▶reasentamiento de granjas); *f* **ferme** [f] **transférée** (Exploitation agricole déplacée en rase campagne ; ▶restructuration d'exploitations agricoles, ▶transfert d'exploitations agricoles) ; *g* **Aussiedlerhof** [m] (Ein im Rahmen einer Aussiedlung aus dem Dorfverband in die freie Feldmark verlegter Bauernhof; ▶Aussiedlung).

relocation [n] *game'man. phyt. zool.* ▶reintroduction; *trans.* ▶road relocation.

relocation [n] **of farm holdings** [UK] *agr. urb.* ▶farmstead resettlement [US].

4960 relocation [n] **of industries** *urb.* (Reestablishment of industrial firms outside the confines of densely populated urban areas or their parts in connection with residential area improvement plans [US]/improvement schemes [UK]; ▶planned industrial development, ▶planned industrial development); *s* **relocalización** [f] **de industrias** (Traslado de industrias afuera de las zonas urbanas residenciales en el marco de la rehabilitación urbana o para darles más espacio para su desarrollo; ▶gestión del desarrollo industrial, ▶industria existente); *f 1* **transfert** [m] **d'un établissement industriel** (Changement d'implantation géographique de tout ou partie des activités d'une entreprise dans le cadre de mesures de réhabilitation urbaine ou d'amélioration de la structure urbaine dans les quartiers à forte densité de population ; ▶établissement industriel, ▶implantation d'activités industrielles) ; *f 2* **relocalisation** [f] **industrielle** (Transfert des activités d'une entreprise sur un autre site, notamment pour réduire les coûts de production ; la relocalisation industrielle est souvent synonyme du transfert à l'étranger des unités de production) ; *syn.* délocalisation [f] industrielle, réimplantation industrielle, déplacement [m] d'un établissement industriel, transfert [m] industriel ; *g* **Aussiedlung** [f] **von Industrie- oder Gewerbebetrieben** (Verlegung von Produktionsbetrieben aus dicht besiedelten Stadtgebieten oder deren Teile im Rahmen von Sanierungs- oder Wohnumfeldverbesserungsmaßnahmen; Industrie- oder Gewerbebetriebe aussiedeln [vb]; ▶geplante Industrieansiedlung, ▶vorhandene Industrieansiedlung).

remain [vb] **vacant** *sociol. urb.* ▶vacancy.

4961 remedial construction claim [n] [US] *constr. contr.* (Owner's claim against a contractor requiring the contractual fulfillment/fulfilment of construction work[s] by correcting ▶defects which have appeared during the ▶defects liability period; ▶surety bond [US]/bond [UK]); *syn.* remedial works claim [n] [UK]; *s* **derecho** [m] **a garantía** (Derecho del comitente hacia el contratista que ha de cumplir con el contrato de obra y, en su caso, ha de corregir los ▶defectos [de obra] que surjan dentro del ▶plazo de garantía; ▶garantía 2); *f* **droit** [m] **à l'obligation de garantie** (Droit du maître de l'ouvrage à exiger la responsabilité d'une personne physique ou morale, en vertu des articles 1792 à 1792-4 du C. civ., de remédier aux ▶imperfections après exécution des travaux, aux ▶malfaçons et ▶vices de construction constatés ou aux désordres signalés pendant le ▶délai de garantie ; ▶garantie de bonne fin, ▶sûretés) ; *g* **Gewährleistungsanspruch** [m] (...sprüche [pl]; Anspruch eines Bauherrn an den Unternehmer nach Erfüllung des Vertrages über die Erbringung von Bauleistungen eine Gewährleistung gem. BGB oder VOB zu übernehmen, demzufolge er ▶Mängel, die in der ▶Gewährleistungsfrist auftreten, zu beseitigen hat; ▶Sicherheitsleistung).

4962 remedial lawn work [n] *constr.* (Repair of failed lawn areas); *s* **resiembra** [f] **de claros de césped** (Nueva siembra de superficies de césped para eliminar las marras); *f* **reprise** [f] **des parties malvenues d'un gazon** (Travaux de restauration des engazonnements pendant le délai de garantie intervenant dès la première période favorable qui suivra le constat de défaut, semis de regarnissage sur les pelades, aération des zones compactées, amendement, et drainage du sol engorgé) ; *g* **Mängelbeseitigung** [f] **an Rasenflächen** (Landschaftsgärtnerische Arbeiten, die Fehlstellen im Rasen durch Nachsaat schließen, Verdichtungen auflockern, Nährstoffmangel durch Düngung und Staunässe durch Dränung beseitigen).

4963 remedial measure [n] *constr. envir. landsc.* (TGG 1984, 157; action taken to improve the environment or to redress environmental damage); *s* **medida** [f] **de mejora** (Medida tomada para remediar un daño al ambiente o mejorar las condiciones

R

ambientales, como p. ej. por medio de protección contra el ruido, contra inmisiones, cambio de suelo contaminado); *syn.* medida [f] de saneamiento; *f* **mesure [f] de valorisation** (Mesures contre les dégradations de l'environnement et du cadre de vie, en faveur des préoccupations de l'environnement, comme celles relatives à la lutte contre les pollutions et les nuisances, aux conditions d'habitabilité d'un ensemble d'immeubles, etc.) ; *syn.* mesure [f] favorable à la défense de l'environnement ; *g* **Verbesserungs-maßnahme [f]** (Maßnahme zur Verbesserung von Umweltbedingungen resp. zur Sanierung von Umweltschäden oder deren Abwendung, z. B. durch Lärmschutz, Immissionsschutz, Gewässerausbau, Bodenaustausch bei kontaminierten Böden); *syn.* Sanierungsmaßnahme [f].

remedial measures [npl] **for groundwater pollution** *envir. wat'man.* ►remediation of groundwater pollution.

remedial work [n] *constr. contr.* ►remedying defects.

remedial works claim [n] [UK] *constr. contr.* ►remedial construction claim [US].

remediation [n] **of contaminated soil** *envir.* ►decontamination of soil.

4964 remediation [n] **of groundwater pollution** *envir. wat'man.* (Measures undertaken to decontaminate polluted water); *syn.* remedial measures [npl] for groundwater pollution; *s* **saneamiento [m] puntual de aguas subterráneas**; *f* **résorption [f] de la pollution de la nappe** (Mesures visant à la dépollution de la nappe) ; *g* **Sanierung [f] der Grundwasserverschmutzung** (Maßnahmen, die dazu dienen, verschmutztes Grundwasser in einen sauberen Zustand zu versetzen).

remediation [n] **of toxic waste sites** [US] *envir. plan.* ►toxic site reclamation.

4965 remedying [n] **defects** *constr. contr.* (Rectification of work executed according to contractual specifications within the scope of ►guarantee; rectify [vb] [US]/make good [vb] [UK], remedy [vb]; ►repair work 1, ►defect, ►replacement planting); *syn.* remedial work [n], making [n] good of defects [also UK]; *s* **corrección [f] de defectos (en obra)** (Subsanación de ►defectos [de obra] en el marco de la ►garantía 1; ►obras de reparación, ►plantación de reemplazo de marras); *syn.* eliminación [f] de defectos, subsanamiento [m] de deficiencias; *f* **réfection [f] des imperfections ou malfaçons** (Travaux de reprise des ouvrages défectueux, des ►imperfections, ►malfaçons, des désordres et défauts apparents dans le cadre des ►garanties ; ►plantation de remplacement, ►remplacement des végétaux, ►travaux de réfection ; *syn.* réfection [f] des ouvrages défectueux, travaux [mpl] de reprise, réparation [f] des imperfections ; *g* **Mängelbeseitigung [f]** (Wiederherstellung von vertraglich zugesicherten Eigenschaften von Bauleistungen im Rahmen der ►Gewährleistung; cf. § 13 VOB Teil B; ►Ausbesserungsarbeiten, ►Mangel, ►Nachpflanzung); *syn.* Beseitigung [f] von Mängeln, Nachbesserung [f], Nacherfüllung [f], Garantiearbeiten [fpl] [CH].

4966 remnant habitat [n] *conserv. ecol. land'man.* (Small remaining area; ►habitat, ►habitat island, ►ecology of habitat islands, ►relict area); *syn.* remnant (habitat) island [n]; *s* **biótopo [m] relíctico** (►área relíctica, ►biótopo, ►ecología de biótopos aislados, ►hábitat residual); *syn.* relicto [m] ecológico, biótopo [m] residual, biótopo [m] relictual; *f* **biotope [m] relictuel** (Biotope qui par suite de bouleversements orogéniques ou climatiques, des actions anthropiques est restreint à une région limitée ou morcelé en îlots disjoints ; habitat souvent considéré comme vestige de milieux disparus ou très rares ; ►aire relictuelle, ►biotope, ►écologie des biocénoses relictuelles, ►habitat relictuel) ; *syn.* biotope [m] vestigial, biotope [m] résiduel ; *g* **Biotoprest [m]** (Teil eines ehemaligen ►Biotopes, der auf

Grund seiner kleinen Größe oder durch Fragmentierung der Landschaft und Ausbreitungsbarrieren für bestimmte Arten keine Überlebensmöglichkeit mehr bietet; ►Habitatinsel, ►Inselökologie, ►Reliktareal); *syn.* reliktartiger Biotop [m], Biotopinsel [f], ökologische Insel [f], Restbiotop [m, *auch* n].

remnant habitat island [n] *conserv. ecol. land'man.* ►remnant habitat.

remnant population [n] *phyt. zool.* ►residual population.

remnant wood patch [n] [UK] *landsc.* ►woodlot [US&CDN] (2).

4967 remodeling [n] [US]/**remodelling** [n] [UK] *arch. constr.* (Modification of, e.g. an existing building, garden, or park with significant alterations to the structure or the existing vegetation; ►partial redesign, ►road redesign); *s* **remodelación [f]** (Modificación de un objeto existente [edificio, jardín, parque, etc]; ►reconfiguración parcial, ►remodelación de calles); *f* **réaménagement [m]** (►Recomposition 1 d'un objet existant, de la voirie, d'un quartier, d'un jardin, d'un parc en procédant à des transformations importantes du bâti ou de l'existant ; ►réaménagement d'une rue, ►remodelage) ; *g* **Umbau [m]** (...bauten [pl]; ►Umgestaltung 1 eines vorhandenen Objektes, z. B. eines Gebäudes, Gartens, einer Parkanlage, mit wesentlichen Eingriffen in Konstruktion oder Bestand; ►Straßenumbau).

remodeling [n] **of an area/terrain** [US] *arch. plan.* ►reconfiguration of an area/terrain.

remodeling [n] **of a site** [US] *arch. plan.* ►reconfiguration [n] of an area/terrain.

remodelling [n] **of an area/terrain** [UK] *arch. plan.* ►reconfiguration of an area/terrain.

remodelling [n] **of a site** [UK] *arch. plan.* ►reconfiguration [n] of an area/terrain.

4968 remontant rose [n] *hort.* (1. Term for a class of roses bred in the 19th century primarily in France. Of the approximate 4,000 cultivars there are some 100 in cultivation today. 2. Hybrid perpetual rose which continues to flower in the same season after the first flush of bloom, though not necessarily continually); *syn.* perpetual flowering rose [n], repeat flowering rose [n], recurrent flowering rose [n] [also US]; *s* **rosa [f] refloreciente** (Rosa que florece dos veces al año); *f* **rosier [m] remontant** (1. Terme désignant une classe de rosiers cultivés principalement en France au 19ème siècle dont il n'est plus cultivé qu'une centaine de variétés sur les quatre mille existant à l'époque. 2. Rosier qui refleurit une deuxième fois dans la période végétative) ; *g* **Remontantrose [f]** (1. Bezeichnung für eine im 19. Jh. vor allem in Frankreich züchterisch entwickelte Rosenklasse. Von den ca. 4000 Sorten gibt es heute noch rund 100 Sorten in Kultur. 2. Bezeichnung für eine zweimal im Jahr blühende Rosensorte); *syn.* remontierende Rose [f].

4969 remote sensing [n] *plan. rem'sens. surv.* (Acquiring of information or data through the use of aerial cameras or other sensing devices that are situated at a distance from the area being investigated; ►scanner, ►terrain analysis); *s* **teledetección [f]** (Técnica que permite obtener información sobre un objeto, área o fenómeno a través del análisis de los datos adquiridos por un instrumento que no está en contacto con el objeto, área o fenómeno bajo investigación. Para ello se utilizan la fotografía aérea y las fotos de satélites; ►análisis de la topografía del terreno, ►escáner; cf. CHUV 1990); *syn.* percepción [f] remota; *f 1* **télédétection [f] (de terrain)** (Détection à distance de la terre utilisant des images non photographiques telles que la gravimétrie, les rayonnements électromagnétiques, la sismographie, la scintillométrie ; ►analyse de la structure terrestre superficielle,

R

▶scanneur) ; *f 2* **télédétection [f] aéroportée** (par avion) ; *f 3* **télédétection [f] satellitaire** (par satellite) ; *g* **Fernerkundung [f]** (Erforschung der Erde aus der Luft mit Hilfe von Luftbildaufnahmen oder [opto]elektronischer Aufnahmeverfahren; ▶Erderkundung, ▶Scanner).

4970 removable post [n] *urb. trans. s* **poste [m] de cierre de paso móvil** ; *f* **poteau [m] d'interdiction de passage amovible** ; *g* **herausnehmbarer Absperrpfosten [m]**.

removal [n] **for replanting** *constr.* ▶vegetation removal for replanting.

removal [n] **of branch with V crotch** *arb.* ▶removal of codominant stem.

4971 removal [n] **of codominant stem** *arb.* (Cutting of a second leader); *syn.* removal [n] of branch with V crotch (ARB 1983, 399); *s* **poda [f] de una de las guías en horquilla** (Poda de una de las dos guías que forman la horquilla en un ejemplar); *f* **défourchage [m] des têtes d'arbres** (Suppression à sa base d'une pousse qui se développe au détriment de la pousse principale) ; *g* **Zwieselentfernung [f]** (Wegschneiden des zweiten Leittriebes als konkurrierenden Stämmling an Jungbäumen); *syn.* Entfernen [n, o. Pl.] des Zwieselwuchses.

4972 removal [n] **of fortifications** *urb.* (In Europe, demolition of mediaeval walls; when levelled, the space was often converted to an urban green belt; ▶greenbelt [US]/green belt [UK]); *s* **desmantelamiento [m] de fortificaciones** (En ciudades de Europa Central práctica de eliminar fortificaciones del medioevo para crear ▶cinturones verdes alrededor de los centros); *f* **arasement [m] de(s) fortifications** (Démantèlement d'ouvrages militaires moyenâgeux suivi en général par la transformation en une ▶ceinture verte urbaine) ; *g* **Schleifen [n, o. Pl.] von Wallanlagen** (Entfernen und Einplanieren von mittelalterlichen Befestigungsanlagen, die meist zu einem innerstädtischen ▶Grüngürtel umgebaut wurden. Mit dem **S.** der Stadtmauern und **W.** verschwand die jahrhundertelange klare Trennung von Stadt und Landschaft. Das **S.** begünstigte zusammen mit der Erfindung der Eisenbahn und des Autos die schnelle Ausbreitung der Städte in die freie Landschaft; Wallanlagen schleifen [vb]).

removal [n] **of tree stumps** *constr. hort.* ▶clearing and removal of tree stumps.

4973 removal [n] **of woody debris** *for. wat'man.* (Disposal of dead trunks, branches and twigs lying on the forest floor or floating in bodies of water; ▶snag 4); *s* **eliminación [f] de restos de leñosas** (Limpia de restos de troncos, ramas y ramillas que yacen sobre el suelo del bosque o flotan sobre el agua de ríos o lagos; ▶obstáculo); *f 1* **enlèvement [m] du bois mort** *for.* (Suppression des arbres morts sur pied et enlèvement du bois mort et des débris ligneux combustibles gisant au sol des forêts ; *f 2* **enlèvement [m] des débris ligneux grossiers** *wat'man.* (Retrait des branches, souches, arbres tombés dans les cours d'eau ou accumulés sur les rives de plans d'eau ; ▶amas de bois, ▶embâcle de bois ; *terme spécifique* enlèvement d'embâcles) ; *g 1* **Totholzentfernung [f] (2)** *for.* (Wegnahme von toten Stämmen, Ästen [Trockenäste/Dürräste] und Zweigen, die auf dem Waldboden liegen); *syn.* Totholzbeseitigung [f]; *g 2* **Entfernung [f] des Treibholzes** *wat'man.* (Entfernung der in Gewässern treibenden ▶Holzhindernisse resp. an Ufern angeschwemmten Hölzer); *syn.* Treibholzentfernung [f], Schwemmholzentfernung [f] [auch A, CH].

4974 remuneration [n] *contr. prof.* (Generic term for the payment of specific work [US]/works [UK] or services; i.e. construction cost, professional services; *specifics terms* ▶bonus payment, ▶lump-sum fee, ▶overtime payment, ▶remuneration of professional fees, ▶time charges); *syn.* payment [n]; *s* **remu-**

neración [f] (Pago de determinados servicios o trabajos; ▶pago de honorarios; *términos específicos* ▶gratificación por entrega adelantada, ▶honorarios globales, ▶pago de horas extras, ▶salario horario); *f* **rémunération [f]** (Règlement des comptes d'un entrepreneur ou d'un maître d'œuvre ; *termes spécifiques* ▶honoraire au forfait, ▶prime d'avance, ▶rémunération des heures supplémentaires, ▶rémunération horaire, rémunération plancher [*en allemand* Grundvergütung] ; HAC 1989, 460 ; ▶rémunération des honoraires) ; *g* **Vergütung [f]** (1. Bezahlung von bestimmten [Arbeits]leistungen wie z. B. Bauleistungen, Architektenleistungen; *UBe* ▶Beschleunigungsvergütung, ▶Honorarvergütung, ▶Pauschalhonorar, ▶Stundenvergütung, ▶Überstundenvergütung. 2. Geldsumme, mit der eine Leistung bezahlt wird. 3. Ein festgelegter Betrag für eine definierte Leistung, der durch zusätzliche Leistungen erhöht werden kann, z. B. durch Leistungsprämien, Mengenprämien, Terminprämien etc., wird **Grundvergütung** genannt); *syn. zu 1.* Bezahlung [f].

remuneration [n] **of professional charges** *contr. prof.* ▶remuneration of professional fees.

4975 remuneration [n] **of professional fees** *contr. prof. syn.* payment [n] of professional fees/charges; *s* **pago [m] de honorarios** *syn.* remuneración [f] de honorarios; *f* **rémunération [f] des honoraires** (Règlement des sommes dues au maître d'œuvre) ; *syn.* règlement [m] des honoraires, règlement [m] des comptes du titulaire, paiement [m] des honoraires ; *g* **Honorarvergütung [f]** (Bezahlung eines Honorares für erbrachte Leistungen); *syn.* Honorierung [f].

4976 remuneration [n] **on a percentage basis** *contr. prof.* (Fees based on a portion of work performed according to contractually agreed upon percentage rates); *s* **remuneración [f] sobre la base porcentual;** *f* **honoraire [m] au pourcentage (du montant des travaux)** (Honoraire calculé sur la base du produit du taux de rémunération par le montant hors taxes des travaux) ; *g* **Honorar [n] nach Prozentsätzen [A]** (Entgelt nach Prozentsätzen der Gebührenordnung [HRLA] oder nach einem vertraglich vereinbarten Prozentsatz der anrechenbaren Baukosten; in der CH spricht man von der Honorarberechnung in Prozentsätzen der Baukosten); *syn.* Honorar [n] in Prozentsätzen der Baukosten [CH].

4977 remuneration [n] **on a time basis** *contr. prof.* (Fees based on certified time taken to perform the work, according to accepted hourly rates; ▶daywork contract); *s* **honorarios [mpl] por tiempo empleado** (▶contrato por horas); *syn.* remuneración [f] por tiempo empleado; *f* **honoraire [m] au temps consacré** (Honoraire déterminé avant le début de la mission et calculé par application d'un prix unitaire au temps consacré par chaque catégorie de personnel ; *terme spécifique* ▶marché au temps passé ; HAC 1989, 482) ; *syn.* honoraire [m] au déboursé, honoraire [m] à la vacation ; *g* **Honorar [n] nach Zeitaufwand** (Honorar, bei dem die erbrachten Leistungen nach dem nachgewiesenen Zeitbedarf auf Grund von vereinbarten Stundensätzen vergütet werden; ▶Stundenlohnvertrag); *syn.* Zeithonorar [n], Zeitgebühr [f] [auch A].

renaturalization [n] [UK] *landsc. land'man. conserv.* ▶restoration (1).

4978 rendzina [n] *pedol.* (Great soil group of the intrazonal order and calcimorphic suborder consisting of soils with brown or black friable surface horizons underlain by light gray to pale yellow calcareous material, developed from soft, highly calcareous parent material in humid to semiarid climates; RCG 1982. In U.S. Soil Classification System, 'rendoll' is approximately equivalent; ▶ranker); *s* **rendsina [f]** (Un gran grupo de suelos calcimagnésicos de color negruzco, gris oscuro hasta gris claro según el contenido en humus, conteniendo caliza hasta extremadamente rico en ella, caracterizado por humatos cálcicos, cuyo horizonte de humus se formó directamente sobre la roca madre que consiste en caliza, mármol, dolomía, magnesita, margas o yeso; KUB 1953; ►ranker); *syn.* rendzina [f]; *f* **rendzine [f]** (Sol intrazonal calcimorphe à profil AC formé sur roche mère calcaire, avec un horizon A1 riche en matière organique et en carbonates, de couleur foncée et de structure grumeleuse ; DIS 1986 ; ►ranker) ; *g* **Rendzina [f]** (Dem ►Ranker 1 entsprechender terrestrischer Boden mit A-C-Profil aus Kalk-, Dolomit- oder Gipsgesteinen. Dem oft humus- und skelettreichen, krümeligen A-Horizont folgt unmittelbar der C-Horizont, in den ersterer häufig in Klüften lappig übergeht; SS 1979),

4979 renewable power source [n] *envir.* (Origin of renewable energy. In comparison to fossil power sources, renewable alternatives are: solar energy, ►wind energy, geothermal energy, tidal energy, biogas, plant oil, which can be exploited at a large scale in perpetuity); *syn.* sustainable power source [n], alternative power source [n]; *s* **fuente [f] de energía renovable** (Son aquéllas que se presentan en la naturaleza de un modo repetitivo y prácticamente inagotable. Se trata de la **e.** solar, la ►energía eólica, la **e.** geotérmica, la **e.** de las mareas, la **e.** térmica de los océanos, la **e.** de las olas, la **e.** térmica del subsuelo y la biomasa; cf. DINA 1987); *syn.* fuente [f] de energía sostenible, fuente [f] de energía alternativa; *f* **source [f] d'énergie renouvelable** (Terme désignant tous les phénomènes naturels réguliers ou constants provoqués par les astres [Soleil, Lune, Terre], à partir desquels il est possible de retirer de l'énergie qui se régénère ou se renouvelle en permanence et assez rapidement [en tenant compte de la vitesse à laquelle elle est consommée] pour être considérée comme inépuisable à l'échelle des temps. Par comparaison avec les énergies fossiles non renouvelables telles que le charbon, le gaz et l'énergie nucléaire, on distingue entre autres énergie renouvelables le vent [►énergie éolienne], l'énergie géothermique, les marées, le biogaz, les huiles végétales et l'énergie solaire dont le développement devrait permettre de limiter les émissions de gaz à effet de serre ; cf. décision du Conseil n° 93/500/CEE du 13 septembre 1993 concernant la promotion des énergies renouvelables [programme ALTENER]) ; *g* **erneuerbare Energiequelle [f]** (Im Vergleich zu fossilen Energieträgern wie Kohle, Erdgas und Kernenergie Quellen, aus denen im großen Rahmen und nach menschlichen Maßstäben zeitlich unbegrenzt Strom gewonnen werden kann wie z. B. Sonnenenergie, Wasserkraft, ►Windenergie, geothermische Wärme, Wellen- und Gezeitenkraft, Biomasse, Photovoltaik. Die Entwicklung **e.r E.n** kann zu einer spürbaren Verringerung der verunreinigenden Emissionen beitragen, die durch den Einsatz fossiler Brennstoffe entstehen und sie hilft, die Emissionen von Treibhausgasen zu reduzieren und der Gefahr der Erwärmung der Erdatmosphäre entgegenzuwirken; cf. Entscheidung 93/500/ EWG des Rates vom 13.09.1993 zur Förderung der erneuerbaren Energieträger in der Gemeinschaft [ALTENER-Programm]). Eine sehr detaillierte Schätzung über die europäischen Potenziale zur Erzeugung von „grünem" Strom einschließlich ihrer regionalen Verteilung durch das Deutsche Zentrum für Luft- und Raumfahrt im Auftrag des Bundesumweltministeriums mit Hilfe seines Geografischen Informationssystems 2007 ergab, dass das wirtschaftlich realisierbare Potenzial mit mehr als 5700 Terrawattstunden wesentlich höher als der heutige und zukünftige Verbrauch von 3700 Terrawattstunden ist [1 Terrawattstunde entspricht einer Milliarde KW/h; zz. fehlt eine auch nur ansatzweise vergleichbare gemeinsame Anstrengung der europäischen Staaten, um den Ausbau **e.r E.n** zu forcieren; cf. SZ 2008; in D. trat 2000 das Erneuerbare-Energien-Gesetz [EEG] in Kraft, das den Ausbau von Energieversorgungsanlagen, die aus regenerativen Quellen Strom erzeugen, vorantreiben soll); *syn.* alternativer Energieträger [m], alternative Energiequelle [f], erneuerbarer Energieträger [m], regenerative Energiequelle [f].

renewal [n] *arb. hort.* ►crown renewal; *plan. urb.* ►village renewal.

renewal [n]**, urban** *urb.* ►center city redevelopment [US]/ urban regeneration [UK].

4980 reno mattress [n] *eng. wat'man.* (**1.** Flat, wire mesh mat packed with loose rock material or gravel installed as a ►gabion to protect stream beds and to secure the base and the foot of the slope of a river bank at the mean water line; **r. m.s** are flexible double twisted woven wire mesh boxes with dimensions 6.0 x 2.0m in plan and are divided into cells by double diaphragms across the width of the unit; the mesh has a special coating protection [Zn-5% AL MM alloy] for durability; with more aggressive environments an additional polymer protective coating is applied; dense, rounded or quarried stone is normally used to fill **r. m.es**; the stone should be weather resistant, non-friable, insoluble and sufficiently hard. The most appropriate size is in the range of 80mm to 0.6 x depth); **2. vegetated gabion** [n] (Bio-engineering method used to stabilize slopes and slope toes where a lot of smaller stones are available. The **v. g.** is constructed with a fine wire mesh filled with coarse gravel or smaller stones, earth, live cuttings and container plants, which should protrude slightly beyond the face of the gabion, but not more than 10cm to prevent desiccation. The wire mesh is pulled together and sewn shut with wire; **v. g.s** are a fast and simple construction to secure wet slopes because they are elastic and improve drainage through plant transpiration; they are laced together and installed at the base of a bank to form a structural toe or sidewall. Vegetation is incorporated by placing live branches, which take root between each layer of rock filled baskets and in the soil behind the structures. **V. g.s** can also be covered with a pre-cultivated, completely rooted, reed mat held in place by wire; the gabion is covered with a coconut or a non-woven geotextile and the cavities of the ballast material are filled with a gravel-sand mixture to promote root growth; ►reed roll); *s 1* **gabión [m] de fondo (±)** (Malla o cesto de alambre plano —p. ej. de 5 x 5 m y 20-30 cm de altura— lleno de grava que se utiliza para proteger el lecho o el pie de taludes en cursos de agua; ►gabión); *s 2* **gabión [m] de carrizo** (*Bioingeniería* gabión de fondo que cierra por arriba con una estera de carrizo precultivada que se sujeta a la malla del gabión. Éste está revestido con un vellón o geotextil y los huecos entre la grava están rellenos de una mezcla de arena y gravilla para favorecer el crecimiento de las raíces; ►fajina de carrizo); *f 1* **gabion [m] matelas** (Structure métallique [►gabion] parallélépipédique de grande surface et de faible épaisseur, fabriquée en grillage métallique à maille hexagonale double torsion type 60 x 80 ou électrosoudée, préfabriquée, remplie de pierres ; les matelas sont compartimentés tous les mètres par des cloisons appelées diaphragmes et fixées à la base ; la fabrication est réalisée par pliage, en une seule nappe de grillage constituant le fond, les côtés et les diaphragmes ; le couvercle est, à priori, indépendant de la cage de base ; tous les bords des éléments grillagés sont renforcés par des fils de plus gros diamètres que ceux du grillage) ; *f 2* **matelas [m] de gabions végétalisés** (Génie biologique matelas utilisé en protection de berges de cours d'eau [►gabion], d'une épaisseur de 20 à 40 cm et d'une dimension de 5 x 2 ou 3 x 8 m ; les cellules au niveau du trait d'eau sont, lors du remplissage, végétalisées par des géonattes en fibre de coco prévégétalisées en aquapépinière et, afin d'assurer un bon développement racinaire, les compartiments empierrés sont remplis d'un mélange gravier-sable ; ►fascine d'hélophytes) ; *g 1* **Gabionenmatratze [f]** (Mit Schotter gefüllter flacher ►Drahtschotterbehälter [z. B. in den Abmessungen 5 x 2 m Höhe 20-30 cm] zur Sicherung der Sohle und des

R

Böschungsfußes im Bereich der Mittelwasserlinie der Fließgewässer); **g 2 Röhrichtgabione [f]** (*Ingenieurbiologie* Gabionenmatratze, die oben mit einer vorkultivierten, flächig durchgewurzelten Röhrichtmatte abschließt. Diese ist mit dem Drahtgeflecht des Schotterbehälters befestigt. Die Matratze ist mit einem Kokosvlies oder einem Geotextil ausgekleidet und die Hohlräume des Schottermaterials sind zur Förderung des Wurzelwachstums mit einem Kies-Sand-Gemisch eingeschlämmt; ▶Röhrichtwalze).

renovation [n], upkeep and *conserv'hist. constr.* ▶maintenance and repair work.

4981 rented garden [n] [UK] *urb.* (Garden leased by the owner to an occupant; no U.S. equivalent; ▶community garden [US]/allotment garden [plot] [UK], ▶tenant garden, ▶leasehold garden); **s jardín [m] alquilado** (Jardín privado que está alquilado a terceros. Al contrario que el ▶jardín de inquilinos no se encuentra directamente junto a la vivienda; ▶huerto recreativo urbano, ▶jardín arrendado, ▶jardín comunitario); *syn.* jardín [m] en alquiler; *f* **jardin [m] loué (±)** (Jardin mis à la disposition d'un locataire dans le cadre d'un bail de location, non attenant au bâtiment mais situé à proximité du lieu d'habitation ; ▶jardin affermé, ▶jardin collectif, ▶jardin locatif, ▶jardin ouvrier) ; *syn.* jardin [m] à jouissance exclusive ; **g Mietgarten [m]** (Garten, den der Vermieter dem Mieter durch Vertrag zum Gebrauch überlässt. Im Vergleich zum ▶Mietergarten ist der **M.** dem Wohngebäude nicht direkt zugeordnet, sondern liegt im wohnungsnahen Umfeld; ▶Kleingarten, ▶Pachtgarten).

4982 reoccurrence [n] of a species *phyt. zool.* (Reappearance of a plant or animal species which was deemed to have become distinct or had not populated a certain area for a long period of time; ▶extinct species, ▶recolonization); **s reaparición [f] de una especie** (Resurgimiento de una especie de flora o fauna que había desaparecido o que no había sido localizada durante mucho tiempo; ▶especie extinta, ▶recolonización natural); *f* **résurgence [f] d'une espèce** (Réapparition d'une espèce animale ou végétale considérée comme disparue [▶espèce éteinte] ou n'étant plus apparue pendant une longue période dans une station donnée ; ▶espèce disparue, ▶recolonisation naturelle) ; **g Wiederauftauchen [n, o. Pl.] einer Art** (Erscheinen einer Pflanzen- oder Tierart, die als verschollen gilt oder lange Zeit in einem bestimmten Gebiet nicht mehr vorkam; ▶verschollene Art, ▶Wiederbesiedelung).

4983 re-organization of a road [n] *trans. urb.* (LD 1988 [9], 59; narrowing or readjustment of traffic lanes to permit planting as well as for widening pedestrian pavement, or for creation of a mall [US]/pedestrian precinct [UK] after ▶street abandonment [US]/extinguishment of rights of way [UK]; ▶right-of-way planting [US]/roadside planting [UK], ▶road redesign, ▶traffic calming); **s recalificación [f] de calles** (Conjunto de medidas como el estrechamiento de la calzada, creación de carriles de bicis y la plantación de árboles y arbustos para mejorar la estética y la funcionalidad de las calles en las que se prioriza el uso peatonal y de bicicletas; ▶apaciguamiento del tráfico [Es]/apaciguamiento del tránsito [AL], ▶cambio de clasificación de calle, ▶remodelación de calles, ▶verde vial); *f* **recalibrage [m] d'une rue** (Réduction de la largeur d'une chaussée au profit des ▶espaces verts d'accompagnement de la voirie ou de l'élargissement des trottoirs ou suite à la création d'une zone à circulation restreinte après le ▶déclassement d'une rue ; ▶réaménagement d'une rue, ▶circulation apaisée) ; **g Straßenrückbau [m]** (...bauten [pl]; Verschmälerung oder bauliche Beseitigung von Fahrbahnstreifen zugunsten von ▶Verkehrsgrün oder breiterer Bürgersteige oder Schaffung von verkehrsberuhigten Bereichen nach ▶Einziehung von Straßen;

▶Straßenumbau, ▶Verkehrsberuhigung); *syn.* Rückbau [m] von Straßen.

4984 re-organisation [n] of land holdings [UK] *agr. leg. urb.* (Rearrangement of property boundaries; **in U.S.,** done only by owners or developers of private property; this is accomplished through mutual agreement of abutting landowners, and when they do not agree through a judicial proceeding known as a "suit to quiet title"; ▶rearrangement of lots [US]/re-organisation of plot boundaries [UK], ▶farmland consolidation [US]/land consolidation [UK]); **s ordenación [f] del suelo** (BEC 1999; nueva ordenación de los límites de propiedades edificadas o no por medio de ▶reordenación de parcelas con el fin de que las parcelas resultantes sean más adecuadas para los fines para los cuales están previstas. Puede realizarse tanto en zonas urbanas como en el campo en el marco de la ▶concentración parcelaria. La adjudicación de las parcelas resultantes a los propietarios debe realizarse en proporción a sus respectivos derechos. La **o. del s.** dentro de los límites de un plan parcial o una unidad de actuación se denomina **reparcelación [f]** y tiene por objeto distribuir justamente los beneficios y cargas de la ordenación urbanística, regularizar la configuración de las fincas y situar su aprovechamiento en zonas aptas para la edificación con arreglo al Plan; art. 97, Texto refundido de la ley sobre régimen del suelo y ordenación urbana, RD 1346/1976); *syn.* regulación [f] del suelo; *f* **réorganisation [f] foncière** (Remembrement, redistribution parcellaire, groupement de parcelles ou modification des limites de propriété réalisés, p. ex. par les associations foncières urbaines afin d'obtenir une meilleure utilisation du sol sur des fonds bâtis ou non dans le cadre d'opérations de construction ou d'ouvrages d'intérêt collectif ; ▶remembrement parcellaire, ▶remembrement rural) ; *syn.* opérations [fpl] de restructuration foncière ; **g Bodenordnung [f]** (OB zu ▶Umlegung und ▶Flurbereinigung; **B.** ist eine hoheitliche Maßnahme einer Gebietskörperschaft, um Grund und Boden zum Zwecke einer bestimmten Nutzung neu zu ordnen; **Umlegung** [§§ 45-79 BauGB] oder vereinfachte Umlegung [§§ 80-84 BauGB] ist die erwirkte Neuordnung bebauter oder unbebauter Grundstücke in einem bestimmten Gebiet mit dem Ziel, dass nach Lage, Form und Größe für die bauliche oder sonstige Nutzung zweckmäßig gestaltete Grundstücke entstehen; **Flurbereinigung** ist das Instrument ganzheitlicher Entwicklung im ländlichen Raum durch Koordination von Maßnahmen zur Erreichung agrar-, umwelt- und raumordnerischer Ziele).

re-organisation [n] of plot boundaries [UK] *leg. urb.* ▶rearrangement of lots [US].

4985 repair [n] *constr. urb.* (Measures which have become necessary when maintenance has been neglected for a long period of time to restore structures and urban complexes to good functional condition and desired appearance; ▶building maintenance, ▶building rehabilitation, ▶modernization 1); **s reparación [f] de edificios** (Obras necesarias en aquellos edificios en los que no se ha hecho ▶mantenimiento de edificios durante mucho tiempo y prácticamente no se pueden utilizar; ▶modernización, ▶rehabilitación de edificios); *syn.* obras [fpl] de reparación de edificios; *f* **remise [f] en état** (Travaux de réparation entrepris lorsque l'▶entretien immobilier a été négligé pendant une période prolongée et nécessite la remise aux normes des conditions d'habitabilité d'un ensemble d'immeubles ; ▶modernisation, ▶réfection, ▶réhabilitation, ▶rénovation immobilière) ; *syn.* réfection [f] ; **g Instandsetzung [f]** (Alle Maßnahmen an Bauwerken, die erforderlich sind, wenn die ▶Instandhaltung von Gebäuden längere Zeit vernachlässigt wurde und der bestimmungsgemäße Gebrauch wieder hergestellt werden soll; ▶Modernisierung, ▶Sanierung); *syn.* Instandsetzen [n, o. Pl.], Reparatur [f].

repair planting [n] [US] *for.* ▶replacement planting.

R

4986 repair work [n] (1) *constr. contr.* (Restore damaged or worn equipment/tools to good working order; ►remedying defects); *s* **obras [fpl] de reparación [f]** (►corrección de defectos [en obra]); *syn.* reparación [f]; *f* **travaux [mpl] de réfection** (Correction des imperfections, travaux de finition en vue d'assurer le parfait achèvement d'un chantier ; ►réfection des imperfections ou malfaçons) ; *syn.* travaux [mpl] de réparation ; *g* **Ausbesserungsarbeiten [fpl]** (Arbeiten, die Schadhaftes reparieren, wiederherstellen, in Stand setzen oder Arbeiten, die dazu dienen, einen abnahmefähigen Zustand herzustellen; ►Mängelbeseitigung).

repair work [n] (2) *conserv'hist. constr.* ►maintenance and repair work.

repeat flowering rose [n] *hort.* ►remontant rose.

replacement biotope [n] *ecol. leg.* ►substitute habitat.

4987 replacement costs [npl] **of habitats/ecosystems** *adm. plan. ecol.* (Financial resources which are necessary for re-creating degraded or destroyed habitats, ecosystems or parts thereof); *syn.* reinstatement costs [npl] of habitats/ecosystems, restoration costs [npl] of habitats/ecosystems; *s* **costos [mpl] de restauración/regeneración (de biótopos o ecosistemas)** (Medios económicos necesarios para restaurar biótopos, paisajes o ecosistemas degradados hasta alcanzar su capacidad funcional); *syn.* coste [m] de restauracion/regeneración (de biótopos o ecosistemas) [Es]; *f* **coûts [mpl] de restauration de biotopes/d'écosystèmes** (Moyens financiers nécessaires à la reconstitution de biotopes dégradés ou détruits afin qu'ils retrouvent leur résilience) ; *g* **Wiederherstellungskosten [pl] von Biotopen/Ökosystemen** (Finanzmittel, die aufgebracht werden müssen, um einen degradierten oder zerstörten Biotop so zu gestalten, dass er seine frühere Funktionsfähigkeit zurückerhält).

4988 replacement guarantee [n] *constr. contr.* (Undertaking to replace plant material which has failed to survive within the maintenance period; ►guarantee, ►project development maintenance [US]/pre-handover maintenance [UK]); *s* **garantía [f] de reposición de marras** (Obligación del contratista de cuidar de las plantaciones de la obra hasta que expire el plazo de ►garantía 1 incluyendo la reposición de la vegetación que no haya arraigado por causas imputables al mismo; ►trabajos de mantenimiento pre-entrega) ; *f 1* **garantie [f] de reprise des végétaux** (Responsabilité pour l'entrepreneur de jardins d'effectuer les travaux d'entretien pour les végétaux plantés et de remplacer à ses frais tous les sujets défectueux et plantés par ses soins ; ►entretien [d'espaces verts] jusqu'à la réception des travaux, ►garanties, ►garantie de parfait achèvement) ; *f 2* **garantie [f] de bonne levée** (Responsabilité pour l'entrepreneur de jardins d'effectuer les travaux d'entretien des gazons semés par ses soins et de réfectionner toutes parties malvenues) ; *g* **Anwachsgarantie [f]** (Verpflichtung des Auftragnehmers während der Garantiezeit den neu gepflanzten Vegetationsbestand zu pflegen und bei Ausfall zu ersetzen; ►Gewährleistung, ►Fertigstellungspflege); *syn.* Anwuchsgarantie [f].

replacement habitat [n] *ecol. leg.* ►substitute habitat.

replacement [n] **of a habitat** *landsc. conserv.* ►habitat restoration.

4989 replacement [n] **of sods** [US] *constr.* (Relaying of sods [US]/turves [UK]); *syn.* replacement [n] of turves [UK], restore [vb] [also US]; *s* **recolocación [f] de tepes** (Extensión de los tepes almacenados provisionalmente después de terminar la obra); *f* **remise [f] en place des plaques de gazon** (Réutilisation des plaques ou des rouleaux de gazon) ; *g* **Wiederandecken [n, o. Pl.] von Rasensoden** (Einbau von zwischengelagerten Rasensoden); *syn.* Wiedereinbau [m] von Rasensoden.

replacement [n] **of turves** [UK] *constr.* ►replacement of sods [US].

4990 replacement planting [n] *constr.* (1. Replanting of or within areas where planting has failed to become properly established or has been destroyed. 2. Replacement of failures in a newly planted site, normally done at yearly intervals after planting; BS 3975: part 5); *syn.* beating [n] [UK], beating up [n] [UK], gapping [n] [UK], *for.* filling [n] fail places [US] (SAF 1983), infill planting [n] [US], *for.* repair planting [n] [US], replantation [n], re-setting [n] [UK]; *s 1* **replantación [f]** (Nueva plantación en el primer o segundo año después de la plantación para sustituir marras); *syn.* reposición [f] de marras (2); *s 2* **plantación [f] de reemplazo de marras** (Replantación en áreas de obra donde las plantas no arraigaron o donde fueron destruidas); *f 1* **remplacement [m] des végétaux** (Substitution des végétaux dans le cadre de la garantie de reprise pour des plantations nouvelles pendant la saison suivant la plantation) ; *f 2* **plantation [f] de remplacement** (Végétaux replantés en remplacement des végétaux morts, détruits ou dont la reprise n'est pas satisfaisante) ; *syn.* végétaux [mpl] de remplacement ; *f 3* **boisement [m] compensateur** (Reconstitution d'une surface boisée équivalente au défrichement exécuté par un propriétaire privé ou une collectivité publique lors de la réalisation de constructions ou d'équipements) ; *syn.* plantation [f] compensatrice ; *g* **Ersatzpflanzung [f]** (1. Ersatzpflanzung von Fehlstellen im ersten oder zweiten Jahr nach der Pflanzung. 2. Neupflanzung von Pflanzen, die nicht angewachsen sind oder zerstört wurden. 3. Wiederherstellung einer Waldfläche an anderer Stelle als Ersatzmaßnahme für eine Fläche, die wegen eines Bauprojektes abgeholzt wurde); *syn. zu 1. und 2.* Nachbesserung [f] von Pflanzarbeiten, Nachpflanzung [f].

4991 replanning [n] *plan.* (New planning measures, e.g. for an existing facility, existing building, or existing area; ►partial redesign); *s* **replanificación [f]** (Nueva planificación de una zona, instalación o edificio; ►reconfiguración parcial); *f* **refonte [f] d'un projet** (En terme de planification, redéfinition du parti général, d'un programme d'aménagement pour un ouvrage de bâtiment, d'infrastructure, un espace public ou privé ; ►recomposition, ►remodelage) ; *g* **Umplanung [f]** (Planerische Neuüberlegung für ein Projekt, z. B. bei einer vorhandenen Grünanlage, einem vorhandenen Gebäude oder einer bereits bestehenden Planung; ►Umgestaltung).

replantation [n] *constr.* ►replacement planting.

4992 replenishment [n] **of sand areas** *constr.* (Addition of sand to sand boxes or sand play areas); *syn.* topping-up [n] of sand areas [also UK]; *s* **relleno [m] de superficies de arena** (Reemplazo de arena en los cajones o en las superficies de arena en áreas de juegos infantiles); *f* **regarnissage [m] des aires de sable** (Complément en sable effectué sur les bacs à sable ou les aires sablées des aires de jeux) ; *g* **Nachfüllen [n, o. Pl] von Sandflächen** (Auftragen von Sand in Sandkästen oder Sandmulden auf Kinderspielplätzen).

report [n] *plan.* ►explanatory report, ►final investigative report, ►final planning report, ►regional plan report; *constr. plan.* ►progress report; *plan.* ►provisional report; *prof.* ►submission of a report.

report [n]**, interim** *plan.* ►provisional report.

report [n] [UK]**, site** *constr.* ►site log.

report [n] [US]**, stopgap** *plan.* ►provisional report.

report [n] [UK]**, submissions** *adm. constr. contr.* ►minutes of bid opening [US].

report [n]**, submittal of a** *prof.* ►submission of a report.

report [n] **of results** *plan.* ►final investigative report.

report [n] **on landscape planning** *landsc.* ▶consultant's report on landscape planning.

report [n] **on noise** *plan.* ▶consultant's report on noise.

representation [n] *plan.* ▶graphic representation.

representative [n] [US]**, public agency** *constr. prof.* ▶site supervisor representing an authority [US].

representative plant [n] *plant.* ▶characteristic plant.

4993 representative poll [n] *sociol.* (Statistical survey to compile data acquired from persons selected within the total population of a sampled area; ▶poll); *syn.* opinion survey [n]; *s* **encuesta** [f] **representativa** (▶encuesta); *syn.* sondeo [m] de opinión representativo; *f* **sondage** [m] **d'opinion** (▶Enquête statistique visant à recueillir des informations sur une population donnée) ; *g* **Repräsentativbefragung** [f] (Statistische Teilerhebung von Daten ausgewählter Personen der Gesamtbevölkerung eines Untersuchungsgebietes; ▶Umfrage); *syn.* Repräsentativerhebung [f].

4994 reprocessing [n] *envir.* (Separation of re-usable fuel elements at a nuclear power plant before ▶final disposal; ▶recycling); *s* **reproceso** [m] **de material nuclear** (Separación de material reutilizable y del destinado al ▶almacenamiento definitivo de residuos radi[o]activos que ha sido utilizado en una central nuclear; ▶planta de tratamiento, ▶reciclado); *syn.* reprocesamiento [m] de material nuclear; *f* **retraitement** [m] (Opération de sélection des combustibles usés issus d'une installation nucléaire en vue du ▶recyclage dans la production de combustible neuf et du ▶stockage définitif des résidus radioactifs de traitement ; ▶installation de traitement) ; *g* **Wiederaufbereitung** [f] (Trennung von wieder nutzbaren und zur ▶Endlagerung bestimmten Bestandteilen von abgebrannten Brennelementen einer kerntechnischen Anlage; ▶Recycling, ▶Aufbereitungsanlage).

reprocessing plant [n]**, fuel** *envir.* ▶nuclear fuel reprocessing plant.

4995 reproduction [n] *hort.* (Sexual multiplication of plants by seeds or vegetative asexual method such as budding, cutting, division, grafting, root cuttings, etc.; *specific terms* sexual reproduction, vegetative reproduction); *syn.* multiplication [n], propagation [n]; *s* **multiplicación** [f] (Regeneración por medio de semillas o vegetativamente por esquejes, rizomas, injertos, etc.); *syn.* reproducción [f], propagación [f]; *f 1* **reproduction** [f] (Processus de propagation des individus d'une espèce dans les règnes végétal et animal par voie sexuée) ; *f 2* **multiplication** [f] **végétative** (Processus de propagation des individus d'une espèce principalement dans le règne végétal, par lequel différentes parties de plantes [stolons, rhizomes, tubercules] peuvent naturellement donner naissance à une plante complète, la multiplication végétative peut être réalisée artificiellement par greffage, bouturage ou marcottage) ; *g* **Vermehrung** [f] (Generative Fortpflanzung durch Samen oder vegetative Fortpflanzung durch Stecklinge, Rhizome, Teilung, Veredlung, Wurzelschnittlinge etc.).

reproduction [n]**, natural** *for. phyt.* ▶natural regeneration (1), ▶natural regeneration (2).

reproduction cutting [n] [UK] *for.* ▶regeneration cutting [US].

4996 request [n] **for qualification** [US] *prof.* (**1.** *Abbr.* RFQ; public announcement requesting submittal by planning and design firms of a record of expertise and experience of each member of the firm for prequalification to design a specific project; ▶design competition 1, ▶public prequalification bidding notice [US]/public prequalification tendering notice [UK]); **2. request** [n] **for proposals** [US] (*Abbr.* RFP; public announcement requesting planning and design firms to submit a technical proposal in a document describing their understanding of the project, the methodology they would use, and the project scheduling, with or without cost—'cost proposal'—and, if not prequalified, a record of expertise and experience of each member of the firm. Several of the firms submitting proposals are then selected for interview and one of them is awarded the contract; ▶aviso de candidaturas a concurso público, ▶concurso de ideas); *s* **demanda** [f] **de proposiciones** (En EE.UU., anuncio público dirigido a gabinetes de paisajismo o planificación de gran experiencia para que se presenten a concurso para determinado proyecto; ▶concurso de ideas); *f* **concours** [m] **restreint** (Concours sur appel public de candidature, réservé à un nombre limité de concurrents sélectionnés, lors d'une deuxième phase par un jury d'après les critères de sélection, compétences, références et moyens des candidats ; ▶appel public de candidatures pour un appel d'offres, ▶concours d'idée) ; *g* **öffentlicher Teilnahmewettbewerb** [m] **für Planungsbüros** (**1. In USA** ist **request for qualification** eine öffentliche Bekanntmachung, in der Planungsbüros für bestimmte Projekte aufgefordert werden, sich mit Nachweis über ihre Berufserfahrung und über die Qualifikation ihrer einzelnen Mitarbeitenden für bestimmte Projekte zu melden. **2. Request for proposals** [US] ist eine öffentliche Aufforderung für Planungsbüros, einen technischen Lösungsvorschlag für eine Planungsaufgabe abzugeben, in dem sie ihr Verständnis über das Projekt darlegen, die Herangehensweise zur Lösung der Aufgabe darstellen, einen Terminplan und eine oder keine Kostenschätzung vorlegen. Wenn die Büros nicht schon durch ein Auswahlverfahren bekannt sind, sind Nachweise zur Berufserfahrung und über die Qualifikation ihrer einzelnen Mitarbeitenden zu erbringen. Nach einer Vorstellungsrunde der eingeladenen Büros wird der Planungsauftrag dann vergeben; ▶Ideenwettbewerb, ▶öffentlicher Teilnahmewettbewerb).

4997 required side yard [n] [US] *leg. urb.* (Space extending from the front yard to the rear yard between the closest point of the principal building and the side lot line; no structure shall be located in this area except as provided in the zoning ordinance; cf. IBDD 1981; no UK equivalent; ▶building setbacks, ▶minimum building spacing, ▶setback line); *s* **androna** [f] (Espaciamiento obligatorio entre el edificio y el límite del predio; ▶distancia legal de construcción, ▶línea de retranqueo, ▶retranqueo legal entre edificios); *f* **recul** [m] **de constructibilité par rapport aux limites séparatives** (≠) (U.S., D., recul de la ▶limite de constructibilité d'une parcelle par rapport à la limite parcellaire la plus proche ; ▶espacement minimal entre les constructions, ▶polygone d'implantation ; aucun équivalent en F. mais disposition se rapprochant de celles prévues dans le règlement du PLU/POS : implantations des constructions par rapport aux limites séparatives ; ▶espace des constructions, ▶gabarit de prospect, ▶marge de reculement, ▶prospect) ; *syn.* recul [m] latéral [B et L] (de tout bâtiment par rapport aux limites de propriété interne ou externe à la zone) ; *g* **Bauwich** [m] (Abstand zwischen bebaubarem Grundstücksteil und seitlicher Nachbargrenze; der **B.** ist grundsätzlich von baulichen Anlagen freizuhalten; cf. § 22 [2] BauNVO; ▶Abstandsflächen, ▶Baugrenze, ▶Mindestabstand von Gebäuden); *syn.* seitlicher Grenzabstand [m].

required yard [n] [US] *leg. urb.* ▶building setbacks.

requirements [npl] **for bidders** [US] *contr.* ▶bidding requirements [US]/requirements for tenderers [UK].

4998 requirements [npl] **for green spaces** *plan. pol. recr. urb.* (Areas necessary for the urban population, with consideration of ecological values, and the physical and mental wellbeing of all age groups; ▶requirements for open spaces); *s* **demanda** [f] **de zonas verdes** (Cantidad de espacios verdes necesarios para cubrir las necesidades de esparcimiento y cumplir

R

las funciones ecológicas; ▶demanda de espacios libres); *f* **besoins [mpl] en espaces verts** (Espaces nécessaires à la population urbaine compte tenu de la valeur écologique et des besoins manifestés par les différentes catégories d'âges indispensables à la santé physique et morale des habitants ; ▶besoin en espaces libres) ; *g* **Grünflächenbedarf [m]** (Erfordernis an vegetationsbestimmten Freiflächen für eine ausreichende Grünflächenversorgung aller Altersgruppen in einem Siedlungsgebiet; ▶Freiraumbedarf); *syn.* Bedarf [m] an Grünflächen.

4999 requirements [npl] **for open spaces** *plan. recr.* (Size of the area necessary for the adequate provision of paved or unpaved open spaces; ▶guideline value, ▶provision of open spaces, ▶requirements for green spaces); *syn.* open space needs [npl]; *s* **demanda [f] de espacios libres** (Superficie de espacios abiertos necesaria para cubrir adecuadamente las necesidades de esparcimiento y actividades al aire libre de la población de un área; ▶demanda de zonas verdes, ▶dotación de espacios libres por habitante, ▶valor estándar); *f* **besoins [mpl] en espaces libres** (permettant de définir les espaces nécessaires à certaines catégories d'usagers ; ▶besoins en espaces verts, ▶dotation en espaces libres, ▶valeur guide) ; *syn.* besoins [mpl] en espaces verts urbains ; *g* **Freiraumbedarf [m]** (Erfordernis an Freiflächen, die für eine ausreichende ▶Freiraumversorgung benötigt werden; ▶Grünflächenbedarf, ▶Richtwert); *syn.* Freiflächenbedarf [m], Bedarf [m] an Freiflächen.

5000 reradiation [n] *met. phys.* (TGG 1984, 76; reflected radiation from prior absorption of sun rays or heat); *s* **rerradiación [f] (solar/de calor)**; *syn.* reflexión [f] de radiación; *f* **radiation [f] réfléchie** (*Terme spécifique* radiation solaire réfléchie) ; *syn.* rayonnement [m] réfléchi ; *g* **Rückstrahlung [f]** (der Sonnenstrahlen, Wärme).

5001 reradiation [n] **of heat** *met.* (cf. TGG 1984, 247; heat reflection/reflexion [also UK] from buildings or hard paved areas, e.g. towards the undersides of a tree's leaves; ▶radiant heat); *s* **rerradiación [f] del calor** (Reflexión del calor de los edificios o de superficies pavimentadas; ▶radiación térmica); *syn.* rerradiación [f] térmica, reflexión [f] del calor, reflexión [f] térmica; *f* **rayonnement [m] thermique des surfaces** (Réflexion de la chaleur par les bâtiments ou les surfaces à revêtement stabilisé ; ▶radiation thermique ; DUV 1984, 56) ; *syn.* radiation [f] thermique des surfaces, radiation [f] thermique de la terre (DUV 1984, 56) ; *g* **Wärmerückstrahlung [f]** (Reflexion der Wärme von Gebäuden oder von befestigten Flächen; ▶Wärmestrahlung).

rerouting [n] *urb.* ▶traffic rerouting [US]/traffic redirection [UK].

resaturation [n] **(of a bog) with water** *conserv.* ▶rewatering of a bog.

rescheduling [n] [UK] *leg. plan.* ▶rezoning [US].

research [n] *envir.* ▶causality research, *ecol. sociol.* ▶environmental research; *plan. pol.* ▶regional research.

research [n] [UK], **background** *contr.* ▶background work [US].

research [n], **field** *ecol. geo. pedol.* ▶field investigation.

research [n], **geographical** *geo.* ▶landscape geographical research.

research [n], **large-scale planning** *plan. pol.* ▶regional research.

research [n] **and knowledge** [n] *geo.* ▶regional research and knowledge.

5002 research area [n] *ecol. pedol.* (Area or subject of scientific research or investigated for a planning project; ▶planning area); *syn.* investigation area [n]; *s* **área [f] de investigación** (Zona que es investigada en el contexto de un planteamiento científico o con fines de planificación; ▶área de planificación); *syn.* zona [f] de investigación; *f* **zone [f] d'étude (1)** (Territoire sur lequel sont effectués des travaux de recherche scientifiques ou des études dans un périmètre limité pour un projet d'aménagement, p. ex. inventaire mené sur un site afin d'identifier ou de caractériser des habitats naturels [relevés phytosociologiques] ; ▶aire d'aménagement) ; *syn.* site [m] d'étude, aire [f] d'investigations, périmètre [m] d'études ; *g* **Untersuchungsgebiet [n]** (Gebiet, das für wissenschaftliche oder planerische Fragestellungen untersucht wird; ▶Planungsraum).

research area [n] [US], **natural** *conserv.* ▶natural area laboratory.

reseeding [n] [US]/**re-seeding** [n] [UK] *constr.* ▶lawn reseeding [US]/lawn re-seeding [UK]; ▶over-seeding [US].

reservation [n] [UK], **motorway central** *trans.* ▶freeway median strip [US].

5003 reservation [n] **of land** *plan.* (Temporary setting aside of public or private land for acquisition or regulation to prevent development on coastlines, hilltops, in urban open spaces, etc.); *s* **reserva [m] de terreno no edificable** (En plan de ordenación territorial o urbana o en plan parcial, superficie que se delimita como libre de construcciones, como p. ej. franja costera, colinas, zonas verdes urbanas, etc); *f* **institution [f] d'une zone non aedificandi** (±) (Mesure consistant à maîtriser le développement urbain en créant des zones inconstructibles, p. ex. sur l'espace littoral, les rives de plans d'eau, les zones de bruit autour des aérodromes, des voies, les zones de risques, les espaces verts en zone urbaine, etc.) ; *g* **Freihaltung [f, o. Pl.] von Bebauung** (Bauleitplanerische Verhinderung von Bauabsichten, z. B. an Ufern, Küsten, auf Bergkuppen, in innerstädtischen Freiräumen).

5004 reserve [n] [US] **(1)** *for.* (Any tree or in U.S. a group of trees, left unfelled in a stand that is being regenerated and kept for part or the whole of the next rotation; SAF 1983; ▶seed bearer); *syn.* standard [n] (SAF 1983), reserved tree [n] [UK]; *s* **árbol [m] padre (2)** (Cualquier árbol productor de semilla y, más concretamente, el que se deja en pie, al cortar el monte a matarrasa, para que lo repueble; DB 1985; ▶portagrano); *f* **réserve [f]** (Arbre choisi pour être maintenu sur pied après que le reste du peuplement aura été abattu. Son maintien au-dessus d'un nouveau peuplement est pour l'abri et l'ensemencement ; cf. DFM 1975 ; ▶semencier) ; *g* **Überhälter [m] (2)** (Ausgesuchter, einzelner, alter Baum oder eine Gruppe besonders geeigneter Bäume, die beim Fällen eines Bestandes in die nächste Generation mit dem Ziel übernommen werden, eine zusätzliche natürliche Verjüngung zu ermöglichen oder als Sonneneinstrahlungsschutz zu dienen; ▶Samenträger).

reserve [n] **(2)** *conserv.* ▶biogenetic reserve; *conserv. ecol. leg.* ▶bioreserve [US]; *conserv. landsc.* ▶coastal reserve [US]/heritage coast [UK]; *conserv. pol.* ▶European Wilderness Reserve; *conserv. game'man. hunt. leg.* ▶game reserve; *conserv. land' man.* ▶nature reserve; ▶strict nature reserve.

reserve [n], **core** *conserv.* ▶core area (2).

reserve [n], **forest** *for. leg.* ▶reserved forest.

reserve [n] [UK], **forest nature** *conserv. leg. nat'res.* ▶forest research natural area [US].

reserve [n] [UK], **natural forest** *conserv. leg. nat'res.* ▶forest research natural area [US].

reserve [n] [UK], **nature** *conserv. leg.* ▶natural area preserve [US]; ▶strict nature reserve.

reserve [n], **scientific** *conserv.* ▶strict nature reserve.

R

reserve [n] [UK]**, shooting** *conserv. hunt.* ▶shooting preserve [US].

reserve [n] [US]**, strict wilderness** *conserv. leg. nat'res.* ▶wilderness area [US].

reserve [n] [UK]**, wildlife** *conserv. leg.* ▶wildlife sanctuary [US].

5005 reserved forest [n] *for. leg.* (An area so designated and constituted under a forest act or ordinance as to give it the necessary legal protection; SAF 1983; ▶forest research natural area [US]/forest nature reserve [UK]); *syn.* (forest) reserve [n], Demarcated Forest [n] [ZA], Permanent State Forest [n] [NZ]; *s* **monte** [m] **catalogado** (Área de bosque registrada en el catálogo de los Montes de Utilidad Pública; cf. art. 13, Ley 43/2003, de Montes; ▶reserva forestal integral); *f* **forêt** [f] **classée** (Zone forestière définie et délimitée comme telle, conformément à un texte législatif ou réglementaire, de façon à lui donner la protection légale nécessaire ; DFM 1975 ; ▶réserve biologique domaniale, ▶réserve biologique forestière) ; *g* **Bannwald** [m] **(3)** (Rechtsbegriff nach Wald- und Forstgesetzen der Bundesländer unterschiedlich definiert. Grundsätzlich ist die Flächenausdehnung des Waldes mit seinen gebietstypischen Waldgesellschaften incl. seiner Tier- und Pflanzenarten zu erhalten und eine Umwandlung in eine andere Nutzungsart untersagt. Ziel ist der Schutz der Lebensräume und -gemeinschaften, die sich in dem betreffenden Wald befinden, sich im Verlauf der eigendynamischen Entwicklung ändern oder durch eine eigendynamische Entwicklung entstehen; ▶Naturwaldreservat).

reserved tree [n] [UK] *for.* ▶reserve [US] (1).

reserve [n] **of nutrients** *pedol. chem.* ▶nutrient pool.

reservoir [n] *wat'man.* ▶flood control reservoir; *hydr.* ▶groundwater reservoir.

reservoir [n]**, detention** *wat'man.* ▶flood control reservoir.

reservoir [n] [UK]**, impounding** *wat'man.* ▶impoundment.

re-setting [n] [UK] *constr.* ▶replacement planting.

5006 resettlement [n] **(1)** *plan. urb.* (Moving of people into a new area; ▶farmstead resettlement [US]/resited farm holding [UK]); *s* **nueva colonización** [f] (Migración de población a un área no o poco habitada hasta la fecha; ▶reasentamiento de granjas); *f* **colonisation** [f] (Conquête territoriale d'une population dans une région nouvelle ; ▶transfert d'exploitations agricoles) ; *g* **Ansiedelung** [f] **(2)** (Das Sesshaftmachen von Leuten in einem Gebiet; ansiedeln [vb]; ▶Aussiedlung); *syn. o. V.* Ansiedlung [f].

5007 resettlement [n] **(2)** *zool.* (**1.** Transport of animal species, e.g. frogs, fish, ants, to a newly created ▶habitat or to a place no longer inhabited by these species—undertaken usually to increase species density of these taxa within the existing area; **2.** *phyt.* ▶salvaging of plants); *s* **reubicación** [f] **de especies** (**1.** Traslado de animales salvajes, p. ej. ranas, peces, hormigas, a áreas resp. ▶hábitats ya no habitados por ellos para conseguir una densidad de población adecuada en toda el área de distribución; **2.** *phyt.* ▶transplante de especies vegetales); *f* **transplantation** [f] **(2)** (**1.** Déplacement d'espèces de faune de leur ▶habitat naturel dans des régions où elles sont en vue de la restauration de populations clairsemées ou menacées ou de la réintroduction d'espèces en voie d'extinction ou disparues ; cf. LR 1983, 7 et s. ; **2.** *phyt.* ▶déplacement de végétaux) ; *g* **Umsiedlung** [f] (**1.** Verfrachtung von Tierarten, z. B. Fröschen, Fischen, Ameisen etc., in neu geschaffene resp. durch diese Arten nicht mehr bewohnte ▶Habitate, meist zwecks einer angemessenen Verdichtung der Vorkommensbestände einer Tierart innerhalb des vorhandenen Areals; **2.** *phyt.* ▶Umsetzen von Pflanzen[arten]); syn. *o. V.* Umsiedlung [f].

reshaping [n] *gard. urb.* ▶redesign (1).

reshaping [n]**, ground** *constr.* ▶ground restoration.

residence building [n] *arch. urb.* ▶residential building.

resident [n] *leg. urb.* ▶abutting resident on a street, ▶adjoining resident, ▶adjoining street resident.

resident bird [n] [US] *zool.* ▶permanent resident.

resident engineer [n] *constr. prof.* ▶design firm's representative [US]/supervisor representing a design practise [UK].

5008 residential area [n] *urb.* (Subdivision which is mainly designated for residential buildings on a ▶zoning map [US]/Proposals Map/Site Allocations Development Plan Document [UK]; ▶low-density residential area, ▶residential community [US], ▶residential zoning district [US]/residential zone [UK]); *syn.* residential subdivision [n] [also US] (HORT 1989 [1], 16); *s* **zona** [f] **residencial** (Término genérico que denomina a la zona de la ciudad que se utiliza primordialmente para la vivienda. **D.,** legalmente se diferencia entre los tipos de ▶suelo residencial 1 y ▶suelo residencial 2 y suelo residencial especial según los usos que están permitido en ellos; ▶barrio residencial ajardinado); *f 1* **zone** [f] **d'habitation** (Zone réservée principalement à la fonction d'habitation ; en matière d'ensembles d'habitations, les urbanistes distinguent le quartier qui groupe 2500 à 4000 logements, l'unité de voisinage qui comporte 800 à 1200 logements et le groupe résidentiel comprenant quelques centaines de logements ; BON 1990, 358) ; *syn.* zone [f] résidentielle [aussi CH] ; *f 2* **zone** [f] **urbaine résidentielle** (*Plan de zonage PLU/POS* zone urbaine mixte réservée à l'habitat résidentiel de moyenne à faible densité et comportant les activités d'accompagnement nécessaires à la vie quotidienne de la population résidente [services, équipements, commerces] ; ▶secteur dense ; *termes spécifiques* secteur résidentiel pavillonnaire, ▶quartier paysager, ▶zone à vocation exclusive résidentielle) ; *f 3* **habitat** [m] **(2)** (Terme générique caractérisant différentes formes architecturales d'habitations comme l'habitat pavillonnaire, ou caractérisant une forme d'occupation du sol comme l'habitat dispersé, l'habitat collinaire, etc.) ; *f 4* **secteur** [m] **d'habitation** (Terme caractérisant la diversification des différentes zones urbaines ou des zones d'urbanisation future dans le cadre du zonage d'un PLU/POS) ; *g* **Wohngebiet** [n] **(2)** (**1.** Ganz oder überwiegend dem Wohnen dienendes Gebiet. **2.** In D. werden gem. Baunutzungsverordnung ▶reine Wohngebiete, ▶allgemeine Wohngebiete und besondere **W.e** unterschieden — abhängig davon, wie stark ein solches Gebiet von anderen Nutzungen durchsetzt sein darf und im ▶Bebauungsplan entsprechend dargestellt; cf. §§ 3 u. 4 BauNVO; ▶durchgrüntes Wohngebiet); *syn.* Wohnquartier [n], Wohnviertel [n], Wohnzone [f] [CH].

5009 residential area traffic [n] *trans. urb.* (Traffic on residential streets in a housing development); *s* **tráfico** [m] **de zona residencial** (Tráfico de vehículos dentro de una zona residencial, sean o no de residentes en la misma); *f* **circulation** [f] **résidentielle** (Trafic des véhicules à l'intérieur d'une zone résidentielle) ; *syn.* trafic [m] résidentiel ; *g* **Anliegerverkehr** [m] **(1)** (Verkehr innerhalb von Straßen eines Wohngebietes).

5010 residential building [n] *arch. urb.* (Building used, serving, or designed as a residence or for occupation by residents; ▶high-rise residential building); *syn.* residence building [n], dwelling building [n]; *s* **casa** [f] **de viviendas** (▶torre de pisos); *syn.* edificio [f] residencial; *f* **maison** [f] **d'habitation** (Maison individuelle utilisée ou prévue à des fins d'habitation ; ▶tour d'habitations) ; *syn.* bâtiment [m] d'habitation ; *g* **Wohnhaus** [n] (Zum Wohnen genutztes oder für die Wohnnutzumng geplantes Gebäude; ▶Wohnhochhaus); *syn.* Wohngebäude [n].

residential building [n] [UK]**, multistorey** *arch. urb.* ▶multistory building for housing [US].

R

5011 residential community [n] [US] *leg. urb.* (Development zone designated for housing in a ►community development plan; in comparison to an area used exclusively for housing; **in U.S.**, there is increasing use of **r. c.** to enable residents to walk from home to work and shopping, etc.; **in U.K.**, no such use class; **in D.**, a zoning term for predominantly residential area in which supporting facilities, such as shops, restaurants, handicrafts, skilled trades, religious, cultural, social and health institutions are also permissible, as specified in a German community development plan; ►residential zoning district [US]/residential zone [UK]); **s suelo [m] residencial (1)** (En la planificación urbana categoría de suelo con fines primordialmente residenciales, pero en los que se permiten usos comerciales y sociales; **en EE.UU.** se define crecientemente este tipo de suelo para posibilitar al vecindario a acudir a pie al trabajo o a los comercios; **en D.** es una categoría de suelo en los ►planes generales de ordenación urbana en el que —en comparación con el ►suelo residencial 2— están permitidos diversos usos como tiendas y restaurantes, talleres de artesanos no causantes de ruidos o molestias e instalaciones culturales, religiosas, sociales y de salud; **en Es.** no existe esta categoría de suelos); **f secteur [m] dense** (F., catégorie du ►zonage urbain : les **s. d.** se divisent en habitations collectives avec espaces verts et en habitation et commerces ; **D.**, surface constructible à vocation générale d'habitation. Par comparaison avec le ►secteur à vocation exclusive résidentielle, il est aussi prévu l'implantation d'équipements divers comme petits commerces, entreprises artisanales non nuisantes ainsi que d'équipements culturels, confessionnels, sociaux et de santé) ; *syn.* secteur [m] à dominante d'habitat, secteur [m] d'agglomération continue ; **g allgemeines Wohngebiet [n]** (**D.**, im Rahmen der ►Bauleitplanung gem. §§ 1 [2] und 4 BauNVO für die Bebauung vorgesehene Fläche [Baugebiet], die vorwiegend dem Wohnen dient. Im Vergleich zum ►reinen Wohngebiet sind die der Versorgung des Gebietes dienenden Läden, Schank- und Speisewirtschaften sowie nicht störende Handwerksbetriebe und Anlagen für kirchliche, kulturelle, soziale und gesundheitliche Zwecke zulässig).

residential community development [n] *urb.* ►housing subdivision [US].

5012 residential complex [n] *arch. urb.* (Development with houses for several families. Such a complex is usually surrounded by a harmoniously-designed green area); *syn.* multifamily housing site [n] [also US]; **s complejo [m] residencial** (Casa multifamiliar, generalmente con zona ajardinada alrededor); **f résidence [f]** (Ensemble d'habitations constitué d'un [immeuble] ou plusieurs bâtiments de standing souvent agrémenté d'espaces verts de conception exigeante) ; *syn.* groupe [m] résidentiel, complexe [m] résidentiel ; **g Wohnanlage [f]** (Mehrfamilienhaus, meist mit gestalteten Außenanlagen).

5013 residential density [n] *sociol. urb.* (Number of inhabitants per hectare or per square mile [US] of ►built-up area 1); **s densidad [f] habitacional** (Número de habitantes por hectárea de área construida; ►zona edificada 1); *syn.* densidad [f] residencial; **f densité [f] résidentielle** (±) (Nombre de logements à l'hectare de ►surface bâtie) ; **g Wohndichte [f] (2)** (Anzahl der Bewohner je Hektar ►bebauter Fläche).

residential environment [n] *sociol. urb.* ►dwelling environs.

residential garden [n] [US] *gard.* ►private garden.

5014 residential green space [n] *urb.* (Generic term for house gardens and courtyards which are directly adjacent to freestanding or attached single-family dwellings as well as for green areas connected with multi-storied residential development. Term also includes directly adjacent weekend gardens and allotment gardens which are used by residents of apartments [US]/flats [UK] which are without gardens; ►community garden); **s zonas [fpl] verdes residenciales** (Término genérico para los espacios libres directamente adyacentes a casas individuales o bloques de pisos, es decir: jardines privados, jardines vecinales o espacios verdes entre bloques, ►huertos reacreativos urbanos o jardines de inquilinos); **f espaces [mpl] verts résidentiels** (Terme générique pour les jardins privatifs et les cours d'habitation attenants à des lotissements de maisons individuelles, pour les espaces verts à usage privé d'accompagnement de résidences et de grands ensembles, les ►jardins ouvriers journaliers et de week-end) ; *syn.* espaces [mpl] verts privés urbains ; **g Wohngrün [n, o. Pl.]** (OB zu Hausgarten und Wohnhof, die unmittelbar an freistehende, gereihte oder flächig aneinandergefügte Einfamilienhäuser grenzen sowie zu Grünfläche im Zusammenhang mit mehrgeschossigem Wohnungsbau, ferner zu ►Kleingarten und Mietergarten als Ergänzung hausgartenloser Wohnungen sowie zu Wochenendgarten; cf. L+S 1969 [2], 60).

5015 residential land use [n] *leg. urb.* (Land use category shown on ►comprehensive plan [US]/Local Development Framework [LDF] [UK] for a municipality; ►mixed use development, ►residential community [US], ►residential zoning district [US]/residential zone [UK], ►zoning map [US]/Proposals Map/Site Allocations Development Plan Document [UK]); *syn.* housing zone [n] [also US]; **s uso [m] residencial** (En D. término genérico de planificación urbana utilizado en los ►planes generales municipales de ordenación urbana para todos los tipos de zonas de vivienda que se diferencian en los ►planes parciales de ordenación en cuatro categorías: suelo residencial rural, ►suelo residencial 1 y ►suelo residencial 2 y suelo residencial especial. En España no existe tal diferenciación; ►zona urbana mixta); **f zone [f] urbaine destinée à l'habitation** (**D.**, terme d'urbanisme utilisé dans le « Flächennutzungsplan »/►schéma de cohérence territoriale [SCOT allemand] ; cette zone d'urbanisme est concrétisée dans le « Bebauungsplan » ►plan local d'urbanisme [PLU/POS allemand] en différenciant le secteur de lotissements ou de groupes d'habitation, ►secteur dense, ►secteur à vocation exclusive résidentielle ou secteur spécialisé ; ►zone à urbaniser à vocation mixte) ; **g Wohnbaufläche [f]** (**D.**, Abk. im ►Flächennutzungsplan W; gemäß § 5 [2] Nr. 1 BauGB im FNP nach der allgemeinen Nutzung darzustellende Fläche, die im weiteren Planungsprozess im ►Bebauungsplan als Kleinsiedlungsgebiet, ►reines Wohngebiet, ►allgemeines Wohngebiet oder besonderes Wohngebiet ausgewiesen wird; cf. auch § 1 BauNVO; ►Mischgebiet).

residential leisure complex [n] *recr. urb.* ►residential leisure development.

5016 residential leisure development [n] *recr. urb.* (General term for holiday village or apartment complex, recreation center [US]/recreation centre [UK], campground [US]/camping ground [UK] with permanent spaces for tents, travel trailers [US]/caravans [UK], mobile homes); *syn.* residential leisure complex [n]; **s 1 urbanización [f] de recreo (≠)** (Término genérico para cualquier tipo de asentamiento o complejo de casas, centro de recreo con posibilidades de alojamiento, zona de camping permanente, etc. utilizado para fines recreacionales); **s 2 urbanización [f] turística** (Barrios de chalets con jardín localizados en las zonas costeras o zonas atractivas desde el punto de vista recreacional y paisajístico, generalmente en áreas turísticas o en los alrededores de las grandes urbes); *syn.* urbanización [f] de «segundas residencias»; **f 1 habitat [m] de loisirs** (Terme générique utilisé pour désigner divers équipements de loisirs tels que villages de vacances, habitations légères de vacances, meublés de tourisme, gîtes de France, terrains de camping-caravaning, parcs résidentiel de loisirs) ; *syn.* habitat [m] de plein air ;

R

f 2 **lotissement [m] de loisirs** (Terrain réservé à l'habitat de plein air prévu pour l'implantation de résidences mobiles ou d'habitations légères de loisirs) ; *g* **Freizeitsiedlung [f]** (Sammelbegriff für Feriendorf, Apartmenthäuser, Ferienzentrum, Campingplatz mit Daueraufstellung von größeren Wohnzelten, Wohnwagen und Mobilheimen; i. d. R. auch mit Freizeiteinrichtungen ausgestattet).

5017 residential street [n] *trans. urb.* (Primarily a ►local street with access only for residents without through traffic in a residential area); *syn.* residential subdivision road [n] [also US]; *s* **calle [f] residencial** (Calle en zona residencial con restricciones para el tráfico rodado; ►calle de vecindario); *f* **rue [f] résidentielle** (Rue à usage des piétons en priorité, pour laquelle la vitesse et le stationnement des véhicules est limité dans un quartier résidentiel ; LEU 1987, 69 ; ►rue riveraine) ; *g* **Wohnstraße [f]** (Weitgehend verkehrseingeschränkte ►Anliegerstraße in einem reinen Wohngebiet).

residential subdivision [n] *urb.* ►residential area.

residential subdivision road [n] [US] *trans. urb.* ►residential street.

5018 residential traffic [n] *trans.* (Limited traffic generated by users along a specific street); *s* **tráfico [m] de residentes** (Tráfico zonal generado por los residentes de un área); *f* **circulation [f] riveraine** (Circulation générée par les habitants d'une rue) ; *syn.* trafic [m] riverain, trafic [m] de desserte ; *g* **Anliegerverkehr [m] (2)** (Auf Anwohner einer Straße beschränkter Verkehr).

residential zone [n] [UK] *leg. urb.* ►residential zoning district [US].

5019 residential zoning district [n] [US] *leg. urb.* (Development zone designated exclusively for housing in a legally-binding land-use plan, such as a ►zoning map [US]/Proposals Map/Site Allocations Development Plan Document [UK]); *syn.* residential zone [n] [UK]; *s* **suelo [m] residencial (2)** (D., en los ►planes parciales de ordenación uso de suelo tipo en el que —en comparación con el ►suelo residencial 1— sólo están permitidas la construcción de viviendas y la instalación de algunos servicios mínimos no causantes de ruidos o molestias); *f* **secteur [m] à vocation exclusive résidentielle** (Zonation d'un ►plan local d'urbanisme [PLU]/plan d'occupation des sols [POS] caractérisant une zone d'extension de l'habitat à densité faible : *termes spécifiques* secteur à faible densité, secteur résidentiel, secteur pavillonnaire aéré, zone d'extension à densité réduite, habitation basse avec jardins ; cf. POS Marseille 1976) ; *syn.* secteur [m] à vocation exclusive d'habitation, zone [f] principalement à caractère résidentiel, zone [f] d'habitation à caractère groupé ; *g* **reines Wohngebiet [n]** (D., Abk. WR; gemäß § 9 [1] BauGB und § 1 [2ff] BauNVO nach der besonderen Art der baulichen Nutzung im ►Bebauungsplan darzustellendes Baugebiet, das ausschließlich dem Wohnen dient; § 3 BauNVO).

residents' underground parking [n] [US] *urb.* ►underground parking garage [US]/underground car park [UK].

residual parcel [n] **of land** *plan. urb.* ►leftover area.

5020 residual pollutants [npl] *envir.* (Toxic and hazardous substances remaining as residue); *s* **residuos [mpl] de sustancias nocivas**; *f* **résidus [mpl] des polluants** *syn.* résidus [mpl] des substances toxiques ou dangereuses, résidus [mpl] des substances nocives ; *g* **Rückstände [mpl] von Schadstoffen** *syn.* Schadstoffrückstände [mpl].

5021 residual population [n] *phyt. zool.* (Remaining individuals of animal and plant species in a defined area); ►relict species); *syn.* remnant population [n]; *s* **población [f] relíctica** (Individuos restantes de especies de flora o fauna de una población antiguamente más grande que se presentan en un área determinada); ►especie relíctica); *syn.* población [f] residual; *f* **population [f] résiduelle** (Restes d'un peuplement d'espèces animales et végétales dans une aire déterminée ; ►espèce relictuelle) ; *syn.* population [f] restante ; *g* **Restbestand [m]** (Rest von Tier- und Pflanzenarten, die in einem bestimmten Gebiet vorkommen; ►Reliktart); *syn.* Restvorkommen [n].

5022 residual woodland [n] *agr. for.* (WT 1980; remaining woodland less than 3 hectares in size in an agrarian landscape; ►coppice [US] 1/copse [UK]); *s* **bosquete [m] residual** (Pequeño bosque de menos de tres hectáreas en zona de uso agrícola; ►bosquecillo 1); *f* **bosquet [m] restant** (Petit massif forestier en paysage rural dont la superficie et inférieur à 3 ha ; ►petit bois) ; *syn.* bosquet [m] résiduel ; *g* **Restgehölz [n]** (Kleiner Waldrest unter 3 ha Größe in einer bäuerlichen Kulturlandschaft; ►Gehölz 1).

r̲e̲sidue mound [n] *min.* ►spoil bank [US] (2)/spoil heap [UK].

r̲e̲sidues [npl] *agr. for. hort.* ►crop residues; *chem. envir.* ►pesticide residues.

5023 resilient layer [n] *constr.* (Layer between wearing surface of a sand-clay sports court [US]/pitch [UK] and the compacted subbase to create an elastic resistance to vertical loading; ►granular surface course [US]/hoggin surface course [UK], ►subbase [US]/sub-base [UK]); *syn.* elastic layer [n]; *s* **capa [f] elástica** (Capa entre el ►suelo de arenilla de canchas/pistas de deporte y la ►capa portante compactada para dar resiliencia elástica a la carga vertical); *f* **couche [f] de souplesse** (Couche intermédiaire entre la couche de finition et la ►couche de fondation pour les ►sols sportifs stabilisés mécaniquement) ; *g* **dynamische Schicht [f]** (Schicht zwischen ►Tennenbelag und ►Tragschicht zur Ermöglichung einer elastischen Nachgiebigkeit bei vertikaler Belastung).

r̲e̲silience [n] *ecol.* ►elasticity.

5024 resistance [n] *ecol. hort.* (Robustness of plants or animals to withstand the effects of injurious organisms, noxious poisons, or climate extremes; ►cold hardiness, ►resistance to air pollution, ►salt resistance); *s* **resistencia [f]** (Capacidad de un individuo para oponerse a las condiciones adversas del medio y a los ataques y enfermedades; DINA 1987; ►resistencia a la contaminación, ►resistencia a las heladas, ►resistencia a la salinidad); *syn.* robustez [f]; *f* **résistance [f]** (Propriété que possèdent certains végétaux et animaux de réagir à un parasite, à un produit, à un facteur climatique ; ►résistance à la pollution, ►résistance au gel, ►résistant à la salinité) ; *g* **Resistenz [f]** (Widerstandsfähigkeit von Pflanzen oder Tieren gegen Schaderreger oder Schadwirkungen [Hitze, Kälte, Gifte, Trockenheit etc.]; es gibt **R.** gegen abiotische Einflussgrößen wie Hitze, Kälte, Dürre, hoher und niedriger pH-Wert, Pflanzenschutzmittel, Luftschadstoffe und Bodenverdichtung sowie gegen biotische Einflussgrößen wie Krankheiten und Schädlinge; *UBe* Frostresistenz [►Frosthärte], ►Immissionsresistenz, Rauchresistenz, ►Salzresistenz, Trockenheitsresistenz).

resistance [n] *constr.* ►frost resistance; *phyt.* ►salt resistance; *constr.* ►weather resistance;

resistance [n]**, ecological** *ecol. recr.* ►environmental stress tolerance.

resistance [n]**, shearing** *constr. stat.* ►shear strength.

5025 resistance against fracture [n] *arb. stat.* (Capacity of trees to withstand breakage); *s* **seguridad [f] contra rotura** (Capacidad de partes de árboles de resistir sin romperse ante presiones externas, como vientos fuertes o carga de nieve); *f* **résistance [f] au bris** (Capacité des arbres à bien résister à la cassure du tronc ou l'arrachement de parties de la couronne lors

R

de tempêtes, sous le poids de la neige) ; *g* **Bruchfestigkeit [f]** (Artspezifische Fähigkeit und Beschaffenheit von Bäumen, dem Bruch von Stamm und Kronenteilen bei äußeren Einflüssen — Sturm, Schneelast, Eisregen — ausreichend zu widerstehen; cf. QBB 1992); *syn.* Bruchsicherheit [f].

5026 resistance [n] **to air pollution** *hort.* (Capability of plants to be insusceptible to air pollution); *s* **resistencia [f] a la contaminación** (Capacidad de algunas plantas de tolerar cierto grado de contaminación atmosférica); *f* **résistance [f] à la pollution** (Capacité de certains végétaux à supporter sans altérations notables les polluants atmosphériques) ; *g* **Immissionsresistenz [f]** (Eigenschaft von Pflanzen, die gegenüber Immissionen unempfindlich sind); *syn.* z. T. Industriehärte [f], Rauchhärte [f], Rauchresistenz [f].

resistance [n] **to skidding** *constr.* ▶nonskid property [US]/non-skid property [UK].

5027 resistance [n] **to wear and tear** *constr.* (Property of surfacing materials to withstand abrasion or attrition); *s* **resistencia [f] al desgaste** (Capacidad de materiales de resistir las inclemencias del uso; resistente [adj] al desgaste, indesgastable [adj]); *f 1* **résistance [f] à l'abrasion** (*Terme générique* capacité des matériaux de revêtement de voirie à résister à l'usure) ; *f 2* **résistance [f] au roulement** (*Terme spécifique* résistance à l'action provoquée par les matériels roulants) ; *f 3* **résistance [f] à l'usure** (*Terme spécifique* résistance à l'action provoquée par les piétons et les véhicules) ; *g* **Verschleißfestigkeit [f]** (Widerstandskraft von Wegebelägen/Deckenbaustoffen gegenüber Abnutzung); *syn.* Abriebfestigkeit [f].

resistant [adj] *hort. plant.* ▶air pollution-resistant; *phyt.* ▶drought-resistant; *agr. bot. constr. hort.* ▶frost-resistant; *phyt.* ▶salt-resistant; *constr.* ▶weather-resistant.

resistant [adj], **deterioration** *biol.* ▶nonrotting [US]/▶non-rotting [UK].

resited farm holding [n] [UK] *agr.* ▶relocated farmstead [US].

5028 resonance [n] *phys.* (Acoustic oscillations at frequencies between roughly 16,000 and 20,000 Hz, which generate sounds or ▶noise sensed by the ear); *s* **sonido [m]** (Ondas acústicas en la frecuencia entre 16 y unos 20 000 Hz que generan sensación sobre el oído; ▶ruido); *f* **son [m]** (Ondes acoustiques d'une fréquence de 16 à environ 20 000 Hz générant une sensation auditive ; ▶bruit) ; *g* **Schall [m]** (Schwingungen im Frequenzbereich von 16 bis ca. 20 000 Hz, welche mit Hilfe des Ohres Ton-, Klang- oder Geräuschempfindungen auslösen; ▶Lärm).

resort [n] *recr.* ▶climatic health resort, ▶health resort, ▶health resort center [US]/health resort centre [UK], ▶seaside resort, ▶tourist resort [US]/torist centre [UK], ▶winter sports resort.

5029 resort community [n] [US] *plan. recr.* (Built development usually in a tourist region consisting of small-scale residential accommodation and communal facilities specifically for vacation [US]/holiday [UK] use; such facilities may include hotel accommodations; ▶family or youth camp); *syn.* holiday village [n] [UK]; *s* **pueblo [m] de veraneo** (Lugar en la costa o en zona rural que se utiliza como lugar de veraneo; ▶colonia de vacaciones); *f* **village [m] de vacances** (Ensemble d'hébergement individuel ou collectif faisant l'objet d'une exploitation commerciale ou non, destiné à offrir des séjours de vacances et de loisirs tout en assurant la restauration et l'usage d'équipements collectifs permettant des activités de loisirs sportifs et culturels ; décret 68-476 ; ▶colonie de vacances) ; *g* **Feriendorf [n]** (Häufig in sich geschlossene, städtebaulich gestaltete Siedlung für Freizeitwohnungen in ein- oder zweigeschossiger Bauweise, die

für begrenzte Zeiträume erholungsbedürftigen Familien gegen Entgelt überlassen werden; ▶Ferienkolonie).

resort park [n] *recr.* ▶health resort park.

resort recuperation [n] *recr.* ▶health resort recuperation.

resource [n] *conserv'hist.* ▶cultural resource [US]/cultural asset [UK]; *ecol.* ▶environmental resource patch; *landsc. plan.* ▶visual resource management.

resource [n], **protected** *conserv'hist.* ▶monument.

Resource Conservation District [n] **(RCD)** [UK] *conserv. leg.* ▶Area of Outstanding Natural Beauty [UK].

resource data management system [n] *adm. geo. landsc. plan.* ▶landscape information system.

resource heritage [n], **landscape** *conserv. land'man.* ▶landscape patrimony.

resource inventory [n] [US], **soil** *geo. pedol.* ▶soil mapping.

resource quality [n], **visual** *landsc. plan. recr.* ▶visual quality of landscape.

resource recovery [n] *envir. nat'res.* ▶recycling.

resources [npl] *hydr.* ▶available groundwater resources; *conserv. land'man. landsc.* ▶conservation of nature and natural resources; *ecol. nat'res. sociol.* ▶consumption of natural resources; *ecol. envir.* ▶depletion of natural resources; *min.* ▶development of mineral resources; *wat'man.* ▶diminution of water resources; *min.* ▶development of mineral resources; *hydr.* ▶groundwater resources; *geo. min.* ▶mineral resources; *nat'res.* ▶natural resources 1; ▶natural resources management; *conserv. zool.* ▶nutritional resources; *conserv'hist.* ▶preservation of cultural resources; *nat'res. pol.* ▶preservation of natural resources; *nat'res. wat'man.* ▶protective forest for water resources; *plan. recr.* ▶recreation resources inventory; *land'man.* ▶spatial landscape characteristics/resources; *conserv. plan.* ▶use capacity of natural resources; *ecol. land'man.* ▶utilization of natural resources; *hydr. nat'res. wat'man.* ▶water resources; *hydr. nat'res.* ▶water resources conservation.

resources [npl], **capacity of natural** *conserv. plan.* ▶use capacity of natural resources.

resources [npl], **exploitation of mineral** *min.* ▶mineral working.

resources [npl], **forest for water** *nat'res. wat'man.* ▶protective forest for water resources.

resources [npl], **landscape** *conserv. landsc. recr.* ▶natural site characteristics; *land'man.* ▶spatial landscape characteristics/resources.

resources [npl] [US], **lessening of water** *wat'man.* ▶diminution of water resources.

resources [npl], **lowering of water** *wat'man.* ▶diminution of water resources.

resources [npl], **spatial landscape** *land'man.* ▶spatial landscape characteristics/resources.

resources information system [n], **natural** *adm. geo. landsc. plan.* ▶landscape information system.

responsibility [n], **maintenance** *leg. trans. wat'man.* ▶maintenance obligation.

resources [npl], **reclamation of used** *envir. nat'res.* ▶recycling.

resources [npl], **recreation** *landsc. recr.* ▶recreation potential.

5030 respreading [n] **of topsoil** *constr.* (Respreading of topsoil after stockpiling [US]/stacking [UK]); *syn.* restore [vb] [also US] (Hort 1989 [1], 16); *syn.* topsoil respreading [n]; **s reextensión [f] de tierra vegetal** (Esparcimiento de la tierra vegetal acopiada provisionalmente en caballones después de terminar la obra); **ƒ remise [f] en place de la terre végétale** (Réutilisation de la terre végétale stockée sur le chantier en début de travaux) ; *syn.* mise [f] en remblai de la terre végétale stockée, mise [f] en remblai de la terre végétale prise au stock sur place) ; **g Wiederandecken [n, o. Pl.] von Oberboden** (Einbau von auf Miete gesetzten Oberboden); *syn.* Wiederandecken [n, o. Pl.] von Mutterboden, Wiedereinbau [m] von Oberboden.

5031 rest area [n] *plan. recr.* (Open-air space reserved for quiet recreation; ▶rest place); *syn.* quiet area [n]; **s zona [f] de descanso** (Parte de zona verde que sirve para el recreo tranquilo; ▶lugar de descanso 1); *syn.* área [f] de descanso; **ƒ zone [f] de silence** (Aire d'un espace libre réservé au repos et à la détente ; ▶aire de repos 2) ; **g Ruhebereich [m]** (Freiraumbereich, der der ruhigen Erholung dient; ▶Rastplatz).

restocking [n] *for.* *▶regeneration (1).

5032 restocking [n] **of fish** *zool.* (Amount of fish released into a given habitat for replenishment purposes; ▶fry); **s 1 suelta [f] de peces** (Introducción de ▶alevines en un curso de agua para aumentar su población); *syn.* suelta [f] de alevines; **s 2 repoblación [f] piscícola** (Introducción de peces en curso de agua para repoblarlo); **ƒ 1 alevinage [m]** (Introduction de jeunes poissons dans un cours d'eau, un plan d'eau ou un bassin ; ▶alevin) ; **ƒ 2 réempoissonnement [m]** (Introduction de poissons adultes) ; **ƒ 3 repeuplement [m] de poissons** (Action de redonner à un écosystème aquatique la population en poissons qu'il pourrait normalement contenir et qu'il a perdue ; AEP 1976, 324) ; **g Fischeinsatz [m]** (Einbringen von Fischarten in Gewässer; ▶Setzling 1).

5033 restocking [n] **of game population** *game'man.* (Augmentation of game population by means of ▶wildlife management); **s repoblación [f] de la fauna cinegética** (▶conservación y gestion de la fauna silvestre); *syn.* repoblación [f] de la caza; **ƒ repeuplement [m] des populations animales sauvages** (Augmentation d'une population d'animaux sauvages par des mesures biologiques ; ▶conservation et gestion durable de la faune sauvage, ▶gestion et conservation de la faune sauvage et des ses habitats) ; **g Aufstockung [f] des Wildbestandes** (Vergrößerung der Wildtierpopulation durch wildbiologische Maßnahmen; ▶Wildlife Management).

5034 restocking [n] **of population** *phyt. zool.* (Successful ▶introduction 1 of new individuals to replenish existing plants or animals); **s reposición [f] de especies de fauna y flora** (En el marco de la protección de especies en peligro de extinción, ▶suelta [masiva] de individuos de una especie de fauna o reintroducción de una especie de flora en una región donde casi están extinguidas); **ƒ renforcement [m] de la population** (▶Réintroduction 2 d'organismes appartenant à des espèces indigènes de la flore et de la faune sauvages dans une partie de leur aire de distribution ou elles étaient déjà présentes, qu'il s'agisse soit du renforcement des effectifs d'une espèce végétale menacée en vue d'un rétablissement dans un état de conservation favorable, soit de lâchers d'animaux appartenant à des espèces de gibier ou de poissons en vue de reconstituer des effectifs suffisamment abondants ou de les renforcer pour que les chasseurs ou les pêcheurs puissent pratiquer leurs activités à partir de populations captives ou de plus grandes populations d'autres localités ; cf. Convention de Berne, Recommandation n° 58 [1997] relative aux réintroductions d'organismes appartenant à des espèces sauvages et aux reconstitutions et renforcements de populations d'organismes appartenant à des espèces sauvages dans l'environnement ; ▶introduction) ; **g Bestandsstützung [f]** (Erfolgreiche

▶Aussetzung von Individuen einer Tier- oder Pflanzenart in einem Gebiet, in dem sie selten oder nur noch lokal vorkommt); *syn.* Bestandsvermehrung [f], *syn. Wildbiologie* Populationsaufstockung [f], Bestandsaufstockung [f].

5035 restoration [n] (1) *conserv. landsc. land'man.* (Attempt to restore natural or near-natural conditions of a landscape unit damaged by man, e.g. worked bog, ▶piped stream, gravel dredging pit, by implementing active or passive measures of habitat management; ▶land reclamation 2, ▶natural river engineering measures); *syn.* regeneration [n] [also UK], rehabilitation [n] (2), recovery [n] (of natural habitat), renaturalisation [n] [also UK]; **s renaturalización [f]** (Intento de reestablecer un estado natural o cuasinatural de un paisaje destruido por usos antropógenos, como ▶arroyos entubados, turberas explotadas, canteras, etc.; ▶recuperación de tierras, ▶renaturalización de cursos de agua) ; **ƒ renaturation [f]** (Intervention qui tente à rétablir l'état naturel ou réhabiliter vers un état proche de son état naturel d'origine un milieu plus ou moins artificialisé grâce à des mesures actives et passives de création de biotopes, p. ex. la remise en eau d'une tourbière en fin d'exploitation, l'ouverture d'une rivière busée, la ▶remise en état des lieux après exploitation d'une carrière dans la nappe, la renaturation des berges d'un cours d'eau ; la renaturation se fixe comme objectif, en tentant de réhabiliter notamment toutes les caractéristiques physiques du milieu, de retrouver toutes les potentialités initiales du milieu en terme de diversité biologique, de capacité d'autoépuration, etc. ; ▶renaturation des cours d'eau) ; *syn.* reconstitution [f] en l'état naturel, restauration [f] en l'état naturel, revitalisation [f], renaturalisation [f] [CDN] ; **g Renaturierung [f]** (Versuch der Wiederherstellung eines natürlichen/naturnahen Zustandes eines durch menschliche Maßnahmen geschädigten Landschaftsraumes durch aktive oder passive Maßnahmen der Biotopgestaltung, z. B. Wiedervernässung eines abgetorften Moores, Offenlegung eines ▶verdolten Baches [▶Gewässerrenaturierung], ▶Rekultivierung einer Nassabbaggerung).

restoration [n] (2) *constr.* ▶ground restoration; *landsc. conserv.* ▶habitat restoration; ▶land restoration (2); *wat'man.* ▶river restoration.

restoration [n], **bog** *conserv.* ▶bog regeneration.

restoration [n], **crown** *arb. hort.* ▶crown renewal.

restoration [n], **land** *agr. for. land'man.* ▶land reclamation (2), ▶land restoration (2).

restoration [n] [US], **toxic site** *envir. plan.* ▶toxic site reclamation.

restoration costs [npl] **of habitats/ecosystems** *adm. plan. ecol.* ▶replacement costs of habitats/ecosystems.

restoration [n] **of contaminated land** *envir. plan.* ▶toxic site reclamation.

restoration [n] **of land** *agr. for. land'man.* ▶land reclamation (2).

restoration [n] **of watercourses** *conserv. landsc. wat' man.* ▶natural river engineering measures.

restore [vb] *constr.* ▶replacement of sods [US]/replacement of turves [UK], replacement of topsoil.

rest period [n] **of grave** [UK] *adm.* ▶use period of grave [US].

5036 rest place [n] [US] *recr.* (Location for taking a break during a walk in a park or rambling in a recreation area; ▶stopping place); *syn.* rest spot [n] [UK] (1); **s lugar [m] de descanso (1)** (Sitio donde se puede descansar durante un paseo por un parque o una caminata en una zona de recreación; ▶área de parada [en zona natural]); *syn.* área [f] de picnic; **ƒ aire [f] de repos (2)** (Lieu aménagé pour le repos des promeneurs dans un parc ou des randonneurs en forêt, dans les zones touristiques, etc. ; ►aire de stationnement en bordure de route ; **g Rastplatz [m]** (**1.** Ort zum Unterbrechen eines Spazierganges in einem Park oder einer Wanderung in einem Erholungsgebiet, um sich auszuruhen. **2.** Platz zum Abstellen eines Fahrzeuges am Rande eines Waldes, Erholungsgebietes, um zu halten, auszusteigen, die Aussicht zu genießen, Hinweistafeln zu lesen oder um zu essen; ►Halteplatz).

restraining order [n] [US] *conserv. leg. plan.* ►temporary restraining order [US]/emergency preservation notice [UK].

restraint [n] *constr.* ►edge restraint.

rest spot [n] [UK] **(1)** *recr.* ►rest place [US].

5037 rest spot [n] **(2)** *recr.* (Resting place in a meadow, forest, etc. for recreation; ►picnic site, #1, ►rest place [US]/rest spot [UK] 1, ►stopping place [US]/rest spot [UK] 3); **s lugar [m] de descanso (2)** (Campa o claro en bosque para descansar, estar, jugar, etc.; ►área de parada [en zona natural], ►área de picnic); **f aire [f] de campement** (Aire aménagée équipé rudimentairement pour permettre pendant une nuit ou pour la restauration pendant le jour d'accueillir les randonneurs de passage ; ►aire de pique-nique, ►aire de repos 2, ►aire de stationnement en bordure de route) ; *syn.* aire [f] de repos (3) ; **g Lagerplatz [m]** (Platz im Freien zum Rasten oder Übernachten; ►Halteplatz, ►Picknickplatz, ►Rastplatz).

rest spot [n] [UK] **(3)** *recr.* ►stopping place [US].

5038 resurfacing [n] *constr.* (Replacing the top pavement layer; ►paving 1); *syn.* surface improvement [n] [also US]; **s pavimentación [f]** (Adición de una capa de asfalto a una carretera, ►adoquinado 2); *syn.* afirmado [m]; **f renforcement [m] (2)** (d'une voie de circulation par la confection d'un nouveau revêtement ; ►pavage 2) ; **g Belagserneuerung [f]** (Erneuern der bestehenden Oberfläche von Wegen, Straßen oder Plätzen durch Aufbringen einer neuen Deckschicht; ►Pflasterung 2); *syn.* Erneuerung [f] der Wegedecke/Straßendecke/des Platzbelages, Wiederherstellung [f] des Belages.

resurgence pool [n] *geo. hydr.* ►spring-fed pool.

retainage [n] [US] *constr. contr.* ►retention fund.

5039 retainer fee [n] *contr. prof.* (Lump-sum minimum payment made in advance by a client for professional services of a designer to ensure his availability for an agreed period of time); **s adelanto [m] de honorarios** (Cantidad mínima de los honorarios que el cliente paga como señal al arquitecto resp. a la arquitecta antes de que sean realizados los servicios profesionales para asegurar su disponibilidad durante el periodo/período de tiempo acordado); **f avance [f] d'honoraires** (Montant forfaitaire versé par le client avant le début d'une mission de travaux pouvant être prévu dans une convention d'honoraires par lequel le client signale à l'architecte l'intérêt qu'il porte à la réalisation d'esquisses ou d'un avant-projet sans s'engager à contracter les missions suivantes ; dans le cas de la poursuite de la mission l'avance sera prise en compte dans le calcul du montant total des honoraires) ; **g Honorarvorauszahlung** (Eine vor Arbeitsbeginn auszuzahlende Pauschalsumme, die vertraglich vereinbart werden kann, um die Ernsthaftigkeit des Kunden für den Auftrag zur Erstellung erster Ideen und Entwürfe zu bekräftigen, ohne sich zu verpflichten, die weiteren Leistungsphasen in Auftrag zu geben; sollte das Projekt zu einem späteren Zeitpunkt fortgeführt werden, wird die Vorauszahlung beim Gesamthonorar angerechnet; dieses Prozedere ist in der HOAI nicht vorgesehen, in den USA aber üblich; *syn.* Honorarvorschuss [m], Vorabhonorar [n].

5040 retaining wall [n] *arch. constr.* (Wall constructed to contain or hold backfilled or cut ground and withstand pressure, as opposed to a ►face wall; ►gravity retaining wall, ►cantilever

wall); **s muro [m] de sostenimiento** (Muro que —al contrario que el ►muro cortina— debe contener un terraplén; *términos específicos* ►muro de gravedad, ►muro de sostenimiento en L); **f mur [m] de soutènement** (Mur construit pour retenir un terrassement en remblais ou en déblais qui, par opposition au ►mur de revêtement, résiste à la poussée des terres ; *termes spécifiques* ►mur-chaise, ►mur poids) ; **g Stützmauer [f]** (Mauer vor aufgeschüttetem oder abgetragenem Boden, die das dahinter liegende Erdreich in seiner Lage hält und im Gegensatz zur ►Futtermauer den Erddruck auffängt; *UBe* ►Schwergewichtsmauer, ►Winkelstützmauer).

retaining wall unit [n] *constr.* ►L-shaped retaining wall unit.

retardant [n] *chem. agr. hort.* ►growth retardant.

5041 retarding basin [n] *hydr. wat'man.* (Generic term covering ►flood control reservoir, ►storm water detention basin, ►rainwater interceptor basin); *syn.* detention basin [n]; **s embalse [m] de retardo** (Embalse con el que se reducen las puntas de crecida de un curso de agua, mediante un almacenaje temporal; WMO 1974; ►embalse de regulación, ►embalse de regulación de aguas pluviales, ►embalse de sobrecarga); *syn.* embalse [m] de regulación; **f bassin [m] de rétention** (*Termes spécifiques* ►bassin de rétention des crues, ►bassin de retenue d'eaux pluviales, ►déversoir d'orage) ; *syn.* bassin [m] de retenue, bassin [m] d'accumulation ; **g Rückhaltebecken [n]** (OB zu ►Hochwasserrückhaltebecken, ►Regenrückhaltebecken, ►Regenüberlaufbecken).

5042 retention [n] *hydr.* (Temporary storage of excess water in flood control reservoirs and ►storm water detention basins to reduce peak flood in watercourse areas, or in sufficiently-wide floodplains [►floodable land]; ►impoundment, ►retention of rainwater); **s retención [f]** (Reducción del caudal máximo en cursos de agua por medio de presas, ►embalses de regulación de aguas pluviales, ►embalses de retención de aguas pluviales o en vegas suficientemente grandes [►terrenos inundables, ►llanuras de inundación]; ►retención de aguas pluviales); **f rétention [f]** (Abaissement de pointes de crues d'un cours d'eau par emmagasinement temporaire des excédents dans des barrages-réservoirs [►lac de retenue], ►bassins de retenue d'eaux pluviales ou des plaines alluviales assez larges [►polder d'inondation ou exutoires de crues] ; ►rétention des eaux pluviales) ; **g Retention [f]** (Abminderung der Hochwasserspitze durch Auffangen des Wassers in Hochwasserrückhaltebecken [►Stausee, ►Regenrückhaltebecken] oder in genügend großen Auenräumen [►Überschwemmungsbereich]; ►Regenwasserrückhaltung).

retention [n], **nutrient** *pedol.* ►nutrient fixation.

retention basin [n] *hydr. wat'man.* ►storm water detention basin.

5043 retention fund [n] *constr. contr.* (Sum of money, usually a percentage of the total construction cost, which is specified in the contract to be deducted from payments to the contractor, and not released until the project has been satisfactorily completed and guarantee period has passed; ►surety bond [US]/bond [UK]); *syn.* retainage [n] [also US]; **s retención [f] de garantía (1)** (Cantidad de dinero que se reduce del pago al contratista hasta que la obra ha sido terminada totalmente; ►garantía 2); **f retenue [f] de garantie (1)** (Somme — en général 5 % — retenue sur les situations comme sûreté et payée à expiration du délai de garantie ; ►sûretés) ; **g Einbehalt [m]** (Betrag, der bei Abschlagszahlungen oder als ►Sicherheitsleistung zurückbehalten wird; cf. § 17 [2] VOB Teil B).

5044 retention [n] **of payment** *constr. contr.* (Contractually-stipulated proportion of payment to contractor or consultant withheld until work is satisfactorily completed and guarantee period has passed); **s retención [f] de garantía (2)** (Proporción de la suma de un contrato retenida hasta que hayan sido terminados los trabajos y el plazo de garantía finalice); *f* **retenue [f] de garantie (2)** (Action de retenir une partie des paiements ou des honoraires jusqu'à la réalisation complète des prestations prévues au marché et de leur règlement) ; *g* **Einbehaltung [f]** (Zurückhaltung eines Teils von Zahlungen, Honorarsummen bis zur vollständigen Erfüllung der vertraglich vereinbarten Leistungen, wenn Teilleistungen mit Mängeln behaftet sind; cf. § 17 [2] VOB Teil B).

5045 retention [n] **of rainwater** *envir. urb. wat'man.* (Any storm drainage technique that retards or detains runoff, such as a ▶storm water detention basin or ▶rainwater interceptor basin, retention basin, parking lot storage, rooftop storage, porous pavement, dry wells or any combination thereof; IBDD 1981; ▶retention; *specific term* on-site retention); *syn.* delaying [n] storm runoff [also US], storm water detention [n]; *s* **retención [f] de aguas pluviales** (Cualquier técnica o instalación para reducir la escorrentía superficial de aguas de lluvia, como plantación de tejados, ▶embalse de regulación de aguas pluviales o ▶embalse de sobrecarga, ▶retención); *f* **rétention [f] des eaux pluviales** (Mesures et dispositifs qui permettent la restitution avec retardement des eaux des précipitations, p. ex. les toitures végétalisées, les ▶déversoirs d'orage, les ▶bassins de retenue d'eaux pluviales, les revêtements de voirie perméables ; ▶rétention) ; *g*°**Regenwasserrückhaltung [f]** (1. Maßnahmen und Einrichtungen, die zur Verzögerung des Niederschlagswasserabflusses dienen, z. B. durch Dachbegrünungssysteme, ▶Regenrückhaltebecken, wasserdurchlässige Wege- und Straßenbeläge, ▶Regenüberlaufbecken; ▶Retention. 2. Wasservolumen, das durch [Hoch]wasserrückhalteräume [Regenrückhaltebecken und Polder] aufgenommen werden kann); *syn. zu 1.* Rückhaltung [f] von Niederschlagswasser; *syn. zu 2.* Rückhaltevolumen [n], Wasserrückhaltemenge [f], Wasserrückhaltevolumen [n].

retention structure [n] *hydr. wat'man.* ▶storm water detention basin.

5046 retirement community [n] *urb.* (Independent residential area, often in an attractive region with a favo[u]rable climate, which is inhabited by senior citizens the whole year round); *syn.* retirement settlement [n] [US]; *s* **colonia [f] de pensionistas** *syn.* asentamiento [m] de pensionistas, colonia [f] residencial de tercera edad; *f* **lotissement [m] pour personnes âgées (1.** Type d'habitat dans certaines régions des États Unis et du pourtour méditerranéen, aux paysages pittoresques et au climat favorable, aménagé principalement pour recevoir les retraités ; **2. F., B., CH.,** ensemble d'habitations spécifiquement conçues pour des personnes âgées ou à mobilité réduite avec entrée de plain-pied, détecteur de présence, interrupteurs situés à hauteur de fauteuil, volets roulants, salle de bain « à l'italienne » avec un léger dénivelé pour l'écoulement de l'eau de la douche, etc.) ; *f* 2 **résidence [f] pour personnes âgées (RPA)** (Logement-foyer qui regroupe des logements autonomes et assure une indépendance de vie identique à celle du domicile classique, assure la sécurité des personnes, par la présence de personnel, grâce à la conception des lieux favorise la convivialité, l'autonomie et répond aux problèmes d'isolement et de sécurité, propose un service de restauration, de lingerie, de soutien ou de maintien à domicile) ; *syn.* résidence [f] pour le troisième âge, résidence [f] troisième âge ; *g* **Rentnersiedlung [f]** (Eigenständiger Siedlungstyp, der vorwiegend von Rentnern und Pensionären ganzjährig in landschaftlich reizvollen und klimatisch begünstigten Regionen der USA und des Mittelmeers bewohnt wird); *g* 2 **Altenwohnanlage [f]** (Gebäude oder Gebäudekomplex, der speziell für die ältere Generation mit einem breitgefächerten Versorgungsangebot wie Restaurationsbetrieb, Wäscherei, Freizeit- und Begegnungsräume, i. d.R. auch mit einer

Pflegestation ausgestattet ist, damit diese ein selbständiges Leben in Gemeinschaft mit den Mitbewohnern führen kann).

retirement settlement [n] [US] *urb.* ▶retirement community.

5047 retractable post [n] *urb. trans.* (Barrier which is typically drawn down below ground surface by hydraulic means to provide vehicular access when required); *s* **poste [m] de cierre de paso escamoteable** *syn.* poste [m] de cierre de paso hundible; *f* **poteau [m] d'interdiction de passage escamotable** ; *g* **versenkbarer Absperrpfosten [m]** (Pfosten, der bei Bedarf hydraulisch für Autozufahrten bodeneben abgesenkt werden kann).

5048 retrogradation [n] **of shoreline** *geo.* (DNE 1978; gradual wearing away of the coastline caused by waves and wind action on the strand, or by cliff subsidence; ▶shoreline erosion [US]/shore line erosion [UK]); *syn.* shoreline retreat [n]; *s* **regresión [f] de la línea de la costa** (Retroceso gradual de la costa causado por la acción del viento y de las olas sobre las playas o sobre las rocas; ▶erosión marina); *syn.* retroceso [m] costero; *f* **recul [m] côtier** (Régression des terres sous l'action de la mer et du vent ; un **r. c.** est aussi possible suite à un affaissement de terrain à l'intérieur des terres derrière le trait de côte ; le **r. c.** est d'une manière générale la conséquence de la transgression marine ; ▶abrasion marine, ▶erosion côtière) ; *syn.* recul [m] du trait de côte ; *g* **Küstenrückgang [m]** (Durch Brandungstätigkeit, Rutschungen am Kliff und Windtätigkeit bewirkter Landverlust an Küsten. Ein **K.** ist auch durch Landsenkung im Hinterland der ursprünglichen Küstenlinie möglich, was letztlich einem Vorrücken des Meeres entspricht; cf. WAG 1984; ▶Küstenerosion).

5049 return migration [n] *zool.* (Seasonal movement of birds back to the home territory which may be modified by climatic conditions, feeding needs, or breeding activity; ▶migratory bird); *s* **migración [f] de contrapasa** (Vuelta de ▶aves migradoras de sus cuarteles de invernada); *syn.* migración [f] prenupcial; *f* **migration [f] de retour** (Déplacement saisonnier périodique ramenant les ▶oiseaux migrateurs de leur aire d'hiver à leur point de départ) ; *g* **Rückzug [m] von Zugvögeln** (Jahreszeitlich bedingtes Zurückkehren von ▶Zugvögeln aus ihren Winterquartieren); *syn.* Frühjahrszug [m], Heimzug [m].

5050 return [n] **of bidding documents** [US] *contr. syn.* return [n] of tendering documents [UK]; *s* **devolución [f] de documentos de concurso**; *f* **remise [f] du dossier d'appel d'offres** *syn.* remise [f] du dossier de consultation des entreprises (DCE) ; *g* **Rückgabe [f] der Ausschreibungsunterlagen.**

return [n] **of tendering documents** [UK] *contr.* ▶return of bidding documents [US].

5051 return seepage [n] *hydr.* (VOLL 1973; groundwater in lowlands which is brought to the surface by the pressure of increasing level of a river); *s* **afloración [f] de agua** (En vegas de ríos, agua subterránea que sale a la superficie al ser empujada por agua de crecida del río); *f* **eau [f] d'exfiltration** (Eaux d'un aquifère dans une plaine basse dont la remontée et l'apparition à la surface du sol sont dues à la pression exercée par les masses d'eau d'un cours d'eau ; DG 1984, 176) ; *g* **Qualmwasser [n, o. Pl.]** (Grundwasser, das in einer Niederung von außen durch steigendes Flusswasser hochgedrückt wird und zu Tage tritt).

5052 reusable material [n] *envir.* (Waste or residue material which can be reused industrially without any further processing; includes, e.g. discarded parts, machine units and whole products; ▶recycling); *syn.* recyclable material [n]; *s* **material [m] reutilizable** (Residuo o resto de producción que puede ser reutilizado sin ser procesado; ▶reciclado); *f* **matériau [m] réutilisable** (Déchets ou résidu effectivement récupérable en raison de leur qualité et de leur propreté, pouvant être directe-

R

ment réintroduit dans le cycle économique ; ►récupération, ►recyclage) ; *syn.* matériau [m] disponible ; *g* **Wertstoff [m]** (Abfallstoff oder Rückstand, der ohne weiteren Aufbereitungsprozess als Input des Wirtschaftssystems verwendet werden kann. Hierzu zählen Stoffe, Teile, Aggregate und komplette Produkte; cf. BIFA 1994; ►Recycling).

5053 reusable waste [n] *envir.* (Byproducts which are suitable for reuse after reprocessing); *s* **residuos [mpl] sólidos reciclables o aprovechables** (Desechos cuyos componentes o partes de ellos pueden ser utilizados nuevamente como materia prima una vez que han sido sido procesados); *f* **produit [m] de récupération** (Produits ou matériaux pouvant être séparé d'autres en vue de leur réemploi ou recyclage) ; *syn.* matériaux [mpl] récupérables ; *g* **verwertbarer Abfall [m]** (Abfall, der nach einer Wiederaufbereitung als Rohstoff wieder verwendet werden kann); *syn.* wieder verwertbarer Abfall [m].

5054 reusable water [n] *wat'man.* (Once-used water suitable for industrial purposes, but unusable as potable water); *s* **agua [f] para uso industrial** (Agua destinada a procesos industriales que no es adecuada para el consumo); *f* **eau [f] à usage industriel** (Eau utilisée dans le cycle de production industrielle et impropre à l'alimentation humaine) ; *syn.* eaux [fpl] de recirculation (DUV 1984, 196) ; *g* **Brauchwasser [n]** (…wässer [pl]; für gewerbliche, industrielle oder landwirtschaftliche Zwecke bestimmtes Wasser [Betriebswasser], das als Trinkwasser ungeeignet ist).

5055 reuse [n] **of building rubble** *constr.* (Renewed usage of material recovered after demolition, excavation, or removal of material from a building site until its replacement or reinstallation; ►reapplication of waste [US]/re-application of waste [UK]); *s* **reutilización [f] de material de derribo** (Uso repetido de materiales de construcción recuperados después del derribo o transformación de un edificio o de excavación de un emplazamiento de obra; ►valorización de residuos [Es]); *syn.* reciclaje [m] de material de derribo; *f* **réutilisation [f] de matériaux de démolition** (Mise en œuvre de matériaux recyclables récupérés sur un chantier lors de travaux de démolition ; ►valorisation des déchets) ; *syn.* réemploi [m] de matériaux conservés ; *g* **Wiedereinbau [m, o. Pl.] von Abbruch-/Aufbruchmaterial** (Wiederverwendung von verwertbaren Baustoffen, die auf einer Baustelle durch Abbruch, Aushub oder Ausbau gewonnen wurden; *UB* **W.** von aufbereitetem Straßenaufbruch und Bauschutt; ►Abfallverwertung); *syn.* Bauschuttrecycling [n], Bauschuttrezyklierung [f], Recycling [n] von Abbruch-/Aufbruchmaterial.

5056 reuse [n] **of effluent** *envir.* (1. Recovery of reusable materials from treated ►effluent. 2. Reuse of untreated ►wastewater suitable for industrial purposes; ►sewage treatment); *s* **aprovechamiento [m] de aguas residuales** (Extracción de ciertas sustancias aprovechables de las ►aguas residuales o reutilización del agua para usos industriales que lo permitan; ►tratamiento de agua residuales); *syn.* reciclaje [m] de vertidos líquidos; *f* **récupération [f] des effluents** (Séparation de certaines matières en suspension ou de substances dissoutes dans les ►effluents urbains à des fins de réemploi, de réutilisation ou de recyclage ainsi que le réemploi de l'effluent dans le système de production ; ►traitement des effluents) ; *g* **Abwasserverwertung [f]** (Nutzbarmachung der im ►Abwasser enthaltenen Stoffe oder Wiederverwendung des Abwassers im Rahmen der Kreislaufführung bei Produktionsprozessen, z. B. als Brauchwasser, um kein zusätzliches Trinkwasser zu benötigen oder für die Landwirtschaft [Rieselfelder]; für manche Produktionszwecke wird auch die Restwärme des Abwassers genutzt; ►Abwasserbehandlung); *syn.* Abwasserwiederverwertung [f].

5057 reuse [n] **of on-site construction material** *constr.* (Relocation of temporarily removed and stockpiled construction material intended for reuse; e.g. stone pavers, subbase [UK]/sub-base [UK] for roads, concrete elements, extracted *in situ* gravel; ►backfilling behind structures); *syn.* reinstallation [n] of stockpiled materials; *s* **reutilización [f] de materiales de construcción** (En obra de construcción, recolocación de materiales o elementos que han sido removidos temporalmente, como adoquines, grava de carreteras, elementos de hormigón; ►terraplenado); *f* **reprise [f] des matériaux stockés** (Réutilisation de matériaux ou d'éléments de construction stockés temporairement sur le chantier, p. ex. la terre végétale, les pavés, le tout-venant routier, des éléments en béton, le sable ou le gravier en provenance de déblai, etc. ; ►remblaiement) ; *g* **Wiedereinbau [m, o. Pl.] von zwischengelagerten Baustoffen** (Im Baubetrieb die Wiederverwendung von ausgebauten und ggf. zwischengelagerten Materialien oder Bauelementen, z. B. Pflastersteine, Straßenschotter, Betonelemente, abgebaggerter anstehender Kies; ►Hinterfüllen); *syn.* Wiederverwendung [f] von zwischengelagerten Baustoffen.

reuse [n] **of recoverable materials** *envir.* ►reuse of waste.

reuse [n] **of recycled material** *envir.* ►reapplication of waste.

5058 reuse [n] **of waste** *envir.* (Re-utilization of materials without any special treatment; e.g. second-hand clothing, worn utensils or equipment, kitchen waste as fodder for livestock, instead of ►recycling into different forms; ►reapplication of used materials, ►waste management); *syn.* reuse [n] of recoverable materials; *s* **reutilización [f] de residuos** (En el marco de la ►gestión de residuos sólidos [urbanos], reuso de materiales, sustancias u objetos recuperables para otros fines sin que sea necesario reprocesarlos, como p. ej. aprovechamiento de restos de cocina para la alimentación animal o de maquinaria en otros países menos industrializados; ►reciclado, ►valorización de residuos); *f* **réemploi [m] des déchets** (1. Réutilisation en l'état des déchets récupérés pour un usage analogue à son usage initial, p. ex. le **r.** des vieux vêtements, des bouteilles consignées, des appareils ménagers, des machines industrielles. 2. Valorisation par réemploi et sans transformation des matériaux usés, p. ex. moellons, briques, sable en place ; ►gestion des déchets, ►recyclage, ►valorisation) ; *syn.* réutilisation [f] de matériaux récupérables, récupération [f] et recyclage [m] de matériaux de construction (DUV 1984, 249) ; *g* **Weiterverwendung [f] von Abfallstoffen** (Bei der ►Abfallwirtschaft der Einsatz von Rückständen in einem neuen Einsatzbereich zu dem gleichen Zweck, für den sie hergestellt wurden, ohne dass hierfür ein spezielles physikalisches, chemisches, thermisches oder biologisches Aufbereitungsverfahren nötig ist, z. B. die **W.** von ausgemusterten Kleidern, Maschinen und Geräten, die Verwendung von Küchenresten als Viehfutter; ►Recycling, ►Abfallverwertung); *syn.* Wiederverwendung [f] von Abfallstoffen.

5059 revegetation [n] *landsc.* (TEE 1980, 42; reestablishment of vegetation in an area without plant growth for a long period of time; ►reintroduction); *s* **revegetalización [f]** (Reestablecimiento de vegetación en un área sin ella durante un largo periodo/período de tiempo; ►reintroducción de especies extintas); *f* **revégétalisation [f]** (Reconstitution du couvert végétal d'un espace resté longtemps sans végétation ; ►réintroduction 2) ; *g* **Wiederbegrünung [f]** (Begrünung einer Fläche, die längere Zeit vegetationslos war; ►Wiederansiedelung).

reverse falls [n] [UK] *constr.* ►reverse slope [US].

5060 reverse slope [n] [US] *constr.* (Reversal of incline in a general downgrade; ►slope [US] 2/fall [UK], ►cross slope

[US]/cross falls [UK]); *syn.* back slope [n] [also US], reverse falls [npl] [UK] (WRP 1974, 117); *s* **contrapendiente [f]** (▶Pendiente contraria a la inclinación predominante de un talud o una superficie; ▶pendiente transversal; *f* **contre-pente [f]** (▶Pente opposée à la pente générale d'une surface, d'un talus ; ▶pente transversale, ▶dévers) ; *g* **Gegengefälle [n]** (Der generellen Neigung einer Fläche oder eines Hanges gegenläufiges ▶Gefälle ; ▶Quergefälle).

5061 reversion [n] *agr.* (Turning-over of the ground by ▶plowing [US]/ploughing [UK]; e.g. for the transformation of grassland to cropland: grassland conversion; ▶plowing up of grassland [US]/ploughing-up of grassland [UK]); *s* **laboreo [m] de un prado** (Transformación de un prado en campo de cultivo; ▶roturación de prados, ▶roturado de la tierra); *f* **labour [m] de défrichement** (Façon culturale de découpage et de retournement de la terre d'une prairie, d'une lande ou de terrains incultes au moyen d'une charrue pour la rendre propre à la culture ; ▶retournement de prairie, ▶retournement du sol) ; *syn.* défrichage [m] d'un pré ; *g* **Umbruch [m, o. Pl.]** (Durch Pflügen bewirktes Wenden einer bewachsenen landwirtschaftlichen Fläche, z. B. die Umwandlung von [Dauer]grünland, Dauerbrache oder Heideflächen in Ackerland; umbrechen [vb] ; ▶Bodenumbruch, ▶Grünlandumbruch); *syn.* Umbrechen [n, o. Pl.].

reverted pasture [n] *agr.* ▶abandoned pasture.

revetment [n], rock *constr. eng.* ▶riprap.

5062 revised version [n] *plan.* (Result of revision of a design; ▶revision); *syn.* amended version [n], up-dated version [n]; *s* **versión [f] revisada** (Resultado de la ▶revisión de un plan; ▶modificación, ▶revisión); *f* **version [f] révisée** (Résultat de la ▶révision d'un plan, d'une planification ; pour la révision d'un document d'urbanisme, on utilisera de préférence l'expression « le plan révisé » ; ▶modification) ; *g* **überarbeitete Fassung [f]** (Ergebnis einer aktualisierten oder neu bearbeiteten Version einer bestehenden Planung; ▶Überarbeitung); *syn.* überarbeitete Version [f].

5063 revision [n] *plan.* (Process of up-dating a design; ▶plan revision, ▶revised version); *syn.* amendment [n], up-date [n]; *s 1* **revisión [f]** (Proceso de actualización de un plan, un esbozo, etc. *con* alteraciones significativas; ▶revisión de un plan, ▶versión revisada); *s 2* **modificación [f]** (Proceso de actualización de un plan, un esbozo, etc. *sin* alteraciones significativas; ▶modificación de un plan); *f 1* **révision [f]** (d'un document d'urbanisme ; procédure analogue à celle de l'élaboration, l'ancien plan continuant de s'appliquer jusqu'à l'approbation du nouveau) ; *terme spécifique* révision simplifiée d'un PLU/POS : procédure simple, et rapide, mais limitée à un seul objet) ; *f 2* **modification [f]** (Processus ou procédure permettant la mise à jour **1.** des plans d'un projet d'aménagement. **2.** Du PLU/POS : modification sur des aspects limités, de forme, de la reformulation des règles de l'actuel règlement, et de l'ajout de nouvelles règles, cette modification n'apporte que des modifications mineures au plan local d'urbanisme [PLU]/plan d'occupation des sols [POS] ; son économie générale n'est pas affectée et les règles d'urbanisme demeurent adaptées au plus près des objectifs définis lors de l'élaboration du PLU, elle ne remet pas en cause les choix généraux faits lors de l'établissement du document initial, elle ne doit réduire un espace boisé classé, une zone agricole [NC dans un POS — A dans un PLU] ou une zone naturelle et forestière [ND dans un POS — N dans un PLU], ou une protection édictée en raison des risques de nuisance, de la qualité des sites, des paysages ou des milieux naturels et ne doit pas comporter de grave risques de nuisance) ; *f 3* **mise [f] à jour** (Modifications destinées à l'actualisation du PLU/POS, p. ex. afin d'intégrer au PLU/POS des éléments réglementaires et carto-graphiques liés à d'autres plans ; ▶modification du plan, ▶version révisée) ; *g* **Überarbeitung [f]** (Verbesserung und Aktualisierung von etwas planerisch oder schriftlich Vorhandenem; das Ergebnis ist eine neue Fassung, eine ▶überarbeitete Fassung, ein Update; ▶Überarbeitung eines Planes).

revision [n] of a plan *plan.* ▶plan revision.

5064 revitalization [n] *sociol. urb.* (PSP 1987, 196; aim and result of measures which restore the vitality of an abandoned landscape, a neglected park, a run-down street, etc. and increase their attractiveness; ▶industrial revitalization [US]/redevelopment of industrial areas [UK]); *s* **revitalización [f]** (Programas y medidas para mejorar la calidad y el atractivo p. ej. de barrios marginales, de paisajes deteriorados, calles, parques, etc.; ▶saneamiento de ruinas industriales); *f* **réanimation [f]** (Action de redonner une âme, de rendre la vie à des monuments désaffectés ou à des ensembles urbains ou ruraux en voie de dépérissement, réalisée au moyen de la réhabilitation, la réimplantation d'anciennes fonctions ou l'implantation de nouvelles ; DUA 1996 ; ▶rénovation de zones industrielles) ; *syn.* revitalisation [f] ; *g* **Wiederbelebung [f]** (Ziel und Maßnahmen, die z. B. ein vernachlässigtes Stadtviertel, eine verödete Landschaft, einen vergessenen Park wieder mit Leben erfüllen und die Attraktivität erhöhen; ▶Sanierung von Industriegebieten); *syn.* Revitalisierung [f].

revitalization [n] [US], neighborhood *plan. urb.* ▶neighborhood improvement [US]/neighbourhood improvement [UK], ▶neighborhood redevelopment [US]/district redevelopment [UK].

revitalization [n] [US], village *plan. urb.* ▶village renewal.

revitalization [n] of street tree pits *constr.* ▶revitalization of tree pits.

5065 revitalization [n] of tree pits *constr.* (Improvement of tree planting sites to encourage growth of trees by changing surface material, and sometimes by installation of technical devices for ventilation, irrigation or fertilization; ▶soil improvement); *syn.* improvement [n] of (street) tree pits; *s* **revitalización [f] (de ubicaciones) de árboles urbanos** (Medidas horticulturales y paisajísticas para mejorar las condiciones de crecimiento de los árboles en la ciudad, p. ej. apertura del alcorque para permitir la aireación, enmiendas de abono, etc.; ▶enmienda del suelo); *syn.* mejora [f] de ubicación de árbol; *f* **amélioration [f] du lieu d'implantation des arbres** (En général en milieu urbain, amélioration des conditions de croissance des arbres en redonnant au sol les qualités indispensables pour assurer un bon développement racinaire du végétal, p. ex. opérations visant à retrouver la perméabilité des supports originels [enlèvement des revêtements imperméables], mise en place de dispositifs d'aération, amélioration des apports hydriques, décompactage pneumatique et nutrition-fertilisation mécanique ; ▶correction de la terre) ; *syn.* revitalisation [f] du trou d'arbre ; *g* **Baumstandortsanierung [f]** (Umfassende Maßnahmen des Garten- und Landschaftsbaues zur Ermöglichung und Förderung der Lebensbedingungen von [Straßen]bäumen in der Stadt, z. B. durch Entsiegelung befestigter Flächen, evtl. durch Einbau technischer Einrichtungen zur Belüftung, Bewässerung oder Nährstoffversorgung; ▶Bodenverbesserung 2); *syn.* Baumstandortverbesserung [f], Baumumfeldsanierung [f], Baumumfeldverbesserung [f], standortverbessernde Maßnahmen [fpl] an Baumscheiben, Standortoptimierung [f] für Bäume, Baumstandortverbesserung [f].

5066 revitalization [n] of woody plants *constr.* (Increasing the vitality of existing trees and shrubs; *specific terms* ▶revitalization of tree pits, ▶tree [treatment] work); *s* **saneamiento [m] de leñosas** (Tratamiento de árboles y arbustos existentes en un parque, jardín o en otras zonas de la ciudad;

R

términos específicos ►saneamiento de árboles, ►revitalización [de ubicaciones] de árboles urbanos); *syn.* revitalización [f] de leñosas; *f* **restauration [f] végétale** (Soins apportés à la végétation existante sur le site d'un aménagement ; ►amélioration du lieu d'implantation des arbres, ►travaux de restauration arboricole) ; *g* **Gehölzbestandssanierung [f]** (Sanierung eines vorhandenen Gehölzbestandes. *UBe* ►Baumsanierung, ►Baumstandortsanierung); *syn.* Bestandssanierung [f] von Gehölzen.

revitalization plan [n] *constr. plan. urb. wat'man.* ►rehabilitation plan.

5067 rewatering [n] **of a bog** *conserv.* (Measures aimed at allowing water to collect in a depleted bog, so that conditions are favo[u]rable for the further generation of peat; ►bog regeneration); *syn.* resaturation [n] (of a bog) with water; *s* **reencharcamiento [m] de turbera** (Medidas de reintroducción de agua con las cuales se pretende alcanzar un grado de humedad de las turberas explotadas o degradadas, de manera que puedan volver a desarrollarse como tales; ►regeneración de turberas); *f* **réengorgement [m] d'une tourbière** (Mesure de conservation des tourbières par remise en eau des fosses de tourbage [dans l'exploitation manuelle traditionnelle] ou des vastes espaces des tourbières exploitées [exploitation industrielle] rétablissant un régime hydrique favorisant la réapparition des stades initiaux de la tourbière ou l'établissement d'une biocénose similaire ; ►régénération de la tourbière) ; *g* **Wiedervernässung [f] eines Moores** (Maßnahmen, mit denen versucht wird, in einem abgetorften oder degradierten Moor soviel Wasser sich ansammeln zu lassen, dass sich für eine Moorentwicklung oder moorähnliche Biozönose ein günstiges Feuchteregime einstellt; ►Moorregeneration).

5068 rezoning [n] **[US]** *leg. plan.* (Change of zoning classification to alter the use by redevelopment of particular plots or areas of land; ►conversion 1, ►rearrangement of lots [US]/reorganization of plot boundaries [UK]); *syn.* rescheduling [n] [UK]; *s* **redefinición [m] de uso(s)** (Modificación del uso del suelo previsto en plan de ordenación municipal o plan del paisaje; ►cambio de uso[s], ►reordenación de parcelas); *f* **déclassement [m]** (Décision administrative souvent accompagnée d'une procédure, par laquelle un catégorie d'usages [documents d'urbanisme], de protection [espaces protégés, monuments historiques], d'appartenance à un certain domaine public [voirie] changent de régime juridique ; ►modification de l'usage, ►remembrement parcellaire) ; *g* **Umwidmung [f]** (Änderung der im Bauleitplan und Landschaftsplan festgesetzten Nutzung oder eine straßenrechtliche Widmungsänderung; ►Nutzungsänderung, ►Umlegung).

rheophilous peatland [n] *geo.* ►fen (1).

rheotrophic peatland [n] *geo.* ►fen (1).

rhithral [n] *limn.* ►salmonid zone [UK].

5069 rhithron [n] *limn.* (Life community of the ►salmonid zone; ►epirhithron, ►potamon); *s* **ritrón [m]** (Biocenosis de la ►región salmonícola; ►epiritrón, ►potamium); *f* **rhithron [m]** (Terme de zonation écologique des eaux courantes caractérisant le cours supérieur à courant modéré et fort dans lequel vit la communauté animale de la ►région salmonicole ; ►épirhithron, ►potamon) ; *g* **Rhithron [n]** (Lebensgemeinschaft der ►Bergbachregion; ►Epirhithron, ►Potamon).

rhizogenesis [n] *biol.* ►root formation.

5070 rhizomatous [adj] *bot.* (LD 1994 [4], 12; descriptive term applied to plants with ►rhizoms or characteristically forming them; ►stoloniferous, ►stem-spreading); *s* **rizomatoso/a [adj]** (Término descriptivo para plantas que desarrolla rizomas; ►estonílfero/a 1, ►estonílfero/a 2, ►rizoma); *f* **rhizomateux, euse [adj]** (Relatif à une plante développant un ►rhizome ;

►traçant, ante ►stolonifère) ; *g* **rhizombildend [ppr/adj]** (Eine Pflanze betreffend, die ►Rhizome bildet; ►Ausläufer treibend, ►Stolonen bildend).

5071 rhizome [n] *bot. hort.* (Creeping underground root-like thickened stem, growing just beneath the surface, consisting of a series of nodes and short internodes. The roots commonly develop from nodes and buds from leaf axils); *s* **rizoma [m]** (Metamorfosis caulinar debida a la adaptación a la vida subterránea, es decir, tallo subterráneo, cuya función es la de defender a la planta contra los rigores del ambiente en la estación desfavorable de zonas frías o extremadamente secas; cf. DB 1985); *f* **rhizome [m]** (Tige souterraine, vivace, gorgée de réserves nutritives, émettant au fil des ans des pousses aériennes, des feuilles écailleuses et des racines adventices chez certaine Ptéridophytes et certains Angiospermes ; DIB 1988, 323) ; *g* **Rhizom [n]** (Unterirdisches Überwinterungsorgan und als Speicher dienende Sprossachse mit kurzen Internodien, die aus mehreren Vegetationsperioden stammen. **R.e** haben sprossbürtige Wurzeln und farblose, schuppenartige Blätter und/oder vergängliche Niederblätter); *syn.* Wurzelstock [m] (1), Erdstamm [m].

5072 rhizosphere [n] *arb. hort.* (Zone of growing medium in which the environment for microbial activity in general is influenced by any root growing in it. The diameter of the **r.** equals that of the cylinder of soil around a root which root hairs explore and into which they may release root exudates; RRST 1983; ►drip line area, ►rooting zone); *s* **rizosfera [f]** (Entorno de las raíces; espacio que influye o es influido, física o biológicamente, por las raíces de las plantas; DINA 1987; ►zona de desarrollo de la raíz, ►zona de penetración [de la raíz]); *f* **rhizosphère [f]** (Volume du sol en contact immédiat avec les racines les plus fines, dans lequel prolifèrent des microorganismes ; TSF 1985 ; ►horizon racinaire, ►zone de développement des racines) ; *syn.* zone [f] rhizosphérique (PED 1979, 180) ; *g* **Rhizosphäre [f]** (Unmittelbar um die Feinwurzeln befindlicher Bodenkörper; ►Wurzelbereich, ►Wurzelraum).

rhythm [n] *biol.* ►circadian rhythm.

rhythm [n], **circannual** *biol.* ►annual periodicity.

rhytidome [n] *arb. bot.* ►outer bark.

ribbon development [n] **[UK]** *urb.* ►strip development [US] (1), ►strip development [US] (2).

rich in forbs [loc] *phyt.* ►forb-rich.

rich in nitrogen [loc] *limn. pedol.* ►nitrogen-rich.

5073 richness [n] *landsc. plan.* (Non-ecological term describing an element of the environment, or a landscape or project with regard to the number of various plants and animals, or design features; ►diversity, ►species richness); *s* **variedad [f]** (Visión no ecológica de una sección de la naturaleza/un paisaje o de un objeto de planificación en cuanto a la cantidad de elementos naturales o de especies de flora y fauna que posee; ►diversidad ecológica, ►riqueza en especies); *syn.* heterogeneidad [f]; *f* **variété [f]** (Approche non écologique de description d'un espace/aire paysagère ou d'un projet d'aménagement ayant pour base le nombre des espèces de faune et de flore ou d'éléments de la composition ; ►diversité, ►richesse en espèces) ; *syn.* hétérogénéité [f], multiplicité [f] ; *g* **Vielfältigkeit [f]** (Nichtökologische Betrachtungsweise eines Natur-/Landschaftsausschnittes oder eines Planungsobjektes hinsichtlich der Anzahl unterschiedlicher Pflanzen- und Tierarten oder Gestaltungselemente; ►Artenvielfalt, ►Diversität); *syn.* Mannigfaltigkeit [f], Vielfalt [f] von Natur und Landschaft.

5074 richness [n] **of food supply** *zool.* (Natural existence or human introduction of sufficient and diversified food for animals); *s* **riqueza [f] de alimentos** (Provisión natural o antropógena de recursos alimenticios para los animales en gran cantidad y diversidad); *f* **richesse [f] en ressources alimentaires** (Aliments présents en abondance naturellement dans le milieu ou apportés par l'homme) ; *g* **reichhaltiges Nahrungsangebot [n]** (Vielfältige, genügend vorhandene natürliche oder durch den Menschen bereitgestellte Nahrungsressourcen).

5075 ridge [n] *geo.* (**1.** Long, narrow, steep-sided upland between valleys. **2.** The term occasionally applied to a range of hills or mountains. **3.** Top of upper part of a hill; a narrow, elongated crest of a hill or mountain; GFG 1997); *syn.* mountain ridge [n]; *s* **cresta [f]** (Cumbre peñascosa de una montaña); *f* **crête [f] de montagne** (Ligne culminante de forme arrondie et continue d'un ensemble montagneux) ; *g* **Bergrücken [m]** (Rundlich, längliche geomorphologische Vollform mit unterschiedlich steilen Hängen; im Vergleich hierzu ist der **Grat** eine scharfkantige Vollform mit durchgehend steilen Abhängen).

ridge [n], scarped *geo.* ▶cuesta.

5076 ridge planting [n] *for. constr.* (**1.** Setting out young woody plants on a long, narrow crest of excavated soil, usually on shallow soils and in regions with high preciptation; ▶mound planting); **2. step planting [n]** (Like ▶ridge planting, but setting out the plants on the shady side of such a ridge); *syn.* side planting [n]; *s* **plantación [f] en caballones** (Plantación de leñosas jóvenes en diques estrechos, usual en zonas con suelos poco profundos y en zonas muy lluviosas; ▶plantación en montículos); *syn.* plantación [f] en camellones; *f 1* **plantation [f] sur bourrelet** (DFM 1975 ; plantation de jeunes végétaux ligneux sur la crête d'un billon, réalisée à l'aide d'une charrue sur les sols de faible épaisseur et dans les régions à grande pluviosité ; ▶plantation sur butte) ; *f 2* **plantation [f] sur ados** (Plantation similaire à la ▶plantation sur bourrelet, mais sur le flanc du bourrelet ; ▶plantation sur butte) ; *g* **Kammpflanzung [f]** (Pflanzen von jungen Gehölzen auf schmalen, langen „Dämmen"; üblich auf flachgründigen Böden und in Gebieten mit hohen Niederschlägen; ▶Hügelpflanzung).

riding [n] [UK], horse *recr.* ▶horseback riding [US].

5077 riding circuit [n] *recr.* (Circular path provided for ▶horseback riding [US]/horse riding [UK]); *s* **itinerario [m] hípico** (Camino preparado para ir a caballo y que conduce de nuevo al lugar de origen; ▶equitación); *syn.* itinerario [m] ecuestre; *f* **circuit [m] équestre** (Piste d'▶équitation, sentier équestre revenant sur son point de départ) ; *syn.* circuit [m] cavalier, parcours [m] équestre ; *g* **Reitrundweg [m]** (Zum ▶Pferdereiten angelegter Weg, der wieder an seinen Ausgangspunkt zurückführt).

riding sport [n] *recr.* ▶horseback riding sport [US]/equestrian sport [UK].

5078 riding stable [n] [US] *recr.* (Building with stalls for shelter and upkeep of riding horses); *syn.* horse riding stable [n] [UK]; *s* **picadero [m]** (Lugar destinado para adiestrar los caballos y aprender a montar; CAS 1985); *f* **ferme [f] équestre** (Équipement touristique en milieu rural élevant des chevaux et constituant une des formes de l'agrotourisme) ; *syn.* centre [m] équestre ; *g* **Reiterhof [m]** (Einrichtung, in der Reitpferde gehalten werden).

riding trail [n] *recr.* ▶bridle path [US]/bridle way [UK].

riding trail network [n] [UK], horse *recr.* ▶bridle path network [US].

right [n] *min. leg.* ▶extraction right.

right [n], hunting *hunt. leg.* ▶hunting laws.

right [n] [UK], pre-emptive *adm. leg.* ▶right of first refusal [US].

right angle parking layout [n] *plan. trans.* ▶ninety-degree parking layout.

5079 right [n] of first refusal [US] *adm. leg.* (Prerogative of a public authority to be the first to be offered a plot of land for sale, for which a legally-binding land-use plan, such as a ▶zoning map [US]/Proposals Map/Site Allocations Development Plan Document [UK] will be drawn up, or a plan will be prepared for ▶re-organization of land holdings; **in U.S.**, right of first refusal may also be the right of a riparian owner to a preference in the acquisition of land under tide waters adjoining his upland; IBDD 1981; ▶eminent domain [power of condemnation], ▶land acquisition); *syn.* preemptive option [US]/pre-emptive right [n] [UK], right of pre-emption [n] [UK]; *s* **derecho [m] de tanteo** (Derecho de la administración o de instituciones públicas y semipúblicas de recibir la primera oferta de venta de terrenos, p. ej. cuando se ha aprobado un ▶plan parcial [de ordenación] o en el marco de ▶reparcelaciones; ▶adquisición de terrenos, ▶expropiación forzosa); *f* **droit [m] de préemption** (Dans le cadre de l'aménagement foncier, droit pour une personne publique ou semi-publique d'exiger le transfert à son profit de la propriété d'un bien, lorsque le propriétaire a l'intention de le céder à un tiers dans les communes dotées d'un ▶plan local d'urbanisme [PLU]/plan d'occupation des sols [POS] rendu public ou approuvé ; cf. C. urb., art. L. 210-1, 211-1, 300-1 ; ▶acquisition foncière, ▶réorganisation foncière) ; *g* **Vorkaufsrecht [n]** (**1.** Das Bürgerliche Gesetzbuch unterscheidet das schuldrechtliche [§§ 463 ff BGB] und das dingliche [§§ 1094 BGB] **V. 2.** In der Stadtplanung ist es das Recht, dass Träger öffentlicher Belange, Grundstücke, die zum Verkauf anstehen, zuerst angeboten bekommen, weil nach dem Baugesetzbuch [BauGB] der Gemeinde ein öffentlich rechtliches **V.** bei Grundstücken zusteht, die bei einem ▶Bebauungsplan als öffentliche Flächen vorgesehen sind, die in einem Umlegungsgebiet [▶Bodenordnung], in einem Sanierungsgebiet oder in einem städtebaulichen Entwicklungsgebiet liegen, die in einem Geltungsbereich einer Erhaltungssatzung oder als künftiges Wohnbauland im Geltungsbereich eines Flächennutzungsplanes liegen oder die als innerörtliche Baulücken für die Wohnbebauung vorgesehen sind; Ziel des **V.**es ist die Verhinderung der Bodenspekulation; ▶Enteignung, ▶Grunderwerb).

right-of-passage [n], vehicular and pedestrian *leg. urb.* ▶vehicular and pedestrian easement.

5080 right-of-passage [n] for fire control (≠) *leg.* (In France, a regulation on private ground as an access right for fire control vehicles; ▶fire lane [US]/fire path [UK]); *s 1* **servidumbre [f] de acceso y gestión para la lucha contra incendios** (≠) (**F.**, norma legal que, tras un procedimiento determinado que culmina en una resolución administrativa y en base a la cual los servicios de extinción de incendios forestales pueden transitar por y aplicar medidas en los montes públicos y privados con el fin de combatir incendios forestales, tomar medidas de prevención de los mismos, así como construir y mantener instalaciones para combatir el fuego que pueden ser ordenadas administrativamente, como la construcción de accesos, cortafuegos y de bocas de incendios); *s 2* **servidumbre [f] de uso para los servicios de prevención y extinción de incendios** (En Es., norma de la Ley de montes para las «zonas de alto riesgo de incendio», según la cual los servicios de lucha contra incendios forestales pueden utilizar las infraestructuras existentes o de nueva creación para cumplir su función. Además, en los Planes de Defensa que deben ser elaborados por las CC.AA. para estas zonas, se puede ordenar —también a los propietarios privados— realizar trabajos de carácter preventivo, incluyendo tratamientos selvícolas, áreas cortafuegos, vías de acceso y puntos de agua; cf. art. 48.2 y 6 Ley 43/2003, de Montes; ▶vía de bomberos); *f* **servitude [f] de**

R

passage et d'aménagement pour assurer la continuité de la lutte contre les incendies (Les bois situés dans les régions particulièrement exposées aux incendies de forêts peuvent faire l'objet d'un classement prononcé par décision administrative après avis des conseils municipaux intéressés et du conseil général. Dans les bois classés, une servitude de passage et d'aménagement est établie par l'Etat à son profit ou au profit d'une autre collectivité publique, d'un groupement de collectivités territoriales ou d'une association syndicale pour assurer exclusivement la continuité des voies de défense contre l'incendie, la pérennité des itinéraires constitués, ainsi que l'établissement des équipements de protection et de surveillance des forêts, pour entreprendre les travaux ayant pour but de protéger ou reconstituer les massifs particulièrement exposés aux incendies, notamment des pare-feu, des voies d'accès, des points d'eau, d'effectuer les travaux de prévention des incendies comme le débroussaillement, le brûlage dirigé des pâturages et des périmètres débroussaillés ; cf. lois n° 85-1273 et n° 92-613 ; ▶voie pompiers) ; *syn.* droit [m] de passage et d'aménagement pour assurer la continuité de la lutte contre les incendies ; *g* **Fahr- und Bewirtschaftungsrecht [n] zur Brandbekämpfung in Wäldern (≠)** (In Frankreich eine gesetzliche Regelung, die per Verwaltungsbeschluss nach Anhörung der Gemeinderäte und des *Conseil général* [Rat eines Départements] in allen brandgefährdeten, öffentlichen und privaten Wäldern ein Fahr- und Nutzungsrecht für Feuerwehreinsätze und brandbekämpfende Maßnahmen sowie der Bau und die ständige Unterhaltung von Einrichtungen für die Bekämpfung des Feuers festgesetzt werden können, insbesondere die Anlage von Brandschneisen, Zufahrtswegen, Hydranten sowie Maßnahmen zur Verhütung der Brandausbreitung wie z. B. das Auf-den-Stock-Setzen von Waldstreifen und das Abbrennen von Wiesen und abgeholzten Flächen; ▶Feuerwehrweg).

right of pre-emption [n] [UK] *adm. leg.* ▶right of first refusal [US].

5081 right [n] **of public access** *leg. recr.* (Legal right to go to or through uncultivated or forested private land for recreational use; ▶vehicular and pedestrian easement); *syn.* "right to roam" legislation [n] [also UK]; *s* **servidumbre [f] de tránsito** (Regulación que garantiza el acceso gratuito del público a la naturaleza en general, fijada en las leyes de protección de la naturaleza de las «Länder» de D. **En Es.** la Ley de Costas fija así mismo la **s. de t.** en una franja de 6 m a partir del límite interior de la ribera del mar para el paso público peatonal, salvo en espacios especialmente protegidos, y la ▶servidumbre de acceso al mar en zonas urbanas o urbanizables tanto para peatones [cada 200 m] como para vehículos [cada 500 m]; cf. art. 27 Ley 22/1988 de 28 de julio); *syn.* servidumbre [f] de paso; *f* **servitude [f] de passage des piétons** (F., on distingue **1.** la servitude de marchepied permettant l'utilisation pour le service de la navigation des bords d'une voie navigable ou flottable, **2.** la ▶servitude de passage à l'usage des pêcheurs le long des cours d'eau et plans d'eau domaniaux ; loi n° 84-512, **3.** la servitude de libre passage sur les berges des cours d'eau non navigables ni flottables ; loi n° 59-96, **4.** la servitude longitudinale et transversale de passage des piétons sur le littoral, permettant la continuité du cheminement des piétons ou leur libre accès au rivage de la mer ; loi n° 76-1285) ; *g* **Betretungsrecht [n]** (In den Naturschutzgesetzen der Bundesländer näher geregeltes Recht, nach dem alle Teile der freien Natur, insbesondere Wald, Bergweide, Fels, Ödland, Brachflächen, Uferstreifen, Moore und landwirtschaftliche Flächen von jedermann auf eigene Gefahr unentgeltlich betreten werden können; cf. Art. 22 BayNatSchG, § 37 NatSchG-BW und Landesforstgesetze; der weiter gefasste Begriff wäre **Zugangsrecht**, womit eindeutig unterschieden werden könnte, dass die

Landschaft auch für alle Formen der sonstigen Sportausübung wie z. B. Radfahren und Mountainbike-Fahren, Reiten, Kanufahren, Segeln, Gleitschirmfliegen und Tauchen nutzbar wäre, sofern diese Aktivitätsarten im konkreten Landschaftsbereich im Sinne des Natur- und Landschaftsschutzes natur- und landschaftsverträglich eingestuft werden; ▶Geh- und Fahrrecht); *syn.* Betretensrecht [n].

5082 right [n] **of use** *leg.* (Right to use the property of another and to draw the profits it produces without wasting its substance; in U.S., use by right, which is conveyed with ownership, as recorded on a land register, and not requiring discretionary approval by an official or administrative agency); *syn.* usufruct [n]; *s* **usufructo [m]** (Derecho a usar de la cosa ajena y aprovecharse de todos sus frutos sin deteriorarla; CAS 1985); *syn.* derecho [m] de uso); *f 1* **droit [m] d'usage** (Droit de jouissance temporaire conférant à une personne déterminée le droit d'utiliser un bien appartenant à autrui et d'en percevoir les fruits dans la limite de ses besoins et de ceux de sa famille) ; *f 2* **droit [m] d'usage et d'habitation** (*Terme spécifique* droit conférant à une personne déterminée la faculté de demeurer dans un bien immobilier mais non celui de le louer ou de le vendre) ; *f 3* **usufruit [m]** (Droit réel d'user, de se servir d'un bien et d'en percevoir les revenus ou de le louer à autruit ; www.paris. notaires.fr); *f 4* **droit [m] d'utilisation** (Droit limité d'utilisation non commerciale d'un bien, d'un produit, de services) ; *g* **Nutzungsrecht [n]** (Berechtigung oder Befugnis, ein fremdes Eigentum, hier ein fremdes Grundstück, in bestimmter Weise zu nutzen).

5083 right-of-way [n] *leg. urb.* (Strip of land acquired by reservation, dedication, forced dedication, prescription or condemnation and intended to be occupied by a road, crosswalk, railroad, electric transmission lines, oil or gas pipeline, water line, sanitary storm sewer, and other similar public uses; IBDD 1981; ▶easement, ▶public right-of-way, ▶utility easement, ▶vehicular and pedestrian easement); *s* **servidumbre [f] de servicios públicos** (En Derecho, privilegio de la sociedad para utilizar terrenos de propiedad privada con el fin de conducir líneas eléctricas, conducciones de abastecimiento, vías de tránsito, etc. a través de ellos en el caso de que no sea posible hacerlo a través de terrenos públicos; ▶servidumbre de acceso, ▶servidumbre de paso de conducciones, ▶servidumbre inmobiliaria); *f* **servitudes [fpl] d'utilité publique** (Terme désignant les mesures imposées par la loi et limitant le droit de propriété au profit de la collectivité, d'établissements publics ou de concessionnaires de services publics ou d'un nombre restreint de personnes et touchant les domaines les plus divers tels que le libre passage, l'alignement, l'écoulement des eaux, les ouvrages du réseau d'alimentation et d'assainissement, etc. ; ▶servitude, ▶servitude de passage ; ▶servitudes relatives à l'utilisation des réseaux d'alimentation et de distribution d'énergie, de canalisation et de télécommunication) ; *g* **Geh-, Fahr- und Leitungsrecht [n]** (Zu Gunsten der Allgemeinheit, eines Erschließungsträgers oder eines beschränkten Personenkreises zu belastende Grundstücksfläche als Durchgang, Durchfahrt oder zum Verlegen von unter- oder oberirdischen Versorgungsleitungen; § 9 [1] 21 BauGB); ▶Geh- und Fahrrecht, ▶Grunddienstbarkeit, ▶Leitungsrecht, ▶Straßen- und Wegerecht)

right-of-way [n] [UK], **railway** *leg. trans.* ▶railroad right-of-way [US].

5084 right-of-way planting [n] [US] *leg. trans. landsc.* (All planted and grassed areas, including slopes in cut or fill, which are part of a public right-of-way; ▶street planting); *syn.* roadside landscaping [n] [also US], roadside planting [n] [UK], street landscaping [n] [also US]; *s* **verde [m] vial** (Conjunto de superficies plantadas o sembradas en la red viaria; ▶plantación a lo largo de la calle); *syn.* plantaciones [fpl] viarias; *f* **espaces [mpl] verts d'accompagnement de la voirie** (Plantations [arbres, pelouses, massifs floraux] d'aspect fonctionnel placées en bordure des voies de circulation, aux abords immédiats des édifices publics, bâtiments et commerces de proximité, contribuant à l'amélioration de la qualité de la vie en milieu urbain ; ►plantations du bord des routes) ; *syn.* espaces [mpl] verts sur voies publiques ; *g* **Verkehrsgrün [n, o. Pl.]** (Alle Pflanz- und Rasenflächen einschließlich der Straßenböschungen im Einschnitt oder Auftrag, wenn sie Bestandteil der öffentlichen Straßen, Wege und Plätze sind; cf. §1 FStrG; ►Straßenbepflanzung); *syn.* Grün [n, o. Pl.] an Straßen, Verkehrsgrünfläche(n) [f(pl)], Straßenbegleitgrün [n, o. Pl.], Straßengrün [n, o. Pl.], Verkehrsbegleitgrün [n, o.Pl.].

rights [n] *leg.* ►neighbourhood rights [UK], ►riparian rights, ►water rights.

rights [npl] **of way** [UK]**, extinguishment of** *adm. leg. urb.* ►street abandonment [US]/extinguishment of rights of way [UK].

"right to roam" legislation [n] [UK] *leg. recr.* ►right of public access.

rigid pavement [n] *constr.* ►concrete pavement.

rill [n] *geo. pedol.* ►erosion rill; *wat'man.* ►streamlet.

5085 rill erosion [n] *geo. pedol.* (Surface erosion process on sloping fields in which numerous and randomly occurring small channels of only several centimeters in depth are formed; occurs mainly on recently cultivated soils; cf. SST 1997; ►gully erosion, ►natural erosion); *s* **erosión [f] en regueros** (Consiste en el arrastre de elementos terrosos al correr el agua por la superficie del suelo, ocasionando la formación de surcos o regueros orientados más o menos normalmente a las curvas de nivel; DINA 1987; ►erosión, ►erosión en cárcavas); *syn.* erosión [f] en surcos; *f* **ruissellement [m] en filets** (Écoulement diffus de l'eau dans un réseau instable de petites rigoles anastomosées en mailles amygdaloïdes ; VOG 1979, 164 ; ►érosion, ►ravinement 1) ; *syn.* érosion [f] en filets ; *g* **Rillenerosion [f]** (**1.** Form der Bodenerosion, bei der durch linear fließendes Niederschlagswasser flache, maximal mehrere Zentimeter tiefe Furchen am Hang entstehen. Diese Rillen verschwinden durch Bodenbearbeitung oder durch natürliche geomorphologische Prozesse an der Erdoberfläche relativ rasch, führen aber zu einer ständigen [schleichenden] Bodenabtragung, was an der Kappung des Bodenprofils erkennbar wird; ►Erosion, ►Grabenerosion 1. **2.** Wenige Dezimeter [15-50 cm] tiefe Erosionslinien ergeben eine **Rinnenerosion**); *syn.* Rillenspülung [f] (WAG 1984).

ring [n] *arb.* ►annual ring; *landsc. urb.* ►green open space ring; *constr.* ►manhole adjustment ring;

ring [n]**, growth** *arb.* ►growth layer.

ring [n]**, manhole adjusting** *constr.* ►manhole adjustment ring.

5086 ring drain [n] *constr.* (Drainage line, collecting water from a gravel filter layer, e.g. around a building or sportsfield); *s* **drenaje [m] circular** (Tubería alrededor de un edificio o de un campo de deportes que recoge el agua de un filtro de gravilla y garantiza su evacuación); *f* **ceinture [f] de drainage** (Conduite drainante assurant la collecte des eaux d'infiltration dans un filtre drainant ou autour d'un bâtiment) ; *g* **Ringdränung [f]** (Im Kreis geführte Dränleitung, die den Wasserabfluss eines Kiesflächenfilters, z. B. eines Sportplatzes oder um ein Gebäude herum, sicherstellt); *syn.* Ringdrainage [f].

ringed bird [n] [UK] *zool.* ►banded bird.

5087 ring road [n] *trans.* (City or suburban road which forms a complete or partial circle around a particular area; ►beltway [US]; ►bypass [US]/by-pass [UK]); *syn.* circumferential road [n]; *s* **anillo [m] periférico** (Carretera principal que forma un círculo total o parcial alrededor de una ciudad; ►autopista periférica, ►carretera de circunvalación); *syn.* cinturón [m] de ronda, ronda [f] [ambos AL]; *f* **rocade [f]** (Voie de communication contournant en totalité ou en partie une zone urbaine définie et souvent reliée au centre par des radiales ou des pénétrantes [routes ou autoroutes] ; ►rocade autoroutière, ►route de contournement) ; *g* **Ringstraße [f]** (Hauptverkehrsstraße, die einen vollständigen oder teilweisen Ring um ein bestimmtes Gebiet einer Stadt beschreibt; in Großstädten gibt es oft innere, mittlere und äußere **R.n**; ►Autobahnring, ►Umgehungsstraße).

Rio de Janeiro Convention [n] **on Biological Diversity** *conserv. leg.* ►Convention on Biological Diversity.

riparial [adj] *phyt. zool.* ►ripicolous.

5088 riparian [n] *leg.* (Land owner with legal ►riparian rights along a watercourse; **in U.S.**, adjacent owner may or may not have the use of adjacent water; law varies in eastern and western states); *syn.* riparian owner [n]; *s* **ribereño/a [m/f]** (Propietario/a de un terreno colindante con un curso de agua o persona con derecho a utilizarlo; ►derecho ribereño); *f* **riverain [m]** (Propriétaire d'un terrain situé le long d'un cours d'eau. Les riverains possèdent les droits de propriété et d'usage des eaux superficielles non domaniales et sont assujettis aux obligations d'entretien et de gestion ; ►droits des riverains) ; *g* **Anlieger [m] (2)** (Eigentümer eines an einem oberirdischen Gewässer angrenzenden Grundstücks oder die zur Nutzung dieses Grundstückes Berechtigten. Eigentümer von Grundstücken, die hinter den **A.**-Grundstücken/Ufergrundstücken liegen, werden **Hinterlieger** genannt; ►Anliegerrechte); *syn.* Anrainer [m], Gewässeranlieger [m], Uferanlieger [m].

5089 riparian alder stand [n] [US] *phyt.* (Natural Area Journal 1991, I, 7; deciduous woodland or copse, mainly in the genus alder *[Alnus]*, on a permanently wet, organic soil; ►fen wood [US]/carr [UK], ►riparian woodland); *syn.* alder carr [n] [UK], alder fen [n] [UK], alder swamp woodland [n] [also UK]; *s* **aliseda [f] turbícola** (Bosque o matorral decíduo, predominantemente del género Alnus, que crece en suelo orgánico saturado de agua permanentemente; ►bosque ripícola, ►bosque turbícola); *syn.* aliseda [f] turfófila, aliseda [f] de ribera (BB 1979, 129); *f* **aulnaie [f] marécageuse** (*Alnion glutinosae* habitat forestier constitué généralement en taillis ou futaie d'arbres moyennement à peu élevés, dominés par l'Aulne glutineux ou Aulne noir *[Alnus glutinosa]*. Boisement typique des marais et plaines marécageuses de basse altitude, aux étages planitiaire, collinéen et sub-montagnard. Il est établi sur des sols marécageux très hydromorphes [gley superficiel] gorgés d'eau une grande partie de l'année et généralement inondés en hiver et au début du printemps, souvent tourbeux, eutrophes, à nappe phréatique stagnante ou peu renouvelée, pauvre en oxygène, proche de la surface. Il occupe les fonds de vallées et plaines alluviales au niveau de dépressions, replats humides et au contact de zones marécageuses ; à ne pas confondre avec les boisements d'aulnes sur les sols très humides à inondables des sols tourbeux [aulnaie tourbeuse] ou les aulnaies humides ripariales mais non marécageuses [aulnaie des cours d'eau et des sources] ; cf. guide des habitats Naturels du département de l'Isère, Conservatoire Botanique National Alpin — Conseil Général de l'Isère — 2007 ; ►forêt tourbeuse ; ►forêt alluviale) ; *syn.* aulnaie [f] turficole, aulnaie [f] tourbeuse, aulnaie [f] à Sphaignes ; *g* **Erlenbruchwald [m]** (Erlenbrücher [mpl], Erlenbruchwälder [pl]; auf ständig nassen, organischen Böden wachsender Laubwald, der hauptsächlich aus Erlen *[Alnus]* besteht; *OB* ►Bruchwald; ►Auenwald).

R

riparian alluvial plain [n] *phyt.* ►upper riparian alluvial plain.

riparian alluvial plain [n] [US]**, lower** *phyt.* ►regularly flooded alluvial plain.

riparian forest [n] *phyt.* ►riparian woodland.

5090 riparian land [n] *leg.* (Property situated along a watercourse or on a waterbody, the owner of which may have ►riparian rights, i.e. right to the use of water on such land for various purposes; ►keeping open of riparian land); *s* **terreno [m] ribereño** (Parcela de tierra situada al borde de un curso o un cuerpo de agua, cuyo/a propietario/a puede poseer el ►derecho ribereño, p. ej. el de utilizar el agua para diferentes usos; ►mantenimiento de zona de servidumbre de acceso a las orillas); *syn.* terreno [m] riparío; *f* **parcelle [f] riveraine** (Terrain situé le long des eaux superficielles domaniales et non domaniales ; les droits de propriété et d'usage des riverains sont réglementés par le Code civil ; ►droits des riverains, ►servitude de libre passage sur les berges) ; *g* **Ufergrundstück [n]** (An ein Gewässer angrenzendes Grundstück. Den Eigentümer- und Anliegergebrauch regeln die Wassergesetze der Bundesländer; ►Anliegerrechte, ►Uferfreihaltung).

5091 riparian meadow [n] *geo.* (Regularly, naturally flooded enduring grassland alongside rivers and streams); *s* **pradera [f] aluvial** (Prado permanente a lo largo de un río que se inunda naturalmente); *syn.* pradera [f] ripícola; *f* **prairie [f] alluviale** (Prairie permanente régulièrement inondée s'étendant sur les alluvions d'une plaine alluviale) ; *syn.* prairie [f] riveraine, prairie [f] inondable ; *g* **Auenwiese [f]** (Regelmäßig, natürlich überflutetes Dauergrünland im Auenbereich); *syn.* Auwiese [f] [A].

riparian owner [n] *leg.* ►riparian.

5092 riparian rights [npl] *leg. wat'man.* (Rights of land owner to the waters on or bordering his property, including the right to make use of such waters and to prevent diversion or misuse of upstream water and to claim title to submerged beachfront lands or dried streams that re-emerge from ocean floor or change in location of a stream when water's edge or stream location is defined as a property boundary; **in U.S.**, the right to use water on or bordering property varies between eastern and western states); *s* **derecho [m] ribereño** (Derecho legal del propietario de un terreno para utilizar el agua de una corriente superficial de agua que bordee su tierra; también se refiere al propietario del suelo existente por debajo de la corriente; DACO 1988, 434); *f* **droits [mpl] des riverains** (Droits définis dans le code rural, livre I, acquis par les propriétaires des deux rives d'un cours d'eau ; les droits des riverains cours d'eau non domaniaux sont réglementés par les articles L215- à 215-6 du Code de l'Environnement) ; *g* **Anliegerrechte [npl]** (Rechte der Eigentümer und Anlieger an Ufergrundstücken sind in den Wassergesetzen der Länder geregelt).

5093 riparian upland woodland [n] *phyt.* (Woodland of the upper riverine zone, occasionally flooded; ►riparian woodland); *syn.* occasionally flooded riparian/riverine woodland [n], upper riparian/riverine woodland [n] [also US]; *s* **bosque [m] ripícola ocasionalmente inundado** (Bosque higrófilo situado en la terraza superior del valle que se inunda muy esporádicamente, en Europa Central está constituído generalmente por el roble carvallo *[Quercus robur]*, los olmos *[Ulmus glabra, Ulmus laevis]* y fresno común *[Fraxinus excelsior]*; en España varía su constitución según la región climática: en la región atlántica su composición de especies es similar a la de Europa Central, en la zona mediterránea están compuestos de fresno *[Fraxinus angustifolia]*, almez *[Celtis australis]*, fresno de flor *[Fraxinus ornus]* y sauzgatillo *[Vitex agnus-castus]*, entre las especies arbustivas se encuentran el tamujo *[Securinega tinctoria]* y la adelfa *[Nerium oleander]*; cf. DINA 1987; ►bosque ripícola); *syn.* bosque [m] de vega ocasionalmente inundado, bosque [m] ribereño ocasionalmente inundado, bosque [m] de ribera ocasionalmente inundado; *f* **forêt [f] alluviale à bois durs** (Formation riveraine occupant la terrasse alluviale humide située sur les partics hautes du lit majeur de grandes vallées occasionnellement inondée ; en Europe centrale peuplée du Chêne pédonculé *[Quercus robur]*, des Ormes *[Ulmus glabra, Ulmus laevis]* et du Frêne *[Fraxinus exselsior]* ; ►forêt alluviale ; ces forêts riveraines peuvent former des galeries ou parfois entrer en contact avec un massif forestier en bordure d'un cours d'eau ; ces formations appartiennent à l'alliance de *l'Alno-Padion*) ; *syn.* forêt [f] riveraine (inondable) à bois durs, forêt [f] humide ripariale, ripisilve [f] haute ; *g* **Hartholzauenwald [m]** (Etwas höher und weiter weg vom Ufer gelegener Auenwald, der nur bei außergewöhnlichen Hochwässern überflutet wird, in Mitteleuropa meist mit Stieleiche *[Quercus robur]*, Ulme *[Ulmus glabra, Ulmus laevis]* und Esche *[Fraxinus excelsior]* bestockt; dieser Wald gehört im gemäßigten Europa zum Verband des Erlen-Ulmenwaldes *[Alno-Ulmion* BB et Tx 1943*/Alno-Padion* Knapp 1948]; ►Auenwald); *syn.* Hartholzauwald [m] [A].

5094 riparian vegetation [n] *phyt.* (**1.** Vegetation growing along the banks of a stream, river or lake. **2.** The term 'littoral', used to describe the coast of a sea or an inland lake, can also be applied to describe their vegetation; whilst in French 'littoral' is only used for a coastal area; in French-speaking Canada, the term is used for the banks of rivers and streams as well as coasts [littoral fluvial]; *general term for both areas* ►littoral vegetation; ►driftline community, ►exposed mud vegetation, ►littoral vegetation, ►planting of riparian vegetation, ►riverine community, ►riverine reed belt); *s* **vegetación [f] ripícola** (Vegetación existente en las orillas de ríos o en la costa; ►asociación lótica, ►carrizal fluvial, ►vegetación de bordes de estanques, ►vegetación de las acumulaciones de restos orgánicos, ►vegetación del litoral y de las riberas de las aguas continentales, ►vegetación halófila de los cúmulos de residuos orgánicos); *f 1* **végétation [f] des grèves** (**1.** Terme générique pour les groupements rencontrés sur les sols alluviaux nitratés et périodiquement inondés en bordures des étangs, lacs et cours d'eau. **2.** Terme spécifique relatif aux cours d'eau végétation des grèves alluviales ; ►flore d'ourlet halonitrophile, ►flore des grèves [alluviales], ►groupement des grèves d'étangs, ►phalaridaie fluviale, ►roselière de rive de cours d'eau, ►végétation du littoral et des rives des cours d'eau) ; *syn.* groupement [m] des grèves alluviales, végétation [f] des rives ; *f 2* **végétation [f] ripicole** (Terme spécifique caractérisant les groupements ou les plantes des rives des cours d'eau ; ►association lotique, ►végétation rivulaire) ; *g* **Uferbewuchs [m, o. Pl.]** (**1.** Vorhandene Vegetation am Ufer eines Fließ- oder Stillgewässers. **2.** Im Deutschen bezieht sich U. nur auf Vegetation am Rande von fließenden und stehenden Gewässern; der Terminus „littoral" deckt im Deutschen und Englischen den Lebensbereich des Meeres und der Binnengewässer ab; im Französischen hingegen wird *littoral* nur für den Küstenbereich verwendet, jedoch im frankophonen Kanada auch für Ufer von Fließgewässern [littoral fluvial]; *OB für beide Bereiche* Litoralvegetation; ►Fließgewässergesellschaft; *UBe* ►Flussröhricht, Seeufervegetation, ►Spülsaumflur, ►Teichuferflur, ►Uferbepflanzung, ►Vegetation entlang von Küsten und Ufern); *syn.* Ufervegetation [f].

5095 riparian wattlework [n] [US]**/riparian wattlework** [n] [UK] *constr.* (*Bioengineering technique* ►wattlework [US]/wattle-work [UK]); *s* **varaseto [m] de ramas en orilla** (Enrejado de ramillas de sauce que se utiliza para evitar daños de erosión en orillas de ríos; ►varaseto de ramas); *f* **tressage [m] de**

berge (WIB 1996 ; technique de génie biologique de confortement d'une berge et de son pied au moyen de ►treillis de branches [de saules] à rejets [tressage oblique]) ; *g* **Uferflechtwerk [n]** (Ingenieurbiologisches ►Flechtwerk zur Ufersicherung).

5096 riparian woodland [n] *phyt.* (Woodland type found alongside rivers and streams, where the species composition is influenced by the effect of flooding and, in contrast to ►fen wood, is always associated with a mineral substrate. In central Europe subdivided into areas which are **regularly flooded** ['softwood' **r. w.**: ►regularly flooded alluvial plain] characterized by willow [Salix], elder [Alnus] and poplar [Populus] species, and those which are only **occasionally flooded** ['hardwood' **r.w.**: ►upper riparian alluvial plain]; ►gallery forest, ►regulary flooded riparian woodland, ►riparian upland woodland); *syn.* wet alluvial woodland [n], riparian forest [n], bottomland hardwood forest [n] [also US] (WET 1993, 32); *s 1* **bosque [m] ripícola** (Tipo de bosque, que se forma en las riberas de ríos y arroyos, cuya composición de especies está influenciada por los efectos de inundación y que, al contrario que el ►bosque turbícola, está siempre asociado a un sustrato mineral. En Europa Central se subdivide en áreas, que son regularmente inundadas caracterizadas por especies como Salix, Alnus, Populus y en aquéllas que sólo son inundadas ocasionalmente; ►bosque de galería, ►bosque ripícola ocasionalmente inundado, ►bosque ripícola regularmente inundado, ►vega aluvial ocasionalmente inundada, ►vega aluvial regurlamente inundada); *syn.* bosque [m] de vega, bosque [m] ribereño, bosque [m] de ribera; *s 2* **soto [m] (ripícola)** (Bosque ribereño o de una vega. Su vegetación suele ser preclímax respecto a la de niveles superiores, sobre todo si ésta es xerofítica; DB 1985); *f* **forêt [f] alluviale** (Terme générique pour le groupement forestier situé dans la zone d'influence des cours d'eau [►plaine alluviale de forêt à bois durs et ►plaine alluviale à bois tendres] et qui, par opposition à la ►forêt tourbeuse, est toujours associé à un substrat minéral ; en Europe centrale on distingue la ►forêt alluviale à bois tendres et la ►forêt alluviale à bois durs ; ►galerie forestière, ►ripisilve basse) ; *syn.* forêt [f] riveraine (sur alluvions), ripisilve [f] alluvionnaire, forêt [f] rivulaire ; *g* **Auenwald [m]** (Waldgesellschaft im Einflussbereich von Fließgewässern, die [in Mitteleuropa] in ►Weichholzaue und ►Hartholzaue untergliedert wird und im Gegensatz zum ►Bruchwald stets an mineralische Sedimente gebunden ist; ►Galeriewald, ►Hartholzauenwald, ►Weichholzauenwald); *syn.* Auwald [m] [A].

riparian woodland [n] [US], **upper** *phyt.* ►riparian upland woodland.

5097 riparian woody species [n] *phyt. syn.* riverine woody species [n]; *s* **leñosa [f] ripícola**; *f* **bois [m] riparial** (Végétal ligneux qui vit sur la rive d'un cours d'eau ou d'un eau stagnante) ; *syn.* bois [m] ripicole ; *g* **Ufergehölz [n]** (Am Ufer eines Fließgewässers oder Sees wachsender Baum oder Strauch).

5098 riparian zonation [n] *phyt.* (Riverbank sequence of plant communities, usually ribbon-like and spreading from water to land); *s* **zonación [f] de la vegetación ribereña** (Secuencia de comunidades vegetales en las riberas de ríos y arroyos, generalmente en forma de banda, extendiéndose desde el agua hacia la tierra); *f* **zonation [f] de la végétation riveraine des cours d'eau** (Juxtaposition de stations transversales typiques [en série] de groupements végétaux le long d'un cours d'eau) ; *g* **Vegetationszonierung [f] an Fließgewässern** (Uferbewuchs als Abfolge verschiedener Pflanzengesellschaften in Abhängigkeit der Strömungsstärke und Wasserspiegelschwankungen [Wassergehalt und Wasserkapazität des Bodens] sowie des Gewässerabschnittes [Oberlauf, Mittellauf oder Unterlauf/küstennahes Tiefland, Brackwasserbereich in Mündungsgebieten] und der Prall- oder Gleituferseite, z. B. im Oberlauf beginnend oberhalb der Niederwasserlinie mit einer Annuellenflur und einem Kriechrasen, ab Mittelwasserstand mit Flussröhricht und ab beginnender Weichholzaue mit Weidengebüsch, Weidenwald und Erlenwald); *syn.* Vegetationsprofil [n] an Fließgewässern (ELL 1978, 797), Vegetationsabfolge [f] an Fließgewässern.

riparian zone [n] **of emergent vegetation** *phyt.* ►riverine reed belt.

riparious [adj] *phyt. zool.* ►ripicolous.

ripe for development [n] *urb.* ►unzoned land ripe for development [US]/white land [UK].

ripicole [adj] *phyt. zool.* ►ripicolous.

5099 ripicolous [adj] *phyt. zool.* (Living on the banks of rivers and streams); *syn.* riparial [adj], riparious [adj], ripicole [adj]; *s* **ripícola [adj]** (Referente a un organismo que vive en las riberas de ríos); *syn.* riparío/a [adj]; *f* **riparial, ale [adj]** (Qui vit sur les berges d'un cours d'eau) ; *syn.* de berge [loc], ripicole [adj] ; *g* **uferbewohnend [ppr/adj]** (Ein Lebewesen betreffend, das an Fließgewässerufern lebt); *syn.* am Ufer lebend [loc], ripikol [adj].

ripping [n] [UK] *constr.* ►scarifying [US].

ripping out [n] **of hedgerows** [UK] *agr. landsc.* ►hedgerow clearance.

5100 riprap [n] *constr. eng. wat'man.* (Broken rock layer placed on outer streambanks for protection against ►streambank erosion, lateral erosion or undermining of foundations; ►bank revetment, ►channel lining, ►intermittent riprap); *syn.* rock revetment [n]; *s* **escollera [f] [Es, RA, YV]** (Capa de rocas partidas colocadas sobre los taludes de las márgenes para protegerlos contra la ►erosión de márgenes de cursos de agua, la erosión lateral o para evitar la socavación de fundamentos; ►escollera de fondo, ►escollera intermitente, ►revestimiento de orillas); *syn.* escollerado [m] [RA], enrocamiento [m] de protección [CO, MEX], enrocado [m] de protección [EC, RCH], riprap [m] [RCH] (MESU 1977); *f* **cailloutage [m] des berges** (Mise en place d'une couche de pierres sur les berges pour la protection contre ►l'affouillement de la berge, l'►érosion des berges des cours d'eaux ; ►empierrement du lit [d'une rivière], ►ouvrage de revêtement de berge, ►plots en enrochements) ; *syn.* empierrement [m] des berges, enrochement [m] des berges (AEP 1976, 253), pose [f] de blocs (ARI 1986, 46) ; *g* **Steinschüttung [f] auf Uferböschungen** (Lage gebrochener Steine auf Uferböschungen zum Schutz gegen ►Erosion an Ufern von Fließgewässern oder gegen Unterspülung von Fundamenten; eine mit Steinen geschüttete Böschung wird **Schüttsteinböschung** genannt; ►Berollung, ►Uferdeckwerk, ►Steinschüttungspakete); *syn.* Steinwurf [m].

rise [n] *pedol.* ►capillary rise; *hydr.* ►natural rise of groundwater level; *constr.* ►stair rise, ►total rise of a flight of steps.

rise [n], **high-** *arch. urb.* ►multistory building for housing [US].

rise [n] **and run** [n] [US] *arch. constr.* ►riser/tread ratio.

rise [n] **of a flight of steps** *constr.* ►total rise of a flight of steps.

rise [n] **of groundwater level** *hydr.* ►natural rise of groundwater level.

riser [n] [US] *constr.* ►overflow pipe.

riser [n], **stair** *constr.* ►stair rise.

riser [n], **step** *constr.* ►stair rise.

riser height [n] *constr.* ►stair rise.

5101 riser step [n] *constr.* (HALAC 1988-II, 226; step constructed of a vertically placed slab [riser] and a paved, asphalted or gravel tread; ►block step, ►flagstone step, ►round timber step); *s* **escalón [m] de tabica** (Peldaño formado por placa colocada verticalmente que se rellena con tierra, se alfalta o pavimenta para formar la huella; ▶escalón de losa, ▶escalón de rodillo de madera, ▶escalón prefabricado); *syn.* escalón [m] de tabla de contrahuella; *f* **emmarchement [m] bloqué par une bordure** (Marche d'escalier dont la dénivellation est bloquée par une dalle verticale ou une bordurette et le palier constitué d'un revêtement réalisé en pavés, en bitume ou en sol stabilisé aux liants hydrauliques ; ▶bloc marche, ▶contremarche en rondins, ▶dalle giron) ; *g* **Stellstufe [f]** (Treppenstufe, bei der die vordere Platte gestellt und der Auftritt ausgepflastert, geteert oder mit einer wassergebundenen Decke versehen wird; ▶Blockstufe, ▶Legstufe, ▶Knüppelstufe).

5102 riser/tread ratio [n] *arch. constr.* (Proportion between vertical rise and horizontal run; ▶step formula); *syn.* rise [n] and run [n] [also US]; *s* **relación [f] entre huella y contrahuella** (▶fórmula de cálculo de peldaño); *f* **proportion [f] d'une marche** (Rapport entre la profondeur du giron de la marche et la hauteur de la marche ; ▶formule de dimensionnement des marches) ; *syn.* dimension [f] d'une marche ; *g* **Steigungsverhältnis [n]** (*Treppenbau* Verhältnis zwischen Stufenhöhe und Breite des Auftritts; ▶Schrittmaßformel).

5103 riser unit [n] [US] *constr.* (Precast concrete ring used in the construction of a drain manhole; ▶cone 1); *syn.* chamber section [n] [UK] (SPON 1986, 171); *s* **anillo [m] de pozo** (Pieza prefabricada de hormigón utilizada en la construcción de pozos de registro; ▶cono asimétrico de pozo de registro); *f* **virole [f]** (Élément cylindrique constitutif de la cheminée d'un regard sur lequel repose le tampon ; ▶hotte de regard) ; *g* **Schachtring [m]** (Vorgefertigter Betonring zum Bau eines Entwässerungsschachtes; ▶Schachthals); *syn.* Zwischenteil [m] (eines Schachtes).

5104 rising spring [n] *geo.* (Issue point of a natural flow of groundwater from the Earth's surface); *syn.* source [n] (of natural water flow); *s 1* **fuente [f] de emergencia** (Fuente de la que el agua fluye a la superficie debido únicamente a que dicha superficie se extiende hasta el nivel freático o por debajo del mismo; WMO 1974); *s 2* **fuente [f]** (Lugar donde fluye agua naturalmente de una roca o suelo sobre la tierra o en una masa de agua superficial; WMO 1974); *syn.* manantial; *f 1* **émergence [f]** (Terme générique qui désigne la zone d'apparition à la surface du sol, sous forme d'écoulement concentré ou diffus, **1.** des eaux d'une nappe aquifère recoupée par la surface topographique, **2.** des eaux souterraines provenant de la perte de cours d'eau en milieu karstique ; DG 1984) ; *syn.* point [m] d'émergence d'eaux souterraines ; *f 2* **source [f]** (Point d'apparition naturelle à la surface du sol, sous forme d'écoulement concentré, des eaux d'une nappe phréatique ; on distingue suivant les conditions hydrogéologiques et le type d'aquifère plusieurs types de sources : a] les sources *de déversement*, *de débordement* ou *de trop plein* [correspondant aux types de nappes souterraines libres de mêmes dénominations], les sources *d'émergence* ou *de dépression* [source d'aquifère à nappe libre non liée à l'affleurement du substratum], les sources *d'étranglement*, b] les sources *artésiennes* ou *jaillissante*s [issue d'une nappe captive], c] les sources *diaclasiennes*, les sources *karstiques* ou *de karst*, les ▶exsurgences [issues d'un aquifère discontinu], les ▶sources vauclusiennes [exutoire d'un conduit karstique ascendant subvertical], d] les ▶résurgences, ou « sources secondaires » [retour en surface d'eau originaire, en tout ou partie, de pertes d'un ou plusieurs cours d'eau dans un aquifère karstique] ; cf. Dictionnaire Français d'hydrologie, Comité National Français des sciences hydrologiques) ; *g* **Quellaustritt [m]** (Stelle an der Erdoberfläche, an der Grundwasser aus dem Untergrund an die Oberfläche tritt und über ein oberirdisches Gewässer abfließt); *syn.* Quelle [f].

R

risk [n] *ecol. plan.* ▶ecological risk; *geo. pedol. plan.* ▶erosion risk.

risk analysis [n] *ecol. landsc. plan.* ▶environmental risk analysis.

5105 risk [n] **of avalanches** *plan.* (Increased probability of massive sliding of snow and ice packs in alpine regions); *s* **riesgo [m] de aludes** (Evaluación del estado de la cubierta de nieve en una área montañosa específica que depende de la altitud, la exposición, la topografía para estimar la probabilidad de que ocurran avalanchas); *f* **risque [m] d'avalanches** (Évaluation de l'état du manteau neigeux en fonction de l'altitude, de l'exposition, de la topographie et estimation du risque de déclenchement provoqué ou spontané d'avalanches dans les régions alpines. L'échelle européenne des risques d'avalanches est graduée de 1 [conditions générales sûres] à 5 [danger aigu]) ; *g* **Lawinengefahr [f]** (Auf einen Standort bezogenes, langfristiges objektives Risiko, dass die Wahrscheinlichkeit eines unerwünschten Lawinenabganges z. B. Gebäude und Verkehrsinfrastrukturen zerstört, Menschen und Tiere unter sich begräbt oder Forstkulturen verwüstet; die **L.** wird durch die Lawinenhäufigkeit und die Größe der Lawinenkraft auf ein Hindernis ausgedückt).

5106 risk [n] **of falling** *leg.* (Peril of accidental fall); *s* **peligro [m] de caída**; *f* **risque [m] de chute accidentelle dans le vide** ; *g* **Absturzgefahr [f]** (Möglichkeit, dass jemand abstürzt).

5107 risk [n] **of groundwater pollution** *envir. hydr.* (Constant or intermittent danger of contamination of groundwater by polluting substances or those which are detrimental to human health); *s* **peligro [m] de contaminación de acuífero** (Posibilidad constante o puntual de polución de la capa freática por medio de sustancias nocivas al medio ambiente o la salud humana); *syn.* peligro [m] de contaminación de las aguas subterráneas; *f* **vulnérabilité [f] de la nappe aux pollutions** (Possibilité de contamination chronique ou accidentelle de la nappe phréatique par des substances ayant des incidences nuisibles sur la santé de la population ; la nappe phréatique est vulnérable aux pollutions par manque de protection naturelle [absence de couverture imperméable]) ; *syn.* vulnérabilité [f] des eaux souterraines aux pollutions, risque [m] de pollution de la nappe phréatique ; *g* **Grundwassergefährdung [f]** (Die Möglichkeit, dass Grundwasser ständig oder hin und wieder mit verschmutzenden oder der Gesundheit abträglichen Stoffen belastet wird).

river [n] *geo.* ▶braided river; *hydr.* ▶clear-water river; *geo.* ▶meandering river; *conserv. leg. nat'res.* ▶wild and scenic river [US].

river [n], **winding** *geo.* ▶meandering river.

river authority [n] [UK] *adm. nat'res. wat'man.* ▶water users agency [US].

5108 riverbank [n] [US]/**river-bank** [n] [UK] *geo.* (**1.** Peripheral zone at the edge/margin of a large body of flowing water, stretching from the shallow water area to the uppermost high-water mark. Various types of vegetation grow linearly according to the level of wetness and the frequency of inundation); **2.** **riverside** [n] (Area comprising the floodplain of a river; ▶riverside park); *s 1* **margen [f] de río** (Límite de la tierra que la separa del río. Legalmente se entiende por **m.** el terreno que linda con el cauce; cf. art. 6 RDL 1/2001); *syn.* orilla [f] de río, margen [f] fluvial, ribera [f]; *s 2* **ribera [f]** (Margen u orilla de río; por extención se denomina también así la tierra cercana a los ríos [▶vega], que suele ser más fértil y puede beneficiarse de sus aguas naturalmente o por la acción humana. Legalmente se entiende por **r.** la faja lateral de un cauce público situada por encima del nivel de aguas bajas; cf. art. 6 RDL 1/2001; ▶parque de ribera); *f 1* **berge [f] d'une rivière** (Bord exhaussé d'un cours d'eau moyen, séparant le lit mineur du lit majeur ; cf. DG 1984) ; *syn.* bord [m] de rivière ; *f 2* **rive [f] d'un fleuve** (Bande de terre qui borde un grand cours d'eau ; ▶parc fluvial, ▶rivage) ; *g* **Flussufer [n]** (Rand eines größeren Fließgewässers von der Flachwasserzone bis zur höchsten Hochwasserlinie — je nach Höhe des Wasserstandes. Entsprechend des Feuchteregimes und der Überflutungshäufigkeit bestehen Streifen unterschiedlicher Vegetationstypen; ▶Flussauenpark).

riverbank cave-in [n] [US]/**river-bank cave-in** [n] [UK] *geo.* ▶bank-caving.

5109 riverbank collapse [n] [US]/**river-bank collapse** [n] [UK] *geo.* (Subsidence caused by ▶streambank erosion; ▶bank-caving, ▶undercut bank); *syn.* riverbank failure [n] [US]/river-bank failure [n] [UK]; *s* **desplome [m] de orillas** (Consecuencia de la ▶socavación de márgenes, es decir de la ▶erosión de márgenes de cursos de agua; ▶arco erosivo de meandro); *f* **effondrement [m] d'une berge** (Résultat de l'▶affouillement de la berge ou de l'▶érosion des berges du cours d'eaux ; ▶berge d'effondrement) ; *syn.* éboulement [m] de berge ; *g* **Uferabbruch [m]** (Herunterbrechen von Erdmassen als Folge der ▶Auskolkung an/von Flussufern resp. der ▶Erosion an Ufern von Fließgewässern; ▶Abbruchufer); *syn.* Uferabriss [m].

5110 riverbank degradation [n] [US]/**river-bank degradation** [n] [UK] *geo.* (Wearing down of banks by erosion and livestock trampling); *s* **deterioro [m] de taludes de márgenes** (Daños en riberas causados por erosión, aprovechamiento ganadero o intrusiones técnicas); *f* **dégradation [f] de berge** (Phénomène occasionné par l'érosion, le piétinement du bétail ou le passage d'animaux fouisseurs) ; *g* **Beschädigung der Uferböschung [f]** (Durch Erosion, Weidenutzung oder technische Eingriffe verursachte Zerstörung der Böschung eines Fließgewässers).

riverbank erosion [n] [US]/**river-bank erosion** [n] [UK] *geo.* ▶streambank erosion.

riverbank failure [n] [US]/**river-bank failure** [n] [UK] *geo.* ▶riverbank collapse [US]/river-bank collapse [UK].

5111 riverbank slippage [n] [US]/**river-bank slippage** [n] [UK] *geo.* (Sliding down of sedimented material on a riverbank caused by changed groundwater regime, whereby the water level of the watercourse is dropping faster than the groundwater level in the slope); *s* **deslizamiento [m] de orillas** (Debido a la reducción del nivel del agua del río, mientras que el nivel de la capa freática de la margen desciende más lentamente); *f* **glissement [m] de berge** (A lieu en présence d'une berge constituée de matériaux alluvionnaires lorsque la ligne d'eau baisse plus vite dans le cours d'eau que le niveau de la nappe dans la berge) ; *g* **Uferabrutschung [f]** (Abgleiten der aus angeschwemmtem Material entstandenen Uferböschung eines Fließgewässers, verursacht durch veränderte Grundwasserverhältnisse, wobei der Wasserspiegel des Fließgewässers schneller absinkt als der im Böschungsbereich; *syn.* Abrutschen [n, o. Pl.] der Uferböschung.

5112 riverbank stabilization [n] [US]/**river-bank stabilization** [n] [UK] *constr.* (Measures to protect against erosion on riverbanks, e.g. ▶riprap on bank slopes, ▶bank revetment, ▶planting of a streambank); *s* **fijación [f] de orillas** (Estabilización de bordes de cursos de agua por medio de medidas de construcción para evitar su erosión; ▶escollera, ▶revestimiento de orillas, ▶vegetalización de riberas); *f* **confortement [m] des berges** (Travaux de protection des berges ; ▶cailloutage des berges, ▶ouvrage de revêtement de berge, ▶végétalisation des berges) ; *syn.* consolidation [f] des berges, stabilisation [f] des berges [CDN] ; *g* **Uferbefestigung [f]** (Erosionsschutzmaßnah-

men an Ufern wie z. B. ▶Steinschüttung an Uferböschungen, ▶Uferdeckwerk, ▶Uferbegrünung); *syn.* Ufersicherung [f], Uferverbau [m], Uferverbauung [f].

river basin [n] [US]/**river-basin** [n] [UK] *geo. hydr. nat'res.* ▶drainage basin.

5113 river basin district [n] *hydr. wat'man. leg.* (The EC Water Framework Directive, 2000/60/EC, which came into force on 22 December 2000, established a new, integrated approach to the protection, improvement and sustainable use of Europe's rivers, lakes, estuaries, coastal waters and groundwater. The introduction of a river basin management planning system is the key mechanism for ensuring the integrated management of all water bodies and the water needs of terrestrial ecosystems that depend on groundwater, such as wetlands. For each river basin district [RBD] a river basin management plan [RBMP] will be prepared, implemented and reviewed on a six year cycle to meet the environmental objectives of the Directive by 2015. England, Scotland and Wales are divided into water districts named after rivers or regions; ▶drainage basin); *s* **demarcación [f] hidrográfica** (Según el art. 2.15 de la Directiva Marco del Agua 2000/60/CE, de 23 de octubre de 2000, que entró en vigor el 22 de diciembre del 2000, una **d. h.** es la zona marítima y terrestre compuesta por una o varias cuencas hidrográficas vecinas y las aguas subterráneas y costeras asociadas, designada como principal unidad a efectos de la gestión de la ▶cuencas hidrográficas. Esta directiva estableció un nuevo planteamiento integrado de la protección, mejora y uso sostenible de los ríos, lagos, estuarios, aguas costeras y subterráneas de Europa. La introducción de un sistema de planificación de la gestión de las cuencas hidrográficas es el mecanismo clave para asegurar el manejo integrado de todos los cuerpos y cursos de agua y de todas las necesidades de agua de los ecosistemas terrestres que dependen del agua subterránea, tal como los humedales. Para cada **d. h.** se ha de preparar, aplicar y revisar cada seis años un plan hidrológico de cuenca [PHC] para alcanzar los objetivos de la directiva hasta el 2015. **En Es.** las demarcaciones hidrográficas cuyo ámbito territorial afecta a más de una CC. AA. o que comparten cuencas con Estados vecinos están fijadas en el RD 125/2007. Según éste se diferencian las **dd. hh.** con cuencas intercomunitarias dentro del territorio español que son 1] **D. H.** del Guadalquivir, 2] **D. H.** del Segura y 3] **D. H.** del Júcar, y se definen las partes españolas de las **dd. hh.** correspondientes a cuencas hidrográficas compartidas con otros países, que son 1] **D. H.** del Miño-Limia, 2] **D. H.** del Norte, 3] **D. H.** del Duero, 4] **D. H.** del Tajo, 5] **D. H.** del Guadiana, 6] **D. H.** del Ebro, 7] **D. H.** de Ceuta y 8] **D. H.** de Melilla; cf. arts. 2 y 3 RD 125/2007. Las **dd. hh.** intracomunitarias son definidas por la correspondiente CC. AA. La cooperación con Portugal en esta materia fue acordada en el Convenio de Albufeira el 30 de noviembre 1998 que entró en vigor el 17 de enero del 2000. La cooperación con Francia se definió en el Acuerdo administrativo de Toulouse firmado el 15 de febrero de 2006 que entró en vigor inmediatamente); *f* **district [m] hydrographique** (Notion définie par la directive cadre européenne sur l'eau [DCE] 2000/60/CE, article 2, du 23 octobre 2000. Zone terrestre et maritime, composée d'un ou plusieurs bassins hydrographiques ainsi que des eaux souterraines et eaux côtières associées, identifiée conformément à l'article identifiée selon la DCE comme principale unité pour la gestion de l'eau ; la liste des districts hydrographiques français a été arrêtée le 16 mai 2005 ; le territoire français compte 7 bassins versants [Adour Garonne, Rhin Meuse, Artois Picardie, Rhône Méditerranée, Corse, Seine Normandie, Loire Bretagne] 3 districts internationaux [Escaut, Meuse, Rhin] et 5 districts des DOM-TOM ; lorsqu'un bassin hydrographique s'étend sur le territoire de plus d'un Etat membre, les Etats veillent à ce qu'il soit intégré à un district hydrographique international ; pour chaque district doivent être établis un état des lieux, un programme de surveillance, un plan de gestion [SDAGE révisé] et un programme de mesures ; ▶bassin versant) ; *g* **Flussgebietseinheit [f]** (In der Richtlinie 2000/60/EG — Wasserrahmenrichtlinie [WRRL] — festgelegtes, nach dem Wasserhaushaltsgesetz zu bewirtschaftendes Land- oder Meeresgebiet, das aus einem oder mehreren benachbarten Einzugsgebieten, dem ihnen zugeordneten Grundwasser und den ihnen zugeordneten Küstengewässern besteht. Für D. sind in der WRRL 10 Einheiten z. T. grenzüberschreitend festgelegt: 1. Donau, 2. Rhein, 3. Maas, 4. Ems, 5. Weser, 6. Elbe, 7. Eider, 8. Oder, 9. Schlei/Trave und 10. Warnow/Peene; cf. § 1b und Anhang 1 WHG ; ▶Abflussgebiet).

5114 river bathing beach [n] *recr.* (Part of an inland stretch of water usually frequented by a large number of people, in which swimming is permitted or not forbidden by the authorities; ▶bathing beach) ; *s* **playa [f] de río** (Tramo de la orilla de un río en el que está permitido o no está prohibido bañarse; ▶playa de mar) ; *f* **zone [f] de baignade d'une rivière** (Bord des eaux douces, courantes sur lequel la baignade est expressément autorisée ou n'est pas interdite, habituellement pratiquée par un nombre important de baigneurs ; Directive du Conseil n° 76/160 CEE du 8. décembre 1975 concernant la qualité des eaux de baignade ; ▶zone de baignade) ; *syn.* aire [f] de baignade d'une rivière ; *g* **Flussbadestrand [m]** (Teil eines fließenden Binnengewässers, an dem das Baden von den zuständigen Behörden ausdrücklich gestattet ist oder nicht untersagt ist und an dem üblicherweise eine große Anzahl von Personen badet; cf. RL 76/160/EWG des Rates vom 08.12.1975; ▶Badestrand) ; *syn.* Badestrand [m] am Flussufer.

5115 riverbed [n] [US]/**river-bed** [n] [UK] *geo.* (Channel in which a river flows; ▶foreland 1, ▶river bottom, ▶stream bed) ; *s 1* **lecho [m] de un río** (Espacio por el que circulan las aguas de un río) ; *s 2* **lecho [m] menor** (Espacio ocupado por las aguas de caudal normal o medio de un río. El espacio ocupado por las aguas de crecida se llama ▶lecho mayor [de rio] ; cf. DGA 1986; ▶fondo del cauce, ▶lecho de arroyo) ; *f 1* **lit [m] d'un cours d'eau** (▶lit d'une rivière) ; *f 2* **lit [m] mineur** (Tracé d'écoulement des eaux d'un fleuve ou d'une rivière, recouvert par les eaux coulant à pleins bords avant tout débordement ; ▶fond du lit d'un fleuve, ▶lit majeur 2) ; *syn.* lit [m] ordinaire (DG 1984, 466) ; *f 3* **lit [m] majeur (3)** (Surface sur laquelle s'étale un cours d'eau en crue ; ▶lit majeur 1) ; *g* **Flussbett [n]** (Oberirdisches Gerinne eines Fließgewässers mit einer Wasserführung von mindestens 10-20 bis 200 m³/s ; größere **F.en** führen bis 2000 m³/s ; das **F.** besteht aus dem **eigentlichen Gewässerbett** und dem flächenmäßig größeren ▶Hochwasserbett, das durch Dämme begrenzt ist; cf. WAG 2001; ▶Bachbett, ▶Flusssohle, ▶Vorland 1).

5116 riverbed improvement [n] [US]/**river-bed improvement** [n] [UK] *eng. wat'man.* (Measures taken for the prevention of river scouring, ▶riverbank collapse [US]/river-bank collapse [UK], ▶bank-caving, sedimentation of riverbeds/river-beds; control of river alignment, inclines and cross-sections at various water levels) ; *s* **corrección [f] del lecho (de un río)** (Medidas tomadas para evitar hondonadas en ríos, ▶desplome de orillas, ▶socavación de márgenes, sedimentación sobre el lecho, etc. y para adaptar la línea del curso, la pendiente y la forma de la sección transversal a los diferentes caudales) ; *f* **correction [f] du lit d'un fleuve** (Les objectifs généraux des travaux de correction sont de protéger les terres contre les inondations et les dévastations, d'empêcher les érosions, les ▶affouillements de berges, les glissements de terrains, les changements de lit et les destructions de zones riveraines, d'as-

surer l'écoulement des eaux provenant des drainages agricoles ou urbains et de rendre possibles le remembrement des terres ; AEP 1976, 251 ; ▶effondrement d'une berge) ; *syn.* restauration [f] d'un cours d'eau, reprofilage [m] et recalibrage [m] d'un cours d'eau ; AEP 1976, 256) ; *g* **Flussbettregelung [f]** (Wasserbauliche Maßnahmen zur Vermeidung von Vertiefungen im Fluss, ▶Auskolkungen an/von Flussufern, ▶Uferabbrüchen, Sohlenerhöhungen durch Anlandung etc., zur Anpassung der Linienführung, des Gefälles und der Querschnittsform bei verschiedenen Wasserständen; cf. VOLL 1973).

river biocenose [n] *limn.* ▶river biocoenosis.

river biocoenose [n] *limn.* ▶river biocoenosis.

5117 river biocoenosis [n] *limn.* (Association of plant and animal species characteristic of a specific ▶river zone according to geological or pedological characteristics and velocity of current) ; *syn.* river community [n], *o.v.* river biocenose [n], river biocoenosis [n], river biocoenose [n]; *s* **biocenosis [f] fluvial** (Agrupación natural de seres vivos animales y vegetales de las diferentes ▶zonas biológicas de los ríos. En Europa Central éstas se denominan según las especies piscícolas características; ▶región ciprinícola, ▶región salmonícola) ; *syn.* biocenosis [f] fluvial ; *f* **biocénose [f] fluviale** (Communauté d'espèces animales et végétales vivant dans les cours d'eau et dépendante de la nature du lit [blocs, graviers, sables, vases] et de la vitesse du courant. La zonation des cours d'eau est établie en tenant principalement compte des poissons caractéristiques vivant sur les divers tronçons ; ▶zone écologique des cours d'eau) ; *syn.* biocénose [f] des cours d'eau, biocénose [f] lotique ; *g* **Flussbiozönose [f]** (Lebensgemeinschaft in ▶Flussregionen auf Grund unterschiedlicher Bodenbeschaffenheit [Fels, Sand, Schlamm] und Strömungsgeschwindigkeit. Die Regionen eines Flusses werden nach dem Überwiegen charakteristischer Fischarten benannt) ; *syn.* Fließwasserbiozönose [f].

5118 river bottom [n] *geo. wat'man.* (▶riverbed [US]/river-bed [UK]) ; *s* **fondo [m] del cauce** (Fondo del ▶lecho de un río; ▶lecho menor) ; *f* **fond [m] du lit d'une rivière** (Le fond du lit d'un cours est caractérisé par des microreliefs d'inégales profondeurs qui dépendent de la dynamique des eaux, la mouille constituant la zone profonde de surcreusement provoquée par une veine de courant érodant le fond du lit ; ▶lit d'un cours d'eau, ▶lit majeur, ▶lit mineur) ; *g* **Flusssohle [f]** (Nicht immer ebener Grund eines ▶Flussbettes, das durch Rinnen und Lockersedimentakkumulationen unregelmäßig ausgeformt sein kann).

5119 rivercliff [n] [US]/**river-cliff** [n] [UK] *geo.* (Steep bank cut by lateral erosion of a river on the outside of a meander; *opp.* river slip bank [US]/river slip-off slope [UK]; *generic term* ▶bluff 1) ; *s* **orilla [f] cóncava de un río** (Orilla abrupta de meandro de un río que en ríos naturales frecuentemente está socavada; *opp.* orilla convexa de un río; *término genérico* ▶orilla cóncava) ; *f* **berge [f] concave d'un fleuve** (Talus abrupt d'un fleuve et en constant recul, sapé par le courant formant des affouillements ; *opp.* berge convexe d'un fleuve ; *terme générique* ▶berge concave) ; *g* **Flussprallhang [m]** (Außenseite eines Flussbogens, der im natürlichen Zustand durch Auskolkung unterhöhlt ist; *opp.* Flussgleithang; *OB* ▶Prallhang).

river community [n] *limn.* ▶river biocoenosis.

5120 river course [n] *geo. s* **curso [m] fluvial**; *f* **cours [m] d'eau** ; *g* **Flusslauf [m]**.

5121 river development [n] *eng. wat'man.* (Changes made to the banks of a natural watercourse, e.g. quays, decks, excavation work; ▶bank revetment) ; *syn.* improvement [n] of a river course; *s* **regulación [f] de ríos** (Medidas de estabilización de cursos de agua naturales por medio de ▶revestimiento de orillas, escolleras, dragado, etc.) ; *f* **correction [f] du cours d'un fleuve**

(Amélioration du cours naturel d'une rivière par des travaux de débroussaillement des berges, de terrassement, de protection des berges comme p. ex. un ▶ouvrage de revêtement de berge ; *syn.* régulation [f] du cours d'un fleuve ; *g* **Flussregelung [f]** (Veränderung eines natürlichen Wasserlaufes durch Ein- oder Uferbauten wie z. B. Buhnen, Deckwerke, Parallelwerke oder Baggerungen zur Verbesserung der Wasserstraße und zum Hochwasserschutz durch Deichbau und Schaffung von Polderflächen; ▶Uferdeckwerk).

5122 river diking [n] *eng. wat'man.* (Method of flood control involving the construction of flood banks along sections of a river subject to flood hazard; ▶damming 2, ▶impounding of a watercourse, ▶river embankment); *s* **construcción [f] de diques fluviales** (Método de control de avenidas en cursos de agua para evitar inundaciones; ▶construcción de presas, ▶dique fluvial, ▶embalse de un curso fluvial); *f* **endiguement [m] fluvial** (Mesure de protection contre les crues par la réalisation d'ouvrages parallèles aux berges d'un cours d'eau ; ▶construction d'oùvrages de rétention, ▶digue fluviale, ▶endiguement d'un bassin fluvial de retenue) ; *syn.* endiguement [m] des rives, canalisation [f] fluviale ; *g* **Eindeichung [f] von Flüssen** (Bau von Deichen entlang eines Fließgewässers zur Verhinderung von Überschwemmungen; ▶Eindeichung eines Gewässeraufstaues, ▶Flussdeich, ▶Stauanlagenbau).

5123 river embankment [n] *wat'man.* (▶levee [US] 2/ flood bank [UK], ▶main dike [US]/main dyke [UK]); *syn.* river levee [n] [also US] (TGG 1984, 130); *s* **dique [m] fluvial** (*Término genérico* ▶dique de contención; ▶dique-guía); *f* **digue [f] fluviale** (*Terme générique* ▶digue ; *terme spécifique* digue de hautes eaux, ▶digue de canalisation des crues) ; *g* **Flussdeich [m]** (Uferdamm eines Flusses; *OB* ▶Deich; ▶Leitdamm).

5124 river engineering measures [npl] *leg. wat'man.* (Generic term for construction work requiring legal approval to make alterations or redirect watercourses into new channels, in order to facilitate flood discharge, improve land use, or lessen maintenance; *specific terms* channelization, ▶conventional river engineering measures, ▶culverting, ▶natural river engineering measures, ▶regulation of watercourses, ▶river straightening, ▶straightening of watercourses, ▶stream improvement, ▶stream straightening); *syn.* improvement [n] of rivers and streams; *s* **ingeniería [f] hidráulica de ríos y arroyos** (Todo tipo de medidas de construcción en cursos de agua aplicadas con el fin de alterar el curso, canalizar, etc. para facilitar el desagüe, mejorar las posibilidades de uso de las riberas o reducir el mantenimiento. Generalmente estas obras hidráulicas deben ser autorizadas; *términos específicos* ▶canalización de cursos de agua, ▶entubado de un arroyo, ▶ingeniería hidráulica de arroyos y torrentes, ▶rectificación de arroyos, ▶rectificación [de ríos y arroyos], ▶rectificación de ríos, ▶regulación del curso de ríos y arroyos, ▶renaturalización de cursos de agua); *syn.* trabajos [mpl] de regulación de cauces de cursos de agua, trabajos [mpl] de ingeniería hidráulica en ríos y arroyos; *f* **travaux [mpl] d'aménagement des cours d'eau** (Ouvrages hydrauliques effectués sur les cours d'eau et leurs berges dans le cadre de la législation en vigueur — enquête d'utilité publique — en vue d'améliorer l'écoulement des eaux, de réguler le débit des crues, de favoriser une meilleure occupation des sols ou pour en faciliter l'entretien ; *termes spécifiques* ▶aménagement des cours d'eau, ▶busage, ▶recalibrage d'une rivière/d'un ruisseau, ▶redressement d'un cours d'eau, ▶régularisation d'un fleuve, ▶régulation des cours d'eau, ▶renaturation d'un cours d'eau, ▶travaux d'aménagement d'une rivière/d'un ruisseau ; AEP 1976, 251 et DFM 1975, 72) ; *g* **Gewässerausbau [m]** (...bauten [pl]; gesetzlich näher bestimmte Baumaßnahmen wie Herstellung, Verlegung oder wesentlichen Umgestaltungen von Gewässerläufen oder

ihrer Ufer zur Verbesserung der Hochwasserabflussverhältnisse, zur Verbesserung der Bodennutzungsbedingungen oder aus unterhaltungstechnischen Gründen; meistens mit einem Planfeststellungs- oder wasserrechtlichen Verfahren verbunden; cf. WaStrG, Landeswassergesetze und § 31 WHG; *UBe* ▶Bachausbau, ▶Bachbegradigung, ▶Begradigung von Fließgewässern, ▶Flussbegradigung, ▶Gewässerrenaturierung, ▶technischer Gewässerausbau, ▶Verdolung, ▶Wasserlaufregelung); *syn.* Ausbau [m, o. Pl.] von Fließgewässern, Gewässerausbaumaßnahme [f], Regulierung [f] von Wasserläufen, Wasserlaufregulierung [f].

5125 river-flow cross section [n] *eng. wat'man.* (Available profile for the flow of a river at highwater, e.g. through bridges, conduits, changes made to the riverbanks, etc.; ▶channel discharge cross section); *s* **perfil [m] transversal de un río** (Sección existente para el paso de caudal máximo como p. ej. bajo puentes, estrechamientos, etc.; ▶capacidad máxima de desagüe); *syn.* perfil [m] fluvial; *f* **profil [m] transversal d'un cours d'eau** (▶section [transversale totale] d'écoulement) ; *g* **Flussprofil [n]** (Das für den Hochwasserabfluss verfügbare Durchflussprofil, z. B. bei Brücken, Durchlässen, Uferprofilveränderungen etc.; cf. VOLL 1973; ▶Abflussquerschnitt 1).

riverine alluvial plain [n] [US]**, lower** *phyt.* ▶regularly flooded alluvial plain.

5126 riverine community [n] *phyt.* (Herbaceous plant community including all wetlands and deepwater habitats of a river or stream bank; wetlands dominated by trees, shrubs, persistent emergents, emergent mosses, or lichens; cf. CWD 1979); *s* **asociación [f] lótica** (Comunidad herbácea de los bordes de aguas corrientes); *f 1* **association [f] lotique** (Groupement herbacé [alliance du *Ranunculion fluitantis*] localisé sur le fond ou le bord de petites rivières à courant vif ; l'optimum de son extension se situe dans la zone à truite ; GGV 1979, 130) ; *syn.* association [f] d'hydrophytes des eaux courantes ; *f 2* **végétation [f] rivulaire** (Terme spécifique désignant la végétation qui croît dans les ruisseaux et ses bords) ; *g* **Fließgewässergesellschaft [f]** (Krautige Pflanzengesellschaft am unmittelbaren Uferbereich der Flüsse und Bäche).

riverine emergent wetland [n] [US] *phyt.* ▶riverine reed belt.

riverine persistent emergent wetland [n] [US] *phyt.* ▶riverine reed belt.

5127 riverine reed belt [n] *phyt.* (Several plant communities living on the banks of slow-flowing watercourses, where rushes and reeds are tolerant of permanently saturated soils: in Europe, plant community of *Phalaridetum arundinaceae*; ▶reed canary grass wamp, ▶reed-swamp [formation]); *syn.* reed-bank zone [n] of rivers or streams [UK], riverine emergent wetland [n] [US], riverine zone [n] of emergent vegetation (CWD 1979, 12), riverine persistent emergent wetland [n] [also US], riparian zone [n] of emergent vegetation; *s* **carrizal [m] fluvial** (1. Diversas comunidades vegetales de helófitos —p. ej. *Phalaridetum arundinaceae*— que crecen en las orillas de cursos de agua de circulación lenta con niveles de agua variables, sobre suelos ricos en nutrientes. 2. Superficie cubierta de carrizos; ▶carrizal de fragmitea, ▶carrizal de hierba cinta); *syn.* carrizal [m] ribereño; *f 1* **phalaridaie [f] fluviale** (Végétation spontanée d'accompagnement des rives de cours d'eau calmes [association du *Phalaridetum arundinaceae*], supportant une variation du niveau d'eau et se développant sur alluvions minérales [graviers, sable et limon] riches en substances nutritives ; ▶phalaridaie, ▶phragmitaie, ▶roselière) ; *syn.* phalaridaie [f] en bordure des cours d'eau ; *f 2* **roselière [f] de rive de cours d'eau** (Localisation du peuplement) ; *syn.* roselière [f] marginale (d'un cours d'eau), roselière [f] de la marge des cours d'eaux, roselière [f] ripariale,

R

roselière [f] rivulaire, roselière [f] fluviale ; *g* **Flussröhricht [n]** (**1.** Diverse Pflanzengesellschaften an Ufern langsam fließender Gewässer, oft mit schwankendem Wasserstand auf nährstoffreichen, sandig-kiesigen bis schluffigen Böden. **2.** Mit **F.** bestandene Fläche; ▶Röhricht, ▶Rohrglanzgrasröhricht); *syn.* Flussröhrichtzone [f].

5128 riverine species [n] *phyt. zool.* (Plant or animal living in a river habitat); *syn.* lotic species [n]; *s* **especie [f] de aguas corrientes** (Organismo que vive en el agua o en las orillas de un curso fluvial); *syn.* especie [f] lótica, especie [f] de ríos y arroyos; *f* **espèce [f] des eaux courantes** (Organisme qui vit dans ou en bordure des eaux courantes) ; *syn.* espèce [f] lotique, espèce [f] rivulaire ; *g* **Fließwasserart [f]** (Organismus in oder im unmittelbaren Einflussbereich eines Fließgewässers).

riverine upland [n] [US] *phyt.* ▶upper riparian alluvial plain.

5129 riverine vegetation [n] *phyt.* (Nitrophilic vegetation along riverbanks which may be inundated occasionally; ▶driftline community, ▶riparian woodland, ▶riverine reed belt); *s* **vegetación [f] ripícola** (Vegetación nitrófila anual de las orillas de los ríos, ocasionalmente inundada, que pertenece a la clase de las Chenopodietea; ▶bosque ripícola, ▶carrizal fluvial, ▶vegetación de las acumulaciones de restos orgánicos); *syn.* vegetación [f] ribereña; *f* **végétation [f] de marge de fleuves** (Végétation nitrophile en bordure des eaux courantes [corridor rivulaire] soumise à une inondation temporaire de faible amplitude ; ▶forêt alluviale, ▶flore des grèves [alluviales], ▶flore d'ourlet halonitrophile, ▶phalaridaie fluviale, ▶roselière de rive de cours d'eau) ; *syn.* végétation [f] ripariale, végétation [f] rivulaire ; *g* **Flussufervegetation [f]** (Nitrophile Vegetation an Flussrändern, die zeitweise überschwemmt wird; ▶Flussröhricht, ▶Au[en]wald, ▶Spülsaumflur); *syn.* Flussuferflur [f].

riverine woodland [n], **occasionally flooded** *phyt.* ▶riparian upland woodland.

riverine woodland [n] [US], **upper** *phyt.* ▶riparian upland woodland.

riverine woody species [n] *phyt.* ▶riparian woody species.

riverine zone [n], **upper** *phyt.* ▶upper riparian alluvial plain.

riverine zone [n] **of emergent vegetation** *phyt.* ▶riverine reed belt.

5130 river landscape [n] *hydr. landsc.* (Landscape characterized by one or more rivers); *s* **paisaje [m] fluvial** (Espacio natural caracterizado por una o más corrientes de agua); *f* **paysage [m] fluvial** ; *g* **Flusslandschaft [f]** (Durch einen oder mehrere Flüsse geprägte Landschaft).

river levee [n] [US] *wat'man.* ▶river embankment.

5131 river maintenance [n] *leg. wat'man.* (Comprehensive measures to maintain navigability, to sustain scenic quality or to conserve river landscapes for recreation, and to enhance biological self-purification capacity and wildlife conditions; ▶management of lakes and streams, ▶river engineering measures); *s* **mantenimiento [m] de cursos de agua (1)** (Conjunto de medidas tomadas en corrientes de agua para garantizar la navegabilidad, mantener la calidad visual como base a su uso recreativo y para mejorar la capacidad biológica de las mismas; ▶gestión de ríos y lagos, ▶ingeniería hidráulica de ríos y arroyos); *f* **entretien [m] des cours d'eau** (Ensemble des mesures [curage, entretien des rives, enlèvement des débris et embâcles] prises dans le but de conserver en l'état conforme le régime hydraulique et la navigation des voies d'eau [maintien de la capacité naturelle d'écoulement du lit et de la navigation], l'aspect et la valeur récréative des paysages tout en prenant en compte la sauvegarde ou l'amélioration de la capacité d'autoépuration biologique et des conditions stationnelles de vie de la flore et de la faune. Contrairement aux mesures d'aménagement, les travaux de maintenance ne constituent pas une atteinte au patrimoine naturel ; ▶gestion des eaux intérieures, ▶travaux d'aménagement des cours d'eau) ; *syn.* travaux [mpl] de maintenance des cours d'eau ; *g* **Gewässerunterhaltung [f, o. Pl.]** (Gesamtheit der Maßnahmen, die dazu dienen, einen ordnungsmäßigen Zustand für den Wasserabfluss, für die Schiffbarkeit, für das Bild oder den Erholungswert der Landschaft unter Berücksichtigung der Sicherung oder Verbesserung der biologischen Selbstreinigungskraft des Wassers und der Verbesserung der Lebensmöglichkeiten für eine artenreiche Pflanzen- und Tierwelt zu erhalten. Unterhaltungsmaßnahmen sind im Gegensatz zu den meisten Gewässerausbaumaßnahmen keine Eingriffe in Natur und Landschaft; cf. § 28 ff WHG, WaStrG und Landeswassergesetze; ▶Gewässerausbau, ▶Gewässerpflege); *syn.* Unterhaltung [f, o. Pl.] von Gewässerstrecken.

river management [n] *land'man. landsc.* **▶management of lakes and streams.

5132 river pollution [n] *envir.* (Contamination of a river by introduction of untreated sewage, toxic substances, fertilizers, etc; ▶water pollution); *s* **contaminación [f] fluvial** (Polución de un río causada por vertido de aguas residuales sin depurar o de sustancias tóxicas, fertilizantes, etc.; ▶contaminación de las aguas continentales); *syn.* contaminación [f] de ríos, polución [f] fluvial, polución [f] de ríos; *f* **pollution [f] des eaux d'une rivière** (Par rejet des eaux usées non traitées, de substances toxiques ou de polluants minéraux et organiques ; ▶pollution des eaux continentales); *g* **Flussverunreinigung [f]** (Durch Einleitung ungeklärter Abwässer oder durch Giftstoffe, Düngemittel etc. verursachte Belastung eines Flusses; ▶Gewässerbelastung); *syn.* Flussverschmutzung [f].

river region [n] *limn.* ▶river zone.

5133 river restoration [n] *wat'man.* (All work[s] which reinstate[s] the effective functioning of a neglected river course, including the riverbed [US]/river-bed [UK] and the stabilization of the banks; ▶river maintenance); *syn.* reconditioning [n] of rivers and streams; *s* **rehabilitación [f] de cursos de agua** (Todos los trabajos de restauración de la solera o las márgenes para recuperar la funcionalidad del curso de agua correspondiente. En D. sólo es necesario informar de ellos a la administración competente; ▶mantenimiento de cursos de agua 1); *f* **restauration [f] d'un cours d'eau** (Tous travaux participant à la remise en l'état d'un cours d'eau perturbé dans son régime hydraulique par le mauvais état de son lit et de ses berges ; ▶entretien des cours d'eau) ; *syn.* réhabilitation [f] d'un cours d'eau ; *g* **Gewässerinstandsetzung [f]** (Alle Planungen und Arbeiten, die die Funktionsfähigkeit eines beeinträchtigten Wasserlaufes an Sohle und Ufer einschließlich deren Befestigung wiederherstellen. Sie bedürfen lediglich einer Benachrichtigung der zuständigen Behörden; ▶Gewässerunterhaltung); *syn.* Instandsetzung [f] von Gewässern.

5134 riverside park [n] *landsc. recr. urb.* (**1.** Public open space or forest along a watercourse for recreational use. **2.** Broad strip of open space along a stream valley with natural vegetation and wildlife; ▶alluvial plain); *syn.* stream valley park [n] [US]; *s 1* **parque [m] de ribera** (Bosque o zona verde pública en ▶vega de río que se utiliza para fines recreativos); *s 2* **corredor [m] verde fluvial (≠)** (Franja ancha a lo largo de un valle con vegetación natural y fauna silvestre); *f* **parc [m] fluvial** (Forêts ou espaces verts publics dans une ▶plaine alluviale gérés et entretenus en vue d'une activité de loisirs ; très souvent il s'agit de bandes de terrain laissées à l'état naturel le long des berges d'un cours d'eau pouvant servir en zone urbaine d'axe de déplacement piéton, constituer un espace de détente, un réseau d'espaces protégés, etc.) ; *g* **Flussauenpark [m]** (Entlang eines Fließgewässers für die Erholung bewirtschafteter Wald oder angelegte [öffentliche] Grünfläche; ►Aue; *OB* Auenpark); *syn.* Grünzug [m] entlang eines Flusses.

Rivers [Prevention of Pollution] Act 1951 [UK] *envir. leg. wat'man.* *►water rights.

5135 river straightening [n] *eng. wat'man.* (Specific term applying to ►river engineering measures); *s* **rectificación [m] de ríos** (Término específico de ►ingeniería hidráulica de ríos y arroyos. Medidas de construcción hidráulica en ríos para evitar inundaciones y poder aprovechar las vegas para usos agrícolas y la edificación, para acelerar el flujo del agua y mejorar la navegabilidad; ►separación de un meandro); *syn.* encauzamiento [m] de ríos; *f* **régularisation [f] d'un fleuve** (Travaux de correction du cours d'un fleuve en vue de protéger les riverains contre les inondations, de gagner de nouvelles terres pour l'agriculture en assurant un meilleur écoulement des crues, de réaliser un chenal navigable par creusement du lit et augmentation de la vitesse des eaux, par ►scindement des méandres et fermeture des bras ; la régulation d'un fleuve a eu très souvent des conséquences fatales [abaissement de la nappe phréatique, fortes inondations avec dommages catastrophiques sur le cours inférieur] sur la zone alluviale naturelle et les habitats de faune et de flore ainsi que sur la culture des sols. Avec la prise de conscience de la portée économique, sociale et écologique que constituent ses effets négatifs sur le milieu naturel, la tendance est à l'heure actuelle à la ►renaturation des cours d'eau ; ►travaux d'aménagement des cours d'eau) ; *syn.* régularisation [f] du lit d'un cours d'eau) ; *g* **Flussbegradigung [f]** (Kappen der Mäander, um einen Fluss, der in mehreren Armen verläuft, auf ein Bett festzulegen, den Hochwasserabfluss zu steuern und häufige Überschwemmungen von Siedlungen und ganzen Talauen zu verhindern, aber auch, um Flüsse durch gleichzeitige Tieferlegung des Flussbettes und Erhöhung der Fließgeschwindigkeit schiffbar zu machen; z. T. dient die **F.** der Eigentumssicherung, indem Grundstücksgrenzen dauerhaft festgelegt werden; mit der **F.** geht meist eine Grundwasserabsenkung mit z. T. fatalen Folgen für die bestehende Tier- und Pflanzenwelt sowie für die Bewirtschaftung der angrenzenden Kulturböden einher; ►Kappen eines Mäanders; *OB* ►Gewässerausbau).

river terrace [n] *geo.* ►gravel river terrace, ►lower river terrace, ►middle river terrace, ►upper river terrace.

5136 river upkeep [n] *wat'man.* (All work[s] on a riverbed [US]/river-bed [UK] and its banks, including stabilization necessary for the unimpeded water flow; ►river maintenance); *s* **mantenimiento [m] de cursos de agua (2)** (Todos los trabajos necesarios para mantener la funcionalidad de cursos de agua, sean en la solera o las márgenes, incluyendo su estabilización; ►mantenimiento de cursos de agua 1); *f* **entretien [m] des eaux courantes** (Tous travaux participant au maintien de la fonction hydraulique d'un cours d'eau dans son lit et sur ses berges ; ►entretien des cours d'eau) ; *g* **Gewässerinstandhaltung [f]** (Alle Planungen und Arbeiten, die zur Erhaltung der Funktionsfähigkeit eines Wasserlaufes an Sohle und Böschungen einschließlich deren Befestigungen durchzuführen sind; TWW 1982; ►Gewässerunterhaltung).

river water authority [n] [UK] *adm. nat'res. wat'man.* ►water users agency [US].

5137 river zone [n] *limn.* (Classification unit of river systems based primarily on the water quality, oxygen content, temperature, current and bed conditions, and named according to the typical fish species. In Central Europe, from the source to the mouth the following regions are distinguished, each with its own characteristic plant and animal communities [►river biocoeno-

sis]: ►trout zone, ►grayling zone [UK], ►barbel zone [UK], ►bream zone [UK] and ►flounder zone [UK]. Outside the European context the two upper regions are known as 'rhithral', the three lower ones as '►potamal'; ►salmonid zone; cf. ILL 1961); *syn.* river region [n]; *s* **zona [f] biológica de un río** (División de los ríos de las zonas templadas según la ►biocenosis fluvial que los caracteriza, que depende principalmente de la pendiente en relación con las materias acarreadas por él, el subsuelo, el grado de oxigenación, la velocidad del agua, la temperatura de la misma, etc. Partiendo de la cabecera se encuentran la ►región salmonícola [►zona de trucha y ►zona de umbra], la ►región ciprinícola [►zona de barbo y ►zona de madrilla] y la ►región salobre); *syn.* región [f] de un río; *f* **zone [f] écologique des cours d'eau** (Classification des différentes zones d'un cours d'eau de la source jusqu'à l'embouchure. Elle est principalement caractérisée par le degré de propreté des eaux et les conditions d'oxygénation, la température, la vitesse du courant et la nature du lit. De la source jusqu'à l'embouchure on distingue la zonation piscicole suivante : ►zone à Truite, ►zone à Ombre, ►zone à Barbeau, ►zone à Brème et la ►zone à Éperlan ; une classification basée sur la composition des communautés d'invertébrés benthiques reconnaît trois zones principales : le crénon, le rhithron [►région salmonicole] et le potamon [région ciprinicole] ; ►biocénose fluviale, ►potamal ; MPE 1983, 36 et s.) ; *syn.* zone [f] piscicole des cours d'eau (AEP 1976, 150) ; *g* **Flussregion [f]** (Flussabschnitt, der in erster Linie durch Reinheitsgrad und Sauerstoffgehalt des Wassers, der Temperatur, Fließgeschwindigkeit und Bodenbeschaffenheit gekennzeichnet ist. In den einzelnen **F.en** leben bestimmte ►Flussbiozönosen. In Mitteleuropa werden von der Quelle bis zur Mündung folgende **F.en** unterschieden: ►Forellenregion, ►Äschenregion, ►Barbenregion, ►Brachsenregion und ►Brackwasserregion. In anderen Regionen der Welt werden Forellen- und Äschenregion auch als ►Bergbachregion [Rhithral], die drei übrigen als Cyprinidenregion oder ►Potamal bezeichnet; cf. ILL 1961).

rivier [n] [ZA] *geo.* ►dry valley.

rivulet [n] *wat'man.* ►streamlet.

road [n] *plan. trans.* ►access road, ►arterial road, ►collector road, ►county road [US]/B-road [UK], ►local feeder road [US]/local distributor road [UK]; ►parking access road, ►relief road; ►re-organization of a road; ►ring road, ►service road; *for.* ►skidding road [US&CDN]/skidding lane [UK]; *constr.* ►sub-base grade of a road/path [US]/formation level of a road/path [UK], ►temporary construction road; *for.* ►timber road [US]/forest path [UK].

road [n] [UK], **B-** *adm. leg. trans.* ►county road [US].

road [n], **circumferential** *trans.* ►ring road.

road [n] [UK], **classified** *adm. trans.* ►classified highway [US].

road [n] [UK], **cloverleaf slip** *plan. trans.* ►cloverleaf ramp [US].

road [n] [US], **connecting** *plan. trans.* ►access road.

road [n] [UK], **elevated** *trans.* ►elevated highway [US].

road [n] [US], **feeder** *plan. trans. urb.* ►local feeder road [US]/local distributor road [UK].

road [n] [US], **frontage** *trans.* ►service road.

road [n] [UK], **local distributor** *plan. trans. urb.* ►local feeder road [US].

road [n], **loop road** *urb. trans.* ►loop street (1).

road [n], **major** *trans.* ►arterial road.

road [n] [UK], **major** *trans.* ►major highway [US] (1).

R

road [n] [UK], **motorway and trunk** *adm. leg. trans.*
▶interstate highway [US].

road [n] [UK], **peripheral** *plan. trans.* ▶peripheral highway
[US].

road [n] [UK], **peripheral distributor** *plan. trans.*
▶peripheral highway [US].

road [n], **residential subdivision road** [US] *trans. urb.*
▶residential street.

road [n], **site** *constr.* ▶temporary construction road.

road [n] [US&CDN], **skid** *for.* ▶skidding road [US&CDN]/
skidding lane [UK].

road [n] [UK], **slip** *plan. trans.* ▶highway ramp [US].

road [n] [UK], **through** *plan. trans.* ▶major cross-town artery
[US].

road [n] [UK], **through traffic** *plan. trans.* ▶major cross-
town artery [US].

road [n] [US&CDN], **travois** *for.* ▶skidding road [US&CDN]/
skidding lane [UK].

road [n], **trunk** *trans.* ▶arterial road.

road [n] [UK], **trunk** *trans.* ▶major highway [US] (1), ▶state
highway [US].

5138 road alignment [n] *trans.* (Plan layout geometry and
both longitudinal and cross-sectional aspects of a road, street or
highway adapted to topography, functional requirements, clear
sighting, design speeds, etc. The plan view is called 'horizontal
alignment', and sectional view is called 'profile' or 'vertical
alignment'; ▶horizontal and vertical alignment, ▶transportation
corridor selection); *s* **alineación** [f] **de carretera** (Resultado de la
proyección de una carretera en cuanto a su adaptación a la topografía, exigencias funcionales, visibilidad en las curvas, velocidad prevista, etc.; ▶alineación horizontal y vertical, ▶selección
de corredor de transporte); *f* **tracé** [m] **routier** (Résultat d'une
planification routière ayant pris en considération l'intégration
dans le site, les différentes fonctions d'accessibilité, la visibilité
dans les virages, la rapidité d'exécution des travaux ; ▶étude de
tracé routier, ▶recherche du tracé optimum) ; *g* **Straßenführung**
[f] (Ergebnis der Planung einer Straße hinsichtlich der Einpassung in die Topographie, der Erfüllung von Erschließungsfunk-
tionen, der Übersichtlichkeit von Kurven, Ausbaugeschwindig-
keit etc.; ▶Trassierung in Lage und Höhe, ▶Trassenfindung).

road base [n] [UK], **granular** *constr.* ▶gravel subbase
[US]/gravel sub-base [UK].

roadbed [n], **shifting of** *trans.* ▶road relocation.

road charges [npl] [UK] *adm. urb.* ∗▶utility connection
charge.

5139 road construction [n] (1) *constr. eng.* (Branch of
▶ground civil engineering concerned with the planning, construction and maintenance of roads; ▶civil engineering works);
syn. road work [n] [also US]; *s* **construcción** [f] **de carreteras**
(Parte de la ▶ingeniería civil que se dedica a planificar, construir
y reparar carreteras; ▶trabajos de obras públicas); *syn.* construcción [f] vial, vialidad [f]; *f* **construction** [f] **routière** (Branche
d'activité des ▶travaux publics se consacrant au domaine de
l'aménagement, la construction et l'entretien des routes ; ▶tra-
vaux de voirie et de réseaux divers) ; *g* **Straßenbau** [m, o. Pl.]
(1. Teilbereich des Bauwesens/▶Tiefbau[e]s, der sich mit der
Planung, dem Bau und der Unterhaltung von Verkehrsstraßen
befasst. 2. Alle Maßnahmen zur Fertigung von Straßen und
Plätzen einschließlich der Nebenanlagen wie Böschungen, Entwässerungsgräben etc.; ▶Tiefbau[arbeiten]).

5140 road construction [n] (2) *eng. trans.* (Finished result
in the construction of a road comprising foundations, subbase

[US]/sub-base [UK] and surface courses with embankments,
incised slopes, traffic intersections, etc.); *s* **estructura** [f] **de
carretera** (Conjunto de estructuras que forman una carretera
[pavimentación, capa portante y firme] incluidos los cortes,
desmontes y terraplenes, puentes, etc.); *syn.* estructura [f] viaria;
f **ouvrage** [m] **routier** (Ensemble des structures formant une
route [partie supérieure des terrassements, couche de forme,
couches de base et de surface], ouvrages en déblai ou en remblai,
ouvrages d'intersection routière, etc.) ; *g* **Straßenbauwerk** [n]
(Gesamtheit des konstruktiven Aufbaus einer Straße [Gründung,
Unterbau und Oberbau] mit Dämmen, Einschnittböschungen,
Kreuzungsbauwerken etc.).

5141 road corridor [n] *plan. trans.* (Line of construction for
a road on an area of land, usually determined by an ▶corridor
approval procedure [US]/approval procedure for route selection
[UK], ▶transportation corridor); *s* **trazado** [m] **de carretera**
(Línea proyectada sobre el terreno a lo largo de la cual se
construye una carretera, que se define generalmente a través de
un ▶procedimiento de fijación de trazado; ▶arteria de
transporte); *f* **tracé** [m] **routier** (Ensemble des lignes sur le
terrain fixant l'emprise d'une route, en général déterminé dans le
cadre d'une ▶procédure de définition de tracé ; ▶couloir de
transport) ; *g* **Straßentrasse** [f] (Linienführung einer Straße im
Gelände, die, wenn sie übergeordnet ist, meist durch ein
▶Linienbestimmungsverfahren festgelegt wird; ▶Verkehrs-
trasse).

5142 road crest [n] *trans.* (Convex-shaped summit of a
vertical road alignment at which the gradient changes; ▶vertical
alignment; *opp.* ▶road sag); *s* **cambio** [m] **de rasante** (Cima
convexa de la alineación vertical de una carretera en la que
cambia la dirección de la pendiente; ▶alineación vertical; *opp.*
▶depresión 2); *f* **courbe** [f] **convexe en profil en long** (Changement d'angle de déclivité [pente/rampe] de forme convexe dans
un ▶profil en long des voies routières ; il est défini par un rayon
de raccordement en angle saillant ; *opp.* ▶courbe concave en
profil en long) ; *g* **Kuppe** [f] (2) (Konvex geformter Neigungswechsel der ▶Gradiente einer Straße; *opp.* ▶Wanne); *syn.*
Straßenkuppe [f].

road crest [n] [UK], **vertical curve of a** *trans.* ▶crest
vertical curve [US].

road dip [n] [UK] *trans.* ▶road sag [US].

road edge [n] (1) *trans.* ▶road limit.

road edge [n] [US] (2) *trans.* ▶road verge.

5143 road edge guide post [n] *trans.* (White posts, mostly
equipped with reflectors, delineating edges of roads in rural areas
and serving as a guide in dark and misty weather, or in snow
removal); *s* **poste-guía** [m] (Poste blanco generalmente equipado
con deflectores para señalar el borde de la carretera y las distancias); *f* **balise** [f] **de signalisation routière** (Poteau blanc de
signalisation de voirie équipé de réflecteurs et placé sur le bas-
côté des routes) ; *g* **Leitpfosten** [m] **am Straßenrand** (Als seitliche Begrenzung einer Landstraße dienender weißer Pfosten,
meist mit Reflektoren ausgestattet).

5144 road embankment [n] *constr. eng.* (Elevated
travelled way [US]/carriage-way [UK]; ▶road structure, ▶trans-
portation, ▶embankment); *syn.* highway embankment [n];
s **terraplén** [m] **para carretera** (▶cuerpo de carretera, ▶talud,
▶terraplén para construcción viaria); *f* **remblai** [m] **routier**
(Ouvrage linéaire constitué d'un terrassement en remblai qui
supporte une chaussée ; *terme spécifique* digue routière ; ▶corps
de chaussée, ▶ouvrage linéaire en remblai, ▶remblai d'infrastructure de transport, ▶talus) ; *syn.* remblai [m] d'infrastructure
routière, remblai [m] d'une route/chaussée, remblai [m] de voirie,
levée [f] d'une route ; *g* **Straßendamm** [m] (Längliches, ge-

R

schüttetes Erdbauwerk als tragfähiger Untergrund für eine Straße; ►geschüttete Böschung, ►Straßenkörper, ►Damm für Verkehrsbauten).

road gully [n] [UK] *constr.* ►catch basin [US].

road gully [n] **with sump** [UK] *constr.* ►catch basin with sump [US].

5145 road limit [n] *trans.* (Edge of ►travelled way [US]/ carriage-way [UK], ►curbstone [US]/kerbstone [UK], ►hard shoulder, ►road verge, ►yellow edge line [US]); *syn.* road edge [n] (1), roadside [n]; *s* **borde** [m] **de carretera** (Límite lateral de una ►carretera; ►arcén, ►berma 2, ►línea lateral de señalización, ►piedra de bordillo); *f 1* **bord** [m] **de route** (Terme courant désignant l'espace en bordure de la chaussée comprenant la bande de guidage et la berme engazonnée lorsqu'elle existe ; ►accotement) ; *f 2* **accotement** [m] (Terme technique désignant suivant la catégorie de ►routes l'espace latéral de la chaussée constitué de la ►bande latérale de signalisation, de la ►bande d'arrêt d'urgence [largeur minimum de 2 m] ou d'une bande dérasée de droite [largeur minimum de 0,50 m] et de la berme engazonnée ; ►accotement non stabilisé, ►accotement stabilisé, ►bordure de trottoir) ; *g* **Straßenrand** [m] (Seitliche Begrenzung einer ►Straße; ►Bankett, ►Bordstein, ►weißer/gelber Randstreifen, ►befestigter Seitenstreifen).

5146 road network [n] *trans.* (Complete street, road, and highway fabric of a given area); *s 1* **red** [f] **vial** (Conjunto de vías de tráfico motor en una región, incluidas las calles de las ciudades); *syn.* red [f] viaria; *s 2* **red** [f] **de carreteras** (Conjunto de vías de tráfico motor fuera de las ciudades, exceptuando las arterias principales que sirven para atravesar éstas); *f* **réseau** [m] **routier** (Ensemble des voies de circulation constituées des routes nationales, des autoroutes concédés ou non, des routes départementales, des routes communales et des rues dans les agglomérations ; le **réseau routier national** est celui qui est placé sous la responsabilité de l'Etat et constitué des routes nationales et des autoroutes, concédées et non-concédées qui accueillent les trafics à longue distance et assurent la desserte des grandes métropoles régionales et des grands pôles économiques ; *terme générique* **réseau des infrastructures de transport** auquel appartiennent aussi les liaisons aériennes, ferroviaires et navigables) ; *g* **Straßennetz** [n] (Gesamtheit der Straßen eines Gebietes; *OB* **Verkehrswegenetz**, zu dem auch Wasserstraßen und Bahntrassen gehören).

5147 road planning [n] *plan. trans.* (**1.** Sectoral planning concerned with the layout of roads necessary for traffic or with an individual road project; ►traffic planning); **2. highway planning** [n] (Layout of main thoroughfares); *s* **planificación** [f] **vial** (Planeamiento sectorial que abarca desde la planificación de toda la red vial para el tráfico de vehículos motores de una región hasta el diseño de un proyecto específico de construcción de carretera; ►planificación de tráfico y transportes); *syn.* planificación [f] de carreteras; *f 1* **étude** [f] **de transport** (Planification sectorielle établissant les programmes spatiaux et financiers résultant de la demande prévisible en réseau de transport routier ; ►planification des transports) ; *f 2* **étude** [f] **routière** (Examen du projet d'aménagement d'une infrastructure routière ; *terme spécifique* étude autoroutière) ; *g* **Straßenplanung** [f] (Fachplanung, die sich je nach Aufgabenstellung mit der Planung des für den Straßenverkehr benötigten Wegenetzes oder mit der Planung eines einzelnen Straßenbauprojektes befasst; ►Verkehrsplanung); *syn. z. T.* Verkehrswegeplanung [f].

5148 road pricing [n] *trans.* (Electronic monitoring of motor vehicles for collection of money to reduce inner-city traffic and increase the ridership of mass transit); *s* **cobro** [m] **de peaje para uso de calles** (Sistema en introducción en algunas ciudades

grandes, según el cual se cobra a los usuarios de vehículos, registrados electrónicamente, si transitan por algunas zonas congestionadas de la ciudad); *syn.* road pricing [m]; *f* **prélèvement** [m] **d'une redevance d'usage des routes** (±) (Système nouveau introduisant le versement d'une indemnité par une vignette ou un procédé électronique ayant pour objectif d'améliorer la qualité de l'air en limitant la circulation dans certains secteurs des centres urbains et en favorisant l'utilisation des transports collectifs) ; *g* **Erhebung** [f] **von Straßenbenutzungsgebühren** (Konzept, das zur Reduzierung des Innenstadtverkehrs eine Gebühr durch Vignetten oder elektronische Erfassung von Wegstrecken mit dem Ziel vorsieht, dass die Verkehrsteilnehmer auf öffentliche Nahverkehrsmittel umsteigen oder Fahrgemeinschaften bilden, um die Luftqualität in der Innenstadt nicht weiter zu verschlechtern oder sie sogar zu verbessern); *syn.* elektronische Erfassung [f] einer Maut, Roadpricing [n].

5149 road redesign [n] *urb. trans.* (New design of an existing street space, e.g. by creating a tree-lined avenue, by a reduction in the width of the travelled way [US]/carriage-way [UK], by a new parking layout, or by the creation of a ►pedestrian zone; ►reorganization of a road, ►traffic calming); *s* **remodelación** [f] **de calles** (Reorganización de los usos de una calle existente por medio de medidas como reducción del ancho de la calzada, plantación de árboles, reorganización de las plazas de aparcamiento o creación de ►zona peatonal; ►apaciguamiento del tráfico, ►recalificación de calles); *f* **réaménagement** [m] **d'une rue** (À la différence de la réfection de la voirie, travaux d'aménagement d'une allée, réduction de la largeur de la chaussée, refonte des parcs à voitures, création d'une ►zone piétonne ; ►recalibrage d'une rue, ►circulation apaisée) ; *g* **Straßenumbau** [m] (Neugestaltung eines vorhandenen Straßenraumes, z. B. durch Schaffung einer Allee mit großen, offenen Baumscheiben, durch Verschmälerung der Fahrbahn und Neuordnung der Parkplätze oder durch die Einrichtung einer ►Fußgängerzone; ►Straßenrückbau, ►Verkehrsberuhigung).

5150 road relocation [n] *trans.* (Moving of an existing traffic route to another place); *syn.* shifting [n] of roadbed (PSP 1987, 199); *s* **desviación** [f] **de una carretera** (Cambio del trayecto de una carretera para mejorar su recorrido o evitar zonas sensibles a los efectos del tráfico); *f* **déviation** [f] **d'une route** (Action de détourner une route de son tracé initial lors de travaux et entrainant un itinéraire de déviation) ; *g* **Verlegung** [f] **einer Straße** (Verlagerung eines bestehenden Straßenkörpers an einen anderen Ort).

5151 road safety [n] *leg. trans.* (Condition obtained by control measures to provide for safe flow of traffic; e.g. design factors, traffic signs and signals; ►public security); *s* **seguridad** [f] **del tráfico** (Condición alcanzada por medio de medidas de control para asegurar el flujo seguro del tráfico; ►seguridad pública); *syn.* seguridad [f] vial; *f* **sécurité** [f] **routière** (Sécurité relative à la circulation routière ; ►sécurité publique) ; *g* **Straßenverkehrssicherheit** [f] (Sicherheit im Straßenverkehr; ►Verkehrssicherheit).

5152 road sag [n] [US] *trans.* (Concave-shaped curve connecting two different grades; ►vertical alignement; *opp.* ►road crest); *syn.* road dip [n] [UK]; *s* **depresión** [f] (2) (En carretera cambio de gradiente cóncavo; ►alineación vertical; *opp.* ►cambio de rasante); *f* **courbe** [f] **concave en profil en long** (±) (Changement d'angle de déclivité [pente/rampe] de forme concave dans un ►profil en long d'un tracé de voies routières ; il est défini par un rayon de raccordement en angle rentrant ; *opp.* ►courbe convexe en profil en long) ; *g* **Wanne** [f] **(einer Straße)** (Konkav geformter Neigungswechsel der Gradiente einer Straße; ►Gradiente einer Trasse; *opp.* ►Kuppe 2).

R

road salt [n] [US] *envir.* ▶de-icing salt.

5153 road salting [n] *trans.* (Removal of snow or ice from roads with the aid of chemical thawing material; ▶de-icing salt, ▶snow clearing); *s* **eliminación [f] de nieve con sal** (Sistema empleado en los países desarrollados de clima de templado a frío para evitar que cuaje la nieve y para que no se hiele la humedad en las carreteras principales y autopistas; ▶sal anticongelante; *opp.* ▶limpia mecánica de nieve); *f* **salage [m]** (Élimination de la neige et du verglas sur la voirie au moyen de produits chimiques de déglaçage ; ▶sel de déneigement ; *opp.* ▶déneigement mécanique) ; *g* **Schwarzräumung [f]** (Entfernung von Schnee oder Eis auf Verkehrsflächen durch den Einsatz von chemischen Auftaumitteln; ▶Streusalz; *opp.* ▶Weißräumung).

roadside [n] *trans.* ▶road limit.

roadside green [n] [UK] *landsc.* ▶travelled way green spaces [US].

roadside green connection [n] [UK] *lands. urb.* ▶urban greenway [US].

roadside landscaping [n] [US] *leg. trans. landsc.* ▶right-of-way planting [US]/roadside planting [UK].

roadside planting [n] [UK] *leg. trans. landsc.* ▶right-of-way planting [US]/roadside planting [UK].

roadside rest area [n] [UK] *recr.* ▶stopping place [US].

road straightening [n] *eng.* ▶street straightening.

5154 road structure [n] *constr. eng.* (Complete construction of a road, street or highway upon subgrade, including embankments where necessary); *s* **cuerpo [m] de carretera** (Todas las capas de una carretera hasta el subsuelo, si lo hubiese incluye el terraplén) ; *f* **corps [m] de chaussée** (Ensemble de couches de matériaux disposées sur le terrain en place pour supporter la circulation des véhicules ; cf. DIR 1977) ; *g* **Straßenkörper [m]** (Gesamtheit des Aufbaus einer Straße über dem anstehenden Boden [Untergrund], ggf. incl. Dammschüttung).

5155 road surfacing [n] *constr. eng.* (Bituminous or concrete paving of a road, which directly bears the traffic load; ▶surface layer, ▶surfacing 3, ▶wearing course); *syn.* road wearing course [n]; *s* **pavimento [m] de carreteras/calles** (Capa que está sometida directamente al desgaste de los vehículos [▶capa de desgaste] que puede ser de asfalto o de hormigón; ▶pavimento 1, ▶revestimiento de superficies); *syn.* firme [m] de carreteras/calles; *f* **revêtement [m] d'une rue/route** (Matériaux en béton de ciment ou en enrobé bitumineux constituant la couche de surface supportant la charge du trafic [couche de roulement] ; ▶couche de surface, ▶revêtement de surface 2) ; *syn.* couche [f] de surface ; *g* **Straßenbelag [m]** (Die der unmittelbaren Belastung durch den Verkehr ausgesetzte Fahrbahndecke [▶Verschleißschicht] als bituminöse oder Betondeckschicht; in vielen Teilen der Welt auch in Form einer Schotterdecke ausgebildet. Das Ergebnis ist dann eine **Schotterpiste** oder **Schotterstraße**; ▶Belag 2, ▶Decke); *syn.* Straßenbefestigung [f].

road tunnel [n] [UK] *trans.* ▶highway tunnel [US].

5156 road verge [n] *trans.* (Unpaved strip of land beside a road; ▶curbstone [US]/kerbstone [UK], ▶emergency lane, ▶hard shoulder, ▶road limit, ▶yellow edge line [US]); *syn.* berm or berme [n], road edge [n] [also US] (2), unpaved shoulder [n] [also US], soft shoulder [n] [also US]; *s* **berma [f] (2)** (▶Borde de carretera no estabilizado; ▶arcén, ▶carril de emergencia, ▶línea lateral de señalización, ▶piedra de bordillo); *syn.* costado [m] no estabilizado, banqueta [f] [Es, MEX] (2), banquina [f] (1) [RA], hombrillo [m] [CO, YZ], acotamiento [m] [MEX], berma [f] lateral [RCH]; *f* **accotement [m] non stabilisé** (Espace non stabilisé entre la chaussée et le fossé en bord de route à l'extérieur des agglomérations ; ▶accotement, ▶accotement non stabilisé, ▶bande d'arrêt d'urgence ▶bande latérale de signalisation, ▶bord de route, ▶élément de bordure) ; *syn.* banquette [f] d'accotement, bas-côté [m] ; *g* **Bankett [n]**, *auch* **Bankette [f]** (Unbefestigter Randbereich der Straßenkrone an Straßen außerhalb geschlossener Ortschaften. Das **B.** liegt unmittelbar neben dem Randstreifen der Fahrbahn und dient der seitlichen Stabilisierung des Straßenkörpers und bietet Platz für die Installation der Leitpfosten, Verkehrsschilder und Leitplanken; ▶Bordstein, ▶weißer/gelber Randstreifen, ▶befestigter Seitenstreifen, ▶Standstreifen, ▶Straßenrand); *syn.* unbefestigter Seitenstreifen [m].

road wearing course [n] *constr. eng.* ▶road surfacing.

5157 road-widening construction [n] [US] *constr. trans. syn.* road-widening works [npl] [UK]; *s* **ensanchamiento [m] de carreteras** (Ampliación de carretera para mejorar el flujo del tránsito); *f* **élargissement [m] d'une route** (Augmentation de la largeur d'une voie ou création d'une seconde chaussée sur une section de route existante afin de faciliter l'écoulement du trafic, d'améliorer la sécurité des usagers ou la qualité environnementale [nuisances sonores, écoulement des eaux, traitement paysager]) ; *syn.* élargissement [m] de la voirie ; *g* **Straßenverbreiterung [f]** (Form des Aus- oder Umbaus einer Straße, um den Verkehrsfluss zu verbessern); *syn.* Ausbau [m, o. Pl.] einer Straße, Straßenausbau [m, o. Pl.], Verbreiterung [f] einer Straße.

road-widening works [npl] [UK] *constr. trans.* ▶road-widening construction [US].

5158 road [n] **with panoramic view** *urb. recr.* (Road on a slope from which a wide-angle view of the surrounding cityscape or landscape is possible); *s* **ruta [f] panorámica** (Carretera que transcurre por zona montañosa lo cual permite una vista panorámica del paisaje circundante); *f* **route [f] de vision panoramique** (C. urb., L. 145-6 ; route située sur un versant à partir de laquelle on a une vue panoramique sur le paysage des alentours) ; *syn.* route [f] de corniche ; *g* **Panoramastraße f]** (Straße in Hanglage, von der man einen Ausblick auf die umgebende [Stadt]landschaft hat); *syn.* Aussichtsstraße [f].

5159 road work [n] [US] *constr. eng.* (Road construction site work; ▶road construction 1); *syn.* road works [npl] [UK]; *s* **obra [f] viaria** (Trabajos de construcción de carretera; ▶construcción de carreteras); *syn.* obra [f] de carretera, obra [f] vial; *f* **travaux [mpl] routiers** (▶construction routière) ; *syn.* chantier [m] routier ; *g* **Straßenbauarbeiten [fpl]** (Straßenneubau oder -reparaturarbeiten; ▶Straßenbau); *syn.* Straßenbaustelle [f].

road works [npl] [UK] *constr.* ▶road work [US].

5160 rock [n] **(1)** *geo.* (Hard material of the Earth's crust, exposed on the surface or underlying the soil; main types are magmatic, sedimentary and metamorphic rock; ▶cleft rock, ▶loose sedimentary rock, ▶sedimentary rock, ▶siliceus rock, ▶solid rock); *s* **roca [f]** (Término genérico para ▶roca dura, ▶roca fisurada, ▶roca sedimentaria, ▶roca sedimentaria no consolidada, ▶roca silícea); *f* **roche [f]** (Matériau minéral ou agrégat de minéraux solide de l'écorce terrestre ; du point de vue pédologique on distingue les roches éruptives [ou ignées], sédimentaires et métamorphiques ; ▶roche fissurée, ▶roche massive/dure, ▶roche sédimentaire, ▶roche sédimentaire meuble, ▶roche siliceuse) ; *g* **Gestein [n]** (Mineralart oder Mineralgemenge der Erdkruste. Als Hauptgesteinsarten werden magmatische, Sediment- und metamorphe **Ge.** unterschieden; ▶Festgestein, ▶klüftiges Gestein, ▶Lockergestein, ▶Sedimentgestein, ▶Silikatgestein).

rock [n] **(2)** *constr.* ▶bedrock [US]/bed-rock [UK], ▶crushed rock, ▶massive bedrock; *geo.* ▶solid rock.

rock [n], **hard** *geo.* ▶solid rock.

R

rock [n], **parent** *geo.* ▶parent material.

rock [n] [UK], **wind** *arb. constr.* ▶wind rocking [US].

rock accumulation [n], **detrital** *geo.* ▶detrital deposit.

rock base [n] [US] *constr.* ▶crushed aggregate subbase [US]/ crushed aggregate sub-base [UK].

rock-cleft association [n] *phyt.* ▶rock crevice plant community.

rock cleft plant [n] *phyt.* rock plant.

5161 rock creep [n] *geo.* (Slow flowage type of ▶mass slippage: downward slope movement of individual rocks or boulders; cf. GGT 1979); *s* **reptación** [f] **de rocas** (Tipo de ▶movimiento gravitacional lento, transporte de bloques individuales de rocas; cf. DINA 1987); *f* **éboulement** [m] **continu** (Chute fractionnée de blocs de petite taille donnant un talus d'éboulis ; DG 1984, 471 ; le résultat est un éboulis actif, une coulée de blocs ou un champ de blocs ; ▶mouvement de masse) ; *syn.* chute [f] des débris (DG 1984, 148) ; *g* **Versatzbewegung** [f] **von Einzelblöcken** (*Form des* ▶*Massenversatzes* langsame Abwärtsbewegung von einzelnen Felsblöcken); *syn.* Steinrutschung [f], Versatz [m] von Einzelblöcken.

rock crevice plant [n] *phyt.* ▶rock plant.

5162 rock crevice plant community [n] *phyt.* (VIR 1982; plants which tolerate strong fluctuations of temperature and water availability on rock or cliff faces from subalpine to nival zones); *syn.* rock-cleft association [n] (BB 1965); *s 1* **comunidad** [f] **fisurícola** (Comunidad de plantas que crece en grietas de rocas y muros, y que está extendida especialmente en las montañas meridionales y alcanza hasta el límite climático de las nieves persistentes. Pertenece a la clase *Asplenietea rupestris*; BB 1979); *syn.* comunidad [f] de fisuras, comunidad [f] de grietas de rocas, comunidad [f] de casmófitos; *s 2* **comunidad** [f] **parietal** (Término específico para un roquedo); *syn.* comunidad [f] de roquedo; *f 1* **association** [f] **rupicole** (*Terme générique* végétation des rochers dont les besoins en substances nutritives sont faibles et capable de résister au mieux à de fortes variations de température et d'humidité sur le « rocher » des étages subalpin à nival ; p. ex. la classe des *Asplenietea rupestris*) ; *syn.* groupement [m] rupestre ; *f 2* **association** [f] **des fissures de rocher** (Terme spécifique) ; *syn.* association [f] de chasmophytes, groupement [m] de chasmophytes ; *f 3* **association** [f] **pariétale** (*Terme spécifique* groupement vivant sur les parois des rochers) ; *f 4* **association** [f] **des corniches rocheuses** (*Terme spécifique* groupement vivant sur la saillie surplombant un escarpement rocheux) ; *g* **Felsspaltengesellschaft** [f] (Pflanzen mit meist oligotrophen Nährstoffansprüchen, die die starken Temperatur- und Wasserhaushaltsschwankungen an Felsen oder Felswänden [im subalpinen bis nivalen Bereich] am besten vertragen können; z. B. die Klasse der Mauer- und Felsspaltengesellschaft *[Asplenietea rupestris]*); *syn.* Felsspaltenflur [f].

rock dam [n] *constr. wat'man.* ▶loose rock dam.

5163 rock debris [n] *geo.* (Coarse angular or rounded rock fragments caused by physical weathering, which create ▶scree at the foot of rock walls and ▶tapered scree at the end of rockfall [US]/rock fall [UK] channels; ▶detrital deposits, ▶gravel river terrace); *syn.* detritus [n]; *s* **detrito** [m] (Fragmentos de roca resultantes de la meteorización física que al pie de paredes rocosas forman ▶derrubios de gravedad, al pie de surcos de caída de rocas ▶conos de derrubios; ▶depósito detrítico, ▶terraza [fluvial] de aluviones); *syn.* detritus [m] (2); *f* **éboulis** [mpl] (**1.** Débris anguleux sous l'action de la météorisation formant un ▶talus d'éboulis au pied de parois rocheuses, un ▶cône d'éboulis en bas d'un couloir de chute de débris ; DG 1984, 148. **2.** Matériaux solides, grossiers, anguleux et meubles arrachés aux

roches par érosion ; ▶accumulation détritique, ▶terrasse alluviale) ; *syn.* débris [m] détritique, détritus [m] (2) ; *g* **Gesteinsschutt** [m] (Durch physikalische Verwitterung entstandene, eckige bis kantengerundete Gesteinsfragmente, die z. B. am Fuße von Felswänden ▶Schutthalden 2, am Ende von Steinschlagrinnen ▶Schuttkegel bilden; ▶Schotterkörper 1, ▶Schotterterrasse); *syn.* Schutt [m], Zersatz [m].

5164 rock face [n] (1) *geo.* (Very steep or vertical surface of rock); *syn.* cliff face [n]; *s 1* **pared** [f] **de roquedo** (Superficie vertical o casi vertical de una ladera rocosa de una montaña); *s 2* **acantilado** [m] (Pared de la costa cortada verticalmente); *f* **paroi** [f] **rocheuse** (Versant rocheux d'une montagne de forte inclinaison comprise entre 30° et la verticale) ; *syn.* falaise [f] rocheuse ; *g* **Felswand** [f] (Sehr steile bis senkrechte Seite eines anstehenden Felsens); *syn.* Felsklippe [f].

rock face [n] (2) *min.* ▶quarry rock face.

rock-faced [pp/adj] *constr.* ▶quarry-faced; *∗*▶tooling of stone [US]/dressing of stone [UK], #2.

5165 rockfall [n] [US]/**rock-fall** [n] [UK] *geo.* (Relatively free falling of rock fragments such as newly detached segment of bedrock from a cliff, steep slope, cave or arch caused by ▶congelifraction; ▶rock slide [US]/rock-slide [UK]); *s* **caída** [m] **de rocas** [Es, EC, MEX] (Caída de fragmentos de rocas resultantes de la ▶gelifracción o de la saturación del material no consolidado de paredes de roquedo; ▶deslizamiento de rocas); *syn.* desprendimiento [m] de piedras, caída [f] de piedras, caído [m] [MEX]; *f* **chute** [f] **de pierres** (Chute de morceaux de pierres qui se détachent des falaises ou des talus rocheux sous l'action de la ▶gélifraction ou de l'eau; ▶glissement de montagne) ; *g* **Steinschlag** [m] (Absturz einzelner Gesteinstrümmer, die sich durch ▶Frostsprengung oder bei Durchfeuchtung aus Felswänden und -hängen lösen; WAG 1984; ▶Bergrutsch).

5166 rock fan [n] *geo.* (Cone of loose, broken stones in various sizes, created by mechanical weathering, at the foot of mountain slopes and exposed rock walls; ▶scree); *s* **abanico** [m] **de derrubios** (Acumulación plana de fragmentos de roca meteorizada resultado de la caída de las piedras por el efecto de la gravedad; ▶derrubio de gravedad); *f* **nappe** [f] **d'éboulis** (Accumulation de blocs rocheux mis en place par gravité et couvrant un versant plan ; ▶talus d'éboulis) ; *g* **Schuttfächer** [m] (Flacher Akkumulationskörper aus eckigen bis kantengerundeten, verwitterten Gesteinsfragmenten; ▶Schutthalde 2).

5167 rock formation [n] *conserv. geo. land'man.* (▶Protected landscape feature in the form of a rock outcrop; ▶natural landmark [US]/natural monument [UK]); *s* **formación** [f] **de rocas** (Espacios naturales que presentan superficies verticales de altura superior a 8 metros, con independencia de la longitud de base y del tipo de roca. Como elementos de la naturaleza pueden ser protegidos según las leyes de protección de naturaleza; ▶componente protegido del paisaje, ▶monumento natural); *f* **formation** [f] **rocheuse** (Formation géologique remarquable, souvent protégée au titre de la loi du 2 mai 1930 ; ▶élément de paysage identifié, ▶monument naturel) ; *g* **Felsenbildung** [f] (Geologische Einzelschöpfung, als Landschaftsbestandteil oder Naturdenkmal häufig unter Schutz gestellt; Art. 9 BayNatSchG; ▶geschützter Landschaftsbestandteil, ▶Naturdenkmal).

5168 rock garden [n] *gard.* (Place where rocks and stones are built into a garden to provide favo[u]rable conditions for a range of rock-loving plants; it is frequently an element in park development; ▶alpine garden 1); *s* **rocalla** [f] (Jardín o parte de tal en el que se utilizan rocas y piedras para proporcionar condiciones favorables a plantas de roquedo o de alta montaña; ▶alpinum); *f* **jardin** [m] **de rocaille** (Jardin aménagé avec des blocs erratiques, des rochers naturels ou des murets en pierres

R

sèches plantés avec une végétation basse rupestre ou alpine ; ►alpinum) ; *g* **Steingarten** [m] (Gärtnerische Anlage mit Findlingen, Felsbrocken oder Mauerbeeten, die ein gutes Gedeihen der vorwiegend niedrigen, langsam wachsenden Fels- und Gebirgspflanzen begünstigt; ►Alpinum).

rock island [n] *geo.* ►meander lobe.

rock-lined ditch [n] *constr.* ►paved rubble drop chute [US]/rough bed channel.

5169 rock plant [n] *phyt.* (Plant growing on or among rocks; *specific terms* **epilithic plant** [n] describes a plant growing on rocks; e.g. algae, lichen; **rock crevice plant** [n] describes a plant growing in rock clefts; ►rupicolous; *syn.* chasmophytic plant [n], rock cleft plant [n]), rupicolous plant [n], petrophyte [n]; *s* **petrófito** [m] (Planta rupícola propiamente dicha que solo se cría en los peñascos; DB 1985; *términos específicos* **litófito** [m]: vegetación que vive sobre la propia piedra, a expensas de ella y atacándola. Se compone de plantas muy elementales: algas, líquenes y musgos o fanerógamas muy pequeñas; *syn.* planta [f] epilítica; **casmófito** [m]: petrófito fisurícola, planta propia de los peñascos, en cuyas grietas hinca sus raíces. La Chasmophytia contiene frecuentemente vegetación de orden superior incluso leñosa y hasta árboles; DB 1985; ►saxícola); *syn.* planta [f] rupícola, planta [f] rupestre, planta [f] lapidícola, planta [f] petrícola; *f* **plante** [f] **saxicole** (*Termes spécifiques* **lithophyte** [m] vivant sur la surface de rochers ; *syn.* épilithe [m] ; **chasmophyte** [m] vivant dans les fissures de rochers ; ►rupicole) ; *syn.* plante [f] rupicole, plante [f] rupestre, pétrophyte [m] ; *g* **Felspflanze** [f] (Auf Felsen oder in Felsspalten lebende Pflanze; *UBe* **Epilith** [m] [auf Felsen haftend, z. B. Algen, Flechten]; *syn.* Lithophyt [m]; **Felsspaltenpflanze** [f] *syn.* Chasmophyt [m] [griech. *chasma* >Öffnung<]; ►felsbewohnend); *syn.* Petrophyt [m].

rock revetment [n] *constr. eng.* ►riprap.

rock salt [n] [US] *envir.* ►de-icing salt.

5170 rockslide [n] [US]/**rock-slide** [n] [UK] *geo.* (Mass of rocks which slides en masse down a hillside over a bedding plane or a fault plane; ►catastrophic rockslide [US]/catastrophic rock-slide [UK], ►earthflow, ►landslide, ►mass slippage [US]/mass movement [UK], ►mud flow, ►rockfall [US]/rock-fall [UK], ►rock creep); *s* **deslizamiento** [m] **de rocas [Es, EC, MEX, YV]** (Masa de rocas que se deslizan pendiente abajo; ►alud de tierras [Es, YV], ►avalancha de rocas, ►caída de rocas, ►colada de barro, ►corrimiento de tierras, ►movimientos gravitacionales, ►reptación de rocas); *f* **glissement** [m] **de montagne** (Mouvement en masse de matériaux meubles et de roches instables sur un versant, sans dérangement considérable de leur agencement dans la partie déplacée ; cf. DG 1984 ; ►chute de pierres, ►coulée de terre, ►éboulement catastrophique, ►éboulement continu, ►mouvement de masse, ►glissement de terrain, ►lave torrentielle) ; *g* **Bergrutsch** [m] (Großflächige Form des ►Massenversatzes an Hängen, wobei der Gleitprozess vorherrscht; bei Lockermaterial spricht man von **Erdrutsch**, bei kleineren Felsmassen von **Felsrutsch**; kleinflächige Rutschungen werden i. d. R. ►Hangrutschungen genannt; ►Bergsturz, ►Erdschlipf, ►Mure, ►Steinschlag, ►Versatzbewegung von Einzelblöcken).

rod [n] [US] *constr.* ►step rod [US]/step iron [UK].

5171 rod bracing [n] *arb. constr.* (*Obsolete technique* fastening with threaded rods); *s* **sujeción** [f] **con cinchas roscadas** (*Técnica obsoleta* reforzamiento de troncos de árboles ahuecados con barras roscadas); *f* **renforcement** [m] **avec des tiges filetées** (*Technique dépassée* renforcement d'un tronc d'arbre dans lequel la pourriture a formé une cavité très importante) ; *g* **Befestigung**

mit Gewindestäben [mpl] (*Obsolete Technik* Verstärkung von ausgehöhlten Baumstämmen durch Einbau von Gewindestangen).

5172 rod bracing [n] **of V crotches** *arb. constr.* (*Obsolete technique* provision of rigid support for weak branch attachments and split crotches with ►threaded rods); *syn.* bolting [n] [also US]; *s* **sujeción** [f] **de horcadura con barra roscada** (*Técnica obsoleta* previsión de soporte rígido para horcaduras flojas de ramas con ►cincha roscada. Actualmente las ►cimas en horquilla se fijan con cinturones); *f* **renforcement** [m] **métallique de l'enfourchement principal** (*Technique dépassée* fourche sur un tronc ou une branche en forme de V entretoisée au moyen de ►tiges filetées pour ne pas créer de point de rupture) ; *g* **Verstärkung** [f] **einer V-Vergabelung durch Stabanker** (*Veraltete Technik* mit ►Gewindestäben verstrebte V-förmige Verzweigung von Stämmen oder Stämmlingen; ►Zwiesel werden heute im Rahmen einer Kronensicherung mit Gurten verspannt).

rodding eye [n] [UK] *constr.* ►access eye [US].

roll [n] *constr.* ►reed roll.

rolled sod [n] [US] *constr. hort.* ►rolled turf.

5173 rolled turf [n] *constr. hort.* (►Sods [US] 1/turves [UK] delivered in rolls; ►sod layer [US]/turf layer [UK]); *syn.* rolled sod [n] [also US]; *s* **césped** [m] **de tepes enrollados** (Bandas de ►césped de tepes de dimensiones especiales para el establecimiento de césped en campos de deportes; ►tepe); *syn.* tepes [mpl] enrollados; *f* **gazon** [m] **en rouleaux précultivé** (Gazon de placage [►gazon précultivé] livré en rouleau de dimensions variables [rouleau de 1 m², 2,50 x 0,40 m ou de 20 m², 27,40 x 0,73 m] ; F., l'expression « gazon en plaques » est couramment utilisée pour les rouleaux de gazon ; ►plaque de gazon) ; *syn.* gazon [m] préfabriqué en rouleau ; *g* **Rollrasen** [m] (Geschälter Rasenstreifen eines ►Fertigrasens mit den Schälmaßen 30 x 167 cm/40 x 250 cm, 1,5-2,5 cm dick; cf. DIN 18 917; für großflächige Sportplatzbegrünungen werden auch Großrollen von 75 x 2000 cm für die maschinelle Verlegung geliefert; ►Rasensode).

5174 rolling [n] *agr. constr. hort.* (1. Rolling in of seeds. 2. Consolidation of the soil or breaking up of clods with a roller. 3. Rotary motion by a smooth roller to consolidate the grass root-zone layer after mowing); *s* **pase** [m] **de rodillo (2)** (1. Operación para integrar las semillas en el suelo. 2. Operación para consolidar el suelo o romper terrones de tierra con un rodillo. 3. Operación para consolidar la capa de enraizamiento del césped después de cortarlo; pasar [vb] el rodillo); *syn.* consolidación [f] con pasada de rodillo, pasada [f] de rodillo; *f* **roulage** [m] (1. Opération effectuée pour raffermir la surface du sol avant [nivellement des mouvements du sol, enfoncement des mottes de terre] ou après le semis au moyen d'un rouleau léger. 2. Opération effectuée pour faciliter le tallage des jeunes gazons ou le déchaussement des plantes lors de la tonte au moyen d'un rouleau léger ou d'une tondeuse à rouleau) ; *syn.* plombage [m] au rouleau ; *g* **Walzen** [n, o. Pl.] (1. Andrücken des Saatgutes mit einer Glattradwalze. 2. Verfestigen des Bodens oder Zerdrücken von Erdschollen mit einer [Ringel]walze. 3. Befestigen der Rasentragschicht mit einer Glattradwalze nach dem Schnitt; walzen [vb]); *syn. zu 1.* Anwalzen [n, o. Pl.].

rolling hills [npl] *constr.* ►creation of rolling hills.

5175 rolling landscape [n] *geo. syn.* hilly landscape [n] (LE 1986, 13), *less hilly* undulating landscape [n]; *s* **paisaje** [m] **ondulado**; *f* **paysage** [m] **vallonné** (Paysage caractérisé par l'alternance de buttes, de reliefs allongés de dimensions modérées [collines] et de vallons) ; *syn.* paysage [m] de collines, paysage [m] collinaire, pays [m] vallonné ; *g* **hügelige Landschaft** [f] (Durch Bergkuppen, flache Täler und niedrige Höhenzüge gekennzeichnete Landschaft; eine **L.** mit sehr flachen Kuppen und Geländerücken wird **sanfthügelig** oder **flachhügelig** genannt).

rollock course [n] [US] *constr.* ►rowlock course [US]/layer of headers-on-edge [UK].

rolok [n] [US] *constr.* ►rowlock course [US]/layer of headers-on-edge [UK].

Romantic Rose [n] *hort.* ►old-fashioned rose.

roof [n] *arch. constr.* ►cold roof, ►flat roof; *constr. urb.* ►grassed roof; *arch. constr.* ►gravel-covered roof, ►low-pitch roof, ►pitched roof; *constr. urb.* ►planted flat roof; ►planted roof; *arch. constr.* ►warm roof.

roof [n], **dead-level** *arch. constr.* ►flat roof.

roof [n] [UK], **inverted** *arch. constr.* ►insulated roof membrane assembly [US].

roof [n], **non-ventilated** *arch. constr.* ►warm roof.

roof [n] [US], **sod** *constr. urb.* ►grassed roof.

roof [n], **ventilated** *arch. constr.* ►cold roof.

5176 roof blockout [n] *arch.* (Hole in the roof surface; ►roof hatch); *syn.* roof penetration [n]; *s* **manguito** [m] **a través de un tejado** (Vano previsto generalmente en tejados planos para claraboya o desagüe; ►escotilla de tejado); *syn.* orificio [m] a través de un tejado; *f 1* **passage** [m] **en traversée de dalle** (Réservation prévue en général dans la dalle d'un édifice, p. ex. orifice d'évacuation des eaux pluviales ou émergences telles lanterneau, cheminée d'aération) ; *syn.* émergence [f] traversant la dalle ; *f 2* **trémie** [f] (Réservation dans une dalle/plancher d'une dimension importante pour lanterneau, édicule de ventilation, d'ascenseur, etc. ; ►tabatière) ; *g* **Dachdurchführung** [f] (In einem [Flach]dach vorgesehene Aussparung, z. B. für Lichtkuppel, Entwässerungsvorrichtung, Lüftungsrohr; ►Dachausstiegsklappe); *syn.* Dachdurchdringung, [f], Dachdurchgang [m] (FLA 1983, 59).

roof deck [n] [US] *constr.* ►concrete slab (2).

roofed vehicular entrance [n] *arch.* ►porte cochère.

5177 roof garden [n] *arch. constr.* (An elaborately designed area with ornamental planting in contrast to ►roof terrace; ►extensive roof planting, ►flat roof); *syn.* rooftop garden [n] [also US]; *s* **azotea** [f] **ajardinada** (Jardín sobre ►tejado plano en el que, al contrario que en la ►terraza en azotea o en el ►revestimiento vegetal extensivo de tejados, generalmente predomina una capa de sustrato con mucha vegetación; ►terraza-jardín); *syn.* jardín [m] en azotea, techo-jardín [m], techo [m] verde, azotea [f] verde, cubierta [f] ajardinada; *f 1* **jardin** [m] **sur dalle** (Terme générique pour tout aménagement implanté sur une dalle en béton, p. ex. garage, habitation, bâtiment industriel) ; *syn.* dalle [f] plantée ; *f 2* **jardin** [m] **sur toiture-terrasse** (*Terme spécifique* tout aménagement d'un jardin réalisé sur la ►toiture à pente nulle d'un immeuble qui, à la différence d'une ►toiture-terrasse ou d'une ►végétalisation extensive des toitures, est caractérisé par une couche de terre d'épaisseur importante ; ►terrasse-jardin) ; *syn.* jardin [m] sur toit-terrasse ; *g* **Dachgarten** [m] (Gärtnerische Anlage auf einem ►Flachdach 1, bei der im Gegensatz zur ►Dachterrasse oder zur ►extensiven Dachbegrünung i. d. R. eine intensiv bepflanzte Erdschicht überwiegt; ►Terrassenhausgarten).

5178 roof garden system [n] *constr.* (Commercially available system for the installation of extensive ►roof plantings; ►planting method on extreme sites); *syn.* eco-roof system [n] [also US] (LA 1998 [5], 51), green-roof construction [n] [also US] (LA 1998 [5], 51); *s* **sistema** [m] **de plantación de tejados** (Técnicas de establecimiento de ►revestimiento vegetal de tejados; ►técnica de revestimiento vegetal para ubicaciones extremas); *f* **procédé** [m] **d'aménagement de jardins sur dalles** ► (Techniques particulières utilisées pour la ►végétalisation des toitures ; ►procédé de végétalisation en milieux extrêmes) ; *g* **Dachgartensystem** [n] (Technische Konstruktionsmethode zur Herstellung von extensiven ►Dachbegrünungen; ►Begrünungssystem für extreme Standorte).

5179 roof hatch [n] *arch.*(Hinged panel unit, providing a watertight means of access to a roof); *syn.* roof scuttle [n] (DAC 1975) [also US]; *s* **escotilla** [f] **de tejado**; *f* **tabatière** [f] (Ouverture en toiture comprenant un châssis dormant incliné et un abattant vitré, permettant l'accès sur la toiture ; cf. DTB 1985) ; *g* **Dachausstiegsklappe** [f].

5180 roofing [n] *arch.* (Process of building a roof or shelter; ►canopy 1, ►overhead shelter, ►park shelter, ►pedestrian shelter); *s* **construcción** [f] **del tejado** (Proceso de construcción o montaje de un tejado o un techado; ►marquesina 1, ►marquesina 2, ►techado, ►techado contra la lluvia); *f 1* **recouvrement** [m] (Processus de couvrir un bâtiment avec un toit ; ►abri couvert, ►auvent, ►couverture, ►hutte) ; *f 2* **mise** [f] **hors d'eau** (Couvrir un bâtiment de manière à le protéger de l'action des éléments naturels ; *contexte* mettre [vb] une construction hors d'eau) ; *g* **Überdachung** [f] (2) (*Vorgang* mit einem Dach versehen; ►Schutzdach 1, ►Schutzdach 2, ►Schutzhütte 2, ►Überdachung 1).

roof installation [n] [US], **green** *arch. constr. urb.* ►roof planting.

5181 roof membrane [n] *constr.* (►Waterproof layer made of plastic, asphaltic mixture, or synthetic rubber sheets laid on a roof; ►thermal stress on roof membrane); *syn.* weatherproofing layer [n] of a roof, waterproofing sheet [n] of roof panels [also US] (LA 1998 [5], 49), waterproof roofing membrane [n] [also US]; *s* **membrana** [m] **impermeabilizante de tejados** (►Capa de impermeabilización de plástico, bitumen o de hojas de goma sintética que se colocan sobre los tejados planos o inclinados para aislarlos de las precipitaciones; ►tensión térmica); *syn.* capa [f] de impermeabilización de tejados, revestimiento [m] impermeabilizante de tejados; *f* **revêtement** [m] **d'étanchéité** (Couche constituée d'une ou de plusieurs feuilles préfabriquées, utilisée pour la réalisation de l'étanchéité des toitures terrasses accessibles ou des toitures inaccessibles ; la membrane de couverture synthétique peut être monocouche, ou pluricouche ; la plupart des membranes sont composées d'un élément étanche, résistant et souple [bitume élastomère, alliage de polyoléfines, mousse de polyuréthane, PVC, Epdm [monomère d'éthylène propylène diène] vulcanisé, etc.] et d'une armature [feutre non tissé, trame/grille de fibre de verre ou de polyester] qui donnent aux membranes leurs propriétés physiques et mécaniques spécifiques ; les membranes de couverture sont fabriquées à partir de matériaux résistants, souples et étanches ; les dernières nées des membranes d'étanchéité intègrent des modules photovoltaïques souples produisant de l'électricité ; ►contrainte thermique ; ►couche d'étanchéité) ; *syn.* complexe [m] d'étanchéité, membrane [f] de couverture [CDN]) ; *g* **Dachhaut** [f] (Oberer Abschluss von Gebäuden auf flachen oder geneigten Dachkonstruktionen in Form einer niederschlagswassersperrenden Schicht. Auf Flachdächern verwendet man Dichtungsbahnen auf Bitumenbasis und auf der Basis von Hochpolymeren aus thermoplastischen elastischen Stoffen wie z. B. Kunststofffolien; weitere Dachabdichtungen sind Polyesterharzbeschichtungen, Polyurethan-Ortschäume oder Flüssigkunststoffe als Bitumen-Latexgemisch; ►Dichtungsschicht, ►thermische Belastung einer Dachhaut); *syn.* Dachabdichtung [f], Dachdichtung [f].

R

roof membrane assembly [n] *arch. constr.* ▶insulated roof membrane assembly [US].

5182 roof parapet [n] *arch.* (Low protective wall at the edge of a rooftop garden); *syn.* low roof wall [n] [also US]; *s* **parapeto [m] de tejado plano** *syn.* pretil [m] de tejado plano; *f* **acrotère [m]** (Remontée verticale autour d'une dalle de toiture-terrasse) ; *syn.* ouvrage [m] de couronnement ; *g* **Attika [f] (2)** (Niedriger waagerechter Aufbau rings um ein Flachdach); *syn.* Dachbrüstung [f].

roof penetration [n] *arch.* ▶roof blockout.

5183 roof planting [n] *arch. constr. urb.* (Installation of a "contained" green space on top of a building, requiring low to high maintenance; ▶roof garden; *specific terms* ▶extensive roof planting, ▶intensive roof planting); *syn.* rooftop planting [n] [also US], green roof installation [n] [also US]; *s* **revestimiento [m] vegetal de tejados** (▶azotea ajardinada; *términos específicos* ▶revestimiento vegetal extensivo de tejados, ▶revestimiento vegetal intensivo de tejados); *syn.* plantación [f] de tejados; *f 1* **végétalisation [f] des toitures** (*Terme générique* création d'un couvert végétal en culture extensive ou d'un aménagement paysager ponctuel sur toitures-terrasses ; *termes spécifiques* ▶végétalisation extensive des toitures, ▶végétalisation intensive des toitures ; ▶jardin sur dalle) ; *syn.* verdissement [m] de dalle ; *f 2* **aménagement [m] de terrasses-jardins** (Création d'un jardin d'agrément ; ▶jardin sur toiture-terrasse) ; *syn.* aménagement [m] de jardins sur dalles ; *g* **Dachbegrünung [f]** (Herstellen von extensiv oder intensiv zu pflegenden Vegetationsflächen oder Vegetationsbestandteilen auf Dächern; ▶Dachgarten; *UBe* ▶extensive Dachbegrünung, ▶intensive Dachbegrünung, Flachdachbegrünung, Schrägdachbegrünung); *syn.* Begrünung [f] unterbauter Flächen.

5184 roof planting method [n] *constr.* (System for roof planting with different types of multilayer construction and growing medium mixtures, depending on the delivery firm); *syn.* green roof technology [n] [also US], green roof system [n]; *s* **técnica [f] de revestimiento vegetal de tejados o azoteas** (Término específico para la plantación de tejados); *f* **procédé [m] de végétalisation des toitures** (*Terme spécifique relatif aux terrasses-jardins* procédé d'aménagement de jardins sur dalles) ; *syn.* procédé [m] de verdissement des toitures ; *g* **Dachbegrünungssystem [n]** (Von Herstellern entwickelte Technologie und angebotenes System für Dachbegrünungen mit unterschiedlicher Aufbaustärke und Substratmischung); *syn.* Begrünungssystem [n] für Dachflächen.

roof scuttle [n] [US] *arch.* ▶roof hatch.

5185 roof structure [n] *arch.* (Part of a building emerging from the roof, such as chimney, elevator tower, skylight, etc.; *opp.* ▶roof blockout); *s* **volumen [m] sobre tejado** (Partes sobresalientes del tejado de un edificio, como chimeneas, claraboyas, etc.; *opp.* ▶manguito a través de un tejado); *f* **émergence [f] de toiture** (Elément, structure qui sort en saillie d'une toiture-terrasse ou d'une couverture de comble, p. ex. souche de cheminées, aérateur, prise d'air, édicule d'ascenseur, lanterneau ; *opp.* ▶passage en traversée de dalle) ; *g* **Dachaufbau [m]** (…aufbauten [pl]; aus einem Dach herausragende Gebäudeteile, z. B. Schornstein, Aufzugsturm, Lichtkuppel; *opp.* ▶Dachdurchführung).

roof system [n]**, green** *constr.* ▶roof planting method.

roof technology [n] [US]**, green** *constr.* ▶roof planting method.

5186 roof terrace [n] *arch. constr.* (Paved surface upon a ▶flat roof, sometimes furnished with ▶planters; ▶roof garden); *s* **terraza [f] en azotea** (▶Tejado plano revestido con losas o azulejos y decorado con ▶jardineras o ▶tiestos de flores; ▶azotea ajardinada); *f* **toiture-terrasse [f]** (Toiture plate aménagée en terrasse d'agrément, comportant dallage ou pavage, ▶bacs de plantation et mobilier de terrasse ; ▶jardin sur dalle, ▶jardin sur toiture-terrasse, ▶toiture à pente nulle) ; *g* **Dachterrasse [f]** (Gepflastertes oder mit Platten verlegtes ▶Flachdach 1, das oft mit ▶Pflanzenkübeln, Pflanzentrögen oder bepflanzten Erdbeeten ausgestattet ist; ▶Dachgarten).

rooftop garden [n] [US] *arch. constr.* ▶roof garden.

rooftop planting [n] [US] *arch. constr. urb.* ▶roof planting.

roof wall [n] [US]**, low** *arch.* ▶roof parapet.

5187 rookery [n] *zool.* (**1.** Most commonly the breeding and nesting place of birds, such as the crow and rook. **2.** The term is also used to describe the breeding grounds of marine animals where mammals, e.g the seal, sea lion, walrus, penguin and other gregarious seabirds breed in general. It may also be the place where they nurse their young, such as a beach or similar location; ▶breeding site; *specific terms* crow rookery or colony of crows, ▶heron rookery [US]/heronry [UK]); *syn.* breeding colony [n]; *s* **colonia [f] de cría** (**1.** Lugar de nidada y cría de aves gregarias, como golondrinas de ribera, grajos, gaviotas y garzas reales, algunas de las cuales, como las dos últimas, anidan en colonias, pero viven aisladas el resto del año. **2.** El término se utiliza también para los lugares de cría de animales gregarios marinos, donde crían mamíferos como focas, leones marinos, morsas y pingüinos, y aves marinas; ▶lugar de cría; *términos específicos* ▶colonia de garzas, colonia de grajos); *f* **colonie [f] nicheuse** (**1.** Site de nidification des oiseaux vivant en groupe. **2.** Organismes individuels appartenant à la même espèce vivant rassemblés en un lieu de vie particulier ; en Français comme en Anglais on désigne ainsi les groupes d'oiseaux nicheurs, de mammifères marins [otaries, phoques, éléphants de mer, baleines], de chauvesouris, d'insectes, certains ne vivant qu'en colonies [p. ex. fourmis et abeilles] ; ▶emplacement de nidification ; *termes spécifiques* corbeautière, ▶héronnière) ; *syn.* colonie [f] nidificatrice, colonie [f] nidifiante ; *g* **Brutgesellschaft [f]** (**1.** Geselliges Zusammenleben von zahlreichen Vögeln einer Art während der Brutzeit, z. B. Graureiher, Kormorane, Möwen, Fregattvögel. **2.** Im Englischen werden auch große Ansammlungen von Säugetieren wie Seehunde und Seelöwen als *rookery* bezeichnet; ▶Brutplatz; *UBe* Krähenkolonie, ▶Reiherkolonie); *syn.* Brutkolonie [f].

5188 roost [n] *conserv. zool.* (**1.** Resting place recurrently visited by migratory birds after a long flight; ▶roost[ing] tree. **2.** A temporary resting place for a bird is a **perch**, ▶nocturnal roost, ▶winter perch); *s* **área [f] de descanso (de aves migratorias)** (Zona en la que descansan aves migratorias después de una etapa larga de migración; ▶árbol de reposo, ▶lugar de invernación, ▶lugar de reposo nocturno); *f* **lieu [m] d'escale** (Aire le long d'une voie de migration sur laquelle les oiseaux migrateurs font escale pour se nourrir, se reposer et muer ; les terres humides servent souvent de haltes migratoires ; ▶arbre de repos, ▶gîte d'hivernation, ▶site de repos des oiseaux) ; *syn.* zone [f] de relais de migration, aire/zone [f] de halte migratoire, halte [f] migratoire, aire [f] de stationnement des oiseaux migrateurs [CDN] ; *g* **Rastplatz [m] für Zugvögel** (Immer wieder während des Vogelzuges angeflogenes Gebiet, in dem Zugvögel nach einer größeren Flugetappe rasten; im Englischen bedeutet *roost* der langjährige, traditionelle **R.** und *perch* der bei Ortsbewegungen zum Verweilen angeflogene Platz; ▶Schlafbaum, ▶Schlafplatz für Vögel, ▶Überwinterungsplatz).

roosting colony [n] *conserv. zool.* ▶roosting population.

5189 roosting population [n] *conserv. zool.* (Group of resting migratory birds; ▶roost); *syn.* roosting colony [n];

s **población [f] de pasada** (Conjunto de aves migratorias que se encuentran en un ►área de descanso); *f* **population [f] de passage** (Ensemble des oiseaux migrateurs regroupés sur une ►aire d'escale) ; *g* **Rastbestand [m]** (Anzahl von Zugvögeln, die sich auf einem ►Rastplatz aufhalten).

5190 roost(ing) tree [n] *zool.* (Tree in which birds sleep regularly; ►roost); *s* **árbol [m] de reposo** (Árbol en el que normalmente duermen pájaros; ►área de descanso de aves migratorias); *f* **arbre [m] de repos** (Arbre sur lequel les oiseaux dorment régulièrement ; ►lieu d'escale) ; *syn.* arbre [m] reposoir ; *g* **Schlafbaum [m]** (Baum, auf dem Vögel regelmäßig schlafen; ►Rastplatz für Zugvögel).

root [n] *bot.* ►absorbing root, ►adventitious climbing root, ►aerial root, ►anchor root; *arb. hort.* ►fibrous root, ►fine root; *bot. hort.* ►heart root; *bot.* ►lateral root; *arb.* ►medium root; *bot. hort.* ►sinker root; *arb. bot.* ►stilt root; *bot. hort.* ►tap root.

root [n], **active** *bot.* ►absorbing root.

root [n], **brace** *arb. bot.* ►stilt root.

root [n], **coarse** *arb.* ►medium root.

root [n], **feeder** *bot.* ►absorbing root.

root [n], **prop** *arb. bot.* ►stilt root.

root [n], **scaffold** *arb.* ►structural root.

root [n], **side** *bot.* ►lateral root.

root [n], **skeletal** *arb.* ►structural root.

root [n], **skeleton** *arb.* ►structural root.

root [n] [US], **to take** *constr. hort. for.* ►become established.

root [n], **tuberous** *bot.* ►root tuber.

5191 root absorbing power [n] *bot.* (Capacity of roots to take up water and mineral nutrients. When the **r. a. p.** declines, the plant wilts, because the amount of water lost due to transpiration can no longer be replaced; ►wilting point); *s* **fuerza [f] de absorción de las raíces** (Capacidad de las raíces de las plantas de absorber agua y nutrientes del suelo. Cuando se reduce la **f. de a.**, de manera que la planta no puede sustituir el agua perdida por transpiración, la planta se marchita; ►punto de marchitez); *f* **pouvoir [m] d'absorption des racines** (Capacité des racines à prélever l'eau et les sels minéraux du sol ; si le **p. d'a. d. r.** vient à diminuer et la transpiration est plus importante que l'apport d'eau par les racines, la plante se flétri ; ►point de flétrissement) ; *g* **Wurzelsaugspannung [f]** (Fähigkeit von Pflanzenwurzeln, Wasser und Nährstoffe aus dem Boden aufzunehmen. Nimmt die **W.** so ab, dass das durch Transpiration abgegebene Wasser nicht mehr nachgeführt werden kann, welkt die Pflanze; ►Welkepunkt).

5192 root anchoring fabric [n] [US] *constr.* (Wire or plastic mesh combined with a plastic filter mat and installed in the ►root-zone layer to improve the penetration of roots and therefore the stability of trees and shrubs, especially on roof gardens); *syn.* nylon anchorage net [n] [US] (LA 1998 [5], 49), root anchoring membrane [n] [UK]; *s* **tejido [m] de anclaje** (Malla metálica o tejido de plástico que se coloca en la capa de ►sustrato de plantación para dar la posibilidad en enraizamiento a los arbustos y árboles, para mejorar la estabilidad. Se utiliza especialmente en azoteas/tejados ajardinados); *syn.* malla [f] de anclaje; *f* **treillis [m] d'ancrage** (Treillis métallique ou filet en polyéthylène tissé de haute résistance intégré à la couche végétale [►support de culture] et permettant aux racines des arbres et arbustes de s'y accrocher et d'augmenter ainsi la stabilité des végétaux plantés, en particulier sur toitures-terrasses) ; *g* **Verankerungsgewebe [n]** (Drahtgeflecht oder Kunststoffgewebe innerhalb der ►Vegetationstragschicht zur Durchdringung von Baum- und Strauchwurzeln, damit die Standsicherheit der

Pflanzen erhöht wird, besonders bei Dachgärten); *syn.* Krallmatte [f].

root anchoring membrane [n] [UK] *constr.* ►root anchoring fabric [US].

5193 root ball [n] *hort.* (Complete mass of roots together with the soil adhering to them of an excavated plant or a plant taken out of a container; ►burlapped root ball, ►frozen root ball, ►root-balled plant, ►soil-root ball, ►straw-wrapped root ball, ►wired root ball); *s* **cepellón [m]** (Toda la masa de raíces de una planta desenterrada y la tierra adherida a ella; ►cepellón congelado, ►cepellón con malla metálica, ►cepellón con paja, ►cepellón de tierra, ►cepellón sujeto con yute, ►planta con cepellón); *f* **motte [f] de racines** (Ensemble du système racinaire et de la terre qui y adhère lors de la transplantation ou du dépotage ; ►motte avec panier métallique, ►motte avec tontine de paille, ►motte de terre, ►motte gelée, ►motte enveloppée en tontine de jute, ►plante en motte) ; *g* **Wurzelballen [m] (2)** (Gesamtheit der ausgegrabenen oder ausgetopften Wurzeln einer Pflanze mit der daran haftenden Erde; ►Ballen mit Strohballierung, ►Ballenpflanze, ►Drahtballen, ►Erdballen, ►Frostballen, ►Juteballen).

root-ball [n] [UK], **earth** *hort.* ►soil root ball.

root ball [n] [UK], **hessian-wrapped** *hort.* ►burlapped root ball.

5194 root-balled plant [n] *constr. hort.* (►container-grown plant, ►nursery stock, ►soil-root ball; *opp.* ►bare-rooted plant; ►quality standards for nursery stock); *syn. pl.* root-balled stock [n]; *s* **planta [f] de cepellón** (Planta desenterrada con raíz y la tierra adherida a ella, la cual se protege con saco o entramado de malla metálica; *opp.* ►planta de raíz desnuda; ►cepellón de tierra, ►estándar de calidad de plantas de vivero, ►planta de vivero, ►planta en maceta); *syn.* planta [f] con mota [C]; *f* **plante [f] en motte** (Végétaux cultivés en pleine terre, arrachés en pépinière en conservant la terre qui entoure les racines et livrés avec un emballage en tontine, sous feuille ou filet plastique, en panier métallique en respectant les ►normes de qualité pour les végétaux de pépinière ; *opp.* ►plante à racines nues ; ►fourniture de pépinière, ►motte de terre, ►plante cultivée en conteneur) ; *g* **Ballenpflanze [f]** (Mit Wurzeln und anhaftender Erde ausgegrabene Pflanze, deren gestochener Ballen zum Verpflanzen meist mit einem Ballentuch oder Drahtgeflecht eingewickelt wird. Der Ballen muss gem. den ►Gütebestimmungen für Baumschulpflanzen der Art/Sorte und der Größe der Pflanze sowie den Bodenverhältnissen entsprechend groß und möglichst gleichmäßig durchwurzelt sein; in D. soll der Ballen von Gehölzen gem. DIN 18 916 mindestens den 8-fachen Durchmesser des Stammes — gemessen 1 m über dem Erdboden — haben. Beim Pflanzen ist die Verknotung am Wurzelhals zu lösen; der Draht wird an 3-4 Stellen aufgekniffen, Ballentuch und Draht werden auf die Pflanzlochsohle heruntergeklappt; ►Baumschulware, ►Containerpflanze, ►Erdballen; *opp.* ►Pflanze ohne Ballen); *syn.* Pflanze [f] mit Ballen, Ballenware [f].

root-balled stock [n] *constr. hort.* ►root-balled plant.

root ball fixing [n] **(with an earth anchor system)** *constr. hort.* ►root bracing.

5195 root balling [n] **with straw** *hort.* (Wrapping of root balls with straw and binding with rope or wire for transport of trees and shrubs; ►straw-wrapped root ball); *s* **embalaje [m] del cepellón con paja** (Envoltura con paja del cepellón de leñosas a transplantar para protegerlo durante el transporte a la ubicación definitiva; ►cepellón con paja); *f* **emballage [m] en tontine** (Emballage en paille utilisé pendant le transport pour la protection de la motte des végétaux à transplanter ; ►motte avec tontine de paille) ; *syn.* tontinage [m] ; *g* **Strohballierung [f]**

R

(Umwickeln mit Stroh von Wurzelballen zu verpflanzender Gehölze für den Transport; ▶Ballen mit Strohballierung).

5196 root bracing [n] *constr. hort.* (Securing of a tree in an upright position by means of wire ropes and boards, or scaffold poles, tensioned across the root-ball below ground level, and fastened to supports buried in the ground. This is done usually with newly transplanted trees which cannot be secured by normal ▶guying methods; BS 3975: part 5 and BS 4043, p. 26s); *syn.* root guying [n] (BS 3975: Part 5), root ball fixing [n] (with an earth anchor system), underground securing [n] (BS 3975: Part 5); *s* **fijación** [f] **del cepellón** (Aseguramiento de árboles recién plantados con ayuda de alambres y estacas de madera que se colocan directamente encima del cepellón y se clavan con pequeñas estacas. Este método se utiliza cuando no se quiere o no se puede utilizar el de ▶fijación con vientos [metálicos]); *f* **haubanage** [m] **de la motte** (Maintien en position verticale d'un arbre nouvellement planté au moyen de haubans métalliques [▶câble de haubanage] fixés à des madriers de bois placés en triangle sur la partie supérieure de la motte et reliés à des piquets fermement enfoncés dans le sol) ; *g* **Wurzelballenverankerung** [f] (Unterirdische Befestigung eines neu gepflanzten Baumes mit Hilfe von Drahtseilen und Holzbohlen oder horizontalen Querhölzern, die direkt über den Wurzelballen niveaugleich eingebracht werden, mit unterirdisch eingeschlagenen Kurzpfählen befestigt werden und nach dem Anwachsen im Boden verbleiben. Diese Methode wird vor allem in windgeschützten Lagen verwendet, wo eine ▶Drahtseilverankerung nicht möglich oder nicht erwünscht ist. Diese Methode setzt einen festen Ballen voraus; Pflanzen mit Kunstballen können so nicht wirkungsvoll verankert werden); *syn.* Ballenverankerung [f], Wurzelballenstützung [f], Unterflurverankerung [f].

5197 root buttress [n] *arb. bot.* (Prominent ▶structural roots which above ground at the base of a tree trunk; cf. RRST 1983); *syn.* root spur [n], root flange [n], buttress flare [n]; *s* **contrafuerte** [f] **del tronco** (Transición trapezoidal de la base del tronco a la ▶raíz primaria); *syn.* arranque [m] de la raíz; *f* **contrefort** [m] **du tronc** (Partie proéminente de forme trapézoïdale d'un arbre située entre le sol et la partie inférieure du fût ; base de l'arbre constituée des ▶racines maîtresses latérales ; DFM, 1975, 69) ; *g* **Wurzelanlauf** [m] (Trapezförmiger Übergang der ▶Starkwurzeln vom Wurzelbereich zum Stamm).

5198 root climber [n] *bot.* (▶Self-clinging climber which climbs by the aid of rootlets developed on the stem; e.g. English ivy [US] *[Hedera helix]*, Virginia creeper *[Parthenocissus tricuspidata]*, many species of Arum family *[Araceae]*, climbing hydrangea *[Hydrangea petiolaris]*; RRST 1983; ▶climbing plant, ▶self-clinging vine with adhesive pads [US]/self-clinging vine with adhesive disks [UK]); *s* **planta** [f] **trepadora con raíces adventicias** (▶Planta autotrepadora con raíces adherentes que crecen del tallo, a menudo de fototropismo negativo, como p. ej. la hiedra *[Hedera helix]*, muchas especies de las araceas *[Araceae]* o la hortensia trepadora *[Hydrangea petiolaris]*; ▶planta zarcillosa con ventosas); *f* **plante** [f] **grimpante à racines-crampons** (▶Plante grimpante à organes d'accrochage sans mouvement préhenseur qui se fixe par de courtes racines adventives adhérant fortement au support tels que p. ex. le Lierre *[Hedera helix]*, certaines vignes-vierges *[Parthenocissus tricuspidata]*, de nombreuses Aracées *[Araceae]*, l'Hydrangea grimpante *[Hydrangea petiolaris]* ; MVV 1977, 196 ; ▶plante grimpante à ventouses) ; *g* **Wurzelkletterer** [m] (▶Selbstklimmer mit sprossbürtigen, oft negativ fototropischen Haftwurzeln, z. B. Efeu *[Hedera helix]*, Falscher Wein *[Parthenocissus tricuspidata]*, viele Aronstabgewächse *[Araceae]*, Kletterhortensie *[Hydrangea anomala ssp. petiolaris]*; ▶Haftscheibenranker); *syn.* Haftwurzelkletterer [m].

5199 root collar [n] *hort.* (RRST 1983; transition point between stem and root; ▶butt of a tree); *syn.* root neck [n] (RRST 1983), root crown [n]; *s* **cuello** [m] **de la raíz** (Zona que separa la radícula del hipocótilo, en las plántulas recién germinadas, generalmente marcada por una ceñidura, y a partir de la cual empiezan a formarse los pelos radicales; DB 1985; el **c. de la r.** y el ▶pie del tronco denominan a la misma parte de las plantas, dependiendo del punto de vista); *f* **collet** [m] **(racinaire)** (Zone de transition entre la partie racinaire et la partie caulinaire [▶base du tronc] d'un végétal supérieur ; DIB 1988, 103) ; *g* **Wurzelhals** [m] (Oberer Teil der [Stark]wurzel im Übergangsbereich zum Stamm/Stängel; W. und ▶Stammfuß bezeichnen den selben Teil einer Pflanze — je nach Betrachtungsstandpunkt).

root competition [n] *phyt.* ▶rooting stress.

5200 root-crop weed community [n] *phyt.* (Vegetation comprising weedy associates of root crops; class: *Chenopodietea*; ▶arable weed community); *syn.* pigweed vegetation [n]; *s* **comunidad** [f] **arvense de los huertos y cultivos de verano** (▶Comunidad arvense de cultivos de la clase *Chenopodietea*); *syn.* vegetación [f] de malas hierbas de los huertos y cultivos de verano; *f* **flore** [f] **de cultures sarclées** (Groupement d'espèces adventices de cultures appartenant à la classe de *Chenopodietea* ; ▶groupement de plantes adventices des cultures) ; *syn.* végétation [f] de cultures sarclées, groupement [m] (d'espèces adventices) de cultures sarclées ; *g* **Hackunkrautflur** [f] (▶Ackerunkrautgesellschaft der Klasse *Chenopodietea*, die vorwiegend auf Hackfruchtfeldern [mit Kartoffeln, Futter- und Zuckerrüben oder Feldgemüse] vorkommt); *syn.* Hackfrucht-Unkrautgesellschaft [f].

root crown [n] *hort.* ▶root collar.

5201 root-crown ratio [n] *arb. bot.* (MET 1985, 4; relationship between the total mass of a plant root system to the total volume of the crown); *syn.* root-top ratio [n] (±) (RRST 1983, 89; MET 1985, 6); *s* **relación** [f] **raíz-copa** (El total de la masa o volumen del sistema radical de una planta dividido por el total de la masa de la copa); *f* **relation** [f] **racines-couronne** (Relation entre le volume de la masse racinaire et celui de la couronne) ; *g* **Wurzel-Kronen-Verhältnis** [n] (Beziehung zwischen der räumlichen Ausdehnung der gesamten Wurzelmasse zum gesamten Kronenvolumen).

5202 root curtain [n] *arb. constr.* (Barrier provided for immediate protection of exposed, or to be exposed, roots adjacent to excavation pits or services trenches. A ditch approximately 50cm wide is excavated next to the affected tree, the roots are cut cleanly pruned and treated with a wound dressing. The ditch is then backfilled with a mixture of compost, peat and nutrients); *s* **cortina** [f] **de raíces** (Medida técnica para proteger las raíces de leñosas expuestas a la intemperie por obras de construcción: Se excava una zanja de unos 50 cm de ancho, se iguala el corte de las raíces dañadas, se les aplica una sustancia protectora y se cierra la zanja con una mezcla de compost, turba y nutrientes); *f* **cernage** [m] **des racines** (Mesure technique de protection et de préparation des racines avant mises ou à mettre à nues sur les bords des parois de fouilles lors de travaux d'affouillement [de préférence une année de végétation avant le début des travaux] ; à cet effet il est effectué une tranchée d'environ 50 cm de large retenue par une séparation en dur [armature métallique ou rangée de pieux] recouverte d'un feutre ou d'un tissu de jute, dans laquelle les racines coupées ou cassées sont soigneusement taillées et, le cas échéant, badigeonnées au moyen d'un produit cicatrisant, la tranchée étant ensuite remplie d'un terreau spécial favorisant le développement de nouvelles racines [rhizogénèse rapide]) ; dans le cas de transplantation de végétaux, cette opération de protection préalable, par creusement d'une tranchée circulaire autour de l'arbre qu'on remplis d'un mélange approprié ayant pour but de stimuler la production de racines fines et de revitaliser le système racinaire, peut être effectuée en plusieurs étapes ; *syn.* rideau [m] à racines [CH] ; *g* **Wurzelvorhang [m]** (Technische Maßnahme zum sofortigen Schutz freigelegter oder freizulegender Wurzelbereiche am Rande von Baugruben oder Leitungsgräben. Dabei wird ein ca. 50 cm breiter Graben ausgehoben, abgesägte und gebrochene Wurzeln sauber bis auf das intakte Holz nachgeschnitten und der Graben mit einem Gemisch aus Komposterde, Torfmull und Nährstoffen wieder verfüllt, um die Wurzelneubildung zu fördern. Bei Großbaumverpflanzungen dient der **W.** dazu, das stammnahe Wurzelwerk [2-3 Jahre vor Pflanztermin] zu konzentrieren und somit die Ausbildung von Feinwurzeln zu fördern.

5203 root cutting [n] *hort.* (Piece of a root, usually 3-5cm long for perennials, and about 15cm long for woody species, which is used for vegetative propagation; ►cutting); *s* **esqueje [m] de raíz** (Fragmento de pequeño tamaño de la raíz de un vegetal que se utiliza para la reproducción vegetativa, en el caso de perennes suele ser de 3-5 cm. de largo, en leñosas de 15 cm; ►esqueje); *f* **bouture [f] de racines** (Méthode de multiplication végétative des végétaux pratiquée à la fin de l'hiver, sous châssis, avec des fragments de 5 cm pour les plantes vivaces et 15 cm pour les végétaux ligneux ; ►bouture) ; *g* **Wurzelschnittling [m]** (Zur vegetativen Vermehrung von einer Wurzel abgeschnittener Teil — 3 bis 5 cm lang bei Stauden und ca. 15 cm bei Gehölzen; ►Steckling).

5204 root damage [n] *arb. hort.* (Mechanically- or chemically-caused injury to roots; ►root treatment); *s* **daño [m] de la raíz** (Daño mecánico o químico causado sobre raíces; ►tratamiento de raíces); *f* **dommage [m] causé aux racines** (Provoqué par une blessure mécanique, chimique ou une attaque de champignons ; ►traitement des racines) ; *g* **Wurzelschaden [m]** (Mechanisch oder chemisch verursachte Verletzung oder durch Pilzbefall bedingte Gewebezerstörung an Wurzeln; ►Wurzelbehandlung).

5205 root decay [n] *biol.* (Disintegration of root tissue or whole roots due to fungi or bacteria); *s* **pudrimiento [m] de raíces** (Descomposición del tejido radical por acción de hongos o bacterias); *f* **pourrissement [m] des racines** (Processus et résultat du dépérissement des tissus causés par les bactéries ou les champignons) ; *g* **Wurzelfäule [f]** (Zersetzung von Wurzelgewebe durch Pilze oder Bakterien).

5206 root development [n] *hort.* (Process or result of a root progressing from earlier to later stages of maturation; RRST 1983; ►root growth); *s* **desarrollo [m] radical** (►crecimiento de la raíz); *syn.* desarrollo [m] de las raíces; *f* **développement [m] racinaire** (Processus et résultat de croissance des racines ; ►croissance [en longeur] des racines) ; *g* **Bewurzelung [f] (1)** (1. Wachstumsfortschritt von Wurzeln oder eines ganzen Systems während einer Zeitspanne. 2. Ergebnis des Wurzelwachstums in einer bestimmten Zeit; ►Wurzelwachstum).

root dip [n] *hort.* ►clay slurry.

root dipping [n] *constr. hort.* ►root puddling.

5207 rooted floating-leaf community [n] *phyt.* (ELL 1967; root-anchored herbaceous plant association with swimming leaves on the water surface in silted-up, standing waterbodies; e.g. water-lily community *[Nymphaeion]*; ►free-floating water plant, ►free-floating freshwater community); *s* **comunidad [f] de hidrófitos radicantes** (Comunidad de plantas acuáticas fijas al fondo y sostenidas por el agua; BB 1979; ►comunidad de hidrófitos libremente flotantes, ►hidrófito libremente flotante); *syn.* comunidad [f] de agua dulce arraigada (UNE 1973);

f **groupement [m] flottant, fixé** (Composé de plantes aquatiques que l'eau supporte structurellement, c.-à-d. qui, à la différence des hélophytes, ne possèdent pas de support propre, p. ex. nupharaie ; ►groupement d'hydrophytes libres et flottants [des eaux calmes], ►plante flottante et libre) ; *syn.* communauté [f] flottante enracinée (UNE 1973), herbier [m] immergé fixé ; *g* **wurzelnde Schwimmblattgesellschaft [f]** (Krautige Pflanzenformation der Wasseroberfläche in verlandenden Stillgewässern auf dem Gewässerboden wurzelnd, z. B. Teichrosengesellschaft *[Nymphaeion]*; ►Schwimmpflanze 2, ►Schwimmpflanzendecke).

rooted hydrophyte [n] *bot. phyt.* ►floating-leaved plant.

rooted water plant [n] *bot. phyt.* ►floating-leaved plant.

root environment [n] *arb. hort.* ►rooting zone.

root exploitation [n] *hort.* ►root penetration (1).

root feeding [n] *arb. constr.* ►tree feeding.

root flange [n] *arb. bot.* ►root buttress.

5208 root formation [n] *biol.* (Genesis in development of a root or a root system); *syn.* rhizogenesis [n]; *s* **rizogénesis [f]** (Proceso de desarrollo de las raíces); *f* **rhizogenèse [f]** (Processus de naissance et de développement d'une racine ou du système racinaire) ; *g* **Wurzelbildung [f]** (Art des Entstehens oder des Sichbildens einer Wurzel oder eines Wurzelsystems).

5209 root growth [n] *bot.* (Permanent increase in volume of a root or root system; ►root development); *syn.* root morphogenesis [n] (RRST 1983); *s* **crecimiento [m] de la raíz** (Resultado de la multiplicación y del alargamiento de las células de la raíz que conllevan el aumento del volumen y de la longitud de las mismas; ►desarrollo radical); *f* **croissance [f] (en longueur) des racines** (Croissance terminale [multiplication cellulaire] et subterminale [élongation cellulaire] sont les facteurs de l'allongement des racines ; MVV 1977, 315 ; ►développement racinaire) ; *g* **Wurzelwachstum [n]** (Ständige Zunahme der Ausdehnung einer Wurzel oder eines Wurzelsystems; ►Bewurzelung).

root guying [n] *constr. hort.* ►root bracing.

5210 root hair [n] *arb. bot.* (Small, tubular outgrowth from an epidermal cell in roots for the uptake of water-dissolved soil nutrients. By exuding certain substances **r. h.** are also able to dissolve nutrient compounds in the soil and make them available to the plant; **r. s.** are thereby able to penetrate the soil; ►absorption zone); *s* **pelo [m] radical** (Cualquiera de las prolongaciones de las células epidérmicas próximas al ápice de las raíces de las plantas terrícolas, de forma cilíndrica más o menos deformada por las irregularidades del terreno en que se desarrollan, de paredes sutiles y aptos para absorber el agua y las sales que ésta lleva en disolución. Tienen vida breve y pierden la turgencia a pocos días de formados, siendo sustituidos por otros nuevos. La porción de la raíz en la que se encuentran se llama ►zona pilífera y se halla siempre inmediata al ápice radical; cf. DB 1985); *syn.* pelo [m] absorbente; *f* **poil [m] absorbant** (Poil unicellulaire, allongé, porté par la partie subterminale de la racine, qui joue un rôle très important dans l'absorption de l'eau et des éléments minéraux contenus dans le sol ; DAV 1984 ; ►zone pilifère) ; *syn.* poil [m] racinaire ; *g* **Wurzelhaar [n]** (Kurzlebiges, schlauchförmiges, ausgezogenes Teil der Saugwurzelepidermis, das der Aufnahme von in Wasser gelösten Nährsalzen im Boden dient und darüber hinaus durch Ausscheidung bestimmter Stoffe Nährstoffverbindungen im Boden löst und somit diese aufschließt; ►Wurzelhaarzone).

root hair zone [n] *bot.* ►absorption zone.

root hardy [adj] [US] *bot.* ►perennial (2).

5211 root herbicide [n] *chem. hort.* (Systemic chemical used to combat weed growth by invading the entire plant and

R

destroying the plant cells; e.g. amino triazine; ►nonselective herbicide [US]/non-selective herbicide [UK], ►selective herbicide); *s* **herbicida [f] de translocación** (Productos químicos así denominados porque son absorbidos por las raíces o las hojas y translocados a los puntos de crecimiento donde los inhiben o causan crecimientos deformes que provocan la muerte de la planta. Algunos de ellos son tóxicos para las dicotiledóneas y tienen gran importancia en el control selectivo de las malas hierbas en las cosechas de cereales y pasto; cf. DINA 1987; ►herbicida selectivo, ►herbicida total); *f* **herbicide [m] racinaire** (Herbicide sélectif ou total résiduaire absorbé uniquement par les racines comme la simazine ou le dimétron empêchant la levée des mauvaises herbes au stade de la germination des graines présentes dans le sol ; ►herbicide sélectif, ►herbicide total) ; *syn.* herbicide [m] systémique ; *g* **Wurzelherbizid [n]** (Chemisches Unkrautbekämpfungsmittel [systemisches Mittel], bei dem die durch Wurzeln aufgenommenen Zellgifte, wie z. B. Aminotriazine, den Tod der Pflanze hervorrufen; ►Teilherbizid, ►Totalherbizid).

5212 rooting stress [n] *phyt.* (TEE 1980, 72; adverse effect of vigorous root activity upon neighbo[u]ring plants or those growing in the drip line area; ►rooting zone); *syn.* root competition [n] (TEE 1980, 94); *s* **competencia [f] edáfica** (SILV 1979, 101; influencia negativa del sistema radical de una planta sobre el de otra vecina o que se encuentra en la misma ►zona de penetración); *syn.* competencia [f] entre raíces; *f* **concurrence [f] des racines** (Influence de racines vigoureuses sur le système racinaire de plantes voisines entraînant la dominance d'une plante sur les sujets proches ; ►horizon racinaire) ; *syn.* compétition [f] racinaire ; *g* **Wurzeldruck [m]** (Einfluss verstärkter Wurzelaktivität auf benachbarte, von benachbarten oder im Wurzelbereich lebenden Pflanzen; ►Wurzelraum); *syn.* Wurzelkonkurrenz [f].

5213 rooting zone [n] *arb. hort.* (Part of the soil invaded by plant roots; ►root penetration 1, ►drip line area, ►rhizosphere); *syn.* root spread [n] (2) (SPON 1974, 87), root environment [n] (MET 1985, 39); *s* **zona [f] de penetración (de la raíz)** (Sección del suelo en el que se desarrolla la raíz de una planta específica; ►enraizamiento, ►rizosfera, ►zona de desarrollo de la raíz); *f* **horizon [m] racinaire** (Partie du sol où se développent les racines des plantes ; DFM 1976 ; ►enracinement, ►rhizosphère, ►zone de développement des racines) ; *g* **Wurzelraum [m]** (Bodenraum, der von Pflanzenwurzeln durchdrungen ist. ►Rhizosphäre bezeichnet den unmittelbar um die Feinwurzeln befindlichen Bodenkörper; ►Durchwurzelung 1, ►Wurzelbereich).

root morphogenesis [n] *bot.* ►root growth.

root neck [n] *hort.* ►root collar.

5214 root nodule [n] *bot.* (Nodular hypertrophy on roots, harbo[u]ring nitrogen-fixing bacteria of the genus *Rhizobium* and occurring especially in leguminous plants *[Leguminosae]* and alder roots *[Alnus]*); *s* **nódulo [m] radical** (Pequeñas protuberancias que crecen en las raíces de las leguminosas *[Leguminosae]* y los alisos *[Alnus]* que hospedan microorganismos del género *Rhizobium* que fijan biológicamente el nitrógeno del aire); *syn.* nódulo [m] de la raíz; *f* **nodosité [f] (racinaire)** (Formation supportée par le système racinaire de certains végétaux [p. ex. légumineuses, Aulne, etc.] née de l'hypertrophie d'une ou plusieurs radicelles sous l'influence de bactéries symbiotiques du genre *Rhizobium* [légumineuses] ou de l'actinomicète symbiotique [Aulne et autres non-papillonacées] hébergés dans la nodosité qui est le siège de la fixation d'azote libre par le couple bactérie/acténomicète — hôte ; DIB 1988, 256) ; *syn.* nodule [m] racinaire (DFM 1975, 187) ; *g* **Wurzelknöllchen [n]** (Durch ein Bakterium der Art *Rhizobium* hervorgerufene Stickstoff bin-

dende, knöllchenartige Verdickung, z. B. bei Leguminosen- und Erlenwurzeln); *syn.* Bakterienknöllchen [n].

5215 root penetration [n] (1) *hort.* (Root growth into a growing medium or the degree to which a material has been penetrated by root growth; cf. GREACEN et al. 1969 cit. in RRST 1983; ►root ball); *syn.* root system development [n], root exploitation [n] (RRST 1983); *s* **enraizamiento [m]** (Penetración del sistema radical en el suelo; ►cepellón); *f* **enracinement [m]** (Développement du système radiculaire d'une plante dans le sol, dans son conteneur, une ►motte de racine) ; *g* **Durchwurzelung [f] (1)** (Durchdringung des Erdreichs/►Wurzelballens 2 mit Wurzeln).

5216 root penetration [n] (2) *constr.* (Perforation of ►waterproofing sheet, drain, sewer or water pipe by aggressive roots); *s* **penetración [f] de las raíces** (Perforación de la ►banda de material impermeabilizante de un tejado plano, en una tubería de agua o desagüe, un dren, etc. causada por raíces agresivas); *f* **pénétration [f] des racines** (à travers le revêtement, le ►lé d'étanchéité ou dans un drain, un tuyau d'assainissement, etc.) ; *syn.* transpercement [m] par les racines ; *g* **Durchwurzelung [f] (2)** (Durchdringung z. B. einer ►Dichtungsbahn auf einem Dach oder von Leitungsrohren durch aggressive Wurzeln); *syn.* Wurzeldurchdringung [f].

rootproof membrane [n] [US] *constr.* ►root protection membrane.

rootproof sheet [n] [US] *constr.* ►root protection membrane.

5217 root protection membrane [n] *constr.* (Non-biodegradable plastic sheet used in the construction of ►roof gardens for the permanent protection of the roof against penetration of aggressive plant roots; e.g. poplar, birch, willow and alder trees or even thistles); *syn.* rootproof sheet [n] (OSM 1999) [also US], rootproof membrane [n] (OSM 1999) [also US], root repellant membrane [n] [also US]; *s* **lámina [f] antirraíces** (Membrana sintética no deteriorable para proteger la membrana impermeabilizante de la agresividad de las raíces de algunas plantas como chopos, abedules, sauces, alisos y cardos; ►azotea ajardinada); *syn.* capa [f] de protección contra raíces; *f* **film [m] (de protection) antiracines** (Membrane synthétique imputrescible utilisée sur les ►jardins sur dalle pour la protection permanente de l'étanchéité d'une toiture contre l'agressivité du système radiculaire de certains végétaux comme le Peuplier, le Bouleau, le Saule, l'Aulne ou le Chardon dont les racines peuvent transpercer l'étanchéité) ; *syn.* membrane [f] d'étanchéité antiracines ; *g* **Wurzelschutzfolie [f]** (Verrottungsfeste Kunstfolie zum nachhaltigen Schutz einer Dachhaut vor aggressiven Wurzeln, z. B. von Pappeln, Birken, Weiden, Erlen oder Disteln, die die Isolierung eines ►Dachgartens durchdringen können); *syn.* wurzelfeste Bahn(en) [f(pl)], Wurzelschutzbahn(en) [f(pl)], Wurzelschutzschicht [f].

5218 root pruning [n] (1) *constr. hort.* (1. Cutting away parts of the root system to stimulate fibrous root growth within a compact ball, or cutting of broken or damaged roots to promote sound growth; ►pruning at planting; 2. ►root pruning 2; 3. ►undercutting of roots); *s 1* **poda [f] de raíces** (1. Corte de raíces dañadas en plantas sin cepellón o durante el transplante y de las puntas antes de la plantación; ►poda de plantación; 2. ►corte lateral [del cepellón]; 3. ►corte de raíces debajo del cepellón); *s 2* **premoteo [m] de raíces** (Corte anual de las raíces de una plántula para que forme cepellón denso); *f 1* **rafraîchissement [m] des racines** (1. Opération qui consiste à supprimer les parties blessées ou à refaire les coupes, avant la plantation ; ADT 1988, 27 ; ►taille de plantation. 2. Coupe avant plantation de l'extrémité des racines d'une plante afin de provoquer le développement des radicelles après la plantation ; ►habillage des parties aériennes) ; *syn.* habillage [m] des racines ; *f 2* **coupe [f] des racines** (1. ▶détourage de la motte. 2. ▶coupe des racines profondes) ; *g* **Wurzelschnitt** (1. Rückschnitt beschädigter Wurzeln ballenloser Pflanzen bei Pflanzarbeiten oder Schneiden gebrochener oder beschädigter Wurzeln bei Verpflanz- oder Aushubarbeiten; Wurzeln schneiden [vb]; ▶Pflanzschnitt. 2. ▶Umstechen eines Wurzelwerks. 3. ▶Unterschneiden der Wurzeln); *syn.* Schneiden [n, o. Pl.] der Wurzeln [m].

5219 root pruning [n] (2) *hort.* (Severing the lateral root system with a spade or machine prior to transplantation of woody plants; ▶root pruning 1, ▶undercutting of roots); *s* **corte [m] lateral (del cepellón)** (Para preparar el transplante; ▶poda de raíces, ▶corte de raíces debajo del cepellón); *f* **détourage [m] de la motte** (Opération culturale consistant à tailler manuellement [bêche] ou par moyens mécaniques les racines de la motte des végétaux en vue de leur transplantation future ; ▶rafraîchissement des racines, ▶coupe des racines profondes) ; *g* **Umstechen [n, o. Pl.]** (Zur Verpflanzung von Gehölzen das Durchschneiden der seitlichen Wurzeln eines Wurzelballens mit dem Spaten oder einer Spatenmaschine, um eine dichtere Durchwurzelung des Ballens zu fördern; ▶Wurzelschnitt, ▶Unterschneiden der Wurzeln).

5220 root puddling [n] *constr. hort.* (Dipping of plant roots into earth/water mixture before planting; ▶clay slurry); *syn.* root dipping [n], coating [n] of roots (RRST 1983); *s* **pralinaje [m] de raíces** (Tratamiento de las raíces de leñosas jóvenes con ▶papilla de barro antes de su plantación); *f* **pralinage [m] des racines** (Traitement des jeunes végétaux ligneux avant plantation consistant à tremper les racines dans une boue d'argile et d'engrais [▶praline] afin d'en favoriser la reprise) ; *g* **Tauchen [n, o. Pl.] der Wurzeln in Lehmschlämme** (Wurzelbehandlung von jungen Gehölzen mit ▶Lehmschlämme/Erdbrei vor dem Pflanzen; einschlämmen [vb] 1); *syn.* Tauchbad [n] der Wurzeln in Lehmbrei.

root repellant membrane [n] [US] *constr.* ▶root protection membrane.

roots [npl] *constr. hort.* ▶undercutting of roots.

roots [npl]**, coating of** *constr. hort.* ▶root puddling.

root shoot [n] *bot.* ▶root sucker.

root spread [n] (1) *bot. constr.* ▶drip line area.

root spread [n] (2) *bot. constr.* ▶rooting zone.

5221 root spreading weed [n] *agr. for. hort.* (MBG 1972; plant regarded as a nuisance because it grows wild amongst other cultivated plants in flower beds etc. and is characterized by its perennial tap roots or rhizomes. Common species in Europe include goutweed [US]/ground elder [UK] *[Aegopodium podagraria]*, couch-grass, witch-grass, quitch-grass *[Agropyron repens]*, chrysanthemum weed, mugwort *[Artemisia vulgaris]*, field and hedge bindweed *[Convolvulus arvensis et C. sepium]*, coltsfoot *[Tussilago farfara]*, dock species *[Rumex spec.]*, etc.; in east-central U.S., common Bermuda grass is an invasive weed); *s* **mala hierba [f] de raíz** (Planta silvestre no deseada porque crece entre otras cultivadas en macizos, arriates de flores, etc. y que se caracteriza por sus raíces axonomorfas o por rizomas. Especies comunes en Europa son p. ej. la podagra *[Aegopodium podagraria]*, la grama *[Agropyron repens]*, la artemisa *[Artemisia vulgaris]*, especies de Convolvulus, la fárfara *[Tussilago farfara]* y especies de la acedera [Rumex spec.], etc.); *f* **mauvaise herbe [f] racinaire** (Caractérise les plantes indésirables, sauvages ou cultivées, possédant une racine pivot ou des rhizomes pérennes [mauvaise herbe à racines traçantes, mauvaise herbe à système racinaire pivotant], poussant spontanément dans les semis ou les parterres de fleurs, telles que l'Aégopode podagraire *[Aegopodium podagraria]* le Chiendent *[Agropyron repens]*, l'Armoise commune *[Artemisia vulgaris]*, le Liseron des champs *[Convolvulus arvensis et C. sepium]*, le Tussilage *[Tussilago farfara]*, les espèces de Patience *[Rumex spec.]*, etc.) ; *g* **Wurzelunkraut [n]** (…kräuter [pl]; aus der Sicht des Nutzers unerwünschte Wildpflanze, die zwischen angebauten Pflanzen, in Schmuckbeeten oder Rabatten wild/spontan wächst und durch ausdauernde Pfahlwurzel oder Rhizome gekennzeichnet ist. Hierzu zählen in Mitteleuropa besonders: Giersch *[Aegopodium podagraria]* Quecke *[Agropyron repens]*, Beifuß *[Artemisia vulgaris]*, Acker- und Zaunwinde *[Convolvulus arvensis et C. sepium]*, Huflattich *[Tussilago farfara]*, Ampferarten *[Rumex spec.]* etc.); *syn.* Dauerunkraut [n].

root sprout [n] [US] *bot.* ▶root sucker.

root spur [n] *arb. bot.* ▶root buttress.

5222 rootstock [n] [US]**/root-stock** [n] [UK] *bot. constr.* (1. Upper part of a ▶root system, especially its central portion, as an underground elongation of the trunk, from which the roots grow; ▶tree stump. 2. *hort.* ▶understock); *s* **cepa [f]** (Parte del tallo de una planta, inmediatamente debajo de la superficie del suelo, desde la que se producen nuevos tallos; DINA 1987; ▶raíz, ▶tocón); *f 1* **souche [f]** (1) (Partie souterraine des végétaux ligneux à la base de laquelle se forment les rejets ; ▶souche 2, ▶système racinaire) ; *f 2* **ensouchement [m]** (L'ensemble des racines d'un peuplement forestier capable ou non se rejeter ; l'anglais « root-stock » est employé dans ce sens exclusivement aux États Unis, et seulement pour l'ensouchement capable de rejeter ; DFM 1975) ; *g* **Wurzelstock [m]** (2) (1. Unterirdische Verlängerung eines Stammes, aus dem Wurzeln herauswachsen. 2. Rest eines gefällten Baumes sowie auf den Stock gesetzten Baumes oder Strauches; ▶Stubben; ▶Wurzelwerk).

5223 root sucker [n] *bot.* (Adventitious shoot arising from a rhizome, as with bamboos, or from a root; cf. SAF 1983; ▶sucker, ▶water sprout); *syn.* root shoot [n], root sprout [n] [also US]; *s* **brote [m] de raíz** (Tallo adventicio en raíces de hierbas o leñosas; ▶chupón, ▶hijuelo); *syn.* retoño [m] de raíz; *f* **drageon [m]** (Rejet qui naît à partir d'un bourgeon de la racine des végétaux ; ▶gourmand, ▶rejet de porte-greffe) ; *syn.* accru [m], surgeon [m] ; *g* **Wurzelschössling [m]** (Adventivspross an Wurzeln von Kräutern oder Gehölzen; ▶Wasserreis, ▶Wildtrieb); *syn.* Wurzelausschlag [m], Wurzelbrut [f], Wurzelspross [m], Wurzel(aus)trieb [m].

root suckers [npl] *bot. hort.* ▶propagation by root suckers.

5224 root system [n] *bot. arb. hort.* (Part of a vascular plant which typically grows within the growing medium and which is the main organ of anchorage and nutrient uptake; ▶fascicular root system, ▶fibrous root system); *s* **raíz [f]** (Órgano de las plantas que crece en dirección inversa a la del tallo, no toma color verde por la acción de la luz, absorbe de la tierra las sustancias nutrientes necesarias para el crecimiento y desarrollo del vegetal y le sirve de sostén; cf. DB 1985; ▶raíz fasciculada, ▶sistema de raíces fibrosas); *syn.* sistema [m] radical, estructura [f] radical; *f* **système [m] racinaire** (L'ensemble des éléments, constituant la partie du corps végétal vasculaire généralement souterraine [racine principale, racines secondaire et radicelles] assurant la fixation au sol de la plupart des Ptéridophytes et des Spermaphytes ; cf. MVV 1977, 301 ; GOR 1985, 43 ; ▶chevelu racinaire, ▶racine fasciculée) ; *syn.* système [m] radiculaire, appareil [m] radiculaire ; *g* **Wurzelwerk [n]** (Gesamtheit des i. d. R. im Boden wachsenden und zur Befestigung und Nahrungsaufnahme dienenden Teils einer höheren Pflanze; ▶Büschelwurzel, ▶Faserwurzelwerk); *syn.* Wurzelsystem [n], Wurzeltracht [f].

root system [n]**, bunched** *bot.* ▶fascicular root system.

R

root system [n], **diffuse** *arb. hort.* ▶fibrous root system.

root system development [n] *hort.* ▶root penetration (1).

root system spread [n] *bot. constr.* ▶drip line area.

5225 root system stage [n] *bot. hort.* (Current amount of root growth; ▶quality standards for nursery stock); *s* **grado [m] de enraizamiento** (Estado actual de desarrollo de las raíces; ▶estándar de calidad de plantas de vivero); *f* **enracinement [m]** (État du développement du système racinaire des végétaux devant répondre ▶norme de qualité pour végétaux de pépinière et p. ex. présenter un enracinement apparent sur les parois de la motte au dépotage ou des racines à travers les parois ajourées des récipients) ; *g* **Bewurzelung [f] (2)** (Zustand/Gesamtheit der erreichten Wurzelausbreitung einer Pflanze; nach den ▶Gütebestimmungen für Baumschulpflanzen muss die **B.** entsprechend der Art/Sorte, dem Alter, den Bodenverhältnissen und der Anzucht entsprechend gut ausgebildet sein).

5226 root tendril [n] *bot.* (Thread-like, branching or non-branching plant organ sprouting from secondary roots used in climbing—often occurs on tropical lianas; e.g. *Vanilla,*—or to be found on some epiphytes; *generic term* ▶tendril); *s* **zarcillo [m] radical** (Raíz epigea con sensibilidad haptotrópica, como en la vainilla; DB 1985; *término genérico* ▶zarcillo); *f* **racine-vrille [f]** (Organe aérien de nature racinaire capable de s'enrouler autour d'un support comme chez certaines lianes des tropiques ou chez certaines orchidées épiphytes, p. ex. *Vanilla*; ▶vrille) ; *g* **Wurzelranke [f]** (Fadenförmiges, unverzweigtes oder verzweigtes Organ aus sprossbürtigen Nebenwurzeln entstanden — oft bei tropischen Lianen, z. B. *Vanilla,* oder bei einigen Epiphyten anzutreffen; *OB* ▶Ranke).

5227 root tip [n] *bot.* (Apical unbranched portion of a root); *s* **ápice [f] de la raíz** (Punto vegetativo de la raíz protegido por la caliptra); *syn.* extremo [m] de la raíz; *f* **extrémité [f] de racine** (Point végétatif de la racine recouvert d'une coiffe ; MVV 1977, 310) ; *syn.* apex [m] radiculaire ; *g* **Wurzelspitze [f]** (Das äußerste [apikale], unverzweigte Stück einer Wurzel).

root-top ratio [n] *arb. bot.* ▶root-crown ratio.

5228 root treatment [n] *arb. hort.* (Cutting back of diseased or dead roots and/or healing of major root wounds; ▶root damage, ▶root curtain); *s* **tratamiento [m] de raíces** (Recorte de raíces enfermas o muertas o tratamiento de ▶daños de la raíz; ▶cortina de raíces); *f* **traitement [m] des racines** (Taille des parties malades ou mortes des racines ainsi que traitement des ▶dommages causés aux racines ; ▶cernage des racines) ; *g* **Wurzelbehandlung [f]** (Rückschnitt kranker, abgestorbener sowie freigelegter Wurzeln oder Behandlung von ▶Wurzelschäden; ▶Wurzelvorhang).

5229 root tuber [n] *bot.* (Swollen adventitious root which serves as a storage organ, e.g. *Dahlia,* lesser celandine *[Ranunculus ficaria],* aconite *[Aconitum napellus],* some terrestrial orchids, etc. *This term is not to be confused with* ▶stem tuber); *syn.* tuberous root [n]; *s* **tuberosidad [f] radical** (En algunas plantas perennes y bianuales, como las dalias *[Dahlia],* el acónito *[Aconitum napellus]* o algunas orquídeas terrestres, raíz adventicia más o menos engrosada que sirve de órgano de reserva; ▶tubérculo caulinar); *syn.* raíz [f] tuberosa; *f* **tubercule [m] radiculaire** (Portion de racine renflée pour le stockage des substances de réserve comme p. ex. chez le *Dahlia,* le Ficaire *[Ranunculus ficaria],* l'Aconit Napel *[Aconitum napellus],* diverses orchidées, etc. ; *ne pas confondre avec* ▶tubercule caulinaire) ; *syn.* racine [f] tubérisée ; *g* **Wurzelknolle [f]** (Bei manchen perennen und biennen Pflanzen ausgebildetes unterirdisches Speicherorgan, das durch eine Verdickung einer begrenzten Zone der Haupt- oder Nebenwurzeln entsteht wie z. B. bei *Dahlia,* beim Scharbockskraut *[Ranunculus ficaria],* Eisenhut

[Aconitum napellus], bei verschiedenen Erdorchideen etc.; ▶Sprossknolle).

root zone [n] *arb. constr.* ▶overfilling of root zone.

5230 root-zone layer [n] *constr.* (In landscape construction, a specially mixed soil for planting [▶growing medium], which because of its stable structure is suitable for installation on natural ground, and on a drainage or filter layer; ▶turf root-zone layer); *s* **sustrato [m] de plantación** (En construcción paisajística, capa de suelo mezclada especialmente y penetrada por las raíces, situada encima de la capa filtrante o de drenaje; ▶sustrato de enraizamiento de césped, ▶sustrato vegetal); *f* **support [m] de culture** (Couche de terre végétale, substrat végétal, substrat de culture ou terreau appropriés au développement racinaire reposant respectivement sur le terrain en place, une couche drainante ou couche filtrante ; ▶mélange terreux, ▶support végétal d'un gazon) ; *g* **Vegetationstragschicht [f]** (Über dem Baugrund resp. auf einer Drän- oder Filterschicht liegende, aufgrund ihrer strukturstabilen Zusammensetzung und ihrer Eigenschaften für Pflanzenbewuchs geeignete Boden□ resp. Substratschicht; cf. DIN 18915; ▶Erdgemisch, ▶Rasentragschicht).

5231 root zone protection [n] *arb.* (Protection of root area against soil compaction or contamination, topsoil stripping or filling; ▶protective fencing around trees); *syn.* drip line protection [n] [also US]; *s* **protección [f] de la zona radical** (Operaciones de escarda contra la compactación del suelo del alcorque; ▶valla protectora de árboles); *f* **protection [f] de trou de plantation** (Protection contre le tassement du sol, la contamination, les déblais et remblais ; ▶barrière de protection d'arbres) ; *syn.* protège-racine [m] ; *g* **Baumschutz [m, o. Pl.] im Wurzelbereich** (Schutz gegen Bodenverdichtung durch Fahrzeuge, gegen Kontamination, Bodenabtrag oder Bodenauftrag; ▶Baumschutzzaun).

ropeway [n] *recr.* *▶ski lift.

rosarium [n] *gard. hort.* ▶rose garden.

5232 rose [n] *bot. hort.* (▶climbing rose, ▶cultivated shrub rose, ▶floribunda rose, ▶ground cover rose, ▶hybrid tea, ▶miniature rose, ▶moss rose, ▶old fashioned rose, ▶park rose, ▶polyantha rose, ▶remontant rose, ▶standard rose, ▶tea rose [US]/tea-rose [UK]); *s 1* **rosa [f]** (Flor del ▶rosal); *s 2* **rosal [m]** (Arbusto sarmentoso y espinoso de la familia de las rosáceas y del género Rosa; DINA 1987; *términos específicos* ▶rosa de té, ▶rosa floribunda, ▶rosa híbrido de té, ▶rosa musgosa, ▶rosa polyantha, ▶rosa refloreciente, ▶rosal antiguo, ▶rosal cubresuelo, ▶rosal de parque, ▶rosal de pie alto, ▶rosal miniatura, ▶rosal trepador); *f 1* **rose [f]** (Fleur du rosier) ; *f 2* **rosier [m]** (Arbrisseau épineux portant des roses ; ▶rosier ancien, ▶rosier arbuste d'ornement, ▶rosier couvre-sol, ▶rosier des parcs [et jardins], ▶rosier floribunda, ▶rosier grimpant, ▶rosier hybride de thé, ▶rosier miniature, ▶rosier mousseux, ▶rosier polyantha, ▶rosier remontant, ▶rosier Thé, ▶rosier tige) ; *g* **Rose [f]** (OB zu Rosenstrauch und Rosenblüte; ▶Bodendeckerrose, ▶Edelrose, ▶Floribundarose, ▶Historische Rose, ▶Kletterrose, ▶Moosrose, ▶Parkrose, ▶Polyantharose, ▶Remontantrose, ▶Rosenhochstamm, ▶Teerose, ▶Zierstrauchrose, ▶Zwergrose).

Rose [n], **English** *hort. plant.* *▶old fashioned rose.

Rose [n], **Historic** *hort.* *▶old-fashioned rose.

Rose [n], **Nostalgic** *hort.* *▶old-fashioned rose.

rose [n], **perpetual flowering** *hort.* ▶remontant rose.

rose [n] [US], **recurrent flowering** *hort.* ▶remontant rose.

rose [n], **repeat flowering** *hort.* ▶remontant rose.

Rose [n], **Romantic** *hort.* ▶old-fashioned rose.

rose [n] [UK]**, weeping standard** *hort.* ▶weeping tree rose [US].

5233 rose bed [n] *gard. landsc.* (Separate plot planted with roses); *s* **macizo** [m] **de rosas**; *f* **massif** [m] **de rosiers** (Surface plantée de rosiers de manière décorative) ; *g* **Rosenbeet** [n] (Eine mit Rosen bepflanzte Fläche).

5234 rose breeder [n] *hort.* (Specialist who propagates new rose cultivars with desired characteristics by selection or cross-breeding; ▶plant breeding); *syn.* rose hybridizer [n]; *s* **cultivador, -a** [m/f] **de rosas** (Persona que se dedica a cultivar rosas por medio de selección o cruce de variedades con características especiales; ▶fitogenética); *f* **sélectionneur, euse** [m/f] **de roses** (Personne ou organisme qui pratique la culture des roses dans le but, par la sélection, de créer des variétés nouvelles en vue de maintenir les caractéristiques spécifiques des variétés obtenues ; ▶sélection végétale) ; *g* **Rosenzüchter/-in** [m/f] (Jemand, der Rosen mit dem Ziel kultiviert, durch Selektion oder Kreuzung Sorten mit besonderen Merkmalen oder Eigenschaften heranzuziehen; ▶Pflanzenzüchtung).

rose bush [n] [UK]**, ornamental** *hort. plant.* ▶cultivated shrub rose.

5235 rose espalier [n] *hort.* (Lattice work fastend to a wall upon which roses are trained to climb; ▶espalier, ▶espalier row); *s* **espaldera** [f] **de rosas** (Soporte en muros o fachadas que sirve de sujeción a rosales; ▶espaldera, ▶hilera de plantas en espaldera); *f* **espalier** [m] **de rosiers** (Armature fixée sur un mur sur laquelle sont palissés les rosiers ; ▶contre-espalier, ▶espalier, ▶plante palissée, ▶treillage) ; *g* **Rosenspalier** [n] (An Mauern oder Gebäudewänden befestigtes Gerüst, an dem Rosen gezogen werden; ▶Spalier 1, ▶Spalier 2).

5236 rose [n] **for the connoisseur** *hort. gard.* (Rose variety or cultivar which is grown only by a few enthusiasts); *s* **rosa** [f] **de aficionado** (±) (Variedad de rosa cultivada por un círculo pequeño de personas amantes de las rosas); *f* **rosier** [m] **d'amateur** (Rosier cultivé pour ses qualités par un cercle restreint de personnes) ; *g* **Liebhaberrose** [f] (Rosensorte, die nur von einem kleinen Interessentenkreis kultiviert wird).

5237 rose garden [n] *gard. hort.* (Garden or part thereof, intended principally for the display of roses); *syn.* rosarium [n], rosetum [n] (DIL 1987); *s* **rosaleda** [f]; *f* **roseraie** [f] (Jardin planté de rosiers permettant aux spécialistes comme aux amateurs de suivre l'histoire de la rose ou d'observer les résultats de nombreuses hybridations ou de jouir de la diversité des parfums et des couleurs) ; *g* **Rosarium** [n] (Garten oder ein Teil davon, in dem Rosen für Fachleute wie für Liebhaber meist nach verschiedenen Kriterien wie z. B. Zuchtergebnis, Duft, Farbe oder Verwendung zur Schau gestellt werden. In D. ist das älteste und größte **R.** in Sangerhausen in Sachsen-Anhalt, das 1898 durch ALBERT HOFFMANN und EWALD GNAU gegründet wurde. In den 1980er-Jahren entwickelte sich dieses **R.** zur größten Rosensammlung der Welt und hatte 2002 auf 12,5 ha ca. 700 Kulturformen und 500 Wildrosenarten in 22 Rosenklassen gegliedert. In F. gab es Anfang des 20. Jhs. bei Paris das **R.** L'Hay-les Roses; cf. DEGA 2002 [32], 27-29); *syn.* Rosengarten [m].

rose hybridizer [n] *hort.* ▶rose breeder.

5238 rosette plant [n] *bot. hort. plant.* (Herbaceous plant whose leaves are spread horizontally from a short axis at ground level with a leafless stalk, bearing flowers; ▶hemicryptophyte); *s* **planta** [f] **en roseta** (*Hemikryptophyta rosulata* ▶hemicriptófito bajo con las hojas basales dispuestas en roseta y tallo sin hojas que soporta la inflorescencia; BB 1979); *f* **plante** [f] **à rosette** (*Hemikryptophyta rosulata* chez les angiospermes acaules, ▶hémicryptophyte développant au printemps des feuilles insérées au niveau du collet, à la base de la tige portant l'inflorescence) ; *g* **Rosettenpflanze** [f] (*Hemikryptophyta rosulata* ▶Hemikryptophyt mit bodenaufliegenden, rosettig angeordneten Grundblättern, darunter bodennah verborgenen Erneuerungs-/Überdauerungsknospen und blattlosem Stängel, der den Blütenstand trägt; z. B. Wegerich *[Plantago]*, Löwenzahn *[Taraxacum]*, Königskerze *[Verbascum]*; *UB* Rosettenstaude).

rosetum [n] *gard. hort.* ▶rose garden.

5239 rotation [n] (1) *for.* (Cycle of planting to felling between the formation or regeneration of a crop or stand and its final cutting at a specified stage of maturity [= rotation length]); *s* **rotación** [f] (Período total de crecimiento de los rodales, desde la plantación [o siembra] hasta la tala); *f 1* **durée** [f] **de renouvellement** (Cycle entier des opérations de récolte [âge d'exploitation] et de régénération totale des peuplements d'une forêt ou d'une série de futaie régulière ; TSF 1985) ; *f 2* **révolution** [f] (Durée séparant deux recépages successifs de taillis ou de taillis sous-futaie ; TSF 1985) ; *f 3* **rotation** [f] (Durée séparant deux passages successifs d'une coupe de même nature dans la même parcelle ; TSF 1985) ; *g* **Umtrieb** [m] (1) (Zeitspanne vom Pflanzen bis zum Ernten eines Bestandes im gewünschten Reifealter); *syn.* Umtriebszeit [f] (1).

rotation [n] (2) *agr.* ▶crop rotation, ▶land rotation.

rotational grazing [n] *agr.* ▶rotation pasture [US].

5240 rotational grazing system [n] *agr.* (HH 1975; type of agriculture involving ▶rotation pasture); *s* **pastoreo** [m] **en rotación** (Sistema de uso pecuario de pastizales divididos en parcelas que, al contrario que en el de pastizales permanentes, sólo se explotan unos días y se dejan regenerar. Permite gran rendimiento, pero necesita gran aporte de abono y de mano de obra; ▶pastizal en rotación); *syn.* pastoreo [m] rotativo, pastoreo [m] de rotación, pastoreo [m] rotacional; *f 1* **pâturage** [m] **rationné** (Forme d'alimentation des animaux récoltant le fourrage sur pied sur des ▶prairies à pâturage rationné) ; *f 2* **pâturage** [m] **en paddocks** (1) (Forme d'alimentation des animaux récoltant le fourrage sur pied sur des ▶prairies à pâturage en paddocks) ; *syn.* système [m] warmbold (LA 1981, 821) ; *g* **Umtriebsweidewirtschaft** [f] (Landwirtschaftliche Betriebsform mit ▶Umtriebsweiden).

5241 rotation pasture [n] [US] *agr.*(Cultivated area utilized as pasture for one or a few years, as part of an agricultural crop rotation; the intensively used pasture is divided into individual paddocks, which in contrast to ▶intensively grazed pasture are always only stocked with cattle for a few days after which the paddocks are allowed to regenerate; high returns with low numbers of species and weeds, with high fertilizer and labor costs; ▶rotational grazing system); *syn.* rotational grazing; *s* **pastizal** [m] **en rotación** (Parcela de pasto utilizada intensamente que al contrario que el ▶pastizal permanente sólo está ocupado pocos días por ganado y se puede recuperar a continuación; ▶pastoreo en rotación); *syn.* pastizal [m] de rotación; *f 1* **prairie** [f] **à pâturage rationné** (Prairie de qualité moyenne sur laquelle la durée et l'aire de pâturage sont fractionnées par l'aménagement de parcelles à l'intérieur d'un enclos au moyen de clôtures électriques mobiles sur lesquelles la pâture ne dure qu'une demi-journée ou une journée ; système utilisé pour augmenter l'efficacité du ▶pâturage tournant ; cf. LA 1981, 821/903 ; ▶pâturage libre) ; *syn.* pâturage [m] rationné ; *f 2* **prairie** [f] **à pâturage en paddocks** (Prairie de qualité moyenne sur laquelle a lieu une combinaison du ▶pâturage tournant et du ▶pâturage rationné ; cf. LA 1981, 821) ; *syn.* pâturage [m] en paddocks (2) ; *g* **Umtriebsweide** [f] (Intensiv genutztes, in Einzelkoppeln aufgeteiltes Weideland, das im Gegensatz zur ▶Standweide immer nur wenige Tage mit Vieh besetzt wird und sich dann regenerieren kann; hohe Erträge bei geringer Artenzahl

R

und Verunkrautung, mit hohem Dünger- und Arbeitsaufwand; ►Umtriebsweidewirtschaft); *syn.* Rotationsweide [f], *auch* Koppelweide [f].

5242 rotation time [n] *agr.* (Cultivation period of one crop or raising period of livestock); *s* **período [m] de rotación** (Duración del aprovechamiento/la explotación de plantas o animales); *f* **période [f] de rotation** (Durée d'utilisation d'une espèce dans la succession des cultures ou cycle de vie du cheptel d'espèces différentes) ; *syn.* durée [f] de rotation ; *g* **Umtrieb [m] (2)** (Nutzungsdauer von Viehbeständen auf einer bestimmten Fläche. *Kontext* wenn z. B. Hennen schon nach einem Legejahr geschlachtet werden, so spricht man vom „einjährigen **U.**"; LFG 1958); *syn.* Umtriebszeit [f] (2).

rotavation [n] [UK] *constr. hort.* ►rototilling [US].

5243 rototilling [n] [US] *constr. hort.* (Tilling the soil with a rotary action by means of a mechanical rototiller [US]/rotavator [UK]; cf. BS 3975: part 5); *syn.* rotavation [n] [UK]; *s* **rotovación [f]** (Arado del suelo con un fresador; SILV 1979, 68; fresar [vb], avellanar [vb]); *f* **fraisage [m]** (Travail du sol en brisant les mottes à la fraise ou au rotavator ; fraiser [vb]) ; *g* **Fräsen [n, o. Pl.]** (Bodenbearbeitung mit rotierenden Schlagmessern).

rotted area [n] *arb.* ►area of rot.

rotting [n] **of refuse/waste** *envir.* ►waste decomposition.

rough bed channel [n] [UK] *constr.* ►paved rubble drop chute [US].

rough cost estimate [n] *contr.* ►preliminary cost estimate.

rough cutting [vb] *agr. constr. hort.* ►scythe.

5244 rough estimate [n] **of earthworks quantities** *constr. eng.* (Preliminary ►earthworks calculation in estimating the amount of cut and fill at the preliminary design stage); *s* **estimación [f] de volúmenes de movimiento de tierras** (Cálculo a grosso modo de volúmenes de terracerías en la fase de ante-proyecto; ►cálculo del movimiento de tierras); *f* **estimation [f] sommaire des cubatures** (Prise en compte dans le calcul des terrassements lors de l'avant-projet de l'équilibre déblai-remblai ; ►évaluation des cubatures) ; *syn.* estimation [f] sommaire des terrassements ; *g* **Erdmassenüberschlag [m]** (Grobe Ermittlung und Gegenüberstellung von Auf- und Abtragsmassen bei Vorentwürfen und für die Ermittlung von Ausschreibungsunterlagen; ►Erdmassenberechnung); *syn.* überschlägige Erdmassenberechnung [f], überschlägige Erdmengenberechnung [f].

rough faced masonry wall [n] *constr.* ►split-face dry masonry.

5245 rough grade [n] *constr.* (Result of grading operations which conforms approximately to proposed lines and levels; ►graded to poper line and levels; ►final grade 1); *syn.* bulk grade [n] [also US]; *s* **nivel [m] de rasante de acabado** (Nivel bruto resultante de trabajos de terracerías o de la modelación del terreno, a continuación se establece la ►superficie refinada; ►nivelado hasta rasante de acabado); *syn.* plano [m] bruto, rasante [f] bruta; *f* **nivellement [m] grosso modo** (Travaux de terrassement pour la mise à niveau générale d'un terrain ►conformément aux plans de profil ; ►règlement final) ; *g* **Grobplanum [n]** (Ergebnis eines mit Maschinen und Geräten durchgeführten Erdbaues, wodurch eine Bodenmodellierung oder ein Erdkörper in seinen geplanten Höhen ►profilgerecht hergestellt wurde; anschließend folgt das ►Feinplanum); *syn.* Rohplanum [n], *obs.* Grobplanie [f], *obs.* Rohplanie [f].

5246 rough grading [n] [US] *constr.* (Grading of subsoil to approximate desired contours, when final levels can only be obtained by removal of topsoil and grading/regulating/excavating into the subsoil beneath; BS 4428: 1969. Stage at which the grade

approximately conforms to the approved plan; ►final grade 1, ►fine grading, ►subbase grade [US]/formation level [UK]); *syn.* bulk grading [n] [US], major grading [n] [UK]; *s* **nivelado [m] bruto** (Nivelar el suelo antes de la ►operación de refino [de superficie]; ►subrasante, ►superficie refinada); *syn.* aplanado [m] bruto; *f* **exécution [f] du nivellement grosso modo** (Réalisation des travaux de terrassement pour la mise à niveau générale d'un terrain ainsi que son ►règlement final ; ►exécution du règlement final, ►fond de forme) ; *g* **Herstellen [n, o. Pl.] des Grobplanums** (Profilgerechtes Herstellen einer Bodenmodellierung oder eines Erdbauwerkes nach Höhenplan oder Angaben der Bauleitung dergestalt, dass ohne zusätzliche Erdbewegungen das anschließende ►Feinplanum hergestellt werden kann; ►Erdplanum, ►Herstellen des Feinplanums); *syn.* Herstellen [n, o. Pl.] des Rohplanums; *obs.* Herstellen [n, o. Pl.] der Grobplanie/Rohplanie.

rough lumber [n] *constr.* ►cut timber.

roughly-hewn [pp/adj] *constr.* *►tooling of stone [US]/dressing of stone [UK], #12.

5247 roughly-hewn block step *constr.* (Natural stone step with split surfaces and broken edges); *s* **escalón [m] de piedras de labra tosca**; *f* **dalle-bloc [f] brute** (Bloc marche de pierre naturelle brut de taille utilisé pour créer créer des escaliers pittoresques, p. ex. les escaliers rustiques des rocailles ; VRD 1994, partie 4, chap. 6.2.2.1, p. 38) ; *syn.* bloc-marche [m] brut ; *g* **grob behauene Natursteinblockstufe [f]** (Blockstufe mit spaltrauen Flächen und angesprengten oder abgeprellten Kanten).

roughly-hewn flag [n] *constr.* ►quarry flagstone.

rough-tooled [pp/adj] *constr.* *►tooling of stone [US]/dressing of stone [UK], #1.

5248 rough-tooled natural stone wall [n] *arch. constr.* (Wall made of rustic stones protruding from the surface for carving in place; ►rusticated masonry); *syn.* embossed natural stone wall [n]; *s* **muro [m] de piedras naturales de labra tosca** (►fábrica de sillarejo); *f* **mur [m] en pierres naturelles bossagées** (Les parements vus sont dressés au burin d'ébauchage, ont un aspect final bombé caractérisé par de gros éclats de formes et de saillies diverses et sont semés irrégulièrement de quelques traces de percussion allongées ; MAÇ 1981, 182 ; ►maçonnerie en pierres bossagées) ; *g* **Natursteinmauer [f] aus bossierten Steinen** (►Bossenwerk).

rounding [n] **the crest of embankment/slope** *constr.* ►rounding the top of an embankment/slope.

5249 rounding [n] **the top of an embankment/slope** *constr.* (►crest of embankment/slope, ►concave slope, ►ogee slope); *syn.* rounding [n] the crest of embankment/slope; *s* **redondeado [m] de talud** (►frente de talud, ►pie de talud cóncavo, ►talud con forma de S); *syn.* concavación [f] de talud; *f 1* **adoucissement [m] de talus** (Réalisation d'un modelé en pied et tête d'un talus pour assurer sa stabilité et son insertion dans le paysage ; adoucir [vb] ; ►épaulement [de talus], ►talus modelé en forme de S) ; *f 2* **arrondi [m] de talus** (Résultat de l'adoucissement de talus ; ►arrondi de pied de talus) ; *g* **Böschungsausrundung [f]** (1. Schaffung eines runden Überganges von der Böschungsfläche zum unteren oder oberen Anschlussgelände. 2. Das Ergebnis im oberen Teil ist ein **ausgerundeter Böschungskopf**, im unteren ein ►**ausgerundeter Böschungsfuß**; ►Böschungsschulter, ►S-förmig ausgerundete Böschung).

5250 round timber step [n] *constr.* (Step constructed of round wooden poles or logs); *s* **escalón [m] de rodillo de madera** (Escalón rústico construido con rodillo de madera como tope que se rellena de tierra); *f* **contremarche [f] en rondins** (Marche en bois utilisée pour un escalier rustique et constituée de rondins retenus par des piquets) ; *g* **Knüppelstufe [f]** (Aus runden Holzstangen gebaute Stufe).

round wood paving [n] *constr.* ▶wood disk paving [US]/ timber disk paving [UK].

route [n]**, communication** *plan. trans.* ▶connecting route.

route [n] **[SCOT], peripheral** *plan. trans.* ▶peripheral highway.

route [n]**, transportation** *plan. trans.* ▶transportation corridor.

5251 route alternative [n] *plan. trans. wat'man.* (Choice of a proposed transportation corridor e.g. for a road, transmission line, pipeline, ship channel, etc. between two or more other possibilities. The word 'alternative' [Latin *alter, -era, erum*] strictly speaking, however, refers only to the selection of one of two possibilities) ; *s* **alternativa [f] de trazado** (Una de las posibilidades disponibles para corredor de transporte, p. ej. para autopista, carretera, ferrocarril, línea de alta tensión, canal u oleo- o gaseoducto); *f* **variante [f] de tracé** (Tracé de principe d'une infrastructure évitant les principales contraintes du site et permettant une comparaison entre plusieurs tracés ; une **solution alternative au tracé** ne prend en compte qu'une des deux solutions possibles) ; *g* **Trassenvariante [f]** (Eine Möglichkeit unter mehreren geplanten Linienführungen einer Infrastrukturlinie, z. B. Straße, Überlandleitung, Pipeline, Schiffskanal. Die **Alternative** hingegen ist strenggenommen nur eine Möglichkeit von zweien [lateinisch *alter, -era, -erum* >der, die, das eine von beiden<]).

route approval procedure [n] [US] *plan. trans.* ▶corridor approval procedure [US].

5252 routine crown pruning [n] *arb. constr.* (Removal mostly of fine branchwork of healthy tree crowns whose branches are not in danger of breakage, as a precautionary measure against ▶dieback, top-heavy crowns and the formation of forked growth as well as the cutting of dead, unhealthy or dying branches, branches competing for space, and crossed or chafing branches; ▶crown reduction, ▶crown thinning, ▶drop-crotching) ; *s* **poda [f] de mantenimiento de la copa** (Operación de aclarado de ramas para mejorar el porte del árbol, darle más aireación y sanearlo. Se eliminan ramas muertas, chupones, ramas entrecruzadas o en paralelo al fuste; ▶aclareo de la copa, ▶ahuecado de la copa, ▶muerte regresiva de plantas, ▶poda de reducción de la copa); *f* **opérations [fpl] d'entretien courant de la cime** (Opérations de taille régulières afin de supprimer les rejets vigoureux qui épuisent l'arbre, de prévenir les accidents, l'▶asphyxie progressive et les maladies et d'aérer la couronne [▶éclaircissage de la couronne] tels que la suppression des gourmands, des branches mortes, dangereuses, mal orientées ou trop rapprochées du tronc, l'élimination des chicots et des branches en surnombres ; ▶éclaircissage 2, ▶réduction de couronne, ▶taille d'allégement) ; *g* **Kronenpflege [f, o. Pl.]** (Vorbeugende Schnittmaßnahmen, überwiegend im Fein- und Schwachastbereich, bei gut versorgten, nicht bruchgefährdeten Kronen zur Verhinderung von Auskahlung, Kopflastigkeit und Zwieselbildung und die Entfernung von toten, kranken, gebrochenen und im ▶Absterben befindlichen Äste sowie das Wegschneiden von Konkurrenzästen, sich kreuzenden und reibenden Ästen; cf. QBB 1992; ▶Kronenentlastungsschnitt, ▶Kronenlichtungsschnitt, ▶Kronenrückschnitt).

row [n] *trans.* ▶angle-parking row; *geo. land'man.* ▶elevated hedgerow [US]/quickset hedge [UK]; *hort.* ▶espalier row; *agr. land'man.* ▶fencerow [US]; *trans.* ▶ninety-degree parking row; *hort.* ▶transplant in nursery row.

row [n] [US]**, in-line parking** *trans.* ▶parking lane.

row [n]**, parking** *trans.* ▶row of parking.

row [n] [US]**, perpendicular parking** *trans.* ▶ninety-degree parking row.

5253 row house [n] *urb.* (Single-family dwelling connected to other, mostly one- and two-stor[e]y houses, in a continuous row); *syn.* terrace house [n] [also UK], town house [n] [also US]; *s* **casa [f] adosada** (Casa unifamiliar de uno o dos pisos construida en hilera, siendo lo característico que las casas vecinas comparten las paredes laterales), *syn.* casa [f] apareada; *f* **maison [f] individuelle construite en ligne** (Maison individuelle attenante, d'architecture identique, en général d'un ou deux étages et disposées en ligne) ; *syn.* construction [f] en file ; *g* **Reihenhaus [n]** (Einfamilienhaus, das in fortlaufender Reihe mit anderen gleichartigen, meist ein-/zweigeschossigen Häusern verbunden ist).

5254 row house garden [n] *gard. urb.* (Usually a long, narrow garden associated with a row house; ▶elongated garden); *s* **jardín [m] de casa adosada** (Jardín situado detrás de una casa adosada que generalmente es estrecho y alargado, aunque no tan desproporcionado como el ▶jardín de bolsillo); *f* **jardin [m] d'une construction en ligne** (Jardin situé à l'arrière d'un ensemble de maisons particulières construites en alignement ; ▶jardin en mouchoir de poche) ; *g* **Reihenhausgarten [m]** (Ein Garten, der hinter einem Reihenhaus angelegt ist. Im Vergleich zum ▶Handtuchgarten muss er nicht überproportional langgestreckt sein).

5255 rowlock course [n] [US] *constr.* (DAC 1975; **1.** layer of masonry units laid on edge. *Specific term* layer of bricks on edge; ▶crown of a wall); *syn.* layer [n] of headers-on-edge [UK] (SPON 1986, 266), rolok [n] [US] (DAC 1975), rollock course [n] (DAC 1975). **2.** Layer of masonry units laid on end is called **soldier course [n]**); *s* **hilada [f] a sardinel** (Hilada de ladrillos sentados de canto y unidos por sus caras mayores; DACO 1988; ▶coronación de un muro); *f* **lit [m] à disposition sur chant** (Assise de pierres de taille, de briques posées sur chant souvent utilisé comme ▶couronnement de mur ; *terme spécifique* galandage [cloison de briques posées sur chant, ou briques sur chape sur chant]) ; *g* **Rollschicht [f]** (Schicht aus quaderförmigen Steinen, die auf der längeren Schmalseite [hochkant] stehen. *UB* Klinkerrollschicht. **R.en** dienen als ▶Mauerkrone beim Sichtmauerwerk aus Klinkern oder Ziegeln, oftmals aus Sonderformaten mit sattelförmigem oder halbrundem Querschnitt; bei Natursteinmauern, traditionell bei trocken aufgesetzten Feldmauern zur Einfriedung von Weideflächen oder Abgrenzung von Feldern, werden oft plattige Materialien verwendet, die meist eine gezackte Oberfläche haben); *syn.* Rollschar [f].

5256 row [n] **of parking** *trans.* (Line of parking spaces located on a street right-of-way adjacent to a lane of traffic, or a line of parking spaces in a car park intended for angled or 90°-parking of cars; ▶aisle, ▶angle-parking row, ▶ninety-degree parking row, ▶parking lane); *syn.* parking row [n]; *s* **banda [f] de aparcamiento** (Franja paralela a la calle o a un ▶pasillo de circulación de un aparcamiento, generalmente para aparcar en diagonal o perpendicular; *términos específicos* ▶carril de aparcamiento, ▶fila de aparcamiento en diagonal, ▶fila de estacionamento en perpendicular); *f* **bande [f] de stationnement (longitudinal)** (Aire de stationnement située le long de la chaussée ou de la ▶voie de circulation sur un parking en général pour un rangement en bataille ou en épi ; *termes spécifiques* ▶aire de stationnement pour stationnement en épi, ▶bande pour le stationnement perpendiculaire, ▶parc à voitures longitudinal (NEU 1983, 365) ; *syn.* bande [f] de parcage (BON 1980, 118) ; *g* **Parkstreifen [m]** (Ein neben der Fahrbahn oder entlang der ▶Fahrgasse eines Parkplatzes verlaufender Streifen, i. d. R. für

R

Schräg- oder Senkrechtaufstellung, der ausschließlich dem ruhenden Verkehr dient; *UBe* ▶Parkspur, ▶Schrägparkstreifen, ▶Senkrechtparkstreifen).

5257 row [n] **of stakes** *constr.* (Line of stakes, e.g. for riverbank protection); *s* **hilera [f] de estacas** (Fila de estacas colocada p. ej. para proteger una orilla); *f* **rideau [m] de pieux** (Alignement de pieux plantés p. ex. pour la protection des berges) ; *syn.* rangée [f] de pieux ; *g* **Pflockreihe [f]** (In einer Linie eingebaute Pfähle, z. B. zur Ufersicherung).

5258 row [n] **of straw bales** *constr.* (Line of bales laid to prevent erosion of soil from construction sites after heavy rain, often used in the U.S.; ▶erosion control facilities); *s* **fila [f] de pacas de paja** (▶Instalación de protección contra la erosión utilizada frecuentemente en EE.UU. para frenar el flujo de tierra procedente de obras de contrucción hacia los cursos de agua); *syn.* hilera [f] de pacas de paja; *f* **rangée [f] de bottes de paille** (**U.S.,** ▶équipement de protection contre l'érosion constitué par une rangée de bottes de paille mis en place sur un chantier afin d'éviter l'entraînement, par ruissellement, du sol dans les cours d'eau ; **F.,** un ou plusieurs alignements de bottes de paille sont parfois utilisés comme filtre pour limiter le départ de matières en suspension lors de travaux en cours d'eau) ; *g* **Strohballenreihe [f]** (In den USA häufig angewendete temporäre ▶Erosionsschutzeinrichtung, bei der Reihen von Strohballen aufgestellt werden, um abgeschwemmten Boden auf der Baustelle zu sichern, bevor er in den ▶Vorfluter gelangt; *syn.* Strohballensperre [f].

5259 row planting [n] *agr. constr. for. hort.* (Method of planting in parallel lines; ▶furrow planting, ▶hole planting [US]/pit planting [UK], ▶trench planting); *s* **plantación [f] en hileras** (▶plantación en hoyos, ▶plantación en surcos, ▶plantación en zanjas); *f* **plantation [f] sur le rang** (Plantation composée de plusieurs alignements de végétaux ; ▶plantation en sillons, ▶plantation en tranchée, ▶plantation sur potets) ; *syn.* plantation [f] d'alignement, plantation [f] en ligne ; *g* **Reihenpflanzung [f]** (Aus mehreren Reihen bestehende Pflanzung; ▶Pflanzung in durchgehendem Pflanzgraben, ▶Lochpflanzung, ▶Riefenpflanzung).

5260 royal hunting forest [n] [UK] *for. hist.* (In Europe, a forest serving as a hunting ground for sovereigns; no U.S. equivalent; ▶game preserve [US]/hunting ground [UK], ▶state forest/national forest [US]/State Forest [UK]); *s* **monte [m] real** (En Europa, antiguamente bosque de dominio exclusivo de los señores feudales o los reyes que se utilizaba predominantemente para la caza; ▶territorio de caza, ▶bosque estatal); *f* **forêt [f] seigneuriale** (Surface forestière sur laquelle les féodaux avaient la souveraineté ; ▶forêt domaniale, ▶territoire de chasse) ; *g* **Bannwald [m] (4)** (Waldfläche, an der die Feudalherren das alleinige Eigentums- und Nutzungsrecht innehatten; ▶Jagdgebiet, ▶Staatswald).

rubbish [n] [US] *envir.* ▶scattered rubbish [US]/litter [UK].

5261 rubble [n] *constr.* (**1.** Irregular, pieces of untooled broken stone, which are delivered from a quarry; ▶cut stone. **2.** *geo.* Loose mass of angular rock fragments, commonly overlying outcropping rock; DOG 1984); *syn.* undressed stone [n]; *s* **piedra [f] sin labrar** (Piedra natural de diferentes tamaños tal como se suministra de la cantera; *opp.* ▶sillar; *término específico* para gran **p. sin l.** que se utiliza para producir hormigones ciclópeos: mampuesto [m]; cf. DACO 1988); *syn.* piedra [f] de cantera; *f* **pierre [f] brute (d'extraction)** (Pierre naturelle, livrée brute d'une carrière, de grosseur et forme variée ; *opp.* matériaux préfabriqués, matériaux artificiels ; *termes spécifiques* moellon, pierre plate, dalle, barrette brute ; ▶matériau de construction minéral, ▶pierre de taille) ; *g* **Bruchstein [m]** (Unbearbeiteter

Stein, der in unterschiedlicher Größe im Steinbruch gebrochen wird; *opp.* ▶Werkstein).

rubble [n]**, coursed** *constr.* ▶squared rubble masonry.

rubble [n]**, ranged** *constr.* ▶squared rubble masonry.

5262 rubble disposal site [n] *envir.* (Dumping area [US]/ tipping area [UK] where ▶building rubble/refuse or ▶demolition material is deposited; ▶building waste dump); *syn.* rubble dumpsite [n] [also US]; *s* **escombrera [f] (2)** (Lugar al aire libre [depresión natural, cantera abandonada] donde se depositan, más o menos ordenadamente, ▶escombros o ▶material de derribo; ▶vertedero de escombros); *f* **dépôt [m] de déchets inertes** (Site [dépression naturelle, carrière abandonnée] sur lequel sont déposés déblais et ▶gravats, ▶matériaux de démolition et décombres ; ▶centre de stockage de déchets inertes) ; *syn.* décharge [f] de matériaux inertes ; *g* **Schuttabladeplatz [m]** (Platz, Geländemulde oder aufgelassener Steinbruch zur Ablagerung von ▶Bauschutt, ▶Abbruchmaterial oder Trümmer; ▶Bauschuttdeponie); *syn.* Schuttabladestelle [f].

rubble dumpsite [n] [US] *envir.* ▶rubble disposal site.

rubble heap [n] [US] *envir.* ▶rubble pile [US]/rubble tip [UK].

rubble masonry [n] *constr. constr.* ▶squared rubble masonry.

5263 rubble masonry [n] **with pinned joints** *constr.* (▶Squared rubble masonry of roughly trimmed stones and joints which are filled with irregular wedge-shaped stones); *s* **mampostería [f] enripiada** (▶Mampostería ordinaria [de piedra partida] en la que se rellenan los huecos con piedrecillas o ripios); *f* **maçonnerie [f] en moellons pleins, grossièrement équarris, à joints rocaillés** (▶Maçonnerie en moellons pleins, grossièrement équarris dans laquelle les gros joints entre les moellons sont garnis d'éclats ; l'opération de garnissage des joints entre les moellons avec des éclats insérés dans le mortier est dénommée rocaillage) ; *g* **ausgezwicktes Bruchsteinmauerwerk [n]** (▶Bruchsteinmauerwerk, bei dem viele Fugen mit Gesteinssplitter ausgefüllt und verkeilt sind).

5264 rubble pile [n] [US] *envir.* (Artificially-created heap composed of building debris); *syn.* rubble heap [n] [US], rubble tip [n] [UK]; *s* **montón [m] de escombros**; *f* **dépôt [m] de gravats** (Exhaussement artificiel constitué de matériaux de construction) ; *g* **Schutthalde [f] (1)** (Künstlicher Berg aus Bauoder Trümmerschutt).

rubble plant [n] *phyt.* ▶ruderal plant.

rubble range ashlar [n]**, random** *constr.* ▶random rubble ashlar masonry.

rubble range work [n]**, random** *constr.* ▶random rubble ashlar masonry.

rubble tip [n] [UK] *envir.* ▶rubble pile [US].

5265 rubble work [n] *constr.* (Stone masonry built of rough stones of irregular sizes and shapes, not laid in courses); *s* **mampostería [f] ordinaria de piedras sin labra** (Obra de fábrica formada por piedras irregulares y sin labra); *f* **maçonnerie [f] en moellons bruts** ; *g* **Bruchsteinmauerwerk [n] aus unbehauenen Steinen** (Mauer aus bruchrauen Natursteinen mit nicht immer horizontalen Lagerfugen).

rub damage [n] [US] *hunt. zool.* ▶rub off the velvet [US]/fray off the velvet [UK].

5266 rub off [vb] **the velvet** [US] *hunt. zool.* (Scraping off antler velvet on tree branches causing damage by game to young, woody plants; *result* rub damage [US]/fray damage [UK]; ▶game damage); *syn.* fray off [vb] the velvet [UK]; *s* **frotar [vb] la borra** (Manera utilizada por el venado para quitarse la piel velluda de la cornamenta frotándola contra árboles jóvenes o contra arbustos; ▶daños causados por la caza); *f* **frotter [vb] le**

velours (De nombreux cerfs et chevreuils [Cervidés] utilisent les branches d'arbres pour gratter le velours de leurs andouillers ou de leurs cornes et provoquent ainsi certaines dégâts aux jeunes plantes [dégâts de frotture] ; ▶dégâts causés par le gibier) ; *g* **Bast fegen [vb]** (*Weidmännische Bezeichnung* Abscheuern der wollig behaarten, eingetrockneten Haut von ausgewachsenen Gehörnen oder Geweihen der Hirsche und Rehe *[Cervidae]* an Jungbäumen [Fegebäumen], Sträuchern, etc.; die behaarten Hautfetzen [der Bast] werden *Gefege*, die Blessuren **Fegeschaden** genannt; ▶Wildschaden).

ruderal community [n] *phyt.* ▶ruderal vegetation.

5267 ruderal habitat [n] *phyt.* (VIR 1982, 313; area frequented by ▶ruderal plants); *s* **estación [f] ruderal** (Ubicación de ▶plantas ruderales); *f* **station [f] rudérale** (Station sur laquelle se développent les ▶plantes rudérales) ; *g* **Ruderalstelle [f]** (Standort, auf dem ▶Ruderalpflanzen wachsen).

5268 ruderalization [n] *phyt.* (Changes to a natural site caused by waste dumping [US]/waste tipping [UK] or other activities which negatively affect soil structure, and lead to colonization by ▶ruderal plants); *s* **ruderalización [f]** (Transformación de una ubicación natural bajo la influencia de residuos de la actividad humana; ▶planta ruderal); *f* **rudéralisation [f]** (Transformation d'une station naturelle sous l'influence des déchets de l'activité humaine ou tout autre activité ayant une influence négative sur la structure des sols et provoquant la colonisation par les ▶plantes rudérales) ; *g* **Ruderalisierung [f]** (Veränderung eines natürlichen Standortes durch Abfallablagerung oder andere die Bodenstruktur negativ beeinflussende Aktivitäten; die Folge ist, dass sich hauptsächlich ▶Ruderalpflanzen ansiedeln).

5269 ruderal plant [n] *phyt.* (In a strict sense, a plant which colonizes building rubble, ruins, refuse dumps [US]/refuse tips [UK], overfertilized path edges or similar dry locations which are strongly influenced by man); *syn.* rubble plant [n]; *s* **planta [f] ruderal** (Planta que habita zonas alteradas por el hombre, ricas en nitrógeno y generalmente bien insoladas. Son **pp. rr.** p. ej. la bolsa de pastor, la ortiga, etc. Algunos autores llaman ruderales a todas las especies adventicias; DINA 1987 y DB 1985); *syn.* especie [f] ruderal; *f* **plante [f] rudérale** (Au sens strict plante vivant sur les ruines, les décombres, les ordures, en bordure de circulation sur les sols riches en nitrates et d'une manière générale sur les lieux dégradés par l'action de l'homme ; au sens large toutes les végétations adventices sont qualifiées de rudérales) ; *syn.* espèce [f] rudérale ; *g* **Ruderalpflanze [f]** (Im engeren Sinne eine Pflanze, die [Bau]schutt, Trümmerplätze, Müll, überdüngte Wegraine oder ähnliche, eher trockene, stark vom Menschen überformte Standorte besiedelt. KRAUSE, W. [1958] zieht es vor, sämtliche Unkrautfluren ‚ruderal' zu nennen, sogar die Flutrasen und Kahlschläge).

5270 ruderal vegetation [n] *phyt.* (VIR 1982, 311; plant community composed primarily of ▶ruderal plants; ▶urban ruderal vegetation, ▶weed community); *syn.* ruderal community [n]; *s 1* **vegetación [f] ruderal** (Vegetación que se presenta en los medios o estaciones creados por la habitación humana y construcciones anejas. Uno de los caracteres de este medio es frecuentemente la elevada proporción de nitrógeno en el suelo, por lo cual muchas de las especies son plantas nitrófilas. Constituye parte de la ▶paranthropophytia, que agrupa a toda la vegetación de los medios modificados por el hombre; ▶comunidad de malas hierbas, ▶planta ruderal, ▶vegetación ruderal urbana); *s 2* **paranthrophytia [f]** (Expresion científica que, en la clasificación ecológica de H. DE VILLAR, se aplica a la vegetación propia de medios modificados por el hombre, por su habitación y sus construcciones o adaptaciones, como ▶tala 2 de bos-

ques, medidas de ▶roturación, cultivos, riegos, etc. Las estaciones que de ello resultan se llaman respectivamente ruderales, viarias, arvenses, etc.; DB 1985; ▶vegetación ruderal urbana); *syn.* vegetación [f] parantrópica; *f* **flore [f] rudérale** (Association végétale résultant de l'activité humaine constituée des ▶plantes rudérales appartenant p. ex. à la classe de *Rudéreto-Secalinetea* ; ▶flore adentice des cultures et des prairies, ▶peuplement rudéral urbain) ; *syn.* végétation [f] anthropique, végétation [f] rudérale ; *g* **Ruderalflur [f]** (Pflanzengesellschaft, die vorwiegend aus ▶Ruderalpflanzen besteht; ▶städtische Ruderalflur, ▶Unkrautflur); *syn.* Ruderalgesellschaft [f], Ruderalvegetation [f].

rumbling strip [n] [UK] *trans.* *▶speed hump.

run [n] (1) *geo.* ▶brook.

run [n] [US] (2) *wat'man.* ▶streamlet.

5271 run-down housing [n] **fit for rehabilitation** *urb.* (Dilapidated dwelling places suitable for restoration; ▶blighted area); *s* **edificación [f] antigua en mal estado** (Edificio que necesita ser rehabilitado; ▶barrio insalubre); *f* **habitat [m] vétuste** (Immeuble dont l'état de l'équipement n'est plus adapté aux règles d'urbanisme en vigueur et dont la remise en état ou la modernisation devient nécessaire ; ▶îlot insalubre) ; *syn.* habitat [m] défectueux ; *g* **Altbebauung [f] (2)** (Veraltete, meist sanierungsbedürftige Bebauung; ▶Baugebiet mit städtebaulichen Missständen).

runlet [n] [US] *wat'man.* ▶streamlet.

5272 runnel [n] *geo.* (**1.** Small natural drainage channel of a raised bog, **2.** in arid regions a small intermittent runoff channel [US]/intermittent run-off channel [UK]); *s* **canal [m] natural de drenaje** (Pequeña zanja natural de drenaje de turbera alta); *f* **chenal [m] d'écoulement** (Canal naturel d'évacuation de l'eau dans une tourbière ombrogène) ; *syn.* dépression [f] naturelle ; *g* **Rülle [f]** (Natürliche Abflussrinne in Hochmooren).

runner [n] *bot.* ▶stolon.

runner up [n] [US] *prof.* ▶honorable mention [US]/honourable mention [UK].

running bond pattern [n] *arch. constr.* ▶stretcher bond pattern.

running costs [npl] *constr.* ▶maintenance costs.

running course [n] *arch. constr.* ▶stretcher course.

5273 running sand/silt [n] *pedol.* (SPON 1986, 155; structureless soil with rough particles caused by aeolian shifting process); *s* **loess [m] arenizo flotante** (Acumulación de partículas grandes del loes causada por procesos de erosión eólica); *f* **lœss [m] flotté** (Sol peu évolué, de structure peu affirmée, formé à la suite de dépôts nivéo-éoliens sous climat froid, constitué d'un mélange de sable fin et de limon ; DIS 1986, 128) ; *g* **Flottsand [m]** (Durch äolische Lagerungsvorgänge hervorgerufene grobkörnige Korngrößensortierung des Lösses); *syn.* Sandlöss [m].

running silt [n] *pedol.* ▶running sand/silt.

running with trailing stems [loc] [US] *bot.* ▶stem-spreading.

runoff/run-off [n] *constr. geo.* ▶direct runoff; *constr. plan.* ▶heavy storm runoff [US]/heavy rainwater run-off [n] [UK]; *constr.* ▶peak runoff; *constr. geo. wat'man.* ▶surface runoff.

runoff [n] [US]**, delaying storm** *envir. urb. wat'man.* ▶retention of rainwater.

runoff [n]**, direct surface** *constr. geo.* ▶direct runoff.

runoff [n] [US]**, farm** *agr. envir.* ▶agricultural wastewater.

runoff [n] [UK]**, heavy rainwater** *constr. plan.* ▶heavy storm runoff [US].

R

runoff [n]**, immediate** *constr. geo.* ▶direct runoff.

runoff [n]**, storm** *constr. geo.* ▶direct runoff.

runoff [n]**, storm water** *constr. geo.* ▶direct runoff.

runoff [n]**, surface water** *constr. geo. wat'man.* ▶surface runoff.

5274 runoff coefficient [n] *constr. hydr.* (Coefficient for calculating the proportion of precipitation lost by surface runoff with multiplier for calculating, depending on the nature of the ground surface and the area involved; ▶runoff rate); *s* **coeficiente** [m] **de escorrentía** (Coeficiente utilizado para calcular la proporción de las precipitaciones que se pierden por la escorrentía superficial que depende del tipo de suelo, de la vegetación y del tamaño de la superficie considerada; ▶caudal); *f* **coefficient** [m] **de ruissellement** (Coefficient utilisé pour la détermination du débit global dans l'évacuation des eaux pluviales recueillies sur les surfaces imperméables — est fonction de la nature et de la taille de la surface de réception ; ▶débit global d'écoulement) ; *g* **Abflussbeiwert** [m] (Dimensionsloser Faktor für die Berechnung des prozentualen Anteils der Regenabflussmenge — jeweils abhängig von der Oberflächenbeschaffenheit und Flächengröße; ▶Abflussspende).

runoff management [n] *ecol. urb.* ▶near-natural storm water runoff management.

5275 runoff rate [n] *constr. wat'man.* (Amount of runoff per unit time discharged from a drainage area; ▶rate of rainfall); *s* **escorrentía** [f] **máxima (2)** (Cantidad máxima de agua superficial que ha de ser descargada por unidad de espacio y tiempo en un sistema de drenaje; ▶intensidad de lluvia); *f* **débit** [m] **global d'écoulement** (Débit d'eau superficielle par unité de temps pour une surface donnée à assainir ; ▶débit d'eaux pluviales) ; *g* **Abflussspende** [f] (Menge des Oberflächenwassers pro Zeiteinheit und Fläche, die entwässert werden muss; ▶Regenspende).

5276 runway [n] *constr. recr.* (Run-up area to high and long jump pits); *s* **carrerilla** [f] (Pista para correr y alcanzar velocidad en instalaciones de salto de altura y de fondo); *f* **aire** [f] **d'élan** (Aire ou piste assurant la préparation d'un saut sur un terrain de sport) ; *syn.* piste [f] d'élan ; *g* **Anlauf** [m] **(2)** (...läufe [pl]; Strecke für das Anlaufen auf Sportplätzen für Sprunganlagen).

rupicoline [adj] *bot. phyt.* ▶rupicolous.

5277 rupicolous [adj] *bot. phyt. zool.* (*Specific term in plant ecology* living or growing on or among rocks); *syn.* epilithic [adj], rupicoline [adj], saxicolous [adj], saxicoline [adj]; *s 1* **saxícola** [adj] (Término genérico); *s 2* **petrícola** [adj] (Término específico en fitosociología); *syn.* rupícola [adj], epilítico [adj], saxícola [adj]; *s 3* **roquero** [adj] *bot.* (Rupícola o fisurícola, es decir, que se cría en las rocas; DB 1985); *f* **rupicole** [adj] (qui vit sur les rochers, le plus souvent dans les fissures) ; *syn.* rupestre [adj], saxicole [adj] ; *g* **felsbewohnend** [ppr] (auf Felsen oder Steinen lebend).

rupicolous plant [n] *phyt.* ▶rock plant.

5278 rural area planning [n] [US] *plan.* (Planning of sparsely settled areas outside of conurbations); *syn.* countryside planning [n] [UK]; *s* **planificación** [f] **rural** *syn.* planeamiento [m] rural; *f* **planification** [f] **rurale** (Planification sur les territoires à vocation rurale) ; *g* **Planung** [f] **im ländlichen Raum** (Planung für meist dünn besiedelte Gebiete außerhalb verstädterter Räume und Agglomerationsgebiete unter Berücksichtigung spezifischer sozioökonomischer Strukturen und sozialer Lebensweisen, die sich von jenen der städtischen Regionen unterscheiden).

rural areas [npl]**, migration from** *agr. sociol.* ▶rural exodus.

5279 rural conservation [n] [US] *conserv. land'man.* (Preservation of the rural cultural landscape); *syn.* countryside conservation [n] [UK]; *s* **preservación** [f] **del paisaje cultural** (≠); *f* **préservation** [f] **des paysages culturels et de récréation** ; *g* **Erhaltung** [f, o. Pl.] **der Kultur- und Erholungslandschaft**.

rural depopulation [n] *agr. sociol.* ▶rural exodus.

5280 rural district [n] [US] *leg. urb.* (**In U.S.,** ▶zoning district category typically restricted to agricultural and other low-density uses predominantly found in rural areas; **in U.K.,** land use class designating an area where barns, agricultural management buildings, part-time agricultural businesses and rural housing is permitted; **in D.,** land use class defined in the zoning code for predominantly farm and forest habitation areas; ▶farmstead); *syn.* agricultural district [n] [US] (2), rural settlement area [n] [UK]; *s* **núcleo** [m] **de población rural** (Según la legislación del suelo supletoria, posible categoría de planificación en zonas de suelo no urbanizable para los asentamientos de población con carácter predominantemente agrícola o ganadero, en los cuales no se podrán permitir utilizaciones que impliquen transformación de su destino o naturaleza o lesionen el valor específico que se quiera proteger; cf. art. 36b RD 2159/1978 y art. 86.2 RD 1346/1976 Texto refundido; **en EE.UU.,** categoría de suelo restringida a usos agrícolas o a otros de baja densidad situados mayormente en zonas rurales; **en GB,** categoría de uso del suelo en la cual están permitidas construcciones para usos agrícolas y viviendas rurales; **en D.** categoría de suelo en los planes de ordenación municipal, que se aplica para los pueblos existentes con dedicación predominantemente agrosilvícola; ▶granja, ▶sector urbano); *f 1* **zone** [f] **agricole (2)** (Zonage du plan local d'urbanisme [PLU]/plan d'occupation des sols [POS], zone équipée ou non, à protéger en raison du potentiel agronomique, biologique ou économique des terres agricoles. Sont susceptibles d'être réalisées en zone A : **1.** les constructions et installations agricoles ainsi que les changements de destination des bâtiments agricoles identifiés par le PLU, **2.** les aménagements accessoires à l'agriculture tels que des gîtes ruraux ou un local de vente sur les lieux de l'exploitation ; si le classement en zone A ne semble pas devoir pas exclure la construction de bâtiments agricoles elle ne permet pas la constitution d'un hameau nouveau. **3.** Les bâtiments agricoles désignés par le PLU qui, en raison de leur intérêt architectural ou patrimonial, peuvent faire l'objet d'un changement de destination, dès lors que celui-ci ne compromet pas l'exploitation agricole. **4.** Les constructions et installations nécessaires aux services publics ou d'intérêt collectif que pour autant qu'elles ne compromettent pas le caractère agricole de la zone ; ▶secteur urbain) ; *f 2* **zone** [f] **urbaine à caractère villageois** (F., en raison de l'étendue du périmètre du zonage et en fonction du caractère rural ou urbain prononcé de la zone, l'affectation d'un tel secteur d'expansion d'installations peut être réalisée en zone urbaine [zone urbaine dense à caractère central d'habitat, de services et d'activités, constituée essentiellement par le village ancien qu'il convient de conserver] ; *f 3* **zone** [f] **naturelle noyaux villageois** (Affectation d'une zone naturelle et forestière au sein d'une zone agricole ou dans un secteur de taille et de capacité d'accueil limitée [hameau et groupements bâtis], à la condition qu'elle ne porte atteinte ni à la sauvegarde des sols agricoles et forestiers, ni à la sauvegarde des sites, milieux naturels et paysages) ; *f 4* **zone** [f] **de noyau villageois (±)** (**D.**, dans le cadre du zonage dans les documents d'urbanisme, zone d'occupation des sols d'urbanisation future prévue principalement pour le développement des exploitations agricoles et forestières ainsi que l'habitat résidentiel correspondant ; ▶ferme ; *g* **Dorfgebiet** [n] (**D.**, im Rahmen der Bauleitplanung gem. §§ 1 [2] und 5 BauNVO für die Bebauung vorgesehene Fläche [▶Baugebiet], die vorwiegend der Unterbringung der Wirtschaftsstellen land- und forstwirtschaftlicher Betriebe und dem dazugehörigen Wohnen sowie auch sonstigem Wohnen dient; ▶Gehöft).

5281 rural exodus [n] *agr. sociol.* (Migration of inhabitants from rural areas, and their resettlement in a conurbation. Mass migration of the rural population into cities and metropolitan areas has occurred since the Industrial Revolution due to increasing differences between the living conditions in the city and the countryside); *syn.* migration [n] from rural areas, rural depopulation [n], flight [n] from the land; *s* **éxodo** [m] **rural** (Fenómeno existente desde la revolución industrial por el cual la población emigra del campo a la ciudad con la consecuencia de un gran crecimiento de ésta; hoy en día el proceso se desarrolla sobre todo en los países del llamado «tercer mundo», dando lugar a la creación de grandes urbes —megápolis— de varios millones de habitantes); *f* **exode** [m] **rural** (Émigration de la population agricole vers les villes, d'un rythme lent à partir le début de la révolution industrielle pour s'accentuer depuis 1945 ; l'**e. r.** a pour cause le déséquilibre entre la pression d'un effectif de population d'une part et la capacité d'emploi et les possibilités d'entretien d'une économie agricole ; le dépeuplement des zone rurales est parfois désigné par le terme de « déruralisation » ; DG 1984, 176) ; *syn.* exodus [m] rural (LA 1981), déprise [f] rurale ; *g* **Landflucht** [f, o. Pl.] (Seit der industriellen Revolution zu beobachtendes massenweises Abwandern der Landbevölkerung hin zu den Städten und Ballungsgebieten durch immer größer werdende Unterschiede zwischen den Lebensbedingungen in der Stadt und auf dem Lande. Oft wird **L.** statt des neutralen Begriffs **Landabwanderung** oder **Stadt-Land-Wanderung** benutzt); *syn.* Landabwanderung [f], Stadt-Land-Wanderung [f].

rural holiday accommodation [n] [UK] *recr.* ▶rural vacation accommodation [US].

5282 rural land [n] (1) *agr. landsc.* (Countryside outside of cities, including villages; ▶conurbation, ▶open field); *s* **zona** [f] **rural** (▶aglomeración urbana, ▶campiña; *syn.* espacio [m] rural, área [f] rural; *f* **espace** [m] **rural** (1. Zone située à l'extérieur des villes et des villages. 2. Catégorie de l'▶aménagement du territoire et de l'urbanisme par opposition à l'espace urbain ou aux ▶zones de concentration urbaine. 3. *Définition INSEE* regroupe l'ensemble des petites unités urbaines et communes rurales n'appartenant pas à l'espace à dominante urbaine [▶pôles urbains, couronnes périurbaines et communes multipolarisées]. Cet espace est très vaste, il représente 70 % de la superficie totale et les deux tiers des communes de la France métropolitaine ; ▶campagne) ; *syn.* milieu [m] rural, zone [f] rurale ; *g* **ländlicher Raum** [m] (1. Landschaft außerhalb der Städte einschließlich der Dörfer mit allen Infrastruktureinrichtungen. 2. Kategorie der ▶Raumordnung für ländliche Kreise höherer und geringerer Dichte, die den verstädterten Räumen und ▶Verdichtungsräumen gegenübersteht; ▶freie Landschaft).

rural land [n] (2) *leg. recr.* ▶public access to rural land.

rural outskirts [npl] [US] *leg. urb.* ▶open land [US]; ▶surrounding region of a city/town.

5283 rural overnight accommodation [n] *recr.* (Provision of lodging for individuals or groups at farmsteads, campgrounds, etc. in rural areas; ▶farm vacation [US]/farmstay holidays [UK]); *s* **hospedaje** [m] **en el campo** (Alojamiento de turistas en granja, camping en zona rural, etc.; ▶agroturismo, ▶vacación en granja); *f* **hébergement** [m] **rural** (Hébergement de vacances à la ferme, sur les terrains de camping en milieu rural etc. ; ▶agrotourisme, ▶vacances à la ferme) ; *g* **Beherbergung** [f] **auf dem Lande** (Unterbringung von Feriengästen auf dem Bauernhof, auf Campingplätzen im ländlichen Raum etc.; ▶Agrotourismus, ▶Ferien auf dem Bauernhof).

5284 rural recreation area [n] *recr.* (Recreation area in a rural district); *s* **zona** [f] **rural de recreo** (Área en zona rural dedicada primordialmente a actividades de recreo); *f 1* **base** [f] **rurale de plein air et de loisirs** (Aménagement relativement peu étendu destiné à satisfaire les besoins d'activités de plein air d'un secteur rural et éventuellement à accueillir une fréquentation estivale complémentaire) ; *f 2* **pays** [m] **d'accueil touristique** [PAT] (Structure d'organisation et de gestion du tourisme rural créée en 1976 ayant pour mission, en partenariat avec les administrations et les collectivités, la mise en place et la réalisation d'un programme de développement des équipements ludiques ainsi que l'organisation, la valorisation et la promotion de l'offre ludique ; cf. LTV 1996, 49) ; *f 3* **station** [f] **verte de vacances** (Communes isolées ou ensembles touristiques offrant un cadre de vacances naturel et agréable, possédant des équipements récréatifs diversifiés, disposant d'un hébergement varié et relativement confortable et s'engageant à assurer l'information et l'encadrement des séjournants) ; *g* **Erholungsgebiet** [n] **im ländlichen Raum**.

5285 rural recreation planning [n] [US] *landsc. plan. pol. recr.* (▶recreation planning); *syn.* countryside recreation planning [n] [UK]; *s* **planificación** [f] **recreacional en zonas rurales** (Planificación de áreas turísticas y de recreo; ▶gestión recreacional 2); *syn.* gestión [f] recreacional en zonas rurales; *f* **aménagement** [m] **des loisirs en milieu rural** (Valorisation des zones rurales par l'▶aménagement touristique) ; *g* **Erholungsplanung** [f] **im ländlichen Raum** (▶Erholungsplanung für ein Gebiet außerhalb verstädterter Räume, das für Freizeit- und Erholungsaktivitäten geeignet ist).

rural settlement area [n] [UK] *leg. urb.* ▶rural district [US].

5286 rural vacation accommodation [n] [US] *recr.* (Place of lodging in the country; e.g. for holidays on a farm, in holiday appartments; ▶hostelry); *syn.* rural holiday accommodation [n] [UK]; *s* **hospedaje** [m] **rural** (Lugar de alojamiento en zona rural, p. ej. para vacaciones en granja, en casa de veraneo, etc.; ▶albergue); *f* **gîte** [m] **rural** (Hébergement de vacances à la campagne, p. ex. à la ferme ; ▶gîte) ; *g* **Ferienunterkunft** [f] **auf dem Lande** (≠) (Unterkunft im ländlichen Raum, z. B. für Ferien auf dem Bauernhof, im Ferienhaus, in Ferienwohnungen, im Appartmenthaus; ▶Beherbergungsstätte).

5287 rush meadow [n] *phyt.* (Permanently moist sites dominated by rush species *[Juncus ssp.]*); *s* **pradera** [f] **de juncos** (Formación vegetal en estaciones permanentemente húmedas con predominio de especies de juncos *[Juncus spp.]*); *f 1* **jonchaie** [f] (Formation herbacée colonisant les milieux humides en permanence et constituée par diverses espèces gazonnantes de *Juncus ssp.*) ; *syn.* jonçaie [f] ; *f 2* **jonchère** [f] (Lieu où poussent les Joncs) ; *syn.* prairie [f] à Joncs ; *g* **Binsenwiese** [f] (Vegetationsformation auf dauernd durchfeuchteten Standorten mit vorherrschend flächendeckenden Binsenarten *[Juncus ssp.]*).

5288 rush swamp [n] *phyt.* (Permanently flooded swamp or marsh dominated by rushes—*Juncus ssp.*); *s* **juncar** [m] **pantanoso** (Zona permanentemente inundada en la que las plantas predominantes pertenecen al género de los juncos *[Juncus ssp.]*); *syn.* pantano [m] de juncos; *f* **marais** [m] **à Joncs** (Zone d'émergence d'eaux résurgentes et constamment humide, couverte par différentes espèces de Joncs *[Juncus ssp.]*) ; *syn.* marais [m] à juncacées ; *g* **Binsensumpf** [m] (Meist quelliger, jedoch stets nasser Standort, vorwiegend mit Binsenarten *[Juncus ssp.]* bewachsen).

rusticated [pp/adj] [UK] *constr.* *▶tooling of stone [US]/ dressing of stone [UK], #1.

5289 rusticated masonry [n] *arch. constr.* (Masonry, which is squared-off and left with a rough surface with deep "V" or square joints or with finished flanking corners that emphasize the edges of each block. The texture of **r. m.** is in strong contrast to smooth ashlar masonry and is often used to visually emphasize the façade of a ground floor with smooth ashlar above; the bold textured look is created by bevelling the edges to form deep-set joints and the central face of the stone is left roughly hewn or carved with various pointed or channelled patterns; ►ashlar masonry, ►natural stone masonry); *syn.* bossage masonry [n], quarry-faced masonry [n] [US]; *s 1* **fábrica** [f] **de sillarejo** (►Sillería de fábrica de mampuestos de cara tosca que están labrados solamente en las juntas; ►mampostería de piedras naturales); *syn.* mampostería [f] de labra tosca; *s 2* **sillería** [f] **almohadillada** (Obra de fábrica de sillares abujardados colocados en hiladas con juntas finas); *f* **maçonnerie** [f] **en pierres bossagées** (►Maçonnerie en moellons à opus quadratum dont la face apparente brute ou traitée par bossage présente un aspect bosselé ; ►maçonnerie en pierres naturelles) ; *syn.* mur [m] en pierres bosselées, mur [m] en moellons de bossage (VRD 1994, partie 4, chap. 6.4.2.3, 15) ; *g* **Bossenwerk** [n] (►Quadermauerwerk, bei dem die Quader an den Vorderseiten buckelig belassen oder geschlagen und an den Kanten meist geglättet sind; ►Natursteinmauerwerk).

rustication [n] [UK] *arch. constr.* ►bossage [US].

S

saber butt [n] *arb.* ►basal sweep.

5290 saddle fitting [n] *constr.* (DAC 1975, 420; fitting for making a connection to a pipe which has already been installed); *s* **accesorio** [m] **con brida** (Pieza de empalme de tuberías que sirve para conectar una a otra ya instalada); *f* **raccord** [m] **de picage à plaquette** (Pièce constituée d'une coque préfabriquée épousant la forme de la canalisation et munie d'une tubulure à emboîtement de même section que le branchement ; réalisation d'une ouverture sur le collecteur et collage du **r. de p. à p.** au droit de celle-ci ; DOFB 1979, 20) ; *syn.* clip [m] ; *g* **Sattelstück** [n] (Formstück für den nachträglichen Anschluss von Zuläufen an anzuschlagende Entwässerungsrohrleitungen).

5291 safeguarding [n] **the effective functioning of natural systems** *conserv. landsc. leg. s* **preservación** [f] **de la capacidad funcional del régimen de la naturaleza** *syn.* protección [f] de la capacidad funcional de la naturaleza; *f* **sauvegarde** [f] **de l'efficacité du système naturel** ; *g* **Erhaltung** [f, o. Pl.] **der Leistungsfähigkeit des Naturhaushaltes** (§ 2 [1] 1 BNatSchG).

safescaping [n] *landsc.* ►firescaping.

safety barrier [n] *urb. trans.* ►guardrail [US] (1)/crash barrier [UK], ►pedestrian guardrail [US]/pedestrian guard-rail [UK].

5292 sag curve [n] [US] *trans.* (TSS 1988; HALAC 1988-II, 29; concave-shaped curve of a vertical road alignment connecting two different grades; *opp.* ►crest vertical curve); *syn.* dip curve [n] [UK]; *s* **redondeo** [m] **de depresión** (En alineación vertical de una carretera, curva cóncava que une dos pendientes de orientación opuesta entes; *opp.* ►curva vertical de culminación); *f* **rayon** [m] **de raccordement concave** (*Techniques d'aména-*

gement des routes ligne verticale qui sur un tracé en long à partir du point le plus bas forme le passage progressif entre la pente et la rampe ; ►rayon de raccordement convexe) ; *syn.* concavité [f] (d'une route) ; *g* **Wannenausrundung** [f] (Bei der Planung einer Wege- oder Straßentrasse die Ausformung einer vertikalen Kurve, die am tiefsten Punkt eines Gradientenabschnittes einen allmählichen Übergang von der fallenden auf die steigende Längsneigung bildet; *opp.* ►Kuppenausrundung).

5293 sag pipe [n] *constr. eng.* (Connecting pipe which loops under a road, watercourse or other obstacle and maintains the level of the carried liquid in accordance with the "principle of common level balance" in a communicative pipe, often but inaccurately called 'inverted siphon'); *s* **acueducto** [m] **sifón** (Conexión de tuberías bajo un curso de agua, una carretera u otro tipo de obstáculo que funciona según el principio de los vasos comunicantes); *syn.* sifón [m] invertido; *f* **siphon** [m] **pour traversée** (ASS 1987, 166 ; canalisation mise en œuvre sur le principe des vases communicants pour traversées enterrées de cours d'eau, voies de circulation ou autres obstacles ; *contexte* passer en siphon) ; *syn.* ouvrage [m] hydraulique en siphon ; *g* **Düker** [m] (Auf dem Prinzip der kommunizierenden Röhren geführte Rohrleitung unter Fließgewässern, Verkehrsanlagen oder anderen Hindernissen).

5294 sailboat harbor [n] [US] *recr.* (Pleasure haven providing secure moorings for sailing boots and yachts; ►marina [US], ►yachting habor [US]/yachting harbour [UK]; *generic term* leisure harbor [US]/leisure harbour [UK]); *syn.* sailing boat harbor [n] [US], sailing-harbour [n] [UK]; *s* **puerto** [m] **de veleros** (►puerto de yates; ►marina; *término genérico* ►puerto de recreo); *f* **port** [m] **de voiliers de plaisance** (F., plan d'eau naturel ou artificiel abrité, le plus souvent créé sur le domaine public maritime, du ressort des communes, aménagé pour les voiliers et bateaux à moteurs pour la pratique de la navigation de plaisance ; ►marina, ►port de navires de plaisance ; *terme générique* ►port de plaisance) ; *g* **Segelboothafen** [m] (Für Segelboote [und Yachten] angelegter Freizeithafen; ►Marina, ►Yachthafen; *OB* ►Freizeithafen).

sailing boat harbor [n] [US] *recr.* ►sailboat harbor [US]/sailing-harbour [UK].

sailing-harbour [n] [UK] *recr.* ►sailboat harbor [US].

Sales of Goods and Services Act [n] [UK] *constr. contr.* ►General Conditions of Contract for Furniture, Furnishings and Equipment [US]/General Conditions of Government Contracts for Building and Civil Engineering Works [UK].

5295 salination [n] (1) *envir.* (Deposition of salt by surface water or spray upon a particular area); *syn.* salt input [n], salt inflow [n]; *s* **salinización** [f] (1) (Introducción de sales en un lugar por medio del agua superficial o por salpicadura del agua marina); *f* **transfert** [m] **de sel** (Sel transporté par les eaux superficielles ou les embruns en un lieu donné) ; *syn.* apport [m] de sel ; *g* **Salzeintrag** [m] (Transport von Salz durch Oberflächenwasser oder durch Gischt auf einen bestimmten Standort oder in ein Gebiet).

5296 salination [n] (2) *limn. pedol.* (Naturally occurring or human induced accumulation of salts in bodies of water or soils; mainly in arid regions, humid coastal areas, or due to the salting of roads in winter; ►salt crust); *syn.* salinization [n] [US] (TGG 1984, 91; LE 1986, 294), salt pollution [n] (1); *s* **salinización** [f] **(2)** (Proceso natural o antropógeno de acumulación de sales en el suelo en zonas áridas por irrigación, en cercanías del mar o por aplicación de sal en inviero para evitar congelamiento en carreteras; ►formación de costra salina); *f* **salinisation** [f] (Processus naturel ou anthropogène d'accumulation de sel dans les eaux ou le sol ; celui-ci a principalement lieu dans les mers fermées des régions sèches, sur les terres irriguées des zones arides, dans les régions humides à proximité des côtes ou sur les zones touchées par les salage d'hiver ; DUV 1984, 260 ; ►croûte salée) ; *g* **Versalzung [f]** (Natürliche oder anthropogene Anreicherung von Salzen in Gewässern oder Böden; die Bodenversalzung findet hauptsächlich in Trockengebieten statt, in humiden Klimaregionen meist in Meeresnähe oder durch winterlichen Streusalzeinsatz auf Straßen; ►Salzverkrustung).

5297 saline site [n] *phyt.* (TEE 1980, 87; salt-rich inland location with halophytic vegetation; ►salt meadow); *s* **ubicación [f] salina** (Lugar del interior rico en sal con vegetación halófila; ►salina); *f* **milieu [m] salé** (Site continental dont le sol contient du sel et recouvert par une végétation halophytique, p. ex. les ►prés salés continentaux ; pour les habitats côtiers on utilise parfois le terme milieu salin) ; *syn.* terrain [m] salifère ; *g* **Salzstelle [f]** (Salzreicher Standort im Binnenland mit halophiler Vegetation; ►Salzwiese).

5298 saline soil [n] *pedol.* (Salty soil primarily characterized by considerable quantities of sulfates, chlorides or nitrates of Ca, Mg or Na); *s* **suelo [m] salino** (Suelo que contiene sales solubles [cloruros o nitratos de Ca, Mg o Na] en cantidades suficientemente elevadas como para que su presencia se manifieste en la vegetación; si el suelo contiene entre 0,2-0,5% de sales disueltas en el agua edáfica se denomina suelo geloide, si tiene entre 0,5-2% suelo haloide y si tiene más de un 2% suelo perhaloide; cf. DB 1985); *f* **sol [m] salsodique** (Sol qui diffère des autres sols par la présence de sels solubles en quantité anormalement élevée. Dans ce milieu salin prédominent les sels d'acides forts et de bases fortes — sulfates, chlorures, nitrates de calcium, de magnésium, de sodium ; cf. PED 1979, 363, 467) ; *syn.* sol [m] halomorphe, sol [m] salé (LA 1981, 1049) ; *g* **Salzboden [m]** (Durch hohe Gehalte an Chloriden und Sulfaten des Na, Mg und K sowie Carbonaten des Na und Mg gekennzeichneter Boden; SS 1979).

5299 salinity [n] *envir.* (Degree of saltiness or quantity of salt found by testing in a particular medium); *s* **salinidad [f]** (Contenido de sal en un medio determinado); *f 1* **salinité [f]** (Teneur en sel; l'ensemble des sels qui sont en solution dans l'eau) ; *f 2* **salure [f]** (Quantité de sel — chlorure de sodium essentiellement — contenu dans l'eau de mer ou l'eau saumâtre ; AEP 1976, 324) ; *g* **Salzgehalt [m]** (Menge an Salz, die in einem untersuchten Umweltmedium vorhanden oder bestimmt wird).

salinization [n] [US] *limn. pedol.* ►salination (2).

5300 salmonid zone [n] [UK] *limn.* (Upper fish zone in the classification of European river systems, encompassing both the ►trout zone and the ►grayling zone [UK] ; ►rhithron, ►river zone); *syn.* rhithral [n]; *s* **región [f] salmonícola** (Zona superior de los ríos [►cabecera] habitada típicamente por salmónidos, que incluye la ►zona de trucha y ►zona de umbra; ►ritrón, ►zona biológica de un río); *f* **région [f] salmonicole** (Terminologie englobant dans un seul terme les deux tronçons supérieurs d'un cours d'eau, la ►zone à truite et la ►zone à ombre ; ►rhithron, ►zone piscicole des cours d'eau) ; *g* **Bergbachregion [f]** (Für Flüsse zusammenfassende Bezeichnung der beiden obersten Fließgewässerabschnitte, der ►Forellenregion und ►Äschenregion; cf. TIS 1975, 46; ►Flussregion, ►Rhithron); *syn.* Rhithral [n], Salmonidenregion [f].

5301 salt accumulation [n] *limn. pedol.* (Increase in salt content in a medium); *s* **acumulación [f] de sal** *syn.* enriquecimiento [m] de sal, acumulación [f] salina; *f* **enrichissement [m] en sels** (Augmentation de la concentration du sel dans un milieu) ; *g* **Salzanreicherung [f]** (Erhöhung des Salzgehaltes in einem Umweltmedium).

5302 salt crust [n] *pedol.* (Surface layer of salt on soils of arid or semiarid regions [US]/semi-arid regions [UK], caused by

evaporation); *s* **formación [f] de costra salina** (Capa superficial de sal en suelos de zonas áridas o semiáridas causada por la evaporación del agua que conlleva la ascensión capilar del agua edáfica y la subsiguiente acumulación de sales en la superficie; este fenómeno natural se agrava en las zonas calurosas en las que se practica la agricultura de regadío sin tomar medidas de precaución); *f* **croûte [f] salée** (Couche de sel formée par évaporation sur les sols des régions arides ou semi-arides) ; *g* **Salzverkrustung [f]** (Durch Verdunstung entstandene Salzschicht auf Böden arider oder semiarider Gebiete).

5303 salt damage [n] *envir.* (Destruction of cellular tissue of plants, e.g. by extreme salination of the soil in arid regions or by road-salting in winter); *s* **efectos [mpl] dañinos de la salinización** (Destrucción del tejido celular de las plantas de cultivo en zonas áridas por alto contenido de sal en el suelo o en la vegetación en zonas frías donde se aplica sal para evitar formación de hielo en las carreteras); *f* **dégâts [mpl] dus à la salinité** (Destruction des tissus cellulaires provoquée p. ex. par la forte salinité des sols dans les régions arides ou causée par le sel de déneigement dans les espaces verts situés à proximité de la voirie) ; *syn.* dégâts [mpl] causés par le sel ; *g* **Salzschaden [m]** (Zellgewebszerstörungen an Pflanzen durch zu hohe Salzkonzentrationen, z. B. durch Versalzung der Böden in ariden Gebieten oder durch Streusalzeintrag in Verkehrsgrünflächen).

salt inflow [n] *envir.* ►salination (1).

salt input [n] *envir.* ►salination (1).

5304 salt marsh [n] *geo. phyt.* (Coastal land along a low-lying shore subject to flooding with salt water. Fine silt and mud are deposited by tides and added to by alluvium brought down by rivers; ►coastal marsh soil, ►tidal mudflat); *syn.* saltwater marsh [n], sea-marsh [n] (GGT 1979, 429), tidal marsh [n] [also US]; *s* **marisma [f] salina** (Tierras saladas del litoral, p. ej. del Mar del Norte, creadas a partir de la colmatación de las ►llanuras de fango o tras la construcción de diques para ganar tierras al mar; ►suelo de marsch); *f* **schorre [m]** (Partie d'estran vaseux souvent séparé de la haute ►slikke par un talus entaillé en microfalaise ; le **s.** est occupé par un tapis dense d'halophytes qui n'est plus recouvert que lors des grandes marées ou des tempêtes ; le sol qui connaît un début d'évolution pédologique y est plus sec, moins salé, de structure plus granuleuse, incorporé de débris végétaux qui contribuent à son exhaussement ; DG 1984, 282 ; ►sol salin à vase marine) ; *syn.* herbu [m], pré [m] salé ; *g* **Salzmarsch [f]** (An Flachküsten mit starker Gezeitenwirkung verbreitete Niederungen, durch Verlandung des ►Watts entstanden; es werden **Vorlandsalzmarschen** und **S. en des Festlandes** unterschieden; an der Nordsee verdanken letztere ihre Entstehung zum größten Teil der systematischen Landgewinnung und wurden durch Eindeichungen dem direkten Meereseinfluss entzogen; ►Marschboden); *syn.* Seemarsch [f] (ELL 1978, 465).

5305 salt meadow [n] *phyt.* (ELL 1988, 366; halophytic grass community growing upon a saline soil on ►coastal marshland; ►salt marsh); *syn.* salt sward [n]; *s 1* **prado [m] salino** (Comunidad de herbáceas halófitas que crece en suelos ricos en sal; ►marisma marítima, ►marisma salina); *s 2* **pasto [m] salado [RA]** (Población del geófito rizomatoso *Distichlis spicata* de 15 a 25 cm de altura, bastante abierta, que va acompañado por algunas especies débilmente halófilas como *Lepidium spicatum*, *L. parodii*, *Spergularia grandis*, *Melilotus indicus* y *Juncus acutus*, que se presenta en Argentina; BB 1979, 367); *s 3* **salina [f]** (Término que se emplea en America Latina en sentido geográfico, equivaliendo, en cuanto al suelo al solonchak, y en cuanto a vegetación a la halophytia; cf. DB 1985); *syn.* salitral [m], saladar [m]; *f* **pré [m] salé** (Végétation herbacée halophile dense située sur la partie la plus haute du ►marais maritime

[►schorre] recouverte par la mer lors des marées et pâturée par les animaux ; la composition de la flore des prés salés dépend de la fréquence de leur recouvrement depuis ceux qui sont inondés à chaque marée haute jusqu'à ceux qui sont recouvert lors des marées de fort coefficient [marée de vives eaux] ; le plus vaste ensemble de prés salés de la côte Nord Amoricaine est situé dans l'anse d'Yffiniac après la baie du Mont St Michel ; le pré salé est considéré comme l'écosystème littoral dont la production primaire est la plus élevée de tous les biotopes du globe ; il assure une fonction d'épuration par lagunage naturel, de protection de la côte contre l'érosion marine et de refuge pour de nombreux oiseaux ; plus de la moitié de la superficie de ces zones humides littorales a été détruite par poldérisation, endiguement ou remblaiement au cours des dernières décennies, la plupart sont maintenant protégées par classement en réserves naturelles ou en site Natura 2000) ; *syn.* prairie [f] salée ; *g* **Salzwiese [f]** (Auf salzreichem Boden wachsende halophytische Rasengesellschaft, die vom Meer periodisch oder in unregelmäßigen Abständen überflutet wird; in D. sind **S.n** in der ►Marsch an der Nordseeküste, vor allem in den Ruhezonen der drei Nationalparke Hamburgisches, Schleswig-Holsteinisches und Niedersächsisches Wattenmeer; ►Salzmarsch); *syn.* Salzrasen [m].

salt pollution [n] (1) *limn. pedol.* ►salination (2).

5306 salt pollution [n] (2) *ecol. envir. limn.* (Accumulation of salt, which has a negative impact upon the environment; ►salination 1, ►salination 2); *s* **hipersalinización [f]** (►salinización 1, ►salinización 2); *f* **salinité [f] excessive** (Augmentation nocive de la salinité dans un milieu ; ►salinisation, ►transfert de sel) ; *g* **Salzbelastung [f]** (Umweltunverträgliche Erhöhung des Salzgehaltes in einem Umweltmedium; ►Salzeintrag, ►Versalzung).

5307 salt resistance [n] *phyt.* (Tolerance of plants to the effects of salinity; ►halophyte; *opp.* ►salt sensitivity); *s* **resistencia [f] a la salinidad** (Característica de ciertas plantas que toleran la presencia de sal en el suelo; ►planta halófila; *opp.* ►sensibilidad a la salinidad); *syn.* resistencia [f] a la sal); *f* **résistance [f] à la salinité** (Propriété de certains végétaux d'être pratiquement insensible à l'action du sel ; ►halophyte ; *opp.* ►sensibilité au sel) ; *g* **Salzresistenz [f]** (Eigenschaft von Pflanzen gegen Salzeinwirkungen [weitgehend] unempfindlich zu sein; ►Salzpflanze; *opp.* ►Salzempfindlichkeit).

5308 salt resistant [adj] *phyt.* (Descriptive term for plants which are mostly insensitive to the effects of salt); *s* **resistente a la salinidad [loc]** (►resistente a la sal [loc]; *f* **résistant, ante à la salinité [loc]** (Plantes pratiquement insensibles aux actions provoquées par le sel) ; *g* **salzresistent [adj]** (Pflanzen betreffend, die gegen Salzeinwirkungen [weitgehend] unempfindlich sind).

5309 salt sensitivity [n] *plant.* (Measure of tolerance of plants to salts in the soil or salt spray; ►salt resistance); *s* **sensibilidad [f] a la salinidad** (Grado de tolerancia de una planta hacia la sal; ►resistencia a la salinidad); *syn.* sensibilidad [f] hacia la sal; *f* **sensibilité [f] au sel** (Taux de tolérance d'une plante à la salinité du sol ou des embruns ; ►résistance à la salinité) ; *g* **Salzempfindlichkeit [f]** (Maß der Verträglichkeit für Pflanzen auf Salze im Boden oder auf salzhaltiges Gischtwasser; ►Salzresistenz).

5310 salt spray [n] *envir. phyt.* (Salt-laden mist caused by wind blowing on land from the sea; ►naturally sprayed area); *s* **salpicadura [f] de sal** (►zona de salpicadura); *f* **embruns [mpl] salés** (SOL 1987, 29 ; bruine salée provenant de l'action du vent sur les vagues déferlantes ; ►zone des embruns) ; *g* **Salzgischt [f]** (Salzhaltiges, aufgesprühtes Wasser; ►Gischtzone).

5311 salt swamp [n] *geo. phyt.* (Marshy area characterized by its high salt content; ►gypsum salt swamp); *s* **pantano [m] salado** (Zona pantanosa caracterizada por un alto contenido de sal; ►pantano gipsáceo); *f* **marais [m] salé** (Zone marécageuse caractérisée par une forte salinité ; ►marais gypsifère) ; *g* **Salzsumpf [m]** (Sumpfgebiet, das durch einen hohen Salzgehalt im Boden gekennzeichnet ist; ►Gipssumpf).

salt sward [n] *phyt.* ►salt meadow.

5312 salt-tolerant [adj] *plant.* (TEE 1980, 86s; ability of plants to grow without any visible damage in spite of ►salination 1 of the soil); *syn.* halophilous [adj]; *s* **tolerante a la salinidad [loc]** (Referente a las plantas cuyo crecimiento no se ve restringido por la ►salinización 1); *f* **tolérant, ante à la salinité [loc]** (Propriété de certains végétaux dont la croissance reste indifférente à une élévation de la salinité dans leur milieu naturel ; ►transfert de sel) ; *syn.* tolérant, ante au sel [loc] ; *g* **salztolerant [adj]** (Pflanzen betreffend, die durch ►Salzeintrag in ihrem Standort keine sichtbaren Wachstumsschäden zeigen); *syn.* salzverträglich [adj], salzvertragend [ppr/adj].

saltwater marsh [n] *geo. phyt.* ►salt marsh.

5313 salvaging [n] **of plants** *constr. conserv.* (Removal/transplantation of plants to another site, in order to save them from · destruction, e.g. during construction; ►resettlement 2); *s* **transplante [m] de especies vegetales** (Traslado de ejemplares p. ej. cuando peligran por obras de construcción, etc.; en el caso de especies de fauna se dice ►reubicación de especies); *f* **déplacement [m] de végétaux** (Transplantation lors des travaux préliminaires sur un chantier de végétaux transplantables d'intérêt écologique ou paysager qui méritent d'être conservés ; ►transplantation 2) ; *g* **Umsetzen [n, o. Pl.] von Pflanzen(arten)** (Sichern und Verpflanzen von Arten, die z. B. durch Baumaßnahmen gefährdet sind; *syn. zool.* ►Umsiedlung); *syn.* Sichern [n, o. Pl.] von Pflanzen, Pflanzensicherung [f].

samphire vegetation [n] *phyt.* ►glasswort vegetation.

5314 sample pit [n] *pedol.* (Hole excavated for the taking of ►soil samples); *s* **calicata [f]** (Perforación o excavación, generalmente de dimensiones métricas, realizada para obtener información sobre el suelo o el sustrato cuando no existe un afloramiento suficientemente claro; ►muestra edáfica); *syn.* hoyo [m] de prospección del suelo; *f* **zone [f] de reconnaissance** (Zone de prélèvement d'►échantillons du sol ou de carottes réalisés par forage ou sondage) ; *syn.* emplacement [m] des prélèvements ; *g* **Bodenentnahmestelle [f]** (Ort, an dem ►Bodenproben für bodenphysikalische oder Nährstoffuntersuchungen entnommen/gezogen werden).

5315 sample plot [n] *phyt.* (Floristically homogeneous area, which characterizes the typical species composition of a plant community, and which is therefore chosen to investigate plant communities over a period of time; ►permanent sample plot, ►sampling 2); *s* **superficie [f] de muestreo** (Área homogénea florísticamente y en sus condiciones mesológicas en la cual crece una comunidad vegetal con composición característica de especies que se utiliza durante un periodo/período largo para estudios fitosociológicos; ►cuadrado permanente, ►inventario fitosociológico); *syn.* área [f] de muestreo; *f* **aire [f] d'échantillonnage floristique** (Station homogène de dimension réduite et choisie en fonction de sa représentativité pour la définition d'une formation végétale prospectée pour une étude phytosociologique de longue durée ; ►carré permanent d'échantillonnage, ►relevé phytosociologique, ►végétation) ; *syn.* aire-échantillon [f], aire [f] du relevé, aire [f] prospectée, placeau [m] d'échantillonnage ; *g* **floristische Probefläche [f]** (Standörtlich und floristisch gleichartige Fläche, die typisch für die Zusammensetzung einer Pflanzengesellschaft ist und deshalb für pflanzensoziologische

Untersuchungen über einen längeren Zeitraum dient; ▶Dauer-quadrat, ▶Vegetationsaufnahme); *syn.* floristische Aufnahme-fläche [f], vegetationskundliche Aufnahmefläche [f].

5316 sampling [n] (1) *constr. pedol.* (Removal of parts of a material, or a single unit of many such items to be furnished in order to test its quality and compliance with specifications; e.g. a concrete test cube; ▶soil sampling); *s* **toma [f] de muestra(s) (1)** (para control de calidad de materiales de obra; ▶toma de mues-tra[s] 2); *f 1* **prélèvement [m] d'échantillons** (Vérification de la qualité d'exécution des travaux par prélèvement de matériaux sur éléments d'ouvrage ou composants de construction et contrôle de la conformité aux normes et spécifications techniques) ; *syn.* prise [f] d'échantillons ; *f 2* **échantillonnage [m]** (Ensemble des opé-rations pour la détermination d'un échantillon ; ▶échantillonnage de sol, ▶prélèvement d'échantillons de sol) ; *g* **Probenahme [f]** (Entnahme von Teilmengen eines Materials zur Prüfung der Qualität hinsichtlich der ausgeschriebenen Bauleistung, z. B. in Form von Probewürfeln bei Beton oder **P.** von einem gelieferten Erdgemisch; ▶Bodenprobenahme); *syn.* Beprobung [f], Proben-entnahme [f].

5317 sampling [n] (2) *phyt.* (*Inventory of vegetation* identifi-cation and compilation of all plant species within comparable plots so that the degree of total vegetative cover and sociability may be determined; ▶field survey 1, ▶landscape inventory, ▶permanent sample plot, ▶sample plot); *syn.* relevé [n] (TEE 1980, 157); *s* **inventario [m] fitosociológico** (Anotación de la composición florística y de los demás caracteres de interés geobotánico que presenta una población vegetal homogénea concreta; DB 1985; ▶cuadrado permanente, ▶inventario paisa-jístico, ▶superficie de muestreo, ▶trabajo de campo); *syn.* inventario [m] florístico; inventario [m] de comunidad; *f* **relevé [m] phytosociologique** (Inventaire floristique caractérisé par des mentions d'ordre sociologiques analytiques ou synthétiques ; ▶aire d'échantillonnage floristique, ▶carré permanent d'échan-tillonnage, ▶étude préliminaire de terrain, ▶relevé des éléments du paysage) ; *syn.* recensement [m], relevé [m] de formations végétales ; *g* **Vegetationsaufnahme [f]** (*Vegetationskundliche Aufnahme, z. B. nach* BRAUN-BLANQUET Erfassung aller Arten einer ▶Minimalfläche und deren tabellarische Auflistung, so dass der Deckungsgrad, die Gesellligkeit jeder einzelnen Art ab-lesbar ist; ▶Aufnahme der Landschaftsbestandteile, ▶Dauerqua-drat, ▶Feldaufnahme, ▶floristische Probefläche); *syn.* Pflanzen-aufnahme [f], vegetationskundliche Aufnahme [f].

5318 sand [n] (**1.** *constr. geo.* ▶*Soil textural class* mineral particles with size—**U.K.**, diameter from > 0.06 to 2mm; **U.S.**, diameter from > 0.05 to 2mm. **2.** *pedol.* 25% or more very coarse, coarse, and medium sand and less than 50% fine or very fine sand; RCG 1982, 161; ▶bedding sand, ▶coarse sand, ▶fine sand, ▶medium sand, ▶washed sand; ▶glacial sand, ▶in-situ sand, ▶quicksand, ▶running sand/silt, ▶wind-blown sand); *s* **arena [f]** (▶Clase textural de suelo mineral con partículas de diámetro entre 0,06 y 2 mm; ▶arena fina, ▶arena gruesa, ▶arena lavada, ▶arena media, ▶arena para adoquinados; ▶arena eólica, ▶arena glaciar, ▶arena in situ, ▶arena movediza); *f* **sable [m]** (**F.**, ▶*classe de texture de sol* en terme de granulométrie, maté-riau meuble formé de quartz [grains de sable] dont les dimensions sont comprises entre 0,05 à 2 mm ; cf. DIS 1986, 192 ; ▶sable de pavage, ▶sable fin, ▶sable grossier, ▶sable lavé, ▶sable moyennement fin, ▶sable moyennement grossier, sable très grossier; ▶sable en place, ▶sable éolien, ▶sable glaciaire, ▶sables mouvants) ; *g* **Sand [m]** (**D.**, mineralische ▶Bodenart im Grobkornbereich von > 0,06 bis 2 mm; *Abk.* S; cf. DIN 4022 und DIN 4023; ▶Feinsand, ▶gewaschener Sand, ▶Grobsand, ▶Mittelsand, ▶Pflastersand; ▶anstehender Sand, ▶Flottsand, ▶Flugsand, ▶Geschiebesand, ▶Treibsand).

sand [n], **aeolian** *pedol.* ▶wind-blown sand.
sand [n], **bedding** *constr.* ▶coarse bedding sand.
sand [n], **running** *pedol.* ▶running sand/silt.
sand [n], **sharp** *constr. eng.* ▶washed sand.

5319 sand accumulation [n] *geo. phyt.* (Piling up of wind-blown sand on dunes leading to the disappearance of, or damage to most of the vegetation); *s* **recubrimiento [m] de arena** (Acumulación de arena sobre dunas con vegetación); *f* **ensable-ment [m]** (Accumulation de sable sur les dunes provoquée par le vent et qui n'est supportée que par un nombre restreint d'espèces végétales ; phénomène naturel d'accumulation de dépôts dans une baie lors des marées) ; *g* **Sandbedeckung [f]** (Durch Wind verursachte Sandanhäufung auf bestockten Dünen, wodurch die meisten Pflanzenarten beeinträchtigt werden); *syn.* Übersandung [f].

5320 sand and gravel pit [n] *min.* (▶borrow pit, ▶extrac-tion site); *s* **hoyo [m] de excavación** (Término común para ▶zona de préstamo de grava, arena o arcilla; ▶zona de extrac-ción); *syn.* foso [m] de excavación; *f* **carrière [f] en fosse** (*Terme générique* ; carrière à ciel ouvert, ▶lieu d'emprunt de sables et graviers alluvionnaires ou autres matériaux meubles dont l'ex-traction est réalisée en fosse ou à flanc de relief ; ▶site d'extrac-tion de matériaux) ; *g* **Baggergrube [f]** (Allgemeiner Begriff für ▶Seitenentnahme und ▶Entnahmestelle 2 für Kies, Sand und Ton); *syn.* Baggerloch [n], Restloch [n].

sand and gravel working [n] *min.* ▶extraction (1).

5321 sand area [n] *plan. recr.* (▶Sand box or ▶sand pit in a children's playground); *s* **superficie [f] con arena** (Término genérico para ▶cajón de arena y ▶hondonada de arena en zonas de juegos infantiles); *f* **aire [f] de sable** (Terme générique pour les ▶bacs à sable ou ▶aires sablées 2 sur les aires de jeux pour enfants) ; *syn.* aire [f] sablée (1) ; *g* **Sandfläche [f]** (OB zu ▶Sandkasten und ▶Sandmulde auf Kinderspielplätzen).

5322 sand arresting trap [n] [US] *constr. hydr.* (Inter-ceptors of various types for the collection of sand or gravel in sewage or drain pipes or in sewage treatment plants; ▶mud trap 1, ▶mud trap 2, ▶sediment basin [US]/silt trap [UK]); *syn.* grit chamber [n], detritus tank [n] (for sewage treatment plants) (DET 1976, 128), sand trap [n] [UK]; *s* **trampa [f] de arena** (Instalación para captar arena o gravilla en tuberías de desagüe o en plantas de depuración de aguas; ▶arqueta de recogida con filtro, ▶colector de fango 1, ▶colector de fango 2); *syn.* atra-padora [f] de arena, desarenador [m]; *f 1* **décanteur [m]** (Dis-positif situé dans une bouche à décantation dont le radier est en contre-pente par rapport à la conduite aval, favorisant la rétention de sable et de gravier ; DOFB 1979, 18 ; ▶débourbeur, ▶dispo-sitif de rétention des boues, ▶radier de décantation) ; *f 2* **cham-bre [f] de dessablement** (Cette installation protège la station d'épuration contre l'intrusion de sable et matières lourdes [gra-viers] de granulométrie supérieure à 0,2 mm) ; *syn.* chambre [f] à sable, bassin [m] de dessablement ; *g* **Sandfang [m]** (Becken unterschiedlicher Ausbildung zum Abfangen von Sand oder Kies bei sandführenden Abwasser-, Dränleitungen oder in Kläran-lagen; bei Abwasserleitungsschächten meist syn. mit ▶Schlamm-fang 1, ▶Schlammfang 2 und ▶Sumpf 1).

5323 sandbank [n] *geo.* (Accumulation of sand in the river or sea, usually exposed at low water; ▶gravel bank); *syn.* sandbar [n], shoal [n]; *s* **banco [m] de arena** (*Términos específicos* **1. barra [f] aluvial:** acumulación detrítica de forma alargada localizada en el cauce de un río y que puede estar estabilizada por vegetación; **2. barra [f] costera/litoral:** acumulación de arena situada paralelamente a la línea de la costa, y que puede estar sumergida o emergida; DGA 1986; ▶banco de grava); *syn.* barra

[f] de arena; *f 1* **banc [m] de sable (élevé)** (Relief en saillie sur un fond marin ou accumulation de sable dans les lits majeurs de fleuves et de rivières chargés d'alluvions ; ►banc de gravier. *Terme spécifique f 2* **levée [f] (2)** (**1.** Forme de relief naturel allongé qui borde le rivage [levée littorale], les rives des chenaux [levée de rive]. **2.** Élévation de terrains dominant la basse plaine le long d'une berge [levée alluviale] ; *syn.* bourrelet [m] de crue ; VOG 1979, 116-117) ; *syn.* levée [f] sablonneuse ; *g* **Sandbank [f]** (*Untiefe als Sandansammlung* Sandinsel in einem Fließgewässer oder Erhebung des Meeresbodens bis nahe unter die Wasseroberfläche; ►Kiesbank).

sandbar [n] *geo.* ►sandbank.

sand base [n] [US] *constr.* ►sand bed.

5324 sand bed [n] *constr.* (Specific term for ►laying course for a paved stone; ►bedding sand); *syn.* sand base [n] [also US] (HALAC 1988-II, 61), sand setting bed [n] [also US], blinding layer [n] of sand [also UK]; *s* **lecho [m] de arena** (Capa de arena sobre la que se tienden adoquines, ladrillos clinker, tuberías, etc.; ►lecho de asiento [de adoquinado]; ►arena para adoquinados); *syn.* capa [f] de arena; *f* **lit [m] de sable** (Couche de pose pour pavés, dalles constituée de ►sable de pavage ; ►lit de pose du pavage) ; *syn.* sablon [m] ; *g* **Sandbett [n]** (Sandschicht, auf die Pflastersteine, Klinker, Rohrleitungen etc. verlegt werden; der Sand für das **S.** wird auch „Bettungssand" [►Pflastersand] genannt; ►Pflasterbett); *syn.* Sandbettung [f].

5325 sandblasted concrete [n] *constr.* (Visible surface of concrete blasted by sand particles after removal of formwork); *s* **hormigón [m] lijado con arena**; *f* **béton [m] sablé** (Béton travaillé sous l'action d'un jet de sable, laissant apparaître les granulats et lui conférant une structure rugueuse) ; *g* **sandgestrahlter Sichtbeton [m]** (Beton, dessen sichtbare Oberfläche nach dem Ausschalen mit einem Sandgebläse bearbeitet wird).

5326 sandblasting [n] *constr.* (Cleaning or roughening of a material's surface with sandblasting equipment); *s* **chorro [m] de arena** (Sistema de limpieza de superficies resp. fachadas con tobera de arena); *f* **sablage [m]** (Projection de sable [soufflé] sur un matériau en vue de le nettoyer ou d'obtenir une rugosité artificielle ; *contexte* dalle sablée, finition/aspect sablé) ; *g* **Sandstrahlen [n, o. Pl.]** (Reinigen oder Aufrauhen einer Materialoberfläche mit einem Sandstrahlgebläse; sandstrahlen [vb], sandgestrahlt [pp]).

sand borrow pit [n] *min.* ►sand extraction site.

5327 sand box [n] *recr.* (Small children's play space composed of sand contained by an edge > 20cm high; ►sand pit); *s* **cajón [m] de arena** (Superficie con borde de > 20 cm de altura rellena de arena para juegos infantiles; ►hondonada de arena); *syn.* pozo [m] de arena, foso [m] de arena); *f* **bac [m] à sable** (Équipement de jeu rempli de sable prévu pour les jeunes enfants, entouré d'une bordure > 20 cm de hauteur ; ►aire sablée 2) ; *g* **Sandkasten [m]** (Spieleinrichtung mit Sandfüllung für Kinder und einer meist >20 cm hohen Einfassung; ►Sandmulde).

sand dump [n] [ZA] *min.* ►sand pile [US]/sand tip [UK].

sand dune [n] *geo.* ►drifting sand dune.

sand dunes [npl] *constr.* ►stabilization of sand dunes.

sand-dwelling [adj] *phyt. zool.* ►psammophilous.

5328 sand extraction [n] *min.* (Removal of sand from sand pits or by dredging; ►mineral extraction); *syn.* quarrying [n] of sand (TGG 1984, 101); *s* **extracción [f] de arena** (Ganancia de arena en arenera o en laguna artificial; ►explotación superficial de recursos minerales); *f* **extraction [f] de sable** (Extraction à ciel ouvert de sable sur un site d'extraction de sable ou dans un étang de fouille ; ►extraction des substances minérales) ; *g* **Sandabbau [m]** (Oberirdische Entnahme von Sand in Sand-gruben oder Baggerseen; ►Materialentnahme); *syn.* Abbau [m] von Sand, Aussandung [f], Sandbaggerung [f], Sandentnahme [f], Sandgewinnung [f].

5329 sand extraction site [n] *min. syn.* sand borrow pit [n]; *s* **lugar [m] de extracción de arenas**; *f* **site [m] d'extraction de sable** ; *syn.* sablière [f] ; *g* **Sandentnahmestelle [f]** (Abbaufläche zur Sandgewinnung); *syn.* Sandgrube [f].

sand-filled joint [n] **topped with liquid cement** [US] *constr.* ►sand-filled joint with cement grout.

5330 sand-filled joint [n] **with cement grout** *constr.* *syn.* sand-filled joint [n] topped with liquid cement [also US]; *s* **junta [f] rellena de arena y cemento líquido**; *f* **joint [m] garni de sable avec applique d'une barbotine de ciment** ; *g* **Fuge [f] mit Sand gefüllt und Zementschlämme ausgegossen**.

5331 sandlot [n] [US] *recr.* (Informal play area, vacant lot or piece of ground with a sandy surface for children esp. as the scene of unorganized sports for boys from city streets; no UK equivalent); *s* **terreno [m] de juegos** (Área de juegos informales, p. ej. béisbol; el término inglés se utiliza en los EE.UU., no hay equivalente para GB); *f* **aire [f] de jeux sans équipements particuliers à l'usage des enfants et des adolescents)** ; *syn.* espace [m] ensablé ; *g* **Sandplatz [m] (±)** (U.S., nicht besonders gestaltete Fläche in der Stadt mit einem wassergebundenen Belag, der für spontane Spielaktivitäten von Kindern und Jugendlichen genutzt wird).

5332 sand pile [n] [US] *min.* (Spoil mound of loose granular tailings resulting from gold mining operations); *syn.* sand dump [n] [ZA], sand tip [n] [UK]; *s* **escombrera [f] de placeres** (Montón de materiales residuales gruesos resultantes de operaciones de minería aurífera); *f* **halde [f] de stériles aurifères (±)** (*Extraction de l'or* accumulation artificielle de matériaux stériles granuleux sur un terrain naturel) ; *g* **Sandbergehalde [f]** (*Goldförderung* künstliche Aufschüttung von feinkörnigem Bergematerial auf unverritztem Gelände).

5333 sand pit [n] *recr.* (Relatively large shallow bowl of sand in a play area with a depth of > 40 cm, often provided with play equipment and an edge element which does not exceed the level of the sand); *s* **hondonada [f] de arena** (Superficie de arena relativamente grande en zona de juegos infantiles, con capa de arena de > 40 cm, a menudo con aparatos de juego y encintada sin que el borde sobresalga a la arena); *f* **aire [f] sablée (2)** (Sur une aire de jeux, surface de sable de 50 à 100 m^2 [épaisseur du sable > 40 cm] comportant souvent diverses installations de jeux [agrès, mobilier de jeux] et ceinturée par une bordure ne dépassant pas en général le niveau du sable) ; *syn.* cuvette [f] de sable ; *g* **Sandmulde [f]** (Größere Sandfläche eines Kinderspielplatzes mit einer > 40 cm hohen Sandfüllung [cf. DIN 18 034], oft mit Spielgeräten versehen und mit einer Einfassung ausgestattet, die die Sandebene meist nicht überragt).

5334 sand playground [n] *recr.* (Sand play area for small children with ►sand box or ►sand pit); *syn.* sand playscape [n]; *s* **zona [f] de juegos con arena** (Area de juegos de niños pequeños con ►cajón de arena u ►hondonada de arena); *f* **aire [f] de jeux sablée** (Place de jeux pour enfants de 3 à 7 ans équipé d'un ►bac à sable ou d'une ►aire sablée) ; *g* **Sandspielplatz [m]** (Spielplatz für Kleinkinder mit ►Sandkasten oder ►Sandmulde).

sand playscape [n] *recr.* ►sand playground.

5335 sand replacement [n] *constr. recr.* (Removal and changing of sand in sand boxes and sand pits in children's playgrounds; ►cleaning of sand areas); *s* **cambio [m] de la arena** (Sustitución periódica de la arena en zonas de juegos infantiles; ►limpieza de superficies con arena); *syn.* reemplazo [m] de la arena; *f* **remplacement [m] du sable** (Opération régulière d'entretien sur les aires de jeux, p. ex. dans les bacs à sable, les fosses de sable, les aires entourant les agrès, etc. ; ▶désinfection des aires sablées) ; *syn.* curage [m] de la fosse sablée ; *g* **Sandaustausch [m]** (Auswechseln des Spielsandes in Sandkästen und Sandmulden auf Kinderspielplätzen; ▶Reinigen von Sandflächen).

sand setting bed [n] [US] *constr.* ▶sand bed.

sand spit [n] *geo.* ▶barrier-beach.

5336 sandstone [n] *geo.* (Sedimentary rock consisting mainly of grains of quartz, often with feldspar [*also* feldspath], mica and other materials, consolidated, cemented and compacted. Cementing material may be calcareous, siliceous, ferruginous and dolomitic. Colo[u]r varies from dark brown or red through yellow to grey and white, mainly due to iron content and its degree of oxidation or hydration; DNE 1978); *s* **arenisca [f]** (Roca sedimentaria detrítica del grupo de las arenitas. Procede de la diagenización de las arenas, por lo que es una roca coherente. La clasificación de las areniscas puede basarse en la naturaleza de los granos [**a.** feldespática, **a.** micácea, **a.** cuarcítica, etc.] o en el tipo de cemento [**a.** calcárea, **a.** silícea, etc.]; DGA 1986); *syn.* gres [m]; *f* **grès [m]** (Roche sédimentaire d'origine détritique formée par des sables très grossiers ou très fins conglomérés. Un grès peut être défini par la nature de son ciment : siliceux, calcaire ou ferrugineux, ainsi que par la nature de ses grains : feldspaths, lamelles de muscovite, glauconie. La couleur jaune à brune de la roche est provoquée par la présence d'hydroxyde de fer, les couleurs rouge et violette par l'oxyde de fer, la couleur verte par la glauconie ; cf. DG 1984) ; *g* **Sandstein [m, o. Pl.]** (Durch Verfestigung von Quarzkörnern, die durch toniges, kaolinisches, kalkiges, mergeliges, kieseliges oder eisenhaltiges Bindemittel verkittet sind, entstandenes Sedimentgestein [Psammit]. Die gelbe bis braune Farbe entsteht durch Eisenhydroxid, die rote und violette durch Eisenoxid, die grüne durch Glaukonit).

5337 sand-swept joints [npl] *constr.* (result); *s* **junta rellenada de arena [loc]**; *f* **joints garnis par balayage de sable [loc]** (résultat) ; *g* **Fugen mit Sand eingefegt [loc]** (Ergebnis); *syn.* mit Sand eingefegte Fugen [loc].

sand tip [n] [UK] *min.* ▶sand pile [US]/sand tip [UK].

sand trap [n] [UK] *constr. hydr.* ▶sand arresting trap [US].

5338 sand trap [n] [US] *constr. recr.* (Hollowed-out sand-filled basin near golf green as a hazard for players); *syn.* bunker [n] [UK]; *s* **bunker [m] de arena** (Obstáculo en un campo de golf); *f* **bunker [m] de sable** (Obstacle sur un parcours de golf) ; *g* **Sandbunker [m]** (Hindernis auf einem Golfplatz).

5339 sandy [adj] *pedol.* (Descriptive term for soils having medium to high sand content); *s* **arenoso/a [adj]** (Término descriptivo para suelo con gran contenido de arena); *f* **sableux, euse [adj]** (Relatif aux sols caractérisés par une proportion en sable moyenne à forte) ; *syn.* sablonneux, euse [adj] ; *g* **sandig [adj]** (Böden betreffend, die einen mittleren bis hohen Sandanteil haben).

5340 sandy soil [n] *agr. hort. pedol.* (▶Easily workable soil, generally with a single particle size, having > 40% up to 80% sand and low humus and nutrient content; ▶clay soil); *s* **suelo [m] arenoso** (▶Suelo ligero con estructura de ceniza, con > 40%, a veces hasta un 80% de arena y con poco contenido de humus y nutrientes; ▶suelo arcilloso); *f* **sol [m] sableux** (▶Terre douce contenant une proportion de sable > 40 % et atteignant souvent 80 %, de faible teneur en matière organique et en substances nutritives ; *opp.* ▶sol argileux) ; *syn.* sol [m] sablonneux ; *g* **Sandboden [m]** (▶Leicht bearbeitbarer Boden, der allgemein ein Einzelkorngefüge aufweist, > 40 %, oft bis 80 % Sandanteil und einen geringen Humus- und Nährstoffgehalt hat; ▶Tonboden).

5341 sanitary landfill [n] (1) *envir.* (Process of controlled dumping [US]/tipping [UK] of industrial or domestic waste material on a landfill site by dumping/tipping in layers, each being sealed with a layer of soil; an impervious plastic liner is placed on the existing ground to prevent seepage into the ground water); ▶sanitary landfill [US] 2/landfill site [UK]); *s* **relleno [m] sanitario** (Proceso de almacenamiento de residuos domésticos o industriales en un ▶vertedero controlado, en el cual los residuos son vertidos en capas que cada vez son revestidas con una capa de tierra aislante; para evitar la lixiviación de líquidos a las aguas subterráneas el fondo del **r. s.** se aisla con material plástico u otro material impermeable); *f* **mise [f] en décharge contrôlée des déchets urbains** (Stockage dans des sites géologiquement adaptés et aménagés et compactage homogène des déchets par un engin mécanique en couches successives d'épaisseur modérée [1,50 à 2,50 m environ], une nouvelle couche n'étant déposée que lorsque la température de la couche précédente s'est abaissée à la température du sol naturel ; les couches sont exactement nivelées et limitées par des talus ; le dépôt doit être compact, ne pas comporter de vides nombreux ou importants ou, en particulier, de vides formant cheminées ; chaque couche après régalage définitif est recouverte de terre ou de matériaux appropriés [sable, gravats], appelés couverture, qui aura 10 à 30 centimètres d'épaisseur dans un délai de 72 heures au maximum et mieux le jour même pour éviter les nuisances et les pollutions qui peuvent être causées au voisinage ; cf. Art 12 du Décret n° 74-338 du 10 avril 1974 réglementant l'évacuation et le dépôt des ordures ménagères ; pour éviter la pollution de la nappe phréatique et des sols voisins le fonds de la décharge est protégé par une étanchéité et les lixiviats sont récupérés et traités ; afin de limiter le rejet de CH_4 les alvéoles d'enfouissement sont recouvertes par une géomembrane permettant la récupération et valorisation du gaz ; ▶décharge contrôlée) ; *g* **lagenweise Abdeckung [f] der/von Müllschichten** (Vorgang der lagenweisen Abdeckung von Müllschichten mit Boden bei der Erstellung einer ▶geordneten Deponie; zur Verhinderung von Sickerwässern wird der gewachsene Boden versiegelt und anfallende Abwässer abgeleitet, gesammelt und gereinigt); *syn.* lagenweise Mülldeponieabdeckung [f].

5342 sanitary landfill [n] [US] (2) *envir.* (Area used for disposal of industrial or domestic waste material, often located in abandoned sand and gravel pits, canyons, excavations or on flat terrain to be landscaped or reclaimed thus creating an ▶artificial landform. The common practice is to spread and compact the refuse as it is delivered and to cover it at the end of each day with a layer of soil, which, in turn, is compacted; ARB 1983; an impervious plastic liner is placed on the existing ground to prevent seepage into the ground water; in modern **s. l.s** the emergent gases [methane] are collected and used to produce energy; ▶unauthorized dumpsite [US]/fly tipping site [UK]); *syn.* controlled landfill site [n] [US] (TGG 1984, 236), landfill site [n] [UK], refuse-landfill site [n] (ARB 1983, 270); *s* **vertedero [m] controlado** (Instalación para depositar residuos sólidos urbanos o residuos industriales, localizada muchas veces en arenales o graveras abandonadas, hondonadas de terreno, excavaciones o en terreno llano a ser recuperado. La técnica actual consiste en aislar el fondo de manera que las aguas subterráneas no sean contaminadas por los líquidos lixiviados e instalar conducciones para recogerlos y tratarlos adecuadamente; además se instalan conducciones para evacuar los gases de putrefacción [metano], que se generan al descomponerse las sustancias orgánicas, y conducirlos a una instalación de aprovechamiento energético; finalmente las capas de residuos son aisladas unas de otras por medio de capas de tierra; *opp.* ▶vertedero clandestino; ▶relleno sanitario); *syn.* confinamiento [m] controlado; *f 1* **décharge [f] contrôlée** (*Terminologie européenne* ; la mise en décharge contrôlée est un

S

procédé d'élimination par lequel les catégories de déchets homogènes sont enfouis en supprimant leur contact direct avec le sol ; ils sont déposés, compactés en couches, en alternance avec des couches de terre et pouvant être réutilisée une fois le site remblayé et reverdi ; la décharge est soumise à des normes strictes : a] récupération des émanations de méthane pour son utilisation comme combustible [valorisation par captage du biogaz] issues de la fermentation, pour les décharges les plus modernes, b] récupération et traitement des lixiviats [jus des déchets polluants], c] étanchéité de la décharge [géomembrane] pour qu'elle ne pollue pas les nappes et sols voisins ; *terme spécifique* décharge contrôlée de résidus urbains ; *opp.* ▶décharge [des déchets] sauvage) ; *syn.* aire [f] de stockage des déchets ; *f 2* **centre [m] de stockage des déchets ultimes [CSDU]** (*Terminologie de la réglementation française, terme générique* lieu de stockage permanent des déchets par dépôt ou enfouissement sur le sol ou dans des cavités artificielles ou naturelles du sol et couverture ultérieure, sans intention de reprise ultérieure, instauré par la loi du 13 juillet 1992 ; on distingue la classe I recevant des déchets industriels spéciaux [DIS], ultimes et stabilisés, la classe II recevant les déchets ménagers et assimilés, la classe III recevant les gravats et déblais inertes) ; depuis le 1er juillet 2002, les centres de stockage ne peuvent accueillir que les déchets ultimes ; *syn. anciennement dénommé* centre [m] d'enfouissement technique [CET] — cf. circulaire ministérielle du 11 mars 1987 ; *g* **geordnete Deponie [f]** (Bauliche und technische Anlage, auf der lagenweise homogene Müllarten mit anschließender, möglichst täglicher, schichtenweiser Abdeckung durch Erdmaterial gelagert werden; zur Verhinderung von Sickerwässern wird der gewachsene Boden versiegelt [mit einem Durchlässigkeitsbeiwert $K_f \leq 1 \times 10^{-9}$] und anfallende Abwässer abgeleitet, gesammelt und gereinigt; **in D.** müssen Deponien gemäß der Deponieverordnung [DepV] betrieben und kontrolliert werden; Grundlagen sind die Technische Anleitung Siedlungsabfall [TASi] und die Technische Anleitung Abfall [TASo]. Es gibt Deponien der Klasse I für Bodenaushub, Bauschutt, Inertabfall; Klasse II für nicht gefährliche Abfälle und zur Deponieklasse III gehören Deponien für gefährliche Abfälle; **in CH** werden die Deponien nach der Technischen Verordnung über Abfälle [TVA] in **Inertstoffdeponien, Reststoffdeponien** und **Reaktordeponien** eingeteilt; **in A.** werden **Bodenaushubdeponien, Baurestmassendeponien, Reststoffdeponien** und **Massenabfalldeponien** unterschieden. Bei modernen **g.n D.n** werden die entstehenden Abgase [Methan] gesammelt und energetisch genutzt; ▶Auffüllberg; *opp.* ▶wilde Deponie); *syn. leg.* Abfallbeseitigungsanlage [f] zur Endablagerung von Abfällen, Abfallentsorgungsanlage [f] zur Endablagerung von Abfällen.; *euphemistisch* Entsorgungspark [m].

sanitary sewage [n] *envir.* ▶domestic sewage.

sanitation department [n] [US] *adm. envir.* ▶waste management department.

5343 sapling [n] *for.* (Young tree which is no longer a ▶seedling, but not yet as long as a ▶pole, approx. 2.5cm in diameter at chest height ▶sapling-stage forest); *s* **árbol [m] nuevo** (Árbol jóven con un diámetro máximo de pocos centímetros; ▶fustal 2, ▶latizal 1, ▶plántula de semilla); *syn.* pimpollo [m] (1), arbolito [m]; *f* **gaule [f]** (Jeune arbre dont le diamètre moyen est de 5 cm dans un jeune peuplement de futaie régulière ; ▶gaulis, ▶perche, ▶semis 2) ; *g* **Gerte [f]** (Junger Baum mit einem Stammdurchmesser von bis 7 cm in einem Jungbestand; ▶Gertenholz, ▶Sämling, ▶Stange); *syn.* Nichtderbholz [n], Reiserholz [n], Stockholz [n].

5344 sapling-stage forest [n] *for.* (developmental stage of a forest; ▶sapling); *s* **latizal [m]** (1) (Masa de brinzales o de pies coetáneos procedentes de semillas de la edad en la que se establece la selección para su futuro desarrollo en diámetro; DB

1985; ▶árbol nuevo); *f* **gaulis [m]** (**1.** Jeune peuplement de futaie régulière dont les brins ont un diamètre moyen de l'ordre de 5 cm et perdent leurs branches basses ; ▶gaule. **2.** Stade développement d'une futaie régulière, qui succède au fourré et précède le perchis ; TSF 1985, 89) ; *g* **Gertenholz [n]** (…hölzer [pl]; Entwicklungsstufe eines Laubhochwaldes — meist Eichen- und Buchenjunghölzer —, die dem Dickicht folgt; ▶Gerte).

sapric peat [n] *pedol.* ▶humified raised bog peat.

saprist [n] [US] *pedol.* ▶anmoor.

5345 saprobic level [n] *limn. wat'man.* (Full range of heterotrophic bioactivity in a body of water. The term is usually limited to the decomposing effects of heterotrophic microorganisms—principally bacteria, fungi and ciliates—and used in the evaluation of organic imbalance in rivers, lakes or the sea; ▶saprobic system, ▶trophic state); *s* **saprobiedad [f]** (Grado de contaminación de las aguas determinado por el tipo de microorganismos indicadores que las habitan y que se dividen en: catarobios, oligosaprobios, mesasaprobios y polisaprobios, según una escala de **s.** creciente. El término **s.** es complementario al de ▶estado trófico; cf. DINA 1987; ▶sistema de saprobiedad); *f* **saprobie [f]** (Somme des activités hétérotrophes dans un milieu aquatique. Communément, dans son acceptation et dans l'évaluation de la qualité des eaux et de leur degré de pollution par la matière organique, le terme de saprobie est limité à l'activité réductrice des micro-organismes hétérotrophes — en premier lieu les bactéries, champignons et paramécies [Ciliates] ; la **s.** est à rapprocher d'un terme complémentaire le ▶degré trophique ; ▶système des saprobies) ; *g* **Saprobie [f]** (Summe der heterotrophen Bioaktivität in einem Gewässer. Gewöhnlich wird der Begriff **S.** auf die Abbautätigkeit der heterotrophen Mikroorganismen — in erster Linie Bakterien, Pilze und Wimpertierchen [Ciliaten] — eingeschränkt und zur Beurteilung eines organisch belasteten Gewässers herangezogen. **S.** ist der Komplementärbegriff zu ▶Trophie; ▶Saprobiensystem); *syn.* Saprobiestufe [f], Saprobität [f].

5346 saprobic system [n] *limn. wat'man.* (System devised by KOLKWITZ und MARSSON [1902, 1908], which classifies the condition of watercourses according to the degree of organic pollution during self-purification processes. Certain organisms are indicators of ▶water quality categories, ▶saprobic level as measured by the extend of their distribution, occurrence and frequency); *s* **sistema [f] de saprobiedad** (Método de clasificación de las aguas continentales según el grado de contaminación orgánica y su capacidad de autodepuración. Para su determinación se utilizan organismos indicadores; ▶clase de calidad de las aguas continentales, ▶saprobiedad); *f* **système [f] des saprobies** (Méthode d'analyse biologique de la qualité des eaux et de leur degré de pollution par les matières organiques — défini par KOLKWITZ ET MARSSON — permettant de caractériser leur degré de pureté ou de souillure, certaines espèces ou groupes d'espèces indicatrices caractérisant un état de qualité donné ; AEP 1976, 324 ; ▶classe de pollution des cours d'eau, ▶niveau de qualité des cours d'eau, ▶saprobie) ; *g* **Saprobiensystem [n]** (Von KOLKWITZ und MARSSON [1902, 1908] beschriebenes stufenweises Zustandsbild von Fließgewässern hinsichtlich des Grades ihrer Belastung mit organischen Inhaltsstoffen im Verlauf des Selbstreinigungsprozesses. Dabei haben bestimmte Organismen auf Grund ihres ökologischen Verbreitungsschwerpunktes — Vorkommen und Häufigkeit — eine Indikatorfunktion für ▶Gewässergüteklassen; ▶Saprobie).

5347 sapropel [n] *pedol. limn.* (Greek *sapros* >rotten<; *pēlos* >loam<; subhydric soil with an A-Gr profile in water extremely deficient in oxygen and aquatic animal life, which has originated under anaerobic conditions); *s* **sapropel [m]** (Sedimento formado a partir de restos orgánicos, básicamente planctónicos, transformados en condiciones reductoras; normalmente bentónicas de poca agitación, lacustres o marinas; DINA 1987); *f* **sapropèle [m]** (Terme qualifiant un humus submergé [famille des vases dans la classification morpho-chimique des humus d'après DELECOUR] se développant dans les milieux aquatiques extrêmement pauvres en oxygène et en organismes aquatiques, à profil A-Gr constitué sous des conditions anaérobes ; cf. DIS 1986, 108) ; *syn.* boue [f] sapropèle, humus [m] submergé (DUV 1984, 193) ; *g* **Sapropel [m]** (Griech. *sapros* >faul<; *pēlos* >Lehm<; subhydrischer Boden in extrem sauerstoff- und wassertierarmen Gewässern mit einem A-Gr-Profil, der unter anaeroben Bedingungen entsteht; cf. SS 1979); *syn. auch* Faulschlamm [m] (2) (GGT 1979, 430).

5348 saprophage [n] *ecol. zool.* (Greek *sapros* >rotten<, *phagein* >eat<; animal which feeds upon decayed organic matter of plant or animal origin; ▶food web); *syn.* saprovore [n]; *s* **saprófago [m]** (Organismo animal que se alimenta de sustancia orgánica muerta sea vegetal o animal; ▶red trófica); *syn.* detritófago [m]; *f* **saprophage [m]** (Organisme qui se nourrit de matières organiques en cours de décomposition ; ▶réseau trophique) ; *g* **Saprophag [m]** (...e [pl]; griech. *sapros* >faul<, *phagein* >essen<; Tier, das sich von toter organischer Substanz pflanzlichen oder tierischen Ursprungs ernährt; saprophag [adj]; ▶Nahrungsnetz).

5349 saprophyte [n] *bot.* (Greek *sapros* >rotten<, *phyton* >plant<; plant which is not capable of assimilation and supplements its nutrient requirement completely or partly with decayed organic matter; ▶decomposer 1); *s* **saprófito [m]** (Término perteneciente a la clasificación de formas de vida de ELLENBERG y MUELLER-DOMBOIS con el que se designa a las plantas heterótrofas que viven y se alimentan de materia orgánica muerta; DINA 1987; ▶descomponedor); *f* **saprophyte [m]** (Organisme végétal se nourrissant de matières organiques en cours de décomposition, p. ex. les champignons du sol ; DEE 1982 ; ▶décomposeur) ; *g* **Saprophyt [m]** (Griech. *sapros* >faul<, *phyton* >Pflanze<; nicht oder nicht ausreichend zur Assimilation befähigte Pflanze, die ihren Nährstoffbedarf ganz oder nur teilweise aus toter organischer Substanz deckt; ▶Destruent); *syn.* Fäulnisbewohner [m].

saprovore [n] *ecol. zool.* ▶saprophage.

sap shoot [n] *arb.* ▶coppice shoot.

5350 sapwood [n] *arb.* (Light colo[u]red, outer and youngest layer of vascular tissue in a woody trunk or stem which conducts and stores plant nutrients, e.g. starch; ▶heartwood); *s* **albura [f]** (Aquella parte de la madera de los árboles vivos que contiene células vivas y materiales de reserva, p. ej. almidón; cf. MGW 1964; ▶duramen); *f* **aubier [m]** (Portion du bois qui renferme dans l'arbre vivant des cellules vivantes et des matières de réserve, p. ex. de l'amidon ; MGW 1964 ; ▶bois parfait) ; *g* **Splintholz [n]** (...hölzer [pl]; der an der Saftleitung beteiligte äußere Teil des Holzes im stehenden Stamm, der lebende Zellen und Reservestoffe [z. B. Stärke] enthält; MGW 1964; ▶Kernholz).

5351 satellite city [n] *urb.* (Independent city/town with its own economic and cultural activities yet within the sphere of influence of a nearby larger community. Such communities serve to ease the pressure upon larger cities, so that the latter do not continue to grow uncontrollably; ▶bedroom community [US]/dormitory suburb [UK]); *syn.* overspill town [n] [also UK] (WAL 1976, 151), satellite town [n]; *s* **ciudad-satélite [f]** (Ciudad pequeña o mediana, próxima a una gran ciudad de la que depende funcionalmente. En los Planes Generales de ámbito metropolitano se asigna la función principal que debe cumplir cada

ciudad-satélite: residencial, industrial, mixta; cf. DGA 1986; ▶ciudad-dormitorio); *f* **ville [f] nouvelle** (*Aménagement du territoire* ville conçue pour freiner la croissance urbaine en tâche d'huile des grandes agglomérations et offrant un cadre de vie urbain équilibré pour la vie familiale, l'emploi, les loisirs et la culture ; ▶cité-dortoir) ; *syn.* ville-satellite [f], cité [f] satellite, village [m] satellite ; *g* **Trabantenstadt [f]** (**1.** Zum Einflussbereich einer größeren Stadt gehörende, aber selbständige Stadt mit eigenen wirtschaftlichen und kulturellen Mittelpunktsfunktionen. **2.** In D. wird eine **T.** im Einflussbereich einer Großstadt mit geringerer Selbständigkeit und geringem Arbeitsplatzangebot auch Satellitenstadt genannt; da die Definition einer *satellite city* im Britischen und Amerikanischen der **T.** entspricht, werden heute beide Termini häufig auch synonym benutzt; ▶Schlafstadt); *syn.* Satellitenstadt [f].

5352 satellite data analysis [n] *plan. rem'sens. surv.* (Evaluation of information contained in satellite photographs; ▶interpretation of aerial photographs); *s* **interpretación [f] de imágenes de satélite** (Conjunto de técnicas y conocimientos con los cuales es posible extraer información fiable sobre el estado de la superficie de la Tierra a partir de las imágenes obtenidas a través de sistemas de ▶teledetección espacial; ▶interpretación de foto[grafías] aéreas); *syn.* interpretación [f] de datos de satélite, interpretación [f] de imágenes de teledetección espacial; *f* **interprétation [f] des images-satellite** (Interprétation scientifique de données provenant de photographies satellite ; ▶interprétation des photos aériennes) ; *g* **Satellitenbildauswertung [f]** (Interpretation von Daten, die von Satellitenbildern stammen; ▶Luftbildauswertung).

5353 satellite photograph [n] *envir. plan. rem'sens. surv.* (Photographs or data taken by satellites in the process of ▶remote sensing); *s* **imagen [f] de satélite** (▶teledetección); *syn.* fotografía [f] de satélite; *f* **photographie [f] satellite** (Photographie prise par satellite et utilisée pour la ▶télédétection [de terrain]) ; *g* **Satellitenaufnahme [f]** (Von einem Satelliten aus aufgenommene Fotos oder Daten als Mittel der ▶Fernerkundung); *syn.* Satellitenfoto [n].

5354 saturated [pp/adj] *pedol.* (Descriptive term applied to a soil when all the interstices between the soil particles are filled with water; ▶waterlogged soil [US]/water-logged soil [UK]); *s* **saturado [pp/adj]** (Estado del suelo en el que todo el volumen de poros está ocupado por agua; ▶suelo saturado); *f* **saturé, ée en eau [loc]** (État d'un sol dont l'ensemble du volume des pores est rempli d'eau ; ▶sol saturé par l'eau) ; *g* **wassergesättigt [pp/adj]** (So beschaffen, dass der gesamte Porenraum eines Bodens mit Wasser gefüllt ist; ▶wassergesättigter Boden).

saturation capacity [n] **of an area** *plan. ecol.* ▶carrying capacity of landscape.

5355 saucer rim [n] [US] *constr. hort.* (Ring of soil around the edge of a ▶planting saucer [US]/watering hollow [UK]); *syn.* watering rim [n] [UK]; *s* **borde [m] (elevado) del hoyo de riego** (▶alcorque de riego); *f* **bord [m] de cuvette d'arrosage** (Anneau de terre ménagé au pied d'un arbre nouvellement planté afin de former une ▶cuvette d'arrosage) ; *g* **Gießrand [m]** (Erhöhter Erdring um die Baumscheibe neu gepflanzter Gehölze zur Schaffung einer ▶Gießmulde).

saum biotope [n] [UK] *phyt. zool.* ▶seam biotope.

5356 savanna(h) [n] *geo. phyt.* (Grassy plain with scattered trees and shrubs in tropical regions with seasonal rains; ▶flood savanna[h], ▶scrub savanna[h], ▶shrubland savanna[h], ▶steppe, ▶tree savanna[h], ▶tropical grassland); *s* **sabana [f]** (Tipo de vegetación, frecuente en los países tropicales cuyo clima comporta una estación seca; su simorfia dominante o característica es un *elatigraminetum*, al que pueden acompañar más o menos abundantemente, hierbas perennes sufrútices, frútices y hasta árboles; DB 1985; ►estepa, ►llanos, ►sabana arbolada, ►sabana arbustiva, ►sabana de espinos, ►sabana de inundación, ►sabana herbácea); *f* **savane** [f] (Mosaïque de formations végétales herbacées, hautes et fermées, dont le cycle biologique est étroitement adapté au climat tropical à deux saisons ; ►savane arborée, ►savane arbustive 1, ►savane arbustive 2, ►savane buissonnante, ►savane galerie, ►savane hallier, ►savane herbacée, ►savane inondable, ►steppe ; DG 1986, 407) ; *g* **Savanne** [f] (Hauptvegetationsformation in tropischen und subtropischen Gebieten mit wechselfeuchtem Klima, d. h. einer ausgeprägten Regenzeit während der heißen Sommerzeit und einer extremen Dürre im kühleren Winterhalbjahr. Die Vegetation besteht aus homogenen Pflanzengesellschaften aus zerstreut stehendem Gehölzbewuchs [Bäume und Sträucher] in einer unterschiedlich geschlossenen Grasschicht mit einigen Kräutern dazwischen; z. B. die **Llano(s)** [m(pl)] nördlich des unteren Orinoco von Venezuela bis nach Kolumbien hinein und die **Campos Cerrados** in Brasilien. Wenn diese **S.n** aus einem Mosaik von Waldinseln in einer offenen Graslandschaft mit wenigen Holzpflanzen bestehen, in welchen der Wald an ökologisch ganz andere Biotope gebunden ist [Flussufer, Talsohlen, Erhebungen] als die Grasflächen, so spricht man von einer ökologisch heterogenen ►Parklandschaft 1; cf. ÖTS 1984; ►Baumsavanne, ►Buschsavanne, ►Dornenstrauchsavanne, ►Grasland, ►Steppe, ►Überschwemmungssavanne).

savanna(h) [n], **grass** *phyt.* ►tropical grassland.

savanna(h) [n], **park** *geo. phyt.* ►parkland.

sawed [pp/adj] *constr.* ∗►tooling of stone [US]/dressing of stone [UK], #3.

saw grass swamp [n] *phyt.* ►saw sedge swamp.

5357 saw [vb] **into pieces** *constr. for.* (Cutting of stems or branches, cutting transversely into lengths/logs); *syn.* bucking off [n] [also US], logging off [n] [AUS], cross-cutting [n]; *s* **serrar** [vb] (Cortar con sierra troncos o ramas de árboles); *f* **débiter** [vb] **à la scie** (Couper le bois de troncs d'arbres abattus ou de branches en morceaux de 1 m ; débitage [m]) ; *g* **zersägen** [vb] (Mit der Säge Baumstämme oder Äste in Stücke zerteilen).

sawn [pp/adj] *constr.* ∗►tooling of stone [US]/dressing of stone [UK], #3.

5358 sawn flagstone [n] *constr.* (►tooling of stone [US]/dressing of stone [UK]); *s* **losa** [f] **aserrada** (►labra de piedra natural o artificial); *f* **dalle** [f] **(en pierre) sciée** (Surface relativement plane, éventuellement striée pouvant comporter de petites ondulations ou décrochements ; MAÇ 1981, 183 ; ►mode de taille des pierres) ; *g* **gesägte Natursteinplatte** [f] (►Steinbearbeitung).

sawn timber [n] *constr.* ►cut timber.

5359 saw sedge swamp [n] *phyt.* (Wetland plant community *[Cladietum mariscii]* in silted-up ponds or lakes, primarily consisting of twig-rush/sawgrass/fen sedge *[Cladium mariscus],* only on calcareous soils in nutient-poor and oxygen-rich water; *generic term* ►reed swamp); *syn.* great fen sedge swamp, saw grass swamp, twig rush swamp [US]; *s* **marisma** [f] **de ciperáceas** (Comunidad vegetal sobre todo de *Cladium mariscus* que se presenta en lagos o lagunas colmatadas pobres en nutrientes y ricas en oxígeno sólo sobre suelos calizos; *término genérico* ►carrizal de fragmitea); *f 1* **cladiaie** [f] (Peuplement souvent rare, d'atterrissement des mares de tourbières alcalines [cladiaephragmitaie], des fosses de tourbage, existant aussi sur tourbe acide [cladiaie turficole], caractérisé par la présence de la Marisque *[Cladium mariscus]* ; *terme générique* ►roselière, ►phragmitaie) ; *f 2* **marais** [m] **à Marisques** (Localisation du peuplement) ; *syn.* roselière [f] à Marisques ; *g* **Schwertried-**

röhricht [n] (**1.** Vorwiegend aus Schneidebinse *[Cladium mariscus]* bestehende Verlandungsgesellschaft *[Cladietum mariscii]* kalkreicher, aber nährstoffärmerer Gewässer, gern an quelligen Stellen mit sauerstoffhaltigem Wasser. **2.** Mit **S.** bestandene Fläche; *OB* ►Röhricht); *syn.* Schneidebinsenried [n], Schneidenried [n].

saxicoline [adj] *bot. phyt. zool.* ►rupicolous.

saxicolous [adj] *bot. phyt. zool.* ►rupicolous.

5360 scaffold branch [n] *arb.* (ARB 1983, 413; branch with a diameter > 100mm; ►sturdy branch, ►tree skeleton, ►upright-growing main branch); *syn.* main branch [n] (ARB 1983, 401), main bough [n] [also UK], main limb [n] [also US]; *s* **rama** [f] **primaria** (Rama de árbol que se desarrolla directamente del tronco; en D., la «rama fuerte» es un término técnico para denominar ramas de diámetro superior a 100 mm; ►esqueleto de un árbol, ►rama mediana, ►rama principal de crecimiento vertical); *syn.* rama [f] de 1er orden; *f* **charpentière** [f] (Branche latérale formant la structure durable d'un arbre ; ►branche de taille moyenne, ►charpente d'arbre, ►charpentière à croissance verticale) ; *syn.* branche [f] maîtresse (DAV 1984, 114 ; LA 1981, 274) ; *g* **Starkast** [m] (Bezeichnung eines Astes mit einem Durchmesser > 100 mm; ►Astgerüst, ►Grobast, ►Stämmling).

scaffold root [n] *arb.* ►structural root.

scale [n] *arch. constr. eng. surv.* ►horizontal scale, ►vertical scale

scale [n] **of charges** [UK] *contr. leg. prof.* ►fee chart.

scale [n] **of fees** [UK] *contr. leg. prof.* ►fee chart.

scale [n] **of professional charges** [UK], **conditions of engagement and** (obs. since 1986) *contr. leg. prof.* ►Guidance for Clients on Fees [UK].

scaling bark [n] *arb. bot.* ∗►outer bark.

5361 scaling damage [n] *game'man.* (SAF 1983, 17; bark peeling of trees and shrubs caused by even-hoofed game, mainly as a consequence of ►overprotection [US]/over-protection [UK]; ►over-browsing damage); *syn.* scarred bark damage [n]; *s* **daño** [m] **de descortezamiento** (Heridas causadas por animal de caza en la corteza de árboles, en general en consecuencia de la ►sobreprotección de la caza; ►recomido); *f* **dégâts** [mpl] **d'écorçage** (Usure ou arrachement localisé de l'écorce des arbres et des arbustes par les ongulés sauvages qui occasionne la perte de jeunes végétaux ligneux souvent dus à une ►réglementation excessive de la chasse et rend impossible, sans mesures de protection particulière, la régénération naturelle des essences en station ; ►abroutissement) ; *syn.* dégâts [mpl] de décortication [CDN] ; *g* **Schälschaden** [m] (Meist durch ►Überhege des Schalenwildes entstandener Schaden an Bäumen und Sträuchern durch Abreißen der Rinde; ►Wildverbis).

5362 scandent plant [n] *bot.* (Type of climbing plant which clings to other plants by means of barbed hook-like shoots *[bittersweet—Solanum dulcamara],* strong clinging hairs *[cleavers—Galium aparine],* prickles *[►climbing rose]* or thorns *[Bougainvillea];* climber with arching branches and supportive hooks, e.g. prickles of rambler roses, blackberries or thorns of *Bougainvillea, Solanum dulcamara;* ►climbing perennial, ►climbing plant); *syn.* hooked climber [n], vine [n] with hooked arching stems [also US]; *s* **planta** [f] **sarmentosa** (►Planta trepadora que se ase a otras plantas por medio de retoños laterales con forma de ganchos *[falso jazmín — Solanum jasminoides],* con fuertes pelos rígidos *[Galium aparine],* con acúleos [►rosal trepador] o con espinas [buganvilla — *Bougainvillea];* ►hemicriptófito trepador); *f* **plante** [f] **sarmenteuse** (►Plante grimpante [ligneuse] sans mouvement préhenseur, à tige souple, relativement grêle, qui s'accroche aux végétaux voisins par des rameaux latéraux courts en forme de grappin comme la Douce-amère *[Solanum dulcamara]*, par des poils rigides comme le Gaillet gratteron *[Galium aparine]* ou par des aiguillons [▶rosier grimpant] ou des épines comme le Bougainvillier *[Bougain-villea]* ; ▶plante grimpante vivace) ; *g* **Spreizklimmer [m]** (▶Kletterpflanze, die mit widerhakenähnlichen Seitensprossen [Bittersüß — *Solanum dulcamara]*, starren Klimmhaaren [Kleb-kraut — *Galium aparine]*, Stacheln [▶Kletterrose, Brombeere *(Rubus agg. fruticosus)]* oder mit Dornen [Bougainvillie — *Bougainvillea]* an anderen Pflanzen hochklettert; cf. LB 1978; ▶Klimmstaude).

5363 scanner [n] *rem'sens. surv.* (1. Optical/mechanical in-strument used to receive and measure sequential, electro-magnetic radiation; scanning planes with scanning cameras and film are employed in ▶remote sensing. 2. Office equipment used to enter information directly into a computer by scanning elec-tronically documents or photographs); *s* **escáner [m]** (1. Equipo sensor que permite explorar secuencialmente la superficie de la Tierra, dividiendo la radiación captada en diversas bandas espectrales; CHUV 1990; ▶teledetección. 2. Aparato para leer electrónicamente textos, representaciones gráficas y fotografías); *syn.* scanner [m]; *syn. de 1.* equipo [m] de barrido (multiespec-tral); *f* **scanner [m]** (1. Appareil optique capable de capter les radiations électromagnétiques émises par des surfaces ; procédé utilisé pour la ▶télédétection [de terrain] à partir d'un vecteur, d'une plate-forme [avions, etc.] ; *syn.* capteur [m]. 2. Appareil servant à réaliser la reproduction d'un document original par balayage électronique) ; *g* **Scanner [m]** (1. Optisch-mechani-sches Gerät zum sequentiellen Empfangen und Registrieren elektromagnetischer Strahlung; Scannerflugzeuge mit entspre-chenden Aufnahmegeräten und geeignetem Filmmaterial werden für die ▶Fernerkundung eingesetzt. 2. Bürogerät zum elektro-nischen Einlesen von Schriftstücken und grafischen Darstel-lungen); *syn.* Abtaster [m].

scare [vb] *conserv. hunt. zool.* ▶frighten away.

5364 scarifying [n] [US] *constr.* (Breaking up of heavily compacted soil, usually by heavy hook-shaped tines mounted on a crawler tractor; ▶deep plowing [US]/deep ploughing [UK], ▶subsoiling); *syn.* ripping [n] [UK]; *s* **escarificado [m]** (Des-compactación de suelos muy compactados, generalmente con escarificador; escarificar [vb]; ▶arado en profundo, ▶desfonde); *f 1* **sous-solage [m]** (1) (Travaux aratoires du sol exécutés sur de grandes superficies par passage d'une sous-soleuse à une profon-deur de 40 à 60 cm ; ▶décompactage, ▶labour profond) ; *syn.* dislocation [f] profonde du sol ; *f 2* **scarifiage [m]** (*Labour superficiel* méthode de préparation d'un terrain par ▶ameublisse-ment du sol très compacté à l'aide d'un scarificateur pour favo-riser la régénération des semences) ; *g* **Aufreißen [n, o. Pl.]** (1) (Tiefes Lockern [> 40 cm] stark verdichteter Böden mit Auf-reißhaken [Heckaufreißer] einer Raupe; ▶Tiefenlockerung, ▶Tiefpflügen).

scarped ridge [n] *geo.* ▶cuesta.

5365 scarp slope [n] *geo.* (Steep slope with ground surface cutting across the underlying strata, usually in connection with a shallower ▶dip slope); *syn.* escarpment [n]; *s* **frente [m] de cuesta** (Vertiente abrupta de una cuesta que consta de un talud en la base, en el que afloran materiales incoherentes, y una cornisa de roca resistente en la cumbre; cf. DGA 1986; ▶reverso de cuesta); *f* **versant [m] anaclinal** (DG 1984, 118 ; versant raide mettant à jour les strates sous-jacentes, en général en relation avec une pente douce sur l'autre versant [▶versant cataclinal] du massif montagneux) ; *syn.* front [m] de cuesta ; *g* **Stufenhang [m]** (Steiler Abhang, der ausstreichende Schichten freilegt — i. d. R. im Zusammenhang mit einem flachen Hang auf der

anderen Seite des Berges; ▶Stufenfläche); *syn.* Stufenstirn [f], Stirn [f], Stufenwand [f], Trauf [m].

scarred bark damage [n] *game'man.* ▶scaling damage.

5366 scattered agricultural holdings [npl] *agr. for.* (Field pattern whereby the individual fields belonging to a farm are dispersed within a rural area); *s* **dispersión [f] parcelaria** (Situación dispersa de los terrenos pertenecientes a una granja); *f* **parcellaire [m] dispersé** (Structure de la propriété foncière agricole dans les régions de petite à moyenne exploitation) ; *g* **Gemengelage [f]** (2) (Flurzustand, bei dem die zu einer Hof-lage gehörenden Grundstücke in vielen Parzellen innerhalb einer Feldmark zerstreut liegen); *syn. o. V.* Gemenglage [f].

5367 scattered rubbish [n] [US] *envir.* (Carelessly-scat-tered [domestic] refuse); *syn.* litter [n] [UK] (3), trash [n]; *s* **basura [f]** (Residuos sólidos tirados por los usuarios y repar-tidos por la naturaleza y por las ciudades como papeles, botellas de plástico o de vidrio, latas de bebidas, etc.); *syn.* papelera [f] (1) [tb YZ]; *f* **détritus [mpl]** (Menus déchets abandonnés, p. ex. emballages, reliefs de pique-nique, pelures quelconques, etc.) ; *g* **Abfall [m]** (1) (Sorglos weggeworfener Kleinmüll, z. B. Ver-packungspapier, Getränkedosen, Obstreste, Plastikflaschen etc.); *syn.* Kleinabfall [m], Kleinmüll [m].

5368 scavenger [n] *plan. sociol.* (People who live near rubbish dumps in the third world and gain an existence from collecting and selling waste matter for ▶recycling, ▶reusable material); *s* **recogedor, -a [m/f] de basuras** (Persona que se gana la vida por medio de la búsqueda de residuos aprovechables en un vertedero y su venta posterior y que vive en el recinto del mismo o en sus alrededores. En las grandes urbes de los países del llamado «tercer mundo» es un fenómeno muy común que afecta a miles de familias; ▶material reutilizable ▶reciclado); *syn.* catador [m] [BR], segregador [m] [CO], catadero [m] [EC], pepenador [m] [MEX]; *f* **ramasseur [m] de déchets** (*Tiers monde* personnes vivant sur les décharges et assurant leur existence grâce à la collecte et la vente de résidus en vue de leur ▶recyclage ; ▶matériau réutilisable) ; *syn.* récupérateur [m] de déchets, ramasseur [m] d'ordures ; *g* **Müllsammler [m]** (Men-schen, die in der dritten Welt auf Mülldeponien leben und vom Sammeln und Verkauf von Müllbestandteilen für die Wieder-verwertung leben; ▶Recycling, ▶Wertstoff).

scenery [n], **natural** *landsc. plan. recr.* ▶visual quality of landscape.

scenery [n], **urban** *urb.* ▶urban setting.

5369 scenic [adj] *landsc. recr.* (Descriptive term applied to beautiful scenery; special term for unusual, but attractive scenes: picturesque); *s 1* **paisajístico/a [adj]** (Término referente al pai-saje o descriptivo de algo cuyas características son similares a las de un [determinado] paisaje); *s 2* **pintoresco/a [adj]** (Término descriptivo para paisajes atractivos, dignos de ser pintados resp. fotografiados); *f 1* **paysager, ère [adj]** (Dont les caractéristiques rappellent un paysage) ; *f 2* **pittoresque [adj]** (Ce qui contribue à d'attirer l'attention par ses composantes frappantes, contrastantes, d'une originalité séduisante) ; *f 3* **paysagique [adj]** (Terme à connotation scientifique parfois utilisé à la place de paysager) ; *f 4* **en terme de paysage [loc]** (En ce qui concerne le paysage) ; *g* **landschaftlich [adj]** (1. So beschaffen, dass es den Eigen-schaften einer [bestimmten] Landschaft entspricht. 2. Im Sinne von landschaftlich reizvoll).

5370 scenic heterogeneity [n] *landsc.* (Diverse wealth of forms, colo[u]rs, and textures in the scenic character of a land-scape; ▶diversity, ▶spatial heterogeneity, ▶spatial landscape characteristics/resources, ▶richness); *s* **riqueza [f] escénica del paisaje** (Variedad de formas, colores y texturas en la ▶organi-zación espacial del paisaje; ▶diversidad ecológica, ▶hetero-

S

genidad espacial, ▶variedad); *syn.* diversidad [f] escénica del paisaje; *f* **diversité f] des paysages** (Pluralité et variété des éléments constitutifs [organisation interne] d'un paysage correspondant à la différenciation de plans, de contrastes de couleurs ou de textures, perspectives des teintes et des volumes, lignes de force et de fuite, etc. ; cf. GEP 1991, 138 ; ▶diversité, ▶hétérogénéité spatiale, ▶organisation spatiale, ▶variété) ; *syn.* diversité [f] optique du paysage ; *g* **Vielfalt [f] des Landschaftsbildes** (Fülle von Formen, Farben und Texturen in der ▶räumlichen Ausstattung einer Landschaft; ▶Diversität, ▶räumliche Vielfalt, ▶Vielfältigkeit).

scenic overlook [n] [US] *recr.* ▶view (1).

scenic quality [n] *landsc. plan. recr.* ▶visual quality of landscape.

scenic quality [n]**, perceived** *plan. recr. sociol* ▶perceived environmental quality.

scenic route [n] [UK] *trans. recr.* ▶parkway [US].

scenic trail [n] *plan. recr.* ▶national scenic trail [US]/long distance footpath [UK].

scenic value [n]**, assessment of** *plan. landsc.* ▶visual landscape assessment.

5371 schedule [n] **of prices** *constr. contr.* (Schedule of rates for the items of work to be carried out and the materials to be used, in which the bidder [US]/tenderer [UK] puts his prices beside the items. In U.K., this system is most commonly used in small contracts for maintenance work and is also referred to as a "Schedule of Rates"; cf. BCD 1990); *s* **lista [f] de precios unitarios** (de salarios y materiales, utilizada generalmente para contratos de poca envergadura); *syn.* relación [f] de precios unitarios; *f* **bordereau [m] des prix unitaires** (Liste de prix de main d'œuvre et de matériaux établie sur la base d'une description précise et détaillée) ; *g* **Einzelpreisverzeichnis [n]** (Auflistung von Lohn- und Materialpreisen, z. B. Stundenlohnsätze, Kilometerpauschalen, Auslösungssätze etc. für kleine Bauvorhaben); *syn.* Kleinvertragspreisliste [f] [HH].

5372 schedule [n] **of quantities** [US] *constr. contr.* (Chart showing amounts of materials and equipment required in progressive stages of construction. Some bids [US]/tenders [UK] are issued in which only the unit rates are required for the bid: a list of quantities is therefore provided for the purpose of pricing, to be read in conjunction with drawings, contract conditions and specifications; the s. of q. may be only approximate or probable and thus liable to alteration; bidders [US]/tenderers [UK] are required to verify the quantities; ▶bill of quantities, ▶calculation of quantities); *s* **lista [f] de cantidades** (Tabla que muestra las cantidades de materiales y equipamiento que se necesitan para las diferentes fases de construcción; ▶cálculo de cantidades, ▶listado de materiales); *f* **établissement [m] du quantitatif** (Définition précise des quantités selon la classification par éléments [CFE] permettant avec les valeurs référentielles d'un descriptif d'établir une estimation détaillée dans l'établissement du devis général à un stade peu avancé d'un projet de construction ; le respect de la norme SIA 416 et de la norme CFE lors de l'établissement du quantitatif est essentiel pour la fiabilité du devis général ; ▶avant-métré, ▶établissement d'un métré) ; *g* **Massenaufstellung [f] (2)** (Liste, in der die Mengen der Baumaterialien, Geräte und Maschinen für den fortlaufenden Bauabschnitt aufgeführt sind; ▶Massenaufstellung 1, ▶Massenberechnung).

schedule [n] **of rates** [UK] *constr. contr.* ▶schedule of prices.

schedule [n] **of tender items** [UK] *constr. contr.* ▶list of bid items and quantities [US].

schematic design [n] *plan. prof.* ▶sketch design.

5373 schematic design phase [n] [US] *contr. plan.* (Stage of a designer's services which includes design and cost studies, and other work necessary at the beginning of a project design; ▶determination of planning context, ▶preliminary cost estimate, ▶preliminary design); *syn.* conceptual design phase [n] [US], outline and sketch scheme proposals [npl] [UK] (work stage C&D); *s* **fase [f] de anteproyecto** (Después del ▶inventario preliminar de un proyecto, fase de elaboración de un concepto de planificación con alternativas, incluida la ▶estimación aproximada de costos según DIN 276 para la preparación de proyectos y planes, y cuyo resultado es el ▶anteproyecto); *f* **avant-projet [m]** (Étape des missions opérationnelles ou de création dite de conception primaire qui suit la phase de ▶recueil des données et définit les caractéristiques essentielles des divers ouvrages, sur la base du programme établi par le maître d'ouvrage et permettant à celui-ci d'apprécier s'il est opportun de poursuivre la réalisation ; faisant suite à la phase de reconnaissance et de recueil des données, elle comprend, l'analyse des informations recueillies, l'élaboration d'une solution d'aménagement éventuellement avec variantes, y compris une ▶estimation sommaire du coût des travaux ; cf. HAC 1989 ; ▶avant-projet sommaire [A.P.S.]) ; *g* **Vorplanung [f]** (Planungsabschnitt nach der ▶Grundlagenermittlung, die u. a. die Analyse der Grundlagen und das Erarbeiten eines Planungskonzeptes, evtl. mit Varianten, incl. ▶Kostenschätzung nach DIN 276 resp. nach dem wohnungsrechtlichen Berechnungsrecht zur Projekt- und Planungsvorbereitung beinhaltet; cf. § 15 [1] 2 HOAI 2002 resp. §§ 33, 38, 42 u. 46 HOAI 2009; das Ergebnis ist der ▶Vorentwurf. In der schweizerischen „Ordnung für Leistungen und Honorare der Landschaftsarchitekten" entspricht diese Leistungsphase dem „Vorprojekt" zusammen mit der „Grobschätzung der Baukosten" als 3. und 4. Leistungsabschnitt der „Vorprojektphase"); *syn.* Vorprojekt [n] [± CH].

scheme [n] *constr. urb.* ▶courtyard rehabilitation scheme.

scheme [n] [UK]**, approved** *leg. urb.* ▶approved plan [US].

scheme [n] [UK]**, back court planting** *constr. urb.* ▶courtyard landscaping.

scheme [n] [UK]**, bikeway** *plan. trans.* ▶approved plan [US].

scheme [n] [UK]**, evaluation** *plan.* ▶evaluation method [US].

scheme [n]**, improvement** *constr. plan. urb. wat'man.* ▶rehabilitation plan.

scheme [n] [UK]**, prototypical planting** *constr. plant.* ▶planting prototype [US].

scheme [n] [UK]**, recommended** *prof.* ▶traffic calming.

scheme [n] [UK]**, traffic restrainment** *plan.* ▶evaluation method [US].

scheme designer [n] [UK] *plan. prof.* ▶preliminary project planner [US].

5374 school garden [n] *landsc.* (▶Demonstration garden 2 usually on school premises used as a teaching aid in horticulture and botany lessons); *s* **jardín [m] escolar pedagógico** (Jardín de escuela que se utiliza para la demostración práctica en clases de biología y horticultura; ▶jardín pedagógico); *f* **jardin [m] scolaire** (▶Jardin pédagogique en général implanté sur un site scolaire comme moyen de sensibilisation et d'illustration du cours de botanique) ; *g* **Schulgarten [m]** (▶Lehrgarten, meist auf dem Schulgelände, zur Veranschaulichung des botanischen Unterrichts).

5375 schoolground planting [n] [US]/**school-ground planting** [n] [UK] *landsc. urb.* (Planted open space within the grounds of a school); *syn.* school landscaping [n]; *s* **patio [m] escolar ajardinado**; *f* **espaces [mpl] verts d'école** (Ensemble des espaces végétalisés à l'intérieur d'un ensemble scolaire) ; *g* **Schulgrün [n, o. Pl.]** (Vegetationsbestimmte Freianlage auf einem Schulgelände); *syn.* Schulfreifläche [f].

school landscaping [n] *landsc. urb.* ►schoolground planting [US]/school-ground planting [UK].

5376 school sports fields [npl] *landsc.* *s* **campos [mpl] de deportes escolares**; *f* **terrain [m] de sport scolaire** ; *g* **Schulsportanlage [f]**.

5377 school yard [n] *landsc.* *s* **patio [m] escolar**; *f* **cour [f] de l'école** *syn.* cour [f] de récréation, cour [f] d'école ; *g* **Schulhof [m]** *syn.* Pausenhof [m].

schwingmoor [n] [EIRE] *phyt.* ►floating sphagnum mat.

sciaphyte [n] *bot.* ►shade plant.

science [n] **of forestry** *for.* ►forest science.

scientific purpose area [n] [CDN] *conserv. leg. nat'res.* ►forest research natural area [US]/forest nature reserve [UK].

scientific reserve [n] *conserv.* ►strict nature reserve.

sciophilous [adj] *phyt. plant.* ►shade-loving.

sciophilous plant [n] *bot.* ►shade plant.

5378 scope [n] **of construction work** *constr. contr.* (Type and extent of work to be executed as defined in the ►technical specifications of a contract. General Technical Requirements and ►standards for construction are also part of a contract); *s* **ámbito [m] de trabajo** (Tipo y amplitud de trabajo que ha de realizarse, tal como se define en las ►especificaciones técnicas de contrato. El ►pliego de normas técnicas generales para obras de construcción es también parte de un contrato); *syn.* conjunto [m] de trabajos a realizar; *f* **consistance [f] et étendue [f] des travaux** (►description des ouvrages fixée dans les ►spécifications techniques détaillées et figurant également dans le ►cahier des clauses techniques générales) ; *syn.* consistance [f] et volume [m] des travaux, nature [f] et consistance [f] des travaux, consistance [f] et masse [f] des travaux ; *g* **Art und Umfang der Leistung [f]** (In der ►Leistungsbeschreibung 2 und dadurch auch im Vertrag genau spezifizierte Leistung oder Funktionsanforderung mit allen notwendigen Mengen, Maßen, Gewichten oder Stückzahlen, so dass dem Auftragnehmer ein klares Bild vom Auftragsgegenstand vermittelt werden kann. Als Bestandteil des Vertrages gelten auch die ►Allgemeinen Technischen Vertragsbedingungen für Bauleistungen; cf. § 1 VOB Teil B); *syn.* Art [f] und Umfang [m] der Bauleistung, Leistungsumfang [m].

5379 scope [n] **of maintenance** *constr.* (Extent of maintenance work in terms of planning, management and financial expense required for a particular project); *s* **gasto [m] de mantenimiento** (Cantidad de trabajo necesario para conservar un objeto); *f* **étendue [f] de la gestion (±)** (Importance prise par les études, l'organisation et les dépenses des travaux d'entretien nécessaires à la conservation d'un ouvrage, d'un équipement, etc.) ; *syn.* importance [f] de la gestion, ampleur [f] de la gestion ; *g* **Pflegeaufwand [m, o. Pl.]** (Planerischer, organisatorischer und finanzieller Umfang der Pflegearbeiten, die nötig sind, um den gewünschten Zustand eines Objektes/einer Anlage zu erhalten).

5380 scope [n] **of professional services** *contr. prof.* (Itemized description of work required for, e.g. project planning, preparing ►general plan for urban green spaces, ►landscape plan, etc. The description of work is subdivided into various ►work phases [US]/work stages [UK]); *syn.* list [n] of professional services, extent [n] of planning services; *s* **descripción [f] de servicios de planificación** (Total de las ►fases de trabajo necesarias para realizar un trabajo profesional específico [proyecto de diseño para edificio o parque, ►plan municipal de zonas verdes y espacios libres, ►plan de gestión del paisaje, etc.]. La descripción se divide punto por punto en fases de trabajo); *f* **forme [f] et étendue [f] de la mission** (Ensemble des contenus des missions normalisées dans un domaine fonctionnel [infrastructure, bâtiment, ►plan paysage, ►étude paysagère, etc.] composées d'éléments normalisés ; lorsque le contenu de la mission est moins important que celui de la mission normalisée on parle de mission partielle ou allégée ; ►étape de la mission de maîtrise d'œuvre) ; *g* **Leistungsbild [n]** (Summe der für ein bestimmtes Leistungsziel [Objektplanung für Gebäude und Freianlagen, ►Grünordnungsplan, ►Landschaftsplan etc.] erforderlichen ►Leistungsphasen. Das jeweilige **L.** wird in Leistungsphasen unterteilt; wenn nicht alle Leistungsphasen komplett beauftragt werden, handelt es sich um ein **reduziertes L.**); *syn.* Leistungsumfang [m], Umfang [m] der Planungsleistungen.

5381 scoping [n] *conserv. leg. nat'res. plan.* (The process of identifying the content and extent of the environmental information to be submitted to the competent authority under ►environmental impact assessment [EIA] [US]/environmental assessment [UK] procedures. **S.** is the first step in the Environmental Impact Study and provides the opportunity for the public and other agencies to be involved. Public sessions and discussions are held in the context of an Environmental Impact Assessment [EIA] whereby the parties involved, chiefly the ►approving authority, as well as other agencies and associations, experts and the public, are informed by the instigator or planner of a project on the subject, scope and methods of the EIA and other relevant issues for its implementation. The final scope of the investigation and the provision of the required documents are then determined by the competent authority. As part of the development planning process, **s.** describes the scope of the investigation, its method and the degree of planning detail in relation to the protection of various resources [fauna, flora, soil, water, climate, air, biodiversity, landscape, human beings, culture and property] and possible interactions. The **scoping opinion** [also Scoping Report] outlines which issues should be addressed by the EIA. It contains the Plan of Study for the EIA, specifying the methodology to be used to assess the potential impacts, and the specialists or specialist reports that are required. An applicant may only proceed with the EIA after the Competent Authority has approved these. **S.** culminates in the submission of the Environmental Impact Assessment Report. **In U.K.**, if an authority advises that a given project is 'EIA development', and the developer does not disagree, an EIA will need to be undertaken and an environmental statement produced before a planning application can be made. In order to establish the topics to be covered by the EIA and the information to be included in the environmental statement, the developer requests the provision of a **scoping opinion** from the relevant planning authority as per Regulation 10 of the EIA Regulations); *s* **consulta [f] sobre los contenidos de un estudio de impacto ambiental (≠)** (En el plazo de diez días después de presentada la memoria-resumen de un proyecto a realizar, el órgano administrativo de medio ambiente puede consultar a las personas, instituciones y administraciones previblemente afectadas por la ejecución del proyecto con relación al impacto ambiental que, a juicio de cada una, se derive de aquél … así como sobre cualquier propuesta que estimen conveniente respecto a los contenidos específicos a incluir en el estudio de impacto ambiental; cf. art. 13 RD 1131/1988; el término «scoping» se refiere en particular al procedimiento de análisis y decisión conjuntas del ámbito de estudio sobre el proyecto a

S

realizar por parte de las administraciones, instituciones, organizaciones y personas afectadas); *f* **cadrage [m] préalable** (Conformément à la directive européenne du 3 mars 1997 et transposée dans le droit français par les décrets du 20 mars 2000 et du 1er août 2003 phase de préparation de l'étude d'impact d'un projet consistant à définir un avant-projet, à sélectionner les composantes de l'environnement à étudier [enjeux environnementaux], à définir les thèmes et les méthodes de l'analyse de l'état initial, à identifier les effets potentiels principaux génériques du projet sur l'environnement et définir les méthodes spécifiques de leur évaluation et déterminer la ou les aires d'études à retenir ; les méthodes de cadrage s'adaptent à la nature du projet et à ses effets prévisibles sur l'environnement ; elles vont de la simple consultation du service instructeur à l'étude de cadrage détaillée, présentée dans un dossier [document de cadrage préalable avec cahier des charges de l'étude d'impact] à l'autorité compétente chargée de l'instruction de la demande ; celle-ci vérifie le sérieux et le caractère complet de l'étude d'impact, contrôle son contenu ainsi que les mesures de compensation des effets dommageables du projet. Le sérieux du cadrage peut être renforcé par la participation du public, des élus et des représentants des associations de défense de l'environnement à l'élaboration du cahier des charges de l'étude d'impact ; MIC, 2002) ; *syn.* scoping [m] ; *g* **Scoping [n]** (Gesprächstermin [Scoping-Termin], bei dem im Rahmen der ▶Umweltverträglichkeitsprüfung [UVP] nach § 5 [1] Satz 1 UVPG die zu beteiligende verfahrensführende Behörde sowie andere Behörden und Verbände, Sachverständige und ggf. die Öffentlichkeit vom Träger eines Vorhabens über Gegenstand, Umfang und Metode der UVP sowie sonstige für ihre Durchführung relevanten Fragen informiert werden. Der endgültige Umfang des Untersuchungsraumes und die beizubringenden Unterlagen werden dann von der Planfeststellungsbehörde festgelegt. Im Rahmen der Bauleitplanung beschreibt **S.** die Festlegung des Untersuchungsumfanges, der Untersuchungsmethode und des Detaillierungsgrades der Planung bezogen auf die verschiedenen Schutzgüter [Tiere, Pflanzen, Boden, Wasser, Klima, Luft, biologische Vielfalt, Landschaft, Mensch, Kultur- und Sachgüter] und mögliche Wechselwirkungen).

scoping opinion [n] *conserv. leg. nat'res. plan.* *▶scoping.

5382 scorch [n] *arb. bot.* (Injury to bark, foliage, flowers or fruit from excessive heat—fire or sunlight—, unbalanced nutrition, or from misapplication of a pesticide; cf. SAF 1983; ▶bark scorch, ▶necrosis, ▶sun scorch); *s* **quemadura [f] (del tronco)** (Daños en el tronco, las hojas u otras partes de árboles causados por el fuego, fuerte insolación o por la aplicación incorrecta de abonos o pesticidas; ▶necrosis; *términos específicos* ▶quemadura de la corteza, ▶socarrado); *syn.* escorche [m]; *f* **brûlure [f]** (Décollement de l'écorce, dessèchement des feuilles ou dépérissement d'autres parties d'une plante causée par le feu, un ensoleillement excessif, des mauvaises conditions d'utilisation d'engrais ou de produits phytosanitaires, le salage des routes ; ▶nécrose ; *termes spécifiques* ▶brûlure solaire, ▶insolation d'écorce) ; *g* **Verbrennungsschaden [m]** (...schäden [pl]; Absterben von Rinde, Blättern oder sonstigen Pflanzenteilen durch Feuer, plötzlich starke Sonneneinstrahlung oder durch unsachgemäße Anwendung von Düngesalzen, Schädlingsbekämpfungsmitteln oder durch Folgen von Streusalzeinsätzen im Straßenraum; ▶Nekrose; *UBe* ▶Rindenbrand, ▶Sonnenbrand); *syn.* Brandschaden [m].

5383 scoria [n] *constr. geo.* (Vesicular cindery material extracted from volcanic lava; processed in various particle sizes; e.g. for roof gardening); *syn.* processed lava [n]; *s* **lava [f] escórica** (Variété de roche volcanique poreuse légère) ; *f* **pierre [f] ponce** (Variété de roche volcanique poreuse légère) ; *syn.* ponce [m], scorie [f] volcanique ; *g* **Schaumlava [f]** (In

verschiedenen Körnungen aufbereitete, feinporige bis blasige, vulkanische Auswurfmasse, z. B. Lavalit).

scoria [n], **volcanic** *constr. envir. geo. min.* ▶cinder.

5384 scoring [n] *plan.* (Award of points to objects or its characteristics in an evaluation procedure); *s* **puntuación [f]** (Asignación de puntos a objetos o a características de ellos en el marco de un procedimiento de evaluación); *f 1* **notation [f] sur une échelle** (Processus d'attribution de points à des objets ou des critères dans le cadre d'un procédé d'évaluation) ; *f 2* **pondération [f]** (Attribution d'une valeur particulière à divers éléments pour leur redonner une place proportionnelle à leur valeur réelle) ; *g* **Punktbewertung [f]** (Zuordnung von Punkten zu Objekten oder deren Eigenschaften im Rahmen eines Bewertungsverfahrens).

scoring [n] **of turves** [UK] *constr. hort.* ▶cutting of sods [US].

5385 scour [n] *geo.* (Natural cutting down of riverbed [US]/river-bed [UK] into the land surface; ▶vertical erosion, ▶watercourse bed erosion); *syn.* channel erosion [n]; *s* **socavación [f]** (Proceso natural de erosión de los cauces de los cursos de agua producida por el agua, excavando y arrastrando materiales del lecho y de las márgenes; ▶erosión del lecho, ▶erosión lineal); *f* **enfoncement [m]** (Résultat de l'▶érosion verticale agissant sur les talwegs, créant un modelé de dissection et provoquant la formation d'atterrissements [bancs alluviaux] qui s'exhaussent et se végétalisent pour devenir des petits îlots, voire des îles ; DG l984, 167 ; ▶érosion du lit d'un cours d'eau) ; *syn.* affouillement [m] du lit d'un cours d'eau ; *g* **Eintiefung [f]** (Natürlicher, senkrechter Erosionsvorgang bei Fließgewässersohlen; ▶Sohlenerosion, ▶Tiefenerosion).

5386 scrapyard [n] *envir.* (Any area or part thereof where discarded metal scrap is collected or stored and sorted for reuse; ▶car salvage yard [US]/car scrap yard [UK]); *s* **depósito [m] de chatarra** (Almacén de restos metálicos resultantes del procesamiento de metales que son clasificados y reutilizados; ▶cementerio de coches); *f* **dépôt [m] de ferraille** (Chantier sur lequel sont effectués, à l'air libre ou sous abri, des dépôts et activités en vue de la récupération, valorisation et recyclage de déchets de métaux ferreux et non ferreux ou d'alliages, résidus métalliques, objets usagés en métal, véhicules hors d'usage, etc. ; circ. du 10 avril 1974 ; ▶cimetière de voitures) ; *g* **Schrottplatz [m]** (Lagerplatz für Metallabfälle aller Art, die entweder als Altmaterialien oder als Abfälle bei der Metallverarbeitung anfallen und sortiert einer Wiederverwendung zugeführt werden; ▶Autofriedhof).

5387 scree [n] *geo.* (Mass of detritus forming a precipitous, stony slope on a mountain side, or at the foot of cliffs, and the material composing such a slope; ▶alluvial fan, ▶rock fan, ▶tapered scree); *syn.* talus [n] (DNE 1978); *s* **derrubio [m] de gravedad** (Depósito detrítico localizado en el talud de una vertiente y procedente de la parte alta de ésta, zona en la que se produce una meteorización de la roca coherente tras lo cual los derrubios se desplazan por gravedad; DGA 1986; ▶abanico aluvial, ▶abanico de derrubios, ▶cono de derrubios); *f* **talus [m] d'éboulis** (Masse de débris anguleux couvrant un versant plan en pente et mis en place par gravité ; l'expression « éboulis de gravité » est un pléonasme ; cf. DG 1984, 148 ; ▶cône d'éboulis, ▶glacis d'épandage, ▶nappe d'éboulis) ; *syn.* pierrier [m], *dans les Alpes* clapier [m], casse [f], *en Roussillon* clapisse [f] ; *g* **Schutthalde [f] (2)** (Fächerförmige, unverfestigte Masse von Gesteinsbrocken verschiedenster Größen, die durch mechanische Verwitterung am Fuß von steilen Berghängen und Felspartien entstanden ist; ▶Schuttfächer, ▶Schuttkegel, ▶Schwemmfächer).

scree [n] [UK], **glacial** *geo.* ▶glacial debris.

S

screed [vb] **a laying course** [US] *constr.* ▶level a laying course.

scree formation [n] *phyt.* ▶scree vegetation.

5388 screen [n] **(1)** *envir. landsc.* (Generic term for a barrier against pollution source, wind, visual intrusion, etc.; ▶antiglare screen [US]/anti-dazzle screen [UK], ▶noise screening facility, ▶protective planting, ▶screen planting 2); *s* **apantallado** [m] (Instalación aislante de agentes contaminantes, viento, intrusión visual, etc.; ▶instalación antirruidos, ▶pantalla vegetal, ▶plantación protectora); *syn.* protección [f] con pantalla; *f* **écran** [m] (Toute installation destinée à protéger contre le vent, le bruit ou à dissimuler des regards ; ▶dispositif antibruit, ▶plantation brise-vue, ▶plantation de protection) ; *syn.* rideau [m] ; *g* **Abschirmung** [f] **(1)** (Einrichtung gegen Emissionen, Wind, Einsicht etc.; ▶Schutzpflanzung, ▶Sichtschutzpflanzung, ▶Lärmschutzeinrichtung).

screen [n] **(2)** *envir.* ▶sewage plant screen.

screen [n] [UK], **anti-dazzle** *trans.* ▶antiglare screen [US].

screen [n], **planting** *gard. landsc.* ▶screen planting (2).

screen [n], **shade-** *arch.* ▶sunscreen.

screen [n], **vegetation** *landsc.* ▶screen planting (2), ▶shelterbelt (1).

screen [n], **visual wood** *constr.* *▶screen wall.

screen [n], **wooden** *constr.* ▶wooden fence.

5389 screened material [n] *constr.* (Rock or soil passed through a sieve); *syn.* sieved material [n]; *s* **material** [m] **cribado** (Piedras o tierra pasadas por una criba); *f* **criblure** [f] (Matière passant à travers un crible ou au tamis ; DTB 1985) ; *g* **Siebgut** [n, o. Pl.] **(1)** (Das bei einem Siebvorgang durch ein Sieb geschüttete Gesteins- oder Bodenmaterial).

5390 screen forest [n] *for. landsc.* (Forest planted to conceal objects which disturb the visual quality of the landscape and to protect against undesirable views); *s* **bosque** [m] **pantalla** (≠) (Bosque cuya principal función es la de tapar estructuras no atractivas para el aspecto escénico del paisaje o evitar la intrusión visual); *syn.* bosque [m] de protección visual (≠); *f* **forêt** [f] **brise-vue** (Forêt contribuant à cacher les nuisances visuelles dans un paysage ou à protéger des regards indésirables) ; *syn.* forêt [f] écran contre les vues ; *g* **Sichtschutzwald** [m] (Wald, der Objekte, die das Landschaftsbild stören, verdecken und vor unerwünschten Einblicken schützen soll).

5391 screening [n] **(1)** *envir. landsc.* (Process of providing protection from light, wind, visual intrusion, etc.; ▶protective planting); *s* **apantallar** [vb] (▶plantación protectora); *syn.* proteger [vb] con pantalla; *f* **préservation** [f] (Processus, se protéger contre quelque chose ; ▶plantation de protection) ; *g* **Abschirmung** [f] **(2)** (Vorgang, sich gegen etwas zu schützen; ▶Schutzpflanzung).

5392 screening [n] **(2)** *conserv. leg. nat'res. plan.* (The process by which the decision is taken on whether or not an EIA is required for a particular project. **In U.K.**, the Town & Country Planning [Environmental Impact Assessment (EIA)] Regulations 1999 for England & Wales include two lists of different types of development projects. Schedule 1 identifies all the types of projects for which EIA is mandatory. Schedule 2 identifies the types of scheme which may require an EIA. To establish whether a project will need to undergo an EIA, a developer applies to the relevant planning authority to provide a **screening opinion**, which has to be prepared and adopted within a given period of time); *s* **evaluación** [f] **preliminar** (Fase de preparación de un proyecto en la que se evalúa si éste debe ser sometido a una evaluación de impacto ambiental, teniendo en cuenta las condiciones concretas de la ubicación en la que está previsto que se realice); *f* **évaluation** [f] **préliminaire** (D., Phase de préparation d'une étude d'impact permettant de déterminer la réalisation ou non de l'étude d'impact pour un projet donné et le champs d'études requis, selon les effets prévisibles potentiels sur l'environnement) ; *syn.* tri [m] préliminaire, screening [m] ; *g* **Screening** [n] (Nach § 3 UVP-Gesetz allgemeine oder standortbezogene Vorprüfung, ob für ein einzelnes Vorhaben eine Umweltverträglichkeitsprüfungspflicht besteht; für die Vorprüfung des Einzelfalls sind die Vorschriften des § 3c [1] und des § 3e [1] 2 und [2] UVPG maßgeblich).

5393 screenings [npl] *envir.* (Coarse material separated at a sewage treatment plant by screening and filtering); *s* **residuo** [m] **de cribado** (Material grueso extraído del agua residual en planta de depuración por medio de criba o trampa de arena); *f 1* **résidus** [mpl] **provenant du traitement mécanique** (Résidus retenus lors du traitement préliminaire des eaux usées urbaines dans une station d'épuration) ; *syn.* refus [mpl] provenant du traitement préliminaire ; *f 2* **résidus** [mpl] **de dessablage** (Matériaux grossiers retenus par la chambre de dessablement en amont d'une station d'épuration) ; *f 3* **résidus** [mpl] **de dégrillage** (Matières volumineuses retenues par les grilles d'entrée dans une station d'épuration) ; *g* **Siebgut** [n, o. Pl.] **(2)** (Grobes Material, das in einer Kläranlage in der ersten Stufe der mechanischen Reinigung durch Grob- und Feinrechen [Rechengut] und Siebanlage [Sandfang] aufgefangen wird; cf. MEL 1971, Bd. 1, 190).

screening [n] **with plants** *landsc.* ▶planting treatment.

screen mound [n] [UK], **acoustic** *constr. envir.* ▶noise attenuation mound.

screen mound [n] [UK], **noise** *constr. envir.* ▶noise attenuation mound.

screen planting [n] **(1)** *agr. for. land'man. landsc.* ▶protective planting.

5394 screen planting [n] **(2)** *gard. landsc.* (Planting which serves a ▶visual screening purpose; a specific linear area of **s. p.** to prevent observation is named **concealment planting**; ▶buffer zone, ▶green separation zone, ▶screen forest); *syn.* vegetation screen [n], screen planting belt [n], planting screen [n]; *s* **pantalla** [f] **vegetal** (Banda de leñosas plantada para proteger contra la intrusión visual; ▶bosque pantalla, ▶protección visual, ▶verde separador, ▶zona de amortiguación); *syn.* plantación [f] en pantalla, plantación [f] de protección visual; *f 1* **plantation** [f] **brise-vue** (Écran végétal, bande, ceinture de plantation contribuant à la ▶protection contre les vues ; *terme spécifique* haie brise-vue ; ▶bande végétale de séparation, ▶forêt brise-vue, ▶zone tampon) ; *syn.* écran [m] brise-vue ; *f 2* **palissade** [f] **(2)** (Rangée d'arbres taillés formant une enceinte de verdure [dans les jardins à la française]) ; *g* **Sichtschutzpflanzung** [f] (Pflanzung, die dem ▶Sichtschutz dient; Pflanzungen, die Gebäude oder Anlagen völlig verstecken sollen, werden im Englischen *concealment planting* genannt; ▶Abstand- und Schutzzone, ▶Sichtschutzwald, ▶Trenngrün); *syn.* Abpflanzung [f], Sichtschutzgrün [n, o. Pl.].

screen planting belt [n] *gard. landsc.* ▶screen planting (2).

5395 screen wall [n] *constr.* (Wall or fence constructed to prevent unwanted views into a property; *specific term* visual wood screen, ▶stockade fence); *s* **pantalla** [f] **de protección visual** (Muro o valla construida para proteger una propiedad de la vista desde el exterior; ▶valla de estacas partidas); *f* **mur** [m] **brise-vue** (Mur protégeant des regards indésirables ; ▶clôture en échalas) ; *syn.* mur [m] de protection contre les vues, mur [m] écran contre les vues ; *g* **Sichtschutzwand** [f] (Mauer, die vor unerwünschten Einblicken schützen soll; ▶Sichtschutzholzzaun aus Latten/Staketen); *syn.* Sichtschutzmauer [f].

S

screen wall [n] [UK]**, acoustic** *constr. envir.* ►noise barrier wall [US].

screen [n] **with horizontal boards** *constr.* ►visual screen with horizontal boards.

screen [n] **with interwoven split boards** *constr.* ►visual screen with interwoven split boards.

screen [n] **with vertical boards** *constr.* ►visual screen with vertical boards.

5396 scree vegetation [n] *phyt.* (Plant communities that grow; **1.** on active scree accumulations, which are continually changing with the introduction of new angular rock-debris. **2.** On 'slipping' screes, which are not active, but still unstable. **3.** On 'stable' scree accumulations at the foot of steep rock buttresses); *syn.* scree formation [n]; *s* **comunidad** [f] **de pedregales** (Comunidad vegetal pionera que crece en acumulaciones de detritos no estables); *syn.* comunidad [f] glareícola; *f* **végétation** [f] **des éboulis** (Associations végétales pionnières occupant les éboulis naturels en formation ou stabilisés situés sous les affleurements rocheux naturels ou artificiels [front de carrière] ou bien des déblais artificiels stables ou en voie d'érosion) ; *syn.* végétation [f] de pierrier, flore [f] de pierrier, végétation [f] d'éboulis, flore [f] des terrils ; *g* **Gesteinsschuttflur** [f] (Pflanzengesellschaften auf **aktiven**, d. h. durch Zufuhr von neuem Material veränderten, auf **rutschenden**, d. h. nicht mehr aktiven aber noch instabilen oder auf **ruhenden** Schutthalden unterhalb verwitternder Felspartien; nach ELL 1978); *syn.* Steinschuttflur [f], Schutthaldenflur [f], Steinschutthaldenbewuchs [m], Felsschuttgesellschaft [f], Felsschutt- und Geröllgesellschaft [f], Geröllhaldenflur [f].

screw rod [n] [US] *arb. constr.* ►threaded rod.

scrub [n] *phyt.* ►brushy area [US], ►garrigue, ►hygrophilic scrub, ►hygrophilic scrub, ►macchia, ►oak scrub; *conserv. land'man.* ►spontaneous colonization by scrub.

scrub [n] [AUS]**, Mallee-** *phyt.* ►brushy area [US].

scrub [n]**, wetland** *phyt.* ►hygrophilic scrub.

scrub community [n] *phyt.* ►woodland edge scrub community.

scrubland [n] *phyt.* ►brushy area.

scrubland [n]**, oak** *phyt.* ►oak scrub [US].

5397 scrub savanna(h) [n] *phyt.* (**1.** Stands of shrubs alternating in various patterns with grassland in the tropics and subtropics; cf. ELL 1967; ►brushy area [US]/scrub, ►shrubland savanna[h]. **2.** In North America, the **oak savanna**, a formation which is composed of moderate to dense thickets of oak sprouts within a prairie matrix, with a few fairly dwarfed open-grown trees, is wide-spread along the west coast and through the central States; cf. CSB 1985); *s* **sabana** [f] **arbustiva** (Vegetación compuesta de grupos de arbustos irregularmente repartidos sobre cubierta herbácea en las regiones tropicales y subtropicales; ►matorral, ►sabana de espinos); *syn.* sabana [f] matorral; *f 1* **savane** [f] **arbustive** (1) (Formation végétale des régions tropicales et subtropicales constituée essentiellement de plantes herbacées et de quelques arbustes disséminés ; on distingue aussi la **savane hallier** [ou savane maquis] avec une strate arbustive développée ; ►brousse, ►fourré 1, ►fruticée, ►savane arborée, ►savane arbustive 2, ►savane buissonnante, ►savane herbacée, cf. DG 1984, 407) ; *f 2* **savane** [f] **buissonnante** (Zone de transition entre la savane et la steppe) ; *g* **Buschsavanne** [f] (*Allgemeinsprachlicher Begriff* unregelmäßig verteilte Gebüsche oder Strauchgruppen auf Grasland in den Tropen oder Subtropen sowie in Teilen der USA; ►Dornstrauchsavanne, ►Gebüsche).

5398 scuba diving [n] *recr.* (Sport diving with breathing equipment); *s* **submarinismo** [m] *syn.* buceo [m] deportivo con

oxígeno; *f* **plongée** [f] **sous-marine** (Activité sportive de descente sous la surface de l'eau muni de matériels de respiration) ; *g* **Sporttauchen** [n, o. Pl.] (1) (Tauchen mit Atmungsgerät).

sculpture [vb] **plant material** *hort.* ►trim (2).

5399 scythe [vb] *agr. constr. hort.* (To cut long grass or other vegetation, either with a scythe or weed eater [US]/trimmer [UK]); *syn.* rough cutting [n]; *s* **guadañar** [vb] (Siega de hierbas altas cortadas con guadaña); *f* **faucher** [vb] **avec une débroussailleuse ou à la faux** (Fauchage d'un pré à la faux, avec une faucheuse ou une débroussailleuse) ; *g* **mit Freischneider oder Sense mähen** [vb] (Schneiden von langem Gras).

sea beach [n] *geo.* *►shore.

sea defences [npl] [UK] *conserv. eng.* ►sea defenses [US].

5400 sea defenses [npl] [US] *conserv. eng.* (Generic term for protective structures [dikes, flood water barriers, etc.] which help to protect the coastline; ►bank revetment, ►coastal dike, ►coastal protection; ►sea wall 1); *syn.* sea defences [npl] [UK], coastal protection structure [n]; *s* **construcción** [f] **de protección del litoral** (Término genérico para construcciones cuyo fin es la ►protección del litoral; ►dique costero, ►dique marítimo, ►revestimiento de orillas); *f* **ouvrage** [m] **de protection côtière** (Terme générique désignant les types d'ouvrages longitudinaux en béton ou autres matériaux de ►protection du littoral [digue en terre, mur vertical en ciment, paroi de soutènement en béton armé, perré et gabion, cordon d'enrochement], transversaux [épis] et **brise-lames**, destinés à diminuer l'action des dynamiques marines sur les secteurs littoraux sensibles ; ►digue côtière ; ►ouvrage de protection des plages, ►ouvrage de revêtement de berge, ►protection du littoral) ; *syn.* construction [f] de protection côtière, ouvrage [m] de défense côtière ; *g* **Küstenschutzbauwerk** [n] (**1.** OB für ein Bauwerk aus Boden, bewehrtem Beton oder anderen Materialien zur Verteidigung der Küstenlinie, wie z. B. Deich, ►Strandmauer, Buhne, Lahnung, ►Uferdeckwerk, Sturmflutsperrwerk, Wellenbrecher in Form von Tetrapoden; ►Küstendeich. **2.** Eine Form des **K.es** ist der **Wellenbrecher**, der die Wucht der Wellen bricht und so die Beschädigung der Strände und Dünen sowie die Zerstörung der Boote in Häfen [weitgehend] verhindert. **Offshore-Wellenbrecher**, die in flacheren Gewässern in einer Reihe angelegt sind und an einer Seite auch mit dem Festland verbunden sein können, dienen beim ►Küstenschutz dem Schutz vor Abtragung des Strandes und dienen der Anlandung von Sand bis zu den Wellenbrechern); *syn.* Küstenschutzwerk [n], Küstenschutzbauten [mpl].

sealant [n] *arb. constr.* ►wound dressing.

5401 sealing [n] *envir.* (Tight surface covering of a refuse dump [US]/refuse tip [UK] with a layer of clayey soil: capping layer [UK]); *syn.* capping [n] [also UK]; *s* **sellado** [m] **de depósito de residuos** (Recubrimiento de vertedero con tierra arcillosa); *syn.* revestimiento [m] de depósito de residuos, revestimiento [m] de vertedero; *f* **couverture** [f] **finale d'une décharge** (Recouvrement après reprofilage final des parties comblées à la fin de l'exploitation d'un dépôt d'ordures ; la couverture est constituée **1.** d'une **couche de fermeture** constituée de limon argileux ou de terre argileuse, destinée à assurer l'isolement du dépôt et de limiter ainsi les infiltrations d'eau dans les déchets et à l'intérieur de l'installation de stockage ainsi que **2.** de la **couche de finition** constituée de terre végétale non compactée, de bonne qualité agronomique pour la revégétalisation du site) ; *g* **Deponieoberflächenabdichtung** [f] (Abdichtung mit mineralischen Materialien wie Tone, Mergel oder Schluffe, die den Deponiekörper nach oben abschließen, um Sickerwasserbildung zu minimieren und den unkontrollierten Austritt von Gasen zu verhindern; im deponierten Abfall laufen biologische, physikalische und chemische Prozesse ab, weshalb ein Deponiebetreiber den

Schrumpfungsprozess und damit einhergehende Rissbildungen in der Abdichtung beobachten muss, um am Ende der Setzungen ein dauerhaftes Dichtungssystem einzurichten; Einzelheiten sind in der 1993 erlassenen Verwaltungsvorschrift „Technische Anleitung Siedlungsabfall [TASi]“ unter Ziffer 10.4.1.4 erläutert); *syn.* Oberflächenabdichtung [f] einer Mülldeponie, Abdeckung [f] der Oberfläche einer Mülldeponie mit Boden.

5402 sealing [n] of soil surface *envir. plan.* (Human actions such as insulation, compaction, filling, building, or surfacing which make the ground impervious [to the atmosphere or ►hydrosphere]; *opp.* ►de-sealing; **s sellado [m] del suelo** (Aislamiento antropógeno del suelo de la atmósfera o ►hidrosfera por medio de material impermeable, compactación o relleno con la consecuencia de la pérdida de la capacidad de desarrollo y de la productividad del suelo; *opp.* ►desimpermeabilización del suelo); *syn.* impermeabilización [f] del suelo); *f* **imperméabilisation [f] du sol** (Résultat ou opération par laquelle le sol est isolé en continu et qui interdit la percolation naturelle des eaux de pluie, consécutif à des travaux de compaction et de tassement, aux activités du bâtiment, à la mise en œuvre de revêtements imperméables à l'air et à l'eau et qui imposent de recueillir ces Eaux pluviales (EP) et de les évacuer dans des réseaux collectifs ; *opp.* ►repérméabilisation du sol) ; *syn.* minéralisation [f] ; *g* **Bodenversiegelung [f]** (*seltener* Versieglung; *Vorgang und Ergebnis* anthropogene Isolierung der Bodenoberfläche von der Atmosphäre resp. ►Hydrosphäre durch Aufschüttungen oder Auffüllungen, Bebauung sowie durch Abdichtung oder Verdichtung des Bodens mit wasser- und luftundurchlässigen Materialien; *opp.* ►Entsiegelung des Bodens); *syn.* Versiegelung [f] des Bodens, *o. V.* Versieglung [f] des Bodens.

sea-marsh [n] *geo. phyt.* ►salt marsh.

5403 seam biotope [n] *phyt. zool.* (Narrow, linear ►habitat 2 between two or more habitats; e.g. between forest and meadow; ►ecotone 1, ►edge effect); *syn.* herbaceous fringe biotope [n], saum biotope [n] [also UK]; **s biótopo [m] de la orla herbácea** (►Biótopo de pequeño tamaño situado en la zona de transición entre dos o más biótopos diferentes, como bosque, prado, etc.; ►ecotono, ►efecto de borde); *f* **biotope [m] d'ourlet herbacé** (►Biotope linéaire de transition, juxtaposé à deux ou plusieurs communautés floristiques, p. ex. entre la forêt et la prairie ; ►écotone, ►effet de lisière) ; *g* **Saumbiotop [m,** *auch* **n]** (►Biotop schmaler Ausdehnung im Grenzbereich zweier oder mehrerer Lebensräume, z. B. zwischen Wald und Wiese; ►Ökoton, ►Randeffekt).

5404 seam community [n] *phyt.* (Herbaceous plant community on the edge of forests, coppices or hedges; ►herbaceous edge, ►woodland edge scrub community, ►forest seam formation); *syn.* herbaceous fringe community [n]; **s comunidad [f] de la orla herbácea** (Comunidad herbácea que crece en los bordes de bosques, matorrales o setos; ►comunidad arbustiva del lindero del bosque, ►comunidad herbácea del lindero del bosque, ►orla herbácea); *f* **peuplement [m] d'ourlet herbacé** (Formation végétale herbacée ou sous-frutescente voisine des groupements ligneux en bordure de forêts ou de haies vives rattachées à l'ordre des *Origanetalia* ; ►groupement de manteau préforestier buissonnant, ►groupement d'ourlet préforestier à mégaphorbes, ►ourlet herbacé) ; *syn.* association [f] d'ourlet (herbacé), groupement [m] d'ourlet (herbacé) ; *g* **Saumgesellschaft [f]** (Krautgesellschaft an Waldrändern, Gebüschen oder Hecken; ►Krautsaum, ►Mantelgesellschaft, ►Waldsaumflur).

seam formation [n] *phyt.* ►forest seam formation.

seashore [n] *geo.* ►shore.

seaside [n] *geo.* ►shore.

5405 seaside resort [n] *recr.* **s lugar [m] de veraneo en la costa** *syn.* estación [f] turística en la costa; *f* **station [f] balnéaire** (Ville ou commune en bord de mer, dont le tourisme littoral est la principale activité économique, spécialement équipée et aménagée [transport, logements, animation, aménagement des plages, qualité de l'environnement] pour l'accueil des touristes) ; *syn.* station [f] littorale ; *g* **Seebad [n]** (Badeort an der Küste).

season [n] [UK], autumnal flowering *hort. plant.* ►fall-flowering period [US].

5406 seasonal aspects [npl] *phyt. plant.* (Changing visual characteristics of a floral habitat, resulting from the apparent seasonal variations in species composition, caused by climatic or edaphic factors and the genetic make-up of the plants themselves. There are seven distinct aspects: prevernal, vernal, preaestival, aestival, serotinal, autumnal, hibernal aspect; for the enhanced enjoyment of a garden, planting design takes **s.a.** into consideration and selects plant species, which create a sequence of a strong visual impact in each of the seasons in which they flower; ►flowering aspect, ►periodicity, ►seasonally changing visual dominance); *syn.* seasonal sequences [npl]; **s sucesión [f] de aspectos estacionales** (Cambios de las características visuales de un hábitat, resultado de los cambios estacionales en la composición de especies causados por los factores climáticos, edáficos y la fenología de las plantas mismas. Se diferencian aspectos prevernal, vernal, estival, serotinal, otoñal [o autumnal], invernal [o hibernal, hiemal]; ►aspecto de floración, ►dominancia visual estacional, ►periodicidad); *syn.* sucesión [f] de aspección, sucesión [f] estacional; *f* **aspects [mpl] saisonniers** (Changement saisonnier de la physionomie d'un ensemble végétal sur un espace homogène correspondant aux étapes successives de la phénologie des individus de cet ensemble spécifique. On distingue en général six aspects successifs : prévernal, vernal, estival, postestival, automnal et hivernal ; ►aspect floral saisonnier, ►composition florale saisonnier, ►périodicité, ►variation saisonnière de la dominance dans la composition florale) ; *syn.* succession [f] des aspects floraux, cycle [m] phénologique floral ; *g* **Aspektfolge [f]** (**1.** Jahreszeitliche, durch Klima- und Standortfaktoren oder durch genetische Veranlagung der Pflanzen bedingte kurzfristige wirkungsvolle Abfolge eines unterschiedlichen Erscheinungsbildes eines Lebensraumes, hervorgerufen durch sich auffällig abwechselnde, visuell dominante Zusammensetzungen gewisser Arten, wahrnehmbar u. a. in großflächigen Wiesengesellschaften mit sich ablösenden „Farbwellen“ [= Blühaspekte], z. B. „Wiesen-Schaumkraut-Aspekt“: weiß, „Löwenzahn-Aspekt“: gelb und folgende]. Es können sieben Hauptaspekte unterschieden werden: Vorfrühlingsaspekt, Frühlingsaspekt, Frühsommeraspekt, Hoch- und Spätsommer-/Frühherbstaspekt, Herbstaspekt und Winteraspekt. **2.** Die für die Erlebnisqualität einer gärtnerischen Pflanzung geplante zeitliche Aufeinanderfolge von ►Blühaspekten kurzfristig visuell dominierender Pflanzenarten oder -sorten; ►Periodizität, ►saisonal wechselnde visuelle Dominanz); *syn.* jahreszeitliche Aspekte [mpl].

5407 seasonal bedding [n] *hort.* (Annual/seasonal rotations of colo[u]rful flowering plants, which provide seasonal plant displays; *specific terms* spring bedding, summer bedding; ►seasonal flower plants, ►summer plant display); *syn.* seasonal bedding display [n]; **s plantación [f] estacional** (Plantación de plantas de floración colorida, anuales o bianuales, que se van alternando año tras año en primavera, verano, otoño y —según el clima del lugar— también en invierno; *término específico* ►plantación de flores de verano; ►flores de estación); *f* **plantation [f] saisonnière** (Plantation de plantes annuelles ou bisannuelles effectuée chaque année au printemps, en été et parfois l'automne participant à l'ornement saisonnier des parcs et jardins ; *termes*

spécifiques plantation printanière, ►plantation florale estivale, ►plantes saisonnières) ; *g* **Wechselbepflanzung [f]** (Pflanzung aus vorwiegend ein- oder zweijährigen Pflanzen, die jedes Jahr für das Frühjahr, den Sommer und ggf. für den Herbst neu durchgeführt werden; *UBe* Frühjahrsbepflanzung, ►Sommerblumenbepflanzung; ►Wechselflor); *syn.* Wechselblumenpflanzung [f].

5408 seasonal climatic conditions [npl] *met.* (Common attributes of a season which influence plants, animals, and humans in various ways); *s* **régimen [m] estacional**; *f* **régime [m] (climatique) saisonnier** (Distribution saisonnière d'un ou plusieurs éléments météorologiques à un endroit donné influençant le comportement des végétaux, des animaux et de l'homme) ; *g* **jahreszeitliche Klimaverhältnisse [npl]** (Charakteristische Eigenschaften einer Jahreszeit, auf die Pflanzen, Tiere und Mensch entsprechend reagieren).

5409 seasonal flower bed [n] *hort.* (Planting bed characterized by annual bedding plants; ►seasonal bedding); *s* **macizo [m] de flores estacionales** (Arriate que se planta según las estaciones; ►plantación estacional); *f* **plate-bande [f] (florale) saisonnière** (Plate-bande caractérisée par une ►plantation saisonnière) ; *syn.* plate-bande [f] de saison ; *g* **Wechselbeet [n]** (Durch ►Wechselbepflanzung gekennzeichnetes Blumenbeet).

5410 seasonal flowering plants [npl] *hort.* (Annuals or biennials which are planted every year for floral displays in spring, summer, and sometimes in autumn; ►bedding plant, ►seasonal bedding); *s* **flores [fpl] de estación** (Plantas anuales o bianuales que se plantan en plantaciones estacionales en la estación en que florecen; ►plantación estacional, ►planta de macizo); *f* **plantes [fpl] saisonnières** (Plantes annuelles ou bisannuelles attrayantes, utilisées chaque année au printemps, en été et à l'automne pour le fleurissement des plates bandes et des massifs de saison ; p. ex. une annuelle, le souci *[Calendula officinalis]*, ou une vivace, la grande capucine, appelée encore le cresson du Pérou *[Tropaeolum majus]* ou encore des arbustes, les pélargoniums d'Afrique du Sud *[Pelargonium]*) ; ►plantation saisonnière, ►plante à massifs) ; *syn.* fleurs [fpl] de saison ; *g* **Wechselflor [m]** (Attraktive ein- oder zweijährige, nicht frostharte ►Beetpflanzen, die jedes Jahr für das Frühjahr, den Sommer und ggf. für den Herbst neu gepflanzt werden; hierzu gehören z. B. die in ihrer Heimat echte einjährige mediterrane Ringelblume *[Calendula officinalis]*, Stauden — die peruanische Kapuzinerkresse *[Tropaeolum majus]* oder Gehölze wie das afrikanische *Pelargonium*; ►Wechselbepflanzung).

seasonally attractive planting composition [n] *phyt. plant.* ►seasonally changing visual dominance.

5411 seasonally changing visual dominance [n] *phyt. plant.* (**1.** Seasonal appearance of an area of plants being admired within a defined space, especially striking, if the vegetation has special characteristics [flower, foliage, autumn color, shapes and colors of fruit and seeds] and is characterized by a visually dominant plant species [aspect creators], e.g. the spring aspect of ground cover vegetation in beech wood communities, **2. s. c. v. d.** and the selection of a corresponding seasonal species composition is an important design tool in creating attractive planting; ►seasonal aspects); *syn.* aspect [n], seasonally attractive planting composition [n]; *s* **dominancia [f] visual estacional** (≠) (**1.** Aspecto estacional llamativo de un área de plantas en un espacio definido de observación, sobre todo cuando el aspecto de la vegetación tiene características llamativas [flor, follaje, colorido otoñal, formas y colores de frutos y semillas] debido a la correspondiente especie de flora visualmente dominante [aspecto dominante], p. ej. estrato herbáceo en comunidades de hayas en primavera. **2.** La **d. v. e.** es un instrumento de diseño muy impor-

tante para crear plantaciones atractivas utilizando la correspondiente **composición estacional de especies**; ►sucesión de aspectos estacionales); *f 1* **variation [f] saisonnière de la dominance dans la composition florale** (Méthode de composition des végétaux très souvent utilisée pour établir contrastes, rythmes et harmonies dans la conception de plantations) ; *syn.* succession [f] saisonnière de la dominance dans la composition florale ; *f 2* **composition [f] florale saisonnière** (Liste des espèces ayant une floraison simultanée et limitée à une certaine période et dont dépendent les ►aspects saisonniers ; l'ensemble de ces plantes constitue la synusie) ; *g 1* **saisonal wechselnde visuelle Dominanz [f]** (**1.** Jahreszeitlich bedingtes Aussehen eines Pflanzenbestandes in einem definierten Betrachtungsraum, insbesondere wenn das Vegetationsbild durch auffällige Merkmale [Blüte, Laubwerk, Herbstfärbung, Formen und Farben von Fruchtständen] jeweils visuell dominierender Pflanzenarten [►Aspektbildner] charakteristisch geprägt wird, z. B. Frühlingsaspekt der Bodenvegetation in Buchenwaldgesellschaften. **2.** Die **s. w. v. D.** durch eine entsprechende **saisonale Artenzusammensetzung** ist in der Pflanzenverwendung ein wichtiges Gestaltungsmittel, um einprägsame Pflanzungen zu schaffen; ►Aspektfolge); *syn.* Aspekt [m], saisonal visuell wirkende Artenzusammensetzung [f]; *g 2* **Aspektbildner [m]** (Pflanzenart, die — bedingt durch einen hohen ►Deckungsgrad oder hohen Mengenanteil und auffällige Merkmale, z. B. Blatt- und Wuchsformen, Farbigkeit der Blätter und Blüten, Fruchtstände — einen Pflanzenbestand in einem betrachteten Raum und Zeitabschnitt [Aspekt] visuell dominiert; **A.** eignen sich in der Pflanzenverwendung für wiesenartige Pflanzungen [Aussaaten, Mischpflanzungen] oder als Frühjahrsaspekt, z. B. in Gruppenpflanzungen. Bei Staudenpflanzungen ist die Anordnung der **A.** im Vergleich zu Leitstauden eher zufällig oder verstreut, werden die Leitstauden jedoch flächendeckend verdichtet, führt dies zur Aspektbildung).

5412 seasonally-flooded soil [n] *pedol.* (Semi-terrestrial soil characterized by regular flooding; e.g. aqualf [US], aquent [US], aquol [US]; ►alluvial soil); *s* **suelo [m] hidromorfo regularmente inundado** (Suelo semiterrestre caracterizado por estar inundado en ciertas épocas del año; ►suelo aluvial); *f* **sol [m] alluvial (2)** (Classification française des sols ; sol peu évolué d'apport, occupant le lit majeur très souvent inondé des rivières et caractérisé par la présence d'une nappe phréatique soumise à de fortes oscillations saisonnières ; PED 1983, 204 ; ►sol alluvial 1 ; *g* **Überflutungsboden [m]** (Durch regelmäßige Überschwemmung geprägter semiterrestrischer Boden; ►Auenboden); *syn.* Überschwemmungsboden [m].

seasonal sequences [npl] *phyt. plant.* ►seasonal aspects.

5413 seasoning check [n] *constr.* (Small split or check, which occurs in the grain of wood when moisture is extracted too rapidly, or from weathering; ►seasoning of wood, ►shrinkage, ►shrinkage crack); *s* **fenda [f] de merma** (Raja en la madera paralela al hilo que se produce principalmente por desecación; cf. DACO 1988; ►grieta de contracción, ►merma, ►merma de la madera); *f* **fente [f] de retrait (2)** (Fissure apparaissant lors du séchage rapide du bois ; ►fissure de retrait [de dessiccation] dans le béton, ►retrait, ►retrait du bois) ; *syn.* fissuration [f] ; *g* **Schwindriss [m] in Holz** (Durch einen schnellen Trocknungsprozess verursachter Schrumpfungsriss in Holz; ►Schwinden, ►Schwinden von Holz, ►Schwindriss in Beton); *syn.* Schwundriss [m].

5414 seasoning [n] **of wood** *constr.* (Reduction in volume due to drying out, either by air or in a kiln); *s* **merma [f] de la madera** *syn.* contracción [f] (de la madera); *f* **retrait [m] du bois** (Désigne la rétractibilité du bois du fait de la diminution du degré d'humidité au séchage ; le retrait apparaît quand l'eau d'imprégnation du bois s'échappe par les ponctuations et lorsque les

parois des cellules s'amincissent ; il se manifeste par une fissuration ou une déformation se produisant principalement dans la largeur et dans l'épaisseur d'un élément, suivant les plans axial et tangentiel ; elles sont pratiquement nulles dans le sens de la longueur ; le phénomène inverse s'appelle le gonflement) ; *syn.* rétraction [f] du bois, contraction [f] du bois ; *g* **Schwinden [n, o. Pl.] von Holz** (Längenveränderung in verschiedene Richtungen, Verkleinerung des Rauminhalts oder Verformung [„Arbeiten" des Holzes], die beim [Aus]trocknen in Abhängigkeit der Holzart [Splintholz schwindet mehr als Kernholz, Laubholz mehr als Nadelholz], Dichte und Wuchseigenschaften [z. B. gerade gewachsen oder Drehwuchs] entsteht; sie findet i. d. R. in den drei anatomischen Richtungen tangential, radial und longitudinal statt; schwinden [vb]); *syn.* Austrocknung [f] von Holz.

seat [n] *gard. landsc.* ►bench 1, ►planter seat, ►wall seat.

seat [n]**, free-standing** *gard. landsc.* ►free-standing bench/seat.

seat [n]**, tree** *gard. landsc.* ►tree bench.

5415 seat bollard [n] *gard. landsc.* (Pillar-shaped post, which can be used to sit upon); *s* **mojón [m] de asiento** *syn.* bolardo [m] de asiento; *f* **borne [f] siège** (Borne de forme tronconique ou cylindre utilisée pour s'asseoir) ; *g* **Sitzpoller [m]** (Säulenförmiger Stein zum Sitzen).

seat step [n] *gard. landsc.* ►sitting step.

seat steps [npl] *gard. landsc.* ►sitting steps.

5416 seatwall [n] *constr. gard. landsc.* (Low retaining wall suitable for sitting); *syn.* sitting wall [n]; *s* **murete [m] asiento** (Muro de contención bajo que se puede utilizar también para sentarse); *f* **muret [m] siège** *syn.* muret [m] banquette ; *g* **Sitzmauer [f]**.

5417 sea wall [n] (1) *constr. wat'man.* (**1.** Protective wall or embankment of stone, reinforced concrete, or other material along a shore to prevent wave erosion and reduce flooding; DOG 1984; ►coastal dike. **2.** *geo.* Long steep-faced embankment of shingle or boulders, built by powerful storm waves along a seacoast at the high-water mark—higher than a ►storm beach [US]/storm-beach [UK]; cf. DOG 1984; ►sea defenses [US]/sea defences [UK]); *s* **dique [m] marítimo** (**1.** Muro protector o terraplén de rocas, hormigón reforzado u otros materiales construido a lo largo de la costa para prevenir la erosión causada por las olas y reducir inundaciones; ►construcción de protección del litoral, ►dique costero. **2.** *geo.* En EE.UU. el término *sea wall* se utiliza también para un ►cordón playero alto y empinado creado por fuertes olas de temporal a lo largo de la costa); *f* **ouvrage [m] de protection des plages** (**1.** Terme générique désignant les types d'ouvrages longitudinaux [digues, enrochement, brise lame, épis, blocs de béton] destinés à absorber l'énergie des vagues et ainsi stabiliser la ligne de plage sur les secteurs littoraux sensibles ; ►digue côtière). **2.** *geo.* US Le terme *sea wall* signifie aussi une ►crête de plage haute et de forte pente ; ►ouvrage de protection côtière) ; *g* **Strandmauer [f]** (**1.** ►Küstenschutzbauwerk als Stützmauer oder wandartiges Bauwerk zur Aufnahme der Wellenkräfte, meist mit konkav vorragendem Oberteil zum Abweisen der Wellen und zur Reduzierung des Wellenüberlaufes; ►Küstendeich. **2.** *geo.* Im US-Amerikanischen bedeutet *sea wall* außer Strandmauer auch ein hoher und steiler ►Strandwall); *syn.* senkrechtes Wellenschutzbauwerk [n], Ufermauer [f].

sea wall [n] [US] (2) *geo.* ►storm beach [US]/storm-beach [UK].

secondary consumer [n] *ecol.* *►consumer.

secondary forest [n] [UK] *phyt. for.* ►second-growth (forest) [US].

5418 secondary production [n] *ecol.* (Total quantity or volume of animal ►biomass incorporated by heterotrophic organisms within a certain period of time and a particular area; ►production 1, ►primary production); *s* **producción [f] secundaria** (Cantidad total de zoomasa creada por organismos heterótrofos en una zona y un tiempo definidos a partir de la sustancia orgánica vegetal o animal por transformación en sustancia corporal propia; ►biomasa, ►producción, ►producción primaria bruta, ►producción primaria neta); *f* **productivité [f] secondaire** (►Biomasse produite par les consommateurs ou les décomposeurs, détritivores ou parasites à partir de l'énergie produite par la biomasse végétale, dans un temps donné et un espace défini ; cf. DEE 1982, 74) ; ►production, ►productivité primaire) ; *g* **Sekundärproduktion [f]** (Gesamtmenge der innerhalb eines definierten Zeitraumes und innerhalb eines definierten Gebietes resp. Volumens von heterotrophen Organismen aus überwiegend organischen Stoffen durch Umformung in körpereigene Substanz gebildeten tierischen ►Biomasse; ►Produktion, ►Primärproduktion).

secondary structure [n] *leg. urb.* ►accessory structure.

5419 second-growth (forest) [n] [US] *phyt. for.* (Natural forest growth after drastic interference in the growth of ►virgin forest, e.g. after ►burn clearing, ►clear felling); *syn.* secondary forest [n] [UK]; *s* **bosque [m] secundario** (Crecimiento natural del bosque después de la intervención humana en el ►bosque primario ; ►tala a fuego, ►tal a matarrasa); *f* **forêt [f] de substitution** (**1.** Peuplement favorisé par l'homme ou ayant conquis des espaces libres provenant de l'érosion ou du défrichement du fait de sa croissance plus rapide et de ses moindres exigences ; VCA 1985, 85. **2.** Peuplement forestier qui s'établit naturellement après quelques interventions importantes sur le peuplement forestier existant antérieurement, p. ex. ►brûlis de défrichement, ►coupe à blanc ; terme souvent utilisé pour caractériser la forêt tropicale et subtropicale pour lesquelles le nouveau peuplement est considérablement différent en composition et en caractère de celui qu'il remplace ; cf. DFM 1975 ; ►forêt vierge) ; *syn.* forêt [f] de seconde venue, forêt [f] secondaire ; *g* **Sekundärwald [m]** (Natürlicher Folgebestand von Bäumen eines vom Menschen beseitigten ►Urwaldes, z. B. durch ►Brandrodung oder ►Kahlschlag).

Secretary of State Approval [n] **(of a plan)** [UK] *adm. leg.* ►design and location approval [US] (2).

5420 section [n] *constr. plan.* (Sectional view of a construction drawing for a site or structure; *specific terms* ►cross section 1, ►longitudinal section); *s* **sección [f]** (*Términos específicos* ►sección longitudinal, ►sección transversal) ; *f* **coupe [f]** (Projection orthogonale sur un plan vertical d'un ouvrage ou d'une partie d'ouvrage coupée par un plan vertical continu ou brisé ; norme P 02-001 ; ►coupe en travers, ►coupe longitudinale) ; *g* **Schnitt [m]** (Zeichnung, die eine Schnittebene zeigt; *UBe* ►Längsschnitt, ►Querschnitt).

sectional completion certificate [n] *contr.* ►issue of a sectional completion certificate.

5421 section of a community area [n] [US] *adm. agr. urb.* (Portion of land within a community; ►agricultural land parcels, ►local community area); *syn.* local district [n] [UK], parish [n] [UK]; *s* **finage [m]** (2) (Territorio sobre el que una comunidad de campesinos se instala para roturarlo y cultivarlo y sobre el que ejerce unos derechos agrarios. El término no se ajusta demasiado a la realidad, ya que una comunidad de campesinos puede poseer propiedades en territorios discontinuos. El concepto de **f.** es válido en pueblos de nueva planta [Plan Badajoz], en el caso de Israel o en ciertas regiones africanas; DGA 1986; ►esquema de parcelación de tierras agrícolas, ►finage 1,

S

▶territorio municipal); *f* **finage [m]** (Au sens strict du terme, territoire administratif et juridique sur lequel la communauté agricole exerce ses droits ; elle comprend en général les champs, les prés, les landes et les bois ; chez les géographes ce terme est passé dans l'usage courant pour désigner des parcelles cultivées par une cellule agricole même si elles ne correspondent pas à une entité administrative ; ▶parcellaire, ▶territoire communal, ▶terroir) ; *g* **Gemarkung [f] (2)** (**1.** Teil der Erdoberfläche, in dem mehrere Fluren zusammengefasst sind. **2.** Begriff im deutschen Südwesten für das Gesamtgebiet einer Gemeinde oder eines Teiles, wobei mehrere ▶Fluren 1 und Gewanne zusammengefasst sein können; ▶Gemeindegebiet); *syn. ugs.* Markung [f].

sectoral landscape planning [n] [UK] *landsc. leg. plan.* ▶functional landscape planning [US]/sectoral landscape planning [UK].

sectoral plan [n] [UK] *plan.* ▶functional plan [US].

sectoral planning [n] [UK] *plan.* ▶functional planning [US].

5422 securing sods [npl] [US] *constr.* (Pinning by wooden pegs or wire pins); *syn.* securing turves [npl] [UK]; *s* **fijación [f] de tepes con clavijas**; *f* **chevillage [m] de gazon précultivé** (Fixation de plaques ou de rouleaux de gazon au moyen d'un long piquet métallique) ; *g* **Sicherung [f] von Fertigrasen** (Befestigen der Rasensoden durch Annageln; DIN 18 917, 5.3.4; Fertigrasen sichern [vb]); *syn.* Annageln [n, o. Pl.] von Rasensoden.

securing turves [npl] [UK] *constr.* ▶securing sods [US].

security [n] *leg.* ▶public security, ▶oil tanker security.

security [n] [US], **bid** *contr.* ▶surety bond [US].

security anchorage point [n] *constr. leg.* ▶protection from falling, #2.

sedentary bird [n] *zool.* ▶permanent resident.

sedge fen [n] *phyt.* ▶sedge swamp.

5423 sedge peat [n] [US] *constr.* (Term used in landscape industry for horticultural peat, extracted from ▶low bogs); *syn.* fen peat [n] [UK]; *s* **turba [f] para jardines** (Tipo de turba minerótrofa empleada en construcción paisajística; ▶turbera baja); *f* **tourbe [f] brune** (Dénomination pour la tourbe utilisée comme amendement dans les travaux d'aménagement de jardins, d'espaces verts et de terrains de sport ; D., la norme DIN 11 542 utilise comme critère de qualité les dénominations de « tourbe fortement décomposée » en provenance de ▶tourbières basses et de transition et « tourbe faiblement décomposée en provenance de tourbières basses et de transition ») ; *g* **Niedermoortorf [m] (2)** (Im Garten-, Landschafts- und Sportplatzbau werden gem. DIN 11 542 als Qualitätskriterien nur noch die Bezeichnungen **stark zersetzter Niedermoor- und Übergangsmoortorf** und **wenig zersetzter Niedermoor- und Übergangsmoortorf** verwendet; ▶Niedermoor).

5424 sedge swamp [n] *phyt.* (Plant communities on periodically inundated margins of standing water bodies or ▶low bogs, primarily covered by sedges; ▶low sedge swamp, ▶saw sedge swamp, ▶tall sedge swamp); *syn.* sedge fen [n] (ELL 1988, 289); *s* **formación [f] de cárices** (Término genérico para comunidades vegetales en zonas de colmatación de aguas estancadas regularmente inundadas o en ▶turberas bajas, compuestas principalmente por especies de la familia de los cárices *[Carex]*; *términos específicos* ▶ciénaga de grandes cárices, ▶formación de pequeños cárices, ▶marisma de ciperáceas); *f 1* **cariçaie [f]** (Peuplements des zones d'atterrissement des eaux stagnantes [base immergée la plus grande partie de l'année] ou sur sols tourbeux [▶tourbière basse] temporairement inondés et principalement occupés par les Laiches/Laîches *[Carex]* ; *termes spécifiques*

▶cariçaie à petites Laiches/Laîches, ▶magnocariçaie, ▶cladiaie) ; *f 2* **prairie [f] à Laiches/Laîches** (station ; ▶prairie à petites Laiches/Laîches) ; *syn.* marais [m] à Carex ; *g* **Seggenried [n]** (Pflanzengesellschaft[en] auf jahreszeitlich überfluteten Verlandungszonen an Stillgewässern oder auf ▶Niedermooren, die hauptsächlich von Seggenarten bewachsen sind; *UBe* ▶Groß seggenried,) ▶Schwertriedröhrricht, ▶Kleinseggenried).

5425 sediment [n] (1) *geo.* (Solid particles that are carried by the action of water, ice or wind and have been deposited on the surface of the land); *syn.* superficial deposits [npl], drift [n], alluvial deposit [n], surface deposit [n]; *s* **sedimento [m]** (*Estado* conjunto de partículas de roca que han sido depositadas por efecto del agua, hielo o viento); *syn.* aluvión [m], llanura [f] aluvial; *f* **sédiment [m]** (Dépôts provenant de l'altération superficielle des roches et due à l'action marine, fluviale, glaciaire, ou éolienne ; p. ex. sédiments fluviaux, sédiments marins, sédiments fluvio-lacustres, sédiments fluvio-lagunaires) ; *syn.* dépôt [m] sédimentaire, dépôt [m] de sédiments ; *g* **Ablagerung [f] (2)** (*Zustand* durch Meer, Fluss, Eis oder Wind umgelagerte Verwitterungsprodukte); *syn. bei Wassertransport* Anschwemmungsboden [m], Anlandung [f] (1), Sediment [n].

5426 sediment [n] (2) *geo. ocean. pedol.* (Deposit of fine-particled soil in alluvial plains, lakes, the sea; ▶gyttja, ▶harbor mud [US]/harbour mud [UK], ▶mud, ▶sapropel, ▶tidal mudflat); *syn. for the sea* tidal mud [n]; *s* **légamo [m]** (Lodo pegajoso de los estuarios, ríos, esteros y otras masas de agua; DGA 1986; ▶gyttja, ▶légamo portuario, ▶llanura de fango, ▶lodo lacustre, ▶sapropel); *f 1* **boue [f] sédimentaire** (Terme générique) ; *f 2* **vase [f] marine** (Dépôt bathyal et abyssal formé d'un mélange de limons argileux et de matières organiques provenant des organismes aquatiques ; on distingue la vase marine déposée en bordure des côtes et des estuaires, la ▶vase lacustre en bordure des berges d'un lac ; la vase salée est un sédiment fin formé de précolloïdes et de colloïdes, ou la fraction sableuse est réduite dans la zone littorale ; DG 1989, 468 ; ▶boue de dragage portuaire, ▶gyttja, ▶sapropèle, ▶slikke) ; *syn.* schlick [m], boue [f] marine ; *g* **Schlick [m]** (Abgelagertes, feinkörniges, schlammartiges Sediment in Auen, Wasserstraßen, Häfen, in Seen und im Meer; ▶Gyttja, ▶Hafenschlick, ▶Sapropel, ▶Schlamm 1, ▶Watt); *syn.* Schlamm [m] (2).

sediment [n], **detrital** *geo.* ▶loose sedimentary deposit.

5427 sediment accumulation belt [n] *geo. phyt.* (Area of siltation on the periphery of a body of water; ▶sediment accumulation zone); *s* **cintura [f] de colmatación** (▶zona de colmatación alrededor de un lago); *f* **ceinture [f] d'atterrissement** (*Végétation des ceintures des bords des eaux* zone périphérique des étangs ou lacs, des rivières ou ruisseaux, des marais ou marécages sur laquelle se succède une végétation amphibie monospécifique de grands hélophytes, inondée une partie de l'année et qui forme une série de peuplements de rives plus ou moins continus en allant du plus profond vers le moins profond. La prolifération de ces peuplements entre dans le processus naturel de comblement progressif et de fermeture des mares, petits étangs et lacs peu profonds, par avancée de saules et autre végétation arbustive qui permettent des atterrissements et la progression successive des différentes ceintures vers l'intérieur du plan d'eau On observe habituellement une succession de différentes ceintures dont la composition varie avec le niveau d'eau, la nature des sols et la qualité de l'eau ; la ceinture externe [côté eau libre] est formée par les roselières [scirpaie lacustre, roselière basse, phragmitaie, typhaie] suivies des peuplements à grandes laîches et de la jonchaie ; côté terrestre, les roselières hautes se mélangent parfois avec les massettes et le peuplement dense de la phalaridaie ; ▶zone d'atterrissement entourant l'eau

libre d'un étang) ; *g* **Verlandungsgürtel [m]** (▶Verlandungs-zone, die sich rund um ein Gewässer aufbaut).

5428 sediment accumulation zone [n] *geo. phyt.* (Lakeside shallow water zone of eutrophic ▶standing waterbody, the vegetation of which is spreading continuously to the middle of the lake according to the silting-up process; various life forms which grow in different water depths are arranged in belts along the lake margin; ▶siltation); *s* **zona [f] de colmatación** (Zona poco profunda de un lago eútrofo que se rellena por ▶entarquina-miento 1 y cuya vegetación va avanzando aguas afuera cuanto más avanza el proceso, formando cinturones según el tipo de plantas que crecen en las diferentes profundidades; ▶aguas estancadas, ▶colmatación); *f* **zone [f] d'atterrissement** (Marge des cours d'▶eaux calmes, étangs et mares, de faible profondeur peuplée par divers groupements végétaux formant une ceinture autour de l'eau libre, les peuplements denses d'hélophytes [roselières] se substituent aux hélophytes dispersés et aux groupements d'hydrophytes et progressent vers le centre du plan d'eau par suite de l'▶atterrissement ; ▶comblement sédimen-taire) ; *g* **Verlandungszone [f]** (Durch Sedimentation sich all-mählich auffüllender Flachwasserbereich eutropher ▶Stillge-wässer, dessen Vegetation mit zunehmendem Verlandungs-prozess fortschreitend zur Mitte des Gewässers vordringt; die einzelnen, in unterschiedlicher Wassertiefe wachsenden Lebens-formen staffeln sich zu Gürteln; ▶Verlandung).

sedimentary deposit [n]**, unconsolidated** *geo.* ▶loose sedimentary deposit.

5429 sedimentary rock [n] *geo.* (Layered rock resulting from the accumulation and consolidation [lithification] of sediment, e.g. **1.** mechanically formed into clastic rock such as sandstone, shale, breccia, conglomerate, **2.** chemically formed into carbonates such as travertine or dolomite, into silicates such as sinter or flint or into ironstone such as haematite or **3.** organi-cally formed into calcareous rock such as coral limestone, shelly limestone or into siliceous rock such as diatomaceous earth and **4.** formed by desiccation into sulphates such as gypsum or chlorides [rock-salt]; **s. r.** is the most common rock exposed on the Earth's surface but is only a minor constituent of the entire crust, which is dominated by igneous and metamorphic rock; ▶loose sedimentary rock); *s* **roca [f] sedimentaria** (Roca de materiales muy diferentes, formada en superficie por acumu-lación de materia procedente de la destrucción de otras rocas preexistentes o de ciertos organismos, cuya transformación en roca coherente se realiza mediante procesos de diagenización. Su génesis requiere una acumulación de materia procedente de la destrucción de otras rocas preexistentes; cf. DGA 1986; ▶roca sedimentaria no consolidada); *syn.* roca [f] exógena (DGA 1986); *f* **roche [f] sédimentaire** (Roche formée à la surface de la terre par **1.** érosion mécanique et sédimentation détritique, p. ex. grès, poudingue ou brèche, **2.** dissolution chimique, érosion et préci-pitation, p. ex. les roches carbonatées [travertin, dolomie], les roches siliceuses [silex, meulière, jaspe], les roches ferrugineuses [hématite], **3.** sédimentation biogène/organogène, p. ex. les roches carbonatées [calcaire corallien, lumachelle], les roches siliceuses [radiolarite, terre à diatomées] ou **4.** dessiccation/vapo-ration, p. ex. les roches sulfatées [chaux] ou les roches salines [sel gemme, gypse] ; ▶roche sédimentaire meuble) ; *g* **Sedi-mentgestein [n]** (Meist ein Schichtgestein, das entstanden ist entweder **1.** durch mechanische Verwitterung und Ablagerung wie z. B. Sandstein, Tonschiefer oder Brekzien, **2.** durch che-mische Lösung, Verwitterung und Ausscheidung wie z. B. Kalk-gestein [Travertin, Dolomit], Silikatgestein [Kieselsinter, Feuer-stein] und eisenhaltiges Gestein [Hämatit], **3.** durch biogene/organogene Ablagerung wie z. B. Kalkgestein [Korallenriff, Schillkalk] oder Silikatgestein [Kieselschiefer, Diatomeen-

schlamm] oder **4.** durch Austrocknung/Eindampfung wie z. B. Sulfatgestein [Gips] oder Salzgestein [Steinsalz, Magnesium-salz]; **S.** ist das Gestein, das am meisten auf der Erdoberfläche vorkommt, obwohl es nur einen geringen Anteil an der Gesamt-erdkruste, die vorwiegend von magmatischen und metamorphen Gesteinen geprägt ist, ausmacht; ▶Lockergestein).

5430 sedimentation [n] *geo.* (Deposition of solid particles after transport by the action of water, ice or wind; ▶aggradation, ▶silting up 2); *syn.* sediment deposition [n]; *s* **sedimentación [f]** (Proceso de acumulación de materiales transportados en diso-lución o por suspensión, saltación, tracción etc., por corrientes de agua [ríos, arroyos, torrentes, corrientes marinas, corrientes de turbidez], de aire o de hielo en la Tierra. Cuando cubre campos de cultivo, obras humanas o ecosistemas más o menos naturales puede denominarse aterramiento; DINA 1987; ▶acumulación, ▶agradación, ▶entarquinamiento); *f* **sédimentation [f]** (Action de transport et d'accumulation de particules ou de roches sous l'action de l'eau, la glace ou le vent avec formation d'un dépôt ; ▶alluvionnement, ▶envasement) ; *g* **Ablagerung [f] (3)** (*Vor-gang* durch Meer, Fluss, Eis oder Wind bewirkte Umlagerung von Bodenmaterial/Verwitterungsprodukten; ▶Anlandung 2, ▶Auflandung 1); *syn.* Sedimentation [f].

5431 sedimentation analysis [n] *constr.* (Determination of the proportion of particle sizes or fines in soil [0.001 to 0.006mm] by applying differential velocities of sinkage in water—rate of settlement; ▶water-graded particle); *syn.* decan-tation test [n] [also UK], sedimentation test [n]; *s* **análisis [m] granulométrico por sedimentación** (Determinación de la granu-lación de las fracciones arcillosas y limosas del suelo; ▶partícula en suspensión decantable en agua); *syn.* prueba [f] de sedimen-tación; *f* **analyse [m] par sédimentation** (Méthode utilisée pour estimer la granulométrie des particules fines du sol [0,001 à 0,06 mm], fondée sur le fait que des grains de diamètre différents sédimentent, dans un milieu liquide au repos, à des vitesses différentes [méthode densimétrique] ; ▶particules en suspension décantables dans l'eau) ; *g* **Schlämmanalyse [f]** (Bestimmung des Massenanteils der Korngrößen oder Feinbestandteile des Bodens [0,001 bis 0,06 mm] mit Hilfe ihrer verschiedenen Sinkgeschwindigkeit im Wasser; ▶Schlämmkorn).

5432 sedimentation basin [n] *envir.* (Tank in which the solid matter is allowed to settle out of the raw sewage at a ▶sewage treatment plant; ▶sediment retention basin); *syn.* settling tank [n] [also UK], sedimentation lagoon [n], settling basin [n]; *s* **decantador [m] primario** (En ▶planta de depu-ración de aguas residuales, tanque de sedimentación y flotación de materias en suspensión o de menor densidad que el agua; ▶decantador de fango/lodos); *syn.* tanque [m] de decantación, sedimentador [m], clarificador [m] primario; *f 1* **cuve [f] de décantation** (Installation de décantation des particules en sus-pension des eaux usées dans une ▶station d'épuration des eaux usées ; ▶bassin de décantation [des boues]); *syn.* décanteur [m], bassin [m] de dessablement, bassin [m] de décantation ; *f 2* **bas-sin [m] de lagunage (2)** (Plan d'eau endigué au fond duquel se déposent les matières décantables qui sont extraites régulièrement et épandues ; on distingue en général 3 sortes de bassins : un bassin de prétraitement [décantation des boues, dégraisseur, déssableur, etc.], un bassin de lagunage à microphytes [plante de petite taille qui agissent avec la photosynthèse] et un bassin de ▶lagunage à macrophytes) ; *g* **Absetzbecken [n] (1)** (Becken in einer ▶Kläranlage, in dem Schwebstoffe der Abwässer langsam absinken; ▶Schlammabsetzbecken).

sedimentation lagoon [n] *envir.* ▶sedimentation basin.

5433 sedimentation tank [n] *envir.* (Tank in which the solid matter is allowed to settle out of the ▶raw sewage at a

►sewage treatment plant; ►raw sewage); *s* **tanque [m] de sedimentación** (de una ►planta de depuración de aguas residuales; ►aguas residuales [sin depurar]); *f* **installation [f] de flottation** (Cuve cylindrique dans la quelle est introduite une suspension boueuse d'eaux résiduaires urbaines dans une ►station d'épuration des eaux usées afin de provoquer l'épaississement par introduction d'air dissous ; ►effluent non épuré) ; *g* **Flotationsbecken [n]** (Absetzbecken in ►Kläranlagen, in dem beim physikalischen Verfahren feine Schmutzpartikel durch Einblasen von Luft entfernt werden; ►ungeklärtes Abwasser).

sedimentation test [n] *constr.* ►sedimentation analysis.

5434 sediment basin [n] [US] *constr.* (Sedimentation space below an outlet pipe at the bottom of a drainage inlet structure. This space serves to collect trash, silt, sand, organic matter, etc., which is periodically removed; *generic term* ►mud trap 1); *syn.* sump [n], silt trap [n] [UK]; *s* **arqueta [f] de recogida con filtro** (*Término genérico* ►colector de fango 1); *f* **radier [m] de décantation** (Partie inférieure d'une bouche d'égout ou d'un regard de visite située en contre-pente par rapport à la conduite aval, de façon à favoriser la rétention du sable ou des boues ; *terme générique* ►dispositif de rétention des boues) ; *g* **Sumpf [m] (1)** (Geschlossener Bodenteil in einem Straßenablauf unterhalb des Muffenteils zum Auffangen von Unrat, Schlamm, Sand oder Kies; *OB* ►Schlammfang 1); *syn.* Bodensumpf [m].

sediment bucket [n] [UK] *constr.* ►silt box [US].

5435 sediment load [n] *hydr.* (Sediment which remains in suspension in flowing water for a considerable period of time without contact and without settling on the streambed; WMO 1974; ►bed load, ►wash load); *s* **carga [f] de sedimentos en suspensión** (Sedimentos que permanecen en suspensión en las aguas durante un periodo/período de tiempo considerable sin sedimentarse; WMO 1974; ►carga de fondo, ►carga en suspensión); *f* **charge [f] sédimentaire** (Sédiments restant en suspension dans l'eau courante [►charge solide en suspension] pendant un temps assez considérable sans toucher ni se déposer sur le fond du lit ; WMO 1974 ; ►charge de fond) ; *syn.* charge [f] solide en suspension ; *g* **Sedimentfracht [f]** (Gesamtgewicht von feinem Gesteinsmaterial, das eine Zeitlang im Schwebezustand vom Fließgewässer mitgeführt wird und je Zeiteinheit an einem Abflussquerschnitt gemessen wird; ►Geschiebe 1, ►Schwebstofffracht).

5436 sediment removal [n] *wat'man.* (Removal of mud, sand and gravel sediments in rivers and streams; ►ditch cleaning; ►excavated muck and plant material; ►river maintenance); *s* **drenaje [m] de cursos de agua** (En el marco del ►mantenimiento de cursos de agua 1, la extracción de fango, arena o cantos rodados; ►limpieza de zanjas de avenamiento, ►material de desbroce); *f* **curage [m] des cours d'eau** (Dans le cadre d'une opération d'►entretien des cours d'eau ou d'étangs et lacs, enlèvement des dépôts vaseux, sablonneux ou de graviers ; ►curage des fossés, ►produit de curage) ; *syn.* curage [m] d'entretien des cours d'eau ; *g* **Gewässerräumung [f]** (Im Rahmen der ►Gewässerunterhaltung das Entfernen von Schlamm-, Sand- und Kiesablagerungen; ►Grabenräumung, ►Räumgut).

5437 sediment retention basin [n] *envir. constr.* (Holding pond for soil erosion control); *s* **decantador [m] de fango/lodos** (Estanque para recoger las partículas deslavadas por la erosión para que no se viertan en el curso receptor); *f* **bassin [m] de décantation (des boues)** (Ouvrage d'épuration permettant dans un procédé de traitement primaire l'élimination par sédimentation des matières en suspension décantables dans l'eau et de ne pas atteindre le milieu naturel, p. ex. la décantation des pluviolessivats des routes) ; *g* **Schlammabsetzbecken [n]** (Be-

cken, das durch Erosion abgeschwemmte Bodenpartikel auffängt, damit diese nicht in die Vorflut gelangen).

5438 seed bank [n] *agr. bot. conserv. hort.* (Repository of seeds collected to preserve genetic diversity; ►soil seed bank); *s* **banco [m] de semillas** (Lugar de almacenamiento de semillas con el fin de preservar la diversidad genética; ►reserva de semillas en el suelo); *f* **banque [f] de gènes et de collections** (Lieu de dépôt de semences en vue de la conservation de la diversité génétique ; LA 1981, 1024 ; ►réserve de semences dans le sol) ; *g* **Samenbank [f]** (...banken [pl]; Aufbewahrungsort für Pflanzensamen, um das Genmaterial der Arten zu erhalten und somit zur Erhaltung der genetischen Vielfalt beizutragen; ►Samenspeicher); *syn.* Saatgutbank [f].

5439 seed bearer [n] *agr. bot. hort. for.* (Any plant retained to produce seed for natural regeneration; *specific terms* seed tree, mother tree; ►reserve [US] 1/reserved tree [UK]); *s* **portagrano [m]** (En silvicultura, seminífero; utilizado para la repoblación; DB 1985; ►árbol padre 2); *f* **semencier [m]** (Arbre produisant des semences, soit d'une façon générale, soit qu'il soit réservé en vue d'obtenir une régénération naturelle ; cf. DFM 1975 ; ►réserve) ; *syn.* porte-graines [m] ; *g* **Samenträger [m]** (Pflanze zur Saatgutgewinnung; *UBe* Samenbaum, Mutterbaum oder Schlaghüter [auf einer Kahlschlagfläche für die Bestandsverjüngung verbleibender Baum]; ►Überhälter 2).

5440 seed-bearing [adj] *bot. hort.* (Pertaining to plants, which have developed seed); *syn.* seminiferous [adj]; *s* **seminífero/a [adj]** (Referente a planta que ha desarrollado semillas); *f* **séminifère [adj]** (Plante sur laquelle se développent les semences et qui contribue à la dispersion de l'espèce) ; *g* **Samen tragend [ppr]** (Pflanze betreffend, die Samen ausgebildet hat).

5441 seed bed [n] *constr. hort.* (Area prepared for seeding; ►seeding of grass/lawn areas); *s* **lecho [m] de siembra** (►siembra de césped 2); *f 1* **surface [f] à ensemencer** (*Terme générique* surface préparée pour l'ensemencement ; ►engazonnement d'une pelouse par semis) ; *syn.* lit [m] de semence ; *f 2* **surface [f] à engazonner** (pour un gazon) ; *g* **Ansaatfläche [f] (1)** (Für die Saat vorbereitete Fläche; ►Herstellung von Rasenflächen durch Ansaat); *syn.* Saatfläche [f], Saatbeet [n].

5442 seed dispersal [n] *phyt. for.* (Spreading of seed by wind [►wind dispersal], water [►seed rain], animals [►animal dispersal of seeds] or humans [►anthropogenic dispersal]; ►barochory, ►dispersal, ►water dispersal); *s 1* **diseminación [f] de semillas** (Dispersión de semillas gracias al viento [►anemocoria], al agua [►hidrocoria], a animales [►zoocoria] o a seres humanos [►antropocoria]; ►nuevo vuelo, ►barocoria, ►dispersión 1); *syn.* dispersión [f] de semillas; *s 2* **policoria [f]** (Proceso de desplazamiento de semillas de una misma planta por diferentes medios); *f* **dissémination [f] des semences** (Libération des diaspores ainsi que leur dispersion active ou passive par le vent [►anémochorie], l'eau [►hydrochorie], les animaux [►dissémination par les animaux] et les êtres humains [►dissémination anthropogène] ; ►barochorie, ►dissémination, ►flux de semences) ; *g* **Samenverbreitung [f]** (Ausbreiten von Samen durch Wind [Anemochorie], Wasser [Hydrochorie], Tiere [Zoochorie] oder durch den Menschen [Anthropochorie]; ►Anflug 2, ►Aufschlag, ►Verdriftung durch Wasser, ►Verdriftung durch Wind, ►kulturbedingte Verschleppung, ►Verschleppung durch Tiere); *syn.* Aussamen [n, o. Pl.], Aussamung [f], Versamung [f].

5443 seeded area [n] *constr. hort.* *s* **superficie [f] sembrada**; *f 1* **surface [f] ensemencée** (Terme générique pour toutes espèces de semences) ; *f 2* **surface [f] engazonnée** (Pour un semis de gazon) ; *g* **Ansaatfläche [f] (2)** (Eingesäte Fläche).

seeded fallow [n] *agr.* ►green fallow.

seed exchange market [n] *constr. hort.* ►seed exchange trade.

5444 seed exchange trade [n] *constr. hort.* (Buying and selling of seed); *syn.* seed exchange market [n]; *s* **comercio [m] de semillas**; *f* **commerce [m] des graines** (Vente et achat de semences) ; *syn.* commerce [m] des semences ; *g* **Samenhandel [m, o. Pl.]** (Kauf und Verkauf von Sämereien).

5445 seeding [n] *constr. hort.* (Act of sowing seed; ►dry seeding, ►establishment of vegetation, ►grass seeding, ►lawn seeding, ►over-seeding [US]/over-sowing [UK], ►seeding of meadows, ►sowing 2, ►sowing 3); *syn.* sowing [n] (1); *s* **siembra [f] (1)** (sembrar [vb]; ►establecimiento de césped/hierba con semillas, ►establecimiento de vegetación, ►inserción de las semillas, ►resiembra de césped 2, ►siembra de césped 1, ►siembra de praderas, ►siembra en seco); *syn.* sembradura [f]; *f* **semis [m] (1)** (Faire [vb] un semis ; ►enfouissement du semences, ►engazonnement par semis, ►enherbement, ►ensemencement à sec, ►semis de regarnissage, ►semis d'un gazon, ►végétalisation) ; *syn.* ensemencement [m] ; *g* **Ansaat [f] (1)** (Ausbringen von Saatgut i. S. v. anbauen; *Kontext z. B.* Gras ansäen [vb]; ►Begrünung, ►Begrünung durch Raseneinsaat, ►Nachsaat, ►Rasenansaat, ►Saatguteinbringung, ►Trockensaat, ►Wiesenansaat).

5446 seeding [n] **of grass/lawn areas** *constr.* (►grass seeding); *s* **siembra [f] de césped (2)** (►siembra de césped 1); *f* **engazonnement [m] d'une pelouse par semis** (►semis d'un gazon) ; *syn.* gazonnage [m] d'une pelouse par semis, établissement [m] d'une pelouse par semis ; *g* **Herstellung [f] von Rasenflächen durch Ansaat** (►Rasenansaat).

5447 seeding [n] **of meadows** *constr. gard. syn.* grass seeding [n]; *s* **siembra [f] de praderas**; *f* **enherbement [m]** (1. Installation d'une couverture végétale semée de façon permanente ou temporaire. 2. Expression désignant le maintien et l'entretien d'un couvert végétal, naturel ou semé, utilisé entre les rangs ou les rangées dans les cultures de la vigne ou d'arbres fruitiers pour limiter le développement des adventices et réduire l'emploi de désherbant, lutter contre l'érosion des sols, le ruissellement et améliorer la portance des sols ; *contexte* enherbement de la vigne, enherbement du vignoble) ; *g* **Wiesenansaat [f]** (Vorgang des Einsäens einer Wiesensaatgutmischung. **W.n** dienen z. B. der Begrünung öffentlicher Grünflächen, um im Vergleich zu den mehrfach geschnittenen Rasenflächen reich blühende ►naturnahe Wiesen zu schaffen, um das Stadtbild mit bunten Blühaspekten im Laufe der Vegetationsperiode zu bereichern und um die Artenvielfalt der Flora und Fauna in der Stadt zu erhöhen. **W.n** werden auch zum Erosionsschutz und zur Unterdrückung des Unkrautes in den Weinbergen vorgenommen).

5448 seeding [n] **of stockpiled topsoil** *constr.* (Sowing seeds for temporary establishment of vegetative cover to control erosion; e.g. ►green manure plants); *syn.* seeding [n] of topsoil heaps [also UK], seeding [n] of topsoil stores, seeding [n] of topsoil piles [also US]; *s* **siembra [f] del caballón de tierra vegetal** (Establecimiento temporal de cubierta de vegetación para evitar la erosion de la tierra acopiada durante una obra); *f* **végétalisation [f] de la terre stockée** (Mesure de protection par ensemencement contre l'érosion de la terre végétale décapée et stockée avant sa réutilisation ; ►engrais vert) ; *syn.* végétalisation [f] du stock ; *g* **Erdmietenbegrünung [f]** (1. Herrichten einer zeitlich befristeten Vegetationsdecke, z. B. aus ►Gründüngungspflanzen, als Erosionsschutz. 2. Pflanzendecke auf Erdmieten als Ergebnis der Begrünungssaat); *syn.* Begrünung [f] von Oberbodenmieten.

seeding [n] **of topsoil heaps** [UK] *constr.* ►seeding of stockpiled topsoil.

seeding [n] **of topsoil piles** [US] *constr.* ►seeding of stockpiled topsoil.

seeding [n] **of topsoil stores** *constr.* ►seeding of stockpiled topsoil.

5449 seedling [n] *hort.* (In nursery practice, a young plant grown from seed that has not been transplanted; ►germinate young plant, ►immature plant); *s* **plántula [f] de semilla** (En vivero, planta joven nacida de semilla que aún no ha sido transplantada; ►plántula, ►plántula germinada, ►postura); *syn.* plantita [f], planta [f] de semilla, planta [f] de semillero (HDS 1987); *f* **semis [m] (2)** (►Jeune plant en cours de germination issu d'un semis 1 ; on distingue : le **semis en place** [plant non repiqué comportant une seule racine principale] et le **semis soulevé** [plant dont la racine principale a été coupée sous la terre du lit de semence et possédant les caractéristiques d'un plant d'un an repiqué] ; ►jeune touffe, ►plantule) ; *g* **Sämling [m]** (Aus Samen entstandene junge Pflanze — im Vergleich zur Pflanze, die aus einem Steckling oder durch Veredlung herangezogen wurde; ►Keimling, ►Jungpflanze).

seed market [n] [UK] *constr. hort.* ►seed store [US] (1).

5450 seed mat [n] *constr.* (Mat of varying manufacture and material measuring 0,5-2m wide and 10-15m long, the undersides of which are coated with seeds in a special adhesive, sometimes containing fertilizer, used on a large scale to protect soils from erosion; ►planting mat, ►substrate mat); *s* **estera [f] de semillas** (para proteger superficies contra la erosion; ►estera de sustrato sintético, ►estera sintética sembrada); *f* **natte [f] préensemencée** (Natte de 0,5 à 2 m de largeur et de 10 à 15 m de longueur fixée par une colle sur la face inférieure, des semences et éventuellement un engrais étant incorporés sur la face supérieure, utilisée pour la protection du sol sur de grandes surfaces contre l'ablation par l'eau ; ►natte de substrat synthétique, ►natte synthétique préensemencée) ; *syn.* natte [f] préensemencée de protection contre l'érosion ; *g* **Saatmatte [f]** (0,5-2,0 m breite, 10-15 m lange Matte unterschiedlicher Herstellung und Materialien, die unterseits mit einem Kleber befestigt wird, auf der Oberseite mit Saatgut und evtl. Dünger beschichtet ist und zum großflächigen Schutz der Bodenoberfläche gegen Abspülung dient; LEHR 1981; ►Begrünungsmatte, ►Substratmatte); *syn.* Erosionsschutzmatte [f].

seed maturation [n] *bot. hort.* ►seed ripening.

5451 seed merchant [n] *hort.* (Person who buys and sells seed); *syn.* seedsman [n]; *s* **vendedor, -a [m/f] de semillas**; *f* **grainetier [m]** (Personne faisant le commerce des graines, des semences) ; *g* **Samenhändler [m]** (Jemand, der Sämereien kauft und verkauft).

seed mix [n] *agr. constr. hort.* ►seed mixture.

5452 seed mixture [n] *agr. constr. hort.* (►grass-seed mixture, ►standard grass-seed mixture); *syn.* seed mix [n] (LD 1994 [4], 13); *s* **mezcla [f] de semillas** (►Mezcla esándar de semillas, ►mezcla de semillas de césped); *f* **mélange [m] de graines** (►Mélange standardise de semences pour gazon [destiné p. ex. à l'engazonnement] ; ►mélange de semences pour gazon) ; *g* **Saatgutmischung [f]** (►Regelsaatgutmischung, ►Rasensaatgutmischung); *syn.* Ansaatmischung   [f].

seed mixture [n] [US]**, turf-** *constr. hort.* ►grass-seed mixture.

seed orchard [n] *for.* *►seed stand.

5453 seed-propagated (stock) [loc] *for. hort.* (Nomenclature for cultivated plants grown from seeds, e.g. in tree, forest, or perennial nurseries; ►seedling); *syn.* propagated [pp] from seed; *s* **criado/a de semillas [loc]** (►plántula de semilla); *f* **obtenu par semis [loc]** (Terme décrivant la provenance de végétaux, p. ex. en pépinière ou en pépinière spécialisée en plantes vivaces ; ▸semis 2) ; *g* aus Samen gezogen [loc] (Kulturbeschreibung für Pflanzen, die durch Aussaat herangezogen wurden, z. B. in Baumschulen, Staudengärtnereien; ▸Sämling).

5454 seed rain [n] *phyt. for.* (Young plants derived from seed dispersal by wind or birds. *Generic term* ▸natural regeneration 1; ▸barochory, ▸natural colonization, ▸natural colonization by seed rain, ▸seed dispersal); *s* **nuevo vuelo [m]** (Proceso de colonización espontánea por plantas de diseminación anemócora; *término genérico* ▸regeneración natural; ▸barocoria, ▸colonización natural de «espacio vacío», ▸colonización natural de «espacio vacío» por nuevo vuelo, ▸diseminación de semillas); *syn.* transporte [m] aéreo de semillas; *f* **flux [m] de semences** (Processus de colonisation spontanée par des plantes dont les diaspores ont été transportées par le vent ou les oiseaux ; *terme générique* ▸régénération naturelle ; ▸barochorie, ▸colonisation spontanée, ▸dissémination des semences, ▸semis naturel par flux de semences) ; *syn.* colonisation [f] naturelle ; *g* **Anflug [m, o. Pl.] (2)** (Vorgang der spontanen Ansiedlung/Besiedlung von Pflanzen durch Wind- oder Vogeleintrag der Diasporen. *OB* ▸Naturverjüngung ; ▸Anflug 1, ▸Aufschlag, ▸Samenverbreitung, ▸spontane Besiedelung); *syn.* Ansamung [f], Samenanflug [m].

5455 seed ripening [n] *bot. hort.* (Point at which seed has developed so that it can germinate or be harvested); *syn.* seed maturation [n]; *s* **maduración [f] de semillas**; *f* **état [m] de maturation de la graine** (État morphologique et physiologique final de la graine apte à germer sitôt semée) ; *syn.* état [m] de maturation des semences ; *g* **Samenreife [f]** (Zustand der fertigen Entwicklung von Samen, so dass sie keimen oder geerntet werden können).

5456 seeds [npl] *agr. constr. for. hort.* (Selected or mixed seeds; ▸seed mixture); *s* **semilla(s) [f (pl)]** (Semillas de una especie o ▸mezcla de semillas prevista para la siembra); *syn.* simientes [fpl]; *f* **semences [fpl]** (Graines destinées à la reproduction, utilisées en semis, en général sélectionnées et certifiées ; ▸mélange de graines) ; *g* **Saatgut [n, o. Pl.]** (Zur Ansaat/Aussaat bestimmte Samen oder -mischung; ▸Saatgutmischung); *syn.* Sämerei [f].

seeds [n] [UK]**, landscape grass** *constr.* ▸low-maintenance grass type.

seed shop [n] [UK] *constr. hort.* ▸seed store [US] (1).

seedsman [n] *hort.* ▸seed merchant.

5457 seed stand [n] *for.* (Stand set aside in a forest and listed in an international directory managed in which seed of genetically high quality forest plants [seed producers] is harvested; in D. four categories are distinguished: **1. source identified:** seeds, only appropriate for the cultivation of trees and shrubs for landscape use, but not for silviculture; **2. selected seed stand:** selected seed for the lowest quality of approved stands used for harvesting seeds for silviculture, *syn.* seed stand with selected material; **3. tested seed stand:** quality seeds for a designated stand of forest used for seed harvesting approved stock, where the young trees are tested according to certain criteria; **4. qualified seed stand:** seeds of the highest quality for silvicultural use, whereby the seeds originate only from specially selected, superior phenotypes [slips on older scions] of the same species or from a collection of clones. These trees must be at a distance of 400m from specimens of the same species in order to be able to control cross-fertilization; a **s. s.** used for harvesting qualified and certified seed is referred to as a **seed orchard**); *s* **rodal [m] semillero** (Tipo de material de base para la producción de material forestal formado por población delimitada de árboles con suficiente uniformidad en su composición que se

aprovecha para cosechar semillas para la reproducción y está registrado en un Catálogo Nacional de Materiales de Base [CNMB], que a su vez forma parte del Catálogo Común Europeo. Este catálogo está constituido por aquellas poblaciones, plantaciones y clones de los que se obtiene material forestal de reproducción [frutos, semillas, partes de plantas y plantas] para utilizar en las repoblaciones. Los tipos de material de base aprobados actualmente son: a] fuentes semilleras, b] rodales selectos, c] huertos semilleros, d] progenitores de familia, e] clones y f] mezcla de clones. Los **materiales forestales de reproducción** se subdividen en 4 categorías: **1. identificados:** m. f. de r. obtenidos de materiales de base que pueden ser bien una fuente semillera, bien un rodal situados dentro de una única región de procedencia y que satisfacen las exigencias establecidas en el anexo II; **2. seleccionados:** m. f. de r. obtenidos de materiales de base que se corresponden con un rodal situado dentro de una única región de procedencia, que hayan sido seleccionados fenotípicamente a nivel de población y que satisfacen las exigencias establecidas en el anexo III; **3. cualificados:** m. f. de r. obtenidos de materiales de base que se correspondan con huertos semilleros, progenitores de familias, clones o mezclas de clones, cuyos componentes han sido individualmente seleccionados fenotípicamente y satisfacen las exigencias establecidas en el anexo I; **4. controlados:** m. f. de r. obtenidos de materiales de base que se correspondan con rodales, huertos semilleros, progenitores de familias, clones o mezclas de clones. La superioridad del material de reproducción debe haber sido demostrada mediante ensayos comparativos o estimada a partir de la evaluación genética de los componentes de los materiales de base. Los materiales de base deberán satisfacer las exigencias establecidas en el anexo V; art. 2, 1) y anexos RD 289/2003; cf. Directiva 1999/105/CE del Consejo, de 22 de diciembre; *término genérico* población [f] portagrano); *f* **peuplement [m] porte-graines** (Population délimitée d'arbres [matériels forestiers admis] de composition uniforme sur une parcelle classée et inscrite sur le registre national des matériels de base des essences forestières ; ce registre regroupe toutes les informations relatives à l'identification des unités d'admission des matériels de base, et notamment **a.** l'identification de référence, **b.** la région de provenance, **c.** la localisation [la zone des latitudes et longitudes pour les catégories identifiée et sélectionnée ; la position géographique précise pour les catégories « qualifiée et testée »], **d.** l'altitude ou la tranche altitudinale des unités d'admission, **e.** le caractère indigène ou non indigène, **f.** l'origine connue ou inconnue ; les matériels forestiers de reproduction sont répartis en quatre catégories **1. identifiée :** si la source de graines est située dans une région de provenance de l'essence considérée ; **2. sélectionnée :** lorsque le matériel de base constitue un peuplement qui est situé dans une seule région de provenance et dont la population a fait l'objet d'une sélection phénotypique ; **3. qualifiée :** un matériel de base peut être admis en catégorie « qualifiée », lorsque le matériel de base constitue un verger à graines, des parents de famille, un clone ou un mélange clonal dont les composants ont fait l'objet d'une sélection phénotypique individuelle ; **4. testée :** lorsque le matériel de base constitue un peuplement, un verger à graines, des parents de famille, un clone ou un mélange clonal pour lequel la supériorité des matériels de reproduction par rapport à des matériels témoins doit avoir été démontrée par des tests comparatifs ou par une estimation établie à partir de l'évaluation génétique des composants des matériels de base ; un **p. p.-g.** avec un matériel de base admis dans les catégories qualifiée et testée est dénommé **verger à graines** [anglais *seed orchard*] ; Code forestier Art. R. 552-1-10 ; *g* **Samenerntebestand [m]** (Im Wald ausgewiesene und in einem internationalen Verzeichnis erfasste Fläche, auf der Saatgut zur bedarfsgerechten Erzeugung von genetisch hochwertigen Forstpflanzen [Saatgutproduzenten] geerntet werden darf. In D. werden vier Kategorien von forstlichem Saatgut gehandelt: **1. quellengesichert:** Saatgut, das nur für die Anzucht von Landschaftsgehölzen, jedoch nicht für den Waldbau geeignet ist, **2. ausgewählt:** Saatgut für die unterste Qualitätsbezeichnung eines für die waldbauliche Samenernte zugelassenen Bestandes, **3. geprüft:** Saatgut für die Qualitätsbezeichnung eines für die waldbauliche Samenernte zugelassenen Bestandes, bei dem die Nachkommenschaft hinsichtlich bestimmter Kriterien geprüft ist und **4. qualifiziert:** Saatgut für die höchste waldbauliche Samenqualität, bei der die Samen nur von besonders ausgelesenen, überdurchschnittlichen Phänotypen [Pfropfreiser auf mehrjährigen Unterlagen] derselben Art oder von einer Klonsammlung stammen. Diese Bäume müssen wegen der Kontrollierbarkeit der Befruchtung über 400 m von Exemplaren derselben Art entfernt sein. Einen **S.** für geprüftes und qualifiziertes Saatgut wird **Samenplantage** [engl. *seed orchard*] genannt).

5458 seed store [n] [US] (1) *constr. hort.* (Place where seed can be obtained); *syn.* seed shop [n], seed market [n] [UK]; *s* **tienda** [f] **de semillas** *syn.* mercado de semillas; *f* **graineterie** **[f]** (Lieu de vente de graines et semences) ; *g* **Samenhandlung** **[f]** (Bezugsstelle für Sämereien).

seed store [n] (2) *bot.* ▶soil seed bank.

seepage [n] (1) *envir.* ▶infiltration.

seepage [n] (2) *envir.* ▶leakage.

5459 seepage [n] (3) *geo. pedol.* (Oozing out of ▶seepage water or ▶groundwater along a fault or joint plane; ▶infiltration, ▶groundwater seepage, ▶return seepage, ▶seepage spring); *s* **rezumo** [m] (▶Agua de percolación o ▶agua subterránea que asciende al exterior de un suelo humedeciéndolo pero sin producir un flujo perceptible; WMO 1974; ▶afloración de agua, ▶filtarción efluente, ▶fuente de percolación, ▶infiltración); *f* **suintement** [m] (3) (▶Eau de gravité ou eau de la ▶nappe phréatique émergeant lentement du sol et humidifiant sa surface sans écoulement perceptible ; WMO 1974 ; ▶eau d'exfiltration, ▶émergence de nappe, ▶infiltration, ▶source de suintement) ; *g* **Heraussickerung** [f] (Langsames Heraustreten von ▶Sickerwasser oder ▶Grundwasser aus dem Boden bei Verwerfungen; ▶diffuser Grundwasseraustritt, ▶Grundwasseraustritt, ▶Qualmwasser, ▶Versickerung).

5460 seepage layer [n] *constr.* (Layer of gravel, 50cm deep, on cut slopes with drainage water or seepage springs); *s* **manto** **[m] de drenaje** (Capa de grava de más de 50 cm de espesor en terraplenes que llevan agua de manantial o de percolación); *syn.* capa [f] de drenaje; *f* **couche** [f] **de drainage** (Large couche de grave de plus de 50 cm d'épaisseur pour le drainage des venues d'eau de surface ou des eaux de suintement, mise en place en pied d'un talus en déblai ou d'un talus de fouille) ; *g* **Sickerschicht** [f] (Über 50 cm dicke, flächige Kiesschüttung an Sickeroder Quellwasser führenden Einschnitt- oder Anschnittböschungen).

seepage pit [n] *constr.* ▶drywell sump [US]/soakaway [UK].

5461 seepage spring [n] *hydr. geo.* (Spring which issues from the permeable rock over a relatively large area: ▶groundwater seepage [US]/ground water seepage [UK]); *syn.* filtration spring [n] (EDH 1969, WMO 1974), outflow [n] of permeable layers (SPON 1974, 107); *s* **fuente** [f] **de percolación** (Fuente que mana de una roca permeable, sobre una extensión relativamente grande; WMO 1974; ▶filtración efluente); *syn.* fuente [f] de filtración; *f* **émergence** [f] **de nappe** (L'eau souterraine qui, pour les nappes libres, s'échappe par les sources, émerge au fond d'un vallon, se déverse à flanc de coteau, suinte le long d'un talus, déborde au contact d'une limite imperméable ou jaillit d'un conduit raccordé à une nappe captive ; ▶source de suintement) ;

g **Grundwasseraustritt** [m] (Wasser, das aus durchlässigem Gestein flächig an die Erdoberfläche tritt; ▶diffuser Grundwasseraustritt); *syn.* Sickerquelle [f], Sickerwasserquelle [f].

5462 seepage water [n] *pedol.* (Water in a soil which moves in a downward direction, as opposed to upward moving; ▶infiltration, ▶percolation, ▶slope seepage water; *opp.* ▶capillary water); *syn.* gravitational water [n]; *s* **agua** [f] **de percolación** (En el suelo no saturado, agua que se mueve libremente hacia las capas inferiores del subsuelo gracias al efecto de la gravedad; ▶filtración efluente en ladera, ▶infiltración, ▶percolación, *opp.* ▶agua capilar); *syn.* agua [f] de gravedad, agua [f] libre, agua [f] móvil; *f* **eau** [f] **de gravité** (Eau entraînée par la pesanteur et circulant dans le sol [infiltration] en général vers le bas par ▶percolation dans les interstices des sédiments et des roches ; PED 1979, 177 ; *opp.* ▶eau capillaire ; ▶eaux d'écoulement hypodermique, ▶infiltration) ; *syn.* eau [f] de percolation ; *g* **Sickerwasser** [n] (2) (…wässer [pl]; Wasser, das sich im Boden von oben nach unten bewegt; ▶Durchsickerung, ▶Hangwasser, ▶Versickerung; *opp.* ▶Kapillarwasser).

seep water [n] [US] *envir.* ▶leachate.

5463 segetal community [n] *phyt.* (Weed vegetation of grain crops, gardens and vineyards, growing only on open ground, belonging to the class of cornfield weeds *[Secalietea]*; ▶arable weed community, ▶weed community); *syn.* vegetation [n] of cornfield weeds (VIR 1982), cereal weed community [n]; *s* **vegetación** [f] **segetal** (Comunidad vegetal silvestre de los campos de cultivo de cereales, los jardines y viñedos que pertenece a la clase *Secali[n]etea*; ▶comunidad arvense de cultivos, ▶comunidad de malas hierbas); *syn.* comunidad [f] arvense de cultivos de cereales, vegetación [f] meseguera (DB 1985); *f* **végétation** **[f] ségétale** (Groupements végétaux adventices des cultures céréalières appartenant à la classe de *Secali[n]etea* ; ▶flore adventice des cultures et des prairies, ▶groupement anthropique, ▶groupement de plantes adventices de cultures) ; *syn.* groupement [m] de moissons, végétation [f] messicole ; *g* **Segetalflur** **[f]** (Wildkräuterbestände der Getreideäcker und Gärten, die zur Klasse der Getreideunkrautflur *[Secali(n)etea]* gehören; ▶Ackerunkrautgesellschaft, ▶Unkrautflur); *syn.* Getreideunkrautflur [f], Getreideunkrautgesellschaft [f], Segetalgesellschaft [f].

segment arc pattern [n] [UK] *constr.* ▶paving in curved pattern.

5464 segregation [n] **(of land uses)** *urb.* (Separation of working places and dwellings in inner cities); *s* **separación** [f] **de usos del suelo** (con fines residenciales, comerciales o industriales); *syn.* segregación [f] de usos del suelo; *f* **séparation** [f] **des usages** (±) (Principe visant à séparer les fonctions de travail et d'habitat dans les centres urbains) ; *g* **Entmischung** [f] (Trennung von Wohnungen und Arbeitsstätten, Versorgungs- und Freizeiteinrichtungen in Innenstädten [Kerngebieten] mit der Folge, dass weitere monofunktionale Nutzungseinheiten von zunehmender Größe [am Stadtrand] entstehen).

5465 selected bidder [n] [US] *contr.* (Bidder [US]/tenderer [UK] who is awarded the contract; ▶bidder offering best value for the price [US]/tenderer offering the best value for the money [UK], ▶awarding of contract [US]/letting of contract [UK]); *syn.* winning bidder [n] [US], selected tenderer [n] [UK]; *s* **licitante** **[m/f] seleccionado/a** (Empresa [▶licitante ganador, -a] que ha presentado la oferta más interesante, que no tiene por que ser la más barata; ▶adjudicación de contrato); *f* **candidat** [m] **attributaire** (L'entreprise retenue [▶candidat retenu] devient attributaire du marché lorsque la personne responsable du marché prend la décision de conclure le marché avec elle ; ▶passation d'un marché, ▶attribution d'un marché de travaux) ; *syn.* entreprise [f] attributaire ; *g* **Bieter** [m], **der den Zuschlag erhält/**

S

Bieterin [f], die den Zuschlag erhält (cf. ▶Bieter des annehmbarsten Angebotes; ▶Auftragserteilung an eine Firma).

selected tenderer [n] [UK] *contr.* ▶selected bidder [US].

selection [n] *prof.* ▶final selection [UK]; *plan.* ▶site selection; *plan. trans. wat'man.* ▶transportation corridor selection, ▶transportation route selection.

selection [n] **and negotiation** [n] [US]**, direct** *contr.* ▶sole source contract award [US]/freely awarded contract [UK].

5466 selection forest [n] *for.* (Uneven-aged high forest with the major concern of management to obtain a sustained yield by selective cutting; i.e. gradation of ages or sizes of stems capable of assuring the regular replacement of the old tree by natural regeneration; ▶natural silviculture); *s* **monte** [m] **irregular** (Masa arbórea compuesta por rodales que contienen pies de todas las posibles clases de edad de forma continua. Este tipo de monte se presenta en las masas naturales inalteradas, y es la estructura más lógica de una población de seres vivos. Es típico de los ecosistemas climácicos y estables; DINA 1987; ▶silvicultura natural); *f* **futaie** [f] **jardinée** (Forme naturelle de la haute futaie à couvert arboré permanent, formée d'arbres ou de bouquets d'arbres d'âges divers présentant un mélange des strates du peuplement ; ▶sylviculture écologique) ; *g* **Plenterwald** [m] (Meist naturnahe gemischte Dauerbestockungsform eines Hochwaldes, in der auf kleinster Fläche ein struktureller Gleichgewichtszustand durch eine Einzel- bis truppweise Mischung von Bäumen einer oder mehrerer Arten aller Wuchs- resp. Altersklassen und Entwicklungsstufen mit kleinlokalen Unterschieden in Ober-, Mittel- und Unterstand[-schicht] erreicht wird. Die Erhaltung eines **P.es** erfordert immer wieder Eingriffe durch den Menschen durch nahezu ausschließlich einzelstammweises Herausnehmen von herrschenden Bäumen, deren Stämme einen bestimmten Zieldurchmesser überschreiten; ▶naturnaher Waldbau).

5467 selection system [n] *for.* (Uneven-aged silvicultural system or multiaged stand in which trees are removed individually, here and there, from a large area each year-ideally over a whole forest or working circle; regeneration mainly natural and crop ideally of all ages; cf. SAF 1983; ▶selection forest); *s* **entresaca** [f] (Tratamiento selvícola consistente en la extracción de entre la masa arbórea de ciertos pies con características concretas, ya sean de diámetro, altura, aspecto, edad, sanidad o vigor; DINA 1987; ▶monte irregular); *syn.* corta [f] por entresaca; *f* **jardinage** [m] (Traitement qui confère ou tend à conférer au peuplement traité en une station donnée une structure dite jardinée équilibrée [▶futaie jardinée], c.-à-d. que le peuplement est constitué d'arbres se répartissant en une suite continue de classes d'âge et de dimensions, avec une distribution dans l'espace, soit pied par pied, soit par bouquets. Les coupes extraient des arbres choisis individuellement ou en petits groupes. En F., on distingue le « **jardinage cultural** » qui répond strictement à la définition précédente et le « **jardinage extensif** » qui correspond à une application étendue à des bouquets de plus grande surface, de l'ordre de grandeur de la parcelle ou même supérieure, appelée parquets ; DFM 1975) ; *syn.* traitement [m] en futaie jardinée, aménagement [m] en futaie jardinée ; *g* **Plenterwaldbetrieb** [m] (Waldbaumethode, bei der eine naturnahe, gemischte mehrstufige Dauerbestockungsform eines Hochwaldes angestrebt wird; bei der Durchforstung gibt es keine Umtriebszeit; es zählt nur der Entwicklungszustand der Einzelbäume. Beim **P.** finden gleichzeitig Verjüngung, Auslese, Pflege, Erziehung, Forstschutz und Ernte statt; die Nutzung findet kontinuierlich und überwiegend in der Oberschicht durch die Entnahme einzelner, starker Bäume statt. Ein **P.** ist nur mit Halbschatten ertragenden Bäumen wie Tanne, Fichte, Buche und

Zirbelkiefer möglich; in Laubwäldern werden die Bäume kurzschaftig; nach dem Umfang der Entnahme wird zwischen **Einzelplenterung** [Entnahme von Einzelstämmen] und **Gruppenplenterung** [Entnahme mehrerer Stämme] unterschieden; cf. WFL 2002, 557; ▶Plenterwald).

5468 selective bidding [n] [US] *constr. contr.* (Contract procedure which involves a limited number of chosen construction or professional firms for a ▶submission of bids [US]/submission of tenders [UK]; ▶bidding [US]/tendering [UK]); *syn.* selective tendering [n] [UK]; *s* **concurso-subasta** [m] **restringido** (▶concurso-subasta de obras en el que sólo se pide a unas pocas empresas que presenten una ▶oferta; ▶entrega de ofertas); *f 1* **appel** [m] **d'offres restreint** (Procédure précédée d'un avis d'appel de candidature et retenant une liste limitée de candidats admis à présenter/remettre une offre ; ▶appel d'offres, ▶remise des offres/soumissions) ; *f 2* **adjudication** [f] **restreinte** (Mise en concurrence d'un nombre restreinte d'entreprises sélectionnées pour rechercher le meilleur prix ; cf. DTB 1985 ; ▶adjudication) ; *g* **beschränkte Ausschreibung** [f] (2) (▶Ausschreibung 1, bei dem nur wenige Firmen zu einer ▶Angebotsabgabe aufgefordert werden; beschränkt ausschreiben [vb]).

5469 selective herbicide [n] *agr. chem. for. hort.* (Chemical used to kill only certain types of plants; ▶non-selective herbicide); *s* **herbicida** [f] **selectivo** (Biocida que sólo aniquila ciertas malas hierbas; ▶herbicida total); *f* **herbicide** [m] **sélectif** (Herbicide permettant de lutter contre certaines adventices tout en respectant les cultures ; ▶herbicide total) ; *syn.* désherbant [m] sélectif ; *g* **Teilherbizid** [n] (Chemisches Unkrautvernichtungsmittel, das nur bestimmte Pflanzen abtötet und von dem vor allem die Kulturpflanzen ausgenommen sind; ▶Totalherbizid); *syn.* selektives Herbizid [n].

5470 selective species [n] *phyt.* (Species found most frequently in a site specific community but also, though rarely, in other communities; ▶grade of fidelity); *s* **especie** [f] **electiva** (Especie con óptimo bien acusado en una comunidad, pero que también se presenta en otras, aunque en éstas es poco abundante y rara o presenta vitalidad disminuída; BB 1979; ▶grado de fidelidad); *f* **espèce** [f] **élective** (Espèce cantonnée surtout dans un groupement déterminé, mais se rencontrant aussi, quoique rarement, dans d'autres groupements ; ▶degré de fidélité ; BB 1928) ; *g* **feste Art** [f] (Art, die hinsichtlich des ▶Treuegrades eine bestimmte Pflanzengesellschaft deutlich bevorzugt, aber auch in anderen Gesellschaften spärlich und seltener oder mit herabgesetzter Vitalität vorkommt).

selective tendering [n] [UK] *constr. contr.* ▶selective bidding [US].

5471 selective waste collection [n] *envir.* (Refuse collection, which differentiates between the various fractions of municipal waste for the purpose of recycling; the separated recyclable garbage is collected in specially marked containers and transported to processing plants; recoverable materials include paper, glass, cans, scrap metal, plastic materials, textiles, garden and organic kitchen waste as well as bulky waste and problem materials such as batteries, oils, medicines, electrical items, etc. **S. w. c.** reduces the volume of waste in garbage disposal by up to 25%; remaining waste is normally deposited at ▶sanitary landfills [US] 2/landfill sites [UK]. Recycling helps in conserving scarce natural resources as well as energy, which would otherwise be required to manufacture products from virgin raw materials); *s* **recogida** [f] **selectiva de residuos urbanos** (Sistema de recogida diferenciada de materiales orgánicos fermentables y de materiales reciclables, así como cualquier otro sistema de recogida diferenciada que permite la separación de los materiales valorizables contenidos en los residuos. Se pueden diferenciar dos sistemas: **1.** Recogida selectiva utilizando contenedores especiales para cada tipo de residuo: vidrio, papel y cartones, envases ligeros, restos orgánicos u otros restos. **2.** Recogida y separación posterior por medio de sistemas de separación automatizada en instalaciones creadas para este fin. **En Es.** desde 2001 los municipios con una población superior a 5000 habitantes están obligados a implantar sistemas de **r. s. de rr.** que posibiliten su reciclado y otras formas de valorización; arts. 3 m] y 20.3 Ley 10/1998, de Residuos; ►eliminación de residuos, ►reciclado, ►reutilización de residuos, ►valorización de residuos, ►vertedero controlado); *f 1* **collecte [f] sélective** (Filière de traitement comprenant l'ensemble des opérations consistant à enlever les déchets mis dans des contenants prévus à cet effet, en vue d'une valorisation ou d'un traitement spécifique pour les acheminer ensuite vers un centre de tri [►centre de tri des déchets], de traitement ou de stockage [►centre de stockage des déchets ultimes]) ; syn. collecte [f] et tri sélectif, *terme pléonasmique* tri [m] sélectif des déchets ; *f 2* **collecte [f] spécifique** (*Terme spécifique* filière de traitement de déchets particuliers tels que les « encombrants », les déchets verts, les déchets de bureaux, etc.) ; *f 3* **collecte [f] sélective des déchets ménagers et assimilés** (Collecte de certains flux de déchets, préalablement séparés par les producteurs, en vue d'une valorisation ou d'un traitement spécifique) ; *f 4* **collecte [f] séparative** (*Terme spécifique* mise à disposition des usagers de plusieurs contenants [deux à quatre selon le niveau de tri] généralement différenciés à l'aide d'un code couleur en vue de des déchets ménagers séparés en plusieurs flux différenciés [recyclables secs, fermentescibles, encombrants, déchets dangereux des ménages et ordures ménagères résiduelles]) ; *f 5* **collecte (en) [f] porte à porte** (Opération de collecte des déchets qui consiste à organiser le passage de bennes collectrices à proximité immédiate du domicile de l'usager ou du lieu de production des déchets) ; *f 6* **collecte [f] en apport volontaire** (*Terme spécifique* conteneurs spécifiques de récupération pour le plastique, les métaux [de 4 m³ à 30 m³] et les papiers-cartons, le verre mis à la disposition des usagers et placés à différents points de regroupement [point-tri] sur le domaine public ; la collecte sélective permet grâce au traitement et la valorisation des déchets d'économiser l'énergie, de préserver les ressources naturelles, de limiter les gaz à effets de serre et de soutenir l'économie et l'emploi) ; *g* **getrennte Müllsammlung [f]** (Trennung von Wertstoffen, die im Müllaufkommen anfallen, in diverse verwertbare Fraktionen und Sammlung in speziell gekennzeichneten Containern, die zu einer Weiterverarbeitungsanlage transportiert werden. Zu den verwertbaren Altstoffen gehören z. B. Papier, Glas, Getränkedosen, Altmetalle, Plastikstoffe, Textilien, Garten- und organische Küchenabfälle sowie Sperrmüll und Problemstoffe wie Batterien, Öle, Medikamente, Elektroschrott etc. Durch die **g. M.** reduziert sich das Abfallaufkommen für die Müllabfuhr bis zu 25 % und wird als **Restmüll** auf eine ►geordnete Deponie gebracht. Die **g. M.** schont durch die Weiterverwertung den Ressourcenverbrauch und spart Energie zur Herstellung von Produkten aus Primärrohstoffen).

5472 self-cleansing [n] *constr.* (Cleaning of sewage pipes by the velocity of the wastewater alone; ►biological self-purification); *s* **autolimpieza [f]** (≠) (Capacidad de un sistema de alcantarillado de mantener las conducciones sin decantaciones y suciedas simplemente por la velocidad de flujo de las aguas residuales sin necesidad de otras medidas; ►autodepuración); *f* **autocurage [m]** (Phénomène de nettoyage d'un réseau d'assainissement sous la seule action de l'écoulement des effluents ; ►auto-épuration) ; *g* **Selbstreinigung [f, o. Pl.]** (Fähigkeit eines Entwässerungssystems, Entwässerungsrohre von Verschmutzungen und Ablagerungen ohne Pflegemaßnahmen durch die Fließ-

geschwindigkeit des Abwassers zu befreien; ►biologische Selbstreinigung).

self-cleansing [n]**, biological** *limn.* ►biological self-purification.

5473 self-cleansing capacity [n] *limn.* (Capacity of a system to breakdown pollution biologically or chemically without external intervention; ►biological self-purification, ►self-cleansing); *s* **capacidad [f] de autodepuración** (Potencia de un sistema de eliminar contaminación orgánica por medios biológicos o químicos; ►autodepuración, ►autolimpieza); *syn.* capacidad [f] autodepurativa; *f* **capacité [f] d'auto-épuration** (Capacité d'un système à éliminer une charge polluante par des processus physiques, chimiques et biologiques naturels ; ►autocurage, ►auto-épuration) ; *syn.* pouvoir [m] d'auto-épuration, pouvoir [m] épurateur (PPC 1989, 41) ; *g* **Selbstreinigungskraft [f]** (Fähigkeit eines Systems, auf biologischem oder chemischem Wege Verunreinigungen abzubauen; ►biologische Selbstreinigung, ►Selbstreinigung); *syn.* Selbstreinigungsvermögen [n].

self-clinger [n] [UK] *hort.* ►self-clinging climber.

5474 self-clinging climber [n] *hort.* (►Climbing plant which maintains its upward growth by the adhesion of modified aerial roots [►grasping branch with adhesive disks] to tree trunks, walls or other structures without the artificial support of a ►climber support; e.g. ►root climbers, ivy [*Hedera helix*], trumpet vine [*Campsis grandiflora* and *C. radicans*], climbing hydrangea [*Hydrangea petiolaris*], as well as ►**self-clinging vines with adhesive pads [US]/self-clinging vines with adhesive disks [UK]** with which they attach themselves to walls; e.g. virginia creeper [*Parthenocissus triscuspidata 'Veitchii'*; ►scandent plant, ►tendril climber, ►twining climber; *syn.* self-clinger [n] [UK], self-clinging vine [n] [also US]; *s* **trepadora [f] autoadherente** (►Planta trepadora que trepa sobre paredes y muros sin necesitar ►enrejado de soporte de trepadoras, sino que se sirve de órganos adhesivos como la ►**planta trepadora con raíces adventicias**, p. ej. la hiedra [*Hedera helix*], la enredadera de trompeta [*Campsis grandiflora, C. radicans*] y la hortensia trepadora [*Hydrangea petiolaris*] así como la ►**planta zarzillosa con ventosas** adhesivas, como p. ej. la parra virgen [*Parthenocissus spp.*]; ►planta sarmentosa, ►zarcillo caulinar con ventosas; *opp.* ►trepadora sobre soporte); *syn.* autotrepadora [f]; *f* **plante [f] grimpante à organes d'accrochage** (►Plante grimpante capable de se fixer d'elle-même à l'aide de racines aériennes, de ventouses ou de vrilles [►vrilles raméale à disques ventouses] sans l'aide d'un support artificiel le long d'un mur ou d'une façade, p. ex. le lierre [*Hedera helix*], la bignone de Chine [*Campsis grandiflora, C. radicans*], l'hortensia grimpant [*Hydrangea petiolaris*] pour les ►plantes grimpantes à racines-crampons, ou la vigne vierge de Veitch [*Parthenocissus triscuspidata 'Veitchii'*] pour les ►plantes grimpantes à ventouses ; ►support de plantes grimpantes ; *opp.* ►grimpante pour treillage) ; *syn.* plante [f] grimpante qui se fixe seule, plante [f] grimpante sans treillage ; *g* **Selbstklimmer [m]** (►Kletterpflanze, die ohne künstliches ►Rankgerüst mit Hilfe von Haftorganen an Mauern, Wänden oder Bäumen hochwächst wie die ►**Wurzelkletterer**, z. B. Efeu [*Hedera helix*], Trompetenwinde [*Campsis grandiflora, C. radicans*] und Kletterhortensie [*Hydrangea petiolaris*] sowie Rankenpflanzen mit Haftscheiben [►**Haftscheibenranker**], z. B. Wilder Wein [*Parthenocissus triscuspidata 'Veitchii'*]; ►Spreizklimmer, ►Sprossranken mit Haftscheiben; *opp.* ►Gerüstkletterer); *syn.* Selbstkletterer [m].

self-clinging vine [n] [US] *hort.* ►self-clinging climber.

self-clinging vine [n] **with adhesive disks** [UK] *bot.* ►self-clinging vine with adhesive pads [US].

5475 self-clinging vine [n] **with adhesive pads** [US] *bot.* (►Self-clinging climber with tendrils equipped with adhesive disks to enable them to cling to walls; e.g. Virginian Creeper *[Parthenocissus]*); *syn.* self-clinging vine [n] with adhesive disks [UK]; *s* **planta** [f] **zarcillosa con ventosas** (►Trepadora autoadherente que se fija al soporte mediante discos adhesivos como es el caso de la enredadera de Virginia *[Parthenocissus quinquefolia]*); *f* **plante** [f] **grimpante à ventouses** (►Plante grimpante à organes d'accrochage sans mouvement préhenseur qui possède des sortes de pelotes adhésives, qui sont de petits rameaux modifiés et permettent de s'accrocher à un mur, p. ex. la vigne vierge *[Parthenocissus]*) ; *syn.* plante [f] grimpante à pelotes adhésives ; *g* **Haftscheibenranker** [m] (►Selbstklimmer, der zum Klettern Sprossranken mit Haftscheiben ausbildet, die sie an Mauern befestigen, z. B. Wilder Wein *[Parthenocissus]*); *syn.* Rankenpflanze [f] mit Haftscheiben.

selfgrown tree [n] [US] *for.* *►woody wildling.

self organization [n] *ecol.* ►self regulation (2).

5476 self-regulation [n] **(1)** *ecol.* (Capacity of an ecosystem to react to changes in environmental conditions such that its own internal equilibrium is restored; ►ecological equilibrium); *syn.* autoregulation [n]; *s* **autoregulación** [f] **(1)** (Capacidad de un ecosistema de reaccionar ante cambios ambientales para alcanzar un ►equilibrio dinámico); *f* **autorégulation** [f] **(1)** (Capacité d'un écosystème à réagir à des changements et tendant à retrouver, par un grand nombre de mécanismes de régulation, son état initial ; ►équilibre écologique) ; *g* **Selbstregulation** [f] (Fähigkeit eines Ökosystems, auf veränderte Umweltbedingungen durch entsprechende Reaktionsweisen in Richtung auf ein Fließgleichgewicht zu reagieren; ►ökologisches Gleichgewicht); *syn.* biologische Selbstregulierung [f].

5477 self regulation [n] **(2)** *ecol.* (Automatic control mechanism for restoration of natural systems, e.g. a forest, a river or a natural area, to their previous ecologically-functioning condition by natural means, after intrusion upon the natural environment or foreign influences or catastrophes, e.g. contamination, fire, flood, etc. This takes the form of a ►cybernetic cycle which operates according to the ►feedback principle); *syn.* self organization [n]; *s* **autoregulación** [f] **(2)** (Mecanismo automático de control de los ecosistemas que sirve para restaurar un estado determinado de los mismos cuando han sufrido perturbaciones. Esto ocurre en forma de ►ciclos cibernéticos que funcionan según el principio del ►retroalimentación); *f* **autorégulation** [f] **(2)** (Rétablissement de l'état antérieur d'un système après intervention de modifications grâce à des mécanismes internes de réaction qui, dans ►système finalisé interviennent d'après le principe de ►rétroaction, c.-à-d. ordonnent sans cesse l'inverse des effets produits par les modifications dans le système ; le système maintient ainsi son comportement) ; *g* **Regelung** [f] (Wiederherstellung des vorhergehenden funktionellen Zustandes [Sollwert, Regelgröße] nach Einwirken einer Änderung [Störgröße] durch bereitliegende Reaktionsmechanismen [Regler], die in Form eines ►Regelkreises nach dem Prinzip der ►Rückkopplung arbeiten, d. h. stets das Gegenteil dessen veranlassen, was in dem System als Störung geschieht).

5478 self-rooted [adj] [US] *hort.* (Descriptive term applied to plants that have been grown from cuttings or layering instead of grafting); *syn.* own-rooted [adj] [UK]; *s* **de raíces propias** [loc] (Denominación de plantas que no han sido injertadas, sino que crecen sobre su propia raíz, se utiliza p. ej. para frutales o rosas); *f* **franc, franche de pied** [loc] (Terme qualifiant les végétaux non greffés se développant par leurs propres racines) ; *syn.* de pied franc [loc] ; *g* **wurzelecht** [adj] (Bezeichnung für nicht veredelte, aus eigener Wurzel wachsende Pflanzen, z. B. bei Obstgehölzen, Rosen); *syn.* auf eigener Wurzel [loc].

self setter [n] [UK] *for.* *►woody wildling.

seller's warranty [n] *contr.* ►guarantee fulfillment obligation [US]/guarantee fulfilment obligation [UK].

semiarid region [n] [US]/**semi-arid region** [n] [UK] *geo.* ►semidesert [US]/semi-desert [UK].

5479 semicircular step [n] [US]/**semi-circular step** [n] [UK] *constr.* *s* **escalón** [m] **semicircular**; *f* **marche** [f] **semi-circulaire** (Marche d'escalier de forme semi-circulaire) ; *g* **Rundstufe** [f] (Halbkreisförmige Treppenstufe).

5480 semidesert [n] [US]/**semi-desert** [n] [UK] *geo.* (Transition area between savanna[h], steppe or desert and characterized by a poor vegetation cover caused by lack of precipitation and soil water); *syn.* semiarid region [n] [US]/semi-arid region [n] [UK]; *s* **semidesierto** [m] (Zona de transición entre la sabana resp. estepa y el desierto); *f* **région** [f] **semi-aride** (Zone de transition entre la savane ou la steppe et le désert, à précipitations saisonnières mais irrégulières et couverte d'une végétation pauvre [steppe ouverte]) ; *syn.* région [f] semi-désertique ; *g* **Halbwüste** [f] (Übergangsbereich zwischen Savanne resp. Steppe und Wüste, der durch Niederschlags- und Bodenwassermangel und somit durch eine entsprechend arme Vegetationsausstattung geprägt ist).

5481 semidetached house [n] *arch. leg. urb.* (Two single-family units attached to each other with a common wall, each on a separate land parcel; ►single-family house); *syn.* duplex [n] [also US], double house [n] [also US]; *s* **casa** [f] **pareada** (Casa de dos viviendas que consiste en dos unidades construidas una junto a la otra con un muro separador común y en la que cada una tiene un terreno propio; ►casa unifamiliar); *syn.* casa [f] para dos familias, casa [f] semiseparada, casa [f] melliza, casas [fpl] gemelas; *f* **maisons** [fpl] **jumelles** (Unité d'habitation indépendante constituée de deux ►maisons individuelles sur deux parcelles distinctes ayant un mur commun ; les **m. j.** sont en général de superficie égale et/ou de plan de masse symétrique) ; *syn.* maison [f] jumelée ; *g* **Doppelhaus** [n] (Im Verständnis des Planungsrechts eine freistehende Gebäudeeinheit, die durch zwei selbständige, aneinander gebaute Häuser zusammengefügt ist, i. d. R. zwei ►Einfamilienhäuser oder zwei Geschosswohnhäuser mit einer gemeinsamen Hauswand, beide auf einem eigenen Grundstück; jede **D.hälfte** hat ein eigenes Erschließungselement, wie Zugang, Eingang und Treppenhaus. Zum Wesensmerkmal eines **D.es** gehört es nicht, dass die beiden das **D.** bildenden Gebäude etwa zur selben Zeit errichtet, gleich groß oder gar symmetrisch sein müssen; cf. BAUR 1999, 479 f).

5482 semidry grassland [n] [US]/**semi-dry grassland** [n] [UK] *phyt.* (Plant community comprising *Mesobromion* growing in shallow soils on rocks, gravel or sand, which has less extreme temperature and dryness requirements than dry grassland; ►dry meadow, ►limestone grassland); *s* **pastizal** [m] **semiseco** (Comunidad herbácea que crece dentro del área de distribución del *Xerobrometum* sobre suelos poco profundos rocosos o arenosos y, en comparación con el ►pastizal seco, con menos predilección por las ubicaciones muy secas y calurosas; ►pastizal calcícola seco); *f* **pelouse** [f] **héliophile dense** (Formation herbacée pâturée ou de fauche, haute et dense, floristiquement diversifiée et riche en espèces herbacées et graminées thermophiles et mésoxérophiles appartenant à l'alliance du *Mesobromion*, résultant de la reconstitution séculaire d'un tapis végétal après l'abandon des cultures sur les sols de la chênaie pubescente [classe des *Querco-Fagetea*], localisée sur les versants et affleurements rocheux, les graviers et sables calcaires, remplaçant ou accompagnant les ►pelouses arides *[Xerobromion]*, dans les

S

zones d'exposition moins favorable ou sur un sol plus profond et une pente plus douce, évoluant vers le pré-bois calcicole puis vers la chênaie-frênaie ; GGV 1979, 306-307 ; ▶pelouse calcicole maigre) ; *syn.* pelouse [f] semi-aride ; *g* **Halbtrockenrasen [m]** (Gras- und krautreiche Pflanzengesellschaft auf überwiegend durch Beweidung oder einschürige Mahd bewirtschafteten Flächen an Stelle anspruchsvoller Gesellschaften der Buchen- und sommergrünen Eichenwälder Europas *[Querco-Fagetea]*. Sie kommt innerhalb des Verbreitungsgebietes des Trespen-Halbtrockenrasens *[Mesobromion erecti]* vor — vor allem im Übergangsbereich vom natürlichen ▶Trockenrasen felsiger Standorte auf flachgründigen Fels-, Kies- oder Sandstandorten mit im Vergleich zum Trockenrasen weniger extremen Wärme- und Trockenheitsbedürfnissen; im Allgemeinen gekennzeichnet durch eine schlechte Wasserversorgung sowie durch Nährstoffarmut oder Einseitigkeit der Nährstoffversorgung; ▶Kalkmagerrasen); *syn.* Halbstrauch-Gebüschtrockenrasen [m].

5483 semi-dwarf fruit tree [n] [US] *arb. hort.* (Clear stem height 120 to 150cm for fruit trees/shrubs, 60cm for standard roses; BS 3936: part 1; in U.S., a **s.-d. f. t.** is measured from 12 to 15 feet of full height; ▶dwarf fruit tree [US]/short standard [UK], ▶standard [UK] 1); *syn.* half standard [n] [UK]; *s* **árbol [m] de pie medio** (Altura de la cruz del tronco de 100-120 cm en frutales, de 60 cm en rosales; ▶árbol de pie alto, ▶árbol de pie bajo); *f* **arbre [m] demi-tige** (Suivant la Norme N.F. V 12 055 terme qui désigne les arbres dont la hauteur du tronc mesurée du collet jusqu'à la première branche est égale à 1,30 m l'écart de formation sur la hauteur de la tige est de ± 10 cm ; ▶arbre courte-tige, ▶arbre mini-tige, ▶arbre tige) ; *syn.* demi-tige [m] ; *g* **Halbstamm [m]** (Stammhöhe von 100-120 cm bei Obstgehölzen, 60 cm bei Stammrosen; ▶Hochstamm 2, ▶Niederstamm).

semi-evergreen [adj] *bot. hort.* *▶winter green.

5484 semiliquid manure [n] [US]/**semi-liquid manure** [n] [UK] *agr. envir.* (Thick liquid mixture of excrements and urine of cattle, hogs or poultry; sometimes diluted with water; ▶liquid manure, ▶straw dung); *s* **licuame [m]** (Mezcla semilíquida de excrementos y orina de animales domésticos [ganado vacuno, porcino o aves de corral] a veces diluída con agua; ▶purín, ▶estiércol [de establo]); *syn.* abono [m] semi líquido, agua [f] de estiércol); *f* **lisier [m]** (Mélange de déjections animales solides et liquides, produites par des animaux [bovins, cochons ou volaille] maintenus sur des aires non paillées ou sur des caillebotis et surtout employé pour la fertilisation des champs et des prairies ; DA 1981 ; ▶purín, ▶fumier) ; *g* **Gülle [f]** (Dickflüssiges Gemisch aus Kot- und Harnausscheidungen von Rindern, Schweinen oder Geflügel, auch vermischt mit Wasser sowie deren natürliche Umwandlungsprodukte; der Begriff **G.** ist im Vergleich zur ▶Jauche ausschließlich auf tierische Ausscheidungen beschränkt, wird jedoch im allemannischen Sprachgebiet oft synonym für Jauche verwendet; ▶Mist).

semi-mature trees [npl] *constr.* ▶transplantation of semi-mature trees.

5485 semi-natural [adj] *ecol.* (Term used to differentiate the degrees of naturalness of ecosystems, meaning those almost exclusively composed of indigenous species, but influenced by extensive human interference, causing the introduction of new characteristic combinations and propotions in numbers; e.g. ▶straw meadows [US]/litter meadows [UK], ▶nutrient-poor grassland, ▶dwarf-shrub plant communities; ▶degree of landscape modification, ▶near natural); *s* **seminatural [adj]** (Para diferenciar los ecosistemas según su ▶grado de modificación [de ecosistemas y paisajes]: Se trata de aquéllos compuestos casi exclusivamente de especies autóctonas [▶poco alterado], pero

donde la combinación característica de especies y la abundancia han sufrido cambios debido al uso extensivo [siega, pastoreo], como p. ej. ▶prados de heno, ▶prados oligótrofo, de camada, muchas turberas bajas, pastizales secos, ▶matorrales bajos, landas de brezal enano en zonas de bosque potencial); *f* **d'artificialisation moyenne [loc]** (Terme utilisé pour caractériser le ▶degré d'artificialisation des écosystèmes presque exclusivement peuplés d'espèces indigènes [▶faiblement artificialisé] dont l'ensemble caractéristique d'un groupement et la densité des individus de chaque espèce ne peuvent rester stable que grâce à une utilisation extensive par l'homme, p. ex. ▶prairie à litière, ▶prairie mésophile de fauche, ▶pelouse oligotrophe, ▶groupement d'arbrisseaux nains [lande à éricacées] ou écosystèmes ayant subis dans le passé une action humaine prolongée et actuellement à l'état semi-naturel) ; *syn.* semi-naturel, elle [adj] ; *g* **halbnatürlich [adj]** (*Zur Unterscheidung von Ökosystemen nach ▶Natürlichkeitsgraden* fast ausschließlich aus einheimischen Arten entstanden [▶naturnah], jedoch zu neuen charakteristischen Artenkombinationen und Mengenverhältnissen vereinigt, die nur bei extensiver menschlicher Nutzung [Mahd, Weide] bestehen, z. B. ▶Streuwiesen, viele Niedermoore, ▶Magerrasen, ▶Zwergstrauchgesellschaften in waldfähigen Gebieten; cf. BUCH 1978, Bd. 2; im Vergleich zur ursprünglichen Vegetation kommt es bei **h.en** Ökosystemen resp. Lebensgemeinschaften zu deutlichen Artenverschiebungen, wobei neben einheimischen Arten meist einzelne alt- oder neuheimische auftreten [Archeo- und Neophyten]. Viele **h.e** Vegetationseinheiten sind typische Vertreter der extensiven Kulturlandschaft früherer Jahrhunderte, enthalten eine Vielzahl gefährdeter Arten [Rote-Listen-Arten] und sind deshalb äußerst schutzwürdig; cf. WIT 1996).

seminiferous [adj] *bot. hort.* ▶seed-bearing.

5486 semiproductive cultivation [n] [US]/**semi-productive cultivation** [n] [UK] *agr. hort.* (Tilling of ground for growing of limited crops on agricultural land, rather than growing intensively for maximum produce, in order to leave the land fallow and provide more landscape diversity; ▶landscape management); *s* **cultivo [m] extensivo** (Método de producción agrícola en el cual la ▶gestión y protección del paisaje es un objetivo declarado al igual que los resultados económicos); *syn. agr.* agricultura [f] extensiva; *f* **culture [f] extensive** (Culture dont le volume de production à l'hectare est faible. Cette forme d'agriculture constitue une occupation des sols ménageant les ressources naturelles et est parfois utilisée comme mesure de ▶gestion des milieux naturels remarquables) ; *g* **extensive Bewirtschaftung [f]** (Landwirtschaftliche Bewirtschaftungsmethode, bei der die ▶Landschaftspflege im Vergleich zur Produktion vorrangig oder zumindest gleichrangig ist); *syn. agr.* extensive Landwirtschaft [f].

semipublic authority [n] *adm. plan.* ▶public or semipublic authority [US]/public or semi-public authority [UK].

5487 semi-shade [n] *phyt. plant.* (Light shade from open foliage trees); *syn.* partial shade [n] [also US], part shade [n] [also US]; *s* **semisombra [f]**; *f* **mi-ombre [f]** (Ombre légère produite par la couronne des arbres dispensant un ombrage léger) ; *syn.* ombre [f] légère ; *g* **Halbschatten [m]** (Durch lichtdurchlässige Baumkronen, Schattiermatten oder durch Gebäude entstehender schwacher Schatten); *syn.* lichter Schatten [m].

5488 semishaded location [n] [US] *phyt. plant.* (Place with partial shade); ▶shady location); *syn.* semi-shaded position [n] [UK]; *s* **exposición [f] semisombra** (▶umbría); *f* **exposition [f] mi-ombragée** (Emplacement d'une strate herbacée ou arbustive recouvert par une légère ombre produite par la couronne des arbres ou par l'ombre projetée d'un bâtiment ; ▶exposition [très] ombragée) ; *syn.* exposition [f] à mi-ombre, situation [f] mi-

S

ombragée, situation [f] à mi-ombre ; **g Halbschattenlage [f]** (Standort einer Kraut- oder Strauchschicht, der im Vergleich zur ▶Schattenlage durch ein lichtdurchlässiges Kronendach oder durch den Schattenwurf von Bauwerken gekennzeichnet ist); *syn.* absonnige Lage [f].

semi-shaded location [n] [US] *hort.* ▶plant in a semi-shaded location [US]/plant in a semi-shaded position [UK].

semi-shaded position [n] [UK] *hort.* ▶plant in a semi-shaded location [US]/plant in a semi-shaded position [UK]; *phyt. plant.* ▶semi-shaded location [US].

semi-shrub [n] *bot.* ▶suffruticose shrub.

5489 senescence [n] (1) *bot.* (Aging process of leaves caused by breakdown of cellular structures leading to abscission at the end of vegetation period; ▶cryptophyte, ▶defoliation, ▶hemycriptophyte, ▶life form); *s* **caída [f] de las hojas** (Proceso periódico de envejecimiento de las hojas causado por la destrucción de las estructuras celulares que llevan a la pérdida de función de las mismas al final del periodo/período vegetativo; ▶criptófito, ▶defoliación, ▶forma biológica, ▶hemicriptófito); *f* **disparition [f] des feuilles** (Processus périodique de jaunissement et flétrissement des feuilles pour stocker des éléments nutritifs dans les organes hivernants et qui provoque la ▶défeuillaison chez les ▶hémicryptophytes et les ▶cryptophytes [▶type biologique]) ; *g* **Einziehen [n, o. Pl.] des Laubes** (Transport der Assimilate aus den Blättern in die Überwinterungsorgane am Ende einer Vegetationsperiode, um bei Gehölzen den ▶Laubfall, bei ▶Hemikryptophyten und ▶Kryptophyten [▶Lebensform] das Absterben oberirdischer Organe und die Winterruhe/Trockenruhe vorzubereiten. Über diesen Zeitpunkt hinaus können die abgetrockneten Sprossachsen [Chinaschilf *Miscanthus*], Blätter [Gräser, Farne] oder Fruchtstände [Brandkraut *Phlomis russeliana*] einiger strukturstabiler Arten den Winteraspekt von Staudenpflanzungen wesentlich bereichern — Rückschnitt je nach Klimaregion und Austriebsverhalten im Februar bis Ende März resp. Mitte April); *syn.* Vergilben [n, o. Pl.] des Laubes.

5490 senescence [n] (2) *arb.* (Advanced aging of trees which is characterized by a decreasing production of biomass); *s* **senescencia [f]** (En las comunidades vegetales, la etapa senescente se caracteriza por disminuir la acumulación de biomas, pérdida de ramas en los árboles viejos, incremento pequeño de la capa superficial del suelo, etc. Se aplica también a los lagos cuyo volumen va disminuyendo por sus propios aportes sedimentarios y en los que se dan procesos de eutrofización; DINA 1987); *syn.* envejecimiento [f], senectud [f]; *f* **sénescence [f]** (Fort vieillissement d'un arbre caractérisé par une croissance faible — formation réduite de la phytomasse aérienne) ; *g* **Vergreisung [f]** (Starke Alterung bei Bäumen, die durch sehr schwaches Wachstum — mangelnde Bildung an Phytomasse — gekennzeichnet ist).

senior landscape architect [n] **in a public authority** [UK] *prof.* ▶chief landscape architect in a public authority/agency [US].

5491 senior landscape architect [n] **in private practice** *prof.* *s* **arquitecto/a [m/f] paisajista jefe/a** (en gabinete de arquitectura paisajista); *f* **architecte [m/f] paysagiste directeur de projet** (dans un bureau d'études) ; *g* **Garten- und Landschaftsarchitekt/-in [m/f] in leitender Stellung (3)** (im freien Planungsbüro).

5492 sensitive area [n] *ecol. plan.* (Identifiable area where the natural ecosystem is so unusual and vulnerable that it could be destroyed, severely altered or irreversibly changed by human intrusion); *syn.* critical area [n] [also US], fragile area [n] [also US]; *s* **zona [f] frágil** (Área identificable de un ecosistema natural que es especialmente sensible a impactos); *f 1* **zone [f] fragile** (Tout ou partie de territoire sujet à être altéré facilement) ;

syn. espace [m] fragile, espace [m] vulnérable ; *f 2* **espace [m] sensible** (Tout ou partie de paysage, site, milieu naturel pour lequel toute atteinte entraîne irrémédiablement une dégradation notable et une altération à long terme de leur caractère ou de leur fonction) ; *f 3* **espaces [mpl] naturels sensibles [ENS]** (Zone de préservation de la qualité des sites, des paysages et des milieux naturels et de sauvegarde des habitats naturels ; cette réglementation qui remplace celle des périmètres sensibles institue le droit de préemption, le classement des espaces boisés, la réglementation de la construction, du camping et caravaning, de la circulation, la taxe départementale d'espaces naturels sensibles ; cf. loi n° 85-729 du 18 juillet 1985 ; cette réglementation remplace celle des périmètres sensibles) ; *syn. obs.* périmètres [mpl] sensibles ; *g* **empfindlicher Raum [m]** (Landschaft oder Teile davon, die keine Eingriffe in Natur und Landschaft ohne nachhaltige Störungen vertragen; in F. ist **e. R.** eine Schutzgebietskategorie); *syn. obs.* umweltempfindlicher Raum [m].

5493 sensitive [adj] **to salt** *plant.* (TEE 1980, 99; descriptive term for plants, which cannot tolerate salt); *s* **halófobo/a [adj]** (Término descriptivo para organismos que no toleran la salinidad); *syn.* sensible a la salinidad [loc]; *f* **sensible au sel [loc]** (Concerne les plantes qui présentent des réactions d'incompatibilité aux sels) ; *g* **salzempfindlich [adj]** (Pflanzen betreffend, die auf Salze unverträglich reagieren).

separate [adj] *urb.* ▶detached.

5494 separate contract maintenance [n] *constr.* (Special maintenance operations, usually during a specified period, which are undertaken beyond the ▶establishment maintenance period in order to ensure the successful establishment of an area of new planting—frequently involving native species planted at a small initial size; ▶maintenance management, ▶maintenance work); *syn.* follow-up care [n] [also US], follow-up maintenance [n] [also US] (MET 1985, 2); *s* **trabajos [mpl] regulares de mantenimiento** (Medidas de mantenimiento después de finalizar los de ▶mantenimiento inicial para asegurar el crecimiento posterior de la vegetación; ▶mantenimiento de zonas verdes, ▶trabajos de mantenimiento); *syn.* trabajos [mpl] generales de mantenimiento; *f* **travaux [mpl] d'entretien courant** (Travaux saisonniers de caractère horticole faisant suite aux ▶travaux d'entretien pendant l'année de garantie afin de prolonger la vitalité des plantations et des surfaces engazonnées, ou travaux réguliers [quotidien, hebdomadaire] afin de maintenir en bon état les ouvrages et circulations ; ▶entretien courant, ▶maintenance) ; *syn.* travaux [mpl] d'entretien après l'année de garantie ; *g* **Erhaltungspflege [f, o. Pl.]** (Pflegearbeiten nach Abschluss der ▶Entwicklungspflege zur Sicherung und bestandstypischen Weiterentwicklung und damit zur Förderung der Funktion der Anpflanzungen; im Deutschen ist der in der DIN 18919 verwendete Terminus *Unterhaltungspflege* sprachlich ungeschickt, da die beiden Begriffe ▶Unterhaltung und ▶Pflege unterschiedliche Bedeutungen haben); *syn.* Bestandspflege [f, o. Pl.].

5495 separate sewerage system [n] *envir.* (Collection and conveyance of sanitary sewage and storm water runoff in different pipe lines; ▶combined sewerage system [US]/combined sewer system [UK]); *s* **sistema [m] de alcantarillado separativo** (Sistema separado de recolectar las aguas residuales y pluviales en la ciudad; ▶sistema de alcantarillado unitario); *f 1* **réseau [m] d'assainissement séparatif** (Système qui implique la présence de deux réseaux, l'un est réservé à l'évacuation des eaux usées et sous certaines conditions, des effluents industriels, le second étant réservé aux eaux pluviales ; ▶réseau [d'assainissement] unitaire) ; *f 2* **réseau [m] pseudo-séparatif** (Système d'assainissement qui se distingue du réseau séparatif par le fait que le réseau d'eaux usées peut collecter les eaux pluviales des pro-

S

priétés riveraines) ; *g* **Trennverfahren** [n] (Getrennte Ableitung von Schmutzwasser und Regenwasser in verschiedenen Leitungssystemen. In Frankreich gibt es ein System *réseau pseudo-séparatif*, bei dem das Regenwasser der Anlieger in Wohngebieten zusammen mit dem Schmutzwasser in die Schmutzwasserkanalisation eingeleitet wird; ►Mischverfahren).

5496 separate specification item [n] *constr. contr.* (Independent item [►specification item number], which is neither related to other items nor limited in scope; ►add-on item, ►alternate specification item, ►list of bid items and quantities [US]/schedule of tender items [UK], ►optional specification item); *s* **partida** [f] **normal** (En el ►resumen de prestaciones, ►número de órdenes independiente que ni se mezcla con otras partidas ni se limita en su envergadura; ►partida alzada, ►partida discrecional, ►partida [de obra] adicional); *f* **numéro** [m] **de prix de prestation courante** (Numérotation d'une prestation courante de travaux dans le ►descriptif-quantitatif ou dans le Bordereau des Prix Unitaires [B.P.U.] ; ►numéro d'ordre, ►numéro de prix pour une variante, ►numéro de prix provisoire, ►numéro de prix pour plus-value) ; *g* **Normalposition** [f] (Im ►Leistungsverzeichnis gekennzeichnete, eigenständige ►Ordnungszahl, die weder mit anderen Positionen im Zusammenhang steht, noch mit einem Vorbehalt verbunden ist; ►Alternativposition, ►Bedarfsposition, ►Zulageposition).

separation [n] *envir. leg.* ►waste separation.

separation plant [n] *envir.* ►waste separation plant.

separation strip [n] *trans.* ►traffic separation strip.

5497 separator [n] *trans.* (Separating element such as a green strip in cross-sections of roads, which divides different types of traffic; e.g. through traffic, residents' traffic, pedestrian and bicycle traffic; ►freeway median strip [US]/motorway central reservation [UK], ►median strip [US]/central reservation [UK], ►traffic separation strip, ►vegetated strip); *s* **banda** [f] **de separación** (Término genérico para todo tipo de instalaciones o demarcaciones para separar el tráfico de diferentes tipos o las direcciones del mismo; ►banda verde central, ►franja divisoria central de autopista, ►franja divisoria central y ►línea de demarcación de las direcciones del tráfico); *syn.* franja [f] de separación; *f* **bande** [f] **de séparation** (Terme générique pour un dispositif séparatif des différentes fonctions sur la voie publique [p. ex. une bande enherbée plantée ou non d'arbres] ou une signalisation horizontale dans le profil en travers d'une chaussée qui matérialise les flux antagonistes dans les différents types de circulations, telles que la circulation riveraine, le trafic de transit, la circulation piétonne et cycliste ; ►îlot directionnelle, ►bande verte d'accompagnement, ►terre-plein central, ►terre-plein central d'une autoroute) ; *syn.* terre-plein [m] séparatif ; *g* **Trennstreifen** [m] (Einrichtung im Straßenraum, um Richtungsfahrbahnen [durch ►Mittelstreifen] oder verschiedene Verkehrsarten wie Durchgangsverkehr, Anliegerverkehr, Fuß- und Radfahrverkehr [z. B. durch begrünte Seitentrennstreifen] zu separieren; *UBe* ►Autobahnmittelstreifen, ►Grünstreifen, ►Richtungstrennstreifen).

septic system [n] [US] *envir.* ►individual sewage disposal system [US].

septic system [n] [US]**, wetland** *envir.* ►wastewater treatment wetland.

septic tank disposal system [n] [US] *envir.* ►individual sewage disposal system [US].

5498 sequence [n] **of operations** *plan.* (Progression of work on a project); *syn.* sequence [n] of work, work procedure [n]; *s* **secuencia** [f] **de trabajo** *syn.* proceso [m] de trabajo; *f* **déroulement** [m] **des travaux** (Succession des phases de travaux) ; *syn.* processus [m] du travail ; *g* **Arbeitsablauf** [m]

(Bestimmte Abfolge von Arbeitsschritten); *syn.* Arbeitsabfolge [f], Arbeitsverlauf [m].

sequence [n] **of work** *plan.* ►sequence of operations.

sequent occupance [n] **of landscape** *landsc.* ►land consumption.

seral plant community [n] *phyt.* ►successional plant community.

5499 sere [n] *phyt.* (Succession of genetically-related plant communities, sequential stages which follow one another. A sere includes at least one pioneer stage, usually several transition stages [seral communities, successional communities; TEE 1980], and a terminal stage [climax community]); *s* **serie** [f] **(de sucesión)** (Conjunto de las etapas de desarrollo de las comunidades, genéticamente relacionadas entre sí, que se van sucediendo con el tiempo; cf. BB 1979, 630); *f* **série** [f] **de végétation** (Concept phytosociologique destiné à faciliter l'étude du tapis végétal, traduisant la succession dynamique lente et spontanée de groupements végétaux depuis le sol nu jusqu'au climax [*série évolutive normale* groupements pionniers, intermédiaires, préforestiers et forestiers] ; ce concept fait intervenir la référence à l'arbre forestier qui domine le climax ainsi qu'à des notations écologiques ou biogéographiques ; *on distingue* les séries progressives, les séries régressives, les séries tronquées DEE 1982, VCA 1985, 71-75) ; *g* **Sukzessionsserie** [f] (Abfolge der genetisch verbundenen, zeitlich sich ablösenden Gesellschaftsstadien. Eine normale Serie beginnt in der Regel auf Neuland, umfasst meist mehrere Übergangsstadien und endet mit einer klimatisch bedingten Schlussgesellschaft); *syn.* Serie [f].

5500 serious intrusion [n] *conserv. plan.* (►intrusion upon the natural environment); *s* **intrusión** [f] **significativa** (►intrusión en el entorno natural); *f* **atteinte** [f] **notable** (Préjudice environnemental portant atteinte à l'intégrité du milieu naturel [destruction de la biodiversité, perte de la jouissance d'un bien environnemental, etc.] et altérant de façon durable la qualité du milieu naturel, un espace protégé, etc. ; ►atteinte au milieu naturel) ; *syn.* agression [f], atteinte [f] grave à l'environnement ; *g* **erheblicher Eingriff** [m] (Vorhaben, das Gestalt oder Nutzung von Flächen verändert und deshalb die Leistungs- und Funktionsfähigkeit der Naturhaushaltes oder das Landschaftsbild sehr stark beeinträchtigen kann; e. E.e sind z. B. der Bau von Straßen, Eisenbahnlinien, Deponien, Kraftwerksbauten und andere die Landschaft stark bestimmende Gebäude, der Ausbau von Gewässern, die Einrichtung von Bodenentnahmestellen und Waldrodungen; ►Eingriff in Natur und Landschaft).

service area [n] [UK]**, motorway** *trans.* ►service area with all facilities.

5501 service area [n] **with all facilities** *trans.* (Stopover place with food dispensers, petrol/gas station, kiosk, and rest rooms [US]/toilet buildings [UK] on a motorway or dual freeway or highway [US]/carriage-way [UK]); *syn.* motorway service area [n] [UK], motorway services [npl] [UK]; *s* **área** [f] **de servicio** (Zona de parada en autopista, equipada con servicios de restaurante, gasolinera, quiosco de venta, servicios sanitarios, etc.); *f* **aire** [f] **de service principale** (Aire de repos assurant le stationnement et le ravitaillement des véhicules dont l'équipement peut être extrêmement développé, tant en ce qui concerne la station service, que les parcs de stationnement et lieux d'accueil des usagers [motel, restaurant, etc. ; ICTA 1985, 45] ; on distingue les **a. de s. normales** et les **a. de s. principales**) ; *g* **bewirtschaftete Rastanlage** [f] (Rastanlage mit Restaurant, Tankstelle oder Verkaufskiosk mit WC an einer Autobahn).

5502 service area [n] **with partial facilities** *trans.* (Stop-over place with toilet facilities, and often with a picnic area; ►pull-off [US]/lay-by [UK]); *s* **área** [f] **de parada (con**

S

servicios) (Área de descanso en autopista equipada solamente con servicios sanitarios e instalaciones básicas como bancos, papeleras, etc.; ▶área de parada sin servicios); *f* aire [f] de service normale (Aire équipée d'installations sanitaires et souvent de tables de pique-nique, le long des routes et autoroutes, permettant l'arrêt et le stationnement des véhicules et destinée au repos et à l'agrément des usagers, offrant généralement des prestations commerciales légères ; ▶aire d'arrêt, ▶aire de repos 1) ; *g* **teilbewirtschaftete Rastanlage [f]** (Rastanlage nur mit WC-Gebäude und oft mit Picknickeinrichtung; ▶unbewirtschaftete Rastanlage).

service charge [n] [UK] *adm. urb.* *▶utility connection charge.

5503 service load [n] *eng.* (UBC 1979, 295; live and dead loads—without load factors—used to proportion construction elements of structures); *s* **carga [f] móvil** (Valor estadístico que indica la capacidad de carga cambiante —p. ej. personas, vehículos, presión del viento— de las piezas de construcción, que ha de añadirse al peso propio de la estructura); *f* **charge [f] d'exploitation** (BON 1990, 156 ; charge déterminée en fonction de la nature des usages divers calculée pour les différentes parties d'un bâtiment [habitation, commerce, restauration, garage et parc de voitures etc.] ; *terme spécifique* **c.** climatique : charge de neige, pluie et vent) ; *g* **Verkehrslast [f]** (Statische Größe bei der Berechnung der veränderlichen Belastbarkeit von Bauteilen, z. B. durch Personen, Fahrzeuge, Winddruck, die zum Eigengewicht des Bauwerkes hinzugerechnet werden muss); *syn.* Nutzlast [f].

5504 service road [n] *trans.* (▶Local street with access only for residents running parallel to a major street, which provides access to the adjoining properties, and which may be separated by a grass or planted strip; ▶delivery access); *syn.* frontage road [n] (WEB 1993) [also US], service street [n] [also US]; *s* **vía [f] de servicio** (▶Calle de vecindario cuya función es la de permitir el abastecimiento de bienes a una zona y que en general no se utiliza para el tránsito de paso; en muchos casos se trata de una vía lateral paralela a la calle principal y separada de ésta por una banda verde o un paseo; ▶acceso de suministro); *f* **contre-allée [f] de service** (▶Rue riveraine en général latérale à la voirie principale à double voie et généralement séparée de celle-ci par une bande plantée, prévue pour la sécurité des usagers, le maintien de la vitesse de référence et la protection contre les nuisances des véhicules [odeurs, bruit]) ; cf. BON 1990 ; *syn.* contre-voie [f], voie [f] de service) ; *g* **Nebenfahrbahn [f]** (▶Anliegerstraße, die der Versorgung der Anlieger mit Gütern dient und im Übrigen dem allgemeinen Straßenverkehr [Durchgangsverkehr] nicht zur Verfügung steht; verläuft häufig seitlich parallel zur mehrspurigen Hauptverkehrsstraße resp. Schnellverkehrsstraße und ist i. d. R. durch einen Grünstreifen abgetrennt; ▶Anlieferungsweg); *syn.* Anliegerfahrbahn [f].

services [npl] *constr. contr.* ▶additional work and services, ▶all additional work and services included; *contr. prof.* ▶basic professional services; *trans. urb.* ▶local post services; *urb.* ▶public utility services; *contr. prof.* ▶special professional services; *constr. contr.* ▶supplemental construction services or supplies.

services [npl], **additional planning services** [UK] *contr. prof.* ▶special professional services.

services [npl], **additional professional services** *constr. contr.* ▶supplemental professional services.

services [npl] [UK], **construction management** *prof.* ▶construction phase—administration of the construction contract [US].

services [npl], **contract for planning** *contr. prof.* ▶planning contract.

services [npl], **designated** [US] *contr. prof.* ▶special professional services.

Services [npl] [US], **Highways and Transportation** *adm.* ▶public works department.

services [npl], **local post and telecommunication** *contr. prof.* ▶local post services.

services [n] [UK], **motorway** *trans.* ▶service area with all facilities.

services [npl] [UK], **post-completion advisory** *prof.* ▶concluding project review.

services [npl] [UK], **site management** *prof.* ▶construction phase—administration of the construction contract [US].

5505 services [npl] **and materials** [npl] **by client/ owner** *constr. contr.* (Construction operations and materials, stipulated in a construction contract, which are to be undertaken or supplied by the client himself); *syn.* construction services [npl] by client/owner [also US], services [npl] and purchases [npl] by client/owner [also US], supplies [npl] (purchased) by client/owner; *s* **servicios [mpl] y materiales [mpl] del comitente** (Servicios previstos por el contratista, pero que son proporcionados por el comitente mismo); *f* **prestations [fpl] et fournitures [fpl] fournies par le maître d'ouvrage** (Prestations et fournitures mises à disposition de l'entrepreneur par le maître de l'ouvrage) ; *g* **bauseitige Leistungen [fpl] und Lieferungen [fpl]** (Im Vertrag des Auftragnehmers ausbedungene Bauleistungen und Materiallieferungen, die vom Auftraggeber selbst übernommen werden; cf. § 2 [4] VOB Teil A und § 13 [3] VOB Teil B); *syn.* Leistungen [fpl] und Bereitstellung [f] von Stoffen durch den Auftraggeber.

services [npl] **and purchases** [npl] **by client/owner** [US] *constr. contr.* ▶services and materials by client/owner.

services [npl] **by client/owner** [US], **construction** *constr. contr.* ▶services and materials by client/owner.

services included [loc] *constr. contr.* ▶all additional work and services included.

services network [n] *plan. trans. urb.* ▶public utility network.

services [npl] **or supplies** [npl], **construction** *constr. contr.* ▶supplemental construction services or supplies.

service street [n] [US] *trans.* ▶service road.

5506 services trench [n] *constr.* (Long narrow excavation for laying utility lines or pipes to a building or area; ▶excavation of service trenches); *s* **zanja [f] de abastecimiento y drenaje** (Excavación lineal para la instalación de tuberías de distribución y recogida de aguas, aguas residuales, gas o líneas de suministro eléctrico); ▶excavación de zanjas [de conducción]); *f* **tranchée [f]** (Excavation linéaire pour la mise en œuvre des réseaux de distribution ou d'évacuation d'eaux, des canalisations de transport de combustibles liquides ou gazeux, des lignes et des câbles électriques, des réseaux de télécommunications, etc. ; CCM 1984, 143 ; ▶ouverture des tranchées) ; *g* **Leitungsgraben [m]** (Graben zum Einbau von Ver- oder Entsorgungsleitungen; ▶Ausheben von Leitungsgräben).

set [n] [UK] *constr.* ▶live picket [US].

set [pp] **apart** [US] *landsc.* ▶isolated (2).

5507 setback [n] [US]/**set-back** [n] [UK] *leg. urb.* (**1.** Distance between a building and front, side and rear lot lines [US]/front, side and rear property lines [UK]; **in U.S.**, in some places front setbacks may be variously measured from the street right-of-way or the middle of the street; ▶required side yard, ▶planting setback [UK]. **2.** *arch. leg. urb.* Making a whole building or upper storeys of a high-rise or parts of a fassade further back than the lower ones for aesthetic, structural, or land-use restriction reasons); *s* retranqueo [m] (Distancia de un edificio al borde del solar; ▶androna, ▶línea de retranqueo, ▶retranqueo de plantación); *syn.* distancia [f] al límite del predio; *f 1* marge [f] de reculement (Implantation des constructions par rapport **a.** aux voies et emprises publiques ou privées, **b.** aux limites séparatives ; cette prescription reportée sur les documents graphiques du PLU, imposant le recul des constructions nouvelles à une certaine distance comptée horizontalement de tout point de la construction imposée à une construction à édifier en bordure d'une voie publique ou privée ou d'une limite séparative [▶limite séparative de fonds de parcelle, ▶limite séparative latérale] ; sa largeur se mesure à partir de l'alignement actuel ou futur si un élargissement de la voie est prévu au plan ; la **m. de r.** répond à un motif de protection acoustique des constructions et/ou à des motifs d'harmonie ou architecturaux ainsi qu'urbanistiques [maintien de perspective, amélioration des constructions existantes, etc.] ; cf. C. urb., art. R 111-24 ; ▶limite de constructibilité, ▶polygone d'implantation, ▶distance d'éloignement) ; *syn.* marge [f] de recul ; *f 2* implantation [f] en retrait *arch. leg. urb.* (de tout ou parties de bâtiments [façade, étages supérieurs] par rapport à l'alignement pour des raisons architecturales, statiques ou de réglementation de la construction) ; *syn.* recul [f] de l'implantation) ; *f 3* marge [f] d'isolement (Distance d'implantation des constructions isolées [▶implantation des constructions en ordre discontinu] par rapport aux ▶limites séparatives définie dans les documents d'un PLU/POS ; la largeur de la marge d'isolement est en général au moins égale à la hauteur du bâtiment pour les parties de constructions comportant des baies de pièces habitables) *g 1* Grenzabstand [m] (Abstand eines Gebäudes von den Grundstücksgrenzen; in der ▶offenen Bauweise werden Gebäude mit seitlichem **G.** [▶Bauwich], bei geschlossener Bauweise ohne **G.** und bei der Blockbebauung i. d. R. an die Straßenbegrenzungslinie/straßenseitige Grundstücksgrenze gebaut; ▶Baugrenze, ▶Grenzabstand von Gehölzen); *g 2* Zurücksetzung [f] *arch. leg. urb.* (Versetzung ganzer Baukörper oder deren Teile, z. B. von oberen Stockwerken oder Fassadenteilen — von der Straßenseite aus betrachtet — aus gestalterischen, statischen oder bauordnungsrechtlichen Gründen weiter nach hinten in das Grundstück).

5508 setback line [n] *leg. urb.* (**1.** A line shown on a legally-binding land-use plan, such as a ▶zoning map [US]/Proposals Map/Site Allocations Development Plan Document [UK], which is fixed at a specific distance from the front, back or side property line of a lot or street right-of-way, beyond which a structure may not extend; **in U.S.**, is shown on a recorded map of a ▶building plot. **2.** *Specific term* **front setback line:** line set at a statutory distance away from and parallel to the front property line indicating the zone within which, no portion of a building or attached ancillary structure may be constructed, including overhanging eaves and canopies. *NOTE* front setback line also establishes the minimum front yard depth requirement. In addition building side and rear setback lines indicate the net zone of permitted construction; ▶buildable area, ▶mandatory building line, ▶required side yard [US], ▶zoning ordinance [US]/use class order [UK]); *syn.* building restriction line [n] [US], yard line [n] [US] (IBDD 1981, 120); *syn. to 2.* building line [n]; *s 1* línea [f] de retranqueo (Línea definida en plan parcial que fija a una distancia específica de los límites delantero, laterales y trasero de la propiedad más allá de los cuales no está permitida la edificación); *s 2* línea [f] de la calle (Línea que limita la zona edificable de la no edificable que servirá como vía de comunicación, por lo que no está permi-

tido levantar edificios, más allá de ella. En EE.UU., GB y D. se define en los planes parciales; ▶androna, ▶línea de fachada, ▶ordenanza de zonificación, ▶planta edificable); *f 1* límite [f] de constructibilité (±) (U.S., U.K., D., ligne ne devant pas être dépassée sur un document d'urbanisme lors de la construction d'un bâtiment sur un terrain à bâtir) ; *f 2* polygone [m] d'implantation (En France règle inscrite aux documents graphiques d'un ▶plan local d'urbanisme [PLU]/plan d'occupation des sols [POS] instituant que les constructions ou parties de construction, travaux ou ouvrages dont la hauteur excède 0,60 mètre à compter du sol naturel, à l'exception des clôtures, doivent être implantées à l'intérieur de la délimitation de l'emprise du polygone ; ▶emprise maximale de construction, ▶ligne d'implantation, ▶marge de reculement, ▶règles générales d'utilisation du sol) ; *syn.* limite [f] de la bande constructible ; *g* Baugrenze [f] (**1. D.**, gem. ▶Baunutzungsverordnung [BauNVO] im ▶Bebauungsplan festgesetzte, bauaufsichtsrechtliche Linie, die bei neu zu bauenden Gebäuden nicht überschritten werden darf. Wird die **B.** über mehrere ▶Baugrundstücke festgelegt, sollte besser von **Bebauungsgrenze** gesprochen werden; ▶Baufenster, ▶Baulinie, ▶Bauwich. **2.** *UB* straßenseitige Baugrenze oder vordere Baugrenze. **3.** Der US-Begriff ▶*front setback line* entspricht dem obs. deutschen Begriff **Fluchtlinie**).

setback plane [n] [US] *leg. urb.* ▶bulk plane [US].

setback regulation [n] *leg. urb* ▶building setback regulation.

set-back requirements [npl] [UK], **minimum** *leg. urb.* ▶minimum building spacing.

setting [n] *urb. landsc.* ▶in a green setting, ▶urban setting.

setting [n] **apart of trees** [US] *for. landsc.* ▶isolation of trees.

5509 setting-aside [n] **of arable land** [UK] *agr. pol.* (*Europe* According to an European Council resolution, 20% of arable land is to be taken out of production for 5 years in order to reduce the amount of agricultural overproduction. **S.-a.** premiums are paid for the ensuing loss of income; **in U.S.**, the Soil Conservation Service is authorized by federal law to make payments, in a different kind of program[me], for not cultivating over-used land in order to restore its fertility; ▶agricultural reduction program [US]/extensification of agricultural production [UK]); *s* abandono [m] de tierras (Política de la Unión Europea llevada a cabo con el fin de reducir la sobreproducción agrícola. Consiste en dejar sin explotar un 20% de las tierras de cultivo durante cinco años. Para compensar las pérdidas de los agricultores se les paga una indemnización por abandono; ▶extensificación de la agricultura); *syn.* plan [m] de abandono de tierras; *f 1* enfrichement [m] (**1.** Mise en jachère pendant cinq ans de plus de 20 % des terres agricoles dans le cadre de la mise en œuvre de la Politique Agricole Commune [PAC] en liaison avec l'institution d'une prime attribuée en compensation de la perte de revenus occasionnée. **2.** Abandon des espaces naturels dans le cadre de la déprise agricole ; p.°ex. l'abandon des pratiques traditionnelles d'entretien par pâturage aboutis à la fermeture des pelouses sèches et à la disparition des espèces qui lui sont inféodées) ; *f 2* gel [m] environnemental (Terme spécifique qui recouvre les parcelles gelées dans le cadre du règlement 2078/92 concernant des méthodes de production agricole compatibles avec les exigences de la protection de l'environnement ; dans ce cadre un gel aidé de 20 ans est possible ; ▶extensivité des activités agricoles) ; *syn.* gel [m] vert ; *g* Flächenstilllegung [f] (**1.** Herausnahme eines Teils der landwirtschaftlichen Flächen aus der ackerbaulichen Bewirtschaftung zur Reduzierung der Überproduktion. Mit der Einführung sollte Ende der 1980er-Jahre in der EU die Getreideproduktion begrenzt werden; ab 1992 wurde die **F.** durch Beschluss des Europäischen Ministerrates obligatorisch

S

und die Landwirte mit flächengebundenen Ausgleichszahlungen im Rahmen der Agrarförderung durch eine sog. Stilllegungsprämie zum Ausgleich von Einkommensverlusten entschädigt. Im Erntejahr 2008 wurden diese Zahlungen ausgesetzt und ab 2009 entfallen sie. **2.** Herausnahme eines Teils der landwirtschaftlichen Flächen für Umweltzwecke, z. B. zur Verringerung der Anwendung von Dünge- und Pflanzenschutzmitteln, zur Erosionsbekämpfung, zur Erhöhung der biologischen Vielfalt, zur Verbesserung des Lebensraumes für Wildtiere in der Feldflur und zur Verminderung des Wildschadensdruckes auf den Wald. **3.** Seit 1992 werden **F.sflächen** auch zum Anbau nachwachsender Rohstoffe, da diese Produkte nicht zur Lebensmittelherstellung dienen, bewirtschaftet; ►Extensivierung der Landwirtschaft); *syn.* Stilllegung [f] landwirtschaftlicher Flächen.

setting bed [n] [US] *constr.* ►laying course for a paved stone surface.

setting bed [n] [US], **mortar** *constr.* ►mortar bed.

setting bed [n] [US], **sand** *constr.* ►sand bed.

setting-out [n] [UK] *constr. surv.* ►staking out [US].

setting-out plan [n] [UK] *constr. surv.* ►staking-out plan.

setting-out [n] **using ranging-poles** [UK] *constr. surv.* ►staking-out with field rods [US].

5510 settled area [n] *plan. urb.* (Generic term for inhabited land); *syn.* settlement area [n]; *s* **zona [f] poblada** (Término de uso común); *syn.* zona [f] habitada, área [f] poblada, área [f] habitada, zona [f] de asentamiento; *f* **zone [f] habitée** (Langage courant) ; *g* **besiedelter Bereich [m] (1)** (*Allgemeiner Sprachgebrauch*); *syn.* besiedelte Fläche [f], besiedeltes Gebiet [n].

5511 settled soil [n] *constr.* (General term for soil which has assumed its practical limit of consolidation and density; *process* ►ground sinkage); *s* **suelo [m] asentado**; *f* **sol [m] tassé naturellement** (Sol, qui après un déplacement ou un ameublissement a repris, au cours d'un processus naturel lent, par son propre poids ses caractéristiques structurales); *g* **gesetzter Boden [m]** (Boden, der nach Schüttung oder Auflockerung durch sein Eigengewicht nachgibt und seine natürliche Lagerungsdichte einnimmt; *Vorgang* ►Bodensetzung).

5512 settled soil material [n] *constr.* (►settled soil); *s* **tierra [f] asentada** (►suelo asentado); *f* **terre [f] tassée** (►sol tassé) ; *g* **gesetzte Bodenmasse [f]** (►gesetzter Boden).

5513 settlement [n] (1) *constr. eng.* (Sinking of buildings or structures; ►ground settlement); *s* **hincamiento [m] de edificios** (►asentamiento [AL], ►subsidencia [Es, CO]); *f* **effondrement [m] de bâtiments** (►subsidence, ►tassement) ; *g* **Absenkung [f] (2)** (Absacken von Bauwerken; ►Senkung).

5514 settlement [n] (2) *urb.* (Human habitation in a particular area); *s* **colonización [f]** (Ocupación de un área anteriormente despoblada para el asentamiento humano); *f* **peuplement [m]** (Processus démographique par lequel un territoire libre est occupé par de nouvelles implantations) ; *g* **Besiedelung [f] (2)** (Das Bebauen und Bewohnen eines [vorher nicht bewohnten] Gebietes); *syn. o. V.* Besiedlung [f].

settlement [n] (3) *envir.* ►dust settlement; *urb.* ►edge of a settlement; *constr. geo. min.* ►ground settlement; *urb. zool.* ►sparse settlement;

settlement [n] [UK], **dispersed** *urb.* ►dispersed building development [US].

settlement [n], **informal** *urb.* ►slum.

settlements [n], **proliferation of** *plan. urb.* ►uncontrolled proliferation of settlements.

settlement [n], **retirement** *urb.* ►retirement community.

settlement [n], **squatter** *urb.* ►shanty town, ►slum.

settlement area [n] *plan. urb.* ►settled area.

settlement area [n] [UK] *plan. urb.* ►developed area [US].

settlement area [n] [UK], **rural** *leg. urb.* ►rural district [US].

settlement density [n] *plan. sociol. urb.* ►population density (1).

settlement pressure [n] [UK] *urb.* ►development pressure [US].

settlement site [n] [US] *urb.* ►building plot.

settling basin [n] *envir.* ►sedimentation basin.

5515 settling solids [npl] *hydr.* (Solid particles borne in flowing water which slowly sink to the bottom; ►wash load); *s* **partículas [fpl] sedimentarias** (Sustancias sólidas que son acarreadas por el agua y se depositan lentamente; ►carga en suspensión); *syn.* carga [f] sedimentaria; *f* **particules [fpl] sedimentaires** (Éléments solides entraînés par le courant, se déposant lentement sur le fond ; ►charge solide en suspension) ; *g* **Sinkstoffe [mpl]** (Vom fließenden Wasser mitgeführte feste Stoffe, die langsam zu Boden sinken; ►Schwebstofffracht).

settling tank [n] [UK] *envir.* ►sedimentation basin.

sett paving [n] [UK] *constr.* ►block paving [US].

sett paving [n] [UK], **mosaic** *constr. eng.* ►mosaic block paving [US].

sett paving [n] [UK], **random** *constr.* ►random cobblestone paving [US].

sett paving [n] [UK], **small** *constr.* ►small stone paving [US].

sett paving [n] [UK] **timber** *constr.* ►wood block paving.

sett paving [n] [UK] **wood** *constr.* ►wood block paving.

severe crown pruning [n] [US] *arb. constr.* ►dropcrotching.

5516 severe disturbance [n] *ecol. envir.* (►disturbance); *syn.* very adverse impact [n]; *s* **perturbación [f] significativa** (►Influencia perturbadora muy negativa para la capacidad funcional de un ecosistema, un espacio natural, etc.); *syn.* daño [m] significativo; *f 1* **nuisance [f] considérable** *syn.* nuisance [f] grave ; *f 2* **inconvénient [m] irrémédiable** *leg.* (►Nuisance grave et persistante qui présente des dangers, pour la commodité du voisinage, pour la santé, la sécurité ou la salubrité publique, pour l'agriculture, pour la protection de la nature et de l'environnement, pour la conservation des sites et des monuments et qui justifie le refus d'autorisation d'une installation classée, la suspension de l'exploitation ou la fermeture ou la suppression d'une installation existante ; loi n° 76-663 du 19 juillet 1976) ; *g* **erhebliche Beeinträchtigung [f]** (Sehr negative Auswirkungen auf die Funktionsfähigkeit des Naturhaushaltes oder dessen Teile sowie auf das Wohlbefinden des Menschen; ►Beeinträchtigung).

severe habitat [n] *ecol.* ►extreme habitat.

5517 severe pruning [n] *hort.* (Heavy cutting back of woody plants without destroying their natural growth habit; *opp.* ►lopping); *s* **poda [f] severa** (HDS 1987; poda fuerte de leñosa que sin embargo mantiene el porte natural de la especie; *opp.* ►trasmocho); *f* **rabattage [m] (d'un végétal ligneux)** (Suppression des branches ou de gros rameaux d'un arbre ou d'une plante dans le but de provoquer le développement de pousses nouvelles ; IDF 1988 ; *opp.* ►rapprochement, ►ravalement) ; *syn.* élagage [m] sévère ; *g* **starker Rückschnitt [m]** (Gehölzschnitt, ohne den natürlichen Habitus zu zerstören; *opp.* ►Stummelschnitt); *syn.* starker Gehölzrückschnitt [m].

5518 sewage [n] (1) *envir.* (1. ►Effluent or ►wastewater from households, industry or agriculture as a suspension of liquid and solid waste, water runoff which is transported by sewers to be disposed of or processed in a ►sewage treatment plant. **2.** The original meaning of **sewage** is a part of wastewater that is contaminated with feces [US]/faeces [UK] or urine; ►mixed effluent); *s* **aguas [fpl] residuales (1)** (Agua contaminada con residuos líquidos o sólidos, generados en actividades comerciales, domésticas, industriales o en operaciones de lavado, etc. que es conducida a una ►planta de depuración de aguas residuales; las ►aguas residuales 2 no incluyen las aguas de escorrentía superficial; ►aguas residuales mixtas, ►efluente); *syn.* aguas [fpl] servidas [AL] (1), agua [f] sucia [C]; *f* **eaux [fpl] usées** (Ensemble des eaux ménagères, et des eaux-vannes, des eaux industrielles, des eaux du service public et de l'agriculture ; ►eaux usées et eaux pluviales, ►effluent, ►effluent urbain, ►station d'épuration des eaux usées) ; *syn.* eaux [fpl] résiduaires ; *g* **Schmutzwasser [n]** (…wässer [pl]; ►Abwasser aus Haushaltungen, Wirtschaftsbetrieben, Gewerbe, Industrie und Landwirtschaft sowie das kommunale Abwasser, das durch die Kanalisation zur Klärung in eine ►Kläranlage fließt; ►abfließendes Abwasser, ►Mischwasser).

sewage [n] (2) *envir.* ►domestic sewage, ►industrial sewage, ►purified sewage, ►raw sewage; *wat'man.* ►storm water sewage.

sewage [n] [UK], foul *envir.* ►toilet wastewater [US].

sewage [n], sanitary *envir.* ►domestic sewage.

sewage [n], untreated *envir.* ►raw sewage.

sewage discharge pipe [n] *envir. urb.* ►effluent discharge pipe.

sewage disposal field [n] *wat'man. agr.* ►wastewater wetland [US]/trickle field [UK].

sewage disposal pipe [n] [UK] *envir. urb.* ►effluent discharge pipe.

sewage disposal plant [n] *envir.* ►sewage treatment plant.

sewage filtering [n] *agr. wat'man.* ►trickle filtering.

5519 sewage plant screen [n] *envir.* (Heavy form of sieve used in ►sewage treatment to separate residual solids from ►wastewater); *s* **reja [f] de cribado** (Dispositivo en planta de depuración para separar residuos sólidos contenidos en las ►aguas residuals 2; ►tratamiento de aguas residuales); *f* **grille [f] d'entrée** (En-tête d'une station d'épuration pour retenir des matières volumineuses des ►effluents urbains ; ►traitement des effluents) ; *g* **Rechen [m] einer Kläranlage** (Gitterähnliche Vorrichtung in einer Kläranlage zum Abfangen groben Materials, das vom ►Abwasser mitgeführt wird; ►Abwasserbehandlung).

5520 sewage sludge [n] *envir.* (Semi-solid residue removed from raw sewage in the treatment of liquid waste; ►digested sludge); *syn.* municipal sludge [n] [also US]; *s* **lodo [m] de depuración** (Resto semisólido separado de las ►aguas residuales 2 por decantación en el proceso de tratamiento de las mismas. El **l. de d.** tiene un 25% de materia seca, el resto es agua. En los **l. de d.** se encuentran miles de compuestos químicos y metales pesados, por lo que no se deben utilizar como material de compostaje para la agricultura. Desde la perspectiva de protección ambiental es mejor incinerarlos, aunque esta variante es bastante más costosa; *término específico* ►lodo digerido); *syn.* fango [m] resultante de la depuración, cieno [m] de depuración; *f* **boues [fpl] d'épuration** (La technologie d'épuration des eaux résiduaires domestiques et industrielles entraîne la production de grandes quantités de boues pendant les différentes phases d'épuration et de traitements primaires et secondaires ; ASS 1987 ; *terme spécifique* boues [fpl] de décantation ; ►boues digérées) ; *g* **Klärschlamm [m]** (…schlamme [pl] u. …schlämme [pl]; bei der Abwasserreinigung in Kläranlagen anfallende

Sinkstoffe im Absetzbecken. **K.** hat 25 % Trockensubstanz und der Rest ist Wasser. In **K.schlämmen** befinden sich eine Reihe von Chemikalien, darunter auch Schwermetalle, weshalb dieses Material nicht zum Bodenverbesserungsstoff für die Agrarwirtschaft kompostiert werden sollte; ökologisch sinnvoller ist die Verbrennung [thermische Verwertung], wenn auch teurer. Die Ausbringung von **K.** wird von den Agrarministern und -senatoren der Bundesländer sehr kritisch gesehen, weshalb ein Konzept zum Ausstieg aus der Klärschlammverbringung erarbeitet wird; *UB* ►Faulschlamm).

5521 sewage sludge digestion [n] *envir.* (Final biochemical reduction stage of organic matter in sewage treatment resulting in the formation of mineral and simpler organic compounds; ►digested sludge); *s* **digestión [f] anaerobia de lodo de depuración** (Tratamiento acelerado de fermentación o putrefacción de fangos, en ausencia de oxígeno, transformando la materia orgánica principalmente en metano y en anhídrido carbónico. El lodo resultante de este proceso tiene alto contenido en nitrógeno, potasio y fósforo por lo que resulta ser un abono de gran valor si no está contaminado con metales pesados; cf. DINA 1987; ►lodo digerido); *f* **digestion [f] anaérobie des boues** (Méthode de stabilisation des boues par fermentation en absence d'oxygène engendrant la fermentation méthanique sans dégagement d'odeurs et éliminant une quantité très importante de matières organiques, produisant un engrais exploitable, riche en azote, sels potassiques et en acides phosphoriques dans la mesure où il est exempt de métaux lourds ; cf. ASS 1987, 99 et s. ; ►boues digérées) ; *g* **Schlammfaulung [f]** (Anaerobe, bakterielle Zersetzung von fäulnisfähigem Frischschlamm. Nach abgeschlossener Schlammzersetzung gibt es einen geruchlosen, ausgefaulten Schlamm, der wegen seines Gehaltes an Stickstoff, Kaliumsalzen und Phosphorsäure als wertvoller Dünger genutzt werden kann, so er nicht schwermetallbelastet ist; ►Faulschlamm).

5522 sewage treatment [n] *envir.* (Generic term for purification by primary, secondary and tertiary measures; ►effluent disposal, ►management of wastewater and sewage treatment, ►reuse of effluent); *s* **tratamiento [m] de aguas residuales** (Proceso de limpieza de las aguas residuales para liberarlas de las sustancias contaminantes o contenidas en exceso, como la materia orgánica. Existen diversas fases de **t.**: mecánica, biológica y química; ►aprovechamiento de aguas residuales, ►eliminación de aguas residuales, ►sistema de tratamiento de aguas residuales); *syn.* depuración [f] de aguas residuales; *f* **traitement [m] des effluents** (Terme générique pour les diverses techniques de mise en œuvre de mesures d'►assainissement des eaux usées urbaines ou résiduaires industrielles telles que les stations d'épuration ; on distingue les procédés d'épuration mécaniques, biologiques et chimiques ; ►élimination des eaux usées, ►récupération des effluents) ; *syn.* dépollution [f] des eaux usées ; *g* **Abwasserbehandlung [f]** (Entfernung schädlicher Abwasserinhaltsstoffe in Kläranlagen — hierzu gehören auch die Kleinkläranlagen zur Behandlung des häuslichen Abwassers. Es werden mechanische, biologische und chemische Abwasserreinigung unterschieden; ►Abwasserbeseitigung, ►Abwasserverwertung, ►Abwasserwesen); *syn.* Klärung [f] der Abwässer, Abwasseraufbereitung [f], Abwasserreinigung [f].

5523 sewage treatment basin [n] *envir.* (Generic term covering primary purification basin [separation/removal of solid sinking material of untreated wastewater] and final treatment basin [separation of activated sludge after chemical and biological treatment]; ►sedimentation basin, ►biochemical sewage treatment basin); *s* **decantador [m]** (Término genérico para ►decantador primario [eliminación de sustancias en suspensión] y decantador secundario [eliminación del fango activado] en planta de depuración de aguas residuales, ►tanque de tratamiento

S

secundario); **f bassin [m] d'épuration** (Terme générique désignant le décanteur primaire [procédé de séparation solide-liquide par gravité] et le décanteur secondaire [utilisation de procédés de traitement chimiques, biologiques naturels et artificiels avec élimination des boues] dans une station d'épuration ; ▶bassin de lagunage 1, ▶cuve de décantation) ; **g Klärbecken [n]** (OB zu Vorklärbecken [Entfernung von Sinkstoffen aus unbehandeltem Abwasser] und Nachklärbecken [Becken zur Behandlung häuslicher und industrieller Abwässer, in dem sie einem natürlichen biochemischen Prozess der Selbstreinigung ausgesetzt sind und Belebtschlamm abscheiden] als Teil einer Kläranlage; ▶Absetzbecken 1, ▶Nachklärbecken); *syn.* Klärbassin [n].

5524 sewage treatment plant [n] *envir.* (Facility for purification of wastewater; ▶wastewater treatment wetland); *syn.* sewage disposal plant [n], sewage works [n] [also UK]; **s planta [f] de depuración de aguas residuales** (Depuradora vegetal [▶filtro verde] de aguas residuales domésticas); *syn.* planta [f] de tratamiento de aguas residuales; **f station [f] d'épuration des eaux usées** (Installation de traitement des eaux usées domestiques et industrielles avant leur rejet dans le milieu naturel ; ▶lagunage à macrophytes) ; **g Kläranlage [f]** (Anlage zur Abwasserreinigung; ▶Pflanzenkläranlage); *syn.* Klärwerk [n].

sewage volume [n] *envir.* ▶effluent volume.

sewage works [n] [UK] *envir.* ▶ sewage treatment plant.

5525 sewer [n] *envir. urb.* (DAC 1975; pipe or conduit for the transport of ▶sewage 1 or other liquid waste; ▶collector sewer, ▶combined sewer, ▶storm sewer [US]/rain-water sewer [UK]); **s alcantarilla [f] [Es, RA, RCH, MEX]** (Canal de desagüe de ▶aguas residuales 1; ▶alcantarilla 1, ▶alcantarilla unitaria, ▶colector de aguas pluviales); *syn.* colector [m] [Es, CO, EC, MEX], drenaje [m] [MEX]; **f égout [m]** (Partie du réseau public d'évacuation des ▶eaux usées ; ▶collecteur d'évacuation d'eaux pluviales, ▶collecteur d'eaux usées et d'eaux pluviales, ▶collecteur unitaire) ; **g Schmutzwasserkanal [m]** (Teil des öffentlichen Entwässerungsnetzes zur Aufnahme und Ableitung des ▶Schmutzwassers; ▶Abwasserkanal, ▶Mischkanal, ▶Regenwasserkanal).

sewer [n] [US]**, trunk** *envir. urb.* ▶collector sewer.

5526 sewerage [n] *envir. urb.* (System of underground pipes for the removal of wastewater from populated areas; ▶drainage 1); *syn.* sewer lines [npl] [also US]; **s canalización [f]** (Sistema de conductos subterráneos para desagüe de las aguas residuales de una población; ▶drenaje); *syn.* red [f] de alcantarillado, alcantarillado [m]; **f réseau [m] (public) d'assainissement** (Système d'▶évacuation des eaux usées d'une zone habitée au moyen de conduites souterraines) ; *syn.* canalisations [fpl] (ASS 1987, 133) ; **g Kanalisation [f]** (Leitungssystem zur Fortleitung der Abwässer eines besiedelten Gebietes durch unterirdische Kanäle; ▶Entwässerung); *syn.* Kanalisationsnetz [n] (TWW 1982, 885).

sewerage system [n] *envir.* ▶combined sewerage system, ▶separate sewerage system.

sewer lines [npl] [US] *envir. urb.* ▶sewerage.

5527 sewer system [n] *envir. urb.* (Drainage system used exclusively for the transport of sewage; ▶combined sewer system); **s canalización [f] de aguas residuales** (▶sistema de alcantarillado unitario); **f réseau [m] d'égouts** (Système de canalisation destiné à recevoir exclusivement les eaux usées domestiques; ▶réseau [d'assainissement] unitaire) ; **g Schmutzwasserkanalisation [f]** (Kanalsystem, das ausschließlich der Schmutzwasserableitung dient; ▶Mischverfahren).

5528 shade bearer [n] *for. phyt.* (SAF 1983; woody plant species which requires shade for optimum growth especially

whilst young; ▶shade plant; *opp.* ▶light-demanding woody species); **s leñosa [f] esciófila** (▶planta de sombra, *opp.* ▶especie leñosa de solana); **f essence [f] d'ombre** (Essence ligneuse qui supporte la lumière ; ▶plante d'ombre ; *opp.* ▶essence de lumière) ; **g Schattholzart [f]** (Gehölzart, die besonders im Jugendstadium für optimales Wachstum Schatten benötigt; ▶Schattenpflanze; *opp.* ▶Lichtholzart); *syn. hort.* Schattengehölz [n].

shade grass [n] *hort. plant.* ▶shade-tolerant grass.

5529 shade-loving [adj] *phyt. plant.* (Descriptive term for a plant which requires shade for optimal growth); *syn.* sciophilous [adj]; **s esciófilo/a [adj]** (Calificativo ecológico de las plantas y sinecias que requieren la sombra; DB 1985); *syn.* umbrófilo/a [adj]; **f sciaphile [adj]** (Se dit d'un végétal ou d'une formation végétale qui préfère l'ombre. Ne pas confondre avec « ombrophile » [= aimant la pluie]) ; **g Schatten liebend [ppr/adj]** (Eine Pflanze betreffend, die zu ihrem optimalen Wachstum Schatten benötigt, im Einzelfall Schatten nur toleriert; *syn.* skiophil [adj].

5530 shade plant [n] *bot.* (Plant which requires a shady location for optimum growth; ▶shade bearer; *opp.* ▶light-demanding plant); *syn.* sciophilous plant [n], sciaphyte, plant [n] for shade; **s planta [f] de sombra** (Planta que necesita una ubicación de umbría para crecer de forma óptima; ▶leñosa esciófila; *opp.* ▶heliófito); *syn.* planta [f] esciófila, esciófito [m], planta [f] de umbría; **f plante [f] d'ombre** (Plante ayant son optimum de développement à une exposition ombragée ; ▶essence d'ombre ; *opp.* ▶espèce héliophile) ; *syn.* espèce [f] d'ombre, espèce [f] sciaphile, sciaphyte [m] ; **g Schattenpflanze [f]** (Pflanze, die einen schattigen Standort für optimales Wachstum benötigt, im Einzelfall Schatten nur toleriert; *for.* ▶Schattholzart; *opp.* ▶Lichtpflanze); *syn.* Skiophyt [m].

shade planting [n] [UK] *plant.* ▶shade-tolerant planting.

shadescreen [n] *arch.* ▶sunscreen.

5531 shade-tolerant [adj] *phyt. plant.* **s que tolera la sombra [loc]**; **f tolérant, ante à l'ombre [loc]** ; **g Schatten vertragend [ppr/adj]** *syn.* schattenverträglich [adj], Schatten ertragend [ppr/adj].

5532 shade-tolerant grass [n] *hort. plant.* (Grass which requires a shady location for optimum growth); *syn.* shade grass [n]; **s gramínea [f] esciófila** *syn.* gramínea [f] de umbría; **f graminée [f] d'ombre** (Graminée convenant particulièrement à une exposition ombragée, p. ex. la laîche japonaise *[Carex morrowii]*, la laîche pendante *[Carex pendula]*) ; *syn.* graminée [f] sciaphile ; **g Schattengras [n]** (Gras, das einen schattigen Standort für optimales Wachstum benötigt, z. B. Japansegge *[Carex morrowii]*, Riesenwaldsegge *[Carex pendula]*).

5533 shade-tolerant lawn [n] *gard.* (Lawn composed of various grass species suitable for a shady location); **s césped [m] de umbría** (Césped que —debido a su composición de especies— puede crecer en la sombra); **f gazon [m] adapté à l'ombre** (Mélange de graminées faisant preuve d'une bonne tolérance vis à vis du manque de lumière) ; *syn.* gazon [m] pour l'ombre ; **g Schattenrasen [m]** (Rasen, der sich durch seine Artenzusammensetzung für schattige Standorte eignet).

5534 shade-tolerant perennial [n] *hort. plant.* (▶Perennial 1 which endures shade for optimal growth); **s planta [f] vivaz esciófila** (▶planta vivaz); *syn.* vivaz [f] esciófila, perenne [f] esciófila; **f vivace [f] des emplacements ombragés** (▶plante vivace) ; *syn.* vivace [f] sciaphile, plante [f] vivace des sous-bois frais, plante [f] vivace supportant l'ombre ; **g Schattenstaude [f]** (Mehrjährige, krautige Pflanze, die einen schattigen Standort für optimales Wachstum benötigt oder toleriert; ▶Staude).

5535 shade-tolerant planting [n] *plant.* (Stand of plants growing in a shady or semi-shaded location); *syn.* planting [n] for shade [also UK], shade planting [n] [also UK]; *s* **plantación [f] de umbría** (Plantación en lugar sombrío o semisombrío); *f* **plantation [f] d'ombre** (Plantation satisfaisant à une situation ou une exposition ombragée, à un ensoleillement réduit) ; *syn.* plantation [f] adaptée à l'ombre ; *g* **Schattenpflanzung [f]** (Angelegter Pflanzenbestand im schattigen oder leicht beschatteten Bereich).

5536 shade-tolerant woody plant [n] *hort. plant.* (►shade bearer); *syn.* woody plant [n] for shade; *s* **leñosa [f] de sombra** (►leñosa esciófila); *syn.* leñosa [f] de umbría; *f* **espèce [f] ligneuse d'ombre** (Terme horticole désignant les ►essences d'ombre) ; *g* **Schattengehölz [n]** (Gärtnerischer Begriff für ►Schattholzart).

5537 shade tree [n] (1) *agr.* (Tree which affords [grazing cattle] shade; ►shade bearer); *s* **árbol [m] da sombra** (Árbol que proporciona sombra al ganado; ►leñosa esciófila); *syn.* árbol [m] para sombra; *f* **arbre [m] d'ombrage** (Arbre qui ombrage [la prairie] ; ►essence d'ombre) ; *syn.* arbre [m] d'ombre ; *g* **Schattenbaum [m]** (Baum, der Weidevieh Schatten spendet; ►Schattholzart).

5538 shade tree [n] [US] (2) *arb. hort.* (In U.S., a standard tree is delimbed to at least 6 ft [1.8 m] in nurseries; in U.K., ornamental tree stock grown in nurseries with minimum 1.8m height from ground level to lowest branch; BS 3936: part 1; ►standard [UK] 1, ►standard street tree [US]/avenue tree [UK]); *syn.* tall standard [n] [UK]; *s* **árbol [m] de pie alto (1)** (Árbol de vivero de un altura mínima de la cruz de 1,8 m para caducifolios; ►árbol de alineación, ►árbol de pie alto 2); *f* **arbre [m] de haute-tige** (Suivant les normes AFNOR N.F. V 12 051 – 054 et 055, terme qui désigne en sylviculture les arbres dont la hauteur du tronc mesurée du sol à la première branche est située entre 1,80 m et 2,50 m selon la vigueur des espèces ou cultivars mis en culture ; ►arbre d'avenue, ►arbre tige) ; *syn.* haute-tige [m] ; *g* **Hochstamm [m] (1)** (In Stamm und Krone gegliederter Baum mit einer Stammhöhe von mindestens 200-250 cm; ►Alleebaum, ►Hochstamm 2).

5539 shading [n] *phyt. urb.* (Generic term for the provision of shade or the screening of plants by other vegetation, structures, such as shade houses, tent roofs, pergolas or adjacent buildings; ►tree canopy shading; ►building setbacks); *s* **sombreado [m]** (*Contexto* unas plantas dan sombra a otras o un edificio da sombra a otro o a un espacio libre adyacente; dar [vb] sombra; ►distancia legal de construcción, ►sombreado por árboles); *f 1* **ombrage [m]** (*Terme générique* ensemble des branches et de feuilles des arbres qui donnent de l'ombre ; par analogie désigne aussi l'ombre projetée [obscurcissement, absence de lumière] par cet ensemble ; ►degré d'ouverture) ; *f 2* **zone [f] d'ombre** (*Terme spécifique urb.* surface projetée sur un bâtiment par un bâtiment voisin ; l'utilisation des distances de ►prospect est justifiée par la volonté d'exposer les bâtiments au soleil ou d'éviter les zones d'ombres sur les bâtiments ; ►espacement minimal entre les constructions) ; *syn.* ombre [f] portée, ombre [f] projetée ; *f 3* **caractère [m] ombragé** (*Terme spécifique* qualité d'un lieu, d'une station) ; *g* **Beschattung [f]** (**1.** Allgemeiner Begriff für Schattenwurf auf Pflanzen durch benachbarte höhere Pflanzen oder Bauwerke; ►Beschirmung. **2.** *urb.* Schattenwurf auf Außenwände von Gebäuden durch benachbarte Gebäude; ►Abstandsfläche); *syn.* Verschattung [f]).

5540 shadow line [n] *constr.* (Dark area beneath the overhang of a step tread caused by the particular design of the front of the step, e.g. by [champfering of] step ►nosing; TSS 1988, 240-6); *s* **línea [f] de sombra** (Área oscura formada por el perfil particular de la contrahuella de una escalera, p. ej. debido al ►mampirlán); *f* **ligne [f] d'ombre (≠)** (Zone d'ombre linéaire formée par le profil particulier de la contremarche d'un escalier, p. ex. par un ►débordement du nez de marche par rapport à la contremarche ou une inclinaison de la contremarche [biseautée] sur un pas de souris) ; *syn.* ombre [f] du nez de marche (≠) ; *g* **Schattenfuge [f] (2)** (Aus gestalterischen Gründen an Treppenstufen vorgesehener linearer Schatten, der durch bestimmte Profilgestaltung der Stirnfläche oder durch Überstand einer Legstufenplatte [►Unterschneidung] erzeugt wird); *syn.* Schattennase [f] (PGL 2005, 219).

5541 shady location [n] *hort. plant.* (►Location 1 of a herb or shrub layer characterized by dense shade caused by the tree canopy above; ►semishaded location [US]/semi-shaded position [UK], ►ubac); *syn.* shady site [n]; *s* **umbría [f]** (►Estación de estrato herbáceo o arbustivo caracterizada por la sombra del dosel del bosque; ►exposición semisombra, ►ladera umbría); *syn.* lugar [m] sombrío; *f* **exposition [f] (très) ombragée** (►Station caractérisée par la présence d'un écran végétal filtrant fortement les rayons solaires ; ►ubac) ; *syn.* milieu [m] ombragé, situation [f] ombragée ; *g* **Schattenlage [f]** (►Standort 1 einer Kraut- oder Strauchschicht, der durch ein lichtundurchlässiges Kronendach gekennzeichnet ist; ►Schattenseite, ►Halbschattenlage); *syn.* schattiger Standort [m], schattige Lage [f].

shady site [n] *hort. plant.* ►shady location.

shady slope [n] *geo.* ►ubac.

5542 shaft [n] *constr.* (Narrow, vertical hollow structure in the ground, mostly of cylindrical or rectangular cross section, with a covered opening; *specific terms* ►cleanout chamber [US]/inspection chamber [UK], ►irrigation equipment manhole, ►manhole, ►drywell sump [US]/soakaway [UK], well shaft, *min.* hoisting shaft); *s* **pozo [m]** (Construcción vertical en el subsuelo de sección circular o cuadrada con diferentes funciones de control de las conducciones subterráneas; *términos específicos* ►pozo de registro 1, ►pozo de riego [subterráneo], ►sumidero); *f* **regard [m]** (Ouvrage généralement vertical de section circulaire ou carrée, constitué suivant sa fonction du bas vers le haut : d'un socle de regard, d'une cheminée pouvant être munie d'échelons et constituée de viroles et de piédroits, d'une hotte et selon la nature du regard d'un dispositif de fermeture ; *termes spécifiques* ►regard d'arrosage, ►regard de visite, ►regard visitable, ►puisard) ; *g* **Schacht [m]** (Schächte [pl]; enger, meist senkrecht-röhrenförmiger oder rechteckiger aus Betonfertigteilen oder Mauersteinen in die Erde gebauter Hohlraum; *UBe min.* Brunnen**sch.**, Förder**sch.**; *constr.* ►Bewässerungsschacht, ►Einstiegsschacht, Kabel**sch.**, ►Kontrollschacht 2, ►Sickerschacht).

shaft lid [n] *constr.* ►manhole cover.

shag [n] *agr.* ►shifting cultivation.

shale [n] *min.* ►coal mine shale, ►red shale.

shale [n] [UK], **colliery** *min.* ►coal mine shale [US].

5543 shallow [adj] (1) *geo. hydr.* (Descriptive term applied to a watercourse or lake with a limited depth of water); *s* **poco profundo/a** [adj] (1) (Término descriptivo aplicable a cursos o masas de agua de poca profundidad); *f* **de faible profondeur [loc]** ; *g* **flachgründig [adj] (1)** (Gewässer mit geringer Tiefe betreffend); *syn.* flach [adj], seicht [adj].

5544 shallow [adj] (2) *geo. pedol.* (Descriptive term applied to a topsoil horizon with a limited depth of soil; *opp.* ►deep [soil]); *s* **poco profundo/a [loc] (2)** (Término descriptivo para suelo con horizontes de muy poca profundidad; *opp.* ►profundo/a); *syn.* superficial [adj]; *f* **superficiel, elle [adj]** (Sol de faible épaisseur ; *opp.* ►profond, onde) ; *syn.* peu profond, onde [loc] ; *g* **flach-**

gründig [adj] (2) (Oberbodenhorizont mit geringer Mächtigkeit betreffend; *opp.* ▶tiefgründig); *syn.* geringmächtig [adj].

5545 shallow bog [n] *geo. min.* (Bog of little depth); *syn.* shallow mire [n], shallow peatland [n]; *s* **turbera [f] poco profunda** *syn.* turbera [f] llana; *f* **tourbière [f] de faible épaisseur** ; *g* **flachgründiges Moor [n]** (Moor mit geringer Torfmächtigkeit); *syn.* gering mächtiges Moor [n].

shallow crown [n] [UK] *arb.* ▶flat crown [US].

shallow fringe [n] **of a water body** *phyt. zool.* ▶shallow water area.

shallow mire [n] *geo. min.* ▶shallow bog.

shallow peatland [n] *geo. min.* ▶shallow bog.

shallow-rooted plant [n] [US] *bot. hort.* ▶shallow rooting plant.

5546 shallow rooting plant [n] *bot. hort.* (Plant with a root system confined or largely confined to upper soil layers; *opp.* ▶deep-rooting plant); *syn.* shallow-rooted plant [n] [also US]; *s* **planta [f] de raíz plana** (Planta cuyo sistema radical se desarrolla en las capas superficiales del suelo; *opp.* ▶planta de radicación profunda); *f* **espèce [f] à racines traçantes** (Plante adaptée à un milieu [sec] avec un système racinaire à extension latérale superficielle ; GOR 1985, 209 ; *opp.* ▶plante à racines profondes) ; *g* **Flachwurzler [m]** (Pflanze ohne Pfahlwurzel — bei gewissen Zweikeimblättlern *[Eudicotyledoneae]* und Nacktsamern *[Gymnospermae]*, die sich mit ihrem Wurzelsystem aus stängelbürtigen Nebenwurzeln oberflächlich im A- und oberen B-Horizont ausbreitet, z. B. die *Wurzelscheibe* der Fichte *[Picea]*; *opp.* ▶Tiefwurzler); *syn.* flach wurzelnde Pflanze [f].

shallows [npl] *phyt. zool.* ▶shallow water area.

shallow soil roof planting [n] [US] *constr.* ▶extensive roof planting.

5547 shallow water area [n] *phyt. zool.* (Shoal water of river and streams with < 3m of water depth); *syn.* shallows [npl], sublittoral zone [n], shallow fringe [n] of a water body; *s* **zona [f] de aguas poco profundas** (Área de aguas continentales de profundidad inferior a 3 m); *f* **zone [f] de faible profondeur** (Zone des rivages lacustres et des rives de cours d'eau dont la profondeur est inférieure à trois mètres ; AEP 1976, 227) ; *syn.* zone [f] (d'eau) peu profonde ; *g* **Flachwasserzone [f]** (Fließ- oder Stillgewässern mit Bereichen unter 3 m Wassertiefe); *syn.* Seichtwasserzone [f].

5548 shanty town [n] *urb.* (1. A sometimes quite extensive gradual development of unsanitary shacks built by squatters using reclaimed materials on the edge of rapidly-growing conurbations in developing countries; ▶overnight land squatting, ▶slum; 2. s. t.s which are occupied by squatters are called ▶squatter settlements; *s* **barrio [m] de chabolas** (Asentamiento urbano ilegal situado generalmente al borde de las grandes urbes en países en desarrollo creado informalmente por la población inmigrante y cuyas casas están construidas de materiales de desecho; ▶paracaidismo, ▶slums); *syn.* barrio [m] insalubre, villa [f] miseria [RA], callampa [f] [RCH], champa [f] [ES], tugurio [m] [ES, HON], asentamiento [m] de paracaidistas [MEX], pueblo [m] joven [PE], cantegril [m] [ROU], rancho [m] endeble [YZ]; *f* **bidonville [m]** (Quartier constitué d'abris de fortune en bois, en tôles dans lequel vit, sous des conditions hygiéniques déplorables, une population des plus misérable en bordure de grandes agglomérations ; le terme **b.,** employé pour la première fois en 1953 à propos du Maroc, désigne des concentrations sauvages des populations utilisant des matériaux de récupération [entre autres, des bidons de kérosène] pour construire des logements précaires sur les friches urbaines de l'agglomération casablancaise, alors en plein développement

fournirent l'idée d'un terme qui allait être appelé à désigner un phénomène en forte expansion das les villes du tiers monde ; ▶quartier informel, ▶quartier misérable, ▶établissement d'un campement sauvage) ; *g* **Barackenstadt [f]** (Illegale Besiedlung obdachloser Menschen, besonders in Entwicklungsländern am Rande großer Ballungsräume, in Form von Notquartieren [Elendsviertel], die aus Wellblech, Kanistern, Brettern, Pappe etc. gebaut sind; ▶Bauen von Notbehausungen über Nacht, ▶Slum, ▶informelle Siedlung); *syn.* Bidonville [f], Hüttensiedlung [f].

shaping [n] **the ground** [US] *constr. plan.* ▶ground modeling [US]/ground modelling [UK].

5549 shared space [n] (Concept introduced by HANS MONDERMAN, Netherlands, and BEN HAMILTON-BAILLIE, Bristol, UK, for inner-city road traffic aimed at eliminating the traditional separation between vehicles and pedestrians by removing road management devices such as kerbs, lines, traffic signs and signals; the intention is to improve road safety by requiring users to drive at slow speeds through such spaces observing simple rules, e.g. yielding to the right. It is believed that an individuals' behaviour in road space is more influenced by a well-planned, built environment and that conventional traffic control devices can have a negative effect. Sponsored by the EU **s. s.** projects are being undertaken as part of road safety policies, to deal with congestion issues and to enhance streets and public open spaces); *s* **espacio [m] compartido** (Concepto introducido por HANS MONDERMAN, Países Bajos, y BEN HAMILTON-BAILLIE, Bristol, GB, para el tráfico dentro de la ciudad que tiene como objetivo eliminar la separación tradicional entre vehículos y peatones quitando instalaciones o dispositivos de regulación como las aceras, líneas de separación de carriles, semáforos y señales de tráfico. La intención es incrementar la seguridad viaria exigiendo a todos/as los/las usuarios/as a conducir a baja velocidad a través de los espacios comunes, cumpliendo reglas simples como el «ceda el paso» a la derecha. Se cree que el comportamiento individual en el espacio-calle está más influenciado por un entorno construido bien planificado y que dispositivos convencionales de regulación del tráfico pueden tener efectos negativos. Con apoyo del *Programa para la Región del Mar del Norte* INTERREG de la UE se han realizado entre 2004 y 2008 proyectos de **e. c.** en siete municipios de Alemania, Bélgica, Dinamarca, Gran Bretaña y los Países Bajos); *f* **espace [m] partagé** (Concept d'apaisement des villes développé par l'urbaniste BEN HAMILTON-BAILLIE en Angleterre et l'ingénieur HANS MONDERMAN en Pays-Bas qui préconisent la responsabilisation des usagers au lieu de leur imposer des règles, la convivialité entre les usagers de la rue, la cohabitation active entre les différents modes de déplacement, entre les besoins des habitants, des passants, des riverains, des commerçants, etc. et propose dans des ▶zones de circulation apaisée l'aménagement de l'espace commun en faveur des flux libres des piétons, des cyclistes, des automobilistes et des transports en commun, c.-à-d. sans marquage au sol [concept de rue/route nue], sans panneaux ni feux rouges, avec trottoir et chaussée au même niveau. Dans le cadre du programme INTERREG de l'Union Européenne a été réalisé entre 2004 et 2008 un programme de développement de nouvelles politiques en matière d'aménagement de l'espace public sur sept communes pilotes en Belgique, au Danemark en Allemagne en Angleterre et aux Pays-Bas) ; *syn.* voirie [f] pour tous, espace [m] pour tous ; *g* **Shared Space [m]** („Gemeinsam genutzter Raum", „Straße für alle", „Gemeinschaftsstraße", „Begegnungszone" [CH]; ein vom niederländischen Verkehrsplaner HANS MONDERMAN entwickeltes und mit den britischen Verkehrsplaner BEN HAMILTON-BAILLIE bekanntgemachtes Konzept, das vorsieht, die traditionelle Trennung zwischen Auto- und Fußgängerverkehr dadurch aufzuheben, dass Bordsteine, Verkehrszeichen, Signalanlagen und Fahrbahnmarkierungen entfernt werden. Durch dieses Konzept soll eine Gleichberechtigung der Verkehrsteilnehmer unter Beibehaltung der Rechts-vor-Links-Vorfahrtsregel hergestellt, die Überbeschilderung und Überregulierung im Straßenraum abgeschafft und der öffentliche Raum für den Menschen unter dem Aspekt „Gegenseitige Rücksichtnahme" wieder aufgewertet werden. Von 2004 bis 2008 wurde **S. S.** testweise im Rahmen des Infrastrukturförderprogramms INTERREG *North Sea Region Programme* der EU in sieben Gemeinden in Belgien, Dänemark, Deutschland, England und den Niederlanden verwirklicht.)

5550 sharp pointed ridge [n] *geo.* (Long, narrow, steepsided ridge between two valleys); *s* **cresta [f] alargada** (Formación geomorfológica entre dos valles); *f* **crête [f]** (d'une montagne) ; *g* **Kamm [m]** (Langgestreckte, schroffe geomorphologische Vollform zwischen zwei Tälern).

sharp sand [n] *constr. eng.* ▶washed sand.

5551 shear development [n] *constr. eng.* (Creation of planes in natural slopes and earthwork[s] by overcoming of the soil's shearing strength, mostly with the occurrence of water in absorptive soils); *syn.* slip-plane development [n] (DNE 1978, 254); *s* **formación [f] de superficie deslizante** (Formación de planos de deslizamiento en cuestas naturales o terraplenes al superarse la resistencia al cizallamiento del suelo. Ocurre generalmente en suelos cohesivos con presencia de agua); *f* **formation [f] d'un plan de glissement** (Formation d'une surface souvent incurvée le long d'un versant naturel ou d'ouvrages de terrassement lorsque sous l'effet de la pesanteur un compartiment de sol perd sa cohésion, décroche et glisse sur la couche sousjacente ; phénomène dû en général à une infiltration d'eau dans un sol plastique) ; *g* **Gleitflächenbildung [f]** (Bildung von Flächen in natürlichen Böschungen, Erd- oder Grundbauwerken bei Überwindung der Scherfestigkeit des Bodens, meist bei Wasserzutritt in bindigen Böden).

5552 shear failure [n] **of embankment** *constr.* (Term usually applied to the failure of an embankment or slope, which due to its own weight, overcomes the shear resistance strength of the soil particles along a definable shear plane. The ground at the foot of the slope is not displaced, in contrast to ▶bearing capacity failure; **s. f. of e.** can be induced by too great an angle of repose, too low a shear strength of the soil, too great a slope height, by vibration, or excessive loads at the crown of the slope or by a change in hydrological conditions in the soil mass; *s* **deslizamiento [m] planar** (Desplazamiento masivo de un talud o una vertiente causado por su propio peso, en el que —al contrario que en el ▶deslizamiento rotacional [Es, RA, COL, EC, MEX, YV]— no se desplaza el terreno al pie del talud o de la vertiente; ▶movimientos gravitacionales); *f* **glissement [m] de terrain (2)** (Descente en masse de matériaux meubles vers le bas versant le long d'une surface de rupture, sans dérangement considérable de leur agencement dans la partie déplacée ; DG 1984 ; ▶glissement rotationnel, ▶mouvement de masse) ; *g* **Geländebruch [m]** (Rutschung der Böschung durch ihr Eigengewicht bei Überwindung der Scherfestigkeit des [aufgeschütteten] Bodens auf einer Gleitfläche, bei dem das Gelände unterhalb des Böschungsfußes im Vergleich zum ▶Grundbruch nicht verlagert wird; der **G.** kann z. B. durch einen zu großen Böschungswinkel, eine zu geringe Scherfestigkeit des Bodens, eine zu große Böschungshöhe, durch Erschütterungen, zu große Belastungen am Böschungskopf oder durch eine Veränderung der hydrologischen Verhältnisse im Erdkörper ausgelöst werden; *syn.* Böschungsbruch [m]).

shearing [n] **of hedges** [US] *constr. hort.* ▶clipping of hedges.

shearing resistance [n] *constr. stat.* ▶shear strength.

5553 shear strength [n] *constr. stat.* (Maximum resistance of a structural member or material to shearing stress, e.g. the resistance of a soil to shearing stress under loading; **s. s.** of soil is measured in terms of two soil parameters: angle of ▶internal friction and cohesion. Practically speaking it is the resistance to mass deformation developed from a combination of particle rolling, sliding, and crushing and is reduced by any pore pressure that exists or develops during particle movement; *syn.* shearing resistance [n]; *s* **resistencia [f] al cizallamiento** (▶frottement interne); *syn.* resistencia [f] al esfuerzo cortante [Es, RA, EC, MEX], resistencia [f] al esfuerzo de corte [RA]; *f* **résistance [f] au cisaillement** (Donnée statistique constituée par les critères de ▶frottement interne et de cohésion sur un plan de chevauchement ; elle est un indicatrice du comportement à la déformation, p. ex. des masses de terre remblayées sur un talus, de leur stabilité face à des contraintes ou une pression) ; *g* **Scherfestigkeit [f]** (Statische Größe aus ▶innerer Reibung und Kohäsion an Gleit- und Schichtflächen [Scherflächen]. Sie ist Indikator für Verformungsverhalten, z. B. von Bodenmassen, die an Böschungen aufgetragen wurden, hinsichtlich ihrer Stabilität bei Belastung oder Druck).

shedding [n] **of leaves** *bot.* ▶defoliation, ▶premature shedding of leaves.

sheep farming [n] [UK] *agr.* ▶sheep raising.

sheep keeping [n] *agr.* ▶sheep raising.

5554 sheep pasture [n] *agr.* (Pasture grazed by sheep); *s* **pasto [m] de ovejas**; *f* **pâturage [m] à ovins** (Prairie utilisée pour la pâture des ovins) ; *g* **Schafweide [f]** (Von Schafen beweidetes Grünland).

5555 sheep raising [n] *agr.* (Branch of lifestock farming concerned with breeding and rearing of sheep); *syn.* sheep keeping [n], sheep farming [n] [also UK]; *s* **explotación [f] de ganado lanar** *syn.* explotación [f] ovina; *f* **élevage [m] des ovins** (Branche d'activité agricole concentrée dans des exploitations spécialisées ayant pour objet la reproduction intentionnelle des ovins pour leur valeur économique) ; *syn.* exploitation [f] ovine, élevage [m] d'ovins ; *g* **Schafhaltung [f]** (Zweig der landwirtschaftlichen Viehwirtschaft, die die Aufzucht und Zucht von Schafen betreibt).

sheet [n] *constr. contr.* ▶daywork sheet; *constr.* ▶metal sheet, ▶plastic sheet, ▶protective sheet, ▶take-off sheet, ▶waterproofing sheet.

sheet [n]**, detailed design** *constr.* ▶construction drawing.

sheet [n] [US]**, rootproof** *constr.* ▶root protection membrane.

sheet [n]**, time** *constr. contr.* ▶daywork sheet.

5556 sheet erosion [n] *geo. pedol.* (Removal of a fairly uniform layer of soil by runoff water; ▶denudation, ▶extreme soil erosion, ▶flash erosion, ▶gully erosion, ▶sheet flow/sheetflow, ▶sheetwash [US]/sheet wash [UK], ▶soil erosion); *s* **erosión [f] laminar** (Consiste en la remoción de delgadas capas de suelo extendida más o menos uniformemente a toda una superficie. Es altamente perjudicial, ya que es la causa de grandes aportaciones de sedimentos a los cursos de agua; además, por afectar a las partículas más finas de tierra, su pérdida significa un notable empobrecimiento de la fertilidad del suelo; DINA 1987; ▶arrastre laminar, ▶denudación, ▶destrucción del suelo, ▶erosión del suelo, ▶erosión en cárcavas 1, ▶escorrentía laminar); *syn.* erosión [f] superficial; *f* **érosion [f] pelliculaire** (Érosion d'une mince couche assez homogène de matériaux fins à la surface du sol : peut être imperceptible, en particulier lorsqu'elle est causée par le vent. Cet enlèvement progressif mécanique de la couche superficielle du sol pouvant avoir des résultats dévasta-

S

teurs ; DG 1984, 471 ; ►ablation, ►dévastation des sols, ►érosion continentale, ►érosion brutale, ►érosion des sols, ►ravinement 1, ►ruissellement en nappes) ; *syn.* érosion [f] en couche (LA 1981, 477 et DUV 1984, 237), érosion [f] en plaque (DFM 1975, 116), érosion [f] en nappe (DFM 1975, 116) ; *g* **Flächenerosion [f]** (Gemäßigte, zunächst kaum bemerkbare, gleichmäßige Abtragung des Oberbodens, ohne dass Erosionsrinnen gebildet oder Gräben eingeschnitten werden, die sich im Laufe der Zeit jedoch verheerend auswirkt; ►Abspülung, ►Abtragung, ►Bodenerosion, ►Bodenverheerung, ►Flächenabfluss 2, ►Grabenerosion 1); *syn.* Schichterosion [f].

sheetflood [n] *geo.* ►sheetwash.

5557 sheetflow [n]/**sheet flow** [n] *geo.* (►Overland flow or downslope movement of surface water runoff in the form of a thin continuous film rather than concentrated in individual channels larger than rills; cf. GFG 1997; ►direct runoff, ►sheet erosion, ►sheetwash [US]/sheet wash [UK]); *s* **escorrentía [f] laminar** (Flujo de una lámina de agua relativamente delgada, de espesor uniforme, sobre la superficie del suelo —generalmente de poca pendiente— en vez de fluir concentrada en regueros; cf. WMO 1974; ►arrastre laminar, ►erosión laminar, ►escorrentía superficial 1, ►escorrentía superficial 2); *f* **ruissellement [m] en nappes** (Écoulement caractéristique des régions arides et semi-arides où les averses, rares mais intenses, donnent naissance à une lame d'eau, relativement mince et d'épaisseur sensiblement uniforme à la surface du sol ; cf. DG 1984 et WMO 1974 ; ►ablation, ►écoulement superficiel, ►érosion laminaire, ►érosion pelliculaire, ►ruissellement) ; *syn.* écoulement [m] en nappe ; *g* **Flächenabfluss [m] (2)** (Abfließen von Niederschlagswasser bei Starkregen oder Dauerregen auf kaum geneigter Erdoberfläche in Form einer dünnen Wasserschicht, statt durch Abfluss über (Spül)rinnen, Gräben oder Fließgewässer; geomorphogenetisch betrachtet ist der F. die Voraussetzung für den Prozess der ►Abspülung; ►Flächenerosion, ►Oberflächenwasserabfluss 1, ►Oberflächenwasserabfluss 2); *syn.* Schichtflut [f] (WAG 2001).

sheet metal [n] *constr.* ►metal sheet.

sheet [n] **of water** *recr. hydr.* ►water surface.

5558 sheetwash [n]/**sheet wash** [n] *geo.* (1. Substantial, broad expanse of moving, storm-borne water that spreads as a thin, continuous film over a large area in arid regions in the form of ►sheet flow/sheetflow; cf. GFG 1997); ►direct runoff, ►gully erosion, ►overland flow, ►sheet erosion; **sw.** usually occur before runoff is sufficient to promote channel flow, or after a period of sudden and heavy rainfall. **2.** Material transported and deposited by the water of a **sw. 3.** Term used as a syn. of sheet flow [a movement] and sheet erosion [a process]); *syn.* to 1. sheetflood [n], sheet water [n] (GFG 1997, 587); *s* **arrastre [m] laminar** (≠) (1. Fuerte ►erosión laminar del suelo causada por flujo de capa de agua de lluvia, que ocurre generalmente después de aguaceros repentinos y fuertes en zonas subtropicales secas o húmedas y en los trópicos húmedos. Los materiales del suelo arrastrados constituyen posteriormente la carga de partículas sedimentarias de los cursos de agua; ►erosión en cárcavas, ►escorrentía laminar, ►escorrentía superficial 1, ►escorrentía superficial 2; **2.** *Resultado* arrancamiento [m] del suelo); *f 1* **ablation [f]** (Forte action mécanique combinée d'entraînement de matériel meuble provoqué par l'impact des gouttes de pluie, par le ruissellement ou par les eaux de fusion glaciaires ; ►écoulement superficiel 1, ►écoulement superficiel 2, ►érosion pelliculaire, ►ravinement 1, ►ruissellement, ►ruissellement en nappes) ; *f 2* **érosion [f] laminaire** (Décapage de la surface du sol provoquée par une lame d'eau de pluie dans les régions subtropicales à tropicales humides) ; *syn.* érosion [f] en nappe ;

g 1 **Abspülung [f]** (Flächenhafte, kräftige Abtragung des Bodens durch eine Regen- oder Schmelzwasserschicht in subtropischen Trockengebieten bis wechselfeuchten Tropen; eine **A.** erfolgt meist nach plötzlichen und heftigen Regenfällen; ►Flächenabfluss 2, ►Flächenerosion, ►Grabenerosion 1, ►Oberflächenwasserabfluss 1, ►Oberflächenwasserabfluss 2); *syn.* Flächenspülung [f] (WAG 2001); *g 2* **abgespülter Boden [m]** (Erosionsmaterial, das die Sinkstofffracht der Fließgewässer bildet); *syn.* abgetragenes Material [n].

sheet water [n] *geo.* ►sheetwash.

5559 shelter [n] *landsc. trans.* (Roof structure which offers protection against the wind, rain, sun, etc.; ►canopy 1, ►overhead shelter, ►park shelter, ►picnic shelter, ►public transit shelter [US]/public transport shelter [UK], ►pedestrian shelter); *s* **refugio [m]** (Instalación de protección contra el viento, la lluvia, etc.; *términos específicos* ►garita de espera, ►marquesina 1, ►marquesina 2, ►techado, ►techado contra lluvia, ►techado para picnics); *syn.* abrigo [m]; *f* **mobilier [m] de protection contre les intempéries** (Construction généralement couverte destinée à protéger le piéton sur la voie publique du soleil, du vent, de la pluie ; ►abri, ►abri couvert, ►abri de pique-nique, ►auvent, ►abribus, ►couverture, ►hutte) ; *g* **Witterungsschutzeinrichtung [f]** (Überdachte, an den Seiten z. T. geschlossene Einrichtung zum Schutz gegen Wind, Niederschlag und Sonne; *UBe* ►Picknickhütte, ►Schutzdach 1, ►Schutzdach 2, ►Schutzhütte 2, ►Überdachung 1, ►Wartehalle); *syn.* Wetterschutz [m].

shelter [n] [UK], **mountain** *recr.* ►mountain lodge [US].

shelter [n] [UK], **public transport** *trans.* ►public transit shelter [US].

5560 shelterbelt [n] (1) *landsc.* (Strip of living trees or shrubs grown and maintained to provide shelter from wind, sun, snowdrift, industrial emissions, salt spray, etc. *Note* in the U.S., the term refers mainly to belts protecting large agricultural areas; those protecting farmsteads are termed ►windbreaks; cf. SAF 1983; ►windbreak planting); *syn.* protective belt [n], vegetation screen [n]; *s* **franja [f] protectora** (Generalmente plantación de árboles y arbustos de varias filas para proteger contra el viento, las emisiones industriales, etc.; ►cortina rompevientos); *f* **rideau [m] forestier** (Plantation linéaire arborescente et arbustive, constituée d'essences variées, disposée sur plusieurs rangées agissant comme protection contre le vent, le soleil, les congères, les émissions industrielles, les embruns etc. ; ►plantation brise-vent, bande boisée brise-vent) ; *syn.* bande [f] boisée ; *g* **Schutzstreifen [m]** (Meist mehrreihige Baum- oder Strauchpflanzung zum Schutz gegen Wind, Sonne, Schneeverwehungen, Industrieimmissionen, Salzgischt etc.; ►Windschutzpflanzung); *syn.* Gehölzschutzstreifen [m].

shelterbelt [n] (2) *landsc.* ►windbreak.

shift [n] *geo. hydr.* ►stream alignment shift.

shift [n] **in floristic species composition, anthropogenic** *phyt.* ►anthropogenic change/alteration of species composition.

5561 shifting cultivation [n] *agr.* (Primitive method of cyclical cultivation, widely practiced in the tropics, whereby cultivators cut some or all of the forest, burn it, and raise agricultural crops for one or two years before moving on to another site and repeating the process; cf. SAF 1983; if the soil is left to itself for a long time without cultivation a secondary forest [►second-growth forest] is often the result; ►land rotation, ►burn clearing); *syn.* shag [n] [also US], slash-and-burn farming [n] [also US]; *s* **agricultura [f] itinerante** (Tipo de actividad agrícola que subsiste en algunas zonas tropicales de América,

África y el Sureste de Asia que consiste fundamentalmente en la ►tala a fuego de zonas boscosas o de matorral para obtener un campo cubierto de cenizas sobre el que se cultiva. El campo conseguido así recibe diversos nombres como «milpa» en México y Centroamérica, «lugan» en África, «landang» en Indonesia. La fragilidad de los suelos tropicales, unida al clima que favorece la erosión y el lavado rápido de los nutrientes del suelo tienen como resultado el agotamiento del mismo a corto plazo. A los pocos años los suelos son abandonados y la vegetación inicia su proceso de reconstrucción natural, dando lugar al ►bosque secundario o a la ►sabana. Este proceso conlleva la destrucción del bosque primario [natural]; cf. DINA 1987; ►cultivo en rotación); *f* **agriculture [f] itinérante (sur brûlis)** (Mode de culture agricole temporaire, pratiquée dans les pays tropicaux, consistant à défricher totalement ou partiellement les formations végétales arborées, à les brûler, puis après un travail sommaire du sol à les cultiver pendant une ou plusieurs années, avant de les abandonner et de se déplacer pour recommencer ailleurs ; les terres abandonnées se régénèrent lentement pour reconstituer une ►forêt de substitution ; DG 1984, 315 ; ►assolement, ►brûlis de défrichement) ; *syn.* culture [f] itinérante, agriculture [f] nomade ; *g* **Wanderfeldbau [m]** (Tropische ►Landwechselwirtschaft bei der nach der Rodung — meist ►Brandrodung — eine ein- oder zweijährige Ackernutzung folgt, wobei auch die Siedlungen in einem gewissen zeitlichen Rhythmus verlegt werden; danach wird der Boden zur Regeneration für einen langen Zeitraum sich selbst überlassen, wobei eine Wiederbewaldung, — die Entstehung eines ►Sekundärwaldes —, stattfindet); *syn.* tropische Landwechselwirtschaft [f].

shifting dune [n] *geo.* ►drifting sand dune.

shifting [n] **of roadbed** *trans.* ►road relocation.

shingle [n] [UK]**, pea** *constr.* ►pea gravel.

shingle bar [n] *geo.* ►storm beach [US]/storm-beach [UK].

shingle beach [n] *geo.* ►storm beach [US]/storm-beach [UK].

5562 shipping [n] **of plants** *constr. hort.* (Loading and transport of plants from nurseries to the construction site); *syn.* plant shipment [n]; *s* **expedición [f] de plantas** (Carga y transporte de plantas del vivero a la obra); *f* **vente [f] et expédition [f] des végétaux** (Chargement et transport des végétaux depuis la pépinière jusqu'au chantier ; *g* **Pflanzenversand [m, o. Pl.]** (Laden und Transport von Pflanzen von der Baumschule, Staudengärtnerei etc. zur Bau- oder Verwendungsstelle).

shoal [n] *geo.* ►sandbank.

5563 shoe-scraper [n] *constr.* (Screened frame placed before a door for wiping mud and dirt from shoes); *syn.* door grate [n] [also US]; *s* **limpiabarros [m]** (Rejilla colocada ante la entrada de las casas que sirve para quitarse el barro de los zapatos); *f* **gratte-pieds [m]** (Grille devant la porte d'entrée d'une habitation ou les portes palières afin d'éliminer la boue collée aux semelles des chaussures) ; *g* **Fußabstreifer [m]** (Rost vor der Haustür zum Abstreifen des Schmutzes von den Schuhen); *syn.* Fußabstreicher [m], Schuhabstreicher [m].

5564 shoot [n] *bot. hort.* (Stem of a young vascular plant; *specific terms* ►adventitious shoot, ►cane, ►coppice shoot, ►epicormic shoot of stem, ►one year's growth, ►sucker, ►water sprout); *syn.* sprout [n]; *s* **brote [m]** (Tallo joven no lignificado; *términos específicos* ►brote adventicio, ►brote de cepa, ►chupón, ►chupón del tronco, ►guía anual, ►hijuelo, ►rama basilar); *syn.* retoño [m], vástago [m]; *f* **pousse [f]** (Terme désignant une jeune tige feuillée, verte, non ligneuse et pourvue d'un bourgeon terminal ; *termes spécifiques* ►gourmand, ►gourmand de tige, ►pousse adventive, ►pousse annuelle, ►rejet de taillis, ►rejet du porte-greffe, ►tige centrale) ; *g* **Trieb [m]** (Bezeichnung für einen jungen, unverholzten Spross; *UBe*

►Adventivspross, ►Grundtrieb, ►Jahrestrieb, ►Stammaustrieb, ►Stockaustrieb 1, ►Wasserreis, ►Wildtrieb).

shoot [n]**, annual** *bot. hort.* ►one years's growth.

shoot [n]**, epicormic** *arb. bot. hort.* ►water sprout.

shoot [n]**, ground** *hort.* ►cane.

shoot [n]**, root** *bot.* ►root sucker.

shoot [n]**, sap** *arb.* ►coppice shoot.

shoot [n]**, stool** *arb.* ►coppice shoot.

5565 shoot elongation [n] *hort.* (Shoot of a plant, which has grown in one year and is not yet woody; growth over several years produces a twig or also a ►cane); *syn.* shoot lengthening [n]; *s* **crecimiento [m] del brote** (Crecimiento de un vástago no lignificado en un año que con el tiempo forma un brote o una ►rama basilar); *syn.* crecimiento [m] del retoño; *f* **allongement [m] de la tige** (Croissance chaque année d'une pousse non ligneuse ; la croissance pendant plusieurs années donnant naissance à un rameau ou une ►branche ; ►tige centrale) ; *g* **Triebzuwachs [m]** (In einem Jahr gewachsener unverholzter Pflanzentrieb; mehrere Zuwachsraten ergeben einen ,Zweig', aber auch einen ►Grundtrieb); *syn.* Trieblänge [f] (±).

shooting out [n] *bot.* ►leafing out.

5566 shooting preserve [n] [US] *conserv. hunt.* (Large area of land enclosed by an ►animal-tight fence [US]/deer-stop fence [UK] on which pen-reared game are kept for hunting purposes under controlled conditions; cf. RCG 1982; ►wild animal enclosure, ►game reserve, ►huntable game); *syn.* game enclosure [n], shooting reserve [n] [UK]; *s* **reserva [f] de caza** (Área especial reservada para la reproducción de la caza, donde se permiten capturas bajo control y que puede cerrarse al público durante periodos; DINA 1987; ►cercado de protección de la caza, ►especie cinegética, ►Refugio de Fauna [Es], ►Reserva Regional de Caza [Es]); *f* **enclos [m] de réserve de chasse** (►clôture gibier, ►enclos d'animaux sauvages, ►espèce chassable, ►réserve de chasse, ►réserve nationale de chasse et de faune sauvage) ; *g* **Wildgehege [n]** (Mit einem ►Wildschutzzaun eingefriedetes Gelände/Revier, in dem ►jagdbare Tiere für jagdliche Zwecke gehalten werden; ►Tiergehege, ►Wildschutzgebiet); *syn.* Gatterrevier [n], Gehege [n], Jagdgatter [n] (§ 21 LJagdG-NW).

shooting reserve [n] [UK] *conserv. hunt.* ►shooting preserve [US].

shooting season [n] *hunt. leg.* ►hunting season.

shoot lengthening [n] *hort.* ►shoot elongation.

5567 shoot tendril [n] *bot.* (Slender, leafless, coiling organ of a climbing plant; e.g. grape *[Vitis vinifera]*, five-leaved Virgina creeper with additional adhesive disks/discs *[Parthenocissus quinquefolia]*; ►tendril); *s* **zarcillo [m] caulinar** (Órgano filamentoso enroscado formado en los extremos de algunos tallos con el cual se agarra una planta en algún otro cuerpo como en la pasionaria o en el género *Parthenocissus*; cf. DB 1985; ►zarcillo); *syn.* zarcillo [m] rameal; *f* **vrille [f] raméale** (Organe aérien de fixation des plantes grimpantes des Vitacées de nature caulinaire constitué par un rameau court volubile, p. ex. la Vigne *[Vitis vinifera]*, la Vigne vierge avec ses pelotes adhésives complémentaires *[Parthenocissus quinquefolia]* ; ►vrille) ; *syn.* vrille [f] caulinaire, rameau-vrille [m] ; *g* **Sprossranke [f]** (Fadenförmig ausgebildeter, kontaktreizbarer, ausschließlich dem Klettern dienender Sprossteil [Achsenfadenranke], z. B. beim Weinstock *[Vitis vinifera]*, Mauerwein mit zusätzlichen Haftscheiben *[Parthenocissus quinquefolia]*; ►Ranke).

5568 shop drawing [n] *constr. plan.* (One of the drawings, diagrams, schedules, illustrations, performance charts, catalog[u]es, other data and installation instructions furnished by the contractor or any subcontractor, manufacturer, or supplier which show how specific portions of the work shall be fabricated and/or installed; ►detail drawing); *syn.* production drawing [n]; **s dibujo [m] de construcción** (Diagrama, dibujo o representación gráfica que indica la forma de montaje o instalación del correspondiente producto; ►dibujo en detalle); *f* **fiche [f] de détail d'un produit** (Dessin, diagramme ou toute représentation graphique produite par le fournisseur ou le producteur présentant un produit et sa mise en œuvre ; ►plan de détail) ; *syn.* détail [m] d'un produit, schéma [m] de construction, schéma [m] d'installation ; *g* **Produktzeichnung [f]** (Zeichnung, Diagramm oder sonstige grafische Darstellung, die vom Hersteller oder Lieferanten eines Produktes geliefert wird, um zu zeigen, wie ein Produkt installiert oder zusammengefügt werden soll; ►Detailplan); *syn.* Konstruktionszeichnung [f].

5569 shore [n] *geo.* (**1.** Narrow strip of land immediately bordering any body of water, especially a sea or large lake; GFG 1997. **2.** The most seaward part of the coast: area between the lowest, low-water spring-tide line and the highest point reached by storm-waves. Sometimes divided into *backshore*, covered with beach material, and *foreshore*, to edge of low-tide terrace; the most typical **sea beach [n]** has a gently concave profile; the landward side is backed by sanddunes, succeeded by shingle, then an area of sand, and rocks covered with seaweed at or about the low-tide mark; DNE 1978; ►flat shore, ►coast; *generic term* ►beach·1) ; *syn. to 2.* seashore [n], seaside [n], *at the ocean* ocean beach [n]; **s playa [f] (costera)** (Franja del litoral formada por acumulación de material arenoso, de grava o limo que está influenciada por el oleaje; en las playas se diferencian varias zonas: La **anteplaya**, la más extensa, que corresponde a la parte siempre sumergida, la **playa baja**, en suave pendiente, que queda cubierta por el agua durante la marea alta, y el **cordón litoral**, la parte más elevada y cuya cumbre se denomina **cresta de playa**; ►orilla del mar llana, ►costa; *término genérico* ►playa); *syn.* playa [f] de mar; *f* **plage [f] côtière** (Zone d'accumulation sédimentaire, de transport ou d'érosion sur le bord de mer de matériaux [galets, sable] située sur l'►estran et comprenant la zone de déferlement des vagues et l'avant-plage [avant-côte] avec les avant-dunes de la zone infralittorale ; les plages s'orientent perpendiculairement à la houle dominante et ont un profil légèrement concave ; *on peut distinguer* le haut de plage, l'avant-plage, le bas de plage, le cordon littoral ; DG 1984 ; ►côte, ►rivage 2, ►zone de baignade ; *contexte* plage océanique, plage méditerranéenne ; *terme générique* ►plage) ; *g* **Meeresstrand [m]** (...strände [pl]; **1.** der aus Sand, Kies, Geröll, Geschiebelehm oder ähnlichem Material bestehende und im Wirkungsbereich der Wellen liegende Küstenstreifen. **2.** Im Englischen wird *shore* je nach Kontext auch als Strand an einem großen See verstanden; ►Badestrand, ►Flachufer 2, ►Küste; *OB* ►Strand).

5570 shoreline [n] [US]/**shore line** [n] [UK] *geo. wat' man.* (Outline of the land at the margin of a lake, river, or the sea. The latter is frequently called a coast line; MAR 1964); **s línea [f] de orilla** (Límite entre un cauce o lecho de agua, o el mar, y sus márgenes; en el caso de este último se habla de línea de costa); *f* **ligne [f] de rive** (Ligne délimitant la rive d'un cours d'eau et une parcelle riveraine et définie par le niveau des moyennes eaux) ; *g* **Uferlinie [f]** (Grenze zwischen einem Gewässer und den Ufergrundstücken, die durch den Mittelwasserstand bestimmt wird; cf. Wassergesetze der Bundesländer).

5571 shoreline erosion [n] [US]/**shore line erosion** [n] [UK] *geo.* (Removal of soil, sand or rock from land adjacent to the sea due to wave action; ►retrogradation of shoreline); *syn.* coastal erosion [n]; **s erosión [f] marina** (Acción destructora del mar sobre las costas; *opp.* [proceso de] acreción; ►regresión de

la línea de la costa); *syn.* erosión [f] del litoral; *f 1* **érosion [f] côtière** (Action de la houle sur les rives côtières ; ►recul côtier) ; *syn.* érosion [f] marine (DG 1984) ; *f 2* **abrasion [f] marine** (Processus d'usure mécanique des roches, lié au frottement des matériaux côtiers brassés par les vagues ; DG 1984, 167) ; *f 3* **ablation [f] marine** (*Terme spécifique pour le* processus d'enlèvement par la mer de matériaux sur la côte) ; *g* **Küstenerosion [f]** (Ausnagende, Hohlkehlen bildende sowie einen Strand abtragende Tätigkeit einer Brandung oder durch Gezeiten und Meeresströmungen ausgelöst; ►Küstenrückgang); *syn.* marine Erosion [f], Brandungserosion [f], Abrasion [f], Erosion [f] an Küsten.

shoreline protection [n] *conserv. land'man.* ►coastal protection.

shoreline retreat [n] *geo.* ►retrogradation of shoreline.

5572 shore oil pollution [n] *envir.* (Pollution of shorelines and their flora and fauna, especially along seacoasts by the leakage [tanker collision, off-shore drillings] or discharge of raw oil from ships [bilge water]; ►oil pollution, ►oil pollution accident); *syn.* black tide [n] [also US] (GE 1977); **s marea [f] negra** (Gran ►contaminación de hidrocarburos de zonas costeras causada por ►accidente con derrame de hidrocarburos o por derrames de instalaciones de perforación en alta mar); *f* **marée [f] noire** (Pollution des rivages due à la diffusion dans la mer de mazout provenant du dégazage des soutes d'un pétrolier au large des côtes ; PR 1987 ; ►pollution par les hydrocarbures, ►accident entraînant une pollution par les hydrocarbures) ; *g* **Ölpest [f]** (Verschmutzung des Meeres und der Meeresküsten samt der dortigen Flora und Fauna durch ausgelaufenes [Tankerhavarien, Offshore-Bohrungen] oder abgelassenes Rohöl [Bilgenwasser der Schiffe]; ►Ölverschmutzung, ►Ölverschmutzungsunfall).

5573 shortcut traffic [n] [US]/**short-cut traffic** [n] [UK] *trans.* (Rush hour traffic whereby so-called 'rat-runners' use residential streets to weave their way across town in order to avoid traffic lights and congestion); *syn.* detouring [n] to avoid traffic lights, cut-through traffic [n] [US], rat run traffic [n] [UK] (TGG 1984, 69); **s tráfico [m] parásito** (Circulación de vehículos por calles residenciales en horas punta causada por aquellos automovilistas que pretenden evitar los atascos de las arterias principales y se buscan una ruta alternativa por calles secundarias generando las consiguientes molestias); *f* **trafic [m] parasite** (Circulation automobile indésirable à l'intérieur d'un quartier résidentiel provoquée par les automobilistes locaux lorsque la circulation sur la voirie principale est embouteillée) ; *g* **Schleichverkehr [m]** (Unerwünschter Autoverkehr in Wohngebieten, vorwiegend verursacht durch Ortskundige, wenn Hauptverkehrsstraßen verstopft sind).

shortening [n] **of trees and shrubs** *constr. hort.* ►cutting back of trees and shrubs.

5574 shortfall [n] *plan.* (Shortage of facilities or land to meet a specific demand/requirement); *syn.* deficit [n], uncovered demand [n], unsatisfied/uncovered requirements [npl]; **s déficit [m]** (Falta de equipamiento o de suelo para cubrir necesidades específicas); *f* **déficit [m]** (Insuffisance des équipements pour certaines activités) ; *g* **Fehlbedarf [m]** (Nichtvorhandensein von/Mangel an Flächen, Ausstattungen und Einrichtungen für Nutzungen); *syn.* Defizit [n].

5575 short list [n] *prof.* (List of construction or planning firms which have prequalified for a project and are invited for interview before the construction or design/planning firm is selected); **s lista [f] selectiva** (Lista de empresas o consultorías que se prestan para realizar un proyecto específico y que son invitadas a entrevistas con este fin); *f* **sélection [f] préalable** (Processus de décision et de nomination par le pouvoir adjudicateur des candidats admis à participer à un concours restreint et sélectionnés selon leurs compétences, références et moyens) ; *g* **engere Wahl [f, o. Pl.] (3)** (Liste der Firmen oder Planungsbüros, die im Bedarfsfalle am besten für ein Projekt qualifiziert sind).

short standard [n] [UK] *hort.* ▶dwarf fruit tree [US].

5576 short-stay campground [n] [US] *recr.* (Temporary campground [US]/camping ground [UK] for stays of short-term visitors in tents or recreation vehicles (RV) [US]/motor home [UK]; *opp.* ▶permanent campground [US]/permanent caravan site [UK]; ▶tent camping [US]/camping [UK]); *syn.* short-stay camping ground [n] [UK]; *s* **camping [m] de tránsito** (Camping o parte del mismo utilizado por campistas que permanecen poco tiempo para pernoctar en tienda de campaña o en ▶autocaravana; *opp.* ▶camping permanente, ▶camping 2); *syn.* camping [m] de paso; *f* **terrain [m] de camping saisonnier** (Terrain aménagé à des fins strictement saisonnières pour lequel est fixée la période d'exploitation en dehors de laquelle tout maintien de la tente ou de caravane est interdit ; ▶camping sous la tente, ▶terrain de camping permanent) ; *syn.* camp [m] de tourisme saisonnier, terrain [m] de camping de transit ; *g* **Durchgangscampingplatz [m]** (Campingplatz oder ein Teil davon, der für Camper vorgesehen ist, die nur kurzfristig bleiben; *opp.* ▶Dauercampingplatz, ▶Zelten); *syn.* Touristikcampingplatz [m].

short-stay camping ground [n] [UK] *recr.* ▶short-stay campground [US].

5577 short-stay recreation [n] *recr.* (Recreation period of up to 4 days; ▶daily recreation, ▶local recreation, ▶vacation [US]/holidays [UK]; *opp.* ▶long-stay recreation); *syn.* short-term recreation [n]; *s 1* **vacaciones [fpl] cortas** (Periodo/período de recreación de hasta 4 días; ▶recreo local, ▶recreación diaria, ▶vacaciones; *opp.* ▶vacaciones largas); *s 2* **puente [m]** (Prolongación de un fin de semana al «saltar» un día laborable gracias a un puente entre éste y un día festivo en martes o jueves. Costumbre española que se está popularizando en Europa Occidental); *f* **tourisme [m] de passage** (Une des premières formes de loisirs en milieu rural, la durée d'un séjour étant au maximum de 4 jours ; ▶détente et récréation quotidienne, ▶loisirs de proximité, ▶vacances ; *opp.* ▶tourisme de séjour) ; *syn.* vacances [fpl] courtes, court séjour [m] ; *g* **Kurzzeiterholung [f, o. Pl.]** (Erholung bis vier Tage; ▶Naherholung, ▶Tageserholung, ▶Urlaub; *opp.* ▶Langzeiterholung); *syn.* Kurzerholung [f, o. Pl.].

short term crop [n] *agr. hort.* ▶green manure catch crop.

5578 shotcrete [n] [US] *constr.* (Concrete pneumatically projected at high velocity onto a surface, e.g. for stabilizing a steep slope); *syn.* gun-applied concrete [n], guncrete [n] [UK]; *s* **gunita [f]** (Nombre comercial de un material de construcción compuesto por cemento, arena o escoria machacada y agua que se mezcla y se proyecta a presión; DACO 1988); *syn.* hormigón [m] bombeado, hormigón [m] proyectado, torcreto [m], hormigón [m] a soplete [CA], mortero [m] lanzado [Mex] (todos BUKSCH 1959); *f* **béton [m] projeté** (Mélange de ciment et de sable mis en place à la machine pneumatique par projection à grande vitesse [opération de gunitage] du mélange frais sur la surface à recouvrir ; DBT 1985) ; *syn.* gunite [f], mortier [m] projeté ; *g* **Spritzbeton [m]** (Mit 2-3 atü durch eine Rohrleitung gepresstes Betongemisch, das in mehreren Lagen aufgespritzt wird und z. B. einer Hangsicherung dient); *syn.* Torkretbeton [m].

shot put area [n] [US] *recr.* ▶shot-putting runway and landing area.

5579 shot-putting runway [n] **and landing area** [n] *recr.* (Competitive sports facility for shot putting delimited by a fixed take-off log at the edge of a 2.135m circle); *syn.* shot put area [n] [also US]; *s* **instalación [f] para lanzamiento de pesas**;

f **installation [f] de lancement du poids** (Installation sportive de lancer du poids de forme circulaire [diamètre 2,135 m] et limitée par une poutre de lancer) ; *syn.* installation [f] de lancer du poids ; *g* **Kugelstoßanlage [f]** (Sportanlage zum Kugelstoßen, die durch einen fest verankerten Abstoßbalken am Rande eines 2,135 m messenden Kreises begrenzt ist).

shoulder [n] [US]**, soft** *trans.* ▶road verge.

shoulder [n] [US]**, unpaved** *trans.* ▶road verge.

5580 shown to scale [loc] *constr. plan.* (Delineated on plan according to a specific scale); *s* **dibujado/a a escala [loc]**; *f 1* à l'échelle [loc]; *f 2* **proportionné, ée [pp/adj]** ; *g* **maßstabsgerecht [adj]** (Einem angegebenen Maßstab genau entsprechend); *syn.* maßstabsgetreu [adj].

showy perennial [n] [US] *hort. plant.* ▶ornamental perennial.

5581 shred [vb] *agr. constr. hort.* (**1.** To cut or tear into thin strips [shreds]. **2.** To cut or break up into fragments [chips]); *syn.* chip [vb] [US]; *s* **triturar [vb]** *syn.* picar [vb], desmenuzar [vb]; *f* **broyer [vb]** (Broiement de la paille, de branches, d'écorces au moyen d'une broyeuse) ; *g* **häckseln [vb]** (Zerkleinern von Stroh, Ästen und Baumrinden mit einem Häcksler [Häckselmaschine]); *syn. constr. hort.* schreddern [vb].

5582 shredded bark humus [n] *constr. hort.* (Bark which has been reduced to small strips, piled in heaps and left to ferment for several months in the open air with the addition of nitrogen, and oxigenated by turning the heap over. The resulting product is used as a ▶soil amendment [US]/soil ameliorant [UK]; ▶bark mulch, ▶bark chips); *s* **humus [m] de corteza** (Corteza de árbol triturada y fermentada en pequeños acopios añadiendo nitrógeno y oxígeno para utilizarla como ▶material de enmienda de suelos; ▶mulch de corteza triturada, ▶astillas de corteza); *f* **humus [m] d'écorce** (Écorce broyée et stockée en tas à l'air libre, soumise à un processus de fermentation avec apport d'azote et d'oxygène [retournement successif des tas] et transformée en ▶amendement ; ▶mulch d'écorce fragmentée, ▶copeaux d'écorce) ; *g* **Rindenhumus [m, o. Pl.]** (Bodensubstrat aus zerkleinerter Rinde, die eine mehrmonatige Fermentation [Gärung] unter Zufügung von Stickstoff und Sauerstoff [Umschichtung der Miete] durchgemacht hat; dieses Substrat wird zu einem ▶Bodenverbesserungsstoff 1 aufbereitet; ▶Rindenmulch, ▶Rindenschrot).

5583 shredded material [n] *agr. constr. hort.* (Straw, branches, bark, etc. cut into small pieces by a shredder for mulching or composting); *s* **material [m] triturado** (Paja, ramas, corteza, etc. cortadas en trozos pequeños con una picadora/trituradora); *f* **déchets [mpl] de broyage** (Produits provenant du broiement de paille, de branches, d'écorces au moyen d'une broyeuse) ; *g* **Häckselgut [n, o. Pl.]** (Von einem Häcksler [Häckselmaschine] zerkleinertes Stroh oder ges[c]hredderte Äste und Baumrinden).

5584 shrinkage [n] *constr.* (*Generic term* volume decrease caused by cooling [metal], drying and chemical changes [wood, concrete]; ▶drying shrinkage of concrete, ▶seasoning check, ▶seasoning of wood, ▶shrinkage crack); *syn.* contraction [n]; *s* **merma [f]** (Término genérico para la contracción de un material causada por enfriamiento [metal], secado o cambios químicos [madera, hormigón]; ▶fenda de merma, ▶grieta de contracción, ▶merma del hormigón, ▶merma de la madera); *f* **retrait [m]** (**1.** Contraction d'un matériau provoquée soit par son refroidissement [métal], soit par un abaissement de taux d'humidité [bois], soit par l'élimination de l'eau de gâchage excédentaire [béton, enduit], soit par l'évaporation d'un solvant [colle, peinture], soit encore par la cuisson [poterie, brique]. **2.** Rétraction que subit un corps ou matériaux au séchage [béton, bois, ciment, mortier, etc.] se manifestant par une fissuration ou

S

une déformation [bois] ; ►fente de retrait 2, ►fissure de retrait [de dessiccation] dans le béton, ►retrait du béton, ►retrait du bois) ; *g* **Schwinden [n, o. Pl.]** (*OB* Längenveränderung oder Verkleinerung des Rauminhaltes von Baumaterialien, z. B. beim Abkühlen [Metall], beim Austrocknen [Beton, Holz, Farbanstrich] oder beim Trocknen und Brennen [Klinker, Kachel]; ►Schwinden von Beton, ►Schwinden von Holz, ►Schwindriss, ►Schwindriss in Holz).

shrinkage coefficient [n] *constr.* ►shrinkage crack.

5585 shrinkage crack [n] *constr.* (Fissure in concrete caused by a rapid, irregular drying process; in building construction, the change in length is defined by the **shrinkage coefficient** or **degree of contraction** in mm/m; the extent of shrinkage depends on the aggregates in the concrete: for cement and concrete, they are given as 0.2-0.8 mm/m: for aerated or gas concrete, 1-3 mm/m. Long concrete and reinforced concrete structures are installed with **contraction joints** to allow specifically for movement during shrinkage and to prevent unexpected ►shrinkages); *s* **grieta [f] de contracción** (Pequeña fisura debida a la reducción de volumen causada por el secado irregular y demasiado rápido del hormigón; en construcciones las modificaciones de largo se indican como **medida de contracción** en mm/m; el tamaño de las medidas de contracción depende de los áridos utilizados: en el cemento y hormigón varían entre 0,2-0,8 mm/m, en el hormigón gaseoso entre 1-3 mm/m. En construcciones de hormigón o de hormigón armado se prevén **juntas de contracción** para posibilitar movimiento en el proceso de retracción y así evitar **g. de c.**; ►merma); *f* **fissure [f] de retrait (de dessiccation) dans le béton** (La fissure liée au retrait de dessiccation, consécutif à l'évaporation de l'eau, peut se manifester quelques minutes après la mise en œuvre du béton, et se poursuivre quelques semaines après ; la **valeur du retrait** est donnée en mm/m, elle dépend des matériaux constituant le béton : elle est dans les conditions courantes de l'ordre de 0,2 à 0,8 mm/m pour le ciment et le béton, de 1 à 3 mm/m pour le béton cellulaire. Pour contrer les effets de retrait sur les dalles ou les éléments en béton brut ou béton armé de grandes longueurs il est recommandé d'introduire des **joints de retrait** afin d'éviter la fissuration aléatoire ou localisée ; ►retrait) ; *syn.* fissuration/ fissure [f] de béton ; *béton et mortier* faïençage [m] ; *g* **Schwindriss [m] in Beton** (Durch einen schnellen, ungleichmäßigen Trocknungsprozess verursachter Schrumpfungsriss in Beton; bei Baukonstruktionen werden die Längenänderungen als **Schwindmaß** in mm/m angegeben; die Größe der Schwindmaße sind abhängig von den Zuschlagstoffen im Beton: bei Zement und Beton liegen sie bei 0,2-0,8 mm/m, bei Gasbeton bei 1-3 mm/m. Bei langen Beton- und Stahlbetonkonstruktionen werden **Schwindfugen** vorgesehen, um Bewegungen während der Schwindung gezielt zu ermöglichen und unerwartete **Sch.e** zu vermeiden; ►Schwinden); *syn.* Schwundriss [m].

shrinkage [n] **of concrete** *constr.* ►drying shrinkage of concrete.

5586 shrub [n] *bot. hort.* (Generic term for *a* woody plant, less than 8m tall, branching below, e.g. hazelnut *[Corylus avellana]* or near ground level, e.g. elder *[Sambucus nigra]* into several main stems without a central trunk; ►dwarf shrub, ►field tree or shrub, ►large shrub, ►nanophanerophyte, ►ornamental shrub, ►prostrate shrub, ►small shrub, ►specimen tree/shrub, ►suffruticose shrub, ►young shrub transplant); *s 1* **arbusto [m]** (Planta leñosa de altura generalmente no superior a los 6 m y de ramificación desde la base; ►arbusto en espaldera, ►arbusto grande, ►arbusto joven para transplante, ►arbusto ornamental, ►espécimen arbóreo/arbustivo, ►leñosa campestre, ►mata, ►nanofanerófito, ►pequeño arbusto, ►sufrútice); *s 2* **subarbusto [m] (1)** (Arbusto de altura entre 50 cm y 2 m. Los arbustos

de una altura de un metro a lo sumo o poco más, se llaman «**matas**» o «**matillas**»; cf. DB 1985); *f 1* **arbrisseau [m]** (*Pour les horticulteurs* végétal ligneux à tige simple et nue à la base, de moins de 7 m de hauteur et présentant pratiquement tous les caractères d'un arbre, p. ex. l'Érable de Mandchourie *[Acer ginnala]*, l'Arbre de Judée *[Cercis siliquastrum]*, le Lagerose *[Lagerstroemia indica]*, le Sureau *[Sambucus nigra]* ou le Cornouiller *[Cornus mas]* ; pour les botanistes ce végétal est nommé ►arbuste nain ; IDF 1988, 19 ; ►arbuste en espalier, ►bosquet champêtre, ►espèce sousfrutescente, ►ligneux champêtre, ►nanophanérophyte, ►petit arbuste) ; *f 2* **arbuste [m]** (*Pour les horticulteurs* végétal ligneux de moins de 4 m de hauteur, formant un ensemble touffu de tiges jaillissant du sol et parmi lesquelles un tronc ne peut être distingué ; on distingue les arbustes buissonnants hauts [4 à 7 m] et les arbustes buissonnants bas [1 à 3 m] ; pour les botanistes ce végétal est nommé ►arbrisseau ; IDF 1988, 19 ; ►arbuste d'ornement, ►nanophanérophyte) ; *syn.* buisson [m] (1) ; *g* **Strauch [m] (2)** (Sträucher [pl]; **1.** bis ca. 8 m hohes, nicht baumartig wachsendes Gehölz, dessen Haupt- und Seitenachsen sich meist aus mehreren basalen, unterirdischen, z. B. bei Haselnuss *[Corylus avellana]* oder oberirdischen Seitenknospen [kahlfüßiger Großstrauch — im Französischen *arbrisseau]* verzweigen, z. B. Holunder *[Sambucus nigra]* oder Kornelkirsche *[Cornus mas]*; **2.** Mindestens 2 x verpflanzte Baumschulware ohne Ballen, die in D. in Angeboten, Ausschreibungen und Rechnungen seit 09/1995 mit den Sortierkriterien ‚Höhe' und ‚Mindesttriebzahl' [mindestens 3 Triebe] spezifiziert werden müssen, wobei nur die Triebe gezählt werden, die die geforderte Mindesthöhe erreichen und die dicht über dem Erdboden ansetzen; *syn. für 2.* verpflanzter Strauch [m]: *Abk.* v. Str.; ►Halbstrauch, ►Kleinstrauch, ►leichter Strauch, ►Solitärgehölz 2, ►Spalierstrauch, ►Zierstrauch, ►Zwergstrauch); *syn.* ►Nanophanerophyt [m].

shrub [n], **flowering tree or** *hort. plant.* ►ornamental woody plant.

shrub [n], **half-** *bot.* ►suffruticose shrub.

shrub [n], **light** *hort.* ►young shrub transplant.

shrub [n], **partially flowering** *hort. plant.* ►ornamental shrub.

shrub [n], **semi-***bot.* ►suffruticose shrub.

shrub [n], **sub-** *bot.* ►suffruticose shrub.

shrub [n], **suffrutescent** *bot.* ►suffruticose shrub.

shrub [n], **tall** *hort. plant.* ►large shrub.

shrub [n], **trailing** *bot.* ►prostrate shrub.

shrub bed [n] *constr.* ►shrub border.

5587 shrubbery [n] *gard.* (**1.** plantation of shrubs. **2.** Growth or group of shrubs); *s* **matorral [m] (2)** (Término popular referente a grupos de matas en jardines o en el campo); *syn.* mata [f]; *f 1* **buisson [m] (2)** (*Terme général désignant un* groupe de plantes ligneuses, buissonnantes) ; *f* **broussaille [f]** (*Terme imprécis utilisé pour désigner une* formation végétale basse de plantes ligneuses, buissonnantes, se développant en sous-bois ou sur des terrains abandonnés et favorable à la propagation du feu ; DG 1984) ; *g* **Gebüsch [n] (1)** (*Allgemeiner Sprachgebrauch* ein Bestand von Sträuchern, z. B. in einer Parkanlage, im Wald).

5588 shrub border [n] *gard.* ([Elongated] bed of shrubs in private gardens or public green areas; ►border with shrubs and trees); *syn.* shrub bed [n], shrub planting area [n] (SPON 1974, 88); *s* **macizo [m] de arbustos** (Arriate alargado en parque público o jardín privado plantado con especies arbustivas; ►macizo de leñosas); *f* **massif [m] d'arbustes** (Groupement dense de végétaux ligneux de faible hauteur dans les jardins privés ou publics ; ►massif d'arbres et d'arbustes [d'ornement]) ; *g* **Strauchrabatte [f]** (Beet in privater oder öffentlicher Freianlage, das vorwiegend mit Sträuchern bepflanzt ist; ►Gehölzrabatte).

5589 shrubby ground cover vegetation [n] *plant.* (LD 1994 [4], 9); *s* **cobertura [f] de arbustos enanos;** *f* **couverture [f] arbustive basse** *syn.* couvert [m] arbustif bas, arbustes [mpl] à végétation étalée, arbustes [mpl] à port étalé, couverture [f] arbustive étalée, végétation [f] arbustive tapissante ; *g* **Strauchbodendecke [f]** *syn.* Bodenbedeckung [f] mit niedrigen Gehölzen, bodendeckende Sträucher [mpl], bodendeckende Kleingehölze [npl].

shrub carpet [n] [US]**, matted dwarf-** *phyt.* ►creeping dwarf shrub carpet [US]/creeping dwarf-shrub thicket [UK].

5590 shrub clearing [n] *constr. for. hort.* (Removal of unwanted scrub or shrub vegetation; ►coppice 2); *syn.* coppicing [n]; *s* **despeje [m] de arbustos** (Limpieza del terreno con corte de la vegetación no deseada; ►poda hasta el tocón); *f* **débroussaillage [m]** (**1.** Élimination des arbustes et broussailles ; **2.** *protection contre l'incendie* opérations dont l'objectif est de diminuer l'intensité et de limiter la propagation des incendies par la réduction des combustibles végétaux en garantissant une rupture de la continuité du couvert végétal et en procédant à l'élagage des sujets maintenus et à l'élimination des rémanents de coupes ; Art. L.321-5-3 du Code forestier ; la zone à débroussailler s'étend sur un rayon de 50 mètres autour des constructions, chantiers travaux, et installations de toute nature ainsi que sur une bande de 10 mètres de part et d'autres de l'emprise des voies privées donnant accès à ces constructions. Les travaux de débroussaillage sont à la charge du propriétaire de la construction ou voie d'accès concernées, y compris si la zone dépasse les limites de son terrain ; les travaux de débroussaillage comportent l'élimination des arbres dont les cimes et les branches basses sont trop proches du toit et des murs, les petits arbustes situés sous des grands arbres, les végétaux très inflammables, les végétaux morts ou très secs, les masses compactes de végétation naturelle souvent très sèches en été, les branches basses des arbres sur une hauteur de 2 m, les herbes sèches ; ►recéper) ; *syn.* débroussaillement [m] ; *g* **Abholzung [f] von Sträuchern** (Entfernen der oberirdischen Teile von Sträuchern auf einer Fläche; ►auf den Stock setzen); *syn.* Abholzen [n, o. Pl.] von Sträuchern, Auf-den-Stock-Setzen [n, o. Pl.] von Sträuchern.

5591 shrub dune [n] *geo. phyt.* (Dune in semiarid regions [US]/semi-arid regions [UK] which is covered with scattered shrubs); *s* **duna [f] arbustiva** (Duna en áreas semiáridas cubierta con arbustos dispersos); *f* **dune [f] arbustive** (**1.** Dune des régions semi-arides, parsemée de végétaux arbustifs. **2.** *Dune côtière* espace dunaire sensible faisant suite à la dune herbacée à oyat sur lequel se développe une végétation arbustive en boule déformée par le vent ; sur les dunes littorales du Nord de la France la dune arbustive est essentiellement composée de l'argousier *[Hyppophae rhamnoïdes]* qui forme d'épaisses broussailles épineuses et impénétrables et généralement côtoyé par le troène vulgaire *[Ligustrum vulgare]* et le sureau noir *[Sambucus nigra]*. La série de végétation évolue ensuite vers le climax, et est représentée par un haut perchis de saules *[Salix aurita, Salix cinerea]*, bouleaux *[Betula verrucosa]*, tremble *[Populus tremula]*, peuplier blanc *[Populus alba]*, grisard *[Populus canescens]*, aulne *[Alnus glutinosa]*, frène *[Fraxinus excelsior]* ; DOR 1979) ; *g* **Heckendüne [f]** (Düne in semiariden Gebieten, die vereinzelt mit Sträuchern bewachsen ist).

5592 shrubland [n] *phyt.* (Land area where mostly individual shrubs are scattered and do not touch each other, often with a low herbaceous undercover; ►coniferous evergreen thicket/shrubland, ►cushion shrubland, ►dwarf-shrub plant community, ►thicket); *s* **matorral [m] claro** (Según la tipología de IONESCO Y SAUVAGE formación vegetal de plantas leñosas de pequeña talla con un grado de recubrimiento entre 25-50%; DINA 1987; ►formación pulvinular, ►matorral achaparrado de coníferas, ►matorral bajo, ►matorral denso, ►monte bravo); *f* **formation [f] buissonneuse** (Peuplement arbustif dans lequel la plupart des sujets ne se touchent pas, le sol étant recouvert d'une végétation herbacée basse ; ►brousse sempervirente de résineux, ►formation d'arbustes nains en coussinets, ►fourré buissonnant, ►groupement d'arbrisseaux nains, ►hallier) ; *g* **Gebüsch [n] (2)** (Bestand von Sträuchern, bei dem sich die meisten der einzelnen Sträucher nicht berühren; der Boden ist oft von einer niedrigen Krautschicht bedeckt; ►Gebüsch 3, ►immergrünes Nadelgebüsch, ►Zwergstrauchgesellschaft, ►Zwergstrauchpolster-Formation); *syn.* Gesträuch [n].

shrubland [n]**, suffruticose** *phyt.* ►suffruticose thicket, #2.

5593 shrubland savanna(h) [n] *phyt.* (Grassy plain with low vegetative cover of 1-3m high thorny shrubs and scattered trees growing in seasonally humid tropic regions having an annual rainfall of 200-700mm and 7-10 dry months. The underlying ►herb layer is formed by an approximately 50cm high grassland, which is green after rainfall); *s* **sabana [f] de espinos** (Tipo de sabana que se presenta en zonas en la que las precipitaciones anuales varían entre 200 y 700 mm con un periodo/período de sequía de ocho a diez meses al año compuesta por arbustos de diferente talla, que suelen tener espinas en sus ramas y troncos. El ►estrato herbáceo está constituido por especies anuales o perennes que pueden alcanzar hasta 75 cm de altura. Se encuentra sobre todo en el norte del Brasil *[caatinga]* y en África; cf. DINA 1987); *syn.* bosque [m] espinoso; *f* **savane [f] arbustive (2)** (Formation végétale de plantes herbacées hautes présentant quelques arbustes épineux isolés de 1 à 3 m de hauteur, spécifique aux régions tropicales à pluviosité annuelle comprise entre 200 et 700 mm et soumise à une période de sécheresse de 7 à 10 mois. La ►strate herbacée constitue un couvert clairsemé vert en période de pluies d'environ 50 cm de hauteur) ; *syn.* savane [f] à buissons ; *g* **Dornstrauchsavanne [f]** (Niedrige Vegetationsformation mit 1-3 m hohen Dornensträuchern, Sukkulenten und wenigen Bäumen in den wechselhaft feuchten Tropen mit Niederschlägen von 200-700 mm bei ca. 7-10 ariden Monaten. Die ►Krautschicht besteht aus einer ca. 50 cm hohen regengrünen, aber nicht geschlossenen Grasdecke; die Böden bilden häufig harte Krusten, der aufliegende lockere Oberboden ist erosionsgefährdet); *syn.* Dornsavanne [f], Dornstrauchsteppe [f].

5594 shrub layer [n] *phyt.* (Vegetation layer of shrubs in a plant community; ►stratification of plant communities); *syn.* shrub stratum [n] (TEE 1980, 279); *s* **estrato [m] arbustivo** (En la ►estratificación de la vegetación de una comunidad vegetal o un rodal capa caracterizada por los arbustos); *syn.* subvuelo [m] (DB 1985); *f* **strate [f] arbustive** (Strate comprenant les espèces de végétaux ligneux de 1 à 7 m dans la structure verticale de la végétation ; ►stratification de la végétation) ; *g* **Strauchschicht [f]** (Die durch Sträucher geprägte Schicht bei der ►Vegetationsschichtung eines Bestandes oder einer Pflanzengesellschaft).

shrub planting area [n] *gard.* ►shrub boder.

shrub pruning [n] *constr. hort.* ►tree and shrub pruning.

5595 shrub rose [n] *hort. plant.* (**1.** Generic term for shrub-like rose types: ►bedding rose, ►climbing rose, ►cultivated shrub rose, ►dwarf standard rose, ►ground cover rose, ►hybrid tea, ►miniature rose, ►moss rose, ►old-fashioned rose, ►park rose, ►standard rose, ►tea rose [US]/tea-rose [UK], ►wild rose, etc. **2.** Hardy, bushy landscape rose which grows from approx. 1 to 3m in height, according to species and variety); *syn.* bush rose

S

[n] [also US]); **s rosal [m] arbustivo** (**1.** Rosal que se desarrolla formando un arbusto; término genérico para ►rosa de macizo, ►rosa de té, ►rosa híbrido de té, ►rosa musgosa, ►rosa silvestre, ►rosal antiguo, ►rosal arbustivo cultivado, ►rosal cubresuelo, ►rosal de parque, ►rosal de pie alto, ►rosal enano de pie alto, ►rosal miniatura, ►rosal trepador, etc. **2.** Arbusto de rosas que en estado adulto puede alcanzar según la variedad entre 1-3 m de altura); **f 1 rosier [m] buisson** (Terme générique pour ►rosier ancien, ►rosier buisson à fleurs groupées, ►rosier couvre-sol, ►rosier grimpant, ►rosier hybride de thé, ►rosier miniature, ►rosier mini-tige miniature, ►rosier mousseux, ►rosier des parcs [et des jardins], ►rosier Thé, ►rosier sauvage, ►rosier arbuste d'ornement, ►rosier tige, ►rosier nain, etc.) ; **f 2 rosier [m] arbustif** (Rose de 1 à 3 m de hauteur selon la variété ; ADT 1988, 209) ; **g Strauchrose [f]** (In der Rosenklassifizierung ein kompakter Wuchstyp, der im ausgewachsenen Zustand je nach Art und Sorte ca. 1-3 m hoch wird und als Solitär in seiner Größe habitusgerecht gepflegt oder als Gruppe resp. für frei wachsende Hecken verwendet wird. Die **Einteilung der Rosen** erfolgt u. a. nach der **Verwendungsart**, z. B. ►Beetrose [Floribundarose, Polyantharose], ►Edelrose, ►Teerose, ►Bodendeckerrose, HECKENROSE, ►Parkrose, Patio-Rose, Topfrose; nach der **Wuchsform** wie Halbstammrose, ►Hochstammrose, ►Kletterrose, ►Zwergrose, ►Zwergstammrose, oder nach dem **Grad der züchterischen Bearbeitung** wie ►Wildrose, ►Edelrose, ►Moosrose, ►Teerose, ►Zierstrauchrose oder nach dem **ersten Erscheinen im Handel**, z. B. ►Historische Rose. Die Einteilung ist nicht immer eindeutig, da sich die Termini wegen unterschiedlicher Kriterien [Abstammung, Blütengröße und -anordnung, Verwendung, Wuchs, züchterische Bearbeitung] z. T. überschneiden; z. B. wird der Unterschied zwischen Strauch- und Wildrose durch die züchterische Veränderung definiert. Die jährlich je nach Pflegekonzept in unterschiedlichen Höhen zurückgeschnittenen Rosen sind dann z. B. Beetrosen [Polyantha und Floribunda], Edelrosen und Teerosen); syn. Buschrose [f].

shrub stratum [n] *phyt.* ►shrub layer.

shrub thicket [n] [UK], **creeping dwarf-** *phyt.* ►creeping dwarf shrub carpet [US].

shrub thicket [n] [UK], **matted dwarf** *phyt.* ►creeping dwarf shrub carpet [US]/creeping dwarf-shrub thicket [UK].

shuttering [n] *constr.* ►formwork.

5596 side dumping [n] *constr.* (Method in earthworks, whereby the transporting vehicles dump [US]/tip [UK] fill material over an exposed, lateral slope of an existing oblong-shaped pile [US]/heap [UK]; ►earth filling 1); **s descargamiento [m] lateral** (En terracerías método de relleno desde vehículo de transporte en el que la tierra se vierte sobre talud lateral de un cuerpo de tierra; ►relleno); **f déversement [m] latéral** (Travaux de terrassement forme de mise en place de terres en les déchargeant latéralement sur talus d'un ouvrage en remblai existant ; ►exécution d'un remblai) ; syn. déchargement [m] latéral ; **g Seitenschüttung [f]** (Erdbau Einbaumethode, bei der Transportfahrzeuge den Boden auf eine seitliche Böschung eines vorhandenen Erdkörpers abkippen; ►Schüttung 1).

5597 side edging [n] **of a flight of steps** *arch. constr.* (Generic term for finishing element at the sides of steps in an outdoor stairway; e.g. ►cheek wall [outdoors]/string [indoors], ►flight of steps with lateral edging); **s muro [m] de caja de escalera** (Límite lateral de escalera; ►escalera incrustada y encintada lateralmente, ►zanca); **f bordure [f] latérale d'un escalier** (Terme générique pour les ouvrages latéraux bordant les escaliers, tels que le ►limon, bordure ou simple raccordement au terrain naturel ; ►escalier sur talus rentrant) ; **g seitliche Treppenbegrenzung [f]** (Seitlicher Treppenabschluss, sei es durch

►Treppenwange, Treppenbekantung oder nur mit seitlichem Erdanschluss; ►Treppe mit Bekantung).

5598 side elevation [n] *plan.* (Vertical view of an object or building from the side; generic term ►elevation); **s alzado [m] lateral** (Representación gráfica de un objeto visto de costado; término genérico ►alzado); syn. alzada [f] lateral, vista [f] lateral; **f vue [f] latérale** (Élément graphique reproduisant un objet de face latérale ; terme générique ►élévation) ; **g Seitenansicht [f]** ([Zeichnerische] Darstellung, die einen Gegenstand oder ein Objekt von der Seite zeigt; OB ►Ansicht).

side-line [n] [UK], **agricultural** *agr.* ►part-time agricultural business.

5599 side lot line [n] [US] *surv.* (►required side yard [US]); syn. side plot boundary [n] [UK]; **s límite [m] lateral de solar** (►androna); syn. límite [m] lateral de lote de terreno/predio; **f limite [f] séparative latérale** (Limites latérales séparant deux propriétés privés et qui donnent sur les voies ou emprises publiques ; ►recul de constructibilité par rapport aux limites séparatives) ; syn. limite [f] parcellaire latérale ; **g seitliche Grundstücksgrenze [f]** (►Bauwich); syn. adm. leg. seitliche Flurstücksgrenze [f].

side planting [n] *constr. for.* ►ridge planting.

side plot boundary [n] [UK] *surv.* ►side lot line [UK].

side root [n] *bot.* ►lateral root.

sidetrack [n] [US]/**side-track** [n] [UK] *trans.* ►railroad siding [US].

5600 sidewalk [n] [US] *trans. urb.* (Paved pedestrian pathway at the side of a road); syn. pavement [n] [UK]; **s acera [f]** syn. vereda [f] [AL]; **f trottoir [m]** (Espace surélevé de circulation des piétons généralement séparé de la chaussée par un caniveau) ; **g Bürgersteig [m]** (Erhöhter Gehweg neben einer Straße); syn. Gehsteig [m], Gehweg [m] z. T., Trottoir [n] ...e und ...s [pl] [CH, in D. obs., nur noch vereinzelt gebräuchlich].

side yard [n] [US] *leg. urb.* ►required side yard [US].

sieved material [n] *constr.* ►screened material.

sight line [n] [UK] *trans.* ►sight triangle.

5601 sight triangle [n] *trans.* (Triangular shaped area clear of sight obstructions at road intersections; IBDD 1981, 20, 171); syn. minimum sight triangle [n] [also US] (HALAC 1988-II, 40), sight line [n] [also UK]; **s triángulo [m] de visibilidad** (Área de forma triangular libre de obstáculos para posibilitar la vista en cruce de carreteras); **f angle [m] de visibilité de voies** (Servitude de visibilité sur les voies frappant les propriétés riveraines ou voisines des voies publiques, situées à proximité de croisements, virages ou points dangereux ou incommodes pour la circulation publique et assurant une meilleure visibilité ; un plan de dégagement détermine, pour chaque parcelle, les terrains sur lesquels peuvent s'exercer les servitudes de visibilité suivantes ; **1.** l'obligation de supprimer les murs de clôtures ou de les remplacer par des grilles, de supprimer les plantations gênantes, de ramener et de tenir le terrain et toute superstructure à un niveau au plus égal niveau qui est fixé par le plan de dégagement. **2.** L'interdiction absolue de bâtir, de placer des clôtures, de remblayer, de planter et de faire des installations quelconques au-dessus du niveau fixé par le plan de dégagement. **3.** Le droit pour l'autorité gestionnaire de la voie d'opérer la résection des talus, remblais et de tous obstacles naturels de manière à réaliser des conditions de vue satisfaisantes — Code de la voirie routière : articles L. 114-1 à L. 114-6, R. 114-4 et R. 114-2) ; syn. espace [m] de dégagement visuel ; **g Sichtdreieck [n]** (Eine von sichtbehindernder Bepflanzung und Bebauung an höhengleichen Knotenpunkten freizuhaltende Fläche, um den Verkehrsteilnehmern eine rechtzeitige gegenseitige Wahrnehmung zu ermöglichen; **S.e** können im

Bebauungsplan festgelegt werden. Grundlage für die Anlage von **S.en** ist die Richtlinie für die Anlage von Straßen [RAS-K 1]).

sight triangle [n]**, minimum** *trans.* ►sight triangle.

sign [n] *urb.* ►commercial sign [US]/advertisement board [UK].

signage [n]**, span structure for** *trans.* ►overhead sign structure.

sign structure [n] *trans.* ►overhead sign structure.

5602 siliceous rock [n] *geo.* (Cristallic rock, e.g. magmatite, gneiss, mica, slate, sandstone, quartz); *s* **roca** [f] **silícea** (Roca cristalina, pobre en bases como magmatita, gneis, mica, arenisca); *syn.* roca [f] silicosa; *f* **roche** [f] **siliceuse** (Roche acide, cristalline, riche en silice, telle que les magmatites, les gneiss, les schistes, le grès) ; *g* **Silikatgestein** [n] (Basenarmes, kristallines Gestein wie z. B. Magmatit, Gneis, Glimmerschiefer, Sandstein).

5603 silt [n] (1) *constr. eng.* (Soil texture class usually classified as consisting of 0.02 to 0.002mm diameter particles in the U.K., and 0.05 to 0.002mm in the U.S.; cf. RCG 1982; ►clay 1, ►coarse silt, ►fine silt, ►medium silt, ►soil textural class); *s* **limo** [m] (2) (►Clase textural de suelos minerales, fracción de las partículas intermedias cuyo grosor va de 0,05 a 0,005 mm de diámetro; cf. DGA 1986; ►arcilla, ►limo fino, ►limo medio, ►limo grueso); *syn.* tarquín [m]; *f* **limon** [m] (►Classe de texture de sol avec les fractions granulométriques dont les dimensions sont comprises entre 0,002 et 0,05 mm ; ►limon fin, ►limon grossier, ►limon moyen, ►argile) ; *g* **Schluff** [m] (...e [pl] u. ...schlüffe [pl]; mineralische ►Bodenart in dem Feinkornbereich von > 0,002 bis 0,06 mm; *Abk.* U; cf. DIN 4022 und DIN 4023; ►Feinschluff, ►Mittelschluff, ►Grobschluff; *Feinkornbereich < 0,002 mm* ►Ton).

silt [n] (2) *pedol.* ►alluvial silt.

silt [n]**, running** *pedol.* ►running sand/silt.

5604 siltation [n] *geo. phyt.* (Gradual deposition of silt and organic debris of organisms in an eutrophic body of water; ►dredging 2, ►sedimentation); *syn.* terrestrialization [n] (WET 1993, 380 ss), filling up [n], silting [n] (1), silting up [n] (1); *s 1* **entarquinamiento** [m] (1) (Proceso de relleno o elevación del lecho de un curso o masa de agua, por deposición de limos o sedimentos; cf. WMO 1974; ►entarquinamiento 2, ►sedimentación); *syn.* aterramiento [m]; *s 2* **colmatación** [f] (Resultado del entarquinamiento de una masa de agua; ►dragado de lodo/fango); *syn.* azolvamiento [m]; *f 1* **comblement** [m] **sédimentaire** (*Terme générique* évolution naturelle d'une vallée, d'un plan d'eau ou d'un système baie-embouchure qui entraîne le comblement de la cuvette ou du chenal par des sédiments [sables, graviers, galets, rochers, boues feuilles branches et tronc d'arbres] apportés par les rivières affluentes, par le ruissellement superficiel [précipitation et fonte des neiges] et par l'accumulation des organismes morts ayant vécu dans leurs eaux ; ce comblement progressif provoque une diminution de la profondeur et de la surface des plans d'eau et le développement d'un marais dans les zones peu profondes ; l'accélération du phénomène peut conduire à la colonisation des berges par les végétaux terrestres et à terme à la disparition de l'eau libre et l'assèchement du milieu ; les activités humaines contribuent souvent à accélérer ce processus qui se déroule normalement à l'échelle des temps géologiques) ; *f 2* **atterrissement** [m] (*Terme spécifique* accumulation limitée basse et plane en bordure d'une étendue d'eau [lac, étang, marais, eaux calmes] réduisant la surface de celle-ci [ceinture/frange/zone d'atterrissement] et favorisant le développement de formations végétales d'atterrissement [roselière d'atterrissement, etc.] ; DG 1984 ; ►sédimentation) ; *f 3* **envasement** [m] (1) (*Terme spécifique* processus d'envasement par les débris végétaux lié à la dégradation de la matière organique du fond des eaux eutrophes stagnantes ou des cours d'eau à faciès lentiques ; ►désenvasement) ; *syn.* colmatage [m] d'un lac, comblement [m] alluvial ; *f 4* **ensablement** [m] (*Terme spécifique* comblement engendré par les dépôts de sable dans la zone d'embouchure des cours d'eau) ; *g* **Verlandung** [f] (Fortschreitende Auffüllung eutropher Gewässer durch Sedimentation eingeschwemmter, anorganischer Substanzen und abgestorbener Organismen; *opp.* ►Schlammräumung; verlanden [vb]; ►Ablagerung 3); *syn.* Verschlammung [f].

5605 silt box [n] [US] *constr.* (Removable mud trap in ►catch basins [US]/road gullies [UK] and ►yard drains [US]/gullies [UK] for separating mud content in storm water runoff); *syn.* sediment bucket [n] [UK]; *s* **cubo** [m] **de fango** (Dispositivo para recoger el fango seco de ►sumideros de calle o ►sumideros de patio); *f* **panier** [m] **(ramasse-boue)** (Dispositif amovible d'un ►regard à grille ou d'un ►siphon de cour destiné à retenir les déchets secs transportés par les eaux de ruissellement) ; *syn.* panier [m] amovible ; *g* **Schlammeimer** [m] (Vorrichtung in ►Straßenabläufen oder ►Hofabläufen zum Abscheiden des Trockenschlammes).

silted-up lake [n] *geo. phyt.* ►raised bog on a silted-up lake.

5606 silt fence [n] [US] *constr.* (Temporary sediment barrier made of timber or synthetic filtration fabric, which is stabilized by steel or hardwood posts, to control erosion by trapping fine soil particles, before they reach collector drains, outfalls or receiving streams; **in U.S.,** temporary plastic barriers are required by law during the construction period to control erosion by trapping fine soil particles on areas undergoing earthworks; ►erosion control facility); *syn.* interceptor [n] [UK]; *s* **interceptor de fango** (En los EE.UU. ►instalación de protección contra la erosión obligatoria para obras de construcción que consiste en láminas de plástico que recogen las partículas finas de suelo para que no se viertan en el curso receptor); *f* **clôture** [f] **(de protection) contre les écoulements de boue (±)** (U.S., ►équipement de protection contre l'érosion réglementaire obligatoire sur les chantiers, constitué par une feuille de plastique afin de retenir les fines particules érodées du sol avant qu'elles ne rejoignent le milieu naturel) ; *g* **Schlammfangzaun** [m] (In den USA für Bauvorhaben gesetzlich vorgeschriebene ►Erosionsschutzeinrichtung in Form einer Plastikfolie, die die erodierten Bodenfeinteile auffängt, damit sie nicht abgeschwemmt werden und nicht in den Vorfluter gelangen können).

5607 silt fencing [n] *constr.* (Often legally required in US, a temporary installation of a sediment barrier made of woven, synthetic filtration fabric or geotextile filter fabrics [►silt fence]; fine soil particles are trapped and allowed to settle in order to prevent the loss of topsoil and the silting up of ►receiving streams, rivers and lakes); *s* **colocación** [f] **de un interceptor de fango** (Medida de protección contra la erosión frecuentemente obligatoria en los EE.UU. en obras de construcción consistente en instalar barrera de tela sintética o tela filtro geotextil para evitar el paso de las partículas finas del suelo y así el entarquinamiento del ►curso receptor; ►interceptor de fango); *f* **mise** [f] **en place d'une clôture (de protection) contre les écoulements de boue (±)** (U.S., mesure temporaire de protection contre l'érosion sur un chantier, constituée en général d'un géotextile, afin de retenir les sols érodés des surfaces nues et d'éviter leur transport dans le milieu naturel) ; *syn.* mise [f] en œuvre de filet de protection contre l'érosion du sol (±), mise [f] en œuvre d'une clôture de retenue des boues (±) ; *g* **Aufstellen [n, o. Pl.] von Schlammfangzäunen** (Temporäre Erosionsschutzmaßnahme in den USA, um den Bodenabtrag von offenen Flächen während Baumaßnahmen in die ►Vorfluter zu stoppen; ►Schlammfangzaun); *syn.* Aufstellung [f] von Schlammfangzäunen, Errichten [n, o. Pl.]/Errichtung [f] von Schlammfangzäunen).

S

silt heap [n] [UK] *min.* ▶silt pile [US].

silting [n] (1) *geo. phyt.* ▶silting up (1).

silting [n] (2) *geo.* ▶silting up (2).

silting up [n] (1) *geo. phyt.* ▶siltation.

5608 silting up [n] (2) *geo.* (Process of deposition of silt on mud flats in watercourses or bodies of water; ▶aggradation, ▶deposition 1, ▶sediment 1, ▶siltation); *syn.* silting [n] (2); *s* **entarquinamiento** [m] (2) (Proceso de acumulación por aguas de crecida de los ríos o de escorrentía de material cenagoarcilloso sobre suelos, rellenando poros, grietas y cavidades, con la consiguiente pérdida de cualidades; DGA 1986; ▶acumulación, ▶agradación, ▶almacenamiento de materia, ▶colmatación, ▶entarquinamiento 1, ▶sedimento); *syn.* sedimentación [f] de limo; *f* **envasement** [m] (2) (Processus et dépôt d'alluvions limoneuses dans les cours d'eau, les lacs, les plaines d'inondation, les baies des estuaires et les cotes littorales résultant de la diminution de la pente ou de la vitesse du courant d'eau ; ▶alluvionnement, ▶atterrissement, ▶comblement sédimentaire, ▶dépôt, ▶ensablement, ▶envasement 1, ▶sédiment) ; *syn. dans la zone de la slikke* extension [f] des vasières littorales ; *processus* accrétion [f] vaseuse ; *résultat* zone [f] d'accrétion ; *g* **Anlandung** [f] (2) (Vorgang und Ergebnis der Schlickanlagerung, z. B. im Wattengebiet oder an Fließgewässerränder; ▶Ablagerung 1, ▶Ablagerung 2, ▶Auflandung 1, ▶Verlandung); *syn. für Wattengebiete* Schlickanlagerung [f], Schlickablagerung [f].

5609 silting up [n] **of soil pores** *pedol.* (Destroying of soil structure by filling of soil voids with fine particles of silt and clay, due to the effects of overwatering or heavy rainfall. *CONTEXT* drainage layer becomes blocked with silt particles; ▶muddying of soil, ▶water-holding capacity); *s* **embarramiento** [m] **del suelo** (Pérdida de la estabilidad estructural del suelo por efecto del agua; ▶riesgo de embarramiento del suelo, ▶capacidad de retención de agua); *syn.* enfangamiento [f] del suelo; *f* **colmatage** [m] **des vides d'un sol** (Modification de la nature d'un sol en y faisant séjourner de l'eau riche en limon et argile, qui s'y dépose ; PED 1979, 264 ; cf. PR 1987 ; ▶mouillabilité, ▶capacité de rétention [maximale] pour l'eau) ; *g* **Verschlämmung** [f] (Zerstörung der Gefügestabilität eines Bodens durch die Wirkung des Wassers, das die Bodenporen mit Ton- und Schluffteilchen verfüllt; ▶Verschlämmbarkeit, ▶Wasseraufnahmefähigkeit).

5610 silt pile [n] [US] *min.* (Spoil mound of sand residue material or dried fine tailings resulting from gold mining operations; ▶tailings slurry); *syn.* silt heap [n] [UK], slimes dam [n] [ZA]; *s* **vaciadero** [m] **de lodos de placeres** (Acumulación de gangas finas resultantes de operaciones de minería aurífera; ▶lodo de minería); *f* **dépôt** [m] **de sable de lavage** (±) (Accumulation de matériaux stériles broyés sous forme de sable fin résultant de l'exploitation de gisements aurifères ; ▶stériles fins) ; *g* **Bergehalde** [f] **aus Spülsand** (Anschüttung von zermahlenem, taubem Gestein in Form von feinem Spülsand bei der Goldförderung; ▶Waschberge); *syn.* Spülsandhalde [f].

silt trap [n] [UK] *constr.* ▶sediment basin [US].

5611 silty [adj] *pedol.* (Descriptive term for a soil with a high percentage of silt); *s* **limoso/a** [adj] (Término descriptivo de un suelo que contiene gran cantidad de limo); *f* **limoneux, euse** [adj] (Terme descriptif pour la texture d'un sol à forte teneur en particules de limon) ; *g* **schluffig** [adj] (Einen Boden mit einem großen Schluffanteil betreffend).

silvicultural management [n] *for.* ▶forest management (2).

silvicultural management system [n] *for.* ▶economic forest management system.

silviculture [n] (1) *for.* ▶forestry (1).

5612 silviculture [n] (2) *for.* (**1.** Generally, the science and art of cultivating, [i.e. growing and tending] forest crops, based on a knowledge of the life history and general characteristics of forest trees and stands, with particular reference to local factors. **2.** More particularly, the theory and practice of controlling the establishment, composition, constitution and growth of forests; SAF 1983; ▶forest management 1, ▶natural silviculture); *s 1* **dasonomía** [f] (*Término genérico* ciencia de los bosques, que trata principalmente de su conservación y aprovechamiento; DB 1985; ▶dasocracia); *s 2* **silvicultura** [f] (Rama de la ▶dasonomía que se ocupa de la regeneración, el establecimiento y desarrollo de los bosques, que fue desarrollada en Europa a partir del Siglo XIX, y puede estar orientada primordialmente al mantenimiento de los ecosistemas forestales [▶silvicultura natural] o dedicarse exclusivamente a la explotación maderera de los bosques [▶silvicultura artificial] sin considerar las demás funciones de éstos; cf. DINA 1987); *syn.* selvicultura [f]; *f* **sylviculture** [f] (2) (**1.** *Sensu lato* la science et l'art de cultiver des peuplements forestiers, c.-à-d., de les créer, de les faire pousser et prospérer en se basant principalement sur la connaissance de l'écologie forestière. **2.** *Sensu stricto* la connaissance théorique et la maîtrise pratique de la création, de la composition, de la structure et de la croissance des peuplements forestiers ; DFM 1975 ; ▶aménagement forestier, ▶sylviculture écologique) ; *g* **Waldbau** [m] (2) (**1.** Wissenschaftliche Erforschung der Grundlagen des forstgerechten Anbaues von Bäumen im Wald zur Produktion von Rohstoffen, zur Erbringung von Schutz- und Sozialleistungen und sonstigen Wirkungen sowie zur Einkommenssicherung. **2.** Planmäßige Begründung, Pflege und Verjüngung von Beständen mit dem Ziel einer optimalen Erreichung der Wirtschafts- und Betriebsziele durch Schaffung von Beständen, die nach Aufbau und Leistung den Betriebszieltypen entsprechen und die Nachhaltigkeit der Bestände und deren Funktionen sichern; cf. WT 1980; ▶Forsteinrichtung, ▶naturnaher Waldbau).

silvopastoralism [n] *agr. conserv.* ▶traditional orchard [UK], #2.

Simplified Planning Zone [n] [UK] *leg. urb.* ∗▶comprehensive plan [US]/Local Development Framework (LDF) [UK].

5613 simply-provided recreation facility [n] *recr.* (Meagerly-provisioned recreation area or facility); *syn.* poorly-provided recreation facility [n]; *s* **instalación** [f] **de recreo extensivo** (Centro de recreo con facilidades sencillas, generalmente en zonas rurales); *f* **aménagement** [m] **touristique léger** (Aménagement favorisant les formes d'activités de loisirs simples ne demandant qu'un équipement à caractère rustique) ; *syn.* équipement [m] touristique léger, installation [f] touristique légère, équipement [m] léger de loisir, équipement [m] récréatif léger [CDN] ; *g* **Erholungseinrichtung** [f] **mit niedrigem Ausstattungsgrad** (Anlage mit geringen, wenig aufwendigen Einrichtungen, besonders im ländlichen Raum).

5614 single crop system [n] *agr.* (Cultivation of only one plant species on the same piece of land; ▶pure stand, ▶monoculture); *syn.* one-crop farming [n]; *s* **cultivo** [m] **monoespecífico** (Cultivo de una sola especie de plantas en una parcela de tierra; ▶monocultura, ▶rodal monoespecífico 1); *syn.* monocultivo [m]; *syn. phyt. for.* rodal [m] monoespecífico (2); *syn.* monocultivo [m]; *syn. phyt. for.* ▶rodal monoespecífico; *f* **culture** [f] **monospécifique** (Culture d'une seule espèce végétale sur une parcelle ; ▶monoculture, ▶boisement monospécifique) ; *g* **Reinkultur** [f] (Anbau nur einer Pflanzenart auf derselben Fläche; ▶Monokultur, ▶Reinbestand).

5615 single digging [n] *hort.* (Digging to the full depth of one spade or fork); *syn.* one spit digging [n] [UK] (BS 3975: Pt.

5); *s* **hacer [vb] cava de un azadón** (Cavar a una profundidad de un azadón); *f* **bêcher [vb] à un fer** (Remuer la terre sur la profondeur de la bêche) ; *g* **spatentief umgraben [vb]**.

single-family detached dwelling [n] [US] *arch. urb.* ▶single-family detached house.

5616 single-family detached house [n] *arch. urb.* (Free-standing house/dwelling; a **s.-f. d. h.** in a remote area is called an 'isolated house'; ▶detached, ▶single-family house); *syn.* single-family detached dwelling [n] [also US]; *s* **casa [f] unifamiliar independiente** (▶casa unifamiliar, ▶separado/a); *f* **maison [f] individuelle isolée** (*Type d'habitat* logement non collectif caractérisé par un bâtiment qui occupe seul une parcelle terrain et abrite une seule famille ; *termes afférents* petit habitat individuel en ville, maison individuelle pavillonnaire ; plus que la maison individuelle isolée, c'est la **maison individuelle pavillonnaire**, qui ces dernières décennies, est responsable de l'étalement des villes et des villages sur la campagne et a, comme jamais dans l'histoire urbaine, démultiplié l'urbanisation dans les espaces périurbains. Avec l'automobile comme unique mode de déplacement, ce modèle d'habitat semble compromis à long terme face à la crise énergétique et environnementale ; ▶maison individuelle, ▶isolé, ée) ; *g* **frei stehendes Einfamilienhaus [n]** (Ein **f. s. E.** in einsamer Lage wird ‚entlegenes Haus' genannt; (▶Einfamilienhaus, ▶frei stehend); *syn.* frei stehendes Eigenheim [n].

single-family dwelling [n] [US] *arch. urb.* ▶single-family house.

5617 single-family house [n] *arch. urb.* (Generic term for ▶single-family detached house, ▶courtyard house, ▶attached dwelling in a cluster [US]/attached single-family house in a cluster [UK]; ▶semidetached house); *syn.* single-family dwelling [n] [also US], detached dwelling [n] [also US]; *s* **casa [f] unifamiliar** (Término genérico para ▶casa unifamiliar independiente, ▶casa de patio l, ▶casa unifamiliar agrupada, ▶casa pareada); *f* **maison [f] individuelle** (Terme générique caractérisant une construction immobilière à usage d'habitation ou mixte (habitation et usage professionnel) comportant au plus deux logements, destinés au même maître de l'ouvrage ; ▶maison individuelle isolée, ▶maison atrium, ▶maison individuelle regroupée, ▶maisons jumelles) ; *syn.* logement [m] individuel, maison [f] unifamiliale [B, CDN, CH, L]; *g* **Einfamilienhaus [n]** (OB zu ▶frei stehendes Einfamilienhaus, ▶Gartenhofhaus, ▶Einfamilienhaus einer Gebäudegruppe; ▶Doppelhaus).

5618 single-grain structure [n] *pedol.* (Loose-bonded aggregation of soil particles of similar size and shape, typified by coarse sandy soils low in organic matter; ▶crumb structure, ▶granular soil structure); *s* **estructura [f] de ceniza** (Estado de no agregación de las partículas del suelo, aunque se encuentren compactadas como en el caso de los suelos arenosos; ▶estado grumoso, ▶estructura grumosa); *f* **structure [f] particulaire du sol** (Sans agglomération des agrégats : le sol est très meuble. Les éléments du squelette ne sont pas liés. C'est surtout le cas des sols sableux ; PED 1979, 235 ; ▶structure fragmentaire, ▶état structural grumeleux) ; *g* **Einzelkorngefüge [n]** (Bodenstruktur, bei der die Einzelteilchen meist in kompakter Lagerung vorliegen; ▶Krümelstruktur 1 und ▶Krümelstruktur 2).

single house [n] *arch. leg. urb.* ▶individual house.

5619 single mowing [n] [US] *agr. constr. hort.* (Mowing of a meadow once a year; ▶late cut; ▶be mown once a year [US]/be cut once a year [UK]); *syn.* annual cut [n] (of a meadow) [UK]; *s* **siega [f] anual** (Contexto prado de siega anual; ▶segado/a anualmente, ▶corte tardío); *f* **fauche [f] exécutée une fois par an** (▶fauché, ée une fois par an, ▶fauche tardive) ;

g **einmalige Mahd [f]** (Einmaliges Mähen von Wiesen; einmal mähen [vb]; ▶einschürig, ▶Reinigungsschnitt).

5620 single row grave [n] *adm.* (▶Burial site occupied by one grave in a row for a specific ▶use period of grave [US]/rest period of grave [UK] in a ▶burial area); *s* **tumba [f] individual en hilera** (▶Tumba situada en un ▶área de tumbas que se van ocupando en orden una tras de otra; ▶periodo/período de descanso de tumbas); *f* **tombe [f] à la ligne** ([Terme courant en CH] Tombe individuelle placée successivement sur une ligne, le temps de la ▶durée d'inhumation, souvent non renouvelable, dans un secteur sans concessions ; ▶sépulture) ; *syn.* tombe [f] individuelle, tombe [f] singulière ; *g* **Reihengrab [n]** (▶Grabstätte für die Aufnahme jeweils eines Verstorbenen, die auf einem ▶Grabfeld anlässlich eines Todesfalles „der Reihe nach" [zeitlich und örtlich] für die Dauer der ▶Ruhefrist zur Verfügung gestellt wird); *syn.* Einzelgrab(stätte) [n/(f)], Einzelgrabstelle [f], Reihengrabstätte [f], Reihengrabstelle [f].

single-storey [adj] [UK] *arch. urb.* ▶single-story [US].

5621 single-story [adj] [US] *arch. urb.* (▶multistory [US]/multistorey [UK]); *syn.* singlestorey [adj] [UK]; *s* **de un piso [loc]** (▶de varias plantas); *f* **d'un seul niveau [loc]** (▶à plusieurs étages) ; *g* **eingeschossig [adj]** (▶mehrgeschossig).

5622 single-stratum planting system [n] *constr.* (▶Roof planting method having only one homogeneous layer of soilmix or a substrate mat; ▶multistrata planting system); *syn.* one-layer planting system [n]; *s* **estructura [f] de plantación monocapa** (▶Técnica de revestimiento vegetal de tejados o azoteas en la que la capa portante de la vegetación sólo consiste en un sustrato de plantación homogéneo o una estera de sustrato; *opp.* ▶estructura de plantación multicapa); *f* **structure [f] monocouche** (▶Procédé de végétalisation des toitures pour lequel le support végétal est constitué d'une seule couche de substrat homogène ou d'une natte préplantée ; *opp.* ▶structure pluricouche) ; *g* **Einschichtaufbau [m]** (...bauten [pl]; ▶Begrünungssystem für Dachflächen, bei dem die Vegetationstragschicht aus nur einer homogenen Pflanzsubstratschicht oder Substratmatte besteht; *opp.* ▶Mehrschichtenaufbau).

single-wythe masonry [n] [UK] *constr.* ▶one-tier masonry [US].

singly [adv] *phyt.* ▶growing singly.

5623 sinker root [n] *bot. hort.* (Any secondary root that descends vertically from an horizontal main root; e.g. Norway Spruce *[Picea abies]*, White Ash *[Fraxinus americana]*, Poplar *[Populus spec.]*, European Mountain Ash [US]/Mountain Ash [UK] *[Sorbus aucuparia (Europe)/Pyrus americana (US)]*); *syn.* dropper [n] [also UK]; *s* **raíz [f] vertical** (Raíz vertical lateral de una raíz principal como p. ej. en el abeto rojo *[Picea abies]*, el fresno *[Fraxinus excelsior]*, el chopo *[Populus spec.]*, el serbal de cazadores *[Sorbus aucuparia]*); *f* **racine [f] verticale** (Racine secondaire à croissance verticale se développant à partir d'une racine principale horizontale, comme p. ex. chez le Pin sylvestre *[Pinus sylvestris]*, l'Épicéa *[Picea abies]*, le Frêne élevé *[Fraxinus excelsior]*, le Sorbier des oiseleurs *[Sorbus aucuparia]*) ; *g* **Senkerwurzel [f]** (Senkrechte Seitenwurzel einer horizontalen Hauptwurzel wie z. B. bei Rotfichte *[Picea abies]*, Esche *[Fraxinus excelsior]*, Pappel *[Populus spec.]*, Vogelbeere *[Sorbus aucuparia]*).

5624 site [n] (1) *plan.* (Place for a specific land use, e.g. industrial plant, refuse dump [US]/refuse tip [UK]; ▶building site); *syn.* location [n] (2); *s* **ubicación [f]** (1. Lugar en el que está localizado o se planea un uso, como fábrica, incineradora de residuos, ▶escombrera, etc. 2. Espacio geográfico [ciudad, región, país, continente] en donde tiene lugar determinada actividad); *syn.* emplazamiento [m]; *f* **site [m]** (1. Lieu affecté à un usage,

S

une utilisation du sol réel ou potentiel, p. ex. pour les diverses infrastructures industrielles, commerciales, urbaines, de transport, de télécommunications ; ▶site industriel. **2.** Espace géographique [ville, région, pays ou continent] assurant ou suscitant des activités économiques) ; *g* **Standort [m] (2) (1.** Ort einer Nutzung, z. B. Industrieanlage, Mülldeponie. **2.** Geografischer Raum [Stadt, Region, Land oder Kontinent], wo bestimmte wirtschaftliche Tätigkeiten stattfinden oder von wo solche Aktivitäten ausgehen; ▶Industriestandort).

site [n] **(2)** *leg. urb.* ▶approved development site; *zool.* ▶breeding site; *adm.* ▶burial site; *recr.* ▶camper with trailer site [US]/camp and caravan site [UK], ▶campground[US]/campsite [UK] 2; *adm.* ▶coffin burial site; *constr.* ▶construction site; *conserv'hist.* ▶cultural site; *constr.* ▶cut site, ▶damaged site; *geo.* ▶deposition site; *constr.* ▶excavation site; *min.* ▶extraction site; *adm.* ▶family burial site; *min.* ▶gravel extraction site; *envir. urb.* ▶hazardous old dump site [US]/contaminated waste site, ▶hazardous waste disposal site; *phyt.* ▶heavy metal contaminated site; *zool.* ▶hibernation site; *conserv'hist.* ▶historic site; *adm. hist.* ▶memorial site; *phyt. zool.* ▶mesic site; *min. plan.* ▶mineral working site; *envir.* ▶orphan contaminated site; *min.* ▶overburden dumping site; *landsc. recr.* ▶permanent site; *recr.* ▶picnic site; *adm. landsc.* ▶prominent urn burial site; *min. plan.* ▶proposed extraction site; *envir.* ▶rubble disposal site; *phyt.* ▶saline site; *min.* ▶sand extraction site; *constr. hort.* ▶sod removal site [US]/turf removal site [UK]; *landsc. plan.* ▶spoil site [US]/tip site [UK]; *min. wat'man.* ▶tap site; *constr. urb.* ▶tree planting site; *adm. landsc.* ▶urn burial site; *conserv. leg. pol.* ▶world heritage site; *phyt. zool.* ▶xeric site; *ecol. phyt.* ▶zinc-contaminated site.

site [n], **abandoned industrial** *plan. urb.* ▶derelict land.

site [n], **brownfield** *envir. urb.* ▶brownfield.

site [n] [UK], **camp and caravan** *recr.* ▶camper with trailer site [US].

site [n], **camping** *recr.* ▶camper with trailer site [US]/camp and caravan site [n] [UK].

site [n], **contaminated** *envir.* ▶orphan contaminated site.

site [n], **contaminated waste** *envir. urb.* ▶hazardous old dumpsite [US].

site [n] [US], **controlled landfill** *envir.* ▶sanitary landfill [US] (2)/landfill site [UK].

site [n], **extreme** *ecol.* ▶extreme habitat.

site [n] [UK], **fly tipping** *envir.* ▶unauthorized dumpsite [US].

site [n] [UK], **gap** *urb* ▶vacant lot [US].

site [n], **grave** *adm.* ▶burial site.

site [n] [UK], **green-field** *leg. urb.* ▶open land [US].

site [n] [US], **holding** *constr. hort.* ▶site nursery.

site [n] [US], **horticultural exhibition** *hort. urb.* ▶horticultural exhibition area.

site [n] [US], **land disposal** *envir.* ▶dumpsite [US]/tipping site [UK].

site [n] [UK], **landfill** *envir.* ▶sanitary landfill [US] (2).

site [n], **marginal agricultural** *agr. (for.)* ▶marginal soil.

site [n], **multifamily housing** *arch. urb.* ▶residential complex.

site [n], **nesting** *zool.* ▶breeding site.

site [n] [UK], **permanent allotment** *leg. urb.* ▶permanent community garden area [US].

site [n] [UK], **permanent caravan** *leg. urb.* ▶permanent campground [US].

site [n] [US], **public and civic** *leg. urb.* ▶site for public facilities.

site [n] [US], **public use** *leg. urb.* ▶site for public facilities.

site [n], **refuse-landfill** *envir.* ▶sanitary landfill [US] (2)/landfill site [UK].

site [n], **rubble dump-** *envir.* ▶rubble disposal site.

site [n] [US], **settlement** *urb.* ▶building plot.

site [n], **shady** *hort. plant.* ▶shady location.

site [n] [US], **static** *landsc. recr.* ▶permanent site.

site [n] [UK], **tip** *landsc. plan.* ▶spoil site [US].

site [n] [UK], **tipping** *envir.* ▶dumpsite [US].

site [n] [UK], **unmanaged tipping** *envir.* ▶unmanaged dumpsite [US].

5625 site advantage [n] *plan.* (Benefits arising from the location of a particular development in a city, region or country); *s* **ventajas [fpl] de ubicación/emplazamiento** (Cualidades de una zona particular para el desarrollo de actividades económicas tales como accesibilidad, presencia de equipamientos técnicos, sociales y culturales, etc. En urbanismo estas cualidades se pueden traducir en la definición de espacios o ejes de desarrollo); *f* **avantage [m] d'un site** (Qualités que présente un site pour le développement d'une activité précise telles que l'accessibilité, la présence d'équipements etc. ; en urbanisme ces qualités se traduisent par la définition de sites géographiques préférentiels, d'axes de développement, d'axes préférentiels d'urbanisation) ; *syn.* atout [m] du/d'un site, rente [f] de situation ; *g* **Standortvorteil [m]** (Der sich für eine geplante Nutzung aus der geografischen Lage, dem wirtschaftlichen, sozialen und infrastrukturellen Umfeld in einer Stadt, Region oder in einem Land ergebende Vorteil); *syn.* Standortgunst [f].

5626 site analysis [n] *plan.* (Study of collected site data for a construction project or to determine design parameters; ▶initial site analysis, ▶landscape analysis); *syn.* site study [n]; *s* **estudio [m] del terreno** (Relevamiento de las condiciones de la ubicación de una obra de construcción; ▶análisis del estado inicial, ▶estudio del paisaje); *syn.* estudio [m] de la ubicación del proyecto; *f* **analyse [f] du site** (Prise en compte des données relatives à un site affecté par un ouvrage ; ▶analyse de l'état initial, ▶étude sitologique) ; *syn.* reconnaissance [f] du terrain, inventaire [m] des données du site ; *g* **Standortanalyse [f]** (Untersuchung der Standortgegebenheiten und des ▶Standortvorteils für ein Bauvorhaben und zur Festlegung der Planungsparameter und planungsbezogenen ersten Schritte; ▶Bestandsanalyse, ▶Landschaftsanalyse).

site assessment [n] *plan.* ▶site evaluation.

site-cast concrete [n] [US] *constr.* ▶poured-in-place concrete [US]/in-situ (cast) concrete [UK].

5627 site clearance [n] *constr.* (Removal of structures and/or vegetation from a project site; ▶preparatory operations, ▶topsoil conservation, ▶site clearance after project completion, ▶topsoil stripping, ▶woody plant clearance); *s* **preparación [f] del terreno** (Trabajos a realizar antes de la construcción: Recogida de material vegetal reutilizable, arranque y transporte de material, suelo o restos inutilizables etc.; ▶despeje del lugar de la obra, ▶despeje y ▶desbroce 2, ▶protección legal de la tierra vegetal, ▶remoción de la capa de tierra vegetal, ▶trabajos preliminares); *f* **nettoyage [m] du terrain** (Travaux préliminaires de la mise en chantier dont les plus courants consistent dans la démolition des constructions existantes et l'évacuation des matériaux non récupérables, l'arrachage, le dessouchage ou abattage des arbres, arbustes et broussailles, le déplacement des végétaux, la protection des végétaux à conserver sur place ; cf. C.C.T.P.-

type, fascicule n° 35 ; ►dégagement du terrain, ►protection de la terre végétale, ►repliement du chantier, ►retroussement de terre végétale, ►travaux préliminaires) ; *g* **Baufeldfreimachung [f]** (*Arbeiten vor Baubeginn gemäß DIN 18 915 und DIN 18 917. Gewinnen von wieder verwendbarem Aufwuchs, Abtransport nicht verwendbaren Aufwuchses und ungeeigneter Bodenarten, störender Stoffe etc.;* ►Entfernen von Gehölzen und Gehölzresten, ►Mutterbodenschutz, ►Oberbodenabtrag 1, ►Räumung der Baustelle nach Fertigstellung, ►Vorarbeiten 2); *syn.* Freimachen [n, o. Pl.] des Baufeldes, Abräumen [n, o. Pl.] des Baufeldes (DIN 18 915, Bl. 3).

5628 site clearance [n] **after project completion** *constr.* *s* **despeje [m] del lugar de la obra** (Desmontaje de instalaciones provisionales y limpieza del terreno alrededor de una obra una vez terminada ésta); *f* **repliement [m] du chantier** (Nettoyage du chantier, évacuation des déchets, contrôle des matériels démontage et repli des installations provisoires à la fin d'un chantier ; *syn.* repliement [m] des installations de chantier ; *g* **Räumung [f] der Baustelle nach Fertigstellung.**

5629 site climate [n] *for. met. phyt.* (Climate, determined by local environmental factors such as soil, vegetation, topography and adjacent buildings, which characterizes a particular location; ►microclimate, ►climate near the ground, ►topoclimate); *syn.* ecoclimate [n]; *s* **microclima [m] de estación** (►Microclima que caracteriza una estación y que es influenciado por el tipo de suelo, la vegetación, el relieve, la exposición, las construcciones existentes, etc.; el **m. de e.** es un término que se solapa con el de ►clima de las capas cercanas a la superficie terrestre; ►mesoclima); *syn.* microclima [m] de residencia ecológica; *f* **climat [m] stationnel** (►Microclimat d'une station ou d'un biotope dont les conditions climatiques sont caractérisées par les facteurs sol, couvert végétal, forme du terrain, exposition, constructions ; *contexte* climat de stations non forestières ; cf. GGV 1979, 51-63 ; ►climat de la couche d'air à proximité du sol, ►climat local) ; *g* **Standortklima [n]** (…ta [pl]; *fachsprachlich* …mate [pl]; ►Mikroklima, das durch Boden, Pflanzenbewuchs, Geländeformen, Exposition, auch durch Bauwerke einen Standort oder Biotop prägt; **S.** und ►Klima der bodennahen Luftschicht sind sich überschneidende Begriffe; ►Geländeklima); *syn.* Bestandsklima [n], Ökoklima [n], Biotopklima [n].

5630 site conditions [npl] **(1)** *plan.* (Necessary requirements for location and implementation of large-scale projects); *s* **condiciones [fpl] de ubicación** (Condiciones previas necesarias para realizar un proyecto [grande]); *syn.* condiciones [fpl] de emplazamiento; *f* **conditions [fpl] d'implantation** (Facteurs nécessaires à la localisation et la réalisation d'un projet) ; *g* **Standortbedingungen [fpl] (2)** (Für die Ansiedlung resp. Realisierung von [Groß]projekten notwendige Voraussetzungen).

site conditions [npl] **(2)** *constr.* ►precontract investigation of site conditions; *landsc. phyt. zool.* ►species suited to site conditions.

site contouring [n] **[UK]** *constr.* ►ground modeling [US]/ ground modelling [UK].

site design [n] *plan. prof.* ►site planning.

5631 site designer [n] *prof.* (Planner who primarily designs projects associated with a particular site as opposed to a town or regional planner or landscape planner or a planner engaged in nature conservation); *syn.* project designer [n]; *s* **diseñador, -a [m/f] de parques y jardines (≠)** (►Arquitecto, -a paisajista especializado/a en el diseño y la realización de parques y jardines); *f* **concepteur, trice [m/f] (paysagiste) d'opération** (Ingénieur qui, par comparaison avec les urbanistes ou les paysagistes spécialistes de la protection de la nature et des paysages, travaillent exclusivement sur des projets isolés de constructions et

d'équipements) ; *g* **Objektplaner/-in [m/f]** (Jemand, der im Vergleich zum Stadt-, Landschafts- oder Naturschutzplaner ausschließlich Einzelbauvorhaben [der Freiraumplanung] bearbeitet).

site engineer [n] *constr. prof.* ►design firm's representative [US]/supervisor representing a design practise [UK].

5632 site evaluation [n] *plan.* (Appraisal and classification of the characteristics of a site by ranking its potential for a particular use); *syn.* site assessment [n]; *s* **evaluación [f] de una ubicación/un emplazamiento;** *f* **analyse [f] du site en terme d'aptitude à un usage** (Évaluation et classification des qualités d'un site pour un usage déterminé) ; *g* **Standortbewertung [f]** (Einschätzung und Einstufung/Skalierung der Gegebenheiten eines Standortes nach der Rangfolge ihrer Wertigkeit für einen bestimmten Nutzungsanspruch).

5633 site facilities installation [n] *constr.* (Erection of temporary construction site facilities; ►dismantlement of building site installations, ►building site facilities and equipment); *s* **montaje [m] de las instalaciones auxiliares de obra** (►instalaciones auxiliares de una obra; *opp.* ►desmontaje de las instalaciones auxiliares de una obra); *syn.* instalación [f] de equipos de obra; *f* **mise [f] en place (des installations de chantier) ;** (►installations de chantier) ; *opp.* ►repli d'installations de chantier ; *g* **Aufbau [m, o. Pl.] von Baustelleneinrichtungen** (Errichtung der für den Betrieb einer Baustelle notwendigen ►Baustelleneinrichtung 1; *opp.* ►Abbau von Baustelleneinrichtungen).

site factor [n] *ecol.* ►abiotic site factor.

5634 site factors [npl] *ecol.* (Environmental conditions prevailing at a particular site, relevant to a plant or plant community; ►ecological factors, ►habitat conditions); *s* **factores [mpl] mesológicos** (Condiciones ambientales que se presentan en una ubicación específica y son relevantes para el crecimiento de las plantas resp. de las comunidades vegetales; ►condiciones mesológicas, ►factores ecológicos); *f* **facteurs [mpl] de la station** (Conditions de l'environnement ou d'une communauté végétale dans un lieu déterminé ; ►conditions de la station, ►facteurs écologiques) ; *syn.* facteurs [mpl] stationnels ; *g* **Standortfaktoren [mpl]** (Messbare Größen einzelner Umweltbedingungen, die an einem bestimmten Standort auf Pflanzen und Tiere wirken; ►Ökofaktoren, ►Standortbedingungen 1).

5635 site fidelity [n] *phyt. zool.* (Preference of a species for belonging continually to a particular community; ►degree of fidelity); *s* **fidelidad [f] (2)** (Constancia de la presencia de una especie en una comunidad. Va desde la exclusividad, cuando la especie sólo aparece en cierta comunidad, hasta la accidentalidad; DINA 1987; ►grado de fidelidad); *f* **fidélité [f] (au site/lieu)** (Rend compte de la manière dont une espèce est localisée exclusivement ou préférentiellement dans les diverses communautés biologiques ; DEE 1982, 42 ; ►degré de fidélité) ; *g* **Standorttreue [f]** (Ausmaß der Beschränkung oder Bevorzugung einer Art, einer bestimmten Gemeinschaft stets anzugehören; ►Treuegrad); *syn.* Fidelität [f].

5636 site [n] **for public facilities** *leg. urb.* (►public facility 1); *syn.* public and civic site [n] [also UK], public use site [n] [also US]; *s* **emplazamiento [m] reservado para equipamiento comunitario** (En Es. la LS Arts. 12 d y 13 d prevé la reserva de áreas para usos de interés público y social en los ►planes generales municipales de ordenación urbana y en los ►planes parciales [de ordenación]; ►instalación de equipamiento comunitario, ►instalación técnica de abastecimiento); *f* **emplacement [m] réservé (aux installations d'intérêt général)** (Prescription édictée par un PLU/POS espaces délimités sur le plan de zonage qui sont réservés aux ►installations d'intérêt général

S

[p. ex. logements sociaux], aux ▶équipements collectifs [p. ex. espaces verts], aux ▶équipements urbains d'intérêt public [voies et ouvrages publics, etc.]) ; *syn.* zone [f] affectée aux équipements collectifs ; *g* **Gemeinbedarfsfläche [f]** (Fläche für ▶Gemeinbedarfseinrichtungen; cf. § 9 [1] 5 BauGB); *syn.* Fläche [f] für den Gemeinbedarf.

site [n] for radioactive waste *envir.* ▶final disposal site for radioactive waste.

site furniture [n] *arch. constr. urb.* ▶outdoor furniture.

5637 site indicator (species) [n] *phyt.* (Specific term for ▶indicator species; ▶location 1); *s* **especie [f] indicadora de estación** (Especie animal o vegetal cuya presencia en una ▶estación muestra —dentro de cierto márgen— la existencia de algún factor o complejo de factores abióticos específicos; *término genérico* ▶especie indicadora); *syn.* especie [f] indicadora de residencia ecológica; *f* **espèce [f] indicatrice de la station** (Espèce animale ou végétale dont la présence ou l'absence, la fréquence ou la quantité sur une ▶station sont caractéristiques d'un ou plusieurs facteurs abiotiques ; *terme générique* ▶espèce indicatrice) ; *g* **Standortzeiger [m]** (Pflanzen- oder Tierart, dessen An- oder Abwesenheit an einem ▶Standort 1 innerhalb gewisser Grenzen einen bestimmten abiotischen Faktor oder einen Faktorenkomplex anzeigt; *OB* ▶Zeigerart).

site inspection [n] [UK] (1) *adm. leg.* ▶building code control [US].

5638 site inspection [n] (2) *landsc.* (Walking or driving on a piece of land or in an area to obtain a general impression); *syn.* site reconnaissance [n] [also UK]; *s* **reconocimiento [m] del terreno** (Visita a pie o en vehículo de una zona específica para obtener una impresión general de ella); *f* **reconnaissance [f] de terrain** (Inspection pédestre ou en véhicule d'un espace naturel afin d'en obtenir sa perception globale) ; *g* **Geländebegehung [f]** (Ablaufen/Abfahren eines Gebietes/einer Fläche, um einen Gesamteindruck zu erhalten).

5639 site inspection [n] (3) *prof.* (Visual examination of existing conditions on a construction site or area; ▶construction site inspection, ▶final site inspection); *s* **inspección [f] del lugar del proyecto (±)** (Examen de las condiciones existentes en la zona de construcción o de otra superficie a planificar; ▶inspección de obra in situ, ▶inspección final de obra); *syn.* inspección [f] de la zona de planificación (±); *f* **reconnaissance [f] des lieux** (*Terme générique* recueil des données sur le terrain d'une zone quelconque, d'un territoire, d'un chantier afin d'en prendre possession ; ▶inspection du site, ▶reconnaissance du chantier) ; *g* **Ortsbesichtigung [f]** (Besichtigung der örtlichen Gegebenheiten einer Baustelle, einer sonstigen Fläche oder eines Gebietes; ▶Baustellenbegehung, ▶Objektbegehung).

site inspector [n] [UK], independent *constr. prof.* ▶client's project representative [US].

site inventory [n] [US] *adm. envir.* ▶toxic site inventory.

site investigations [npl] *landsc. plan.* ▶comprehensive site survey and analysis.

5640 site log [n] *constr.* (**1.** Continual record of site work and labo[u]r etc. compiled by the contractor's site agent. **2. Site report [n] [UK]:** summary containing information abstracted from the site log and submitted regularly to the supervising consultant); *s* **libro [m] de obra** (PGT 1987, 28; registro continuo de los trabajos realizados en una obra que es llevado por el jefe de obra); *f* **journal [m] de chantier** (CCM 1984, 150 ; rapport journalier des opérations sur le chantier — planning, matériaux, construction) ; *syn.* carnet [m] de chantier ; *g* **Bautagebuch [n]** (Heft/Buch mit lückenlosen, täglichen Eintra-

gungen über das Baugeschehen durch den örtlichen Bauleiter); *syn.* Baubuch [n] [A].

site management [n], contractor's *contr. constr.* ▶supervision by the person-in-charge.

site management services [npl] [UK] *prof.* ▶construction phase—administration of the construction contract.

site manager [n], supervision by the *contr. constr.* ▶supervision by the person-in-charge.

5641 site mapping [n] *pedol.* (Delineation of site conditions by ▶great soil groups [US]/soil types [UK] with respect to their suitability for plant cultivation or building projects; ▶soil mapping; *generic term* ▶mapping); *s* **inventario [m] de suelos (de una ubicación)** (Registro en un mapa de los grandes grupos de suelos de una zona para determinar su idoneidad para uso agrícola o para construcción; ▶cartografía de los suelos, ▶[gran] grupo de suelos [US], ▶tipo de suelos; *término genérico* ▶inventario cartográfico); *f* **cartographie [f] de(s) station(s)** (Représentation sur une carte des ▶groupes de sols après détermination de leur aptitude à la production végétale, l'assainissement pluvial, la construction, etc. ; ▶cartographie des sols; *terme générique* ▶cartographie) ; *g* **Standortkartierung [f]** (Erfassung und Darstellung von ▶Bodentypen 1 auf Karten als Ergebnis bodenkundlicher Untersuchungen hinsichtlich der Eignung für bestimmte Kulturpflanzen oder Bauprojekte; ▶Bodenkartierung; *OB* ▶Kartierung).

5642 site nursery [n] *constr. hort.* (Temporary storage area on construction site where plant material is heeled-in between delivery and planting out in its final position; ▶heeling-in on site); *syn.* holding site [n] [also US], temporary site nursery [n]; *s* **depósito [m] de material de plantación** (Almacén provisional del material de plantación en obra hasta que es plantado en su lugar definitivo; ▶enterrado de plantas in situ); *f* **jauge [f]** (Aire de stockage provisoire des végétaux à racines nues qui sont recouverts de terre aussi longtemps que leur plantation définitive n'a été effectuée ; ▶jauge d'attente) ; *g* **Einschlagplatz [m]** (Schattiges und windgeschütztes Zwischenlager von Pflanzgut, dessen Wurzeln solange mit Boden abgedeckt sind bis sie endgültig gepflanzt werden können; cf. DIN 18 916; ▶Baustelleneinschlag).

5643 site [n] of community importance *conserv. ecol.* (*Abbr.* SCI; according to Directive 92/43/EEC, dated 21 May 1992, concerning the conservation of natural habitats, fauna and flora, a site which, in the biogeographical region or regions to which it belongs, contributes significantly to the preservation or restoration of a favourable ▶conservation status 2 of a natural habitat type (Annex I) or of a species (Annex II) and may also contribute significantly to the network 'Natura 2000' and/or contributes significantly to the maintenance of biological diversity within the biogeographic region or regions concerned; for animal species ranging over wide areas, sites of community importance correspond to the places within the natural range of such species, where the physical or biological factors essential to their life and reproduction exist); *s* **lugar [m] de importancia comunitaria** (Según el art. 1k de la Directiva 92/43/CEE, de 21 de mayo, de Hábitats, un lugar que, en la región o regiones biogeográficas a las que pertenece, contribuya de forma apreciable a mantener o restablecer un tipo de hábitat natural de los que se citan en el Anexo I o una especie de las que se enumeran en el Anexo II en un ▶estado de conservación favorable y que pueda de esta forma contribuir de modo apreciable a la ▶coherencia ecológica de Natura 2000 tal como se contempla en el artículo 3, y/o contribuya de forma apreciable al mantenimiento de la diversidad biológica en la región o regiones biogeográficas de que se trate. Para las especies animales que ocupan territorios extensos, los lugares de importancia comunitaria corresponderán a las ubicaciones concretas dentro de la zona de reparto natural de dichas especies que presenten los elementos físicos o biológicos esenciales para su vida y su reproducción); *f* **site [m] d'importance communautaire [SIC]** (Site sélectionné, sur la base des propositions des Etats membres, par la Commission Européenne pour intégrer le réseau Natura 2000 en application de la directive « Habitats, faune, flore » ; site qui, dans la ou les régions biogéographiques auxquelles il appartient, contribue de manière significative à maintenir ou à rétablir un type d'habitat naturel de l'Annexe 1 ou une espèce de l'Annexe 2 dans un état de conservation favorable et peut aussi contribuer de manière significative à la ►cohérence du réseau « Natura 2000 » et/ou contribue de manière significative au maintien de la diversité biologique dans la ou les régions biogéographiques concernées ; pour les espèces animales qui occupent de vastes territoires, les SIC correspondent aux lieux, au sein de l'aire de répartition naturelle de ces espèces, qui présentent les éléments physiques ou biologiques essentiels à leur vie et reproduction ; la liste nominative de ces sites est arrêtée par la Commission Européenne pour chaque région biogéographique ; ces sites sont ensuite désignés en ►Zones Spéciales de Conservation [ZSC] par arrêtés ministériels) ; *g* **Gebiet [n] von gemeinschaftlicher Bedeutung** (*Richtlinie 92/43/EWG des Rates vom 21. Mai 1992* ein von der EU-Kommission im Rahmen eines Bewertungsverfahrens festgelegtes Gebiet, das in der oder den biogeographischen Region[en], zu welchen es gehört, in signifikantem Maße dazu beiträgt, einen natürlichen Lebensraumtyp des Anhangs 1 oder eine Art des Anhangs II in einem günstigen ►Erhaltungszustand zu bewahren oder einen solchen wiederherzustellen und auch in signifikantem Maße zur ►Kohärenz des in Artikel 3 genannten Netzes „Natura 2000" und/oder in signifikantem Maße zur biologischen Vielfalt in der biogeographischen Region beitragen kann. Bei Tierarten, die große Lebensräume beanspruchen, entsprechen die Gebiete von gemeinschaftlichem Interesse den Orten im natürlichen Verbreitungsgebiet dieser Arten, welche die für ihr Leben und ihre Fortpflanzung ausschlaggebenden physischen und biologischen Elemente aufweisen; cf. Art. 1 der Richtlinie 92/43/EWG des Rates vom 21. Mai 1992 zur Erhaltung der natürlichen Lebensräume sowie der wild lebenden Tiere und Pflanzen).

Site [n] of Special Scientific Interest (S.S.S.I.) [UK] *conserv. leg.* ►Area of Outstanding Natural Beauty [UK]; ►natural landmark [US].

5644 site plan [n] *plan.* (Scaled development plan of a property parcel or particular area, showing existing conditions or proposed development; ►illustrative site plan, ►overall plan 1, ►overall site plan, ►survey plan 2); *syn.* layout plan [n]; *s* **plano [m] de situación** (Representación cartográfica a muy gran escala de un área determinada en la que se muestran las condiciones existentes o la planificación prevista para ella; ►plano de agrimensura, ►plano de ordenación general, ►plano general, ►plano general [de ubicación]); *f* **plan [m] de situation (du terrain)** (Représentation graphique ou cartographique des caractéristiques existantes ou projetées d'un espace donné ; ►plan d'ensemble, ►plan de principe, ►plan général, ►plan régulier) ; *g* **Lageplan [m]** ([Karto]grafische Darstellung, die die Situation einer realen örtlichen Gegebenheit oder die Planung für eine Fläche oder einen bestimmten Raum wiedergibt; ►Gesamtlageplan, ►Gestaltungsplan, ►Übersichtsplan, ►Vermessungsplan).

5645 site planning [n] *plan. prof.* (**1.** Process of design and preparation of construction drawings and specifications for the external space around a building project; ►construction project; **2.** In a broad sense, **s. p.** is an interdisciplinary activity. The scientific aspects encompass environmental science and engineering while the visually creative aspects of site planning require the perception and creation of outdoor spaces, building masses, formulation of land use through economic evaluation and the orchestration of spatial human activities and circulation; cf. IISP 2009); *syn.* site design [n], design [n] of landscape around buildings [also UK]; *s* **planificación [f] de un proyecto paisajístico** (Proceso de diseño y preparación de ►proyecto de construcción paisajista de los exteriores de una obra arquitectónica, incluidas todas las especificaciones necesarias para realizar la obra); *syn.* planificación [f] de parques y jardines; *f* **mission [f] d'études particulières et de maîtrise d'œuvre du paysagiste** (Conception [primaire, secondaire et tertiaire] et maîtrise d'œuvre de ►projets d'aménagement de construction ou de l'espace extérieur en particulier dans les domaines suivants : **1.** aménagements urbains ou périurbains : places, rues piétonnières, aires de jeux, parkings, habitat collectif, entrées de villes et de bourgs, squares et jardins de villes, politique de réhabilitation et d'intégration urbaine, pré-verdissement, jardins familiaux, **2.** aménagements ruraux : mise en valeur [aménagements fonciers et hydrauliques], remembrement, reboisements, friches, aménagements de cours d'eau ou plans d'eau, bases de loisirs, parcs d'attraction, valorisation du patrimoine : jardins et parcs historiques, sites, espaces naturels protégés, littoral, montagne, **3.** réhabilitation de lieux dégradés : friches industrielles, carrières, gravières, terrils, décharges, **4.** aménagements de zones industrielles, portuaires, **5.** aménagements de jardins privés ; cf. les missions du concepteur paysagiste, Fédération Française du Paysage, 2008) ; *g* **Objektplanung [f] für Freianlagen** (Entwurfs- und baureife Planung vegetationsbestimmter Außenanlagen eines ►Bauvorhabens im städtischen und außerstädtischen Bereich incl. der dazu gehörenden Bauwerke wie z. B. Lärmschutzwälle, Stützbauwerke und Geländeabstützungen, Stege und Brücken, ohne tragwerksplanerische Leistungen, die einer gesonderten Fachplanung unterliegen, Wegebau, Teichbau etc. Je nach Umfang der Aufgabenstellung und der dafür erforderlichen Leistungen ist diese Planung eine interdisziplinäre Tätigkeit, die gestalterisch-ästhetische Gesichtspunkte mit verschiedenen wissenschaftlichen Aspekten der Landschaftsökologie, Pflanzensoziologie und des Ingenieurbaus zusammenführt).

site plan review [n] *leg. urb.* *►directive on planting requirements [US].

site plant [n] [UK] *constr.* ►machinery stock.

5646 site productivity [n] *agr. ecol. for.* (Potential biomass production of a defined area in a given period of time dependent upon environmental conditions/site factors; ►potential yield, ►soil fertility); *s* **productividad [f] de una ubicación** (Productividad alcanzable en un emplazamiento específico que depende de las condiciones ecológicas y factores mesológicos; ►capacidad de rendimiento, ►fertilidad del suelo); *f* **productivité [f] de la station** (Quantité de matière vivante pouvant être produite sur une station en fonction des facteurs écologiques existants ; ►fertilité du sol, ►productivité 1, ►rendement) ; *g* **Standortproduktivität [f]** (Realisierbare Produktivität an Biomasse eines Standortes in einem definierten Zeitraum in Abhängigkeit von ökologischen Bedingungen/Standortfaktoren; ►Bodenfruchtbarkeit, ►Ertragsfähigkeit); *syn. for.* Standortleistungsfähigkeit [f].

site reconnaissance [n] [UK] *landsc.* ►site inspection (2).

site recycling [n] *plan. urb.* ►recycling of derelict sites.

site report [n] [UK] *constr.* ►site log.

5647 site requirements [npl] (1) *phyt. zool.* (An organism's prerequisites for existence at a particular location); *s* **exigencias [fpl] mesológicas** (Condiciones que deben reinar en el lugar de crecimiento para que un organismo pueda desarrollarse); *syn.* necesidades [fpl] mesológicas; *f* **exigences [fpl] stationnelles** (Facteurs environnementaux indispensables au développement des organismes sur un site particulier) ; *g* **Standortansprüche [mpl] (1)** (Die zum Leben von Organismen notwendigen Umweltfaktoren an einem Ort).

5648 site requirements [npl] (2) *plan.* (Necessary preconditions at the location for a project); *s* **exigencias [fpl] de ubicación** (Condiciones con las que debe cumplir un emplazamiento para que se pueda realizar un proyecto industrial, de infraestructura, etc. en él); *syn.* exigencias [fpl] de emplazamiento; *f* **exigences [fpl] du site** (Contraintes à prendre en compte à l'égard des ressources fondamentales comme conditions préalables requises en vue de l'implantation d'une infrastructure) ; *g* **Standortansprüche [mpl] (2)** (Forderungen für notwendige Voraussetzungen zur Ansiedlung oder Durchführung von Projekten).

site road [n] *constr.* ▶temporary construction road.

5649 site selection [n] *plan.* (Choice of a site, e.g. for industrial development, refuse dump [US]/refuse tip [UK]); *s* **selección [f] de ubicación/emplazamiento** *syn.* determinación [f] de ubicación/emplazamiento, elección [f] de ubicación/emplazamiento; *f* **choix [m] (en faveur) du/d'un site** (Décision en faveur d'un site d'implantation d'une activité industrielle, d'une grande infrastructure, d'une aire de stockage, etc.) ; *syn.* localisation [f] ; *g* **Standortwahl [f, o. Pl.]** (Entscheidung für einen bestimmten Standort, z. B. für eine Industrieansiedlung, Mülldeponie; einen Standort wählen [vb]).

5650 site soil [n] *constr. pedol.* (Natural ground in undisturbed state; ▶consolidated subgrade, ▶undisturbed subgrade [US]/undisturbed subsoil [UK], ▶volume of on-site soil); *syn.* in situ soil [n]; *s* **suelo [m]** *in situ* (Suelo natural inalterado; ▶subsuelo estable, ▶subsuelo inalterado [Es, RCH], ▶volumen de tierra compacta); *f 1* **sol [m] en place** *pedol.* (Sol naturel rencontré sur un site ; ▶terrain naturel) ; *f 2* **terrain [m]** *in situ constr.* (terre non foisonnée ; VRD 1986, 48 ; ▶sol stable, ▶bon sol, ▶volume en place) ; *syn.* terrain [m] en place, terre [f] en place, terre [f] en site, sol [m] compact ; *g* **anstehender Boden [m]** (Natürlicher Boden, der ungestört in einem Gelände vorhanden ist; ▶fester Baugrund, ▶gewachsener Boden; ▶feste Bodenmasse).

site study [n] *plan.* ▶site analysis.

5651 site suitability [n] *plan.* (Existence of certain functional attributes, which are prerequisites for the realization of a proposed land-use); *s* **idoneidad [f] de una ubicación** *syn.* idoneidad [f] de un emplazamiento; *f* **aptitude [f] d'un site** (Présence des caractéristiques fonctionnelles nécessaires d'un site, d'une region, favorables à la localisation d'un usage) ; *syn.* aptitude [f] du milieu, vocation [f] d'un site ; *g* **Standorteignung [f]** (Vorhandensein der erforderlichen, zweckentsprechenden Eigenschaften eines Ortes, einer Region für die Verwirklichung eines Nutzungsanspruches).

5652 site supervision [n] *constr.* (Direction and surveillance of the proper execution of a project on the construction site carried out by a ▶client appointed architect or by the contractor himself aimed at ensuring that the project is executed according to planning approval, the contract documents including construction drawings and specifications as well as state-of-the-art technology and relevant building regulations; ▶supervision of works); *s* **dirección [f] en pie de obra** (Control de la ejecución de los trabajos y verificación del avance de la obra por parte de arquitecto/a encargado/a por la ▶propiedad de obra o por el mismo contratista para asegurar que la obra se realiza de acuerdo con el permiso recibido y los documentos del contrato, incluyendo los planos y especificaciones constructivas, así como que

se respetan las normas de construcción relevantes y se utiliza la mejor tecnología asequible; ▶dirección de obra); *f* **surveillance [f] de chantier** (Contrôle de l'exécution des travaux, vérification de l'avancement des travaux, direction des réunions de chantier ; ▶direction de l'exécution du contrat de travaux, ▶maître d'ouvrage) ; *g* **örtliche Bauleitung [f]** (Leitung und Überwachung der ordnungsgemäßen Ausführung eines Objektes auf der Baustelle seitens des durch den ▶Bauherrn beauftragten Planers oder des Unternehmers mit dem Ziel, dass entsprechend der Baugenehmigung, den Ausführungsplänen und den Leistungsbeschreibungen sowie nach den allgemein anerkannten Regeln der Technik und den einschlägigen Vorschriften gebaut wird; ▶Bauleitung 3).

site supervisor [n] **representing a design practise** [UK] *constr. prof.* ▶design firm's representative [US].

5653 site supervisor [n] **representing an authority** *constr. prof.* (**1.** Representative of a [public] client or authority responsible for the supervision and controlling of construction or maintenance work [▶supervision of works]; tasks include the monitoring of the execution in compliance with the time schedule and the budget as well as with state-of-the-art technology and relevant regulations; he must also ensure that the contractor's project representative and foreman comply with the necessary regulations for safety, health and environmental protection on the construction site. The **s. s. r. an a.** is responsible for the fulfillment/fulfilment of all statutory, governmental and trade association conditions; monitoring of deadlines in relation to budget expenditures with monthly reports; as a rule the **s. s. r. an a.** is a civil engineer, landscape engineer, or architect. **2.** In D., local authorities employ a **chief s. s. r. an a.** who is a very experienced civil or landscape engineer responsible for the administration of construction projects at the municipal, state and federal level; his tasks include the uniform quality management of all construction projects in a department and he is the arbitrator of disputes during construction; in addition he must check bills of quantities to ensure they are complete and to avoid contractor claims and monitor the time schedules of all projects under construction; furthermore, the **chief s. s. r. an a.** monitors the flow of available financial resources and must deliver monthly reports to the department head on deadlines and the financial status of all major projects; ▶public works inspector); *syn.* public contract manager [n] [US], public agency representative [n] [US], building control officer [n] [UK]; *s 1* **interventor, -a [m/f] en pie de obra** (Representante de una administración comitente de una obra pública y responsable de la ▶dirección de obra del proyecto. Entre sus tareas se encuentra la vigilancia de la realización de los trabajos puntual, económica y con mejor tecnología de contrucción asequible. Es responsable de que el jefe de obra o el capataz de la empresa contratista ponga en práctica las normas de seguridad y de protección de la salud y las medidas de protección ambiental. El **i. en p. de o.** también es responsable del cumplimiento de las prescripciones legales, administrativas y de la mutua de accidentes de trabajo, de la vigilancia del calendario y del flujo de los fondos a disposición, y debe presentar informes mensualmente. Generalmente es ingeniero de caminos [Es]/civil [AL], arquitecto, arquitecto paisajista o maestro de obras; ▶inspector de obra pública); *s 2* **arquitecto/a [m/f] interventor, -a** (*Término administrativo para* ingeniero de caminos [Es]/civil [AL] o arquitecto [paisajista] con mucha experiencia responsable del control de calidad de todos los proyectos de una administración específica en proceso de construcción. Es la instancia de consulta con facultad decisoria en el caso de que se planteen divergencias en el proceso de construcción. Le corresponde además controlar el resumen de prestaciones para evitar ofertas complementarias, vigilar el cumplimiento del calendario de todas

las obras y el flujo de los fondos a disposición y tiene además la obligación de presentar informes mensuales ante la dirección administrativa sobre el estado de avance y de gasto de todos los proyectos importantes); *f 1* **représentant [m] du maître d'ouvrage public** (Technicien d'une administration territoriale responsable de l'assistance du maître d'ouvrage dans le planning de consultation, de préparation et de rédaction de l'avis d'appel public à la concurrence [AAPC] et du règlement de la consultation [RC] ; il a pour responsabilité d'analyser les dossiers de candidatures et d'établir un rapport à l'attention du jury, de préparer et rédiger avec la maîtrise d'ouvrage les pièces du marché de maîtrise d'œuvre [AE, CCAP, CCTP], de présenter le programme [par le programmiste] aux candidats retenus, d'animer la commission technique et rédiger le rapport, de suivre administrativement le choix du lauréat [rédaction des courriers, de la mise au point du marché, de la rédaction de l'avis d'attribution, etc.], assurer le suivi du travail du maître d'œuvre en veillant au respect du programme, des délais et de l'enveloppe budgétaire, en participant à toutes les réunions de conception, d'analyser les différents documents [APS, APD, PRO], de soumettre les rapports d'analyse au maître d'ouvrage à chaque étape, de transmettre et de s'assurer de la prise en compte des avis du coordonnateur SPS et Bureau de Contrôle Technique ; préparer et faire le suivi des marchés de CSPS, CT, OPC, élaborer les cahiers des charges CSPS, Contrôle Technique et OPC, d'assister le maître d'ouvrage pour la passation des marchés, d'organiser les consultations des marchés CSPS, Contrôle Technique et OPC, de suivre l'exécution des prestations, d'analyser les documents produits, de vérifier et d'émettre des avis sur les notes d'honoraires, d'animer et suivre les prestations du coordonnateur SPS et du Bureau de Contrôle Technique, de s'assurer pour la gestion administrative de l'obtention des autorisations administratives, de la communication autour du projet et de la traçabilité de l'opération, de s'assurer pour la gestion financière du respect de l'enveloppe budgétaire, de contrôler et de valider les avenants et de faire le suivi financier de l'opération ; marchés publics, www.envt.fr ; ▶personne responsable des travaux [et de la maintenance]) ; *f 2* **conducteur [m] d'opération** (F., personne à laquelle peut recourir le maître de l'ouvrage pour une mission d'assistance générale à caractère administratif, financier et technique ayant pour tâche de traduire les orientations financières dans un programme et d'apporter son aide dans le choix des modes de réalisations, des intervenants, de la gestion des marchés, etc. D'une manière générale, le conducteur d'opération est, sur le plan des relations contractuelles, l'unique interlocuteur « public » des architectes, contrôleurs techniques, des ingénieurs et des entrepreneurs ; la mission de conduite d'opération est exclusive de toute mission de maîtrise d'œuvre, de contrôle technique ou de travaux ; ▶direction de l'exécution du contrat de travaux) ; *g 1* **Bauleiter/-in [m/f] (3)** (Vertreter eines öffentlichen Auftraggebers/einer Baubehörde, der für die ▶Bauleitung 3 eines Projektes verantwortlich ist. Zu den Aufgaben gehört die termingerechte, wirtschaftliche und nach den allgemein anerkannten Regeln der Baukunst durchzuführende Abwicklung der Arbeiten; er hat dafür zu sorgen, dass der Bauleiter/Polier oder Vorarbeiter der bauausführenden Firma die nötigen Vorschriften für die Sicherheit und den Gesundheitsschutz und Maßnahmen zum Umweltschutz auf der Baustelle umsetzt. Der **B.** ist für die Erfüllung der gesetzlichen, behördlichen und berufsgenossenschaftlichen Auflagen verantwortlich; Überwachung des Terminplanes und des Abflusses der zur Verfügung stehenden Finanzmittel mit monatlicher Berichtspflicht; der **B.** ist im Regelfall ein Bauingenieur, Gartenbauingenieur, Architekt oder Baumeister; ▶Bauaufseher); *g 2* **Oberbauleiter/-in [m/f]** (*Verwaltungstechnischer Begriff für die kommunale, Landes- und Bundesebene*

sehr erfahrener Bauingenieur oder Gartenbauingenieur für den Landschaftsbau, der für ein einheitliches Qualitätsmanagement aller im Bau befindlichen Projekte einer Dienststelle zuständig und ein zentraler Ansprechpartner bei Meinungsverschiedenheiten in der Bauausführung und mit entsprechender Entscheidungskompetenz ausgestattet ist; zu seinen weiteren Aufgaben gehört die Prüfung der Leistungsverzeichnisse auf Vollständigkeit zur Vermeidung von Nachtragsangeboten; die zentrale Terminüberwachung aller im Bau befindlichen Vorhaben; Überwachung des Abflusses der zur Verfügung stehenden Finanzmittel; monatliche Berichtspflicht gegenüber der Dienststellenleitung über den terminlichen und finanziellen Stand aller wichtigen Vorhaben).

5654 site survey [n] (1) *eng. surv.* (Measuring a site to produce a detailed plan of existing conditions and features on a small site; ▶topographic survey 3); *s* **medición [f] del terreno** (Medición de un solar para hacer un plan de detalle de las condiciones existentes antes de comenzar a planear; ▶medición topográfica); *syn.* medida [f] (del terreno); *f* **métré [m]** (Mesure de tous les éléments rencontrés sur le terrain caractérisant l'état des lieux avant l'établissement d'un projet et reporté sur un document constituant le fond de plan ; ▶levé, ▶arpentage) ; *g* **Aufmaß [n] (2)** (Vermessen eines Grundstücks vor Planungsbeginn für die Bestandspläne; ▶Vermessung).

site survey [n] (2) *constr. contr.* ▶method of site survey [US]/ method of measurement [UK]; *constr. landsc.* ▶topographic survey (1).

site survey [n] **and analysis** [n] *landsc. plan.* ▶comprehensive site survey and analysis.

5655 siting [n] *landsc.* (Exact location of a project and its individual features); *s* **emplazamiento [m]** (Localización exacta de elementos individuales); *syn.* localización [f], ubicación [f]; *f* **emplacement [m]** (Lieu occupé par quelque chose ou propice à la construction, ou encore, la localisation exacte d'un édifice) ; *syn.* lieu [m], endroit [m], position [f] ; *g* **Anlage [f] (2)** (Genaue Lokalisierung eines Bauobjektes; anlegen [vb]); *syn.* Platzierung [f], *f. S.* Plazierung [f].

5656 sitting area [n] *landsc.* (TGG 1984, 72; area equipped with seats or benches); *s* **lugar [m] de descanso (3)** (Sitio equipado con sillas o bancos); *syn.* lugar [m] para sentarse, sitio [m] de descanso; *f* **emplacement [m] de repos (équipé de bancs)** (Aire équipée de bancs ou de chaises ; *terme spécifique* coin de repos) ; *g* **Sitzplatz [m]** (Mit Bänken oder Stühlen ausgestatteter Platz).

5657 sitting step [n] *gard. landsc. syn.* seat step [n]; *s* **grada [f] de asiento**; *f* **gradin [m]** (Marche d'environ 45 cm de hauteur servant de siège, p. ex. aux spectateurs dans les stades, théâtres, amphithéâtres) ; *g* **Sitzstufe [f]** (Ca. 45 cm hohe Stufe einer Sitztreppe).

5658 sitting steps [npl] *gard. landsc.* (Flight of steps used as seats from which events etc. can be viewed); *syn.* seat steps [npl]; *s* **gradería [f] de asiento**; *f* **escalier [m] en gradins** (Escalier en accompagnement de talus constitué de marches en forme de gradins) ; *g* **Sitztreppe [f]** (In einen Hang oder eine Böschung eingebaute Treppenanlage aus Sitzstufen).

sitting wall [n] *constr. gard. landsc.* ▶seatwall.

situation [n] **before planning starts [UK], existing** *ecol. plan.* ▶existing condition before planning starts [US].

situation [n] **of practical completion [UK]** *constr. contr.* ▶condition for final acceptance [US].

5659 size [n] **of planting stock** *constr. hort.* (Height and spread of a plant at the time of planting. Trees are defined by trunk caliper measurement. Herbaceous plants are selected for

planting according to the diameter of the plant container; ►plant size); *s* **tamaño [m] del material de plantación** (Altura o, en el caso de los árboles, perímetro del tronco en el momento de ser plantados; ►tamaño de planta); *f* **taille [f] des végétaux** (Dimension définie soit par le diamètre ou la contenance du pot, soit par la hauteur et la largeur de la plante ou pour les arbres par la circonférence du tronc ; ►taille des végétaux) ; *g* **Pflanzgröße [f]** (Topfdurchmesser oder Höhe und Breite einer Pflanze zum Zeitpunkt der Pflanzung, bei Bäumen auch durch den Stammumfang definiert; ►Pflanzengröße).

5660 sizing [n] of woody plants *hort.* (Standarized system to facilitate the trade of nursery stock by classifying height, crown width, stem circumference or diameter, number of shoots, as well as age in the case of young plants and diameter in the case of ►understocks); *s* **calibradura [f] de leñosas** (Clasificación de las leñosas de vivero según la altura y la envergadura, el diámetro del tronco, número de ramas, en las plantas jóvenes según la edad y en ►patrón según el diámetro); *f* **calibrage [m] des plantes ligneuses** (Classement des produits ligneux de pépinières commercialisés d'après les méthodes de culture [multiplication], leurs formes [arbres fruitiers palissés ou non], les caractéristiques d'aspect [sans tête formée, touffe et cépées, sujets fastigiés etc.], les caractéristiques dimensionnelles [hauteur, largeur, diamètre de la tige, circonférence du tronc], les caractéristiques d'âge pour les jeunes plants ; ►jeune plant portegreffe issu de semis, ►porte-greffe) ; *g* **Sortierung [f] von Gehölzen** (Unterscheidung von verkaufsfähiger Baumschulware nach Höhe oder Breite, Stammumfang, Triebzahl sowie Alter bei Jungpflanzen, Wildlinge nach Durchmesser; ►Wildling 2).

skeletal root [n] *arb.* ►structural root.

5661 skeletal soil [n] *pedol.* (Weak soils with very shallow humus layer [►organic horizon] above unweathered ►parent material; ►ranker [UK], ►rendzina, ►truncated soil profile); *syn. obs.* lithosol [n] [US]; *s* **suelo [m] esquelético** (Suelo muy superficial con una capa de humus y ►mantillo muy fina directamente sobre la ►roca madre; ►ranker, ►rendsina, ►suelo decapitado); *f* **sol [m] squelettique** (Sol d'érosion peu évolué caractérisé par une couche peu épaisse et pourvue d'une faible quantité de matière organique, en place sur la ►roche-mère inaltérée ; ►horizon organique [Ao], ►ranker, ►rendzine, ►sol décapé) ; *g* **Gesteinsrohboden [m]** (Gering mächtiger Boden mit sehr wenig Humusauflage über unverwittertem ►Ausgangsgestein; ►geköpftes Bodenprofil, ►Ranker 1, ►Rendzina); *syn.* Skelettboden [m], Schuttboden [m].

skeleton root [n] *arb.* ►structural root.

5662 sketch [vb] *plan.* (To produce a rough rapid drawing, outlining the most important characteristics; ►preliminary design); *s* **bosquejar [vb]** (Hacer un dibujo rápido que muestra las características más importantes de un proyecto; ►anteproyecto); *syn.* hacer [vb] un boceto, esbozar [vb], trazar [vb]; *f* **esquisser [vb]** (Représenter sommairement les traits principaux de la conception d'un projet ; ►avant-projet sommaire [A.P.S.]) ; *g* **aufreißen [vb]** (2) (1. Schnelles zeichnerisches Darstellen einer Idee. 2. Zeichnen der Vorder- oder Seitenansicht eines Gebäudes oder Bauteils; ►Vorentwurf); *syn. zu 1.* skizzieren [vb].

5663 sketch design [n] *plan. prof.* (Drawing showing the first general rough outline of a planning concept, usually made before the ►preliminary design; ►schematic design phase [US]/outline and sketch scheme proposals [UK]); *syn.* schematic design [n], concept (site) design [n], sketch proposal [n] [also UK]; *s* **esbozo [m] de diseño preliminar** (Dibujo que muestra esquemáticamente los planteamientos preliminares del concepto de un proyecto que se realiza generalmente antes del ►anteproyecto; ►fase de anteproyecto); *f* **esquisse [f] d'avant-projet**

sommaire (Propositions graphiques permettant au maître d'ouvrage de fixer son choix sur un parti général et sur la base desquelles l'architecte/paysagiste concrétise le parti général de composition sous forme d'►avant-projet sommaire [A.P.S.] ; ►avant-projet) ; *g* **Vorentwurfsskizze [f]** (Zeichnerische Darstellung über erste Überlegungen zu einem Projektplanungskonzept, die i. d. R. dem ►Vorentwurf vorausgehen; ►Vorplanung).

sketch proposal [n] [UK] *plan. prof.* ►sketch design.

5664 ski [vb] *recr.* *s* **esquiar [vb]**; *f* **faire [vb] du ski** *syn.* skier [vb] ; *g* **Ski laufen [vb]** *syn. o. V.* Schi laufen [vb].

skidding [n], resistance to *constr.* ►nonskid property [US]/non-skid property [UK].

skidding damage [n] [US&CDN] *for.* ►felling and logging damage.

skidding lane [n] [UK] *for.* ►skidding road [US&CDN].

5665 skidding road [n] [US&CDN] *for.* (Roughly prepared track over which logs are dragged after felling; ►timber road [US]/forest path [UK]); *syn.* skidding lane [n] [UK]; skid road [n] [US&CDN], snig track [n] [AUS], travois road [n] [US&CDN]; *s* **camino [m] de arrastre** (DFM 1975; ►camino forestal 2 estrecho por el cual se arrastran los troncos tras la tala); *syn.* picada [f]; *f* **chemin [m] de débardage** (Un itinéraire tracé dans une forêt non ou très peu aménagée, utilisé par un tracteur pour la vidange des bois ; cf. DFM 1975 ; ►chemin forestier, ►laie) ; *syn.* layon [m], piste [f] de débardage ; *g* **Rückeweg [m]** (Schmaler, meist nicht befestigter Weg in einem Waldbestand, auf dem das abzuführende Holz abtransportiert wird; ►forstwirtschaftlicher Weg).

skidfree [adj] [US]/skid-free [adj] [UK] *constr.* ►nonslip [US]/non-slip [UK].

skid road [n] [US&CDN] *for.* ►skidding road [US&CDN]/skidding lane [UK].

5666 skiing [n] *recr.* (Generic term for ►back-country skiing, ►cross-country skiing, ►down mountain skiing, ►ski touring); *s* **esquí [m]** (Término genérico para ►esquí alpino, ►esquí de fondo, ►esquí de travesía y ►esquí de travesia en alta montaña); *f* **ski [m]** (Terme générique pour la pratique du ►ski de piste, ►ski de fond, ►ski de randonnée, de la ►randonnée à ski en altitude) ; *g* **Skilauf [m, o. Pl.]** (OB zu ►Abfahrtslauf, ►Skilanglauf, ►Skiwandern, ►Tourenskilauf; Ski laufen [vb]); *syn. o. V.* Schilauf [m].

skiing [n], downhill *recr.* ►down mountain skiing.

5667 skiing area [n] *plan. recr.* (Hilly or mountainous area suitable for skiing); *s* **zona [f] de esquí**; *f* **domaine [m] skiable** (Région de montagne appropriée à la pratique du ski) ; *syn.* domaine [m] de ski ; *g* **Skigebiet [n]** (Gebirgsregion, die sich zum Skifahren eignet); *syn. o. V.* Schigebiet [n].

5668 ski lift [n] *recr.* (Power-driven conveyor for carrying skiers up or down a mountain slope. Tow-lift, chair lift and ropeway are distinguished); *s* **remonte [m] mecánico** (*Términos específicos* telearrastre, telesilla, teleférico, telecabina); *syn.* remonte [m] de esquí; *f* **remontée [f] mécanique** (Équipement technique permettant au skieur d'atteindre le haut des pistes ; *termes spécifiques* remonte-pente/tire-fesses/téléski, téléférique/ téléphérique, télésiège) ; *g* **Skilift [m]** (Technische Einrichtung [Schlepplift, Sessellift oder Seilbahn], die Skiläufer bergauf befördert); *syn. o. V.* Schilift [m].

5669 skin diving [n] *recr.* (Sport diving without breathing equipment); *s* **snórkeling [m]** *syn.* buceo [m] deportivo con respirador; *f* **plongée [f] sous-marine en apnée** (Activité sportive de descente sous la surface de l'eau sans matériel de respiration) ; *g* **Sporttauchen [n, o. Pl.]** (2) (1. Tauchen ohne

Atmungsgerät. **2.** Mit Hilfe eines Schnorchels tauchen, um das Leben unter Wasser zu beobachten; schnorcheln [vb]); *syn. zu 2.* Schnorcheltauchen [n, o. Pl.], Schnorcheln [n, o. Pl.].

skip [n] [UK] *constr.* ►dumpster [US].

5670 ski run [n] *recr.* (Ski slope specially prepared for downhill skiing, with artificial snow from snow canons, if necessary; ►downhill ski run); *s* **pista [f] de esquí** (Pista preparada para ►descenso en esquís); *f* **piste [f] de ski alpin** (Piste enneigée naturellement ou préparée avec de la neige artificielle [canon à neige] pour une ►descente à ski) ; *g* **Skipiste [f]** (Für ►Skiabfahrten besonders angelegte und, wenn nötig, mit Kunstschnee aus Schneekanonen präparierte Schneebahn); *syn. o. V.* Schipiste [f].

5671 ski touring [n] *recr.* (Skiing travel in mountainous terrain with synthetic self-sealing 'skins' for ascent and with overnight accommodations in cabins; ►back-country skiing, ►cross-country skiing, ►down mountain skiing); *s* **esquí [m] de travesía en alta montaña** (Excursión con esquís alpinos, incluso de varios días, al margen de las pistas preparadas; ►esquí alpino, ►esquí de fondo, ►esquí de travesía); *f* **randonnée [f] à ski en altitude** (Ascension avec des skis équipés de peaux de phoque et des descentes hors-pistes en poudreuses ; ►randonnée à ski, ►ski de fond, ►ski de piste, ►ski de randonnée) ; *syn.* ski [m] de randonnée en altitude, raid [m] à ski ; *g* **Tourenskilauf [m, o. Pl.]** (Alpiner Skisport abseits der Pisten mit Steigfellen und Tiefschneeabfahrten sowie mit Übernachtungen in Hütten bei mehrtägigen Touren; ►Abfahrtslauf, ►Skilanglauf, ►Skiwandern); *syn. o. V.* Tourenschilauf [m, o. Pl.], Hochtourenskilauf [m, o. Pl.], *o. V.* Hochtourenschilauf [m, o. Pl.].

sky exposure plane [n] [US] *leg. urb.* ►bulk plane [US].

5672 skyline [n] *urb.* (Outline of hills, building, etc., defined against the sky on the visible horizon of a city); *s* **silueta [f] de la ciudad** (Perfil de una ciudad proyectado hacia el horizonte); *syn.* skyline [m], perfil [m] de la ciudad; *f* **silhouette [f] de la ville** (Profil d'une ville qui se détache sur l'horizon) ; *g* **Stadtsilhouette [f]** (Ansicht einer Stadt gegen den Horizont); *syn.* Skyline [f].

slab [n] *constr.* ►concrete slab (1), ►concrete slab (2), ►cover slab, ►footpath paving slab; *adm.* ►grave slab; *constr.* ►natural stone slab, ►patterned concrete paving slab, ►paving slab, ►poured-in-place concrete slab [US]/in-situ concrete slab [UK].

slab [n] [UK]**, bg-** *constr.* ►grass paver.

slab [n] [UK]**, chamber cover** *constr.* ►cover slab.

slab [n] [US]**, concrete protective** *constr.* ►protective screed.

slab [n]**, concrete roof** *constr.* ►concrete deck.

slab [n] [UK]**, garden paving** *constr.* ►natural garden stone flag.

slab [n] [UK]**, grass concrete** *constr.* ►grass paver.

slab [n] [UK]**, in-situ concrete** *constr.* ►poured-in-place concrete slab [US].

slab [n] [UK]**, planting** *constr.* ►planting mat.

slab [n] [UK]**, precast concrete paving** *constr.* ►precast paving block [US].

slab [n]**, shaft cover** *constr.* ►cover slab.

slab [n]**, structural** *constr.* ►concrete slab (1).

slab [n] **on riser** [UK] *arch. constr.* ►flagstone step.

5673 slab paving [n] *constr.* (Blocks of stone shaped or selected for use in a pavement surface; *e.g. with natural stones* flagstone paving; *with concrete paving slabs* paving of precast concrete slabs; ►surface layer); *s* **enlosado [m]** (Pavimento de losas de hormigón o de piedra natural; ►pavimento 1); *syn.* embaldosado [m]; *f* **dallage [m]** (Revêtement de sols ou de voirie réalisé avec des dalles de béton ou de pierres naturelles ; ►revêtement de sol) ; *syn.* revêtement [m] avec dalles préfabriquées ; *g* **Plattenbelag [m]** (...beläge [pl]; Boden- oder Wegebelag aus Beton- oder Natursteinplatten; ►Decke); *syn.* Plattendecke [f].

5674 slab paving pattern [n] *constr.* (Distinctive joint design of slabs laid for surfacing; ►bond); *s* **aparejo [m] de losas** (Disposición de losas de pavimento siguiendo un diseño específico; ►aparejo); *f* **appareillage [m] de dalles** (Disposition suivant un dessin donné d'un revêtement de dalles, l'aspect de l'**a.** étant donné par l'assemblage des joints ; ►appareillage) ; *g* **Plattenverband [m]** (Art und Weise wie Platten auf einer zu befestigenden Fläche angeordnet sind; ►Verband 2).

"slack" space [n] [UK] *recr.* ►teenage playground [US].

slag [n] *constr. envir. min.* ►cinder.

slag heap [n] [UK] *envir. min.* ►slag pile [US].

5675 slag pile [n] [US] *envir. min.* (Pile [US]/tip [UK] of blast furnace waste resulting from smelting of ores; ►spoil bank [US] 2/spoil heap [UK]); *syn.* slag heap [n] [UK]; *s* **escorial [m]** (Depósito para residuos de la industria siderúrgica; ►vaciadero de gangas); *f* **crassier [m]** (Amoncellement des scories de hauts fourneaux ; ►terril) ; *g* **Schlackenhalde [f]** (Deponie für Abfallprodukte der Hüttenindustrie; ►Bergehalde).

slanting stake [n] *arb. constr.* ►angled stake.

5676 slash [n] *for.* (►Crop residues left on the ground after tree felling, or material which has accumulated there as a result of storm, fire, or girdling; *specific term* ►windbreakage: all such material blown down by wind; SAF 1983); *syn.* brash [n] [also UK]; *s* **residuo [m] forestal** (Restos de material ligno-celulósico procedentes de los aprovechamientos selvícolas del monte como entresacas, clareos, podas, etc.; cf. DINA 1987; ►residuos de cosecha); *syn.* restos [mpl] de corta/poda; *f* **rémanent(s) [m/(pl)] de coupe** (DFM 1975 : résidus laissés sur le sol après l'exécution d'une coupe, d'une opération d'amélioration ou qui viennent s'y ajouter à la suite d'une tempête, d'un feu, etc. On distingue les **gros rémanents** [grumes non utilisées, souches déracinées, grosses branches] et des **petits rémanents** [petites branches, rameaux, feuilles, écorces, chips] ; n'inclut pas les ►bris de vent qui, contrairement au concept anglais « slash », seront exploités ultérieurement ; ►déchets de récolte) ; *g* **Schlagabraum [m]** (Am Boden liegende Reste von Bäumen nach Baumfällarbeiten oder Verjüngungshieben im Forst; ►Ernterückstand).

slash-and-burn farming [n] *agr.* ►shifting cultivation.

slate mound [n] *min.* ►spoil bank [US] (2)/spoil heap [UK].

sleeping policemen [n] *trans.* ►speed hump.

5677 slender branch [n] [US] *arb.* (Branch with diameter > 30-50mm; ►small branch); *syn.* feather (branch) [n] [UK]; *s* **rama [f] delgada** (Rama con un diámetro de > 30-50 mm; ►rama fina); *f* **branche [f] de faible diamètre** (Branche d'un diamètre supérieur à 30-50 mm ; ►rameau) ; *g* **Schwachast [m]** (Ast mit einem Durchmesser von > 30-50 mm; ►Feinast).

slide [n] *geo.* ►catastrophic rockslide [US]/catastrophic rock-slide [UK], ►debris slide; ►flow slide; *constr. geo. pedol.* ►landslide; *geo.* ►rockslide [US]/rock-slide [UK].

slide [n]**, mud-** *constr.* ►mud flow.

5678 sliding snow [n] *land'man. conserv.* (Snowmass slippage on slopes of over 25°, which can cause extensive damage to woodlands at high altitudes by sliding downwards); *s* **nieve [f] reptante** (Nieve que se desliza lentamente en cuestas de más de 25° y puede producir grandes daños a las masas de bosque); *f* **neige [f] reptante** (Glissement lent de neige accumulée le long

S

d'un versant de pente supérieure à 25° et dont le déplacement occasionne des dommages importants dans les massifs forestiers en haute altitude) ; *g* **Gleitschnee [m]** (An über 25° steilen Hängen auftretender Schnee, der durch Abwärtsbewegungen erhebliche Schäden an Waldbeständen in Hochlagen verursachen kann).

slimes dam [n] [ZA] *min.* ▶silt pile [US], ▶tailings pond.

5679 slip [n] (1) *hort.* (Cutting pulled from a woody or herbaceous plant); *s* **esqueje [m] arrancado** (Parte de una planta [tallo, rama, raíz, etc.] arrancada para reproducirla vegetativamente); *f* **bouture [f] arrachée** (Bouture détachée d'une plante ligneuse ou herbacée) ; *g* **Rissling [m]** (Zur Stecklingsvermehrung abgerissener Teil einer Pflanze).

slip [n] (2) *constr. geo.* ▶landslip; *constr. eng.* ▶rotational slip.

slip [n], **translational** *constr. geo. pedol.* ▶landslide, #2.

5680 slip bank [n] [US] *geo.* (WRP 1974, 12; slump bank on the inside curve of a stretch of flowing water; ▶flat bank, *opp.* ▶bluff 1); *syn.* slip-off slope [n] [UK]; *s* **orilla [f] convexa** (Curva interna de un tramo de río; *opp.* ▶orilla cóncava; ▶orilla llana); *syn.* orilla [f] interna; *f* **berge [f] convexe** (Talus d'un cours d'eau en pente douce sur laquelle se déposent les alluvions ; ▶berge à pente douce; *opp.* ▶berge concave) ; *syn.* rive [f] convexe ; *g* **Gleithang [m]** (Innenkurve eines Fließgewässers; ▶Flachufer 1; *opp.* ▶Prallhang); *syn.* Gleitufer [n].

slip-off slope [n] [UK] *geo.* ▶slip bank [US].

slippage [n] *geo.* ▶mass slippage [US]/mass movement [UK], ▶riverbank slippage [US]/river-bank slippage [UK]; *geo.* ▶soil slippage.

5681 slippage hazard [n] *constr.* (Soil slipping danger on a steep slope); *s* **peligro [m] de deslizamiento** (Peligro de que el suelo de una pendiente fuerte o de partes de ella se ponga en movimiento vertiente abajo); *f* **risque [m] de glissement** (Caractérise l'éventualité d'un glissement de talus ou de versant) ; *syn.* menace [m] de glissement ; *g* **Rutschgefährdung [f]** (Die Möglichkeit, dass eine Böschung, ein Hang oder Teile davon abrutschen können).

5682 slip plane [n] *constr. eng.* (Small or large area, mostly concave-shaped, which occurs by sliding or slump of soil/rock material; ▶shear failure of embankment, ▶shear development); *syn.* slip surface [n] (DOG 1984); slump plane [n]; *s* **plano [m] de deslizamiento** (Plano neto por el que se produce un deslizamiento. Si el plano es arqueado, de manera que la masa de tierra se desliza según un eje de círculo imaginario, se trata de un deslizamiento rotacional; cf. DGA 1986; ▶deslizamiento planar, ▶formación de superficie deslizante); *f 1* **plan [m] de glissement** (1. Surface formée sur un terrain en pente ou par une couche pédologique plastique lorsque des ouvrages de terrassement ou des fondations de bâtiment surpassent la ▶résistance au cisaillement des matériaux sur lesquels ils ont été mis en place et commencent à glisser. 2. Plan incurvé, concave lors d'un ▶rupture de terrain ; ▶formation d'un plan de glissement) ; *syn.* surface [f] de rupture, face [f] de glissement ; *f 2* **plan [m] de faille** (Surface approximativement plane le long de laquelle ont glissé deux compartiments de terrain ; DG 1984, 182) ; *g* **Gleitfläche [f]** (1. Abbruchfläche auf geneigten Flächen [Hängen] oder geneigten bindigen Bodenschichten, die entsteht, wenn Erdbauwerke oder Fundamente für Hochbauten die Scherfestigkeit des Bodenmaterials, auf dem sie gebaut wurden, überwinden und abrutschen. 2. Konkave Abbruchfläche beim ▶Geländebruch, die durch Überwindung der ▶Scherfestigkeit von zu hohen Lasten im Erdkörper selbst entsteht; ▶Gleitflächenbildung); *syn.* Abbruchfläche [f] (eines Geländebruches).

slip-plane development [n] *constr. eng.* ▶shear development.

slip road [n] [UK] *plan. trans.* ▶highway ramp [US].

slip road [n] [UK], **cloverleaf** *plan. trans.* ▶cloverleaf ramp [US].

slip surface [n] *constr. eng.* ▶slip plane.

5683 slit drain [n] *constr.* (Very narrow trench excavated by a drainage plow [US]/plough [UK], filled with ▶pea gravel 2-8mm; a series of such drains at close intervals is used for the drainage of an athletic field; ▶slot trench drain); *s* **canalillo [m] de drenaje** (En campos de deportes de hierba, zanja muy estrecha y poco profunda rellena de ▶garbancillo cuya función es el drenaje del suelo. Generalmente se excavan muchos **cc. de d.** paralelos con poca distancia entre ellos; ▶conducto de desagüe con ranura); *f* **fente [f] de drainage** (Tranchée de drainage très étroite réalisée sur une pelouse de sport par une trancheuse et remplie avec du ▶gravillon de granularité 2/8 cm ; ▶caniveau à fente) ; *g* **Sickerschlitz [m]** (Sehr schmaler Entwässerungsschlitz, in kurzen Abständen über ein ganzes Rasensportfeld aneinandergereiht, mit einer Schlitzfräse hergestellt und mit ▶Gartenriesel 2/8 mm verfüllt; ▶Schlitzrinne); *syn.* Schlitzdrän [m].

5684 slit planting [n] [US] *constr. for.* (Prizing open a cut made, e.g. by a spade, mattock or planting bar, inserting a young tree, then closing the cut on the latter by pressure on the soil around it; cf. SAF 1983; ▶hole planting [UK]/pit planting [UK], ▶mound planting, ▶T-notching); *syn.* notch planting [n] [UK]; *s* **plantación [f] en raja** (Método de plantación en repoblaciones forestales o en la ▶construcción biotécnica; ▶plantación en hoyos, ▶plantación en montículos, ▶plantación en T); *f* **plantation [f] en fentes** (Mode de plantation de jeunes plants racinés : ouverture d'une fente par un coup de bêche, de pioche, de barre à planter, etc. et introduction du plant et fermeture de la fente sur ce dernier par pression du pied ; cf. DFM 1975 ; ▶plantation à la bêche en T, ▶plantation sur butte, ▶plantation sur potets) ; *g* **Klemmpflanzung [f]** (Pflanzmethode, bei der bei Aufforstungen oder ▶ingenieurbiologischen Bauweisen in einen mittels einer Wiedehopfhaue geschaffenen Bodenspalt Pflanzen eingebracht werden; ▶Hügelpflanzung, ▶Lochpflanzung, ▶Winkelpflanzung); *syn.* Spaltpflanzung [f], Schrägpflanzung [f].

slitting [n] *constr.* ▶lawn aeration, #2.

5685 slope [n] (1) *geo.* (Inclined part of a hill or mountain; ▶natural slope 1); *syn.* incline [n]; *s* **ladera [f]** (Lado inclinado de una montaña o colina; ▶talud natural); *syn.* falda [f], vertiente [f]; *f* **versant [m]** (Plan incliné d'une montagne ou d'une vallée ; ▶levée, ▶talus, ▶talus naturel) ; *syn.* coteau [m], côte [f] ; *g* **Hang [m]** (Geneigte Seite eines Berges oder Hügels; ▶natürliche Böschung); *syn.* Halde [f] (2).

5686 slope [n] [US] (2) *constr.* (Road or ground incline; ▶gradient 1; *specific terms* ▶cross slope [US]/cross fall [UK], ▶reverse slope [US]/reverse falls [UK], ▶longitudinal gradient, ▶slopes [US] 2/falls [UK], ▶pipe gradient); *syn.* fall [n] [UK], pitch [n] [US]; *s* **pendiente [f]** (En la construcción de carreteras y caminos, grado de inclinación de una superficie; ▶pendiente en subida; *términos específicos* ▶contrapendiente, ▶gradiente de tubería, ▶pendiente de una superficie, ▶pendiente longitudinal, ▶pendiente transversal); *f* **pente [f]** (Inclinaison d'une surface de voirie ; ▶déclivité en rampe ; *termes spécifiques* ▶contre-pente, ▶dévers, ▶forme de pente, ▶pente de la conduite/du collecteur, ▶pente longitudinale, ▶pente transversale) ; *syn.* gradient [m] ; *g* **Gefälle [n]** (Grad der Neigung einer Fläche im Wegebau; ▶Steigung 1; *UBe* ▶Gegengefälle, ▶Längsgefälle, ▶Leitungsgefälle, ▶Oberflächengefälle, ▶Quergefälle).

S

slope [n] (3) *constr.* ►concave slope, ►create a slope, ►crest of embankment/slope, ►cross slope [US]/cross fall [UK], ►cut slope; *geo.* ►dip slope, ►east-facing slope; *constr.* ►fill slope, ►flatten a slope, ►incised slope, ►lay to slope [US]/lay to falls [UK], ►natural slope (1), ►natural slope (2); *geo.* ►natural terrace on a slope, ►north-facing slope; *constr.* ►ogee slope, ►reverse slope [US]/reverse falls [UK]; *geo.* ►scarp slope,. ►south-facing slope; *constr. eng.* ►standard slope; *geo.* ►standard slope; *constr.* ►steepen an existing slope; *geo.* ►steep slope, ►terraced slope; *constr. min.* ►terracing of a slope; *constr.* ►transition slope, ►unstable embankment slope; *geo.* ►unstable slope; ►west-facing slope.

slope [n] [US], **back** *constr.* ►reverse slope [US]/reverse falls [UK].

slope [n], **brow of** *constr.* ►crest of embankment/slope.

slope [n], **change in** *constr.* ►change in gradient.

slope [n], **change of** *constr.* ►gradient change.

slope [n], **construct a** *constr.* ►construct an embankment/slope.

slope [n] [US], **drop of a** *constr.* ►slope gradient.

slope [n] [US], **equator-facing** *geo. met.* ►south-facing slope.

slope [UK], **fall** [n] **of a** *constr. geo.* ►slope gradient.

slope [n], **foot of a** *constr.* ►foot of an embankment/slope.

slope [n] [US], **grade of a** *constr. geo.* ►slope gradient.

slope [n] [US], **gradient of a** *constr. geo.* ►slope gradient.

slope [n], **hazardous** *geo. pedol.* ►unstable slope.

slope [n], **lessen a** *constr.* ►flatten a slope.

slope [n], **rounding the crest of** *constr.* ►rounding the top of an embankment/slope.

slope [n], **shady** *geo.* ►ubac.

slope [n] [UK], **slip-off** *geo.* ►slip bank [US].

slope [n] [US], **soil-slippage-prone** *geo. pedol.* ►unstable slope.

slope [n], **sunny** *geo.* ►adret.

slope [n], **toe of a** *constr.* ►foot of an embankment/slope.

5687 slope analysis map [n] *plan.* (Topographical depiction showing varying degrees of average slopes in an area for evaluation of erosion susceptibility; ►erosion risk); *s* **mapa [m] de pendientes** (Mapa topográfico en el que se representan las pendientes medias. Sirve para evaluar el ►peligro de erosión); *f* **carte [f] des pentes** (Carte topographique représentant en fonction de l'altitude les pentes moyennes d'une région ; le système de pente est défini avec une courbe clinographique, la valeur des pentes se mesurant en degré d'angle, plus rarement en grade ou en pourcentage ; elle sert de document de base à l'évaluation du ►risque d'érosion des sols ; DG 1984, 341) ; *g* **Karte [f] der mittleren Hangneigung** (Topografische Karte, auf der die mittleren Hangneigungen in Gradstufungen dargestellt sind. Sie dient als Grundlage zur Einschätzung der ►Erosionsgefahr).

5688 slope angle [n] *constr.* (Angle of inclination: slopes may be expressed in degrees or as a ratio; ►slope ratio, ►slope gradient); *syn.* angle [n] of slope, embankment gradient [n]; *s* **ángulo [m] de talud** (Ángulo de inclinación de una ladera o talud; el **á. de t.** se expresa en grados, la pendiente en relación 1:n ►pendiente de una ladera, ►razón de pendiente); *syn.* inclinación [f] de talud, pendiente [f] de talud; *f* **angle [f] d'un/du talus** (L'angle de pente d'un talus est exprimé en degré, en pourcentage ou dans le rapport 1:n [►rapport élévation-longueur] ; ►pente d'un versant) ; *syn.* inclinaison [f] d'un/du talus, gradient

[m] de pente d'un/du talus ; *g* **Böschungswinkel [m]** (Ein **B.** wird in Grad, ein Neigungsverhältnis in 1:n ausgedrückt; ►Neigungsverhältnis einer Böschung, ►Hanggefälle); *syn.* Böschungsneigung [f].

5689 slope aspect [n] *geo. met.* (TEE 1980, 335; direction in which a slope faces with respect to points of a compass with reference to possible amounts of sunshine; ►adret, ►ubac); *syn.* slope exposure [n] [also US], slope orientation [n]; *s* **exposición [f] de una ladera** (Situación de una pendiente respecto a los puntos cardinales del horizonte; ►ladera solana, ►ladera umbría); *f* **exposition [f] d'un versant** (Désigne l'orientation d'un sol en pente, d'un talus, d'un versant, par rapport à la lumière ou son orientation par rapport à un pont cardinal ; ►adret, ►ubac) ; *g* **Hangexposition [f]** (Neigung eines Hanges zur Himmelsrichtung; ►Schattenseite, ►Sonnenseite); *syn.* Exposition [f] eines Hanges, Neigungsrichtung [f] eines Hanges, Hangausrichtung [f], Hangrichtung [f].

5690 sloped [pp] *constr. syn.* inclined [pp]; *s* **con declive [loc]** *syn.* con pendiente [loc]; *f* **avec une pente [loc]** ; *g* **mit Gefälle [loc]**.

5691 slope damage [n] *constr. geo.* (Erosion damage on an incline; ►bearing capacity failure, ►shear failure of embankment, ►soil slippage); *s* **daños [m] en ladera por erosión** (►deslizamiento del suelo, ►deslizamiento planar, ►deslizamiento rotacional [Es, RA, COL, EC, MEX, YV]); *f* **dégâts [mpl] d'érosion de versant** (Dommage engendré par l'érosion sur les terrains en pente ; ►foirage, ►rupture de terrain ; ►glissement rotationnel) ; *syn.* érosion [f] de coteau ; *g* **Hangschaden [m]** (Erosionsschaden an einem Hang; ►Geländebruch, ►Grundbruch, ►Rutschung); *syn. constr.* Schaden [m] an Erdbauwerken.

slope exposure [n] *geo. met.* ►slope aspect.

5692 slope failure [n] *constr.* (Generic term for the [partial] collapse of an embankment or hillside through the development of a shear plane, resulting in mass downslope movement of material of which the slope is composed, mostly caused by incorrect earthwork construction; ►land slide, ►rotational slip); *s* **caída [f] de un talud [Es]** (Término genérico para derrumbe [parcial] de un talud o una ladera, generalmente causado por fallos de construcción; ►corrimiento de tierras [Es, RA], ►deslizamiento rotacional [Es, RA, COL, EC, MEX, YV]); *syn.* rotura [f] de un talud [RA], falla [f] de talud [RA, CO, RCH, EC, MEX, YV] (MESU 1977); *f* **glissement [m] de talus (±)** (Terme caractérisant le mouvement souvent profond de masses de terre importantes, de couches de terrains gonflées par l'humidité et glissant sur un soubassement imperméable, une couche de sol plastique ; phénomène en général provoqué par un drainage insuffisant lors de travaux de terrassement ; ►glissement de terrain, ►glissement rotationnel) ; *g* **Hanggleiten [n, o. Pl.]** (Meist durch unsachgemäßen Erdbau verursachtes, oft tiefgründiges Abrutschen von Bodenmassen eines Hanges entlang einer natürlich geneigten bindigen Bodenschicht, die bei unzureichender Dränung wie eine Gleitschicht wirkt; ►Grundbruch, ►Hangrutschung); *syn.* Gleiten [n, o. Pl.] (LEHR 1981, 166).

5693 slope gradient [n] *constr. geo.* (Drop [US]/fall [UK] of a slope); *syn.* fall [n] of a slope [also UK], drop [n] of a slope [also US], grade [n] of a slope [also US], gradient [n] of a slope; *s* **pendiente [f] de una ladera** (Grado de inclinación de una ladera); *syn.* inclinación [f] de una ladera, gradiente [m] de una ladera [RA, RCH, EC]; *f* **pente [f] d'un versant** (Terme utilisé pour exprimer différence de hauteurs d'un versant de montagne ou d'un versant de toiture [la/le dénivelée/dénivelé] rapportée à l'unité de longueur horizontale [p. ex. 10 cm par mètre], ou par leur rapport en pourcentage [pente de 35 %] ; ne pas confondre la

S

pente avec l'inclinaison qui est la mesure de l'angle formé avec le plan horizontal, exprimée en degrés) ; *g* **Hanggefälle [n]** (Grad der Neigung einer Berg- oder Hügelseite); *syn.* Gefälle [n] eines Hanges, Hangneigung [f], Neigung [f] eines Hanges.

5694 slope [n] **of a flight of steps** *arch. constr.* (Incline dependant upon the height of the riser [▶stair rise] and the depth of the tread); *syn.* steepness [n] of a stairway *(only indoors)*, steepness [n] of a flight of steps; *s* **inclinación [f] de escalera** (Relación entre huella y ▶contrahuella que determina cuan empinada es una escalera); *f* **raideur [f] d'un escalier** (Ligne générale de pente d'un escalier définie par le rapport entre la ▶hauteur de la contremarche et la profondeur du giron et qui détermine son confort ; selon que ce rapport dépasse 1, est compris entre 3/4 et 1 ou est inférieur à 3/4, l'escalier est raide, moyen ou confortable, la hauteur de marche devant être de l'ordre de 17 cm et le giron de profondeur suffisante pour bien poser le pied à la descente [si possible supérieur à 24 cm] ; *g* **Treppenneigung [f]** (Die durch das Verhältnis von ▶Steigungshöhe und Auftrittstiefe bestimmte Steilheit einer Treppe).

5695 slope [n] **of tread** *constr.* (Inclination of each individual step in a flight of stairs/steps); *s* **pendiente [f] de peldaño** (Inclinación de cada peldaño individual de una escalera); *f* **pente [f] de la marche** (Pente de chacune des marches d'un escalier à respecter en extérieur [1,5 %] et orientée vers l'avant pour permettre l'écoulement de l'eau) ; *g* **Stufengefälle [n]** (1,5 % Neigung jeder einzelnen Stufe in einer Treppenanlage, damit das Regenwasser abfließen kann).

slope orientation [n] *geo. met.* ▶slope aspect.

5696 slope ratio [n] *constr.* (Proportion of the vertical to the horizontal or vice versa used to define the gradient of a slope mostly referred to as 1 in n or n in1; e.g. a slope that has a rise of 10 meters for every 100 meters of run would have a **s. r.** of 1 in 10 ; in U.K., gradient is usually expressed as a ratio, but also as a percentage, on <u>road signs</u>, <u>maps</u> and in <u>construction</u> work; ▶slope angle); *s* **razón [f] de pendiente (≠)** (Relación expresada en 1:n ó n:1 para indicar la inclinación de una ladera o un talud; ▶ángulo de talud); *f* **rapport [m] élévation-longueur** (Angle de pente d'un talus exprimé dans le rapport de l'élévation verticale sur la distance horizontale de 1:n ou inversement de n:1 ; ▶angle d'un/du talus ; *g* **Neigungsverhältnis [n] einer Böschung** (Der Winkel einer Böschung, ausgedrückt in einer Relation 1:n oder n:1, d. h. eine vertikale Einheit zu n horizontalen Messeinheiten resp. n horizontale Messeinheiten zu einer vertikalen; ▶Böschungswinkel).

5697 slopes [npl] [US] *constr.* (Downgrade of ground surface); *syn.* falls [npl] [UK]; *s* **pendiente [f] de una superficie;** *f* **forme [f] de pente** (d'une surface, d'un revêtement) ; *g* **Oberflächengefälle [n]** (Neigung einer Fläche).

5698 slope seepage water [n] *constr.* (Diffused discharge of groundwater to the surface of a slope; ▶groundwater seepage [US]/ground water seepage [UK]); *s* **filtración [f] efluente en ladera** (Salida difusa de agua subterránea a la superficie en la falda de un monte; ▶filtración efluente); *f* **eaux [fpl] d'écoulement hypodermique** (Eau de gravité provenant de la descente des eaux à travers le sol des versants en voie de ressuyage ; DG 1984, 152 ; ▶source de suintement) ; *g* **Hangwasser [n]** (…wässer [pl]; aus einem Hang austretendes Wasser; ▶diffuser Grundwasseraustritt).

5699 slope stabilization [n] *constr. eng.* (Erosion control on slopes using ▶biotechnical construction techniques or ▶conventional engineering methods); *s* **consolidación [f] de taludes** (Estabilización de laderas/pendientes por métodos de ▶construcción convencional o ▶construccion biotécnica); *syn.* fijación [f] de laderas/pendientes; *f* **stabilisation [f] de versant**

(Fixation des terres d'un versant par des techniques paysagères [▶technique de génie biologique] ou par des ▶techniques d'ouvrages en dur) ; *syn.* stabilisation [f] de pente ; *g* **Hangsicherung [f]** (Festlegung eines Hanges durch ▶ingenieurbiologische Bauweisen oder ▶Hartbauweisen); *syn.* Hangverbauung [f].

5700 slope wind [n] *met.* (*Specific terms* ▶downslope wind and ▶upslope air flow); *syn.* hillside breeze [n] [also US] (TGG 1984, 77); *s* **viento [m] de ladera** (Término genérico para el ▶viento anabático y el ▶viento catabático); *f* **vent [m] de versant** (Circulation thermique caractérisant les vents journaliers en montagne [vent de versant et vent de vallée] qui montent pendant la journée [▶vent anabatique] et descendent pendant la nuit [▶vent catabatique] ; ce phénomène local est dû au réchauffement différentiel entre le fond de la vallée et les versants. Plus le volume d'air en présence est petit, plus il est rapide à se réchauffer ou se refroidir ; ainsi, pendant la journée, l'air du fond de la vallée se réchauffe plus vite que celui des versants et crée un mouvement de convection ascendant ; la nuit, le phénomène s'inverse, le fond de la vallée se refroidit plus vite que les versants et le mouvement de convection est descendant) ; *g* **Hangwind [m]** (Luftbewegung an einem Hang; *UBe* ▶Hangaufwind und ▶Hangabwind).

5701 sloping [n] **down** *envir. constr.* (Lessening the slope of an embankment); *s* **reducción [f] de la pendiente de un talud** (reducir [vb] la pendiente de un talud); *f* **adoucissement [m] de la pente d'un talus** (Réduction du profil d'un talus imposée par des contraintes géotechniques ou dans un but d'insertion paysagère ; *opp.* raidissement de talus ; *syn.* réduction [f] de la pente d'un talus ; *g* **Abflachen [n, o. Pl.] von Böschungen** (Die Umwandlung von steilen Böschungen in flachere; Böschungen flacher machen [vb], abflachen [vb]).

sloping ground [n] *constr.* ▶gradually sloping ground.

5702 slot trench drain [n] *constr.* (Drainage channel enclosed on the surface except for a 8-25mm wide slit, which collects surface water runoff from paved areas; ▶slit drain, ▶trench drain [US]/channel drain [UK]); *syn.* slit drain [n] [also UK]; *s* **conducto [m] de desagüe con ranura** (▶Canal de drenaje [superficial] con perfil cerrado, utilizado para drenar caminos y plazas, que en la parte superior tiene una ranura abierta continua o discontinua de 8 a 25 mm; ▶canalillo de drenaje); *f* **caniveau [m] à fente** (▶Caniveau couvert utilisé pour l'assainissement des places et des cheminements dans lequel les eaux pluviales s'engouffrent par une fente continue ou discontinue, de largeur comprise entre 8 et 25 mm ; ▶fente de drainage) ; *syn.* satujo [m] ; *g* **Schlitzrinne [f]** (▶Entwässerungsrinne mit einem geschlossenen Profil für die Wege-, Platz- oder Fassadenentwässerung, die an der Oberseite einen durchgehenden oder unterbrochenen Einlaufschlitz aufweist; Schlitzweiten 8-25 mm; cf. DIN 19 580; je nach Fabrikat haben manche **S.n** 2-3 Einlaufschlitze; ▶Sickerschlitz).

slough [n] [US], **tidal** *geo.* ▶tidal creek.

5703 slow-growing [adj] *agr. for. hort.* (*opp.* ▶fast growing); *s* **de crecimiento lento [loc]** (*opp.* ▶de crecimiento rápido); *f* **à croissance lente [loc]** *opp.* ▶à croissance rapide ; *g* **langsam wüchsig [adj]** (Pflanzen betreffend, die langsam wachsen; *opp.* ▶schnell wüchsig); *syn.* träg wüchsig [adj].

5704 slow-moving stream [n] *geo. limn. wat'man.* *s* **río [m] de flujo lento;** *f* **eau [f] calme** (à débit lent) ; *g* **langsam fließendes Gewässer [n]**

5705 slow-release fertilizer [n] *agr. constr. for. hort.* (Mineral fertilizer which slowly sets free inorganic substances necessary for the nutrition of plants); *syn.* controlled-release fertilizer [n] [also US]; *s* **fertilizante [m] de liberación contro-**

lada (Abono que contiene nutrientes vegetales en alguna forma tal que su disponibilidad para la absorción y uso por las plantas se demora luego de su aplicación, o que su disponibilidad para las plantas se extiende por un tiempo suficiente en comparación a los «nutrientes de rápida disponibilidad», de otros fertilizantes tales como nitrato de amonio ó urea. Tal demora en su disponibilidad inicial, extendida en el tiempo o disponibilidad continua puede ocurrir por una variedad de mecanismos. Éstos incluyen la solubilidad controlada del material en agua [por una cobertura semipermeable, oclusión, o por la inherente insolubilidad en agua de polímeros, sustancias orgánicas nitrogenadas, materiales proteicos u otras formas químicas], la lenta hidrólisis de compuestos solubles de bajo peso molecular u otros mecanismos desconocidos; AAPFCO 2005); *syn.* fertilizante [m] de acción controlada, fertilizante [m] de disolución lenta, fertilizante [m] de efecto retardado, fertilizante [m] de liberación lenta; *f* **engrais [m] à décomposition lente** *syn.* engrais [m] à libération lente, engrais [m] à libération progressive, engrais [m] à libération contrôlée, engrais [m] longue durée ; *g* **langsam fließender Dünger [m]** (Mineralischer Dünger, der die zur Ernährung der Pflanzen notwendigen anorganischen Stoffe langsam abgibt); *syn.* Langzeitdünger [m], Vorratsdünger [m].

5706 slow vehicle lane [n] *trans.* (Additional traffic lane for slow-moving vehicles on steep ascents or long, downward slopes); *syn.* truck lane [n] [also US, Vt]; *s* **carril [m] lento** (Carril adicional en tramos de carretera o autopista con pendiente fuerte o con pendiente suave pero prolongada, destinado a vehículos lentos); *f* **voie [f] pour véhicules lents** (Voie supplémentaire aménagée dans les sections en rampe lorsque leur longueur et leur déclivité sont telles que la vitesse des véhicules lents est réduite à moins de 50 km/h ; ICTA 1985, 25) ; *g* **Kriechspur [f]** (Zusätzliche Fahrspur bei großen oder langen Steigungen/Neigungen für langsam fahrende Fahrzeuge).

5707 sludge [n] *envir.* (Fine-particled, fluid mixture containing organic waste substances from municipal or industrial processing; ►activated sludge, ►digested sludge, ►sewage sludge); *s* **fango [m] residual orgánico** (Residuo semilíquido resultante de procesos industriales formado por partículas finas mezcladas con sustancia orgánica; ►lodo activado, ►lodo de depuración, ►lodo digerido); *f* **boues [fpl] organiques** (Mélange fin de matières organiques en suspension et d'eau, résidu d'installations industrielles qui se développent sous forme de flocons ; ►boue activée, ►boues d'épuration, ►boues digérées) ; *g* **Schlamm [m] (3)** (Feinkörniges, fließfähiges Gemisch mit organischen Beimengungen als Abfallprodukt von Industrieanlagen; ►belebter Schlamm, ►Faulschlamm, ►Klärschlamm).

sludge [n] [US], **municipal** *envir.* ►sewage sludge.

5708 slum [n] *sociol. urb.* (Urban area in which residents live under very poor social, economic and physical conditions. **1.** Inner-city area, characterized by its run-down buildings, and correspondingly low living standards and infrastructure: inhabitants are, therefore, socially marginalized populations; in developing countries **s.**s are mostly poor, immigrant neighbourhoods on the outskirts of a city; the term was originally coined around the year 1820 in London and was used for a dwelling of a low standard. **S.**s are inhabited by low-income groups; they include urban core areas created, mainly at the beginning of industrialization, for the accommodation of workers and have not been redeveloped since. Low income housing areas with relatively dilapidated buildings are usually not designated as such; ►slum clearance); **2. squatter settlement** [n] (New, unplanned 'living quarters' on the fringes of metropolitan areas in the Third World, but also in some industrialized countries [e.g. agglomeration of Madrid], which have been illegally occupied by the poor and dispossessed; these areas are not designated in any urban development plan, the residents usually have no connection to public infrastructure; it is estimated that today approximately 1 billion people [1 in 6 persons] live in **s. s.**s—with a rising trend, since the rural exodus into the larger cities continues to increase; ►shanty town); *syn.* poverty neighborhood [n], informal settlement [n]; *s 1* **slums [m]** (Zona urbana en la que generalmente las viviendas están deterioradas, el nivel de hacinamiento es elevado, el equipamiento social es deficitario y la población residente está socialmente marginada. El término **s.** fue acuñado hacia 1820 en Londres para denominar viviendas de estándar bajo; ►barrio de chabolas); *syn.* barrios [mpl] bajos, barrio [m] pobre; *s 2* **barrio [m] marginal** (Término genérico para asentamiento urbano con malas condiciones de habitabilidad y servicios, en muchos casos creado informalmente por la población inmigrante, y cuyas casas están construidas de materiales de desecho o de materiales estables pero que están sin terminar, porque a la población frecuentemente le faltan los medios económicos para construir de una vez. Estos barrios constituyen una gran parte de las zonas de vivienda de las grandes aglomeraciones de los países en vías de desarrollo; ►barrio de chabolas); *syn.* barrio [m] informal, villa [f] miseria [RA], favela [f] [BR], campamento [m], población [f] [RCH], barrio [m] [CO], barriada [f] [CO] [PE], ciudad [f] perdida, colonia [f] proletaria [MEX], asentamiento [m] [NIC]; *s 3* **arrabal [m]** (En Es. barrio popular con vivienda de baja calidad situado en las afueras de una ciudad); *f 1* **quartier [m] misérable** (Zone occupée par une population vivant dans des conditions misérables ; ►bidonville, ►suppression des bidonvilles) ; *f 2* **quartier [m] pauvre** (Terme générique désignant les zones urbaines d'habitat précaire ou insalubre en général dans les pays en voie de développement possédant des infrastructures [eau potable, assainissement et collecte des déchets, électricité, transports] désuètes et dans lesquelles les populations vivent souvent dans des conditions d'hygiène précaire) ; *f 3* **habitat [m] indigne** (F., *terme spécifique* concept politique et non juridique qui recouvre l'ensemble des situations d'habitat qui sont un déni au droit au logement et portent atteinte à la dignité humaine tels que logements, immeubles et locaux insalubres, locaux où le plomb est accessible [risque saturnin], immeubles menaçant ruine, hôtels meublés dangereux, habitat précaire et dont la suppression ou la réhabilitation relève des pouvoirs de police administrative exercés par les maires et les préfets ; l'habitat indigne ne recouvre ni les logements inconfortables, c'est-à-dire ne disposant pas à la fois d'une salle d'eau, de toilettes intérieures, et d'un chauffage central, ni les logements vétustes — notion qui renvoie à l'entretien — ni les logements non « décents » au sens de la loi « SRU » et de son décret d'application du 30 janvier 2001 ; *termes spécifiques* ►bidonville, ►quartier informel, ►quartier misérable) ; *f 4* **quartier [m] informel** (Terme récent désignant un zone urbaine à la périphérie des grandes agglomérations des pays du tiers monde occupée par une population misérable qui n'est ni propriétaire, ni locataire du terrain et dont les constructions sont juridiquement illégales ; n'apparaissant sur aucun plan d'urbanisme, les occupants ne peuvent bénéficier des aménagements urbains tels que les réseaux routiers, les égouts, l'adduction en eau, l'électricité ; on estime aujourd'hui à un milliard les personnes vivant dans les quartiers informels, soit un habitant sur six ou un tiers de la population urbaine dans le monde, tendance croissante due à un taux de mortalité en baisse et à un fort exode des populations rurales vers les villes ; ►bidonville) ; *syn.* habitat [m] informel ; *g 1* **Slum [m]** (*Meist* Slums [pl]; innerstädtisches Wohngebiet, das durch eine heruntergekommene Bausubstanz gekennzeichnet ist, entsprechend niedrig ist der Wohnstandard und die Ausstattung mit Infrastruktur. Dort wohnen sozial marginalisierte Bevölkerungsgruppen; in Entwicklungsländern sind **S.**s meist Auffangquartiere für mittellose Zuwanderer am Stadtrand; der Begriff **S.** stammt ursprünglich aus London um das Jahr 1820 und bedeutete damals eine Wohnung mit niedrigem Standard; ▶Slumbeseitigung); *g 2* **Elendsviertel [n]** (OB für Siedlungstypen oder Wohnviertel, meist in Entwicklungsländern, von besonders niedrigem Standard. Diese werden von einkommensschwachen Bevölkerungsschichten bewohnt; cf. WAG 2001; zu den **E.n** gehören auch die zu Beginn der Industrialisierung entstandenen Kernbereiche einer Stadt, die vorwiegend für die Unterbringung von Arbeitern konzipiert waren und seitdem nicht saniert wurden. Viertel des sozialen Wohnungsbaus mit vergleichsweise heruntergekommener Bausubstanz werden i. d. R. nicht als **E.** bezeichnet. Die ***Favelas*** in Rio de Janeiro, Brasilien, z. B., sind durch Eigeninitiative der Bewohner, die zuvor aus dem Zentrum der Stadt resp. aus den auf der Ebene liegenden Stadtgebieten [sog. *asfalto*] in die Berge um die Stadt herum [sog. *morro* — Buckel, Hügel] verdrängt wurden, entstanden. Erst später wurden einige Infrastrukturmaßnahmen vom Staat bzw. von der Stadtverwaltung durchgeführt. Bis heute tauchen auf Stadtplänen [für Touristen] die *Favelas* in Rio nicht auf. Auf den Karten sind die *morros* grün gezeichnet, als ob es sich um grüne Hügel und nicht um dicht besiedelte, dürftig ausgestattete Wohnquartiere handele; *UBe* ▶Barackenstadt, ▶informelle Siedlung, ▶Slum); *syn.* Marginalsiedlung [f], Armutsviertel [n], Elendsquartier [n]; *g 3* **informelle Siedlung [f]** (Neue, ungeplante „Wohnviertel" am Rande von Ballungsräumen in der Dritten Welt, aber auch in einigen Industrieländern [z. B. Ballungsraum Madrid], die von den Armen und Besitzlosen illegal errichtet werden; diese Gebiete sind in keinem Bauleitplan vorgesehen; die Bewohner haben i. d. R. keinen Anschluss an die öffentliche Infrastruktur; man schätzt, dass heute ca. 1 Milliarde Menschen [jeder Sechste] in **i.n S.en** leben — mit steigender Tendenz, da die Landflucht zu den großen Städten zunimmt).

5709 slum clearance [n] *sociol. urb.* (Removal and replacement of slum housing through program[me]s for better dwellings; ▶slum); *s* **eliminación [f] de slums** (Derribo y sustitución de bloques de viviendas en slums por medio de programas de rehabilitación de barrios pobres); *f* **suppression [f] des bidonvilles** (Évacuation des populations des zones d'habitat insalubres, destruction des ▶quartiers misérables et construction de nouvelles cités modernes dans le cadre de programmes de réhabilitation urbaine ; *syn.* élimination [f] des bidonvilles, résorption [f] des bidonvilles, éradication [f] des bidonvilles, traitement [m] des slums ; *g* **Slumbeseitigung [f]** (Abriss der heruntergekommenen Wohngebiete und Wiederaufbau von neuen Siedlungen im Rahmen von Stadtsanierungsprogrammen; ▶Slum); *syn.* Beseitigung [f] der Slums.

5710 slum housing [n] *urb.* (Housing which, from a structural and sanitary point of view, is regarded as no longer being fit for human habitation; ▶slum); *s* **edificación [f] insalubre y ruinosa** (Edificio considerado inadecuado para la vivienda humana por su estado ruinoso y sus malas condiciones sanitarias; ▶slums); *f* **habitat [m] insalubre** (Immeuble devenu ou rendu inhabitable lorsque la substance du bâtiment ou l'état de l'équipement rendent l'utilisation des locaux dangereuse ou impossible ; ▶habitat indigne, ▶quartier misérable) ; *syn.* bâtiment [m] menaçant ruine ; *g* **Altbebauung [f] (3)** (Bebauung in baufälligem Zustand; ▶Slum).

slump [n] *constr. geo.* ▶landslip.

slump plane [n] *constr. eng.* ▶slip plane.

5711 slurry [n] (1) *envir.* (Fine-particled mixture of water and any insoluble material, such as clay, cement, etc.; ▶sludge); *s* **fango [m] residual mineral** (Residuo semilíquido resultante de procesos industriales formado por partículas finas mezcladas con sustancias anorgánicas como cal, cemento, etc.; ▶fango residual orgánico); *f* **boues [fpl] minérales** (Mélange fin de matières en suspension et d'eau retenu, p. ex. dans les chambres de dessablement ou provenant de l'épuration physico-chimique ; cf. ASS 1987, 80 ; ▶boues organiques) ; *g* **Schlamm [m] (4)** (Feinkörniges, fließfähiges Gemisch mit anorganischen Beimengungen als Ausfallstoffe, z. B. Kalk; ▶Schlamm 3).

slurry [n] (2) *hort.* ▶clay slurry; *constr.* ▶concrete slurry; *min.* ▶tailings slurry.

slurry pond [n] [UK] *min.* ▶tailings pond.

slurry spreading [n] *eng.* ▶topsoil slurry spreading.

5712 small branch [n] *arb.* (Branch with a diameter of 10-30mm; ▶sturdy branch, ▶twig); *syn.* branchlet [n]; *s* **rama [f] fina** (Rama con un diámetro de 10-30 mm; ▶rama mediana, ▶rama muy fina); *f* **rameau [m]** (Branche d'un diamètre compris entre 10 et 30 mm ; ▶branche de taille moyenne, ▶ramille); *g* **Feinast [m]** (Ast mit einem Durchmesser von 10-30 mm; ▶Feinstast, ▶Grobast).

5713 small cultural features [npl] *land.man. landsc.* (Minor buildings or artifacts e.g. chapels, shrines, historic objects or structures, which have been erected or preserved for cultural or other purposes); *s* **microestructuras [fpl] culturales** (Pequeñas construcciones u objetos como ermitas, cruces religiosas, monumentos conmemorativos, etc. creados y preservados por razones culturales); *f 1* **petit patrimoine [m] culturel** (Petits édifices tels que p. ex. chapelles, croix, calvaires, lavoirs, moulins et autres vestiges naturels ou artificiels, témoignages de la vie économique, sociale, religieuse, etc.) ; *f 2* **micro-patrimoine [m] [architectural]** (Petits éléments à mi chemin entre l'architecture et le mobilier, qui, comme les puits, les auges, les fontaines, les lavoirs, les chasse-roues, les barrières, les quais à vendange, les colonnes, les porches, les portes, jardinets, enseignes, les statues et statuettes, horloges, cadrans solaires, girouettes, marquises, bas-reliefs, etc. ponctuent le territoire urbain ou rural et l'enrichissent. Ces éléments protégés sont répertoriés sur la carte du bâti remarquable dans les ▶zones de protection du patrimoine architectural et paysager [ZPPAUP], dans lesquelles sont instituées des prescriptions particulières quant à leur protection, conservation et restauration) ; *g* **kulturelle Kleinstrukturen [fpl]** (Kleine Bauwerke in der Landschaft, z. B. Kapelle, Bildstock, Sühne- und Gedenksteine sowie andere Gegenstände, die aus natürlichen oder künstlichen Materialien von Menschen für kulturelle oder andere Zweckbestimmungen erstellt wurden).

5714 small fauna [n] *zool.* (There is no generally accepted definition of this term; it depends on the context: with regard to ecological aspects in central and northern Europe, the term includes primarily invertebrates, amphibia, reptilia, most families of singing birds *[Oscines]*, Macrochires, kingfisher *[Halcyones]* and ▶small mammals; *in general* small animals found in a particular area, specific environment, or period of time); *s* **fauna [f] menor** (No existe definición generalmente aceptada de esta categoría; depende del contexto: en el de ecología del paisaje en Europa Central y Occidental comprenden los siguientes grupos de invertebrados *[Invertebrata]*, amfibios *[Amphibia]*, reptiles *[Reptilia]*, la mayor parte de las familias de pájaros cantores *[Oscines]*, alondras *[Macrochires]*, martines pescadores *[Halcyones]* y ▶micromamíferos); *f* **monde [m] des invertébrés et des petits mammifères** (Terme d'acceptation limitée dépendant du contexte dans lequel il est utilisé ; dans le domaine de la protection de la nature en Europe du Nord et en Europe Centrale il inclut en particulier les invertébrés *[Invertebrata]*, les amphibiens *[Amphibia]*, les reptiles *[Reptilia]*, la plupart des familles des oiseaux chantants *[Oscines]*, les voiliers *[Macrochires]*, le martin-pêcheur *[Halcyones]* et les ▶petits mammifères) ; *g* **Kleintierlebewelt [f, o. Pl.]** (Es gibt keine anerkannte Abgrenzung dieses Begriffes; er ist abhängig vom

Kontext: bei landschaftsökologischen Fragestellungen in Mittel- und Nordeuropa umfasst er vor allem Wirbellose *[Invertebrata]*, Lurche *[Amphibia]*, Kriechtiere *[Reptilia]*, die meisten Familien der Singvögel *[Oscines]*, Seglerartige *[Macrochires]*, Eisvögel *[Halcyones]* und ▶Kleinsäuger).

5715 small landscape features [npl] *land'man. landsc.* (Minor elements in the landscape, e.g. hedges, groups of trees, little wetland meadows, sunken paths, mounds as well as ▶small cultural features); *syn.* field features [npl] [also UK]; *s* **microestructuras [fpl] del paisaje** (Pequeños componentes del paisaje, como setos, grupos de árboles, prados húmedos, caminos hondos, etc. asi como ▶microestructuras culturales); *f* **microstructures [fpl] du paysage** (Structures de grandeur limitée, éléments isolés ayant un rôle structurant dans le paysage telles que les haies, groupes d'arbres, prairies humides, chemins creux, mamelons, ▶micro patrimoine [architectural], ▶petit patrimoine culturel) ; *g* **Kleinstrukturen [fpl] in der Landschaft** (Kleine Landschaftsbestandteile, wie z. B. Hecke, Baumgruppe, Feuchtwiese, Hohlweg, Kuppe sowie ▶kulturelle Kleinstrukturen); *syn.* landschaftliche Mikrostrukturen [fpl].

5716 small mammal [n] *zool.* (An ill defined generic term in zoological literature for certain Mammalia in the relevant zoological literature, which can include the following animal groups within the scope of ▶nature conservation and landscape management: insectivores *[Insectivora]*, bats *[Chiroptera]*, Lagomorpha without hares, rodents *[Rodentia]* without beavers] and from the carnivores the family of martens *[Mustelidae]* without badgers *[Meles]* and otters *[Lutra]*); *s* **micromamífero [m]** (Término no definido claramente: en la ▶protección de la naturaleza se incluyen generalmente los siguientes grupos: insectívoros *[Insectivora]*, murciélagos *[Chiroptera]*, Lagomorpha sin la liebre, roedores *[Rodentia]* sin el castor *[Castor]* y entre los carnívoros *[Carnivora]* la familia de los mustélidos *[Mustelidae]* sin el tejón *[Meles]* ni la nutria *[Lutra]*); *syn.* mamífero [m] menor; *f* **petits mammifères [mpl]** (Terme générique à la définition incertaine qui dans le domaine de la protection de la nature peut englober les ordres et sous ordres d'espèces animales suivantes : les insectivores *[Insectivora]*, les chiroptères *[chiroptera]*, chez les rongeurs les sous-ordres des lagomorphes *[lagomorpha]* sans le lièvre et des rodentiés *[rodentia]* sans le castor *[Castor]* et chez les carnivores *[Carnivora]* la famille des mustélidés *[mustelidae]* sans le blaireau *[Meles]* et la loutre *[Lutra]*) ; *syn.* micromammifère [m] ; *g* **Kleinsäuger [mpl]** (In der Literatur nicht eindeutig abgegrenzter OB, der im Bereich von ▶Naturschutz und Landschaftspflege folgende Tierordnungen und Unterordnungen umfassen kann: Insektenfresser *[Insectivora]*, Fledermäuse *[Chiroptera]*, Hasenartige *[Lagomorpha* ohne Hasen], Nagetiere *[Rodentia* ohne Biber *(Castor)]* und von den Raubtieren *[Carnivora]* die Familie der Marderartigen *[Mustelidae* ohne Dachse *(Meles)]* und Fischotter *[Lutra]*).

small paving sett [n] [UK] *constr.* ▶small paving stone [US].

5717 small paving stone [n] [US] *constr.* (**In U.S.**, typical **s. p. s.s** range from 80-100mm on a side and vary from square to rectangular; Referred to as setts **in U.K.**, small natural stone paving blocks are usually, but not exclusively, granite or basalt, and are often referred to as "cubes"; they are more or less regularly sized, usually 100x100x100mm or 125x125x125mm. Both new and reclaimed igneous setts are finished to one of the Classes given in BS EN 1342:2000 *Specification for dressed natural stone kerbs, channels, quadrants and setts*: fine picked, fair picked or rough punched; sizes of granite cubes are 50x50x 50mm, 80x80x80mm, 100x100x100mm, 150x150x 150mm; sizes for granite setts: 100x100x50mm, 200x75x150mm, 200x 150x100mm, 300x100x200mm,and sizes for gritstone setts:

150x125x100mm, 100x150x150mm, 275x100x200mm; **in D.**, according to DIN 18 502, natural stone pavers with sides measuring 80-100mm on a side; ▶mosaic paver); *syn.* small paving sett [n] [UK], small stone paver [n] [US]; *s* **adoquín [m] pequeño** (Según la norma alemana DIN 18 502, piedra natural para pavimentos con caras vistas laterales entre 80-100 mm; en Es. este tamaño de adoquín corresponde aprox. al ▶adoquín mosaico; cf. PGT 1987, 118); *f* **petit pavé [m]** (Classe de pavé en roche naturelle dont les dimensions et les tolérances dimensionnelles sont normalisées ; **F.**, conforme à la norme N.F. P 98 401 — pavés et bordures de trottoir ; **D.**, à la norme DIN 18 502 ; ▶pavé mosaïque en pierre naturelle) ; *g* **Kleinpflasterstein [m]** (Nach DIN 18 502 beschriebene Güteklassen und Größeneinteilungen für Natursteinpflastersteine mit Kantenlängen von 80-100 mm; ▶Mosaikpflasterstein).

5718 small playing field [n] *recr.* (Paved or grassed sports ground for various ball games with a standard size of 20 x 40m in D.); *s* **campo [m] de deportes pequeño** (Superficie pavimentada o de césped para juegos de pelota, en D. con un tamaño estándar de 20 x 40 m); *f* **terrain [m] de jeux réduit** (Espace de jeux aux dimensions réduites, stabilisé ou engazonné [**D.**, dimension standard 20 x 40 m] prévu pour divers jeux de ballon ; *en F.* terminologie variée p. ex. espace multisports, espace sportif, espace sportif polyvalent, etc.) ; *g* **Kleinspielfeld [n]** (**D.**, befestigter oder Rasenplatz für diverse Ballspiele mit einer Normgröße von 20 x 40 m).

small-sedges fen [n] [UK] *phyt.* ▶low sedge swamp.

small sett paving [n] [UK] *constr.* ▶small stone paving [US].

5719 small shrub [n] *hort. plant.* (Shrub usually shorter than 1.50m [5 ft]; ▶dwarf shrub; *opp.* ▶large shrub); *s* **pequeño arbusto [m]** (Término no claramente definido; ▶mata; *opp.* ▶arbusto grande); *f* **petit arbuste [m]** (Terme non clairement défini ; en général arbuste pouvant atteindre une hauteur de 1,5 m ; ▶arbuste nain ; *opp.* ▶grand arbrisseau) ; *g* **Kleinstrauch [m]** (...sträucher [pl]; kein klar definierter Begriff; i. d. R. ein Strauch bis zu einer Höhe von 1,50 m; ▶Zwergstrauch; *opp.* ▶Großstrauch).

small stone paver [n] [US] *constr.* ▶small paving stone [US].

5720 small stone paving [n] [US] *constr.* (Paving consisting of small stone pavers [US]/stone setts [UK] [natural stones] or concrete blocks [US]); *syn.* small sett paving [n] [UK]; *s* **adoquinado [m] pequeño** (Superficie revestida con adoquines pequeños de piedra natural u hormigón); *f* **pavage [m] en petits pavés** (Revêtement réalisé avec des petits pavés en béton ou en pierre naturelle) ; *g* **Kleinpflaster [n]** (Mit Kleinpflastersteinen aus Naturstein oder Beton verlegter Belag).

small stump [n] [UK] *arb. for. hort.* ▶snag [US&CDN] (2).

5721 small waterbody [n] **or watercourse** [n] *geo. land'man.* (Generic term covering small stream, pond, ▶manmade pond, ▶natural pond, pool, fishing pond, small flooded gravel pit, etc.); *s* **aguas [fpl] continentales menores** (Término genérico para arroyo, ▶estanque [artificial], ▶estanque natural, charca, etc.); *f* **petits cours [mpl] d'eau et petits plans d'eau** (±) (**D.**, terme générique désignant les rus, ruisseaux, mares, ▶étangs naturels et les ▶étangs artificiels, petits lacs de gravières, etc.) ; *g* **Kleingewässer [n]** (OB zu Bach, ▶Teich, Tümpel, ▶Weiher, kleiner Baggersee etc.).

small wood [n] *landsc.* ▶woodlot [US&CDN] (2).

5722 smog [n] *envir.* (*Portmanteau word combining the two words 'smoke' and 'fog'*; heavy air pollution from gaseous, liquid and solid particles over an urban or industrial area which occurs during ▶inversion weather; e.g. **London smog**: primarily soot

S

and sulphur dioxide. **Los Angeles smog**: photochemically produced and called 'photochemical smog'; ▶summer smog); **s smog [m]** (Tipo de contaminación atmosférica caracterizada por la formación de nieblas que contienen sustancias tóxicas para la salud. El término proviene de la contracción de dos palabras inglesas «smoke» [humo] y «fog» [niebla]. Su formación se debe a la combinación de procesos meteorológicos [▶condiciones de inversión térmica] y geográficos además de la presencia de gran cantidad de contaminantes en la atmósfera. Se diferencian el **smog ácido** o **londinense**, caracterizado por una alta presencia de SO_2 y de cenizas, que es típico de ciudades frías y con tendencia a la niebla, y el ▶**smog oxidante** o **de Los Ángeles**, que se produce cuando hay una gran cantidad de oxidantes fotoquímicos en la atmósfera, como el NO_2 causado principalmente por las emisiones de los vehículos, que conllevan la formación de ozono); *syn.* niebla [f] sucia; *f* **smog [m]** (Formation nuageuse toxique engendrée par des gouttelettes d'eau et des poussières industrielles ou urbaines [gaz de combustion] formées au-dessus des grandes agglomérations ou des zones industrielles lors de ▶conditions d'inversion thermique ; on distingue le **smog londonien** [formé principalement à partir de fumées de combustion et dioxyde de soufre] et le ▶smog d'été appelé **smog de Los-Angeles** [smog photochimique]) ; *g* **Smog [m]** (Kurzwort aus dem Englische *smoke* = Rauch und *fog* = Nebel. Starke Luftverunreinigung aus gasförmigen, flüssigen und festen Bestandteilen über städtischen oder industriellen Ballungsräumen bei ▶Inversionswetterlagen. Es werden der **London-** [vorwiegend Ruß und Schwefeldioxid] und der **Los-Angeles-S.** [fotochemischer S. oder ▶Sommersmog] unterschieden. Der Begriff wurde erstmals 1905 von H. A. Des Voeux benutzt; cf. ENCYCLOPEDIA BRITANNICA 1905).

5723 smoldering [n] *envir.* (Combustion without flame in, e.g. waste dumps [US]/waste tips [UK]); *syn.* smouldering [n]; **s combustión [f] sin llamas** (Combustión lenta e incompleta a baja temperatura que no produce llamas, pero a veces gran cantidad de humo, p. ej. en depósito de residuos); *f* **feu [m] couvant** (Feu brûlant sans flamme [avec fumée importante], p. ex. incendie de tourbière, sur une décharge) ; *g* **Schwelbrand [m]** (1. Ohne offene Flamme, langsam dahin glimmende, unvollständige Verbrennung [z. T. mit großer Rauchentwicklung] bei niedriger Temperatur und ungenügender Sauerstoffzufuhr, z. B. in Deponien. 2. Das [kontrollierte] Verbrennen von Weidegras der Graslandvegetation in der Brandwirtschaft).

smonitza [n] *pedol.* ∗▶vertisol.

5724 smooth-faced concrete [n] *constr.* (Concrete finish created by non-textured metal forms, metal troweling, or burnishing prior to curing); *syn.* fair-faced concrete [n] [also UK]; **s hormigón [m] alisado** *syn.* hormigón [m] allanado; *f* **béton [m] banché** (Béton coulé dans des coffrages verticaux [banches] pour former à son emplacement définitif dans la construction un mur en béton banché) ; *syn.* béton [m] lisse ; *g* **glatter, strukturloser Sichtbeton [m]**.

smouldering [n] *envir.* ▶smoldering.

5725 snag [n] (1) *arb. for. hort.* (**1. U.K.**, jagged point of a broken, small, thin branch, often dead; BS 3975: part 5; ▶branch stub [US]/branch stump [UK]; *syn.* spike [n] [UK], spine [n] (1). **2. U.S.**, the projecting base of a broken or cut branch on a tree stem); **s muñón [m]** (Resto de una ramilla rota, frecuentemente muerta; HDS 1987; ▶tocon de rama); *syn.* garrón [m] (HDS 1987); *f* **moignon [m]** (Reste d'une branche d'environ 1 m, coupée dans un deuxième temps afin d'éviter l'éclatement du bois du tronc lors de la coupe de grosses branches ; ▶chicot 1) ; *g* **dünner Aststummel [m]** (Am Stamm herausragender Rest

eines abgebrochenen oder nicht fachgerecht abgeschnittenen Kleinastes; ▶Aststummel).

5726 snag [n] [US&CDN] (2) *arb. for. hort.* (Standing section of a tree stem, broken off at a height of more than 6m [20 ft], from which the leaves and most of the branches have fallen. *NOTE* if less than 6m [20 ft], it is called a **stub**; cf. SAF 1983; ▶snag 3 [US&CDN]); *syn. for stub* small stump [n] [UK]; **s tocón [m]** (1) (Resto de tronco de árbol, desgarrado a una altura de más de 6 m, del cual se han caido la mayor parte de las ramas y hojas; en EE.UU. y Canadá si el tocón [snag] mide menos de 6 m se le llama «stub»); *f* **chicot [m]** (2) (Arbre ou partie d'un arbre rompu sous l'action du vent et resté sur pied et qui se décompose ; en général, important comme habitat de certaines espèces sauvages, comme les oiseaux qui nichent dans des cavités ; CDN, on distingue **a] les chicots durs** — gros et grands arbres morts ayant encore beaucoup de branches au sommet et dont l'état de décomposition est peu avancé, **b] les chicots mous** — arbres morts depuis plusieurs années et leur état de décomposition est plus avancé, **c] les chicots vivants** — arbres de taille variable qui partiellement vivants et dont une bonne partie de la cime est dégarnie ; ▶arbre mort debout, ▶arbre mort par terre) ; *syn.* tronc [m] brisé ; *g* **abgebrochener Baumstamm [m]** (Durch Windbruch geschädigter Baum, dessen Krone in einer Höhe von mehreren Metern abgedreht wurde oder schlicht abbrach; ▶Totbaum).

5727 snag [n] [US&CDN] (3) *ecol.* (**1.** Standing dead tree from which the leaves and most of the branches have fallen; SAF 1983. **2.** Dead tree trunk laying down); *syn. for* 1. dead (standing) tree [n], dotard [n] [UK]; **s 1 árbol [m] muerto de pie** (Árbol sin ramas o con ramas tronchadas); **s 2 árbol [m] tumbado** (Árbol muerto que yace sobre el suelo y se encuentra en proceso de descomposición); *f 1* **arbre [m] mort debout** (Arbre mort érigé dont il ne reste plus que le tronc et quelques branches maîtresses) ; *syn.* arbre [m] mort sur pied ; *f 2* **arbre [m] mort par/à terre** (Arbre mort couché sur le sol) ; *g* **Totbaum [m]** (**1.** Abgestorbener Baum, von dem nur noch der Stamm mit einigen Starkästen steht. **2.** Umgestürzter toter Baum/Totbaum; alter, starker Baumstamm, der ungenutzt im Wald liegt und verfault, wird auch *Urholz [n]* oder *Rohne [f] (obs.)* genannt; cf. WFL 2002, 601, 747); *syn.* toter Baum [m].

5728 snag [n] (4) *wat. man.* (Tree or branch, log, or stump embedded and held fast in a lake or stream bed in such a manner that projecting parts often form a hazard to boats); **s obstáculo [m]** (±) (Árbol, rama, tronco o tocón incrustado en el fondo de un río o lago de manera que partes del mismo sobresalen y pueden ser un peligro para la navegación; ▶madera flotante); *f 1* **amas [m] de bois** (Bois abandonné sur les berges d'un cours d'eau, d'un lac ou sur l'estran ; ▶bois flottant ; *termes spécifiques pisc.* amas de bois flotté, amas de bois noyé) ; *f 2* **embâcle [m] de bois** (Débris ligneux grossiers [branches, troncs, arbres tombés dans le lit, bouchon de végétaux morts] encombrant le lit mineur d'un cours d'eau, qui se bloquent sur les branches basses ou se déposent sur un atterrissement, constituant un barrage ou un frein à l'écoulement et pouvant provoquer une gène pour la circulation fluviale ou des dégâts exceptionnels par rupture d'embâcle) ; *syn.* bois [m] formant embâcle ; *g* **Holzhindernis [n]** (Auf dem Wasser treibender oder aus dem Wasser eines Gewässerlaufes herausragender Ast, Baum, Holzstamm oder Stubben, der für die Schifffahrt eine Gefahr darstellen kann; ▶Treibholz).

snagging list [n] [UK] *constr. contr.* ▶punch list [US].

snecked rubble wall [n] [UK] *arch. constr.* ▶broken range work.

snig track [n] [AUS] *for.* ▶skidding road [US&CDN]/skidding lane [UK].

snowbank [n] *landsc. trans.* ▶snowdrift.

5729 snowbed community [n] *phyt.* (Alpine plant association in small snowpockets with a longer growing period than a ▶snowland community); *syn.* snowpocket association [n]; *s* **formación** [f] **de nevero** (*En geobotánica* vegetación fanerogámica cuyos gérmenes se conservan en invierno bajo la nieve y cuyos brotes asoman y florecen en la época en que ésta se funde; DB 1985; ▶comunidad de ventisqueros); *syn.* vegetación [f] de nevero; *f* **groupement** [m] **de combe à neige** (Peuplement de l'étage nival qui par comparaison avec les ▶groupements de mode nival subit l'influence d'un enneigement court ; appartient aux alliances de l'*Arabidion caeruleae* [substrat carbonaté] et du *Salicion herbaceae* [substrat acide]) ; *g* **Schneetälchengesellschaft** [f] (Pflanzengesellschaft, die im Vergleich zur ▶Schneebodengesellschaft eine längere Vegetationsperiode hat, da kleine Schneeflächen der Schneetälchen nicht so lange liegen bleiben).

5730 snowbreak [n] *landsc. trans.* (Shelterbelt, windbreak, or other barrier—natural or artificial—constructed to trap blowing snow and thereby prevent excessive accumulations on roadways, etc.; SAF 1983); *s* **paranieves** [m] (Plantación protectora o valla paranieves a lo largo de las carreteras para evitar la acumulación de nieve en éstas); *syn.* protección [f] paranieves; *f* **protection** [f] **contre les congères** (*Terme spécifique* plantations ou dispositifs de protection contre la neige [p. ex. filet pare-neige, barrière de retenue de neige, barrière pare neige à claire-voie] placés le long des routes afin d'amasser la neige du côté sous le vent et d'éviter la formation de congères) ; *g* **Schneeschutz** [m] (Schutzpflanzung oder seitlich an Straßen aufgestellte Zäune und Gatter zur Verhinderung von Schneeverwehungen); *syn.* Verwehungsschutz [m].

5731 snowbreakage [n] *arb. for.* (Breaking of trees or branches by snow; ▶snowbreak); *s* **rotura** [f] **por la nieve** (Daños en leñosas causados por el peso excesivo de la nieve; ▶paranieves); *f* **bris** [m] **de neige** (Toute cassure d'un arbre ou d'une partie d'arbre, provoquée par une accumulation de neige ; ▶protection contre les congères) ; *g* **Schneebruch** [m] (Bruchschäden an Bäumen durch übermäßige Schneebelastung; ▶Schneeschutz).

5732 snow clearing [n] *trans.* (Mechanical removal of snow from traffic areas without resorting to the use of thawing substances; *opp.* ▶road salting; ▶de-icing salt); *syn.* snowplowing [n] [also US]; *s* **limpia** [f] **mecánica de nieve** (Eliminación mecánica de la nieve de las carreteras y calles, sin utilizar ▶sal anticongelante; *opp.* ▶eliminación de la nieve con sal); *f* **déneigement** [m] **mécanique** (Enlèvement mécanique de la neige sur la voirie sans utilisation de produits de salage ; *opp.* ▶salage, ▶sel de déneigement) ; *g* **Weißräumung** [f] (Mechanische Entfernung von Schnee auf Verkehrsflächen, ohne anschließenden Auftaumitteleinsatz; *opp.* ▶Schwarzräumung; ▶Streusalz); *syn.* Schneeräumung [f].

5733 snowdrift [n] *landsc. trans.* (Mass of snow piled up by wind); *syn.* snowbank [n]; *s 1* **remolino** [m] **de nieve** (Proceso); *syn.* torbellino [m] de nieve; *s 2* **acumulación** [f] **de nieve** (Resultado); *syn.* montón [m] de nieve; *f 1* **tourmente** [f] **de neige** (Tourbillon violent et prolongé de neige) ; *syn.* tourbillons [mpl] de neige ; *f 2* **congère** [f] (Accumulation importante de neige provoquée par le vent derrière un obstacle) ; *g* **Schneeverwehung** [f] (1. Vorgang der Schneeanhäufung durch Windverfrachtung. 2. Durch Wind aufgehäufte, große Menge Schnee. 3. Eine Schneeanhäufung im Gebirge auf der Leeseite einer Geländekante oder eines Kammes heißt **(Schnee)wechte**, *frühere Schreibweise* Wächte); *syn. zu 2.* Schneewechte [f] [A], Schneewehe [f].

5734 snow fence [n] *landsc. trans.* (Temporary, light-weight lath and wire barrier erected to prevent drifting of snow, e.g. onto a road; ▶snowbreak); *s* **valla** [f] **paranieves** (Valla colocada temporalmente a lo largo de carreteras para evitar acumulaciones de nieve; ▶paranieves); *f* **clôture** [f] **pare-neige** (Clôture en bois ou filet plastique de ▶protection contre les congères) ; *g* **Schneezaun** [m] (Temporärer Zaun, meist aus Kunststoffnetzen, der dem ▶Schneeschutz dient; aufgestellte Lattenroste heißen **Schneegatter**); *syn.* Schneefangzaun [m].

5735 snowland community [n] *phyt.* (Plant association which grows in surface depressions in alpine regions, where snow often covers large areas until July or the beginning of August, thus allowing only a short growing period; ▶snowbed community); *s* **comunidad** [f] **de ventisqueros** (Comunidad vegetal del piso alpino superior que se encuentra cubierta por la nieve de 8 a 10 meses al año y está constituida principalmente por musgos, hepáticas, hemicriptófitos y caméfitos reptantes; BB 1979; ▶formación de nevero); *f* **groupement** [m] **de mode nival** (Association de l'étage alpin se développant dans les combes d'ubac mal ensoleillées, des petites dépressions ou en pied de versant et soumis à un long enneigement ; ces groupements sont soumis à des contraintes extrêmes : courte saison végétative, températures basses qui ralentissent la croissance, poids de la neige, sols froids, humides ou hydromorphes en profondeur ; *terme spécifique* pelouse de mode nival ; ▶groupement de combes à neige) ; *g* **Schneebodengesellschaft** [f] (Pflanzengesellschaft, die in Rinnen oder Hangfußmulden der alpinen Stufe, in denen der Schnee oft bis Juli/Anfang August großflächig liegen bleibt, wächst und somit im Vergleich zur ▶Schneetälchengesellschaft eine kürzere Vegetationszeit hat).

5736 snow line [n] *geo.* (Lowest altitudinal limit of the region of perpetual snow; SAF 1983); *s* **límite** [m] **de la nieve** (Desde el punto de vista climático, es la menor altitud a la que puede subsistir la capa de nieve en la alta montaña durante el verano; orográficamente, es la menor altitud en la que pueda persistir, en verano, la nieve sobre la superficie terrestre, en forma de manchas aisladas que deben su persistencia a condiciones orográficas locales favorables; DM 1986); *syn.* cota [f] de la nieve; *f* **limite** [f] **des neiges persistantes** (Ligne idéale désignant l'altitude la plus basse à laquelle la neige persiste en été sous forme de plaques isolées [neige pérenne] et qui défini théoriquement l'étage nival ; VOG 1979 ; *syn.* limite [f] inférieure des neiges, limite [f] des neiges permanentes) ; *g* **Schneegrenze** [f] (1. Grenze zwischen schneefreiem und auch im Sommer schneebedecktem Gebiet; die **Sch.** wird durch die Mitteltemperaturen und Niederschlagsverhältnisse einer jeweiligen Klimazone bestimmt; es wird zwischen **temporärer [Schneefallgrenze]** und **klimatischer Sch.** unterschieden. Oberhalb der klimatischen **Sch.** fällt im Mittel mehr Schnee als abtaut. 2. Die **Firngrenze** [Firn ist alter, mehrjähriger Schnee] auf Gletschern liegt wegen der abkühlenden Wirkung des Eises ca. 100 m tiefer als die **Sch.** in unvergletscherten Nachbarbereichen).

5737 snow load [n] *arb. constr.* (Additional pressure caused by the weight of snow on trees, roofs, etc.); *s* **carga** [f] **de nieve** (Acción estática de la nieve sobre ramas de árboles, tejados, etc.); *f* **surcharge** [f] **de neige** (Action de la neige tombée sur des arbres, des toitures, etc., statiquement, en grande masse) ; *g* **Schneelast** [f] (Durch aufliegenden Schnee entstehende Last, z. B. auf Bäumen, Dächern).

snowplowing [n] [US] *trans.* ▶snow clearing.

snowpocket association [n] *phyt.* ▶snowbed community.

soak [n] [EIRE] *geo. phyt.* ▶bog lake.

soakaway [n] [UK] *constr.* ▶drywell sump [US].

soaking [n] *hort.* ▶watering.

S

soap [n] *arch. constr.* ►queen closer.

social benefits [npl] **of open areas** *recr.* (Positive effects which green areas exert upon the physical and psychological well-being of humans by; e.g. shaded areas, dust-filtering capacity, beauty of flowers, verdancy, opportunities for sport, play, and relaxation); *s* **función** [f] **beneficiosa (de zonas verdes)** (Efecto positivo de los espacios libres con vegetación debido a su influencia favorable sobre el estado físico y [p]síquico de las personas, p. ej. a través de la filtración del polvo, el gozo estético de la vegetación y de las flores, posibilidades de descanso y de juego); *f* **fonction** [f] **sanitaire et sociale des espaces verts** (Effets bénéfiques des espaces de végétation sur le bien-être psychique et physique de l'homme p. ex. les aires d'ombrage, la filtration des poussières, la richesse florale, la verdure, les activités ludiques, le repos et la détente) ; *syn.* rôle [m] social des espaces verts ; *g* **Wohlfahrtsfunktion** [f] **von Grünanlagen** (Positive Wirkung von vegetationsbestimmten Freiräumen auf das psychische und physische Befinden der Menschen, z. B. durch Schattenplätze, Staubfilterung, Blütenpracht, grüne Vegetation, Spiel- und Ruhemöglichkeiten); *syn.* Sozialfunktion [f] von Grünanlagen.

5739 social benefits [npl] **of the forest** *land'man.* (All of the positive influences which a forest exerts upon the ►landscape ecosystem. These benefits include the effects of a forest upon physical living conditions and upon the well-being of humans, besides its being a source of raw materials; ►amenity benefits of a forest, ►long-term benefits); *syn.* beneficial effects [npl] of the forest, beneficial influences [npl] of the forest; *s* **efectos** [mpl] **beneficiosos del bosque** (Todos aquellos efectos más allá de los productivos, sobre todo los referentes al ►régimen ecológico del paisaje y al desarrollo del paisaje y los que influyen en las condiciones de vida de la población; ►efecto a largo plazo, ►efectos [p]sicológicos del bosque); *syn.* efectos [mpl] saludables del bosque; *f* **fonction** [f] **sanitaire et sociale de la forêt** (Au-delà de la fonction de production, ensemble des multiples usages de la forêt qui influencent les ►écosystèmes paysagers, la santé corporelle et morale de l'homme ; ►effet à long terme, ►rôle de la forêt sur la santé corporelle et l'équilibre psychique) ; *syn.* rôle [m] bénéfique de la forêt (LA 1981, 319), utilité [f] sociale et récréative de la forêt ; *g* **Wohlfahrtswirkungen** [fpl] **des Waldes** (Alle über die Rohstoffproduktion hinausgehenden Wirkungen des Waldes, die besonders den ►Landschaftshaushalt und die Landschaftsentwicklung [landeskulturelle Wirkungen] sowie die materiellen Lebensbedingungen und die Bewusstseinsbildung der Menschen positiv beeinflussen; ►Langzeitwirkung, ►umweltpsychologische Wohlfahrtswirkungen des Waldes).

5740 social obligation [n] **of private property** *leg. pol. sociol.* (Legal principle limiting property rights on private land for public benefit; ►easement, ►utility easement); *s* **obligación** [f] **social de la propiedad** (En D. según art. 14 [1] 2 de la Constitución la propiedad privada conlleva responsabilidad de velar por el bien común. En Es. la Constitución Española de 1978, que entró en vigor el 29.12.1978, reconoce el derecho a [p]spropiedad privada y a la herencia [art. 33.1], determina que la función social de estos derechos delimitará su contenido, de acuerdo con las Leyes [art. 33.2] y que nadie podrá ser privado de sus bienes y derechos sino por causa justificada de utilidad pública o interés social, mediante la correspondiente indemnización y de conformidad con lo dispuesto por las Leyes [art. 33.3]; ►servidumbre de paso de conducciones, ►servidumbre inmobiliaria); *syn.* compromiso [m] social de la propiedad; *f* **responsabilité** [f] **sociale de la propriété** (±) (**D.**, principe selon lequel la propriété est tenu de servir le bien-être de la communauté et ne peut donc être utilisé comme pouvoir de domination absolue ;

l'article 14 de la constitution allemande actuelle de 1947, fait de la propriété une institution sociale imposant à tout propriétaire d'exercer son droit conformément à la nature de la chose qui lui appartient en respectant l'intérêt général de la société. Les restrictions au droit de propriété lui sont inhérentes ; cf. Art. 4 [1] 2 de la loi fondamentale de RFA ; **F.**, l'article 17 de la constitution proclame la propriété comme un droit inviolable et sacré et que nul ne peut en être privé, sauf si une nécessité publique, légalement constatée, l'exige et sous réserve d'une juste et préalable indemnité ; dans la pratique, notamment en matière d'urbanisme [contrôle de la propriété privée par l'intérêt général] le caractère absolu du droit de propriété est limité par le droit positif qui tend de plus en plus à organiser un partage de l'usage du sol entre le propriétaire et la collectivité. Le principe constitutionnel allemand rejoint la notion nouvelle de **responsabilité sociale de l'entreprise** [RSE] en relation avec la propriété privée des moyens de production, traduction pour l'entreprise des concepts de développement durable, intégrant les trois piliers environnementaux, sociaux, et économiques ; ►servitude, ►servitudes relatives à l'utilisation des réseaux d'alimentation et de distribution d'énergie, de canalisation et de télécommunication ; *syn.* en L. obligations [fpl] sociales découlant de la propriété, fonction [f] sociale de la propriété) ; *g* **Sozialpflichtigkeit** [f, o. Pl.] **des Eigentums** (**1.** Nach deutschem Recht ist das Eigentum dem Wohl der Allgemeinheit verpflichtet und kann somit nicht als uneingeschränktes Herrschaftsrecht aufgefasst werden; cf. Art. 14 [1] 2 GG. **2.** Im regionalplanerischen Sinne kann die gemeindliche Selbstverwaltung der Kommunen eingeschränkt werden, wenn diese auf Grund ihrer standortgünstigen geografischen Lage als Vorrangstandort für z. B. großindustrielle Anlagen einer gewissen **Sozialgebundenheit** unterliegen; cf. BAUR 1999, 959; ►Grunddienstbarkeit, ►Leitungsrecht); *syn.* Sozialbindung [f] des Eigentums, soziale Bindung [f] des Eigentums.

5741 social service facilities [npl] *leg. urb.* (Public facilities defined in a legally-binding land-use plan, such as a ►zoning map [US]/Proposals Map [UK] for ►communal use; e.g. school, hospital); *syn.* communal facilities [npl]; *s* **equipamientos** [mpl] **sociales** (En Es. como norma supletoria, emplazamiento a determinar en ►plan parcial [de ordenación] reservado para centros asistenciales y sanitarios y demás servicios de interés público y social; art. 13d RD 1346/1976 Texto refundido); *f* **bâtiments et équipements** [mpl] **à vocation sociale** (±) (**D.**, dans le cadre du plan local d'urbanisme allemand, aire affectée aux équipements collectifs et aux ►services collectifs) ; *g* sozialen Zwecken dienende Gebäude [npl] und Einrichtungen [fpl] (Im ►Bebauungsplan auszuweisende Flächen für den ►Gemeinbedarf; cf. Anlage zur PlanzV 90, Kap. 4).

5742 socio-economic factors [npl] *plan. sociol.* (Background information used in interpreting, e.g. behavio[u]ral patterns of recreation or traffic); *s* **datos** [mpl] **socioeconómicos** (Conjunto de informaciones económicas y sociales necesarias para conocer el comportamiento resp. las necesidades de la población en campos como recreación, vivienda, transportes, etc.); *f* **facteurs** [mpl] **socio-économiques** (Données économiques ou sociologiques prises en compte pour déterminer les comportements relatifs de loisirs, d'habitat, de transports, etc.) ; *g* **sozioökonomische Größen** [fpl] (Ökonomische oder soziologische Daten, die zur Erklärung für Freizeit-, Wohn-, Verkehrsverhalten etc. herangezogen werden).

socket pipe [n] [UK] *constr.* ►bell pipe [US].

5743 sod [n] [US] (1) *constr. hort.* (**1.** Pre-grown piece of rooted grass mat, approximately 600 x 1500 x 20mm [2 ft x 5 ft x ¾ in], pared off for transporting to a landscape construction site; **in U.K.**, the turf is generally cut to the size of one square metre [1,640mm long and 610mm wide]; thickness of uncompressed thatch should not exceed 10mm; the soil layer beneath the thatch should be between 5 and 15mm deep; the most common type of cultivated turf is the roll); *syn.* turf [n] [UK]; **2. thick-cut sod [n]** or **extra thick sod [n]** (Sod with a thickness of more than 30mm, is used for sports fields; due to the weight of the sods, they can be played upon immediately, even without their rooting; the turf is delivered in extra-large rolls; widths vary according to the supplier from 60cm to 200cm wide and in lengths of 8-14m, or as required; weighing up to 800 kg, the jumbo rolls are installed with very close joints by special machines; ►replacement of sods [US]/replacement of turves [UK], ►rolled turf, ►sod layer [US]/turf layer [UK], ►turfing 1); *s 1* **tepe [m]** (Porción de tierra cubierta de césped, muy trabada por las raíces, que se corta en forma generalmente rectangular para colocarla en otro sitio; RA 1970; ►césped de tepes, ►césped de tepes enrollados, ►recolocación de tepes o tierra vegetal, ►revestimiento con tepes); *syn.* pan [m] [C], alfombra [f] [C]; *s 2* **tepe [m] grueso** (Tipo de tepe de grosor superior a 30 mm que se usa principalmente para campos de deporte. Debido al propio peso pueden utilizarse inmediatamente, antes de enraizar. Este tipo de tepe se suministra en grandes rollos, cuyo ancho varía según la empresa sumistradora de 60-200 cm y el largo de 8-14 m o según se necesite. El peso puede ascender hasta 800 kg. Estos grandes rollos son instalados con juntas muy estrechas con ayuda de máquinas especiales); *f 1* **plaque [f] de gazon** (Gazon de ►placage, en général de dimension 0,40 x 0,40 m, d'épaisseur comprise entre 2,5 et 4 cm ; ►gazon précultivé ; **D.**, conformément à la norme DIN 18 917 ; ►gazon en rouleaux précultivé, ►remise en place des plaques de gazon) ; *f 2* **rouleau [m] épais** (*Terme spécifique* gazon de placage utilisé pour la création, la réparation et le renouvellement de moyennes et grandes surfaces ; avec une épaisseur de 4-5 cm le gazon a une excellente résistance au piétinement et est mis en place pour les terrains de sport de haut niveau ; la préparation est sommaire, la mise en place rapide et l'utilisation immédiate ; les caractéristiques dimensionnelles varient selon les fournisseurs de 60 à 120 cm de largeur, de 10 à 20 m de longueur avec un poids allant jusqu'à 500 kg selon l'hygrométrie) ; *f 3* **pavé [m] (de) gazon** (*Terme spécifique* gazon de placage utilisé pour la réparation rapide de certaines zones très dégradées de terrains de sport comme les devants de but, le rond central, etc. permettant de limiter la durée d'immobilisation du terrain ; selon les fournisseurs l'épaisseur est de 3 à 5 cm de support, la largeur de 0,40 à 0,80 cm, la longueur de 100 à 120 cm et le poids de 30 à 100 kg par pavé) ; *g 1* **Rasensode [f]** (Geschältes Stück eines ►Fertigrasens mit den Schälmaßen ca. 30 x > 30 cm, 2,5-4 cm dick; cf. DIN 18 917; im Handel sind auch andere Maße: 40 x 250 cm, 1,5-2,0 cm dick); *syn.* Rasenziegel [m], *syn. obs.* Rasenplatte [f]; *g 2* **Dicksode [f]** (Rasensode für Sportplätze mit einer Sodenstärke je nach Lieferant von 35-42 mm, die durch ihr hohes Eigengewicht auch ohne sofortige Verwurzelung mit der Tragschicht für den Spielbetrieb sofort belastbar ist; Lieferung in Großrollen [Rollrasengroßrolle], je nach Firma, in 60 cm, 100 cm, 120 cm und 200 cm Breite und in Längen von 8-14 m oder je nach Bedarf; die bis zu 800 kg schweren Großrollen werden mit Spezialverlegemaschinen abgerollt und fugeneng angerückt; ►Andecken von Fertigrasen, ►Rollrasen, ►Wiederandecken von Rasensoden).

sod [n] (2) *agr. pedol.* ►heath sod.

sod [n], reed *constr.* ►reed clump.

sod [n] [US], rolled *constr. hort.* ►rolled turf.

sod [n], thick-cut *constr. hort.* ►sod [US] (1), #2.

sod [n], turf *agr. pedol.* ►heath sod.

sodding [n] [US] *constr.* ►turfing (1) and ►turfing (2), #1&2.

5744 sodium chloride soil [n] *pedol.* (Soil primarily characterized by its NaCl-content. Such soils also contain a significant amount of $CaCO_3$, $MgCO_3$ and, in certain areas, considerable quantities of sulphates; ►saline soil); *s* **suelo [m] sódico** (►suelo salino); *f* **sol [m] salin à complexe sodique** (Sol à forte salinité, à sodisation et alcalinisation modérées, à profil AC, rencontré en bordure de mer ou dans les lagunes côtières ; PED 1983, 469-480 ; ►sol salsodique) ; *g* **Kochsalzboden [m]** (Boden, der vorwiegend durch NaCl geprägt ist; außerdem enthalten diese Böden auch einen bedeutenden Anteil an $CaCO_3$, $MgCO_3$ und in vielen Gebieten Sulfate in wechselnder Menge; ►Salzboden).

5745 sod layer [n] [US] *constr. hort.* (Industrially grown layer of turf grass for transporting to a development site; ►plugging, ►rolled turf, ►securing sods [US]/securing turves [UK], ►seed mat, ►sod [US] 1/turf [UK], ►sod nursery [US]/turf nursery [UK], ►turfing 1); *syn.* turf layer [n] [UK]; *s 1* **césped [m] para transplante** (Capa de césped cultivada industrialmente para ser transportada al lugar de implantación; ►césped de tepes enrollados, ►estera de semillas, ►fijación de tepes con clavijas, ►plantación de césped con tepes o con macollas, ►plantel de césped, ►revestimiento con tepes, ►tepe, ►tepe grueso); *s 2* **césped [m] de tepes** (Resultado de 1.); *f* **gazon [m] précultivé** (Gazon implanté et prélevé pour être transporté et mise en place sur une surface à engazonner. Cette technique est employée pour un engazonnement rapide, p. ex. création de gazon sur talus, décoration d'expositions, jardins d'agrément ; ►chevillage de gazon précultivé, ►engazonnement par bouturage, ►gazon en rouleaux précultivé, ►gazonnière, ►natte préensemencée, ►placage, ►plaque de gazon) ; *syn.* gazon [m] en plaques ; *g* **Fertigrasen [m]** (1. Vorkultivierter und zum Transport geeigneter Rasenbelag zur schnellen Begrünung und zur Sicherung erosionsgefährdeter Böschungen; **F.** gibt es für alle Rasentypen, wobei der Anteil des rhizombildenden Grases *Poa pratensis* i. d. R. höher gewählt wird als bei Saatrasen, damit die 15-20 mm starken Soden und Rollen fest zusammenhalten; für den Hausgarten werden üblicherweise handliche, ein qm große Rollen im Format 2,50 m x 0,40 m verwendet; für große Flächen und Sportplätze gibt es 60-220 cm große Rollen, die auf Grund ihres hohen Gewichtes nur mit Spezialmaschinen verlegt werden können; für Rasensportplätze für den Profifußball, bei denen die Anwachszeit wegen der engen Spieltermine oft nicht reicht, werden ►Dicksoden von 3-4 cm Stärke verwendet, die bereits nach wenigen Tagen bespielbar sind. **2.** Mit Rasensoden oder Rollrasen angedeckte Fläche; ►Andecken von Fertigrasen, ►Anzuchtfläche für Fertigrasen, ►Herstellung von Rasenflächen durch Auspflanzen von Rasen- und Grasteilen, ►Rasensode, ►Rollrasen, ►Saatmatte, ►Sicherung von Fertigrasen).

5746 sod laying [n] **and seeding techniques** [npl] [US] *constr.* (Generic term for laying of pre-cultivated grass strips, ►seed mats, and for ►hydroseeding; ►sod layer [US]/turf layer [UK]); *syn.* laying [n] of turves and seeding techniques [npl] [UK]; *s* **técnica [f] de aplicación de césped** (*Construcción biotécnica* término genérico para todo tipo de maneras de establecer una superficie de hierba o césped como ►césped de tepes, ►esteras de semillas o establecimiento de vegetación por siembra fluída; ►césped para transplante, ►riego de semillas por emulsión); *f* **technique [f] d'enherbement** (*Termes spécifiques en génie biologique* végétalisation au moyen de ►gazon précultivé, de ►nattes préensemencées, ►procédé d'enberbement par projection hydraulique) ; *g* **Rasenbau [m, o. Pl.]** (*Ingenieurbiologische Bauweise* OB zu Verlegen von ►Fertigrasen, ►Saatmatten und für die Begrünung im ►Anspritzverfahren).

5747 sod nursery [n] [US] *hort.* (Cared for area of ground where turf is cultivated and maintained until required to be transferred to permanent positions; BS 3975: part 5; ►sod layer [US]/turf layer [UK]); *syn.* turf nursery [n] [UK]; *s* **plantel [m] de césped** (Superficie de tierra en donde se cultiva césped hasta su traslado al lugar definitivo en forma de tepes; ►césped de tepes); *f* **gazonnière [f]** (Aire de culture de ►gazon précultivé destiné au placage) ; *g* **Anzuchtfläche [f] für Fertigrasen** (Fläche für die Vorkultivierung von ►Fertigrasen; je nach Rasentyp dauert die Produktion 10-14 Monate; ein Betrieb, der Rasensoden herstellt, wird **Rasenschule** genannt); *syn.* Anbaufläche [f] für Fertigrasen.

5748 sod removal site [n] [US] *constr. hort.* (►sod nursery [US]/turf nursery [UK]); *syn.* turf removal site [n] [UK]; *s* **zona [f] de préstamo para tepes** (►plantel de césped); *f* **lieu [m] de prélèvement de gazon** (►gazonnière) ; *g* **Entnahmestelle [f] für Fertigrasen** (►Anzuchtfläche für Fertigrasen).

sod roof [n] [US] *constr. urb.* ►grassed roof.

sods [npl] [US] (1) *constr. hort.* ►cutting of sods [US]/scoring of turves [UK], ►replacement of sods [US]/replacement of turves [UK], ►securing sods [US]/securing turves [UK].

sods [npl] (2) *agr.* ►cutting of heath sods.

5749 soften [vb] **a pathway's hard edge** *gard.* (Allowing plants to grow loosely over the edge to give a pathway a less rigid and softer character; ►edge design); *syn.* interrupt [vb] a hard-edged pathway [also US]; *s* **suavización [f] de bordes de caminos** (≠) (Configuración de los bordes duros de caminos por medio de plantas que crecen más allá de los mismos para darles una impresión óptica menos rígida; ►delimitación); *f* **adoucir [vb] les angles des cheminements** (Conception utilisant des végétaux débordant sur les cheminements piétons afin d'en réduire les contours rigides ; *substantif* adoucissement [m] des angles ; ►aménagement des lisières, ►délimitation) ; *g* **Brechen [n, o. Pl.] der harten Wegekante** (Gestaltung der Wegekanten in der Art, dass durch überwachsende Pflanzen der optische Eindruck einer „harten" Kontur aufgelöst wird; ►Randausbildung).

soft landscape [n] *constr.* ►softscape.

5750 soft landscaping [n] *constr. plan.* (Planning and execution of earthwork[s], soil preparation, sowing and planting as well as the maintenance of vegetation in open spaces; ►softscape, ►hard landscaping, ►hardscape); *s* **diseño [m], ejecución [f] y plantación [f] de componentes vivos en espacios libres** (≠) (Planificación, ejecución y plantación de los componentes vivos en espacios libres incluyendo el movimiento de tierras, la siembra y plantación, así como el mantenimiento de la vegetación; ►componentes duros de espacios libres, ►diseño y construcción de componentes duros/inertes en espacios libres, ►superficie blanda); *f* **conception [f] des éléments vivants du jardin** (Élaboration et exécution des terrassements et travaux de préparation du sol, des semis et des plantations par comparaison avec l'emploi des matériaux inertes ; ►conception des éléments inertes de l'espace, ►éléments inertes d'un espace libre, ►éléments vivants d'un espace libre) ; *g* **vegetationsbestimmte Gestaltung [f] des Außenraumes** (≠) (Planung und Ausführung von Erdbauarbeiten, Bodenvorbereitungen, Ansaaten und Pflanzungen zur Unterscheidung des technisch-baulichen Teils einer Freianlage; ►technisch-bauliche Gestaltung des Außenraumes, ►Technisch-Bauliches einer Freianlage, ►vegetationsbestimmte Freianlage).

5751 softscape [n] *constr.* (Components of open spaces, covering vegetation and ground model[l]ing, in comparison with ►hardscape; ►soft landscaping); *syn.* soft landscape [n]; *s* **superficie [f] blanda** (≠) (Partes de los espacios libres determinadas por la vegetación o por estructuras blandas como modelado de superficie, al contrario que los ►componentes duros de

espacios libres; ►diseño, ejecución y plantación de componentes vivos en espacios libres); *f* **éléments [mpl] vivants d'un espace libre** (≠) (Composantes d'un espace libre caractérisées par les sols, eaux, engazonnements et plantations, par opposition aux ►éléments inertes d'un espace libre ; ►conception des éléments vivants d'un espace libre) ; *g* **vegetationsbestimmte Freianlage [f]** (Der Teil des Außenraumes, der durch Erdmodellierungen, Ansaaten und Pflanzungen geprägt ist — zur Unterscheidung befestigter Flächen, Steinbauten und sonstiger technisch-baulicher Einrichtungen; ►Technisch-Bauliches einer Freianlage, ►vegetationsbestimmte Gestaltung des Außenraumes); *syn.* vegetationsbestimmte Außenanlage [f], begrünter Freiraum [m], grünbestimmte Freianlage [f], nicht versiegelte Freianlage [f], vegetationsbestandene Freianlage [f], vegetationsbestandene Freifläche [f], vegetationsgeprägte Freianlage/Freifläche [f].

soft shoulder [n] [US] *trans.* ►road verge.

soft timber [n] *for.* ►softwood.

5752 softwood [n] *arb. for.* (**1.** Conventional term for both the timber and the trees belonging to the botanical group *Gymnospermae*. *Note* commercial timbers of this group are practically confined to the class *Coniferopsida*; cf. SAF 1983); *syn. for.* soft timber [n]. **2.** In German-speaking countries, the term 'softwood' [Weichholz] also includes alder *[Alnus]*, linden *[Tilia]*, poplar *[Populus]*, willow *[Salix]* and chestnut *[Aesculus]*; ►density of wood; *opp.* ►hardwood); *s 1* **madera [f] blanda** (Término convencional que denomina la madera de coníferas [clase *Coniferopsida*], excepto el tejo *[Taxus]*; en países de habla alemana también se incluyen algunas frondosas como el aliso *[Alnus]*, tilo *[Tilia]*, álamo *[Populus]*, sauce *[Salix]* y castaño de indias *[Aesculus]*; en inglés el término sólo incluye las coníferas; ►densidad de la madera; *opp.* ►madera dura); *s 2* **madera [f] de resinosas** *for.* (Término convencional en silvicultura para denominar la madera de las especies coníferas [clase *Coniferopsida*]; ►especie conífera); *f* **bois [m] tendre** (Terme conventionnel désignant tous les bois de résineux à l'exception de l'If *[Taxus]* ou les **bois très légers** de feuillus avec une densité de 0,4 à 0,5 g/cm^3] : le peuplier *[Populus]*, le saule *[Salix]*, le tilleul *[Tilia]*, le tremble *[Populus tremula]* ; les **bois légers** [densité 0,5 à 0,65 g/cm^3] : l'aulne *[Alnus]*, le bouleau *[Betula]*, le Marronnier *[Aesculus]* ; *opp.* ►bois dur ; en anglais le terme **b. t.** ne désigne que les bois de résineux [conifères] ; ►densité du bois) ; *g* **Weichholz [n]** (Bezeichnung für alle Nadelhölzer, außer Eibe *[Taxus]* und für Laubhölzer wie z. B. Erle *[Alnus]*, Linde *[Tilia]*, Pappel *[Populus]*, Weide *[Salix]* und Rosskastanie *[Aesculus]*; *opp.* Hartholz [►Laubholz]; im deutschsprachigen Raum gehören im waldbaulichen Sinne nur die o. g. Laubholzarten [Weichlaubhölzer] und die Birke *[Betula]* zum **W.**, das ein Trockenraumgewicht bis zu 0,55 g/cm³ [Darrdichte] hat. Im englischen Sprachgebrauch werden nur Nadelhölzer [Klasse der *Coniferopsida*] als **W.** bezeichnet; ►Holzdichte).

softwood [n], **brush mat of** *constr.* ►brush mat.

5753 soil [n] (1) *agr. constr. hort. pedol.* (Upper unconsolidated layer of the Earth's crust, which develops by physical, chemical, and biological processes; *specific terms* ►subsoil, ►topsoil 1); *syn. generally speaking* earth [n], ground [n]; *s* **suelo [m]** (Producto de la alteración, la reestructuración y la organización de las capas superiores de la corteza terrestre bajo la acción de la vida, la atmósfera y de los intercambios de energía que en ellas se manifiestan; EDAFO 1982; ►subsuelo 2, ►tierra vegetal); *syn. agr. hort.* tierra [f]; *f* **sol [m]** (Formation naturelle de surface à structure meuble, d'épaisseur variable, résultant de la transformation de la roche mère sous-jacente, sous l'influence de divers processus physiques, chimiques et biologiques ; *termes spécifiques* ►horizon structural, ►sous-sol, ►terre végétale) ;

g **Boden [m]** (Obere unverfestigte Schicht der Erdkruste [als Grundlage des Pflanzenwachstums], die durch physikalische, chemische und biologische Vorgänge entstanden ist; *UBe* ▶Oberboden, ▶Unterboden); *syn. allgemeinsprachlich* Erde [f], Erdreich [n].

soil [n] (2) *pedol.* ▶acid soil, ▶alkaline soil, ▶alluvial soil, ▶anmoor, ▶anmoor soil, ▶aquepts [US]/gley soil [UK], ▶arable soil, ▶bog soil; *agr. hort.* ▶cemented soil; *pedol. chem.* ▶chemical properties of soil; *agr. for. hort. pedol.* ▶clay soil; *pedol.* ▶coastal marsh soil; *constr. pedol.* ▶colloidal soil; *geo. pedol.* ▶creeping soil, ▶creeping soil on frozen ground; *agr. hort.* ▶cultivated soil; *envir.* ▶decontamination of soil; *pedol.* ▶deep plowed soil [US]/deep ploughed soil [UK], ▶degradation of soil; *agr. hort.* ▶easily workable soil, ▶fine-textured soil, ▶firm soil; *agr. for. hort.* ▶friable soil; *constr.* ▶frostproof soil, ▶frost-susceptible soil; *pedol.* ▶gleyed anmoor, ▶granular soil [US]/non-cohesive soil [UK], ▶gray-brown podzolic soil, ▶groundwater soil [US]/ground water soil [UK], histosol [US] [▶organic soil], ▶humosic soil [US]/humose soil [UK], ▶hydromorphic soils; *constr. eng. pedol.* ▶liquid limit of soil; *pedol.* lithosol [US]/ ▶skeletal soil; *ecol.* ▶living in the soil; *constr.* ▶load-bearing soil; *pedol.* ▶loosening (of soil); *agr. hort.* ▶loose soil; *pedol.* ▶man-made humic soil; *agr. (for.)* ▶marginal soil; *pedol.* ▶mellow soil; *pedol.* ▶mineral soil, ▶muddying of soil; ▶organic soil, ▶peat soil (1), ▶peat soil (2); *constr.* ▶placement of soil; *pedol.* ▶plaggen soil; *constr. hort.* ▶planting soil; *constr. eng. pedol.* ▶plastic limit (of soil); *pedol.* ▶podzol; *constr.* ▶poor load-bearing soil; *agr. hort.* ▶poorly drained soil; *pedol.* ▶rendzina, ▶saline soil; *agr. hort. pedol.* ▶sandy soil, ▶seasonally-flooded soil; *constr.* ▶settled soil, ▶settled soil material; *constr. pedol.* ▶site soil; *pedol.* ▶skeletal soil, ▶sodium chloride soil; *chem. pedol.* ▶soil reaction; *pedol.* spodosol [US] [▶podzol]; *constr.* ▶stabilized soil; *pedol.* ▶structurally stable soil; *pedol.* ▶structural stability of soil; ▶subhydric soil; *constr. pedol.* ▶subsoil; *pedol.* ▶swelling soil; *agr. constr. (pedol).* ▶topsoil (1); *constr.* ▶volume of on-site soil; *agr. hort. pedol.* ▶waterlogged soil [US]/water-logged soil [UK]; *constr. pedol.* ▶well-drained soil; *pedol.* ▶zinc-contaminated soil.

soil [n], alkali *pedol.* ▶alkaline soil.

soil [n] [US], altered *agr. constr. hort. pedol.* ▶man-made soil.

soil [n], anthropic *agr. constr. hort. pedol.* ▶man-made soil.

soil [n], brown forest *pedol.* ▶brown earth.

soil [n] [US], cohesionless *pedol.* ▶granular soil [US]/non-cohesive soil [UK].

soil [n] [UK], cohesive *pedol.* ▶colloidal soil.

soil [n], consolidated *constr.* ▶stabilized soil.

soil [n] [US], crumbly *agr. for. hort.* ▶friable soil.

soil [n] [UK], deep ploughed *pedol.* ▶deep plowed soil [US].

soil [n], drainage of *pedol.* ▶water permeability.

soil [n], drainage-poor *agr. hort.* ▶poorly drained soil.

soil [n], expansive *pedol.* ▶swelling soil.

soil [n] [US], field check of *constr.* ▶on-site soil investigation.

soil [n], forest *for. pedol.* ▶forest floor.

soil [n], frost-resistant *constr.* ▶frostproof soil.

soil [n] [US], garden *pedol.* ▶man-made humic soil.

soil [n] [UK], gley *pedol.* ▶aquepts [US].

soil [n], hard *agr. hort.* ▶firm soil.

soil [n], heavy *obs. agr. hort.* ▶fine-textured soil.

soil [n] [UK], humose *pedol.* ▶humosic soil [US].

soil [n] [US], humous *pedol.* ▶humosic soil [US]/humose soil [UK].

soil [n] [UK], immature *pedol.* ▶entisol [US] (1).

soil [n], in situ *constr. pedol.* ▶site soil.

soil [n], installation of *constr.* ▶placement of soil.

soil [n] [US], light *agr. hort.* ▶easily workable soil.

soil [n] [UK], non-cohesive *pedol.* ▶granular soil [US].

soil [n] [US], nonplastic *pedol.* ▶granular soil [US]/non-cohesive soil [UK].

soil [n], peaty *constr.* ▶peat soil (2).

soil [n], permeable *constr. pedol.* ▶well-drained soil.

soil [n], pH value of *chem. pedol.* ▶soil reaction.

soil [n] [US], plastic *pedol.* ▶colloidal soil.

soil [n], podzolic *pedol.* ▶podzol.

soil [n] [US], subaqueous raw *pedol.* ▶subhydric soil.

soil [n] [US], very firm *agr. hort.* ▶fine-textured soil.

soil [n], water-saturated soil *agr. hort. pedol.* ▶waterlogged soil [US]/water-logged soil [UK].

soil acidity [n] *pedol.* ▶increase in soil acidity.

soil additive [n] *agr. constr. hort. leg.* ▶soil amendment [US].

5754 soil [n] affected by impeded water *pedol.* (▶Hydromorphic soils with an impermeable sublayer causing water confinement; ▶aqualf [US]/pseudogley [UK]); *s* **suelo [m] de agua estancada** (▶Suelo semiterrestre hidromórfico cuya evolución se debe o ha debido al encharcamiento. Este conlleva un déficit de aireación que lleva consigo la reducción de iones con la consecuencia de colores particulares azul verdoso o rojizo; ▶pseudogley); *f* **sol [m] à nappe temporaire perchée** (Sol caractérisé par des phénomènes de réduction ou de ségrégation locale du fer, liés à une saturation temporaire ou permanente des pores par l'eau, provoquant un déficit prolongé en oxygène ; la présence d'une nappe d'eau libre temporaire ou permanente à propriétés réductrices est mise en évidence dans les pseudogley, stagnogley, gley et tourbes ; PED 1983, 374 ; ▶sol à pseudogley, ▶sols hydromorphes) ; *g* **Stauwasserboden [m]** (Durch temporäres oder dauerndes Stauwasser und Sauerstoffmangel stark veränderter Landboden, z. B. ▶Pseudogley, Stagnogley; ▶Grundwasser- und Stauwasserböden).

soil ameliorant [n] [UK] *agr. constr. hort. leg.* ▶soil amendment [US].

soil amelioration [n] [UK] *constr. pedol.* ▶soil improvement.

5755 soil amendment [n] [US] *agr. constr. hort. leg.* (Organic and inorganic material added to a soil in order to enhance its physical and chemical characteristics; e.g. structure, moisture, friability; ▶water-holding capacity, ▶soil improvement, ▶soil conditioning 1, ▶workability of a soil); *syn.* soil additive [n], soil ameliorant [n] [UK]; *s* **material [m] de enmienda de suelos** (Todo tipo de sustancias orgánicas e inorgánicas que se aplican a los suelos para mejorar sus características físicas y químicas como ▶capacidad de retención de agua, ▶enmienda del suelo, ▶friabilidad del suelo, pH; ▶medidas de mejora estructural de suelos); *syn.* enmienda [f] de suelos; *f* **amendement [m]** (*Terme générique* substances organiques ou inorganiques, naturelles ou synthétiques incorporées dans le sol comme l'intercalation de cultures d'engrais verts, pour en améliorer les propriétés physiques et chimiques telle que l'augmentation de la ▶capacité de rétention [maximale] pour l'eau, de la résistance au roulement, des ▶possibilités de travail du sol, la correction du pH, l'allège-

S

ment des terres lourdes. Certains auteurs utilisent toutefois l'expression **a.** chimique pour désigner les engrais, qu'ils distinguent de l'**a.** physique [apport de sable, etc.] et de l'**a.** mécanique [travail d'irrigation, etc.] ; ►amélioration de la structure du sol, ►correction de la terre) ; *g* **Bodenverbesserungsstoff [m] (1)** (**1.** Organischer und nicht organischer Stoff natürlicher oder synthetischer Herkunft, der Böden beigemengt wird sowie Saatgut für Voranbau oder Zwischenbegrünung, um die physikalischen und chemischen Eigenschaften zu verbessern, z. B. ►Wasseraufnahmefähigkeit, Erweiterung des Bereiches zwischen Ausroll- und Schrumpfgrenze, ►Bearbeitkeit des Bodens, Veränderung des pH-Wertes; cf. DIN 18 915, Bl. 2; ►Bodenverbesserung 2, ►Bodenstrukturverbesserung). **2.** *leg.* Stoff, der als Produkt für den Endverbraucher zur Verwendung im Garten verkauft und dem Boden zugeführt wird, um zumindest dessen physikalische und biologische Beschaffenheit zu verbessern, ohne nachteilige Auswirkungen zu haben; cf. Art. 1 Entscheidung 98/488/EG der Kommission v. 07.04.1998); *syn.* Bodenhilfsstoff [m] (N+L 1994 [12], 901), *leg.* Bodenverbesserungsmittel [n] (Entscheidung 98/488/EG), Bodenverbesserer [m] (Art. 1 [2] lit. b der Verordnung [EWG] Nr. 94/92 der Kommission).

5756 soil analysis [n] *pedol.* (Determination of particle-size distribution and chemical properties of a soil; ►soil investigation); *s* **análisis [m] del suelo** (Análisis físico-químico de laboratorio para determinar las propiedades de un suelo; ►prospección del suelo); *syn.* análisis [m] edáfico; *f* **analyse [f] du sol (au laboratoire)** (Analyse de caractérisation du sol, analyse granulométrique, analyse physique, analyse de fertilité par étude physico-chimique en laboratoire ; ►prospection du sol) ; *g* **Bodenanalyse [f]** (Chemisch-physikalische Laboruntersuchung einer Bodenprobe; ►Bodenuntersuchung).

5757 soil and water conservation district [n] [US] *leg. wat'man.* (Officially-declared area for the protection of water resources for future public water supply against lasting harmful effects and the accumulation of water in aquifers or reservoirs, as well as the prevention of damaging surface water runoff. **In U.S.**, a subdivision of a state or territory organized pursuant to the State soil conservation district law; *specific term* ►protective forest for water resources); *syn.* water conservation area [n] [UK]; *s* **perímetro [m] de protección de acuífero** (En el marco del aprovechamiento de aguas subterráneas, la delimitación de **p. de p. de a.** tiene como fin someter ciertos usos, como realización de obras de infraestructura o extracción de áridos, a la autorización de los organismos de cuenca para proteger el acuífero correspondiente; art. 54 [3] Ley de Aguas, 29/1985; ►bosque protector de acuífero); *syn.* área [f] de protección hidrogeológica, zona [f] de protección de un acuífero; *f* **périmètre [m] de protection d'un captage d'eau potable** (Zone protégée autour des points de prélèvement d'eau souterraine et des sources destinée à l'alimentation des collectivités humaines dont les périmètres de protection [immédiate, rapprochée et éloignée] sont déterminés par une procédure de déclaration d'utilité publique ; *terme spécifique* ►forêt de protection des ressources en eau) ; *g* **Wasserschutzgebiet [n]** (Im Rahmen eines förmlichen Verfahrens festgelegtes Gebiet, um die künftige öffentliche Wasserversorgung vor nachhaltigen Einwirkungen zu schützen, um das Grundwasser anzureichern oder um das schädliche Abfließen von Niederschlagswasser zu verhüten; cf. § 19 WHG; *UB* ►Wasserschutzwald).

5758 soil-bearing capacity [n] *constr. eng.* (Civil engineering term referring to the carrying power of a soil or subbase to support a load, usually expressed as kp per cm^2 or pounds per square foot; ►load-bearing capacity); *syn.* bearing capacity [n] of soil; *s* **capacidad [f] de soporte del suelo** (►capacidad de

carga); *syn.* resistencia [f] del terreno, valor [m] portante/soporte, capacidad [f] de asiento [RA]; *f* **portance [f] d'un sol** (Caractérise l'aptitude d'un sol à supporter des charges ; ►capacité de charge [d'exploitation]) ; *syn.* capacité [f] portante ; *g* **Tragfähigkeit [f, o. Pl.] eines Bodens** (Standsicherheit gewährende Beschaffenheit eines Bodens, Untergrundes etc.; ►statische Belastbarkeit); *syn.* Belastbarkeit [f, o. Pl.] eines Baugrundes.

5759 soil cementing [n] *constr.* (Process of ►soil consolidation [for road construction] to stabilize and solidify with lime or cement; ►cemented soil); *s* **estabilización [f] del suelo con mortero** (Término específico de ►estabilización del suelo; ►suelo cimentado); *f* **traitement [m] de sol avec un liant** (Stabilisation du sol avec un liant [ciment, chaux, cendres volantes, etc. — liants de traitement de sol] qui augmente la résistance du matériau face aux variations de l'humidité pour le rendre apte à constituer une couche de forme ; *termes spécifiques* traitement du sol à la chaux, stabilisation du sol au ciment ; ►sol-ciment ; *terme générique* ►stabilisation du sol) ; *g* **Vermörtelung [f] des Bodens** (Verfestigungsmethode bei bindigen Böden mit Kalken und bei Sanden, Kiesen und gut durchmischbaren Schluffböden mit Zement, da diese Böden sonst nicht gut verdichtet werden können; ►vermörtelter Boden; *OB* ►Bodenverfestigung); *syn.* Bodenvermörtelung [f].

5760 soil characteristics [npl] *pedol.* (Any soil condition which influences organisms; RCG 1982; ►physical properties of soils); *syn.* soil properties [npl], edaphic properties [npl]; *s* **características [fpl] del suelo** (►características físicas del suelo); *syn.* propiedades [fpl] del suelo; *f* **propriétés [fpl] du sol** (Propriétés physiques et chimiques du sol dans ses rapports [facteurs édaphiques] avec la végétation [structure, texture, pH, salinité, aération] ; ►caractéristiques physiques du sol) ; *syn.* édaphisme [m], propriétés [fpl] édaphiques ; *g* **Bodeneigenschaften [fpl]** (Physikalische und chemische Beschaffenheit des Bodens [Bodenstruktur, Textur, pH-Wert, Salz- und Nährstoffgehalt, Wassergehalt, Belüftung] in Bezug zur Vegetation; ►bodenphysikalische Eigenschaften); *syn.* Bodenbeschaffenheit [f], edaphische Eigenschaften [fpl].

soil class [n] [UK] *constr.* ►class of soil materials [US].

5761 soil classification [n] **(1)** *constr.* (Division of loose soils into soil groups according to particle size and distribution, texture, plasticity, and organic composition for construction purposes: sand, gravel, silt, clay, humus admixtures, peat, mud, lime; ►class of soil material [US]/soil class [UK], ►construction soil classification, ►soil suborder); *syn.* use classification [n] of soil materials [also US]; *s* **clasificación [f] granulométrica** (Clasificación de suelos sueltos para fines constructivos según norma DIN 18 196 en ►grupos de suelos según la granulometría, las características plásticas y las enmiendas orgánicas; ►clase de suelos, ►clasificación de suelos 1, ►clasificación de suelos 2); *f* **classification [f] granulométrique des sols** (F., classification des sols en fonction de la dimension des grains qui les composent ; par ordre décroissant de diamètre on distingue : rochers, cailloux, graviers, sables grossiers, sables fins, limons, argiles, ultra argiles ; D., classification des sols meubles en fonction de la granulométrie, la plasticité, et la présence de matière organique ; ►catégorie de sols, ►classe de terrains, ►classification des sols, ►classification des terrains, ►sous-classe pédologique) ; *g* **Bodenklassifikation [f] (2)** (**1.** *constr.* D., Untergliederung der Lockerböden gem. DIN 18 196 in Bodengruppen nach Korngrößenbereichen, Korngrößenverteilung, plastischen Eigenschaften und organischen Beimengungen für bautechnische Zwecke: Kies, Sand, Schluff, Ton, organische Beimengungen, Torf [Humus], Mudde [Faulschlamm], Kalk; ►Bodengruppe, ►Bodenklassifikation 1. **2.** *pedol.* ►Bodenklassifikation 3).

5762 soil classification [n] (2) *pedol.* (Systematic arrangement of soils into scientific orders, groups, and series on the basis of such characteristics as moisture, texture and organic components. The U.S. Comprehensive Soil Classification System has been used in the U.S. since 1966; ►soil suborder [US]); *s* **clasificación [f] de suelos** (2) (Ordenación científica sistemática de los suelos en clases, grupos y series sobre la base de su génesis, sus características fisico-químicas y su contenido de materia orgánica; ►clase de suelos, ►suborden de suelos); *f* **classification [f] des sols** (2) (La classification utilisée par les pédologues français a été mise sur pied par ALBERT DEMOLON et V. Oudin. Elle comporte la répartition des sols en **classes** et **sous-classes** en fonction des conditions physiques ou climatiques, physico-chimiques et chimiques d'évolution des sols, qui s'expriment par un certain nombre de caractères essentiels **1.** le degré d'évolution du sol et développement du profil, **2.** le mode d'altération climatique, **3.** le type et la répartition de la matière organique susceptible d'influer sur l'évolution du sol et la différenciation des horizons du profil, **4.** certains phénomènes fondamentaux d'évolution ; les classes et sous-classes sont subdivisées en **groupes de sol**, définis par des caractères morphologiques du profil correspondant à des processus d'évolution de ces sols ; les groupes comprennent en général plusieurs **sous-groupes** dont les profils sont différenciés soit par une intensité du processus fondamental d'évolution caractéristique du groupe, soit par la manifestation d'un processus secondaire ; dans certaines études assez détaillées on peut être amené à définir dans les sous-groupes des « **faciès** » de sols, correspondant à des stades d'évolution ou à des types intermédiaires entre deux sous-groupes ; à l'intérieur des sous-groupes on peut distinguer des **familles de sols**, en fonction des caractères pétrographiques de leur roche-mère ou de leur matériau original ; les **séries** correspondent à des différenciations de détail du profil ; dans certains cas, les séries sont subdivisées en **types de sols** en fonction des caractères précis de la texture de leurs horizons supérieurs ; cf. AUB 1962 ; ►sous-classe pédologique) ; *syn.* classification [f] pédologique ; *g* **Bodenklassifikation [f]** (3) (In D. setzte sich das nach WALTER KUBIENA [1897-1970] und von EDUARD MÜCKENHAUSEN [1907-2005] verfeinerte „natürliche System" durch, das auf der genetischen Bodenentwicklung basiert und den gesamten Profilaufbau eines Bodens berücksichtigt. Diese Bodensystematik wird von der *Deutschen Bodenkundlichen Gesellschaft [DBG]* ständig fortgeschrieben: Die Arbeitsgruppe Bodensystematik, eine AG der Kommission V der BDG, berät und beschließt **1.** die Gliederung und Definition der Bodenhorizonte, **2.** die systematische Gliederung der Böden, **3.** die systematische Gliederung der Bodengesellschaften und **4.** die systematische Gliederung der bodenbildenden Substrate; ►Bodenklasse 2).

5763 soil compaction [n] *constr.* (Physical treatment of soil as a result of ►trampling or by mechanical equipment; ►soil consolidation for road construction); *syn.* soil compression [n]; *s* **compactación [f] del suelo** (Resultado del ►pisoteo o del uso de maquinaria pesada; ►estabilización del suelo); *f 1* **compactage [m] du sol** (Opération d'augmentation de la compacité d'un sol par utilisation d'engins de compactage ; ►piétinement, ►stabilisation du sol) ; *f 2* **tassement [m] du sol** (Tassement de la couche superficielle du sol dû au piétinement excessif) ; *g* **Bodenverdichtung [f]** (**1.** Vorgang der Verdichtung eines Bodens durch Maschinen bei Baumaßnahmen oder durch ständige Trittbelastung. Die Neigung von Böden zur mechanischen Verdichtung wächst mit steigendem Tongehalt, abnehmendem Strukturgefüge [Sand, Kies] und zunehmender Durchnässung [zerstörtes Bodengefüge ohne Krümelstruktur]. **2.** Ergebnis durch Einsatz von Verdichtungsgeräten oder von Übernutzung durch ►Trittbelastung; ►Bodenverfestigung).

soil compression [n] *constr.* ►soil compaction.

soil condition [n]**, friable** *pedol.* ►granular soil structure.

5764 soil conditioner [n] *agr. constr. hort.* (*Specific term for a* ►*soil amendment [US]/soil ameliorant [UK]* synthetic linear organic polymers added to stabilize soil structure); *s* «**soil conditioner**» [m] (Término específico para los polímeros sintéticos organo-lineares, aportados para estabilizar la estructura del suelo; ►material de enmienda de suelos); *f* **amendement [m] synthétique** (Produit de synthèse organique [p. ex. à partir du pétrole] ; *terme générique* ►amendement) ; *g* **Bodenverbesserungsstoff [m]** (2) (Synthetische, lineare organische Polymere zur Strukturverbesserung des Bodens; ►Bodenverbesserungsstoff 1).

5765 soil conditioning [n] (1) *pedol.* (Improving the physical structure of a soil by the addition of [artificial] organic polymers [soil conditioners] to facilitate drainage and increase productivity/fertility; ►soil amendments [US]/soil ameliorants [UK], ►soil improvement); *s* **medidas [fpl de mejora estructural de suelos** (Medidas aplicadas para mejorar las propiedades físicas del suelo, añadiéndole materiales [artificiales = «soil conditioners»] para mejorar el drenaje e incrementar la fertilidad; ►enmienda del suelo, ►materiales de enmienda de suelos); *syn.* mejora [f] estructural de suelos; *f* **amélioration [f] de la structure du sol** (Mesures tendant à corriger les propriétés physiques du sol ; ►amendement, ►amélioration du sol) ; *syn.* amélioration [f] physique des potentialités du sol (VRD 1994, partie 4, chap. 8.2.1, 1) ; *g* **Bodenstrukturverbesserung [f]** (Maßnahmen, die dazu beitragen, die physikalischen Eigenschaften eines Bodens zu verbessern; ►Bodenverbesserung 2, ►Bodenverbesserungsstoff 1, ►Bodenverbesserungsstoff 2).

soil conditioning [n] (2) *constr. hort. agr.* ►soil preparation.

5766 soil conservation [n] *conserv. land'man.* (**1.** Policies and measures to prevent the loss of topsoil through erosion; ►soil erosion. **2.** Measures against contamination of soils or to remove impervious surfaces; ►decontamination of soil); *s* **protección [f] del suelo** (**1.** Conjunto de medidas preventivas de la ►erosión del suelo. **2.** Medidas contra la contaminación del suelo y su destrucción y su sellado excesivo por usos urbanos; ►saneamiento de suelos); *f* **protection [f] du sol** (**1.** Mesures de stabilisation de la terre végétale — drainage, ancrage des terrains, entretien, boisement — contre les phénomènes d'►érosion du/des sol[s]. **2.** Mesures de protection de la terre arable existante lors de l'étude d'avant-projet. **3.** Mesures favorables à une gestion économe dans l'utilisation des sols dans le cadre de l'établissement de documents d'urbanisme [prescriptions du règlement de PLU/POS] ; ►décontamination d'un sol pollué) ; *syn. de f 1.* maintien [m] du sol ; *g* **Bodenschutz [m, o. Pl.]** (**1.** Gesamtheit der Maßnahmen zur Erhaltung der Bodenertragsfähigkeit durch Sicherung des Oberbodens vor Abtragungsvorgängen; ►Bodenerosion. **2.** Maßnahmen gegen Kontamination und Versiegelung des Bodens; cf. § 1a BauGB u. § 5 BBodSchG; in D. wurde seit 1998 der **B.** als öffentlicher Belang in das Baurecht aufgenommen [cf. § 35 (3) 5 BauGB] und seit dem 17.03. 1998 gibt es das Bundes-Bodenschutzgesetz — BBodSchG; ►Bodensanierung).

5767 soil conservation crop [n] *conserv. land'man.* (Usually short-lived crop planted to minimize ►soil erosion, also for ►green manuring; ►protective forest for soil conservation purposes, ►soil conservation); *s* **cultivo [m] para proteger el suelo** (Plantación generalmente de corta duración para el ►abonado verde o que sirve para proteger el suelo de la erosión del mismo; ►bosque protector del suelo, ►erosión del suelo, ►protección del suelo); *syn.* plantación [f] para la protección del suelo; *f* **culture [f] de protection du sol** (Opération culturale

S

utilisant des plantes à rythme végétatif court, souvent employées comme culture intercalaire ou engrais vert [►fertilisation par (apport) d'engrais vert], et devant limiter l'►érosion du/des sol[s] ; le maintien de la jachère et de la culture des légumineuses dans les assolements sont des remèdes au ruissellement et au ravinement ; ►forêt de protection contre l'érosion, ►protection du sol) ; *g* **Bodenschutzkultur [f]** (Meist kurzlebige Pflanzenkulturen, oft als Zwischenkultur oder zur ►Gründüngung verwendet, die ►Bodenerosion vermindern sollen; ►Bodenschutz, ►Bodenschutzwald).

soil conservation purposes [n] *conserv. land'man.* ►protective forest for soil conservation purposes.

5768 soil consolidation [n] **for road construction** *constr. eng.* (Chemical treatment of soil in connection with road work; e.g. by mixing in lime, cement, or bitumen, to increase stability and improve bearing capacity; ►soil cementing, ►soil compaction, ►stabilized soil); *syn.* soil stabilization [n] for road construction; *s* **estabilización [f] del suelo** (Tratamiento químico y mecánico del suelo para incrementar su estabilidad y su capacidad de carga, de manera que resulte accesible en todo momento; ►compactación del suelo, ►estabilización del suelo con mortero, ►suelo estabilizado); *f* **stabilisation [f] du sol** (Augmentation ou maintien de la résistance du sol dans les travaux de voirie par épandage de liants hydrauliques, chaux grasse, ciment ou par imprégnation de liants bitumineux ; DIR 1977 ; ►compactage du sol, ►sol stabilisé, ►traitement de sol avec un liant) ; *g* **Bodenverfestigung [f]** (Erhöhung resp. Erhaltung der Widerstandsfähigkeit [Tragfähigkeit] des Bodens im Wegebau gegen äußere Belastung [Erhöhung der Tragfähigkeit] und Witterung durch Beimengungen von Kalk, Zement, bituminösen Bindemitteln etc.; ►Bodenverdichtung, ►verfestigter Boden, ►Vermörtelung des Bodens); *syn.* Bodenstabilisierung [f], Bodenverbesserung [f] (1).

soil-covered cultural monument [n] *conserv'hist. land' man.* ►buried cultural monument.

5769 soil creep [n] **(1)** *geo. pedol.* (Downslope movement of water-saturated or dry soils; ►creeping soil, ►mass slippage [US]/mass movement [UK], ►soil creep 2, ►solifluction); *s* **solifluxión [f]** (Tipo de ►movimiento gravitacional lento de suelo o materiales alterados sobre laderas y pendiente abajo resultado de la combinación de la gravedad y fenómenos de expansión y contracción. Puede deberse a ciclos de hielo-deshielo [►selifluxión] o humectación-desecación. Al contrario que en alemán e inglés, la solifluxión incluye también los movimientos causados por el hielo-deshielo; cf. DINA 1987; ►plancha de solifluxión, ►reptación hídrica del suelo); *syn.* reptación [f] del suelo; *f* **solifluxion [f]** (Descente lente sur un versant, de matériaux boueux ramollis par l'augmentation de leur teneur en eau ; ►boue reptante, ►gélifluxion, ►mouvement de masse, ►reptation d'un versant) ; *g* **Bodenfließen [n, o. Pl.]** (Langsame Abwärtsbewegung von Wasser gesättigtem oder trockenem Boden auf geneigten Hängen; ►Gekriech, ►Hangkriechen, ►Massenversatz, ►Solifluktion).

5770 soil creep [n] **(2)** *geo. pedol.* (Imperceptibly slow flowage type of ►mass slippage [US]/mass movement [UK]): gradual viscous movement downhill of slope-forming soil lubricated by rainwater [US]/rain-water [UK], under the influence of gravity; movement is caused by shear stress sufficient to produce permanent deformation, but too small to produce shear failure; **s. c.** is indicated by curved tree trunks, bent fences or retaining walls, tilted poles or fences, and small soil ripples or ridges; cf. DNE 1978, NAUS 2009; ►landslide); *s* **reptación [f] hídrica del suelo** (Tipo de transporte en vertientes que afecta a la parte más superficial de éstas, frecuente en suelos capaces de absorber

agua, con el consiguiente aumento de volumen, y de perderla durante los períodos de desecación; DGA 1986. Es uno de los tipos de ►movimiento gravitacionales lentos, ►corrimiento de tierras); *f* **reptation [f] d'un versant** (Déplacement lent et discontinu dans le temps de la pellicule superficielle de matériaux meubles d'un versant ; DG 1984, 471 et VOG 1979 ; ►glissement de terrain, ►mouvement de masse) ; *g* **Hangkriechen [n, o. Pl.]** (Langsames Rutschen der Hangoberfläche; kann auch bei geschlossener Vegetationsdecke zusammen mit einer Hangneigung > 2° auftreten, wenn plastische, quellfähige Bodenmassen anstehen; ►Hangrutschung, ►Massenversatz).

soil crumbing [n] *pedol.* ►crumb structure.

5771 soil cultivation [n] *agr. constr. hort.* (**1.** Steady working and loosening of soil by creating pores for aeration and water and thus more favo[u]rable conditions for plant growth; **2. in tillage [loc]:** under cultivation ; *context* 50 ha in tillage; ►soil mellowness); *syn.* tillage [n], tilling [n]; *s* **cultivo [m] de la tierra** (Laboreo mecánico continuo del suelo para mantener las condiciones óptimas de crecimiento de las plantas; ►sazón); *syn.* arado [m] de la tierra, laboreo [m] de la tierra; *f 1* **travail [m] du sol** (Travail mécanique d'un sol ; ►bon état de structure grumeleuse) ; *f 2* **façon [f] superficielle** (Travail ou série de travaux effectués sur une terre labourée pour créer les conditions de structure et d'ameublissement satisfaisantes en vue d'un semis ou d'une plantation ; LA 1981, 499 ; ►bon état de structure grumeleuse) ; *façon* obs. façon [f] culturale ; *agr. syn.* culture [f] du sol (PED 1983, 148, LA 1981, 1125) ; *g* **Bodenbearbeitung [f]** (Brechen und Lockern von Verdichtungs- und Verschlämmungshorizonten zur Herstellung von Luft und Wasser führenden Hohlräumen, das Bekämpfen des Unkrautes durch Zerkleinern und Unterarbeiten, Heraufholen ausgewaschener Bodenteilchen, die Förderung der Stickstoffmineralisation und der Krümelung des Bodens zur Pflanzflächenvorbereitung und zur Voraussetzung für optimales Pflanzenwachstum; cf. DIN 18 915; ►Bodengare).

5772 soil decontamination [n] **with plants** *envir.* (Soft technology which purifies contaminated soil with microorganisms and specific plants which are suitable for soil decontamination; ►decontamination plant); *s* **fitosaneamiento [m] de suelos (≠)** (Tecnología alternativa que utiliza ►plantas descontaminantes y microorganismos para limpiar suelos contaminados con sustancias orgánicas); *f* **décontamination [f] du sol par les plantes** (Technologie douce d'épuration des sols contaminés par des matières organiques polluantes par l'intermédiaire de ►plantes décontaminantes et des micro-organismes) ; *syn.* réhabilitation [f] biologique des sols pollués, dépollution [f] du sol par les plantes, phytorémédiation [f] ; *g* **Phytosanierung [f]** (Sanfte Technologie, die mit Hilfe von ►Repositionspflanzen und Mikroorganismen Böden reinigt, die mit organischen Schadstoffen kontaminiert sind); *syn.* grüne Dekontamination [f] (S+G 1995 [9], 643), Phytoremediation [f].

5773 soil depletion [n] *agr. hort.* (Reduced ►natural soil fertility caused by monoculture over many years with the result of impoverishment in trace elements, disease-causing agents, or increase in harmful metabolic substances produced by plants); *s* **empobrecimiento [m] del suelo** (Reducción de la ►fertilidad natural del suelo causada generalmente por el mismo cultivo durante muchos años que puede deberse a la falta de oligoelementos, a la contaminación con bacterias o a la alta presencia de sustancias tóxicas resultantes del metabolismo de las plantas en el suelo); *f* **épuisement [m] du sol** (Diminution importante de la ►fertilité naturelle du sol consécutive à une rotation des cultures inadaptée se traduisant par un appauvrissement en oligo-éléments ou en éléments fertilisants, le développement des parasites, l'envahissement par certaines mauvaises herbes) ; *syn.* fatigue [f]

S

du sol ; *g* **Bodenmüdigkeit [f]** (Meist durch jahrelange, einseitige Fruchtfolge bedingte verminderte ►Bodenfruchtbarkeit, die auf einer Verarmung an Spurenelementen, auf Verseuchung durch Krankheitserreger oder durch vermehrtes Vorhandensein schädlicher pflanzlicher Stoffwechselprodukte im Boden beruht).

soil development [n] *pedol.* ►soil formation.

5774 soil distribution pattern [n] *pedol.* (Spatial arrangement of soils within a geographical area shown on ►soil maps); *syn.* soil pattern [n]; *s* **distribución [f] espacial de los suelos** (►mapa cartográfico); *syn.* pattern [m] de suelos; *f* **couverture [f] pédologique** (Répartition spatiale des sols ; ►carte pédologique, ►carte des pédopaysages) ; *g* **Bodenverbreitung [f]** (Räumliche Verteilung von Böden in einem Gebiet; ►Bodenkarte).

5775 soil drainage [n] (1) *constr.* (Generic term for removal of excess water from the soil profile to prevent deterioration of plant growth or damage to structures; ►field drainage, ►land drainage, ►subsurface drainage); *s* **drenaje [m] del suelo** (Desagüe de las capas superiores de suelo; ►drenaje a gran escala, ►drenaje agrícola subterráneo, ►drenaje subterráneo); *f* **drainage [m] de sol** (Processus de collecte des eaux d'infiltration ; réseau de drains posés dans le sens de la plus grande pente et reliés à un collecteur — pelouse, terrain de sport, périphérie des bâtiments — afin d'en corriger les mauvaises propriétés physiques ; *terme spécifique* drainage de sol en pied de mur ; ►drainage [agricole] à ciel ouvert, ►drainage agricole souterrain, ►drainage souterrain) ; *g* **Bodenentwässerung [f]** (Vorgang des Abzugs überschüssigen Wassers im Bodenprofil im Sinne der ►Wasserdurchlässigkeit und ggf. Ableitung in Sauger und Sammler; eine gute **B.** ist die Voraussetzung für die Funktionsfähigkeit von technischen Anlagen und anspruchsvollen Bepflanzungen; ►Dränung, ►Dränung landwirtschaftlicher Flächen, ►Wasserbaumaßnahme zur Entwässerung landwirtschaftlicher Flächen).

soil drainage [n] (2) *agr. constr.* ►subsurface drainage.

soil drainage [n] (3) *pedol.* ►water permeability.

5776 soil erosion [n] *geo. pedol.* (Generic term covering all aspects of ►degradation of soil [►denudation, erosion, accumulation] resulting from the action of wind or water; usually understood to mean the **s. e.**, which results from human intervention above and beyond that resulting from natural processes alone. This man-made **s. e.** may be normal wearing away of the land surface used by man, not greatly exceeding ►natural erosion, or may be 'accelerated erosion', which is much more rapid than normal or natural erosion and may be caused by grazing animals on sloping ground, on bare surfaces exposed by fire or by large construction projects; cf. RCG 1982; ►extreme soil erosion, ►truncated soil profile); *s* **erosión [f] del suelo** (Incluye todos los procesos que llevan a la decapitación total o parcial del suelo. Puede ser de origen natural o causada por la acción humana [erosión antrópica/acelerada]; ►erosión, ►denudación, ►destrucción del suelo, ►degradación del suelo [vegetal], ►suelo decapitado); *f* **erosion [f] du/des sol(s)** (Terme générique caractérisant le processus de décapement total ou partiel d'un profil suite à un phénomène naturel ; lorsque ce processus est engendré par un facteur humain on parle d'érosion anthropique ; ►érosion, ►érosion continentale, ►dégradation [du sol], ►dévastation des sols, ►sol décapé) ; *g* **Bodenerosion [f]** (Alle jene Erscheinungen der Abtragung [Denudation, ►Erosion u. Akkumulation] durch Wasser oder Wind bewirkt, die den Haushalt der Landschaft durch menschlichen Einfluss über die natürlichen Abtragungsvorgänge hinaus verändern. Der Begriff ist im deutschen Sprachraum nicht immer identisch mit dem englischen *soil erosion*. Nur durch den Menschen direkt

oder indirekt durch Maßnahmen der Bodenbewirtschaftung verursachte **B.** heißt auch ,anthropogene Bodenabtragung'; ►Abtragung, ►Bodenverheerung, ►Degradation [des Bodens], ►geköpftes Bodenprofil).

soil examination [n] [US] *pedol.* ►soil investigation.

5777 soil excavation [n] *constr.* (Removal of soil/rock by digging, scraping, chiseling, or blasting, etc.); *s* **excavación [f] de suelos**; *f* **extraction [f] des terres** (Enlèvement des terres après décapage, désagrégation ou éclatement de roches) ; *syn.* prélèvement [m] des terres ; *g* **Lösen [n, o. Pl.] von Bodenmassen** (Abschieben, Ausheben, Meißeln oder Sprengen von Boden/Fels).

soil fauna [n] *pedol. zool.* ►species of soil fauna.

5778 soil fertility [n] *agr. hort. pedol.* (Generic term for either human influenced or natural productive capacity of a soil to support plant growth; ►enhancement of soil fertility, ►natural soil fertility, ►potential yield, ►site productivity, ►soil productivity); *s* **fertilidad [f] del suelo** (Capacidad natural o influenciada antropógenamente del suelo de servir como base para el crecimiento de las plantas [de cultivo]; ►capacidad de rendimiento, ►fertilidad natural del suelo, ►mejora de la fertilidad del suelo, ►productividad del suelo, ►productividad de una ubicación); *syn.* fertilidad [f] de la tierra; *f* **fertilité [f] du sol** (Capacité du sol à procurer aux plantes un support et à assurer leur développement ; ►fertilité naturelle du sol, ►mise en valeur du sol par la culture, ►productivité de la station, ►rendement, ►productivité du sol); *g* **Bodenfruchtbarkeit [f]** (Fähigkeit des natürlich gewachsenen Bodens, den Pflanzen langfristig als Standort zu dienen und Pflanzenwachstum zu ermöglichen; im rein gärtnerischen Sinne, nachhaltig hohe Erträge guter Qualität zu erwirtschaften; ►Bodenpflege, ►Bodenproduktivität, ►natürliche Bodenfruchtbarkeit, ►Ertragsfähigkeit, ►Standortproduktivität).

soil filter [n] [US] *constr.* ►filter mat.

soil flora [n] *bot. pedol.* ►species of soil flora.

5779 soil formation [n] *pedol.* (Transformation of the upper layer of the rock and soil of the earth's surface, influenced by the prevailing climate, the particular type of vegetation and its litter as well as the population of soil organisms; the soil-forming processes are characterized by weathering and mineral formation, decomposition and humification, structural stability, and various biochemical cycles. **S. f.** begins on the surface of the parent rock and penetrates deep down over a long period of time; hereby, layers known as ►soil horizons are formed, which can be clearly distinguished from one another by their properties); *syn.* soil development [n], pedogenesis [n]; *s* **edafogénesis [f]** (Proceso de transformación de las capas superiores de las rocas sobre la superficie de la tierra, influenciado por el clima predominante, la vegetación existente y el litter que produce así como por la fauna edáfica. Estos procesos están caracterizados por la meteorización de las rocas y la transformación de los minerales, la descomposición y humificación, la estabilidad estructural de los agregados y varios ciclos bioquímicos. La **e.** comienza en la superficie de la roca madre y penetra en su interior a través de un largo periodo/período de tiempo, formando los llamados ►horizontes edáficos, que pueden ser bien diferenciados entre sí por sus propiedades; ►horizonte edáfico); *syn.* pedogénesis [f], formación [f] de suelos; *f* **pédogenèse [f]** (Mode de formation des sols défini par des processus fondamentaux liés à l'humification, à l'altération géochimique prolongée, aux conditions physico-chimiques de la station ou conditionnés par les forts contrastes saisonniers et concourant, à partir des matériaux parentaux, à la formation et à l'évolution des couvertures pédologiques au cours du temps ; PED 1984, 121, INRA ; ►horizon) ; *syn. v.o.* pédogénèse [f],

S

formation [f] du sol ; *g* **Bodenbildung** [f] (Umformung der obersten Gesteins- und Bodenschicht der Erdoberfläche, beeinflusst durch das jeweils vorherrschende Klima, eine bestimmte streuliefernde Vegetation und Population von Bodenorganismen; die bodenbildenden Prozesse sind durch Verwitterung und Mineralbildung, Zersetzung und Humifizierung, Gefügebildung und verschiedene Stoffumlagerungen gekennzeichnet. Die **B.** beginnt an der Oberfläche des Ausgangsgesteins und setzt sich im Laufe der Jahrhunderte und Jahrtausende in die Tiefe fort und es entstehen Schichten, die sich mit ihren Eigenschaften deutlich abgrenzen lassen und als ▶Bodenhorizonte bezeichnet werden); *syn.* Bodenentwicklung [f], Pedogenese [f].

soil granule [n] *pedol.* ▶crumb.

soil group [n] *pedol.* ▶great soil group [US]/soil type [UK].

soil heap [n] [UK] *constr.* ▶soil pile [US].

soil heaps [npl] [UK], **stacked in** *constr.* ▶stacked in soil piles [US].

5780 soil horizon [n] *pedol.* (Soil layer roughly parallel to the land surface and separate from adjacent genetically-related layers with different physical, chemical and biological properties; **in U.S.** soil classification system [soil taxonomy], capital letters, such as O, P, A, B, E, C, D, R are used for master horizons and layers; lower case letters indicate specific characteristics and arabic numerals are used for vertical subdivisions within a horizon or layer, as well as for indicating discontinuities; cf. SST 1997); *syn.* horizon [n]; *s* **horizonte** [m] **edáfico** (Cada una de las diferentes capas separadas por superficies de contacto sensiblemente horizontales, y originadas en los suelos al evolucionar progresivamente el proceso de la pedogénesis. Constituyen conjuntamente el suelo, al mismo tiempo que por sus diferencias reflejan su anisotropía vertical. Los **h. e.** se designan mediante las letras A, B y C; cf. DINA 1987 y EDAFO 1982); *syn.* horizonte [m] del suelo; *f* **horizon** [m] (Couche du sol plus ou moins épaisse et distincte, définie par des caractères morphologiques, biologiques et physico-chimiques. Les horizons principaux pour la classification FAO sont : [B], B, C, G, R ; les horizons diagnostics principaux pour la classification américaine sont : épidon = horizon de surface, horizons [B] ou B = horizon de profondeur) ; *g* **Bodenhorizont** [m] (Bei der Bodenentwicklung aus dem anstehenden Gestein durch Verwitterung, Zersetzung und Humifizierung, Mineralbildung, Gefügebildung, Stoffumlagerungen und Bodenbearbeitung parallel zur Erdoberfläche entstehende relativ gleichförmige Schicht. Im Deutschen werden die Horizonte eines Bodenkörpers von oben nach unten mit den großen Buchstaben O, A, B, C etc. bezeichnet und durch Zahlen- oder Buchstabenindizes unterschieden; cf. SS 1979); *syn.* Horizont [m].

soil horizons [npl] *pedol.* ▶exposed soil horizons.

5781 soil impoverishment [n] **and structural degeneration** [n] *pedol.* (Degradation of soil, usually brought about by the destruction of vegetative cover; ▶truncated soil profile); *s* **empobrecimiento** [m] **y degradación** [f] **del suelo** (Evolución regresiva de los suelos causada en general al ser destruida la cubierta vegetal, debido a incendio, tala indiscriminada, erosión pluvial, etc.; ▶suelo decapitado); *f* **appauvrissement** [m] **du sol** (Évolution régressive des sols due en général à la disparition progressive du couvert végétal à la suite d'incendie, d'érosion pluviale, de surpâturage, etc. ; ▶sol décapé) ; *syn.* épuisement [m] du sol ; *g 1* **Aushagerung** [f] (1. *pedol.* Qualitätsminderung des Bodens, meist ausgelöst durch Abtragung oder Verwehung des Bodens in Form feinsten Sandes oder Staubes, z. B. Lössboden, oder durch Verlust der Vegetationsdecke, Überweidung etc.; aushagern [vb]; ▶geköpftes Bodenprofil); **2.** *pedol. land' man.* Bewusste Verminderung des Nährstoffgehaltes in Böden

durch Entfernen des Bewuchses ohne nachzudüngen; in der Landschaftspflege wird aus naturschutzfachlichen Gründen durch wiederholtes Mähen und Abfuhr des Schnittgutes erreicht, dass die Artenvielfalt durch Veränderung der Vegetationszusammensetzung zunimmt, besonders an Trockenstandorten und auf Feuchtwiesen); *g 2* **Verhagerung** [f] (Verringerung der Produktivität einer Biozönose auf Grund standörtlicher Bedingungen in Relation zur unmittelbaren Umgebung; **V.** beinhaltet das je unterschiedliche Zusammenspiel verschiedener ▶Ökofaktoren, z. B. Bodenfeuchte, Nährstoffgehalt des Bodens, Strahlung, Wind, mit unterschiedlicher räumlicher Ausdehnung; im Vergleich zur ▶Aushagerung handelt es sich bei der **V.** um ungeplante und ungerichtete Prozesse; cf. KNO 2003).

5782 soil improvement [n] *constr. pedol.* (General term for measures taken to enhance ▶soil structure and fertility in order to provide more favo[u]rable growing conditions; ▶agricultural land improvement, ▶fertilizing, ▶liming, ▶revitalization of tree pits, ▶soil conditioner, ▶soil conditioning 1); *syn.* soil amelioration [n] [also UK]; *s* **enmienda** [f] **del suelo** (Término genérico para la aportación al suelo de materiales de enmienda como compost, arena, arcilla o materiales sintéticos, para mejorar la ▶textura del suelo vegetal y con ello su fertilidad; ▶abonado cálcico, ▶fertilización, ▶medidas de mejora estructural de suelos, ▶mejoramiento de suelos, ▶revitalización [de ubicaciones] de árboles urbanos, ▶«soil conditioner»); *f* **amélioration** [f] **du sol (2)** (Terme générique caractérisant l'apport et le mélange uniforme de substances servant à l'amélioration de la ▶structure du sol [▶amendement synthétique] comme le sable, l'argile, les substances synthétiques et la fertilité du sol [apport nutritif], etc. ; ▶amélioration de la structure du sol, ▶amélioration du lieu d'implantation des arbres, ▶amélioration du sol, ▶chaulage, ▶compostage, ▶fertilisation, ▶fumure, ▶limonage, ▶terreautage 1) ; *syn.* correction [f] du sol ; *g* **Bodenverbesserung** [f] **(2)** (Einbringung und gleichmäßige Vermischung von ▶Bodenverbesserungsstoffen wie z. B. Kompost, Sand, Ton, synthetische Stoffe in den Oberboden zur Verbesserung der ▶Bodenstruktur; ▶Bodenstrukturverbesserung, ▶Düngung, ▶Kalkung, ▶Melioration, ▶Baumstandortsanierung).

5783 soil improvement [n] **by working in a leaf mold/sand mixture** [US]/**soil improvement** [n] **by working in a leaf-mould/sand mixture** [UK] *constr. hort.* (▶topdressing [US] 1/top-dressing [UK], ▶mulch); *s* **enmienda** [f] **de mantillo** (Mejora del suelo por incorporación de mantillo; ▶cubierta de mantillo, ▶mulch); *f* **terreautage** [m] **(1)** (Épandage d'une couche de terreau sur un semis ou incorporation d'un mélange de terreau et sable ; ▶épandage en couverture, ▶mulch, ▶terreautage 2) ; *g* **Bodenverbesserung** [f] **mit Lauberde-Sandgemisch** (Einarbeitung eines Lauberde-Sand-Gemisches in den Boden oder Abdecken des ausgebrachten Saatgutes; ▶Bodenabdeckung, ▶Mulch).

5784 soil improvement [n] **with clay** *constr. s* **enmienda** [f] **de arcilla** (Mejora del suelo arenoso por incorporación de arcilla o limo); *f* **limonage** [m] (Enrichissement du sol sablonneux avec des matériaux limoneux) ; *g* **Bodenverbesserung** [f] **mit Ton** (Verbesserung von Sandböden mit Ton).

5785 soil improvement [n] **with compost** *constr. hort. s* **abonado** [m] **con compost** (Mejora del suelo por incorporación de compost); *syn.* enmienda [f] de compost, fertilización [f] con compost; *f* **fertilisation** [f] **(du sol) au compost** (Amendement organique — enrichissement de la terre végétale par apport de compost qui améliore la structure du sol, développe l'activité biologique, augmente la capacité à retenir l'eau et les éléments nutritifs, tue certaines graines de plantes indésirables et certains germes pathogènes durant sa fermentation) ; *syn.* amendement

[m] en/de compost, fertilisation [f] (du sol) par compostage de surface, amendement [m] (du sol) par compostage de surface ; *g* **Bodenverbesserung [f] mit Kompost** (Einarbeitung von Kompost in Böden zur Ertragssteigerung und zur Verbesserung der Bodenstruktur und der Mikrobiologie sowie in leichten sandigen Böden zur Erhöhung des Wasserhaltevermögens).

soiling [n] of holes *constr.* ►filling of paver holes [US].

soiling [n] of voids [UK] *constr.* ►filling of paver holes [US].

5786 soil inventory [n] *pedol.* (Record of ►soil survey; ►soil map); *s* **inventario [m] edafológico** (Resultado de la ►investigación edafológica; ►mapa de suelos); *syn.* inventario [m] de suelos; *f* **inventaire [m] pédologique** (Résultat d'un ►relevé pédologique en vue de la représentation cartographique des unités de sol et de leurs caractéristiques [►carte pédologique, ►carte des pédopaysages] ; F., le programme national « Inventaire Gestion et Conservation des Sols » [IGCS] est coordonné par le Groupement d'Intérêt Scientifique SOL [INRA, centre d'Orléans] et a pour objectif la réalisation d'un inventaire exhaustif des sols de France, par région ou par département. Cet inventaire sert de support à la réalisation d'un référentiel pédologique basé sur la constitution d'une base de données géographique pilotée par un Système d'Information Géographique) ; *syn.* inventaire [m] des sols ; *g* **Bodenaufnahme [f] (1)** (Ergebnis der ►Bodenaufnahme 2; ►Bodenkarte).

5787 soil investigation [n] *pedol.* (Detailed study of soil horizons; ►soil analysis, ►on-site soil investigation); *syn.* soil examination [n] [also US]; *s* **prospección [f] del suelo** (Estudio detallado de los suelos *in situ* para elaborar mapas edáficos o conocer el perfil de un suelo específico; ►análisis del suelo, ►análisis del suelo in situ); *f* **prospection [f] du sol** (Examen sur le terrain en vue de la constitution de cartes pédologiques et de l'établissement de profils de sols ; ►analyse du sol [au laboratoire], ►analyse du sol sur site) ; *syn.* reconnaissance [f] du sol ; *g* **Bodenuntersuchung [f]** (Untersuchung im freien Gelände zur Anfertigung von Bodenkarten oder zur Bodenprofilerstellung; ►Bodenanalyse, ►Bodenbeurteilung).

5788 soil loosening [n] *agr. constr. hort.* (Working up of soil, mechanically or by hand: ►scarifying [US]/ripping [UK], ►rototilling [US]/rotovation [UK], plowing [US]/ploughing [UK], ►subsoiling, ►deep plowing [US]/deep ploughing [UK], ►loose soil, shallow tillage [US]); *s* **descompactación [f] del suelo** (Operación mecánica o manual de laboreo del suelo con ayuda de herramientas; desapelmazar [vb] el suelo, mullir [vb] el suelo; ►arado en profundo, ►desfonde, ►escarificado, ►rotovación, ►subsolar en profundo, ►suelo mullido); *syn.* mullido [m] del suelo, desapelmazar [vb] el suelo, mullir [vb] el suelo; *f* **ameublissement [m] du sol** (Opérations culturales, **1.** mécaniques profondes : ►décompactage, ►défonçage, ►labour profond, ►sous-solage ou **2.** mécaniques superficielles : ►fraisage, amélioration de la texture du sol ; résultat : ►sol ameubli) ; *g* **Bodenauflockerung [f] (2)** (Lockern des Bodens [Baugrund oder Vegetationstragschicht] mit Bodenbearbeitungsgeräten: ►Aufreißen 1, ►Fräsen, Pflügen, ►Tiefenlockerung, ►Tiefpflügen, ►gelockerter Boden; cf. DIN 18 915); *syn.* Bodenlockerung [f], Lockerung [f] (des Baugrundes oder der Vegetationstragschicht).

5789 soil loosening [n] of planting areas *constr. hort.* (Making earth less compact with the use of a cultivating tool; ►soil loosening); *s* **descompactación [f] del terreno para plantación** (Operación mecánica o manual de laboreo del suelo antes de la plantación con ayuda de herramientas; ►descompactación del suelo); *syn.* mullido [m] del terreno para plantación, esponjado [m] del suelo para plantación; *f* **griffage [m] de sol (des plantations)** (Ameublissement superficiel du sol au moyen d'une griffe ; ►ameublissement du sol) ; *g* **Lockern [n, o. Pl.] der Pflanzflächen** (Auflockerung der Bodenoberfläche mit einem Bodenbearbeitungsgerät zur Vorbereitung einer Pflanzung; auflockern [vb], lockern [vb], locker machen [vb]; ►Bodenauflockerung 2).

5790 soil management [n] *agr. for.* (Productive use of soils while simultaneously conserving and enhancing their fertility); *s* **cultura [f] del suelo (≠)** (En alemán término tradicional para denominar un uso óptimo del suelo que contribuye además a su conservación y a la mejora de su fertilidad); *f* **mise [f] en valeur des sols** (Utilisation du sol en tenant compte simultanément de la préservation et de l'amélioration de la fertilité) ; *g* **Bodenkultur [f]** (Nutzung des gewachsenen Bodens bei gleichzeitiger Erhaltung und Förderung der Bodenfruchtbarkeit wie z. B. in Baumschulen, Staudengärtnereien, im Obst-, Gemüse- und Weinbau, bei bestimmten Pflanzenarten auch im Zierpflanzenbau [Anzucht von Wechselflor und Frühjahrsgeophyten im Freiland).

5791 soil map [n] *pedol.* (Cartographic representation of the soil occurring in a particular area. ►Soil types are shown on maps in scales 1:25,000 to 1:50,000 and at smaller scales, groups of soils and soils both vulnerable and worthy of protection are shown; on maps of larger scales, 1: 5000 to 1:10,000, additional properties such as soil textural class, groundwater conditions, and ground relief are also illustrated; ►great soil group, ►soil inventory); *s* **mapa [m] de suelos** (Representación cartográfica de los suelos que se presentan en un área específica. Los tipos de suelos se muestran en mapas con escalas de 1:25.000 hasta 1:50.000 e inferiores; los grupos de suelos y los suelos vulnerables o protegibles en mapas de mayor escala, de 1:5.000 a 1:10.000, en los que representan además otras propiedades como la clase textural, las condiciones del agua subterránea y el relieve topográfico; ►inventario edáfico, ►tipo de suelo); *syn.* mapa [m] edafológico; *f 1* **carte [f] pédologique** (Représentation graphique de la délimitation de zones homogènes de ►groupes de sols définis par leurs caractéristiques morphologiques et génétiques sur un territoire donné ; la carte pédologique de la France est à l'échelle du 1 : 1 000 000 ; F., la couverture pédologiques des régions françaises à grande et moyenne échelle n'est pas exhaustive et très inégale selon les départements ; les échelles sont très variées [du 5000e au 200 000e] et les principes de leur établissement sont très divergents ; dans le cadre du Programme Pédologique de la France et la création du Service d'Étude des Sols et de la Carte Pédologique en 1968, 24 cartes avec un fond topographique au 100 000e ont été publiées et 15 cartes sont en voie d'achèvement ; elles représentent environ 15 % du territoire français ; ►inventaire pédologique) ; *syn.* carte [f] des sols ; *f 2* **carte [f] des pédopaysages** (Représentation graphique d'un ensemble litho-géomorpho-pédologique cohérent résultat de la combinaison des éléments du paysage [roche mère, topographie, eaux de surface, occupation du sol] et des caractéristiques des sols [types de sol, profils et horizons pédologiques] ; ce découpage géographique est réalisé dans le cadre du programme national IGCS [inventaire, gestion et conservation des sols] de l'INRA qui fait partie du programme européen de constitution de base de données géoréférencées des sols d'Europe au 1/250 000 proposé en 1990 par le Bureau Européen des Sols) ; *g* **Bodenkarte [f]** (Kartografische Darstellung der in einem Gebiet vorkommenden Böden. Auf Karten in den Maßstäben 1:25 000 bis 1:50 000 werden ►Bodentypen und in kleineren Maßstäben i. d. R. Vergesellschaftungen von Böden und schutzwürdige sowie schutzbedürftige Böden gezeigt; auf Karten größerer Maßstäbe 1:5000 bis 1:10 000 werden zusätzliche Eigenschaften wie Bodenart, Grundwasserverhältnisse, Reliefneigungen etc. dargestellt; ►Bodenaufnahme 1).

S

5792 soil mapping [n] *geo. pedol.* (▶soil survey, ▶soil inventory); *syn.* soil resource inventory [n] [also US]; *s* **cartografía [f] de suelos** (▶inventario edafológico, ▶investigación edáfica); *syn.* cartografía [f] edáfica; *f* **cartographie [f] des sols** (A pour but de définir les unité-sols d'une zone déterminée et d'en préciser leur extension géographique ; ▶inventaire pédologique, ▶relevé pédologique) ; *syn.* cartographie [f] pédologique ; *g* **Bodenkartierung [f]** (▶Bodenaufnahme 1 u. ▶Bodenaufnahme 2).

5793 soil mechanics [npl] *constr. eng.* (*Term with a singular connotation* Science of the physical properties of soils with regard to their structural and load-bearing characteristics, and as a building material); *s* **mecánica [f] de suelos** (Ciencia de las características físicas de los suelos como soporte de fundación); *f* **mécanique [f] des sols** (Étude minéralogique, physique et géologique des sols de fondation) ; *g* **Bodenmechanik [f]** (Lehre von den physikalischen Eigenschaften der Böden als Baustoff und Baugrund).

5794 soil mellowness [n] *agr. hort. pedol.* (**1.** State of maturity of a soil at which it is at maximum productivity, being characterized by a highly developed and stable ▶crumb structure; ▶mellow soil); **2. tilth** [n] (Physical condition of fine/good soil as related to its ease of ▶soil cultivation, fitness as a seed bed, and encouragement to seedling emergence and root penetration; cf. RCG 1982); *s* **sazón [f]** (*Contexto* Sazón que adquiere la tierra con la lluvia; *término específico* tempero; DIS 1986, 99; ▶estructura grumosa, ▶suelo con sazón); *f* **bon état [m] de structure grumeleuse** (Aptitude culturale maximale des sols agricoles, forestiers et horticoles, reconnaissable à la stabilité de la structure des agrégats, d'une bonne aération, d'un fort pouvoir de rétention de l'eau ; ▶sol à forte stabilité structurale, ▶structure fragmentaire) ; *syn.* bon état [m] structural ; *g* **Bodengare [f]** (Zustand höchster Leistungsfähigkeit der Acker-, Wald- und Gartenböden, erkenntlich an der Beständigkeit der ▶Krümelstruktur 1, der guten Durchlüftung, dem hohen Wasserhaltevermögen; die Teilgare eines Bodens, die durch Bodenbedeckung entsteht, wird **Schattengare [f]** genannt; ▶garer Boden); *syn.* Gare [f], Garezustand [m] des Bodens, Dauerkrümelstruktur [f].

5795 soil moisture [n] *pedol.* (Retained soil water which is not lost as a result of drainage; ▶field capacity, ▶seepage water, ▶soil wetness); *s* **humedad [f] del suelo** (Humedad que retiene el suelo en la zona de aireación, que se equilibra con el vapor de agua de la atmósfera; WHO 1974; ▶agua de percolación, ▶alta humedad del suelo, ▶capacidad de campo); *syn.* agua [f] adsorbida, agua [f] higroscópica, agua [f] ligada, humedad [f] del suelo; *f* **humidité [f] du sol** (Eau retenue par le sol au cours de l'infiltration des pluies lorsque toute l'eau susceptible d'être naturellement drainée a été éliminée ; on distingue l'eau capillaire absorbable par les racines et l'eau liée [ou eau d'absorption] non absorbable par les racines ; ▶capacité [de rétention] au champ, ▶eau de gravité, ▶engorgement du sol ; cf. WHO 1974, n° 536 et PED 1979 II, 272) ; *syn.* eau [f] retenue, eau [f] de rétention ; *g* **Bodenfeuchte [f]** (Das im Boden verbleibende, oberhalb des Grundwassers verteilte Wasser; cf. LES 1976; ▶Bodennässe, ▶Feldkapazität, ▶Sickerwasser 2); *syn.* Haftwasser [n].

soil obliteration [n] [US] *conserv. pedol.* ▶extreme soil erosion.

soil [n] **on frozen ground** *geo. pedol.* ▶creeping soil on frozen ground.

5796 soil order [n] [US] *pedol.* (Category at the highest level of generalization in a soil classification system to reflect the degree of horizon development and the kinds of horizons present; main subdivision in soil classification after AVERY 1973: litho-

morphic soils, brown soils, ▶podzols, pelosols, ▶aquepts [US], ▶man-made soils, ▶peat soils 2; **in U.S.**, a soil classification system was continually developed. In 1928 C. F. MARBUT differentiated the soils primarily according to morphological and chemical characteristics. In the 1938 classification system, the three soil orders were zonal soil, intrazonal soil, and azonal soil. In 1975 there were 10, and since 1994, according to the USDA classification scheme, eleven **s. o.**, differentiated by the presence or absence of diagnostic horizons: Alfisols, Andisols, Aridisols, Entisols, Histosols, Inceptisols, Mollisols, Oxisols, Spodosols, Ultisols, and Vertisols; orders are divided into Suborders and the Suborders are farther divided into Great Groups; cf. SST 1997; ▶hydromorphic soils, ▶peatland [US]/moor [UK], ▶soil affected by impeded water, ▶subhydric soil, ▶terrestrial soil); *syn.* major (soil) group [n] [UK]; *s 1* **grupos [mpl] principales (de suelos)** (En D., primera subdivisión en la clasificación de suelos sobre la base del régimen acuático que diferencian entre los ▶suelos terrestres, los suelos hidromorfos o ▶suelos semiterrestres [▶suelos semiterrestres de inundación y ▶suelos de agua estancada], los ▶suelos subacuáticos no turbosos, los ▶suelos subacuáticos turbosos, las ▶turberas y los ▶cultosoles; la inventariación de suelos se basa en el sistema «natural» propuesto por KUBIËNA [1950] y modificado por MÜCKENHAUSEN, en el cual los suelos se ordenan primordialmente según características genéticas; cf. BB 1979, 438); *s 2* **orden [m] de suelos** (Según la taxonomía de suelos del *Soil Survey Staff* de los EE.UU. [1960], nivel jerárquico más general de clasificación que utiliza como criterio las propiedades o características más condicionantes para el uso del suelo y los subdivide en 11: alfisoles, andisoles, aridisoles, entisoles, histosoles, inceptisoles, mollisoles, oxisoles, spodosoles, ultisoles y vertisoles); *f* **classe [f] pédologique** (F., unité principale de classification de sols utilisée par le service de la carte des sols de France et basée sur les processus biogéochimiques ; on distingue 12 classes : sols peu évolués, sols peu différenciés humifères désaturés, sols calcimagnésiques, sols isohumiques, vertisols, sols brunifiés à profil A[B]C ou ABC, sols podzolisés, sols fersiallitiques, sols ferrugineux, sols ferrallitiques, sols hydromorphes, sols salsodiques ; cf. PED 1983, 196 et *s* ; U.S., la classification américaine est basée sur l'identification d'horizons diagnostics fondamentaux [de surface et de profondeur] et distingue 11 ordres fondamentaux : alfisols, andisols, aridisols, entisols, histosols, inceptisols, mollisols, oxisols, spodosols, ultisols, vertisols ; PED 1983, 186 et SST 1997 ; D., la classification allemande est établie d'après le critère de l'hydromorphie dans laquelle on distingue les ▶sols de culture, ▶sols hydromorphes, ▶sols semi-terrestres, ▶sols terrestres, ▶sols subaquatiques [PED 1983, 371], ▶tourbe) ; *syn. pour la classification américaine* **ordre [m] pédologique** ; *syn. pour la classification allemande* **groupe [m] pédologique** ; *g* **Bodenabteilung [f]** (D., oberste pedologische Klassifizierungskategorie der Böden Mitteleuropas: es wird nach dem Wasserregime zwischen den terrestrischen oder ▶Landböden, den hydromorphen oder ▶Grundwasser- und Stauwasserböden, den subhydrischen oder ▶Unterwasserböden, den ▶Mooren 2 und Kultosolen [▶anthropogene Böden] unterschieden; die Bodenkartierung in D. basiert auf dem von WALTER KUBIENA [1897-1970] vorgeschlagenen „natürlichen" und von EDUARD MÜCKENHAUSEN [1907-2005] modifizierten System, in dem die Böden vorrangig nach genetisch bedingten Merkmalen und ihrem genetischen Zusammenhang geordnet sind. Hierbei wird der gesamte Profilaufbau eines Bodens, in dem sich die Auswirkungen aller Faktoren der Bodenentwicklung widerspiegeln resp. seine Horizontkombination in den Mittelpunkt gestellt; SS 1979, 314; **U.S.**, das Klassifikationssystem der Böden, das seit 1928 nach CURTIS F. MARBUT [1863-1935] primär auf morphologischen und chemischen Eigenschaften der Böden beruhte, wurde ständig weiterentwickelt. 1938 bestand die Gliederung aus drei Ordnungen *[Orders]*: **1.** *zonale Böden*, die in erster Linie durch den Einfluss des Klimas entstanden, **2.** *intrazonale Böden*, die vorwiegend durch andere Faktoren als das Klima gestaltet wurden und **3.** nur wenig entwickelte ‚azonale Böden'. 1975 gab es zehn und seit 1994 gibt es nach dem USDA-Klassifikationssystem elf Ordnungen. Die Ordnungen gliedern sich in Unterordnungen *[Suborders]* und diese in *Great Groups*; cf. SST 1997).

soil organic matter [n] *pedol.* ▶humus.

5797 soil organism [n] *pedol.* (Generic term for a flora or fauna species living in the soil; ▶soil organisms, ▶species of soil fauna, ▶species of soil flora); *s* **organismo** [m] **edáfico** (Término genérico para las especies de flora y fauna del suelo; ▶animal edáfico, ▶edafon, ▶especie de flora edáfica); *f* **organisme** [m] **édaphique** (Terme générique pour l'▶espèce édaphique et le ▶pédobionte ; ▶édaphon) ; *g* **Bodenlebewesen** [n] (*UBe* ▶pflanzliches Bodenlebewesen und ▶tierisches Bodenlebewesen, ▶Bodenlebewelt); *syn.* Bodenorganismus [m].

5798 soil organisms [npl] *biol. pedol.* (Term covering both soil fauna and soil flora); ▶species of soil fauna); *s* **edafon** [m] (Conjunto de organismos que viven en el suelo; ▶animal edáfico); *syn.* edaphon [m]; *f* **édaphon** [m] (Ensemble des organismes vivant dans le sol ; *terme spécifique* animaux édaphiques ; ▶pédobionte) ; *g* **Bodenlebewelt** [f] (Gesamtheit der nur im Boden lebenden Organismen: Bodenfauna und Bodenflora; ▶tierisches Bodenlebewesen); *syn.* Bodenorganismen [mpl], Mikroorganismen [mpl] im Boden, Bodenleben [n, o. Pl.], Edaphon [n].

soil pattern [n] *pedol.* ▶soil distribution pattern.

soil phase [n] [US] *pedol.* ▶soil series [US], #2.

5799 soil pile [n] [US] *constr.* (Stockpile of stripped topsoil or excavated subsoil for later reuse; ▶storage of topsoil, ▶topsoil pile); *syn.* soil heap [n] [UK], soil store [n]; *s* **depósito** [m] **de tierra** (Montón ordenado de tierra vegetal o de otras capas de suelo excavadas para su posterior reutilización; ▶acopio de tierra vegetal, ▶depósito de tierra vegetal); *f* **dépôt** [m] **de terre** (Stockage en tas non tassé de la terre végétale décapée et des autres terres provenant des terrassements pour utilisation ultérieure en remblai ; ▶dépôt de terre végétale, ▶stockage de la terre végétale) ; *g* **Bodenmiete** [f] (Geordnetes Lager zur Wiederverwendung von abgetragenem Ober- oder Unterboden; ▶Oberbodenlagerung, ▶Oberbodenmiete); *syn.* Erdmiete [f].

soil piles [npl] *constr.* ▶stacked in soil piles [US]/stacked in soil heaps [UK].

5800 soil pocket [n] *constr.* (Wide masonry joint between two adjacent stones or a recess in a concrete or stone wall, for the insertion of a plant; ▶dry masonry, ▶paving joint vegetation); *s* **junta** [f] **de plantación** (Junta ancha en fábrica de mampostería o hueco en muro de hormigón para insertar plantas; ▶mampostería en seco, ▶vegetación de las juntas de adoquinado); *f* **niche** [f] **de plantation** (Réservation effectuée pendant la construction d'un mur par décalement de pierres pour créer un évidement qui sera rempli de terreau pour le fleurissement d'un mur ; ▶maçonnerie à sec, ▶végétation d'interstices de pavage) ; *syn.* poche [f] de terre, buse [f] de plantation, poche [f] de plantation ; *g* **Pflanzfuge** [f] (Breiter Spalt im Mauerwerk, oft in Trockenmauern, zwischen zwei nebeneinander liegenden Steinen zum Einsetzen von Pflanzen. Eine Aussparung im Betonmauerwerk heißt **Pflanzöffnung**; ▶Pflastergrün, ▶Trockenmauerwerk).

5801 soil preparation [n] *constr. hort. agr.* (Measures undertaken to provide optimum conditions for planting or sowing; ▶soil cultivation); *syn.* ground preparation [n], soil conditioning [n] (2); *s* **tratamiento** [m] **de superficies** (Medidas tomadas en el suelo para crear condiciones óptimas para la plantación o siembra; ▶cultivo de la tierra); *f* **travaux** [mpl] **de préparation du**

sol (Pour la plantation ou l'engazonnement ; ▶travail du sol) ; *syn.* mise [f] en état de la terre végétale ; *g* **Bodenvorbereitung** [f] (Lockerung und Verbesserung der Bodenstruktur sowohl des Unterbodens wie des Oberbodens für die Bepflanzung oder Aussaat gem. DIN 18 915 „Bodenarbeiten"; die Bearbeitkeitsgrenzen, z. B. Maschineneinsatz bei Nässe, der unterschiedlichen Bodenarten sind besonders wegen irreversibler Schädigungen des Bodengefüges mit entsprechenden negativen Auswirkungen auf das Pflanzenwachstum zu berücksichtigen; ▶Bodenbearbeitung).

soil pressure [n] *constr. eng.* ▶earth pressure.

5802 soil productivity [n] *agr. hort. pedol.* (▶potential yield); *s* **productividad** [f] **del suelo** (▶capacidad de rendimiento); *f* **productivité** [f] **du sol** (Capacité, aptitude d'un sol à la production d'une espèce ou d'une variété dans des conditions optimales de culture ; la productivité s'exprime par le ▶rendement) ; *g* **Bodenproduktivität** [f] (Ergiebigkeit des landwirtschaftlich oder gärtnerisch genutzten Bodens, gemessen in Rohertrag je Hektar; ▶Ertragsfähigkeit).

5803 soil profile [n] *pedol.* (Vertical section of soil from its surface through the major horizons A, B and C; ▶expose a soil profile, ▶exposed soil horizons, ▶outcrop, ▶truncated soil profile); *s* **perfil** [m] **del suelo** (Sección vertical de un suelo desde la superficie hasta la roca madre; ▶abrir una calicata, ▶afloramiento natural, ▶horizonte edáfico, ▶perfil aflorado, ▶suelo decapitado); *f* **profil** [m] **pédologique** (Coupe plane du sol distinguant un ensemble d'▶horizons successifs caractérisant l'évolution du sol ; dans la classification américaine le concept du profil est remplacé par celui du pédon ; ▶affleurement, ▶établir une fosse pédologique, ▶profil pédologique tranché, ▶sol décapé) ; *g* **Bodenprofil** [n] (Vertikale Abfolge der ▶Bodenhorizonte durch einen Bodenkörper von der Bodenoberfläche bis zum unverwitterten Ausgangsgestein; ▶Aufschluss, ▶Bodenaufschluss, ▶Bodenprofil freilegen, ▶geköpftes Bodenprofil).

soil properties [npl] *pedol.* ▶soil characteristics.

5804 soil quality analysis [n] *plan.* (Evaluation of a soil for a specific purpose; e.g. land use requirements, soil-bearing capacity for structures; ▶land capability classification [US], ▶on-site soil investigation, ▶soil analysis); *s* **valoración** [f] **del suelo** (Evaluación de las características de un suelo según demanda de usos, fertilidad, capacidad de carga, etc.; ▶análisis del suelo, ▶análisis del suelo in situ, ▶clasificación de aptitud de suelos); *syn.* evaluación [f] del suelo; *f* **évaluation** [f] **de la qualité du sol** (Pour son aptitude aux usages, pour sa fertilité ; ▶analyse du sol [au laboratoire], ▶analyse du sol sur site, ▶classification des aptitudes culturales et pastorales des sols) ; *g* **Bodenbewertung** [f] (2) (Bewertung des Bodens nach Nutzungsansprüchen, Fruchtbarkeit; ▶Bodenanalyse, ▶Bodenbeurteilung, ▶Bodenschätzung).

5805 soil reaction [n] *chem. pedol.* (Degree of acidity or alkalinity, usually expressed as a pH value; ▶acid soil, ▶alkaline soil); *syn.* pH value [n] of soil; *s* **reacción** [f] **(química) del suelo** (▶suelo alcalino, ▶suelo ácido); *syn.* pH [m] del suelo; *f* **réaction** [f] **du sol** (Degré d'acidité ou d'alcalinité d'un sol, habituellement exprimé par le pH ; ▶sol alcalin, ▶sol acide) ; *syn.* pH [m] du sol ; *g* **Bodenreaktion** [f] (▶alkalischer Boden, ▶saurer Boden); *syn.* pH-Wert [m] des Bodens.

5806 soil replacement [n] *constr.* (Exchange of unstable for load-bearing, or contaminated for uncontaminated, soil); *s* **recambio** [m] **del suelo** (Sustitución de suelo no portante por portante o de tierra contaminada por no contaminada); *syn.* sustitución [f] (de una capa) del suelo; *f* **remplacement** [m] **du sol** (Substitution d'un sol instable par un sol stable ou d'un sol contaminé par un sol non contaminé) ; *g* **Bodenaustausch** [m] (Auswechseln eines Bodens, der für eine vorgesehene Nutzung

S

nicht geeignet ist, durch einen geeigneten; z. B. einen nicht tragfähigen durch tragfähiges Material, einen verseuchten durch einen nicht verseuchten Boden oder einen stark mit Wurzelunkräutern belasteten durch einen unkrautfreien Boden).

soil resource inventory [n] [US] *geo. pedol.* ▶soil mapping.

5807 soil root ball [n] *hort.* (MET 1985, 4; ball-shaped solid mass of topsoil or other growing medium containing roots *in situ*, lifted undisturbed, as distinguished from a root ball with loose soil which is called an 'artificial' root ball; ▶bare-rooted plant, ▶root ball, ▶root-balled plant); *syn.* earth root-ball [n] [also UK] (SPON 1986, 236), root ball [n]; *s* **cepellón** [m] **de tierra** (Al contrario que la ▶planta con raíz desnuda, ▶cepellón de planta con sistema radical bien insertado en tierra o sustrato de plantación, que ha sido preparado para el trasplante; ▶planta de cepellón); *syn.* mota [f] [C]; *f* **motte** [f] **de terre** (Système radiculaire des végétaux de pleine terre préparé en vue de sa transplantation ou des plants élevés en conteneur ; ▶motte de racines, ▶plante en motte par opposition à ▶plante à racines nues ; les mottes façonnées avant la vente sont dénommées **mottes reconstituées** ou **fausses mottes** ; cette opération qui consiste à réduire de manière importante le système racinaire [absence de chevelu] et de l'entourer d'un substrat lors de son conditionnement dans une tontine, un filet grillagé ou un conteneur est en général responsable de la mauvaise reprise des végétaux après plantation et est à proscrire) ; *syn.* motte [f] ; *g* **Erdballen** [m] (Mit einem festen, gut durchwurzelten Erdkörper oder Pflanzsubstrat umschlossenes Wurzelsystem, das zum Verpflanzen ausgegraben wurde oder in einem Pflanzgefäß kultiviert wurde und ohne Ballenleinen fest zusammenhält; *vergleiche dazu* ▶Pflanze ohne Ballen. Ballen aus losem Erdoder Pflanzsubstrat werden **Kunstballen** genannt; ▶Ballenpflanze); *syn.* Wurzelballen [m] (1).

soils [npl] *pedol.* ▶hydromorphic soils, ▶physical properties of soils, ▶terrestrial soils.

soils [npl]**, hydric** *pedol.* ▶hydromorphic soils.

5808 soil sample [n] *constr. pedol.* (Representative core of soil specimen from a specific site, used for testing its chemical and physical properties; ▶sample pit, ▶sampling 1, ▶soil sampling); *s* **muestra** [f] **edáfica** (▶calicata, ▶toma de muestra[s] 1, ▶toma de muestra[s] 2); *syn.* muestra [f] de un suelo; *f* **échantillon** [m] **du sol** (Petite quantité de sol prélevée lors d'un ▶échantillonnage de sol en vue d'une analyse ; ▶prélèvement d'échantillons de sol, ▶prélèvement d'échantillons, ▶zone de reconnaissance) ; *g* **Bodenprobe** [f] (Entnommene Menge eines Bodens oder Schüttgutes; ▶Bodenentnahmestelle, ▶Bodenprobenahme, ▶Probenahme).

5809 soil sampling [n] *constr. pedol.* (Removal of representative soil from an area to determine its physical and chemical properties or to identify the ▶great soil group [US]/soil group [UK] or a ▶soil series [US]); *s* **toma** [f] **de muestra(s)** (2) (Extracción de una porción representativa de un suelo para determinar las características físicas del mismo o para determinar el ▶tipo de suelo; ▶serie de suelos); *f 1* **prélèvement** [m] **d'échantillons de sol** (Afin d'effectuer des analyses ou essais en laboratoire et de déterminer le ▶groupe de sol et les caractéristiques physiques et mécaniques des terrains rencontrés ; ▶sous-groupe de sol) ; *syn.* prise [f] d'échantillons de sol ; *f 2* **échantillonnage** [m] **de sol** (Ensemble des opérations pour la détermination d'un échantillon de sol) ; *g* **Bodenprobenahme** [f] (Entnehmen von Boden aus einem Schüttgut oder mit einem Pürckhauer-Bohrstock aus einer repräsentativen Fläche zur Untersuchung bodenphysikalischer Eigenschaften, der Nährstoffversorgung oder zur Bestimmung des ▶Bodentyps 1; ▶Bodentyp 2); *syn.* Boden-

probenentnahme [f], Entnahme [f] von Bodenproben, Ziehen [n, o. Pl.] von Bodenproben.

5810 soil science [n] *pedol.* (Scientific study of the origin and classification of soils); *syn.* pedology [n]; *s* **edafología** [f] (Ciencia que estudia la génesis, las características y la clasificación de los suelos); *syn.* ciencia [f] del suelo, pedología [f]; *f* **pédologie** [f] (Étude des constituants et des propriétés physiques, chimiques et biologiques des sols ainsi que de la pédogenèse et la formation des profils) ; *syn.* édaphologie [f], agrologie [f] ; *g* **Bodenkunde** [f] (Wissenschaft vom Boden, die die Bodenentstehung und Entwicklung [Bodengenese: spezielle **B.**], die im Boden ablaufenden physikalischen, chemischen und biologischen Prozesse erforscht und die Zusammensetzung der Böden [allgemeine **B.**] und ihre land- und forstwirtschaftliche Nutzungseigenschaften und Verbesserungsmöglichkeiten untersucht und aufzeigt [angewandte **B.**]); *syn.* Pedologie [f], *selten* Edaphologie [f].

5811 soil seed bank [n] *bot.* (TEE 1980, 60; seeds existing in the soil, which have yet to germinate); *syn.* seed store [n] (2); *s* **reserva** [f] **de semillas en el suelo** (Conjunto de semillas existentes en el suelo, aun sin germinar); *f* **réserve** [f] **de semences dans le sol** (Totalité des graines dans le sol susceptibles de germer) ; *g* **Samenspeicher** [m] (Summe der im Boden vorhandenen keimfähigen Samen).

soil separate group [n] [US] *constr. (pedol.)* ▶graded sediment group [US]/particle-size group [UK].

soil separator [n] *constr.* ▶filter mat.

soil separator sleeve [n]**, drain pipe with** *constr.* ▶perforated drain pipe with woven fabric soil separator sleeve.

5812 soil series [n] [US] *pedol.* (1. Lowest class of soils in the United States Comprehensive Soil Classification System, which have horizons similar in differentiating characteristics and arrangement in the soil profile; RCG 1982, WPG 1976); *syn.* soil taxonomic unit [n] (RCG 1982, 161), soil taxon [n]; **2. soil phase** [n] [US] (Subdivision of a *soil taxonomic unit*, usually a *soil series* or other unit of classification based on characteristics that affect the use and management of the soil, the degree of slope, degree of erosion, content of stones, etc.; WPG 1976); *s* **serie** [f] **de suelos** (En la clasificación de los EE.UU., la unidad taxonómica inferior del sistema); *syn.* unidad [f] taxonómica pedológica; *f* **sous-groupe** [m] **de sols** (F., dernière subdivision dans la nomenclature de classification des sols ; U.S., la plus petite unité taxinomique de sol pouvant être cartographiée, dénommée d'après le lieu où elle est caractérisée) ; *syn. classification US* série [f] de sol ; *g* **Bodentyp** [m] (2) (U.S., Lokale Bodenformen der *Great Soil Groups* als unterstes Glied der Bodenklassifikation).

5813 soil slippage [n] *constr.* (1. Type of erosion in ▶earthworks whereby part of the topsoil slips off the ▶slip plane parallel to the slope; frequently caused by ▶impeded water in top soil or saturation of the subsoil; ▶bearing capacity failure, ▶shear failure of embankment, ▶slope failure, ▶soil creep 1. **2.** In case of fluid state of the material the slippage is called ▶flow slide); *s* **deslizamiento** [m] **del suelo** (1. Forma de erosión en ▶terracerías por la que parte de la tierra vegetal resbala sobre un ▶plano de deslizamiento frecuentemente paralela al talud, debido al encharcamiento del suelo o la saturación del subsuelo; ▶caída de un talud, ▶deslizamiento planar, ▶deslizamiento rotacional, ▶solifluxión. **2.** En el caso de estado fluído del material el d. se llama ▶corrimiento por flujo); *f* **foirage** [m] (Forme d'érosion consécutive à des ▶terrassements ; mise en mouvement d'une partie des terres rapportées [matériaux meubles] sur un ▶plan de glissement généralement parallèle à la pente du talus [▶glissement de talus] ; phénomène provoqué par un engorgement de la couche superficielle ou un ramollissement de la partie supérieure des terrassements ; *terme spécifique* **coup de cuillère** [mouvement de masse rapide sur un talus, la partie supérieure forme une petite niche de foirage suivie d'un chenal légèrement creux et à l'extrémité duquel s'étalent les matériaux ; VOG 1979] ; ▶fluage, ▶glissement rotationnel, ▶rupture de terrain, ▶solifluxion) ; *g* **Rutschung [f]** (Erosionsform beim ▶Erdbau, bei dem Teile des angedeckten Oberbodens auf einer ▶Gleitfläche meist parallel zur Böschungsebene abrutschen; häufig durch Staunässe im Oberboden resp. Aufweichen des Untergrundes ausgelöst; ▶Bodenfließen, ▶Geländebruch, ▶Grundbruch, ▶Hanggleiten, ▶Schlammrutschung).

soil-slippage-prone slope [n] [US] *geo. pedol.* ▶unstable slope.

5814 soil stabilization [n] *constr.* (Measures taken to keep soil in place and prevent erosion; ▶brushwood thatching, ▶ground cover, ▶soil consolidation for road construction); *s* **fijación [f] del suelo** (Medidas de protección del suelo contra la erosión; ▶capa protectora de desbrozo y leña menuda, ▶cobertura vegetal del suelo, ▶estabilización del suelo, ▶revestimiento [del suelo] con ramas secas); *f* **fixation [f] du sol** (Mesures de protection contre l'érosion ; ▶couverture de branchage, ▶couvert végétal du sol, ▶fixation du sol au moyen de ramilles, ▶stabilisation du sol) ; *syn.* maintien [m] du sol en place, consolidation [f] du sol ; *g* **Bodenfestlegung [f]** (Erosionsschutzmaßnahmen; cf. DIN 18 915, Bl. 2; ▶Abdeckung mit einem Rauwehr, ▶Bodenbedeckung, ▶Bodenverfestigung); *syn.* Bodenbefestigung [f], Bodenstabilisierung [f], Oberflächenfestlegung [f] (DIN 18 918, Kap. 2.2.4).

soil stabilization [n] **for road construction** *constr. eng.* ▶soil consolidation for road construction.

soil stabilization [n] **with brushwood** [US] *constr.* ▶brushwood thatching.

soil stockpile [n] [US] *constr.* ▶soil storage.

5815 soil storage [n] *constr.* (Temporary storage of soil during construction operations; ▶keep soil beside an excavation [US]/leave soil lying by side of an excavation [UK]; ▶soil pile [US]/soil heap [UK], ▶topsoil pile [US]/topsoil heap [UK], ▶storage of topsoil); *syn.* soil stockpile [n] [also US]; *s* **acopio [m] de tierra** (Almacenamiento provisional del suelo durante una obra o en la extracción minera a cielo abierto; ▶acopio de tierra vegetal, ▶acopio de tierra vegetal [en obra], ▶depósito de tierra, ▶depósito de tierra vegetal); *f* **stockage [m] (sur place) du sol** (Mise en réserve temporaire de terre pendant la durée du chantier ; ▶dépôt de terre, ▶dépôt de terre végétale, ▶stockage de la terre végétale, ▶terre jetée sur le côté) ; *syn.* mise [f] en dépôt de la terre ; *g* **Bodenlagerung [f]** (Bodenzwischenlagerung — getrennt nach Ober- und Unterboden — während des Baubetriebes; ▶Boden seitlich lagern, ▶Bodenmiete, ▶Herstellung von Oberbodenmieten, ▶Oberbodenmiete); *syn.* Lagerung [f] von Boden.

soil store [n] *constr.* ▶soil pile [US]/soil heap [UK].

5816 soil structure [n] *pedol.* (Spatial arrangement and aggregation of individual solid particles within the soil; ▶crumb structure, ▶granular soil structure); *s* **textura [f] del suelo** (Conjunto de propiedades físicas que son el resultado directo de la dimensión de sus constituyentes y está relacionada con la composición granulométrica; EDAFO 1982, 24; ▶estado grumoso, ▶textura grumosa); *f* **structure [f] du sol** (Arrangement spécial des particules minérales du sol et leur éventuelle liaison par des matières organiques, des hydroxydes de fer ou d'alumine ou des deux. L'assemblage de ces particules solides forme des unités structurales qui sont séparées par des surfaces de moindre résistance ; ces unités sont appelées agrégats élémentaires ; DIS

1986 ; ▶état structural grumeleux, ▶structure fragmentaire) ; *g* **Bodenstruktur [f]** (Räumliche Anordnung und Verbindung der festen Einzelteile eines Bodens; ▶Krümelstruktur 1, ▶Krümelstruktur 2); *syn.* Bodengefüge [n].

5817 soil suborder [n] [US] *pedol.* (**In U.S.**, next inferior category following the 'orders' in the USDA soil classification system to reflect the degree of horizon development and the kinds of horizons present; the **s. s.** name has two syllables. The first syllable connotes something of the diagnostic properties of the soil, and the second syllable is the formative element or syllable for the order; e.g. Aqualfs: aqu is derived from Latin *aqua*, meaning water, and alf from the order Alfisols; cf. FOS 1990; **in U.K.**, extensive groups of soils, having similar internal characteristics, are distinguished: **great soil groups** include tundra soils, desert soils, podzol soils, chernozem soils; ▶great soil group [US]/soil type [UK], ▶hydromorphic soils, ▶man-made soil, ▶peatland [US]/moor [UK], ▶soil order [US]/major soil group [UK], ▶terrestrial soils); *syn.* soil group [n] (±) [UK]; *s 1* **clase [f] de suelos** (En D., categoría de clasificación de suelos que sigue a la de los ▶grupos principales de suelos, en la cual la clase de los ▶suelos terrestres se subdividen según el grado de desarrollo resp. el grado de diferenciación de los horizontes, la clase de los ▶suelos semiterrestres con los suelos de agua estancada, los suelos de inundación y los suelos salinos y la clase de los ▶suelos subacuáticos con los suelos subacuático no turbosos y los suelos subacuáticos turbosos; ▶cultosol, ▶orden de suelos, ▶tipo de suelo, ▶turbera); *s 2* **suborden [m] de suelos** (Según la taxonomía de suelos del *Soil Survey Staff* de los EE.UU. [1960], segundo nivel jerárquico de clasificación de los suelos según los criterios como hidromorfismo, influencia del clima y la vegetación, texturas extremas, etc. —indicados en la primera sílaba de la denominación como p. ej. «Aqua»— y el elemento formativo del orden —indicado en la segunda sílaba como p. ej. «Alf», es decir del orden de los alfisoles: Aqualfs); *f* **sous-classe [f] pédologique** (**F.**, catégorie pédologique inférieure dans la classification des sols, caractérisant des variantes du processus de bases induites par de conditions physico-chimiques, un processus secondaire modifiant une ▶classe pédologique, un degré d'altération. *Contexte* classe de sols brunifiés, sous classe : sol vertique ; PED 1983, 195 ; **U.S.**, la classification est définie par le pédoclimat traduisant le régime d'humidité et le régime thermique ; **D.**, catégorie pédologique inférieure établie sur la base du degré d'évolution du profil, l'altération climatique, les mouvements de matière, les cycles d'évolution ; ▶groupe de sols, ▶sols anthropogènes, ▶sols hydromorphes, ▶sols subaquatiques, ▶sols terrestres, ▶tourbière) ; *g* **Bodenklasse [f] (±) (2)** (**D.**, pedologische Klassifizierungskategorie, die der ▶Bodenabteilung direkt nachgeordnet ist: z. B. wird die Klasse der ▶Landböden nach dem Entwicklungsstand resp. nach dem Grad der Differenzierung der Horizonte untergliedert; die Klassen der **hydromorphen Böden** [▶Grundwasser- und Stauwasserböden] sind die Stauwasserböden, die Grundwasserböden [Gleye] sowie die mit Süßwasser überfluteten Auenböden [▶Unterwasserboden], die mit Salzwasser überfluteten Marschen und die Klassen der ▶anthropogenen Böden; **U.S.**, im amerikanischen Bodenklassifizierungssystem entspricht hinsichtlich der Hierarchiestufe die **B.** der Unterordnung *[suborder]*; das *USDA Soil Classification System* ist mit dem deutschen nicht direkt vergleichbar: siehe dazu die Ausführungen unter ▶Bodenabteilung; ▶Bodentyp 1, ▶Moor).

5818 soil survey [n] *pedol.* (Systematic examination of ▶great soil groups in the field and in laboratories, their description and classification, mapping, interpretation for growing purposes, behavio[u]r under use and treatment for plant production and productivity under different management systems; cf.

S

RCG 1982; ▶soil mapping); *s* **investigación [f] edáfica** (Estudio sistemático de los ▶tipos de suelos *in situ* y en laboratorio para describirlos y clasificarlos para fines productivos; ▶cartografía de suelos); *syn.* estudio [m] de suelos; *f* **relevé [m] pédologique** (Reconnaissance et prospection systématique des ▶groupes de sols sur le terrain afin de définir les unités-sols, de dresser l'inventaire des sols pour une région déterminée et analyses physico-chimiques en laboratoires en vue de l'établissement de cartes des sols ; ▶cartographie des sols) ; *syn.* relevé [m] des sols ; *g* **Bodenaufnahme [f] (2)** (*Vorgang* Erfassung der ▶Bodentypen 1 eines Untersuchungsgebietes zur Erstellung einer Bodenkarte; **B.n** erfolgen je nach Fragestellung: **1.** Zur Untersuchung der Bodengenese werden horizontbezogene Proben mittels Stechzylinder, i. d. R. bis zum Ausgangsgestein, gezogen, die dann nach physikalischen, chemischen und biologischen Parametern untersucht werden. **2.** Bei Schadstoffen im Boden werden je nach erwartungsmäßiger Tiefe der Verunreinigungen Rammkernsondierungen durchgeführt und die Proben chemisch und biologisch untersucht. **3.** Bei der Erkundung der Standsicherheit eines Bodens hinsichtlich seiner Eignung als Baugrund werden Rammkernsondierungen nach physikalischen Größen wie Bodendichte, Korngrößenverteilung, Wassergehalt etc. untersucht. **4.** Bei der Frage nach pflanzenverfügbaren Nährstoffen werden mit einem Bohrzylinder mit definiertem Volumen im durchwurzelbaren Raum Proben gezogen und nach chemischen, biologischen und physikalischen Parametern untersucht; ▶Bodenkartierung); *syn.* Bestandsaufnahme [f] von Böden.

soil taxon [n] *pedol.* ▶soil series [US].

soil taxonomic unit [n] *pedol.* ▶soil series [US].

5819 soil textural class [n] *constr. pedol.* (Classification of soil according to the particle size distribution of the three main mineral components, expressed in terms of the percentages of clay, silt and sand; ▶particle-size class, ▶soil texture diagram); *s* **clase [f] textural** (Clasificación de suelos según la composición granulométrica dominante que depende de los tres componentes sólidos principales y sus porcentajes: arcilla, limo y arena; ▶clase granulométrica, ▶diagrama de granulometría); *f* **classe [f] de texture de sol** (Sol à texture argileuse, limoneuse, sableuse ou équilibrée : classement des minéraux des sols établi sur la base des 4 principales fractions de particules classées en fonction de leur diamètre : en F. on distingue la fraction dite de terre fine telle que les ultra argiles < 0,2 microns, les argiles 0,2 < D < 2 microns, les limons 2 < D < 20 microns, les sables fins 20 < D < 200 microns, les sables grossiers 0,2 < D < 2 mm et la fraction grossière telle que les graviers 2 < D < 20 mm, les cailloux 20 mm < D < 200 mm et les rochers > 200 mm ; ▶classe granulométrique, ▶diagramme de texture des sols) ; *g* **Bodenart [f]** (**1.** *Mineralische Bodenart* Unterscheidung nach Korngrößen der drei Hauptbodenarten Ton, Schluff und Sand, die im Deutschen auch als Körnungsklassen bezeichnet und in ▶Bodenartendiagrammen übersichtlich dargestellt werden. **2.** Im Erdbau wird unter **B.** die Einteilung in „nicht bindiger" und „bindiger" Boden verstanden; ▶Korngrößenbereich); *syn.* Körnungsklasse [f].

5820 soil texture diagram [n] *pedol.* (Three-coordinate graph for showing the textural class of a soil according to the relative content of clay, silt, and sand); *s* **diagrama [m] de granulometría** (Representación gráfica con tres coordinadas para determinar la clase de textura de un suelo según los porcentajes de arcilla, limo y arena en la composición mineral del mismo); *syn.* esquema [m] de granulometría; *f* **diagramme [m] de texture des sols** (Représentation graphique de la classification des textures matérialisée à l'aide d'un triangle dont les côtés correspondent respectivement aux pourcentages de sable, de limon et d'argile) ; *syn.* triangle [m] des textures, diagramme [m] triangulaire des textures, triangle [m] textural ; *g* **Bodenartendia-**

gramm [n] (Grafische Dreiecksdarstellung, mit der die mineralische Zusammensetzung eines Bodens entsprechend den Körnungsklassen mit den Anteilen an Ton, Schluff, Lehm und Sand bestimmt werden kann); *syn.* Dreiecksdiagramm [n] der Körnungsklassen.

5821 soil transport [n] *constr.* (Movement of excavated soil on a building site or removal to a storage area); *s* **transporte [m] de tierra (vegetal)** (Traslado del suelo excavado dentro del terreno de obra o a almacén provisional o a vertedero); *f* **transport [m] de la terre** (Déplacement de la terre à l'intérieur du chantier ou chargement et mise en dépôt sur une aire de stockage) ; *g* **Fördern [n, o. Pl.] des Bodens** (Transport des gelösten Bodens innerhalb der Baustelle zur Einbaustelle oder Abtransport auf ein Zwischenlager resp. auf eine Absatzkippe; LEHR 1981); *syn.* Fördern [n, o. Pl.] von Boden.

soil type [n] [UK] *pedol.* ▶great soil group [US].

5822 soil water [n] *pedol.* (All water that exists in the soil, including groundwater and humidity; *specific terms* ▶available soil water, ▶capillary water, ▶groundwater [US]/ground water [UK], ▶seepage water, ▶soil moisture, ▶superheated soil water); *s* **agua [f] edáfica** (Toda el agua existente en el suelo, incluida el agua subterránea y la humedad; *términos específicos* ▶agua capilar, ▶agua de percolación, ▶agua edáfica realmente combinada, ▶agua subterránea, ▶humedad del suelo); *syn.* holardía [f], agua [f] del suelo; *f* **eau [f] du sol** (Eau de ruissellement captée par le sol ; *termes spécifiques* ▶eau capillaire, ▶eau de gravité, ▶eau liée, ▶humidité du sol, ▶nappe phréatique) ; *g* **Bodenwasser [n]** (Alles im Boden vorkommende Wasser, ▶Grundwasser und ▶Bodenfeuchte. *UBe* ▶Kapillarwasser, ▶Kristallwasser, ▶Sickerwasser).

5823 soil wetness [n] *pedol.* (MET 1985, 5; condition of soil with impeded drainage having a high level of ▶soil moisture, where most of the soil air is replaced by water; ▶impeded water, ▶poorly drained soil, ▶waterlogged soil [US]/water-logged soil [UK], ▶waterlogging); *s* **alta humedad [f] del suelo** (▶agua estancada, ▶encharcamiento del suelo, ▶humedad del suelo, ▶suelo encharcado, ▶suelo saturado); *f* **engorgement [m] du sol** (Très fort taux d'humidité du sol lorsque la majorité des macropores est occupée par l'eau ; ▶humidité du sol, ▶imbibition [capillaire], ▶nappe perchée, ▶processus d'engorgement du sol, ▶sol engorgé, ▶sol saturé par l'eau) ; *g* **Bodennässe [f]** (Sehr hohe ▶Bodenfeuchte, bei der der größte Anteil der Bodenluft durch Wasser ersetzt ist; ▶Bodenvernässung, ▶Staunässe, ▶wassergesättigter Boden, ▶staunasser Boden).

solar exposure [n] *met.* ▶insolation.

5824 solar-generated current [n] *envir. ecol.* (Electricity generated directly with the use of the sun's energy stored in the cells of a solar energy system, whereby sunlight is used to convert hydrogen into helium and heat is released); *syn.* solar-generated electricity [n]; *s* **electricidad [f] solar** (Corriente eléctrica producida por medio de células fotovoltaicas instaladas en panel solar gracias a la transformación de la luz del sol); *syn.* corriente [f] eléctrica solar; *f* **courant [m] solaire** (Courant électrique produit par les cellules des panneaux solaires résultant de la conversion directe de la lumière du soleil libérée lors de la transformation de l'hydrogène en hélium) ; *g* **Solarstrom [m]** (Erneuerbare elektrische Energieform, die mit Hilfe von Solarzellen, die in Photovoltaikmodulen zusammengefasst sind, in einer Solaranlage als Ergebnis der direkten Nutzung des Sonnenlichtes erzeugt wird [Photovoltaikenergie]. **S.** kann durch Fotovoltaikanlagen und in Sonnenwärmekraftwerken [z. B. Solarfarmkraftwerk, Aufwindsolarkraftwerk] gewonnen werden).

5825 solar orientation [n] *gard.* (TGG 1984, 80; slope which directly faces the midday sun: in the northern hemisphere

S

this is synonymous with ▶south orientation; in the southern hemisphere the corresponding term is ▶north orientation; *opp.* ▶shady location); *syn.* orientation [n] to sun (TGG 1984, 80); *s* **solana [f]** (Lugar orientado al sol, donde da el sol más tiempo y más plenamente; en el hemisferio norte es sinónimo de ▶orientación sur, en el hemisferio sur de ▶orientación norte; *opp.* ▶umbría; *f* **exposition [f] ensoleillée** (Versant exposé directement au soleil de midi ; *syn. dans l'hémisphère nord* ▶exposition au sud, dans l'hémisphère austral correspondant à l'▶orientation vers le nord ; *opp.* ▶exposition [très] ombragée) ; *g* **Sonnenlage [f]** (Der Mittagssonne direkt ausgesetzte Hanglage. Auf der nördlichen Halbkugel syn. mit ▶Südlage, korrespondiert in der südlichen Hemisphäre mit ▶Nordlage; *opp.* ▶Schattenlage).

5826 solar power station [n] *envir.* (Power plant which produces energy from sunlight by means of solar-thermic or solar-electric cells/units); *s 1* **central [f] solar** (Instalación para producir energía a partir de la radiación solar utilizando dispositivos de conversión térmica o conversión fotovoltaica); *syn.* central [f] de energía solar, horno [m] solar; *s 2* **central [f] termosolar** (Instalación para producir energía eléctrica a partir de la radiación solar utilizando dispositivos de conversión térmica. En Es., en Sanlúcar la Mayor, Sevilla, está en funcionamiento desde finales de febrero de 2007 la primera **c. t.** de torre comercial del mundo. Por medio de helioestatos se capta la energía fotovoltaica, que es concentrada en el receptor solar. Éste genera vapor saturado con el que se alimentan las turbinas para producir energía termoeléctrica; cf. wikipedia.es consulta 1.11.2009); *syn.* central [f] térmica solar; *f* **centrale [f] solaire** (Installation produisant de l'énergie à partir de la lumière solaire au moyen de cellules thermiques ou électriques) ; *g* **Sonnenkraftwerk [n]** (Anlage zur Energiegewinnung aus Sonnenlicht durch Zusammenschaltung von solarthermischen oder solarelektrischen Einheiten; in Europa werden zz. in Spanien die ersten kommerziellen thermischen Solarkraftwerke betrieben; cf. SZ 2008); *syn.* Solarkraftwerk [n].

5827 solar radiation [n] *met.* *s* **radiación [f] solar** (Energía radiante procedente del sol); *f* **radiation [f] solaire** ; *g* **Sonnenbestrahlung [f]**.

soldier course [n] *constr.* ▶rowlock course [US]/layer of headers-on-edge [UK].

5828 sole source contract award [n] [US] *contr.* (Awarding of a contract without previous ▶bidding [US]/tendering [UK] procedure); *syn.* direct selection [n] and negotiation [n] [US], freely awarded contract [n] [UK]; *s* **adjudicación [f] negociada** (Procedimiento de ▶adjudicación de contratos sin previo ▶concurso subasta de obras por el que la Administración elige justificadamente al empresario, previa consulta y negociación de los términos del contrato con uno o varios empresarios; cf. art. 74.4 Ley 13/1995); *syn.* adjudicación [f] directa, contratación [f] directa/negociada, adjudicación [f] a/de mano libre; *f* **passation [f] du/des marché(s) sous forme de marché négocié** (La personne responsable du marché attribue librement le marché au candidat qu'elle a retenu [sans mise en concurrence préalable] ; ▶appel d'offres) ; *syn.* passation [f] des marchés négociés ; *g 1* **freihändige Vergabe [f]** (Direkte Vergabe eines Auftrages an einen Auftragnehmer ohne förmliches Verfahren [§ 3 (1) 3 VOB Teil A], wenn die öffentliche oder beschränkte Ausschreibung unzweckmäßig ist [§ 3 (4) VOB Teil A]); *g 2* **Verhandlungsverfahren [n]** (Im **V.**, das an Stelle der freihändigen Vergabe tritt, wendet sich der Auftraggeber an ausgewählte Unternehmer und verhandelt mit einem oder mehreren dieser Unternehmer über den Auftragsinhalt, die Qualität der Ausführung, Preise und Termine; cf. § 3b (1) lit. c VOB Teil A; ▶Ausschreibung).

solid board fence [n] [US] *constr.* ▶close-boarded fence.

solid bracing [n] [UK] *arb. constr.* ▶cavity bracing.

solid newel stair [n] *arch.* ▶spiral stair(case).

5829 solid rock [n] *geo.* (Generic term covering magmatic, solidified sediments or metamorphic rocks often underlying superficial strata; *specific term* bedrock; ▶loose sedimentary rock); *syn.* hard rock [n] [also US]; *s* **roca [f] dura** (Término genérico que cubre las rocas eruptivas [magmáticas], las metamórficas o los sedimentos sólidos; ▶roca sedimentaria no consolidada); *f* **roche [f] massive** (Terme générique relatif aux roches en général non altérées endogènes, exogènes consolidées ou métamorphiques ; ▶roche sédimentaire meuble) ; *syn.* roche [f] dure ; *g* **Festgestein [n]** (Im Gegensatz zum ▶Lockergestein festes Gestein [Fels], das insbesondere zu den Hauptgesteinsgruppen Magmatite, Metamorphite und Sedimentite gehört).

solid wall [n] [US] *constr.* ▶one-tier masonry [US]/one-leaf masonry [UK].

5830 solifluction [n] *pedol.* (Slow form of downward ▶mass slipage [US]/mass movement [UK] of waterlogged soil, normally 0.5-5.0 m/yr above frozen ground; ▶creeping soil on frozen ground, ▶soil creep 1, ▶soil creep 2; *syn.* solifluxion [n]; *s* **sɛlifluxión [f]** [sic!] (Tipo de ▶movimiento gravitacional lento del suelo o materiales alterados sobre las laderas y pendiente abajo, resultado de combinación de gravedad y fenómenos de expansión y contracción debidos a ciclos de hielo-deshielo; ▶reptación hídrica del suelo, ▶solifluxión, ▶tierra fangosa reptante); *f* **gélifluxion [f]** (▶Mouvement de masse lent et descente le long d'un versant de matériaux imbibés d'eau provenant du dégel, en général sous climat froid et sur sous-sol gelé ; DG 1984, 419 ; ▶reptation d'un versant, ▶solifluxion, ▶terre fluente) ; *g* **Solifluktion [f]** (Langsame Form des ▶Massenversatzes: ▶Hangkriechen über gefrorenem Untergrund [0,5-5,0 m/Jahr, meist in Regionen mit Dauerfrostböden; ▶Bodenfließen, ▶Fließerde).

solifluxion [n] *pedol.* ▶solifluction.

soligenous peatland [n] *geo.* ▶fen (1).

solitary [adj] *landsc.* ▶isolated (2).

solubility, of low [loc] *chem. pedol.* ▶poorly soluble.

soluble [adj] *chem. pedol.* ▶easily soluble, ▶poorly soluble, ▶water soluble.

5831 sorrel-top [n] [US&CDN] *arb. for.* (Initial phase of ▶dieback of conifers, caused by pests [e.g. bark beetles] or pathogens; the advanced phase is called **red-top**, and the final phase, when all the needles have fallen, ▶black-top [US&CDN] 2, ▶inital phase of top-kill, ▶opening up of tree crowns; cf. SAF 1983); *s* **puntiseco [m] incipiente en conífera** (≠) (Fase inicial de la muerte de coníferas [▶puntiseco incipiente], causada por plagas o organismos patógenos que consiste en el enrojecimiento y enmarronamiento de la parte superior de la copa; en EE.UU. la fase avanzada se denomina «red-top» y la fase final, cuando se han caido todas las agujas, «black-top»; ▶aclarado de la copa, ▶muerte regresiva de plantas, ▶puntiseco en conífera); *f* **rougissement [m] et brunissement [m] des conifères** (Phénomène provoqué par les agents pathogènes [p. ex. le rouge], la dessiccation hivernale ou la canicule estivale entraînant le brunissement ou le rougissement des aiguilles accompagné de la chute prématurée de celles-ci ; la dessiccation hivernale [surévaporation hivernale] se produit en hiver ou au début du printemps, alors que des journées tièdes, ensoleillées et venteuses provoquent une forte évaporation alors que le sol est encore gelé ; la canicule estivale provoque une rupture complète d'alimentation en eau du houppier, l'embolie des vaisseaux ligneux, suivie de brunissement ou rougissement

[▶phase initiale du dépérissement terminal] puis la chute des aiguilles en cime, le dépérissement et la mort progressive de l'arbre [▶asphyxie] ; la température élevée, combinée avec une sécheresse prolongée peut entraîner une mort très rapide des résineux ; ▶éclaircie de cime) ; *g* **Rötungen und Verbraunungen [f] bei Koniferen** (Beginnendes ▶Absterben der obersten Koniferenkrone; bei Schadensfortschritt tritt die sog. **Nadelröte** ein, Endstadium ist die völlige **Entnadelung** oder völlige ▶Verlichtung 2; ▶beginnende Wipfeldürre).

sorting [n], **waste** *envir. leg.* ▶waste separation.

5832 sound-absorbing [adj] *envir.* (Capable of reducing the effect or intensity of noise); *s* **absorbente [adj] acústico** (Capacidad de materiales de amortiguar el ruido y las vibraciones); *f* **absorbant, ante le bruit [loc]** (Constitué de telle manière que le bruit est absorbé) ; *g* **schallmindernd [ppr/adj]** (So beschaffen, dass Schall absorbiert wird); *syn.* schallabsorbierend [ppr/adj], schallschluckend [ppr/adj].

5833 sound absorption [n] *envir.* (Reduction of spread and the effect or intensity of noise); *s* **absorción [f] acústica** (Reducción de la transmisión y del efecto o la intensidad de ruidos y vibraciones por medio de medidas constructivas adecuadas); *f 1* **atténuation [f] du niveau sonore** (Réduction à la source ou par ouvrages interposés du niveau sonore) ; *f 2* **affaiblissement [m] phonique** (Isolation phonique à l'intérieur d'un logement ; réduction de la différence de niveau sonore entre une pièce dans laquelle sont émises des ondes sonores et la pièce voisine ; DTB 1985) ; *g* **Schallminderung [f]** (Reduzierung der Schallausbreitung); *syn. ugs.* Lärmminderung [f].

sound barrier [n] *envir.* ▶noise screening facility.

sound control [n] [US] *envir.* ▶noise control.

5834 sound pressure level [n] *constr. envir. leg.* (Measurement of environmental noise nuisance in decibels); *syn.* decibel level [n] [also UK]; *s* **nivel [m] de intensidad de ruido** (Grado de contaminación acústica medido en decibelios [dB(A)]); *syn.* nivel [m] de intensidad de sonido; *f* **niveau [m] de pression acoustique** (Grandeur exprimée en décibels du bruit perçu dans un lieu donné ou émis par une source) ; *g* **Schalldruckpegel [m]** (Kennwert für den von z. B. einer Maschine am zugeordneten [Arbeits-/Betriebs]ort erzeugten Schall, der in Dezibel [dB(A)] das Niveau der akustischen Umweltbelästigung ausdrückt).

5835 sound wood [n] *arb. hort.* (Wood which is not infected or decayed); *s* **madera [f] sana** (Tejido vegetal leñoso no infectado ni podrido); *f* **bois [m] sain** (Bois non endommagé ou tissu de végétaux ligneux non infecté par une maladie) ; *g* **gesundes Holz [n, o. Pl.]** (Unbeschädigtes und durch keine Krankheit beeinträchtigtes Gewebe eines Gehölzes).

source [n] *envir.* ▶noise source, ▶nonrenewable power source [US]/non-renewable power source [UK], ▶pollution source, ▶power source.

source [n], **alternative power** *envir.* ▶renewable power source.

source [n], **emission** *envir.* ▶pollution source.

source [n], **energy** *envir.* ▶power source.

source [n], **nutritional** *conserv. zool.* ▶nutritional resources.

source [n], **sustainable power** *envir.* ▶renewable power source.

source area [n] *geo. plan.* ▶avalanche source area, *met.* ▶cold air source area.

source [n] **of natural water flow** *geo.* ▶rising spring.

5836 south-facing slope [n] *geo. met.* (Slope headed south; *syn. for the northern hemisphere* equator-facing slope [n]

[also US] (ARB 1983, 118; ▶solar orientation); *s* **vertiente [f] sur** (En el hemisferio norte, ladera de montaña orientada al sur o al suroeste, que recibe la mayor cantidad de sol posible en la ubicación, por lo que es más luminosa y caliente que las otras vertientes; en el hemisferio sur corresponde a la vertiente norte; ▶solana); *syn.* ladera [f] sur, vertiente [f] meridional; *f* **adret [m]** (Versant d'une montagne exposé vers le sud dans l'hémisphère nord ; ▶exposition ensoleillée) ; *syn.* versant [m] sud ; *syn. dans l'hémisphère nord* versant [m] ensoleillé, versant [m] du soleil, versant [m] exposé au soleil ; *g* **Südhang [m]** (Hangfläche eines Berges, die gegen Süden geneigt ist; ▶Sonnenlage); *syn. auf der Nordhalbkugel* sonnenexponierter Hang [m].

5837 south orientation [n] *geo. urb.* (TGG 1984, 77; south-facing direction of slopes, building elevations, walls, etc.; in the northern hemishere the corresponding term is ▶solar orientation); *s* **orientación [f] sur** (▶solana); *syn. en el hemisferio norte* exposición [f] de solana; *f* **exposition [f] au sud** (Versant, façade, mur, etc. exposé vers le sud) ; *syn.* orientation [f] au sud, *syn. dans l'hémisphère nord* ▶exposition ensoleillée ; *g* **Südlage [f]** (Gen Süden gerichteter Hang, gerichtete Gebäudeseite, Mauer etc.); *syn. auf der nördlichen Halbkugel* ▶Sonnenlage [f].

sovereignty [n] [UK], **planning** *leg. plan.* ▶planning jurisdiction.

sowing [n] (1) *constr. hort.* ▶seeding.

5838 sowing [n] (2) *constr. hort.* (Scattering seed, fruits, or shredded shoots; ▶hand sowing, ▶mechanical sowing, ▶seeding, ▶sprigging 1); *syn.* broadcasting [n]; *s* **siembra [f] (2)** (Proceso de esparcir semillas, frutos o retoños triturados en el suelo; sembrar [vb]; ▶inserción de las semillas, ▶siembra con retoños triturados, ▶siembra manual, ▶siembra mecánica); *syn.* sembradura [f]; *f* **semis [m] (3)** (Action de semer les graines ; ▶enfouissement des semences, ▶semis effectué à la machine, ▶semis manuel, ▶semis par jet de jeunes pousses) ; *syn.* ensemencement [m] ; *g* **Aussaat [f]** (Ausstreuen von Samen, Früchten oder Sprossen und in die Erde bringen. Im Englischen spricht man bei ‚breitwürfiger Aussaat' von *broadcasting*; aussäen [vb]; ▶Handsaat, ▶Maschinensaat, ▶Saatguteinbringung, ▶Sprossenaussaat); *syn.* Ansaat [f] (2).

5839 sowing [n] (3) *constr. hort.* (Embedding of seeds in the soil, e.g. with a roller or rake; ▶seeding); *s* **inserción [f] de las semillas** (Introducción de las semillas en la tierra, p. ej. con un rodillo o rastrillo; ▶siembra); *f* **enfouissement [m] des semences** (Mise en terre des semences au moyen d'une griffe, d'un râteau ou d'un rouleau ; ▶semis 1) ; *g* **Saatguteinbringung [f]** (Einarbeitung eines Saatgutes in die Erde, z. B. mit Igelwalze oder Vierzahn; Saatgut einbringen [vb]; ▶Ansaat 1); *syn.* Einarbeiten [n, o. Pl.] des Saatgutes.

5840 sowing rate [n] *constr. hort.* (Amount of seed, fruits, or shredded shoots broadcast per unit area, in landscape construction work usually measured in g/m^2); *s* **cociente [m] de siembra** (Cantidad de semillas sembradas por unidad de suelo en kg/a o g/m^2); *syn.* dosificación [f] de semillas; *f* **densité [f] de semis** (Quantité de semis par unité de surface exprimée soit au kg/are ou en g/m^2) ; *syn.* dosage [m] ; *g* **Aussaatmenge [f]** (Bei der Ansaat ausgebrachte Saatgutmenge; Angabe im Garten- und Landschaftsbau meist in g/m^2); *syn.* Saatgutmenge [f], Saatmenge [f].

5841 spa [n] *recr.* (Place characterized by the occurrence of mineral water springs and with special facilities for the health care of visitors; *generic term* ▶health resort; ▶seaside resort); *s* **estación [f] hidrotermal** (Lugar en el que existen manantiales de aguas minerales calientes y que está equipado para su aprovechamiento medicinal; ▶lugar de veraneo en la costa;

término genérico ►estación balnearia); *syn.* estación [f] termal; *f* **établissement** [m] **thermal** (Lieu caractérisé par l'existence de sources thermales naturelles et possédant les infrastructures permettant le traitement médical et le repos des curistes ; *terme générique* ►station thermale ; ►station balnéaire) ; *syn.* thermes [mpl] ; *g* **Heilbad** [n] (Ort für Sanatorienaufenthalte, der sich durch Vorkommen natürlicher Heilquellen auszeichnet und über besondere Einrichtungen für die Pflege und Erholung von Kurgästen verfügt; ►Seebad; *OB* ►Kurort).

space [n] *constr.* ►clearance space; *plan.* ►common open space; *urb.* ►covered parking space; *constr. urb.* ►external space; *landsc. urb.* ►green space; *envir.* ►intermediate storage space; *leg. phyt. zool.* ►living space; *plan.* ►open space; *gard.* ►outdoor living space [US]/garden living space [UK]; *plan.* ►outdoor space; *trans. plan.* ►parking maneuver space [US]/parking manoeuvre space [UK]; *trans. urb.* ►parking space; *leg. urb.* ►private green space; *plan.* ►private green space; *urb.* ►proximity green space; *leg. urb.* ►public green space; *plan. sociol.* ►public open space, ►public space; *plan. recr.* ►recreation space; *urb.* ►residential green space; *urb.* ►street space; *trans. urb.* ►turnaround space; *urb.* ►urban space; *leg. urb.* ►zoned green space.

space [n] [UK], **community open** *envir.* ►common open space.

space [n] [UK], **garden living** *gard.* ►outdoor living space [US].

space [n], **green open** *landsc. urb.* ►green space.

space [n], **landscaped open** *landsc. urb.* ►green space.

space [n], **neighbourhood green** *urb.* ►proximity green space.

space [n] [US], **offsetting open** *plan.* ►open space compensation area.

space [n], **play** *recr.* ►playground.

space [n], **playable** *recr.* ►playground.

space [n] [US], **polluted air** *envir.* ►contaminated airspace.

space [n], **pore** *pedol.* ►pore volume.

space [n], **shared** *trans. urb.* ►traffic calming, #2.

space [n] [UK], **"slack"** *recr.* ►teenage playground [US].

space [n], **vertical green** *gard. urb.* ►façade planting.

5842 space [n] **covered by a parked vehicle** *trans.* (Area covered by a parked vehicle in a ►parking space); *s* **superficie** [f] **ocupada por un vehículo** (►aparcamiento 2); *f* **surface** [f] **occupée par un véhicule** ; (►emplacement de stationnement) ; *g* **Stellfläche** [f] (Fläche, die von einem stehenden Fahrzeug überdeckt wird; ►Parkplatz 3).

spaced [pp] **extra wide** *hort.* ►widely-spaced plant.

5843 space net [n] *constr. recr.* (Play equipment constructed of a network of ropes for climbing, acrobatics and trapeze exercises); *s* **juego** [m] **de cuerdas (≠)** (Instalación de juego en parque infantil consistente en red de cuerdas gruesas colgada de una estructura que sirve para trepar, balancearse, suspenderse, etc.); *f* **jeu** [m] **de cordes** (Cordage mobile pour grimper, se balancer, se tenir en équilibre, se suspendre, équipant les pyramides de cordes, les chapiteaux, les filets d'escalade, etc.) ; *syn.* équipement [m] de jeux en cordages ; *g* **Seilspielgerät** [n] (Kletternetz aus beweglichen Seilen zum Klettern, Schaukeln, Turnen, Hangeln etc.); *syn.* Spielraumnetz [n].

5844 spacing [n] **(1)** *constr. hort.* (Distance between individual plants, fence posts, etc.; ►regular spacing, ►spacing of lateral drains; ►square planting grid, ►triangular planting grid); *s* **espaciamiento** [m] (Distancia entre plantas, estacas de vallas, etc.; ►distancia entre drenes secundarios, ►esquema de plan-

tación rectangular, ►esquema de plantación triangular); *f* **espacement** [m] (Distance entre les plantes, les poteaux de clôture, etc. ; ►disposition en quinconce, ►disposition rectangulaire, ►écartement des adducteurs) ; *g* **Abstand** [m] (Zwischenraum zwischen Pflanzen, Pfosten, Saugern etc.; ►Dreiecksverband, ►Saugerabstand, ►Vierecksverband).

spacing [n] **(2)** *leg. urb.* ►minimum building spacing.

spacing [n], **even** *constr.* ►regular spacing.

spacing [n] [US], **plant** *constr. hort.* ►planting distance.

5845 spacing [n] **between rows** *constr. hort. for.* (Interval between two rows; ►spacing of individuals within a row, ►planting distance); *syn.* spacing [n] of rows; *s* **espaciamiento** [m] **entre hileras** (►espaciamiento en hilera de plantación, ►distanciamiento de plantación); *syn.* espacio [m] entre surcos; *f* **espacement** [m] **entre les lignes** (Distance séparant deux lignes de plantation ; ►espacement, ►distance de plantation, ►intervalle de plantation) ; *g* **Reihenabstand** [m] (Abstand zwischen zwei Reihen; ►Abstand in der Reihe; ►Pflanzabstand).

spacing [n] **of drainage lines** *constr.* ►spacing of lateral drains.

5846 spacing [n] **of individuals within a row** *constr. hort.* (►widely-spaced plant; ►spacing 1); *s* **espaciamiento** [m] **en hilera de plantación** (►planta de gran espaciamiento; ►espaciamiento); *f* **intervalle** [m] **de plantation** (Espace entre les végétaux dans la même rangée ; ►plante à grand écartement ; ►espacement) ; *syn.* espacement [m] en alignement ; *g* **Abstand** [m] **in der Reihe** (►Pflanze aus extra weitem Stand; ►Abstand).

5847 spacing [n] **of lateral drains** *constr.* (Distance between lateral drains; ►spacing 1); *syn.* spacing [n] of drainage lines; *s* **distancia** [f] **entre drenes secundarios** (►espaciamiento); *f* **écartement** [m] **des adducteurs** (Distance comprise entre deux adducteurs ; PED 1979, 302 ; ►espacement) ; *g* **Saugerabstand** [m] (Entfernung zwischen zwei Saugern; ►Abstand); *syn.* Dränabstand [m].

spacing [n] **of rows** *constr. hort. for.* ►spacing between rows.

span structure [n] **for signage** *trans.* ►overhead sign structure.

spare time [n] *plan. recr.* ►leisure.

5848 sparse foliage [n] **of woody plants** *bot. hort.* (Widely-spaced leaves of shrubs and trees); *s* **follaje** [m] **hueco** *syn.* follaje [m] ralo; *f* **frondaison** [f] **légère** (Caractéristique du feuillage de certaines espèces de végétaux ligneux chez lesquels la répartition de la ramure et la disposition des feuilles font que la couronne est transparente à la lumière) ; *syn.* frondaison [f] claire ; *g* **lockere Belaubung** [f, o. Pl.] (**1.** Laubwerk eines Gehölzes, bei dem die Blätter arttypisch nicht dicht zusammen sind und die Krone somit lichtdurchlässig ist. **2.** Durch Vitalitätsverlust lückiges Blattwerk eines Gehölzes; *syn.* lockeres Laubwerk [n], *syn. zu 2.* schwache Belaubung [f, o. Pl.], schüttere Belaubung [f, o. Pl.], schütteres Laub [n, o. Pl.].

5849 sparse settlement [n] *urb. zool.* (Thin human or animal population of an area); *s* **poblamiento** [m] **excaso** (Bajo grado de ocupación de una zona por parte humana o animal); *f* **peuplement** [m] **faible** (Faible densité de population) ; *g* **dünne Besiedelung** [f] *syn. o. V.* dünne Besiedlung [f].

spatial articulation [n] *landsc.* ►structuring of landscape.

spatial configuration [n] *plan.* ►spatial distribution (2).

spatial configuration [n] **of a town** *landsc. urb.* ►structuring of townscape.

spatial distribution [n] **(1)** *phyt.* ►dispersion (1).

5850 spatial distribution [n] **(2)** *plan.* (LE 1986, 204s; division of a landscape according to specific criteria, e.g. soil

S

types with pedogenetic characteristics or geological attributes or ecological factors; ►ecological distribution of spatial patterns, ►physiographic division); *syn.* spatial structuring [n], spatial configuration [n]; *s* **estructuración [f] del territorio** (Zonificación del territorio según criterios científicos sectoriales, como p. ej. según el tipo de suelo o condiciones geológicas, o según criterios ecológicos; ►estructuración ecológica del territorio, ►tipología del paisaje); *syn.* zonificación [f] normativa del territorio; *f* **division [f] spatiale** (Division de l'espace paysager selon des critères scientifiques spécifiques [p. ex. classification en groupes de sols selon leur caractéristiques génétiques] ou suivant des données écologiques [►division de l'espace en territoires naturels, ►typologie physiographique des paysages]) ; *syn.* division [f] de l'espace, définition [f] d'unités spatiales ; *g* **Raumgliederung [f]** (Räumliche Gliederung der Landschaft nach fachwissenschaftlichen Kriterien [z. B. Gliederung in Bodentypen nach genetischen Merkmalen oder nach geologischen Gegebenheiten] oder nach ökologischen Gesichtspunkten [►ökologische Raumgliederung]; ►naturräumliche Gliederung).

spatial diversity [n] *plan.* ►spatial heterogeneity.

5851 spatial heterogeneity [n] *plan.* (Diverse spatial constituents, characterized by specific vegetational [biogenic], edaphic, geological [geogenic] features, i.e. moraines, valleys, watercourses, lakes, woods, hedges, farmholdings, cropland or grassland in a defined landscape unit; ►diversity); *syn.* spatial diversity [n]; *s* **heterogenidad [f] espacial** (Presencia de gran variedad de estructuras en un espacio natural, caracterizada por sus elementos bióticos [vegetación], edáficos, geológicos y estructurales del paisaje como valles, cursos de agua, lagos, bosques, setos, diferentes usos agrícolas, etc.; ►diversidad ecológica); *syn.* heterogenidad [f] territorial; *f* **hétérogénéité [f] spatiale** (Existence d'une multitude de structures spatiales, caractérisées par des éléments végétaux, édaphiques, géologiques et/ou constitutifs, p. ex. l'existence de moraines, de vallées, cours d'eau, forêts, haies, fermes, prairies dans un espace naturel donné ; l'**h. s.** constitue avec la variabilité et l'équitabilité un élément de la ►diversité au sein de l'écosystème ; DUV 1984, 103) ; *g* **räumliche Vielfalt [f]** (Vorhandensein einer Fülle unterschiedlicher Strukturen in einem bestimmten Naturraum, geprägt durch pflanzliche [biogene], edaphische, geologische [geogene] und die Landschaft bildende Elemente, z. B. Moränen, Täler, Wasserläufe, Seen, Wälder, Hecken, Hoflagen, Acker- und Grünland; ►Diversität).

5852 spatial landscape characteristics/resources [npl] *land'man.* (Landscape components including water, vegetation, and constructed development; ►scenic heterogeneity); *syn.* landcover [n] [also US], spatial organisation [n]; *s* **organización [f] espacial del paisaje** (Grado de existencia de diferentes componentes con efectos de conformación espacial del paisaje, como elementos geomorfológicos, acuáticos, asentamientos humanos, obras de infraestructura, etc.; ►riqueza escénica del paisaje); *f* **organisation [f] spatiale** (Éléments caractéristiques pour l'ensemble d'un espace paysager, p. ex. phénomènes géomorphologiques, eau, végétation ; ►diversité des paysages) ; *syn.* caractéristiques [fpl] spatiales, caractéristiques [fpl] paysagères, composantes structurantes [fpl] du paysage, ressources [fpl] paysagères ; *g* **räumliche Ausstattung [f]** (Ausstattung eines Landschaftsraumes, z. B. mit geomorphen Erscheinungen, Vegetationselementen, Gewässern, Siedlungen, Verkehrsbauten; ►Vielfalt des Landschaftsbildes).

spatial organisation [n] *land'man.* ►spatial landscape characteristics/resources.

spatial patterns [npl] *ecol.* ►ecological distribution of spatial patterns.

spatial planning [n] *plan.* ►physical planning.

5853 spatial requirements [npl] *plan.* (TGG 1984, 223; Demand for open space established by planning authorities for public use when developing land; ►land consumption, ►land requirements); *s* **exigencias [fpl] de espacio** (Necesidades de espacio físico para un uso determinado en una región o parte de ella; ►demanda de suelo, ►gasto de paisaje); *f* **exigence [f] spatiale** (Exigences exprimées par des aménageurs sur un secteur défini en vue d'un usage public ; ►besoins fonciers, ►consommation d'espaces) ; *g* **Raumanspruch [m]** (Ansprüche von Planungsträgern an eine Region oder deren Teile für gesellschaftliche Nutzungen; ►Flächenanspruch, ►Landschaftsverbrauch).

spatial structuring [n] *plan.* ►spatial distribution (2).

5854 spatial unit [n] *plan.* (LE 1986, 7; extent of an area defined within the context of ►spatial distribution 2; ►ecological spatial unit, ►spatial unit of the landscape); *s* **unidad [f] estructural del territorio** (Parte delimitada del territorio en el marco de ►estructuraciones del territorio; ►unidad ambiental, ►unidad espacial del paisaje); *f* **unité [f] spatiale** (Espace de base dans le cadre de la définition d'unités spatiales dans l'étude des paysages ; GEP 1991, 29 ; ►espace paysager, ►écotope territorial naturel, ►division spatiale) ; *syn.* unité [f] structurale paysagère ; *g* **Raumeinheit [f]** (Abgegrenzter Raum im Rahmen von ►Raumgliederungen; ►Landschaftsraum, ►ökologische Raumeinheit).

5855 spatial unit [n] **of a landscape** *geo. landsc.* (**1.** Section of the geosphere which is demarcated or may be demarcated according to certain planning or research criteria, e.g. ecological, structural, physiognomical, genetic or socio-economic aspects. **2.** Area of distinct, but not homogeneous, visual character which is spatially enclosed at ground level; a visually identifiable place or "outdoor room"; useful for visual assessment and management; ►ecological spatial unit, ►natural landscape unit/region); *s* **unidad [f] espacial de paisaje** (Sección de la geosfera demarcada o demarcable según criterios de investigación o planificación, p. ej. ecológicos, estructurales, fisionómicos, histórico-genéticos o socioeconómicos; ►unidad ambiental, ►unidad de espacio natural); *syn.* unidad [f] paisajística/del paisaje; *f* **espace [m] paysager** (Partie définie de la géosphère caractérisée par différents critères — écologiques, structurels, physiogéographiques, historiques et génétiques, socio-économiques ; ►écotope territorial naturel, ►espace naturel) ; *syn.* espace [m] paysagique ; *g* **Landschaftsraum [m]** (**1.** Je nach Aufgabenstellung nach bestimmten Kriterien — ökologischen, strukturellen, physiognomischen, historisch-genetischen oder sozioökonomischen Aspekten — zielorientiert abgegrenzter und abgrenzbarer Ausschnitt der Geosphäre; cf. BUCH 1978. **2.** Gebiet, das ökologisch aus ähnlichen geologisch-morphologischen Komplexen zusammengesetzt ist und traditionell von ähnlichen Nutzungsmosaiken geprägt wird, die diese Komplexe widerspiegeln; z. B. sind die traditionellen Muster der Primärproduktion oft auch heute noch — trotz vielfältiger Veränderungen — in modernen landwirtschaftlichen Anbausystemen und der Landschaftsstruktur erkennbar; ►Naturraum, ►ökologische Raumeinheit).

spatial units [npl] *plan.* ►definition of spatial units.

5856 spawning area [n] *zool.* (Area in which fish and amphibians lay eggs; ►fish spawning area, ►zona de freza protegida); *s* **frezadero [m]** (Lugar donde ponen sus huevos los peces y anfibios; ►área de freza, ►área de freza protegida); *syn.* lugar [f] de desove; *f* **frayère [f]** (Lieu où la femelle du poisson et des amphibiens dépose ses œufs, et où le mâle les recouvre de semence : la notion de frayère peut être élargie aux endroits où se reproduisent les mollusque et les crustacés marins ou dulçaquicoles Les frayères se trouvent souvent sur les fonds sableux ou sablo-vaseux des rivières, des étangs, des lacs, des marais, des estuaires, voire des zones marines très profondes ; AEP 1976, 88 et s. ; ►réserve d'élevage, ►réserve statique, ►cours d'eau de fraye) ; *syn.* aire [f] de ponte, aire [f] de fraie [CDN], habitat [m] de fraie [CDN] ; *g* **Laichplatz [m]** (Eiablegestelle für Fische und Amphibien ; ►Fischlaichgewässer, ►Laichschonbezirk).

5857 spawning period [n] *zool.* (Season during which fish and amphibians lay eggs); *s* periodo/período [m] de freza *syn.* periodo/período [m] de desove; *f* période [f] de fraie (Période de reproduction pendant laquelle les poissons pondent les œufs dans des frayères) ; *syn.* époque [f] de ponte ; *g* **Laichzeit [f]** (Zeit, in der Fische oder Amphibien Eier ablegen [laichen]).

5858 spawning water [n] *zool.* (Water in which fish and amphibians lay eggs); *s* **aguas [fpl] de freza** (Aguas en las que ponen sus huevos los peces y anfibios); *syn.* aguas [fpl] de desove; *f* **eaux [fpl] de fraie/fraye** (Plans et cours d'eau naturels ou artificiels dans lesquels les poissons se reproduisent) ; *g* **Laichgewässer [n]** (Gewässer, in dem Fische oder Amphibien ablaichen).

5859 special area [n] **of conservation (SAC)** *conserv. ecol.* (Site of community importance designated by EU Member States through a statutory, administrative and/or contractual act where the necessary conservation measures are applied for the maintenance or restoration, at a favourable ►conservation status 2, of the natural habitats and/or the populations of the species for which the site is designated); *syn.* special protection area [n] (SPA); *s* **zona [f] especial de conservación** (Según el art. 1 de la Directiva 92/43/CEE, de Hábitats, es un lugar designado por los Estados miembros mediante un acto reglamentario, administrativo y/o contractual, en el cual se apliquen las medidas de conservación necesarias para el mantenimiento o el restablecimiento, en un estado de conservación favorable, de los hábitats naturales y/o de las poblaciones de las especies para las cuales se haya designado el lugar); *f* **zone [f] spéciale de conservation [ZSC]** (Site d'importance désigné par les États membres par un acte réglementaire, administratif et/ou contractuel où sont appliquées les mesures de conservation nécessaires au maintien ou au rétablissement, dans un état de conservation favorable, des habitats naturels et/ou des populations des espèces pour lesquels le site est désigné) ; *g* **besonderes Schutzgebiet [n]** (Ein national kann resp. Länderrecht rechtsverbindlich ausgewiesenes besonderes Schutzgebiet im Sinne der ►Fauna-Flora-Habitat-Richtlinie, in dem Maßnahmen, die zur Wahrung oder Wiederherstellung eines günstigen Erhaltungszustandes der natürlichen Lebensräume und/oder Populationen der Arten, für die das Gebiet bestimmt ist, erforderlich sind und durchgeführt werden; in D. werden die besonderen Schutzgebiete als **FFH-Gebiete** bezeichnet; cf. Art. 1 der Richtlinie 92/43/EWG des Rates vom 21. Mai 1992 zur Erhaltung der natürlichen Lebensräume sowie der wild lebenden Tiere und Pflanzen; *syn.* FFH-Gebiet [n].

5860 special conditions [npl] **of (a) contract** *constr. contr.* (Contract conditions which cover special project-related prerequisite terms, e.g. ►liquidated damages, use of access roads, availability of utility connections; ►additional contractual conditions, ►contractual terms, ►Standard Form of Building Contract [UK]); *syn.* special contractual terms [npl] [also UK]; *s* **pliego [m] de prescripciones técnicas particulares** (Condiciones fijadas en un contrato de obra que reflejan exigencias específicas del proyecto; ►condiciones de contrato, ►multa por retraso, ►pliego de cláusulas administrativas generales para la contratación de obras, ►pliego de prescripciones administrativas particulares); *f 1* **cahier [m] des clauses administratives particulières**

(C.C.A.P.) (Clauses contractuelles qui, par comparaison avec le ►cahier des clauses administratives générales [des marchés publics], définit les dispositions administratives propres à chaque marché et précise clairement, p. ex. les prestations fournies par le maître de l'ouvrage et propre au chantier telles que les fournitures d'eau, d'électricité, le prix et mode d'évaluation des ouvrages, le règlement des travaux, les délais d'exécution, ►pénalités pour retard, assurances, contrôle de l'avancement, hygiène, sécurité, etc.) ; *f 2* **cahier [m] des clauses administratives particulières — type (C.C.A.P. — type)** (Selon les termes prévus par les diverses directives réglementaires ; ►clauses administratives complémentaires, ►clauses du contrat) ; *g* **besondere Vertragsbedingungen [fpl]** (Vertragsbedingungen, die im Vergleich zu den ►allgemeinen Vertragsbedingungen für die Ausführung von Bauleistungen und ►zusätzlichen Vertragsbedingungen besondere projektbezogene Erfordernisse wie z. B. ►Vertragsstrafen, Benutzung von Zufahrtswegen, Wasseranschlüssen etc. regeln; cf. § 10 VOB Teil A; ►Vertragsbedingungen).

5861 special construction requirements [npl] *contr. constr.* (Additional work [US]/works [UK], either included in the contract specifications or in an addendum, which the contractor is required to calculate as a separate cost item; e.g. protection provisions for bad weather; ►additional work and services, ►bid documents, ►supplemental construction services or supplies); *s* **servicios [mpl] especiales de construcción** (Partidas adicionales incluidas en los ►documentos de contrato de las administraciones públicas, como p. ej. protección contra la intemperie, vigilancia del trabajo de terceros, etc., ya que en el ►pliego de prescripciones técnicas generales no están consideradas como parte de los servicios y para que se tengan en cuenta su costo; ►servicios adicionales, ►trabajos adicionales [de construcción]); *f* **sujétions [fpl] particulières inhabituelles** (Sujétions d'exécution rencontrées lors de l'établissement des prix et indiquées par l'entrepreneur à l'état de prix forfaitaire ou au bordereau des prix unitaires afin de faciliter les comparaisons de prix entre les divers marchés, p. ex. protection particulière contre les intempéries, surveillance des travaux d'autres corps de métiers, etc. ; CCM 1984, 114 ; ►cahier des clauses techniques générales — documents types, ►pièces contractuelles constitutives du marché, ►sujétions d'exécution, ►travaux supplémentaires) ; *g* **besondere Leistung [f] (1)** (Vom Unternehmer zusätzlich zu erbringende Leistung, z. B. Schutz gegen Witterungseinflüsse, Überwachung der Leistungen anderer etc., die in den ►Verdingungsunterlagen besonders anzugeben ist, da sie nach den ►Allgemeinen Technischen Vertragsbedingungen für Bauleistungen oder der gewerblichen Verkehrssitte nicht unmittelbar zur Leistung gehört und damit sie bei der Preiskalkulation für die Angebotsabgabe berücksichtigt werden kann; cf. § 9 [6] VOB Teil A und § 2 [1] VOB Teil B; ►Nebenleistungen, ►zusätzliche Leistung 1).

special contractual terms [npl] [UK] *constr. contr.* ►special conditions of (a) contract.

5862 special garden [n] *gard. hort.* (Particular kind of garden devoted to specific plants, e.g. ►alpine garden 1, ►rock garden, ►rose garden); *s* **jardín [m] especializado** (Jardín dedicado a determinadas plantas como p. ej. ►jardín alpino, ►rocalla, ►rosaleda); *f* **jardin [m] spécialisé** (Jardin réservé à l'emploi ou à la culture de plantes spécifiques, p. ex. un ►alpinum, un ►jardin de rocaille, une ►roseraie) ; *g* **Sondergarten [m]** (Sonderform eines Gartens, der allein auf spezifische Pflanzen ausgerichtet ist, z. B. ein ►Alpinum, ►Rosarium, ►Steingarten).

5863 specialist [n] **(1)** *prof.* (Person devoted to one subject or to one particular branch, professional field or pursuit; e.g. consultant for a special planning task; in good English, specialist and ►expert are not synonymous!); *s* **especialista [m/f] (2)** (Profe-

S

sional de otra profesión a quien se consulta p. ej. para solucionar un problema específico en la planificación; ▶especialista 1); *syn.* experto/a [m/f] (3), perito/a [m/f] (3); *f* **spécialiste [m/f]** (Personne de métier spécialisée dont l'activité professionnelle déterminée est p. ex. nécessaire à la solution de problèmes particuliers, limités dans le domaine de l'aménagement ; ▶professionnel) ; *g* **Sonderfachmann [m]** (...fachleute [pl]; an einem Bauvorhaben beteiligter Fachplaner, Gutachter oder Wissenschaftler, der z. B. zur Bewältigung einer Planungs- oder Bauaufgabe notwendig ist und gesondert hinzugezogen wird, z. B. Bodengutachter, Tragwerksplaner, Bauphysiker, Pflanzensoziologe, Denkmalpfleger; der Planer muss den Bauherren über die Notwendigkeit der Einschaltung von Sonderfachleuten umfassend beraten; fällt die Leistung des **S.es** in den vertraglichen Aufgabenbereich des Planers, so muss er selbst den **S.** beauftragen; ▶Fachmann).

specialist [n] (2) *arb.* ▶tree nursery specialist.

specialist firm [n] *constr. hort.* ▶speciality contractor.

5864 specialist trade [n] *constr.* (Part of construction project [US]/part of construction works [UK] awarded separately; e.g. landscaping, masonry, carpentry, paving; ▶construction project section); *s* **lote [m] de oficio** (Partida de trabajos de obra que es ejecutada por un ramo de artesanos u oficio y que es adjudicada independientemente, p. ej. trabajos de jardinería, mampostería, carpintería, etc.; ▶tramo de obra); *f* **lot [m] technique** (Partition des travaux dans un appel d'offre correspondant à une spécialité précise, p. ex. travaux d'espaces verts, travaux de maçonnerie, mise en œuvre des enrobés, travaux de charpente ; CCM 1984, 47 ; ▶tranche de travaux) ; *syn.* lot [m] de travaux ; *g* **Fachlos [n]** (Zu einem Bauvorhaben gehörende Bauleistungen eines Handwerks-, Fachgebietes oder Gewerbezweiges, die zusammengefasst und getrennt vergeben werden, z. B. landschaftsgärtnerische Arbeiten, Maurerarbeiten, Asphaltbelagsarbeiten, Zimmermannsarbeiten; ▶Baulos); *syn.* Einzelgewerk [n].

5865 speciality contractor [n] *constr. hort.* (Company, e.g. for landscape construction, which is specialized in a particular field. In France there are landscape construction companies, which have been awarded certificates by the Ministry of Agriculture after qualifying for the execution of certain works); *syn.* specialist firm [n]; *s* **empresa [f] especializada** (Empresa, p. ej. de construcción paisajística, que se ha especializado en un área determinada. En Francia el Ministerio de Agricultura, a través del Comité Nacional Interprofesional de Horticultura, emite certificados de cualificación a las empresas que se han especializado en una determinada área de trabajo); *f 1* **entreprise [f] spécialisée** (dans un secteur d'activité particulier) ; *f 2* **entreprise [f] qualifiée** (Entreprise paysagiste possédant un titre de qualification attribué par le Comité National Interprofessionnel de l'Horticulture [CNIH] pour le compte du ministère de l'agriculture pour la compétence dans un secteur d'activité tel que la création et l'entretien, plantations autoroutes, élagage, terrains de sport, golf, forêts, arrosage ; p. ex. avec le Titre de Qualification des entreprises de reboisement et d'arrosage intégré ; LEU 1987) ; *g* **Fachfirma [f]** (Unternehmen [z. B. des Garten- und Landschaftsbaues], das sich auf ein bestimmtes Tätigkeitsfeld spezialisiert hat. In F. gibt es GaLaBau-Firmen, die vom Landwirtschaftsministerium für ein bestimmtes Arbeitsgebiet eine Qualifikationsbescheinigung erhalten haben); *syn.* Fachbetrieb [m].

5866 special professional services [npl] *contr. prof.* (Specified work carried out by a planning professional in excess of the ▶basic professional services); *syn.* designated services [npl] [also US], additional planning services [npl] [also UK]; *s* **servicios [mpl] complementarios de planificación** (Prestaciones adicionales más allá de los ▶servicios profesionales

básicos, que se pueden definir como encargo específico en el caso de que los trabajos previstos vayan más allá de lo normal); *f* **prestation [f] complémentaire** (Prestations d'un concepteur caractérisant des éléments d'une ▶mission de maîtrise d'œuvre normalisée complète lorsque les exigences imposées à la réalisation de la mission sont supérieures à celle contenues dans les ▶prestations élémentaires ; il s'agit d'une mission étendue comprenant des éléments complémentaires ; HAC 1989, 367) ; *g* **besondere Leistung [f] (2)** (D., Leistung eines Planers, die zu den ▶Grundleistungen [HOAI 2002]/Leistungen [HOAI [2009] oder an deren Stelle treten kann, wenn besondere Anforderungen an die Ausführung eines Auftrages gestellt werden, die über die Grundleistungen hinausgehen; cf. § 2 HOAI 2002 u. § 1 [3] HRLA; **b. L.** können wie bisher frei vereinbart werden, sind jedoch im Hauptteil der HOAI 2009 nicht mehr erwähnt, sondern in einer Anlage 2 als Orientierungsrahmen aufgeführt).

special protection area [n] *conserv. ecol.* ▶special area of conservation (SAC).

5867 special technical requirements [npl] **of contract** *constr. contr.* (Particular stipulations of a construction contract, which may be negotiated individually by the client; ▶standards for construction [US]/Codes of practice (C.P.) [UK]); *s* **prescripciones [fpl] técnicas particulares** (Normas técnicas a acordar individualmente, más allá de las incluidas en el ▶pliego de prescripciones técnicas generales; cf. Art. 124 [1c], Lib. II, Tít. I, LCAP); *f* **prescriptions [fpl] techniques particulières** (Normes en vigueur réglementant l'exécution des travaux publics fixées dans les documents techniques unifiés [D.T.U.] et qui font sans cesse partie intégrante des conditions applicables aux marchés de travaux publics ; ▶cahier des clauses techniques générales — documents types [C.C.T.G. — documents types]) ; *g* **Besondere Technische Vorschriften [fpl]** (*Abk.* BTV; Vorschriften, die neben den ▶Allgemeinen Technischen Vertragsbedingungen für Bauleistungen vom jeweiligen Auftraggeber individuell vereinbart werden können).

5868 special waste [n] *envir. leg.* (In U.S., items such as household hazardous waste, bulky wastes [refrigerators, pieces of furniture, etc.], tires, and used oil; EPA 1994; ▶hazardous waste); *s* **residuos [mpl] especiales (2)** (En EE.UU., término utilizado para residuos domésticos peligrosos, basura voluminosa [refrigeradores, muebles, etc.], llantas y aceites usados; ▶residuos industriales peligrosos); *f* **déchets [mpl] encombrants et déchets dangereux** (≠) (**US.**, Catégorie de déchets englobant les déchets ménagers spéciaux, les encombrants, les déchets de pneumatiques et les huiles usées ; ▶déchets dangereux) ; *g* **Sperr- und Sondermüll [m]** (≠) (In den USA eine Sammelbezeichnung für Haushaltssondermüll, Sperrmüll, Autoreifen und Altöl. In D. wird unter Sondermüll — *special waste* — ▶besonders überwachungsbedürftiger Abfall verstanden).

5869 species [n] (1) *biol. leg.* (Entire population of individuals, which reproduce amongst themselves naturally without restrictions and mutually correspond in all typical characteristics and with their offspring; ▶subspecies); *s* **especie [f]** (1. Categoría sistemática base del catálogo de los seres vivos. Reúne a aquellos con la misma morfología hereditaria, los mismos caracteres biofísicos y fisiológicos y un género de vida en común. Los individuos que la componen son interfecundos, presentando una proporción normal de sexos. Es la categoría inferior al género; cf. DGA 1986. 2. *leg.* Especie, ▶subespecie o población parcial de una especie o subespecie; cf. Ordenanza CC.EE. N° 338/97 del Consejo del 09.12.1996); *f* **espèce [f]** (1. Ensemble de tous les individus capables d'engendrer entre eux naturellement des individus féconds et présentant avec leurs descendants une identité des caractères. 2. *leg.* Espèce, ▶sous-espèce et population d'une espèce ou d'une sous-espèce ; cf. art. 2 du Règlement du Conseil n° 338/97/CE) ; *g* **Art** [f] (**1.** Gesamtheit der Individuen, die sich auf natürliche Weise untereinander uneingeschränkt fortpflanzen und in allen typischen Merkmalen untereinander und mit ihren Nachkommen übereinstimmen; cf. ANL 1984. **2.** Grundlegende taxonomische Einheit/Kategorie des biologischen Ordnungssystems. **3.** *leg.* Juristisch wird der Begriff **A.** nicht als taxonomische Rangstufe gesehen, sondern umfassender unter Berücksichtigung auch von ►Subspecies und Teilpopulationen einer Art oder Unterart; cf. Art. 2 Verordnung EG Nr. 338/97 des Rates v. 09.12.1996 und § 10 BNatSchG); *syn.* Spezies [f].

species [n] (2) *phyt.* ►acidophilous species; *zool.* ►animal species; *biol.* ►anthropophilous species; *agr. for. hort.* beneficial species; *bot. for.* broad-leaved woody species; *phyt.* ►calcareous indicator species, ►calcicolous species, ►casual species, ►character species; *phyt. zool.* ►character species of a habitat, ►character species of an alliance; *bot. zool.* ►cosmopolitan species; *conserv. phyt. zool.* ►critically endangered species; *phyt. zool.* ►density of species, ►differential species, ►dominant species; *arb. hort.* ►dwarf woody species; *conserv. phyt. zool.* ►endangered species, ►endangerment of species, ►endemic species; *phyt.* ►exclusive species; *phyt. zool.* ►exotic species; *conserv. phyt. zool.* ►extinct species; *phyt. zool.* ►indigenous plant species; *phyt. zool.* ►feral species, ►fluctuation of species; *for. phyt.* ►forest species; *agr. hort.* ►fruit species; *phyt. zool.* ►generalist species; *agr. phyt.* ►grassland species; *phyt.* ►hemerophobic species; *ecol. envir.* ►indicator species; *phyt.* ►indifferent species; *phyt. plant.* ►indigenous plant species; *phyt. zool.* ►indigenous species, ►introduced species; *hort. phyt. plant.* ►invasive species; *for. phyt.* ►light-demanding woody species; *phyt. zool.* ►lignicolous species, ►lime-avoiding species; *phyt. zool.* ►list of species; *conserv. zool.* ►migratory species; *phyt. zool.* ►naturalized species; *conserv. phyt. zool.* ►near threatened species; *bot. for.* needle-leaved species; *phyt.* ►nitrogen indicator (species), ►nitrophilous species; *leg. hunt.* ►nongame species [US]/non-game species [UK]; *phyt. zool.* ►nonindigenous species [US]/non-indigenous species [UK]; *phyt.* ►peat-forming species, ►pioneer species; *bot. phyt.* plant species; *phyt.* preferential species; *conserv. phyt. zool.* ►relict species; *phyt.* ►riparian woody species; *phyt. zool.* ►riverine species; *phyt.* ►selective species, ►strange species; *biol.* ►subspecies; *agr. conserv. for. hort.* ►target species; *zool.* ►terrestrial species; *conserv. phyt. zool.* ►threatened species; *phyt. zool.* ►ubiquitous species; *agr. for. hort.* ►useful plant species; ►use of beneficial animal species; *conserv. phyt. zool.* ►vulnerable species; *plant. hort.* ►wild woody species; *landsc. phyt.* ►woody pioneer plant/species; *bot. phyt.* woody species (1).

species [n], **alien** *phyt. zool.* ►introduced species.

species [n], **autochthonous** *phyt. zool.* ►indigenous species.

species [n], **autochthonous plant** *phyt. plant.* ►indigenous plant species.

species [n], **basophilous** *phyt.* ►calcicolous species.

species [n], **calcicole** *phyt.* ►calcicolous species.

species [n], **calcifuge** *phyt.* ►lime-avoiding species.

species [n], **calciphile** *phyt.* ►calcicolous species.

species [n], **calciphobe** *phyt.* ►lime-avoiding species.

species [n], **candidate** *conserv. phyt. zool. obs.* ►near threatened species.

species [n] [US], **cavity-nester** *zool.* ►cavity-nesting bird [US]/hole-nesting bird [UK].

species [n], **characteristic** *phyt.* ►character species.

species [n], **coniferous** *phyt.* ►conifer.

species [n], **constant** *phyt.* ►exclusive species.

species [n], **critical** *obs. phyt.* ►critically endangered species.

species [n], **disappeared** *conserv. phyt. zool.* ►extinct species, #3.

species [n], **economic plant** *agr. for. hort.* ►useful plant species.

species [n], **extinct in the wild** *conserv. phyt. zool.* ►extinct species, #2.

species [n], **faithful** *phyt.* ►exclusive species.

species [n], **fall-blooming** *hort. plant.* ►autumn-blooming plant.

species [n], **game** *hunt.* ►game, ►huntable game.

species [n], **heliophilous** *for. phyt.* ►light-demanding plant.

species [n], **heliophilous woody** *for. phyt.* ►light-demanding woody species.

species [n], **hemerophilous** *biol.* ►anthropophilous species.

species [n], **lime-loving** *phyt.* ►calcicolous species.

species [n], **lotic** *phyt. zool.* ►riverine species.

species [n], **mesophil(ic)** *phyt.* ►moisture indicator plant.

species [n], **native** *phyt. zool.* ►indigenous species.

species [n], **native plant** *phyt. plant.* ►indigenous plant species.

species [n], **naturally occurring** *phyt. zool.* ►indigenous species.

species [n] [UK], **non-native** *phyt. zool.* ►nonindigenous species [US].

species [n] [US], **nonnative** *phyt. zool.* ►nonindigenous species [US].

species [n], **pest-eating animal** *agr. for. hort.* ►beneficial species.

species [n], **riverine woody** *phyt.* ►riparian woody species.

species [n], **suppressing** *hort. phyt.* ►invasive species.

species [n], **synanthropic** *biol.* ►anthropophilous species.

species [n], **ubiquitous** *phyt. zool.* ►ubiquist.

5870 species abundance [n] *phyt. zool.* (Number of individuals of a single species per unit area; ►degree of species cover, ►degree of total vegetative cover, ►population density 3, ►species density); *syn.* number [n] of individuals; *s* **abundancia** [f] **(de una especie)** (Número de individuos de una especie por unidad de superficie; ►densidad de especies por área, ►densidad de población 3, ►dominancia); *f* **abondance** [f] (Appréciation relative du nombre des individus de chaque espèce entrant dans la constitution d'une population ou d'un peuplement sur un territoire étudié ; ►degré de recouvrement, ►densité des espèces, ►densité des individus, ►dominance) ; *g* **Abundanz** [f] (Häufigkeit von Organismen in Bezug auf eine Flächen- oder Raumeinheit; es werden **1.** absolute Individuen-A., **2.** absolute Arten-A. und **3.** relative Arten-A. unterschieden; A. wird meistens nur im Sinne von ►Individuendichte gebraucht; cf. ÖKO 1983; ►Artendichte, ►Deckungsgrad, Deckungsgrad einer Art); *syn.* Individuenzahl [f].

5871 species area curve [n] *phyt.* (TEE 1980, 158; graphic method of determining the area/species relationship and ►minimal area where all species of a plant community are adequately represented); *s* **curva** [f] **de especies y área** (Representación gráfica de la relación entre el tamaño de la superficie y la cantidad de especies y que se utiliza para determinar el ▶área mínima de una comunidad vegetal); *f* **courbe [f] d'aire** (Représentation graphique reproduisant la relation entre le nombre d'espèces d'un peuplement avec la superficie de l'aire explorée et servant à définir ▶l'aire minimale de ce groupement) ; *g* **Arealkurve [f]** (Grafische Darstellung, die das Verhältnis zwischen Flächengröße und Artenzahl zeigt und zur Bestimmung der ▶Minimalfläche einer Pflanzengesellschaft dient); *syn.* Art-Arealkurve [f].

5872 species composition [n] *phyt. zool.* (Combination of different species living in a defined area; ▶characteristic combination of species, ▶species inventory, ▶species pattern, ▶synusia; *specific term* floristic composition [TEE 1980, 130]); *s* **composición [f] de especies de una comunidad** (*Términos específicos* composición florística, composición faunística, ▶combinación característica de especies, ▶estructura específica [de especies], ▶inventario de especies, ▶sinusia); *f* **composition [f] des espèces d'une communauté (animale ou végétale)** (Liste des espèces qui constituent la population animale ou l'association végétale d'un peuplement ; *termes spécifiques* composition floristique, composition faunistique, ▶ensemble caractéristique, ▶ensemble spécifique, ▶inventaire des espèces, ▶synusie) ; *syn. phyt.* cortège [m] floristique ; *g* **Artenzusammensetzung [f]** (Bestand unterschiedlicher Arten auf einer definierten Fläche, abhängig von Boden Klima, Relief; *UB* floristische/faunistische Zusammensetzung; ▶Artengefüge, ▶Arteninventar, ▶charakteristische Artenkombination, ▶Synusie); *syn.* Artenspektrum [n], Artenkombination [f].

species composition [n]**, anthropogenic shift in animal** *zool.* ▶anthropogenic alteration of the genetic fauna pool.

species composition [n]**, anthropogenic shift in floristic** *phyt.* ▶anthropogenic alteration of flora composition.

5873 species conservation [n] *conserv. leg.* (Objective of nature conservation involving preservation and development of the total populations of animal and plant species within their natural distribution ranges and in their existing variety, so as to safeguard their continued evolution. Original objective of ▶nature conservation, involving the protection of populations, of individuals, usually endangered and often conspicuous plant or animal species, is now largely superseded by the broader concept of ▶habitat conservation; ▶animal species conservation, ▶Birds Directive, ▶Convention on Biological Diversity, ▶range 1, ▶Endangered Species Act [E.S.A.] [US]/Conservation of Wild Creatures and Wild Plants Act 1975 [UK], ▶Habitats Directive, ▶integrated species conservation, ▶protection of conservation areas, ▶protection of natural habitats and of wild fauna and flora, ▶conservation of flora and fauna, ▶protected area, ▶Washington Convention, ▶wildlife conservation); *s* **protección [f] de especies (de flora y fauna)** (Uno de los objetivos de la ▶protección de la naturaleza que tiene como fin salvaguardar las poblaciones de flora y fauna salvajes en sus zonas naturales de distribución [▶área de distribución] de manera que se mantenga la diversidad de especies y la capacidad evolutiva de las mismas; ▶conservación de los hábitats naturales y de la fauna y flora silvestres, ▶Convención de Washington, ▶Convención sobre la Diversidad Biológica, ▶Directiva de Hábitats, ▶Directiva de las Aves, ▶espacio natural protegido 2, ▶ley de protección de especies, ▶protección de especies de fauna en peligro, ▶protección de la fauna salvaje, ▶protección de zonas naturales, ▶protección integral de flora y fauna, ▶protección de biótopos); *f 1* **protection [f] des espèces (animales et végétales)** (Terme générique qualifiant un des domaines de la ▶protection de la nature ayant pour objectif de préserver et de reconstituer les populations animales et végétales ou de leurs habitats de telle manière que soit assuré le développement naturel des espèces ; ▶aire de répartition, ▶arrêté réglementant la protection de la faune et de la flore sauvages, ▶conservation de la faune et de la flore sauvage, ▶conservation des habitats naturels ainsi que de la faune et de la flore sauvages, ▶conservation intégrée des espèces animales et végétales, ▶Convention de Washington, ▶convention sur la diversité biologique, ▶Directive Habitat Faune Flore, ▶Directive Oiseaux, ▶préservation des biotopes, ▶protection de la faune sauvage, ▶protection des espèces animales, ▶protection de zones) ; *f 2* **protection [f] des espèces de faune et de flore sauvages** (Terminologie utilisée dans la ▶Convention de Washington du 3 mars 1973 sur le commerce international des espèces de faune et de flore sauvages menacées d'extinction connue sous l'acronyme anglais : CITES — Convention on International Trade of Endangered Species of Wild Fauna and Flora) ; *g* **Artenschutz [m, o. Pl.]** (Aufgabenbereich des Naturschutzes mit dem Ziel, den Gesamtbestand an wild lebenden Tier- und Pflanzenarten innerhalb ihres natürlichen Verbreitungsgebietes [▶Areal] in ihrer gegebenen Vielfalt so zu erhalten und zu fördern, dass die Evolution der Arten gesichert bleibt; ANL 1984; in D. ist der **A.** einerseits über Rechtsvorschriften wie das ▶Bundesnaturschutzgesetz, die Landesnaturschutzgesetze und die Bundesartenschutzverordnung [▶Artenschutzverordnung] verankert. Andererseits werden viele Arten über den Schutz ihrer Lebensräume im Rahmen des ▶Biotopschutzes oder durch Ausweisung von ▶Schutzgebieten geschützt. Internationale Artenschutzbestimmungen und –abkommen wie z. B. das ▶Washingtoner Artenschutzabkommen, das ▶Übereinkommen über die biologische Vielfalt [Biodiversitätskonvention]. Seit den 1990er-Jahren hat sich ein komplexes Gefüge von nationalen, europäischen und internationalen Rechtsvorschriften wie z. B. die ▶Fauna-Flora-Habitat-Richtlinie [FFH-RL], ▶Vogelschutzrichtlinie und EU-Artenschutzverordnung [Verordnung EG 338/97] entwickelt. So gehen die Anforderungen des Europäischen Naturschutzrechtes weit über die nationalen Bestimmungen hinaus — insbesondere für die europäischen Vogelarten und die Arten des Anhanges IV der FFH-RL; in Europa natürlich vorkommende Vogelarten werden durch Artikel 1 der Vogelschutzrichtlinie sowie in einer Rechtsverordnung nach § 52 BNatSchG, z. B. Bundesartenschutzverordnung, geschützt; Tier- und Pflanzenarten von gemeinschaftlichem Interesse sind die in den Anhängen II, IV und V der FFH-RL aufgeführt. Der **A.** vor Ort wird mit Artenschutzmaßnahmen resp. Artenhilfsmaßnahmen, die den Erhalt bestimmter Populationen und Vegetationsstrukturen fördern sollen, durchgeführt. Dies sind z. B. die Schaffung von Biotopverbundnetzen, Anlage von Ackerrandstreifen, die Wiederansiedlung verdrängter wild lebender Arten innerhalb ihres natürlichen Areals. Der **A.** erstreckt sich auch außerhalb der natürlichen Lebensräume, z. B. in Botanischen und Zoologischen Gärten oder in Genbanken; ▶Erhaltung der natürlichen Lebensräume sowie der wild lebenden Tiere und Pflanzen, ▶Erhaltung der Tier- und Pflanzenwelt, ▶Faunenschutz, ▶Gebietsschutz, ▶integrierter Artenschutz, ▶Naturschutz, ▶Tierartenschutz).

species cover [n] *phyt.* ▶degree of species cover.

5874 species density [n] *ecol.* (Frequency of plant or animal species per unit area or volume; ▶diversity, ▶population density 3, ▶population dynamics, ▶species abundance); *s* **densidad [f] de especies** (Cantidad de especies de flora y fauna que se presentan en un área determinada; ▶abundancia [de una especie], ▶densidad de población 3, ▶dinámica de poblaciones, ▶diversidad); *f* **densité [f] des espèces** (Rapport exact du nombre des individus d'une même espèce observée sur un certain territoire avec l'étendue de ce territoire ; ▶abondance, ▶densité des individus, ▶diversité, ▶dynamique de population) ; *g* **Arten-**

dichte [f] (Häufigkeit von Tier- oder Pflanzenarten bezogen auf eine bestimmte Fläche; ▶Abundanz, ▶Diversität, ▶Individuendichte, ▶Populationsdynamik); *syn.* absolute Artenabundanz [f].

species diversity [n] *ecol.* ▶taxonomic diversity.

species diversity [n], **decline in** *phyt. zool.* ▶decline in number of species.

species diversity [n], **reduction in** *phyt. zool.* ▶decline in number of species.

species equilibrium [n] *ecol.* ▶dynamic species equilibrium.

5875 species inventory [n] *phyt. zool.* (Recorded list of species known to occur within a specific area); *s* **inventario [m] de especies** (Conjunto de especies en un área de muestreo o en un hábitat determinado); *f* **inventaire [m] des espèces** (Somme totale des espèces d'un peuplement pour une aire de prospection ou une aire caractéristique de la communauté) ; *g* **Arteninventar [n]** (Artenbestand einer bestimmten Aufnahmefläche oder eines bestimmten Lebensraumes.

species list [n] *landsc. plant.* ▶plant list, ▶list of undesirable plants.

species [n] **living on wood** *phyt. zool.* ▶lignicolous species.

species [n] **of a habitat** *phyt. zool.* ▶character species of a habitat.

species [n] **of an alliance** *phyt.* ▶character species of an alliance.

5876 species [n] **of soil fauna** *pedol. zool.* (▶soil organism, ▶living in the soil); *s* **animal [m] edáfico** (▶edafon, ▶hipogeo/a); *f* **pédobionte [m]** (organisme édaphique ; ▶édaphon, ▶endogé, ée) ; *syn.* animal [m] édaphique ; *g* **tierisches Bodenlebewesen [n]** (Im Boden lebendes Tier; ▶Bodenlebewelt, ▶im Boden lebend); *syn.* Bodentier [n].

5877 species [n] **of soil flora** *bot. pedol.* (Fungus or other vegetal microorganism); *s* **especie [f] de flora edáfica** (Hongos u otros microorganismos vegetales que viven en el suelo); *syn.* especie [f] de flora hipogea; *f 1* **espèce [f] édaphique** (Terme générique qui englobe les bactéries, les actinomycètes et les champignons vivant dans le sol ; PED 1984, 37s) ; *f 2* **microorganisme [m] édaphique** (Bactéries, champignons mycorrhiziens, protozoaires vivant dans le sol et dont les fonctions sont d'assurer la biodisponibilité en azote minéral d'origine essentiellement microbiologique, l'altération des minéraux primaires dont dépend la fertilité des sols, le recyclage du carbone, et le fonctionnement des cycles biogéochimiques et plus particulièrement la minéralisation de la matière organique ; PED 1984, 37s ; *terme spécifique* flore bactérienne édaphique) ; *syn.* microorganisme [m] du sol, microflore [f] édaphique ; *g* **pflanzliches Bodenlebewesen [n]** (Mikroorganismus, der den Bakterien [Actinomycetaceae], Archaeen [Urbakterien], Pilzen oder Algen zuzuordnen ist).

species [n] **of wild animals, migratory** *conserv.* ▶effective management of migratory species of wild animals.

Species [n] **of Wild Animals, Convention on the Conservation of Migratory** *conserv.* ▶Bonn Convention.

5878 species pattern [n] *biol. ecol.* (Spatial distribution of plant or animal species within a specific area; ▶species composition); *s* **estructura [f] específica (de especies)** (Distribución espacial de especies animales o vegetales dentro de un área determinada; ▶composición de especies de una comunidad); *syn.* «pattern» [m] de especies; *f* **ensemble [m] spécifique** (Population animale ou végétale à l'optimum de son développement naturel pour une aire donnée ; ▶composition des espèces d'une communauté [animale ou végétale]) ; *g* **Artengefüge [n]** (Natür-

lich gewachsene Gemeinschaft von Pflanzen- oder Tierarten in einem Lebensraum; ▶Artenzusammensetzung).

5879 species-poor [adj] *ecol.* (Containing a relatively low number of species; *opp.* ▶species-rich); *s* **baja diversidad de especies [loc]** (Población o zona en la que se presenta un número relativamente bajo de especies; *opp.* ▶alta diversidad de especies); *syn.* pobre en especies [loc]; *f* **pauvre en espèces [loc]** (Se dit d'un milieu, d'une communauté, d'un peuplement qui regroupe ou n'abrite qu'un petit ou très petit nombre d'espèces ; *opp.* ▶riche en espèces) ; *syn.* paucispécifique [adj] ; *g* **artenarm [adj]** (So beschaffen oder ausgestattet, dass nur wenige Arten vorkommen; *opp.* ▶artenreich).

5880 species-rich [adj] *biol. ecol. opp.* ▶species-poor; *s* **alta diversidad [f] de especies** (Población o zona en la que se presenta un número relativamente alto de especies; *opp.* ▶baja diversidad de especies); *syn.* rico/a [adj] en especies; *f* **riche [adj] en espèces** (Se dit d'un milieu, d'un peuplement qui abrite un grand nombre d'espèces ; *opp.* ▶pauvre en espèces) ; *syn.* plurispécifique [adj] ; *g* **artenreich [adj]** (So beschaffen oder ausgestattet, dass viele Arten vorkommen; *opp.* ▶artenarm).

5881 species richness [n] *ecol.* (Variety of species within an area of study. The richness of species does not necessarily mean a ▶diversity!; ▶taxonomic diversité); *s* **riqueza [f] en especies** (Número de especies que aparecen en una comunidad de organismos; DINA 1987, 314; ▶diversidad ecológica, ▶espectro de diversidad); *syn.* diversidad [f] de especies, diversidad [f] biológica; *f* **richesse [f] en espèces** (Nombre d'espèces vivant sur un territoire étudié. Ne pas confondre avec la diversité biologique ; ▶diversité, ▶diversité des espèces) ; *syn.* richesse [f] des peuplements ; *g* **Artenvielfalt [f, o. Pl.]** (Menge an unterschiedlichen Arten einer Artenzusammensetzung in einem Untersuchungsraum; ▶Artendiversität, ▶Diversität); *syn.* Artenmannigfaltigkeit [f], Artenreichtum [m].

5882 species-specific [adj] *phyt. zool.* (TEE 1980, 116; appropriate to the character of a particular species; ▶species suited to site conditions); *s* **específico/a de una especie [loc]** (Que corresponde a las exigencias y reacciones naturales de una especie; ▶especie adaptada a la residencia ecológica); *syn.* adecuado/a para una especie [loc]; *f* **conforme à la personnalité écologique de l'espèce [loc]** (Répondant aux exigences naturelles, aux spécificités et réactions d'une espèce ; ▶espèce bien adapté aux conditions de la station) ; *syn.* spécifique de l'espèce [loc], spécifique à une espèce [loc], typique d'une espèce [loc], typique pour une espèce [loc] ; *g* **artgerecht [adj]** (Den natürlichen Ansprüchen, Eigenschaften und Reaktionen einer Art entsprechend; ▶standortgerechte Art); *syn.* artgemäß [adj], artspezifisch [adj], arttypisch [adj].

5883 species [n] **suited to site conditions** *landsc. phyt. zool.* (Species which exist or has settled in its typical habitats within its ▶natural range or ▶potential range; this is not always syn. with ▶indigenous plant species); *s* **especie [f] adaptada a la residencia ecológica** (▶área natural de distribución, ▶área virtual [de distribución], ▶especie vegetal autóctona); *syn.* especie [f] adaptada a la estación, especie [f] adaptada al hábitat; *f* **espèce [f] bien adaptée aux conditions de la station** (Espèce qui à l'intérieur de son aire naturelle ou de son aire potentielle se développe dans ses habitats naturels ; ▶aire naturelle d'une espèce, ▶aire potentielle d'une espèce, ▶espèce végétale indigène ; *syn.* espèce [f] adaptée aux conditions du milieu) ; *g* **standortgerechte Art [f]** (Art, die innerhalb ihres ▶natürlichen Areals oder ▶potentiellen Areals die für sie typischen Habitate bewohnt resp. dort angesiedelt wurde; muss nicht mit ▶bodenständiger Pflanzenart syn. sein. Eine nicht **s. Pflanzenart** erkennt man z. B. auch daran, dass diese Pflanze schlecht an-

S

wächst, einen hohen Pflegeaufwand benötigt, kümmert oder sogar abstirbt); *syn.* standortgemäße Art [f], standortspezifische Art [f].

5884 specifically agreed rate [n] *contr. prof.* (Accepted sum of money for the payment of work per man and time, or for executed construction work); *s* **tarifa [f] (de pago) acordada** (Monto de dinero acordado para el pago de la mano de obra por hora o de los servicios realizados); *f* **taux [m] de rémunération** (Valeur monétaire par unité de temps établie pour la rémunération des travaux effectués au métré ou en régie) ; *syn.* tarif [m] de rémunération ; *g* **Verrechnungssatz [m]** (Vereinbarter Geldbetrag für die Vergütung von Arbeitskräften je Zeiteinheit oder von erbrachten Leistungen; cf. § 7 Nr. 1 VOB Teil A).

specification clause [n] [UK]**, basic** *contr.* ▶basic specification item.

specification item [n] *contr.* ▶alternate specification item, ▶basic specification item, ▶collective specification item, ▶optional specification item, ▶separate specification item.

specification item [n]**, provisional** *contr.* ▶optional specification item.

5885 specification item number [n] *constr. contr.* (Additional computer code number attached to items in a ▶list of bid items and quantities [US]/schedule of tender items [UK] which includes a specific description of construction work, used in standardized specifications such as Uniform Construction Specifications [US] of the Construction Specification Institute [CSI]/National Building Specifications [UK]; ▶bid item [US]/tender item [UK]); *s* **número [m] de órdenes** (Desde la introducción de los pliegos de normas técnicas de construcción estandarizados, códigos adicionales a los de las ▶partidas de oferta que contienen la descripción de un tipo trabajo de construcción; ▶resumen de prestaciones); *f* **numéro [m] d'ordre** (Depuis l'existence de logiciels de métré et de pièces écrites [logiciel de descriptif] désigne la numérotation des ▶articles désignant une prestation de travaux dans un ▶descriptif quantitatif ou un métré ; ▶numéro de prix décomposé) ; *syn.* poste [m] d'ordre, code [m] ; *g* **Ordnungszahl [f]** (Seit der Einführung von standardisierten Leistungsbüchern zusätzlich zur Positionsnummer eingeführte Codenummer im ▶Leistungsverzeichnis, die eine spezifizierte Beschreibung einer Bauleistung beinhaltet; ▶Position).

specifications [npl] *constr. contr.* ▶outline specifications; *constr.* ▶standard specifications; *constr. contr.* ▶technical specifications; *constr. eng.* ▶Uniform Construction Specifications [US]/National Building Specification (NBS) [UK].

specific land-use area [n] [D] *leg. urb.* ▶zoning district [US].

5886 specimen grass [n] *hort. plant.* (Very decorative grass suitable for planting on its own for aesthetic effect as an accent plant; ▶ornamental grass); *syn.* accent grass [n]; *s* **gramínea [f] solitaria** (Tipo de hierba grande apropiada para plantar sola con fines decorativos; *término genérico* ▶hierba ornamental); *f* **graminée [f] (en position) isolée** (Critère de qualité caractérisant les graminées décoratives; *terme générique* ▶graminée ornementale) ; *g* **Solitärgras [n]** (*Terminus für das Wuchsverhalten eines Grases hinsichtlich seiner ästhetischen Wirkung und gärtnerischen Verwendung* 1. Gras in Einzelstellung. 2. Zur Einzelpflanzung geeignetes, besonders dekoratives Gras; *OB* ▶Ziergras).

specimen nursery stock [n] *hort. plant.* ▶specimen tree/shrub.

5887 specimen perennial [n] *hort. plant.* (**1.** Individually-planted perennial. **2.** Tall perennial species specially suitable for aesthetic effect); *s* **vivaz [f] solitaria** (**1.** Planta vivaz que crece sola. **2.** Planta vivaz muy decorativa criada para plantarse sola); *syn.* perenne [f] solitaria; *f* **plante [f] vivace isolée** (Plante décorative par ses fleurs et/ou son feuillage, en général de haute végétation, plantée en évidence en groupe de un, trois, cinq sujets) ; *g* **Solitärstaude [f]** (*Terminus für das Wuchsverhalten einer Staude hinsichtlich ihrer gestalterischen Wirkung und gärtnerischen Verwendung* 1. Standfeste, formprägnante Großstaude in Einzelstellung. 2. Zur Einzelpflanzung besonders geeignete dekorative Großstaude).

5888 specimen plant [n] *hort. plant.* (Generic term for ▶individual woody plant, ▶specimen tree/shrub, ▶specimen perennial, or ▶specimen grass planted for a particular trait [flower, colo(u)r texture, structure], to create a focal point within a design composition); *syn.* accent plant [n]; *s 1* **planta [f] solitaria** (Término genérico para ▶espécimen arbóreo/arbustivo, ▶gramínea solitaria, ▶leñosa solitaria y ▶vivaz solitaria); *s 2* **espécimen [m]** (Planta criada para plantarse como ▶ejemplar); *f 1* **plante [f] isolée** (Plante en position isolée ; ▶arbre/arbuste isolé, ▶graminée [en position] isolée, ▶plante vivace isolée) ; *syn.* plante [f] à isoler, solitaire [m] ; *f 2* **spécimen [m]** (**1**) (Critère de qualité caractérisant les plantes de pépinière de très grande force cultivées pour prendre une position isolée dans un espace vert) ; *syn.* spécimen [m] remarquable, sujet [m] exceptionnel ; *g* **Solitärpflanze [f]** (*Terminus für das Wuchsverhalten einer Pflanze hinsichtlich ihrer gestalterischen Wirkung und gärtnerischen Verwendung;* Pflanze in Einzelstellung; *UBe* ▶Solitärgehölz 1, ▶Solitärgehölz 2, ▶Solitärgras und ▶Solitärstaude. 2. Für Einzelstellung in Baumschulen oder Staudengärtnereien herangezogene Pflanze); *syn.* Solitär [m].

5889 specimen tree/shrub [n] *hort. plant.* (Exceptional, nursery-grown tree or shrub suitable for accent planting, transplanted three or more times, finally at extra-wide spacing for shapeliness. In order to keep such woody plants transplantable, it is necessary either to root-prune the plant in place or to continue the transplanting process); *syn.* specimen nursery stock [n], advanced nursery-grown stock [n] [also UK], large nursery stock [n]; *s 1* **espécimen [m] arbóreo/arbustivo** (Árbol o arbusto grande cultivado en vivero que es adecuado para plantación individual. Es transplantado tres o más veces, finalmente con espaciamiento muy grande para dejar desarrollar su porte típico. Para que estas leñosas sigan siendo transplantables es necesario cortar lateralmente el cepellón o transpartarlas periódicamente); *s 2* **ejemplar [m]** (Árbol o arbusto grande con forma especialmente bonita); *f* **spécimen [m]** (**2**) (Critère de qualité caractérisant les gros sujets de pépinière de très grande force, transplantés trois fois et plus, cultivés comme plante à grand écartement) ; *g* **Solitärgehölz [n]** (**2**) (Zur Einzelpflanzung geeignetes, in der Baumschule kultiviertes Gehölz: drei- oder mehrmals verpflanzt, beim dritten Mal in „extra weitem Stand". Nach dem 3. Verpflanzen wird es gegebenenfalls durch weiteres, wiederholtes Verpflanzen oder Umstechen des Wurzelballens verpflanzfähig gehalten; es gibt „Solitärsträucher", „Solitärstammbüsche" und „Solitärhochstämme"); *syn.* Einzelgehölz [n].

spectator sport [n] *recr.* ▶participatory sport [US], #2.

speed [n] *eng. trans.* ▶design speed.

speed bump [n] [US] *eng. trans.* ▶speed hump.

speed check ramp [n] [UK] *trans.* ▶speed hump.

5890 speed hump [n] *trans.* (Traffic management device for discouraging speeding in residential areas; a smaler device in parking areas is a **speed bump [US]/rumbling strip [UK]**); *syn.* speed check ramp [n] [also UK] (LD 1988 [9]); *in the vernacular* sleeping policeman [n] [also UK]; *s* **frena-coches [m]** (Montadura para desacelerar el tráfico rodado); *syn.* tope [m] para el tráfico, badén [m], policía [m] acostado [AL], lomo [m] de toro

[RCH]; *f* **dispositif** [m] **dos d'âne** (Surélévation placée en travers de la chaussée mis en place dans certaines zones urbaines, dont le franchissement désagréable contraint les automobilistes à réduire leur vitesse ; *terme générique* limitateur de vitesse [*syn.* ralentisseur]) ; *syn.* bande [f] rugueuse ; *g* **Schwelle** [f] (Quer zur Fahrtrichtung in eine Verkehrsfläche eingebaute Erhöhung mit dem Ziel, dass Fahrer ihre Geschwindigkeit verlangsamen. Wegen der Lärm- und Auspuffbelastung für Anwohner, Rettungs- und Einsatzfahrzeuge und der Schadenshaftung des Straßenbaulastträgers werden in D. **S.n.** kaum noch eingebaut); *syn.* Straßenschwelle [f].

sphagnum [n] *phyt.* ▶peat moss.

sphagnum mat [n] *phyt.* ▶floating sphagnum mat.

5891 sphagnum peat [n] *min. pedol.* (Generic term for ▶humified raised bog peat and ▶fibric peat 1); *syn.* raised bog peat [n]; *s* **turba** [f] **de esfagnos** (Término genérico para ▶turba negra y ▶turba blanca); *f* **tourbe** [f] **de Sphaigne** (Résidu organique naturel plus ou moins décomposé qui provient d'une accumulation de plantes dans les tourbières parmi lesquelles la sphaigne prédomine ; *terme générique* pour la ▶tourbe noire, ▶tourbe blanche et la ▶tourbe fibreuse) ; *g* **Hochmoortorf** [m] (OB zu ▶Schwarztorf und ▶Weißtorf 1); *syn.* ombrogener Torf [m].

sphere [n] **of influence** *envir. plan.* ▶zone of influence.

spicatum opus [n] [UK] *constr.* ▶herringbone pattern.

5892 spices garden [n] *gard.* (Garden for growing aromatic or pungent vegetable substances used to flavo[u]r food; ▶culinary herb, ▶herb garden); *syn.* condiment garden [n]; *s* **jardín** [m] **de hierbas condimenticias** (Jardín o parte de uno que se utiliza para cultivar ▶plantas condimenticias; ▶jardín de hierbas finas); *f* **jardin** [m] **d'herbes condimentaires** (Tout ou partie de jardin utilisé pour la culture de ▶plantes condimentaires ; ▶jardin d'herbes) ; *g* **Gewürzgarten** [m] (Garten oder ein Teil davon, der zur Anzucht von ▶Gewürzpflanzen dient; ▶Kräutergarten); *syn.* Gewürzpflanzengarten [m].

spike [n] [UK] *arb. for. hort.* ▶snag (1).

spike-top [n] [US] *arb. for.* ▶dead crown, ▶top-kill [US].

spike-topped [pp/adj] [US] *arb. for.* ▶stag-headed, #2.

spiking [n] *constr.* ▶lawn aeration, #3.

spill [n] [US]**, oil** *envir.* ▶oil pollution accident.

5893 spillway [n] *eng. wat'man.* (Water passage in or around a dam or other hydraulic structure for the escape of excess flood waters. An **emergency spillway** will be used in the event of floods exceeding the capacity of the main spillway; cf. WMO 1974); *syn.* overfall [n]; *s* **aliviadero** [m] (Desagüe de una presa u obra estructura hidráulica, en forma de canal abierto o conducto cerrado. Un **a.** de emergencia es utilizado cuando las crecidas exceden de la capacidad del **a.** principal; cf. WMO 1974); *syn.* rebosadero [m]; *f* **déversoir** [m] **de crue** (Ouvrage par lequel s'écoule le trop-plein des eaux d'un cours d'eau ou permettant de contrôler, dériver, régler l'écoulement des eaux excédentaires d'une retenue d'eau ; un déversoir de secours est utilisé lorsque les crues dépassent la capacité du déversoir principal) ; *syn.* évacuateur [m] ; *g* **Hochwasserentlastungsanlage** [f] (Anlage zur Verhinderung des Überströmens von Staudämmen bei ungünstigen meteorologischen Verhältnissen); *syn.* Überlauf [m].

spine [n] (1) *arb. for. hort.* ▶snag (1).

5894 spine [n] (2) *bot.* (Modified leaf in the form of a sharp, rigid structure such as of a cactus, or a projection from the margins of a leaf blade, e.g. *Ilex* y los cactus; ▶thorn); *s* **espina** [f] **foliar** (Hoja lignificada, endurecida y puntiaguda que se presenta p. ej. en *Ilex*, los cactus; ▶espina); *f* **épine** [f] **foliaire** (Organe acéré, sclérifié et donc piquant, né de la spécialisation de l'apex d'une feuille, p. ex. *Berberis* et *Cactus* ou d'un partie de feuille, p. ex. *Ilex* ; GOR 1985, 205 ; ▶épine) ; *g* **Blattdorn** [m] (Zu spitzen, harten, holzigen ▶Dornen umgebildete — metamorphosierte — Blätter, z. B. bei Berberitze *[Berberis]* und Kaktus oder bei Teilen der Blätter, z. B. bei Stechpalme *[Ilex]*).

5895 spiral stair(case) [n] *arch.* (Flight of steps, the treads of which wind around a central newel); *syn.* circular stair [n], caracole [n], cockle stair [n], corkscrew stair [n], helical stair [n], solid newel stair [n], spiral stairway [n] [also US]; *s* **escalera** [f] **de caracol** (Tipo de escalera helicoidal, sin descansillos; DACO 1988); *f* **escalier** [m] **à vis** (Escalier de forme hélicoïdale, tournant en spirale autour d'un axe [= escalier à vis à noyau], qui soutient toutes les marches) ; *syn.* escargot [m] (DTB 1985, escalier [m] en colimaçon) ; *g* **Wendeltreppe** [f] (Treppe, die schraubenförmig um eine Achse von einem Stockwerk zum anderen führt).

spiral stairway [n] [US] *arch.* ▶spiral stair(case).

5896 splash erosion [n] *geo. pedol.* (Displacement of soil particles by the impact of large raindrops, particularly under intense convectional precipitation and bare earth conditions; ▶sheetwash [US]/sheet wash [UK]); *syn.* raindrop erosion [n] [also UK]; *s* **erosión** [f] **causada por gotas de lluvia** (Desplazamiento de la tierra por la acción de grandes gotas de agua individuales; en tormentas fuertes se pueden soltar hasta 200 t de tierra/ha y así dar comienzo al ▶arrastre laminar); *f* **érosion** [f] **provoquée par l'action de la pluie** (Détachement de particules d'un sol découvert provoqué par l'impact des grosses gouttes de pluie lors d'un fort orage ; jusqu'à 200 t/ha de terre peuvent être ainsi détachées pour être ensuite déplacées par ▶ablation, ▶érosion laminaire) ; *g* **Spritzerosion** [f] (Abtragung des vegetationslosen Oberbodens durch den Aufprall großer Regentropfen; bei einem heftigen Gewitterregen können bis zu 200 t Boden/ha aufgespritzt und für die ▶Abspülung aufbereitet werden).

splayed kerb [n] [UK] *constr. eng.* ▶mountable curb [US].

split-faced [pp/adj] *constr.* ▶quarry-faced; ∗▶tooling of stone [US]/dressing of stone [UK], #2.

5897 split-face dry masonry [n] *constr.* (Coursed ashlar masonry composed of roughly split, rough-hewn natural stone with variable joint widths); *syn.* rough faced masonry wall [n]; *s* **mampostería** [f] **de piedras de corte natural** (Paredón de piedra de corte natural sin labrar o de labra tosca con juntas de diferentes anchuras); *f* **maçonnerie** [f] **de moellons de taille rugueuse** (Mur en pierres naturelles assisées constitué de moellons tranchés, éclatés ou lités) ; *g* **Raumauerwerk** [n] (Lagerhaftes Natursteinmauerwerk aus spaltrauen, grob bossierten Steinen mit unterschiedlich breiten Fugen aufgesetzt); *syn.* f. S. Rauhmauerwerk [n].

split-level house garden [n] [UK] *arch. constr.* ▶terraced townhouse garden [US].

splitting-up [n] **of an area** *conserv. plan. urb.* ▶fragmentation of an area.

spodosol [n] [US] *pedol.* ▶podzol.

spoil [n] [US] *min.* ▶coal mine spoil [US]/colliery spoil [UK], ▶coarse spoil, ▶disposal of mining spoil, ▶fine spoil, ▶mining spoil, ▶underground disposal of mining spoil.

spoil [n] [UK]**, colliery** *min.* ▶coal mine spoil [US].

spoil area [n] *envir.* ▶dumpsite [US]/tipping site [UK], *envir. min.* ▶large spoil area [US]/large tip [UK].

spoil bank [n] [US] (1) *envir.* ▶dumpsite [US]/tipping site [UK].

5898 spoil bank [n] [US] (2) *min.* (Mound of deposited mining spoil, originating from coal, iron ore or mineral exploitation; ▶slag pile [US]/slag heap [UK]); *syn.* mine dump [n] [US], spoil heap [n] [UK], spoil pile [n] [US], spoil tip [n] [UK], coal bing [n] [also SCOT], residue mound [n], *in the case of slatelike mining spoil* slate mound [n]; *s* **vaciadero** [m] **de gangas** (Depósito de materiales de desecho de minas; ▶escorial); *f 1* **terril** [m] (Ouvrage de dépôts de résidus [stériles francs excavés lors du creusement des galeries, stériles minéralisés provenant du triage ou stériles non minéralisés provenant du lavage] au voisinage d'une mine [charbon, fer, potasse, aluminium, etc.] ; le terril d'une mine de charbon est en général essentiellement composé de matériaux schisteux, ainsi que de grès houillers et des fragments de charbon ce qui lui confère donc sa couleur noire ; les produits stockés sur certains terrils datant d'une époque où les méthodes de séparation de triage et de lavage du charbon n'était pas aussi efficace qu'aujourd'hui, sont relavés et servent de combustible aux centrales électriques, d'autres sont utilisés comme matériaux de construction [briques, surschistes], pour les travaux publics [emploi des schistes houillers brûlés en technique routière], les espaces verts de loisirs [revêtement d'allées] et de sports [terre battue des courts de tennis] ; certains terrils ont été réaménagés en base de loisirs [VTT, ski], d'autres, présentant un intérêt écologique, ont été classés en zones naturelles protégées ; ▶crassier) ; *syn.* terri [m], verse [f], terril [m] minier ; *f 2* **halde** [f] (*Terminologie régionale* grand tas de déblais formé par les résidus de recherche ou d'exploitation au voisinage d'une mine de fer) ; *syn.* halde [f] de stériles ; *g* **Bergehalde** [f] (Anschüttung tauben Gesteins bei der Kohle- und Eisenerzförderung; ▶Schlackenhalde); *syn.* Bergbauhalde [f].

spoil heap [n] [UK] *min.* ▶spoil bank [US] (2)/spoil heap [UK].

spoil pile [n] [US] *min.* ▶high spoil pile [US]/high tip [UK], ▶spoil bank [US] (2)/spoil heap [UK].

5899 spoil plateau [n] [US] *landsc.* (Flat top of spoil area [US]/tip [UK]); *syn.* tip plateau [n] [UK]; *s* **esplanada** [f] **de escombrera** (≠) (Superficie grande y llana de una escombrera de zafras o de estériles); *f 1* **plateau** [m] **d'un dépôt de matériaux** (Large plate-forme supérieure plane d'un dépôt de matériaux minéraux) ; *f 2* **plateau** [m] **d'un terril** (Surface relativement plane ou légèrement ondulée d'un dépôt de stériles réalisé à l'horizontale) ; *syn.* plateau [m] d'un dépôt de stériles ; *g* **Haldenplateau** [n] (Große, ebene Oberfläche einer Abraum- oder Bergehalde).

5900 spoil reclamation [n] [US] *land'man. landsc.* (Reclaiming with vegetation established on spoil areas [US]/tips [UK]); *syn.* tip reclamation [n] [UK]; *s* **plantación** [f] **de escombreras** (Vegetalización de antiguas escombreras); *f* **végétalisation** [f] **de terril** (Technique de retraitement des anciens terrils par installation d'une végétation par ensemencement ou plantation ; en fonction de la nature des résidus déposés et les caractéristiques des sites et de l'utilisation ultérieure différentes méthodes de végétalisation sont mises en œuvre : étanchement-végétalisation ou végétalisation absorbante pour les terrils potassiques, ensemencement au canon hydraulique, filet biodégradable, fixation des pentes par des pieux en bouture de saule, plantation de diverses variétés d'épineux et de plantes spécifiques aux terrils, telles le tussilage ou le pavot cornu, pour les terrils houillers, végétalisation dépolluante avec la mise en place de métallophytes sur les terrils calaminaires) ; *syn.* plantation [f] de terril, revégétalisation [f] de(s) terrils ; *g* **Haldenbegrünung** [f] (Schaffung einer Pflanzendecke auf einer Abraum- oder Bergehalde); *syn.* Haldenrekultivierung [f].

5901 spoil site [n] [US] *landsc. plan.* (Location where a spoil area [US]/tip [UK] exists or is being developed); *syn.* tip site [n] [UK]; *s* **ubicación** [f] **de escombrera** (Lugar en el que existe o está previsto emplazar una escombrera); *f 1* **site** [m] **d'un dépôt de matériaux** (Emplacement sur lequel se trouve ou est prévu l'implantation d'un dépôt de stériles) ; *syn.* site [m] d'un dépôt de stériles ; *f 2* **site** [m] **d'un terril** ; *g* **Haldenstandort** [m] (Fläche, auf der eine Abraum- oder Bergehalde entstehen soll, entsteht oder entstanden ist).

spoil tip [n] [UK] *min.* ▶spoil bank [US] (2)/spoil heap [UK].

sponsor [n] [UK]**, tree** *conserv.* ▶voluntary caretaker of a tree [US].

5902 sponsor [n] **of a competition** *prof.* (Individual or company responsible for originating and financing a business awards competition, including the provision of prize or money; ▶awarding of prizes, ▶initiating authority); *syn.* promoter [n] of a competition; *s* **promotor, -a** [m/f] **de concurso** (2) (Persona, fundación o empresa privada que convoca un concurso de ideas o de realización y financia y otorga los premios o las menciones; ▶entrega de premios, ▶promotor de concurso); *f* **promoteur** [m] **du concours** (Personne physique, institution ou société privé qui prépare le programme d'un concours, en organise et assure financièrement la bonne marche, nomme les membres du jury et attribue les prix, primes, indemnités et récompenses ; ▶pouvoir adjudicateur d'un concours, ▶remise des prix) ; *g* **privater Auslober** [m]/**private Ausloberin** (Einzelperson, Stiftung oder Firma, die einseitig verspricht, dass sie einen Wettbewerb ausschreibt, durchführt und für die besten Ergebnisse Preise in Geld vergibt und Auszeichnungen und Belobigungen verleiht; ▶Preisverleihung, ▶auslobende Behörde/Organisation).

spontaneous colonization [n] *phyt. plant.* ▶natural colonization.

5903 spontaneous colonization [n] **by scrub** *conserv. land'man.* (Natural growth of low woody species, e.g. in heaths or boggy areas, on agricultural land which has been left uncultivated for a period of time; ▶clearance of unwanted spontaneous woody vegetation); *s* **invasión** [f] **de leñosas** (Colonización de una turbera por especies leñosas; ▶desbroce 1); *f* **embroussaillement** [m] (Recolonisation spontanée par des espèces ligneuses, p. ex. sur une tourbière, sur les pâturages alpins abandonnés ; ▶débroussaillage) ; *syn.* envahissement [m] des broussailles, invasion [f] par des broussailles, colonisation [f] spontanée par les broussailles, *peu utilisé* embroussaillage [m] ; *g* **Verbuschung** [f] (Natürlicher, niedriger Aufwuchs von Gehölzen, z. B. in Heide- oder Moorflächen, auf länger brachliegenden landwirtschaftlichen Flächen; wenn verbuschte Flächen sich selbst überlassen werden und zu Wald werden, spricht man auch von **Verwaldung**; N+L 1999, 411); ▶Entkusselung).

5904 spontaneous growth [n] *hort.* (Plants which have grown without human intervention); *s* **crecimiento** [m] **espontáneo** (de la vegetación sin la intervención humana. *Contexto* favorecer el **c. e.** de especies raras); *f* **croissance** [f] **spontanée** (des végétaux sans l'action de l'homme ; *contexte* favoriser la **c. s.** d'espèces rares) ; *g* **Wildwuchs** [m, o. Pl.] (1) (Vom Menschen nicht beeinflusstes Wachsen von Pflanzen. *Kontext* den **W.** seltener Pflanzen fördern).

5905 spontaneous plants [n] *hort.* (Plants which have established themselves spontaneously; ▶natural colonization, ▶weed 1, ▶weed tree); *s* **vegetación** [f] **espontánea** (1) (Plantas que se han establecido espontáneamente; ▶colonización natural de «espacio vacío», ▶leñosa adventicia, ▶planta advenediza); *f* **végétation** [f] **spontanée** (1) (Espèces non introduite pouvant se développer sans l'action de l'homme ; ▶colonisation spontanée, ▶plante adventice, ▶végétation arbustive adventice) ; *syn.* plantes [fpl] spontanées ; *g* **Wildwuchs** [m] (2) (*Fachsprachlich* ...wüchse [pl], *sonst ohne Pl.*; Pflanzen, die sich spontan angesiedelt haben; gärtnerischer/allgemeiner Begriff für ▶spontane Besiedelung; ▶Fremdaufwuchs, ▶Unkraut); *syn.* Fremdbewuchs [m].

spontaneous regeneration [n] *for. phyt.* ▶natural regeneration (1).

5906 spontaneous vegetation [n] *phyt.* (Young plant cover resulting from ▶seed rain or ▶barochory on exposed sites, previously lacking in vegetation; ▶recolonization, ▶weed tree); *s* **vegetación** [f] **espontánea (2)** (Cubierta vegetal joven en superficie anteriormente sin vegetación resultante de regeneración natural por ▶nuevo vuelo o ▶barocoria; ▶brinzal natural [de barocoras], ▶leñosa adventicia, ▶recolonización natural); *f* **végétation** [f] **spontanée (2)** (Jeune peuplement végétal dont le développement résulte de ▶flux de semences, d'apport barochore ou de la régénération naturelle sur des milieux ouverts, en général pauvres en végétation ; ▶barochorie, ▶recolonisation barochore, ▶recolonisation naturelle, ▶végétation arbustive adventice) ; *g* **Aufwuchs** [m] **(2)** (*Fachsprachlich* ...wüchse [pl], *sonst ohne Pl.*; junger Pflanzenbestand als Ergebnis von ▶Anflug 2 oder ▶Aufschlag [Naturverjüngung] auf offenen, meist vegetationsarmen Flächen; ▶Wiederbesiedelung, ▶Fremdaufwuchs); *syn.* spontane Vegetation [f] (2), Spontanvegetation [f].

spontaneous woody vegetation [n] *conserv. land'man.* ▶clearance of unwanted spontaneous woody vegetation.

spoon drain [n] [AUS] *constr.* ▶dished channel unit.

sport [n] *recr.* ▶horseback riding sport [US]/horse riding sport [UK], ▶participatory sport [US]/sport enjoyed by the masses [UK].

sport [n] [UK], **equestrian** *recr.* ▶horseback riding sport [US]/horse riding sport [UK].

sport [n] [US], **mass** *recr.* ▶participatory sport [US]/sport enjoyed by the masses [UK].

sport [n] [US], **spectator** *recr.* ▶participatory sport [US], #2.

sport [n] **enjoyed by the masses** [UK] *recr.* ▶participatory sport [US].

5907 sport fishing [n] *recr.* (Angling done in leisure time); *syn.* angling [n]; *s 1* **pesca** [f] **recreativa** (Actividad de pesca de anzuelo desarrollada por aficionados en su tiempo libre); *s 2* **pesca** [f] **deportiva** (Actividad de pesca desarrollada por aficionados con fines competitivos); *f* **pêche** [f] **(de loisirs)** *syn.* pêche [f] à la ligne, pêche [f] sportive, *syn. scientifique* sport [m] halieutique, loisirs [mpl] halieutiques ; *g* **Angelsport** [m] (In der Freizeit betriebene Angelfischerei); *syn.* Angeln [n, o. Pl.], Freizeitangeln [n, o. Pl.].

sports [npl] **and physical activities** [npl] [US] *plan. recr. urb.* ▶overall master planning for sports and physical activities [US]/strategic planning for sports and physical activities [UK].

5908 sports area [n] *urb.* (Zoning use category for sports activities); *s* **suelo** [m] **para usos deportivos** (En D. categoría de planificación urbanística fijada en los planes municipales de ordenación o en los planes parciales); *f* **sols** [mpl] **sportifs** (Catégorie d'usage utilisée dans le zonage des plans d'urbanisme) ; *g* **Sportflächen** [fpl] (In Bauleitplänen ausgewiesene Nutzungskategorie für Sport).

sports area construction [n] [US] *constr.* ▶sports ground construction.

5909 sports facility [n] *recr.* *s* **instalación** [f] **de deportes**; *f* **équipement** [m] **sportif** ; *g* **Sporteinrichtung** [f].

sports field [n] [US] **(1)** *recr.* ▶sports ground.

sports field [n] **(2)** *constr. recr.* ▶grass sports field.

sports fields [npl] *landsc.* ▶school sports fields.

sports fields [npl] [US], **master plan for** *plan. recr. urb.* ▶overall master plan for sports and physical activities [US]/strategic planning for sports and physical activities [UK].

5910 sports fields complex [n] *recr.* (Enclosed area of sports fields and structures; ▶sports ground); *s* **complejo** [m] **de deportes** (Conjunto de áreas e instalaciones de cualquier tipo y tamaño que sirve para realizar actividades deportivas; ▶campo de deporte); *f 1* **installation** [f] **sportive** (Toutes formes d'aires utilisées pour une activité sportive) ; *f 2* **complexe** [m] **sportif** (Ensemble des différentes installations sportives [▶terrains de sport, aires de sport, bâtiments et installations annexes]) ; *g* **Sportanlage** [f] (**1.** Jegliche Anlage, groß oder klein, die dem Sport dient. **2.** Gesamtheit sportlich unterschiedlich nutzbarer Flächen und Gebäude, in einem Komplex arrondiert; ▶Sportplatz); *syn.* Sportstätte [f].

5911 sports grass mixture [n] *constr.* (Especially hardwearing grass-seed mixture used for lawn regularly played upon. A **regeneration mix** is sown to repair damaged or bare patches of grass on sportsfields); *s* **mezcla** [f] **de semillas para campos de deporte** (Mezcla de semillas de especies especialmente resistentes al pisoteo. Para reparar manchas dañadas o calvas existe la llamada **mezcla de regeneración**); *f* **mélange** [m] **de semences pour gazon pour terrains de sport** (Mélange de semences ou semences destinées à l'engazonnement de terrains de jeux, de forte résistance au piétinement, à l'arrachement et à la sécheresse ; pour la rénovation des surfaces dégradées des terrains de sport on utilise un **mélange de semences pour gazon de regarnissage** [gazon de regarnissage pour pelouses de sport]) ; *g* **Sportrasenmischung** [f] (Besonders strapazierfähige Saatgutmischung für regelmäßig bespielte Rasenflächen. Für die Ausbesserung von Fehlstellen in Sportrasenflächen gibt es sog. „Regenerationsmischungen").

5912 sports ground [n] *recr.* (Unroofed sports area with a ball field, which is used for organized competition sports as well as informal recreation); *syn.* sports pitch [n] [also UK], sports field [n] [also US]; *s* **campo** [m] **de deporte** (▶Complejo de deportes al aire libre con campo de juegos de pelota [campo de fútbol]); *f* **terrain** [m] **de sport** (Installation de plein air dotée d'un terrain de jeu, utilisée pour la compétition ou toutes autres formes d'activités corporelles ; *terme spécifique* terrain de sport sommaire ; ▶complexe sportif, ▶installation sportive) ; *g* **Sportplatz** [m] (Nicht überdachte ▶Sportanlage mit einem Ballspielplatz, die sowohl dem organisierten Wettkampfsport als auch der nicht wettkampforientierten, spielerisch-sportlichen Freizeitbeschäftigung dient).

5913 sports ground construction [n] *constr.* (Branch of the construction industry concerned with the planning and execution of sports facilities used for competitions and other physical activities); *syn.* sports area construction [n] [also US]; *s* **construcción** [f] **de instalaciones deportivas** (Parte del ramo de la construcción que se dedica a diseñar y construir equipamiento de deportes); *f* **construction** [f] **d'équipements sportifs** (Branche de l'activité de construction relative à la réalisation d'installations sportives en vue d'épreuves de compétition ou à toute autre activité corporelle) ; *g* **Sportstättenbau** [m] (Teilgebiet des Bauwesens, das die Planung und Errichtung von Sportanlagen zur Durchführung von Wettkämpfen und sonstigen körperlichen Betätigungen umfasst).

sports ground turf [n] [UK] *constr.* ▶playfield turf [US].

sports pitch [n] [UK] *recr.* ▶sports ground.

sports pitches [npl] [UK], **city plan for** *plan. recr. urb.* ▶overall master plan for sports and physical activities [US].

5914 spot elevation [n] *surv. constr.* (Existing or proposed level on a plan or on site); *s* **cota** [f] (Nivel existente o previsto en

S

un plan o una ubicación); *f* **cote [f] en altimétrie** (Hauteur ou altitude d'un point existant ou proposé par rapport à une cote de référence) ; *g* **Kote [f]** (Geplante oder vorhandene Höhe auf einem Plan oder vor Ort); *syn.* Höhenangabe [f].

5915 spread [n] (1) *phyt. zool.* (Extended distribution of a plant or animal species; ►range 1, ►distribution); *syn.* dispersal [n]; *s* **expansión [f]** (Proceso de ampliación del ►área de distribución de una especie de flora o fauna; ►distribución biogeográfica); *f* **dispersion [f]** (Processus d'extension de l'►aire de répartition des espèces végétales ou animales ; ►répartition [des espèces]) ; *syn.* diffusion [f], extension [f] ; *syn. phyt.* envahissement [m] végétal ; *g* **Ausbreitung [f]** (Prozess der Ausdehnung des Besiedlungsgebietes von Pflanzen- oder Tierarten; ausbreiten [vb]; ►Areal, ►Verbreitung 2); *syn.* Ausbreiten [n, o. Pl.].

5916 spread [vb] (2) *hort. phyt. plant.* (Process of outward growth of plant species. Where other plants are suppressed by strong growth, they are ►overrun; ►invasive species); *s* **extenderse [vb/refl]** (Fuerte crecimiento de especies de plantas que reprimen el crecimiento de otras; ►invadir, ►especie invasora); *syn.* expandirse [vb/refl]; *f* **s'étendre [vb]** (Extension naturelle de plantes vigoureuses ; pour les végétaux ayant tendance à étouffer leurs voisins, à l'envahissement ; ►espèce envahissante, ►envahir) ; *g* **sich ausbreiten [vb]** (Sich verbreiten von Pflanzen durch starke Wüchsigkeit; überhandnehmen [vb]; *bei Unterdrückung anderer Pflanzen* ►überwachsen; ►sich stark ausbreitende Art).

5917 spreading [n] **of soil to true contours** *constr.* (SPON 1986, 105; laying of soil material to finish level; ►make a finished grade flush with [US]/marry finish grade with [a structure] [UK], ►graded to proper line and levels); *syn.* grading [n] of finished level to all falls and gradients [also UK] (SPON 1986, 156), grading [n] of finished level to all gradients [also US]; *s* **relleno [m] de tierra hasta nivel de rasante** (►equiparar alturas, ►nivelado hasta rasante de acabado); *f* **profilage [m] aux pentes indiquées** (►conforme/conformément aux plans de profil, ►se raccorder au niveau fini) ; *g* **profilgerechter Einbau [m, o. Pl.]** (von Bodenmassen; ►höhengerecht anschließen, ►profilgerecht).

5918 sprigging [n] (1) *constr.* (Means of vegetation establishment appropriate to certain species involving spreading of shredded shoots, e.g. mossy stonecrop *[Sedum acre]*, on a prepared bed; ►establishment of grass/lawn areas, ►hydroseeding, ►plugging, ►stolonizing); *s* **siembra [f] con retoños triturados** (Medio de establecer vegetación en azoteas ajardinadas adecuado para ciertas especies como p. ej. *Sedum*; ►establecimiento de césped con estolones, ►establecimiento de superficies de césped, ►plantación de césped con tepes o con macollas, ►riego de semillas por emulsión); *f* **semis [m] par jet de jeunes pousses** (Technique de végétalisation par épandage de fragments de jeunes pousses, p. ex. *Sedum*, souvent employée pour la végétalisation extensive de toitures-terrasses ; ►engazonnage par bouturage, ►engazonnement d'une pelouse, ►engazonnement par jet de stolons de graminées, ►procédé d'enherbement par projection hydraulique) ; *syn.* semis [m] de fragments ; *g* **Sprossenaussaat [f]** (Ausbringen von zerkleinerten Pflanzensprossen zur Begrünung, z. B. von Mauerpfeffer *[Sedum acre]* auf Dachgärten; die **S.** kann auch im ►Anspritzverfahren erfolgen; Sprossen aussäen [vb]; ►Begrünung mit Grasstolonen, ►Rasenflächenherstellung, ►Herstellung von Rasenflächen durch Auspflanzen von Rasen- und Grasteilen); *syn.* Ausstreuen [n, o. Pl.] von Sprossschnittlingen/Triebschnittlingen, Sprossensaat [f].

5919 sprigging [n] (2) *constr.* (Planting by machine of warm season grass stolon or rhizome pieces in shallow trenches or small holes or by tilling up the ground and spreading sprigs across the surface followed by rolling to mash the sprigs into the soil. Method often employed in semi-arid and arid regions; this method is common for establishing Bermuda Grass *[Cýnodon dáctylon]*, Zoysia-Grass *[Zoysia]* and Bahiagras *[Paspalum notatum]*; shredded shoots, stolons and rhizomes can also be broadcast in a slurry as in ►hydroseeding); *s* **siembra [f] con estolones o rizomas triturados** (Sistema de plantación en zonas áridas o semiáridas en las que las semillas germinan mal, por lo que se utiliza el método de sembrar trozos de estolones o rizomas en zanjas poco profundas o en pequeños hoyos, o de arar el suelo y esparcir ramitas y pasar rodillo para mezclarlas con la tierra; ►riego de semillas por emulsión); *f* **engazonnement [m] par repiquage de stolons de graminées** (Réalisation de gazons par enfouissement en bandes ou par repiquage de tiges racinées ou de stolons de Chiendent pied de poule *[Cýnodon dáctylon]* et de gazon de Mascareignes *[Zoysia]* dans les régions arides et semi-arides ou le semis a peu de chance d'être efficace ; ►procédé d'enherbement par projection hydraulique) ; *syn.* enherbement [m] par repiquage de stolons de graminées ; *g* **Flächenbegrünung [f] mit zerteilten Grasstolonen oder Rhizomen** (Pflanzmethode in ariden und semiariden Gebieten, in denen Saaten sehr schlecht keimen, bei der geschnittene Grasstolone, z. B. vom Bermudagras *[Cýnodon dáctylon]*, Zoysiagras *[Zoysia]* und Bahiagras *[Paspalum notatum]* in die Bodenoberfläche, in flache Furchen oder kleine Löcher maschinell eingearbeitet werden; zerhackte Triebsprossen, Grasstolone und Rhizome können auch durch ►Anspritzverfahren aufgebracht werden); *syn.* Begrünung [f] von Flächen mit zerteilten Grasstolonen oder Rhizomen, Rasenflächenherstellung [f] mit zerteilten Grasstolonen oder Rhizomen, Herstellung [f] von Rasenflächen mit zerteilten Grasstolonen oder Rhizomen.

spring [n] *hydr.* ►dimple spring; *geo.* ►rising spring; *geo. hydr.* ►seepage spring; *hydr.* ►tapping of a spring; *geo. hydr.* ►vauclusian spring.

spring [n]**, filtration** *hydr. geo.* ►seepage spring.

spring [n]**, pool** *geo. hydr.* ►vauclusian spring.

spring bloom [n] *gard. hort. plant.* ►spring flower.

spring bloomer [n] *gard. hort. plant.* ►spring-blooming plant.

5920 spring-blooming plant [n] *gard. hort. plant.* (Herbaceous or woody species which flowers in March, April or May; ►spring flower); *syn.* spring bloomer [n] [also US], spring flowerer [n] [also UK], spring-flowering plant [n]; *s* **especie [f] de floración primaveral** (►flor de primavera); *syn.* especie [f] de floración prever20al; *f* **plante [f] à floraison printanière** (Terme horticole pour une plante herbacée ou ligneuse qui fleurit entre mars et mai ; ►fleur printanière) ; *g* **Frühjahrsblüher [m]** (Gärtnerische Bezeichnung für eine krautige oder Gehölzpflanze, die je nach Pflanzenart und geografischer Lage im März, April oder Mai blüht; ►Frühjahrsblume); *syn.* Frühlingsblüher [m].

spring bud break [n] [US] *bot.* ►leafing out.

5921 spring circulation period [n] *limn.* (Mixing period of stratified water masses in a body of water during the early spring: [1] melting of ice cover, [2] warming of surface waters, [3] density changes in surface waters producing convection currents from top to bottom, [4] circulation of the total water volume by wind action, and [5] vertical temperature equality; RCG 1982; ►summer stagnation phase); *syn.* spring overturn [n] [also US]; *s* **circulación [f] vertical prevernal** (Mezcla vertical del agua, producida por los vientos fuertes y posibilitada por la variación del perfil de densidades, que depende de los cambios de temperatura en relación con el balance térmico anual. A la **c. v. p.** le sigue la ►estratificación estival; cf. MARG 1977); *f* **brassage [m] printanier** (Processus de mélange de l'ensemble de la masse d'eau d'un lac lors du réchauffement des eaux superficielles au printemps ; l'échange surface/fond amène une égalisation des

températures ; au **b. p.** succèdent la ►stagnation estivale, le ►brassage automnal et la ►stagnation hivernale) ; **g Frühjahrszirkulation [f]** (Vorgang und Zustand der Durchmischung der gesamten Wassermasse eines Sees [nach dem Abschmelzen der Eisdecke], bedingt durch kräftige Winde im Frühjahr. Der **F.** folgt die ►Sommerstagnation).

5922 spring discharge [n] *hydr.* (Flow rate of a spring in units of time; in U.K., measured in litres [UK]/liters [US] per second [l/s]; in U.S., measured in gallons per minute [g.p.m.]); *s* **descarga [f] de una fuente** (Volumen de agua por unidad de tiempo); *f* **débit [m] d'une source** (Volume d'eau d'une source dans l'unité de temps à travers une section déterminée) ; **g Quellschüttung [f]** (Austretendes Quellwasser je Zeiteinheit, gemessen in l/s — Liter pro Sekunde); *syn.* Schüttung [f] einer Quelle.

5923 spring-fed pool [n] *geo. hydr.* (Small, open stretch of water, primarily fed from an underground source; ►vauclusian spring); *syn.* resurgence pool [n]; *s* **laguna [f] de manantial** (Pequeña superficie de agua en hondonada generalmente alimentada por un manantial situado en el fondo de ella; ►fuente vauclusiana); *f* **étang [m] de résurgence** (Petit plan d'eau localisé dans une dépression en forme d'entonnoir ou de cuvette principalement alimenté par la nappe phréatique et dont les eaux s'écoulent par un des bord ; ►résurgence, ►mare de résurgence, ►source vauclusienne) ; **g Quellteich [m]** (Quelle als offene, kleine Wasserfläche in einer trichter- oder schalenförmigen Vertiefung, die mit überwiegend unterirdischem Zufluss gespeist wird und über den Teichrand abfließt; ►Tümpelquelle); *syn.* Quelltopf [m].

5924 spring flower [n] *gard. hort. plant.* (Horticultural designation of an herbaceous plant which blooms in spring; ►spring-blooming plant); *syn.* spring bloom [n]; *s* **flor [f] de primavera** (►especie de floración primaveral); *f* **fleur [f] printanière** (Terme horticole caractérisant une plante herbacée qui fleurit entre mars et mai ; ►plante à floraison printanière) ; *syn.* fleur [f] de printemps ; **g Frühjahrsblume [f]** (Gärtnerische Bezeichnung für eine krautige Pflanze, die je nach Pflanzenart und geografischer Lage im März, April oder Mai blüht; ►Frühjahrsblüher); *syn.* Frühlingsblume [f].

spring flowerer [n] [UK] *gard. hort. plant.* ►spring-blooming plant.

5925 spring-flowering period [n] *hort. plant.* (►flowering period); *s* **floración [f] preverval/vernal** (►floración); *f* **floraison [f] printanière** (►floraison de mars, avril à fin mai) ; *g 1* **Blütezeit [f] im Frühjahr** (►Blütezeit im März, April und/oder Mai; nicht zu verwechseln mit **Frühjahrsflor [m]:** gärtnerischer Ausdruck für die Gesamtheit der Pflanzen, die für die Blühsaison im Frühjahr kultiviert und — meistens im Herbst — ausgepflanzt und bis zum Wechsel zum Sommerflor gepflegt werden); *g 2* **Frühjahrsblüte [f]** ([Visuell] wahrnehmbare Fülle einer Pflanzenart oder mehrerer blühender Arten im Garten oder in der Landschaft im Frühling); *syn.* Frühlingsblüte [f].

spring-flowering plant [n] *gard. hort. plant.* ►spring-blooming plant.

5926 spring foliage [n] *bot. hort.* (Young leaves which appear in spring); ►emerging leaves); *syn.* early foliage [n] [also US]; *s* **foliación [f] preverval** (1. Periodo/período de aparición de las hojas de las leñosas en el hemisferio norte entre febrero y mayo, dependiendo de la latitud. 2. *Por extensión* el follaje mismo; ►foliación); *syn.* follaje [m] preverval; *f* **frondaison [f] printanière** (1. Période d'apparition du feuillage des végétaux ligneux entre mars et mai. 2. *Par extension* le feuillage lui-même ; ►feuillaison) ; **g Frühjahrsbelaubung [f]** (Frische

Blätter von Gehölzen, die je nach Pflanzenart und geografischer Lage im März, April oder Mai gebildet werden; ►Laubaustrieb).

5927 spring horizon [n] *geo. hydr.* (Groundwater-bearing layer on a slope, brought to the surface in a linear seepage spring or a series of springs); *syn.* spring line [n] (DNE 1978); *s* **horizonte [m] de rezumo** (Capa freática en una pendiente caracterizada por percolación lineal o por una serie de manantiales); *f* **zone [f] d'émergence (de nappe)** (Apparition linéaire permanente des eaux de la nappe phréatique quand elle est recoupée par la surface topographique sous la forme d'une ligne de sources ou d'une ligne de suintement soit sur les versants ou au bas de ceux-ci, soit sur des replats lorsque la nappe affleure) ; *syn.* aire [f] d'émergence (de nappe), source [f] d'affleurement ; **g Quellhorizont [m]** (Durch linienhaften Grundwasseraustritt oder durch eine Quellenreihe gekennzeichneter Ausstrich einer Grundwasser führenden Schicht, eines sog. Grundwasserleiters, an einem Hang).

springing [n] *constr.* ►lawn aeration, #5.

spring line [n] *geo. hydr.* ►spring horizon.

spring overturn [n] [US] *limn.* ►spring circulation period.

5928 spring protection area [n] *leg. nat.res.* (Area legally designated for the protection of a recognized medicinal spring; ►soil and water conservation district [US]/water conservation area [UK]); *s* **perímetro [m] de protección de aguas minerales y termales** (Declaración de ►perímetro de protección de acuífero por medio de procedimiento formal según la legislación específica; cf. art. 99 bis RD Legislativo 1/2001); *f* **périmètre [m] de protection des sources d'eau minérale** (Zone de protection des sources et gisements d'eaux minérales naturelles à l'intérieur de laquelle sont réglementées toutes activités, dépôts ou installations pouvant nuire directement ou indirectement sur la qualité des eaux ; Code de la santé publique, art. L. 736 ; ►périmètre de protection d'un captage d'eau potable) ; **g Quellenschutzgebiet [n]** (Durch förmliches Verfahren festgesetztes ►Wasserschutzgebiet zum Schutz staatlich anerkannter Heilquellen; cf. § 49 WG-BW).

5929 spring swamp [n] *geo. phyt.* (Marshy area caused by the seepage of a spring and usually colonized by various plant communities in concentric belts around the wettest spots); *s* **ciénaga [f] de manantial** (Zona pantanosa originada por emergencia de un manantial, generalmente con diferentes comunidades vegetales ordenadas en orlas concéntricas alrededor de la zona más húmeda); *syn.* pantano [m] de manantial; *f* **marais [m] de suintement** (Marais localisé autour de la zone d'infiltration des eaux de suintement, présentant une zonation de la végétation en fonction de l'humidité du sol) ; *syn.* marais [m] de pente ; **g Quellsumpf [m]** (Durch Sickerquelle entstandener Sumpf mit gewöhnlich verschiedenen Pflanzengesellschaften, die sich gürtel- oder streifenartig um die nassesten Stellen ansiedeln; ELL 1978).

5930 spring vegetation [n] *phyt.* (Plant community around a spring fed by water of even temperature; in Western and Central Europe belonging to the spring vegetation class *[Montio-Cardaminetea]*; cf. VIR 1982); *s* **vegetación [f] fontinal** (Comunidades de altitudes medias y elevadas que aparecen junto a aguas muy movidas, ricas o pobres en cal, pertenecientes a la clase *Montio-Cardaminetea*); *f* **végétation [f] fontinale** (VCA 1985, 248 ; association végétale hygrophile vivant dans la zone immédiatement limitrophe des sources et ruisselets appartenant à la classe des *Montio-Cardaminetea*) ; *syn.* groupement [m] fontinal (VCA 1985, 322), groupement [m] de source et de suintement (VCA 1985, 255) ; **g Quellflur [f]** (Feuchtigkeitsliebende Pflanzengesellschaften im Quellbereich eines Fließ-

S

gewässers, die in West- und Mitteleuropa zur Klasse der Kalt-wasserquellfluren *[Montio-Cardaminetea]* gehören).

5931 spring water [n] *wat'man.* (Water of a rising spring as a source of a ▶headwater stream; ▶protection of spring water); *syn.* headwater [n]; *s* **agua** [f] **de manantial** (Agua que sirve de fuente a la ▶cabecera [de río]; ▶laguna de manantial, ▶protección de aguas de manatial); *syn.* agua [f] de fuente; *f* **eau** [f] **de source** (Eau en provenance d'une source qui coule dans le ▶cours supérieur près de la source ou dans un ▶étang de resurgence ; ▶protection des eaux de source) ; *syn.* eau [f] vive ; *g* **Quellwasser** [n] (Wasser aus einer Quelle, das in einen ▶Quelllauf oder ▶Quellteich fließt; ▶Quellwasserschutz).

5932 spring water bog [n] **[US]/spring-water bog** [n] **[UK]** *geo. phyt.* (Small bog created by a rising spring in a depression or a cavity on a slope); *s* **turbera** [f] **de manantial** (Pequeña turbera originada por emergencia de un manantial en una hondonada o en una cavidad de una pendiente); *f* **tourbière** [f] **de source** (Tourbière de faible étendue localisée sur une pente faible à moyenne ou dans une petite dépression de terrain au niveau de sources ou d'émergences diverses qui dépendent d'un écoulement permanent [ruissellement ou suintement]) ; *syn.* tourbière [f] de pente, tourbière [f] soligène ; *g* **Quellmoor** [n] (Kleines Moor, das durch einen Quellwasseraustritt in einer Geländevertiefung oder Hangnische entstanden ist).

spring wood [n] *arb. bot.* ▶early wood.

5933 sprinkler [n] *agr. constr.* (Irrigation device used for distributing freshwater or wastewater; ▶pop-up sprinkler); *s* **aspersor** [m] (Dispositivo de irrigación; ▶aspersor escamoteable); *f* **arroseur** [m] (Appareil d'arrosage pour l'aspersion d'eau ou des eaux usées ; ▶arroseur escamotable) ; *syn.* asperseur [m] ; *g* **Regner** [m] (Bewässerungsgerät, das zum Verregnen von Wasser oder Abwasser dient; ▶Versenkregner).

5934 sprinkler irrigation [n] *agr. hort.* (Method of mechanical watering from lateral lines laid either above or below ground; ▶irrigation); *s* **irrigación** [f] **por aspersión** (Método de regadío mediante aparatos que distribuyen el agua imitando a la lluvia. Generalmente se utiliza en climas con grado de aridez no demasiado elevado. Tiene importantes ventajas: consumo de agua reducido, mejor control de la cantidad y frecuencia de los riegos y distribución más uniforme de éstos sobre el terreno. Sin embargo, su instalación supone costes iniciales elevados y se necesita combustible para su aplicación; cf. DINA 1987; ▶irrigación); *f* **arrosage** [m] **par aspersion** (Forme d'arrosage consistant à diffuser sous forme de pluie fine et régulière de l'eau claire ou usée sur pelouses ou cultures à l'aide d'appareils à basse pression [asperseur ou rampe d'arrosage] ou à haute pression [canon à eau, etc.] ; ▶arrosage, ▶irrigation) ; *g* **Beregnung** [f] (Dosierte, regenähnliche ▶Bewässerung von Kulturpflanzen oder Rasenanlagen durch gleichmäßige, Klarwasser- oder Abwasserzufuhr mit Hilfe von Einzel- oder Reihenregnern).

sprout [n] *bot. hort.* ▶shoot.

sprout forest [n] **[US]** *for.* ▶coppice forest.

sprouting capacity [n] *for. hort.* ▶capacity of making new shoots.

sprouting vigo(u)r [n] *for. hort.* ▶capacity of making new shoots.

5935 spun concrete [n] *constr.* (Concrete compacted by centrifugal action, e.g. in prefabricating pipes or columns); *s* **hormigón** [m] **(vibro-)centrifugado** (Hormigón compactado por acción centrifugante); *syn.* concreto [m] (vibro-)centrifugado [AL]; *f* **béton** [m] **centrifugé** (Béton compacté par la force centrifuge) ; *g* **Schleuderbeton** [m] (Mit einem langen Löffel in eine rotierende, einwandige Schalung gebrachter Beton, der

durch die Zentrifugalkraft an die Schalung gedrückt und dabei unter Aussonderung des Überschusswassers stark verdichtet wird).

5936 spur [n] *bot. hort.* (Stubby lateral twig with short internodes, e.g. of fruit trees, *Ginkgo*, pine trees; ▶thorn); *s* **vástago** [m] **corto** (▶espina); *syn.* brote [m] corto; *f 1* **pousse** [f] **courte** (Terme générique désignant une pousse dont la croissance en longueur des entre-nœuds est limitée ; ▶épine) ; *f 2* **lambourde** [f] (Terme spécifique pour un rameau court correspondant à plusieurs années de végétation et entouré d'une rosette de feuilles chez certains arbres fruitiers) ; *g* **Kurztrieb** [m] (Seitentrieb mit sehr kurzen Internodien, z. B. bei Obstbäumen, *Ginkgo*, Kiefern; ▶Dorn); *syn.* Stauchspross [m] (LB 1978, 144).

5937 square [n] *urb.* (Urban open space either paved, unpaved or predominantly planted, which may be enhanced with ▶slab paving patterns, water features or sculptures; ▶urban square); *s* **plaza** [f] **(1)** (Esplanada no edificada generalmente de forma cuadrada, rectangular o circular situada entre edificios y —según el contexto arquitectónico y la zona urbana en que se encuentre— decorada con fuentes, macizos de flores, árboles, ▶aparejos de losas, esculturas y con mobiliario urbano; en los centros históricos de ciudades mediterráneas las **p.** están predominantemente caracterizadas por la arquitectura circundante más que por los elementos decorativos contenidos en ellas); *syn. para plazas con predominio de la vegetación* parque [m] [AL]; *f 1* **place** [f] (Large espace découvert public urbain généralement entouré de constructions organisé autour de fontaines de statues et structuré par différents revêtements de sols, ▶appareillage de dalles ou d'éléments paysagers ; ▶place publique) ; *f 2* **square** [m] (Espace libre urbain de dimension réduite, en moyenne de quelques centaines de m² à 2-3 ha, créé à la fin du XIXème siècle, souvent de forme carrée, utilisé dans les quartiers urbains comme lieu de promenade et de détente de proximité) ; *g* **Platz** [m] (Unbebaute, ebene Fläche in einem unbebauten Gebiet, z. T. durch Verlegemuster oder besondere ▶Plattenverbände, Brunnen oder Skulpturen gegliedert, oft grünplanerisch gestaltet; ▶Stadtplatz).

square [n], **city** *urb.* ▶urban square.

square [n], **public** *urb.* ▶urban square.

square [n], **town** *urb.* ▶urban square.

5938 squared rubble masonry [n] *constr.* (Wall of roughly-trimmed stones, not always laid in courses, usually bedded in mortar; ▶ashlar masonry, ▶broken range work, ▶coursed dressed ashlar masonry, ▶cyclopean wall, ▶dry masonry, ▶natural stone masonry, ▶polygonal masonry); *syn.* coursed rubble [n], ranged rubble [n], rubble masonry [n]; *s* **mampostería** [f] **ordinaria (de piedra partida)** (Obra hecha con piedras en bruto o con labra irregular o de mampuestos de piedra partida, generalmente sobre tortada de mortero de 2 ó 3 cm de espesor, y con aparejo irregular de hiladas de diferentes alturas; ▶mampostería ciclópea, ▶mampostería de piedras naturales, ▶mampostería en hiladas, ▶mampostería en seco, ▶mampostería poligonal, ▶sillería de fábrica, ▶sillería ordinaria); *f* **maçonnerie** [f] **en moellons pleins, grossièrement équarris** (Mur en pierres brutes et rugueuses, dont l'appareillage ne présente pas toujours des lits horizontaux avec des joints en général exécutés au mortier ; ▶maçonnerie à appareillage à l'anglaise, ▶maçonnerie à appareillage cyclopéen, ▶maçonnerie à appareillage polygonal ▶maçonnerie à sec, ▶maçonnerie en moellons à opus quadratum, ▶maçonnerie en moellons assisés, ▶maçonnerie en pierres naturelles) ; *syn.* mur [m] en pierre de carrière, mur [m] en pierres de taille ; *g* **Bruchsteinmauerwerk** [n] (Mauer aus wenig bearbeiteten bruchrauen Steinen, nicht immer in allen Lagerfugenbereichen horizontal, lagerhaft im Verband, meist mit Mörtel; das Mauerwerk ist spätestens in

1,50 m Höhe in seiner ganzen Dicke waagerecht auszugleichen, d. h. mit durchlaufender Lagerfuge zu horizontieren; ▶Natursteinmauerwerk, ▶Polygonalmauerwerk, ▶Quadermauerwerk, ▶Schichtenmauerwerk, ▶Trockenmauerwerk, ▶Wechselmauerwerk, ▶Zyklopenmauerwerk).

square grid [n] *constr. hort.* ▶square planting grid.

square pattern [n] [US] *constr. hort.* ▶square planting grid.

5939 square planting grid [n] *constr. hort.* (Planting layout used in mass plantings; ▶planting grid, ▶triangular planting grid); *syn.* square grid [n], square pattern [n] [US]; *s* **esquema [f] de plantación rectangular** (Ordenación de las plantas en cuadros o rectángulos en un ▶esquema de plantación 1; ▶esquema de plantación triangular); *f* **disposition [f] rectangulaire** (Arrangement selon un ordre rectangulaire des sujets lors d'une plantation sur plusieurs rangs ; ▶disposition en quinconce, ▶disposition des végétaux) ; *g* **Vierecksverband [m]** (Rechteckige oder quadratische Anordnung von Pflanzen in einem Pflanzschema; *UB* Rechtecksverband; ▶Dreiecksverband, ▶Pflanzenverband).

squatter camping [n] [ZA] *urb.* ▶overnight land squatting.

squatter settlement [n] *urb.* ▶shanty town, ▶slum.

S.S.S.I. *conserv. leg.* ▶natural landmark [US].

stability [n] *ecol.* ▶ecological stability; *constr. eng.* ▶structural stability; *pedol.* ▶structural stability of soil.

stability [n] [US]**, ped** *pedol.* ▶structural stability of soil.

stability [n] **of soil, aggregate** *pedol.* ▶structural stability of soil.

stability [n] **of trees** *arb.* ▶structural stability of trees.

5940 stabilization [n] *constr.* (Protective measures on the ground surface to keep it in place and prevent soil erosion; ▶bank revetment); *s* **estabilización [f]** (Medidas de protección de taludes, dunas, orillas, etc.; ▶revestimiento de orillas); *syn.* afianzamiento [m], fijación [f]; *f* **stabilisation [f]** (Mesures de protection d'un talus, de berges, de dunes, etc. contre l'érosion ; ▶ouvrage de revêtement de berge) ; *syn.* confortement [m], fixation [f] (des dunes) ; *g* **Befestigung [f] (1)** (Schutzmaßnahmen an Böschungen, Hängen, Dünen, Ufern etc. gegen Erosionseinwirkungen; ▶Uferdeckwerk); *syn.* Festlegung [f], Sicherung [f].

stabilization [n]**, dune** *constr.* ▶stabilization of sand dunes.

5941 stabilization [n] **of sand dunes** *constr.* (Measures to hold sand dunes in place; ▶brushwood thatching, ▶dune planting); *syn.* dune stabilization [n]; *s* **fijación [f] de dunas** (Medidas para proteger la superficie de las dunas de la erosión; ▶capa protectora de desbrozo y leña menuda, ▶plantación de dunas, ▶revestimiento [del suelo] con ramas secas); *syn.* estabilización [f] de dunas; *f* **fixation [f] des dunes** (Mesures de protection des dunes ou de reconquête de terrain sur la mer grâce à la plantation de végétaux aréneux conservateurs des dunes tels que l'oyat, le pin maritime, l'aulne, etc. ; ▶couverture de branchage, ▶fixation du sol au moyen de ramilles, ▶plantation dunaire) ; *syn.* stabilisation [f] des dunes ; *g* **Festlegung [f] von Dünen** (Maßnahmen, die die Dünenoberfläche gegen Erosion widerstandsfähig machen; z. B. durch ▶Abdeckung mit einem Rauwehr; ▶Bodenabdeckung mit Reisig, ▶Dünenbepflanzung); *syn.* Befestigung [f] von Dünen, Dünenbefestigung [f], Dünenfestlegung [f].

5942 stabilized soil [n] *constr.* (▶soil consolidation for road construction); *syn.* consolidated soil [n]; *s* **suelo [m] estabilizado** (Resultado de la ▶estabilización del suelo o por compactación); *syn.* suelo [m] consolidado; *f* **sol [m] stabilisé** (▶stabilisation du sol) ; *g* **verfestigter Boden [m]** (durch ▶Bodenverfestigung).

5943 stabilized subgrade [n] [US]/**stabilized subgrade** [n] [UK] *constr.* (Existing ground made stable by compaction or by addition of granular material, chalk, or cement or by blinding with weak concrete to fill voids; ▶cemented soil); *s* **subsuelo [m] estabilizado** (Resultado de estabilización de subsuelo mecánicamente por compactación o por adición de mortero; ▶suelo cimentado); *syn.* subsuelo [m] mejorado; *f* **fond [m] de forme stabilisé** (par addition de chaux grasse ou de ciment ou par compactage avec des agrégats anguleux ; ▶solciment) ; *g* **verbesserter Untergrund [m] (2)** (Mechanisch durch Verdichtung oder durch Vermörtelung mit Bindemitteln verfestigter Untergrund; ▶vermörtelter Boden).

stabilized surface [n] [US] *constr.* ▶water-bound surface.

5944 stable [adj] *constr.* (Descriptive term applied to ▶consolidated subgrade); *syn.* buildable [adj]; *s* **firme [adj]** (Término descriptivo referente a subsuelo, suelo, fundación, etc. con estabilidad estructural; ▶subsuelo estable); *syn.* estable [adj], resistente [adj]; *f* **portant, ante [ppr/adj]** (Caractéristique de la couche de fondation, d'un sol, etc. ; ▶sol stable) ; *g* **tragfähig [adj]** (Baugrund, Boden, Unterbau etc. betreffend, der standsicher ist; ▶fester Baugrund).

stable manure [n] *agr. hort.* ▶straw dung.

stable soil level [n] [US] *constr. eng.* ▶consolidated subgrade.

5945 stack bond [n] *constr.* (Pattern in which brickwork or pavement units of a single size are set with continuous vertical and horizontal joints; ▶bond, ▶stack bond joint); *s* **aparejo [m] de juntas cruzadas** (Patrón de colocación de losas o adoquines caracterizado por ▶juntas cruzadas; ▶aparejo); *f* **appareillage [m] à joints croisés** (Mode d'assemblage de dalles, de pavés selon lequel les côtés perpendiculaires au point de rencontre de deux dalles de section carrée ou rectangulaire sont dans le prolongement les uns des autres ; ▶appareillage, ▶joint croisé) ; *g* **Kreuzfugenverband [m]** (...verbände [pl]) durch ▶Kreuzfugen gekennzeichnetes Fugenbild eines Plattenbelages und Pflasters oder einer Plattenverblendung; ▶Verband 2).

5946 stack bond joint [n] *arch. constr.* (Joint pattern in masonry and paving, comprising a ▶coursing joint and two butt joints located vertically above each other; ▶heading joint, ▶stack bond); *s* **junta [f] cruzada** (Tipo de junta en mampostería y pavimentación que se compone de una ▶junta de asiento y dos ▶juntas a tope colocadas verticalmente una sobre otra; ▶aparejo de juntas cruzadas); *f* **joint [m] croisé** (Joint formé par la rencontre d'un joint filant avec deux joints montants dans le prolongement l'un de l'autre ; ▶appareillage à joints croisés, ▶joint délit, ▶joint montant) ; *syn.* joint [m] coupé, joint [m] droit ; *g* **Kreuzfuge [f]** (Fugenbild aus ▶Lagerfuge und zwei senkrecht übereinander stehenden ▶Stoßfugen im Mauerwerk, analog im Plattenverband; ▶Kreuzfugenverband).

stacked [pp] **in soil heaps** [npl] [UK] *constr.* ▶stacked in soil piles [US].

5947 stacked [pp/adj] **in soil piles** [US] *constr. syn.* stacked [pp] in soil heaps [UK]; *s* **tierra [f] depositada en caballón** *syn.* tierra [f] acopiada en caballón; *f* **terre [f] en dépôt** ; *g* **in Mieten gelagerter Boden [loc]**.

stacking [n] **of topsoil** [UK] *constr.* ▶storage of topsoil.

5948 stadia rod [n] *surv.* (Graduated surveyor's rod used for optical measurement of distances and levels); *s* **regla [f] dividida** (Listón de 2 a 4 m de largo con escala de medición utilizado para mediciones topográficas de distancia óptica y de alturas); *f* **mire [f]** (Règle de 2 à 4 m de hauteur, graduée en cm, utilisée en altimétrie et en nivellement) ; *g* **Messlatte [f]** (Zur Gelände-

S

vermessung skalierter Messstab für die optische Distanz und Höhenmessung).

staff [n]**, administrative** *adm. contr.* *▶ office worker.

staff member [n] [UK]**, technical** *adm. contr.* ▶ technical employee [US].

5949 staff person [n] *adm. constr. contr.* (Generic term); *syn.* employee [n] [also US], member [n] of operatives [also UK]; *s* **fuerza** [f] **de trabajo** (Término genérico); *syn.* empleado/a [m/f]; *f* **main** [f] **d'œuvre** (Terme générique) ; *g* **Arbeitskraft** [f] **(2)** (*Abk.* AK; Arbeit leistender Mensch).

5950 stage [n] **of a watercourse** *hydr.* (Elevation of the water surface of a river or stream relative to a datum; WMO 1974; ▶ flow rate, ▶ water level); *s* **nivel** [m] **de una corriente** (Altura de la superficie de las aguas en una corriente, en un punto determinado; WMO 1974; ▶ caudal, ▶ nivel de agua 2); *f* **hauteur** [f] **d'eau** (Hauteur d'une surface d'eau au-dessus d'un niveau de référence préalablement choisi ; WMO 1974 ; ▶ débit, ▶ niveau de l'eau) ; *syn.* hauteur [f] limnimétrique ; *g* **Wasserführung** [f] (Wassermenge, die ein Wasserlauf zu einer bestimmten Zeit führt — als Erscheinungsbild, weniger als gemessene ▶ Abflussmenge; ▶ Wasserspiegel).

5951 staggered joint [n] *constr.* (Joint in a ▶ jointing pattern where each stone, brick or paver is offset vertically from the one below, in comparison with ▶ stack bond joint; ▶ heading joint); *s* **junta** [f] **alternada** (▶ Junta a tope que, al contrario que la ▶ junta cruzada, está desplazada verticalmente de la inferior en el ▶ patrón de juntas); *f* **joint** [m] **décalé** (par opposition au ▶ joint croisé, dallage à l'antique formé d'une succession de lignes parallèles et d'une ligne de joint droite discontinue d'au moins la largeur d'un pave; ▶ calepinage, ▶ joint montant) ; *g* **versetzte Fuge** [f] (Im Vergleich zur ▶ Kreuzfuge diejenige ▶ Stoßfuge, die im ▶ Fugenbild zur unteren Steinreihe [mindestens um Steinhöhe] seitlich verschoben ist).

staggered joints [npl] *constr.* ▶ lay with staggered joints.

5952 stag-headed [pp/adj] *arb. for.* (**1.** Descriptive term applied to the dead top of a crown as a result of injury, disease, old age, deficiency, excess of moisture or nutrient; cf. BS 3975: part 5; ▶ top-kill [US]/top drying [UK]; **2. spike-topped** [pp/adj] **[US]** (Descriptive term applied to stag-headed conifers; ▶ blacktop [US&CDN] 2); *s* **puntiseco** [adj] (Término descriptivo referente a árboles cuya punta está muerta como resultado de enfermedades, deficiencias o excesos de nutrientes o humedad; ▶ puntiseco); *syn.* seco [adj] en la punta; *f* **sec, sèche en cime** [loc] (Caractérise les arbres dont la cime est en train de mourir ou est desséchée ; ▶ descente de cime ; pour les conifères on utilise le terme « noir en cime ») ; *syn.* à cime desséchée [loc] ; *g* **wipfeldürr** [adj] (Einen Baum betreffend, dessen oberstes Astwerk sichtlich abstirbt oder vertrocknet ist; ▶ Wipfeldürre); *syn.* zopftrocken [adj].

5953 stagnation phase [n] *limn.* (Generic term for ▶ summer stagnation phase and ▶ winter stagnation phase); *s* **estagnación** [f] (Término genérico para ▶ estagnación estival, ▶ estagnación invernal); *f* **stagnation** [f] (Dans le cycle thermique d'un plan d'eau désigne la période pendant laquelle s'effectue une stratification sans circulation des masses d'eau ; *termes spécifiques* ▶ stagnation estivale et ▶ stagnation hivernale) ; *g* **Stagnation** [f] (OB zu ▶ Sommerstagnation und ▶ Winterstagnation).

5954 stagnicolous [adj] *biol.* (Descriptive term for organisms living in stagnant water; *opp.* ▶ torrenticolous); *s* **estagnícola** [adj] (Término descriptivo para organismos que habitan en aguas dulces estancadas; *opp.* ▶ torrentícola); *f* **vivant, ante en eaux stagnantes** [loc] (Concerne les organismes vivant dans les eaux douces stagnantes ; *opp.* torrenticole) ; *syn.* vivant, ante en eaux calmes [loc], stagnicole [adj] ; *g* **stagnikol** [adj] (Orga-

nismen betreffend, die nur in stehenden Süßgewässern leben; *opp.* ▶ torrentikol); *syn.* stehendes Gewässer bewohnend [loc], in Stillgewässern lebend [loc].

stair [n] *arch. constr.* *▶ steps.

stair [n]**, circular** *arch.* ▶ spiral stair(case).

stair [n]**, cockle** *arch.* ▶ spiral stair(case).

stair [n]**, corkscrew** *arch.* ▶ spiral stair(case).

stair [n]**, helical** *arch.* ▶ spiral stair(case).

stair [n]**, solid newel** *arch.* ▶ spiral stair(case).

5955 stair axis [n] *constr.* (Centre line of a flight of steps used in calculating the riser/tread ratio; ▶ step formula); *s* **eje** [m] **de escalera** (Línea central de una escalera por medio de la cual se calcula la altura de los escalones; ▶ fórmula de cálculo de peldaño); *f* **ligne** [f] **de foulée** (Axe de l'emmarchement d'un escalier pris en compte pour définir la proportion des marches d'un escalier, d'une gradine ; ▶ formule de dimensionnement des marches) ; *g* **Lauflinie** [f] (Mittelachse einer Treppe, für die das Steigungsverhältnis der Stufen berechnet wird; ▶ Schrittmaßformel); *syn.* Treppenlauflinie [f].

staircase [n] *arch. constr.* ▶ stairway, #3.

5956 staircase step [n] *arch. constr.* (Indoor step; ▶ stair rise, ▶ step in a flight of steps; ▶ tread length, width of a flight of steps); *s* **peldaño** [m] **(1)** (Travesaño de una escalera interior; las medidas de un **p.** se definen por la ▶ huella de escalón y la ▶ contrahuella; ▶ anchura de escalera, ▶ escalón 2); *syn.* escalón [m] (1); *f* **marche** [f] **d'escalier (1)** (Pour escalier d'intérieur ; degré, pièce horizontale sur laquelle on pose le pied ; la marche se définit dimensionnellement par son ▶ giron et la ▶ hauteur de la contremarche ; formellement une marche peut être *droite* ou *carrée* [si elle est rectangulaire], *balancée* ou *dansante* si les deux extrémités ont des largeurs différentes, *biaise* si, sans être balancée, elle n'est pas perpendiculaire au limon ; WIK 2007 ; ▶ emmarchement, ▶ marche d'escalier 2) ; *g* **Treppenstufe** [f] **(1)** (Stufe im Innern eines Gebäudes; die Abmessungen einer **T.** sind durch ▶ Auftrittstiefe und ▶ Steigungshöhe definiert; ▶ Treppenstufe 2, ▶ Treppenbreite).

stair landing [n] *constr.* ▶ landing.

5957 stair rise [n] *constr.* (Vertical measurement between two stair treads; ▶ staircase step, ▶ step in a flight of steps); *syn.* riser height [n], stair riser [n], step riser [n]; *s* **contrahuella** [f] (Altura de un ▶ peldaño 1 de escalera; ▶ escalón 2); *f 1* **hauteur** [f] **de la contremarche** *syn.* hauteur [f] d'une marche ; *f 2* **dénivelé(e)** [m/f] **d'une marche** (Différence de hauteur entre deux marches ou hauteur de la marche ; elle prend en compte la hauteur de la contremarche et la pente du giron, calculée de pied de contremarche à pied de contremarche ; ▶ marche d'escalier 1, ▶ marche d'escalier 2) ; *g* **Steigungshöhe** [f] (Höhe einer ▶ Treppenstufe 1, ▶ Treppenstufe 2); *syn.* Steigung [f] (2).

stair riser [n] *constr.* ▶ stair rise.

stair run [n] *arch. constr.* ▶ tread length.

stairs [npl] *arch. constr.* ▶ stairway, #3.

5958 stairway [n] *arch. constr.* (HALAC 1988-II, 224; **1.** Single flight or multiple **flight of steps** with intermediate landings; ▶ open stairway [US]/flight of steps without cheek walls [UK], ▶ steps, ▶ total run of a stairway [US]/length of a flight of steps); **2. flight** [n] (Uninterrupted series of steps; a series of steps in a building is also called a **staircase** [n], and is often used interchangeably with **stairs** [npl] and **stairway** [n]); *syn.* flight [n] of stairs; *s* **tramo** [m] **de escalera** (▶ escalera, ▶ escalera abierta, ▶ longitud de tramo [de escalera]); *syn.* tramo [m] de escalones (DACO 1988); *f* **volée** [f] **d'escalier** (Ouvrage constitué d'un ensemble de marches successives ; ▶ escalier,

►escalier sans limon, ►longueur d'escalier) ; **g Treppenlauf [m]** (Zusammenhängende Folge aus mehreren Stufen [mindestens drei] und nach geltenden Vorschriften höchstens 18 Treppenstufen zur Verbindung unterschiedlich hoher Ebenen; ►Treppe, ►Treppenlauflänge, ►wangenlose Treppe).

stairway [n] [US], spiral *arch. constr.* ►spiral stair(case).

stairway [n], steepness of a *arch. constr.* ►slope of a flight of steps.

5959 stake [n] (1) *constr. hort.* (Thick, bottom-pointed, wooden stake for guying newly-planted trees; ►angled stake, ►tree stake); **s tutor [m] (1)** (Estaca gruesa afilada por uno de los extremos que se emplea para dar apoyo a árboles recién plantados; ►estaca oblicua); *syn.* estaca [f]; **f tuteur [m] (1)** (Support constitué d'un pieu écorcé et épointé sur lequel sont attachés les arbres en formation ; ►tuteur oblique, ►tuteur 2) ; **g Pfahl [m]** (Pfähle [pl]; unten zugespitzter, dicker Stab, i. d. R. aus geschältem, nicht imprägniertem Holz, zum Befestigen neu gepflanzter Gehölze wie Hochstämme, Stammbücke, Heister, Solitärsträucher sowie Stämmchen von Rosen und anderen Ziersträuchern; es werden Senkrechtpfahl, ►Schrägpfahl und Doppelpfähle unterschieden; für die Verankerung von Bäumen gibt es die Einzel-, 2- und 3-Pfahlsicherung resp. Verankerungsböcke aus drei oder vier Pfählen; *UB* ►Baumpfahl); *syn.* Stützpfahl [m].

5960 stake [n] (2) *constr.* (In bioengineering, a short, straight piece of wood sharpened at one end for driving into the ground in a row of stakes to prevent erosion. Such stakes have a diameter of > 3cm and length > 50cm; ►row of stakes); *syn.* picket [n]; **s estaca [f] muerta** (Palo delgado, aguzado en su borde superior, utilizado en vallas o empalizadas; ►hilera de estacas); **f pieu [m] en bois mort** (VRD 1994, 2/1.3.2.1.4, 23 ; branche droite, d'un diamètre minimum de 3 cm et d'une longueur d'environ 50 cm, utilisée pour les travaux de génie biologique ; ►rideau de pieux) ; **g toter Pflock [m]** (*Ingenieurbiologie* gerader Stangenabschnitt, der einen Durchmesser von > 3 cm und eine Länge von > 50 cm hat. Es können auch gespaltene Hölzer verwendet werden; cf. DIN 18 918; ►Pflockreihe).

5961 stake [vb] (3) *constr. hort.* (Staking is used to guy and anchor newly-planted trees or large shrubs; *specific terms* double staking, triple staking; stake placement [also US]); **s entutorado [m]** (Proceso de afianzar árboles o incluso arbustos recién plantados con tutores de madera); *syn.* fijación [f] de plantas con tutor, afianzamiento [m] (de plantas) con tutor; **f 1 tuteurer [vb]** (Maintenir [un arbre] par un tuteur) ; **f 2 tuteurage [m]** (Action d'attacher une plante en formation à des pieux de support en bois [tuteur]) ; **g pfählen [vb]** (Mit Holzpfählen nach dem Pflanzen befestigen, z. B. Bäume oder große Sträucher); *syn.* anpfählen [vb], mit Holzpfählen befestigen [vb].

stake [n] (4) *arb. constr.* ►angled stake; *constr. surv.* ►grade stake [US]/level stake [UK]; *constr.* ►live stake, ►row of stakes; *surv.* ►surveyor stake; *constr. hort.* ►tip of a stake, ►tree stake.

stake [n] [UK], level *constr. surv.* ►grade stake [US].

stake [n], slanting *arb. constr.* ►angled stake.

stake [n], surveying [US] *surv.* ►surveyor stake.

stake placement [n] [US] *constr. hort.* *►stake [vb] (3).

staking [n], double *constr. hort.* *►stake [vb] (3).

staking [n], triple *constr. hort.* *►stake [vb] (3).

5962 staking out [n] [US] *constr. surv.* (Process used in marking surveyed lines and proposed contours with short wooden stakes; ►staking-out with fields rods); *syn.* setting-out [n] [UK]; **s replanteo [m] de obra sobre el terreno** (Dibujo de la planta de una obra ya proyectada en el suelo con pequeñas estacas; ►jalonamiento); *syn.* trazado [m]; **f piquetage [m]** (Sur le chantier il

s'agit de toute opération de repérage des fouilles générales en plan et en altitude jusqu'à établissement des profils y compris toutes les opérations topographiques pour la réalisation des ouvrages dus à l'entrepreneur ; ►jalonnement) ; **g Absteckung [f]** (Markierung in einem Gelände von geplanten Höhen und eingemessenen Wegeachsen, Pflanzbeeten etc. mit Holzpflöcken o. Ä.; ►Absteckung mit Fluchtstäben).

5963 staking out [n] before construction begins *surv.* (Setting out of a defined area to determine the extent of the design and its corresponding levels on site in preparation for construction); **s replanteo [m] del terreno** (Nivelado y jalonamiento de una superficie antes del comienzo de una obra); **f levé [m] (de plan) d'état des lieux** (Exécution d'un relevé topographique d'un site en planimétrie et altimétrie avant le début des travaux de dresser un état des lieux, un état référent) ; *syn.* (re)levé [m] topographique d'état des lieux, lever [m] de terrain ; **g Geländeabsteckung [f] vor Baubeginn** (Nivellieren und Ausstecken einer bestimmten Fläche vor Beginn der Bauarbeiten).

5964 staking-out plan [n] *constr. surv.* (Plan showing spot elevations and contours for driving stakes on the construction site to define the designed layout and carry out the ►grading design); *syn.* setting-out plan [n] [also UK]; **s plan [m] de replanteo** (Acta de comprobación del replanteo de una obra); ►estudio de nivelación); **f plan [m] de piquetage** (Représentation planimétrique du tracé des futurs espaces verts suivant des repères et des figures géométriques cotées en valeur réelle ; ►cotation d'un projet) ; **g Absteckungsplan [m]** (Plan, der die Vermaßung von Objekten oder eines Layouts für die Arbeiten vor Ort darstellt sowie zur Übertragung von Koordinaten dient; ►Planung der Höhenabwicklung); *syn.* Plan [m] zur Absteckung (von), Absteckplan [m].

5965 staking-out [n] with field rods [US] *constr. surv.* (Traditional process used in ground surveying); *syn.* setting-out [n] using ranging-poles [UK]; **s jalonamiento [m]** *syn.* piquetaje [m], estacado [m], estaquillado [m], picada [f]; **f jalonnement [m]** (Repérage sur le terrain permettant de matérialiser le tracé au moyen de jalons) ; **g Absteckung [f] mit Fluchtstäben** (Markierung von einzumessenden Höhen und Objekten mit Fluchtstäben).

stand [n] *phyt.* ►mangrove stand; *for.* ►tending of a forest stand.

5966 standard [n] [UK] (1) *arb. hort.* (In U.K., nursery stock tree with an upright clear stem supporting a head; plants in standard form are described, in descending order according to the height of the clear stem, as **tall standard, ¾ standard,** and **½ standard.** ["Intermediate standard" and "short standard" are depreciated]; BS 3975: Part 4: 1966; ►shade tree [US] 2/tall standard [UK], ►standard rose, ►standard street tree [US]/avenue tree [UK]); **s árbol [m] de pie alto (2)** (Árbol de vivero de un altura mínima de la cruz de 1,8 m para caducifolios y de 1,6-1,8 m para frutales; ►árbol de alineación, ►árbol de pie alto, ►rosal de pie alto 1); **f arbre [m] tige** (*Terme générique* arbre constitué du tronc dépourvu de branches latérales et de la couronne comprenant la flèche et les ramifications ; selon les caractéristiques dimensionnelles on distingue les arbres mini-tige, courte-tige, demi-tige, haute-tige ; ►arbre d'avenue, ►arbre de haute-tige, ►rosier tige) ; **g Hochstamm [m] (2)** (*Kulturform in der Baumschule* in Stamm und Krone gegliedertes Gehölz; Stammhöhe für Laubbäume mindestens 200-250 cm, für Obstgehölze 160-180 cm; veredelte Hochstämme dürfen als Kronen- oder Fußveredelungen herangezogen werden; die Krone muss artspezifische Stammverlängerungen haben [durchgehender Mitteltrieb] und artspezifischen Aufasten, insbesondere bei Hochstämmen für Alleebäume zulassen. Hiervon ausgenommen sind Kronenveredelungen, Kugel- und Hängeformen; zweimal ver-

S

pflanzte Hochstämme [2xv H] werden **leichte Hochstämme**, H.stämme für Straßenbepflanzungen **Alleebäume** und 4xv H.stämme **Solitärhochstämme** genannt; cf. FLL-Gütebestimmungen für Baumschulpflanzen; ►Alleebaum, ►Hochstamm 1, ►Rosenhochstamm).

standard [n] (2) *adm. envir. pol.* ►air quality standard; *envir. leg.* ►emission standard; *urb. recr.* ►open space standard; *for.* ►reserve [US] (1)/reserved tree [UK].

standard [n]**, air pollution** *adm. envir. pol.* ►air quality standard.

standard [n]**, conservation** *conserv'hist. leg.* ►conservation criterion.

standard [UK]**, half** *hort.* ►semi-dwarf fruit tree [US].

standard [n] [UK]**, short** *hort.* ►dwarf fruit tree [US].

standard [UK]**, tall** *hort.* ►shade tree [US] (2).

standard building regulations [npl] [ZA] *leg. urb.* ►uniform building code [US].

Standard Codes [npl] [US]**, Form** *constr. contr.* ►standards for construction [US].

Standard Codes [npl] [US]**, Performance** *constr. contr.* ►standards for construction [US]/Codes of Practice (C.P.) [UK].

Standard Codes [npl] of Practice (C.P.) [UK]**, British** *constr. contr.* ►standards for construction [US].

standard concrete paving block [n] [UK] *constr.* ►standard non-interlocking concrete paver [US].

5967 standard concrete paving block [n] **without exposed aggregate** *constr.* *s* **piedra** [f] **estándar sin árido visto**; *f* **pavé** [m] **en béton de masse** (Pavé sans parement) ; *g* **Betonpflasterstein** [m] **ohne Natursteinvorsatz**.

5968 standard contract form [n] [US] *contr. prof.* (Preprinted document with standard text, used as the basis for the wording of a contract); *syn.* form [n] of agreement [UK]; *s* **formulario** [m] **de contrato**; *f* **cadre** [m] **d'acte d'engagement** (Document type utilisé par le maître d'œuvre dans le dossier de consultation et définissant les termes du marché concernant le contractant, les prix, les délais et les paiements) ; *g* **Vertragsformular** [n] (Vorgeschriebener, typisierter Vertragstext als Grundlage für die sprachliche Gestaltung eines Vertrages).

5969 standard data form [n] *adm. conserv. ecol.* (Standardized sheet for compilation of information on special protection areas and special areas of conservation according to the EU ►Habitats Directive and ►Birds Directive, published in the *Official Journal of the European Union* (OJ); *s* **formulario** [m] **estándar de datos** (Lista estandarizada para recopilar información sobre las ►zonas especiales de conservación y las ►zonas de protección especial, de acuerdo a la ►Directiva de las Aves y la ►Directiva de Hábitats, publicado en el boletín oficial de la Unión Europea); *f* **formulaire standard de données (FSD)** (Formulaire standardisé accompagnant la décision de transmission d'un projet de site ou l'arrêt désignant un site, élaboré pour chaque site Natura 2000 et transmis à la Commission européenne par chaque État membre ; il liste sur la base des codes nationaux expliqués dans la clé de lecture les espèces animales ou végétales et les habitats naturels qui justifient la désignation du site à la Commission européenne conformément à la ►Directive Habitat Faune Flore et la ►Directive Oiseaux ; document officiel, publié au journal officiel de la Communauté européenne) ; *g* **Standarddatenbogen** [m] (Standardisiertes Meldeformular für die EU-Kommission zur Übermittlung von gebietsbezogenen Informationen zu Schutzgebieten nach ►Fauna-Flora-Habitat-Richtlinie [FFH-RL] resp. ►Vogelschutzrichtlinie [VSchRL]; offizielles, im Amtsblatt der Europäischen Gemeinschaften veröffentlichtes Dokument).

Standard Form [n] **of Agreement between Owner and Architect** [US] *contr. leg. prof.* ∗►Guidance for Clients on Fees [UK].

5970 Standard Form [n] **of Building Contract** [UK] *constr. contr.* (In U.K., there are three main standard forms of contract used: The Association of Consultant Architects Form of Building Agreement 1984 [ACA], The General Conditions of Government Contracts for Building and Civil Engineering Works 1989 [GC/Works], and Joint Contracts Tribunal Standard Form of Building Contract 1980 [JTC 80]; ►General Conditions of Contract; In D., the portion of contract documents setting forth non-technical rights, responsibilities and relationships of parties involved, invitation for bidders [US]/tenderers [UK], instruction to bidders/tenderers, bonds, supplementary conditions and other special conditions is called **general conditions** [npl] **of contract for project execution**; *s* **pliego** [m] **de cláusulas administrativas generales para la contratación de obras** (Conjunto de normas que determinan las disposiciones contractuales generales de contratos para la ejecución de obras públicas; cf. art. 49 LCAP; ►condiciones generales de contratos de las administraciones públicas); *f* **cahier** [m] **des clauses administratives générales (des marchés publics) (C.C.A.G.)** (Applicable aux marchés publics, ce document préétabli contractuel fixe les dispositions administratives applicables à tous marchés de travaux, de fournitures et de services d'un maître d'ouvrage ; pour les marchés de maîtrise d'œuvre est appliqué le **cahier des clauses administratives générales applicables aux marchés publics de prestations intellectuelles, [C.C.A.G.P.I.]** ; il n'existe pas de documents légalement imposés pour les marchés privés, mais d'une manière générale les maîtres d'ouvrage privés ont recours au C.C.A.G. constitué par la norme N.F. P 03 001 ; ►clauses générales de passation des marchés) ; *g* **allgemeine Vertragsbedingungen** [fpl] **für die Ausführung von Bauleistungen** (DIN 1961: **V. für die A. von B.**, die regelmäßig nur durch ausdrückliche Vereinbarung zum Vertragsinhalt werden; in der VOB Teil B wird das Werkvertragsrecht des Bürgerlichen Gesetzbuches [BGB] durch spezielle, auf die besonderen Bedingungen des Bauens abgestellte Regelungen mit dem Ziel ergänzt, für beide Vertragsparteien ein ausgewogenes Vertragswerk zu schaffen, wie z. B. Festlegungen über Vergütung, Ausführung, Ausführungsfristen, Behinderungen, Kündigungen, Haftung, Abnahme, Gewährleistung, Abrechnung, Sicherheitsleistungen, Streitigkeiten; cf. VOB Teil B; ►Verdingungsordnung).

standard form rose [n] [US] *hort.* ►standard rose.

5971 standard fruit tree [n] *arb. hort.* (Type of fruit-bearing tree with a structured crown on a straight trunk; usual trunk height, 160-180cm; a grafted **s. f. t.** may be grafted at the crown or base of a trunk; ►fruit tree, ►standard [UK] 1; in U.S., a standard fruit tree is measured from 18-25 feet of full height); *s* **árbol** [m] **frutal de pie alto** (Frutal de vivero de una altura mínima de la cruz de 1,6-1,8 m; ►árbol de pie alto 1, ►árbol frutal); *syn.* frutal [m] de pie alto; *f* **arbre** [m] **fruitier (de) haute tige** (Arbre de plein vent composé d'un tronc d'environ 2 m [la greffe est en général effectuée à 1,80 m du sol] de hauteur surmonté d'une couronne de cinq à six branche ; à l'âge adulte l'arbre peut atteindre une hauteur de 8 à 10 m ; la distance de plantation est de 10 à 12 m ; les arbres haute tige portent en général 2 greffes ; ►arbre fruitier, ►arbre tige) ; *syn.* arbre [m] plein vent ; *g* **Obsthochstamm** [m] (In Stamm und Krone gegliederter ►Obstbaum; Stammhöhe für Obstgehölze 160-180 cm;

veredelte Hochstämme dürfen als Kronen- oder Fußveredelungen herangezogen werden; ▶Hochstamm 2).

5972 standard grass-seed mixture [n] *constr.* (**D.,** standardized composition of grass seeds for a specific ▶lawn type. The seed is limited to cultivars which are contained in a list compiled according to their suitability for a particular lawn use; ▶grass-seed mixture); *s* **mezcla** [f] **estándar de semillas** (**D.,** mezcla estandarizada de semillas de gramíneas para diferentes ▶tipos de césped; cf. Norma DIN 18 719; ▶mezcla de semillas de césped); *f* **mélange** [m] **standardisé de semences pour gazon** (**F.,** composition du mélange de semences d'un gazon établie pour des usages définis ; il n'existe pas en France de standardisation généralisée des mélanges de gazon les sélectionneurs de gazons commercialisant des produits portant leur dénomination propre [p. ex. composition « MDG » pour la maison des gazons, mélange « club-vert » pour Rhône-poulenc] ; ▶mélange de semences pour gazon ; **D.,** compositions standards de graminées pour mélanges de semences pour gazon correspondant à différents ▶types de gazons ; les semences proviennent d'espèces à gazon sélectionnées par le « service fédéral de contrôle des espèces » qui atteste la qualité génétique, sanitaire et technologique des semences et leur aptitude à répondre aux types d'usage proposés sur la base de l'analyse des caractères des espèces ; norme DIN 18 719) ; *syn.* composition [f] du mélange de gazon ; *g* **Regelsaatgutmischung** [f] (*Abk.* RSM; standardsierte Zusammenstellung von Gräsern zu Grassamenmischungen für unterschiedliche ▶Rasentypen. Die Grassamen stammen nur von solchen Sorten, die nach der beschreibenden Sortenliste für Rasengräser des Bundessortenamtes in der Bewertung ihrer Eigenschaften eine auf den Verwendungszweck bezogene besondere Eignung aufweisen; cf. DIN 18 917; ▶Rasensaatgutmischung).

5973 standard non-interlocking concrete paver [n] [US] *constr.* (Simple square, rectangular or hexagonal concrete paving block/paver without vertical or horizontal interlocking features; ▶interlocking paver [US]/interlocking paving block [UK]); *syn.* standard concrete paving block [n] [UK]; *s* **piedra** [f] **estándar de hormigón** (Piedra ordinaria sin función de ensamblaje; ▶adoquín de ensamblaje); *f* **pavé** [m] **en béton ordinaire** (Pavé simple de forme carrée, rectangulaire ou hexagonale sans liaison verticale ou horizontale ; ▶pavé autobloquant) ; *g* **Betonpflasterstein** [m] (2) (Rechteckiger, quadratischer oder sechseckiger Normalstein ohne horizontale oder vertikale Verbundwirkung; ▶Verbundpflasterstein).

standard paver [n] *constr.* ▶edge restraint, #1.

5974 standard rose [n] *hort.* (Rose grown in standard form, i.e. grafted on a single tall stem; ▶dwarf standard rose, ▶weeping tree rose [US]/weeping standard rose [UK]); *syn.* standard form rose [n] [also US], tree rose [n] [also US]; *s* **rosal** [m] **de pie alto** (Rosal en arbolito generalmente con cuatro alturas diferentes: dependiendo del vigor de crecimiento de la variedad de rosa comercialmente se diferencia entre pies de 40, 60, 90 y 140 cm; ▶rosal llorón, ▶rosal enano de pie alto); *f* **rosier** [m] **tige** (Forme de rosier dont les caractéristiques végétatives et dimensionnelles correspondent à deux catégories de comercialisation et dont la hauteur de tige mesurée de la racine supérieure jusqu'à la greffe inférieure est inférieure à 70 cm [rosier courte-tige, ▶rosier mini-tige miniature], de 70 à 90 cm [rosier demi-tige], de 90 à 120 cm [rosier tige] et supérieure à 140 cm [▶rosier pleureur, rosier haute-tige]) ; *g* **Rosenhochstamm** [m] (Aus Unterlagen-Wildlingen gezogene und veredelte Rosenwuchsform mit i. d. R. vier unterschiedlichen Stammhöhen: je nach Wuchsstärke der Rosensorte werden im Handel 40 cm [Zwergstamm], 60 cm [Halbstamm], 90 cm [Hochstamm]

und 140 cm hohe Hochstämme [für ▶Trauerrosen] angeboten; ▶Zwergstammrose); *syn.* Hochstammrose [f], Stammrose [f].

standards [npl] *adm. constr.* ▶American Standards [US]; *for.* ▶coppice with standards; *envir. pol.* ▶environmental quality standards; *urb.* ▶land use intensity standards [US]; *wat'man.* ▶water quality standards.

5975 standards [npl] **for construction** [US] *constr. contr.* (Technical documents setting forth standards of good construction for various materials and trades; DAC 1975; **in U.S., Form Standard Codes** contain exact kinds of construction materials and methods of installation, etc. **Performance Standard Codes** designate the manner in which a construction design must work; **in U.K.,** British Standard Codes of Practice; ▶additional contractual conditions, ▶special technical requirements of contract); *syn.* Codes [npl] of Practice (C.P.) [UK]; *s* **pliego** [m] **de prescripciones técnicas generales** (Conjunto de normas técnicas para contratos de obras públicas, fijadas según las indicaciones del art. 53 de la Ley de Contratos de las Administraciones Públicas [LCAP]; éstas pueden ser ampliadas por ▶prescripciones técnicas particulares; ▶pliego de prescripciones administrativas particulares; además las construcciones se rigen por las normas fijadas en el ▶código técnico de la edificación); *syn.* prescripciones [fpl] técnicas generales; *f* **cahier** [m] **des clauses techniques générales — documents types (C.C.T.G. — documents types)** (Spécifications techniques générales définissant les normes de construction ou de réalisation d'ouvrages ainsi que les stipulations concernant la qualité des matériaux. Le C.C.T.G. fait partie des pièces constitutives du marché et est complété par le cahier des clauses techniques particulières — C.C.T.P. ; ▶clauses administratives complémentaires, ▶prescriptions techniques particulières) ; *g* **Allgemeine Technische Vertragsbedingungen** [fpl] **für Bauleistungen** (*Abk.* ATV; **D.,** der Verdingungsordnung für Bauleistungen [VOB Teil C: DIN 18 299-18451] festgelegte Normen für die Erstellung von Bauleistungen, die stets unverändert Bestandteil der Vertragsbedingungen sind. Sie können durch ▶zusätzliche Vertragsbedingungen ergänzt werden; ▶Besondere Technische Vorschriften); *syn.* Allgemeine Technische Vorschriften [fpl].

5976 standard slope [n] *constr. eng.* (Slope gradient laid down by planning guidelines for a specific excavation or fill of a road or river embankment); *s* **talud** [m] **estándar** (Desmonte o terraplén para obra de infraestructura cuya inclinación está regulada en directriz de planificación); *syn.* terraplén [m] standard; *f* **profil** [m] **type de talus** (Talus en déblai ou en remblai d'ouvrages projetés dont l'angle de pente est fixé dans le cadre d'un règlement ou de spécifications techniques) ; *g* **Regelböschung** [f] (Auftrag- oder Einschnittböschung für bautechnische Vorhaben, z. B. im Straßen- oder Deichbau, mit in Planungsrichtlinien festgelegten Böschungsneigungen).

5977 standard specifications [npl] *constr.* (Sample text describing the work to be performed in building or in landscape construction and used as an aid in compiling particular specifications; ▶Uniform Construction Specifications); *s* **modelo** [m] **de resumen de prestaciones** (Texto descriptivo de los servicios estándar en construcción civil y paisajística como base para elaborar resúmenes de prestaciones específicos para ofertas o contratos de obra; ▶Registro General del Código Técnico de Edificación [Es]); *syn.* pliego [m] de prescripciones técnicas tipo; *f 1* **cahier** [m] **de prescriptions techniques-type** (±) (**D.,** propositions de textes écrits standardisés, décrivant les dispositions techniques et règles de l'art nécessaires à l'exécution des prestations de travaux du bâtiment et d'espaces verts et utilisés dans la rédaction des spécifications techniques ; ▶documents techniques unifiés) ; *f 2* **Cahier** [m] **de clauses techniques** [CCT] (**F.,** textes qui indiquent les conditions techniques à respecter

S

pour le choix et la mise en œuvre de matériaux) ; *g* **Musterleis-tungsverzeichnis [n]** (Textvorschläge zur Beschreibung von Standardleistungen im Bauingenieurwesen sowie im Garten- und Landschaftsbau als Hilfe zur Aufstellung von örtlichen Leistungsverzeichnissen; ►Standardleistungsbuch für das Bauwesen); *syn.* Musterausschreibungstext [m].

5978 standard street tree [n] [US] *hort.* (In U.S. and U.K., bid [US]/tender [UK] specifications for ►street trees should specify the height to which the tree is to be free of branching; ►quality standards for nursery stock, ►shade tree [US] 2/tall standard [UK] delimbed to >6 ft [US]); *syn.* avenue tree [n] [UK]; *s* **árbol [m] de alineación** (*Forma de cultivo de vivero* ►árbol de pie alto con comienzo muy alto de la copa que se utiliza para plantar en calles; ►árbol de calle, ►estándar de calidad de plantas de vivero); *f* **arbre [m] d'avenue** (►Arbre de haute-tige dont la base de la couronne est particulièrement haute [au moins 4,20 mètres — Direction Départementale de l'Équipement (D.D.E.)], p. ex. le long des voies urbaines ; ►arbre d'alignement, ►norme de qualité pour végétaux de pépinière) ; *syn.* arbre [m] en allée ; *g* **Alleebaum [m]** (**1.** *Kulturform in der Baumschule* mindestens 3x verpflanzter, in extra weitem Stand kultivierter ►Hochstamm mit gerader Stammverlängerung und besonders hohem Kronenansatz, der später an der Verwendungsstelle, z. B. an Straßen oder für Reihenpflanzungen, bis auf seine Endstammhöhe/seinen Endkronenansatz von 4,50-6,00 m oder noch höher, aufgeastet werden muss; Stammhöhe der Baumschulware bis 25 cm Stammumfang mindestens 220 cm, über 25 cm StU mindestens 250 cm; ►Gütebestimmungen für Baumschulpflanzen; ►Straßenbaum. **2.** Baum innerhalb einer Allee); *syn. zu 1.* Hochstamm [m] für die Straßenbepflanzung.

stand density [n] *for.* ►forest stand density.

stand density [n] **of forest/woodland** *for.* ►forest stand density.

standing biomass [n] *phyt. zool.* ►biomass.

Standing Conference [n] **of local planning authorities** [UK] *adm. plan.* ►Metropolitan Council of Governments (COG) [US].

standing crop biomass [n] *bot. ecol.* ►phytomass.

5979 standing forest crop [n] *for.* (Large area of trees, bamboo, etc. grown especially for harvesting timber; ►forest cover); *syn.* forest stand [n]; *s* **población [f] forestal** (Masa arbórea de edad mixta en una zona específica; ►cobertura forestal); *f* **peuplement [m] forestier (1)** (Ensemble de la végétation arborescente de composition floristique, d'âge et de structure homogène, faisant partie de la ►couverture forestière. En USA et CDN on désigne sous le terme *forest crop* tout ce qui est susceptible d'être récolté ; DFM 1975) ; *g* **Waldbestand [m] (2)** (*Sensu stricto* gesamter Baumbestand aller Altersstufen, der einen Teil der ►Waldbedeckung auf größerer Fläche ausmacht. Im US-Englischen wird mit *forest crop* der für die Holznutzung bestimmte Teil bezeichnet).

standing tree [n], **dead** *ecol.* ►snag [US&CDN] (3).

5980 standing waterbody [n] *geo. limn* (Pond, pool, fishpond, lagoon or lake with motionless water; e.g. pond, ►lagoon, ►man-made pond, ►natural pond, lake; *opp.* ►watercourse 1; *generic term* ►waterbody 1); *s* **aguas [fpl] estancadas** (Cuerpo de aguas quietas como estanque, charco, laguna, lago, etc.; ►albufera, ►estanque [artificial], ►estanque natural; *opp.* ►curso de agua; *término genérico* ►aguas continentales); *syn.* aguas [fpl] quietas; *f 1* **eaux [fpl] calmes** (Eaux superficielles intérieures stagnantes telles que lacs, étangs, mares, etc. ; ►étang artificiel, ►étang naturel, ►lagune, *opp.* ►cours d'eau ; *terme générique* ►eaux) ; *syn.* eaux [fpl] stagnantes ; *f 2* **eaux [fpl]**

closes (Plan d'eau privé ou non qui ne communique pas d'une manière continue avec des ruisseaux ou canaux, sans communication amont ou en communication exceptionnelle [crues] avec les eaux libres ; *contexte* pêcher en eaux closes) ; *g* **Stillgewässer [n]** (Stehende Wasseransammlung wie ►Teich, Tümpel, ►Weiher, Haff, ►Lagune, See etc.; *opp.* ►Fließgewässer; *OB* ►Gewässer); *syn.* stehendes Gewässer [n], Stehgewässer [n], Stillwasser [n].

5981 stand [n] **of rushes** *phyt.* (►rush meadow); *s* **juncar [m]** (►pradera de juncos); *f* **peuplement [m] de Joncs** (►jonchaie, ►jonchère) ; *syn.* peuplement [m] à *Juncus* ; *g* **Binsenbestand [m]** (cf. Art. 21 NHG; ►Binsenwiese).

5982 stand [n] **of timber** *for.* (A silvicultural or forest management unit; ►mixed forest, ►pure stand); *syn.* timber stand [n], tree crop [n], timber crop [n]; *s* **rodal [m]** (Unidad de gestión forestal; ►bosque mixto, ►rodal monoespecífico); *f* **peuplement [m] forestier (2)** (Ensemble d'arbres ou végétaux arborescents poussant sur un terrain forestier ; ►boisement monospécifique, ►forêt mélangée, ►forêt mixte) ; *g* **Baumbestand [m] (1)** (Bestand an Holz in einem Forstbetrieb oder Teilen davon; ►Mischwald, ►Reinbestand).

stand structure [n] *for. phyt.* ►stratified stand structure.

5983 stand thinning [n] *constr. for.* (Removal of individual trees to reduce the density of a stand, leaving the remainder with space to develop; ►alternate tree removal, ►isolation of trees); *syn.* chopping-out [n] [also UK]; *s 1* **limpia [f]** (Extracción de individuos para reducir la densidad de un ►brinzal 2 y así dejar más espacio de crecimiento a los individuos restantes); *s 2* **raleo [m]** (Extracción de individuos para reducir la densidad de un ►latizal y así dejar más espacio de crecimiento a los individuos restantes; ►aclareo esquemático, ►corta de deliberación); *syn.* aclareo [m] (2), ralear [vb], apertura [f] de un rodal; *s 3* **corta [f] de mejora** (Extracción de individuos en bosque natural descuidado); *f* **éclaircissage [m] (1)** (Opération d'arrachage de jeunes plants en vue de favoriser la croissance des plants restant ; ►démariage, ►isolation [d'arbres]) ; *g* **Auslichten [n, o. Pl.] im Bestand** (**1.** Herausnahme einiger Stämme aus einem jungen Bestand, damit die übrigen Bäume bessere Wuchsbedingungen haben); **2.** ►Freistellung von Bäumen; ►Herausnahme eines jeden zweiten Baumes; auslichten [vb]); *syn.* Bestandsauflockerung [f].

stand [vb] **vacant** *sociol. urb.* ►vacancy.

5984 start [n] **of construction** [US] *constr. contr. syn.* commencement [n] of works [UK]; *s* **comienzo [m] de una obra** *syn.* comienzo [m] de los trabajos de obra; *f* **ouverture [f] des travaux** ; *g* **Aufnahme [f] der Bauarbeiten** (Beginn der Arbeiten auf der Baustelle); *syn.* Baubeginn [m].

state [n] *bot. zool.* ►dormant state, *ecol.* ►trophic state.

state [n]**, contracting member** *contr. pol.* ►party to a contract.

state [n]**, steady** *ecol.* ►ecological equilibrium.

state college [n] [US] *prof.* ►polytechnic institute [US]/university of applied sciences [UK].

5985 state development [n] *plan.* (Complete socio-economic development of a [federal] State); *s* **desarrollo [m] territorial socioeconómico**; *f 1* **développement national** (Résultat des actions et programmes menés au niveau national assurant et une croissance harmonieuse et une amélioration qualitative des activités socio-économiques au niveau national) ; *f 2* **développement régional** (Résultat des actions et programmes menés au niveau national assurant et une croissance harmonieuse et une amélioration qualitative des activités socio-économiques au

niveau régional) ; *g* **Landesentwicklung [f]** (Sozioökonomische Gesamtentwicklung eines Staates oder Bundeslandes).

5986 state development plan [n] [US] *leg. plan.* (A plan which lays down the objectives of ▶regional planning policy and policy for the comprehensive development of local or regional areas according to state laws; **in U.S.**, comprehensive land use planning is sometimes done by state governments, but is more often authorized by state enabling legislation for planning, land use regulation and zoning at the local level; **in U.K.**, a regional plan for an Economic Planning Region; **in D.**, a plan which is revised every 10 years at the latest and lays down the objectives of ▶state regional planning and policy for the comprehensive development of local or regional areas according to state laws); *syn.* regional policy/strategy plan [n] [UK ±]; *s* **plan [m] de desarrollo territorial** (En D. plan revisado como mínimo cada 10 años que define los objetivos del desarrollo regional integrado de una región o un estado federado de acuerdo con las leyes vigentes; **en Es.** corresponde al ▶plan director territorial de coordinación; ▶planificación regional); *f* **schéma [m] régional d'aménagement et de développement du territoire** (**F.**, Document de la collectivité territoriale de la Région fixant, à travers un rapport et des cartes, les orientations fondamentales en matière d'environnement, de développement durable, de grandes infrastructures de transport, de grands équipements et de services d'intérêt régional et prenant en compte les orientations du schéma national ; cf. loi 95-115 du 4 février 1995 ; ▶aménagement du territoire, ▶planification régionale ; *syn. anciennement* schéma [m] d'orientation et d'aménagement ainsi que schéma [m] régional d'aménagement et de développement durable du territoire ; *syn. pour la Guadeloupe, la Guyane, la Martinique et la Réunion* schéma [m] régional d'aménagement) ; *g* **Landesentwicklungsplan [m]** (**D.**, Plan, der spätestens alle 10 Jahre fortgeschrieben und erneut aufgestellt wird und gemäß den Landesplanungsgesetzen der Bundesländer unter Berücksichtigung der Grundsätze des § 2 ROG die Ziele der ▶Raumordnung und ▶Landesplanung für die Gesamtentwicklung eines Bundeslandes festlegt).

5987 state forest/national forest [n] [US]/**State Forest** [n] [UK] *adm. for.* (Publicly-owned area providing for multiple uses of timber production, mineral extraction, watershed protection, wildlife management, scenic and wilderness preservation as well as recreational development; ▶reserved forest. **In U.K.**, the term 'State' is used in the national sense and not in the Federal sense as in the US: Large-scale forests are managed by the Forestry Commission, set up in 1918. In former times the term 'Crown Forest' was employed for large areas of royal land used for hunting, but also as a source of timber for ship building; ▶royal hunting forest; **in U.S.**, state and national forests can only be created by US Congress and state legislatures in each state); *s 1* **monte [m] de dominio público [Es]** (Aquellos montes que están incluidos en el Catálogo de Montes de Utilidad Pública, los comunales, pertenecientes a las entidades locales, en tanto su aprovechamiento corresponda al común de los vecinos y aquellos otros montes, que sin reunir las características anteriores, hayan sido afectados a un uso o servicio público. Todos ellos integran el **dominio público forestal**; cf. art. 12.1 Ley 43/2003); *syn.* monte [m] demanial; *s 2* **monte [m] patrimonial [Es]** (Bosques de propiedad pública que no son demaniales; art. 12.2. Ley 43/2003; ▶monte real); *s 3* **bosque [m] estatal** (Bosque que se encuentra en manos del Estado, es gestionado por la correspondiente administración pública, cuya función no se reduce a la producción de madera y otros productos forestales, sino que se gestiona de tal manera que sirva para cumplir funciones ecológicas, protectoras y recreativas); *f* **forêt [f] domaniale** (Forêt appartenant à l'État dont la gestion et l'équipement sont assurés par l'Office National

des Forêts [ONF], établissement public, national, à caractère industriel et commercial ; l'ONF est en outre responsable de la mise en œuvre du régime forestier dans les forêts appartenant aux départements, communes, établissements publics ainsi que de la gestion cynégétique et du développement des équipements récréatifs destinés à l'accueil du public ; ▶forêt seigneuriale) ; *g* **Staatswald [m]** (Wald, der sich im Staatseigentum befindet, von Forstbehörden bewirtschaftet und verwaltet wird und nicht nur der reinen Holzproduktion, sondern auch der Erholungsnutzung und anderen Wohlfahrtswirkungen dient. In D. werden **Bundes-** und **Landesforsten** unterschieden. Der **S.** ist aus dem ▶Bannwald 1 und den Wäldern der Landesfürsten hervorgegangen); *syn.* Staatsforst [m].

State Forest [n] [NZ]**, Permanent** *adm. for.* ▶reserved forest.

5988 state government agency [n] [US] *adm.* (State department/division responsible for executing government program[me]s; ▶federal government agency [US]); *syn.* central government body [n] [UK], state government authority [n] (±); *s 1* **autoridad [f] estadual [AL]** (Instituciones de administración regional de los países con estados federados, como Brasil, México y EE.UU.; ▶autoridad estatal [federal]); *syn.* administración [f] estadual [AL]; *s 2* **administración [f] autónoma [Es]** (Instituciones de administración de las Comunidades Autónomas del Estado Español); *s 3* **administración [f] foral [Es]** (Instituciones de administración de las provincias de España con derecho foral Araba, Bizkaia, Gipuzkoa y Navarra); *f* **autorité [f] régionale** (Administration d'État d'une région ; ▶autorité centrale [fédérale]) ; *syn.* autorité [f] administrative régionale ; *g* **Landesbehörde [f]** (Staatliche Behörde eines Bundeslandes; ▶Bundesbehörde).

state government authority [n] *adm.* ▶state government agency [US].

5989 state government construction project [n] *plan.* (Project of a federal, state or local authority/agency [with national significance]; ▶federal government construction project); *s* **proyecto [m] de obras públicas estatal (2)** (Proyecto de infraestructura o de construcción realizado por una autoridad estadual, de CC.AA. o de un estado federado; ▶proyecto de obras públicas 1); *f* **projet [m] immobilier de l'État (2)** (Opération de construction d'un ensemble immobilier de logements ou d'équipements publics réalisée et financée par l'État ; ▶projet immobilier de l'État 1) ; *g* **staatliches Bauvorhaben [n] (2)** (Vom Finanz- oder Bauministerium eines Bundeslandes [Staatsbauamt resp. Landesbauamt] durchgeführtes Bauprojekt; ▶staatliches Bauvorhaben 1); *syn.* Bauvorhaben [n] eines Bundeslandes.

5990 state highway [n] [US] *adm. leg. trans.* (**1. In U.S.**, major thoroughfare in a state-wide system of highways, which are funded on a fifty-fifty, federal/state basis; ▶interstate highway, ▶maintenace obligation); **2. trunk road [n] [UK] (3)** (**In U.K.**, there are A-roads [main trunk roads] and B-roads [minor roads]. Responsible for planning and maintenance are county highway departments which also receive funds from the national government); *s* **carretera [f] provincial** (En España carretera de conexión regional cuya responsabilidad recae en la administración provincial; ▶carretera estatal, ▶carretera federal, ▶carretera interestatal, ▶obligación de mantenimiento); *syn.* vía [f] interurbana; *f 1* **route [f] départementale** (**F.**, route classée appartenant au domaine public et dont l'▶obligation d'entretien est à la charge d'un département ; ▶grand itinéraire, ▶réseau routier national) ; *syn.* chemin [m] départemental ; *f 2* **route [f] régionale (≠)** (**D.**, route classée appartenant au domaine public et à la charge d'un Bundesland ; de par sa fonction et l'importance

S

du réseau correspond à *f 1*) ; *g* **Landstraße [f]** (In D. eine klassifizierte Straße, bei der die ▶Unterhaltungslast vom jeweiligen Bundesland getragen wird; ▶Bundesfernstraße).

5991 State Historic Preservation Office [n] **(SHPO)** [US] *adm. conserv'hist.* (**In U.S.,** an office or bureau created by law in each state to administer the national historic preservation program, carry out state historic preservation responsibilities, and assist local governments. SHPOs focus on several activities: **1.** statewide surveys and inventories; **2.** a special form of environmental review of federal projects having an impact on National Register/Register-eligible projects where federal grants or permits are involved, and **3.** helping the owners of National Register commercial properties obtain federal income tax credits where appropriate under federal law. Some states by state law ask or require the SHPO office to make recommendations to the local ▶historic district commissions regarding the enforcement of local historic district or landmarks ordinances); **in U.K.,** a public authority concerned with the finding, protection or restoration and administration of ▶cultural monuments); *syn.* Historic Buildings and Monuments Commission [n] [UK]; *s 1* **organismo [m] responsable del Patrimonio Histórico Español** (En España la competencia de protección de Patrimonio está compartida, siendo exclusiva del Estado la «defensa del patrimonio cultural, artístico y monumental español contra la exportación y la expoliación; museos, bibliotecas y archivos de titularidad estatal», mientras que las CC.AA. tienen derecho a asumir competencia sobre el «patrimonio monumental de interés de la comunidad autónoma»; arts. 148.16° y 149.1.28° Constitución Española; ▶monumento, ▶Patrimonio Histórico Español); *s 2* **administración [f] regional de protección de monumentos** (En D. la **a. de p. de m.** es competencia de los estados federados y su nombre varía entre ellos, es la administración que supervisa las ▶administraciones locales de protección de monumentos); *f* **Conservation [f] régionale des monuments historiques** (F., administration responsable chargée de la protection, la conservation, la réutilisation et la mise en valeur du patrimoine historique [archéologique, historique et ethnologique] public ou privé de la région et qui dispose pour accomplir ces missions d'une équipe administrative, scientifique et technique, [cellule administrative, cellule recensement-documentation, cellule ZPPAUP et Urbanisme, cellule travaux-marchés] dirigée par le conservateur régional des monuments historiques qui élabore les programmes techniques et financiers et conduit les travaux de restauration sur les ▶monuments historiques 1 et instruit les dossiers de demandes de protection ; celui-ci assure le secrétariat de la Commission régionale du patrimoine et des sites [CRPS] qui est consultée pour l'inscription des monuments et l'opportunité de leur classement ou les créations de ZPPAUP ; la conservation régionale des monuments historiques a pour mission de coordonner l'action du ▶service départemental de l'architecture et du patrimoine [SDAP] et contribue ainsi à la prise en compte du patrimoine et de la qualité architecturale dans l'aménagement du territoire et la politique de la ville, notamment par l'action du conseiller pour l'architecture, et de l'architecte-conseil dont chaque région est dorénavant dotée. Toutes ces activités sont gérées en F. par l'administration des affaires culturelles [Direction régionale des affaires culturelles — DRAC] ; cf. www. culture.gouv.fr) ; *g* **Landesdenkmalamt [n]** (...ämter [pl]; Fachbehörde, die mit der Auffindung, Erhaltung oder Wiederherstellung und Verwaltung von ▶Kulturdenkmälern befasst ist; sie berät die untere und höhere Denkmalschutzbehörde in allen Fachfragen. Das **L.** ist der unteren ▶Denkmalschutzbehörde als Fachbehörde übergeordnet resp. zugeordnet. *In D. gibt es in den einzelnen Bundesländern unterschiedliche Synonyma* Landeskonservator [m] (1) [Berlin], Denkmalschutzamt [n] [HH], Landesamt [n] für Denkmalpflege [Bremen, Westfalen, SH, HS], Staatliches Konservatoramt [n] [SL], Bayerische Verwaltung [f] der Staatlichen Schlösser, Gärten und Seen; *in CH* Eidgenössische Kommission [f] für Denkmalpflege).

5992 State Historic Preservation Officer [n] **(SHPO)** [US] *conserv'hist.* (Official title for the head of a public authority concerned with historic landmark preservation [US]/conservation of historic monuments and sites [UK] at the state level. **In U.S.,** official title of the head of a ▶State Historic Preservation Office—concerned with historic landmark preservation [US] at the state level. The officer is appointed by the Governor of each state and administers the federal historic preservation program[me], as well as being responsible for the state's historic preservation program[me]s; **in England,** the authority is known as the Historic Buildings and Monuments Commission [commonly called English Heritage] and advises the Secretary of State for the Environment, who in case of dispute makes the final decision); *syn.* Chief Executive [n] of English Heritage [UK]; *s 1* **director, -a [m/f] general de Bellas Artes y Archivos [Es]** (En Es. es el presidente del Consejo del Patrimonio Histórico cuya función esencial es facilitar la comunicación e intercambio de programas de actuación e información relativos al Patrimonio Histórico Español entre las Administraciones del Estado y de las Comunidades Autónomas; cf. art. 1 RD 111/1986, de 10 de enero, de desarrollo parcial de la Ley 16/1985, del Patrimonio Español; ▶organismo responsable del Patrimonio Histórico Español); *s 2* **director, -a [m/f] del departamento de protección de monumentos** (Persona que dirige una administración de protección del patrimonio histórico resp. de monumentos de un estado federado u otra entidad geográfico-administrativa de un país; en alemán el término se utiliza también para científicos/as que trabajan en tal institución; ▶administración regional de protección de monumentos); *f* **conservateur [m] régional des monuments historiques** (Titre officiel désignant le responsable de la direction régionale des monuments historiques ; ▶conservation régionale des monuments historiques) ; *g* **Landeskonservator [m] (2)** (Amtsbezeichnung des Leiters oder auch eines wissenschaftlichen Mitarbeiters eines ▶Landesdenkmalamtes); *syn.* Landesdenkmalpfleger [m].

statement [n] *conserv. leg. nat'res. plan.* ▶environmental impact statement (EIS); *plan.* ▶explanatory statement; *leg. plan.* ▶position statement [US]/information notice [UK].

statement [n]**, written** *plan.* ▶explanatory report.

statement [n] [US]**, written award** *contr.* ▶written notice of contract award.

statement [n] **of the project** [US] *plan. prof.* ▶project brief.

state [n] **of a landscape** *landsc.* ▶current state of a landscape.

5993 state-of-the-art [n] **(1)** *arch. constr.* (Phrase used to describe a method or type of building construction, which has been determined empirically in the current state of development, or carried out according to the latest written standards and is universally accepted as being the most appropriate, both in theory and practice. The method is well-known and applied in design and construction, is in compliance with the laws of mathematics and physics, and is proven to be durable and efficient; ▶technical state-of-the-art 2); *s* **reglas [fpl] de construcción reconocidas y aceptadas** (Forma o método empírico y fijado en norma para realizar un trabajo arquitectónico específico con materiales y construcciones reconocidos que cumplen con las exigencias tanto en cuanto a las bases matemático-físicas como desde el punto de la teoría arquitectónica, que son conocidos y utilizados por profesionales destacados y han dado buenos resultados en la práctica; ▶mejor tecnología asequible, ▶técnica reconocida y aceptada); *f* **Règles [fpl] de l'Art** (Forme ou méthode d'exécution de travaux de construction déterminée empiriquement ou fixant par écrit les règles d'aptitude à l'emploi et de mise en œuvre de procédés, éléments ou équipements utilisés dans la construction ; ces règles qui ont été reconnues sur des bases physico-mathématiques et selon les théories de l'architecture font dans la pratique l'objet d'un avis technique, d'une normalisation, d'une certification ou d'une codification dans le cadre des documents techniques unifiés [D.T.U.] mais restent dans beaucoup de cas à l'appréciation des hommes de l'art ; ▶principes de la technique courante, ▶suivant les règles de l'art) ; *g* **allgemein anerkannte Regeln [fpl] der Baukunst** (Empirische und schriftlich festgelegte Art oder Methode des Vorgehens zur Lösung bestimmter Bauaufgaben mit anerkannten Materialien und Konstruktionen, die in der Praxis, nach mathematisch-physikalischen Grundlagen sowie nach bauhandwerklichen und nach Grundsätzen der Architekturtheorie als richtig anerkannt sind, den maßgeblichen Fachleuten bekannt sind, von ihnen angewendet werden und sich in der Praxis bewährt haben. Die **a. a. R. d. B.** werden begrifflich weiter gefasst als die ▶**allgemein anerkannten Regeln der Technik**, da sich das Anforderungsprofil der ersteren über die Konstruktionsgrundsätze, die Statik und die Materialkunde hinaus auch auf nicht technische bauhandwerkliche und architektonische Grundsätze erstreckt, was in bauordnungsrechtlicher Hinsicht bei Baugenehmigungen oft von Bedeutung ist; cf. BAUR 2000 [6]; ▶Stand der Technik).

5994 state-of-the-art [n] (2) *arch. eng. envir.* (**1.** TGG 1984, 122; applied technology recognized as being the most up-to-date by the majority of specialists in the professional fields of science, engineering, architecture, etc.; ▶best practice, ▶state-of-the-art 1, ▶technical state-of-the-art); **2. Best Practicable Means [npl] (BPM)** (Essentially pragmatic philosophy for the control of emissions from scheduled processes. The term was first used in the 1906 Alkali Act and was implemented by the superseded *UK Alkali and Clean Air Inspectorate* who were responsible for controlling emissions from scheduled industrial premises [oil refineries, chemical works, cement plants, etc.; cf. DES 1991); **3. best practicable environmental option [n] [BPEO]** (BPEO in addition to controlling atmospheric emissions, includes "appropriate measures to deal with any harmful discharges to water and for the treatment or disposal of other solid and liquid wastes to land. A BPEO should take into account the risk of transfer of pollutants from one medium to another"—DES 1991, 32; term was first used by the *Royal Commission on Environmental Pollution*, 5th Report [1976] in order to take account of the total pollution from a process and the technical possibilities for dealing with it); *s* **mejor tecnología [f] asequible** (Procedimientos o métodos para realizar servicios, fabricar productos o bienes de equipo, desarrollar procesos productivos y de construcción que utilizan los descubrimientos científicos y desarrollos técnicos más recientes, sin que se hayan generalizado en la práctica y que —al contrario que en el caso de la ▶técnica reconocida y aceptada y de las ▶reglas de construcción reconocidas y aceptadas— aún no han sido incluidos en las correspondientes normas técnicas; ▶mejor práctica posible); *f* **suivant les règles de l'art [loc]** (**1.** Pratique reconnue par la majorité des spécialistes d'une profession qui s'engagent à exercer leur mission conformément à l'ensemble de la réglementation et de la pratique en vigueur ; ▶Règles de l'Art. **2.** Le terme « ▶**bonne pratique** » désigne, dans un milieu professionnel donné et en particulier dans le domaine environnemental, un ensemble de démarches qualité et de comportements personnels et collectifs faisant consensus et considérés comme indispensables en vue de concourir à l'amélioration des pratiques visant à minimiser l'impact d'une activité sur

l'environnement ; p. ex. **bonnes pratiques environnementales [BPE]** ; ▶principes de la technique courante) ; *syn.* dans les règles de l'art [loc] ; *g* **Stand [m, o. Pl.] der Technik** (**1.** Fundierte berufliche Meinung der Mehrzahl von Fachleuten in Wissenschaft und Praxis. **2.** Im technischen Umweltschutz, z. B. beim Immissionsschutz, wird unter **S. d. T.** der aktuelle „Entwicklungsstand fortschrittlicher Verfahren, Einrichtungen und Betriebsweisen" verstanden; cf. § 3 [6] BImSchG. **3.** Im deutschen Recht gibt es noch den Terminus ▶**gute fachliche Praxis [GfP]**, der die Einhaltung gewisser Grundsätze des Tier- und Umweltschutzes in der Land-, Forst- und Fischereiwirtschaft umfasst. **Im U.K.** sind im Immissionsschutz die Termini **best practicable environmental option [BPEO]** resp. **best practicable means [BPM]** üblich, die nur teilweise mit den deutschen Begriffsinhalten übereinstimmen; ▶allgemein anerkannte Regeln der Baukunst, ▶allgemein anerkannte Regeln der Technik).

state-of-the-art technique [n] *arch. constr.* ▶technical state-of-the-art.

state regional planner [n] *prof.* ▶regional planner.

5995 State regional planning [n] [US] *plan.* (**In U.S.**, preparation of plans to meet socio-economic, cultural and environmental needs of separate regions within a State; **in U.K.**, preparation and implementation of plans for an Economic Planning Region, to fulfill socio-economic, cultural and environmental requirements; **in D.**, comprehensive planning for socio-economic, cultural and environmental requirements according to regional policies for a State [in D., Land Government]; ▶regional planning); *syn.* regional planning [n] (2) [(±) UK]; *s* **planificación [f] regional (2)** (En EE.UU., elaboración de planes para satisfacer necesidades socioeconómicas, culturales y ambientales de regiones individuales dentro de un estado de la unión; **en GB**, preparación e implementación de planes para cada Región de Planificación Económica para cumplir exigencias socioeconómicas, culturales y ambientales; **en D.**, planificación del territorio a nivel de estado federado; **en Es.**, la **p. r.** corresponde a las CC.AA. a través del instrumento ▶plan director territorial de coordinación; ▶ordenación del territorio; ▶planificación regional 1); *f* **planification [f] régionale** (Actions d'encadrement, décidées au niveau national ou résultant de décisions prises et appliquées dans le cadre des régions, par les pouvoirs publics du développement économique et social à l'aide d'un plan, grâce à la mise en œuvre d'instruments spécifiques ; DUA 1996 ; ▶aménagement du territoire, ▶aménagement du territoire au niveau régional) ; *g* **Landesplanung [f]** (**D.**, übergeordnete, zusammenfassende Planung für soziale, kulturelle, ökologische und wirtschaftliche Erfordernisse auf der Ebene eines Bundeslandes durch die Aufstellung von Programmen und Plänen sowie die Abstimmung raumbedeutsamer Planungen und Maßnahmen von Bundes- und Landesbehörden mit den Erfordernissen der ▶Raumordnung; cf. Landesplanungsgesetze der Länder und § 3 [1] und § 4 ROG 2008; ▶Regionalplanung).

state surveillance system [n] **for air pollution** *envir.* ▶national or state surveillance system for air pollution.

static caravan ground [n] [UK] *recr.* ▶permanent campground [US].

5996 statics [npl] and dynamics [npl] *constr. eng. phys.* (Field of mechanical sciences involving the calculation of forces exerted upon an object or structure and the measures required to assure their stability. *Context* to work out the structural stability of a building; important for the structural calculations of the building [LA 1998 [5], 49]); *syn.* structural engineering [n] (1); *s* **estática [f]** (Parte de la ciencia mecánica referente al cálculo de las fuerzas que actúan sobre los objetos o estructuras y las medidas requeridas para asegurar su estabilidad); *f* **statique [f]**

(Partie de la mécanique qui étudie les systèmes de points matériels soumis à l'action de forces, quand elles ne créent aucun mouvement ; PR 1987) ; *syn.* stabilité [f] d'une construction ; *g* **Statik [f]** (*Teilgebiet der Mechanik* Lehre vom Gleichgewicht der an einem ruhenden Körper angreifenden Kräfte und den dabei zu erfüllenden Gleichgewichtsbedingungen).

static site [n] [US] *landsc. recr.* ▶permanent site.

station [n] *envir.* ▶air quality control station, ▶hydroelectric power station, *ecol.* ▶location (1); *envir.* ▶nuclear power station; *conserv. zool.* ▶ornithological station; *envir.* ▶solar power station.

station [n], **power** *envir.* ▶power plant.

5997 stationary traffic [n] *plan. trans.* (Generic term covering 'stopped' traffic [= short-term stop for getting in or out of a vehicle], 'delivery and haulage' traffic [= short-term stop of vehicles for loading and unloading], 'parked' traffic [▶parked vehicles], and 'traffic standing' in a traffic jam); *s* **tráfico [m] quieto (≠)** (Término genérico para los vehículos parados al borde de la calle/carretera para que se suban o bajen personas, para cargar y descargar bienes y los aparcados; ▶vehículos aparcados); *f* **circulation [f] intermittente (±)** (Terme générique regroupant le trafic des véhicules effectuant un arrêt occasionnel, temporaire sur voies et emplacements aménagés sur le domaine public routier [arrêt court pour monter et descendre du véhicule], le trafic de livraison et le trafic lié au stationnement ; ▶véhicules en stationnement) ; *syn.* trafic [m] de stationnement intermittent (±) ; *g* **ruhender Verkehr [m]** (OB zu ,haltender' [= kurzfristiges Stehenbleiben von Fahrzeugen am Fahrbahnrand zum Ein- und Aussteigen], ,arbeitender' [= kurzfristiges Stehenbleiben von Fahrzeugen am Fahrbahnrand zum Be- und Entladen von Gütern] und ,parkender' V.; *nicht zu verwechseln mit* ,stehendem' V. [Verkehrsstau]; ▶parkender Verkehr).

5998 stationing [n] *constr. eng.* (Marking out of road alignment continuously along the centerline); *s* **jalonamiento [m] de carretera** (Marcado con jalones a distancias regulares del trazado y alineación de una carretera, generalmente a lo largo del eje central); *f* **jalonnement [m] d'un tracé** (Détermination du tracé d'un projet de route en plan et en altimétrie pour des espaces réguliers en général en prenant l'axe médian) ; *g* **Stationierung [f]** (Punktgenaue Bestimmung einer geplanten Wegetrasse in Lage und Höhe in regelmäßigen Abständen, i. d. R. in der Mittelachse).

5999 status [n] **of endangerment** *conserv. phyt. zool.* (Classified condition of disappearing faunistic and floristic species; ▶Blue List, ▶Red List, ▶threatened species); *s* **categoría [f] de amenaza de extinción de especies** (Diferenciación de la situación de las especies de flora y fauna en diversos grados según el peligro de extinción que sufran. En Es. según la legislación básica estatal existen dos: a] en peligro de extinción: taxones o poblaciones cuya supervivencia es poco probable si los factores causantes de su actual situación siguen actuando, b] vulnerable: taxones o poblaciones que corren el riesgo de pasar a la categoría anterior en un futuro inmediato si los factores adversos que actúan sobre ellos no son corregidos; las CC.AA. pueden establecer en sus respectivos ámbitos territoriales otras categorías específicas y, en su caso, incrementar el grado de protección de las especies del ▶Catálogo Español de Especies Amenazadas en sus catálogos autonómicos, incluyéndolas en una categoría superior de amenaza; art. 55, Ley 42/2007; ▶especie amenazada, ▶Lista Azul [US], ▶lista roja); *f* **catégorie [f] de régression des populations et de distribution des espèces** (Hiérarchisation traduisant la situation de patrimoine faunistique et floristique. En F. on distingue les cinq catégories suivantes : **1.** espèce disparue, **2.** espèce amenée par leur régression à un

niveau critique des effectifs, **3.** espèce affectée d'une régression forte et continue et a qui a déjà disparue de nombreuses régions, **4.** espèce dont la population n'a pas sensiblement diminuée, **5.** espèce dont une régression s'est manifestée sans qu'il soit possible de définir dans laquelle mesure. Pour la faune existent trois catégories supplémentaires : **1.** espèce remarquable, **2.** espèce endémique et **3.** espèce migratrice ; ▶espèce menacée, ▶Liste bleue [US], ▶Liste rouge, ▶Livre rouge des espèces menacées) ; *g* **Gefährdungskategorie [f]** (Einheit bei der Einteilung der faunistischen und floristischen Bestandssituation hinsichtlich des Verbreitungs- und Bestandsrückganges; die internationale Klassifizierung der IUCN sieht für die ▶Rote Liste weltweit gefährdeter Arten sieben Kategorien vor: ▶bedrohte Art; ▶Blaue Liste [US]).

statute [n] *adm. leg. pol.* ▶law, #2.

statute [n] [UK], **enabling** *adm. leg.* ▶enabling act [US].

statutes [n] [UK], **city** *adm. leg. urb.* ▶city ordinance [US].

statutes [n] [UK], **town** *adm. leg. urb.* ▶city ordinance [US].

6000 statutory consultation [n] **with public agencies** [US] *plan.* (Legal obligation of [line or staff] agencies to seek the views of interested public or official bodies on a planning proposal before approving that proposal; ▶public participation); *syn.* coordinated public agency review [n] [also US], statutory consultation [n] with public authorities [UK]; *s 1* **informe [f] a las corporaciones locales y a los departamentos ministeriales** (En España, según la legislación supletoria los planes territoriales se someten a información ante las corporaciones locales a cuyo territorio afectaren. En el caso de los planes directores territoriales de coordinación también a los departamentos ministeriales que no hayan intervenido en su elaboración y a los que puedan interesar por razón de su competencia; cf. arts. 39-41 LS Texto refundido; ▶participación pública en la planificación); *s 2* **audiencia [f] de autoridades e instituciones públicas y semipúblicas** (D., en el marco del procedimiento de aprobación de proyectos públicos, de aprobación de planes generales de ordenación urbana y de planes sectoriales se consultan a las autoridades e instituciones afectadas como pueden ser las empresariales, sindicatos, iglesias, etc.); *f* **consultation [f] des services de l'État et des personnes publiques associées** (**1.** Cette consultation a lieu au sein du groupe de travail [G.T.] du PLU/POS et de la commission locale d'aménagement et d'urbanisme [CLAU] du SCOT/SDAU dans lesquels les plans d'urbanisme sont élaborés conjointement. **2.** Transmission pour avis du projet de plan aux personnes associées à l'élaboration des plans d'urbanisme et non représentées au sein des groupes de travail ; ▶participation du public) ; *syn.* participation [f] ; *g* **Beteiligung [f] (der) Träger öffentlicher Belange** (Frühzeitige Mitwirkung von Behörden und sonstigen Stellen, die Träger öffentlicher Belange sind und von Planungen, z. B. Bauleitplanungen, Fachplanungen oder im Rahmen von Planfeststellungsverfahren berührt werden können, indem diese angehört und von diesen Stellungnahmen eingeholt werden; cf. §§ 4, 4a, 13, 22 Abs. 9 BauGB; ▶Beteiligung der öffentlichkeit 2); *syn.* Anhörung [f] (der) Träger öffentlicher Belange, Anhörung [f] der Behörden und sonstiger Träger öffentlicher Belange, Anhörung [f] der Planungspflichtigen [CH], Beteiligung [f] der Behörden, Behördenbeteiligung [f]).

statutory consultation [n] **with public authorities** [UK] *plan.* ▶statutory consultation with public agencies [US].

6001 statutory designation [n] *leg. plan.* (**In U.S.,** authorization by Acts of Congress for federal facilities, and by state enabling laws and local ordinances, except in 'Home Rule' states having constitutional authority for such local responsibilities. This term **s. d.** is only one of dozens of legally-binding

S

forms of restriction regarding land use; **in U.K.**, authorization by Acts of Parliament [statutes], bye-laws or legally-binding plans and compliance by local communities; ▶design ordinance [US]/ design regulations [UK], ▶statutory land use specification); *syn.* legal designation [n] [also US]; *s* **prescripción [f] de planificación** (Todo tipo de regulaciones [decretos, planes u otros instrumentos legales] que determinan las obligaciones a las que está sujeto el uso del suelo; ▶asignación de usos pormenorizados del suelo, ▶reglamento de diseño); *syn.* designación [f] legal, fijación [f] legal; *f* **prescriptions [fpl] d'urbanisme** (Dispositions spécifiques écrites et graphiques fixées par un PLU/POS et destinées à préciser l'implantation des construction, la hauteur des bâtiments les espaces libres à conserver, etc. ; ▶prescriptions communales architecturales et paysagères, ▶zonage) ; *syn.* prescription [f] édictée par un document d'urbanisme ; *g* **Festsetzung [f]** (Bestimmung durch Gesetze, Verordnungen oder durch rechtsverbindliche Pläne von Flächen, Grunddienstbarkeiten, Gestaltungsvorschriften, Pflanzbindungen und -zwängen etc.; ▶Ausweisung von Flächen, ▶Gestaltungssatzung); *syn.* förmliche Festlegung [f].

6002 statutory designation [n] **of landscape planning requirements** *leg. landsc.* (Legal specification of landscape characteristics which recognize conservation of nature and requirements for recreation area development which are shown, e.g. on ▶zoning maps [US]/Proposals Map [UK]); *s* **prescripción [f] de norma paisajística** (Fijación de medidas de ordenación o diseño paisajístico en el plan general de gestión del paisaje o en el ▶plan parcial [de ordenación]); *syn.* ordenación [f] paisajística; *f* **prescription [f] paysagère** (Mode de protection et de mise en valeur des espaces forestiers, des sites et des paysages prescrit aux documents d'urbanisme et déterminant l'utilisation de l'espace, la forme et l'intensité des usages) ; *g* **landschaftsplanerische Festsetzung [f]** (Festsetzungen in Landschafts- oder ▶Bebauungsplänen, die die Belange des Naturschutzes und der Landschaftspflege durchsetzen sollen).

6003 statutory floodplain zone [n] [US]/**statutory flood-plain zone** [n] [UK] *leg. plan.* (**In U.S.**, legally-established area to protect against flood damage and contain floodwater on land which can be used only for agriculture; **in Europe**, area established by law to protect against flood damage and contain floodwater on land which cannot be built upon or planted and which can only be used as permanent grassland); *s* **zona [f] inundable** (Zona de la márgen de ríos en la que el Gobierno, por Real Decreto, podrá establecer limitaciones en el uso de las **zz. ii.** que estime necesarias para garantizar la seguridad de las personas y bienes. Los Consejos de Gobierno de las CC.AA. podrán establecer, además, normas complementarias de esta regulación; art. 11.3 RD Legislativo 1/2001); *f* **surface [f] submersible** (Zone exposée aux risques d'inondations définie par un plan d'exposition aux risques naturels prévisibles publié, dans laquelle ne pourra être établi aucun remblai, digue, dépôt de matières encombrantes, clôture, plantation, construction ou ouvrage pour assurer le libre écoulement des eaux ou la conservation des champs d'inondations ; cf. art. 5-1 du loi n° 82-600 du 13 juillet 1982 ; les **s. s.** sont définies par le plan d'exposition aux risques naturels prévisibles) ; *syn. obs.* plan [m] des surfaces submersibles ; *g* **gesetzliches Überschwemmungsgebiet [n]** (Durch Rechtsverordnung festgesetztes Gebiet, das im Interesse des Hochwasserschutzes und der Regelung des schadlosen Wasserabflusses von Bebauung, Anpflanzungen und anderen Hindernissen freizuhalten ist und in D. nur als Dauergrünland genutzt werden darf [cf. § 32 WHG]. Ein **g. Ü.** wird gem. § 5 [2] 7 u. [6] und § 9 [1] 16 u. [6] BauGB im Bauleitplan nachrichtlich übernommen).

statutory instrument [n] [UK] *leg.* ▶order [US] (2).

6004 statutory land use specification [n] *adm. leg. plan.* (Legal identification of precise land uses on a community development plan; ▶designation by zoning map [US]/designation by a Proposals Map/Site Allocations Development Plan Document [UK], ▶statutory designation); *syn.* designation [n] of land uses; *s* **asignación [f] de usos pormenorizados del suelo** (Determinación precisa de la función del suelo en el ▶plan parcial [de ordenación], incluida la delimitación de las zonas en las que se divide el territorio planeado por razón de los usos pormenorizados previstos y, en su caso, la división en polígonos o unidades de actuación; cf. art. 45 RD 2159/78; ▶prescripción de planificación, ▶prescripción en plan parcial); *syn.* zonificación [f] [AL], fijación [f] de usos del suelo; *f* **zonage [m]** (**1.** Représentation graphique de la répartition d'un territoire en zones sur le parcellaire d'un document de planification urbaine. **2.** Déterminer l'affectation dominante des sols par zones en précisant l'usage ; cf. C. urb., art. R. 123-21 ; ▶prescription d'urbanisme, ▶prescription édictée par un plan local d'urbanisme/plan d'occupation des sols. **3.** *recr.* Technique consistant à définir des espaces en fonction de leur fragilité, de leur difficulté d'accès, de leur importance culturelle, écologique ou scientifique et de leur capacité à recevoir des équipements et des aménagements touristiques) ; *g* **Ausweisung [f] von Flächen** (Parzellenscharfe und rechtsverbindliche Nutzungsdarstellung in ▶Bauleitplänen oder Festsetzung von ▶Schutzgebieten; Flächen ausweisen [vb], Nutzungen ausweisen [vb]; ▶Festsetzung durch den Bebauungsplan); *syn.* Flächenausweisung [f], Ausweisung [f] von Nutzungen.

6005 statutory provision [n] **for participation** *adm. leg.* (Legal obligation to obtain views as soon as possible of both public authorities/agencies and private persons affected by planning proposals; ▶public announcement, ▶publication of planning proposals, ▶public participation); *s* **prescripción [f] legal de participación pública** (La normativa de Es., D. y F. obliga a las autoridades responsables a informar y permitir la participación de la población afectada y de las instituciones públicas y semipúblicas en el proceso de planificación; ▶participación pública en la planificación, ▶anuncio preceptivo, ▶publicación de un plan, ▶sumisión a información pública); *f* **principe [m] de la participation publique** (à l'occasion de projets ou d'opérations d'intérêt public lors de ▶mise à la disposition du public ; intervention des associations agréées à l'élaboration, p. ex. d'un PLU/POS [facultative et à posteriori], d'une Z.E.P. [obligatoire et préalable] ; ▶enquête publique, ▶participation du public, ▶publication) ; *syn.* principe [m] de l'intervention du public ; *g* **Beteiligungspflicht [f]** (Verpflichtung eines Planungsträgers, Behörden und Stellen als Träger öffentlicher Belange und Bürger, die von der Planung berührt werden können, möglichst frühzeitig durch öffentliche Darlegung [▶Auslegung eines Planentwurfes] und Anhörung zu beteiligen; cf. z. B. §§ 3 u. 4 BauGB; ▶Beteiligung der Öffentlichkeit, ▶öffentliche Bekanntmachung).

statutory regional plan [n] *plan.* ▶regional plan report.

stay [n] *recr. zool.* ▶period of stay.

steady state [n] *ecol.* ▶ecological equilibrium.

6006 steel grate [n] *constr.* (Cover for drainage channels or cleanout sumps; ▶grille cover); *syn.* steel grill(e) [n]; *s* **rejilla [f] de acero** (▶Rejilla de cubierta de canales de desagüe; *f* **grille [f] maille métallique** (Grille métallique à maillage rectangulaire recouvrant les caniveaux d'évacuation des eaux ; ▶grille) ; *g* **Stahlgitterrost [m]** (Abdeckrost aus sich kreuzenden Stahlstäben für Entwässerungsrinnen; ▶Gitterrost); *syn.* Stahlmaschenrost [m], Stahlgitterabdeckung [f].

S

steel grill(e) [n] *constr.* ▶steel grate.

steel plan [n] *constr.* ▶reinforcing steel plan [US].

6007 steepen [vb] **an existing slope** *constr.* (▶construct an embankment/slope, ▶create a slope); *s* **igualar [vb] terreno con talud** (En un terreno dado adaptar el nivel a la cota planificada por medio de un talud; ▶ataluzar, ▶taluzar); *f* **taluter [vb] en déblai** (Confection d'un talus lors du raccordement du terrain naturel à un plate-forme inférieure ; ▶dresser un talus ; ▶taluter en remblai) ; *g* **abböschen [vb]** (Eine vorhandenes Gelände auf eine niedriger gelegene Planungshöhe durch eine Böschung angleichen; ▶Böschung herstellen; ▶anböschen).

steeping treatment [n] [UK] *constr.* ▶open tank treatment [US&CDN].

steepness [n] **of a flight of steps** *arch. constr.* ▶slope of a flight of steps.

steepness [n] **of a stairway** *arch. constr.* ▶slope of a flight of steps.

6008 steep slope [n] *geo.* *s* **pendiente [m] fuerte**; *f* **forte pente [f]** (LA 1981, 477) ; *g* **steiler Hang [m]**.

stele [n] *adm. conserv'hist.* ▶burial stele.

stem [n] *arb.* ▶removal of codominant stem.

stem [n], **base** *hort.* ▶cane.

stem [n], **short** *hort.* *▶fruit tree.

6009 stem apex [n] *arb.* (Upper part of a tree trunk from which ▶upright-growing main branch and horizontal main branches begin to grow at the point where the tree crown starts; also the point where twin stems begin their ▶forked growth; *opp.* ▶butt of a tree); *syn.* apex [n] of a stem; *s* **ápice [m] del tronco** (Extremo superior del tronco del cual salen las ▶ramas principales de crecimiento vertical y las ramas horizontales principales, es decir donde empieza la copa; también es el punto donde en el caso de una ▶cima en horquilla el tronco se subdivide en dos ramas principales; *opp.* ▶pie del tronco); *f* **apex [m] du fût** (Partie supérieure du tronc qui se divise en branches de fortes taille ou partie du tronc à partir de laquelle, après la ▶fourche, se développent les ▶charpentières ; *opp.* ▶base du tronc) ; *g* **Stammkopf [m]** (Oberer Teil eines Baumstammes, bei dem die Kronenbeastung beginnt resp. beim ▶Zwiesel sich der Stamm in ▶Stämmlinge verzweigt; *opp.* ▶Stammfuß).

stem damage [n] [UK] *arb. hort.* ▶stem injury.

6010 stem flow [n] *phyt.* (Precipitation that accumulates on a tree and flows down the trunk to the ground); *s* **escurrido [m] del agua por el tronco** (Flujo del agua de lluvia a lo largo del tronco hacia el suelo); *syn.* escurrimiento [m] del agua por el tronco; *f* **écoulement [m] le long du tronc** (Portion des précipitations atmosphériques qui est interceptée par le couvert végétal et qui s'écoule le long des branches principales et des troncs) ; *syn.* écoulement [m] sur écorce (DFM 1975), ruissellement [m] sur les troncs [CDN] (DFM 1975) ; *g* **Stammabfluss [m]** (Am Baumstamm herunterfließendes Regenwasser).

6011 stem injury [n] *arb. hort.* (Mechanically-caused tree trunk damage; ▶bark wound, ▶trunk wound [US]/stem wound [UK]); *syn.* stem damage [n] [UK], trunk injury [n] [also US]; *s* **lesión [f] del tronco** (Daño mecánico causado en un tronco; ▶herida de la corteza, ▶herida del tronco); *f* **blessure [f] du tronc** (Dégât mécanique occasionné au tronc, p. ex. par des véhicules ; ▶blessure de l'écorce, ▶plaie du tronc) ; *syn.* plaie [f] du tronc ; *g* **Stammverletzung [f]** (Mechanischer Schaden an einem Baumstamm; ▶Rindenverletzung, ▶Stammwunde); *syn.* Stammschaden [m].

6012 stem-spreading [adj] *bot.* (Having the nature of forming horizontal stems growing above ground [runner or ▶stolon] or those growing underground [▶rhizome]. From nodes on stolons and rhizomes, roots and upright shoots develop, the latter from axillary buds; cf. BOT 1990, 101; ▶rhizomatous, ▶stoloniferous); *syn.* running with trailing stems [loc] [also US], strong creeping [ppr]; *s* **estolonífero/a [adj] (1)** (Se aplica a las plantas, los ▶rizomas, que producen ▶estolones. Éstos nacen de la base del tallo y pueden arrastrarse por la superficie del suelo [estolones epígeos] o arrastrarse debajo de ésta [estolones subterráneos]. En ambos casos, enraizando y muriendo en las porciones intermedias, engrendran nuevos individuos y propagan vegetativamente la planta; ▶planta vivaz cespitosa, ▶rizomatoso/a, ▶estolonífero/a 2); *f* **traçant, ante [ppr/adj]** (Concerne les végétaux développant des tiges latérales aériennes [▶stolon] ou souterraines [▶rhizome] présentant des unités de végétation allongées ainsi que des écailles foliaires et qui s'enracinent à distance de la plante mère afin de former de nouveaux individus avec la mort de la partie intermédiaire ; ▶rhizomateux, euse, ▶stolonifère) ; *g* **Ausläufer treibend [ppr/adj]** (Krautige, sich vegetativ ausbreitende Pflanze betreffend, die stets horizontal verlaufende, teils ober- [▶Stolo(n)], teils unterirdische Seitenzweige/Sprosse [▶Rhizom] mit stark verlängerten Internodien und reduzierten Blättern bildet und in einiger Entfernung von der Mutterpflanze bewurzelt und durch Absterben des dazwischen liegenden Stückes neue Individuen bildet; hierzu gehören auch niedrige Bodendeckstauden [z. B. Stachelnüsschen — *Acaena*] und hochwüchsige Schaftstauden [z. B. Goldrute *[Solidago canadensis]*; ▶rhizombildend, ▶Stolone bildend); *syn.* Ausläufer bildend [ppr/adj].

6013 stem thorn [n] *bot.* (Modified, sharply pointed, short spine growing from an ▶axillary bud, which may be simple, e.g. sloe *[Prunus spinosa]* or branched as in the case of the honeylocust tree *[Gleditsia triacanthos]*; ▶thorn); *s* **espina [f] caulinar** (Extremo de tallo endurecido y puntiagudo que puede ser simple como en la endrina *[Prunus spinosa]* o ramificado como en la acacia de tres púas *[Gleditsia triacanthos]*; ▶espina, ▶yema lateral); *f* **rameau-épine [m]** (Désigne une pousse courte, dure, à pointe acérée, formée à partir de la transformation d'un rameau simple, p. ex. sur le Prunellier *[Prunus spinosa]* ou un rameau ramifié sur le Févier à trois épines *[Gleditsia triacanthos]* ; GOR 1985, 14 ; ▶bourgeon latéral, ▶épine) ; *syn.* dard [m] ; *g* **Sprossdorn [m]** (Blattachselständiger Kurztrieb mit reduzierten Blättern, die einfach, z. B. bei Schlehe *[Prunus spinosa]* oder verzweigt sind, z. B. beim Lederhülsenbaum, fälschlich Christusdorn, *[Gleditsia triacanthos]*; ▶Dorn, ▶Seitenknospe).

6014 stem tuber [n] *bot.* (Swollen part of a stem [e.g. Kohlrabi—*Brassica oleracea var. gongylodes*, hypocotyl of *Cyclamen*] or underground stem [e.g. potato—*Solanum tuberosum* or **corms** of *Crocus*, Autumn Crocus—*Colchicum*, Winter Aconite—*Eranthis*] serving as an organ both of vegetative reproduction and of storage; *specific term* hypocotylar tuber; ▶root tuber); *s* **tubérculo [m] caulinar** (Porción del tallo engrosada en mayor o menor grado, generalmente subterránea, como la patata, rica en sustancias de reserva [almidón, inulina, etc.], o bulbo sólido o macizo del azafrán; DB 1985; ▶tuberosidad radical); *f* **tubercule [m] caulinaire** (Organe aérien renflé, né de la croissance primaire d'une partie de la tige au niveau duquel la plante stocke des réserves, p. ex. la tubercule hypocotylée du Cyclamen et du Chou-rave *[Brassica oleracea var. gongylodes]*, les tubercules de la Pomme de terre *[Solanum tuberosum]* ; ▶tubercule radiculaire) ; *syn.* tige [f] tubérisée, tige [f] tubéreuse ; *g* **Sprossknolle [f]** (Ober- oder unterirdisches, fleischiges Speicherorgan, das durch primäres Dickenwachstum eines Sprossteils entsteht, z. B. die typische **S.** beim Kohlrabi *[Brassica oleracea var. gongylodes]*, das Hypocotyl [Hypocotylknolle] beim Alpenveilchen *[Cyclamen]*, die unterirdische **S.** beim *Crocus* oder die unterirdischen Seitenzweige [Ausläufer] der Kartoffel *[Solanum tuberosum]*; ▶Wurzelknolle).

stem wound [n] [UK] *arb. hort.* ▶trunk wound [US].

stenecious [adj] *ecol.* ▶stenoecious.

stenoecic [adj] *ecol.* ▶stenoecious.

6015 stenoecious [adj] *ecol.* (Descriptive term for organisms capable of withstanding only very limited fluctuations in vital environmental factors: such organisms only occur in certain habitats; ▶euryecious, ▶stenotopic); *syn. o.v.* stenecious [adj], stenoecic [adj]; *s* **estenoico/a** [adj] (Especie que habita en lugares de ambiente muy precisamente definido, ya que para su normal desarrollo necesitan condiciones ambientales especiales sin las cuales no pueden vivir. El que una especie sea **e.** no implica necesariamente que su área geográfica de distribución sea pequeña; *términos específicos* estenofótico, estenohalino, estenoiónico, estenotérmico; DINA 1987; ▶estenótopo/a, ▶eurioico); *f* **sténoèce** [adj] (Ne pouvant supporter que de faibles amplitudes des facteurs écologiques ; ▶sténotope ; *opp.* ▶euryèce) ; *syn.* sténoapte [adj], sténoécologique [adj], sténoïque [adj] ; *g* **stenök** [adj] (Schwankungen lebenswichtiger Umweltfaktoren nur innerhalb enger Grenzen ertragend; meist Organismen betreffend, die nur in bestimmten Biotopen vorkommen; ▶stenotop; *opp.* ▶euryök); *syn.* standortgebunden [adj]).

6016 stenotopic [adj] *ecol.* (Descriptive term for organisms living only in few habitats with relatively similar environmental factors; ▶stenoecious; *opp.* ▶euryecious); *s* **estenótopo/a** [adj] (*En geobotánica* expresión del grado máximo de fidelidad geográfica; o sea, calificativo aplicado a las sinecias que sólo viven naturalmente en un área geográfica reducida o limitada; el término se aplica también a los organismos que cumplen las mismas características; cf. DB 1985; ▶estenoico/a, ▶eurioico); *f* **sténotope** [adj] (Désigne les organismes qui vivent sur un territoire restreint, sur un biotope limité ; ▶sténoèce ; *opp.* ▶euryèce) ; *g* **stenotop** [adj] (Organismen betreffend, die nur in wenigen, relativ gleichartigen Lebensräumen vorkommen. *MERKE,* **s.** ist nicht syn. mit ▶stenök; *opp.* ▶euryök).

step [n] *arch. constr.* ▶block step, ▶block step with nosing, ▶diminishing step, ▶diminishing step, ▶flagstone step, ▶ramp step, ▶riser step, ▶round timber step, ▶sitting step, ▶semi-circular step [US]/semi-circular step [UK], ▶staircase step.

step [n], **bottom** *arch. constr.* ▶step in a flight of steps.

step [n], **seat** *gard. landsc.* ▶sitting step.

step [n], **top** *arch. constr.* ▶step in a flight of steps.

6017 step formula [n] *constr.* (Mathematical method of calculating the relationship of riser to tread [▶riser/tread ratio]. According to the French architectural theorist JACQUES-FRANÇOIS BLONDEL [1616-1686] the formula is: length of pace = twice the riser + once the tread [61-64cm]. The German landscape architect ALWIN SEIFERT [1890-1972] modified this formula for outdoor use and suggested that for the comfortable climbing of step risers less than 15cm, longer treads are more appropriate, e.g. 13/42, 10/54. Because people today are getting taller and taller, the measure of pace has increased accordingly: for women between 58 and 74cm and for men 58 to 81; cf. DEGA 2002 [45], 11. *NOTE* **US standard** is derived from a 19th Century study of European staircases to arrive at a formula of: 2 x Riser + Tread = 685mm for exterior stairs [TSS/LA]; this formula lowers the angle of ascent for exterior stairs); *s* **fórmula** [f] **de cálculo de peldaño** (Fórmula para calcular la ▶relación entre huella y contrahuella de un escalón de escalera o la distancia entre dos losas de un camino. De acuerdo al teórico de arquitectura francés JACQUES-FRANÇOIS BLONDEL [1616-1686] la fórmula es: largo

de paso = 2 x contrahuella + 1 huella [61-64cm]. El arquitecto paisajista alemán ALWIN SEIFERT [1890-1972] modificó esta fórmula para el uso al aire libre y propuso que para la subida cómoda de escalones de menos de 15 cm de contrahuella, huellas más largas son más apropiadas, p. ej. 13/42, 10/54. Como las personas hoy en día son cada vez más altas, el largo de paso aumenta correspondientemente: para mujeres entre 58 y 74 cm y para hombres entre 58 y 81; cf. DEGA 2002 [45], 11); *f* **formule** [f] **de dimensionnement des marches** (Formule définissant la proportion des marches d'escalier [▶proportion d'une marche], c.-à-d. le rapport entre la profondeur du giron et la hauteur de la contremarche et déterminée par JACQUES-FRANÇOIS BLONDEL [1616-1686], architecte français, qui étudia le calcul des marches d'un escalier pour en rendre son utilisation agréable et sécurisante et constata qu'à chaque fois qu'on s'élève d'un pouce, la valeur de la partie horizontale se trouve réduite de deux pouces et que la somme de la hauteur doublée de la marche et de son giron doit demeurer constante et être de deux pieds ; la formule, dite « **formule de Blondel** » est la suivante : M = 2h + g, où M est le pas [entre 60 et 64 cm], *h* la hauteur de la marche, et *g* son giron [distance entre deux nez de marche consécutifs mesurée sur la ligne de foulée] ; cette formule a été modifiée par l'architecte paysagiste allemand ALWIN SEIFERT [1890-1972] qui proposa pour le confort d'un escalier en espaces extérieurs de prévoir, pour une hauteur de marche inférieure à 15 cm, un giron plus profond dans un rapport de 13/42, 10/54 ; l'augmentation de la taille des individus entraîne naturellement l'augmentation da la longueur d'un pas, entre 58 et 74 cm chez les femmes et 58 et 81 cm chez les hommes ; *syn.* formule [f] de calcul des marches, formule [f] de détermination de la proportion des marches ; *g* **Schrittmaßformel** [f] (Formel zur Bemessung des ▶Steigungsverhältnisses von Treppenstufen. Nach dem französischen Architekturtheoretiker JACQUES-FRANÇOIS BLONDEL [1616-1686] lautet die Steigungsformel: Schrittmaß = 2 Steigungen + 1 Auftritt [61-64 cm]. Der deutsche Gartenarchitekt ALWIN SEIFERT [1890-1972] modifizierte diese Formel für den Außenbereich und schlug wegen des bequemeren Gehens vor, bei Steigungen unter 15 cm längere Auftritte vorzusehen, z. B. 13/42, 10/54. Dadurch dass die Menschen heute immer größer werden, vergrößert sich auch das Schrittmaß entsprechend: bei Frauen zwischen 58 und 74 cm und bei Männern zwischen 58 und 81; cf. DEGA 2002 [45], 11); *syn.* Steigungsformel [f].

6018 step [n] **in a flight of steps** *arch. constr.* (The individual riser/tread element in an outdoor stairway: the lowest is the **bottom step**, the highest is the **top step**; *specific terms* ▶block step, ▶flagstone step, ▶riser step, ▶round timber step, ▶staircase step); *s* **escalón** [m] (2) (Cada uno de los peldaños de una escalera exterior; *términos específicos* ▶escalón de losa, ▶escalón de rodillo de madera, ▶escalón de tabica, ▶escalón prefabricado; ▶peldaño 1); *syn.* peldaño [m] (2); *f* **marche** [f] **d'escalier (2)** (Pour escalier d'extérieur ; la marche du bas est la **marche de départ**, celle du haut est la **marche d'arrivée** ou **marche palière** ; *terme spécifiques* ▶bloc-marche, ▶contremarche en rondins, ▶dalle giron, ▶emmarchement bloqué par une bordure ; ▶marche d'escalier 1) ; *g* **Treppenstufe** [f] (2) (Das einzelne Steigungselement einer Treppe; in einem Treppenlauf heißt die unterste Stufe **Antrittsstufe**, die oberste **Austrittsstufe**; *UBe* ▶Blockstufe, ▶Knüppelstufe, ▶Legstufe, ▶Stellstufe, ▶Treppenstufe 1).

step iron [n] [UK] *constr.* ▶step rod [US].

6019 step lip [n] *constr.* (front edge of a step); *s* **arista** [f] **de peldaño** (En escalera, borde delantero de la huella); *f* **rive** [f] **(1)** (L'arête formée par le giron et la contremarche) ; *syn.* nez [m] de marche ; *g* **Stufenvorderkante** [f] (einer Treppenstufe).

6020 steppe [n] *geo. phyt.* (**1.** Mid-latitude dry grassland extending in ▶arid regions across Eurasia from Ukraine to Manchuria; treeless, and according to the amount of precipitation, either with complete or patchy vegetation cover, principally comprising drought resistant grasses, solitary suffruticose shrubs, perennials and annuals; shrubs may occur where there is sufficient moisture; ▶savanna[h], ▶tall grass steppe. **2.** In North America, the **prairie** is an extensive area of level or rolling land that was originally treeless and grass-covered, and has a temperate climate, late summer drought and winter frost season. *In South America south of the Amazon* **pampa**); *s* **estepa** [f] (Paraje abierto ocupado por especies vegetales herbáceas más o menos amacolladas y ausencia casi total de individuos arbóreos. Existen diferentes formas como la ▶estepa de gramíneas sub-xerofítica [las **Pampas** de Sudamérica, la **pradera** de Norteamérica], la **e.** de gramíneas xerofítica [Norte de África], la **e.** leñosa, la **e.** salina *[Halophytetum]*, la **e.** suculenta, la **e.** alpina y la **e.** desértica; cf. DB 1985; ▶zona árida, ▶sabana); *f* **steppe** [f] (Mosaïque de formations végétales ouvertes et basses, constituées de plantes xérophiles, herbacées ou sous-frutescentes, souvent en touffes espacées. Cette formation est climacique dans les zones arides ou subarides à climat tempéré ou continental saisonnièrement froid et sec. « La Prairie » nord-américaine occidentale est une steppe ainsi que la partie occidentale de la « Pampa » sud-américaine. Par contre, la partie de la « Steppe » russe à tapis végétal fermé est une prairie ; les vastes étendues du nord de l'Amérique du sud [Llanos] qui couvrent presque 30 % du territoire vénézuélien sont une ▶savane ; cf. DG 1984 ; ▶prairie à hautes herbes, ▶région aride) ; *g* **Steppe** [f] (In außertropischen, kontinentalen ▶Trockengebieten [z. B. Ukraine bis zur Mandschurei] vorherrschende, baumlose, je nach Niederschlagsmenge geschlossene oder lückige Vegetationsformation, die hauptsächlich aus dürreharten Gräsern und einzelnen Halbsträuchern, Stauden und Annuellen besteht — bei ausreichender Feuchtigkeit können auch Sträucher vorkommen; ▶Hochgrassteppe, ▶Savanne); *syn.* Trockengrasland [n], *Nordamerika* **Prärie** [f], *südlich des Amazonas* **Pampa(s)** [f(pl)].

stepped building garden [n] [UK] *arch. constr.* ▶terraced townhouse garden [US].

6021 stepped foundation [n] *constr.* (Foundation cut in a series of steps on sloping ground); *s* **cimentación** [f] **escalonada** (Fundación dividida en escalones para construir sobre una ladera; cf. DACO 1988); *f* **fondation** [f] **à redans** (Fondation dont le profil préfigure celui des marches d'un escalier) ; *syn.* fondation [f] à redents ; *g* **Stufenfundament** [n] (Abgetrepptes Gründungsbauwerk in hängigem Gelände).

6022 stepped house [n] *arch.* (Dwelling in which the floors are at different levels creating roof terraces or ▶terraced townhouse gardens [US]/stepped building garden at each floor—a construction method on sloping sites; ▶terraced housing block[s]); *syn.* terraced dwelling [n], terraced house [n]; *s* **casa** [f] **aterrazada** (Casa construida en escalones de manera que cada piso tiene una ▶terraza-jardín; ▶edificación aterrazada); *syn.* casa [f] escalonada; *f* **maison** [f] **en terrasse** (Habitation dont les étages sont aménagés en palier de sorte que chaque étage dispose d'une toiture-terrasse ou d'une ▶terrasse-jardin ; ▶habitat collinaire) ; *syn.* habitation [f] en terrasse ; *g* **Terrassenhaus** [n] (Haus an einem Hang, dessen Geschosse stufenförmig so versetzt sind, dass die nächsthöhere Wohnung die darunter liegende teilweise überdeckt und dadurch jede Wohneinheit eine besonnte Dachterrasse oder einen ▶Terrassenhausgarten hat; ▶Terrassenbebauung).

6023 stepped pathway [n] *gard. constr.* (Hillside path in which the slope is broken by steps at intervals; i.e. a series of ramps interconnected by steps; ▶step ramp); *syn.* stepped ramp [n]; *s* **camino** [m] **escalonado** (Camino empinado que en intervalos está interrumpido por tramos de escalera; ▶rampa escalonada); *f* **chemin** [m] **aménagé de marches d'escalier** (±) (Chemin raide dont la pente en long est entrecoupée par une série de marches ; ▶gradine) ; *g* **Treppenweg** [m] (Steiler Weg, bei dem das Längsgefälle abschnittsweise durch Stufenpakete unterbrochen ist; ▶Stufenrampe).

stepped ramp [n] *trans.* ▶stepped pathway.

6024 steppe-heath [n] *geo. phyt.* (ELL 1988, 173ss; treeless community of shrubs which favo[u]r warm limestone, xeric sites with grasses and perennial herbs, e.g. ▶garrigue in the Mediterranean region of France; ▶garide, ▶heathland, ▶macchia); *s* **landa** [f] **esteparia** (Comunidad vegetal de estaciones secas y ricas en calizas formada sobre todo por gramíneas y perennes y pocas leñosas; en las zonas cálidas del Mediterráneo se presenta en forma de la ▶garriga; ▶matorral [de sustitución], ▶landa, ▶maquis); *f* **landes** [fpl] **steppiques** (Complexe de groupements xérophiles de ▶garides, herbacés et épineux, bas et discontinus, à faciès faiblement arbustif, des plateaux calcaires des étages montagnard et subalpin de la série du Pin sylvestre, landes à Genévriers, p. ex. landes à *Juniperus sabina*, landes basses à Astragale épineux et lavandaies d'altitude, comme survivance de la végétation de steppe froide du tardiglaciaire ; ▶garrigue, ▶lande 1, ▶maquis) ; *syn.* landes [fpl] xérophiles de la série subalpine altiméditerranéenne (cf. VCA 1985, 220 et s.) ; *g* **Steppenheide** [f] (Wärme liebende strauch- und baumarme Pflanzengesellschaft auf kalkreichen Trockenstandorten mit oft flachgründigen Böden, vorwiegend aus Gräsern und Stauden bestehend; begrenzender Standortfaktor sind geringe Niederschläge. In Mitteleuropa Reliktstandorte für kontinentale Steppenflora einer nacheiszeitlichen Wärmeperiode mit z. B. Federgras *[Stipa capillata]*; im warmen Mittelmeergebiet in Form der ▶Gariden [*in Frankreich* ▶Garrigue, *in Italien* ▶Macchia] vorhanden; ▶Heide 2).

6025 steppe-like grassland [n] *phyt.* (Plant communities composed of heliophytic species, e.g. limestone grassland *[Festuco-Brometea]* with many tussock grasses occurring in warm, dry areas on shallow, humus-poor and mostly alkaline soils; ▶dry meadow); *s* **pastizal** [m] **xérico** **(eurosiberiano)** (BB 1979; comunidad vegetal rica en especies heliófilas como los pastizales calcícolas secos *[Festuco-Brometea]*; ▶pastizal seco); *syn.* pastizal [m] seco hemicriptófito; *f* **pelouse** [f] **steppique** (Formations végétales herbacées ouvertes et basses, formées de graminées à caractère xérophile, p. ex. les pelouses calcicoles de la classe *Festuco-Brometea*, en touffes espacées, d'une grande richesse biologique, qui se développe sur des sols généralement calcaires, superficiels, secs et parfois arides, pauvres en matières organiques, sous un climat caractérisé par une sécheresse estivale prononcée et un très grand froid hivernal ; ▶pelouse aride) ; *g* **Steppenrasen** [m] (Artenreiche, Wärme und Trockenheit ertragende Pflanzengesellschaften der Trocken- und Steppenrasen, z. B. Kalkmagerrasen *[Festuco-Brometea]* aus Licht liebenden Artenverbindungen mit vielen Horstgräsern auf flachgründigen, humusarmen, meist kalkreichen Böden; ▶Trockenrasen).

6026 stepping stone [n] *constr.* (Slab mostly laid in lawns or in wide planting borders according to the length of stride); *s* **losa** [f] **para pasos** (Piedra plana tendida generalmente en céspedes o macizos a distancia de un paso); *f* **dalle** [f] **d'un pas japonais** (Posée en général sur une pelouse, cette dalle est séparée des suivantes par une longueur de pas ; ▶pas japonais en dalle) ; *syn.* dalle [f] pour pas japonais ; *g* **Schrittplatte** [f] (Meist in Rasen oder Pflanzbeeten in Schrittlänge verlegte Platte).

6027 stepping stone path [n] *constr.* (Series of slabs spaced apart for walking on lawns or in wide planting borders; ►stepping stone); *s* **caminito [m] de losas espaciadas** [►losa para pasos]; *f* **pas [m] japonais** (Cheminement de jardin discontinu, régulier ou irrégulier, souvent établi sur une pelouse, composé de dalles posées au sol, non solidaires entre elles, disposées légèrement en quinconce et dont l'espacement correspond à une longueur de pas — l'espacement des centres de dalle est de 65 cm ; ►dalle d'un pas japonais) ; *g* **Schrittplattenweg [m]** (Befestigter Pfad im Rasen oder in breiten Rabattenpflanzungen mit auf Lücke verlegten Platten; ►Schrittplatte).

step planting [n] *constr. for.* ►ridge planting, #2.

6028 step ramp [n] *constr.* (HALAC 1988-II, 225; ramp into which steps are introduced to reduce its steepness; ►ramp step); *s* **rampa [f] escalonada** (Conjunto de escalones construidos siguiendo una pendiente; DACO 1988; ►escalón de rampa); *f* **gradine [f]** (Escalier constitué d'une succession de marches de faible hauteur et très espacées les unes des autres, autrefois accessible aux chevaux, dont la pente générale est comprise entre 12 et 16 % ; ►marche d'une gradine) ; *syn.* escalier [m] à pas d'âne, escalier [m] à pas de mule ; *g* **Stufenrampe [f]** (Rampe oder steiler Weg, dessen Steilheit durch weit auseinander liegende Einzelstufen reduziert ist; bei kürzeren Auftrittstiefen spricht man von **Rampentreppe**; ►Rampenstufe).

step riser [n] *constr.* ►stair rise.

6029 step rod [n] [US] *constr.* (Part of a manhole ladder) ; *syn.* step iron [n] [UK] (SPON 1986, 171); *s* **peldaño [m] de hierro** (Pieza de una escalera en pozo de registro); *f* **pose-pied [m] en acier** (Échelons ou échelles en acier galvanisé ou en fonte fixés à la paroi des cheminées des regards de visite ; les échelons placés à la partie supérieure de la cheminée sont munis de trous permettant la mise en place d'une crosse) ; *syn.* échelon [m] ; *g* **Steigeisen [n]** (1. In die Wandung eines besteigbaren Kontrollschachtes eingelassene[r] Metallbügel als Kletterhilfe zum Hinab- und Heraussteigen. 2. An Fuß und Unterschenkel anschnallbare Kletterhilfe zum Besteigen von zu fällenden Bäumen, um diese in der Krone stückweise abzusetzen; der Gebrauch von **S.** wird notwendig, wenn der Einsatz eines Hubsteigers unmöglich ist).

6030 steps [npl] *arch. constr.* (Open air structure for change in walking levels; ►broad flight of steps, ►cantilever steps, ►construction of steps, ►foundation of a flight of steps, ►natural stone steps, ►sitting steps, ►stairway, ►stepped pathway; *syn.* outdoor steps [npl], flight [n] of steps; *s* **escalera [f]** (Construcción consistente en una serie de escalones que une dos niveles de diferente altura. Dependiendo de la localización dentro o fuera de edificios se diferencia entre **escalera interior** y **escalera exterior** o **escalinata**. En inglés el término *steps* denomina solamente una escalera exterior, si está dividida en tramos, cada uno de éstos se denomina *flight of steps*; ►camino escalonado, ►cimentación de escalera, ►construcción de escaleras, ►escalera rústica en piedra, ►escalera volante, ►gradería de asiento, ►tramo de escalera); *f* **escalier [m]** (Ouvrage constitué d'une suite de marches et reliant deux niveaux ; suivant leur localisation par rapport à un bâtiment on différencie l'**escalier intérieur** et l'**escalier extérieur** ; la première marche est dénommée **marche de départ** et la dernière marche de plain-pied avec le palier la **marche palière** ; *termes spécifiques* ►gradine, ►escalier suspendu, ►perron; ►construction d'escalier, ►escalier en gradins, ►escalier rustique, ►fondation d'un escalier, ►volée d'escalier) ; *g* **Treppe [f]** (Aufgang aus mehreren Stufen, der unterschiedlich hoch liegende Ebenen verbindet; im Gebäudeinnern als **Innentreppe**, im Freien in Verbindung mit Gebäuden als **Außentreppe** oder ►**Freitreppe**. Die erste Stufe einer Treppe heißt **Antrittsstufe**, die letzte **Austrittsstufe**; im Englischen wird eine einzelne Treppe im Gelände oder am Gebäude als *steps* bezeichnet; gibt es mehrere „Treppenpakete" hintereinander, so heißen die einzelnen Einheiten *flight of steps*; ►freitragende Treppe, ►Natursteintreppe, ►Sitztreppe, ►Treppenbau, ►Treppengründung, ►Treppenlauf, ►Treppenweg).

step slabs on risers [loc] *arch. constr.* ►flight of step slabs on risers.

6031 step strip foundation [n] *constr.* (Footing at the lower and upper end of a flight of steps); *s* **cimiento [m] de escalera** *syn.* fundación [f] de escalera; *f* **fondation [f] d'ancrage** (Fondation en tête et en pied d'escalier) ; *g* **Streifenfundament [n] (1)** (Gründungsstreifen am unteren und oberen Ende einer Treppe).

6032 steps [npl] **with adjacent soil** *constr.* (Flight of steps without cheek walls in which the sloping ground at the sides of the steps is graded to meet either the front or the back of each tread); *s* **escalera [f] incrustada en pendiente** (Escalera sin zancas incrustada en talud); *f* **escalier [m] sur talus saillant** (Escalier posé sur un talus, dont la ligne de pente passe par les pieds des contremarches) ; *g* **Treppe [f] mit seitlichem Erdanschluss** (Wangenlose Treppe, bei der der seitliche Erdanschluss über die vordere oder über die hintere Auftrittsflächenbegrenzung geführt werden kann).

6033 stewardship [n] *conserv.* (Responsibility of a citizen or group of citizens who take care of public trees, streams, etc. in cooperation with a public authority/agency; *context* to adopt a shade tree [US]; ►stewardship of trees [US]/tree adoption of trees [UK], ►stream stewardship [US]); *s* **apadrinamiento [m]** (Responsabilidad tomada por parte de persona[s] de cuidar árboles públicos, arroyos, etc. en cooperación con la administración competente; ►apadrinamiento de árboles, ►apadrinamiento de arroyos); *f* **parrainage [m]** (Prise en charge par une personne ou un groupe de personnes de l'entretien d'arbres, de massifs de fleurs, de ruisseaux etc. sur le domaine public ; ►parrainage d'un ruisseau, ►parrainage des arbres) ; *g* **Patenschaft [f]** (Mitverantwortung eines Bürgers oder einer Gruppe in Form einer ehrenamtlichen Betreuung von öffentlichen Bäumen und Pflanzbeeten, von Kinderspielplätzen, Bächen etc.; Patenschaft übernehmen [vb]; ►Bachpatenschaft, ►Baumpatenschaft).

6034 stewardship [n] **of trees** [US] *conserv.* (Feeling of responsibility by one or more citizens to take care of street trees on a public site out of civic pride; ►voluntary caretaker of a tree); *syn.* tree adoption [n] [UK]; *s* **apadrinamiento [m] de árboles (≠)** (Responsabilidad tomada por parte de persona[s] de cuidar uno o más árboles de la calle en la que vive[n]; ►padrino/madrina voluntario/a de un árbol); *f* **parrainage [m] des arbres** (Politique municipale menée avec la collaboration du service des espaces verts tendant à confier le soin des arbres des espaces publics aux habitants d'une commune ; ►parrain d'un arbre) ; *g* **Baumpatenschaft [f]** (Ehrenamtliche Mitverantwortung einer oder mehrerer Privatpersonen [►Baumpate/Baumpatin] für das Wohlergehen eines oder mehrerer Bäume im öffentlichen Raum).

6035 stilling basin [n] *wat'man.* (Pool constructed at the foot of a dam to prevent undercutting of the structure and abutting banks); *syn.* stilling pool [n] [also US]; *s* **tanque [m] amortiguador** (MESU 1977; estanque construido aguas abajo para reducir el efecto de socavación y erosión de la caída del agua de una presa); *f* **bassin [m] de tranquillisation** (Bassin construit à la base d'ouvrages de rétention des eaux afin de réduire l'énergie des eaux déferlantes et de protéger le pied de l'ouvrage ou les enrochements des berges contre le battement des vagues) ; *syn.* bassin [m] de dissipation d'énergie (ARI 1986, 58), bassin [m] d'amortissement (WIB 1996), fosse [f] de dissipation d'énergie ;

S

g **Tosbecken [m]** (Am Fuße von Stauanlagen gebautes Becken zum Abbau der Bewegungsenergie des stürzenden oder schießenden Wassers, um Unterspülungen von Bauwerk und Uferbefestigung zu verhindern).

stilling pool [n] [US] *wat'man.* ►stilling basin.

6036 stilt root [n] *arb. bot.* (Adventitious aerial root that develops from the butt of a tree above ground level and supports the stem); *syn.* brace root [n], prop root [n] (SAF 1983, 258); *s* **raíz [f] fúlcrea** (Tipo de raíz epigea que brota de la parte inferior del tronco y sirve de sostén a la planta, a modo de fulcro, como las raíces del género *Pandanus*; cf. DB 1985); *f* **racine-échasse [f]** (Racine adventive jouant le rôle de support du tronc, p. ex. du Palétuvier dans la mangrove ou du maïs) ; *syn.* racine [f] en chasse ; *g* **Stelzwurzel [f]** (Am unteren Stammteil entspringende Luftwurzel, die zum Stützen des Stammes dient; BOR 1958).

stock [n] *for.* ►growing stock; *hort.* ►container-grown stock; *zool.* ►fish stock; *agr.* ►livestock; *constr.* ►machinery stock; *hort.* ►nursery stock; *constr. hort.* ►quality of nursery stock; *bot. constr.* ►rootstock [US]/root-stock [UK]; *hort.* ►understock.

stock [n] [UK]**, advanced nursery-grown** *hort. plant.* ►specimen tree/shrub.

stock [n] [US]**, bare-rooted** *hort. plant.* ►bare-rooted plant.

stock [n]**, large nursery** *hort. plant.* ►specimen tree/shrub.

stock [n] [US]**, nursery-bred** *hort.* ►nursery stock.

stock [n] [US]**, nursery-grown** *hort.* ►nursery stock.

stock [n] [US]**, planting** *constr. for. hort.* ►planting material.

stock [n]**, plant species** *phyt.* ►existing area vegetation.

stock [n] [UK]**, poor quality housing** *urb.* ►poor quality housing [US].

stock [n]**, root-** *hort.* ►understock.

stock [n]**, root-balled** *constr. hort.* ►root-balled plant.

stock [n]**, specimen nursery** *hort. plant.* ►specimen tree/shrub.

6037 stockade fence [n] [US] *constr.* (Fence constructed of abutting pointed and split wooden stakes intended to provide privacy; ►picket fence, ►visual screen with vertical boards); *s* **valla [f] de estacas partidas** (Valla de rollizos de madera partidos longitudinalmente por la mitad; ►valla de listones verticales, ►valla de protección visual de tablones verticales); *f* **clôture [f] en échalas** (Clôture constituée de lattes éclatées en demi-rond, espacées ou fixées côte à côte sur des ►barreaux ; ►clôture à lattes, ►clôture [en bois] non jointive à lattes verticales) ; *g 1* **Sichtschutzholzzaun [m] aus Latten/Staketen** (Gartenzaun aus gesägten Hölzern/Latten, die ohne Lücke auf Querhölzern [►Riegel] befestigt sind; ►Lattenzaun); *syn.* Sichtschutz-Staketenzaun [m] (aus gesägten Hölzern); *g 2* **Spaltzaun [m]** (Zaun mit unterschiedlich breiten Latten, die mit Beil und Keil längs gespalten wurden; ►Sichtschutzwand mit senkrechter Lattung); *syn.* Sichtschutzspaltzaun [m].

stockbroker belt [n] [US] *recr. urb.* ►origination area [US].

stocking density [n] *agr.* ►increase in stocking density.

6038 stocking level [n] *agr. game'man.* (Existing population of game or domestic animals expressed as numbers living off a given area; *specific terms* high/low stocking level; ►animal population, ►game population, ►overstocking, ►restocking of fish); *s* **carga [f] animal** (Cabezas de ganado existentes en relación a la superficie de pastoreo a disposición; *términos específicos* carga animal alta, carga animal baja; ►población animal, ►población cinegética, ►repoblación piscícola, ►sobrecarga de ganado, ►suelta de peces); *f* **densité [f] d'animaux** (*Termes*

spécifiques forte densité, faible densité ; ►alevinage, ►peuplement animal, ►population cynégétique, ►réempoissonnement, ►surcharge pastorale) ; *g* **Besatz [m] (2)** (Zahl der vorhandenen Tiere in Bezug auf die zur Verfügung stehende Weide-/Äsungsfläche; *UBe* starker Besatz, geringer Besatz; ►Fischeinsatz, ►Tierbesatz, ►Überbesatz, ►Wildbesatz).

6039 stocking [n] **with conifers** *for.* (Planting of conifers in an area; ►natural colonization by seed rain, ►natural seeding of conifers; ►seed rain); *syn.* planting [n] of conifers; *s* **repoblación [f] de coníferas** (Colonización natural o plantación de coníferas en un área talada que anteriormente estaba poblada de ellas; ►colonización natural de «espacio vacío» por nuevo vuelo, ►nuevo vuelo, ►siembra natural de coníferas); *f* **enrésinement [m] (naturel ou artificiel)** (Semis naturel, ensemencement artificiel ou plantation de résineux sur une station qui en était exempt ; ►ensemencement naturel des conifères, ►flux de semences, ►semis naturel par flux de semences) ; *g* **Bestockung [f] mit Nadelgehölzen** (Natürliche Ansaat oder Anpflanzung von Nadelgehölzen auf einem Standort, der vorher nicht mit Koniferen bewachsen war; ►Anflug 1 u. ►Anflug 2, ►natürliche Versamung von Koniferen).

6040 stockpile area [n] *constr.* (Place where stripped topsoil is kept during construction; ►dumpsite [US]/tipping site [UK], ►storage of topsoil, ►topsoil pile [US]/topsoil heap [UK]); *s* **área [f] de acopio de tierra vegetal** (Lugar de almacenamiento provisional de la tierra durante la obra; ►acopio de tierra vegetal, ►depósito de residuos sólidos, ►depósito de tierra vegetal); *f* **lieu [m] de dépôt de la terre végétale** (Emplacement de stockage pour une utilisation ultérieure pendant la durée des travaux ; ►décharge publique, ►stockage de la terre végétale, ►dépôt de terre végétale) ; *syn.* aire [f] de dépôt de la terre végétale, aire [f] de stockage provisoire de terre végétale ; *g* **Lagerplatz [m] für Oberboden** (Zwischenlagerungsort für Mutterboden während der Bauzeit; ►Deponie, ►Oberbodenlagerung, ►Oberbodenmiete); *syn.* Lagerfläche [f] für Oberboden, Oberbodenlagerplatz [m].

stockpiled materials [npl]**, reinstallation of** *constr.* ►reuse of on-site construction material.

stockpiling [n] **of topsoil** [US] *constr.* ►storage of topsoil.

6041 stock plant [n] [US] *hort.* (Plant from which young plants may be propagated vegetatively); *syn.* parent plant [n] [UK]; *s* **planta [f] madre** (Planta que sirve para la reproducción vegetativa); *f* **plante [f] mère** (Végétal utilisé pour la multiplication végétative ; LA 1981, 764) ; *syn.* pied [m] mère (LA 1981, 74, 389) ; *g* **Mutterpflanze [f]** (Pflanze, die der vegetativen Vermehrung dient).

stock ratio [n]**, increase in** *agr.* ►increase in stocking density.

6042 stolon [n] *bot.* (Horizontal stem or branch growing above ground which takes root at nodes and forms new plants; ►stem-spreading, ►stoloniferous); *syn.* runner [n]; *s* **estolón [m]** (Vástago rastrero que nace de la base del tallo y echa a trechos raíces que producen nuevas plantas; DINA 1987; ►estolonífero/a 1, ►estolonífero/a 2); *f* **stolon [m]** (Tige provenant d'un bourgeon axillaire, qui croît couché sur le sol et s'enracine en produisant de nouveaux individus ; ►stolonifère, ►traçant, ante) ; *g* **Stolo(n) [m]** (Stolonen [*meist* pl]; oberirdischer Ausläufer mit stark verlängerten Internodien, der sich in einiger Entfernung von der Mutterpflanze bewurzelt und eine neue Pflanze bildet; ►Ausläufer treibend, ►Stolonen bildend).

6043 stoloniferous [adj] *bot. hort.* (►stolon); *s* **estolonífero/a** [adj] **(2)** (►estolón); *f* **stolonifère** [adj] (►stolon) ; *g* **Stolonen [mpl] bildend** *bot. hort.* (►Stolo[n]).

6044 stolonizing [n] *constr.* (Spreading and planting of shredded grass stolons or runners over a moist, prepared lawn bed; method often employed in semiarid [US]/semi-arid [UK] and arid regions on large areas where irrigation is readily available, such as golf courses, polo fields and parklands; ▶sprigging 2); *s* **establecimiento** [m] **de césped con estolones** (Método de creación de superficies herbáceas utilizado en regiones áridas y semiáridas, en el que se esparcen e introducen mecánicamente en la tierra estolones de hierbas triturados; se aplica en áreas extensas con posibilidad de irrigación como en campos de golf, de polo o en parques; ▶siembra con estolones o rizomas triturados); *f* **engazonnement** [m] **par jet de stolons de graminées** (Méthode de multiplication végétative des graminées en particulier utilisée dans les régions arides et semi-arides dans les quelles les semis germent difficilement, p. ex. l'herbe des Bermudes [*Cýnodon dactylon*] et le gazon de Mascareignes [*Zoysia*] dont les stolons et rhizomes sont semés par jet et ensuite et enfouis machinellement dans le sol ; cette méthode est particulièrement adaptée pour augmenter la production fourragère en prairie permanente ou pour l'aménagement de terrains de sport [terrains de polo, de golf] ou de parcs lorsque peut être assuré un minimum d'arrosage ; ▶engazonnement par repiquage de stolons de graminées) ; *syn.* enherbement [m] par jet de stolons de graminées ; *g* **Begrünung** [f] **mit Grasstolonen** (Begrünungsmethode in ariden und semiariden Gebieten, in denen Saaten sehr schlecht keimen, bei der zur Rasenflächenherstellung zerteilte Grasstolonen, z. B. vom Bermudagras [*Cýnodon dactylon*] und Zoysiagras [*Zoysia*], breitwürfig ausgebracht und in den Boden eingearbeitet werden. Diese Methode eignet sich gut für große Flächen, für die eine regelmäßige Bewässerung gesichert ist, wie z. B. für Golfplätze, Polosportplätze und große Parkanlagen; ▶Flächenbegrünung mit zerteilten Grasstolonen oder Rhizomen); *syn.* Anlegen [n, o. Pl.] von Rasenflächen mit Grasstolonen.

stone [n] *constr.* ▶cobblestone [US]/natural stone sett [UK], ▶curbstone [US]/kerbstone [UK]; *arch. constr.* ▶cut stone; ▶edging stone; *adm. conserv'hist.* ▶gravestone; *constr.* ▶large-sized paving stone [US]/large-sized concrete paving sett [UK]; *geo. pedol.* ▶limestone, ▶marly limestone; *constr.* ▶natural stone, ▶natural stone flag, ▶paving stone [US]/paving sett [UK], ▶quarry flagstone, ▶ragstone; *geo.* ▶sandstone; *constr.* ▶sawn flagstone, ▶small paving stone [US]/small paving sett [UK], ▶stepping stone.

stone [n] **[UK], concrete paving** *constr.* ▶concrete paver.

stone [n], **crushed** *constr.* ▶crushed rock.

stone [n], **dressed flag-** *constr.* ▶flag cut to shape.

stone [n] **[UK], flat undressed** *constr.* ▶ragstone [US].

stone [n] **[US], jumbo cobble-** *constr.* ▶large-sized paving stone [US]/large-sized concrete paving sett [UK].

stone [n], **large-sized natural paving** *constr.* ▶large-sized paving stone [US].

stone [n], **marl-** *geo. pedol.* ▶marl.

stone [n], **precut paving** *constr.* ▶natural stone flag.

stone [n], **tomb-** *adm. conserv'hist.* ▶gravestone.

stone [n], **tooled natural** *arch. constr.* ▶cut stone, #2.

stone [n], **undressed** *constr.* ▶rubble.

stone block [n], **concrete** *constr.* ▶cut stone, #3.

stone chippings [n] **[UK]** *constr.* ▶fine aggregate [US].

stone collecting [n] **[US]** *constr. hort.* ▶stone picking.

6045 stone construction [n] *constr.* (Generic term for wall and step construction as well as paving in landscape construction; ▶hardscape); *s* **obras** [fpl] **de fábrica y de revestimiento de suelos** (En construcción paisajística, cualquiera de los trabajos relacionados con la mampostería y albañilería; ▶componentes duros de espacios libres); *f* **maçonnerie** [f] **paysagère** (Terme générique pour les travaux d'ouvrages techniques comme les murets, les emmarchements, dallages, etc. ; ▶éléments inertes d'un espace libre) ; *g* **Steinbau** [m] **im Garten- und Landschaftsbau** (OB zu Mauerbau, Treppenbau und Erstellung von Pflastersteinbelägen im Garten- und Landschaftsbau; ▶Technisch-Bauliches einer Freianlage).

stone course [n] *constr.* ▶stone layer.

stone crumbling [n] *envir.* ▶stone decay.

6046 stone decay [n] *envir.* (Desintegration of stone in buildings caused by gaseous emissions or mechanical forces [frost action]; ▶weathering); *syn.* stone weathering [n], stone crumbling [n]; *s* **desagregación** [f] **de la piedra** (Deterioro de la piedra u otros materiales inertes de construcción o edificio causada por la contaminación o por procesos de helamiento; ▶meteorización); *syn.* descomposición [f] de la piedra; *f* **désagrégation** [f] **de la pierre** (Fragmentation provoquée par une action mécanique [gel] ou des émissions de gaz entraînant la dégradation ou la destruction des pierres sur les bâtiments ; ▶météorisation) ; *g* **Steinzerfall** [m] (Durch Abgase und mechanische Sprengung [Frost] verursachter Schaden resp. bedingte Zerstörung von Steinen an Bauwerken; ▶Verwitterung).

stone flag [n] *constr.* ▶natural stone flag, ▶natural garden stone flag.

stone formation [n] *geo. pedol.* ▶exposed stone formation.

6047 stone layer [n] *constr.* (Row or course of stones in a wall; ▶header course, ▶stretcher course); *syn.* stone course [n]; *s* **hilada** [f] (Conjunto de mampuestos que forman una fila horizontal de un muro o tabique; DACO 1988; ▶hilada de sogas, ▶hilada de tizones); *f* **assise** [f] (Rang des pierres, briques, etc. posées horizontalement et sensiblement de même hauteur ; DTB 1985 ; ▶assise de panneresse, ▶lit de boutisse) ; *syn.* cours [m] d'assise ; *g* **Steinschicht** [f] (Reihe oder Lage von Steinen in einer Mauer; ▶Binderschicht, ▶Läuferschicht); *syn.* Steinlage [f].

stone path [n] *constr.* ▶stepping stone path.

stone pavement [n] *constr.* ▶paved stone surface.

stone pavement [n] **[US], natural** *constr.* ▶natural stone paving.

stone paving [n] *constr.* ▶natural stone paving.

6048 stone picking [n] *constr. hort.* (Removal of stones after ▶topsoil spreading, and before sowing); *syn.* stone collecting [n] [also US]; *s* **despeje** [m] **de piedras** (Remoción de las piedras después de la ▶extensión de tierra vegetal y antes de la siembra); *f* **épierrage** [m] (Enlèvement des pierres après la ▶mise en place de la terre végétale, avant le semis ; *contexte description des travaux d'espaces verts*, épierrage fin et complet des granulats de taille supérieure à 5 cm, épierrage des éléments > 5 cm de diamètre) ; *syn.* épierrement [m] ; *g* **Entfernen** [n, o. Pl.] **von Steinen** (Aufsammeln/Auflesen von Steinen nach dem Oberbodeneinbau, vor der Einsaat; ▶Andecken von Oberboden).

stone surface [n] *constr.* ▶paved stone surface.

6049 stone tooling wastage [n] *constr.* (Loss of stone material, e.g. during cutting to size of ▶cut stones or ▶paving slabs); *s* **pérdida** [f] **de material pétreo** (Residuos de mampuestos, ▶sillares o ▶losas por rotura al cortarlos a medida); *f* **chutes** [fpl] **de taille** (Déchets produits lors du débit des blocs bruts de ▶pierres de taille afin de les ajuster à l'ouvrage requis ; *terme générique* **chutes** ; *termes spécifiques* **chutes de sciage** [lors du sciage des pierres, des ▶dalles ; *syn.* sciage [m] perdu], **perte de coupe** [p. ex. perte de dalles lors des coupes et découpes pour un calepinage d'appareil irrégulier]) ; *syn.* perte [f] de taille ; *g* **Verschlag [m, o. Pl.]** (Verlust von Steinmaterial, z. B. beim Zuschlagen von ▶Werksteinen oder ▶Platten; Verlust/Abfall durch Schneiden von Platten oder Werksteinen wird **Verschnitt** genannt); *syn.* Materialverschlag [m].

stoneware pipe [n] [UK]**, glazed** *constr.* ▶vitrified clay pipe [US].

stone weathering [n] *envir.* ▶stone decay.

stonework [n] *arch. constr.* ▶masonry construction.

6050 stool [n] *constr. for. hort.* (Living tree stump, capable of producing sprouts for regeneration [▶coppice shoot] or cuttings, layers, etc.); *s* **cepa** [f] (Parte del tallo de una planta, inmediatamente debajo de la superficie del suelo, desde la que se producen nuevos tallos; DINA 1987; ▶brote de cepa); *f* **souche** [f] **vivante** (Partie souterraine des végétaux ligneux à la base de laquelle se forment les rejets ; ▶rejet de taillis) ; *g* **ausschlagfähiger Stubben** [m] (Der nach dem Fällen eines Baumes im Boden verbleibende Wurzelstock, der neue Triebe bilden kann; ▶Stockaustrieb); *syn.* ausschlagfähiger Baumstumpf [m], ausschlagfähiger Stock [m], ausschlagfähiger Wurzelstock [m].

stool shoot [n] *arb.* ▶coppice shoot.

stopgap report [n] [US] *plan.* ▶provisional report.

6051 stopover bird [n] *zool.* (Bird, which uses an area [stopover site] for a few days or more as a temporary resting place during its migration; ▶bird of passage, ▶visitor bird); *s* **ave** [f] **en etapa migratoria** (Especie de ave que utiliza un territorio de forma temporal para descansar y reponer fuerzas, que se encuentra de camino en su desplazamiento hacia el lugar final de la ruta migratoria, pero donde no cría; ▶ave de paso, ▶ave visitante); *f* **oiseau** [m] **en étape migratoire** (Oiseau qui effectue des déplacements périodiques, généralement saisonniers ; par rapport à une région considérée, espèce qui y fait halte au cours de sa migration sans s'y reproduire ; ▶oiseau de passage, ▶oiseau visiteur) ; *syn.* oiseau [m] en halte migratoire , *g* **Rastvogel** [m] (Vogel, der auf seinem Zug ein Gebiet zur zwischenzeitlichen Rast aufsucht; ▶Durchzügler, ▶Gastvogel).

stopped traffic [n] *plan. trans.* *▶stationary traffic.

stopping area [n] [UK] *recr.* ▶stopping place [US].

6052 stopping place [n] [US] *recr.* (Relatively small space along highways or roads, on the edge of a forest or in a recreation area providing opportunity for travellers to halt or leave a vehicle, to relax, enjoy scenic views, read historical markers or eat; ▶pull-off [US]/lay-by [UK]); *syn.* rest spot [n] [UK] (3), roadside rest area [n] [UK], stopping area [n] [UK], stopping point [n] [UK], wayside area [n] [US]; *s* **área** [f] **de parada (en zona natural)** (Pequeño espacio para estacionar vehículos a lo largo de carreteras, al borde de un bosque o una zona de recreación; ▶área de parada [sin servicios]); *f* **aire** [f] **de stationnement en bordure de route** (Aire aménagée pour le stationnement des véhicules en bordure d'une route, en forêt ou dans une zone de loisirs ; ▶aire d'arrêt, ▶aire de repos 1) ; *syn.* aire [f] d'arrêt en bordure de route, aire [f] d'accueil en bordure de route, parking [m] en bordure de route ; *g* **Halteplatz** [m] (Platz am Rande eines Waldes oder in einem Erholungsgebiet, um zu halten, auszusteigen, die Aussicht zu genießen, Hinweistafeln zu lesen oder zu picknicken; ▶unbewirtschaftete Rastanlage).

stopping point [n] [UK] *recr.* ▶stopping place [US].

storage [n] *envir.* ▶long-lasting storage; *hydr.* ▶recharged groundwater storage; *constr.* ▶soil storage; *urb.* ▶trash storage [US]/bin store [UK].

storage [n]**, groundwater** *hydr.* ▶groundwater reservoir.

6053 storage capacity [n] *pedol.* (Capacity of a soil to retain nutrients and water necessary for plant growth; ▶waterholding capacity); *s* **capacidad** [f] **de retención** (Característica del suelo de almacenar agua y nutrientes para ponerlos a disposición de las plantas; ▶capacidad de retención de agua); *f* **capacité** [f] **de rétention** (Quantité d'eau et de substances nutritives qu'un sol peut absorber et restituer dans des conditions de pH bien définies ; ▶capacité de rétention [maximale] pour l'eau) ; *syn.* pouvoir [m] de rétention , *g* **Speicherfähigkeit** [f] (Eigenschaft des Bodens, Wasser und Nährstoffe für Pflanzenwachstum vorhalten zu können; ▶Wasseraufnahmefähigkeit); *syn.* Speichervermögen [n].

storage [n] **of recyclable materials** *constr.* ▶interim storage of recyclable materials.

6054 storage [n] **of topsoil** *constr.* (Interim deposit of stripped topsoil at a distance from construction operations until it can be respread/reused; ▶topsoil pile [US]/topsoil heap [UK] *generic term* ▶soil storage); *syn.* stacking [n] of topsoil [also UK], stockpiling [n] of topsoil [also US], storing [n] of topsoil; *s* **acopio** [m] **de tierra vegetal** (Almacenamiento provisional de tierra vegetal hasta su reutilización; ▶depósito de tierra vegetal; *término genérico* ▶acopio de tierra); *syn.* almacenamiento [m] provisional de tierra vegetal; *f* **stockage** [m] **de la terre végétale** (Mise en dépôt provisoire de la terre végétale décapée en début des travaux en vue d'une réutilisation ultérieure ; ▶dépôt de terre végétale ; *terme générique* ▶stockage [surplace] du sol) ; *syn.* mise [f] en dépôt de la terre végétale, constitution [f] d'un dépôt de terre végétale ; *g* **Oberbodenlagerung** [f] (**1.** Aufsetzen von Mutterbodenmieten zur Oberbodensicherung als Zwischenlager. **2.** Aufbewahrung des abgetragenen Oberbodens abseits des Baubetriebes bis zur Wiederverwendung; ▶Oberbodenmiete; *OB* ▶Bodenlagerung); *syn.* Aufsetzen [n, o. Pl.] von Oberbodenmieten, Herstellen [n, o. Pl.] von Oberbodenmieten.

storage space [n] *envir.* ▶intermediate storage space.

store [n]**, topsoil** *constr.* ▶topsoil pile [US].

stored automobiles [npl] [US] *trans.* ▶parked vehicles.

stored coppice [n] [UK] *for.* ▶coppice with standards.

storey [n] [UK] *leg. urb.* ▶story [US].

storing [n] **of topsoil** [UK] *constr.* ▶storage of topsoil.

6055 stork's nest growth [n] *envir. for.* (Typical deformation of the crown of fir trees *[Abies]* in old age or by disease due to the ceasing of top growth; ▶dieback of firs [US]/die-back of firs [UK]); *s* **nido** [m] **de cigüeña** (Conformación plana típica de la punta de la copa del abeto *[Abies]* causada por la reducción del crecimiento por la edad o por enfermedad; ▶muerte de abetos); *f* **nid** [m] **de cigogne** (Forme typique de la cime d'un Sapin adulte ou atteint par une maladie, présentant des zones d'interruption de croissance ; ▶dépérissement des Sapins) ; *g* **Storchennest** [n] (Typische, flache Ausformung der Kronenspitze einer Tanne *[Abies]* durch stockenden Höhenzuwachs im Alter oder bei Erkrankung; ▶Tannensterben).

6056 storm beach [n] [US]**/storm-beach** [n] [UK] *geo.* (Accumulation of shingle, cobbles and boulders, formed during exceptionally powerful storm-waves; ▶shore); *syn.* shingle beach [n], shingle bar [n] (GGT 1979), sea wall [also US] (2); *s* **cordón** [m] **playero** (Acumulación alargada de arena formada por la acción del oleaje; ▶playa [costera]); *syn.* resalte [m] playero; *f* **crête** [f] **de plage** (Bourrelet d'accumulation situé sur la partie la plus élevée de la plage avec une pente plus raide vers le large et édifié par les vagues qui y jettent les matériaux les plus grossiers ; ▶plage côtière) ; *g* **Strandwall** [m] (Durch Brandungstätigkeit aus Sand und Strandgeröllen aufgebauter niedriger Wall, der sich jahreszeitlich verlagern kann; ▶Meeresstrand).

6057 storm damage [n] *agr. for. hort.* (Physical destruction due to severe wind shear conditions that snap large limbs and causes extensive structural damage; term usually refers—in comparison to ▶wind damage—to a much more severe outcome, typified by wind velocities exceeding No. 9 on the BEAUFORT Scale—> 75 km/h [47.2 mph]; a **blasted tree** is one usually damaged in a heavy storm by blasts of wind or is struck by lightning; ▶windbreakage); *syn.* blast damage [n], wind blast [n] [UK]; *s* **daños [mpl] causados por tormenta** (Destrucción física considerable de plantas, edificios, infraestructuras y vehículos debida a vientos fuertes [fuerza del viento superior a 9 de la escala de BEAUFORT, de 13 grados (de 0 a 12) y dada la velocidad media del viento durante un mínimo de 10 minutos de 75-88 km/h]; ▶daños causados por el viento, ▶rotura por viento); *f* **dégâts [mpl] causés par la tempête** (Dommages mécaniques très importants provoqués par un vent très fort [vent de force > 9 de l'échelle de BEAUFORT, comportant 13 degrés [de 0 à 12] et donnant la vitesse moyenne du vent sur une durée de dix minutes — 75-88 km/h] et provoquant la destruction des bâtiments et de la végétation ; les deux tempêtes de décembre 1999 ont causé, en France, la destruction d'environ 500 000 hectares de forêts ; ▶bris de vent, ▶dommage causé par le vent) ; *syn.* destruction [f] par la tempête de vent, dégâts [mpl] occasionnés par la tempête ; *g* **Sturmschaden [m]** (...schäden [pl]; durch sehr heftigen, bis zur Orkanstärke aufkommender Wind [Windstärke > 9 der 12-stufigen BEAUFORT-Skala — > 75 km/h] hervorgerufene mechanische Beschädigungen resp. Zerstörungen an Pflanzen, Gebäuden, baulichen Anlagen und Fahrzeugen; ▶Windbruch, ▶Windschaden); *syn.* Orkanschaden [m].

storm drain [n] [UK] *urb. wat'man.* ▶storm sewer [US].

storm drainage service charge [n] [US] *adm. urb.* *∗*▶utility connection charge.

storm drainage system [n]**, underground** [US] *envir. urb.* ▶storm water drainage system.

6058 storm drain grate [n] [US] *constr.* (Cover to a catch basin inlet [US]/gutter inlet [UK], usually of cast iron; ▶grille cover); *syn.* drainage grate [n] [US], drain grate [n] [US], gully grating [n] [UK]; gutter grate [n] [US], inlet grate [n] [US]; *s* **rejilla [f] de entrada (de alcantarilla)** (▶rejilla de cubierta); *syn.* parrilla [f] de entrada, emparrillado [m] de entrada; *f* **grille [f] de regard d'évacuation** (Couvercle généralement en fonte pour regards d'évacuation des eaux pluviales, etc. ; *terme spécifique* grille-avaloir ; ▶grille) ; *syn.* couvercle [m] de regard ; *g* **Einlaufrost [m]** (Meist gusseiserne Abdeckung auf Straßen- und Hofabläufen; ▶Gitterrost).

storm flow [n] *constr. geo.* ▶direct runoff.

storm runoff [n] *constr. geo.* ▶direct runoff; *constr. plan.* ▶heavy storm runoff [US]/heavy rainwater run-off [UK].

storm runoff [n] [US]**, delaying** *envir. urb. wat'man.* ▶retention of rainwater.

6059 storm sewer [n] [US] *urb. wat'man.* (TGG 1984, 129; sewer used for conveying rainwater or other similar discharges, but not sewage or industrial waste, to a point of disposal or to a ▶receiving stream; DAC 1975; ▶sewer); *syn.* rain-water sewer [n] [UK], storm drain [n]; *s* **colector [m] de aguas pluviales** (▶alcantarilla [Es, RA, RCH, MEX]); *syn.* alcantarilla [f] para aguas pluviales; *f* **collecteur [m] d'évacuation d'eaux pluviales** (Canalisation d'évacuation des eaux superficielles urbaines vers une station d'épuration ou de rejet dans le ▶milieu récepteur ; ▶égout) ; *syn.* canalisation [f] d'eaux pluviales ; *g* **Regenwasserkanal [m]** (Kanal, der das städtische Oberflächenwasser in eine Kläranlage oder in einen ▶Vorfluter leitet; ▶Schmutzwasserkanal); *syn.* Tagwasserkanal [m].

storm water [n] *met. wat'man.* ▶rainwater (1).

storm water [n]**, heavy** [US] *met.* ▶heavy rainfall.

storm water curb opening [n] [US] *constr.* ▶curb inlet [US]/kerb-inlet [UK].

storm water detention [n] *envir. urb. wat'man.* ▶retention of rainwater.

6060 storm water detention basin [n] *hydr. wat'man.* (TGG 1984, 150; impoundment area constructed to collect storm runoff from a management system for the purpose of reducing peak flow and controlling rate of flow. A **retention basin** may be defined as having a permanent pool, while a **detention basin** is normally dry; ▶rainwater interceptor basin, ▶retention); *syn.* detention structure [n], retention structure [n], storm water management pond [n] [also US], storm water detention pond [n] (TGG 1984, 150); *s* **embalse [m] de regulación de aguas pluviales** (Embalse artificial construido para almacenar las aguas de lluvias fuertes o torrenciales y así evitar la descarga inmediata en el sistema de alcantarillado; ▶embalse de sobrecarga, ▶retención); *syn.* embalse [m] de retardo de aguas pluviales; *f* **bassin [m] de retenue d'eaux pluviales** (Bassin destiné à stocker les pointes d'orages en prévenance des zones amont en restituant les eaux pluviales par un débit compatible avec la capacité du réseau ou du sol ; ▶déversoir d'orage, ▶rétention) ; *syn.* bassin [m] de rétention d'eaux pluviales, bassin [m] d'accumulation des eaux pluviales ; *g* **Regenrückhaltebecken [n]** (*Abk.* RRB; Becken zur vorübergehenden Aufnahme von Starkregenwassermengen, um Abflussspitzen der Siedlungsentwässerung zu reduzieren; ▶Regenüberlaufbecken, ▶Retention); *syn.* Regenwasserrückhaltebecken [n].

storm water detention pond [n] *hydr. wat'man.* ▶storm water detention basin.

storm water drainage [n] [US] *constr. urb. wat' man.* ▶surface water drainage.

6061 storm water drainage system [n] *envir. urb.* (TGG 1984, 130; network containing all the components [catch basin inlet, conduits, outfall pipes, etc.] that convey storm water runoff into ▶receiving streams or lakes and, in combined systems, into sewage treatment plants); *syn.* underground storm water system [n], underground storm drainage system [n] [also US]; *s* **alcantarillado [m] (separativo) para aguas pluviales** *syn.* sistema [m] de drenaje de aguas pluviales; *f* **réseau [m] d'évacuation des eaux pluviales** (Ensemble des canalisations d'évacuation des eaux superficielles urbaines vers une station d'épuration ou de rejet dans le ▶milieu récepteur) ; *g* **Regenwasserkanalisation [f]** (Kanalnetz zur Ableitung des städtischen Oberflächenwassers in eine Kläranlage oder in einen ▶Vorfluter).

storm water management pond [n] [US] *hydr. wat'man.* ▶storm water detention basin.

storm water runoff [n] *constr. geo.* ▶direct runoff.

storm water runoff management [n] *ecol. urb.* ▶near-natural storm water runoff management.

6062 storm water sewage [n] *wat'man.* (Rainwater flowing in storm sewers [▶separate sewerage system] or in combined sewers—▶combined sewer system—to a point of disposal); *s* **aguas [fpl] pluviales vertidas** (Agua de precipitación que fluye al ▶sistema de alcantarillado separativo o al ▶sistema de alcantarillado unitario para ser conducidas en este caso a una depuradora); *f* **eaux [fpl] pluviales collectées** (Eaux de pluies provenant des précipitations naturelles recueillies par les toitures, les chaussées et les diverses surfaces stabilisées contenant une charge polluante dissoute ou non dissoute et évacuées vers la station d'épuration par un système de collecte un égout d'eaux pluviales

S

[▶réseau d'assainissement séparatif] ou dans un égout collecteur [▶réseau (d'assainissement) unitaire]) ; **g abgeführtes Regenwasser [n, o. Pl.]** (Niederschlagswasser, das i. d. R. gelöste und ungelöste Verunreinigungen von Dächern, Straßen und sonstigen befestigten Flächen mitführt und in einem Regenwasserkanal [▶Trennverfahren] oder in einem Schmutzwasserkanal meist einer Kläranlage zugeleitet wird [▶Mischverfahren]); *syn.* eingeleitetes Niederschlagswasser [n].

storm water system [n], underground *envir. urb.*
▶storm water drainage system.

6063 story [n] [US] *leg. urb.* (That portion of a building included between the upper surface of any floor and the upper surface of the floor next above, except that the topmost stor[e]y is that portion of a building included between the upper surface of the topmost floor and the ceiling or roof above. If the finished floor level directly above a basement or unused under-floor space is more than 6 feet [1.8m] above grade as defined herein for more than 50% of the total perimeter or is more than 12 feet [3.6m] above grade as defined herein at any point, such basement or unused under-floor space shall be considered as a stor[e]y; UBC 1979, 45; ▶permitted number of stories [US]/permitted number of storeys [UK]); *syn.* floor [n], storey [n] [UK]; **s planta [f] completa** (Planta de un edificio que se encuentra sobre nivel y cuya altura cumple los requisitos legales mínimos de altura libre para servir de vivienda; ▶número de plantas completas); *syn.* piso [m] completo; **f niveau [m] d'une construction (±)** (Surface de plancher prise en compte dans le calcul de la surface hors œuvre nette [SHON] d'une construction, en projet ou existante, à l'exception des surfaces de plancher de combles dont la hauteur sous toiture [calcul à partir de la sous face de la toiture] est inférieure à 1,80 mètre et les surfaces de plancher de sous-sols dont la hauteur sous plafond est inférieure à 1,80 mètre [faux plafonds non compris] ; les surfaces de plancher non aménageables pour l'habitation ou d'autres activités en raison de leur usage [locaux techniques affectés au fonctionnement technique, comme les chaufferies, les machineries d'ascenseur, les pièces de stockage des ordures ménagères], les surfaces de plancher des caves sans ouvertures sur l'extérieur [aération exceptée], les surfaces de plancher non aménageables en raison de leur impossibilité à supporter des charges liées à des usages d'habitation ou d'activité [particularité devant être clairement exprimée sur les plans et les coupes du dossier de permis de construire], les surfaces de plancher non aménageables en raison de l'encombrement de la charpente [certaines charpentes anciennes, fermettes, etc.], les surfaces de plancher non closes, au niveau des rez-de-chaussée — les surfaces de balcons [surfaces de coursives et de bow-windows non déductibles], les surfaces des loggias non closes ; — les surfaces des toitures-terrasses, les surfaces de plancher affectées au stationnement des véhicules [automobiles, caravanes, remorques, deux-roues, voitures de personnes à mobilité réduite] ; **D.,** plan situé en totalité au-dessus du sol et sur lequel la hauteur des pièces habitables atteint la hauteur libre sur les deux tiers de la surface d'implantation de la construction ; ▶nombre des niveaux de la construction ; **F.,** dans les règles d'urbanisme on utilise le terme de « surface de plancher ») ; **g Vollgeschoss [n]** (Geschoss, das vollständig über der festgelegten Geländeoberfläche liegt und über mindestens zwei Drittel seiner Grundfläche die für Aufenthaltsräume notwendige lichte Höhe hat. In D. richtet sich die Definition des **V.es** nach dem jeweiligen Landesrecht. In F. wird das Maß der baulichen Nutzung nicht durch **V.e** definiert, sondern nur durch GRZ und GFZ; ▶Zahl der Vollgeschosse).

6064 straggler [n] *zool.* (Individual of a bird species starting its migratory journey later than the rest of the population); **s ave [f] migratoria tardía** (Ave que emigra más tarde que los otros

individuos de su especie); **f migrateur [m] tardif** (Oiseau effectuant ses déplacements réguliers annuels plus tard que les autres membres de l'espèce) ; **g Nachzügler [m]** (Vogel, der sein jahreszeitlich bedingtes Fortziehen später als der Rest der Population beginnt).

straightening [n] *eng. wat'man.* ▶river straightening; *eng.* ▶street straightening; *wat'man.* ▶stream straightening.

straightening [n]**, road** *eng.* ▶street straightening.

6065 straightening [n] of watercourses *eng. wat'man.* (Reduction or removal of bends in watercourses for faster flood water discharge and for better navigability, by shortening the channel, avoiding flooding, and making urban development as well as agricultural use possible on alluvial plains; ▶river engineering measures, ▶stream straightening); **s rectificación [f] (de ríos y arroyos)** (Medidas de construcción en ríos y arroyos para evitar inundaciones y poder aprovechar las vegas para usos agrícolas y la edificación, para acelerar el flujo del agua y mejorar la navegabilidad; ▶ingeniería hidráulica en ríos y arroyos, ▶rectificación de arroyos); *syn.* encauzamiento [m] de ríos y arroyos; **f redressement [m] d'un cours d'eau** (Travaux hydrauliques sur les cours d'eau dont le but est d'aménager ceux-ci comme voie d'eau ou d'augmenter dans les plaines alluviales l'occupation des sols à des fins agricoles ou pour une urbanisation future, par réduction des risques d'inondation en augmentant la vitesse de l'écoulement ou par la diminution de la longueur de certains tronçons, le scindement des méandres ou l'endigage ; ▶recalibrage d'une rivière/d'un ruisseau, ▶travaux d'aménagement des cours d'eau) ; *syn.* rectification [f] du tracé d'un cours d'eau ; **g Fließgewässerbegradigung [f]** (Baumaßnahmen an Flüssen oder Bächen zur Vermeidung von Überschwemmungen und zur zusätzlichen Nutzung der Auenbereiche als Siedlungs- und landwirtschaftliche Nutzfläche, zur Beschleunigung des Hochwasserabflusses durch Reduzierung der Fließgewässerstrecke, indem Mäander abgeschnitten und Deiche gebaut werden und zur Verbesserung der Nutzbarkeit als Wasserstraße; ▶Bachbegradigung, ▶Gewässerausbau); *syn.* Begradigung [f] von Fließgewässern, Gewässerkorrektion [f] [CH].

6066 strainer [n] *constr.* (Sieve installed over an ▶overflow pipe opening in pools and fountains for the removal of floating litter/for debris collection); *syn.* trash rack [n] [also US] (WRP 1974, 94); dome grate [n] [also US] (OSM 1999, 169); **s rejilla [f] de cribar** (Dispositivo de metal que permite el paso del agua colocado en el orificio de salida de tubería en piscinas o surtidores para eliminar basura flotante); *syn.* cesta [f] de cribado; **f 1 crapaudine [f] à emboîtement** (Filtre grossier placé en tête de l'orifice supérieur d'un tuyau de descente d'eaux pluviales ou d'une ▶bonde de surverse ; cf. DTB 1985) ; **f 2 crépine [f]** (Tôle perforée servant à arrêter les corps étrangers placée devant l'ouverture d'un tuyau ou d'une ▶bonde de surverse ; cf. PR 1987) ; **g Siebkorb [m]** (Wasserdurchlässige Metallvorrichtung auf einem Standrohr [▶Anstaurohr] in einem Wasserbecken oder an einem Fallrohr zum Abfangen von Schwimmgut).

strand [n] *geo.* ▶beach (1).

strandline community [n] *phyt.* ▶driftline community.

6067 strange species [n] *phyt.* (Species that is a rare and accidental intruder from another plant community or relict of a preceding community; ▶accidental, ▶degree of fidelity, ▶relict species); **s especie [f] extraña** (Especie procedente de otras comunidades o reliquias de otras comunidades que habían ocupado el mismo lugar; BB 1979; ▶accidental, ▶especie relíctica, ▶grado de fidelidad); **f espèce [f] étrangère** (Espèce provenant d'un ensemencement de hasard ou constituant une relique d'un stade évolutif inférieur n'apparaissant qu'accidentellement dans un groupement déterminé ; ▶degré de fidélité, ▶espèce acciden-

telle, ▶espèce relictuelle) ; *g* **fremde Art [f]** (Art, die hinsichtlich des ▶Treuegrades bei Pflanzengesellschaften als zufälliges Einsprengsel aus anderen Pflanzengesellschaften oder als Relikt früher vorhandener Gesellschaften vorkommt; ▶Reliktart, ▶Zufällige).

strata [npl], overlying *geo. min.* ▶overburden (1).

strategic plan [n] [US], regional *leg. plan.* ▶regional plan.

strategic plan [n] for play spaces [UK] *plan. recr. urb.* ▶overall master plan for play areas [US].

strategic planning [n] [UK] *plan.* ▶general planning [US]/strategic planning [n] [UK].

strategic planning [n], play spaces [UK] *plan. recr. urb.* ▶overall master planning for play areas.

strategic planning [n] for sports and physical activities [UK] *plan. recr. urb.* ▶overall master planning for sports and physical activities [US].

strategic planning [n] of play areas [UK] *plan. recr. urb.* ▶overall master planning for play areas.

strategy [n] *biol. ecol.* ▶survival strategy.

strategy [n] [UK], green space *landsc. urb.* ▶general urban green space planning.

strategy [n] [US], land assembly *plan. pol.* ▶land assembly policy [US]/land acquisition policy [UK].

strategy plan [n] [UK], green space *landsc. urb.* ▶general plan for urban open spaces [US].

strategy plan [n] [UK], landscape *landsc.* ▶landscape plan.

strategy plan [n] [UK], regional *leg. plan.* ▶state development plan [US]/regional policy/strategy plan [UK].

6068 stratification [n] (1) *geo.* (Accumulation of sedimentary rocks in layers or strata); *s* **estratificación [f]** (Parte de la geología, que estudia la disposición de capas o estratos del subsuelo); *f* **stratification [f]** (Disposition et succession des strates ou des couches sédimentaires de la croûte terrestre) ; *g* **Schichtenaufbau [m] (2)** (In der Geologie die Abfolge von aufeinander folgenden Ablagerungen sowohl bei Lockergesteinen als auch bei Festgesteinen); *syn.* Schichtung [f] (1).

stratification [n] (2) *limn.* ▶thermal stratification.

6069 stratification [n] of plant communities *phyt.* (Profile of layers with characteristic life forms; generally four principal layers are recognized: ▶moss layer [ground layer], ▶herb layer, ▶shrub layer, ▶tree layer; it is also possible to distinguish between a higher and lower tree layer); *syn.* layering [n] of plant communities; *s* **estratificación [f] de la vegetación** (Disposición vertical de capas de vegetación: ▶estrato muscinal, ▶estrato herbáceo, ▶estrato arbustivo, ▶estrato arbóreo); *f* **stratification [f] de la végétation** (Structure verticale de l'appareil aérien des végétaux par strates superposées ; on peut distinguer quatre strates principales : ▶strate muscinale, ▶strate herbacée, ▶strate arbustive et ▶strate arborescente ; on peut distinguer encore, s'il y a lieu, une strate arborescente supérieure et une inférieure, etc., tout comme la stratification aérienne et la stratification souterraine) ; *syn.* structure [f] verticale de la végétation ; *g* **Vegetationsschichtung [f]** (Vertikaler Aufbau eines Lebensraumes. Es werden in der Regel vier Hauptvegetationsschichten unterschieden: ▶Moosschicht [Bodenschicht], ▶Krautschicht [Feldschicht], ▶Strauchschicht, ▶Baumschicht; man kann auch manchmal eine höhere, niedrigere etc. Baumschicht unterscheiden); *syn.* Schichtung [f] (2), Schichtenaufbau [m] (3), Stratifikation [f].

6070 stratified [pp] in one, two, three or several layers *phyt.* (e.g. moss-, forb-, shrub-, tree layer; ▶stratification of plant communities); *s* **formación [f] monoestratificada, biestratificada, triestratificada, etc.** (▶estratificación de la vegetación); *f* **formation [f] unistrate, bistrate, tristrate, pluristrate** (Organisation verticale d'une plantation ou d'un groupement végétal ; ▶stratification de la végétation) ; *g* **Aufbau [m, o. Pl.] einer Pflanzung (2)** (Einstufiger, zweistufiger, dreistufiger oder mehrstufiger Aufbau eines Pflanzenbestandes: z. B. Krautschicht, Strauchschicht, Baumschicht 1. Wuchsgröße und Baumschicht 2. Wuchsgröße; ▶Vegetationsschichtung); *syn.* stufiger Aufbau [m] eines Bestandes.

6071 stratified stand structure [n] *for. phyt.* (Vertical layers of a forest stand or plant community according to height graduations; ▶forest structure); *s* **estructura [f] estratificada** (de un rodal o de una comunidad vegetal; ▶estructura de un bosque); *f* **structure [f] stratifiée** (Distribution et structure verticale d'un peuplement végétal définies d'après les variations de hauteur ; ▶structure d'une forêt) ; *g* **stufiger Bestandsaufbau [m]** (Vertikale Anordnung und Struktur eines Vegetationsbestandes hinsichtlich der Höhenstaffelung; ▶Waldaufbau).

straw [n] *agr.* ▶fertilizing with straw; *pedol.* ▶neddle straw [US]/needle litter [US]; *hort.* ▶root balling with straw; *constr.* ▶row of straw bales; *constr. hort.* ▶wrapping with straw ropes.

6072 straw baling [n] *constr.* (Installation of rows of straw bales, often used in the U.S.; ▶erosion control facility); *s* **colocación [f] de pacas de paja** (Instalación de filas de pacas de paja, método utilizado frecuentemente en EE.UU. para proteger provisionalmente obras de construcción contra la erosión; ▶instalación de protección contra la erosión); *f* **mise [f] en place d'une rangée de bottes de paille** (U.S., mesure de protection contre l'érosion sur les chantiers ; ▶équipement de protection contre l'érosion) ; *g* **Aufstellen [n, o. Pl.] von Strohballenreihen** (In den USA häufig angewendete, temporäre Erosionsschutzmaßnahme auf Baustellen; Strohballenreihe(n) aufstellen [vb]; ▶Erosionsschutzeinrichtung); *syn.* Errichten [n, o. Pl.] einer Strohballensperre [f].

6073 straw dung [n] *agr. hort.* (Animal excrements mixed with straw bedding); *syn.* barn manure [n] [also US], stable manure [n], straw manure [n]; *s* **estiércol [m] (de establo)**; *f* **fumier [m]** (Mélange de déjections animales avec une litière, utilisé comme fertilisant) ; *g* **Mist [m, o. Pl.]** (Mit Stroh vermischte tierische Exkremente [Kot-Urin-Einstreugemisch der Stalltiere], meist als Dünger verwendet); *syn.* Stallmist [m, o. Pl.], Stalldung [m, o. Pl.].

straw manure [n] *agr. hort.* ▶straw dung.

6074 straw meadow [n] [US] *agr. land'man. phyt.* (Meadow cut once a year for farm animal bedding, e.g. moorgrass meadow [*Molinion* alliance], low-sedge swamp [*Caricion* alliance]); *syn.* litter meadow [n] [UK]; *s* **prado [m] de heno** (Prado de siega anual que se corta para henificar); *f* **prairie [f] à litière** (Prairie non fertilisée, de faible valeur agronomique, fauchée une fois en automne quand les herbes sont sèches, celles-ci étant utilisées comme litière hivernale pour les animaux de la ferme, p. ex. les moliniaies [alliance du *Molinion*] ou les prairies à petites Laiches/Laîches [alliance du *Caricion*]) ; *syn.* surface [f] à litière ; *g* **Streuwiese [f]** (Einschürige Wiese zur Streugewinnung für die Tierhaltung, z. B. Pfeifengraswiese [*Molinion*-Verbände], Kleinseggenried [*Caricion*-Verbände]); *syn. o. V.* Streuewiese [f]; Streuefläche [f] [CH].

6075 straw-wrapped root ball [n] *hort.* *s* **cepellón [m] con paja** (Cepellón protegido por cubierta de paja); *f* **motte [f] avec tontine de paille** (Motte entourée par une enveloppe de paille de seigle et fermée par un lien en osier à la base du collet et

parfois renforcée au milieu de la motte par une ficelle supplémentaire) ; *g* **Ballen [m] mit Strohballierung** (Zum Verpflanzen gestochener Wurzelballen eines Gehölzes, der für den Transport zur Baustelle/Pflanzstelle mit einem Strohgeflecht umwickelt ist; diese in F. und den USA praktizierte Methode wird in D. nicht angewandt).

6076 stream [n] (1) *geo.* (Medium-sized natural watercourse); *s* **arroyo [m]** *syn.* riachuelo [m]; *f 1* **ruisseau [m]** (Cours d'eau naturel de moindre importance de largeur moyenne de 1 m) ; *f 2* **rivière [f]** (Cours d'eau naturel de moindre importance de largeur moyenne de 8 m ; *termes spécifiques* petite rivière, rivière flottable, rivière navigable) ; *g* **Bach [m]** (Kleines Fließgewässer mit einer Wasserführung von maximal 10-20 m³/s, dessen Verlauf dem Kleinrelief einer Landschaft angepasst ist und oft ein unregelmäßiges Längsprofil aufweist).

stream [n] (2) *geo.* ►bed of a continually flowing stream; *wat' man. wat'man.* ►canalized stream; *geo. limn. wat'man.* ►fast-moving stream; *geo.* ►headwater stream [US]/headstream [UK]; *geo. hydr.* ►intermittent stream; *wat'man.* ►piped stream [US]/culverted stream [UK]; *constr. eng. wat'man.* ►receiving stream; *geo. limn. wat'man.* ►slow-moving stream.

stream [n] [UK], **air** *plant.* ►traffic-caused wind [US].

stream [n], **braided** *geo.* ►braided river.

stream [n], **channelized** *wat'man.* ►canalized stream.

stream [n] [UK], **culverted** *wat'man.* ►piped stream [US].

stream [n], **outlet** *constr. eng. wat'man.* ►receiving stream.

6077 stream alignment shift [n] *geo. hydr.* (Change in the course of a river due to lateral erosion); *s* **desplazamiento [m] del curso de un río** (Cambio del curso de un río por erosión de las orillas); *f* **divagation [f] d'un cours d'eau** (Changement du tracé d'un lit par l'érosion latéral ; *syn.* déplacement [m] d'un cours d'eau ; DG 1984, 167); *g* **Flusslaufverschiebung [f]** (Lageveränderung des Flussbettes durch Ufererosion); *syn.* Flussbettverlagerung [f].

streambank [n] *landsc.* ►planting of a streambank.

6078 streambank erosion [n] *geo.* (Scouring of material and undercutting of channel banks by running water; ►bank-caving, ►bank erosion by surface runoff, ►riverbank collapse [US]/river-bank collapse [UK], ►riverbank degradation [US]/river-bank degradation [UK], ►riverbank slippage [US]/river-bank slippage [UK], ►undercut bank); *syn.* riverbank erosion [n] [US]/river-bank erosion [n] [UK], undercutting [n] of streambanks/riverbanks; *s* **erosión [f] de márgenes de cursos de agua** (Socavación lateral y de fondo de las márgenes de ríos sobre todo con el choque de la corriente contra la orilla cóncava de los meandros; ►arco erosivo de meandro, ►deslizamiento de orillas, ►desplome de orillas, ►deterioro de taludes de márgenes, ►erosión de orillas por escorrentía, ►socavación de márgenes); *f* **érosion [f] des berges des cours d'eaux** (Enlèvement naturel des matériaux d'une berge ; ►affouillement de la berge, ►berge d'effondrement, ►dégradation de berge, ►effondrement d'une berge, ►érosion de berge par ruissellement, ►glissement de berge) ; *syn.* érosion [f] latérale ; *g* **Erosion [f] an Ufern von Fließgewässern** (Abtragender Vorgang an Uferböschungen; ►Abbruchufer, ►Auskolkung an/von Flussufern, ►Beschädigung der Uferböschung, ►Uferabbruch, ►Ufererosion durch Oberflächenabfluss, ►Uferabrutschung); *syn.* Seitenerosion [f], Seitenschurf [f].

6079 stream bed [n] *geo. hydr. syn.* stream channel [n]; *s* **lecho [m] de arroyo**; *f* **lit [m] d'une rivière** (Espace occupé, en permanence ou temporairement, par un cours d'eau. On distingue le ►lit majeur 1 qui est l'espace occupé par le cours d'eau lors de ses plus grandes crues et le ►lit mineur qui est la zone limitée par les berges) ; *syn.* lit [m] d'un ruisseau ; *g* **Bachbett [n]** (Schmales oberirdisches Gerinne im niedrigsten Bereich einer Landschaft, das ständig oder zeitweilig Wasser führt; es besteht aus dem eigentlichen Gewässerbett und dem flächenmäßig größeren ►Hochwasserbett, das je nach landschaftlicher Situation durch Dämme begrenzt ist).

streambed erosion [n] *geo.* ►watercourse bed erosion.

stream-bed rampart [n] [UK] *wat'man.* ►stream ramp.

stream channel [n] *geo. hydr.* ►stream bed.

stream channelization [n] *wat'man.* ►stream straightening.

6080 stream corridor deciduous vegetation [n] *land'man.* (Strip of woody species of a ►regularly-flooded alluvial plain; primarily of such species of a riparian woodland as: willow *[Salix]*, alder *[Alnus]*, ash *[Fraxinus]* and poplar *[Populus]*; ►gallery forest); *s* **franja [f] de leñosas al borde de cursos de agua** (Banda de árboles y/o arbustos en la ►vega aluvial regularmente inundada, generalmente de especies como el sauce *[Salix]*, el aliso *[Alnus]*, el fresno *[Fraxinus]* y el álamo *[Populus]*); ►bosque de galería); *f* **boisement [m] de berge des cours d'eau** (Liseré végétal en pied de berge des cours d'eau actifs dans la ►plaine alluviale à bois tendres principalement colonisée par les Saules *[Salix]*, l'Aulne *[Alnus]*, le Frêne *[Fraxinus]*, le peuplier *[Populus]* ; ►galerie forestière); *syn.* cordon [m] végétal riverain [CH], ripisylve [f] [CH], bande [f] riveraine arbustive ou arborescente des cours d'eau [CDN], boisement [m] rivulaire, lisière [f] boisée riveraine [CDN] ; *g* **Gehölzsaum [m] an Fließgewässern** (Gehölzstreifen der ►Weichholzaue entlang von regelmäßig über die Ufer gehenden Bächen und [kleinen] Flüssen, der vorwiegend aus Arten wie Weide *[Salix]*, Erle *[Alnus]*, Esche *[Fraxinus]* und Pappel *[Populus]* besteht; ►Galeriewald).

6081 stream course [n] *geo. syn.* stream-course [n] [also UK]; *s* **curso [m] de un arroyo** *syn.* recorrido [m] del cauce (de un arroyo); *f* **cours [m] de la rivière/du ruisseau** ; *g* **Bachlauf [m]**.

6082 stream erosion [n] *geo.* (This washing away occurs mainly in arid and semiarid regions [US]/semi-arid regions [UK], when massive heavy storms cause severe removal of soil by creating deep channels; ►extensive erosion, ►flash erosion, ►rill erosion); *s* **erosión [f] en cárcavas (2)** (Tipo de erosión de las regiones áridas y semiáridas, en las cuales pueden presentarse precipitaciones muy fuertes que —junto con la falta de vegetación— conllevan el arrastre masivo del suelo; ►erosión aerolar, ►erosión en regueros, ►erosión repentina brusca); *syn.* erosión [f] en barrancos (2); *f* **ravinement [m] (2)** (Important phénomène d'ablation des sols sans couvert végétal provoqué par des pluies torrentielles dans les régions arides et semi-arides ; ►érosion aréolaire, ►érosion brutale, ►ruissellement en filets) ; *g* **Grabenerosion [f] (2)** (±) (Besonders in ariden und semiariden Gebieten, in denen durch schwere Niederschläge und durch fehlende Vegetation massive Bodenabspülungen in Form von bis zu mehreren Metern tiefen, unregelmäßigen Gräben entstehen; ►massive Bodenabspülung, ►massive, großflächige Erosion, ►Rillenerosion).

6083 stream flow [n] *geo. wat'man.* (Water movement down a watercourse); *s* **flujo [m] de un cauce** (Flujo de agua en un cauce natural o artificial [corriente de agua] con una superficie libre); *f* **écoulement [m] fluviatile** (Résultat de la concentration dans les talwegs des eaux de ruissellement superficiel direct, des eaux de l'écoulement hypodermique et des eaux souterraines arrivant à émergence ; DG 1984, 152) ; *g* **Abfluss [m] (2)** (Das Abfließen eines Fließgewässers; Messdaten über die Schwankungen des **A.es** liefern wichtige Informationen für viele wasserwirtschaftliche und wasserbauliche Planungen, insbesondere für die Hochwasservorhersage und Planungen für Hochwasserschutzmaßnahmen).

6084 stream improvement [n] *wat'man.* (Channelization involving straightening, paving or confining of streams purely on the basis of river engineering considerations; ►culverting, ►river engineering measures, ►stream straightening); *s* **ingeniería [f] hidráulica en arroyos y torrentes** (Encauzamiento, adoquinado o incluso ►entubado de un arroyo realizado teniendo en cuenta exclusivamente los principios de ingeniería hidráulica; ►ingeniería hidráulica de ríos y arroyos, ►rectificación de arroyos); *syn.* regulación [f] de arroyos y torrentes; *f* **travaux [mpl] d'aménagement d'une rivière/d'un ruisseau** (Travaux de terrassements lourds comportant le calibrage, l'enrochement ou la canalisation de ruisseaux réalisés selon les pratiques courantes [techniques dures] de l'ingénierie hydraulique ; ►busage, ►recalibrage d'une rivière/d'un ruisseau, ►travaux d'aménagement des cours d'eau) ; *g* **Bachausbau [m, o. Pl.]** (Begradigung, Auspflasterung oder sogar Verrohrung von Bächen nach rein wasserbaulichen Gesichtspunkten; ►Bachbegradigung, ►Gewässerausbau, ►Verdolung); *syn.* Ausbau [m, o. Pl.] von fließenden Kleingewässern, Bachregulierung [f].

6085 streamlet [n] *wat'man.* (Very small stream); *syn.* rivulet [n], brooklet [n], rill [n], run [n] [also US] (2), runlet [n] [also US]; *s* **arroyuelo [m]** *syn.* regato [m]; *f* **ruisselet [m]** ; *g* **sehr kleiner Bach [m]** *syn.* Rinnsal [n].

6086 stream ramp [n] *wat'man.* (Structure in a watercourse to control ►watercourse bed erosion; ►ground sill); *syn.* streambed rampart [n] [also UK]; *s* **rampa [f] de solera** (Estructura construida en cauce de río o torrente para evitar la ►erosión del lecho; ►solera de fondo); *f* **rampe [f] en enrochement(s)** (ARI 1986, 43 ; ouvrage de stabilisation du profil en long d'un cours d'eau par suppression de l'érosion du fond du lit ; ►érosion du lit d'un cours d'eau, ►seuil transversal d'un cours d'eau) ; *g* **Sohlenrampe [f]** (In Fließgewässern eingebaute Absturzrampe zum Schutz vor ►Sohlenerosion; ►Sohlenschwelle).

streamside flood area [n] [US] *geo.* ►flood area of a stream/brook.

6087 stream stewardship [n] *conserv.* (Feeling of responsibility by one or more citizens to maintain and care for a stream); *s* **apadrinamiento [m] de arroyos** (≠) (Política llevada en algunos países por parte de las administraciones de protección de la naturaleza de integrar a la población en el cuidado de secciones específicas de arroyos naturales o cuasi-naturales en zonas urbanizadas); *f* **parrainage [m] d'un ruisseau** (Participation de la collectivité, des propriétaires riverains, des utilisateurs à l'entretien d'une rivière) ; *syn.* parrainage [m] de rivières ; *g* **Bachpatenschaft [f]** (Mitverantwortung eines Bürgers oder einer Gruppe von Bürgern für die Pflege und Unterhaltung eines Baches).

6088 stream straightening [n] *wat'man.* (Reduction or removal of the bends in a natural stream by excavation, realignment, lining or other means to accelerate the flow of water; ►stream improvement, ►straightening of watercourses); *syn.* stream channelization [n]; *s* **rectificación [f] de arroyos** (Término específico de ►ingeniería hidráulica en arroyos y torrentes, ►rectificación [de ríos y arroyos]); *syn.* encauzamiento [m] de arroyos; *f* **recalibrage [m] d'une rivière/d'un ruisseau** (Terme spécifique pour les ►travaux d'aménagement d'une rivière/d'un ruisseau; ►redressement d'un cours d'eau) ; *g* **Bachbegradigung [f]** (UB zu ►Bachausbau und ►Begradigung von Fließgewässern).

stream valley corridor [n] [US] *landsc. urb.* ►linear open space pattern/system.

stream valley park [n] [US] *landsc. recr. urb.* ►riverside park.

street [n] *trans. urb.* ►local street with access only for residents; *trans. urb.* ►loop street (1); *trans.* ►major street; *trans. urb.* ►residential street.

street [n]**, abutter on a** [US] *leg. urb.* ►adjoining street resident.

street [n]**, abutting owner on a** [US] *leg. urb.* ►adjoining street resident.

street [n]**, abutting resident on a** [US] *leg. urb.* ►adjoining street resident.

street [n] [US]**, collector** *trans. urb.* ►collector road.

street [n] [US]**, cross-town** *trans.* ►major cross-town artery.

street [n] [US]**, dead-end** *trans. urb.* ►cul-de-sac.

street [n] (1)**, loop** *trans. urb.* ►loop street (1).

street [n] [US] (2)**, loop** *trans.* ►bypass [US]/by-pass [UK].

street [n]**, play-** *recr. urb.* ►home zone [UK].

street [n] [US]**, service** *trans.* ►service road.

6089 street abandonment [n] [US] *adm. leg. urb.* (Procedure according to law taken for removal of road pavement and conversion to pedestrian use or open green space executed by the governing body; complex legal proceedings are required for such action; ►street closure, ►traffic calming); *syn.* extinguishment [n] of rights of way [UK] (TCPA 1971, 214); *s* **cambio [m] de clasificación de calle** (≠) (Procedimiento legal de cambiar la función de una calle pública, p. ej. de cerrarla al tráfico de paso y convertirla en una calle peatonal; ►cierre de calle al tráfico de paso, ►tranquilización del tráfico); *f* **déclassement [m] de rue** (Décision administrative par laquelle une voie routière change d'utilisation ou de catégorie juridique, en général lui fait perdre son caractère de voie publique et la soustrait au régime juridique du réseau auquel elle se trouvait incorporée ; ►fermeture à la circulation de la rue, ►circulation apaisée) ; *g* **Straßeneinziehung [f]** (Ein vom Gemeinderat zu beschließender Verwaltungsakt, der eine öffentliche Straße in eine andere Nutzungsart umwidmet; ►Straßenschließung, ►Verkehrsberuhigung); *syn.* Auflassung [f] von Straßen, Einziehung [f, o. Pl.] von Straßen.

street amenities [npl]**, placement of** *constr.* ►installation of street furniture.

street amenity [n] [US] *arch. urb.* ►piece of street furniture.

street appearance [n] [US] *arch. urb.* ►streetscape.

6090 street canyon [n] *urb.* (TGG 1984, 67; ►street space between facing high-rise buildings); *s* **cañón [m] entre calles** (≠) (Espacio de calle entre edificios de gran altura; ►espacio calle); *f* **couloir [m] de la rue** (►Emprise de la rue/route située entre deux alignements de bâtiments) ; *g* **Straßenschlucht [f]** (Raum zwischen lückenlos aneinandergereihten Hochhäusern oder hohen Häuserzeilen auf beiden Seiten der Straße; derselbe ►Straßenraum mit weniger hohen Häusern wird **Straßenzug** genannt).

street cleaning [n] *urb.* ►hydrant for street cleaning.

6091 street closure [n] *adm. leg. urb.* (Act of closing a road to through traffic, e.g. by erection of traffic barriers; involving complex legal proceedings; ►street abandonment [US]/extinguishment of rights of way [UK]); *s* **cierre [m] de calle al tráfico de paso** (Decisión administrativa de evitar el tráfico de paso en una calle, tomando medidas que dificultan el tránsito rápido, como construcción de barreras o cierre de un extremo de la calle. Esta medida implica el ►cambio de clasificación de calle); *f* **fermeture [f] à la circulation de la rue** (Disposition administrative qui vise temporairement ou définitivement à

S

empêcher l'accès d'une ou partie de rue à la circulation par la mise en place d'un dispositif de barrage ; ►déclassement de rue) ; *syn.* barrage [m] de rue, mise [f] en barrage de rue ; *g* **Straßenschließung [f]** (Ein vom Gemeinderat zu beschließender Verwaltungsakt mit dem Ziel, den Durchgangsverkehr in einer Straße, z. B. durch Querbauwerke, zu unterbinden; wird eine Straße an einem Ende geschlossen, weil sie in der bisherigen Form für den durchgehenden Verkehr entbehrlich ist, für den Anliegerverkehr jedoch weiterhin nutzbar bleibt, spricht man von einer **Teileinziehung**; ►Einziehung von Straßen); *syn.* Schließung [f] von Straßen, Aufhebung [f] einer Straße [auch CH].

street design element [n] *arch. urb.* ►piece of street furniture.

street frontage [n] *urb.* ►frontage.

6092 street furniture [n] *arch. constr. urb.* (Equipping of a street, road or public square with ►pieces of street furniture elements, such as lighting, signage, benches, waste containers, fences, planters, etc.; ►installation of street furniture, ►outdoor furniture); *s* **mobiliario [m] de calles** (►elemento de mobiliario urbano, ►instalación de mobiliario urbano, ►mobiliario urbano); *f* **mobilier [m] de voirie** (Équipement utilitaire ou décoratif d'un espace de circulation ou d'une place publique ; ►élément du mobilier de voirie, ►implantation de mobilier urbain, ►mobilier urbain [d'espaces verts]) ; *g* **Straßenmöblierung [f]** (Ausstattung eines Straßenraumes oder öffentlichen Platzes mit ►Straßenmöblierungselementen; ►Aufstellung von Straßenmöblierungselementen, ►Möblierung für Außenanlagen); *syn.* Ausstattung [f] von Straßen, Straßenausstattung [f].

6093 street hardware [n] [US] *arch. urb.* (Elements connected with mechanical and utility systems within a street right-of-way such as hydrants, manhole covers, traffic lights and directional signs, utility poles and lines, parking meters; ►piece of street furniture); *s* **equipamiento [m] técnico en calles** (≠) (Elementos presentes en calles que son parte de sistemas técnicos o de servicios como bocas de agua, cajas de distribución, farolas, líneas aéreas de electricidad o teléfono, postes indicadores, semáforos, tapas de registro, etc.; ►elemento de mobiliario urbano); *f* **mobilier [m] (urbain) de voirie** (Terme générique pour les équipements fonctionnels d'une rue comme p. ex. le mobilier de confort [abris-bus, bancs], le mobilier d'information et de publicité [panneaux d'affichage libre, relais d'information service et touristique, horloges], le mobilier de propreté et d'hygiène [corbeilles à papiers, borne de propreté, conteneur pour collecte sélective, compacteur à déchets, sanitaires], le mobilier de protection et de sécurité [bornes, potelet, garde-corps, barrières de protection, grille d'arbres, arceau de protection des arbres, ralentisseur], le mobilier d'exploitation et de service [cabine téléphonique, borne d'appel de taxis, coffres relais pour le courrier, armoire à feux, armoire EDF, PTT, vidéo, détente gaz, poteau d'incendie, horodateurs, lampadaires, supports de feux, caméras, caisses automatiques, contrôle d'accès des « parcs à enclos »], etc. ; ►élément de mobilier de voirie) ; *g* **Straßenausstattungselement(e) [n(pl)]** (OB für eine ortsgebundene funktionsnotwendige Einrichtung einer Straße wie z. B. Hydrant, Schachtdeckel, Straßenlaterne, Verkehrsampel, Wegweiser, Schaltkasten, oberirdische Versorgungsleitung, Parkautomat ; ►Straßenmöblierungselement).

street landscaping [n] [US] *leg. trans. landsc.* ►right-of-way planting [US]/roadside planting [UK].

6094 street planting [n] *landsc.* (►seeding, ►right-of-way planting [US]/roadside planting [UK]); *s* **plantación [f] a lo largo de la calle** (Cualquier tipo de plantación en calles, aunque en alemán la ►siembra 1 de hierba o césped no se considera tal; *término genérico* ►verde vial); *f* **plantations [fpl] du bord des routes** (IDF 1988, 294 ; toute forme de plantation d'accompagnement de la voirie ; *terme générique* ►espaces verts d'accompagnement de la voirie) ; *syn.* végétalisation [f] des bords de routes, aménagement [m] paysager routier ; *g* **Straßenbepflanzung [f]** (Jegliche Art von Bepflanzung im Straßenraum; dazu gehören i. d. R. auch die Straßennebenflächen, soweit sie liegenschaftlich zur Straße gehören; im Deutschen wird ►Ansaat [Rasen- oder Wiesenansaat] nicht als Bepflanzung gedacht; *OB* ►Verkehrsgrün); *syn.* Straßenpflanzung [f].

street resident [n] *leg. urb.* ►adjoining street resident.

6095 streetscape [n] *arch. urb.* (Visual appearance of a street, its spatial definition and features; ►cityscape, ►townscape); *syn.* street appearance [n] [also US]; *s* **imagen [f] de la calle** (Conjunto de elementos perceptibles visualmente, generadores de espacios y articulantes de la calle; ►aspecto escénico urbano, ►fisionomía de la ciudad); *f* **image [f] de la rue** (Ensemble des composantes perceptibles par l'observateur qui limitent, compartimentent ou cadrent l'espace de la voirie urbaine ; ►aspect d'une agglomération, ►morphologie urbaine; *syn.* représentation [f] de la rue, image [f] de la voirie urbaine ; *g* **Straßenbild [n]** (Vom Menschen optisch wahrnehmbare Gesamtheit der raumbegrenzenden und gliedernden Elemente einer Straße; ►Ortsbild, ►Stadtgestalt).

streetscape element [n] [US] *arch. urb.* ►piece of street furniture element.

6096 street space [n] *urb.* (Ground and air space across a street between facing buildings or avenue trees; ►tree-lined avenue [US]/avenue [UK], ►street canyon; *s* **espacio [m] calle** (Espacio físico entre los edificios de una calle como lugar de usos urbanos; ►avenida 2, ►cañón entre calles); *f* **emprise [f] de la rue/route** (Espace de voirie bordé en agglomération de bâtiments aux façades alignées et en rase campagne parfois par une allée comprenant, outre la chaussée, le trottoir, le fossé, le talus, la piste cyclable, la plantation d'accompagnement ; ►allée d'arbres, ►couloir de la rue) ; *syn.* espace [m] de voirie ; *g* **Straßenraum [m]** (Senkrecht über der Straße liegender Luftraum, der im Siedlungsgebiet durch gegenüberliegende Häuserzeilen definiert ist oder im Außenbereich durch Baumalleen begrenzt sein kann; ►Allee, ►Straßenschlucht).

6097 street straightening [n] *eng.* (Reduction or removal of bends to shorten travelled ways [US]/carriage-ways [UK] and make traffic use safer and more efficient); *syn.* road straightening [n]; *s* **rectificación [f] de carreteras** (Reducción de la cantidad y peligrosidad de las curvas en carreteras para hacer más seguro el tráfico); *f* **redressement [m] de la voirie** (Réaménagement de la voie publique avec modification de l'emprise par déplacement de l'axe de la plate-forme et changement corrélatif des caractéristiques géométriques pour obtenir un tracé plus rectiligne) ; *g* **Begradigung [f] von Straßen** *syn.* Straßenbegradigung [f].

6098 street tree [n] *landsc. urb.* (In the "American Standard for Nursery Stock" street trees are specified with branching height, 6 to 7 ft [1.83 to 2.13 m] or more; ►standard street tree [US]/avenue tree [UK]. The British term for specifying street trees is ►tall-standard); *s* **árbol [m] de calle** (1. Árbol que crece en una calle. 2. *hort.* Árbol adecuado para ser plantado en calles o plazas pavimentadas de zonas urbanas; en diferentes países existen normas de la altura mínima del tronco y lista de árboles que se adaptan bien a las condiciones mesológicas de la ciudad; ►árbol de alineación); *f* **arbre [m] d'alignement** (1. Arbre planté le long de la chaussée. 2. *hort.* Arbre utilisé pour l'ornement des rues, des places en milieu urbain, dont les essences sélectionnées sont reconnues pour leur forme esthétique, leur résistance à la pollution atmosphérique, leur entretien aisé ; les sujets cultivés en pépinières ont en général les caractéristiques dimensionnelles suivantes : haute tige, circonférence minimum de la tige à 1 m du sol 18/20 cm et une hauteur de tige comprise entre 2,50 et 3 m sur circulations piétonnières et supérieure à 4 m sur circulation routière ; ►arbre d'avenue) ; *g* **Straßenbaum [m]** (1. Baum, der in einer Straße steht. 2. *hort.* Für die Anpflanzung an Straßen und auf überwiegend befestigten Plätzen im städtischen Bereich geeignete Baumart, die je nach Verkaufsgröße in Baumschulen als Hochstamm mit einer Stammhöhe von mindestens 2,0-2,50 m herangezogen wird; besonders geeignet sind vor allem anspruchslose Arten in Bezug auf Boden, Nährstoffe und Klima; in D. gibt es einen Arbeitskreis „Stadtbäume" der Ständigen Konferenz der Gartenamtsleiter beim Deutschen Städtetag, der alle paar Jahre eine aktualisierte „Straßenbaumliste der Gartenamtsleiter", herausgibt; *UB* ►Alleebaum); *syn. zu 2.* Hochstamm [m] für eine Straßenbepflanzung.

6099 street tree setback [n] (≠) *leg. urb.* (Under French planning law the distance from buildings to street tree line; **in U.S. and U.K.**, not a legal requirement according to planning law; ►setback [US]/set-back [UK]); *s* **retranqueo [m] frontal de plantación** (En Francia, distancia mínima entre árboles de alineación y la fachada en el frente de la calle; ►retranqueo); *f* **distance [f] de plantation par rapport aux façades** (Distance réglementaire [3 m] de l'axe de la plantation des arbres d'alignement par rapport à la façade d'un édifice ; ►marge de reculement) ; *syn.* écartement [m] des arbres par rapport aux façades ; *g* **Pflanzabstand [m] von der Hausfassade (≠)** (In D. kein planungsrechtlicher Begriff; im franz. Planungsrecht ein einzuhaltender Mindestabstand von Straßenbäumen zur Häuserfront; ►Grenzabstand).

street [n] **with access only for residents** *trans. urb.* ►local street with access only for residents.

strength [n] *constr. stat.* ►compressive strength, ►shear strength, ►tensile strength.

strength [n]**, testing the structural** *eng. stat.* ►structural testing.

strength [n] **of trees, structural** *arb. stat.* ►structural stability of trees.

stress [n] *ecol. envir. recr.* ►environmental stress; *ecol. envir.* ►environmental stress degree; *boil. ecol.* ►environmental stress indicator; *ecol. recr.* ►environmental stress tolerance; *agr.* ►grazing stress; *biol.* ►heat stress; *limn. pedol.* ►nutrient stress; *phyt.* ►rooting stress; *constr.* ►thermal stress on roof membrane.

stress [n]**, competition** *ecol.* ►pressure of competition.

stress [n]**, electro-** *envir.* ►electromagnetic radiation.

stress degree [n] *ecol. envir* ►environmental stress degree.

6100 stress level [n] *biol. ecol. envir.* (Amount of strain on the environment; ►disturbance, ►environmental stress); *s* **nivel [m] de contaminación** (Grado de impacto desde el punto de vista cuantitativo; ►estrés ambiental, ►influencia perturbadora); *syn.* nivel [m] de polución; *f 1* **niveau [m] de charge polluante** (Quantité d'éléments polluants contenus dans le milieu naturel et contribuant à la détérioration de sa qualité ; ►nuisance) ; *f 2* **degré [m] de pollution** (Concentration d'un polluant dans l'air ambiant ou son dépôt sur les surfaces en un temps donné ; art. 2 dir. n° 96/62/CE du 27 septembre 1996 concernant l'évaluation et la gestion de la qualité de l'air ambiant ; ►stress environnemental) ; *g* **Belastungsgrad [m] (2)** (Ausmaß der ►Beeinträchtigung bestimmt durch die Menge; oberhalb des kritischen **B.es** [Grenzwert] können direkte Schadeffekte an Menschen, Pflanzen, Ökosystemen, Materialien] auftreten; in der Rahmenrichtlinie 96/62/EG vom 27.09.1996 wurde für die EU ein Zeitplan für die Definition und Festlegung von Luftqualitätszielen mit Hilfe der Grenzwerte für die wichtigsten Luftschad-

stoffe vorgegeben; mit der RL 2008/50 vom 21.05.2008 sind zum Schutze der menschlichen Gesundheit im Anhang XI und XIV die Grenzwerte der Belastung und in Anhang XII die Alarmschwelle für Ozon festgelegt; ►Belastung).

stress [n] **on roof membrane** *constr.* ►thermal stress on roof membrane.

stress tolerance [n] *ecol. recr.* ►environmental stress tolerance.

6101 stretcher [n] *constr.* (Masonry unit laid horizontally with its length parallel to the face of a wall; ►header); *s* **soga [f]** (Término que se utiliza para indicar que un mampuesto está apoyado sobre su longitud mayor; DACO 1988; ►tizón); *syn.* ladrillo [m] al hilo; *f* **panneresse [f]** (Matériau de construction [pierre, brique] parallélépipédique dont la face la plus longue est disposée parallèlement à l'alignement du mur ; ►boutisse) ; *g* **Läufer [m]** (Der mit seiner Längsseite parallel zur Mauerflucht vermauerte Stein; ►Binder); *syn.* Läuferstein [m].

6102 stretcher bond pattern [n] *arch. constr.* (Masonry pattern, whereby all courses are laid as ►stretchers with the vertical joint of one course falling midway between the joints of the courses above and below; ►masonry bond, ►header bond pattern); *syn.* running bond pattern [n], stretching bond [n] [also US]; *s* **aparejo [m] a soga** (En albañilería, típico ►aparejo de mampuestos, en el que las hiladas de ►sogas se asientan sobre su superficie mayor, uniéndose por sus caras intermedias sin que las llagas de cada dos hiladas consecutivas coincidan en vertical; cf. DACO 1988; ►aparejo a tizón); *f* **appareillage [m] à panneresses** (►Appareillage d'une maçonnerie dont les assises réglées ne sont constituées que de ►panneresses et tel que les joints verticaux d'un lit tombent au niveau du milieu de la pierre ou de la brique du lit immédiatement inférieur [déharpement par demi-brique un rang sur deux] ; ►appareillage en boutisses ; *termes spécifiques* appareil à quart-panneresse, appareillage à assises réglées « à la grecque ») ; *syn.* appareil [m] à demi-brique en long, appareillage [m] harpé, appareillage [m] en quinconce ; *g* **Läuferverband [m]** (►Mauersteinverband, bei dem die ►Läufer i. d. R. um eine halbe Steinlänge gegeneinander versetzt sind; ►Binderverband); *syn.* Längsverband [m].

6103 stretcher course [n] *arch. constr.* (Course in masonry whereby stones/bricks are laid with their lengths parallel to the horizontal alignment of the wall; ►header course); *syn.* running course [n]; *s* **hilada [f] de sogas** (En mampostería, serie horizontal de ladrillos o sillares colocados a soga, ►hilada de tizones); *f* **assise [f] de panneresses** (En maçonnerie, lit de pierres ou de briques dont la face la plus longue est parallèle à l'alignement du mur, p. ex. les ouvrages par assises réglées ; ►lit de boutisses) ; *syn.* lit [m] de panneresses ; *g* **Läuferschicht [f]** (Schicht im Mauerwerk, bei der die Steine mit der Längsseite parallel zur Mauerflucht verlegt werden; ►Binderschicht).

stretching bond [n] [US] *arch. constr.* ►stretcher bond pattern.

6104 strict nature reserve [n] *conserv.* (1. Area of land and/or sea possessing some outstanding or representative ecosystems, geological or physiological features and/or species, available primarily for scientific research and/or environmental monitoring for the *in situ* conservation of geomorphological and biological diversity. 2. Large area of unmodified or slightly modified land, and/or sea, retaining its natural character and influence, without permanent or significant habitation, which is protected and managed to preserve its natural condition; BAS 1996, 3. 3. Category proposed by the *Commission on Natural Parks and Protected Areas* of the IUCN in the context of defining ►biosphere reserves [UNESCO Programm "Man and Biosphere"] whereby the protected area is to be kept free of

S

human interference except for scientific research purposes and environmental monitoring—as far as this is possible under existing environmental conditions; ►natural area preserve [US]/nature reserve [UK]); *syn.* scientific reserve [n]; *s* **reserva [f] natural (científica)** (Área natural en la que el uso económico está restringido o prohibido, para la protección de los caracteres naturales y que sirve para fines científicos; cf. DINA 1987; ►espacio natural protegido 1, ►reserva de la biosfera); *f* **réserve [f] naturelle intégrale** (Catégorie de protection proposée par la Commission des parcs nationaux et des aires protégées de l'UICN [Union internationale pour la conservation de la nature] dans le cadre de la désignation de ►réserves de la biosphère [Programme MAB — Man and Biosphère — lancé par l'UNES-CO], prévoyant, à l'exclusion des observations scientifiques, l'interdiction des activités humaines sur les territoires protégés ; ►site classé) ; *g* **Vollnaturschutzgebiet [n]** (Von der *Commission on Natural Parks and Protected Areas* der IUCN vorgeschlagene Schutzgebietskategorie im Rahmen der Ausweisung von ►Biosphärenreservaten [UNESCO-Programm MAB — Man and Biosphere], bei der Schutzgebiete außer durch wissenschaftliche Forschung und Umweltüberwachungstätigkeiten frei von menschlicher Beeinflussung gehalten werden sollen — soweit dies bei der heutigen Umweltsituation überhaupt noch möglich ist; ►Naturschutzgebiet 2); *syn.* wissenschaftliches Schutzgebiet [n].

strict wilderness reserve [n] [US] *conserv. leg. nat'res.* ►wilderness area [US].

6105 string bog [n] *geo.* (Type of boreal-subarctic bog, found particularly in Fenno-Scandinavia, with longitudinal parallel rows of deep ►bog hollows, caused by frequent frost and thaw and formed within the overall bog area); *syn.* Aapa moor [n], Aapa mire [n]; *s* **turbera [f] acordonada** (Tipo de turbera subártica-boreal que se presenta en especial en el área fenoescandinavia, está formada por hileras paralelas de ►pocinas de turbera profundas y alargadas, resultado de procesos de helamiento y deshielo frecuentes, dentro de la superficie de turbera cerrada); *syn.* turbera [f] tipo «Aapa»; *f* **tourbière [f] cordée** (Tourbière des pays froids à sols gelés en permanence en profondeur [pergélisol] constituée d'une alternance de lanières de végétation [bourrelets saillants de tourbe] et de dépressions [►gouilles profondes] habituellement remplies d'eau, due à la ségrégation de la glace » ; cf. DG 1984, 451) ; *syn.* tourbière [f] réticulée de type « Aapa », tourbière [f] structurée en lanières ; *g* **Strangmoor [n]** (Boreal-subarktischer Moortyp, vornehmlich im Fennoskandia [Baltischer Schild], mit langgestreckten und in parallelen Reihen durch häufiges Gefrieren und Wiederauftauen geordneten, tiefen ►Schlenken, die mit der sonst geschlossenen Moorfläche abwechseln); *syn.* Aapa-Moor [n].

strip [n] *conserv. landsc. urb.* ►buffer strip; *geo.* ►coastal strip; *limn.* ►drawdown strip; *trans.* ►freeway median strip [US]/motorway central reservation [UK]; *constr.* ►grass strip (1), ►meadow strip; *trans.* ►median strip [US]/central reservation [UK]; *constr.* ►mowing strip; *agr. constr. ecol.* ►pathside strip [US]/wayside [UK]; *constr.* ►paving strip, ►planting strip; *trans.* ►traffic separation strip; *constr.* ►turf strip (1), ►turf strip (2); *landsc. trans.* ►vegetated strip.

strip [n] [US]**, bus pull-off** *trans.* ►bus pullout [US]/bus lay-by [UK].

strip [n]**, grass** *constr.* ►grass strip (1), ►turf strip (1).

strip [n]**, grassed** *constr.* ►grass strip (1).

strip [n] [US]**, pull-off** *trans.* ►pullout [US]/lay-by [UK].

strip [n] [UK]**, reed** *phyt.* ►reed canary grass swamp [US].

strip [n] [UK]**, rumbling** *trans.* ►speed hump.

6106 strip development [n] (1) *urb.* (**1.** Linear development of a community, caused by geomorphological conditions; e.g. in valleys. **2.** Long narrow rows of buildings and facilities, most often commercial in nature, along the edges of urban highways and between city cores [US]/town centres [UK]; **in U.K.**, ribbon development was common between the two World Wars); *syn.* ribbon development [n] [UK] (1); *s* **desarrollo [m] lineal de la ciudad** (**1.** Orientación lineal de una ciudad debida a condiciones geomorfológicas, p. ej. a lo largo de valles. **2.** Alineación de edificios y construcciones a lo largo de vías de comunicación); *f* **extension [f] urbaine linéaire** (**1.** Forme d'urbanisation liée à des caractères géomorphologiques ayant pris naissance dans le fond d'une vallée encaissée, le long d'un cours d'eau. **2.** Processus de développement urbain le long des axes routiers en milieu rural ou à l'entrée des villes) ; *g* **lineare Stadtentwicklung [f]** (**1.** Geomorphologisch bedingte streifenförmige städtebauliche Ausrichtung, z. B. in Tälern. **2.** Aneinanderreihung von Gebäuden und baulichen Anlagen, oft gewerblicher und industrieller Art, entlang von Straßen).

6107 strip development [n] [US] (2) *urb.* (In U.S., commercial or retail development, usually one-store deep, that fronts onto a ►major street or service road; generally associated with intensive use of billboards [US]/advertisements to attract passersby. **S. d.** is contrary to the basic of good urban planning: it consumes open space and depletes natural resources, impedes pedestrian and non-motorized traffic, and grows outward from the limits of existing development; **s. d.** is a precursor to ►urban sprawl; cf. MOL 2009); *syn.* commercial strip development [n] [US], commercial corridor [n] [US], ribbon development [n] [UK] (2); *s* **crecimiento [m] desordenado de la periferia de la ciudad** (Desarrollo caótico a lo largo de las ►arterias [principales] de tráfico debido a la instalación de hipermercados, centros comerciales o sedes de empresas, que generalmente ocupan mucho espacio); *f* **développement [m] urbain périphérique désordonné en bordure des axes d'accès** (Un des objectifs de l'urbanisme consiste dans la maîtrise de l'urbanisation le long des axes de communication ; ►voie primaire) ; *g* **Ortsrandauswucherung [f] entlang von Ausfallstraßen** (Ungeordnete, bandartige Stadtentwicklung entlang einer ►Hauptverkehrsstraße 2, die vorwiegend aus großflächigen, Gewerbebetrieben, Supermärkten und sonstigen Firmenniederlassungen besteht).

strip development [n] [US]**, commercial** *urb.* ►strip development [US] (2).

strip drain [n] **with a metal grille** [US] *constr.* ►cast iron drain with grating [US].

strip farming [n] *agr. land'man.* ►contour strip cropping.

strip footing [n] [US] *constr.* ►strip foundation.

6108 strip foundation [n] *constr.* (Narrow, longitudinal footing as opposed to a ►post foundation [US]/spot foundation [UK]; ►step strip foundation); *syn.* strip footing [n] [also US] (TSS 1997); *s* **zapata [f] corrida** (En comparación con la placa de fundamentación o la ►cimentación puntual una cimentación alargada; ►cimiento de escalera); *f 1* **semelle [f] filante** (Fondation filante sous un mur, escalier, etc.) ; *f 2* **longrine [f] en béton** (Poutre en infrastructure reposant sur des points d'appui liés à une fondation ; ►fondation d'ancrage, ►fondation sur pieux) ; *g* **Streifenfundament [n]** (2) (Im Vergleich zur Fundamentplatte oder zum ►Punktfundament eine streifenförmige Gründung; ►Streifenfundament 1).

strong creeping [ppr] *bot.* ►stem-spreading.

strong-growing [adj] *hort. plant.* ►vigorous.

6109 strong light exposure [n] *bot. syn.* high level [n] of light exposure; *s* iluminación [f] **fuerte**; *f* forte exposition [f] à la lumière ; *g* starke Belichtung [f].

6110 strongly branched liner [n] [US] *arb. hort.* (Immature plant of a ▶ground branching tree [US]/feathered tree [UK], once transplanted, in sizes of 60-80 to 125-150cm); *syn.* one-year-old feathered tree [n] [UK]; *s* leñosa [f] **joven ramificada desde la base (≠)** (**D.**, ▶leñosa ramificada desde la base, transplantada una vez, y de tamaños entre 60-80 y 125-150 cm); *f* jeune baliveau [m] (Végétal de genre et d'espèce ayant les qualités d'arbre à l'état adulte, élevé en pépinière à des distances permettant un développement harmonieux du système racinaire et foliaire et livré aux dimensions normalisées ; dimension généralement de 60/90 jusqu'à 120/150 ; N.F. V 12 051 ; ▶baliveau) ; *g* leichter Heister [m] (**D.**, Jungpflanze eines ▶Heisters, 1 x verpflanzt, in Größen zwischen 60-80 bis 125-150 cm sortiert. Bei Angeboten, Ausschreibungen und Rechnungen gilt seit 09/1995 als Sortierkriterium nur noch die Höhenangabe).

6111 strong upright branch [n] *arb.* (Large vertical branch which develops mostly at the end of a pruned branch or highest part of a drooping branch); *s* flecha [f] (Rama primaria o secundaria que crece en vertical a menudo al final de una rama podada o en la parte superior de una rama con inclinación hacia el suelo); *f* rameau [m] **vertical** (Forte branche verticale se développant en particulier après la coupe à l'aisselle d'une ramification, en amont d'un œil supérieur ou au point haut d'une branche courbée) ; *g* Ständer [m] (Senkrecht wachsender starker Ast, der sich vor allem an Kappstellen, auf waagerechten Stämmen und Ästen oder im Scheitelpunkt geneigter Äste entwickelt).

6112 structural change [n] *plan.* (Long-term and fundamental change, e.g. in the economy of a country and the associated economic process; ▶old industrial region); *s* cambio [m] **estructural** (Transformación profunda de la estructura económica de una región o un país, debido al avance tecnológico, a la integración en el mercado mundial y a las interdependencias internacionales; ▶zona industrial en declive); *f* changement [m] **structurel** (Transformation profonde et à long terme de la structure économique d'une région ou d'un pays en général causée par l'internationalisation des processus de production dans l'économie mondiale ; ▶région industrielle en déclin) ; *g* Strukturwandel [m] (Langfristige und grundsätzliche Änderungen, z. B. der Wirtschaftsstruktur und die damit verbundenen Wirtschaftsprozesse in einem Land oder Teilen davon; ▶altindustrialisierte Region); *syn.* Strukturänderung [f].

6113 structural characteristics [npl] **of a plant community** *phyt.* (Phytosociological grouping and ranking of species and their composition according to ▶life forms within a ▶plant association); *syn.* structure [n] of plant community; *s* estructura [f] **de una comunidad** (Agrupación fitosociológica e importancia de las especies así como su composición según ▶formas biologicas dentro de una ▶asociación vegetal); *f* organisation [f] **des groupements végétaux** (*Phytosociologie* caractères analytiques et synthétiques définissant les espèces et individus d'espèces entrant dans la constitution d'une ▶association [végétale] selon l'importance accordée à leurs ▶types biologiques) ; *syn.* structure [f] des groupements végétaux ; *g* Gesellschaftsstruktur [f] (Gruppierung und soziologische Einordnung der Arten sowie ihre Zusammensetzung nach ▶Lebensformen innerhalb einer ▶Assoziation); *syn.* Gesellschaftsorganisation [f].

6114 structural diversity [n] *ecol.* (Multiplicity of three-dimensional spatial elements in an ecosystem or part thereof, including the organisms living there; ▶diversity, ▶taxonomic diversity); *s* diversidad [f] **estructural** (Gran variedad de elementos estructurales espaciales en un paisaje o ecosistema o en partes de los mismos, incluidos los organismos que habitan en ellos; ▶diversidad ecológica, ▶espectro de diversidad); *f* diversité [f] **structurale** (Multiplicité des éléments spatiaux structuraux dans tout ou partie d'écosystème/paysage, y compris les organismes vivant en son sein ; ▶diversité, ▶diversité des espèces) ; *g* Strukturdiversität [f] (Vielfältigkeit von Raumstrukturelementen in einem Landschaftsraum/Ökosystem oder in Teilen davon, inklusive der darin lebenden Organismen; ▶Artendiversität, ▶Diversität).

6115 structural engineer [n] *arch. eng.* (Engineer engaged in the structural calculations of buildings and other structures); *s* ingeniero/a [m/f] **de estructuras** (Ingeniero de caminos [Es]/ civil [AL] que se dedica a calcular la estática de construcciones); *syn.* especialista [m/f] en cálculos estáticos; *f* ingénieur [m] **(en) structure(s)** (Ingénieur établissant le calcul des structures dans les projets de construction) ; *syn.* ingénieur [m] spécialisé en structures ; *g* Statiker/-in [m/f] (Bauingenieur, der sich mit statischen Berechnungen von Bauwerken befasst); *syn.* Tragwerksplaner/-in [m/f].

structural engineering [n] (1) *constr. eng.* ▶statics and dynamics.

6116 structural engineering [n] (2) *eng.* (Branch of civil engineering concerned primarily with the design and construction of structures [bridges, buildings, dams, walls, etc] and the structural calculation of these structures as well as of materials; ▶statics and dynamics); *s* ingeniería [f] **de construcción** (Rama de la ingeniería civil que estudia las características de los materiales de construcción y el comportamiento estructural y estático; ▶estática); *f* ingénierie [f] **du bâtiment** (Catégorie de l'ingénierie du génie civil regroupant les professions de l'ingénieur relatives aux travaux de gros œuvre, de second œuvre et d'équipements techniques de constructions ; ▶statique) ; *g* Bauingenieurwesen [n, o. Pl.] (2) (Arbeitsgebiet des Bauingenieurs und Statikers im konstruktiven Hochbau und Ingenieurbau; ▶Statik); *syn.* Ingenieurbau [m, o. Pl.].

6117 structural engineering office [n] *eng. stat.* (Planning practice concerned with the structural calculations of buildings and other structures); *s* oficina [f] **de ingeniería estructural** (Gabinete de ingeniería que se dedica a realizar los cálculos estáticos de construcciones); *f* bureau [m] d'études **(en) structure(s)** (Bureau d'ingénierie établissant le calcul des structures dans les projets de construction) ; *g* Statikbüro [n] (Ingenieurbüro, das sich mit statischen Berechnungen von Bauwerken und baulichen Anlagen befasst); *syn.* Büro [n] für Tragwerksplanung.

structural fill [n] *constr.* ▶fill material, ▶land filling.

6118 structural load(ings) [n(pl)] *eng. stat.* (Force or system of forces which impact a structure or parts thereof; ▶service load); *s* sobrecarga [f] (Carga variable que soporta un elemento constructivo, además de su peso; DACO 1988; ▶carga móvil); *f 1* charges [fpl] **(appliquées à un bâtiment)** (*Terme générique* ensemble des charges [▶charge statique et charge dynamique] qui doivent être prises en compte lors de la construction d'un bâtiment afin d'éviter le tassement, l'effondrement ou la déformation permanente de la construction ; ▶charge d'exploitation) ; *f 2* charge [f] **statique** (Charge appliquée à un bâtiment comprenant le poids du bâtiment lui-même, ainsi que tous ses éléments) ; *f 3* surcharge [f] (*Terme générique* décrivant au delà des charges permanentes [charges statiques] les charges temporaires telles **1.** les ▶charges d'exploitation qui prennent en compte l'occupation humaine et des mobiliers sur un plancher, **2.** les charges d'entretien sur les couvertures d'un bâtiment [charpente, terrasse accessible ou non], **3.** les charges dues a la neige, à la glace et à la pluie, **4.** les charge dues au vent, **5.** les charges dues au séismes) ; *syn.* charge [f] dynamique ; *g* **statische Belastung [f]** (Last, die auf ein Bauwerk oder auf dessen Teile wirkt; ▶Verkehrslast).

6119 structurally stable soil [n] *pedol.* (▶structural stability of soil); *s* **suelo [m] de estructura estable** (Suelo con gran ▶estabilidad de la grumosidad); *f* **sol [m] à structure stable** (▶stabilité structurale du sol) ; *g* **strukturstabiler Boden [m]** (Boden mit grosser ▶Gefügestabilität).

6120 structural root [n] *arb.* (RRST 1983; one of the anchor roots of a tree with diameter > 50mm; ▶medium root, ▶tap root); *syn.* skeletal root [n], skeleton root [n], scaffold root [n]; *s* **raíz [f] primaria** (Raíz de > 50 mm de diámetro cuya función principal es la de soporte del árbol, aunque también sirve para transportar agua y nutrientes; ▶raíz axonomorfa, ▶raíz secundaria); *syn.* raíz [f] principal; *f* **racine [f] maîtresse** (Racine de dimension importante [diamètre > 50 mm] à la base d'un tronc [souche] permettant essentiellement la fixation de l'arbre au sol ; *terme spécifique* racine contrefort ; GOR 1985, 47 ; ▶racine pivotante, ▶racine secondaire) ; *g* **Starkwurzel [f]** (Vorwiegend der Verankerung eines Baumes, aber auch dem Wasser- und Nährstofftransport dienende Wurzel mit einem Durchmesser > 50 mm; ▶Grobwurzel, ▶Pfahlwurzel).

structural slab [n] *constr.* ▶concrete slab (1).

6121 structural stability [n] *constr. eng. stat.* (Intrinsic strength of a structure, e.g. due to a stable foundation, which prevents it from subsiding or collapsing); *s* **estabilidad [f] estática** (Seguridad intrínseca de edificios, taludes, diques, etc. gracias a una cimentación estable que previene subsidencia o colapso); *f* **stabilité [f] des ouvrages** (Qualité caractérisant la structure, l'ossature d'un ouvrage permettant de transmettre aux fondations et donc au sol, les sollicitations dues au poids de l'ouvrage, aux charges d'occupation, aux actions exercées par divers agents tels que le vent, les secousses sismiques, etc.) ; *g* **Standsicherheit [f, o. Pl.] von (Erd)bauwerken** (Sicherheit eines Baukörpers gegen Setzungen oder [Um]kippen); *syn.* Stabilität [f] von (Erd)bauwerken.

6122 structural stability [n] **of soil** *pedol.* (Firmness of soil with resistance to changes in its structure caused by, e.g. precipitation, pedestrian or vehicular traffic or infiltration; ▶structurally stable soil); *syn.* aggregate stability [n] of soil, ped stability [n] [also US]; *s* **estabilidad [f] de la grumosidad** (BB 1979, 444; resistencia de los grumos o agregados del suelo a deshacerse o disgregarse en condiciones de humedad. Ésta depende del tipo y cantidad de arcilla, de materia orgánica y de cualquier otro agente cementante; DINA 1987, 908; ▶suelo de estructura estable); *syn.* estabilidad [f] de terrones, estabilidad [f] estructural, estabilidad [f] de los agregados; *f* **stabilité [f] structurale du sol** (Capacité d'un sol à résister aux modifications de sa structure et aux facteurs de dégradation comme les orages violents, le tassement dû au piétinement ou au roulement des véhicules. La destruction de la structure se traduit par une perte de porosité, de perméabilité, par une prise en masse, un état de dispersion et par une croûte de battance ; DIS 1986 ; ▶sol à structure stable) ; *syn.* résistance [f] des agrégats, stabilité [f] de la structure ; *g* **Gefügestabilität [f]** (Widerstandsfähigkeit eines Bodengefüges gegen Veränderungen infolge erhöhter Beanspruchung, z. B. durch Regenschlag, Betreten, Befahren oder Sickerwasserfronten; SS 1979; ▶strukturstabiler Boden); *syn.* Aggregatstabilität [f], Strukturstabilität [f].

6123 structural stability [n] **of trees** *arb. stat.* (stable tree conditions); *syn.* structural strength [n] of trees; *s* **estabilidad [f] de árboles** (Capacidad natural de los árboles de mantenerse en pie tras tormentas); *f* **stabilité [f] d'un arbre** (Capacité naturelle d'un arbre à résister aux intempéries) ; *g* **Standsicherheit [f, o. Pl.] von Bäumen** (Natürliche Fähigkeit eines Baumes, bei normalen äußeren Einflüssen, auch bei starkem Wind, nicht umzustürzen); *syn.* Standfestigkeit [f] von Bäumen.

structural strength [n]**, testing the** *eng. stat.* ▶structural testing.

structural strength [n] **of trees** *arb. stat.* ▶structural stability of trees.

6124 structural testing [n] *eng. stat.* (Examination of the ▶load-bearing capacity of bridges, roofs, slabs, walls, etc.); *syn.* testing [n] the structural strength; *s* **análisis [m] de la capacidad de carga** (Estudio de la ▶capacidad de carga de puentes, azoteas, cubiertas de aparcamientos subterráneos, muros, etc.); *syn.* examen [m] de la capacidad de carga; *f* **examen [m] de la capacité de charge** (Étude de la ▶capacité de charge [d'exploitation] de ponts, toitures, dalles armées, etc.) ; *g* **Prüfung [f] der Belastbarkeit** (Untersuchung der ▶statischen Belastbarkeit von z. B. Brücken, Dachflächen, Tiefgaragendecken, Mauern und deren Gründungen).

6125 structure [n] **(1)** *leg. urb.* (Any facility on or permanently attached to the ground and not classified as a building, constructed according to building regulations); *s* **construcción [f]** (Término genérico que incluye edificios y otras instalaciones construidas de materiales inertes y definidas como tal por la correspondiente legislación); *f* **construction [f]** (Ouvrage érigé conformément aux règles relatives à l'acte de construire et aux prescriptions régissant les modes d'utilisation des sols) ; *g* **bauliche Anlage [f]** (D., im Sinne der Landesbauordnungen mit dem Erdboden verbundene, aus Baustoffen und Bauteilen hergestellte Anlage; eine Verbindung mit dem Erdboden besteht auch dann, wenn die Anlage durch eigene Schwere auf dem Erdboden ruht oder auf ortsfesten Bahnen begrenzt beweglich ist oder wenn die Anlage nach ihrem Verwendungszweck dazu bestimmt ist, überwiegend ortsfest benutzt zu werden; zu **b.n A.n** gehören auch Aufschüttungen und Abgrabungen [auch zur Gewinnung von Steinen, Erden und anderen Bodenschätzen], Lager-, Abstell- und Ausstellungsplätze, Camping- und Dauerwohnwagenplätze sowie Kraftfahrzeugstellplätze).

6126 structure [n] **(2)** *constr. eng.* (1. Combination of units constructed and so put together spatially as to provide rigidity between its elements. 2. Composite of individual elements assembled in accordance with rules and laws); *s* **estructura [f]** (Distribución y orden espacial de elementos que tiene como resultado una construcción estable); *f* **structure [f] (1)** (1. Agencement, arrangement, disposition des parties d'un ensemble présentant un ordre spatial stable. 2. Ensemble complexe dont l'organisation des éléments essentiels qui le compose répond aux règles et aux lois qui le définissent) ; *g* **Struktur [f] (1)** (Lat. *structura* >Bau, Zusammenfügung< zu lat. *struere* >bauen<; 1. Anordnung von Teilen eines Ganzen zueinander als räumlicher, in sich stabiler Aufbau. 2. Ein aus Einzelelementen nach Regeln und Gesetzmäßigkeiten zusammengefügtes oder aufgebautes, komplexes, konstruktives Ganzes); *syn.* Aufbau [m], Bau [m] (2).

6127 structure [n] **(3)** *plant.* (Lat. *structura* >building, assembly< from lat. *struere* >to build<; an organizing principle, which facilitates the perception of various individual elements and contours within a clear overall structure. **S.s** require transparency, i.e. the last or previous element must be at least partly identifiable between the individual elements, so that an incoherent overall image [loose structure] does not occur. *NOTE* an image, or the interaction of various objects within a space, is always a **s.**, and not the occurrence of individual elements. In *Landscape Architecture:* [transparent] **spatial s.s** occur, for example, in a landscape park, where individual trees and groups of shrubs present a scene of both transparency and view corridors, or in planting where grasses and tree branches present their own **s.s.**, e.g. trees and shrubs in winter have **s.s** from the very coarse [e.g. the thick tree branches of Kentucky coffee tree *(Gymnocladus)*] to the thin branches of the Japanese pagoda tree *[Sophora]*. Allees and hedges also represent a spatially linear structure. **Graphic s.s** are refered to when branches clearly stand out against a bright sky or a wall. **Surficial s.s** become clear in the top view, for example, when a leaf pattern is observed from above, see Hosta species. **S.s** also occur as colors or tree bark patterns, e.g. hornbeam *[Carpinus]* or in various ground cover plants. In planting terms, **open s.s** indicate the transition between existing species and take changes in their development into account on the basis of long term maintenance goals. **Enclosed spatial s.s** have defined borders with neighboring species, for example, in block planting or large areas of planting, and are intended to be long-term planting structures; cf. W. BORCHARDT and K. H. RÜCKER Garden Praxis 2005, H. 4, pp. 31-37; ▶degree of total vegetative cover, ▶texture); *s* **estructura [f] (3)** (Lat. *structura* >construcción, ensamblaje< del lat. *struere* >construir<; un principio de ordenación que facilita claramente la percepción de varios elementos individuales y sus contornos dentro de un todo. Las **ee.** requieren transparencia, es decir, entre los elementos individuales debe ser visible —por lo menos parcialmente— lo que está detrás para que no surja una imagen general incoherente [desestructurada]. *NOTA:* La imagen, o sea la interacción de varios objetos dentro de un espacio de percepción, siempre refleja una **e.**, no el elemento individual participante en su creación. En *arquitectura paisajista* **ee. espaciales** [transparentes] se manifiestan p. ej. en un parque inglés, en el que ejemplares de árboles o grupos de arbustos dejan percibir una escena con vistas y corredores visuales, o en las plantaciones donde las hierbas y las ramas de leñosas presentan sus propias **ee.** que van de «muy burdas» [como las ramas gruesas del *Gymnocladus* a «muy finas» como las finas ramas del árbol de la pagoda *(Sophora)*]. También avenidas y setos son **ee.** espaciales lineales. De **ee. gráficas** se habla cuando el conjunto de los vástagos y ramas resalta claramente ante el trasfondo del cielo o de un muro de color claro. **Estructuras superficiales** son perceptibles en la vista de planta, p ej. en los dibujos de las hojas del llantén *[Hosta 'Crispula']*, en los patterns de colores de la corteza del carpe *[Carpinus]* o entre diferentes tipos de plantas tapizantes. **Ee. superficiales abiertas** muestran transiciones entre macizos monoespecíficos y plantaciones cambiantes según estación y toleran también cambios temporales de las cantidades relativas de las diferentes especies, según los objetivos de cuidado perseguidos. **Ee. superficiales cerradas** tienen límites bien definidos hacia las especies vecinas, como en el caso de plantaciones en macizos homogéneos o en grandes áreas, y tienen como objetivo crear imágenes de vegetación estáticas y de largo plazo; ▶expansión horizontal, ▶textura); *f* **structure [f] (2)** (lat. *structura*, de *struere* « construire » principe d'ordonnancement, favorisant la perception visuelle de différents éléments isolés et de leurs contours qui se dégagent clairement dans un espace ; les structures ont besoin de transparence, l'élément en arrière plan étant encore en partie reconnaissable entre les éléments isolés afin d'éviter la formation d'un tableau d'ensemble astructuré ; *dans l'architecture des paysages* : les **structures spatiales transparentes** s'observent dans les parcs paysagers dans lesquels des arbres isolés ou des groupes de végétaux ligneux laissent percevoir un décors constitué de vues traversantes et de couloirs de vue ; en hiver les arbres et arbustes laissent discerner des structures allant de « très grossières » [p. ex. les branches fortes du chicot du Canada *(Gymnocladus)*] jusqu'à « très fines » [p. ex. les fines branches de l'arbre des pagodes *(Sophora)*], les allées et les haies constituant elles aussi des structures spatiales. On parle

de **structure graphique** lorsque l'ensemble pousses/branches se détache clairement du ciel ou d'un mur de couleur claire ; une **structure de surface** est reconnaissable par vue en plan tel le graphisme d'une feuille de Hosta *[Hosta 'Crispula']*, verte bordée de blanc ou celui de l'écorce du charme *[Carpinus]* ; une **structure de surface ouverte** peut aller des transitions harmonieuses entre les différents massifs d'espèces de plantes jusqu'à la plantation en structure liée ; elle tolère une modification des pourcentages entre les espèces et de leurs ▶degré de recouvrement conformément aux objectifs d'entretien fixés au planning annuel. La **structure de surface fermée** dégage des limites bien définies par rapport aux espèces d'une plantation voisine, comme par exemple la plantation en massifs homogènes ou la plantation en structure de surface et ont pour objectif le maintien d'un aspect végétal statique et de long terme ; ▶texture) ; *g* **Struktur [f] (2)** (Lat. *structura* >Bau, Zusammenfügung< zu lat. *struere* >bauen<; ein Ordnungsprinzip, das die Wahrnehmung erleichtert, nach dem verschiedene Einzelelemente mit ihren Konturen in einem Gesamtgefüge deutlich werden. **S.en** bedürfen der Transparenz, d. h. zwischen den Einzelelementen muss mindestens das dahinter Liegende teilweise erkennbar sein, damit kein ungegliedertes, [strukturloses] Gesamtbild entsteht. *MERKE*, stets weist das Bild, also das Zusammenwirken von Objekten in einem Wahrnehmungsraum, eine **S.** auf, nicht das an der Entstehung beteiligte Einzelelement. *In der Landschaftsarchitektur*: [transparente] **Raum-S.en** zeigen sich z. B. im Landschaftspark, in dem Einzelbäume und Gehölzgruppen eine Szenerie mit Durchsichten und Blickkorridoren wahrnehmen lassen, oder beim Spross-/Zweiggefüge bei Gräsern und Gehölzen. Z. B. lassen Bäume und Sträucher im Winter **S.en** erkennen, die von „sehr grob" [z. B. starke Äste beim Geweihbaum *(Gymnocladus)*] bis „sehr fein" [z. B. dünne Äste beim Schnurbaum *(Sophora)*] reichen. Auch Alleen und Hecken sind räumliche **S.en**. Von **grafischen Strukturen** spricht man, wenn das Spross-/Zweiggefüge sich deutlich gegen den Himmel oder eine helle Wand abhebt. **Flächen-S.en** werden in der Draufsicht z. B. beim Blattmuster bei der Weißrandfunkie *[Hosta 'Crispula']*, beim farblich abgesetzten Rindenmuster der Hainbuche *[Carpinus]* oder bei verschiedenartiger Bodenvegetation erkennbar. **Offene Flächen-S.en** zeigen Übergänge zwischen Artbeständen bis hin zur Verlaufspflanzung und tolerieren auch zeitliche Veränderungen der Artmengenanteile/▶Deckungsgrade entsprechend den im Pflegeplan festgelegten Pflegezielen. **Geschlossene Flächen-S.en** haben definierte Grenzen zu benachbarten Artbeständen, z. B. bei der Blockpflanzung oder Flächenstrukturpflanzung, und zielen auf statische, langfristig zu erhaltende Vegetationsbilder; cf. W. BORCHARDT and K. H. RÜCKER in Gartenpraxis 2005, H. 4, pp. 31-37; ▶Textur).

structure [n] **(4)** *leg. urb.* ▶accessory structure; *landsc. urb.* ▶advertisement structure; *constr.* ▶control structure; *pedol.* ▶crumb structure; *constr.* ▶drop structure; *agr. land'man.* ▶farmland structure; *for.* ▶forest structure; *pedol.* ▶granular soil structure; *conserv'hist.* ▶historic structure; *agr. plan.* ▶improvement of agrarian structure; *landsc.* ▶landscape structure; *plan.* ▶land use structure; *trans.* ▶overhead sign structure; *constr. eng.* ▶pavement structure; *constr. trans.* ▶planted noise attenuation structure; *constr. landsc.* ▶planting structure; *constr. eng.* ▶road structure; *arch.* ▶roof structure; *pedol.* ▶single-grain structure; *constr. eng.* ▶surface inlet [US]; *pedol.* ▶soil structure; *for. phyt.* ▶stratified stand structure.

structure [n], **aggregate** *pedol.* ▶crumb structure.

structure [n] [UK], **ancillary** *leg. urb.* ▶accessory structure.

structure [n], **coastal protection** *conserv. eng.* ▶sea defenses [US]/sea defences [UK].

S

structure [n], **detention** *hydr. wat'man.* ►storm water detention basin.

structure [n] [US]**, drainage inlet** *constr. eng.* ►surface inlet [US]/gully [UK].

structure [n]**, improvement of agricultural** *agr. plan.* ►improvement of agrarian structure.

structure [n] [UK]**, land ownership** *agr. plan.* ►land ownership pattern [US].

structure [n]**, load-bearing** *constr.* ►load-bearing construction.

structure [n]**, noise attenuating earth** *constr. envir.* ►noise attenuation mound.

structure [n] [US]**, outfall** *constr. envir.* ►drop structure.

structure [n] [UK]**, paving** *constr. eng.* ►pavement structure.

structure [n]**, retention** *hydr. wat'man.* ►storm water detention basin.

structure [n]**, secondary** *leg. urb.* ►accessory structure.

structure [n]**, sub-** *constr.* ►footing.

structure [n] [US]**, supportive permanent** *leg. urb.* ►accessory structure.

structure [n]**, tree** *arb.* ►skeleton.

structured landscape [n] *landsc.* *►structuring of landscape.

structure [n] **for signage, span** *trans.* ►overhead sign structure.

structure [n] **of a forest stand** *for.* ►forest structure.

structure [n] **of plant community** *phyt.* ►structural characteristics of a plant community.

structure plan [n] (1) *landsc. plan.* ►landscape structure plan.

structure plan [n] [UK] (2) *leg. urb. obs.* *►community development plan [US]/urban development plan [UK].

structure plan [n]**, green open space** [± UK] *landsc. urb.* ►general plan for urban open spaces [US].

structure planning [n] *landsc.* ►landscape structure planning; *agr. plan.* ►preliminary agrarian structure planning.

6128 structure planting [n] *gard. landsc.* (Tall, space-defining planting in open spaces; ►division of spaces); *syn.* divisional planting [n] [also US]; *s* **plantación [f] estructurante** (Conjunto de plantas que sirven para crear y estructurar espacios en zonas verdes; ►creación de espacios); *f* **plantation [f] structurante** (Élément de végétation utilisé pour définir la structure d'un espace vert ; ►structuration spatiale) ; *syn.* plantation [f] de structuration paysagère ; *g* **Rahmenpflanzung [f]** (Hohe, raumbildende Pflanzung in Freianlagen; ►Raumbildung).

6129 structuring [n] **of landscape** *landsc.* (Spatial distribution of landscape elements/features, e.g. the change from woodlands to agricultural areas or the planting of field hedges in re-organized agricultural areas; *result* **landscape mosaic pattern, structured landscape**); *syn.* spatial articulation [n]; *s* **estructuración [f] del paisaje** (Distribución espacial de los elementos del paisaje, como bosques, campos de cultivo o setos interparcelarios); *f* **structuration [f] d'un paysage** (Configuration spatiale, aspect général d'un paysage dû à la présence de haies dans un paysage bocagé ou à l'alternance de surfaces agricoles et forestières) ; *syn.* structure [f] spatiale ; *g* **räumliche Gliederung [f] der Landschaft** (Dreidimensionale Aufteilung der Landschaft durch den Wechsel einzelner Landschaftselemente, z. B. durch den Wechsel von Wald- und landwirtschaftlichen Flächen, durch Gebäudegruppen, Feldhecken oder Gehölzstreifen entlang von Fließgewässern).

6130 structuring [n] **of townscape** *landsc. urb.* (Spatial arrangement of elements which create the visual appearance of a town, e.g. high-rise buildings, parks and open spaces, rivers); *syn.* spatial configuration [n] of a town; *s* **trama [f] urbana** (Forma de agruparse y ordenarse las edificaciones en el plano de la ciudad. Según se dispongan los edificios y según la proporción del espacio edificado con respecto al abierto se distinguen **t.** cerrada, **t.** en orden abierto; DGA 1986); *syn.* tejido [m] urbano; *f* **tissu [m] urbain** (Ensemble des éléments physiques du cadre urbain défini par la structuration architecturale, le réseau de voirie, les espaces verts, etc. ; expression pouvant être rapprochée de celle de la composition urbaine) ; *syn.* structure [f] urbaine, structuration [f] urbaine ; *g* **städtebauliche Gliederung [f]** (G. des Stadtbildes z. B. durch Grünflächen, abwechselnd hohe Gebäude, Wechsel von dichter und lockerer Bebauung, Hauptverkehrsstraßen etc.); *syn.* Gliederung [f] des Stadtbildes.

strutting [n] *constr.* ►planking and strutting.

stub [n] *arb. for. hort.* ►branch stub [US]/branch stump [UK], ►snag [US&CDN] (2)/small stump [UK].

stubbing [n] [US] *arb. constr.* ►lopping.

stubbing [n] **of main scaffolds** *arb. hort.* ►heading back of tree scaffolds.

stub out [vb] [US&CND] *constr. for. hort.* ►stump out.

study [n] *conserv. leg. nat'res. plan.* ►environmental impact study (EIS) [US]/environmental assessment [UK]; *plan. trans.* ►feasibility study; *ecol. landsc. plan.* ►pilot study; *prof.* ►university study; *envir. phyt.* ►vegetation impact study.

study [n]**, demonstration** *ecol. landsc. plan.* ►pilot study.

study [n]**, site** *plan.* ►site analysis.

study [n]**, suitability** *plan.* ►suitability evaluation.

6131 study area [n] *plan.* (Defined area for planning work or obtaining an expert's opinion; ►research area); *syn.* planning area [n] (2); *s* **zona [f] de estudio** (►área de investigación); *syn.* área [f] de estudio; *syn. parcial* área [f] a planificar, zona [f] a planificar; *f* **zone [f] d'étude** (2) (Espace défini pour l'élaboration d'études ou de travaux d'aménagement, p. ex. délimitation d'une zone d'étude et justification de ses limites en fonction de la description des composantes des milieux biophysique et humain pertinentes au projet dans une étude d'impact ; ►zone d'étude 1) ; *syn.* zone [f] à étudier, zone [f] étudiée ; *g* **Bearbeitungsgebiet [n]** (►Untersuchungsgebiet für Planungen oder gutachtliche Stellungnahmen); *syn.* z. T. Planungsgebiet [n].

6132 study [n] **of leisure activities** *plan. recr.* (Quality research on leisure requirements); *s* **estudio [m] de las actividades de tiempo libre** (Investigación sobre el comportamiento y las necesidades de la población para el aprovechamiento de su tiempo libre); *f* **étude [f] d'activités de loisirs** (Étude réalisée afin de connaître le comportement et les besoins des acteurs pendant leurs loisirs) ; *g* **Freizeituntersuchung [f]** (Studie, die das Freizeitverhalten und die Ansprüche der Menschen an ihre Freizeit untersucht).

study trail [n] *recr.* ►forest study trail.

stump [n] *constr. for. hort.* ►tree stump.

stump [n] [UK]**, branch** *constr. for. hort.* ►branch stub [US].

stump [n] [UK]**, small** *arb. for. hort.* ►snag [US&CDN] (2).

6133 stump-chipping [n] *constr.* (Destruction of stumps *in situ* by a special machine; BS 3975: Part 5); *syn.* stump shredding [n] [also US]; *s* **trituración [f] de tocones** (Destrucción de tocones de árboles *in situ* con ayuda de una máquina especial); *f* **broyage [m] de la souche** (Destruction d'arbres par moyens mécaniques des souches d'arbres sur le site) ; *g* **Stubben fräsen**

[vb] (Maschinelles Zerkleinern von Baumstümpfen an Ort und Stelle; *Substantiv* Stubbenfräsen [n, o. Pl.]); *syn.* Baumstumpf fräsen [vb], Wurzelstock fräsen [vb].

6134 stump out [vb] *constr. for. hort.* (Extraction of ►tree stumps by digging, lifting, or by using explosives; stump extraction, grubbing up; ►clearing and removal of tree stumps); *syn.* stub out [vb] [also US&CND], grub up [vb] [UK], grub out [vb] [US]; *s* **extraer** [vb] **tocones** (Arranque de tocones de árboles con raíz con pala o con excavadora; *sustantivo* extracción [f] de tocones; ►destoconado); *f* **dessoucher** [vb] (Extraction de la souche d'un arbre abattu manuellement au moyen d'une bêche ou avec une dessoucheuse ; *substantif* dessouchage, dessouchement ; ►élimination de la souche) ; *g* **Stubben roden** [vb] (Entfernung eines Wurzelstocks mit Handspaten oder Baggerschaufel, ►Stubbenbeseitigung; *syn.* Stubben ausgraben [vb], Wurzelstock ausgraben [vb], Stockrodung [f], Stubbenrodung [f], Wurzelstockrodung [f].

stump [n] **removal** *constr. for. hort.* ►tree clearing and stump removal (operation) [US]/grub felling [UK].

stump shredding [n] [US] *constr.* ►stump-chipping.

stunted wood [n] [US] *phyt.* ►krummholz.

stunting [ppr] **(plant growth)** *biol. hort.* ►inhibiting (plant growth).

6135 sturdy branch [n] *arb.* (Lateral branch with diameter of 5-10cm; size between ►small branch and ►scaffold branch); *s* **rama** [f] **mediana** (Rama lateral con un diámetro entre 5 y 10 cm; tamaño entre la ►rama fina y la ►rama primaria); *f* **branche** [f] **de taille moyenne** (Terme qualifiant une branche d'un diamètre compris entre 5 et 10 cm ; ►charpentière, ►rameau) ; *g* **Grobast** [m] (Bezeichnung eines Astes mit einem Durchmesser von 5-10 cm; in der Größenklassifizierung zwischen ►Feinast und ►Starkast).

subalpine belt [n] *geo. phyt.* ►subalpine zone.

6136 subalpine zone [n] *geo. phyt.* (►Vegetation altitudinal zone between ►montane zone and ►alpine zone); *syn.* subalpine belt [n]; *s* **piso** [m] **subalpino** (En la ►zonación altitudinal ►piso de vegetación inmediatamente inferior al alpino y por encima del montano, que se escalona entre los 900 y los 2100 a 2400 m de altitud; cf. DGA 1986; ►área cacuminal, ►piso montano); *f* **étage** [m] **subalpin** (►Étage de végétation de 1700 à 2400 m environ situé entre l'►étage alpin et l'►étage montagnard correspondant souvent à l'étage forestier climax de forêts arborées claires de conifères, associées à la lande ; cf. DG 1984 ; ►étagement de la végétation) ; *g* **subalpine Stufe** [f] (Hinsichtlich der ►Höhenstufung der Bereich zwischen ►montaner Stufe und ►alpiner Stufe; die Höhenangaben variieren, je nach Quelle und in Abhängigkeit von der Höhe über NN, zwischen > 1500 m in den Mittelgebirgen und zwischen 1500-2500 m in den Alpen ; ►Vegetationsstufe).

subaqueous carpet [n] **of chandelier algae** *phyt.* ►chara vegetation.

subaqueous meadow [n] *phyt.* ►chara vegetation.

subaqueous raw soil [n] *pedol.* ►subhydric soil.

sub-area [n] [UK] *plan.* ►planning district [US].

6137 subassociation [n] *phyt.* (Plant community which does not have ►character species. A **s.** is distinguished from an ►association plant by ►differential species); *s* **subasociación** [f] (Comunidades que difieren del tipo de la asociación pero a las que faltan las ►especies características específicas. Se diferencian de la composición típica de la ►asociación por la presencia de ►especies diferenciales; BB 1979, 118); *f* **sous-association** [f] (Groupement végétal caractérisé par l'absence de certaines ►espèces caractéristiques [exclusives]. La distinction par rapport à l'►association [végétale] s'effectue par l'intermédiaire des ►espèces différentielles) ; *g* **Subassoziation** [f] (Vom Typus der ►Assoziation abweichende Gesellschaft, der spezifische ►Kennarten fehlen; von der typischen Assoziation unterscheidet sie sich durch ►Trennarten).

6138 subbase [n] [US]**/sub-base** [n] [UK] *constr.* (Frostresistant layer composed of sand, gravel or crushed rock with particle sizes of 1-100mm, for homogeneous distribution of traffic load on the subgrade; *specific terms* ►crushed aggregate subbase [US]/crushed aggregate sub-base [UK], ►frost-resistant subbase [US]/frost-resistant sub-base [UK], ►gravel subbase [US]/gravel sub-base [UK], ►hand-laid stone subbase [US]/hand-pitched stone sub-base [UK], ►hard base [US]/hardcore [UK], ►root-zone layer, ►turf root-zone layer); *syn.* base course [n] (2), bearing course [n] [UK]; *s* **capa** [f] **portante** (*Construcción viaria* capa de grava o piedra partida de 1-100 mm de diámetro resistente a la congelación cuya función es repartir resp. transmitir el peso de la carga móvil a las capas inferiores y al subsuelo; *términos específicos* ►base de grava, ►capa de asiento de piedra partida, ►capa de protección contra las heladas, ►cimiento de Telford, ►subbase de piedra compactada; ►sustrato de enraizamiento de césped, ►sustrato de plantación); *syn.* capa [f] de asiento, base [f] de asiento, base [f] portante, capa [f] solera (de hormigón); *f* **couche** [f] **de fondation** (*Travaux de voirie* couche non gélive constituée de graviers ou de matériaux concassés de diamètre compris entre 1 et 100 mm permettant de répartir uniformément les charges sur la couche de forme ou la partie supérieure des terrassements ; en fonction de la classe de trafic, de la nature de la plate-forme support de chaussée et de la couche de roulement peut être mise en œuvre une couche de fondation ; *termes spécifiques* ►couche antigel, ►couche de base en grave, ►couche de fondation en grave non traité, ►empierrement, ►empierrement manuel, ►support de culture, ►support végétal d'un gazon) ; *g* **Tragschicht** [f] (1. *Wegebau* frostfreie Schicht aus Kies oder Schotter in den Größen von 1-100 mm zur Aufnahme und gleichmäßigen Verteilung/Ableitung der Verkehrslast auf den Untergrund/Unterbau; bei ungünstigen Untergrundverhältnissen können auch *Schroppen* mit größeren Fraktionen von 32/200 und 65/x mm verwendet werden. Man unterscheidet **T.en ohne Bindemittel**, **T.en mit hydraulischen Bindemitteln** und **Asphalttragschichten**; *UBe* ►Kiestragschicht, ►Schottertragschicht, ►Schüttpacklage, ►Rasentragschicht, ►Setzpacklage, ►Vegetationstragschicht. 2. Als Unterbau wird heutzutage meist eine **kombinierte Frostschutz-Tragschicht [KFT]** in den Körnungen 0-45 und 0-56 mit entsprechender Sieblinie eingebaut. 3. Bei begrünbarem Pflaster und Plattenbelägen gibt es auch die **T. mit vegetationstechnischen Eigenschaften**. Dies ist eine Verknüpfung zwischen bautechnischen und vegetationstechnischen Eigenschaften in einer **T.** Ein ausreichendes Porenvolumen in dieser **T.** sorgt für die Einwurzelung und einen schnellen Wasserabzug; cf. FFL-Empfehlungen für die Planung, Ausführung und Unterhaltung von Flächen aus begrünbaren Pflasterdecken und Plattenbelägen).

subbase [n], **granular** *constr.* ►crushed aggregate subbase [US]/crushed aggregate sub-base [UK].

sub-base [n] [UK], **hand-pitched stone** *constr.* ►hand-laid stone subbase [US].

6139 subbase grade [n] [US] *constr.* (Surface grade of the subbase to true level before installation of topsoil or base course for hard surfacing; ►excavation to subbase grade [US]/excavation to formation grade [UK], ►subbase grade of a road/path [US]/formation level of a road/path [UK]); *syn.* formation level [n] [UK]; *s* **subrasante** [f] (Nivel de superficie de la subbase preparada para el pavimento con las características mecánicas requeridas; ►lecho de carretera/camino); *syn.* rasante [f], plataforma [f], cancha [f] [RA] (BU 1959); *f* **fond** [m] **de forme**

(Surface du terrain préparée pour les bâtiments, couche de forme pour les encaissements de voirie et présentant le profil et les qualités mécaniques requises ; ►arase de terrassement) ; *syn.* plate-forme [f], forme [f] (DIR 1977, 57) ; *g* **Erdplanum [n]** (Technisch bearbeitete Oberfläche des Baugrundes oder Unterbaues mit festgelegten Merkmalen wie Gefälle, Höhenlage und Ebenheit [profilgerechte Herstellung], die vor der Lockerung auf einer 4-m-Messstrecke nicht mehr als 5 cm von der Ebenheit, bei Anschlüssen nicht mehr als 3 cm von der Nennhöhe abweicht; ►Koffersohle); *syn.* Planum [n].

subbase grade [n] [US]**, excavation for** *constr.* ►subbase grade preparation [US]/grading of formation level [UK].

6140 subbase grade [n] **of a road/path** [US] *constr.* (Final grade prior to subbase course preparation for paved road and other paved surface construction; ►pavement structure, ►pathway excavation); *syn.* subgrade [n] of a road/path [US], formation level [n] of a road/path [UK]; *s* **lecho [m] de carretera/camino** (Nivel definitivo de la subrasante de construcción viaria; ►estructura del cuerpo de carreteras y caminos, ►vaciado para carretera); *syn.* fondo [m] de carretera/camino; *f* **arase [m] de terrassement** (Plate-forme constituée des sols en place ou des matériaux rapportés formant la ►partie supérieure des terrassements [PST] ; ►décaissement des circulations, ►structure de la voirie) ; *g* **Koffersohle [f]** (Planum beim Wegebau; ►Wegeaufbau, ►Wegekoffer).

6141 subbase grade preparation [n] [US] *constr.* (Grading and consolidation operations [i.e.; with vibrating roller for granular soils or sheeps foot roller for clay soils] for establishing the subgrade level for proposed final elevations and to achieve the specified bearing capacity and linear deflection tolerances for slabs, road beds, etc.; excavate [vb] to formation [UK]); *syn.* excavation [n] to formation level [UK], grading [n] of formation level [UK]; *s* **preparación [f] de la (sub)rasante** (Operaciones de enrasado y consolidación [p. ej. con vibradora para suelos sueltos o con pisón para suelos arcillosos] para establecer el nivel de subrasante de la elevación final y para conseguir la capacidad de carga especificada y las tolerancias de desviación lineal para placas, lechos de carreteras, etc.; excavar [vb] hasta nivel de [sub]rasante); *syn.* preparación [f] de la plataforma/cancha; *f* **dressement [m] du fond de forme** (Opération comprenant compactage ou décompactage et profilage) ; *syn.* traitement [m] du fond de forme ; *g* **Herstellen [n, o. Pl.] des Erdplanums** (Profilgerechtes Herrichten des Baugrundes, Ausbildung und Verdichtung der Oberfläche nach bestimmten bautechnischen Vorgaben; beim Wegebau spricht man auch vom **Herstellen der Koffersohle**); *syn.* Herstellen [n, o. Pl.] des Rohplanums.

6142 subcontract [n] *constr. contr.* (Contract which is negotiated by a contractor with a subcontractor); *s* **contrato [m] de subcontrata** (Acuerdo negociado entre un contratista y un subcontratista); *f* **contrat [m] de sous-traitance** (CCM 1984, 67 ; contrat passé entre une entreprise et un sous-traitant en vue de l'exécution de certaines parties d'un marché) ; *g* **Nachunternehmervertrag [m]** (Vertrag, den ein Unternehmer mit einem Nachunternehmer abschließt); *syn.* Subunternehmervertrag [m].

6143 subcontract construction [n] [US] *constr. contr.* (►subcontractor); *syn.* subcontract works [npl] [UK]; *s* **subcontrata [f]** (Operación confiada a un ►subcontratista por el contratista principal de la obra); *f* **sous-traitance [f]** (Opération confiée à un ►sous-traitant suivant les directives de l'entrepreneur principal : travaux sous-traités, prestations sous-traitées) ; *g* **Nachunternehmertätigkeit [f]** (Umfang der Bauleistungen, die ein ►Nachunternehmer im Auftrage des Hauptunternehmers durchführt); *syn.* Subunternehmertätigkeit [f].

6144 subcontractor [n] *constr. contr.* (Contractor who carries out construction [US]/works [UK] under contract to the main contractor); *s* **subcontratista [m]** (Empresa que realiza partes de una obra por encargo de la empresa contratista principal); *f* **sous-traitant [m]** (Sous-entrepreneur qui est chargé d'une partie du travail concédé à un entrepreneur principal) ; *g* **Nachunternehmer [m]** (Unternehmer, der im Auftrage des Hauptunternehmers Bauleistungen ausführt); *syn.* Subunternehmer [m], Unterauftragnehmer [m].

subcontract works [npl] [UK] *constr. contr.* ►subcontract construction [US].

6145 subdivision [n] [US] *urb.* (**1.** Tract of land surveyed and divided into residential lots for purposes of sale; *specific terms* ►housing subdivision [US]/housing estate [UK], ►residential area. **2.** Established housing or commercial development. **3. In U.K.,** a subdivision is referred to as an estate which can be either a housing estate or a business park, formerly known as an industrial estate); *syn.* real estate development [n]; *s* **unidad [f] de ejecución urbanística (±)** (Superficie de terreno claramente delimitado y dividido en parcelas destinadas a la venta para fines residenciales. El terreno puede estar ya parcial o totalmente construido; ►parcelación urbanística, ►polígono residencial, ►zona residencial); *f* **lotissement [m]** (**1.** Opération de division d'une propriété foncière [en plusieurs lots] en vue de l'implantation de bâtiments qui a pour objet ou qui, sur une période de moins de dix ans, a eu pour effet de porter à plus de deux le nombre de terrains issus de ladite propriété ; Art R 315-1 Code de l'urbanisme ; cette opération d'urbanisme peut être réalisée par une collectivité locale ou un promoteur privé qui réalisent la voirie et les réseaux divers [routes, réseaux d'assainissement, d'adduction d'eau potable et d'éclairage public, etc.] les espaces verts et qui se rémunèrent en vendant les parcelles aménagées ; **2.** le terme **l.** désigne aussi l'ensemble des habitations construites sur le terrain loti et, par extension, le quartier ainsi aménagé ; on parle, par exemple, de lotissement pavillonnaire ; ►ensemble résidentiel, ►secteur d'habitation, ►zone d'habitation, ►zone urbaine résidentielle) ; *g* **Wohnbaugebiet [n]** (**1.** Durch Umlegung so zugeschnittenes Gebiet, das für eine Wohnbebauung vorgesehen ist, dessen Parzellen zum Verkauf angeboten werden oder schon mit Wohngebäuden bebaut ist; ►Wohngebiet 2, ►Wohnsiedlung; **2.** Im Amerikanischen bedeutet *subdivision* auch der Vorgang einer bodenordnenden Aufteilung eines Baugebietes, sowohl für Wohngebiete als auch Gewerbegebiete; ►Umlegung).

6146 subdivision [n] **(of land)** *leg. plan.* (Act of dividing a lot/tract or parcel of land into two or more lots, parcels, or tracts for sale, development or lease; cf. IBBD 1975. **In U.S.,** land may be subdivided by provision in a will, voluntary partition among heirs, or by court order; ►rearrangement of lots [US]/reorganization of plot boundaries [UK], ►farmland consolidation [US]/land consolidation [UK]); *s 1* **parcelación [f] rústica** (División de un terreno en uno o más lotes, parcelas o zonas para redistribución con finalidad exclusivamente vinculada a la explotación agraria de la tierra, p. ej. en el caso de la ►concentración parcelaria; cf. DICURB 2008); *syn.* división [f] de fincas; *s 2* **parcelación [f] urbanística** (División o segregación simultánea o sucesiva de terrenos en dos o más lotes, que tiene por finalidad permitir o facilitar la realización de actos de edificación o uso del suelo o del subsuelo sometidos a licencia urbanística; cf. DICURB 2008; ►parcelación, ►reordenación de parcelas); *syn.* lotificación [f] [AL], loteo [m] [AL], loteamiento [m] [AL], reparto [m] [AL], fraccionamiento [m] [MEX]; *f* **fixation [f] de l'état parcellaire nouveau** (Modifications des limites de propriété lors de la redistribution parcellaire et transcrit sur un plan de remembre-ment, p. ex. dans le cadre d'un ►remembrement

parcellaire effectué par une association foncière urbaine [cf. C. urb., art. R. 322, 10-15] ou d'un ▶remembrement rural ; *syn.* délimitation [f] des fonds ; *g* Festsetzung [f] von Grundstücks-grenzen (Parzellenscharfe Bestimmung von Grundstücken, z. B. bei einer ▶Umlegung oder ▶Flurbereinigung. In den USA erfolgt die **F. v. G.** auf einem *subdivision plan*, aber auch durch Willenserklärungen sowie durch einvernehmliche Grundstücks-teilung der beteiligten Erben oder durch Gerichtsbeschluss); *syn.* Festsetzen [n, o. Pl.] von Grundstücksgrenzen.

subdivision [n] [US]**, new housing** *urb.* ▶new housing development.

subdivision [n] [US]**, residential** *urb.* ▶residential area.

subdivision road [n] [US]**, residential** *trans. urb.* ▶residential street.

6147 subgrade [n] [US] **(1)/sub-grade** [n] [UK] *constr.* (Existing ground or embankment, installed for leveling [US]/levelling [UK] or to improve the load-bearing capacity, upon which subbase [US]/sub-base [UK] and upper paving layers are constructed; ▶pavement structure, ▶subgrade [US] 2/sub-grade [UK], ▶subgrade preparation); *s* **subbase** [f] (▶Subsuelo 1 existente o talud de tierra, arena y grava levantado para mejorar la capacidad de carga sobre el que se tienden las capas superiores de una carretera o un camino; ▶composición estratificada de superficies revestidas, ▶estructura del cuerpo de carreteras y caminos, ▶preparación de la [sub]rasante); *syn.* cimiento [m] de firme, cimiento [m] de pavimento, apisonado [m] [YV], sub-base [f] [PE]; *f* **couche** [f] **de forme** (Couche située sur la ▶partie supérieure des terrassements sur laquelle on vient répandre les ▶éléments constructifs de la voirie ; ▶réalisation de la couche de forme, ▶structure de la voirie) ; *syn.* assise [f], sous-couche [f] ; *g* **Unterbau** [m] **(1)** (Auf einem ▶Untergrund zum Höhenaus-gleich oder zur Verbesserung der Tragfähigkeit geschütteter Erdkörper, auf dem die Schichten des ▶Oberbaus eingebaut werden; ▶Herstellen des Unterbaues, ▶Wegeaufbau); *syn.* Koffer [m].

6148 subgrade [n] [US] **(2)/sub-grade** [n] [UK] *constr. eng.* (Existing ground with established formation level, upon which the foundations for traffic routes and building structures are constructed; ▶consolidated subgrade, ▶pavement construc-tion courses, ▶stabilized subgrade [US]/stabilized sub-grade [UK], ▶subgrade [US] 1/sub-grade [UK], ▶subsoil level, ▶un-disturbed subgrade [US]/undisturbed subsoil [n] [UK]); *s* **sub-suelo** [m] **(1)** (Superficie del terreno nivelado que soporta la ▶subbase y las capas superiores de una carretera o los cimientos de una construcción; ▶composición estratificada de superficies revestidas, ▶subsuelo 3, ▶subsuelo estable, ▶subsuelo estabi-lizado, ▶subsuelo inalterado [Es, RCH]); *syn.* suelo [m] de cimentación, subrasante [f] [EC, CO, MEX, PA, RA, RCH, YV] (MESU 1977); suelo [m] de fundación, terreno [m] de cimen-tación (BU 1959); *f* **partie** [f] **supérieure des terrassements (PST)** (Zone supérieure des terrains en place [cas des profils en déblai] ou des matériaux rapportés [cas des profils en remblai] ; STCN 1993 ; ▶bon sol, ▶couche de fondation, ▶couche de forme, ▶éléments constructifs de la voirie, ▶fond de forme stabilisé, ▶sol stable, ▶terrain naturel) ; *syn.* terrain [m] sous-jacent, forme [f] de support ; *g* **Untergrund** [m] (Im Vergleich zum ▶Unterbau der anstehende Boden, der mit dem Planum abschließt und im Erdbau die Schichten des ▶Oberbaus und die Fundamente für Bauwerke trägt; ▶Baugrund, ▶fester Baugrund, ▶gewachsener Boden, ▶verbesserter Untergrund).

subgrade [n] **of a road/path** [US] *constr.* ▶subbase grade of a road/path [US]/formation level of a road/path [UK].

6149 subgrade preparation [n] [US]**/sub-grade pre-paration** [n] [UK] *constr.* (Placement and grading to proper

line and levels of specified aggregate on the prepared ▶subsoil level at the specified thickness and density consolidation stan-dards; ▶subgrade [US] 1/sub-grade [UK]); *s* **preparación** [f] **de la subbase** (Colocación y compactación del material específico de relleno hasta nivel de rasante sobre el subsuelo preparado, según estándares técnicos descritos en el resumen de presta-ciones; ▶subbase, ▶subsuelo 3); *f* **réalisation** [f] **de la couche de forme** (Mise en place des matériaux élaborés [sols, matériaux rocheux, sous produits industriels] utilisés dans la construction des plates-formes rigides et stables du domaine routier qui assureront portance, insensibilité à l'eau et au gel ainsi que le passage du trafic de chantier, selon les prescriptions techniques définies dans le descriptif ; ▶couche de fondation, ▶couche de forme) ; *g* **Herstellen** [n, o. Pl.] **des Unterbau(e)s** (Profilge-rechter Einbau des zu verdichtenden Schüttgutes auf den ▶Bau-grund nach technischen Vorgaben, die im Leistungsverzeichnis beschrieben sind; ▶Unterbau 1).

6150 subhydric soil [n] *pedol.* (Deposits of soil in bodies of still water, usually having a characteristic type of humus which is populated by soil organisms; *specific terms* ▶dy, ▶gyttja, ▶sapropel); *syn.* subaqueous raw soil [n] (GGT 1979, 452); *s* **suelo** [m] **subacuático** (Según el sistema de clasificación alemán, clase de suelos formados bajo aguas estancadas. Están poblados de microorganismos y tienen generalmente una forma característica de humus. La clase se subdivide en suelos turbosos y suelos no turbosos. *Términos específicos* ▶dy, ▶gyttja, ▶sa-propel); *syn.* suelo [m] lacustre; *f* **sol** [m] **subaquatique** (*Classi-fication allemande* groupe des sols formé sous l'eau constitué des sols à nappe stagnante et des sols à nappe phréatique profonde ; PED 1983, 180, 371 ; *termes spécifiques* ▶dy, ▶gyttja, ▶sapro-pèle) ; *g* **Unterwasserboden** [m] (Gemäß des deutschen Klassifi-kationssystems ein Boden, der durch Ablagerungen unterhalb des Wasserspiegels nicht fließender Gewässer entsteht. Er ist durch Organismen besiedelt und besitzt auch meist eine charakteristi-sche Humusform; *UBe* ▶Dy, ▶Gyttja, ▶Sapropel); *syn.* subhy-drischer Boden [m].

subjective quality [n] **of the environmental expe-rience** *plan. recr. sociol* ▶perceived environmental quality.

sublittoral zone [n] *phyt. zool.* ▶shallow water area.

6151 submerged aquatic plant [n] *phyt.* (Either rooted or non-rooted ▶hydrophyte which lies entirely beneath the water surface, except for flowering parts in some species; CWD 1979, 43); *syn.* submergent plant [n] [also US] (CWD 1979, 43), submersed aquatic [n] [also US] (WET 1993, 344); *s* **planta** [f] **acuática sumergida** (▶hidrófito); *syn.* hidrófito [m] sumergido, hidatófito [m] sumergido; *f* **plante** [f] **immergée** (Plante hydro-phile, enracinée **[plante submergée fixée]** ou non dont l'appareil végétatif est en permanence totalement en immersion ; ▶hydro-phyte) ; *syn.* plante [f] nageante, hydrophyte [m] enraciné et immergé, plante [f] submergée (VOC 1979) ; *g* **Unterwasser-pflanze** [f] (Völlig untergetauchte Wasserpflanze, wurzelnd oder nicht wurzelnd; ▶Hydrophyt); *syn.* submerser Hydrophyt [m], submerse Schwimmpflanze [f], submerse (Wasser)pflanze [f], untergetauchte (Wasser)pflanze [f].

submergent plant [n] [US] *phyt.* ▶submerged aquatic plant.

submersed aquatic [n] [US] *phyt.* ▶submerged aquatic plant.

6152 submersible pump [n] *constr.* *s* **bomba** [f] **sumer-gible**; *f* **pompe** [f] **immergée** ; *g* **Tauchpumpe** [f] (Pumpe, die als Ganzes samt Elektromotor im Wasser arbeitet); *syn.* Unter-wasserpumpe [f].

submission [n]**, closing date for design** *plan. prof.* ▶submission deadline for design competition.

S

submission date [n] [UK] *contr.* ▶bid opening date [US].

submission date [n] [UK]**, minutes of** *adm. constr. contr.* ▶minutes of bid opening [US].

submission deadline [n] *constr. contr.* ▶date for submission of bids [US]/date for submission of tenders [UK].

6153 submission deadline [n] **for design competition** *plan. prof.* (Final date for receipt of competition entries); *syn.* closing date [n] for design submission; *s* **fecha** [f] **tope de entrega de proposiciones a concurso** *syn.* cierre [m] de concurso; *f* **date** [f] **limite d'envoi** (Dernière date de remise d'un projet dans le cadre d'un concours) ; *g* **Abgabetermin** [m] **bei einen Planungswettbewerb** (Zeitpunkt, bis zu dem eine Wettbewerbsarbeit abgegeben werden muss).

submission material [n] [UK] *prof.* ▶competition entry.

6154 submission [n] **of a report** *prof. syn.* submittal of a report; *s* **entrega** [f] **de un informe;** *f* **remise** [f] **d'une expertise** *syn.* remise [f] d'un avis d'expert ; *g* **Abgabe** [f] **eines Gutachtens.**

6155 submission [n] **of bids** [US] *constr. contr.* (Delivery of proposals before opening of bids [US]/tenders [UK] at the ▶bid opening date); *syn.* bid submission [n] [US], submission [n] of tenders [UK]; *s* **entrega** [f] **de ofertas** (Acto de entregar una propuesta antes de la ▶fecha de apertura de pliegos); *f* **remise** [f] **des offres/soumissions** (dans un délai compté à partir de la date d'envoi de l'avis pour publication jusqu'à la ▶date d'ouverture des plis ; de l'avis d'adjudication jusqu'à l'ouverture de l'adjudication) ; *syn.* dépôt [m] d'une offre, dépôt [m] d'une soumission ; *g* **Angebotsabgabe** [f] (Das Einreichen eines Angebotes bis zum ▶Eröffnungstermin); *syn.* Abgabe [f] eines Angebotes, Abgeben [n, o. Pl.] eines Angebotes.

submission [n] **of bids** [US]**, date for** *constr. contr.* ▶date for submission of bids [US]/date for submission of tenders [UK]

submission [n] **of bids** [US]**, deadline for** *constr. contr.* ▶date for submission of bids [US]/date for submission of tenders [UK].

6156 submission [n] **of planning comments** *urb. plan.* (Suggestions and opinions submitted by local authority departments on planning proposals as a result of ▶statutory consultation with public agencies [US]/statutory consultation with public authorities [UK]); *s* **entrega** [f] **de comentarios o proposiciones** (En el marco de la participación de las autoridades e instituciones públicas y semipúblicas en la elaboración de planes de ordenación urbana posibilidad de colaborar en la planificación; ▶audiencia de autoridades e instituciones públicas y semipúblicas, ▶audiencia de corporaciones locales y a los departamentos ministeriales; *f* **présentation** [f] **d'un rapport** (Dans le cadre de la participation des services publics, des collectivités locales et des organismes socioprofessionnels lors de l'élaboration conjointe des documents d'urbanisme ; ▶consultation des services de l'État et des personnes publiques associées) ; *syn.* émission [f] d'un avis ; *g* **Abgabe** [f] **einer Stellungnahme** (im Rahmen der ▶Beteiligung [der] Träger öffentlicher Belange, z. B. bei der Aufstellung von Bauleitplänen [cf. §§ 4 (2) und 4a (3-6) BauGB] und der Beteiligung der Öffentlichkeit [§ 3 (2) BauGB]); *syn.* Abgeben [n, o. Pl.] einer Stellungnahme.

submission [n] **of tenders** [UK] *constr. contr.* ▶submission of bids [US].

submission [n] **of tenders** [UK]**, date for** *constr. contr.* ▶date for submission of bids [US].

submissions report [n] [UK] *adm. constr. contr.* ▶minutes of bid opening [US].

submittal of a report [n] *prof.* ▶submission of a report.

6157 submittal requirements [npl] *prof.* (All material requested for a competition entry to be delivered on a specified submission date, such as plans, reports, cost estimate, time schedule, model; technical and financial offers are delivered in separate envelopes); *s* **requisitos** [mpl] **de concurso** (Todos los documentos o productos que deben ser entregados por los participantes en concurso de ideas o de realización en la fecha determinada, como planos, informes, calendario, modelos, cálculos de costes [Es]/costos [AL]; las ofertas técnicas y financieras se entregan en sobres separados); *f* **rendus** [mpl] **de concours** (Ensemble des pièces remis par les concurrents à une date définie pouvant comporter selon la nature du concours dans une enveloppe anonyme une note synthétique, des plans, coupes, élévations, axonométries et vues perspectives, une maquette, un planning des travaux une ébauche de cahier des charges ; le projet de marché avec son enveloppe financière prévisionnelle ainsi que la proposition d'honoraires seront remis dans une deuxième enveloppe) ; *syn.* prestations [fpl] demandées aux concurrents ; *g* **geforderte Wettbewerbsleistungen** (Die für einen Ideen- oder Realisierungswettbewerb zu einem festgesetzten Termin abzugebenden/einzureichenden Unterlagen, z. B. Pläne, Erläuterungsbericht, Kostenschätzung, Zeitplan, Modell; das technische Arbeitsprogramm und die Honorarkalkulation werden in gesonderten Umschlägen abgegeben).

submittee [n] **of a bid** [US] *contr.* ▶bidder [US] (1)/tenderer [UK].

submitter [n] **of a bid** [US] *contr.* ▶bidder [US] (1)/tenderer [UK].

6158 subregional plan [n] *plan.* (Plan for a demarcated portion of a region; ▶regional plan, ▶regional planning policy); *s* **plan** [m] **director territorial de coordinación (2) (de una comarca)** (En Es. según la legislación supletoria los **p. d. t. de c.** establecen las directrices para la ▶ordenación del territorio, el marco físico y el modelo territorial en el que se han de desarrollar las previsiones del plan para una comarca, provincia o varias de ellas de acuerdo con los principios de la planificación económica y social y las exigencias del desarrollo regional; arts. 9 y 10, RD 2159/1978; ▶plan director territorial de coordinación 1, ▶plan regional); *f* **schéma** [m] **directeur régional d'aménagement et d'urbanisme** (Document fixant, à travers un rapport et des cartes, les orientations, spatialisées de la politique à long terme d'aménagement et d'urbanisme d'une région urbaine [zone d'urbanisation, secteurs à préserver, implantations des activités, projets d'infrastructures et de grands équipement, etc.] ; DUA 1996 ; ▶schéma régional d'aménagement du territoire, ▶aménagement du territoire) ; *g* **Gebietsentwicklungsplan** [m] **(D.,** ▶Regionalplan; ▶Raumordnung).

subsequent use [n] *plan.* ▶adaptive use [US].

sub-shrub [n] *bot.* ▶suffruticose shrub.

subsidence [n] *min.* ▶area subject to mining subsidence; *land'man.* ▶bog subsidence; *min.* ▶damage due to mining subsidence; *constr. geo. min.* ▶ground settlement; *eng. landsc. min.* ▶reclamation after subsidence.

subsidence [n]**, mining** [n] *min.* ▶damage due to mining subsidence.

6159 subsidiary bid [n] [US] *constr. contr.* (Ancillary tender/bid which may or may not be submitted together with the main ▶bid [US] 1/tender [UK]); *syn.* subsidiary tender [n] [UK], supplementary bid [n] [US]; *s* **oferta** [f] **subsidiaria** (▶Oferta adicional que se puede o no añadir a la oferta principal); *f* **offre** [f] **subsidiaire** (▶Offre [de prix] remise avec l'offre principale lorsque celle-ci est souhaitée ou proposée dans le dossier de consultation, ou offre non prévue dans le règlement particulier de l'appel d'offres) ; *g* **Nebenangebot** [n] (▶Angebot, das zum

Hauptangebot oder in Verbindung mit diesem erwünscht oder ausdrücklich zugelassen wird resp. nicht abgegeben werden darf; cf. § 17 Nr. 4 [3] VOB Teil A).

subsidiary tender [n] [UK] *constr. contr.* ▶subsidiary bid [US].

6160 subsidized holiday [n] *recr.* (Subsidized recreation for low income groups); *syn.* subsidized vacation [n] [also US]; *s* **turismo [m] social** (Oferta subvencionada de vacaciones para familias de recursos reducidos, personas de la tercera edad, etc.); *syn.* turismo [m] popular; *f* **tourisme [m] social** (Activités de loisirs gérées par des organismes reconnus d'intérêt public et financées pour partie par les collectivités territoriales en faveur des couches de population défavorisées) ; *g* **Ferienreisen [fpl] für einkommensschwache Bevölkerungsschichten (≠)** (Urlaubsmöglichkeit für Familien oder Senioren mit geringem Einkommen, die von gemeinnützigen Organisationen geplant und z. T. finanziell getragen werden); *syn.* Erholungsreisen [f, o. Pl.] für Bevölkerungsschichten mit niedrigem Einkommen (≠).

subsidized vacation [n] [US] *recr.* ▶subsidized holiday.

subsidy [n] *econ.* *∗*▶public grant.

6161 subsoil [n] *constr. pedol.* (Weathered soil layer beneath the ▶topsoil 1; synonymous with B horizon in ▶soil science, according to the development of the soil profile); *s* **subsuelo [m] (2)** (Capa de suelo inferior a la ▶tierra vegetal constituida por material meteorizado; en ▶edafología sinónimo del horizonte B); *f 1* **sous-sol [m]** *constr. agr.* (**1.** Couche de sol de profondeur située sous la couche de terre végétale superficielle et constituant une couche de transition de la roche-mère ; ▶terre végétale. **2.** *agr.* Partie du sol non remuée par le labour ordinaire) ; *f 2* **horizon [m] structural** *pedol.* (▶*Pédologie* Horizon d'altération du sol situé sous l'▶horizon de surface et constituant un horizon de transition du matériau originel ; ▶sol superficiel) ; *syn.* horizon [m] B, horizon [m] cambique ; *g* **Unterboden [m]** (Die unter dem ▶Oberboden liegende verwitterte Bodenschicht, die bei den meisten Landböden in das Ausgangsgestein [C-Horizont] überleitet. In der ▶Bodenkunde je nach Bodenentwicklung syn. mit B-Horizont. U. kann im Landschaftsbau durch entsprechende Maßnahmen für Vegetationszwecke verbessert und verwendbar gemacht werden; cf. DIN 18 915); *syn.* B-Horizont [m].

subsoil [n] [UK]**, undisturbed** *constr.* ▶undisturbed subgrade [US]/undisturbed subsoil [UK].

6162 subsoil compaction [n] *constr.* (Consolidation of underlying soil); *s* **compactación [f] del subsuelo**; *f* **compaction [f] du fond de forme** ; *g* **Unterbodenverdichtung [f]**.

6163 subsoiling [n] *constr.* (Tillage of subsurface soil below the normal depth of cultivation without disturbing the sequence of the soil profile for the purpose of breaking up heavily compacted subsoil which restricts water and root penetration; ▶deep plowing [US]/deep ploughing [UK], ▶scarifying); *syn.* subsoil tilling [n] [also US], deep ripping [n] [also UK] (SPON 1986, 154); *s* **desfonde [m]** (Consiste en dar a la tierra una labor profunda, de 50 cm o más, con la finalidad de romper la compacidad del suelo sin voltearlo; RA 1970; ▶arado en produndo, ▶escarificado); *f 1* **décompactage [m]** (Terme générique pour l'ameublissement en profondeur de sols compactés sans retournement ni enfouissement, pour maintenir la matière organique à la surface du sol ; ▶labour profond, ▶scarifiage, ▶sous-solage) ; *f 2* **sous-solage [m] (2)** (Travaux aratoires du sol exécutés sur de grandes superficies par passage d'une sous-soleuse [outil à dents droites] à une profondeur de 50 à 80 cm, but est de régénérer la structure des horizons de sol situés sous le fond de labour ; terme est parfois utilisé au sens large pour un ▶décompactage ; ▶labour profond) ; *syn.* dislocation [f] profonde du sol ; *g* **Tie-**

fenlockerung [f] (*Vorgang und Ergebnis* Lockerung des stark verdichteten Untergrundes durch Aufreißhaken in einer Tiefe über 50 cm; die Heckaufreißer werden von Planierraupen hydraulisch in den Boden gedrückt und gezogen; ▶Tiefpflügen, ▶Aufreißen 1); *syn.* Unterbodenauflockerung [f], Untergrundlockerung [f].

6164 subsoil level [n] [US] *constr. stat.* (**1.** Existing ground or filled consolidated subsoil at the base of an excavated site for architectural or engineering construction. **2.** In landscape construction, an existing ground or filled consolidated subsoil on which top soil, drainage or filter layer is laid; ▶consolidated subgrade); *syn.* formation level [n] [UK]; *s* **subsuelo [m] (3)** (En la construcción paisajística el suelo natural o artificial bajo la vegetación o la subbase bajo la capa de drenaje o filtrante; ▶subsuelo estable); *syn.* terreno [m] de cimentación; *f* **couche [f] de fondation** (Sur un chantier terrain en place ou couche rapportée situé sous la couche végétale ou la couche drainante/filtrante ; ▶sol stable) ; *syn.* sol [m] de fond ; *g* **Baugrund [m, o. Pl.]** (**1.** Natürlich anstehender Boden [Lockergestein] resp. Fels [Felsgestein] oder aufgetragener, standfest verdichteter Boden, auf dem ein Bauwerk errichtet werden soll. **2.** Im Landschaftsbau der natürlich anstehende oder aufgetragene Boden als Untergrund unter der Vegetationstragschicht resp. der Unterbau [evtl. Aufschüttung auf dem Untergrund] unter der Drän- oder Filterschicht; ▶fester Baugrund).

subsoil tilling [n] [US] *constr.* ▶subsoiling.

6165 subspecies [n] *biol.* (In taxonomy a systematic group or category classified under ▶species 1, which may be distinguished from its nearest related group by special important characteristics, but is associated with it by unmistakable, nonhybrid intermediate forms. The International Committee on the Nomenclature and Registration of Plants decides upon whether to class a plant as a species, subspecies, varietas, cultivar/variety, or form *[forma]*; species); *s* **subespecie [f]** (Raza geográfica de una ▶especie cuyo areal es muy grande; DINA 1987, 387); *f* **sous-espèce [f]** (Unité taxonomique de rang immédiatement inférieur à l'▶espèce, distinguable des autres sous-espèces de la même espèce par des caractères discriminants (morphologiques, hydrologiques, etc.) constants. La commission internationale d'horticulture — nomenclature et enregistrement — décide du classement d'une plante au rang d'espèce *[species]*, de sous-espèce *[subspecies]*, de variété *[varietas]*, de cultivar ou de forme *[forma]* ; TSF 1985) ; *g* **Subspecies [f]** (*Abk.* subsp.; in der Taxonomie eine systematische Gruppe [hierarchische Kategorie] innerhalb einer ▶Art, die von der nächstverwandten Gruppe zwar durch einzelne wichtige Merkmale abweicht, aber mit ihr durch unverkennbare, nicht hybride Zwischenformen verbunden ist; cf. BOR 1958. Im Gartenbau entscheidet der Internationale Ausschuss für Gartenbau-Nomenklatur und Registrierung darüber, ob eine Pflanze als Art *[species]*, Unterart *[subspecies]*, Varietät *[varietas]*, Kulturvarietät [Cultivar/Sorte] oder Form *[forma]* einzuordnen ist); *syn.* Unterart [f], o. V. Subspezies [f].

substance [n] *envir.* ▶noxious substance, *envir.* ▶water-polluting substance.

substitute association [n] *phyt.* ▶substitute community.

substitute biotope [n] *ecol. leg.* ▶substitute habitat.

6166 substitute community [n] *phyt.* (Secondary plant community, that has developed under the direct or indirect influence of man, often as a result of agricultural activity, and replaces the natural climax vegetation; ▶economic land management, ▶potential natural vegetation); *syn.* substitute association [n]; *s* **comunidad [f] (secundaria) sustituyente** (Comunidad vegetal originada por modificaciones de las condiciones naturales de la vegetación provocadas sobre todo por la explotación agrí-

S

cola y que sustituye a la ▶vegetación natural potencial; cf. BB 1979, 505; ▶gestión agrícola); *f* **association [f] de substitution** (Peuplement végétal secondaire qui pour des raisons de culture des sols ou de phénomènes d'érosion occupe l'espace revenant à un groupement climax ; *terme spécifique* forêt de substitution ; cf. VCA 1985, 85 ; ▶gestion agricole, ▶végétation naturelle potentielle) ; *syn.* peuplement [m] de substitution ; *g* **Ersatzgesellschaft [f]** (Anthropo-zoogene Pflanzengesellschaft, die insbesondere durch Landbewirtschaftungsmaßnahmen an die Stelle der ursprünglichen ▶potentiellen natürlichen Vegetation tritt; ▶Landbewirtschaftung).

6167 substitute habitat [n] *ecol. leg.* (Defined spatial unit serving to replace a ▶habitat 2, which was destroyed or severely disturbed, and which can perform the same ecological functions); *syn.* replacement habitat [n], substitute biotope [n], replacement biotope [n]; *s* **biótopo [m] sustituyente** (Espacio delimitado y correspondientemente acondicionado para reemplazar un ▶biótopo natural destruido por acción humana); *f* **biotope [m] de substitution** (Aire limitée utilisée pour le remplacement d'un ▶biotope détruit dans une zone voisine et présentant des fonctions écologiques identiques) ; *syn.* habitat [m] de substitution ; *g* **Ersatzbiotop [m,** *auch* **n]** (Abgrenzbarer Raumabschnitt, der auf Grund der Vernichtung oder Zerstörung eines intakten ▶Biotopes an anderer Stelle wiederhergestellt wird oder worden ist und dort den gleichen oder einen gleichwertigen ökologischen Funktionszusammenhang übernehmen kann).

substitute plant [n] *gard. plant.* ▶lawn substitute plant.

substrate [n] *constr.* ▶noncompactible substrate [US]/non-compactible substrate [UK].

6168 substrate mat [n] *constr.* (Artificial growing medium in the form of lightweight mats used in roof gardens instead of conventional growing media; *specific terms* foam substrate slab, ▶planting mat; ▶seed mat); *s* **estera [f] de sustrato sintético** (Estera artificial utilizada en el revestimiento vegetal de tejados en vez de mezcla convencional de tierra cuando se necesita ahorrar peso; *términos específicos* estera de gomaespuma, ▶estera de semillas, estera de sustrato precultivada, ▶estera sintética sembrada); *f* **natte [f] de substrat synthétique** (Natte synthétique utilisée pour la végétalisation extensive de toitures en remplacement de mélanges terreux traditionnels ; *termes spécifiques* plaque de substrat synthétique en mousse, ▶natte préensemencée, ▶natte synthétique préensemencée) ; *g* **Substratmatte [f]** (Zur Dachbegrünung verwendete, gewichtssparende Kunststoffmatte als Vegetationsschicht anstelle konventioneller Erdmischungen; *UBe* Schaumstoffmatte, vorkultivierte Substratmatte, ▶Begrünungsmatte; ▶Saatmatte).

substructure [n] *constr.* ▶footing.

subsurface contamination [n] [US] *envir. urb.* ▶hazardous old dumpsite [US].

6169 subsurface drainage [n] *agr. constr.* (Removal of excess subsurface water through a connected series of clay pipes or perforated pipes as underground lines, or through a drainage layer; ▶agricultural drainage, ▶field drainage, ▶land drainage, ▶drainage 1, ▶soil drainage 1, ▶subsurface drainage system); *syn.* soil drainage [n] (2); *s* **drenaje [m] subterráneo** (Desagüe de las capas superiores de suelo por medio de conjunto de tubos de arcilla, de hormigón poroso o de plástico perforado colocados en filas regulares o por medio de capas de material filtro; ▶drenaje a gran escala, ▶drenaje agrícola, ▶drenaje agrícola subterráneo, ▶drenaje del suelo, ▶sistema de drenaje subterráneo); *f* **drainage [m] souterrain** (**1.** Processus de collecte des eaux d'infiltration, constituée d'un réseau de drains posés dans le sens de la plus grande pente et reliés entre eux par un collecteur — pelouse, terrains de sports, périphérie des bâtiments — afin d'en

corriger les mauvaises propriétés physiques. **2.** Mesures d'assainissement des sols agricoles par suppression des excès d'eau pour une amélioration des conditions de croissance des cultures, LA 1981, 113. **F.**, le terme « drainage » désigne l'élimination naturelle et artificielle des eaux excédentaires sur et dans les sols. Le terme allemand « Dränung » désigne seulement l'élimination naturelle et artificielle des excès d'eau dans le sol ; *termes spécifiques* ▶drainage agricole, ▶drainage [agricole] à ciel ouvert [*syn.* drainage par fossé], ▶drainage agricole souterrain, drainage par canalisations enterrées [*syn.* drainage enterré], drainage par techniques associées ; ▶drainage de sol, ▶évacuation des eaux, ▶système de drainage souterrain) ; *syn.* drainage [m] du sol ; *g* **Dränung [f]** (Entwässerung zur Beseitigung bauten- oder pflanzenschädlicher Bodennässe durch unterirdische Dränleitungen oder durch eingebaute Dränschichten; ▶Bodenentwässerung, ▶Dränsystem, ▶Dränung landwirtschaftlicher Flächen, ▶Entwässerung landwirtschaftlicher Flächen, ▶Wasserbaumaßnahmen zur Entwässerung landwirtschaftlicher Flächen; *s. auch Anmerkungen zu dem englischen Begriff* ,drainage' *unter* ▶Entwässerung; dränieren [vb]; *obs. o. V.* drainieren [vb]); *syn. obs.* Drainage [f].

6170 subsurface drainage line [n] *constr.* (▶subsurface drainage); *syn.* subsurface drain line [n]; *s* **dren [m]** (▶drenaje subterráneo); *syn.* línea [f] de dren; *f* **drain [m]** (**1.** *Terme générique* canalisation souterraine constituée de tuyaux perforés ou poreux permettant l'évacuation des eaux de drainage. **2.** Terme aussi très souvent utilisé pour désigner le tuyau de ▶drainage souterrain) ; *syn.* ligne [f] de drains ; *g* **Dränleitung [f]** (*O. V.* Drainleitung; unterirdische, perforierte oder poröse Leitung zur Aufnahme und Abführung von Überschusswasser im Boden; **D.en** werden i. d. R. mit 0,5 % Gefälle mit Rohren von DN 100 Nennweite verlegt; ▶Dränung); *syn.* Dränstrang [m], *o. V.* Drainstrang [m].

6171 subsurface drainage system [n] *constr.* (Pattern of connected series of pipes for removal of excess surface water or groundwater from land areas; ▶drainage 1, ▶grid pattern drainage, ▶herringbone pattern drainage); *syn.* drainage piping pattern [n] [also US]; *s* **sistema [m] de drenaje subterráneo** (Conjunto de tubos conectados entre sí para evacuar agua excedente del suelo en superficies agrícolas; ▶drenaje, ▶drenaje longitudinal/paralelo a la pendiente, ▶drenaje en forma de espina de pez); *syn.* sistema [m] de desagüe [m], sistema [m] de avenamiento [m]; *f* **système [m] de drainage souterrain** (Terme générique ; *termes spécifiques* ▶drainage longitudinal, ▶drainage transversal ; selon la position du collecteur par rapport aux drains le drainage peut être axial, radial ou latéral et suivant la répartition des drains drainage [à disposition] en peigne, épi ou arêtes de poisson, pattes d'araignée ; ▶évacuation des eaux) ; *syn.* réseau [m] de drainage ; *g* **Dränsystem [n]** (Mehrsträngige Entwässerungsanlage, die durch eine profilgerechte Anordnung von Saugern und Sammlern gekennzeichnet ist; *UBe* ▶Entwässerung, ▶Längsdränung, ▶Querdränung; *o. V.* Drainsystem); *syn. obs.* Drainagesystem [n].

subsurface drain line [n] *constr.* ▶subsurface drainage line.

subterranean construction [n] *urb.* ▶below-grade construction.

6172 subterranean plant part [n] *bot.* (▶bulb, ▶geophyte, ▶rhizome, ▶stem tuber); *syn.* underground plant part [n]; *s* **órgano [m] subterráneo** (Parte de una planta que crece bajo tierra, como raíz, ▶rizoma, ▶tubérculo caulinar, ▶tuberosidad radical o ▶bulbo; ▶geófito); *syn.* parte [f] subterránea (de una planta); *f* **organe [m] souterrain** (DAV 1984, 4 ; partie d'un végétal croissant dans le sol ; ▶bulbe, ▶géophyte, ▶rhizome, ▶tubercule caulinaire) ; *g* **unterirdischer Pflanzenteil [m]** (Im

Boden wachsendes, pflanzliches Organ, z. B. Wurzel, ▶Rhizom, unterirdische ▶Sprossknolle [Kartoffel], Wurzelknolle [Steppenkerze], Erdknolle [Crocus, Winterling] und ▶Zwiebel; ▶Geophyt).

subterranean water [n] *hydr. wat'man.* ▶groundwater [US]/ ground water [UK].

suburb [n] *urb.* ▶garden suburb; *plan.* ∗▶vicinity.

suburb [n]**, bedroom** *urb.* ▶bedroom community [US]/ dormitory suburb [UK].

suburb [n]**, dormitory** *urb.* ▶bedroom community [US]/ dormitory suburb [UK].

suburban area [n] *plan.* ▶vicinity.

6173 suburban forest [n] *urb.* (Woodland, appropriate for recreational use on the outskirts of large cities, providing walking paths, parking and picnic areas, horseback riding and bike trails, and resting areas, as well as fulfilling climatic and ecological objectives; *syn.* suburban wood [n] [UK], suburban woods [npl] [US]; *s* **bosque** [m] **suburbano** (Bosque cercano a la ciudad utilizado especialmente para fines recreativos y de higiene ambiental); *f* **forêt** [f] **suburbaine** (Forêt située à proximité des grands centres urbains et accessible aux citadins grâce à la réalisation d'aménagements légers en vue de l'accueil du public : terrains de pique-nique, camping, parkings, de sentiers de promenade, pistes cavalières, zone de silence, etc.) ; *g* **stadtnaher Wald** [m] (Der Naherholung zugeeigneter Wald, der am Rande großer Städte z. B. Spazierwege, Park- und Picknickplätze, Reitwege und Ruhebereiche vorhält und auch klimaökologischen Zielen dient).

6174 suburban green areas [npl] *landsc. urb.* (**1.** Green open spaces outside of the central core of a city. **2.** Abiotic and biotic phenomena in close vicinity of cities or agglomerations, which are either man-made or have been little influenced by man); *syn.* suburban green open spaces [npl], peripheral urban green spaces [npl], nature [n] close to the city; *s* **espacios** [mpl] **verdes suburbanos** (Conjunto de zonas verdes y áreas naturales situadas en las cercanías de las ciudades, que pueden ser de creación humana o de origen natural); *syn.* zonas [fpl] verdes suburbanas; *f 1* **espaces** [mpl] **verts périurbains** (Espaces ouverts qui s'étendent à la périphérie des agglomérations entre les marges de la ville et les frontières de l'espace rural) ; *f 2* **espace** [m] **naturel suburbain** (Ensemble des phénomènes biotiques et abiotiques naturels ou influencés par l'homme à la périphérie des villes ou des agglomérations) ; *g 1* **stadtnahes Grün** [n, o. Pl.] (Grünflächen an der Peripherie von Städten); *g 2* **stadtnahe Natur** [f, o. Pl.] *recr. plan.* (Gesamtheit der vom Menschen geschaffenen, naturnahen, belebten und unbelebten Erscheinungen in der Nähe von Städten oder Ballungsräumen).

suburban green open spaces [npl] *landsc. urb.* ▶suburban green areas.

6175 suburban park [n] *urb. recr.* (Developed public recreation area on the outskirts of a city); *s* **parque** [m] **suburbano** (Zona verde en las afueras de una ciudad); *f* **parc** [m] **périurbain** (Espace vert public de plusieurs centaines d'hectares dans une banlieue urbaine et nécessitant un moyen de transport) ; *syn.* parc [m] suburbain ; *g* **Vorortpark** [m] (Öffentliche Grünanlage im Vorort einer größeren Stadt).

suburban recreation area [n] *plan. recr.* ▶local recreation area.

suburban sprawl [n] *urb.* ▶urban sprawl.

suburban wood [n] [UK] *urb.* ▶suburban forest.

suburban woods [n] [US] *urb.* ▶suburban forest.

suburbia [n] *plan.* ▶vicinity.

suburbs [npl] [UK]**, inner** *urb.* ▶central city environs [US].

successful bidder [n] [US] *contr.* ▶bidder offering best value for the price [US]/tenderer offering the best value for the money [UK].

successful tenderer [n] [UK] *contr.* ▶bidder offering best value for the price [US]/tenderer offering the best value for the money [UK].

6176 succession [n] *phyt. zool.* (Sequence of plant or animal communities which follow each other on one site, caused by changes of environmental conditions, such as lowering of groundwater table, silting up of lakes, climatic change, etc., finally culminating in a ▶climax community; ▶climax, ▶ecological factors, ▶old-field succession, ▶zonation of vegetation); *s* **sucesión** [f] **ecológica** (Sucesiones naturales, en las cuales un organismo o grupo de organismos reemplaza a otro en un hábitat con el paso del tiempo; existe la sucesión primaria y la secundaria; cf. DINA 1987; ▶asociación climácica, ▶clímax, ▶cliserie, ▶factores ecológicos, ▶sucesión secundaria); *f* **succession** [f] (Remplacement successif des communautés biologiques au cours du temps en un lieu donné et qui constituent une série évolutive ; DEE 1982 ; lorsque les ▶facteurs écologiques restent stables la biocénose du ▶climax devient un ▶peuplement climax ou un ▶peuplement paraclimacique ; ▶succession sur friche agricole, ▶zonation de la végétation) ; *g* **Sukzession** [f] (Zeitliche [nicht räumliche!] Entwicklungsreihe von natürlichen Organismengemeinschaften bis zur völligen Ablösung der Ausgangsgemeinschaft durch eine andere am gleichen Ort, ausgelöst durch Veränderung der Umweltverhältnisse wie z. B. Klimaveränderung, Grundwasserabsenkung, Verlandung von Seen, Brachfallen von Kulturland, Kahlschlag, Brand, Überschwemmungen etc.; das Endstadium einer **S.**, das, wenn die ▶Ökofaktoren konstant bleiben, relativ stabil und dauerhaft ist, heißt ▶Klimax, die Pflanzengesellschaft ▶Klimaxgesellschaft; ▶Sukzession auf Ackerbrachland, ▶Vegetationszonierung).

successional meadow [n] [US] *gard. landsc.* ▶wildflower meadow.

6177 successional plant community [n] *phyt.* (TEE 1980, 193, 145; **1.** transitional stage of a plant community during succession). **2.** Result of a natural plant ▶succession process; ▶ecotonal association); *syn.* seral plant community [n] (SAF 1983, 265; TEE 1980, 145), transient community [n] (TEE 1980, 145), transitional community [n]; *s* **comunidad** [f] **sucesora** (**1.** Comunidad vegetal resultado del desarrollo de la ▶sucesión ecológica a partir de una comunidad pionera. **2.** Estado de transición de comunidad vegetal en el proceso de ▶sucesión ecológica; ▶communidad de ecotono); *syn.* comunidad [f] de transición, comunidad [f] transitoria; *f* **association** [f] **successive** (**1.** Groupement végétal succédant à un autre groupement ou au groupement initial [p. ex. végétation pionnière] dans une même station [relation évolutive]. **2.** Peuplement traduisant un stade d'évolution dans la dynamique de la végétation caractéristique du remplacement successif dans le temps de groupements végétaux constituant la série évolutive [évolution progressive ou régressive] ; ▶groupement de transition 1, ▶succession) ; *syn.* groupement [m] de transition (2), association [f] de transition, groupement [m] transitoire, groupement [m] transitionnel ; *g* **Folgegesellschaft** [f] (**1.** Pflanzengesellschaft, die sich aus einer bestehenden Gesellschaft oder Ausgangsgesellschaft [Pionierpflanzengesellschaft] im Rahmen der ▶Sukzession entwickelt. **2.** Pflanzengesellschaft als Übergangsstadium in der Sukzessionsabfolge auf dem selben Standort; ▶Übergangsgesellschaft 1); *syn.* Übergangsgesellschaft [f] (2), Sukzessionsgesellschaft [f], transitorische Gesellschaft [f].

S

6178 successional stage [n] *phyt.* (Clearly recognizable species change caused by modification of environmental site conditions, whereby a particular combination of species develops); *s* **etapa [f] de sucesión** (Cambio de especies claramente reconocible en una comunidad en desarrollo; se distinguen etapas iniciales, intermedias y finales; BB 1979, 628s); *f* **stade [m] d'une succession** (Modification perceptible de la composition floristique d'une station suite à une évolution des conditions écologiques locales) ; *syn.* stade [m] d'une série évolutive ; *g* **Sukzessionsstadium [n]** (Im Rahmen der Sukzessionsfolge ein deutlich erkennbarer Artenwechsel durch Änderung der Umweltverhältnisse oder durch Veränderung der Standortverhältnisse, wodurch sich eine bestimmte Artenkombination entwickelt).

6179 succession area [n] *conserv. land'man. phyt.* (Area on which ►succession occurs. Such an area may appear as an ►environmental resource patch on abandoned farmland or urban site and becomes an altered habitat for flora and fauna, thus serving to invigorate the rural or urban landscape); *s* **área [f] de sucesión** (Territorio en donde está teniendo lugar una ►sucesión ecológica; en el paisaje humanizado medida para desarrollar nuevos espacios vitales para la flora y la fauna a partir de ►islote de vegetación natural); *f* **aire [f] de succession** (Emplacement sur lequel a lieu une ►succession. Dans le paysage culturel ou urbain celle-ci constitue souvent un ►îlot de ressources naturelles favorisant la constitution d'un nouvel habitat pour la faune et la flore et participant à la diversification du système naturel et de l'aspect du paysage) ; *g* **Sukzessionsfläche [f]** (Fläche, auf der eine ►Sukzession stattfindet. In der Kultur- oder Stadtlandschaft oft eine ►Ökozelle in Form einer Brachfläche zur Schaffung von neuem Lebensraum für Flora und Fauna, der zur Diversifizierung des Landschaftshaushaltes und zur Belebung des Landschaftsbildes beiträgt).

6180 sucker [n] *hort.* (Vigo[u]rous shoot that arises below the graft union; ARB 1983, 20; ►root sucker, ►water sprout); *s* **chupón [m]** (Retoño que crece por debajo del punto de esqueje, por lo que es indeseado; ►brote de raíz, ►hijuelo); *syn.* hijo [m], mamón [m], pimpollo [m], serpollo [m] (HDS 1987); *f* **rejet [m] du porte-greffe** (Pousse vigoureuse, indésirable en horticulture, prenant naissance en dessous de la greffe ; ►drageon, ►gourmand) ; *g* **Wildtrieb [m]** (Gärtnerisch unerwünschter Trieb unterhalb einer Vered[e]lungsstelle; ►Wasserreis, ►Wurzelschössling); *syn.* Wildling [m] (1).

suckers [npl] *bot. hort.* ►propagation by root suckers.

suffrutescent shrub [n] *bot.* ►suffruticose shrub.

6181 suffruticose shrub [n] *bot.* (Plant with woody and hardy lower shoots; the upper herbaceous shoots die back and new shoots are produced; e.g. Common Sage *[Salvia officinalis]*, Thyme *[Thymus vulgaris]*, Hyssop *[Hyssopus officinalis]*); *syn.* semi-shrub [n], suffrutescent shrub [n], sub-shrub [n], half-shrub [n] (TEE 1980, 61); *s* **sufrútice [m]** (Planta semejante a un arbusto, generalmente pequeña y sólo lignificada en la base, como el tomillo; DB 1985); *syn.* subarbusto [m]; *f* **espèce [f] sousfrutescente** (Espèce végétale dont les parties inférieures de la tige se lignifient et persistent l'hiver pendant que leurs organes supérieurs herbacés meurent en général après la floraison) ; *syn. hort.* sous-arbrisseau [m] ; *g* **Halbstrauch [m]** (*Chamaephyta suffruticosa* Pflanze, deren untere Sprossteile verholzen und ausdauern, während die oberen krautig bleiben und am Ende einer Vegetationsperiode absterben, z. B. Gartensalbei *[Salvia officinalis]*, Gartenthymian *[Thymus vulgaris]*, Ysop *[Hyssópus officinális]*; die Erneuerungsknospen befinden sich an den ausdauernden, meist verholzten oberirdischen Organen); *syn.* Hemiphanerophyt [m], Suffrutex [m].

suffruticose shrubland [n] *phyt.* ►suffruticose thicket, #2.

6182 suffruticose thicket [n] *phyt.* (**1.** Close-grown stand of somewhat shrubby, semi-lignified nanophanerophytes [►suffruticose shrubs] which in dry years may shed part of their shoot systems; ELL 1967); **2. suffruticose shrubland** [n] (Open stand of semi-lignified nanophanerophytes that in dry years may shed part of their shoot systems; ELL 1967); *s* **formación [f] sufruticosa** (Formación vegetal de ►sufrútices semilignificados que puede perder sus brotes en los años secos); *f* **formation [f] sousfrutescente** (Peuplement végétal de sous-abrisseaux pouvant perdre tout ou parties de leurs tiges pendant les années sèches ; ►espèce sousfrutescente) ; *syn.* sous-frutiçaie [f] ; *g* **Halbstrauch-Gebüsch [n]** (Vegetationsformation von ►Halbsträuchern, die in trockenen Jahren Teile ihres Sprosssystems abstoßen können).

suit [n] [US] *conserv. leg.* ►non-profit organization suit.

6183 suitability evaluation [n] *plan.* (Evaluation of an area of land to assess its potential for human use; ►suitability for human use); *syn.* suitability study [n]; *s* **evaluación [f] de la aptitud intrínseca** (Evaluación de las potencialidades de un paisaje para diversas formas de aprovechamiento; ►aptitud intrínseca para usos); *syn.* evaluación [f] de la vocación (del suelo/de un paisaje), evaluación [f] de la idoneidad de un paisaje; *f* **évaluation [f] de l'aptitude** (Prise en compte de l'aptitude paysagère d'un milieu à diverses utilisations de l'espace ; ►aptitude à un/aux usage[s]) ; *g* **Eignungsbewertung [f]** (Festtellen der natürlichen Eignung einer Landschaft für geplante Nutzungen; ►Nutzungseignung).

6184 suitability [n] **for human use** *plan.* (Potential of a piece of land for its use in, e.g. agriculture, forestry, for development or for building upon); *s* **aptitud [f] intrínseca para usos** (Capacidad del territorio de acoger diversos usos, debido a sus condiciones idóneas para ellos); *syn.* vocación [f] del suelo para usos, idoneidad [f] para usos; *f 1* **aptitude [f] à un/aux usage(s)** *agr.* (Capacité des fonds pour différents usages tels que l'agriculture, la sylviculture, l'habitat ; *termes spécifiques* aptitude culturale, aptitude des terrains à la fondation) ; *f 2* **aptitude [f] paysagère (du milieu)** (Disposition naturelle du milieu définie par des descripteurs, p. ex. la sensibilité, la qualité des vues) ; *syn.* vocation [f] paysagère ; *g* **Nutzungseignung [f]** (Eignung von Grund und Boden für Nutzungen wie z. B. Landwirtschaft, Forstwirtschaft, Siedlungserweiterung, Eignung als Baugrund; von diesem Begriff zu unterscheiden ist die „Nutzungsverträglichkeit", die die **N.** aufheben kann).

suitability study [n] *plan.* ►suitability evaluation.

sulphur emissions [npl] *envir.* ►dehydration of sulphur emissions.

6185 summer annual [n] *bot. hort.* (Annual herbaceous plant which germinates, flowers and seeds within the same growing season); *s* **planta [f] anual estival** (Planta herbácea que germina, florece y produce semillas en un mismo año); *f* **annuelle [f] estivale** (Plante herbacée annuelle qui germe, fleurit et fructifie pendant la même période de végétation) ; *syn.* plante [f] annuelle estivale, plante [f] annuelle à floraison estivale, thérophyte [m] vernal ; *g* **Sommerannuelle [f]** (Einjährige krautige Pflanze, die in derselben Vegetationsperiode keimt, blüht und fruchtet, d. h. im Frühjahr keimt und bis zum Herbst einzieht); *syn.* sommerannuelle Pflanze [f], sommerannuelle Art [f], Sommereinjährige [f], Sommerpflanze [f].

6186 summer bloom display [n] *gard. hort. plant.* (Wide group of ►summer flowers; ►summer-flowering period, ►summer-flowering plant); *s* **flores [fpl] estivales** (Conjunto de plantas estivales en floración; ►flor de verano, ►floración estival, ►planta de floración estival); *f* **fleurs [fpl] estivales** (Ensemble des plantes molles en fleurs ; ►fleur à floraison estivale, ►floraison estivale, ►plante à floraison estivale) ; *g* **Sommerflor [m]** (**1.** Gärtnerischer Ausdruck für die Gesamtheit der Pflanzen [i. d. R. Annuelle und mehrjährige, nicht frostresistente exotische Stauden], die für die Sommerblüte in Blumenbeeten kultiviert und in Zentraleuropa in spätfrostgefährdeten Regionen meist nach dem 15. Mai ausgepflanzt werden. **2.** Blütenfülle im Sommer; ►Blütezeit im Sommer, ►Sommerblüher, ►Sommerblume).

summer bloomer [n] *gard. hort. plant.* ►summer-flowering plant.

summer-blooming plant [n] *gard. hort. plant.* ►summer-flowering plant.

6187 summer dike [n] *eng. wat'man.* (Embankment designed to prevent negative effects of summer storm floods in order to guarantee use of foreland on the shoreline and to improve coast[al] protection; ►levee [US] 2/flood bank [UK]); *s* **dique [m] de verano** (Obra de protección ante las tormentas de verano para permitir el aprovechamiento de las tierras situadas detrás del ►dique de contención); *f* **digue [f] d'été** (Ouvrage de protection de la côte ou d'une plaine alluviale retenant les grandes marées ou les crues d'été ; ►digue) ; *g* **Sommerdeich [m]** (**1.** Kleines Deichschutzwerk, das eine vor Sommersturmfluten sichere Vorlandnutzung als Weideland ermöglicht und den Küstenschutz verbessert. **2.** Ein **S.** wird auch zur Landgewinnung vor den höheren Winterdeich errichtet; ►Deich 2).

6188 summer flower [n] *gard. hort. plant.* (*Gardening term* precultivated, exotic bedding plant or border annual, which usually flowers in summer and dies off before winter, e.g. *Ageratum, Begonia, Lobelia*. Large areas of these are collectively termed a 'summer flower display'; ►summer annual, ►summer-flowering plant); *s* **flor [f] de verano** (Herbácea ornamental que florece en verano; ►planta anual estival, ►planta de floración estival); *f* **fleur [f] à floraison estivale** (*Terme horticole* plante herbacée annuelle de décoration florale estivale, précultivée [sous châssis ou sous serre], souvent exotique, en général à grande floribondité, très employée pour la décoration des jardins en été, p. ex. les variétés de l'Agérate *[Ageratum]*, du Bégonia *[Begonia]*, du Lobelia/de la Lobélie *[Lobelia]* ; ►annuelle estivale, ►plante à floraison estivale) ; *syn.* fleur [f] d'été, plante [f] molle ; *g* **Sommerblume [f]** (*Gärtnerische Bezeichnung* oft exotische, krautige Pflanze, die meist im Sommer blüht und in kalten Klimagebieten vor dem Winter abstirbt, z. B. *Ageratum,* Begonie *[Begonia]*, Lobelie *[Lobelia]*; die Summe der heranzuziehenden oder herangezogenen Pflanzen wird auch **Sommerflor** genannt; ►Sommerannuelle, ►Sommerblüher).

summer flowerer [n] [UK] *gard. hort. plant.* ►summer-flowering plant.

6189 summer-flowering period [n] *hort. plant.* (►flowering period, ►summer bloom display); *s* **floración [f] estival** (Periodo/período de verano en el que florecen una o varias especies de plantas o una plantación entera; ►floración, ►flores estivales); *f* **floraison [f] estivale** (Période d'épanouissement des fleurs pendant la ►floraison entre juin et août ; ►fleurs estivales) ; *g* **Blütezeit [f] im Sommer** (Jahreszeitlicher Abschnitt von Juni bis August, in dem eine Pflanzenart, eine Gruppe von Arten oder eine Bepflanzung blüht; ►Blütezeit; nicht zu verwechseln mit ►Sommerflor); *syn.* Sommerblüte [f].

6190 summer-flowering plant [n] *gard. hort. plant.* (Generic gardening term for a plant which flowers in the summer, thus a ►summer flower as well as an ►perennial 1 or a woody plant); *syn.* summer bloomer [n], summer-blooming plant [n], summer flowerer [n] [also UK]; *s* **planta [f] de floración estival** (*Término hortícola para* una planta que florece en verano que puede ser una ►flor de verano, una ►planta vivaz o una leñosa);

f **plante [f] à floraison estivale** (*Terme horticole désignant* un végétal fleurissant en été ; il s'agit aussi bien de ►fleurs à floraison estivale, de ►plantes vivaces ou de végétaux ligneux) ; *g* **Sommerblüher [m]** (*Gärtnerische Bezeichnung* eine Pflanze, die im Sommer blüht; dies kann sowohl eine ►Sommerblume, eine ►Staude oder ein Gehölz sein).

6191 summer green [adj] *bot. hort.* (Characteristic of plants which adapt to the seasons of the year in temperate regions by bearing their green leaves only during the summer growing season; term is used to emphasize the appearance of the plant in summer; ►deciduous, ►defoliation, ►evergreen, ►winter green); *s* **verde [adj] vernal** (*Término horticultor de las zonas templadas para designar a las plantas de caducifolias que sólo tienen hojas en el periodo/período vegetativo; syn. para todas las latitudes* ►caducifolio/a, ►verde invernal; *opp.* ►perennifolio/a; ►defoliación); *f* **à feuillage vert pendant l'été [loc]** (Terme caractérisant les végétaux dont les feuilles [caduques] se détachent chaque année à la fin de la période de végétation [►défeuillaison]. Il est syn. du terme ►à feuilles caduques mais souligne en particulier l'aspect vert du feuillage pendant l'été ; *opp.* ►à feuilles persistantes ; ►semi-persistant) ; *syn.* caduc, uque [adj] ; *g* **sommergrün [adj]** (Eigenschaft der Pflanzen, die auf Grund der Anpassung an die periodische Jahreszeitenabfolge ihre grünen Blätter nur während der Vegetationsperiode haben. Der Begriff **s.** ist syn. mit ►Laub abwerfend, betont jedoch den grünen Aspekt im Sommer, und gilt auch für Stauden; der ►Laubfall eines **s.en** Gehölzes erfolgt konzentriert mit den ersten leichten Frösten im Herbst [Herbstlaubfall]; ►wintergrün; *opp.* ►immergrün).

6192 summer holiday traffic [n] *plan. recr. sociol.* *s* **tráfico [m] turístico de verano** *syn.* tráfico [m] de vacaciones de verano; *f* **circulation [f] touristique estivale** ; *g* **Sommerreiseverkehr [m]** (Urlaubsverkehr im Sommer); *syn.* Hauptreiseverkehr [m].

summerhouse [n] *gard.* ►gazebo.

6193 summer plant display [n] *gard. hort.* (Plantation of ornamental flower beds with annual or biennial plants which bloom in summer and die off before winter; *generic term* ►seasonal bedding); *s* **plantación [f] de flores de verano** (Plantación de macizos ornamentales con plantas anuales o bianuales que florecen en verano; también se utilizan vivaces exóticas que en las latitudes templadas no pueden sobrevivir en invierno; *término genérico* ►plantación estacional); *syn.* plantación [f] de flores estivales; *f* **plantation [f] florale estivale** (Décoration estivale de plates-bandes en utilisant des plantes molles, annuelles ou bisannuelles florifères, fleurissant en été et mourant avant l'hiver dans les régions de climat froid ; *terme générique* ►plantation saisonnière) ; *syn.* décoration [f] florale estivale ; *g* **Sommerblumenbepflanzung [f]** (Bepflanzung von Schaubeeten mit ein- oder zweijährigen Pflanzen, die im Sommer blühen und in kalten Klimagebieten vor dem Winter absterben; es werden auch exotische perenne Pflanzen verwendet, die in gemäßigten Breiten nicht überwintern können; *OB* ►Wechselbepflanzung); *syn.* Sommerbepflanzung [f], Sommerflorbepflanzung [f].

summer pruning [n] *agr.* ►green pruning.

6194 summer smog [n] *envir.* (►Smog occurring in summer during ►inversion weather and containing man-made pollutants arising from urban or industrial concentrations. It is characterized by photochemical processes as well as the increase in ozone in air layers close to the ground); *s* **smog [m] oxidante** (Uno de los dos tipo de ►smog causado por la fuerte presencia de NO_2 y una fuerte radiación solar, por lo que se produce ozono que es muy irritante; ►condiciones de inversión térmica); *syn.*

S

smog [m] fotoquímico; *f* **smog [m] d'été** (▶Smog provoqué par une forte pollution de l'air constituée de particules gazeuses, liquides et prenant naissance dans les agglomérations urbaines ou industrielles sous des ▶conditions d'inversion thermique en été ; le smog est caractérisé par des processus photochimiques et se manifeste entre autre par une augmentation de la concentration d'ozone dans les couches de l'atmosphère proches du sol) ; *syn.* smog [m] photochimique, smog [m] de Los Angeles ; *g* **Sommersmog [m]** (▶Smog, der durch starke Luftverunreinigung aus gasförmigen, flüssigen und festen Bestandteilen über städtischen oder industriellen Ballungsräumen bei ▶Inversionswetterlagen in den Sommermonaten regelmäßig entsteht und durch fotochemische Vorgänge sowie unter anderem durch eine Erhöhung der Konzentration von Ozon und Fotooxidantien in den bodennahen Luftschichten gekennzeichnet ist); *syn.* fotochemischer Smog [m], *Kurzform* Fotosmog, [m], *o. V.* photochemischer Smog [m], *Kurzform o. V.* Photosmog [m].

6195　summer stagnation phase [n] *limn.* (Stabilization of the thermic layers of a lake in summer, followed by ▶autumn circulation period; ▶thermal stratification); *s* **estratificación [f] estival** (Estado de ▶estratificación térmica fuerte de un lago templado dimíctico en dos capas: una superior relativamente caliente [> 4°C] y otra inferior de < 4°C; DINA 1987; ▶circulación vertical autumnal/otoñal); *f* **stagnation [f] estivale** (État de la ▶circulation thermique ; en été la température d'un lac stagne dans l'épilimnion et décroît lentement à partir du thermocline jusqu'au fond ; à ce profil de température saisonnier suit la ▶brassage automnal) ; *g* **Sommerstagnation [f]** (Zustand der stabilen thermischen ▶Schichtung 3 eines Sees; der Sommerstagnation folgt die ▶Herbstzirkulation).

6196　summer visitor [n] *zool.* (Bird, which lives in summer in a defined area; they may be **regular s. v.s** [visiting every year], **irregular s. v.s.** [not visiting every year] and **occasional s. v.s.** [visiting now and again]; some s. v.s breed in the area, some do not; they can be vagrants, species that breed in the southern hemisphere, or birds that are too young to breed; ▶stopover bird, ▶visitor bird, ▶winter visitor); *syn.* summer visitor bird [n]; *s* **visitante [m] estival** (Especie de ave migratoria de primavera o prenupcial que visita una zona en verano. Puede ser para reproducirse y críar a sus poyuelos, marchando después en otoño a invernar a otra zona más cálida, o puede tratarse de aves divagantes, de especies que se reproducen en el hemisferio sur [resp. en el norte, desde el punto de vista del hemisferio sur] o aves demasiado jóvenes para reproducirse; ▶ave visitante, ▶ave en etapa migratoria, ▶visitante invernal); *syn.* ave [f] estival; *f* **visiteur [m] d'été** (*En Europe* oiseau qui hiverne totalement en dehors de l'Europe dans les contrées plus chaudes et, dans son déplacement annuel, y revient en été pour s'y reproduire ; parmi les ▶oiseaux visiteurs d'été on distingue les visiteurs d'été réguliers [apparition chaque année], les visiteurs d'été irréguliers [apparition irrégulière ou rare] et les visiteurs d'été accidentels [apparition très peu connues] ; ▶oiseau en étape migratoire, ▶visiteur d'hiver) ; *syn.* oiseau [m] estivant ; *g* **Sommergast [m]** (Vogelart, die sich während des Sommers in einem bestimmten Gebiet aufhält, dort seine Jungen aufzieht und im Herbst in wärmere Gebiete zieht; es werden **regelmäßige S.gäste** [treten jedes Jahr auf], **unregelmäßige S.gäste** [treten nicht jedes Jahr auf] und **gelegentliche S.gäste** [treten hin und wieder auf] unterschieden; einige **S.gäste** brüten, andere nicht; es können Vögel sein, die in wärmeren, südlicheren Gebieten brüten oder solche, die noch nicht geschlechtsreif sind; ▶Gastvogel, ▶Rastvogel, ▶Wintergast); *syn.* Übersommerer [m].

summer visitor bird [n] *zool.* ▶summer visitor.

summer wood [n] *arb. bot.* ▶late wood.

sump [n] *constr.* ▶sediment basin [US]/silt trap [UK].

6197　sunbathing area [n] *recr.* (Space where leisure seekers can relax lying flat on lawns, beaches, etc.; ▶sunbathing lawn); *syn.* relaxation area [n]; *s* **área [f] de reposo** (Zona en campas o playas donde la gente puede relajarse y tomar el sol; ▶campa para reposo); *f* **zone [f] de repos** (Espace d'accueil de touristes, randonneurs, etc. favorable à la détente, la relaxation, le bain de soleil, p. ex. sur un pré en bordure de rivière et de lac, en forêt, etc. ; ▶pelouse pour la détente) ; *g* **Liegeplatz [m] (2)** (Fläche, auf dem Erholungsuchende sich der Länge nach ausgestreckt entspannen und ggf. sonnen können — auf Wiesen, an Stränden etc.; ▶Liegewiese).

6198　sunbathing lawn [n] *recr.* (Lawn for passive recreation; ▶sunbathing area); *s* **campa [f] para reposo** (Área de césped utilizable para usos recreativos pasivos; ▶área de reposo); *f* **pelouse [f] pour la détente** (Espace engazonné favorable à la détente, la relaxation, le bain de soleil ; ▶zone de repos) ; *syn.* pré [m] de repos, prairie [f] de repos ; *g* **Liegewiese [f]** (Zum Liegen, Ruhen und Sichsonnen geeignete Wiese; ▶Liegeplatz 2).

sunburn [n] *arb. bot.* ▶sun scorch.

sunken driveway [n] [UK] *constr. urb.* ▶driveway cut [US].

6199　sunken garden [n] *gard.* (Part of a garden situated at a lower level than its surroundings); *s* **jardín [m] hundido** (Parte de un jardín que se encuentra a menor altura que el resto); *f* **jardin [m] creux** (Jardin ou partie de jardin situé plus bas que le terrain environnant) ; *g* **Senkgarten [m]** (Gartenteil, der tiefer als das ihn umgebende Gelände liegt); *syn.* abgesenkter Garten [m], vertiefter Garten [m].

sunken lane [n] *agr. for.* ▶sunken path.

6200　sunken path [n] *agr. for.* (Pathway below the level of adjacent ground on both sides, mostly in regions with soils, susceptible to erosion; e.g. loess, sandstone); *syn.* sunken lane [n]; *s* **camino [m] hondo** (Camino encajado entre taludes o paredes de roca); *f* **chemin [m] creux** (Chemin encaissé entre deux talus parfois empierrés, voie traditionnelle de circulation avant le 20e siècle, reliant les villages entre eux, souvent caractéristique du paysage de bocage, considéré à l'heure actuelle comme élément de paysages protégé et faisant partie du patrimoine culturel) ; *g* **Hohlweg [m]** (**1.** Weg in einem Geländeeinschnitt, meist in Gebieten mit leicht erodierbaren Böden [Löss, Buntsandstein]. Je nach Alter entsteht dieses Kulturlandschaftselement durch jahrzehnte- oder jahrhundertelange Benutzung eines Weges. Mehrere **H.e** nebeneinander [Hohlwegbündel] entstehen, wenn ein Weg zu tief eingeschnitten und nicht mehr benutzbar ist. **2.** Von steilen Felspartien begrenzter Weg).

6201　sunny day [n] *met. recr.* (Day of sunny weather); *s* **día [m] soleado** *syn.* día [m] con sol; *f* **journée [f] d'ensoleillement** (Jour par temps ensoleillé) ; *syn.* journée [f] ensoleillée ; *g* **Sonnentag [m]** (Tag mit sonnigem Wetter).

6202　sunny location [n] *plant.* (location in full sun; *opp.* ▶shady location); *s* **exposición [f] de solana** (*opp.* ▶umbría); *syn.* solana [f], exposición [f] a pleno sol; *f* **exposition [f] en plein soleil** (*opp.* ▶exposition [très] ombragée) ; *syn.* exposition [f] très ensoleillée ; *g* **vollsonnige Lage [f]** (*opp.* ▶Schattenlage).

sunny slope [n] *geo.* ▶adret.

sunscald [n] *arb. bot.* ▶sun scorch.

6203　sun scorch [n] *arb. bot.* (Excessive damage to bark and cambium, caused by sudden, strong exposure of a stem or branch to the sun; ▶bark scorch, ▶scorch); *syn.* sunburn [n], sunscald [n]; *s* **socarrado [m]** (▶Quemadura [del tronco] que afecta la corteza y el cambium causada por fuerte radiación solar; ▶quemadura de la corteza); *f* **brûlure [f] solaire** (**1.** *Sensu lato* ensemble des désordres causés par une exposition excessive au soleil. **2.** *Sensu stricto* Blessure localisée de l'écorce et du cambium causée p. ex. par un ensoleillement intensif ; DFM 1975 ; ▶brûlure, ▶insolation d'écorce) ; *g* **Sonnenbrand [m]** (Ein ▶Verbrennungsschaden an Borke und Kambium, der durch plötzlich starke Sonneneinstrahlung resp. durch Witterungsextreme [plötzlicher Temperaturunterschied von +20°C tagsüber und nur wenige Minusgrade nachts] an Stamm und Ästen entsteht und ein flächiges Absterben der Rinde auf der südlich bis südwestlich exponierten Seite zur Folge hat. Der Schaden wird meist erst später bemerkt, wenn Rindenpartien, die offenbar seit längerer Zeit abgestorben sind, rissig werden, sich vom Stamm lösen oder herunterfallen. Da diese Rindennekrosen sowohl im Sommer als auch im Winter entstehen können, werden sie **Sommer-Sonnennekrosen** resp. **Winter-Sonnennekrosen** genannt. Um in Baumschulen bei der Anzucht Sonnennekrosen zu vermeiden, ist es erforderlich, für eine ausreichende Wasserversorgung zu sorgen und auf eine optimale stickstoffreduzierte Ernährung zu achten. Im Straßenraum ist um junge Hochstämme ein geeigneter Sonnenschutz anzubringen, z. B. Schilfmatten, weißer Farbanstrich, Lehm-Jute-Verband; ▶Rindenbrand); *syn.* Frostplatte [f] (NL 2002, H. 9, p. 24), Sonnennekrose [f]).

6204 sunscreen [n] *arch.* (Shading device for protection against direct sunlight); *syn.* shadescreen [n]; *s 1* **dispositivo [m] protector contra el sol** (Techado u otro tipo de construcción cuya función es proteger contra la radiación solar directa); *s 2* **umbráculo [m]** (*Término específico* cobertizo de ramaje o de otro material que deja pasar el aire. Se usa para dar sombra y para proteger las plantas de la fuerza del sol; cf. CAS 1985); *f* **brise-soleil [m]** (Dispositif de protection contre le soleil) ; *syn.* pare-soleil [m] ; *g* **Sonnenschutzeinrichtung [f]** (Vorrichtung zum Schutze gegen direkte Sonneneinstrahlung).

sunshade [n] *phyt.* ▶protection from light.

sunshield [n] *phyt.* ▶protection from light.

6205 superelevation [n] *eng. trans.* (Raising of the outer edge of highway pavement or railroad rails along a horizontal curve to counteract centrifugal force); *s* **peralte [m] de curva** (En carreteras, vías férreas, etc., mayor elevación de la parte exterior de una curva en relación con la interior para contrarrestar la fuerza centrífuga; cf. CAS 1985); *f* **dévers [m] de la courbe** (Inclinaison transversale de la voie dans les courbes pour combattre la force centrifuge) ; *g* **Kurvenüberhöhung [f]** (Hebung des äußeren Randes eines Verkehrsbauwerkes [Straße oder Gleiskörper] in der horizontalen Kurve, um die Fliehkraft eines Fahrzeuges beim Durchfahren der Kurve aufzunehmen).

superficial deposits [npl] *geo.* ▶sediment (1)

6206 superheated soil water [n] *pedol.* (Water which remains in the soil despite heating above the boiling point. Its volume is added to that of the soil mass; ▶soil water, ▶soil moisture); *s* **agua [f] edáfica realmente combinada** (▶Agua edáfica que permanece en el suelo en estado de absorción aún cuando la temperatura supera el punto de ebullición. Su volumen se considera parte del volumen del suelo; ▶humedad del suelo); *syn.* agua [f] edáfica realmente ligada; *f* **eau [f] liée** (Eau d'absorption formant une fine pellicule à la surface des particules du sol et qui n'est pas absorbable par les racines ; ▶eau du sol, ▶humidité du sol) ; *syn.* eau [f] d'absorption, eau [f] hygroscopique, eau [f] pelliculaire ; *g* **Kristallwasser [n]** (Wasseranteil, der im Boden trotz 105°C Erhitzung verbleibt. Er wird der festen Bodenmasse zugeordnet; ▶Bodenwasser, ▶Bodenfeuchte).

superhighway [n] [US] *trans.* ▶freeway [US]/motorway [UK].

6207 superimposition [n] *plan* (Evaluation technique used in ▶environmental impact studies, whereby the various functions, land uses, or environmental stresses occurring in the study area are superimposed cartographically, so that sensitivity and stresses/impacts of individual spatial units may be compared; ▶layering of different uses); *s* **técnica [f] (gráfica) de superposición** (En el proceso de elaboración de ▶estudio de impacto ambiental, método de representar en mapas los diferentes tipos de hábitats y usos en una misma superficie para reconocer zonas especialmente sensibles y analizar posibles conflictos con el proyecto estudiado; ▶superposición de usos); *f* **superposition [f]** (Technique utilisée dans les ▶études d'impact opérant par chevauchement d'informations cartographiques traduisant différents facteurs biologiques, usages du sol ou nuisances sur un espace donné afin de pouvoir comparer la sensibilité et les conflits de portions d'espaces entre eux ; *terme spécifique* ▶superposition des usages) ; *g* **Überlagerung [f]** (In der Bewertung bei ▶Umweltverträglichkeitsstudien gebrauchte Technik, die dazu dient, verschiedene Funktionen, Nutzungen oder Belastungen, die in demselben Raum vorkommen, kartografisch übereinanderzulegen, so dass Empfindlichkeit und Belastung der Einzelräume untereinander vergleichbar werden; *UB* ▶Nutzungsüberlagerung).

6208 supervised children's playground [n] *recr.* *s* **área [f] de juegos educativos**; *f* **terrain [m] de jeux éducatifs** ; *g* **betreuter Kinderspielplatz [m]** *syn.* Spielpark [m].

supervising officer [n] [US] *constr. prof.* ▶design firm's representative [US]/supervisor representing a design practise [UK].

supervision [n] *constr.* ▶site supervision.

6209 supervision [n] **by the person-in-charge** *contr. constr.* (Oversight by the [general] ▶contractor's agent who is responsible for coordinating the work, in executing the contract documents including sub-contractor trades); *syn.* contractor's site management [n], supervision [n] by the site manager; *s* **jefatura [f] de obra** (Vigilancia por parte del ▶jefe/a de obra del contratista que es responsable de coordinar y controlar la actuación de cada oficio participante en la obra); *f* **maîtrise [f] de chantier** (Tâche d'ordonnancement et le pilotage correspondant à la coordination des travaux confiée à l'entreprise générale ou au mandataire d'un groupement dans le cas d'un marché unique, à des organismes spécialisés ou au ▶maître d'œuvre dans le cas de marchés séparés ; ▶conducteur de travaux) ; *g* **Bauleitung [f] (2)** (Tätigkeit des Generalunternehmers oder desjenigen Unternehmers, an den das Hauptgewerk zur Koordinierung und Überwachung des termingerechten Einsatzes aller am Bau beteiligten Gewerke beauftragt wurde; ▶Bauleiter/-in 1); *syn.* Bauführung [f] (2).

supervision [n] **of artistic quality** *constr. prof.* ▶observation of artistic quality.

supervision [n] **of the building works** [UK], **contractor's** *constr.* ▶contractor's supervision of the project [US].

6210 supervision [n] **of works** *constr. prof.* (General term for the overseeing and inspection of construction by a design firm's representative especially with regard to the interpretation of the contract drawings; **s. o. w.** is carried out in accordance with the contract schedule, cost-control and state-of-the-art technology and other standards and construction regulations; the term is applied whether carried out by the client as commissioning agent, the planner appointed by the client or the construction company as contractor, in accordance with assigned powers and functions; **s. of w.s** includes coordinating and monitoring of those professionally involved, especially the checking for consistency and approval of plans by third parties; compilation and monitoring of the time schedule; notice of delays caused by the contractor; acceptance of the work[s] and supplies in cooperation with the supervision team and others involved in the planning;

S

monitoring and issue of certificates of completion; applications to the authorities for final and partial acceptance; handing-over of the completed project, including compilation and submission of necessary documents, e.g. inspection records and test protocols; compilation of the maintenance requirements for the project; listing the limitation periods for warranty claims; establishment of costs; cost control by reviewing the performance of the contractor and his invoices in comparison with contract prices and the updating of final costs; ►building code control, ►contractor's agent, ►contractor's supervision of the project, ►design firm's representative [US]/supervisor representing a design practise [UK], ►observation of artistic quality, ►project management, ►site supervisor representing an authority, ►supervision by the person-in-charge; *syn.* inspection [n] of the works [UK]; *s* **dirección [f] de obra** (Término genérico para la vigilancia e inspección de las obras por parte del ►arquitecto director de las obras, sobre todo en lo que se refiere a la interpretación de los planos del contrato. La **d. de o.** incluye la vigilancia del cumplimiento del calendario del contrato, el control de costes [Es]/costo [AL] y la aplicación de la mejor tecnología asequible así como de otras regulaciones y estándares de construcción. El término se utiliza tanto por parte de la propiedad [comitente] y del arquitecto encargado por ella [adjudicatario de planificación] como por parte de la empresa de construcción [contratista] con las competencias y tareas correspondientes a cada caso; ►gestión de proyecto, ►inspección de la construcción, ►interventor, -a en pie de obra, ►jefatura de obra, ►jefe/a de obra, ►supervisión de la calidad artística, ►supervisión de obra); *f* **direction [f] de l'exécution du contrat de travaux (F.**, prestations intellectuelles ayant pour objet le contrôle des documents d'exécution, la direction et le contrôle de l'exécution et de l'avancement des travaux, la vérification des décomptes mensuels et l'établissement du décompte général ; **D.**, la **d. de l'é. du c. de t.** comprend en outre en Allemagne l'**ordonnancement, la coordination et le pilotage du chantier** [détermination de l'enchaînement des travaux ainsi que des délais d'exécution, harmonisation des interventions des divers corps de métiers, mise en application des mesures d'ordonnancement et de coordination] et l'**assistance au maître de l'ouvrage lors des opérations de réception** [opérations préalables à la réception, suivi et levée des réserves, constitution du dossier des ouvrages exécutés ; ►conducteur d'opération, ►contrôle technique des ouvrages, ►direction de chantier responsable de la qualité artistique des travaux, ►direction du chantier, ►maîtrise de chantier, ►maîtrise d'ouvrage, ►représentant du maître d'ouvrage public, ►représentant du maître d'œuvre) ; *syn.* direction [f] des travaux (HAC 1989, 464, direction [f] de l'exécution des travaux ; *g 1* **Bauleitung [f] (3)** (Überwachung der gemäß Vertrag termingerecht, wirtschaftlich und nach den allgemeinen Regeln der Technik und sonstigen Vorschriften und Auflagen auszuführenden Bauarbeiten; der Begriff **B.** wird sowohl seitens des Bauherrn [Auftraggeber] und des von ihm beauftragten Planers [Auftragnehmer Planung] als auch vom Bauunternehmen [Auftragnehmer Bau] mit entsprechend zugeordneten Kompetenzen und Aufgaben verwendet; ►Bauleiter/-in 1, ►Bauleiter/-in 2, ►Bauleitung 1, ►Bauleitung 2, ►künstlerische Bauoberleitung, ►Projektmanagement); *g 2* **Bauoberleitung [f]** (*Terminus, der nur in der HOAI verwendet wird [cf. § 55 (2) 8 HOAI 2002 und §§ 42 (1) 8 und 46 (1) 8 HOAI 2009]* Aufsicht über die örtliche Bauüberwachung bei Ingenieurbauwerken und Verkehrsanlagen, soweit die **B.** und **örtliche Bauüberwachung** getrennt vergeben werden; zu den weiteren Aufgaben gehören z. B. das Koordinieren der an der Objektüberwachung fachlich Beteiligten, insbesondere das Prüfen auf Übereinstimmung und Freigeben von Plänen Dritter; Aufstellen und Überwachen eines Zeitplanes, Inverzugsetzen der ausführenden Unternehmen; Abnahme von

Leistungen und Lieferungen unter Mitwirkung der örtlichen Bauüberwachung und anderer an der Planung und Objektüberwachung fachlich Beteiligter unter Fertigung einer Abnahmeniederschrift; Antrag auf behördliche Abnahmen und Teilnahme daran; Übergabe des Objekts einschließlich Zusammenstellung und Übergabe der erforderlichen Unterlagen, z. B. Abnahmeniederschriften und Prüfungsprotokolle; Zusammenstellung der Wartungsvorschriften für das Objekt; Auflistung der Verjährungsfristen der Gewährleistungsansprüche; Kostenfeststellung; Kostenkontrolle durch Überprüfung der Leistungsabrechnung der bauausführenden Firmen im Vergleich zu den Vertragspreisen und der fortgeschriebenen Kostenberechnung); *g 3* **Oberbauleitung [f]** *adm.* (*Verwaltungstechnischer Begriff für die kommunale, Landes- und Bundesebene* Aufsichts- und Überwachungstätigkeit eines ►Oberbauleiters; ►Bauaufsicht); *syn. zu 1. für den Planer* Bauüberwachung [f], *für den Bauunternehmer* Bauführung [f] (3); *syn. zu 2.* Oberleitung [f] der Bauausführung.

supervisor [n] [UK], contractor's construction *constr. prof.* ►contractor's agent.

6211 supervisory authority [n] *adm.* (Authority/agency which oversees subordinate administrative units and the implementation of laws and regulations; **in U.K.**, this is done directly by the ministry concerned); *s* **autoridad [f] inspectora** (Autoridad que controla a instancias administrativas subalternas sobre todo en cuanto a la aplicación de la normativa vigente); *syn.* autoridad [f] supervisora, autoridad [f] fiscalizadora [AL]; *f* **autorité [f] de tutelle** (Administration contrôlant les services subordonnés ainsi que les collectivités locales et veillant à l'application des lois) ; *g* **Aufsichtsbehörde [f]** (A. und D., Behörde, die nachgeordnete Verwaltungseinheiten kontrolliert und die Ausführung der Gesetze überwacht; im UK wird die Aufsicht durch das jeweilige Fachministerium unmittelbar ausgeübt).

6212 supplemental construction services [npl] or supplies [npl] *constr. contr.* (Additional work, which is neither included in the contract specifications nor indicated on the construction drawings but which has to be executed due to unforeseen circumstances. For such services or supplies an additional amount of money has to be negotiated); *s* **trabajos [mpl] adicionales (de construcción)** (Servicios u obras adicionales que ha de realizar el contratista, aunque no estaban previstos en el contrato, pero que —por presentarse condiciones imprevistas— son necesarios para llevar a cabo la obra encargada. Los costos de estos trabajos deben ser abonados adicionalmente); *f* **travaux [mpl] supplémentaires** (Prestations consécutives à un changement constaté, en cours d'exécution, dans l'importance ou la consistance des diverses natures d'ouvrages [modification du programme] à la suite d'un ordre de service ou de circonstances qui ne sont ni de la faute ni du fait de l'entrepreneur et lui donnant droit au versement d'une indemnité ; CCM 1984) ; *g* **zusätzliche Leistung [f] (1)** (Leistung, die von einem Unternehmer gegen zusätzliche Vergütung erbracht werden muss, obwohl sie weder in seinem Vertrag noch in dem ihm zu Grunde liegenden Entwurf enthalten ist, jedoch wegen eines unvorhergesehenen Ereignisses zur Ausführung der im Hauptauftrag beschriebenen Leistung erforderlich ist; cf. § 3 [6] VOB Teil A); *syn.* zusätzliche Bauleistungen [fpl].

6213 supplemental professional services [npl] *contr. prof.* (Additional planning or supervision services, which are usually provided together with ►basic professional services, and may lead to a reduction in building and running costs under certain circumstances; **in U.S.**, services which are not included in 'basic services' or ►special professional services in the "Standard Form of Agreement between Owner and Architect for

Designated Services", written down in AIA document B161/CM); *syn.* additional professional services [npl]; *s* **servicios [mpl] suplementarios** (Servicios especiales de planificación y control que se desarrollan generalmente además de los ►servicios profesionales básicos y que en algunos casos pueden ayudar a reducir los costos de construcción y de servicio); *f* **prestations [fpl] supplémentaires** (Intervention accrue d'un concepteur [mission étendue] réalisant des charges supplémentaires plus importantes que celles des ►prestations élémentaires prévues dans les missions normalisées de base dans le but d'une optimisation des ouvrages ; la rémunération prévue peut être **1. en F.** forfaitaire, avec majoration de la note de complexité, par application du coût d'objectif à l'élément normalisé supplémentaire ; cf. HAC 1989, 304. **2. En D.**, forfaitaire, en dépenses contrôlées, avec incitation au respect du coût d'objectif définitif, à la réduction du coût réel ou à la qualité et au délai d'exécution des ouvrages) ; *g* **zusätzliche Leistung [f] (2)** (Spezielle Planungs-, Entwicklungs- oder Überwachungsleistung, die im Allgemeinen begleitend zu den ►Grundleistungen erbracht wird und unter besonderen Optimierungsgesichtspunkten zu einer Senkung der Bau- und Nutzungskosten führen kann sowie rationalisierungswirksame besondere Leistungen, Projektsteuerung und Leistungen für den Winterbau [cf. §§ 28-32 HOAI]. Diese **L.en** konnten nach der HOAI 2002 als Erfolgs-, Pauschal- oder Zeithonorar vereinbart werden; nach der neuen HOAI 2009 gibt es die **z. L.en** nicht mehr; diese können im Rahmen der Honorarvereinbarung frei vereinbart werden).

supplementary bid [n] [US] *constr. contr.* ►subsidiary bid [US]/subsidiary tender [UK].

Supplementary Planning Documents [npl] [UK] *leg. urb.* *►comprehensive plan [US]/Local Development Framework (LDF) [UK].

supplies [npl] **(purchased) by client/owner** [US] *constr. contr.* ►services and materials by client/owner.

supply [n] *plan.* ►electric power supply, ►energy supply; *pedol. chem.* ►nutrient supply; *envir. limn. pedol.* ►oxygen supply; *plan. urb.* ►public water supply; *zool.* ►richness of food supply.

supply [vb] **and install** [vb] *constr. contr.* ►furnish and install.

supply line [n] *constr.* ►feed pipe.

6214 supply [n] **of developed green spaces** *landsc. urb.* (Quantity of developed green spaces available in urban areas for public recreation, and for climatic improvement; ►green space); *s* **oferta [f] de zonas verdes** (Total de ►zonas verdes existentes en áreas urbanas cuya función principal es servir para la recreación de la población); *syn.* oferta [f] de espacios verdes; *f* **offre [f] en espaces verts** (►Espaces verts existants mis à la disposition de la population, agissant favorablement sur la santé physique le psychisme et les rapports sociaux des habitants et assurant parallèlement une fonction hygiénique) ; *g* **Grünflächenangebot [n]** (Im Siedlungsbereich zur öffentlichen Nutzung vorhandene ►Grünflächen, die der Freizeit und Erholung, aber auch stadthygienischen Erfordernissen dienen); *syn.* Grünflächenbestand [m].

6215 supply [n] **of food** *zool.* (Natural existence or manmade provision of food for consumption by animals); *s* **disponibilidad [f] de alimentos** (Provisión natural o antropógena de alimentos para los animales); *f* **ressources [fpl] alimentaires** (Aliments présents naturellement dans le milieu ou apportés) ; *g* **Nahrungsangebot [n]** (Natürlich vorhandene oder künstlich bereitgestellte Nahrung für Tiere); *syn.* Angebot [n] an Nahrung.

6216 supply [n] **of materials** *constr. contr.* (**1.** Provision of building materials and prefabricated compound units used during construction. **2. In U.K.**, the term **s.** is often used for both supply and **delivery** to the site; ►delivery of plants, ►services and materials by the client/owner); *s* **suministro [m] de materiales** (►servicios y materiales del comitente, ►suministro de plantas); *syn.* provisión [f] de materiales; *f* **approvisionnement [m] de matériaux** (Action de livraison sur le chantier des matériaux nécessaires à l'évacuation des travaux ; ►fourniture de végétaux, ►prestations et fournitures fournies par le maître d'ouvrage) ; *g* **Baustofflieferung [f]** (Beschaffung und Anlieferung von Baustoffen und Bauteilen zur Verwirklichung einer Bauaufgabe; ►bauseitige Leistungen und Lieferungen, ►Pflanzenlieferung); *syn.* Materiallieferung [f], Lieferung [f] von Baustoffen, Lieferung [f] von Materialien.

6217 supply [n] **of sustenance** *ecol.* (Availability, movement and absorption of water and nutrients in plants; ►nutrient supply); *s* **avituallamiento [m] de la vegetación** (Disponibilidad de agua y nutrientes para el crecimiento normal de las plantas; ►suministro de nutrientes); *syn.* abasto [m] de la vegetación; *f* **alimentation [f]** (Apport et consommation, p. ex. d'eau potable, d'énergie, de substances nutritives pour les végétaux ; ►nutrition minérale [des plantes]) ; *g* **Versorgung [f] der Pflanzen** (Vorhandensein, Aufschließungsmöglichkeit und Zuführung von Wasser, Spurenelementen und Nährstoffen für Pflanzen; ►Nährstoffversorgung).

supportive permanent structure [n] [US] *leg. urb.* ►accessory structure.

suppressing species [n] *hort. phyt.* ►invasive species.

6218 supralocal [adj] [US]/**supra-local** [adj] [UK] *adm. plan.* (Descriptive term applied to significance beyond a local area); *s* **supralocal [adj]** (De importancia más allá de los límites locales); *f* **dépassant, ante les limites locales [loc]** (Dont l'importance dépasse les limites d'une localité ; *par extension* régional) ; *g* **überörtlich [adj]** (Über den Ort hinaus von Bedeutung); *syn.* gemeindeübergreifend [ppr/adj].

6219 supraregional [adj] [US]/**supra-regional** [adj] [UK] *adm. plan.* (Descriptive term applied to significance beyond a regional area); *s* **supraregional [adj]** (De importancia más allá de los límites regionales); *f* **dépassant, ante les limites régionales [loc]** (Dont l'importance dépasse les limites d'une région ; *par extension* national) ; *g* **überregional [adj]** (Über eine Region hinaus von Bedeutung).

6220 surety bond [n] [US] *contr.* (Generic term for a sum of money usually < 5% of the contract value, withheld from the construction contractor as a guarantee for the fulfillment/fulfilment of contractual obligations or for the remediation of any defects; ►bid bond [US]/tender bond [UK], ►retention fund, ►performance bond); *syn.* bid security [n] [US] (DAC 1975, 52), bond [n] [UK]); *s* **garantía [f] (2)** (Término genérico que denomina una cantidad de dinero que el contratista tiene que depositar como fianza para garantizar el cumplimiento de las obligaciones del contrato o para remediar posibles defectos; en Es. según la Ley de Contratos de las Administraciones Públicas, las garantías varían entre un máximo del 2% [garantías provisionales], un máximo del 4% [garantías definitivas y especiales] y un máximo del 6% [garantías complementarias] del presupuesto del contrato adjudicado; las garantías pueden ser constituidas a] en metálico, en valores públicos o en valores privados debidamente avalados, b] mediante aval prestado, en la forma y condiciones reglamentarias, por alguno de los Bancos, Cajas de Ahorros, Cooperativas de Crédito y Sociedades de Garantía Recíproca autorizados para operar en España y c] por contrato de seguro de caución celebrado en la forma y condiciones que reglamentariamente se establezcan, con entidad aseguradora autorizada para operar en el ramo de caución; cf. arts. 36 y 37 Ley 13/1995; ►aval, ►garantía de cumplimiento de contrato, ►retención de garantía); *syn.*

constitución [f] de garantía, depósito [m] de fianza; *f 1* sûretés **[fpl]** (*Terme générique* garanties financières fournies par une entreprise pour l'exécution d'un marché de travaux) ; *f 2* **garantie [f] de bonne fin** (*Terme spécifique* sûretés constituées en vue d'assurer une bonne exécution des travaux et couvrant le maître de l'ouvrage pour le risque de défaillance de l'entrepreneur, exigée de la part des entreprises ou de l'entreprise générale, le plus souvent sous forme de cautionnement ou de ►caution personnelle et solidaire [5 % du montant du marché lorsque a été fixé un délai de garantie], fournie à la signature du marché et restituée en une fois à la date de réception des travaux ; ►retenue de garantie, ►garantie de bonne fin des travaux) ; *g* **Sicherheitsleistung [f]** (Geldbetrag, der meist < 5 % der Auftragssumme bei Auftragnehmern betragen kann, die keine genügende Gewähr für die Erfüllung vertragsgemäßer Leistungen oder für die Beseitigung etwa auftretender Mängel bieten; cf. § 14 VOB Teil A. Die Sicherheitsleistung kann auch ein Geldinstitut in Form einer ►Bürgschaft erbringen; ►Einbehalt, ►Vertragserfüllungsbürgschaft).

surface [n] *constr.* ►all-weather surface, ►below ground surface, ►granular playing surface [US]/hoggin playing surface [UK]; *bot.* ►leaf surface; *constr.* ►natural ground surface, ►paved stone surface, ►water-bound surface, ►tree pit surface; *hydr. recr.* ►water surface.

surface [n] [UK], **coated macadam** *constr.* ►macadam [US].

surface [n] [US], **compacted granular** *constr.* ►water-bound surface.

surface [n], **existing ground** *constr. plan.* ►natural ground surface.

surface [n] [UK], **flexible** *constr.* ►water-bound surface, ►surface layers, #12.

surface [n], **groundwater** *constr.* ►watertable.

surface [n], **hard** *constr. urb.* ►paved area.

surface [n] [UK], **hoggin** *constr.* ►water-bound surface.

surface [n] [UK], **hoggin playing** *constr.* ►granular playing surface [US].

surface [n] [US], **pervious** *constr.* ►water-bound surface.

surface [n] [UK], **processed aggregate** *constr.* ►water-bound surface.

surface [n], **slip** *constr. eng.* ►slip plane.

surface [n] [US], **stabilized** *constr.* ►water-bound surface.

surface [n], **tar macadam** *constr.* ►macadam [US]/coated macadam surface [UK].

surface [n] [US], **water-bound playing** *constr.* ►granular playing surface [US]/hoggin playing surface [UK].

6221 surface articulation [n] *arch. constr.* (Design of surfaces using various paving materials and jointing patterns); *s* **articulado [m]** (Disposición sistemática, p. ej. de un pavimento; cf. DACO 1988); *f* **agencement [m] des surfaces** (p. ex. de revêtements de sols par l'utilisation de différents matériaux ou de leur appareillage ou calepinage) ; *g* **Oberflächengliederung [f]** (Gestaltung von Bodenbelägen, z. B. durch Verwendung unterschiedlicher Materialien, Verlegerichtungen oder durch eine differenzierte Höhenstaffelung der Teilflächen; *syn.* Gliederung [f] von Oberflächen.

surface composting [n] *agr. constr. hort.* *►mulching.

surface course [n] *constr.* ►granular surface course [US]/hoggin surface course [UK], ►surface layer.

surface course [n] [UK], **hoggin** *constr.* ►granular surface course [US].

6222 surface cultivation [n] *constr.* (Form of soil surface treatment; e.g. ►rototilling [US]/rotavation [UK], cultivating or hoeing); *s* **tratamiento [m] de superficie** (Forma de preparación del suelo para su mejor aprovechamiento, como subsolado, rastrillado, ►rotovación, etc.); *f* **traitement [m] superficiel** (Travail de préparation du sol lors de la mise en forme définitive, p. ex. défoncement, ►fraisage, binage) ; *syn.* traitement [m] de surface ; *g* **Oberflächenbehandlung [f] des Bodens** (Form der Bodenbehandlung, z. B. ►Fräsen, Grubbern, Hacken); *syn.* Behandlung [f] der Bodenoberfläche, Bodenoberflächenbehandlung [f].

6223 surface finish [n] *constr.* (Exterior structure and texture of natural stones, pavers, flags, wall faces, plastered walls, etc.); *syn.* surface texture [n]; *s* **acabado [m] de la superficie** (Características y estructura de la superficie de pavimentos de adoquines o losas, muros, enfoscados, etc.); *syn.* textura [f] de la superficie; *f* **aspect [m] de surface** (Structure d'un revêtement tel que pavage, dallage, mur, etc.) ; *syn.* texture [f] de la surface, texture [f] de la couche de surface, texture [f] de la couche superficielle (MAÇ 1981, 215) ; *g* **Oberflächenbeschaffenheit [f]** (Beschaffenheit und Struktur der Oberfläche von Steinen, Pflaster- und Plattenbelägen, Maueroberflächen, eines Verputzes etc.); *syn.* Oberflächenstruktur [f].

surface improvement [n] [US] *constr.* ►resurfacing.

6224 surface infiltration [n] *constr. ecol. wat'man.* (Percolation of surface water runoff over a wide surface area into the soil, thus recharging the groundwater; in **s. i.** systems, where rainwater is to penetrate the soil quickly, a sandy soil should be installed; ►swale infiltration, ►near-natural storm water runoff management); *s* **infiltración [f] superficial** (Percolación de las aguas pluviales en el suelo sobre un área grande de superficie, que favorece la recarga de acuíferos; en sistemas de **i. s.** en los que el agua de lluvia debe penetrar rápidamente en el suelo, se deben instalar suelos arenosos; ►infiltración por hoyos, ►gestión natural de las aguas pluviales); *f* **infiltration [f] superficielle** (Gestion des eaux pluviales non polluées permettant de recharger les nappes phréatiques ; on distingue deux modes d'infiltration : **1.** avec passage au travers d'une couche d'humus [cuvettes d'infiltration, basins d'infiltration diffuse, etc.] ou **2.** souterraine diffuse à faible profondeur sous la surface du sol sans transit sur sol humifère [galerie d'infiltration, cuves/citernes/réservoirs d'infiltration, massif graveleux dans la couche de couverture] ; ►gestion naturelle des eaux de pluie, ►infiltration par noues) ; *g* **Flächenversickerung [f]** (Breitflächige Sickerbewegung von anfallendem Oberflächenwasser in den Untergrund zur Grundwasserneubildung; in **F.sanlagen**, bei denen das Wasser von Regenereignissen schnell in den Boden eindringen soll, sind sandige Böden einzubauen; ►Muldenversickerung, ►naturnahe Regenwasserbewirtschaftung).

6225 surface inlet [n] [US] *constr. eng.* (Entry for storm runoff from a pavement or flat roof to enter the drainage system; generic term for ►yard drain and ►catch basin [US]/road gully [UK]); *syn.* drainage inlet structure [n] [US] (TSS 1988, 330-15), gully [n] [UK]; *s* **sumidero [m]** (Término genérico de ►sumidero de patio y ►sumidero de calle); *f* **regard [m] d'évacuation** (Terme générique pour tout élément de canalisation destiné à récolter les eaux de surface ; ►regard à grille, ►siphon de cour) ; *syn.* orifice [m] d'évacuation des eaux pluviales, orifice [m] d'écoulement, sac [m] d'eau pluviale [CH] ; *g* **Ablauf [m]** (Entwässerungsöffnung in einer Straßen-, Wege-, Platz- oder Flachdachoberfläche; *UBe* ►Hofablauf und ►Straßenablauf); *syn.* Gully [m].

6226 surface layer [n] *constr.* (Generic term for the top finish layer of roads, paths and other paved areas; ►pavement

construction courses, ▶surfacing 3; *specific terms* ▶crushed rock top course, ▶paved stone surface, ▶slab paving, ▶road surfacing, ▶path surfacing, ▶water-bound surface, ▶wearing course); *syn.* pavement layer [n], surface course [n], surfacing [n] (1); *s 1* **firme [m]** (En la estructura de carreteras, caminos y otras superficies pavimentadas, la capa superior; ▶capa de desgaste, ▶composición estratificada de superficies revestidas, ▶revestimiento compactado); *syn.* revestimiento [m]; *s 2* **pavimento [m]** (1) (Término genérico para todo tipo de superficie que calles, caminos y plazas; ▶estabilización de plazas y caminos; *términos específicos* ▶adoquinado, ▶enlosado, ▶capa superior de grava, ▶pavimento de carreteras/calles, ▶pavimento de caminos, ▶revestimiento compactado); *syn.* revestimiento [m]; *f 1* **couche [f] de surface** (*Structure de la chaussée* terme générique désignant la ▶couche de roulement et la couche de liaison [▶éléments constructifs de la voirie] ; ▶surface stabilisée aux liants hydrauliques) ; *syn.* revêtement [m] de surface, revêtement [m] superficiel ; *f 2* **revêtement [m] de sol** (Terme générique pour les différents types de sols stables ; ▶consolidation ; *termes spécifiques* ▶pavage, ▶dallage, ▶revêtement en grave, ▶revêtement d'une rue/route, revêtement de cheminement, ▶revêtement de cheminement) ; *g* **Decke [f]** (**1.** *Wegeaufbau* Schicht des ▶Oberbaus: oberste, verdichtete oder eingebaute/verlegte Schicht auf einer Bodenoberfläche. **2.** OB zu Oberfläche von Straßen-, Wege- und Platzflächen; ▶Befestigung 2; *UBe* Asphaltbelag, Asphaltdecke, ▶Pflaster, ▶Plattenbelag, ▶Schotterdecke, Steinbelag, ▶Straßenbelag, ▶Verschleißschicht, ▶wassergebundener Belag, ▶Wegebelag); *syn.* Belag [m] (1), Bodenbelag [m], Deckschicht [f], Oberflächenschutzschicht [f].

6227 surface layers [npl] *constr.* (Compilation of specific terms for paved surfaces using different materials; ▶bond, ▶paving pattern); 1. bituminous surfacing [n] [UK], asphaltic concrete (a.c.) pavement [n] [US]; *syn.* bituminous pavement [n] [UK], bituminous concrete pavement [n] [US], flexible pavement [n] (with cold asphalt) [US]; 2. ▶pavement [n] of concrete pavers [US]/concrete block pavement [n] [UK], precast Portland cement (p.c.) concrete paving [n] [US]; 3. in-situ concrete paving [n] [UK], poured-in-place Portland cement (p.c.) concrete paving [n] [US]; *syn.* rigid pavement [n] [US]; 4. concrete slab paving [n] [UK], precast Portland cement (p.c.) concrete paving [n] [US]; 5. ▶wood paving [n]; 6. ▶cobble paving [n]; 7. ▶brick paving [n]; 8. natural stone flag paving [n] [UK], natural stone slab paving [n] [UK]; *syn.* flagstone paving [n] [US]; 9. ▶paved stone surface [n]; 10. crazy paving [n] [UK], irregular flagstone paving [n] [US]; *syn.* opus [n] incertum; 11. porous pavement [n]; 12. ▶water-bound surface [n], flexible surface [n] [UK], 13. ▶wood block paving [n]; 14. ▶wood disk paving [n] [US]/timber disk paving [UK]; *s* **revestimiento [m] de superficies** (Términos específicos caracterizados por el material utilizado o por el ▶aparejo; ▶patrón de aparejo); *syn.* pavimento [m] (2); 1. pavimento [m] de asfalto/asfáltico; 2. ▶adoquinado [m] de piedras hormigón; *syn.* pavimento [m] de adoquín prefabricado en hormigón; 3. pavimento [m] de cemento; 4. pavimento [m]/enlosado [m] de piezas prefabricadas de cemento (baldosa, loseta, granito artificial); 5. ▶adoquinado [m] de madera; 6. ▶pavimento [m] de guijarros; *syn.* pavimento/revestimiento [m] de garbancillo; 7. ▶pavimento [m] de ladrillo clinker; 8. adoquinado [m]/pavimento [m] de losas de piedra natural; 9. ▶adoquinado [m]; 10. opus [m] incertum; *syn.* enlosado [m] irregular; 11. revestimiento [m] permeable; 12. ▶revestimiento [m] compactado; 13. ▶pavimento [m] de madera en bloques; 14. ▶adoquinado [m] de discos de madera; *f* **revêtement [m] de surface** (2) (Termes spécifiques caractérisant les différents types de revêtements par l'utilisation de matériaux divers et de différentes formes d'▶appareillage des matériaux ; ▶calepinage) ; 1. revête-

ment [m] d'enrobé ; *syn.* revêtement [m] en asphalte ; 2. ▶pavage [m] en pavés en béton ; 3. dallage [m] en béton/en ciment ; 4. revêtement [m] avec dalles préfabriquées en béton/en ciment ; 5. ▶pavage [m] en bois, revêtement [m] en bois ; 6. ▶pavage [m] en galets ; 7. ▶pavage [m] en terre cuite ; 8. revêtement [m] avec dalles naturelles ; 9. ▶pavage [m] ; 10. dallage [m] en tout-venant ; *syn.* opus [m] incertum ; 11. revêtement [m] perméable ; 12. ▶sol [m] stabilisé aux liants hydrauliques ; 13. ▶revêtement [m] en pavés de bois équarri ; 14. ▶dallage [m] en tranches de bois ; *syn.* revêtement [m] superficiel ; *g* **Belag [m] (2)** (Übersicht über UBe, die den **B.** durch Verwendung bestimmter Baustoffe oder ▶Verlegemuster kennzeichnen; ▶Verband 2); 1. Asphaltbelag [m], Asphaltdecke [f]; 2. ▶Betonsteinpflaster [n]; 3. Betonwegebelag [m], Betonwegedecke [f]; 4. Betonplattenbelag [m], Gehwegplattenbelag [m]; 5. ▶Holzpflaster [n]; 6. ▶Kieselsteinpflaster [n]; 7. ▶Klinkerpflaster [n]; 8. Natursteinplattenbelag [m]; 9. ▶Pflaster [n]; 10. unregelmäßiger Plattenbelag [m]; *syn.* Opus incertum [n]; 11. wasserdurchlässiger Belag [m]; 12. ▶wassergebundener Belag [m]; 13. ▶Kantholzpflaster [n]; 14. ▶Rundholzpflaster [n].

surface mining [n] [US] *min.* ▶open-pit mining [US]/opencast mining [UK].

6228 surface plant [n] *bot.* (Life form classification of the *Chamaephyta* after RAUNKIAER: renewal buds are above the surface of the earth; the plants enjoy only such protection as is afforded by the plant itself, either through protective mechanisms on the bud or by dense growth or dead shoots; ▶cushion plant); *syn.* chamaephyte [n]; *s* **camófito [m]** (División de primer orden en la clasificación biotípica y simorfial de RAUNKIAER: conjunto de formas cuyas yemas de reemplazo se elevan en el aire a menos de 25 cm, de modo que pueden quedar protegidas en la estación desfavorable por un manto de nieve o de hojarasca; DB 1985; ▶planta pulviniforme); *f* **chaméphyte [m]** (Classification des végétaux vasculaires selon RAUNKIAER désignant un végétal nain dont les bourgeons hivernaux situés au-dessus du sol ne bénéficient que de la protection assurée par l'appareil aérien de la plante mère ; ▶plante [vivace] en coussin[et]) ; *g* **Oberflächenpflanze [f]** (Lebensformklasse der *Chamaephyta* — nach RAUNKIAER: ausdauernde, holzige oder krautige Pflanze, bei der die Erneuerungs-/Überdauerungsknospen bis 50 cm über der Erdoberfläche liegen und daher nur den Schutz genießen, den die Pflanze selbst zu gewähren vermag wie z. B. Knospenschutz, dicht gedrängter Wuchs oder auf eine hohe Schneedecke angewiesen ist. Hierzu gehören z. B. Kriechstauden wie Kaukasus-Fetthenne [*Sedum spurium*], ▶Polsterpflanzen, Halbsträucher, Zwergsträucher und kriechende Holzpflanzen der Hochgebirge und nordischen Tundren, z. B. Zwergbirke [*Betula nana*], viele *Ericaceae* der ozeanischen Heiden, z. B. Besenheide [*Calluna*], Glockenheide [*Erica tetralix*], sowie niedrige, immergrüne Sukkulente, z. B. Hauswurz [*Sempervivum*]); *syn.* Chamaephyt [m], Zwergpflanze [f].

6229 surface runoff [n] *constr. geo. wat'man.* (That part of the precipitation which cannot be absorbed by the soil and moves over the ground surface to the nearest ▶receiving stream; it does not include stream channels affected by artificial diversions, storage, or other human works; ▶direct runoff, ▶overland flow, ▶surface water drainage); *syn.* runoff [n], *o. v.* run-off [n] (GGT 1979), surface water runoff [n]; *s* **escorrentía [f] superficial (3)** (Parte de la precipitación que se desplaza en la superficie del terreno; WMO 1974; ▶escorrentía superficial 1, ▶escorrentía superficial 2, ▶drenaje superficial); *syn.* escurrimiento [m] superficial; *f* **eaux [fpl] superficielles** (Toutes les eaux qui s'écoulent et qui ne sont pas infiltrées dans le sol ; ▶écoulement superficiel 2, ▶évacuation superficielle, ▶ruissellement) ; *g* **Oberflächenwasser [n]** (**1.** Der Teil des Niederschlagswassers, der

vom Boden nicht aufgenommen werden kann und oberflächig zum nächsten ▶Vorfluter und in letzter Konsequenz ins Meer oder in abflusslose Senken abfließt. **2.** Der Teil des Wassers bei Starkregen, Schneeschmelze oder Bewässerung, der in den Boden nicht einsickern kann und durch Entwässerungsmaßnahmen abgeleitet werden muss; ▶Oberflächenwasserabfluss 1, ▶Oberflächenwasserabfluss 2, ▶Oberflächenentwässerung).

surface runoff [n]**, direct** *constr. geo.* ▶direct runoff.

surface texture [n] *constr.* ▶surface finish.

6230 surface treatment [n] *constr.* (Surfacing of manufactured stone products, metals, wood, etc. *Specific term for surface treatment of stones with hammer and chisel* tooling); *syn.* surfacing [n] (2); *s* **tratamiento [m] de superficie visible de revestimientos** (Modelado de las superficies de elementos de piedra manufacturados, metales, madera, etc.; *términos específicos para modelado de piedra con martillo y cincel* estampación en seco, fileteado); *f* **traitement [m] des parements** (MAÇ 1981, 214 ; des pierres de taille, de la couche de finition des revêtements en béton, des enduits intérieurs ou extérieurs ou de différents matériaux tels que le métal, le bois, etc.) ; *syn.* surfaçage [m] (VRD 1994, tome II, chap. 4,6.4.2.2.1, 10) ; *g* **Oberflächenbehandlung [f] von Materialien** (Bearbeitung der Ansichtsflächen von Werksteinen, Metall, Holz etc.).

surface vegetation [n] *phyt.* ▶ground cover.

6231 surface water [n] *constr. hydr. wat'man.* (Water which remains or runs off from the ground after a period of precipitation; ▶flowing surface water); *s* **agua [f] superficial** (Aquella que circula por los cauces naturales, la que transitoriamente lo hace por la superficie de la tierra hasta encontrar un cauce natural y la almacenada en lagos y embalses; DINA 1987); *f* **précipitations [fpl] superficielles quotidiennes** (Eaux de pluie tombées pendant une journée qui stagnent ou s'écoulent sur un site ; ▶eau d'écoulement de surface) ; *g* **Tagwasser [n]** (Auf der Bodenoberfläche stehendes oder abfließendes Niederschlagswasser; ▶Vorflut 1).

6232 surface waterbody [n] *geo.* (Watercourse, pond, lake, reservoir, etc. on the Earth's surface); *s* **aguas [fpl] continentales (superficiales)** (Término genérico para todas las superficies de la litosfera cubiertas de agua); *f 1* **eaux [fpl] superficielles** (Terme générique pour toutes les eaux libres, courantes et stagnantes à la surface de la lithosphère) ; *syn.* eaux [fpl] de surface ; *f 2* **eaux [fpl] superficielles domaniales** (Eaux appartenant à l'État. Il s'agit des cours d'eau navigables ou flottables, des lacs navigables ou flottables, canaux, étangs, lagunes, lagons, etc.) ; *g* **Oberflächengewässer [n]** (Gewässer, das an die Oberfläche der Erde tritt).

6233 surface water drainage [n] *constr. urb. wat' man.* (Generic term for drainage of storm runoff from the Earth's surface, first by sheet or non-point drainage, then into open ditches, depressions or channels [channel drainage] or a system of pipes to a main collector drain or to a receiving stream, and then into a large watercourse [point drainage]; ▶drainage 1); *syn.* storm water drainage [n] [also US], conduction [n] of surface water/storm water; *s* **drenaje [m] superficial** (Evacuación del agua de áreas pavimentadas por medio de zanjas abiertas, hondonadas o canales o por un sistema de tuberías hasta un collector; ▶drenaje); *syn.* evacuación [f] de aguas pluviales; *f* **évacuation [f] des eaux superficielles** (Écoulement des eaux de pluie superficielles vers un exutoire par l'intermédiaire de fossés, caniveaux, rigoles ou canalisations ; ▶évacuation des eaux) ; *syn.* évacuation [f] des eaux pluviales ; *g* **Oberflächenentwässerung [f]** (Abführung des Regenwassers von der Erdoberfläche über offene Gräben, Mulden oder Rinnen oder über geschlossene Abflussleitungen; ▶Entwässerung); *syn.* Entwässerung [f] des Oberflächen-

wassers, Abführung [f] des Oberflächenwassers/Regenwassers, Ableitung [f] des Oberflächenwassers.

surface water runoff [n] *constr. geo. wat'man.* ▶surface runoff.

surfacing [n] (1) *constr.* ▶surface layer.

surfacing [n] (2) *constr.* ▶surface treatment.

6234 surfacing [n] (3) *constr.* (Stabilizing of pavement by applying the top layer; ▶path surfacing, ▶resurfacing, ▶surface layer, ▶wearing course); *s* **estabilización [f] de plazas y caminos** (Cobertura de aceras, caminos y plazas con adoquines, asfalto, losas, etc.; ▶capa de desgaste, ▶firme, ▶pavimentación, ▶pavimento 1, ▶pavimento de caminos); *syn.* pavimentación [m] de plazas y caminos; *f* **consolidation [f]** (Travaux de réfection du corps de chaussée d'une route, d'un chemin, d'une place par remplacement du pavage, application d'une couche d'asphalte, etc. ; contexte consolidation de la chaussée, consolidation de la plate-forme autoroutière ; ▶couche de roulement, ▶couche de surface, ▶renforcement 2, ▶revêtement, ▶revêtement de cheminement) ; *syn.* renforcement [m] (3) : *contexte* travaux de renforcement de la chaussée et renouvellement de la couche de roulement ; *g* **Befestigung [f] (2)** (Befestigen von Straßen, Wegen, Plätzen durch Pflastern, Asphaltieren, Legen von Platten etc.; ▶Belagserneuerung, ▶Decke, ▶Verschleißschicht, ▶Wegebelag); *syn.* Herstellung [f] eines Belages/einer Decke.

surfacing [n] **of lattice concrete blocks** [US] *constr.* ▶grid pavement [US]/pavement of grass pavers [UK].

surgery [n] *arb. hort.* ▶tree surgery.

6235 surplus excavated material [n] *constr. eng.* (SPON 1974, 101; material which must be disposed of off-site; ▶excavated material, ▶final disposal of surplus material, ▶off-site disposal of surplus material); *s* **material [m] sobrante** (Material de excavación que no se utiliza en la obra por lo que ha de ser trasladado a y depositado en otro lugar; ▶depósito definitivo de material sobrante, ▶disposición de excedentes de terracerías, ▶material de desmonte); *f* **déblais [mpl] excédentaires** (▶Matériaux extraits d'une fouille ne pouvant pas être réutilisés et donc évacués ; ▶dépôt définitif de déblais excédentaires, ▶enlèvement des excédents) ; *syn.* matériaux [mpl] excédentaires ; *g* **Überschussmasse [f]** (Aushub- oder Abtragsmaterial, das bei einem Bauvorhaben nicht wieder eingebaut werden kann und deshalb abgefahren werden muss; ▶Aushub 1, ▶Endlagerung von Überschussmassen, ▶Massenverbringung).

surplus material [n] *constr.* ▶off-site disposal of surplus material.

surplus material [n]**, disposal of** *constr.* ▶final disposal of surplus material.

6236 surrounding area [n] *landsc. urb.* (General term for an area situated around a building, a traffic route, an urban square, a city etc.; ▶preservation of the surrounding area); *s* **alrededores [mpl]** (Área que rodea a un edificio, una ciudad, etc.; ▶protección de zonas periféricas); *f* **abords [mpl]** (**1.** Espace entourant un bâtiment, une voie de circulation, une place, une localité. **2.** Dans la pratique du patrimoine, ce terme en est venu à désigner par extension les immeubles bâtis ou non formant le cadre d'un monument historique ; ▶protection des abords) ; *syn.* alentours [mpl], environs [mpl] ; *g* **Umfeld [n]** (...felder [pl]; Bereich, der ein Gebäude, eine Verkehrstrasse, einen Platz, eine Stadt etc. umgibt; ▶Umgebungsschutz); *syn.* Umgebung [f].

6237 surrounding region [n] **of a city/town** *urb.* (**1.** Inexactly defined built-up area surrounding a city and its suburbs, which is connected to the city socio-economically and culturally; often used to describe the area from which commuter traffic emanates. **2.** Area of land which surrounds and depends upon an urban center [US]/urban centre [UK] for its socio-economic and cultural existence, sometimes used in the sense of ►originating area [US]/catchment area [UK]; ►vicinity); *syn.* rural outskirts [npl] [US] (TGG 1984, 83), burgeon [n] [also US], *o.v.* bourgeon [n] [also US]; **s espacio [f] periurbano** (Espacio rural situado en la periferia de una ciudad y de su banlieue. Lugar de transformaciones profundas en el plano demográfico y social. La emigración de personas de la ciudad a este espacio como lugar de residencia, sin dejar de trabajar en la ciudad, produce cambios formales en el hábitat y en las comunicaciones, así como transformaciones sociales. El espacio se altera y pasa de ser considerado solo como un espacio productivo a convertirse en un espacio demandado por usos diferentes [residencial, industrial]; DGA 1986; ►área de influencia; ►área de influencia inmediata); *syn. coloquial* alrededores [mpl] de la ciudad; **f 1 espace [m] périurbain** (Terme souvent imprécis désignant les espaces situés à autour des villes, soumis à leur influence directe et susceptibles d'être significativement touchés par les processus enclenchés par cette proximité [forte pression foncière, conflits d'usages, etc.] ; TUP 1998 ; le rayon de cette zone varie de 15 à 30 kilomètres en fonction de l'importance de l'agglomération centrale ; ce territoire est situé au voisinage immédiat d'une ville-centre mais, par ses activités et ses modes de vie, en fait fonctionnellement partie et est en continuité physique du bâti ; le périurbain correspond aux espaces d'interpénétration ville — campagne dans lesquels, suite à une extension non maîtrisée, sans plan d'ensemble ou de politique de développement, s'observent des phénomènes de mitage, de friches industrielles, ferroviaires ; l'occupation des sols souvent très hétéroclites accueille des activités diverses résultat du desserrement urbain [surfaces commerciale, entrepôts, centres de recherche, etc.] ainsi que des formes d'habitat diversifiées [collectifs populaires, lotissements pavillonnaires, espaces résidentiels de populations aisées, jusqu'au cas limite des quartiers ou îlots clôturés [*gated communities*] et des enclaves résidentielles fermées, l'ensemble mêlé à des infrastructures routières envahissantes ; *syn.* zone [f] périurbaine ; **f 2 périphérie [f] urbaine (2)** (Espace constituant la ceinture située dans la zone d'influence directe d'une agglomération, présentant une spécialisation fonctionnelle [résidence, activités de commerce et culturelles] ; ►zone d'attraction, ►zone d'influence immédiate) ; **f 3 zone [f] de prolifération suburbaine (±)** (*D. jargon d'aménageur* [litt. ceinture de lard] limite brouillée d'une certaine forme de ville-banlieue en forte expansion, sans aménagement rationnel des espaces autour des grandes agglomérations, reliant les zones rurales à la ville-centre et ses équipements dont elle profite) ; **g 1 Stadtumland [n, o. Pl.]** (Nicht eindeutig definierter, relativ stark urbanisierter Raum im Umkreis einer politisch selbständigen Gemeinde/Stadt [suburbane Umgebung einer Stadt], der durch sozioökonomische und kulturelle Verflechtungen mit der Stadt verbunden ist und sich auf diese Weise durch den Prozess der Suburbanisierung zu einer Stadtregion entwickelt hat; oft wird der Bereich der Pendlerbewegungen und das zentralörtliche ►Einzugsgebiet gemeint; ►Nahbereich); **g 2 Speckgürtel [m]** (*Umgangssprachlich* Planerjargon für sich stark ausbreitende Bereiche um Großstädte und Ballungsräume, die von den zentralen Einrichtungen des naheliegenden Zentrums profitieren); *syn.* Umland [n, o. Pl.], Vorortgürtel [m] (±).

6238 surveillance [n] *adm. envir.* (Continual or periodic checking for compliance with environmental protection regulations; ►monitoring, ►national or state surveillance system for air pollution); **s vigilancia [f] (ambiental)** (Control regular o puntual del cumplimiento de normas ambientales; ►monitoreo ambiental, ►sistema español de información, vigilancia y prevención de la

contaminación atmosférica [Es]); **f surveillance [f]** (Contrôle permanent ou ponctuel des prescriptions réglementaires pour la protection de l'environnement ; ►monitorage, ►réseau de surveillance, ►dispositif de surveillance) ; **g Überwachung [f]** (Kontinuierliche oder punktuelle Kontrolle von Auflagen des Umweltschutzes; ►Monitoring; ►Überwachungsnetz).

surveillance [n]**, technical** *leg. envir.* ►technical monitoring.

surveillance system [n] **for air pollution** *envir.* ►national or state surveillance system for air pollution.

survey [n] *plan.* ►area study survey; *surv.* ►boundary survey; *landsc. plan.* ►comprehensive site survey and analysis; *ecol. plan.* ►environmental survey; *biol. geo. landsc. phyt. surv.* ►field survey (1); *adm. landsc. urb.* ►green space survey; *plan.* ►landscape survey, ►landscape survey; *constr. contr.* ►method of site survey [US]/method of measurement [UK]; *eng. surv.* ►site survey (1); *pedol.* ►soil survey; *constr. landsc.* ►topographic survey (1); *landsc.* ►topographic survey (2); *surv.* ►topographic survey (3); *constr. landsc.* ►tree survey; *phyt.* ►vegetation survey.

survey [n] **[UK], enumeration** *for. obs.* ►forest stand inventory [US].

survey [n]**, field** *surv.* ►topographic survey (3).

survey [n]**, lot** *surv.* ►boundary survey.

survey [n] **[US], metes and bounds** *surv.* ►boundary survey.

survey [n]**, opinion** *sociol.* ►representative poll.

survey [n]**, plant ecological** *phyt.* ►vegetation survey.

survey [n]**, plot** *surv.* ►boundary survey.

survey [n]**, property** *surv.* ►boundary survey.

surveyed to proposed falls and gradients [loc] *constr.* ►graded to proper line and levels.

surveying stake [n] **[US]** *surv.* ►surveyor stake.

6239 survey [n] **of a built-up area** *plan. urb.* (Investigation and mapping of settlement conditions in a defined urban area, identified by on-site inspection and/or aerial photographs); **s inventario [m] urbano** (Levantamiento de datos y mapeo de las condiciones de asentamiento en una zona urbana determinada, que puede realizarse sobre terreno o por medio de fotos aéreas); **f relevé [m] du patrimoine bâti existant** (L. 145-3 ; *résultat* inventaire du patrimoine bâti existant) ; **g städtebauliche Bestandsaufnahme [f]** (Erfassung und grafische Darstellung aller für die Planung relevanten städtebaulichen Gegebenheiten durch Aufnahmen vor Ort und/oder durch Luftbildauswertung).

survey [n] **of existing site conditions** *landsc. plan.* ►comprehensive site survey and analysis.

6240 survey [n] **of existing uses** *plan.* (Delineated record on a map of the actual land uses within a planning area at a given time; *result* existing use plan [US]; cf. IBDD 1981, 82; *generic term* ►mapping); *syn.* mapping [n] of existing land uses; **s inventario [m] cartográfico de usos** (Registro de usos reales del suelo en una zona; *termino genérico* ►inventario); *syn.* inventario [m] de usos del suelo; **f cartographie [f] des usages** (Inventaire et représentation cartographique de l'utilisation existante du sol dans une région déterminée ; *terme générique* ►cartographie) ; **g Realnutzungskartierung [f]** (Erfassung und kartografische Darstellung der tatsächlich vorhandenen Flächennutzungen in einem Planungs- oder Erhebungsgebiet; *OB* ►Kartierung).

6241 surveyor [n] *surv.* (Professional who identifies and measures the boundaries and topographic features of a site); **s geodeta [m/f]** (Profesional que se dedica a estudiar y medir la

S

topografía del terreno con fines científicos, agrarios y catastrales); *syn.* agrimensor, -a [m/f], apeador, -a [m/f] (BU 1959); *f* **géomètre-expert [m]** (Technicien réalisant les études et travaux topographiques dans les domaines foncier et géographique ; *syn.* géomètre [m], arpenteur-géomètre [m] ; *g* **Vermessungsingenieur/-in [m/f]** (Jemand, der als Ingenieur/-in die Landes-, Kataster- und Agrarvermessung betreibt; **V.e** werden an technischen Fachhochschulen [Master- und Bachelor-Studienabschluss, *vormals* Dipl.-Ing. (FH)] oder Universitäten [Master- und Bachelor-Studienabschluss, *vormals* Dipl.-Ing., A. *auch* DI] ausgebildet); *syn. obs.* Feldmesser/-in [m/f], *obs.* Geodät/-in [m/f], *obs.* Geometer/-in [m/f], *obs.* Landmesser/-in [m/f], *obs.* Landvermesser/-in [m/f].

6242 surveyor stake [n] *surv.* (Wooden or metal stake, which is set out to mark survey lines, usually 2.5m tall; ▶staking out with field rods [US]/setting-out using ranging poles [UK]); *syn.* surveying stake [n] [also US]; *s* **jalón [m]** (Estaca de madera o metal de unos 2 m utilizada para marcar líneas en terreno que se está estudiando; ▶jalonamiento); *f* **jalon [m]** (Bâton en bois ou en acier avec revêtement PVC de 1,50 à 2 mètres de hauteur, pointu à une extrémité, peint de couleurs rouges et blanches, planté en terre comme repère, pour déterminer un alignement, une direction, une distance ou une limite lors de travaux de nivellement, de terrassement ; ▶jalonnement) ; *g* **Fluchtstab [m]** (2 m langer Stab mit 3 cm Durchmesser und regelmäßigen roten und weißen Farbabschnitten zum Abstecken von Linien und zur Markierung von Punkten im Gelände; ▶Absteckung mit Fluchtstäben).

6243 survey phase [n] *landsc. plan. prof.* (First part of the landscape planning process, which presents the results of basic investigations, the landscape analysis and the evaluation of the planning area; ▶comprehensive site survey and analysis, ▶landscape planning proposals; ▶landscape survey); *s* **capítulo [m] de análisis de un plan** (D., resultado de la primera fase de elaboración de plan general del paisaje, en la cual se registra, analiza y valora el estado de la naturaleza y del paisaje; ▶capítulo de desarrollo de un plan general del paisaje, ▶elaboración de inventario ambiental, ▶inventariación paisajística); *f* **étude [f] préliminaire d'un plan de paysage** (D., premier élément normalisé d'une mission normalisée d'étude paysagère ou d'environnement présentant dans un document écrit et cartographique les résultats du recueil des données du milieu naturel [▶relevé cartographique], de l'état du milieu initial, de l'analyse du paysage et du diagnostic paysager ; cf. § 45 a HOAI 2002 et § 23 HOAI 2009 ; ▶étude de reconnaissance, ▶rapport d'orientation du plan de paysage) ; *syn.* recherches [fpl] et études ; *g* **Grundlagenteil [m]** (Erster Abschnitt bei landschaftsplanerischen Leistungen, der die Ergebnisse der ▶Grundlagenuntersuchung und ▶Bestandsaufnahme einer Landschaft, der Landschaftsanalyse und die Bewertung des Planungsgebietes [Landschaftsbewertung] in einem Erläuterungstext und in Karten wiedergibt, entsprechend der Leistungsphase 2 § 45 a HOAI 2002 resp. § 23 i. V. m. Anlage 6 Leistungsphase 2 HOAI 2009; ▶Entwicklungsteil einer Landschaftsplanung); *syn.* Analyseteil [m].

6244 survey plan [n] (1) *plan.* (Cartographic presentation of existing conditions on development or land treatment area; ▶tree survey plan); *s* **plano [m] de inventario** (Representación cartográfica de las condiciones existentes en zona a planificar o desarrollar al comienzo de la planificación; ▶plano de muestreo de árboles); *syn.* plano [m] de datos básicos; *f* **plan [m] de situation** (Représentation cartographique des données locales d'une zone au début de la procédure d'aménagement ; ▶plan d'implantation des arbres d'ornement) ; *g* **Bestandsplan [m] (2)** (Kartografische Darstellung der örtlichen Gegebenheiten eines Planungsgebietes bei Planungsbeginn; ▶Baumbestandsplan); *syn.* Bestandskarte [f].

6245 survey plan [n] (2) *surv.* (Map of an area with all existing objects shown to scale as a result of a ▶topographic survey 3; ▶site plan); *s* **plano [m] de agrimensura** (Mapa de un área que muestra todos los objetos existentes en escala como resultado de la ▶medición topográfica; ▶plano de situación); *f* **plan [m] régulier** (Document graphique ou numérique établi par un géomètre résultat d'un ▶levé de points [levé planimétrique] représentant la planimétrie et permettant la reproduction ; on distingue les plans périmétriques de propriété, plans de masse, plan de lotissement, plans parcellaires, etc. ; cf. HAC 1989, 531 ; ▶plan de situation [du terrain], ▶relevé planimétrique d'un terrain) ; *syn. langage courant* plan [m] de géomètre ; *g* **Vermessungsplan [m]** (Durch einen Vermessungsingenieur erstellter maßstabsgerechter ▶Lageplan eines Geländes mit den darauf befindlichen Objekten, z. B. Gebäude, Mauern, Wege, Bäume und sonstige Pflanzenstrukturen als Ergebnis einer ▶Vermessung).

survey work [n] *surv.* ▶topographic survey (3).

survival conditions [npl] *ecol. plan. sociol.* ▶living conditions.

6246 survival group [n] *zool.* (Zoological term used to describe a community characterized by the fundamental absence of any family connection, whilst not excluding the possibility that members of the same family or tribe are also in the same group; all members of the same species may be members of the group and are as individuals replaceable); *syn.* aggregation [n]; *s* **grupo [m] de supervivencia** (Término de la zoología que denomina una comunidad caracterizada por la falta de relaciones parentescas entre sus miembros, aunque no excluye la posibilidad de que miembros de una misma familia sean parte del mismo g. de s.; todos los miembros de una especie pueden ser miembros del grupo y son sustituibles como individuos); *f 1* **agrégat [m]** (Regroupement d'animaux entre individus différents, sans liens familiaux, attirés par un endroit particulier résultant souvent d'une préférence dans l'habitat, de la présence d'une source de nourriture ; le phénomène de regroupement se dénomme l'« agrégation » ; LAF 1990) ; *f 2* **grégarisme [m]** (Tendance des animaux à former des groupes sociaux comme les bandes d'oiseaux, les bancs de poissons ou les troupeaux de mammifères) ; *g* **Gesellungsverband [m]** (Verband, der durch das grundsätzliche Fehlen familiärer Bindungen gekennzeichnet ist, was nicht ausschließt, dass Glieder derselben Familie oder Sippe ihm angehören. Alle Artgenossen können Verbandsgenossen sein und sind als Individuen auswechselbar); *syn.* Gesellung [f].

6247 survival strategy [n] *biol. ecol.* (Adaption of organisms to unfavo[u]rable living conditions); *s* **estrategia [f] de supervivencia** *syn.* estrategia [f] de sobrevivencia [AL]; *f* **stratégie [f] de survie** (Adaptation des organismes à des conditions de vie défavorables afin de survivre) ; *g* **Überlebensstrategie [f]** (Anpassung von Organismen an ungünstige Lebensbedingungen, um zu überleben).

susceptibility [n], **erosion** *geo. pedol. plan.* ▶erosion risk.

susceptibility [n], **environmental** *envir. land'man.* ▶environmental sensitivity.

6248 suspended particulate matter [n] *hydr.* (LE 1986, 146; finely-distributed/divided organic or inorganic particles in water without forming deposition); *syn.* suspended particulates [npl]; *s* **materias [fpl] en suspensión** (Sustancias contenidas en las masas de agua que no se depositan); *f* **matières [fpl] en suspension (M.E.S.)** (LA 1981, 474 ; matières organiques et minérales en suspension dans les cours d'eau ne se déposant pas, constituant la partie non dissoute de la pollution ;

terme utilisé comme paramètre pour définir le degré de pollution des eaux résiduaires domestiques) ; *g* **Schwebstoffe [mpl]** (Im Gewässer vorhandenes Bodenmaterial und organische Stoffe, die sich nicht absetzen).

suspended particulates [npl] *hydr.* ▶suspended particulate matter.

suspension [n] of construction works [UK] *constr. contr.* ▶suspension of project [US].

6249 suspension [n] of project [US] *constr. contr.* (Contractor's decision to stop construction work [US]/construction works [UK] because of bad weather, or to stop the work/works completely); *syn.* suspension [n] of construction works [UK]; *s* **suspensión [f] de trabajos (de obras)** (Interrupción de más de tres meses, p. ej. por mal tiempo); *f* **interruption [f] des travaux** (Décision de l'entrepreneur d'arrêter temporairement [p. ex. pour cause d'intempéries] ou définitivement les travaux) ; *g* **Einstellung [f] der Bauarbeiten** (Unternehmerische Entscheidung, den Fortgang der Arbeiten [vorübergehend] zu unterbrechen, z. B. wegen schlechten Wetters, oder überhaupt nicht mehr weiterzuführen); *syn.* Einstellen [n, o. Pl.] der (Bau)arbeiten.

sustainability principle [n] *agr. ecol. for.* ▶principle of sustained yield.

6250 sustainable development [n] *envir. nat'res. sociol.* (**1.** Overriding goal first internationally recognized and ratified at the UN conference on the environment in Rio de Janeiro in 1992 and proclaimed in the Agenda 21 as the basic principle for economic development for the 21st century. According to this ideal all of mankind's actions are to be based on the premise that the fulfil[l]ment of today's human needs will not jeopardize the fulfil[l]ment of the requirements of future generations and shall not adversely affect the earth's limited ability to continue to provide natural resources and absorb wastes. The principle should not only apply to individuals, but also to every nation, enterprise, family, etc. In order to insure that life on earth has equal chances of existence, social and cultural development opportunities must be co-ordinated in a long-term strategy. **2.** The term also implies the maintenance of a dynamic balance between the needs of an urban population and the even distribution of natural resources required to support those needs in all forms of anthropogenic activity. **3.** Furthermore, successful **sustainability** means living more harmoniously with nature by minimizing environmental interventions and damage, reducing energy consumption, using replenishable resources and recycling already extracted and processed resources, even if they are replenishable; i.e. the rate of consumption of renewable sources of raw materials, water and energy should not be greater than the rate of their regeneration and also that non-renewable resources should not be depleted faster than the rate at which they can be substituted for by sustainable, renewable resources. This in turn means that the emission of pollutants must not be permitted to be greater than the capacity of air, water and soil to bind or biodegrade contaminating substances. Chapter 28 of Agenda 21 pays great attention to the necessity for action plans to ensure **s. d.**: these are to be realized by a collaboration of involved members of the public together with the authorities. In all multifaceted debates on the subject it is undisputed, that industrial and resource-intensive methods of production and ways of life cannot be practised in all countries and regions of the world, because of the earth's limited capacity for resource regeneration. Four basic strategies for the fulfil[l]ment of **s. d.** may be pursued: **1. efficiency:** minimization of the use of resources, whilst achieving the same output; **2. sufficiency:** changes in lifestyles and patterns of consumption; **3. compatibility:** adaptation to natural cycles and processes; e.g. substitutes for non-regenerative resources; **4. participation:**

democratization of decision-making processes [clarification of requirements, goals and strategies]; cf. ÖKO 2003; ▶principle of sustained yield, ▶sustained growth); *s* **desarrollo [m] sostenible** (El desarrollo social y ambientalmente sustentable es aquel que satisface las necesidades de las generaciones presentes, sin comprometer la capacidad de las generaciones futuras para satisfacer sus propias necesidades. Además prioriza la equidad en la distribución de los recursos, considerando las necesidades de las personas, pero al mismo tiempo impone límites al crecimiento, garantizando los principios del bien común y el mejoramiento de la calidad de vida. El **d. s.** promueve así mismo la profundización de la democracia, garantizando a la sociedad civil ser actora en la definición de su propio desarrollo; cf. Chile Sustentable 1999; ▶crecimiento sostenido, ▶sostenibilidad); *syn.* desarrollo [m] sustentable [CS]; *f* **développement [m] durable** (**1.** Innovations de nature à mettre en permanence à la disposition des hommes les ressources naturelles et d'éliminer sans dommages les produits rejetés, celles-ci couvrant les besoins actuels des hommes sans que soit compromise la faculté de répondre aux besoins des générations futures et la mise en péril de la capacité de régénération restreinte de l'écosystème terre. **2.** Maintien d'un équilibre dynamique entre les besoins des régions à forte densité de population et la mise à leur disposition des ressources naturelles nécessaires. **3.** Un développement durable réussi implique une plus grande harmonie entre la nature et l'homme, qui se doit de limiter la destruction des espaces naturels et des paysages et de réduire la consommation d'énergie, d'utiliser les ressources naturelles primaires renouvelables et de recycler/réutiliser les ressources produites même si celles-ci sont renouvelables ; ▶croissance durable, ▶principe d'exploitation durable) ; *syn.* développement [m] soutenable (**1.** Im Vergleich zur ▶Nachhaltigkeit der Nutzung ist der Begriff **n. E.** jüngeren Datums und die deutsche Übersetzung des englischen Terminus *sustainable development*. Er wurde 1987 von der BRUNDTLAND-Kommission definiert — der „Weltkommission für Umwelt und Entwicklung", die von der UNO eingesetzt und von der ehemaligen norwegischen Ministerpräsidentin GRO HARLEM BRUNDTLAND geleitet wurde. Mit der Veröffentlichung des BRUNDTLAND-Berichtes begann eine weltweite Expertendiskussion. 1992 wurde auf der Konferenz über Umwelt und Entwicklung der UN in **Rio de Janeiro** in der Agenda 21 **n. E.** zum proklamierten Leitbild für das 21. Jahrhundert. Danach soll das Handeln der Menschen prinzipiell so organisiert und bestimmt sein, dass die heutigen ökologischen, ökonomischen und sozialen Bedürfnisse der Menschen abgedeckt werden, ohne Ansprüche zukünftiger Generationen zu beeinträchtigen und ohne die begrenzte Fähigkeit des Ökosystems Erde zunichte zu machen. Dies gilt sowohl für den einzelnen Menschen als auch für jeden Staat, jedes Unternehmen, jede Familie etc. Um gerechte Lebenschancen heute und für die Zukunft zu möglichen, müssen ökologische, ökonomische, soziale und kulturelle Entwicklungsmöglichkeiten langzeitstrategisch miteinander verknüpft werden. Es folgte 1997 die Klimaschutzkonferenz in Kyoto mit dem so genannten **Kyoto-Protokoll**, das als Abkommen bis zum Jahre 2012 verbindliche Ziele für den Ausstoß von Treibhausgasen, welche als die wichtigste Ursache der globalen Erwärmung gelten, festschreibt. **2. N. E.** bedeutet auch die Aufrechterhaltung eines dynamischen Gleichgewichtes zwischen den Bedürfnissen der Bevölkerung in dicht besiedelten Gebieten und dem gleichmäßig verteilten Nachliefern der benötigten natürlichen Lebensgrundlagen bei allen anthropogenen Aktivitäten. **3.** Erfolgreiche **n. E.** bedeutet ferner, dass der Mensch mehr im Einklang mit der Natur leben muss, indem er die Zerstörung von Natur und Landschaft und den Energieverbrauch dadurch einschränkt, dass er primär erneuerbare Naturgüter verwendet und schon gewonnene und verarbeitete Ressourcen rezykliert/wiederverwendet,

S

auch wenn sie erneuerbar sind; d. h. dass die Verbrauchsrate von erneuerbaren Rohstoff-, Wasser- und Energieressourcen nicht höher ist als die Neubildungsrate, und dass nicht erneuerbare Ressourcen nicht schneller verbraucht werden, als sie durch dauerhafte, erneuerbare Ressourcen ersetzt werden können; das bedeutet auch, dass der Ausstoß von Schadstoffen nicht größer sein darf als die Fähigkeit von Luft und Wasser und Boden, diese Schadstoffe zu binden und abzubauen. Im Kapitel 28 der Agenda 21 wird besonders auf die Umsetzung von Aktionsplänen für eine **n. E.** hingewiesen, die von Behörden in Zusammenarbeit mit betroffenen Bürgern erstellt und umgesetzt werden sollen. Bei allen facettenreichen Diskussionen ist es unstrittig, dass die industrielle, ressourcenintensive Lebens- und Produktionsweise wegen der begrenzten Tragfähigkeit der Erde nicht auf alle Staaten und Regionen übertragen werden kann. Vier Basisstrategien können zur Realisierung der **n. E.** verfolgt werden: **a. Effizienz:** Minimierung des Einsatzes von Ressourcen bei vergleichbarer Leistung; **b. Suffizienz:** Änderung der Lebensstile und Konsummuster; **c. Konsistenz:** Anpassung an natürliche Kreisläufe und Prozesse, z. B. Ersatz für nicht regenerative Ressourcen; **d. Partizipation:** Demokratisierung von Entscheidungsprozessen [Klärung von Bedürfnissen, Zielen, Strategien]; cf. ÖKO 2003. Der Begriff **zukunftsfähige Entwicklung** [1991 von UDO E. SIMONIS geprägt] verfolgt einmal einen sozial-ökologischen Strukturwandel im eigenen Land und zum anderen, beim anstehenden Wandel die globalen Zusammenhänge systematisch mitzubedenken und die Beziehungen zwischen Industrie- und Entwicklungsländern auf eine gerechtigkeitsorientierte Basis zu stellen; HOL 2006]; ►nachhaltiges Wachstum); *syn.* dauerhafte Entwicklung [f], dauerhaft umweltverträgliche Entwicklung [f], tragfähige Entwicklung [f], zukunftsfähige Entwicklung [f].

6251 sustainable management [n] *ecol. econ. nat'res.* (Economic method of balancing profitable human use of natural resources with ecological values, so that those resources are not exhausted; ►principle of sustained yield); *s* **gestión [f] económica sostenible** (Actuación en empresas con el objetivo de equiparar los aspectos económicos y ecológicos, para así ahorrar recursos naturales y preservar la naturaleza para generaciones futuras; ►sostenibilidad); *syn.* gestión [f] económica sustentable; *f* **gestion [f] durable** (Démarche économique des entreprises dont l'objectif est la prise en compte des intérêts économiques et écologiques de manière à prévenir l'épuisement des ressources naturelles ; ►principe d'exploitation durable) ; *g* **nachhaltiges Wirtschaften [n, o. Pl.]** (Betriebswirtschaftliches Handeln mit dem Ziel, wirtschaftliche und ökologische Aspekte so abzustimmen, dass natürliche Ressourcen nicht erschöpft werden; ►Nachhaltigkeit der Nutzung); *syn.* zukunftsfähiges Wirtschaften [n, o. Pl.].

sustainable power source [n] *envir.* ►renewable power source.

sustainable tourism [n] *recr.* ►eco-tourism.

6252 sustainable use [n] *agr. ecol. for.* (►principle of sustained yield, ►sustainable development); *s* **aprovechamiento [m] sostenible** (►desarrollo sostenible, ►sostenibilidad, ►turismo ecológico); *syn.* aprovechamiento [m] sustentable; *f* **usage [m] durable** (►développement durable, ►principe d'exploitation durable ; *termes spécifiques* ►tourisme durable, urbanisme durable, ►écotourisme) ; *g* **nachhaltige Nutzung [f]** (►nachhaltige Entwicklung, ►Nachhaltigkeit der Nutzung, ►Ökotourismus).

sustainably [adv] *ecol.* ►use sustainably.

6253 sustained growth [n] *econ. nat'res. sociol.* (Growth which permits renewability and restoration of natural resources for economic development, in comparison with unlimited growth

which causes environmental destruction; ►sustainable development); *s* **crecimiento [m] sostenido** (En teoría, crecimiento económico en el marco de un desarrollo sostenible, es decir que permite la renovación y restauración de los recursos naturales para producción y consumo y, por lo tanto, no causa daños al medio ambiente; ►desarrollo sostenible); *f* **croissance [f] durable** (Par opposition à une croissance illimitée, face aux dangers que représentent les déséquilibres écologiques globaux et ayant conscience que l'utilisation et la capacité de charge de l'écosystème terre sont limitées, un développement économique soutenable ne peut être assuré qu'avec la mise en œuvre de mesures favorables au renouvellement et à la restauration des ressources dans le système de production, ceci avec le souci du respect des équilibres naturels fondamentaux ; ►développement durable) ; *syn.* croissance [f] soutenable ; *g* **nachhaltiges Wachstum [n]** (Im Vergleich zum unbegrenzten **W.** sind durch die globalen ökologischen Gefahren und im Bewusstsein nicht übersteigbarer Grenzen der Nutz- und Belastbarkeit des [Gesamt]ökosystems Erde nur noch ein **W.** möglich, das eine dauerhafte wirtschaftliche Entwicklung unter Berücksichtigung der Erneuerbarkeit resp. Wiederherstellung der Produktions- und Lebensgrundlagen gewährleistet; ►nachhaltige Entwicklung).

swale [n] *constr. eng.* ►drainage swale, ►infiltration swale.

swale [n], **check** *agr. constr.* ►berm ditch.

6254 swale infiltration [n] *constr. ecol. wat'man.* (Percolation of surface water runoff, which retained temporarily in swales, depressions, channels, etc. to induce slow penetration into the soil thus helping to recharge groundwater; ►surface infiltration, ►near-natural storm water runoff management, ►infiltration swale); *s* **infiltración [f] por hoyos** (≠) (Técnica alternativa de ►gestión natural de aguas pluviales en zonas urbanas que consiste en promover la infiltración lenta del agua pluvial en el lugar de caída por medio de ►acequias de infiltración, depresiones del terreno, canales, etc. y así ayudar a recargar los acuíferos; ►infiltración superficial); *f* **infiltration [f] par noues** (Technique alternative de ►gestion naturelle des eaux de pluies en milieu urbain visant à infiltrer les eaux pluviales au plus près de leur source au moyen de noues engazonnés ou parfois plantées, larges fossés peu profond de ruissellement [►noue d'infiltration] ; cette technique présente l'intérêt de ralentir le ruissellement des eaux pluviales et de réapprovisionner les nappes d'eaux souterraines par infiltration, la dépollution des eaux de ruissellement, une intégration d'aménagements paysagers dans le tissu urbain et un faible coût ; ►cuvette d'infiltration, ►gestion naturelle des eaux de pluie, ►infiltration superficielle) ; *g* **Muldenversickerung [f]** (Sickerbewegung von anfallendem Oberflächenwasser, das, bevor es in den Untergrund zur Grundwasserneubildung gelangt, in einer oberirdischen Vertiefung zwischengespeichert wird; ►Flächenversickerung, ►naturnahe Regenwasserbewirtschaftung, ►Versickerungsmulde).

6255 swamp [n] (1) *geo. phyt.* (Area saturated with water throughout much of the year but with the surface of the soil usually not deeply covered; this area commonly has herbaceous, tree or shrub vegetation; cf. RCG 1982; ►fen wood [US]/carr [UK]; ►swamp forest); *s* **pantano [m]** (2) (Hondonada donde se recogen de forma natural las aguas de la lluvia y de escorrentía, que suele ser de fondo cenagoso y estar más o menos cubierta de vegetación; DGA 1986; ►bosque turbícola, ►bosque pantanoso); *syn.* ciénaga [f], aguazal [m]; *f* **marécage [m]** (Étendue de terrain périodiquement inondée et recouverte par une végétation herbacée hydrophile, arbustive ou arborescente ; ►forêt tourbeuse, ►forêt marécageuse, ►forêt humide) ; *syn.* mouille [f], mouillère [f] ; *g* **Sumpf [m]** (2) (Im Allgemeinen mit Kräutern, Sträuchern und Bäumen bewachsenes Gebiet, das im Jahresverlauf überwiegend durch wassergesättigten Boden gekennzeichnet ist, gewöhnlich jedoch keine mächtige Bodenschicht hat; ▶Bruchwald, ▶Sumpfwald); *syn.* Sumpfgebiet.

swamp [n] (2) *phyt.* ▶bulrush swamp [US]/club-rush swamp [UK], ▶cattail swamp [US]/reedmace swamp [UK]; *geo.* ▶gypsum salt swamp; *phyt.* ▶low sedge swamp, ▶reed canary grass swamp [US]/reed strip [UK], ▶reed swamp, ▶rush swamp, ▶salt swamp, ▶saw sedge swamp, ▶sedge swamp, ▶spring swamp, ▶tall sedge swamp.

swamp [n], **great fen-sedge** *phyt.* ▶saw sedge swamp.

swamp [n] [UK], **reedmace** *phyt.* ▶cattail swamp [US]/reedmace swamp [UK].

swamp [n], **saw grass** *phyt.* ▶saw sedge swamp.

swamp [n] [UK], **tall forb reed** *phyt.* ▶tall forb reed marsh [US].

swamp [n] [UK], **tall herb reed** *phyt.* ▶tall forb reed marsh [US]/tall forb reed swamp [UK].

swamp [n] [US], **twig rush** *phyt.* ▶saw sedge swamp.

6256 swamp forest [n] *geo. phyt.* (Forest on soils with abundant moisture, particularly along rivers in palustrine or estuarine regions. This forest possesses an overstor[e]y of trees, an understor[e]y of young trees or shrubs, and an herbaceous layer; cf. CWD 1979. *Specific term for s. f. in low bogs* fen woodland [EIRE]; ▶fen wood [US]/carr [UK], ▶mangrove stand); *syn.* hydric forest [n] (TEE 1980, 215); *s* **bosque** [m] **pantanoso** (Bosque que crece en suelo siempre encharcado e inundado varias veces al año, perteneciente a las alianzas *Alno-Ulmion* [bosques mixtos de alisos y olmos] y *Salicion albae*, a veces sobre suelos de turbera éutrofos o mesótrofos, perteneciente a las alianzas *Salicion cinereae* [matorrales de sauces y ▶bosques turbícolas de abedules de turbera] y *Alnion glutinosae* [alisedas turbícolas]; bosque de ▶manglares); *f 1* **forêt** [f] **marécageuse** (Forêt sur sols mouillés en permanence, asphyxiant, des sources et petits marais de pente, des vallées tourbeuses ou tourbières eutrophes ou mésotrophes [alliances du *Salicion cinereae* et de l'*Alnion glutinosae*] ; GGV 1979 ; ▶forêt tourbeuse, ▶mangrove) ; *syn.* forêt [f] hygrophile ; *f 2* **forêt** [f] **humide** (Forêt sur sols engorgés temporairement avec faible circulation de la nappe, des grandes vallées marécageuses en bordure des cours d'eau [alliances de l'*Alno-Padion* et du *Salicion albae*] ; GGV 1979) ; *g* **Sumpfwald** [m] (Wald, der auf ständig nassen oder jährlich häufig überfluteten Böden vorkommt [Verband der Grauerlen-, Ulmen- und Eschen-Mischwälder *(Alno-Ulmion)* und des *Salicion albae*], zuweilen auf eutrophen oder mesotrophen Moorböden [Verband der Grauweidengebüsche und Moorbirken-Bruchwälder (*Salicion cinereae*) und der Erlenbruchwälder (*Alnion glutinosae*)]; cf. OBER 1992; ▶Bruchwald, ▶Mangrove).

swamp formation [n], **reed** *phyt.* ▶reed swamp.

swamp plant [n] [US] *bot.* ▶helophyte.

6257 swamp vegetation [n] *phyt.* (Vegetation occurring on permanently saturated or frequently flooded soils; ▶swamp forest); *s* **helophytia** [f] (Vegetación característica de lagos y pantanos que crece en suelos encharcados o frecuentemente inundados; ▶bosque pantanoso); *syn.* vegetación [f] helófita; *f* **végétation** [f] **des marais** (Végétation de zone humide établie sur des sols constamment engorgés ; la composition de cette végétation herbacée hydrophile émergeante varie avec la hauteur de l'eau, l'importance des périodes d'assèchement, et selon le taux de salinité ; la végétation hélophyte du marais est variée, avec des prairies humides, des roselières, de la végétation aquatique, des friches, des boisements ; ▶forêt marécageuse) ; *syn.* végétation [f] marécageuse ; *g* **Sumpfvegetation** [f] (Vege-

tation, die auf ständig nassen oder jährlich häufig überfluteten Böden vorkommt; ▶Sumpfwald).

swamp vegetation [n], **reed** *phyt.* ▶reed swamp.

swamp wood [n] *phyt.* ▶fen wood.

swamp woodland [n] [UK], **alder** *phyt.* ▶riparian alder stand [US].

6258 sward [n] *constr.* (Ground covered with thick, closely mown or cropped grass; ▶turf sward); *syn.* turf [n]; *s* **tapiz** [m] **herbáceo** (Capa superior del suelo cubierto de herbáceas densamente enraizadas, generalmente de gramíneas; ▶tapiz de césped); *f* **tapis** [m] **herbacé** (Couverture végétale homogène d'un sol principalement constituée de graminées ; ▶tapis de graminées) ; *g* **Grasnarbe** [f] (Die oberste Bodenschicht dicht überziehende und durchwurzelnde Krautschicht, die vorwiegend aus Gräsern besteht. G. hat im Vergleich zu ▶Rasennarbe einen weiteren Begriffsumfang, wird jedoch je nach Zusammenhang oft synonym mit Rasennarbe verwendet. Bei *Rasennarbe* wird i. d. R. die kurze Halmlänge durch häufiges Mähen mitgedacht).

sward [n], **alpine** *phyt.* ▶alpine grassland.

sweep circle [n] [UK] *trans.* ▶compound curve [US].

swelling [n] *constr.* ▶bulking.

6259 swelling soil [n] *pedol.* (TGG 1984, 98; in US classification there are five classes of expansion: very low, low, medium, high, very high); *syn.* expansive soil [n]; *s* **suelo** [m] **expansivo** (Suelos que, como p. ej. los vertisoles, debido al alto contenido de arcilla hinchable, primordialmente montmorillonita, aumentan claramente de volumen cuando absorben agua); *f* **sol** [m] **gonflant** (Sol contenant une forte proportion d'argiles gonflantes, telles que les montmorillonites ; LA 1981, 425) ; *g* **quellfähiger Boden** [m] (Tonreicher, montmorillonitischer Boden, z. B. Vertisole in Gebieten mit warmem Klima und ausgeprägter Wechselfeuchte, der bei Wiederbefeuchtung einen hohen Quellungsdruck erzeugt).

swimming area [n] [US] *recr.* ▶bathing area.

6260 swimming center [n] [US]/**swimming centre** [n] [UK] *recr.* (Generic term for ▶outdoor bathing complex [US]/open-air swimming baths [UK] and ▶indoor bathing complex [US]/indoor swimming baths [UK]); *syn.* swimming pool complex [n] [US], bathing complex [n] [US]; *s* **instalación** [f] **de natación** (Término genérico para ▶piscina cubierta y ▶piscina al aire libre); *f* **piscine** [f] (Terme générique englobant la ▶piscine de plein air et la ▶piscine couverte) ; *g* **Schwimmbad** [n] (OB zu ▶Freibad und ▶Hallenbad).

swimming facilities [npl], **competition** *recr.* ▶competition swimming facilities.

swimming meet installation [n] [US] *recr.* ▶competition swimming facilities.

6261 swimming pool [n] (1) *recr.* (1. Generic term); **2. natatorium** [n] (Swimming pool, especially one in a gymnasium or other building); *s* **piscina** [f] *syn.* alberca [f] [MEX], pileta [f] [RA]; *f* **bassin** [m] **de natation** ; *g* **Schwimmbecken** [n].

6262 swimming pool [n] (2) *recr.* (Pool for swimmers, frequently one structure with descending depth which may include a diving section; ▶swimming center [US]/swimming centre [UK]; *opp.* ▶wading pool [US]/pool for non-swimmers [UK]); *s* **piscina** [f] **para nadadores/as** (En ▶instalación de natación, piscina de suficiente profundidad como para nadar; *opp.* ▶piscina para niños/as); *syn.* alberca [f] para nadadores/as [MEX], pileta [f] para nadadores/as [RA]; *f* **bassin** [m] **nageurs** (Bassin dans une ▶piscine d'une profondeur prévue pour les nageurs ; *opp.* ▶bassin non-nageurs) ; *g* **Schwimmerbecken** [n] (Im ▶Schwimmbad

S

für Schwimmer angelegtes Becken mit einer entsprechenden Mindesttiefe; *opp.* ▶Nichtschwimmerbecken).

swimming pool complex [n] [US] *recr.* ▶swimming center [US]/swimming centre [UK].

6263 swimming pool hall [n] [US] *recr.* (That part of an ▶indoor bathing complex [US]/indoor swimming baths [UK], which is equipped with a swimming pool); *syn.* indoor pool [n] [UK]; *s* **pabellón** [m] **de piscina cubierta** (Parte de una ▶instalación de natación donde se encuentra una ▶piscina cubierta); *f* **bassin** [m] **couvert** (Zone de la ▶piscine couverte aménagée avec un bassin de natation) ; *g* **Schwimmhalle** [f] (Der Teil eines ▶Hallenbades, der mit einem Schwimmbecken ausgerüstet ist).

synanthropic species [n] *biol.* ▶anthropophilous species.

6264 synchorology [n] *phyt.* (Science of the geographical distribution of vegetation units); *s* **sincorología** [f] (Ciencia que estudia la distribución de las unidades vegetales en el espacio); *syn.* biosociología [f] corológica (DB 1985); *f* **synchorologie** [f] (Science qui étudie la répartition géographique des associations végétales ; DEE 1982) ; *syn.* distribution [f] des communautés végétales ; *g* **Synchorologie** [f] (Wissenschaft von der Verbreitung der Vegetationseinheiten auf der Erde); *syn.* Gesellschaftsverbreitung [f].

6265 syndynamics [npl] *phyt.* (Subdivision of ecological research dealing with the spread and development of vegetation units in particular locations; ▶succession); *s* **sindinamismo** [m] (Parte de la ecología que estudia la distribución y el desarrollo de las asociaciones en determinadas ubicaciones; ▶sucesión); *f* **syndynamique** [f] (Science qui étudie l'évolution des associations végétales ; DEE 1982 ; ▶succession) ; *g* **Syndynamik** [f] (Teilgebiet der ökologischen Forschung, das sich mit der Ausbreitung und Entwicklung von Vegetationseinheiten an bestimmten Standorten beschäftigt; ▶Sukzession); *syn.* Gesellschaftsentwicklung [f].

6266 synecology [n] *ecol.* (Subdivision of ecology that deals with the study of relationships of life communities to each other in their habitat and with related environmental factors; it is subclassified in bioc[o]enology, ecosystem research, production biology, and ecogeography; ▶phytosociology; *opp.* ▶autecology); *s* **sinecología** [f] (Término de ▶geobotánica, cuando se emplea sobreentendiendo la especificación de «vegetal». En sentido lato, sinónimo de ▶fitosociología. En sentido estricto parte de la ecología que estudia las relaciones entre las colectividades vegetales, o sinecias, y su medio. En este sentido fue aceptado por la Ponencia de Nomenclatura del Congreso Internacional de Bruselas, de 1910; cf. DB 1985; ▶ecología de poblaciones; *opp.* ▶autoecología); *f* **synécologie** [f] (Partie de l'écologie qui étudie les relations entre les facteurs écologiques et l'ensemble des individus d'une communauté biologique [biocénoses] ainsi que les rapports entre organismes ; DEE 1982 ; la **s.** se divise en la biocénotique et la recherche sur les écosystèmes, la biologie des biomasses et la géographie écologique ; en ▶phytosociologie on entend par **s.** l'enseignement de l'écologie d'une association végétale ; *sensu lato* ▶demécologie ; *opp.* ▶autécologie) ; *g* **Synökologie** [f] (Teilgebiet der Ökologie, das die Beziehungen von Lebensgemeinschaften [= Biozönosen] im jeweiligen Lebensraum und die einzelnen Umweltfaktoren untersucht. Sie unterteilt sich in Biozönologie und Ökosystemforschung, Produktionsbiologie und Ökogeographie. In der ▶Pflanzensoziologie versteht man unter **S.** nur die Lehre vom Gesellschaftshaushalt. *Sensu lato* ▶Demökologie; *opp.* ▶Autökologie).

6267 synusia [n] *phyt.* (Group of plants of a particular ▶life form, as defined by RAUNKIAER, which grows under relatively uniform site conditions. It can be composed of various species,

which, to a certain extent, can represent plants [vicarious species] of geographically, widely separated areas or similar habitats, or which may come from different parts of the world, if their climates are relatively similar); *s* **sinusia** [f] (Comunidad constituida por especies pertenecientes a un biotipo determinado, de exigencias ecológicas uniformes. En algunos casos, la sinusia se confunde con la asociación, pero ordinariamente en la estructura de una asociación intervienen cierto número de sinusias; DB 1985; ▶forma biológica); *syn.* simorfia [f]; *f* **synusie** [f] (Groupe d'organismes ayant les mêmes exigences écologiques [▶type biologique] et constituant une sous-unité au sein d'une biocénose, p. ex. les organismes qui vivent sur les écorces ou ceux qui vivent dans le feuillage constituent dans un écosystème forestier des synusies particulières ; DEE 1982) ; *g* **Synusie** [f] (Gruppe von Pflanzen einer bestimmten ▶Lebensform im Sinne RAUNKIAERS, die unter relativ einheitlichen Standortbedingungen wächst. Sie kann von verschiedenen Species gebildet werden, die z. T. miteinander vikariieren resp. aus verschiedenen Teilen der Erde stammen, wenn deren Klimata ziemlich einheitlich sind); *syn.* Verein [m].

system [n] *envir.* ▶air quality monitoring network/system; *plan. urb.* ▶circulation system; *for.* ▶clear-cutting system; *envir.* ▶closed recycling system, ▶combined sewer system; *landsc. urb.* ▶combined open space pattern/system, ▶concentric open space pattern/system; *urb.* ▶district heating system; *constr.* ▶drainage system (1); *geo.* ▶dune system; *for.* ▶economic forest management system; *ecol.* ▶ecosystem; *envir. pol.* ▶emissions trading system; *bot.* ▶fascicular root system; *arb. hort.* ▶fibrous root system; *phyt. ecol.* ▶forest ecosystem; *adm. geo. landsc. plan.* ▶geographic information system (GIS); *envir.* ▶individual sewage disposal system [US]/private sewage treatment plant [UK]; *agr. constr. hort.* ▶irrigation system; *ecol.* ▶landscape ecosystem; *adm. geo. landsc. plan.* ▶landscape information system; *landsc. urb.* ▶linear open space pattern/system; *trans.* ▶local public transportation system [US]/local transport system [UK]; *constr.* ▶multistrata planting system; *envir.* ▶national or state surveillance system for air pollution; *envir.* ▶open recycling system; *landsc. urb.* ▶open space pattern/system of peninsular interdigitation; *landsc. urb.* ▶open space system; *trans. urb.* ▶park-and-ride system; *agr. for.* ▶pollard system; *landsc. urb.* ▶radial open space pattern/system; *constr.* ▶roof garden system; *bot.* ▶root system; *agr.* ▶rotational grazing system; *limn. wat'man.* ▶saprobic system; *for.* ▶selection system; *envir.* ▶separate sewerage system; *envir. urb.* ▶sewer system; *agr.* ▶single crop system; *constr.* ▶single-stratum planting system; *envir. urb.* ▶storm water drainage system; *constr.* ▶subsurface drainage system; *plan. trans.* ▶transportation system.

system [n], **agroforestry** *agr. conserv.* ▶traditional orchard [UK], #2.

system [n], **agropastoral** *agr. conserv.* ▶traditional orchard [UK], #2.

system [n], **agrosilvopastoral** *agr. conserv.* ▶traditional orchard [UK], #2.

system [n] [US], **bicycle trail** *trans. urb.* ▶bikeway network.

system [n], **bunched root** *bot.* ▶fascicular root system.

system [n] [UK], **circular pathway** *plan.* ▶circular walk system [US].

system [n] [US], **circular trail** *plan.* ▶circular walk system [US]/circular pathway system [UK].

system [n] [UK], **clear felling** *for.* ▶clear cutting system [US].

system [n], **diffuse root** *arb. hort.* ►fibrous root system.

system [n] [UK], **drainage** *geo.* ►drainage pattern.

system [n], **ecological** *ecol.* ►ecosystem.

system [n] [US], **eco-roof** *constr.* ►roof garden system.

system [n], **green roof** *constr.* ►roof planting method.

system [n] [US], **gridiron** *constr.* ►grid pattern drainage.

system [n] [UK], **hard** *constr. eng.* ►conventional engineering.

system [n], **interconnected** *conserv. ecol.* ►network (1).

system [n], **interfingering open space pattern/** *landsc. urb.* ►open space pattern/system of peninsular interdigitation.

system [n], **land information** *adm. geo. landsc. plan.* ►landscape information system.

system [n], **land management information** *adm. geo. landsc. plan.* ►landscape information system.

system [n] [UK], **local transport** *trans.* ►local public transportation system [US].

system [n], **natural resources information** *adm. geo. landsc. plan.* ►landscape information system.

system [n], **one-layer planting** *constr.* ►single-stratum planting system.

system [n] [US], **pavement** *constr.* ►pavement structure.

system [n], **resource data management** *adm. geo. landsc. plan.* ►landscape information system.

system [n] [US], **septic** *envir.* ►individual sewage disposal system [US]/private sewage treatment plant [UK].

system [n] [US], **septic tank disposal** *envir.* ►individual sewage disposal system [US]/private sewage treatment plant [UK].

system [n], **silvicultural management** *ecol.* ►economic forest management system.

system [n] [US], **underground storm drainage** *envir. urb.* ►storm water drainage system.

system [n] [US], **wetland septic** *envir.* ►wastewater treatment wetland.

system [n] **of green areas** *landsc. urb* ►open space system.

system [n] **of peninsular interdigitation** *landsc. urb.* ►open space pattern/system of peninsular interdigitation.

System [n] **of Protected Areas Natura 2000** *conserv. leg. pol.* ►Natura 2000 Protection Area Network.

systems [npl] *ecol.* ►natural systems.

systems [npl], **effective functioning of natural** *ecol.* ►natural potential.

systems [npl] [UK], **hard** *constr. eng.* ►conventional engineering.

T

tag [n] [US] *hort.* ►label.

tagging [n] [US] *hort.* ►labeling.

6268 taiga [n] (Type of landscape in the boreal, coniferous, virgin forest biome with tree species such as spruce, larch, pine, and fir. The **t.** of central and eastern Siberia is the largest contiguous forest area in the world); *s* **taiga** [f] (Bioma en el cual la formación vegetal dominante es el bosque de coníferas, en el que se presentan las mayores masas arboladas continuas de la tierra; DINA 1987); *f* **taiga** [f] (Paysage des régions continentales froides et boréales de l'hémisphère nord peuplé par une forêt sempervirente constituée de conifères [Epicéa, Mélèze, Pin sylvestre et Sapin]. Par extension ce terme est devenu synonyme de l'étendue forestière la plus importante sur le globe) ; *g* **Taiga** [f] *geo.* (Landschaftstyp des borealen Nadelwaldbioms in Eurasien, welches urwaldartige Nadelwälder aus Fichte, Sibirischer Lärche, Kiefer und Tanne darstellt. Die **T.** Mittel- und Ostsibiriens ist das größte flächenhaft zusammenhängende Waldgebiet der Erde).

tailings lagoon [n] *min.* ►tailings pond.

6269 tailings pond [n] *min.* (Artificial settling pond dammed for the sedimentation of ►mining spoil; ►site pile); *syn.* tailings lagoon [n], slurry pond [n] [also UK], slimes dam [n] [also ZA]; *s* **estanque** [m] **de decantación** (Cerrado por terraplén de tierra en el que se descargan los lodos de minería; ►vaciadero de lodos de placeres, ►zafras); *syn.* laguna [f] de decantación; *f* **bassin** [m] **de décantation** (1. Bassin entouré d'un remblai de terre dans lequel les stériles de charbon [< 2 m] sont transportés comme sable de lessivage. 2. Bassin peu profond dans lequel séjournent les effluents provoquant la sédimentation des fines particules < 2 mm provenant de l'exploitation minière ; ►dépôt de sable de lavage, ►stérile) ; *g* **Absetzbecken** [n] **(2)** (Mit einem Erdwall eingefasstes Becken, in das Bergematerial [< 2 mm] als Spülsand zum Ablagern transportiert wird; ►Berge 3, ►Bergehalde aus Spülsand); *syn.* Flotationsbecken [n], Flotationsteich [m].

6270 tailings slurry [n] *min.* (Mineral refuse [< 2mm] from milling operations hydraulically transported in pipes to the spoil dump [US]/spoil heap [UK]; ►mining spoil, ►silt pile [US]); *s* **lodo** [m] **de minería** (Residuos de minería obtenidos por medio de sedimentación; ►vaciadero de lodos de placeres, ►zafras); *f* **stériles** [mpl] **issus du lavage** (Fraction broyée [< 2 mm] de charbon non minéralisé récupérée après lavage au fond des appareils à liqueur dense qui est ensuite évacuée et déposée en terrils ; ►dépôt de sable de lavage, ►stérile) ; *syn.* stériles [mpl] de flottation, stériles [mpl] de lavage ; *g* **Waschberge** [pl] (1. *Erzabbau* zerkleinertes Bergematerial [< 2 mm] bei der Erzgewinnung, das als Spülsand durch Leitungen zu Erdauffüllungen transportiert wird. 2. *Kohleförderung* das nach dem Brechvorgang in einer Schwereflüssigkeitswäsche von der Kohle getrennte Nebengestein; ►Berge 3, ►Bergehalde aus Spülsand); *syn. zu 1.* Flotationsschlamm [m], Flotationsbergematerial [n].

take [vb] *plant.* ►become established.

take-off [n], **quantity** *contr.* ►calculation of quantities.

6271 take-off sheet [n] *constr.* (Form for recording quantities from drawings); *s* **formulario** [m] **de cómputo de cantidades** (Formulario para calcular las cantidades de servicios y obras); *f* **formulaire** [m] **d'avant-métré** (En vue du calcul du volume des quantités des travaux permettant l'établissement d'un devis) ; *g* **Massenermittlungsformblatt** [n] (Formular zur Berechnung der Massen von Bauleistungen).

take [vb] **root** [US] *plant.* ►become established.

taking [n] [US] *adm. leg.* ►land assembly policy [US]/land acquisition policy [UK].

taking [n] **effect** [US] *leg.* ►coming into effect.

taking [n] **of species** *conserv. phyt. zool.* ►collection of wild plants or animals.

6272 tall forb [n] *phyt.* (Tall-growing, herbaceous perennial, except grasses, which usually forms a dense cover in moist locations with soils rich in nutrients; ▶perennial 1); **s mega-forbia** [f] (Cualquier ▶planta vivaz de gran porte; DB 1985); **f mégaforbe** [f] (Plante en général hémicryptophyte, constituée surtout de dicotylédones et notamment de composées, à l'exception des graminées, croissant sur des sols humides et formant des peuplements hauts et denses ; cf. VCA 1985, 208 ; ▶plante vivace) ; **g Hochstaude** [f] **(1)** (Massereiche, oft mehr als mannshohe, mehrjährige krautige Pflanze [Großstaude meist als Schaftstaude], die gewöhnlich an frische bis nasse, nährstoffreiche Böden sonniger Standorte gebunden ist. In der Landschaft treten **H.n** als **H.n**-Säume [an Gehölzrändern und Ufern] oder **H.n**-Fluren [Offenflächen] auf und können ebenso Thema einer gärtnerischen Pflanzung sein; ▶Staude).

6273 tall forb community [n] *phyt.* (Vegetation of mostly large-leaved, lush-growing perennials on moist soils, rich in nutrients, at sub-alpine and alpine levels as well as lowland; ▶perennial forb community); *syn.* tall herb community [n]; **s formación** [f] **megafórbica** (Formación vegetal caracterizada por el predominio de hierbas vivaces —▶hemicriptófitos, ▶geófitos— vigorosas, de gran talla y follaje exuberante; DB 1985; ▶vegetación de herbáceas perennes); *syn.* megaforbio [m]; **f mégaphorbiaie** [f] (Formation dense et haute de plantes vivaces à grandes feuilles se développant sur les sols humides et riches en substances nutritives des étages subalpins et alpins ainsi que dans les plaines ; VCA 1985, 208 ; ▶végétation herbacée pérenne) ; *syn.* flore [f] herbacée géante, groupement [m] de hautes herbes, végétation [f] de hautes herbes ; **g Hochstaudenflur** [f] (Überwiegend großblättrige, meist üppige Staudengesellschaften auf nährstoffreichen und feuchten Böden in subalpinen und alpinen Lagen sowie in Niederungen; ▶Staudenflur).

6274 tall forb reed marsh [n] [US] *phyt.* (Flora covering wet, often marshy riverbanks of streams and rivers or estuaries, rich in soil nutrients and with a lush growth of ▶tall forbs; ▶tall sedge swamp); *syn.* tall herb reed swamp [n] [UK], tall forb reed swamp [n] [UK]; **s pantano** [m] **megafórbico** (Zona ribereña de cursos de agua y estuarios, muy húmeda y rica en nutrientes, cubierta de gran cantidad de ▶megaforbias; ▶ciénaga de grandes cárices); *syn.* megaforbio [m] húmedo; **f marais** [m] **à mégaphorbes** (Zone riveraine des cours d'eau et estuaires, très humide et riche en substances nutritives, à sol paratourbeux souvent recouverte de plantes vivaces hautes [▶mégaforbe] ; le peuplement se développant sur ces stations est appelé « mégaphorbiaie de marais » ; ▶magnocariçaie) ; **g Hochstaudenried** [n] (Sehr feuchter, nährstoffreicher, oft anmooriger Uferbereich von Fließgewässern oder Ästuaren mit üppigen ▶Hochstauden 1 bewachsen; ▶Großseggenried).

tall forb reed swamp [n] [UK] *phyt.* ▶tall forb reed marsh [US].

tall grass prairie [n] [US] *phyt.* ▶tall grass steppe.

6275 tall grass steppe [n] *phyt.* (Extensive, relatively treeless plain more humid than a ▶savanna[h] and covered with caespitose grasses taller than 1m; the western prairie in North America, the eastern center of the South American pampa, and parts of the russian steppe are **t. g. s.s**); *syn.* tall grass prairie [n] [US]; **s estepa** [f] **de gramíneas subxerofítica** (Estepa de la zona templada, con precipitaciones todo el año, e incluso un máximo estival, como las del sur de Rusia, interior-oeste de Norteamérica y las Pampas sudamericanas; DB 1985; ▶sabana); **f prairie** [f] **à hautes herbes** (Plaine recouverte d'une formation végétale climacique fermée de graminées tapissantes à caractère hygrophile dont certaines peuvent atteindre une hauteur supérieure à 1 m comme p. ex. dans la partie occidentale de la prairie nord-américaine, le centre-est de la pampa sud-américaine ou une partie de la steppe russe ; ▶savane) ; *syn.* prairie [f] à hautes

graminées, steppe [f] à hautes graminées, haute prairie [f] [CDN] ; **g Hochgrassteppe** [f] (Steppe in einem mehr humiden Klima mit Rasen bildenden Gräsern, von denen viele größer als 1 m sind; die westliche Prärie in Nordamerika, das östliche Zentrum der südamerikanischen Pampa und Teile der russischen Steppe sind **H.n**; ▶Savanne); *syn. Südamerika* Pampa(s) [f(pl)], *Nordamerika* Prärie [f].

tall helophyte [n] *bot.* ▶emergent aquatic plant (2).

tall herb community [n] *phyt.* ▶tall forb community.

tall herb reed swamp [n] [UK] *phyt.* ▶tall forb reed marsh [US]/tall forb reed swamp [UK].

6276 tall perennial [n] *plant.* (Tall-growing persistent herbaceous plant cultivated in nurseries for use in planting design; ▶perennial 1); **s vivaz** [f] **alta** (En horticultura, ▶planta vivaz de crecimiento alto cultivada en vivero de vivaces para usos paisajísticos); **f plante** [f] **vivace haute** (Vivace de croissance vigourreuse cultivée dans les pépinières de ▶plantes vivaces pour l'utilisation comme élément de décoration dans les jardins d'agrément ou les compositions des espaces publics) ; **g Hochstaude** [f] **(2)** (In Staudengärtnereien kultivierte, hochwüchsige ▶Staude zur Verwendung in der Gestaltung von Freianlagen); *syn.* Großstaude [f], hohe Staude [f], hochwüchsige Staude [f].

6277 tall sedge formation [n] *phyt.* (▶tall sedge swamp); **s comunidad** [f] **de grandes cárices** (BB 1979, 28; ▶ciénaga de grandes cárices); **f peuplement** [m] **à grands Carex** (▶magnocariçaie) ; *syn.* flore [f] à grandes Laiches/Laîches ; **g Großseggenflur** [f] (Pflanzengesellschaft, die vorwiegend aus Großseggen besteht; ▶Großseggenried).

6278 tall sedge swamp [n] *phyt.* (Vegetation dominated by large sedges *[Magnocaricion]* in eutrophic to mesotrophic water on soft organic or mineral substrates; often in a zone around open water behind reed swamps; in Central Europe character species are e.g. Tufted Sedge *[Carex elata]* which tolerates high oscillations of flooding, Cyperus Sedge *[C. pseudocyperus]*, Pond Sedge *[C. riparia]*; ▶tall sedge formation, ▶tall forb reed marsh [US]/tall forb reed swamp [UK]); *syn.* large-sedges fen [n] [also UK] (ELL 1988, 573); **s ciénaga** [f] **de grandes cárices** (Zona de colmatación de superficies de agua en ubicaciones de húmedas a permanentemente estancadas, en la que crecen sobre todo especies de grandes cárices *[Magnocaricetalia]*; ▶comunidad de grandes cárices, ▶pantano megafórbico); **f magnocariçaie** [f] (Zone d'atterrissement des lacs et des cours d'eau recouverte d'une végétation herbacée dominée par les espèces de grandes Laiches/Laîches, p. ex. *Carex elata,* appartenant à l'ordre de *Magnocaricetalia* ; ▶peuplement à grands Carex, ▶marais à mégaphorbes) ; *syn.* prairie [f] à grandes Laiches/Laîches ; **g Großseggenried** [n] (Vegetationsbestand im Verlandungsbereich von Gewässern landeinwärts auf feuchten bis überstauten Standorten, der vorwiegend aus Großseggenarten — *Magnocaricetalia* — besteht; ▶Kennarten in Mitteleuropa sind z. B. Steife Segge *[Carex elata]*, die die größten Wasserspiegelschwankungen erträgt, Scheinzypergras-Segge *[C. pseudocyperus]*, Ufersegge *[C. riparia]*; ▶Großseggenflur; ▶Hochstaudenried).

tall shrub [n] *hort. plant.* ▶large shrub.

tall standard [n] [UK] *arb. hort.* ▶shade tree [US] (2), ▶fruit tree.

talus [n] *geo.* ▶scree.

6279 talus creep [n] *geo.* (Slow downslope movement of talus or ▶scree; ▶mass slippage [US]/mass movement [UK]); **s reptación** [f] **de detritus de talud** (Tipo de ▶movimiento gravitacional lento por el cual son transportados derrubios pendiente abajo; ▶derrubio de gravedad); **f reptation** [f] **d'éboulis** (Mouvement superficiel et lent de débris sur un ▶talus d'éboulis sous l'action gravitaire ou par un processus de solifluction ; ►mouvement de masse) ; *g* **Schuttgleitung [f]** (Langsame Abwärtsbewegung von Schutt auf ►Schutthalden 2 durch gravitative oder solifluidale Prozesse; cf. WAG 1984; ►Massenversatz).

talweg [n] *geo.* ►thalweg.

tanker accident [n] *envir.* ►oil tanker accident.

tanker security [n] *envir. leg.* ►oil tanker security.

tapered pipe [n] [UK] *constr.* ►increaser [US].

6280 tapered scree [n] *geo.* (Conically shaped, unconsolidated accumulation of broken rock of various sizes, which has arisen due to mechanical weathering at the end of rockslide corridors [US]/rock slide corridors [UK] ; ►rock fan, ►scree) ; *s* **cono [m] de derrubios** (Montón de forma cónica de fragmentos de roca no consolidada resultado de la caída de rocas sueltas por la acción de la meteorización mecánica; ►abanico de derrubios, ►derrubio de gravedad); *f* **cône [m] d'éboulis** (Masse de débris anguleux couvrant un versant et mis en place par gravité au bas de couloirs qui canalisent la chute des matériaux de météorisation ; l'expression « éboulis de gravité » est un pléonasme ; cf. DG 1984, 148 ; ►nappe d'éboulis, ►talus d'éboulis) ; *syn.* cône [m] d'accumulation, cône [m] rocheux ; *g* **Schuttkegel [m]** (Dreiecksförmige, unverfestigte Masse von Gesteinsbrocken verschiedenster Größen, die durch mechanische Verwitterung am Ende von Steinschlagrinnen entstanden ist; ►Schuttfächer, ►Schutthalde 2).

taper elbow [n] *constr.* ►reducing elbow.

taper unit [n] *constr.* ►cone (1).

6281 tapping [n] **of a spring** *hydr.* (VOLL 1973; gathering of outflow from springs or permeable layers in a structure above or below ground); *s* **captación [f] de una fuente** (Recogida de las aguas de una fuente por obras de drenaje en tuberías o canales; WMO 1974); *syn.* captación [f] de un manantial; *f* **captage [m] de source** (Récupération d'une eau de source par des ouvrages de drainage et d'amenée dans des canalisations ; WMO 1974) ; *g* **Quellfassung [f]** (Zur Sammlung und Ableitung des Quellwassers an der Austrittstelle eingerichtete ober- oder unterirdische Vorrichtung).

6282 tap root [n] *bot. hort.* (Main descending root of a plant; e.g. oak *[Quercus spec.]*, Scotch Pine [US]/Scots Pine [UK] *[Pinus sylvestris]*, Black Locust [US]/Common or False Acacia [UK] *[Robinia pseudoacacia]*, Balsam Fir *[Abies balsámea]*; ►anchor root, ►heart root, ►structural root); *s* **raíz [f] axonomorfa** (La raíz cuyo eje es preponderante, ramificada de manera racemosa, con los ejes secundarios, etc., poco desarrollados en comparación del principal; DB 1985; ►raíz de anclaje, ►raíz fasciculada, ►raíz primaria); *syn.* raíz [f] pivotante [AL]; *f* **racine [f] pivotante** (Racine principale, qui conserve sa prédominance sur les racines secondaires comme p. ex. chez le Chêne *[Quercus spec.]*, le Pin sylvestre *[Pinus sylvestris]*, le Robinier faux-acacia *[Robinia pseudoacacia]*, le Sapin blanc *[Abies alba]* ; ►racine d'ancrage, ►racine fasciculée, ►racine maîtresse) ; *syn.* pivot [m], racine [f] pivot ; *g* **Pfahlwurzel [f]** (Senkrecht nach unten verlaufende Hauptwurzel, z. B. bei Eiche *[Quercus spec.]*, Waldkiefer *[Pinus sylvestris]*, Robinie *[Robinia pseudoacacia]*, Weißtanne *[Abies alba]*; ►Ankerwurzel, ►Herzwurzel; ►Starkwurzel; *OB* Vertikalwurzel).

6283 tap-rooted plant [n] *bot. hort.* (Plant with a root system characterized by a ►tap root); *s* **planta [f] con raíz axonomorfa** (►raíz axonomorfa); *syn.* planta [f] con raíz pivotante [AL]; *f* **plante [f] à racine pivotante** (Plante caractérisée par un système radiculaire à ►racine pivotante) ; *syn.* espèce [f] à racine pivotante ; *g* **Pfahlwurzler [m]** (Pflanze, deren Wurzelsystem durch eine ►Pfahlwurzel gekennzeichnet ist).

6284 tap site [n] *min. wat.man.* (Deep hole or shaft sunk or drilled into the earth to tap an underground supply of water, gas, or oil); *syn.* well [n]; *s* **punto [m] de alumbramiento** (Lugar en el que se perfora para extraer agua, petróleo o gas); *f* **point [m] de prélèvement** (PPC 1989, 101 ; puits ou lieu de forage destiné à l'exploitation de l'eau potable, du pétrole ou du gaz) ; *g* **Entnahmestelle [f] (3)** (Ort, an dem durch Bohrungen Trinkwasser, Öl oder Gas gewonnen wird).

6285 target biotope [n] *ecol. plan.* (Biotope which is to be protected and sustained by appropriate long-term measures); *s* **biótopo [m] meta** (Biótopo a preservar y desarrollar a largo plazo por medio de medidas específicas de protección); *f* **biotope [m] prioritaire** (Biotope dont la conservation est assurée par des mesures de gestion et de protection particulières ; *syn.* habitat [m] naturel prioritaire ; *g* **Zielbiotop [m,** *auch* **n]** (Biotop, der langfristig durch geeignete Managementmaßnahmen nachhaltig gesichert und entwickelt werden soll).

6286 target planning [n] *plan.* (Planning process which defines a long-term program[me], and outlines the steps toward[s] reaching its goals); *s* **planificación [f] a largo plazo** (Proceso de planeamiento que define los objetivos a largo plazo y los pasos a seguir para alcanzarlos); *f* **planification [f] stratégique** (Processus définissant des objectifs prospectifs ou prévisionnels de développement) ; *g* **Zielplanung [f]** (Planung, die für eine zu erwartende oder angestrebte Entwicklung Vorgaben erarbeitet).

6287 target species [n] *agr. conserv. for. hort. phytopath.* (**1.** Organism which damages crops or animals to be exterminated by pest control. The various groups are indicated by the names of the products: fungicides, insecticides, molluscicides, etc. **2.** Animals and plants selected for nature conservation, which are representtative of certain communities and biotopes and may be used to monitor the effectiveness of conservation methods); *s* **especie [f] meta** (**1.** En la lucha antiplaguicida organismo que se desea eliminar por su efecto dañino en los cultivos o hacia el ganado. Los diferentes grupos de **ee. mm.** se indican en los nombres de los productos: fungicidas, insecticidas, herbicidas, etc. **2.** En la conservación de la naturaleza, especie animal o vegetal representativa de determinada comunidad vegetal o biótopo que es seleccionada para controlar el efecto de medidas de protección); *f 1* **espèce [f] cible** (Organisme nuisible combattu dans le cadre de la lutte contre les parasites ; sous la dénomination des produits phytopharmaceutiques doit figurer la fonction des substances actives, p. ex. acaricide, bactéricide, fongicide, herbicide etc. ; *opp.* espèce non cible) ; *syn.* espèce [f] ciblée ; *f 2* **espèce [f] contrôle** (Organisme végétal ou animal sélectionné comme représentatif de certaines formes de vie et de types de biotopes et utilisé comme contrôle de l'efficience des mesures de protection) ; *syn.* espèce [f] témoin ; *g* **Zielart [f]** (**1.** In der Schädlingsbekämpfung der zu eliminierende Organismus, der eine Nutzpflanzenkultur oder Nutztierzucht schädigt. In den Namen der Bekämpfungsmittel finden sich die **Z.en** oder Zielartengruppen wieder: Fungizide, Insektizide, Molluskizide etc. **2.** Im Naturschutz ausgewählte Tiere und Pflanzen, die als Repräsentanten für bestimmte Lebensformen und Biotoptypen fungieren und anhand derer die Wirksamkeit von Naturschutzmaßnahmen kontrolliert werden kann).

tarmac [n] *constr.* ►macadam [US]/coated macadam surface [UK].

tar macadam surface [n] *constr.* ►macadam [US]/coated macadam surface [UK].

tax [n] *adm. envir. leg. pol.* ►environmental tax.

tax [n]**, emissions** *adm. envir. leg.* ►air pollution fee.

6288 taxonomic diversity [n] *ecol.* (Variety of life forms found in a particular place, including all the plants, birds, mammals, reptiles, and invertebrates; *diversity can be distinguished as follows* **alpha d.** is the number of different species in a local area, often called species richness; **beta d.** is a measure of the degree of change in species composition, or vertical and longitudinal structure of communities along an environmental gradient—the **d.** occurring between habitats or ecosystems; **gamma d.** is the species **d.** across a variety of habitats in a large area. It depends upon alpha **d.** in each habitat and beta **d.** among them; **delta d.** is the **d.** of entire plant communities within the vegetational mosaic of management units at a landscape level; cf. NRM 1996; ▶biodiversity, ▶species richness); *syn.* species diversity [n]; *s* **espectro [m] de diversidad [f]** (Número de especies y abundancia relativa de las mismas en un área específica. El término **e. de d.** se utiliza en la investigación ecológica sobre la diversidad de diferentes tipos de ecosistemas. Dependiendo de la taxocenosis y del tamaño del espacio que se estudien, la diversidad puede variar mucho. Por ello es preferible hablar de **e. de d.**, en el que la diversidad es función del tamaño de la muestra, que de diversidad a secas, sin referencia espacial alguna. PIELOU considera al espectro como algo discontinuo y diferencia entre diversidad de especies o **d. alfa** refieriéndose al nivel inferior de una muestra pequeña o normal y diversidad de motivo o **d. beta** [«pattern diversity»] con referencia a la que deriva del hecho de que pequeñas estructuras aparezcan diferentes unas de otras y combinadas en estructuras mayores; cf. MARG 1977; ▶biodiversidad, ▶riqueza en especies); *f* **diversité [f] des espèces** (La **d. d. e.** dans une communauté floristique ou faunistique dépend de l'importance numérique [abondance] des espèces dans un site donné et rend compte de sa richesse biologique ; DEE 1982 ; la diversité au sein d'un écosystème est caractérisée par la richesse spécifique des espèces : **1.** la variabilité spécifique du nombre d'espèces par unité de surface, **2.** l'équitabilité ou répartition des individus entre les diverses espèces ; DUV 1984 ; la conservation des espèces et de leurs milieux fait l'objet de la Convention de Rio sur la diversité biologique ; ▶biodiversité, ▶richesse en espèces) ; *syn.* diversité [f] taxonomique ; *g* **Artendiversität [f]** (*Artenmannigfaltigkeit* Vielfalt von Arten in einer Lebensgemeinschaft unter Berücksichtigung der relativen Abundanz der Arten [Arten-Individuen-Relation]. *Es werden unterschieden* Diversität als Anzahl unterschiedlicher Arten innerhalb einer Lebensgemeinschaft; Diversität als Maß der Veränderung in der Artenzusammensetzung entlang eines Umweltgradienten von einem Lebensraum in einen nächsten; Diversität als A. in einer Reihe von Biotopen einer größeren Landschaft; cf. ÖKO 1983; ▶Artenvielfalt, ▶Biodiversität); *syn.* taxonomische Diversität [f].

teacup syndrome [n] [US] *constr. arb.* ▶bathtub effect [US]/pot-binding effect [UK].

6289 tea rose [n] [US]/**tea-rose** [n] [UK] *hort. plant.* (Evergreen or semi-evergreen rose of unknown origin cultivated in China which grows in warm climates—*Rosa odorata, Rosa indica odorata, Rosa indica fragrans, Rosa thea.* Introduced into England from China in 1810 in the pink-flowering form and then in 1824 with yellow flowers. **T. r.** is the common name today for several roses with a tea-like scent; ▶hybrid tea); *s* **rosa [f] de té** (Variedad de rosa siempreverde o semisiempreverde de clima cálido cultivada en China —*Rosa odorata, Rosa indica odorata, Rosa indica fragrans, Rosa thea*— e importada a Europa vía Inglaterra en 1810 en variedad con flores rosas y en 1824 en variedad con flores amarillas claras. La **r. de t.** es hoy en día una denominación general para diferentes rosas con aroma de té; ▶rosa híbrido de té); *f* **rosier [m] Thé ou rosier [m] thé** (Rosier à feuillage persistant ou semi-persistant cultivé en Chine et d'origine inconnue *[Rosa odorata, Rosa indica odorata, Rosa indica fragrans, Rosa thea]*, importé de Chine et d'Angleterre — à fleur rose en 1810, à fleur jaune en 1824. Le rosier-thé constitue aujourd'hui la désignation générale pour les différents rosiers à odeur de thé et les rosiers grimpants à petites fleurs en bouquets ; ▶rosier hybride de thé) ; *syn.* rosier [m] à odeur de thé ; *g* **Teerose [f]** (In China kultivierte immergrüne oder halbimmergrüne Rose unbekannten Ursprungs *[Rosa odorata, Rosa indica odorata, Rosa indica fragrans, Rosa thea]*, die 1810 in einer rosa blühenden [Rosa indica odorant] und 1824 in einer hellgelb blühenden Form aus China und England eingeführt wurde, die *"Parks Yellow Tea-scented China"*-Rose. Weil diese neuen Rosen einen an echten Tee erinnernden Duft hatten — andere Autoren berichten, weil sie auf Teeschiffen der East India Company nach England transportiert wurden, wiedere andere schreiben der Gärtnerei Fan Tee oder Fan Ti in Kanton die Namengebung zu —, nannte man sie und die aus ihnen gezüchteten Rosen Teerosen. Die Rosenzüchtung bekam durch die Importe aus China neue entscheidende Impulse, vor allem, weil **T.n** im Sommer öfter blühen. Die **T.** ist eine der Ausgangsformen der ▶Edelrose oder Teehybride; cf. ROS 2006).

6290 technical agricultural methods [npl] *agr.* (Measures taken for land reclamation and preservation of crop land, soil improvement, and thus the increase in production yields); *s* **medida [f] agrotécnica** (Conjunto de medidas para mantener y acrecentar la productividad de suelos agrícolas); *f* **techniques [fpl] culturales** (Interventions prises en faveur de la sauvegarde et de la mise en valeur des terres agricoles ayant pour objectif l'amélioration des sols et par conséquence l'augmentation de la productivité agricole) ; *g* **kulturtechnische Maßnahme [f]** (Maßnahme, die die Gewinnung und Erhaltung landwirtschaftlicher Nutzflächen, die Bodenverbesserung und damit die Steigerung landwirtschaftlicher Erträge zum Ziel hat).

technical contract conditions [npl] **for tree work** *arb. constr. contr.* ▶additional technical contract conditions for tree work.

6291 technical employee [n] [US] *adm. contr.* (Single person on a technical staff; a group of such persons is called by the collective term 'technical staff', **in U.S.**, also by 'technical personnel'); *syn.* technical staff member [n] [UK]; *s* **personal [m] técnico**; *f* **personnel [m] technique** ; *g* **technische Arbeitskraft [f]**.

technical journal [n] *prof.* ▶professional journal.

6292 technical monitoring [n] *leg. envir.* (Regular investigations and inspection tests imposed by law or local regulations; e.g. of pollutants or emissions caused by industrial plants); *syn.* technical surveillance [n]; *s* **inspección [f] técnica** (Sistema de control técnico de instalaciones industriales y productos del cumplimiento de normas ambientales y para la seguridad de los usuarios como p. ej. la inspección técnica de vehículos); *syn.* vigilancia [f] técnica; *f* **surveillance [f] technique** (Contrôles de pollution effectués sur les installations industrielles dans le cadre des lois et des directives en vigueur) ; *g* **technische Überwachung [f]** (Durch Gesetze oder Verordnungen festgelegte Überprüfung von z. B. Schadstoffe emittierenden Anlagen); *syn.* technisches Monitoring [n].

technical personnel [n] [US] *adm. contr.* *∗*▶technical employee [US]/technical staff member [UK].

6293 technical protection [n] **of the environment** *conserv. pol.* (Measures and technical procedures which prohibit injury or impairment to natural elements and the health of living organisms by economic activities. Protective measures are mainly concerned with ▶air pollution control, ▶noise abatement, ▶radiation protection, ▶soil conservation, ▶water quality management, ▶sewage treatment, ▶waste disposal, etc.); *s* **protección**

[f] técnica del medio ambiente (Medidas y procesos técnicos que sirven para evitar daños y distorsiones de componentes o del conjunto de la naturaleza o para la salud de los organismos en el proceso productivo y consuntivo de la sociedad. Estas medidas conciernen sobre todo la ►eliminación de residuos sólidos, ►lucha contra el ruido, ►protección contra las radiaciones, ►protección del ambiente atmosférico, ►protección de las aguas continentales y subterráneas, ►protección del suelo y el ►tratamiento de aguas residuales); **ƒ génie [m] environnemental** (Ensemble des activités qui produisent des biens ou des services visant à mesurer, prévenir, limiter ou corriger les atteintes à l'environnement par la mise en œuvre de solutions technologiques concernant en particulier les domaines de la gestion des ressources naturelles, la ►gestion des eaux superficielles, la ►lutte contre le bruit, la ►protection du sol, la ►protection, surveillance et contrôle de la qualité de l'air, la ►radioprotection, le ►traitement des effluents, le traitement, la gestion et l'►élimination des déchets, etc.); **g technischer Umweltschutz [m, o. Pl.]** (Maßnahmen und technische Verfahren, die Schäden und Beeinträchtigungen einzelner Elemente im Naturhaushalt und an der Gesundheit der Lebewesen durch die ökonomische Tätigkeit des Menschen gar nicht erst aufkommen lassen. Schutzmaßnahmen betreffen hauptsächlich die ►Abfallbeseitigung, ►Abwasserbehandlung, den ►Bodenschutz, ►Gewässerschutz, die ►Lärmbekämpfung, ►Luftreinhaltung, den ►Strahlenschutz etc.); *syn.* technologisch-hygienischer Umweltschutz [m].

technical requirements [npl] **of contract** *constr. contr.*
►special technical requirements of contract.

6294 technical specifications [npl] *constr. contr.* (*Abbr.* specs; written description of materials, equipment, construction systems, standards, and workmanship contained in contract documents; ►outline specifications, ►list of bid items and quantities [US]/schedule of tender items [UK]); **s especificaciones [fpl] técnicas** (Instrucciones escritas del constructor para dar la información acerca de los materiales, uso y aplicación de cada uno de aquéllos; ►resumen de prestaciones, ►resumen de especificaciones técnicas); **ƒ 1 description [fpl] des ouvrages** (Description précise et détaillée des ouvrages dans le cadre du descriptif-quantitatif indiquant leurs dispositions d'ensemble et de détail, les sollicitations auxquelles ils doivent répondre ainsi que les spécifications techniques imposées pour les matériaux et équipements ; CCM 1984) ; **ƒ 2 spécifications [fpl] techniques détaillées** (Document définissant les travaux des divers corps d'état concurremment avec les plans d'exécution des ouvrages ; fait parties de l'élément normalisé spécifications techniques détaillées (S.T.D.) comprenant en outre le programme général, le devis quantitatif [►descriptif quantitatif] et l'estimation détaillée des dépenses ; ►description sommaire des ouvrages au stade de l'avant-projet, ►description sommaire des ouvrages au stade du projet) ; *syn.* descriptif [m] des travaux ; **g Leistungsbeschreibung [f] (2)** (Beschreibung des Umfangs der einzelnen Bauleistungen/Teilleistungen in einem ►Leistungsverzeichnis; ►verkürzte Leistungsbeschreibung).

technical staff [n] [US] *adm. contr.* *►technical employee [US]/technical staff member [UK].

technical staff member [n] [UK] *adm. contr.* ►technical employee [US].

6295 technical state-of-the-art [n] *arch. constr.* (Method or procedure carried out according to latest written standards for the up-to-date provision of certain services or products, which is in compliance with the laws of mathematics and physics, is well-known and applied by leading experts and has proven to be durable and efficient in practice; ►state-of-the-art 2); *syn.* state-of-the-art technique [n]; **s técnica [f] reconocida y aceptada** (Procedimientos o métodos estandarizados en normas para realizar servicios, fabricar productos o bienes de equipo y desarrollar procesos productivos que han sido reconocidos por la ciencia, son conocidos y utilizados por profesionales destacados y han dado buenos resultados en la práctica; ►mejor tecnología asequible); **ƒ principes [mpl] de la technique courante** (GEN 1982, 56 ; forme ou méthode d'exécution de travaux de construction fixant par écrit les règles d'aptitude à l'emploi et de mise en œuvre de procédés, éléments ou équipements utilisés dans la construction ; ces règles reconnues par les hommes de l'art et les organismes spécialisés font dans la pratique l'objet d'un avis technique, d'une normalisation, d'une certification ou d'une codification dans le cadre des documents techniques unifiés [D.T.U.] ; ►suivant les règles de l'art) ; **g allgemein anerkannte Regeln [fpl] der Technik** (Schriftlich [z. B. in DIN-Normen] festgelegte Art oder Methode des Vorgehens zur Erstellung bestimmter Leistungen oder Produkte und andere Verfahrensweisen, die in der Wissenschaft als theoretisch richtig anerkannt sind, den maßgeblichen Fachleuten bekannt sind, von ihnen angewendet werden und sich in der Praxis bewährt haben; im Bauwesen wird statt des ►Standes der Technik üblicherweise die Einhaltung der **a. a.n R. d. T.** vertraglich gefordert).

technical surveillance [n] *leg. envir.* ►technical monitoring.

technique [n] *constr. land'man.* ►avalanche control technique; *constr.* ►biotechnical construction technique, ►planting technique.

technique [n]**, network planning** *constr. plan.* ►critical path method.

technique [n]**, open cut** *eng.* ►cut and cover excavation [US]/cut and cover tunnelling [UK].

technique [n]**, state-of-the-art** *arch. constr.* ►technical state-of-the-art.

technique [n]**, vegetating** *constr.* ►planting technique.

techniques [npl] *for.* ►forestry techniques; *constr.* ►landscape construction techniques; *constr.* ►sod laying and seeding techniques [US]/laying of turves and seeding techniques [UK].

technology [n] *envir.* ►clean technology, ►environmental protection technology, ►low-waste technology.

technology [n] [US]**, green roof** *constr.* ►roof planting method.

teenage hangout [n] [UK] *recr.* ►teenage playground [US].

6296 teenage playground [n] [US] *recr.* (Play area with facilities such as basketball boards, kick-about pitch or teen shelter); *syn.* teenage hangout [n] [UK]; *syn. in the vernacular* "slack" space [n] [UK]; **s terreno [m] para juegos juveniles** (Área de juegos para adolescentes equipada con cestas de baloncesto, cancha de juegos de pelota, cobijos cubiertos, etc.); **ƒ terrain [m] de jeux pour les adolescents** *syn.* terrain [m] de jeux pour les plus grands, terrain [m] de jeux pour les plus de 14 ans ; **g Spielplatz [m] für Jugendliche** (Spielbereich, der für Jugendliche mit Einrichtungen, z. B. Basketballständern, Bolzfläche oder Lümmelecke mit Schutzdach, ausgestattet ist); *syn.* Jugendspielplatz [m].

telecommunications services [n]**, local postal and** *trans. urb.* ►local post services.

telephone traffic [n]**, local** *trans. urb.* ►local traffic, #2.

temperature [n]**, raising the** *envir. limn.* ►heating up.

temporary allotment land [n] [UK] *urb.* ►temporary community garden area [US].

temporary allotment plot [n] [UK] *leg. urb.* ►temporary community garden plot [US].

T

6297 temporary community garden area [n] [US] *urb.* (Garden area ploughed for annual renting of individual plots with short-term agreements negotiated with a local authority. This area may become later on ▶permanent community garden areas [US]/permanent allotment sites [UK] or may be used for building development); *syn.* temporary allotment land [n] [UK]; *s* **área** [f] **de huertos urbanos temporales** (En D. área de terreno dividido en parcelas que se alquilan por un año de la autoridad local hasta la fijación del uso definitivo del terreno. Al contrario que en las zonas de ▶huertos recreativos urbanos, no tienen apenas infraestructura; ▶zona permanente de huertos recreativos urbanos); *f* **terres [fpl] affermées temporairement** (Parcelles régies par le statut du fermage à courte durée [souvent bail annuel renouvelable] à des fins de jardinage ou d'activités récréatives, dont la location est le plus souvent contractée avec la commune en vue d'une affectation future en jardins familiaux ou en zone urbaine ; ▶lotissement de jardins [ouvriers]) ; *g* **Grabeland [n]** (Kleinpachtland, das mit kurzfristigen, meist mit der Gemeinde abgeschlossenen Rechtsverträgen, in der Freizeit bewirtschaftet werden kann, später im Rahmen städtebaulicher Entwicklungsmaßnahmen entweder in baurechtlich abgesicherte ▶Dauerkleingartenanlagen oder in Bauflächen umgewidmet wird. **G.flächen** dienen oft Nutzern, die wenig Interesse am Vereinsleben haben oder sich ungern auf Dauerpachtverträge festlegen wollen sowie solchen, meist ausländischen Mitbürgern, die andere Vorstellungen von Gartenkultur haben).

6298 temporary community garden plot [n] [US] *leg. urb.* (Plot which can be used on the basis of a short-term rental agreement; ▶temporary community garden area [US]/temporary allotment land [UK], ▶community garden [US]/allotment garden [plot] [UK]); *syn.* temporary allotment plot [n] [UK]; *s* **parcela [f] de huerto urbano temporal** (▶área de huertos urbanos temporales, ▶huerto recreativo urbano); *syn.* jardín [m] de hortalizas [C]; *f* **parcelle [f] affermée temporairement** (Propriété foncière privée dont l'usage temporaire s'effectue sur la base du paiement d'une redevance ; ▶terres affermées temporairement, ▶jardin ouvrier) ; *g* **Grabelandparzelle [f]** (Privat für gärtnerische Tätigkeiten nutzbarer Freiraum, der durch kurzfristige Pachtabschlüsse zeitlich begrenzt in Besitz genommen werden kann; ▶Grabeland, ▶Kleingarten); *syn.* Zeitgarten [m].

6299 temporary construction road [n] *constr.* (Road built to serve a building or other construction project and then removed when the work has been completed); *syn.* site road [n]; *s* **carretera [f] provisional de acceso a obra** (Vía construida exclusivamente para servir de acceso a una obra que se desmonta una vez terminada ésa); *f* **voie [f] de chantier** (Voie provisoire exécutée pour la circulation des engins sur le chantier) ; *syn.* piste [f] de chantier (VRD 1986, 149) ; *g* **Baustraße [f]** (Provisorisch ausgebaute Straße zur Andienung oder inneren Erschließung von Baustellen); *syn.* Baustellenstraße [f].

6300 temporary grassland [n] [US] *agr.* (Farmland used in rotation for crops and as ▶pasture; ▶crop rotation); *syn.* ley [n] [UK]; *s* **pradera [f] temporal** (Área agrícola sembrada con especies forrajeras y de trébol de uso temporal que, al contrario que los ▶pastizales, se siega con regularidad para hacer heno y para ensilaje; ▶rotación de cultivos); *syn.* pradera [f] agronómica; *f* **prairie [f] temporaire** (Surface cultivée, ensemencée avec des graminées ou avec un mélange de graminées et de légumineuses, et destinée à être pâturée, fanée ou ensilée et qui occupe une sole pendant une durée variable ; c'est donc une prairie assolée ; cf. DAV 1984 et DA 1977 ; ▶pâturage, ▶rotation des cultures) ; *g* **zeitweiliges Grünland [n]** (Durch Aussaat von Futtergräsern und Kleearten angelegtes temporäres Grünland, das im Gegensatz zur ▶Weide regelmäßig gemäht und der Heugewinnung oder zu Erzeugung von Silagefutter dient. Es

unterliegt dem Wechsel von Acker- und Grünlandnutzung; ▶Fruchtfolge); *syn.* Feldgraswiese [f], Wechselgrünland [n].

6301 temporary restraining order [n] [US] *conserv. leg. plan.* (Court order to protect an area of land for nature conservation or historical importance or to delay another change in land use prior to adjudication at a hearing, or to prevent immediate demolition or alteration of unlisted historic buildings for conservation reasons; ▶preliminary injunction on alteration); *syn.* emergency preservation notice [n] [UK]; *s 1* **declaración [f] de régimen de protección preventiva** (Medida preventiva de protección cautelar de zonas bien conservadas, amenazadas por un factor de perturbación que potencialmente pudiera alterar tal estado; art. 23 Ley 42/2007; ▶prohibición provisional de alteración); *s 2* **anotación [f] preventiva** (Protección cautelar de los bienes declarados de interés cultural después de la incoación del correspondiente expediente hasta que recaiga la resolución definitiva de declaración; art. 12 [1] Ley 16/1985); *syn.* régimen [m] de protección preventiva; *f* **classement [m] d'office** (Décision juridique empêchant provisoirement la démolition ou une quelconque modification d'immeubles pour des raisons de protection du patrimoine historique et esthétique et ordonnant la protection provisoire d'espaces naturels en vue de leur classement définitif ; ▶sursis à statuer) ; *g* **einstweilige Sicherstellung [f]** (Vorläufige gerichtliche oder behördliche Verfügung/Anordnung, die den sofortigen Abbruch oder Veränderungen an Gebäuden aus Gründen des Denkmalschutzes solange verhindert oder wertvolle Flächen für Naturschutz und Landschaftspflege solange schützt, bis die endgültige Sicherung/Sicherstellung verfügt ist; ▶Veränderungssperre).

temporary site nursery [n] *constr.* ▶site nurserey.

tenant agricultural land [n] *agr.* ▶agricultural leasehold (property).

tenant farm [n] [US] *agr.* ▶agricultural leasehold (property).

6302 tenant garden [n] *urb.* (Individual area of garden usually made available only to the occupants of ground-floor apartments [US]/ground-floor flats [UK]; ▶rented garden [UK], ▶leasehold garden); *s* **jardín [m] de inquilinos** (Zona parcelada dentro del recinto de casas de pisos para el uso individual de los inquilinos como jardín. En el marco de la remodelación de barrios *en D.* fueron creados en las zonas ajardinadas entre bloques producto de la distancia legal entre ellos; ▶jardín alquilado, ▶jardín arrendado); *f* **jardin [m] locatif** (Jardin attenant à l'habitation dans l'habitat collectif, directement accessible par un escalier ou une terrasse dont le locataire de l'appartement situé en rez-de-jardin a la jouissance dans le cadre du bail de location ; ▶jardin loué, ▶jardin affermé) ; *g* **Mietergarten [m]** (Parzelliertes, meist den Erdgeschosswohnungen von Geschosswohnungsbauten direkt mit der Wohnung über Treppe oder Terrasse zugeordnetes Gartenland, das den Bewohnern von Mietwohnungen zur individuellen Nutzung per Mietvertrag zur Verfügung steht. Für Mieter der oberen Geschosse über Haus- oder Kellertür erreichbarer Garten, der wie ein in unmittelbarer Nähe liegender Kleingarten anzusehen ist. Heute im Rahmen der Wohnumfeldverbesserung planerisch wieder häufig berücksichtigt; ▶Mietgarten, ▶Pachtgarten); *syn.* Erdgeschossgarten [m].

tender [n] [UK] *constr. contr.* ▶bid [US] (1).

tender [n] [UK]**, accept a** *constr. contr.* ▶accept a bid [US].

tender [n] [UK]**, acceptance of a** *constr. contr.* ▶acceptance of a bid [US] (1), ▶acceptance of a bid [US] (2).

tender [n] [UK]**, invitation for a** *constr. contr.* ▶invitation for a bid [US].

tender [n] [UK]**, subsidiary** *constr. contr.* ▶subsidiary bid [US].

tender accepted [loc] [UK]**, have a** *constr. contr.* ►have a bid accepted [US].

tender action [n] **and contract preparation** [n] [UK] *contr. prof.* ►bidding and negotiation phase [US].

tender analysis [n] [UK] *constr. contr.* ►bid analysis [US].

tender bond [n] [UK] *contr.* ►bid bond [US].

tender documents [npl] [UK] *constr. contr. leg.* ►bid documents [US].

tender documents [npl] [UK]**, inspection of additional** *constr. contr.* ►inspection of additional bid documents [US].

tender documens [npl] [UK]**, priced** *constr. contr.* ►priced bidding documents [US].

tenderer [n] [UK] (1) *contr.* ►bidder [US] (1).

tenderer [n] [UK] (2) *contr.* ►bidder [US] (2).

tenderer [n]**, highest** [UK] *contr.* ►highest bidder [US].

tenderer [n]**, lowest** [UK] *contr.* ►lowest bidder [US].

tenderer [n]**, selected** [UK] *contr.* ►selected bidder [US].

tenderer [n]**, successful** [UK] *contr.* ►bidder offering best value for the price [US]/tenderer offering the best value for the money [UK].

tenderer [n] **offering the best value for the money** [UK] *contr.* ►bidder offering best value for the price [US].

tenderer's confirmation [n] [UK] *contr.* ►bidder's affidavit [US].

tendering [n] [UK] *constr. contr.* ►bidding [US].

tendering [n] [UK]**, compulsory competitive** *adm. constr. contr.* ►open bidding [US].

tendering [n] [UK]**, open** *adm. constr. contr.* ►open bidding [US].

tendering n] [UK]**, selective** *adm. constr. contr.* ►selective bidding [US].

tendering documents [npl] [UK] *constr. contr. leg.* ►bid documents [US].

tendering documents [npl] [UK]**, priced** *constr. contr.* ►priced bidding documents [US].

tendering documents [npl] [UK]**, return of** *constr. contr.* ►return of bidding documents [US].

tendering procedure [n] [UK] *adm. contr.* ►bidding procedure [US].

tender item n] [UK] *constr. contr.* ►bid item [US].

tender items [npl] [UK]**, comparative analysis of** *constr. contr.* ►comparative analysis of bid items [US].

tender items [npl] [UK]**, schedule of** *constr. contr.* ►list of bid items and quantities [US].

tender negotiation [n] [UK] *constr. contr.* ►bid negotiation [US].

tender negotiations [n] [UK] *constr. contr.* ►bid negotiations [US].

tender notice [n] [UK] *adm. constr.* ►bid advertisement [US].

tender period [n] [UK] *constr. contr.* ►bidding period [US].

tender sum [n] [UK] *constr. contr.* ►bid price [US].

tender summary [n] [UK] *constr. contr.* ►bid summary [US].

6303 tending [n] **of a forest stand** *for.* (Generally, any silvicultural operation carried out for the benefit of a forest crop or an individual thereof, at any stage of its life; cf. SAF 1983); *s* **tratamiento** [m] **silvicultural (de mantenimiento, mejora y protección de rodales)** (En general todo tipo de operaciones silviculturales realizadas para promover el crecimiento de rodales o de árboles individuales dentro de los mismos) *syn.* sistema [m] silvicultural de mantenimiento; *f* **soins** [mpl] **culturaux d'un peuplement forestier** (Opérations culturales [tailles, coupes, dégagement, désherbage, travail du sol] effectuées en fonction de l'origine et la nature [naturelle, artificielle], des essences, de l'âge d'un peuplement en vue d'assurer la mise en bonnes conditions de croissance [équilibre, résistance], d'améliorer la qualité du bois produit ainsi que d'assurer sa pérennité) ; *syn.* soins [mpl] sylvicoles, entretiens [mpl] culturaux ; *g* **Waldbestandspflege** [f, o. Pl.] (Waldbauliche Maßnahmen, die günstige Wachstumsvoraussetzungen für die Weiterentwicklung von wuchskräftigen und qualitativ hochwertigen Bäumen schaffen und nachhaltig sichern, um Kronen- und Zuwachsentwicklung zu fördern. Bei der Pflege werden schwache, kranke, abgestorbene und bedrängende Bäume entnommen. Es werden drei verschiedene Hauptpflegemaßnahmen unterschieden: Jungwuchspflege, Läuterung und Durchforstung); *syn.* Bestandspflege [f, o. Pl.], Pflege [f, o. Pl.] eines (Wald)bestand(e)s.

6304 tendril [n] *bot.* (Usually an elongated, curling, filamentlike stem or leaf, which enables a plant to climb; *specific terms* ►leaf tendril, ►shoot tendril, and ►root tendril; ►tendril climber); *s* **zarcillo** [m] (Hilo, por lo común enroscado, con el cual se agarra la planta en algún otro cuerpo; como en la vid. En sentido estricto, se define hoy como tal cualquier órgano filamentoso y haptotrópico que la planta utiliza *exclusivamente* para trepar. Según el origen de los *z.* se diferencian los ►zarcillos caulinares, los ►zarcillos foliares y los ►zarcillos radicales; cf. DB 1985; ►planta zarcillosa); *f* **vrille** [f] (Organe aérien des végétaux d'origine foliaire ou raméale capable de s'enrouler en spirale autour d'un support, caractéristique des plantes grimpantes ; ►plante grimpante à vrilles, ►racine-vrille, ►vrille foliaire, ►vrille raméale) ; *g* **Ranke** [f] (Meist fadenförmig verlängertes, kontaktreizbares, ausschließlich dem Klettern dienendes Organ. Es wird zwischen ►Blattranke, ►Sprossranke und ►Wurzelranke unterschieden; cf. BOR 1958; ►Rankenpflanze).

6305 tendril climber [n] *bot.* (►Climbing plant which maintains its upward growth by means of tendrils, e.g. sweet pea; ►liana, ►root climber, ►scandent plant, ►self-clinging climber, ►twining climber); *syn.* tendril-climbing plant [n]; *s* **planta** [f] **zarcillosa** (Trepadora que utiliza zarcillos [caulinares o foliares] para asirse a un soporte; cf. DB 1985; ►bejuco, ►planta autotrepadora, ►planta sarmentosa, ►planta trepadora, ►planta trepadora con raíces adventicias, ►planta voluble); *f* **plante** [f] **grimpante à vrilles** (Végétal se développant en hauteur en s'appuyant sur un support avec des mouvements préhenseurs et dont les organes de fixation n'assurent que cette seule fonction ; ►liane, ►plante grimpante, ►plante volubile, ►plante grimpante à organes d'accrochage, ►plante grimpante à racines-crampons, ►plante sarmenteuse) ; *g* **Rankenpflanze** [f] (Im Boden wurzelnde, krautige ►Kletterpflanze, die durch Windebewegungen im Kampf um das Licht mit dünnen Greiforganen [Blatt- oder Sprossranken] auf Berührungsreiz hin Stützen in ein- bis mehrfachen Windungen umwächst, z. B. Waldrebe *[Clematis]*, Weinrebe *[Vitis vinifera]*, Wilder Wein *[Parthenocissus quinquefolia]*; ►Liane, ►Selbbstklimmer, ►Spreizklimmer, ►Windepflanze, ►Wurzelkletterer); *syn.* Rankenkletterer [m], Ranker [m] (2).

tendril-climbing plant [n] *bot.* ►tendril climber.

tennis building [n] [US] *recr.* ►indoor tennis facility [US]/indoor tennis courts [UK].

tennis court [n] *recr.* ►tournament tennis court.

6306 tensile strength [n] *constr. eng. stat.* (Resistence of a material to rupture from ever-increasing tension; tensile strength is evaluated by a tension test); *s* **resistencia** [f] **a la tracción**

(Resistencia de los materiales a la rotura por tensión creciente, que se mide en prueba de tracción); *f* **résistance [f] à la traction** (Caractéristique mécanique des matériaux, valeur calculée lors d'essais normalisés de stabilité sur des matériaux soumis une tension progressive provoquant une extension importante [variations dimensionnelles], le déchirement [résistance à la déchirure] ou la rupture [allongement à la rupture]); *g* **Zugfestigkeit [f]** (Werkstoffkenngröße, die bei einem Zugversuch ermittelt wird, wobei auf den zu prüfenden Werkstoff eine stetig zunehmende Zugbeanspruchung ausgeübt wird, die zu einer starken Dehnung resp. zum Zerreißen des Materials führt).

6307 tent and motor camping [n] [US] *recr.* (General term for all types of outdoor temporary living space used in recreation travel; ▶tent camping [US]); *syn.* camping [n] and caravaning [n] [UK]; *s* **campismo [m]** (Término genérico para la actividad de tiempo libre de pernoctar en tienda de campaña, caravana u otro vehículo adaptado a ese fin, en general en zona acotada y equipada con instalaciones y servicios; ▶camping 1 y ▶camping 2); *f 1* **hôtellerie [f] de plein air** (Terme générique désignant le mode de séjour estival englobant l'hébergement sous tente ou en caravane mobile, les caravanes ou mobile homes installés dans les ▶parcs résidentiels de loisirs ou sur des parcelles privées, les ▶habitations légères de loisirs ; LTV 1996, 68-69 ; ▶camping sous la tente) ; *syn.* camping [m] ; *f 2* **camping-caravaning [m]** (*Terme spécifique* **1.** activité touristique qui prône comme moyen d'hébergement l'utilisation d'un abris mobile tel d'une caravane routière, d'un motorhome ou de tout autre abri analogue, non conçus pour servir d'habitation permanente. **2.** Terrain aménagé avec des emplacements de passage pour les caravanes et pourvus des équipements communs et sanitaires nécessaires ; ▶camping sous la tente) ; *syn.* caravanage [m] (PR 1987), terrain [m] de camping-caravaning ; *g* **Campingwesen** ▶camping sous la tente) ; *syn.* caravanage [m] (PR 1987) ; *g* **Campingwesen [n, o. Pl.]** (Sämtliche juristischen, planerischen, technischen und organisatorischen [Betriebs]maßnahmen, die die Ausstattung und den Betrieb von Anlagen für das Freizeitwohnen in transportablen Unterkünften [Caravaning], z. B. Zelt, Wohnwagen, Reise- oder Wohnmobil, auf besonders ausgestatteten Campingplätzen regeln; das **C.** hatte sich zwischen den Weltkriegen als Urlaubsform aus dem ▶Zelten der Jugendbewegung entwickelt und seit den 1950-Jahren durch die zunehmende Motorisierung in Form des Campingtourismus rasch an Bedeutung zugenommen).

6308 tent camping [n] [US] *recr.* (Overnight lodging in a tent; *context* to go camping; ▶camp out); *syn.* camping [n] [UK]; *s* **camping [m] (2)** (Término genérico para acampada con tienda de campaña; *contexto* ir de camping; ▶acampar por libre); *syn.* campin [m], acampe [m] [AL]; hacer [vb] camping, acampar [vb] en tienda; *f* **camping [m] sous la tente** (Séjourner sous la tente ; ▶camping sauvage) ; *syn.* camping [m] sous toile ; *g* **Zelten [n, o. Pl.]** (Aufschlagen eines Zeltes, um darin zu übernachten und sich vor den Unbilden des Wetters zu schützen; ▶wildes Lagern und Zelten).

6309 terminal bud [n] *bot. hort. s* **yema [f] terminal** *syn.* yema [f] apical; *f* **bourgeon [m] terminal** ; *g* **Terminalknospe [f]** (Knospe am Ende eines Sprosses); *syn.* Endknospe [f], Gipfelknospe [f].

6310 terminal moraine [n] *geo.* (Mound of ▶till [US]/boulder-clay [UK] deposited at the end of a glacier); *syn.* end moraine [n], frontal moraine [n] (GFG 1997), *obs.* marginal moraine [n] (GFG 1997); *s* **morrena [f] frontal** (Producida cuando los materiales transportados por el hielo hasta el frente de fusión, donde son depositados. *Términos específicos* **m.** de estabilización, **m.** de empuje, **m.** de ablación; cf. DINA 1987; ▶till); *syn.* morrena [f] terminal; *f* **moraine [f] frontale** (Amas de débris rocheux transportés et déposés à l'extrémité de la langue d'un glacier ; ▶argile à blocaux) ; *syn.* moraine [f] terminale ; *g* **Endmoräne [f]** (Das an der Stelle des weitesten Gletschervorstoßes abgelagerte und aufgehäufte Geschiebematerial; ▶Geschiebemergel); *syn.* Stirnmoräne [f].

termination [n] **of a commission** [UK] *contr. prof.* ▶termination of a contract [US].

6311 termination [n] **of a contract** [US] *contr. prof.* (Client's act of invalidating the agreement [US]/engagement [UK] awarded to a designer/planner); *syn.* cancellation [n] of contract [US], termination [n] of an agreement [US], termination [n] of a commission [UK], termination of engagement/appointment [n] [UK]; *s* **resolución [f] de contrato (de servicios)** (Finalización anticipada de un contrato de servicios de planificación por parte del comitente; cf. arts. 112-114 LCAP); *syn.* derogación [f] de un contrato de trabajo profesional; *f* **résiliation [f] du marché d'études par le maître d'ouvrage** (Décision de mettre fin à l'exécution d'un marché avant son terme prise soit du fait du maître de l'ouvrage, soit aux torts du maître d'œuvre) ; *g* **Entziehung [f] des Auftrages (2)** (Widerrufen von übertragenen Aufgaben an einen Planer durch den Auftraggeber); *syn.* Auftragsentziehung [f] (2), Auftragsentzug [m] (2).

termination [n] **of an agreement** [US] *contr. prof.* ▶termination of a contract [US].

termination [n] **of appointment** [UK] *contr. prof.* ▶termination of a contract [US].

termination [n] **of engagement** [UK] *contr. prof.* ▶termination of a contract [US].

terrace [n] **(1)** *arch.* ▶building terrace; *agr.* ▶cultivated terrace; *geo.* ▶gravel river terrace, ▶lower river terrace, ▶middle river terrace; *constr.* ▶narrow plant terrace; *geo.* ▶natural terrace on a slope; *arch. constr.* ▶roof terrace; ▶upper river terrace.

terrace [n] [US] **(2)** *constr. eng.* ▶ledge (1).

6312 terrace cultivation [n] *agr.* (GGT 1979; system of cultivation in artificial terraces on the sides of hills and mountains; e.g. terraced vineyards in Central Europe or paddy terraces in rice cultivation regions of SE-Asia; ▶cultivated terrace, ▶terraced slope); *s* **cultivo [m] en terrazas** (Método de cultivo en pendientes fuertes por medio del abancalado, p. ej. en Europa Central utilizado para el cultivo de la vid y en el sureste de Asia en ▶campos en bancales para cultivo de arroz y té; ▶vertiente en bancales); *f* **culture [f] en terrasse** (Pratique agricole pour laquelle les versants d'une colline ou d'une montagne sont remodelés en parcelles planes [terrasses], p. ex. les vignobles en Europe centrale et méridionale, la riziculture et la culture du thé en Asie du sud-est ; ▶terrasse de culture, ▶versant en terrasses) ; *syn.* culture [f] en étages (PR 1987) ; *g* **Terrassenanbau [m, o. Pl.]** (Landwirtschaftliche Bodennutzung, bei der steile Hänge in weniger geneigte oder ebene Parzellen [Terrassen] umgestaltet werden; z. B. in Mitteleuropa beim Weinbau in Weinbergen, in Südostasien in Berghänge eingebaute ▶Terrassenfelder für Reis und Teekulturen; ▶Terrassenhang).

terraced dwelling [n] *arch.* ▶stepped house.

terraced house [n] *arch.* ▶stepped house.

6313 terraced housing block(s) [n(pl)] *urb.* (Layout of housing units with terraces at different levels on a sloping site. The staggered juxtaposition of each housing unit creates a series of open spaces, each of which may be contained by walls of adjacent units. Each apartment thus has its own open terrace with a sunny aspect; ▶courtyard house, ▶stepped house); *s* **edificación [f] aterrazada** (▶casa aterrazada, ▶casa de patio); *syn.* edificación [f] escalonada; *f* **habitat [m] collinaire** (Superposition d'appartements d'un ensemble immobilier le long d'un versant et assurant un bon ensoleillement de chaque unité d'habitation grâce à leur disposition en terrasse ; ►maison atrium, ►maison en terrasse) ; *g* **Terrassenbebauung [f]** (Übereinanderschichtung von Gartenhofwohneinheiten an einem Geländehang, so dass durch seitliches und rückwärtiges Verschieben der Grundrisse gegeneinander jede Wohneinheit ihren besonnten Freiraum [Atrium] erhält ; ►Gartenhofhaus, ►Terrassenhaus); *syn.* Terrassenhäuser [npl], Wohnhügel [m].

6314 terraced slope [n] *agr. geo.* (Hill or mountain slope characterized by ►cultivated terraces); *s* **vertiente [m] en bancales** (►campo en bancales); *syn.* pendiente [f] aterrazada; *f* **versant [m] en terrasses** (DG 1984, 446 ; pente forte d'une montagne aménagée par l'homme en terrasses étagées ; ►terrasse de culture) ; *g* **Terrassenhang [m]** (Von ►Terrassenfeldern geprägter Berghang).

6315 terraced townhouse garden [n] [US] *arch. constr.* (Patio garden on the roof terrace of a dwelling unit [one above another] in a building constructed on a slope; ►roof garden, ►roof terrace); *syn.* stepped building garden [n] [UK], split-level house garden [n]; *s* **terraza-jardín [f]** (Jardín en tejado de casa aterrazada; ►azotea ajardinada, ►terraza en azotea); *f* **terrasse-jardin [f]** (Jardin implanté sur la toiture de l'appartement inférieur, p. ex. dans un habitat collinaire ; ►jardin sur dalle, ►jardin sur toiture-terrasse, ►toiture-terrasse) ; *syn.* jardin [m] sur terrasse, jardin [m] sur immeuble en terrasses ; *g* **Terrassenhausgarten [m]** (Garten auf einem Haus am Hang, dessen Geschosse so versetzt sind, dass ab dem zweiten Geschoss jede Wohneinheit eine Dachterrasse auf der unterliegenden Wohnung hat; ►Dachgarten, ►Dachterrasse).

terrace house [n] [UK] *urb.* ►row house.

6316 terrace [n] **of border perennials** *gard.* (Level change in a garden or park, planted predominantly with ►perennials 1); *s* **terraza [f] de vivaces/perennes (±)** (Escalón en el terreno de un parque o jardín, plantado mayormente con ►plantas vivaces); *f* **terrasse [f] de vivaces** (Plantation constituée en majorité de ►plantes vivaces sur une levée de terrain, dans un jardin ou un parc) ; *syn. en Provence* restanque [f] de vivaces, replat [m] de plantes vivaces ; *g* **Staudenterrasse [f]** (Geländeabsatz in einem Garten oder Park, der vorwiegend mit ►Stauden bepflanzt ist).

6317 terracing [n] *agr. constr. landsc.* (Construction of terraces on the sides of hills and mountains for cultivation purposes, e.g. paddy terraces in SE-Asia and terraced vineyards in Central Europe as well as in the Mediterranean region; ►cultivated terrace, ►narrow plant terrace, ►terrace cultivation, ►terracing of a slope); *s* **abancalado [m]** (*Términos específicos* **a.** de escalonamiento, **a.** circular, **a.** por talud; DGA 1986; ►aterrazamiento de talud, ►berma de plantación, ►campo en bancales, ►cultivo en terrazas); *syn.* aterrazamiento [m]; *f* **aménagement [m] en terrasses** (LA 1981, 1167 ; installation de terrasses sur des pentes fortes en vue de l'exploitation agricole intensive, p. ex. la riziculture en Asie, les vignobles en Europe centrale et autour de la Méditerranée ; ►crantage d'un talus, ►culture en terrasse, ►redan, ►terrasse de culture) ; *syn.* crantage [m] d'une pente ; *g* **Hangterrassierung [f]** (Schaffung ebener Flächen in einem Hang: In der Landwirtschaft werden für manche Kulturen steile Hänge in ►Terrassenfelder umgewandelt, z. B. für den Reisanbau in Südostasien, für den Weinbau in Mitteleuropa und im Mittelmeergebiet; ►schmale Pflanzterrasse, ►Terrassenanbau, ►Terrassierung von Böschungen); *syn.* Geländeterrassierung [f].

6318 terracing [n] **of a slope** *constr. min.* (Construction of ►ledges 1 or ►working benches on steep slopes or rock faces);

s **aterrazamiento [m] de talud** (Construcción de ►berma en un talud, ►talud escalonado); *f* **crantage [m] d'un talus** (Exécution d'une ►risberme sur un talus ; ►banquette d'exploitation) ; *g* **Terrassierung [f] von Böschungen** (Bau von Bermen in steilen Böschungen oder in Tagebauflächen; ►Berme 1, ►Berme 2).

terrain [n] *landsc.* ►flat terrain; *for. leg.* ►forest terrain; *geo.* ►nature of the terrain; *arch. plan.* ►reconfiguration of an area/terrain.

terrain [n]**, natural** *constr. plan.* ►natural ground surface.

terrain [n]**, open** *landsc.* ►open country; ►open space.

6319 terrain analysis [n] *plan. rem'sens. surv.* (Evaluation of remote sensing data with respect to the geomorphological characteristics of the land surface; ►remote sensing); *s* **análisis [m] de la topografía del terreno** (Evaluación de los datos de satélite en cuanto a la estructura y textura de la superficie terrestre; ►teledetección); *f* **analyse [f] de la structure terrestre superficielle** (Interprétation des données de la ►télédétection de terrain relatives à la structure et la texture de la surface de la terre) ; *g* **Erderkundung [f]** (Auswertung der Fernerkundungsdaten bezüglich der Struktur und Textur der Erdoberfläche; ►Fernerkundung); *syn.* Analyse [f] der Erdoberflächenbeschaffenheit.

6320 terrain relief [n] *geo. ecol.* (Description of an area showing much or little variation in elevation: rough country, or flat country); *s* **relieve [m] del terreno** (Grado de contraste del relieve en una zona determinada); *f* **énergie [f] de relief** (Degré de variation des formes superficielles du terrain dans une unité paysagère donnée — différence d'altitude entre le point le plus haut et le plus bas d'une surface, p. ex. un sur relief mou, un relief très marqué, un relief très accentué, un relief accidenté) ; *g* **Reliefenergie [f]** (Grad der potentiellen Abtragungsenergie eines Landschaftsausschnittes in Abhängigkeit des Wechsels der Höhenverhältnisse — Höhenunterschied zwischen niedrigstem und höchstem Punkt einer Flächeneinheit).

Terres Noires [n] *pedol.* *►vertisol.

terrestrialization [n] *geo. phyt.* ►siltation.

terrestrialization plant community [n] *phyt.* ►plant community in silting up ponds or lakes.

6321 terrestrial radiation [n] *met.* (Emergent warmth from the ground, specially at night; ►night of ground radiation); *s* **radiación [f]** (Pérdida de calor nocturna, en especial en las noches despejadas; ►noche radiativa); *f* **rayonnement [m] réfléchi** (Perte nocturne de l'énergie du sol emmagasinée pendant le jour ; ►nuit en situatio radiative) ; *g* **Ausstrahlung [f]** (Wärmeabgabe der Erdoberfläche, besonders in ►Strahlungsnächten; ausstrahlen [vb]).

6322 terrestrial soils [npl] *pedol.* (*German classification system* soil divisions found above groundwater and unaffected by man, which are differentiated according to their state of development or degree of variation in the horizons and are divided into the following soil classes: ►entisol [US] 1/immature soil [UK], AC-horizon soils, aridisols, inceptisols, ►podzols, terrae calcis and "plastosols"; in the U.S. soil classification system of soil taxonomy this category does not exist; ►brown earth, ►soil orders; ►soil suborder, ►subhydric soils); *s* **suelos [mpl] terrestres** (En la clasificación alemana de suelos, ►grupo principal [de suelos] determinados por su evolución sin influencia ni del agua freática ni de agua estancada y tampoco humana que, según el grado de desarrollo y de diferenciación de los horizontes, se dividen en las siguientes ►clases de suelos: ►suelos minerales brutos, suelos rankeriformes, suelos rendsiniformes, suelos de estepa, pelosoles, ►tierras pardas, ►podsoles, terrae calcis y plastosoles; en la clasificación de los EE.UU. no existe la

T

clase de los **ss. tt.**; ▶suborden de suelos, ▶suelos subacuáticos); *f* **sols [mpl] terrestres** (PED 1983, 18O, 371 ; *classification des sols allemande* ▶classe pédologique n'ayant pas été formés sous l'eau et subi d'influences anthropogènes et qui suivant leur cycle d'évolution et le degré de différenciation des horizons présentent les classes suivantes: ▶sols peu évolués, sols à profil peu différencié AC, le sols isohumiques, les pélosols, les ▶sols brunifiés, les ▶podzols, Terrae calcis et les plastosols ; PED 1983, 180 ; ▶sol subaquatique, ▶sous-classe pédologique) ; *g* **Landböden [mpl]** (*Klassifikationssystem in D.* Bodenabteilung außerhalb des Grund- und Stauwasserbereiches und nicht anthropogene Böden, die nach ihrem Entwicklungsstand resp. dem Grad der Differenzierung in Horizonte in folgende [Boden]klassen unterschieden werden: ▶Rohböden, A-C-Böden, Steppenböden, Pelosole, ▶Braunerden, ▶Podsole, Terrae calcis und Plastosole. In dem USDA-Klassifizierungssystem gibt es keine Bodenabteilung ‚L.' Dort gibt es eine völlig andere Einteilung der Böden, die mit der deutschen kaum vergleichbar ist; siehe dazu die Ausführungen unter ▶Bodenabteilung; ▶Bodenklasse 2, ▶Unterwasserböden); *syn.* terrestrische Böden [mpl].

6323 terrestrial species [n] *zool.* (▶living in the soil); *syn.* epigaeic species [n]; *s* **terrícola [f]** (▶hipogeo/a); *syn.* especie [f] terrícola, especie [f] epigea; *f* **espèce [f] terricole** (▶endogé) ; *syn.* espèce [f] épigéique ; *g* **auf dem Boden lebende Art [f]** (▶im Boden lebend); *syn.* epigäische Art [f].

territorial behaviour [n] [UK] *zool.* ▶territoriality [US].

6324 territoriality [n] [US] *zool.* (GE 1977; pattern of behavior by which an animal or a group of animals appropriates and demarcates a clearly defined territory to ensure that not too many individuals live and breed in the same area; ▶individual territory); *syn.* territorial behaviour [n] [UK]; *s* **territorialidad [f]** (Comportamiento de animales o grupos de los mismos derivado de la defensa de un territorio que se considera como propio; DINA 1987; ▶territorio de un individuo); *f* **comportement [m] territorial** (Prise de possession d'une surface de terrain précise par un individu ou un groupe d'individus qui le défend contre des intrus appartenant le plus souvent à la même espèce ; ▶territoire exclusif ; LAF 1990, 883) ; *syn.* territorialité [f] ; *g* **Territorialverhalten [n]** (Die Inbesitznahme eines bestimmten Raumausschnittes durch ein Individuum oder eine Gruppe von Individuen und die Verteidigung dieses Raumes gegen Artgenossen; ▶Individualterritorium); *syn.* Revierverhalten [n].

6325 territorial waters [npl] *geo. leg. pol.* (Waters under the jurisdiction of a nation, especially that part of the sea within a stated distance of the shore, formerly accepted as being three nautical miles in width, then widened to 12 miles, and recently by some countries to 200 miles for control of fishing ground; ▶inland waters); *syn.* coastal waters [npl]; *s* **aguas [fpl] territoriales** (Concepto creado a mediados del siglo XVI por los holandeses, referido al sector marino sobre el cual el Estado ejercía su soberanía y era considerado parte integrante de su territorio. A finales del siglo XVIII se acordó parcialmente delimitar las **aa. ta.** a una franja de tres millas marinas. En la 1ª Conferencia sobre el Derecho del Mar [Ginebra 1958] la mayoría de los Estados acordaron el principio de las tres millas y ya en la 2ª Conferencia [Ginebra 1960] se aceptó el límite de las 12 millas. En 1952 los Estados sudamericanos decretaron unilateralmente las 200 millas como reacción a la declaración del presidente Truman de 1945 que proclamaba el derecho de los EE.UU. a la explotación de su plataforma continental. A partir del 1977 los países del Mercado Común adoptaron también esta postura. España decretó en 1978 la ley 15/1978 sobre zona económica. Hoy en día las 200 millas se consideran parte del patrimonio económico de cada Estado, sobre todo con el fin de controlar mejor los recursos pesqueros y minerales, cf. DGA 1986; ▶aguas interiores); *f* **eaux [fpl] terri-**toriales (Eaux situées en bordure de la côte, à l'intérieur desquelles l'État riverain exerce tous les pouvoirs découlant de sa souveraineté ; espaces maritimes faisant partie du territoire d'un État souverain situés en deçà des lignes de base à partir desquelles est mesurée la largeur de la mer territoriale comprenant: port maritime, mer intérieure, embouchure, baie, rade ; *terme spécifique* eaux maritimes intérieures ; cf. Convention des Nations Unies du 10 décembre 1990 ; ▶eaux intérieures); *g* **Territorialgewässer [n]** (Vor einer Küste liegende, die maritimen ▶Binnengewässer begrenzenden Meeresteile, die nicht zur hohen See, sondern zum Staatsgebiet des Anrainerstaates gehören und herkömmlich eine Breite von drei Seemeilen haben [Dreimeilenzone], bei manchen Ländern bis zu 12 sm erweitert ist und neuerdings von einigen Ländern auf 200 sm erweitert wurde; cf. Übereinkommen der Vereinigten Nationen vom 10.12.1990. Die Grenze zwischen dem Staatsgebiet der Anrainerstaaten und der hohen See verläuft nicht dort, wo Land und Meer zusammenstoßen, sondern im Meer selbst; zur Gebietshoheit gehören auch die Luftsäule über und der Meeresgrund unter den **T.n**); *syn.* Küstengewässer [n], Küstenmeer [n].

6326 territory [n] *zool.* (Marked area usually including the nesting or denning site and variable foraging space that is preempted by an animal species [bird or mammal] and defended against the intrusion of rival individuals; ▶breeding territory, ▶courtship territory, ▶feeding territory, ▶home range 1, ▶home range 2, ▶individual territory, ▶living space, ▶mating territory); *s* **territorio [m]** (Área que incluye el lugar del nido o la guarida de un animal y un ámbito variable de vivienda, defendido por él; ▶área de actividad, ▶área de habitación, ▶área trófica, ▶dominio vital, ▶territorio de cría, ▶territorio de reproducción, ▶territorio de un individuo); *f* **territoire [m]** (Surface délimitée, partie de l'▶aire d'habitation d'une espèce animale [portion du ▶domaine vital] pouvant varier de quelques millimètres à plusieurs kilomètres et marquée et défendue contre d'autres individus ; ▶domaine d'existence, ▶habitat naturel, ▶territoire alimentaire, ▶territoire d'activité, ▶territoire de nidification, ▶territoire de reproduction, ▶territoire exclusif) ; *g* **Territorium [n]** (Begrenztes Gebiet, Teil des Wohngebietes einer Tierart, das von dieser markiert und gegen Artgenossen verteidigt wird; **T.** und ▶Streifgebiet können sich überschneiden, was hinsichtlich von möglichen Begattungspartnern sinnvoll ist. Durch Revierkämpfe oder nach dem Ableben eines Revierinhabers können sich die Grenzen von **T.** und Streifgebiet verändern; je besser die naturräumliche Ausstattung eines **T.s** in Bezug auf Nahrung, Sicherheit und paarungswilligen Artgenossen, desto erfolgreicher ist der Revierinhaber; ▶Balzrevier, ▶Brutrevier, ▶Fortpflanzungsrevier, ▶Individualterritorium, ▶Lebensraum, ▶Nahrungsrevier, ▶Wohngebiet 1); *syn.* Revier [n].

tertiary consumer [n] *ecol.* *∗*▶consumer.

tessera [n] *geo.* ▶ecotope.

tesserae [npl] [US], **pattern of** *ecol.* ▶pattern of ecotopes.

test [n] *constr.* ▶plate load-bearing test, ▶waterproofing test.

test [n] [UK], **decantation** *constr.* ▶sedimentation analysis.

test [n], **sedimentation** *constr.* ▶sedimentation analysis.

6327 test area [n] *agr. for. hydr. phyt.* (Piece of land used for research into scientific, production or technical issues; ▶permanent sample plot, ▶sample plot); *syn.* experimental area [n]; *s* **campo [m] experimental** (▶cuadrado permanente, ▶superficie de muestreo); *syn.* campo [m] de ensayo; *f* **champ [m] d'essais comparatifs** (Parcelle de terrain délimitée en vue de l'étude et du suivi d'observations scientifiques ; ▶aire d'échantillonnage floristique, ▶carré permanent d'échantillonnage) ; *g* **Versuchsfläche [f]** (Freilandfläche, die der Erforschung wissenschaftlicher, produktions- oder pflege- und unterhaltungstechnischer

Fragestellungen dient; ▶Dauerquadrat, ▶floristische Probefläche).

6328 test area [n] **of paving surface** [US] *constr.* (Small, temporarily-selected area; e.g. for the testing and sampling of various paving surfaces); *syn.* trial area [n] of paving surface [UK]; *s* **superficie** [f] **de prueba** (para estudiar diferentes tipos de pavimentos); *syn.* área [f] de prueba; *f* **échantillons** [mpl] **de revêtement du sol** (Aire de surface réduite et mise en place temporairement afin de faire un choix parmi différents échantillons de revêtement de sol) ; *g* **Probefläche** [f] **eines Belag(e)s** (Kleine, temporäre Fläche, die z. B. zur Prüfung und Bemusterung unterschiedlicher Wegebeläge angelegt wird).

6329 test borehole [n] [US] *geo. min.* (Hole made by boring for investigation purposes; ▶test boring); *syn.* trial borehole [n] [UK] (LD 1991 [11], 27); *s* **pozo** [m] **de sondeo de prueba** (Agujero perforado en el subsuelo con fines exploratorios; ▶perforación de prueba); *syn.* agujero [m] de perforación de prueba, taladro [m] de prueba; *f* **trou** [m] **de sondage** (Puits creusé mécaniquement destiné à la reconnaissance du sous-sol ; ▶sondage) ; *syn.* puit [m] de sondage ; *g* **Probebohrloch** [n] (Für eine Untersuchung niedergebrachtes Bohrloch; ▶Probebohrung).

6330 test boring [n] *geo. min.* (Drilling of holes in order to investigate subsoil or bedrock; ▶test borehole [US]/trial borehole [UK], ▶test trenching); *syn.* prospective drilling [n]; *s* **perforación** [f] **de prueba** (**1.** Realizar una perforación del subsuelo para investigar las características del mismo; ▶pozo de sondeo de prueba, ▶zanja de prospección. **2.** Agujero de perforación resultante); *syn.* perforación [f] de exploración, sondeo [m] de prueba; *f* **sondage** [m] (Reconnaissance des sols — essai de sol en place ; prélèvement pour une étude en laboratoire d'échantillons de sol par forage, par percussion ou par rotation dans le cadre de la campagne de reconnaissance du site afin de définir la nature et l'épaisseur des couches [études des caractéristiques physiques et mécaniques], de préciser les caractéristiques de la nappe aquifère dans la zone d'influence des fondations ; COB 1984, 14 ; ▶reconnaissance par tranchées, ▶trou de sondage) ; *g* **Probebohrung** [f] (Niederbringen von Bohrlöchern, um den Untergrund in größeren Tiefen zu erkunden; ▶Probebohrloch, ▶Schlitzsondierung).

6331 test garden [n] *hort.* (Garden in which plant breeders, botanists and other scientists can make comparisons of new cultivars, plant novelties, plant communities or techniques of vegetation establishment, as well as test new cultivation and maintenance methods; ▶perennials test garden); *syn.* trial garden [n] [also UK]; *s* **jardín** [m] **experimental** (Jardín que sirve a jardineros y botánicos para comparar las plantas y comunidades y experimentar diferentes métodos de cultivo y de composición; ▶jardín experimental de vivaces/perennes); *f* **jardin** [m] **d'essai** (Jardin dans lequel des sélectionneurs, des botanistes ou des chercheurs peuvent pratiquer des expérimentations sur des nouvelles sélections végétales, sur des compositions floristiques d'un groupement ou sur des procédés de végétalisation ainsi qu'éprouver de nouvelles méthodes culturales ou d'acclimatation ainsi que l'évaluation comparative des plantes ; cf. DIC 1993 ; *termes spécifiques* ▶jardin d'essai comparatif des plantes vivaces, jardin de présentation de la flore alpine) ; *g* **Sichtungsgarten** [m] (Garten, in dem Züchter, Botaniker oder andere Wissenschaftler vergleichende Untersuchungen an Pflanzenzüchtungen, Pflanzengemeinschaften oder Begrünungstechniken vornehmen sowie neue Kultur- und Bewirtschaftungsmethoden erproben können; ▶Staudensichtungsgarten).

testing [n] *eng. stat.* ▶structural testing.

6332 testing [n] **of perennials** *hort.* (Systematic observation and evaluation of perennials in a perennials ▶test garden, according to certain criteria, with field log entries being made of the inspections; ▶perennial plant breeding); *s* **evaluación** [f] **comparativa de vivaces/perennes** (Observación y evaluación sistemáticas de perennes en ▶jardín experimental de vivaces utilizando determinados criterios; ▶cultivo de vivaces/perennes); *f* **évaluation** [f] **comparative des plantes vivaces/pérennes** (Évaluation des plantes herbacées pérennes dans un ▶jardin d'essais comparatifs des plantes vivaces ; ▶sélection de plantes vivaces) ; *g* **Staudensichtung** [f] (Systematische Beobachtung und Bewertung von Stauden nach bestimmten Kriterien in einem ▶Sichtungsgarten; über die Beobachtungen und Bonitierungen werden Feldbücher/Sichtungsbücher geführt. Die wichtigsten Bewertungskriterien sind Krankheitsanfälligkeit, Standfestigkeit, Wuchskraft, Winterhärte, Wuchshöhe mit und ohne Blüten, Blattfarbe und -form, Blütenform und Blütezeit, Remontieren, Wetterempfindlichkeit, Selbstreinigung, Ähnlichkeit mit anderen Sorten, Standort, Boden und Klima. Auf Grund der Sichtung lassen sich Stauden in ein Hauptsortiment/Standardsortiment und ein Ergänzungssortiment einteilen; Standardsortimente werden jeweils nach 5-10 Jahren überprüft und mit inzwischen neuen Sorten verglichen. Bereits in den 1920er-Jahren wurde der Ruf nach Sichtung der Staudensortimente laut, um die Wahl aus der Fülle der kultivierten Sorten zu erleichtern und um den Staudenverwendern eine fundierte Entscheidungshilfe zur Wahl geeigneter Sorten für die Park- und Gartengestaltung zu geben. Einer der eifrigsten Verfechter der Sichtungsidee war KARL FOERSTER [1874-1970], der dafür sorgte, dass in den 1930er-Jahren die ersten **S.en** an unterschiedlichen Standorten stattfanden; anlässlich der Reichsgartenschau entstand 1939 im Höhenpark Killesberg in Stuttgart der erste Schau- und Sichtungsgarten. Heute werden die in den verschiedenen **S.sgärten** erfassten Ergebnisse an die Staatliche Forschungsanstalt für Gartenbau in Weihenstephan weitergeleitet und dort zentral ausgewertet. Die strengen Bewertungsmaßstäbe führen zum international üblichen und leicht verständlichen Sternchensystem: 1. ausgezeichnete Sorte [***], 2. sehr gute Sorte [**], 3. gute Sorte [*], 4. Liebhabersorte [Li] — nur für Sonderverwendungszwecke zu empfehlen; 5. Lokalsorte [Lo] — eine günstige Entwicklung der Sorte ist an bestimmte klimatische Bedingungen geknüpft; 6. entbehrliche Sorte [Ø]; ▶Staudenzüchtung).

testing [n] **the structural strength** *eng. stat.* ▶structural testing.

test nursery [n] [US], **perennials** *hort.* ▶perennials test garden.

6333 test pit [n] [US] *constr. pedol.* (Hole excavated for investigation of the upper soil profile; ▶test trenching); *syn.* trial pit [n] [UK]; *s* **calicata** [f] [Es, RA] (Perforación o excavación, generalmente de dimensiones métricas, realizada para obtener información sobre el suelo o el sustrato; DINA 1987; ▶zanja de prospección); *syn.* pozo [m] a cielo abierto [también RA], apique [m] [CO], pozo de prueba [EC, MEX]; *f* **tranchée** [f] **de prospection** (Ouverture d'une tranchée en vue de procéder à une reconnaissance ou des essais de sol ; ▶reconnaissance par tranchées) ; *syn.* galerie [f] de prospection ; *g* **Schürfgrube** [f] (Aufgrabung zur Erkundung der oberflächennahen Bodenschichten; ▶Schlitzsondierung); *syn.* Probegrabung [f].

6334 test trenching [n] *geo. stat.* (Excavation of narrow ditches for the investigation of subsoil conditions; ▶test boring); *s* **zanja** [f] **de prospección** (▶perforación de prueba); *f* **reconnaissance** [f] **par tranchées** (Reconnaissance du sol pratiquée par ouverture de tranchées très étroites, solution très utilisée pour des petites constructions individuelles ; COB 1984 ; ▶sondage) ;

T

g **Schlitzsondierung** [f] (**1.** Ausheben von schmalen Gräben zur Untersuchung der Untergrundverhältnisse, z. B. einer zu bebauenden Fläche; ▶Probebohrung; **2.** Ausheben von schmalen Gräben in historischen Anlagen zur Auffindung Reste alter Wegeführungen, Fundamente etc.).

6335 texture [n] *plant.* (Term used to describe the visual perception and characteristic, tactile qualities or structures of a fabric, material or object, which are sometimes relatively unobtrusive, but readily appreciated, when repeated within a certain space or when distributed over a large area, e.g. the surface of a plant's leaf or leaves. **T.** is also related to a particular arrangement or proportion; in gardens, **t.** is an essential aspect in planting design, e.g. it can be created by planting numerous or regularly occurring elements such as leaves, groups of plants, and shoots with quickly recognizable characteristics. Categories of **t.** range from 'very fine' [small-scale, fine-leaved, and densely flowering, such as *Gypsophila*] to 'very rough' [large-leaved like *Bergenia*]; trees are described as having 'dense' crowns [e.g. sycamore, horse chestnut] or 'open' crowns [birch, *Robinia*]. In contrast to the short-lived colours of flower **t.**s tend to have a much longer effect—depending upon how long the foliage or seed capsules remain on a plant, for example, during winter or if the plant is evergreen. **T.**s dispersed over a large area are known as ▶structures 3); *s* **textura** [f] (Lat. *textum* >tejido< de lat. *texere* >tejer<; término utilizado para describir la percepción y las características visuales, la calidad táctil o las estructuras de un tejido, material u objeto, que a veces son relativamente poco visibles, pero sí apreciables si se repiten en un cierto espacio o si están distribuidas dentro de un área grande, p. ej. la superficie de las hojas de una planta. La **t.** también está relacionada con un arreglo o unas proporciones particulares; en jardines, la **t.** es un aspecto esencial en el diseño de las plantaciones y puede ser creada plantando elementos numerosos o regularmente recurrentes como aquéllos con hojas o retoños o grupos de plantas con características rápidamente reconocibles. Las categorías de **t.** van de «muy fina» [fragmentada, de hojas finas, de floración densa como la *Gypsophila*] hasta «muy burda» [de hojas grandes como la bergenia]. En el caso de los árboles se diferencia entre «densa» o «cerrada» [arce blanco, castaño de Indias] y «ligera» o «abierta» [abedul, falsa acacia]. En comparación con el colorido efímero de las flores, las **tt.** tienen la ventaja de tener efecto a mucho más largo plazo, dependiendo de la persistencia del follaje, incluso en el invierno en el caso de las perennifolias o las plantas con verde invernal. Con la atenuación progresiva de las **tt.** pasa a predominar visualmente la ▶estructura); *f* **texture** [f] (Lat. *textum*, de *texere* « tisser », principe d'ordonnancement, favorisant la perception visuelle ou tactile par la répétition d'éléments isolés de formes identiques ou semblables, relativement discrets et répartis dans un espace ou sur une surface. Dans la conception des jardins la texture est une composante essentielle dans l'utilisation des végétaux tels que p. ex. les feuilles, les groupes de feuilles, les tiges ; les échelles de texture varient de « fine » [fragmenté, à fine feuille, resserré comme la gypsophile en fleurs *(Gypsophila paniculata)* dont la texture ne se différencie qu'à faible distance et qu'on utilise dans des massifs amenés à être observés de près] jusqu'à « grossier » [à feuilles larges et épaisses comme la bergénie *(Bergenia)*] ; chez les arbres on fait la distinction entre une texture « dense » ou « touffue » [érable, châtaignier] et la texture « ouverte » ou « légère » [bouleau, acacia] ; par comparaison avec les couleurs éphémères des fleurs les textures ont l'avantage d'avoir un effet nettement prolongé en fonction de la capacité du feuillage à persister au-delà d'une période de végétation ; avec l'atténuation progressive de sa spécificité la texture prend le caractère d'une ▶structure 2) ; *g* **Textur** [f] (Lat. *textum* >Gewebe< zu lat. *texere* >weben<, >flechten<; ein Ordnungsprinzip, das die Wahrnehmung erleichtert, nach dem formal identische oder ähnliche, vergleichsweise unauffällige Einzelelemente durch die Wiederholung in einem Raum oder auf einer Fläche verteilt sind. In der Gartengestaltung ist die **T.** ein wesentlicher Aspekt in der Pflanzenverwendung, z. B. durch sich genügend zahlreich und regelmäßig wiederholende Einzelelemente wie Blätter, Blattgruppen, Triebe. **T.**stufen reichen von „sehr fein" [kleinteilig, feinblättrig, dicht; wie blühendes Schleierkraut *(Gypsophila paniculata)*] bis „sehr grob" [großblättrig wie Bergenie *(Bergenia)*]; bei Bäumen wird zwischen „dicht geschlossen" [Bergahorn, Rosskastanie] und „offen" oder „leicht" [Birke, Robinie] unterschieden. **T.**en haben im Vergleich zu kurzlebigen Blütenfarben den Vorzug der deutlich längeren Wirkung — je nach Dauerhaftigkeit des Laubwerks auch über eine Vegetationsperiode hinaus, z. B. Winter- und Immergrüne. Mit zunehmender Auflockerung gehen **T.**en in ▶Strukturen 2 über; cf. W. BORCHARDT und K. RÜCKER in GAP 2005, H. 4, p. 31-37 und W. BORCHARDT in GAP 2005, H. 6, pp. 35-41).

texture [n]**, surface** *constr.* ▶surface finish.

6336 thalweg [n] *geo.* (**1.** Line joining the lowest points of a valley from the source of a watercourse to its mouth. **2.** Line of continuous maximum descent from any point on a land surface, e.g. the line crossing all contours at right angles. **3.** Groundwater stream percolating beneath and in the same direction as a surface stream. **4.** Deepest or best navigable channel defining water boundaries between states); *syn.* longitudinal valley profile [n], valley line [n], talweg [n]; *s* **talweg** [m] (Voz alemana que significa «camino del valle» y que se aplica en hidrología fluvial a la línea que dibuja un curso de agua siguiendo los puntos de mayor profundidad del lecho. Con un sentido menos preciso a veces se identifica este término con los de lecho y cauce; DGA 1986); *syn.* thalweg [m]; *f* **talweg** [m] (*Orthographie ancienne* thalweg. Ligne joignant les points les plus bas d'une vallée) ; *g* **Talweg** [m] (Verbindungslinie der tiefsten Punkte eines Flusslaufes von der Quelle bis zur Mündung).

6337 thatch [n] *constr.* (SPON 1974, 361; interwoven grass mat as a result of an accumulation of dead grass blade remains in the sward); *s* **fieltro** [m] (Resultado de la acumulación en la base del césped de restos de vegetales. Este **f.** llega a alcanzar un alto grado de compactación, impidiendo la libre circulación de agua y aire a través del perfil del suelo, provocando asfixia radicular y favoreciendo la aparición de enfermedades fúngicas); *f* **feutre** [m] **de gazon** (Accumulation de matière morte du système racinaire des graminées à chaque cycle végétal ; le **f. d. g.** lorsqu'il devient trop abondant constitue un écran à la pénétration de l'air dans le sol et un obstacle à l'absorption de l'eau et des éléments nutritifs ; il favorise en outre le développement des maladies cryptogamiques) ; *syn.* bourre [f] ; *g* **Rasenfilz** [m] (Ergebnis einer starken Anreicherung von abgestorbenen Pflanzenresten in der Grasnarbe).

6338 thatching [n] *hort.* (SPON 1974, 361; process of dense accumulation of vegetation debris—clippings, etc.—in the grass sward; ▶dethatching, ▶thatch); *s* **formación** [f] **de fieltro** (Fuerte presencia de restos muertos de plantas en capa de hierba densa; ▶escarificación, ▶fieltro); *f* **feutrage** [m] **(d'un gazon)** (Processus d'accumulation de déchets de gazon en décomposition ; ▶feutre de gazon, ▶scarification, ▶verticutage) ; *syn.* bourrage [m] ; *g* **Verfilzen [n, o. Pl.]** (Starke Durchsetzung der Grasnarbe mit abgestorbenen Pflanzenresten; ▶Rasenfilz, ▶Vertikutieren); *syn.* Verfilzung [f].

thatching [n] **with brushwood** *constr.* ▶brushwood thatching.

6339 theme garden [n] *hort. gard.* (Demonstration garden on a particular subject in ►horticultural exhibitions; ►garden design theme); *s* **jardín** [m] **temático** (Jardín presentado generalmente en ►exhibición de horticultura cuyo diseño está orientado a un tema especial; ►tema de diseño de jardín); *f* **jardin** [m] **thématique** (Jardin ayant une vocation déterminée ou étant organisé autour d'un thème particulier, très souvent lors d'►expositions horticoles ; cf. DIC 1993 ; ►thème d'un jardin) ; *syn.* jardin [m] à thèmes ; *g* **Themengarten** [m] (Garten, meist in ►Gartenbauausstellungen, dessen Gestaltung durch ein spezielles Thema besonders geprägt ist; ►Gartenthema).

theme [n] **of garden design** *hort. gard.* ►garden design theme.

theme perennial [n] *hort. plant.* ►accent perennial.

6340 The Nature Conservancy [n] [US] *conserv. leg.* (**In U.S.**, an independent international membership organization committed to global preservation of natural diversity with the mission of funding, protecting and maintaining the best examples of communities, ecosystems and endangered species in the natural world. **T. N. C.** is a publicly-supported, non-profit corporation which owns and manages over 1,000 preserves and has chapters in each state; **in U.K.**, exists an independent advisory body, the **Nature Conservancy Council**, which gives advice to the government on the development and implementation of policies affecting nature conservation as well as managing nature reserves, conducting research and disseminating knowledge about nature conservation; founded in U.K. in 1973; ►conservation/recreation agency [US]/nature conservation body [UK]); *syn.* Nature Conservancy Council [n] [UK]; *s 1* **La Conservación** [f] **de la Naturaleza** (En EE.UU., organización internacional independiente comprometida con la preservación global de la biodiversidad que tiene como misión financiar, proteger y mantener los mejores ejemplos de comunidades vegetales, ecosistemas y especies en peligro en el medio natural. **L. C. N.** es una corporación sin fines de lucro que recibe financiamiento público, es propietaria y gestora de más de 1000 reservas y tiene delegaciones en todos los estados); *s 2* **Consejo** [m] **Estatal para el Patrimonio Natural y la Biodiversidad** (**En Es.** a través de la nueva Ley del Patrimonio Natural y la Biodiversidad se crea el **C. E.** como órgano de participación pública en el ámbito de la conservación y el uso sostenible de patrimonio natural y la biodiversidad, en el que deben estar representadas las organizaciones profesionales, científicas, empresariales, sindicales y ecologistas más representativas, así como las CC.AA. y una representación de las entidades locales, aunque éstas últimas con voz pero sin voto. La composición y funciones de este gremio se determinarán reglamentariamente previa consulta de las CC.AA.; cf. art. 8 Ley 42/2007); *s 3* **Consejo** [m] **para la protección de la naturaleza** (En D. y GB, cuerpo asesor de la ►administración de conservación de la naturaleza sobre cuestiones relacionadas con la protección, formado por expertos reconocidos y conservacionistas, en GB el **Nature Conservancy Council**, fundado en 1973, se dedica además a la investigación y a la propagar conocimiento sobre la protección de la naturaleza); *f* **conseil** [m] **de la protection de la nature** (±) (Organe de consultation pour la protection de la nature ; **F.**, ces organes de consultation et de concertation sont très nombreux et ont en général une vocation très sectorielle ; pour la protection de la nature on dénombre : **1.** au niveau central : les **comités interministériels de l'environnement, des parcs nationaux, de la mer**, le **conseil national de la protection de la nature**, la **commission supérieure des sites**, etc. **2.** Au niveau régional : le **comité régional de l'environnement**, les **conseils de rivage**, le **comité régional du patrimoine et des sites**, etc. **3.** Au niveau départemental : le **comité départemental de l'environnement**, la **commission départementale des** sites et de l'environnement, le **conseil départemental de la chasse et de la faune sauvage**. Les associations agréées de protection de l'environnement sont appelées à participer à l'action de certains de ces organismes et des organismes publics concernant l'environnement ; **D.**, organisme consultatif de concertation rattaché à l'►administration pour la protection de la nature constitué de personnalités bénévoles en raison de leur compétence et de leur responsabilité dans le domaine de la protection de l'environnement) ; *g* **Naturschutzbeirat** [m] (...beiräte [pl]; **1. D.**, den ►Naturschutzbehörden zugeordnetes Gremium aus ehrenamtlich tätigen sachverständigen Personen zur wissenschaftlichen und fachlichen Beratung in Fragen des Naturschutzes und der Landschaftspflege; cf. Art. 41 BayNatSchG, § 49 NatSchG-BW etc.; **2. U.S.**, *The Nature Conservancy* ist eine unabhängige internationale Mitgliederorganisation, die sich dem weltweiten Schutz der natürlichen Diversität verpflichtet und die Aufgabe hat, dafür Gelder bereitzustellen und die besten Beispiele von Lebensgemeinschaften, Ökosysteme und bedrohte Arten in ihrer natürlichen Umwelt zu schützen); *syn.* Naturschutzrat [m] (§ 45 HmbNatSchG).

6341 thermal balance [n] *ecol.* (Equilibrium induced by thermal flows/fluxes, heat accumulation and heat reduction in organisms, plant populations, larger ecosystems and in the biosphere); *syn.* thermal energy budget [n]; *s* **balance** [m] **térmico** (Equilibrio existente en los organismos y ecosistemas causado por los flujos, el almacenamiento y la producción de calor así como la regulación térmica); *f* **bilan** [m] **d'énergie** (Énergie provenant du rayonnement total [radiation solaire, radiation thermique] réfléchie, interceptée et transmise par les organismes et les surfaces terrestres ; DUV 1984, 56 et s. ; à ne pas confondre avec le bilan énergétique qui est un diagnostic de performance permettant de mettre en évidence la perte ou le gain en énergie d'un système) ; *g* **Wärmehaushalt** [m] (Gleichgewicht, das sich durch Wärmeflüsse, Wärmespeicherung, Wärmeproduktion und Regulierung in Organismen, Pflanzenbeständen, größeren Ökosystemen und in der Biosphäre einstellt); *syn.* Wärmebilanz [f].

6342 thermal belt [n] *met.* (Climatically mild area on clear and calm nights on upper slopes, close to the crest and above an accumulation of cold air below; ►frost pocket); *s* **cinturón** [m] **térmico** (≠) (En zona de inversión térmica en noches de invierno sin viento parte superior de las laderas situadas por encima de la ►hondonada de aire frío donde la temperatura es más alta que en aquélla); *f* **ceinture** [f] **chaude** (Phénomène provoqué par l'écoulement de l'air froid le long d'un versant pendant les nuits calmes et claires et qui conduit à l'établissement d'un profil thermique d'inversion [►lac d'air froid] et de la formation d'une zone altitudinale d'accumulation d'air chaud située le long du versant d'une vallée ; l'importance de ce phénomène varie avec la différence d'altitude entre le bord supérieur et le fond de la vallée, sa connaissance permettant de définir les zones dans lesquelles les risques de gelées printanières sont plus réduits ; cette zone est aussi propice à l'urbanisation en raison de son microclimat plus chaud et moins brumeux que celui du fond de vallée ainsi que pour sa bonne qualité de l'air) ; *g* **warme Hangzone** [f] (*Geländeklimatologie* in klaren und windstillen Nächten ein Bereich oberhalb von ►Kälteseen an oberen Hangpartien in der Nähe der Hangschulter. Je größer die Höhendifferenz zwischen Talsohle und oberer Talrand ist, desto deutlicher prägt sich die **w. H.** mit ihrer längeren Vegetationszeit aus und ist auch als Standort für Wohnsiedlungen auf Grund des trockeneren und nebelärmeren Klimas sowie der guten Luftqualität besser geeignet als im Talgrund).

thermal energy budget [n] *ecol.* ►thermal balance.

6343 thermal image [n] *plan. rem'sens.* (Colo[u]r-coded grey-scale value photograph, which is created by thermal radiation from objects/areas, when taken in medium infrared range/spectrum; ►infrared color aerial photograph [US]/infrared colour aerial photograph [UK]); *s* **imagen [f] térmica infrarrojo** (Imagen con valores de grises codificados que corresponden al espectro lejano de 7,0 a 15,0 micras; CHUV 1990; ►fotografía aérea infrarrojo color); *syn.* imagen [f] de infrarrojo térmico, imagen [f] de infrarrojo lejano; *f* **photographie [f] infrarouge thermique** (En télédétection, technique photographique permettant d'obtenir des photographies de différents tons codés de gris émis par réflexion du rayonnement des corps ou des surfaces correspondant au spectre compris entre 7,0 et 15,0 μ ; ►photographie aérienne infrarouge couleur) ; *g* **thermales Infrarotbild [n]** (Bild mit farbkodierten Grauwerten, das auf Grund des Rückstrahlungsverhaltens von angestrahlten Körpern/Flächen entsteht: im mittleren Infrarot aufgenommen; ►Infrarotluftbild in Farben).

thermal insulation [n] *constr.* ►heat insulation (1).

thermal load [n] *envir. limn.* ►thermal pollution of watercourses.

6344 thermal pollution [n] **of watercourses** *envir. limn.* (Artificial heating of rivers or streams, e.g. by the discharge of cooling water from nuclear power stations, which causes harmful effects on aquatic life); *syn.* thermal load [n]; *s* **contaminación [f] térmica de cursos de agua** (Calentamiento artificial de ríos, p. ej. por descarga de aguas de refrigeración de centrales nucleares, que tiene efecto negativo considerable sobre la flora y fauna acuáticas); *f* **pollution [f] thermique des cours d'eau** (Réchauffement artificiel des eaux courantes causé p. ex. par la décharge des eaux de refroidissement des centrales nucléaires) ; *syn.* charges [fpl] thermiques (des cours d'eau), pollution [f] thermique des eaux courantes ; *g* **thermische Belastung [f] von Fließgewässern** (Aufheizung von Flüssen durch Einleitung von Abwärme, z. B. von Kühlwasser aus Kernkraftwerken, wodurch das aquatische Leben erheblich beeinträchtigt oder sogar zerstört wird); *syn.* thermische Fließgewässerbelastung [f], Wärmebelastung [f] von Fließgewässern, künstliche Aufheizung [f] der Fließgewässer.

6345 thermal stratification [n] *limn.* (Sequence of various water temperatures at different depths in still water during ►summer stagnation phase and ►winter stagnation phases); *s* **estratificación [f] térmica de las masas de agua dulce** (Es el fenómeno que ocurre en algunos lagos y embalses durante ciertas épocas del año, en las que la circulación vertical está impedida por el establecimiento de distintas capas de agua con diferentes densidades debido a grandes diferencias en su temperatura; DINA 1987; ►estratificación estival, ►estratificación invernal); *f* **stratification [f] thermique** (Profils de température saisonniers dans la répartition des couches des eaux calmes pendant la ►stagnation estivale et la ►stagnation hivernale) ; *g* **Schichtung [f] (3)** (Abfolge von unterschiedlich temperierten Wasserschichten im Stillgewässer während der ►Sommerstagnation und ►Winterstagnation); *syn.* Temperaturschichtung [f], thermische Schichtung [f].

6346 thermal stress [n] **on roof membrane** *constr.* (Strain caused by intensive heating up of a roof membrane by sun exposure); *s* **tensión [f] térmica** (de una membrana impermeabilizante de tejados por exceso de insolación; cf. DACO 1988); *f* **contrainte [f] thermique** (Sujétions à une membrane d'étanchéité causée par l'exposition au rayonnement solaire et à la différence de température) ; *g* **thermische Belastung [f] einer Dachhaut** (Starke Aufheizung einer Dachhaut durch Sonneneinstrahlung).

6347 thermocline [n] *limn. ocean.* (Horizontal layer of water in lakes or seas with steep temperature gradients: during the ►summer stagnation phase a layer is formed underneath the warm surface layer; in the ►winter stagnation phase a layer is formed under the surface ice); *syn.* discontinuity layer [n]; *s* **termoclina [f]** (En los océanos o lagos grandes, capa de agua que separa otras dos de temperaturas notablemente diferentes, produciéndose a través de ella un salto brusco. Durante la ►estratificación estival está situada debajo de la capa superficial caliente, durante la ►estratificación invernal debajo de la capa de hielo o capa fría de la superficie; cf. DGA 1986); *syn. limn.* metalimnión [m]; *f* **thermocline [f]** (Zone de transition dans la mer ou dans un lac où la température diminue brusquement [1 °C au moins par mètre] apparaissant pendant la ►stagnation estivale en dessous des eaux superficielles chaudes [épilimnion] et pendant la ►stagnation hivernale sous la glace à la surface de l'eau) ; *syn. limn.* métalimnion [m] ; *g* **Sprungschicht [f]** (Horizontale Wasserschicht im See oder Meer mit besonders steilem Temperaturgradienten, d. h. die Schicht, die die unterschiedlichen Temperarturbereiche in einem geschichteten Wasserkörper voneinander trennt. Während der ►Sommerstagnation unterhalb der warmen Oberflächenschicht, während der ►Winterstagnation unter dem Eis an der Oberfläche ausgebildet; cf. EIL 1977); *syn.* Thermokline [f], *limn.* Metalimnion [n].

6348 thermophilic [adj] *biol.* (Descriptive term for organisms which require high temperatures—up to about 70 °C—for development); *s* **termófilo/a [adj]** (Término descriptivo para organismos cuyo crecimiento óptimo se da a altas temperaturas, cercanas a la coagulación de algunas proteínas [> 70°C]; *f* **thermophile [adj]** (Qualifie un organisme dont l'optimum de croissance se situe à des températures particulièrement élevées, p. ex. pour les bactéries thermophiles au voisinage de la température de coagulation [> 70 °C] de certaines protéines) ; *g* **thermophil [adj]** (Lebewesen betreffend, deren Wachstumsoptimum bei hohen Temperaturen liegt, d. h. bei extrem thermophilen Bakterien in der Nähe der Koagulationstemperatur [>70 °C] gewisser Eiweißkörper); *syn.* Wärme liebend [ppr/adj].

therophyte [n] *bot. hort. phyt. plant.* ►annual.

thick-cut sod [n] *constr. hort.* ►sod [US] (1), #2.

6349 thicket [n] *phyt. for.* (**1.** Dense growth of small trees, shrubs, bamboos, canes, etc.; SAF 1973; ►coniferous evergreen thicket, ►shrub layer, ►suffruticose thicket, ►thicket stage, ►woodlot [US&CDN] 2/wood patch [UK]; *syn.* coppice [n] (4). **2.** Interlocked individual shrubs; ►shrubland); *s 1* **espesura [f] frondosa** (Término genérico; ►bosquecillo, ►bosquete, ►estrato arbustivo, ►formación sufruticosa, ►latizal); *s 2* **monte [m] bravo** (►Espesura frondosa de origen natural); *s 3* **matorral [m] denso** (Formación vegetal de leñosas de pequeña talla que, según la tipología de IONESCO Y SAUVAGE, tiene un grado de recubrimiento de más de un 75%; DINA 1987; ►matorral achaparrado, ►matorral claro); *f 1* **fourré [m] buissonnant** (Terme générique pour une ►strate arbustive très dense ; ►boqueteau 2, ►bosquet 1, ►brousse sempervirente de résineux, ►formation sousfrutescente, ►fourré 2, ►parquet) ; *f 2* **hallier [m]** (Réunion de buissons très touffus ; ►formation buissonneuse) ; *g 1* **Dickicht [n]** (*Allgemeiner Begriff für* eine sehr dichte, undurchdringliche ►Strauchschicht; ►Dickung, ►Gebüsch 2, ►Halbstrauchgebüsch, ►immergrünes Nadelgebüsch, ►Wäldchen); *g 2* **Gebüsch [n] (3)** (Bestand von dicht ineinander gewachsenen Sträuchern; ►Gebüsch 2).

thicket [n] [UK], **creeping dwarf-shrub** *phyt.* ►creeping dwarf shrub carpet [US].

thicket [n] [UK], **matted dwarf-shrub** *phyt.* ►creeping dwarf shrub carpet [US]/creeping dwarf-shrub thicket [UK].

T

thicket [n] [UK]**, shrub** *phyt.* ►creeping dwarf shrub carpet [US].

6350 thicket stage [n] *for.* (Stage in the growth of a young plantation during which the lower branches of young trees meet and interlace; BS 3975: part 5); *s* **latizal** [m] **(2)** (Masa de brinzales o de pies coetáneos procedentes de semillas de la edad en que se establece la selección para su futuro desarrollo en diámetro; DB 1985); *syn.* rodal [m] joven frondoso; *f* **fourré** [m] **(2)** (Stade de développement d'une futaie, formation arborescente dense, difficilement pénétrable, constituée de jeunes arbres [entre 0 et 10 ans et composé de brins de moins de 2,50 m de hauteur] issus des semis après la coupe d'ensemencement à laquelle succède le ►gaulis, le ►perchis et la jeune futaie) ; *g* **Dickung** [f] (Junger Baumbestand nach Eintritt des Bestandsschlusses).

6351 thickness [n] *geo. pedol.* (Height of bed or soil horizons; SOT l976, 13); *s* **espesor** [m] (Altura de capa geológica u horizonte edáfico); *f* **épaisseur** [f] (Hauteur d'un horizon pédologique ou d'une couche géologique) ; *g* **Mächtigkeit** [f] (Höhe von Boden- oder Gesteinsschichten).

6352 thickness [n] **of a layer** *constr.* (Depth of a stratum of growing medium, e.g. on roof gardens; ►installation depth); *syn.* depth [n] of a layer; *s* **profundidad** [f] **de capa** (Espesor de una capa de sustrato, p. ej. en revestimiento vegetal de tejados; ►espesor a poner en obra); *f* **épaisseur** [f] **de couche** (Épaisseur minimum de couche[s], p. ex. de la terre végétale dont l'épaisseur dépend du type de végétation prévue mise en œuvre sur la couche filtrante sur une toiture-terrasse-jardin, requise pour garantir des propriétés physiques et mécaniques conformément à l'Avis Technique ; ►épaisseur de mise en œuvre, ►épaisseur de pose) ; *g* **Schichtstärke** [f] **(2)** (Höhe einer Substratschicht, z. B. bei Dachbegrünungen; ►Einbaustärke); *syn.* Schichtdicke [f].

thick sod [n]**, extra** *constr. hort.* ►sod [US] (1), #2.

6353 thinning [n] *arb. hort.* (Pruning to reduce the number of upper branches of an individual woody plant, to improve visual or productive functions; ►crown lifting, ►crown thinning, ►dead wooding 1); *syn.* thinning-out [n] [also US] (ARB 1983); *s* **aclareo** [m] **de la copa (2)** (Poda seca consistente en eliminar las ramas desde la base para reducir las ramas inútiles de una leñosa y aligerar la copa y así favorecer las funciones productivas y estéticas; abrir [vb] la copa; ►aclareo de la copa 1, ►escamonda, ►limpia del tronco); *syn.* poda [f] de limpieza (de la copa); *f* **éclaircissage** [m] **(2)** (Opération de suppression des branches inutiles ou de repousses en vue de favoriser la croissance des plantes ; éclaircir [vb] ; ►éclaircissage de la couronne, ►élagage 1, ►élagage en sec) ; *syn.* réduction [f] de la couronne ; *g* **Auslichten** [n, o. Pl.] (Reduzierung des Astwerkes eines Gehölzes; auslichten [vb]; ►Aufasten, ►Kronenlichtungsschnitt; *syn.* Astolzentfernung]; *syn.* Ausasten [n, o. Pl.], Astung [f], Lichtungsschnitt [m].

thinning-out [n] [US] *arb. hort.* ►thinning.

6354 thin planting [n] *constr.* (Scattered planting in an area; *opp.* ►intensive planting 1); *s* **plantación** [f] **rala** (*opp.* ►plantación densa); *syn.* plantación [f] hueca; *f* **plantation** [f] **clairsemée** (*opp.* ►plantation dense) ; *syn.* plantation [f] extensive ; *g* **lockere Anpflanzung** [f] (Ergebnis einer transparenten Pflanzung; *opp.* ►dichte Anpflanzung); *syn.* lockere Bepflanzung [f], lockere Pflanzung [f].

6355 thorn [n] *bot.* (Generic term for a modified short shoot growing from an axillary bud, and terminating in a sharp-pointed hard protective structure; a modified leaf is called ►spine 2 and is not to be confused with ►prickle [!]; ►stem thorn); *s* **espina** [f] (Órgano o parte orgánica axial [p. ej. raíz] o apendicular [hoja, estípula, etc.] endurecida y puntiaguda. La **e.** está lignificada y posee tejido vascular, no así el ►acúleo, que es de origen super-ficial; cf. DB 1985; ►espina caulinar, ►espina foliar); *f* **épine** [f] (Terme générique désignant une pousse courte dure et à pointe acérée dont la croissance s'est arrêtée peu après le printemps, formée à partir de la métamorphose d'un rameau [►rameau épine] ou des feuilles [►épine foliaire] ; à ne pas confondre avec l'►aiguillon [!]) ; *g* **Dorn** [m] (OB zu starrer, spitzer, pfriemförmiger Kurztrieb, entstanden durch Umwandlung aus Zweigen [►Sprossdorn] oder Blättern [►Blattdorn]; nicht zu verwechseln mit ►Stachel).

6356 thorny hedge [n] *landsc. plant.* (Impenetrable hedge composed of thorny and prickly shrubs; ►hedge); *s* **seto** [m] **espinoso** (►Seto impenetrable de arbustos espinosos); *f* **haie** [f] **défensive** (►Haie impénétrable plantée d'arbustes épineux) ; *g* **Dornenhecke** [f] (Undruchdringliche ►Hecke aus dornigen oder stacheligen Sträuchern).

thoroughfare [n] *urb. plan.* ►pedestrian thoroughfare.

thoroughfare [n]**, main** *trans.* ►major street.

6357 threaded rod [n] *arb. constr.* (Metal rod used, for example, in ►tree surgery for bracing hollowed-out trunks. This method, however, has been shown to have no significant effect and is no longer carried out; ►cavity bracing); *syn.* screw rod [n] [also US]; *s* **cincha** [f] **roscada** (Barra metálica utilizada en la ►cirugía arbórea para sujetar los troncos ahuecados; ►reforzamiento del tronco con barras metálicas); *f* **tige** [f] **filetée** (Tige métallique utilisée en ►chirurgie arboricole et mise en place pour le rétablissement de la cohésion des troncs creux mais n'ayant aucune influence sur la stabilité de l'arbre ; ►renforcement métallique des troncs) ; *g* **Gewindestab** [m] (...stäbe [pl]; Metallstange mit Gewinde zur Stabilisierung von baumchirurgisch ausgehöhlten Stämmen [►Baumchirurgie]. Diese Arbeiten sind jedoch statisch i. d. R. ohne Bedeutung und werden im Zuge der Weiterentwicklung der Baumpflege seit Ende der 1980er-Jahre nicht mehr durchgeführt, da dynamische Windlasten durch starre **G.stäbe** nicht aufgefangen werden können; ►Stammverstärkung durch Stabanker); *syn.* Stabanker [m].

6358 threatened species [n] *conserv. phyt. zool.* (Generic term for ►status of endangerment of faunistic and floristic species at high risk of global extinction, classified by the IUCN in different threatened categories of the ►Red List according to their extinction risk: 1. Extinct [EX], 2. Extinct in the Wild [EW], 3. Critically endangered [CR], 4. Endangered [EN], 5. Vulnerable [VU], 6. Near Threatened [NT], 7. Least Concern [LC], Data Deficient [DD], Not Evaluated [NE]; cf. IUCN Red List Categories and Criteria, Version 3.1, 2001. In U.S., the term is defined as any species which is likely to become an endangered species within the forseeable future throughout all or a significant portion of its range; in U.K., a new list of 1,149 threatened UK species was published in 2007 that will be the focus of future government conservation action under the Biodiversity Action Plan [BAP]. All species listed are declining rapidly in the UK, are internationally endangered, or are under some other extreme threat. In addition 65 habitats have been listed as being highly threatened and in desperate need of conservation action; ►critically endangered species, ►endangered species, ►extinct species, ►vulnerable species); *s* **especie** [f] **amenazada** (Término genérico para especie de flora o fauna en alto riesgo de extinción global, clasificada por la UICN en diferentes categorías de amenaza de la ►lista roja según su riesgo de extinción; cf. UICN 2001. **En Es.** especie cuya población ha sufrido una reducción significativa debida a la influencia humana directa o indirecta o que han desaparecido regionalmente, de manera que se incluye en el correspondiente registro de **ee. aa.**; en Es. existen dos categorías de **ee. aa.**: a] en peligro de extinción y b] vulnerables; art. 55 Ley 42/2007; ►categoría de amenaza de extinción

T

de especies, ►Catálogo Español de Especies Amenazadas, ►especie desaparecida, ►especie en peligro [EN], ►especie en peligro de crítico, ►especie en peligro de extinción, ►especie extincta, ►especie extincta en estado silvestre, ►especie vulnerable); *f* **espèce [f] menacée** (►*Catégorie de régression des populations et de distribution des espèces* terme générique qualifiant les taxons de flore et de faune qui risquent de s'éteindre à l'échelle mondiale et pour lesquels il est nécessaire de prendre des mesures de conservation visant leur protection dans le cadre d'un système de catégories pour les ►listes rouges ; 1. Extinct [EX] **éteint :** ►espèce éteinte, ►espèce éteinte à l'état sauvage, 2. Extinct in the Wild [EW] — **éteint à l'état sauvage :** ►espèce disparue, 3. Critically endangered [CR] – **en danger critique d'extinction :** ►espèce amenée par sa régression à un niveau critique des effectifs, 4. Endangered [EN] – **en danger :** ►espèce affectée d'une régression forte et continue et qui a déjà disparue de nombreuses régions, 5. Vulnerable [VU] – **vulnérable :** ►espèce dont la population n'a pas sensiblement diminué, mais dont les effectifs sont faibles, donc en danger latent, 6. Near Threatened [NT] – **quasi menacé,** 7. Least Concern [LC] – **préoccupation mineure,** 8. Data Deficient [DD] – données insuffisantes, 9. Not Evaluated [NE] – non évalué ; cf. Catégories et Critères de l'UICN pour le Liste Rouge, Version 3.1, 2001 ; ►liste des espèces menacées, livre rouge [des espèces menacées]); *g* **bedrohte Art [f]** (*Faunistische und floristische* ►*Gefährdungskategorie* OB für Taxa von Flora und Fauna, die weltweit in ihrem Bestand bis hin zum Aussterben gefährdet sind; die *International Union for Conservation of Nature and Natural Resources [IUCN]* empfiehlt für die ►Roten Listen die Einteilung in folgende Gefährdungskategorien: 1. Extinct [EX] **ausgestorben:** ►ausgestorbene Art, 2. Extinct in the Wild [EW] — **in freier Wildbahn ausgestorben:** ►verschollene Art, 3. Critically endangered [CR] — **vom Aussterben bedroht:** ►vom Aussterben bedrohte Art, 4. Endangered [EN] — **stark gefährdet:** ►stark gefährdete Art, 5. Vulnerable [VU] — **gefährdet:** ►gefährdete Art, 6. Near Threatened [NT] — **gering gefährdet,** 7. Least Concern [LC] — **nicht gefährdet,** 8. Data Deficient [DD] — keine ausreichenden Daten vorhanden; cf. IUCN Red List Categories and Criteria, Version 3.1, 2001.

6359 threatened with extinction [loc] *conserv. phyt. zool.* (►Red List, ►status of endangerment); *s 1* **amenazado de extinción [loc]** (►*Categoría de amenaza de extinción de especies* según la ►lista roja de la Unión Mundial para la Naturaleza UICN, un taxón está amenazado si está registrado en una de las categorías siguientes: en peligro crítico [CR], en peligro [EN] o vulnerable [VU]. Éste es el caso cuando cumple o la evidencia disponible indica que cumple cualquiera de los criterios "A" a "E" [reducción del tamaño de la población; distribución geográfica; tamaño de la población (en dos escalas por cada categoría) y probabilidad de extinción en estado silvestre, definida en porcentaje o número de generaciones] especificados para cada categoría y, por consiguiente, se considera que está enfrentando un riesgo extremadamente, muy alto resp. alto de extinción en estado silvestre; cf. UICN 2001); *s 2* **en peligro de extinción [loc]** (Taxón o subtaxón cuya supervivencia es poco probable si los factores causantes de su actual situación siguen actuando; art. 55 a] Ley 42/2007; ►Catálogo Español de Especies Amenazadas); *f* **en voie d'extinction [loc]** (►*Catégorie de régression des populations et de distribution des espèces,* espèce animale ou végétale dont les effectifs sont en très forte régression qui tend à s'aggraver et faisant face à un risque élevé d'extinction à l'état sauvage dans un avenir proche [définition IUCN]) ; *syn.* en voie de disparition [loc] ; *g* **vom Aussterben bedroht [loc]** (*Höchste faunistische und floristische* ►*Gefährdungskategorie* gemäß der ►Roten Liste gefährdeter Arten der IUCN betrifft es diejenige

Population, deren beobachtete, geschätzte, hergeleitete oder vermutete Größe sich in den letzten 10 Jahren oder in drei Generationen über 80[-90] % verringert hat).

threat [n] **of extinction** *conserv. phyt. zool.* ►endangerment of species.

threat [n] **of trees toppling over** *arb. leg.* ►danger of trees falling down.

three-layered planting [n] [UK] *constr. plant.* ►three-level planting [US].

6360 three-level planting [n] [US] *constr. plant.* (Vertical stratification of a shrub layer and two different heights of tree layers; e.g. windbreak plantings; ►planting structure, ►stratification of plant communities, ►stratified in one, two, three or several layers); *syn.* three-layered planting [n] [UK]; *s* **plantación [f] en tres estratos** (Estructura vertical de plantaciones [p. ej. de protección contra el viento] con un estrato arbustivo y dos arbóreos [1ª y 2ª clase]; ►estratificación de la vegetación, ►estructura de plantación protectora, ►formación monoestratificada, biestratificada, triestratificada, etc.); *syn.* plantación [f] en tres capas, plantación [f] triestratificada; *f* **plantation [f] à trois strates** (Disposition verticale de plantations, par exemple écran pare-vent, sous la forme d'une strate arbustive et de deux strates arborescentes ; ►formation unistrate, bistrate, tristrate, pluristrate, ►stratification de la végétation, ►structure d'une plantation) ; *g* **dreistufige Pflanzung [f]** (Vertikaler Aufbau von Gehölzpflanzungen in Form einer Strauchschicht und zweier Baumschichten [1. und 2. Wuchsklasse], z. B. bei Windschutzpflanzungen; ►Aufbau einer Pflanzung 1, ►Aufbau einer Pflanzung 2, ►Vegetationsschichtung).

threshold [n]**, adulteration** *adm. envir. leg.* ►adulteration limit.

threshold [n]**, injury** *agr. for. hort.* ►injury limit.

6361 threshold level [n] **of air pollution** *envir. leg.* (Statutory upper limit of the air pollution load, permissible within the area of influence of the source of emission; ►adulteration limit, ►emission standard, ►pollution impact); *s* **nivel [m] de referencia de calidad del aire** (Son los valores de inmisión individualizados por contaminante y período de exposición, a partir de los cuales se determinarán las situaciones ordinarias, las de zona de atmósfera contaminada y las de emergencia; cf. Orden 18/10/76, Anexo 1; ►inmisión, ►valor límite, ►valor límite de emisión); *syn.* concentraciones [fpl] de referencia, valores [mpl] de referencia; *f* **valeur [f] limite de la qualité de l'air** (Niveau maximal de concentration de substances polluantes dans l'atmosphère, fixé sur la base des connaissances scientifiques dans le but d'éviter, de prévenir ou de réduire les effets nocifs de ces substances pour la santé humaine ou pour l'environnement dans son ensemble, à atteindre dans la mesure du possible et à ne pas dépasser une fois atteint ; art. 3 de la loi n° 96/1236 du 30 décembre 1996 sur l'air et utilisation rationnelle de l'énergie ; cf. dir. n° 96/62/CE du 27 septembre 1996 ; ►immission, ►valeur limite, ►valeur limite d'émission) ; *syn.* seuil [m] de nocivité ou de gêne ; *g* **Immissionsgrenzwert [m]** (Gesetzlich festgelegter Höchstwert an Immissionsbelastung, der im Einwirkungsbereich einer emittierenden Anlage zulässig ist; ►Emissionsgrenzwert, ►Grenzwert, ►Immission).

6362 threshold value [n] *envir. conserv.* (Measurable limit above or below which a new environmental quality is created by the degree of pollution or environmental stress; e.g. noise, amount of noxious substances; *s* **valor [m] umbral** (Valor límite de tolerancia ambiental); *f* **seuil [m]** (Limite d'utilisation, de concentration dont le dépassement vers le haut ou vers le bas entraîne irrémédiablement une nouvelle qualité dans l'aggravation de la pollution de l'environnement, p. ex. pour le bruit, les substances toxiques ; *terme spécifiques* seuil de dépassement, seuil d'alerte à la population [dir. n° 92/72/CEE], seuil de pollution, seuil de toxicité, seuil thermique) ; *syn.* intensité [f] plafond, limite [f] de tolérance ; *g* **Schwellenwert [m]** (Stufe der Erheblichkeit, ausgedrückt in messbaren Werten, oberhalb oder unterhalb derer eine neue Umweltqualität durch Belastungsintensität, z. B. Lärm, Schadstoffmenge, besteht; *UBe* Schwellenwertüberschreitung, Schwellenwert für die Auslösung des Warnsystems [Art. 1 (2) RL 92/72/EWG]).

6363 throat [n] *arch. constr.* (MEA 1985; groove cut in the underside of an exterior projecting piece, such as a sill or coping, to prevent water from running down the wall face; cf. MEA 1985); *syn.* water drip [n] [also US]; *s* **goterón [m]** (Canalillo del plano inferior de una cornisa que escurre el agua de la lluvia; DACO 1988); *f* **larmier [m]** (Cannelure en quart de cercle effectuée en sous face du profil d'un couronnement de mur ou d'une corniche, pour éviter l'écoulement des eaux contre la sous-face ; DTB 1985) ; *syn.* goutte [f] d'eau (VRD 1994) ; *g* **Tropfnase [f]** (An der Unterseite von Gesimsen und anderen überstehenden Bauteilen eingearbeitete Nut, die das Herunterlaufen von Niederschlagswasser an der darunter liegenden Maueransichtsfläche verhindert); *syn.* Tropfkante [f], Tropfleiste [f], Wassernase [f] (NAL 1986).

through road [n] [UK] *plan. trans.* ▶major cross-town artery [US].

6364 through traffic [n] *trans.* (Non-local traffic having origin and destination outside the area in question; ▶destination traffic, ▶origin traffic); *syn.* thru traffic [n] [also US]; *s* **tráfico [m] de tránsito** (Flujo de vehículos cuyo origen y destino están fuera del espacio considerado; ▶tráfico de destino, ▶tráfico de origen); *syn.* tráfico [m] de paso; *f* **trafic [m] de transit** (Circulation fluide dont la provenance et la destination sont situées à l'extérieur de la zone considérée ; ▶circulation de destination, ▶circulation de provenance) ; *g* **Durchgangsverkehr [m]** (Fließender Verkehr, dessen Fahrtquellen und Fahrtziele außerhalb des Betrachtungsbietes liegen; erfolgt die Durchfahrt auf direktem Wege und ohne Aufenthalt, so nennt man dies **echten D.**; wird ein Umweg oder eine kurze Fahrtunterbrechung gemacht, so wird dies ein **gebrochener D.** genannt; ▶Quellverkehr, ▶Zielverkehr); *syn.* Transitverkehr [m].

through traffic road [n] [UK] *trans.* ▶major cross-town artery [US].

throughway [n] [US] *trans.* ▶major cross-town artery [US].

6365 throw-away-society [n] *sociol* (Society which continually discards used products without recycling and consumes non-renewable natural resources); *s* **sociedad [f] del despilfarro** (Sociedad de consumo que cada vez gasta más recursos no renovables y que no recicla los productos después de ser utilizados); *f* **société [f] de gaspillage** (Société qui consomme de manière excessive les ressources naturelles non renouvelables et ne recycle pas les produits dont ils sont issus) ; *g* **Wegwerfgesellschaft [f]** (Gesellschaft, die immer mehr nicht erneuerbare Ressourcen verbraucht und die aus ihnen gewonnenen Produkte nach Gebrauch nicht rezykliert).

thru traffic [n] [US] *trans.* ▶through traffic.

thruway [n] [US] *trans.* ▶major cross-town artery [US].

tidal channel [n] *geo.* ▶tidal creek.

6366 tidal creek [n] *geo.* (Sluggish channel running through a ▶tidal mudflat); *syn.* tidal channel [n], tidal slough [n] [US]; *s* **canal [m] de marea baja** (En la ▶llanura de fango de la costa del Mar del Norte canal sinuoso por donde corre agua en la marea baja y que si es ancho y profundo se utiliza para la navegación); *f* **chenal [m] de marée** (Canalisant les eaux du flux ou du reflux

sur l'estran du marais maritime ; DG 1984 ; ▶slikke) ; *g* **Priel [m]** (Relativ ortsstete Wasserrinne für den Gezeitenstrom im ▶Watt; große **P.e**, die als Fahrrinne dienen, werden **Balje** [ndt.] oder **Balge** [franz.-ndt.] genannt).

6367 tidal flats [npl] *geo.* (Flat shore of a sea coast between ▶driftline and low-water surface of the sea); *s* **zona [f] litoral** (En oceanografía es el espacio que se extiende entre los niveles de pleamar y bajamar. Debido a la variación de éstos, se divide en: zona supralitoral, litoral media, infralitoral y sublitoral. Algunos autores incluyen también la zona de salpicadura; DINA 1987; ▶orla de cúmulos de residuos); *f* **estran [m]** (Partie du littoral située entre les niveaux des hautes et basses mers, zone de balancement des marées sur laquelle aucune plante ne parvient à s'installer ; ▶haute de plage, ▶hourlet halonitrophile) ; *syn.* zone [f] intertidale, zone [f] de balancement des marées, zone [f] de marnage ; *g* **Tidebereich [m]** (Flacher Uferbereich einer Meeresküste zwischen ▶Spülsaum und Wasserfläche des Meeres).

tidal marsh [n] *geo.* ▶coastal marshland; *geo. phyt.* ▶salt marsh.

tidal mud [n] *ocean. pedol.* ▶sediment (2).

6368 tidal mudflat [n] *geo. ocean* (Stretch of muddy land left uncovered at low tide. In Europe, tidal marshes lying between the mainland coasts, e.g. of Denmark, North Germany and the Netherlands, and the protective line of the offshore N. and E. Friesian Islands. At low tide a maze of creeks and channels separates the sheets of mud. The term **t. m.** is synonymous with 'wadden' in the Netherlands. In tropical waters **t. m.** are commonly colonized by ▶mangrove stands; cf. DNE 1978. In U.S., tidal areas usually have **t. m.** adjacent to coastlines and tidal estuaries during low tides; ▶Wadden Sea); *syn.* coastal mudflat [n], watten [n] [also UK] (DNE 1978); *s* **llanura [f] de fango** (Extensión de limos y arcillas que puede contemplarse durante la bajamar, y que queda sumergida en la pleamar. Su origen se debe a los aportes de limo y arcilla que, llevados en suspensión por las corrientes de marea, son depositados tras la floculación que se produce donde el agua dulce se junta con el agua salada, en las bahías y estuarios. En Europa se presenta solamente en la costa del Mar del Norte de Holanda, Alemania y Dinamarca; cf. DINA 1987; ▶manglar, ▶mar de watt); *syn.* watt [m] (litoral); *f* **slikke [f]** (Terme d'origine hollandaise [= vase salée] ; partie inférieure de l'estran formée de vases plus ou moins sableuses, sans évolution pédologique, recouverte plusieurs heures à chaque marée, découvrant à marée basse, peu ou pas colonisée, la végétation halophyte commençant seulement à coloniser la haute slikke ; la **s.** est parcourue par des chenaux de marée encaissés et ramifiés caractéristiques de la dynamique du relief de cette zone d'estran ; la **s.** des tropiques est l'aire de développement de la ▶mangrove ; ▶mer des wadden) ; *syn.* estran [m] vaseux, vasière [f] molle, wadden [m] (*du hollandais*, VOG 1979), vasière [f] salée ; *g* **Watt [n]** (Watten [pl]; an flachen Gezeitenküsten, z. B. an der Nordseeküste, der zweimal täglich vom Meerwasser überspülte Bereich, der bei Ebbe ganz oder teilweise trockenfällt [Eulitoral] und auf dem im Falle von Misch- und Schlickwatt Quellergesellschaften wachsen. Das **W.** wird von vielverzweigten Prielen durchzogen, die als Zu- und Abflussrinnen für die Gezeitenströme dienen und somit dem sehr differenzierten Feinrelief eine große Dynamik verleihen. Das **W.** der Tropen ist der Lebensraum der ▶Mangrove; ▶Wattenmeer); *syn.* Küstenwatt [n].

tidal slough [n] [US] *geo.* ▶tidal creek.

6369 ties [npl] *constr. hort.* (Rope or twine for tying trees or shrubs to stakes or shoots to climber supports; materials such as twisted coconut fibre, plastic tape or belt, osier twigs are used); *s* **material [m] ligante** (Material como cuerda, cinta de plástico,

etc. utilizado para sujetar leñosas a los tutores o ramas de enredaderas a las estructuras portantes); *syn.* material [m] de sujeción; *f* **matériel [m] de ligature** (Bande de plastique souple, gaine plastique, chanvre torsadé, rameau de bambou ou de saule utilisés pour attacher ou palisser une plante sur un support fixe [pieux, treillage, etc.] ; *contexte* attache en chanvre torsadé, collier pour arbre en plastique) ; *g* **Bindegut [n, o. Pl.]** (Material wie z. B. Kokosstrick, Plastikband, Kunststoffgurt, Weidenrute zum Befestigen von neu gepflanzten Gehölzen an Pfählen, von Trieben an Rankhilfen etc.; *UB* Baumbindematerial); *syn.* Bindematerial [n].

tight board fence [n] [US] *constr.* ►close-boarded fence.

tight-butted [adj/adv] *arch. constr.* ►butt-jointed.

tight crotch [n] *arb.* ►forked growth.

tight-jointed [adj/adv] *arch. constr.* ►butt-jointed.

6370 till [n] [US] *geo. pedol.* (Unstratified glacial drift deposited directly by the ice and consisting of clay, silt, sand, gravel, and boulders intermingled in any proportion; RCG 1982); *syn.* boulder-clay [n] [UK]; *s* **till [m]** (Depósitos de origen glaciar generalmente caracterizados por la falta de estratificación y la presencia de partículas de todos los tamaños; DINA 1987); *syn.* tillita [f] (DGA 1986); *f* **argile [f] à blocaux** (*Alluvion glaciaire* formation morainique de fond des glaciations inlandsissiennes à très forte hétérométrie, alliant des argiles plus ou moins sableuses, avec des cailloutis pouvant être de très forte taille, irrégulièrement emballés dans les formations fines ; cf. DG 1984, 24, 49, 143) ; *g* **Geschiebemergel [m]** (Lockersediment aus kalkhaltigem Mischgestein [Ton, Schluff, Sand, Kies, Geröll], das vom Inlandeis oder von einem Gletscher als Grundmoräne abgelagert wurde; nach MEL 1979, Bd. 10).

tillage [n] *agr. constr. hort.* ►soil cultivation.

tillage pan [n] [US] *agr. pedol.* ►plow pan [US]/plough pan [UK].

6371 tillering [n] *bot.* (Formation of new shoots at the base of a grass or similar type plant; BS 3975, Part 5; ►tussock formation); *s* **macollamiento [m]** (Formación de vástagos nacidos de la base del mismo pie [= macolla], que ocurre particularmente en las gramíneas y plantas graminoides; cf. DB 1985; amacollar [vb]; ►formación de fascículos); *syn.* ahijamiento [m], amacollado [m], macollaje [m]; *f* **tallage [m]** (Développement de talles [tiges adventices et racines] formés au voisinage du collet d'une plante, en particulier chez les graminées ; ►formation en touffes) ; *g* **Bestockung [f] (2)** (Bildung von Seitensprossen und Wurzeln an oberirdischen, bodennahen Knoten des Hauptsprosses, besonders bei Gräsern und krautartigen Pflanzen; ►Horstbildung); *syn.* Bestaudung [f].

tilling [n] *agr. constr. hort.* ►soil cultivation.

tilth [n] *agr. hort. pedol.* ►soil mellowness, #2.

6372 timber [n] *for.* (Commercially exploitable wood other than firewood); *s* **madera [f]** (Todo tipo de madera aprovechable comercialmente. No incluye la leña); *f* **bois [m] d'œuvre et d'industrie** (Tous les bois ronds autres que les bois de feu [aptes au sciage, tranchage ou déroulage] et les bois utilisés dans l'industrie [bois à pâte, distillation, etc.]) ; *syn.* bois [m] débité brut ; *g* **Nutzholz [n]** (...hölzer [pl]; alles technisch verwertbare Holz im Gegensatz zu Brennholz).

timber [n], **growing** *for.* ►growing stock.

timber [n] [UK], **hedgerow** *agr. land'man.* *►field tree.

timber [n] [US], **overhead** *constr.* ►hanging beam.

timber connection [n] [UK] *constr.* ►lumber connection [US].

6373 timber crib wall [n] *constr.* (Retaining wall: framework of round timber logs erected by alternate laying of transverse and perpendicular rows to create an anchored retaining wall; ►concrete crib wall); *s* **muro [m] de entramado de rollizos (≠)** (Muro de contención construido con rollizos de madera entrecruzados que se rellenan de tierra; ►muro de entramado de piezas de hormigón); *f* **mur [m] de/en caissons en bois** (Mur de soutien constitué par des pieux ou travées de bois posés à angle droit en alternance les uns sur les autres et rempli de terre végétale ; ►système de construction en caissons) ; *g* **Krainerwand [f]** (Aus Rundhölzern gebaute Wand oder Schwelle, bei der abwechselnd runde [Querhölzer und im rechten Winkel dazu Rundpfähle [Zangen] übereinander gesetzt und bergseitig mit Boden eingefüllt werden; ►Raumgitterkonstruktion).

timber crop [n] *for.* ►stand of timber.

timber disk paving [n] [UK] *constr.* ►wood disk paving [US].

6374 timber extraction [n] *for.* (Removing forest products, particularly timber, from its place of growth to a delivery point. The English term 'logging' covers 'felling' and 'extraction' of timber); *s* **arrastre [m]** (Extracción de los troncos de árboles talados hasta el camino más próximo desde donde se recogen en camiones); *f* **débardage [m]** (Déplacement des bois des exploitations forestières jusqu'à un emplacement où ils sont groupés pour être transportés à un autre dépôt ou à l'usine ; cf. DFM 1975) ; *g* **Rücken [n, o. Pl.]** (In der Forstwirtschaft das Verlagern der gefällten Stämme an den nächstgelegenen Abfuhrweg); *syn.* Holzrücken [n, o. Pl.].

timber fence [n] [UK] *constr.* ►wooden fence.

6375 timberline [n] *for. geo. phyt.* (Altitude or geographical latitude beyond which dense forests no longer occur due to the severe site conditions; ►timberline ecotone, ►tree line); *s* **límite [m] del bosque** (Altitud o latitud pasada la cual debido a las condiciones mesológicas extremas no existen bosques —aunque sí árboles individuales—; *términos específicos* límite altitudinal del bosque, límite polar del bosque; ►límite de los árboles, ►zona de cumbres); *syn.* frontera [f] del bosque, límite [m] de arbolado [MEX]; *f* **limite [f] des forêts** (Limite latitudinale ou altitudinale naturelle et potentielle au-delà de laquelle il n'y a pas de peuplement arboré. *Remarques* en F. le terme **l. d. f.** est souvent appliqué au sens large, c.-à-d. aux arbres isolés [*en englais* tree line] ; ►limite supérieure de la végétation arborescente, ►zone de combat) ; *syn.* limite [f] forestière ; *g* **Waldgrenze [f]** (Geografische Breite oder klimatisch und waldgeschichtlich bedingte Höhenzone, über die hinaus auf Grund der Standortbedingungen keine geschlossenen Waldbestände mehr vorkommen. Im Französischen wird oft der Begriff **W.** für ►Baumgrenze benutzt); ►Kampfwaldzone).

6376 timberline ecotone [n] *phyt.* (Transitional zone at the ►tree line, in which closely-growing stands of trees become more and more sparse; ►timberline); *s* **zona [f] de cumbres** (En la zonación altitudinal de la vegetación, en el ámbito del ►límite del bosque, zona en la que comienzan a desaparecer los bosques cerrados; ►limite de los árboles); *f* **zone [f] de combat** (Tranche altitudinale [200 à 500 m de largeur] de transition naturelle entre la lisière supérieure de la forêt et les landes ou pelouses alpines sur laquelle les parties boisées prennent d'abord un aspect discontinu pour ne plus former, avant de disparaître, que des îlots d'arbres d'aspect chétif et tourmenté ; ►limite des forêts, ►limite supérieure de la végétation arborescente) ; *g* **Kampfwaldzone [f]** (Höhengürtel im Hochgebirge, dessen Untergrenze die örtlich anzunehmende Grenze des geschlossenen Baumbewuchses und dessen Obergrenze die natürliche ►Baumgrenze

bildet, an der sich die geschlossenen Baumbestände zunehmend auflösen; ▶Waldgrenze); *syn.* Kampfzone [f] an der Baumgrenze.

timber pile [n] *constr.* ▶palisade.

6377 timber research area [n] *ecol. for.* (Forested area for research into timber production: natural forest patch for research and growth behavio[u]r into the performance and risks of exotic trees whereby certain ecological factors or changes in the structure of a forest stand are scientifically investigated over a long period of time); *s* **reserva [f] forestal experimental** (Territorio forestal para observar y estudiar a largo plazo la evolución de las condiciones ecológicas o las transformaciones de la estructura de la población forestal principales); *f* **parcelle [f] permanente d'observation** (Aire d'observation scientifique sur laquelle sont étudiés certains facteurs écologiques ainsi que l'évolution, sur le long terme, de la structure d'un peuplement forestier, p. ex. dans une réserve biologique, aire permettant l'étude de la croissance, des performances et des risques encourus par les espèces arborescentes étrangères ; *syn.* parcelle [f] expérimentale ; *g* **Weiserfläche [f]** (Wissenschaftliche Beobachtungsfläche, auf der bestimmte ökologische Verhältnisse oder Veränderungen der Bestandsstruktur langfristig untersucht warden, z. B. Naturwaldzelle, Fläche zur Erforschung des Wuchsverhaltens, der Leistungsfähigkeit und der Risiken fremdländischer Baumarten).

6378 timber road [n] [US] *for.* (▶skidding road [US&CDN]/ skidding lane [UK], ▶forest track); *syn.* forest path [n] [UK]; *s* **camino [m] forestal (2)** (Vía estabilizada, a veces asfaltada, utilizada para fines de explotación forestal; ▶camino forestal 1, ▶camino de arrastre); *syn.* pista [f] forestal, trocha [f] [C]; *f 1* **laie [f]** (laie [f] (Chemin rectiligne tracé dans une forêt et aménagé pour transporter le bois des différentes divisions de coupes ; une **laie sommière** est une voie plus large servant de base au réseau de laies et de layons ; TSF 1985, 109 ; ▶chemin de débardage, ▶chemin forestier) ; *syn.* piste [f] forestière ; *f 2* **layon [m]** (Petite laie, sentier étroit ne dépassant pas en principe 2 mètres de largeur ; DIC 1993, 421) ; *g* **forstwirtschaftlicher Weg [m]** (Für die Bewirtschaftung des Waldes ausgebauter Weg, manchmal asphaltiert und dann **Forststraße** genannt; ▶Rückeweg, ▶Waldweg); *syn.* Forstweg [m].

timber sett paving [n] [UK] *constr.* ▶wood block paving.

timber stand [n] *for.* ▶stand of timber.

timber stand improvement [n] *constr. for.* ▶improvement cut [US], #2/improvement felling [UK].

time [n] *constr. contr.* ▶contract time; *plan. recr.* ▶organized use of leisure time; *agr.* ▶rotation time.

time [n], **free** *plan. recr.* ▶leisure.

time [n], **spare** *plan. recr.* ▶leisure.

time basis [n] *prof. contr.* ▶payment on a time basis, ▶remuneration on a time basis.

6379 time charges [npl] *contr. prof.* (SPON 1974, 44; payment for time spent by a principal and other professional staff of an architectural or engineering practice; ▶hourly billing rate, ▶overtime payment); *syn.* professional fee [n] based on an hourly rate; *s* **salario [m] horario** (Remuneración de arquitectos, ingenieros o expertos según ▶tasas de honorarios por hora; ▶pago de horas extras); *syn.* pago [m] por hora; *f* **rémunération [f] horaire** (Mode de rémunération des prestations, d'architectes, d'ingénieurs ou d'experts calculée par application d'un ▶prix horaire ; ▶rémunération des heures supplémentaires) ; *g* **Stundenvergütung [f]** (Vergütung von Architekten-, Ingenieur- oder Gutachterleistungen nach ▶Stundensätzen; ▶Überstundenvergütung).

6380 time [n] **of completion** *contr.* (Contractually agreed upon time for the execution of the work [US]/works [UK]); *s* **plazo [m] de terminación** (Tiempo previsto para la terminación de una obra); *f* **délai [m] d'exécution** (Temps nécessaire ou délai contractuel pour la réalisation d'un ouvrage) ; *g* **Fertigstellungszeit [f]** (Zeit, die für die Bauabwicklung nötig resp. vertraglich vereinbart ist); *syn.* Ausführungszeit [f].

time sheet [n] *constr. contr.* ▶daywork sheet.

6381 tinted concrete paver [n] [US] *constr. syn.* coloured paving block [n] [UK]; *s* **piedra [f] de hormigón coloreada**; *f* **pavé [m] en béton coloré dans la masse** (Pavé décoratif en béton coloré au moyen de colorants [en général pigments inorganiques] sur toute leur épaisseur ou uniquement dans la couche de revêtement sur une épaisseur d'au moins 4 mm ; l'utilisation de granulats de couleur dans la couche de revêtement permet de garantir la stabilité des teintes) ; *g* **eingefärbter Betonpflasterstein [m]** (Betonstein oder Betonplatte, meist im zweistufigen Verfahren hergestellt: die Vorsatzbetonschicht oder Mörtelschicht [mindestens 4 mm] muss mit lichtund wetterfesten Farbpigmenten gemischt sein, damit die Farbbeständigkeit im Laufe der Jahre trotz Sonnenlichteinstrahlung und Witterung gewährleistet ist; bei Verwendung farbiger Vorsatzsplitt- oder Granulatstoffe aus Naturstein ist ein Verblassen der Farbe ausgeschlossen).

tip [n] *envir.* ▶dumpsite [US]/tipping site [UK], ▶rubble pile [US]/rubble tip [UK]; *bot.* ▶root tip.

tip [n] [UK], **construction waste** *constr. eng.* ▶building waste dump].

tip [n], **decomposition** *envir.* ▶decomposition dump [US]/ decomposition tip [UK].

tip [n] [UK], **fly** *envir.* ▶unauthorized dumpsite [US].

tip [n] [UK], **high** *envir. min.* ▶high spoil pile [US].

tip [n] [UK], **large** *envir. min.* ▶large spoil area [US].

tip [n] [UK], **refuse** *envir. min.* ▶refuse dump [US].

tip [n] [UK], **rupple** *envir.* ▶rubble pile [US].

tip [n] [UK], **sand** *envir.* ▶sand pile [US].

tip [n] [UK], **spoil** *min.* ▶spoil bank [US] (2).

6382 tip [n] **of a stake** *constr. hort.* (Top end of a wooden stake); *s* **punta [f] de una estaca** (Extremo superior de una estaca de madera); *f* **tête [f] d'un pieu** (Partie supérieure d'un pieu en bois) ; *g* **Zopf [m] (1)** (Oberster Teil eines Holzpfahles).

tip [n] **of a tree** *for. hort.* ▶top of a tree.

tipping certificate [n] [UK] *constr.* ▶dumping certificate [US].

tipping layer [n] [UK] *min.* ▶dumping layer [US].

tipping [n] **of earth** [UK] *constr. eng.* ▶dumping of earth [US].

tipping [n] **of refuse** [UK] *envir. min.* ▶dumping of refuse [US].

tipping record [n] [UK] *constr.* ▶dumping record [US].

tipping site [n] [UK] *envir.* ▶dumpsite [US].

tip plateau [n] [UK] *landsc.* ▶spoil plateau [US].

6383 tip pruning [n] *hort.* (ARB 1983, 387; selective heading of shoot terminals that may or may not still be growing); *s* **despunte [m]** (Intervención de poda en verde, consistente en la eliminación de la porción apical de las ramas jóvenes, cuando tiene una longitud de 20 o 30 cm, con el fin de favorecer el brote de ramificaciones laterales, que aumentan la frondosidad. El **d.** es necesario para los setos y las matas y arbustos de flor; cf. PODA 1994); *syn.* despuntado [m]; *f* **raccourcissement [m] des

T

extrémités des branches (Couper les extrémités des branches des jeunes plants) ; *syn.* rapprochement [m] des extrémités des branches, taille [f] d'épointement, épointement [m] des branches ; *g* **Rückschnitt [m] der Astspitzen** (Einkürzen der Astspitzen bei jungen Gehölzen); *syn.* Einkürzen [n, o. Pl.] der Astspitzen.

tip reclamation [n] [UK] *land'man. landsc.* ►spoil reclamation [US].

tip site [n] [UK] *landsc. plan.* ►spoil site [US].

tissue [n] *arb.* ►callus, ►outer bark, ►phloem, ►underprovided plant tissue, ►xylem.

6384 title block [n] *plan.* (Information such as project name, location of project, name of client, name of design firm, drawing title, drawing number, scale and date, normally arranged in the lower right-hand corner of a drawing sheet); *s* **cajetín [m]** (Espacio que se deja en la parte inferior derecha de un plan; contiene el nombre de la compañia, título de la obra, del dibujo, escala, fecha y cualquier otro dato necesario, así como las firmas de los que colaboraron en él); *f* **cartouche [m]** (Partie délimitée d'un plan [ou d'une carte], généralement disposée dans l'angle droit pour mentionner le titre [nom et le lieu d'un projet], le nom du maître d'œuvre et de l'auteur du plan, le contenu et le numéro du plan, l'échelle et la date) ; *g* **Schriftfeld [n]** (Informationshinweis in der rechten unteren Ecke eines Planes über Name und Lage des Projektes, Name des Bauherrn, Name des Planungsbüros, Planinhalt, Plannummer, Maßstab und Datum); *syn.* Planstempel [m].

6385 T-jointed pipe fitting [n] *constr.* (Standardized 90° connecting fitting for pipelines); *syn.* pipe tee [n] [US]; *s* **tubo [m] bifurcado a 90°**; *f* **té [m] de raccordement à 90°** (Accessoire d'une canalisation permettant un changement de direction de 90°) ; *syn.* jonction [f] à 90°; *g* **Abzweig [m] 90°** (Teilstück einer Rohrleitung, das die Richtung um 90° ändert); *syn.* T-Stück [n], 90°-Abzweig [m].

6386 T-notching [n] *constr. for.* (Slant planting method similar to ►slit planting [US]/notch planting [UK] whereby a T-shaped or triangular opening on the ground is made into which young plants for afforestation or bioengineering methods are planted; cf. SAF 1983; ►biotechnical construction technique, ►hole planting [US]/pit planting [UK]); *s* **plantación [f] en T** (Método de plantar según el cual se crea con una azada un surco en el suelo en el que se colocan las plántulas. Se utiliza en la reforestación y en la ►construcción biotécnica. La **p. en T** es una forma de ►plantación en rajas; ►plantación en hoyos); *f* **plantation [f] à la bêche en T** (Mode de plantation de jeunes plants racinés en ouvrant une fente d'un coup de bêche en T en y introduisant le plant, et en refermant la fente sur ce dernier par pression du pied ; ►plantation en fente, ►plantation sur potets, ►technique de génie biologique) ; *g* **Winkelpflanzung [f]** (Pflanzmethode, bei der mittels einer Haue ein T-förmiger oder dreiecksförmiger Bodenspalt geschaffen wird, um in diesen eine Jungpflanze für Aufforstungen oder ►ingenieurbiologische Bauweisen zu pflanzen. Die **W.** ist eine Form der ►Klemmpflanzung; ►Lochpflanzung).

toddlers play area [n] [UK] *recr.* ►tot lot [US].

6387 toddlers wading pool [n] [US] *recr.* (Shallow pool of water in which small children can play; ►wading pool [US]/pool for non-swimmers [UK]); *syn.* paddling pool [n] [UK]; *s* **estanque [m] para juegos infantiles** (►piscina para niños/as); *f* **pataugeoire [f]** (Petit bassin plat rempli d'eau pour les jeux d'enfants ; ►bassin non-nageurs) ; *g* **Planschbecken [n]** (Flaches Becken mit stehendem Wasser für Kinderspiele; ►Nichtschwimmerbecken).

toe [n] **of an embankment/slope** *constr.* ►foot of an embankment/slope.

6388 toilet wastewater [n] [US] *envir.* (Sewage containing only human excrements and liquid bathroom waste); *syn.* foul sewage [n] [UK]; *s* **aguas [fpl] negras** (Término obsoleto); *f* **eaux [fpl] vannes** (Eaux en provenance d'installations sanitaires) ; *g* **Toilettenabwasser [n]**.

token payment [n] [US] *prof.* ►honorarium.

tolerance [n] *ecol.* ►ecological tolerance; *ecol. recr.* ►environmental stress tolerance; *constr. hort.* ►mowing tolerance; *agr. for. hort.* ►plant tolerance; *envir.* ►pollution level tolerance of a water body.

tolerance [n] [US]**, environmental** *ecol.* ►ecological tolerance.

tolerant [adj] *hort. plant.* ►air pollution-tolerant; *phyt.* ►heavy metal-tolerant, ►heavy metal-tolerant plant, ►heavy metal-tolerant vegetation; *plant.* ►salt-tolerant; *phyt. plant.* ►shade-tolerant; *hort. plant.* ►shade-tolerant grass; *gard.* ►shade-tolerant lawn; *hort. plant.* ►shade-tolerant perennial; *plant.* ►shade-tolerant planting; *hort. plant.* ►shade-tolerant woody plant; *phyt.* ►zinc-tolerant plant, ►zinc-tolerant plant community.

6389 tomb [n] *adm.* (Burial structure or chamber; ►family burial site); *syn.* burial vault [n]; *s* **panteón [m]** (Monumento funerario dedicado a la sepultura de varias personas, generalmente de una familia o un grupo social, o a alguien especialmente ilustre; ►panteón familiar); *f* **caveau [m]** (Construction abritant une sépulture souterraine, généralement en béton avec un revêtement intérieur [pierre, marbre, carrelage, granit, etc.], pouvant comporter une à quatre places, réalisée sur l'emplacement destiné à recevoir le cercueil ou une urne dans un cimetière ou une église ; *terme spécifique* ►caveau de famille) ; *g* **Gruft [f]** (Ausgemauertes Grab oder Grabgewölbe — ober- oder unterirdisch —, in das ein Sarg oder eine Urne beigesetzt wird; i. d. R. eine alte ►Familiengrabstätte).

tombstone [n] *adm. conserv'hist.* ►gravestone.

6390 tongue [n] **and groove** [n] *constr.* (Connection of boards by insertion of the tongue of one member into the corresponding groove of another, forming a T&G joint [US]; cf. DAC 1975); *syn.* dressed and matched boards [npl]; *s* **machihembrado [m]** (Ensambladura con las tablas a ranura y lengüeta; DACO 1988); *f 1* **rainure [f] et languette [f]** (Type d'assemblage utilisé en menuiserie qui permet de solidariser des de pièces de bois [en général des planches], le chant d'une des pièces comportant une rainure dans laquelle s'emboîte exactement le chant de l'autre pièce comportant une languette ; l'emboitement de deux pièces de bois à rainures et languettes se dénomme **embrèvement**) ; *f 2* **tenon [m] et mortaise [f]** (Type d'assemblage utilisé en menuiserie qui permet d'ajuster la partie saillante [tenon] à l'extrémité d'une pièce, dans l'entaille [mortaise] correspondante de l'autre pièce) ; *g* **Nut [f] und Feder [f]** (Holzverbindung bei Brettern, wobei die „Nut" eine längliche Vertiefung [Schlitz] am Brettrand [Nutbrett] und die „Feder" entweder eine angefräste Brettleiste ist [Federbrett], die in den Schlitz des Nutbrettes eingreift oder eine besondere Leiste, die beide Nuten der anstoßenden Bretter formstabil verbindet).

6391 tool [n] *hort. constr.* (Gerneric term for an implement used to carry out mechanical functions by hand or machine; ►garden tool); *s* **herramienta [f]** (►herramienta de jardinería/ ►equipamiento de jardinería); *f* **outil [m]** (►outil de jardinage) ; *g* **Gerät [n] (2)** (►Gartengerät); *syn.* Werkzeug [n].

6392 tooled concrete [n] *constr.* (Mechanically finished concrete, e.g. wire brushed, bush-hammered, striated, jackhammered); *s* **hormigón [m] trabajado con herramientas (±)** (Hormigón cuya superficie es modelada con herramientas como bujarda, escorda o martillo de puntas); *f* **béton [m] à parement éclaté** (Aspect de surface d'un béton travaillé à l'état durci au moyen d'outils de taille tels que la boucharde [béton bouchardé], la gradine ou le poinçon afin de faire apparaître l'ensemble des constituants avec la cassure des gros granulats et lui donner une structure plus naturelle ; *termes spécifiques* béton pointé, béton fendu) ; *syn.* béton [m] à surface éclatée ; *g* **Sichtbeton [m] mit Schlagbearbeitung** (Beton, dessen sichtbare Oberfläche nach dem Ausschalen z. B. mit Stockhammer, Kröneleisen oder Spitzmeißel bearbeitet wird; *UBe* gestockter Beton, Kratzbeton, Spitzbeton); *syn.* Beton [m] mit Schlagbearbeitung.

tooled finish [n] *constr.* *►tooling of stone [US]/dressing of stone [UK].

tooled natural stone [n] *arch. constr.* ►cut stone, #2.

tooled stones [npl] [US] *constr.* to *►work stones.

tooling [n] *constr.* *►surface treatment.

6393 tooling [n] **of stone** [US] *constr.* (Working on stone with hammer and chisel cutting tool; the result is a **tooled finish** on the face of a stone with corrugations; ►workability of construction material: 1. rusticated [pp/adj] [UK]; *syn.* embossed [pp/adj], rough-tooled [pp/adj]; 2. split-faced [pp/adj]; *syn.* rock-faced [pp/adj], quarry-faced [pp/adj]; 3. sawn [pp/adj]; *syn.* sawed [pp/adj], 4. pointed [pp/adj]; 5. bush-hammered [pp/adj]; 6. hammered [pp/adj]; 7. polished [pp/adj]; 8. broached [pp/adj]; 9. undressed [pp/adj], untooled [pp/adj]; 10. dummy-jointed [pp/adj]; 11. flamed [pp/adj]; 12. roughly-hewn [pp/adj]); *syn.* dressing [n] of stone [UK]; *s* **labra [f] de piedra natural o artificial** (Trabajado de la superficie de la piedra con cincel y martillo; ►manejabilidad de un material de construcción; 1. de labra tosca [loc], 2. partidas [adj], *syn.* con superficie partida [loc]; 3. aserrada [adj], 4. apuntada [adj]; 5. escodada [adj]; 6. tallada con martillo [loc]; 7. pulida [adj]; 8. cincelada [adj]; 9. sin labrar [vb]; 10. con ranuras [loc]; 11. flameada [adj], *syn.* pulida al fuego [loc], 12. desbastada [adj]); *syn.* labrado [m] de piedra natural o artificial; *f* **mode [m] de taille des pierres** (travailler la pierre ; ►maniabilité des matériaux : 1. bosselé, ée [pp/adj] ; *syn.* en bossage [loc], bossagé, ée [pp/adj] ; 2. brut, brute de taille [loc] ; 3. scié, ée [pp/adj] ; 4. pointé, ée [pp/adj] ; 5. bouchardé, ée [pp/adj] ; 6. têtue [adj] ; 7. poli, ie [pp/adj] ; 8. layé, ée [pp/adj] ; 9. lité, ée [pp/adj] ; 10. rainuré, ée [pp/adj] ; 11. flammé, ée [pp/adj] ; 12. éclaté, ée [pp/adj]) ; *g* **Steinbearbeitung [f]** (Grobe bis feine Oberflächenbearbeitung von Natur- oder Kunststeinen; ►Bearbeitbarkeit der Baumaterialien: 1. bossiert [pp/adj]; 2. bruchrau [adj]; *syn.* spaltrau [adj]; 3. gesägt [pp/adj]; 4. gespitzt [pp/adj]; 5. gestockt [pp/adj]; 6. hammerrecht [adj]; 7. poliert [pp/adj]; 8. scharriert [pp/adj]; 9. unbearbeitet [pp/adj]; 10. mit Scheinfuge; 11. beflammt [pp/adj]; *syn.* flammgestrahlt [pp/adj]; 12. grob bearbeitet [pp/adj]); *syn.* grob zugerichtet [pp/adj].

6394 toolshed [n] *hort.* (Shelter for tools and equipment); *s* **caseta [f] de herramientas**; *f* **resserre [f] à outils** (Construction légère en général en bois dans laquelle sont entreposés les outils de jardinage) ; *syn.* abri [m] à outils, cabane [f] à outils ; *g* **Gerätehaus [n]** (I. d. R. kleines Holzhaus, in dem Gartengeräte, Blumentöpfe und Gartenmöbel aufbewahrt werden); *syn.* Geräteschuppen [m].

top aspect [n] *plan.* ►top view.

6395 top bog layer [n] [US] *pedol.* (Upper layer of raised bog overlaying peat deposits, which often may be used as cultural soil after peat extraction); *syn.* top spit [n] [UK]; *s* **turba [f] bruta** (Capa superior de turbera alta, entrelazada con raíces, que se utiliza como tierra vegetal para cultivo después de la explotación de la turbera); *f* **tourbe [f] brute** (Couche superficielle de tourbe peu décomposée et parcourue d'un lacis plus ou moins

serré de rhizomes et de racines ; l'enlèvement de la couche tourbeuse superficielle s'appelle « étrépage ») ; *g* **Bunkerde [f]** (**1.** Obere Torfschicht der Hochmoore, die nach der Torfgewinnung meist als Kulturboden verwendet wird. **2.** Material der oberen Torfschicht von Hochmooren); *syn.* Bunkmaterial [n, o. Pl.], Bunkschicht [f].

6396 top consumer [n] *ecol.* (Animal species in the food chain which has no natural enemies; ►trophic level); *s* **consumidor [m] final** (Especie animal que no tiene enemigos naturales en la red alimentaria; ►nivel trófico); *f* **consommateur [m] final** (Espèce animale située en bout de chaîne alimentaire et qui ne possède plus d'ennemis naturels ; ►niveau trophique) ; *g* **Endkonsument [m]** (In der Konsumentenfolge diejenige Tierart, die von keinem Feind erlegt und gefressen wird; ►Trophiestufe).

6397 topdressing [n] [US] **(1)/top-dressing** [n] [UK] *constr. hort.* (Thin surface application of ►fertilizer or ►soil amendment [US]/soil ameliorant [UK]; *specific terms* dressing with sand, dressing with compost; ►mulching); *s* **cubierta [f] de mantillo** (Aplicación al suelo de compost u otros materiales fertilizantes descompuestos; ►fertilizante, ►material de enmienda de suelos, ►mulching; *syn.* arrope [m] vegetal (paja, hojarasca, aserrín) [C]; *f 1* **épandage [m] en couverture** (*Terme générique* apport de sable, de terreaux, de compost, etc. avec incorporation d'►engrais et ►amendements ; ►mulching, ►terreautage 1) ; *f 2* **terreautage [m] (2)** (*Terme spécifique* recouvrement du sol d'une mince couche de terreau ou de tourbe après le semis) ; *g* **Bodenabdeckung [f]** (Bedeckung des Bodens mit einer dünnen Schicht Sand, Torf, Komposterde etc. Das englische *top-dressing* bedeutet auch das bloße Ausstreuen von Düngergaben ohne Einarbeitung in den Boden; ►Bodenverbesserungsstoff 1, ►Dünger, ►Mulchen); *syn.* Abdeckung [f] des Bodens.

6398 topdressing [n] [US] **(2)/top-dressing** [n] [UK] *agr. hort.* (Spreading of manure or fertilizer over a growing crop, meadow, or lawn); *s* **abonado [m] de/en cobertura** (HDS 1987); *syn.* abono [m] de cobertura (HDS 1987), fertilización [f] de cobertura; *f* **fumure [f] de couverture** (LA 1981, 555 ; AD 1992 ; amélioration de la fertilité du sol par l'épandage de matières fertilisantes sur une culture déjà installée ou par l'apport d'engrais sur une prairie ou pelouse) ; *g* **Kopfdüngung [f]** (Düngung bereits wachsender Pflanzenbestände mit leicht aufnehmbaren Düngesalzen).

6399 topdressing [n] **for turfing** [US]/**top-dressing** [n] **for turfing** [UK] *constr.* (On completion of turf/sod laying, the whole area is top-dressed to a depth of 10mm and joints brushed in well with finely sieved soil); *s* **relleno [m] de juntas de tepes** (con tierra); *f* **terreautage [m] des joints** (Remplissage des joints d'un gazon en plaques au moyen de terre végétale ou d'un mélange terreux approprié) ; *syn.* garnissage [m] des joints ; *g* **Verfüllen [n, o. Pl.] der Rasensodenfugen** (Füllen der Fugen nach dem Verlegen von Rasensoden mit gesiebtem Oberboden oder einem Erdsubstratgemisch).

top drying [n] [UK] *arb. for.* ►top-kill [US].

top drying [n] [UK]**, initial phase of** *arb. for.* ►initial phase of top-kill [US].

top dying [n] *arb. for.* ►top-kill [US].

6400 topiary [n] *hort.* (Practice of training and shearing woody plants into various formal shapes, geometric or mimetic; e.g. trees trained into boxheads; the result is a **topiary specimen**; *syn.* topiary work [n]); *s* **poda [f] de topiaria** (Práctica de dar una forma artificial a árboles y arbustos, p. ej. en jardines barrocos); *f* **art [m] topiaire** (Lat. *ars* « art », grec *topos* « lieu » ; taille ornementale d'un arbre ou d'un arbuste, p. ex. en forme de boule, de cône, d'animal, etc. exécutée par un jardinier décorateur

T

utilisant la malléabilité d'espèces persistantes ; l'opus topiarium est le travail qui dans l'antiquité comprenait la culture, l'entretien des arbres et des arbrisseaux, la formation des bosquets, des tonnelles et principalement la taille ornementale des arbres ; DIC 1993) ; *syn.* topiaria [f] (Art du jardinier décorateur) ; *g* **Formschnitt [m]** (Gehölzschnitt, der einem Baum, Strauch oder einer Gehölzgruppe eine künstliche Form gibt, z. B. Kugel, Kegel, Tiergestalt, geschnittene Hecke, Dach- oder Kastenform; eine besondere Form ist die in manchen Landschaften vorkommende, in Spalierform geschnittene Linde, die auch als Kastenlinde, Spalierlinde oder Lindenhochhecke bezeichnet wird. Das Ergebnis des **F.es** ist ein *Formgehölz*; *obs.* UB Baumverschnitt [m] [GOT 1926-II, 449 f]); *syn.* Ars [f] topiaria (Lat. *ars* >Kunst<, griech. *topos* >Ort<); Gehölzformschnitt [m], Opus topiarum [m] (GDG 1965, 25), *obs.* Verschnitt [m] von Hecken und Bäumen (GOT 1926-II, 449 f).

topiary specimen [n] *hort.* *▶topiary.

topiary work [n] *hort.* ▶topiary.

6401 top-kill [n] [US] *arb. for.* (Death or ▶dieback of the upper part of a crown which creates a stag head. *NOTE* with hardwoods, the appearance in the earlier stages of **t. k.** is termed ▶stag-headed; in **U.S.**, the dead tip of conifers is called **spike-top** [n]; ▶dead crown, ▶initial phase of top-kill [US]/initial phase of top drying [UK], ▶open crown, ▶opening up of tree crowns, ▶stag-headed); *syn.* top drying [n] [UK], top dying [n], dead top [n] [AUS] (SAF 1983); *s* **puntiseco** [m] (DB 1985; muerte progresiva de la parte superior de la ▶copa de un árbol; ▶aclarado de la copa, ▶copa hueca, ▶copa muerta, ▶muerte regresiva de plantas, ▶puntiseco, ▶puntiseco incipiente); *f* **descente [f] de cime** (Éclaircissement général de la couronne par perte de feuilles suivi par un désordre progressif de la cime, du haut vers le bas et de l'extérieur vers l'intérieur [diminution de croissance radiale et apicale], d'où peut résulter la mort d'une partie ou de la totalité de l'arbre ; DFM 1975 ; ▶asphyxie, ▶couronne d'arbre, ▶éclaircie de cime, ▶houppe au port aéré, ▶houppe morte, ▶phase initiale du dépérissement terminal, ▶sec, sèche en cime) ; *syn.* mort [f] en cime, dessèchement [m] de la cime, dépérissement [m] de la cime, dépérissement [m] terminal ; *g* **Wipfeldürre [f]** (Absterbende oder abgestorbene obere Teile einer ▶Baumkrone; ▶abgestorbene Baumkrone, ▶Absterben, ▶beginnende Wipfeldürre, ▶Verlichtung 2, ▶wipfeldürr); *syn.* Zopfdürre [f], Zopftrockenheit [f], Zopftrocknis [f] (MEY 1982).

6402 top level [n] *arch. constr.* (Upper surface of a structure; e.g. top of wall, top of stair, pipe; ▶finished floor level; *opp.* ▶bottom level); *syn.* upper edge [n]; *s* **borde [m] superior** (p. ej. de un muro, una escalera; ▶nivel final del suelo, *opp.* ▶borde inferior); *f* **arase [f]** (*Contexte* niveau arase mur, niveau arase terrassement, ▶niveau plancher ; *opp.* ▶bord inférieur) ; *g* **Oberkante [f]** (*Abk.* OK; oberster Teil eines Bauwerks, einer Oberfläche, z. B. OK Mauer, OK Gelände, OK Rohr, ▶Oberkante Fußboden; *opp.* ▶Unterkante).

6403 top level [n] **of wall** *arch. constr.* (*abbr.* T.W.; ▶top level); *syn.* TW-elevation [n] [also US]; *s* **borde [m] superior de un muro** (▶borde superior); *f* **arase [f] de mur** (1. Face supérieure d'un mur correctement mise de niveau ; *contexte* niveau fini arase de mur ; 2. dans un mur en pierres, dernière ligne de pierres de faible épaisseur servant à mettre de niveau la face supérieure du mur ; *terme générique* ▶arase) ; *g* **Maueroberkante [f]** (*Abk.* MOK; OB ▶Oberkante); *syn.* Oberkante [f] Mauer, *Abk.* OKM.

6404 topoclimate [n] *met.* (Characteristic climate of an area based on land configuration, soil and vegetative cover as well as by ▶slope aspect; ▶microclimate); *syn.* topographic climate [n];

s **mesoclima [m]** (Componentes en que se diferencia un macroclima cuando aparecen modificaciones locales en algunas de sus características; DINA 1987; ▶exposición de una ladera, ▶microclima); *f 1* **climat [m] local** (Ensemble des phénomènes météorologiques locaux faisant référence à une échelle climatique très réduite située entre le ▶macroclimat et le ▶microclimat, caractérisant l'état moyen de l'atmosphère et son évolution en un point donné, pour plusieurs centaines de mètres jusqu'à quelques dizaines d'hectares, par des informations météorologiques dépendant de la topographie locale, de la nature du sol ou du couvert végétal) ; *f 2* **mésoclimat [m]** (Terme très souvent utilisé en sylviculture et en viticulture. Le mésoclimat délimite les éléments topographiques Pour une simple colline, par exemple, ses différentes parties [sommet, pente, pied et ▶exposition d'un versant] seront toutes décrites en fonction de leur propre mésoclimat. En viticulture le mésoclimat peut souvent s'appliquer à un vignoble particulier, voire à une parcelle de vignoble. Il permet en particulier de décrire les influences climatiques sur le caractère de vins issus de parcelles ou de l'ensemble d'un vignoble) ; *syn.* mésoclimat [m] ; *g* **Geländeklima [n]** (...ta [pl]; *fachsprachlich* ...mate [pl]; Klima an einem Geländepunkt, das von den Reliefgegebenheiten, der Bodenart und von der Pflanzendecke bestimmt wird; ▶Hangexposition, ▶Mikroklima); *syn.* Topoklima [n], Mesoklima [n].

top [n] **of an embankment/slope** *constr.* ▶rounding the top of an embankment/slope.

6405 top [n] **of a tree** *for. hort.* (Uppermost part of a trunk and crown of a tree); *syn.* tip [n] of a tree; *s* **punta [f] de un árbol** (Extremo superior del tronco de un árbol); *f* **houppier [m] (2)** (Partie du tronc non comprise dans le fût et portant les ramifications d'un arbre) ; *g* **Zopf [m] (2)** (Oberster Stammteil einer Baumkrone).

6406 top [n] **of crown** *arb.* (Upper part of a tree crown); *s* **punta [f] de un árbol**; *f* **cime [f]** (La partie la plus élevée d'un arbre) ; *g* **Wipfel [m]** (Oberer Teil einer Baumkrone); *syn.* Kronenspitze [f].

top [n] **of the curb** [US] *constr.* ▶curb level [US]/kerb level [UK].

topographic climate [n] *met.* ▶topoclimate.

topographic layout [n] *arch. constr.* ▶configuration of an area.

6407 topographic mapping [n] *rem'sens. surv.* (Geodetic survey method, especially used for mountain topography and engineering ground survey; ▶aerial photogrammetry, ▶photogrammetry); *s* **cartografía [f] topográfica** (Método geodésico utilizado sobre todo para mediciones de altas montañas o en ingeniería para mediciones de la superficie del subsuelo; ▶fotogrametría, ▶fotogrametría aérea); *syn.* fotogrametría [f] terrestre; *f* **phototopographie [f]** (En géodésie, technique d'élaboration de plans pratiquée par recouvrement stéréoscopique de photos et effectuée en particulier pour les levés topographiques en haute montagne ; ▶photogrammétrie, ▶photogramétrie aérienne) ; *g* **Erdbildmessung [f]** (Im geodätischen Bereich eingesetzte Methode der ▶Photogrammetrie, besonders zur Vermessung der Topographie im Hochgebirge und bei Ingenieurvermessungen, wobei die Messbilder von erdfesten Standpunkten aus aufgenommen werden; ▶Luftbildmessung); *syn.* terrestrische Photogrammetrie [f].

6408 topographic survey [n] **(1)** *constr. landsc.* (Mapping of a construction site by surveyor's transit [US]/tilting level [UK] and stadia rods, or other means of delineating existing conditions and elevations of a site area before planning and construction commences; ▶site inspection 2) *syn.* site survey [n] (2); *s* **levantamiento [m] de planos** (BU 1959; medición e inventario de un terreno antes de comenzar su edificación; ▶reconocimiento del terreno); *syn.* topometría [f]; *f* **reconnaissance [f] du site** (Visite du site d'un chantier afin d'établir un état des lieux avant le début des travaux ; ▶reconnaissance du terrain, ▶levé) ; *syn.* reconnaissance [f] d'un chantier ; *g* **Geländeaufnahme [f] (1)** (Bestandsaufnahme einer Fläche vor Planungs- und Baubeginn bezüglich der Höhen, baulichen Elemente, Wege, Vegetation etc.; ▶Geländebegehung).

6409 topographic survey [n] (2) *landsc.* (Mapping of landscape elements on an extensive area); *s* **levantamiento [m] topográfico** (Registro de los elementos del paisaje de un área extensa); *f* **relevé [m] topographique** (Prise en compte des éléments du paysage d'un grand espace naturel) ; *g* **Geländeaufnahme [f] (2)** (Aufnahme eines Gebietes oder großen Raumes hinsichtlich seiner Landschaftselemente).

6410 topographic survey [n] (3) *surv.* (Exact measurement of metes and bounds, contours, dimensions and positions of plots, structures, roads, paths and landscape conditions on a tract of land for their delineation on a topographic map; ▶site survey 1, ▶geodesy); *syn.* survey work [n], field survey [n] (2); *s* **medición [f] topográfica** (Medición exacta de superficie y altitud de terrenos, dimensión de parcelas, estructuras y áreas de vegetación para elaborar mapas topográficos; esto se realiza hoy en día frecuentemente con métodos de teledetección; ▶geodesia, ▶medición del terreno); *syn.* levantamiento [m] topográfico, agrimensura [f]; *f 1* **levé [m]** (Travaux consistant à relever sur le terrain l'implantation planimétrique et altimétrique des éléments au sol tels que limites de propriété, bâtiments, circulations, etc. ; ▶arpentage, ▶métré) ; *syn.* lever [m], relevé [m] topographique ; *f 2* **arpentage [m]** (Ensemble des procédés utilisés en planimétrie par le géomètre pour représenter graphiquement des données topographiques ; ▶géodésie, ▶levé) ; *g* **Vermessung [f]** (Das genaue Festlegen in Größe und Lage vor Ort von Grundstücken, Gebäuden, baulichen Anlagen, Wegetrassen, Höhenabwicklungen etc. sowie die grafische Darstellung; **V.** wird heute i. d. R. satellitengestützt mit GPS-Verfahren durchgeführt [GPS = Global Positioning System]; ▶Aufmaß 2, ▶Geodäsie); *syn.* topografische Aufnahme [f], Feldvermessung [f], Landvermessung [f], Vermessungsarbeit [f].

topography [n] *geo.* ▶local topography.

topomap [n] [US] *geo.* ▶contour map (1).

topping [n] [UK] (1) *constr.* ▶first mowing [US].

topping [n] (2) *arb. hort.* ▶pollarding.

6411 topping [n] (3) *arb.* (Removing all or most of the crown of a tree; ▶lopping, ▶pollarding); *syn.* heading [n] [also US], dehorning [n] [also US], de-horning [n] [also UK], crowning off [n] [AUS], heading off [n] [NZ]; *s* **descopado [m]** (Poda de toda la copa de un árbol sin tener en cuenta el porte o las necesidades fisiológicas; en algunos casos se justifica como medida de recuperación de árboles viejos o descuidados; descabezar [vb]; ▶desmoche, ▶trasmocho); *syn.* descopadura [f]; *f* **éhoupage [m]** (Opération controversée consistant à étêter l'ensemble du houppier d'un arbre ; ▶étêtage, ▶rapprochement, ▶ravalement) ; *syn.* éhouppement [m], écimage [m] (±), étêtement [m] (±) ; *g* **Kronenkappung [f]** (Nicht fachgerechter Baumschnitt, bei dem die gesamte Krone ohne Rücksicht auf Habitus oder physiologische und biomechanische Erfordernisse im Stark- und Grobastbereich abgesetzt wird; das Ergebnis ist eine „gekappte Krone"; **K.en** bedeuten — die kulturhistorischen Kopfweiden an Gewässern und Weinbergen mal ausgenommen — immer einen Totalschaden, so dass die Bäume auf Grund der Schnittverletzungen im internodialen Bereich nicht zu überwallende Rindenverletzungen und Kambiumschäden bekommen und in Folge davon sich oft nicht sichtbare zentrale Morschungen

entwickeln. Durch diese Verstümmelungen wird die Standzeit erheblich verkürzt, ferner muss der Eigentümer erhebliche Folgekosten tragen, die aus einer ständigen, alle drei- bis fünfjährigen Nachbehandlung resultiert; ▶auf den Kopf setzen, ▶Stummelschnitt); *syn.* umfangreiches, baumzerstörendes Absetzen [n, o. Pl.] der Krone [cf. ZTV Baumpflege 2001], Absetzen [n, o. Pl.] der Krone, Abwerfen [n, o. Pl.] der Krone, Kappen [n, o. Pl.] der Krone, Kappschnitt [m], Kappung [f] der Krone.

topping-up [n] **of sand areas** [UK] *constr.* ▶replenishment of sand areas.

6412 top-size bulb [n] *hort.* (ANSI 1986; large bulb that flowers in the following season; ▶bulbil); *s* **bulbo [m] apto para florecer** (Producto horticultural desarrollado por siembra o por multiplicación a partir de ▶bulbillo); *f* **bulbe [f et m] apte à fleurir** (Specimen horticole cultivé par semis ou obtenus à partir de ▶bulbilles) ; *g* **blühfähige Blumenzwiebel [f]** (Gärtnerisches Erzeugnis, das durch Aussaat oder Vermehrung aus ▶Brutzwiebeln entstanden ist).

6413 topsoil [n] (1) *agr. constr. (pedol.)* (Mostly humus-rich, uppermost layer of soil [A horizon] created by physical, chemical and biological processes; in agriculture the permanently-plowed/ ploughed topsoil is 15-35cm deep, and grassland soil has a root-penetration depth of 7-10cm; ▶cultivated soil, ▶subsoil); *syn. pedol.* A horizon [n]; *s* **tierra [f] vegetal** (Capa superior del suelo, generalmente rica en humus y bien aireada, resultado de procesos físico-químicos y biológicos [meteorización y mineralización, descomposición y humificación, formación de grumos y diferentes procesos de transporte de sustancias]; en la agricultura la **t. v.** es la capa de 15-35 cm que se ara, en praticultura la de 7-10 cm en la que enraizan las plantas herbáceas; ▶subsuelo, ▶suelo de cultivo); *syn.* horizonte [m] A, suelo [m] vegetal [RA], capa [f] vegetal [CO, EC]; *f 1* **sol [m] superficiel** *agr.* (Couche supérieure de terre définie par sa nature, ses propriétés physiques et chimiques et sa granulométrie et travaillée par les instruments aratoires ; à la différence du sol profond, couche meuble exploitée par les racines ; ▶sol cultivé) ; *syn.* horizon [m] ; *f 2* **horizon [m] de surface** *pedol.* (Horizon contenant de la matière organique, souvent appauvri en éléments fins ou en fer par lessivage ; PED 1983) ; *syn.* horizon A [m] ; *f 3* **terre [f] végétale** *constr.* (À la différence du ▶sous-sol **1.** couche arable supérieure de sol préexistante sur un chantier récupérée par décapage superficiel ou retroussement sur les emprises du chantier et mise en stock pour réutilisation sur le chantier. **2.** Terre d'apport dont la composition devrait tendre vers les normes argile 20 à 25 %, calcaire 5 à 10 %, sable 55 à 60 %, humus et limon 10 %, pH neutre voisin de 7) ; *g* **Oberboden [m] (1.** Meist humusreiche, gut durchlüftete, durch physikalische und biologische Vorgänge [Verwitterung und Mineralisierung, Zersetzung und Humifizierung, Gefügebildung und verschiedene Stoffumlagerungen] entstandene oberste Schicht der Erdkruste [A-Horizont]; in der Landwirtschaft ist der **O.** der ständig 15-35 cm tief bearbeitete Ackerboden und bei Grünlandböden der 7-10 cm stark durchwurzelte Horizont. **2.** *leg.* In dt. Gesetzen wird nicht vom **O.**, sondern vom Synonym ‚Mutterboden' [§ 202 BauGB] oder vom OB ‚Boden' in seiner vielfältigen Funktion gesprochen [cf. § 2 (1 u. 2) BBodSchG]; ▶Kulturboden, ▶Unterboden); *syn.* Mutterboden [m], *pedol.* A-Horizont [m], *agr.* Ackerkrume [f], Kulturerde [f] [CH].

topsoil [n] (2) *constr.* ▶replacement of topsoil, ▶respreading of topsoil, ▶seeding of stockpiled topsoil, ▶storage of topsoil, ▶volume of stripped topsoil.

topsoil [n]**, covering with** *constr.* ▶topsoil spreading.

topsoil [n]**, regulating of** *constr.* ▶fine grading.

topsoil [n] [UK]**, stacking of** *constr.* ▶storage of topsoil.

topsoil [n] [US]**, stockpiling of** *constr.* ▶storage of topsoil.

6414 topsoil conservation [n] *leg. nat'res.* (Measures for the protection of topsoil, e.g. required stripping and stockpiling during construction and grading work; in D., this is prescribed by legal enactment; ▶storage of topsoil, ▶topsoil 1, ▶topsoil stripping); **s protección [f] legal de la tierra vegetal** (▶acopio de tierra vegetal, ▶remoción de la capa de tierra vegetal, ▶tierra vegetal); *syn.* protección [f] del suelo vegetal, protección [f] de la capa vegetal superior, protección [f] del capote [CO], protección [f] del migajón [MEX], protección [f] de la tierra húmica [YV]; *f* **protection [f] de la terre végétale** (Mesures légales prises en vue de la protection et la conservation des sols concernés par des travaux de génie civil afin de conserver leurs caractéristiques physiques et chimiques, de les protéger contre les pollutions et de développer une gestion prévoyante et parcimonieuse de cette ressource ; il n'existe pas en France de législation spécifique à la protection des sols ; ▶retroussement de terre végétale, ▶stockage de la terre végétale, ▶terre végétale) ; *syn.* conservation [f] de la terre arable, protection [f] du sol (sur les chantiers [CH]) ; *g* **Mutterbodenschutz [m, o. Pl.]** (Durch Gesetze geregelter Schutz des ▶Oberbodens bei der Errichtung und Änderung baulicher Anlagen sowie bei wesentlichen anderen Veränderungen der Erdoberfläche, um ihn in nutzbarem Zustand zu erhalten und vor Vernichtung oder Vergeudung zu schützen; in D. gelten z. B. § 202 BauGB und das Bundes-Bodenschutzgesetz — BBodSchG; ▶Oberbodenabtrag 1, ▶Oberbodenlagerung); *syn.* Oberbodenschutz [m], Schutz [m] des Mutterbodens.

topsoil heap [n] [UK] *constr.* ▶topsoil pile [US].

topsoil heaps [npl] [UK]**, seeding of** *constr.* ▶seeding of stockpiled topsoil.

topsoiling [n] *constr.* ▶topsoil spreading.

6415 topsoiling [n] **and cultivation** [n] *constr.* (Preliminary 'soft landscape' operations for the preparation of planting and seeding areas, including soil stripping, storage, placement, soil amelioration, ground preparation, fertilizing, etc; ▶earthworks, ▶enhancement of soil fertility, ▶rototilling [US]/rotovation [UK], ▶soil preparation); **s modificación [f] de suelos** (Trabajos para preparación de plantaciones como excavación de tierra, relleno de tierra, acopio de tierra vegetal, descompactación del suelo, preparación del suelo, abonado e introducción de enmiendas; ▶mejora de la fertilidad del suelo, ▶terracerías, ▶tratamiento de superficies, ▶rotovación); *f* **travaux [mpl] de jardinage** (Travaux d'espaces verts comprenant les fosses des plantations la reprise, la mise en place et réglage des terres végétales, la formation grosso-modo, fertilisants et produits phytosanitaires, façons culturales, consolidation de talus, travaux de plantation et d'engazonnement ; cf. C.C.T.G. fascicule n° 35 ; ▶fraisage, ▶mise en valeur du sol par la culture, ▶terrassements, ▶travaux de préparation du sol) ; *g* **Bodenarbeiten [fpl]** (Arbeiten für vegetationstechnische Zwecke wie Bodenabtrag, Bodenlagerung, Bodeneinbau, Bodenlockerung, ▶Bodenvorbereitung, ▶Bodenpflege, Düngung, Bodenverbesserung etc. gem. DIN 18 915 u. DIN 18 320; ▶Erdbau, ▶Fräsen).

topsoil mix(ture) [n] *constr. hort.* ▶growing medium.

6416 topsoil pile [n] [US] *constr.* (Well-ordered stack of stripped topsoil for later reuse, which should be protected against saturation, the proliferation of weeds and other contaminants; ▶storage of topsoil; *generic term* ▶soil pile; *syn.* topsoil heap [n] [UK], topsoil store [n]; **s depósito [m] de tierra vegetal** (Montón ordenado de tierra vegetal protegido contra encharcamiento, crecimiento de malas hierbas u otro tipo de contaminación, para garantizar la posibilidad de reutilización; ▶acopio de tierra vegetal; *término genérico* ▶depósito de tierra); *f* **dépôt**

[m] de terre végétale (Résultat du ▶stockage de la terre végétale soigneusement effectué [sans compactage] en tenant compte d'un bon équilibre des talus et des protections contre le développement des mauvaises herbes, les eaux de ruissellement ; *terme générique* ▶dépôt de terre) ; *syn.* stock [m] de terre végétale ; *g* **Oberbodenmiete [f]** (Geordnetes Lager zur Wiederverwendung des abgetragenen Oberbodens, der gegen Vernässung, Verunkrautung und sonstige Verunreinigung zu schützen ist. Entgegen alten Festlegungen über Mietenabmessungen, gibt es nach der VOB keine Beschränkungen mehr in Breite und Höhe; ▶Oberbodenlagerung; *OB* ▶Bodenmiete); *syn.* Mutterbodenmiete [f], Oberbodenlager [n].

topsoil piles [npl] [US]**, seeding of** *constr.* ▶seeding of stockpiled topsoil.

6417 topsoil replacement [n] *constr.* (Substitution of good topsoil for poor or contaminated soil; ▶respreading of topsoil); **s sustitución [f] de la tierra vegetal** (Reemplazo de la tierra donde el suelo es pobre o está contaminado; ▶reextensión de tierra vegetal); *syn.* sustitución [f] de tierra vegetal; *f* **remplacement [m] de la terre végétale** (Substitution de terres de mauvaise qualité ou de sols contaminés par une terre saine, exempte de matières toxiques, de racines, de mottes d'argile, de cailloux de plus de 25 mm de diamètre, de mauvaises herbes, de branchages ni d'autres détritus ; ▶remise en place de la terre végétale) ; *g* **Oberbodenaustausch [m]** (Ersatz von schlechtem, total mit Wurzelunkräutern durchsetztem oder verseuchtem Boden; ▶Wiederandecken von Oberboden); *syn.* Austausch [m] von Oberboden.

topsoil respreading [n] *constr.* ▶respreading of topsoil.

6418 topsoil slurry spreading [n] *eng.* (Wet process of spreading soil with organic matter; ▶hydraulic filling); **s relleno [m] hidráulico de tierra vegetal** (Método técnico de relleno de superficies con material que contiene sustancia orgánica; ▶relleno hidráulico); *f* **remblaiement [m] hydraulique de terre végétale** (Opération de refoulement des produits de dragage avec introduction de matières végétales [terre végétale] ou de matériaux d'emprunts mis en place sous forme liquide par des tuyaux sous pression en vue de la création de terre-pleins dans le cadre de travaux de génie civil ; *syn.* mise [f] en place de remblai hydraulique; ▶remblayage hydraulique) ; *g* **Auflandung [f] (3)** (Technische Bodenablagerungsmethode, bei der Boden, der organische Stoffe enthält [Oberboden], mit Druckwasserleitungen in tiefer liegende oder eigens dafür vorbereitete Flächen gespült wird; ▶Auflandung 2).

6419 topsoil spreading [n] *constr.* (Placing of topsoil to designed lines and contours and, on dumpsites [US]/tipping sites [UK], ▶capping with soil material; ▶placement of soil); *syn.* topsoiling [n], covering [n] with topsoil; **s extensión [f] de tierra vegetal** (Extensión de tierra hasta la cota final en depósito de residuos o escombrera; ▶cobertura con tierra vegetal, ▶relleno de tierra); *syn.* esparcir [vb] tierra vegetal, extender [vb] tierra vegetal; *f* **mise [f] en place de la terre végétale** (▶couverture de terre végétale, ▶remblaiement de/en terre) ; *syn.* épandage [m] de terre végétale, mise [f] en remblai de terre végétale ; *g* **Andecken [n, o. Pl.] von Oberboden** (Profilgerechte Oberbodenandeckung; *bei Halden auch* ▶Übererdung; ▶Einbau von Boden); *syn.* Andecken [n, o. Pl.] von Mutterboden, Einbau [m] von Kulturerde [CH], Einbau [m] von Oberboden, Mutterbodenauftrag [m], Oberbodenandeckung [f], Oberbodenauftrag [m], Oberbodeneinbau [m].

topsoil store [n] [UK] *constr.* ▶topsoil pile [US].

6420 topsoil stripping [n] *constr.* (Removing of topsoil layer); **s remoción [f] de la capa de tierra vegetal** (Proceso de retirar la capa de tierra superior del suelo para saneamiento o acopio); *syn.* extracción [f] de la capa de tierra vegetal; *f* **retroussement [m] de terre végétale** (*Processus.* Récupération de la terre végétale avant les travaux de terrassement dans les emprises du chantier susceptible d'être réutilisée ultérieurement et mise en stock) ; *syn.* décapage [m] de la terre végétale ; *g* **Oberbodenabtrag [m] (1)** (*Vorgang* das Entfernen der Oberbodenschicht); *syn.* Abschieben [n, o. Pl.] des Oberbodens, Abschieben [n, o. Pl.] des Mutterbodens, Abtrag [m] von Mutterboden, Abtrag [m] von Oberboden, Mutterbodenabtrag [m]).

top spit [n] [UK] *pedol.* ▶top bog layer [US].

top step [n] *arch. constr.* ▶step in a flight of steps.

6421 top view [n] *plan.* (Graphic representation of an object, structure or site from above; ▶plan view); *syn.* top aspect [n]; *s* **vista [f] de planta** (Dibujo o plano que representa una estructura o un proyecto paisajístico visto desde arriba; ▶planta); *f* **vue [f] en plan** (Représentation graphique d'un ▶plan de masse ; ▶dessin en plan) ; *g* **Draufsicht [f]** (Zeichnerische Darstellung eines Gegenstandes, Bauwerkes oder eines Geländes von oben; ▶Grundriss); *syn.* Aufsicht [f].

6422 torrent [n] *geo.* (Small, cold, rushing watercourse in a mountainous area, confined parts of which may suddenly contain very high levels of water, and which drain away or transport sediments with great fluctuation, dependent upon the roughness of the stream bed and the sudden thrust or particle size of the bed load; ▶fast-moving stream); *syn.* white water [n], wild water [n] [also US]; *s 1* **torrente [m]** (Corriente natural de agua cuyas crecidas son súbitas e irregulares, mientras que el caudal de estiaje es muy reducido o incluso nulo; un **t.** pequeño o una rama fluente de otro principal se llama «**torrentera**» [f]; *término específico* **rambla [f]**: término local que se da a los arroyos de las áreas mediterráneas. Generalmente se trata de cursos esporádicos de gran caudal, que discurren por cauces de tipo artesa frecuentemente desbordados; cf. DINA 1987 y DGA 1986); *s 2* **aguas [fpl] vivas** (Arroyo en alta montaña de flujo rápido y caudal variable dependiendo de las lluvias o del deshielo; ▶arroyo de flujo rápido); *syn.* aguas [fpl] bravas, arroyo [m] de alta montaña; *f 1* **torrent [m]** (Cours d'eau irrégulier et excessif des régions montagneuses avec une charge en boue et cailloux importante ; ▶eau vive) ; *syn.* ruisseau [m] torrentiel (AEP 1976, 253) ; *f 2* **gave [m]** (Cours d'eau montagnard dans les Pyrénées occidentales et par extension tout cours d'eau montagnard non torrentiel, c.-à-d. à lit rocheux ou encombré de blocs, à cours rapide mais à débit permanent ; cf. DG 1984) ; *g* **Wildbach [m]** (Kleiner, nicht regulierter, zeitweilig oder streckenweise steil abwärts führender, kalter, schnell fließender Gebirgswasserlauf mit heftigen und plötzlichen hohen Wasserständen, dessen Wasserablauf resp. Transport fester Bestandteile durch die Rauigkeit des Flussbettes, durch plötzliche Geschiebeeinstöße und durch die Korngröße des Geschiebes große Schwankungen aufweist; ▶schnell fließendes Gewässer); *syn.* Gebirgsbach [m], Wildwasser [n].

6423 torrent control [n] *constr.* (Stabilization of banks or construction of ▶checkdams in order to regulate accumulation of bed load and to restrain water flowing with great velocity or turbulence, as after a heavy rainfall or down a mountain ravine or watercourse); *s* **corrección [f] de torrentes** (Estabilización de las orillas o construcción de ▶diques transversales en el lecho de los arroyos para regular el flujo de agua y para controlar la sedimentación de la carga de fondo); *syn.* corrección [f] de torrenteras; *f* **stabilisation [f] d'un torrent** (Confortement des berges ou construction d'▶ouvrages de confortement transversaux dans le lit de torrents en vue de la régulation du débit et des accumulations grossières) ; *syn.* confortement [m] d'un torrent ; *g* **Wildbachverbauung [f]** (Befestigung der Ufer oder Bau von ▶Quer-

werken im Wildbachbett zur Regelung des Wasserabflusses und zur kontrollierten Ablagerung von Geschieben); *syn.* Wildbachverbau [m].

torrential downpour [n] [US] *met.* ▶heavy rainfall.

torrential pour [n] [US] *met.* ▶heavy rainfall.

6424 torrenticolous [adj] *biol.* (Descriptive term for organisms which only live in very turbulent water; e.g. in white water or ocean surf; *opp.* ▶stagnicolous); *s* **torrentícola [adj]** (Término descriptivo para organismos que habitan en aguas muy turbulentas; *opp.* ▶estagnícola); *f* **torrenticole [adj]** (Organismes habitant dans les torrents ou par extension vivant dans les eaux très mouvementées comme dans la zone de ressac de la mer ; ▶vivant, ante en eaux stagnantes) ; *g* **torrentikol [adj]** (Organismen betreffend, die nur in heftig bewegtem Wasser, z. B. im Wildbach, in der Meeresbrandung leben/vorkommen; *opp.* ▶stagnikol); *syn.* wildbachbewohnend [adj/ppr].

torrent zone [n] *limn.* ▶trout zone.

6425 total construction cost estimate [n] [US] *contr.* (Accurate calculation of the costs expected to be incurred by the client, ascertained before contract award negotiations and based upon prices contained in previous bids [US]/tenders [UK]; ▶estimate of probable construction cost, ▶preliminary cost estimate); *syn.* priced bills [npl] of quantities [UK] (WAL 1976, 145), final construction cost estimate [n] [UK]; *s* **presupuesto [m] de gastos** (Cálculo exacto de los costes [Es]/costos [AL] de construcción que va a contraer el propietario de obra, determinados antes de las negociaciones del contrato y sobre la base de los precios unitarios y globales de ofertas anteriores; ▶estimación aproximada de costos, ▶estimación del costo); *syn.* cómputo [m] presupuesto [AL]; *f 1* **détermination [f] du coût prévisionnel des travaux** (Partie financière de l'analyse des offres dans la mission d'assistance apportée au maître de l'ouvrage pour la passation d'un contrat de travaux ; détermination exacte du coût total des travaux établi sur la base des prix unitaires et forfaitaires figurant dans les offres lors de la consultation et remis au maître d'ouvrage avant les négociations avec le candidat retenu et faisant partie ; arrêté du 21 déc. 1993, annexe III ; ▶estimation détaillée des dépenses, ▶estimation sommaire du coût des travaux, ▶évaluation détaillée du coût des travaux) ; *syn.* établissement [m] du coût prévisionnel des travaux ; *f 2* **devis [m]** (Offre de prestations de travaux établie par un prestataire sur la base d'une description technique détaillée des travaux comprenant nature, dimensions et qualités des ouvrages et des matériaux [devis descriptif] ainsi qu'un état détaillé des prix évalués pour l'exécution des ouvrages [devis estimatif]) ; *g* **Kostenanschlag [m]** (...schläge [pl]; genaue Ermittlung der tatsächlich zu erwartenden Kosten nach DIN 276 oder nach dem wohnungsrechtlichen Berechnungsrecht resp. aus Einheits- oder Pauschalpreisen der Angebote für den Bauherrn vor den Vergabeverhandlungen; §§ 10 [2] und 15 [2] Nr. 7 HOAI 2002 resp. §§ 33 und 38 HOAI 2009 mit Anlage 11; ▶Kostenberechnung, ▶Kostenschätzung).

6426 total estimate [n] *phyt.* (Combined estimation of plant ▶species abundance and ▶degree of species cover, ignoring density, according to a seven-degree scale: **r** = very sparsely present; **+** = sparsely present, cover very small; **1** = plentiful but of small cover value; **2** = very numerous, or covering at least 1/20 of the area; **3** = any number of individuals covering ¼ to ½ of the area; **4** = any number of individuals covering ½ to ¾ of the area; **5** = any number of individuals covering more than ¾ of the area); *syn.* relative coverage [n]; *s* **estima [f] (global) del grado de cobertura** (Estimación combinada de la ▶abundancia [de una especie] y del grado de cobertura de las plantas [▶dominancia], sin considerar la densidad, según una clasificación de siete escalas: **r** =presencia esporádica; **+** = planta escasa con un valor de cobertura muy pequeño; **1** = abundante pero con un valor de cobertura bajo, o bien bastante escaso pero con un valor de cobertura mayor; **2** = muy abundante con cobertura escasa o cubriendo entre 1/10 y ¼ parte de la superficie investigada; **3** = cubriendo entre ¼ y ½ de la superficie investigada, número de individuos cualquiera; **4** = cubriendo entre ½ y ¾ de la superficie investigada, número de individuos cualquiera; **5** = cubriendo más de ¾ de la superficie investigada, número de individuos cualquiera; BB 1979, 37-38); *f* **estimation [f] globale** (Estimation globale de la répartition de chaque espèce sur une aire donnée, utilisant le coefficient d'abondance-dominance de BRAUN-BLANQUET, en tenant compte de l'►abondance [nombre d'individus] et de la ►dominance [recouvrement] : **r** = individus très rares ; **+** = individus disséminés/rares, recouvrement très faible ; **1** = individus abondants mais recouvrement faible ; **2** = individus très abondants ou recouvrant au moins 25 % de la surface étudiée ; **3** = nombre d'individus quelconque recouvrant de 25 à 50 % de la surface étudiée ; **4** = nombre d'individus quelconque recouvrant de 50 à 75 % de la surface étudiée ; **5** = nombre d'individus quelconque recouvrant plus de 75 % de la surface étudiée) ; *g* **Gesamtschätzung [f]** (Kombinierte Schätzung von ►Abundanz und ►Deckungsgrad einer Art unter Verzicht auf die Dichtigkeitsbestimmung mit Hilfe einer siebenteiligen Skala: **r** = nur ganz vereinzelt vorkommend; **+** = spärlich mit sehr geringem Deckungswert; **1** = reichlich, aber mit geringem Deckungswert oder ziemlich spärlich, aber mit großem Deckungswert; **2** = sehr zahlreich oder mindestens 1/10 bis ¼ der Aufnahmefläche deckend; **3** = ¼ bis ½ der Aufnahmefläche deckend, Individuenzahl beliebig; **4** = ½ bis ¾ der Aufnahmefläche deckend, Individuenzahl beliebig; **5** = mehr als ¾ der Aufnahmefläche deckend, Individuenzahl beliebig); *syn.* Artmächtigkeit [f].

total fixed costs [npl] *contr. prof.* ►overall fixed costs.

totally cleared agrarian landscape [n] *land'man.* ►extensively cleared land (for cultivation).

6427 totally covering [ppr] *phyt.* (Descriptive term applied to vegetation, which covers completely a defined area); *s* **cubriendo [ger] totalmente** (Término descriptivo aplicado a la vegetación que cubre completamente un área definida); *f* **sur l'ensemble de/du [loc]** (Terme appliqué à une surface, un territoire, un site, une aire précis, p. ex. en totalité recouvert[e] par des végétaux) ; *g* **flächendeckend [ppr/adj] (2)** (Eine bestimmte Fläche betreffend, die vollständig bedeckt ist, z. B. mit Pflanzen, Bodenstreu).

6428 total [n] **of separate spaces** *plan.* (Sum of the distributed parts of a defined area, e.g. measured in hectares or acres); *s* **balance [m] de superficie** (Visión general de la distribución de usos en un área definida); *f* **bilan [m] spatial** (Résultat et représentation de la répartition des espaces sur un territoire donné) ; *g* **Flächenbilanz [f]** (Ergebnis und Übersicht der Flächenverteilung in einem Gebiet).

6429 total parking spaces [n] *plan. trans.* (Number of vehicles that may be parked in a designated area); *s* **espacio [m] de aparcamiento** (Suma de plazas de estacionamiento existentes en una zona dada); *f* **aires [fpl] de stationnement** (Somme des emplacements de stationnement disponibles dans une zone donnée) ; *syn.* espaces [mpl] de stationnement ; *g* **Parkraum [m]** (Summe der Parkmöglichkeiten in einem Betrachtungsgebiet).

total porosity [n] *pedol.* ►pore volume.

6430 total range [n] *zool.* (Zoogeographical term for the area covered by animal movements of the same species during different seasons—spring, summer, autumn, and winter habitats, molting, breeding, spawning grounds, migration routes, etc.; ►home range 2, ►territory); *s* **territorio [m] vital total** (Tér-

mino de la geografía animal que designa el ámbito de actividad de una especie faunística que varía según las estaciones del año [hábitats de primavera, verano, otoño e invierno, de muda, de cría o freza, rutas de migración, etc.]; ►área de habitación, ►territorio); *syn.* dominio [m] vital); *f* **domaine [m] vital** (Terme de géographie animale désignant la surface sur lequel un animal vit pendant toute l'année et constitué en général de territoires différents suivant les activités et les modes de dispersion des espèces considérées : territoire d'accouplement, aire de reproduction, de nidification, de ponte, routes de migration, etc.] ; ►aire d'habitation, ►territoire) ; *g* **Jahreslebensraum [m]** (Tiergeografische Bezeichnung für den Aktivitätsbereich einer Art während eines ganzen Jahres, der meist im Jahresgang unterschiedliche Landschaftsteile erfasst [Frühlings-, Sommer-, Herbst- und Winterstreifgebiet, Mausergebiet, Brut- und Laichplätze, Wanderrouten etc.]; ►Territorium, ►Wohngebiet).

total redevelopment [n] [US] *plan. urb.* ►comprehensive redevelopment.

6431 total rise [n] **of a flight of steps** *constr.* (Height of a stairway; ►stair rise[r]); *s* **altura [f] de tramo (de escalera)** (►contrahuella); *f* **hauteur [f] d'un escalier** (Hauteur totale d'un escalier ; ►dénivelé[e] d'une marche, ►hauteur de la contremarche) ; *g* **Treppenhöhe [f]** (Gesamthöhe eines Treppenlaufes; ►Steigungshöhe).

6432 total run [n] **of a stairway** [US] *arch. constr.* (DAC 1975, 417; horizontal dimension of a flight of steps); *syn.* length [n] of a flight of steps; *s* **longitud [f] de tramo (de escalera)**; *f* **longueur [f] de l'escalier** (Distance horizontale projetée au sol séparant la contremarche inférieure de la contremarche supérieure d'un escalier, d'une volée d'escalier c.-à-d. son encombrement ou son emprise) ; *syn.* développé [m] d'un escalier, développement [m] d'un escalier, reculement [m] d'un escalier ; *g* **Treppenlauflänge [f]** (In der Horizontalen gemessene Baulänge einer Treppe); *syn.* Steigungslänge [f].

6433 tot lot [n] [US] *recr.* (Equipped play area for small children, usually up to elementary school age; cf. IBDD 1981); *syn.* play lot [n] [US], toddlers play area [n] [UK]; *s* **área [f] de juegos para niños y niñas pequeñas** (En D., equipamiento colectivo obligatorio en la construcción de bloques de viviendas; zona de juego generalmente equipada con cajón de arena, aparatos sencillos y bancos situada en el recinto de complejos de viviendas o en patios interiores); *f 1* **aire [f] de jeux pour petits enfants** (Aménagement de jeux pour petits enfants pouvant comprendre, bac à sable, pataugeoire, petites installations pour grimper, glisser, se balancer, abris et petite hutte, bancs pour adultes, etc.) ; *syn.* enclos [m] de tout petits ; *f 2* **enclos [m] bac à sable** (Petite espace de jeu clos, prévue pour les enfants de moins de trois ans, composé d'un bac à sable accompagné de mobilier pour adultes) ; *g* **Spielplatz [m] für Mutter und Kind** (In D. ist in der LBO eines jeweiligen Bundeslandes näher geregelt, bei welchen neu zu errichtenden baulichen Gegebenheiten — z. B. in BW bei Gebäuden mit mehr als zwei Wohnungen — auf dem Grundstück ein Kinderspielplatz anzulegen ist, sofern nicht in unmittelbarer Nähe eine Gemeinschaftsanlage geschaffen wird oder vorhanden ist; cf. § 9 LBO-BW); *syn.* Kleinkinderspielplatz [m].

6434 tourism [n] **(1)** *recr. trans.* (**1.** Organization and operation of tours, especially as a commercial enterprise, to accommodate stays and journeys of persons away from home for purposes of leisure, health and study; business trips are not included. *Specific terms* vacation tourism [US]/holiday tourism [UK], pleasure trip tourism, educational tourism. **2.** Economic activities, guidance and management associated with and dependent upon tourists as a business or a governmental function.

3. Generic term for any economic and social phenomenon related to the travel industry); *syn.* tourism enterprise [n], tourism industry [n], tourist business [n], tourist trade [n]; *s* **turismo [m]** (Movimiento de población muy ligado a la satisfacción del ocio. Por ello, se relaciona con apetencias humanas muy variadas, tales como el descanso, el deporte, el termalismo o el deseo de ampliar conocimientos culturales. No es un fenómeno nuevo, aunque sí de auge reciente. Los cambios socioeconómicos y los de los medios de transporte fueron los factores decisivos para el desarrollo masivo del t. *Términos específicos* t. de vacaciones, t. de salud, t. educativo, t. internacional, t. científico, t. cultural); *f* **tourisme [m]** (Ensemble des activités liées au déplacement et au séjour des personnes — à l'exception des voyages d'affaires — à des fins de repos, thérapeutiques, de loisirs, d'études etc. ; *termes spécifiques* tourisme culturel, tourisme de loisirs, tourisme de santé, climatisme, tourisme scientifique, tourisme religieux) ; *g* **Fremdenverkehr [m, o. Pl.]** (Die in einer modernen Industriegesellschaft ermöglichten Reisen und Aufenthalte Ortsfremder zu Erholungs-, Heil- und Studienzwecken auf internationaler, nationaler, regionaler und örtlicher Ebene, gefördert durch Ausdehnung von Freizeit, Lebensalter, Bildung und Konsum sowie durch organisierte Reiseangebote in Gebiete mit entsprechenden Freizeit- und Erholungsinfrastruktur. Geschäftsreisen werden nicht mitgerechnet. *UBe* Bildungstourismus, Erholungstourismus, Heiltourismus, Radtourismus, Städtetourismus, Vergnügungstourismus, Wissenschaftstourismus); *syn.* Tourismus [m].

tourism [n] **(2)** *recr.* ▶cycling tourism, ▶eco-tourism; *recr. trans.* ▶mass tourism, ▶overseas tourism; *econ. plan. recr.* ▶promotion of national tourism.

tourism [n] **[US], agricultural** *recr.* ▶agritourism [US]/agro-tourism [UK].

tourism [n] **[UK], agro-** *recr.* ▶agritourism [US].

tourism [n]**, bike** *recr.* ▶cycling tourism.

tourism [n]**, promotion of foreign** *econ. plan. recr.* ▶promotion of national tourism, #2.

tourism [n]**, sustainable** *recr.* ▶eco-tourism.

tourism [n] **[UK], waste** *envir.* ▶transboundary waste shipment [UK].

tourism enterprise [n] *recr. trans.* ▶tourism (1).

tourism industry [n] *recr. trans.* ▶tourism (1).

6435 tourist [n] *recr.* (**1.** Person who visits other places or countries for pleasure or culture. **2.** Person who stays overnight at an inn or motel); *s* **turista [m/f]**; *f* **touriste [m/f]** ; *g* **Tourist/-in [m/f]** (Jemand, der fremde Orte oder Länder bereist); *syn.* Urlaubsreisende/-r [f/m].

6436 tourist area [n] *plan. recr.* (Area heavily frequented by vacationists [US]/holiday-makers [UK]; ▶recreation resort community); *syn.* tourist region [n] [US], holiday area [n] [UK]; *s* **zona [f] turística** (Lugar muy frecuentado por personas con fines recreativos; ▶lugar receptor de recreacionistas); *syn.* zona [f] de veraneo, área [f] turística; *f* **zone [f] de tourisme** (Espace à forte fréquentation touristique ; ▶collectivité réceptrice) ; *syn.* site [m] touristique, zone [f] touristique, espace [m] vacancier ; *g* **Ferienerholungsgebiet [n]** (Von Erholungsuchenden stark frequentiertes Gebiet; ▶Erholungszielgebiet 2); *syn.* Fremdenverkehrsgebiet [n].

tourist business [n] *recr. trans.* ▶tourism (1).

tourist centre [n] **[UK]** *recr.* ▶tourist resort [US].

6437 tourist flow [n] *recr. plan.* (Number of recreation travellers); *syn.* flood [n] of tourists; *s* **flujo [m] turístico** (Cantidad de personas que viajan por actividades de turismo en una determinada área); *f* **flux [m] touristique** (Mouvement important

de touristes vers une région touristique) ; *g* **Touristenstrom [m]** (Starkes Aufkommen von Urlaubern, die in ein bestimmtes Gebiet reisen).

6438 tourist resort [n] **[US]** *recr.* (Place heavily frequented by holidaying/vacationing visitors); *syn.* tourist centre [n] [UK]; *s* **ciudad [f] turística** (Lugar muy frecuentado por turistas); *syn.* lugar [m] turístico; *f 1* **station [f] touristique** (Ensemble fonctionnel à vocation prédominante [balnéaire, nautique, sports d'hiver, tourisme d'été, thermale, climatique, etc.] et caractérisé par la combinaison variable d'équipements d'accueil et d'hébergement touristiques, de loisirs et de culture ainsi que de services généraux ; DG 1984, 426) ; *f 2* **unité [f] touristique nouvelle (U.T.N.)** (*Terme spécifique pour* une opération de développement touristique importante [urbanisation, équipement, aménagement touristique] en zone de montagne ; cf. directive nationale d'aménagement du 22 novembre 1977 et loi montagne du 9 janvier 1985) ; *f 3* **base [f] littorale de loisirs et de nature (B.L.L.N.)** (*Terme spécifique pour* des équipements touristiques de loisirs et de culture dans l'espace littoral dans le respect des exigences écologiques et des paysages ; cf. circ. du 25 mars 1974) ; *g* **Fremdenverkehrsort [m]** (Von Erholungsuchenden stark frequentierter Ort).

tourists [npl]**, flood of** *recr. plan.* ▶tourist flow.

tourist trade [n] *recr. trans.* ▶tourism (1).

6439 tourist traffic [n] *plan. recr. trans.* (Movement of people and conveyances from or to a holiday resort; ▶tourism 1); *syn.* holiday traffic [n], vacation traffic [n] [also US]; *s* **tráfico [m] turístico** (Tráfico de o a un centro turístico; ▶turismo); *f* **trafic [m] des vacances** (Volume de la circulation en direction ou en provenance du lieu de séjour en vacances ; ▶tourisme) ; *syn.* trafic [m] des vacanciers, migration [f] vacancière ; *g* **Urlaubsreiseverkehr [m]** (Verkehr vom oder zum Urlaubsort; ▶Fremdenverkehr).

6440 tournament tennis court [n] *recr.* (Court designed for championship matches); *s* **pista [f] de tenis de campeonato** (*Términos históricos* palenque [m], liza [f]); *syn.* cancha [f] de tenis de campeonato [AL]; *f* **court [m] (de tournoi) de tennis** (Aire de jeu de dimension très précise pour la pratique du sport de tennis entourée d'une tribune constituée de gradins fixes ou mobiles) ; *syn.* terrain [m] de tennis de compétition ; *g* **Tennisturnierplatz [m]** (Für den Tennissport mit Zuschauertribünen ausgestatteter Platz).

tower [n] *arch. urb.* ▶office tower.

tower [n] **[US], apartment** *arch. urb.* ▶high-rise residential building.

tower block [n] **[UK]** *arch. urb.* ▶high-rise residential building.

6441 tower building [n] *arch. urb.* (Narrow, high-rise edifice, in which all living apartments or offices are served by a central bank of elevators; ▶office tower); *s* **torre [f] de viviendas** (Edificio de viviendas muy alto y aislado; ▶torre de oficinas); *f* **tour [f] à noyau central** (Immeuble de grande hauteur [I.G.H.] qui concentre en un noyau central l'ensemble des circulations verticales (escaliers, cages d'ascenseurs, etc.) autour desquelles se développent les planchers courants ; ▶tour à usage de bureaux) ; *syn.* immeuble [m] de grande hauteur à noyau central ; *g* **Punkthaus [n]** (Einzeln stehendes Hochhaus, bei dem alle Wohnungen oder Büros von einem zentral gelegenen Treppenhaus und Aufzug erreichbar sind; ▶Bürohochhaus).

towing path [n] **[US]/towing-path** [n] **[UK]** *recr. trans.* ▶towpath [US].

tow-lift [n] *recr.* *▶ski lift.

town [n] *urb.* ▶old part of a town, ▶shanty town, ▶surrounding region of a city/town.

town [n] [US]**, down-** *urb.* ▶inner city.

town [n]**, fringe of a city/** *urb.* ▶urban fringe.

town [n]**, old** *urb.* ▶historic core.

town [n]**, outskirts of a** *urb.* ▶urban fringe.

town [n] [UK]**, overspill** *urb.* ▶satellite city.

town [n]**, satellite** *urb.* ▶satellite city.

town [n]**, spatial configuration of a** *landsc. urb.* ▶structuring of townscape.

town [n]**, visual picture of a** *urb.* ▶townscape.

town and country planning [n] [UK] *adm. leg. urb.* ▶community development planning [US]/urban land-use planning [UK].

6442 Town and Country Planning Act [n] [UK] *leg. urb.* (**In U.K.**, Government Act existing since 1909 and continually revised, gives comprehensive powers to central and local authorities for planning control and land development; **in U.S.**, no equivalent terms: There are state planning enabling acts authorizing local communities to prepare their own planning ordinances and zoning codes. **In D.**, the **Federal Building Code**, a federal law regulates the use of land in the ▶community development planning [US]/urban land-use planning [UK]; ▶zoning and building regulation); *s* **Ley [f] sobre Régimen del Suelo y Valoraciones [Es]** (Ley 6/1998, de 13 de abril; el Tribunal Constitucional sentenció el 20 de marzo de 1997 la inconstitucionalidad de la última reforma legislativa de ordenación urbana, recogida en el texto refundido de la Ley sobre Régimen del Suelo y Ordenación Urbana, aprobada por RD legislativo 1/1992, de 26 de junio. La Constitución Española atribuye la competencia en materia de urbanismo y de ordenación del territorio a las CC.AA. En aplicación de la Constitución, el tribunal atribuyó al Estado las tareas de «regular las condiciones básicas que garanticen la igualdad en el ejercicio del derecho de propiedad del suelo en todo el territorio nacional, así como regular otras materias que inciden en el urbanismo como son la expropiación forzosa, las valoraciones, la responsabilidad de las Administraciones públicas o el procedimiento administrativo común»; Exposición de motivos, para. 1 [3] Ley 6/1998. A raíz de esta sentencia se aprobó la **L. sobre R. del S. y V.** que diferencia entre las competencias exclusivas del Estado y las regulaciones de carácter básico y deroga la mayor parte de la ley anterior, que hasta esa fecha había sido la base general del planeamiento urbanístico y territorial de todo el Estado Español; cf. Disposición derogatoria única, ▶legislación urbanística, ▶planeamiento urbanístico); *s 2* **código [m] urbanístico** (Término genérico para la legislación que regula la ordenación del territorio, el uso del suelo y la contrucción); *f* **Code [m] de l'urbanisme [F]** (Ensemble des textes législatifs et réglementaires [six livres] traitant des règles générales d'aménagement et d'urbanisme, de la préemption et des réserves foncières, de l'aménagement foncier, des règles relatives à l'acte de construire et à divers modes d'utilisation des sols, de l'implantation des services, établissements et entreprises, des dispositions relatives aux contentieux de l'urbanisme ; ▶droit de construire, ▶planification urbaine) ; *g* **Baugesetzbuch [n] [D]** (*Abk.* BauGB; aktualisiertes, öffentlich-rechtliches Städtebaurecht, das als Bundesgesetz die Nutzung des Grund und Bodens in der ▶Bauleitplanung regelt; es löste am 01.07.1987 das Bundesbaugesetz [BBauG] und das Städtebauförderungsgesetz [StFG] ab und wurde 1998 wesentlich geändert; 2004 nochmals auf Grund des Artikels 6 des Europarechtsanpassungsgesetzes Bau vom 24. Juni 2004 neu gefasst; ▶Baurecht).

Town and Country Planning (Assessment of Environmental Effects) Regulations [npl] **1988** [UK] *conserv. leg. nat'res. plan.* ▶National Environmental Policy Act [US].

town beautification [n] [UK] *landsc.* ▶Garden City Movement [UK].

town centre clearance [n] [UK] *plan. urb.* ▶comprehensive redevelopment.

town centre redevelopment [n] [UK] *urb.* ▶center city redevelopment [US]/urban regeneration [UK].

town climate [n] [UK] *met. urb.* ▶urban climate.

6443 town commons [npl] [US] *agr. for.* (*leg. hist.*) (Pasture or woodland owned and managed by a local community for joint use; in U.S., 'town commons' of colonial America; *specific term* grazing commons [US]/grazing common [land] [UK]); *syn.* common (land) [n] [UK]; *s* **tierras [fpl] comunales** (Conjunto de pastos, bosques o tierras yermas que son aprovechados conjuntamente por los vecinos de un pueblo); *syn.* terreno [m] patrimonial, ejido [m] [MEX]; *f* **communaux [mpl]** (*Article 542 du Code civil* biens à la propriété ou au produit desquels les habitants d'une ou de plusieurs communes ont un droit acquis, une utilisation commune ; il s'agit le plus souvent de bois, de pâturages, de terres incultes, de carrières, de marais, etc. dont les habitants, pour des raisons historiques très anciennes, ont la jouissance directe à la différence des biens du domaine privé proprement dit de la commune dont celle-ci demeure seule propriétaire et dont elle dispose librement) ; *syn.* biens [mpl] communaux ; *g* **Allmende [f]** (Wirtschaftstyp gemeinschaftlicher Nutzung von Ländereien wie Weiden, Wald oder Ödland durch Mitglieder einer Gemeinde; *UB* Allmendweide); *syn. leg. hist.* Heide [f].

town expansion [n] [UK] *urb.* ▶community expansion [US].

6444 town forest [n] [UK] *urb.* (In Europe, area of woodland in public ownership within the boundaries of a town or city directly adjoining the built-up area. **In U.S.**, town forests exist in small New England Towns and are managed by appointed committees; such a publicly-owned urban woodland is sometimes called a 'park'; ▶recreation forest); *syn.* municipal woodland [n], urban forest [n], park [n] [also US]; *s* **bosque [m] urbano** (Superficie grande de bosque público situada dentro de los límites de una ciudad y cerca de zonas edificadas; ▶monte recreativo); *f* **forêt [f] urbaine** (Espace boisé d'au moins 100 ha souvent inclus dans l'agglomération ou tout au moins en contact direct avec celle-ci et remplissant pour les habitants du voisinage le rôle d'espaces verts urbains de quartier ; ▶forêt récréative) ; *g* **Stadtwald [m]** (Größerer, öffentlicher Wald innerhalb der Gemarkungsgrenze einer Stadt mit unmittelbarem Bezug zur Bebauung; ▶Erholungswald).

town fortification [n] [UK] *urb.* ▶earthen wall fortification [US].

town house [n] [US] *urb.* ▶row house.

town house garden [n] [US] *gard. plan.* ▶elongated garden; *arch. constr.* ▶terraced town house garden [US].

town ordinance [n] [US] *adm. leg. urb.* ▶city ordinance [US].

town planner [n] [UK] *prof.* ▶urban planner [US].

town planning consultant [n] [UK] *prof.* ▶planning consultant [US].

6445 townscape [n] *urb.* (Visual impression/appearance of a town or village, or parts thereof, based upon its particular design characteristics; ▶characteristic of the townscape, ▶cityscape, ▶structuring of townscape, ▶visual quality of landscape); *syn.* visual picture [n] of a town; *s* **aspecto [m] escénico urbano**

T

(Impresión visual dada por una ciudad o un pueblo, basada en sus características específicas; ►aspecto escénico del paisaje, ►característico/a del aspecto escénico urbano, ►fisionomía de la ciudad, ►trama urbana); *syn.* calidad [f] visual urbana; *f* **aspect [m] d'une agglomération** (Impression visuelle générale que dégage une agglomération à travers ses caractéristiques architecturales, les matériaux utilisés, l'implantation des masses construites, l'organisation des voies de communication et l'importance de la végétation caractéristique d'un paysage urbain ; ►caractéristique de l'aspect de l'agglomération, ►image paysagère, ►morphologie urbaine, ►tissu urbain) ; *syn.* caractère [m] d'une agglomération, caractère [m] du/des lieu(x) ; *g* **Ortsbild [n]** (Zusammenfassender optischer Eindruck einer Ortschaft oder deren Teile auf Grund bestimmter Gestaltungsmerkmale; bei einer Stadt spricht man von Stadtbild oder ►Stadtgestalt; im Vergleich zum **O.** resp. Stadtbild kann der Begriff ►Landschaftsbild für den besiedelten und unbesiedelten Bereich verwendet werden; ►ortsbildbestimmend, ►städtebauliche Gliederung).

6446 townscape analysis [n] *arch. urb.* (Visual analysis of the appearance and characteristics of a city or town which may be incorporated in a written document; ►cityscape, ►townscape); *syn.* community appearance analysis [n] [also US]; *s* **análisis [m] visual del paisaje urbano** (Estudio de la calidad escénica de la ►fisionomía de la ciudad con fines de planificación y determinación de medidas para su mejora en ►plan parcial [de ordenación]; ►aspecto escénico urbano); *f* **analyse [f] du paysage urbain** (Analyse des caractéristiques visuelles et esthétiques du milieu urbain ; ►aspect d'une agglomération, ►morphologie urbaine) ; *syn.* analyse [f] de l'aspect de la ville ; *g* **Stadtbildanalyse [f]** (1. Analyse der ►Stadtgestalt; ►Ortsbild; **2.** Besondere planerische Leistung bei der Ermittlung der Vorgaben für die Aufstellung von ►Bebauungsplänen; cf. § 40 [2] Nr. 2 HOAI 2002).

townscape feature [n] [UK]**, green** *landsc. urb.* ►green space feature [US].

town square [n] *urb.* ►urban square.

town statutes [n] [UK] *adm. leg. urb.* ►city ordinance [US].

6447 towpath [n] [US]/**tow-path** [n] [UK] *recr. trans.* (Pathway immediately adjacent to rivers or canals upon which barges were formerly pulled upstream by horses or mules); *syn.* towing path [n] [US]/towing-path [n] [UK]; *s* **camino [m] de sirga** (Sendero paralelo a ríos o canales desde el cual antiguamente se remolcaban las chalanas río arriba); *syn.* camino [m] de remolque; *f* **chemin [m] de halage** (Chemin aménagé le long de certains fleuves, rivières ou canaux à partir duquel étaient autrefois tirées les péniches) ; *g* **Leinpfad [m]** (Unmittelbar an Flüssen oder Kanälen entlang führender Weg, von dem aus früher Kähne stromaufwärts gezogen wurden); *syn.* Treidelpfad [m].

toxic accident [n] *envir. leg.* ►industrial plant toxic accident.

toxic agent [n] *envir.* ►noxious substance.

toxic dumpsite [n] [US] *envir. urb.* ►hazardous old dumpsite [US].

6448 toxic site inventory [n] [US] *adm. envir.* (Comprehensive survey and list compiled on a computer of all known or suspected ►hazardous waste materials within a particular area, together with records of previous finds and details of how the contamination was treated. Location and, if possible, type and date of the dumping [US]/tipping [UK] of the waste are also documented. Transferred or excavated waste is registered as a disposed item with details of its whereabouts; **in U.S.,** a list of Superfund sites is kept by the Environmental Protection Agency in sequence according to the severity of contamination); *syn.* contaminated land registry [n] [UK]; *s* **inventario [m] de suelos potencialmente contaminados** (Registro cartográfico y descrip-

tivo de todos los antiguos depósitos de residuos peligrosos y zonas industriales contaminadas o potencialmente contaminadas [►suelos contaminados] en una región dada. En Es. el primer inventario se elaboró en la CC.AA. del País Vasco a principios de los 1990); *f* **inventaire [m] (national) des sites et sols pollués** (F., recensement faisant l'objet d'un traitement automatisé d'informations relatives aux sites et sols pollués : deux programmes permettent de dresser l'inventaire des sites et sols pollués en France : **1. ba**se de données sur les sites et **sol**s pollués [base de données **BASOL**] établie à partir de 1994, qui recense les sites et sols pollués [ou potentiellement pollués] appelant une action des pouvoirs publics, à titre préventif ou curatif ; les sites recensés dans BASOL sont répartis en 4 catégories : **a]** sites traités et libres de toute restriction, **b]** sites traités avec restriction, **c)** sites en activité doivent encore faire l'objet d'un diagnostic, **d]** sites en cours d'évaluation ou de travaux ; l'inventaire est librement consultable sur Internet : http://basol.environnement.gouv.fr ; il a vocation à être actualisé tous les trois mois ; en janvier 2007, environ 3 900 sites pollués sont recensés en France, plus d'un tiers des sites recensés sont situés dans le Nord-Pas-de-Calais [538], en Rhône-Alpes [544] et en Ile-de-France [330, dont 3 à Paris, 42 dans le Val-de-Marne, 58 dans les Hauts de Seine et 40 en Seine-Saint-Denis]. **2. BASIAS** [**b**ase de données d'**a**nciens **s**ites **i**ndustriels et **a**ctivités de **s**ervice] qui dresse l'inventaire des anciens sites industriels et activités de service pouvant éventuellement être à l'origine d'une pollution, mis en place en 1998 ayant pour vocation de reconstituer le passé industriel d'une région ; il devrait être achevé en 2007 et contenir environ 300 000 sites ; *f* 2 inventaire [m] historique régional d'ancien sites industriels et activités de service [IHR] (Les résultats de l'IHR sont engrangés dans la base de données BASIAS, élaborée par le **B**ureau de **r**echerches **g**éologiques et **m**inières [BRGM], régulièrement enrichies et accessibles sur Internet ; le recensement a lieu à l'échelle d'une région, l'inventaire est mené au niveau du département, le recensement et l'actualisation permanente par les Préfectures et les DRIRE qui rendent compte des travaux d'évaluation, de décontamination et de remise en état ; il est effectué à partir d'une analyse critique des archives locales [administrations, archives départementales et communales, Institut Géographique National, chambres de commerce et d'industrie, directions régionales des affaires culturelles, etc.], cf. circ. du 3 décembre 1993 ; le fichier est tenu à jour par la Direction de la Prévention des Pollutions et des Risques du Ministère chargé de l'environnement ; l'objectif principal de cet inventaire est d'apporter une information concrète aux propriétaires de terrains, aux exploitants de sites et aux collectivités, pour leur permettre de prévenir les risques que pourraient occasionner une éventuelle pollution des sols en cas de modification d'usage ; **D.,** l'établissement du fichier des ►déchets historiques est réalisé à l'échelon de la commune) ; *g* **Altlastenkataster [m,** *in A nur so — oder* **n]** (Rechnergestützte, flächendeckende historische Erhebung und grafische und tabellarische Darstellung aller erfassten ►Altlasten, Altablagerungen und altlastverdächtigen Flächen in einem Gemeindegebiet oder Landkreis sowie die Archivierung der bisherigen Altlasten- und Schadensfallbearbeitung. Es werden der Ort, wenn möglich, Art und Zeitpunkt der Ablagerung dokumentiert. Umgelagerte oder ausgelagerte Altlasten werden als Abgänge unter Angabe des Verbleibs vermerkt).

6449 toxic site reclamation [n] *envir. plan.* (**1.** Decontamination to remove or minimize noxious substances on contaminated soils on former industrial and commercial sites, former military bases, or abandoned and illegal waste disposal sites, in order to prevent the spread of noxious substances; or **2.** sealing of noxious substances on site. **In U.S.,** the Resource, Conservation and Recovery Act [RCRA] of 1976, as amended, authorizes management of ▶hazardous waste from generation to ultimate disposal; the Comprehensive Environmental Resource Compensation and Liability Act [CERCLA] of 1980, as amended, authorizes funding through annual appropriations for the 'Superfund'. An 87.5% portion of this fund is financed by a special tax on raw oil and raw materials used by chemical industries, and 12.5% from unidentifiable polluting companies through general tax revenues [▶polluter-pays principle]. In 35 states additional funds come from extra payments per ton of hazardous waste); *syn.* reclamation [n] of hazardous waste sites (TGG 1984, 233), decontamination [n] of toxic waste sites, remediation [n] of toxic waste sites [also US], treatment [n] of contaminated land [also UK], restoration [n] of contaminated land, toxic site restoration [n] [also US]; *s* **recuperación [m] de suelos contaminados** (PDPS 1994, Tomo I; ▶principio de causalidad, ▶residuos industriales peligrosos); *syn.* saneamiento [m] de suelos contaminados, saneamiento [m] de «viejas cargas»; *f* **réhabilitation [f] d'un site pollué (par les déchets historiques)** (Opérations de traitement des friches industrielles polluées, des sites contaminés orphelins consistant en général dans l'évacuation ou l'élimination des produits dangereux ainsi que des déchets présents sur le site, la décontamination et le démantèlement des bâtiments, la dépollution des sols ou des eaux souterraines éventuellement polluées, l'insertion du site de l'installation da ns son environnement, la surveillance de l'impact de l'installation sur son environnement ; cf. art. 34-1 du décret n° 77-1133 du 21 septembre 1977 pris pour l'application de la loi n° 76-663 du 19 juillet 1976 relative aux installations classées pour la protection de l'environnement ; dans le cas de la solvabilité de l'exploitant et du propriétaire du terrain la réhabilitation est réalisée en vertu de la loi du 19 juillet 1976 sur les installations classées ; dans le cas contraire elle est confiée à l'ADEME, le financement étant assuré dans le cas d'espèce **1.** par le produit de la taxe sur les déchets spéciaux pour les sites industriels pollués par des déchets et devenus orphelins ou **2.** par le produit des sommes consignées ; ▶principe pollueur-payeur, ▶déchets dangereux) ; *syn.* restauration [f] des sites contaminés par les déchets, traitement [m] des sites et sols pollués, remise [f] en état des sites pollués ; *g* **Altlastensanierung [f] (1.** Maßnahmen [a] zur Beseitigung oder Verminderung von Schadstoffen [Dekontaminationsmaßnahmen] in verseuchten Böden ehemaliger Industrie- und Gewerbestandorte, ehemaliger Militärgelände sowie stillgelegter und wilder Mülldeponien, [b] die eine Ausbreitung der Schadstoffe langfristig verhindern oder vermindern, ohne die Schadstoffe zu beseitigen [Sicherungsmaßnahmen] und [c] zur Beseitigung oder Verminderung schädlicher Veränderungen der physikalischen, chemischen oder biologischen Beschaffenheit des Bodens. **2.** In den USA wird der US-Bundesfonds [‚Superfund'] für die Altlastensanierung zu 87,5 % von einer Sondersteuer auf Rohöl und Grundstoffe der chemischen Industrie gespeist, zu 12,5 % aus allgemeinen Steuermitteln [▶Verursacherprinzip auf Branchenebene]. Weitere Fonds werden in 35 Bundesstaaten über Sonderabgaben pro Tonne von ▶besonders überwachungsbedürftigem Abfall finanziert; cf. § 2 BBodSchG, BROCK 1986, 444); *syn.* Altstandortsanierung [f].

toxic site restoration [n] [US] *envir. plan.* ▶toxic site reclamation.

6450 toxic substance *chem. envir.* (Poisonous element causing pollution; ▶hazardous material); *syn.* contaminant [n] [also US]; *s* **sustancia [f] tóxica** (Producto o elemento químico que envena a seres humanos, animales o el medio ambiente; ▶sustancia tóxica y peligrosa); *syn.* sustancia [f] nociva; *f* **substance [f] toxique** (Substances et préparations qui, par inhalation, ingestion ou pénétration cutanée en petites quantités, entraînent la mort ou des risques aigus ou chroniques ; directive du Conseil n° 67/548/CEE du 27 juin 1967 ; sous cette dénomination sont souvent comprises par simplification les substances très toxiques ou les substances nocives ; ▶substance dangereuse) ; *g* **Giftstoff [m]** (Chemisches Element und seine Verbindung in natürlicher Form oder hergestellt durch Produktionsverfahren, das beim Eindringen in den menschlichen oder tierischen Organismus Vergiftungen hervorruft. Es werden Umwelt**g.e** [z. B. Quecksilber, Pestizide] und gewerbliche **G.e** [z. B. Benzol, Blei, Cadmium] unterschieden; ▶Gefahrstoff); *syn.* giftiges Gut [n].

toxic waste [n] *envir. leg.* ▶hazardous waste.

toxic waste [n]**, incineration plant for** *envir.* ▶incinerator for toxic/hazardous waste [US].

6451 toxic waste disposal [n] *envir. leg.* (Legally-required disposal of contaminated soil, hazardous material and structures, by excavation, interim storage [if necessary], and final disposition; ▶hazardous waste material); *s* **eliminación [f] de suelo contaminado y materiales tóxicos** (Procedimiento regido por normas legales para eliminar sustancias venenosas o suelos contaminados en depósito o en incineradora de residuos peligrosos; ▶suelo contaminado); *f* **dépollution [f] d'un site pollué (par les déchets historiques)** (Obligation faite à l'exploitant d'une installation classée de procéder à l'élimination contrôlée [extraction, évacuation, stockage de transit et définitif] des sols contaminés et des déchets présentant des nuisances et laissés sur place après cessation ou interruption des activités ; art 1 de la loi du 19 juillet 1976 relative aux installations classées pour la protection de l'environnement ; ▶déchets historiques) ; *g* **Altlastenentsorgung [f]** (Nach gesetzlichen Bestimmungen geordnete Beseitigung [Aushub, ggf. Zwischenlagerung und Endlagerung] von kontaminierten Böden, Gebäudeteilen oder nicht ordnungsgemäß gelagerten, gefährlichen Stoffen; in D. wird die **A.** nach §§ 13 ff BBodSchG i. V. m. § 27 [1] des KrW-/AbfG geregelt; ▶Altlast).

toys [npl] *recr.* ▶play equipment.

6452 trace element [n] *chem.* (Chemical element which, in only very low amounts, is indispensable for human, animal and plant nourishment; e.g. iron, manganese, copper, fluorine, iodine, boron, molybdenum, chlorine); *s* **oligoelemento [m]** (Elemento imprescindible para el desarrollo de un organismo, aunque se precise sólo en cantidad muy pequeña; DINA 1987, 356); *syn.* elemento [m] traza; *f* **oligo-élément [m]** (Éléments chimiques sous forme anionique [bore, molybdène, chlore] ou cationique [fer, manganèse, cuivre, fluor, iode, zinc] présents dans le sol ou dans les végétaux à l'état de trace et indispensable à la croissance des végétaux ou à l'alimentation ; leur insuffisance engendre une carence, leur excès, une toxicité) ; *syn.* élément [m] trace ; *g* **Spurenelement [n]** (Chemisches Element, das in nur sehr geringer Menge für die menschliche, tierische [z. B. Eisen, Mangan, Kupfer, Fluor, Jod] und pflanzliche Ernährung [z. B. Bor, Molybdän, Chlor] unentbehrlich ist); *syn.* Mikroelement [n].

trachiotis [n] [Cyprus] *phyt.* *▶garide.

track [n] *geo.* ▶avalanche track [US]; *agr.* ▶field track [US]; *landsc. urb.* ▶footpath; *for.* ▶forest track [US]/forest ride [UK].

track [n] [US]**, bicycle** *trans.* ▶bikeway [US].

track [n] [US]**, farm** *agr.* ▶field track.

track [n] [US]**, pull-off** *trans.* ▶railroad siding [US].

track [n] [AUS]**, snig** *for.* ▶skidding road [US&CDN].

trackage [n] [US] *leg. trans.* ▶railroad right-of-way [US]/railway right-of-way [UK].

6453 tracks [npl] **of grazing animals** *agr.* (Narrow path made by animals moving along steep slopes; *specific terms* sheep tracks, cattle tracks); *s* **camino [m] de ganado** (Senda estrecha creada por animales en laderas de montañas pendientes; *término* *específico* camino de cabras); *syn.* sendero [m] de ganado; *f* **pieds [mpl] de vaches** (Série de replats étroits [gradin] et dénudés, souvent parallèles, entaillant un versant couvert de pelouse, prairie ou pré-bois, façonnés par le parcours des troupeaux ; cf. DG 1984, 347) ; *syn.* stries [fpl] de parcours, sentiers [mpl] de vaches ; *g* **Viehgangeln [fpl]** (Durch Weidevieh verursachte schmale Trittterrassen an steilen [Gras]hängen).

6454 tractive force [n] *hydr.* (WMO 1974, 1114; energy exerted upon ▶bed load by flowing water); *s* **fuerza [f] de tracción** (Fuerza ejercida por la corriente de agua sobre las partículas sedimentadas [▶carga de fondo] que permanecen en el lecho del río y que tiende a desplazarlas; WMO 1974); *f* **force [f] tractive** (Force que le courant exerce sur une particule de sédiment immobile sur le lit d'un cours d'eau pour la faire se déplacer ; cf. WMO 1974 ; ▶charge de fond) ; *g* **Schleppkraft [f]** (Die von fließendem Wasser auf das ▶Geschiebe 1 im Gerinnebett ausgeübte Kraft).

6455 trade [n] *constr.* (Skilled craft or piece of work concerned with providing specialized manual or mechanical services, e.g. as a sub-contractor for a construction project; there may be main and subordinate trades; ▶specialist trade); *s* **oficio [m]** (Ramo de construcción que interviene en una obra al servicio del contratista principal haciendo p. ej. trabajos de adoquinado, de carpintería, etc.; ▶lote de oficio); *f* **corps [m] d'état** (Entreprise engagée pour un ou plusieurs lots de travaux. On distingue les corps d'état principaux et les corps d'état secondaires ; ▶lot technique) ; *g* **Gewerk [n]** (Gewerbe oder Handwerk, das für ein Bauprojekt einen Teilauftrag erhält, z. B. für Pflasterarbeiten, Zimmermanns-arbeiten. Es wird unter Haupt- und Nebengewerken unterschieden; ▶Fachlos).

trade [n], **nursery** [US] *hort.* ▶commercial horticulture, #1.

trade [n], **tourist** *recr. trans.* ▶tourism (1).

traditional agriculture [n] *agr.* ▶traditional farming.

6456 traditional farming [n] *agr.* (Conventional form of agriculture characterized by the often considerable use of ▶plant care products, ▶plant protection agents, chemically produced fertilizers and pesticides in contrast to ▶organic farming); *syn.* traditional agriculture [n]; *s* **cultivo [m] tradicional** (Agricultura convencional que emplea grandes cantidades de ▶productos fitosanitarios, biocidas y abonos químicos y, generalmente, está especializada en unos pocos cultivos; ▶cultivo alternativo); *syn.* agricultura [f] tradicional; *f* **agriculture [f] traditionnelle** (Par rapport à la ▶culture biologique, forme de culture pratiquant l'apport d'engrais, de ▶produits de traitement phytosanitaire et de produits antiparasitaires d'origine chimique) ; *syn.* agriculture [f] conventionnelle ; *g* **konventioneller Landbau [m, o. Pl.]** (Herkömmliche Landbewirtschaftung, die sich im Gegensatz zum ▶alternativen Landbau durch oft erheblichen Einsatz von ▶Pflanzenbehandlungsmitteln und chemisch produzierten Handelsdüngern auszeichnet).

6457 traditional orchard [n] [UK] *agr. conserv.* (**1.** Existing area or linear plantation of widely-spaced ▶standard fruit trees of different species and cultivars on grassland [▶open orchard grassland (US)/open orchard meadow (UK)], on arable fields [▶plowed land with fruit trees (US)/ploughed land with fruit trees (UK)], along field tracks or roads [fruit tree-lined road], or planted as an orchard belt around villages; **in U.K., t. o.s** are variously protected either as trees [under **T.P.O.s**], as landscape features or as habitats with biodiversity value; under the U.K. Biodiversity Action Plan [BAP], **t. o.s** are not defined as national habitats, but as a sub-group in lowland wood pasture habitat. They are concentrated in the southern counties; e.g. the BAP for Gloucester defines "old orchards" as 'sites with a continuous presence since before 1950 of fruit or nut trees on

vigorous rootstocks and at traditional spacing, with a grass sward usually either grazed by livestock or cut for hay'; *syn.* old orchard [n] [UK], dispersed fruit tree planting [n] [US], agrosilvopastoral system [n], silvoarable agroforestry [n]; **2. agroforestry system [n]** (Since the early 1980s silvoarable agroforestry—the traditional practice of using agricultural land planted with widely-spaced trees for the annual production of arable crops or pasture has become popular again. It is made possible by either thinning woodlots or tree planting on agricultural parcels and various forms may be distinguished: **silvopastoralism** [trees and undergrowth grazed by domestic animals, forest grazing]; **agroforestry systems** [forage crops and livestock] and **agropastoral systems** [forestry, crop and livestock fodder]. In support of the European Common Agricultural Policy, the European project Silvoarable Agroforestry For Europe [SAFE] combines trees and crops and is encouraged by E.U. governments. The project builds on recent findings, which indicate that growing high quality trees in association with arable crops in European fields may improve the sustainability of farming systems, diversify farmers' incomes, provide new products for the wood industry, and create novel landscapes of high value. In D. conservationists, farmers, the public sector and vineyards are endeavoring to increase and protect the approximately 400,000 ha of **d. f. t. p**. Due to its diversity **d. f. t. p**. is a formative part of the Central and Western European cultural landscape, like agroforestry farming systems in southern Europe, such as olive, walnut and almond groves. In the Iberian *dehesas* and *montados*, meadows under trees were previously used as forage pastures for animals and the fruit trees served the livestock for shade. Also of economic interest are oak trees under which wild pigs forage on acorns [high quality ham production] as well as under cork-producing oak trees *[Quercus suber]*; in France, walnut trees *[Juglans regia]*); *s* **plantación [f] dispersa de árboles frutales** (Plantación lineal o en superficie de ▶árboles frutales de pie alto de diferentes especies y variedades en pradera [▶prado de huerta frutal], en tierras de cultivo [▶campo cultivado con frutales intercalados], a lo largo de caminos rurales, carreteras locales, como orlas arbóreas al borde de campas, zanjas o como orlas de frutales alrededor de asentamientos); *f 1* **peuplement [m] dispersé d'arbres fruitiers** (Plantation linéaire ou étendue de différentes variétés d'▶arbres fruitiers [de] haute tige sur les prairies permanentes [▶prairie complantée d'arbres fruitiers], sur les champs [▶champ complanté d'arbres fruitiers], le long des chemins ruraux et des routes [allée d'arbres fruitiers], comme ourlet en lisière des champs, des fossés ou ceinture d'arbres fruitiers entourant les agglomérations rurales) ; *f 2* **agroforesterie [f]** (Mot à la mode depuis le début des années 80 pour désigner une pratique ancestrale d'exploitation des terres agricoles associant sur les mêmes parcelles une production à long terme par la plantation d'arbres d'une part, et une production annuelle de cultures ou pâturages d'autre part ; elle s'effectue soit par éclaircie des parcelles boisées soit par plantation des parcelles agricoles. On distingue le sylvopastoralisme [arbres et sous bois pâturés par les animaux domestiques, ▶pâturage en forêt] ou système sylvopastoral, l'**agrisylviculture** [arbres et cultures] ou système agrosylvicole, l'**agropastoralisme** [cultures fourragères et élevage pastoral] ou système agropastoral et l'**agrosylvopastoralisme** [forêt, culture fourragères et élevage pastoral] ou système agrosylvopastoral. Le projet européen « Des Systèmes Agroforestiers pour les Fermes Européennes » [SAFE] qui s'inscrit dans le cadre de l'évolution de la **Politique Agricole Commune [PAC]** concerne les systèmes agroforestiers associant arbres et cultures intercalaires et sont encouragés par les pouvoirs publics) ; *syn.* système [m] agroforestier ; *g 1* **Streuobstbestand [m]** (**1.** Gesamtheit der vorhandenen Streuobstpflanzungen in einem definierten Gebiet. **2.** Flächenhafte oder lineare vorhandene Anpflanzung von meist hochstämmigen Obstbäumen

T

[►Obsthochstamm] verschiedener Arten und Sorten auf Grünland [►Streuobstwiese] — bei manchen Kernobstarten und vor allem bei Kirschen und Zwetschgen auch unter 1,60 m Stammhöhe—, auf ackerbaulich genutzten Flächen [►Baumacker, Baumfeld], entlang von Feld-, Fahrwegen und Straßen [Obstallee, Straßenobst], als Saum an Feldrainen, Gräben, Feldrändern oder als Obstgartengürtel um Ortschaften, die mit umweltverträglichen Bewirtschaftungsmethoden extensiv genutzt und gepflegt werden. In D. bemühen sich seit Anfang der 1980er-Jahre Naturschützer, Landwirte, die öffentliche Hand und Keltereien vermehrt um die Förderung und den Schutz der ca. 400 000 ha Streuobstbestände. In ihrer Vielfalt der Anbauformen sind Streuobstbestände prägender Bestandteil der mittel- und westeuropäischen Kulturlandschaft, vergleichbar agroforstwirtschaftlichen Anbausystemen in Südeuropa wie Oliven-, Walnuss- und Mandelhaine und die iberischen *Dehesas* und *Montados*; cf. RÖS 2003, www.Streuobst.de und N+L 2008 [7], [8]. Früher wurden die Wiesen als Weiden für die Grasfuttergewinnung genutzt und die Obstbäume dienten dem Vieh als Schatten. Aus dem Obst wurde Most gemacht oder es wurde als Tafelobst geerntet); *syn.* Streuobstpflanzung [f], Streuobstanlage [f], Obstwald [m] (N+L 2008 [6], 166); *g 2* **Agroforstwirtschaft [f]** (Landnutzungssystem, bei dem eine baumbestandene Fläche gleichzeitig landwirtschaftlich genutzt wird; die iberischen *Dehesas* und *Montados* sind vitale Nutzungssysteme, die sich über mehrere 100 000 ha ausdehnen. Von ökonomischem Interesse sind auch die Eichelmast für überwiegend frei lebende Schweine [hochwertige Schinkenproduktion — *Jamón ibérico*] und die Korkeichenbestände *[Quercus suber]* zur Korkgewinnung; in Frankreich gibt es entsprechende **A.en** zur Walnussproduktion *[Juglans regia]*. Eine spezielle ackerbauliche Variante ist eine streifenartige Ackernutzung, die sich mit streifenartigen Baumpflanzungen abwechselt [engl. *alley cropping*]; cf. N+L 2009 [2], 47-52); *syn.* Agroforstsystem [n], Wertholzpflanzung [f], Holzwiese [f].

6458 traditional orchard meadow [n] [UK] *agr. conserv.* (Grassland with widely and regularly-spaced fruit tree planting, mostly unfenced, poorly maintained and predominantly on slopes or heavy soils, established with frost- and disease-resistant varieties of different ages, scattered or isolated in the agrarian landscape. Especially characteristic of the landscape of Southwest Germany and Northern France; **o. o. m./o. o. g.** are amongst the most species-rich habitats of Central Europe. In U.K., more than half the country's **traditional orchards** on grassland have declined or disappeared in the last fifty years: existent orchards have been added to the government's priority list of protected habitats; a pilot two-year study for nine counties has been completed and an inventory of **t. o.s** will be carried out in the next three years; in U.S., a landscape typology referring to old orchards that are out of production and minimally maintained through grazing or mowing for either aesthetic or commercial purposes. **T. o. m.s** are usually planted with single species fruit trees [apple, pear, peach, etc.]; such landscapes occur in New England, upstate New York and southern Canada); *syn.* old orchard grassland [n], open orchard grassland [n] [≠ US], open orchard meadow [n] [≠ US]; *s* **prado [m] de huerta frutal** (≠) (Plantación espaciada de frutales de diferentes edades, resistentes a heladas y a plagas, generalmente aprovechada y cuidada extensivamente, que se da en especial en el sur de Alemania y el norte de Francia en ubicaciones en pendiente o sobre suelos pesados con cobertura de herbáceas. Estos prados se encuentran entre los hábitats de Europa Central más ricos en especies); *f* **prairie [f] complantée d'arbres fruitiers** (Forme traditionnelle en voie de disparition de la culture fruitière extensive implantée sur les coteaux bien exposés ou sur les terres marginales ; surface composée d'une prairie de fauche ouverte parsemée d'arbres fruitiers de haute-tige et d'âge divers ; DG 1984, 75 ; les **p. c. d'a. f.** constituent un des types d'espaces vitaux les plus riches en Europe ; *opp.* verger industriel) ; *syn.* pré [m] complanté d'arbres fruitiers, pré-verger [m], verger [m] haute tige ; *g* **Streuobstwiese [f]** (**1.** Anthropogener, mit umweltverträglichen Bewirtschaftungsmethoden extensiv genutzter Lebensraum, vorwiegend in Hanglage oder auf schweren Böden, bestehend aus robusten, widerstandsfähigen, hochstämmigen Obstbäumen unterschiedlichen Alters, die im Unterschied zu den geschlossenen Blöcken der Intensivobstplantagen gemischt, locker und meist uneingezäunt in der freien Landschaft verstreut sind, besonders landschaftsprägend in Südwestdeutschland und Nordfrankreich. **S.n** zählen mit über 5000 Tier- und Pflanzenarten sowie über 3000 Obstsorten zu den artenreichsten Lebensraumtypen und naturschutzfachlich bedeutendsten Kulturlandschaften Mitteleuropas bis nach Slowenien und Rumänien; **S.n** sind Lebensräume für anspruchsvolle, auf halboffene Landschaften angewiesene Vogelarten wie z. B. Wendehals *[Jynx torquilla]*, Rotkopfwürger *[Lanius senator]*, Gartenrotschwanz *[Phoenicurus phoenicurus]*, Halsbandschnäpper *[Ficedula albicollis]* und Grauspecht *[Picus canus]*. Streuobstwiesen waren lange Zeit keine *per se* gesetzlich geschützten Flächen; heute stehen viele Streuobstflächen im Zusammenhang mit ►Natura 2000 unmittelbar unter dem Schutz der EU-Vogelschutzrichtlinie, bei einem Vorhandensein artenreicher Wiesen oder entsprechender Arten oftmals auch unter dem Schutz der ►Fauna-Flora-Habitat-Richtlinie [FFH-RL] **2.** Der Begriff **S.** rührt vom Landschaftsbild her, da das „gestreute" Vorkommen im Vergleich zu den einheitlichen Anpflanzungen geschlossener Niederstamm-Dichtpflanzungen ein sehr auffälliges Unterscheidungsmerkmal ist; ►Streuobstbestand); *syn.* Obstbaumwiese [f].

traffic [n] *trans.* ►destination traffic, ►interregional traffic, ►interurban traffic, ►local recreational traffic, ►local telephone traffic, ►local traffic, ►origin traffic, ►origin and destination traffic, ►planting for traffic control [US]/planting for traffic guidance [UK], ►recreational traffic, ►regional traffic, ►residential area traffic, ►residential traffic, ►shortcut traffic [US]/short-cut traffic [UK], ►tourist traffic, ►shortcut traffic [US]/short-cut traffic [UK], ►summer holiday traffic, ►stationary traffic, ►through traffic, ►tourist traffic, ►weekend traffic.

traffic [n] [US], **cut-through** *trans.* ►shortcut traffic [US]/short-cut traffic [UK].

traffic [n], **delivery and haulage** *plan. trans.* ►stationary traffic.

traffic [n], **holiday** *plan. recr. trans.* ►tourist traffic, ►summer holiday traffic.

traffic [n], **local telephone** *trans. urb.* *►local traffic.

traffic [n] [US], **O** *trans.* ►origin traffic.

traffic [n] [US], **O&D** *trans.* ►origin and destination traffic.

traffic [n], **parked** *plan. trans.* ►stationary traffic.

traffic [n], **pedestrian** *urb. trans.* ►pedestrian circulation.

traffic [n] [UK], **rat run** *trans.* ►shortcut traffic [US]/short-cut traffic [UK].

traffic [n], **stopped** *plan. trans.* *►stationary traffic.

traffic [n] [US], **thru** *trans.* ►through traffic.

traffic [n] [US], **vacation** *plan. recr. trans.* ►tourist traffic.

traffic artery [n] [UK] *trans.* ►arterial road.

6459 traffic axis [n] *plan. trans.* (Major traffic route connecting two communities or regions); *s* **eje [m] de tráfico [Es]/eje [m] de tránsito [AL]** (Conexión importante para el tráfico entre dos ciudades o regiones); *syn.* eje [m] de transporte;

f **axe [m] de transport** (Grande liaison d'aménagement du territoire ou d'intérêt régional, entre deux pôles d'activités importants) ; *g* **Verkehrsachse [f]** (Für den Verkehr wichtige Hauptverbindungslinie zwischen zwei Orten oder Regionen).

6460 traffic bollard [n] *urb. trans.* (Short post made of metal, wood, concrete or natural stone, installed on pathways, squares and pedestrian areas to prevent vehicular traffic from encroaching on designated areas; ▶bollard); *s* **mojón [m] obstáculo** (Término genérico para todo tipo de postes pequeños de diferentes materiales colocados para evitar la circulación o el aparcamiento de vehículos en zonas específicas como calles peatonales, plazas, aceras, etc.; *término genérico* ▶mojón); *f* **borne [f] de sécurité** (Terme générique désignant le mobilier en métal, en pierre ou en bois conçu pour protéger, délimiter efficacement les trottoirs, les zones de circulation des piétons ; *termes spécifiques* ▶borne d'interdiction de passage placée, p. ex. à l'entrée de zones piétonnes pour en empêcher l'accès aux véhicules [*syn.* borne obstacle], borne d'interdiction de stationner placée en général le long des trottoirs) ; *g* **Absperrpoller [m]** (Niedriger Pfosten aus Holz, Metall, Beton oder Naturstein, z. B. auf Gehwegen, Plätzen, am Beginn von Wohnstraßen, Fußgängerzonen, um die Durch- oder Überfahrt zu verhindern; *OB* ▶Poller).

6461 traffic calming [n] *trans. urb.* (Measures taken to reduce the speed of vehicles. By giving precedence to pedestrians and cyclists, for example by narrowing the width of streets or introducing obstacles, e.g. speed bumps [US]/speed check ramps [UK], as well as introducing low speed limits, the quality of life in residential areas is improved; ▶shared space); *syn.* traffic restrainment scheme [n] [also UK], traffic-restraint program [n] [also US] (PSP 1987, 322); *s* **apaciguamiento [m] del tráfico [Es]/apaciguamiento [m] del tránsito [AL]** (Medidas de restricción o de reducción de la velocidad del tráfico por medio de estrechamiento de la calzada, colocación de obstáculos [p. ej. frena-coches] así como implantando límites de velocidad bajos, que dan preferencia a peatones y ciclistas. De esta manera se mejora la calidad de vida en las ciudades y aumenta la seguridad de los niños y las niñas; ▶espacio compartido); *syn.* tranquilización [f] del tráfico [Es], tranquilización [f] del tránsito [AL]; *f* **circulation [f] apaisée** (Ensemble de mesures visant à améliorer la qualité de la vie dans les quartiers résidentiels par le détournement du trafic de transit, la réduction du flux des véhicules [rue à circulation restreinte ou rue à fréquentation automobile réduite], la limitation de vitesse et divers aménagements de la chaussée pour donner la priorité au confort et à la sécurité des usagers vulnérables dans l'espace de circulation ; LEU 1987, 59-69 ; ▶espace partagé) ; *syn.* vitesse [f] apaisée, apaisement [m] de la circulation [CDN] ; *g* **Verkehrsberuhigung [f]** (1. Befreiung von allzu starkem Durchgangsverkehr. 2. Maßnahmen zur Verbesserung der Lebensqualität in Wohngebieten, in denen durch Geschwindigkeitsbegrenzungen für Kraftfahrzeuge Fußgänger und Radfahrer gegenüber dem Autoverkehr mit Hilfe von Rück- oder Umbau von Straßen bevorrechtigt werden; das Ergebnis sind **verkehrsberuhigte Straßen**; ▶Shared Space).

6462 traffic-caused wind [n] [US] *plant.* (Current of wind following a fast line of moving traffic, which is caused by air displacement from vehicle passage); *syn.* air stream [n] [UK]; *s* **viento [m] de marcha** (Viento causado por la circulación de vehículos); *f* **déplacement [m] d'air des véhicules** (Déplacement d'air provoqué par la circulation des véhicules) ; *syn.* vent [m] engendré par le déplacement d'un véhicule ; *g* **Fahrtwind [m]** (Durch Fahrzeugverkehr entstehender Wind).

6463 traffic control [n] [US] *plan. trans.* (1. Regulated flow of vehicular traffic which is pre-determined in traffic planning. 2. Visual help by means of plantations and technical devices at the edge of travelled ways [US]/carriage-ways [UK] which give car drivers a better orientation on curved or winding routes; ▶planting for traffic control [US]/planting for traffic guidance [UK]); *syn.* traffic guidance [n] [UK]; *s* **guía [f] directiva del tráfico [Es]/guía [f] directiva del tránsito [AL] (±)** (1. Gestión del flujo del tráfico [Es]/tránsito [AL] a través de una ciudad o zona determinada. 2. Ayuda óptica por medio de plantaciones o dispositivos técnicos al borde de las carreteras, sobre todo en tramos con muchas curvas; ▶plantacíon guía [del tráfico]); *f* **guidage [m] de la circulation** (Flux de déplacement automobile organisé pour un itinéraire prévu/établi dans le cadre d'un plan de circulation ; ▶plantation directionnelle) ; *g* **Verkehrsführung [f]** (1. Planerisch festgelegter Verlauf des Fahrzeugverkehrs. 2. Optische Hilfe in Form von Bepflanzungen oder technischen Einrichtungen am Straßenrand, die den Autofahrern zur besseren Orientierung, besonders auf kurvigen Strecken, dienen; ▶Leitpflanzung); *syn.* Verkehrslenkung [f].

6464 traffic convenience [n] *plan. trans.* (Ease of access to a transport network from a residence or working place—good connection[s]—depending upon the reason for the trip; ▶accessibility); *s* **buena accesibilidad [f] (en vehículo)** (Ventaja de ciertas zonas de estar situadas en lugares bien comunicados, es decir accesibles por autopista o autovía, por ferrocarril y por avión, según la función a la que estén destinadas; ▶accesibilidad); *f* **niveau [m] de desserte** (Qualité d'une zone en fonction de la présence ou de l'éloignement des équipements de transports publics ; ▶accessibilité) ; *g* **Verkehrsgunst [f]** (Merkmal, durch das die ▶Erreichbarkeit eines Verkehrsmittels im Verkehrsnetz [gute Verkehrsverbindung] — abhängig vom Fahrtzweck — beschrieben wird); *syn.* Lagegunst [f].

6465 traffic density [n] *trans.* (Number of vehicles in origin and destination traffic ["O&D" traffic] within a defined area and specified unit of time); *syn.* traffic intensity [n]; *s* **densidad [f] de tráfico [Es]/densidad [f] de tránsito [AL]** (Cantidad de vehículos de origen y destino en un lugar determinado); *syn.* intensidad [f] de tráfico, intensidad [f] de tránsito [AL]; *f* **densité [f] du trafic** (Flux de déplacement [nombre de véhicules en transit ou en déplacement interne à une zone] pour un temps donné établi lors d'une enquête de circulation) ; *syn.* intensité [f] de la circulation ; *g* **Verkehrsdichte [f]** (Messgröße für die Quell- und Zielverkehr bestimmende Anzahl von Fahrzeugen in einem bestimmten Bereich in einer definierten Zeiteinheit); *syn.* Verkehrsaufkommen [n].

traffic derouting [n] [US] *trans. urb.* ▶traffic rerouting [US]/traffic redirection [UK].

6466 traffic facility [n] *trans. urb.* (Equipment or installation which facilitates the movement of traffic on roads, railways, water, or in the air; ▶transportation infrastructure [US]/traffic infrastructure [UK]); *s* **instalación [f] de tráfico y transportes [Es]/instalación [f] de tránsito y transportes [AL]** (Equipamiento que hace posible el tránsito en un sistema específico [carretera, vía de ferrocarril, agua, aire]; la suma de todas las **ii. de t. y t.** en una región conforman la ▶infraestructura de tráfico y transportes [Es]/instalación de tránsito y transportes [AL]); *f* **équipement [m] de transport et de communication** (Installation, ouvrage, réseau, permettant d'assurer les déplacements sur les réseaux routiers, ferroviaires et aériens ; ▶infrastructure de transport) ; *g* **Verkehrsanlage [f]** (Einrichtung, die den Verkehr auf einem Verkehrsträger, z. B. Straße, Schiene, Wasser oder in der Luft ermöglicht; die Summe aller **V.n** in einem bestimmten Gebiet ist die ▶Verkehrsinfrastruktur).

traffic fumes [npl] *envir.* ▶exhaust gas.

traffic guidance [n] [UK] *plan. trans.* ▶traffic control [US].

traffic guidance [n] [UK]**, planting for** *landsc.* ▶planting for traffic control [US].

6467 traffic guidance facility [n] *trans.* (Any installation serving to direct vehicular traffic, visually or mechanically); *s* **dispositivo** [m] **de guía del tráfico** [Es]/**dispositivo** [m] **de guía del tránsito** [AL] (Cualquier equipo óptico o mecánico [técnico] que sirve para dirigir la circulación viaria); *f* **dispositif** [m] **directionnel** (Équipement, mobilier ou accessoire de voirie servant à dirigé visuellement ou mécaniquement sur certaines parties ou itinéraires de l'espace urbain la circulation des véhicules, p. ex. un îlot directionnel, signalétique, etc.) ; *g* **Leiteinrichtung** [f] (Jede Verkehrseinrichtung, die der optischen oder mechanischen Führung des Straßenverkehrs dient).

traffic infrastructure [n] [UK] *plan. trans.* ▶transportation infrastructure [US].

traffic intensity [n] *trans.* ▶traffic density.

6468 traffic intersection [n] *trans.* (Generic term for road intersection in the form of either ▶at-grade junction or ▶grade-separated junction); *syn.* vehicular intersection [n]; *s* **intersección** [f] **de tráfico** [Es]/**intersección** [f] **de tránsito** [AL] (Término genérico para cruce de carreteras que puede ser ▶cruce a nivel o ▶nudo sin cruces a nivel); *f* **ouvrage** [m] **d'intersection routière** (Ouvrage d'▶intersection routière à un niveau [carrefour] ou à plusieurs niveaux [▶échangeur]) ; *g* **Kreuzungsbauwerk** [n] (Verkehrskreuzung als ▶plangleicher Knotenpunkt oder ▶planfreier Knotenpunkt).

traffic jam [n]**, standing in a** *plan. trans.* ＊▶stationary traffic.

6469 traffic linkage [n] *plan. trans. urb.* (Means of connecting tracts or whole subdivisions to an existing transportation system; ▶provision of access for the public); *syn.* connection [n] to a transportation system; *s* **urbanización** [fpl] **viaria** (Resultado de la conexión de una zona al sistema de circulación; ▶planificación viaria); *f* **desserte** [f] (Résultat du raccordement du réseau de transport d'un lieu à un autre ; le niveau de desserte influe fortement sur le niveau des valeurs foncières ; *termes spécifiques* desserte routière, desserte ferroviaire ; ▶viabilisation) ; *g* **Verkehrserschließung** [f] (Ergebnis der Anbindung eines Gebietes an ein bestehendes Verkehrsnetz; ▶Erschließung für den Verkehr); *syn.* Verkehrsanbindung [f].

6470 traffic load [n] *envir. plan. trans.* (Actual number of vehicles within a certain section of road containing traffic); *s* **carga** [f] **de tráfico** [Es]/**carga** [f] **de tránsito** [AL] (Cantidad de vehículos que transitan por una zona determinada); *f* **intensité** [f] **du trafic** (Nombre de véhicules dans un secteur du réseau, fréquence des convois sur une ligne) ; *g* **Verkehrsbelastung** [f] (Aufkommen von Fahrzeugen in einem Verkehrsabschnitt).

6471 traffic noise [n] *envir.* (Sound caused by moving vehicles; ▶noise pollution); *s* **ruido** [m] **de tráfico** [Es]/**ruido** [m] **de tránsito** [AL] (Emisión acústica causada por la circulación de vehículos; ▶contaminación acústica); *f* **bruit** [m] **de circulation** (Bruit causé par les véhicules sur une infrastructure de transport ; ▶nuisance sonore) ; *syn.* bruit [m] des infrastructures routières ; *g* **Verkehrslärm** [m] (Durch Fahrzeuge verursachter Lärm; ▶Lärmbelastung).

6472 traffic noise dispersion [n] *envir.* (▶noise corridor, ▶noise level contouring); *s* **difusión** [f] **del ruido en calles y carreteras** (▶banda de ruido, difusión del ruido); *f* **émissions** [fpl] **sonores aux abords d'infrastructures routières** (▶corridor d'étude du bruit, ▶propagation du bruit) ; *g* **Lärmausbreitung** [f] **an Straßen** (▶Lärmausbreitung, ▶Lärmband); *syn.* Lärmausbreitung [f] durch Verkehr, Lärmbelastung [f] an Straßen.

6473 traffic planning [n] *adm. plan. pol. trans.* (Specialist planning discipline, aimed at achieving optimal infrastructural and organisational parameters for the smooth functioning of traffic at national, state, regional, or local levels; ▶long-range transportation plan [US]/long-range transport plan [UK]); *syn.* transportation planning [n]; *s* **planificación** [f] **de tráfico y transportes** [Es]/**planificación** [f] **de tránsito y transportes** [AL] (Planeamiento sectorial de las condiciones infraestructurales y organizativas para permitir un tráfico lo más efectivo posible; ▶plan general de tráfico y transportes [Es]/plan general de tránsito y transportes [AL]); *f* **planification** [f] **des transports** (Établissement de programmes, spatiaux [aux échelons national, régional, départemental et communal] et économiques, déterminant la demande prévisible à l'horizon temporel prévisible, les investissements à réaliser pour les satisfaire, leur échelonnement dans le temps et leurs conséquences prévisibles, en particulier sur le développement urbain et la localisation des activités et des équipements ; DUA 1996, 587 ; ▶plan de circulation) ; *g* **Verkehrsplanung** [f] (Im Vergleich zur Gesamtplanung die Fachplanung, die sich zum Ziel gesetzt hat, die infrastrukturellen und organisatorischen Rahmenbedingungen zur Ermöglichung eines reibungslosen Verkehrs auf Bundes-, Landes-, Kreis- oder Gemeindeebene zu erarbeiten; ▶Generalverkehrsplan); *syn.* Verkehrsentwicklungsplanung [f].

traffic redirection [n] [UK] *trans. urb.* ▶traffic rerouting [US].

6474 traffic reduction [n] *plan. trans.* (Lessening of traffic levels by re-routing or construction of ▶relief roads; ▶bypass [US]/by-pass [UK], ▶traffic calming); *s* **reducción** [f] **del tráfico** [Es]/**reducción** [f] **del tránsito** [AL] (Disminución del paso de vehículos por una zona construyendo una variante o desviándolo por otras calles o carreteras; ▶apaciguamiento del tráfico [Es]/apaciguamiento del tránsito [AL], ▶calle de desvío, ▶carretera de circunvalación); *f* **délestage** [m] **de la circulation** (Réduction du trafic dans une zone donnée au moyen d'un changement d'itinéraire ou de la construction de routes nouvelles ; ▶itinéraire de délestage, ▶circulation apaisée, ▶route de contournement, ▶route de délestage) ; *syn.* délestage [m] d'une voie routière/du trafic routier ; *g* **Entlastung** [f] **des Verkehrs** (Verringerung des Verkehrs in einem bestimmten Gebiet durch Umleitung oder Neubau zusätzlicher Straßen; ▶Entlastungsstraße, ▶Verkehrsberuhigung, ▶Umgehungsstraße); *syn.* Verkehrsentlastung [f].

6475 traffic rerouting [n] [US] *trans. urb.* (1. Measures for re-routing vehicles to reduce the amount of traffic in residential or commercial areas permanently. **2. traffic detouring** [n] [US] (Measures for re-routing vehicles temporarily); ▶pedestrian zone, ▶traffic calming); *syn.* traffic redirection [n] [UK]. **3.** A special measure of **t. r.** is the ▶pedestrianization); *s* **restricción** [f] **del tráfico de paso** [Es]/**restricción** [f] **del tránsito de paso** [AL] (1. Medida de ▶apaciguamiento del tráfico [Es]/apaciguamiento del tránsito [AL] en zonas residenciales de las ciudades y sus suburbios o en centros comerciales. 2. Una forma especial de la **r. del t.** es la ▶peatonalización; ▶zona peatonal); *syn.* restricción [f] de la circulación automovilística; *f* **restriction** [f] **de la circulation automobile** (1. Mesure de ▶circulation apaisée [limitation de la pollution, de l'encombrement et de l'insécurité] dans les quartiers résidentiels ou les zones commerçantes des centres-villes ; ▶création de zone piétonne, ▶zone piétonne. 2. Transformation de la voirie pour la circulation automobile pour la rendre aux piétons) ; *g* **Herausnahme** [f] **des Verkehrs** (1. Maßnahme der ▶Verkehrsberuhigung in Wohngebieten der Städte und deren Vororte oder in den Einkaufsbereichen der Innenstädte, bei der der der Autoverkehr auf ein Minimum beschränkt wird. **2.** Eine besondere Form der **H. d. V.** ist die ▶Schaffung von Fußgängerzonen; ▶Fußgängerzone).

traffic restrainment scheme [n] [UK] *trans. urb.* ▶traffic calming.

traffic-restraint program [n] [US] *trans. urb.* ▶traffic calming.

traffic route [n] (1) *trans.* ▶transportation corridor.

traffic route [n] [UK] (2) *trans.* ▶travelled way [US].

6476 traffic separation strip [n] *trans.* (Painted lines or ▶median strip [US]/central reservation [UK] separating traffic lanes in which traffic flows in opposite directions; ▶separator); *s* **línea** [f] **de demarcación de las direcciones del tráfico [Es]/línea** [f] **de demarcación de las direcciones del tránsito [AL]** (▶Banda de separación en forma de línea sencilla o doble o de ▶franja divisoria central entre las dos direcciones del tráfico de vehículos motores); *syn.* línea [f] de separación de las direcciones del tráfico [Es]/línea [f] de separación de las direcciones del tránsito [AL]; *f* **îlot** [m] **directionnel** (Séparation matérielle des deux sens de la circulation, dans le cas où les tracés en plan et les profils en long des deux chaussées ne sont pas indépendants, servant à canaliser la circulation et constitué soit d'un marquage au sol [bandes blanches], soit d'une surélévation de la chaussée [séparateur de chaussée], soit des deux ; ▶bande de séparation de la voirie ; *terme spécifique* ▶terre-plein central) ; *syn.* îlot [m] séparateur ; *g* **Richtungstrennstreifen** [m] (▶Trennstreifen als Linie/Doppellinie oder ▶Mittelstreifen zwischen Fahrbahnen, die in entgegengesetzten Fahrtrichtungen befahren werden).

trail [n] *recr.* ▶fitness trail; *landsc. urb.* ▶footpath; *recr.* ▶forest study trail; *game'man. hunt.* ▶game trail; *recr.* ▶hiking trail [US]/hiking footpath [UK]; *recr.* ▶loop trail (1); *plan. recr.* ▶national scenic trail [US]/long distance footpath [UK]; *recr.* ▶nature trail [US]/nature study path [UK].

trail [n] [US]**, alignment of a** *plan. trans.* ▶alignment of a pathway.

trail [n] [US]**, bike** *trans.* ▶bikeway [US]/bicycle path [UK].

trail [n] [US]**, circular** *plan. recr.* ▶circular path [US]/circular pathway [UK].

trail [n]**, equestrian** *recr.* ▶bridle path [US]/bridle way [UK].

trail [n] [US]**, exercise** *recr.* ▶fitness trail.

trail [n] [US] (2)**, loop** *plan. recr.* ▶circular path [US]/circular pathway [UK].

trail [n]**, migratory** *game'man. hunt.* ▶game trail.

trail [n]**, raised** *recr.* ▶boardwalk.

trail [n]**, riding** *recr.* ▶bridle path [US]/bridle way [UK].

trail [n]**, scenic** *plan. recr.* ▶national scenic trail [US]/long distance footpath [UK].

trail [n]**, trim** *recr.* ▶fitness trail.

trail [n]**, walking** *recr.* ▶hiking trail [US]/hiking footpath [UK].

trail hut [n] [US] *recr.* ▶mountain lodge [US].

trailing shrub [n] *bot.* ▶prostrate shrub.

trailing stems [npl] [US]**, running with** *bot.* ▶stemspreading.

trail maintenance [n] [US] *constr. for.* ▶pathway maintenance.

trail network [n] [UK]**, horse riding** *recr.* ▶bridle path network [US].

trail system [n] [US]**, bicycle** *trans. urb.* ▶bikeway network.

trail system [n] [US]**, circular** *plan.* ▶circular walk system [US]/circular pathway system [UK].

training [n] *prof.* ▶practical training.

training [n]**, on-the-job** *prof.* ▶continuing education [US], #2.

training [n]**, professional** *prof.* ▶professional education.

6477 training [n] **of young trees** *hort.* (Pruning of young trees to further the development of the desired branching structure; cf. BS 3975: part 5 et ARB 1983; ▶tree and shrub pruning, ▶tip pruning, ▶cutting back of trees and shrubs); *syn.* formative pruning [n]; *s* **poda** [f] **de formación** (Poda moderada de árboles jóvenes para formar un esqueleto regular y favorecer el desarrollo rápido de la planta; cf. PODA 1994; ▶despunte, ▶poda de leñosas, ▶rebaje); *syn.* poda [f] de mejora; *f* **taille** [f] **de formation de la charpente** (▶Taille des végétaux ligneux effectuée afin de favoriser la croissance en épaisseur du tronc, d'obtenir une charpente solide et la forme désirée ; ▶élagage 2, ▶raccourcissement des extrémités des branches) ; *syn.* taille [f] charpentière (DAV 1984, 114) ; *g* **Erziehungsschnitt** [m] **1.** ▶Gehölzschnitt bei Jungbäumen, der das Dickenwachstum des Stammes fördert, einem habitusgerechten Kronenaufbau dient, d. h. Konkurrenztriebe an Jungbäumen, insbesondere Zwiesel, Drehäste, sich reibende und kreuzende Äste sowie instabile Triebe entfernt und Astquirle vereinzelt, oder einer beabsichtigten Krone für eine vorgesehene Funktion oder Form dienen soll. **2.** Regelmäßige Schnittmaßnahmen an alten Bäumen zur Förderung des Aufbaus habitusgerechter Kronen nach Kappung. Wenn dieser **Kronenerziehungsschnitt** unterbliebe, entstünden „Besenkronen" mit labilen Stammköpfen. **3.** Das Herausarbeiten der Leittriebe von Jungbäumen, die zwischen alten Alleebäumen gepflanzt wurden. Durch behutsames Ausdünnen der bedrängenden Seitenäste der Altbäume wird ein schräg zum Licht wachsender Leittrieb verhindert; ▶Rückschnitt der Astspitzen, ▶Rückschnitt von Gehölzen); *syn.* Aufbauschnitt [m], Kronenerziehungsschnitt [m].

trampled areas [npl] *phyt.* ▶plant community of trampled areas.

6478 trampling [n] *agr. constr. phyt.* (Frequent treading on soil or vegetation; ▶cattle trampling, ▶recoverability from trampling, ▶vulnerable to trampling); *s* **pisoteo** [m] (Deterioro de la vegetación por pisadas, sea de seres humanos o de animales; ▶pisoteo de animales, ▶resistencia al pisoteo, ▶sensible al pisoteo); *f* **piétinement** [m] (Dégradation du couvert végétal ou du sol provoquée par le piétinement de l'homme ou des animaux ; *résultat* végétation piétinée, sol piétiné, sol surpiétiné ; *termes spécifiques* piétinement diffus, piétinement occasionnel ; *opp.* piétinement répété ; ▶capacité au piétinement, ▶piétinement de bétail, ▶vulnérable au piétinement) ; *g* **Trittbelastung** [f] (Beeinträchtigung der Bodenvegetation durch Betreten, sei es durch Mensch oder Tier; ▶Belastbarkeit durch Tritt, ▶trittempfindlich, ▶Viehtritt).

6479 trampling damage [n] *agr. constr. phyt.* (Treading harm to vegetation; ▶trampling); *s* **daño** [m] **por pisoteo** (Destrucción de la vegetación debida al ▶pisoteo); *syn.* destrucción [mpl] por pisoteo; *f* **dégâts** [mpl] **de piétinement** (Dommages causés par le ▶piétinement) ; *g* **Trittschaden** [m] (Mechanische Vegetationszerstörung durch Tritt; ▶Trittbelastung).

6480 transboundary air pollution [n] *envir. leg.* (Air contamination crossing jurisdictional boundaries; *generic term* long-range transboundary air pollution; cf. art. 1, Council decision 81/462/EEC); *syn.* transfrontier air pollution [n]; *s* **contaminación** [f] **atmosférica transfronteriza** (Polución del aire que atraviesa fronteras interestatales; *término genérico* contaminación atmosférica transfronteriza a larga distancia); *syn.* polución [f] transfronteriza; *f* **pollution** [f] **atmosphérique transfrontière** (Transport à longue distance des polluants atmosphériques ; *terme générique* pollution atmosphérique transfrontière à longue distance ; cf. art. 1 de la Décision n° 81/462/CEE) ; *g* **grenzüberschreitende Luftverunreinigung** [f] (Ein weiter gehender Begriff ist die **weiträumige g. L.** gemäß Art. 1 des Beschlusses 81/462/EWG des Rates vom 11.06.1981); *syn.* grenzüberschreitende Luftverschmutzung [f].

6481 transboundary movements [npl] **of hazardous wastes and their disposal** *envir. leg.* (▶Basel Convention; ▶transboundary waste shipment [US]/waste tourism [UK]); *s* **movimientos** [mpl] **transfronterizos de residuos tóxicos y peligrosos y su eliminación** (▶Convención de Basilea; ▶turismo de residuos); *f* **(contrôle des) mouvements** [mpl] **transfrontières de déchets dangereux et de leur élimination** (▶Convention de Bâle ; ▶transfert [transfrontalier] de déchets) ; *g* **grenzüberschreitende Verbringung** [f] **gefährlicher Abfälle und ihre Entsorgung** (▶Basler Konvention; ▶Mülltourismus).

transboundary planning [n] *plan.* ▶transfrontier regional planning.

6482 transboundary waste shipment [n] [US] *envir.* (Legal or illegal transportation of waste material to other federal states or export to foreign countries caused by the non-existence of sufficient dumpsites [US]/tipping sites [UK], deficiency of suitable waste incineration and waste processing plants; ▶Basle Convention); *syn.* waste tourism [n] [UK]; *s* **turismo** [m] **de residuos** (Fenómeno que se presenta con cierta frecuencia en los países industrializados a partir de los años 1980 de exportar legal o ilegalmente residuos a países más pobres, debido a la falta de capacidad de tratamiento y/o deposición o al alto coste de eliminación de residuos peligrosos en aquellos países. Esto llevó en 1989 a la firma de la ▶Convención de Basilea, un acuerdo internacional que regula el control del transporte y la eliminación transfronterizos de residuos tóxicos y peligrosos); *f* **transfert** [m] **(transfrontalier) de déchets** (Transport légal et/ou illégal sur de grandes distances de déchets à l'intérieur d'un État ou mouvements transfrontaliers et transferts de déchets à l'intérieur des États membres de la Communauté ou vers des pays tiers à l'Union Européenne par suite d'insuffisances de capacités de stockage ou de traitement ou de réduction des coûts de traitement des déchets ; ▶Convention de Bâle) ; *syn.* exportation [f] de déchets, mouvement [m] transfrontalier de déchets, tourisme [m] des déchets ; *g* **Mülltourismus [m, o. Pl.]** (Durch fehlende Mülldeponiekapazitäten, Mangel an geeigneten Müllverbrennungs- und Müllverarbeitungsanlagen sowie durch Reduzierung von Entsorgungskosten verursachter legaler und/oder illegaler Abfallexport in andere Bundesländer oder ins Ausland; ▶Basler Konvention); *syn.* Abfallexport [m], *juristischer Begriff nach dem Abfallverbringungsgesetz [AbfVerbrG]* grenzüberschreitende Verbringung [f] von Abfällen.

6483 transboundary water [n] *geo. pol.* (DED 1993, 9; water legally defined as river, lake or other water that flows across or forms a part of state or international boundaries; RCG 1982; *specific terms* interstate water [also US], transnational water [also US]); *s* **aguas [fpl] continentales interestatales** (Río o lago que fluye o se encuentra entre dos Estados o hace frontera entre ellos); *syn.* aguas [fpl] continentales transfronterizas; *f* **eaux [f] continentales transfrontalières** (Fleuve, lac ou toutes autres eaux continentales traversant ou appartenant à différents États) ; *g* **grenzüberschreitendes Gewässer [n]** (Fluss, See oder anderes Gewässer, das über eine Staatsgrenze fließt oder einen Teil dieser bildet).

transfer [n] *pedol.* ▶nutrient transfer.

6484 transferring [n] **of soil piles** [US] *constr. hort.* (Moving of a stockpile of soil from one place to another on a building site, within a plant nursery or a compost area); *syn.* turning-over [n] of soil heaps [UK]; *s* **traslado** [m] **de acopios de tierra**; *f* **déplacement** [m] **de terre mise en dépôt** (Déplacement à l'intérieur d'un chantier ou d'une pépinière ou d'une aire de compostage) ; *syn.* déplacement [m] des terres stockées ; *g* **Umsetzen [n, o. Pl.] von Erdmieten** (Umlagerung von Erdmieten an einen anderen Platz einer Baustelle, Gärtnerei oder auf einen Kompostplatz).

6485 transfers [npl] **of real estate** [US] *leg. urb.* (Buying and selling of parcels of land; ▶land register); *syn.* land transactions [npl] [UK]; *s* **transacciones [fpl] inmobiliarias** (▶catastro); *syn.* mercado [m] de terrenos, operaciones [fpl] inmobiliarias; *f* **opérations [fpl] foncières et immobilières** (Modifications apportées au droit de propriété et à la réglementation foncière ; ▶cadastre) ; *syn.* mutations [fpl] immobilières ; *g* **Bodenverkehr [m]** (Eigentumsänderungen an Grundstücken; cf. §§ 19 ff BauGB; ▶Grundbuch); *syn.* Handänderung [f] [CH].

transfrontier air pollution [n] *envir. leg.* ▶transboundary air pollution.

6486 transfrontier regional planning [n] *plan.* (Cross border zonal planning at a municipal, state, or international level; *specific term* multistate regional planning); *syn.* transboundary planning [n] (DED 1993, 9), transnational planning [n] (±); *s* **planificación [f] transfronteriza** (Planificación que se realiza en concertación con los vecinos, p. ej. otras corporaciones locales, comunidades autónomas, estados federados o Estados); *f 1* **planification [f] intercommunale** (Touchant le territoire de plusieurs communes) ; *f 2* **planification [f] interrégionale** (Touchant le territoire de plusieurs régions) ; *f 3* **planification [f] transfrontalière** (Touchant le territoire de plusieurs États) ; *syn.* planification [f] transfrontière) ; *g* **grenzüberschreitende Planung [f]** (Planung, die in Abstimmung mit Nachbarn erfolgt, z. B. auf kommunaler, Länder- oder internationaler Ebene).

transient [n] *zool.* ▶bird of passage.

transit [n] *trans.* ▶public transit [US]/public transport [UK]; ▶terrain de camping de transit.

transit [n] [US]**, mass** *trans.* ▶local public transportation system [US]/local transport system [UK].

transitional area [n] [US] *conserv. landsc. urb.* ▶buffer zone.

6487 transitional biotope [n] *ecol.* (GE 1977, 22; narrow overlap zone [LE 1986, 60] between two different habitats, e.g. field-forest edge, edge of a bog, hedgerow, etc., with intermediate environmental conditions, generally with a more diversified bioc[o]enosis than in either adjoining system, and more intense biological activity; ▶ecotone 1, ▶edge effect); *syn.* habitat transition line/zone [n], transitional habitat [n]; *s* **cinturón** [m] **de transición** (Espacio limítrofe entre dos tipos de biótopos como p. ej. entre bosque y pradera; ▶ecotono, ▶efecto de borde); *syn.* zona [f] de transición, zona [f] de unión; *f* **ligne [f] de démarcation** (Espace limitrophe plus ou moins marqué entre deux ou plusieurs formations végétales distinctes voisines, p. ex. forêt/champ, pré/haies ; ▶écotone, ▶effet de bordure, ▶effet de lisière) ; *syn.* zone [f] de contact, ceinture [f] de transition ; *g* **Grenzlinienbereich [m]** (Bereich, in dem zwei unterschiedliche Biotoptypen aneinandergrenzen, z. B. Feld-/Waldgrenze, Moorrand, wenig gedüngte Wegraine in der Agrarlandschaft, Heckenreihe; ▶Randeffekt, ▶Ökoton); *syn.* Randzonenbereich [m], Biotoprand [m].

transitional habitat [n] *ecol.* ▶transitional biotope.

6488 transitional regulation [n] *adm. leg.* (Legal requirements partially modifying current regulations during an adaptation period after a new law has come into effect); *s* **norma [f] transitoria** (Reglamentación de las condiciones y los plazos de adaptación a una norma legal nueva); *f* **dispositions [fpl] (réglementaires) transitoires** (Textes réglementaires définissant les modalités d'application des nouvelles dispositions jusqu'à la date de leur entrée en vigueur, portant modification de la réglementation existante) ; *g* **Übergangsvorschrift [f]** (Gesetzliche Bestimmung/Vorschrift zur Anpassung an bisher gültige Bestimmungen nach In-Kraft-Treten der neuen Rechtsvorschrift).

6489 transition bog [n] *pedol.* (Wetland, often called 'poor fen', which is intermediate between mineral-nourished [= ►low bog] and precipitation-dominated [= ►raised bog] peatlands; cf. WET 1983, 374); *syn.* mesotrophic peatland [n] (WET 1983, 374), transition peatland [n]; *s* **humedal [m] de transición** (Turbera intermedia entre la ►turbera alta y la ►turbera baja 2 caracterizada por una turba formada en agua neutra o débilmente ácida. Puede ser una evolución de la turbera baja, por acidificación del agua); *syn.* turbera [f] mesótrofa; *f* **tourbière [f] de transition** (Tourbière intermédiaire entre la ►tourbière haute et la ►tourbière basse, caractérisée par une tourbe formée en eau neutre ou très faiblement acide, évolution de la tourbière basse par acidification des eaux) ; *syn.* tourbière [f] mésotrophe, marais [m] de transition ; *g* **Übergangsmoor [n]** (Moor, das in seiner Entstehung zwischen ►Hochmoor und ►Niedermoor liegt).

transition peatland [n] *pedol.* ►transition bog.

6490 transition period [n] *leg. envir.* (Interval of time specified during which existing regulations are adjusted to new environmental provisions after a new law has come into effect); *s* **plazo [m] de adecuación (progresiva)** (En las normas transitorias de nuevas regulaciones, p. ej. a nivel ambiental, periodo/período concedido una vez entrada en vigor la regulación para adaptar las instalaciones o equipamientos a la nueva norma); *f* **délai [m] d'application** (Période de transition fixée s'écoulant entre le vote de textes réglementaires et l'entrée en vigueur des nouvelles dispositions) ; *g* **Übergangsfrist [f]** (In der Übergangsbestimmung/-vorschrift für neue Vorschriften, z. B. Umweltauflagen, festgesetzter Zeitraum, der der Anpassung an bisher gültige Bestimmungen nach In-Kraft-Treten der neuen Rechtsvorschrift dient); *syn.* Anpassungsfrist [f].

6491 transition radii [npl] *trans. plan.* (Curve with continuously increasing or decreasing radii in horizontal and vertical alignment used in creating a gradual change in direction or elevation of a traffic route); *s* **curva [f] de transición** (BU 1959; en el trazado de rutas de tránsito, curva con radio creciente o decreciente en alineación horizontal o vertical empleada para posibilitar un cambio gradual de dirección o de elevación); *syn.* curva [f] de acuerdo, curva [f] de enlace, curva [f] amplia, arco [m] de transición; *f* **courbe [f] de raccordement progressif** (Dans un tracé de route, courbe à rayon croissant ou décroissant engendrant horizontalement [rayon inférieur à une fois et demi la valeur du rayon normal déversé] un changement de direction du tracé, ou verticalement le passage d'une pente sur un plat et inversement) ; *syn.* courbe [f] de transition ; *g* **Übergangsbogen [m] (2)** (Kurve mit mehrfach oder stetig zu- oder abnehmendem Halbmesser im Grundriss oder Aufriss einer Verkehrslinie zur Herbeiführung einer allmählichen Richtungsänderung).

6492 transition slope [n] *constr.* (►Embankment overcoming differences in level between two adjoining areas); *s* **talud [m] de unión** (►Talud que sirve para superar diferencias entre el nivel del terreno y el planeado); *syn.* talud [m] de transición; *f* **talus [m] de raccordement** (Dans le but de niveler des différences de niveau sur le terrain ; ►talus) ; *g* **Anschlussböschung [f]** (►geschüttete Böschung, die zum Ausgleich unterschiedlicher Gelände- resp. Planungshöhen dient).

6493 transition zone [n] *plan.* (Intermediate area; e.g. between a city and the countryside); *syn.* middle landscape [n] [also UK] (LD 1995 [4], 6); *s* **zona [f] de transición** (Área intermedia, p. ej. entre el campo y la ciudad); *f* **zone [f] de transition** (Espace caractéristique du changement des formes l'utilisation du sol ou de la morphologie des paysages ; *urbanisme* zone d'un plan local d'urbanisme appelée à évoluer p. ex. **1.** entre le tissu urbain existant et les secteur naturels et agricoles, **2.** vers des secteurs plus denses pour favoriser la construction de petits collectifs ou entre les quartiers pavillonnaires et les quartiers d'habitat collectif) ; *g* **Übergangsbereich [m]** (Bereich des Wechsels von einem Bodennutzungsbereich zum anderen, z. B. Zone zwischen Stadt und Land).

transit lane [n] *trans.* ►offside lane, #2.

transit-mixed concrete [n] [US] *constr.* ►ready-mixed concrete.

translational slip [n] *constr. geo. pedol.* *►landslide.

6494 translocation [n] **of clay** *pedol.* (Downward movement of clay particles, especially fine clay fractions < 0.002mm such as clay minerals, fine-grained oxides of Fe, Al and Si as well as humic material bound with mineral particles; ►claypan, ►illuvial horizon); *s* **eluviación [f] de arcilla** (Remoción mecánica de arcilla en suspensión por las aguas de percolación; ►horizonte de iluviación 1, ►horizonte iluvial); *f* **processus [m] d'accumulation d'argile** (Migration de la fraction argileuse et des hydroxydes de Fe, d'Al et de Si qui lui sont liés par les eaux de gravité vers les horizons profonds ; ►horizon d'accumulation d'argile, ►horizon illuvial) ; *syn.* lessivage [m] d'argile, entraînement [m] d'argile (cf. PED 1983, 90), éluviation [f] ; *g* **Tonverlagerung [f]** (Abwärtsbewegung von Bestandteilen der Tonfraktion, vor allem der Feintonfraktion [< 0,2 μm] wie Tonminerale, feinkörnige Oxide des Fe, Al und Si sowie mit Mineralteilchen verbundene Huminstoffe; SS 1979; ►Tonband, ►Anreicherungshorizont); *syn.* Lessivierung [f], Illimerisation [f].

6495 transmission [n] *envir.* (Transport of pollutant matter through the air to the place of precipitation/input; ►polluter); *s* **transmisión [f]** (Transporte de contaminantes en la atmósfera desde el ►emitente hasta el lugar de disposición); *f* **transfert [m]** (DUV 1984 ; déplacement dans l'air des matières polluantes de la source d'émission jusqu'à la zone de dépôt ; ►émetteur) ; *g* **Transmission [f]** (Transport von Stoffen durch die Luft vom ►Emittenten zum Eintragsgebiet).

transmission line [n] [US]**, electric power** *constr. envir.* ►high tension power line.

transmission line [n] [US]**, high voltage** *constr. envir.* ►high tension power line.

transnational planning [n] *plan.* ►transfrontier regional planning.

6496 transpiration [n] *bot.* (Photosynthetic and physiological process by which living plants release water into the air in the form of water vapo[u]r; RCG; ►evapotranspiration, ►protection against transpiration); *s* **transpiración [f]** (Transferencia de vapor de agua a la atmósfera a través de los estomas de las plantas; DINA 1987; ►evapotranspiración, ►protección contra la transpiración); *f* **transpiration [f]** (Fonction d'élimination d'eau sous forme de vapeur que réalisent les végétaux au niveau de leurs stomates ; DIB 1988 ; ►évapotranspiration, ►protection contre la transpiration) ; *g* **Transpiration [f]** (Form der Verdunstung durch Abgabe von Wasserdampf aus lebenden Pflanzenteilen; ►Evapotranspiration, ►Verdunstungsschutz).

6497 transplant [n] (1) *hort.* (Seedling after it has been lifted and replanted one or several times in a nursery; cf. SAF 1983);

T

s **plantón** [m] (Planta joven cultivada en vivero con suficiente tamaño para ser plantada en una ubicación definitiva); *f* **jeune plant** [m] **(repiqué)** (Végétal au début de son développement résultant de semis, marcotte, bouture, éclat, greffe ou tout autre mode de reproduction ou de multiplication, ayant en général subi un repiquage en pépinière et ayant les qualités requises pour être proposé à la vente pour plantation) ; *g* **Setzling** [m] **(2)** (Ausreichend große, gärtnerisch kultivierte Jungpflanze, die an einen endgültigen Standort gepflanzt wird); *syn.* Setzpflanze [f].

transplant [n] (2) *for. hort.* ►whip.

transplant [n] (3) *hort.* ►young shrub transplant.

6498 transplantation [n] **of semi-mature trees** *constr.* (Moving of tall trees, mostly with a trunk diameter of more than 20 cm, usually done with tree movers; ►mature tree, ►root curtain); *s* **transplante** [m] **de árboles maduros** (Procedimiento de transplantación de árboles grandes, generalmente de un diámetro de tronco de más de 20 cm, que se realiza en general con ayuda de máquina de transplante. Para preparar el transplante hay que analizar, por medio de zanjas, el desarrollo del sistema radical y —por lo menos dos años antes de llevarlo a cabo— crear una ►cortina de raíces para que se formen raicillas en la proximidad de la raíz principal. Para eliminar daños mecánicos de las raíces y promover su cicatrización, se podan hasta las zonas sanas y se aplica sustancia protectora. Se debe realizar aclareo de la copa de forma adecuada para la especie. En el lugar de plantación, se planta el árbol —sin almacenamiento intermedio— en hoyo de plantación no descompactado para evitar que el árbol se hunda; ►árbol maduro); *syn.* transplante [m] de árbol grande; *f* **transplantation** [f] **de gros végétaux** (Opération horticole de déplacement d'arbres adultes [en général ►arbre mature avec tronc d'un diamètre > 60 cm], réalisée par moyens mécaniques avec une transplanteuse pour le replanter ailleurs ; ►cernage des racines) ; *syn.* déplacement [m] d'arbre de grande taille, transplantation [f] d'arbre de haute-tige ; *g* **Großbaumverpflanzung** [f] (Das Verpflanzen eines großen Baumes, meist mit einem Stammumfang > 60 cm, i. d. R. mit Verpflanzungsmaschinen. Zur Vorbereitung sollte durch Vorgrabungen die Ausbildung des Wurzelsystems ermittelt und mindestens zwei Jahre vor Umsetzung mit einem ►Wurzelvorhang darauf vorbereitet werden, um ein konzentriertes stammnahes Feinwurzelwerk auszubilden. Um mechanische Schäden an Wurzeln zu beseitigen und die Abschottung gegen Pilzbefall zu fördern, sind sie bis auf das intakte Holz nachzuschneiden und mit Pflanzenschutzmitteln zu behandeln. Die Krone muss artgerecht ausgelichtet und nicht durch Kappen von Starkästen verstümmelt werden. Am Pflanzort sind die Bäume ohne Zwischenlagerung in ungelockerte Pflanzgruben zu setzen, um ein Sacken zu verhindern; ►Großgehölz); *syn.* Verpflanzen [n, o. Pl.] von Großgehölzen.

6499 transplanting [n] **(1)** *hort.* (Lifting and replanting of woody plants or perennials in another place; ►transplantation of semi-mature trees; to ►transplant in nursery row); *s* **trasplante** [m] **(1)** (Trasplante de leñosas o vivaces a otro lugar; ►transplante de árboles maduros, ►trasplantar en hilera de vivero; trasplantar [vb]); *f* **transplantation** [f] **(3)** (Opération horticole consistant à déterrer des végétaux ligneux ou des plantes vivaces pour les replanter dans un autre lieu ; ►contre-planter, ►repiquer, ►transplantation de gros végétaux) ; *syn.* déplacement [m] des végétaux ; *g 1* **Verpflanzung** [f] **(1)** (Umpflanzen von Gehölzen oder Stauden an einen anderen Ort; ►Großbaumverpflanzung, ►verschulen); *g 2* **Auspflanzen** [n, o. Pl.] (Das Verpflanzen junger Pflanzen aus dem Gewächshaus ins Freiland oder aus der Anzucht an die Stelle der weiteren Kultur); *syn.* Umpflanzen [n, o. Pl.], Umpflanzung [f] (2), Verpflanzen [n, o. Pl.].

6500 transplanting [n] **(2)** *constr.* (Reuse of plants at another on-site location); *syn.* moving [n] (BS 4428, 2.3.3); *s* **trasplante** [m] **(2)** (Arranque y plantación de planta en obra; RA 1970; trasplantar [vb]); *f* **récupération** [f] **et transplantation** [f] **des végétaux à conserver** (Travaux préliminaires de la mise en chantier, conservation des végétaux pour une réutilisation après déplacement) ; *g* **Verpflanzung** [f] **(2)** (Wiederverwendung ausgegrabener Pflanzen aus einem Bestand, der umgestaltet wird); *syn.* Umpflanzen [n, o. Pl.] ausgegrabener Pflanzen, Umpflanzung [f] ausgegrabener Pflanzen, Wiederverwendung [f] ausgegrabener Pflanzen.

6501 transplant [vb] **in nursery row** *hort.* (ANSI 1986, 22; to move ►nursery stock from one part of a nursery to another, essentially so as to improve its development before planting in the final position; to ►plant trees or shrubs in nursery row); *s* **trasplantar** [vb] **en hilera de vivero** (Trasplante de leñosas en vivero sobre todo para mejorar el desarrollo del cepellón antes de ser plantadas a su lugar definitivo; ►planta de vivero, ►plantar leñosas en hileras); *f 1* **repiquer** [vb] (Opération horticole par laquelle **1.** un plan provenant d'une planche de semis ou de bouture ou de marcotte ou tout autre mode de reproduction est remis en pleine terre à l'emplacement définitif ou s'achèvera leur croissance ; AFNOR 1991. **2.** Soit pour écarter les jeunes plants afin qu'ils aient assez de place, d'air et de lumière pour se développer de manière satisfaisante) ; *f 2* **contre-planter** [vb] (Transplantation d'espèces ligneuses âgées effectuée à racines nues ou en motte dans une pépinière en vue de fortifier le système radiculaire pour une bonne formation de la motte et une amélioration de la croissance ; dans les catalogues allemands le nombre de transplantations effectuées figure toujours mentionné avec la taille et le conditionnement des végétaux, p. ex. 3 x v. [transplanté 3 fois] ; ►fourniture de pépinière, ►planter des végétaux ligneux sur des lignes de culture) ; *g* **verschulen** [vb] (Verpflanzen von Gehölzen in Baumschulen, damit sie einen gut durchwurzelten Ballen erhalten; im Deutschen wird im Vergleich zum Französischen der Begriff ‚verschulen' sowohl für Jungpflanzen als auch für ältere Pflanzen benutzt; in Baumschulkatalogen hingegen wird nur der Begriff ‚verpflanzt' verwendet, z. B. 2 xv [zweimal verpflanzt]; zu verschulende Pflanzen werden *Verschulpflanzen* genannt; ►aufschulen, ►Baumschulware).

transport [n] *envir. limn. pedol.* ►nutrient transport; *trans.* ►private transport; *constr.* ►soil transport.

transport [n] [UK], **public** *trans.* ►public transit [US].

transportation [n], **public** *trans.* ►public transit [US]/public transport [UK].

6502 transportation corridor [n] *plan. trans.* (TGG 1984, 223; major traffic route which often combines various modes of transport, e.g. a road, railroad [US]/railway [UK], river, canal; *specific term* [rail]road corridor [US]); *syn.* transportation route [n] (TGG 1984, 248), traffic route [n] (1); *s* **arteria** [f] **de transporte** (Vía principal de tráfico en una zona; bajo el término **a. de t.** a menudo se entiende una concentración de diferentes infraestructuras de tránsito [carretera, río, canal, ferrocarril]); *f* **couloir** [m] **de transport** (Espace linéaire, souvent caractérisé par une concentration de divers modes de transport — route, fleuve, canal, chemin de fer) ; *g* **Verkehrstrasse** [f] (Hauptverkehrsweg/-linie in einem Gebiet, oft auch als Bündelung unterschiedlicher Verkehrseinrichtungen — Straße, Fluss, Kanal, Schiene — verstanden); *syn.* Verkehrsschiene [f], Trasse [f].

6503 transportation corridor selection [n] *plan. trans. wat'man.* (Process of evaluating various options for the most environmentally-friendly transportation route; e.g. canal, highway, railroad [US]/railway [UK]; ►transportation route selection); *s* **selección [f] de corredor (de transporte) (≠)** (Proceso de decisión sobre la banda de terreno por donde ha de realizarse un proyecto de carretera, vía de ferrocarril o canal teniendo en cuenta los aspectos funcionales, técnico-constructivos, financieros, geográficos y ecológicos; ►selección de trazado); *f* **recherche [f] du tracé optimum** (Phase de décision dans une étude de faisabilité permettant la recherche et l'identification d'un corridor de transport optimal [p. ex. route, chemin de fer, canal] ; ►définition du tracé) ; *syn.* étude [f] des tracés ; *g* **Trassenfindung [f]** (Planerischer Entscheidungsprozess bei der Auswahl von Trassenkorridoren [z. B. für Bahn, Straße, Kanal] unter Abwägung funktionaler, bautechnischer, finanzieller geografischer und ökologischer Gegebenheiten und Zielsetzungen; ►Trassenfestlegung).

6504 transportation embankment [n] *eng. trans.* (Artificial, linear mound of compacted soil and other fill material for carrying transportation facilities; *specific terms* ►road embankment, railroad embankment [US]/railway embankment [UK]); *syn.* carriage-way embankment [n] [also UK]; *s* **terraplén [m] para construcción viaria** (Formación lineal de tierra o material compactado que sirve de lecho a una carretera, una vía de ferrocarril o a otro sistema de transporte; ►terraplén para carretera); *f* **remblai [m] d'infrastructure de transport** (Ouvrage linéaire en remblai en terre compactée ou en pierres, de coupe transversale en général trapézoïdale servant de fondation aux voies de communication ; *termes spécifiques* ►remblai routier, remblai ferroviaire ; *syn.* remblai [m] d'un ouvrage de transport) ; *g* **Damm [m] für Verkehrsbauten** (Linienhafte, verdichtete Aufschüttung als tragfähiger Untergrund für den Verkehrswegebau; *UBe* Bahndamm, ►Straßendamm, [aufgeschütteter] Straßenkörper/Bahnkörper).

6505 transportation infrastructure [n] [US] *plan. trans.* (All facilities and installations used by airway, highway, railway and waterway traffic within a particular area, community, or region; ►infrastructure, ►traffic facility); *syn.* traffic infrastructure [n] [UK]; *s* **infraestructura [f] de tráfico y transportes [Es]/infraestructura [f] de tránsito y transportes [AL]** (Conjunto de instalaciones y vías de tránsito de todo tipo para el tráfico y transporte de personas y bienes; ►infraestructura, ►instalacion de tráfico y transportes [Es]/instalacion de tránsito y transportes [AL]); *f* **infrastructure [f] de transport** (Ensemble des équipements ou installations publics de communication au niveau d'un espace défini — commune, région ; ►équipement de transport et de communication, ►infrastructure) ; *g* **Verkehrsinfrastruktur [f]** (Gesamtheit der Verkehrswege [Straßen, Gleistrassen, Wasserstraßen und Flughäfen] und die dazugehörenden Einrichtungen in einem Raum — Gemeinde oder Region; ►Infrastruktur, ►Verkehrsanlage); *syn.* Verkehrsanlagen [fpl].

transportation network [n] *plan. trans.* ►transportation system.

6506 transportation [n] **of soil material** *geo. pedol.* (Generic term covering all forms of downslope [US]/down-slope [UK] movements of rock and soil material—►mass movement— as well as natural transport of soil by wind, water, ice or gravity; ►soil erosion); *s* **transporte [m] de material en la naturaleza** (Movimiento de rocas y de materiales de suelos por diversas causas: gravitacional, agua, viento, etc.; cf. DINA 1987; ►erosión del suelo, ►movimientos gravitacionales); *f 1* **transport [m] de matériaux** (Déplacement de matériaux sous l'effet de la pesanteur [►mouvement de masse], mouvement de matériaux enlevés par un processus d'érosion [vent et eau] jusqu'au lieu d'accumulation ou de sédimentation ; DG 1984 ; ►érosion du/des sol[s]) ; *f 2* **transit [m] de matériaux** (Situation momentanée des matériaux dans un transport discontinu ; VOG 1979) ; *g* **Umlagerung [f] in der Landschaft** (Bewegung von ganzen

Bodenkörpern [►Massenversatz] durch Schwerkrafteinwirkung oder Bewegung von Bodenmaterial durch Wasser oder Wind; ►Bodenerosion).

transportation plan [n] *trans. pol.* ►long-range transportation plan [US]/long-range transportat plan [UK].

transportation planning [n] *adm. plan. pol. trans.* ►traffic planning.

transportation route [n] *plan. trans.* ►transportation corridor.

6507 transportation route selection [n] *plan. trans.* (Process for selecting from several possible transportation alignments within a wide corridor, based on civil engineering criteria, and then evaluating the environmental impact of each alignment on a comparative basis; cf. TAN 1975, 136; *specific term* selection of highway location; ►road alignment, ►transportation corridor selection); *s* **selección [f] de trazado** (Decisión sobre el trazado exacto con el que se ha de construir una vía de tránsito [carretera, vía de ferrocaril, canal] teniendo en cuenta los aspectos funcionales, técnico-constructivos, financieros, geográficos y ecológicos; ►alineación de carretera, ►selección de corredor [de transporte]); *f* **définition [f] du tracé** (Décision sur le choix d'un variante [une parmi plusieurs variantes ou sous-variantes] ou d'un tracé alternatif dans les limites de l'aire d'étude [p. ex. pour les grands projets d'infrastructures de transport] sur la base de l'appréciation des données de l'état initial et des partis envisagés amenant à retenir la solution offrant le meilleur compromis entre les différentes contraintes techniques, financières, environnementales, économiques, sociales etc. ; ►recherche du tracé routier optimum, ►tracé routier) ; *g* **Trassenfestlegung [f]** (Planerische Entscheidung bei der Auswahl einer Trassenvariante [einer von vielen] oder einer Trassenalternative [einer von zweien] innerhalb eines breiten Korridors [z. B. für Bahn, Straße, Kanal] unter Abwägung bautechnischer, finanzieller, geographischer und ökologischer Gegebenheiten und Zielsetzungen; ►Straßenführung, ►Trassenfindung).

6508 transportation system [n] *plan. trans.* (General term for a comprehensive network of various transportation routes either proposed or existing within a particular, extensive area; transport hubs or interchanges connect the various routes ►circulation system, ►local public transportation system [US]/local transport system [UK]); *syn.* transportation network [n] (TGG 1984: 244); *s* **red [f] de transportes** (Conjunto de medios de tránsito de personas y bienes en una zona determinada. La **r. de t.** se refiere a los medios, mientras que la ►red de circulación lo hace a las infraestructuras que se necesitan para ellos, ►transporte público urbano); *syn.* sistema [m] de transportes; *f* **réseau [m] de transport** (Ensemble des lignes assurant la liaison et la desserte des aires d'emploi, des principales villes et des zones rurales sur un territoire donné [voies routières, ferroviaires et navigables ; ►réseau de voirie, ►transports suburbains) ; *g* **Verkehrssystem [n]** (Gesamtheit der in einem bestimmten Raum geplanten oder zur Verfügung stehenden und an Verkehrsknotenpunkten miteinander verbundenen Verkehrswege [Straßen, Bahntrassen und Wasserstraßen]; ►Erschließungssystem, ►öffentlicher Personennahverkehr); *syn.* Verkehrsnetz [n].

transportation system [n], **connection to a** *trans. urb. plan.* ►traffic linkage.

transport plan [n] [UK], **long-range** *trans. pol.* ►long-range transportation plan [US].

transport system [n] [UK], **local** *trans.* ►local public transportation system [US].

6509 transport [n] **to the construction site** *constr.* (►delivery at the construction site free of charge); *s* **transporte [m] a la obra** (►suministro a la obra sin recargo); *f* **achemine-**

T

ment [m] sur le chantier (Apport de matériaux ; ►fourniture à pied d'œuvre) ; *g* **Transport [m] zur Baustelle** (**1.** Beförderung/Lieferung von Baumaterialien und Arbeitskräften zur Verwendungs-/Einsatzstelle; **2.** für den **T.** zusammengestellte Materialien, Maschinen, Geräte und ggf. zu befördernde Arbeitskräfte; ►Lieferung frei Baustelle); *syn.* Baustellenandienung [f].

transverse section [n] *constr. trans.* ►cross section (1).

trap [n] *envir.* ►gasoline trap [US]/petrol trap [UK]; *constr.* ►mud trap (1), ►mud trap (2); *constr. hydr.* ►sand arresting trap [US]; *constr. recr.* ►sand trap [US]

trap [n], **air** *constr.* ►water seal, #2.

trap [n], **grease** *constr.* ►grease separator.

trap [n] [US], **odo(u)r** *constr.* ►water seal.

trap [n] [UK], **petrol** *envir.* ►gasoline trap [US].

trap [n] [UK], **silt** *constr.* ►sediment basin [US].

trash [n] *envir.* ►scattered rubbish [US]/litter [UK].

trash [n] [US], **bulk** *envir.* ►bulky waste.

6510 trash can [n] [US] (1) *envir.* (Container for collection of household waste which is located in or outside of a building); *syn.* dust bin [n] [UK], waste bin [n] [UK]; *s* **contenedor [m] de basura** (Recipiente localizado en la calle o en un patio interior destinado para recoger los residuos sólidos en un bloque o una casa); *f* **poubelle [f]** (Récipient destiné aux déchets solides ménagers placé à proximité ou à l'intérieur d'un immeuble ou d'un appartement) ; *syn.* bac [m] roulant ; *g* **Mülltonne [f]** (Im Gebäude oder außerhalb eines Gebäudes aufgestellte Tonne zum Sammeln fester Haushaltsabfallstoffe); *syn.* Müllcontainer [m].

6511 trash can [n] [US] (2) *envir.* (Small container installed in parks and open spaces for paper and small refuse; ►waste container); *syn.* trash receptacle [n] [US], litter bin [n] [UK]; *s* **papelera [f]** (2) (Pequeño recipiente instalado en calles, parques u otros espacios públicos para recoger residuos de viandantes; ►recipiente para residuos sólidos); *syn.* pipote [m] de basura [YZ]; *f* **corbeille [f] à ordures** (*Terme générique* récipient de collecte des papiers et petits résidus dans les espaces libres publics ou privés ; le terme « corbeille à papiers » est principalement utilisé dans la terminologie de la Bureautique ; ►récipient de collecte [des ordures ménagères]) ; *syn.* corbeille [f] de propreté ; *g* **Abfallkorb [m]** (Im Freien aufgestellter kleinerer Behälter zur Aufnahme von Kleinmüll/Kleinstabfällen); *OB* ►Abfallbehälter); *syn.* Papierkorb [m], Abfallkübel [m].

6512 trash can support [n] **with bag liner** [US] *envir.* (Free-standing container on metal or precast concrete support with disposable paper or plastic sack/bag; *generic term* ►waste container); *syn.* litter bin support [n] with sackholder [UK]; *s* **soporte [m] de sacos de basura** (Dispositivo metálico o de hormigón para sujetar sacos de basura; *término genérico* ►recipiente para residuos sólidos); *f* **support [m] sacs poubelles** (Équipement en général métallique recevant les sacs poubelles en polyéthylène de 80 à 110 litres ; *terme générique* ►récipient de collecte [des ordures ménagères]) ; *g* **Müllsackständer [m]** (Haltevorrichtung mit Plastiktüte zum Sammeln von Müll; *OB* ►Abfallbehälter).

trash dump [n] [US] *envir.* ►refuse dump [US]/refuse tip [UK].

trash platform [n] [US] *urb.* ►trash storage [US]/bin store [UK].

trash rack [n] [US] *constr.* ►strainer.

trash receptacle [n] [US] *envir.* ►trash can [US] (2)/litter bin [UK].

6513 trash storage [n] [US] *urb.* (Area in which trash cans [US]/dust bins [UK] are placed near a street or integrated in a multistor[e]y building); *syn.* trash platform [n] [US], bin store [n]

[UK]; *s* **emplazamiento [m] de cubos de basura** (Lugar perteneciente a un solar o en la calle en donde se encuentran los cubos o contenedores de basura para el edificio o el tramo de calle correspondiente); *f* **logette [f] vide-ordures** (Circulaire n° 77-127 du 25 août 1977 ; espace réservé aux conteneurs poubelles mobiles sur des espaces libres ou encastrés le long d'un bâtiment) ; *syn.* emplacement [m] pour poubelles ; *g* **Standplatz [m] für Abfallbehälter** (Standplatz mit beweglichen Mülltonnen auf einem Grundstück oder an ein Gebäude an- oder in dieses eingebaut); *syn.* Mülltonnenstandplatz [m], Müllanlage [f], Standort [m] für die Müll- und Abfallsammlung, Standort [m] für Müllcontainer/Müllbehälter.

travel costs [npl] *contr.* ►travel expenses.

6514 travel expenses [npl] *contr.* (Costs incurred during a business trip, such as travel, accommodation and food in connection with a design or construction contract); *syn.* travel costs [npl]; *s* **gastos [mpl] de viaje** (Incluyen los gastos de transporte, alojamiento y las dietas); *syn.* gastos [mpl] de desplazamiento; *f* **frais [mpl] de déplacement** (Frais particuliers à une opération constitués par les temps de déplacements par tous moyens, les temps immobilisés, l'hébergement et la restauration) ; *g* **Reisekosten [pl]** (Auf einer [Geschäfts]reise anfallende Kosten für z. B. Fahrt, Unterkunft und Verpflegung); *syn.* Reisespesen [pl].

6515 travelled way [n] [US] *trans.* (Generic term for street, road or highway which is funded and constructed by local, state, or federal authorities to meet traffic needs; ►interstate highway [US], ►navigable waterway, ►state highway [US]); *syn.* carriage-way [n] [UK], traffic route [n] [also UK] (2); *s 1* **carretera [f]** (Vía pavimentada para la circulación de vehículos fuera de la ciudad. Según qué nivel administrativo se haga cargo de su construcción y mantenimiento y según su importancia se clasifican en ►carreteras estatales, ►carreteras provinciales, **c.** locales, etc.; en **D.** el término «Straße» incluye las ►hidrovías; ►carretera federal, ►carretera interestatal); *s 2* **calle [f]** (Vía de tránsito de vehículos y personas dentro de la ciudad); *f 1* **route [f]** (Chaussée prévue pour la circulation terrestre des véhicules automobiles ; F., selon l'autorité administrative on distingue le réseau routier national [autoroutes et routes nationales], les ►routes départementales et les voies communales) ; *f 2* **voie [f]** (Terme spécialement utilisé pour désigner les eaux continentales navigables [►voie navigable (ou flottable)] ou le réseau ferroviaire [voie ferroviaire]) ; *g* **Straße [f]** (**1.** Für den Fahrzeugverkehr besonders hergerichteter befestigter Weg; nach dem Träger der Straßenbaulast unterscheidet man in Deutschland ►Bundesfernstraßen [Bundesautobahnen und Bundesstraßen], ►Landstraßen, Kreisstraßen und Gemeindestraßen. **2.** Im Dt. auch auf schiffbare Binnengewässer [►Wasserstraße] angewandt; *OB* Verkehrsweg).

6516 travelled way green spaces [npl] [US] *landsc.* (Plantings and grassed areas along roads or streets; ►right-of-way planting [US]/roadside planting [UK]); *syn.* roadside green [n] [UK]; *s* **verde [m] de acompañamiento** (≠) (Plantaciones y superficies de hierba generalmente a lo largo de carreteras o en zonas industriales; ►verde vial); *f* **plantations [fpl] et espaces [mpl] verts d'accompagnement** (Plantations et zones enherbées situées en général en bordure des axes de circulation ; ►espaces verts d'accompagnement de la voirie) ; *g* **Begleitgrün [n, o. Pl.]** (Bepflanzungen und Wiesenflächen, i. d. R. entlang von Verkehrswegen; ►Verkehrsgrün).

travellers [npl] [US], **accommodation-seeking holiday** *recr. plan.* ►accommodation-seeking vacationers [US]/accommodation-seeking holiday-makers [UK].

6517 travel trailer [n] [US] *recr.* (Trailer vehicle for camping purposes; ►camping vehicle, ►recreation vehicle [RV]

[US]/motor home [UK]); *syn.* camping trailer [n] [US], caravan [n] [UK]; *s* **remolque [m] (de camping)** (Vehículo sin motor adecuado para vivir durante las vacaciones; ►autocaravana, ►remolque-vivienda); *syn.* roulotte [f], caravana [f]; *f* **caravane [f]** (Véhicule ou élément de véhicule qui, équipé pour le séjour ou l'exercice d'une activité, conserve en permanence des moyens de mobilité lui permettant de se déplacer par lui-même ou être déplacé par simple traction ; C. urb., art. R. 443-2) ; ►camping-car, ►remorque de camping) ; *g* **Caravan [m]** (►Wohnwagen zur Freizeitnutzung in Form eines Anhängers; ►Wohnmobil); *syn.* Wohnanhänger [m].

travois road [n] [US&CDN] *for.* ►skidding road [US& CDN].

tread [n] *constr.* ►slope of tread.

tread [n] **and riser** [n] [US]**, flagstone** *constr.* ►flagstone step.

tread [n] [UK]**, landing** *constr.* ►landing.

6518 tread length [n] *arch. constr.* (Horizontal part of a step; includes the ►nosing; the width of a single stair tread excluding the nosing is called **stair run [n]**); *s* **huella [f] de escalón** (Porción horizontal de un escalón; ►mampirlán); *f* **giron [m]** (Constitue la largeur de la marche, c.-à-d. sa profondeur ; ►débordement) ; *g* **Auftrittstiefe [f]** (Breite einer Treppenstufe; ►Unterschneidung).

treatment [n] *envir.* ►industrial effluent treatment; *wat'man.* ►drinking water treatment; *envir.* ►management of wastewater and sewage treatment; *constr.* ►open tank treatment [US&CDN]/ steeping treatment [UK]; *landsc.* ►planting treatment; *constr.* ►pressure treatment [US]/pressure impregnation [UK]; *arb. hort.* ►root treatment; *envir.* ►sewage treatment; *constr.* ►surface treatment; *envir.* ►waste treatment, ►waste treatment and disposal [US]/public cleansing and waste disposal [UK]; ►wetland wastewater treatment; *arb. constr. hort.* ►wound treatment.

treatment [n] [US]**, biotechnical** *constr.* ►biotechnical construction technique.

treatment [n]**, landscape** *landsc.* ►planting treatment.

treatment [n] [UK]**, steeping** *constr.* ►open tank treatment [US&CDN].

treatment [n] **of contaminated land** [UK] *envir. plan.* ►toxic site reclamation.

6519 treatment plant [n] *envir. wat'man.* (Installation which process waste materials; *specific terms* ►drinking water extraction and treatment plant, ►sewage treatment plant, ►waste recycling plant, ►water treatment plant); *s* **planta [f] de tratamiento** (Término genérico para instalación de tratamiento de sustancias para prepararlas para un uso determinado; *términos específicos* ►planta de depuración de aguas residuals, ►planta de extracción y potabilización de agua, ►planta de reciclaje de residuos solidos, ►planta de tratamiento de aguas); *f* **installation [f] de traitement** (*Terme générique* tout ouvrage de traitement des déchets ; *termes spécifiques* ►station d'épuration des eaux usées, ►station de pompage et de traitement d'eau potable, ►usine de traitement des eaux, ►usine de traitement des résidus) ; *syn.* station [f] d'épuration ; *g* **Aufbereitungsanlage [f]** (Anlage für die Behandlung von Stoffen, um sie mit bestimmten Eigenschaften für bestimmte Verwendungszwecke anzupassen; *UBe* ►Kläranlage, ►Müllverwertungsanlage, ►Trinkwassergewinnungs- und -aufbereitungsanlage, ►Wasseraufbereitungsanlage).

6520 tree [n] (1) *bot. arb.* (Wood species usually with a single vertical main trunk, which develops a crown in the upper part. Big trees [> 20m], medium trees [> 15m] and small trees [> 7m] are the main categories. For ►fruit trees, other criteria are used according to the vigo[u]r of their understocks); *syn.* macro-

phanerophyte; *s* **árbol [m]** (Vegetal leñoso generalmente con único tronco recto y con una copa en su extremo superior; se clasifican en árboles grandes [> 20 m], medianos [> 15 m] y pequeños [> 7 m]; en el caso de los ►árboles frutales se utiliza otro tipo de clasificación); *syn.* macrofanerófito [m]; *f* **arbre [m]** (Végétal ligneux présentant un fut nu cylindrique ou à peine conique [tige ou tronc], surmonté d'un ensemble de plusieurs branches appelé tête ou couronne ; les arbres sont caractérisés selon leur port, leur utilisation et leur développement ; D., on distingue les grands arbres [> 20 m], les arbres de taille moyenne [> 15 m] et les arbres de petite taille [> 7 m] ; on peut aussi classer les arbres selon leur pouvoir de croissance ou de développement, la classe 1 qualifiant les arbres ayant une croissance supérieure à 20 m et la classe 2 une croissance supérieure à 7 m ; ►arbre fruitier) ; *syn.* macrophanérophyte [m] ; *g* **Baum [m] (2)** (Holzgewächs mit i. d. R. einem einzigen aufrechten Stamm, der im oberen Teil eine Krone entwickelt; es werden Großbäume [> 20 m], mittelgroße Bäume [> 15 m] und Kleinbäume [> 7 m] unterschieden; es gibt auch die Einteilung: Bäume 1. Wuchsklasse [> 20 m], Bäume 2. Wuchsklasse [> 7 m]; bei ►Obstbäumen, die gärtnerisch nach Stammhöhen und auf unterschiedlich starken Unterlagen gezogen werden, gelten andere Unterscheidungen); *syn.* Makrophanerophyt [m].

tree [n] (2) *arb.* ►accidental damage to a tree; *hort.* ►ball wiring (of a tree); *arb.* ►butt of a tree; *landsc. urb.* ►city tree; *constr. for.* ►crop tree; *bot.* ►deciduous tree, ►coniferous tree; *constr. for.* ►crop tree; *hort.* ►dwarf fruit tree [US]/short standard [UK], ►espaliered fruit tree; *agr.* ►field tree, *agr.* ►fodder tree; *hort.* ►fruit tree; *arb. hort.* ►ground-branching tree [US]/feathered tree [UK]; *arb. constr.* ►groundfill around a tree; *arb.* ►high crowned tree, ►large-caliper tree/large-calliper tree, ►mature tree; *hort.* ►miniature fruit tree [US] (2); *conserv. leg.* ►monarch tree [US]; *hort. gard. plant.* ►ornamental tree; *agr. conserv.* ►pollarded tree; *hort.* ►pyramidal dwarf fruit tree [US]/pyramidal fruit bush [UK], ►pyramidal fruit tree; *zool.* ►roost(ing) tree; *arb. hort.* ►semi-dwarf fruit tree [US]/half standard [UK]; *agr.* ►shade tree (1); *arb. hort.* ►shade tree [US] (2)/tall standard [UK]; *hort.* ►specimen tree/shrub, ►standard fruit tree, ►standard street tree [US]/avenue tree [UK]; *landsc. urb.* ►street tree; *for. hort.* ►top of a tree; *conserv.* ►voluntary caretaker of a tree [US]; *constr. hort.* ►weed tree; *arb.* ►wind-blown tree [US]/wind-trimmed tree [UK].

tree [n] [UK]**, avenue** *hort.* ►standard street tree [US].

tree [n]**, blasted** *agr. for. hort.* ►storm damage.

tree [n]**, broadleaf** *bot.* ►deciduous tree.

tree [n]**, broad-leaved** *bot.* ►deciduous tree.

tree [n]**, canopy edge of a** *arb. hort.* ►drip line.

tree [n] [US]**, caretaker of a** *conserv.* ►voluntary caretaker of a tree.

tree [n] [US]**, champion** *conserv. leg.* ►monarch tree [US].

tree [n]**, dead (standing)** *ecol.* ►snag [US&CDN] (3).

tree [n]**, earth fill around a** *arb. constr.* ►groundfill around a tree.

tree [n]**, fan-trained fruit** *hort.* ►palmette espalier.

tree [n] [UK]**, feathered** *arb. hort.* ►ground-branching tree [US].

tree [n] [US]**, hedgerow** *agr. land'man.* ►field tree.

tree [n] [US]**, high branched** *arb.* ►high crowned tree.

tree [n] [US]**, landmark** *conserv. leg.* ►monarch tree [US].

tree [n]**, lopped** *agr.* ►fodder tree.

T

tree [n] [US] **(1), miniature fruit** *hort.* ▶dwarf fruit tree [US]/short standard [UK].

tree [n] [US]**, multiple-stemmed** *arb.* ▶multistem tree [US]/multistemmed tree [UK].

tree [n]**, needle-leaved** *bot. for.* ▶coniferous tree.

tree [n] [UK]**, one-year-old feathered** *arb. hort.* ▶strongly branched liner [US].

tree [n] [UK]**, reserved** *for.* ▶reserve [US] (1).

tree [n] [US]**, selfgrown** *for.* ▶woody wildling.

tree [n]**, tip of a** *for. hort.* ▶top of a tree.

tree [n]**, wind-deformed** *arb.* ▶wind-blown tree [US]/wind-trimmed tree [UK].

tree [n]**, wind-swept** *arb.* ▶wind-blown tree [US]/wind-trimmed tree [UK].

tree [n] [UK]**, wind-trimmed** *arb.* ▶wind-blown tree [US].

tree adoption [n] [UK] *conserv.* ▶stewardship of trees [US].

6521 tree and shrub evaluation [n] *arb. conserv.* (Assessment of existing woody plants in monetary terms for the public good and to determine their significance for a particular site. In U.S., various methods are used according to the laws of each State: the *Replacement Cost Method* [natural restitution: the value of a tree up to a trunk circumference of 1m comprises the cost of replacing a tree of the same size, including maintenance operations and company profits], the *Compounded Replacement Cost Method* [a kind of cost accounting] and the *Trunk Formula Method* [trunk sectional area method: the value of large trees (trunk circumference >1m) is calculated by combining the Replacement Cost Method with the basic value per cm² sectional area of the trunk; the sectional area to be applied results from the difference between that of the tree under evalaution and that of its substitute, whereby the condition and location of the new tree are also taken into account; *s* **cálculo** [m] **del valor de leñosas** (Estima del valor económico de poblaciones de leñosas o de ejemplares individuales existentes para la sociedad o para valorar un terreno privado específico que se aplica p. ej. en casos de expropiación. En D. y EE.UU. existen diferentes métodos de valoración); *f* **calcul** [m] **de la valeur des végétaux ligneux** (Mesure de la valeur financière de peuplements ligneux existants en fonction des différentes utilités que peut en retirer la société — agrément, utilité sociale ou récréative. *Terme spécifique* calcul de la valeur des arbres) ; *syn.* directive [f] pour le calcul de la valeur des arbres [CH], taxation [f] de la valeur des végétaux ligneux ; *g* **Gehölzwertermittlung** [f] (Berechnung des monetären Wertes bei Schadenersatz oder im Falle der Enteignungsentschädigung von vorhandenen Gehölzbeständen, gemessen an der Bedeutung der Gehölze für die Allgemeinheit oder für das betreffende Grundstück; in D. entspricht gängiger Rechtsprechung die „Methode Koch", die den Geldwert von Gehölzen in Analogie der „Verordnung über Grundsätze für die Ermittlung der Verkehrswerte von Grundstücken [WertV]" berechnet. Alle Kosten einer wieder herzustellenden Neupflanzung, einschließlich der Pflegekosten bis zur Funktionserfüllung oder bis zur vorgefundenen Gehölzgröße, werden dabei zu Grunde gelegt. In den USA werden je nach den Rechtsverhältnissen in den einzelnen Bundesstaaten unterschiedliche Methoden bevorzugt: die *Replacement Cost Method* [Naturalrestitution: der Wert von Bäumen bis zu einem StU von 1 m errechnet sich aus den Kosten für eine Neupflanzung in gleicher Größe incl. Fertigstellungspflege und Unternehmergewinn], die *Compounded Replacement Cost Method* [eine Art Kostenrechnung] und die *Trunk Formula Method* [Stammquerschnittsverfahren: der Wert größerer Bäume (StU >1 m) wird aus einer Kombination von Replacement Cost Method und Basiswerten je cm² Stammquerschnittsfläche er-

rechnet; die maßgebliche Stammquerschnittsfläche resultiert aus der Differenz der Stammquerschnitte des zu bewertenden Baumes und des Ersatzgehölzes, wobei Zustand und Lage des neu zu pflanzenden Gehölzes als Korrekturfaktoren die **G.** beeinflussen]; cf. FLL 2000); *syn.* Wertermittlung [f] von Gehölzen.

6522 tree and shrub planting [n] *constr. landsc.* (**1.** Area occupied by planted woody vegetation. **2.** Planting process of woody species); *syn.* woody species planting [n]; *s* **plantación** [f] **de leñosas** (Conjuntos de árboles y arbustos plantados); *f 1* **plant** [m] (Ensemble de végétaux d'une même espèce planté sur un même terrain) ; *f 2* **plantation** [f] **de végétaux ligneux** (**1.** Action de planter des arbres et des arbustes. **2.** Espace planté avec des arbres et des arbustes) ; *syn.* plantation [f] d'arbres et d'arbustes, plantation [f] de ligneux) ; *g* **Gehölzpflanzung** [f] (**1.** Tätigkeit des Pflanzens von Bäumen und Sträuchern. **2.** Mit Bäumen und Sträuchern bepflanzte Fläche).

6523 tree and shrub pruning [n] *constr. hort.* (Cutting away of parts of a woody plant to control shape and size, to remove dead or diseased wood, to improve flowering, fruiting or the structural stability, as well as to induce vigo[u]rous growth; ▶cutting back of trees and shrubs, ▶dead wooding [US] 1/cleaning out [UK], ▶drop-crotching, ▶pollarding, ▶pruning at planting, ▶thinning[-out], ▶topiary, ▶topping 3, ▶training of young trees); *s* **poda** [f] **de leñosas** (Corte de ramificaciones de árboles y arbustos o porciones de éstas con el fin de regular su tamaño, y forma, favorecer el crecimiento y la floración, eliminar ramas muertas o enfermas y mejorar la estática; podar [vb] leñosas; *términos específicos* ▶aclareo de la copa, ▶descopado, ▶desmoche, ▶poda de formación, ▶poda de plantación, ▶poda de reducción de la copa, ▶poda de topiaria, ▶ramoneo, ▶rebaje); *f* **taille** [f] **des végétaux ligneux** (Opération culturale consistant à réduire la longueur des rameaux aériens d'arbustes et arbrisseaux dans le but d'améliorer l'aspect esthétique et la croissance, d'augmenter l'éclat de la floraison, d'améliorer ou de régulariser la fructification ; *termes spécifiques* ▶art topiaire, ▶éclaircissage 2, ▶éhoupage, ▶élagage, ▶étêtage, ▶réduction de couronne, ▶taille d'entretien courant, ▶taille de formation de la charpente, ▶taille de plantation) ; *g* **Gehölzschnitt** [m] (Wegschneiden von oberirdischen Teilen eines Baumes oder Strauches, um z. B. Größe, Form und Wüchsigkeit zu regulieren, bei Bäumen die Standsicherheit zu verbessern, Totholz und kranke Astpartien zu entfernen, das Blühen und den Fruchtansatz zu verbessern etc.; *UBe* ▶auf den Kopf setzen, ▶Auslichten, ▶Erhaltungsschnitt, ▶Erziehungsschnitt, ▶Formschnitt, ▶Kronenkappung, ▶Kronenrückschnitt, ▶Pflanzschnitt, ▶Rückschnitt von Gehölzen).

tree assessment [n]**, visual** *adm. arb. leg.* ▶tree inspecttion.

tree [n] [UK]**, avenue** *hort.* ▶standard street tree [US].

tree barrier [n] *urb.* ▶protective tree barrier.

tree base [n] *constr. hort.* ▶tree pit surface.

tree bed [n] *constr. hort.* ▶tree pit surface.

6524 tree bench [n] *gard. landsc.* (Bench encircling a tree); *syn.* tree seat [n]; *s* **banco** [m] **alrededor de un árbol** ; *f* **banc** [m] **autour d'un arbre** ; *g* **Baumbank** [f] (Um einen Baumstamm aufgestellte Sitzbank).

6525 tree canopy [n] *for. phyt.* (Continuous cover of branches and foliage formed by the crown of a single tree or the space occupied by a number of adjacent tree crowns; ▶woodland canopy); *syn.* crown canopy [n], leaf canopy [n], *for.* forest canopy [n]; *s* **dosel** [m] **de árbol** (Cobertura continua de ramas y follaje formada por la copa de un árbol o espacio ocupado por copas de un grupo de árboles adyacentes; ▶dosel del bosque);

f **couvert [m] de la cime** (Écran formé par l'ensemble continu des branches et du feuillage d'un arbre ou d'un peuplement forestier ; ►couvert continu des cimes ; *termes spécifiques* couvert [m] de la cime d'un arbre, couvert [m] d'un peuplement forestier, voûte [f] forestière) ; *g* **Kronenraum [m]** (Raum in einer einzelnen Baumkrone oder der Raum, der in einem Waldbestand von der Summe der Baumkronen eingenommen wird; ►Kronendach des Waldes).

tree canopy level [n] *phyt.* ►tree layer.

6526 tree canopy shading [n] *phyt.* (Protection from or reduction of sunlight by the tree crown; ►canopy closure, ►canopy cover, ►shading, ►woodland canopy); *s* **sombreado [m] por árboles** (Protección o reducción de la intensidad de la luz por el ►cierre del dosel [del bosque], ►cobertura del dosel de un rodal, ►dosel del bosque); *f* **degré [m] d'ouverture** (Expression caractérisant la limitation ou l'arrêt complet de la lumière par la strate arborescente ; ►couvert continu des cimes, ►couvert d'un peuplement forestier, ►fermeture du couvert) ; *g* **Beschirmung [f]** (Schutz vor oder Reduzierung der Belichtung der/des Lichteinfalls in die Kraut- oder Strauchschicht durch das ►Kronendach eines Waldes; ►Beschattung, ►Kronenschluss 1 und ►Kronenschluss 2).

tree care [n] [US] *arb. constr.* ►tree maintenance.

tree circle edging [n] [UK] *constr.* ►tree pit edging [US].

6527 tree clearing [n] **and stump removal (operation)** [n] [US] *constr. for. hort.* (Tree removal by first cutting the trunk and then digging up the stump; ►deforestation, ►stump out); *syn.* grub felling [n] [UK], tree clearing work [n] [US]; *s 1* **despeje [m] de árboles y tocones** (►extraer de tocones); *syn. constr.* descuajo [m], descuaje [m], descuajar [vb] (HDS 1987), trabajos [mpl] de descuaje/despeje; *s 2* **roturación [f]** *for.* (Preparación de terreno inculto o no laborado durante mucho tiempo para su siembra. Si el terreno está cubierto de vegetación herbácea, se corta y extiende ésta y se aplica cal para neutralizar la acidez. Si la vegetación es arbustiva se realiza una **r. con rozas:** Se prende fuego y después se procede en la forma indicada anteriormente. Si el campo está arbolado se hace la **r. con descuaje:** Se cortan los árboles y luego se arrancan o descuajan los tocones a mano, con arado o explosivos y finalmente se procede a la **r.** de la forma ya descrita; cf. DGA 1986; ►de[s]forestación); *f* **abattage [m] par extraction de souche** (Coupe effectuée en laissant la souche ou une partie de la souche, attenante à l'arbre abattu ; cf. DFM 1975 ; ►déforestation, ►défrichement, ►dessoucher) ; *syn.* abattage [m] à culée noire, abattage [m] après déchaussement, arrachage [m] des arbres, travaux [mpl] d'arrachage des arbres (GEN 1982-I, 111) ; *g* **Rodung [f]** (Bäume mit Wurzeln aus dem Boden entfernen resp. die beim Fällen zurückgebliebenen Stubben ausgraben, ausheben oder heraussprengen; ►Entwaldung, ►Stubben roden); *syn.* Rodungsarbeiten [fpl]; *for.* Ausreutung [f] [CH].

tree clearing work [n] [US] *constr. for.* ►tree clearing and stump removal (operation) [US]/grub felling [UK].

6528 tree cover [n] *for. landsc.* (Total amount and kind of trees covering a specific land area; ►tree data bank); *s* **cobertura [f] arbórea** (Conjunto de áreas de bosque, grupos e hileras de árboles en una superficie dada; ►inventario de árboles urbanos); *syn.* población [f] forestal; *f* **peuplement [m] arborescent** (Ensemble des massifs forestiers, allées et groupes d'arbres sur une aire donnée ; ►fichier informatique de description des arbres) ; *g* **Baumbestand [m] (2)** (Summe von Waldflächen, Baumreihen, Baumgruppen und Einzelbäumen auf einer definierten Fläche; ►Baumdatei); *syn.* Baumbewuchs [m], Bestockung [f], *for.* stehender Bestand [m].

tree crop [n] *for.* ►stand of timber.

6529 tree crown [n] *arb. hort.* (Upper part of a tree carrying the main branch system and foliage/needles; ►crown form); *syn.* tree head [n] [also UK]; *s* **copa [f] de un árbol** (►forma de la copa); *f* **couronne [f] d'arbre** (Ensemble des branches, des rameaux et des feuilles d'un arbre ; ►forme de la cime) ; *syn.* cime [f], *for.* houppier [m] (IDF 1988, 291) ; *g* **Baumkrone [f]** (Gesamtheit des Astwerkes mit den Blättern/Nadeln eines Baumes; ►Kronenform).

tree crowns [npl] *arb. for.* ►opening up of tree crowns.

6530 tree data bank [n] *adm. arb.* (Computer record of tree species and their mapping or numbering according to age, health, condition and decline of individual trees within the scope of a city ►tree survey; ►cadastral map of trees, ►green space data bank, ►tree inspection, ►tree maintenance); *s* **inventario [m] de árboles urbanos** (Inventario informatizado de la ubicación, la edad y el estado de los árboles individuales que sirve de base para organizar el ►mantenimiento de los árboles 2; ►catastro de árboles, ►inspección de árboles, ►inventario de zonas verdes y de árboles, ►muestro de árboles urbanos); *f* **fichier [m] informatique de description des arbres** (Inventaire informatisé comprenant diverses informations sur les arbres en milieu urbain, le ►relevé des essences [arborescentes], leur état phytosanitaire, éventuellement constaté par des images infrarouges, ainsi que les travaux d'entretien effectués ou à réaliser. Ce fichier permet de programmer les interventions nécessaires aux ►travaux d'entretien des arbres, d'entretien [personnel et matériels] et d'évaluer les coûts pour l'établissement du budget ; ►cadastre des arbres plantés, ►contrôle visuel des arbres, ►fichier informatique de description des espaces verts) ; *g* **Baumdatei [f]** (Rechnergestützte Datensammlung, die Daten der ►Baumaufnahme und zusätzlich den Gesundheitszustand durch ►Baumschau oder Infrarotluftbilder ermittelt sowie notwendige und durchgeführte Pflegemaßnahmen enthält. Sie dient als Grundlage für die Organisation der ►Baumpflege [Personal- und Maschineneinsatz] und für die Haushaltsplanung. In manchen Städten beschränkt sich die **B.** auf Daten der Baumaufnahme mit unterschiedlich vielen Erfassungskriterien; ►Baumkataster, ►Grünflächendatei); *syn.* Inventur [f] des Baumbestandes, Bauminventur [f]).

6531 tree-fall gap community [n] [US] *phyt.* (LE 1986, 97; *specific term* forb formation in opened-up forests; ►clearance herb formation, ►clear felling); *syn.* forest clearing community [n] [UK]; *s* **vegetación [f] de claros** (Comunidad vegetal, al principio herbazales megafórbicos, posteriormente matorrales de claros, que surge en manchas de luz directa originados por la muerte natural de árboles maduros, en claros creados por ►tala a matarrasa o debidos a incendios sobre todo de las copas en los que no se destruyó totalmente la capa de humus bruto; ►vegetación de bosques talados); *syn.* formación [f] de claros; *f* **flore [f] de clairière** (Associations végétales, en général à l'origine une végétation de plantes herbacées hautes, suivie d'une végétation buissonnante, localisées dans les clairières naturelles, ►coupes à blanc, brûlis dans lesquels le feu n'a pu détruire la totalité de la couche d'humus ; ►flore des coupes à blanc[-étoc]) ; *syn.* végétation [f] de clairière, formation [f] clairiérée ; *g* **Lichtungsflur [f]** (Pflanzengesellschaft — zunächst meist ►Hochstaudenfluren, später Waldlichtungsgebüsche — in natürlichen Waldlichtungen, ►Kahlschlägen oder auf Waldbrandflächen, auf denen durch Kronenfeuer nicht der gesamte Auflagehumus zerstört wurde; cf. ELL 1978; *UBe* Lichtungskrautflur, [Wald]lichtungsgebüsche, ►Kahlschlagflur); *syn.* Waldlichtungsflur [f].

6532 tree feeding [n] *arb. constr.* (Application of fertilizer to the soil around the roots of a tree; ►fertilizing hole); *syn.* root feeding [n], tree fertilizing [n] [also US]; *s* **fertilización [f] de árboles** (Aporte de abono y agua a árboles maduros por medio de

►hoyo de alimentación); *syn.* abono [m] de árboles; *f* **nutrition [f] des arbres** (Apport d'engrais et d'eau autour de sujets adultes ou en voie de vieillissement à l'aide d'►orifices d'alimentation espacés sur une circonférence précise dans la zone du système racinaire d'un arbre) ; *syn.* fertilisation [f] des arbres ; *g* **Baumfütterung [f]** (Versorgung älterer oder alter Bäume mit Nährstoffen und Wasser durch ►Fütterungslöcher im Wurzelbereich); *syn.* Tiefenvorratsfütterung [f].

tree felling [n] *constr.* ►tree felling work.

6533 tree felling work [n] *constr.* (Complete or partial removal of trees from an area; ►felling 1, ►tree clearing and stump removal [operation] [US]/grub felling [UK]); *syn.* tree felling [n]; *s* **trabajos [mpl] de tala y despeje de árboles** (Conjunto de trabajos necesarios para talar árboles individuales o un grupo de ellos para preparar el terreno o como medida de mantenimiento en un jardín, parque o en otro tipo de zona verde; ►despeje de árboles y tocones, ►roturación, ►tala 1); *f* **travaux [mpl] d'abattage d'arbres** (►abattage, ►abattage par extraction de souche) ; *g* **Baumfällarbeiten [fpl]** (Gärtnerische Leistung, die das Umsägen von Einzelbäumen oder eines Baumbestandes zur Baufeldfreimachung oder als Pflegemaßnahme in einer Grünanlage oder in einem anderen Freiraum mit allen dazu gehörenden Arbeiten beinhaltet; ►Fällen, ►Rodung).

tree fertilizing [n] [US] *arb. constr.* ►tree feeding.

6534 tree grate [n] [US] *constr.* (*Specific terms* cast iron tree grate [US]/cast iron tree grille [UK], ►concrete tree grate [US]/concrete tree grid [UK]; *generic term* ►tree pit cover); *syn.* tree grill(e) [n], tree grid [n] [UK]; *s* **rejilla [f] (metálica) para alcorque** (Término específico para ►cobertura de alcorque [prefabricada] de hormigón; ►cobertura de alcorque); *f* **grille [f] (de protection de pied) d'arbre** (*Mobilier urbain* Terme spécifique pour tout ►revêtement de protection de pied d'arbre en fonte, en acier galvanisé ou en béton, de forme demi-circulaire, circulaire monobloc profilé ou non, hexagonale, carré monobloc profilé ou non, rectangulaire monobloc, implanté sur trottoirs et zones piétonnes non circulées ; ►panneau de béton perforé pour la protection de pied d'arbre) ; *g* **Baumscheibenrost [m]** (Gitterelemente aus hochfestem Gusseisen oder Metall zur Abdeckung von Baumscheiben im Straßenraum, um den Wurzelraum vor Verdichtung zu schützen und Luft- und Wasserzufuhr zu ermöglichen; *OB* ►Baumscheibenabdeckung; ►Betonbaumscheibe); *syn.* Gitterrost [m] für Baumscheiben, Gitterabdeckung [f] für Baumscheiben.

tree grid [n] [UK] *constr.* ►tree grate [US].

tree grid [n] [UK]**, concrete** *constr.* ►concrete tree grate [US].

tree grill [n] *constr.* ►tree grate [US]/tree grid [UK].

tree grille [n] *constr.* ►tree grate [US]/tree grid [UK].

tree grille [n] [UK]**, concrete** *constr.* ►concrete tree grate [US].

tree grove [n] [US]**, fruit** *hort.* ►orchard.

6535 tree guard [n] *arb. constr.* (Generally a metal or wire cage enclosing a tree trunk to protect it from mechanical damage); *s* **rejilla [f] de protección del tronco** (Enrejado metálico o de alambre para proteger el tronco de daños mecánicos); *f* **corset [m] de protection (métallique)** (Corset composé d'éléments verticaux, en général en feuillard galvanisé et reliés par des cerces soudées pour la protection du tronc contre les chocs occasionnés par la circulation de véhicules) ; *g* **Baumschutzgitter [n]** (Aus Metallstäben gefertigter Schutz für Baumstämme gegen mechanische Schäden).

tree guard [n]**, wire netting** *arb. constr. for.* ►wire mesh tree guard.

tree head [n] [UK] *arb. hort.* ►tree crown.

6536 tree inspection [n] *adm. arb. leg.* (Annual or otherwise regular inspection of public trees to ascertain their health and safe condition); *syn.* visual tree assessment [n]; *s* **inspección [f] de árboles** (Control anual o regular de los árboles públicos para comprobar su estado de salud y su estabilidad con el fin de evitar daños a terceros y tomar medidas de saneamiento en caso de que se detecte alguna enfermedad); *syn.* inspección [f] arborícola; *f* **contrôle [m] visuel des arbres** (Opération de contrôle annuel ou semestriel du patrimoine arborescent public afin d'en établir l'état sanitaire et la nature des interventions nécessaires) ; *syn.* inspection [f] arboricole ; *g 1* **Baumschau [f]** (Jährliches Betrachten des [öffentlichen] Baumbestandes zur Feststellung und Dokumentation des Gesundheitszustandes und der Standsicherheit [Verkehrssicherungspflicht des Eigentümers als eine gesetzliche Pflichtaufgabe]. Die Kontrolle kann sich auf „fußläufiges Ansehen von unten" ohne Zuhilfenahme von Werkzeugen oder anderen Hilfsmitteln beschränken, wenn keine Auffälligkeiten zu erkennen sind); *g 2* **Visual-Tree-Assessment-Methode** (Neuere und rechtlich anerkannte Methode der Baumschau [VTA-Baumdiagnose] zur Erkennung des Gesundheitszustandes und der Standsicherheit/ Verkehrssicherheit von Bäumen. Die Protokollierung beinhaltet vier Teilschritte: **1.** die **Symptomerkennung** an Hand der Körpersprache der Bäume [z. B. Ausbildung des Astwerkes und der Belaubung, Zustand der Rinde, Vorkommen von Pilzfruchtkörpern], **2.** die genaue **Beschreibung der Schäden** und deren Lokalisation, **3. Schadensbewertung** und **4.** die **Festlegung der zu treffenden Maßnahmen**); *syn.* Baumkontrolle [f] durch Inaugenscheinnahme, optische Baumbeurteilung [f], Sichtkontrolle [f] an Bäumen, Sichtprüfung [f] an Bäumen, visuelle Baumbewertung [f], VTA-Methode [f].

6537 tree layer [n] *phyt.* (Vegetation layer of a forest; according to the situation there may be a lower, medium, and upper **t. l.**; ►stratification of plant communities); *syn.* tree canopy level [n] (of different heights) [also US]; *s* **estrato [m] arbóreo** (Capa de vegetación de un bosque, que —dependiendo del tipo de éste— puede diferenciarse en **e. a.** superior, medio e inferior; ►estratificación de la vegetación); *syn.* vuelo [m] (DB 1985); *f* **strate [f] arborescente** (*d'une forêt* il est possible — p. ex. dans la forêt tropicale — de différencier une strate arborescente supérieure, moyenne et inférieure ; ►stratification de la végétation) ; *g* **Baumschicht [f]** (Oberste Vegetationsschicht eines Waldes. Je nach Erfordernis können eine untere, mittlere und obere **B.**, z. B. im Tropenwald, unterschieden werden; ►Vegetationsschichtung).

tree-like [adj] *arb. phyt.* ►arborescent.

tree limit [n] *for. phyt.* ►tree line.

6538 tree line [n] *for. phyt.* (Limit created by climatic, topographical, or environmental factors, beyond which trees do not grow. 'Tree line' is more generally used for the altitudinal boundary and 'tree limit' for the latitudinal boundary. A distinction may be drawn between 'tree line' and ►timberline, the latter being roughly the limit of timber rather than isolated trees; cf. SAF 1983; ►timberline ecotone); *syn.* tree limit [n]; *s* **límite [m] de los árboles** (Línea altitudinal o latitudinal más allá de la cual no pueden crecer los árboles; ►límite del bosque, ►zona de cumbres); *f* **limite [f] supérieure de la végétation arborescente** (Limite d'altitude ou de latitude au-delà de laquelle s'achève la végétation arborescente. En F. le terme ►limite des forêts est souvent appliqué au sens large, c.-à-d. aux arbres isolés [*en anglais* tree line] ; cf. DFM 1975 ; ►zone de combat) ; *g* **Baumgrenze [f]** (Höhenzone oder geografische Breite, über die hinaus auf Grund der Standortbedingungen keine Bäume mehr wachsen können. Im Französischen wird oft der Begriff ►Waldgrenze für **B.** benutzt; ►Kampfwaldzone).

6539 tree-lined avenue [n] [US] *gard. landsc. urb.* (Broad road or walk bordered by rows of trees; *syn.* allee lane [n] [US] (DIL 1987), avenue [n] [UK]; *s* **avenida [f] (2)** (Paseo arbolado a ambos lados. En alemán el término «Chaussee» se utiliza para una **a.** recta; *término específico* alameda [f]); *f* **allée [f] d'arbres** (Double rangée d'arbres d'espacement régulier, bordant une route ou un chemin) ; *g 1* **Allee [f]** (Verkehrsweg, der mit parallel zueinander verlaufenden Baumreihen mit i. d. R. gleichen Baumabständen gesäumt ist. Eine **A.** muss im Vergleich zur ‚Chaussee' nicht geradlinig verlaufen. Die **A.** ist aus der barocken Parkallee entstanden und wurde zum Boulevard. **A.n** prägten die Städte des 18. und 19. Jhs., besonders gut in Paris nachzuvollziehen, wo der Stadtgärtner ADOLPH ALPHAND in seiner Amtszeit seit 1845 in den Pariser Boulevards 82 000 Bäume pflanzen ließ; cf. CHEV 2001. In Berlin wurden bereits 1647 die ersten öffentlichen stadtbildprägenden Straßenbäume am Boulevard „Unter den Linden" gepflanzt; cf. UB 2004, 3; es gibt nirgend so viele **A.n** wie in Europa; sie sind Teil der europäischen Kulturgeschichte und deshalb dauerhaft zu erhaltende Landschaftselemente); *g 2* **Festonallee [f]** (Aus dem Ital. *festone* >girlandenförmiges Ornament aus Zweigen, Blättern und Blüten, an zwei Enden aufgehängt<; eine in D. einzigartige Allee ist die **F.** aus holländischen Linden im Bothmerschen Schlosspark in Mecklenburg-Vorpommern, bei der die spalierartig gezogenen Bäume durch regelmäßige Schnittmaßnahmen girlandenförmig gezogen werden).

6540 tree maintenance [n] *arb. constr.* (Regular operations undertaken to maintain and conserve the health and safety of trees; e.g. crown maintenance, treatment of minor wounds and areas of decay, cultivation, fertilizing and watering, provision of protective structures for the trunk or root area, etc.; ►cabling, ►clearance space of trees, ►crown pruning, ►crown pruning for public safety, ►crown reduction, ►crown thinning, ►dead wooding [US] 1/cleaning out [UK], ►dead wooding [US] 2/ cutting out [UK], ►revitalization of tree pits, ►tree treatment work); *syn.* tree maintenance practice [n], tree maintenance work [n], tree management [n], tree care [n] [also US]; *s* **mantenimiento [m] de árboles (2)** (Actividades regulares de mantenimiento y conservación de la salud de los árboles urbanos y de su seguridad hacia terceros, como aireación, fertilización y riego, poda [►aclareo de la copa 1, ►ahuecado de la copa, ►escamonda, ►poda de corrección de la copa, ►poda «en copa»], cura de pequeñas heridas y zonas podridas e instalación de estructuras de protección y sujeción [►sujeción de las ramas principales (o primarias)]; creación de ►espacio libre para vehículos; ►revitalización [de ubicaciones] de árboles urbanos; ►saneamiento de árboles); *syn.* trabajos [mpl] de mantenimiento de árboles; *f* **travaux [mpl] d'entretien des arbres** (Opérations renouvelées régulièrement ayant pour but la maintenance et la protection de la vitalité et de la santé des arbres et nécessaires à la sécurité du public, p. ex. travaux des sols, arrosage et fertilisation exécutés aux pieds des arbres, taille, élagage et chirurgie arboricole ; opérations de protection du tronc et du trou de plantation etc. ; ►opérations de restauration arboricole, ►amélioration du lieu d'implantation des arbres, ►éclaircissage de la couronne, ►élagage en sec, ►haubanage de la charpente, ►hauteur sous couronne, ►taille d'allégement, ►taille de correction, ►taille en cime, ►travaux de restauration arboricole) ; *syn.* entretien [m] des arbres ; *g* **Baumpflege [f, o. Pl.]** (Regelmäßige Leistungen an Baum und Baumumfeld zur Erhaltung und Förderung der Funktion des Baumes, d. h. zur Vermeidung von Fehlentwicklungen, zur Förderung der Vitalität und Gesundheit sowie zur Erhaltung der Verkehrssicherheit. Hierzu gehören Maßnahmen der Kronen-,

Stamm- und Wurzelpflege, wie z. B. ►Baumsanierung, ►Baumstandortsanierung, Behandlung und Betreuung kleiner Wunden und Faulstellen am Stamm und an Starkästen, Einbau von technischen Schutzvorrichtungen für den Stamm- und Wurzelbereich, Herstellung des ►Lichtraumprofils, ►Kronenlichtungsschnitt ►Kronenschnitt, ►Kronenentlastungsschnitt, ►Kronensicherungsschnitt, ►Kronenverankerung, Lockern, Düngen und Wässern des Wurzelbereichs, ►Totholzentfernung etc.); *syn.* Baumpflegearbeiten [fpl], Baumpflegemaßnahmen [fpl].

tree maintenance practice [n] *arb. constr.* ►tree maintenance.

tree maintenance work [n] *arb. constr.* ►tree maintenance.

tree management [n] *arb. constr.* ►tree maintenance.

6541 tree mapping [n] *landsc.* (Delineating trees on a map; ►mapping, ►tree survey); *s* **cartografía [f] de árboles** (Representación de árboles o grupos de árboles en un mapa; ►inventario cartográfico, ►muestreo de árboles urbanos) *syn.* inventario [m] cartográfico de árboles; *f* **cartographie [f] d'arbres** (►cartographie, ►relevé des essences [arborescentes]) ; *syn.* relevé [m] cartographique des arbres ; *g* **Baumkartierung [f]** (Darstellung eines eingemessenen Baumes oder Baumbestandes in seiner räumlichen Lage auf eine Karte/einen Plan oder als ►Baumaufnahme in eine elektronische Datei; ►Kartierung).

6542 tree nursery [n] *hort.* (Place where ornamental ►woody species 1 are grown for sale; ►forest tree nursery, ►plant nursery); *s* **vivero [m] (de árboles)** (►Empresa de horticultura donde se cultivan ►especies leñosas ornamentales; en alemán el término «Baumschule» se utiliza para vivero que además de leñosas también cultiva vivaces; ►vivero forestal); *f* **pépinière [f]** (Terrain consacré à la multiplication de plantes ligneuses ou herbacées qu'on élève jusqu'à ce qu'elles puissent être transplantées ; les pépinières sont généralement spécialisées dans un type de production [jeunes plants, gros sujets, cultures spécialisées, gamme large] ; on distingue les pépinières d'élevage et de multiplication [pépinière de production] et les pépinières de négoce ; ►entreprise horticole, ►pépinière forestière) ; *g* **Baumschule [f]** (►Gartenbaubetrieb zur Anzucht und zum Verkauf von Freilandgehölzen, oft auch von Stauden. Es gibt Spezial-**B.n** für Zier- und Obstgehölze, Unterlagen, Rosen, Alleebäume, Bambus, Rhododendron und Forstpflanzen; die deutsche Baumschulwirtschaft ist mit 1400 Mitgliedern und den bedeutendsten Unternehmen im „Bund deutscher Baumschulen [BdB] e. V., gegründet 1907, als gemeinsame Interessenvertretung organisiert; ►Forstbaumschule, ►Gehölz 3).

tree nurseryman [n] *arb.* ►tree nursery specialist.

6543 tree nursery specialist [n] *arb. hort. prof. syn.* tree nurseryman [n]/tree nursery woman [n]; *s* **arboricultor, -a [m/f] de vivero** (Especialista en el cultivo de leñosas en un vivero); *f* **pépiniériste [m/f]** (Personne spécialisée dans la culture des plantes ligneuses ; le terme « arboriste » désignait autrefois la personne qui s'occupait de la culture des arbres ; DIC 1993) ; *syn.* arboriculteur [m], *obs.* arboriste [m] ; *g* **Gärtner/-in [m/f] der Fachrichtung „Baumschule"** (Fachkraft, die Gehölze in einer Baumschule anzieht); *syn.* Baumschuler/-in [m/f].

tree nursery woman [n] *arb.* ►tree nursery specialist.

6544 tree of seedling origin [n] *for.* (Tree which has never been cut back or coppiced; such a young tree is called 'maiden'); *s 1* **árbol [m] generado de semilla** (Arbolillo obtenido por germinación natural que nunca ha sido podado hasta el tocón); *s 2* **brinzal [m] (1)** (Plantita leñosa que nace en los rodales de los montes procedente de semillas caídas naturalmente de los árboles); *f* **brin [m] de franc-pied** (Arbre obtenu par germination d'un semis sans avoir été soumis à une coupe) ; *g* **Kernwuchs [m]** (...wüchse [pl]; aus einem Keimling oder Steckling

T

entstandener Laubbaum, der nicht auf den Stock gesetzt wurde. Bäume aus Wurzelbrut sind auch keine Kernwüchse).

6545 tree pit [n] *constr. hort.* (Excavated hole for tree planting; ►planting hole); *s* **foso** [m] **de plantación** (►Hoyo de plantación grande para plantar árbol); *f* **fosse** [f] **de plantation** (Grand trou creusé pour la plantation d'un arbre ou d'un gros arbuste en motte ; ►trou de plantation) ; *g* **Pflanzgrube** [f] (Ausgehobenes, großes Loch zur Pflanzung eines Baumes oder Großstrauches; ►Pflanzloch).

6546 tree pit cover [n] *constr.* (Generic term covering cast iron tree grate [US]/cast iron tree grille [UK], ►tree grate [US]/ tree grid [UK], ►concrete tree grate [US]/concrete tree grid [UK]; ►tree pit surface); *s* **cobertura** [f] **de alcorque** (Término genérico para ►cobertura de alcorque [prefabricada] de hormigón, cobertura de hierro con ranuras y agujeros para posibitar la aireación y el paso de agua a las raíces, ►rejilla [metálica] para alcorque; ►alcorque); *f* **revêtement** [m] **de protection de pied d'arbre** (Terme générique pour ►grille [de protection] d'arbre, ►panneau de béton perforé pour la protection de pied d'arbre, protections en fonte, acier, éléments en béton alvéolés préfabriqués ou en briques perforées utilisés comme recouvrement des ►tours d'arbre et permettant le passage de l'air et de l'eau) ; *syn.* revêtement [m] de protection du trou de plantation, revêtement [m] protège-racine, disque [m] protège-racine ; *g* **Baumscheibenabdeckung** [f] (OB zu Baumscheibe aus Gusseisen, ►Baumscheibenrost, ►Betonbaumscheibe oder Betonfertigteile mit Schlitzen oder Löchern sowie hart gebrannte Hochlochklinker zur Befestigung von ►Baumscheiben bei gleichzeitiger Möglichkeit des Luftaustausches und der Wasserzufuhr für den Wurzelraum).

6547 tree pit edging [n] [US] *constr.* (Border separating pavement from a ►tree pit surface); *syn.* tree circle edging [n] [UK]; *s* **recintado** [m] **de alcorque** (Borde de ladrillos o adoquines alrededor de un árbol que lo separa del resto del pavimento; ►alcorque); *f* **entourage** [m] **d'arbre** (Ouvrage de protection délimitant le ►tour d'arbre sur des surfaces imperméables) ; *g* **Baumscheibeneinfassung** [f] (Umrandung einer ►Baumscheibe im Pflaster-, Platten- oder Asphaltbelag).

tree pits [npl]**, improvement of (street)** *constr.* ►revitalization of tree pits.

6548 tree pit surface [n] *constr. hort.* (ULD 1992, 300) (**1.** Area of soil around the trunk of a tree maintained initially free of other vegetation. **2.** Generally planted or grass-covered area around the trunk of a tree in paved areas; ►planting saucer [US]/watering hollow [UK]); *syn.* tree (planting) bed [n], tree base [n] (PSP 1987, 202); *s* **alcorque** [m] (**1.** Círculo de suelo alrededor de un árbol que se mantiene sin vegetación. **2.** En zona asfaltada, área no pavimentada generalmente cubierta de hierba o de otras plantas; ►alcorque de riego); *f* **tour** [m] **d'arbre** (**1.** Espace laissé nu autour d'un tronc d'arbre dans un rayon allant jusqu'à 3 m sur une surface engazonnée. **2.** Surface ouverte plantée ou engazonnée correspondant à la fosse de plantation d'un arbre entouré d'un revêtement de sol imperméable ; ►cuvette d'arrosage) ; *syn.* pied [m] d'arbre ; *g* **Baumscheibe** [f] (**1.** Von Bewuchs freigehaltene, kreisförmige Bodenfläche um den Baumstamm in einer Rasenfläche, die nach der Pflanzung mit einem Gießrand ausgebildet wird. **2.** In versiegelten Flächen offene, meist bepflanzte oder mit Gras angesäte Fläche eines Baumstandortes; ►Gießmulde; *OB* Pflanzscheibe).

tree planting [n] [US] *agr. conserv.* ►ditch corridor tree planting [US].

tree planting [n] [US]**, dispersed fruit** *agr. conserv.* ►traditional orchard [UK].

tree planting bed [n] *constr. hort.* ►tree pit surface.

tree planting grid [n] *constr. hort.* ►tree planting pattern.

6549 tree planting pattern [n] *constr. hort.* (Arrangement of trees in, e.g. a ►square planting grid or ►triangular planting grid); *syn.* tree planting grid [n]; *s* **esquema** [m] **de plantación (de árboles)** (►esquema de plantación rectangular, ►esquema de plantación triangular); *f* **groupement** [m] **d'arbres** (Disposition des arbres regroupés en général en nombre impair ; ►disposition en quinconce, ►disposition rectangulaire) ; *g* **Baumverband** [m] (Stellung der Bäume im ►Dreiecksverband oder ►Vierecksverband).

6550 tree planting site [n] *constr. urb.* *s* **lugar** [m] **de plantación de un árbol**; *f* **lieu** [m] **d'implantation d'un arbre** ; *g* **Baumstandort** [m] (Lebensort eines Baumes samt seiner ihn bestimmenden Umwelteinflüsse); *syn. sensu lato* Baumumfeld [n].

tree preservation order [n] [UK] *arb. leg. conserv.* ►tree preservation ordinance [US].

6551 tree preservation ordinance [n] [US] *arb. leg. conserv.* (In U.S., local ordinance [►city ordinance (US)/city bye-laws (UK)] for control of tree cutting and replacement; in U.K., order made by a local planning authority in accordance with town and country planning legislation for the preservation of individual trees, groups of trees or areas of woodland in the interest of amenity; cf. TCPA 1971, pt. IV, §§ 59-62); *syn.* tree protection ordinance [n] [US], tree preservation order [n] [UK]; *s* **ordenanza** [f] **de protección de árboles** (En EE.UU., D. y F. por medio de ►ordenanza municipal. En GB sobre la base de la legislación de urbanismo, prohibe talar árboles a partir de un determinado tamaño. En Es. según la legislación supletoria, que se aplica en el caso de que no exista regulación en la correspondiente CC.AA., está prohibida la corta de arbolado en los predios sometidos a algún tipo de plan urbanístico, a no ser que éste indique lo contrario; art. 58 [1] RD 1346/1976); *f* **réglementation** [f] **concernant la protection des arbres** (En Allemagne, Angleterre et en France, réglementation communale [►arrête municipal] ayant pour objectif la protection, l'entretien et l'extension des plantations [arbres isolés, groupes d'arbres, massifs forestiers] publiques et privées) ; *syn.* code [m] de l'arbre urbain ; *g* **Baumschutzverordnung** [f] (Verordnung von Behörden [in D. Gemeinden, Kreise] zum Schutze von Einzelbäumen, Baumgruppen und Waldbeständen. In D., A. und CH erfolgt die Baumschutzregelung meist per Erlass einer Baumschutzsatzung durch den Gemeinderat. Somit können kraft ►Ortssatzung alle Bäume eines Gemeindegebietes oder eines definierten Teiles der Gemarkung, die über einen festgelegten Stammumfang — meist 60 oder 80 cm — oder Stammdurchmesser hinaus gewachsen sind, unter Schutz gestellt werden. In D. liegen die Anfänge des städtischen Baumschutzes 1. bei der Berliner Gassenordnung von 1660, in der frevelhafte Baumbeschädigungen mit dem Abschlagen einer Hand bestraft wurde, 2. in dem Patent zur Konservation und Fortpflanzung von „Unter den Linden" durch Friedrich I. [1707], 3. im Preußischen Gesetz zur Erhaltung des Baumbestandes und Erhaltung und Freigabe von Uferwegen im Interesse der Volksgesundheit [29.07.1922] und 4. im Reichsnaturschutzgesetz von 1935. Im UK werden hingegen die zu schützenden Bäume oder Waldbestände im Einzelnen von der örtlichen Behörde bestimmt; cf. TCPA 1971, Art. IV, §§ 59-62); *syn.* Baumschutzsatzung [f], Baumsatzung [f], Baumschutzgesetz [n] [A, CH].

6552 tree prop [n] *arb. hort.* (**1.** Post or pole to hold up a heavy tree limb; ►guying, ►scaffold branch, ►tree stake); **2. propping** [n] (Supporting measure for heavy tree limbs); *s* **rodrigón** [m] (Estaca fuerte colocada bajo una ►rama primaria lateral de árbol maduro con porte ancho para evitar su rotura;

►fijación con vientos [metálicos], ►tutor); *syn.* estaca [f] (de árbol), rodriga [f] (HDS 1987); *f* **support [m] de branche** (Mesure empêchant la rupture d'une charpentière sur de vieux arbres au port très étalé ; ►charpentière, ►haubanage, ►tuteur) ; *g* **Astunterstützung [f]** (Sondermaßnahme der Kronensicherung [dicker Stützpfahl] zur Verhinderung des Abbrechens eines ►Starkastes bei ausladenden Kronen alter Bäume, wenn Trag-/Haltesicherungen nicht ausreichen; zur Aufnahme seitlich wirkender Belastungen sind A-förmige Stützen erforderlich; bei einer Abstützung von schräg stehenden Altbäumen, bei der der Stützpfahl meist in einem Betonfundament eingebunden ist, spricht man von „**Baumstütze**". Bei der Wahl der Dimensionierung sind die auf die Stützen wirkenden Lasten [und die damit verbundenen Wandstärken der Stahlrohre] zu berücksichtigen; ►Baumpfahl, ►Drahtseilverankerung); *syn.* Aststütze [f].

tree protection [n] *arb. constr.* ►protection of trees.

tree protection ordinance [n] [US] *arb. leg. conserv.* ►tree preservation ordinance [US]/tree preservation order [UK].

6553 tree pruning [n] *hort.* (Cutting of tree branches for training of young trees, appearance and health; ►fruit tree pruning, ►rejuvenation of trees; *generic term* ►tree and shrub pruning); *s* **poda [f] de árboles** (►poda de [árboles frutales], ►rejuvenecimiento de árboles; *término genérico* ►poda de leñosas); *f* **taille [f] d'arbre** (►rajeunissement d'un arbre, ►taille d'arbres fruitiers ; *terme générique* ►taille des végétaux ligneux) ; *g* **Baumschnitt [m]** (►Baumverjüngung, ►Obstbaumschnitt; *OB* ►Gehölzschnitt).

tree removal [n] *constr. hort.* ►alternate tree removal.

tree rose [n] [US] *hort.* ►standard rose, *hort.* ►weeping tree rose [US]/weeping standard rose [UK].

6554 trees [npl] *urb. landsc.* (All trees in e.g. a given garden, park, parking area, or throughout a community; ►group of trees); *s* **arbolado [m]** (Grupo pequeño de árboles en jardines, parques, etc.; ►bosquete, ►grupo de árboles); *f* **patrimoine [m] arborescent** (L'ensemble des arbres des parcs et jardins, des espaces verts, de la voirie, etc. dans une agglomération ; ►groupe d'arbres, ►massif d'arbres) ; *g* **Baumbestand [m] (3)** (Summe der Einzelbäume in Gärten, Parkanlagen, Straßen etc.; ►Baumgruppe).

trees [npl] [UK]**, block of** *agr. land'man.* ►coppice [US] (1).

trees [npl]**, clump of** *landsc.* ►group of trees.

trees [npl]**, cluster of** *landsc.* ►group of trees.

trees [npl]**, existing** *arb. constr. landsc. urb.* ►preservation of existing trees.

trees [npl]**, forest border** *for.* ►edge trees.

trees [npl]**, fruit** *agr. conserv.* ►plowed land with fruit trees [US]/ploughed land with fruit trees.

trees [npl]**, map of** *adm. landsc. surv.* ►cadastral map of trees.

trees [npl]**, semi-mature** *constr.* ►transplantation of semi-mature trees.

trees [npl] [US]**, setting apart of** *for. landsc.* ►isolation of trees.

trees [npl]**, stability of** *arb.* ►structural stability of trees.

trees [npl]**, structural strength of** *arb. stat.* ►structural stability of trees.

trees [npl]**, young** *hort.* ►training of young trees.

trees [npl] **and shrubs** [npl] *land'man.* ►band of trees and shrubs; *constr.* ►border with shrub and trees, ►cutting back of trees and shrubs; *arb. conserv.* ►expert opinion on the value of trees and shrubs; ►valuation chart for trees and shrubs.

trees [npl] **and shrubs** [npl]**, shortening of** *constr.* ►cutting back of trees and shrubs.

trees [npl] **of a forest, marginal** *for.* ►edge trees.

6555 tree savanna(h) [n] *phyt.* (Grassy plain with isolated trees dispersed more or less regularly over it; ►savanna[h]); *s* **sabana [f] arbolada** (►Sabana salpicada de grandes árboles, como el baobab *[Adansonia digitata]*, varias especies de acacia, etc. Según el elemento arbóreo va aumentando en importancia sinecológica, se pasa insensiblemente al bosque sabanero, en la que la sabana es el sotobosque del arboretum dominante; DB 1985); *syn.* sabana [f] arbórea; *f* **savane [f] arborée** (Association herbacée des régions tropicales et subtropicales caractérisée par un couvert arborescent dispersé irrégulier ; ►savane) ; *g* **Baumsavanne [f]** (Grasland mit unregelmäßig verteilten einzelnen Bäumen in den Tropen oder Subtropen; ►Savanne).

tree seat [n] *gard. landsc.* ►tree bench.

tree seedling [n] *for. hort.* ►forest tree seedling.

6556 tree skeleton [n] *arb.* (Structural branch system; ►branching 1, ►branching 2, ►scaffold branch); *syn.* tree structure [n]; *s* **esqueleto [m] de un árbol** (Estructura básica de un árbol, es decir el tronco y las ramas principales; ►rama primaria, ►ramaje, ►ramas colgantes); *f* **charpente [f] d'arbre** (Ensemble des branches constituant le squelette de l'arbre ; ►branchage, ►charpentière, ►ramure) ; *syn.* ossature [f] d'arbre ; *g* **Astgerüst [n, o. Pl.]** (Statisches Grundgerüst eines Baumes; ►Astbehang, ►Astwerk, ►Starkast).

tree sponsor [n] [UK] *conserv.* ►voluntary caretaker of a tree [US].

6557 tree stake [n] *arb. constr. hort.* (Support for a newly planted tree; ►angled stake, ►stake 3, ►tree prop); *s* **tutor [m] (2)** (Apoyo para árbol recién plantado; ►entutorado, ►estaca oblicua, ►rodrigón); *f* **tuteur [m] (2)** (Perche/pieu de châtaignier, de chêne ou acacia servant de ►support de branche lors de nouvelles plantations d'arbres ; ►tuteur oblique, ►tuteurage) ; *g* **Baumpfahl [m]** (Zur Baumverankerung neu gepflanzter Bäume verwendete Stütze; ►Astunterstützung, ►pfählen, ►Schrägpfahl).

tree steward [n] [US] *conserv.* ►voluntary caretaker of a tree [US].

tree structure [n] *arb.* ►skeleton.

6558 tree stump [n] *constr. for. hort.* (Woody base of a tree, left in the ground after felling; SAF 1983; ►stool); *syn.* stump [n]; *s* **tocón [m] (2)** (Parte del tronco de un árbol que queda unida a la raíz cuando lo cortan por el pie; DB 1985; ►cepa); *f* **souche [f] (2)** (Partie souterraine d'un arbre restant dans le sol après son abattage ; ►souche vivante) ; *g* **Stubben [m]** (Der nach dem Fällen eines Baumes im Boden verbleibende Wurzelstock; ►ausschlagfähiger Stubben); *syn.* Baumstrunk [m], Baumstumpf [m], Stock [m], Wurzelstock [m] (3).

tree stumps [npl] *constr. hort.* ►clearing and removal of tree stumps.

6559 tree surgery [n] *arb. hort.* (Art and practice of treating injuries, deformities and damage of trees by major cutting, filling, ►cabling [US]/cable-bracing [UK] and similar operations; BS 3975: part 5; ►area of rot, ►tree treatment work); *s* **cirugía [f] arbórea** (Tratamiento de daños en árboles para evitar la rotura de ramas mediante cortes o ►sujeción de las ramas principales [o primarias], eliminando la ►caries húmeda y aplicando productos curativos para sanearlos; ►saneamiento de árboles); *f* **chirurgie [f] arboricole** (Ensemble des soins à donner aux arbres en voie de dépérissement, aux troncs en voie de ►pourrissement, aux charpentières menaçant de se rompre ; ►haubanage de la charpente, ►pourriture, ►travaux de restauration arboricole) ;

T

g **Baumchirurgie** [f] (Behandlung wertvoller, durch Holzschäden gefährdeter Bäume sowie Verankerung ausladender, bruchgefährdeter Stämme und Äste; ▶Baumsanierung, ▶Kronenverankerung, ▶Morschung).

6560 tree survey [n] *constr. landsc.* (Process of on-site computer listing of tree species and their mapping and numbering according to age, health, damage, and decline of individual trees as surveyed in an entire area within a specified period of time. The action taken is to make a ▶tree survey plan or set up a ▶tree data bank; do [vb] a tree survey [US]/carry out [vb] a tree survey [UK]); *s* **muestreo** [m] **de árboles urbanos** (Registro del estado de los árboles individuales según especie, tamaño, características de la ubicación, daños visibles para crear los ▶planos de muestreo de árboles y los ▶inventarios de árboles urbanos); *f* **relevé** [m] **des essences (arborescentes)** (Inventaire des arbres isolés, des caractéristiques de leur lieu de croissance et de leur état phytosanitaire et répertoriés sur des fiches de travaux. Ce relevé constitue la base du ▶plan d'implantation des arbres d'ornement ou des peuplements forestiers en milieu urbain ; ▶fichier informatique de description des arbres) ; *syn.* inventaire [m] des arbres existants, recensement [m] des arbres existants ; *g* **Baumaufnahme** [f] (Erfassung von Einzelbäumen mit Baumart, Standortgröße, Standorteigenschaften, Schadensmerkmalen vor Ort an Hand von Erfassungsbelegen oder EDV-gestützt mit einem Palm-Top. Oft wird auch der Gesundheitszustand durch Augenschein bestimmt [▶Baumschau, ▶Visual-Tree-Assessment-Methode]. Die **B.** ist die Grundlage für den ▶Baumbestandsplan und für die ▶Baumdatei); *syn.* Baumerfassung [f], Aufnahme [f] des Baumbestandes, Erfassung [f] des Baumbestandes.

6561 tree survey plan [n] *adm. constr. landsc.* (Plan showing trees that specifies crown spread as well as trunk diameter, and indicates the altitude above sea level at the foot of the trunk; ▶tree survey); *s* **plano** [m] **de muestreo de árboles** (Plano en el que reunen datos sobre los árboles urbanos como diámetro de copa, perímetro de tronco y altura; ▶muestreo de árboles); *f* **plan** [m] **d'implantation des arbres d'ornement** (Plan représentant les arbres et leurs caractéristiques principales telles le diamètre du houppier, la circonférence du tronc et la hauteur du pied de l'arbre cotation NGF ; ▶relevé des essences [arborescentes]) ; *g* **Baumbestandsplan** [m] (Plan, der für eine bestimmte Fläche Bäume mit Angaben über deren Kronendurchmesser, Stammumfang und Stammfußhöhe über NN darstellt; ▶Baumaufnahme).

tree technician [n] *arb. hort.* ▶arborist.

6562 tree treatment work [n] *arb. constr. hort.* (One-time operations [US]/one-off operations [UK] to reestablish the health and public safety of a tree and its planting site; ▶preservation of existing trees, ▶revitalization of tree pits, ▶revitalization of woody plants, ▶tree maintenance, ▶tree surgery); *s* **saneamiento** [m] **de árboles** (Operaciones excepcionales para reestablecer la salud de los árboles y garantizar la seguridad pública a largo plazo; ▶cirugía arbórea, ▶mantenimiento de árboles 1, ▶mantenimiento de árboles 2, ▶revitalización [de ubicaciones] de árboles urbanos, ▶saneamiento de leñosas); *syn.* trabajos [mpl] de saneamiento de árboles; *f* **travaux** [mpl] **de restauration arboricole** (Opérations exceptionnelles exécutées sur un sujet ou sur son lieu de croissance en vue de rétablir sa santé et d'assurer la sécurité du public ; ▶amélioration du lieu d'implantation des arbres, ▶chirurgie arboricole, ▶conservation des arbres existants, ▶restauration végétale, ▶travaux d'entretien des arbres) ; *syn.* opération [f] de restauration arboricole ; *g* **Baumsanierung** [f] (Einmalige Leistungen am Baum und Standort zur langfristigen Wiederherstellung seiner Gesundheit/Vitalität und Verkehrssicherheit; ▶Baumchirurgie, ▶Baumpflege, ▶Baum-

standortsanierung, ▶Bestandssanierung von Gehölzen, ▶Erhaltung von Bäumen); *syn.* Baumsanierungsarbeiten [fpl].

6563 tree trunk [n] *hort.* *s* tronco [m] (de árbol); *f* tronc [m] d'arbre ; *g* Baumstamm [m].

tree trunk [n]**, raised soil level near a** *arb. constr.* ▶groundfill around a tree.

tree trunks [npl] *constr.* ▶boarding up of tree trunks.

tree vault [n] [US]**, concrete** *constr.* ▶precast concrete tree vault.

6564 tree water hollow [n] *arb.* (Pit at the base of ▶scaffold branches [stem apex] or at the fork of big branches, which retains water); *s* **cavidad** [f] **arbórea con agua** (Hundimiento en el ápice del tronco o en las bifurcaciones de las ramas en donde se puede acumular agua de la lluvia; ▶rama primaria); *f* **cavité** [f] **remplie d'eau** (Creux naturel formé, pendant la croissance, à la partie supérieure du tronc ou au point de ramification des branches d'un arbre ; ▶charpentière) ; *g* **Wassermulde** [f] (Wachstumsbedingte Vertiefung im Stammkopf oder in Astgabeln von Bäumen; ▶Starkast).

6565 tree well [n] *constr.* (Hole inside a retaining wall built to protect the root zone of an existing tree within an area of filled ground; ▶groundfill around a tree, ▶precast concrete tree vault [US]/precast concrete tree well [UK]); *s* **murete** [m] **de protección de árbol** (Cerco mamposteado alrededor del hoyo de un árbol para proteger el tronco contra posible ▶relleno con tierra [alrededor del tronco], ▶jardinera de hormigón para árbol); *f* **muret** [m] **de protection d'arbre** (±) (Élément maçonné construit autour d'un arbre afin d'éviter l'▶enfouissement du tronc lors de travaux de mise en remblai ; ▶anneau en béton de protection d'arbre) ; *g* **Baumschacht** [m] (Gemauerter Schacht im Bereich des Wurzeltellers eines Baumes zur Verhinderung einer ▶Stammeinschüttung bei Auffüllarbeiten; ▶Betonbaumschacht);

tree work [n] *arb. constr. contr.* ▶additional technical contract conditions for tree work.

tree worker [n] [UK] *arb. hort.* ▶arborist.

tree wrapping [n] *arb. constr. hort.* ▶trunk wrapping.

trellis [n] *constr.* ▶lattice.

trellis [n]**, wooden** *constr.* ▶wooden lattice.

6566 trellis climber [n] *hort. plant.* (▶Climbing plant which needs the support of a trellis structure in order to climb, e.g. the ▶tendril climber *Clematis*, the ▶twining climber *Wisteria sinensis* and honeysuckle - *Lonicera henryi, L. tellmanniana* and ▶scandent plants such as climbing roses, *rambler* roses, blackberry [*Rubus fruticosus*]); *syn.* climber for façade planting; *syn.* wall-covering vine [n] [US], house wall climber [n] [UK]; *s* **trepadora** [f] **sobre soporte** (▶Planta trepadora que necesita una estructura de apoyo para trepar [▶enrejado de soporte de trepadoras], como p. ej. la ▶planta zarcillosa clemátide [*Clematis spp.*], las ▶plantas volubles glicinia [*Wisteria sinensis*] y madreselva [*Lonicera periclymenum, L. tellmanniana*] y las ▶plantas sarmentosas como el rosal trepador y la mora [*Rubus fruticosus*]); *f* **grimpante** [f] **pour treillage** (▶Plante grimpante qui n'est capable de recouvrir un mur ou une façade qu'avec l'aide d'un support tel un treillage, etc., p. ex. le *Clematis* [▶plante grimpante à ventouses], la glycine [*Wisteria sinensis*] et les chèvrefeuille [*Lonicera henryi, L. tellmanniana*] [▶plante volubile] ou les rosiers sarmenteux [▶plantes sarmenteuses] et les rosiers liane ; ▶support de plantes grimpantes) ; *syn.* grimpante [f] sur treillage, grimpante [f] pour/sur treillis ; *g* **Gerüstkletterer** [m] (▶Kletterpflanze, die eine Rankhilfe [▶Rankgerüst] benötigt, z. B. die ▶Rankenpflanze *Clematis*, die ▶Windepflanze Glyzinie [*Wisteria sinensis*] und das Geißblatt [*Loni-

cera henryi, L. tellmanniana] und der ▶Spreizklimmer Kletter-rose, *Rambler*-Rose, Brombeere *[Rubus fruticosus]*); *syn.* Gerüst-kletterpflanze [f].

trellis-covered walk [n] *arch. gard.* ▶tunnel-arbor [US]/ tunnel-arbour [UK].

6567 trellis fence [n] *constr.* (Low fence constructed of pointed, half-round pickets fixed diagonally to one another); *s* **valla** [f] **trenzada** (Cercado de troncos partidos longitudinal-mente y unidos diagonalmente entre sí); *syn.* valla [f] al estilo de los pioneros; *f* **clôture** [f] **à treillis** (Échalas entrecroisés pour former en général un panneau à maille losangée renforcé par 2 lices demi-rondes) ; *syn.* clôture [f] « type croisillon » (GEN 1982-II, 155) ; *g* **Scherenzaun** [m] (Gitterzaun aus zugespitzten, diagonal miteinander befestigten Halbrundhölzern); *syn.* Jäger-zaun [m], Polygonzaun [m], Spriegelzaun [m].

trelliswork [n] *constr.* ▶wooden lattice.

trench [n] *conserv'hist.* ▶exploratory trench; *constr.* ▶founda-tion trench, ▶planting trench, ▶services trench.

trench [n], **drainage** *agr. constr.* ▶drainage ditch,

6568 trench bottom [n] *constr.* *s* **fondo** [m] **de zanja**; *f* **fond** [m] **de tranchée** *syn.* fond [m] de fouille en tranchée ; *g* **Grabensohle** [f] (Boden eines Grabens).

6569 trench bracing [n] *constr.* (Stabilizing of trench walls against collapse by erection of metal or wooden braces; ▶planking and strutting); *s* **arriostramiento** [m] **de una zanja** (Estabilización lateral de zanjas con riostras de metal o madera para evitar que se derrumben los costados; ▶encofrado y apuntalado); *f* **étaiement** [m] **d'une fouille** (*Terme générique* mise en place d'un ensemble d'étais pour soutenir la façade d'un bâtiment, les parois d'une fouille, etc. ; *terme spécifique* ▶blin-dage (Étaiement de parois ayant recours à un coffrage de planches, un platelage en bois, un voile de palplanches, des panneaux préfabriqués) ; *g* **Grabenverbau** [m] (Sicherung der Grabenwände gegen Einsturz, z. B. mit verspießten Holzbohlen oder Verbaukorb; ▶Verbau 2); *syn.* Grabenaussteifung [f].

6570 trench drain [n] [US] *constr.* (Prefabricated drain for draining hard surfaces; ▶cast iron drain with grating [US]/cast iron channel with grating [UK], ▶slot trench drain; *specific term for a very shallow trough-shaped gutter* dished channel [UK]); *syn.* channel drain [n] [UK]; *s* **canal** [m] **de drenaje (super-ficial)** (ALB 1996; conjunto de piezas de drenaje prefabricadas colocadas una junto a otra para formar un conducto para evacuar el agua de lluvia de superficies pavimentadas; ▶canalón de desagüe cubierto con rejilla, ▶conducto de desagüe con ranura); *f* **caniveau** [m] (Élément de voirie de conduite des eaux pluviales provenant de la chaussée, des chemins et surfaces à sol stabilisé ; on distingue le caniveau simple pente [CS] et double pente [CC], les caniveaux de pavés, ▶caniveaux à grille, ▶caniveaux à fente ou satujos) ; *g* **Entwässerungsrinne** [f] (Aneinander gereihte, vorgefertigte einzelne Rinnenelemente an Rändern von oder auf befestigten Wegen oder Plätzen zur linienförmigen Weiterleitung des Niederschlagswassers; ▶Kastenrinne, ▶Schlitzrinne).

trench drain [n] [US], **covered** *constr.* ▶cast iron drain with grating [US].

6571 trench drain infiltration [n] *constr.* (Percolation of precipitation from a trench filled with coarse gravel, often with a drainage pipe, to the groundwater table. The trench acts as an intermediate reservoir until the water can percolate through the pores of the subsoil; ▶infiltration, ▶infiltration swale); *s* **perco-lación** [f] **por acequia de infiltración** (Percolación de aguas desde zanja rellena de grava gruesa, a menudo con tubería de drenaje, hasta la capa freática, actuando la zanja como almacén intermedio hasta que el agua pueda infiltrarse a través de los poros del suelo; ▶acequia de infiltración, ▶infiltración); *f* **infiltration** [f] **par fossé drainant (±)** (Infiltration des eaux pluviales à travers un revêtement superficiel perméable [p. ex. chaussée drainante] dans un fossé rempli de matériaux drainants sur le fond duquel est souvent placé un drain collecteur ; cette technique permet le stockage des eaux pluviales jusqu'à ce que celles-ci s'évacuent lentement par les pores du sol pour rejoindre la nappe phréatique ; ▶cuvette d'infiltration, ▶infiltration, ▶noue d'infiltration) ; *g* **Rigolenversickerung** [f] (Nieder-schlagsversickerung in einem mit Grobkies verfüllten und oft mit einem Dränrohr ausgestatteten Graben [Rohrrigole], der als Zwischenspeicher dient, bis das Wasser durch die Poren des umliegenden Bodens dem Grundwasser zufließen kann. Bei der Nutzung von Kies als Speicher kann von einem Hohlraum bis ca. 35 % ausgegangen werden. Bei Einbau eines Vollsickerrohres wird der Speicher je nach Rohrdurchmesser entsprechend erhöht; ▶Versickerung, ▶Versickerungsmulde).

trench frame [n] **with grated or solid cover** [US] *constr.* ▶cast iron drain with grating [US].

6572 trenching [n] *agr. hort.* (Cultivation of soil to a depth of two or three spits [spade depths], as necessary, by digging so as to invert the position of the layers; e.g. 2-spit trenching, 3-spit trenching; BS 3975: part 5; ▶deep plowing [US]/deep ploughing [UK], ▶dig, ▶double digging); *s* **subsolado** [m] **en profundo** (Proceso de volteo del suelo hasta más de 50 cm de profundidad que se realiza generalmente con un arado especial o manualmente a tres palas; ▶arado en profundo, ▶cavar, ▶desfonde a dos palas); *syn.* subsolado [m] a tres palas; *f* **défonçage** [m] (Travail profond du sol effectué par retournement et permettant d'ameublir à la fois la zone normalement travaillée par les labours ou les bêchages ordinaires à trois fers, et le sous-sol, jusqu'à une profondeur de 50 à 60 cm ; ▶bêcher, ▶bêcher à deux fers, ▶labour profond) ; *syn.* défoncement [m] ; *g* **Rigolen [n, o. Pl.]** (Vorgang der tiefgründigen Bodenschichtung [> 50 cm tief] mit einem Spezialpflug; früher geschah dies durch aufwendige Spatenarbeit: drei Spaten tief; ▶holländern, ▶Tief-pflügen, ▶umgraben).

trenching [n] [UK], **trial** *conserv'hist.* ▶exploratory trench-ing.

6573 trench planting [n] *constr. hort.* (Setting out hedge plants or young trees in a trench; ▶furrow planting); *s* **plan-tación** [f] **en zanjas** (▶plantación en surcos); *f* **plantation** [f] **en tranchée** (Plantation de plantes pour haie dans une fouille en longueur sur le terrain de plantation ; ▶plantation en sillons) ; *syn.* plantation [f] en rigole (DFM 1975) ; *g* **Pflanzung** [f] **in durchgehendem Pflanzgraben** (▶Riefenpflanzung); *syn.* Grabenpflanzung [f].

trial area [n] **of paving surface** [UK] *constr.* ▶test area of paving surface [US].

trial borehole [n] [UK] *geo. min.* ▶test borehole [US].

trial garden [n] [UK] *hort.* ▶test garden.

trial garden [n] [UK], **perennials** *hort.* ▶perennials test garden.

trial pit [n] [UK] *constr. pedol.* ▶test pit [US].

trial trenching [n] [UK] *conserv'hist.* ▶exploratory trenching.

triangular grid [n] *constr.* ▶triangular planting grid.

6574 triangular planting grid [n] *constr.* (Planting layout used for mass plantings in which plant centers [US]/plant centres [UK] form equilateral triangles; ▶planting grid, ▶square plant-ing grid); *syn.* quincunx planting grid [n] (cf. BS 3975: part 5, 54071), triangular grid [n]; *s* **esquema** [m] **de plantación triangular** (▶Esquema de plantación 1 en el que los centros de las plantas forman triángulos equiláteros; ▶esquema de plantación rectangular); *f* **disposition [f] en quinconce** (Disposition triangulaire des sujets lors d'une plantation sur plusieurs rangs ; ►disposition des végétaux, ►disposition rectangulaire) ; *g* **Dreiecksverband [m]** (Auf Lücke versetzte Pflanzenanordnung bei mehrreihigen Pflanzungen; ►Pflanzenverband, ►Vierecksverband).

trickle field [n] [UK] *wat'man. agr.* ►wastewater wetland [US].

6575 trickle filtering [n] *agr. wat'man.* (Gradual percolation of sewage on a ►wastewater wetland [US]/trickle field [UK]); *syn.* sewage filtering [n]; *s* **irrigación [f] con aguas residuales** (Irrigación lenta de aguas residuales en ►campo de regado de aguas residuales); *syn.* filtración [f] de aguas residuales, percolación [f] de aguas residuales; *f* **épandage [m] par aéro-aspersion** (Aspersion lente et continue d'effluents urbains sur des ►champs d'épandage) ; *g* **Verrieselung [f]** (Langsames Aufbringen von Abwasser auf ►Rieselfelder); *syn.* Verregnung [f] von Abwässern.

trickle irrigation [n] [UK] *agr. hort.* ►drip irrigation.

trim [n] [UK] **(1)** *constr.* ►flush edging.

6576 trim [vb] **(2)** *hort.* (To clip any plant material frequently to align edge or shape; BS 3975: part 5; ►topiary); *syn.* sculpture [vb] plant material; *s* **recortar [vb]** (Poda ligera de árboles o arbustos para mantener la forma deseada; ►poda de topiaria); *f* **entretenir [vb] par des tontes régulières** (ADT 1988, 186 ; tailler des arbres et arbustes formés pour leur donner une forme en cône, en pyramide, en boule, etc. ; ►art topiaire) ; *g* **in Formschnitt halten [vb]** (Die gewünschte künstliche Form eines Gehölzes durch mehrmaliges Nachschneiden während eines Jahres erhalten; ►Formschnitt).

trimmed hedge [n] *constr. hort.* ►clipped hedge.

trimmings [npl] *agr. constr. hort.* ►prunings.

trim trail [n] *recr.* ►fitness trail.

trophic factors [npl] *ecol.* ∗►ecological factors.

6577 trophic level [n] *ecol.* (Level in a nutritive series of an ecosystem in which organisms, acquiring their food in the same way, are grouped together in one step in the ►food chain. **1. chain:** the first or lowest **t. l.** consists of primary producers [green plants]; the second level of herbivores; the third level of secondary carnivores; the last level of ►top consumers. **2. chain:** the first level consists of dead, organic material; the second level of ►saprophages; the succeeding levels are carnivores; cf. RCG 1982); *s* **nivel [m] trófico** (Cada uno de los diferentes eslabones de las ►cadenas tróficas: productores, herbívoros, carnívoros. Aquellos organismos que obtienen su alimento de los vegetales a través del mismo número de eslabones se dice que pertenecen al mismo nivel trófico; DINA 1987; ►saprófago); *f* **niveau [m] trophique** (Ensemble des organismes appartenant à la même séquence de la ►chaîne trophique. **Première chaîne :** les plantes vertes [producteurs primaires] forment le premier **n. t.**, les herbivores [consommateurs de premier ordre] le deuxième **n. t.**, les carnivores se nourrissant d'herbivores [consommateurs de deuxième ordre] le troisième **n. t.**, les carnivores se nourrissant de carnivores [consommateurs de troisième ordre] formant le dernier **n. t. Deuxième chaîne :** la matière morte, les substances organiques forment le premier **n. t.**, les ►saprophages le deuxième et les carnivores de premier, deuxième et troisième ordre, les niveaux trophiques suivants) ; *g* **Trophiestufe [f]** (Gesamtheit der Organismen, die demselben Glied der ►Nahrungskette angehören. **1. Kette:** grüne Pflanzen [Primärproduzenten] bilden die erste Trophiestufe, Pflanzenfresser die zweite [Stufe der Primärkonsumenten], Fleischfresser, die Pflanzenfresser fressen, die dritte Stufe [Sekundärkonsumenten]; die letzte Stufe bilden die Endkonsumenten. **2. Kette:** tote, organische Substanzen bilden die erste Stufe, ►Saprophage die zweite und Fleischfresser 1., 2. und 3. Ordnung die folgenden Stufen; cf. ODUM 1980); *syn.* Nahrungsstufe [f], Trophieebene [f], trophische Ebene [f].

6578 trophic level [n] **of a waterbody** *limn.* (►Water quality category of standing waterbodies depending on the degree of ►nutrient stress. **In D.,** four levels are distinguished: ►oligotrophic, ►mesotrophic, ►eutrophic and polytrophic; ►eutrophication, ►mesotrophic lake, ►polytrophic lake); *s* **nivel [m] trófico de masa de agua** (►Clase de calidad de las aguas continentales de lago o laguna dependiendo del grado de ►eutrofización; generalmente se diferencian cuatro niveles ►oligótrofo, ►mesótrofo, ►eutrófico y polítrofo; ►eutrofización, ►lago mesótrofo, ►lago polítrofo; ►exceso de nutrientes); *f* **niveau [m] trophique des eaux douces** (Classe de pollution des eaux calmes caractérisée par le niveau de ►charge en éléments nutritifs ; **F.,** on distingue différentes classifications [selon l'interprétation des analyses chimiques : pH, matières en suspension — MeS —, dureté, alcalinité etc.] ; ►niveau de qualité des cours d'eau ; **D.,** on distingue quatre classes d'interprétation de la pollution par les matières organiques : ►oligotrophe, ►mésotrophe, ►eutrophe, polytrophe ; la productivité primaire est généralement exprimée par la fixation de C par le phytoplancton [mg C/m²/jour] ; ►eutrophisation, ►lac mésotrophe, ►lac polytrophe) ; *syn.* niveau [m] de consommation ; *g* **Trophiestufe [f] eines stehenden Gewässers** (►Gewässergüteklasse bei Stillgewässern in Abhängigkeit der ►Nährstoffbelastung. **In D.** werden vier Stufen unterschieden: ►oligotroph, ►mesotroph, ►eutroph, polytroph/hypertroph; ►Eutrophierung, ►mesotropher See, ►polytropher See); *syn.* Trophiegrad [m] eines stehenden Gewässers.

6579 trophic state [n] *ecol.* (Level of available nutrients in an ecosystem; ►eutrophy, ►mesotrophy, ►oligotrophy); *s* **estado [m] trófico** (Grado de suministro de nutrientes de un ecosistema; ►eutrofia, ►mesotrofia, ►oligotrofia); *f* **degré [m] trophique** (Richesse d'un écosystème en nutriments minéraux et en matière organique ; ►eutrophie, ►mésotrophie, ►oligotrophie) ; *syn.* richesse [f] en éléments nutritifs ; *g* **Trophie [f]** (Grad der Versorgung eines Ökosystems mit verfügbaren Nährstoffen; ►Eutrophie, ►Mesotrophie, ►Oligotrophie).

6580 tropical grassland [n] *phyt.* (Large open area practically without woody phanerophytes, in general due to anthropogenic influence; normally called '**t. g.**', but the grass cover is physiognomically identical with tree and scrub savanna[h]s; ELL 1967; ►savanna[h], ►shrubland savanna[h], ►tree savanna[h]); *syn.* grass savanna[h] [n]; *s* **sabana [f] herbácea** (Vegetación de gramíneas sin árboles existente en zonas tropicales y subtropicales; *subtipos* **s. h.** alta, **s. h.** baja; ►sabana, ►sabana arbolada, ►sabana de espinos); *syn.* sabana [f] de gramíneas; *f* **savane [f] herbacée** (Formation végétale herbacée dépourvue de végétaux ligneux, spécifique aux régions tropicales ; ►savane, ►savane arborée, ►savane arbustive 2) ; *g* **Grasland [n, o. Pl.]** (Natürliche Grasvegetation ohne Baumbestand in den Tropen; ►Baumsavanne, ►Dornstrauchsavanne, ►Savanne); *syn.* tropisches Grasland [n, o. Pl.].

tropical ombrophilous cloud forest [n] *for. phyt.* ►cloud forest.

trough garden [n] [UK] *gard.* ►container garden [US].

6581 trout zone [n] *limn.* (Uppermost zone of a river system with a steep gradient and fast flowing current; in Europe the characteristic fish species are trout *[Salmo trutta]* and Salmon *[Salmo salar]*. The unpolluted water is very clear and well oxygenated; ►epirhithron, ►river biocoenosis, ►river zone);

syn. torrent zone [n]; *s* **zona [f] de trucha** (Zona superior de la región salmonícola con pez característico trucha común *[Salmo trutta]*; ►biocenosis fluvial, ►epiritrón, ►zona biológica de un río); *f* **zone [f] à truite** (Tronçon supérieur d'un cours d'eau caractérisé par la truite *[Salmo trutta]*, espèce adoptée à un fort courant des eaux bien oxygénées et assez froides. La zone à truite et la zone à ombre forment la région salmonicole ou rhithron ; ►biocénose fluviale, ►épirhithron, ►zone piscicole des cours d'eau) ; *syn.* zone [f] de l'épirhithron ; *g* **Forellenregion [f]** (Durch Vorkommen der Forelle *[Salmo trutta]* gekennzeichneter oberster Abschnitt eines Fließgewässers. Er hat eine kräftige bis sehr starke Strömung, ist klar, sauerstoffreich und konstant kalt; ►Epirhithron, ►Flussbiozönose, ►Flussregion); *syn.* Epirhithral [n].

truck gardening [n] [US] *trans.* ►market gardening.

truck lane [n] [US, Vt] *trans.* ►slow vehicle lane.

6582 true to name [loc] *hort.* (Descriptive term correctly applied to a plant which corresponds genetically to the indicated species or cultivar/variety; *noun* trueness to variety); *s* **de variedad certificada [loc]** (Garantía de que una planta corresponde genéticamente a la especie o variedad indicada); *syn.* con garantía de identificación [loc]; *f* **en conformité spécifique et variétale [loc]** (Conforme aux caractéristiques génétiques de l'espèce et de la variété) ; *syn.* en authenticité des variétés [loc] ; *g* **arten- und sortenecht [adj]** (So beschaffen, dass eine Pflanze genetisch der bezeichneten Art resp. der angegebenen Sorte entspricht; Sortenechtheit [f]).

6583 truncated soil profile [n] *pedol.* (Exposed lower [B and/or C] horizons, caused by soil erosion or excavation of the upper [A and/or B] horizons; ►denudation, ►exposed stone formation, ►extreme soil erosion); *s* **suelo [m] decapitado** (Suelo erosionado hasta dar luz al horizonte C o a roca madre; ►denudación, ►destrucción del suelo, ►formación de suelo decapitado); *f* **sol [m] décapé** (Sol dont les horizons de surface ont été enlevés sous l'action de l'érosion et laissant apparaître en surface les horizons B ou C ; DIS 1986, 63 ; ►dévastation des sols, ►érosion continentale, ►formation de pavage) ; *syn.* profil [m] tronqué, profil [m] érodé ; *g* **geköpftes Bodenprofil [n]** (Bis auf den C-Horizont zerstörtes Bodenprofil, bei dem sich Bodenbildung und ►Abtragung etwa die Waage halten; ►Bodenverheerung; ►Steinpflasterbildung); *syn.* gekapptes Bodenprofil [n], Scherbenboden [m] (WAG 1984).

trunk [n] *constr.* ►boarding up of tree trunks; *hort.* ►tree trunk.

trunk [n]**, raised soil level near a tree** *arb. constr.* ►groundfill around a tree.

trunk cal(l)iper [n] [US] *hort. constr.* ►trunk diameter.

6584 trunk diameter [n] *hort. constr.* (Straight line passing from side to side through the center [US]/centre [UK] of a tree stem, usually stated as **diameter breast height [d.b.h.]**; ►caliper growth, ►girth); *syn.* trunk caliper/trunk calliper [n] [also US]; *s* **diámetro [m] de tronco** (En leñosas se mide a 1 m del suelo; ►circunferencia de tronco, ►crecimiento en espesor); *syn.* espesor [m] de tronco; *f 1* **diamètre [m] (à hauteur d'homme)** (Hauteur à laquelle on a coutume de mesurer le diamètre ou la circonférence des arbres. Il est généralement admis que cette hauteur est de 1,50 m au-dessus du sol ; DFM 1975, 146 ; ►circonférence du tronc, ►croissance en épaisseur) ; *f 2* **diamètre [m] de la tige** (Caractéristique dimensionnelle des jeunes baliveaux, baliveaux, touffes et cépées d'arbres d'alignement, d'ornement fruitiers mesurée à 100 cm du collet) ; *g* **Stammdurchmesser [m]** (Bei Gehölzen in 1 m Höhe über Grund durch die Mitte des Stammes verlaufende Messlinie; Stämmlinge mehrstämmiger Bäume werden einzeln gemessen; in angelsächsischen Ländern beziehen sich die Maße auf Brusthöhe

[**Brusthöhendurchmesser** bei 1,30 m Stammhöhe]; ►Dickenwachstum, ►Stammumfang); *syn.* Durchmesser [m] eines Stammes.

trunk highway [n] [US] *trans.* ►major highway [US] (1)/major road [UK].

trunk injury [n] [US] *arb. hort.* ►stem injury.

6585 trunk protection [n] *arb. constr.* (Measures to prevent the effects on tree trunks of drying by wind or sun, and mechanical damage; ►trunc wrapping, ►wrapping with burlap [US]/wrapping with Hessian strips [UK], ►wrapping with straw ropes); *s* **protección [f] del tronco** (Medida para prevenir contra quemaduras y daños mecánicos del tronco p. ej. con ►saco de protección del tronco, ►envoltura del tronco, ►envoltura de paja o lechada de cal); *f* **protection [f] du tronc** (Mesures de protection du tronc d'arbre contre les effets des rayons solaires, du vent ainsi que contre les dégâts mécaniques ; ►paillage de tronc, ►tampon de protection en jute, ►tampon protecteur de tronc) ; *syn.* protection [f] contre les agressions ; *g* **oberirdischer Baumschutz [m, o. Pl.]** (Schutz gegen Wind- und Sonneneinwirkung, Rindenbrand und mechanische Schäden; ►Lehm-Juteverband, ►Stammumwicklung, ►Strohbandage); *syn.* Stammschutz [m].

trunk road [n] (1) *trans.* ►arterial road.

trunk road [n] [UK] (2) *adm. leg. trans.* ►major highway [US] (1).

trunk road [n] [UK] (3) *adm. leg. trans.* ►state highway [US].

trunk sewer [n] [US] *envir. urb.* ►collector sewer.

6586 trunk wound [n] [US] *arb. hort.* (Result of a ►stem injury); *syn.* stem wound [n] [UK]; *s* **herida [f] del tronco** (Resultado de una ►lesión del tronco); *f* **plaie [f] du tronc** (Résultat de la ►blessure du tronc) ; *g* **Stammwunde [f]** (Ergebnis einer ►Stammverletzung).

6587 trunk wrapping [n] *arb. constr. hort.* (ZIO 1968; enfolding of a trunk for protection against sun scorch and transpiration; e.g. by ►wrapping with burlap [US]/wrapping with Hessian strips [UK], ►wrapping with straw ropes, wrapping with prepared paper [US]; ►trunk protection); *syn.* tree wrapping [n]; *s* **envoltura [f] del tronco** (Cobertura de paja, tela o papel especial para proteger el tronco contra quemaduras; ►saco de protección del tronco, ►envoltura de paja, ►protección del tronco); *f* **tampon [m] protecteur de tronc** (Protection contre le soleil et le dessèchement des gros végétaux fraîchement transplantés, p. ex. au moyen d'un ►tampon de protection en jute ou du ►paillage de tronc ; ►protection du tronc) ; *g* **Stammumwicklung [f]** (1. Sonnen- und Verdunstungsschutz an neu gepflanzten Großgehölzen, z. B. mit einer ►Lehm-Juteverband, ►Strohbandage oder Phragmites-Manschette. 2. Rindenschutz bei Transport von einer Baumschule zur Pflanzstelle); *syn.* oberirdischer Baumschutz); *syn.* Bandage [f].

tuber [n] *bot.* *►geophyte, ►stem tuber, ►root tuber.

tuber [n]**, ground** *bot.* *►geophyte.

tuber [n]**, hypocotylar** *bot.* *►stem tuber.

tuberous root [n] *bot.* ►root tuber.

6588 tuberous-rooted plant [n] *bot. hort.* (Plant with fleshy, swollen tubers, known as ►root tuber, ►stem tuber or hypocotylar tuber; ►bulbous plant); *s* **planta [f] tuberosa** (►planta de cebolla, ►tubérculo caulinar, ►tuberosidad radical); *f* **plante [f] à tubercule** (Plante qui persiste grâce à un organe souterrain, dans lequel sont accumulées des réserves ; ►plante à bulbe, ►tubercule caulinaire, ►tubercule radiculaire) ; *syn.* plante [f] tubéreuse ; *g* **Knollengewächs [n]** (Pflanze mit fleischig verdickten Sprossen als ►Wurzelknolle, ►Sprossknolle oder Hypokotylknolle; ▶Zwiebelgewächs); *syn.* Knollengeophyt [m].

tubing size [n] [US] *constr.* ▶internal diameter.

6589 tuft [n] (1) *phyt. plant. hort.* (Clump of several perennials or grasses, mostly of the same species); *syn.* clump [n], bunch [n] [also US]; *s* **macolla** [f] (≠) (Grupo de vivaces o gramíneas, generalmente de la misma especie, que crece en un cuadro de flores); *f* **touffe** [f] (Assemblage de plusieurs plantes vivaces ou de graminées souvent de la même espèce dans une plantation) ; *syn.* groupe [m] ; *g* **Tuff** [m] (Gruppe mehrerer Stauden oder Gräser, meist einer Art, auf einer Fläche).

tuft [n] (2) *bot.* ▶tussock.

tuft [n] **divided for propagation** *hort.* ▶tussock divided for propagation.

tuft formation [n] *bot. hort.* ▶tussock formation.

6590 tuft-forming [adj] *phyt. plant. hort.* (▶tussock formation); *s* **que forma macollas** [loc] (▶formación de fascículos); *f* **formant, ante une touffe** [loc] (Espèce herbacée croissant en touffe ; ▶formation en touffe) ; *syn.* formant, ante touffe [loc] ; *g* **tuffbildend** [ppr/adj] (Krautige Pflanze einer Art betreffend, die in Gruppen wächst; ▶Horstbildung).

tumbling bay [n] [UK] *constr.* ▶drop drain [US].

6591 tundra [n] *geo. phyt.* (Treeless, species-poor, vegetation type occurring on flat land, characteristic of northern arctic regions, marking the limit of arborescent vegetation on black mucky soil with permanently frozen subsoil—permafrost soil); *syn.* arctic prairie [n] (GGT 1979, 495); *s* **tundra** [f] (1. Tipo de vegetación circumpolar definido por las características mesológicas siguientes: período vegetativo inferior a 3 meses, precipitación entre 200 y 300 mm al año, temperatura media en el mes más cálido 5-10°C, suelo mal drenado [permafrost], pobre en nitrógeno, de poca profundidad; cf. DINA 1987. 2. Designación geográfica del paisaje de los países circumpolares. Aparte de su carácter climático, encierra un sentido geobotánico y otro edafológico; cf. DB 1985); *f* **toundra** [f] (1. Formations végétales, buissonnantes, frutescentes et herbacées, généralement fermées et non arborées, des régions subpolaires. Correspondent à des climats froids à été court et à des sols gelés en profondeur ; cf. DG 1984, 451. 2. Terme géographique désignant les écosystèmes couvrant les régions circumpolaires semi-désertes) ; *g* **Tundra** [f] (1. Baumloser, artenarmer Vegetationstyp jenseits der polaren Baumgrenze auf Dauerfrostböden — Permafrostböden, die im Sommer nur kurzzeitig gering mächtig auftauen. 2. Geografische Bezeichnung für zirkumpolare Landschaften. Neben der klimatischen Bedeutung umfasst **T.** den pflanzensoziologischen und bodenkundlichen Aspekt).

6592 tunnel [n] *eng. trans.* (Underground passageway; *specific terms* ▶bored tunnel, ▶cut and cover tunnel, ▶highway tunnel [US]/road tunnel [UK], ▶open cut tunnel); *s* **túnel** [m] (Ruta subterránea de tráfico; *términos específicos* ▶túnel construido a zanja abierta, ▶túnel construido a cielo abierto y cubierto, ▶túnel de carretera, ▶túnel perforado); *f* **tunnel** [m] (Ouvrage souterrain de circulation ; *termes spécifiques* ▶tunnel de montagne, ▶tunnel en tranchée ouverte, ▶tunnel réalisé en fouille ouverte, ▶tunnel routier) ; *g* **Tunnel** [m] (Künstliches unterirdischer Verkehrsbauwerk; *UBe* ▶Bergtunnel, ▶Straßentunnel, ▶Tunnel in Schlitzlage, ▶Unterpflastertunnel); *syn.* Unterflurtrasse [f] [auch A].

tunnel [n] [UK], **road** *trans.* ▶highway tunnel [US].

6593 tunnel-arbor [n] [US]/**tunnel-arbour** [n] [UK] *arch. gard.* (Tunnel-shaped trellis or a series of arches over which climbing plants are trained to cover a walk; ▶grapevine arbor [US]/grapevine arbour [UK]); *syn.* trellis-covered walk [n];

s **arcada** [f] (Pequeña construcción de enrejado en forma de arco para apoyo de trepadoras hasta conseguir un túnel; ▶emparrado); *f 1* **tonnelle** [f] (Petite construction circulaire à sommet arrondi, faite de lattes en treillis soutenues par des cerceaux, sur laquelle s'accroche des plantes grimpantes qui sert d'abri ; PR 1987 ; ▶treille) ; *syn.* berceau [m], treillage [m] en voûte (PR 1987) ; *f 2* **allée** [f] **taillée** (Arbres taillés de façon rectiligne en palissade de manière à présenter une surface plane, un mur de verdure) ; *f 3* **charmille** [f] (*Terme spécifique* allée plantée de charmes [*Carpinus betulus*] très rapprochés et souvent taillés en forme de berceau]) ; *g* **Laubengang** [m] (1. Ein meist mit Kletterpflanzen bewachsener, offener Bogengang in einem Garten oder Park [in der Fachsprache *Berceau artificiel* oder *B. de treillage*]; ▶Weinlaube). 2. Geformte Hecken aus Heckenpflanzen oder Bäumen, die einen Weg überspannen, wobei die Pflanzen selbst das Gerüst bilden [in der Fachsprache *Berceau naturel* oder *B. de verdure*; cf. GAR 2003; H. 7, 17 ff]); *syn.* Berceau [m].

tunnel construction [n] *constr. eng.* ▶tunneling.

6594 tunnel [n] **for amphibians** *constr. ecol. zool.* (Underpass of a travelled way [US]/carriage-way [UK] which facilitates safe passage of migrating amphibians; ▶game pass); *s* **túnel** [m] **para anfibios** (Paso subterráneo bajo una carretera para permitir que la crucen sin peligro los anfibios y sobre todo los sapos; ▶cruce para animales salvajes); *f* **tunnel** [m] **pour amphibiens** (Passage inférieur spécifique réalisé sous une infrastructure routière et emprunté par la faune amphibienne dans ses pérégrinations vers les lieux de reproduction ; ▶passage à faune) ; *syn.* crapauduc [m], batrachoduc [m] ; *g* **Amphibientunnel** [m] (Unterführung einer Verkehrsstraße, die Amphibien, vor allem Kröten, das gefahrenfreie Queren ermöglichen sollen; *UB* Krötentunnel [m]; ▶Wildtierpassage); *syn.* Amphibiendurchlassanlage [f], Amphibienschutzanlage [f] (L+S 2008, 248 ff).

6595 tunneling [n] *constr. eng.* (*Also* tunnelling [n]; construction of an underground passageway, e.g. driving a roadway underneath a watercourse, airport, mountain ridge, channel, urban development, etc.; ▶tunnel); *syn.* tunnel work [n], tunnel construction [n]; *s* **construcción** [f] **de túnel** (Construcción de un pasaje subterráneo para cruzar por debajo de un río, monte, un estrecho de mar, etc.; ▶túnel); *f* **passage** [m] **sous-terrain** (Aménagement d'une infrastructure de transport souterraine, p. ex. le passage sous un cours d'eau, un canal, un aéroport, un massif montagneux, une zone bâtie, un détroit maritime, etc. ; *termes spécifiques* passage sous ouvrage, passage sous voirie, passage sous piste, passage sous-fluvial ; ▶tunnel) ; *syn.* passage [m] en souterrain ; *g* **Untertunnelung** [f] (Anlegen eines unterirdischen Verkehrsbauwerkes, z. B. zur Unterfahrung eines Gewässerlaufes, Flughafens, Bergmassivs, einer Bebauung, Meerenge etc.; ▶Tunnel); *syn.* Tunnelbau [m].

tunnelling [n] *constr. eng.* ▶tunneling.

tunnel work [n] *constr. eng.* ▶tunneling.

turf [n] *constr.* ▶playfield turf [US]/sportsground turf [UK], ▶reinforced turf, ▶rolled turf, ▶sod [US] 1/turf [UK], ▶sward.

turf [n], **hard-wearing** *constr.* ▶hard-wearing lawn.

turf [n] [UK], **sports ground** *constr.* ▶playfield turf [US].

6596 turf area [n] *constr. recr.* (Grass area such as a ▶lawn; ▶turfing 2, ▶turf root-zone layer, ▶turf sward); *syn.* lawn area [n], grass area [n]; *s* **superficie** [f] **de césped** (Terreno cubierto de vegetación densa de gramíneas segadas; ▶césped, ▶establecimiento de césped con tepes, ▶sustrato de enraizamiento de césped, ▶tapiz de césped); *f* **pelouse** [f] (Terrain couvert de plantes herbacées, le plus souvent de graminées, enchevêtrées les unes dans les autres et constituant un tapis ras par des tontes successives ; ▶engazonnement par placage, ▶gazon, ▶support végétal d'un gazon, ▶tapis de graminées) ; *g* **Rasenfläche** [f]

(Mit kurz gehaltenen Gräsern bestockte Fläche, besonders in Ziergärten, öffentlichen Grünanlagen und in Sportanlagen; beim Sportplatzbau besteht der Aufbau der **R.** aus ►Rasennarbe, ►Rasentragschicht, Baugrund und ggf. Entwässerungseinrichtungen; zum Oberbau gehören Rasendecke, Rasentragschicht, ggf. Dränschicht; ►Herstellung von Rasenflächen durch Fertigrasen, ►Rasen).

turf block [n] [US] *constr.* ►grass paver.

turf corning [n] [UK] *constr.* ►lawn aeration, #6.

turf density [n] *constr. hort.* ►density of sward (1).

turfed pitch [n] [UK] *recr.* ►play lawn [US] (2).

6597 turf-filled joint [n] *constr.* (Wide gap in ►grass-filled modular paving [US]/grass setts paving [UK] that is left between pavers and sown with hard-wearing grasses); *syn.* grass-filled joint [n]; *s* **junta** [f] **de césped** (Junta ancha entre las piezas de ►pavimento de adoquines con césped que se siembra con mezcla de semillas de especies resistentes); *f* **joint** [m] **gazonné** (Joint large dans certaines surfaces pavées prévu pour le semis d'un gazon résistant au piétinement ; ►pavage engazonné) ; *syn.* joint [m] gazon, joint [m] engazonné ; *g* **Rasenfuge** [f] (In Belägen aus Großpflastersteinen o. Ä. vorgesehene breite Fuge für die Einsaat strapazierfähiger Gräser resp. Fuge, die mit strapazierfähigen Gräsern bestockt ist; ►Rasenpflaster).

6598 turf formation [n] *hort.* (Growing together of individual grasses to form a ►turf sward); *s* **formación** [f] **de tapiz de césped** (Crecimiento de las plantas de hierba individuales hasta formar un ►tapiz de césped); *f* **formation** [f] **de la couche de gazon** (Développement des différents graminées pour former un gazon à feuillage dense ; ►tapis de graminées) ; *g* **Narbenbildung** [f] (Das Zusammenwachsen der einzelnen Gräser zu einer dichten ►Rasennarbe).

6599 turf grid [n] *constr.* (*Bioengineering technique* network of ►turf strips 2 laid diagonally at an angle of 45° and at spacings of 100-200cm; ►rolled turf, ►sod layer [US]/turf layer [UK]); *s* **entramado** [m] **de tepes** (*Construcción biotécnica* bandas diagonales de césped prefabricado —generalmente ►césped de tepes enrollados— que se colocan para fijar laderas en cruz a 45°; ►banda de tepes, ►césped para transplante); *f* **maillage** [m] **de gazon** (±) (*Génie biologique* bandes de gazon préfabriqué [en général gazon en rouleaux] parallèles, obliques et croisées formant un angle de 45°, espacées régulièrement tous les 1 à 2 m, mises en place sur les talus ; ►bande de gazon, ►gazon en rouleaux précultivé, ►gazon précultivé) ; *g* **Rasengitter** [n] (*Ingenieurbiologische Bauweise* am Hang aus ►Fertigrasen [meist ►Rollrasen] kreuzweise hergestellte Schrägstreifen in 100-200 cm Abstand im 45° Winkel; ►Rasenband).

turficolous [adj] *phyt.* ►growing on peat.

6600 turfing [n] (1) *constr.* (Covering an area with sod [US]/ covering an area with turfs [UK]; ►plugging, ►replacement of sods [US]/replacement of turves [UK], ►rolled turf); *syn.* sodding [n] [also US] (1); *s* **revestimiento** [m] **con tepes** (Cubrir un área con césped por medio de planteles de césped precultivados; ►césped de tepes enrollados, ►plantación de césped con tepes o con macollas, ►recolocación de tepes o tierra vegetal); *f* **placage** [m] (Mise en place sur une aire d'engazonnement d'une pelouse cultivée en ►gazonnière ; ►engazonnement par bouturage, ►remise en place des plaques de gazon, ►rouleau de gazon précultivé) ; *syn.* mise [f] en place d'un gazon en plaque, engazonnement [m] par placage ; *g* **Andecken** [n, o. Pl] **von Fertigrasen** (Bedecken einer Bodenfläche mit vorkultivierten Rasensoden; ►Herstellung von Rasenflächen durch Auspflanzen von Rasen- oder Grasteilen, ►Rollrasen, ►Wiederandeckung von Rasensoden); *syn.* Begrünung [f] mit Rasensoden, Verlegen [n, o. Pl.] von Fertigrasen/Rasensoden.

6601 turfing [n] (2) *constr.* (**1.** Creating a lawn with turf; ►sod layer [US]/turf layer [UK], ►turfing 1); *syn.* sodding [n] [also US] (2); **2. sodding** [n] [US] (3) (May also include placement of pregrown mats of herbaceous perennials, ground covers, ornamental grasses, or mosses); *s* **establecimiento** [m] **de césped con tepes** (►césped para transplante, ►revestimiento con tepes); *syn.* plantación [f] de tepes; *f* **engazonnement** [m] **par placage** (Réalisation d'une surface engazonnée par mise en place d'un gazon de placage récolté sur un lit de sable, livré prêt à être déroulé sur le sol. Le travail du sol s'effectue comme pour un semis : enlèvement des cailloux et débris végétaux, incorporation d'engrais racinaire, nivellement et roulage léger du sol [100-400 kg/m²] sans le compacter. La pose des plaques s'effectue bord à bord, de préférence en quinconce pour éviter les joints croisés ; roulage ou damage des plaques suivi d'un arroser copieux [2 x jour pendant 10 jours, matin et soir en été] ; ►gazon précultivé, ►placage) ; *g* **Herstellung** [f] **von Rasenflächen durch Fertigrasen** (Bei der Bodenvorbereitung ist ein erdfeuchtes Planum herzustellen und eine ausreichende Grundversorgung von Nährstoffen sicherzustellen; der Fertigrasen wird engfugig ohne Kreuzfugen verlegt; mit einer leichten Walze [100-400 kg/m²] wird für den nötigen Bodenschluss gesorgt und anschließend mit 15 l/m² kräftig gewässert; ►Andecken von Fertigrasen, ►Fertigrasen).

turf layer [n] [UK] *constr. hort.* ►sod layer [US].

turf nursery [n] [UK] *hort.* ►sod nursery [US].

turf paver [n] [US] *constr.* ►grass paver.

turf perforation [n] *constr.* ►lawn aeration, #1.

turf playing field [n] *constr. recr.* ►grass sports field.

turf removal site [n] [UK] *constr. hort.* ►sod removal site [US].

6602 turf root-zone layer [n] *constr.* (Bed of soil or other growing medium for grass root growth); *s* **sustrato** [m] **de enraizamiento de césped** (Capa de tierra vegetal o sustrato creado artificialmente, bien aireado e intensamente enraizado, situado sobre la capa filtrante; almacena parte del agua de percolación y deja pasar aquélla que sobra hacia el subsuelo o a la instalación de drenaje); *syn.* sustrato [m] de siembra de césped; *f* **support** [m] **végétal d'un gazon** (Sous-sol bien drainant et bien préparé sur lequel s'effectue la pose de gazon de placage) ; *syn.* assise [f] végétale d'un gazon ; *g* **Rasentragschicht** [f] (Für Rasen und seine zukünftige Nutzung durchlässige, belastbare und intensiv durchwurzelbare Boden- resp. Substratschicht, die auf dem Baugrund resp. auf einer Dränschicht liegt; sie speichert einen Teil des einsickernden Oberflächenwassers und gibt das Überschusswasser an den Baugrund oder eine Entwässerungseinrichtung ab; cf. DIN 18 035 Teil 4); *syn.* Vegetationstragschicht [f] für Rasen.

turf-seed mixture [n] [US] *constr. hort.* ►grass-seed mixture.

turf sod [n] *agr. pedol.* ►heath sod.

6603 turf strip [n] (1) *constr.* (Narrow, grassed area often between sidewalk and roadway, mowed 6 to 10 times during the growth period; ►grass strip 1, ►meadow strip, ►turf strip 2); *syn.* grass strip [n] (2); *s* **franja** [f] **de césped** (►Franja de hierba estrecha en carretera o calle que se corta entre 6 y 10 veces en cada periodo/período vegetativo; ►banda de herbáceas, ►banda de tepes, ►franja de hierba); *f* **accotement** [m] **engazonné** (Espace linéaire aménagé le long une chaussée sur lequel sont effectuées entre 6-10 tontes par an ; ►accotement enherbé, ►bande de gazon, ►bande engazonnée) ; *g* **Rasenstreifen** [m] (1) (Schmaler, oft straßenbegleitender, kurz geschnittener ►Grasstreifen, der im Laufe der Vegetationsperiode 6-10-mal gemäht wird; ►Rasenband, ►Wiesenstreifen).

T

6604 turf strip [n] (2) *constr.* (*Bioengineering* diagonal strips of ►sods [US] 1/turves [UK] which are laid at spacings of 100-200cm and at an angle of about 35° to the slope; on steep slopes, crosswise at, 45°; ►sod layer [US]/turf layer [UK], ►turf grid); *s* **banda** [f] **de tepes** (*Construcción biotécnica* bandas diagonales de ►césped de tepes —generalmente ►césped de tepes enrollados— colocadas en las laderas a 100-200 cm de distancia en ángulos de 35° o —para proporcionar una buena protección contra la erosion— cruzadas a 45°; ►entramado de tepes); *f* **bande** [f] **de gazon** (*Génie biologique* méthode de protection des versants contre l'érosion réalisée au moyen de ►gazon précultivé [►gazon en rouleau précultivé] disposé en bandes parallèles obliques d'un angle de 35° avec un espacement de 1 à 2 m ou en bandes croisées obliques d'un angle de 45° ; ►maillage de gazon) ; *g* **Rasenband** [n] (*Ingenieurbiologische Bauweise* aus ►Fertigrasen [meist ►Rollrasen] hergestellter, im Winkel von ca. 35°, in 100-200 cm Abständen auf Böschungen zur Bodensicherung verlegter Schrägstreifen. Bei starker Erosionsgefahr werden die **R.bänder** kreuzweise im 45° Winkel eingebaut; ►Rasengitter); *syn.* Rasenstreifen [m] (2).

6605 turf sward [n] *constr.* (Ground covered with dense, closely mown or cropped grass; ►sward); *s* **tapiz** [m] **de césped** (Capa superior del suelo de césped cuidado intensivamente con denso enraizamiento de la vegetación generalmente de gramíneas; ►tapiz herbáceo); *f* **tapis** [m] **de graminées** (Couverture végétale d'un sol maintenue rase, constituée principalement de graminées sélectionnées ; ►tapis herbacé) ; *g* **Rasennarbe** [f] (Oberste Bodenschicht eines intensiv gepflegten Nutz- oder Zierrasens, die mit einer dichten, vorwiegend aus Gräsern bestehenden Vegetationsdecke durchwurzelt ist; ►Grasnarbe deckt einmal als OB jede mit Gräsern bestandene Fläche ab, wird deshalb auch syn. für intensiv gepflegte Rasenflächen gebraucht; bei **R.** wird i. d. R. die kurze Halmlänge durch häufiges Mähen mitgedacht); *syn.* Rasendecke [f], Grasnarbe [f].

turn [n] *trans. urb.* ►turnaround.

6606 turnaround [n] *trans. urb.* (Generic term covering all forms of turning areas at the end of ►cul-de-sacs; ►hammerhead turnaround, ►loop turnaround, ►turnaround space); *syn.* turn [n]; *s* **revuelta** [f] (Término genérico para todos los tipos de lugares de viraje al final de un ►fondo de saco; ►área de [re]vuelta, ►lazo de [re]vuelta, ►martillo de [re]vuelta); *syn.* zona [f] de giro; *f* **dispositif** [m] **de retournement** (Élément de voirie aménagé à l'extrémité de toutes les voies en impasse ; ►voie sans issue, ►T de retournement, ►aire de retournement, ►boucle de retournement ; BON 1990, 359) ; *g* **Wendeanlage** [f] (Fahrbahnausweitung am Ende einer ►Sackgasse für das Wenden von Fahrzeugen; *UBe* ►Wendehammer, ►Wendeplatz, ►Wendeschleife).

6607 turnaround space [n] *trans. urb.* (Widening at the end of a ►cul-de-sac without a central island which allows vehicles to turn); *s* **área** [f] **de (re)vuelta** (≠) (Ampliación de la calzada sin isla central al final de un ►fondo de saco para que los vehículos puedan dar la vuelta; *término genérico* ►revuelta); *syn.* área [f] de giro (≠); *f* **aire** [f] **de retournement** (Espace permettant à un véhicule de faire demi-tour dans les voies en impasse ; il existe deux familles d'aires de retournement : en giratoire et en T ; ►dispositif de retournement, ►voie sans issue) ; *syn.* aire [f] de manœuvre ; *g* **Wendeplatz** [m] (Fahrbahnausweitung am Ende einer ►Sackgasse ohne Mittellinsel zum Wenden der Fahrzeuge; *OB* ►Wendeanlage); *syn.* Wendeplatte [f] [BW], Umkehrplatz [m] [auch A].

turnaround [n] **with central island** [UK] *trans. urb.* ►loop turnaround.

turnaround [n] **with open centre** [UK] *trans. urb.* ►loop turnaround.

turning circle [n] *constr. plan.* ►vehicle turning circle.

turning clearance circle [n] [UK] *trans.* ►compound curve [US].

turning-over [n] **of grassland** [US] *agr.* ►plowing up of grassland [US]/ploughing up of grassland [UK].

turning-over [n] **of soil heaps** [UK] *constr. hort.* ►transferring of soil piles [US].

turnover [n] *ecol.* ►organic turnover.

turves [npl] [UK], **replacement of** *constr. hort.* ►replacement of sods [US].

turves [npl] [UK], **scoring of** *constr. hort.* ►cutting of sods [US].

turves [npl] [UK], **securing** *constr.* ►securing sods [US].

6608 tussock [n] *bot.* (Thick clump of grass or similar plants as opposed to a ►sward); *syn.* tuft [n] (2); *s* **macolla** [f] (Conjunto de vástagos nacidos de la base de un mismo pie, sobre todo tratándose de gramíneas y plantas graminoides; DB 1985; ►tapiz de césped); *f 1* **touffe** [f] (Tiges basales d'égale force et dépourvues de rameaux latéraux formant un bouquet compact ; ►tapis de graminées) ; *f 2* **touradon** [m] (Terme spécifique pour la forme élevée de certaines graminées adultes, p. ex. la Canche bleue *[Molinia cærulea]*, ou la touffe de certaines Laîches comme la Laîche paniculée *[Carex paniculata]*, pouvant dépasser 1,5 m et qui résulte de la persistance et de l'accumulation, au cours des années, de la souche et des restes des feuilles basales sèches) ; *g* **Horst** [m] (1) (Grasbüschel aus gleich starken, unverzweigten Achselknospen; ►Rasennarbe).

6609 tussock [n] **divided for propagation** *hort.* (►tussock plant); *syn.* tuft [n] divided for propagation; *s* **macolla** [f] **divida** (para la multiplicación de ►plantas cespitosas); *f* **éclat** [m] **d'une plante cespiteuse** (Fragment d'une ►plante cespiteuse pourvu de racines et de bourgeons obtenu lors de la multiplication végétative artificielle d'une touffe) ; *g* **geteilter Horst** [m] (Zur vegetativen Vermehrung in zwei oder mehrere Teile zertrennte ►Horstpflanze).

6610 tussock formation [n] *bot. hort.* (Clusters of [grass] leaves growing together from the lowest bud axils, to form clumps with equally strong, non-branching axils, e.g. grasses without rhizomes or stolons); *syn.* tuft formation [n]; *s* **formación** [f] **de fascículos**; *f* **formation** [f] **en touffe** (Croissance sur la périphérie des bourgeons situés au niveau du sol et formant une touffe serrée de nouvelles pousses d'égale force chez les graminées) ; *g* **Horstbildung** [f] (Auswachsen der untersten Achselknospen, so dass Büschel mit gleich starken, unverzweigten Achsen entstehen, z. B. bei Gräsern; cf. BOR 1958).

6611 tussock plant [n] *bot. hort.* (Category of *Hemikryptophyta caespitosa* herbaceous plant where renewal buds are surrounded and protected by a thick strawlike cover of old leaf sheaths); *syn.* bunch plant [n] [also US]; *s* **planta** [f] **cespitosa** (Tipo de ►hemicriptófito radicante cuyas yemas de renovación están recubiertas y protegidas por una envoltura pajiza formada por las vainas de las hojas en descomposición. Son frecuentes en las turberas, en el norte y en las montañas por encima del límite superior del bosque; cf. BB 1979, 143); *syn.* planta [f] fasciculada; *f* **plante** [f] **cespiteuse** (Catégorie d'►hémicryptophytes qui croît en touffes compactes chez lesquels les bourgeons sont protégés par la gaine d'une feuille) ; *g* **Horstpflanze** [f] (*Hemikryptophyta caespitosa* **1.** horstig wachsende ►Staude oder Staudengras, die/das sich — im Gegensatz zur Schaftstaude — von Grund an verzweigt [cf. ELL 1967a]. Die Erneuerungsknospen sind von verwitternden Blattscheiden umhüllt und geschützt. Neben anderen Stauden, z. B. Prachtstorchschnabel *[Geranium x magnificum]*, bei zahlreichen Gräsern — Horstgräser, z. B. Blauschwingel *[Festuca cinerea]*. **2.** *Gärtnerischer Sprachgebrauch* hier wird unter einer **H.** eine wenig aggressive, „verträgliche" Staude ohne oder mit nur kurzen Ausläufern verstanden, z. B. Golderdbeere *[Waldsteinia geoides]*, und entsprechend Stauden mit längeren Ausläufern gegenüber gestellt); *syn.* Horst [m] (2), Horststaude [f], Horsthemikryptophyt [m].

TW-elevation [n] [US] *arch. constr.* ▶top level of wall.

6612 twig [n] *hort. arb.* (Small branch with diameter < 10mm; ▶osier twig, ▶slender branch [US]/feather [branch] [UK]); *syn.* branchlet [n]; *s* **rama [f] muy fina** (Rama con un diámetro de < 10 mm; ▶rama delgada); *syn.* ramilla [f]; *f* **ramille [f]** (Branche menue d'un diamètre inférieur à 10 mm ; ▶branche de faible diamètre) ; *syn.* brindille [f] ; *g* **Feinstast [m]** (Ast mit einem Durchmesser < 10 mm; ▶Schwachast); *syn.* Zweig [m].

twig rush swamp [n] [US] *phyt.* ▶saw sedge swamp.

6613 twining climber [n] *bot.* (Climbing plant which maintains its upward growth by twining around a vertical support, e.g. honeysuckle; BS 3975: part 4, 1966; ▶climbing plant, ▶liana, ▶tendril climber); *s* **planta [f] voluble** (Tipo de ▶planta trepadora con tallo voluble que —en comparación con la ▶planta zarcillosa— se enreda en matas o arbustos o en un soporte cualquiera, como la madreselva *[Lonicera periclymenum]* y la maravilla; cf. DB 1985; ▶bejuco); *f* **plante [f] volubile** (▶Plante grimpante dont les tiges s'enroulent autour d'un support avec des mouvements préhenseurs ; ▶liane, ▶plante grimpante à vrilles) ; *g* **Windepflanze [f]** (▶Kletterpflanze, die im Vergleich zur ▶Rankenpflanze mit ihrem ganzen Spross eine Stütze umschlingt [▶Gerüstkletterer] und sich aufwärts dreht, z. B. Blauregen *[Wisteria sinensis]*, Schlingknöterich *[Fallopia aubertii,* früher *Polygonum aubertii]*, Waldgeißblatt *[Lonicera periclymenum]* und Pfeifenwinde *[Aristolochia macrophylla,* früher *A. durior]*; ▶Liane); *syn.* Schlinger [m], Schlingpflanze [f], Winder [m].

twin stems [npl] *arb.* ▶forked growth.

twisted growth [n] *arb. for. hort.* ▶contorted growth.

twisted wood [n] *phyt.* ▶krummholz.

two-leaf masonry [n] [UK] *constr.* ▶two-tier masonry [US].

two-storey grave [n] [UK] *adm.* ▶bi-level grave [US].

6614 two-tier masonry [n] [US] *constr.* (Masonry whereby the facing bond is constructed in front of the backing wall without bonding through the entire thickness of the wall. Therefore, only the backing wall may have structural strength. A wythe [withe] or tier is a continuous vertical section of wall one brick thick; ▶veneer masonry, ▶one-tier masonry [US]/one-leaf masonry [UK]); *syn.* two-wythe masonry [n] [UK], composite wall [n], two-leaf masonry [n] [UK]; *s* **mampostería [f] de dos capas** (Tipo de fábrica en la que la cubierta [▶mampostería enchapada] está colocada delante del muro en sí sin conexión, de manera que solo éste último puede aguantar carga; ▶mampostería compacta); *f* **maçonnerie [f] à double paroi** (Mur constitué d'un habillage de briques pleines ou perforées, d'un mur de pierre pelliculaire [dalles, plaquettes, etc.] en placage de façades, ménageant une lame d'air contre le corps de mur assumant la ▶résistance mécanique de l'ouvrage; ▶maçonnerie à parement, maçonnerie à paroi simple) ; *syn.* mur [m] double ; *g* **zweischaliges Mauerwerk [n]** (Mauerwerk, bei dem der Vormauerverband [▶Verblendmauerwerk] als schmale Scheibe ohne Verbandseinheit vor die Hintermauerung vorgeblendet wird. Deshalb ist nur die Hintermauerung statisch beanspruchbar; ▶einschaliges Mauerwerk).

two-wythe masonry [n] [UK] *constr.* ▶two-tier masonry [US].

6615 tying [n] **to a climber support** *constr. gard.* (Fastening of climbing plants to supporting wires or trellis; ▶espalier row); *s* **sujeción [f] a enrejado de soporte de trepadoras** (Operación de atar plantas trepadoras a un soporte; ▶hilera de plantas en espaldera); *f* **palissage [m]** (Opération consistant à fixer un rameau d'un végétal [arbre fruitier, plant grimpante] sur un support ; ▶espalier, ▶contre-espalier) ; *g* **Befestigen [n, o. Pl.] an ein Rankgerüst** (Befestigung von Kletterpflanzen an eine Kletterhilfe; ▶Spalier 2).

type [n] *constr.* ▶construction type; *pedol.* ▶humus type; *landsc. geo.* ▶landscape type; *constr.* ▶lawn type, ▶low-maintenance grass type; *conserv. ecol.* ▶natural habitat type; *conserv. ecol.* ▶priority natural habitat type; *plan. recr.* ▶recreation type.

type [n] [UK]**, soil** *pedol.* ▶great soil group [US].

6616 type [n] **of building development** *leg. urb.* (Generic term covering ▶attached building development and ▶detached building development within a dense urban area; ▶zero lot line development [US]); *s* **tipo [m] de ordenación de edificios** (En D. término del reglamento de la construcción que se refiere al ordenamiento de los edificios entre sí y en relación a la calle. Se diferencia entre la ▶edificación discontinua y la ▶edificación continua); *f* **type [m] d'implantation des constructions** (Terme utilisé dans les règles d'utilisation du sol dans l'urbanisme, caractérisant les ▶règles de prospect pour les constructions sur un même terrain, par rapport à la voie publique et aux propriétés voisines, p. ex. dans le centre-ville, la protection du caractère et des formes urbaines veut qu'en général, dans la bande constructible en ordre continu, l'interruption de façade ne soit pas autorisée ; ▶implantation des constructions en ordre continu, ▶implantation des constructions en ordre discontinu) ; *syn.* genre [m] d'implantation des constructions ; *g* **Bauweise [f] (2)** (Begriff der Bauordnung, der die Anordnung der Gebäude zueinander, besonders entlang einer Straße, bezeichnet. Man unterscheidet ▶offene Bauweise 2, ▶geschlossene Bauweise und ▶abweichende Bauweise mit einseitiger Grenzbebauung).

6617 type [n] **of leisure** *plan. recr.* (Specific free time pursuit, covering the entire range of leisure-time activities; e.g. cultural events, nature walks, sports, games, etc.; ▶recreation time category, ▶recreation type); *s* **tipo [m] de actividad de tiempo libre** (p. ej. deportes, juegos al aire libre, paseos por el campo, visita a museo; ▶categoría recreativa, ▶tipo de recreo); *f* **type [m] d'activités de loisirs** (Différents domaines de la pratique d'activités de loisirs tels que sport et jeux en plein air, découverte de la nature, activités à caractère culturel et éducatif ; ▶catégorie de loisirs, ▶type d'activités de détente) ; *g* **Freizeittyp [m]** (Freizeittätigkeitsbereiche, z. B. Sport und Spiel, Natur und Landschaft erleben, Kennenlernen von kulturellen Einrichtungen, Basteln; ▶Erholungsform, ▶Erholungstyp).

6618 type [n] **of use** *adm. agr. for. plan. urb.* (Kind of use for land or a building shown on zoning maps [US]/in land registers [UK]; ▶zoning district category [US]/use class [UK]); *s* **tipo [m] de uso** (Naturaleza del uso principal del suelo o de edificios [p. ej. residencial, industrial, servicios comunitarios, parques y jardines] que se registra en el catastro; ▶categoría de zonificación); *f 1* **type [m] d'occupation** (▶Nature de l'affectation des sols déterminée selon l'usage principal ou la nature des activités dominantes et fixée par les plans d'occupation des sols dans le cadre des orientations des schémas directeurs ou des schémas de secteur) ; *syn.* type [m] d'utilisation ; *f 2* **type [m] d'usage** (Terme en général utilisé pour définir la nature de l'affectation de bâtiments) ; *g* **Nutzungsart [f]** (**1.** Im Liegenschaftskataster dargestellte und beschriebene Art, wie Boden oder Gebäude genutzt werden; z. B. Gebäudefläche, Hof- und Wegefläche, Unland. **2.** In D. wird die ▶Art der baulichen Nutzung

T

und sonstiger Nutzung detailliert in den Bauleitplänen gemäß der Baunutzungsverordnung und Planzeichenverordnung dargestellt); *syn.* Art [f] der Nutzung.

6619 typical cross-section [n] *constr. eng.* (Section at right angles to the longitudinal axis which corresponds with the given technical specifications; ▶standard slope); *s* **corte [m] transversal estándar/standard** (Sección transversal de una obra de infraestructura cuyas dimensiones corresponden a normas reguladas; ▶talud estándar); *syn.* corte [m] transversal normado; *f 1* **profil [m] type** (Coupe transversale d'un ouvrage correspond-dant aux principes généraux ou aux spécifications techniques figurant comme dessin standard dans un dossier ; ARI 1986, 45 ; ▶profil type de talus) ; *syn.* structure [f] type ; *f 2* **coupe [f] de principe** (Représentation graphique d'une partie d'un ouvrage tranché sur un plan vertical et servant de référence pour l'exécution de l'ensemble de l'ouvrage ; ▶profil type de talus) ; *g* **Regelquerschnitt [m]** (Querschnitt eines bautechnischen Vor-habens, der in seinen Bestandteilen den gegebenen Richtlinien entspricht; ▶Regelböschung); *syn.* Regelprofil [n].

U

6620 ubac [n] *geo.* (Steep sloping side of a valley or a ▶north-facing slope, esp. in high mountains, which receives hardly any or no sun at all in the northern hemisphere; usually a forest; *opp.* ▶adret); *syn.* shady slope [n]; *s* **ladera [f] umbría** (Vertiente norte [en el hemisferio norte] o sur [en el hemisferio sur] que apenas recibe los rayos del sol, en las latitudes templadas generalmente cubierta de bosque; ▶ladera norte; *opp.* ▶ladera solana); *f* **ubac [m]** (Versant montagnard exposé au nord [dans l'hémisphère nord] et restant à l'ombre pendant une grande partie de la journée ; *opp.* ▶adret); *g* **Schattenseite [f]** (Hangfläche/Talseite, die kaum oder gar nicht besonnt wird; auf der N-Halbkugel N- und NO-Hänge; ▶Nordhang; *opp.* ▶Sonnenseite); *syn.* z. T. Schatthang [m].

6621 ubiquist [n] *phyt. zool.* (Wide-ranging species occurring in various habitats without a recognizable relation to a particular biotope; ▶cosmopolitan species); *syn.* ubiquitous species [n]; *s* **especie [f] ubicuista** (Calificativo aplicado por la generalidad de los geobotanistas a las especies que no están adscritas a deter-minadas agrupaciones. Se puede entender desde dos puntos de vista: **u.** de formación y **u.** de asociación. Éste es el más emplea-do y en este sentido es el opuesto al de ▶especie característica. En cambio, ubicuista de formación resulta sinónimo de hetero-córico; DB 1985; ▶especie cosmopolita); *f* **espèce [f] ubiquiste** (Espèce à grande plasticité écologique capable de vivre dans les milieux très différents ; *ne pas confondre avec* ▶espèce cosmo-polite) ; *g* **Ubiquist [m]** (In verschiedenen Lebensräumen ohne erkennbare Bindung an bestimmte Biotoptypen vorkommende Pflanzen- oder Tierart; ▶Kosmopolit); *syn.* ubiquitäre Art [f], Allerweltsart [f].

6622 ubiquitous [adj] *phyt. zool.* (Descriptive term applied to species which occur commonly in most latitudes; ▶euryecious, ▶cosmopolitan species); *syn.* wide-ranging [adj] (TEE 1980, 42), omnipresent [adj]; *s* **ubicuo/a [adj]** (Término descriptivo apli-cado a especies que se presentan en la mayoría de las latitudes; ▶eurioico, ▶eurítopo, ▶especie cosmopolita); *syn.* ubicuista [adj], omnipresente [adj], ubiquista [adj]; *f* **ubiquiste [adj]** (Se

dit d'un organisme à large amplitude écologique, capable de vivre dans des milieux très différents ; DEE 1982, 91 ; ▶euryèce, ▶espèce cosmopolite) ; *g* **ubiquitär [adj]** (Eine Art betreffend, die fast überall vorkommt; ▶euryök, ▶Kosmopolit).

ubiquitous species [n] *phyt. zool.* ▶ubiquist.

U crotches [npl] *arb.* ▶formation of U or V crotches.

udalf [n] [US] *pedol.* ▶gray-brown podzolic soil.

6623 unauthorized dumped waste [n] *envir.* (Illegally deposited, scattered waste on plots which are not intended as disposal sites; ▶unauthorized dumpsite [US]/fly tipping site [UK]); *s* **basura [f] clandestina** (Residuos sólidos desechados ilegalmente en solares que no tienen oficialmente la función de vertedero; ▶vertedero clandestino); *f* **détritus [m] abandonné** (Déchets dont le dépôt est légalement interdit dans les lieux publics ou privés impropres au stockage et causant des pollutions physiques, olfactives et visuelles ; ▶décharge [des déchets] sauvage) ; *g* **wilder Müll [m]** (Abfall, der verbotswidrig auf Grundstücken, die nicht für eine ordnungsgemäße Lagerung vorgesehen sind, abgelegt wird; ▶wilde Deponie); *syn.* wilde Müllablagerung [f].

6624 unauthorized dumpsite [n] [US] *envir.* (Illegal de-posit of refuse on places where this is not permitted; ▶unauthor-ized dumped waste); *syn.* fly-tip [n] [UK], fly tipping site [n] [UK]; *s* **vertedero [m] clandestino** (Terreno u otro lugar utili-zado como vertedero sin que exista ni autorización ni control; ▶basura clandestina); *syn.* vertedero [m] ilegal, basurero [m] clandestino/ilegal, tiradero [m] clandestino [MEX]; *f* **décharge [f] (des déchets) sauvage** (Dépôt d'ordures non autorisée ; ▶détritus abandonné) ; *syn.* dépôt [m] clandestin (DUV 1984, 311), dépotoir [m] sauvage, décharge [f] illégale ; *g* **wilde Deponie [f]** (Verbotswidriger Müll- und Schuttabladeplatz auf Grundstücken, die nicht für eine ordnungsgemäße Lagerung vorgesehen sind; ▶wilder Müll).

unbuilt plot area [n] [UK] *plan. urb.* ▶unbuilt yard [US].

unbuilt site area [n] [UK] *plan. urb.* ▶unbuilt yard [US].

6625 unbuilt yard [n] [US] *plan. urb.* (That part of a plot of land which is not built upon: open space lying between the principle or accessory buildings and the nearest lot line [US]/plot line [UK]; this space is unoccupied and unobstructed from the ground upward except as may be specifically provided in building codes or zoning ordinances; cf. UBC 1979); *syn.* unbuilt plot area [n] [UK], unbuilt site area [n] [UK], curtilage [n]; *s* **superficie [f] no edificada** (Parte de un solar que no está edi-ficada, pueden ser zonas verdes entre el edificio principal y los accesorios. Este espacio está desocupado a no ser que las regulaciones de construcción prevean otra cosa; *opp.* superficie edificada); *f* **surface [f] non construite** (Espace non occupé par les constructions) ; *g* **nicht überbaute Fläche [f]** (Der Teil eines Grundstückes, auf dem kein Gebäude oder keine sonstige bauliche Anlage steht); *syn.* nicht bebaute Grundstücksfläche [f]

unclear water [n] *limn.* ▶cloudy water.

unconsolidated sedimentary deposit [n] *geo.* ▶loose sedimentary deposit.

uncontrolled grass growth [n] *for. hort.* ▶uncontrolled grass intrusion.

6626 uncontrolled grass intrusion [n] *for. hort.* (Inva-sive grass development, e.g. on forest floors, perennial borders, etc.); *syn.* uncontrolled grass growth [n]; *s* **intrusión [f] incontrolada de hierba** (Crecimiento muy fuerte de hierba, p. ej. en suelo forestal, macizo de vivaces, etc.); *f* **envahissement [m] par les herbes** (Développement incontrôlé d'un couvert herbeux, p. ex. en zone de forêt, dans un massif de plantes vivaces etc.) ;

syn. invasion [f] par les herbes ; **g Vergrasung [f]** (Starker Grasbewuchs, z. B. auf Waldbodenflächen, Staudenflächen etc.).

6627 uncontrolled proliferation [n] **of settlements** *plan. urb.* (Unplanned scattering of housing developments in rural areas, and, in some places, of vacation homes in the countryside; ►landscape of proliferated housing, ►urban sprawl); **s desfiguración [f] del paisaje** (1. Consecuencia del desarrollo urbanístico desordenado en zonas rústicas con proliferación de edificios y urbanizaciones dispersas; ►urbanización caótica, ►paisaje desfigurado. 2. Expansión desordenada de una población en zona rural con el resultado de un ►paisaje desfigurado; ►urbanización caótica); **f mitage [m] du paysage** (Développement anarchique de l'urbanisation [maisons individuelles, pavillonnaire, petit lotissement isolé, habitat de week-end, terrain de camping, etc.] dans les espaces ruraux et les paysages naturels périurbains qui est responsable de la fragilisation des équilibres naturels et de la dégradation du patrimoine esthétique et paysager ; ►accroissement désordonné d'une agglomération, ►paysage mité) ; *syn.* diffusion [f] anarchique des constructions, éparpillement [m] des constructions dans le paysage ; **g Zersiedelung [f] (der Landschaft)** (Ungeordnete Streuung von Einzelgebäuden, Industriebetrieben, Einkaufszentren, Wohngebieten [auch Ferienwohngebieten, Dauercampingplätzen etc.] in der freien Landschaft mit der Folge der Zerstörung des Landschaftsbildes; ►ungeordnete Ausbreitung einer Stadt, ►zersiedelte Landschaft); *syn. o. V.* Zersiedlung [f] (der Landschaft).

uncovered demand [n] *plan.* ►shortfall.

uncovered requirements [n] *plan.* ►shortfall.

uncultivated [pp] *agr.* ►leave uncultivated.

6628 uncultivated farmland [n] *plan. agr.* (Previously worked land, left uncultivated for longer than one year; ►fallow land, ►derelict land, ►abandoned land); **s parcela [f] de cultivo abandonada** (Tierra de cultivo dejada más de un año sin cultivar; ►ruina industrial, ►tierra abandonada, ►tierra en barbecho); **f terre [f] en friche** (Terre restée non cultivée pendant plus d'un an par suite de l'interruption de la culture ; ►friche industrielle, ►friche sociale, ►terre en jachère) ; *syn.* friche [f] agricole, terre [f] inculte ; **g Brachfläche [f]** (Länger als ein Jahr nicht mehr durch die Landwirtschaft bewirtschaftete Fläche; ►Brachland, ►Industriebrache, ►Sozialbrache).

under-canopy [n] *for. phyt.* ►understory [US]/understorey [UK].

6629 undercut [n] *geo.* (Result of erosion of banks or cliffs by water; ►blow-out); **s retroceso [m] del acantilado** (Resultado de la acción de socavado del acantilado; ►excavación eólica); **f encoche [f] littorale** (Résultat de l'action érosive d'affouillement des vagues dans les zones de faiblesse au pied des falaises ; ►creux de déflation) ; *syn.* encoche [f] d'abrasion, encoche [f] de surcreusement, encoche [f] côtière ; **g Auskolkung [f] an der Küste (1)** (Ergebnis der Brandungserosion an Kliffs der [Steil]küsten; auskolken [vb]; ►Windanriss); *syn.* Auswaschung [f] an [Steil]küsten, Erosionskessel [m] an [Steil]küsten, Kolk [m] an [Steil]küsten.

6630 undercut bank [n] *geo.* (Undermined outside bend of river or stream which has been eroded by action of the current; ►bluff 1, ►riverbank collapse [US]/river-bank collapse [UK], ►streambank erosion); *syn.* eroding bank [n]; **s arco [m] erosivo de meandro** (DINA 1987, 449; curva externa de un río que se ve afectada por la fuerza erosiva del agua; ►desplome de orillas, ►erosión de márgenes de cursos de agua, ►orilla cóncava); *syn.* arco [m] externo de meandro (DINA 1987), orilla [f] exterior de meandro; **f berge [f] d'effondrement** (Rive extérieure du coude d'un cours d'eau sapée par la poussée de l'eau ; ►berge concave, ►effondrement d'une berge, ►érosion des berges des cours

d'eaux) ; *syn.* berge [f] d'éboulement ; **g Abbruchufer [n]** (Außenkurve eines Fließgewässerbogens, die durch die Prallwirkung des Wassers unterspült wird; ►Erosion an Ufern von Fließgewässern, ►Prallhang, ►Uferabbruch).

6631 undercutting [n] **of coastal cliffs** *geo.* (Removal of material at the base of a steep slope, overfall, or ►cliff by falling water, a stream, wind erosion, or wave action, which steepens the slope or produces an overhanging cliff; RCG 1982); **s 1 socavado [m] del acantilado** (Proceso de destrucción de la costa por acción de las olas del mar); *syn.* socavación [f] del acantilado; **s 2 abrasión [f] marina** (Erosión superficial producida sobre las rocas del litoral por diversos agentes, fundamentalmente el hielo, las olas y el viento, en combinación con los materiales que transportan dando lugar a la creación de plataformas de abrasión [= rasas]; cf. DGA 1986; ►acantilado); **f 1 affouillement [m] de la côte** (Action de creusement des eaux sur les côtes à falaises et la plate-forme littorale) ; *syn.* creusement [m] de la côte ; **f 2 affouillement [m] d'une falaise** (Creusement par la mer du pied de falaise suivi de l'effondrement brutal de pans entiers de l'abrupt ; ►falaise) ; *syn.* sapement [m] d'une falaise ; **g Erosion [f] an Steilküsten** (1. Brandungserosionstätigkeit in Locker- oder Festgesteinen; ►Kliff; *syn.* Steilküstenerosion [f]. 2. Durch Erosionsvorgänge verursachte Unterspülung von Küstenteilen; auskolken [vb]); *syn.* Auswaschung [f] an Steilküsten, Auskolkung [f] an der Küste (2), Steilküstenaushöhlung [f].

undercutting [n] **of riverbanks** *geo.* ►streambank erosion.

6632 undercutting [n] **of roots** *constr. hort.* (Operation of severing downward growing plant roots in situ, to control root development or, e.g. to prepare semi-mature trees for transplanting operations; cf. BS 3975: part 5; ►transplantation of semi-mature trees); **s corte [m] de raices debajo del cepellón** (Corte de las raíces largas para controlar el crecimiento de un árbol o para preparar su transplante; ►transplante de árboles maduros); **f coupe [f] des racines profondes (±)** (Coupe effectuée sur les racines profondes afin de contrôler le développement du système racinaire ou en vue de la préparation de la contre plantation des arbres ou la ►transplantation de gros végétaux ; opération en général effectuée par une machine comportant une lame en forme de demi-cercle) ; **g Unterschneiden [n, o. Pl.] der Wurzeln** (Durchtrennen von tief streichenden Wurzeln, um das Wurzelwachstum von Kulturen unterschiedlichen Alters in Baumschulen zu kontrollieren oder zur Vorbereitung für ►Großbaumverpflanzungen; geschieht meist maschinell mit einem halbkreisförmigen Unterschneidemesser eines Ballenschneiders [Unterschneidepflug] oder mit einem Seilzug; früher wurden Wurzeln mit Handspaten unterstochen/unterschnitten; unterstechen [vb], unterschneiden [vb]).

undercutting [n] **of streambanks/riverbanks** *geo.* ►streambank erosion.

underground car park [n] [UK] *urb.* ►underground parking garage [US].

underground construction [n] *urb.* ►below-grade construction.

underground disposal [n] **of mining gob** [US] *min.* ►underground disposal of mining spoil.

6633 underground disposal [n] **of mining spoil** *min.* (Return of ►mining spoil to the galleries after exploitation of mineral resources; ►disposal of mining spoil); *syn.* underground disposal [n] of mining gob [also US]; **s relleno [m] de minas (subterráneas) con zafras** (Devolución de las ►zafras a los filones explotados después de sacar el mineral; ►depositar zafras); *syn.* atibación [f]; **f stockage [m] souterrain de stériles** (Remise en place des ►stériles dans les galeries après exploi-

U

tation d'un gisement ; ▶stockage de stériles) ; *syn.* comblement [m] de galeries de mines, comblement [m] de cavités minières ; *g* **Bergeversatz [m]** (Rückführung der ▶Berge 3 nach Abbau der Bodenschätze in die Stollen; ▶Bergeverbringung); *syn.* Untertageversatz [m].

underground hose bib [n] [UK] *urb.* ▶underground hydrant [US].

6634 underground hydrant [n] [US] *urb.* (Water supply device installed under a paved surface with an attachment of a stand pipe from which water can be extracted; *opp.* ▶hydrant 2); *syn.* underground hose bib [n] [UK]; *s* **boca [f] de agua subterránea** (Boca de agua instalada bajo el pavimento de la calle de la cual con ayuda de una tubería se puede extraer agua; *opp.* ▶boca de agua superficial); *syn.* hidrante [m] subterráneo; *f* **bouche [f] d'alimentation en eau** (Prise d'eau installée dans un regard dans le trottoir ou dans sa bordure ; *on distingue* la bouche d'incendie, la bouche d'arrosage, la bouche de lavagem ; VRD 1986, 222 et s. ; ▶borne d'alimentation en eau) ; *g* **Unterflurhydrant [m]** (Unter der Straßendecke eingebauter ▶Hydrant, an dem mit Hilfe eines Standrohres Wasser entnommen werden kann; *opp.* ▶Überflurhydrant).

6635 underground irrigation [n] *constr.* (Trapping of water, e.g. in a drainage layer of a roof garden, or automatic irrigation devices for street trees; ▶drip irrigation); *syn.* underground watering [n] [also UK] (BS 3957: Part 5); *s* **irrigación [f] subterránea** (Riego subterráneo p. ej. en azotea ajardinada [agua estancada en la capa de drenaje] con dispositivos automáticos de riego para árboles en las calles; ▶irrigación por goteo); *syn.* riego [m] subterráneo; *f* **irrigation [f] souterraine** (Apport artificiel d'eau par un réseau souterrain, p. ex. pour les jardins sur dalles — réserve d'eau dans la couche drainante et arrosage par remontée capillaire — ou pour les arbres en sols urbains ; ▶arrosage au goutte à goutte) ; *syn.* arrosage [m] par irrigation souterraine ; *g* **Unterflurbewässerung [f]** (Unterirdische Bewässerung, z. B. für Dachgärten ein Wasserstau in der Dränschicht oder mit Bewässerungsautomaten für Bäume im Straßenraum sowie durch Regulierung des Grundwasserspiegels bei Ausnutzung der Kapillarität; ▶Tröpfchenbewässerung).

underground mining [n] [US] *min.* ▶deep mining.

underground parking [n] [US], **residents'** (TGG 1984, 194) *urb.* ▶underground parking garage [US]/underground car park [UK].

6636 underground parking garage [n] [US] (TGG 1984, 194) *urb.* (**1.** Underground parking structure with numerous parking spaces, usually for cars; **u. p. g.s** are constructed in urban areas with high land costs and scarce land reserves [inner cities, or large enterprises and transportation centers]); *syn.* underground car park [n] [UK] (HUL 1978, 233); **2. residents' underground parking** [n] (In Spain, publicly funded parking garage, built beneath urban plazas/squares or road intersections, to relieve the cities of surface-parked vehicles; the garage spaces are leased at a reasonable price to local residents; ▶communal garage); *s 1* **garaje [f] subterráneo** (Construcción subterránea que acoge plazas de aparcamiento [Es]/estacionamiento [AL], generalmente situada bajo los edificios residenciales o comerciales); *s 2* **aparcamiento [m] de residentes** (En España, garaje subterráneo construido por la administración municipal bajo plazas o cruces de calles de la ciudad para descongestionarla. Las plazas de los **aa. de rr.** se arriendan a los residentes del vecindario a un precio relativamente bajo; ▶garaje de comunidad de vecinos); *f 1* **garage [m] souterrain** (Endroit public servant d'abri aux véhicules édifié sous un édifice) ; *syn.* parking [m] enterré, parking [m] souterrain ; *f 2* **parking [m] souterrain résidentiel** (Parking construit à proximité des voies principales

sous les places, et carrefours et réservé au stationnement résidentiel des riverains ; ▶garage collectif) ; *syn.* parc [m] de stationnement souterrain résidentiel ; *g 1* **Tiefgarage [f]** (Unterirdisches Parkierungsbauwerk mit mehreren Stellplätzen, i. d. R. für PKW; **T.n** werden in Gebieten [Innenstädte, Großbetriebe und Verkehrszentren] mit hohen Grundstückskosten und knappen Flächenreserven angelegt); *g 2* **Anwohnertiefgarage** (In Spanien öffentlich geförderte Tiefgarage, die unterhalb von innerstädtischen Plätzen oder Straßenkreuzung gebaut wird, um die Städte von ruhendem Verkehr zu entlasten und deren Plätze für einen günstigen Preis an die Anwohner verpachtet werden; ▶Gemeinschaftsgarage).

6637 underground power line [n] *envir.* (Cable used for electric power transmission below ground level; *opp.* ▶overhead power line); *s* **cable [m] eléctrico subterráneo** (Línea de conducción eléctrica tendida en el subsuelo; *opp.* ▶línea aérea); *syn.* conducción [f] eléctrica subterránea; *f* **ligne [f] électrique souterraine** (Câble électrique posé en souterrain. L'enfouissement [mise en souterrain] des circuits à haute tension aériens constitue une mesure de protection des paysages et des milieux naturels ; *opp.* ▶ligne aérienne de distribution d'électricité) ; *syn.* câble [m] électrique souterrain ; *g* **unterirdisches Stromkabel [n]** (In die Erde verlegtes elektrisches Kabel; *opp.* ▶Freileitung).

underground securing [n] *constr. hort.* ▶root bracing.

underground storm water system [n] *envir. urb.* ▶storm water drainage system.

underground water [n] *hydr. wat'man.* ▶groundwater [US]/ground water [UK].

underground watering [n] [UK] *constr.* ▶underground irrigation.

undergrowth [n] *for. phyt.* ▶understory [US]/understorey [UK].

6638 underplanting [n] (1) *for.* (Setting out young trees, or sowing tree seed under an existing stand. The trees themselves are termed 'underplants'; SAF 1983; ▶understory [US]/understorey [UK]); *s* **plantación [f] en sotobosque** (Cultivo de una segunda capa de rodal bajo un estrato más viejo para proteger el suelo y los troncos; ▶sotobosque); *syn.* plantación [f] bajo cubierta (HDS 1987); *f* **plantation [f] en sous-étage** (Plantation de jeunes plants, ou semis direct, à l'intérieur et sous le couvert d'un peuplement forestier existant ; ▶sous-étage) ; *syn.* sousplantation [f] [CDN] ; *g* **Unterbau [m]** (2) (Anzucht einer zweiten Bestandsschicht unter einem älteren Baumbestand zur Boden und Stammpflege; WT 1980; ▶Unterschicht); *syn.* Unterpflanzung [f] (1).

6639 underplanting [n] (2) *constr. plant.* (Lower layer of planted vegetation, e.g. shrubs beneath trees, perennials beneath shrubs; ▶ground cover, ▶understory [US]/understorey [UK]); *s* **capa [f] inferior de plantas** (En plantación antropógena, capa de plantas que se encuentra debajo de una superior, p. ej. arbustos debajo de los árboles, las vivaces bajo arbustos; ▶cobertura vegetal del suelo, ▶sotobosque); *f* **plantation [f] sous couvert végétal** (Strate plantée sous une strate sus-jacente, p. ex. les arbustes sous les arbres, les vivaces sous les arbustes ; ▶couvert végétale du sol, ▶sous-étage) ; *syn.* plantation [f] sous couverture végétale ; *g* **Unterpflanzung [f]** (2) (Bepflanzung, die unter einer höheren angelegt wurde, z. B. Sträucher unter Bäumen, Stauden unter Sträuchern; ▶Bodenbedeckung 2, ▶Unterschicht).

6640 underprovided plant tissue [n] *arb.* (Poorly developed plant tissue of trees and shrubs caused by obstruction of the flow of sap, especially above and below the forking of branches, callouses and injured parts); *s* **zona [f] no suministrada de savia** (Tejido vegetal mal desarrollado causado por la obstrucción del flujo de la savia sobre todo alrededor de heridas,

bifurcaciones, etc.); *f* **zone [f] sous-alimentée en sève (±)** (Tissu végétal des arbres et arbrisseaux sujet à une sous-alimentation en sève, en particulier au-dessus et en dessous du point de ramification des branches, des bourrelets, des plaies) ; *syn.* tissu [m] sous-alimenté en sève ; *g* **Versorgungsschatten [m]** (Durch Behinderung oder Unterbrechung des Saftstromes unterversorgtes Gewebe an Gehölzen, insbesondere ober- oder unterhalb von Abzweigungen, Wülsten oder Schadstellen sowie infolge falscher Schnittführung bei Astableitungen; führt auf Dauer zu lokal verringertem Dickenwachstum resp. zu Trockenstellen ggf. mit Pilzbefall).

6641 underseed [n] *agr. hort. landsc.* (Slow-growing, cultivated plant species which requires the protection of a faster-growing plant [= cover crop] to improve its growth; this plant also serves as protection against erosion where spacing of plant rows is exceptionally large; *s* **siembra [f] de especies de germinación tardía** (Especie vegetal de crecimiento lento que necesita la protección de otras especies de crecimiento rápido para mejorar el suyo. Esta planta también sirve para proteger contra la erosión cuando el espaciamiento entre las filas es muy grande); *syn.* siembra [f] bajo abrigo, siembra [f] bajo cubierta, siembra [f] bajo cobertura; *f* **semis [m] sous couverture** (Espèce culturale à développement lent, requérant la protection de végétaux à croissance rapide ; plante utilisée comme protection contre l'érosion dans les cultures caractérisées par un grand espacement des lignes de plantation) ; *syn.* semis [m] sous couvert végétal ; *g* **Untersaat [f]** (Sich langsam entwickelnde Kulturpflanzenart, die zum besseren Wachstum den Schutz von schneller wachsenden Kulturpflanzen [= Decksaat] benötigt; dient auch dem Erosionsschutz bei Kulturen mit größerem Reihenabstand).

6642 understanding [n] of planning *plan.* (**1.** Comprehension of planning principles and procedures. **2.** Basic attitude of a planner; who e.g. may dominate or participate in the planning process); *s* **enfoque [m] ideológico de la planificación** (Visión sociopolítica de la profesión desde la que parte un/a profesional al realizar su trabajo; *contexto* e. i. de p. elitista, e. i. de p. participativo, etc.); *f 1* **compréhension [f] des processus de planification** (Capacité à saisir par la pensée les principes et processus de l'aménagement) ; *f 2* **conception [f] en matière de planification** (Attitude personnelle de l'aménageur pouvant traduire un comportement « élitaire » ou « participateur ». *Contexte* conception élitaire/participatrice en matière de planification) ; *g* **Planungsverständnis [n]** (**1.** Verständnis über Planungsprinzipien und -abläufe. **2.** Gedankliche Grundhaltung eines Planers, z. B. ein ‚elitäres' oder ‚partizipatorisches' P.).

6643 understock [n] *hort.* (ANSI 1986, 28; **1.** stem or root in which a graft is inserted. **2.** Specially propagated plant upon which a scion is grafted; ▶grafting); *syn.* stock [n] (HLT 1971, 33), rootstock [n]; *s* **patrón [m]** (Planta sobre la cual se realiza un ▶injerto); *syn.* portainjerto [m] (HDS 1987), hospedador [m]; *f 1* **porte-greffe [m]** (Plante sur laquelle on fixe le greffon ; ▶greffage) ; *syn.* sujet [m] (LAH 1983) ; *f 2* **jeune plant [m] porte-greffe issu de semis** (Jeune plant d'espèces fruitières ou ornementales obtenu par semis destiné à la ▶greffe) ; DINA 1987; *g 1* **Unterlage [f]** (Bei der ▶Veredelung von Gehölzen der untere Pflanzenteil, z. B. Wurzel, Stamm); *syn.* Veredelungsunterlage [f], Wurzelpflanze [f]; *g 2* **Wildling [m] (2)** (Durch Aussaat entstandene Jungpflanze als Unterlage zur Veredelung. Im Deutschen wird der Begriff **W.** nur für Gehölze verwandt).

6644 understory [n] [US]/understorey [n] [UK] *for. phyt.* (**1.** Generally, trees [underwood (US)] and other woody species, growing under an overstor[e]y. **2.** Any plants growing under the canopy formed by others—more particularly, herbaceous and shrub vegetation under a brushwood or tree canopy; SAF 1983); *syn.* undergrowth [n], understory(e)y plants [npl]

(TEE 1980, 96), under-canopy [n] (GP 2003, Dec., p. 35); *s* **sotobosque [m]** (Vegetación que se desarrolla bajo las copas de los árboles del bosque, generalmente en los estratos arbustivo, subarbustivo y herbáceo; DINA 1987); *syn.* subpiso [m] (DFM 1975); *f* **sous-étage [m]** (Les arbres, grands ou petits, poussant sous un étage dominant) ; *g* **Unterschicht [f]** (Vegetation unter dem Schirm größerer Pflanzen); *syn.* Unterwuchs [m].

understory plants [npl] [US]/understorey plants [npl] [UK] *for. phyt.* ▶understory [US]/understorey [UK].

6645 underwater effluent discharge pipe [n] into the sea *envir.* (▶water polluting firm [US]); *s* **emisario [m] submarino** (Conducción que transporta los efluentes líquidos a cierta distancia de la orilla y que dispone al final de una sección dotada con difusores, con el fin de diluir mejor los residuos con el agua del mar; ▶causante de vertidos directos); *f* **émissaire [m] en mer** (ASS 1987, 267 ; conduite utilisée pour l'évacuation dans les eaux marines des eaux usées en provenance des agglomérations littorales ; ▶auteur d'un déversement direct [des effluents]) ; *g* **Abwasserleitung [f] ins Meer** (Leitung für den Transport ungeklärter Haushaltsabwässer ins Meer; ▶Direkteinleiter).

undesirable plant list [n] *landsc.* ▶list of undesirable plants.

6646 undeveloped land [n] *plan. urb.* (That part of an area, which has not been developed; ▶open land [US]/green-field site [UK], ▶undeveloped zoned land); *syn.* unimproved property [also US]); *s* **área [f] no edificada** (Zona de un municipio en la que aún no se ha construido ningún edificio; ▶suelo rústico, ▶suelo urbanizable); *f 1* **espace [m] non urbanisé** (Espace qui n'est pas situé dans les parties actuellement urbanisées d'une commune ; ▶zone non urbanisable, ▶zone agricole) ; *syn.* zone [f] non urbanisée ; *à ne pas confondre avec f 2* **terrain [m] non construit** (Terrain non [encore] construit et situé dans un espace actuellement urbanisé ; cf. circulaire n° 96-32 du 13 mai 1996 ; ▶zone constructible, ▶terrain [constructible] non viabilisé) ; *syn.* terrain [m] non bâti ; *g* **unbebaute Fläche [f]** (Der Teil eines Gemeindegebietes, der [noch] nicht durch eine Bebauung erschlossen ist; ▶Außenbereich, ▶Rohbauland).

undeveloped outskirts area [n] *leg. urb.* ▶open land [US].

undeveloped peripheral area [n] [UK] *leg. urb.* ▶open land [US].

6647 undeveloped zoned land [n] *leg. urb.* (Piece of land designated in a ▶comprehensive plan [US]/local development framework [UK] for future specific building use but not yet developed; ▶approved development site, ▶unzoned land ripe for development [US]/white land [UK], ▶zoning district [US]/development zone [UK]); *s* **suelo [m] urbanizable** (Suelo declarado apto para urbanizar en un Plan General Municipal de Ordenación Urbana, que ha de ser urbanizado según el programa del propio plan; ▶sector urbano, ▶suelo edificable no urbanizado, ▶suelo urbanizado 1); *f* **terrain [m] (constructible) non viabilisé** (Zone constructible prescrite dans un plan local d'urbanisme [PLU]/ plan d'occupation des sols [POS] et définies comme ▶zone d'urbanisation future dans le schéma directeur ; les parcelles ne sont pas encore pré-équipées en réseaux divers : assainissement, eau potable, électricité, gaz, téléphone ; ▶terrain viabilisé, ▶zone urbaine 3) ; *syn.* parcelle [f] (constructible) non équipée, terrain [m] (constructible) non équipé ; *g* **Rohbauland [n, o. Pl.]** (Nach §§ 30, 33 und 34 BauGB für eine bauliche Nutzung festgesetztes, noch unerschlossenes Gebiet, dessen Erschließung aber noch nicht gesichert ist oder das nach Lage, Form oder Größe für eine bauliche Nutzung unzureichend gestaltet ist und das vor Aufstellung eines Bebauungsplanes als Bauland im Flächennut-

U

zungsplan dargestellt ist; cf. § 4 [3] WertV. Bei der Grundstücksbewertung wird dies als ▸Bauerwartungsland bezeichnet; ▸Baugebiet 1, ▸fertiges Bauland).

6648 undissected area [n] **with low traffic intensity/density** (≠) *conserv. plan.* (Term used in Germany since 1978 for areas compiled on maps by the Federal Research Institute for Nature Conservation and Landscape Ecology in Bonn with the aim of determining which areas are least affected by traffic and thus important in terms of recreation. Updated in 1987 and 1999 and extended to include concerns such as land conservation and landscape management the detailed map shows areas with a minimum size of 100 km². In addition, all railway tracks and all roads [from highways to the category of district roads] with an average daily traffic volume of more than 1,000 vehicles in 24 hours are included. In the whole of D. there are 480 areas with a total area of 80,062 km², equivalent to 22.4% of the federal territory; cf. N + L 76, 11/2001, 481ss); *syn.* unfragmented habitats [npl] in an area with few traffic corridors (≠), unfragmented area [n] poor in linear infrastructure (≠); **s área [f] no fragmentada con poca densidad de tráfico** (Zona natural de gran tamaño que no sufre las consecuencias de la división por grandes infraestructuras de transporte. En D. en 1978 por primera vez la Agencia Federal de Protección de la Naturaleza elaboró un mapa de las áreas no fragmentadas superiores a 100 km² con el objetivo de registrar los espacios grandes poco sometidos a la contaminación acústica del tráfico, que son importantes para la recreación al aire libre. Este mapa fue reelaborado y ampliado en contenido a otros usos del espacio, como la protección de la naturaleza, en 1987 y en 1999 incluyó a los nuevos estados federados; según este último mapa en D. existen 480 áreas no fragmentadas con una superficie total de 80 062 km², que corresponden al 22,4% del territorio de ese país); **f zone [f] naturelle pauvre en infrastructures** (D., carte établie en 1978 par la « *Bundesforschungsanstalt für Naturschutz und Landschaftsökologie* » [Institut fédéral de recherche sur la protection de la nature et de l'écologie des paysages] pour le territoire de la République Fédérale d'Allemagne avec l'objectif de recenser les espaces naturels non fragmentés par les infrastructures de transport [réseau ferroviaire et axes routiers avec un trafic journalier moyen supérieur à 1000 véh./24 h], espaces d'importance pour les loisirs et le tourisme ; le document mis à jour en 1987 et 1999 avec la prise en compte des territoires de la RDA s'est enrichi des préoccupations d'autres occupations des sols comme p. ex. la protection de la nature et a permis la délimitation d'espaces d'une superficie supérieure à 100 km² ; ils sont au nombre de 480 et couvrent une superficie de 80.062 km², soit 22,4 % du territoire fédéral) ; *syn.* espace [m] non fragmenté à faible densité d'infrastructures de transport ; **g unzerschnittener verkehrsarmer Raum** [m] (*Abk.* UZV-Raum; **D.**, 1978 wurde von der Bundesforschungsanstalt für Naturschutz und Landschaftsökologie zu Bonn erstmalig eine Karte der UZV-Räume der Bundesrepublik Deutschland mit dem Ziel erstellt, großflächige, unzerschnittene und damit vom Verkehrslärm gering belastete Räume zu ermitteln, welche für die Erholungsvorsorge von Bedeutung sind. Diese 1987 und 1999 fortgeschriebene, um die neuen Bundesländer erweiterte und auf weitere Belange der Raumnutzung wie z. B. Naturschutz und Landschaftspflege eingehende Karte weist Flächen mit einer Mindestgröße von 100 km² aus. Ferner werden als zerschneidend alle ein- und mehrgleisigen Bahnstrecken, alle Straßen [von Autobahnen bis zur Kategorie der Kreisstraßen] mit einer durchschnittlichen täglichen Verkehrsmenge von mehr als 1000 Kfz/24 h berücksichtigt. In ganz D. ergeben sich somit 480 UZV-Räume mit einer Gesamtfläche von 80 062 km², ent-

sprechend 22,4 % des Bundesgebietes; cf. N+L 76, 11/2001, 481 ff).

undisturbed earth [n] [US] *constr.* ▸undisturbed subgrade [US]/undisturbed subsoil [UK].

6649 undisturbed subgrade [n] [US] *constr.* (Unaltered, consolidated *in situ* soil that is suitable for foundations; ▸consolidated subgrade, ▸site soil); *syn.* undisturbed subsoil [n] [UK], undisturbed earth [n] [also US]; **s subsuelo [m] inalterado [Es, RCH]** (Capa de subsuelo que no ha sufrido cambios antropógenos y puede servir para la cimentación de construcciones; ▸subsuelo estable, ▸suelo *in situ*); *syn.* suelo [m] de cimentación inalterado [Es], subrasante [f] inalterada [AL] (cf. MESU 1977); **f 1 terrain [m] naturel** (Sol en place inaltéré ; surface du terrain d'un chantier avant le commencement des travaux ; abréviation cartographique : T.N. ; *contexte constr.* hauteur sur T. N. ; ▸sol stable, ▸bon sol) ; **f 2 bon sol [m]** (Couche géologique homogène ferme et dure, de capacité portante pour des fondations normales ; elle détermine la profondeur de la fouille de fondation liée à l'atteinte du bon sol) ; **g gewachsener Boden [m]** (Vor Ort vorhandener, nicht veränderter Untergrund für die Erstellung von Bauwerken; ▸fester Baugrund, ▸anstehender Boden).

undisturbed subsoil [n] [UK] *constr.* ▸undisturbed subgrade [US].

undressed [pp/adj] *constr.* ∗▸tooling of stone [US]/dressing of stone [UK], #9.

undressed stone [n] *constr.* ▸rubble.

undressed stone [n] [UK]**, flat** *constr.* ▸ragstone [US].

undulating landscape [n] *geo.* ▸rolling landscape.

unfolding foliage [n] *bot.* ▸leafing out.

6650 unforeseen construction work [n] *constr. contr.* (Unexpected project requirements, not awarded in the contract, which have to be executed for an agreed payment); **s trabajos [mpl] imprevistos** (Trabajos de construcción a los cuales no está obligado el contratista pero que se abonan por separado si el comitente los acepta); **f travaux [mpl] non prévus** (Ouvrages ou travaux dont la réalisation ou la modification est décidée par ordre de service et pour lesquels le marché ne prévoit pas de prix ; art. 14, règlement du prix des ouvrages ou travaux non prévus, décret n° 76-87 du 21 janvier 1976 du Code des marchés publics) ; *syn.* travaux [mpl] nouveaux, ouvrage [m] nouveau, prestation [f] non prévue ; **g nicht vorhergesehene Leistung [f]** (Bauleistung, zu der der Auftragnehmer nach dem Vertrag nicht verpflichtet ist, die jedoch nach Anerkennung durch den Auftraggeber besonders vergütet werden muss; cf. § 2 [6] VOB Teil B); *syn.* außervertragliche Leistung [f], unvorhergesehene Leistung [f].

unfragmented area [n] **poor in linear infrastructure** *conserv. plan.* ▸undissected area with low traffic intensity/density.

unfragmented habitats [npl] **in an area with few traffic corridors** *conserv. plan.* ▸undissected area with low traffic intensity/density.

6651 uniform building code [n] [US] *leg. urb.* (**In U.S.**, the Uniform Building Code [UBC] is a document dedicated to the development of better building construction and greater safety to the public by uniformity in building laws. The purpose is to provide minimum standards to safeguard life or limb, health, property and public welfare by regulating and controlling the design, construction, quality of materials, use and occupancy, location and maintenance of all buildings and structures within this jurisdiction and certain equipment specifically regulated in the Code; UBC 1979. Some building codes in the U.S. deviate

from the UBC; **in U.K.**, exist the **building regulations** which are the basis of the system of building control in the form of functional requirements; ►zoning and building regulations); *syn.* building regulations [npl] [UK], standard building regulations [npl] [ZA]; *s* **Código [m] Técnico de la Edificación (CTE) [Es]** (Instrumento normativo, aprobado con el objetivo de mejorar la calidad de la edificación y promover la innovación y la sostenibilidad, que fija las exigencias básicas de calidad de los edificios y sus instalaciones, incluyendo los requisitos básicos de seguridad y bienestar de las personas, que se refieren tanto a la seguridad estructural y de protección contra incendios, como a la salubridad, la protección contra el ruido, el ahorro energético o la accesibilidad para personas con movilidad reducida. El código es por una parte, la modernización de las Normas Básicas de la Edificación de 1977, por otra la armonización de las normas nacionales con las disposiciones de la UE vigentes en esta materia, tal como la directiva relativa a los productos de construcción y la relativa a la eficiencia energética de los edificios. El **CTE** se divide en dos partes: **1.** Disposiciones de carácter general [ámbito de aplicación, estructura, clasificación de usos, etc.] y exigencias que deben cumplir los edificios para satisfacer los requisitos de seguridad y habitabilidad. **2.** Documentos Básicos cuya adecuada utilización garantiza el cumplimiento de las exigencias básicas y que contienen procedimientos, reglas técnicas y ejemplos de soluciones, que no tienen carácter excluyente. Por ello se crea también el ►Registro General del Código Técnico de Edificación para registrar aquéllos Documentos Reconocidos que no son parte integrante del código; cf. RD 314/2006; ►legislación urbanística); *syn.* código [m] de construcción [Es, EC], código [m] de edificación [RA], ordenanza [f] de construcción [RCH], reglamento [m] de construcción [MEX] (MESU 1977); *f* **Code [m] de la construction et de l'habitation (C.C.H.)** (**1. F.**, ensemble hétérogènes de textes législatifs et réglementaires ordonné en six livres traitant des règles générales de la construction, du statut des constructeurs, des aides à la construction d'habitations et à l'amélioration de l'habitat, des habitations à loyers modérés, des bâtiments menaçant ruine et bâtiments insalubres et des mesures tendant à remédier à des difficultés exceptionnelles de logement ; Code de l'urbanisme, ►droit de construire. **2. D.**, règles d'urbanisme régissant les prescriptions relatives aux terrains constructibles ainsi qu'à la construction, la modification des ouvrages et de leur utilisation, l'entretien de constructions et d'ouvrages divers ; chaque Land possède un Code de la construction propre : « Landesbauordnung — LBO ») ; *g* **Bauordnung [f]** (*Abk.* BauO; **D.**, öffentlich-rechtliche Vorschriften über die Anforderungen an das einzelne Baugrundstück sowie an die Errichtung, bauliche Änderung, Nutzungsänderung, Instandhaltung etc. der einzelnen baulichen Anlagen. Die **B.** eines einzelnen Bundeslandes heißt Landesbauordnung [LBO]; Landesbauordnungen enthalten nur Bauordnungsrecht, nicht jedoch Bauplanungsrecht, das im ►Baugesetzbuch abgehandelt wird; **in CH.** wird das öffentliche Baurecht durch mehrere Verordnungen ergänzt. Im Kanton Zürich gibt es z. B. das Planungs- und Baugesetz [2005] mit die *Allgemeine Bauverordnung* [Verordnung über die nähere Umschreibung der Begriffe und Inhalte der baurechtlichen Institute sowie über die Mess- und Berechnungsweisen v. 22.06. 1977 (LS 700.2)], geändert 14.05.2003, die *Besondere Bauverordnung I* [Verordnung über die technischen und übrigen Anforderungen an Bauten, Anlagen, Ausstattungen und Ausrüstungen v. 06.06.1981 (LS 700.21), zuletzt geändert am 31.03.2009, seit 01.07.2009 in Kraft getreten] und die *Besondere Bauverordnung II* [Verordnung über die Verschärfung oder die Milderung von Bauvorschriften für besondere Anlagen (LS 700.22), geänderte Fassung seit 01.11. 2009 in Kraft getreten]; cf. RBU 1999; ►Baurecht); *syn.* Bauverordnung [f] [CH].

6652 Uniform Construction Specifications [npl] [US] *constr. eng.* (**In U.S.**, a collection of computer-coded standard texts published by the Construction Specification Institute [CSI], and covering basic specification clauses set out in work sections or divisions. Guide specification databases are being produced for planners to use in landscape, civil, and site construction work; **in U.K.**, the NBS is a computer-based, comprehensive library of specification clauses arranged in work sections, available on subscription either as the 'Full Version' for major works or the 'Small Jobs Version' for minor works. Clauses are optional and a regular updating service is included; ►standard specifications); *syn.* National Building Specification [n] (N.B.S.) [UK]; *s* **Registro [m] General del Código Técnico de Edificación [Es]** (≠) (Registro de carácter público e informativo, adscrito a la Dirección General de Arquitectura y Política de Vivienda del Ministerio de Vivienda, que tiene como fin inscribir entre otros los Documentos Reconocidos del CTE [documentos técnicos, sin carácter reglamentario, que cuenten con el reconocimiento del Ministerio de Vivienda], las marcas, sellos u otros distintivos de calidad voluntarios de los productos, equipos o sistemas que se incorporen a los edificios y ayuden a cumplir la exigencias básicas, los sistemas de certificación que fomenten la mejora de la calidad de la edificación, y los organismos autorizados por las Administraciones Públicas competentes para concesión de evaluaciones técnicas de la idoneidad de productos o sistemas innovadores; cf. art. 4 RD 314/2006; ►modelo de resumen de prestaciones); *f* **documents [mpl] techniques unifiés (DTU)** (Ensemble des textes écrits fixant les dispositions techniques et règles de l'art nécessaires à l'exécution des prestations de travaux du bâtiment prévues au marché et utilisés dans la rédaction des spécifications techniques détaillées ; les DTU comprennent les Cahiers de Clauses Techniques [CCT] qui indiquent les conditions techniques à respecter pour le choix et la mise en œuvre de matériaux, les Cahiers des Clauses Spéciales [CCS] qui définissent les clauses administratives et les Règles de Calcul qui permettent de dimensionner les ouvrages ; les DTU sont publiés par le centre scientifique et technique du bâtiment [C.S.T.B.] ; il y a actuellement une centaine de DTU, ceux-ci devenant depuis 1990, dans le cadre de l'harmonisation européenne, les normes-DTU ; ►cahier de prescriptions techniques-type) ; *g 1* **Standardleistungsbuch [n] für das Bauwesen** (*Abk.* StLB; eine in den 1970er-Jahren vom **G**emeinsamen **A**usschuss für **E**lektronik im **B**auwesen [GAEB] nach Leistungsbereichen gegliederte Sammlung standardisierter, datenverarbeitungsgerechter Texte zur Beschreibung von Standardleistungen als Hilfe zur Zusammenstellung von Bauleistungsverzeichnissen. Den Ausschreibungstexten liegen die Vertragsbedingungen der Vergabe- und Vertragsordnung für Bauleistungen [VOB] zugrunde. Das Textsystem StLB-Bau als Fachwerkzeug für die rationelle Beschreibung von Bauleistungen wurde als Version 04/1998 mit Erlass des Bundesministeriums für Raumordnung, Bauwesen und Städtebau am 03.08.1998 eingeführt und wurden durch die überarbeitete und ergänzte Fassung 10/1999 mit den neuesten technischen Regeln und Erkenntnissen aktualisiert. Seit 1999 gibt es auf dem Markt parallel dazu eine mit Normen, Richtlinien und Expertenwissen sowie Kalkulationsmöglichkeiten versehene Version als **StLB-Bau — Dynamische Baudaten**; ferner gibt es das **StLB-Standardleistungsbuch für das Bauwesen Zeitarbeitsverträge [Z]**; ►Musterleistungsverzeichnis); *g 2* **Standardleistungskatalog [m]** (*Abk.* StLK; eine nach Leistungsbereichen gegliederte Sammlung standardisierter, datenverarbeitungsgerechter Texte im Bauingenieurwesen sowie für den Garten- und Landschaftsbau [StLK-GaLaBau] als Anhalt für die Erstellung örtlicher Leistungsverzeichnisse. In D. gibt es mehrere **S.e**, da

U

das Standardleistungsbuch [StLB] noch nicht überall angewendet wird, z. B. für Erdbau, Landschaftsbau, Baugruben, Leitungsgräben, Pflaster, Platten, Borde, Rinnen); *syn.* Normpositionenkatalog [m] (N.P.K.) der Schweizerischen Zentralstelle für Baurationalisierung [CH].

unimproved property [n] [US] *plan. urb.* ▶undeveloped land.

uninhabited area [n] *plan. urb* ▶unpopulated area.

unique character [n] **of a (scenic) landscape** *conserv. landsc. plan. recr.* ▶landscape character.

6653 unique character [n] **of natural features** *conserv. syn.* uniqueness [n] of natural features [also US]; *s* **particularidad** [f] **de los elementos naturales**; *f* **caractère** [m] **particulier d'un objet naturel remarquable** (Caractère d'un espace ponctuel naturel présentant un intérêt particulier remarquable) ; *g* **Eigenart** [f] **von Einzelschöpfungen der Natur**.

6654 unique natural character [n] **of a landscape** *conserv. s* **particularidad** [f] **de un paisaje**; *f* **caractère** [m] **d'un milieu naturel présentant un intérêt particulier** *syn.* caractère [m] d'un milieu naturel présentant des qualités remarquables ; *g* **Eigenart** [f] **von Natur und Landschaft** *syn.* Charakter [m] von Natur und Landschaft.

unique natural feature [n] *conserv.* ▶outstanding natural feature.

6655 uniqueness [n] *conserv.* (Special quality according to its singularity, e.g. of an ▶outstanding natural feature); *s* **singularidad** [f] (Característica notable e irreproducible de cualquier parte de la naturaleza. La **s.** de los rasgos bióticos o abióticos de un territorio es una de las razones importantes para su conservación; cf. DINA 1987; ▶elemento natural singular); *f* **singularité** [f] (Propriété caractérisant l'unicité ou l'incomparabilité, p. ex. de peuplements faunistiques, floristiques, de paysages ou de formations géologiques ; ▶objet naturel remarquable) ; *g* **Einzigartigkeit** [f] (Charaktereigenschaft hinsichtlich der Einmaligkeit oder Unvergleichbarkeit, z. B. von Bestandteilen der Fauna, Flora oder geologischen Formationen; ▶Einzelschöpfung der Natur); *syn.* Unverwechselbarkeit [f].

uniqueness [n] **of a landscape** *conserv. landsc. plan. recr.* ▶landscape character.

uniqueness [n] **of natural features** [US] *conserv.* ▶unique character of natural features.

unit [n] *constr.* ▶dished channel unit; *ecol. land'man. plan.* ▶ecological spatial unit; *constr.* ▶edging unit; *land'man. plan.* ▶landscape unit; *constr.* ▶living construction unit; *constr.* ▶L-shaped retaining wall unit; *geo. landsc.* ▶natural landscape unit; *constr.* ▶precast concrete unit; *constr.* ▶riser unit [US]/chamber section [UK]; *plan.* ▶spacial unit.

unit [n] [US], **brick paving** *constr.* ▶paver brick [US]/clinkerbrick for paving [UK].

unit [n], **concrete** *constr.* ▶precast concrete unit.

unit [n] [UK], **concrete paving** *constr.* ▶concrete paver [US].

unit [n] [AUS], **interlocking concrete paving** *constr.* ▶interlocking paver [US].

unit [n], **multifamily dwelling** *arch. urb* ▶multi-family housing.

unit [n] [US], **paving** *constr.* ▶paving stone [US]/paving sett [UK].

unit [n] [US], **picnic** *recr.* ▶picnic site, #2.

unit [n] [UK], **prefab(ricated) concrete compound** *constr.* ▶precast concrete unit.

unit [n], **soil taxonomic** *pedol.* ▶soil series [US].

unit [n], **taper** *constr.* ▶cone (1).

6656 unit area [n] *constr. hort.* (Defined portion of an area); *s* **unidad** [f] **de territorio** (Tamaño definido de un área); *f* **unité** [f] **spatiale** (Superficie définie d'une aire) ; *syn.* aire [f] donnée ; *g* **Flächeneinheit** [f] (definierte Flächengröße).

unitary development plan [n] [UK] *leg. urb. obs.* ∗▶comprehensive plan [US]/Local Development Framework (LDF) [UK].

unit [n] **of a landscape** *geo. landsc.* ▶spatial unit of a landscape.

6657 unit [n] **of computation** *contr. prof.* (**In some European countries**, ratio per unit of area which is the basis for remuneration of professional services; e.g. in preparing ▶general plans for urban open spaces [US]/green open space structure plans [UK]); *s* **unidad** [f] **de cálculo** (En reglamento de honorarios en D., unidad que se utiliza para liquidar servicios en el contexto de elaboración de ▶planes municipales de zonas verdes y espacios libres); *f 1* **unité** [f] **d'intervention** (Un des modes de détermination des honoraires à la superficie pour les missions d'urbanisme réglementaire [Schéma directeur, POS], certaines études préalables [schéma de secteur], préopérationnelles et opérationnelles [PAZ, étude de lotissement] en matière d'urbanisme ; la valeur de l'**UI** est fixée dans certaines limites et varie dans le temps suivant l'indice SYNTEC ; HAC 1989, 459-461) ; *f 2* **unité** [f] **de compte** (Un des modes de détermination des honoraires à la superficie pour les travaux et prestations des géomètres-experts ; la valeur monétaire de l'**U.C.** était fixée jusqu'en 1986 par le Conseil supérieur de l'Ordre et est désormais fixée par le cabinet de géomètre-expert en fonction des paramètres internes liées à son activité ; HAC 1989, 525) ; *g* **Verrechnungseinheit** [f] (In § 29 HOAI 2009 angegebene Einheit, nach der landschaftsplanerische Leistungen für ▶Grünordnungspläne abgerechnet werden; aus der Summe der Einzelansätze je Hektar zu bearbeitender Fläche ergibt sich die aus der Honorartabelle zu interpolierende Vergütung); *syn.* Einzelansatz [m].

unit paver [n] [US], **concrete** *constr.* ▶concrete paver [US]/concrete paving unit [UK].

6658 unit price [n] *constr. contr.* (Calculated construction cost stated in the bid [US]/tender [UK] as a price per unit of measurement for materials and labo[u]r per construction item, including delivery and installation; *context* item cost of materials and unit or hourly cost of labo[u]r); *s* **precio** [m] **unitario** (Precio por unidad de obra o por unidad de cantidad de material suministrado incluidos el suministro a obra y la mano de obra que se emplea en ofertas); *f* **prix** [m] **unitaire** (Prix par unité de quantité mentionné dans le cadre du détail estimatif ou du bordereau des prix unitaires ; *contexte* rémunération sur prix unitaires) ; *g* **Einheitspreis** [m] (*Abk.* EP; Preis für Bauleistungen oder Lieferungen je Mengeneinheit, z. B. im Kostenanschlag oder im Angebot; *Kontext* im EP enthaltene Materialkosten).

6659 unit price [n] **for an alternative** *constr. contr.* (Calculated cost of a construction alternative for materials and labo[u]r); *s* **precio** [m] **unitario para partida alzada** (Coste calculado para una partida alzada en una oferta); *f* **prix** [m] **pour mémoire (P.M.)** (pour une variante de construction) ; *g* **Einheitspreis** [m] **für eine Alternative** (Kalkulierter Angebotspreis für eine Alternativposition).

6660 unit price list *constr. contr.* (Usual method in France of entering only the ▶unit prices for specified items in a bill of quantities and listing of total prices at the end of the tender document [UK]); *s* **listado** [m] **de precios unitarios** (Método

común en F. de incluir en el resumen de prestaciones sólo los ▶precios unitarios. Los precios totales se añaden al final del resumen de prestaciones en forma de una lista); *f* **bordereau [m] des prix unitaires [B.P.U.]** (Description précise, claire, détaillée et complète de la définition et du mode d'évaluation de chaque nature d'ouvrage ou d'élément d'ouvrage permettant d'inscrire les ▶prix unitaires de chaque prestation de travaux, la liste des prix totaux étant jointe à la fin du document ; pièce faisant partie du dossier de consultation des entreprises) ; *g* **Verzeichnis [n] der Einheitspreise** (In F. übliche Methode, im Leistungsverzeichnis neben den Leistungsbeschreibungen nur die ▶Einheitspreise einzutragen. Die Gesamtpreise werden am Schluss des Leistungsverzeichnisses als Übersicht aufgeführt).

6661 unit price reduction [n] *constr. contr.* (Lowering of unit price with an increase in estimated volume of required material); *s* **rebaja [f] de precio unitario** (Reducción de un precio unitario en el caso de aumento del volumen de material necesitado); *f* **moins-value [f]** (Réduction sur un prix unitaire souvent consentie lors d'une augmentation dans la masse des travaux) ; *g* **Abschlag [m]** (*Planerjargon* Reduzierung eines Einheitspreises bei Massenmehrung); *syn.* Nachlass [m].

6662 universal design [n] *leg. plan. pol. sociol.* (Planning, shaping and functional arrangement, organization and layout of environments and products to be usable by everyone, to the greatest extent possible, without the need for adaptation or specialized design. The intent of **u. d.** is to simplify life for the benefit of people of all ages and abilities, as many as possible, by making the built environment, products, and communications more usable at little or no extra cost. **U. d.** is easy to understand regardless of the user's experience, knowledge, language skills, or current concentration levels; the design communicates necessary information effectively to the user, regardless of ambient conditions or the user's sensory abilities; it minimizes hazards and the adverse consequences of accidental or unintended actions; appropriate size and space is provided for approach, reach, manipulation, and use regardless of user's body size, posture, or mobility); *s* **diseño [m] universal** (Planificación, diseño y arreglo funcional, organización y estructuración de ambientes y productos de manera que sean utilizables por todas las personas, en el grado máximo posible, sin necesidad de adaptación o ayuda especial. El objetivo del **d. u.** es simplificar la vida de la gente de todas las edades y capacidades haciendo el medio construido, los productos y los medios de comunicación más fáciles de utilizar y sin o con pocos costes [Es]/costos [AL] adicionales. El **d. u.** es fácil de entender, independientemente de la experiencia, conocimiento, habilidad lingüística o grado de concentración del/de la usuaria. El diseño comunica al usuario/a la usuaria la información necesaria con efectividad, independientemente de las condiciones ambientales o de las capacidades sensoriales de las personas, y minimiza peligros y consecuencias adversas de acciones accidentales o desintencionadas. Tamaño y espacio apropiados son previstos para acceder, alcanzar, manipular y utilizar los objetos o servicios, independientemente de la altura, postura y mobilidad del/de la usuaria); *f* **conception [f] universelle** (Concevoir des produits, des services, des environnements qui soient utilisables pour un éventail d'usagers le plus large possible, sans nécessiter d'adaptation ou de conception spéciale ; cette conception s'organise autour des principes suivants : **1.** utilisation équitable [la conception est utile et commercialisable auprès de personnes ayant différentes capacités], **2.** flexibilité d'utilisation [la conception peut être conciliée à une vaste gamme de préférences et de capacités individuelles], **3.** utilisation simple et intuitive [l'utilisation de la conception est facile à comprendre, indépendamment de l'expérience, des connaissances, des compétences linguistiques de l'utilisateur ou de son

niveau de concentration au moment de l'utilisation], **4.** information perceptible [la conception communique efficacement à l'utilisateur l'information nécessaire, quelles que soient les conditions ambiantes ou les capacités sensorielles de la personne], **5.** tolérance pour l'erreur [la conception réduit au minimum les dangers et les conséquences adverses des accidents ou des actions involontaires], **6.** effort physique minimal [la conception permet une utilisation efficace et confortable, générant une fatigue minimale], **7.** dimensions et espace libre pour l'approche et l'utilisation [la conception prévoit une taille et un espace adéquats au moment de s'approcher, de saisir, de manipuler et d'utiliser, quelles que soient la taille, la posture ou la mobilité de l'utilisateur]) ; *syn.* conception [f] pour tous, aménagement [m] adapté aux handicapés/personnes à mobilité réduite ; *g* **behindertengerechte Planung [f]** (Planerische Bearbeitung, Formgebung und Ausstattung, funktionale Anordnung und Gliederung der gebauten Umwelt und Gestaltung technischer Gebrauchsgegenstände, die von jedermann in jedweder Hinsicht ohne besondere Erschwernis und grundsätzlich ohne fremde Hilfe zugänglich und nutzbar resp. benutzbar sind. **B. P.** beabsichtigt, das Leben aller Menschen jeden Alters und jeder Fähigkeit in der gestalteten Umwelt, mit den Kommunikationseinrichtungen und mit dem Umgang aller vorhandenen Gebrauchsgegenstände so angenehm wie möglich zu machen — dies zu geringen oder ohne Extrakosten. Das Ergebnis der **b.n P.** ist unabhängig der individuellen Erfahrung, Kenntnisse und Sprachfertigkeiten oder der aktuellen Konzentrationsfähigkeit leicht verständlich und ablesbar. **B. P.** vermittelt dem Nutzer leicht nötige Informationen, unabhängig von den umgebenden Bedingungen oder von seinen Wahrnehmungsfähigkeiten; sie reduziert Gefahren und die negativen Folgen zufälliger oder unbeabsichtigter Handlungen; für Zugänge, Aktionsradien und sonstige Bewegungsabläufe wird genügend Platz bereitgestellt, unabhängig von Körpergröße, Gestalt und Beweglichkeit); *syn.* barrierefreie Planung [f].

6663 university [n] *prof.* (Institution of higher education for students in many branches of learning, which confers undergraduate and post-graduate degrees in various faculties. An **u.** often includes colleges for general instruction and other institutions, as well as a graduate school for specialized studies, which is empowered to confer various advanced academic degrees); **in U.S.**, institution of higher education, including a college for general instruction, and graduate school for specialized studies, which is empowered to confer various advanced academic degrees); *s* **universidad [f]** (Centro de educación superior con orientación científico-teórica, en el que los estudios duran generalmente 5 años); *f* **université [f]** (**F.**, établissement public d'enseignement supérieur constituant une communauté d'enseignants, de chercheurs et d'étudiants et qui offrent des programmes diversifiés ; **D.**, établissement de formation scientifique attribuant un diplôme universitaire [ingénieur diplômé] après des études scientifiques théoriques d'une durée minimale de quatre ans ; *terme spécifique* unité d'enseignement et de recherche [U.E.R.] — anciennement facultés ; DUA 1996) ; *g* **Universität [f]** (Hochschule, an der ein mindestens 4-jähriges [achtsemestriges], wissenschaftlich-theoretisches Studium absolviert wird).

university [n] **of applied sciences** [UK] *prof.* ▶polytechnic institute [US].

6664 university study [n] *prof.* (Courses of higher education); *s* **carrera [f] universitaria** *syn.* estudios [mpl] universiarios; *f* **études [fpl] universitaires** ; *g* **Hochschulstudium [n]**.

6665 unmanaged dumpsite [n] [US] *envir.* (Open refuse dump [US]/refuse tip [UK] with uncontrolled rotting of waste, spontaneous combustion, and emitting of malodorous gases;

U

►unauthorized dumpsite [US]/fly tipping site [UK]); *syn.* unmanaged tipping site [n] [UK]; *s* **vertedero [m] incontrolado** (Vertedero abierto en el que se pudre la basura incontroladamente, lo que conlleva frecuentemente combustión espontánea y malos olores causados por gases de putrefacción; ►vertedero clandestino); *syn.* basurero [m], tiradero [m] [MEX]; basural [m] [RCH], botadero [m] [RCH, COL]; *f* **décharge [f] non contrôlée** (Aire sur laquelle la décomposition incontrôlée des ordures a souvent pour effet l'embrasement spontané du dépôt et la constitution de gaz nauséabonds ; ►décharge [des déchets] sauvage) ; *syn.* dépôt [m] non contrôlé ; *g* **ungeordnete Deponie [f]** (Offene Müllkippe mit unkontrollierter Verrottung der Abfälle; dadurch häufig Selbstentzündung und Geruchsbelästigung durch übelriechende Gase; ►wilde Deponie); *syn.* Müllhalde [f].

unmanaged tipping site [n] [UK] *envir.* ►unmanaged dumpsite [US].

unnatural [adj] *ecol.* ►man-developed.

6666 unpaved area [n] *constr. plan. urb.* (Area which has not been surfaced with a pavement; ►waterbound surface; *opp.* ►paved area); *s* **superficie [f] no pavimentada** (►revestimiento compactado; *opp.* ►superficie pavimentada); *f* **surface [f] semi-stabilisée** (Espace qui n'a pas reçu de revêtement particulier ; ►sol stabilisée aux liants hydrauliques ; *opp.* ►surface stabilisée) ; *syn.* terrain [m] semi-stabilisé ; *g* **unbefestigte Fläche [f]** (Fläche, die nicht mit einem Belag versehen ist; ►wassergebundener Belag; *opp.* ►befestigte Fläche).

unpaved shoulder [n] [US] *trans.* ►road verge.

6667 unpopulated area [n] *plan. urb* (Area devoid of habitation); *syn.* uninhabited area [n]; *s* **zona [f] deshabitada** *syn.* zona [f] no poblada, zona [f] desierta; *f* **zone [f] non urbanisée** *syn.* zone [f] inhabitée, zone [f] déserte ; *g* **unbesiedelter Bereich [m]** *syn.* unbesiedeltes Gebiet [n], nicht besiedeltes Gebiet [n], nicht besiedelter Bereich [m], unbebauter Bereich [m] (cf. § 2 [1] 2 BNatSchG), unbebaute Flächen [fpl] (cf. § 2 [1] 9 BNatSchG).

unsatisfied requirements [n] *plan.* ►shortfall.

6668 unserved land [n] [US] *plan. urb.* (Area not served by public utilities or roads); *s* **área [f] no urbanizada** (Zona en donde no se ha construido ninguna infraestructura); *f* **terrain [m] non équipé** (Terrain ne possédant aucune desserte par les équipements de réseaux publics) ; *g* **unerschlossene Fläche [f]** (Gebiet, das noch durch keine Erschließungsmaßnahmen zugänglich gemacht wurde); *syn.* unerschlossenes Gebiet [n].

6669 unsolicited prequalification document [n] *prof.* (Unasked brochure submitted by a planning and design firm, containing a record of the expertise and experience of each member of the firm and the firm's work to prequalify for selection to perform future projects; ►request for proposals [US], ►request for qualification [US]); *s* **documentación [f] de referencia (de empresa paisajista)** (Presentación de un gabinete de paisajismo o planificación con descripciones de los proyectos realizados; ►demanda de proposiciones); *f* **brochure [f] de prospection** (Document [maquette, plans] présentant les qualités et capacités d'un bureau d'études dans le cadre des démarches de prospection du responsable ; ►concours restreint) ; *g* **Bewerbungsunterlagen [fpl] eines Planungsbüros** (Broschüre über abgewickelte Projekte, Leistungsfähigkeit und berufliche Erfahrung eines Planungsbüros zur Akquisition für neue Aufträge; ►öffentlicher Teilnahmewettbewerb für Planungsbüros); *syn.* Akquisitionsmappe [f].

6670 unstable embankment slope [n] *constr.* (Artificial slope liable to failure; e.g. ►soil slippage; ►incised slope); *s* **talud [m] inestable (1)** (Pendiente de un talud artificial con peligro de deslizamiento; ►cara de corte, ►deslizamiento del

suelo); *f* **talus [m] instable** (État d'un talus menacé de ►foirage ; ►talus de fouille) ; *g* **instabile Böschung [f]** (Rutschgefährdete Böschung an Erdbauwerken; ►Rutschung); *syn. bei großen* ►*Einschnittböschungen* Rutschhang [m].

6671 unstable slope [n] *geo. pedol.* (Slope in danger of slipping due to soil properties, precipitation, freezing and thawing, or other natural causes; ►landslide, ►unstable embankment slope); *syn.* hazardous slope [n] (TGG 1984, 272), soil-slippage-prone slope [n] [also US]; *s* **talud [m] inestable (2)** (Pendiente con peligro de deslizamiento debido a las características del suelo, a la lluvia, a heladas y deshaladas u otras causas naturales; ►corrimiento de tierras [Es, RA], ►talud inestable 1); *f* **versant [m] instable** (Versant exposé aux risques de glissement ou de solifluxion ; ►glissement de terrain, ►talus instable) ; *g* **Rutschhang [m]** (Durch Bodengleit- oder Erdrutschvorgänge gefährdeter Hang; ►Hangrutschung, ►instabile Böschung).

unstructured play area [n] [US] *recr.* ►free play area.

untooled [n] *constr.* *►tooling of stone [US]/dressing of stone [UK], #9.

untrained hedge [n] [UK] *gard. landsc.* ►untrimmed hedge [US].

untreated sewage [n] *envir.* ►raw sewage.

6672 untrimmed hedge [n] [US] *gard. landsc.* (Linear planting of closely-growing shrubs and trees which develop according to their natural habit; *specific term* ►hedgerow); *syn.* untrained hedge [n] [UK]; *s* **seto [m] vivo** (Plantación lineal de arbustos que no se podan y pueden desarrollarse naturalmente; *término específico* ►seto vivo interparcelario); *f 1* **haie [f] vive** (Plantation arbustive linéaire bocagère ou champêtre non taillée et en mélange ; *terme spécifique* ►haie champêtre) ; *f 2* **haie [f] libre** (Plantation arbustive très modérément taillée, de hauteur libre en général constituée de plusieurs espèces et utilisée en clôture ou comme cloisonnement) ; *g* **freiwachsende Hecke [f]** (Linear angeordnete Anpflanzung von Sträuchern und stockausschlagfähigen Bäumen, die nicht in Form geschnitten werden und sich habitusgerecht entwickeln können; *UB* ►Feldhecke); *syn.* Naturhecke [f], Strauchhecke [f].

unzoned land [n] (±) *leg. urb.* ►open land [n] (±).

6673 unzoned land [n] **ripe for development** [US] *urb.* (An area that can actually be expected to be built upon in the foreseeable future because of its properties and location, as well as the fact that it has been designated in the development planning process for land use or by the resolved will of the local community or on account of the general urban development of the municipal area; *opp.* ►land zoned for development; ►undeveloped area, ►undeveloped zoned land); *syn. obs.* white land [n] [UK]; *s* **suelo [m] edificable no urbanizado** (Superficie que por sus características y su localización probablemente será edificada en un plazo breve; ►área no urbanizada, ►suelo edificable, ►suelo urbanizable, ►suelo urbanizado); *f* **zone [f] d'urbanisation future** (Zone naturelle peu ou non équipée définie par un PLU/POS et destinée à être urbanisée en respectant un équilibre entre l'aménagement et la protection ; ►zone constructible, ►terrain [constructible] non viabilisé, ►espace non urbanisé) ; *syn.* zone [f] ouverte à l'urbanisation, terrains [mpl] en réservation, terrain [m] potentiellement constructible ; *g* **Bauerwartungsland [n, o. Pl.]** (Flächen, die in ihrer Eigenschaft, ihrer sonstigen Beschaffenheit und ihrer Lage eine bauliche Nutzung in absehbarer Zeit tatsächlich erwarten lassen, da sie vom Planungsprozess der Bauleitplanung erfasst sind [Ausweisung im Flächennutzungsplan] oder ein entsprechendes Verhalten der Gemeinde oder die allgemeine städtebauliche Entwicklung des Gemeindegebietes dies begründet; cf. § 4 [2] WertV; ►Bauland, ►Rohbauland, ►unbebaute Fläche).

U

up-date [n] *plan.* ▶revision.

up-dated version [n] *plan.* ▶revised version.

6674 updating [n] *adm. plan.* (Bringing a policy, plan or report in line with conditions prevailing at a current date); *s* **actualización [f] periódica** (1. Puesta al día de datos estadísticos. 2. Revisión de planes para adaptarlos a las condiciones actuales); *syn.* puesta [f] al día; *f* **actualisation [f]** (1. Mise à jour de données statistiques, de listes de monuments naturels classés, des prix d'un marché selon l'indice BT [bâtiment] dans le cadre de la révision des prix. 2. Mise à jour d'une planification existante) ; *syn.* mise [f] à jour, tenue [f] à jour ; *g* **Fortschreibung [f]** (1. Fortlaufende Ergänzung einer bestehenden Statistik; auf den neuesten Stand bringen [vb], fortschreiben [vb]. 2. Eine bestehende Planung weiterbetreiben und an veränderte Gegebenheiten anpassen; auf den neuesten Stand bringen [vb], fortschreiben [vb]); *syn.* Aktualisierung [f], Nachführung [f] [CH].

upkeep [n] *constr. hort.* ▶maintenance work; *wat'man.* ▶river upkeep.

upkeep [n] [UK]**, garden** *constr. hort.* ▶garden maintenance [US].

upkeep [n] **and renovation** [n] *conserv'hist. constr.* ▶maintenance and repair work.

upland [n] [US]**, riverine** *phyt.* ▶upper riparian alluvial plain.

upland afforestation [n] [UK] *for. land'man.* ▶high-altitude afforestation.

uplift wind pressure [n] [US] *met. constr.* ▶wind suction.

upper edge [n] *arch. constr.* ▶top level.

6675 upper reaches [npl] *geo.* (Part of a watercourse in the higher region of the ▶drainage basin; WMO 1974; ▶headwater stream [US]/headstream [UK], ▶middle reaches; *opp.* ▶lower reaches); *syn.* headwaters [npl] [also US]; *s* **tramo [m] superior (de un curso de agua)** (Parte del cauce situada en la zona superior de la cuenca de drenaje; WMO 1974; ▶cabecera [de río], ▶tramo medio [de un curso de agua]; *opp.* ▶tramo inferior [de un curso de agua]); *f* **cours [m] supérieur** (Partie d'un cours d'eau dans la région haute du bassin versant ; cours moyen ; ▶cours supérieur près de la source, ▶cours moyen; *opp.* ▶cours inférieur) ; *g* **Oberlauf [m]** (Oberer Teil eines Fließgewässers; ▶Quelllauf, ▶Mittellauf; *opp.* ▶Unterlauf).

6676 upper riparian alluvial plain [n] *phyt.* (Upper terrace of middle and lower reaches of a river with a vegetation cover dominated by trees, incapable of withstanding continued flooding; ▶regularly flooded alluvial plain); *syn.* upper riverine zone [n], riverine upland [n] [also US]; *s* **vega [f] aluvial ocasionalmente inundada** (Terraza superior en los valles de cursos bajo y medio de ríos en la que crecen árboles que no toleran inundaciones continuas; *opp.* ▶vega aluvial regularmente inundada); *f* **plaine [f] alluviale de forêt à bois durs** (*Terme peu utilisé.* Terrasse alluviale humide située sur les parties hautes du lit majeur de grandes vallées occasionnellement inondée dans le cours moyen et inférieur des fleuves et recouverte par la frênaie-ormaie typique et/ou la chênaie alluviale à Chêne pédonculé ; *opp.* ▶plaine alluviale à bois tendres) ; *syn.* zone [f] alluviale à forêt de bois durs, plaine [f] alluviale occasionnellement inondée, terrasse [f] alluviale irrégulièrement inondée ; *g* **Hartholzaue [f]** (Höchste Stufe innerhalb des Überschwemmungsbereiches am Mittel- und Unterlauf der Flüsse mit kräftigen, dauerhaften Baumarten; *opp.* ▶Weichholzaue); *syn.* harte Au [f] [auch A].

upper riparian/riverine woodland [n] [US] *phyt.* ▶riparian upland woodland.

upper riverine zone [n] *phyt.* ▶upper riparian alluvial plain.

6677 upper river terrace [n] *geo.* (Upper terrace of gravel bordering a river, indicating a former floodplain formed at different stages of erosion or deposition during the last Pleistocene age; ▶lower river terrace, ▶middle river terrace); *s* **terraza [f] fluvial alta** (Llanura aluvial superior a ambos lados de los cursos fluviales originada por la acción de los procesos fluviales de carácter erosivo; cf. DINA 1987, 939; ▶terraza fluvial baja, ▶terraza fluvial media); *f* **terrasse [f] (fluviatile) supérieure** (Étendue plane supérieure représentant la partie du lit ancien d'un cours d'eau entaillée par suite de l'enfoncement continu du lit mineur résultant des alternances de biostasie et de rhéxistasie ; ▶terrasse [fluviatile] inférieure, ▶terrasse [fluviatile] moyenne) ; *syn.* terrasse [f] alluviale supérieure ; *g* **Hochterrasse [f]** (Die obere Schotterterrasse in Flusstälern, die durch mehrfachen Wechsel von Akkumulation und Erosion während der letzten pleistozänen Kaltzeit entstanden ist. Unterhalb der **H.** liegen ▶Mittelterrasse und ▶Niederterrasse).

6678 upright-growing main branch [n] *arb.* (ARB 1983, 402; ▶scaffold branch growing vertically from the ▶stem apex or main lateral branch); *syn.* upright scaffold branch [n] (ARB 1983, 405); *s* **rama [f] principal de crecimiento vertical** (Rama principal que crece derecha hacia arriba del ▶ápice del tronco o de una rama lateral contribuye a dar forma a la copa; ▶rama primaria); *f* **charpentière [f] à croissance verticale (±)** (▶Charpentière se développant verticalement à partir de l'▶apex du fût ou d'une forte branche latérale) ; *syn.* charpentière [f] centrale ; *g* **Stämmling [m]** (Ein vom ▶Stammkopf oder vom kräftigen Seitenast steil aufrecht wachsender, kronenbildender ▶Starkast).

upright scaffold branch [n] *arb.* ▶upright-growing main branch.

upsiloidal dune [n] *geo.* ▶parabolic dune.

6679 upslope air flow [n] *geo. met.* (Wind moving upward sloping terrain; ▶valley wind; *opp.* ▶downslope wind); *syn.* anabatic wind [n]; *s* **viento [m] anabático** (Viento ascendiente a lo largo de las laderas previamente calentadas por la acción de la radiación solar. Es mucho menos frecuente que el ▶viento catabático; ▶viento de valle); *f* **vent [m] anabatique** (Vent ascendant le long d'une pente. Ce vent est associé à un échauffement de la surface de la pente ; ▶brise de vallée ; *opp.* ▶vent catabatique) ; *g* **Hangaufwind [m]** (Den Hang hinauf wehender Wind; ▶Talwind, *opp.* ▶Hangabwind); *syn.* anabatischer Wind [m].

urban agglomeration [n] *urb.* ▶ ▶urban concentration.

6680 urban area [n] *urb.* (▶urban space); *s* **área [f] urbana** (▶espacio urbano); *syn. leg.* término [m] municipal; *f* **espace [m] urbain** (▶milieu urbain) ; *syn.* territoire [m] urbain ; *g* **Stadtgebiet [n]** (Zweidimensionale Vorstellung einer Stadt — im Vergleich zum ▶Stadtraum).

6681 urban area recreation planning [n] *landsc. plan. pol. recr.* (▶recreation area planning); *s* **planificación [f] de zonas urbanas de recreo** (▶planificación de áreas turísticas y de recreo); *syn.* gestión [f] de zonas urbanas de recreo; *f* **planification [f] du tourisme urbain** (Réorganisation physique, paysagère et fonctionnelle de l'espace urbain [mise en tourisme des espaces urbains] en vue de sa mise en valeur touristique par la prise en compte de l'identité et de l'originalité culturelle des villes ; ▶étude d'un schéma de développement des loisirs) ; *syn.* planification [f] touristique en milieu urbain, aménagement [m] touristique urbain, aménagement [m] urbain touristique ; *g* **Erholungsplanung [f] im städtischen Raum** (▶Erholungsplanung für ein Gebiet).

urban clearance [n] **and redevelopment** [n] [US] *plan. urb.* ▶comprehensive redevelopment.

U

6682 urban climate [n] *met. urb.* (Climate of a town or city, which is influenced by the density of its buildings, and is primarily characterized by higher air temperatures than the surrounding countryside, low fresh air exchange, lower temperature differences between day and night, high concentrations of condensation nuclei and aerosols due to air pollution. These conditions cause the occurrence of ►smog over large cities, leading to frequent rain and fog; ►heat island); *syn.* town climate [n] [also UK]; *s* **clima** [m] **urbano** (Clima condicionado por la densa edificación y pavimentación de áreas urbanizadas así como por las emisiones de la industria, el tráfico, etc. que se caracteriza por temperaturas más altas que en los alrededores, menores diferencias de temperatura entre el día y la noche, renovación del aire reducida, mayor concentración de contaminantes atmosféricos y la formación de capas de calima [►smog]; ►isla de calor); *syn.* clima [m] de la ciudad; *f* **climat** [m] **urbain** (Ambiance atmosphérique en milieu urbain caractérisée par rapport aux milieux ruraux environnant par un échauffement de l'air dû à une augmentation de la réception des radiations solaires, un faible renouvellement de l'air et une réduction des différences de température pendant la nuit, une forte concentration des noyaux de condensation et des aérosols due à la pollution ; tous ces phénomènes participent à l'augmentation de l'effet de serre urbain [►smog], les précipitations et le brouillard ; ►îlot thermique) ; *syn.* microclimat [m] urbain ; *g* **Stadtklima** [n] (…ta [pl]; fachsprachlich …mate [pl]; das durch die dichte Bebauung einer Stadt bedingte Klima; es ist vornehmlich geprägt durch eine stärkere Erhöhung der Lufttemperatur [Sommerhitze und relative Wintermilde] gegenüber der freien Landschaft, durch geringere Lufterneuerung und geringere Temperaturunterschiede während der Nacht, hohe Konzentration an Kondensationskernen und Aerosolen durch Luftverschmutzung. Über Großstädte entstehen ‚Dunstglocken' [►Smog] und mit vermehrtem Regen und Nebel ist zu rechnen; ►Wärmeinsel).

6683 urban climatology [n] *met.* (Branch of science concerned with research into the ►urban climate); *s* **climatología** [f] **urbana** (Ciencia que estudia el ►clima urbano); *f* **climatologie** [f] **urbaine** (Science qui étudie les conditions météorologiques des villes ; ►climat urbain) ; *g* **Stadtklimatologie** [f] (Wissenschaft, die das ►Stadtklima erforscht).

6684 urban concentration [n] *urb.* (1. Crowding of dwelling units or floor area within a defined area; ►infill development); **2. urban agglomeration** [n] (1. Process in which cities and urban developments grow together; 2. result of 1.; ►conurbation, ►metropolitan region); *s* **concentración** [f] **urbana** (1. Crecimiento de construcciones en un área determinada de una ciudad; ►área metropolitana, ►densificación urbana. 2. Proceso de crecimiento de dos ciudades vecinas hasta formar una ►aglomeración urbana); *f* **concentration** [f] **urbaine** (1. Processus d'urbanisation, d'augmentation des constructions sur une aire donnée ; ►aire métropolitaine, ►construction dans le bâti existant, ►politique de la ville compacte. 2. Resserrement d'agglomérations pour former des zones de concentration urbaine) ; *g 1* **Verdichtung** [f] **von Siedlungsgebieten** (Zunahme der auf eine bestimmte Grundfläche bezogenen Wohnungseinheiten oder Geschossflächen; ►Nachverdichtung); *g 2* **Zusammenwachsen** [n, o. Pl.] **von großen Städten** (Planung und Durchführung von Infrastrukturmaßnahmen, die einzelne Städte verbinden sowie die planerische Festlegung und Realisierung neuer Baugebiete in den Bereichen zwischen den Städten, so dass im Laufe der Zeit eine Stadtregion entsteht; ►Ballungsraum); *syn.* Agglomeration [f], Ballung [f], Zusammenwachsen [n, o. Pl.] großer Städte.

6685 urban controlling [n] *adm. urb.* (Intervention and guidance of the development of a town or city by means of ►community development planning [US]/urban land-use

planning [UK], due to socioeconomic change); *s* **gestión** [f] **urbana** (Intervención gestora en el desarrollo urbano por medio del ►planeamiento urbanístico para reaccionar ante los cambios socioeconómicos); *f* **maîtrise** [f] **de l'urbanisme** (Préoccupation des collectivités publiques d'assurer un développement urbain en recourant aux instruments de ►planification urbaine et aux acquisitions foncières en vue d'aménager le cadre de vie, d'assurer sans discrimination aux populations résidentes et futures des conditions d'habitat, d'emploi, de services et de transports répondant à la diversité de ses besoins et de ses ressources, de gérer le sol de façon économe, de réduire les émissions de gaz à effet de serre, de réduire les consommations d'énergie, d'économiser les ressources fossiles d'assurer la protection des milieux naturels et des paysages, la préservation de la biodiversité notamment par la conservation, la restauration et la création de continuités écologiques, ainsi que la sécurité et la salubrité publiques et de promouvoir l'équilibre entre les populations résidant dans les zones urbaines et rurales et de rationaliser la demande de déplacements, les collectivités publiques harmonisent, dans le respect réciproque de leur autonomie, leurs prévisions et leurs décisions d'utilisation de l'espace. Leur action en matière d'urbanisme contribue à la lutte contre le changement climatique et à l'adaptation à ce changement ; Code de l'Urbanisme, art. L 110) ; *g* **Steuerung** [f] **der Stadtentwicklung** (Planerisches Eingreifen in die städtebauliche Entwicklung durch die ►Bauleitplanung auf Grund sozioökonomischer Veränderungen).

6686 urban design [n] *arch. urb.* (City planning process increasingly incorporating aesthetic aspects of qualtity in urban development after the completion of a phase of growth in urban areas with all its negative concomitants. Residents of towns are taking the design of their surroundings more and more into their own hands; ►village enhancement); *s* **diseño** [m] **urbano** (Conjunto de medidas de planificación orientadas a mejorar la calidad estética y vivencial de la ciudad. En creciente grado se entiende como una tarea de la comunidad en su conjunto de conformar su propio entorno residencial; ►mejora de pueblos); *f 1* **composition** [f] **urbaine** (Figuration tridimensionnelle d'une ville ou partie de ville, conçue et dessinée de façon suffisamment précise pour permettre la construction, et correspondant à une implantation sur un site réel ou décrit comme tel, compte tenu de ses accidents et particularités ; DUA 1996, 183) ; *f 2* **urban design** [m] (Expression utilisée aux États-Unis et en Angleterre désignant le processus de conception et de réalisation d'arrangements physiques permettant de maîtriser l'organisation formelle de la croissance urbaine à travers permanences et changements ; champ d'activité tentant de résoudre les questions posées par la mauvaise répartition et l'usage défectueux des ressources foncières ainsi que la destruction inutile du tissu historique, afin d'intégrer cohérence et beauté dans le domaine bâti ; DUA 1996, 813 ; ►mise en valeur d'un village) ; *g* **Stadtgestaltung** [f] (Planungsaufgabe, die verstärkt qualitative Gesichtspunkte in die Stadtentwicklung einbezieht, nachdem die Phase des unbedingten Wachstums im Städtebau mit all seinen negativen Begleiterscheinungen abgeschlossen ist. Wird vermehrt auch als Selbstgestaltung der eigenen Umwelt durch die Bewohner begriffen; ►Dorfgestaltung).

6687 Urban Design Framework Plan [n] *urb.* (An U.D.F.P. creates a vision for the development of an area consistent with the principles of an adopted spatial strategy. This illustrated development plan outlines the future sustainable development of a city and promotes regeneration, improvement and redevelopment, where appropriate. It gives a physical form to proposed land use and infrastructure, based upon design guidelines, and aims to secure high standards of design and development; to create investor confidence, by establishing a shared vision and consensus for action and providing a strategy for implementation. An **U.D.F.P.** draws on the skills of several professional disciplines including urban planning and design, architecture and landscape planning. The plan involves the preparation of realistic design concepts based on consultation, research and analysis. **In UK.**, an **U.D.F.P** is drawn up as part of Supplementary Guidance within the Local Development Framework. It may also include site development briefs and is important in the determination of planning application. The term is used in most English-speaking countries); *s* **programa [m] marco de desarrollo urbanístico** (≠) (En GB y D., documento adicional de planificación sin estatus legal definido, pero a menudo parte importante de un ▶plan general municipal de ordenación urbana. Constituye un instrumento estratégico del desarrollo urbanístico sostenible/sustentable de una ciudad y promociona la rehabilitación, mejora y reestructuración donde se considere adecuado. El programa proporciona un marco de diseño general para las áreas existentes y las de expansión del centro de la ciudad y una estrategia para guiar el desarrollo futuro de forma planeada y coordinada con diseño de alta calidad y creatividad, integrando áreas periféricas al marco propuesto. A su elaboración contribuyen expertos/as de diferentes disciplinas, como urbanismo, arquitectura y planificación del paisaje, desarrollando ideas y conceptos realistas de diseño basados en análisis, investigación y audiencias públicas); *f* **schéma [m] directeur de zones de développement urbain** (≠) (U.K./D., document cadre informel d'urbanisme servant de référence pour la définition et la programmation d'actions de développement soutenable dans un ou plusieurs quartiers, un centre-ville ou pour l'ensemble du territoire d'une commune dans le but de prendre en compte aussi bien la promotion de la qualité du paysage urbain, la préservation ou l'amélioration de la qualité de l'habitat, les fonctions climatiques et paysagères, que la modération de la circulation automobile et le développement des transports urbains ou la mise en place d'infrastructures sociales et culturelles ; dans le cadre de cette démarche a lieu un examen des règles d'urbanisme existantes afin de garantir leur compatibilité avec les exigences futures ; celle-ci est accompagnée d'une vaste et intensive participation des acteurs locaux, élus, propriétaires, artisants et commerçants au processus de planification) ; *g* **städtebaulicher Rahmenplan [m]** (Informelles Planwerk der Stadtplanung, das eine Gemeinde als Ganzes, quartiersbezogen oder z. B. zur Innenstadtentwicklung oder für definierte Planungs- und Entwicklungsprobleme erfasst und dabei auf Grund von Analysen und Bewertungen der an der Planung zu beteiligenden Fachdisziplinen Aussagen zu bestehenden und zukünftigen städtischen Strukturen neu trifft, darstellt und weiterentwickelt, um insgesamt für voraussehbare Bedürfnisse und Entwicklungen eine nachhaltige Lösung zu erreichen. Die Qualität des Stadtbildes, Erhaltung oder Verbesserung der Wohnqualität, Klima- und Landschaftsfunktionen werden genauso berücksichtigt wie Ziele der Verkehrsentwicklung und eine ausreichende Versorgung mit kulturellen und sozialen Infrastrukturen. In diesem Zusammenhang werden bestehende planungsrechtliche Festsetzungen überprüft, ob sie die zukünftigen Erfordernisse abdecken. Eine umfassende und intensive Beteiligung der Bürgerschaft, Grundstückseigentümer und Gewerbetreibenden begleitet den Planungsprozess).

6688 urban development [n] *urb.* (Generic term for ▶center city redevelopment [US]/town centre redevelopment [UK], ▶community expansion [US]/town expansion [UK], and ▶urban redevelopment dependent upon demographic, economic and cultural factors and requirements of a city); *s* **desarrollo [m] urbano** (1. Término genérico para ▶expansión urbana ▶rehabilitación urbana y ▶remodelación urbana; proceso que depende de las condiciones demográficas, económicas y culturales de una ciudad. 2. Planificación integrada para el desarrollo espacial coordinado de una ciudad o municipio); *f* **développement [m] urbain** (Terme générique englobant l'▶expansion urbaine, l'▶extension urbaine 1, le ▶remodelage urbain, la ▶rénovation urbaine et la ▶restauration urbaine, dépendant des données et nécessités démographiques, économiques et culturelles d'une agglomération) ; *g* **Stadtentwicklung [f]** (1. OB zu ▶Stadterweiterung, ▶Stadtsanierung und ▶Stadtumbau, abhängig von demografischen, wirtschaftlichen, soziologischen und kulturellen Gegebenheiten und Erfordernissen einer Stadt. Im Städtebaulichen Bericht der Bundesregierung vom Dezember 2004 wird **S.** als „ein dynamisches Zusammenwirken von Stadterneuerung, Stadtumbau und Stadterweiterung" beschrieben. 2. **S.** ist die zusammenfassende, gemeindliche Planung im Interesse einer koordinierten räumlichen Entwicklung); *syn. zu 2.* Stadtentwicklungsplanung [f].

urban development plan [n] [UK] *leg. urb.* ▶community development plan [US].

6689 urban ecology [n] *ecol. urb.* (Science of ecological functions and interrelationships within the urban habitat; ▶ecology); *s* **ecología [f] urbana** (Ciencia que estudia las sociedades urbanas utilizando el modelo de ecosistema, es decir considerando a aquéllas como un todo interrelacionado, en el que el ser humano es tratado como cualquier otra especie natural, aunque más activa, debido a su capacidad de crear cultura, que habita en un contexto de factores físicos [ambientales, energéticos y mecánicos] que codeterminan su desarrollo y la calidad ambiental reinante en ellas; cf. DINA 1987; ▶ecología); *f* **écologie [f] urbaine** (Science étudiant les relations entre l'homme et son cadre de vie dans l'espace urbain. **F.**, terme aussi utilisé pour définir une politique visant à promouvoir des démarches globales pour une gestion environnementale de l'espace urbain dans le cadre d'une charte pour l'environnement [circ. n° 94-48 du 11 mai 1994] ; ▶écologie) ; *g* **Stadtökologie [f]** (Wissenschaft von den ökologischen Funktionszusammenhängen im Lebensraum Stadt; ▶Ökologie); *syn.* Urbanökologie [f].

6690 urban environment [n] *urb.* (▶urban setting); *syn.* urban milieu [n]; *s* **medio [m] ambiente urbano** (▶escenario urbano); *syn.* entorno [m] urbano; *f* **environnement [m] urbain** (▶environnement bâti) ; *syn.* milieu [m] urbain (±) ; *g* **städtische Umwelt [f]** (▶städtebauliches Umfeld).

6691 urban expansion [n] *urb.* (Extension of land development; ▶urban sprawl); *s* **expansión [f] urbana (2)** (Extensión de la superficie de una ciudad; ▶urbanización caótica); *syn.* crecimiento [m] urbano; *f* **expansion [f] urbaine (2)** (Extension affirmée hors des frontières de la ville au détriment des zones rurales ; ▶accroissement désordonné d'une agglomération) ; *g* **Ausbreitung [f] einer Stadt** (Ausdehnung einer Stadt/eines Siedlungsgebietes; sich ausbreiten [vb]; ▶ungeordnete Ausbreitung einer Stadt); *syn.* Erweiterung [f] einer Stadt, Vergrößerung [f] einer Stadt.

urban fabric [n] *arch. urb.* ▶cityscape.

urban forest [n] *urb.* ▶town forest [UK].

6692 urban forestry [n] *adm. for. hort. landsc. urb.* (1. Science, art and practice of managing and using natural resources on forest lands for human benefit in urban areas as well as trees on roadsides and all kind of public green open spaces, such as parks and gardens; it advocates the role of trees as a critical part of urban infrastructure. U. f. is practiced by municipal and commercial arborists, municipal and utility foresters; it has its origins in the **U.S.**, where federal policy is overseen by the USDA Forest Service; work is funded by non-profit organisations, private donations and government grants; some of the

U

earliest urban forestry and forest conservation legislation was carried out in New England at the end of the nineteenth century with the introduction of tree warden laws); **2. near-natural management [n] of green spaces** (Care of individual green areas, or parts thereof, whereby the maintenance is highly differentiated according to the use of an area. As a result of this, for example, not all lawn and meadow areas are mown at the same time and same intervals any more, but rather divided into different mowing periods, once per year, twice per year or several times. Ecologically such a system of mowing encourages biodiversity in dense urban areas and allows the population to experience nature and its rich diversity of flora and fauna at close hand. This change in the principles of maintaining grassed areas was first practised in Germany in 1977, on the occasion of the Stuttgart Federal Garden Show in that year. It represented a huge break in the conventional maintenance of public parks, but it was gradually adopted in the 1980s and with the need to curb public spending in the 1990s, the principles became widespread. The switching of intensive cultivation of many green areas to more extensive maintenance, has had a long-lasting effect on the appearance of German towns and cities); *s 1* **gestión [f] de zonas verdes públicas** (Aplicación de todos los métodos y todas las medidas necesarias para planificar, construir, cuidar y mantener los espacios verdes públicos, optimizando el empleo de recursos, para alcanzar el mayor grado posible de disponibilidad de zonas verdes y de recreo cercanas a las áreas residenciales como una red de diferentes tipos de espacios libres [parques, zonas de juegos (infantiles), campos de deporte, huertos recreativos urbanos, cementerios, superficies de agua, bosques, campos agrícolas, espacios naturales protegidos, etc.] para el beneficio público, que responda a las necesidades de uso ecológicas, económicas y sociales así como a las exigencias higiénicas de las ciudades. A la **g. de zz. vv. pp.** corresponde además realizar medidas de educación de la población sobre el valor de las áreas verdes para la sociedad y establecer una administración de zonas verdes autónoma dentro de la administración municipal, incluido su financiamiento. El término angloamericano **silvicultura [f] urbana** se ha establecido en los países escandinavos y se utiliza a menudo como sinónimo de planificación de zonas verdes urbanas en general y de gestión de zonas verdes en particular. En ciudades de EE.UU. donde no se había desarrollado tan diferenciadamente la gestión de espacios libres como en Europa, los ingenieros forestales, basándose en sus conocimientos sobre los árboles, ampliaron su campo de actividad a las ciudades); *s 2* **gestión [f] cuasi-natural de zonas verdes** (Gestión de zonas verdes en la cual los espacios verdes resp. partes de ellos son cuidados diferenciadamente teniendo en cuenta los usos. Esto supone que p. ej. no se cortan los céspedes y las campas de hierba con la misma regularidad, sino que se diferencian entre superficies con un, dos, tres o más cortes anuales. Con este sistema de corte orientado a criterios ecológicos se promueve la biodiversidad en áreas urbanas de gran densidad, y se posibilita a la población la vivencia de la naturaleza en su diversidad florística y faunística. Este cambio de paradigma en el cuidado de zonas verdes se puso en práctica por primera vez en 1977 en Stuttgart, Alemania, en el contexto de la Exposición Federal de Horticultura y supuso un cambio sustancial en el cuidado y mantenimiento convencional de zonas verdes públicas. En los años 1980 se extendió paulatinamente y en los 1990 se amplió su aplicación debido a la necesidad de consolidar los presupuestos públicos. Gracias a la reclasificación del estatus de muchas zonas verdes de cuidado intensivo a cuidado extensivo, se ha cambiado notablemente el aspecto de muchas ciudades); *f 1* **gestion [f] des espaces verts publics** (Modes d'actions touchant la conception, la réalisation et l'entretien des espaces verts urbains avec l'objectif de proposer aux citadins un large éventail d'espaces verts de proximité et d'espaces de détente constituant un réseau varié de différents types d'espaces [parcs, squares et autres jardins, aires ludiques, sportives, familiales, festives, pédagogiques, de calme et de repos, jardins ouvriers, cimetières, espaces aquatiques (bords de rivières, ruisseaux, lacs, mares) et le patrimoine arboré et les espaces naturels — arbres d'alignement, fourrés, haies bocagères, zones boisées, prairies, — territoires agro-pastoraux] tenant compte des besoins écologiques, économiques et sociaux ainsi que du bien-être des utilisateurs ainsi que de l'hygiène urbaine ; le terme de « **foresterie urbaine** » n'est pas couramment utilisé dans la terminologie d'une grande partie des pays européens ; ce terme angloaméricain s'est établi dans les pays scandinaves ou au Canada, pays fortement marqués par la forêt ou encore dans les pays en développement où l'urbanisation croît à un taux accéléré dans des sociétés en grande partie rurale [terme utilisé p. ex. dans les documents de la FAO]. **U.S.,** les villes nordaméricaines ne possédant pas une aussi grande variété d'espaces verts qu'en Europe, la gestion des espaces verts urbains est réalisée par les forestiers qui grâce à leur savoir en matière d'arboriculture ont trouvé là un champ d'action supplémentaire) ; *syn.* gestion [f] des arbres et espaces boisés en milieu urbain [CDN], foresterie [f] urbaine [CDN] ; *f 2* **gestion [f] différenciée des espaces verts publics** (Mode d'actions diversifiées et coordonnées d'entretien des espaces verts selon leur spécificité paysagère et leur fonction, usage et fréquentation dans l'espace urbain dont l'objectif fait une part importante au principe écologique d'entretien des espaces verts ; mode de gestion adopté dans les années 1980 par un grand nombre de villes européennes ; dans les années 1990 avec l'émergence du concept de « ville durable » cette gestion a évolué vers une naturalisation de l'espace vert public dans la ville intégrant des techniques diverses en vue de conserver ou d'augmenter la biodiversité ; la gestion différenciée est devenue un moyen de favoriser le développement d'une diversité faunistique et floristique, le rétablissement des équilibres biologiques et la protection de la biodiversité ordinaire au sein même des espaces bâtis ; cette gestion est en rupture avec les pratiques traditionnelles d'entretien des espaces verts, fortes consommatrices d'engrais, de produits phytosanitaires et d'eau ; dans le contexte d'une gestion différenciée, un espace vert et les éléments, qui le composent, sont entretenus selon une classe précise de gestion ; exemple de la commune de Meylan par intensification décroissante : classe I — espaces de « prestige » [parterres de fleurs en centre-ville, haies et massifs arbustifs soignés, parcs historiques, etc.], classe II — espaces à entretien soigné [pelouses tondues régulièrement, fleurissement par plantes annuelles en taches à géométrie contrôlée, abords d'avenues, etc.], classe III — espaces à pratiques horticoles [p. ex. parcs de loisirs, de promenade, terrains d'entraînement], classe IV — espaces à pratiques extensives : espaces de jeux de découvertes, [prairies, bosquets éclaircis, etc.], classe V — Zone d'intervention limitée [talus subnaturels, abords de sentiers, promenades de sous-bois, etc.], classe VI — zones de simple inventaire des milieux, pouvant être classées en Zone Naturelle d'Intérêt Floristique et Faunistique [p. ex. zone humide, ripisylve] ; cf. COT 2006) ; *g 1* **Management [n] öffentlicher Grünflächen** (Unter Optimierung des Ressourceneinsatzes die Anwendung aller erforderlichen Methoden und Maßnahmen zum Planen, Bauen sowie zur Pflege und Unterhaltung des öffentlichen städtischen Grüns, um einen möglichst hohen Versorgungsgrad nutzbarer wohnungsnaher Grün- und Erholungsflächen als Netz unterschiedlicher Freiraumtypen [Parkanlagen, Spielflächen, Sportstätten, Kleingärten, Friedhöfe, Wasserflächen, Wald, landwirtschaftliche Flächen, Natur- und Landschaftsschutzgebiete etc.] für die Daseinsvorsorge vorzuhalten, der den ökologischen, wirtschaftlichen und sozialen Nutzungsbedürfnissen sowie stadthygienischen Anforderungen entspricht. Dazu gehören ferner alle Maßnahmen der Verankerung des Wertes städtischen Grüns in unserer Gesellschaft und der Etablierung einer eigenständigen Grünflächenverwaltung innerhalb kommunaler Ämterstrukturen und ihrer Finanzierung. Im dt. Sprachgebrauch werden im Vergleich zur *urban forestry* städtische Waldflächen bei öffentlichen Grünflächen nicht mitgedacht; dieser angloamerikanische Begriff hat sich auch in den skandinavischen Ländern etabliert und wird häufig mit städtischer Grünplanung im Allgemeinen und Grünflächenmanagement im Besonderen gleichgesetzt. In nordamerikanischen Städten war ein Freiflächenmanagement nicht so differenziert entwickelt wie in Europa, weshalb die Förster auf Grund ihres Wissens um Bäume ein erweitertes Betätigungsfeld entdeckt und ausgebaut haben); *syn.* Freiflächenmanagement [n], Grünflächenbewirtschaftung [f], Grünflächenmanagement [n], Management [n] von Stadtgrün, Planung [f], Bau [m], Pflege [f, o. Pl.] und Unterhaltung [f, o. Pl.] öffentlicher Grünflächen, Management [n] des städtischen Baum- und Waldbestandes sowie öffentlicher Grünflächen, Stadtgrünmanagement [n]; *g 2* **naturnahe Grünflächenbewirtschaftung [f]** (Management, bei dem die einzelnen Grünflächen resp. deren Teile nutzungsorientiert sehr differenziert gepflegt werden mit der Folge, dass z. B. nicht alle Rasen- und Wiesenflächen zur gleichen Zeit in gleichen Zeitabständen gemäht werden, sondern in einschürige, zweischürige, dreischürige und mehrschürige Wiesen und Rasenflächen eingeteilt werden. Durch dieses mit ökologischen Gesichtspunkten durchgeführte Mähregime wird die Biodiversität in dichten Siedlungsgebieten gefördert und für die Bewohner Natur in ihrer floristischen und faunistischen Vielfalt stärker unmittelbar erlebbar gemacht. Dieser Paradigmenwechsel in der öffentlichen Grünpflege wurde erstmals 1977 in der Bundesgartenschau in Stuttgart vorgeführt und stellte einen gewaltigen Bruch mit der konventionellen Pflege und Unterhaltung öffentlicher Grünanlagen dar; sie setzte sich allmählich in den 1980er-Jahren durch und wurde in den 1990er-Jahren unter dem Gesichtspunkt der Haushaltskonsolidierung öffentlicher Haushalte weiter ausgebaut. Durch die Herunterzonung vieler intensiv gepflegter Grünflächen auf extensive Bewirtschaftung hat sich das Stadtbild in vielen Gemeinden nachhaltig verändert).

6693 urban fringe [n] *urb.* (Particular term for an area of land on the boundary of an urban area, which forms a transition zone or an interface between the built-up area and the open landscape or countryside. Viewed as a landscape in itself, it is often an area in which the interaction of urban and rural land uses leads to conflicts; ►edge of a settlement); *syn.* fringe [n] of a city/town, *in the vernacular* outskirts [npl] of a city/town; *s* **franja [f] rururbana** (Término acuñado por WEHREIN [1942] para describir un área de transición entre los usos del suelo urbano bien diferenciados y el área dedicada a la agricultura. PRYOR define este espacio como área de transición en el uso del suelo y en las características demográficas y sociales, entre el espacio urbano edificado de la ciudad central y el *umland* rural, caracterizado por la casi ausencia de habitantes y usos de suelo agrarios, por una incompleta gama de servicios y por la falta de planificación y zonificación coordinadas. PRYOR distingue una **franja urbana** de mayor densidad que el conjunto de la franja, con fuertes movimientos pendulares de los trabajadores, y una **franja rural**, con menor densidad y crecimiento de la población, menor proporción de tierras sin cultivar y menores cambios en el uso del suelo; cf. DGA 1986; ►borde de la ciudad); *syn.* franja [f] perimetral de la ciudad; *f 1* **frange [f] urbaine** (Composante de l'►espace périurbain désignant une zone aux frontières floues, espace de mitage où la friche est présente d'une façon dispersée, paysage en processus d'urbanisation, territoire mi-urbain mi-rural où les activités de type urbain dominent les activités rurales ; cette zone est bordée par **la frange [f] rurale** composée de zones

aux limites difficiles à tracer et où les activités de type rural dominent les activités urbaines) ; *f 2* **limite [f] du périmètre urbain [CDN]** (Délimite le territoire municipal entre **1.** l'aire urbanisée qui comprend une forte concentration de la population, du bâti, des infrastructures et des activités économiques et **2.** l'aire rurale qui comprend une faible concentration de la population et qui est caractérisée par des activités extensives, c.-à-d. réalisées sur de grandes surfaces, comme l'agriculture, la foresterie, l'extraction des ressources naturelles, les activités récréatives ; cf. Loi sur l'aménagement et l'urbanisme qui oblige les municipalités régionales de comté [MRC] à définir un périmètre urbain dans leur schéma d'aménagement et de développement ; ►limite de l'agglomération, ►limite du site construit) ; *g* **Übergangsbereich [m] einer Stadt zur offenen Kulturlandschaft** (≠) (Bereich beiderseits der Stadtgrenze zweier Nachbargemeinden, der sich deutlich von den Ortsteilen, die im Zusammenhang bebaut sind, dadurch unterscheidet, dass er wesentlich dünner besiedelt ist und oft Konflikte zwischen land- und forstwirtschaftlichen Nutzungen und Vorhaben für Siedlungserweiterungen verursacht; ►Siedlungsrand); *syn.* Übergangsbereich [m] einer Stadt zum ländlichen Raum.

urban gardener [n] [US] *urb. sociol.* ►community gardener [US].

urban green space planning [n] *landsc. urb.* ►general urban green space planning.

6694 urban green spaces [npl] *landsc. urb.* (Sum of public and private open spaces primarily characterized by vegetation within an urban area; ►green space, ►public green space); *s* **tejido [m] vegetal urbano** (Conjunto de áreas verdes públicas y privadas en una zona urbanizada; ►zona verde, ►parques y jardines públicos); *syn.* verde [m] urbano; *f* **tissu [m] végétal urbain** (IDF 1988, 294 ; ensemble des ►espaces verts privés et publics contribuant à définir l'apparence végétale de la ville ; ►espace vert public) ; *g* **Stadtgrün [n, o. Pl.]** (Gesamter vegetationsbestimmter, öffentlicher und privater Freiraumbestand im Siedlungsbereich. ‚Städtisches Grün' bedeutet i. d. R. ‚öffentliche Grünflächen' als Abgrenzung zu privaten Grünflächen; ►Grünfläche, ►öffentliche Grünfläche); *syn.* Grün [n, o. Pl.] einer Stadt.

6695 urban greenway [n] [US] *lands. urb.* (**1.** On a local level, a narrow green area with wide building setbacks from an avenue or walk, within a community for recreation or pleasure, for mitigating the effects of overdeveloped hard landscape elements, such as buildings and roads, or connecting two or more urban centers. **2. In U.S.**, urban greenways are designed to connect ecological systems or networks, cultural institutions, economic markets, and inter-modal transportation networks providing pedestrian and cycling access to green corridors that enhance natural habitat interconnectivity, while serving cultural needs as well); *syn.* roadside green connection [n] [UK]; *s* **conexión [f] verde (2)** (Espacio verde estrecho y alargado que conecta dos ciudades o se encuentra dentro de una situado a lo largo de calles o avenidas y se utiliza para fines recreativos y para mitigar el efecto de la excesiva construcción con materiales inertes, como edificios y superficies asfaltadas); *f* **espaces verts [mpl] (lineaires) de liaison** (Espaces verts d'accompagnement [allées, bandes engazonnées ou plantées] ou alignement de jardins de façade le long de la voirie reliant entre eux plusieurs quartiers et pouvant former une partie d'un réseau structurant hiérarchisé urbain d'espaces verts) ; *g* **Grünverbindung [f] entlang von Straßen** (Verkehrsgrünflächen und/oder breite Vorgärten, die städtische Wohngebiete miteinander verbinden oder einen Teil eines innerstädtischen Grünverbundnetzes darstellen).

U

6696 urban habitat mapping [n] *conserv. phyt. zool.* (Process of defining and delineating the extent and distribution of all significant urban natural areas with the intention of creating a habitat network; e.g. vacant lots, abandoned quarries, steep slopes of transportation corridors, ▶derelict land; ▶habitat mapping; *generic term* ▶mapping); *syn.* mapping [n] of urban wild sites [also US], urban wilds inventory [n] [also US]); *s* **inventario [m] de biótopos urbanos** (Registro cartográfico de los biótopos dentro del tejido urbano, en solares, ▶ruinas industriales, terraplenes de vías de ferrocarril, etc., con el fin de protegerlos y crear una red de zonas protegidas en la ciudad; ▶inventario cartográfico de biótopos; *término genérico* ▶inventario cartográfico); *f* **cartographie [f] des biotopes urbains** (Relevé et représentation cartographique des biotopes en milieu urbain, p. ex. les pelouses urbaines extensives, marges des chaussées de routes, talus de voies ferrées, terrains vagues ou abandonnés à la périphérie de l'agglomération, anciens jardins, ▶friches industrielles afin de définir un réseau de biotopes ; ▶cartographie des biotopes, ▶cartographie écobiocénotique ; *terme générique* ▶cartographie) ; *g* **Stadtbiotopkartierung [f]** (Kartenmäßige Erfassung von Biotopen eines Stadtgebietes zur Erarbeitung eines Biotopverbundnetzes im innerstädtischen Bereich, z. B. in Baulücken, aufgelassenen Steinbrüchen, auf Standorten von ▶Industriebrachen, auf Böschungen an Eisenbahndämmen oder in Böschungseinschnitten; ▶Biotopkartierung; *OB* ▶Kartierung).

6697 urban improvement [n] *urb.* (Upgrading the overall quality of the urban environment, e.g. through the enhancement of parks and open spaces, infrastructure, etc.; ▶center city redevelopment [US]/urban regeneration [UK], ▶landscape urbanism); *s* **mejora [m] urbanística** (Mejora de la situación urbanística por medio de creación de espacios verdes, de nuevas infraestructuras, etc.; ▶rehabilitación urbana, ▶urbanismo paisajista); *syn.* revalorización [f] urbanística; *f* **revalorisation [f]** (Amélioration de la qualité de la structure urbaine par exemple en favorisant le développement et la création d'espaces verts ; ▶rénovation urbaine, ▶restauration urbaine, ▶urbanisme paysager) ; *syn.* mise [f] en valeur ; *g* **Aufwertung [f] einer Stadt** (Verbesserung der städtebaulichen Gesamtsituation, z. B. durch Ausbau der Grünflächensysteme, der Infrastruktur etc.; ▶grünraumorientierte Stadtplanung, ▶Stadtsanierung).

6698 urbanization [n] *urb. plan.* (TGG 1984, 105; continual increase in the size of cities and their amalgamation into a ▶metropolitan region or ▶conurbation); *s* **urbanización [f]** (Proceso de crecimiento continuo de las ciudades debido a la concentración de la población en las áreas urbanas; en los países en desarrollo sobre todo debido a la migración de gran cantidad de población a las megaciudades. Este proceso puede conllevar la unión de dos o varias ciudades; ▶aglomeración urbana; ▶área metropolitana); *s 2* **rururbanización [f]** (Distribución de la población con actividades económicas y mentalidades urbanas sobre espacios rurales. En la geografía sajona se habla de *dispersed city* para referirse a este fenómeno. Este proceso se ve facilitado por la creciente accesibilidad motivada por los medios de transporte públicos y privados que permiten movimientos migratorios pendulares superiores a los 100 km de distancia entre el lugar de residencia [rural] y el lugar de trabajo [ciudad central de un área metropolitana]. El proceso de **r.** se produce en los espacios próximos a las ciudades. La extensión territorial del mismo depende del tamaño de la ciudad central, con su dinamismo y el grado de desarrollo del sistema de transporte; cf. DGA 1986); *f 1* **urbanisation [f]** (Concentration croissante de la population dans un ensemble d'agglomérations urbaines pour former une ▶zone de concentration urbaine dans laquelle les limites entre la ville et l'espace rural finissent par s'effacer ; *termes spécifiques* suburbanisation, [urbanisation continue de

l'espace autour des villes], périurbanisation [urbanisation continue des franges des agglomérations] métropolisation ; LA 1981, 473 ; ▶aire urbaine) ; *f 2* **rurbanisation [f]** (Transformation des zones périurbaines procédant d'une nouvelle répartition des populations citadines, marquée par un exode vers la campagne [autour des noyaux de l'habitat rural] au détriment du centre) ; *g* **Verstädterung [f]** (Immer größere Ausbreitung von Städten und deren Zusammenwachsen zu ▶Ballungsgebieten; *Kontext* die **V.** des ländlichen Raumes nimmt ständig zu; die Landschaft verstädtert immer mehr; ▶Verdichtungsraum); *syn.* Urbanisierung [f], Urbanisation [f].

6699 urban landscape [n] *urb.* (Densely populated urban area with a high rate of sealing of the ground in comparison with the open countryside. The quality of **u. l.** is a function of a combination of the diverse social, ecological, aesthetic and cultural requirements imposed upon the built area in association with public and private green areas; ▶open country, ▶urban environment); *s* **paisaje [m] urbano** (Paisaje determinado por la densa urbanización de ciudades y grandes aglomeraciones caracterizado por la gran cantidad de construcciones e infraestructuras, una gran proporción de suelo sellado y, al contrario que en el paisaje natural, presencia reducida de zonas verdes y vegetación; la calidad del **p. u.** está determinada por la integración óptima de las funciones sociales, económicas, ecológicas y estéticas que deben cumplir el conjunto de las edificaciones e infraestructuras junto con las áreas verdes públicas y privadas; ▶campiña, ▶medio ambiente urbano); *f* **paysage [m] urbain** (Espaces fortement urbanisés des villes et grandes agglomérations caractérisés par une importante imperméabilisation du sol par comparaison ou opposition à la ▶campagne 2 ; ▶environnement urbain) ; *g* **Stadtlandschaft [f]** (Siedlungsgebiet mit einem ständig wechselnden Nebeneinander von Gebäuden, Gebäudekomplexen und Freiräumen, das im Vergleich zur ▶freien Landschaft mit einem sehr hohen Versiegelungsgrad gekennzeichnet ist; die Qualität von **S.** zeigt sich in der optimalen Integration der vielfältigen sozialen, ökologischen, ästhetisch-künstlerischen Anforderungen, die an das Gebaute im Verbund mit städtischen und privaten Grün- und Freiflächen gestellt werden; ▶städtische Umwelt).

6700 urban land use category [n] *leg. urb.* (Generic term covering each of the following land uses; e.g. ▶residential land use, commercial land use, ▶industrial land use, ▶recreational land use, ▶institutional land use, and ▶mixed land use 1, which are shown on a ▶comprehensive plan [US]/Local Development Framework [LDF] [UK]); *syn.* zoning category [n] [also US], general land-use area [n] [D] (FBC 1993); *s* **categoría [f] de suelo urbano** (Término genérico para los usos del suelo según la categoría de zonificación: ▶zona residencial, ▶suelo residencial, ▶suelo urbano de uso mixto, ▶suelo industrial, ▶reserva de terreno para equipamiento público, etc. En D. es un término legal utilizado a nivel del ▶plan general municipal de ordenación urbana); *f* **zone [f] urbaine (2)** (F., *Plan local d'urbanisme [PLU]/Plan d'occupation des sols [POS]* on distingue deux catégories de zones pour lesquelles s'appliquent les règles d'urbanisme : les zones urbaines [zones U] et les zones naturelles [zones N] ; à la différence de l'ancien schéma directeur d'aménagement et d'urbanisme [SDAU], le ▶schéma de cohérence territoriale [SCOT] ne comporte pas de carte de destination générale des sols ; néanmoins, l'article R.122-3 permet au SCOT de délimiter, dans des documents graphiques, les espaces et sites naturels ou urbains à protéger ou des zones pouvant être ouvertes à l'urbanisation en constituant l'enveloppe des zones U et AU des PLU ; cette délimitation facultative peut renforcer le caractère opposable des dispositions prises dans le document d'orientations générales en lui apportant une traduction géographique plus précise ; **D.**, terme générique pour les quatre catégories de zones prévues dans le « Flächennutzungsplan » [schéma de cohérence territoriale allemand] ; ▶zone urbaine destinée à l'habitation, ▶zone d'activités mixtes, ▶zone à vocation d'activités économiques, ▶emplacement réservé pour ouvrage public, installation d'intérêt général ou espace vert) ; *g* **Baufläche [f]** (OB zu den vier im ▶Flächennutzungsplan dargestellten Bauflächenarten gemäß BauNVO: ▶Wohnbauflächen [W], ▶gemischte Bauflächen [M], ▶gewerbliche Bauflächen [G], ▶Sonderbauflächen [S]).

urban land use planning [n] [UK] *adm. leg. urb.* ▶community development planning [US].

urban milieu [n] *urb.* ▶urban environment.

6701 urban park [n] *recr. urb.* (**1.** Large park within a city, serving the recreational needs of neighbo(u)ring residents; ▶park 1, ▶neighborhood park [US]/district park [UK], ▶inner city park, ▶public park, ▶suburban park. **2.** An extensiv park in an big agglomeration with a large service area [US]/catchment area [UK]; e.g. Central Park in New York, is called **metropolitan park**); *s* **parque [m] urbano** (Gran zona verde situada en el interior de una ciudad cuya función principal es el recreo de la población; ▶parque, ▶jardín público, ▶jardín de barrio, ▶parque público, ▶parque suburbano); *f 1* **parc [m] urbain** (*Terme générique* grand espace vert urbain composé de plusieurs espaces en majorité créés pour offrir des activités de repos, de détente et de loisirs spécialisés ; selon leurs dimensions et leur aire d'influence on distingue les ▶jardins publics, les ▶parcs de voisinage, les squares, les ▶parcs de quartiers, les parcs centraux, les ▶parcs publics, les ▶parcs périurbains ; DUA 1996, 540) ; *f 2* **parc [m] central** (Parc de quelques dizaines d'hectares accessible dans un rayon de 15 minutes de marche ; ▶parc) ; *g* **Stadtpark [m]** (Großflächige Grünanlage innerhalb einer Stadt, die der wohnungsnahen, mehrstündigen bis Ganztageserholung dient; ▶Park, ▶Stadtgarten, ▶Stadtteilpark, ▶Volkspark, ▶Vorortpark).

6702 urban planner [n] [US] *prof.* (Planner concerned with the spatial development of a town or city, i.e. its land use, as well as the development of a town's character; ▶urban planning); *syn.* city planner [n] [US], community planner [n] [US], town planner [n] [UK]; *s* **urbanista [m/f]** (Profesional que se dedica a estudiar, diseñar y gestionar los procesos de desarrollo urbano; ▶urbanismo); *f* **urbaniste [m]** (Aménageur réalisant des missions d'▶urbanisme concernant l'urbanisme réglementaire, les études préalables et opérationnelles en milieu urbain et rural) ; *g* **Stadtplaner/-in [m/f]** (Jemand, der durch seine planerische Arbeit die räumliche Entwicklung einer Gemeinde, d. h. die Bodennutzung wie auch die bauliche Entwicklung lenkt und die Gestaltung des Ortsbildes mitbestimmt; ▶Städtebau).

6703 urban planning [n] *urb.* (**1.** Comprehensive planning activity of a municipality in which the spatial development, i. e., land use, as well as the aesthetic values [townscape and visual quality] are determined within the area of a city's authority with the intention of advocating environmentally appropriate policies while coordinating local community or district plans and initiatives. **In U.S., u. p.** aims to promote social, economic, environmental and spatial equity using public and private resources at federal, state and local levels of government); **2. community planning [n]** (Planning which focuses on a district level of a city or town, and typically serves the needs of a local population within the district); *syn.* city planning [n] [also US], community planning [n] [also US]; *s* **urbanismo [m]** (Conjunto de medios destinados a prefigurar el desarrollo futuro de una ciudad. En un sentido general se aplica tanto a los trabajos de ingeniería, como a los planes de las ciudades o a las formas urbanas de cada época; DGA 1986. *Términos específicos* **u.** concertado, **u.** culturalista,

u. marginal, **u.** progresista); *f* **urbanisme [m]** (Ensemble de mesures techniques, administratives, économiques et sociales à la disposition des communes leurs permettant de définir une utilisation harmonieuse, rationnelle et humaine de l'espace communal avec le souci de l'amélioration du cadre de vie et du respect de l'environnement) ; *g* **Städtebau [m]** (**1.** Zusammenfassende Planungstätigkeit und Gestaltung der baulichen Umwelt einer Gemeinde, mit der sie ihre *räumliche* Entwicklung, d. h. die Bodennutzung wie auch die Gestaltung des Orts- und Landschaftsbildes innerhalb ihres Hoheitsgebietes bestimmt. Der traditionelle Begriff ‚S.' stammt aus historischer Zeit, als Landesfürsten mit Hilfe von Fachleuten der „Stadtbaukunst" ganz neue Städte oder deren Teile planten und bauen ließen. Heute spricht man mehr von **Stadtplanung/Ortsplanung [f]**, weil in demokratischen Staaten außer bloß *baulichen* Gesichtspunkten ökologische und sozioökonomische gleichermaßen berücksichtigt werden müssen. **2.** Jahrhunderte waren urbane Zentren vom Wachstum bestimmt, weshalb sich Stadtplanung als Disziplin mit dem drängenden Problem der Wachstumsbewältigung entwickelte. Bei den seit vielen Jahren sichtbar werdenden Leerständen und der weiter fortschreitenden Suburbanisierung der Städte muss beim **S.** eine Konzentration der städtebaulichen Projekte auf die Innenstadt und eine Attraktivitätssteigerung öffentlicher Räume als neue Wertschätzung des *Urban Designs* erfolgen. **3.** Die Tätigkeiten eines Planers im Rahmen der ▶Bauleitplanung werden **städtebauliche Leistungen** [Teil V HOAI 2002] oder **bauleitplanerische Leistungen** [Teil 2 HOAI 2009] genannt); *syn.* Stadtplanung [f], Ortsplanung [f].

urban reconstruction [n] *urb.* ▶urban redevelopment.

urban recreation [n] *plan. recr.* ▶outdoor urban recreation.

urban recreation facilities [npl] *plan. recr* ▶concentration of urban recreation facilities.

6704 urban redevelopment [n] *urb.* (**1.** Reorganization of the structure of a city through compulsory acquisition of blighted buildings, their clearance, and replacement according to established guidelines; e.g. the mixing or separation of places of employment and residential areas, or the transformation of central areas into residential zones to combat land speculation. **2.** The exodus of inhabitants of towns and cities leading to a decline in the urban population, and thus an increase in the number of vacant homes and commercial properties, requires new strategies and types of open spaces on land, which has become empty after demolition. Such strategies have to consider not only the appearance of a city, the actual use of open spaces, as well as ecology and sustainability, but also the limited financial capacity of many municipalities to maintain the additional land. Strategies have to be adopted to prevent the abandonment and dereliction of land in many urban areas; ▶neighborhood redevelopment [US]/district redevelopment [UK]. **3** For centuries urban centers have been characterized by growth, which is why urban planning developed as a discipline aimed at combating the pressing problems of growth in cities. With the visible concentration of empty properties in downtown areas, **u. r.** is nowadays concerned with discouraging the further progressive suburbanization of cities by concentrating projects in the center of cities and increasing the attractiveness of public spaces, in which there is a new appreciation for urban design); ▶center city redevelopment [US]/urban regeneration [UK]); *syn.* urban reconstruction [n]; *s* **remodelación [f] urbana** (Transformación de las funciones y del mosaico de usos del suelo de una ciudad orientándose en determinado modelo urbanístico con el fin de adaptarla a las necesidades y demandas contemporáneas; ▶rehabilitación de barrios, ▶rehabilitación urbana); *f* **remodelage [m] urbain** (Réorganisation des espaces urbanisés en tenant compte notamment de l'équilibre entre emploi, services et habitat ;

U

▶réaménagement des quartiers, ▶réhabilitation des quartiers, ▶rénovation urbaine, ▶restauration urbaine) ; *syn.* renouvellement [m] urbain, recomposition [f] urbaine ; *g* **Stadtumbau [m]** (**1.** Entsprechend dem städtebaulichen Leitbild die Umstrukturierung des Funktions- und Nutzungsmusters in einer Stadt zum Wohl der Allgemeinheit; z. B. Mischung oder Entmischung von Arbeitsstätten/Dienstleistungsbetrieben und Wohnungen oder zur Bekämpfung der Bodenspekulation die Umnutzung von Teilen der Citybereiche in Wohnbereiche resp. die Erhöhung des Wohnanteils zur Verbesserung der Wohn- und Arbeitsverhältnisse, Rückbau baulicher Anlagen, die keiner anderen Nutzung zugeführt werden können; ▶Stadtsanierung. **2.** Durch Stadtflucht verursachter Bevölkerungsrückgang und damit verbundener zunehmender Wohnungs- und Geschäftsraumleerstand — besonders in vielen ostdeutschen Städten — erfordern neue Strategien der Freiraumtypologien auf abzureißenden Flächen, die nicht nur die Stadtgestalt und Freiraumnutzung, den Naturhaushalt und die Nachhaltigkeit, sondern auch die engen finanziellen Rahmenbedingungen vieler Kommunen bei der Pflege und Unterhaltung zusätzlicher Flächen berücksichtigen, um der Verbrachung und Verödung vieler Stadtquartiere entgegenzuwirken; cf. hierzu §§ 171a-171d BauGB; ▶Stadtteilsanierung).

6705 urban refuse [n] *envir.* (Generic term for household rubbish, kitchen waste, rubbish-like waste from light industry, bulky household waste material, market refuse, road sweepings, mud from road gully mudtraps, etc.); *syn.* urban waste [n], urban trash [n] [also US]; *s* **residuos [mpl] sólidos urbanos** (Término genérico para los residuos domésticos, los de talleres y oficinas de tipo doméstico, los de mercados, las barreduras de las calles, etc.); *syn.* desechos [mpl] urbanos; *f* **déchets [mpl] urbains** (Terme générique regroupant les ordures ménagères, les déchets d'origine commerciale, artisanale ou industrielle assimilés aux déchets ménagers, les déchets volumineux ou encombrants, les déchets des marchés, les déchets provenant du nettoyage de la voirie etc.) ; *syn.* résidus [mpl] urbains, déchets [mpl] municipaux ; *g* **Siedlungsabfall [m]** (OB zu Hausmüll, Büro- und hausmüllähnliche Gewerbeabfälle, Sperrmüll, Marktabfälle, Straßenkehricht, Garten- und Parkabfälle, Bauabfälle, Klärschlamm, Fäkalien, Fäkalschlamm, Rückstände aus Abwasseranlagen, Wasserreinigungsschlämme und Sinkkastenschlämme sowie solche Abfälle, die auf Grund ihrer Beschaffenheit oder Zusammensetzung den Abfällen von Haushalten ähnlich sind); *syn.* Siedlungsmüll [m] (RL 89/369/EWG).

urban regeneration [n] [UK] *urb.* ▶center city redevelopment [US].

6706 urban region [n] *plan. urb.* (Area of influence of a city, which is interconnected with other surrounding towns and cities forming an agglomeration of communities. The population is primarily occupied in non-agricultural professions, and to a great extent, work for their living directly within the city or as commuters. An urban region can be divided into the following zones, starting at the center [US]/centre [UK]: *main urban center/centre, and downtown districts, suburban area* [communities in the proximity of the center], *peripheral zone, outskirts* [of other towns and cities]; ▶conurbation, ▶metropolitan region); *s* **region [f] urbana** (Zona de asentamiento interrelacionada formada por una ciudad y los pueblos de los alrededores, cuyos habitantes se dedican mayormente a actividades laborales no agrícolas que son desarrolladas sobre todo en aquélla. La estructura de una **r. u.** está formada por el centro de la ciudad, los barrios residenciales céntricos, los suburbios residenciales y la periferia; ▶aglomeración urbana, ▶área metropolitana); *f 1* **région [f] urbaine** (*Terme générique* espace constitué d'un ensemble de villes [villes-centres] et de leurs banlieues étroitement liées les unes aux autres, morphologiquement par leur

voisinage et leurs contacts, structurellement par leurs relations qui dominent et organisent un espace régional ; DUA 1996 ; la région urbaine se différencie d'une ▶agglomération et d'une conurbation par son polycentrisme et par son étendue ; DG 1984 ; ▶aire métropolitaine, ▶communauté urbaine, ▶métropole d'équilibre, ▶métropole régionale) ; *f 2* **agglomération [f]** (Ensemble administratif d'une ville et du territoire urbanisé qui l'entoure et dépend de ses services centraux et de son appareil de gestion économique: la ville et sa banlieue ; DG 1984 ; ▶communauté urbaine, ▶zone de concentration urbaine) ; *g* **Stadtregion [f]** (Verflechtungsbereich einer Stadt mit dem Umland, dessen Bewohner überwiegend nicht landwirtschaftliche Berufe ausüben und überwiegend oder zu einem erheblichen Teil ihre wirtschaftliche Existenz unmittelbar in den Arbeitsstätten der Stadt finden [Pendler]. Die Stadtregion gliedert sich von innen nach außen in *Kernstadt, Ergänzungsgebiet* [die unmittelbar angrenzenden Gemeinden], *verstädterte Zone* [Nahbereich der Umlandgemeinden] und *Randzone* [übrige Umlandgemeinden]; ▶Ballungsraum, ▶Verdichtungsraum).

urban renewal [n] *urb.* ▶center city redevelopment [US]/ urban regeneration [UK].

urban renewal area [n] *leg. urb.* ▶neighborhood improvement area [US]/neighbourhood improvement area [UK].

6707 urban ruderal vegetation [n] *phyt.* (Plant communities developing on wasteland, unoccupied building lots, etc., or in heat islands of towns and cities; ▶plant association of pavement joints, ▶ruderal vegetation, ▶wall community); *s* **vegetación [f] ruderal urbana** (Comunidad de plantas que se desarrolla en áreas yermas, terrenos abandonados o en islas de calor en ciudades; ▶comunidad de las grietas de muros, ▶comunidad de rendijas de pavimentos, ▶paranthrophytia, ▶vegetación ruderal); *syn.* vegetación [f] espontánea urbana; *f* **peuplement [m] rudéral urbain** (Terme générique englobant les groupements végétaux, p. ex. *Agropyretea, Artemisietalia, Onopordion, Sisymbrietalia*, vivant spontanément en milieu urbain sur les murs, dans les rues et terrains vagues. *Termes spécifiques* peuplement urbain de terrain vague, flore urbaine des terrains vagues ; ▶flore rudérale, ▶groupement des interstices des pavés, groupement muricole, ▶groupement saxicole) ; *syn.* végétation [f] rudérale urbaine, flore [f] rudérale urbaine ; *g* **städtische Ruderalflur [f]** (OB zu Pflanzengesellschaft auf innerstädtischen Brachflächen oder Baulücken — im Wärmearchipel Großstadt wie z. B. halbruderale Pioniertrockenrasen *[Agropyretea]*, Beifußgesellschaft *[Artemisietalia vulgaris]*, [Esels]distelgesellschaft *[Onopordion]*, Raukengesellschaft *[Sisymbrietalia]*; ▶Mauerfugengesellschaft, ▶Pflasterritzengesellschaft, ▶Ruderalflur); *syn.* städtische Ruderalgesellschaft [f], städtische Ruderalvegetation [f], spontane Stadtvegetation [f].

urban scenery [n] *urb.* ▶urban setting.

6708 urban setting [n] *urb.* (LD 1988 [9], 61; *genius loci* of a city, the identity of which is perceived due to its spatial structures and is influenced by socio-psychological factors; ▶urban environment, ▶urban space); *syn.* urban scenery [n]; *s* **escenario [m] urbano** (±) (Especificidad de una ciudad que se percibe debido a sus estructuras espaciales y arquitectónicas, y está influenciada por factores sociosicológicos; ▶medioambiente urbano); *f* **environnement [m] bâti** (**1.** Environnement architectural perçu par un observateur comme l'ensemble des éléments structurants et limitants de l'espace bâti ; **2.** organisation structurelle [habitation, travail, loisirs] d'une ville qui influence les liens physiques et psychologiques noués par une collectivité ; l'acception de ce terme a évolué et inclut aujourd'hui la notion de la qualité de la vie et d'impact sur la santé ; ▶environnement urbain) ; *g* **städtebauliches Umfeld [n]** (**1.** Bereich einer Stadt,

U

der auf einen Betrachter als Gesamtheit der raumgliedernden und raumbegrenzenden Gegebenheiten wirkt. **2.** Bereich einer Stadt, der auf Grund seiner räumlichen Strukturen durch wahrnehmungs- und sozialpsychologische Faktoren in Wechselbeziehungen zu seinem Betrachter steht; ▶ städtische Umwelt).

6709 urban space [n] *urb.* (Aspect of an urban setting or place defined by building façades, canopies, and/or by topographic features and vegetative masses that contribute to the perception of three dimensional enclosure structured by floor, wall, and canopy elements); *s* **espacio** [m] **urbano** (Imagen tridimensional de una ciudad, en comparación con el término «área urbana» que se entiende bidimensionalmente); *f* **milieu** [m] **urbain** (Ensemble des éléments caractérisant la structure spatiale de la ville) ; *syn.* espace [m] urbain ; *g* **Stadtraum** [m] (Dreidimensionale Vorstellung von der Raumstruktur einer Stadt, d. h. die klare Ablesbarkeit unterschiedlicher geometrischer Grundrissmuster, die durch Abgrenzungen mit Hausfassaden und Überdachungen [baulich und pflanzlich] sowie mit raumbildenden Vegetationsbeständen und topografischen Oberflächenformen in Stadt- und Ortsteilen definiert sind — im Vergleich zum Begriff ▶ Stadtgebiet).

6710 urban sprawl [n] *urb.* (Haphazard spreading of urban development: linear development along main arteries, or outlying pockets of development; ▶ housing proliferation); *syn.* suburban sprawl [n]; *s* **urbanización** [f] **caótica** (Crecimiento desordenado de una ciudad; ▶ desfiguración del paisaje); *f* **accroissement** [m] **désordonné d'une agglomération** (Croissance galopante [rapide, chaotique, incontrôlée] d'une ville dont le tissu urbain s'étale sur le territoire et engendre une urbanisation désordonnée avec ses déséquilibres spatiaux et sociaux, une dilution radiale vers la périphérie par opposition à l'homogénéité du bâti ; ▶ mitage du paysage) ; *syn.* développement [m] désordonné d'une agglomération, développement [m] urbain désordonné ; *g* **ungeordnete Ausbreitung** [f] **einer Stadt** (Planlose Ausdehnung einer Stadt; ▶ Zersiedelung [der Landschaft]); *syn.* Siedlungsbrei [m] *(Ergebnis).*

6711 urban square [n] *urb.* (**1.** Public open space surrounded by buildings or blocks of buildings, which depending on its dimensions may be provided with, e.g. with a public monument, sitting area, play area, planting areas, individual trees, fountains); **2. plaza** [n] (*Of Spanish origin* formally-designed open urban public space in a town or city with a public building, often with a monument and water features, which is a center of community life) *syn.* city square [n], public square [n], town square [n], plaza [n] [also US]; *s* **plaza** [f] **(2)** (Espacio libre público delimitado por construcciones por los lados; según su tamaño y su ubicación en la ciudad, está plantado con árboles o parterres, decorado con fuentes, estatuas u otros elementos decorativos y equipado con mobiliario urbano, instalaciones de juego, etc. En Es. las plazas centrales de las ciudades [plaza mayor] son generalmente cuadradas y están rodeadas por edificios con arcadas); *f 1* **square** [m] (Espace libre urbain de dimension réduite, en moyenne de quelques centaines de m² à 2-3 ha, créé à la fin du XIXᵉ siècle, souvent de forme carrée, utilisé dans les quartiers urbains comme lieu de promenade et de détente de proximité) ; *f 2* **plaza** [f] (Terme espagnol utilisé en français pour désigner les places d'importances dans les pays de langue espagnole telles que la Plaza Mayor à Madrid, la Plaza Baquedano à Santiago du Chili, la Plaza de Mayo à Buenos-Aires, etc. ; *g 1* **Stadtplatz** [m] (Öffentlicher Freiraum durch Baublockseiten oder Bauwerke begrenzt oder geprägt, der je nach Größe mit einem Denkmal, mit Sitz-, Spiel-, Pflanzflächen, Einzelbäumen, Wasseranlagen o. Ä. ausgestattet ist); *g 2* **Plaza** [f] (*Im Amerikanischen aus dem Spanischen übernommen* bedeutender städtischer, formal und repräsentativ gestalteter, öffentlicher Platz im Zentrum einer Stadt,

i. d. R. mit Denkmal oder Brunnenanlage und angrenzendem öffentlichen Gebäude).

urban trash [n] [US] *envir.* ▶ urban refuse.

urban waste [n] *envir.* ▶ urban refuse.

urban wilds inventory [n] [US] *conserv. phyt. zool.* ▶ urban habitat mapping.

6712 urn burial [n] *adm.* (Interment of ▶ cremation ashes); *s* **entierro** [m] **en urna** (Última fase de entierro por ▶ cremación); *f* **inhumation** [f] **cinéraire** (*Opération funéraire* dernier acte de l'▶ incinération au cours duquel l'urne cinéraire est déposée dans une sépulture) ; *g* **Urnenbeisetzung** [f] (Letzter Akt der ▶ Feuerbestattung, bei dem der eingeäscherte Tote in einem Gefäß in ein Urnengrab beigesetzt wird); *syn.* Urnenbegräbnis [n], Urnenbestattung [f].

6713 urn burial site [n] *adm. landsc.* (**1.** Grave site for cremation ashes container; ▶ burial site, ▶ individual urn grave, ▶ prominent urn burial site. **2.** A particular type is the ▶ columbarium); *syn. to 1.* urn grave [n]; *s* **tumba** [f] **para urnas** (▶ Tumba para depositar o enterrar las urnas cinerarias, que es de tamaño mucho menor que las de entierro en ataúd; *términos específicos* ▶ columbario, ▶ tumba conmemorativa para urna, ▶ tumba individual para urna); *f* **jardin** [m] **d'urnes** (Sépulture dans laquelle est déposée une urne cinéraire qui, à la différence de la fosse prévue pour un cercueil, est de faible dimension, environ un tiers ou la moitié de la taille d'une tombe ; *terme spécifique* tombe cinéraire individuelle ; ▶ columbarium, ▶ jardin d'urnes familial, ▶ jardin d'urnes individuelles, ▶ sépulture) ; *f 2* **jardin** [m] **du souvenir** (Espace dans l'enceinte d'un cimetière sur lequel, à la demande des familles des défunts, peuvent être répandues les cendres pulvérisées des corps incinérés) ; *g* **Urnengrabstätte** [f] (**1.** ▶ Grabstätte für Feuerbeisetzung[en], die im Vergleich zur Erdgrabstelle für Särge nur eine geringere Größe hat, etwa ein Drittel oder die Hälfte eines Erdgrabes; **U.n** können Reihen- oder Wahlgräber sein; *UBe* ▶ Urneneinzelgrabstätte, ▶ Urnensondergrabstätte. **2.** Eine Sonderform ist das ▶ Kolumbarium); *syn. zu 1.* Urnengrab [n], Urnengrabstelle [f].

urn grave [n] *adm. landsc.* ▶ individual urn grave, ▶ urn burial site.

urn grave [n] [UK]**, class one** *adm. landsc.* ▶ prominent urn burial site.

6714 urn grave [n] **in a row** *adm. landsc.* (▶ Single row grave for ▶ cremation ashes container; ▶ urn burial site); *s* **panteón** [m] **para urnas en hilera** (▶ Tumba individual en hilera para enterrar urnas de ▶ cremación; ▶ tumba para urnas); *f* **tombe** [f] **cinéraire à la ligne** (▶ Tombe à la ligne prévue pour l'inhumation par ▶ incinération ; ▶ jardin d'urnes, ▶ jardin du souvenir) ; *syn.* jardin [m] d'urnes en ligne ; *g* **Urnenreihengrabstätte** [f] (▶ Reihengrab für ▶ Feuerbestattung[en]; ▶ Urnengrabstätte); *syn.* Urnenreihengrab [n], Urnenreihengrabstelle [f].

6715 usability [n] *plan. syn.* usefulness [n]; *s* **utilidad** [f] *syn.* posibilidad [f] de uso; *f* **capacité** [f] **d'utilisation** *syn.* possibilité [f] d'utilisation ; *g* **Benutzbarkeit** [f] (Geeignetsein zum Benutzen); *syn. o. V.* Benützbarkeit [f] [südd., A, CH].

6716 use [n] (1) *plan.* (Purpose of activity for which land and buildings are designed, arranged or intended, or for which they are occupied; cf. IBDD 1981. Generally, if there is a coincidence of several land uses on the same plot[s], this is called **mixed land use** [US]**/overlapping of land uses** [UK]; ▶ layering of different uses); *s* **uso** [m] (Función asignada a una superficie como residencial, industrial, comercial, recreativa, de tránsito, etc., o a edificios o instalaciones. Cuando se mezclan varios usos en un lugar se habla de mezcla de usos o ▶ superposición de usos); *syn.*

utilización [f]; *f 1* **usage** [m] (*Terme générique* activités humaines déterminées telles que se loger, travailler, se détendre, se déplacer, communiquer pratiquées sur des territoires géographiques déterminés ; ►superposition des usages) ; *f 2* **utilisation** [f] (Action d'utiliser un espace, un objet ; ►utilisation du sol) ; *syn.* occupation [f], affectation [f] ; *g* **Nutzung** [f] (1. OB zu menschliche Tätigkeit wie z. B. Wohnen, Arbeiten, Erholung, Verkehr, die auf Flächen oder in Gebäuden oder sonstigen Einrichtungen stattfindet. Finden verschiedene **N.en** gleichzeitig auf einer Fläche statt, so spricht man von **Nutzungsmischung** oder ►Nutzungsüberlagerung. 2. Gebrauch einer immobilen Sache oder eines Rechtes und die daraus resultierenden Vorteile; cf. auch Ausführungen bei ►Nutzer); *syn.* Nutzen [n, o. Pl.].

use [n] (2) *plan.* ►adaptive use [US]; *agr.* ►agricultural land use; *agr. plan.* ►agricultural use, ►arable use; *recr.* ►area intended for general recreational use; *agr.* ►communal use; *plan.* ►density of use (1); *recr.* ►frequency of use, ►heavy recreation use; *leg. urb.* ►industrial land use, ►industrial use zone, ►institutional land use, *agr. for. hort.* ►intensity of agricultural and forest land use; *agr. plan. urb.* ►land use (1); *plan.* ►land use pattern; *plan.* ►land use structure; *plan. recr.* ►level of recreation use, ►limited recreation use; *leg. urb.* ►mixed land use (1), ►mixed land use (2), ►mixed use development, *plan.* ►multiple use; *plan. recr.* ►peak visitor use; *conserv. leg. plan.* ►prohibition of a land use; *leg.* ►public; *recr.* ►recreational land use, ►recreational use, ►reduction in recreation use; *leg. urb.* ►residential land use; *leg.* ►right of use; *adm. leg. plan.* ►statutory land use specification; *plan.* ►suitability for human use; *agr. ecol. for.* ►sustainable use; *adm. agr. for. plan. urb.* ►type of use.

use [n] [UK], **after-** *plan.* ►adaptive use [US].

use [n], **area for industrial** *leg. urb.* ►industrial use zone.

use [n], **degree of** *recr.* ►frequency of use.

use [n] (2), **density of** *recr.* ►frequency of use.

use [n], **institutional** *leg. urb.* ►institutional land use.

use [n], **pesticide** *agr. envir. for. hort. phytopath.* ►pesticide application.

use [n], **recreation** *plan. recr.* ►recreational use.

use [n], **subsequent** *plan.* ►adaptive use [US].

use [n] **as defined by Use Class Order** [UK] *leg. urb.* ►building use category [US].

6717 use capacity [n] **of natural resources** *conserv. plan.* (Potential of natural resources for sustainable human use without imposing restrictions on their renewability); *syn.* exploration capacity [n]; *s* **aprovechamiento** [m] **sostenible de los recursos naturales** (Potencial de los recursos naturales de ser utilizados a largo plazo sin perder su capacidad de reproducción); *f* **capacité** [f] **de gestion (durable) des ressources naturelles** (Capacité que possède l'homme d'utiliser sans limites les ressources naturelles renouvelables selon le principe d'exploitation durable) ; *g* **Nutzungsfähigkeit** [f] **der Naturgüter** (Das Vermögen, Naturgüter durch den Menschen ohne Einschränkung der Erneuerbarkeit im Sinne der Nachhaltigkeit zu nutzen; cf. § 1 [1] BNatSchG).

use category [n] *leg. urb.* ►building use category [US]/use as defined by Use Class Order [UK], ►urban land use category,

use class [n] [UK] *leg. urb.* ►zoning district category [US].

use classification [n] **of soil materials** [US] *constr.* ►soil classification (1).

use class order [n] [UK] *leg. urb.* ►zoning ordinance [US].

Use Class Order [n] [UK], **use as defined by** *leg. urb.* ►building use category [US].

use conflict [n] *plan.* ►land use conflict.

use density [n] [US], **land** *urb.* ►density of development.

used materials [n] *envir.* ►reapplication of used materials.

usefulness [n] *plan.* ►usability.

6718 usefulness [n] **for cut flowers** *hort.* (Quality of perennials or annuals with regard to their longevity as cutflowers in a vase; useful [adj] as cut flowers); *s* **aptitud** [f] **de flores para el corte** *syn.* idoneidad [f] de flores para el corte; *f* **qualités [fpl] requises pour les fleurs à couper** (Qualités recherchées du point de vue de la durée, de la couleur et de la forme de floraison chez une fleur à couper) ; *syn.* aptitude [f] à être cultivée pour la fleur à couper, qualité [f] comme fleur à couper ; *g* **Schnittwert** [m] (Qualität von Stauden und Sommerblumen hinsichtlich ihrer Haltbarkeit als Schnittblumen).

6719 useful plant species [n] *agr. for. hort.* (Generic term for ►cultivated plant 1 and useful ►wild plant); *syn.* economic plant species [n]; *s* **planta** [f] **útil** (Término genérico para planta que se puede aprovechar económicamente, sea ►planta de cultivo o ►planta silvestre aprovechable); *syn.* planta [f] de aprovechamiento económico; *f* **plante [f] utile** (Terme générique désignant les ►plantes de culture et les ►plantes sauvages utiles) ; *g* **Nutzpflanze** [f] (OB zu ►Kulturpflanze 1 und nutzbare ►Wildpflanze).

use intensity standards [npl] [US] *urb.* ►land use intensity standards.

6720 use [n] **of beneficial animal species** *agr. for. hort.* (Application of pest-eating species for biological pest control); *s* **aprovechamiento** [m] **de animales beneficiosos** (Aplicación de especies animales para combatir plagas en la agricultura); *f* **utilisation [f] d'auxiliaires** (Introduction d'un ennemi spécifique [micro-organismes, parasites, prédateurs] du parasite qui est à détruire dans la lutte biologique) ; *syn.* emploi [m] d'auxiliaires ; *g* **Nützlingseinsatz** [m] (Verwendung von Nützlingen zur biologischen Schädlingsbekämpfung); *syn.* Einsatz [m] von Nützlingen.

use [n] **of leisure time** *recr.* ►organized use of leisure time.

6721 use [n] **of open space** *plan. recr.* (Active or passive use of public and semi-public open spaces for leisure and recreation, usually characterized by its neighbo[u]rhood location, open spaces include all non-roofed surfaces in an urban area, whether with or without vegetation); *s* **uso** [m] **de espacios libres** (Aprovechamiento activo o pasivo de los espacios libres públicos o semipúblicos para fines recreativos o de tiempo libre, generalmente caracterizado por su localización en el vecindario. Espacios libres incluyen todas las superficies sin techo, tengan o no vegetación); *f* **utilisation [f] des espaces libres** (Usage actif ou passif des espaces publics, semi-publics ainsi que des espaces publics des lieux bâtis de droit privé à des fins de loisirs et de récréation ; les espaces libres sont constitués de toutes les surfaces extérieures ouvertes publiques, végétalisées [espaces verts] ou non, aménagées [espaces minéraux] ou non, situées en zone urbanisée) ; *g* **Freiraumnutzung** [f] (Meist quartiersbezogene, aktive oder passive Nutzung öffentlicher und halböffentlicher Freiräume für Freizeit und Erholung; zu den Freiräumen gehören alle nicht überdachten, vegetations- oder nicht vegetationsbestimmten Flächen in einem Siedlungsgebiet).

use [n] **of plant material** *gard. landsc.* ►planting design.

use [n] **of the environment** *conserv. nat'res.* ►beneficial use of the environment.

use pattern [n] *plan.* ►land use pattern, ►pattern of interrelated land uses.

6722 use period [n] **of grave** [US] *adm.* (Minimum period of time in which an individual ▶burial site may be occupied. The period must be no less than the time necessary for the corpse to decay); *syn.* rest period [n] of grave [UK]; *s* **periodo/período [m] de descanso de tumbas** (Tiempo mínimo de ocupación de ▶tumba individual que tiene que superar el tiempo necesario para la descomposición del cadáver); *f* **durée [f] d'inhumation** (Période minimum d'utilisation d'une ▶sépulture, devant correspondre au temps minimum de décomposition du corps) ; *syn.* durée [f] de concession ; *g* **Ruhefrist [f]** (Mindestnutzungsdauer einer einzelnen ▶Grabstätte, die mindestens der Verwesungsdauer einer Leiche entsprechen muss); *syn.* Grabnutzungsdauer [f], Ruhezeit [f].

use plan [n] [US]**, existing** *plan.* ▶survey of existing uses.

use planning [n] [UK]**, urban land** *adm. leg. urb.* ▶community development planning [US].

use pressure [n] *plan.* ▶competing land-use pressure.

6723 user [n] *adm. plan.* *s* **usuario/a [m/f]** (Persona que utiliza una instalación, servicio o dispositivo. En alemán el término «Nutzer» se utiliza solo para personas jurídicas que utilizan p. ej. tierras o edificios); *f* **usager [m]** (Individu [usager particulier] ou personne morale qui a un droit réel d'usage et est bénéficiaire des prestations d'un service public déterminé ou utilise un bien du domaine public en rapport à essentiellement trois usages possibles : l'usage domestique, l'usage industriel, et l'usage agricole ; terme à ne pas confondre avec l'« utilisateur » qui se sert de produits acquis ou non à titre personnel) ; *g* **Nutzer [m]** (Juristische Person, die z. B. eine Liegenschaft oder ein Gebäude nutzt; *Kontext* N. eines Gebäudes; gewerblich genutztes Grundstück, landwirtschaftlich genutzter Boden; *nicht zu verwechseln mit* ‚Benutzer': jemand, der einen beweglichen Gegenstand benutzt).

6724 use regulation [n] *conserv. plan.* (Use control, e.g. of areas which are significant for nature conservation purposes, by zoning maps, by legally-binding land-use and landscape plans or by private agreements); *s* **regulación [f] de usos** (Determinación de los posibles usos, p. ej. en un espacio natural protegido, por medio de ordenanza, plan vinculante o acuerdo en base al derecho civil); *f* **maîtrise [f] d'usage** (Mesures réglementaires — documents d'urbanisme — administratives ou contractuelles instituées afin de gérer ou maintenir des activités humaines favorables ou compatibles avec la protection des milieux naturels remarquables) ; *g* **Nutzungsregelung [f]** (Bestimmung der Nutzung, z. B. auf für Naturschutz und Landschaftspflege bedeutsamen Flächen, durch Festsetzungen im Bebauungsplan, durch rechtsverbindliche Landschaftspläne, durch privatrechtliche Vereinbarungen oder durch gesetzliche Vorschriften [Betretungsrecht] und Auflagen im Genehmigungsverfahren); *syn.* Regelung [f] der Nutzung.

6725 use restriction [n] *conserv. leg. plan.* (Limitation of use possibilities specified by 'restrictive covenant' usually running with ownership of the land, by zoning, or other local regulations); *s* **restricción [f] del uso** (Limitación fijada en ley u ordenanza de las posibilidades de usar determinados terrenos, p. ej. por sus funciones en la protección de la naturaleza); *f* **restriction [f] d'usage** (F., il existe différentes formes de restrictions d'usage : **1.** la servitude d'utilité publique [SUP] — Code de l'Environnement, articles L. 515-8 à L. 515-12 et R.515-24 à R 515-31, **2.** la restriction d'usage conventionnelle au profit de l'Etat [RUCPE], **3.** la restriction d'usage entre parties [RUP], **4.** le porter à connaissance [PAC] - Code de l'Urbanisme, articles L. 121-2 et R. 121-1, **5.** le projet d'intérêt général [PIG] - Code de l'Urbanisme, articles L. 121-9 et R. 121-3 ; ces restrictions touchent l'utilisation du sol, du sous-sol, de la nappe phréatique, des eaux superficielles, etc.) ; *syn.* limitation [f] de l'usage ;

g **Nutzungsbeschränkung [f]** (Einengung der Nutzungsmöglichkeit von Grund und Boden durch Gesetze oder Vorschriften [cf. § 32 BauGB u. § 3 b BNatSchG], z. B. **N.en** durch Rechtsvorschriften oder Anordnungen der für Naturschutz und Landschaftspflege zuständigen Behörden, die erhöhte Anforderungen in der Land- und Forstwirtschaft zur Verwirklichung der Ziele des Naturschutzes und der Landschaftspflege festsetzen und damit die ausgeübte forst-, land- und fischereiwirtschaftliche Bodennutzung über die Anforderungen der guten fachlichen Praxis hinaus beschränken; cf. § 3 b [1] BNatSchG; die **Nutzungsunterbrechung** [§ 3 b (2) BNatSchG] ist eine Form der **N.**); *syn.* Nutzungseinschränkung [f], Einschränkung [f] der Nutzung, Beschränkung [f] der Nutzung.

use site [n] [US]**, public** *leg. urb.* ▶site for public facilities.

use structure [n] *plan.* ▶land use structure, ▶land use pattern.

6726 use [vb] **sustainably** *ecol.* *s* **usar [vb] sosteniblemente** *syn.* usar [vb] sustentablemente, utilizar [vb] sosteniblemente/sustentablemente; *f* **utiliser [vb] durablement** ; *g* **nachhaltig nutzen [vb]**.

use zone [n] *plan.* ▶industrial use zone.

usufruct [n] *leg.* ▶right of use.

utilities [npl]**, laying of** *constr.* ▶installation of utilities.

6727 utilities plan [n] *trans. urb.* (Plan showing the location of ▶public supply utility lines and ▶effluent discharge pipe); *s* **plano [m] de conducciones** (Plano en el que están registradas todas las ▶conducciones de abastecimiento y los ▶conductos de desagüe); *syn.* plano [m] de servicios; *f* **plan [m] des réseaux existants** (Plan sur la situation des câbles, canalisations, ouvrages souterrains, etc. *Terme spécifique* plan des canalisations enterrées existantes ; ▶conduite d'alimentation 2, ▶conduite d'assainissement) ; *g* **Leitungsplan [m]** (Plan, in dem ▶Versorgungsleitungen 2 und ▶Entsorgungsleitungen eingetragen sind); *syn.* Spartenplan [m].

6728 utility connection charge [n] *adm. urb.* (Payment made by landowner as a contribution towards the cost of providing infrastructural facilities; *specific term* storm drainage service charge [US] (TGG 1984, 158); road charges [UK]); *syn.* utility connection payment [n] [US]; *in the vernacular* hookup charge [n] [US]/hook-up charge [n] [UK]; *s* **costo [m] de urbanización** (Porcentaje de los costes [Es]/costos [AL] de infraestructura que corren a cargo de los propietarios de los predios); *syn.* coste [m] de urbanización [Es]; *f* **taxe [f] locale d'équipement (T.L.E.)** (Recette de nature fiscale instituée par la loi d'orientation foncière [LOF] n° 67-1253, destinée à faire participer les constructeurs et les lotisseurs aux dépenses d'équipements collectifs revenant aux communes et généralement induite par les nouvelles constructions ; DUA 1996) ; *g* **Erschließungsbeitrag [m]** (Der von der Gemeinde zu erhebende und von einem Grundstückseigentümer resp. Erbbauberechtigten zu tragende Kostenanteil für die Realisierung von Infrastruktureinrichtungen. Der Erschließungsaufwand umfasst die Kosten für den Grunderwerb und die erstmalige Herstellung oder die Übernahme von Anlagen. Dieser Aufwand wird gemäß Regelung der Gemeindesatzung auf die durch die Anlage erschlossenen Grundstücke nach Art und Maß der baulichen oder sonstigen Nutzung nach Grundstücksflächen oder nach der Grundstücksbreite an der Erschließungsanlage verteilt); *syn.* Anliegerbeitrag [m] für Erschließungsmaßnahmen, Anliegerleistung [f] für Erschließungsmaßnahmen).

6729 utility easement [n] *leg. urb.* (Legal grant of limited property rights to lay public utility lines. **In U.S.**, such 'easements' are recorded to define property ownership in county land records, deed registers, maps, property plats, etc., and may be shown on zoning maps only as information for regulatory purposes; ▶easement, ▶vehicular and pedestrian easement; **in**

U

Europe, such 'easements' are shown on an legally-binding land-use plan and recorded in a land register [UK]); *s* **servidumbre [f] de paso de conducciones** (Derecho público de instalar/tender tuberías de abastecimiento público, líneas eléctricas, etc. a través de una propiedad privada; ▶servidumbre de acceso, ▶servidumbre inmobiliaria); *f* **servitudes [fpl] relatives à l'utilisation des réseaux d'alimentation et de distribution d'énergie, de canalisation et de télécommunication** (Servitudes d'utilité publique affectant l'utilisation du sol en réglementant sur le domaine privé l'accès des fonds et le passage des différents réseaux d'alimentation et de distribution d'énergie, de canalisation et de télécommunication ; C. urb., art. A 126-1 ; ▶servitudes relatives à l'utilisation de certaines ressources et équipements ; ▶servitude de passage) ; *g* **Leitungsrecht [n]** (In Bebauungsplänen gem. § 9 [1] 21 BauGB ausgewiesenes und im Grundbuch eingetragenes Recht [▶Grunddienstbarkeit] über ein Grundstück zugunsten der Allgemeinheit oder eines beschränkten Nutzerkreises, Ver- oder Entsorgungsleitungen zu legen; ▶Geh- und Fahrrecht).

utility hole [n] [US] *constr.* ▶manhole.

utility line [n] *trans. urb.* ▶public supply utility line.

utility services [npl] *urb.* ▶public utility services.

6730 utilization [n] **of natural resources** *ecol. land' man.* (Use of renewable and non-renewable natural resources; ▶exploitation); *s* **utilización [f] de los recursos naturales** (Aprovechamiento de los recursos naturales renovables y no renovables; ▶explotación); *syn.* aprovechamiento [m] de los recursos naturales; *f* **utilisation [f] des ressources naturelles** (▶surexploitation) ; *g* **Nutzung [f] natürlicher Ressourcen** (Nutzung der erneuerbaren und nicht erneuerbaren Naturgüter; ▶Ausbeutung).

6731 vacancy [n] *sociol. urb.* (Sum of unused homes or business premises in a defined area, which are empty; *verbs* to be vacant, to stand empty, to remain empty; *specific terms* apartment vacancies, business space vacancy, apartment building vacancy [rate], vacancy at office buildings); *s* **desocupación [f]** (Total de viviendas o locales de oficinas no utilizados en un área determinada; estar [vb] vacía/estar [vb] desocupada una vivienda/una oficina; *términos específicos* desocupación [f] de viviendas, desocupación [f] de locales de oficinas); *f* **taux [m] de vacances** (Proportion de logements ou de bureaux vacants dans un périmètre donné ; être vacant/inoccupé [vb] ; *termes spécifiques* taux de vacance/d'inoccupation de logements locatifs/d'immeubles/ de bureaux) ; *syn.* taux [m] d'inoccupation ; *g* **Leerstand [m]** (Summe nicht genutzter Wohnungen oder Geschäftsräume in einem definierten Gebiet; leer stehen [vb]; *UBe* Wohnungsleerstand, Geschäftsraumleerstand).

vacant [adj] *sociol. urb.* ∗▶vacancy.

6732 vacant lot [n] [US] *urb.* (TGG 1984, 199; unbuilt plot in the frontage of a block; ▶infill development, ▶undeveloped land, ▶odd-lot development [US]/gap-stopping [UK]); *syn.* empty plot [n] [UK], vacant parcel [n] [US] (PSP 1987, 200), gap site [n] [UK], gap plot [n] [UK]; *s* **solar [m] desocupado (entre edificios)** (Terreno no edificado en una manzana construida; ▶área no urbanizada, ▶densificación urbana, ▶relleno de solares desocupados); *f* **dent [f] creuse** (Parcelle ou une unité

foncière non construite appartenant à un même propriétaire, entourée de parcelles bâties, de voiries existantes, en limite d'une voirie ou d'une zone inconstructible [zone agricole, zone naturelle, espace boisé classé, etc.] ; ▶politique de la ville compacte, ▶terrain non construit, ▶reconstitution du front bâti) ; *syn. en zone urbaine* interruption [f] de façade, trouée [f] dans le bâti, terrain [m] libre, terrain [m] nu ; *g* **Baulücke [f]** (Unbebautes Baugrundstück zwischen bebauten Grundstücken; **B.n** gibt es in Baugebieten mit ▶geschlossener Bauweise und ▶offener Bauweise 2; in Europa besteht zunehmend die Tendenz, **B.n** zu bebauen, um den Landschaftsverbrauch zu verringern; ▶Nachverdichtung, ▶unbebaute Fläche, ▶Schließen von Baulücken).

vacant parcel [n] [US] *urb.* ▶vacant lot [US]/empty plot [UK].

6733 vacation [n] [US] *recr.* (Period of relaxation, usually more than 3 weeks per annum if complete rest is to be achieved; according to tourism statistics, the average length is more than 5 days; ▶farm vacation [US]/farmstay holidays [UK], ▶short-stay recreation); *syn.* holidays [npl] [UK]; *s* **vacaciones [fpl]** (Periodo/período largo de descanso del trabajo, generalmente de más de tres semanas al año; en la estadística turística se consideran periodos de más de cinco días; ▶puente, ▶vacación en granja, ▶vacaciones cortas); *f* **vacances [fpl]** (F., période de repos d'environ trois semaines pour toute personne qui, dans les statistiques des loisirs touristiques, quitte son domicile pour au moins quatre jours [D., au moins cinq jours] consécutifs pour des raisons autres que professionnelles, d'étude ou de santé ; LTV 1996 ; ▶tourisme de passage, ▶vacances à la ferme) ; *syn.* congés [mpl] ; *g* **Urlaub [m]** (Erholungszeitraum, i. d. R. >3 Wochen; in der Fremdenstatistik > 5 Tage; ▶Ferien auf dem Bauernhof, ▶Kurzzeiterholung).

vacation [n] [US]**, backpacking** *recr.* ▶hiking holiday.

vacation [n] [US]**, subsidized** *recr.* ▶susidized holiday.

6734 vacation accommodation [n] [US] (1) *recr.* (Generic term for dwelling in cottages and other residential complexes while engaged in leisure time living; e.g. condominium [US]/holiday village flat [UK], apartment in large recreation centers [US]/centres [UK] or resort community [US], ▶vacation house [US]/holiday house [UK]); *syn.* holiday accommodation [n] [UK]; *s* **segunda residencia [f]** (Vivienda, casa o plaza de camping permanente utilizada durante las vacaciones o en los fines de semana; *término específico* ▶casa de veraneo); *f* **habitation [f] de loisirs** (1. D'une manière générale résidence secondaire constitutive d'une résidence de week-end, de bâtiments à usage de loisirs ou d'immeubles en jouissance à temps partagé. 2. *Terme générique* les logements ou appartements constitutifs de résidences de tourisme, de parcs résidentiels de loisirs ; loisirs ; ▶résidence de vacances, ▶habitation légère de loisirs) ; *g* **Freizeitwohnung [f]** (In erster Linie eine Zweitwohnung in Wochenendhäusern und sonstigen Wohngebäuden zur Freizeitnutzung. OB zu z. B. Feriendorfwohnung, Appartmenthaus der großen Ferienzentren, Zweitwohnung in Wochenendhäusern; ▶Ferienhaus).

6735 vacation accomodation [n] [US] (2) *recr.* (Holiday accomodation for families which is run by an organization; ▶rural vacation accommodation [US]); *syn.* holiday home [n] [UK] (1), *in the vernacular* getaway [n] [US]; *s* **centro [m] vacacional** (Término genérico para un centro de hospedaje en lugar turístico o de recreo regido por la seguridad social o de otras instituciones no lucrativas; ▶hospedaje rural; *término específico* ciudad [f] infantil de las Cajas de Ahorros); *f 1* **centre [m] d'hébergement** (*Terme générique* lieu de séjour temporaire, utilisé en majorité pour les vacances, ou pour une période courte) ; *syn.* centre [m] de vacances ; *f 2* **séjour [m] vacances** (*Terme spécifique,* anciennement « centre de vacances et de loisirs » ; formule d'accueil des jeunes pour les vacances en séjour court ou séjour spécifique [séjours sportifs linguistiques artistiques et culturels]) ; *f 3* **maison [f] familiale de vacances** *(Terme spécifique* équipement communautaire de vacances géré par la fédération nationale des maisons familiales de vacances ; ►gîte rural) ; *g* **Erholungsheim [n]** (Von einer gemeinnützigen Organisation getragenes Heim, in dem Familien ihren Urlaub/ihre Ferien gegen Entgelt verbringen können; *UBe* Familienerholungsheim, Walderholungsheim; ►Ferienunterkunft auf dem Lande); *syn.* Ferienheim [n].

6736 vacation activity [n] [US] *recr.* (Physical, mental or handicraft activities during a vacation [US]/holiday [UK]; ►recreation activity); *syn.* holiday activity [n] [UK], vacation occupation [n] [US]/holiday occupation [n] [UK]; *s* **actividad [f] de vacaciones** (Cualquier tipo de actividad física, mental o manual que se realiza durante unas vacaciones; ►actividad de recreo); *f* **pratique [f] vacancière** (Toutes formes d'activités corporelles, intellectuelles et manuelles pendant la période des vacances ; ►activité de récréation) ; *g* **Urlaubsbetätigung [f]** (Jede körperliche, geistige oder handwerkliche Tätigkeit während der Urlaubszeit; ►Erholungsaktivität).

vacation development [n] [US] *recr.* ►park-like vacation development.

vacationer [n] [US] *recr. sociol.* ►vacationist [US]/holidayer [UK].

vacationers [npl] [US] *recr. plan.* ►accommodation-seeking vacationers.

6737 vacation frequency [n] [US] *plan. recr. sociol.* (Number of times a vacation [US]/holiday [UK] of five days or longer is repeated during one calender year); *syn.* holiday frequency [n] [UK]; *s* **frecuencia [f] de viajes (de vacaciones)** (Porcentaje de la población mayor de 14 años que realiza uno o varios viajes de recreo de 5 o más días de duración al año); *f* **taux [m] de départ en vacances** (F., part [en %] de la population âgée de plus de quatorze ans qui quitte son domicile à l'occasion des vacances pour une durée supérieure à quatre jours [**D.,** cinq jours]) ; *g* **Urlaubsreiseintensität [f]** (Größe, die den prozentualen Anteil der über 14 Jahre alten Bevölkerung misst, der im Laufe eines Kalenderjahres mindestens eine Urlaubsreise von fünf Tagen Dauer oder länger unternommen hat); *syn.* Reiseintensität [f].

6738 vacation house [n] [US] *recr.* (House used for stays away from home; in U.S., this may be called a 'second home' which is used primarily for short-term recreation stay); *syn.* holiday cottage [n] [UK], holiday home [n] [UK] (2); *s* **casa [f] de veraneo** (Casa habilitada para residir durante las vacaciones —sean en verano o en otras épocas del año— y durante estancias cortas de esparcimiento y que normalmente no se utiliza como residencia permanente); *f 1* **résidence [f] de vacances** (Construction à usage d'habitation pendant les périodes de vacances ; *terme spécifique* ►maison familiale de vacances) ; *syn.* logement [m] de loisirs, maison [f] de vacances ; *f 2* **habitation [f] légère de loisirs** (Construction à usage non professionnel, démontable ou transportable et à laquelle s'appliquent des règles particulières de construction [cf. art. 10 décret 92-273 du 23 mars 1992 relative aux plans de zones sensibles aux incendies de forêts], souvent implantée sur un terrain de camping-caravaning) ; *g* **Ferienhaus [n]** (Für Ferienaufenthalte bestimmtes Wohnhaus).

6739 vacationist [n] [US] *recr. sociol.* (Person, who takes a vacation [US]/holiday [UK] away from home; ►recreation user; *syn.* holidayer [n] [also UK], holiday-maker [n] [UK], vacationer [n] [US]; *s* **vacacionista [m/f]** (*Término específico* veraneante [m/f]; ►recreacionista); *f* **vacancier [m]** (Personne en vacances à l'extérieur de son domicile principal ; ►séjournant) ; *g* **Urlau-**

ber/-in [m/f] (Person, die Urlaub außerhalb ihres ständigen Wohnsitzes macht; ►Erholungsuchende/-r).

vacationists [npl]**, accommodation-seeking vacationists** [US] *recr. plan.* ►accomodation-seeking vacationers [US].

vacation occupation [n] [US] *recr.* ►vacation activity [US]/holiday activity [UK].

6740 vacation residence [n] [US] *recr.* (Second home [US] for recreational use); *syn.* holiday residence [n] [UK]; *s* **segundo domicilio [m]** (*Término específico* domicilio [m] de veraneo); *f* **résidence [f] de tourisme** (Lieu d'une résidence secondaire utilisée à des fins touristiques ; on parle alors de tourisme de résidences secondaires) ; *g* **Freizeitwohnsitz [m]** (Ort einer Zweitwohnung, die der Freizeitnutzung dient).

vacation traffic [n] [US] *plan. recr. trans.* ►tourist traffic.

valley [n] *geo.* ►dry valley; *landsc.* ►meadow valley.

valley [n]**, dead** *geo.* ►dry valley.

6741 valley bridge [n] *trans.* (Bridge structure which crosses a valley); *s* **viaducto [m]** (Puente largo que atraviesa un valle); *f* **viaduc [m]** (Édifice de grande longueur traversant une vallée) ; *g* **Talbrücke [f]** (Ein Tal querendes Verkehrsbauwerk); *syn.* Viadukt [m, *auch* n].

valley flat [n] *geo.* ►valley floor.

valley floodplain [n] *geo.* ►alluvial plain.

6742 valley floor [n] *geo.* (Usually low flat land lying between ranges of hills or mountains; ►thalweg); *syn.* bottom [n] of the valley, valley flat [n] (DOG 1984); *s* **vaguada [f] fluvial** (Línea que marca la parte más honda de un valle; DGA 1986; ►talweg); *syn.* fondo [m] del valle; *f* **fond [m] de vallée** (Partie plane la plus basse d'une vallée, bordée par des versants ; ►talweg) ; *g* **Talboden [m]** (Von Talhängen begrenzter, niedrigster und i. d. R. flacher Teil eines Tales; ►Talweg); *syn.* Talgrund [m], Talsohle [f].

valley line [n] *geo.* ►thalweg.

valley park [n] [US]**, stream** *landsc. recr. urb.* ►riverside park.

valley profile [n]**, longitudinal** *geo.* ►thalweg.

6743 valley wind [n] *met.* (Flow of air up a valley during the day: air movement due to difference between heating of mountain slopes and plains below. This frequently alternates with nocturnal ►mountain wind; ►upslope air flow); *s* **viento [m] de valle** (Viento local, provocado por la influencia directa de las formas del terreno; ►viento anabático, *opp.* ►brisa de montaña); *f* **brise [f] de vallée** (Vent anabatique soufflant le jour le long des pentes d'une vallée vers la montagne [brise montante] ; si la vallée est en pente, on peut observer « l'effet de canyon », les vents montent et descendent le long des versants qui entourent la vallée ; CILF 1978 ; ►vent anabatique, *opp.* ►brise de montagne) ; *g* **Talwind [m]** (Während des Tages talaufwärts gerichtete Luftströmung im Bergland/Gebirge; ►Hangaufwind; *opp.* ►Bergwind).

6744 valuable [adj] **for nature conservation** *conserv. pol.* (Descriptive term applied to a landscape feature or to a whole area which should be awarded protection status); *s* **digno/a de protección [loc]** (Dícese de un componente del paisaje o de toda un área que debería tener estatus de protección legal); *syn.* protegible [adj]; *f* **digne d'être protégé, ée [loc]** (Dont le caractère d'un site ou d'un vaste territoire permet le classement en espace protégé) ; *syn.* apte à être protégé [loc] ; *g* **naturschutzwürdig [adj]** (So beschaffen, dass ein Landschaftsbestandteil oder ein ganzes Gebiet rechtsverbindlich unter Schutz gestellt werden sollte).

V

6745 valuation chart [n] **for trees and shrubs** *arb. conserv.* (Calculation table on relative values of woody plants; ►tree and shrub evaluation); *s* **tabla [f] de evaluación del valor de leñosas** (Tabla de ►cálculo del valor de leñosas); *syn.* baremo [m] de evaluación del valor de leñosas; *f* **barème [m] d'évaluation de la valeur des végétaux ligneux** (►calcul de la valeur des végétaux ligneux) ; *g* **Gehölzwerttabelle [f]** (Tabelle zur ►Gehölzwertermittlung).

value [n] *plan.* ►benefit-value analysis; *contr.* ►bidder offering best value for the price [US]/tenderer offering the best value for the money [UK]; *plan.* ►guideline value; *phyt.* ►indicator value; *plan. recr.* ►leisure value; *envir.* ►maximum value; *chem. limn. pedol.* ►pH value, ►increasing of pH value, ►lowering of pH value; *plan. recr.* ►recreation value; *envir. conserv.* ►threshold value.

value [n]**, aesthetic** *recr. plan.* ►visual quality.

value [n]**, assessment of scenic** *plan. landsc.* ►visual landscape assessment.

value [n] **[UK], computation of full contract** *constr.* ►determination of final contract amount [US].

value [n]**, raising of pH** *chem. linm. pedol.* ►increasing of pH value.

value [n]**, reducing of pH** *chem. limn. pedol.* ►lowering of pH value.

value [n] **of a forest, amenity** *recr.* ►amenity benefits of a forest.

value [n] **of a landscape, aesthetic** *landsc. plan. recr.* ►visual quality of landscape.

value [n] **of soil, pH** *chem. pedol.* ►soil reaction.

value [n] **of the environmental experience** *plan. recr. sociol.* ►perceived environmental quality.

value [n] **of trees and shrubs** *arb. conserv.* ►expert opinion on the value of trees and shrubs.

valve [n] *constr.* ►gate valve.

6746 vandalism [n] *sociol.* (Wil[l]ful or ignorant destruction of works of art, planting, facilities, street furniture, etc. in public areas. Term first coined by HENRI GRÉGOIRE, a French theologist, in 1794); *s* **vandalismo [m]** (Destrucción o deterioro sin sentido de objetos o instalaciones en lugares públicos o semi-públicos); *f* **vandalisme [m]** (Tendance à détruire stupidement, à détériorer par ignorance, des œuvres d'art, des installations dans les espaces publics) ; *g* **Vandalismus [m]** (Blinde Zerstörungswut im öffentlichen Raum. Der Begriff wurde 1794 von HENRI GRÉGOIRE, franz. kath. Theologe, geprägt).

vantage point [n] *recr.* ►viewpoint [US].

6747 vapor barrier [n] **[US]/vapour barrier** [n] **[UK]** *arch.* (A moisture-impervious layer or coating, which prevents or is highly resistant to the passage of moisture or vapo[u]r into a material or structure); *syn.* vapor retarder [n] [US], vapour control layer [n] (VCL) [UK]; *s* **barrera [f] antihumedad** (Cualquier material que forme una película impermeable que retrase el paso de vapor o humedad al interior de los muros y, por tanto, prevenga la condensación dentro de ellos. También usual en tejados aislados; cf. DACO 1988); *f* **écran [m] pare-vapeur** (Membrane utilisée p. ex. sur les toitures « chaudes » et destinée à empêcher la pénétration de la vapeur d'eau dans les matériaux d'isolation) ; *syn.* barrière [f] de vapeur ; *g* **Dampfsperre [f]** (Folie, z. B. bei Warmdächern, die verhindert, dass Diffusionsfeuchte der Raumluft in den Dachaufbau eindringt und zu Wirkungsverlusten der Dämmstoffe führt).

vaporizing [n] **of nitrogen** *pedol.* ►volatilization of nitrogen.

vapor retarder [n] **[US]** *arch.* ►vapor barrier [US]/vapour barrier [UK].

vapour barrier [n] **[UK]** *arch.* ►vapor barrier [US].

vapour control layer [n] **[UK]** *arch.* ►vapor barrier [US]/vapour barrier [UK].

6748 variance [n] **[US]** *adm. leg. urb.* (Permission to depart from the literal requirements of a zoning ordinance where strict enforcement would create hardship amounting to confiscation of the property; IBDD 1981); *syn.* relaxation [n] of building regulations [UK]); *s* **exención [f]** (Permiso para no cumplir normas legales o restricciones y condiciones administrativas); *syn.* relevación [f]; *f* **dérogation [f]** (Autorisation d'être libéré de prescriptions juridiques ou de décisions administratives) ; *g* **Befreiung [f]** (Dispens von gesetzlichen Normen und Regelungen seitens einer Behörde; von behördlichen Auflagen gibt es keine **B.**, diese können zurückgenommen werden [Zurücknahme behördlicher Auflagen]).

variation [n] **[UK]** *constr. contr.* ►altered construction segment [US].

variety [n] *hort.* ►cultivar.

6749 vauclusian spring [n] *geo. hydr.* (Naturally occurring small pool-forming underground spring, mostly found in limestone areas; ►spring-fed pool); *syn.* pool spring [n]; *s* **fuente [f] vauclusiana** (En terrenos calizos estanque alimentado por manantial de caudal variable determinado por la presencia de lluvias. El término tiene su origen en la región francesa de Vaucluse en la que existen este tipo de fuentes que se toman como prototipo; cf. DGA 1986; ►laguna de manantial); *f 1* **résurgence [f]** (*Terme générique* point de réapparition des eaux souterraines qui proviennent de la restitution des eaux d'un écoulement superficiel ou souterrain qui se perd en un endroit précis d'un massif calcaire, à sa surface ou dans sa profondeur [grotte ou aven] ; *f 2* **mare de [f] résurgence** (*Terme spécifique* petite dépression dans laquelle réapparaissent les eaux souterraines, ►étang de résurgence) ; *f 3* **exsurgence [f]** (*Terme spécifique* source karstique dont l'eau ne provient pas d'une perte, mais de condensations et d'infiltrations cavernicoles ; cf. Convention de Ramsar, résolution VII.13 — lignes directrices pour l'identification et l'inscription de systèmes karstiques et autres systèmes hydrologiques souterrains sur la Liste des zones humides d'importance internationale) ; *f 4* **source [f] vauclusienne** (*Terme spécifique* source généralement à gros débit issue d'une rivière souterraine en pays karstique ; les résurgences en terrain karstique sont très souvent commandées par un siphon et fonctionnent de façon intermittente. La « Fontaine de Vaucluse » a donné le nom à ce type de résurgence et se situe au 5ème rang mondial dans l'importance des sources) ; *g* **Tümpelquelle [f]** (Natürlich entstandenes Kleingewässer mit unterirdischem Quellwasserzufluss, meist in Kalksteingebieten; ►Quellteich); *syn.* Limnokrene [f].

V crotch [n]**, removal of branch with** *arb.* ►removal of codominant stem.

V crotches [npl] *arb. constr.* ►formation of U or V crotches, ►rod bracing of V crotches.

vegetable garden [n] **[US]** *gard.* ►kitchen garden.

vegetal cover [n] **[US]** *phyt.* ►plant cover.

vegetated area [n] *constr.* ►area of vegetation.

vegetated gabion [n] *eng. wat'man.* ►reno mattress.

vegetated rock sill [n] *constr. wat'man.* ►brush and rock dam.

6750 vegetated strip [n] *landsc. trans.* (Linear strip, planted or grassed, separating pedestrian and vehicular traffic or

V

opposite car traffic lanes; a **v. s.** in car parking areas are called ▶planting island; ▶grass strip 1, ▶median strip [US]/central reservation [UK], ▶planting strip); *s* **banda [f] verde central** (BU 1959; franja de separación de calzadas en autopistas, carreteras o calles; ▶banda de plantación, ▶banda verde, ▶franja de hierba, ▶franja divisoria central); *f* **bande [f] verte d'accompagnement** (Espace linéaire végétalisé séparant les voies d'une chaussée ou différents types de circulation ; ▶bande engazonnée, ▶bande plantée, ▶îlot de plantation, ▶terre-plein central) ; *g* **Grünstreifen [m]** (Mit Pflanzenbewuchs ausgestattete, lineare Fläche zur Trennung von Fahrbahnen oder Verkehrsarten; ▶Grasstreifen, ▶Mittelstreifen, ▶Pflanzstreifen 1 u. ▶Pflanzstreifen 2).

vegetating technique [n] *constr.* ▶planting technique.

6751 vegetation [n] *phyt.* (**1.** Total plant cover of an area, soil, forest, etc.; ▶chara vegetation, ▶existing vegetation, ▶wetheath vegetation, ▶xeric vegetation. **2.** Character of plant growth; e.g. lush vegetation, sparse vegetation); *s* **vegetación [f]** (**1.** Tapiz vegetal que resulta de la disposición en el espacio de los diferentes tipos de vegetales presentes en una porción cualquiera del territorio; DINA 1987; ▶vegetación real, ▶vegetación xerófita. **2.** Forma de crecimiento de las plantas, p. ej. vegetación exuberante/rala); *f* **végétation [f] (3)** (**1.** Ensemble des végétaux recouvrant le sol ; ▶végétation réelle, ▶végétation xérique. **2.** Mode de croissance des végétaux, p. ex. végétation luxuriante) ; *g* **Vegetation [f]** (**1.** Gesamtheit der Pflanzen, die den Boden bedecken; ▶reale Vegetation, ▶Trockenvegetation. **2.** Pflanzenwuchs, z. B. üppige/spärliche Vegetation); *syn.* Bewuchs [m].

vegetation [n]**, chasmophytic** *phyt.* ▶vegetation of rock clefts.

vegetation [n]**, existing** *contr. constr.* ▶damage to existing vegetation; *constr. leg.* ▶preservation of existing vegetation.

vegetation [n]**, reed swamp** *phyt.* ▶reed swamp.

6752 vegetation altitudinal zone [n] *geo. phyt.* (Generic term for such ▶altitudinal belts as ▶planar zone, ▶colline zone, ▶montane zone, ▶subalpine zone, ▶alpine zone, ▶nival zone); *syn.* vegetation belt [n]; *s* **piso [m] de vegetación** (En la ▶zonación altitudinal de la vegetación cada uno de los niveles en los que se subdivide aquélla como ▶piso planar, ▶piso colino, ▶piso montano, ▶piso subalpino, ▶area cacuminal, ▶piso nival, y en los que predominan comunidades vegetales características para cada uno de ellos); *f* **étage [m] de végétation** (Unité de végétation correspondant à une amplitude altitudinale, généralement caractérisée par une espèce dominante, expression de l'▶étagement de la végétation en altitude comme l'▶étage de plaine, l'▶étage collinéen, l'▶étage montagnard, l'▶étage subalpin, l'▶étage alpin, l'▶étage nival ; DEE 1982) ; *g* **Vegetationsstufe [f]** (▶Höhenstufung wie ▶planare Stufe, ▶kolline Stufe, ▶montane Stufe, ▶subalpine Stufe, ▶alpine Stufe, ▶nivale Stufe, die jeweils durch charakteristische Pflanzengesellschaften geprägt ist).

vegetation belt [n] *phyt. geo.* ▶vegetation altitudinal zone.

6753 vegetation climatic zone [n] *geo. phyt.* (Climatic region of the world, usually corresponding to a latitude, in which particular ▶plant formations grow, e.g. evergreen rain forests, boreal coniferous forests, humid, warm-temperate forests, tundra); *syn.* vegetation zone [n]; *s* **zona [f] de vegetación** (Amplias zonas geográficas con un tipo de vegetación característico que depende en primer lugar del clima, pero también de otras condiciones mesológicas, como desierto helado, tundra, taiga, bosque húmedo templado, selva tropical, etc.; cf. DINA 1987, 984; ▶formación vegetal); *f* **zone [f] de végétation** (Région géographique dont le climat et la nature des ▶formations végétales sont

déterminés par la latitude, p. ex. les forêts équatoriales, tropicales, méditerranéennes sempervirentes, tempérées caducifoliées, boréales, toundra ; DEE 1982, 93) ; *g* **Vegetationszone [f]** (Den Klimazonen der Erde zugeordnete, meist breitenkreisparallel verlaufende Gebiete, in denen bestimmte ▶Pflanzenformationen wachsen wie z. B. immergrüne Regenwälder, feuchte, warmtemperierte Wälder, boreale Nadelwälder, Tundra).

6754 vegetation cover [n] *landsc. urb.* (Existing vegetation in a defined settlement area); *s* **cubierta [f] de vegetación** (Toda la vegetación existente en una zona urbana dada); *f* **couvert [m] végétal existant** (Ensemble du couvert végétal d'une unité spatiale en milieu urbain) ; *g* **Grünbestand [m]** (Gesamtheit des Pflanzenbewuchses eines Betrachtungsraumes im Siedlungsbereich).

vegetation ecology [n] *phyt.* ▶phytosociology, ▶plant ecology.

6755 vegetation establishment [n] **on entisol** [US] *constr. land'man.* (▶establishment of vegetation, ▶entisol [US] 1/immature soil [UK]); *syn.* vegetation establishment [n] on immature soil [UK]; *s* **establecimiento [m] de vegetación en suelo mineral bruto** (▶establecimiento de vegetación, ▶suelo mineral bruto); *syn.* plantación [f] de suelos minerales brutos; *f* **végétalisation [f] de sol brut** (*Insertion paysagères d'ouvrages* Méthode de ▶végétalisation de ▶sols peu évolués caractérisés par l'absence de support de végétation [roches mères affleurantes, sols à teneur en matières organiques inférieure à 1 % ou à activité biologique faible ou inexistante] ou par un support de végétation superficiel, compacté ; la végétalisation vise à installer un peuplement initial assuré par les espèces semées constituant un milieu d'accueil pour la flore spontanée locale qui, suivant une succession de communautés transitoires, remplacera progressivement les espèces semées) ; *g* **Rohbodenbegrünung [f]** (Begrünungsmethode von [nahezu] oberbodenlosen Flächen, die eine oft sehr langsam anlaufende Initialphase der natürlichen Sukzession beschleunigt, um schneller eine erosionsstabile Vegetationsentwicklung zu ermöglichen; ▶Begrünung, ▶Rohboden); *syn.* oberbodenlose Begrünung [f].

vegetation establishment [n] **on immature soil** [UK] *land'man.* ▶vegetation establishment on entisol [US].

6756 vegetation geography [n] *geo.* (Research branch of geography concerned with the differing character of vegetation in various regions of the world, whereby individual plants and their communities are not the major objects of interest, but rather whole landscapes with their vegetation patterns; ▶plant geography); *s* **biogeografía [f]** (Ciencia con el mismo carácter de amplitud conceptual como la geografía que estudia los paisajes biológicos desde múltiples puntos de vista: corológico, ecológico-fisonómico, biocenológico, sociológico y evolutivo; pero siempre haciéndolo estructural, funcional y genéticamente así como en sus distribuciones espaciales; cf. DGA 1986; ▶fitogeografía); *f* **biogéographie [f]** (Courant dans la conception de la géographie se libérant des éléments prépondérants de la géomorphologie et se référant à une appréhension globale des systèmes de paysages [domaine, faciès, unités géographiques globales] et une approche perceptive des paysages ; cf. GEP 1991, 78 ; ▶phytogéographie) ; *syn.* géographie [f] de la végétation ; *g* **Vegetationsgeographie [f]** (Erdkundlicher Forschungszweig, der sich mit dem unterschiedlichen Charakter der Vegetation in den verschiedenen Gegenden der Erde befasst, wobei nicht primär die einzelnen Pflanzen der Pflanzengesellschaften das eigentliche Forschungsobjekt sind, sondern ganze Landschaften mit ihrer Vegetation; cf. ÖKO 1983; ▶Pflanzengeografie).

6757 vegetation impact study [n] *envir. phyt.* (Monitoring procedure to assess the continuity or change of ecological

V

factors in an area by mapping the existing vegetation before and after an ▶intrusion upon the natural environment or encroachment which may have an adverse impact on the environment, especially by altering the water table or nutrient regime; ▶existing condition before planning starts); *s* **procedimiento [m] de comprobación de efectos** (≠) (D., proceso jurídico para demostrar las consecuencias causadas en la vegetación por un impacto en el medio producido por una obra de construcción o por el uso de un recurso; ▶estado cero, ▶intrusión en el entorno natural); *f* **procédure [f] de suivi environnemental par relevé de végétation** (≠) (D., procédure juridique utilisée pour mettre en évidence la continuité ou la variation des facteurs écologiques dans un milieu ; à cet effet on procédé à un inventaire floristique et est établi un relevé des unités phytosociologiques selon le principe BACI [Before-After Control-Impact/contrôle des impacts d'une atteinte portée sur le milieu naturel par comparaison avec l'état initial], en particulier lorsqu'on peut s'attendre à une perturbation du cycle de l'eau et des substances nutritives ; ▶atteinte au milieu naturel, ▶état initial du site et de l'environnement) ; *g* **Beweissicherungsverfahren [n] durch Vegetationskartierung** (Verfahren, das die Kontinuität oder den Wandel von ökologischen Faktoren in einem Gebiet nachvollziehbar macht, indem die tatsächlich vorhandenen Vegetationseinheiten vor und nach einem ▶Eingriff in Natur und Landschaft aufgenommen werden, insbesondere bei zu erwartenden Änderungen des Wasser- und Nährstoffhaushaltes; ▶Zustand vor Planungsbeginn).

6758 vegetation layer [n] *phyt.* (Component of the ▶stratification of plant communities); *s* **estrato [m] de vegetación** (Una de las capas en la ▶estratificación de la vegetación); *syn.* capa [f] de vegetación; *f* **strate [f] de végétation** (caractérisant la ▶stratification de la végétation); *g* **Vegetationsschicht [f]** (Bestandteil der ▶Vegetationsschichtung).

6759 vegetation management [n] *for. hort. land'man.* (Horticultural measures to improve the growth of existing vegetation by creating favo[u]rable site conditions; ▶maintenance work, ▶park management program [US]/park management programme [UK]); *s* **mantenimiento [m] de la vegetación** *hort.* (Medidas horticulturales para promover el crecimiento sano de la vegetación; ▶plan de conservación de jardín patrimonial, ▶trabajos de mantenimiento); *f* **soins [mpl] culturaux (d'un peuplement forestier)** (Opérations culturales [tailles, coupes, dégagement, désherbage, travail du sol] effectuées en fonction de l'origine et la nature [naturelle, artificielle], des essences, de l'âge d'un peuplement en vue d'assurer la mise en bonnes conditions de croissance [équilibre, résistance], d'améliorer la qualité du bois produit ainsi que d'assurer sa pérennité ; ▶entretien courant, ▶programme de gestion et d'entretien d'un parc) ; *syn.* soins [mpl] sylvicoles, entretiens [mpl] culturaux ; *g* **Bestandspflege [f, o. Pl.]** (1. Maßnahmen, die auf den Entwicklungsstand von Vegetationsbeständen mit dem Ziel abgestimmt werden, den Bestand zu sichern und günstige Wachstumsvoraussetzungen für die Weiterentwicklung zu schaffen; ▶Parkpflegewerk, ▶Pflege. **2.** Bei der **waldbaulichen B.** sind es Maßnahmen, die günstige Wachstumsvoraussetzungen für die Weiterentwicklung von wuchskräftigen und qualitativ hochwertigen Bäumen schaffen und nachhaltig sichern, um Kronen- und Zuwachsentwicklung zu fördern. Bei der Pflege werden schwache, kranke, abgestorbene und bedrängende Bäume entnommen. Es werden drei Hauptpflegemaßnahmen unterschieden: Jungwuchspflege, Läuterung und Durchforstung); *syn.* Pflege [f, o. Pl.] eines Bestand(e)s; *syn. zu 2.* Waldbestandspflege [f, o. Pl.], Waldpflege [f, o. Pl.].

6760 vegetation map [n] *phyt.* (Cartographic representation documenting the distribution of plants upon the Earth's surface or parts thereof, according to certain themes; e.g. plant communities, plant formations, zonation of vegetation); *s* **mapa [m] de vegetación** (Mapa temático en el que está representada la vegetación de la tierra o parte de ella según determinados criterios como comunidad, formación vegetal, etc.); *f* **carte [f] de la végétation** (Représentation cartographique thématique du couvert végétal selon divers critères définis ; on distingue les cartes de groupements végétaux, les cartes de groupements physiologiques, les cartes des séries de végétation) ; *g* **Vegetationskarte [f]** (Thematische Karte, die das Pflanzenreich nach bestimmten Einteilungskriterien [Gesellschaften, Formationen, zonenmäßige Abfolge] für die Erdoberfläche oder deren Teile darstellt).

6761 vegetation [n] **of animal rest area** *phyt.* (Community of perennials mostly of anthropogenic species composition, e.g. *Rumicetum alpinii*, on fertile soil which has been created over decades or centuries with the accumulation of animal droppings during grazing or in places where wild animals have their lairs; ▶pasture weeds vegetation; *specific term* ▶vegetation of cattle grazing areas; *generic term* ▶weed community); *syn.* vegetation [n] of animal resting area; *s* **vegetación [f] de lugares de reposo** (Comunidad herbácea de especies ruderales, p. ej. *Rumicetum alpinii*, formada en los lugares que han servido como pastos o reposaderos de animales salvajes o de ganado durante largo tiempo, con la correspondiente acumulación de excrementos; cf. BB 1979; ▶comunidad arvense de pastizales; *término genérico* ▶comunidad de malas hierbas; *términos específicos* ▶vegetación de lugares de reposo de ganado, majada [vegetación de reposaderos de ganado a 1600-1900 m de altitud especializada en tolerar el pisoteo], vegetación de lugares de reposo de animales salvajes; cf. BB 1979); *syn.* vegetación [f] de reposadero; *f* **végétation [f] de reposoir** (Groupement herbacé composé d'espèces rudérales, p. ex. *Rumicetum alpinii*, formé sur les stations enrichies pendant des décennies ou des siècles par les déjections des animaux d'embouche ou sauvages ; ▶végétation des prairies pacagées négligées ; *termes spécifiques* ▶végétation de reposoir à bétail, végétation de reposoir de gibier ; *terme générique* ▶flore adventice de cultures et de prairies) ; *g* **Lägerflur [f]** (Staudengesellschaft aus vorwiegend anthropogenen Pflanzenkombinationen, z. B. Alpenampferlägerflur *[Rumicetum alpinii]*, die auf nährstoffreichen Böden wachsen und durch jahrzehnte- oder jahrhundertelange Kotanhäufung durch Weidebetrieb oder Wildliegeplätze entstanden sind; nach ELL 1978; ▶Weideunkrautflur; *UBe* ▶Viehlägerflur, Wildlägerflur; *OB* ▶Unkrautflur).

vegetation [n] **of animal resting area** *phyt.* ▶vegetation of animal rest area.

6762 vegetation [n] **of cattle grazing areas** *phyt.* (Community of perennial plants in a meadow, which are not grazed upon by cattle; ▶vegetation of rest areas); *s* **vegetación [f] de los lugares de reposo de ganado** (Vegetación de perennes que el ganado no consume; *término genérico* ▶vegetación de los lugares de reposo); *f* **végétation [f] de reposoir à bétail** (Peuplement fortement piétiné dominé en général par un pourcentage plus ou moins important de mégaphorbiaies que l'on rencontre sur les pâturages intensifs, les pâturages continus avec fourrage supplémentaire ou les pâturages de montagne ; *terme générique* ▶végétation de reposoir) ; *g* **Viehlägerflur [f]** (Durch Weidebetrieb entstandene Staudengesellschaft, die vom Weidevieh nicht gefressen wird; *OB* ▶Lägerflur).

vegetation [n] **of cornfield weeds** *phyt.* ▶segetal community.

6763 vegetation [n] **of perennial halophytes** *phyt.* (Plant community existing between land and sea, principally composed of salt-tolerant perennials; ▶salt meadow); *s* **vegetación [f] halófila perenne** (Comunidad vegetal en la zona de

V

transición entre la tierra y el mar, constituida mayormente de vivaces que toleran la sal; ►prado salino); *f* **végétation [f] halophile herbacée vivace** (Groupement végétal dans la zone de transition entre la terre et la mer, principalement consistant de plantes vivaces ; ►pré salé) ; *g* **Salzstaudenflur [f]** (Pflanzengesellschaft im Kampfbereich zwischen Land und Meer, hauptsächlich aus salztoleranten perennierenden Pflanzen bestehend; ►Salzwiese); *syn.* Salzpflanzengesellschaft [f].

6764 vegetation [n] of rock clefts *phyt.* (Plants of the ►rock crevice plant community); *syn.* vegetation [n] of rock crevices, chasmophytic vegetation [n]; *s* **vegetación [f] fisurícola** (Vegetación que vive en las quiebras de las peñas, sobre lo poco de tierra que hay en ellas. Puede contener, y contiene frecuentemente, vegetación de orden superior, incluso leñosa y hasta árboles; DB 1985; ►comunidad fisurícola); *syn.* vegetación [f] de casmófitos; *f* **végétation [f] chasmophytique** (Végétaux saxicoles appartenant à l'►association rupicole des fissures de rocher ; ►association des fissures de rocher, ►association pariétale) ; *syn.* florule [f] chasmophytique, végétation [f] des fissures de rochers ; *g* **Felsspaltenbewuchs [m, o. Pl.]** (Pflanzen der ►Felsspaltengesellschaft); *syn.* Felsspaltenflur [f].

vegetation [n] of rock crevices *phyt.* ►vegetation of rock clefts.

6765 vegetation [n] of wall joints, rock crevices or paving joints *phyt.* (►rock crevice plant community, ►plant association of pavement joints, ►wall community); *s* **vegetación [f] de rendijas** (Crecimiento de vegetación en las rendijas de muros, entre rocas o entre losas del pavimento; ►comunidad fisurícola, ►comunidad de las grietas de muros, ►comunidad de rendijas [de pavimentos]); *f* **végétation [f] saxicole et muricole** (*Terme générique* la végétation des murs, des interstices de pavés et de rochers telles que le ►groupement saxicole, l'►association des fissures de rochers, l'►association pariétale, l'►association des corniches rocheuses, le groupement muricole et le ►groupement des interstices des pavés) ; *g* **Fugenbewuchs [m, o. Pl.]** (Pflanzenwuchs in Mauer- und Pflasterfugen oder Felsspalten; ►Felsspaltengesellschaft, ►Mauerfugengesellschaft, ►Pflasterritzengesellschaft).

6766 vegetation removal [n] for replanting *constr.* (Careful digging up of desired plants to save for relocation on the project site; e.g. before construction; ►heeling-in); *s* **arranque [m] de leñosas con cepellón** (Extracción de plantas aprovechables antes de comenzar una obra para utilizarlas posteriormente; cf. PGT 1987; ►enterrado de plantas); *f* **enlèvement [m] des végétaux réutilisables** (Extraction des végétaux destinés à être réutilisés ; *syn.* ►mise en jauge) ; *syn.* arrachage [m] des plantes réutilisables ; *g* **Herausnahme [f] des verwertbaren Pflanzenbestandes** (Ausgraben von wieder verwendbaren Pflanzen auf einer Baustelle vor Baubeginn; ►Einschlag).

vegetation science [n] *phyt.* ►phytosociology.

vegetation screen [n] *landsc.* ►screen planting (2), ►shelterbelt (1).

6767 vegetation survey [n] *phyt.* (Large-scale inventory and mapping of vegetation patterns; ►sampling 2); *syn.* plant ecological survey [n]; *s* **inventario [m] de los tipos de vegetación** (►Inventario fitosociológico a gran escala de las formaciones de vegetación); *f* **relevé [m] de formations végétales** (Étude de la composition floristique de stations d'un périmètre important ; ►relevé phytosociologique) ; *g* **großräumige Vegetationsaufnahme [f]** (Großräumige Erhebung von Vegetationsstrukturen; ►Vegetationsaufnahme).

vegetation zone [n] *phyt. geo.* ►vegetation altitudinal zone; ►vegetation climatic zone.

vegetative cover [n] *phyt.* ►degradation of vegetative cover, ►degree of total vegetative cover, ►density of vegetative cover, ►plant cover.

vegetative cover [n], percentage of *phyt.* ►degree of total vegetative cover.

6768 vegetative mat [n] *constr. hort.* (Compact mass of leaf and stem tissue resulting from herbaceous or ground cover plants growing so closely together that stems, foliage and roots are entwined to an undesirable extent; cf. BS 3975: part 5; ►thatch); *s* **cobertura [f] herbácea densa** (Masa compacta de tejido de hojas y tallos resultante de plantas herbáceas o tapizantes que crecen tan juntas que los tallos, las hojas y las raíces se entremezclan hasta un extremo no favorable; ►fieltro); *f* **couverture [f] végétale dense** (►feutre de gazon) ; *g* **dichter Bewuchs [m, o. Pl.]** (Dicht gewachsene Pflanzendecke; ►Rasenfilz).

vehicle barrier [n] [UK] *trans.* ►guardrail [US] (1)/crash barrier [UK].

vehicle crossing [n] [UK] *urb.* ►driveway.

vehicle entrance [n] [UK] *urb.* ►driveway.

6769 vehicle turning circle [n] *constr. plan.* (SPON 1986, 300; SAF 1983, 313; circle described by the farthest reaching part of a vehicle turning at its maximum steering lock, and other circles for different sized vehicles, measured by turning radii; ►compound curve [US]/sweep circle [UK]); *s* **radio [m] de giro** (Círculo descrito por la parte más externa de un vehículo al girarse su volante al máximo; ►curva de arrastre); *f* **rayon [m] (extérieur) de braquage** (Cercle décrit par les roues extérieures des véhicules et définissant les dimensions minimales de certaines circulations [aire de retournement, rayon de courbure des voies] ; VRD 1986, 181 ; ►épure de giration) ; *g* **Wendekreis [m]** (Derjenige Kreis, der beim stärksten Lenkeinschlag eines Kraftfahrzeugs durch seine am weitesten nach außen stehenden Fahrzeugteile beschrieben wird; ►Schleppkurve).

vehicle turning radii [npl] [US] *trans.* ►compound curve [US]/sweep circle [UK].

vehicular access [n] *trans.* ►limitned vehicular access.

6770 vehicular and pedestrian easement [n] *leg. urb.* (**In Europe**, provision for permanent right of passage to or through private property, or on private property along a public right-of-way, shown on a legally-binding land-use plan, such as a ►Proposals Map [UK], and recorded in a land register [UK]. **In U.S.**, such an easement has legal validity only when recorded with county deed registers, maps, property plats, etc. Representation on a ►zoning map [US] has nothing to do with the legal validity of a vehicular or pedestrian right-of-way. Furthermore, in many if not most states, legally valid pedestrian rights of way may be acquired by the public through a specialized application of the law of adverse possession known as 'prescriptive rights', which would rarely, if ever, be indicated on a map. They are acquired through open, notorious use by the public in a manner adverse to the interest of the owner of the fee simple estate for a specified period, which may be as long as 21 years. In short, vehicular and pedestrian easements and other less-than-fee interests are typically, but not always, acquired through registered land transactions. Current practice in the subdivision of land leans towards requiring the developer to dedicate a fee simple interest in vehicular and pedestrian areas, rather than a mere right-of-way or easement; ►easement, ►right of public access, ►vehicular easement); *syn.* vehicular and pedestrian right-of-passage [n]; *s* **servidumbre [f] de acceso** (Derecho permanente de un grupo de personas o de la población en general a transitar por un terreno privado para llegar a otro. En

V

Es. la Ley de Costas fija la **s. de a.** al mar en zonas urbanas o urbanizables tanto para peatones [cada 200 m] como para vehículos [cada 500 m]; cf. art. 27, Ley 22/1988, de 28 de julio; ▶plan parcial [de ordenación], ▶servidumbre de acceso para vehículos, ▶servidumbre de tránsito, ▶servidumbre inmobiliaria); ***f*** **servitude [f] de passage** (Droit constituant une des servitudes d'utilité publique affectant l'utilisation du sol et apparaissant sur les documents graphiques du ▶plan local d'urbanisme [PLU]/plan d'occupation des sols [POS] et sur les registres du cadastre et réglementant l'accès ou le passage d'un groupe d'utilisateurs ou du public sur le domaine privé ou public. La servitude de passage permettant à un propriétaire dont les fonds n'ont pas d'issue sur une voie publique d'assurer la desserte de ses fonds sur ceux de ses voisins est réglementée par l'art. 682 du C. civil et la loi n° 67-1253 ; ▶servitude, ▶servitude de passage des engins mécaniques, ▶servitude de passage des piétons, ▶servitude de passage des véhicules ; *terme spécifique* **servitude [f] de marchepied**, qui oblige tout propriétaire, locataire, fermier ou titulaire d'un droit réel, riverain d'un cours d'eau ou d'un lac domanial à ne pas planter ni clore les rives d'un cours d'eau ou d'un lac domanial) ; *syn.* droit [m] de passage ; ***g*** **Geh- und Fahrrecht [n]** (In ▶Bebauungsplänen gem. § 9 [1] 21 BauGB ausgewiesenes und im Grundbuch eingetragenes Recht [▶Grunddienstbarkeit] über ein Grundstück zu Gunsten der Allgemeinheit oder eines beschränkten Nutzerkreises eine Wegeerschließung/Zuwegung zu ermöglichen; ▶Betretungsrecht, ▶Fahrrecht).

6771 vehicular and pedestrian infrastructure [n] *trans. urb.* (Sum of facilities and installations used by motor-vehicles, cyclists and walkers; i.e. roadways, pathways, cycleways and sidewalks; ▶transportation infrastructure [US]/traffic infrastructure [UK]); ***s*** **superficies [fpl] viarias** (Conjunto de vías de tránsito en una zona determinada que, dependiendo del contexto, pueden incluir las vías de ferrocarril; ▶infraestructura de tráfico y transportes [Es]/infraestructura de tránsito y transportes [AL]); ***f 1*** **circulations [fpl]** (Ensemble des espaces de circulation ou de déplacement utilisés par les piétons, les cycles et les véhicules automobiles ; ▶infrastructure de transport) espaces [mpl] de circulation ; ***f 2*** **voirie [f]** (Ensemble des voies de circulation pour les cycles et les véhicules automobiles) ; ***g*** **Verkehrsflächen [fpl]** (Gesamtheit der Verkehrswege eines Betrachtungsraumes; je nach Erfordernis werden auch Flächen für den Eisenbahnverkehr hinzugerechnet; ▶Verkehrsinfrastruktur).

vehicular and pedestrian right-of-passage [n] *leg. urb.* ▶vehicular and pedestrian easement.

6772 vehicular easement [n] *leg. urb.* (**In Europe**, permanent public right of vehicular passage to or through private property, or on private property along a public ▶right-of-way, shown on a ▶zoning maps [US]/Proposals Map [UK], and recorded in a land register [UK]. **In U.S.**, a **v. e.** has legal validity only when recorded on county deed registers, maps, property plats, etc. This is a very complex area of the law, which varies greatly from state to state; there is also an unrecorded **prescriptive** ▶**easement** which is acquired by adverse use for a certain period which varies from state to state); ***s*** **servidumbre [f] de acceso para vehículos** (Derecho permanente del público en general de acceder con vehículo a través de o a una propiedad privada; ▶servidumbre de servicios públicos, ▶servidumbre inmobiliara); ***f 1*** **servitude [f] de passage de véhicules** (Servitude permettant à un propriétaire dont les fonds n'ont pas d'issue sur la voie publique d'assurer la desserte de ses fonds sur ceux de ses voisins ; cf. C. civil, art. 682, loi d'orientation foncière n° 67-1253. **F.**, on distingue entre autres la ▶servitude de passage des engins mécaniques, la servitude de passage et d'aménagement

pour assurer la continuité de la lutte contre les incendies, etc. ; ▶servitude d'utilité publique) ; ***f 2*** **servitude [f] de passage des engins mécaniques** (Servitude permettant l'entretien des canaux d'irrigation et d'assainissement ; cf. loi n° 92-1283) ; *syn.* droit [m] de passage des engins mécaniques ; ***g*** **Fahrrecht [n]** (Im ▶Bebauungsplan festgesetzte ▶Grunddienstbarkeit, mit der gem. § 9 [1] 21 BauGB private Grundstücksflächen zugunsten der Allgemeinheit oder eines beschränkten Benutzerkreises belegt werden können; ▶Geh-, Fahr- und Leitungsrecht).

vehicular entrance [n], roofed *arch.* ▶porte cochère.

vehicular intersection [n] *trans.* ▶traffic intersection.

velocity [n] of groundwater flow *hydr.* ▶groundwater velocity.

veneer concrete [n] *constr.* ▶facing concrete.

veneer(ed) construction [n] *arch. constr.* ▶veneering.

veneered wall [n] *arch. constr.* ▶veneer masonry.

6773 veneering [n] *arch. constr.* (Masonry facing fastened to a wall, but not structurally bonded to it; ▶veneer masonry, ▶veneering with panels); *syn.* veneer(ed) construction [n] (DAC 1975); ***s*** **enchapado [m]** (Paramento exterior en ▶mampostería enchapada; ▶paramento de losas); ***f*** **parement [m]** (Revêtement vertical d'un mur constituant l'habillage du ▶maçonnerie à parement ; ▶placage) ; *syn.* revêtement [m] vertical (MAÇ 1981) ; ***g*** **Verblendung [f]** (Verkleidung als Vormauerverband bei einem ▶Verblendmauerwerk; verblenden [vb] ▶Plattenverkleidung); *syn.* Vormauerschale [f].

6774 veneering [n] with panels *arch. constr.* (Facing of concrete walls with a non-load-bearing external layer of, e.g. ▶natural stone slabs, exposed aggregate slabs or other facing material; ▶veneering); *syn.* cladding [n] with panels [also UK]; ***s*** **paramento [m] de losas** (Enchapado p. ej. de un muro de hormigón con una capa exterior no portante de ▶losas de piedra [natural], losas de hormigón o de otro material de cara vista; ▶enchapado); *syn.* chapado [m] de losas; ***f*** **placage [m]** (Revêtement en extérieur ou en intérieur de surfaces verticales exécuté au moyen d'éléments de 3 à 8 cm d'épaisseur en pierre [plaque de parement], en béton, en bois posé, scellé ou accroché ; ▶parement) ; *syn.* habillage [m] par plaques ; ***g*** **Plattenverkleidung [f]** (Vorhängen von meist Naturstein- oder Betonplatten, z. B. vor eine Fassade; ▶Verblendung).

6775 veneer masonry [n] *arch. constr.* (Any wall having an attached though not bonded or loadbearing facing; *specific terms* brick veneer wall, brick-facing wall; ▶two-tier masonry, ▶wall face); *syn.* veneered wall [n]; ***s*** **mampostería [f] enchapada** (▶Mampostería de dos capas con núcleo de hormigón o de mampuestos y una ▶cara vista de un muro de piedras naturales o artificiales de calidad; *término específico* paramento de ladrillo; cf. DACO 1988); ***f*** **maçonnerie [f] à parement** (▶Maçonnerie à double paroi constituée d'un corps principal de mur en béton et d'un habillage décoratif de grande valeur épais de pierres naturelles ou artificielles ; lorsque les deux faces du mur sont parementées, on parle d'un mur à ▶parement double ; *terme spécifique* maçonnerie à parement en briques) ; *syn.* mur [m] parementé ; ***g*** **Verblendmauerwerk [n]** (▶Zweischaliges Mauerwerk mit Betonkern oder Hintermauerung und einer Sichtfläche [▶Mauerhaupt] mit hochwertigen natürlichen oder künstlichen Steinen verkleidet, ‚verblendet'. Sind beide Seiten mit Vormauerverbänden versehen, so spricht man von einem ‚zweiseitigen Verblendmauerwerk'; *UB* Ziegelverblendmauerwerk); *syn.* Verblendmauer [f].

venery [n] *hunt.* ▶game hunting.

ventilated roof [n] ▶cold roof.

V

6776 ventilation [n] *met. urb.* (Exchange or movement of fresh air in a room or urban settlement; ▶air exchange); *s* **ventilación** [f] **(2)** (Provisión de aire fresco a la ciudad o en espacios cerrados; ▶ventilación 1); *syn.* aireación [f]; *f* **ventilation** [f] (Alimentation en air frais des quartiers urbains ; ▶brassage de l'air) ; *g* **Durchlüftung** [f] (Versorgung von Stadtteilen oder Räumen mit Frischluft; ▶Luftaustausch); *syn.* Belüftung [f] (2), Ventilation [f].

ventilation corridor [n] *landsc. met.* ▶fresh air corridor.

6777 venturi-effect [n] *landsc.* (Acceleration in wind speed [or water flow rates] caused by a narrowing of the ▶wind corridor; e.g. between high-rise blocks of buildings or windbreaks, etc.); *s* **efecto** [m] **venturi** (Aceleración de la velocidad del viento causada por el estrechamiento del canal de flujo, p. ej. entre bloques de casas; ▶canal de viento); *f* **effet** [m] **venturi** (Accélération de la vitesse du vent due à un écoulement dans un passage étroit, p. ex. entre des bâtiments, des haies de protection, une gorge, etc. ; *terme spécifique* **effet de vallée** lorsqu'une vallée à versants prononcés canalise le vent, pouvant provoquer un changement du vent dominant et une accélération conséquente de sa vitesse ; ▶couloir de vent) ; *syn.* effet [m] entonnoir ; *g* **Düseneffekt** [m] (Verengung des Winddurchlasses und dadurch bedingte Beschleunigung der Windgeschwindigkeit bei Windschutzpflanzungen, zwischen Häuserblocks etc.; ▶Windschneise).

verified delivery notes [npl] [UK]**, on proof of** *constr. contr.* ▶according to certified delivery [US].

version [n]**, amended** *plan.* ▶revised version.

version [n]**, up-dated** *plan.* ▶revised version.

vertebrates [npl]**, native** *conserv.* ▶wildlife.

6778 vertical aerial photograph [n] *envir. plan. rem' sens. surv.* (Picture taken from an aeroplane with the camera positioned at a 90° angle to the ground; ▶oblique photograph); *s* **fotografía** [f] **aérea vertical** (Foto aérea tomada desde un avión en ángulo de 90° con respecto a la superficie del terreno; ▶fotografía oblicua); *f* **photographie** [f] **verticale** (Photo prise avec un angle de 90° à partir d'un avion ; ▶photographie oblique) ; *g* **Senkrechtluftbild** [n] (Fotografie, die in einem 90°-Winkel aus einem Flugzeug gemacht wird; ▶Schrägaufnahme); *syn.* Senkrechtaufnahme [f].

6779 vertical alignment [n] *trans.* (Delineated longitudinal road profile showing proposed downhill and uphill slope with marked changes in elevation; ▶vertical profile); *syn.* gradient profile [n] [also US]; *s* **alineación** [f] **vertical** (Delineación longitudinal del perfil de una vía de transporte que muestra cuestas y declives y los cambios de altitud; ▶trazado en perfil longitudinal); *f* **profil** [m] **en long d'un tracé** (Ligne représentant sur un plan le tracé en hauteur d'une voie routière ; ▶cotation altimétrique et planimétrique d'un tracé) ; *g* **Gradiente** [f] **einer Trasse** (Linie, die den Höhenverlauf der Längsachse [Längsschnitt] einer Straße im Plan festlegt; anhand der **G.** können ▶Steigungen 2 und ▶verlorene Steigungen berechnet werden; ▶Höhenabwicklung einer Verkehrstrasse); *syn.* Nivellette [f] [A].

vertical curve [n] **of a road crest** [UK] *trans.* ▶crest vertical curve [US].

6780 vertical erosion [n] *geo.* (Scouring of a streambed or river bottom due to waterflow, bed load, and large-scale changes in the level of the earth's crust—epeirogenesis; ▶scour, ▶watercourse bed erosion); *s* **erosión** [f] **lineal** (Profundización del lecho de cursos de agua por levantamiento de la corteza terrestre; ▶erosión del lecho, ▶socavación); *f* **érosion** [f] **verticale** (DG 1984, 167 ; phénomène d'▶enfoncement des talwegs résultant d'un vaste soulèvement de la croûte terrestre ou de l'abaissement général du niveau de la mer [épirogénèse] ; ▶érosion du lit d'un cours d'eau) ; *syn.* érosion [f] linéaire ; *g* **Tiefenerosion** [f] (▶Eintiefung der Fließgewässersohle durch großräumige Anhebung der Erdkruste [Epirogenese]; ▶Sohlenerosion); *syn.* rückschreitende Erosion [f], Tiefenschurf [m].

vertical garden [n] [CDN] *gard. urb.* ▶façade planting.

vertical greenery [n] *gard. urb.* ▶façade planting.

vertical green space [n] *gard. urb.* ▶façade planting.

vertical joint [n] *arch. constr.* ▶heading joint.

vertical landscaping [n] *gard. urb.* ▶façade planting.

6781 vertical profile [n] *plan. trans.* (Stationing of levels in the form of vertical curves determined along the longitudinal axis in the planning of a travelled way [US]/carriage-way [UK]; these parapolic curves are used to change the slope of a road; ▶horizontal alignment; ▶road crest, ▶road sag [US]/road dip [UK], ▶vertical alignement); *s* **trazado** [m] **en perfil longitudinal** (En la planificación del trazado de vía de transporte, determinación de las cotas de altura del eje longitudinal y representación en forma de curvas verticales; ▶alineación horizontal, ▶alineación vertical, ▶cambio de rasante, ▶depresión 2, ▶trazado en planta); *syn.* trazado [m] en alzado; *f* **cotation** [f] **altimétrique et planimétrique d'un tracé** (Détermination de la cotation en altimétrie et en planimétrie de l'axe d'un projet d'infrastructure de circulation ; ▶courbe concave en profil en long, ▶courbe convexe en profil en long, ▶définition des caractéristiques géométriques principales du tracé, ▶profil en long d'un tracé) ; *syn.* dimensionnement [m] altimétrique du profil en long ; *g* **Höhenabwicklung** [f] **einer Verkehrstrasse** (Planung des Längsprofils und der Querprofile mit allen Höhenangaben für ein Verkehrswegeprojekt; ▶Kuppe 2, ▶Lagetrassierung, ▶Wanne, ▶Gradiente einer Trasse).

6782 vertical scale [n] *arch. constr. eng. surv.* (Proportional representation of heights shown on a cross section; ▶horizontal scale); *s* **escala** [f] **vertical** (En un plano o mapa relación entre las alturas representadas y las reales; ▶escala horizontal); *f* **échelle** [f] **des hauteurs** (Rapport de la hauteur mesurée sur un document et sa hauteur réelle qui est indiqué sur une coupe ou un profil ; ▶échelle des longueurs) ; *g* **Höhenmaßstab** [m] (In einem gezeichneten Profil oder einem Querschnitt benutztes Verhältnis von Höhen in der Darstellung und denen in der Wirklichkeit; ▶Längenmaßstab).

vertical slatted fence [n] *constr.* ▶picket fence.

6783 vertisol [n] *pedol.* (Soil with high shrink-swell potential, that has wide, deep cracks when dry, containing predominantly $CaCO_3$-rich sediments or weathered, clayey and Ca-silicious-rich, solid rock. This soil is formed in depressions with poor drainage or on vast plains in alternately humid warm climates, having 3-4 dry months and an annual precipitation of 300-1,300mm. These soils also include the *Terres Noires* of North Africa and the *smonitza* of the Balkan); *s* **vertisol** [m] (Suelo de textura arcillosa pesada que se agrieta notablemente cuando se seca. Son difíciles de labrar, pero adecuados para diversos cultivos, siempre y cuando se controle la cantidad de agua para que no se inunden ni sequen. Son buenos para pastos. *Términos específicos* **v.** pélico [suelo de color negro en la superficie], **v.** crómico [suelo de color gris en la superficie, generalmente de manejo más fácil que los anteriores]; cf. MEX 1983); *f* **vertisol** [m] (Classification des sols ; sols à argiles gonflantes caractérisant les climats chauds avec une période sèche marquée de 3 à 4 mois situés dans les dépressions mal drainées et en présence de roches basiques riches en fer, en zone tropicale [argiles noires d'Afrique], subtropicale [terre à coton des Indes] ou continentale du sud-est européen [Smonitza sur les Balkans]) ; *syn.* tirs [m] [Maroc], régur [m]

V

[Inde] ; *g* **Vertisol [m]** (Boden mit starken Schrumpfrissen in Trockenzeiten aus tonreichen, vorwiegend CaCO$_3$-haltigen Sedimenten oder tonreich verwitterten Ca-silicatreichen Festgesteinen, der sich bevorzugt in abflussträgen Senken oder weiten Ebenen wechselfeuchter Warmklimate mit 3-4 Trockenmonaten und Jahresniederschlägen zwischen 300 und 1300 mm bildet. Zu diesen Böden gehören z. B. auch die *Terres Noires* Afrikas sowie die *Smonitza* auf der Balkanhalbinsel); *syn.* Tirs [m], tropische Schwarzerde [f] (MEL 1978, Bd. 23, 520), *obs.* Grumosol [m].

very adverse impact [n] *ecol. envir.* ▶severe disturbance.

very firm soil [n] [US] *agr. hort.* ▶fine-textured soil.

6784 vicinity [n] *plan.* (Generic term for areas close to a centrally-located town or city which provide them with basic short-term needs and services; such areas are generally referred to as the suburbs; in Australia, one refers to inner and outer suburbs; ▶surrounding region of a city/town); *syn.* suburban area [n], suburbia [n], *one specific area* suburb [n]; *s* **área [f] de influencia inmediata** (Área de influencia de un lugar central para la cual éste pone a disposición su infraestructura social y cultural y su oferta de servicios; ▶espacio periurbano); *f* **zone [f] d'influence immédiate** (Zone située autour d'une ville centre couvrant l'approvisionnement et les services de base ; ▶périphérie urbaine) ; *syn.* zone [f] d'attraction immédiate ; *g* **Nahbereich [m]** (Einzugsbereich eines zentralen Ortes, für den dieser den kurzfristigen Grundbedarf sowie Dienstleistungen bereitstellt; ▶Stadtumland); *syn.* Versorgungsnahbereich [m].

6785 view [n] (1) *recr.* (1. Visible scene or prospect. 2. Extent or range of vision; ▶vista, ▶panoramic view); *syn.* (scenic) overlook [n] [also US]; *s* **vista [f]** (Término genérico; ▶corredor visual, ▶vista panorámica); *f* **vue [f]** (1. porter son regard vers le lointain. 2. Vers tous les côtés ; *contexte* avoir une vue étendue sur ... ; ▶couloir de vue, ▶vue panoramique) ; *g* **Aussicht [f]** (1. Blick in die Ferne. 2. Blick nach verschiedenen Seiten; *Kontext* eine schöne Aussicht auf etwas haben; ▶Sichtachse; ▶Panoramablick).

view [n] (2) *plan.* ▶plan view; *urb. recr.* ▶road with panoramic view; *plan.* ▶top view.

view [n]**, perspective** *plan. recr.* ▶prospect.

view corridor [n] [UK] *hist. landsc. gard.* ▶vista.

viewing platform [n] [UK] *recr.* ▶viewpoint [US].

6786 view obstruction [n] *urb.* (Hindrance of a view, e.g, by buildings); *s* **obstrucción [f] de la vista por edificación** *syn.* intercepción [f] de la vista por edificación; *f* **obstruction [f]** (Obstacle au champ de visibilité, à l'éclairement ou à l'ensoleillement occasionné p. ex. par un bâtiment ou par son ombre projetée ; *contexte* les baies éclairant les pièces principales ne doivent être masquées par une partie de bâtiment) ; *syn.* masque [m] ; *g* **Verbauung [f] (2)** (Behinderung des Sichtfeldes, z. B. durch Baukörper; verbauen [vb]).

6787 viewpoint [n] [US] *recr.* (Viewing site from which one has an attractive scenic view); *syn.* vantage point [n], viewing platform [n] [UK] (LD 1991 [11], 29), vista point [n] (TGG 1984, 243); *s* **plataforma [f] de vista panorámica** *syn.* lugar [m] con vista panorámica; *f* **belvédère [m]** (Lieu élevé permettant une vue panoramique exceptionnelle) ; *syn.* point [m] de vue, plateforme [f] panoramique ; *g* **Aussichtsplatz [m]** (Stelle mit schönem Ausblick); *syn.* Aussichtspunkt [m].

viewshed [n] [US] *plan. recr.* ▶prospect.

6788 vigor [n] [US]/**vigour** [n] [UK] *hort. plant.* (Growth characteristic of a vigorous plant; ▶slow-growing, ▶fast-growing, ▶vitality); *s* **vigor [m]** (Característica del crecimiento de una planta, de la vegetación; ▶de crecimiento lento, ▶de crecimiento rápido, ▶vitalidad); *f* **vigueur [f]** (Propriété d'une plante, de la

végétation caractérisant sa capacité de croissance ; ▶à croissance lente, ▶à croissance rapide, ▶vitalité) ; *g* **Wüchsigkeit [f]** (Eigenschaft einer Pflanze, die ihr Wachstum beschreibt; ▶langsam wüchsig, ▶schnell wüchsig, ▶Vitalität); *syn.* Wuchskraft [f].

vigor [n]**, sprouting** *for. hort.* ▶capacity of making new shoots.

6789 vigorous [adj] *hort. plant.* (Term applied to understock with strong vigo[u]r on which cultivars are grafted; ▶vigorous-growing, *opp.* ▶dwarfish); *syn.* strong-growing [adj]; *s* **de crecimiento vigoroso [loc]** (Término horticultor aplicado p. ej. a patrón sobre el que ha sido injertada una planta; *opp.* ▶de crecimiento débil, ▶vigoroso/a); *syn.* de crecimiento fuerte [loc]; *f* **à croissance rapide [loc]** (Capacité des végétaux à très forte vitalité ; ▶vigoureux ; *opp.* ▶à croissance lente) ; *syn.* très vigoureux, euse [loc] ; *g* **stark wüchsig [adj]** (Eine Pflanze betreffend, die in ihrem Wachstum im Vergleich zu anderen [der selben Art oder Gattung] mit starker Vitalität ausgestattet ist und deshalb entsprechend groß wird, z. B. eine **s. w.e** Veredlungsunterlage, **s. w.es** Ziergehölz; ▶wüchsig; *opp.* ▶schwach wachsend); *syn.* stark wachsend [ppr/adj].

6790 vigorous-growing [adj] *hort. plant.* (Strong; descriptive term applied to the strength and health of plants or understock; *context* to be a vigorous grower; ▶dwarfing, ▶intermediate, ▶vigorous); *s* **vigoroso/a** [adj] (Término horticultor aplicado a las plantas que presentan un crecimiento fuerte y sano; ▶de crecimiento débil, ▶de crecimiento medio, ▶de crecimiento vigoroso); *f* **vigoureux, euse** [adj] (Terme désignant une plante à forte croissance ; ▶à croissance lente, ▶à croissance moyenne, ▶vigoureux) ; *g* **wüchsig [adj]** (*Allgemeiner Begriff* so beschaffen, dass eine Pflanze ein gesundes und kräftiges Wachstum aufweist, ▶schwach wüchsig, ▶mittelstark wüchsig, ▶stark wüchsig).

vigour [n] [UK] *hort. plant.* ▶vigor [US]/vigour [UK]; *bot. phyt. zool.* ▶vitality.

vigour [n]**, sprouting** *for. hort.* ▶capacity of making new shoots.

villa garden [n] [UK] *gard.* ▶estate garden [US].

village [n] *recr. urb.* ▶marina village, *urb.* ▶old part of a city/town/village.

village [n] [UK]**, holiday** *plan. recr.* ▶resort community [US].

village competition [n] [UK]**, best kept** *hort. plan. urb.* ▶All-America City [US].

village edge [n] *urb.* ▶city periphery.

6791 village enhancement [n] *plan. urb.* (Planning measures aimed primarily at improving the appearance of village buildings and open spaces); *s* **mejora [f] de pueblos (≠)** (Medidas de planificación tomadas principalmente para mejorar el aspecto visual de las casas, plazas y calles en zonas rurales); *f* **mise [f] en valeur d'un village** (Projet d'aménagement prenant en compte les composantes qualitatives dans le développement d'un village, en particulier le caractère remarquable et la physionomie interne du village ainsi que la préservation de la structure villageoise) ; *syn.* aménagement [m] d'un village ; *g* **Dorfgestaltung [f]** (Planerische Aufgabe, die verstärkt qualitative Gesichtspunkte in die Dorfentwicklung einbezieht, besonders um die Eigenart des Dorfbildes und der Dorfstruktur zu erhalten).

village improvement [n] [US] *plan. urb.* ▶village renewal.

6792 village redevelopment planning [n] *plan. urb.* (Planning measures undertaken to sustain a rural settlement, in response to departure of a growing number of inhabitants, according to altered demographic, economic and cultural needs and desires; ▶urban planning, ▶village renewal); *s* **planificación [f] de pueblos (≠)** (Término genérico que cubre medidas de

V

desarrollo tomadas para reformar los asentamientos rurales como respuesta a los cambios estructurales demográficos y económicos y de acuerdo con las precepciones del correspondiente plan director territorial de coordinación; ▶renovación rural, ▶urbanismo); *f* **développement [m] de l'habitat rural** (Mesures et actions d'initiative locale, départementale et régionale de transformation et de restructuration des zones villageoises avec prise en compte des particularités démographiques, et économiques et des pratiques historiques et culturelles dans le cadre de l'aménagement du territoire [schémas d'orientation et d'aménagement régionaux] ; ▶rénovation de village, ▶urbanisme) ; *g* **Dorfentwicklung [f]** (Planerische Veränderung eines Dorfgebietes gemäß den demografischen, wirtschaftlichen und kulturellen Wünschen und Erfordernissen unter Berücksichtigung raumordnerischer Vorgaben [Regionales Raumordnungsprogramm]; ▶Dorferneuerung, ▶Städtebau).

6793 village renewal [n] *plan. urb.* (Physical improvements to structures and the environment of rural settlements; ▶village redevelopment planning); *syn.* village improvement [n] [also US], village revitalization [n] [also US]; *s* **renovación [f] rural** (Rehabilitación de pueblos para adecuarlos a las necesidades modernas; ▶planificación de pueblos); *syn.* remodelación [f] de pueblos, revitalización [f] de pueblos; *f* **rénovation [f] de village** (Opération de transformation du tissu villageois touchant en général les équipements [élargissement et aménagement des trottoirs, parc de stationnement, entretien de la voirie, éclairage et signalisation] et les espaces publics, réalisée souvent en relation avec l'extension ou la réfection des équipements commerciaux et communaux dans le respect du patrimoine culturel et architectural et paysager ; *termes génériques* rénovation rurale, revitalisation rurale ; *termes spécifiques* réhabilitation de l'habitat rural, réhabilitation du bâti rural ; ▶développement de l'habitat rural) ; *g* **Dorferneuerung [f]** (Sanierung eines Dorfes, um es zeitgemäßen Ansprüchen anzupassen; ▶Dorfentwicklung); *syn.* Dorfsanierung [f].

village revitalization [n] [US] *plan. urb.* ▶village renewal.

vine [n] [US] *bot.* ▶climbing plant.

vine [n] [US], **self-clinging** *bot.* ▶self-clinging climber.

vine-covered façade [n] [US] *gard. urb.* ▶planted façade.

6794 vine-covered wall [n] [US] *landsc. urb.* (TGG 1984, 75; ▶planted façade, ▶façade planting); *syn.* climber-covered wall [n] [UK]; *s* **muro [m] verde** (Fachada o muro cubierto de plantas trepadoras; ▶fachada vegetalizada, ▶plantación de fachadas); *f* **mur [m] végétalisé** (Avec des plantes grimpantes ; ▶façade tapissée, ▶végétalisation des façades) ; *g* **begrünte Mauer [f]** (Mit Kletterpflanzen bewachsene Wand; ▶begrünte Fassade; ▶Fassadenbegrünung); *syn.* begrünte Wand [f].

vine [n] **with adhesive disks, self-clinging** [UK] *bot.* ▶self-clinging vine with adhesive pads [US].

vine [n] **with adhesive pads** [US] *bot.* ▶self-clinging vine with adhesive pads [US]/self-clinging vine with adhesive disks [UK].

vine [n] **with hooked arching stems** [US] *bot.* ▶scandent plant.

6795 vineyard [n] *agr.* (Grapevine plantation, usually on terraced slopes); *syn.* winery [n] [also US]; *s* **viñedo [m]** (Plantación de viñas, en los países septentrionales generalmente en laderas aterrazadas de orientación sur o suroeste en el hemisferio norte); *f* **vignoble [m]** (Territoire — zone géographique, région ou pays — planté de vignes) ; *g* **Weinberg [m]** (Meist in Terrassen gestufter, mit Weinreben bepflanzter Hang); *syn.* Rebhang [m], Wengert [m] [BW].

6796 viral disease [n] *hort.* (ARB 1983, 584); (Infection of a plant caused by viruses); *s* **enfermedad [f] viral** *syn.* enfermedad [f] de virus. *f* **maladie [f] virale** (LA 1981, 1175 ; maladie due à un virus pathogène) ; *syn.* virose [f], maladie [f] à virus (LA 1981, 1174) ; *g* **Viruskrankheit [f]** (Durch Viren hervorgerufene Infektionskrankheit bei Pflanzen; eine Form der **V.** ist die Mosaikkrankheit); *syn.* Virose [f].

6797 virgin [adj] *ecol.* (Used to distinguish the ▶degree of landscape modification: an absence of any direct human influence and originating without any changes caused by human intervention; ▶pristine; *opp.* ▶man-developed); *syn.* primeval [adj]; *s* **virgen [adj]** (Para diferenciar los ecosistemas según el ▶grado de modificación [de ecosistemas y paisajes], uno que no ha sufrido influencia directa de la acción humana; ▶primigenio/a; *opp.* ▶antropógeno/a); *f* **d'artificialisation nulle [loc]** (Différenciation du ▶degré d'artificialisation d'un écosystème qui s'est développé en l'absence d'une action directe de l'homme ; ▶vierge ; *opp.* ▶fortement artificialisé) ; *g* **natürlich [adj] (2)** (*Zur Unterscheidung von Ökosystemen nach* ▶*Natürlichkeitsgraden* ohne direkten menschlichen Einfluss und ohne vom Menschen ausgehende Veränderung entstanden; in Mitteleuropa gäbe es nach o. g. Definition angesichts des allgegenwärtigen menschlichen Einflusses eigentlich keine natürlichen Ökosysteme und Lebensgemeinschaften. Deshalb schlägt DIERSCHKE vor, von einem fehlenden bis schwachen menschlichen Einfluss zu sprechen, der in jedem Fall so gering ist, dass die Vegetation zumindest auf der Ebene der Formation unverändert bleibt und einheimische Arten [Indigene] gegenüber nicht heimischen [Archaeophyten und Neophyten] deutlich überwiegen; cf. DIER 1984; ▶ursprünglich; *opp.* ▶naturfern).

6798 virgin area [n] *landsc.* (Landscape of virtually undisturbed natural vegetation, which is the same as ▶potential natural vegetation; ▶wilderness, ▶wilderness area, ▶wild land; *opp.* ▶cultural landscape); *s* **paisaje [m] virgen** (Área natural no o mínimamente influenciada por la sociedad, cuya vegetación no coincide con la ▶vegetación potencial natural; ▶paraje silvestre, ▶reserva natural 2, ▶zona virgen; *opp.* ▶paisaje humanizado); *syn.* paisaje [m] natural; *f 1* **espace [m] vierge** (Terme utilisé par les géographes anglo-saxons pour décrire un paysage non ou faiblement influencé par l'homme ; la végétation existante est identique à la ▶végétation naturelle potentielle ; ▶paysage sauvage, ▶zone vierge ; *opp.* ▶paysage culturel) ; *f 2* **espaces [mpl] naturels** *conserv. plan.* (Terme utilisé dans l'aménagement du territoire et la protection de la nature pour désigner les territoires à très fort degré de naturalité et en général soumis à une protection spécifique, p. ex. les paysages exceptionnels dits patrimoniaux ou les ▶espaces naturels sensibles, lieux privilégiés pour la conservation, la gestion et la découverte des richesses naturelles ; ▶espaces naturels sensibles [ENS], ▶paysage naturel, ▶réserve naturelle de flore et faune sauvage) ; *g* **Naturlandschaft [f] (1)** (Ein vom Menschen nicht oder kaum beeinflusster Zustand der Landschaft. Die reale Vegetation ist in diesem Falle identisch mit der ▶potentiellen natürlichen Vegetation; zu diesem Landschaftstyp gehören z. B. die Hochlagen der Alpen, Kernzonen vieler Nationalparke, weite Teile der Tundra; Landschaften, die bis vor Jahrzehnten noch genutzt wurden und dann sich selbst überlassen wurden, wie z. B. ▶Bergbaufolgelandschaften, können auch dazu gezählt werden; *opp.* ▶Kulturlandschaft; ▶Naturlandschaft 2, ▶Wildnis, ▶Wildnis[schutz]gebiet); *syn. z.* T Urlandschaft [f].

6799 virgin forest [n] *phyt.* (Natural forest virtually uninfluenced by human activity; ▶second-growth [forest] [US]/secondary forest [UK]); *syn.* prim(ev)al forest [n], primordial forest [n], pristine forest [n]; *s* **bosque [m] primario** (Aquel que se presenta como etapa de máxima autoorganización en una localidad, originado exclusivamente por las fuerzas naturales, y

V

libre de toda alteración humana. El **b. p.**, en una región determinada, tendrá una composición particular en función de la flora del territorio, del clima general, del relieve y del sustrato litológico; DINA 1987; ▶bosque secundario); *syn.* bosque [m] virgen, selva [f] virgen; *f* **forêt [f] vierge** (Forêt naturelle pratiquement non influencée par les activités humaines ; ▶forêt de substitution) ; *syn.* forêt [f] primitive (CAT 1987, 128), forêt [f] primaire (TSF 1985, 144) ; *g* **Urwald [m]** (Naturwald ohne jeden menschlichen Einfluss in Vergangenheit und Gegenwart; ▶Sekundärwald); *syn.* Primärwald [m].

6800 virgin landscape [n] *geo. nat'res.* (Condition of a landscape before human intervention affects the environment; ▶virgin area, ▶wild land [US], ▶wilderness area [US]); *s* **paisaje [m] virgen** (Estado de un paisaje antes de la intervención del hombre; ▶paisaje virgen, ▶paraje silvestre, ▶reserva natural 2); *f* **paysage [m] vierge** (État d'un paysage exempt de toute influence humaine ou précédant toute intervention humaine ; ▶espaces naturels, ▶espace vierge, ▶paysage naturel, ▶paysage sauvage, ▶réserve naturelle de flore et faune sauvage) ; *g* **Urlandschaft [f]** (Landschaftszustand vor den umweltwirksamen Eingriffen des Menschen; ▶Naturlandschaft 1, ▶Naturlandschaft 2, ▶Wildnis[schutz]gebiet).

6801 visibility [n] **into an area** *plan.* (Seeing into or within a landscape or portion thereof; e.g. more **v.** in the opening up of a vista/view, or less **v.** in the blocking out or screening of a garden); *s* **visibilidad [f]** (Situación/estado en un lugar específico que permite ver p. ej. un jardín privado o un parque público desde el exterior); *f* **soumission [f] à la vue** (État d'un lieu [p. ex. dans un jardin] visible de l'extérieur) ; *g* **Einsehbarkeit [f]** (Situation/Zustand an einem bestimmten Ort, die/der den Einblick in ein Objekt, in einen öffentlichen oder privaten Freiraum ermöglicht und eine [soziale] Kontrolle darstellt).

6802 visible [adj] *plan.* (able to be seen); *s* **visible [adj]** (Relativo a un espacio libre privado o público, que permite visibilidad desde el exterior); *syn.* abierto/a a la vista [loc]; *f* **offert, erte à la vue [loc]** (*opp.*à l'abri des vues/regards) ; *syn.* à la vue de [loc], exposé à la vue/aux vues [loc], exposé aux regards [loc] ; *g* **einsehbar [adj]** (Einen öffentlichen oder privaten Freiraum betreffend, der Einblick gewährt oder zulässt).

visitor [n] *zool.* ▶summer visitor, ▶winter visitor.

visitor [n], **passage** *zool.* ▶bird of passage.

visitor [n] [US], **recreation** *recr. sociol.* ▶recreation user.

6803 visitor bird [n] *zool.* (Bird, which stays for one season of the year at the most in a particular area, to breed or not, or visits an area for a short stay [▶summer visitor, ▶winter visitor]; typical birds are those visiting from Nordic breeding grounds, which migrate from breeding to wintering areas and stop on the way for a rest [▶stopover bird] to gain energy [nutrition guest] or spend the winter [▶overwintering (migratory) bird]; moulting birds or non-breeders are also guest birds in the summer months in the respective areas; ▶breeding bird, ▶local migrant, ▶migratory bird, ▶permanent resident); *s* **ave [f] visitante** (Especie de ave que visita una región por un periodo/período limitado, en la que en algunos casos se reproduce, o para pasar el invierno. Hay 2 tipos distintos: ▶visitante estival y ▶visitante invernal; ▶ave de dispersión posgenerativa, ▶ave en etapa migratoria, ▶ave migradora, ▶ave nidificante, ▶ave sedentaria); *f* **oiseau [m] visiteur** (Espèce d'oiseau qui séjourne dans une région considérée pendant une période courte sans s'y reproduire [p. ex. un oiseau en étape migratoire], ou pendant la période d'hivernage [▶oiseau hivernant] ; ▶oiseau migrateur, ▶oiseau nicheur, ▶oiseau sédentaire, ▶visiteur d'été, ▶visiteur d'hiver) ; *g* **Gastvogel [m]** (Vogel, der sich in einem Bezugsraum nur einen kurzen Zeitraum, längstens eine Saison lang aufhält; typische

Gastvögel sind Vögel nordischer Brutgebiete, die auf dem Zug von den Brut- in die Überwinterungsgebiete unterwegs rasten [▶Rastvogel], um Energiereserven aufzufüllen [Nahrungsgast] oder den Winter verbringen [▶Überwinterer]; auch mausernde Vögel oder Nichtbrüter im Sommerhalbjahr sind in den jeweiligen Gebieten Gastvögel [▶Sommergast]; ▶Brutvogel, ▶Standvogel, ▶Strichvogel, ▶Zugvogel).

6804 visitor number [n] *plan. recr.* (Persons present per unit of time or performance event; ▶peak visitor use); *s* **afluencia [f] de visitantes** (Cantidad de personas presentes por unidad de tiempo o en una actividad; ▶cantidad máxima de visitantes, ▶presión recreacional); *syn.* cantidad [f] de visitantes, cantidad [f] de usuarios/as; *f* **affluence [f] touristique** (Nombre de visiteurs par unité de temps ou par manifestation ; ▶pointe de fréquentation [touristique]) ; *syn.* afflux [m] touristique, fréquence [f] de visite, fréquentation [f] touristique ; *g* **Besucheraufkommen [n]** (Anzahl der Besucher je Zeiteinheit oder Veranstaltung; ▶Spitzenbesucheraufkommen); *syn.* Besuchermenge [f].

6805 visitor pressure [n] *plan. recr.* (Cumulative stress of visitors on an area; ▶peak visitor use, ▶recreation load, ▶visitor number); *s* **presión [f] turística** (Perturbación causada a un paisaje o partes del mismo por la gran ▶afluencia de visitantes; ▶cantidad máxima de visistantes, ▶presión recreacional); *syn.* presión [f] de usuarios/as; *f* **pression [f] exercée par les visiteurs** (Nuisance exercée par les flux touristiques sur un paysage ou parties de celui-ci ; ▶affluence touristique, ▶pointe de fréquentation [touristique], ▶pression récréative, ▶pression touristique) ; *g* **Besucherdruck [m]** (Beeinträchtigung einer Landschaft oder Teile davon durch ▶Besucheraufkommen; ▶Erholungsdruck, ▶Spitzenbesucheraufkommen); *syn.* Belastung [f] durch Besucher, Inanspruchnahme [f] durch Besucher.

visitor use [n] *plan. recr.* ▶peak visitor use.

6806 vista [n] *hist. landsc. gard.* (Long, narrow, and usually framed view; ▶panoramic view, ▶prospect); *syn.* visual corridor [n], view corridor [n] [also UK]; *s* **corredor [m] visual** (En parques y jardines, pasillo libre de estructuras o grandes plantas para posibilitar la vista en la distancia; ▶cuenca visual, ▶vista panorámica); *syn.* vista [f] axial; *f 1* **couloir [m] de vue** (Espace laissé dégagé afin de permettre une vue libre, p. ex. dans un parc ; ▶champ de vision, ▶vue panoramique) ; *syn.* axe [m] de vue, corridor [m] visuel [CDN] ; *f 2* **cône [m] de visibilité** (Espace laissé dégagé à partir de lieux ou d'itinéraires privilégiés d'appréhension du paysage assurant la protection de panoramas même lointains et pouvant concerner des territoires vastes [plusieurs communes] ; cf. directive paysagère, circ. n° 94-88 du 21 novembre 1994) ; *syn.* cône [m] de vue ; *g* **Sichtachse [f]** (Für den Ausblick von Sichthindernissen freigehaltener Korridor, z. B. in Parkanlagen; ▶Panoramablick, ▶Sichtfeld); *syn.* Sichtschneise [f], Blickschneise [f].

6807 vista fan [n] *plan.* (Several vistas towards separate features from the same park viewpoint); *s* **abanico [m] de visión** (Conjunto de varios corredores visuales perceptibles desde un punto, p. ej. en un parque); *syn.* abanico [m] de vistas; *f* **éventail [m] d'axes de vue** (Plusieurs axes de vues perceptibles à partir d'un point donné, p. ex. au carrefour de plusieurs allées dans un parc paysager) ; *syn.* éventail [m] de couloirs de vue ; *g* **Sichtenfächer [m]** (Mehrere Blickachsen, die von einem Standpunkt aus wahrzunehmen sind, z. B. in einem Landschaftspark).

vista point [n] *recr.* ▶viewpoint [US].

visual amenities [npl] *urb.* ▶preservation of local visual amenities.

visual appearance [n] **of a landscape** *plan. landsc.* ▶analysis of the visual appearance of a landscape.

V

visual assessment [n] **of a landscape** *plan. landsc.* ►visual landscape assessment.

visual character [n] **of a landscape** *conserv. landsc. plan. recr.* ►landscape character.

6808 visual connection [n] *plan.* (►prospect); *syn.* visual relationship [n]; *s* **relación [f] visual** (►cuenca visual); *f* **relation [f] visuelle** (Rapport entre un observateur et un objet éloigné ; ►champ de vision) ; *g* **Blickbeziehung [f]** (Verhältnis zwischen Betrachter und entferntem Objekt, das eine visuelle Kommunikation ermöglicht; ►Sichtfeld); *syn.* Blickverbindung [f], Sichtbeziehung [f], Sichtverbindung [f].

visual corridor [n] *hist. landsc. gard.* ►vista.

visual defacement [n] [US] *landsc.* ►detraction from visual quality.

visual detriment [n] *landsc.* ►detraction from visual quality.

6809 visual disturbance [n] *landsc.* (Negative change in visual effect caused by a development project and viewer response thereto; ►detraction from visual quality, ►impairment of landscape); *syn.* visual impairment [n]; *s* **desfiguración [f] del paisaje (rural o urbano)** (Alteración visual negativa de un paisaje o de la imagen de una ciudad causada p. ej. por un proyecto de construcción; ►desfiguración [visual], ►deterioro del paisaje); *syn.* desvalorización [f] visual del paisaje (rural o urbano); *f* **altération [f] visuelle du caractère (des sites et des paysages)** (Modification visuelle de l'état, de l'aspect esthétique d'un paysage ou d'un site urbain ; ►défiguration des sites et des paysages, ►nuisance visuelle) ; *syn.* pollution [f] esthétique (des sites et des paysages), altération [f] de l'aspect esthétique ; *g* **visuelle Beeinträchtigung [f] (2)** (Verschlechterung des Erscheinungsbildes einer Landschaft oder eines Stadtbildes, z. B. durch unsensibel in die Umgebung eingepasste Bauwerke; ►Verunstaltung der Landschaft, ►visuelle Beeinträchtigung 1).

visual dominance [n] *phyt. plant.* ►seasonally changing visual dominance.

visual element [n] *conserv. land'man. landsc.* ►landscape feature.

visual (impact) assessment [n] **of a landscape** *plan. landsc.* ►visual landscape assessment.

visual impairment [n] *landsc.* ►visual disturbance.

visual landscape [n] *landsc. plan. recr.* ►visual quality of landscape.

6810 visual landscape assessment [n] *plan. landsc.* (Aesthetic evaluation of a landscape's visual impact; ►visual disturbance); *syn.* aesthetic landscape assessment [n], visual (impact) assessment [n] of a landscape (LD 1985 [6]), assessment [n] of scenic value; *s* **evaluación [f] del aspecto escénico del paisaje** (Evaluación del aspecto escénico del paisaje en cuanto al grado de desfiguración visual del mismo; ►desfiguración del paisaje [rural o urbano]); *syn.* análisis [m] visual del paisaje, evaluación [f]/valoración [f]/estimación [f] de la calidad visual del paisaje; *f* **évaluation [f] de l'image paysagère** (Évaluation de l'impact visuel sur qualité esthétique d'un paysage ; ►altération du caractère des sites et des paysages) ; *syn.* évaluation [f] de la qualité visuelle du paysage ; *g* **Landschaftsbildbewertung [f]** (Bewertung eines Landschaftsbildes hinsichtlich des Grades der ►visuellen Beeinträchtigung 2); *syn.* Bewertung [f] des Landschaftsbildes.

visual landscape management [n] *landsc. plan.* ►visual resource management.

visual orientation [n] *landsc. trans.* ►optical guidance.

visual picture [n] **of a town** *urb.* ►townscape.

6811 visual quality [n] *recr. plan.* (Visual excellence of form and shape; ►perceived environmental quality); *syn.* aesthetic quality [n], aesthetic value [n]; *s* **calidad [f] visual** (Valor beneficioso de forma y diseño de un objeto o un espacio; ►calidad vivencial del ambiente); *syn.* calidad [f] estética, composición [f] visual, composición [f] estética; *f* **valeur [f] esthétique** (Caractéristiques traduisant la perception visuelle des formes et de leurs associations ; ►qualité perceptuelle) ; *syn.* qualité [f] visuelle [CDN] ; *g* **Gestaltqualität [f]** (Empirische oder verstandesmäßige Verknüpfung visuell oder taktil wahrnehmbarer Einzelelemente zu Gestalten oder Formkomplexen; bei landschaftlichen **G.en** werden figürliche und räumliche unterschieden; cf. NOHL 1997; ►Erlebnisqualität).

visual quality [n]**, reduction in** *landsc.* ►detraction from visual quality.

6812 visual quality [n] **of landscape** *landsc. plan. recr.* (Collective impression made by a landscape with relatively similar characteristics, e.g. relief, vegetation, land use patterns, forms of settlement. The dividing line between one landscape and the next is marked by the change in characteristic features. Their interrelationships can be described in terms of dominance, scale, diversity, and continuity. Individual judgements of quality are affected by the values and activity of the viewer; ►visual disturbancce); *syn.* aesthetic value [n] of a landscape, physiognomy [n] of a landscape (LE 1986, 8), natural scenery [n], visual landscape [n], scenic quality [n], visual resource quality [n]; *s* **aspecto [m] escénico del paisaje** (Aspecto característico de un paisaje con elementos configurantes específicos, como p. ej. relieve, estructura de uso del suelo, tipo[s] de vegetación, presencia de agua, etc. Las «escenas» se diferencian al cambiar los elementos configurantes característicos; ►desfiguración [visual]); *syn.* calidad [f] escénica del paisaje, «escena» [f], calidad [f] visual del paisaje; *f* **image [f] paysagère** (Représentation synoptique des traits spécifiques à plusieurs secteurs paysagers présentant un caractère global relativement similaire et intervenant dans la perception immédiate de l'organisation paysagère tels que la géomorphologie, la couverture végétale, les modes d'occupation du sol, les formes d'urbanisation ; la différenciation des paysages est mise en évidence par une variation des composantes perceptuelles caractéristiques ; ►altération du caractère des sites et des paysages) ; *syn.* qualité [f] visuelle du paysage, aspect [m] esthétique du paysage, traits [mpl] du paysage ; *g* **Landschaftsbild [n]** (Zusammenfassende Vorstellung von mehreren sinnlich wahrnehmbaren Landschaftsausschnitten relativ gleichartiger Ausprägung bestimmter Gestaltmerkmale wie z. B. Oberflächenform, Vegetationsausstattung, Flächennutzungsmuster, Siedlungsformen. Die Abgrenzung zu anderen Landschaftsbildern ist durch den Wechsel der charakteristischen Gestaltmerkmale gekennzeichnet; das **L.** wird bei jedem einzelnen durch strukturell-objektive und ästhetisch-subjektive [emotionale] Elemente geprägt; es ist nach HABER nicht intersubjektiv oder objektivierbar, obwohl die grundlegenden Züge der Landschaftswahrnehmung stammesgeschichtlich und damit biologisch bestimmt sind; cf. G+L 1999 [12]; ►visuelle Beeinträchtigung 1).

visual relationship [n] *plan.* ►visual connection.

6813 visual resource management [n] *landsc. plan.* (Inventory, planning and governmental or legal actions taken to identify values and establish objectives for managing those values, and the management actions necessary to achieve those objectives; NRM 1996; ►visual quality of landscape); *syn.* visual landscape management [n], landscape aesthetics management [n]; *s* **preservación [f] del aspecto escénico del paisaje** (Conjunto de medidas de planificación e implementación que sirven para

V

registrar, valorar y proteger las características visuales típicas de un paisaje o la belleza de la naturaleza y el paisaje; ►aspecto escénico del paisaje); *syn.* preservación [f] del carácter visual del paisaje, gestión [f] de la estética del paisaje; *f* **protection [f] et mise en valeur de la qualité esthétique (des sites et) des paysages** (Mesures et programmes d'action en faveur de la préservation de l'image paysagère des territoires remarquables par leur intérêt paysager ou de la reconquête de paysages dégradés, de la réhabilitation du paysage rural et urbain ; ►image paysagère) ; *g* **Landschaftsbildpflege [f, o. Pl.]** (Sämtliche planerischen Maßnahmen und Handlungen, die den Wert eines ►Landschaftsbildes erfassen, bewerten und dazu beitragen, die charakteristischen Eigenarten und Gegebenheiten einer Landschaft oder die Schönheit von Natur und Landschaft zu fördern, zu erhalten und zu verbessern); *syn.* Pflege [f, o. Pl.] des Landschaftsbildes.

visual resource quality [n] *landsc. plan. recr.* ►visual quality of landscape.

6814 visual screening [n] *gard. landsc.* (LD 1985 [6]; measures or installations to conceal unattractive buildings, to provide a barrier to reduce such nuisances as glare or strong sunlight, or to discourage visual intrusion by inquisitive persons; ►anti-glare protection); *s* **protección [f] visual** (Conjunto de medidas para bien tapar un edificio poco atractivo o para proteger contra la intrusión visual; ►medidas [de protección] antideslumbramiento); *f* **protection [f] contre les vues** (Toute mesure ou dispositif de protection [écran visuel] contre des bâtiments inesthétiques, contre les nuisances lumineuses, contre les regards indiscrets, etc. ; ►protection contre l'éblouissement) ; *g* **Sichtschutz [m, o. Pl.]** (Maßnahmen oder Einrichtungen zur Abdeckung unschöner Bauwerke sowie Schutzeinrichtungen vor störender Lichtwirkung, vor Einblicken Neugieriger etc.; ►Blendschutz).

6815 visual screen [n] **with horizontal boards** *constr.* (Wooden screen with planks installed tightly together on one side of the supporting frame or on alternate sides of the frame with a gap in between; ►visual screen with interwoven split boards); *s* **valla [f] de protección visual de tablones horizontales** (Pantalla formada por tablones de madera colocados en horizontal estrechamente unidos a un lado del marco de apoyo o alternando entre los lados del marco con un hueco intermedio; ►valla de protección visual de tablas entrecruzadas); *f* **mur [m] brise-vue à lattis horizontal** (La disposition des lattes peut être jointive ou à claire-voie [mur écran visuel à lattis à claire-voie] ; ►écran brise-vue à lattis horizontal) ; *syn.* mur [m] écran à lattis horizontal ; *g* **Verbretterung** (►Sichtschutzwand mit geschlossener oder „auf Lücke" montierter Brettanordnung).

6816 visual screen [n] **with interwoven split boards** *constr.* *s* **valla [f] de protección visual de tablas entrecruzadas;** *f* **écran [m] brise-vue à lattis horizontal** (Clôture de protection contre les vues, constituée de panneaux dont les lattes de bois entrelacées possèdent une ondulation horizontale) ; *syn.* palissade [f] à lattis horizontal ; *g* **Sichtschutzwand [f] mit Spaltbretterverflechtungen.**

6817 visual screen [n] **with vertical boards** *constr.* (**1.** Wooden screen with planks installed tightly together on one side of the supporting frame or on alternate sides of the frame with a gap in between); **2. close board panel fence** [n] (Fence constructed with prefabricated elements; ►close-boarded fence, ►picket fence); *s* **valla [f] de protección visual de tablones verticales** (Pantalla formada por tablones de madera colocados en vertical estrechamente unidos a un lado del marco de apoyo o alternando entre los lados del marco con un hueco intermedio; ►valla de listones, ►valla de listones verticales, ►valla de

tablones); *f* **mur [m] brise-vue à lattis vertical** (Mur de protection contre les vues constitué de lattes disposées à joint serrés ou écartés ; ►clôture à lattes, ►clôture en planches jointives, ►clôture [en bois] non jointive à lattis verticales) ; *syn.* mur [m] écran à lattis vertical ; *g* **Sichtschutzwand [f] mit senkrechter Lattung** (Blickdichte Sichtschutzwand mit geschlossener oder auf „Lücke" montierter, senkrechter Brettanordnung; die senkrechte Lattung kann auch beidseitig der Riegel angebracht sein; ►Bretterzaun, ►Lattenzaun); *syn.* Plankenzaun [m] mit senkrechter Brettanordnung/Lattung.

visual tree assessment [n] *adm. arb. leg.* ►tree inspection.

6818 vitality [n] *phyt. zool.* (**1.** Capacity of a species to survive in given environmental circumstances. **2.** Characteristic of a roadside tree to provide environmental benefits because of its genetic potential and adaptability to thrive under difficult conditions and without costly maintenance measures; ►degree of vitality, ►vigor [US]/vigour [UK]); *s* **vitalidad [f]** (Por v. se entiende la capacidad de una planta de fructificar, de formar semillas y su fuerza de germinación. El término se aplica también en la valoración de los árboles en la ciudad, es decir de su capacidad de desarrollarse en ese hábitat hostil; ►grado de vitalidad, ►condiciones mesológicas, ►vigor); *f* **vitalité [f]** (**1.** Capacité vitale d'une espèce à s'adapter aux conditions du milieu. **2.** Lorsqu'il s'agit d'un arbre en milieu urbain ce terme désigne sa capacité, pour un entretien économiquement et écologiquement acceptable, à surmonter des conditions stationnelles hostiles ; ►degré de vitalité, ►vigueur) ; *g 1* **Vitalität [f]** (Grad der Lebenskraft einer Art, ihre ►Standortfaktoren zu ertragen — bedingt durch Genpotenzial und Umweltbedingungen); *syn.* Lebenskraft [f], Lebensfähigkeit [f]; *g 2* **Vitalität [f] eines Straßenbaumes** (Fähigkeit eines Baumes, durch Wüchsigkeit und Widerstand gegen Krankheiten sowie auf Grund seines genetischen Potenzials trotz schwieriger Standortbedingungen spezifische Wohlfahrtswirkungen bei ökonomisch und ökologisch vertretbarem Pflegeaufwand zu leisten; ►Vitalitätsgrad, ►Wüchsigkeit).

6819 vitrified clay pipe [n] [US] *constr.* (Pipe manufactured of an earthenware material which is glazed and has a high resistance to chemical corrosion; it is used for house sanitary pipes; ►clay pipe); *syn.* glazed stoneware pipe [n] [UK]; *s* **tubería [f] de drenaje vitrificada** (Tubo cerámico cocido en exceso y, por lo tanto, vitrificado hasta llegar a ser impermeable; DACO 1988; ►tubo de arcilla); *f* **tuyau [m] en grès** (*Matière constitutive* grès vitrifié et émaillé intérieur et extérieur ; fabriqué à partir d'un mélange d'argiles, avec apport de déchets d'argiles cuites, émaillé, passage en filières, séché et cuit dans un four tunnel ; tuyau utilisé pour les canalisation d'assainissement ; ►drain en poterie) ; *g* **Steinzeugrohr [n]** (Glasiertes, bis zur Sinterung gebranntes ►Tonrohr mit glatter Oberfläche für erdverlegte Abwasserleitungen).

6820 volatilization [n] **of nitrogen** *pedol.* (TEE 1980, 297); *syn.* mobilization [n] of nitrate (TEE 1980, 296), vaporizing [n] of nitrogen; *s* **mineralización [f] del nitrógeno** (Transformación del nitrógeno orgánico contenido en los desechos de plantas y cadáveres de animales en nitrógeno anorgánico [amonio, nitrato] por medio de los procesos de proteólisis, amonificación y nitrificación, después de los cuales puede ser nuevamente asimilado por las plantas): *syn.* mobilización [f] del nitrógeno; *f* **minéralisation [f] de l'azote** (Transformation biologique [protéolyse, ammonification, nitrification] des formes organiques de l'azote pour devenir directement assimilable par les végétaux ; cf. PED 1984, 91) ; *g* **Stickstofffreisetzung [f]** (**1.** Umwandlung von organischen Stickstoffverbindungen durch Bakterien in anorganische Formen wie Ammonium oder Nitrat, primäre Nährstoffe, die für das Pflanzenwachstum unabdingbar sind. **2.** Umwandlung von Ammonium in atmosphärischen Stickstoff durch Bakterien. Der Prozess läuft *ohne* molekularen Sauerstoff ab; die anaerobe Ammonium-Oxidation wird von den gleichnamigen Ammonium-Bakterien geleistet; cf. ORF 2009.

volcanic scoria [n] *constr. envir. geo. min.* ▶cinder.

volume balance [n] **of cut and fill** *constr.* ▶balance of cut and fill.

6821 volume [n] **of cut and fill** *constr.* (Amount of soil involved in earth-moving operations; ▶cut and fill); *s* **volumen [m] de desmonte y terraplén** (Cantidad de tierra que se mueve en operaciones de ▶desmonte y terraplén); *f* **déblai [m] de masse** (Matériau à déplacer ; ▶déblai et remblai) ; *g* **Boden-masse [f]** (Zu bewegendes resp. bewegtes Erdmaterial; ▶Auftrag und Abtrag).

6822 volume [n] **of on-site soil** *constr.* (Bulk of existing consolidated soil before excavation that is used as a basis for payment to the contractor; ▶consolited subgrade, ▶site soil); *s* **volumen [m] de tierra compacta** (Volumen de tierra en el lugar de la obra antes de ser excavada que se utiliza como base para la facturación; ▶subsuelo estable, ▶suelo *in situ*); *syn.* volumen [m] de suelo compacto; *f* **volume [f] en place** (Volume à prendre en compte dans la facturation et désignant le volume de terre décapé ou extrait sur l'épaisseur prescrite ou le profil imposé par opposition au volume de terre transporté foisonné ; ▶sol stable, ▶terrain *in situ*) ; *syn.* mètre [m] cube en place ; *g* **feste Bodenmasse [f]** (Begriff für die Abrechnung von abgetragenem/ausgekoffertem Boden, im Vergleich zum losen, zu transportierenden Boden; ▶anstehender Boden, ▶fester Baugrund).

6823 volume [n] **of spread soil** *constr.* (Bulk of excavated soil that has been graded to proposed levels and contours, and used as a basis for payment to the contractor); *s* **volumen [m] de relleno** (Cantidad de tierra utilizada hasta llegar a la cota necesitada, que se liquida por cubos); *f* **volume [m] de terre en œuvre** (Cubature mise en place sur le chantier conformément aux plans de profils et prise en compte dans la facturation) ; *syn.* cubature [f] de terre en œuvre ; *g* **eingebaute Bodenmasse [f]** (Kubatur, die in profilgerecht eingebautem Zustand abgerechnet wird).

6824 volume [n] **of stripped topsoil** *constr.* (Amount of removed topsoil; ▶topsoil pile [US]/topsoil heap [UK]); *s* **volumen [m] de acopio de tierra vegetal** (Masa de tierra resultante de la remoción de la capa de tierra vegetal; ▶depósito de tierra vegetal); *f* **retroussis [m] de terre végétale** (Quantité de sol excavé et pouvant être stocké pour réutilisation, p. ex. pour le reprofilage en fin de travaux ; ▶dépôt de terre végétale) ; *syn.* terre [f] végétale décapée, terre [f] végétale excavée, terre [f] de découverte ; *g* **Oberbodenabtrag [m] (2)** (*Ergebnis des Abtrages* abgetragene Bodenmasse; ▶Oberbodenmiete); *syn.* abgetragene Bodenmasse [f].

volumetrics [npl] [US] *constr. plan.* ▶calculation of quantities.

6825 voluntary caretaker [n] **of a tree** [US] *conserv.* (Citizen who feels responsibility to care for one or more trees on a public site out of civic pride; ▶stewardship of trees [US]); *syn.* tree steward [n] [US]; tree sponsor [n] [UK]; *s* **padrino [m]/madrina [f] voluntario/a de un árbol** (Ciudadano/a que se hace responsable de cuidar un árbol situado en un espacio público, generalmente en la calle donde está su vivienda; ▶apadrinamiento de árboles); *f* **parrain [m] d'un arbre** (Citadin en général riverain prenant la responsabilité de l'entretien d'un arbre planté dans un espace public — place, rue, avenue ; ▶parrainage des arbres) ; *g* **Baumpate [m]/Baumpatin [f]** (Jemand, der ehrenamtlich Mitverantwortung für die Pflege und Unterhaltung eines

öffentlichen Baumes, z. B. für einen Straßenbaum, meist in seiner unmittelbaren Nachbarschaft, übernimmt; ▶Baumpatenschaft).

6826 voluntary land exchange [n] *agr. leg.* (Interchange of [p]lots for the purpose of land consolidation; ▶transfers of real estate [US]/land transactions [UK]); *syn.* voluntary land swap [n] [also US]; *s* **intercambio [m] voluntario de tierras** (Forma simple y eficaz de reagrupación de las parcelas de tierras de cultivo en el contexto de la concentración parcelaria; ▶transacciones inmobiliarias); *f* **échange [m] amiable de terrain** (Formule simple et efficace de regroupement des terres dans le cadre du remembrement rural ; ▶opérations foncières et immobilières) ; *syn.* échange [m] parcellaire amiable ; *g* **freiwilliger Landtausch [m]** (Schnelles und einfaches Flurbereinigungsverfahren zur Verbesserung der Agrarstruktur; cf. § 103 a FlurbG; ▶Bodenverkehr).

voluntary land swap [n] [US] *agr. leg.* ▶voluntary land exchange.

volunteer growth [n] [US] *for. phyt.* ▶natural regeneration (2).

6827 vulnerable species [n] *conserv. phyt. zool.* (*Faunistic and floristic* ▶*status of endangerment* animal or plant species which has either significantly decreased in its natural range due to direct or indirect human interference or which has regionally disappeared; *according to IUCN categories of threat* VU [Vulnerable]; ▶Red List, ▶threatened species). *s* **especie [f] vulnerable (VU)** (Según las categorías de la ▶lista roja de la UICN, un taxón es vulnerable [Vulnerable] cuando la mejor evidencia disponible indica que cumple cualquiera de los cinco criterios para esta categoría y, por consiguiente, se considera que está enfrentando un alto riesgo de extinción en estado silvestre; cf. UICN 2001. **En Es.** una de las dos ▶categorías de amenaza de extinción, utilizada para taxones o poblaciones que corren el riesgo de pasar a la categoría de ▶especie en peligro de extinción en un futuro inmediato si los factores adversos que actúan sobre ellos no son corregidos; cf. art. 55 Ley 42/2007; ▶Catálogo Español de Especies Amenazadas); *f 1* **espèce [f] vulnérable (VU)** (Une des neufs catégories du système de la ▶Liste Rouge de l'UICN : espèce remplissant un des critères suivants [texte non exhaustif] : **a.** réduction des effectifs constatée, estimée, déduite ou supposée depuis 10 ans ou 3 générations égale ou supérieure à 50 % lorsque les causes de la réduction sont clairement réversibles et ont cessé ou de 30 % lorsque les causes de la réduction n'ont pas cessé ou ne sont peut-être pas réversibles…, **b.** espèce dont la répartition géographique dans la zone d'occurrence [estimée inférieure à 20 000 km²] et la zone d'occupation [estimée à moins de 2000 km²] indique un déclin continu, constaté, déduit ou prévu ainsi qu'une fluctuation extrême de la zone d'occurrence, de la zone d'occupation, de l'étendue et/ou de la qualité de l'habitat, du nombre de localités ou de sous-populations, du nombre d'individus matures, **c.** une population estimée à moins de 10 000 individus matures dont le déclin continu est estimé à 10 % au moins en dix ans ou trois générations, **d.** une population très petite et limitée, estimée à moins de 1000 individus matures exposés aux impacts des activités anthropiques en une très brève période de temps et **e.** taxon confronté à une probabilité d'extinction à l'état sauvage d'au moins 10 % en l'espace de 100 ans ou de cinq générations ; cf. Catégories et Critères de l'UICN pour la Liste Rouge, Version 3.1, UICN 2001 ; ▶espèce menacée); *f 2* **espèce [f] dont la population n'a pas sensiblement diminué, mais dont les effectifs sont faibles, donc en danger latent** (▶*Catégorie de régression des populations et de distribution des espèces* toute espèce animale ou végétale dont le peuplement ou l'habitat est directement ou indirectement menacé par l'homme et dont les effectifs sont réduits par suite de leur régression régionale ou disparition locale ; également espèce végétale caractéri-

V

sée par une alternance de l'habitat ; ►liste des espèces menacées, ►Livre rouge [des espèces menacées]) ; *syn.* espèce [f] en danger ; *g* **gefährdete Art [f]** (*Faunistische und floristische* ►*Gefährdungskategorie* jede durch direkte oder indirekte menschliche Einwirkungen in ihren Beständen regional oder vielerorts lokal zurückgehende oder lokal verschwundene Tier- oder Pflanzenart, auch eine Pflanzenart mit wechselnden Wuchssorten; in D. zur Gefährdungskategorie 3, gemäß IUCN zur Kategorie VU [Vulnerable] gehörig: diejenige Population, deren beobachtete, geschätzte, hergeleitete oder vermutete Größe sich [lt. einem von fünf Kriterien] in den letzten 10 Jahren oder in drei Generationen über 50 % verringert hat oder sich in den nächsten 10 Jahren über 30 % verringern wird; ►Rote Liste, ►bedrohte Art).

6828 vulnerable [adj] **to trampling** *agr. constr. phyt.* (DEN 1971, 13; *opp.* resistant [adj] to trampling); *s* **sensible al pisoteo [loc]** (Referente a plantas que no toleran los daños mecánicos causados por el pisoteo; *opp.* resistente al pisoteo); *f* **vulnérable au piétinement [loc]** (Végétaux ne résistant pas à l'action du piétinement ; *opp.* résistant) ; *g* **trittempfindlich [adj]** (Pflanzen betreffend, die mechanische Schädigungen durch Tritt nicht vertragen; *opp.* trittfest, trittunempfindlich).

W

6829 Wadden Sea [n] *geo. ocean.* (Tidal area extending from Den Helder in the Netherlands [30%] along Germany's coast [60%] and the islands to Esbjerg in Denmark [10%]. This is the largest unbroken stretch of ►tidal mudflats in the world, encompassing approximately 900,000ha. Various parts of it have been placed under comprehensive national and international legal protection); *s* **mar [m] de watt** (La mayor ►llanura de fango litoral del mundo de hasta 30 km de ancho y una superficie total de unas 900 000 ha está situada en las costas holandesas, alemanas y danesas del Mar del Norte. Varias partes de ella están protegidas nacional e internacionalmente, en D. gracias a la declaración como Parque Nacional Wattenmeer); *f* **mer [f] des wadden** (D., Bande littorale de la mer du Nord pouvant atteindre 30 km de large, située entre la côte et l'alignement des îles accompagnant celle-ci, la côte hollandaise à partir de Den Helder représentant 30 %, la côte allemande 60 % et la côte danoise jusqu'à Esbjerg 10 % du plus grand marais maritime au monde, avec une superficie totale d'environ 900 000 ha ; ►slikke) ; *g* **Wattenmeer [n]** (D., bis zu 30 km breiter Meeresbereich der Nordsee zwischen Küste und vorgelagerten Inseln, d. h. vom holländischen Den Helder [30 %] entlang der deutschen Küste [60 %] bis hin zu den dänischen Inseln vor Esbjerg [10 %]. Es ist das längste zusammenhängende Küstenwatt der Welt mit einer Gesamtfläche von ca. 900 000 ha und wurde deshalb zum Nationalpark erklärt: Schleswig-Holsteinisches **W.** [ca. 285 000 ha], 1985 eingerichtet zw. dänischer Grenze und Elbemündung; Niedersächsisches **W.** [ca. 240 000 ha], 1986 eingerichtet zw. Ems- und Elbemündung, und Hamburgisches **W.** [ca. 117 000 ha], 1990 eingerichtet; ►Watt).

6830 wader [n] *zool.* (Relatively tall-legged bird which inhabits the shores of flat freshwater and salt-water bodies, swamps and other wetland areas. The protected status of these birds *[Limicolae]* was particularly emphasized by the ►Ramsar Convention; ►waterfowl); *syn.* wading bird [n]; *s* **ave [f] limícola** (Ave generalmente con patas muy largas que vive en zonas

húmedas dulces y saladas o salobres. A través de la ►Convención de Ramsar se ha creado un instrumento de protección de estas aves y sus hábitats a nivel mundial; ►aves acuáticas); *syn.* limícola [f]; *f* **oiseau [m] limicole** (Oiseaux de rivage des grands cours d'eau, des lacs et des zones littorales, à pattes assez longues, la plupart des espèces nichent dans les zones marécageuses, prairies humides, landes tourbeuses, tourbières fangeuses, lagunes d'eau douce, rives herbeuses, plages de sable, de gravier, vasières, marais salants ; les limicoles sont protégés dans le cadre de la ►Convention de Ramsar relative aux zones humides ; ►sauvagine) ; *syn.* limicole [m] ; *g* **Watvogel [m]** (Meist ziemlich hochbeiniger Vogel, der an flachen Süß- und Salzgewässern, in Sümpfen und in anderen feuchten Landschaften lebt. Die Schutzwürdigkeit dieser Vögel *[Limicolae]* wurde durch die ►Ramsar-Konvention besonders herausgestellt; ►Wasser- und Watvögel); *syn.* Limikole [f].

6831 wadi [n] *geo.* (Arabic term for dry watercourse in a desert or semiarid area [US]/semi-arid area [UK], especially in North Africa, which sometimes contains a torrent for a short time after heavy rain; *s 1* **wadi [m]** (Voz árabe aplicada a los cauces o valles secos de las áreas desérticas por los que sólo corre el agua ocasionalmente; DGA 1986); *syn.* uadi [m]; *s 2* **rambla [f]** (*Término regional* arroyo de las áreas mediterráneas. Generalmente se trata de cursos esporádicos pero de gran caudal, que discurren por cauces de tipo artesa frecuentemente desbordados; DINA 1987; *f* **oued [m]** (*Mot arabe « ouadi »* cours d'eau dont l'écoulement est spasmodique dans les régions arides, temporaire ou saisonnier dans les régions semi-arides, comme le sont les précipitations elles-mêmes ; DG 1984 ; le terme désigne aussi bien l'écoulement de crue que le lit lui-même) ; *g* **Wadi [n]** (*Arab.* ouadi >Bach<; sporadisch oder nur saisonal fließender Bach- oder Flusslauf in ariden oder semiariden Gebieten Nordafrikas und des Vorderen Orients. Oft jahrelang wasserlos; bei Regenereignissen werden plötzlich große Wassermassen transportiert. Wegen des ungleichmäßigen Fließens haben **W.s** nur abschnittsweise ein ausgeglichenes Gefälle).

wading bird [n] *zool.* ►wader.

6832 wading pool [n] [US] *recr.* (Pool with a shallow water depth for non-swimmers; ►toddlers wading pool [US]/paddling pool [UK]); *syn.* pool [n] for non-swimmers [UK]; *s* **piscina [f] para niños/as** (Piscina muy poco profunda generalmente para niños y niñas pequeñas que aún no saben nadar; ►estanque para juegos infantiles; *opp.* ►piscina para nadadores/as); *syn.* piscina [f] para no nadadores/as; *f 1* **bassin [m] non-nageurs** (NEU 1983 ; petit bassin utilisé pour s'habituer à l'eau et à apprendre à nager) ; *f 2* **bassin [m] récréatif** (Petit bassin en général nonnageur doté d'attractions aquatiques et d'équipements de jeu [pentaglisse, jeux d'eau, bain tourbillon, vagues, etc.] ; ►pataugeoire ; *opp.* ►bassin nageurs) ; *syn.* bassin [m] pour nonnageurs ; *g* **Nichtschwimmerbecken [n]** (Schwimmbecken mit geringer Wassertiefe; ►Planschbecken; *opp.* ►Schwimmerbecken).

waldsterben [n] [UK] *envir. for.* ►forest decline.

walk [n] *landsc. urb.* ►footpath.

walk [n], **raised** *recr.* ►boardwalk.

6833 walking distance [n] *plan. recr.* (Length of space expressed in terms of time needed to cover it on foot; e.g. five minutes walking distance from home); *s* **distancia [f] a pie** (Medida utilizada en el contexto de la planificación de equipamiento y de espacios libres que indica el tiempo en minutos necesario para acceder a un servicio, p. ej. de transporte público, o a una zona verde); *f* **distance [f] (d'une) minute à pied** (Mode d'évaluation du rayon d'influence d'un équipement de loisirs en milieu urbain) ; *g* **Gehminuten [fpl]** (Zeitlicher Entfernungsmaßstab für wohnungsnahe Infrastruktureinrichtungen, z. B. fünf Gehminuten zum öffentlichen Nahverkehrsmittel, zum Park).

walking holiday [n] [UK] *recr.* ▶hiking holiday.

walking path [n] [US] *recr.* ▶hiking trail [US]/hiking footpath [UK].

walking trail [n] [US] *recr.* ▶hiking trail [US]/hiking footpath [UK].

6834 wall [n] *arch. constr.* (Elongated structure constructed with bricks, stone, concrete and other natural or man-made materials; the base of a wall is ▶bottom level of a wall; the top is the ▶crown of a wall, the visible, narrow, side elevation is the ▶wall head; the depth of a wall is the ▶wall thickness and the alignment is the▶extension of the wall line. Depending on the location and function of the wall, the following distinctions are made: ▶bearing wall, ▶cheek wall, ▶common boundary wall, ▶concrete crib wall, ▶double-sided masonry wall, ▶dry masonry [masonry without mortar], ▶face wall, ▶free-standing concrete crib wall, ▶free-standing wall, ▶gravity retaining wall, ▶low wall, ▶masonry foundation wall, ▶natural stone masonry, ▶noise barrier wall [US]/acoustic screen wall [UK], ▶perimeter wall, ▶property boundary wall, ▶retaining wall, [to contain earth], ▶screen wall, ▶seatwall, ▶sea wall (1), ▶timber crib wall, ▶veneer wall [cladding with high quality material to simulate masonry], ▶wing wall; ▶battered wall, ▶cyclopean wall, ▶masonry bond, ▶pierced wall, ▶rough-tooled natural stone wall); *s* **muro** [m] (Estructura vertical de piedra, ladrillo u hormigón construida para cerrar una espacio; ▶aparejo de mampuestos, ▶mampostería ciclópea, ▶mampostería de piedras naturales, ▶mampostería en hiladas horizontales, ▶mampostería en seco, ▶murete asiento, ▶muro calado, ▶muro con desplome, ▶muro cortina, ▶muro de cimentación, ▶muro de contención, ▶muro de cercamiento, ▶muro de deslinde, ▶muro de entramado de piezas de hormigón, ▶muro de entramado de rollizos, ▶muro de gravedad, ▶muro de piedras naturales de labra tosca, ▶muro de sostenimiento, ▶muro de sostenimento en L, ▶muro en ala, ▶muro libre, ▶muro libre de entramado de piezas de hormigón, ▶muro medianero, ▶sillería almohadillada, ▶zanca); *f* **mur** [m] (Ouvrage réalisé avec des éléments solides ou constitué d'un aggloméré de matériaux inertes mélangés à un liant, d'origine naturelle ou artificielle ; ▶appareillage d'une maçonnerie, ▶claustra, ▶limon, ▶maçonnerie à appareillage cyclopéen, ▶maçonnerie à parement, ▶maçonnerie à sec, ▶maçonnerie en pierres naturelles, ▶mur à appareillage cyclopéen, ▶mur à appareillage en assises, ▶mur [à parement] incliné, ▶mur brise-vue, ▶mur-chaise, ▶mur de caissons en béton, ▶mur de/en caissons en bois, ▶mur de cave, ▶mur de clôture, ▶mur de revêtement, ▶mur de soutènement, ▶mur en aile, ▶mur en pierres naturelles bossagées, ▶muret siège, ▶mur libre, ▶mur mitoyen, ▶mur poids, ▶mur portant, ▶mur séparatif de propriété) ; *g* **Mauer** [f] (Aus dem Lateinischen *murus*; ein aus Lehm, Werk-, Natursteinen, Beton und anderen natürlichen oder künstlichen Materialien hergestellter, meist langgestreckter Baukörper; die Standfläche einer **M.** wird ▶**Mauersohle**, der obere Abschluss ▶**Mauerkrone**, die sichtbare Schmalseite **Haupt**, die Dicke **Mauerstärke** und die Richtung ▶**Mauerflucht** genannt; je nach Lage und Funktion unterscheidet man u. a. ▶**Grundmauer** oder **Fundamentmauer**, **Sockelmauer**, ▶**Umfassungsmauer** und ▶**Tragmauer**; zum Abstützen des Erdreiches dient die ▶**Stützmauer**, ▶**Schwergewichtsmauer**, ▶**Winkelstützmauer** und bedingt auch eine ▶**Futtermauer**, das Wasserreservoir in einem Tal reguliert die **Staumauer** [▶Talsperre], eine **Blendmauer** [▶Verblendmauerwerk] täuscht ein Mauerwerk aus einem höherwertigen Material vor; eine **M.**, die ohne Mörtel aufgesetzt wird, ist eine **Trockenmauer** [▶Trockenmauerwerk]; ▶dossierte Mauer, ▶Flügelmauer, ▶freistehende Mauer, ▶Grenzmauer, ▶Grundstücksmauer, ▶Mauer aus durchbrochenen Steinen, ▶Mauerwerk, ▶Natursteinmauer aus bossierten Steinen, ▶Strandmauer, ▶Zyklopenmauerwerk).

wall [n] [UK], **acoustic screen** *constr. envir.* ▶noise barrier wall [US].

wall [n], **breast** *constr.* ▶face wall.

wall [n], **composite** *constr.* ▶two-tier masonry [US]/two-leaf masonry [UK].

wall [n], **dry (stone)** *arch. constr.* ▶dry masonry.

wall [n], **embossed natural stone** *arch. constr.* ▶rough-tooled natural stone wall.

wall [n], **enclosing** *arch.* ▶perimeter wall.

wall [n] [US], **faced** *constr.* ▶one-sided masonry.

wall [n] [US], **green** *gard. urb.* ▶planted façade.

wall [n] [US], **low roof** *arch.* ▶roof parapet.

wall [n], **natural stone** *arch. constr.* ▶natural stone masonry.

wall [n] [US], **party** *constr. leg.* ▶common boundary wall.

wall [n], **rough faced masonry** *constr.* ▶split-face dry masonry.

wall [n], **sitting** *constr. gard. landsc.* ▶seatwall.

wall [n] [UK], **snecked rubble** *arch. constr.* ▶broken range work.

wall [n] [US], **solid** *constr.* ▶one-tier masonry [US]/one-leaf masonry [UK].

wall [n], **veneered** *arch. constr.* ▶veneer masonry.

wall base [n], **masonry** *arch. constr.* ▶bottom of wall base.

6835 wall base course [n] *constr.* (Lowest part of a wall, with a wider cross section than the upper part, which is distinct from the foundation; ▶bottom of wall base); *syn.* plinth [n] of a wall; *s* **zapata** [f] **de muro** (Cimentación o ensanchamiento de la base de un soporte, que tiene como cometido repartir las cargas sobre el terreno; DACO 1988; ▶pie de muro); *f* **soubassement** [m] **d'un mur** (Partie basse d'un mur [▶pied de mur] en général constituée de matériaux plus durs que le reste du mur ou faisant saillie sur le nu du mur et pouvant être constitué lui-même de trois parties : la plinthe, le tableau et le cordon ; MAÇ 1981, 316) ; *syn.* socle [m] de mur ; *g* **Mauersockel** [m] (Unterer Teil einer Mauer, der sich durch einen meist breiteren Querschnitt oder durch ein anderes Baumaterial von dem oberen Teil abhebt; ▶Mauerfuß).

6836 wall base [n] **of a fence** *constr.* *s* **zapata** [f] **de valla**; *f* **mur** [m] **bahut** (Mur épais et de faible hauteur servant de banc ou d'assise pour un garde-corps, une clôture) ; *syn.* soubassement [m] d'une clôture ; *g* **Mauersockel** [m] **zur Errichtung eines Zaunes** *syn.* Sockelmauer [f].

6837 wall-climbing plant [n] *hort. plant.* (▶Climbing plant for greening façades which uses various strategies, either self-clinging [▶self-clinging climber] or growing up a support attached to a wall [▶trellis climber]; *syn.* wall-covering vine [n] [US], house wall climber [n] [UK]; *s* **trepadora** [f] **cubremuros** (▶Planta trepadora que crece sobre un muro utilizando diferentes estrategias, bien como ▶trepadora autoadherente bien como ▶trepadora sobre soporte); *syn.* enredadera [f] cubremuros, planta [f] trepadora cubremuros; *f* **plante** [f] **grimpante couvre-mur** (▶Plante grimpante utilisant différentes stratégies pour grimper le long d'un mur ; on distingue les ▶plantes grimpantes à organes d'accrochage et les ▶grimpantes pour treillage) ; *syn.* grimpante [f] couvre-mur, plante [f] couvre-mur, couvre-mur [m] ; *g* **Kletterpflanze** [f] **zur Wandbegrünung** (▶Kletterpflanze, die mit unterschiedlicher Kletterstrategie entweder als

W

▶Selbstklimmer oder als ▶Gerüstkletterer an einer Wand hochwächst); *syn.* Kletterpflanze [f] zur Fassadenbegrünung.

6838 wall community [n] *phyt.* (Association of plants rooted in the cement and mortar of wall joints and rock crevices; ▶rock crevice plant community, ▶vegetation of wall joints, rock crevices or paving joints); *s* **comunidad [f] de las grietas de muros** (Comunidad vegetal que crece en las fisuras de los muros o rocas; ▶comunidad fisurícola, ▶vegetación de rendijas); *f* **groupement [m] saxicole** (Associations végétales très spécialisées se développant sur les murs [groupement muricole] et sur les rochers [▶association rupicole] — fixation à la surface et dans les fissures — ; *termes spécifiques* groupement des fissures de mur, groupement de sommet de mur, groupement de base de mur, ▶association des corniches rocheuses, ▶association des fissures de rocher, ▶association pariétale ; ces peuplements forment les classes des *Asplenietea rupestris* pour les chasmophytes et lithophytes des rochers et murs secs ou frais, des *Adianthetea* pour les lithophytes et pseudochasmophytes des parois très humides ; ▶végétatoin saxicole et muricole) ; *g* **Mauerfugengesellschaft [f]** (1. Stickstoff- und feuchtigkeitsliebende Pflanzengesellschaft von bestimmter floristischer Zusammensetzung in Mauer- und Felsspalten wintermilder Gebiete mit einem Schwerpunkt der Verbreitung im mediterranen und atlantischen Europa; z. B. Klasse der Mauer-Unkrautgesellschaften *[Parietarietea judaicae]*. 2. Oligotrophe Nährstoffansprüche und schwankende Wasser- und Temperaturverhältnisse vertragende Pflanzengesellschaft, die überwiegend aus Kleinfarnen und Moosen besteht, z. B. Klasse der Mauer- und ▶Felsspaltengesellschaft *[Asplenietea rupestris]*; cf. OBER 1977; ▶Fugenbewuchs); *syn.* Mauerritzengesellschaft [f].

6839 wall core [n] *constr.* (Central bearing portion of a wall; e.g. of a veneered wall); *s* **cuerpo [m] de un muro** (Parte central portante p. ej. de un muro de cara vista); *f* **corps [m] principal du mur** (Partie d'un mur assumant la résistance mécanique dans le cas d'un mur double dans les maçonneries de brique pleine ou perforée, d'un mur de pierre banche ou pelliculaire en placage de façades) ; *syn.* âme [f] maçonnée d'un mur ; *g* **Mauerkern [m]** (Innerer, tragender Teil einer Mauer, z. B. eines verblendeten Mauerwerks).

wall-covering vine [n] [US] *hort.* ▶wall-climbing plant [US]/house wall climber [UK].

wall covering [n] **with vines** [US] *gard. urb.* ▶façade planting.

6840 wall face [n] *arch. constr.* (Outer surface of a wall; ▶one-side masonry, ▶double-side masonry wall); *s* **cara vista [f] de un muro** (Superficie exterior visible de un muro; ▶mampostería de cara vista [unilateral], ▶mampostería de cara vista bilateral); *f* **parement [m]** (Surface apparente d'un mur ; ▶maçonnerie à un parement, ▶maçonnerie à deux parements) ; *syn.* nu [m] d'un mur, parement [m] vu d'un mur ; *g* **Mauerhaupt [n]** (Äußere Ansichtsfläche einer Mauer; ▶einhäuptiges Mauerwerk, ▶zweihäuptiges Mauerwerk).

6841 wall footing [n] *constr.* (Usually frost-free foundation of a wall of concrete, stone or fill material on undistained subgrade, constructed to bear the load of the wall and the forces acting upon it, e.g. soil pressure on retaining walls, or the subsoil as well as to protect against cracks caused by settling. In non-rigid construction methods, such as dry stone walls, rigid, frost-free footings are not necessary, because a compacted layer of crushed rock, slag or gravel is sufficient; *syn.* wall foundation [n]; *s* **cimentación [f] de muro** (Fundación de muro —en zonas que lo exijan a nivel inferior al de helada— de hormigón, piedra o material de relleno en subsuelo inalterado, construida para soportar la carga del muro y las fuerzas que actúan sobre él, como

p. ej. la presión del suelo sobre un muro de sostenimiento, o sobre el subsuelo, así como para proteger contra fisuras causadas al asentarse. En métodos de construcción no rígidos, como en muros de mampostería en seco, no es necesaria la cimentación rígida e inferior al nivel de helada, porque una capa compactada de roca partida, escoria o grava es suficiente); *syn.* asiento [m] de muro; *f* **fondation [f] d'un mur** (Partie élargie en pierre ou en béton servant de fondation d'une maçonnerie et en général mise en œuvre hors gel) ; *syn.* semelle [f] d'un mur, semelle [f] de fondation ; *g* **Mauerfundament [n]** (In der Regel frostfreie Gründung einer Mauer aus Beton, Steinen oder Schüttmaterial auf gewachsenem Boden zur Ableitung der Mauerlast und der auf sie einwirkenden Kräfte, z. B. den Erddruck bei Stützmauern in den Untergrund, sowie zum Schutz vor Setzungsrissen. Bei unstarren Bauweisen, z. B. bei Trockenmauern, kann auf ein starres und frostfreies Fundament verzichtet werden, da ein verdichtetes Lager aus Schotter, Schlacke oder Kiessand ausreicht); *syn.* Mauergründung [f].

wall foundation [n] *constr.* ▶wall footing.

6842 wall head [n] *arch. constr.* (End side of a [freestanding] wall); *syn.* end face [n] of a wall [also US]; *s* **cabecera [f] de un muro** (Límite lateral de un muro libre); *f* **tête [f] de mur** (Face d'extrémité de mur isolé ; DTB 1985) ; *g* **Mauerkopf [m]** (Stirnseite einer Mauer; bei Natursteinmauerköpfen müssen besonders große Steine mit exakt geschlagenem Eckwinkel und vorgegebener Dossierung vermauert werden).

6843 wall joint [n] *constr.* (▶pointing, ▶soil pocket; *specific terms* ▶butt joint, ▶coursing joint, ▶heading joint); *s* **junta [f] de un muro** (Intervalo entre los mampuestos o ladrillos de un muro; ▶junta de plantación, ▶rejuntado; *términos específicos* ▶junta a tope, ▶junta de asiento, ▶junta prensada); *f* **joint [m] d'un mur** (Intervalle laissé libre ou rempli de mortier entre les éléments d'une maçonnerie; ▶jointoiement ; ▶niche de plantation ; *termes spécifiques* joint filant, ▶joint délit, ▶joint montant, ▶joint serré) ; *g* **Mauerfuge [f]** (Schmaler Zwischenraum zwischen Mauersteinen — mit Mörtel verfugt oder beim Trockenmauerwerk unverfugt; ▶Verfugen; ▶Pflanzfuge; *UBe* ▶Lagerfuge, ▶Pressfuge, ▶Stoßfuge).

6844 wall offset [n] *arch. constr.* (Parallel setback of a ▶extension of a wall line); *s* **retallo [m] de muro** (Resalto que queda en un muro cuando disminuye su espesor; ▶alineación de muro); *f* **décrochement [m] de mur** (Saillie ou retrait dans l'▶alignement d'un mur) ; *g* **Mauerversatz [m]** (Meist paralleles Zurückspringen einer ▶Mauerflucht).

6845 wall seat [n] *gard. landsc.* (Sitting place attached to a wall); *s* **banco [m] adosado a muro**; *f* **banc [m] mural** ; *g* **Sitzbank [f] an einer Mauer** (Bank an der Mauer befestigt oder direkt vor ihr fest installiert).

warming [n] *met.* ▶global warming.

6846 warm roof [n] *arch. constr.* (Non-ventilated single-ply flat roof, whereby the insulation is located above, or above and between, rafters. A typical single ply roof system comprises structural support [generally not installed by the roofing contractor], a deck providing continuous support, a vapour control layer [▶vapor barrier (US)/vapour barrier (UK), if required], thermal insulation [if required] and a waterproof ▶roof membrane covering with aesthetic traffic or load resistant finish, if required. The principal thermal insulation is placed immediately below the roof covering, resulting in the structural deck and support being at a temperature close to that of the interior of the building; ▶inverted roof; *opp.* ▶cold roof); *syn.* non-ventilated roof [n]; *s* **tejado [m] caliente** (Tejado plano de una hoja, no ventilado, en el que la capa impermeabilizante está colocada arriba, consistente generalmente de una capa de cartón

W

alquitranado enarenado. En general la secuencia de capas, de abajo a arriba, es la siguiente: **1.** Capa portante [vigas y viguetas del techo]. **2.** Capa de nivelación/capa intermedia. **3.** ▶Barrera antihumedad. **4.** Capa de aislamiento térmico y capa de control del vapor. **5.** ▶Membrana impermeabilizante de tejado. **6.** Protección de superficie [p. ej. gravilla, capa portante de vegetación]; ▶tejado invertido; *opp.* ▶tejado frío); *syn.* tejado [m] no ventilado; *f* **toiture [f] chaude** (Toit en pente à simple ventilation, avec circulation d'air entre couverture et sous toiture, mais sans ventilation entre isolation thermique et sous toiture ; la **t. c.** est la composition de toiture aujourd'hui la plus fréquemment rencontrée et recommandée ; l'isolation est posée sur le support de toiture, un ▶écran pare vapeur est posé sur le support de toiture ; le revêtement d'étanchéité est ensuite mis en œuvre sur l'isolation ; le support de toiture est protégé des fortes variations de température par l'isolant et le lestage éventuel, et ne subit donc que de faible chocs thermiques ; par contre les ▶revêtements d'étanchéité subissent directement les actions climatiques [variations de températures, rayons solaires UV, etc.] ; on peut néanmoins réduire les effets de ces actions climatiques en plaçant un lestage [protection lourde telle que graviers, dalles, etc.] sur l'étanchéité en traitant celle-ci au moyen d'un surfaçage adéquat [protection légère telle que paillettes ardoisées, peinture, etc.] ou en la protégeant par une végétalisation ; ▶toiture froide, ▶toiture inversée ; *syn.* toit [m] chaud ; *g* **Warmdach [n]** (Unbelüftetes einschaliges Flachdach; diese einschalige Dachaufbauform mit der oberseitigen Abdichtung aus Teer- und Bitumenprodukten ist für D. seit Ende der 1920er-Jahre ein Import aus dem Klimagebiet südlich der Alpen; cf. FLA 1983, 27; die Schichtenfolge besteht i. d. R. von unten nach oben: **1.** Tragschicht [Dachdecke], **2.** Ausgleichsschicht/Zwischenlage, **3.** ▶Dampfsperre, **4.** Wärmedämmschicht, Dampfdruckausgleichsschicht, **5.** ▶Dachhaut, **6.** Oberflächenschutz [z. B. Bekiesung, Vegetationstragschicht]; ▶Umkehrdach; *opp.* ▶Kaltdach); *syn.* nicht belüftetes Dach [n].

warranty [n], seller's *contr.* ▶guarantee fulfillment obligation [US]/guarantee fulfilment obligation [UK].

6847 washed gravel [n] *constr.* (Gravel used to prevent ▶frost heave under pavement; ▶frost-resistant subbase); *syn.* clean gravel [n]; *s* **grava [f] contra heladas** (Gravilla colocada o por colocar en la ▶capa de protección contra las heladas; ▶levantamiento por congelación); *syn.* grava [f] anticongelante, gravilla [f] contra heladas, gravilla [f] anticongelante; *f* **grave [f] antigel** (Granulat [grave tout venant] constituant la ▶couche antigel d'une chaussée; ▶foisonnement par le gel) ; *syn.* grave [f] non gélive ; *g* **Frostschutzkies [m]** (In der ▶Frostschutzschicht eingebauter oder einzubauender Kies; ▶Frosthebung).

6848 washed sand [n] *constr. eng.* (Flushed out sand without silt and clay particles); *syn.* sharp sand [n]; *s* **arena [f] lavada** (Arena que no contiene ni limo ni arcilla); *f* **sable [m] lavé** (Sable exempt des fraction limoneuses et argileuses) ; *g* **gewaschener Sand [m]** (Sand ohne Schluff- und Tonanteile); *syn.* scharfer Sand [m].

6849 wash [vb] fine material *constr. pedol.* (Washing down of clay or silt particles into an underlying drainage or soil layer); *syn.* wash [vb] out fine particles; *s* **colmatación [f]** (Penetración de partículas finas en la capa de drenaje o una capa inferior del suelo); *syn.* deslavado [m] de partículas finas; *f* **colmatage [m]** (Introduction de particules fines colmatantes dans une couche drainante ou dans l'épaisseur d'un feutre filtrant anticontaminant qui vont lier les granulats ou les fibres, ceux-ci perdant progressivement leur efficacité et devenant complètement imperméable ; par des, pour palier à ce désagréments on utilise des dispositifs anti-colmatage ; colmater [vb]) ; *g* **Einschlämmen [n, o. Pl.] von Feinanteilen** (Eindringen von Schluff- und Tonteilchen in eine Drän- oder tiefer liegende Bodenschicht; ein-

schlämmen [vb] (2), zuschlämmen [vb]); *syn.* Zuschlämmen [n, o. Pl.] (mit Feinanteilen), Einschlämmen [n, o. Pl.] von Bodenfeinteilen.

6850 Washington Convention [n] *conserv. leg. pol. zool.* (Convention on International Trade in Endangered Species of Wild Fauna and Flora [CITES] which was agreed to in 1973 and became operational in the U.S. in 1975; ▶species conservation); *s* **Convención de Washington** (Convención sobre comercio internacional de fauna y flora silvestres, de 3 de mayo de 1973. España se adhirió por Instrumento del 16 de mayo de 1986; ▶protección de especies [de flora y fauna]); *f* **Convention [f] de Washington** (Convention signée en 1973 sur le commerce international des espèces de faune et de flore sauvages [C.I.T.E.S. — Convention on International Trade in Endangered Species of Wild Fauna and Flora] menacées d'extinction instaurant une procédure de délivrance et de présentation de permis pour l'exportation et l'importation de spécimen des espèces couvertes par la convention ; ▶protection des espèces [animales et végétales], ▶protection des espèces de faune et de flore sauvages) ; *syn.* convention [f] C.I.T.E.S. ; *g* **Washingtoner Artenschutzabkommen [n]** (Am 03.03.1973 beschlossenes Abkommen zur Beschränkung des internationalen Handels mit gefährdeten Pflanzen und Arten frei lebender Tiere. In D. seit 1976 in Kraft; im internationalen Sprachgebrauch CITES *[Convention on International Trade in Endangered Species of Wild Fauna and Flora]* genannt; ▶Artenschutz); *syn.* Washingtoner Artenschutzübereinkommen [n].

6851 wash load [n] *hydr.* (Particulate matter suspended almost permanently in a stream, which is transported entirely throughout the system without deposition; ▶sediment load, ▶settling solids); *s* **carga [f] en suspensión** (Partículas detríticas finas, en suspensión casi permanente transportadas en un curso de agua, sin depositarse; WMO 1974; ▶carga de sedimentos en suspensión, ▶partículas sedimentarias); *f* **charge [f] solide en suspension** (Matériaux de très petites dimensions, en suspension de façon quasi permanente et transportés en un cours d'eau sans se déposer ; WMO 1974 ; ▶charge sédimentaire, ▶particules sédimentaires) ; *g* **Schwebstofffracht [f]** (Gesamtgewicht des Gesteinsmaterials und der organischen Stoffe, das in aufgeschwemmter Form, ohne sich abzusetzen, von fließendem Wasser mitgeführt wird, und das je Zeiteinheit einen Abflussquerschnitt durchläuft; ▶Sedimentfracht, ▶Sinkstoffe).

6852 washoff [n] of fine particles *landsc. pedol.* (RCG 1982, 184); (Carrying of clayey and silty soil fractions from a land surface; ▶fine particles, ▶silting up 2); *s* **arrastre [m] de partículas finas del suelo** (Transporte pendiente abajo de las fracciones del suelo de limo o arcilla por escorrentía superficial; ▶entarquinamiento 2, ▶partículas finas); *f* **entraînement [m] des particules fines** (Transport superficiel en solution de ▶particules fines du sol par les eaux de ruissellement ; *terme spécifique* lessivage [m] latéral ; PED 1983, 143 ; *opp.* ▶envasement 2) ; entraîner [vb] des particules fines ; *g* **Abschwemmung [f]** (Abtransport der ▶Feinanteile eines Bodens durch Wasser; abschwemmen [vb], abschlämmen [vb]; *opp.* ▶Anlandung 2).

wash [vb] out fine particles *constr. pedol.* ▶wash fine material.

6853 waste [n] *envir. leg.* (Generic term for both ▶reusable waste, and for all unwanted, movable domestic and industrial solid material, produced directly or indirectly as unusable byproducts from household activities, as well as from factories; ▶biodegradable waste, ▶bulky waste, ▶green waste, ▶household waste, ▶kitchen waste, ▶hazardous waste, ▶industrial waste, ▶organic kitchen waste, ▶radioactive waste, ▶reappli-

W

cation of waste [US]/re-application of waste [UK], ►reuse of waste, ►special waste, ►unauthorized dumped waste, ►urban refuse; ►disposal of radioactive waste, ►final disposal site for radioactive waste, ►waste disposal); *syn.* waste material [n], refuse [s, no pl.]; *s* **residuos [mpl] sólidos** (Término genérico para todo tipo de restos que genera la sociedad sean de origen industrial, doméstico, comercial, sanitario o agrario, y que han de ser reutilizados, reciclados o eliminados de forma segura para el medio ambiente; ►basura clandestina, ►basura voluminosa, ►eliminación de residuos sólidos, ►residuos biodegradables, ►residuos domésticos, ►residuos industriales, ►residuos industriales peligrosos, ►residuos radioactivos/nucleares, ►residuos sólidos urbanos, ►residuos verdes, ►restos de cocina, ►restos orgánicos de cocina, ►valorización de residuos; ►cementerio nuclear, ►disposición de residuos radi[o]activos/nucleares); *syn.* desechos [mpl]; *f* **déchets [mpl]** (Terme générique ; selon la loi n° 75-633 du 15 juillet 1975 est considéré comme déchet « tout résidu d'un processus de production, de transformation ou d'utilisation, toute substance, tout matériau, produit ou plus généralement, tout bien meuble abandonné ou que son détenteur destine à l'abandon » ; voir aussi Directive CEE n° 75/442 du 15 juillet 1975 ; ►biodéchets, ►déchets dangereux, ►déchets de cuisine, ►déchets industriels, ►déchets organiques de cuisine, ►déchets radioactifs, ►déchets urbains, ►déchets verts, ►détritus abandonné, ►élimination des déchets, ►encombrants, ►ordures ménagères [OM], ►réemploi des déchets, ►valorisation des déchets ; ►centre de stockage de déchets radioactifs, ►élimination des déchets radioactifs) ; *g* **Abfall [m] (2)** (...fälle [pl]; OB zu allen beweglichen, unerwünschten, stofflichen Sachen, die durch menschliche Tätigkeiten unmittelbar oder mittelbar entstehen und deren sich der Besitzer entledigt, entledigen will oder entledigen muss und für die eine geordnete Beseitigung zur Wahrung des Wohls der Allgemeinheit erforderlich ist; cf. § 3 [1] KrW-/AbfG. Nach dem Kreislaufwirtschafts- und Abfallgesetz [KrW-/AbfG] wird in nicht überwachungsbedürftigen, überwachungsbedürftigen und ►besonders überwachungsbedürftigen Abfall unterschieden; **A. zur Beseitigung [**auch **Beseitigungsabfall]** wird i. d. R. auf Deponien gelagert oder in Abfallverbrennungsanlagen entsorgt; ►Abfallbeseitigung, ►Abfallverwertung, ►Bioabfall, ►Gewerbemüll, ►Grünabfall, ►Hausmüll, ►Küchenabfall, ►organischer Küchenabfall, ►radioaktiver Abfall, ►Siedlungsabfall, ►Sperrmüll, ►verwertbarer Abfall, ►Weiterverwendung von Abfallstoffen, ►wilder Müll, ►Endlager für radioaktive Stoffe, ►Entsorgung des Atommülls); *syn.* Müll [m], Abfallstoff [m].

waste [n], biological *envir. leg.* ►biodegradable waste.

waste [n], dangerous *envir. leg.* ►hazardous waste.

waste [n], domestic *envir.* ►household waste.

waste [n], incineration plant for hazardous *envir.* ►incinerator for toxic/hazardous waste [US].

waste [n], toxic *envir. leg.* ►hazardous waste.

waste [n], urban *envir.* ►urban refuse.

6854 waste avoidance [n] *envir. sociol.* (Strategy of focussing on the elimination of waste through source reduction and recycling. The complete **w. a.** is the ideal case of ►waste reduction); *s* **evitación [f] de residuos** (Estrategia orientada a reducir la cantidad de residuos en procesos productivos o de consumo. La **e. de r.** es el caso ideal de ►reducción de residuos); *f* **prévention [f] de la production de déchets** (Stratégie commerciale visant, grâce à la mise en œuvre de technologies propres, à concevoir un produit dans une approche globale permettant de limiter de manière significative la génération de déchets ou de résidus [réduction à la source vers un niveau 0], et ce en prenant compte tout son cycle de vie jusqu'à son élimina-

tion ; la prévention absolue constitue en la matière le cas idéal de ►réduction de la production de déchets ; le meilleur déchet est celui qui n'est pas produit. Dans le domaine des déchets ménagers la prévention s'attache à réduire les flux de déchets; on parle alors de « flux évités » lorsque les déchets, grâce à un changement d'usage, ne sont pas mis à la collecte et de « flux détournés » lorsque les déchets sont collectés et reconditionnés) ; *g* **Müllvermeidung [f]** (Strategie, die darauf ausgerichtet ist, unerwünschte Abfälle/Rückstände gar nicht erst entstehen zu lassen. Vollständige **M.** ist der Idealfall der ►Müllverminderung); *syn.* Abfallvermeidung [f].

waste bin [n] [UK] *envir.* ►trash can [US] (1).

waste collection [n] *envir.* ►household waste collection [US], ►selective waste collection.

waste collection [n] [UK]**, domestic** *envir.* ►household waste collection [US].

waste collection charges [npl] [UK] *adm. envir. leg.* ►garbage collection charges [US].

waste compost [n] *envir. pedol.* ►refuse compost.

waste compost production plant [n] [US] *envir.* ►refuse compost production plant.

6855 waste container [n] *envir.* (Generic term for refuse receptacles, such as ►dumpster [US]/skip [UK], ►trash can [US] 1/dust bin [UK], ►trash can [US] 2/litter bin [UK], ►trash storage [US]/bin store [UK], ►trash can support with bag liner [US]/litter bin support with sackholder [UK]); *s* **recipiente [m] para residuos sólidos** (Término genérico para todo recipiente utilizado para recoger desechos sólidos; *términos específicos* ►contenedor, ►contenedor de basura; ►emplazamiento de cubos de basura, ►papelera 2, ►soporte de sacos de basura); *syn.* basurero [m] [YZ]; *f* **récipient [m] de collecte (des ordures ménagères)** (Terme générique pour tout récipient servant à la collecte des déchets des ménages ; *termes spécifiques* ►bac, ►corbeille à ordures, ►logette vide-ordures, ►poubelle, ►support sacs poubelles) ; *g* **Abfallbehälter [m]** (Behälter, der der Abfallaufnahme dient; *UBe* ►Abfallkorb, ►Mulde 2, ►Mülltonne; ►Standplatz für Abfallbehälter, ►Müllsackständer); *syn.* Müllbehälter [m].

6856 waste decomposition [n] *envir.* (Method of waste removal whereby a special storage system allows the rotting of a great part of organic matter found in waste by way of oxygenation; ►refuse composting); *syn.* rotting [n] of refuse/waste; *s* **descomposición [f] de residuos urbanos** (Proceso de eliminación de residuos domésticos por medio de un almacenamiento especial por el cual la mayor parte de la materia orgánica se descompone en condiciones aeróbicas; ►compostaje de residuos orgánicos domésticos); *f* **fermentation [f] des ordures ménagères** (Procédé d'élimination des déchets d'origine domestique utilisant la fermentation dirigée et contrôlée des ordures ménagères étant l'œuvre de micro-organismes aérobies faisant évoluer après 3 ou 4 brassages [réoxygénation] le compost à sa maturité ; ►compostage de déchets ménagers) ; *g* **Müllverrottung [f]** (Verfahren der Hausmüllbehandlung, bei dem durch eine spezielle Art der Lagerung ein großer Teil der im Abfall vorhandenen organischen Substanz unter Zutritt von Luftsauerstoff zersetzt wird; ►Müllkompostierung).

6857 waste disposal [n] *envir.* (Discarding of all non-usable solid waste from the economic cycle, reprocessing of reusable material or interim storage and hauling to a disposal site for public health and convenience; ►effluent disposal, ►household waste collection [US]/domestic waste collection [UK], ►refuse composting, ►waste treatment, ►waste treatment and disposal [US]/public cleansing and waste disposal [UK]); *syn.* refuse disposal [n]; *s* **eliminación [f] de residuos sólidos** (Todo

W

procedimiento dirigido, bien al vertido de los residuos o bien a su destrucción total o parcial, realizado sin poner en peligro la salud humana y sin utilizar métodos que puedan causar perjuicios al medio ambiente; art. 3 l] Ley 10/1998; ►compostaje de residuos orgánicos domésticos, ►eliminación de aguas residuales, ►recogida de residuos sólidos urbanos, ►recogida y eliminación de residuos, ►tratamiento de residuos [sólidos]); *f* **élimination [f] des déchets** (Terme désignant les opérations de collecte, de transport, de stockage transitoire, de tri, de traitement et de dépôt ou rejet définitif des déchets ultimes issu du processus de production, de transformation ou d'utilisation ; ►assainissement et élimination des déchets, ►collecte des ordures ménagères, ►compostage des déchets ménagères, ►élimination des eaux usées, ►traitement des déchets) ; *g* **Abfallbeseitigung [f]** (Gemeinwohlverträgliches Konzept, das darauf gerichtet ist, nicht wieder verwertbare feste Rückstände aus dem Wirtschaftsprozess bereitzustellen, zu überlassen, einzusammeln und abzutransportieren, zu behandeln, ggf. zwischenzulagern und anschließend zu deponieren [ablagern zur Beseitigung]; cf. § 10 KrW-/AbfG; ►Abfallbehandlung, ►Abwasserbeseitigung, ►Entsorgung, ►Müllkompostierung, ►Müllsammlung); *syn.* Abfallentsorgung [f], Müllbeseitigung [f], Müllentsorgung [f].

waste disposal [n] [UK]**, public cleansing and** *envir.*
►waste treatment and disposal [US].

waste disposal authority [n] [UK] *adm. envir.* ►waste disposal site authority [US].

waste disposal company [n] [UK] *adm. envir.* ►waste site contractor [US].

6858 waste disposal ordinance [n] [US] *envir. leg.* (**In U.S.**, a federal law does not exist. Local governments have extensive authority to regulate waste disposal by local ordinances, which are typically authorized by the state enabling legislation. Specific authorization may not be required in 'Home Rule' states having constitutional authority for such local responsibilities; **in U.K.**, the C.O.P.A. was enacted in 1974; the Deposit of Poisonous Waste Act 1972 was repealed and replaced by the Control of Pollution [special waste] Regulations/UK Special Wastes Regulations in 1980; cf. DES 1991, 359; **in D.**, there is a federal law with which states have to comply); *syn.* Control [n] of Pollution Act [UK]; *s* **ley [f] de residuos** (En Es. rigen la ley 10/1998, de 21 de abril, de residuos, el reglamento para la ejecución de la ley 20/1986, aprobado por RD 833/1988, de 20 de julio, exceptuando los arts. 50, 51 y 56 derogados por la ley 10/1998, y el RD 952/1997, de 20 de junio, por el que se modifica el citado reglamento. La ley 10/1998 incorpora a la normativa española la concepción moderna de la política de residuos definida en la Directiva Comunitaria 91/156/CEE, del Consejo, de 18 de marzo de 1991, y pretende contribuir también a las políticas económica, industrial y territorial, con el objetivo de incentivar la reducción de residuos en origen y dar prioridad a la reutilización, reciclado y valorización de residuos sobre otras técnicas de gestión. En cuanto al ejercicio efectivo de las competencias, la ley respeta el reparto constitucional entre el Estado y las CC.AA., al tiempo que garantiza las competencias que tradicionalmente han venido ejerciendo las Entidades locales. La ley consta de seis títulos: **1.** Normas generales. **2.** De las obligaciones nacidas de la puesta en el mercado de productos generadores de residuos. **3.** De la producción, posesión y gestión de residuos. **4.** Instrumentos económicos en la producción y gestión de residuos. **5.** Suelos contaminados y **6.** Inspección y vigilancia. Responsabilidad administrativa y régimen sancionador. Además consta de un anexo con las categorías de residuos); *f* **loi [f] relative à l'élimination des déchets et la récupération des matériaux** (**F.**, réglementation hétérogène relevant de différents codes, en particulier du code de la santé publique, du règlement sanitaire départe-

mental et de divers arrêtés et circulaires ; les grands principes de réglementation sont cependant formulés dans la loi n° 75-633 du 15 juillet 1975 modifiée et la loi du 19 juillet 1976 sur les installations classées pour la protection de l'environnement ; voir aussi Directive CEE n° 75/442 du 15 juillet 1975 relative aux déchets) ; *g* **Kreislaufwirtschafts- und Abfallgesetz [n]** (*Abk.* KrW-/AbfG; **D.**, das Rahmengesetz des Bundes vom 11.06.1972 [BGBl. I 873], das „Abfallbeseitigungsgesetz", wurde erst durch das Abfallgesetz von 1986 und dann durch das „Kreislaufwirtschafts- und Abfallgesetz" vom 27.09.1994 [BGBl. I 2705], zuletzt geändert am 22.12.2008 [BGBl. I 2986], abgelöst. Das Abfallrecht des Bundes wird durch zahlreiche zusätzliche spezifizierende Regelungen ergänzt, wie z. B. durch das Tierkörperbeseitigungsgesetz, die abfallrechtlichen Vorschriften des Tierseuchengesetzes, das Pflanzenschutzgesetz, das BBergG, das BImSchG und das Atomgesetz. Alle Bundesländer haben Landesabfallgesetze erlassen, die nicht nur Ausführungsvorschriften zum Bundesrecht, sondern auch eigenständige Regelungen enthalten, insbesondere zur Abfallvermeidung. Diese Regelungen müssen an das **KrW-/AbfG** angepasst werden. Auf kommunaler Ebene gibt es Abfallsatzungen, die Regelungen über den Anschluss- und Benutzungszwang für die öffentliche Abfallentsorgung, den Ausschluss bestimmter Abfallprodukte von der kommunalen Entsorgung sowie die Sammlung des häuslichen Mülls enthalten).

6859 waste disposal site authority [n] [US] *adm. envir.* (Public authority or agency which stores delivered refuse material in a regular and legal way [►sanitary landfill (US) 2/landfill site (UK)] and ensures that no detrimental environmental influences and other dangers, considerable disadvantages and serious disturbances for the public and the neighbo[u]rhood of the landfill may occur; *syn.* waste disposal authority [n] [UK]; *s* **gestor [m] público de depósito de residuos** (Entidad pública que tiene bajo su responsabilidad el funcionamiento y la gestión correctas de un ►vertedero controlado); *f* **établissement [m] public de traitement des déchets** (Personne morale de droit public, collectivités, établissements publics de coopération intercommunale, syndicats de communes assurant une ou plusieurs formes de traitement multifilières des déchets collectés faisant appel au stockage, recyclage ou à différentes formes de valorisation et réalisant ces opérations dans des conditions propres à éviter les nuisances de nature à produire des effets nocifs sur le sol, la flore et la faune, à dégrader les sites ou les paysages, à polluer l'air ou les eaux, à engendrer des bruits et des odeurs et d'une façon générale à porter atteinte à la santé de l'homme et à l'environnement ; cf. loi n° 75-633 du 15 juillet 1975 relative à l'élimination des déchets et à la récupération des matériaux, loi n° 92-646 du 13 juillet 1992 relative à l'élimination des déchets ainsi qu'aux installations classées pour la protection de l'environnement) ; *g* **öffentlicher Deponiebetreiber [m]** (Behörde oder Eigenbetrieb, die/der angelieferte Abfälle ordnungsgemäß lagert und u. a. sicherstellt, dass schädliche Umwelteinwirkungen und sonstige Gefahren, erhebliche Nachteile und erhebliche Belästigungen für die Allgemeinheit und die Nachbarschaft nicht hervorgerufen werden können; cf. § 5 [1] BImSchG); *syn.* öffentlicher Betreiber [m] einer Deponie.

6860 waste disposer [n] *envir.* (Firm or agency which collects waste material, and which will process, recycle or haul it to a final disposal site); *s* **gestor, -a [m/f] de residuos** (Administración o empresa que presta servicios en el campo de la eliminación de residuos; *término específico* gestor, -a [m/f] de residuos tóxicos y peligrosos [Es]); *f* **éliminateur [m]** (Un service de l'administration municipale, un syndicat de communes, une entreprise privée ou une personne morale chargés de l'élimination des déchets) ; *syn.* établissement [m] éliminateur de déchets ;

W

g **Entsorger** [m] (Ein Amt für Abfallwirtschaft einer Gemeinde, eine Person oder eine Firma, an die Abfälle versandt werden, die Abfallstoffe abholt, ggf. verbrennt oder behandelt und weiterverwertet oder endlagert).

6861 waste exchange service [n] [UK] *envir.* (In Europe, private firm which gathers data about industrial waste and coordinates its transfer for ▶reuse of waste, ▶reapplication of waste [US]/re-application of waste [UK], recycling and disposal; **in U.S.**, no comparable firm which coordinates the disposal of industrial waste); *s* **bolsa** [f] **de gestión de residuos tóxicos y peligrosos** (Centro de información de datos relativos a las materias primas contenidas en los residuos tóxicos y peligrosos susceptibles de su aprovechamiento por terceros; ▶reutilización de residuos, ▶valorización de residuos [Es]); *f* **bourse** [f] **de déchets (industriels)** (Établissement d'initiative industrielle permettant d'échanger toute information concernant des marchés de déchets industriels à valoriser ou à récupérer ; ▶réemploi des déchets, ▶valorisation des déchets) ; *g* **Abfallbörse** [f] (Organisatorische Einrichtung, die mit Hilfe einer bundesweiten und europäischen Datenbank unentgeltliche Informationen über wieder verwertbare Produktionsrückstände resp. Reststoffe aus Industrie und Gewerbe sammelt und den Transfer zur ▶Weiterverwendung von Abfallstoffen oder ▶Abfallverwertung organisiert; **in D.** gibt es seit 1974 von den Industrie- und Handelskammern [IHK] zusammen mit dem Deutschen Industrie- und Handelstag [DIHT] bundesweit eingerichtete **A.n**. Seit 1980 koordiniert der Deutsche Industrie- und Handelstag [DIHT] die europäische **A.**, v. a. in Frankreich, der Schweiz, Italien, Österreich und den Niederlanden; die eigentlichen Transaktionsbedingungen handeln Anbieter und Nachfrager unter sich aus. Die irreführende Bezeichnung **A.** in z. T. in **Recyclingbörse** umbenannt, da nicht Abfall schlechthin vermittelt wird, sondern Reststoffe als Wirtschaftsgüter. **In A.** gibt es seit 1974 die Bundes-Abfall- und Recyclingbörse in Linz; seit 1978 fördert in Genf eine Expertengruppe der UN-Wirtschaftskommission für Europa [ECE] auf Vorschlag Österreichs den internationalen Austausch von wieder verwertbaren Abfallstoffen sowie Initiativen zum Einsatz neuer Technologien für die Einsparung von Rohstoffen und für die Energiegewinnung; cf. WKO 2009; **in CH** gibt es die *abfallbörse schweiz.ch AG*, **in F.** die *Bourse des déchets industriels*); *syn.* Recyclingbörse [f].

6862 wasteful exploitation [n] *agr. for. land'man. min. nat'res.* (Ruthless use of resources aimed at achieving maximum short-term financial gains whilst ignoring the need for conservation and up-keep to permit long-term production; ▶consumption of natural resources, ▶exhaustion, ▶sustainable management); *syn.* overexploitation [n] (BEX 1991, 13); *s* **expoliación** [f] **de los recursos naturales** (▶agotamiento, ▶consumo de los recursos naturales; *opp.* ▶gestión económica sostenible); *syn.* sobreexplotación [f] de los recursos naturales; *f* **pillage** [m] **des ressources naturelles** (Forme d'économie destructive en agriculture, sylviculture et dans la pêche dont l'objectif est la recherche à court terme du profit maximal au dépend de la conservation et du développement à long terme de la production ; exploitation extrême des ressources minières ; ▶épuisement des ressources naturelles, ▶consommation des ressources naturelles ; *opp.* ▶gestion durable) ; *syn.* déprédation [f] des ressources naturelles, exploitation [f] effrénée des ressources ; *g* **Raubbau** [m, o. Pl.] (**1** Wirtschaftsweise in der Landwirtschaft, Forstwirtschaft und Fischerei mit dem Ziel, kurzfristig einen maximalen Ertrag auf Kosten der Erhaltung und Förderung einer langfristigen Erzeugungsgrundlage zu erzielen; **2.** im Bergbau die extreme Ausbeutung der Bodenschätze; ▶Abbau 2, ▶Ressourcenverbrauch; *opp.* ▶nachhaltiges Wirtschaften).

6863 waste heat [n] *envir.* (Heated air or water discharged into the environment, e.g. from power stations; ▶thermal pollution of watercourses); *s* **calor** [m] **residual** (Emisión de calor p. ej. de una central de electricidad; ▶contaminación térmica de cursos de agua); *syn.* calor [m] de escape, calor [m] perdido; *f* **rejet** [m] **thermique** (Perte ou rejet de chaleur dans l'atmosphère ou le milieu aquatique occasionné par exemple par l'exploitation de centrales thermiques ; *termes spécifiques* effluent thermique, chaleur résiduaire, ▶pollution thermique des cours d'eau) ; *g* **Abwärme** [f] (Wärmeabgabe an Luft oder Gewässer durch den Betrieb von z. B. Kraftwerken; ▶thermische Belastung von Fließgewässern).

6864 waste incineration [n] *envir.* (Thermal method of waste disposal in a ▶waste incineration plant; ▶incinerator for toxic/hazardous waste); *s* **incineración** [f] **de residuos sólidos urbanos** (▶planta de incineración de residuos peligrosos, ▶planta de incineración de residuos sólidos urbanos); *f* **incinération** [f] **des déchets** (Procédé thermique d'élimination **1.** des ordures ménagères dans une ▶usine d'incinération de résidus urbains, **2.** des déchets industriels spéciaux dans une ▶installation d'incinération ou dans une ▶installation spécialisée d'incinération) ; *g* **Müllverbrennung** [f] (Thermisches Verfahren der Abfallbeseitigung in einer ▶Müllverbrennungsanlage; ▶Sondermüllverbrennungsanlage); *syn.* thermische Abfallbehandlung [f].

6865 waste incineration plant [n] *envir.* (Plant for the processing of household waste by incineration with or without recycling of the heat which is released by the incineration process. Not included are facilities which incinerate on land or at sea sewage waste, chemical, poisonous and hazardous material, medicinal waste from hospitals or other special wastes, even though it might be possible to incinerate domestic waste in those plants); *syn.* (garbage) incinerator [n] [also US], waste incinerator [n] [also US] (TGG 1984, 253); *s* **planta** [f] **de incineración de residuos sólidos urbanos** (Instalación de procesamiento térmico de residuos urbanos con o sin aprovechamiento del calor residual; en esta categoría no se incluyen las incineradoras de residuos peligrosos, residuos de hospitales, las incineradoras de fango de depuración en alta mar o en tierra o incineradoras de otros tipos de residuos tóxicos; cf. art. 1, Directiva 89/369/CEE); *syn.* incineradora [f] de residuos sólidos urbanos; *f* **usine** [f] **d'incinération de résidus urbains** (Installation classée d'élimination des déchets ménagers et assimilés par incinération avec ou sans récupération de la chaleur de combustion produite, à l'exclusion des installations spéciales affectées sur terre et en mer, à l'incinération des boues d'épuration, des déchets chimiques, toxiques et dangereux, de déchets d'établissements hospitaliers et autres déchets spéciaux, même si ces installations peuvent également incinérer des déchets urbains ; directive du Conseil n° 89/429 du 21 juin 1989 ; la pollution de l'air et le traitement des résidus de fumées d'incinération d'ordures ménagères [REFIOM] constituent les inconvénients majeurs de ce procédé d'élimination) ; *syn.* incinérateur [m] de résidus urbains, unité [f] d'incinération de déchets ménagers, installation [f] d'incinération des déchets municipaux [CEE]) ; *g* **Müllverbrennungsanlage** [f] (Technische Betriebsstätte, die der Behandlung von Siedlungsabfällen durch Verbrennung mit oder ohne Rückgewinnung der bei der Verbrennung frei werdenden Wärme dient. Ausgenommen sind Anlagen, in denen an Land oder auf See Klärschlamm, chemische, giftige und gefährliche Stoffe, medizinische Abfälle aus Krankenhäusern oder andere Sonderabfälle verbrannt werden, selbst dann, wenn in diesen Anlagen auch Siedlungsmüll verbrannt werden kann; cf. Art. 1, RL 89/369/EWG; *OB* thermische Entsorgungsanlage); *syn.* Abfallverbrennungsanlage [f], Kehrrichtverbrennungsanlage [f] [CH], Verbrennungsanlage [f] für Siedlungsmüll.

W

waste incinerator [n] [US] *envir.* ►waste incineration plant.

6866 waste land [n] *agr. for.* (Uncultivated agricultural or forest land on ►marginal soil with no profitable use. *Specific term in U.K.* heath and rough grassland; **2.** ►open country [n], #2.; ►abandoned land, ►barren land, ►derelict land); *s* **tierra [f] yerma** (Tierra que no se cultiva en la actualidad, improductiva y con aspecto de abandono; ►ruina industrial, ►suelo marginal, ►tierra abandonada, ►terreno improductivo); *syn.* erial [m], eriazo [m]; *f* **terre [f] improductive** (Terre abandonnée, espace improductif qui n'est plus cultivé, à très faible valeur agricole, pastorale ou forestière dont la récupération ou la mise en valeur n'est souvent plus praticable en particulier en zones de montagnes ; ►friche industrielle, ►friche sociale, ►terre inculte, ►terre marginale) ; *syn.* contrée [f] déserte/pauvre, lande [f] improductive, terrain [m] impropre à la culture ; *g* **Ödland [n]** (**1** Land- und forstwirtschaftlich ertragloses Land [ödes Land] auf leistungsschwachen Standorten; **2.** die britische Begriffskategorie ►*open country, Ziff. 2*, gibt es in D. nicht, da mehr in Nutzungskategorien gedacht wird. Da *open countries* einer Bewirtschaftung nicht unterliegen, können sie z. T. als Ö. bezeichnet werden; ►Grenzertragsboden, ►Industriebrache, ►Sozialbrache, ►Unland).

waste land [n], **industrial** *plan. urb.* ►derelict land.

6867 waste management [n] *envir. urb.* (**1.** Waste treatment and disposal handling. This term is similar to ►waste disposal. **2.** Treatment and disposal of waste by waste-producing companies, which operate from a commercial standpoint and wish to economize on materials, whereby waste may be either directly reused [►reuse of waste] to create the same product again without any particular treatment, or reused after reprocessing [►reapplication of waste (US)/re-application of waste (UK)] to create a new product; ►recycling. Waste, which cannot be reused, is taken to a ►waste exchange service or to a waste disposal facility; ►waste treatment and disposal industry); *s* **gestión [f] de residuos sólidos (urbanos)** (**1.** Tratamiento y deposición de residuos. En sentido general el término puede utilizarse como sinónimo de ►eliminación de residuos sólidos. **2.** Una empresa productora de residuos reutiliza al máximo sus desechos para ahorrar recursos, bien directamente [►reutilización de residuos] o después del correspondiente tratamiento [►valorización de residuos] de manera que puedan ser utilizados en la misma empresa o vendidos en una ►bolsa de gestión de residuos tóxicos y peligrosos. Si no se pueden aprovechar deben ser eliminados a través de la correspondiente ►industria de tratamiento y eliminación de residuos); *f* **gestion [f] des déchets** (Mesures permettant la prévention, la réduction, la valorisation maximale des déchets [►valorisation des déchets], par leur ►réemploi des déchets immédiat ou leur réutilisation après traitement dans l'entreprise ; les déchets restants non réutilisables étant mis à la disposition de la ►bourse de déchets [industriels] ou transportés dans les installations de recyclage, de valorisation et d'élimination ; ►élimination des déchets, ►industrie de la récupération) ; *g* **Abfallwirtschaft [f]** (**1.** Im allgemeinen Sprachgebrauch meist syn. mit ►Abfallbeseitigung. **2.** Abfallbehandlung und Entsorgung des Abfalls von Abfall produzierenden Firmen, die aus betriebswirtschaftlicher Sicht eine kostenorientierte Materialwirtschaft betreiben, bei der Reststoffe unmittelbar [►Weiterverwendung von Abfallstoffen] oder nach entsprechender Behandlung [►Abfallverwertung] im eigenen Unternehmen wieder eingesetzt werden. Diejenigen zum Abfall gewordenen Reststoffe, die nicht wiederverwendet werden können, werden der ►Abfallbörse oder entsprechenden Abfallbeseitigungseinrichtungen zugeführt; ►Entsorgungswirtschaft); *syn.* Abfallbewirtschaftung [f].

6868 waste management department [n] *adm. envir.* (Municipal authority responsible for the disposal of solid refuse, toxic waste, and may include street cleaning and snow clearing); *syn.* public hygiene department [n], municipal hygiene department [n], city hygiene department [n], division [n] of solid waste, collecting and recycling [also US], sanitation department [n] [also US]; *s* **servicio [m] municipal de limpieza** (Administración municipal responsable de la eliminación de residuos sólidos urbanos, de la limpieza de las calles y espacios públicos y, en países fríos, de la eliminación de la nieve en las calles); *f* **service [m] de nettoyage municipal** (Administration municipale responsable de l'organisation et de l'exécution de l'élimination des déchets, du nettoyage de la voirie, du déverglaçage en hiver) ; *g* **Stadtreinigungsamt [n]** (Behörde einer Stadt, die für die Entsorgung von festen Abfallstoffen, die Straßenreinigung und je nach Ortsstatut für den Winterdienst auf städtischen Straßen und Wegen verantwortlich ist. In D. werden seit Mitte der 1990er-Jahre im Rahmen der Umorganisation von städtischen Behörden viele **S.ämter** in kostenrechnende Eigenbetriebe umgewandelt oder privatisiert); *syn.* Amt [n] für Abfallwirtschaft und Stadtreinigung, *obs.* Fuhramt [n].

waste material [n] *envir. leg.* ►hazardous waste material, ►waste.

waste material [n] [UK], **tip of** *envir. leg.* ►large spoil area [US].

6869 waste oil [n] *envir.* (Oil which is left over and no longer needed after it has been used to drive machinery); *s* **residuos [mpl] de hidrocarburos** *syn.* aceites [mpl] residuales; *f* **résidus [mpl] d'hydrocarbures** (Huiles et lubrifiants usagés provenant de l'exploitation normale de machines) ; *syn.* huiles [fpl] résiduaires ; *g* **Ölrückstände [mpl]** (Öl, das nach Gebrauch zum Antrieb von Maschinen nicht mehr benötigt wird); *syn.* Ölabfall [m].

6870 waste producer [n] *envir.* (**1.** A person or firm who generates waste in product fabrication. **2.** Owner of unwanted movable refuse material; ►waste disposer); *s* **productor [m] de residuos** (Cualquier persona física o jurídica cuya actividad, excluida la derivada del consumo doméstico, produzca residuos o que efectúe operaciones de tratamiento previo, de mezcla o de otro tipo que ocasionen un cambio de naturaleza o de composición de esos residuos. Tendrá también carácter de productor el importador de residuos o adquirente en cualquier Estado miembro de la UE; art. 3 d] Ley 10/1998; ►gestor de residuos); *f* **producteur [m] de déchets** (Personne produisant des déchets pendant le processus de production ; ►éliminateur) ; *g* **Abfallproduzent [m]** (**1.** Natürliche oder juristische Person, die bei der Herstellung von Produkten Abfall erzeugt; *syn.* Abfallerzeuger [m]. **2.** Besitzer von beweglichen Sachen, deren er sich entledigen möchte; ►Entsorger).

6871 waste recycling plant [n] *envir.* (Commercial enterprise dealing with the separation of raw materials from waste and their reuse; *specific terms* ►waste separation plant, ►refuse compost production plant, ►waste incineration plant); *syn.* waste reprocessing plant [n] [also US]; *s* **planta [f] de reciclaje de residuos sólidos** (Planta en la que se extraen materias primas de los residuos para su reutilización; *términos específicos* ►planta de compostaje de residuos orgánicos, ►planta de incineración de residuos sólidos urbanos, ►planta de separación de residuos sólidos); *syn.* planta [f] de aprovechamiento de residuos sólidos; *f* **usine [f] de traitement de résidus** (Installation pratiquant la récupération d'énergie ou de matières secondaires pour être réintroduit [valorisation, recyclage] dans le cycle industriel ; *termes spécifiques* ►centre de tri de déchets, ►installation de compostage des ordures ménagères, installation de récupération

W

des déchets, installation de valorisation des déchets, installation/unité de compostage, ▶usine d'incinération de résidus urbains) ; *syn.* station [f] de traitement des déchets, unité [f] de traitement ; *g* **Müllverwertungsanlage [f]** (Gewerblicher Betrieb, der aus Abfall die Rohstoffe trennt, die einer Wiederverwendung zugeführt werden; *UBe* ▶Abfallsortieranlage, ▶Müllkompostwerk, ▶Müllverbrennungsanlage); *syn.* Abfallverwertungsanlage [f].

6872 waste reduction [n] *envir. sociol.* (Strategy organizing production processes and waste treatment with the aim of reducing the amount of waste and residues to a minimum; ▶waste avoidance); *s* **reducción [f] de residuos** (Medidas encaminadas a reducir la generación de residuos en la producción y en el consumo humano, como p. ej. evitando envases no retornables, bolsas de plástico, etc. ▶evitación de residuos); *f* **réduction [f] de la production de déchets** (Stratégie économique [et souvent de marketing] consistant à organiser les processus de production, de distribution et de récupération/valorisation des déchets de telle manière que soit assurée une réduction exemplaire des volumes de déchets générés ; ▶prévention de la production de déchets) ; *g* **Müllverminderung [f]** (Strategie, die darauf ausgerichtet ist, den allgemeinen Konsum, Produktionsprozesse und Abfallbehandlung so zu organisieren, dass unerwünschte Abfälle/Rückstände auf ein Minimum reduziert werden; ▶Müllvermeidung); *syn.* Abfallverminderung [f].

waste reprocessing plant [n] [US] *envir.* ▶waste recycling plant.

6873 waste separation [n] *envir. leg.* (Division of refuse material into various categories for recycling; ▶waste separation plant); *syn.* waste sorting [n]; *s* **separación [f] de residuos** (División de los residuos según los materiales de los que están compuestos: vidrio, papel y cartones, envases ligeros, restos orgánicos u otros restos para posibilitar su aprovechamiento posterior. Puede realizarse a nivel doméstico o en ▶planta de separación de residuos sólidos); *f* **tri [m] des déchets** (Terme générique désignant la séparation dans un centre de tri ou à la source [▶collecte sélective] des déchets recyclable des ordures ménagères ; ▶centre de tri des déchets) ; **Abfalltrennung [f, o. Pl.]** (Sortierung des Abfalls in unterschiedlich verwertbare Abfallstoffe [Fraktionen]; ▶Abfallsortieranlage); *syn.* Mülltrennung [f].

6874 waste separation plant [n] *envir.* (Technology centre, where waste is sorted into separate kinds of reusable material, with disposal of the remainder. Such centres include buildings for ▶waste separation, recycling, composting of green kitchen waste and a percolation and anaerobic digestion facility to treat household waste. There are also laboratories on site controlling the treatment of municipal waste; ▶selective waste collection); *s* **planta [f] de separación de residuos sólidos** (▶recogida selectiva de residuos, ▶recogida selectiva de residuos urbanos, ▶separación de residuos); *f* **centre [m] de tri des déchets** (Installation industrielle ayant vocation de réception, sélection industrielle et de conditionnement des fractions valorisables des différentes catégories de déchets ménagers et industriels ou commerciaux assimilés pour une valorisation ultérieure dans des filières appropriées ; les déchets concernés sont les corps creux [verres, plastiques, acier, aluminium], et les corps plats [journaux et magazines, papiers, cartons, verres, plastiques, acier, aluminium] ; fait partie intégrante de la filière de la ▶collecte sélective ; ▶tri des déchets) ; *g* **Abfallsortieranlage [f]** (Anlage, in der angelieferter Abfall/Müll nach bestimmten Abfallgruppen sortiert wird; die Sortierung nennt man ▶Abfalltrennung oder Mülltrennung; ▶getrennte Müllsammlung).

waste shipment [n] *envir.* ▶transboundary waste shipment [US]/waste tourism [UK].

6875 waste site contractor [n] [US] *adm. envir.* (Firm which operates a sanitary landfill and is charged to ensure that detrimental environmental consequences and danger to the public or serious nuisances for the neighbo[u]rhood of the landfill do not occur); *syn.* waste disposal company [n] [UK]; *s* **gestor [m] de depósito de residuos** (Empresa que opera un depósito de residuos sólidos de tal manera que evita impactos negativos sobre el medio ambiente y cualquier otro peligro o molestia considerable para la sociedad en general); *f* **entreprise [f] de traitement des déchets** (Personne physique ou morale assurant une ou plusieurs formes de traitement multifilières des déchets collectés faisant appel au stockage, recyclage ou à différentes formes de valorisation et réalisant ces opérations dans des conditions propres à éviter les nuisances de nature à produire des effets nocifs sur le sol, la flore et la faune, à dégrader les sites ou les paysages, à polluer l'air ou les eaux, à engendrer des bruits et des odeurs et d'une façon générale à porter atteinte à la santé de l'homme et à l'environnement ; cf. loi n° 75-633 du 15 juillet 1975 relative à l'élimination des déchets et à la récupération des matériaux, loi n° 92-646 du 13 juillet 1992 relative à l'élimination des déchets ainsi qu'aux installations classées pour la protection de l'environnement ; *termes spécifiques* entreprise de traitement d'ordures ménagères, exploitant de centre d'enfouissement de déchets, exploitant d'usine d'incinération) ; *syn.* exploitant [m] d'une installation d'élimination ; *g* **privater Deponiebetreiber [m]** (Natürliche oder juristische Person, die angelieferte Abfälle ordnungsgemäß lagert und u. a. sicherstellt, dass schädliche Umwelteinwirkungen und sonstige Gefahren, erhebliche Nachteile und erhebliche Belästigungen für die Allgemeinheit und die Nachbarschaft nicht hervorgerufen werden können; cf. § 5 [1] BImSchG); *syn.* privater Betreiber [m] einer Deponie.

waste sorting [n] *envir. leg.* ▶waste separation.

waste technology [n] *envir.* ▶low-waste technology.

waste tourism [n] [UK] *envir.* ▶transboundary waste shipment [UK].

6876 waste treatment [n] *envir.* (In ▶waste management, the processing of waste by separating, dehydrating, composting or incinerating of refuse material; ▶refuse composting, ▶waste); *s* **tratamiento [m] de residuos (sólidos)** (El conjunto de operaciones encaminadas a la eliminación de los residuos o al aprovechamiento de los recursos contenidos en ellos; *término específico* ▶gestión de residuos sólidos [urbanos], tratamiento avanzado/terciario; ▶compostaje de residuos orgánicos domésticos, ▶residuos solidos); *f* **traitement [m] des déchets** (Dans le cadre de la ▶gestion des déchets, opérations d'élimination consistant dans le broyage, le compactage, le drainage, le compostage ou l'incinération des ▶déchets ; ▶compostage de déchets ménagers) ; *g* **Abfallbehandlung [f]** (Im Rahmen der ▶Abfallwirtschaft umfasst die Behandlung die Sammlung, den Transport, das Zerkleinern, Verdichten, Entwässern, Kompostieren oder Verbrennen der Abfallstoffe; ▶Abfall 2, ▶Müllkompostierung).

6877 waste treatment and disposal [n] [US] *constr. envir. leg.* (Generic term for collecting, conveying, treatment and disposal of liquid and solid waste, or for recycling; ▶effluent disposal, ▶recycling, ▶reuse of waste, ▶waste treatment and disposal industry); *syn.* public cleansing and waste disposal [n] [UK]; *s* **recogida [f] y eliminación [f] de residuos** (Conjunto de operaciones necesarias para eliminar los desechos urbanos sólidos o líquidos, como recogida, transporte, almacenamiento provisional, clasificación, tratamiento, incineración, deposición final o reciclaje; ▶eliminación de aguas residuales, ▶eliminación de residuos sólidos, ▶industria de tratamiento y eliminación de residuos, ▶reciclado, ▶reutilización de residuos); *syn.* recolección [f], manejo [m] y disposición [f] de desechos

W

(urbanos sólidos o líquidos); *f* **assainissement [m] et élimination [f] des déchets** (Terme désignant les opérations de collecte, de transport, de stockage transitoire, de tri, d'épuration ou de traitement des déchets et des eaux usées en vue de la ▶récupération d'énergie ou de matières primaires et secondaires réutilisables ; ▶élimination des déchets, ▶élimination des eaux usées, ▶industrie de la récupération, ▶réemploi des déchets, ▶recyclage) ; *syn.* assainissement [m] collectif et élimination [f] des déchets ; *g* **Entsorgung [f]** (Alle erforderlichen logistischen Maßnahmen des Einsammelns, Beförderns, Ableitens, Behandelns und Lagerns sowie die Aufbereitung von festen Abfallstoffen und Abwässern hinsichtlich der Gewinnung von Stoffen oder Energie zur Wahrung des Wohls der Allgemeinheit. Für den Baubetrieb sind in den Leistungsbeschreibungen nach den Erfordernissen des Einzelfalles insbesondere 1. Art, Zusammensetzung und Menge der aus dem Bereich des Auftraggebers zu entsorgenden Böden, Stoffe und Bauteile, 2. Art der Verwertung resp. bei Abfall die Entsorgungsanlage, 3. Anforderungen an die Nachweise über Transporte, Entsorgung und 4. die vom Auftraggeber zu tragenden Entsorgungskosten aufzuführen; ▶Abfallbeseitigung, ▶Abwasserbeitigung, ▶Entsorgungswirtschaft, ▶Recycling, ▶Weiterverwendung von Abfallstoffen).

6878 waste treatment and disposal industry [n] *envir.* (Branch of industry dealing with ▶waste treatment and disposal [US]/public cleansing and waste disposal [UK]; ▶waste management); *s* **industria [f] de tratamiento y eliminación de residuos** (Rama industrial que se dedica a la ▶recogida y eliminación de residuos; ▶gestión de residuos sólidos [urbanos]); *f* **industrie [f] de la récupération** (Branche de l'industrie spécialisée dans l'élimination des déchets ; ▶assainissement et élimination des déchets, ▶gestion des déchets) ; *g* **Entsorgungswirtschaft [f]** (Wirtschaftssektor, der die ▶Entsorgung betreibt; ▶Abfallwirtschaft).

6879 wastewater [n] *envir.* (Contaminated water arising from domestic, commercial and industrial facilities, which is discharged into sanitary sewer or river systems; includes polluted water such as storm water runoff from vehicular areas and roofs, landfill leachate, ▶mixed effluent and coolant; *specific terms* ▶agriculture wastewater, ▶sewage 1, ▶domestic sewage and industrial effluents, ▶industrial sewage and storm water discharged into sewers, ▶toilet wastewater [US]/foul sewage [UK]); *s* **aguas [fpl] residuales (2)** (Agua contaminada con residuos líquidos o sólidos, generados en actividades comerciales, domésticas, industriales o en operaciones de lavado, etc. que incluye también las aguas de escorrentía de tejados y superficies asfaltadas, las aguas mixtas y las de refrigeración y que se vierten en una ▶planta de depuración de aguas residuales o en un ▶curso receptor; *términos específicos* ▶aguas residuales domésticas, ▶aguas negras, ▶aguas residuales industriales, ▶aguas residuales mixtas, ▶aguas residuales 1); *syn.* aguas [fpl] servidas [AL] (2), aguas [fpl] de desecho [C]; *f* **effluent [m] urbain** (Ensemble des déchets liquides en provenance d'usines de traitement des eaux usées domestiques et des stations de lavage industrielles [eaux vannes], de déversoirs d'égout unitaire [eaux pluviales en provenance des toits et de la voirie] et de déversoirs d'orage rejeté par le réseau d'assainissement dans l'environnement ; aujourd'hui encore les eaux domestiques et les eaux pluviales sont souvent transportées par un même égout unitaire ; *terme générique* ▶effluent ; *termes spécifiques* ▶eaux ménagères et eaux vannes, eaux pluviales collectées, ▶eaux usées, ▶eaux usées et eaux pluviales, ▶eaux vannes, ▶effluent industriel) ; *g* **Abwasser [n]** (…wässer [pl]; das durch häuslichen, gewerblichen und industriellen Gebrauch verunreinigte Wasser, dessen sich die Erzeuger durch Kanalisation oder ▶Vorfluter entledigen. Zu diesem Wasser gehören auch Was-

sergemische wie das durch Verkehrsflächen und Dächer verschmutzte Niederschlagswasser, Deponiesickerwasser, ▶Mischwasser und Kühlwasser; *UBe* ▶landwirtschaftliches Abwasser, ▶Schmutzwasser, ▶Haushaltsabwasser und ▶Industrieabwasser sowie das in die Kanalisation eingeleitete Regenwasser).

wastewater [n]**, industrial** *envir.* ▶industrial sewage.

wastewater and sewage treatment [n] *envir. plan.* ▶management of wastewater and sewage treatment.

wastewater treatment field [n] *wat'man. agr.* ▶wastewater wetland [US]/trickle field [UK].

6880 wastewater treatment wetland [n] *envir.* (WWT 1989, 14; constructed water area for sewage treatment with plants which clean domestic wastewater in the root system of, e.g. reed canaris grass *[Phalaris arundinácea]*, rushes *[Juncus]* and reeds *[Phragmites]* of a wetland); *syn.* wetland wastewater treatment system [n], wetland septic system [n] (LAND 1998 [5], 5) [also US]; ▶sewage treatment plant, ▶wastewater wetland); *s* **filtro [f] verde** (Instalación de ▶tratamiento de aguas residuales por filtros verdes que consta generalmente de un tanque de decantación, un depósito de varias cámaras o de putrefacción para el pretratamiento y una superficie vegetal como principal fase de tratamiento biológico. Lo más importante de la superficie de vegetación es la capacidad filtrante del suelo y de las plantas para reducir el contenido de sustancia orgánica en el agua. Como generalmente se utilizan plantas de humedales como carrizo *[Phragmites communis]*, calamagrostis *[Phalaris arundinácea]* y juncos *[Juncus]* se denominan en alemán «plantas de depuración de carrizo»; ▶planta de depuración de aguas residuales); *f* **lagunage [m] à macrophytes** (Système décentralisé de traitement biologique par certains végétaux des effluents domestiques [▶lagunage par roselière] ; station composée d'une série de bassins de décantation ou décanteurs primaires pour la séparation des matières solides couplés à un bassin planté, la fixation de la biomasse ayant lieu par filtration et assimilation par les plantes aquatiques telles que le Roseau à balai *[Phragmites communis]*, l'Alpiste *[Phalaris arundinácea]* et les Joncs *[Juncus]* ; ▶station d'épuration des eaux usées) ; *g* **Pflanzenkläranlage [f]** (Dezentrale Abwasserreinigungsanlage [Kleinkläranlage] i. d. R. für einzelne Wohnhäuser oder Streusiedlungen, die häusliche Abwässer mit Hilfe der ▶Wurzelraumentsorgung reinigt. Auch wenn es unterschiedliche Ausführungen gibt, besteht eine solche Anlage grundsätzlich aus Absetzbecken, Mehrkammergruben oder Rottebehältern für eine Vorreinigung von Grobstoffen und einem Pflanzenbeet als biologische Hauptreinigungsstufe. Die wichtigsten Teile des Pflanzenbeetes sind der Bodenfilter und die darauf wachsenden Pflanzen. Weil vor allem feuchtgebietstypische Pflanzen wie Schilf *[Phragmites communis]*, Rohrglanzgras *[Phalaris arundinácea]* und Binsen *[Juncus]* auf dem Bodenkörper angepflanzt werden, werden solche Anlagen landläufig **Schilfkläranlagen** genannt; ▶Kläranlage); *syn.* bewachsener Bodenfilter [m] (N+L 1999 [6], 389), Schilfkläranlage [f], Wurzelraumentsorgungsanlage [f], Wurzelraumkläranlage [f].

6881 wastewater wetland [n] [US] *wat'man. agr.* (WET 1993, 594; fields of grassland prepared for biological wastewater treatment which are systematically irrigated or flooded with domestic, agricultural or industrial wastewater, as well as urban and rural runoff, which is filtered through the soil; ▶wetland wastewater treatment); *syn.* wastewater treatment field [n], sewage disposal field [n] (FAO 1972), trickle field [n] [UK]; *s* **campo [m] de regado con aguas residuales** (Sistema de ▶tratamiento de aguas residuales por filtros verdes); *f* **champ [m] d'épandage** (Milieu récepteur de terrains labourables ou prairies utilisés pour l'épuration naturelle par le sol des effluents épandus ; ▶lagunage par roselière) ; *g* **Rieselfeld [n]**

W

(Hergerichtete Feldfläche für ein natürliches, biologisches Reinigungsverfahren von Abwässern durch planmäßige Verrieselung; ▶Wurzelraumentsorgung).

6882 water [vb] (1) *hort. for.* (Manual watering such as sprinkling or soaking using a hose-pipe; ▶irrigation); *s* **regar** [vb] (Proporcional agua a las plantas manualmente; ▶irrigación); *f* **arroser** [vb] (manuellement ; ▶arrosage 1, ▶irrigation) ; *syn.* effectuer [vb] des arrosages ; *g* **gießen** [vb] (Manuell bewässern; ▶Bewässerung).

water [n] (2) *pedol.* ▶available soil water; *geo.* ▶backwater; *recr. leg.* ▶bathing water; *pedol. phys.* ▶capillary water; *limn.* ▶cloudy water; *wat'man.* ▶drinking water (1); *hydr.* ▶flowing surface water; *geo. limn.* ▶freshwater; *pedol.* ▶impeded water; *met.* ▶precipitation water; *met. wat'man.* ▶rainwater (1); *wat'man.* ▶reusable water; *pedol.* ▶seepage water; *constr.* ▶slope seepage water; *pedol.* ▶soil water; *zool.* ▶spawning water; *wat'man.* ▶spring water; *pedol.* ▶superheated soil water; *constr. hydr. wat'man.* ▶surface water; *geo. pol.* ▶transboundary water;

water [n] [US], **fast** *geo. limn. wat'man.* ▶fast-moving stream.

water [n], **gravitational** *pedol.* ▶seepage water.

water [n] [US], **heavy storm** *met.* ▶heavy rainfall.

water [n], **phreatic** *hydr. wat'man.* ▶groundwater [US]/ ground water [UK].

water [n] [US], **potable** *wat'man.* ▶drinking water (1).

water [n], **receiving** *constr. eng. wat'man.* ▶receiving stream.

water [n] [US], **seep** *envir.* ▶leachate.

water [n], **sheet** *geo.* ▶sheetwash/sheet wash.

water [n], **storm** *met. wat'man.* ▶rainwater (1).

water [n], **subterranean** *hydr. wat'man.* ▶groundwater [US]/ground water [UK].

water [n], **unclear** *limn.* ▶cloudy water.

water [n], **underground** *hydr. wat'man.* ▶groundwater [US]/ ground water [UK].

water [n], **white** *geo.* ▶torrent.

water [n] [US], **wild** *geo.* ▶torrent.

Water Act [n] **1945** [UK] *envir. leg.* *▶water rights.

6883 water authority [n] *adm. hydr. wat'man.* (**In U.K.**, a national Water Authority draws up strategies and advises upon water resource policy and gives advice to ▶water users agencys and to the Department of the Environment. Water Boards deal with the local distribution of water; cf. Water Resources Act 1963; **in U.S.**, no equivalent term, because control of water resources is spread among several federal agencies; e.g. Department of Agriculture [soil and water conservation]; U.S. Department of the Interior [recreation, fish and wildlife conservation, water reclamation, designated National Rivers]; U.S. Environmental Protection Agency [water quality]; U.S. Department of Defense, Army Corps of Engineers, Civil Works [navigable waters, etc.]. U.S. Coast Guard. Potable water is distributed from reservoirs in a system of pipelines controlled by a local water authority or water board; ▶drinking water distribution); *syn.* Water Resources Board [n] [UK]; *s 1* **administración** [f] **de aguas** (Término genérico; en Es. la competencia administrativa y de planificación y gestión del ▶dominio público hidráulico está repartida entre la administración del Estado, de las CC.AA. y las instituciones creadas específicamente para ese fin: el ▶Consejo Nacional del Agua, las confederaciones de cuencas hidrográficas, dentro de éstas las Asambleas de Usuarios, las Comisiones de Desembalse, las Juntas de Explotación y las Juntas de Obras, y los ▶Consejos del Agua de las cuencas hidrográficas; cf. arts. 17-24 Ley de Aguas 29/1985); *s 2.1* **Consejo** [m] **Nacional del**

Agua [Es] (Órgano consultivo superior en la materia de ▶suministro de agua [potable], en el que están representados la Administración del Estado, las de las CC.AA., los ▶organismos de cuencas hidrográficas, organizaciones profesionales y económicas representativas, de ámbito nacional, relacionadas con los distintos usos del agua. Su función es desarrollar el ▶Plan Hidrológico Nacional e informar sobre y ordenar y titular su realización; cf. arts. 17-18, Ley de Aguas, 29/1985); *s 2.2* **Consejo** [m] **del Agua** (de cuenca hidrográfica) (Órgano de planificación a nivel de cuenca hidrográfica cuya función principal es elevar al Gobierno el ▶Plan Hidrológico de cuenca; arts. 24 [3] y 33); *f 1* **administration** [f] **dans le domaine de l'eau** (F., la compétence administrative dans le domaine des eaux continentales est dévolue **1.** *au niveau central* au ministère de l'environnement [organisme de coordination : direction de l'eau, direction de la prévention des pollutions et des risques — organisme consultatif : comité national de l'eau, etc.], **2.** *au niveau du bassin* au préfet coordinateur de bassin et des six ▶agences de l'eau, **3.** *au niveau régional* aux directions régionales de l'environnement [D.I.R.E.N.], **4.** *au niveau départemental* aux préfets de département assistés des directions départementales compétentes de différents ministères ; ▶alimentation en eau potable, ▶gestion de la resource en eau) ; *f 2* **police** [f] **des eaux** (Police administrative exercée par les services départementaux et régionaux du ministère chargé de l'environnement ayant pouvoir de surveillance sur les eaux superficielles et souterraines) ; *g* **Wasserbehörde** [f] (In D. wird zwischen **W.** und **Wasserwirtschaftsbehörde** unterschieden. **W.n** sind nach dem Wasserhaushaltsgesetz [WHG] und den Wassergesetzen der Bundesländer die zuständigen Landesbehörden für den Erlass wasserrechtlicher Hoheitsakte. I. d. R. sind dies die Landratsämter bzw. die Verwaltungen oder Magistrate der kreisfreien Städte. Die **Wasserwirtschaftsverwaltung** umfasst mehrinstanzlich gegliederte **W.n** [z. B. Landwirtschaftsministerium, Regierungspräsidium, Landkreis resp. kreisfreie Stadt]. In einzelnen Bundesländern wirken Wasserwirtschaftsämter [WWA] als technische Fachbehörden beim Vollzug der Wassergesetze mit; TWW 1982; zum 01.01. 2000 fusionierten der Deutsche Verband für Wasserwirtschaft und Kulturbau [DVWK] und die Abwassertechnische Vereinigung [ATV] zum europaweit stärksten Verband mit ca. 16 000 Mitgliedern; ▶Flussgebietsverband, ▶öffentliche Wasserversorgung, ▶Wasserwirtschaft; *syn. für WWA* Gewässerdirektion [f] [BW]).

6884 water authority directives [npl] *wat'man.* (Regulations related to water as the basis of life and its importance for water and land management as well as for navigable waterways; ▶water rights); *s* **regulaciones** [fpl] **del dominio hidráulico** (Conjunto de normas que regulan el uso del agua como elemento vital y de los cursos y masas de agua en sus diversas funciones; ▶legislación de las aguas); *f* **règles** [fpl] **générales relatives à la ressource en eau** (Dispositions réglementaires ayant pour objet général la gestion équilibrée de l'eau et en particulier la protection, développement, valorisation de l'eau comme ressource économique dans le respect des équilibres naturels ; cf. loi n° 92-3 du 3 janvier 1992 ; ▶droit de l'eau) ; *g* **wasserrechtliche Vorschriften** [fpl] (Regelungen, die das Wasser als Lebensgrundlage und seine Bedeutung für Wasserwirtschaft, Landeskultur und als Wasserstraße betreffen; ▶Wasserrecht).

6885 water balance [n] *biol. pedol.* (Water storage and changes thereof in organisms, in the soil, in an ecosystem or parts thereof, measured in terms of the difference between water uptake and transpiration); *syn.* water budget [n]; *s* **balance** [m] **hídrico** (Balance de las entradas y salidas de agua en una zona hidrológica bien definida, tal como una cuenca, un lago, etc., teniendo en cuenta el déficit o el superávit de agua acumulada;

W

WMO 1974); *syn.* régimen [m] hídrico, economía [f] hídrica (BB 1979, 257); *f 1* **bilan [m] d'eau** (Variations de la teneur en eau dans les organismes, le sol, dans tout ou partie d'un écosystème comme expression du rapport apports/pertes entre des différentes phases du mouvement de l'eau à la surface du globe telles que l'interception, la pénétration, l'évaporation, la transpiration, l'infiltration, le ruissellement et le drainage ; DUV 1984, 58) ; *f 2* **bilan [m] hydrique** (Bilan des apports et des sorties d'eau a l'intérieur d'une zone hydrologique bien définie, telle qu'un bassin, un lac, etc., compte tenu des variations nettes de l'emmagasinement ; WMO 1974) ; *g* **Wasserbilanz [f]** (Vorräte und Vorratsveränderungen des Wassers in Organismen oder deren Teile, im Boden, im Ökosystem oder Teilen davon als Differenz zwischen Wasseraufnahme und Transpiration).

water-bearing stratum [n] *hydr.* ▶aquifer.

waterbirds [npl] *conserv. ecol. pol.* ▶Convention on the Conservation of African-Eurasian Migratory Waterbirds [AEWA].

6886 waterbody [n] (1) *geo. limn.* (Large body of water; e.g. ▶inland waters, ▶natural waterbody, ▶small waterbody or watercourse, ▶standing waterbody, ▶surface waterbody, ▶territorial waters, ▶watercourse 1); *syn.* body [n] of water; *s* **aguas [fpl] continentales** (Término genérico para todas las aguas corrientes [arroyos, ríos, canales] y las ▶aguas estancadas [lagos, lagunas, estanques], ▶aguas continentales en estado natural, ▶aguas continentales [superficiales]; en la definición alemana se incluyen también las aguas costeras intraestatales; ▶aguas interiores, ▶aguas territoriales, ▶curso de agua); *f 1* **eaux [fpl]** (*Termes spécifiques* les eaux maritimes ou continentales, les ▶cours d'eau [ruisseau, rivière, fleuve, canal], les ▶eaux calmes [lac, plan d'eau artificiel, étang] et souterraines, les ▶eaux territoriales ; ▶eaux intérieures ; ▶eaux superficielles) ; *f 2* **eaux [fpl] continentales** (*Termes spécifiques* les eaux courantes, stagnantes et souterraines ; ▶eaux superficielles, ▶eaux continentals naturelles ; *g* **Gewässer [n]** (Eine natürliche Ansammlung von Wasser wie ▶Fließgewässer [Bach, Fluss, Kanal] und ▶Stillgewässer [See, Teich]; *UBe* ▶Binnengewässer, ▶Kleingewässer, ▶natürliches Gewässer, ▶Oberflächengewässer, ▶Territorialgewässer).

waterbody [n] (2) *limn. ocean.* ▶biological dying of a waterbody; *limn.* ▶trophic level of a waterbody; *envir.* ▶water purification of a waterbody.

waterbody [n], **water cleanup of a** [US] *envir* ▶water purification of a waterbody.

waterbody [n] **for navigability** *trans.* ▶canalization of a waterbody for navigability.

water-bound playing surface [n] [US] *constr.* ▶granular playing surface [US]/hoggin playing surface [UK].

water-bound porous court [n] *constr.* ▶granular playing court [US]/hoggin playing surface [UK].

6887 water-bound surface [n] *constr.* (Dampened and compacted sand, gravel, well-graded crushed rock or other loose aggregate forming a top layer of a path or road; *specific terms* gravel surface, pea gravel surface, ▶crushed rock top course, fine aggregate surface [US]/stone chippings surface [UK]; ▶granular surface course [US]/hoggin surface course [UK]); *syn.* compacted granular surface [n] [also US], flexible surface [n] [also UK], hoggin surface [n] [also UK], pervious surface [n] [also US], processed aggregate surface [n] [also UK], stabilized surface [n] [also US]; *s* **revestimiento [m] compactado** (Capa superior de caminos permeable formada por materiales ligantes como arena limosa; ▶suelo de arenilla. Otros revestimientos permeables, aunque no compactados son: 1. revestimiento [m] de grava, 2. revestimiento [m] de cantos rodados, 3. revestimiento [m] de

garbancillos); *f* **sol [m] stabilisé aux liants hydrauliques** (Couche supérieure d'une voirie perméable à l'eau exécutée mécaniquement à partir de matériaux plastiques, p. ex. grave, sable limoneux ; ▶sol sportif stabilisé mécaniquement ; *termes spécifiques* revêtement en gravier, revêtement en tout-venant, revêtement gravillonné, sol gravillonné) ; *syn.* revêtement [m] perméable, sol [m] stabilisé, surface [f] non revêtue ; *g* **wassergebundener Belag [m]** (Wasserdurchlässige Straßen-, Wegeoder Platzoberfläche aus bindigem Material; ▶Tennenbelag; *wasserdurchlässige, aber nicht wassergebundene Beläge* Gartenrieselbelag, Kiesbelag, Schotterdecke, Splittbelag); *syn.* Deckschicht [f] ohne Bindemittel, (z. T.) Erdstraße [f], wassergebundene Decke [f].

water budget [n] *biol. pedol.* ▶water balance.

water cleanup [n] **of a waterbody** [US]/**water cleanup** [n] **of a waterbody** [UK] *envir* ▶water purification of a waterbody.

water conservation area [n] [UK] *leg. wat'man.* ▶soil and water conservation district [US].

6888 water content [n] *bot. pedol.* (Amount of water in a defined mass of plants or in a defined soil volume); *s* **contenido [m] de agua** (Cantidad de agua en una masa definida de plantas o en volumen definido de suelo); *f* **teneur [f] en eau** (Quantité d'eau [volumique ou massique] pour une masse végétale déterminée ou pour un volume de sol défini) ; *g* **Wassergehalt [m]** (Wassermenge in einem definierten Pflanzenanteil oder in einem definierten Bodenvolumen).

6889 watercourse [n] (1) *geo. hydr. limn.* (rivers and streams [pl]; route of a waterway; *opp.* ▶standing waterbody); *s* **curso [m] de agua** (Aguas continentales fluyentes: ríos y arroyos; *opp.* ▶aguas estancadas); *syn.* aguas [fpl] lóticas, aguas [fpl] corrientes, corriente [f] fluvial; *f* **cours [m] d'eau** (Terme générique désignant l'écoulement naturel des eaux douces des torrents, ruisseaux, rivières, fleuves ; *opp.* ▶eaux calmes) ; *syn.* eaux [fpl] courantes ; *g* **Fließgewässer [n]** (Oberirdischer Wasserlauf — Bach, Fluss, Strom —, der ständig oder zeitweilig Wasser führt; *opp.* ▶Stillgewässer); *syn.* fließendes Gewässer [n], Gewässerlauf [m], lotisches Gewässer [n], Wasserlauf [m].

watercourse [n] (2) *wat'man.* ▶high-water (level) of watercourses; *eng. wat'man.* ▶impounding of a watercourse; *geo. hydr.* ▶intermittent watercourse; *limn.* ▶oligotrophic lake or watercourse ; *wat'man.* ▶lowering of watercourses; *wat'man.* ▶redirection of a watercourse; *hydr.* ▶regulation of watercourses; *geo. land'man.* ▶small waterbody or watercourse; *hydr.* ▶stage of a of watercourse; *eng. wat'man.* ▶straightening of watercourses; *envir. limn.* ▶thermal pollution of watercourses.

watercourse [n], **meandering** *geo.* ▶meandering river.

watercourse [n], **restoration of** *conserv. landsc. wat'man.* ▶natural river engineering measures.

6890 watercourse bed erosion [n] *geo.* (Scouring of a stream or riverbed due to the erosive action of water. The term applies chiefly to ▶vertical erosion of the bed and therefore does not usually include lateral erosion; ▶scour); *syn.* streambed erosion [n]; *s* **erosión [f] del lecho** (▶Socavación del lecho de un curso fluvial por efecto de la acción del agua. El término **e. del l.** implica también la posibilidad de socavación lateral del lecho, por lo que no es sinónimo de ▶erosión lineal); *f* **érosion [f] du lit d'un cours d'eau** (Mise en mouvement des matériaux constitutifs et approfondissement du fond du lit d'un cours d'eau provoqués par le dépassement de la vitesse d'écoulement critique ; ▶enfoncement, ▶érosion verticale) ; *g* **Sohlenerosion [f]** (▶Eintiefung einer Fließgewässersohle durch die abtragende Tätigkeit des Wassers. Der Begriff ,Sohlenerosion' impliziert auch die

W

Möglichkeit einer lateralen Aushöhlung der Flusssohle, so dass **S.** nicht vollständig syn. mit ▶Tiefenerosion ist; cf. WAG 1984).

6891 watercourse bed protection [n] *wat'man.* (Reinforcement of a stream or riverbed with a layer of stone or paving in order to prevent ▶watercourse bed erosion; ▶channel lining); *s* **defensa** [f] **del fondo** (Estabilización del lecho de un curso de agua con pacas de piedra o adoquinado para evitar la ▶erosión del lecho; ▶escollera de fondo); *syn.* protección [f] del fondo; *f* **confortement** [m] **du fond du lit (d'un cours d'eau)** Mesures ou ouvrages de stabilisation du profil en long d'un cours d'eau [enrochements en vrac, pavage] par diminution de l'érosion du fond du lit ; ▶empierrement du lit [d'une rivière], ▶érosion du lit d'un cours d'eau) ; *g* **Sohlensicherung** [f] (Befestigung der Fließgewässersohle durch Steinschüttung oder Pflasterung zur Verhinderung der ▶Sohlenerosion; ▶Berollung).

6892 watercourse segmentation [n] *wat'man.* (Classification of segments of streams and rivers according to their degree of purity, expressed by ▶water quality categories; ▶water quality of rivers and lakes; ▶lake and river water quality map); *s* **clasificación** [f] **de los cursos de agua** (En D. según su grado de contaminación orgánica en ▶clases de calidad de las aguas continentales; ▶calidad de las aguas continentales, ▶mapa de calidad de las aguas continentales); *f* **classification** [f] **des eaux** (Classement de cours d'eau, sections de cours d'eau selon la ▶qualité des eaux continentales ; la **c. d. e.** est différenciée sur la base de divers critères d'utilisation des eaux tels que, la production d'eau alimentaire, les eaux de baignade, les eaux piscicoles ; ▶carte départementale d'objectifs de qualité, ▶carte de pollution des cours d'eau, ▶classe de pollution des cours d'eau, ▶niveau de qualité des cours d'eau) ; *g* **Gütegliederung** [f] **der Fließgewässer** (Einteilung der Fließgewässer oder deren Teile hinsichtlich ihres Reinheitsgrades nach ▶Gewässergüteklassen; ▶Gewässergüte, ▶Gewässergütekarte).

6893 water demand [n] *hydr. envir. nat'res. syn.* water need [n] (WMO 1974); *s* **demanda** [f] **de agua** *syn.* necesidades [fpl] de agua (WMO 1974); *f* **besoins** [mpl] **en eau** (WMO 1974) ; *g* **Wasserbedarf** [m] (In einer bestimmten Zeitspanne benötigte Trink- und Betriebswassermenge, die von einer Wasserversorgungsanlage in ausreichender Menge in ein Versorgungsgebiet zu liefern ist; cf. TWW 1982, 765).

6894 water discharge [n] *wat'man.* (Release of water from a stream into a river or into the sea; ▶flow rate, ▶stream flow, ▶surface water drainage); *s* **caudal** [m] (Volumen de fluído que pasa en la unidad de tiempo a través de una superficie [sección transversal de una corriente], en una corriente determinada; ▶caudal, ▶drenaje superficial, ▶flujo de un cauce); *f* **circulation** [f] **des eaux** (Mouvement des eaux dans une canalisation, un fossé, un cours d'eau ; ▶débit, ▶écoulement fluviatile], ▶évacuation des eaux superficielles) ; *g* **Abfluss** [m] **(3)** (1. Wassermenge, die von einem Gewässer höherer Ordnung in ein Gewässer niederer Ordnung fließt. 2. Wassermenge, die durch Drän- oder Kanalisationsleitungen fließt; ▶Abfluss 2, ▶Abflussmenge, ▶Oberflächenentwässerung); *syn.* Durchfluss [m].

6895 water dispersal [n] *phyt.* (Transport of diaspores by water); *syn.* hydrochory [n]; *s* **hidrocoria** [f] *syn.* dispersión [f] por el agua, diseminación [f] por el agua; *f* **hydrochorie** [f] (Dissémination des semences par l'eau ; espèce hydrochore) ; *g* **Verdriftung** [f] **durch Wasser** (Verbreitung von Diasporen [verbreitungsbiologisch-funktionelle Einheit einer Pflanze, z. B. Samen, Sporen] durch Wasser); *syn.* Hydrochorie [f]; Verbreitung [f] durch Wasser, Verfrachtung [f] durch Wasser.

water distribution [n]**, drinking** *plan. urb.* ▶public water supply.

water distribution line [n] [US] *constr.* ▶water main.

water drainage [n] *constr. urb. wat'man.* ▶surface water drainage.

water drip [n] [US] *arch. constr.* ▶throat.

6896 water entrapment [n] [US] *constr.* (Planned containment of water, e.g. on rooftops; technique used for rooftop planting; cf. TGG 1984, 217); *syn.* water impoundment [n] [UK]; *s* **estancamiento** [m] **de agua** (Acumulación deseada de agua en tejado plano; técnica utilizada para la plantación de tejados); *f* **réserve** [f] **d'eau pluviale** (Stockage des eaux pluviales p. ex. dans la couche drainante d'un aménagement sur toiture-terrasse) ; *syn.* réservoir [m] d'eau pluviale ; *g* **Wasseranstau** [m] (Geplante Ansammlung von Wasser, z. B. für eine Dachbegrünung auf Flachdächern).

6897 water extraction [n] *hydr. nat'res. wat'man.* (Capture/removal of water for potable, agricultural, or industrial use; ▶groundwater extraction [US]/ground water extraction [UK]); *s* **captación** [f] **de agua** (Extracción de agua para el suministro de agua potable o para la industria o la agricultura de cualquiera de las posibles fuentes: aguas subterráneas, embalses, filtrado de aguas de río, etc.; ▶extracción de aguas subterráneas); *syn.* extracción [f] de agua; *f* **prélèvement** [m] **des eaux** (Puisage d'eau dans le milieu naturel à l'usage de l'alimentation humaine et des activités agricoles et industrielles ; ▶captage d'eau souterraine, ▶prélèvement d'eau souterraine ; *terme spécifique* captage d'alimentation en eau potable) ; *syn.* prise [f] d'eau ; *g* **Wassergewinnung** [f] (Förderung des Wassers für den Trinkwasser- und Brauchwasserbedarf durch Entnahme aus dem Grundwasser, aus Speicherseen, aus dem Uferfiltrat oder aus Meerwasserentsalzungsanlagen sowie das Sammeln von Regenwasser; ▶Grundwasserförderung); *syn.* Wasserentnahme [f].

water extraction and treatment plant [n] *envir. wat'man* ▶drinking water extraction and treatment plant.

6898 water features [npl] *arch. gard.* (Devices which produce special water effects such as cascades, fountains, water sculptures, etc.; ▶composition of water features, ▶water theater [US]/water theatre [UK]); *s* **hidroarte** [m] (Instalación construida para producir efectos especiales con el agua, como p. ej. cascadas artificiales, surtidores, ▶teatro de aguas, esculturas de agua, etc.; ▶juego de aguas); *f* **art** [m] **des jeux d'eau** (Installation produisant des effets d'eau et réalisée pour des ▶jeux d'eau tels que les cascades artificielles, jets d'eau, ▶théâtre d'eau, fontaines monumentales, fontaines musicales, etc.) ; *g* **Wasserkunst** [f] (Gebaute Anlage für Wasserspiele, z. B. künstliche Kaskaden, Springbrunnen, ▶Wassertheater, Wasserchoreografie; ▶Wasserspiel).

6899 waterfowl [n] *zool.* (All birds ecologically dependent on ▶wetlands; ▶Ramsar Convention, ▶waders); *s* **aves** [f] **acuáticas** (Nombre colectivo de la avifauna que depende de las ▶zonas húmedas; ▶ave limícola, ▶Convención de Ramsar); *f* **sauvagine** [f] (Nom collectif donné aux oiseaux sauvages dépendants écologiquement des ▶zones humides ; ▶oiseau limicole) ; *g* **Wasser- und Watvögel** [mpl] (Vögel, die von ▶Feuchtgebieten im Sinne der ▶Ramsar-Konvention abhängig sind; ▶Watvogel); *syn.* Wassergeflügel [n].

Water Framework Directive [n] *wat'man. pol.* ▶EU Water Framework Directive.

waterfront [n] *geo. urb.* ▶water's edge.

6900 waterfront promenade [n] *urb.* (Wide, specially laid out pathway for taking walks in a pedestrian environment along the edge of a large body of water such as a lake, river or canal as well as the sea-facing parts of a city, e.g. the Riverbank Promenade along the Rhine in Düsseldorf, Germany, the Promenade, Brighton on the English Channel, the Sliema Prome-

W

nade, Malta on the Mediterranean; contemporary urban planning refers simply to a *waterfront*, which is generally used for wharves that have been converted into modern residential and commercial areas and includes attractive promenades, e.g. the Victoria and Alfred Waterfront in Cape Town, South Africa, the Belfast Waterfront, Northern Ireland, Baltimore Harborplace [a harbour attraction in Maryland]; the word **esplanade** is sometimes used as a synonym for **promenade**, but it usually also comprises a boulevard or avenue which runs parallel to the promenade (cf. **corniche**, e.g. in Beirut, Lebanon); ▶beach promenade; *specific term* **River Walk [n]** [a network of walkways along the banks of the San Antonio River in San Antonio, Texas], London Docklands [a new business district with promenades on the River Thames, London], Fisherman's Wharf [a tourist attraction in San Francisco]; *s* **paseo [m] ribereño** (Amplia vía peatonal a lo largo de la orilla de ríos, lagos o del mar diseñada para pasear y disfrutar del paisaje; ▶paseo marítimo); *f* **promenade [f] le long de la berge** (Cheminement aménagé pour les promeneurs le long des berges d'un lac ; ▶promenade de bord de mer) ; *g* **Uferpromenade [f]** (**1.** Eigens zum Spazierengehen angelegter Weg entlang eines großen Gewässers wie See, Fluss oder Kanal, z. B. die *Rheinuferpromenade* in Düsseldorf, die *Promenade in Brighton* am Ärmelkanal, die *Sliema-Promenade* auf Malta. **2.** Der englische Begriff *waterfront* beinhaltet heute in der Stadtplanung auch zum Meer hin ausgerichtete Stadtteile, die durch Umwandlung alter Hafenanlagen zu modernen Wohn- und Geschäftsgebieten mit attraktiven Promenaden aufgewertet wurden, z. B. die *Victoria and Alfred Waterfront* in Kapstadt, Südafrika, die *Belfast Waterfront*, Nordirland, *Baltimore Harborplace* [eine Attraktion am Hafen in Maryland]; ▶Strandpromenade).

6901 water-graded particle [n] *constr.* (Generic term for settled particles of such fine soil as ▶clay 1 or ▶silt 1. The gradation of such particle is determined by ▶sedimentation analysis); *s* **partícula [f] en suspensión decantable en agua** (Término genérico para aquellas partículas del suelo de ▶arcilla o ▶limo 2 que se sedimentan al hacer un ▶análisis granulométrico por sedimentación); *f* **particules [fpl] en suspension décantables dans l'eau** (Particules fines d'▶argile et de ▶limon déterminées lors de l'▶analyse par sédimentation) ; *g* **Schlämmkorn [n]** (Bodenfeinbestandteil [▶Ton oder ▶Schluff], der bei der ▶Schlämmanalyse bestimmt wird).

6902 water-holding capacity [n] *pedol.* (Capacity of soil to retain water in its pores, and store hygroscopic water for plants, making it available to plants; ▶available soil water, ▶field capacity); *syn.* water storage capacity [n]; *s* **capacidad [f] de retención de agua** (Capacidad de un suelo de almacenar agua disponible para las plantas y agua higroscópica; ▶cresardía, ▶capacidad de campo); *syn.* capacidad [f] de retención capilar (de agua), capacidad [f] de absorción (de agua); *f* **capacité [f] de rétention (maximale) pour l'eau** (Capacité d'un sol en place bien réhumecté et réssuyé à former une réserve d'▶eau utile ; cet état correspond à son humidification maximale après cessation du drainage naturel ; cf. DIS 1986 ; ▶capacité [de rétention] au champ) ; *syn.* pouvoir [m] de rétention de l'eau, capacité [f] de stockage en eau (d'un sol) (PED 1979) ; *g* **Wasseraufnahmefähigkeit [f]** (Vermögen eines Bodens, ▶pflanzenverfügbares Wasser in Poren und Adsorptionswasser zu speichern; ▶Feldkapazität); *syn.* Wasserhaltekraft [f], Wasserhaltevermögen [n], Wasserhaltung [f] und Sorptionsvermögen [n], Wasserspeicherfähigkeit [f], Wasserspeicherkapazität [f], Wasserspeichervermögen [n].

water hollow [n] *arb.* ▶tree water hollow.

water impoundment [n] [UK] *constr.* ▶water entrapment [US].

6903 water-in [vb] *constr. hort.* (Thorough ▶watering of new planting to ensure establishment); *s* **embarrar [vb]** (▶Regar abundantemente después de la plantación para asegurar el enraizamiento); *f* **plomber [vb] par arrosage** (**1.** Arrosage copieux jusqu'à saturation du substrat après la plantation. **2.** Tassement de la terre autour des arbres et arbustes par un ▶arrosage 2 abondant) ; *g* **einschlämmen [vb]** (**3**) (Frisch gesetzte Pflanzen gründlich angießen, damit der Boden sich setzt und ein guter Kontakt zwischen Wurzeln und Bodensubstrat hergestellt wird; ▶Wässern); *syn.* durchdringend wässern [vb], kräftig angießen [vb].

6904 water influence zone [n] [US] *hydr. landsc.* (*U.S. Forest Service usage* areas comprising significant existing or anticipated public outdoor recreational occupancy, use, and enjoyment along streams and rivers and around lakes, reservoirs, and other bodies of water, areas in which uses and activities are oriented to overwater travel and outdoor recreation; WPG 1976; ▶lake landscape, ▶waterscape); *s* **zona [f] de recreo acuático** (≠) (En EE.UU. áreas del Servicio de Bosques existentes o previstas para uso recreativo público, situadas a lo largo de ríos o arroyos y alrededor de lagos y otros cuerpos de agua. Estas áreas se utilizan para actividades náuticas y de recreo al aire libre; ▶paisaje acuático, ▶paisaje lacustre); *f 1* **base [f] de loisirs aquatique** (Espace naturel aménagé de loisirs des bords de l'eau [rivière, lac] proposant la détente au bord de l'eau, des activités aquatiques et nautiques diverses [parcours de découverte, baignade, la pêche, le canoë-kayak de course en ligne, kayaks et canoës d'initiation, kayaks de descente, kayaks de slalom, canoës de randonnée, canotage, pédalos, vélos d'eau, voiliers, etc.] équipé d'aires de pique nique et de jeux [tobogan aquatique, beach volley, roller skate, football, parcours de santé, tennis, tir à l'arc, minigolf, volley, jeux de boules, etc.] ; ▶paysage aquatique, ▶paysage lacustre); *syn.* parc [m] de loisirs aquatique ; *f 2* **base [f] de loisirs nautique** (*Terme spécifique* espace proposant spécifiquement des activités de navigation [stages de voile, planche à voile, dériveur, canoë, ski nautique, parachute ascensionnel) ; *syn.* base [f] nautique, base [f] nautique et de loisirs ; *g* **Erholungslandschaft [f] an Gewässern** (±) (Landschaft, die durch Oberflächengewässer geprägt ist und auf Grund ihrer Vielfalt, Eigenart und Schönheit besonders der Freizeit- und Erholungsnutzung dient; ▶Gewässerlandschaft, ▶Seenlandschaft, ▶Seenplatte); *syn.* Gewässerlandschaft [f] für Freizeit und Erholung, gewässergeprägte Erholungslandschaft [f].

6905 watering [n] *constr. hort.* (Application of water to plants in open ground, pots or containers to the extent of temporary saturation of the soil; BS 3975: part 5; water [vb]; the thorough watering of a flower pot is achieved by **soaking**; i.e. standing the pot in water; ▶water-in); *s* **riego [m] (2)** (Aplicación de agua a las plantas para empapar temporalmente el sustrato de crecimiento, sea éste el suelo o tierra en macetas o contenedores; ▶embarrar); *syn.* empapado [m]; *f* **arrosage [m] (2)** (Pratique destinée à humidifier le substrat des végétaux en pleine terre, en pots ou en conteneurs ; arroser [vb] ; ▶plomber par arrosage) ; *g* **Wässern [n, o. Pl.]** (Wasserzuführung an Pflanzen im Boden, in Töpfen oder Containern, so dass das Erdsubstrat durchnässt ist; dies nennt man auch **durchdringendes Wässern**; ▶einschlämmen 3).

watering depression [n] *constr. hort.* ▶planting saucer [US]/watering hollow [UK].

watering hollow [n] [UK] *constr. hort.* ▶planting saucer [US].

watering rim [n] [UK] *constr. hort.* ▶saucer rim [US].

6906 water level [n] *constr. hydr.* (**1.** Surface of a waterbody. **2.** Altitude/height of water surface above sea level; ▶average

W

high-water level/mark, ►fluctuating water level, ►lowest low-water level, ►low-water level, ►mean-water level, ►measured water level, ►record high-water level/mark, ►watertable); *s* **nivel [m] del agua (2)** (**1.** Superficie de un lago, río, etc. **2.** Altitud sobre el nivel del mar; ►crecida máxima posible, ►estiaje, ►nivel de crecida media, ►nivel del agua 1, ►nivel de estiaje mínimo registrado, ►nivel freático, ►nivel medio [del agua], ►nivel variable del agua); *syn. de 1.* superficie [f] del agua; *f* **niveau [m] de l'eau** (**1.** d'une surface d'eau libre ; ►crue maximale, ►niveau d'eau à l'échelle des eaux, ►niveau des moyennes eaux, ►niveau d'étiage, ►niveau d'étiage absolu, ►niveau du débit moyen caractéristique, ►plan variable du niveau d'eau ; **2.** de la hauteur des eaux par rapport à l'altitude absolue ; **3.** de la ►niveau de la nappe phréatique) ; *g* **Wasserspiegel [m]** (**1.** Gewässeroberfläche. **2.** Höhe einer Gewässeroberfläche über NN; ►Grundwasserspiegel, ►höchstes Hochwasser, ►mittleres Hochwasser, ►Mittelwasserstand, ►niedrigster Niedrigwasserstand, ►Niedrigwasser[stand], ►Wasserstandshöhe, ►wechselnder Wasserspiegel); *syn. zu 1.* Wasseroberfläche [f], *syn. zu 2.* Wasserstand [m].

water level [n], normal *leg. wat'man.* ►mean-water level.

water level [n], oscillating *hydr.* ►fluctuating water level.

6907 waterlogged soil [n] [US]/water-logged soil [n] [UK] *agr. hort. pedol.* (Soil saturated or nearly saturated with water; ►impeded water, ►soil wetness); *syn.* water-saturated soil [n] (TGG 1984, 173); *s* **suelo [m] saturado** (Suelo cuyos poros están completamente llenos de agua; ►agua estancada, ►alta humedad en el suelo); *f* **sol [m] saturé par l'eau** (Sol dans lequel tous les vides sont entièrement occupés par l'eau ; ►engorgement du sol, ►nappe perchée) ; *syn.* sol [m] gorgé d'eau, sol [m] engorgé, sol [m] détrempé ; *g* **wassergesättigter Boden [m]** (Boden, dessen gesamter Porenraum mit Wasser gefüllt ist; ►Bodennässe, ►Staunässe).

6908 waterlogging [n] *pedol.* (Soil saturation caused by poor drainage, with the result of poorly-drained soil in which water accumulates to the exclusion of air; becoming waterlogged [US]/water-logged [UK], ►impeded water, ►paludification, ►soil wetness, ►waterlogged soil [US]/water-logged soil [UK], ►water saturation); *s* **encharcamiento [m] del suelo** (Gran presencia de agua en el suelo debida a capas de suelo cuasi impermeables; ►agua estancada, ►alta humedad del suelo, ►empantamiento, ►saturación de agua, ►suelo saturado); *f 1* **processus [m] d'engorgement en eau du sol** (Excès d'eau permanent ou temporaire dans le sol dû à la présence de couches peu perméables ; ►engorgement du sol, ►nappe perchée, ►paludification, ►saturation en eau, ►sol saturé par l'eau) ; *f 2* **imbibition [f] (capillaire)** (Humectation progressive que subit un sol ou un horizon par ascension capillaire, par arrosage ou sous l'influence des pluies ; DIS 1986, 114) ; *f 3* **humectation [f]** (Processus naturel d'accroissement de l'humidité dans le sol lorsque la conductivité hydraulique est faible ; PED 1979, 248) ; *g* **Bodenvernässung [f]** (Übermäßige Wasseranreicherung im Boden auf Grund schwer wasserdurchlässiger Schichten; ►wassergesättigter Boden, ►Bodennässe, ►Staunässe, ►Versumpfung, ►Wassersättigung); *syn.* Vernässung [f].

6909 water main [n] *constr.* (Major pipeline in water distribution system; ►water supply line); *syn.* main water supply line [n] [also UK], water distribution line [n] [also US]; *s* **tubería [f] principal de agua** (Conducción enterrada en la calle que sirve para abastecer de agua potable a una comunidad o una zona de una ciudad y es alimentada con agua a presión desde la central de abastecimiento o desde un depósito de agua; la tubería secundaria de distribución suministra las subzonas y la ►tubería de suministro de agua a los usuarios finales); *syn.* conducción [f] principal de agua, cañería [f] principal de agua; *f* **canalisation [f] d'eau** (►conduite de raccordement d'eau) ; *syn.* conduite [f] d'eau ; *g* **Hauptwasserleitung [f]** (*Abk.* HW-Leitung; für die kommunale Wasserversorgung frostfrei verlegte Druckrohrleitung als Hauptstrang, der vom Wasserwerk/Wasserbehälter gespeist wird; die nachgeordnete **Verteilwasserleitung [VW-Leitung]** versorgt die einzelnen Siedlungseinheiten bis zu den ►Anschlusswasserleitungen der Endabnahmestellen).

6910 water management [n] *adm. hydr. nat'res. wat'man.* (Administration, operation and distribution of water resources; ►public water supply, ►water management plan); *s 1* **Plan [m] Hidrológico Nacional** (En Es., plan a nivel nacional elaborado por el Ministerio del Medio Ambiente, conjuntamente con los Departamentos ministeriales relacionados con el uso de los recursos hidráulicos, con el fin de alcanzar los objetivos de la ►planificación hidrológica. Se debe aprobar por ley y contener en todo caso: a] las medidas necesarias para la coordinación de los diferentes planes hidrológicos de cuenca; b] la solución para las posibles alternativas que aquéllos ofrezcan; c] la previsión y las condiciones de las transferencias de recursos hidráulicos entre ámbitos territoriales de distintos planes hidrológicos de cuenca y d] las modificaciones que se prevean en la planificación del uso del recurso y que afecten a aprovechamientos existentes para abastecimiento de poblaciones o regadíos; art. 45 [1 y 2] RD Legislativo 1/2001, de 20 de julio, por el que se aprueba el texto refundido de la Ley de Aguas); *s 2* **plan [m] hidrológico de cuenca** (En Es., plan a nivel de ►cuenca hidrográfica elaborado con el fin de alcanzar los objetivos de la ►planificación hidrológica. Los **pp. hh. de c.** comprenden los siguientes contenidos: a] Descripción general de la demarcación hidrográfica; b] descripción general de los usos, presiones e incidencias antrópicas significativas sobre las aguas; c] identificación y mapas de las zonas protegidas; d] redes de control establecidas para el seguimiento del estado de las aguas superficiales, subterráneas y de las zonas protegidas y los resultados de este control; e] lista de objetivos medioambientales para las aguas; f] resumen del análisis económico del uso del agua; g] resumen de los programas de medidas adoptados para alcanzar los objetivos previstos; h] registro de los programas y planes hidrológicos más detallados relativos a subcuencas, cuestiones específicas o categorías de aguas, así como las determinaciones pertinentes derivadas del Plan Hidrológico Nacional; i] resumen de medidas de información pública y de consulta tomadas, sus resultados y los cambios consiguientes efectuados en el plan; j] lista de autoridades competentes designadas y k] los puntos de contacto y procedimientos para obtener la documentación de base y la información requerida por las consultas públicas; art. 42 [1], RDL 1/2001); *syn.* gestión [f] de recursos hídricos, manejo [m] de recursos hídricos [MEX]; *f* **gestion [f] de la ressource en eau** (Protection, mise en valeur et développement de la ressource utilisable de l'eau pour l' ►alimentation en eau potable dans le respect des équilibres naturels ; ►schéma directeur d'aménagement et de gestion des eaux) ; *syn.* gestion [f] de l'eau, gestion [f] des ressources en eau ; *g* **Wasserwirtschaft [f]** (Gesamtheit der Maßnahmen zur ►Wasserversorgung und zur Regulierung des Wasserhaushaltes; ►wasserwirtschaftlicher Rahmenplan); *syn.* Bewirtschaftung [f] der Wasserressourcen.

6911 water management plan [n] *leg. wat'man.* (Management plan for the water economy of a region, e.g. the ►drainage basin of a river or an economic region, which effectively ensures the availability of water necessary for sustainable economic development); *s* **plan [m] hidrológico** (En España según la Ley de Aguas, planes a nivel nacional —Plan Hidrológico Nacional— y a nivel de ►cuenca hidrográfica elaborados con el fin de «conseguir la mejor satisfacción de las demandas de agua y equilibrar y armonizar el desarrollo regional y sectorial incrementando las disponibilidades del recurso, protegiendo su calidad, economizando su empleo y racionalizando sus usos en armonía con el medio ambiente y los demás recursos naturales»; art. 38 [1], Ley 29/1985, de 2 de Agosto, de Aguas); *f 1* **schéma [m] directeur d'aménagement et de gestion des eaux (S.D.A.G.E.)** (Le S.D.A.G.E. fixe pour chaque bassin ou un groupement de bassins les orientations fondamentales d'une gestion équilibrée de la ressource en eau ; art. 5 de la loi n° 92-3 du 3 janvier 1992 ; ►bassin versant) ; *f 2* **schéma [m] d'aménagement et de gestion des eaux (S.A.G.E.)** (Le S.A.G.E. fixe les objectifs généraux d'utilisation, de mise en valeur et de protection quantitative et qualitative des ressources en eau superficielle et souterraine et des écosystèmes aquatiques, ainsi que de préservation des zones humides, dans un groupement des sous-bassins ou un sous-bassin correspondant à une unité hydrographique ou à un système aquifère ; art. 5 de la loi n° 92-3 du 3 janvier 1992) ; *g* **wasserwirtschaftlicher Rahmenplan [m]** (Plan, der die Lenkung und Ordnung des Wasserhaushalts für eine Region [Niederschlagsgebiet/►Abflussgebiet oder Wirtschaftsraum] regelt, damit die für die Entwicklung der Lebens- und Wirtschaftsverhältnisse notwendigen wasserwirtschaftlichen Voraussetzungen nachhaltig gesichert sind; cf. § 36 WHG. Durch die europäische RL 2000/60/EG v. 23.10.2000 und das siebte Gesetz zur Änderung des Wasserhaushaltsgesetzes vom 18.06. 2002 tritt an Stelle des **wasserwirtschaftlichen R.es** der **Bewirtschaftungsplan**).

6912 water management planning [n] *adm. hydr. wat' man.* (Planning which deals with the administration, operation and distribution of water resources, either yet to be determined or in use, as well as regulations and control measures, required to protect the hydrological cycle of a particular area; ►water management plan); *s* **planificación [f] hidrológica** (En Es. la **p. h.** tiene por objetivos generales «conseguir el buen estado y la adecuada protección del dominio público hidráulico y de las aguas objeto de esta ley, la satisfacción de las demandas de agua, el equilibrio y armonización del desarrollo regional y sectorial, incrementando las disponibilidades del recurso, protegiendo su calidad, economizando su empleo y racionalizando sus usos en armonía con el medio ambiente y los demás recursos naturales». Se realiza mediante el ►Plan Hidrológico Nacional y los ►planes hidrológicos de cuenca; art. 40 [1 y 3] RDL 1/2001; ►plan hidrológico); *f* **établissement [m] d'un programme d'une agence de l'eau** (En F. fait partie de la mission des agences de l'eau ; études ayant pour objectif d'assurer l'équilibre des ressources et des besoins en eau, de développer les ressources en eau, de coordonner la réalisation de toutes les actions et projets d'intérêt commun aux bassins ou groupements de bassins pour un territoire défini ; ►schémadirecteur d'aménagement et de gestion des eaux) ; *g* **wasserwirtschaftliche Planung [f]** (Planung, die die Bewirtschaftung des festzustellenden und nutzbaren Wasserdargebots, die Ordnung und Abstimmung aller wasserwirtschaftlichen Maßnahmen unter Sicherung des Wasserkreislaufes in einem bestimmten Gebiet vorsieht; ►wasserwirtschaftlicher Rahmenplan).

water meadow [n] [UK] *agr.* ►inundated meadow [US].

water need [n] *hydr. envir. nat'res.* ►water demand.

water-parting [n] [UK] *geo. hydr. wat'man.* ►watershed.

6913 water permeability [n] *pedol.* (Property of a porous medium to allow downward movement of water into deeper strata by gravity; *context* soil with good/bad drainage; ►hydraulic conductivity); *syn.* drainage permeability [n] of soil, drainage [n] of soil, soil drainage [n] (3); *s* **permeabilidad [f] (para el agua)** (Propiedad del suelo, de una capa del subsuelo o de un medio poroso de permitir el movimiento de líquidos a

través de él bajo la acción de la gravedad; ►conductividad hidráulica); *f* **perméabilité [f] à l'eau** (Faculté d'un sol [perméabilité d'interstice] ou d'une roche [perméabilité de fissures] d'un milieu poreux à permettre l'infiltration de l'eau ; ►conductivité hydraulique) ; *g* **Wasserdurchlässigkeit [f]** (**1.** Eigenschaft eines Bodens oder einer Gesteinsschicht, Wasser in tiefere Schichten [bis zum Grundwasser] sickern zu lassen. **2.** Fähigkeit eines Bodens zur Ableitung von Oberflächenwasser durch eine Vegetationstrag-, Drän- oder Filterschicht, im Regelfall in den Baugrund; DIN 18 915; ►Wasserleitfähigkeit); *syn.* Wasserabzug [m] im Boden.

water plant [n] *bot. phyt.* ►free-floating water plant.

water plant [n]**, rooted** *bot. phyt.* ►floating-leaved plant.

6914 water polluting firm [n] [US] *envir. leg.* (Firm or local authority which discharges untreated ►effluent directly into the sea or a ►receiving stream; ►point source pollution); *syn.* person [n] responsible for direct discharge [UK]; *s* **causante [m] de vertidos directos** (Fábrica, entidad o municipio que vierte sus aguas residuales —depuradas o no— directamente en un curso de agua o en el mar; ►contaminación focal, ►curso receptor, ►efluente); *f* **auteur [m] d'un déversement direct (des effluents)** (Particuliers, industriels ou collectivités publiques rejetant directement un ►effluent [traité ou non] dans un milieu recapteur ; ►émissaire, ►milieu récepteur, ►pollution ponctuelle) ; *g* **Direkteinleiter [m]** (Betrieb oder Gemeinde, die ►abließendes Abwasser direkt in den ►Vorfluter ungeklärt einleitet; ►punktuelle Verschmutzung).

6915 water-polluting substance [n] *envir.* (Material dangerous to the environment, which must not be allowed to come into contact with surface water or to percolate into groundwater); *s* **sustancia [f] tóxica para el agua** (Materia peligrosa para el medio ambiente, sobre todo en disolución, por lo que no debería entrar en contacto ni con las aguas superficiales ni con las subterráneas); *syn.* sustancia [f] contaminante para el agua; *f* **substance [f] polluante pour les eaux** (Matière solide ou liquide dangereuse ne devant pas entrer en contact des eaux superficielles ou souterraines) ; *syn.* produit [m] polluant pour les eaux ; *g* **wassergefährdender Stoff [m]** (Für die Umwelt gefährlicher fester, flüssiger oder gasförmiger Stoff, der geeignet ist, nachhaltig die physikalische, chemische oder biologische Beschaffenheit des Wassers nachteilig zu verändern; cf. § 19g [5] WHG 2009).

6916 water pollution [n] *envir.* (Presence in water of enough harmful or objectionable material to adversely effect its usefulness or quality; cf. RCG 1982; ►biological self-purification, ►thermal pollution of watercourses, ►water quality standards); *s* **contaminación [f] de las aguas continentales** (Impacto negativo sobre las aguas o sus funciones ecológicas debido a la presencia de sustancias nocivas, al aumento de la temperatura [►contaminación térmica de cursos de agua] o a la presencia de gran cantidad de materia orgánica, que al ser descompuesta biológicamente tiene como consecuencia la reducción del contenido de oxígeno, la pérdida de capacidad de ►autodepuración y el deterioro de las condiciones ecológicas del hábitat; ►calidad del agua); *syn.* polución [f] de las aguas continentales; *f* **pollution [f] des eaux continentales** (Modification des caractéristiques physiques, chimiques, biologiques ou bactériologiques des eaux superficielles et souterraines dont la qualité est remise en cause par les différents usages qui en sont faits ; la pollution peut être de nature thermique [►pollution thermique des eaux], bactérienne, mécanique, chimique et radioactive, celle-ci pouvant remettre en cause le pouvoir d'►auto-épuration et menacer les communautés vivantes aquatiques ; ►qualité de l'eau) ; *g* **Gewässerbelastung [f]** (Beeinträchtigung der Reinheit oder

W

des ökologischen Funktionsgefüges oberirdischer Gewässer durch Schmutz- und Schadstoffe, durch Temperaturerhöhung [▶thermische Belastung von Fließgewässern] oder durch organische Stoffe, die unter starker Sauerstoffzehrung biologisch abgebaut werden mit der Folge, dass die Selbstreinigungskraft nicht mehr gegeben ist und somit Flora und Fauna im Wasser wie auch in den Uferzonen erheblich gefährdet sind; ▶ Wassergüte); *syn.* Gewässerverschmutzung [f], Gewässerverunreinigung [f], Gewässerkontamination [f].

6917 waterproofing [n] *constr.* (Treatment to make material watertight; ▶dampproofed mortar [US]/damp-proof mortar [UK]); *s* **sellado** [m] (Tratamiento para evitar la entrada o salida de agua; ▶mortero impermeabilizante); *syn.* impermeabilización [f]; *f* **étanchéisation** [f] (Mesure de protection contre la pénétration ou la fuite de l'eau ; ▶mortier hydrofuge) ; *syn.* imperméabilisation [f], isolation [f] hydrofuge, étanchéification [f] ; *g* **Dichtung** [f] (Maßnahmen zur Verhinderung des Wasserzu- oder -austritts; ▶Sperrmörtel); *syn.* Abdichtung [f], Wasserdichtung [f] (FLA 1983, 65).

waterproofing layer [n] **of a roof** *constr.* ▶roof membrane.

6918 waterproofing membrane [n] *constr.* (Plastic material installed to form a ▶waterproof layer; ▶waterproofing sheet); *syn.* liner [n]; *s* **membrana** [f] **de impermeabilización** (Hoja de material sintético que se coloca en tejados o en estanques como ▶capa de impermeabilización; ▶banda de material impermeabilizante); *f* **membrane** [f] **d'étanchéité** (▶lé d'étanchéité; ▶couche d'étanchéité) ; *syn.* membrane [f] d'étanchéification ; *g* **Dichtungsfolie** [f] (Kunststofffolie zur Isolierung von Mauern gegen aufsteigende Feuchtigkeit, zur Abdichtung von Becken und Teichen, als Dampfsperre etc.; ▶Dichtungsbahn, ▶Dichtungsschicht).

6919 waterproofing sheet [n] *constr.* (Usually lengths of synthetic material which, when rolled out, are bonded together to provide an homogeneous membrane); *s* **banda** [f] **de material impermeabilizante** (Material generalmente sintético (p. ej. polivinilo) enrollado. Una vez colocadas las **b.** se sueldan entre sí para impermeabilizar superficies formando una membrana); *f* **lé** [m] **d'étanchéité** (Membrane PVC ou bitumineuse soudée pour l'étanchéisation des toitures-terrasses) ; *g* **Dichtungsbahn** [f] (Meist homogen zu verschweißende Kunststoffbahnen, z. B. zur Abdichtung von Flachdächern).

waterproofing sheet [n] **of roof panels** [US] *constr.* ▶roof membrane.

6920 waterproofing test [n] *constr.* (Trial to check the water impermeability of roof membranes and pond liners); *s* **prueba** [f] **de impermeabilidad** (Control de membranas de impermeabilización de tejados, estanques, etc.); *f* **essai** [m] **d'étanchéité** (Contrôle de l'imperméabilité des membranes d'étanchéité de toitures-terrasses, de pièces d'eau, etc.) ; *g* **Wasserdichtigkeitsprobe** [f] (Prüfung der Wasserundurchlässigkeit von Dachdichtungen, Teichdichtungen etc.); *syn.* Wasserdichtigkeitsprüfung [f].

6921 waterproof layer [n] *constr.* (Protective installation to prevent water penetration, which may be made of plastic, asphaltic, or synthetic rubber sheets, of asphaltic mastics or coatings, of special cementitious plasters, ▶dampproof mortar US]/damp-proof mortar [UK], or of bentonite clay; ▶vapor barrier [US]/vapour barrier [UK], ▶waterproofing membrane); *s* **capa** [f] **de impermeabilización** (Capa de protección contra humedad en edificios o de impermeabilización de estanques por medio de rollos de material bituminoso, láminas sintéticas termoplásticas o ▶mortero impermeabilizante; ▶barrera antihumedad, ▶membrana de impermeabilización); *syn.* capa [f] de sellado; *f* **couche**

[f] **d'étanchéité** (Couche de protection contre les dommages des eaux sur les édifices ou de rétention de l'eau pour les étangs artificiels constituée par les matériaux suivants : film PVC, membrane de bitume élastomère, couche d'argile, béton hydrofuge, etc. ; ▶écran pare-vapeur, ▶membrane d'étanchéité, ▶mortier hydrofuge) ; *g* **Dichtungsschicht** [f] (Feuchteschutzeinrichtung wassergefährdeter Bauwerke, z. B. mittels bituminöser Dichtungsbahnen, thermoplastischer Kunststofffolien [▶Dichtungsfolie], ▶Sperrmörtel sowie tonigen Lehms zur Dichtung von Teichen; ▶Dampfsperre); *syn.* Abdichtung [f].

waterproof roofing membrane [n] [US] *constr.* ▶roof membrane.

water purification [n] *envir. wat'man.* ▶water treatment.

6922 water purification [n] **of a waterbody** *envir.* (Cleansing of lakes and watercourses of contamination, and removal of rubbish, pollutants, and other waste material; ▶biological self purification); *syn.* water cleanup [n] of a waterbody [US]/water clean-up [n] of a waterbody [UK]; *s* **limpieza** [f] **de masas de agua** (Eliminación de residuos sólidos y de materias contaminantes de lagos o ríos; ▶autodepuración); *f* **dépollution** [f] **des eaux** (Enlèvement des embâcles, des détritus et autres dans un cours d'eau, un étang, un lac ; ▶auto-épuration) ; *g* **Reinigung** [f] **eines Gewässers** (Entfernung von Unrat, Schmutzstoffen o. Ä. aus einem Fließgewässer oder See; ▶Selbstreinigung).

water purification plant [n] *envir. wat'man.* ▶water treatment plant.

6923 water quality category [n] *wat'man.* (Degree of nutrient in rivers or streams with seven possible ▶saprobic levels, or as in the case of stagnant water, degrees of nutrient stress with four possible categories as shown on a ▶lake and river water quality map; ▶saprobic system, ▶trophic level); *s* **clase** [f] **de calidad de las aguas continentales** (D., nivel de ▶saprobiedad de un curso de agua [entre 7 posibles] o ▶nivel trófico de un lago [entre 4 posibles] que se representan en el ▶mapa de calidad de las aguas continentales; ▶sistema de saprobiedad); *f 1* **classe** [f] **de pollution des cours d'eau** (D., taux de pollution des eaux courantes [selon sept niveaux de saprobie] ou ▶niveau trophique des eaux calmes superficielles [selon quatre niveaux de pollution] représenté sur la ▶carte de pollution des cours d'eau ; ▶saprobie, ▶système des saprobies ; *f 2* **niveau** [m] **de qualité des cours d'eau** (F., permettent de caractériser l'aptitude du milieu à satisfaire les différents usages [six niveaux pour les eaux superficielles destinées à la production d'eau alimentaire, 2 niveaux pour les eaux de baignade ; cf. Directives du Conseil du 16 juin 1975 et du 8 décembre 1976) ; *syn.* catégorie [f] de qualité des eaux superficielles ; *g* **Gewässergüteklasse** [f] (D., Verschmutzungsgrad eines Fließgewässers [von sieben möglichen Saprobiestufen: I = unbelastet bis gering belastet (oligosaprob); I-II = gering belastet (oligosaprob bis betamesosaprob); II = mäßig belastet (betamesosaprob); II-III = kritisch belastet (betamesosaprob bis alphamesosaprob); III = stark verschmutzt (alphamesosaprob); III-IV = sehr stark verschmutzt (alphamesosaprob bis polysaprob); IV = übermäßig verschmutzt (polysaprob)] oder ▶Trophiestufe eines stehenden Gewässers [von vier möglichen Belastungsstufen: I nährstoffarm (oligotroph); II mittlere Nährstoffversorgung (mesotroph); III nährstoffreich (eutroph); IV hohe Nährstoffbelastung (polytroph oder hypertroph)], dargestellt in ▶Gewässergütekarten. Die Gewässergüte wird nach Saprobiestufen klassifiziert, die von der „Länderarbeitsgemeinschaft Wasser" [LAWA] erarbeitet wurden; die Güteklassen der Fließgewässer sind nicht mit denen der Stillgewässer vergleichbar, da Fließgewässer nach dem Verschmutzungsgrad und Stillgewässer nach dem Nährstoffgehalt bewertet

W

werden; ►Saprobie, ►Saprobiensystem); *syn.* Wassergütestufe [f].

water quality control [n] [UK] *wat'man.* ►best management practices (BMP) [US].

6924 water quality improvement measures [npl] *pol. wat'man.* (Measures to improve the water quality of waterbodies with a high pollution load; ►water pollution); *s* **saneamiento [m] integral de las aguas** (Conjunto de medidas orientadas a mejorar la calidad del agua de cursos muy contaminados, p. ej. a través de los «planes generales de limpieza de arroyos y cauces» o «planes de saneamiento integral»; ►contaminación de las aguas continentales); *f* **restauration [f] des milieux naturels aquatiques** (Mesures visant l'amélioration de la qualité des eaux des cours d'eau fortement pollués, souvent mises en œuvre dans le cadre des schémas d'aménagement des eaux et des schémas départementaux de vocation piscicole ; cf. circ. n° SM/GP PN/84-1243 du 9 novembre 1983 ; ►pollution des eaux continentales) ; *g* **Gewässersanierung [f]** (Maßnahmen zur Verbesserung der Gewässergüte von stark belasteten Gewässern; ►Gewässerbelastung).

6925 water quality management [n] *conserv. nat'res. wat'man.* (Comprehensive measures for the public benefit, which are designed to preserve and improve the water balance and visual quality of landscapes characterized by rivers and lakes. They include the conservation and reinstatement of biological self-purification capacity, the improvement of life conditions for a species-rich flora and fauna by landscape and watershed management as well as by the statutory designation of water protection zones; **in U.S., w. q. m.** for water balance and visual landscape quality is a function of land management agencies governed by their legal enactments; ►water management plan); *s* **protección [f] de las aguas continentales** (Conjunto de medidas destinadas a mantener y mejorar el régimen hidrológico y la calidad visual los paisajes acuáticos como las que sirven para incrementar la capacidad de autodepuración de las aguas y mejorar las condiciones de vida de la flora y fauna naturales o la designación de perímetros de protección de acuíferos y la fijación de otras normas de protección; ►plan hidrológico); *f* **gestion [f] des eaux superficielles** (Ensemble des mesures prises en faveur du public dans le but d'une gestion équilibrée des ressources en eau superficielle, de sauvegarder ou d'améliorer, l'aspect des paysages aquatiques, de préserver les zones humides. Celles-ci comprennent la sauvegarde ou le rétablissement de la capacité d'autoépuration biologique, l'amélioration des conditions stationnelles de vie de la flore et de la faune aquatiques et terrestres par des mesures de protection de l'environnement et des paysages, par des dispositions réglementaires concernant les périmètres de protection de zones de captage et la qualité des eaux ; ►schéma directeur d'aménagement et de gestion des eaux) ; *g* **Gewässerschutz [m, o. Pl.]** (Gesamtheit der Maßnahmen zum Wohl der Allgemeinheit, die dazu dienen, einen geordneten Wasserhaushalt und das Landschaftsbild von Gewässerlandschaften zu erhalten oder zu verbessern. Dazu gehören die Erhaltung oder Wiederherstellung der biologischen Selbstreinigungskraft, die Verbesserung der Lebensmöglichkeiten für eine artenreiche Pflanzen- und Tierwelt durch technische Umweltschutz- und Landschaftspflegemaßnahmen sowie die Festsetzungen von Wasserschutzgebieten und der Erlass von Gewässerreinhalteordnungen; cf. auch WHG und Landeswassergesetze; ►wasserwirtschaftlicher Rahmenplan).

water quality map [n] *wat'man.* ►lake and river water quality map.

6926 water quality [n] **of rivers and lakes** *wat'man.* (Physical, chemical and biological properties of flowing water

primarily expressed by the amount of organic content being biologically decomposed by deoxygenization processes; the water quality of lakes is measured by the amount of nutrients and their influence upon the oxygen balance; ►lake and river quality map, ►saprobic level, ►water quality category); *s* **calidad [f] de las aguas continentales** (Calidad del agua desde el punto de vista biológico. En D. los ríos se clasifican según la contaminación con sustancias orgánicas que consumen oxígeno en su degradación, los lagos por la carga de nutrientes y sus consecuencias sobre el contenido de oxígeno; ►clase de calidad de las aguas continentales, ►mapa de calidad de las aguas continentales, ►saprobiedad); *f* **qualité [f] des eaux continentales** (Propriétés physiques, chimiques et biologiques des eaux caractérisant le degré de pollution des eaux et mesuré principalement au moyen de la teneur en substances organiques biodégradables pour les cours d'eau et au moyen de la concentration en éléments nutritifs ainsi que de leurs effets sur le bilan d'oxygène pour les eaux stagnantes ; ►carte de pollution des cours d'eau, ►classe de pollution des cours d'eau, ►niveau de qualité des cours d'eau, ►saprobie) ; *syn.* qualité [f] des eaux de surface, qualité [f] des cours d'eau ; *g* **Gewässergüte [f]** (Wasserqualität, die bei einem Fließgewässer in erster Linie durch die Belastung mit organischen, unter Sauerstoffzehrung biologisch abbaubaren Inhaltsstoffen definiert wird; bei einem stehenden Gewässer durch die Nährstoffbelastung und deren Auswirkung auf den Sauerstoffhaushalt. D., seit 1975 wird alle fünf Jahre die biologische G. mit den von der „Länderarbeitsgemeinschaft Wasser" [LAWA] aufgestellten Saprobiestufen bewertet; durch Abwasserreinigungsmaßnahmen in den alten Bundesländern und Änderung von Produktionsprofilen, Stilllegungen bedeutender Industriebetriebe und Neubau von Kläranlagen in den neuen Bundesländern, die in den 1990er-Jahren verstärkt durchgeführt wurden, hat sich die biologische G. so verbessert, dass 1995 bereits 48 % des Gewässernetzes [ca. 30 000 Flusskilometer] die angestrebte Güteklasse II [mäßig belastet] und besser aufwiesen; cf. R+U 2001 [4], 50. Im Dezember 2000 verpflichteten sich die Staaten der EU in der Wasserrahmenrichtlinie, ihre Oberflächengewässer bis 2015 in einen guten ökologischen Zustand mit einer natürlichen Zusammensetzung der Lebensgemeinschaften zu versetzen; das Umweltbundesamt bilanzierte 2004, dass das Erreichen dieses Zieles für 62 % der deutschen Flüsse und für 38 % der dt. Seen unwahrscheinlich sei, nicht zuletzt wegen der Veränderungen der Gewässermorphologie; ein hoher Prozentsatz an Gewässerabschnitten ist erheblich verändert; cf. UBA 2007; ►Gewässergütekarte, ►Gewässergüteklasse, ►Saprobie).

6927 water quality standards [npl] *wat'man.* (Minimum requirements of water purity for intended uses with respect to the physical, chemical, and biological characteristics; cf. RCG 1982); *s* **calidad [f] del agua** (Condiciones mínimas de las características físicas, químicas y biológicas del agua con respecto a los usos previstos); *f* **qualité [f] de l'eau** (Paramètres physiques, chimiques et microbiologiques caractérisant l'état de propreté d'une eau par rapport aux exigences de certains usages, tels que potabilité, etc.) ; *g* **Wassergüte [f]** (Eigenschaft eines Wassers hinsichtlich seiner Sauberkeit, Nutzbarkeit etc.); *syn.* Wasserqualität [f].

6928 water recreation [n] *recr.* (Leisure-time activities for enjoyment in connection with streams, rivers, lakes, reservoirs, and other bodies of water); *s* **esparcimiento [m] asociado al agua** *syn.* recreación [f] asociada al agua, recreo [m] asociado al agua; *f* **loisirs [mpl] à caractère spécifique aquatique** (Activité récréative autour des plans d'eau, des étangs et lacs, en zone humide, le long des torrents et rivières ; *termes spécifiques* pratique balnéaire [liée à la mer], loisirs et tourisme liés à l'eau douce, tourisme fluvial, loisirs halieutiques [liés à la pêche]) ;

W

g **gewässerbezogene Erholung [f, o. Pl.]** (Erholung an Gewässerufern, auf ufernahen Gewässerflächen oder auf Gewässern in Form von Lagern, Sonnen, Spielen, Aussicht genießen, Angeln, Schwimmen, mit manuell-, wind- oder motorgetriebenen Booten fahren etc.); *syn.* Erholung [f, o. Pl.] an Gewässern/am Wasser.

6929 water resources [n] *hydr. nat'res. wat'man.* (**1.** The total quantity of water existing on Earth is 1.38 billion cu km, of which only 35 million cu km [2.53%] is fresh water, and only 46,800cu km [0.13%] of the latter is available for man's use. **2.** Supply of available surface and underground water in a given catchment area or basin; ▶diminution of water resources); *s 1* **recursos [mpl] hídricos** (Cantidad de agua dulce a disposición para usos humanos que supone una parte muy reducida [0,13%] del agua dulce total existente en la tierra; ▶dominio público hidráulico); *syn.* recursos [mpl] hidráulicos; *s 2* **dominio [m] público hidráulico [Es]** (Término jurídico que incluye las aguas continentales superficiales y subterráneas, los cauces de las corrientes naturales, los lechos de los lagos y lagunas y los de los embalses superficiales en cauces públicos, los acuíferos subterráneos y las aguas procedentes de la desalación del agua del mar; art. 2, RDL 1/2001; ▶degradación de los recursos hídricos); *f* **ressources [fpl] hydrauliques** (**1.** la quantité totale des eaux de la terre est de 1,38 milliards de km^3 ; seulement 35 millions de km^3 soit 2,53 % sont des eaux douces, dont seulement 46 800 km^3 soit 0,13 % sont à la disposition de l'homme. **2.** Quantité des eaux superficielles et souterraines aptes à la consommation humaine ; ▶altération de la qualité des resources en eau) ; *g* **Wasserdargebot [n]** (**1.** Die gesamte, auf der Erde vorhandene Wassermenge beträgt 1,38 Mrd. km^3, wovon nur 35 Mio. km^3 [2,53 %] Süßwasser sind, und davon auch nur 46 800 km^3 [0,13 %] der menschlichen Nutzung zugänglich ist. **2.** Nutzbare Grund- und Oberflächenwässer nach Menge und Qualität in einem Betrachtungsraum; ▶Beeinträchtigung der Wasserressourcen); *syn.* Wasserdargebotspotenzial [n], *o. V.* Wasserdargebotspotential [n], Wasserressourcen [fpl], Wasservorkommen [n], Wasservorrat [m].

water resources [npl] **[US], lessening of** *wat'man.* ▶diminution of water resources.

water resources [npl]**, lowering of** *wat'man.* ▶diminution of water resources.

Water Resources Act [n] **[UK]** *envir. leg.* *water rights.

Water Resources Board [n] **[UK]** *adm. hydr. wat'man.* ▶water authority [UK].

6930 water resources conservation [n] *hydr. nat'res.* *s* **preservación [f] de los recursos hídricos** *syn.* preservación [f] de los recursos de agua; *f* **protection [f] de la ressource en eau** *syn.* préservation [f] des ressources hydrauliques, préservation [f] des ressources en eau ; *g* **Erhaltung [f, o. Pl.] des Wasservorkommens.**

6931 water resources management [n] *adm. leg. wat' man.* (Economic use of water from a natural source, aimed at **1.** preserving or restoring the ecological equilibrium of water resources, **2.** guaranteeing water supply for a population or industry, and **3.** ensuring the long-term use of water for the public benefit; in U.S., Federal Water Pollution Control Act, as amended [Clean Water Act] of 1972 and 1982, Drinking Water Act 1973; ▶EU Water Framework Directive, ▶water users agency); *s* **gestión [f] de recursos hídricos** (Es el manejo ahorrativo del agua con el objetivo de **1.** mantener o reestablecer las funciones ecológicas de las aguas en la naturaleza, **2.** garantizar el suministro de la población y las actividades económicas y **3.** posibilitar los diversos usos a largo plazo de acuerdo con los fines de la sociedad; ▶Directiva Marco del Agua, ▶organismo de cuenca hidrográfica [Es]); *f* **gestion [f] du cycle de l'eau** (Recherche d'un équilibre durable dans l'utilisation, la mise en valeur et la protection quantitative et qualitative de la ressource en eau, assurant d'une manière globale la protection des milieux naturels ainsi que la satisfaction des différents usages économiques ; ▶agence de l'eau, ▶Directive pour la protection des eaux intérieures de surface, des eaux de transition, des eaux côtières et des eaux souterraines) ; *g* **Bewirtschaftung [f] der Wasserressourcen** (Bewirtschaftung der Gesamtheit des in der Natur in einem definierten Gebiet vorhandenen Wassers mit dem Ziel, **1.** das ökologische Gleichgewicht der Wasserressourcen zu erhalten oder wiederherzustellen, **2.** eine gesicherte Wasserversorgung der Bevölkerung und der Wirtschaft [Industrie, Gewerbe und Landwirtschaft] zu gewährleisten und **3.** die dem Gemeinwohl dienenden Wassernutzungen langfristig sicherzustellen. Zur Gesamtheit des Wassers gehören alle oberirdischen Gewässer, Küstengewässer [Küstenmeer] und das Grundwasser; cf. § 1 WHG. Zur rechtlichen Regelung der Materie Wasser hat der Bund unter dem Gesichtspunkt der Wasserwirtschaft und Landeskultur auf Grund seiner Rahmenkompetenz [Art. 75 I Nr. 4 GG a. F.] das Wasserhaushaltsgesetz [WHG] erlassen, das in seiner jetzigen Fassung durch die ▶Wasserrahmenrichtlinie der EU [RL 2000/60/EG] wesentlich geprägt wurde; durch die Aufhebung des Art. 75 GG durch die Föderalismusreform wird die Gesetzregelungskompetenz jetzt durch Art. 72 III Nr. 5 und 74 I Nr. 32 GG geregelt; das WHG gilt zunächst nach Art. 125 b GG fort; die Bundesländer können abweichende Regelungen treffen).

6932 water rights [npl] *envir. leg.* (Laws and ordinances which govern the classification of surface water bodies, groundwater and coastal waters, their ownership, use, and conservation, as well as the establishment of water conservation areas, river engineering measures, planning for economic use of water and the powers/responsibilities of water authorities; in U.S., Clean Water Act; in U.K., the following acts apply: Rivers [Prevention of Pollution] Act 1951, Water Resources Act 1963, Clean Rivers [Estuaries and Tidal Waters] Act 1960, Water Act 1945); *s* **legislación [f] de las aguas** (Conjunto de normas que regulan el uso de las aguas continentales, las costeras, las construcciones hidráulicas, etc.; **en Es.** RD Legislativo 1/2001, de 20 de julio, por el que se aprueba el texto refundido de la Ley de Aguas y Ley 22/1988, de 28 de julio, de Costas, y sus correspondientes reglamentos); *f* **droit [m] de l'eau** (Ensemble des textes réglementaires régissant le régime juridique des eaux douces et de la mer [propriété et usage], le régime général de prélèvement et de rejets, l'action des administrations et organismes compétents en matière d'eau, la responsabilité et sanctions pénales en matière de pollution des eaux) ; *syn.* législation [f] sur l'eau ; *g* **Wasserrecht [n]** (**1.** Gesamtheit der gesetzlichen Vorschriften unter dem Gesichtspunkt der Wasserwirtschaft und Landeskultur, die die Klassifizierung der oberirdischen Gewässer, Küstengewässer und des Grundwassers, das Eigentum daran, die Nutzung, den Gewässerschutz, die Festsetzung der Wasserschutzgebiete, den Gewässerausbau, die wasserwirtschaftliche Planung, die Zuständigkeit der Wasserbehörden etc. regeln. **In D.** ist das W. teils Bundesrecht und wird nach dem Wasserhaushaltsgesetz [WHG] geregelt, teils Landesrecht; **in A.** wird das W. bundeseinheitlich nach dem Wasserrechtsgesetz von 1959 [i. d. F. BGBl. I Nr. 123/2006] geregelt; in der **Schweiz** hat der Bund die Oberaufsicht über die Wasserbaupolizei und die Gesetzgebungskompetenz über die Erhaltung und Erschließung der Wasservorkommen sowie die Benutzung der Gewässer zur Energieerzeugung und für Kühlzwecke sowie andere Eingriffe in den Wasserkreislauf; cf. MEL 1979, Bd. 25. **2.** Rechtliche Regelung der Materie Wasser hinsichtlich seiner Bedeutung als Wasserstraße und Verkehrsweg; cf. WaStrG).

W

water runoff [n], **surface** *constr. geo. wat'man.* ▶surface runoff.

waters [npl] *geo. leg. pol.* ▶inland waters, ▶territorial waters.

waters [npl], **coastal** *geo. leg* ▶territorial waters.

waters [npl] [US], **head-** *geo.* ▶upper reaches.

water-saturated glei [n] *pedol.* ▶water-saturated gley.

6933 water-saturated gley [n] *pedol.* (Hydromorphic soils permanently saturated by groundwater which almost reaches the soil's surface; horizons AG$_0$-Gr); *syn. o.v.* water-saturated glei [n]; *s* **gley** [m] **de moder** (Suelo casi siempre temporalmente encharcado, o no cubierto por agua estancada superficialmente, pero siempre con humedad de agua subterránea; cf. KUB 1953, 158); *f* **gley** [m] **oxydé humifère** (Sol hydromorphe à nappe phréatique profonde, le niveau de la nappe étant en permanence très proche de la surface du sol [oscillation moyenne de la nappe] ; profil AG$_0$-Gr ; PED 1983, 393) ; *g* **Nassgley** [m] (Grundwasserboden [hydromorpher Boden] mit permanent bis nahe zur Bodenoberfläche reichendem Grundwasser; Horizontfolge AG$_0$-Gr).

water-saturated soil [n] [UK] *agr. hort. pedol.* ▶waterlogged soil [US]/water-logged soil [UK].

6934 water saturation [n] *pedol.* (Degree to which voids between soil particles are filled with water; ▶impeded water, ▶waterlogging); *s* **saturación** [f] **de agua** (Estado de un suelo cuyos huecos están llenos de agua; ▶agua estancada, ▶encharcamiento del suelo); *f* **saturation** [f] **en eau** (État de remplissage permanent ou temporaire par l'eau des vides entre les particules solides d'un sol ; ▶humectation, ▶imbibition, ▶nappe perchée, ▶processus d'engorgement en eau du sol ; PED 1983, 368 ; DIS 1986, 206) ; *syn.* degré [m] de saturation en eau ; *g* **Wassersättigung** [f] (Grad des Ausgefülltseins aller Hohlräume eines Bodens mit Wasser; ▶Staunässe, ▶Bodenvernässung); *syn.* Wassersättigungsgrad [m].

6935 water saturation [n] **of a slope** *constr.* (▶Waterlogging of a slope caused by ▶seepage water; ▶slope seepage water); *s* **encharcamiento** [m] **de una ladera** (Fuerte ▶encharcamiento del suelo de una pendiente causado por ▶agua de percolación; ▶filtración efluente en ladera); *f* **engorgement** [m] **du sol d'un versant** (Forte concentration d'eau dans le sol d'un versant par suite de ressuyage ; ▶eau de gravité, ▶eaux d'écoulement hypodermique, ▶processus d'engorgement en eau du sol) *g* **Hangvernässung** [f] (Wasseraustritt an einem Hang, der durch Wasseranreicherung über Schichten geringer hydraulischer Leitfähigkeit entsteht; ▶Bodenvernässung, ▶Hangwasser, ▶Sickerwasser 2).

6936 waterscape [n] *hydr. landsc.* (GGT 1979; landscape characterized by rivers, lakes, and other bodies of water; ▶lake district, ▶lake landscape, ▶river landscape); *s* **paisaje** [m] **acuático** (Paisaje caracterizado primordialmente por la presencia de ríos, lagos, etc.; *términos específicos* ▶paisaje fluvial, ▶paisaje lacustre); *f* **paysage** [m] **aquatique** (**1.** Paysage caractérisé par la présence de grands ou nombreux plans et cours d'eau et dont la structure spatiale est conditionnée par une mosaïque d'habitats aquatiques, p. ex. le ▶paysage fluvial, le ▶paysage lacustre. **2.** *recr.* Paysage dont certains éléments donnent l'impression, l'ambiance d'un environnement paysager naturel [piscines naturelles, plans d'eau et bassins à épuration biologique]) ; *g 1* **Gewässerlandschaft** [f] (Landschaft, die durch Oberflächengewässer geprägt ist, z. B. ▶Flusslandschaft, ▶Seenlandschaft); *g 2* **Erholungslandschaft** [f] **an Gewässern** *recr.* (Eine Landschaft, die durch Oberflächengewässer geprägt ist und auf Grund ihrer Vielfalt, Eigenart und Schönheit besonders der Freizeit- und Erholungsnutzung dient); *syn.* Gewässerlandschaft [f] für Freizeit und Erholung, gewässergeprägte Erholungslandschaft [f].

6937 water seal [n] *constr.* (**1.** Water standing in a drain trap to prevent foul air from rising); *syn.* odo(u)r trap [n]; **2. air trap** [n] (U-shaped pipe filled with water that prevents the escape of foul air or gas from such systems as drains); *s 1* **cierre** [m] **hidráulico** (En fontanería, presencia de agua en la parte inferior de la U de un sifón con objeto de bloquear el paso de los gases y/u olores dentro de la habitación; DACO 1988); *s 2* **sifón** [m] **obturador** (Tubo doblemente acodado en el que el agua detenida dentro de él impide la salida de aire viciado y malos olores, producidos en lavabos, tuberías de drenaje y colectores; DACO 1988); *syn.* sifón [m] inodoro; *f* **coupe-odeur** [m] (Dispositif d'occlusion hydraulique [siphon] assurant une garde d'eau permanente et évitant le retour de gaz malodorants) ; *syn.* siphon [m] disconnecteur ; *g* **Geruchsverschluss** [m]. (Mit Wasser gefülltes, mehrfach gekrümmtes Rohr bei Abwassereinlaufstellen, das verhindert, dass der Kanalisationsgeruch nach oben dringt); *syn.* Siphon [m].

6938 water's edge [n] *geo.* (**1.** Transition area between land and standing or flowing water—in contrast to ▶coast. **W. e.** extends from a shallow water depth to the highest flood line; *specific terms* ▶lakeshore, ▶riverbank [US]/river bank [UK]); **2. waterfront** [n] *urb.* (Part of a city orientated to a body of water [river, lake or ocean]; in the past few decades some cities have transformed their run-down areas on the **w. e.**, e.g. obsolete harbo[u]r facilities, such as Victoria & Alfred Waterfront in Cape Town and London Docklands, into modern developments with a touristic appeal); **3. lakefront** [n] (Lot or part of an area which is orientated to a lake); ▶lakeshore); *s* **orilla** [f] (Borde de un curso o masa de agua o del mar; ▶costa; *términos específicos* ▶margen de río, ▶orilla de un lago, ▶ribera); *f* **berge** [f] (Bande de terre relevée bordant un cours d'eau ou un plan d'eau fermé [lac, étang] ; ▶côte ; *termes spécifiques* ▶berge, ▶rivage lacustre, ▶rive d'un fleuve) ; *syn.* rive [f] (2), rivage [m] d'un lac (PR 1987) ; *g 1* **Ufer** [n] (Übergangsbereich zwischen dem Land und einem stehenden oder fließenden Gewässer — im Gegensatz zur ▶Küste. Das **U.** dehnt sich vom Flachwasserbereich bis zur höchsten Hochwasserlinie aus; *UBe* ▶Flussufer, ▶Seeufer); *syn.* Gewässerrand [m], *hort.* Wasserrand [m]; *g 2* **Waterfront** [f] (*Aus dem Englischen übernommener Begriff, z. B. Waterfront Bremen zum Wasser [Fluss, See oder Meer] hin ausgerichteter Stadtteil; in den letzten Jahrzehnten wurden in einigen Städten zu sanierende Bereiche, häufig Hafenanlagen, in moderne Baugebiete umgewandelt, die oft touristisch sehr attraktiv sind, z. B. London Docklands, Victoria & Alfred Waterfront in Kapstadt*).

6939 watershed [n] *geo. hydr. wat'man.* (Topographic boundary, usually a high point, that marks the dividing line ['divide'] separating headstreams which flow into different ▶drainage basins. The term 'continental divide' is applied to the separation between major drainage systems flowing into oceans or seas); *syn.* water-parting [n] [UK] (DNE 1978, 313), divide [n] [also US]; *s* **divisoria** [f] **de aguas** (Límite topográfico entre dos ▶cuencas hidrográficas); *syn.* divisoria [f] de cuencas, divisoria [f] de drenaje, parteaguas [m]; *f* **ligne** [f] **de partage des eaux** (Ligne de part et d'autre de laquelle les eaux superficielles se dirigent vers des ▶bassins versants contigus) ; *g* **Wasserscheide** [f] (Auf einem Höhenrücken oder Gebirgskamm verlaufende Grenzlinie zwischen zwei ▶Abflussgebieten).

water shoot [n] [UK] *arb. bot. hort.* ▶water sprout.

6940 water soluble [adj] *pedol.* (Descriptive term applied to solubility of nutrients in soil ▶easily soluble; ▶poorly soluble); *s* **soluble en agua** [loc] (Capacidad de los nutrientes de disolverse en el agua del suelo; ▶de gran solubilidad, ▶poco soluble);

W

f **soluble dans l'eau [loc]** (Capacité qu'ont les éléments organiques ou minéraux du sol à se solubiliser dans l'eau libre ou l'eau liée ; ►difficilement soluble, ►facilement soluble) ; *g* **wasserlöslich [adj]** (Nährstoffe betreffend, die im Boden mit Hilfe von Wasser gelöst werden können; ►leicht löslich, ►schwer löslich).

6941 water sports [npl] *recr. syn.* aquatic sports [npl]; *s* **deporte [m] acuático** *syn.* sólo para los deportes marinos deporte [m] náutico; *f* **sport [m] nautique** (pratiqué sur l'eau) ; *g* **Wassersport [m]** (Auf dem Wasser ausgeübte Sportart).

6942 water sprout [n] *arb. bot. hort.* (Shoot arising spontaneously from an adventitious or dormant bud on a stem or branch of a woody plant; *generic term* 'secondary growth'; cf. SAF 1983; ►coppice shoot); *syn.* epicormic shoot [n], epicormic [n] [also US], water shoot [n] [also UK]; *s* **hijuelo [m]** (Retoño largo, a menudo con hojas más grandes de lo normal, que crece de yemas durmientes en la base del tronco o de las ramas; ►brote de cepa); *syn.* brote [f] epicórmico; *f* **gourmand [m]** (Rameau se développant directement sur le fût d'un arbre à partir d'un bourgeon préexistant qui se réveille après que le tronc de l'arbre ait été isolé et mis en lumière par une coupe forte ; TSF 1985 ; ►rejet de taillis) ; *g* **Wasserreis [n]** (...reiser [pl]; junger Langtrieb [Schössling] mit oft vergrößerten Laubblättern, der nach Lichtstellung oder Verletzung aus einem schlafenden Auge aus Stämmen oder Ästen hervorbricht. Besonders junge **W.er**, die wegen ihrer schlechten Einbindung in den Stamm ‚kleben', werden **Klebäste** genannt; ►Stockaustrieb); *syn.* Wasserschoss [m], Wasserschössling [m].

water storage capacity [n] *pedol.* ►water-holding capacity.

water supply [n] *plan. urb.* ►public water supply.

6943 water supply connection [n] [US] *constr.* (Water tap on construction sites or outside faucet for maintenance from a ►water supply line); *syn.* water supply point [n] [UK]; *s* **acometida [f] de agua** (Conexión a red de suministro de agua en obra de construcción o para trabajos de mantenimiento; ►tuberia de suministro de agua); *f* **raccordement [m] à un point d'eau** (à une vanne d'alimentation en eau sur le chantier ; ►conduite de raccordement d'eau) ; *g* **Wasseranschluss [m]** (Wasserzapfvorrichtung auf einer Bau- oder Pflegestelle, an einem Gebäude ode rein Hydrant; cf. § 10 [4] lit. b VOB Teil A; ►Anschlusswasserleitung); *syn.* Wasserzapfstelle [f].

6944 water supply line [n] *constr.* (Connecting pipe supplying water to specific user; ►water main); *s* **tubería [f] de suministro de agua** (Conducción enterrada en la calle para abastecer de agua potable a los usuarios finales; ►tubería principal de agua); *syn.* conducción [f] de suministro de agua, cañería [f] de suministro de agua; *f* **conduite [f] de raccordement d'eau** (Canalisation d'alimentation reliant l'usager au réseau public d'eau potable ou d'assainissement ; pour les tuyaux de faible diamètre on parle plutôt de **conduit de raccordement** ; ►canalisation principale d'alimentation en eau [potable]) ; *syn.* canalisation [f] de raccordement d'eau ; *g* **Anschlusswasserleitung [f]** (*Abk.* AW-Leitung; Stichleitung für die Trinkwasserversorgung von der Verteilerwasserleitung in den Straßen zum Endabnehmer; ►Hauptwasserleitung).

water supply point [n] [UK] *constr.* ►water supply connection [US].

6945 water surface [n] *hydr. recr.* (**1.** Top of a body of water. **2.** *constr.* Puddle on a paved area); *syn.* sheet [n] of water; *s* **superficie [f] del agua** (de un lago o río); *f 1* **surface [f] de l'eau** (d'un cours d'eau) ; *f 2* **plan [m] d'eau** (Surface d'une eau calme naturelle ou artificielle) ; *syn.* pièce [f] d'eau (d'un bassin) ; *g* **Wasserfläche [f] (2)** (**1.** Oberfläche eines Gewässers als

Erscheinungsbild. **2.** *constr.* Große Pfütze auf befestigten Flächen).

6946 watertable [n] *hydr.* (Level of groundwater, measured in meters [US]/metres [UK] below ground surface, variable in elevation and fluctuating somewhat from dry to wet season, and located at the meeting point of porous and impermeable rock strata; ►water level); *syn.* groundwater surface [n], groundwater level [n] (DOG 1984); *s* **nivel [m] freático** (Profundidad bajo la superficie del suelo en la que se encuentra la superficie de las aguas subterráneas y que se mide en metros; ►nivel de agua); *syn.* superficie [f] de agua subterránea, superficie [f] freática; *f* **niveau [m] de la nappe phréatique** (Position altitudinale de la surface de la nappe par rapport au niveau de la surface au sol ; ►niveau de l'eau) ; *g* **Grundwasserspiegel [m]** (Höhe des Grundwassers, gemessen in Metern unter Flur; ►Wasserspiegel); *syn.* Grundwasserstand [m].

6947 watertable contour [n] *hydr.* (Line connecting the points which represent the same levels of groundwater table; ►contour); *syn.* watertable isohypse [n]; *s* **isohipsa [f] freática** (Línea que une todos los puntos de un nivel freático que tienen la misma cota en relación a una cota de referencia; WMO 1974; ►curva de nivel); *syn.* contorno [m] freático; *f* **courbe [f] hydroisohypse (de la nappe phréatique)** (Ligne joignant les points d'égale altitude de la surface de la nappe phréatique ; ►courbe de niveau) ; *syn.* courbe [f] isohypse ; *g* **Grundwasserhöhenlinie [f]** (Linie, die Punkte gleicher Grundwasserspiegelhöhen miteinander verbindet; ►Höhenlinie); *syn.* Grundwassergleiche [f], Isohypse [f].

watertable isohypse [n] *hydr.* ►watertable contour.

6948 water theater [n] [US]/water theatre [n] [UK] *arch. gard. gard'hist.* (Deliberately composed arrangement of ►water features and fountains designed for dramatic effect; e.g. Villa Aldobrandini and Villa Mondragone in Italy, Longwood Gardens in Delaware, U.S.; ►composition of water features); *s* **teatro [m] de aguas** (Composición coreográfica de fuentes y surtidores con efecto dramático, a menudo con iluminación, existente p. ej. en la Villa Aldobrandini, Villa Mondragone en Italia, en los jardines Longwood en Delaware, EE.UU.; ►hidroarte, ►juego de aguas); *f* **théâtre [m] d'eau** (Arrangement chorégraphique constitué de jets d'eau et de statues-fontaines créant une ambiance dramatique comme p. ex. dans les jardins des villas d'Aldobrandini ou de Mondragone ; ►art des jeux d'eau, ►jeu d'eau) ; *g* **Wassertheater [n]** (Choreografische Komposition von Wasserstrahlfiguren [Fontänen] in Springbrunnen, Teichen und Seen mit dramatischer Wirkung, oft mit Beleuchtung in den Abendstunden wie z. B. in Villa Aldobrandini, Villa Mondragone in Italien, Longwood Gardens in Delaware, USA; ►Wasserkunst, ►Wasserspiel).

6949 watertightness [n] *constr.* (e.g. of a ►roof membrane); *syn.* imperviousness [n]; *s* **impermeabilidad [f]** (Calidad o estado de un material, p. ej. de una ►membrana de impermeabilización de tejado o de estanque); *f* **étanchéité [f]** (Qualité ou état d'une ►revêtement d'étanchéité) ; *syn.* imperméabilité [f] ; *g* **Dichtigkeit [f] (eines Materials)** (Eigenschaft oder Zustand eines Materials, Flüssigkeiten — insbesondere Wasser, nicht ein-/ausdringen oder durchzulassen, z. B. **D.** einer Teichfolie, einer ►Dachhaut, eines Lehmschlages, eines Rohres); *syn.* Dichtheit [f] eines Materials.

6950 water treatment [n] *envir. wat'man.* (Production of usable water from groundwater or surface water by chemical and physical treatment processes; 'potable water treatment' and 'reusable water treatment' are differentiated; ►drinking water treatment, ►public water supply); *syn.* water purification [n]; *s* **tratamiento [m] del agua** (Preparación de agua subterránea o

W

superficial para el ►suministro de agua [potable] utilizando procesos químicos o físicos de tratamiento que se diferencian según sean para producir agua potable o agua para uso industrial; ►potabilización del agua); *f* **traitement [m] des eaux** (Dans le cadre de l'alimentation de la population en eau potable [►alimentation en eau potable], clarification [élimination des algues et des particules en suspension par décantation ou filtration] et stérilisation [élimination des éléments vivants au moyen du chlore ou de l'ozone] des eaux souterraines ou superficielles avant leur transport dans le réseau d'eau potable ; à ne pas confondre avec l'épuration des eaux qui est le traitement des eaux usées avant leur rejet dans le milieu naturel ; ►traitement d'eau potable) ; *g* **Wasseraufbereitung [f]** (Gewinnung von nutzbarem Wasser aus Grund- oder Oberflächenwasser durch chemisch-physikalische Behandlungsverfahren im Rahmen der ►öffentlichen Wasserversorgung; es wird zwischen Trinkwasser- und Brauchwasseraufbereitung unterschieden; ►Trinkwasseraufbereitung); *syn.* Wasserreinigung [f].

6951 water treatment plant [n] *envir. wat'man.* (Technical installation for ►water treatment); *syn.* water purification plant [n]; *s* **planta [f] de tratamiento de aguas** (Instalación técnica para el ►tratamiento del agua); *f* **usine [f] de traitement des eaux** (Installation technique de ►traitement des eaux) ; *syn.* installation [f] de traitement des eaux ; *g* **Wasseraufbereitungsanlage [f]** (Technische Einrichtung, die Wasser den speziellen Anforderungen der Verbraucher mit mechanischen, physikalischen und/oder chemischen Verfahren anpasst; **W.n** dienen der Trinkwasser und Abwasseraufbereitung. Das Wasser wird je nach gewünschtem Endprodukt in mehreren Verfahrensschritten behandelt: mechanische Vorreinigung bei Fluss- und Oberflächenwasser, Belüftung, Entsäuerung, Flockung, Filterung, Enteisenung, Enthärtung für Haushaltsgeräte, Entsalzung bei Meerwasser etc.; ►Wasseraufbereitung).

6952 water users agency [n] [US] *nat'res. wat'man.* (Alliance of public and private users of river water in a particular river catchment area or part thereof, formed in order to preserve the regional water quality and its supply by implementing planning measures for water economy); *syn.* river authority [n] [UK], river water authority [n] [UK], water users association [n] [US]; *s* **organismo [m] de cuenca hidrográfica [Es]** (Entidad de derecho público con personalidad jurídica propia y distinta de la del Estado, adscrita a efectos administrativos a la/s Consejería/s competente/s de la/s correspondiente/s CC.AA. y con plena autonomía funcional, cuyas funciones principales son, entre otras, elaborar el ►plan hidrológico de cuenca; cf. art. 41 RDL 1/2001); *f* **agence [f] de l'eau** (Établissement public administratif doté de la personnalité civile et de l'économie financière dont la mission est de faciliter les diverses actions de gestion de l'eau d'intérêt commun aux bassins ou groupements de bassins) ; *syn. obs.* agence [f] de bassins ; *g* **Flussgebietsverband [m]** (Zusammenschluss der Wassernutzer eines Flussgebietes oder Teile davon als Körperschaft des öffentlichen Rechts zur Sicherung der regionalen wassergüte- und mengenwirtschaftlichen Versorgung durch wasserwirtschaftliche Planung und Maßnahmen; ►wasserwirtschaftlicher Rahmenplan); *syn.* Wasser- und Bodenverband [m].

watten [n] [UK] *geo. ocean* ►tidal mudflat.

6953 wattlework [n]/**wattle-work** [n] [also UK] *constr.* (In bioengineering, a generic term for a vertical framework of live interwoven willows, alders or the like, which are usually capable of rooting, and installed on slopes and banks for erosion control; *specific terms* ►diagonal wattlework [US]/diagonal wattle-work [UK], horizontal **w.**); *syn.* wickerwork [n] [US]/wicker-work [n] [UK]; *s* **varaseto [m] de ramas** (Enrejado de ramillas de sauce que se utiliza para fijar el suelo de taludes o las

orillas de ríos contra la erosión; ►malla reticular de ramas); *syn.* trenzado [m] de ramas; *f 1* **treillis [m] de branches (de saules) à rejets** (*Génie biologique* ouvrage de stabilisation de talus constitué par un entrecroisement de branchages à rejets) ; *syn.* treillage [m] de branches à rejets ; *f 2* **claie [f] de branches à rejets** (Panneau en treillis constitué de branchages à rejets tressés ; *termes spécifiques* ►treillis à maillage croisé, treillis à maillage horizontal, treillis à maillage en losange ; *une petite claie* clayon [m]) ; *syn.* claie [f] vive ; *g* **Flechtwerk [n]** (*Ingenieurbiologie* aus ausschlagfähigen Weidenruten erstelltes Geflecht zur Sicherung von Oberbodenrutschungen und zur Ufersicherung; cf. DIN 18 918 und DIN 19 657; *UBe* ►Diagonalgeflecht, Längsgeflecht); *syn.* Geflecht [n] (LEHR 1981).

6954 wave action [n] *geo.* (Impact of waves upon a shore caused by rhythmic wave movement); *syn.* wave attack [n] (VCS 1981, 378), wave-wash energy [n] (RIW 1984, 171); *s* **embate [m] de las olas** (Impacto de las olas en la costa, causado por el movimiento rítmico de las mismas); *f* **jet [m] de rive** (Mouvement rythmique montant d'une vague s'abattant sur le rivage lors de son déferlement) ; *g* **Wellenschlag [m]** (Anprall der Wasserwellen ans Ufer durch rhythmische Wellenbewegungen).

wave attack [n] *geo.* ►wave action.

wave bath(s) [n(pl)] [UK] *recr.* ►wave pool.

6955 wave pool [n] *recr.* (Indoor or outdoor swimming pool in which waves are created artificially at periodic intervals); *syn.* wave bath(s) [n(pl)] [also UK]; *s* **piscina [f] con olas** (Piscina interior o exterior que tiene integrado sistema de producción de olas); *f* **piscine [f] à vagues** (Piscine en plein air ou couverte dans laquelle l'eau est mise en mouvement à intervalles réguliers pour former artificiellement des vagues) ; *g* **Wellenbad [n]** (Ein Hallen- oder Freibad, in dem das Wasser im periodischen Rhythmus künstlich bewegt wird).

wave-wash energy [n] *geo.* ►wave action.

way [n] ►natural drainage way [US]/natural hillside drainage channel [UK]; *trans.* ►travelled way [US]/carriage-way [UK].

way [n] [UK]**, bridle** *recr.* ►bridle path [US].

way [n] [UK]**, extinguishment of rights of** *adm. leg. urb.* ►street abandonment [US].

wayside [n] [UK] *agr. constr. ecol.* ►pathside strip [US].

wayside area [n] [US] *recr.* ►stopping place [US]/rest spot [UK] (3).

weak light exposure [n] *bot.* ►poor light exposure.

6956 wearing course [n] *constr.* (Top layer in paving construction which is resistant to traffic wear; ►surface layer); *s* **capa [f] de desgaste** (Capa superior de revestimientos de calles o caminos de hormigón o de material bituminoso; ►firme); *syn.* superficie [f] de rodadura; *f* **couche [f] de roulement** (Couche superficielle de la chaussée, servant à absorber les efforts tangentiels et le cisaillement importants provoqués par la circulation dans la partie haute de la chaussée ; DIR 1977 ; ►couche de surface, ►revêtement de sol) ; *syn.* couche [f] d'usure ; *g* **Verschleißschicht [f]** (Gegen Abnutzung widerstandsfähige, oberste Schicht bei Wege-, Platz- oder Straßenbelägen aus Beton oder bituminösen Decken; ►Decke); *syn.* Oberflächenschutzschicht [f].

6957 weather damage [n] *landsc. met.* (Destruction caused by the influence of adverse weather); *s* **daño [m] causado por la intemperie** (Efecto destructivo del tiempo meteorológico sobre la naturaleza o los objetos); *f* **dommage [m] causé par des intempéries** (Dégâts provoqués par les agents atmosphériques) ; *g* **Witterungsschaden [m]** (Schaden, der durch Wettereinflüsse entstanden ist).

W

6958 weathered material [n] *geo. pedol.* (Decomposed or disintegrated mineral material caused by weathering processes; ▶rock debris); *s* **material** [m] **meteorizado** (Mezcla de minerales resultante de la meteorización; ▶detrito); *f* **produit** [m] **d'altération** (Matériau secondaire résultant de la transformation partielle par des processus physico-chimiques des roches, minéraux, sols ou sédiments meubles ; ▶éboulis) ; *g* **Verwitterungsprodukt** [n] (Durch Verwitterungsvorgänge entstandenes Mineralgemenge; ▶Gesteinschutt); *syn.* Verwitterungsmaterial [n].

6959 weathering [n] *geo. pedol.* (Disintegration and decay of rock caused by atmospheric and climatic influences, thereby producing an *in situ* mantle of waste); *s* **meteorización** [f] (Los cambios físicos y químicos producidos en las rocas por los agentes atmosféricos; MEX 1983); *syn.* alteración [f] de la roca; *f 1* **météorisation** [f] (*Terme générique* ensemble des processus mécaniques, physico-chimiques ou biochimiques qui fragmentent, désagrègent, ameublissent et altèrent les roches ; VOG 1979, 124) ; *f 2* **altération** [f] (2) (Désagrégation d'une roche sous l'effet d'agents chimiques, physico-chimiques ou biochimiques ; VOG 1979, 7) ; *g* **Verwitterung** [f] (Durch atmosphärische und klimatische Einflüsse laufende Veränderung, Umwandlung und Zerstörung von Gesteinen und Mineralen).

weathering [n]**, stone** *envir.* ▶stone decay.

weatherproof [adj] [US]/**weather-proof** [adj] [UK] *constr.* ▶weather-resistant.

6960 weatherproofing [n] *constr.* (Measures taken to protect materials against exposure to the weather and to ensure durability; ▶weather resistance, ▶weather-resistant); *s* **protección** [f] **contra la intemperie** (Medidas tomadas para proteger los materiales ante las diferentes influencias de la intemperie [humedad, cambios bruscos de temperatura, viento con efecto mecánico de presión, aspiración y abrasión] para asegurar su función a largo plazo; ▶resistencia a la intemperie, ▶resistente a la intemperie); *f* **protection** [f] **contre les intempéries** (Mesures prises contre l'altération des matériaux due à l'action multiple des intempéries afin d'assurer une protection satisfaisante [contre l'humidité, les fortes variations de température, le vent et ses effets mécaniques de pression, d'aspiration et d'abrasion] et de préserver à long terme la fonction des ouvrages ; *terme spécifique dans la préservation des monuments historiques et des œuvres d'art* restauration, consolidation des matériaux, conservation de la pierre, conservation de surface [hydrofugation] ; ▶résistance aux intempéries, ▶résistant, ante aux intempéries) ; *g* **Schutz** [m] **gegen Witterungseinflüsse** (Maßnahmen und Vorkehrungen für einen ausreichenden Schutz auf Materialien gegenüber vielseitigen Einwirkungen der Witterung [Feuchtigkeit, starke Temperaturschwankungen, Wind mit seinen mechanischen Wirkungen von Winddruck, Windsog und Schliff], um deren Leistung und Funktion langfristig zu sichern; *UB bei Oberflächenschutz von Kulturdenkmalen* Oberflächensicherung [f], Konservierung [f] gegen Witterungseinflüsse; ▶Wetterfestigkeit, ▶witterungsbeständig).

weatherproofing layer [n] **of a roof** *constr.* ▶roof membrane.

6961 weather resistance [n] *constr.* (Property of materials which are unaffected by the deteriorating effects of weather, including rain, sun, frost, etc.; **in U.S., 'w. r.'**, rather than 'weather proofing', is a term used to protect manufacturer against a legal suit for damages by the user; ▶weather-resistant); *s* **resistencia** [f] **a la intemperie** (Característica de los materiales, p. ej. de un revestimiento de tejado, de resistir las inclemencias del tiempo sin deteriorarse; ▶resistente a la intemperie); *f* **résistance** [f] **aux intempéries** (Caractéristiques d'un matériau résistant aux influences des agents atmosphériques ; ▶résistant, ante aux

intempéries) ; *syn.* résistance [f] aux agents atmosphériques ; *g* **Wetterfestigkeit** [f] (Eigenschaft von Materialien, die gegenüber Witterungseinflüssen unempfindlich sind; ▶Schutz gegen Witterungseinflüsse, ▶witterungsbeständig).

6962 weather-resistant [adj] *constr.* (Descriptive term applied to surfaces of walls, roofs, pavements, wood, etc. exposed to the weather, which have powers of resistance to effects of weather; ▶nonrotting [US]/non-rotting [UK], ▶weather resistance); *syn.* weather proof [adj]; *s* **resistente a la intemperie** [loc] (Término descriptivo aplicado a los materiales expuestos a las inclemencias del tiempo; ▶no degradable, ▶resistencia a la intemperie); *f* **résistant, ante aux intempéries** [loc] (Caractéristique de certains matériaux comme la pierre, le bois, etc. capables de résister aux différentes conditions atmosphériques — vent, pluie, froid, chaleur, sécheresse, etc. ; ▶imputrescible, ▶résistance aux intempéries) ; *g* **witterungsbeständig** [adj] (So beschaffen, dass Materialien gegenüber Einwirkungen des Wetters unempfindlich sind; ▶verrottungsfest, ▶Wetterfestigkeit); *syn.* wetterbeständig [adj], wetterfest [adj].

weather side [n] *met.* ▶windward side.

6963 weed [n] (1) *agr. for. hort.* (From the cultivator's point of view, an unwanted ▶wild plant or spontaneous growth of a ▶cultivated plant 1 within a crop; ▶root spreeding weed); *s* **planta** [f] **advenediza** (▶Planta silvestre no deseada o problemática que crece en los cultivos en detrimento de las especies cultivadas; cf. DINA 1987; ▶mala hierba de raíz, ▶planta de cultivo); *syn.* mala hierba [f]; *f* **plante** [f] **adventice** (Plante herbacée ou par extension, une plante ligneuse qui, à l'endroit où elle se trouve, est indésirable. Le terme « adventice » est admis comme synonyme, bien que son sens botanique soit différent : il désigne une plante introduite accidentellement à l'insu de l'homme ; ▶mauvaise herbe racinaire, ▶plante de culture, ▶plante sauvage) ; *syn.* mauvaise herbe [f], adventice [f] ; *g* **Unkraut** [n] (…kräuter [pl]; aus der Sicht des Nutzers unerwünschte ▶Wildpflanze sowie spontan aufwachsende ▶Kulturpflanze in einem Nutzpflanzenbestand, die das Wachstum der Kulturpflanzen behindert und i. d. R. bei Produktionskulturen den Ertrag mindert; bei der Definition der Begriffe ‚Unkraut' und ‚Wildkraut' in städtischen Grünanlagen sowie auf und an Verkehrsflächen gibt es ideologische Meinungsverschiedenheiten: Wildkräuter sollen sich wegen ihrer Nahrungsgrundlage für zahlreiche tierische Organismen und wegen ihres nicht geringen Beitrages zur Verbesserung des kleinräumigen Stadtklimas [Staubbindung, Sauerstoffproduktion, Abgabe von Feuchtigkeit an die Luft] möglichst unbeeinträchtigt entwickeln, solange nicht Gründe der Verkehrssicherheit dagegen stehen. Der Terminus **Problemunkraut** bezeichnet eine krautige Pflanze, die in Kulturen trotz Bekämpfung immer wieder massenhaft vorkommt oder ein ▶Wurzelunkraut, das auf Grund seiner tief liegenden Rhizome, z. B. Schachtelhalm *[Equisetum arvense]*, schwer zu bekämpfen ist; ▶Rasenkräuter; *OB* unerwünschter Aufwuchs [DIN 18 919]).

6964 weed [vb] (2) *constr. hort.* (To remove unwanted or harmful plants; ▶root spreading weed); *s* **escardar** [vb] (Acción de quitar las malas hierbas manualmente; ▶mala hierba de raíz); *syn.* desyerbar [vb], desherbar [vb]; *f* **désherber** [vb] (Action d'enlever manuellement [binage, sarclage] les mauvaises herbes ; le **désherbage manuel sélectif** consiste à enlever certaines espèces adventices et à en favoriser d'autres afin d'éviter la levée des autres semences consécutive à la mise à du sol ; ▶mauvaise herbe racinaire) ; *g 1* **Unkraut jäten** [vb] (Manuelles Entfernen von Unkraut durch Ausreißen, Ausstechen, Schuffeln, Hacken, Ausgraben etc.); *g 2* **selektives Jäten** [n, o. Pl.] (Eine besondere Form des Jätens, das nur einzelne Unkräuter entfernt und dabei

W

die übrige Bodenoberfläche nicht lockert, um den Samenspeicher im Boden nicht zur Keimung zu aktivieren; ►Wurzelunkraut).

weed [n]**, lawn** *gard. plant.* *►wildflowers in the lawn.

6965 weed community [n] *phyt.* (Generic term covering ►arable weed community, ►pasture weeds vegetation, ►root crop weed community, ►segetal community, ►vegetation of rest areas; ►ruderal vegetation); *s* **comunidad [f] de malas hierbas** (Término genérico de comunidades que se dividen en dos grandes clases: *Secalinetea* [cereales] y *Chenopodietea* [huertos, etc.], la primera con 140 especies características y acompañantes, y los órdenes *Chenopodietalia, Onopordetalia* y *Paspalo-Heleochloetalia* con 390 especies en total; BB 1979, 478s; ►comunidad arvense de cultivos, ►comunidad arvense de los huertos y cultivos de verano, ►comunidad arvense de pastizales, ►vegetación de lugares de reposo, ►vegetación ruderal, ►vegetación segetal); *syn.* comunidad [f] advenediza (DB 1985); *f 1* **flore [f] adventice des cultures et des prairies** (Végétation des mauvaises herbes spontanées ou étrangères introduites accompagnant les lieux cultivés ou incultes ; flore d'adventices de friches, ►flore de culture sarclées, ►flore rudérale, ►groupement de plantes adventices des cultures, ►végétation de reposoir, ►végétation des prairies pacagées négligées, ►végétation ségétale) ; *syn.* flore [f] de mauvaises herbes ; *f 2* **groupement [m] anthropique** (Association phytosociologique résultant de l'activité humaine constituée des ►flores rudérales appartenant p. ex. à la classe de *Rudéreto-Secalietea* ; ►groupement rudéral) ; *g* **Unkrautflur [f]** (Pflanzengesellschaft, die aus der Sicht desjenigen, der eine Produktionskultur betreibt, vorwiegend aus Unkräutern besteht; OB zu ►Ackerunkrautgesellschaft, ►Segetalflur, ►Hackunkrautflur, ►Lägerflur, ►Ruderalflur, ►Weideunkrautflur und ggf. für Brachlandvegetation); *syn.* Wildkrautflur [f].

weed community [n]**, cereal** *phyt.* ►segetal community.

6966 weed control [n] *agr. for. hort.* (Method of maintenance carried out to restrict or prevent the spreading of weeds; ►biological pest control, ►integrated pest control, ►pest and weed control); *s* **control [m] de las malas hierbas** (Medida de mantenimiento para restringir o evitar el crecimiento de hierbas advenedizas, que se realiza por medios mecánicos, químicos y térmicos; ►control biológico, ►control integrado de plagas, ►control antiplaguicida); *syn.* lucha [f] contra las malas hierbas; *f* **désherbage [m]** (Méthode culturale [rotation, drainage, amendement calcique, désinfection par la chaleur etc.], enlèvement mécanique [sarclage], thermique ou chimique [produits herbicides sélectifs ou non] visant à limiter ou réduire l'extension des plantes adventices ou plus simplement de les éliminer ; LA 1981 ; ►lutte biologique, ►lutte intégrée, ►protection des végétaux) ; *g* **Unkrautbekämpfung [f]** (Pflegemaßnahme, die die Ausbreitung von Unkräutern/Wildkräutern einschränkt oder verhindert; es wird zwischen mechanischer, chemischer und thermischer **U.** unterschieden; ►biologischer Pflanzenschutz, ►integrierter Pflanzenschutz, ►Pflanzenschutz); *syn.* Wildkrautkontrolle [f], Unkrautkontrolle [f].

weed-free [adj] *agr. constr. for. hort.* ►weedless.

6967 weed growth [n] *agr. for. hort.* (Luxuriant growth of ►weeds 1/►wild plants in cultivated areas or public green spaces); *s* **crecimiento [m] de malas hierbas** (Fuerte crecimiento de ►plantas advenedizas o ►plantas silvestres en un cultivo o en zona verde); *f* **envahissement [m] par les adventices** (Implantation, développement et extension importante des ►plantes adventices/►plantes sauvages dans les cultures ou dans les espaces verts) ; *g* **Verkrautung [f]** (Üppiges Wachstum von ►Unkräutern/►Wildpflanzen in Kulturbeständen oder Grünflächen; von Unkraut überwuchert werden); *syn.* Überwucherung [f] von Unkraut, Verkrauten [n, o. Pl.], Verunkrautung [f].

6968 weed growth [n] **in bodies of water** *limn.* (Luxuriant growth of vegetation in eutrophic waters); *s* **crecimiento [m] fuerte de hidrófitos** (Desarrollo incontrolado de vegetación en cuerpos de agua causado por eutrofización); *f* **prolifération [f] des plantes aquatiques** (Développement incontrôlé des végétaux hydrophytes en milieu eutrophe) ; *syn.* invasion [f] par les plantes aquatiques ; *g* **Verkrautung [f] von Gewässern** (Üppiges Pflanzenwachstum in eutrophen Gewässern).

weeding [n] **on planted areas** [UK]**, picking up litter and** *constr.* ►cleaning up planted areas [US].

weedkiller [n] *agr. chem. envir. for. hort.* ►herbicide.

6969 weedless [adj] *agr. constr. for. hort.* (free from weeds); *syn.* weed-free [adj], without weeds [loc]; *s* **libre de malas hierbas [loc]**; *syn.* sin malas hierbas [loc]; *f* **dépourvu, ue de mauvaises herbes [loc]** (Caractéristique d'une surface cultivée ou plantée sur laquelle ne poussent aucunes plantes adventices) ; *g* **unkrautfrei [adj]** (So beschaffen, das kein Unkraut auf einer Kultur- oder bepflanzten Fläche wächst).

6970 weed tree [n] *constr. hort.* (Undesirable woody plant which has begun to grow adventitiously in a cultivated area; ►natural colonization by ►seed rain, ►spontaneous plants, ►weed 1); *s* **leñosa [f] adventicia** (Árbol o arbusto que crece espontáneamente en una plantación; ►colonización natural, ►colonización natural del «espacio vacío», ►nuevo vuelo, ►planta advenediza, ►vegetación espontánea); *f* **végétation [f] arbustive adventice** (Espèces ligneuses étrangères à la ►végétation spontanée, installées accidentellement sur des surfaces cultivées ; ►colonisation spontanée, ►flux de semences, ►plante adventice) ; *syn.* ligneux [mpl] adventices ; *g* **Fremdaufwuchs [m]** (*Fachsprachlich* ...wüchse [pl]; Wildgehölzsämling, der unerwünscht in Kulturflächen aufwächst; ►Anflug 2, ►spontane Besiedelung, ►Unkraut, ►Wildwuchs 2); *syn.* unerwünschter Gehölzaufwuchs [m], unerwünschter Gehölzsämling [m], wild aufgegangenes Gehölz [n].

weekend break [n] [UK] *plan. recr.* ►recreational weekend [US].

weekend cottage [n] [UK] *recr. urb.* ►weekend house [US].

6971 weekend house [n] [US] *recr. urb.* (Domicile on the periphery of a city, or further away therefrom, which is separate from the home and serves as a recreational retreat on a weekend, mainly used only by the owner and his friends or relatives); *syn.* weekend cottage [n] [UK]; *s* **casa [f] de fin de semana** (Domicilio en la periferia de la ciudad que se utiliza para retirarse durante los fines de semana; en Es. es más usual la casa de veraneo); *f* **résidence [f] de fin de semaine** (Habitation [résidence secondaire] située à la périphérie ou en dehors des agglomérations occupée de façon intermittente par le propriétaire, sa famille et ses proches) ; *syn.* résidence [f] de week-end, maison [f] de fin de semaine, habitation [f] de fin de semaine, habitation [f] de week-end, maison [f] de week-end ; *g* **Wochenendhaus [n]** (Ein nicht mit der Hauptwohnung in Verbindung stehendes Haus, das dem Freizeitwohnen auf einem Grundstück am Stadtrand oder außerhalb der Stadt am Wochenende dient und meist nur dem Eigentümer und seinem Bekannten- und Verwandtenkreis zugänglich ist).

6972 weekend house development [n] *recr. urb.* (Residential area comprising weekend and holiday homes); *s* **urbanización [f] de casas de fin de semana** (Zona residencial de casas de fin de semana o de veraneo); *f* **habitat [m] de fin de semaine** (Groupe d'habitations utilisées pendant les vacances et en fin de semaine situé dans un secteur de résidences secondaires ; *syn.* habitat [m] de week-end, ensemble [m] résidentiel de

W

week-end, lotissement [m] de vacances ; *g* **Wochenendhaus-siedlung [f]** (Komplex von Wochenend- und Ferienhäusern in einem Wochenendhausgebiet); *syn.* Ferienhaussiedlung [f].

weekend recreation [n] [UK] *plan. recr.* ►recreational weekend [US].

6973 weekend traffic [n] *plan. trans.* (Traffic which occurs on a weekend; this term usually refers to Sunday afternoon traffic of vehicles returning between the hours of 3pm and 7pm after a weekend outing, outside the main holiday period, taking into account the fluctuations in the actual numbers of vehicles during the course of the year; ►recreational traffic); *s* **tráfico [m] de fin de semana** (Tránsito durante los fines de semana. En D. se entiende sobre todo el de retorno a la ciudad los domingos por la tarde fuera de la época principal de vacaciones; ►tráfico recrea-cional); *f* **trafic [m] de fin de semaine** (Circulation durant la fin de la semaine ; terme souvent utilisé pour décrire l'abondance du ►trafic touristique de départ le vendredi soir ou de retour le dimanche soir entre 15 et 20 heures) ; *g* **Wochenendverkehr [m]** (Verkehr am Wochenende; speziell wird hierunter außerhalb der Hauptreisezeit der sonntags nachmittags rückflutende Verkehr in der Zeit von 15-19 Uhr verstanden — unter Berücksichtigung des Schwankens der absoluten Menge im Laufe des Jahres; ►Erho-lungsverkehr).

6974 weep hole [n] *constr.* (Small opening left in lower levels of retaining walls, aprons, linings or foundations to permit drainage, thereby preventing the build-up of pressure from ►seepage water or groundwater); *syn.* drain hole [n]; *s* **mechinal [m]** (En los muros huecos o revestidos, agujeros que se practican en las juntas de mortero a intervalos regulares para permitir el drenaje de la humedad condensada. En los muros de sosteni-miento se llama **m.** a las aberturas practicadas, corrientemente, al tresbolillo para dar salida al agua del terreno sostenido; DACO 1988; ►agua de percolación); *syn.* agujero [m] de drenaje, ori-ficio [m] de drenaje, ojada [f] [CO]; *f* **barbacane [f]** (Ouverture étroite dans un mur de soutènement aménagée pour faciliter l'écoulement des eaux ; DTB 1985; ►eau de gravité) ; *syn.* chantepleure [f] (DTB 1985) ; *g* **Entwässerungsöffnung [f] im unteren Stützmauerbereich (≠)** (►Sickerwasser 2); *syn.* Mauer-öffnung [f] zur Entwässerung, Wasserabzugsloch [n].

weeping standard rose [n] [UK] *hort.* ►weeping tree rose [US].

6975 weeping tree rose [n] [US] *hort.* (►Standard rose with a stem roughly 1.40m high, on which usually small-flower-ing, climbing roses are grafted); *syn.* weeping standard rose [n] [UK]; *s* **rosal [m] llorón** (►Rosal de pie alto de aprox. 1,40 m de altura en el que generalmente se injertan rosales trepadores de flores pequeñas); *f* **rosier [m] pleureur** (►Rosier tige dont la hauteur de tige mesurée de la racine supérieure jusqu'à la greffe inférieure est supérieure à 140 cm sur lequel est en général greffé un rosier grimpant à petites fleurs) ; *syn.* rosier [m] retombant ; *g* **Trauerrose [f]** (1,40 m hoher ►Rosenhochstamm, auf den meist kleinblütige Kletterrosen gepfropft werden); *syn.* Kaska-denrose [f], Hängerose [f].

6976 weigh bill [n] *contr.* (Written and signed list of weighed load of loose material being transported on a vehicle; ►bill of lading [US]/delivery note [UK]); *s* **certificado [m] de peso** (Formulario en el que se anota el peso de materiales a granel; ►recibo de entrega); *f* **bon [m] de pesage** (Preuve de la livraison de matériaux en général en vrac, pesés sur la bascule ; ►borde-reau de livraison) ; *g* **Wiegekarte [f]** (Liefernachweis über die Tonnage von Schüttgütern; ►Lieferschein); *syn.* Wiegeschein [m].

6977 welded joint [n] *constr.* (Watertight joint of, e.g. ►roof protection membrane; ►hot air welding, ►joint sealing,

►welded joint); *s* **junta [f] soldada** (Línea de unión de dos piezas por medio de soldadura, p. ej. de ►láminas antirraíces; ►junta de solape, ►sellado de juntas, ►termosoldadura); *f* **lèvre [f] de soudure** (Bande étroite de liaisonnement de deux parties d'un produit soudé, p. ex. à la lisière du lé d'un ►film [de pro-tection] antiracines ; ►garnissage des lèvres de soudure, ►jonc-tion des lés, ►thermosoudure) ; *g* **Schweißnaht [f]** (Stelle, an der zwei Stücke durch Schweißen miteinander verbunden werden, z. B. ►Wurzelschutzfolien; ►Nahtstelle, ►Nahtversiegelung, ►Verschweißen der Fügestellen mit Heißluft).

6978 welded lap joint [n] *constr.* (Result of joining sheets of PVC liner, ►waterproofing membranes or ►root protection membranes; ►welded joint); *s* **junta [f] de solape** (Línea de encuentro en soldaduras de capas de cloruro de polivinilo [CPV] de ►láminas antirraíces o de ►membranas de impermeabili-zación; ►junta soldada); *f* **jonction [f] des lés** (Ligne de soudure constituée par le jointoiement des lés de la ►membrane d'étan-chéité ou du ►film [de protection] antiracines ; ►lèvre de soudure) ; *g* **Nahtstelle [f]** (Stelle, an der z. B. PVC-Folien, ►Dichtungsfolien, ►Wurzelschutzfolien zusammengefügt sind; ►Schweißnaht); *syn.* Nahtverbindung [f] (DAB 1992 [1], 110).

welding [n] *constr.* ►hot air welding.

well [n] *arch.* ►basement light well [US]/light well [UK]; *constr.* ►drywell sump [US]/soakaway [UK]; *min. wat'man.* ►tap site; *constr.* ►tree well

well [n] [UK], **light** *arch.* ►basement light well [US].

well [n] [UK], **precast concrete tree** *constr.* ►precast concrete tree vault [US].

well [n] [UK], **window** *arch.* ►basement light well [US].

6979 well-drained soil [n] *constr. pedol. syn.* permeable soil [n], drained granular material [n] [also US]; *s* **suelo [m] bien drenado** *syn.* suelo [m] permeable; *f* **sol [m] perméable** *syn.* sol [m] bien drainé, sol [m] de bon drainage (naturel) (LA 1981, 424) ; *g* **wasserdurchlässiger Boden [m]** *syn.* Boden [m] mit gutem Wasserabzug.

6980 well-graded [adj] *constr.* (Descriptive term applied to a mixture of soil particle sizes for its adequate structural and loadbearing capacity; ►graded sediment group); *s* **bien gra-duado [adj]** (Denominación de suelos y mezclas de áridos que están compuestos de granos de diferentes tamaños; ►grupo granulométrico); *f* **de granulométrie étalée [loc]** (Terme dési-gnant les sols ou les matériaux de remblaiement dont la granulo-métrie présente plusieurs classes de granulation, les grains du granulat s'étendent sur plusieurs tamis successifs ; DIR 1977, 62 ; ►fraction granulométrique) ; *syn.* à granulométrie étalée [loc] ; *g* **kornabgestuft [pp/adj]** (Bodenphysikalische Bezeich-nung für Böden und Schüttgüter, die aus mehreren ►Korn-gruppen zusammengesetzt sind und eine entsprechende Kör-nungskurve ergeben).

well grate [n] *constr.* ►window well grate [US]/light well grid [UK].

well grid [n] [UK], **light** *constr.* ►window well grate [US]/light well grid [UK].

6981 well-provided recreation facility [n] *recr.* (Well-provisioned recreation installation and equipment); *s* **instalación [f] de recreo intensivo** (Equipamiento recreacional de gran variedad y de gran calidad); *f* **équipement [m] de loisirs lourd** (Équipements ayant une incidence sur le milieu naturel, p. ex. équipements sportifs et annexes, plans d'eau artificiels, centre équestres, bases de plein-air, station de ski, golf, etc. ; *opp.* équipement de loisirs légers) ; *syn.* aménagement [m] de loisirs lourd ; *g* **Erholungseinrichtung [f] mit hohem Ausstattungs-**

W

grad (Anlage mit einer großen freizeitrelevanten Infrastrukturvielfalt und einem hohen Ausstattungsniveau).

western hemisphere [n] *conserv. pol.* ►Convention on nature protection and wild life preservation in the western hemisphere.

6982 west-facing slope [n] *geo. met.* *s* **ladera** [f] **oeste** (Ladera de un monte orientada al oeste); *f* **versant** [m] **ouest** (Versant d'une montagne exposé à l'ouest) ; *g* **Westhang** [m] (Hang eines Berges, der nach Westen exponiert ist).

wet alluvial woodland [n] *phyt.* ►riparian woodland.

6983 wet gravel pit [n] *min.* (Extraction site for gravel in groundwater or running water; ►dredging 1, ►flooded gravel pit, ►gravel dredging); *s* **gravera** [f] **subacuática** (Lugar de ►extracción de grava por dragado en agua subterránea o en un curso de agua; ►dragado subacuático, ►laguna artificial); *f* **gravière** [f] **en nappe** (**1.** Lieu d'►extraction de matériaux alluvionnaires en nappe, dans le lit mineur des cours d'eau ou en plan d'eau ; ►étang de fouille ; **2.** par extension la gravière est aussi le plan d'eau d'origine artificielle créé par extraction de granulats et alimenté essentiellement par la nappe souterraine ; ►dragage de gravier) ; *syn.* gravière [f] en eau ; *g* **Kiesentnahmestelle** [f] (**2**) (Abbaustelle für Kies im Grundwasserbereich oder im Fließgewässer; ►Baggersee, ►Nassauskiesung, ►Nassbaggerung); *syn.* Kiesgrube [f] im Grundwasserbereich.

wet gravel workings [npl] [UK] *min.* ►gravel dredging.

6984 wet-heath vegetation [n] *phyt.* (VIR 1982, 338; ►dwarf-shrub plant community on waterlogged, minerotrophic peatland [pH < 4.0]; e.g. class of bog and wet-heath *[Oxycocco-Sphagnetea]*); *s* **landa** [f] **pantanosa** (►Matorral bajo que crece sobre suelos de tipo anmoor o podsoles gleyizados ácidos [pH < 4,0], ricos en turba, con la capa freática cercana a la superficie y pertenece a la clase de las comunidades de turbera alta ricas en matas *[Oxycocco-Sphagnetea]*); *f* **lande** [f] **marécageuse** (Végétation des landes à éricacées [lande rase dépassant rarement 30 cm] sur un sol tourbeux très acide à engorgement permanent [podzol à gley superficiel], appartenant à la classe des *Oxycocco-Sphagnetea* au sein de l'alliance de l'*Ericion tetralicis* ; lorsque le sol reste fortement minéral on parle plutôt d'une lande humide appartenant à l'alliance de l'*Ulicion nani* ; lande d'arbrisseaux nains) ; *syn.* lande [f] hygrophile, lande [f] tourbeuse, lande [f] humide ; *g* **Sumpfheide** [f] (►Zwergstrauchgesellschaft auf baumarmen, sauren, torfreichen Standorten [pH < 4,0] mit stagnierendem Grundwasser, z. B. auf Anmoor- und Gleypodsolböden; Klasse der zwergstrauchreichen Hochmoor-Torfgesellschaften *[Oxycocco-Sphagnetea]*); *syn.* Feuchtheide [f].

6985 wetland [n] *pedol.* (Lands which are transitional between terrestrial and aquatic systems where the water table is usually at or near the surface, or the land is covered at least periodically by shallow water; CWD 1979; ►impeded water, ►Ramsar Convention, ►soil affected by impeded water); *s* **zona** [f] **húmeda** (Zonas de transición entre la tierra y las aguas continentales o marinas determinadas por la presencia de agua o de un nivel freático próximo a la superficie. Según el proyecto 5° del programa MAB se dividen en ríos, lagos y embalses, terrenos húmedos y zonas costeras; cf. DINA 1987; ►agua estancada, ►Convención de Ramsar, ►suelo de agua estancada); *syn.* humedal [m]; *f* **zone** [f] **humide** (Espaces exploités ou non, constamment ou temporairement gorgés d'eau douce, salée ou saumâtre de façon permanente ou temporaire ; la végétation quand elle existe, y est dominée par des plantes hygrophiles pendant au moins une partie de l'année ; cf. art. 1 loi n° 92-3 sur l'eau du 3 janvier 1992 ; ►Convention de Ramsar, ►nappe perchée, ►sol à nappe temporaire perchée) ; *syn.* milieu [m] humide ; *g* **Feuchtgebiet** [n] (Flächen mit subhydrischen Böden

[►Unterwasserboden] und Mooren, ►Stauwasserböden mit unterschiedlichem Entwässerungszustand, Bruchwälder, Seenlandschaften, Auenlandschaften, Feuchtheiden, Riede, etc., die mit feuchtigkeitsliebenden [hygrophilen] Pflanzen bestockt sind; das größte deutsche Feuchtgebiet ist das Wattenmeer; **F.e** internationaler Bedeutung sind nach der ►Ramsar-Konvention besonders geschützt; ►Staunässe).

6986 wetland boardwalk [n] [US] *recr. zool.* (Slightly raised wooden walkway built through wetlands for bird observation and identification); *syn.* observation boardwalk [n] [UK]; *s* **pasarela** [f] **de observación** (En humedales, camino elevado para permitir a naturistas, ornitólogos/as, etc. acercarse a observar a la fauna acuática); *f* **passerelle** [f] **pour la découverte de la nature** (Passage en bois légèrement surélevé construit dans les milieux littoraux ou dans les zones humides destiné à l'information du public ou aux observations scientifiques) ; *g* **Beobachtungssteg** [m] (Leicht erhöhter Holzsteg an Gewässern oder in Feuchtgebieten für naturkundliche Erkundungen).

6987 wetland habitat [n] *ecol.* (Grouping of ►biocenoses, dependent upon the ecological conditions of a ►wetland; ►habitat 2, ►mesic, ►mesic site); *syn.* mesic habitat [n] (TEE 1980, 72); *s* **biótopo** [m] **húmedo** (Hábitat de ►biocenosis que dependen de las condiciones ecológicas de una ►zona húmeda; ►biótopo, ►mesófilo/a, ►ubicación mesofítica); *f* **biotope** [m] **humide** (Aire de répartition de ►biocénoses liées aux facteurs caractérisant une ►zone humide ; ►biotope, ►mésophile, ►station humide) ; *syn.* biotope [m] mésophile, biotope [m] mésohygrophile ; *g* **Feuchtbiotop** [m, *auch* n] (Lebensraum von ►Biozönosen, die an Bedingungen eines ►Feuchtgebietes gebunden sind; ►Biotop, ►mesophil, ►feuchter Standort).

6988 wetland [n] **of international significance** *conserv. nat.res.* (Areas of marsh, fen, peatland or water, whether natural or artificial, permanent or temporary, with water that is stagnant or flowing, fresh, brackish or salt, including areas of marine water the depth of which at low tide does not exceed six meters [US]/metres [UK]; habitat with a uniquely-broad variety of plant and animal species; cf. art. 1 ►Ramsar Convention 1971; *s* **zona** [f] **húmeda de importancia internacional** (Conjunto de marismas, pantanos, turberas o aguas rasas, naturales o artificiales, permanentes o temporales, de aguas remansadas o corrientes, dulces, salobres o salinas, con inclusión de las aguas marinas cuya profundidad en marea baja no rebase los seis metros; cf. ►Convención de Ramsar, sobre la conservación de zonas húmedas y aves acuáticas, art. 1); *syn.* humedal [m] de importancia internacional; *f* **zone** [f] **humide d'importance internationale** (Étendues de marais, de tourbières ou d'eaux naturelles ou artificielles, permanentes ou temporaires, où l'eau est statique ou courante, douce, saumâtre ou salée, y compris des étendues d'eau marine dont la profondeur à marée basse n'excède pas six mètres pour lesquelles les fonctions écologiques fondamentales en tant que régulateur des eaux et qu'habitats d'une flore et d'une faune caractéristiques sont protégées ; cf. art. 1 ►Convention de Ramsar 1971) ; *syn.* zone [f] Ramsar ; *g* **Feuchtgebiet** [n] **internationaler Bedeutung** (Feuchtwiesen, Moor- und Sumpfgebiete oder Gewässer, die natürlich oder künstlich, dauernd oder zeitweilig, stehend oder fließend und durch Süß-, Brack- oder Salzwasser gekennzeichnet sind. Dazu gehören auch Meeresgebiete, die eine Tiefe von sechs Metern bei Niedrigwasser nicht übersteigen; Art. 1 ►Ramsar-Konvention 1971).

wetland scrub [n] *phyt.* ►hygrophilic scrub.

wetland septic system [n] [US] *envir.* ►wastewater treatment wetland.

6989 wetland wastewater treatment [n] *envir.* (TGG 1984, 150; a near-natural method of purifying domestic wastewater by horizontal permeation through a settling pond, planted with rushes and sedges where organic and, to a certain extent, chemical impurities are retained in the root area or are decomposed by soil bacteria. The water is then purified to enter a watercourse or lake; ▶wastewater treatment wetland); *s* **tratamiento** [m] **de aguas residuales por filtros verdes** (Técnica de tratamiento de aguas residuales que aprovecha la capacidad de las raíces de algunas plantas, como p. ej. la caña o la enea, de retener la carga orgánica que es descompuesta con ayuda de bacterias del suelo; ▶filtro verde); *f* **lagunage** [m] **par roselière** (Procédé d'épuration naturelle biologique utilisé en milieu rural consistant à faire écouler les eaux usées dans des bassins peu profonds plantés de végétaux épurateurs tels que les Roseaux *[Phragmites]*, les Laîches *[Carex]* ou les Joncs *[Juncus]* sur un support perméable ; les substances organiques et chimiques sont retenues par le système racinaire des plantes aquatiques et dégradées par les bactéries aérobies du sol, les effluents ainsi traités étant ensuite rejetés directement dans le milieu naturel ; ▶lagunage à macrophytes) ; *g* **Wurzelraumentsorgung** [f]. (Naturnahe Klärtechnik von [Haushalts]abwässern, bei der das Schmutzwasser den Wurzelraum von ausgesuchten Pflanzen, z. B. Schilf *[Phragmites]*, Sumpfsegge *[Carex acutiformis]* oder Binse *[Juncus]*, über eine längere Entfernung in einem durchlässigen Bodenkörper horizontal durchströmt [Hauptreinigungsstufe]. Dabei werden organische und teilweise auch chemische Verunreinigungen im Wurzelbereich zurückgehalten resp. von Bodenbakterien abgebaut, so dass das Wasser danach in ein Gewässer eingeleitet werden kann; ▶Pflanzenkläranlage); *syn.* Wurzelraumverfahren [n].

wetland wastewater treatment system [n] *envir.*
▶wastewater treatment wetland.

6990 wet meadow [n] *agr. phyt.* (TEE 1980, 335; grassland for grazing or hay production on slightly wet to very wet soils with mostly poor aeration: communities; e.g. *Calthion, Molinion, Rumici-Alopecuretum,* etc.; ▶damp meadow, ▶dry meadow); *s* **prado** [m] **higrófilo** (Prado de diente o de siega sobre suelos húmedos o encharcados donde predominan comunidades de *Molinion, Calthion, Rumici-Alopecuretum, Lolio-Cynosuretum lotetosum.* En comparación con el ▶prado húmedo, está encharcado incluso en verano; *opp.* ▶pastizal seco); *syn.* prado [m] muy húmedo; *f* **prairie** [f] **hygrophile** (Prairie naturelle marécageuse de fauche ou pacagée de faible productivité, à engorgement fréquent, rattachée aux alliances du *Molinion, Calthion* et aux alliances du *Bromion* et de l'*Agropyron-rumicion* lorsqu'elle subit un assèchement temporel estival ; *opp.* ▶pelouse aride ; ▶prairie mésophile) ; *syn.* prairie [f] [permanente] inondable, prairie [f] marécageuse ; *g* **Nasswiese** [f] (Grünlandpflanzengesellschaft auf mäßig nassen bis nassen, meist schlecht durchlüfteten Böden, die im Vergleich zur ▶Feuchtwiese sogar im Sommer meist vernässt sind; z. B. Gesellschaften der Pfeifengraswiesen *[Molinion]*, Sumpfdotterblumenwiesen *[Calthion]*, Knickfuchsschwanzwiesen *[Rumici-Alopecuretum]*, Weidelgrasweiden *[Lolio-Cynosuretum lotetosum]*; *opp.* ▶Trockenrasen); *syn. agr.* Nassweide [f].

6991 wetness indicator plant [n] *phyt.* (Plant living with optimum growing conditions in slightly wet or wet soils; ▶indicator plant, ▶moisture indicator plant; *opp.* ▶aridity indicator plant); *s* **planta** [f] **indicadora de encharcamiento** (Especie vegetal cuyas condiciones ecológicas óptimas se encuentran en suelos encharcados, por lo que se utiliza de ▶planta indicadora; ▶especie indicadora de humedad; *opp.* ▶planta indicadora de sequía); *syn.* indicadora [f] de humedad alta; *f* **plante** [f] **indicatrice de forte humidité** (Espèce végétale qui possède son opti-

mum écologique sur un sol à saturation en eau temporaire ou permanente ; ▶espèce indicatrice d'humidité, ▶plante indicatrice ; *opp.* ▶plante indicatrice de sécheresse) ; *g* **Nässezeiger** [m] (Bodenpflanze, die auf mäßig nassen bis nassen Böden ihr ökologisches Optimum resp. den geringsten Konkurrenzdruck hat; Feuchtezahlen als Zeigerwert: ▶Feuchtigkeitszeiger; ▶Zeigerpflanze; *opp.* ▶Trockenheitszeiger).

6992 wetwood [n] *arb. for.* (Dark brown-colo[u]red, decaying ▶heartwood in the middle of trunks and main scaffolds caused by the infiltration of bacteria; ▶area of rot, ▶decayed area); *s* **duramen** [m] **podrido** (Madera muerta del centro del tronco [▶duramen] en estado de descomposición producida por ▶caries húmeda; ▶zona podrida); *syn.* corazón [m] (de tronco) podrido; *f* **cœur** [m] **mouillé** (▶Bois parfait ayant un taux d'humidité anormalement élevé et une apparence vitreuse qui se développant dans l'arbre vivant et non pas au cours d'un séjour dans l'eau — phénomène apparaissant en particulier chez les Peupliers, les Saules et les Sapins et qui au stade ultime se termine par le ▶pourrissement ; DFM 1975, 63 ; ▶point de pourriture [sur le tronc], ▶pourriture) ; *syn.* bois [m] parfait pourri ; *g* **Nasskern** [m] (Im mittleren Bereich von Stamm und Starkästen durch eingedrungene Bakterien dunkelbraun verfärbtes und in Zersetzung befindliches ▶Kernholz — tritt besonders bei Pappeln, Weiden und Tannen auf und führt im Spätstadium zur ▶Morschung, insbesondere dann, wenn der **N.** austrocknet; ▶Faulstelle).

6993 wheel-chair-accessible [adj] *plan. sociol.* (Descriptive term for design and construction of buildings and sidewalks [US]/pavements [UK] to meet the needs of [motorized] wheel chair and electric cart users as well as of users with any other type of vehicle for the handicapped; accessability is achieved by proper slope and spatial requirements; ▶developed for the handicapped/disabled); *s* **accesible con silla de ruedas** [loc] (Término descriptivo para el diseño y la construcción de edificios y aceras [Es]/veredas [AL] para cumplir las necesidades de personas que transitan en sillas de ruedas [motorizadas] así como permitir el uso a cualquier otro vehículo utilizado por personas con movilidad reducida. La accesibilidad se consigue planificando las pendientes y los accesos adecuados, espacios suficientemente amplios y utilizando materiales de pavimentación que no causen vibraciones; ▶accesible para discapacitados/as); *f* **accessible aux fauteuils roulants** [loc] (Conçu de telle manière à répondre aux besoins des personnes handicapées circulant en fauteuil roulant, ▶adapté, ée aux personnes handicapées) ; *syn.* conçu, ue en fonction des fauteuils roulants [loc], accessible aux personnes circulant en fauteuil roulant [loc] ; *g* **rollstuhlgerecht** [adj] (So gestaltet und gebaut, dass es den Bedürfnissen von Rollstuhlfahrern entspricht; z. B. müssen Wege leicht und erschütterungsarm befahrbar sein, die Verwendung von Rasengittersteinen und Betonpflaster mit breiten Fugen ist unzulässig; ▶behindertengerecht).

6994 whip [n] *for. hort.* (Very young tree grown from a seedling, transplanted once, 50-100cm tall); *syn.* transplant [n] (2) (SAF 1983, 281), liner [n]; *s* **renuevo** [m] (Generalmente una planta joven transplantada sólo una vez); *syn.* retoño [m], plántula [f] de renuevo; *f* **plant** [m] **repiqué** (Jeune semis après qu'on l'a extrait et transplanté dans la pépinière ; DFM 1975, 210) ; *syn.* fouet [m] ; *g* **Lode** [f] *obs.* (Meist einmal verpflanzte Jungpflanze von Baumarten. Je nach Höhe wurde zwischen Halblode [< 50 cm] und Starklode [< 100 cm] unterschieden); *syn. o. V.* Lohde [f].

6995 white dune [n] *geo. phyt.* (Steep-crested dune lying between ▶embryo dune and ▶gray dune [US]/grey dune [UK], stabilized by invasion or introduction of a grass species; e.g. in Europe Marram Grass *[Ammophila arenaria]*; in U.S., Beach or

Dune Grass *[Ammophila breviligulata]*); *s* **duna [f] blanca** (Duna que se encuentra entre la ▶duna embrionaria y la ▶duna gris, en Europa cubierta frecuentemente por el barrón *[Ammophila arenaria]*); *f* **dune [f] blanche** (Rehaussement de dune à sommet escarpé sur lequel le sable de la plage se trouve projeté chaque fois qu'il y a du vent, située entre la ▶dune embryonnaire et la ▶dune grise, en partie fixée par un couvert d'Oyat *[Ammophila arenaria]*) ; *syn.* dune [f] mobile, dune [f] vive, dune [f] claire ; *g* **Weißdüne [f]** (Zwischen ▶Primärdüne und ▶Graudüne gelegene steilkuppige Düne, die oft durch Strandhaferbewuchs *[Ammophila arenaria]* geprägt ist).

white land [n] [UK] *obs. leg. urb.* ▶open land [US], ▶unzoned land ripe for development [n] [US].

white water [n] *geo.* ▶torrent.

6996 white-water boating [n] [US] *recr.* (Watersport with foldboats, kayaks and [Canadian] canoes in river rapids/torrents; in U.S., there is also a popular watersport of 'white-water rafting' for running of rapids by a group of persons); *syn.* white-water canoeing [n] [UK]; *s* **piragüismo [m] de aguas bravas** (Deporte acuático con piraguas y kayaks en rápidos de arroyos o torrentes); *f* **canoë-kayak [m]** (Sport nautique léger pratiqué en eaux vives au moyen de canoë, de kayak ; la descente des rapides sur un bateau à fond plat [rafting] gagne en particulier aux USA en popularité) ; *syn.* canoëisme [m] ; *g* **Wildwasserkanusport [m, o. Pl.]** (Auf Wildwasser ausgetragener Kanusport für Kajaks und Kanadier; in den USA wird die Freizeitbeschäftigung „whitewater rafting", bei dem eine Gruppe von Leuten mit einem großen flachen Boot Wildwasserfahrten durchführt, immer beliebter); *syn.* Wildwassersport [m, o. Pl.].

white-water canoeing [n] [UK] *recr.* ▶white-water boating [US].

6997 wicker fence [n] *constr.* (*In bioengineering a type of* ▶wattlework a low line of thin willow branches [osier twigs] capable of rooting and woven between stakes; ▶interwoven wood fence, ▶wickerfence construction); *s* **valla [f] de trenzado** (*Bioingeniería* forma del ▶varaseto de ramas en el que se entrelazan ramillas vegetativas de sauces alrededor de postes; ▶construcción de valla trenzada, ▶valla de enrejado de madera); *f* **palissade [f] tressée de branchages à rejets** (*Génie biologique* forme de treillis constitué de pieux entre lesquels sont tressés des branchages de saules ; ▶claie de branches à rejets, ▶clayonnage, ▶clôture à lamelles tressées, ▶treillis de branches [de saules] à rejets) ; *syn.* grille [f] de branchages à rejets [CH] ; *g* **Flechtzaun [m]** (*Ingenieurbiologie* Form des ▶Flechtwerks, bei dem um Pfähle ausschlagfähige Weidenruten geflochten werden; ▶Flechtzaunbau, ▶Holzflechtzaun).

6998 wicker fence construction [n] *constr.* (1. Erosion barrier provided for slope protection. 2. Barrier provided for wind-blown snow deflection); *s* **construcción [f] de valla trenzada** (*Bioingeniería* construcción para retener la tierra en taludes); *f* **clayonnage [m]** (Préparation et pose d'une claie destinée à retenir les terres sur une pente) ; *g* **Flechtzaunbau [m]** (*Ingenieurbiologie* Bau eines Flechtzaunes zur Böschungssicherung).

wickerwork [n]/**wicker-work** [n] [also UK] *constr.* ▶wattlework/wattle-work [also UK].

wickerwork [n] [US]**, diagonal** *constr.* ▶diagonal wattlework [US]/diagonal wattle-work [UK].

6999 wide ecological range [n] *ecol.* (e.g. wide ranging species; TEE 1980, 51; ▶ecological tolerance; *opp.* ▶narrow ecological range); *s* **valencia [f] ecológica amplia** (Tolerancia ambiental amplia de una especie, por lo que su área de distribución es superior a la media; ▶tolerancia ecológica; *opp.* ▶valencia ecológica estrecha); *syn.* amplitud [f] ecológica amplia; *f* **valence [f] écologique large** (Espèce à large ▶plasti-

cité écologique ; *opp.* ▶faible valence écologique) ; *syn.* amplitude [f] écologique large ; *g* **breite ökologische Amplitude [f]** (Großer Bereich, in dem eine Art zwischen zwei Werten eines Standort-/Umweltfaktors auf Grund ihrer ▶ökologischen Potenz leben kann; *opp.* ▶enge ökologische Amplitude).

7000 widely-spaced plant [n] *hort.* (Tree or shrub grown in a nursery row for specimen use; ▶specimen tree/shrub, ▶spacing 1); *syn.* spaced [pp] extra wide; *s* **planta de gran espaciamiento [loc]** (Criterio de calidad para árbol o arbusto cultivado en vivero que está destinado a plantarse en solitario; ▶espaciamiento, ▶espécimen arbóreo/arbustivo); *f* **plante à grand écartement [loc]** (Critère de qualité d'un produit de pépinière caractérisant les plantes d'ornement, à grand développement, cultivées à distance avec contretransplantation régulière et utilisées comme sujets isolés ; ▶espacement, ▶spécimen 2) ; *g* **Pflanze aus extra weitem Stand [loc]** (Qualitätsbeschreibung einer Baumschulware, die für Solitärpflanzen geeignet ist; ▶Abstand, ▶Solitärgehölz 2).

wide-ranging [adj] *phyt. zool.* ▶ubiquitous.

wide-scale erosion [n] *geo.* ▶extensive erosion.

7001 wide spacing [n] **in a housing development** *urb.* (Spreading of fewer houses per hectare/acre to have more green area between them for lower housing density; ▶low-density building development, ▶low-density housing); *syn.* reduction [n] in housing density; *s* **descongestión [f] urbana** (Reducción de la densidad de construcción en un barrio urbano para mejorar las condiciones de vida; ▶edificación de baja densidad, ▶edificación residencial de baja densidad); *f* **aération [f] de l'habitat** (Diminution de la densité de l'habitat en vue de l'amélioration de la qualité de la vie dans un quartier urbain ; ▶habitat dispersé, ▶urbanisation aérée) ; *g* **Auflockerung [f] einer Bebauung** (Realisierung von [zusätzlichen] Freiflächen zwischen Gebäuden einer Bebauung zur Verbesserung der Wohnqualität eines Stadtteiles; ▶lockere Bebauung, ▶aufgelockerte Wohnbebauung).

width [n] *constr.* ▶clearance width; *leg. surv. urb.* ▶plot width.

width [n]**, lot** *leg. surv. urb.* ▶plot width.

7002 width [n] **of a flight of steps** *constr.* *s* **anchura [f] de escalera** *syn.* ancho [f] de escalera; *f* **emmarchement [m]** (Largeur des marches d'escalier et correspond à la largeur du passage) ; *g* **Treppenbreite [f]** (Breite einer Treppenanlage).

wild [n] [UK]**, in the** *hunt.* ▶in an open hunting area.

7003 wild and scenic river [n] [US] *conserv. leg. nat'res.* (Federal legislation authorizes and directs land-administering government agencies to cooperate in establishing stretches of rivers, which are comparatively undisturbed by man, as Wild and Scenic Rivers for conservation and limited recreation; some states are doing likewise); *s* **tramo [m] de río protegido (≠)** (En EE.UU. la legislación federal permite y orienta a las agencias federales de gestión territorial de cooperar para establecer zonas de ríos poco influenciadas antropógenamente y declararlas como *«wild and scenic rivers»* [literalmente: ríos salvajes y escénicos] para su preservación y usos recreativos limitados; algunos estados también están aplicando este instrumento de protección. En Es. no existe esta categoría de protección); *f 1* **plaine [f] alluviale classée (≠)** (1. US., Catégorie de protection des plaines alluviales gérées par l'administration fédérale des forêts [U.S. Forest Service] et l'administration centrale du cadastre [Bureau of Land Management] instaurée en vue de la conservation d'un patrimoine rare en raison de la richesse et de la rareté de ces milieux et tentant à limiter certaines activités de loisirs. 2. F., les plaines alluviales ou les rivières sauvages sont protégées dans le cadre [a] de la Convention de Ramsar par le classement des zones humides de vallées alluviales ; ces mesures de protection

W

sont les mêmes que celles qui concernent les milieux naturels ; environ 58 % des superficies des zones humides « d'importance majeure » sont protégées à un titre ou à un autre [réserves naturelles nationales et régionales] ; **[b]** dans le cadre de la loi sur l'eau de 1992 et des contrats de rivière, p. ex. **rivière réservée** ; **[c]** dans le cadre du réseau européen Natura 2000, les zones de protections spéciales [ZPS] et les propositions de sites d'intérêt communautaire couvrant en effet 40 % des zones humides concernées ; **[d]** dans le cadre de la protection des sites, p. ex. **rivière classée**) ; *f 2* zone **[f] alluviale d'importance nationale [CH]** (Protection des zones alluviales par l'ordonnance fédérale du 15 novembre 1992 prévoyant l'inscription sur l'inventaire fédéral des zones alluviales d'importance nationale et visant à conserver intactes les principales zones alluviales de Suisse) ; *g* **Auenschutzgebiet [n] (≠)** (**1.** U.S., eine durch amerikanische Bundesgesetzgebung unter Schutz gestellte Auenlandschaft, die von der Bundesforstverwaltung *[U.S. Forest Service]* und Bundesliegenschaftsbehörde *[Bureau of Land Management]* verwaltet wird, um den ursprünglichen Naturcharakter und die einzigartige Schönheit einer solchen Landschaft nachhaltig zu sichern und um sie der Öffentlichkeit in beschränktem Maße zur Erholung bereit zu stellen. **2. In D.** gibt es solch eine spezifizierte Schutzkategorie nicht. Derartige Gebiete werden den [Natur]schutzgebieten zugeordnet oder können als geschützte Landschaftsbestandteile ausgewiesen werden; cf. § 12 BNatSchG; *weiterer Übersetzungsvorschlag* Flussauennaturschutzgebiet [n]).

7004 wild animal [n] *conserv. leg.* (Feral creature with freedom of movement from place to place; ►game, ►wildlife); *s* **animal [m] salvaje** (Animal que mora en la naturaleza y puede transitar libremente de un lugar a otro; ►caza, ►vertebrados endémicos); *f 1* **espèce [f] animale non domestique** (cf. Code rural, art. L. 211-1 ; ►gibier, ►animaux vertébrés indigènes ; *syn. terme courant* animal [m] sauvage ; *f 2* **faune [f] sauvage** (Ensemble des espèces animales sauvages) ; *g* **wild lebendes Tier [n]** (In freier Natur lebendes, gefangenes oder gezüchtetes und nicht herrenlos gewordenes sowie totes Tier einer wild lebenden Art; cf. § 20 ff BNatSchG; ►Wild, ►heimische Wirbeltiere; *syn.* frei lebendes Tier [n] [CH] (cf. Art. 19 NHG).

7005 wild animal enclosure [n] *recr.* (Usually a fenced area of a park or forest as part of a recreation facility in which wild animals are kept in the open and partially under shelter; ►game park 2, ►shooting preserve); *s* **parque [m] de animales salvajes (≠)** (Terreno cercado dentro de un bosque con funciones generalmente recreativas en el cual habitan, más o menos libremente, animales salvajes que pueden ser observados por los turistas; ►parque de animales salvajes, ►reserva de caza); *f* **enclos [m] d'animaux sauvages** (Zone entourée d'une clôture généralement installée en forêt ou dans un parc dans laquelle les animaux sauvages sont gardés en liberté ou en semi-liberté ; ►enclos de réserve de chasse, ►parc animalier) ; *g* **Tiergehege [n]** (Eingezäunter Bereich in einem Park oder Wald als Teil einer Freizeiteinrichtung, in der wild lebende Tiere ganz oder teilweise im Freien gehalten werden; ►Wildgehege, ►Wildpark).

Wild Creatures and Wild Plants Act [n] 1975 [UK], Conservation of *conserv. leg.* ►Endangered Species Act [E.S.A.] [US].

7006 wilderness [n] *conserv. nat'res.* (Any area with a primeval character which has been neither directly nor indirectly influenced by human activities; ►natural area preserve, ►nature reserve, ►wilderness area); *syn.* virgin area [n] (2); *s* **zona [f] virgen** (Área natural que no ha sufrido ninguna injerencia humana; ►espacio natural protegido 1, ►reserva natural 1, ►reserva natural 2); *f* **zone [f] vierge** (Toute zone qui, de mémoire d'homme, a été exempte d'influences humaines tendant à en

dégrader la végétation ; DFM 1975 ; ►réserve naturelle de flore et faune sauvage, ►réserve naturelle nationale, ►site classé) ; *g* **Wildnis [f]** (Gebiet, das vom direkten Einfluss des Menschen unberührt oder weitgehend unberührt geblieben ist; ►Naturreservat, ►Naturschutzgebiet 2, ►Wildnis[schutz]gebiet).

7007 wilderness area [n] [US] *conserv. leg. nat'res.* (Untouched federal land retaining its primeval character and influence, without permanent improvements or human habitation, which is protected and managed so as to preserve its natural conditions and which **1.** generally appears to have been affected primarily by the forces of nature, with the imprint of man's work substantially unnoticeable, **2.** has outstanding opportunities for solitude or a primitive and unconfined type of recreation, and **3.** may also contain ecological, geological, or other natural features of scientific, educational, scenic, or historical value; cf. Wilderness Act 1964 and counterpart state legislative enactments; such an area is established by federal legislation and administered by federal agencies; e.g. U.S. Forest Service, National Park Service and Bureau of Land Management; ►European wilderness reserve, ►forest research natural area [US]/forest nature reserve [UK]); *syn.* strict wilderness reserve [n] [US]; *s* **reserva [f] natural (2) (≠)** (En EE.UU., categoría de protección para áreas en estado primigenio, en las que las actividades humanas no han dejado huella visible, que están prácticamente despobladas, que ofrecen oportunidades de recreo solitario y que pueden tener valores ecológicos, geológicos o componente naturales de interés científico, educativo, escénico o histórico; ►reserva forestal integral, ►reserva natural europea); *f* **réserve [f] naturelle de flore et de faune sauvage (≠)** (*USA* paysages naturels protégés par la législation des États fédéraux, gérés par l'administration fédérale des forêts [U.S. Forest Service] et les services fédéraux des affaires foncières [Bureau of Land Management] ; *objectif* gestion orientée vers la protection et la mise en valeur du point de vue récréatif de paysages remarquables et originels ; cette catégorie de protection n'existe pas en F. mais serait à rapprocher des ►réserves biologiques domaniales et des ►réserves biologiques forestières ; ►réserve européenne) ; *g* **Wildnis(schutz)gebiet [n]** (Eine durch amerikanische Bundesgesetzgebung unter Schutz gestellte Naturlandschaft, die von der Bundesforstverwaltung *[U.S. Forest Service]* und Bundesliegenschaftsbehörde *[Bureau of Land Management]* verwaltet wird, um den ursprünglichen Naturcharakter und die einzigartige Schönheit einer solchen Landschaft für die Öffentlichkeit nachhaltig zu sichern; die erste Ausweisung von Wildnisschutzgebieten erfolgte in den USA 1924. Die Richtlinien der Nationalparkkommission der IUCN 1994 für Management-Kategorien von Schutzgebieten definieren **Wildnis** in der deutschen Übersetzung wie folgt: „Ein ausgedehntes ursprüngliches oder leicht verändertes Landgebiet und/oder marines Gebiet, das seinen natürlichen Charakter bewahrt hat, in dem keine ständigen oder bedeutenden Siedlungen existieren und dessen Schutz und Management dazu dienen, seinen natürlichen Zustand zu erhalten". **In D.** ist weder diese Schutzkategorie noch ein solches Gebiet. Die ca. 635 bundesweit eingerichteten Naturwaldreservate können noch am ehesten mit dem amerikanischen „Wildnis"-Konzept in Verbindung gebracht werden, obwohl in D. die Durchschnittsgröße nur ca. 30 ha beträgt; ►Europareservat, ►Naturwaldreservat).

wilderness recreation [n] *recr.* ►recreation in the natural environment.

Wilderness Reserve [n] *conserv. pol.* ►European Wilderness Reserve.

wild herb [n] *hort.* ►wild flower.

7008 wild flower [n] *hort.* (Indigenous, herbaceous plant growing in the wild; ▶weed 1); *syn.* wild herb [n]; *s* **hierba [f] silvestre** (Planta herbácea indígena creciendo en estado silvestre; ▶planta advenediza); *f* **herbe [f] sauvage** (Spécimen d'espèces herbacées se développant spontanément ; ▶plante adventice) ; *syn.* herbacée [f] sauvage, plante [f] herbacée sauvage ; *g* **Wildkraut [n]** (...kräuter [pl]; heimische, krautige, züchterisch nicht veränderte, wild wachsende Pflanze; aus der Sicht des Nutzers unerwünschtes W. sowie spontan aufwachsende Kulturpflanze in einem Nutzpflanzenbestand, die das Wachstum der Kulturpflanze behindert und i. d. R. den Ertrag mindert, wird ▶Unkraut genannt).

7009 wildflower meadow [n] *gard. landsc.* (TGG 1984, 196; floristically rich meadow which may occur naturally or be developed as a plant community appropriate to the site, with only 1 to 2 cuts per year, or no cutting at all and without additional fertilizing; ▶wildflower meadow gardening); *syn.* flowery mead [n] [also UK], successional meadow [n] [also US] (TGG 1984, 198); *s* **pradera [f] silvestre** (Prado rico florísticamente que puede ser de origen natural o haberse creado artificialmente, que generalmente no se siega o como máximo dos veces al año y no se fertiliza, de manera que se puede desarrollar la comunidad pratense que corresponde a las condiciones mesológicas de la ubicación; ▶cuidado de praderas silvestres); *f* **prairie [f] extensive ensemencée** (Prairie ensemencée avec un mélange de semences florales indigènes ou réalisée à partir d'un épandage d'herbe à semences, fauchée une à deux fois l'an, non amendée qui après plusieurs années présente l'aspect d'une formation prairiale floristiquement diversifiée et dont la composition botanique se rapproche de celle des prairies semi-naturelles ; selon la localisation et la composition des sols ces prairies peuvent évoluer vers des prairies à Molinia, des pelouses calcaires, des ▶pelouses héliophiles denses ; ▶entretien de prairies extensives) ; *syn.* pelouse [f] fleurie ; *g* **naturnahe Wiese [f]** (Mit einer reichhaltigen, standortgerechten Wildblumenmischung ausgesäte Wiese, die durch gezielte Schnitte — 1-2-mal je Vegetationsperiode — oder ohne Schnitt und ohne zusätzliche Düngung eine vielfältige, dem Standort entsprechende stabile, kulturbetonte Wiesengesellschaft entwickeln kann. Aus der Sicht der Pflanzensoziologie gehören unterschiedliche Graslandgesellschaften mit Kennarten von gedüngten und ungedüngten Wiesen dazu; in der gärtnerischen Praxis erhält man solche Wiesen durch Umstellung der Pflege von 6- bis 10-fachem Schnitt auf ein- bis zweimalige Mahd erst nach vielen Jahren; das Ergebnis ist meist eine Wiese der Pfeifengras-Glatthaferwiesengesellschaft *[Molinio-Arrhenatheretea]*. Erst nach weiterem Nährstoffentzug kann auf entsprechendem Standort ein bunt blühender ▶Halbtrockenrasen entstehen, wenn der Boden nicht zu nährstoffreich ist; ▶Pflege naturnaher Wiesen); *syn.* blumenbunte Wiese [f], Blumenmatte [f], Blumenwiese [f], Ökowiese [f], Naturwiese [f].

7010 wildflower meadow gardening [n] [US] *constr. landsc.* (Horticultural and maintenance measures, which are necessary for encouraging growth conditions of ▶wildflower meadows; e.g. by cutting once or twice per year; ▶prairie gardening [US], ▶wildflower); *syn.* wildflower meadows management [n] [UK]; *s* **cuidado [m] de praderas silvestres** (Todas las medidas necesarias para mantener las condiciones de crecimiento óptimas para ▶praderas silvestres; ▶hierba silvestre, ▶jardinería de las praderas norteamericanas); *f* **entretien [m] de prairies extensives ensemencées** (1. Travaux horticoles nécessaires au bon développement des ▶prairies extensives ensemencées ; ▶herbe sauvage. 2. U.S., ▶aménagement et entretien des pelouses sèches de la Prairie) ; *g* **Pflege [f, o. Pl.] naturnaher Wiesen** (1. Alle gärtnerischen Maßnahmen, die notwendig sind, um die Wuchsbedingungen und die Funktion ▶naturnaher Wiesen dauerhaft zu erhalten; ▶Wildkraut. **2.** U.S., ▶Anlage, Pflege und Unterhaltung naturnaher Prärie-Trockenrasen).

wildflower meadows management [n] [UK] *constr. landsc.* ▶wildflower meadow gardening [US].

wildflower perennial [n] [US] *hort. plant.* ▶wild perennial.

7011 wildflowers [npl] **in the lawn** *gard. plant.* (Herbaceous plants which have seeded themselves in grass areas or have been deliberately introduced. Undesirable plants in lawns are termed 'lawn weeds'; ▶meadow flower); *s* **flor [f] silvestre de césped** (Plantas herbáceas que se han propagado espontáneamente o han sido sembradas deliberadamente en un césped. A las no deseadas se les denomina despectivamente «malas hierbas de césped»; ▶flor de prado); *f* **plantes [fpl] herbacées des pelouses** (Végétaux herbacés qui croissent spontanément sur un gazon ou qui sont associées à un mélange de gazon pour installer une pelouse fleurie ; les plantes indésirables, les adventices dicotylédones, sont dénommées « mauvaises herbes du gazon » ; ▶fleur des prés) ; *g* **Rasenkräuter [npl]** (Krautige Pflanzen, die sich auf Rasenflächen spontan einstellen oder zusätzlich eingebracht werden. Unerwünschter Krautwuchs wird als **Rasenunkraut** oder in der DIN 18 919 als **unerwünschter Fremdartenbesatz** bezeichnet; ▶Wiesenblume).

7012 wild land [n] [US] *landsc.* (Uncultivated land, other than fallow, which may be neglected altogether or maintained for such purposes as wood or forage production, wildlife, recreation, or protective plant cover; cf. SAF 1983; ▶virgin landscape); *s* **paraje [m] silvestre** (≠) (En EE.UU. término utilizado para área natural no cultivada, pero no yerma, que puede estar mal cuidada o simplemente mantenida en estado natural con fines de producción maderera o forrajera; ▶paisaje virgen); *f* **1 paysage [m] naturel** (Paysage sans empreinte apparente de l'homme ; *opp.* ▶paysage culturel) ; *f* **2 paysage [m] sauvage** (**US.**, terme utilisé par les géographes anglo-saxons pour décrire des terres n'ayant pas ou presque pas subi l'influence de l'homme, restées soit à l'état naturel et non cultivées ou soit gérées au travers de la production de bois ou de fourrage comme habitat pour la faune vivant à l'état sauvage, le tourisme ou la protection contre l'érosion ; ▶paysage vierge) ; *syn.* espace [m] sauvage, nature [f] sauvage ; *g* **Naturlandschaft [f] (2)** (**U.S.**, ein vom Menschen nicht oder kaum beeinflusster Zustand der Landschaft, der nicht durch Brachfallen entstand, sondern vom Menschen als Ganzes unbewirtschaftet oder zum Zwecke der Holz- oder Futterproduktion, als Lebensstätte für wild lebende Tiere, zur Erholung oder für den Erosionsschutz entsprechend gepflegt wird; ▶Urlandschaft).

7013 wildlife [n] *conserv.* (All ▶wild animals, birds, reptiles, and amphibians, not including fish, which are living in a natural environment. This term covers game and nongame species; cf. WPG 1976, 233; in the literature, *wildlife* has various definitions and usually includes all vertebrate animals living in the wild within the area under observation [cf. WEB 1993, 2616 and NRM 1996, 332]; some definitions include plants, fungi, algae and bacteria; cf. DED 1993, 264 and NRM 1996, 332]; others exclude fish [WPG 1976, 233, RCG 1982, 187]); ▶animal kingdom); *syn.* native vertebrates [npl]; *s* **vertebrados [mpl] endémicos** (El término inglés *«wildlife»* se utiliza para los ▶animales salvajes [mamíferos, aves, reptiles, anfibios] exceptuando a los peces; ▶reino animal); *f* **animaux [mpl] vertébrés indigènes** (Ensemble des ▶espèces animales non domestiques comprenant tous les organismes possédant une colonne vertébrale osseuse ou cartilagineuse. Le terme anglais *«wildlife»* a plusieurs définitions ; en règle générale il englobe tous les invertébrés vivant à

W

l'état sauvage sur une aire déterminée [cf. WEB 1993, 2616 et NRM 1996, 332] ; certaines définitions prennent en compte la flore, les champignons, les algues et les bactéries [cf. DED 1993, 264 et NRM 1996, 332], d'autres excluent les poissons [WPG 1976, 233, RCG 1982, 187] ; ▶règne animal) ; *g* **heimische Wirbeltiere [npl]** (±) (Gesamtheit der ▶wild lebenden Tiere eines Untersuchungsraumes. Der englische Begriff *wildlife* wird in der Literatur unterschiedlich definiert: I. d. R. umfasst er alle wild lebenden Wirbeltiere eines Betrachtungsraumes [cf. WEB 1993, 2616 und NRM 1996, 332]; manche Definitionen schließen auch Pflanzen, Pilze, Algen und Bakterien mit ein [cf. DED 1993, 264 und NRM 1996, 332]; andere Definitionen schließen Fische aus [WPG 1976, 233, RCG 1982, 187]; ▶Tierreich).

wildlife [n]**, protection of** *conserv. zool.* ▶wildlife conservation.

7014 wildlife biologist [n] *game'man.* (Scientist who studies ▶wild animals and their environment; ▶game biologist); *s* **biólogo/a [m/f] de fauna salvaje** (Científico/a que estudia los ▶animales salvajes y su entorno; ▶biólogo/a cinegético/a); *syn.* biólogo/a [m/f] de fauna cinegética; *f* **biologiste [m/f] spécialiste de la faune sauvage** (Scientifique spécialisé[e] dans l'étude des ▶espèces animales non domestiques ; ▶cynégète) ; *g* **Wildbiologe [m]/Wildbiologin [f]** (Wissenschaftler, der sich mit der Erforschung der ▶wild lebenden Tiere in ihrer Umwelt befasst; ▶Jagdwissenschaftler/-in).

7015 wildlife conservation [n] *conserv. leg. zool.* (Legal measures for protection, care and development of a population of wild or undomesticated animals within a specific landscape area; ▶animal species conservation, ▶conservation of flora and fauna, ▶domestic animal protection, ▶nature conservation, ▶special conservation, ▶wildlife); *syn.* protection [n] of wildlife; *s* **protección [f] de la fauna salvaje** (Medidas legales de conservación de especies de ▶vertebrados endémicos en el contexto de la protección del patrimonio biológico y la diversidad genética, independientemente de si aquéllas están amenazadas de extinción o no; ▶preservación de la flora y fauna silvestres, ▶protección de animales, ▶protección de especies de fauna en peligro, ▶protección de especies [de flora y fauna], ▶protección de la naturaleza); *syn.* conservación [f] de la fauna salvaje; *f* **protection [f] de la faune sauvage** (**1.** Conservation des espèces animales non domestiques, vivant à l'état sauvage, dans le cadre de la préservation du patrimoine biologique ; cf. Code rural, art. L. 211-1 à 211-4. **2.** Mesures de conservation, préservation, maintien et de restauration des populations d'espèces animales sauvages ; fait partie de la politique de protection des espèces animales et végétales ; ▶animaux vertébrés indigènes, ▶conservation de la faune et de la flore sauvage, ▶protection de la nature, ▶protection des animaux domestiques et d'expérience, ▶protection des espèces animales, ▶protection des espèces [animales et végétales], ▶protection des espèces de faune et de flore sauvages) ; *g* **Faunenschutz [m]** (Schutz, Pflege und Förderung der wild lebenden Tierwelt für einen definierten Landschaftsraum; die Maßnahmen tragen in signifikanter Weise dazu bei, dass das Gedeihen und die Entwicklung der Tiere gefördert wird, um sie in einem günstigen Erhaltungszustand zu bewahren; Aufgabengebiet des Artenschutzes; Gesetzliche Grundlage für den **F.** bilden das Bundesnaturschutzgesetz resp. die Landesnaturschutzgesetze, die ▶Fauna-Flora-Habitat-Richtlinie — [FFH-Richtlinie] und die ▶Vogelschutzrichtlinie der Europäischen Union. *OB* ▶Artenschutz; zum englischen Begriff ▶*wildlife*; ▶Erhaltung der Tier- und Pflanzenwelt, ▶Naturschutz, ▶Tierartenschutz, ▶Tierschutz); *syn.* Schutz [m, o. Pl.] wild lebender Tiere.

wildlife corridor [n] *ecol.* ▶native plants and wildlife corridor.

7016 wildlife management [n] *conserv.* (Term used for the planning and administration of species conservation and game reserve program[m]s. ased on an American concept, **w. m.** comprises both planning and practical implementation with particular emphasis on the sustainable management and regulation of wild stocks on a scientific basis, e.g. biological research into wildlife; cf. WFL 2002. **W. m.** also includes planning and measures taken to control biocenoses of plants, predators, diseases and parasites so that the living conditions of animals are brought into equilibrium with the demands of human beings. Originally **w. m.** was concerned with game management, i.e. care and regulation of animal populations exclusively for the sport hunting—for example, big game hunting in Africa and India, but in the early 1980s new thought processes began in African countries. Under the aegis of the World Wildlife Fund for Nature [WWF] and the International Union for the Conservation of Nature [IUCN], concepts for the preservation of wild animal diversity and use have been developed. **W. m.** has gained great importance for the world's population, since for many people wild animals are economically important, on the one hand, and because of its influence on nature conservation on the other, which aims at the preservation and sustainable management of wild animals especially in the management and shaping of large National Parks. **W. m.** also takes into account the concerns of tourism, with hunting trips and animal observation. Even trophy hunting can be a sustainable pursuit, which protects resources and regulates animal populations, if, in order to maintain the biological equilibrium, mostly older male animals are killed, and their culling does not negatively impact on the stock and the reproductive capacity of a population; cf. KGJ 1999, 923 s; ▶Bonn Convention, ▶hunting management); *s* **conservación [f] y gestión [f] de la fauna silvestre** (En inglés el término *«wildlife management»* se utiliza para denominar la gestión de programas de conservación de especies salvajes y de reservas de caza e incluye tanto la planificación como la implementación práctica con especial énfasis en la gestión y regulación sostenible/sustentable de las poblaciones de fauna basándose en principios científicos. También incluye medidas de control de biocenosis vegetales, de depredadores, de enfermedades y parásitos para lograr un equilibrio entre las condiciones de vida de la fauna y las necesidades humanas. Originalmente esta actividad estaba dirigida a la gestión de la caza como deporte, como p. ej. en el caso de África y la India, pero a partir de los años 1980 comenzaron a surgir nuevos planteamientos en los países africanos. Bajo el auspicio del WWF y de la UICN se desarrollaron conceptos para la preservación de la biodiversidad y el aprovechamiento de la fauna salvaje. Este tipo de gestión ha cobrado gran importancia para la población a nivel mundial, ya que, por un lado, para muchas personas la **c. y g. de la f. s.** es importante económicamente y, por otro, por su influencia en la conservación de la naturaleza, especialmente en la creación y gestión de grandes Parques Nacionales. Esta política tiene en cuenta los intereses turísticos, ofreciendo safaris de caza y de observación de animales, dándose el caso de que incluso la caza de trofeo puede ser sostenible, ya que sirve para proteger recursos y regular las poblaciones cinegéticas, si se matan sobre todo animales machos viejos, de manera que su muerte no amenaza la capacidad de reproducción de las poblaciones; ▶Convención de Bonn, ▶gestión de la caza 2); *f 1* **conservation [f] et gestion durable de la faune sauvage** (Terme désignant les projets et réalisations nécessaires à la conservation des espèces de la faune et de la flore sauvage au niveau mondial ; pour les forêts tropicales d'Afrique centrale [Bassin du Congo] protection de la biodiversité remise en cause par la chasse de subsistance et commerciale. La pratique de prélèvement de viande en forêt [viande de brousse] contribue à l'équilibre alimentaire des populations vivant dans la forêt. Les mécanismes de gestion traditionnelle garantissant la durabilité de cette pratique se trouvent fortement perturbés par **1.** l'ouverture de nouvelles routes à l'intérieur des forêts, **2.** l'augmentation de la population dans la zone forestière, **3.** le développement du marché de viande de brousse. Pour certaines espèces les taux de prélèvement exercés par des chasseurs provenant de l'extérieur du territoire de chasse ne semblent pas soutenables et pénalisent les communautés locales. Avec la création en mai 2000 du Réseau des Aires Protégées d'Afrique Centrale [RAPAC] suite à la déclaration de Yaoundé de 1999, les États d'Afrique Centrale se sont engagés dans une politique de gestion soutenable de leurs forêts [protection intégrale de 10 % de la surface forestière], la conservation de la biodiversité intégrant la participation des populations locales dans la gestion durable des forêts. Cette engagement a été soutenu en 2002 par le lancement de l'initiative pour le patrimoine forestier en Afrique Centrale [Central African World Heritage Forest Initiative — CAWHFI] soutenu par les ONG de conservation, initiative visant à améliorer la gestion durable des paysages écologiques clefs dans les bassins du Congo en coopération avec UNESCO-Centre du Patrimoine Mondial ; cf. CGDFS ; ▶Convention de Bonn) ; *syn.* gestion [f] de la faune sauvage et des aires protégées [programme FAO]) ; *f 2* **gestion [f] et conservation de la faune sauvage et des ses habitats (F.,** Mesures qui permettent de préserver la biodiversité à l'échelle de tout le territoire français ; elles sont mises en place dans le cadre des Orientations Régionales de Gestion et de Conservation de la Faune Sauvage et de ses Habitats [ORGFH], stratégie nationale adoptée en 2004 pour la biodiversité dont l'objectif est d'enrayer la perte de biodiversité d'ici 2010, conformément à la convention de Rio de Janeiro sur la préservation de la biodiversité adoptée en 1992. L'ORGFH a donc établi une liste d'orientations pour les différents régions françaises dans quatre principaux domaines concernant la gestion de la faune sauvage et des mieux naturels : préservation des habitats, gestion des espèces animales, prise en compte de la faune et de ses habitats dans les activités humaines et diffusion des connaissances de la faune sauvage et de ses habitats. Ces orientations se traduisent sur le terrain par la mise en place de corridors écologiques, la meilleure prise en compte de la faune dans les documents d'urbanisme et les projets d'infrastructures, la lutte contre les espèces envahissantes, la conciliation des sports et des loisirs de nature avec l'environnement, le développement des jachères « environnement et faune sauvage », la mise en place d'un observatoire régional de la biodiversité ; ▶gestion cynégétique) ; *g* **Wildlife Management [n]** (Aus dem Amerikanischen stammender Begriff, der die Planung und praktische Ausführung von Artenschutz und Wildhege unter besonderer Berücksichtigung der nachhaltigen Bewirtschaftung und Regulierung von Wildtierbeständen auf wissenschaftlicher Grundlage beinhaltet [wildbiologische Forschung]; cf. WFL 2002. Zum **W. M.** gehören auch Planungen und Maßnahmen, die die gesamte Biozönose mit ihren Pflanzen, Räubern, Krankheiten und Parasiten so steuern, dass die Lebensbedingungen der Tiere mit den Ansprüchen des Menschen in Einklang gebracht werden können. Ursprünglich war **W. M.** ein *game management*, d. h., die Pflege und Regulierung von Tierbeständen ausschließlich für die sportliche Jagd — z. B, Großwildjagd in Afrika und Indien. Anfang der 1980er-Jahre fand jedoch in den Ländern Afrikas ein Umdenken statt. Unter Mithilfe des *World Wide Fund For Nature [WWF]* und der Internationalen Naturschutz-Union [IUCN] wurden Konzepte zur Bewahrung der Vielfalt der Wildtiere und deren Nutzung entwickelt. **W. M.** hat zum einen eine große Bedeutung für die Bevölkerung erhalten, da für sie der Wildbestand eine ökonomische Bedeutung bekommt, und zum anderen für den Natur-

schutz, der mit seinem Einfluss den Erhalt und die nachhaltige Bewirtschaftung der wild lebenden Tiere verfolgt und für die Betreuung und Gestaltung großer Nationalparke von Bedeutung ist; **W. M.** berücksichtigt auch die Belange des Tourismus mit Jagdreisen und Tierbeobachtung. Auch die Trophäenjagd kann eine nachhaltige, die Ressourcen schonende Nutzung und Regulierung sein, wenn, um das biologische Gleichgewicht zu erhalten, vorwiegend ältere männliche Tiere erlegt werden, deren Entnahme keinen negativen Einfluss auf den Bestand und die Reproduktionsfähigkeit einer Population haben; cf. KGJ 1999, 923 f; ▶Bonner Konvention, ▶jagdliche Hege); *syn.* nachhaltige Wildbestandslenkung [f], Pflege- und Entwicklungsmaßnahmen [fpl] für faunistische Schutzgebiete (±).

wildlife preservation [n] *conserv. pol.* ▶Convention on nature protection and wildlife preservation in the western hemisphere.

wildlife reserve [n] [UK] *conserv. leg.* ▶wildlife sanctuary [US].

7017 wildlife sanctuary [n] [US] *conserv. leg.* (Area set aside for the complete protection of all forms of wildlife, save perhaps specific predators and parasites doing excessive harm; SAF 1983); *syn.* refuge [n] [US], wildlife reserve [n] [± UK]; *s 1* **santuario** [m] **de fauna salvaje (1.** En EE.UU. área dedicada exclusivamente a la protección de todas las formas de vida silvestre, exceptionando quizás depredadores y parásitos que causan daños excesivos. **2.** En general, término usado para denominar a reserva natural, parque nacional, etc. donde existen especies importantes o una alta biodiversidad. Implica una gestión del espacio para la conservación de la vida salvaje, pero el nombre tiene más que ver con la gestión del uso público, para hacer el lugar más atractivo. Hay muchos ejemplos de parques nacionales o reservas que usan este término, aunque uno de sus principales objetivos es el turismo, por lo que —entre otros— frecuentemente ofrecen servicios de educación ambiental. *s 2* **santuario** [m] **de ballenas** (*Término específico;* en Chile zona libre de caza de cetáceos en aguas chilenas, que tiene como fin promocionar el rescate, la rehabilitación, reinserción y observación de mamíferos marinos, reptiles y aves acuáticas, creado por ley 20.293, que entró en vigor el 25.10.2008, y que será efectivo en toda la Zona Económica Exclusiva; cf. Ley 20.293. En otros países de AL también existen **ss. de bb.** pero no tienen ni el mismo estatus de protección ni cubren un área tan extensa como en el caso chileno); *f* **sanctuaire** [m] **(de faune) sauvage (1.** U.S., espace classé en vue de la protection de la faune vivant à l'état sauvage. **2.** F., terme utilisé pour désigner des lieux refuge protégés [p. ex. sous forme de réserve naturelle] qui par leur caractère inviolable [p. ex. îles au large des côtes bretonnes] sont de grandes réserves d'animaux ; le 25 novembre 1999, la France, l'Italie et la Principauté de Monaco signent un Accord relatif à la création en Méditerranée d'un sanctuaire [87 500 km² d'espace maritime] pour les mammifères marins ; en 2001 le sanctuaire est inscrit sur la liste des Aires Spécialement Protégées d'Importance Méditerranéenne [ASPIM] et un plan de gestion est adopté en 2004 ; dans divers pays du monde des espaces de conservation et de sauvegarde de la faune sauvage ont pris la dénomination de sanctuaire) ; *g* **Naturschutzgebiet [n] [US] (3)** (In den USA ausgewiesener Landschaftsraum zum vorwiegenden Schutz der frei lebenden Tiere).

wildlife viaduct [n] *constr. ecol.* *▶game pass.

wildling [n] *for.* ▶woody wildling.

7018 wild perennial [n] *hort. plant.* (LD 1994 [4], 10; ▶perennial 1 which has been taken from its natural site for cultivation in nurseries, as compared to horticulturally-bred ▶ornamental perennials; ▶wildflower); *syn.* wildflower perennial [n]

W

[also US]; *s* **perenne [f] silvestre** (►Planta vivaz que, al contrario que la ►vivaz ornamental cultivada, ha sido sustraída de una ubicación natural para ser reproducida en vivero; ►hierba silvestre); *syn.* vivaz [f] silvestre; ƒ**plante [f] vivace sauvage** (Par comparaison avec les ►plantes vivaces d'ornement sélectionnées en pépinière, ►plantes vivaces sauvages ramassées dans leur milieu naturel pour être ensuite cultivées et commercialisées ; ►herbe sauvage, ►plante vivace à feuilles ornementales) ; *g* **Wildstaude [f]** (*Terminus für das Wuchsverhalten und die Wirkung einer Staude hinsichtlich ihrer gärtnerischen Verwendung* im Vergleich zur gärtnerisch gezüchteten ►Schmuckstaude, natürlichen Standorten entnommene ►Staude, die in Gärtnereien so für den Verkauf weiterkultiviert wird, dass sie ihren ursprünglichen Charakter beibehält. Es gibt nur wenige, durch Auslesen entstandene Sorten. Die Lebensbereiche der **W.n** wurden zum Vorbild ökologisch begründeter, standortgerechter, naturnaher gärtnerischer Pflanzungen; **W.n** mit dekorativem Blattwerk [►Blattschmuckstauden] steigern in ihrer Schlichtheit die Wirkung der Blütenstauden; cf. AKP 1999; ►Wildkraut).

7019 wild plant [n] *plant.* (Generic term for any plant which has grown without direct or indirect influence of man over a long period of time and has established itself in an area containing both indigenous and naturalized plant species; ►wild woody species); *s* **planta [f] silvestre** (Especie vegetal que vive y se reproduce de forma natural y sin cultivo en campos y selvas; *opp.* planta doméstica; cf. DINA 1987; ►leñosa silvestre); ƒ**plante [f] sauvage** (Spécimen sauvage des espèces se développant spontanément sans l'influence directe ou indirecte de l'homme et capable de subsister pendant une longue période sur son lieu de croissance ; parmi les espèces sauvages on distingue les espèces indigènes, naturalisées et subspontanées ; ►espèce ligneuse sauvage) ; *syn.* espèce [f] végétale sauvage ; *g* **Wildpflanze [f]** (Pflanze, die sich ohne direkten oder indirekten Einfluss des Menschen über einen längeren Zeitraum hinweg entwickelt und in einem bestimmten Bereich auf Dauer Fuß gefasst hat. Es werden heimische und eingebürgerte **W.n** unterschieden; ►Wildgehölz); *syn.* wild wachsende (Pflanzen)art [f].

Wild Plants Act [n] **1975** [UK]**, Conservation of Wild Creatures and** *conserv. leg.* ►Endangered Species Act [E.S.A.] [US].

wild plants [npl] **or animals** [npl] *conserv. phyt. zool.* ►collection of wild plants or animals.

7020 wild rose [n] *bot. hort.* (Wild species of the rose genus; e.g. in Europe, apple rose *[Rosa pomifera]*, French rose *[R. gallica]*, field rose *[R. arvensis]*, hedge rose *[R. corymbifera]*, dog rose *[R. canina]*, eglanteria rose *[R. eglanteria]*, sweet briar *[R. rubiginosa]*; in U.S., smooth rose *[R. blanda]*, swamp rose *[R. palustris]*, dwarf wild rose *[R. virginiana]*, pasture rose *[R. carolina]*, shining rose *[R. nitida]*, sweetbriar *[R. eglanteria]*, rugosa rose *[R. rugosa]*; *opp.* ►cultivated shrub rose; ►park rose); *s* **rosa [f] silvestre** (Especie silvestre del género de las rosas, como p. ej. en Europa *[Rosa pomifera]*, *[R. gallica]*, *[R. arvensis]*, *[R. corymbifera]*, escaramujo o rosal perruno *[R. canina]*, egalanteria rosa o roja *[R. eglanteria, R. rubiginosa]*; ►rosal arbustivo cultivado, ►rosal de parque); ƒ**rosier [m] sauvage** (Espèce sauvage de la famille des rosacées p. ex. le Rosier de France *[R. gallica]*, le Rosier des champs *[R. arvensis]*, le Rosier des chiens *[R. canina]*, le Rosier à feuilles odorantes *[R. eglanteria, R. rubiginosa]* qui appartient à la famille des ►rosiers des parcs [et jardins] ; *opp.* ►rosier arbuste d'ornement ; ADT 1988, 203) ; *syn.* rosier [m] botanique ; *g* **Wildrose [f]** (In der Natur wild lebende resp. natürlich vorkommende Art der Gattung der Rosengewächse mit einfachen Blüten und kleinen Blättern, die je nach Standort in den unterschiedlichsten

Erscheinungsformen vorkommt, z. B. als bodendeckende Rose, z. B. Feldrose *[Rosa arvensis]* und Dünenrose *[Rosa spinosissima, syn. R. pimpinellifolia]*, als Strauchrose, z. B. Apfelrose *[Rosa pomifera]*, Essigrose *[R. gallica]*, Heckenrose *[R. corymbifera]*, Hundsrose *[R. canina]*, Weinrose *[R. eglanteria, R. rubiginosa]* und als Kletterform wie z. B. Vielblütige Rose *[Rosa multiflora]*, Wichuras Rose *[Rosa wichuraiana]*; Züchtungen von Strauchrosen mit Wildrosencharakter gehören zu den ►Parkrosen; *opp.* ►Zierstrauchrose).

wild water [n] [US] *geo.* ►torrent.

7021 wild woody species [n] *plant. hort.* (Wild tree or shrub which has grown without direct or indirect human interference over a long period of time, and which has become established permanently in an area containing both indigenous and naturalized types of those species; ►indigenous plant species, ►weed tree, ►wild plant); *s* **leñosa [f] silvestre** (*Término genérico* ►planta silvestre; árbol o arbusto que durante un largo periodo/período de tiempo ha crecido sin interferencia humana directa o indirecta, de manera que se ha establecido permanentemente en un área determinada. Se diferencian la **l.** indígena y la **l.** naturalizada; ►especie naturalizada, ►especie vegetal autóctona, ►leñosa adventicia); ƒ**espèce [f] ligneuse sauvage** (*Terme spécifique pour la* ►plante sauvage ; spécimen sauvage d'arbres ou d'arbustes se développant spontanément sans l'influence directe ou indirecte de l'homme et capable de subsister pendant une longue période sur son lieu de croissance ; parmi les espèces ligneuses sauvages on distingue les espèces indigènes, naturalisées et subspontanées ; ►espèce naturalisée, ►espèce subspontanée, ►espèce végétale indigène, ►végétation arbustive adventice) ; *g* **Wildgehölz [n]** (*OB* ►Wildpflanze; Baum oder Strauch, der ohne Zutun des Menschen sich über einen längeren Zeitraum hinweg entwickelt und ohne direkten oder indirekten Einfluss des Menschen in einem bestimmten Bereich auf Dauer Fuß gefasst hat. Es werden heimische und eingebürgerte **W.e** unterschieden; ►bodenständige Pflanzenart, ►Fremdaufwuchs, ►neuheimische Art); *syn.* Landschaftsgehölz [n].

willow [n] *agr. conserv.* ►pollarded willow.

willow mattress [n] *constr.* ►brush mat.

7022 wilting point [n] *bot.* (Degree of dryness in soil, at which soil absorption power is greater than ►root absorbing power so that the roots can no longer retain moisture and the plant begins to wilt. Permanent **w. p.** occurs when the soil absorbing power reaches -15 bar, a figure based upon agricultural practice); *s* **punto [m] de marchitez** (Límite inferior de humedad del suelo en que se manifiesta la marchitez en una planta cultivada en el mismo y que es en mayor cuantía función de la capacidad hídrica del terreno que de la especie vegetal de que se trate; DB 1985; ►fuerza de absorción de las raíces); ƒ**point [m] de flétrissement** (Point caractérisant l'état hydrique d'un sol qui correspond à la teneur en eau pour laquelle le pouvoir de rétention du sol est supérieur au pouvoir d'absorption des végétaux. À ce moment-là les plantes se fanent. On a défini conventionnellement une valeur de pression [-15 atmosphères] qui permet d'évaluer le **p. d. f.** d'un sol indépendamment de la végétation ; DEE 1982 ; ►pouvoir d'absorption des racines) ; *syn.* point [m] de fanaison ; *g* **Welkepunkt [m]** (Trockenheitsgrad im Boden, bei dem die Bodensaugspannung größer als die ►Wurzelsaugspannung ist, so dass die Wurzeln Wasser an den Boden abgeben müssen und die Pflanzen welken. Die landwirtschaftliche Praxis nimmt einen **permanenten W.** bei -15 bar Saugspannung des Bodens an).

W

wind [n] *geo. met.* ►downslope wind; *met.* ►mountain wind, ►prevailing wind direction, ►slope wind; *plant.* ►traffic-caused wind [US]/air stream [UK]; *met.* ►valley wind.

wind [n]**, anabatic** *geo. met.* ►upslope air flow.

wind [n]**, katabatic** *geo. met.* ►downslope wind.

wind blast [n] [UK] *agr. for. hort.* ►storm damage.

wind-blown [pp/adj] *landsc.* ►wind-swept.

7023 wind-blown sand [n] *pedol.* (Fine granular soil material accumulated through aeolian shifting process); *syn.* aeolian sand [n]; *s* **arena** [f] **eólica** (Arena transportada por el viento); *f* **sable** [m] **éolien** (Sable transporté sous l'action du vent) ; *g* **Flugsand** [m] (vom Wind transportierter Sand).

7024 wind-blown tree [n] [US] *arb.* (Tree with one-sided growth due to the long-term influence of prevailing winds; ►wind deformation, ►wind-swept); *syn.* wind-deformed tree [n], wind-swept tree [n], wind-trimmed tree [n] [UK]; *s* **árbol** [m] **con copa de bandera** (Copa de un árbol deformada por la acción del viento en una dirección predominante; *término específico* **matorral** [m] **en forma de duna**: arbusto deformado por acción del viento; ►anemorfosis, ►deformado por el viento); *f* **arbre** [m] **en drapeau** (Couronne d'un arbre soumise à l'influence constante du vent ou à l'action conjointe de cristaux de glace et ne présentant de ramure que du côté abrité du vent ; ►anémomorphose, ►couché, ée par le vent) ; *g* **zur Fahne geschorener Baum** [m] (Durch ständigen Wind oder zusätzlich noch durch Eiskristallgebläse verformte Baumkrone, die meist nur einen leeseitigen Astbehang hat; ►windgeschoren, ►Windschur); *syn.* windgeschorener Baum [m], windverformter Baum [m].

7025 windbreak [n] *landsc.* (Generic term for *a* barrier serving to break the force of the wind and to protect soil and plants from wind damage, which is often done by planting a strip of several rows of trees and shrubs; **w.** and 'shelterbelt' are often used synonymously; the former is more appropriate for a fence or screen, or a single narrow row of plants, while the latter refers to several rows of trees or shrubs; cf. ARB 1983, 125; in midwestern U.S., the term 'shelterbelt' applies specifically to a wide **w.** in extensive agricultural areas; ►windbreak hedge, ►windbreak planting); *syn.* shelterbelt [n] (2); *s* **protección** [f] **contra el viento** (**1.** Medidas de protección del suelo y la vegetación contra daños causados por el viento, generalmente en forma de pantallas artificiales [muros] o naturales [bandas de leñosas de varias hileras, setos] para reducir la velocidad del viento. **2.** Instalación protectora contra el viento; ►cortina rompevientos, ►seto rompevientos); *f* **protection** [f] **contre le vent** (Mesures de protection au moyen de dispositifs [brise-vent, abrivent] artificiels [écran, grillage, paillasson, mur] ou naturels [►haie brise-vent, ►plantation brise-vent] destinés à réduire la vitesse du vent et de protéger ainsi le sol et les cultures) ; *g* **Windschutz** [m, o. pl] (**1.** Maßnahmen zum Schutze des Bodens und der Pflanzen gegen Windschäden — meist in Form von mehrreihigen Gehölzstreifen, Hecken, Mauern oder ortsfesten Zäunen zur Bremsung der Windgeschwindigkeit. **2.** Windschutzeinrichtungen als Ergebnis der unter Ziff. 1 genannten Maßnahmen; ►Windschutzhecke, ►Windschutzpflanzung).

windbreak [n]**, forest** *landsc.* ►forest wind shield.

windbreak [n]**, porous** *landsc.* ►wind-penetrable plantation.

7026 windbreakage [n] *for. hort.* (Breaking of tree stems or branches caused by strong wind or storm; ►storm damage, ►wind damage, ►windthrow); *syn.* wind slash [n] (SAF 1983); *s* **rotura** [f] **por viento** (Rotura de troncos, ramas, etc. causada por vientos fuertes; ►daños causados por el viento, ►daños causados por tormenta, ►derribamiento de árbol [por el viento]); *syn.* rotura [f] por tormenta; *f* **bris** [m] **de vent** (Parties d'arbres,

branches, brisés ou abattus par le vent ; les deux tempêtes de décembre 1999 ont causé, en France, la destruction d'environ 500 000 hectares de forêts ; ►chablis déraciné [par le vent], ►dégâts causés par la tempête, ►dommage causé par le vent) ; *syn.* bois [m] chablis ; *g* **Windbruch** [m] (Durch starken Wind oder Sturm verursachter Schaden an Baumstämmen oder Ästen; im Falle eines Sturmes spricht man von **Sturmbruch**; das Ergebnis ist ,Sturmholz'. Der Orkan „Lothar" vom 26.12.1999 verursachte allein in Baden-Württemberg ca. 30 Millionen Festmeter Sturmholz und vernichtete ca. 6 % des stehenden Holzvorrates des Bundeslandes — ein doppelt so hoher Schaden wie bei den Orkanen „Vivian" und „Wibke" im Februar 1990; während des Orkans „Kyrill" am 18.-19.01.2007, der noch heftiger als „Lothar" war, fielen in Deutschland ca. 37 Mio. Kubikmeter Holz in den Forstbeständen; cf. Wikipedia; ►Sturmschaden, ►Windschaden, ►Windwurf).

7027 windbreak hedge [n] *landsc.* (*Generic term* ►windbreak planting); *s* **seto** [m] **rompevientos** (*Térmimo genérico* ►cortina rompevientos); *f* **haie** [f] **brise-vent** (LA 1981, 201 ; obstacle végétal en forme d'une haie qui filtre et réduit, sans formation de turbulences, la vitesse du vent sur une distance égale à 10-15 fois sa hauteur ; elle protège en outre contre le froid, le soleil et l'érosion, améliore les rendements des cultures, dans la lutte contre l'érosion, elle contribue au développement des espèces animales sauvages et constitue dans le paysage agricole un élément du réseau d'habitats naturels maintien l'équilibre entre les espèces animales. La haie brise-vent se compose de 3 éléments : **1. les arbres de haut-jet** se caractérisent par un tronc sans branche sur une longueur de 4 à 5 mètres et un houppier composé de branches au-delà. Le houppier constitue la protection haute de la haie ; **2. les arbres conduits en cépée** taillés très court dès la seconde année de plantation afin de se développer tel un taillis et qui assurent la protection de la haie dans sa partie intermédiaire ; **3. les arbustes buissonnants** assurent quant à eux l'imperméabilité de la base de la haie. Ils apportent aussi couvert et nourriture aux animaux sauvages et aux oiseaux ; cf. www.planfor.fr , ►plantation brise-vent) ; *g* **Windschutzhecke** [f] (Je nach Situation eine ein- bis dreireihige Pflanzung, meist aus Bäumen und Sträuchern, zur Abschirmung von Gebäuden und Gärten gegen Wind; in der Agrarlandschaft zur Minderung der Windgeschwindigkeit und somit zur Verbesserung der Kulturbedingungen und örtlichen Umweltfaktoren sowie zum Ausbau eines Biotopverbundes; *OB* ►Windschutzpflanzung, die oft auch synonym mit **W.** gebraucht wird).

7028 windbreak planting [n] *landsc.* (One or several rows of trees and shrubs to reduce wind velocity and thereby to ameliorate local site conditions; ►shelterbelt 1, ►windbreak, ►windbreak hedge); *s* **cortina** [f] **rompevientos** (Plantación de una o varias filas de árboles y arbustos para reducir la velocidad del viento y mejorar las condiciones ambientales; ►franja protectora, ►protección contra el viento, ►seto rompevientos); *syn.* faja [f] rompevientos (SILV 1979, 85); *f 1* **plantation** [f] **brise-vent** (Alignement de plusieurs rangées d'arbres et d'arbustes constituant un écran végétal destiné à protéger les habitations, cultures, etc. contre le vent ; ►haie brise-vent, ►protection contre le vent, ►rideau forestier) ; *f 2* **bande** [f] **boisée brise-vent** (Peuplement forestier installé sur le côté le plus exposé d'une parcelle afin de protéger du froid et du vent la végétation située derrière ; TSF, 1985, 41) ; *g* **Windschutzpflanzung** [f] (Mehr linien- als flächenhafte Pflanzung, mit i. d. R. geringer Breite [ein- bis mehrreihig], meist aus Bäumen und Sträuchern, zur Minderung der Windgeschwindigkeit und somit zur Verbesserung der örtlichen Umweltfaktoren und zum Ausbau eines Biotopverbundes, besonders in ausgeräumten Agrarlandschaften; **W.en** dienen dem Schutze vor Windschäden,

W

der Verminderung der Flugerdebildung, der Erhaltung der Bodenfeuchtigkeit und in manchen Gebieten der Verringerung der Temperaturgegensätze, in schneereichen Regionen auch der Schneebindung; ▶Windschutz, ▶Windschutzhecke, ▶Schutzstreifen); *syn.* Windschutzstreifen [m], Windschutzanlage [f].

7029 wind corridor [n] *landsc.* (Narrow longitudinal area without obstacles, through which wind currents can flow); *s* **canal [m] de viento** (Franja de terreno libre de obstáculos para facilitar el desplazamiento de las masas de aire que ventilan la ciudad); *syn.* corredor [m] de viento; *f* **couloir [m] de vent** (Zone étroite, dépourvue d'obstacles le long de laquelle s'écoule le vent) ; *syn.* canal [m] de vent ; *g* **Windschneise [f]** (Langgestreckter, verengter, hindernisfreier Bereich, durch den der Wind strömen kann).

7030 wind damage [n] *agr. for.* (Injury, essentially mechanical, to twigs, foliage, flowers or fruits, usually of a tree by strong winds exceeding No. 8 on the BEAUFORT Scale—> 62 km/h [39,1 mph]; ▶storm damage; ▶windbreakage); *s 1* **daños [mpl] causados por el viento** (Daños mecánicos en las ramas, hojas o flores, hasta la destrucción de la vegetación causados por vientos fuertes [fuerza del viento superior a 8 de la escala de BEAUFORT, de 13 grados (de 0 a 12) y dada la velocidad media del viento durante un mínimo de 10 minutos de 62-74 km/h]; ▶daños causados por tormenta, ▶rotura por viento); *s 2* **agostamiento [m] por el viento** (DFM 1975; desecación y muerte de órganos vegetales jóvenes o de plantas enteras debida a la acción del viento excesivo); *f* **dommage [m] causé par le vent** (Ensemble des dégâts mécaniques causés par un vent très violent, par un coup de vent, [vent de force > 8 de l'échelle de Beaufort, comportant 13 degrés [de 0 à 12] et donnant la vitesse moyenne du vent sur une durée de dix minutes — 62-74 km/h] sur les bâtiments et la végétation ; les dégâts plus importants sont causés par les tempêtes de vent ; ▶bris de vent, ▶dégâts causés par la tempête) ; *syn.* dégâts [m] du vent (SOL 1987, 5) ; *g* **Windschaden [m]** (...schäden [pl]; durch starken Wind [Windstärke > 8 der 12-teiligen BEAUFORT-Skala] hervorgerufene Beschädigung an Gebäuden und Vegetation; noch größere Zerstörungen werden durch Stürme verursacht: ▶Sturmschaden, ▶Windbruch).

7031 wind deformation [n] *phyt.* (Disfigurement caused by continual strong winds against trees and shrubs; ▶life form, ▶wind-blown tree, ▶wind-swept); *s* **anemorfosis [f]** (Crecimiento asimétrico de la copa debido a fuertes vientos predominantes en una dirección; ▶árbol con copa de bandera, ▶deformado por el viento, ▶forma biológica; cf. SILV 1979, 88); *syn.* abrasión [f] eólica (BB 1979, 286); *f* **anémomorphose [f]** (Modifications morphologiques des végétaux ligneux causée par l'action du vent, p. ex. port prostré, en drapeau, en coussinet ; ▶arbre en drapeau, ▶forme naturelle, ▶couché, ée par le vent, ▶type biologique) ; *g* **Windschur [f]** (Durch ständigen, starken Wind entstehendes Deformationsbild an Gehölzen; ▶Lebensform, ▶windgeschoren, ▶zur Fahne geschorener Baum).

wind-deformed [pp/adj] *landsc.* ▶wind-swept.

wind-deformed tree [n] *arb.* ▶wind-blown tree [US]/wind-trimmed tree [UK].

7032 wind direction [n] *met.* (Direction from which wind blows; ▶prevailing wind direction); *s* **dirección [f] del viento** (Dirección de donde viene el viento; ▶dirección del viento predominante); *f* **direction [f] des vents** (▶direction principale des vents) ; *g* **Windrichtung [f]** (Richtung, aus der ein Wind weht; ▶Hauptwindrichtung).

7033 wind dispersal [n] *phyt.* (Transport of diaspores by wind); *syn.* anemochory [n]; *s* **anemocoria [f]** *syn.* dispersión [f] por el viento, diseminación [f] por el viento; *f* **anémochorie [f]** (Dissémination des semences par le vent ; espèce anémochore) ;

g **Verdriftung [f] durch Wind** (Verbreitung von Diasporen oder Insekten durch Wind); *syn.* Anemochorie [f], Verbreitung [f] durch Wind, Verfrachtung [f] durch Wind, Windverbreitung [f].

7034 wind energy [n] *envir.* (Renewable source of electricity, which is generated by wind-powered rotor blades, traditionally in a windmill, but increasingly by more complicated designs including turbines. It is regarded as environmentally friendly, because there are no polluting emissions except for noise; ▶wind energy plant); *syn.* aeolian energy [n]; *s* **energía [f] eólica** (Energía renovable producida por el viento. Por medio de aeroturbinas se transforma la energía cinética en energía eléctrica. Se considera poco impactante al medio ambiente ya que aparte de ruidos no produce emisiones; ▶aerogenerador de energía eléctrica); *f* **énergie [f] éolienne** (Ressource renouvelable utilisée pour la production d'électricité générée par des ▶éoliennes; l'énergie éolienne couvre en France fin 2007 1 %, en Allemagne 7 % de la consommation électrique nationale et avec 2 455 MW la France était le 6ème producteur d'énergie éolienne en Europe; depuis 2005, la loi de programme n° 2005-781 du 13 juillet 2005 fixe les orientations de la politique énergétique et modifie le régime d'obligation d'achat de l'électricité éolienne ; elle introduit le principe de **zones de développement de l'éolien** [Z.D.E.], définies par le préfet sur proposition des communes concernées, et qui permettent aux installations éoliennes qui y sont situées de bénéficier de l'obligation d'achat. Ces zones sont définies en fonction du potentiel éolien, des possibilités de raccordement aux réseaux électriques et de la protection des paysages, des monuments historiques et des sites remarquables et protégés) ; *syn.* éolien [m] ; *g* **Windenergie [f]** (Erneuerbare Energie aus bewegten Luftmassen der Atmosphäre, die durch windbetriebene Rotoren in elektrischen Strom umgewandelt wird; sie wird als umweltfreundlich angesehen, da sie außer Lärm keine Emissionen verursacht. Der Anteil der **W.** am gesamten Stromverbrauch der Bundesrepublik Deutschland lag im Jahre 2007 bei ca. 7 %; cf. wind-energie.de/de/statistiken; ▶Windenergieanlage); *syn.* Windkraft [f, o. Pl.].

7035 wind energy plant [n] *envir.* (Installation which generates electricity from natural energy of wind currents with the help of rotor blades, wind wheels, turbines, etc.); *syn.* wind turbine [n]; *s* **aerogenerador [m] de energía eléctrica** (Instalación de producción de energía con ayuda del viento); *syn.* máquina [f] eólica, aeroturbina [f], *obs.* molino [m] de viento; *f* **éolienne [f]** (Installation de production d'électricité à partir de l'énergie naturelle du vent au moyen de d'hélices géantes ; cette forme d'énergie propre ne produit pas de nuisances sur l'environnement à l'exception du bruit, de la projection d'ombre ou le passage du reflet du soleil [effet stroboscopique] induits par la rotation des pales, du flash lumineux du balisage nocturne ainsi que de l'impact sur les chauves-souris ou sur les oiseaux lorsque les éoliennes sont éclairées de nuit ou situées sur un corridor de migration) ; *g* **Windenergieanlage [f]** (*Abk.* WEA; Anlage zur Gewinnung elektrischer Energie aus der natürlichen Energie der Windströmung mit Hilfe von Rotoren, Windrädern, Turbinen etc. In D. sind bis zum 31.12.2007 19 460 **W.n** mit einer installierten Leistung von ca. 22 247 MW in Betrieb; cf. wind-energie.de/de/statistiken. Seit der Novellierung des BauGB 1997 sind Anlagen zur Nutzung der Windenergie im ▶Außenbereich privilegiert, d. h., sie sind i. d. R. zulässig, wenn keine öffentlichen Belange entgegen stehen; cf. N+L 2004 [11], 507); *syn.* Windanlage [f], Windkonverter [m], Windkraftanlage [f], Windkraftwerk [n].

7036 wind erosion [n] *geo.* (Generic term covering ▶deflation and corrasion; removal of material from a beach, desert or other land surface by wind action and smoothing of stones by abrasive action wind-blown sand [corrasion]; cf. GGT 1979);

s **erosión [f] eólica** (Proceso de barrido, abrasión y arrastre de partículas del suelo por la acción del viento; DINA 1987; término genérico de ▶deflación y corrasión); *f* **érosion [f] éolienne (2)** (Terme générique caractérisant le transport éolien de matériaux meubles et secs arrachés par le vent [▶déflation éolienne] et l'érosion exercée par la dérive des grains [corrasion] ; *terme spécifique* vannage : déflation sélective selon la granulométrie ; DG 1984, 163) ; *g* **Windabtragung [f]** (Entfernung des Oberbodens durch starke Winde [Bodendegradation]; größere Bodenteilchen mit einer Korngröße von 0,1-1,0 mm werden vom Wind rollend und springend fortgetragen; intensive sichtbare **W.** findet in den großen Trockengebieten der Erde statt; weitere Folgen der **W.** sind Geländedeformationen durch sog. Ausblasungskavernen, Wanderdünenbildung und die Überdeckung von Böden mit erodierten Feinteilen an anderer Stelle; OB zu ▶Deflation und Korrasion — i. S. v. Windschliff oder Sandschliff); *syn.* Winderosion [f].

windfall [n] *landsc.* ▶windthrow.

7037 windfall area [n] *landsc.* (Any area on which trees have been thrown or broken by the wind; SAP 1983; ▶windthrow); *syn.* blowdown area [n] [also US]; *s* **zona [f] de daños causados por tormenta/huracán** (Cualquier zona en la que ha habido una tormenta o un huracán y en la que han ocurrido daños considerables; ▶derribamiento de árbol [por viento]); *f* **zone [f] de chablis** (Une étendue de terrain où les chablis sont nombreux ; ▶chablis déraciné [par le vent]) ; *syn.* zone [f] de bris de vent (DFM 1975) ; *g* **Sturmschadensgebiet [n]** (Durch Einwirkung von Sturm verwüstetes Gebiet; ▶Windwurf).

7038 wind farm [n] *envir.* (Area of land on which several ▶wind energy plants have been installed; a w. f. on the sea is called **offshore w. f.**, on the waterfront **onshore w. f.**); *s* **parque [m] eólico** (Terreno en el que están instalados varios ▶aerogeneradores de energía eléctrica. Un **p. e.** en el mar es un **parque eólico marítimo**); *f* **ferme [f] éolienne** (Aire sur laquelle sont installées plusieurs ▶éoliennes ; **F.**, le plan de développement des énergies renouvelables, a pour objectif de porter à au moins 23 % la part des énergies renouvelables dans la consommation d'énergie à l'horizon 2020. A cet effet, il prévoit notamment une accélération du développement de l'éolien en mer [**parc éolien maritime**] ; *syn.* parc [m] éolien) ; *g* **Windenergiepark [m]** (Gelände, auf dem mehrere ▶Windenergieanlagen installiert sind; nach dt. Recht müssen gem. Nr. 1.6 der Anlage zum UVPG und der Nr. 1.6 des Anhangs zur 4. BImSchV mindestens drei Windkraftanlagen einander räumlich so zugeordnet sein, dass sich ihre Einwirkungsbereiche überschneiden oder wenigstens berühren; cf. BVerwG-Urteil vom 30.06.2004 — 4 C 9.03; einen **W.** auf dem Meer nennt man **Offshore-Windpark** oder **maritime Windfarm**); *syn.* Windfarm [f], Windmühlenpark [m], Windpark [m].

7039 wind funnel [n] *met.* (Narrowing of a wind opening, e.g. in a windbreak planting or between blocks of buildings; ▶venturi-effect); *s* **tolva [f] de viento** (Estrechamiento de un canal de flujo del viento, p. ej. por medio de plantaciones rompevientos o entre los bloques de casas; ▶efecto venturi); *syn.* embudo [m] de viento; *f* **goulet [m] d'étranglement** (Étranglement formé par le rétrécissement d'un couloir à vent provoquant une accélération de la vitesse d'écoulement du vent, p. ex. en météorologie le couloir rhodanien, entre le Massif central et les Alpes, peut être comparé, à grande échelle, à un goulet d'étranglement dans la zone de circulation des masses d'air venues du nord et qui est responsable de la grande vitesse du Mistral en Provence ; ▶effet venturi) ; *syn.* goulet [m] de vent, goulot [m] de vent, couloir [m] de vent, canal [m] de vent, canal [m] venturi ; *g* **Winddüse [f]** (Verengung eines Winddurchlasses, z. B. bei

Windschutzpflanzungen oder zwischen Häuserblöcken; ▶Düseneffekt).

winding river [n] *geo.* ▶meandering river.

windlean [adj] *landsc.* ▶wind-swept.

7040 wind lessening [n] *met. landsc.* (Reduction in the velocity of wind by obstacles; e.g. by ▶windbreak planting); *s* **frenado [m] del viento** (Reducción de la velocidad del viento, p. ej. por medio de ▶cortinas rompevientos); *f* **freinage [m] du vent** (Diminution par des obstacles de la vitesse du vent : vent freiné ; SOL 1987, 5 ; ▶plantation brise-vent) ; *g* **Windbremsung [f]** (Verlangsamung der Windgeschwindigkeit durch Hindernisse, z. B. durch eine ▶Windschutzpflanzung).

7041 wind load [n] *arb. constr. stat.* (ARB 1983, 69; wind pressure on an object from the structural engineering point of view; e.g. wind pressure upon the crown of a tree or against a screen or structure); *syn.* wind pressure [n]; *s* **presión [f] del viento** (Presión total estáticamente relevante causada sobre un objeto por la acción del viento; *syn.* acción [f] del viento, empuje [m] debido al viento (ambos BU 1959); *f* **pression [f] du vent** (Pression totale exercée par le vent sur une surface exposée à son action) ; *g* **Winddruck [m]**. (Durch Wind hervorgerufene, statisch relevante Last auf ein Objekt); *syn.* Windbelastung [f], Windlast [f].

7042 window box [n] *gard.* (Long narrow plant container for hanging outside a windowsill or a railing; ▶planter); *s* **jardinera [f] para balcón** (Recipiente para plantas largo y estrecho para colocar en un balcón o un petril; ▶jardinera); *f* **jardinière [f] pour balcon** (▶bac de plantation) ; *syn.* bac [m] pour balcon ; *g* **Balkonkasten [m]** (Auf oder an der Brüstung eines Balkons befestigter Kasten zum Bepflanzen mit [Sommer]blumen, Kleinstauden oder Kleingehölzen; *OB* Blumenkasten; ▶Pflanzenkübel).

window well [n] [US] *arch.* ▶basement light well [US].

7043 window well grate [n] [US] *constr.* (Grill-like metal cover for a basement light well [US]/light well [UK]); *syn.* light well grid [n] [UK]; *s* **parrilla [f] de pozo de luz** (Cobertura de enrejado de metal para cubrir un pozo de luz); *f* **grille [f] de cour anglaise** (Treillis métallique placé sur une cour anglaise devant un soupirail ou une fenêtre ; MAÇ 1981, 460) ; *syn.* grille [f] de puits de lumière ; *g* **Lichtschachtrost [m]** (Gitterartige Metallabdeckung für einen Lichtschacht).

7044 wind-penetrable [adj] *landsc. land'man.* (Descriptive term for an obstacle which allows wind to penetrate; e.g. a ▶windbreak hedge); *s* **permeable al viento [loc]** (Término descriptivo para un obstáculo que deja pasar al viento, p. ej. ▶seto rompevientos poroso); *f* **perméable au vent [loc]** (LA 1981, 478 ; propriété d'un obstacle de laisser passer le vent, p. ex. les ▶haies brise-vent claires) ; *g* **winddurchlässig [adj]** (So beschaffen, dass Wind ein Hindernis durchdringen kann, z. B. lockere ▶Windschutzhecken).

7045 wind-penetrable plantation [n] *landsc.* (Area of tree and shrub planting not dense enough to block the passage of wind; ▶wind penetrable, ▶windbreak planting); *syn.* porous windbreak [n] (LE 1986, 325); *s* **plantación [f] permeable al viento** (Grupo de leñosas que no impide el paso del viento aunque sí frena su velocidad; ▶cortina rompevientos, ▶permeable al viento); *syn.* plantación [f] hueca; *f* **plantation [f] perméable au vent** (Plantation ligneuse espacée qui ne retient pas le passage du vent mais le laisse traverser ; ▶perméable au vent, ▶plantation brise-vent) ; *g* **durchblasbare Pflanzung [f]** (Locker aufgebaute Gehölzpflanzung, die den Wind beim Abbremsen nicht staut, sondern durchstreichen lässt; ▶winddurchlässig, ▶Windschutzpflanzung).

W

wind pressure [n] *arb. constr. stat.* ►wind load.

wind pressure [n] [US]**, uplift** *met. constr.* ►wind suction.

wind rock [n] [UK] *arb. constr.* ►wind rocking [US].

7046 wind rocking [n] [US] *arb. constr.* (Loosening of the root ball of a tree or other plant through the oscillation of the stem by wind; BS 3975: Part 5); *syn.* wind rock [n] [UK]; *s* **desarraigue [m] debido al viento** (Despeje del suelo de la raíz de un árbol o de otra planta causado por el movimiento provocado por el viento); *f* **déchaussement [m] dû au vent** (Désolidarisation du système racinaire [léger déracinement] provoquée par le balancement du tronc sous l'action du vent) ; *g* **Lockerung [f] des Wurzelstocks durch Windeinwirkung** (Vom Winddruck verursachtes allmähliches Lösen des Wurzelwerks aus seiner Verankerung durch starke Stammschwankungen).

7047 wind rose [n] *met.* (Diagram with 8, 16 or 32 marks showing for one place the relative frequency and strength of winds from different directions at one particular location); *s* **rosa [f] de los vientos** (Diagrama con 8, 16 o 32 marcas que muestra la relativa frecuencia y fuerza de los vientos de diferentes direcciones para un lugar específico); *syn.* rosa [f] de frecuencias del viento, rosa [f] de velocidades del viento; *f* **rose [f] des vents** (À une station et pour une période donnée, diagramme étoilé [8, 16 ou 32 directions] indiquant les fréquences relatives des diverses directions du vent, en tenant compte, éventuellement, de groupes de vitesses du vent) ; *g* **Windrose [f]** (Grafische Darstellung der Windrichtungen für einen bestimmten Ort mit 8, 16 oder 32 Strichen, an denen die relative Häufigkeit und Stärke der Winde für einen gemessenen Zeitraum angetragen werden).

wind shield [n] *landsc.* ►forest wind shield.

wind slash [n] *for. hort.* ►windbreakage.

7048 wind suction [n] *met. constr.* (Severe atmospheric turbulence capable of lifting soil, plants, and even small structures from roof gardens; ►wind funnel); *syn.* wind uplift [n] [also US], uplift wind pressure [n] [also US] (UBC 1979, 125); *s* **acción [f] ascendente de torbellinos** (Remolino ascendente causado por el viento alrededor de edificios altos; ►tolva de viento); *f* **action [f] ascendante des vents tourbillonnants** (Fortes turbulences pouvant arracher des éléments non fixés sur les toitures ; ►goulet d'étranglement) ; *g* **Windsog [m]** (Starke Luftturbulenz, die nicht fest verankerte Einrichtungen wegreißen kann, z. B. Vegetationssubstrat oder Pflanzen auf Dächern; ►Winddüse).

7049 wind-swept [pp/adj] *landsc.* (ARB 1983, 132; descriptive term applied to trees or shrubs deformed on one side by continual strong winds; ►wind-blown tree [US]/wind-trimmed tree [UK], ►wind deformation); *syn.* wind-deformed [pp/adj], wind-trimmed [pp/adj], windlean [adj], wind-blown [pp/adj]; *s* **deformado por el viento [loc]** (Leñosa individual o población de leñosas deformadas físicamente por el viento; ►anemorfosis, ►árbol con copa de bandera); *f* **couché, ée par le vent [loc]** (Physionomie particulière de la végétation ligneuse déformée sous la forte action du vent ; DFM 1975, 6688 ; ►anémomorphose, ►arbre en drapeau) ; *syn.* déformé, ée par le vent [loc], torturé, ée par le vent [loc] ; *g* **windgeschoren [pp/adj]** (Durch ständigen, starken Windeinfluss [einseitig] deformiert, z. B. Einzelgehölz, Gehölzbestand; ►Windschur, ►zur Fahne geschorener Baum); *syn.* windverformt [pp/adj].

wind-swept tree [n] *arb.* ►wind-blown tree [US]/wind-trimmed tree [UK].

7050 windthrow [n] *landsc.* (Uprootal of a tree or trees by strong wind, or one or those which have been so uprooted; ►windbreakage); *syn.* windfall [n]; *s* **derribamiento [f] de árbol (por el viento)** (Caída de árbol causada por vientos fuertes;

►rotura por viento); *syn.* derrumbe [m] (por viento); *f* **chablis [m] déraciné (par le vent)** (Arbre naturellement renversé, déraciné ou rompu par le vent ; ►bris de vent) ; *syn.* déracinement [m] par le vent (DFM 1975, 48) ; *g* **Windwurf [m]** (Durch Sturm oder Windbö umgerissener Baum; ►Windbruch); *syn. bei sehr starkem Wind* Sturmwurf [m].

wind-trimmed [pp/adj] *landsc.* ►wind-swept.

wind-trimmed tree [n] [UK] *arb.* ►wind-blown tree [US].

wind turbine [n] *envir.* ►wind energy plant.

wind uplift [n] [US] *met. constr.* ►wind suction.

7051 windward side [n] *met.* (Side toward or facing into the wind; ►leeward side); *syn.* weather side [n]; *s* **barlovento [m]** (Lado en dirección al viento; *opp.* ►sotavento); *syn.* lado [m] de barlovento; *f* **face au vent [loc]** (dans la direction du vent ; ►côté sous le vent) ; *syn.* (côté) au vent [loc], lof [m] ; *g* **Luv [f u. n]** (*Meist ohne Artikel gebräuchlich* dem Wind zugewandte Seite; *opp.* ►Lee).

wine bower [n] *gard.* ►grapevine arbor [US]/grapevine arbour [UK].

wine bowery [n] *gard.* ►grapevine arbor [US]/grapevine arbour [UK].

winery [n] [US] *agr.* ►vineyard.

7052 wing wall [n] *constr.* (Short retaining wall at an angle to a bridge abutment); *s* **muro [m] en ala** (Muro de contención corto construído generalmente en ángulo con el correspondiente estribo de puente); *f* **mur [m] en aile** (Mur lié à la culée d'un pont et biais par rapport à celui-ci ; DTB 1985) ; *syn.* mur [m] en éventail ; *g* **Flügelmauer [f]** (Seitliche, meist abgewinkelte Stützmauer an einem Brückenwiderlager).

winning bidder [n] [US] *contr.* ►selected bidder [US]/selected tenderer [UK].

7053 winter bare [adj] *bot. hort.* (Horticultural term applied to woody plants which have no leaves in winter; ►deciduous; *opp.* ►evergreen); *syn.* bare [adj] in winter; *s* **desnudo/a [adj] en invierno [loc]** (Término horticultor para las plantas que no tienen hojas en invierno; ►caducifolio/a; *opp.* ►perennifolio/a); *f* **dépourvu, ue de feuillage pendant l'hiver [loc]** (±) (D., terme horticole appliqué aux végétaux qui ne portent pas de feuillage pendant l'hiver ; ►à feuilles caduques ; *opp.* ►à feuilles persistantes) ; *syn.* dénudé en hiver [loc] ; *g* **winterkahl [adj]** (*Gärtnerische Bezeichnung* eine Pflanze betreffend, die im Winter kein Laub hat; ►Laub abwerfend; *opp.* ►immergrün).

winter burn [n] [US] *agr. bot. for. hort.* ►frost-desiccation.

winter desiccation [n] *agr. bot. for. hort.* ►frost-desiccation.

7054 winter dike [n] *wat'man.* (Dike above the winter highwater line as opposed to a ►summer dike); *s* **dique [m] de invierno** (Dique para proteger contra las mareas altas de invierno y, por lo tanto, más alto que el ►dique de verano); *f* **digue [f] d'hiver** (Par comparaison avec la ►digue d'été, protection contre les grandes marées d'hiver) ; *g* **Winterdeich [m]** (Im Vergleich zum ►Sommerdeich ein besonders hoher Küstendeich zum Schutz gegen winterliche Sturmfluten).

winter drying [n] *agr. bot. for. hort.* ►frost-desiccation.

7055 winter flower [n] *hort. gard. plant.* (Flower of herbaceous or woody plants blooming in winter; e.g. winter aconites *[Eranthis]*, witchhazel *[Hamamelis]*, Fragrant Viburnum *[Viburnum farreri; syn. V. fragrans]*); *s* **flor [f] invernal** (Flor de leñosa o herbácea que se presenta en invierno, como en el matalobos de invierno *[Eranthis hyemalis]*, el olmo escocés *[Hamamelis]* o el viburno *[Viburnum farreri; syn. V. fragrans]*); *f 1* **fleur [f] hivernale** (Terme de jardinage désignant la fleur ou l'infloraison des végétaux ligneux et herbacés qui fleurissent en hiver, p. ex.

l'éranthe d'hiver [*Eranthis hyemalis*] jaune dès le début février, le noisetier de sorcière [*Hamamelis*] en variétés jaune, orange, rouge de janvier à mars, la viorne parfumée d'hiver [*Viburnum farreri* ; *syn. V. fragrans*] blanc à rose carmin, de longue floraison de décembre à mars) ; *f* 2 **floraison [f] hivernale** (Terme de jardinage désignant la période de floraison des végétaux ligneux et herbacés qui fleurissent en hiver) ; *g* **Winterblüte [m]** (1. Gärtnerische Bezeichnung für eine einzelne Blume oder einen Blütenstand einer krautigen oder Gehölzpflanze, die im Winter blüht, z. B. Winterling [*Eranthis*], Zaubernuss [*Hamamelis*], Duftender Schneeball [*Viburnum farreri; syn. V. fragrans*]. 2. Im Deutschen bedeutet **W.** auch der Blühzeitraum während des Winters).

7056 winter flowerer [n] *hort. gard. plant.* (Generic term covering herbaceous and woody species which bloom in winter [▶winter flower]; e.g. witchhazel [*Hamamelis*], Christmas Rose [*Helléborus niger*], winter aconites [*Eranthis*]; *s* **planta [f] de floración invernal** (Leñosa o herbácea que florece en invierno [▶flor invernal], en Europa Central p. ej. el olmo escocés [*Hamamelis*], el acónito de invierno [*Eranthis*]); *f* **plante [f] à floraison hivernale** (Végétaux herbacés ou ligneux qui fleurissent pendant l'hiver [▶fleur hivernale], p. ex. l'hamamelis [*Hamamelis*], l'aconit d'hiver [*Eranthis*]) ; *g* **Winterblüher [m]** (Gärtnerische Bezeichnung für eine krautige oder Gehölzpflanze, die im Winter blüht, z. B. Zaubernuss [*Hamamelis*], Christrose [*Helléborus niger*], Winterling [*Eranthis*]); eine einzelne Blume oder ein Blütenstand einer krautigen oder Gehölzpflanze, die im Winter blüht, ist eine ▶Winterblüte).

7057 winter green [adj] *bot. hort.* (Horticultural term for plants which keep their green leaves during the winter and lose them in spring before the new leaves appear; to a certain extent, ▶defoliation often already occurs in the fall/autumn; **w. g.** leaves often serve as storage organs for the new leaves. Due to the capacity of some leaves to last longer than others, there are various types of **w. g.**: spontaneous loss of leaves after the first strong frost, e.g. coral berry [*Symphoricarpos x chenaultii*], semi-evergreen: leaf-loss before new leaves appear, usually in March, e.g. honeysuckle [*Lonicera fragrantissima*] and **w. g.** leaves, which are shed with the appearance of new leaves with e.g. Burkwood viburnum [*Viburnum x burkwoodii*] and privet [*Ligustrum vulgare*]; ▶evergreen, ▶perennial 2, ▶summer green; *syn.* hiemvirent [adj]); *s* **verde [m] invernal** (Término horticultor para designar a las plantas de hoja caduca que las mantienen también en invierno y las pierden en primavera cuando crecen las nuevas hojas, p. ej. el aligustre [*Ligustrum vulgare*]; ▶defoliación, ▶perennifolio/a, ▶planta vivaz, ▶verde vernal]; *f* **semi-persistant, ante [adj]** (Terme horticole désignant les végétaux ligneux ▶à feuilles caduques et les ▶plantes vivaces qui gardent une partie de leur feuillage pendant l'hiver la ▶défeuillaison ayant lieu au plus tard au début du printemps à l'apparition des nouvelles feuilles ; **D.**, selon les conditions abiotiques on désigne par **persistant hors gel** [semi-persistant sous climat doux] les végétaux qui perdent leurs feuilles en cas de grand froid tel que la symphorine [*Symphoricarpos x chenaultii*], ou par **partiellement semi-persistant** les végétaux qui perdent leurs feuilles avant la pousse de printemps en général en mars tels que le chèvrefeuille odorant [*Lonicera fragrantissima*] et par **s-p.** ceux qui perdent leur feuillage lors de la pousse des nouvelles feuilles tels que le viorne du Burkwood [*Viburnum x burkwoodii*] et le Troën [*Ligustrum vulgare*] ; ▶à feuilles persistantes, ▶à feuillage vert pendant l'été) ; *g* **wintergrün [adj]** (Gärtnerische Bezeichnung für Laub abwerfende Gehölze und ▶Stauden, die auch während der Winterzeit ihr grünes Laub behalten, dieses dann spätestens im Frühjahr mit dem Laubaustrieb vollständig abwerfen. Oft erfolgt der ▶Laubfall teilweise

bereits im Herbst. Die **w.en** Blätter dienen oft als Reserveorgane für den Neuaustrieb. Aufgrund der unterschiedlichen Ausdauerfähigkeit der Blätter lassen sich Zwischenstufen unterscheiden: **fakultativ wintergrün** bei Laubfall nach den ersten stärkeren Frösten, z. B. Korallenbeere [*Symphoricarpos x chenaultii*], **halbwintergrün** bei Laubfall vor dem Neuaustrieb, meist im März, z. B. Wohlriechende Heckenkirsche [*Lonicera fragrantissima*] und **w.** bei Laubfall mit dem Neuaustrieb, z. B. Burkwoods Schneeball [*Viburnum x burkwoodii*] und Liguster [*Ligustrum vulgare*]; ▶immergrün, ▶sommergrün); *syn.* hiemvirent [adj]).

wintering [n] *bot.* ▶overwintering (1).

wintering ground [n] *zool.* ▶hibernation site.

7058 winterkill [n] *hort.* (Frost death of whole plants caused by lack of hardiness; ▶cold hardiness, ▶frost desiccation); *s* **muerte [f] por helada** (Destrucción de una planta causada por la ausencia de su ▶resistencia a las heladas; ▶desecación invernal); *f* **mort [f] par le gel** (d'une plante par manque de ▶résistance au gel ; ▶dessèchement par le gel) ; *g* **Erfrieren [n, o. Pl.]** (Absterben einer gesamten Pflanze wegen mangelnder ▶Frosthärte; ▶Frosttrocknis).

7059 winter perch [n] *zool.* (Place where a migrating bird sits during winter periods); *s* **lugar [m] de invernación** (Sitio en el cual pasan el invierno las aves migradoras); *syn.* cuartel [m] de invernada (DINA 1987); *f* **gîte [m] d'hivernation** (Site sur lequel les oiseaux migrateurs passent l'hiver) ; *g* **Überwinterungsplatz [m]** (Ort, an dem Zugvögel überwintern); *syn.* Winterrastplatz [m]).

winter resting place [n] *zool.* ▶hibernation site.

7060 winter seed stalk [n] *bot. hort.* (Dried, woody stem, with or without seed heads that normally remain on the plant during winter; *specific term also* dried seedheads; OEH 1990, 214; ▶perennial with lignified seed stalks); *s* **infructescencia [f] invernal** (Tallo seco lignificado con o sin cabezuelas que se conserva durante el invierno; ▶perenne con infructescencia invernal); *syn.* inflorescencia [f] fructífera de invierno; *f* **infrutescence [f] en hiver** (Ensemble des fruits regroupés sur le même axe qui succèdent aux fleurs et qui subsiste en général pendant l'hiver ; DIB 1988, 213 ; ▶vivace à infrutescence [lignifiée] pérenne) ; *syn.* infrutescence [f] hivernale ; *g* **Fruchtstand [m] im Winter** (Verholzter Blütenstand, mit oder ohne Fruchtkapseln, der nach der Fruchtreife meist als Ganzes überwintert; überwinternde Fruchtstände werden auch ‚trockene Samenstände‘ genannt; ▶Wintersteher); *syn.* trockener Samenstand [m] im Winter.

7061 winter sleep [n] *zool.* (Condition similar to normal sleep into which certain mammals fall at the beginning of winter, whereby all vital functions are reduced to a minimum; ▶hibernation, ▶hibernation area/region); *s* **hibernación [f] (2)** (Letargo invernal causado por descenso del metabolismo, que permite evitar el periodo más frío del invierno y las carencias alimenticias a la mayor parte de los invertebrados y a un gran número de vertebrados; DINA 1987; ▶área de invernación, ▶invernación 1); *f* **sommeil [m] hibernal** (État de vie ralentie de certains vertébrés pendant la saison froide ; DG 1984, 142 ; ▶hibernation, ▶territoire d'hibernation) ; *g* **Winterschlaf [m]** (Zu Beginn des Winters einsetzender schlafähnlicher Zustand einiger Säugetiere, in dem alle Lebensfunktionen auf ein Minimum reduziert sind; ▶Überwinterung 1, ▶Überwinterungsgebiet).

7062 winter sport(s) [n(pl)] *recr.* (All kinds of sports pursued on snow or ice in winter); *s* **deportes [mpl] de invierno** (Término genérico para todo tipo de deportes que se practican sobre la nieve o el hielo); *f* **sports [mpl] d'hiver** (Terme générique pour les sports pratiqués sur la neige ou sur la glace en

W

hiver) ; *g* **Wintersport** [m, o. Pl.] (Sammelbegriff für alle, besonders im Winter auf Schnee oder Eis betriebene Sportarten).

7063 winter sports resort [n] *recr.* (Community which provides and maintains accommodations and facilities for various types of winter sport activities); *s* **lugar** [m] **de deportes de invierno** (Municipio que ofrece hospedajes y facilidades para practicar diferentes tipos de deportes de invierno); *syn.* estación [f] de esquí; *f* **station** [f] **de sports d'hiver** (Commune dotée d'équipement permanents relevant des sports d'hiver) ; *syn.* station [f] d'altitude ; *g* **Wintersportort** [m] (Gemeinde, die Einrichtungen für Wintersportarten baut, pflegt und unterhält).

7064 winter stagnation phase [n] *limn.* (State of a lake or pond with stable thermic strata and two temperature zones: cold surface water and deep water with 4 °C. The **w. s. p.** is followed by the ▶spring circulation period); *s* **estratificación** [f] **invernal** (Estado de estratificación térmica débil de un lago templado dimíctico, con dos zonas, una superior fría y otra inferior de unos 4°C; cf. DINA 1987; ▶circulación vertical prevernal); *syn.* estagnación [f] invernal; *f* **stagnation** [f] **hivernale** (Profil de température hivernal d'un lac caractérisé par une couche superficielle dont la température peut atteindre 0 °C et d'une couche de température homogène de 4 °C ; après la **s. h.** suit le ▶brassage printanier) ; *g* **Winterstagnation** [f] (Zustand der stabilen thermischen Schichtung eines Sees mit zwei Temperaturzonen, dem kalten Oberflächenwasser und dem 4 °C warmen Tiefenwasser. Der **W.** folgt die ▶Frühjahrszirkulation).

7065 winter visitor [n] *zool.* (Bird, which overwinters in a certain area [▶overwintering (migratory) bird] or rests for a short while during its autumn migration; Central European **w.v.s** are mostly breeding birds of the far north, e.g. bean goose, common golden eye [duck], merlin [falcon]; *syn.* winter visitor bird [n]; *s* **visitante** [m] **invernal** (Especie de ave migratoria de otoño o postnupcial que descansa en la zona de referencia en su migración otoñal o que hiverna en la misma, donde viaja en busca de mejores condiciones meteorológicas y de alimento [▶ave invernante]; no se reproduce en esa zona que visita en invierno sino en otra área donde cría en verano; *opp.* ▶visitante estival); *f* **visiteur** [m] **d'hiver** (Espèce qui hiverne dans la région considérée [▶oiseau hivernant] et espèces de passage présentes durant leurs migrations de l'automne) ; *g* **Wintergast** [m] (Vogelart, die in einem bestimmten Gebiet überwintert [▶Überwinterer] oder die während ihres Zuges im Herbst gen Süden eine kurze Rast einlegt; mitteleuropäische Wintergäste sind meist Brutvögel des hohen Nordens, z. B. Saatgans, Schellente, Merlin).

winter visitor bird (species) [n] *zool.* ▶winter visitor.

7066 winter weather construction [n] *constr.* (Work executed by a contractor during periods of cold weather); *s* **trabajo** [m] **en invierno** (BU 1959; trabajos de construcción realizables en periodos/períodos de invierno con lluvias permanentes o heladas); *syn.* construcción [f] en invierno (BU 1959); *f* **travaux** [mpl] **d'hiver** (±) (Travaux pouvant être exécutés par l'entrepreneur en période d'hiver) ; *g* **Winterbau** [m] (Leistungen eines ausführenden Betriebes, die im Zeitraum von Anfang November bis Ende März bei witterungsbedingten Erschwernissen durchgeführt werden können; cf. auch § 32 [1] HOAI 2002).

7067 wired root ball [n] *hort.* (Root ball of a large woody plant stabilized with wire mesh for shipment; ▶ball wiring [of a tree]); *s* **cepellón** [m] **con malla metálica** (Cepellón estabilizado para el transporte; ▶embalaje en malla metálica [del cepellón]); *f* **motte** [f] **avec panier métallique** (Motte des végétaux ligneux renforcée au moyen d'un treillis ou grillage métallique dégradable en vue de leur transport ; ▶emballage en panier metallique) ; *syn.* motte [f] grillagée ; *g* **Drahtballen** [m] (Ein für den

Transport mit einem unverzinkten, möglichst geglühten Drahtgeflecht [Maschendrahtsicherung oder Drahtkorb] stabilisierter Wurzelballen eines Großgehölzes; ▶Drahtballierung); *syn.* Ballen [m] mit Drahtballierung.

7068 wire fence [n] *constr.* (Generic term for ▶chain-link fence and ▶barbed-wire fence); *s* **valla** [f] **metálica** (Término genérico de ▶valla de malla metálica, ▶valla de alambre de púas); *syn.* alambrada [f], cerca [f] de alambres); *f* **clôture** [f] **métallique** (*Terme générique* clôture métallique telle que la ▶clôture grillagée et ▶clôture en fil de fer barbelé) ; *g* **Drahtzaun** [m] (*UBe* ▶Maschendrahtzaun, ▶Stacheldrahtzaun).

7069 wire guy [n] *arb. constr. hort.* (Metal rope or cable securing newly planted large trees; ▶deadman, ▶guying); *s* **viento** [m] **de anclaje** (Cable de metal para asegurar árboles grandes transplantados; ▶fijación con vientos [metálicos], ▶rollizo de anclaje); *syn.* viento [m] de fijación, cable [m] de fijación; *f* **hauban** [m] (Câbles en fer galvanisés et en général torsadés, reliant un piquet solidement enfoncé dans le sol à un collier sur le tronc pour retenir les gros arbres plantés ; ▶câble de haubanage, ▶rondin d'ancrage) ; *g* **Drahtanker** [m] (Draht oder dünnes Drahtseil, das zur Befestigung von verpflanzten Großbäumen von einem im Boden versenkten [„toter Mann"] oder eingeschlagenen Holzpflock bis zum Kronenansatz gespannt wird; ▶Drahtseilverankerung, ▶Rundholzanker).

7070 wire mesh tree guard [n] *arb. constr. for.* (Protective wire fencing around a young tree trunk to prevent wild animal browsing damage); *syn.* wire netting tree guard [n]; *s* **malla** [f] **metálica de protección del tronco** (Dispositivo mecánico de protección de árboles jóvenes para evitar descortezamiento causado por animales); *f* **manchon** [m] **de protection** (Grillage ou filet en plastique pour la protection des jeunes plants contre les morsures des animaux); *g* **Drahthose** [f] (Mechanische Schutzvorrichtung am Stamm frisch gepflanzter Jungbäume gegen Wildverbiss und Fegeschäden).

wire netting tree guard [n] *arb. constr. for.* ▶wire mesh tree guard.

with legally-binding effect [loc] *adm. leg.* ▶legally-binding.

without weeds [loc] *agr. constr. for. hort.* ▶weedless.

witness [n] *prof.* ▶expert witness [US]/expert adviser [UK].

wood [n] [UK] (1) *landsc.* ▶woodlot [US&CDN] (2).

wood [n] [UK] (2) *ecol. for. phyt.* ▶woods [US].

wood [n] (3) *for. phys.* ▶density of wood; *envir.* ▶driftwood; *arb. bot.* ▶early wood; *phyt.* ▶fen wood; *for.* ▶hardwood; *arb.* ▶heartwood; *arb. bot.* ▶late wood; *phyt.* ▶pinewood; *arb.* ▶sapwood; *arb. for.* ▶softwood; *arb. hort.* ▶sound wood; *arb. for.* ▶wetwood; *landsc.* ▶woodlot [US&CDN] (2).

wood [n], **brush-** *arb. bot.* ▶brushwood thatching.

wood [n], **elfin-** *phyt.* ▶krummholz community.

wood [n], **gnarled** *phyt.* ▶krummholz.

wood [n], **mangrove** *phyt.* ▶mangrove stand.

wood [n] [US], **pinus-dominated elfin-** *phyt.* ▶pinus-dominated krummholz formation.

wood [n], **small** *landsc.* ▶woodlot [US&CDN] (2).

wood [n], **species living on** *phyt. zool.* ▶lignicolous species.

wood [n], **spring** *arb. bot.* ▶early wood.

wood [n] [US], **stunted** *phyt.* ▶krummholz.

wood [n] [UK], **suburban** *urb.* ▶suburban forest.

wood [n], **summer** *arb. bot.* ▶late wood.

wood [n], **swamp** *phyt.* ▶fen wood.

wood [n], **twisted** *phyt.* ▶krummholz.

7071 wood block paving [n] *constr.* (Hard surface covering with pavers cut from squared timber; ▶wood paving); *syn.* wood sett paving [n] [also UK], timber sett paving [n] [also UK]; *s* **pavimento** [m] **de madera en bloques** (▶Adoquinado de madera de elementos cuadrados o rectangulares); *f* **revêtement** [m] **en pavés de bois équarri** (Terme caractérisant un ▶pavage en bois utilisant une tranche de bois équarri) ; *g* **Kantholzpflaster** [n] (Belag aus rechteckigen oder quadratischen Holzelementen; ▶Holzpflaster).

wood board finish [n] *constr.* ▶concrete with wood board finish.

7072 wood disk paving [n] [US] *constr.* (Stretch of wooden cylinders—often varying in diameter—laid upright to form an even surface, and gaps brushed with sand or gravel; ▶wood paving); *syn.* round wood paving [n], timber disk paving [n] [UK]; *s* **adoquinado** [m] **de discos de madera** (Pavimento de piezas cilíndricas de madera que pueden variar en el diámetro; ▶adoquinado de madera); *f* **dallage** [m] **en tranches de bois** (*Terme spécifique pour un* ▶pavage en bois constitué de rondelles ou de tranches plus importantes [10 cm et plus], souvent de différents diamètres) ; *syn.* pavage [m] en tranches de bois ; *g* **Rundholzpflaster** [n] (Aus runden Holzscheiben oder –abschnitten bestehender Bodenbelag; ▶Holzpflaster).

wooded patch [n] [US] *agr. land'man.* ▶woody clump.

wooden deck module [n] [US] *constr.* ▶wooden grid.

7073 wooden edging [n] *constr.* (Border made of wooden boards; e.g. of a path); *s* **encintado** [m] **de madera** (Borde lateral de camino realizado en madera); *f* **bordurette** [f] **de voliges** (VRD 4/5.4, 4 ; fines planchettes de bois mise en place pour délimiter le bord de chemins) ; *g* **Holzeinfassung** [f] (Seitliche Begrenzung von Fußwegen mit Holzbrettern als kostensparende Alternative zu Pflastergurten).

7074 wooden fence [n] *constr.* (Generic term for an upright wooden barrier or enclosure; *specific terms* ▶close-boarded fence, ▶decorative board fence, ▶fence, ▶interwoven wood fence, ▶palisade fence, ▶picket fence, ▶wooden rail fence, ▶trellis fence); *syn.* timber fence [n] [also UK], wooden screen [n]; *s* **valla** [f] **de madera** (*Término genérico* ▶valla; *términos específicos* ▶cerca de parcela de pasto, ▶valla de empalizada, ▶valla de enrejado de madera, ▶valla de listones verticales, ▶valla de tablas perfiladas, ▶valla de tablones, ▶valla trenzada); *f* **clôture** [f] **en bois** (*Terme générique* ▶clôture 2 ; *termes spécifiques* ▶clôture à lamelles tressées, ▶clôture à lattes, ▶clôture à treillis, ▶clôture de lices/lisses en bois, ▶clôture de planches ajourée, ▶clôture [en bois] non jointive à lattes verticales, ▶clôture en planches jointives, ▶clôture jointive en rondins) ; *g* **Holzzaun** [m] (Aus Holz gefertigte Einfriedigung; *OB* ▶Zaun; *UBe* ▶Bretterzaun, ▶Holzflechtzaun, ▶Koppelzaun, ▶Lattenzaun, ▶Palisadenzaun, ▶Profilbretterzaun, ▶Scherenzaun, ▶Sichtschutzholzzaun aus Latten/Staketen).

7075 wooden grid [n] *constr.* (Framework composed of boards or laths and used as a low profile deck on grade; modules put together to form a 'wooden deck'; cf. TSS 1988, 460-30); *syn.* wooden deck module [n] [also US]; *s* **parrilla** [f] **de madera** (Paneles de madera entrelazada que se utilizan de pavimento); *f* **claie** [f] **en bois** (Lattes de bois fixées sur un cadre et formant les éléments d'un revêtement de sol) ; *syn.* caillebotis [m] en bois ; *g* **Holzrost** [m] (Auf Riegeln befestigte Latten oder Bretter als Bodenbelag für die Außengestaltung).

7076 wooden lattice [n] *constr.* (Wood lath construction for the support of climbing plants; *generic term* ▶climber support); *syn.* trelliswork [n], wooden trellis [n]; *s* **enrejado** [m] **de madera** (Estructura de listones para apoyar las plantas trepadoras; *término genérico* ▶enrejado de soporte de trepadoras); *f* **treillage** [m] **en bois** (Assemblage de bois fabriqué avec de fines lattes de bois, [autrefois le plus souvent avec des perches de châtaignier refendues de couleur bleu-vert, due à la bouillie bordelaise] ; le treillage à l'ancienne a des mailles rectangulaires assez larges (24 x 30 cm), très élégantes. En ville, la forme est plus dense avec des mailles carrées ou en losange de 15 cm de côté. Il a de multiples fonctions telles que 1. d'habiller des murs nus et peu esthétiques, dissimuler les défauts de construction, 2. d'accueillir comme support les plantes grimpantes sur les murs, 3. de former des séparations et moduler ainsi l'espace, 4. en jouant sur le dessin du treillage et sur la taille des mailles de profiter des effets d'optique pour donner plus de profondeur à un jardin et offrir des perspectives à l'infini [trompe-l'œil], etc. ; *terme générique* ▶support de plantes grimpantes ; *g* **Lattengerüst** [n] (Hilfskonstruktion aus Holzleisten als Rankhilfe für Kletterpflanzen; *OB* ▶Rankgerüst); *syn.* Lattenwerk [n] (GOT 1926-II).

7077 wooden rail fence [n] *constr. agr.* (Fence with wooden rails, 2-4cm thick, used to enclose pastures); *syn.* post and rail fence [n]; *s* **cerca** [f] **de parcela de pasto** (Valla con estacas horizontales de 2-4 cm de espesor para cerrar pastizales); *syn.* palizada [f] de parcela de pasto; *f* **clôture** [f] **de lices/lisses en bois** (HEC 1985, 65 ; clôture en bois constituée de 2 à 4 éléments horizontaux [p. ex. dosses] fixés sur des pieux/piquets en général pour délimiter des pâturages) ; *syn.* clôture [f] en barreaux (GEN 1982-II, 155) ; *g* **Koppelzaun** [m] (Holzzaun mit meist 2-4 cm dicken Querhölzern zum Einzäunen von Weiden); *syn.* Weidezaun [m].

wooden screen [n] *constr.* ▶wooden fence.

wooden trellis [n] *constr.* ▶wooden lattice.

wood fence [n] *constr.* ▶interwoven wood fence.

wood fence [n], **woven** *constr.* ▶interwoven wood fence.

woodland [n] *phyt.* ▶dune woodland community; *ecol.* ▶field-woodland edge [US]; *agr. for.* ▶grazed woodland; *conserv. land'man.* ▶protective coastal woodland; *phyt.* ▶regularly flooded riparian woodland; *agr. for.* ▶residual woodland; *phyt.* ▶riparian upland woodland; *phyt.* ▶riparian woodland; *landsc.* ▶woodlot [US&CDN] (2)/woodland [UK]; *ecol. for. phyt.* ▶woods [US]/wood [UK].

woodland [n] [UK], **alder swamp** *phyt.* ▶riparian alder stand [US].

woodland [n] [UK], **amenity** *for. plan. recr.* ▶recreation forest.

woodland [n], **broad-leaved** *for. phyt.* ▶deciduous forest.

woodland [n] [EIRE], **fen** *geo. phyt.* ▶swamp forest.

woodland [n] [US], **fen** *phyt.* ▶fen wood.

woodland [n], **lower riparian/riverine** *phyt.* ▶regularly flooded riparian woodland.

woodland [n], **municipal** *urb.* ▶town forest [UK].

woodland [n], **occasionally flooded riparian** *phyt.* ▶riparian upland woodland.

woodland [n], **pasture** *agr. for.* ▶grazed woodland.

woodland [n], **stand density of** *for.* ▶forest stand density.

woodland [n] [US], **upper riparian/riverine** *phyt.* ▶riparian upland woodland.

woodland [n], **wet alluvial** *phyt.* ▶riparian woodland.

7078 woodland canopy [n] *phyt.* (Sum of all tree crowns of a forest; ▶canopy closure, ▶crown cover, ▶tree canopy); *s* **dosel** [m] **del bosque** (Conjunto de las copas de un bosque; ▶cierre del

W

dosel del bosque, ▶cubierta de la copa, ▶dosel de árbol); *f* **couvert [m] continu des cimes** (Ensemble des houppiers d'un peuplement ; ▶couvert de la cime, ▶couvert vertical au sol, ▶fermeture du couvert) ; *syn.* canopée [f], couvert [m] forestier ; *g* **Kronendach [n] des Waldes** (Summe aller Baumkronen eines Waldes; ▶Kronenraum, ▶Kronenschirmfläche, ▶Kronenschluss 1); *syn. for.* Schirm [m].

woodland clearing [n] *phyt.* ▶forb vegetation in woodland clearing

woodland clearing community [n] *phyt.* ▶clearance herb formation.

woodland community [n] *phyt.* ▶dune woodland community.

woodland cover [n] *for. geo. phyt.* ▶forest cover.

woodland edge [n] *for.* ▶forest edge.

woodland edge [n] [US]**, prairie-** *ecol.* ▶field-woodland edge [US]/forest-meadow boundary [UK].

7079 woodland edge scrub community [n] *phyt.* (Woody plants forming the ▶forest mantle; ▶seam community); *syn.* woodland mantle scrub community [n] (ELL 1988), woody mantle community [n] [also US] (LE 1986, 108); *s* **comunidad [f] arbustiva del lindero del bosque** (Formación vegetal lineal formada sobre todo por arbustos que constituyen el ▶abrigo del bosque; ▶comunidad de la orla herbácea); *f* **groupement [m] de manteau préforestier buissonnant** (Formation végétale linéaire essentiellement arbustive formant le ▶manteau forestier ; ce peuplement constitue un faciès de dégradation des groupements forestiers qu'il précède et est rattaché à l'ordre de *Prunetalia spinosae* auquel se rattache également les haies vives ; ▶peuplement d'ourlet herbacé) ; *g* **Mantelgesellschaft [f]** (Licht liebende Laubholzgebüsche, die den ▶Waldmantel 1, aber auch frei wachsende Hecken bilden und in Mitteleuropa zur Ordnung der Hecken- und Gebüsche — Schlehen-Gesellschaften *[Prunetalia spinosae]* — gehören; am artenreichsten sind **M.en** bei Eichenmischwäldern, da sich hier im Vergleich zu Buchen-, Fichten- und Tannenbeständen keine bis zum Boden reichende Astschleppen entwickeln; ▶Saumgesellschaft).

woodland mantle scrub community [n] *phyt.* ▶woodland edge scrub community.

woodland patches [npl]**, hedgerows and** *agr. land' man.* ▶planting of hedgerows and woodland patches.

woodlot [n] [US] **(1)** *agr. land'man.* ▶coppice [US] (1).

7080 woodlot [n] [US&CDN] **(2)** *landsc.* (**1.** in U.S., segment of a woodland or forest; **2.** *in commercial terms* **w.** would be used to harvest, e.g. maple syrup or firewood; **3.** *in general terms* small area of woodland in an agricultural landscape; ▶coppice [US] 1, ▶woods [US]); *syn. to 3.* small wood [n], remnant wood patch [n], wood [n] [UK] (1), woodland [n] [UK], coppice [n] [UK] (5); *s* **bosquecillo [m] (2)** (Pequeña superficie de bosque en un paisaje agrícola; ▶bosque 3); *f 1* **bosquet [m] (1)** (Au sens de l'▶inventaire forestier national, toute surface, d'une largeur d'au moins 15 m, comprise entre 5 et 50 ares, où l'état boisé est acquis ; TSF 1985) ; *syn. langage courant* petit bois [m] ; *f 2* **boqueteau [m] (2)** (Au sens de l'▶inventaire forestier national, toute surface, d'une largeur d'au moins 25 m, comprise entre 50 ares et 5 ha, où l'état boisé est acquis ; TSF 1985 ; ▶bois) ; *f 3* **parquet [m]** (Grand bouquet, d'une surface supérieure à une dizaine d'ares) ; *g* **Wäldchen [n]** (Kleines Waldstück in der Agrarlandschaft — kleiner als *woods* [▶ Wald 2]).

wood pasture [n] *agr. for.* ▶forest pasture.

wood patch [n]**, remnant** *landsc.* ▶woodlot [US&CDN] (2).

7081 wood paving [n] *constr.* (Generic term for ▶paved stone surface made of wood; *specific terms*; ▶laying course for a paved stone surface, ▶wood block paving and ▶wood disk paving [US]/timber disk paving [UK]); *s* **adoquinado [m] de madera** (*Término genérico* ▶adoquinado 1; *términos específicos* ▶pavimento de madera en bloques, ▶adoquinado de discos de madera; ▶lecho de asiento [de adoquinado]); *syn.* pavimento [m] de madera, entarugado [m] de madera, firme [m] de madera; *f* **pavage [m] en bois** (Tranches de bois débitées à façon ronde carrée ou rectangulaire pour former un ▶pavage 1 disposé sur un lit de sable [ou de mortier] ; ▶lit de pose du pavage ; *termes spécifiques* ▶dallage en tranches de bois, ▶revêtement en pavés de bois équarri) ; *g* **Holzpflaster [n]** (Mit rechteckigen/quadratischen oder runden Holzelementen, meist 10 cm hoch, befestigte Bodenoberfläche, die auf wasserdurchlässigem ▶Pflasterbett verlegt und mit scharfem Sand oder Splitt ausgefugt wird; *OB* ▶Pflaster; *UBe* ▶Kantholzpflaster und ▶Rundholzpflaster).

7082 wood preservation [n] *chem. constr.* (All protective treatments to make wood more resistant to decay or against fire. There are two methods: **1.** dipping, painting or spraying, and **2.** pressure treatment); *s* **protección [f] de la madera** (Todo tipo de tratamiento de protección para hacerla más resistente a la intemperie o contra el fuego. En general se utilizan dos métodos: **1.** Baño, pintura o pulverización y **2.** tratamiento a presión); *syn.* preservación [f] de la madera; *f* **protection [f] du bois** (Tous les traitements du bois contre les agents biologiques et le feu. On distingue l'imprégnation en surface — trempage, application ou pulvérisation —, l'imprégnation à cœur en autoclave et l'ignifugation avec des sels bicarbonatés) ; *syn.* préservation [f] du bois ; *g* **Holzschutz [m, o. Pl.]** (Alle Maßnahmen, die dazu dienen, Holz gegen Verrottung und Feuer widerstandsfähig zu machen. Man unterscheidet den oberflächlichen **H.** durch Tauchbad, Anstrich oder Anspritzen des **H.mittels** und die Druckimprägnierung).

7083 wood preservative [n] *chem. constr. envir.* (Chemical solution used for wood protection); *s* **conservante [m] de la madera** (Sustancia química que sirve para evitar daños en la madera causados por insectos, humedad, fuego, etc.); *f* **produit [m] de protection du bois** (Produit chimique permettant de rendre un bois imputrescible, insensible à l'eau, aux insectes, aux champignons, aux rayons UV ou au feu) ; *syn.* produit [m] de traitement du bois, produit [m] de préservation du bois, agent [m] de conservation du bois ; *g* **Holzschutzmittel [n]** (Chemisches Mittel, das dazu bestimmt ist, tierische oder pflanzliche Schädlinge vom Holz fernzuhalten, gegen Witterungseinflüsse, Feuchtigkeit oder Feuer zu schützen); *syn.* Holzkonservierungsmittel [n].

7084 woods [npl] [US] *ecol. for. phyt.* (**1.** Plant community, composed principally of trees, with a smaller extent than a ▶forest 1; smaller than woods is a ▶woodlot [US&CDN] 2. **2.** U.K., area of woodland bearing a local name; SAF 1983, 6704); *syn.* wood [n] [UK] (2), woodland [n]; *s* **bosque [m] (3)** (**1.** Comunidad vegetal compuesta principalmente de árboles, de menor extensión que un ▶monte; de menor extensión aún es un ▶bosquecillo. **2.** En GB bosque con nombre propio local); *f* **bois [m]** (Espace arboré de taille moyenne ; terme parfois utilisé comme syn. de ▶forêt avec l'idée d'une surface moindre ; souvent accolé au terme forêt : nos bois et forêts) ; *g* **Wald [m] (2)** (Im Amerikanischen wird ein kleinflächiger Wald *woods*, ein großflächiger grundsätzlich *forest* genannt; ▶Wald 1).

woods [n] [US]**, suburban** *urb.* ▶suburban forest.

wood sett paving [n] [UK] *constr.* ▶wood block paving.

woody bee forage plant [n] *hort. landsc.* ▶woody host plant for bees.

W

woody bee plant [n] *hort. landsc.* ►woody host plant for bees.

7085 woody clump [n] *agr. land'man.* (►Group of trees and shrubs/small woody spinney [also UK] in a rural landscape); *syn.* wooded patch [n] [also US] (LE 1986, 100); *s* **bosquecillo** [m] **(3)** (Grupo pequeño de árboles y arbustos en paisaje campestre; ►bosquete, ►grupo de árboles); *f 1* **bosquet** [m] **(2)** (Petit massif forestier isolé dans le paysage rural d'une superficie de 10 à 50 ares ; au sens de l'►inventaire forestier national, toute surface, d'une largeur d'au moins 15 m, comprise entre 5 et 50 ares, où l'état boisé est acquis ; TSF 1985 ; ►groupe d'arbres, ►massif d'arbres ; *f 2* **bouquet** [m] (Groupe d'arbres de dimension et d'âge sensiblement voisin s'étendant sur quelques ares ; TSF 1985) ; *g* **Gehölz** [n] **(2)** (Kleine Gruppe von Bäumen und Sträuchern von 10-50 Ar in der bäuerlichen Kulturlandschaft; ►Baumgruppe); *syn.* Horst [m] (3).

woody food plant [n] **for birds** *landsc. plant.* ►woody host plant for birds.

woody growth [n] *landsc.* ►regenerated woody growth.

7086 woody heathland [n] *geo. land'man.* *s* **landa** [f] **arbolada**; *f* **lande** [f] **boisée** ; *g* **bewaldete Heidefläche** [f] *syn.* bewaldete Heide [f].

7087 woody host plant [n] *hort. landsc.* (Woody plant providing a food source for insects or birds. *Specific terms* ►woody host plant for bees, ►woody host plant for birds); *s* **leñosa** [f] **útil para insectos y aves** (Leñosa que ofrece gran cantidad de alimentos a las aves y a los insectos; *términos específicos* ►leñosa apícola, ►leñosa útil para las aves); *f* **ligneux** [m] **nourricier** (Végétaux sources de nourriture pour les insectes et les oiseaux ; *termes spécifiques* ►végétal ligneux mellifère, ►ligneux nourricier pour les oiseaux) ; *g* **Nährgehölz** [n] (Gehölz, das Insekten oder Vögeln reichlich Nahrung bietet; *UBe* ►Bienennährgehölz, ►Vogelnährgehölz).

7088 woody host plant [n] **for bees** *hort. landsc.* (Woody plant providing a food source for bees; ►food plant, ►host plant for bees, ►woody host plant); *syn.* woody bee forage plant [n], woody nectar plant [n], woody pollen plant [n], woody bee plant [n]; *s* **leñosa** [f] **apícola** (Árbol o arbusto que ofrece alimento para abejas; ►leñosa útil para insectos y aves, ►planta apícola, ►planta útil para animales salvajes); *f* **végétal** [m] **ligneux mellifère** (Se dit d'une plante ligneuse dont le nectar est récolté par les abeilles pour faire du miel ; *terme générique* ►plante nourricière ; ►ligneux nourricier, ►plante mellifère) ; *g* **Bienennährgehölz** [n] (Holzige ►Bienennährpflanze; ►Nahrungspflanze; *OB* ►Nährgehölz).

7089 woody host plant [n] **for birds** *landsc. plant.* (Tree or shrub providing birds with an important source of food; ►bird refuge plant); *syn.* woody food plant [n] for birds, bird forage plant [n]; *s* **leñosa** [f] **útil para las aves** (Árbol o arbusto que ofrece alimento a los pájaros; ►leñosa escondite para aves); *f* **ligneux** [m] **nourricier pour les oiseaux** (Plante ligneuse source de nourriture aux oiseaux sauvages ; ►plante ligneuse refuge pour les oiseaux) ; *g* **Vogelnährgehölz** [n] (Gehölz, das wild lebenden Vögeln Nahrung bietet; ►Vogelschutzgehölz).

woody mantle community [n] [US] *phyt.* ►woodland edge scrub community.

woody nectar plant [n] *hort. landsc.* ►woody host plant for bees.

7090 woody pioneer plant/species [n] *landsc. phyt.* (First woody plants to colonize an area; ►pioneer species); *s* **leñosa** [f] **pionera** (►especie pionera); *syn.* especie [f] leñosa pionera, planta [f] leñosa pionera; *f* **essence** [f] **ligneuse pionnière** (►Espèce pionnière ligneuse qui apparaît dans les premiers stades de recolonisation d'un milieu) ; *g* **Pioniergehölz** [n] (Holzige ►Pionierart, die sich am Anfang einer Bestockung in großen Stückzahlen ansiedelt und im Laufe der Sukzession anderen Arten Platz macht); *syn.* Pionierholzart [f].

woody plant [n] *bot.* ►broad-leaved woody species; *arb. hort.* ►cultivation of woody plants; *hort. plant.* ►flowering woody plant; *landsc. plant.* ►individual woody plant; *hort. plant.* ►ornamental woody plant, ►shade-tolerant woody plant; *bot. hort.* ►woody species (1).

7091 woody plant clearance [n] *constr. for.* (Removal of woody growth for ►site clearance or in preparation for new planting or afforestation; *specific term* ►clearance of unwanted spontaneous woody vegetation, scrub clearance, ►vegetation removal for replanting; ►stump out, ►tree clearing and stump removal [operation] [US]/grub felling [UK]); *s 1* **despeje** [m] (Operación de ►preparación del terreno que consiste en quitar impedimento u obstrucción para la realización de las obras. Su objeto son principalmente los árboles; y también los postes y elementos de algún tamaño que no queden comprendidos en la demolición; RA 1970; ►arranque de leñosas con cepellón, ►desbroce 1, ►despeje de árboles y tocones, ►extraer de tocones); *s 2* **desbroce** [m] **(2)** (Operación consistente en quitar la broza de la superficie y del interior del suelo; RA 1970); *f* **dégagement** [m] **du terrain** (Enlèvement de la végétation arbustive — avec ou sans souche — dans le cadre des travaux préliminaires de mesures de reforestation ; *termes spécifiques* enlèvement des gros bois du chantier, évacuation des produits d'abattage, nettoyage d'une coupe, défrichage [enlèvement de la végétation d'une friche agricole] ; ►abattage par extraction de souche, ►débroussaillage, ►dessoucher, ►enlèvement des végétaux réutilisables, ►nettoyage du terrain) ; *g* **Entfernen** [n, **o. Pl.] von Gehölzen oder Gehölzresten** (Abräumen von Gehölzen — mit oder ohne Stubben — im Rahmen der ►Baufeldfreimachung oder zur Vorbereitung von Bepflanzungs- oder Aufforstungsmaßnahmen; ►Entkusselung, ►Herausnahme des verwertbaren Pflanzenbestandes, ►Rodung, ►Stubben roden); *syn.* Abräumen [n, o. Pl.] der Gehölzvegetation.

woody plant for shade [n] *hort. plant.* ►shade-tolerant woody plant.

woody plants [npl] *arb. hort.* ►cultivation of woody plants.

woody pollen plant [n] *hort. landsc.* ►woody host plant for bees.

7092 woody species [n] **(1)** *bot. hort.* (**1.** Generic term for tree and shrub; general term for the following ►life forms: ►phanerophytes, ►surface plants [chamaephytes], and partially epiphytes; **2.** in plant nursery catalog[ue]s, w. s. are categorized according to specialities, local or regional markets, however, they are generally organized as follows: 1. Deciduous trees and shrubs, 2. Evergreen trees and shrubs, 3. Bamboo and grasses, 4. Roses, 5. Vines, and 6 Ground covers. Coniferous trees, Rhododendrons and fruit trees are sometimes listed separately; **in U.K.**, sizes for nursery stock of **w. s.** are: **Standard trees:** 8-10cm – 20-25cm girth; **Semi-mature trees:** 20-25cm – 40cm girth ; **Super semi-mature trees:** 40-80cm+ girth; **Container grown trees:** 8-80cm girth; **Specimen shrubs:** 10/15/45 liter container); *syn.* woody plant [n]; *s* **especie** [f] **leñosa** (**1.** Término genérico de árbol y arbusto; término general para las siguientes ►formas biológicas: ►caméfito, ►fanerófito y en parte epífitos. **2.** En viveros de plantas en GB se clasifican en las siguientes categorías: 1. Árboles y arbustos deciduos, 2. Árboles y arbustos siempreverdes, 3. Bambús y gramíneas, 4. Rosas, 5. Enredaderas y 6. Tapizantes); *syn.* planta [f] leñosa, leñosa [f]; *f* **espèce** [f] **ligneuse** (Terme générique pour l'arbre et l'arbrisseau rassemblant les ►types biologiques suivants : les ►phanérophytes, les

W

►chaméphytes et en partie les épiphytes [p.°ex. le gui]) ; *syn.* végétal [m] ligneux ; *g* **Gehölz** [n] (3) (1. OB zu Baum und Strauch; Sammelbegriff für folgende ►Lebensformen: ►Phanerophyt und Chamaephyt [►Oberflächenpflanze], z. T. auch Epiphyt [z. B. Mistel — *Viscum album*]. **2.** Nach der standardisierten „Namensliste der Gehölze" vom Bund deutscher Baumschulen [BdB] werden Baumschulgehölze in sechs Kapitel untergliedert: Laubgehölze, Bambus, Rhododendron, Nadelgehölze, Rosen und Obstgehölze).

woody species [n] (2) *bot. for.* ►broad-leaved woody species; *arb. hort.* ►dwarf woody species; *for. phyt.* ►light-demanding woody species; *phyt.* ►riparian woody species; *hort. plant.* ►wild woody species.

woody species [n], **heliophilous** *for. phyt.* ►light-demanding woody species.

woody species [n], **riverine** *phyt.* ►riparian woody species.

woody species planting [n] *constr. landsc.* ►tree and shrub planting.

woody vegetation [n] *conserv. land'man.* ►clearance of unwanted spontaneous woody vegetation.

7093 woody wildling [n] *for.* (Spontaneously-growing tree or shrub which has become established by natural seeding/regeneration; *specific term for trees* selfgrown tree [US]/self setter [UK]; ►spontaneous plants, ►weed tree); *s* **plántula** [f] **de renuevo** (Leñosa joven resultado de la regeneración natural; ►leñosa adventicia, ►vegetació espontánea 1); *syn.* arbolillo [m] de rejuvenecimiento, arbolillo [m] de renuevo; *f 1* **jeune plant** [m] **de semis** (Jeune plant ligneux obtenu par semis destiné à être proposé à la vente ou à être replanté à distance en pépinière d'élevage) ; *f 2* **semis** [m] **naturel** (Jeunes plants d'arbres, d'arbustes, de vivaces ou de graminées, issus de l'ensemencement naturel ; **D.**, ce terme n'est utilisé que pour les jeunes plants issus de végétaux ligneux ; ►végétation arbustive adventice, ►végétation spontanée 1) ; *g* **Wildling** [m] (3) (Durch natürliche Aussaat entstandener Baum oder Strauch; im Deutschen wird der Begriff **W.** nur für Gehölze verwandt; ►Fremdaufwuchs, ►Wildwuchs 2).

work [n] *contr.* ►accounting of executed work [US]/accounting of executed works [UK]; *arb. constr. contr.* ►additional technical contract conditions for tree work; *constr. contr.* ►additional work and services, ►agreed measurement of completed work [US]/agreed measurement of completed works [UK], ►all additional work and services included; *contr.* ►background work [US]/background research [UK]; *constr.* ►building work; *arch. constr.* ►broken range work; *constr. eng.* ►civil engineering works; *constr. contr.* ►completed work [US]/completed works [UK], ►daywork, ►determination of executed work [US]/assessment of executed works [UK]; *constr.* ►earthworks, ►execution of building works; *constr. contr.* ►extra work; *hort.* ►floristry work; *constr.* ►formwork; *constr. contr.* ►interim agreed measurement of completed work [US]/works [UK]; *constr.* ►landscape construction and maintenance work, ►machine work; *conserv'hist. constr.* ►maintenance and repair work; *constr. hort.* ►maintenance work; *constr. contr.* ►measurement of completed work [US]/measurement of completed works [UK]; *constr.* ►packed fascine-work; *plan.* ►planning work; *contr.* ►progress of work; *constr. contr.* ►proof of executed work; *adm.* ►public works department; *constr.* ►ragwork, ►remedial lawn work; *constr. contr.* ►repair work 1; *constr. eng.* ►road work [US]/road works [UK]; *constr.* ►rubble work; *constr. contr.* ►scope of construction work; *constr.* ►tree felling work; *arb. constr. hort.* ►tree treatment work; *constr. contr.* ►unforeseen construction work; *constr.* ►wattlework/wattle-work [UK].

work [n] [US], **completion of** *constr.* ►project completion [US]/practical completion [UK].

work [n], **execution of piece of** *constr.* ►execution of construction items.

work [n], **fence of lattice** *constr.* ►interwoven wood fence.

work [n] [US], **hand** *constr.* ►manual labor [US]/manual labour [UK].

work [n] [UK], **instruction to rectify defective** *constr. contr.* ►rectification directive [US].

work [n] [UK], **itemised** *constr. contr.* ►construction segment [US].

work [n], **laying** *constr.* ►laying of prefabricated stones.

work [n] [UK], **manual** *constr.* ►manual labor [US]/manual labour [UK].

work [n] [US], **random** *arch. constr.* ►broken range work.

work [n] [US], **random rubble range** *arch. constr.* ►random rubble ashlar masonry.

work [n], **remedial** *constr. contr.* ►remedying defects.

work [n] [US], **road** *constr. eng.* ►road construction (1).

work [n], **sequence of** *plan.* ►sequence of operations.

work [n], **survey** *surv.* ►topographic survey (3).

work [n], **topiary** *hort.* ►topiary.

work [n] [US], **tree clearing** *constr. for.* ►tree clearing and stump removal (operation) [US]/grub felling [UK].

work [n], **tree maintenance** *arb. constr.* ►tree maintenance.

work [n], **tunnel** *constr. eng.* ►tunneling.

7094 workability [n] **of a soil** *agr. hort.* (►easily workable soil, ►friable soil); *s* **friabilidad** [f] **del suelo** (Aptitud del suelo al desmenuzamiento; ►suelo friable, ►suelo ligero); *f* **possibilités** [fpl] **de travail du sol** (*Selon le contexte* amélioration des possibilités de travail du sol ; difficultés de travail du sol ; facilités de travail du sol ; être [vb] facile à travailler ; ►sol à structure grumeleuse, ►terre douce) ; *g* **Bearbeitbarkeit** [f] **des Bodens** (Eigenschaft eines Bodens auf Grund seiner Struktur [Korngefüge, ►Krümelstruktur 2] und plastischen Eigenschaften mit Geräten [leicht oder schwer] bearbeitet werden zu können; ►leicht bearbeitbarer Boden, ►krümeliger Boden).

7095 workability [n] **of construction material** *constr.* *s* **manejabilidad** [f] **de un material de construcción** (Característica de materiales de construcción en cuanto a la facilidad de trabajarlos o de incorporarlos a la obra); *f* **maniabilité** [f] **des matériaux** (État de consistance et de plasticité d'un matériau traduisant sa capacité à être transporté, manipulé et mis en œuvre ; on dira p. ex. qu'un béton est d'autant plus maniable qu'il est d'autant plus aisé à mettre en place dans les coffrages ou mis en œuvre que ce soit par gravité, par projection, par talochage ou en jeté à la truelle) ; *syn.* ouvrabilité [f] d'un matériau ; *g* **Bearbeitbarkeit** [f] **der Baumaterialien** (Eigenschaft von Baustoffen hinsichtlich ihres Aufwandes beim Einbau, beim Zuschlagen der Werksteine oder Einpassen von Einzelelementen); *syn.* Bearbeitbarkeit [f] von Werkstoffen, Verarbeitbarkeit [f] von Baumaterialien.

work and services [npl] *constr. contr.* ►additional work and services.

work and services included [loc] *constr. contr.* ►all additional work and services included.

7096 worked bog [n] *conserv. land'man.* (Bog destroyed by extraction of peat; ►cut-over bog, ►peat-cut area, ►peat cutting [US]/peat-cutting [UK]); *syn.* cutaway bog [n] [also UK] (VIR

1982, 341); *s* **turbera** [f] **explotada** (Turbera destruida por extracción de la turba; ►área de extracción de turba, ►extracción de turba, ►turbera alta explotada); *f* **tourbière** [f] **entièrement exploitée** (Tourbière mise en valeur par ►exploitation de la tourbe ; il s'agit d'une tourbière détruite ; ►fosse de tourbage, ►tourbière partiellement exploitée) ; *g* **abgetorftes Moor** [n] (Durch vollständigen ►Torfabbau gekennzeichnetes Moor; ►Leegmoor, ►Torfstich).

work [vb] **free of charge** [UK] *contr. prof.* ►work pro bono [US].

work [vb] **gratis** *contr. prof.* ►work pro bono [US].

7097 working bench [n] *min.* (Broad horizontal ledge created as a working platform during surface mining operations, or in the construction of a steep embankment; ►ledge 1, ►terracing of a slope); *syn.* working shelf [n] [also US]; *s* **talud** [m] **escalonado** (En explotación a cielo abierto, terraza ancha utilizada como plataforma de trabajo; ►aterrazamiento de talud, ►berma 1); *syn.* pared [f] escalonada (DINA 1987, 801); *f* **banquette** [f] **d'exploitation** (Large surface horizontale sur laquelle circulent les engins d'abattage, de chargement et de transport dans un exploitation à ciel ouvert ; ►crantage d'un talus, ►risberme) ; *g* **Berme** [f] **(2)** (Breiter, horizontaler Absatz in Böschungen beim Tagebau; ►Berme 1, ►Terrassierung von Böschungen).

working drawing [n] *constr.* ►construction drawing.

7098 working plan [n] *min. envir.* (Plan indicating the intended course of mineral extraction, the chronological and spatial coverage of the work [US]/works [UK] as well as subsequent reclamation of the land); *s* **plan** [m] **de explotación** (Plan indicativo del proceso propuesto de extracción mineral); *syn.* plan [m] de aprovechamiento; *f* **plan** [m] **d'exploitation** (Plan définissant dans le temps et l'espace les modalités d'exploitation, précisant la nature des réaménagements et la remise en état des lieux après exploitation, en général consigné dans un cahier des charges ; cf. loi n° 76-663 modifiée par la loi n° 93-3 du 4 janvier 1993) ; *g* **Abbauplan** [m] (Plan, der die Gewinnung von Bodenschätzen regelt, insbesondere Angaben über das räumliche und zeitliche Ausmaß des beabsichtigten Arbeitsprogrammes und zur anschließenden Wiedernutzbarmachung der Oberfläche enthält; cf. § 50 BBergG); *syn.* Betriebsplan [m], Gewinnungsbetriebsplan [m] [A], *z. T.* Abgrabungsplan [m].

workings [npl] [UK], **open-cast** *min.* ►open-pit mining [US].

working shelf [n] [US] *min.* ►working bench.

working site [n] *min. plan.* ►mineral working site.

work item [n] *constr.* *►item (1).

workpeople [n] [UK] *constr. contr.* ►construction worker.

7099 work phase [n] [US] *contr. prof.* (HALAC 1985; distinct portion of related ►basic professional services of a planner which is required to be completed for each section of a project; conditions of engagement and scale of charges are to be described for each work stage; ►Guidance for Clients on Fees); *syn.* work stage [n] [UK] (LPP 1988); *s* **fase** [f] **de trabajo** (Conjunto de ►servicios profesionales básicos necesarios para completar una parte de un proyecto; en el ►reglamento de honorarios para arquitectos e ingenieros alemán se diferencian p. ej. nueve fases de trabajo); *f* **étape** [f] **de la mission de maîtrise d'œuvre** (Ensemble de prestations constituant un élément normalisé ou mission élémentaire de la mission de maîtrise d'œuvre. F., on distingue 12 missions élémentaires normalisées ; art. 5 de l'Arrêté du 29 juin 1973, Annexe 1 ; ►prestations élémentaires, ►rémunération des prestations de maîtrise d'œuvre [exercée pour le compte de maîtres d'ouvrage publics]) ; *syn.* tranche [f] [B] ; *g* **Leistungsphase** [f] (Summe der sachlich zusammengehörenden ►Grundleistungen eines Planers/einer Planerin, die erforderlich sind, um in sich abgeschlossene Leistungsabschnitte [Teilleistungen] zu erzielen. In der ►Honorarordnung für Architekten und Ingenieure [HOAI] werden z. B. bei der Objektplanung neun Leistungsphasen unterschieden, in den österreichischen Honorarrichtlinien für Landschaftsarchitekten wird das Gesamtleistungsbild in sechs Teilleistungen gegliedert; cf. § 22 HRLA); *syn. obs. in D.* Einzelleistung [f], Teilleistung [f] [A].

7100 work preparation [n] [US]/**works preparation** [n] [UK] *constr. contr.* (Organizing activities of a contract holder which are necessary in order to begin the work [US]/works [UK] on the date specified in the contract; ►preparatory operations); *s* **preparación** [m] **de las obras** (Actividades organizativas necesarias para poder comenzar a construir que tiene que realizar la empresa constructora después de haberle sido adjudicada una obra; ►trabajos preliminares); *f* **préparation** [f] **du chantier** (Phase d'organisation des travaux avant l'ouverture de chantier, coordination des moyens matériels, de la main-d'œuvre et des besoins en matériaux pour une mise en route du chantier dans des délais raisonnables ; ►travaux préliminaires) ; *g* **Bauvorbereitung** [f] (Organisatorische Tätigkeiten des Auftragnehmers nach Auftragserteilung, die nötig sind, um in einer angemessenen Frist die Bauarbeiten zu beginnen; ►Vorarbeiten 2).

7101 work [vb] **pro bono** [US] *contr. prof.* (To perform a professional service without remuneration); *syn.* work [vb] gratis, work [vb] free of charge [UK]); *s* **trabajar** [vb] **por amor al arte**; *f* **travailler** [vb] **sans honoraires** ; *g* **ohne Honorar arbeiten** [vb].

work procedure [n] *plan.* ►sequence of operations.

works [npl] *contr.* ►accounting of executed works [UK]; *constr.* ►civil engineering works, ►execution of building works, ►public works, ►public works inspector; *constr. prof.* ►supervision of works.

works [npl] [UK], **agreed measurement of completed** *constr. contr.* ►agreed measurement of completed work [US].

works [n] [UK], **clerk of the** *constr. prof.* *►client's project representative [US].

works [npl] [UK], **commencement of** *onstr. contr.* ►start of construction [US].

works [npl] [UK], **completed** *constr. contr.* ►completed work [US].

works [n] [UK], **completion of** *constr.* ►project completion [US].

works [npl] [UK], **contractor's supervision of the building** *constr.* ►contractor's supervision of the project [US].

works [n] [UK], **earth shaping** *constr.* ►ground modeling [US]/ground modelling [UK].

works [npl] [UK], **ground-** *constr. eng.* ►earthworks.

works [n] [UK], **inspection of the** *constr. prof.* ►supervision of works.

works [npl] [UK], **measurement of completed** *constr. contr.* ►measurement of completed work [US].

works [npl] [UK], **road** *constr.* ►road work [US].

works [npl] [UK], **road-widening** *constr. trans.* ►road-widening construction [US].

works [npl] [UK], **sewage** *envir.* ►sewage treatment plant.

works [npl] [UK], **subcontract** *constr. contr.* ►subcontract construction [US].

W

works [npl] [UK]**, suspension of construction** *constr. contr.* ▶suspension of project [US].

works claim [n] [UK]**, remedial** *constr. contr.* ▶remedial construction claim [US].

works siding [n] [UK] *trans.* ▶railroad siding [US]/railway siding [UK].

work stage [n] [UK] *contr. prof.* ▶work phase [US].

work stage [n] **"L"** [UK] *prof.* ▶concluding project review.

7102 work [vb] **stones** *constr.* (To shape stones with hammer and chisel; *result* tooled stones [US]; ▶tooling of stone [US]/dressing of stone [UK]); *s* **trabajar** [vb] **la(s) piedra(s)** (Dar forma a las piedras con martillo y cincel; ▶labra de piedra natural o artificial); *f* **travailler** [vb] **la pierre** (*Résultat* travail de la pierre ; il concerne les opérations de découpe, de taille ou de surfaçage de la pierre à l'aide d'outils portatifs et recouvre une grande diversité de métiers faisant appel à des méthodes et des outillages spécifiques pour l'usinage de pierres de différentes natures pour la restauration de monuments, la sculpture, la gravure, la fabrication de bordures ou de pierres tombales, de cheminées, d'éléments de décoration, etc. ; ▶mode de taille des pierres) ; *g* **Bearbeiten** [n, o. Pl.] **von Steinen** (Das Zurichten von Steinen mit Steinwerkzeugen [z. B. Hammer und Meißel]; Steine bearbeiten/zurichten [vb]; *Ergebnis* behauener Stein; *OB* bearbeiteter Stein; ▶Steinbearbeitung); *syn.* Zurichten [n, o. Pl.] von Steinen.

7103 world heritage site [n] *conserv. leg. pol.* (Protected area or place of outstanding international significance designated pursuant to the 1972 Paris Convention of UNESCO; normally without any international legal status; such areas may be protected under an individual country's conservation law); *s* **sitio** [m] **del patrimonio mundial (natural)** (Espacio protegido de significado internacional declarado de acuerdo a la Convención de París de la UNESCO. En Es. el Convenio de 23 de noviembre de 1972 para la Protección de Patrimonio Mundial, Cultural y Natural fue aceptado por instrumento de 18 de marzo de 1982; BOE 156 de 1 de julio de 1982); *f* **site** [m] **du patrimoine mondial** (Espace protégé par la convention pour la protection du patrimoine mondial culturel et naturel adoptée par la conférence de l'UNESCO le 16 novembre 1972 à Paris et entrée en vigueur le 19 décembre 1975. Elle concerne les monuments et les ensembles architecturaux ou paysagers et d'une manière générale les biens culturels ayant une valeur universelle exceptionnelle du point de vue historique, esthétique, ethnologique ou anthropologique, ceux-ci étant inscrits sur la liste du patrimoine mondial et parmi ceux inscrits, ceux inclus sur la liste du patrimoine mondial en péril) ; *g* **Schutzgebiet** [n] **für das Welterbe** (Zu schützendes Gebiet — in Deutschland ohne legalen Status —, das durch die Konvention zum Schutze des Weltkultur- und Weltnaturerbes [Paris 1972] geschaffen wurde und Bereiche von hervorragender, globaler Bedeutung erfassen soll; diese Gebiete können in Deutschland z. B. durch flächenhafte Kulturdenkmale oder Nationalparke legal gesichert werden); *syn.* Gebiet [n] zum Schutze des Weltkultur-/Weltnaturerbes.

worship [n] [UK]**, place of public** *leg. urb.* ▶church [US].

worthwhile [adj] **to protect** [US]**/worth-while** [adj] **to protect** [UK] *conserv. leg.* ▶worthy of protection.

7104 worthy of protection [loc] *conserv. leg.* (Having value and importance of protection for conservation; ▶place under legal protection, ▶protection suitability); *syn.* worthwhile [adj] to protect [US]/worth-while [adj] to protect [UK]; *s a* **proteger** [loc] (Espacio natural u objeto construido que reúne características como para ser declarado como espacio protegido o bien de interés cultural; ▶declarar protegido, ▶idoneidad para la protección); *syn.* merecedor de protección [loc]; *f* **qui mérite d'être conservé** [loc] (Dont les caractéristiques exigent la protection ; ▶aptitude à la protection, ▶classer) ; *syn.* qui mérite conservation [loc], qui mérite d'être protégé [loc], apte à être classé[e] [loc] ; *g* **schützenswert** [adj] (So beschaffen, dass es wert ist, unter Schutz gestellt zu werden; ▶Schutzwürdigkeit, ▶unter Schutz stellen); *syn.* schutzwürdig [adj].

wound [n] *arb. hort.* ▶bark wound, ▶pruning wound, ▶trunk wound [US]/stem wound [UK].

wound [n] [UK]**, stem** *arb. hort.* ▶trunk wound [US].

7105 wound area [n] *arb.* (Place of damage or decay on bark or outer sapwood; ▶area of rot, ▶decayed area, ▶wound treatment); *s* **herida** [f] (Daño causado en la corteza y/o en la madera externa de una leñosa; ▶caries húmeda, ▶tratamiento de heridas, ▶zona podrida); *f* **plaie** [f] (Résultat de la blessure ou du ▶pourrissement de l'écorce ou de l'écorce et de l'aubier ; ▶point de pourriture [sur le tronc], ▶traitement des plaies) ; *g* **Wundstelle** [f] (Mechanisch verursachte Verletzung oder Fäulnis von Rinde oder von Rinde und äußerem Splintholz; ▶Faulstelle, ▶Morschung, ▶Wundbehandlung); *syn.* Wunde [f], Wundfläche [f].

7106 wound closure [n] *arb.* (ARB 1983, 506; **1.** Closing of a wound with ▶wound dressing. **2.** Process of healing due to ▶callusing); *s* **cierre** [m] **de heridas** (**1.** Aplicación de ▶pintura de sellado de heridas sobre una herida. **2.** Proceso de ▶cicatrización de herida); *f 1* **badigeonnage** [m] **des plaies** (**1.** Pratique controversée par de nombreux spécialistes consistant dans l'adjonction d'un produit protecteur [▶produit de badigeonnage] sur une plaie afin d'éviter le développement des pourritures ; l'application d'un produit fongistatique et contenant des hormones favorisant le développement du cal semble tout aussi efficace ; IDF 1988, 114. **2.** ▶cicatrisation) ; *f 2* **badigeon** [m] (Résultat du badigeonnage) ; *g* **Wundverschluss** [m] (**1.** Versorgung einer Wunde mit einem ▶Wundverschlussmittel. **2.** Vorgang des Schließens einer Wunde durch ▶Wundheilung); *syn. zu 2.* Schließung [f] der Wunde.

7107 wound dressing [n] *arb. constr.* (Sealant [latex emulsion], which is dressed in a thin coat on the edge of a cleaned and shaped wound in order to prevent decay by wood-invading fungi, or a sealant over a clean cavity to prevent further destruction by wood-eating insects, and to stimulate callus growth; *specific terms* fungicidal sealant, insecticidal sealant); *syn.* sealant [n], wound paint [n] (ARB 1983, 506); *s* **pintura** [f] **de sellado de heridas** (Emulsión de latex que se aplica en capa fina sobre una herida limpia para prevenir la putrición de la madera por invasión de hongos y para estimular el crecimiento del callo cicatricial); *f* **produit** [m] **de badigeonnage** (Le produit protecteur doit être appliqué immédiatement après la coupe d'une branche ou après le parement d'une plaie pour éviter toute attaque de la plaie par les microorganismes et leur emprisonnement sous le produit ; IDF 1988, 114) ; *syn.* baume [m] protecteur pour les plaies, baume [m] cicatrisant, matériel [m] de badigeonnage (IDF 1988, 114) ; *g* **Wundverschlussmittel** [n] (Ein auf den Rand einer geglätteten, größeren Gehölzschnittwunde verstrichenes, dauerelastisches [pilztötendes] Mittel, das dazu dient, eine schnelle Verheilung zu gewährleisten und Holz zerstörende Pilze fernzuhalten).

wound paint [n] *arb. constr.* ▶wound dressing.

7108 wound treatment [n] *arb. constr.* (Generic term covering the treatment of ▶bark wound and infected or dead wood—sap or heart wood); *s* **tratamiento** [m] **de heridas** (Término genérico que cubre el tratamiento de ▶herida de la corteza y de infecciones o partes muertas de la madera); *f* **traitement** [m] **des plaies** (Terme générique pour les travaux

W

arboricoles de protection et de cicatrisation des ►blessures de l'écorce ou des parties infectées ou pourries de branches ou du tronc ; on distingue le parement, le badigeonnage et le curage des plaies) ; *syn.* curage [m] des plaies (IDF 1988, 116) ; *g* **Wundbehandlung [f]** (Behandlung von ►Rindenverletzungen und von befallenen oder abgestorbenen Splint- resp. Kernholzteilen).

7109 woven jute mesh [n] *constr.* (Fabric used for wrapping root balls, and for holding seeds and soil on slopes and ditches); *s* **tejido [m] de yute** (Material textil utilizado para fijar el suelo y las semillas en terraplenes hasta su crecimiento); *f* **filet [m] de jute** (Produit textile constitué de fils de jute *[Corchorus]* dégradables utilisé pour la fixation des sols sur talus et dans les fossés de drainage ou comme tontine pour les végétaux de pépinière) ; *g* **Jutegewebe [n]** (Aus Bastfasern der Jute *[Corchorus]* hergestelltes Textil zur Befestigung des Bodens auf Böschungen und in Entwässerungsrinnen bis zur eingewachsenen Begrünung sowie Ballentuch für Baumschulware).

woven wood fence [n] *constr.* ►interwoven wood fence.

wrapping [n] *arb. constr. hort.* ►trunk wrapping.

wrapping [n]**, hessian** *hort.* ►burlapping.

wrapping [n]**, tree** *arb. constr. hort.* ►trunk wrapping.

7110 wrapping of perforated drain pipe with soil separator [n] *constr.* *s* **enfundado [m] de dren** *syn.* revestimiento [m] de dren; *f* **enrobage [m] d'un drain** (p. ex. avec de la fibre de coco, du feutre) ; *g* **Dränrohrummantelung [f]** (Umwick[e]lung von Dränrohren, z. B. mit Kokosstricken; *o. V.* Drainrohrummantelung).

7111 wrapping [n] **with burlap** [US] *constr. arb.* (Measure to prevent excessive transpiration from the bark of newly-transplanted trees. Burlap is seldom used anymore in the U.S.: preferred are 'prepared paper strips'; ►bark scorch, ►protection oftrees, ►trunk protection, ►trunk wrapping); *syn.* wrapping [n] with Hessian strips [UK]; *s* **saco [m] de protección del tronco** (Vendaje de tela de saco para la protección del tronco de árboles [►quemadura de la corteza] recién transplantados; ►envoltura del tronco, ►protección de árboles [en obras]); *f* **tampon [m] de protection en jute** (Protection de l'écorce du tronc contre les rayons solaires [►insolation d'écorce] et le dessèchement par le vent ; ►protection des arbres, ►tampon protecteur de tronc) ; *g* **Lehm-Juteverband [m]** (Baumrindenschutz gegen schädigende Wind- und Sonneneinwirkung [►Rindenbrand]; ►Baumschutz, ►Stammumwicklung); *syn.* Lehm-Juteumwicklung [f], Lehm-Jutebandage [f].

wrapping [n] **with Hessian strips** [UK] *constr. arb.* ►wrapping with burlap [US].

7112 wrapping [n] **with straw ropes** *constr. hort.* (Covering of a tree trunk and main branches with straw, depending on trunk size, to protect against transpiration and the sun during transplanting and establishment of mature trees; ►trunk protection, ►trunk wrapping); *s* **envoltura [f] de paja** (Cobertura de paja del tronco y de las ramas principales de un árbol para protegerlas contra quemaduras; ►envoltura del tronco, ►protección del tronco); *f* **paillage [m] de tronc** (Méthode de protection du tronc et des charpentières d'un arbre fraîchement transplanté au moyen d'un manchon de paille ; ►protection du tronc, ►tampon protecteur de tronc) ; *g* **Strohbandage [f]** (Umwicklung des Baumstammes sowie der Grob- oder Starkäste mit Stroh als Verdunstungs- und Sonnenschutz bei zu verpflanzenden oder verpflanzten Großbäumen; ►oberirdischer Baumschutz, ►Stammumwicklung).

written award statement [n] [US] *constr.* ►written notice of contract award.

7113 written notice of contract award [n] *contr.* (In U.S., if the contract is made with a public authority/agency, it must be widely published); *syn.* written award statement [n] [also US], written notification [n] of contract award [also UK]; *s* **formalización [f] de contrato por escrito** (Firma de contrato ante gabinete de planificación o empresa constructora. Según los casos, los contratos se formalizan en documento administrativo o se elevan a escritura pública dentro del plazo de treinta días después de la notificación de la adjudicación; cf. art. 55 Ley 13/1995, de Contratos de las Administraciones Públicas [LCAP]); *f* **notification [f] d'un marché** (Attribution écrite d'un marché à un entrepreneur ou bureau d'études) ; *syn.* notification [f] d'un contrat ; *g* **schriftliche Auftragserteilung [f]** (an ein Planungsbüro oder an eine Baufirma) ; *syn.* Auftragsschreiben [n], Zuschlagsschreiben [n] (an eine Baufirma).

written notification [n] **of contract award** [UK] *contr.* ►written notice of contract award.

written statement [n] *plan.* ►explanatory report.

xeric habitat [n] *phyt. zool.* ►dry habitat.

7114 xeric site [n] *phyt. zool.* (TEE 1980, 72; dryland site; ►dry habitat); *s* **estación [f] xerofítica** (Ubicación en la cual los factores topográficos [vertiente, pendiente, altitud o exposición], climáticos [insolación, pluviosidad, temperatura, exposición al viento, etc.] y edafológicos [suelo, textura, profundidad, etc.] tiene carácter xerofítico; ►biótopo seco, *opp.* ►estación mesofítica); *syn.* residencia [f] ecológica xerofítica; *f* **station [f] sèche** (Milieu naturel dont les facteurs topographiques [versant, pente, altitude ou exposition], climatiques [ensoleillement, pluviométrie, température, exposition au vent, etc.] et pédologiques [sols, texture, profondeur, etc.] ont un caractère xérique ; ►biotope sec ; *opp.* station humide) ; *syn.* station [f] xérique ; *g* **Trockenstandort [m]** (Standort, der durch trockenheitsbedingte Umweltbedingungen wie exponierte Topografie [Hang, Höhen- oder Südlage], Klima [z. B. starke Besonnung, geringe Niederschlagsmenge, Temperatur, Windexposition] und Boden [z. B. Textur, hohe Wasserdurchlässigkeit, geringe Mächtigkeit] geprägt ist; *opp.* feuchter Standort; ►Trockenbiotop).

7115 xeric vegetation [n] *phyt.* (Vegetation occurring in very dry areas, e. g, on the lee side—'rain shadow'—of high mountains, or in arid regions); *s 1* **vegetación [f] xerófila** (Plantas y sinecias que se presentan en medios secos [por el clima o por las condiciones edáficas]; DB 1985); *syn.* vegetación [f] de los climas secos; *s 2* **vegetación [f] xerófita** (Plantas resistentes a la sequía); *f* **végétation [f] xérique** (Végétation croissant dans un milieu très sec [à l'abri de la pluie] ou dans les régions arides) ; *g* **Trockenvegetation [f]** (Vegetation, die in sehr trockenen Lagen [im Regenschatten] oder in ariden Gebieten vorkommt).

xerophile plant [n] *bot.* ►xerophyte.

7116 xerophilous [adj] *phyt. zool.* (Preferring or adapted to a dry site); *syn.* xerophytic [adj] (TGG 1984, 179); *s* **xeròfilo/a [adj]** (Calificativo que se aplica a las plantas y sinecias que, por el clima o por las condiciones edáficas, viven en los medios secos; cf. DB 1985); *f* **xérophile [adj]** (Se dit d'une plante susceptible de vivre dans un milieu aride ou adaptée à un milieu

X

sec); *g* **xerophil** [adj] (Trockene Lebensräume bevorzugend oder an diese angepasst); *syn.* Trockenheit liebend [ppr/adj].

xerophilous plant [n] *bot.* ▶xerophyte.

7117 xerophyte [n] *bot.* (A plant that can subsist on dry sites [▶heathland, ▶steppe-heath, etc.]; ▶geophyte); *syn.* xerophile plant [n], xerophilous plant [n]; *s* **xerófito** [m] (Vegetal adaptado a la sequedad, propio de los climas secos [▶landa, ▶landa esteparia, etc.] o con un periodo/período de sequía más o menos largo; DB 1985); *syn.* planta [f] xerófila; *f* **xérophyte** [m] (Plante vivant dans un milieu sec [déserts, steppes arides, ▶lande 1, ▶landes steppiques, etc.] présentant généralement des adaptations morphologiques, anatomiques ou physiologiques, xérophytiques pour résister à la sécheresse ; ▶géophyte) ; *syn.* plante [f] xérophile ; *g* **Xerophyt** [m] (Griech. *xeros* >trocken< und *phyton* >Pflanze<: Pflanze eines klimatisch oder edaphisch trockenen Standortes [Wüste, Halbwüste, ▶Heide, ▶Steppenheide etc.] mit typischen morphologischen [xeromorphen] und ökophysiologischen Anpassungen, die z. T. deren Physiognomie prägen: Reduktion der Blattfläche, reflektierende und silbergrau erscheinende Haarüberzüge, versenkte Spaltöffnungen [Stomata], tief reichende Pfahlwurzeln oder intensiv verzweigtes Wurzelsystem [Steppengräser], verschiedene Formen der Wasserspeicherung im Blatt-/Wurzel- oder Sprossgewebe [Sukkulenz], Standfestigkeit auch bei Trockenheit durch faserreiche Sprossachsen und Blütenteile [Zellulose, Lignin], Verdunstungsschutz durch Abgabe ätherischer Öle [Aerosolglocke]. Zu den **X.en** gehören auch ▶Geophyten); *syn.* Trockenpflanze [f], xerophile Pflanze [f].

xerophytic [adj] *phyt. zool.* ▶xerophilous.

7118 xerothermous meadow [n] *phyt.* (Grass community growing under extremely dry and warm conditions; ▶dry meadow); *s* **pastizal** [m] **xerotérmico** (Comunidad herbácea que crece en ubicaciones extremadamente secas y calurosas; ▶pastizal seco); *f* **pelouse** [f] **xérophile et thermophile** (Formation ouverte colonisant souvent les sols calcaires superficiels orientés au secteur sud sur pentes accusées [gradins de falaises] ; Code NATURA 2000 : 6210 : pelouses sèches semi-naturelles et faciès d'embroussaillement sur calcaire [Festuco-Brometalia] [sites d'orchidées remarquables] ; ▶pelouse aride) ; *g* **Xerothermrasen** [m] (Rasengesellschaft, die auf extrem trockenwarmen, südexponierten Standorten mit oft flachgründigen Kalkböden oder auf Felsabsätzen wächst; ▶Trockenrasen).

7119 xylem [n] *arb. bot.* (Vascular bundle consisting of tracheids, vessels, parenchyma cells and fibres, which transports water within the cambium, and forms the woody tissue of a plant); *s* **xilema** [m] (Término morfológico para los hacecillos conductores, conjunto formado por los vasos o las traqueidas, el parénquima xilemático y las fibras leñosas; a veces se extiende la denominación a todas las partes leñosas de la planta, tanto primarias como secundarias sean conductoras o tengan funciones mecánicas; cf. DB 1985); *f* **xylème** [m] (Tissu conducteur de la sève brute des racines vers les organes aériens résultant de l'activité du cambium pour le xylème secondaire et du méristème apical pour le xylème primaire, constitué d'éléments conducteurs [trachéides], d'éléments de soutien [fibres ligneuses] et des cellules d'un parenchyme ligneux) ; *syn.* bois [m] ; *g* **Xylem** [n] (Aus dem Kambium hervorgehendes, Wasser führendes Leitbündel [Gefäßbündel], das aus Tracheen [weitlumige Gefäßzellen] und Tracheiden [englumige Zellen] besteht und den Gefäß- oder Holzteil bildet).

Y

7120 yachting habor [n] [US]/**yachting harbour** [n] [UK] *recr.* (Anchorage and moorings on lakes or sea coasts for [fast-travelling] motorboats and sailboats; ▶sailboat harbor [US]/sailing-harbour [UK]); *syn. o.v.* yachting harbour [n] [UK]; *s* **puerto** [m] **de yates** (▶puerto de veleros); *f* **port** [m] **de navires de plaisance** (Équipement nautique [loisirs touristique] de mouillage des bateaux de plaisance sur les lacs ou les côtes) ; ▶port de voiliers de plaisance) ; *g* **Yachthafen** [m] (Anker- und Anlegeplatz an Binnenseen oder am Meer für schnell fahrende Sport- und Freizeitschiffe; ▶Segelboothafen); *syn. o. V.* Jachthafen [m].

yard [n] *gard.* ▶private garden, ▶garden courtyard.

7121 yard drain [n] [US] *constr.* (Surface opening to a closed drain or pipe in courtyards, squares, etc. with smaller dimensions than those of catch basins [US]/road gullies [UK]; it usually employs a 150-300mm grate attached to plastic or masonry drain structure. In U.S., a 'gully' is a channel or ravine cut in the earth by running water; ▶catch basin inlet [US]/gutter inlet [UK], ▶storm drain grate [US]/gully grating [UK]); *syn.* drop inlet [n] [US], gully [n] [UK]; *s* **sumidero** [m] **de patio** (Punto de drenaje de patios, plazas, etc. de dimensiones menores que los ▶sumideros de la calle; ▶rejilla de entrada [de alcantarilla]); *syn.* desagüe [m] de patio; *f* **siphon** [m] **de cour** (Dispositif d'évacuation des eaux dans les cours et placettes, de section plus petite que les ▶regards à grille ; ▶grille de regard d'évacuation) ; *g* **Hofablauf** [m] (Punktuelle Entwässerungsvorrichtung für Höfe oder Plätze, die sich vom ▶Straßenablauf durch einen kleineren Einlaufquerschnitt unterscheidet; ▶Einlaufrost); *syn.* Hofeinlauf [m].

yard equipment [n] *constr.* ▶machinery stock.

yard line [n] [US] *leg. urb.* ▶setback line.

yard waste [n] [US] *envir.* *▶green waste.

7122 yellow edge line [n] [US] *trans.* (**1.** Paved or painted strip along the outside road lane, to clearly demarcate the edge of the road; ▶hard shoulder. **2. U.K.**, a white edge line marks the left-hand side of the carriageway; a single yellow line on the edge of the carriageway indicates no waiting or parking during times shown on a nearby sign; a double yellow line indicates no waiting or parking at anytime); *s* **línea** [f] **lateral de señalización** (Banda de pintura blanca o amarilla u hormigón pintado para señalizar ópticamente el borde de la carretera; ▶arcén); *f* **bande** [f] **latérale de signalisation** (Ligne blanche continue peinte sur la couche de roulement, matérialisant le bord de la voie de circulation ; ▶accotement stabilisé) ; *syn.* ligne [f] blanche d'accotement ; *g* **weißer/gelber Randstreifen** [m] (In Fahrbahnhöhe liegender und die Fahrbahn seitlich begrenzender, befestigter, weiß [in den USA gelb] gefärbter Streifen, der den Fahrbahnrand optisch kennzeichnet; ▶befestigter Seitenstreifen); *syn.* Leitstreifen [m], weiße/gelbe Randmarkierung [f].

7123 yerma [n] *pedol.* (Spanish word for extremely poor and dusty soil in desert regions); *s* **yerma** [f] (Suelo bruto de desierto seco existente en regiones pobres en vegetación con clima de veranos cálidos y precipitaciones extremadamente reducidas; cf. KUB 1953); *f* **yermosol** [m] (Unité de sols rencontrés en régime hydrique aridique ; DIS 1986, 238) ; *g* **Yerma** [f] (*Aus dem Spanischen* extrem humusarmer, staubreicher, lockerer Boden in Wüstengebieten); *syn.* Wüstenstaubboden [m].

Y

yield capabilities [npl] *agr. for.* ▶potential yield.

7124 yield predictability [n] *agr. for.* (Expected capacity of managed ecosystems to balance out the effects of external influences, and thereby to produce broadly constant yields; ▶environmental stress tolerance); *syn.* yield reliability [n] [also UK]; *s* **previsibilidad [f] de cosechas (≠)** (Capacidad de los sistemas agrícolas de equilibrar las influencias externas y así producir resultados más o menos regulares; ▶resiliencia de un ecosistema); *f* **prévisibilité [f] de la productivité (±)** (Aptitude de certains écosystèmes à minimiser les influences négatives afin d'assurer un rendement régulier ; ▶capacité de charge d'un écosystème) ; *g* **Ertragssicherheit [f]** (Fähigkeit von Nutzungssystemen, störende Einwirkungen so zu puffern, dass möglichst gleichmäßige Erträge gewährleistet sind; ANL 1984; ▶Belastbarkeit 1).

yield reduction [n]**, agricultural** *agr. pol.* ▶agricultural reduction program [US]/extensification of agricultural production [UK].

yield reliability [n] [UK] *agr. for.* ▶yield predictability.

7125 young shrub transplant [n] *hort.* (Young ▶shrub plant, which has been grown under average soil and climatic conditions in nursery beds and fields, and then transplanted once, root pruned and trimmed according to regular nursery practice. Grades in American Standard for Nursery Stock relate heights to root spread or to minimum earth ball, and need not be referenced); *syn.* light shrub [n] (ENA-E 1996); *s* **arbusto [m] joven para transplante** (Arbusto joven de vivero, cultivado en suelo y bajo condiciones climáticas normales, transplantado una vez, con un mínimo de vástagos estipulado en normas nacionales; ▶arbusto, ▶subarbusto); *f 1* **jeune touffe [f] (d'arbustes et d'arbrisseaux)** (F., végétal provenant d'un jeune plant ayant subi un repiquage en pleine terre, en pot ou en toute autre récipient dont les caractéristiques qualitatives sont fixées dans la norme N.F. V 12 031 ; les caractéristiques dimensionnelles minimales des **j. t. d'a. et d'a.** sont définies selon un échelonnement qui est fonction du diamètre moyen en centimètres [12/15, 51/20, 20/25, 25/35] ou du nombre de branches [3 ou 4] et de leur longueur selon les échelles en centimètres [15/20, 20/25, 25/35, 35/45, 45/65] ; cf. norme N.F. V 12 037 ; **D.**, produit de pépinière ayant subi une transplantation, à racines nues, avec 2-3 branches ; pour la commercialisation de ces produits les spécifications particulières mentionnent la hauteur [longueur de la branche] et le nombre minimum de branches [2 branches], n'étant pris en compte que les branches situées à ras du sol et ayant la longueur requise ; ▶arbrisseau, ▶arbuste) ; *f 2* **touffe [f] (d'arbustes)** (Plante présentant un ensemble d'au moins trois branches fortes, dont la plus basse part au raz du sol ou à ras de la greffe ; cf. norme N.F. V 12-051 ; les caractéristiques dimensionnelles des arbustes à feuilles caduques et à feuilles persistantes sont définies par des classes de diamètre moyen ou des classes de hauteur ; cf. norme N.F. V 12 057) ; *f 3* **touffette [f]** (Jeune plant fort, ayant au moins deux ans, et ayant été au moins une fois transplanté ou cerné ; ENA-F 1996, 6) ; *g* **leichter Strauch [m]** (*Abk.* l. Str.; 1 x verpflanzte Baumschulware ohne Ballen, mit zwei bis drei Basistrieben, die in D. in Angeboten, Ausschreibungen und Rechnungen seit 09/1995 mit den Sortierkriterien ‚Höhe' und ‚Mindesttriebzahl' [2 Triebe] spezifiziert werden muss, wobei nur die Triebe gezählt werden, die die geforderte Mindesthöhe erreichen und die dicht über dem Erdboden ansetzen; ▶Strauch 2).

7126 young stand [n] *for.* (Group of young trees, as the product of planting or ▶natural regeneration 2, that have not yet grown together to form a closed canopy); *s* **brinzal [m] (2)** (Grupo de árboles jóvenes producto de plantación o de

▶regeneración natural que aún no han llegado al estado de ▶latizal 1); *syn.* plantación [f] joven; *f* **jeunes plants [mpl]** (Jeunes arbres provenant d'un semis, d'une transplantation ou de la ▶régénération [forestière] ; ▶rajeunissement de peuplement) ; *g* **Aufwuchs [m] (3)** (*Fachsprachlich* ...wüchse [pl], *sonst ohne Pl.*; junge gepflanzte, durch Aussaat oder ▶Naturverjüngung 2 entstandene Bäume, die noch nicht zur Dickung zusammen gewachsen sind).

youth camp [n] *recr.* ▶family or youth camp.

Y turn [n] [US] *trans. urb.* ▶hammerhead turnaround.

Z

7127 zero lot line building [n] [US] *urb.* (Construction of a building sited directly on a property line; not always allowed by zoning ordinances in the U.S.); *syn.* boundary line building [n] [UK]; *s* **construcción [f] sin retranqueo lateral** (Construcción hasta el borde lateral de la parcela); *syn.* construcción [f] en el límite del predio; *f* **bâtiment [m] mitoyen** (Édifice construit sur la limite de propriété, sans retrait par rapport à une limite séparative) ; *g* **Grenzanbau [m]** (Errichtung eines Gebäudes unmittelbar an der Nachbargrenze — ohne Einhaltung eines Grenzabstandes).

7128 zero lot line development [n] [US] *leg. urb.* (Specific building arrangement of detached building approach in which the location of a building on a lot is sited in such a manner that one or more of the building's sides rest directly on a lot line; the intent is to allow more flexibility in site design and to increase the amount of usable green space on the lot; cf. IBDD 1981; ▶detached building development); *s* **edificación [f] discontinua sin retranqueo (≠)** (Disposición específica de los edificios en área de casas individuales, mellizas o grupos de casas, por la cual —para dar más flexibilidad de diseño y aprovechar mejor los espacios libres— se construyen uno o dos lados de las casas sobre el borde del terreno, sin retranqueo lateral y/o frontal, ▶edificación de baja densidad); *f* **exception [f] aux règles d'implantation des constructions (par rapport aux limites séparatives) avec édification en limite latérale (±)** (Afin d'obtenir une meilleure utilisation des espaces verts ou un meilleur ensoleillement, exception faite aux règles d'urbanisme d'implantation des constructions par rapport aux limites séparatives du terrain avec non application de la marge d'isolement, la construction pouvant être édifiée sur l'une des deux limites séparatives latérales ; ▶implantation des constructions en ordre discontinu ; en général, les règles d'urbanisme applicable localement prévoient que lorsque la construction est faite en limite de propriété, elle est faite sans ouverture sur la propriété voisine) ; *syn.* implantation [f] en limite séparative latérale ; *g* **abweichende Bauweise [f] mit einseitiger Grenzbebauung** (Bei dieser besonderen Form der ▶offenen Bauweise 2 liegen die Gebäude in einem Baugebiet, um z. B. eine günstigere Ausnutzung oder eine bessere Sonnenlage der Gartenflächen zu erhalten, mit einer Gebäudeseite direkt auf der seitlichen Grundstückslinie und müssen zur gegenüberliegenden Grundstücksgrenze mindestens den doppelten seitlichen Grenzabstand haben).

7129 zinc-contaminated site [n] *ecol. phyt.* (Site on which only ▶zinc-tolerant plants will grow; *generic term* ▶heavy-metal contaminated site); *s* **estación [f] calaminar** (Lugar en el que sólo crecen plantas que toleran la toxicidad del

Z

cinc/zinc; ▶planta calaminar; *término genérico* ▶ubicación contaminada con metales pesados); *syn.* estación [f] contaminada por cinc/zinc; *f* **station [f] calaminaire** (Station caractérisée par une toxicité en zinc, sur laquelle ne se développe que des ▶plantes calaminaires ; *terme générique* ▶site métallicole) ; *syn.* site [m] naturel calaminaire, halde [f] calaminaire [aussi B] ; *g* **Galmeistandort [m]** (Standort, der durch Zinktoxizität des Bodens geprägt ist und auf dem deshalb nur ▶Galmeipflanzen wachsen; *OB* ▶Schwermetallstandort).

7130 zinc-contaminated soil [n] *pedol.* (Soil with a very high concentration of zinc due to its proximity to zinc-producing factories or to zinc outcrops; it is toxic to most plants; ▶zinc-tolerant plant); *s* **suelo [m] calaminar** (Suelo con gran concentración de cinc/zinc debido a la vecindad con fábrica de producción de cinc/zinc que es tóxico para la mayor parte de las plantas; ▶planta calaminar); *f* **sol [m] calaminaire** (Sol situé à proximité d'établissements industriels producteurs de zinc caractérisé par une très forte concentration en zinc et en général toxique pour la plupart des végétaux ; ▶plante calaminaire) ; *syn.* sol [m] zincifère ; *g* **Galmeiboden [m]** (In der Nähe von Zink produzierenden Industriebetrieben oder an Zinkausbissen vorkommender Boden mit sehr hohen Zn-Konzentrationen, die für die meisten Pflanzen toxisch sind; ▶Galmeipflanze).

zinc-enduring plant [n] *phyt.* ▶zinc-tolerant plant.

7131 zinc-tolerant plant [n] *phyt.* (*Generic term* ▶heavy-metal-tolerant plant); *syn.* zinc-enduring plant [n]; *s* **planta [f] calaminar** (Planta que crece en suelos que contienen cinc/zinc; *término genérico* ▶planta metalofita); *f* **plante [f] calaminaire** (Plante qui pousse sur sols calaminaires ; *terme générique* ▶espèce métallophyte) ; *syn.* zincophyte [m], plante zincicole ; *g* **Galmeipflanze [f]** (Pflanze, die auf zinkhaltigen Böden wächst, z. B. Galmeiveilchen *[Viola calaminaria]*; *OB* ▶Schwermetallpflanze).

7132 zinc-tolerant plant community [n] *phyt.* (Herbaceous vegetation of areas rich in heavy metals, mainly zinc; e.g. class of *Violetea calaminariae* Tx. 1961; *specific term* zinc-violet vegetation; *generic term* ▶heavy-metal-tolerant vegetation); *s* **vegetación [f] calaminar** (Vegetación característica del cinc/zinc; *término genérico* ▶vegetación matalofita); *f* **groupement [m] calaminaire** (Association végétale se développant sur des sols toxiques zincifères comme p. ex. la pelouse calaminaire du Violetalia calaminariae [habitat d'intérêt communautaire, code Natura 2000 6130] ; sur les pelouses calaminaires on trouve la pensée, le tabouret, la silène, la fétuque, l'arméria calaminaires ; *terme générique* ▶groupement des sols contaminés par les métaux lourds) ; *syn.* flore [f] calaminaire, végétation [f] calaminaire ; *g* **Galmeiflur [f]** (Pflanzengesellschaft auf zink-toxischen Böden, z. B. die Galmeiveilchenflur *[Violetea calaminariae* Tx. 1961]; *OB* ▶Schwermetallflur); *syn.* Galmeivegetation [f].

7133 zonal biocenosis [n] *phyt. zool.* (Term introduced by BALOGH in 1958 to describe life communities or parts of them, which are zonally separated into linear structures, which frequently merge with one another. Their order is influenced by continuous or discontinuous fluctuations in environmental factors. Species, which remain separated from one another, possess low adaptability in relation to one or more environmental factors); *s* **biocenosis [f] de zonación** (Término introducido por BALOGH en 1958 que denomina a las comunidades o partes de ellas que están subdivididas en zonas formando estructuras lineales y frecuentemente con zonas de transición en las que se mezclan, siendo la fluctuación de los factores ambientales la que determina su ordenación espacial); *f* **biocénose [f] zonale (±)** (Terme introduit par BALOGH en 1958 décrivant tout ou parties de communautés vivantes dont la répartition est caractérisée par

un zonage ou une structure en bandes et provoquée par une alternance continue ou discontinue des facteurs environnementaux. Cette différenciation spatiale concerne tout particulièrement les espèces à faible plasticité écologique par rapport à un ou plusieurs facteurs du milieu) ; *g* **Zonationsbiozönose [f]** (Von BALOGH [1958] eingeführter Begriff, der Lebensgemeinschaften oder Teile davon bezeichnet, die mit bandartiger Struktur und häufig fließenden Übergängen zonenartig auseinander gelagert sind und deren Abfolge durch den kontinuierlichen oder diskontinuierlichen Wechsel von Umweltfaktoren gesteuert wird. Es sind vor allem solche Arten voneinander getrennt, die in Bezug auf einen oder mehrere Umweltfaktoren eine geringe ökologische Plastizität besitzen).

7134 zonation [n] **of vegetation** *phyt.* (Regular, belt-like distribution of different plant communities according to their environmental requirements for growth; e.g. ▶altitudinal belts of mountain ranges, latitudes, etc.); *s* **cliserie [f]** (Sucesión de distintas asociaciones vegetales que se corresponden con las variaciones del clima o el ambiente derivadas de la existencia de gradientes latitudinales y altitudinales; DINA 1987; ▶zonación altitudinal); *syn.* zonación [f] de la vegetación; *f* **zonation [f] de la végétation** (Distribution des séries végétales selon la latitude) ; le terme de zonation est souvent improprement utilisé pour désigner un étagement ; ▶étagement de la végétation) ; *g* **Vegetationszonierung [f]** (Den Standortansprüchen entsprechende, regelmäßige und zonenartige Abfolge verschiedener Pflanzengesellschaften, z. B. auf Grund der ▶Höhenstufung im Gebirge, der Breitengrade, unterschiedlicher Feuchtigkeitsregime an Gewässerrändern etc.); *syn.* Vegetationszonation [f], Zonation [f].

zone [n] *bot.* ▶absorption zone; *ocean.* ▶abyssal zone; *geo. phyt.* ▶alpine zone, ▶altitudinal zone; *limn.* ▶barbel zone [UK]; *ocean.* ▶bathyal zone; *limn. ocean.* ▶benthic zone; *limn.* ▶bream zone [UK]; *conserv. landsc. urb.* ▶buffer zone; *geo. phyt.* ▶cloud level zone, ▶colline zone; *limn.* ▶flounder zone [UK], ▶grayling zone [UK]; *plan. landsc.* ▶green separation zone; *recr. urb.* ▶home zone [UK]; *leg. urb.* ▶industrial use zone; *phyt.* ▶lacustrine reed zone; *limn.* ▶limnetic zone; ▶littoral zone [US]/littoral [UK]; *geo. phyt.* ▶montane zone; *leg.* ▶natural danger zone; *geo. phyt.* ▶nival zone; *leg. plan.* ▶noise zone [US]/noise abatement zone [UK]; *geo. phyt.* ▶nival zone; *arb. constr.* ▶overfilling of root zone; *urb.* ▶pedestrian zone; *limn. ocean.* ▶pelagic zone; *geo. phyt.* ▶planar zone; *limn.* ▶profundal zone; *plan.* ▶raw material protection zone; *limn.* ▶river zone; *arb. hort.* ▶rooting zone; *limn.* ▶salmonid zone [UK]; *geo. phyt.* ▶sediment accumulation zone, ▶subalpine zone; *plan.* ▶transition zone; *limn.* ▶trout zone; *geo. phyt.* ▶vegetation altitudinal zone, ▶vegetation climatic zone; *hydr. landsc.* ▶water influence zone [US].

zone [n], **absorbing** *bot.* ▶absorption zone.

zone [n] [US], **air pollution** *envir.* ▶contaminated airspace.

zone [n], **capillary** *pedol.* ▶capillary fringe.

zone [n], **chub** *limn.* ▶barbel zone [UK].

zone [n] [US], **day-use recreation** *plan. recr.* ▶day-use recreation area [US] (2)/day-trip recreation area [UK].

zone [n] [UK], **development** *leg. urb.* ▶zoning district [US].

zone [n], **elevational** *geo. phyt.* ▶altitudinal zone.

zone [n] [US], **floodplain** *leg. wat'man.* ▶preservation of floodplain zone [US]/preservation of flood-plain zone [UK], ▶statutory floodplain zone [US]/statutory flood-plain zone [UK].

zone [n], **habitat transition line/** *ecol.* ▶transitional biotope.

zone [n] [US], **housing** *leg. urb.* ▶residential land use.

Z

zone [n], **light industry** *leg. urb.* ▶commercial and light industry area.

zone [n], **mullet** *limn.* ▶flounder zone [UK].

zone [n] [UK], **noise abatement** *leg. plan.* ▶noise zone [US].

zone [n] [US], **noise reduction** *leg. plan.* ▶noise zone [US].

zone [n], **perimeter protection** *conserv. landsc.* ▶marginal protection area.

zone [n], **peripheral** *leg. urb.* ▶city periphery.

zone [n] [UK], **residential** *urb.* ▶residential zoning district [US].

zone [n], **root hair** *bot.* ▶absorption zone.

zone [n], **simplified planning** *leg. urb.* ▶comprehensive plan [US]/Local Development Framework [UK].

zone [n], **sublitteral** *phyt. zool.* ▶shallow water area.

zone [n], **torrent** *limn.* ▶trout zone.

zone [n], **upper riverine** *phyt.* ▶upper riparian alluvial plain.

zone [n], **vegetation** *geo. phyt.* ▶vegetation altitudinal zone; ▶vegetation climatic zone.

zoned for development [n] *urb.* ▶land zoned for development.

zoned green area [n] [US] *leg. urb.* ▶zoned green space.

7135 zoned green space [n] *leg. urb.* (Green space designated as public or private on a ▶comprehensive plan [US]/Local Development Framework [LDF] [UK] or a legally-binding land-use plan, such as a ▶zoning map [US]/Proposals Map/Site Allocations Development Plan Document [UK]; *syn.* zoned green area [n] [also US]; *s* **zona [f] verde clasificada** (Superficie delimitada en ▶Plan General Municipal de Ordenación Urbana o en ▶plan parcial [de ordenación] para uso como zona verde privada o pública); *syn.* espacio [m] verde clasificado; *f* **espace [m] vert réservé** (Espaces sous le régime des règles de construction et d'urbanisme affectés dans les ▶schéma de cohérence territoriale/schémas directeurs et les ▶plan local d'urbanisme/ ▶plan d'occupation des sols pour l'aménagement ou le maintien d'espaces verts publics ou privés) ; *g* **ausgewiesene Grünfläche [f]** (Im ▶Bebauungsplan festgelegte oder im ▶Flächennutzungsplan dargestellte öffentliche oder private Grünfläche).

zoned land [n] *leg. urb.* ▶undeveloped zoned land.

zone [n] **of emergent vegetation, lacustrine** *phyt.* ▶lacustrine reed zone.

zone [n] **of emergent vegetation, riverine** *phyt.* ▶riverine reed belt.

7136 zone [n] **of influence** *envir. plan.* (Area of a program[me] or project in which ▶causal factors occur, the boundaries of such an area are defined by those factors, which are relevant in determining the conservation objectives of a protected area, taking into account the specific sensitivities of the relevant habitats and species); *syn.* sphere [n] of influence; *s* **zona [f] de influencia** (Área de realización de un programa o proyecto en la que pueden influir ▶factores causantes y cuyos límites se definen basándose en los aspectos relevantes para los objetivos de conservación, teniendo en cuenta la fragilidad específica de los hábitats y las especies importanes); *f* **zone [f] d'effet** (Espace à l'intérieur duquel les effets induit par un PPTOA sont détectables ; la délimitation de cet espace est effectuée sur la base des effets pertinents en rapport aux objectifs de conservation du site Natura 2000 ; pour cela il convient de prendre en compte les sensibilités spécifiques des habitats et espèces concernés ; ▶effet) ; *syn.* zone [f] d'influence ; *g* **Wirkraum [m]** (Gebiet, in dem vorhabenbedingte Wirkprozesse auftreten können; für die Abgrenzung sind diejenigen Wirkprozesse zugrunde zu legen, die für die Erhaltungsziele eines Schutzgebietes relevant sind — unter Berücksichtigung der spezifischen Empfindlichkeiten der für die Erhaltungsziele maßgebenden Lebensräume und Arten; cf. LEIT 2004; ▶Wirkfaktor).

7137 zone [n] **with an easement limiting the height of a building** (≠) *leg. urb.* (Building regulations are often part of a zoning plan in order to maintain views, preserve the appearance of a building ensemble or to protect the countryside; ordinances are also enacted to limit the height of structures adjacent to airports. In France, a designation in a zoning plan for urban development in which the height of all buildings is restricted; **non altius tollendi** [Latin] means literally that the right of a property owner to raise the height of a building is denied; *syn.* development zone [n] with maximum building height (≠); *s* **zona [f] de altura de construcción limitada** (≠) (F., sector urbano en el que está limitado el número de plantas que se pueden construir); *f* **zone [f] non altius tollendi** (F., zone dans laquelle aucune construction ne doit s'élever au-dessus d'une hauteur donnée ; cette servitude vise souvent à protéger le paysage ou à intégrer les constructions au site) ; *g* **Baugebiet [n] mit festgesetzter Gebäudehöhe** (≠) (F., im französischen „Bebauungsplan" [PLU, POS oder ZAC] ausgewiesenes Baugebiet mit festgesetzter Gebäudehöhe, die nicht überschritten werden darf; *altius tollendi* ist das Recht eines Hauseigentümers höher zu bauen; **in D.** kann im Bebauungsplan die Höhe baulicher Anlagen festgesetzt werden, wenn ohne diese Festsetzung öffentliche Belange, insbesondere das Orts- und Landschaftsbild beeinträchtigt werden könnte. Die Festsetzung erfolgt über Firsthöhe, Traufhöhe oder ein zwingend einzuhaltendes Maß, bezogen auf eine Bezugshöhe im Gelände resp. ü. NN).

zone [n] **with maximum building height** [UK], **development** *leg. urb.* ▶zone with an easement limiting the height of a building.

7138 zoning and building regulations [npl] *leg. urb.* (Entirety of legal requirements governing urban zoning of land uses in community and building construction on individual plots; in D., an owner is legally limited in order to provide harmonious development for the benefit of a whole community; ▶uniform building code [US]/building regulations [UK]); *syn.* building law(s) [n(pl)]; *s* **legislación [f] urbanística** (Conjunto de normas jurídicas que regulan la construcción y la planificación urbana; ▶Código Técnico de la Edificación); *f* **droit [m] de construire** (1. *Droit public* ensemble des lois et dispositions légales et réglementaires régissant l'urbanisme et la construction, relevant du ▶Code de l'urbanisme et du ▶Code de la construction et de l'habitation. 2. *Droit privé* droit traditionnellement lié au droit de propriété et largement limité puisqu'il s'exerce dans le respect des dispositions législatives et réglementaires relatives à l'utilisation du sol) ; *syn. obs.* droit [m] relevant du Code de l'urbanisme et de l'habitation ; *g* **Baurecht [n]** (1. **Öffentliches B.:** Aus dem Baupolizeirecht des 19. Jahrhunderts entstandene Gesamtheit der öffentlich-rechtlichen Gesetze, Durchführungsverordnungen und Vorschriften, die die städtebauliche Planung [Bauleitplanung] — als ,Städtebaurecht' oder ,Bauplanungsrecht' bezeichnet — und das Bauen auf dem Grundstück durch das ,Bauordnungsrecht' oder ,Bauaufsichtsrecht' regeln; ▶Bauordnung. 2. **Privatrechtliches B.** wird v. a. durch das Nachbarrecht und die nachbarschützenden Vorschriften des öffentlichen **B.s**, besonders aber durch die §§ 906-923 des BGB abgedeckt. **In D.** ist die **Baufreiheit** — das Recht des Eigentümers, nach Belieben auf seinem Grundstück zu bauen — im Vergleich zu den U.S.A. im Interesse einer organischen baulichen Entwicklung der Gemeinden gesetzlich beschränkt. Die Beschränkungen können

Z

entschädigungslos zumutbar sein oder, wenn sie die verfassungsmäßig geschützten Schranken des Eigentums gem. Art. 14 GG überschreiten, eine Entschädigung begründen); *syn.* Baugesetzgebung [f].

zoning category [n] [US] *leg. urb.* ▶urban land use category.

7139 zoning district [n] [US] *leg. urb.* (**1. In U.S.**, a specifically delineated area in a municipality within which regulations and requirements uniformly govern the use, placement, spacing and size of land and buildings; IBDD 1981; there are different types of **z. d.s**; e.g. housing, commercial, industrial; some zoning districts are conservation or agricultural zones, so building development is not always allowed; **2. in U.S.**, a **development zone** is a designated area in a zoning map for various land use types; ▶urban land use category; **in U.K.** a **development area** is designated within the Local Development Framework planning process, which *inter al* contains **Local Development Orders** [made by a planning authority in order to extend permitted rights for certain forms of development, with regard to a relevant local development document] and **Simplified Planning Zones** [areas in which a local planning authority wishes to stimulate development and encourage investment]; specifically delineated areas for development are shown on a ▶Proposals Map or Site Allocations Development Plan Document, which give broad indications of the type of use); *syn. general term* building area; *syn. for D.* specific land-use area [n] (FBC 1993); *s 1* **distrito** [m] **de zonificación** (≠) (En EE.UU. área delimitada de un municipio dentro de la cual rigen normas y exigencias uniformes en cuanto al tipo de uso, localización, espaciamiento y tamaño de las parcelas de terreno y de los edificios. Hay diferentes categorías de **d. de z.** como residenciales, comerciales, industriales. Algunos **d. de z.** son área protegidas o agrícolas, de manera que no está siempre permitida la construcción); *s 2* **sector** [m] **urbano** (≠) (D., término legal utilizado a nivel de ▶plan parcial [de ordenación] para diferenciar los tipos de uso del suelo en 10 categorías; ▶categoría de suelo urbano); *f 1* **zone** [f] **urbaine** (3) (F., *plan local d'urbanisme [PLU]* on distingue **1.** les ▶**zones urbaines 2 [zones U] :** secteurs déjà urbanisés et secteurs où les équipements publics existants ou en cours de réalisation ont une capacité suffisante pour desservir les constructions à implanter ; cf. art. R.123-5 — code de l'urbanisme, **2.** les **zones à urbaniser [zones AU] :** secteurs à caractère naturel de la commune destinés à être ouverts à l'urbanisation ; cf. art. R.123-6 — code de l'urbanisme, **3.** les **zones agricoles [zones A] :** secteurs de la commune, équipés ou non, à protéger en raison du potentiel agronomique, biologique ou économique des terres agricoles ; cf. art. R.123-7 — code de l'urbanisme], **4.** les **zones naturelles et forestières [zones N] :** secteurs de la commune, équipés ou non, à protéger en raison soit de la qualité des sites, des milieux naturels, des paysages et de leur intérêt, notamment du point de vue esthétique, historique ou écologique, soit de l'existence d'une exploitation forestière, soit de leur caractère d'espaces naturels ; [art. R.123-8 — code de l'urbanisme] ; *f 2* **secteur** [m] **urbain** (D., terminologie spécifique du « Bebauungsplan » [plan local d'urbanisme [PLU]/plan d'occupation des sols allemand] subdivisant la zone urbaine en dix catégories distinctes ; est comparable au secteur du règlement d'une zone ZAC) ; *g* **Baugebiet** [n] (D., UB zu ▶Baufläche. Durch die besondere Art der baulichen Nutzung [10 Möglichkeiten], soweit erforderlich, gem. BauNVO im Bebauungsplan festgesetzte Flächen).

7140 zoning district category [n] [US] *leg. urb.* (**In U.S.**, a specifically delineated zoning class within which regulations and requirements uniformly govern the use, placement, spacing and size of land and buildings; **in U.K.**, there are 18 different categories of uses to which buildings and the land can be put according to the Use Class Order, 1972; **in D.**, the use class is defined by zoning ordinance for ▶community development plans [US]/urban development plans [UK]; e.g. residental, commercial, light industries, heavy industries, etc.; ▶building use category, ▶development plan, ▶urban land use category, ▶zoning map [US]/Proposals Map [UK]/Site Allocations Development Plan Document [UK]); *syn.* use class [n] [UK]; *s* **categoría** [f] **de zonificación** (≠) (En EE.UU. tipo de distrito de zonificación definido por normas y exigencias uniformes en cuanto al tipo de uso, localización, espaciamiento y tamaño de las parcelas de terreno y de los edificios. En GB existen 18 categorías diferentes. En D. el tipo de uso del suelo fijado en los ▶planes urbanísticos [▶plan general municipal de ordenación urbana y ▶plan parcial (de ordenación)], p. ej. residencial, comercial, industrial, etc.; ▶categoría de suelo urbano, ▶volumen edificable. En Es. no existe este tipo de delimitación como categoría de planificación aplicable a todo el Estado); *f* **nature** [f] **de l'affectation des sols** (cf. C. urb., L. 123-1 ; F., déterminée selon l'usage principal ou la nature des activités dominantes l'affectation des sols est déterminée par le Règlement applicable aux diverses zones délimitées sur les documents graphiques d'urbanisme ; ▶conditions de l'occupation du sol [et de l'espace], ▶document de planification urbaine, ▶utilisation des sols pour la construction, ▶zone urbaine 2 ; D., la nature ou l'usage exclusifs des sols portés sur les documents d'urbanisme [Flächennutzungsplan et Bebauungsplan] sont réglementés par les § 5 [2] alinéa 1 et § 9 [1] alinéa 1 de la loi fédérale sur l'urbanisme [BauGB] ainsi que les §§ 1 à 15 de l'arrêté sur l'occupation des sols [BauNVO]) ; *syn.* destination [f] de l'affectation des sols ; *g* **Art** [f] **der baulichen Nutzung** (D., in Bauleitplänen gemäß § 5 [2] Nr. 1 und § 9 [1] Nr. 1 BauGB sowie §§ 1-15 BauNVO festzulegende Nutzungsart von ▶Bauflächen resp. Baugebieten; ▶bauliche Nutzung, ▶Maß der baulichen Nutzung, ▶Bauleitplan).

7141 zoning map [n] [US] *leg. urb.* (Map or series of maps, which implement a ▶zoning ordinance [US]/use class order [UK] and delineate the boundaries of zoning districts; **in U.S.**, **z. m.s** provide detailed guidance for the future pattern of land use in an area, and cover issues such as housing, employment, transport, shopping and the environment. A **z. m.** is not as detailed as a German "Bebauungsplan" [legally-binding land-use plan]. **In U.K.**, there is no exact equivalent; an approximate term is the **Proposals Map**, which shows the location of proposed schemes in all current Development Plan Documents, on an Ordnance Survey base map. Local authorities are required to include an **P. M.** in their Local Development Frameworks to explain geographically the adopted policies and proposals of the Development Plan Documents. The Map reflects the most up-to-date spatial plan for the District and is revised when new policies and proposals are adopted. Other documents, which specify more detailed information, are **1.** the **Area Action Plan** [AAP], which is a development plan document focused upon a specific location or an area subject to conservation or significant change. This could include a major regeneration project or a growth area. The AAP focuses on implementation, providing an important mechanism for ensuring development of an appropriate scale, mix and quality for key areas of opportunity, change or conservation; **2.** the **Site Allocations Development Plan Document** allocates land for new development such as housing, retail, economic development, recreation and community uses, and identifies the characteristics and requirements of such development. It will also include the definition of development boundaries or settlement limits for those places where some further growth may take place; ▶community development plan [US]/urban development plan [UK]. **D.**, the *"Bebauungsplan"* is the **legally-binding land-use plan**, which lays down statutory regulations for the development and organisation of parts of an urban area. As a

Z

Zoning Map, it categorizes land-use areas, e.g. purely residential areas, general residential areas, mixed use areas, commercial areas, industrial areas. It determines the category of use and degree of building coverage, ►type of building development [attached or detached], and lot coverage. The degree of building coverage is determined, for example, by setting the plot ratio, floor-space index, height of structures, and number of full storeys. Permissible lot coverage can be set by means of building lines, set-back lines, or coverage depths. The plan may earmark sites for mitigation measures to offset intrusions and ►directives on planting requirements [US]/enforcement notices on planting requirements [UK] for planting of trees, shrubs and roof gardens; ►zoning ordinance [US]/use class order [UK], ►comprehensive plan [US]/Local Development Framework [LDF] [UK]); *s 1* **plan [m] parcial (de ordenación)** (En Es. según la legislación supletoria tipo de plan de ordenación que tiene por objeto en el suelo calificado como urbanizable programado, desarrollar, mediante la ordenación detallada de una parte de su ámbito territorial, el Plan General y, en su caso, las Normas Complementarias y Subsidiarias del Planeamiento; y en el suelo clasificado como urbanizable no programado, el desarrollo de los Programas de Actuación Urbanística; art. 13 [1] Texto refundido LS RD 1346/1976; ►ordenanza sobre la intensidad de ocupación del suelo y sobre el tipo de uso, ►plan general municipal de ordenación urbana, ►plan urbanístico); *s 2* **programa [m] de actuación urbanística** (En Es. según la legislación supletoria tipo de plan de ordenación que tiene por objeto la ordenación y urbanización de terrenos clasificados como suelo edificable no urbanizado en los que se han de realizar unidades urbanísticas integradas; art. 16 [1] Texto refundido LS RD 1346/1976; *f 1* **plan [m] local d'urbanisme (PLU) (±)** (F., depuis le 1er avril 2001, le PLU remplace le ►plan d'occupation des sols [POS] selon la loi 2000-1208 du 13 décembre 2000 relative à la solidarité et au renouvellement urbains, [loi SRU]. Le PLU fixe dans le cadre des orientations prévues par les documents de rang supérieur [compatibilité] élaborés par l'État [lois Montagne et Littoral, directives territoriales d'aménagement [DTA] ►document de gestion de l'espace agricole et forestier] ou des collectivités territoriales [►schémas de cohérence territoriale (SCOT), programme local de l'habitat les ►schémas de secteurs, etc.] les règles générales et les servitudes d'utilisation des sols pour le territoire d'une commune ou d'une structure intercommunale ; il réglemente et planifie le développement urbain à brève échéance et exprime le projet urbain de la collectivité locale à l'horizon de 10 à 20 ans dans le cadre du projet d'aménagement et de développement durable [PADD] ; le PLU est opposable au tiers, c.-à-d. à toute personne publique ou privée). **D.**, le « Bebauungsplan » [plan d'occupation des sols allemand] fixe les règles d'aménagement et d'urbanisme pour un secteur géographique limité du territoire communal ; documents de planification urbaine, ►règles générales d'utilisation du sol) ; *f 2* **plan [m] d'occupation des sols (POS) (±)** (F., le plan d'occupation des sols fixe dans le cadre des orientations prévues par les schémas directeurs et les ►schémas de secteurs les règles générales et les servitudes d'utilisation des sols pour tout ou partie le territoire d'une commune ; il réglemente et planifie le développement urbain à brève échéance ; le POS est opposable au tiers, c.-à-d. à toute personne publique ou privée. Il remplace, depuis l'institution de la L.O.F. en 1967, les plans d'urbanisme directeur ou de détails et les plans sommaires de la réglementation de 1958 ; *f 3* **carte [f] communale** (Document d'urbanisme sommaire précisant pour les communes non couvertes par un plan d'occupation des sols les règles locale d'application du règlement national d'urbanisme [RNU] ; *syn.* modalités [fpl] d'application du règlement national d'urbanisme [M. A. R. N. U.] ; DUA 1996 ; *g* **Bebauungsplan [m] (±)** (**D.**, Plan, der die Einzelheiten der städtebaulichen Ordnung für einen genau abge-

grenzten Teilbereich des Gemeindegebietes rechtsverbindlich gegen jedermann festsetzt; der **B.** wird als verbindlicher ►Bauleitplan von der Gemeindevertretung — je nach Bundesland als Gemeinderat, Stadtrat, Stadtverordnetenversammlung oder Rat bezeichnet — als Satzung [►Ortssatzung] beschlossen; das Verfahren zur Aufstellung eines **B.es** bis hin zur Genehmigung durch das Regierungspräsidium oder die Bezirksregierung ist im Baugesetzbuch geregelt. **In F.** hingegen deckt ein PLU die gesamte Gemarkungsfläche einer Gemeinde resp. eines Zusammenschlusses mehrerer Gebietskörperschaften mit rechtsverbindlichen Festsetzungen gegen jedermann ab; ►Baunutzungsverordnung, ►Flächennutzungsplan); *syn.* verbindlicher Bauleitplan [m], Gestaltungsplan [m] [CH] (2), Überbauungsplan [m] [CH].

7142 zoning ordinance [n] [US] *leg. urb.* (Regulations on the use of building land; **in U.S.**, legal control is exercised by local governments on the use, placement, spacing and size of land and buildings; **in U.K.**, there are 18 different categories of uses to which buildings and the land can be put according to the Use Class Order, 1972; **in D.**, "**Federal Land Utilization Ordinance**"; cf. FBC 1993; ►Town and Country Planning Act [UK]); *syn.* use class order [n] [UK]; *s* **ordenanza [f] de zonificación** (En EE.UU. ley municipal que regula el uso del suelo con fines constructivos. En D., norma legal que desarrolla la ley federal de construcción en cuanto a las posibilidades de uso constructivo de los predios que se encuentran dentro del ámbito de los ►planes generales municipales de ordenación urbana y de ►planes parciales. En Es. existen leyes a nivel de CC.AA. que desarrollan la legislación urbanística en aquellos ámbitos en los que tienen competencia exclusiva); *syn.* ordenanza [f] sobre la intensidad de ocupación del suelo y sobre el tipo de uso; *f* **règles [fpl] générales d'utilisation du sol** (Règles générales d'urbanisme ne s'appliquant pas sur les territoires couverts par un PLU/POS rendu public ou approuvé ou d'un document d'urbanisme en tenant lieu [elles sont alors remplacées par les articles correspondants du règlement du PLU/POS] et définissent principalement la localisation et la desserte des constructions, l'implantation et le volume des constructions, l'aspect des constructions ; dans les communes du littoral et riveraines des lacs et étangs, les **r. g. d'u. du s.** sont définies par la **directive d'aménagement national** ; ►Code de l'urbanisme) ; *syn. obs.* règlement [m] national d'urbanisme (R.N.U.) ; *g* **Baunutzungsverordnung [f]** (*Abk.* BauNVO; **D.**, eine 1962 vom Bundesminister für Wohnungswesen und Städtebau auf Grund des damaligen Bundesbaugesetzes [BBauG], seit 1987 ►Baugesetzbuch [BauGB], erlassene Verordnung — zuletzt 1993 geändert — über die bauliche Nutzungsmöglichkeit von Grundstücken, die im Geltungsbereich von Flächennutzungsplänen und rechtsverbindlichen Bebauungsplänen liegen).

zoochory [n] *phyt. zool.* ►animal dispersal of seeds.

7143 zoological garden [n] *recr.* (Public garden or park with a collection of animals and birds in structures, yards and aviaries for exhibition and scientific study, as well as conveniences for visitor use and, in modern form, with simulated natural habitats and protective enclosure; *syn.* zoological park [n]; *s* **parque [m] zoológico** (Instalación pública o privada, dirigida científicamente, en la que se mantienen diferentes especies de animales, muchas veces exóticas, para el disfrute y la ilustración de la población. El término alemán «Tiergarten» se aplica a los zoológicos pequeños); *syn.* zoológico [m]; *f 1* **jardin [m] zoologique** (Équipement de loisirs public ou privé conçu pour servir de refuge à des animaux déjà acclimatés et de permettre à un grand public de venir les observer) ; *f 2* **jardin [m] d'acclimatation** (Équipement de loisirs public ou privé conçu pour servir de refuge à des animaux exotiques en vue de leur acclimatation et de permettre à un grand public de venir les

Z

observer ; ce jardin s'efforce de reproduire les conditions de vie respectives des animaux, la forme paysagère de parc, une végétation inhabituelle, des rivières, des étangs, d'immenses volière, etc. ayant pour but de faire évoluer les animaux dans un cadre naturel ; JAR 1975, 101) ; *g* **zoologischer Garten [m]** (Öffentliche oder private, wissenschaftlich geleitete Einrichtung zur Haltung und zur Schaustellung von vor allem exotischen Tierarten in Käfigen, Freigehegen, Volieren und Gebäuden. Alle Einrichtungen sind in eine parkartig gestaltete, großflächige Gesamtanlage eingefügt, oft mit naturgetreu nachgestalteten Lebensräumen. Ein **Tiergarten** ist ein kleiner **z. G.**); *syn.* Tierpark [m], Zoo [m].

zoological park [n] *recr.* ▶zoological garden.

7144 zoomass [n] *ecol.* (Living or dry weight of a population of animals present at any given time. **Z.** may be above or below ground; *generic term* ▶biomass); *s* **zoomasa [f]** (Cantidad de materia viva de una o varias especies animales en una unidad dada del espacio; *término genérico* ▶biomasa); *syn.* biomasa [f] animal; *f* **zoomasse [f]** (▶Biomasse animale constituée par les animaux d'une zoocénose. On distingue la **z.** aérienne ou épigée et la **z.** souterraine ou hypogée) ; *g* **Zoomasse [f]** (Gewicht des zu einem bestimmten Zeitpunkt vorhandenen Tierbestandes je Flächen- oder Volumeneinheit einer Lebensstätte als Frisch- oder Trockengewicht. Es wird zwischen oberirdischer und unterirdischer **Z.** unterschieden; *OB* ▶Biomasse).

zoome [n] *ecol.* * ▶biome.

zoophenology [n] *phyt.* * ▶phenology.

Z